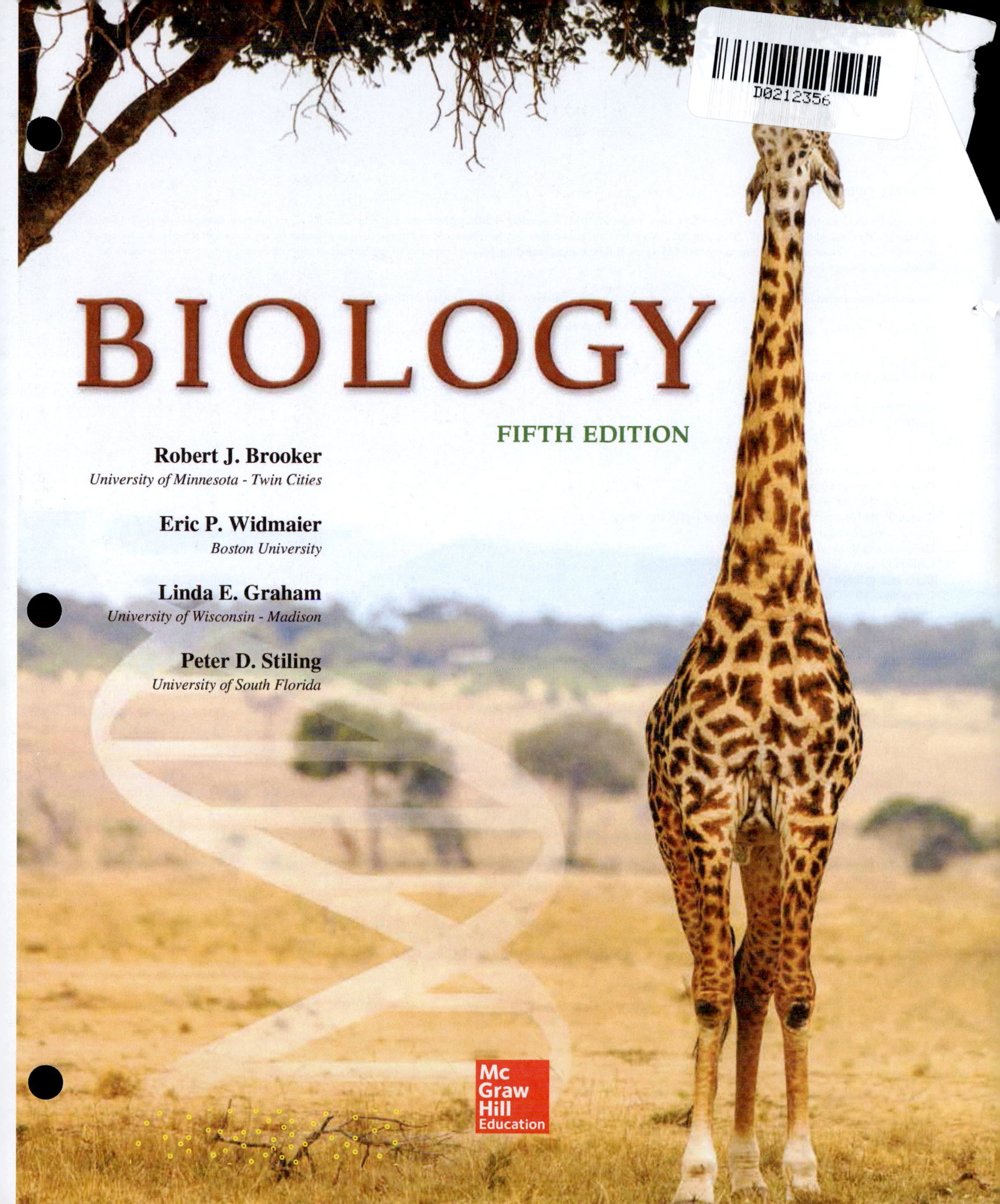

# BIOLOGY

## FIFTH EDITION

**Robert J. Brooker**
*University of Minnesota - Twin Cities*

**Eric P. Widmaier**
*Boston University*

**Linda E. Graham**
*University of Wisconsin - Madison*

**Peter D. Stiling**
*University of South Florida*

Mc
Graw
Hill
Education

BIOLOGY, FIFTH EDITION

Published by McGraw-Hill Education, 2 Penn Plaza, New York, NY 10121. Copyright © 2020 by McGraw-Hill Education. All rights reserved. Printed in the United States of America. Previous editions © 2017, 2014, and 2011. No part of this publication may be reproduced or distributed in any form or by any means, or stored in a database or retrieval system, without the prior written consent of McGraw-Hill Education, including, but not limited to, in any network or other electronic storage or transmission, or broadcast for distance learning.

Some ancillaries, including electronic and print components, may not be available to customers outside the United States.

This book is printed on acid-free paper.

4 5 6 7 8 9  LWI 21 20 19

ISBN 978-1-260-16962-1
MHID 1-260-16962-6

Portfolio Manager: *Andrew Urban*
Product Developer: *Elizabeth M. Sievers*
Marketing Manager: *Kelly Brown*
Content Project Managers: *Jessica Portz/Brent Dela Cruz/Sandra Schnee*
Buyer: *Laura M. Fuller*
Design: *David W. Hash*
Content Licensing Specialist: *Lori Hancock*
Cover Image: *©BlueOrange Studio/Shutterstock*
Compositor: *MPS Limited*

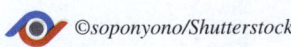 *©soponyono/Shutterstock*

All credits appearing on page are considered to be an extension of the copyright page.

**Library of Congress Cataloging-in-Publication Data**

Brooker, Robert J., author.
  Biology / Robert J. Brooker, University of Minnesota - Twin Cities,
  Eric P. Widmaier, Boston University, Linda E. Graham, University of
  Wisconsin - Madison, Peter D. Stiling, University of South Florida.
  Fifth edition. | New York, NY : McGraw-Hill Education, [2020] |
  Includes index.
  LCCN 2018023793 | ISBN 9781260169621
  LCSH: Biology—Textbooks.
  LCC QH308.2 .B564445 2020 | DDC 570—dc23 LC record
  available at https://lccn.loc.gov/2018023793

The Internet addresses listed in the text were accurate at the time of publication. The inclusion of a website does not indicate an endorsement by the authors or McGraw-Hill Education, and McGraw-Hill Education does not guarantee the accuracy of the information presented at these sites.

mheducation.com/highered

# Brief Contents

# About the Authors

## Robert J. Brooker

Rob Brooker (Ph.D., Yale University) received his B.A. in biology at Wittenberg University, Springfield, Ohio, in 1978, and studied genetics while a graduate student at Yale. For his postdoctoral work at Harvard, he studied lactose permease, the product of the *lacY* gene of the *lac* operon. He continued working on transporters at the University of Minnesota, where he is a Professor in the Department of Genetics, Cell Biology, and Development and the Department of Biology Teaching and Learning. At the University of Minnesota, Dr. Brooker teaches undergraduate courses in biology, genetics, and cell biology. In addition to many other publications, he has written two undergraduate genetics texts published by McGraw-Hill: *Genetics: Analysis & Principles*, 6th edition, copyright 2018, and *Concepts of Genetics*, 3rd edition, copyright 2019.

## Eric P. Widmaier

Eric Widmaier received his B.A. degree in biological sciences at Northwestern University in 1979, where he performed research in animal behavior. In 1984, he earned his Ph.D. in endocrinology from the University of California at San Francisco, where he examined hormonal actions and their mechanisms in mammals. As a postdoctoral fellow at the Worcester Foundation for Experimental Research and later at The Salk Institute, he continued his focus on the cellular and molecular control of hormone secretion and action, with a particular focus on the brain. His current research focuses on the control of body mass and metabolism in mammals, the hormonal correlates of obesity, and the effects of high-fat diets on intestinal cell function. Dr. Widmaier is currently Professor of Biology at Boston University, where he teaches undergraduate human physiology and recently received the university's highest honor for excellence in teaching. Among other publications, he is lead author of *Vander's Human Physiology: The Mechanisms of Body Function*, 15th edition, published by McGraw-Hill, copyright 2019.

## Linda E. Graham

Linda Graham earned an undergraduate degree from Washington University (St. Louis), a master's degree from the University of Texas, and Ph.D. from the University of Michigan, Ann Arbor, where she also did postdoctoral research. Presently Professor of Botany at the University of Wisconsin-Madison, her research explores the evolutionary origins of algae and land-adapted plants, focusing on their cell and molecular biology as well as microbial interactions. In recent years Dr. Graham has engaged in research expeditions to remote regions of the world to study algal and plant microbiomes. She teaches undergraduate courses in microbiology and plant biology. She is the coauthor of, among other publications, *Algae*, 3rd edition, copyright 2016, a textbook on algal biology, and *Plant Biology*, 3rd edition, copyright 2015, both published by LJLM Press.

Left to right: Eric Widmaier, Linda Graham, Peter Stiling, and Rob Brooker

*The authors are grateful for the help, support, and patience of their families, friends, and students, Deb, Dan, Nate, and Sarah Brooker, Maria, Caroline, and Richard Widmaier, Jim, Michael, Shannon, and Melissa Graham, and Jacqui, Zoe, Leah, and Jenna Stiling.*

## Peter D. Stiling

Peter Stiling obtained his Ph.D. from University College, Cardiff, United Kingdom. Subsequently, he became a postdoctoral fellow at Florida State University and later spent two years as a lecturer at the University of the West Indies, Trinidad. Dr. Stiling was formerly Chair of the Department of Integrative Biology at the University of South Florida (USF) at Tampa, where he is currently an Assistant Vice Provost for Strategic Initiatives and Professor of Biology. His research interests include plant-animal relationships and invasive species. He currently teaches biology to students in the USF in London summer program which he established in 2015. Dr. Stiling was elected an AAAS Fellow in 2012. He is also the author of *Ecology: Global Insights and Investigations*, 2nd edition, published by McGraw-Hill.

## A Message from the Authors

As active teachers and writers, one of the great joys of this process for us is that we have been able to meet many more educators and students during the creation of this textbook. It is humbling to see the level of dedication our peers bring to their teaching. Likewise, it is encouraging to see the energy and enthusiasm so many students bring to their studies. We hope this book and its digital resources will serve to aid both faculty and students in meeting the challenges of this dynamic and exciting course. For us, this remains a work in progress, and we encourage you to let us know what you think of our efforts and what we can do to serve you better.

*Rob Brooker, Eric Widmaier, Linda Graham, Peter Stiling*

# Acknowledgements

The lives of most science-textbook authors do not revolve around an analysis of writing techniques. Instead, we are people who understand science and are inspired by it, and we want to communicate that information to our students. Simply put, we need a lot of help to get it right.

Editors are a key component who help the authors modify the content of this textbook so it is logical, easy to read, and inspiring. The editorial team for this *Biology* textbook has been a catalyst that kept this project rolling. The members played various roles in the editorial process. Andrew Urban and his predecessor Justin Wyatt, Portfolio Managers (Majors Biology), have done an excellent job overseeing the 5th edition. Elizabeth Sievers, Senior Product Developer, has been the master organizer. Liz's success at keeping us on schedule is greatly appreciated. We would also like to acknowledge our copy editor, Jane Hoover, for her thoughtful editing that has contributed to the clarity of this textbook.

Another important aspect of the editorial process is the actual design, presentation, and layout of materials. It's confusing if the text and art aren't on the same page, or if a figure is too large or too small. We are indebted to the tireless efforts of Jessica Portz, Content Project Manager, and David Hash, Senior Designer at McGraw-Hill. Likewise, our production company, MPS Limited, did an excellent job with the paging, revision of existing art, and the creation of new art for the 5th edition. Their artistic talents, ability to size and arrange figures, and attention to the consistency of the figures have been remarkable. We would also like to acknowledge the ongoing efforts of the superb marketing staff at McGraw-Hill. Special thanks to Kelly Brown, Executive Marketing Manager, whose effort intensifies when this edition comes out.

Finally, other staff members at McGraw-Hill Higher Education have ensured that the authors and editors were provided with adequate resources to achieve the goal of producing a superior textbook. These include G. Scott Virkler, Senior Vice President, Products & Markets; Michael Ryan, Vice President, General Manager, Products & Markets; and Betsy Whalen, Vice President, Production and Technology Services.

## Reviewers for *Biology*, 5th edition

- Lubna Abu-Niaaj *Central State University*
- Joseph Covi *University of North Carolina at Wilmington*
- Art Frampton *University of North Carolina at Wilmington*
- Brian Gibbens *University of Minnesota*
- Judyth Gulden *Tulsa Community College*
- Alexander Motten *Duke University*
- Melissa Schreiber *Valencia College*
- Madhavi Shah *Raritan Valley Community College*
- Jack Shurley *Idaho State University*
- Om Singh *University of Pittsburgh at Bradford*
- Michelle Turner-Edwards *Suffolk County Community College*
- Ryan Udan *Missouri State University*
- D. Alexander Wait *Missouri State University*
- Kimberly Wallace *Texas A & M University San Antonio*
- Megan Wise de Valdez *Texas A & M University San Antonio*

# A Modern Vision for Learning: Emphasizing Core Concepts and Core Skills

Over the course of five editions, the ways in which biology is taught have dramatically changed. We have seen a shift away from the memorization of details, which are easily forgotten, and a movement toward emphasizing core concepts and critical thinking skills. The previous edition of *Biology* strengthened skill development by adding two new features, called CoreSKILLS and BioTIPS (described later), which are aimed at helping students develop effective strategies for solving problems and applying their knowledge in novel situations. In this edition, we have focused our pedagogy on the five core concepts of biology as advocated by "Vision and Change" and introduced at a national conference organized by the American Association for the Advancement of Science (see www.visionandchange.org). These core concepts, which are introduced in Chapter 1 (see Figure 1.4) include the following:

1. *Evolution:* The diversity of life evolved over time by processes of mutation, selection, and genetic exchange.
2. *Structure and function:* Basic units of structure define the function of all living things.
3. *Information flow, exchange, and storage:* The growth and behavior of organisms are activated through the expression of genetic information.
4. *Pathways and transformations of energy and matter:* Biological systems grow and change via processes that are based on chemical transformation pathways and are governed by the laws of thermodynamics.
5. *Systems:* Living systems are interconnected and interacting.

In addition to core concepts, "Vision and Change" has strongly advocated the development of core skills (also called core competencies). Those skills that are emphasized in this textbook are as follows:

- The ability to apply the process of science
- The ability to use quantitative reasoning
- The ability to use models and simulation (each chapter in *Biology, 5e,* contains a new feature called Modeling Challenge that asks students to create their own model or interpret a model provided)
- The ability to tap into the interdisciplinary nature of science
- The ability to communicate and collaborate with professionals in other disciplines
- The ability to understand the relationship between science and society

A key goal of this textbook is to bring to life these five core concepts of biology and the core skills. These concepts and skills are highlighted in each chapter with a "Vision and Change" icon, which indicates subsections and figures that focus on one or more of them. This approach will serve two purposes. First, the icon will help students to see how the various topics in this textbook are connected to each other by the five core concepts of biology. Second, the icon will allow students to appreciate the important skills they are developing as they progress through the text.

## KEY PEDAGOGICAL FEATURES OF THIS EDITION

The author team is dedicated to producing the most engaging and current text available for undergraduate students who are majoring in biology. We have listened to educators and reviewed documents, such as *Vision and Change, A Call to Action,* which includes a summary of recommendations made at a national conference organized by the American Association for the Advancement of Science. We want our textbook to reflect core concepts and skills and provide a more learner-centered approach. To achieve these goals, *Biology,* 5th edition, has the following pedagogical features.

- *NEW!* Core Concepts: As mentioned, the five core concepts are introduced in Chapter 1 (see Figure 1.4). Throughout Chapters 2 through 60, these core concepts are emphasized by a Vision and Change icon, placed next to headings of particular subsections and beneath certain figure legends.

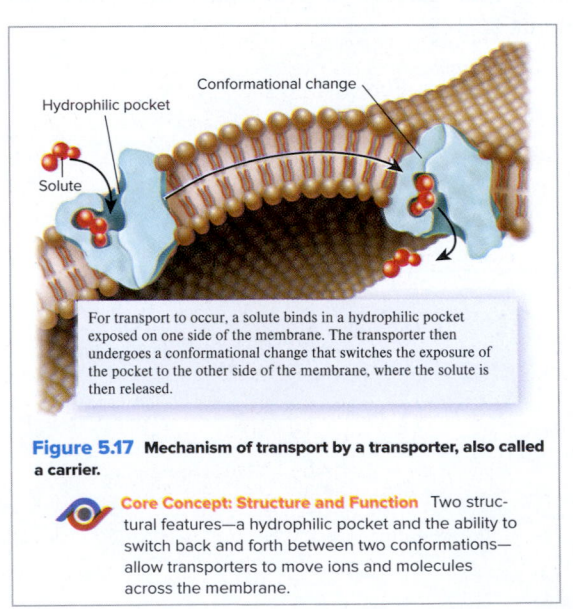

For transport to occur, a solute binds in a hydrophilic pocket exposed on one side of the membrane. The transporter then undergoes a conformational change that switches the exposure of the pocket to the other side of the membrane, where the solute is then released.

**Figure 5.17** Mechanism of transport by a transporter, also called a carrier.

**Core Concept: Structure and Function** Two structural features—a hydrophilic pocket and the ability to switch back and forth between two conformations—allow transporters to move ions and molecules across the membrane.

- **NEW! Core Skills:** Six core skills are also introduced in Chapter 1 (see Section 1.6). In Chapters 2 through 60, these core skills are emphasized by a Vision and Change icon, , placed next to headings of particular subsections, such as Feature Investigations, and beneath certain figure legends. To distinguish them from the **Core Concepts,** the **Core Skills** are highlighted in blue type. In addition, the designator **CoreSKILLS** has been added to certain learning outcomes and end-of-chapter questions that emphasize skills needed in the study of biology.

| Genotype | *PP* | *Pp* | *pp* |
|---|---|---|---|
| Amount of functional protein P produced | 100% | 50% | 0% |
| Phenotype | Purple | Purple | White |
| Only 50% of the functional protein is needed to produce the purple phenotype | | | |

Colorless precursor molecule          Protein P          Purple pigment

**Figure 17.16  How genes give rise to traits during simple Mendelian inheritance.** In the heterozygote, the amount of protein encoded by a single dominant allele is sufficient to produce the dominant phenotype. In this example, the gene encodes an enzyme that is needed to produce a purple pigment. A plant with one or two copies of the dominant allele makes enough pigment to produce purple flowers. In a *pp* homozygote, the complete lack of the functional protein (enzyme) results in white flowers.

 **Core Skill: Quantitative Reasoning** In a simple dominant/recessive relationship, even though the heterozygote may produce less of a functional protein compared to the homozygote that has two copies of the dominant allele, the amount made by the heterozygote is sufficient to yield the dominant phenotype.

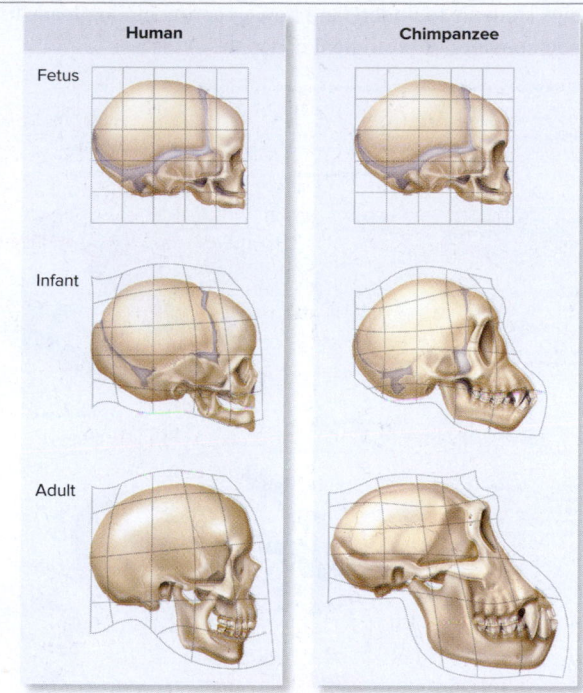

**Figure 24.16  Heterochrony.** Due to heterochrony, one region of the body may grow faster than another during development in different species. For example, the skulls of adult chimpanzees and humans have different shapes even though their fetal skull shapes are quite similar.

**Core Skill: Modeling** The goal of this modeling challenge is to make a series of models that show the differences in limb lengths among orangutans, chimpanzees, and humans.

**Modeling Challenge:** Search the Internet and look at photos of orangutans, chimpanzees, and humans. Even though these species look similar, one noticeable difference is the relative lengths of their limbs. Although the limbs in an early fetus look similar in all three species, the limbs in the adults show significant differences in their relative lengths. Draw models, similar to those in **Figure 24.16**, that show an early fetus, infant, and adult for all three species. Include an explanation of how heterochrony affects limb development.

- **NEW! Modeling Challenges:** A growing trend is the use of models in biology education. Students are asked to interpret models and to create models based on data or a scenario. Furthermore, using models and simulations is one of the core skills that is emphasized by "Vision and Change." The author team has added a new feature called **Modeling Challenge** that asks students to create a model or to interpret a model they are given. Possible answers to the **Modeling Challenges** are provided in Connect.

- **Feature Investigations:** The emphasis on skill development continues in the Feature Investigations, which provide complete descriptions of experiments. These investigations begin with background information in the text that describes the events that led to a particular study. The study is then presented as an illustration that begins with the hypothesis and then describes the experimental protocol at the experimental and conceptual levels. The illustration also includes data and the conclusions that were drawn from the data. This integrated approach helps students to understand how experimentation leads to an understanding of biological concepts.

Figure 5.16  The discovery of water channels (aquaporins) by Agre. (4) Courtesy Dr. Peter Agre

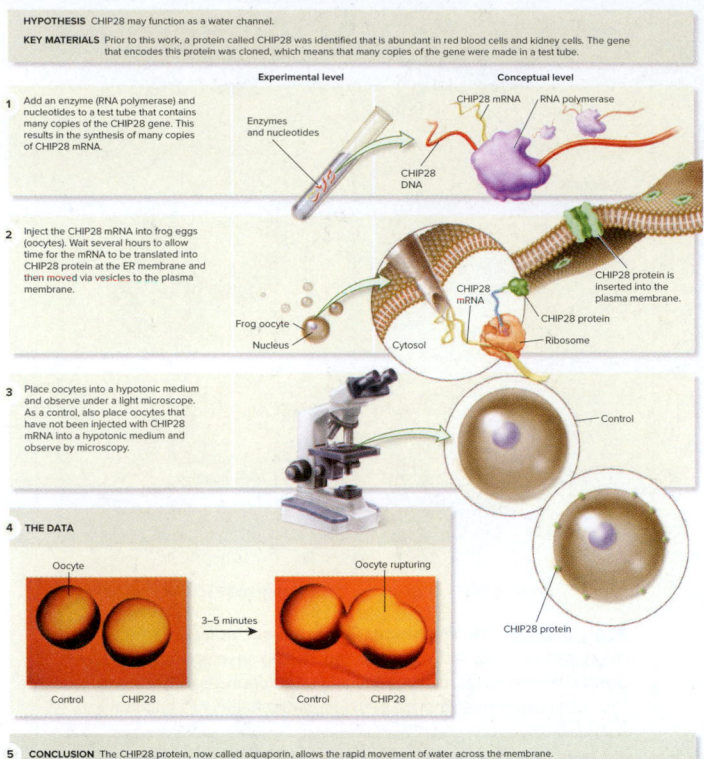

HYPOTHESIS  CHIP28 may function as a water channel.

KEY MATERIALS  Prior to this work, a protein called CHIP28 was identified that is abundant in red blood cells and kidney cells. The gene that encodes this protein was cloned, which means that many copies of the gene were made in a test tube.

Experimental level / Conceptual level

1  Add an enzyme (RNA polymerase) and nucleotides to a test tube that contains many copies of the CHIP28 gene. This results in the synthesis of many copies of CHIP28 mRNA.

CHIP28 mRNA  RNA polymerase
Enzymes and nucleotides
CHIP28 DNA

2  Inject the CHIP28 mRNA into frog eggs (oocytes). Wait several hours to allow time for the mRNA to be translated into CHIP28 protein at the ER membrane and then moved via vesicles to the plasma membrane.

CHIP28 protein is inserted into the plasma membrane.
Frog oocyte
Nucleus
CHIP28 mRNA
CHIP28 protein
Cytosol
Ribosome

3  Place oocytes into a hypotonic medium and observe under a light microscope. As a control, also place oocytes that have not been injected with CHIP28 mRNA into a hypotonic medium and observe by microscopy.

Control
CHIP28 protein

4  THE DATA

Oocyte                     Oocyte rupturing

3–5 minutes →

Control   CHIP28         Control   CHIP28

5  CONCLUSION  The CHIP28 protein, now called aquaporin, allows the rapid movement of water across the membrane.

6  SOURCE  Preston, G. M., Carroll, T. P., Guggino, W. B., and Agre, P. 1992. Appearance of water channels in Xenopus oocytes expressing red cell CHIP28 protein. Science 256: 385–387.

- **BioTIPS:** A feature that was added to the previous edition is aimed at helping students improve their problem-solving skills. Chapters 2 through 60 contain solved problems called **BioTIPS**, where "TIPS" stands for Topic, Information, and Problem-Solving Strategy. These solved problems follow a consistent pattern in which students are given advice on how to solve problems in biology using 11 different problem-solving strategies: Make a drawing. Compare and contrast. Relate structure and function. Sort out the steps in a complicated process. Propose a hypothesis. Design an experiment. Predict the outcome. Interpret data. Use statistics. Make a calculation. Search the literature.

BIO:TIPS  **THE QUESTION**  *A diploid cell has 12 chromosomes, or 6 pairs. In the following diagram, in what phase of mitosis, meiosis I or meiosis II, is this cell?*

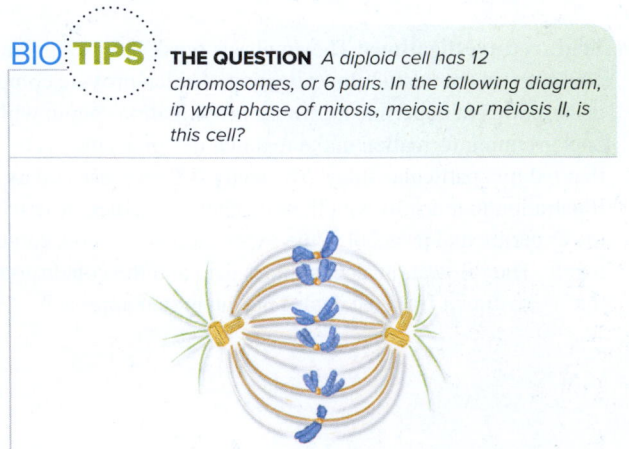

**T**OPIC  *What topic in biology does this question address?* The topic is cell division. More specifically, the question is asking you to be able to look at a drawing and discern which phase of cell division a particular cell is in.

**I**NFORMATION  *What information do you know based on the question and your understanding of the topic?* In the question, you are given a diagram of a cell at a particular phase of the cell cycle. This cell is derived from a mother cell with 6 pairs of chromosomes. From your understanding of the topic, you may remember the various phases of mitosis, meiosis I, and meiosis II, which are described in Figures 16.8 and 16.13. If so, you may initially realize that the cell is in metaphase.

**P**ROBLEM-SOLVING **S**TRATEGY  *Sort out the steps in a complicated process.* To solve this problem, you may need to describe the steps, starting with a mother cell that has 6 pairs of chromosomes. Keep in mind that a mother cell with 6 pairs of chromosomes has 12 chromosomes during $G_1$, which then replicate to form 12 pairs of sister chromatids during S phase. Therefore, at the beginning of M phase, this mother cell will have 12 pairs of sister chromatids. During mitosis, the 12 pairs of sister chromatids will align at metaphase. During meiosis I, 6 bivalents will align along the metaphase plate in the mother cell. During meiosis II, 6 pairs of sister chromatids will align along the metaphase plate in the two cells.

ANSWER  *The cell is in metaphase of meiosis II. You can tell because the chromosomes are lined up in a single row along the metaphase plate, and the cell has only 6 pairs of sister chromatids. If it were mitosis, the cell would have 12 pairs of sister chromatids. If it were in meiosis I, bivalents would be aligned along the metaphase plate.*

- **Formative Assessment:** A trend in biology education is to spend more class time engaging students in active learning. While this is a positive approach that fosters learning, a drawback is that instructors have less time to explain the material in the textbook. When students are expected to learn textbook material on their own, it is imperative that they are regularly given formative assessment—feedback regarding their state of learning while they are engaging in the learning process. This allows students to gauge whether they are mastering the material. Formative assessment is a major feature of this textbook and is bolstered by Connect—a state-of-the art digital assignment and assessment platform. In *Biology*, 5th edition, formative assessment is provided in multiple ways.

  - First, many figure legends have Concept Check questions that focus on key concepts of a given topic.

  - Second, questions in Assess and Discuss at the end of each chapter explore students' understanding of concepts and mastery of skills. Core Concepts and Core Skills are again addressed under the Conceptual Questions. The answers to the Concept Checks and the end-of-chapter questions are in Appendix B, so students can immediately see if they are mastering the material.

## Conceptual Questions

1. The Earth's atmosphere consists of 78% nitrogen. Why is nitrogen a limiting nutrient?

2. Why does maximum sustainable yield occur at the midpoint of the logistic curve and not where the population is at carrying capacity?

3. **Core Skill: Science and Society** In one family, parents, who were born in 1900, have twins at age 20 but then have no more children. Their children, grandchildren, and so on behave in the same way. In another family, parents, who were also born in 1900, delay reproduction until age 33 but have triplets. Their children and grandchildren behave in the same way. Which family has the most descendants by 2000? What can you conclude?

- In Connect, a particularly robust type of formative assessment is SmartBook, which guides a student through the textbook. SmartBook is an adaptive learning tool that is described later in this Preface.

- **Unit openers:** Each unit begins with a unit opener that provides an overview of the chapters within that unit. This overview allows the student to see the big picture of the unit. In addition, the unit openers draw attention to the core concepts and core skills of biology that will be emphasized in each unit.

- **Learning Outcomes:** As advocated in *Vision and Change*, educational materials should have well-defined learning goals. Each section of every chapter begins with a set of Learning Outcomes. These outcomes inform students of the key concepts they will learn and the skills they will acquire in mastering the material. They also provide a tangible indication of how student learning will be assessed. The assessments in Connect were developed using these Learning Outcomes as a guide in formulating online questions, thereby linking the learning goals of the text with the assessments in Connect.

### 13.1 Overview of Non-coding RNAs

**Learning Outcomes:**

1. Describe the ability of ncRNAs to bind to other molecules and macromolecules.
2. Outline the general functions of ncRNAs.
3. Define ribozyme.
4. List several examples of ncRNAs, and describe their functions.

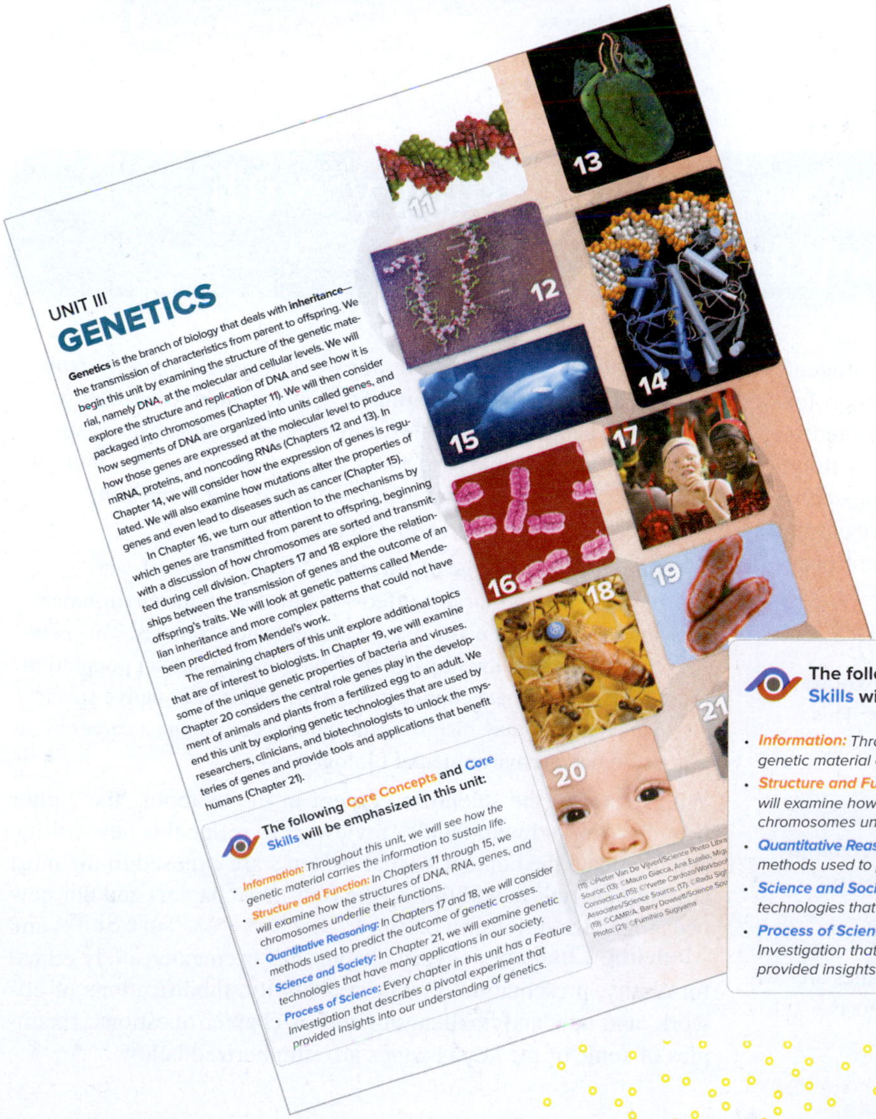

## UNIT III
# GENETICS

**Genetics** is the branch of biology that deals with inheritance—the transmission of characteristics from parent to offspring. We begin this unit by examining the structure of the genetic material, namely DNA, at the molecular and cellular levels. We will explore the structure and replication of DNA and see how it is packaged into chromosomes (Chapter 11). We will then consider how segments of DNA are organized into units called genes, and how those genes are expressed at the molecular level to produce mRNA, proteins, and noncoding RNAs (Chapters 12 and 13). In Chapter 14, we will consider how mutations alter the properties of genes and even lead to diseases such as cancer (Chapter 15).

In Chapter 16, we turn our attention to the mechanisms by which genes are transmitted from parent to offspring, beginning with a discussion of how chromosomes are sorted and transmitted during cell division. Chapters 17 and 18 explore the relationships between the transmission of genes and the outcome of an offspring's traits. We will look at genetic patterns called Mendelian inheritance and more complex patterns that could not have been predicted from Mendel's work.

The remaining chapters of this unit explore additional topics that are of interest to biologists. In Chapter 19, we will examine some of the unique genetic properties of bacteria and viruses. Chapter 20 considers the central role genes play in the development of animals and plants from a fertilized egg to an adult. We end this unit by exploring genetic technologies that are used by researchers, clinicians, and biotechnologists to unlock the mysteries of genes and provide tools and applications that benefit humans (Chapter 21).

**The following Core Concepts and Core Skills will be emphasized in this unit:**

- **Information:** Throughout this unit, we will see how the genetic material carries the information to sustain life.
- **Structure and Function:** In Chapters 11 through 15, we will examine how the structures of DNA, RNA, genes, and chromosomes underlie their functions.
- **Quantitative Reasoning:** In Chapters 17 and 18, we will consider methods used to predict the outcome of genetic crosses.
- **Science and Society:** In Chapter 21, we will examine genetic technologies that have many applications in our society.
- **Process of Science:** Every chapter in this unit has a Feature Investigation that describes a pivotal experiment that provided insights into our understanding of genetics.

**The following Core Concepts and Core Skills will be emphasized in this unit:**

- **Information:** Throughout this unit, we will see how the genetic material carries the information to sustain life.
- **Structure and Function:** In Chapters 11 through 15, we will examine how the structures of DNA, RNA, genes, and chromosomes underlie their functions.
- **Quantitative Reasoning:** In Chapters 17 and 18, we will consider methods used to predict the outcome of genetic crosses.
- **Science and Society:** In Chapter 21, we will examine genetic technologies that have many applications in our society.
- **Process of Science:** Every chapter in this unit has a Feature Investigation that describes a pivotal experiment that provided insights into our understanding of genetics.

(11) ©Peter Van De Vijver/Science Photo Library; Source: (13) ©Mauro Giacca, and Eusebio, Mgo, Connecticut; (15) ©Yvette Cardozo/Newscom; Associate/Science Source; (17) ©Radui Sigheti/ (19) ©CAMR/A. Barry Dowsett/Science Source; Photo (21) ©Fumihiko Sugiyama

## USING STUDENT USAGE DATA TO MAKE IMPROVEMENTS

To help guide the revision for the 5th edition, the authors consulted student usage data and input, which were derived from thousands of SmartBook® users of the 4th edition. SmartBook "heat maps" provided a quick visual snapshot of chapter usage data and the relative difficulty students experienced in mastering the content. These data directed the authors to evaluate text content that was particularly challenging for students. These same data were also used to revise the SmartBook probes.

- If the data indicated that the subject was more difficult than other parts of the chapter, as evidenced by a high proportion of students responding incorrectly to the SmartBook questions, the authors revised or reorganized the content to be as clear and illustrative as possible, for example, by rewriting the section or providing additional examples or revised figures to assist visual learners.

- In other cases, one or more of the SmartBook questions for a section was not as clear as it should have been or did not appropriately reflect the content in the chapter. In these cases the question, rather than the text, was edited.

Below is an example of one of the heat maps from Chapter 8. The color-coding of highlighted sections indicates the various levels of difficulty students experienced in learning the material, topics highlighted in red being the most challenging for students.

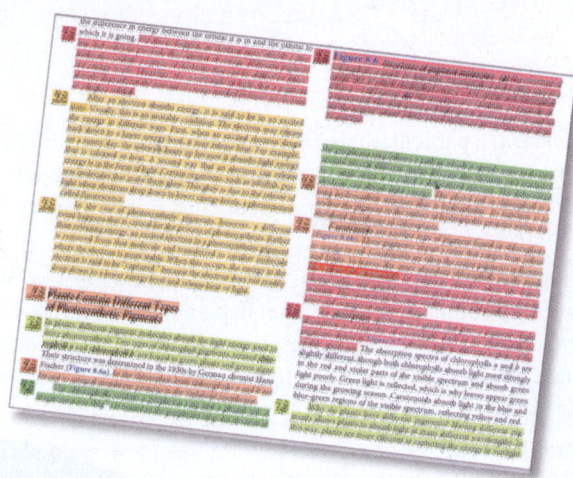

# Preparing Students for Careers in Biology with *NEW* Cutting-Edge Content

A key purpose of a majors biology course is to prepare students for biology-related careers, including those in the health professions, teaching, and research. The author team has reflected on the direction of biology and how that direction will affect future careers that students may pursue. We are excited to announce that *Biology,* 5th edition, has four new chapters that reflect current trends in biology research and education. These trends are opening the doors to exciting new career options in biology.

- ***Chapter 13. Gene Expression at the Molecular Level II: Non-coding RNAs.*** The past decade or so has seen an explosion in the discovery of different types of non-coding RNAs. This work has revealed a variety of roles of non-coding RNAs at the molecular level, as well as roles in human diseases and plant health.

- ***Chapter 30. Microbiomes: Microbial Systems On and Around Us.*** Recent research has revealed the staggering complexity and biological importance of microbiomes—assemblages of microbes that are associated with a particular host or environment. This new chapter explores how microbiomes are analyzed and describes their interactions with diverse hosts, including humans, protists, and plants.

- ***Chapter 53: Integrated Responses of Animal Organ Systems to a Challenge to Homeostasis.*** Systems biology has been a recent trend in biological research and education. This chapter takes systems biology to a new level by exploring how multiple organs systems respond in a coordinated way to the same threat—a challenge to homeostasis.

- ***Chapter 59: The Age of Humans.*** We face a tug-of-war between the undesirable effects of humans on the environment and the efforts of ecologists to prevent such changes. This new chapter surveys the impacts that the growing human population has had on climate change and on the survival of native species. This material may inspire some students to pursue a career as an ecologist or environmental biologist.

With regard to the scientific content in the textbook, the author team has worked with faculty reviewers to refine this new edition and to update the content so that students are exposed to the most current material. In addition to the four new chapters and our new pedagogical additions involving Core Concepts, Core Skills, and Modeling Challenges, every chapter has been extensively edited for clarity, presentation, layout, readability, modifications of artwork, and new and challenging end-of-chapter questions. Examples of some of the key changes are summarized below.

- **Chapter 1. An Introduction to Biology.** Chapter 1 provides a description of the **Core Concepts** (see Figure 1.4) and the **Core Skills** (see Section 1.6) that are advocated by *Vision and Change*.

## Chemistry Unit

- **Chapter 2. The Chemical Basis of Life I: Atoms, Molecules, and Water.** The topics of pH and buffers have been placed in their own section (see Section 2.4).

## Cell Unit

- **Chapter 4. Evolutionary Origin of Cells and Their General Features.** This chapter now begins with a discussion of the evolutionary origin of cells (see Section 4.1). It also discusses a new topic, droplet organelles, which are organelles that are not surrounded by a membrane (see Section 4.3).
- **Chapter 6. An Introduction to Energy, Enzymes, and Metabolism.** For the topic of how cells use ATP as a source of energy, a revised subsection compares the **Core Concept: Information** to the **Core Concept: Energy and Matter.**
- **Chapter 7. Cellular Respiration and Fermentation.** A **Modeling Challenge** asks students to predict the effects of a mutation on the function of ATP synthase (see Figure 7.12).
- **Chapter 10. Multicellularity.** Four figures have been revised to better depict the relative locations of cell junctions between animal cells.

## Genetics Unit

- **Chapter 11. Nucleic Acid Structure, DNA Replication, and Chromosome Structure.** Figure 11.8b has a **Modeling Challenge** that asks students to predict how the methylation of a base would affect the ability of that base to hydrogen bond with a base in the opposite strand.
- **Chapter 13.** *NEW!* **Gene Expression at the Molecular Level II: Non-coding RNAs.** This new chapter begins with an overview of the general properties of non-coding RNAs and then describes specific examples in which non-coding RNAs are involved with chromatin structure, transcription, translation, protein sorting, and genome defense.
- **Chapter 16. The Eukaryotic Cell Cycle, Mitosis, and Meiosis.** The **Core Concept: Evolution** is highlighted in a subsection that explains how mitosis in eukaryotes evolved from binary fission in prokaryotic cells (see Figure 16.10).
- **Chapter 17. Mendelian Patterns of Inheritance.** The organization of this chapter has been revised to contain the patterns of inheritance that obey Mendel's laws.
- **Chapter 18. Epigenetics, Linkage, and Extranuclear Inheritance.** This chapter now covers inheritance patterns that violate Mendel's laws. The topic of epigenetics has been expanded from one section in the previous edition to four sections in the 5th edition (see Sections 18.1 through 18.4).
- **Chapter 19. Genetics of Viruses and Bacteria.** Discussion of the Zika virus has been added to this chapter.
- **Chapter 21. Genetic Technologies and Genomics.** The use of CRISPR-Cas technology to alter genes is now discussed (see Figure 21.10).

## Evolution Unit

- **Chapter 22. An Introduction to Evolution.** This chapter has been moved so that it is the first chapter in this unit on evolution.
- **Chapter 23. Population Genetics.** After learning about the Hardy-Weinberg equation, students are presented with a **Modeling Challenge** that asks them to propose a mathematical model that extends the Hardy-Weinberg equation to a gene that exists in three alleles (see Figure 23.2).
- **Chapter 25. Taxonomy and Systematics.** The topic of taxonomy is related to the **Core Concept: Evolution** through an explanation of how taxonomy is based on the evolutionary relationships among different species.
- **Chapter 26. History of Life on Earth and Human Evolution.** The topic of human evolution has been moved from the unit on diversity to this unit. The expanded version of this topic describes recent examples of human evolution and discusses the amount of genetic variation between different human populations (see Section 26.3).

## Diversity Unit

- **Chapter 27. Archaea and Bacteria.** This chapter has been reorganized to provide essential background for new Chapter 30 (an exploration of microbiomes). The **Core Skill: Connections** is illustrated by linking electromagnetic sensing in bacteria with that in certain animals.
- **Chapter 29. Fungi.** An overview of fungal phylogeny has been updated to reflect new research discoveries. Coverage of plant root-fungal associations (mycorrhizae) and lichens has been moved to new Chapter 30.
- **Chapter 30.** *NEW!* **Microbiomes: Microbial Systems On and Around Us.** This new chapter integrates information about microbial diversity (Chapters 27 through 29) with material on genetic technologies that is introduced in Chapter 21 to explain the evolutionary, medical, agricultural, and environmental importance of microbial associations.
- **Chapter 31. Plants and the Conquest of Land.** The diagrammatic overview of plant phylogeny has been updated to reveal challenges in understanding the pattern of plant evolution.
- **Chapter 33. An Introduction to Animal Diversity.** Figure 33.3, animal phylogeny, has been redrawn to reflect the idea that ctenophores, rather than sponges, are now considered to be the earliest diverging animals. Section 33.2 on animal classification has been largely revised.
- **Chapter 34. The Invertebrates.** Following the new themes introduced in Chapter 33, this chapter has been reorganized to discuss ctenophores as the earlier evolving animals, followed by sponges, cnidria, jellyfish, and other radially symmetrical animals.

## Flowering Plants Unit

- **Chapter 36. An Introduction to Flowering Plant Form and Function.** A new chapter opener links the economic importance of plants, represented by cotton, to the significance of plant structure-function relationships.

- **Chapter 37. Flowering Plants: Behavior.** A **Modeling Challenge** links plant responses to conditions on Earth to those experienced in space.
- **Chapter 38. Flowering Plants: Nutrition.** In a **Modeling Challenge** related to plant-microbe interaction process, students infer how specific mutations might affect an important nutritional feature.
- **Chapter 40. Flowering Plants: Reproduction.** This chapter explores intriguing parallels between the reproductive processes of animals and those of plants.

## Animals Unit

- **Chapter 41. Animal Bodies and Homeostasis.** A section entitled "Homeostatic Control of Internal Fluids" (Section 41.4) now follows the section "General Principles of Homeostasis," providing students with an understanding of body fluid compartments, osmolarity, and how animal bodies exchange ions and water with their environments. These concepts are important to students' understanding of subsequent chapters.
- **Chapter 42. Neuroscience I: Cells of the Nervous System.** The **Core Skill: Science and Society** is featured numerous times in the unit on animals, including in Figure 42.18 which describes the use of magnetic resonance imaging in modern medicine.
- **Chapter 43. Neuroscience II: Evolution, Structure, and Function of the Nervous System.** The **Core Skill: Connections** is also featured throughout the unit on animals, including in Figure 43.1 in which students are asked to identify the defining features of animals by referring to Chapter 33.
- **Chapter 44. Neuroscience III: Sensory Systems.** New research demonstrating a correlation between the types of locomotion of vertebrates and the relative sizes of their semicircular canals is described.
- **Chapter 46. Nutrition and Animal Digestive Systems.** A **Modeling Challenge** was added in which students are tasked with creating models of hypothetical alimentary canals of two species with different diets, eating patterns, and teeth.
- **Chapter 47. Control of Energy Balance, Metabolic Rate, and Body Temperature.** The meaning of body mass index and its usefulness and limitations are more fully elucidated, and data on obesity statistics in the United States have been updated to reflect current trends.
- **Chapter 48. Circulatory and Respiratory Systems.** These topics were formerly addressed in two chapters but are now integrated into a single chapter that streamlines the presentation and emphasizes important connections between the two systems.
- **Chapter 49. Excretory Systems.** The chapter has been more narrowly focused on excretory systems by moving the material on osmoregulation and body fluids earlier in the unit, to Chapter 41.
- **Chapter 51. Animal Reproduction and Development.** Formerly two chapters, this material is now covered in one chapter, which eliminated redundancy in coverage. For example, the topic of fertilization (Section 51.2) is now covered in its entirety in the same section as the topic of gametogenesis, rather than being split between two chapters.
- **Chapter 52. Immune Systems.** Exciting new information has been added that describes the evolution of toll-like receptors and the presence of a TLR-domain in bacterial genes associated with immune defenses.
- **Chapter 53. *NEW!* Integrated Responses of Animal Organ Systems to a Challenge to Homeostasis.** This new chapter integrates material from virtually the entire unit on animals, using a classic challenge to homeostasis as an example. It includes a compelling case study of a young athlete that begins and concludes the chapter.

## Ecology Unit

- **Chapter 54. An Introduction to Ecology and Biomes.** The section on aquatic biomes as been expanded with a new figure and explanation of the annual cycle of temperate lakes, as well as new information on tide formation and waves.
- **Chapter 57. Species Interactions.** This chapter has been reduced in length by the deletion of four figures and streamlined for easier understanding.
- **Chapter 58. Communities and Ecosystems: Ecological Organization at Large Scales.** This chapter has been reorganized to include both community ecology and ecosystems ecology.
- **Chapter 59. *NEW!* The Age of Humans.** This new chapter synthesizes information concerning the effects of humans on the natural environment. It contains discussions of human population growth (previously covered in Chapter 56), the effect of global warming on climate change (previously covered in Chapter 54), and human effects on biogeochemical cycles and biomagnification (previously covered in Chapter 59), and new information on habitat destruction, overexploitation, and invasive species.
- **Chapter 60. Biodiversity and Conservation Biology.** The coverage of the value of biodiversity to human welfare, detailed in Section 60.3 has been updated and expanded.

# Strengthening Problem-Solving Skills and Key Concept Development with Connect®

## Detailed Feedback in Connect®

Learning is a process of iterative development, of making mistakes, reflecting, and adjusting over time. The question and test banks in Connect® for *Biology*, 5th edition, are more than direct assessments; they are self-contained learning experiences that systematically build student learning over time.

For many students, choosing the right answer is not necessarily based on applying content correctly; it is more a matter of increasing the statistical odds of guessing. A major fault with this approach is students don't learn how to process the questions correctly, mostly because they are repeating and reinforcing their mistakes rather than reflecting and learning from them. To help students develop problem-solving skills, all higher-level Bloom's questions in Connect are supported with hints, to help students focus on important information needed to answer the questions, and detailed feedback that walks students through the problem-solving process, using Socratic questions in a decision-tree framework to scaffold learning, in which each step models and reinforces the learning process.

The feedback for each higher-level Bloom's question (Apply, Analyze, Evaluate) follows a similar process: Clarify Question, Gather Content, Consider Alternatives, Choose Answer, Reflect on Process.

Rather than leaving it up to the student to work through the detailed feedback, we present a second version of the question in a stepwise format. Following the problem-solving steps, students need to answer questions about the problem-solving process, such as "What is the key concept addressed by the question?" before answering the original question. A professor can choose which version of the question to include in the assignment based on the problem-solving skills of the students.

## Graphing Interactives

To help students develop analytical skills, Connect® for *Biology*, 5th edition, is enhanced with interactive graphing questions. Students are presented with a scientific problem and the opportunity to manipulate variables, producing different results on a graph. A series of questions follows the graphing activity to assess if the student understands and is able to interpret the data and results.

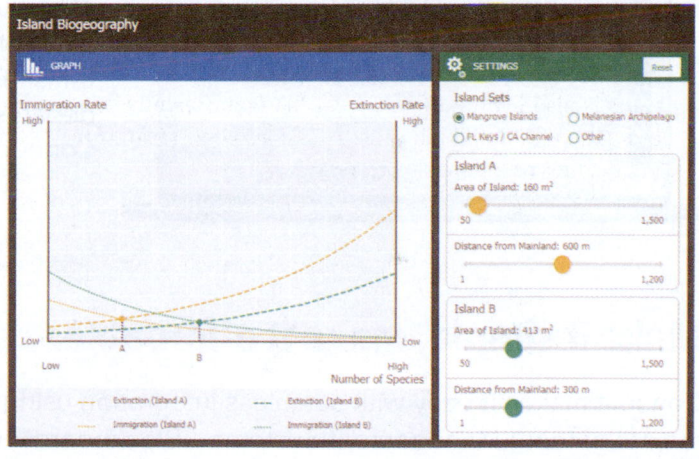

## Unpacking the Concepts

We've taken problem solving a step further. In each chapter, two higher-level Bloom's questions in the question and test banks are broken down according to the steps in the detailed feedback.

 **connect**®

Students—study more efficiently, retain more and achieve better outcomes. Instructors—focus on what you love—teaching.

# SUCCESSFUL SEMESTERS INCLUDE CONNECT

## FOR INSTRUCTORS

### You're in the driver's seat.

Want to build your own course? No problem. Prefer to use our turnkey, prebuilt course? Easy. Want to make changes throughout the semester? Sure. And you'll save time with Connect's auto-grading too.

**65%**

**Less Time Grading**

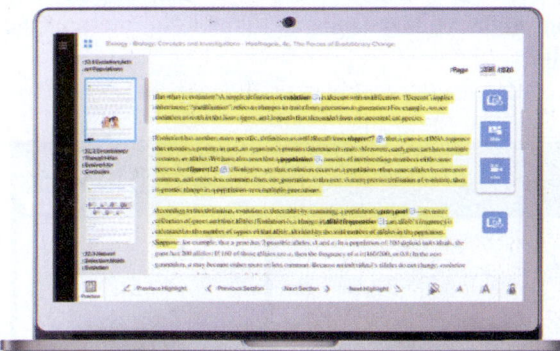

### They'll thank you for it.

Adaptive study resources like SmartBook® help your students be better prepared in less time. You can transform your class time from dull definitions to dynamic debates. Hear from your peers about the benefits of Connect at **www.mheducation.com/highered/connect**

### Make it simple, make it affordable.

Connect makes it easy with seamless integration using any of the major Learning Management Systems—Blackboard®, Canvas, and D2L, among others—to let you organize your course in one convenient location. Give your students access to digital materials at a discount with our inclusive access program. Ask your McGraw-Hill representative for more information.

### Solutions for your challenges.

A product isn't a solution. Real solutions are affordable, reliable, and come with training and ongoing support when you need it and how you want it. Our Customer Experience Group can also help you troubleshoot tech problems—although Connect's 99% uptime means you might not need to call them. See for yourself at **status.mheducation.com**

## Effective, efficient studying.

Connect helps you be more productive with your study time and get better grades using tools like SmartBook, which highlights key concepts and creates a personalized study plan. Connect sets you up for success, so you walk into class with confidence and walk out with better grades.

©Shutterstock/wavebreakmedia

> **"** I really liked this app—it made it easy to study when you don't have your textbook in front of you. **"**
>
> - Jordan Cunningham,
> Eastern Washington University

## Study anytime, anywhere.

Download the free ReadAnywhere app and access your online eBook when it's convenient, even if you're offline. And since the app automatically syncs with your eBook in Connect, all of your notes are available every time you open it. Find out more at **www.mheducation.com/readanywhere**

## No surprises.

The Connect Calendar and Reports tools keep you on track with the work you need to get done and your assignment scores. Life gets busy; Connect tools help you keep learning through it all.

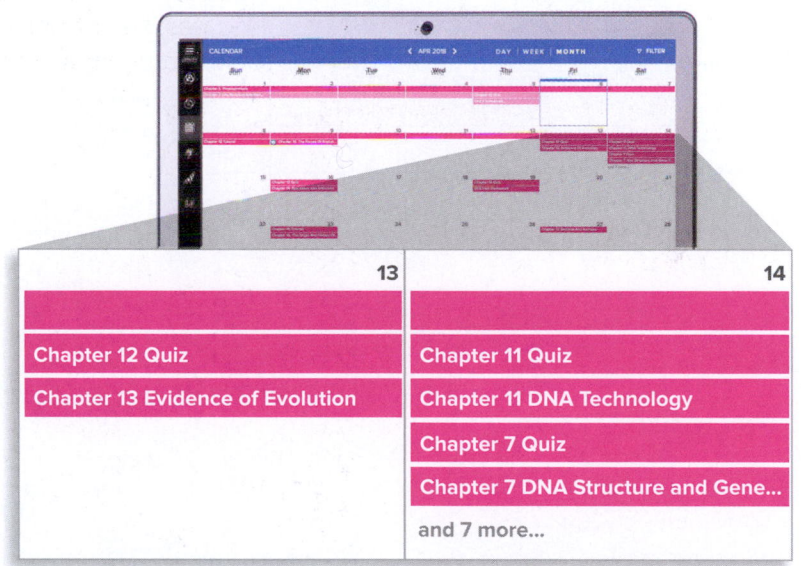

| | 13 | | 14 |
|---|---|---|---|
| Chapter 12 Quiz | | Chapter 11 Quiz | |
| Chapter 13 Evidence of Evolution | | Chapter 11 DNA Technology | |
| | | Chapter 7 Quiz | |
| | | Chapter 7 DNA Structure and Gene... | |
| | | and 7 more... | |

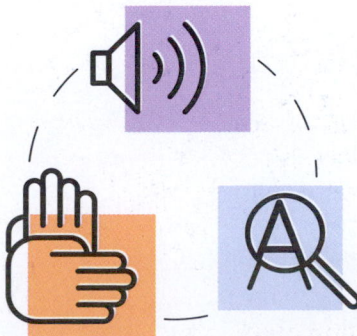

## Learning for everyone.

McGraw-Hill works directly with Accessibility Services Departments and faculty to meet the learning needs of all students. Please contact your Accessibility Services office and ask them to email accessibility@mheducation.com, or visit **www.mheducation.com/about/accessibility.html** for more information.

# Contents

## UNIT I   Chemistry

©Dr. Parvinder Sethi

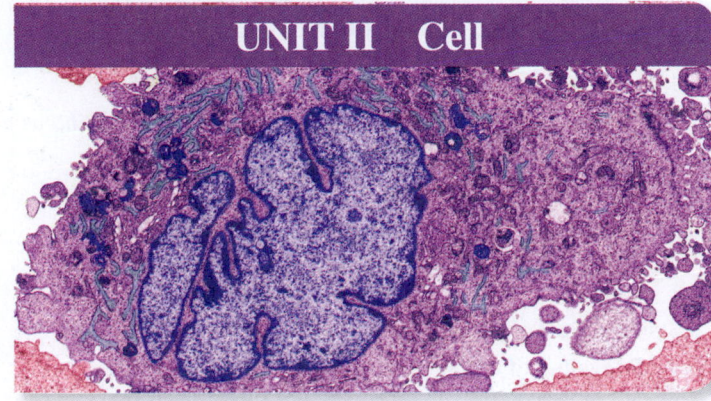

## UNIT II   Cell

©Steve Gschmeissner/Science Source

## UNIT III   Genetics

©Pieter Van De Vijverl/Science Photo Library/Corbis

## Chapter 11

### Nucleic Acid Structure, DNA Replication, and Chromosome Structure 220

### UNIT IV  Evolution

©Mark Dadswell/Getty Images

## UNIT V   Diversity

©Dr. Jeremy Burgess/SPL/Science Source

## Chapter 27

### Archaea and Bacteria   561

## Chapter 28

### Protists   581

## Chapter 29

### Fungi   605

## Chapter 30

### Microbiomes: Microbial Systems On and Around Us   622

## Chapter 31

### Plants and the Conquest of Land   641

## Chapter 32

### The Evolution and Diversity of Modern Gymnosperms and Angiosperms  664

## Chapter 33

### An Introduction to Animal Diversity  686

## Chapter 34

### The Invertebrates  701

## Chapter 35

### The Vertebrates  734

UNIT VI  Flowering Plants

©Linda Graham

## Chapter 36

### An Introduction to Flowering Plant Form and Function  760

©John Rowley/Getty Images

**UNIT VII   Animals**

## UNIT VIII  Ecology

©Dante Fenolio/Science Source

## Chapter 55

## Behavioral Ecology  1180

## Chapter 56

## Population Ecology  1201

## Chapter 57

## Species Interactions  1217

## Chapter 58

## Communities and Ecosystems: Ecological Organization on Large Scales  1236

## Chapter 59

## The Age of Humans  1257

## Chapter 60

## Biodiversity and Conservation Biology  1280

# An Introduction to Biology

# 1

**The giraffe, genus *Giraffa*.** Giraffes, which are found in Africa, are the tallest living terrestrial animals. They are members of the genus *Giraffa*. Until recently, biologists thought that all giraffes belonged to a single species. As discussed later in this chapter, that view may be changing as a result of analyses of genetic features of giraffes from different regions of Africa. ©Robert Muckley/Getty Images

**B**iology is the study of life. The diverse forms of life found on Earth provide biologists with an amazing array of organisms to study. In many cases, the investigation of living things leads to discoveries that no one would have imagined. For example, researchers determined that the venom from certain poisonous snakes contains a chemical that lowers blood pressure in humans. By analyzing that chemical, scientists developed drugs to treat high blood pressure (**Figure 1.1**).

Biologists have discovered that plants can communicate with each other. For example, the beautiful umbrella thorn acacia (*Vachellia tortillis*), shown in **Figure 1.2**, emits volatile organic molecules when it is attacked by herbivores. These molecules warn other nearby acacia trees that herbivores are in the area, and those trees release toxins to protect themselves.

Another interesting example of a biological discovery is a seemingly bizarre phenomenon known as **zombie parasites**. As you may know, zombies are fictional creatures featured in some horror and fantasy novels and movies, where they appear as dead

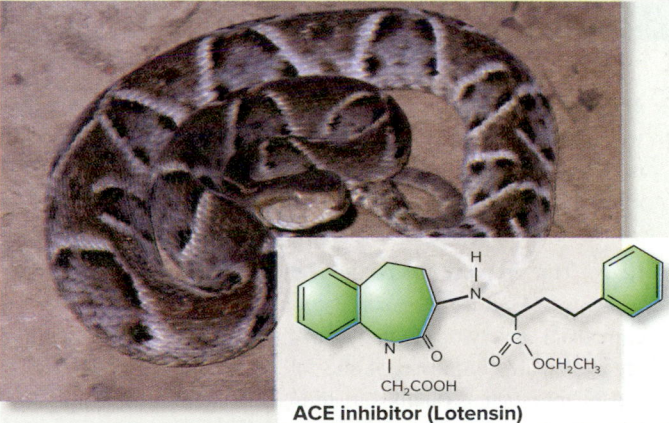

**ACE inhibitor (Lotensin)**

**Figure 1.1**  **The Brazilian arrowhead viper and an inhibitor of high blood pressure.** Derivatives of a chemical, called an angiotensin-converting enzyme (ACE) inhibitor, are found in the venom of the Brazilian arrowhead viper and are commonly used to treat high blood pressure. ©Francois Gohier/Science Source

**Figure 1.2**  **Plant communication.** If attacked by herbivores, this acacia tree will emit molecules that will warn other acacia trees in the area. ©Mark Snodgrass/Getty Images

| Table 1.1 | Examples of Zombie Parasites | |
|---|---|---|
| **Host** | **Parasite** | **Description** |
| House cricket (*Acheta domesticus*) | Horsehair worm (*Paragordius varius*) | A horsehair worm larva infects a cricket and grows inside it. The cricket is terrestrial, but the adult stage of the horsehair worm is aquatic. When the larva matures into an adult, it alters the behavior of the cricket, causing it to jump into the nearest body of water! As the cricket drowns, an adult horsehair worm emerges. |
| Spider (*Plesiometa argyra*) | Wasp (*Hymenoepimecis argyraphaga*) | A female wasp glues an egg onto a spider's body. After the egg develops into a larva, the larva pokes a few holes in the spider's abdomen, which allows it to suck the spider's blood and also to transfer chemicals into the spider, which control its behavior. The spider stops building its normal orb-shaped web and starts building a web whose geometry is strikingly different: The new web is designed to suspend the larva's cocoon in the air, where it will be protected from predators. |
| Various vertebrates, including mice and rats | Protozoan (*Toxoplasma gondii*) | *Toxoplasma gondii* is a parasite whose life cycle involves more than one vertebrate host. The definitive host is the cat, which is where *T. gondii* becomes mature and reproduces sexually. An intermediate host can be any of a variety of vertebrates, including mice and rats, which can ingest the parasite from cat feces. In the intermediate host, the parasite develops and reproduces asexually. To escape an intermediate host, such as a mouse or rat, and move to the definitive host, *T. gondii* dramatically alters the host's behavior. The infected animal becomes attracted to the smell of cat urine! This makes it more likely to be eaten by a cat and thereby allows *T. gondii* to enter its definitive host and mature. |

creatures that are able to move because of some magical force. A zombie parasite is a parasite that infects its host and is then able to control the host's behavior. A relatively small group of researchers have begun to investigate this phenomenon, and their work has spawned a new field called **neuroparasitology**—the study of how parasites control the nervous systems of their hosts. During the past few decades, researchers have discovered many examples of zombie parasites. A few are described in **Table 1.1**.

These are but a few of the many discoveries that make biology an intriguing discipline. The study of life not only reveals the fascinating characteristics of living species but also leads to the development of medicines and research tools that benefit the lives of people.

To make new discoveries, biologists view life from many different perspectives: What is the composition of living things? How is life organized? How do organisms reproduce? Sometimes the questions posed by biologists are fundamental and even philosophical in nature: How did living organisms originate? Can we live forever? What is the physical basis for memory? Can we save endangered species?

Future biologists will continue to make important advances. Biologists are scientific explorers looking for answers to some of life's most enduring mysteries. Unraveling these mysteries presents exciting challenges to the best and brightest minds. The rewards of a career in biology include the excitement of forging into uncharted territory, the thrill of making discoveries that can improve the health and lives of people, and the satisfaction of trying to preserve the environment and protect endangered species. For these and many other compelling reasons, students seeking challenging and rewarding careers may wish to choose biology as a lifelong pursuit.

In this chapter, we will begin by examining the levels of biology and the core concepts that are common to all forms of life. One of those core concepts is evolution, which is discussed in greater depth in Section 1.3. We then explore the general approaches that scientists follow when making new discoveries. Finally, we will consider the skills that students need to develop as they pursue careers in this exciting discipline and the ways in which this textbook fosters those skills.

## 1.1 Levels of Biology

### Learning Outcome:

1. Explain how life can be viewed at different levels of biological complexity.

Let's begin our journey through the wonderful world of biology by considering how life is organized. The term **organism** can be applied to all forms of life. Organisms maintain an internal order that is separated from the environment. The complexity of living organisms can be analyzed at different levels, starting with the smallest level of organization and progressing to levels that are physically much larger and more complex. **Figure 1.3** depicts a biologist's view of the levels of biological organization.

1. *Atoms.* An **atom** is the smallest unit of an element that has the chemical properties of the element. All matter is composed of atoms.

2. *Molecules and macromolecules.* As discussed in Unit I, atoms bond with each other to form **molecules**. A polymer such as a polypeptide is formed of many molecules bonded together and is called a **macromolecule**. Carbohydrates, proteins, and nucleic acids (DNA and RNA) are important macromolecules found in living organisms.

3. *Cells.* The simplest unit of life is the **cell**, which we will examine in Unit II. A cell is surrounded by a membrane and contains a variety of molecules and macromolecules. Unicellular organisms are composed of one cell, whereas multicellular organisms, such as plants and animals, contain many cells.

4. *Tissues.* In multicellular organisms, many cells of the same type associate with each other to form **tissues**. An example is muscle tissue.

5. *Organs.* In complex multicellular organisms, an **organ** is composed of two or more types of tissue. For example, the heart is composed of several types of tissues, including muscle, nervous, and connective tissue.

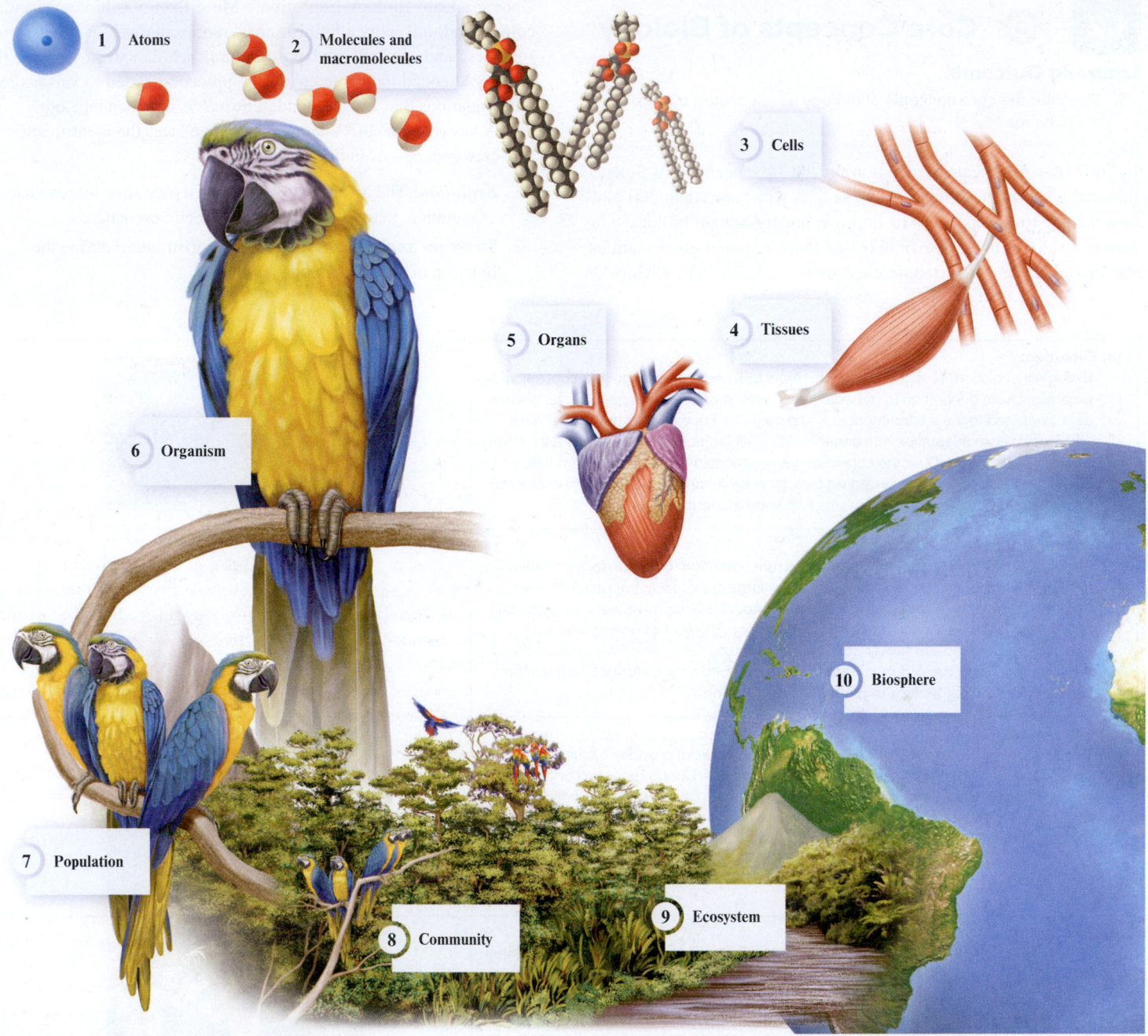

**Figure 1.3** The levels of biological organization.

Concept Check: *At which level of biological organization would you place a herd of buffalo?*

6. *Organism.* All living things can be called **organisms**. Biologists classify organisms as belonging to a particular **species**, which is a related group of organisms that share a distinctive form and set of attributes in nature. The members of the same species are closely related genetically. In Units VI and VII, we will examine plants and animals at the level of cells, tissues, organs, and complete organisms.

7. *Population.* A group of organisms of the same species that occupy the same environment is called a **population**.

8. *Community.* A biological **community** is an assemblage of populations of different species. The types of species found in

a community are determined by the environment and by the interactions of species with each other.

9. *Ecosystem.* Researchers may extend their work beyond living organisms and also study the physical environment. Ecologists analyze **ecosystems**, which are formed by interactions of a community of organisms with their physical environment. Unit VIII considers biological organization from populations to ecosystems.

10. *Biosphere.* The **biosphere** includes all of the places on the Earth where living organisms exist. Life is found in the air, in bodies of water, on the land, and in the soil.

## 1.2  Core Concepts of Biology

**Learning Outcome:**

1. Describe the core concepts of biology as advocated by "Vision and Change."

In 2007, the American Association for the Advancement of Science initiated a series of regional conversations with more than 200 biology faculty to discuss how to improve undergraduate biology education. In 2009, using the findings of these regional conversations, the organization held a conference called "Vision and Change in Undergraduate Biology Education." More than 500 biology faculty, college and university administrators, representatives of professional societies, and students and postdoctoral scholars from around the country attended the conference. The proceedings led to various recommendations that can be found at http://visionandchange.org.

A key outcome of "Vision and Change" was the identification of five core concepts of biology (**Figure 1.4**):

1. **Evolution:** The diversity of life evolved over time by processes of mutation, natural selection, and genetic exchange.

2. **Structure and function:** Basic units of structure define the function of all living things.

| | |
|---|---|
| **(a) Evolution:**<br>Biological evolution, or simply evolution, refers to a heritable change in a population of organisms from generation to generation. As a result of evolution, populations become better adapted to the environment in which they live. For example, the long snout of an anteater is an adaptation that enhances its ability to obtain food, namely ants, from hard-to-reach places. Over the course of many generations, the fossil record indicates that the long snout occurred via biological evolution in which modern anteaters evolved from populations of organisms with shorter snouts. |  |
| **(b) Structure and function:**<br>Biologists often say "structure determines function." This core concept pertains to very tiny biological molecules and to very large biological structures. The feet of different birds provide a striking example. Aquatic birds have webbed feet that function as paddles for swimming. By comparison, the feet of nonaquatic birds are not webbed and are better adapted for grasping food, perching on branches, and running along the ground. The structure of a bird's feet, webbed versus non-webbed, is a critical feature that affects their function. |  |
| **(c) Information:**<br>Genetic material composed of DNA (deoxyribonucleic acid) provides a blueprint for the organization, development, and function of living things. During reproduction, a copy of this blueprint is transmitted from parents to offspring. DNA is heritable, which means that offspring inherit DNA from their parents. A key feature of reproduction is that offspring tend to have characteristics that greatly resemble those of their parent(s). As seen here, this mother dolphin and her offspring have strikingly similar features. |  |
| **(d) Energy and matter:**<br>All living organisms acquire energy and matter from the environment and use them to synthesize essential molecules and maintain the organization of their cells and bodies. These sunflower plants carry out photosynthesis in which they capture light energy and acquire carbon dioxide and water, thereby allowing them to make carbohydrates. This process provides energy and organic molecules so the plants can grow and produce beautiful flowers. |  |
| **(e) Systems:**<br>When the parts of an organism interact with each other or with the external environment to create novel structures and functions, the resulting characteristics are called emergent properties. For example, the human eye is composed of many different types of cells that are organized to sense incoming light and transmit signals to the brain. Our ability to see is an emergent property of this complex arrangement of different cell types. Biologists use the term systems biology to describe the study of how new properties of life arise by complex interactions of its individual parts. |  |

 **Figure 1.4  Core concepts of biology, as advocated by "Vision and Change."**  These core concepts will be emphasized throughout this textbook. a: ©Lucas Leuzinger/Shutterstock; b: ©G.K. & Vikki Hart/Getty Images; c: ©Image Source/Getty Images; d: Source: Photo by Bruce Fritz, USDA-ARS; e: ©Maria Teijeiro/Getty Images

3. *Information flow, exchange, and storage:* The growth and behavior of organisms are activated through the expression of genetic information.

4. *Pathways and transformations of energy and matter:* Biological systems grow and change via processes that are based on chemical transformation pathways and are governed by the laws of thermodynamics.

5. *Systems:* Living systems are interconnected and interacting.

A key goal of this textbook is to bring to life these five core concepts of biology. These concepts will be highlighted in each chapter with a "Vision and Change" icon, , which indicates subsections and figures that focus on one or more of these five core concepts.

## 1.3  Biological Evolution

**Learning Outcomes:**

1. Explain two mechanisms by which evolutionary change occurs: vertical descent with mutation and horizontal gene transfer.
2. Describe how changes in genomes and proteomes underlie evolutionary changes.

Unity and diversity are two words that often are used to describe the living world. All modern forms of life display a common set of characteristics that distinguish them from nonliving objects. In this section, we will explore how this unity of common traits is rooted in the phenomenon of **biological evolution**, or simply **evolution**, which is a heritable change in a population of organisms from one generation to the next. Life on Earth is united by an evolutionary past in which modern organisms have evolved from populations of pre-existing organisms. This unity is a core concept of biology.

However, evolutionary unity does not mean that organisms are exactly alike. The Earth has many different types of environments, ranging from tropical rain forests to salty oceans, hot and dry deserts, and cold mountaintops. Diverse forms of life have evolved in ways that help them prosper in the different environments the Earth has to offer. In this and the following section, we will begin to examine the unity and diversity that exists within the biological world.

### Modern Forms of Life Are Connected by an Evolutionary History

Life began on Earth as primitive cells about 3.5–4 billion years ago (bya). Since that time, populations of living organisms have undergone evolutionary changes that ultimately gave rise to the species we see today. Understanding the evolutionary history of species can provide key insights into the structure and function of an organism's body, because evolutionary change frequently involves modifications of characteristics in pre-existing populations. Over long periods of time, populations may change so that structures with a particular function become modified to serve a new function. For example, the wing of a bat is used for flying, and the flipper of a dolphin is used for swimming. Evidence from

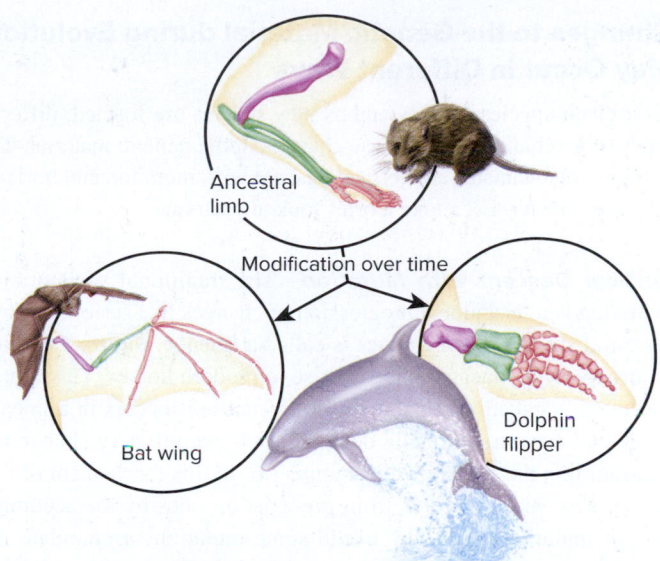

**Figure 1.5  An example of a modification that has occurred as a result of biological evolution.** The wing of a bat and the flipper of a dolphin are modifications of a limb that was used for walking in a pre-existing ancestor.

**Core Concepts: Evolution, Structure and Function**
Via evolution, the different structures of the front limbs seen here result in functions that are best suited for these organisms.

the fossil record indicates that both structures were modified from a front limb that was used for walking in a pre-existing ancestor (**Figure 1.5**).

### Evolutionary Change Involves Changes in the Genetic Material

The example shown in Figure 1.5 represents evolution at the macroscopic level. At the molecular level, evolution involves changes in the genetic material, which is composed of **DNA (deoxyribonucleic acid)**. DNA provides a blueprint for the organization, development, and function of living things. During reproduction, a copy of this blueprint is transmitted from parent to offspring. DNA is **heritable**, which means that offspring inherit DNA from their parents.

As discussed in Unit III, **genes**, which are segments of DNA, govern the characteristics, or traits, of organisms. Most genes are transcribed into a type of **RNA (ribonucleic acid)** molecule called messenger RNA (mRNA), which is then translated into a **polypeptide** with a specific amino acid sequence. A **protein** is composed of one or more polypeptides. The structures and functions of proteins play a key role in determining the traits of organisms.

On relatively rare occasions, changes may occur in DNA. A **mutation** is a heritable change in the genetic material—one that can be passed from cell to cell or from parent to offspring. Mutations can alter the properties of genes and thereby affect the characteristics of the offspring that inherit them. With regard to survival, mutations can be beneficial, detrimental, or neutral. As described next, changes in the genetic material underlie the process of evolution.

## Changes to the Genetic Material during Evolution May Occur in Different Ways

As a given species evolves and as new species are formed, different types of mechanisms may cause changes in the genetic material. Two common mechanisms are vertical descent with mutation and horizontal gene transfer. Let's take a brief look at each one.

***Vertical Descent with Mutation***    The traditional way to study evolution is to examine a progression of changes in a series of related ancestral species. Such a series is called a **lineage**. **Figure 1.6** shows a portion of the lineage that gave rise to modern horses. This type of evolution is called **vertical evolution** because it occurs in a lineage. Biologists have traditionally depicted such evolutionary change in a diagram like the one shown in Figure 1.6. In this mechanism of evolution, new species evolve from pre-existing ones by the accumulation of mutations. But why would some mutations accumulate in a population and eventually change the characteristics of an entire species? One reason is that a mutation may alter the traits of organisms

in a way that increases their chances of survival and reproduction. When a mutation causes such a beneficial change, the frequency of the mutation may increase in a population from one generation to the next, a process called **natural selection**. This topic is discussed in Units IV and V. Evolution also involves the accumulation of neutral changes that do not benefit or harm a species, and evolution sometimes involves rare changes that may be harmful.

With regard to the horses shown in Figure 1.6, the fossil record has revealed adaptive changes in various traits such as size and tooth morphology. The first horses were the size of dogs, whereas modern horses typically weigh more than a half ton. The teeth of *Hyracotherium* were relatively small compared with those of modern horses. Over the course of millions of years, horses' teeth have increased in size, and a complex pattern of ridges has developed on the molars. How do evolutionary biologists explain these changes in horse characteristics? They can be attributed to natural selection, in which changing global climates favored the survival and reproduction of horses with certain types of traits. Over North America, where much of horse evolution occurred, large areas changed from dense forests to grasslands.

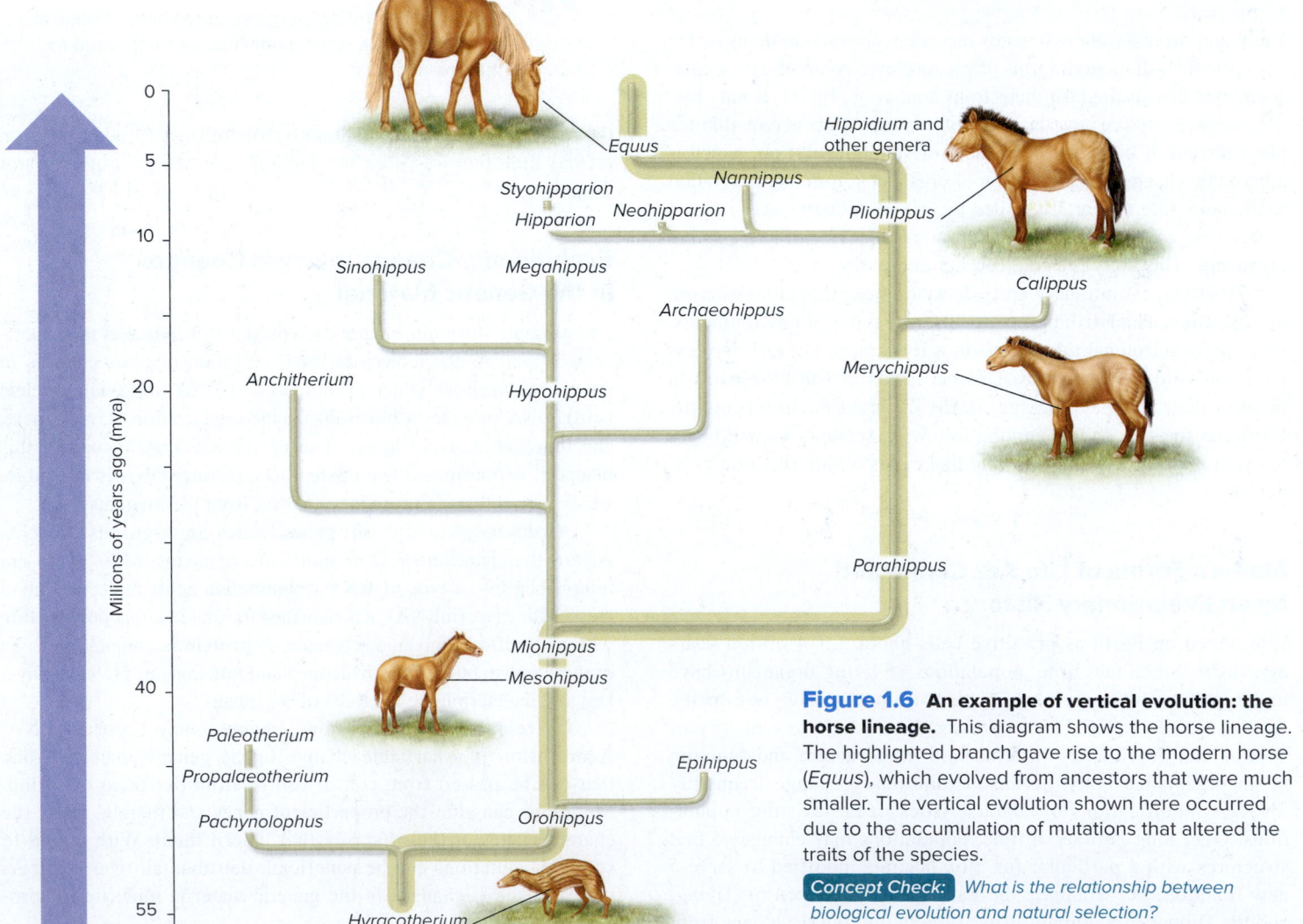

**Figure 1.6  An example of vertical evolution: the horse lineage.**  This diagram shows the horse lineage. The highlighted branch gave rise to the modern horse (*Equus*), which evolved from ancestors that were much smaller. The vertical evolution shown here occurred due to the accumulation of mutations that altered the traits of the species.

**Concept Check:**  *What is the relationship between biological evolution and natural selection?*

Horses with genetic variation that made them larger were more likely to escape predators and to be able to travel greater distances in search of food. The changes seen in horses' teeth are consistent with a dietary shift from eating tender leaves to eating grasses and other types of vegetation that are more abrasive and require more chewing.

***Horizontal Gene Transfer***   The most common way for genes to be transferred is in a vertical manner. This can involve the transfer of genetic material from a mother cell to daughter cells, or it can occur via gametes—sperm and egg—that unite to form a new organism. However, as discussed in later chapters, genes are sometimes transferred between organisms by other mechanisms. These other mechanisms are collectively known as **horizontal gene transfer**, which is the transfer of genetic material from one organism to another organism that is not its offspring. In some cases, horizontal gene transfer can occur between members of different species. For example, you may have heard in the news media that resistance to antibiotics among bacteria is a growing medical problem. As discussed in Chapter 19, genes that confer antibiotic resistance are sometimes transferred between different bacterial species (**Figure 1.7**).

Genes transferred horizontally may be subject to natural selection and promote changes in an entire species. This has been an important mechanism of evolutionary change, particularly among bacterial species. In addition, during the early stages of evolution, which occurred a few billion years ago, horizontal gene transfer was an important part of the process that gave rise to all modern species.

Traditionally, biologists have described evolution using diagrams that depict the vertical evolution of species on a long time scale. This type of evolutionary tree was shown earlier in Figure 1.6. For many decades, a simplistic view held that all living organisms evolved from a common ancestor, resulting in a "tree of life" that depicted the vertical evolution that gave rise to all modern species. Now that we understand the great importance of horizontal gene transfer in the evolution of life on Earth, biologists have re-evaluated the way evolution has occurred over time. Rather than a tree of life, a more appropriate way

to view the unity of living organisms is as a "web of life," as shown in **Figure 1.8**, which accounts for both vertical descent and horizontal gene transfer. In a lineage in which the time scale is depicted on a vertical axis, horizontal gene transfer between different species is shown as a horizontal line.

## Core Concept: Evolution

### The Study of Genomes and Proteomes Provides an Evolutionary Foundation for Our Understanding of Biology

As we have seen, evolutionary unity is a core concept of biology. We can understand the unity of modern organisms by realizing that all living species evolved from an interrelated group of ancestors. However, from an experimental perspective, this realization presents a dilemma—we cannot take a time machine back over the course of 4 billion years to carefully study the characteristics of extinct organisms and fully appreciate the series of changes that have led to modern species. Fortunately, though, evolution has given biologists some wonderful puzzles to study, including the fossil record and the genomes of modern species.

The term **genome** refers to the complete genetic composition of an organism or species (**Figure 1.9a**). The genomes of bacteria and archaea usually contain a few thousand genes, whereas those of eukaryotes may contain tens of thousands. A genome is critical to life because it performs these functions:

- *Stores information in a stable form:* The genome of every organism stores information that provides a blueprint for producing that organism's characteristics.

- *Provides continuity from generation to generation:* The genome is copied and transmitted from generation to generation.

- *Acts as an instrument of evolutionary change:* Every now and then, the genome undergoes a mutation that may alter the characteristics of an organism. In addition, a genome may acquire new genes by horizontal gene transfer. The accumulation of genome changes from generation to generation produces the evolutionary changes that alter species and produce new species.

An exciting advance in biology over the past couple of decades has been the ability to analyze the DNA sequence of genomes, a technology called **genomics**. For example, a researcher can compare the genomes of a frog, a giraffe, and a petunia and discover intriguing similarities and differences. These comparisons help us to understand how new traits evolved. All three types of organisms have the same kinds of genes needed for the breakdown of nutrients such as sugars. In contrast, only the petunia has genes that allow it to carry out photosynthesis. Also, genomics helps us to understand evolutionary relationships. As discussed later in this chapter, researchers analyzed the genomes of giraffes across Africa and concluded that they constitute four distinct species.

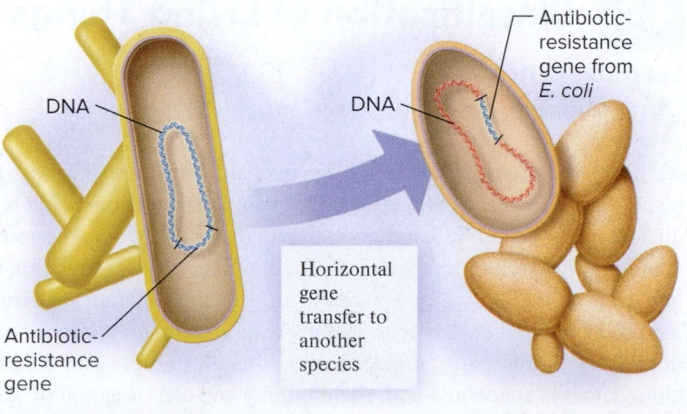

DNA

Antibiotic-resistance gene

DNA

Antibiotic-resistance gene from *E. coli*

Horizontal gene transfer to another species

Bacterial species such as *Escherichia coli*

Bacterial species such as *Streptococcus pneumoniae*

**Figure 1.7  An example of horizontal gene transfer: antibiotic resistance.** One bacterial species may transfer a gene, such as a gene that confers resistance to an antibiotic, to another bacterial species.

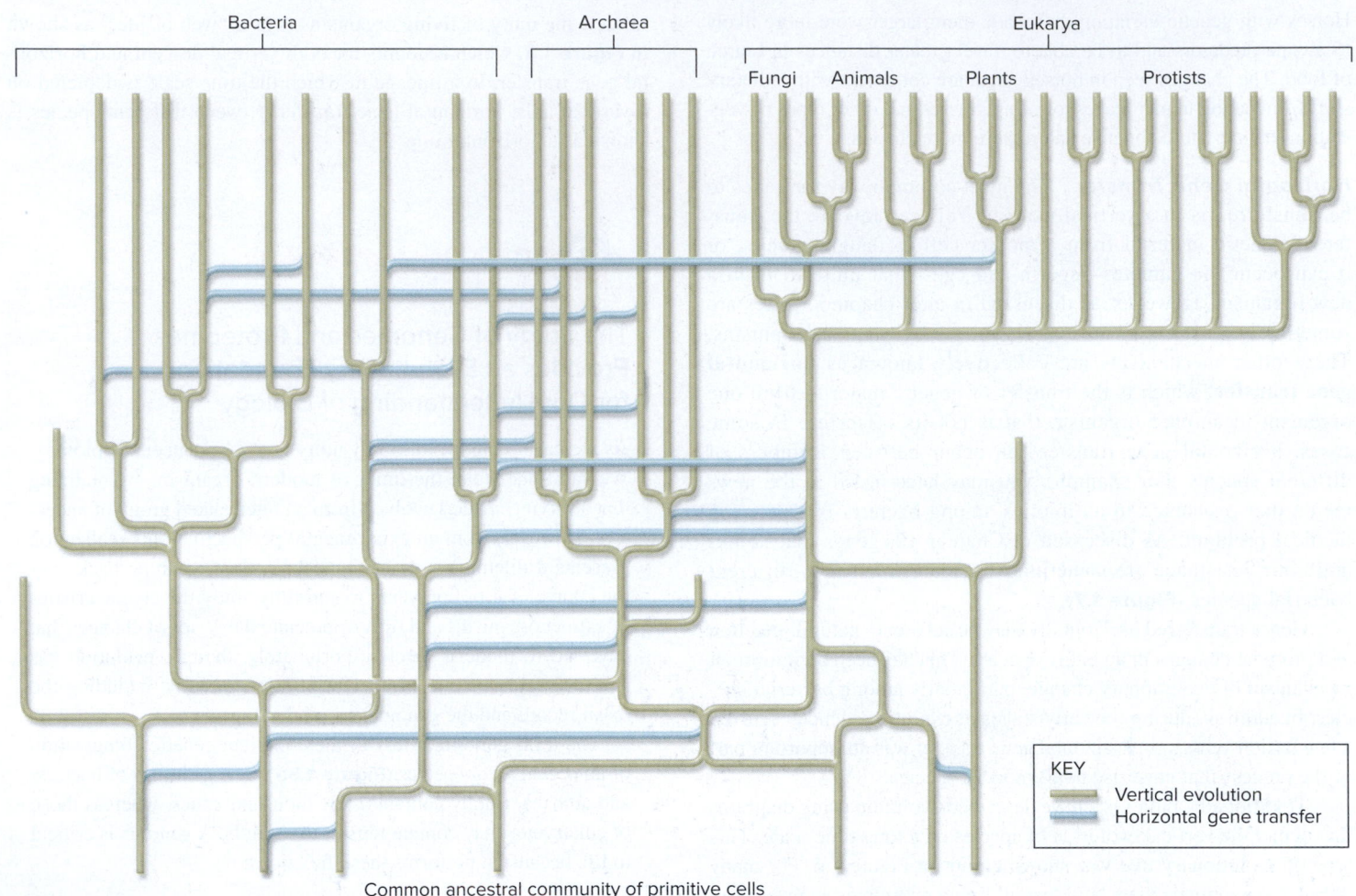

**Figure 1.8**  **The web of life, showing both vertical descent and horizontal gene transfer.**  This diagram includes both of these important mechanisms in the evolution of life on Earth. Note: Archaea are unicellular species that are similar in cell structure to bacteria.

*Concept Check:*  *How does the concept of a tree of life differ from that of a web of life?*

An extension of genome analysis is the study of the **proteome**, which refers to all of the proteins that a cell or organism makes. The function of most genes is to encode polypeptides that become units in proteins. As shown in **Figure 1.9b**, these include proteins that form a cytoskeleton and proteins that function in cell organization and as enzymes, transport proteins, cell-signaling proteins, and extracellular proteins. The genome of each species carries the information to make its proteome—the hundreds or thousands of proteins that each cell of that species makes. Proteins are largely responsible for the structures and functions of cells and organisms. The set of techniques known as **proteomics** allows researchers to analyze the proteome of a single species and compare the proteomes of different species. Proteomics helps us understand how the various levels of biology are related to one another, from the molecular level—at the level of protein molecules—to higher levels, such as how the functioning of proteins produces the characteristics of cells and organisms and affects the ability of populations of organisms to survive in their natural environments.

## 1.4  Classification of Living Things

### Learning Outcome:

1. Outline how organisms are classified.

As biologists study species and discover new species, they try to place them into groups based on their evolutionary history. This is a difficult task because researchers estimate that the Earth has between 5 and 50 million different species! The rationale for classification is based on vertical descent. Species with a recent common ancestor are grouped together, whereas species whose common ancestor was in the very distant past are placed into different groups. The field of biology that is concerned with the grouping and classification of species is termed **taxonomy**.

Why is taxonomy useful? First, taxonomy allows use to appreciate the amazing diversity of life on Earth. Also, because taxonomy is based on evolution, it provides a view of the evolutionary relationships among living species, and between living and extinct species.

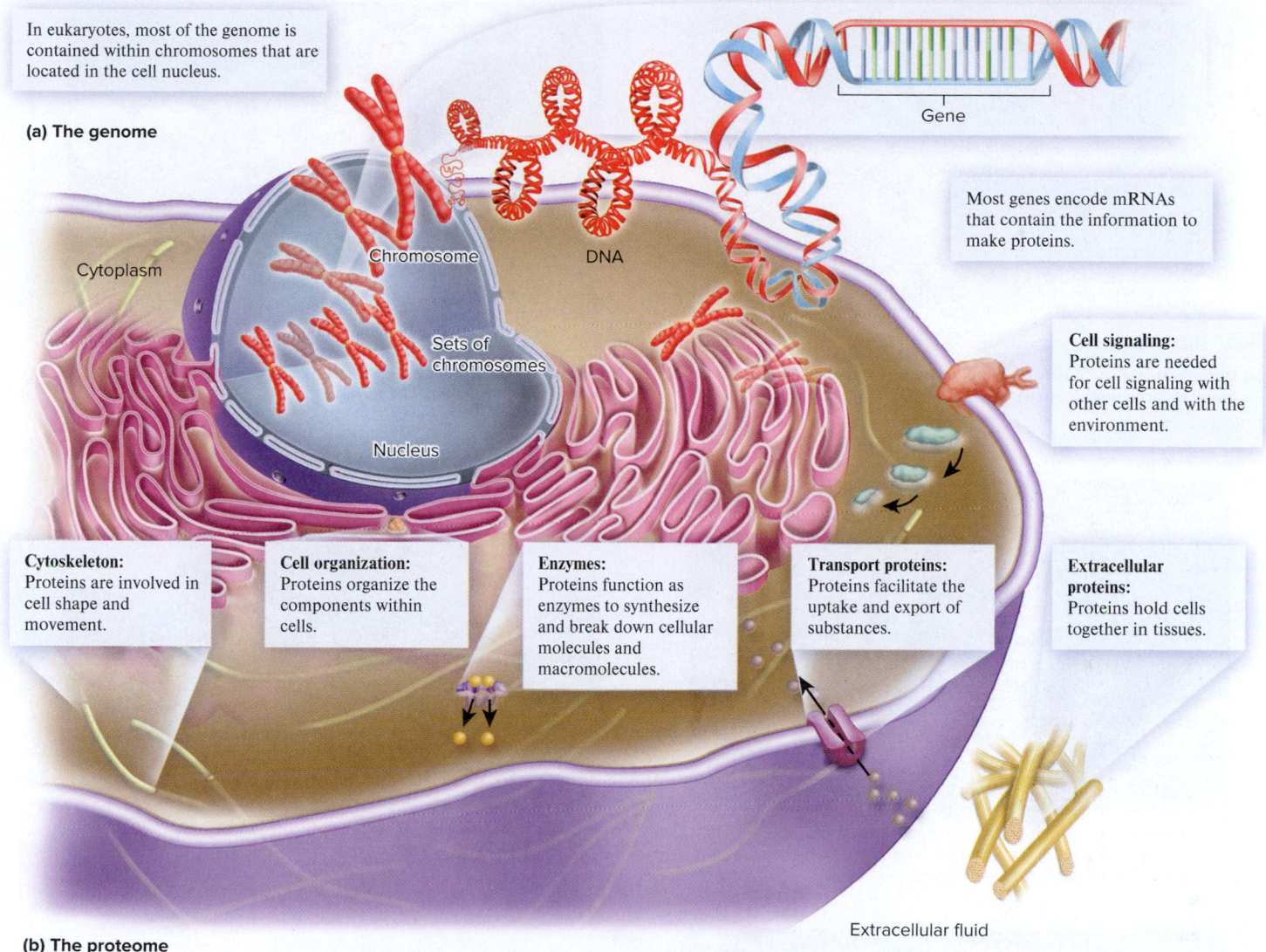

In eukaryotes, most of the genome is contained within chromosomes that are located in the cell nucleus.

**(a) The genome**

Gene

Most genes encode mRNAs that contain the information to make proteins.

Cytoplasm

Chromosome     DNA

Sets of chromosomes

Nucleus

**Cell signaling:** Proteins are needed for cell signaling with other cells and with the environment.

**Cytoskeleton:** Proteins are involved in cell shape and movement.

**Cell organization:** Proteins organize the components within cells.

**Enzymes:** Proteins function as enzymes to synthesize and break down cellular molecules and macromolecules.

**Transport proteins:** Proteins facilitate the uptake and export of substances.

**Extracellular proteins:** Proteins hold cells together in tissues.

**(b) The proteome**

Extracellular fluid

**Figure 1.9   Genomes and proteomes.   (a)** The genome, which is composed of DNA, is the entire genetic composition of an organism. Most of the genetic material in eukaryotic cells is found in the cell nucleus. The primary function of the genome is to encode the proteome **(b)**, which is the entire protein complement of a cell or organism. Six general categories of proteins are illustrated. Proteins are largely responsible for the structure and function of cells and organisms.

**Concept Check:** *Biologists sometimes say that the genome is the storage unit of life, whereas the proteome is largely the functional unit of life. Explain this statement.*

## The Classification of Living Organisms Allows Biologists to Appreciate the Unity and Diversity of Life

Let's first consider taxonomy on a broad scale. You may have noticed that Figure 1.8 showed three main groups of organisms. From an evolutionary perspective, all forms of life can be placed into those three large categories, or domains, called **Bacteria**, **Archaea**, and **Eukarya** (**Figure 1.10**). Bacteria and archaea are microorganisms that are also termed **prokaryotic** because their cell structure is relatively simple. At the molecular level, bacterial and archaeal cells show significant differences in their compositions. By comparison, organisms in the domain Eukarya are **eukaryotic** and have cells with internal compartments that

serve various functions. A defining distinction between prokaryotic and eukaryotic cells is that eukaryotic cells have a **cell nucleus** in which the genetic material is surrounded by a membrane.

The organisms in domain Eukarya were once subdivided into four major categories, or kingdoms, called Protista (protists), Plantae (plants), Fungi, and Animalia (animals). However, as discussed in Chapter 25 and Unit V, this traditional view became invalid as biologists gathered new information regarding the evolutionary relationships of these organisms. We now know that the protists do not form a single kingdom but instead are divided into several broad categories called supergroups.

Taxonomy involves multiple levels in which particular species are placed into progressively smaller and smaller groups whose

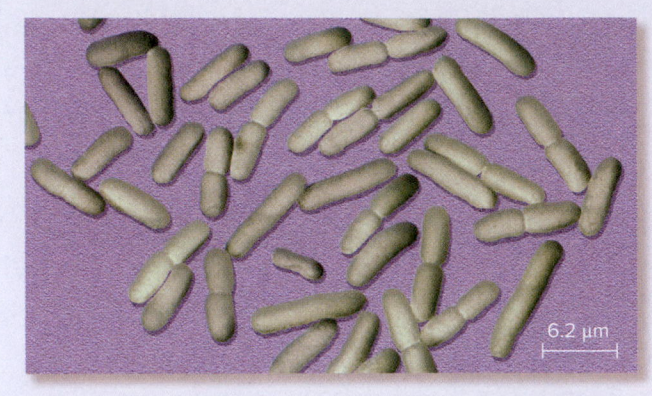

**(a) Domain Bacteria:** Mostly unicellular prokaryotes that inhabit many diverse environments on Earth.

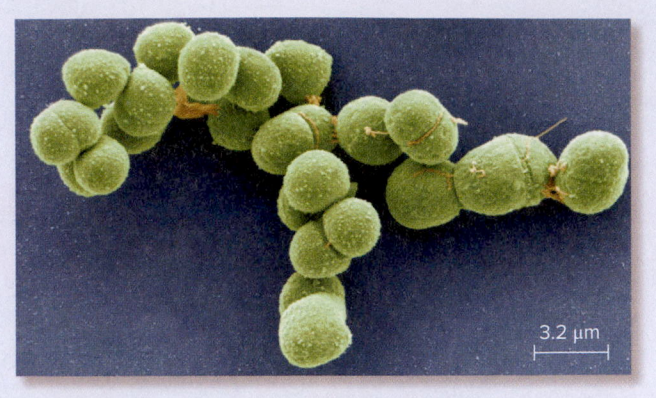

**(b) Domain Archaea:** Unicellular prokaryotes that often live in extreme environments, such as hot springs.

**Protists:** Unicellular and small multicellular organisms that are now subdivided into seven broad groups based on their evolutionary relationships.

**Plants:** Multicellular organisms that can carry out photosynthesis.

**Fungi:** Unicellular and multicellular organisms that have a cell wall but cannot carry out photosynthesis. Fungi usually survive on decaying organic material.

**Animals:** Multicellular organisms that usually have a nervous system and are capable of locomotion. They must eat other organisms or the products of other organisms to live.

**(c) Domain Eukarya:** Unicellular and multicellular organisms having cells with internal compartments that serve various functions.

**Figure 1.10** **The three domains of life.** Two of these domains, **(a)** Bacteria and **(b)** Archaea, consist of species with prokaryotic cells. The third domain, **(c)** Eukarya, comprises species that are eukaryotes. a: ©BSIP/age fotostock; b: ©Eye of Science/Science Source; c (protists): ©Jan Hinsch/Getty Images; c (plants): ©Kent Foster/Science Source; c (fungi): ©Carl Schmidt-Luchs/Science Source; c (animals): ©Ingram Publishing/age fotostock

 **Core Skill: Connections** Look ahead to Figure 25.1. Are fungi more closely related to plants or animals?

| Taxonomic group | The ocellaris clownfish is found in | Approximate time when the common ancestor for this group arose | Approximate number of modern species in this group | Examples |
|---|---|---|---|---|
| **Domain** | Eukarya | 2,000 mya | > 5,000,000 | |
| **Supergroup** | Opisthokonta | 2,000 mya | > 1,000,000 | |
| **Kingdom** | Animalia | 600 mya | > 1,000,000 | |
| **Phylum** | Chordata | 525 mya | 50,000 | |
| **Class** | Actinopterygii | 420 mya | 30,000 | |
| **Order** | Perciformes | 80 mya | 7,000 | |
| **Family** | Pomacentridae | ~ 40 mya | 360 | |
| **Genus** | *Amphiprion* | ~ 9 mya | 28 | |
| **Species** | *ocellaris* | < 3 mya | 1 | |

**Figure 1.11** **Taxonomic classification of the ocellaris clownfish.**

**Concept Check:** *Why is it useful to place organisms into taxonomic groupings?*

members are more closely related to each other evolutionarily. Such an approach emphasizes the unity and diversity of different species. As an example, let's consider the clownfish, a popular saltwater aquarium fish (**Figure 1.11**). Several species of clownfish have been identified.

One species of clownfish, which is orange with white stripes, has several common names, including ocellaris clownfish. The broadest grouping for this clownfish is the domain, namely, Eukarya, followed by progressively smaller divisions, from supergroup (Opisthokonta), to kingdom (Animalia), and eventually to species. In the animal kingdom, clownfish are part of a phylum, Chordata, the chordates, which is subdivided into classes. Clownfish are in a class called Actinopterygii, which includes all ray-finned fishes.

The common ancestor that gave rise to ray-finned fishes arose about 420 million years ago (mya). Actinopterygii is subdivided into several smaller orders. The clownfish are in the order Perciformes (bony fish). The order is, in turn, divided into families; the clownfish belong to the family of marine fish called Pomacentridae, which are often brightly colored. Families are divided into genera (singular, genus). The genus *Amphiprion* is composed of 28 different species; these are various types of clownfish. Therefore, the genus contains species that are very similar to each other in form and have evolved from a common (extinct) ancestor that lived relatively recently on an evolutionary time scale.

Biologists use a two-part description, called **binomial nomenclature**, to provide each species with a unique scientific name. The

**Figure 1.12**  **A proposal that giraffes constitute four distinct species.**  Each species has its own distinctive coat pattern. From left to right: northern giraffe (*Giraffa camelopardalis*), reticulated giraffe (*G. reticulata*), Masai giraffe (*G. tippelskirchi*), and southern giraffe (*G. giraffa*). (northern giraffe): ©NSP-RF/Alamy Stock Photo; (reticulated giraffe): ©McGraw-Hill Education; (Masai giraffe): ©iStock/Getty Images; (southern giraffe): ©Egmont Strigl/Westend61/ Getty Images

 **Core Skill: Process of Science**  The gathering and analysis of new data suggest that giraffes, which were once thought to be a single species, constitute four different species.

scientific name of the ocellaris clownfish is *Amphiprion ocellaris*. The first word is the genus, and the second word is the specific epithet, or species descriptor. By convention, the genus name is capitalized, whereas the specific epithet is not. Both names are italicized. Scientific names are usually Latinized, which means they are made similar in appearance to Latin words. The origins of scientific names are typically Latin or Greek, but they can come from a variety of sources, including a person's name.

### Taxonomy Changes as Researchers Gather Evidence Regarding the Characteristics and Genetic Composition of Organisms Located in Different Places

How do we judge if a gray squirrel living in Minnesota and another one living in California are members of the same species? As discussed in Chapter 24, biologists use different criteria to decide if similar organisms living in different places are the same species or different species. For example, they may analyze morphological features or study DNA samples. Science is a work in progress. As biologists gather new information and conduct experiments, their views often change. An interesting example involves the classification of giraffes, shown on the cover of this textbook.

Giraffes are currently classified as a single species, *Giraffa camelopardalis,* but recent studies are challenging that conclusion. In 2016, a study by Axel Janke and colleagues suggested that there are four distinct species of giraffes. This work was based on a genetic analysis of DNA samples taken from 190 giraffes across Africa. By comparing these DNA samples, the researchers concluded that giraffes should be classified as four distinct species (**Figure 1.12**).

While not all experts agree on this conclusion, this work illustrates how our perception of biological diversity can change as we gather more information.

## 1.5 Biology as a Scientific Discipline

**Learning Outcomes:**

1. Explain how researchers study biology at different levels, ranging from molecules to ecosystems.
2. **CoreSKILL »** Distinguish between discovery-based science and hypothesis testing, and describe the steps of the scientific method.

What is science? Surprisingly, the definition of science is not easy to state. Most people have an idea of what science is, but actually articulating that idea proves difficult. In biology, we can define **science** as the observation, identification, experimental investigation, and theoretical explanation of natural phenomena.

Science is conducted in different ways and at different levels. Some biologists study the molecules that compose life, and others try to understand how organisms survive in their natural environments. Experimentally, researchers often focus their efforts on **model organisms**—organisms studied by many different researchers so they can compare their results and determine scientific principles that apply more broadly to other species. Examples of model organisms include *Escherichia coli* (a bacterium), *Saccharomyces cerevisiae* (a yeast), *Drosophila melanogaster* (fruit fly), *Caenorhabditis elegans* (a nematode worm), *Mus musculus* (mouse), and *Arabidopsis thaliana* (a flowering plant). Model organisms offer experimental advantages over other species. For example, *E. coli* is a very simple organism that can be easily grown in the laboratory. By limiting their work to a few model organisms, researchers can gain a deeper understanding of these species. Importantly, the discoveries made using model organisms help us to understand how biological processes work in other species, including humans.

In this section, we will examine how biologists follow a standard approach, called the **scientific method**, to test their ideas. We will explore how scientific knowledge makes predictions that can be experimentally tested. However, not all discoveries are the result of researchers following the scientific method. Some discoveries are made simply by gathering new information. For example, as illustrated earlier, in Figure 1.1 the characterization of many living organisms has led to the development of important medicines. In this section, we will also consider how researchers often set out on "fact-finding missions" aimed at uncovering new information that may eventually lead to important discoveries in biology.

## Biologists Investigate Life at Different Levels of Organization

In Figure 1.3, we examined the various levels of biological organization. The study of these different levels depends not only on the scientific interests of biologists but also on the tools available to them. The study of organisms in their natural environments is a branch of biology called **ecology**, which considers populations, communities, and ecosystems (**Figure 1.13a**). Some researchers examine the structures and functions of plants and animals; these subjects form the disciplines called **anatomy** and **physiology** (**Figure 1.13b**). With the advent of microscopy, **cell biology**, which is the study of cells and their interactions, became an important branch of biology in the early 1900s and remains so today (**Figure 1.13c**). In the 1970s, genetic tools became available for studying single genes and the proteins they encode. This genetic technology enabled researchers to study individual molecules, such as proteins, in living cells and thereby spawned the field of **molecular biology**. Together with biochemists and biophysicists, molecular biologists focus their efforts on the structure and function of the molecules of life (**Figure 1.13d**). Such researchers want to understand how biology works at the molecular and even atomic levels. Overall, the 20th century saw a progressive increase in the number of biologists who used an approach to understanding biology called **reductionism**—reducing complex systems to simpler components as a way to understand how the system works. In biology, reductionists study the parts of a cell or organism as individual units.

In the 1990s, the pendulum began to swing in the other direction. Scientists have invented new tools that allow them to study groups of genes (genomic techniques) and groups of proteins (proteomic techniques). Biologists now use the term **systems biology** to describe research aimed at understanding how emergent properties arise. This term is often applied to the study of cells. In this context, systems biology may involve the investigation of groups of genes that encode proteins with a common purpose (**Figure 1.13e**). For example, a systems biologist may conduct experiments that try to characterize an entire cellular process, which is driven by dozens of different proteins. However, systems biology is not new. Animal

Ecologists study species in their native environments.

Anatomists and physiologists study how the structures of organisms are related to their functions.

**(a) Ecology—population/ community/ecosystem levels**

**(b) Anatomy and physiology— tissue/organ/organism levels**

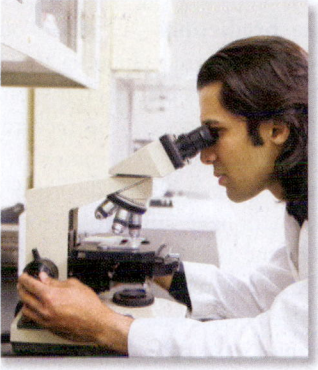
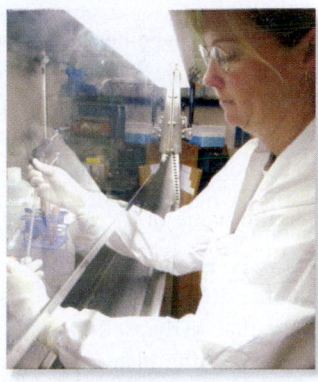

Cell biologists often use microscopes to learn how cells function.

Molecular biologists and biochemists study the molecules and macromolecules that make up cells.

**(c) Cell biology—cellular levels**

**(d) Molecular biology— atomic/molecular levels**

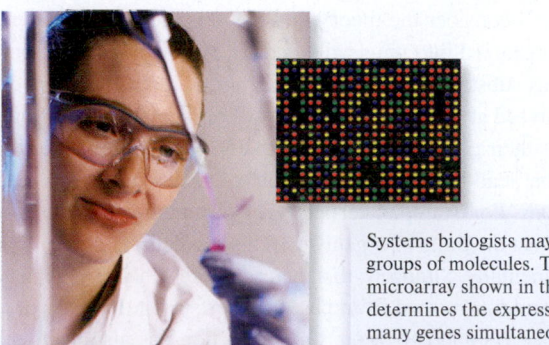

Systems biologists may study groups of molecules. The microarray shown in the inset determines the expression of many genes simultaneously.

**(e) Systems biology—all levels, shown here at the molecular level**

**Figure 1.13  Biological investigation at different levels of organization.** a: ©Diane Nelson; b: ©Purestock/SuperStock; c: ©Erik Isakson/Blend Images; d: ©Northwestern, Shu-Ling Zhou/AP Images; e: ©Andrew Brookes/Corbis/Getty Images; e (inset): ©Alfred Pasieka/Science Source

and plant physiologists have been studying the functions of complex organ systems for centuries. Likewise, ecologists have been characterizing ecosystems for a very long time. The excitement surrounding systems biology in recent years has been the result of new experimental tools that allow biologists to study complex interactions at the molecular level.

## A Hypothesis Is a Proposed Idea, Whereas a Theory Is a Broad Explanation Backed by Extensive Evidence

Let's now consider the process of science. In biology, a **hypothesis** is a proposed explanation for a natural phenomenon. It is a proposition based on previous observations or experimental studies. For example, with knowledge of seasonal changes, you might hypothesize that maple trees drop their leaves in the autumn because of the shortened amount of daylight. An alternative hypothesis might be that the trees drop their leaves because of lower temperatures. In biology, a hypothesis requires more work by researchers to evaluate its validity.

A useful hypothesis must make **predictions**—expected outcomes that can be shown to be correct or incorrect. In other words, a useful hypothesis is **testable**. If a hypothesis is incorrect, it should be **falsifiable**, which means that it can be shown to be incorrect by additional observations or experimentation. Alternatively, a hypothesis may be correct, so further work will not disprove it. In such cases, we say that the researchers have failed to reject the hypothesis. Even so, in science, a hypothesis is never really proven but rather always remains provisional. Researchers accept the possibility that perhaps they have not yet conceived of the correct hypothesis. After many experiments, biologists may conclude that a hypothesis is consistent with known data, but they should never say the hypothesis is proven.

By comparison, the term **theory**, as it is used in biology, is a broad explanation of some aspect of the natural world that is substantiated by a large body of evidence. Biological theories incorporate observations, hypothesis testing, and the laws of other disciplines such as chemistry and physics. Theories are powerful because they allow us to make many predictions about the properties of living organisms. As an example, let's consider the theory that DNA is the genetic material and that it is organized into units called genes. An overwhelming body of evidence has substantiated this theory. Thousands of living species have been analyzed at the molecular level. All of them have been found to use DNA as their genetic material and to express genes that produce the proteins that lead to their characteristics. This theory makes many valid predictions. For example, certain types of mutations in genes are expected to affect the traits of organisms. This prediction has been confirmed experimentally. Similarly, this theory predicts that genetic material is copied and transmitted from parents to offspring. By comparing the DNA of parents and offspring, this prediction has also been confirmed. Furthermore, the theory explains the observation that offspring resemble their parents. Overall, two key attributes of a scientific theory are (1) consistency with a vast amount of known data and (2) the ability to make many correct predictions.

The meaning of "theory" is sometimes muddled because the word is used in different situations. In everyday language, a theory is often viewed as little more than a guess. For example, a person might say, "My theory is that Professor Simpson did not come to class today because he went to the beach." However, in biology, a theory is much more than a guess. A theory is an established set of ideas that explains a vast amount of data and offers valid predictions that can be tested. Like a hypothesis, a theory can never be proven to be true. Scientists acknowledge that they do not know everything. Even so, biologists would say that theories are extremely likely to be true, based on all known information. In this regard, theories are viewed as **knowledge**, which is the awareness and understanding of information.

## Discovery-Based Science and Hypothesis Testing Are Scientific Approaches That Help Us Understand Biology

The path that leads to an important discovery is rarely a straight line. Rather, scientists ask questions, make observations, ask modified questions, and may eventually conduct experiments to test their hypotheses. The first attempts at experimentation may fail, and new experimental approaches may be needed. To suggest that scientists follow a rigid scientific method is an oversimplification of the process of science. Scientific advances often occur as scientists dig deeper and deeper into a topic that interests them. Curiosity is the key phenomenon that sparks scientific inquiry. How is biology actually conducted? As discussed next, researchers typically follow two general types of approaches: discovery-based science and hypothesis testing.

***Discovery-Based Science*** The collection and analysis of data without the need for a preconceived hypothesis is called **discovery-based science**, or simply **discovery science**. Why is discovery-based science carried out? The information gained from discovery-based science may lead to the formation of new hypotheses and, in the long run, may have practical applications that benefit people. Researchers, for example, have identified and begun to investigate previously unknown genes within the human genome without already knowing the function of those genes. The goal is to gather additional clues that may eventually allow them to propose a hypothesis that explains a gene's function. Discovery-based science often leads to hypothesis testing.

***Hypothesis Testing*** In biological science, the scientific method, also known as **hypothesis testing**, is usually followed to formulate and test the validity of a hypothesis. This strategy may be described as a five-step method:

1. Observations are made regarding natural phenomena.
2. These observations lead to a hypothesis that tries to explain the phenomena. A useful hypothesis is one that is testable because it makes specific predictions.
3. Experimentation is conducted to determine if the predictions are correct.
4. The data from the experiment are analyzed.
5. The hypothesis is considered to be consistent with the data, or it is rejected.

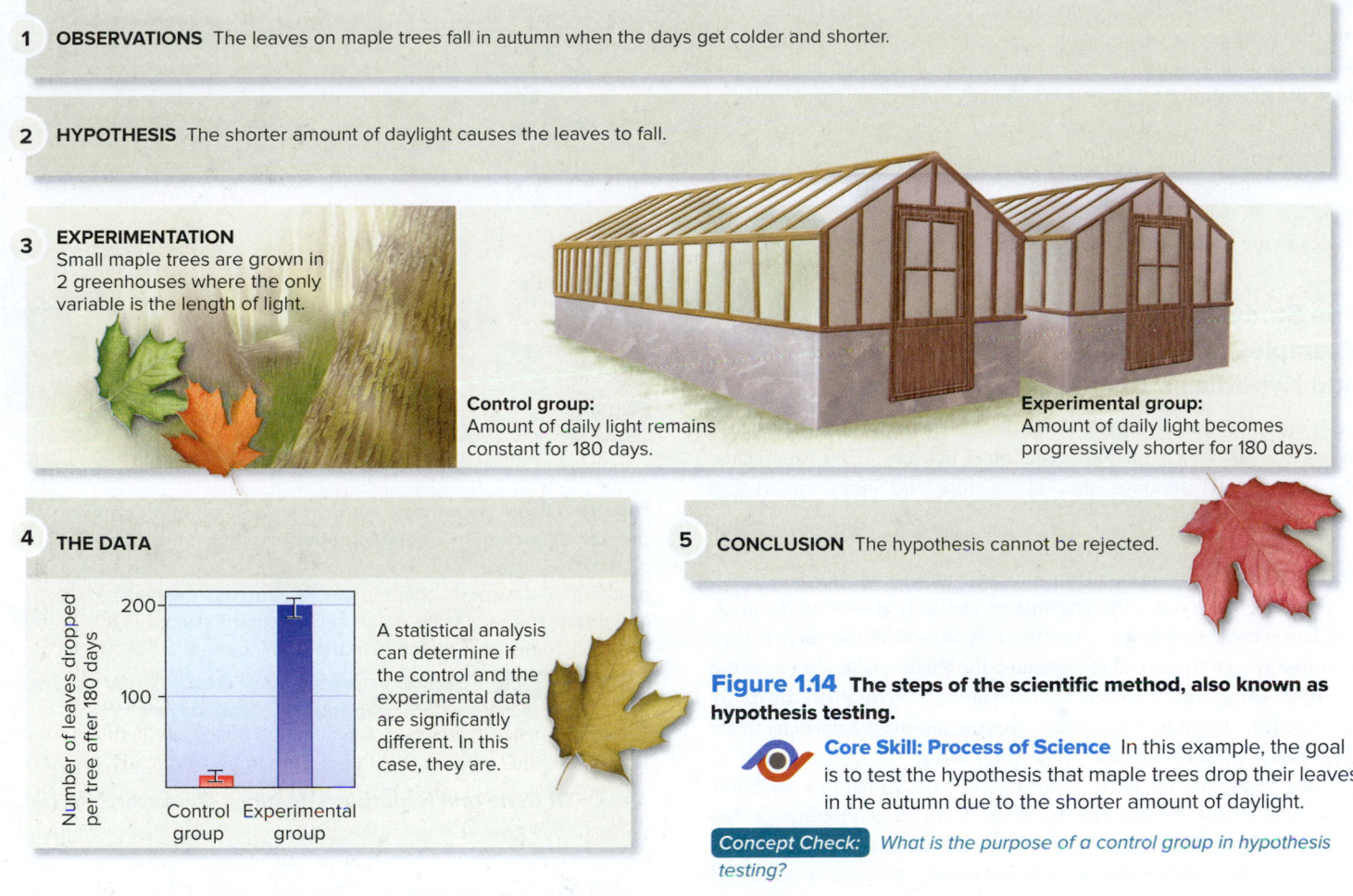

**1  OBSERVATIONS** The leaves on maple trees fall in autumn when the days get colder and shorter.

**2  HYPOTHESIS** The shorter amount of daylight causes the leaves to fall.

**3  EXPERIMENTATION**
Small maple trees are grown in 2 greenhouses where the only variable is the length of light.

**Control group:**
Amount of daily light remains constant for 180 days.

**Experimental group:**
Amount of daily light becomes progressively shorter for 180 days.

**4  THE DATA**

A statistical analysis can determine if the control and the experimental data are significantly different. In this case, they are.

**5  CONCLUSION** The hypothesis cannot be rejected.

**Figure 1.14  The steps of the scientific method, also known as hypothesis testing.**

**Core Skill: Process of Science** In this example, the goal is to test the hypothesis that maple trees drop their leaves in the autumn due to the shorter amount of daylight.

**Concept Check:** *What is the purpose of a control group in hypothesis testing?*

The scientific method is intended to be an objective way to gather knowledge. As an example, let's return to the question of why maple trees drop their leaves in autumn. By observing the length of daylight throughout the year and comparing that data with the time of the year when leaves fall, one hypothesis might be that leaves fall in response to a shorter amount of daylight (**Figure 1.14**). This hypothesis makes a prediction—exposure of maple trees to shorter periods of daylight will cause their leaves to fall. To test this prediction, researchers would design and conduct an experiment.

How is hypothesis testing conducted? Although hypothesis testing may follow many paths, certain experimental features are common to this approach. First, data are often collected in two parallel ways. One set of experiments is done on the **control group**, while another set is conducted on the **experimental group**. In an ideal experiment, the control and experimental groups differ by only one factor. For example, an experiment could be conducted in which two groups of trees are observed, and the only difference between their environments is the length of light each day. To conduct such an experiment, researchers would grow small trees in a greenhouse where they could keep other factors such as temperature, water, and nutrients the same between the control and experimental groups, while providing the two groups with different amounts of light via artificial lighting. In the control group, the number of hours of light

provided is kept constant each day, whereas in the experimental group, the amount of light provided each day becomes progressively shorter to mimic seasonal light changes. The researchers would then record the number of leaves dropped by the two groups of trees over a certain period of time.

Another key feature of hypothesis testing is data analysis. The result of experimentation is a set of data from which a biologist tries to draw conclusions. Biology is a quantitative science. When experimentation involves control and experimental groups, a common form of analysis is to determine if the data collected from the two groups are truly different. Biologists apply statistical analyses to their data to determine if the outcomes from the control and experimental groups are likely to differ because of the single variable that is different between the two groups. When differences between the control and experimental data are statistically significant, they are not likely to have occurred as a matter of random chance.

In our example in Figure 1.14, the trees in the control group dropped far fewer leaves than did those in the experimental group. A statistical analysis could determine if the data collected from the two greenhouses are significantly different from each other. If the two sets of data are found not to be significantly different, the hypothesis will be rejected. Alternatively, if the differences between the two sets of data are significant, as shown in Figure 1.14, biologists can conclude that the hypothesis is consistent with the data, though it is not proven. A hallmark of science is that valid experiments are **repeatable**, which means that similar results are obtained when an experiment is conducted on multiple occasions. For our example in Figure 1.14, the data would be valid only if the experiment was repeatable.

As described next, discovery-based science and hypothesis testing are often used together to learn more about a particular scientific topic. As an example, let's look at how both approaches led to successes in the study of the disease called cystic fibrosis.

## The Study of Cystic Fibrosis Provides Examples of Discovery-Based Science and Hypothesis Testing

Let's consider how biologists made discoveries related to the disease cystic fibrosis (CF), which affects about 1 in every 3,500 Americans. Persons with CF produce abnormally thick and sticky mucus that obstructs the lungs and leads to life-threatening lung infections. The thick mucus also blocks ducts in the pancreas, which prevents the digestive enzymes this organ produces from reaching the intestine. Without these enzymes, the intestine cannot fully absorb amino acids and fats, which can cause malnutrition. Persons with this disease may also experience liver damage because the thick mucus can obstruct the liver. On average, people with CF in the United States currently live into their late 30s. Fortunately, as more advances have been made in treatment, this number has steadily increased.

Because of its medical significance, many scientists are interested in CF and are conducting studies aimed at gaining greater information regarding its underlying cause. The hope is that knowing more about the disease may lead to improved treatment options, and perhaps even a cure. As described next, discovery-based science and hypothesis testing have been critical to gaining a better understanding of this disease.

*The CFTR Gene and Discovery-Based Science*    In 1935, American physician Dorothy Andersen determined that cystic fibrosis is a genetic disorder. Persons with CF have inherited two faulty *CFTR* genes, one from each parent. (We now know this gene encodes a protein named the cystic fibrosis transmembrane regulator, abbreviated *CFTR*.) In the 1980s, researchers used discovery-based science to identify this gene. Their search for the *CFTR* gene did not require any preconceived hypothesis regarding the function of the gene. Rather, they used genetic strategies similar to those described in Chapter 21. Research groups headed by Lap-Chee Tsui, Francis Collins, and John Riordan identified the *CFTR* gene in 1989.

The discovery of the *CFTR* gene made it possible to devise diagnostic testing methods to determine if a person carries a faulty version of that gene. In addition, the characterization of the *CFTR* gene provided important clues about its function. Researchers observed striking similarities between the *CFTR* gene and other genes that were already known to encode proteins that function in the transport of substances across membranes. Based on this observation, as well as other kinds of data, the scientists hypothesized that the function of the normal *CFTR* gene is to encode a transport protein. In this way, the identification of the *CFTR* gene led them to conduct experiments aimed at testing a hypothesis about its function.

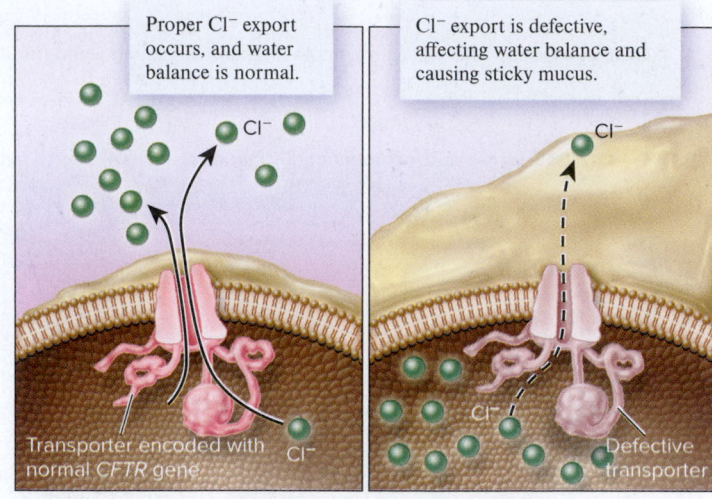

Lung cell with normal *CFTR* gene          Lung cell with faulty *CFTR* gene

**Figure 1.15**    **A hypothesis suggesting an explanation for the defective function of a gene in patients with cystic fibrosis.** The normal *CFTR* gene, which does not carry a mutation, encodes a protein that transports chloride ions (Cl⁻) across the plasma membrane to the outside of the cell. In persons with CF, this protein is defective due to a mutation in the *CFTR* gene.

*Concept Check:*  *Explain how discovery-based science helped researchers to hypothesize that the CFTR gene encodes a transport protein.*

*The CFTR Gene and Hypothesis Testing*    Researchers interested in the *CFTR* gene also considered studies showing that patients with CF have an abnormal regulation of salt balance across their plasma membranes. They hypothesized that the normal *CFTR* gene encodes a protein that functions in the transport of chloride ions (Cl⁻) across the membranes of cells (**Figure 1.15**). This hypothesis led to experimentation that tested normal cells and cells from CF patients for their ability to transport Cl⁻. The CF cells were found to be defective in chloride transport. In 1990, scientists successfully transferred the normal *CFTR* gene into cells from CF patients in the laboratory. The introduction of the normal gene corrected the cells' defect in chloride transport. Overall, the results showed that the *CFTR* gene encodes a protein that transports Cl⁻ across the plasma membrane. A mutation in this gene causes it to encode a defective protein, leading to a salt imbalance that affects water levels outside the cell, which explains the thick and sticky mucus in CF patients. In this example, hypothesis testing provided a way to evaluate a hypothesis about how a disease is caused by a genetic change.

## Biology Is a Social Discipline

Finally, it is worthwhile to point out that biology is a social as well as a scientific discipline. Several laboratories often collaborate on scientific projects. After performing observations and experiments, biologists communicate their results in different ways. Most importantly, papers are submitted to scientific journals. Following submission, a paper usually undergoes a **peer-review process** in which other scientists, who are experts in the area, evaluate the paper and make comments regarding its quality. As a result of peer review, a paper is either accepted for publication or rejected, or the authors of the paper

**Figure 1.16**  **One of the social aspects of science.**
©Dita Alangkara/AP Images

 **Core Skill: Communication and Collaboration**  At scientific meetings, researchers from various disciplines gather together to discuss new data and discoveries. Research that is conducted by professors, students, lab technicians, and industrial participants is sometimes hotly debated.

may be given suggestions for how to revise the work or conduct additional experiments to make it acceptable for publication.

Another social aspect of research is that biologists often attend meetings where they report their most recent work to the scientific community (**Figure 1.16**). They comment on each other's ideas and results, eventually putting together the information that builds into scientific theories over many years. As you develop your skills at scrutinizing experiments, it is helpful to discuss your ideas with other people, including fellow students and faculty members. Importantly, you do not need to "know all the answers" before you enter into a scientific discussion. Instead, a more realistic way to view science is as an ongoing and never-ending series of questions.

## 1.6  Core Skills of Biology

**Learning Outcomes:**

1. **CoreSKILL »** Describe the core skills of biology as identified by "Vision and Change."
2. **CoreSKILL »** Explain the process of science.
3. **CoreSKILL »** Describe what a model is in biology, and explain why models are useful.
4. **CoreSKILL »** List the types of problem-solving skills you will develop by completing BioTIPS.

In addition to the five core concepts of biology (see Section 1.2), the participants in "Vision and Change" also identified certain skills that students should develop so they can become successful in careers in biology. Educators need to focus on these skills, which are also referred to as core competencies:

- The ability to apply the process of science
- The ability to use quantitative reasoning

- The ability to use models and simulation
- The ability to tap into the interdisciplinary nature of science
- The ability to communicate and collaborate with professionals in other disciplines
- The ability to understand the relationship between science and society

In this section, we will consider the features of this textbook that will help you to develop these skills. These features are summarized below:

- Each chapter has a Feature Investigation that allows you to apply the process of science. Likewise, the BioTIPS features are aimed at helping you refine and apply your problem-solving skills.

- Quantitative reasoning is also a key component of the Feature Investigations. It is involved in answering many of the questions at the end of Feature Investigations, as well as many end-of-chapter questions and BioTIPS questions.

- A new feature of the fifth edition, introduced later in this section, is the Modeling Challenges. After learning about a particular topic in biology, you will be asked to either interpret a given model or propose your own model based on a scenario or data.

- The interdisciplinary nature of science is highlighted in features titled "Connections" that follow some figure legends.

- Another new feature of the fifth edition is the addition of a core skill called "Science and Society" following some of the figure legends.

The "Vision and Change" icon, , that highlights core concepts throughout the text, also highlights material that promotes the core skills.

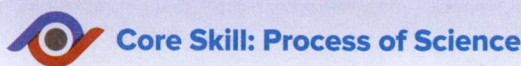

# Feature Investigation | Observation and Experimentation Form the Core of Biology

Biology is largely about the process of discovery. Therefore, a recurring theme of this textbook is how scientists design experiments, analyze data, and draw conclusions. Although each chapter contains many examples of data collection and experiments, a consistent element is a Feature Investigation—which presents an actual study by current or past researchers. Some of these involve discovery-based science, in which biologists collect and interpret data in an attempt to make discoveries that are not hypothesis driven. Most Feature Investigations, however, involve hypothesis testing in which a hypothesis is stated and the experiment and resulting data are presented. Figure 1.14, illustrating the experiment with maple trees, shows the general form of Feature Investigations.

The Feature Investigations allow you to appreciate the connection between science and scientific theories. As you read a Feature Investigation, you may find yourself thinking about different approaches and alternative hypotheses. Different people can view the same data and arrive at very different conclusions. As you progress through the experiments in this textbook, we hope you will try to develop your own skills at formulating hypotheses, designing experiments, and interpreting data.

### Experimental Questions

1. Discuss the difference between discovery-based science and hypothesis testing.
2. What are the steps in the scientific method, also called hypothesis testing?
3. **CoreSKILL »** In an experiment, explain how a control group and an experimental group differ from each other.

## Model-Based Learning Will Enhance Your Understanding of Biological Concepts and Improve Your Critical-Thinking Skills

What is a model? A **scientific model**, or simply a model, is a conceptual, mathematical, or physical depiction of a real-world phenomenon. A model is a simplification and abstraction of a researcher's perception of reality. In biology, models are testable ideas that are usually derived from observations and experiments. Because of the vast amount of complexity and variation found in nature, all but the simplest models are imperfect depictions of living things, their working parts, and their interactions with the environment. The majority of figures in this textbook are models, based on the ideas of biologists and drawn by professional illustrators.

Why are models useful? One reason is they promote communication. Models allow scientists to convey their ideas in a relatively simple way. For example, a model of the human heart depicts how the parts of the heart work together to pump blood (look ahead to Figure 48.6). Another useful aspect of a model is that it can be used as a working hypothesis that helps researchers visualize or explain biological phenomena. Such models form the basis for conducting further experiments. Models are evaluated by their consistency with experimental data, which enables researchers to accept, reject, or refine them. Likewise, models allow biologists to make meaningful predictions. Such predictions can be refuted or supported via experimentation. A model for gene regulation in Chapter 14 predicts that a repressor protein inhibits gene expression (look ahead to Figure 14.7). This prediction was verified by experimentation, as shown in Figure 14.9. Finally, models can lead to conceptual frameworks. For example, the concept of a species niche, which is described in Chapter 57, was derived from species competition models and has subsequently become one of the most important concepts in ecology.

Models take on many different forms. Let's consider some common categories of models that you will see.

- *Structural models.* A structural model shows the physical structures of components that make up living organisms. Biochemists and biologists have proposed many different models that depict biological structures at the cellular and molecular level. Figure 3.13 is a collection of 20 models of the structures of amino acids that are found in proteins.

- *Mechanistic models.* A mechanistic model (also called a physiological model) describes the workings of the individual parts of a complex system, and the manner in which they interact. As an example, plant biologists have proposed two models, called symplastic and apoplastic transport, which describe two possible pathways by which minerals are taken into the root of a plant (look ahead to Figure 39.7).

- *Mathematical models.* A mathematical model is a description of a process or a system using mathematical concepts, symbols, and diagrams. Many mathematical models are presented as one or more equations. For example, ecologists use equations to describe two different modes of population growth, termed exponential and logistic growth. Such equations allow biologists to make predictions about population growth, which can be illustrated graphically (look ahead to Figure 56.10).

- *Temporal models.* A temporal model depicts a biological process as it occurs over a short or long period of time. In cell biology, some processes occur very quickly. For example, the absorption of light energy during photosynthesis occurs in less than a second (look ahead to Figure 8.11). In contrast, the evolution of new groups of species may occur on a timescale of millions of years (look ahead to Figure 26.4).

- *Hierarchical models.* In a hierarchical model, organisms, parts of organisms, or observations fall into nested levels. For example, the field of taxonomy organizes species into progressively smaller groups, such as kingdom, family, and genus (plural, genera). One or more genera are found within a family, and many different families are found within a kingdom (see Figure 1.11).

Some models incorporate two or more of these categories. Take a look at the model for DNA replication in Figure 11.17, which is a combination of a structural model, a mechanistic model, and a temporal model.

**Model-based learning** is an educational approach in which students evaluate or generate models as a way to enhance their understanding of scientific concepts and improve their critical-thinking skills. In this textbook, you will be engaged in this strategy via figures that present a modeling challenge. Each of these figures shows a model that pertains to a particular topic in biology. After you study this model, your modeling skills will be challenged in one of two different ways. In some cases, you will be given a second model and asked to explain it or describe what types of predictions can be made based on it (**Figure 1.17**). In other cases, you will be given a scenario and asked to generate your own model that is consistent with the scenario (**Figure 1.18**). Even though explaining and

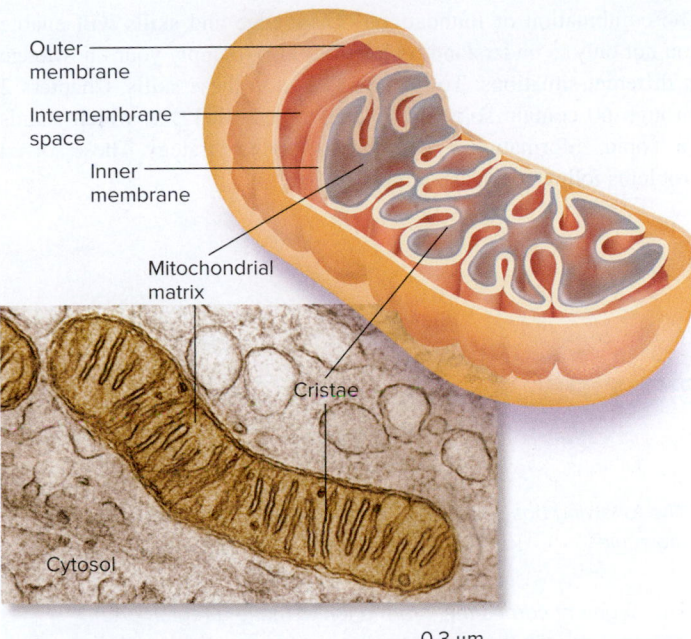

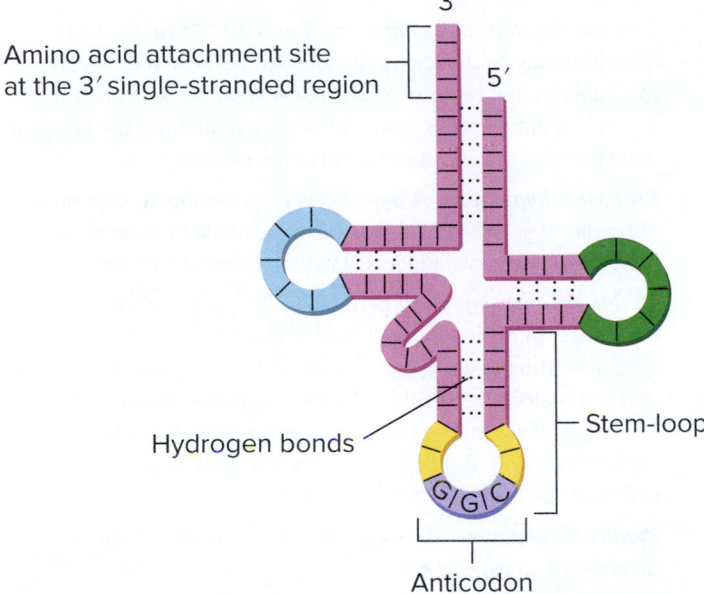

**Figure 1.17** **A modeling challenge to explain a revised model.** This figure shows a model for the structure of a tRNA molecule, which is described in Chapter 12. The stem regions are regions where the RNA is double stranded as a result of complementary base pairing, in which A hydrogen-bonds to U, and G hydrogen-bonds with C. The modeling challenge below involves an alteration in this model.

**Core Skill: Modeling** In this modeling challenge, you are asked to explain how the model below differs from the one in Figure 1.17.

**Modeling Challenge:** In a tRNA molecule, four of the bases were changed. One A was changed to a G, and three C's were changed to U's. A model of the secondary structure of this altered tRNA is shown to the right. Explain where the altered bases are located and how the alteration affects the structure of the tRNA.

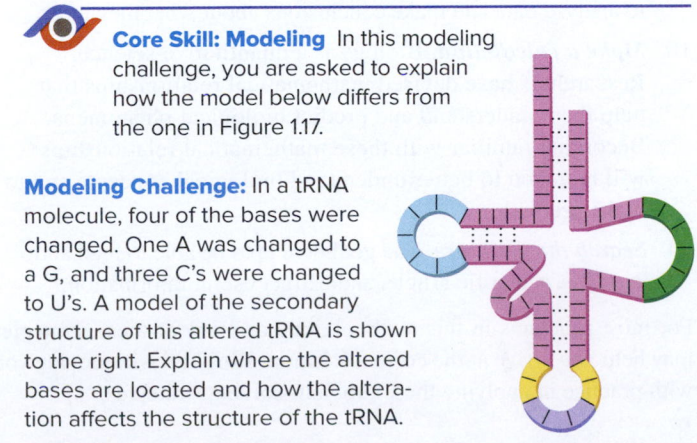

**Figure 1.18** **A modeling challenge to make a prediction.** This figure shows the structure of a mitochondrion. It emphasizes the membrane organization of the mitochondrion, which has outer and inner membranes. The invaginations (infoldings) of the inner membrane, which are called cristae, occur because of the large surface area of that membrane. The modeling challenge below involves proposing an altered model. ©Don W. Fawcett/Science Source

**Core Skill: Modeling** This modeling challenge asks you to propose a model for the structure of a mitochondrion in the presence of a drug that decreases the surface area of the inner membrane.

**Modeling Challenge:** Let's suppose a cell is exposed to a drug that decreases the surface area of the inner mitochondrial membrane, but has no effect on the outer mitochondrial membrane. Draw a model of the structure of the mitochondrion in the presence of this drug.

generating models can be a challenge, the educational benefits are worth it. Give the modeling challenges a try.

## BioTIPS Will Help You Improve Your Problem-Solving Skills

As you progress through this textbook, your learning will involve two general goals:

- You will gather foundational knowledge. In other words, you will be able to describe basic ideas and discoveries in biology. For example, you will be able to explain how photosynthesis works.

- You will develop skills that will allow you to apply that foundational knowledge in different ways. For example, you will learn how to use statistics to determine if a hypothesis is consistent with experimental data.

The combination of foundational knowledge and skills will enable you not only to understand biology but also to apply your knowledge in different situations. To help you develop these skills, Chapters 2 through 60 contain solved problems called **BioTIPS**, which stands for **T**opic, **I**nformation, and **P**roblem-Solving **S**trategy. These solved problems follow a consistent pattern.

 **THE QUESTION** *All of the **BioTIPS** begin with a question. As an example, let's consider the following question:*

*The following base sequence is found within a messenger RNA molecule:*

AUG GGC CUU AGC

*This segment carries the information to make a region of a polypeptide with the amino acid sequence methionine-glycine-leucine-serine. What would be the consequences if a mutation in the gene that encodes this mRNA changed the second cytosine (C) in the base sequence to an adenine (A)?*

**T**OPIC *What topic in biology does this question address?* The topic is gene expression. More specifically, the question is about the relationship between a base sequence and the genetic code.

**I**NFORMATION *What information do you know based on the question and your understanding of the topic?* In the question, you are given the base sequence of a short segment of an mRNA and told that one of the bases has been changed. From your understanding of the topic, you may remember that a polypeptide sequence is determined by reading the mRNA (transcribed from a gene) in groups of three bases called codons.

**P**ROBLEM-SOLVING **S**TRATEGY *Compare and contrast. Predict the outcome.* One strategy to begin to solve this problem is to compare the mRNA sequence before and after the mutation:

Original: AUG GGC CUU AGC
Mutant: AUG GGC AUU AGC
↑

**ANSWER** *The mutation has altered the sequence of bases in the mRNA, changing the third codon from CUU to AUU (see the arrow). Because codons specify amino acids, this may change the third amino acid in the polypeptide to something else. Note: If you look ahead to Table 12.1, you will see that CUU specifies leucine, whereas AUU specifies isoleucine. Therefore, you can predict that the mutation will change the third amino acid from leucine to isoleucine.*

Though many different problem-solving strategies exist, **BioTIPS** will focus on 11 strategies that will help you solve problems. You will see these strategies over and over again as you progress through this textbook:

1. *Make a drawing*. Biology problems are often difficult to solve in your head. Making a drawing may make a big difference in your ability to see the solution.

2. *Compare and contrast*. Making a direct comparison between two biological structures or processes may help you understand how they are similar and how they are different.

3. *Relate structure and function*. A recurring theme in biology is that structure determines function. This relationship holds true at many levels of biology, including the molecular, microscopic, and macroscopic levels. For some questions, you will need to understand how certain structural features are related to their biological functions.

4. *Sort out the steps in a complicated process*. At first, some questions may be difficult to understand because they involve mechanisms that occur in a series of several steps. Sometimes, if you sort out the steps, you will be able to identify the key step that you need to understand to solve the problem.

5. *Propose a hypothesis*. A hypothesis is an attempt to explain an observation or data. Hypotheses may be made in many forms including statements, models, equations, and diagrams.

6. *Design an experiment*. Experimental design lies at the heart of science. In many cases, an experiment begins with some type of starting material(s), such as strains of organisms or purified molecules, and then the starting materials are subjected to a series of steps. The Feature Investigations throughout the textbook will also help you refine the skill of designing experiments.

7. *Predict the outcome*. Biologists may want to predict the outcome of an experiment.

8. *Interpret data*. Experimentation involves the analysis of data. Such an analysis often involves the use of statistics to determine if the experimental and control data show significant differences. The interpretation of data allows scientists to propose models that describe what the data may mean.

9. *Use statistics*. A variety of different statistical methods are used to analyze data and make conclusions about what they mean.

10. *Make a calculation*. Biology is a quantitative science. Researchers have devised mathematical relationships that help them understand and predict biological phenomena. Becoming familiar with these mathematical relationships will help you to better understand biological concepts and to make predictions.

11. *Search the literature*. The goal here is to be able to read and explain a scientific article, and extract useful information.

For most problems in this textbook, one or more of these strategies may help you arrive at the correct solution. **BioTIPS** will provide you with practice in applying these problem-solving strategies.

# Summary of Key Concepts

- Biology is the study of life. Discoveries in biology help us understand how life exists, and they also have many practical applications, such as the development of drugs to treat human diseases (Figures 1.1, 1.2, Table 1.1.).

## 1.1 Levels of Biology

- Living organisms can be viewed at different levels of biological organization: atoms, molecules and macromolecules, cells, tissues, organs, organisms, populations, communities, ecosystems, and the biosphere (Figure 1.3).

## 1.2 Core Concepts of Biology

- "Vision and Change" has identified five core concepts in biology (Figure 1.4). These are evolution; structure and function; information flow, exchange, and storage; pathways and transformations of energy and matter; and systems.

## 1.3 Biological Evolution

- Changes in species often occur as a result of modification of pre-existing structures (Figure 1.5).

- During vertical evolution, mutations in a lineage alter the characteristics of species from one generation to the next. Individuals with greater reproductive success are more likely to contribute to future generations, a process known as natural selection. Over the long run, this process alters species and may produce new species (Figure 1.6).

- Horizontal gene transfer is the transfer of genetic material from one organism to another organism that is not its offspring. Along with vertical descent with mutation, it is an important process in biological evolution, producing a web of life (Figures 1.7, 1.8).

- An analysis of genomes and proteomes helps us to understand how information at the molecular level relates to the characteristics of individuals and how they survive in their native environments (Figure 1.9).

## 1.4 Classification of Living Things

- Taxonomy is the grouping of species according to their evolutionary relatedness to other species. Going from broad to narrow groups, each species is placed into a domain, supergroup, kingdom, phylum, class, order, family, and genus (Figures 1.10, 1.11).

- The classification of species changes as biologists gather new information (Figure 1.12).

## 1.5 Biology as a Scientific Discipline

- Biological science is the observation, identification, experimental investigation, and theoretical explanation of natural phenomena.

- Biologists study life at different levels, ranging from ecosystems to the molecular components in cells (Figure 1.13).

- A hypothesis is a proposal to explain a natural phenomenon. A useful hypothesis makes a testable prediction. A biological theory is a broad explanation that is substantiated by a large body of evidence.

- Discovery-based science is an approach in which researchers conduct experiments and analyze data without a preconceived hypothesis.

- The scientific method, also called hypothesis testing, is a series of steps to formulate and test the validity of a hypothesis. The experimentation often involves a comparison between control and experimental groups (Figure 1.14).

- The study of cystic fibrosis provides an example in which both discovery-based science and hypothesis testing led to key insights regarding the nature of the disease (Figure 1.15).

- Biology is a social discipline in which scientists often work in teams. To be published, a scientific paper is usually subjected to a peer-review process in which other scientists evaluate the paper and make suggestions regarding its quality. Advances in science often occur when scientists gather and discuss their data (Figure 1.16).

## 1.6 Core Skills of Biology

- "Vision and Change" recognized the need to focus on the development of certain skills in students: the ability to apply the process of science; the ability to use quantitative reasoning; the ability to use models and simulation; the ability to tap into the interdisciplinary nature of science; the ability to communicate and collaborate with professionals in other disciplines; and the ability to understand the relationship between science and society.

- Each chapter in this textbook has a "Feature Investigation," an actual study by current or past researchers that highlights the experimental approach and helps you appreciate how science has led to key discoveries in biology.

- Biologists use models to convey their ideas, evaluate experiments, and make predictions that apply to their research studies. Modeling challenges will help you to understand and propose models (Figure 1.17, Figure 1.18).

- BioTIPS are intended to develop your problem-solving skills.

# Assess & Discuss

## Test Yourself

1. Which of the following is *not* a core concept of biology, as advocated by "Vision and Change"?
   a. Evolution
   b. Information flow, exchange, and storage
   c. Structure and function
   d. Taxonomy
   e. Pathways and transformation of energy and matter

2. Populations of organisms change over the course of many generations. Many of these changes are the result of greater reproductive success. This phenomenon is
   a. evolution.          d. genetics.
   b. homeostasis.        e. metabolism.
   c. development.

3. A biologist is studying the living organisms in a valley in western Colorado. She is studying
   a. an ecosystem.       d. a viable land mass.
   b. a community.        e. a population.
   c. the biosphere.

4. Which of the following is an example of horizontal gene transfer?
   a. the transmission of an eye color gene from father to daughter
   b. the transmission of a mutant gene causing cystic fibrosis from father to daughter

c. the transmission of a gene conferring pathogenicity (the ability to cause disease) from one bacterial species to another

d. the transmission of a gene conferring antibiotic resistance from a mother cell to its two daughter cells

e. all of the above.

5. The scientific name for humans is *Homo sapiens*. The name *Homo* is the _____ to which humans are classified.

a. kingdom
b. phylum
c. order
d. genus
e. species

6. The complete genetic makeup of an organism is called its

a. genus.
b. genome.
c. proteome.
d. genotype.
e. phenotype.

7. After observing certain desert plants in their native environment, a researcher proposes that they drop their leaves to conserve water. This is an example of

a. a theory.
b. a law.
c. a prediction.
d. a hypothesis.
e. an experiment.

8. In science, a theory should

a. be viewed as knowledge.
b. be supported by a substantial body of evidence.
c. provide the ability to make many correct predictions.
d. do all of the above.
e. b and c only.

9. Conducting research without a preconceived hypothesis is called

a. discovery-based science.
b. the scientific method.
c. hypothesis testing.

d. a control experiment.
e. none of the above.

10. What is the purpose of using a control group in a scientific experiment?

a. A control group allows the researcher to practice the experiment first before actually conducting it.

b. A researcher can compare the results in the experimental group and control group to determine if a single variable is causing a particular outcome in the experimental group.

c. A control group provides the framework for the entire experiment so the researcher can recall the procedures that should be conducted.

d. A control group allows the researcher to conduct other experimental changes without disturbing the original experiment.

e. all of the above.

## Conceptual Questions

1. Of the five core concepts of biology described in Figure 1.4, which apply to individuals and which apply to populations?

2. Explain how it is possible for evolution to result in unity among different species yet also produce amazing diversity.

3. **Core Concept:** In your own words, describe the five core concepts of biology that are detailed at the beginning of this chapter in Figure 1.4.

## Collaborative Questions

1. Discuss whether or not you think that theories in biology are true. Outside of biology, how do you decide if something is true?

2. In certain animals, such as alligators, sex is determined by temperature. When alligator eggs are exposed to low temperatures, most alligator embryos develop into females. Discuss how this phenomenon is related to genomes and proteomes.

# UNIT I
# CHEMISTRY

Living organisms are composed of chemicals, which are altered via chemical reactions. These reactions occur between atoms and molecules and may require, or in some cases release, energy. Chemical reactions and interactions between molecules play a role in virtually all aspects of a cell's activities. In order to understand how living organisms function, grow, develop, behave, and interact with their environments, therefore, we first need to understand some basic principles of atomic and molecular structure and the forces that allow atoms and molecules to interact with each other. We begin this unit with an overview of **inorganic chemistry**—that is, the nature of atoms and molecules, with the exception of those that contain rings or chains of carbon. Such carbon-containing molecules form the basis of **organic chemistry** and are covered in Chapter 3.

 **The following Core Concepts and Core Skills will be emphasized in this unit:**

- *Energy and matter:* We will see how the chemical energy stored in the bonds of molecules, such as sugars and fats, can be released and used by living organisms to perform numerous functions that support life, including growth, digestion, and locomotion.

- *Structure and function:* As described in Chapter 3, the three-dimensional structure of molecules is critical in enabling them to carry out their function.

- *Information:* Nucleic acids, the basis of inherited genetic material, are first introduced in Chapter 3.

- *Systems:* You will learn in this unit how simple molecules are joined to create a more complex molecule with new biological properties. The newly created molecule has properties that are different from those of its component atoms.

- *Science and society:* In Chapter 2, we will see how an understanding of chemistry has transformed the ability of physicians to diagnose disease in humans. One example of an application of chemistry to medicine is the PET scan.

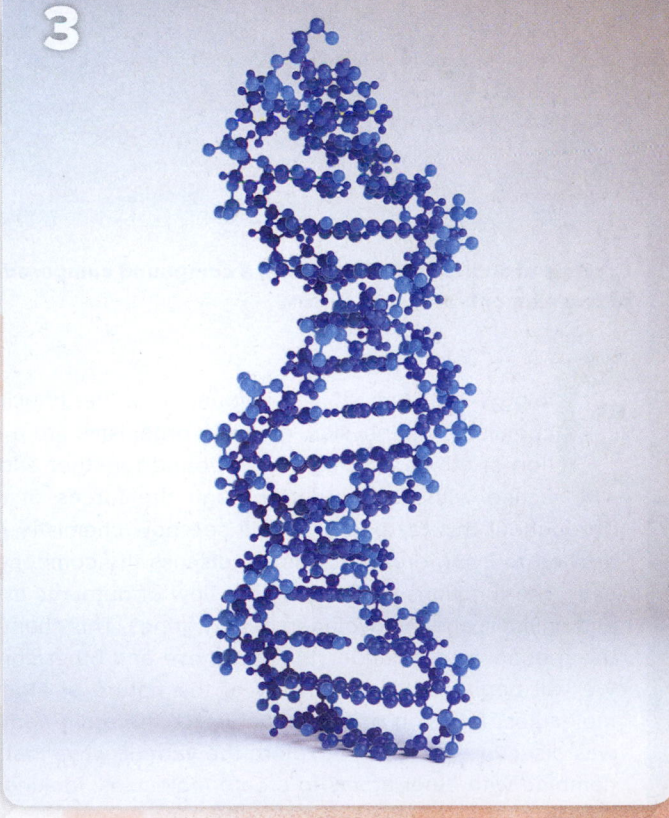

# The Chemical Basis of Life I: Atoms, Molecules, and Water

# 2

**Crystals of sodium chloride (NaCl), a compound composed of two elements.** ©Dr. Parvinder Sethi

**B**iology—the study of life—is founded on the principles of chemistry and physics. All living organisms are a collection of atoms and molecules bound together and interacting with each other through the forces of nature. Throughout this textbook, we will see how chemistry can be applied to living organisms as we discuss the components of cells, the functions of proteins, the flow of nutrients in plants and animals, and the evolution of new genes. This chapter lays the groundwork for understanding these and other concepts. We will begin with an overview of the nature of atoms and molecules, focusing on the structure of the atom and how it was discovered. We next explore the various ways that atoms combine with other atoms to create molecules, looking at the different types of chemical bonds between atoms, how these bonds form, and how they determine the structures of molecules. We end with an examination of the water molecule and the properties that make it a crucial component of living organisms and their environment.

## 2.1 Atoms

**Learning Outcomes:**

1. Describe the general structure of atoms.
2. **CoreSKILL »** Interpret the results of an experiment indicating that most of an atom is empty space.
3. Define orbital and electron shell.
4. Relate atomic structure to the periodic table of the elements.
5. **CoreSKILL »** Quantify atomic mass using units of daltons and moles.
6. Explain how a single element may exist in two or more forms, called isotopes, and how certain isotopes have importance in human medicine.
7. List the elements that make up most of the mass of all living organisms.

All life-forms are composed of **matter**, which is defined as anything that contains mass and occupies space. In living organisms, matter may exist in any of three states: solid, liquid, or gas. All matter is composed of **atoms**, which are the smallest functional units of matter that form all chemical substances and ultimately all organisms; they cannot be further broken down into other substances by ordinary chemical or physical means. Atoms, in turn, are composed of different types of smaller, subatomic particles. Chemists study the properties of atoms and **molecules**, which are two or more atoms bonded together. A major interest of the physicist, by contrast, is to uncover the properties of subatomic particles. Chemistry and physics merge in attempts to understand the mechanisms by which atoms and molecules interact. When atoms and molecules are studied in the context of a living organism, the science of biochemistry emerges. In this section, we will explore the physical properties of atoms so we can understand how atoms combine to form molecules of biological importance.

### Atoms Are Composed of Subatomic Particles

The chemicals within living organisms are composed of many different types of atoms. The simplest atom, hydrogen, is approximately

0.1 nanometer (nm) in diameter, roughly one-millionth the diameter of a human hair. Each specific type of atom—nitrogen, hydrogen, oxygen, and so on—is called an **element** (or chemical element), which is defined as a pure substance made up of only one kind of atom.

Three subatomic particles—**protons** ($p^+$), **neutrons** ($n^0$), and **electrons** ($e^-$)—are found within atoms (**Figure 2.1**). The protons and neutrons are confined to a very small volume at the center of an atom, the **atomic nucleus**, whereas the electrons are found in regions at various distances from the nucleus. In most atoms, the numbers of protons and electrons are identical, but the number of neutrons may vary. Each of the subatomic particles has a different electric charge. Protons have one unit of positive charge, electrons have one unit of negative charge, and neutrons are electrically neutral. Like charges repel each other, and opposite charges attract each other. The positive charges in the nucleus attract the negatively charged electrons.

Because the protons are located in the atomic nucleus, the nucleus has a net positive charge equal to the number of protons it contains. The entire atom has no net electric charge, however, because the number of negatively charged electrons around the nucleus is equal to the number of positively charged protons in the nucleus.

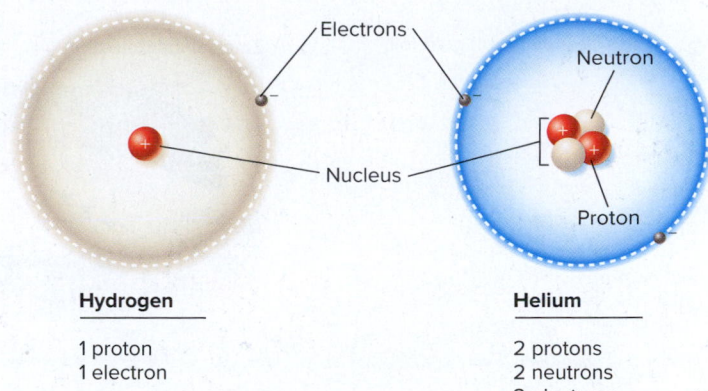

**Hydrogen**

1 proton
1 electron

**Helium**

2 protons
2 neutrons
2 electrons

**Figure 2.1  Diagrams of two simple atoms.** These are models of the two simplest atoms, hydrogen and helium. The nucleus consists of protons and neutrons, whereas electrons are found outside the nucleus. Note: In all figures of atoms, the sizes and distances are not to scale.

This basic concept of the structure of the atom was not established until a landmark experiment conducted by Ernest Rutherford during the years 1909–1911, as described next.

## Core Skill: Process of Science

# Feature Investigation | Rutherford Determined the Modern Model of the Atom

Nobel laureate Ernest Rutherford was born in 1871 in New Zealand, but he did his greatest work at McGill University in Montreal, Canada, and later at the University of Manchester in England. At that time, scientists knew that atoms contained charged particles but had no idea how those particles were distributed. Neutrons had not yet been discovered, and many scientists, including Rutherford, hypothesized that the positive charge and the mass of an atom were evenly dispersed throughout the atom.

In a now-classic experiment, Rutherford aimed a fine beam of positively charged α (alpha) particles at an extremely thin sheet of gold foil only 400 atoms thick (**Figure 2.2**). α particles consist of two

protons and two neutrons and are thus identical to the nuclei of helium atoms; you can think of them as helium atoms without their electrons (see Figure 2.1). Surrounding the gold foil were zinc sulfide screens that registered any α particles passing through or bouncing off the foil, much like film in a camera detects light. Rutherford hypothesized that if the positive charges of the gold atoms were uniformly distributed, many of the positively charged α particles would be slightly deflected, because one of the most important features of electric charge is that like charges repel each other. Due to their much smaller mass, he did not expect electrons in the gold atoms to have any effect on the ability of an α particle to move through the metal foil.

**Figure 2.2  Rutherford's gold foil experiment, demonstrating that most of the volume of an atom is empty space.**

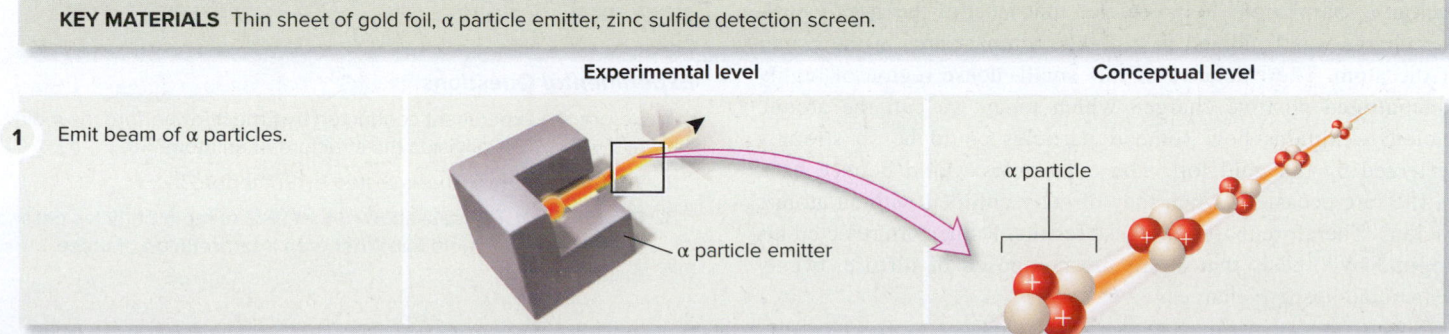

**HYPOTHESIS** Atoms in gold foil are composed of diffuse, evenly distributed positive charges that should usually cause α particles to be slightly deflected as they pass through.

**KEY MATERIALS** Thin sheet of gold foil, α particle emitter, zinc sulfide detection screen.

Experimental level          Conceptual level

1   Emit beam of α particles.

α particle

α particle emitter

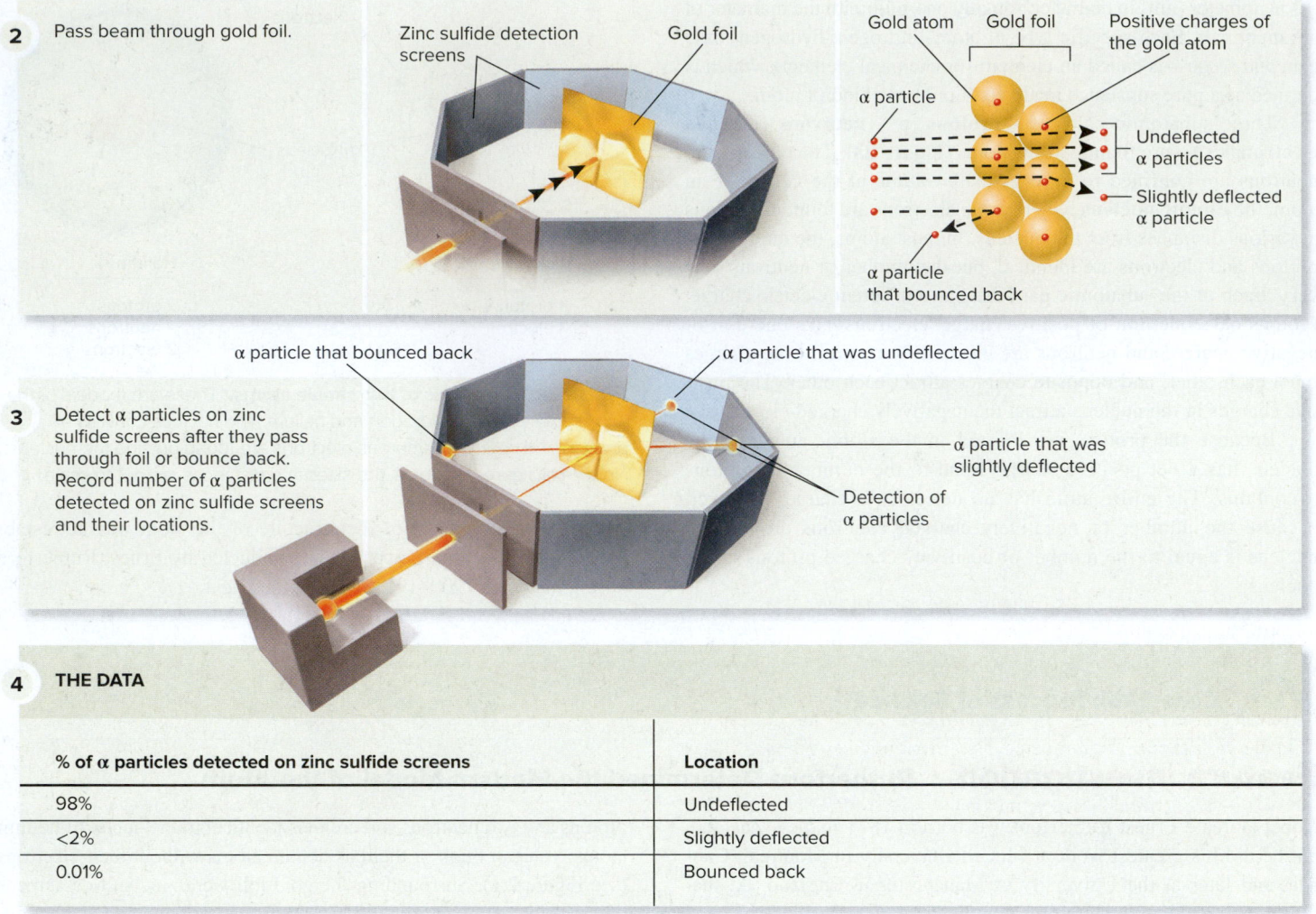

**2** Pass beam through gold foil.

Zinc sulfide detection screens

Gold foil

Gold atom　Gold foil　Positive charges of the gold atom

α particle

Undeflected α particles

Slightly deflected α particle

α particle that bounced back

**3** Detect α particles on zinc sulfide screens after they pass through foil or bounce back. Record number of α particles detected on zinc sulfide screens and their locations.

α particle that bounced back

α particle that was undeflected

α particle that was slightly deflected

Detection of α particles

**4** THE DATA

| % of α particles detected on zinc sulfide screens | Location |
| --- | --- |
| 98% | Undeflected |
| <2% | Slightly deflected |
| 0.01% | Bounced back |

**5** **CONCLUSION** The hypothesis is rejected. Most of the volume of an atom is empty space, with the positive charges concentrated in a small volume.

**6** **SOURCE** Rutherford, E. 1911. The scattering of α and β particles by matter and the structure of the atom. *Philosophical Magazine* 21: 669–688.

Surprisingly, Rutherford discovered that more than 98% of the α particles passed right through the foil as though it was not there, and only a small percentage was slightly deflected; a few even bounced back at a sharp angle! To explain the 98% that passed right through, Rutherford concluded that most of the volume of an atom is empty space. To explain the few α particles that bounced back at a sharp angle, he postulated that most of the atom's positive charge was localized in a highly compact area at the center of the atom. The existence of this small, dense region of highly concentrated positive charge—which today we call the atomic nucleus—explains how some α particles could be so strongly deflected by the gold foil. The α particles would bounce back on the rare occasions when they directly collided with an atomic nucleus. Therefore, based on these results, Rutherford rejected his original hypothesis that atoms are composed of diffuse, evenly distributed positive charges.

From this experiment, without being able to actually visualize an atom, Rutherford proposed a model of an atom, with its small, positively charged nucleus surrounded at relatively great distances by negatively charged electrons. Today we know that more than 99.99% of an atom's volume is outside the nucleus. The nucleus accounts for only about 1/10,000 of an atom's diameter—most of an atom is empty space!

### Experimental Questions

1. Before the experiment conducted by Ernest Rutherford, how did many scientists envision the structure of an atom?

2. What was the hypothesis tested by Rutherford?

3. **CoreSKILL »** The data showed that 98% of the α particles passed right through the gold foil. What is an interpretation of these results?

## Electrons Occupy Orbitals Around an Atom's Nucleus

At one time, scientists visualized an atom as a mini–solar system, with the nucleus being the Sun and the electrons traveling in clearly defined orbits around it. Diagrams of the two simplest atoms—hydrogen and helium—which have the smallest numbers of protons, were shown in Figure 2.1. This model of the atom is now known to be an oversimplification, because as described shortly, electrons do not actually orbit the nucleus in defined paths like planets around the Sun. However, this depiction of an atom remains a convenient way to diagram atoms in two dimensions.

For complex reasons associated with the physics of subatomic particles, it is impossible to precisely predict the exact location of a given electron. We can only describe the region of space surrounding the atomic nucleus in which there is a high probability of finding that electron. Such a region is called an **orbital**. A better model of an atom, therefore, is a central nucleus surrounded by cloudlike orbitals. Some orbitals are spherical, called *s* orbitals, whereas others assume a shape that is often described as similar to a propeller or dumbbell and are called *p* orbitals (**Figure 2.3**). An orbital can contain a maximum of two electrons. Consequently, any atom with more than two electrons must contain more than one orbital.

Orbitals are found within **electron shells**, or energy levels. **Energy** can be defined as the capacity to do work or cause a change. Electrons have kinetic energy, that is, the energy of moving matter. The electron shells are numbered, with shell number 1 being closest to the nucleus. Different electron shells may contain one or more orbitals, each orbital holding up to two electrons. The innermost electron shell of all atoms has room for only two electrons, which spin in opposite directions within a spherical *s* orbital (1*s*). The second electron shell is composed of one spherical *s* orbital (2*s*) and

three dumbbell-shaped *p* orbitals (2*p*). Therefore, the second shell can hold up to four pairs of electrons, or eight electrons altogether (see Figure 2.3).

Electrons vary in the amount of energy they have. The shell closest to the nucleus fills up with the lowest energy electrons first, and then each subsequent shell fills with higher and higher energy electrons, one shell at a time. Within a given shell, the energy of electrons can also vary among different orbitals. In the second shell, for example, the *s* orbital has lower energy, whereas the three *p* orbitals have slightly higher and roughly equal energies. In that case, two electrons fill the *s* orbital first. Any additional electrons fill the *p* orbitals one electron at a time.

Although electrons are found in orbitals of varying shapes, as shown in Figure 2.3, scientists often use more simplified models when depicting the electron shells of atoms. **Figure 2.4a** presents an example of such a depiction of an atom of the element nitrogen. An atom of this element has seven protons and seven electrons. Two electrons fill the first shell, and five electrons are found in the outer shell. Two of these

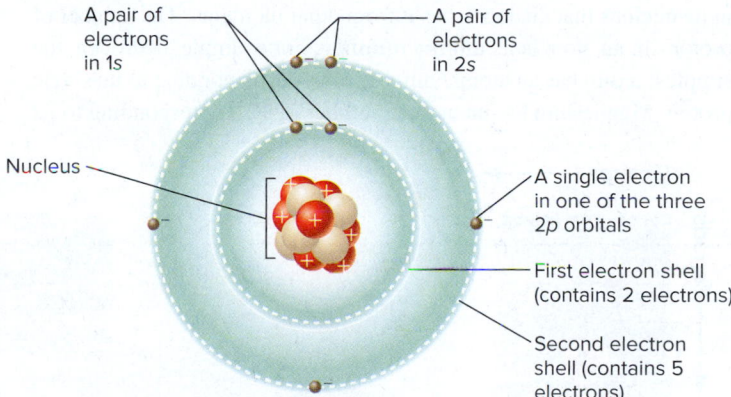

**(a) Simplified depiction of a nitrogen atom**

**(b) Nitrogen atom showing electrons in orbitals**

**Figure 2.4  Diagrams showing the multiple electron shells and orbitals of a nitrogen atom.** The nitrogen atom is shown **(a)** in a simplified depiction and **(b)** with all of its orbitals and shells. An atom's shells fill up one by one. In shells containing more than one orbital, the orbital with lowest energy fills first. Subsequent orbitals gain one electron at a time, shown schematically in boxes, where e⁻ represents an electron. Atoms of heavier elements contain additional shells and orbitals.

*Concept Check:* Explain the difference between an electron shell and an orbital.

| Orbital name | 1s | 2s | 2p |
|---|---|---|---|
| | Nucleus | | |
| Number of electrons per electron shell | 2 | 2 per orbital; 8 total | |
| Orbital shape | Spherical | First orbital: spherical | Second to fourth orbital: dumbbell-shaped |

**Figure 2.3  Diagrams of individual electron orbitals.** Electrons are found outside the nucleus in orbitals that may resemble spherical or dumbbell-shaped clouds. The orbital cloud represents a region in which the probability is high of locating a particular electron. For this illustration, only two shells are shown; the heaviest elements contain a total of seven shells.

fill the 2s orbital and are shown as a pair of electrons in the second shell. The other three electrons in the second shell are found singly in each of the three p orbitals. The diagram in Figure 2.4a makes it easy to see whether electrons are paired within the same orbital and whether the outer shell is full. **Figure 2.4b** is a more accurate model of a nitrogen atom, showing how the electrons occupy orbitals with different shapes.

Most atoms have outer shells that are not completely filled with electrons. Nitrogen, as we just saw, has a first shell filled with two electrons and a second shell with five electrons (see Figure 2.4a). Because the second shell can actually hold eight electrons, the outer shell of a nitrogen atom is not full. As discussed later in this chapter, atoms that have unfilled electron shells tend to share, release, or obtain electrons to fill their outer shell. Those electrons in the outermost shell are called the **valence electrons**. As you will learn shortly, such electrons allow atoms to form chemical bonds with each other, a process in which two or more atoms become joined together to create a new substance.

## Each Element Has a Unique Number of Protons

Each chemical element has a specific and unique number of protons in its nucleus that distinguishes it from other elements. The number of protons in an atom is its **atomic number**. For example, hydrogen, the simplest atom, has an atomic number of 1, corresponding to its single proton. Magnesium has an atomic number of 12, corresponding to its

12 protons. With the exception of ions, which are described later, the number of protons and electrons in a given atom are identical. Therefore, the atomic number is also equal to the number of electrons in the atom, resulting in a net electric charge of zero.

**Figure 2.5** shows the first three rows of the periodic table of the elements, which arranges the known elements according to their atomic numbers and electron shells (see Appendix A for the complete periodic table). A one- or two-letter symbol is used as an abbreviation for each element. The rows (known as periods) indicate the number of electron shells. For example, hydrogen (H) has one shell, lithium (Li) has two shells, and sodium (Na) has three shells. The columns (called groups) indicate the numbers of electrons in the outer shell. As you move along the columns from left to right, the outer shell of lithium (Li) has one electron, beryllium (Be) has two, boron (B) has three, and so forth. This organization of the periodic table tends to arrange elements based on similar chemical properties. The similarities of elements within a group occur because they have the same number of valence electrons, and therefore, they have similar chemical bonding properties.

## Atoms Have a Small but Measurable Mass

Atoms are extremely small and thus have very little mass. A single hydrogen atom, for example, has a mass of about $1.67 \times 10^{-24}$ g (grams). Protons and neutrons are nearly equal in mass, and each has

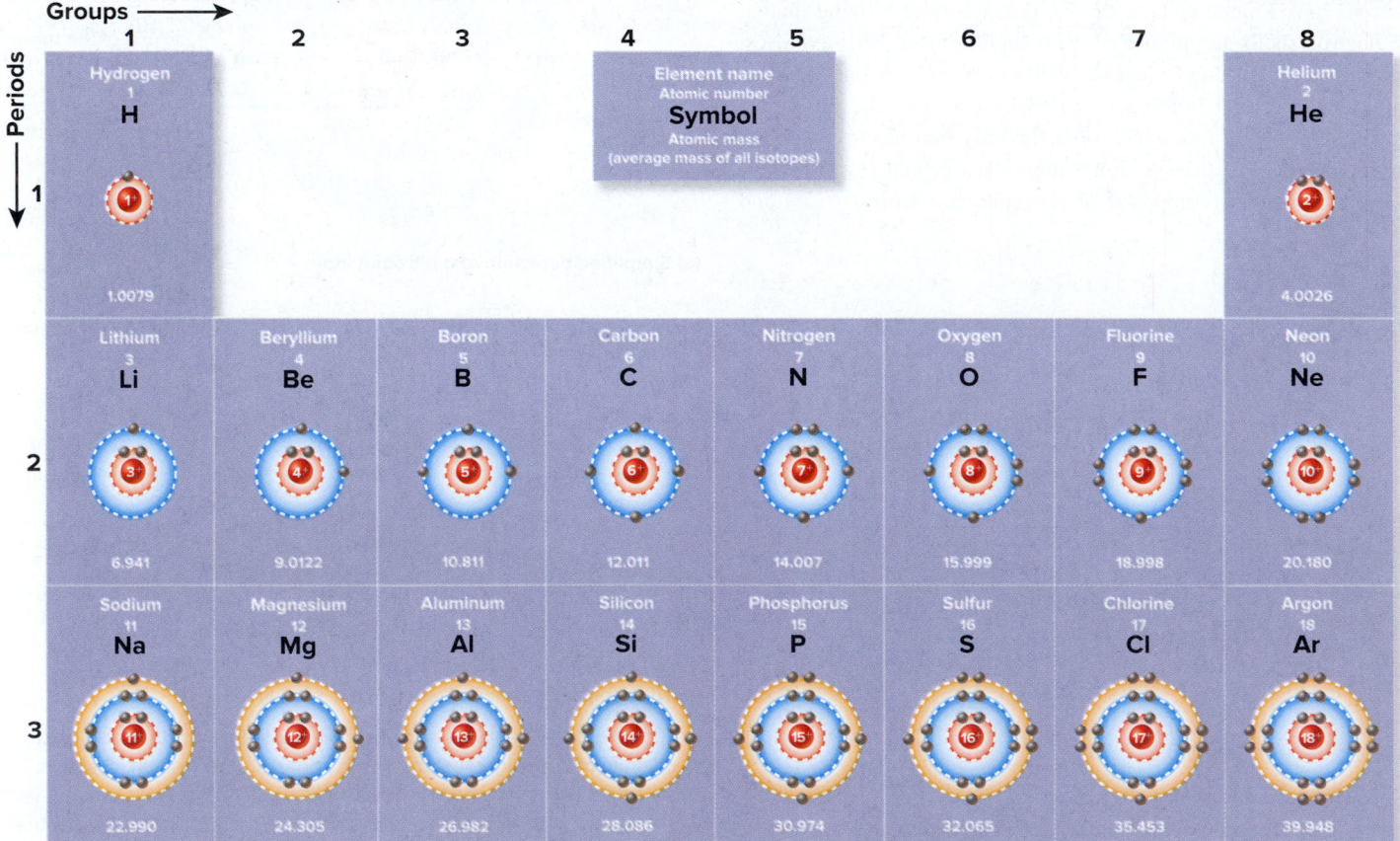

**Figure 2.5  The first three rows of the periodic table of the elements.** The elements are shown in models that depict the electron shells in different colors and show the total number of electrons in each shell. The occupancy of orbitals is that of the elements in their pure state. The red sphere represents the nucleus of the atom, and the numerical value with the $^+$ symbol represents the number of protons and, therefore, the positive charge of the nucleus. Elements are arranged in groups (columns) and periods (rows). For the complete periodic table, see Appendix A.

| Table 2.1 | Characteristics of Major Subatomic Particles | | |
|---|---|---|---|
| Particle | Location | Charge | Mass relative to electron |
| Proton | Nucleus | +1 | 1,836 |
| Neutron | Nucleus | 0 | 1,839 |
| Electron | Around the nucleus | −1 | 1 |

more than 1,800 times the mass of an electron (**Table 2.1**). Because of their tiny size relative to protons and neutrons, the mass of the electrons in an atom is ignored in calculations of atomic mass.

The **atomic mass** indicates an atom's mass relative to the masses of other atoms. By convention, the most common type of carbon atom, which has six protons and six neutrons, is assigned an atomic mass of exactly 12. On this scale, a hydrogen atom has an atomic mass of 1, indicating that it has 1/12 the mass of a carbon atom. A magnesium atom, with an atomic mass of 24, has twice the mass of a carbon atom.

The term mass is sometimes confused with weight, but these two terms refer to different features of matter. Weight is derived from the gravitational pull on a given mass. For example, a man who weighs 154 pounds on Earth would weigh only 25 pounds if he were standing on the Moon, and he would weigh 21 trillion pounds if he could stand on a neutron star. However, his mass is the same in all locations because he has the same amount of matter.

Atomic mass is measured in units called daltons, after the English chemist John Dalton, who postulated that matter is composed of tiny indivisible units he called atoms and laid the groundwork for atomic theory. One **dalton (Da)**, also known as an atomic mass unit (amu), equals 1/12 the mass of a carbon atom, or about the mass of a proton or a hydrogen atom. Therefore, the most common type of carbon atom has an atomic mass of 12 Da.

Because atoms such as hydrogen have a small mass, but atoms such as carbon have a larger mass, 1 g of hydrogen contains more atoms than 1 g of carbon. A **mole** (mol) of any substance contains the same number of particles as there are atoms in exactly 12 g of carbon. Twelve grams of carbon equals 1 mol of carbon, and 1 g of hydrogen equals 1 mol of hydrogen. As first described by Italian physicist Amedeo Avogadro, 1 mole of any element contains the same number of atoms—$6.022 \times 10^{23}$. For example, 12 g of carbon contains $6.022 \times 10^{23}$ atoms, and 1 g of hydrogen, whose atoms have 1/12 the mass of a carbon atom, also contains $6.022 \times 10^{23}$ atoms. This number is known as Avogadro's number. To visualize the enormity of this number, imagine that people could pass through a turnstile at a rate of 1 million people per second. Even at that incredible rate, it would require almost 20 billion years for $6.022 \times 10^{23}$ people to go through that turnstile!

## Isotopes Vary in Their Number of Neutrons

Many elements can exist in multiple forms, called **isotopes**, that differ in the number of neutrons they contain. For example, the most abundant form of the carbon atom, $^{12}C$, contains six protons and six neutrons, and thus has an atomic number of 6 and an atomic mass of 12 Da. The superscript placed to the left of $^{12}C$ is the sum of the protons and neutrons. The rare carbon isotope $^{14}C$, however, contains six protons and eight neutrons. Although $^{14}C$ has an atomic number of 6, it has an atomic mass of 14 Da. Nearly 99% of the carbon in living organisms is $^{12}C$. Consequently, the average atomic mass of carbon is slightly greater than 12 Da because of the existence of a small amount of heavier isotopes. This explains why the atomic masses given in the periodic table do not add up exactly to the predicted masses based on the atomic number and the number of neutrons of a given atom (for example, see carbon in Figure 2.5).

Isotopes of an atom have similar chemical properties but may have very different physical properties. For example, many isotopes found in nature are inherently unstable; the length of time they persist is measured in half-lives—a half-life is the time it takes for 50% of an isotope to decay. Such unstable isotopes are called **radioisotopes**. They emit radiation, which converts them to a stable form. At the very low amounts found in nature, radioisotopes usually pose no serious threat to life, but exposure of living organisms to high amounts of radioactivity can result in the disruption of cellular function, cancer, and even death.

Modern medical treatment and diagnosis make use of the special properties of radioactive compounds in many ways. For example, beams of high-energy radiation can be directed onto cancerous parts of the body to kill cancer cells. In another example, one or more atoms in the metabolically important sugar molecule glucose can be chemically replaced with a radioactive isotope of fluorine ($^{18}F$) to create a molecule called fluorodeoxyglucose (FDG). $^{18}F$ has a half-life of about 110 minutes. When a solution containing such modified radioactive glucose is injected into a person's bloodstream, the organs of the body take up the molecules from the blood just as they would ordinary glucose. Special imaging techniques, such as the positron-emission tomography (PET) scan shown in **Figure 2.6**, can detect the amount of the radioactive FDG in the body's organs. In this way, it is possible to visualize whether organs such as the lungs or brain are functioning normally, or at an increased or decreased rate. For example, cancer cells tend to take up much more glucose than normal cells do. Therefore, PET scans can reveal the presence of cancer—a disease characterized by uncontrolled cell growth. The scan of the individual shown in Figure 2.6, for example, identified numerous regions of high activity, suggestive of cancer.

## The Mass of All Living Organisms Is Largely Composed of Four Elements

Just four elements—oxygen, carbon, hydrogen, and nitrogen—account for the vast majority of atoms in living organisms (**Table 2.2**). These elements typically make up about 95% of the mass of living organisms. Much of the oxygen and hydrogen occur in the form of water, which accounts for approximately 60% of the mass of most animals and up to 95% or more in some plants. Carbon is a major building block of all living matter, and nitrogen is a vital element in all proteins. Note in Table 2.2 that although hydrogen accounts for about 63% of the atoms in the body, it makes up only a small percentage of the mass of the human body. That is because the atomic mass of hydrogen is so much smaller than that of heavier elements such as oxygen.

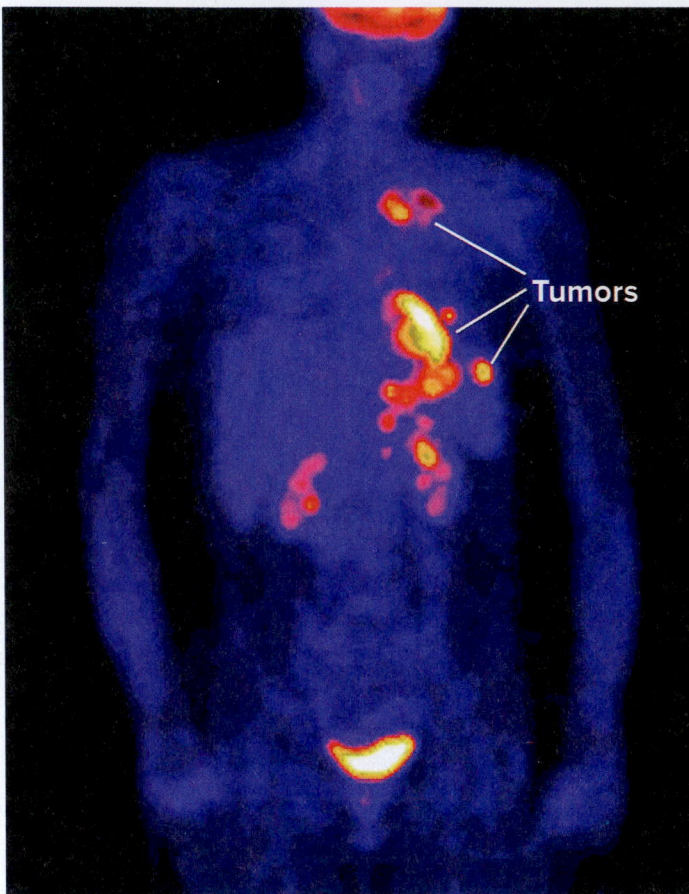

**Figure 2.6** **Diagnostic image of the human body using radioisotopes.** An imaging technique called positron-emission tomography (PET) highlights regions of the body that are actively using glucose, the body's major energy source. Radioactivity in this image shows up as a color. The bright patches are regions of extremely intense activity, some of which were later determined to be cancer in this patient. Note: In this PET scan, the brain, kidneys, and bladder are not cancerous. The brain naturally uses a high amount of glucose. The kidneys filter the blood, and the bladder accumulates urine, which contains some glucose. ©Steven Needell/Science Source

 **Core Skill: Science and Society** Applying an understanding of chemistry to biology has transformed the ability of physicians to diagnose diseases in humans. In the United States alone, between 1 and 2 million PET scans such as this one are performed each year, helping to localize the sites and extent of diseased structures, greatly facilitating subsequent drug or surgical treatments.

Other important elements in living organisms include the mineral elements. Calcium and phosphorus, for example, are important constituents of the skeletons and shells of animals. Minerals such as potassium and sodium are key regulators of water movement and of electric currents that occur across the surfaces of many cells.

In addition, all living organisms require **trace elements**. These elements are present in extremely small quantities but are essential for normal growth and function. For example, iron plays an important role in how vertebrates transport oxygen in their blood, and copper serves a similar role in some invertebrates.

| Table 2.2 | Chemical Elements Essential for Life in Many Organisms* | | |
|---|---|---|---|
| **Most abundant in living organisms (approximately 95% of total mass)** | | | |
| Element | Symbol | % Human body mass | % All atoms in human body |
| Oxygen | O | 65 | 25.5 |
| Carbon | C | 18 | 9.5 |
| Hydrogen | H | 9 | 63.0 |
| Nitrogen | N | 3 | 1.4 |
| **Mineral elements (less than 1% of total mass)** | | | |
| Calcium | Ca | Potassium | K |
| Chlorine | Cl | Sodium | Na |
| Magnesium | Mg | Sulfur | S |
| Phosphorus | P | | |
| **Trace elements (less than 0.01% of total mass)** | | | |
| Boron | B | Manganese | Mn |
| Chromium | Cr | Molybdenum | Mo |
| Cobalt | Co | Selenium | Se |
| Copper | Cu | Silicon | Si |
| Fluorine | F | Tin | Sn |
| Iodine | I | Vanadium | V |
| Iron | Fe | Zinc | Zn |

*Although these are the most common elements in living organisms, many other trace and mineral elements have reported functions. For example, aluminum is believed to be a cofactor for certain chemical reactions in animals, but it is generally toxic to plants.

## 2.2 Chemical Bonds and Molecules

### Learning Outcomes:

1. Compare and contrast the types of atomic interactions that lead to the formation of molecules.
2. Explain the concept of electronegativity, and describe how it contributes to the formation of polar and nonpolar covalent bonds.
3. Describe how a molecule's shape is important to its ability to interact with other molecules.
4. Relate the concepts of a chemical reaction and chemical equilibrium.

The linkage of atoms with other atoms serves as the basis for life and also gives life its great diversity. Two or more atoms bonded together make up a molecule. For example, two oxygen atoms can combine to form one oxygen molecule, represented as $O_2$. This representation is called a **molecular formula**. It consists of the chemical symbols for all of the atoms that are present (here, O for oxygen) and a subscript that tells you how many of those atoms are present in the molecule (in this case, two). The term **compound** refers to a molecule composed of two or more different elements. Examples include water ($H_2O$), with two hydrogen atoms and one oxygen atom, and the sugar glucose ($C_6H_{12}O_6$), which has 6 carbon atoms, 12 hydrogen atoms, and 6 oxygen atoms.

One of the most important features of compounds is their emergent properties. This means that the properties of a compound differ greatly from those of its elements. Let's consider sodium as an example. Pure sodium (Na) is a soft, silvery white metal that can be cut with a knife. When sodium forms a compound with chlorine (Cl), table salt (NaCl) results. NaCl is a white, relatively hard crystal (as seen in the chapter opening photo) that dissolves in water. Thus, the properties of a compound can be dramatically different from the properties of the elements that combined to form it.

The atoms in molecules are held together by chemical bonds. In this section, we will examine the different types of chemical bonds, how those bonds form, and how they determine the structures of molecules.

## Covalent Bonds Are Formed When Atoms Share Electrons to Fill Their Outer Shells

A **covalent bond** is a chemical bond in which two atoms share a pair of electrons. Covalent bonds can occur between atoms whose outer shells are not full. A fundamental principle of chemistry is the following:

*Atoms tend to be most stable when their outer shells are filled with electrons.*

**Figure 2.7** shows this principle as it applies to the formation of hydrogen fluoride (HF), a molecule with many important industrial and medical applications, such as petroleum refining and pharmaceutical production. The outer shell of a hydrogen atom is full when it contains two electrons, though a hydrogen atom has only one electron. The outer shell of a fluorine atom is full when it contains eight electrons, though a fluorine atom has only seven electrons in its outer shell. In the HF molecule, the two atoms share a pair of electrons, which spend time in the outer shells of both atoms. This allows both of the outer shells to be full. Covalent bonds are strong chemical bonds, because the shared electrons behave as if they belong to each atom.

Chemists sometimes depict molecules with a **structural formula** in which each covalent bond is represented by a line indicating a pair of shared electrons. For example, HF is depicted as

$$H—F$$

A molecule of water ($H_2O$) can be represented as

$$H—O—H$$

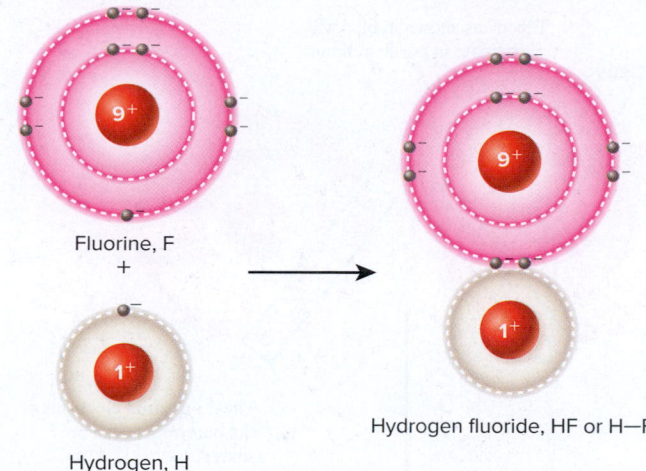

Fluorine, F
+

Hydrogen, H

Hydrogen fluoride, HF or H—F

**Figure 2.7  The formation of a covalent bond.** In a covalent bond, electrons from the outer shells of two atoms are shared with each other in order to complete both outer shells. This simplified illustration shows hydrogen forming a covalent bond with fluorine.

The structural formula of water indicates that the oxygen atom forms a covalent bond with both hydrogen atoms.

Each atom forms a characteristic number of covalent bonds, which depends on the number of electrons required to fill the outer shell. The atoms of some elements important for life, notably carbon, form more than one covalent bond and become linked simultaneously to two or more other atoms. Figure 2.8 shows the number of covalent bonds formed by several atoms commonly found in the molecules of living cells—hydrogen, oxygen, nitrogen, and carbon.

For many types of atoms, their outermost shell is full when it contains eight electrons, an octet. The **octet rule** states that many atoms are most stable when they have eight electrons in their outermost electron shell. This rule applies to most atoms found in living organisms, including oxygen, nitrogen, carbon, phosphorus, and sulfur. These atoms form a characteristic number of covalent bonds to make an octet in their outermost shell (see **Figure 2.8**). However, the octet rule does not always apply. For example, hydrogen has an outermost shell that can contain only two electrons, not eight.

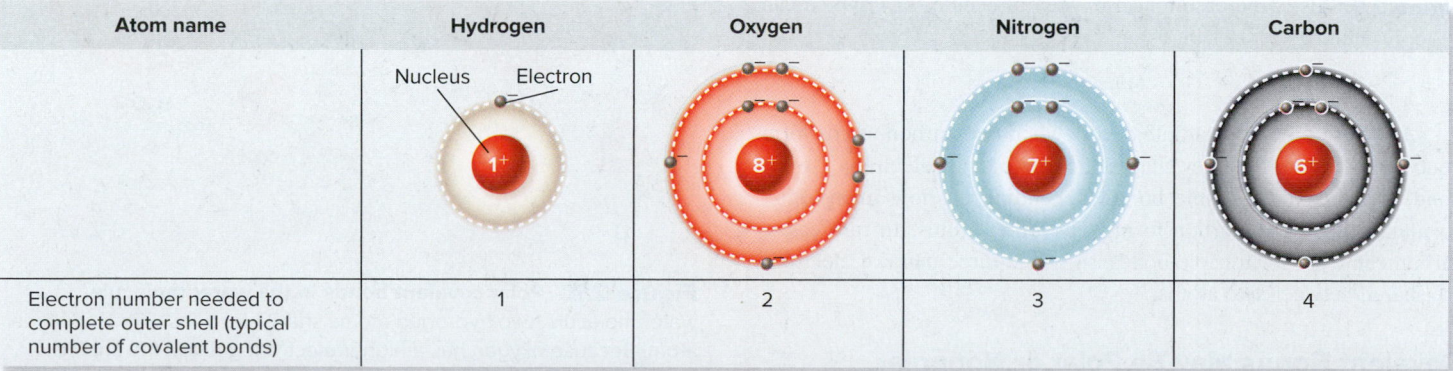

| Atom name | Hydrogen | Oxygen | Nitrogen | Carbon |
|---|---|---|---|---|
| Electron number needed to complete outer shell (typical number of covalent bonds) | 1 | 2 | 3 | 4 |

**Figure 2.8  The number of covalent bonds formed by common elements found in living organisms.** These elements form different numbers of covalent bonds due to the electron configurations in their outer shells.

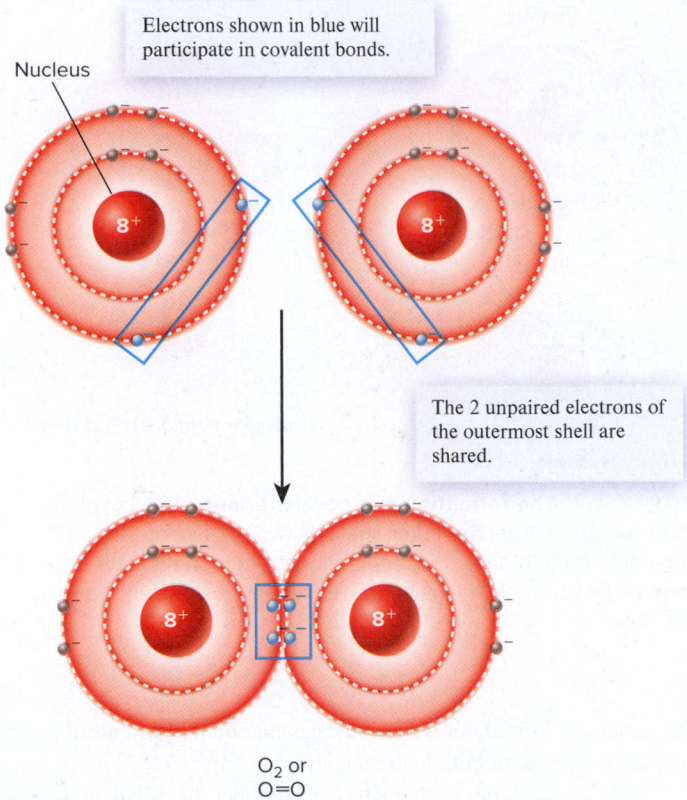

Electrons shown in blue will participate in covalent bonds.

Nucleus

The 2 unpaired electrons of the outermost shell are shared.

$O_2$ or
$O=O$

**Figure 2.9** A double bond between two oxygen atoms.

 **Core Skill: Modeling** The goal of this modeling challenge is to apply your understanding of molecular structure and propose a model for carbon dioxide.

**Modeling Challenge:** Carbon dioxide ($CO_2$) is a colorless and odorless gas that plays a key role in photosynthesis, as discussed in Chapter 8. In a $CO_2$ molecule, each oxygen forms two covalent bonds with the carbon atom. Draw a model for $CO_2$, using the same format as shown for the model of $O_2$ in this figure.

In some molecules, a **double bond** occurs when atoms share two pairs of electrons (four electrons) rather than one pair. As shown in **Figure 2.9**, this is the case for an oxygen molecule ($O_2$), which can be represented as

$$O=O$$

Another common example occurs when two carbon atoms form bonds in compounds. They may share one pair of electrons (single bond) or two pairs (double bond), depending on how many other covalent bonds each carbon forms with other atoms. In rare cases, carbon can even form triple bonds, in which three pairs of electrons are shared between two atoms.

## Covalent Bonds May Be Polar or Nonpolar

Some atoms attract shared electrons more strongly than do other atoms. The **electronegativity** of an atom is a measure of its ability to attract

electrons in a bond with another atom. When two atoms with different electronegativities form a covalent bond, the shared electrons are more likely to be closer to the nucleus of the atom of higher electronegativity than to the nucleus of the atom of lower electronegativity. Such bonds are called **polar covalent bonds**, because the distribution of the shared electrons around the nuclei creates a polarity, or difference in electric charge, across the molecule. Water is a classic example of a molecule containing polar covalent bonds. Because oxygen is much more electronegative than hydrogen, the shared electrons tend to be pulled closer to the oxygen nucleus than to either of the hydrogens. This unequal sharing of electrons gives the molecule a region of partial negative charge (indicated by the Greek letter δ and a minus sign, δ⁻) and two regions of partial positive charge (δ⁺) (**Figure 2.10**).

Atoms with high electronegativity, such as oxygen and nitrogen, have a relatively strong attraction for electrons. These atoms form polar covalent bonds with hydrogen atoms, which have low electronegativity. Examples of polar covalent bonds include O—H and N—H. In contrast, bonds between atoms with similar electronegativities, for example, between two carbon atoms (C—C) or between carbon and hydrogen atoms (C—H), are called **nonpolar covalent bonds**. Polar molecules usually have one or more polar covalent bonds, whereas nonpolar molecules tend to have bonds that are mostly nonpolar covalent. A single molecule may have different regions with nonpolar bonds and polar bonds. As we will explore later, the physical characteristics of polar and nonpolar molecules, especially their solubility in water, are quite different.

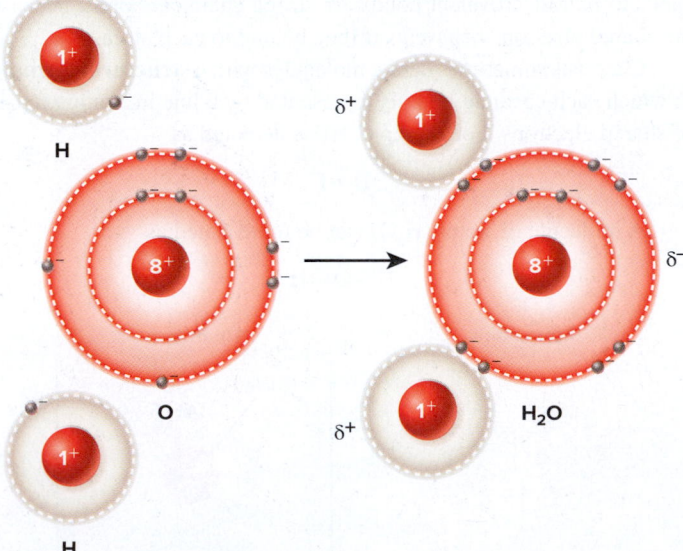

In water, the shared electrons tend to be closer to the oxygen atom. This gives oxygen a partial negative charge (δ⁻) and each hydrogen a partial positive charge (δ⁺).

H

O

H

δ⁺

δ⁻

δ⁺

$H_2O$

**Figure 2.10** Polar covalent bonds in the water molecule. In a water molecule, two hydrogen atoms share electrons with an oxygen atom. Because oxygen has a higher electronegativity, the shared electrons tend to be closer to the oxygen nucleus. This gives oxygen a partial negative charge, designated δ⁻, and each hydrogen a partial positive charge, designated δ⁺.

## Hydrogen Bonds and van der Waals Dispersion Forces Promote Interactions Between and Within Molecules

An important effect of certain polar covalent bonds is the ability of one molecule to loosely associate with another molecule through a weak interaction called a **hydrogen bond**. A hydrogen bond forms when a hydrogen atom in one polar molecule becomes electrically attracted to an electronegative atom, such as an oxygen or nitrogen atom, in another polar molecule. Hydrogen bonds, like those between water molecules, are represented in diagrams by dashed or dotted lines to distinguish them from covalent bonds (**Figure 2.11a**). A single hydrogen bond is very weak. The strength of a hydrogen bond is only a small percentage of the strength of the polar covalent bonds linking the hydrogen and oxygen within a water molecule.

Hydrogen bonds can also occur within a single large molecule. Large molecules may have many hydrogen bonds within their structure. Collectively, many hydrogen bonds may provide a strong force that helps maintain the three-dimensional structure of a molecule. This is particularly true in deoxyribonucleic acid (DNA)—the molecule that makes up the genetic material of living organisms. DNA exists as two long, twisting strands of many millions of atoms. The two strands are held together along their length, in part, by hydrogen bonds between different portions of the molecule (**Figure 2.11b**). Due to the large number of hydrogen bonds, it takes considerable energy to separate the strands of DNA.

In contrast to the cumulative strength of many hydrogen bonds, the weakness of the individual bonds is also important. When an interaction between two molecules involves relatively few hydrogen bonds, such an interaction tends to be weak and readily disrupted. The reversible nature of hydrogen bonds allows molecules to interact and then to become separated again. For example, small molecules may bind to proteins called enzymes via hydrogen bonds. **Enzymes** are molecules that catalyze many biologically important chemical reactions. The small molecules are later released, after the enzymes have changed their structure.

In addition to hydrogen bonds, another type of weak molecular attraction is due to **van der Waals dispersion forces**. These van der Waals dispersion forces arise because electrons are located within orbitals in a random way, as described previously. At any moment, the electrons in the outer shells of the atoms in a nonpolar molecule may be evenly distributed or unevenly distributed. In the latter case, a fleeting electrical attraction to other nearby molecules may arise. As with hydrogen bonds, the collective strength of these temporary attractive forces between molecules can be quite strong.

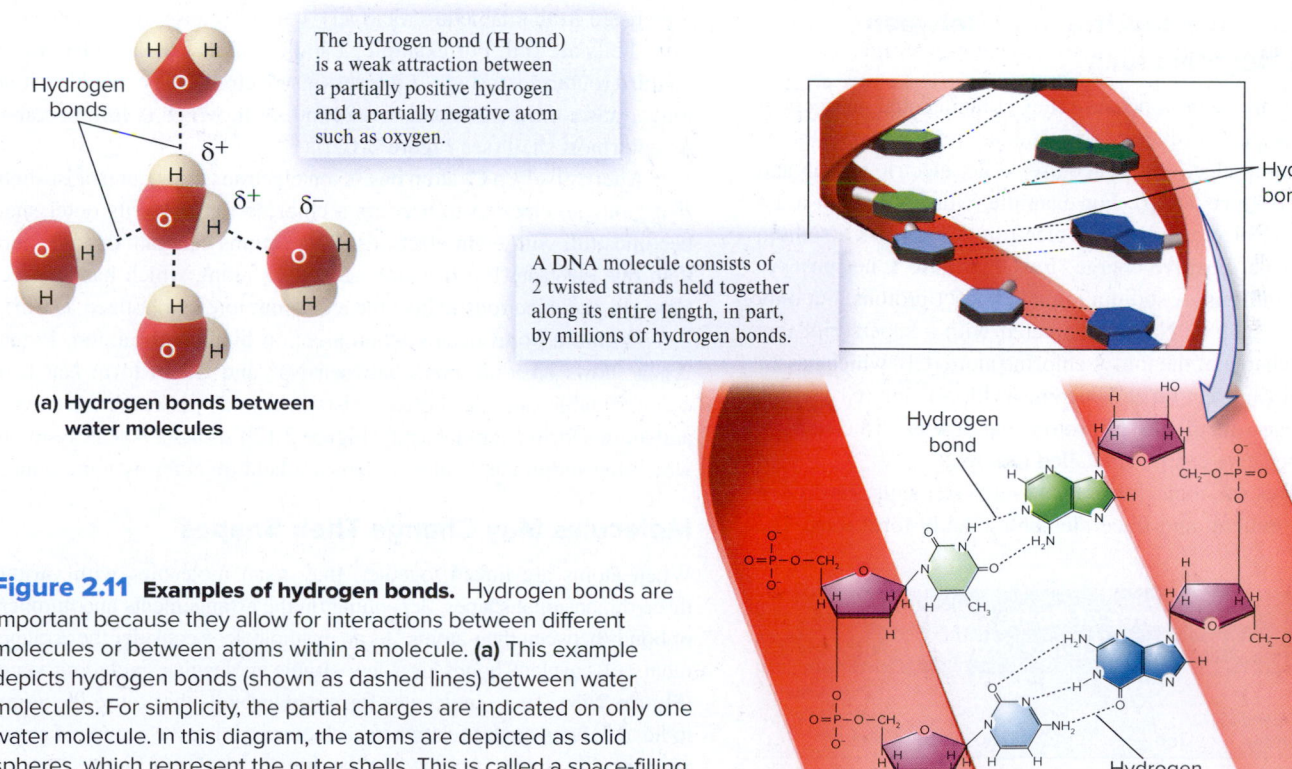

**Figure 2.11  Examples of hydrogen bonds.** Hydrogen bonds are important because they allow for interactions between different molecules or between atoms within a molecule. **(a)** This example depicts hydrogen bonds (shown as dashed lines) between water molecules. For simplicity, the partial charges are indicated on only one water molecule. In this diagram, the atoms are depicted as solid spheres, which represent the outer shells. This is called a space-filling model of an atom. **(b)** A DNA molecule is composed of two twisting strands connected to each other by hydrogen bonds (dashed lines). Although each individual bond is weak, the sum of all the hydrogen bonds in a large molecule like DNA imparts considerable stability to the molecule.

**Concept Check:** *In Chapter 11, you will learn that the two DNA strands must first separate into two single strands for DNA to be replicated. Do you think the process of strand separation requires energy, or do you think the strands can separate spontaneously?*

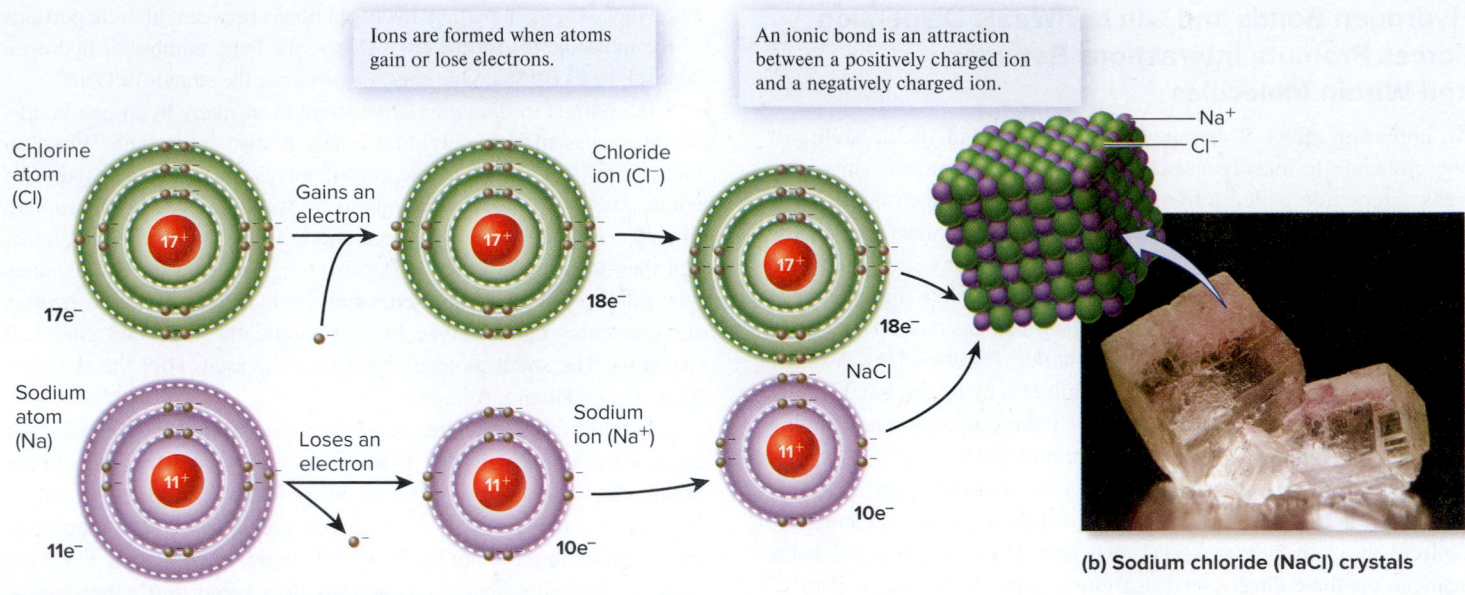

**Ions are formed when atoms gain or lose electrons.**

**An ionic bond is an attraction between a positively charged ion and a negatively charged ion.**

Chlorine atom (Cl) · 17e⁻ → Gains an electron → Chloride ion (Cl⁻) · 18e⁻ → · 18e⁻

Sodium atom (Na) · 11e⁻ → Loses an electron → Sodium ion (Na⁺) · 10e⁻ → · 10e⁻

NaCl · Na⁺ · Cl⁻

**(b) Sodium chloride (NaCl) crystals**

**(a) Formation of ions and an ionic bond**

**Figure 2.12  Ionic bonding in table salt (NaCl). (a)** When a sodium atom loses an electron, a sodium ion is formed, and when a chlorine atom gains an election, a chloride ion is formed. The resulting ions are attracted to each other via an ionic bond. **(b)** In a salt crystal, a lattice is formed in which the positively charged sodium ions ($Na^+$) are attracted to negatively charged chloride ions ($Cl^-$). Note: Electrons do not exist free in solution; when an electron is removed from one atom, it must be transferred to some other atom. b: ©Charles D. Winters/Science Source

## Ionic Bonds Involve an Attraction Between Positive and Negative Ions

Atoms are electrically neutral because they contain equal numbers of negative electrons and positive protons. If an atom or molecule gains or loses one or more electrons, it acquires a net electric charge and becomes an **ion** (**Figure 2.12a**). For example, when a sodium atom (Na), which has 11 electrons, loses 1 electron, it becomes a sodium ion ($Na^+$) with a net positive charge. Ions that have a net positive charge are called **cations**. A sodium ion still has 11 protons, but only 10 electrons. Ions such as $Na^+$ are depicted with a superscript that indicates the net charge of the ion. A chlorine atom (Cl), which has 17 electrons, can gain an electron and become a chloride ion ($Cl^-$) with a net negative charge—it still has 17 protons but now has 18 electrons. Ions with a net negative charge are called **anions**.

**Table 2.3** lists the ionic forms of several elements. Hydrogen atoms and most mineral and trace elements readily form ions. The ions listed in this table are relatively stable because their outer electron shells are full. For example, a sodium atom has one electron in its third (outermost) shell. If it loses this electron to become $Na^+$, it no longer has a third shell, and the second shell, which is full, becomes its outermost shell (see Figure 2.12a).

Alternatively, a Cl atom has seven electrons in its outermost shell. If it gains an electron to become a chloride ion ($Cl^-$), its outer shell becomes full with eight electrons. Some atoms can gain or lose more than one electron. For instance, a calcium atom, which has 20 electrons, loses 2 electrons to become a calcium ion, symbolized as $Ca^{2+}$.

An **ionic bond** occurs when a cation binds to an anion. Figure 2.12a shows an ionic bond between $Na^+$ and $Cl^-$ to form NaCl, or common table salt. NaCl often exists as crystals in which the cations and anions form a regular array. Figure 2.12b shows a NaCl crystal, in which the sodium and chloride ions are held together by ionic bonds.

## Molecules May Change Their Shapes

When atoms are linked together, they form molecules with various three-dimensional shapes, depending on the arrangements and numbers of bonds between their atoms. As an example, let's consider the arrangements of covalent bonds in a few simple molecules, including water (**Figure 2.13**). These molecules form new orbitals that cause the atoms to lie at defined angles relative to each other, giving the groups of atoms very specific shapes, as shown in the three examples of Figure 2.13.

Molecules containing covalent bonds are not rigid, inflexible structures. Think of a single covalent bond as an axle around which the joined atoms can rotate. Within certain limits, the shape of a molecule can change without breaking its covalent bonds. As illustrated in **Figure 2.14a**, a molecule of six carbon atoms bonded together can assume a number of shapes as a result of rotations around various covalent bonds. The three-dimensional, flexible shape of molecules

| Table 2.3 | Ionic Forms of Some Common Elements in Living Organisms | | | |
|---|---|---|---|---|
| Atom | Chemical symbol | Ion | Ion symbol | Electrons gained or lost |
| Calcium | Ca | Calcium ion | $Ca^{2+}$ | 2 lost |
| Chlorine | Cl | Chloride ion | $Cl^-$ | 1 gained |
| Hydrogen | H | Hydrogen ion | $H^+$ | 1 lost |
| Magnesium | Mg | Magnesium ion | $Mg^{2+}$ | 2 lost |
| Potassium | K | Potassium ion | $K^+$ | 1 lost |
| Sodium | Na | Sodium ion | $Na^+$ | 1 lost |

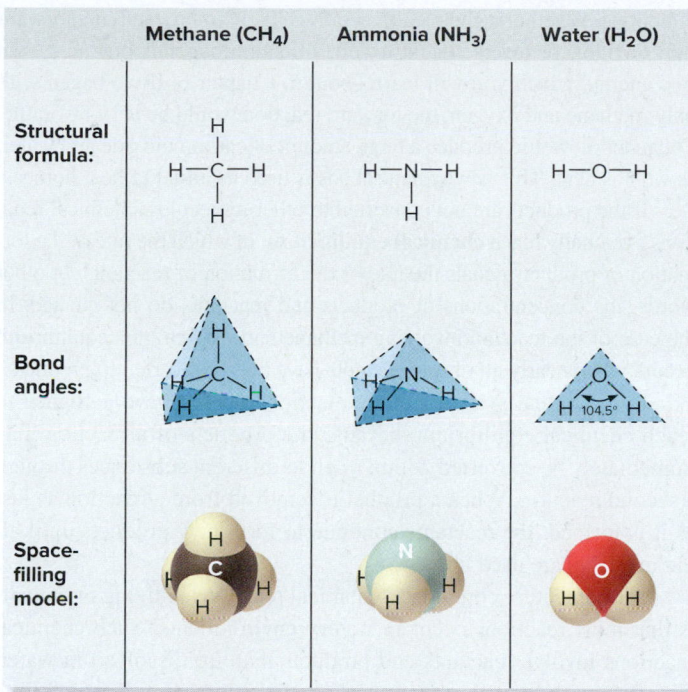

|  | Methane (CH$_4$) | Ammonia (NH$_3$) | Water (H$_2$O) |
|---|---|---|---|
| Structural formula: | | | |
| Bond angles: | | | |
| Space-filling model: | | | |

**Figure 2.13   Shapes of molecules.**  Molecules may assume different shapes, depending on the types of bonds between their atoms. The angles between the bonds are well defined. For example, in liquid water at room temperature, the angle formed by the covalent bonds between the two hydrogen atoms and the oxygen atom is approximately 104.5°. This bond angle can vary slightly, depending on the temperature and degree of hydrogen bonding between adjacent water molecules.

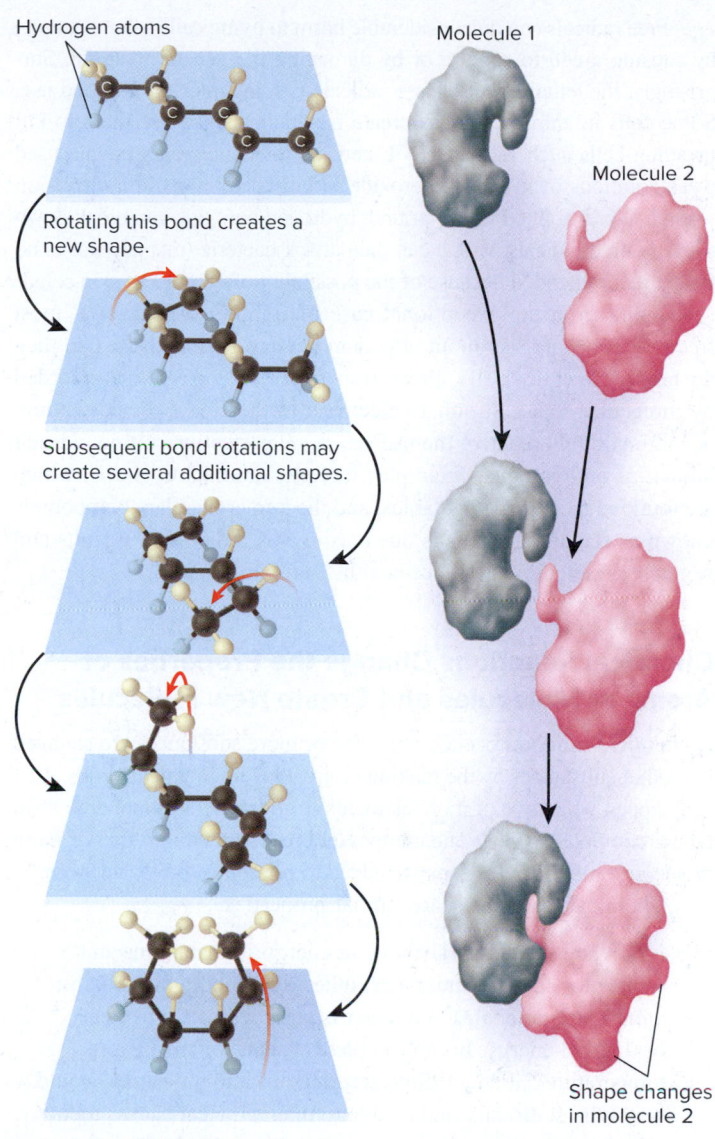

Hydrogen atoms

Rotating this bond creates a new shape.

Subsequent bond rotations may create several additional shapes.

Molecule 1

Molecule 2

Shape changes in molecule 2

**(a) Bond rotation in a small molecule**

**(b) Noncovalent interactions that may alter the shape of molecules**

**Figure 2.14   Shape changes in molecules.**  A single molecule may assume different three-dimensional shapes without breaking any of the covalent bonds between its atoms, as shown in **(a)** for a six-carbon molecule. Hydrogen atoms above the blue plane are shown in white; those below the blue plane are blue. **(b)** Two molecules are shown schematically as having complementary shapes that permit them to interact. During the interaction, the flexible nature of the molecules causes molecule 2 to twist sufficiently to assume a new shape. This change in shape is often an important mechanism by which one molecule influences the activity of another.

 **Core Concept: Structure and Function**  The three-dimensional structure of the two molecules in Figure 2.14b is critical in ensuring that they are capable of specifically interacting with each other and not with other molecules. The function of the molecules is defined by their ability to bind to each other, which in turn depends on their unique structures.

contributes to their biological properties. As shown in Figure 2.14b, the binding of one molecule to another may affect the shape of one of the molecules. A person can smell food, for instance, because odor molecules interact with special proteins called receptors in their nose (see Figure 44.24). When an odor molecule encounters a receptor, the two molecules recognize each other by their unique shapes, somewhat like a key fitting into a lock. As molecules in the food interact with the receptor, the shape of the receptor changes.

## Free Radicals Are a Special Class of Highly Reactive Molecules

Recall that an atom or an ion is most stable when each of its orbitals is occupied by a full complement of electrons. A molecule containing an atom with a single, unpaired electron in its outer shell is known as a **free radical**. Free radicals can react with other molecules to "steal" an electron from one of their atoms, thereby filling the orbital in the free radical. In the process, a new free radical may be created from the donor molecule, setting off a chain reaction.

Free radicals can be formed in several ways, including exposure of cells to radiation and toxins. They are depicted with a dot (representing the unpaired electron) next to the atomic symbol. Examples of biologically important free radicals are superoxide anion, $O_2^{-}$; hydroxyl radical, ·OH; and nitric oxide, NO·. Note that free radicals can be either charged or neutral.

Free radicals can do considerable harm to living cells—for example, by causing a cell to rupture or by damaging the genetic material. Surprisingly, the lethal effect of free radicals is sometimes put to good use. Some cells in animals' bodies create free radicals and use them to kill invading cells such as bacteria. Likewise, for many years people used weak solutions of hydrogen peroxide to kill bacteria, as in a dirty skin wound. When applied to the wound, hydrogen peroxide can break down to create free radicals, which can then attack bacteria (this practice is no longer recommended because of the possibility of damage to skin cells).

Aside from the exceptional case of fighting off bacteria, most free radicals that arise in an organism need to be inactivated so they do not kill healthy cells. Protection from free radicals is afforded by molecules that can donate electrons to the free radicals without becoming highly reactive themselves. Such protective compounds are known as antioxidants. Examples include vitamins C and E, which are found in fruits and vegetables, and the numerous plant compounds known as flavonoids. This is one reason why a diet rich in fruits and vegetables is beneficial to our health.

## Chemical Reactions Change the Properties of Atoms or Molecules and Create New Molecules

A **chemical reaction** occurs when one or more substances are changed into other substances by the making or breaking of chemical bonds. This can happen when two or more elements or compounds combine to form a new compound, when one compound breaks down into two or more molecules, or when electrons are added to or removed from an atom.

Chemical reactions share similar properties.

- First, they all require a source of energy so that atoms and molecules can encounter each other. The energy required for atoms and molecules to interact is provided partly by heat or thermal energy. In the complete absence of any heat (a temperature called absolute zero), atoms and molecules would be totally stationary and unable to interact. Heat causes them to vibrate and move, a phenomenon known as Brownian motion.

- Second, chemical reactions that occur in living organisms often require more than just Brownian motion to proceed at a reasonable rate. Such reactions need to be catalyzed. As discussed in Chapter 6, a **catalyst** is an agent that speeds up the rate of a chemical reaction. Enzymes are proteins that are found in all cells and catalyze most chemical reactions.

- Third, chemical reactions tend to proceed in a particular direction but eventually reach a state of equilibrium.

To understand what we mean by "direction" and "equilibrium," let's consider a chemical reaction between methane (a component of natural gas) and oxygen. When a single molecule of methane reacts with two molecules of oxygen, one molecule of carbon dioxide and two molecules of water are produced:

$$CH_4 + 2\,O_2 \;\rightleftharpoons\; CO_2 + 2\,H_2O$$
(methane) (oxygen)    (carbon dioxide) (water)

As it is written here, methane and oxygen are the starting materials, or **reactants**, and carbon dioxide and water are the **products**. The bidirectional arrows indicate that this reaction can proceed in both

directions. Whether a chemical reaction is likely to proceed in a forward (left to right) or reverse (right to left) direction depends on changes in free energy, which you will learn about in Chapter 6. If we began with only methane and oxygen, the forward reaction would be very favorable. The reaction would produce a large amount of carbon dioxide and water, as well as heat. This is why natural gas is used as a fuel to heat homes.

If the products are not converted to other molecules, chemical reactions eventually reach **chemical equilibrium**, in which the rate of the formation of products equals the rate of the formation of reactants. In other words, the concentrations of products and reactants do not change. In the case of the reaction involving methane and oxygen, this equilibrium occurs when nearly all of the reactants have been converted to products.

In biological systems, many reactions do not have a chance to reach chemical equilibrium, because the products of a reaction may immediately be converted within a cell to different substances through a second reaction. When a product is removed from a reaction as fast as it is formed, the reactants continue to form new products until all the reactants are used up.

A final feature common to chemical reactions in living organisms is that most reactions occur in watery environments. Such chemical reactions involve reactants and products that are dissolved in water. Next, we will examine the properties of this amazing liquid and its importance to biology.

## 2.3 Properties of Water

**Learning Outcomes:**

1. Describe how hydrogen bonding determines many properties of water.
2. List the properties of water that make it a good solvent, and distinguish between hydrophilic and hydrophobic substances.
3. **CoreSKILL »** Calculate the molarity of a solution, and explain its meaning.
4. Discuss the properties of water that are critical for the survival of living organisms.

It would be difficult to imagine life without **water**, which is the liquid form of $H_2O$. People can survive for a month or more without food but usually die in less than a week without water. The bodies of all organisms are composed largely of water; most of the cells in an organism's body are filled with water and surrounded by it. Up to 95% of the weight of certain plants comes from water. In humans, typically 60–70% of body weight is due to water. The brain is roughly 70% water, blood is about 80% water, and the lungs are nearly 90% water. Even our bones are about 20% water! In addition, water is an important liquid in the surrounding environments of living organisms. For example, many species are aquatic organisms that live in watery environments.

Thus far in this chapter, we have considered the features of atoms and molecules and the nature of bonds and chemical reactions between atoms and molecules. In this section, we will turn our attention to issues related to the liquid properties of living organisms and the environment in which they live. Most of the chemical reactions that occur in nature involve molecules that are dissolved in water, including those reactions that happen inside the cells of living organisms and in the spaces that surround the cells (**Figure 2.15**).

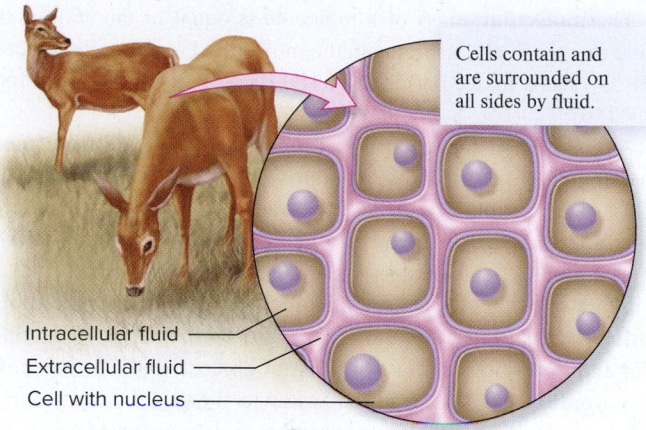

**Figure 2.15 Fluids inside and outside of cells.** Water is found in the intracellular fluid and in the extracellular fluid. Chemical reactions are always ongoing in both fluids.

In this section, we will examine the properties of chemicals that influence whether they dissolve in water and consider how biologists measure the amounts of dissolved substances. In addition, we will examine some of the other special properties of water that make it a vital component of living organisms and their environments.

## Ions and Polar Molecules Readily Dissolve in Water

Substances dissolved in a liquid are known as **solutes**, and the liquid in which they are dissolved is the **solvent**. In all living organisms, the solvent for nearly all chemical reactions is water, which is the most abundant solvent in nature. Solutes dissolve in a solvent to form a **solution**. Solutions made with water are called **aqueous solutions**.

To understand why a substance dissolves in water, we need to consider the chemical bonds in the solute molecule and those in water. As discussed earlier, the covalent bonds linking the two hydrogen atoms to the oxygen atom in a water molecule are polar. Therefore, the oxygen in water has a slight negative charge, and each hydrogen has a slight positive charge. To dissolve in water, a substance must be electrically attracted to water molecules. For example, table salt (NaCl) is a solid crystalline substance because of the strong ionic bonds between positive sodium ions (Na⁺) and negative chloride ions (Cl⁻). When a crystal of sodium chloride is placed in water, the partially negatively charged oxygens of water molecules are attracted to Na⁺, and the partially positively charged hydrogens are attracted to Cl⁻ (**Figure 2.16**). Clusters of water molecules surround the ions, allowing Na⁺ and Cl⁻ to separate from each other and enter the water—that is, to dissolve.

Generally, molecules that contain ionic and/or polar covalent bonds dissolve in water. Such molecules are said to be **hydrophilic**, which literally means "water-loving." In contrast, molecules composed predominantly of carbon and hydrogen are relatively insoluble in water, because carbon-carbon and carbon-hydrogen bonds are nonpolar. These molecules do not have partial positive and negative charges and, therefore, are not attracted to water molecules. Such molecules are **hydrophobic**, or "water-fearing." Oils are a familiar

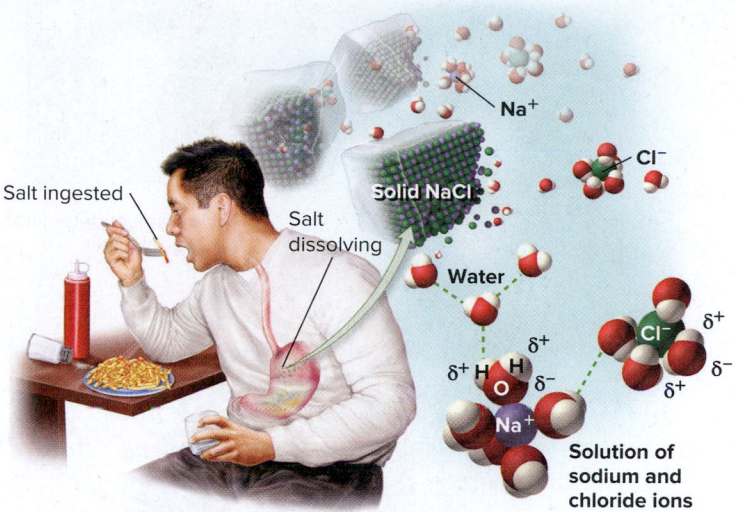

**Figure 2.16 Table salt (NaCl crystals) dissolving in water.** The ability of water to dissolve sodium chloride crystals depends on the electrical attraction between the polar water molecules and the charged sodium (Na⁺) and chloride ions (Cl⁻). Water molecules surround each ion as it becomes dissolved. For simplicity, the partial charges are indicated for only two water molecules.

example of hydrophobic molecules. Try mixing vegetable oil with water and observe the result. The two liquids separate into an oil layer and a water layer, with very little oil dissolving in the water.

## Some Molecules Have Both Hydrophilic and Hydrophobic Regions

Molecules that have both hydrophilic regions at one or more sites and hydrophobic regions at other sites are called **amphipathic** (or amphiphilic, from the Greek for "both loves"). When mixed with water, long amphipathic molecules may aggregate into spheres called **micelles**, with their polar (hydrophilic) regions at the surface of each micelle, where they are attracted to the surrounding water molecules. The nonpolar (hydrophobic) ends are oriented toward the interior of the micelle (**Figure 2.17**). Such an arrangement minimizes the interaction between water molecules and the nonpolar ends of the amphipathic molecules, which face inward. Nonpolar molecules can dissolve in the central nonpolar regions of these clusters and thus exist in an aqueous environment in far higher amounts than would otherwise be possible based on their low solubility in water. Familiar examples of amphipathic molecules are those in detergents, which can form micelles that help to dissolve the oils and nonpolar molecules found in dirt. The detergent molecules found in soap have polar and nonpolar ends. Oils on your skin dissolve in the nonpolar regions of the detergent micelles, and the polar ends help the detergent rinse off in water, taking the oil with it.

In addition to micelles, amphipathic molecules may form structures consisting of double layers of molecules called bilayers. Such bilayers have two hydrophilic surfaces facing outside, in contact with water, and a hydrophobic interior facing away from water. As you will learn in Chapter 5, bilayers play a key role in cell membrane structure (look ahead to Figure 5.1).

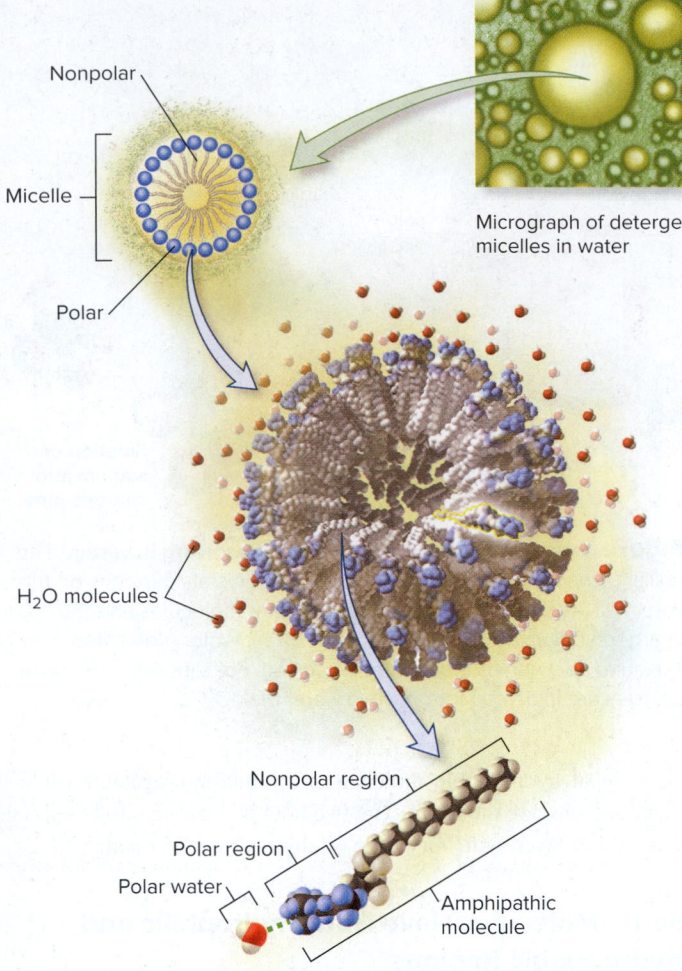

Micrograph of detergent micelles in water

**Figure 2.17   The formation of a micelle by amphipathic molecules.** In water, amphipathic molecules tend to arrange themselves so their nonpolar regions are directed away from water molecules and their polar regions are directed toward the water and can form hydrogen bonds with it. (top right): ©Jeremy Burgess/Science Source

**Concept Check:** *When oil dissolves in a soapy solution, where is the oil found?*

## The Amount of a Dissolved Solute per Unit Volume of Liquid Is Its Concentration

Solute **concentration** is defined as the amount of a solute dissolved in a unit volume of solution. For example, if 1 gram (g) of NaCl was dissolved in enough water to make 1 liter (L) of solution, we would say that its solute concentration is 1 g/L.

A comparison of the concentrations of two different substances on the basis of the number of grams per liter of solution does not directly indicate how many molecules of each substance are present. For example, let's compare 10 g each of glucose ($C_6H_{12}O_6$) and sodium chloride (NaCl). Because the individual molecules of glucose have more mass than those of NaCl, 10 g of glucose contains fewer molecules than 10 g of NaCl. Therefore, another way to describe solute concentration is according to the moles of dissolved solute per volume of solution. To make this calculation, we must know three things: the amount of dissolved solute, the molecular mass of the dissolved solute, and the volume of the solution.

The **molecular mass** of a molecule is equal to the sum of the atomic masses of all the atoms in the molecule. For example, glucose ($C_6H_{12}O_6$) has a molecular mass of 180 ([6 × 12] + [12 × 1] + [6 × 16] = 180).

As mentioned earlier, 1 mole (abbreviated mol) of a substance is the amount of the substance in grams equal to its atomic or molecular mass. The **molarity** of a solution is defined as the number of moles of a solute dissolved in 1 L of solution. A solution containing 180 g of glucose (1 mol) dissolved in enough water to make 1 L is a 1 **molar** solution of glucose (1 mol/L). By convention, a 1 mol/L solution is usually written as 1 M, where the capital M stands for molar and is defined as mol/L. If 90 g of glucose (half its molecular mass) were dissolved in enough water to make 1 L, the solution would have a solute concentration of 0.5 mol/L, or 0.5 M.

The concentrations of solutes dissolved in the fluids of living organisms are usually much less than 1 M. Many have concentrations in the range of millimoles per liter (1 mM = 0.001 M = $10^{-3}$ M), and others are present in even smaller concentrations—micromoles per liter (1 µM = 0.000001 M = $10^{-6}$ M), nanomoles per liter (1 nM = 0.000000001 M = $10^{-9}$ M), picomoles per liter (1 pM = 0.000000000001 M = $10^{-12}$ M), or even less.

### Core Skill: Quantitative Reasoning

**BIO TIPS**

**THE QUESTION** *Insulin is a hormone that regulates the uptake of glucose into muscle and fat cells. It is composed of 51 amino acids and has a molecular mass of approximately 5,808 g/mol. In healthy individuals, insulin levels in the bloodstream rise sharply after eating a meal, which increases the ability of muscle and fat cells to take up glucose. Two individuals, Alfonzo and Gordan, had their blood insulin levels tested before and after a meal.*

|         | Before meal (ng/L) | After meal (ng/L) |
|---------|---------|---------|
| Alfonzo | 475     | 2,850   |
| Gordan  | 399     | 789     |

*Calculate the molarity of insulin in each individual's bloodstream before and after the meal, expressed in units of picomoles/L (pM). (Note: 1 mole = $10^{12}$ picomoles.) Which person do you think may be diabetic because he does not release enough insulin into his bloodstream?*

**TOPIC** *What topic in biology does this question address?* The topic is insulin levels in the bloodstream. More specifically, the question is about calculating the concentration of insulin (in pM) before and after a meal. You are also asked to consider if either individual may be diabetic.

**INFORMATION** *What information do you know based on the question and your understanding of the topic?* From the question, you have learned that insulin increases the ability of muscle and fat cells to take up glucose and that diabetic individuals have a diminished ability to produce insulin. You are given the blood insulin levels of two individuals before and after a meal. From your understanding of the topic, you may remember that molarity refers to the number of moles of a substance dissolved in a liter of solution.

**P**ROBLEM-SOLVING **S**TRATEGY *Make a calculation. Compare and contrast.* To begin to solve this problem, you first need to calculate the molarities. After this is done, you can compare the insulin concentrations between Alfonzo and Gordan.

Let's go through the calculation for Alfonzo's insulin level before a meal. We begin with his blood insulin level, which is 475 ng/L.

To convert this value to g/L, remember that $1 g = 10^9$ ng.

$$475 \text{ ng/L} \times \frac{1 \text{ g}}{10^9 \text{ ng}} = 475 \times 10^{-9} \text{ g/L}$$

We then divide this value by the molecular mass of insulin, which is 5,808 g/mol.

$$\frac{475 \times 10^{-9} \text{ g/L}}{5,808 \text{ g/mol}} = 0.082 \times 10^{-9} \text{ mol/L}$$

The question asks you to express your value in pM, where 1 mole = $10^{12}$ picomoles:

$$0.082 \times 10^{-9} \text{ mol/L} \times \frac{10^{12} \text{ pmol}}{\text{mol}} = 82 \text{ pmol/L, or 82 pM}$$

**ANSWER**

|          | Before meal (pM) | After meal (pM) |
|----------|------------------|-----------------|
| Alfonzo  | 82               | 492             |
| Gordan   | 69               | 136             |

*Alfonzo's blood insulin level after the meal is six times higher than before the meal, whereas Gordan's is only about twice as high. Gordan may be diabetic, but this would need to be substantiated by more extensive testing of his insulin and glucose levels.*

## H₂O Exists in Three States

$H_2O$ is an abundant compound on Earth that exists in all three states of matter—solid (ice), liquid (water), and gas (water vapor). At the temperatures found over most regions of the planet, $H_2O$ is found primarily as a liquid in which the weak hydrogen bonds between molecules are continuously being formed, broken, and formed again. If the temperature rises, the rate at which hydrogen bonds break increases, and molecules of water escape into the gaseous state, becoming water vapor. If the temperature falls, hydrogen bonds are broken less frequently, so larger and larger clusters of water molecules are formed, until at 0°C water freezes into a crystalline matrix—ice. The $H_2O$ molecules in ice tend to lie in a more orderly and open arrangement, that is, with greater intermolecular distances, which makes ice less dense than water. This is why ice floats on water (**Figure 2.18**). Compared with water, ice is also less likely to participate in most types of chemical reactions.

Changes in state, such as changes between the solid, liquid, and gaseous states of $H_2O$, involve an input or a release of energy. For example, when energy is supplied to make water boil, it changes from the liquid to the gaseous state—a process called vaporization. The heat required to vaporize 1 mole of any substance at its boiling point is called the substance's **heat of vaporization**. For water, this value

Ice: Hydrogen bonds are more stable.

Liquid water: Hydrogen bonds continually break and reform.

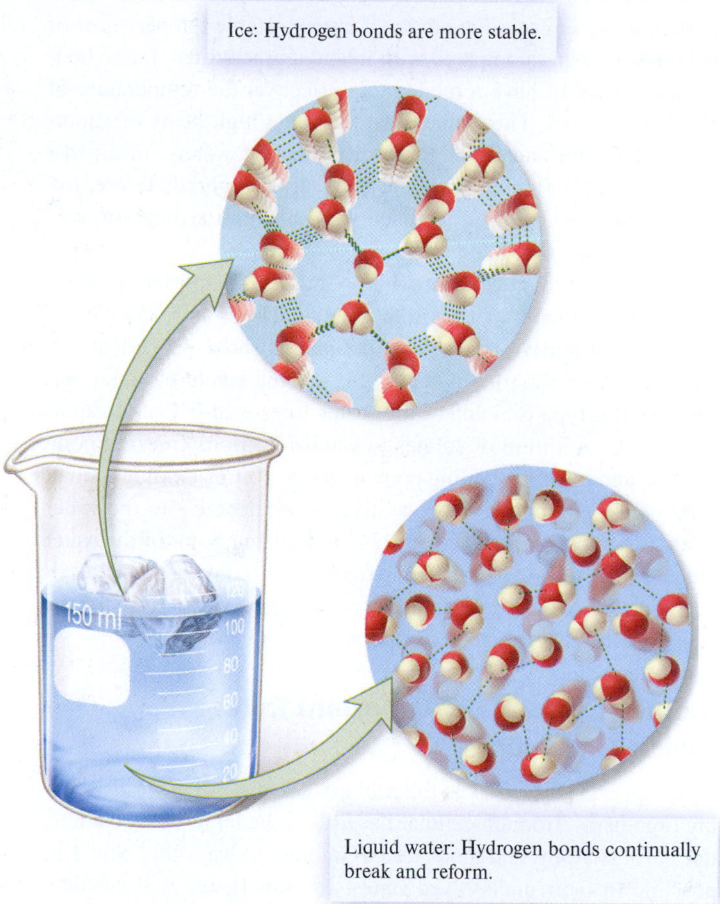

**Figure 2.18  Structure of liquid water and ice.** In the liquid form of $H_2O$, the hydrogen bonds between molecules continually form, break, and re-form, resulting in a changing arrangement of molecules from instant to instant. At temperatures at or below its freezing point, $H_2O$ forms a crystalline matrix called ice. In this solid form, hydrogen bonds are more stable. Ice has a hexagonal crystal structure. The greater space between $H_2O$ molecules in this crystal structure causes ice to have a lower density than liquid water. For this reason, ice floats on water.

is very high, because of the high number of hydrogen bonds between the molecules. It takes more than five times as much heat to vaporize water than it does to raise the temperature of water from 0°C to 100°C. In contrast, energy is released when water freezes to form ice. A substance's **heat of fusion** is the amount of heat that must be withdrawn or released from a substance to cause it to change from the liquid to the solid state. For water, this value is also high.

Another important feature for living organisms is that water has a very high **specific heat**, defined as the amount of heat required to raise the temperature of 1 gram of a substance by 1°C (or conversely, the amount of heat that must be lost to lower the temperature by 1°C). The high specific heat means that it takes considerable heat to raise the temperature of water. A related concept is **heat capacity**, which refers to the amount of heat required to raise the temperature of an entire object or amount of substance. A large beaker of distilled water has a greater heat capacity than a small beaker of distilled water, but the water in both beakers has the same specific heat. These properties of water contribute to the relatively stable temperatures of large bodies of water compared with inland temperatures. Large bodies of water tend to have a moderating effect on the temperature of nearby land masses. These three features—the high heats of vaporization and fusion and the high specific heat of water—mean that water is extremely stable as a liquid. Not surprisingly, therefore, living organisms have evolved to function best within a range of temperatures consistent with the liquid phase of water.

The temperature at which a solution freezes or vaporizes is influenced by the amounts of dissolved solutes. These are examples of a solution's **colligative properties**, defined as those properties that depend strictly on the total number of dissolved solute particles, not on the specific type of solute. Pure water freezes at 0°C and vaporizes at 100°C. Addition of solutes to water lowers its freezing point below 0°C and raises its boiling point to above 100°C. Adding a small amount of the compound ethylene glycol—antifreeze—to the water in a car's radiator, for instance, lowers the freezing point of the water and consequently prevents it from freezing in cold weather. Similarly, the presence of large amounts of solutes partly explains why the oceans do not freeze when the temperature falls below 0°C.

## Water Performs Many Important Roles in Living Organisms

As noted previously, water is the primary solvent in the fluids of all living organisms, from unicellular bacteria to the largest sequoia tree. Water permits atoms and molecules to interact in ways that would be impossible in their undissolved states. In Unit II, we will consider many ions and molecules that are solutes in living cells.

However, it is important to recognize that in addition to acting as a solvent, water serves many other remarkable functions that are critical for the survival of living organisms. For example, water molecules participate in many chemical reactions of this general type:

$$R_1—R_2 + H—O—H \rightarrow R_1—OH + H—R_2$$

R is a general symbol used in this case to represent a group of atoms. In this reaction, $R_1$ and $R_2$ are distinct groups of atoms. On the left side of the reaction, $R_1—R_2$ is a compound in which these groups of atoms are connected by a covalent bond. To be converted to products, a covalent bond is broken in each reactant, $R_1—R_2$ and H—O—H, and OH and H (from water) form covalent bonds with $R_1$ and $R_2$, respectively. Reactions of this type are known as **hydrolysis reactions** (from the Greek *hydro*, meaning water, and *lysis*, meaning to break apart), because water is used to break apart another molecule (**Figure 2.19a**). As discussed in Chapter 3 and later chapters, many large molecules are broken down into smaller, biologically important units by hydrolysis reactions.

Another property of water is that it is incompressible—its volume does not significantly decrease when subjected to high pressure. This has biological importance for many organisms that use water to provide force or support. For example, water supports the bodies of worms and some other invertebrates in a structure called a hydrostatic skeleton, and it provides turgidity (stiffness) and support for plants (**Figure 2.19b**).

Water is also the means by which unneeded and potentially toxic waste compounds are eliminated from an animal's body (**Figure 2.19c**). In mammals, for example, the kidneys filter out

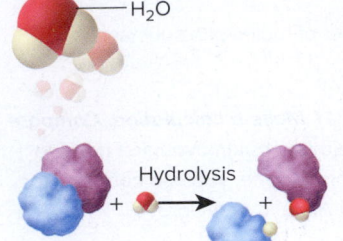

**(a) Water participates in chemical reactions.**

Hydrolysis

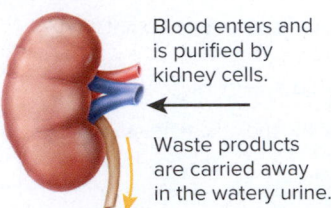

Blood enters and is purified by kidney cells.

Waste products are carried away in the watery urine.

**(c) Water is used to eliminate soluble wastes.**

**(e) The cohesive force of water molecules aids in the movement of fluid through vessels in plants.**

**(g) The surface tension of water explains why this water strider doesn't sink.**

**(b) Water provides support.** The plant on the right is wilting due to lack of water.

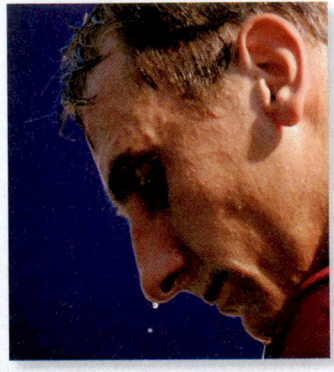

**(d) Evaporation helps some animals dissipate body heat.**

**(f) Water in saliva serves as a lubricant during—or as shown here, in anticipation of—feeding.**

**Figure 2.19   Some of the amazing functions of water.** In addition to acting as a solvent, water serves many crucial functions in nature. b: ©Aaron Haupt/Science Source; d: ©Chris McGrath/Getty Images; e: ©Dana Tezarr/Getty Images; f: ©Gallo Images-Anthony Bannister/DigitalVision/Getty Images; g: ©Matti Suopajarvi/mattisj/Getty Images

**Core Concept: Structure and Function**  The structure of water allows it perform a variety of functions as illustrated in this figure.

soluble waste products derived from the breakdown of proteins and other compounds. The filtered products remain in solution in a watery fluid, which eventually becomes urine and is excreted.

Recall from our earlier discussion of the three states of $H_2O$ that it takes considerable energy in the form of heat to convert water from a liquid to a gas. Although you may be familiar with the phenomenon of boiling water being converted to water vapor, water can also vaporize into the gaseous state even at ordinary temperatures. This process is known as **evaporation**. The simplest way to understand this is to imagine that in any volume of water at any temperature, some vibrating water molecules have higher energy than others. Those with the highest energy break their hydrogen bonds and escape into the gaseous state. During this process, energy in the form of heat is released into the environment. Evaporation is an important mechanism by which many animals cool themselves on hot days (**Figure 2.19d**).

The hydrogen-bonding properties of water affect its ability to form droplets and to adhere to surfaces. The phenomenon of water molecules attracting each other is called **cohesion**. Water exhibits strong cohesion due to hydrogen bonding. Cohesion aids in the movement of water through the vessels of plants (**Figure 2.19e**). A property similar to cohesion is **adhesion**, which refers to the ability of water to be attracted to, and thus adhere to, a surface that is not electrically neutral. Water tends to cling to surfaces to which it can hydrogen bond. For example, the adhesive properties of water allow it to coat the surfaces of the digestive tract of animals and act as a lubricant for the passage of food (**Figure 2.19f**).

**Surface tension** is a measure of the attraction between molecules at the surface of a liquid. In the case of water, the attractive force between hydrogen-bonded water molecules at the interface between water and air is what causes water to form droplets. The surface water molecules attract each other into a configuration (roughly that of a sphere) that reduces the number of water molecules in contact with air. You can see this by slightly overfilling a glass with water; the water forms a rounded bulge above the rim. Likewise, surface tension allows certain insects, such as water striders, to walk on the surface of a pond without sinking (**Figure 2.19g**).

## 2.4  pH and Buffers

### Learning Outcomes:

1. **CoreSKILL »** Calculate the concentrations of hydrogen and hydroxide ions at a given pH.
2. Give examples of how buffers maintain a stable environment in an animal's body fluids.

As we have seen, water is an essential solvent that is needed by all living organisms. In this section, we will examine the factors that determine the relative concentrations of $H^+$ and $OH^-$ ions in water, and see how those concentrations are calculated. We will also consider how buffers are used by living organisms to minimize fluctuations in the $H^+$ and $OH^-$ concentrations.

### Hydrogen Ion Concentrations Are Changed by Acids and Bases

Pure water has the ability to ionize to a very small extent into **hydroxide ions (OH–)** and hydrogen ions ($H^+$). In pure water, the concentrations of $H^+$ and $OH^-$ are both $10^{-7}$ mol/L, or $10^{-7}$ M. An inherent property of water is that the product of the concentrations of $H^+$ and $OH^-$ is always $10^{-14}$ M at 25°C. Therefore, in pure water, $[H^+][OH^-] = [10^{-7} M][10^{-7} M] = 10^{-14}$ M. (The brackets around the symbols for the hydrogen and hydroxide ions indicate that we are considering their concentrations.)

When certain substances are dissolved in water, they may release or absorb $H^+$ or $OH^-$, thereby altering the relative concentrations of these ions. Substances that release hydrogen ions in solution are called **acids**. Two examples are hydrochloric acid and carbonic acid:

$$HCl \quad \rightarrow \quad H^+ \quad + \quad Cl^-$$
(hydrochloric acid)                         (chloride ion)

$$H_2CO_3 \quad \rightleftharpoons \quad H^+ \quad + \quad HCO_3^-$$
(carbonic acid)                         (bicarbonate ion)

Hydrochloric acid is called a **strong acid** because it completely dissociates into $H^+$ and $Cl^-$ when added to water (which is why the arrow is not bidirectional in that reaction). By comparison, carbonic acid is a **weak acid** because some of it remains in the $H_2CO_3$ state when dissolved in water (note the bidirectional arrow, $\rightleftharpoons$).

Compared with an acid, a **base** has the opposite effect when dissolved in water—it decreases the $H^+$ concentration. This can occur in different ways. Some bases, such as sodium hydroxide (NaOH), release $OH^-$ when dissolved in water:

$$NaOH \quad \rightarrow \quad Na^+ \quad + \quad OH^-$$
(sodium hydroxide)    (sodium ion)

Recall that the product of $[H^+]$ and $[OH^-]$ is always $10^{-14}$ M. When a base such as NaOH raises the $OH^-$ concentration, some of the hydrogen ions bind to these hydroxide ions to form water. Therefore, increasing the $OH^-$ concentration lowers the $H^+$ concentration.

Let's consider another example. Ammonia reacts with water to produce ammonium ion:

$$NH_3 \quad + \quad H_2O \rightleftharpoons NH_4^+ \quad + \quad OH^-$$
(ammonia)            (ammonium ion)

In this case, $NH_3$ increases the $OH^-$ concentration by removing $H^+$ from $H_2O$. Both sodium hydroxide and ammonia have the same effect—they lower the concentration of $H^+$. NaOH achieves this by directly increasing the $OH^-$ concentration, whereas $NH_3$ reacts with water to produce $OH^-$.

### The pH Is a Measure of the $H^+$ Concentration of a Solution

The addition of an acid or base to water can greatly change the $H^+$ and $OH^-$ concentrations over a very broad range. Therefore, scientists use a log scale to describe the concentrations of these ions. The $H^+$ concentration is expressed as the solution's **pH**, which is defined as the negative logarithm to the base 10 of the $H^+$ concentration. A logarithmic scale is used because the concentrations of hydrogen ions can vary over a very wide range.

$$pH = -\log_{10} [H^+]$$

To understand what this equation means, let's consider a few examples. A solution with an $H^+$ concentration of $10^{-7}$ M has a pH of 7. A concentration of $10^{-7}$ M is the same as 0.1 μM. A solution in which $[H^+] = 10^{-6}$ M has a pH of 6. A concentration of $10^{-6}$ M is the same as 1.0 μM. A solution at pH 6 is said to be **acidic**, because it contains more $H^+$ ions than $OH^-$ ions. Note that as the acidity increases, the pH decreases. A solution in which the pH is 7 is said to be neutral because $[H^+]$ and $[OH^-]$ are equal. A solution with a pH above 7 is considered to be **alkaline**. **Figure 2.20** considers the pH values of some familiar fluids. Keep in mind that each change of 1 pH unit represents a 10-fold difference in $H^+$ concentration.

Why is pH of importance to living organisms? The answer lies in the observation that $H^+$ and $OH^-$ can readily bind to many kinds of ions and molecules. For this reason, the pH of a solution can affect:

- the shapes and functions of molecules,
- the rates of many chemical reactions,
- the ability of two molecules to bind to each other, and
- the ability of ions or molecules to dissolve in water.

Due to the various effects of pH, many biological processes function best within very narrow ranges of pH, and even small shifts can have a negative effect. In living cells, the pH ranges from about 6.5 to 7.8 and is carefully regulated to avoid major shifts. The blood of the human body has a normal range of about pH 7.35–7.45 and is therefore slightly alkaline. Certain diseases, such as kidney disease, can decrease or increase blood pH by a few tenths of a unit. When this happens, the enzymes in the body that are required for normal metabolism can no longer function optimally, leading to additional symptoms. As described next, living organisms have buffers to help prevent such changes in pH.

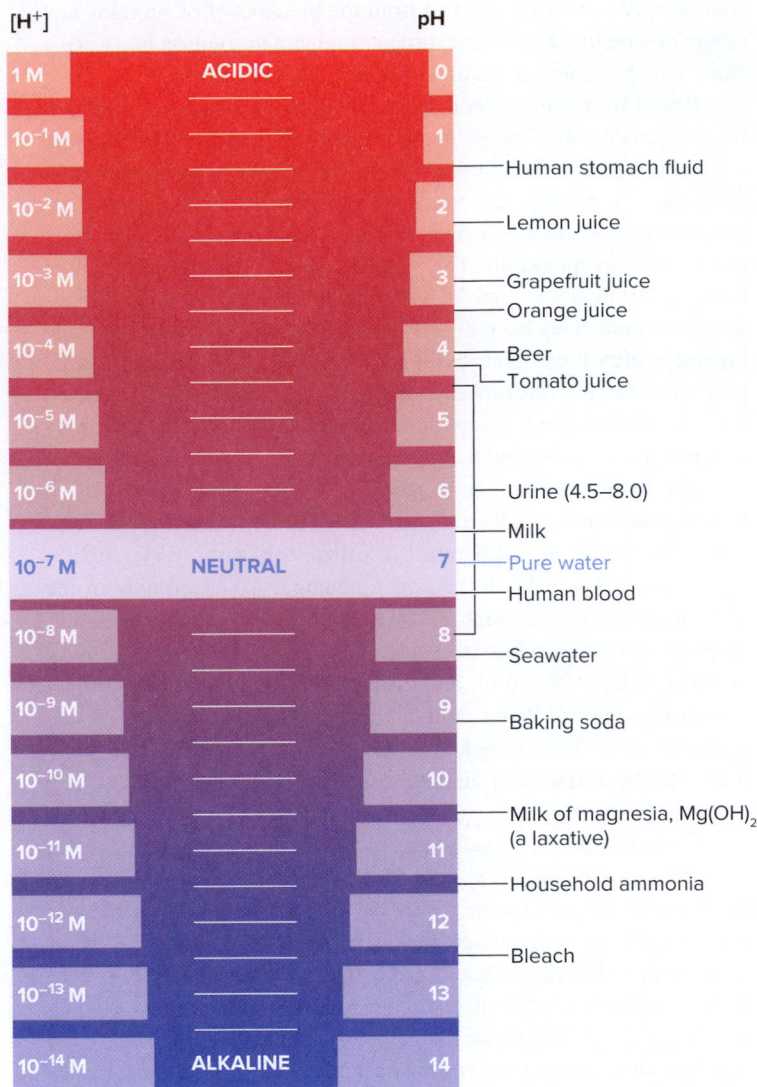

**Figure 2.20**   **The pH scale and the relative acidities of common substances.**

 **Core Skill: Connections**  Look ahead to Figure 54.14. The plant life shown growing in part (b) of that figure is sparse because the soil is very acidic. If the pH of the soil were 5.0, what would the $H^+$ concentration be?

## Buffers Minimize Fluctuations in pH

What factors might alter the pH of an organism's fluids? In plants, external factors such as acid rain and other forms of pollution can reduce the pH of water entering the roots. In animals, exercise generates lactic acid, and certain disease states can raise or lower the pH of blood.

Organisms have several ways to cope with changes in pH. In mammals, for example, the kidneys secrete acidic or alkaline compounds into the bloodstream when the blood pH becomes imbalanced. Another mechanism by which pH balance is regulated in diverse organisms involves the actions of acid-base buffers. A **buffer** is usually a pair of substances, an acid and its related base, that minimizes pH fluctuations in the fluids of living organisms. For example, carbonic acid ($H_2CO_3$) and bicarbonate ions ($HCO_3^-$) function to keep the pH of an animal's body fluids within a narrow range:

$$CO_2 + H_2O \rightleftharpoons H_2CO_3 \rightleftharpoons H^+ + HCO_3^-$$
$$\text{(carbonic acid)} \quad \text{(bicarbonate ion)}$$

This buffer can affect pH in both directions. For example, if the pH of an animal's body fluids increases (that is, the $H^+$ concentration decreases), the reaction proceeds from left to right. Carbon dioxide combines with water to make carbonic acid, and then the carbonic acid dissociates into $H^+$ and $HCO_3^-$. This increases the $H^+$ concentration and thereby decreases the pH. Alternatively, when the pH of an animal's blood decreases (that is, the $H^+$ concentration increases), the reaction runs in reverse. Bicarbonate combines with $H^+$ to make $H_2CO_3$, which then dissociates to $CO_2$ and $H_2O$. This process removes $H^+$ from the blood, restoring it to its normal pH, and the $CO_2$ is exhaled from the lungs. Many buffers exist in nature. Buffers found in living organisms function most efficiently at the normal range of pH values found in that organism.

## Summary of Key Concepts

### 2.1 Atoms

- Atoms are the smallest functional units of matter that form all chemical elements and cannot be further broken down into other substances by ordinary chemical or physical means. Atoms are composed of protons ($p^+$, positive charge), electrons ($e^-$, negative charge), and (except for hydrogen) neutrons ($n^0$, electrically neutral). Electrons are found in orbitals around the atomic nucleus (Figures 2.1, 2.2, 2.3, 2.4).

- Each element contains a unique number of protons—its atomic number. The periodic table organizes all known elements by atomic number and electron shells (Figure 2.5).

- Each atom has a small but measurable mass, measured in daltons (Da). The atomic mass scale indicates an atom's mass relative to the mass of other atoms (Table 2.1).

- Many atoms exist as isotopes, which differ in the number of neutrons they contain. Some isotopes are unstable radioisotopes and emit radiation (Figure 2.6).

- Four elements—oxygen, carbon, hydrogen, and nitrogen—account for the vast majority of atoms in living organisms. In addition, living organisms require mineral and trace elements that are essential for growth and function (Table 2.2).

### 2.2 Chemical Bonds and Molecules

- A molecule is two or more atoms bonded together. The properties of a molecule are different from the properties of the atoms that combined to form it. A compound is a molecule composed of two or more different elements.

- Atoms tend to form bonds that fill their outer shell with electrons. Covalent bonds, in which atoms share electrons, are strong chemical bonds. Atoms form two covalent bonds—a double bond—when they share two pairs of electrons (Figures 2.7, 2.8, 2.9).

- The electronegativity of an atom is a measure of its ability to attract electrons in a bond with another atom. When two atoms with different electronegativities combine, they form a polar covalent bond because the distribution of electrons around the atoms creates a difference in electric charge across the molecule. Polar molecules, such as water, typically have one or more polar covalent bonds, whereas nonpolar molecules tend to have mostly nonpolar covalent bonds (Figure 2.10).

- A hydrogen bond is a weak interaction between a hydrogen atom and an electronegative atom such as oxygen or nitrogen. The van der Waals dispersion forces are weak electrical attractions that arise between molecules due to variations in the locations of electrons in atoms (Figure 2.11).

- If an atom or molecule gains or loses one or more electrons, it acquires a net electric charge and becomes an ion. The strong attraction between two oppositely charged ions forms an ionic bond (Table 2.3, Figure 2.12).

- The three-dimensional, flexible shapes of molecules allow them to interact and contribute to their biological properties (Figures 2.13, 2.14).

- A free radical is an unstable molecule that can cause cellular damage by taking electrons away from other molecules.

- A chemical reaction occurs when one or more substances are changed into different substances. All chemical reactions eventually reach an equilibrium, unless the products of the reaction are continually removed.

### 2.3 Properties of Water

- Water is the solvent for most chemical reactions in all living organisms, both inside and outside of cells. Atoms and molecules dissolved in water interact in ways that would be impossible in their undissolved states (Figure 2.15).

- A solute dissolves in a solvent to form a solution. Solute concentration refers to the amount of a solute dissolved in a unit volume of solution. The molarity of a solution is defined as the number of moles of a solute dissolved in 1 L of solution (Figure 2.16).

- Molecules with ionic and polar covalent bonds are hydrophilic, whereas nonpolar molecules, composed predominantly of carbon and hydrogen, are hydrophobic. Amphipathic molecules, such as detergents, have both hydrophilic and hydrophobic regions (Figure 2.17).

- $H_2O$ exists as ice, liquid water, and water vapor (gas) (Figure 2.18).

- The colligative properties of water depend on the number of dissolved solute particles and allow it to function as an antifreeze in certain organisms.

- Water's high heat of vaporization and high heat of fusion make it very stable in its liquid form.

- Water molecules participate in many chemical reactions in living organisms. Hydrolysis reactions break down large molecules into smaller units. In living organisms, water provides support, is used to eliminate wastes, dissipates body heat, aids in the movement of liquid through vessels, and serves as a lubricant; also, its surface tension allows certain insects to walk on water (Figure 2.19).

### 2.4 pH and Buffers

- The pH of a solution is the negative logarithm to the base 10 of the $H^+$ concentration. The pH of pure water is 7 (neutral). Alkaline solutions have a pH higher than 7, and acidic solutions have a pH lower than 7 (Figure 2.20).

- A buffer is usually a pair of substances, an acid and its related base, that minimizes pH fluctuations in the fluids of living organisms. Buffers in living cells or body fluids can raise or lower pH to keep its value within a narrow range.

## Assess & Discuss

### Test Yourself

1. _____ make(s) up the nucleus of an atom.
   a. Protons and electrons
   b. Protons and neutrons
   c. DNA and RNA
   d. Neutrons and electrons
   e. DNA only

2. Living organisms are composed mainly of which atoms?
   a. calcium, hydrogen, nitrogen, and oxygen
   b. carbon, hydrogen, nitrogen, and oxygen
   c. hydrogen, nitrogen, oxygen, and helium
   d. carbon, helium, nitrogen, and oxygen
   e. carbon, calcium, hydrogen, and oxygen

3. The ability of an atom to attract electrons in a bond with another atom is termed its
   a. hydrophobicity.
   b. electronegativity.

c.  solubility.

d.  valence.

e.  both a and b.

4.  Hydrogen bonds differ from covalent bonds in that

a.  covalent bonds can form between any type of atom, and hydrogen bonds form only between H and O.

b.  covalent bonds involve sharing of electrons, and hydrogen bonds involve the complete transfer of electrons.

c.  covalent bonds result from equal sharing of electrons, but hydrogen bonds involve unequal sharing of electrons.

d.  covalent bonds involve sharing of electrons between atoms, but hydrogen bonds are the result of weak attractions between a hydrogen atom of a polar molecule and an electronegative atom of another polar molecule.

e.  covalent bonds are weak bonds that break easily, but hydrogen bonds are strong links between atoms that are not easily broken.

5.  A free radical

a.  is a positively charged ion.

b.  is an atom with one unpaired electron in its outer shell.

c.  is a stable atom that is not bonded to another atom.

d.  can cause considerable cellular damage.

e.  both b and d.

6.  Chemical reactions in living organisms

a.  require energy to begin.

b.  usually require a catalyst to speed them up.

c.  are usually reversible.

d.  occur in liquid environments, such as water.

e.  are all of the above.

7.  Solutes that easily dissolve in water are said to be

a.  hydrophobic.

b.  hydrophilic.

c.  polar molecules.

d.  all of the above.

e.  b and c only.

8.  The molecular mass of glucose is about 180 g/mol. If 45 g of glucose is dissolved in water to make a final volume of 0.5 L, what is the molarity of the solution?

a.  0.125 M

b.  0.25 M

c.  0.5 M

d.  1.0 M

e.  2.0 M

9.  The sum of the atomic masses of all the atoms of a molecule is its

a.  atomic weight.

b.  molarity.

c.  molecular mass.

d.  concentration.

e.  polarity.

10. Reactions in which water is used to break apart other molecules are known as _____ reactions.

a.  hydrophilic

b.  hydrophobic

c.  dehydration

d.  anabolic

e.  hydrolysis

## Conceptual Questions

1.  Compare and contrast the different types of bonds commonly found in biological molecules.

2.  What is the significance of molecular shape, and what may change the shapes of molecules?

3.  **Core Concept: Systems**  A core concept of biology is that systems are interconnected and interacting. As mentioned in Figure 1.4e, emergent properties arise from complex interactions within systems. How is this core concept of biology related to chemical reactions that make molecules from other starting materials? What examples can you cite from this chapter of emergent properties of molecules, in which atoms with one type of property combine to form molecules with completely different properties?

## Collaborative Questions

1.  Discuss the properties of the three subatomic particles of atoms.

2.  Discuss several properties of water that make it possible for life to exist.

# The Chemical Basis of Life II: Organic Molecules

# 3

In Chapter 2, we learned that all life is composed of atoms, which, in turn, combine to form molecules. Molecules may be simple in atomic composition, such as water ($H_2O$) or hydrogen gas ($H_2$), or may bind with other molecules to form larger ones. Of the countless possible molecules that can be produced from the known elements in nature, certain types contain carbon and are found in all forms of life. These carbon-containing molecules are collectively referred to as **organic molecules**, so named because they were first discovered in living organisms. Organic molecules include lipids and large, complex compounds called **macromolecules**, which can be carbohydrates, proteins, or nucleic acids. In this chapter, we will survey the structures of these molecules and examine their main functions. We begin with the element whose chemical properties are fundamental to the formation of biologically important molecules: carbon. This element provides the atomic scaffold on which life is built.

**A model showing the structure of DNA—a type of organic macromolecule that stores genetic information.** ©Zoonar GmbH/Alamy Stock Photo

## 3.1 The Carbon Atom

### Learning Outcomes:

1. Explain the properties of carbon that make it the chemical basis of all life.
2. Describe the variety and chemical characteristics of common functional groups of organic compounds.
3. Compare and contrast different types of isomers.

The science of carbon-containing molecules is known as **organic chemistry**. A long time ago, the study of organic molecules was considered a fruitless endeavor because of a concept called vitalism, which persisted into the 19th century. Vitalism held that organic molecules were created by, and therefore imparted with, a vital life force that was contained within a plant or an animal's body. Supporters of vitalism argued that chemists could not synthesize an organic compound, because such molecules could arise only through the intervention of mysterious qualities associated with life.

Vitalism was disproved by Friedrich Wöhler, a German physician and chemist interested in the properties of inorganic and organic compounds. He spent some time studying urea (($NH_2$)$_2$CO), a natural organic molecule formed from the breakdown of proteins in an animal's body. In mammals, urea accumulates in the urine formed by the kidneys, and then is excreted from the body. During the course of his studies, Wöhler purified urea from the urine of mammals. He noted the color, size, shape, and other characteristics of the urea crystals.

In 1828, while exploring the reactive properties of ammonia and cyanic acid, Wöhler attempted to synthesize an inorganic molecule, ammonium cyanate ($NH_4OCN$), which is not found in living organisms. Instead, to his surprise, Wöhler discovered that ammonia and cyanic acid reacted to produce a third compound, which, when heated, formed familiar-looking crystals. After careful analysis, he concluded that these crystals were urea. No mysterious life force was required to make this organic molecule. Other scientists, such as Hermann Kolbe, would soon demonstrate that organic compounds such as acetic acid ($CH_3COOH$) could also be synthesized directly from simpler molecules. These studies refuted the concept of vitalism, and so began the field of organic chemistry.

Central to Wöhler's and Kolbe's reactions is the carbon atom. Urea and acetic acid, like all organic compounds, contain carbon atoms bonded to other atoms. In this section, we will consider the chemical features of carbon that make it such an important element in living organisms.

## Carbon Forms Four Covalent Bonds with Other Atoms

A key property of the carbon atom is its ability to form four covalent bonds with other atoms, including other carbon atoms. This occurs because carbon has four electrons in its outer (second) shell, and it requires eight electrons, or four additional electrons, to fill this shell (**Figure 3.1**). In living organisms, carbon atoms most commonly form covalent bonds with other carbon atoms and with hydrogen, oxygen, nitrogen, and sulfur atoms. Bonds between two carbon atoms, between carbon and oxygen, or between carbon and nitrogen can be single or double, or in the case of C≡C and C≡N bonds, triple. The variation in bonding of carbon with carbon and other atoms allows a vast number of organic compounds to be formed from only a few chemical elements. Carbon and other atoms may be bonded together in configurations that are linear, ringlike, or highly branched. Such molecular shapes can produce molecules with a variety of functions.

Carbon and hydrogen have similar electronegativities (see Chapter 2). Therefore, carbon-carbon and carbon-hydrogen bonds are nonpolar. As a consequence, molecules with a high proportion of hydrogen-carbon bonds, called **hydrocarbons**, are hydrophobic and poorly soluble in water. In contrast, when carbon forms polar

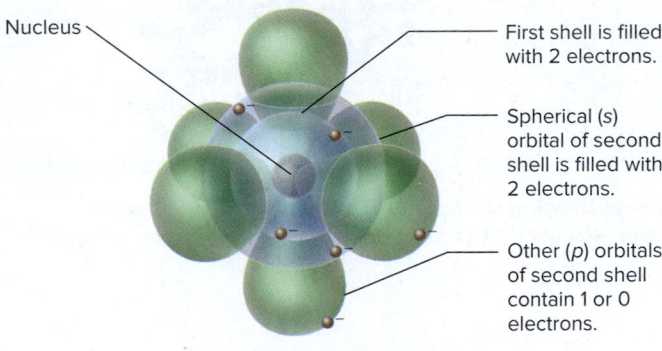

Nucleus

First shell is filled with 2 electrons.

Spherical (s) orbital of second shell is filled with 2 electrons.

Other (p) orbitals of second shell contain 1 or 0 electrons.

**(a) Electron orbitals in carbon**

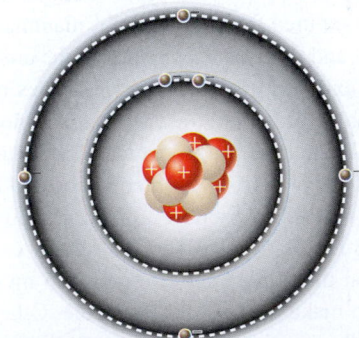

**(b) Simplified depiction of carbon's electron shells**

**Figure 3.1  Models for the electron orbitals and shells of carbon.** A carbon atom has four electrons in its outer (second) electron shell. When a carbon atom forms four covalent bonds with other atoms, its outer shell is full with eight electrons, which is a stable condition.

C–C and C–H bonds are electrically neutral and nonpolar.

Oxygen is more electronegative than carbon; thus, C–O and C=O bonds are polar.

Propionic acid

**Figure 3.2  Nonpolar and polar bonds in an organic molecule.** Carbon can form both nonpolar and polar bonds, and both single and double bonds, as shown here in the molecule propionic acid, a common food preservative.

covalent bonds with more electronegative atoms, such as oxygen or nitrogen, the resulting molecule is much more soluble in water because of its electrical attraction to polar water molecules. The ability of carbon to form both polar and nonpolar bonds (**Figure 3.2**) contributes to its ability to serve as the backbone for an astonishing variety of biologically important molecules.

Another feature of carbon that is important to living organisms is that carbon bonds are stable within the large range of temperatures associated with life. This property arises in part because the carbon atom is small relative to most other atoms. Therefore, the distance between carbon atoms forming a carbon-carbon bond is quite short. Shorter bonds tend to be stronger and more stable than longer bonds, which form between two large atoms. For this reason, carbon bonds are compatible with what we observe about life-forms today; namely, living organisms can inhabit environments with a range of temperatures, from the Earth's frigid icy poles to the superheated water of deep-sea vents.

## Carbon Atoms Bond to Several Biologically Important Functional Groups

Aside from the simplest hydrocarbons, most organic molecules and macromolecules contain **functional groups**—groups of atoms with characteristic chemical structures and properties. Each type of functional group exhibits similar chemical properties in all molecules in which it occurs. For example, the amino group (—NH$_2$) acts like a base. In the pH range found in living organisms, an amino group readily binds H$^+$ to become NH$_3$$^+$, thereby removing H$^+$ from an aqueous solution and raising the pH. As discussed later in this chapter, amino groups are found in proteins and also in other types of organic molecules. **Table 3.1** describes examples of functional groups found in many different types of organic molecules. We will discuss each of these groups at numerous points throughout this textbook.

## Carbon-Containing Molecules May Exist in Multiple Forms Called Isomers

Wöhler was surprised to discover that urea ((NH$_2$)$_2$CO) and ammonium cyanate (NH$_4$OCN) contained the exact same ratio of carbon, nitrogen, hydrogen, and oxygen atoms, yet they were different molecules with distinct chemical and biological properties. Two (or more)

| Table 3.1 | Some Biologically Important Functional Groups That Bond to Carbon | | |
|---|---|---|---|
| **Functional group\*** (with shorthand notation) | **Formula†** | **Examples of where the group is found** | **Properties** |
| Amino (–NH$_2$) | $R-N$ with H, H | Amino acids (proteins) | Weakly basic (can accept H$^+$); polar; forms part of peptide bonds |
| Carbonyl (–CO)‡ Ketone | O ‖ R–C–R' | Steroids, waxes, and proteins | Polar; highly chemically reactive; forms hydrogen bonds |
| Aldehyde (–CHO) | O ‖ R–C–H | Linear forms of sugars and some odor molecules | |
| Carboxyl (–COOH) | R–C with O (double) and OH | Amino acids, fatty acids | Acidic (gives up H$^+$ in water); forms part of peptide bonds |
| Hydroxyl (–OH) | R–OH | Steroids, alcohol, carbohydrates, some amino acids | Polar; forms hydrogen bonds with water |
| Methyl (–CH$_3$) | H \| R–C–H \| H | May be attached to DNA, proteins, and carbohydrates | Nonpolar |
| Phosphate (–PO$_4^{2-}$) | O ‖ R–O–P–O$^-$ \| O$^-$ | Nucleic acids, ATP, phospholipids | Polar; weakly acidic and negatively charged at typical pH of living organisms |
| Sulfate (–SO$_4^-$) | O ‖ R–O–S–O$^-$ ‖ O | May be attached to carbohydrates, proteins, and lipids | Polar; negatively charged at typical pH of living organisms |
| Sulfhydryl (–SH) | R–SH | Proteins that contain the amino acid cysteine | Polar; forms disulfide bridges in many proteins |

\*This list contains many of the functional groups that are important in biology. However, many more functional groups have been identified by biochemists.
†R and R' represent the remainder of the molecule.
‡A carbonyl group is C=O. In a ketone, the carbon of this group forms covalent bonds with two other carbon atoms. In an aldehyde, the carbon is bonded to a hydrogen atom.

molecules with the same chemical formula but different structures and characteristics are called **isomers**.

**Figure 3.3** depicts three ways in which isomers may occur. **Structural isomers** contain the same atoms but in different bonding relationships. Urea and ammonium cyanate fall into this category. A simpler example of structural isomers (isopropyl alcohol and propyl alcohol) is illustrated in Figure 3.3a.

**Stereoisomers** have identical bonding relationships, but the spatial positioning of their atoms differs. Two types of stereoisomers are *cis-trans* isomers and enantiomers. In ***cis-trans* isomers**, like those shown in Figure 3.3b, the two hydrogen atoms linked to the two carbons of a C=C double bond may be on the same side of the carbons, in which case the C=C bond is called a *cis* double bond. If the hydrogens are on opposite sides, it is a *trans* double bond. *Cis-trans* isomers may have very different chemical properties from each other, most notably their stability and sensitivity to heat and light. For instance, the light-sensitive region of your eye contains a molecule called retinal, which exists in either a *cis* or *trans* form. In darkness, the *cis*-retinal form predominates.

The energy of sunlight, however, causes retinal to isomerize to the *trans* form. The *trans*-retinal activates the light-capturing cells in the eye.

A second type of stereoisomer, called an **enantiomer**, exists as one of a pair of molecules that are mirror images. Four different atoms can bind to a single carbon atom in two possible ways, designated as a left-handed and a right-handed structure. The resulting structures are not identical, but instead are mirror images of each other (Figure 3.3b). A convenient way to visualize the contrasting structures of enantiomers is to consider a pair of gloves. No matter which way you turn or hold a left-hand glove, it cannot fit properly on your right hand. Any given pair of enantiomers shares identical chemical properties, such as solubility and melting point. However, due to the different orientation of their atoms in space, their ability to noncovalently bond to other molecules can be strikingly different. For example, **enzymes** are molecules that catalyze (speed up) the rates of many biologically important chemical reactions. An enzyme that recognizes one enantiomer usually does not recognize the other.

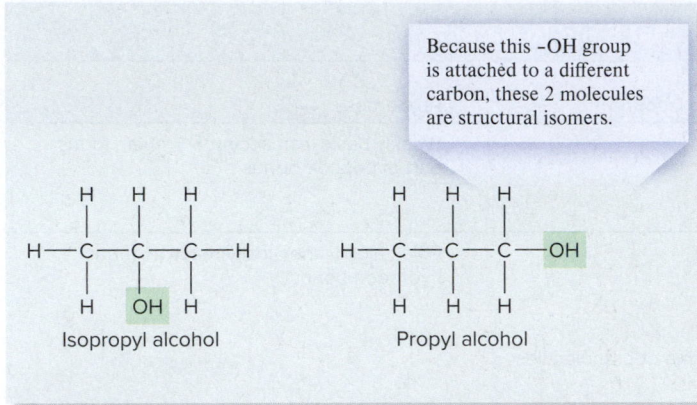

**(a) Structural isomers**

Because this –OH group is attached to a different carbon, these 2 molecules are structural isomers.

Isopropyl alcohol

Propyl alcohol

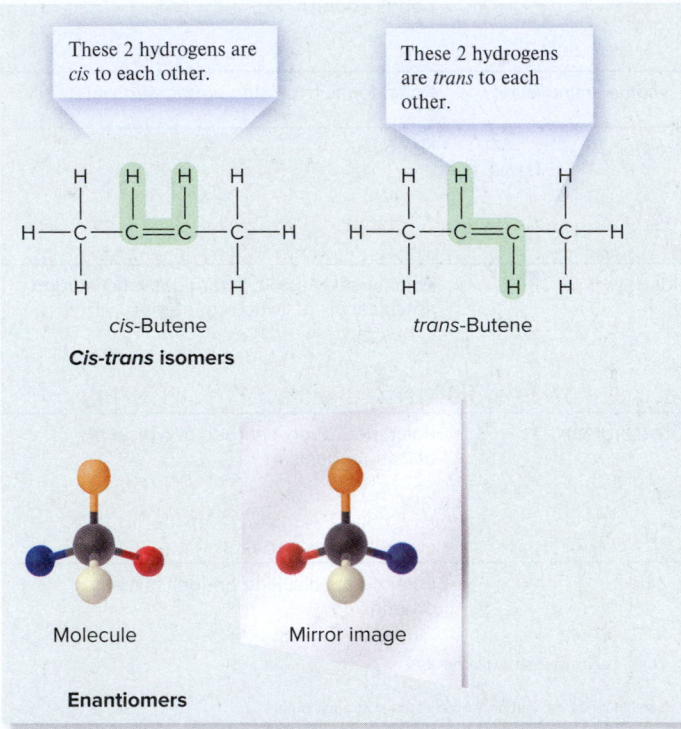

These 2 hydrogens are *cis* to each other.

These 2 hydrogens are *trans* to each other.

*cis*-Butene

*trans*-Butene

**Cis-trans isomers**

Molecule

Mirror image

**Enantiomers**

**(b) Two types of stereoisomers**

**Figure 3.3   Types of isomers.** Isomers are molecules with the same chemical formula but different structures. The differences in structure result in different biological properties. Isomers can be grouped into **(a)** structural isomers and **(b)** stereoisomers.

## 3.2   Formation of Organic Molecules and Macromolecules

**Learning Outcome:**

**1.** Explain how small molecules are assembled into larger ones by dehydration reactions and how hydrolysis reactions reverse this process.

As we have seen, organic molecules have various shapes due to the bonding properties of carbon. During the past two centuries,

biochemists have studied many organic molecules found in living organisms and determined their structures at the molecular level. Many of these are relatively small molecules. However, some organic molecules are extremely large macromolecules composed of thousands or even millions of atoms. Such large molecules are formed by linking together many smaller molecules called **monomers** (meaning one part) and are known as **polymers** (meaning many parts). When a polymer is being formed, two smaller molecules combine by a **condensation reaction**, which produces a larger organic molecule plus a water molecule. This specific type of condensation reaction is also called a **dehydration reaction**, because a molecule of water is removed when a monomer is added to a growing polymer.

The mechanism of a dehydration reaction is illustrated in **Figure 3.4a**. The length of a polymer is extended with each dehydration reaction. Some polymers reach great lengths by this mechanism. For example, during the synthesis of DNA, which is described in Chapter 11, dehydration reactions produce linear strands of DNA that contain millions of monomers called nucleotides.

Although DNA is a stable polymer, other polymers, such as large carbohydrates, are often broken down. As discussed later in this chapter, a large carbohydrate in plants, which is called starch, plays a role in storing energy. When a plant cell needs that stored energy, the starch is broken down into its constituent monomers. The process by which a polymer is broken down into monomers is called a **hydrolysis reaction**, because a molecule of water is added back each time a monomer is released (**Figure 3.4b**).

## 3.3   Overview of the Four Major Classes of Organic Molecules Found in Living Cells

**Learning Outcome:**

**1.** Compare and contrast the structures and functions of carbohydrates, lipids, proteins, and nucleic acids.

By analyzing the cells of many different species, researchers have determined that all forms of life have organic molecules and macromolecules that fall into four broad categories, based on their chemical and biological properties: carbohydrates, lipids, proteins, and nucleic acids. **Table 3.2** outlines the general structures and functions of these molecules and provides some examples. In the next sections, we will examine them in greater detail.

## 3.4   Carbohydrates

**Learning Outcomes:**

**1.** Distinguish among different forms of carbohydrate molecules, including monosaccharides, disaccharides, and polysaccharides.

**2.** Relate the functions of plant and animal polysaccharides to their structure.

**Carbohydrates** are organic molecules composed of carbon, hydrogen, and oxygen atoms in or close to the proportions represented by the general formula $C_n(H_2O)_n$, where $n$ is a whole number. This

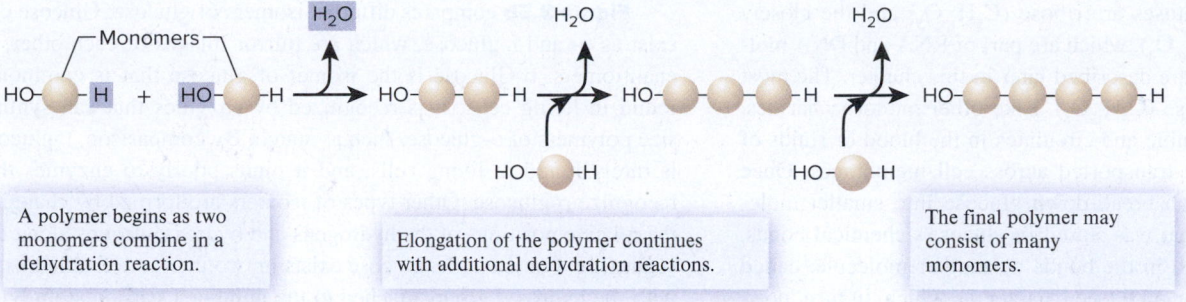

A polymer begins as two monomers combine in a dehydration reaction.

Elongation of the polymer continues with additional dehydration reactions.

The final polymer may consist of many monomers.

**(a) Polymer formation by dehydration reactions**

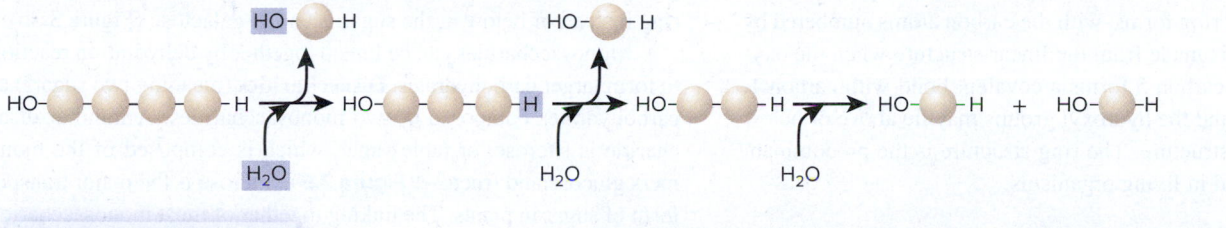

Polymers are broken down one monomer at a time by hydrolysis reactions.

**(b) Breakdown of a polymer by hydrolysis reactions**

**Figure 3.4  Formation and breakdown of polymers. (a)** Monomers combine to form polymers in living organisms by dehydration reactions, in which a molecule of water is removed each time a new monomer is added to the growing polymer. **(b)** Polymers can be broken down into their constituent monomers by hydrolysis reactions, in which a molecule of water is added each time a monomer is released.

formula gives carbohydrates their name—carbon-containing compounds that are hydrated, that is, contain water. Most of the carbon atoms in a carbohydrate are linked to a hydrogen atom and a hydroxyl functional group. However, other functional groups, such as amino and carboxyl groups, are also found in certain carbohydrates. As discussed next, sugars are relatively small carbohydrates, whereas polysaccharides are large macromolecules.

## Sugars Are Small Carbohydrates That Usually Taste Sweet

Sugars are small carbohydrates that usually taste sweet. The simplest sugars are monomers known as **monosaccharides** (from the Greek, meaning single sugars). The most common types of monosaccharides contain either five carbons (pentoses) or six carbons

| Table 3.2 | A Comparison of the Four Types of Organic Molecules Found in Living Organisms | | |
|---|---|---|---|
| **Type** | **Structure** | **Key functions** | **Examples** |
| Carbohydrates | The general formula is $C_n(H_2O)_n$, where $n$ is a whole number. | Simple carbohydrates are broken down to make ATP, which is used as a source of energy. Larger carbohydrates store energy or may play a structural role, as in plant cell walls. Some carbohydrates function as molecular tags allowing recognition of specific cells and molecules. | Simple sugars, such as glucose; larger polymers, such as starch and cellulose |
| Lipids | Lipids are nonpolar molecules that are primarily composed of carbon and hydrogen, with some oxygen. | Lipids are a key part of cell membranes and also function as hormones and in energy storage; in animals, they act as insulators and shock absorbers. | Phospholipids, estrogen, testosterone, triglycerides |
| Proteins | A polypeptide is a structural unit composed of a linear sequence of amino acids. A protein is a functional unit composed of one or more polypeptides. | Proteins play a key role in cell structure and carry out a diverse array of cellular functions; for example, there are proteins involved with gene expression and regulation, motor proteins, defense proteins, cell-signaling proteins, metabolic enzymes, structural proteins, and transporters. | See Table 3.3 |
| Nucleic acids | A nucleic acid is a linear sequence of nucleotides; DNA is double-stranded. | DNA stores genetic information in units called genes. RNA is made from DNA and provides access to that information. | DNA and RNA |

(hexoses). Important pentoses are ribose ($C_5H_{10}O_5$) and the closely related deoxyribose ($C_5H_{10}O_4$), which are part of RNA and DNA molecules, respectively, and are described later in this chapter. The most common hexose is glucose ($C_6H_{12}O_6$). Like other monosaccharides, glucose is very water-soluble and circulates in the blood or fluids of animals, where it can be transported across cell membranes. Once inside a cell, enzymes can break down glucose into smaller molecules, releasing energy that was stored in glucose's chemical bonds. This energy is then stored in the bonds of another molecule, called adenosine triphosphate, or ATP (see Chapter 7), which, in turn, powers a variety of cellular processes. In this way, sugar is often used as a source of energy by living organisms.

**Figure 3.5a** depicts the bonds between atoms in a monosaccharide in both linear and ring forms, with the carbon atoms numbered by convention. The ring is made from the linear structure when the oxygen atom attached to carbon 5 forms a covalent bond with carbon 1. The hydrogen atoms and the hydroxyl groups may lie above or below the plane of the ring structure. The ring structure is the predominant type of structure found in living organisms.

**Figure 3.5b** compares different isomers of glucose. Glucose can exist as D- and L-glucose, which are mirror images of each other, or enantiomers. D-Glucose is the isomer of glucose that is commonly found in living cells. It is recognized by enzymes that can synthesize polymers of D-glucose, such as starch. By comparison, L-glucose is rarely found in living cells, and it binds poorly to enzymes that recognize D-glucose. Other types of isomers are formed by changing the relative positions of the hydrogens and hydroxyl groups along the sugar ring. For example, glucose exists in two interconvertible forms, with the hydroxyl group attached to the number 1 carbon atom lying either above (the β form of glucose, Figure 3.5b) or below (the α form, Figure 3.5a) the plane of the ring. As another example, if the hydroxyl group on carbon atom number 4 of glucose is above the plane of the ring instead of below it, the sugar is called galactose (Figure 3.5b).

Monosaccharides can be linked together by dehydration reactions to form larger carbohydrates. **Disaccharides** (meaning two sugars) are carbohydrates composed of two monosaccharides. A familiar disaccharide is sucrose, or table sugar, which is composed of the monomers glucose and fructose (**Figure 3.6**). Sucrose is the major transport form of sugar in plants. The linking together of most monosaccharides involves the removal of a hydroxyl group from one monosaccharide and a hydrogen atom from the other, giving rise to a molecule of water and covalently bonding the two sugars together through an oxygen atom. The bond formed between two sugar molecules by such a dehydration reaction is called a **glycosidic bond**. Other disaccharides frequently found in nature are maltose, formed in animals during the digestion of large carbohydrates in the intestinal tract, and lactose, present in the milk of mammals. Maltose is α-D-glucose linked to α-D-glucose, and lactose is β-D-galactose linked to β-D-glucose.

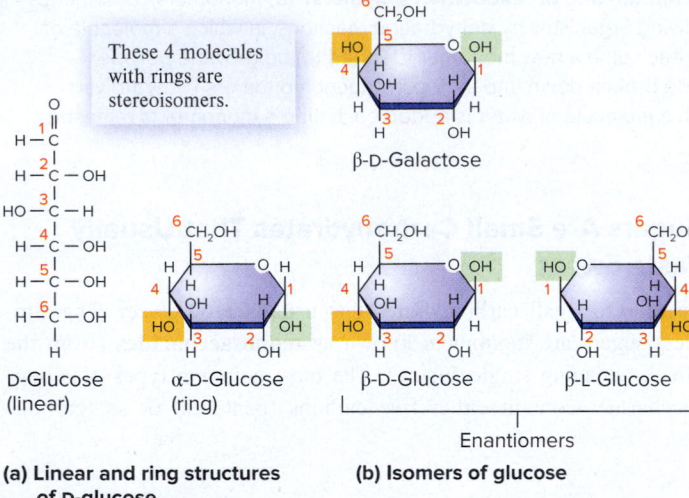

**(a) Linear and ring structures of D-glucose**

**(b) Isomers of glucose**

**Figure 3.5 Monosaccharide structure. (a)** A comparison of the linear and ring structures of glucose. In solution, such as in the fluids of organisms, nearly all glucose is in the ring form. **(b)** Isomers of glucose. The locations of the hydroxyl groups on carbon 1 and carbon 4 are emphasized with green and orange boxes, respectively. Glucose exists as stereoisomers designated α- and β-glucose, which differ in the position of the —OH group attached to carbon atom number 1. Glucose and galactose differ in the position of the —OH group attached to carbon atom number 4. Enantiomers of glucose, called D-glucose and L-glucose, are mirror images of each other. D-Glucose is the form found in living cells.

 **Core Concept: Energy and Matter  Living organisms use energy.** The chemical energy stored in the bonds of glucose molecules can be harnessed by living organisms. This energy is used to perform numerous functions that support life, including the synthesis of new molecules, growth, digestion, locomotion, and many others.

*Concept Check:*  *Why do enantiomers such as D- and L-glucose differ in their ability to bind to enzymes?*

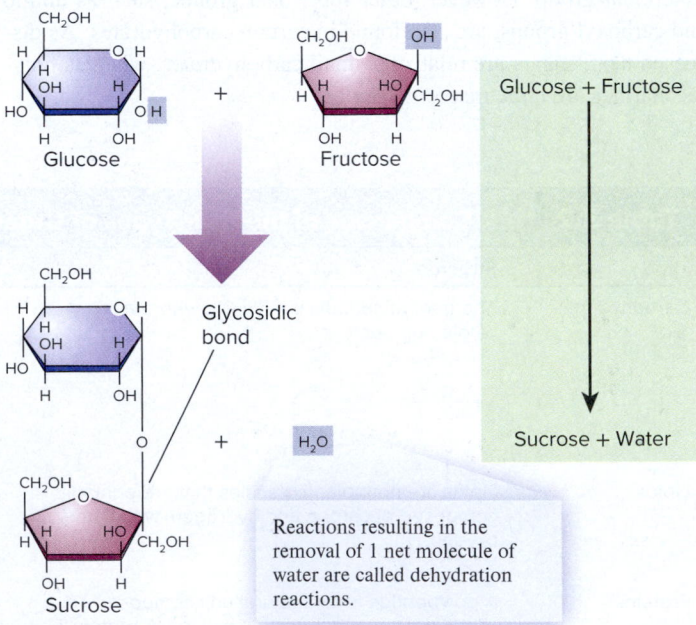

**Figure 3.6  Formation of a disaccharide.** Two monosaccharides can bond to each other to form a disaccharide, such as sucrose, maltose, or lactose, by a dehydration reaction.

*Concept Check:*  *What type of reaction is the reverse of the one shown here, in which a disaccharide is broken down into two monosaccharides?*

## Polysaccharides Are Carbohydrate Polymers That Include Starch, Glycogen, and Cellulose

When many monosaccharides are linked together to form long polymers, the products are **polysaccharides** (meaning many sugars). **Starch**, found in plant cells, and **glycogen**, found in animal cells, are examples of polysaccharides (**Figure 3.7**). Both of these polysaccharides are composed of thousands of α-D-glucose molecules linked together in long, branched chains, differing only in the extent of branching along the chain. The bonds that connect the monomers are very specific. In starch and glycogen, the bonds form between carbons 1 and 4 and between carbons 1 and 6. The high degree of branching in glycogen contributes to its solubility in animal tissues,

such as muscle tissue, because the extensive branching creates a more open structure in which many hydrophilic hydroxyl (—OH) side groups have access to water and can hydrogen-bond with it. Starch is less branched and less soluble, which contributes to the properties of plant structures (think of a potato or a kernel of corn).

Some polysaccharides, such as starch and glycogen, store energy in cells. Like disaccharides, these polysaccharides can be hydrolyzed to yield monosaccharides, which are broken down to produce ATP, a common energy source for cells. Starch and glycogen, the polymers of α-glucose, provide an efficient means of storing energy for those times when a plant or animal cannot obtain sufficient energy from its environment or diet for its metabolic requirements.

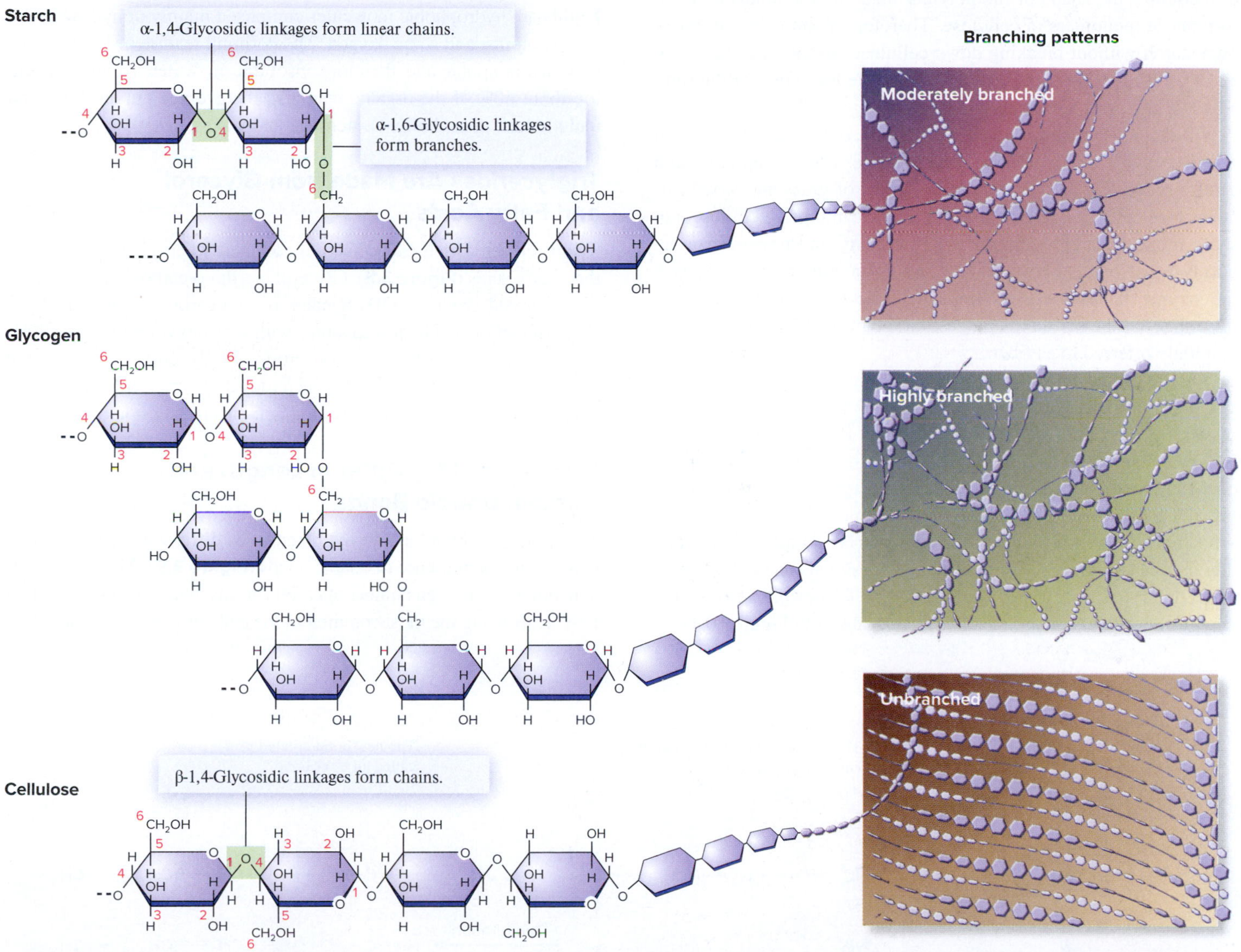

**Starch**

α-1,4-Glycosidic linkages form linear chains.

α-1,6-Glycosidic linkages form branches.

**Glycogen**

**Cellulose**

β-1,4-Glycosidic linkages form chains.

**Branching patterns**

Moderately branched

Highly branched

Unbranched

**Figure 3.7  Polysaccharides that are polymers of glucose.** These polysaccharides differ in their arrangement, extent of branching, and type of glucose isomer. In cellulose, the bonding arrangements cause every other glucose to be upside-down with respect to its neighbors.

 **Core Skill: Connections**  Look ahead to Figures 10.5 and 10.6 for the role of cellulose in plant structure and to Figures 36.10 and 36.11 for its role in plant growth. Considering the amount of plant life on Earth, what might you conclude about the abundance of cellulose on the planet?

Other polysaccharides play a structural role, rather than storing energy. For example, cellulose is a major constituent of plant cell walls. **Cellulose** is a polymer of β-D-glucose, with a linear arrangement of carbon-carbon bonds and no branching (see Figure 3.7). Each glucose monomer in cellulose is in an opposite orientation from its adjacent monomers (flipped over), forming long chains of several thousand glucose monomers.

Linear chains of cellulose can form hydrogen bonds with each other and thereby arrange themselves in a parallel pattern (see the lower right panel in Figure 3.7). These sheets provide great strength to plant cell walls. The bond orientations in β-D-glucose prevent cellulose from being hydrolyzed in most types of organisms. The enzymes that break the bonds between monomers of α-D-glucose in starch do not recognize the shape of the polymer made by the bonds between β-D-glucose monomers in cellulose. Therefore, plant cells can break down starch without breaking down cellulose. In this way, cellulose can be used for other functions, notably in the formation of the rigid cell walls characteristic of plants.

Unlike most animals and plants, some organisms do have an enzyme capable of breaking down cellulose. For example, certain bacteria present in the gastrointestinal tracts of grass and wood eaters, such as cows and termites, respectively, can digest cellulose into usable monosaccharides because they contain an enzyme that can hydrolyze the bonds between β-D-glucose monomers. Humans lack this enzyme. Therefore, we eliminate in the feces most of the cellulose ingested in our diet. Undigestible plant matter we consume is commonly referred to as fiber.

Other polysaccharides also play structural roles. **Chitin**, a tough, structural polysaccharide, forms the external skeleton of insects and crustaceans (shrimp and lobsters) as well as the cell walls of fungi. The sugar monomers within chitin have nitrogen-containing groups attached to them. **Glycosaminoglycans** are large polysaccharides that play a structural role in animals. For example, they are abundantly found in cartilage, the tough, fibrous material found in joints and other animal structures. Glycosaminoglycans are also abundant in the extracellular matrix that provides a structural framework surrounding many of the cells in an animal's body (look ahead to Figure 10.4).

## 3.5 Lipids

### Learning Outcomes:

1. List the classes of lipid molecules important in living organisms.
2. Diagram the structure of a triglyceride, and explain how it is affected by the presence of saturated and unsaturated fatty acids.
3. Explain why some fats are solid at room temperature and others are liquid.
4. Discuss how fats function as energy-storage molecules.
5. Explain why phospholipids form a bilayer when dissolved in water.
6. Describe the chemical nature of steroids, and give an example of their biological importance.

**Lipids** are hydrophobic molecules composed mainly of hydrogen and carbon atoms, and some oxygen. The defining feature of lipids is that they are nonpolar and therefore insoluble in water. Lipids account for about 40% of the organic matter in the average human body and include fats, phospholipids, steroids, and waxes.

### Triglycerides Are Made from Glycerol and Fatty Acids

**Triglycerides** (often called fats) are formed when glycerol bonds to three fatty acids (**Figure 3.8**). Glycerol is a three-carbon molecule with one hydroxyl group (—OH) bonded to each carbon. A fatty acid is a chain of carbon and hydrogen atoms with a carboxyl group (—COOH) at one end. Each of the hydroxyl groups in glycerol is linked to the carboxyl group of a fatty acid by the removal of a molecule of water by a dehydration reaction. The resulting bond is an ester bond.

### Fatty Acids May Differ in Length and Contain Double Bonds

The fatty acids found in fats and other lipids differ with regard to their lengths and the presence of double bonds (**Figure 3.9**). Most fatty acids in nature have an even number of carbon atoms, with 16- and 18-carbon fatty acids being the most common in the cells of plants and animals.

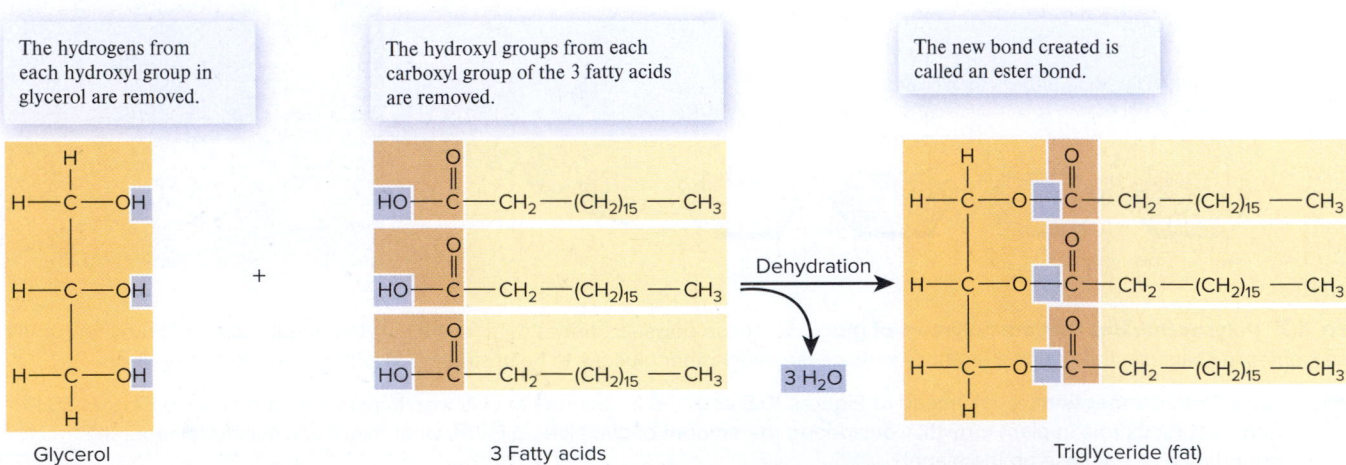

**Figure 3.8 The formation of a triglyceride.** The formation of a triglyceride occurs via three dehydration reactions in which fatty acids are bonded to glycerol. Note in this figure and in Figure 3.9 a common shorthand notation used for depicting fatty acid chains, in which a portion of the $CH_2$ groups forming the chain are represented as $(CH_2)_n$, where $n$ is 2 or greater.

Carboxyl group

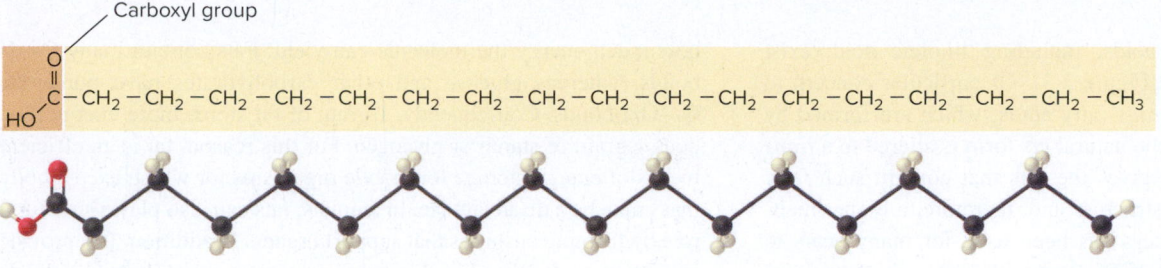

Saturated fatty acid
(Stearic acid)

**Figure 3.9** **Examples of fatty acids.** Fatty acids are hydrocarbon chains that have a carboxyl functional group at one end and contain either no double bonds between carbons (saturated) or one or more double bonds (unsaturated). Stearic acid, for example, is an abundant saturated fatty acid in animals, whereas linoleic acid is an unsaturated fatty acid found in plants. Note that the presence of two C=C double bonds introduces two kinks into the chain structure of linoleic acid. As a consequence, unsaturated fatty acids are not able to pack together as tightly as saturated fatty acids.

Double bonds deform the linear chain and give the fatty acid a kinked 3-dimensional structure.

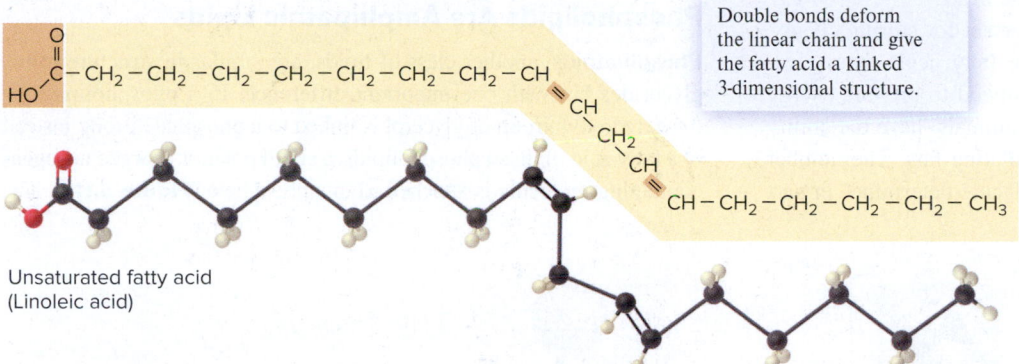

Unsaturated fatty acid
(Linoleic acid)

Fatty acids also differ with regard to the presence of double bonds. When all of the carbons in a fatty acid are linked by single covalent bonds, the fatty acid is said to be a **saturated fatty acid**, because all of the carbons are saturated with covalently bonded hydrogens. Alternatively, some fatty acids contain one or more C=C double bonds. These fatty acids are known as **unsaturated fatty acids**. The C=C double bond introduces a kink into the linear shape of a fatty acid. A fatty acid with one C=C bond is a monounsaturated fatty acid, whereas a fatty acid with two or more C=C bonds is a polyunsaturated fatty acid.

In organisms such as mammals, certain fatty acids are necessary for good health but cannot be synthesized by the body. Such fatty acids are called essential fatty acids, because they must be obtained in the diet; one example is linoleic acid (see Figure 3.9).

Fats (triglycerides) that contain high amounts of saturated fatty acids pack together tightly, resulting in numerous intermolecular interactions that stabilize the fat. Saturated fats have a high melting point and tend to be solid at room temperature. Animal fats generally contain a high proportion of saturated fatty acids. For example, beef fat contains high amounts of stearic acid, a saturated fatty acid with a melting point of 70°C (see Figure 3.9). When you cook a hamburger on the stove, the stearic acid and other saturated animal fats melt, and liquid grease appears in the frying pan (**Figure 3.10**). When allowed to cool to room temperature, however, the liquid grease in the pan returns to its solid form.

Because of kinks in their chains, unsaturated fatty acids do not pack together as tightly as saturated fatty acids. Fats high in unsaturated fatty acids usually have low melting points and are liquid at room temperature. Such fats are called oils. Fats derived from plants generally contain unsaturated fatty acids. For example, olive oil contains high amounts of oleic acid, a monounsaturated fatty acid with a melting point of 16°C. Fatty acids with additional double bonds have even lower melting points. Linoleic acid (see Figure 3.9) has two double bonds and melts at −5°C. Safflower and sunflower oils contain high amounts of linoleic acid.

High temperature converts solid, saturated fats to liquid.

After cooling, saturated fats return to their solid form.

Unsaturated fats have low melting points and are liquid at room temperature.

**Figure 3.10** **Fats at different temperatures.** Saturated fats found in animals tend to have higher melting points than do the unsaturated fats found in plants. a (left, right): ©Tom Pantages; b: ©Felicia Martinez Photography/PhotoEdit

**(a) Animal fats at high and low temperatures**

**(b) Vegetable fats at low temperature**

**Concept Check:** *Certain types of fats used in baking are called shortenings. Shortenings are often made from vegetable oils by a process called hydrogenation, in which the addition of hydrogens causes double bonds to become single bonds. How do you think hydrogenation affects the melting point of the resulting fat?*

Most unsaturated fatty acids, including linoleic acid, exist in nature in the *cis* form (see Figure 3.3). Of particular concern to human health, however, are *trans* fatty acids, which are formed by an artificial process in which the natural *cis* form is altered to a *trans* configuration. This alteration gives the fats that contain such fatty acids a more compact, linear structure and, therefore, a higher melting point. Although this process has been used for many years to produce fats with a longer shelf life and with better characteristics for baking, research has revealed that *trans* fats are linked to human diseases. Notable among these is coronary artery disease, caused by a narrowing of the blood vessels that supply the heart.

Like starch and glycogen, fats are important for storing energy. The hydrolysis of triglycerides releases the fatty acids from glycerol, and these products can then be metabolized to provide energy to make ATP. Certain organisms, such as mammals, have the ability to store large amounts of energy by accumulating fats. The number of C—H bonds in a fat or carbohydrate molecule determines in part how much energy the molecule can yield. Fats contain many C—H bonds, whereas glucose and other carbohydrates have numerous C—OH bonds. Consequently, 1 gram of fat stores more energy than does 1 gram of starch or glycogen. For this reason, fat is an efficient means of energy storage for mobile organisms for which excess body mass may be a disadvantage. In animals, fats can also play a structural role by forming cushions that support organs. In addition, fats provide insulation under the skin that helps protect many terrestrial animals during cold weather and marine mammals in cold water.

## Phospholipids Are Amphipathic Lipids

**Phospholipids**, another class of lipids, are similar in structure to triglycerides but with one important difference. In a phospholipid, the third hydroxyl group of glycerol is linked to a phosphate group instead of a fatty acid. In most phospholipids, a small polar or charged nitrogen-containing molecule is attached to this phosphate (**Figure 3.11a**). The

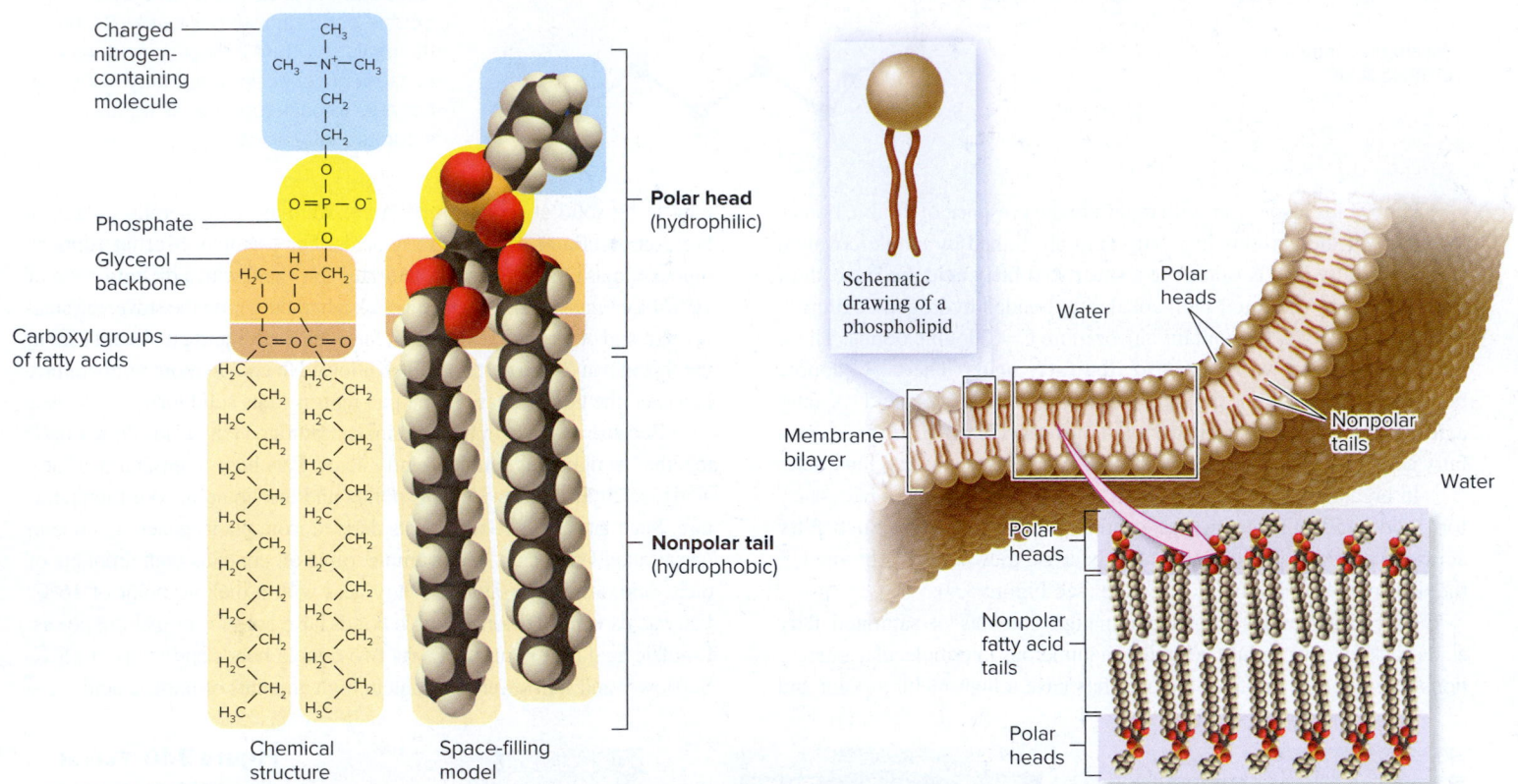

(a) **Structure and model of a phospholipid**

(b) **Arrangement of phospholipids in a bilayer**

**Figure 3.11 Structure of phospholipids. (a)** Chemical structure and space-filling model of phosphatidylcholine, a common phospholipid found in living organisms. Phospholipids contain both polar and nonpolar regions, making them amphipathic. The fatty-acid tails are nonpolar. The rest of the molecule is polar. **(b)** Arrangement of phospholipids in a biological membrane, such as the plasma membrane that encloses cells. The polar regions of the phospholipids face the watery environment, whereas the nonpolar regions associate with each other in the interior of the membrane, forming a bilayer.

 **Core Skill: Modeling** The goal of this modeling challenge is to propose a model for a lipid droplet based on a description of its components.

**Modeling Challenge:** Within human cells, lipids are stored in structures called lipid droplets. The surface of such a droplet has a monolayer of phospholipids and the interior is composed of neutral lipids, such as triglycerides. Some proteins are also attached to the polar head groups of the phospholipids. Draw a model showing the structure of a lipid droplet. In your model, draw phospholipids schematically as shown in part (b) of Figure 3.11, depict neutral lipids with blue dots, and depict proteins as green blobs.

glycerol backbone, phosphate group, and a charged molecule (in this case, choline) constitute a polar (hydrophilic) head at one end of the phospholipid, whereas the fatty acid chains form nonpolar (hydrophobic) tails at the opposite end. Recall from Chapter 2 that molecules with polar and nonpolar regions are called amphipathic molecules.

In water, phospholipids become organized into bilayers, with their polar heads interacting with the water molecules and their nonpolar tails facing the interior, where they are shielded from water. As you will learn in Chapter 5, this bilayer arrangement of phospholipids is critical for determining the structure of cellular membranes, as shown in **Figure 3.11b**.

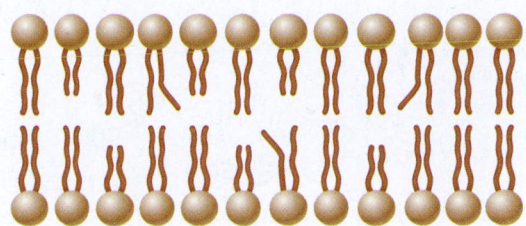

A mixture of long and short nonpolar tails, with double bonds in some of the long tails

**ANSWER** *When you compare the structures in the two drawings, you see that the lipids are more tightly packed in the membrane that is composed primarily of lipids with long nonpolar tails. As you can see, there is less open space in this membrane. It would be less fluid than the membrane containing a mixture of long and short nonpolar tails with double bonds in some of the long tails.*

## BIO TIPS

**THE QUESTION** *Biological membranes contain a phospholipid bilayer. As discussed in Chapter 5, membranes are somewhat fluid. For example, lipids move laterally within a membrane. When the lipid molecules are packed tightly together, the membrane tends to be less fluid than when they are more loosely packed.*

*Let's suppose that one membrane is composed primarily of lipids with long nonpolar tails that do not have any double bonds. A second membrane has some lipids with long nonpolar tails and others with short nonpolar tails. Also, some of the longer tails in this second membrane contain double bonds. Which of these two membranes would you expect to be less fluid?*

**T** OPIC *What topic in biology does this question address?* The topic is how lipid structure may affect the fluidity of a membrane.

**I** NFORMATION *What information do you know based on the question and your understanding of the topic?* From the question, you have learned that membranes with tightly packed lipids are less fluid than membranes in which the lipids are loosely packed. From your understanding of the topic, you may recall that the hydrophobic lipid tails associate with each other in the nonpolar region of the membrane (see Figure 3.11b). You may also remember that phospholipids vary with regard to the lengths of their nonpolar tails and that a double bond introduces a kink into a tail.

**P** ROBLEM-SOLVING **S** TRATEGY *Make a drawing. Compare and contrast.* One strategy to solve this problem is to make a drawing that shows the structures of the two membranes described in the question. Then compare the two structures, and decide which membrane is more tightly packed.

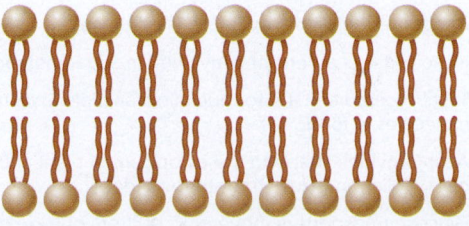

All long nonpolar tails

## Steroids Contain Ring Structures

**Steroids** have a distinctly different chemical structure from the other types of lipid molecules discussed thus far. Four fused rings of carbon atoms form the general structure of all steroids. One or more polar hydroxyl groups are attached to the fused ring structure, but they are not numerous enough to make a steroid highly water-soluble. For example, steroids with a hydroxyl group are known as sterols—one of the most well known being cholesterol (**Figure 3.12**, top). Cholesterol is found in the blood and cellular membranes of animals.

In steroids, minor differences in chemical structure result in profoundly different biological properties. For example, all steroid hormones are derived from cholesterol and share similarities in structure, but with some important differences. Estrogen is a steroid hormone found in high amounts in female vertebrates. Estrogen differs from testosterone, a steroid hormone found largely in males, by having one less methyl group, a hydroxyl group instead of a ketone group (see Table 3.1), and additional double bonds in one of its rings (compare the structures in Figure 3.12). However, these small differences are sufficient to make these two molecules largely responsible for whether an animal exhibits male or female characteristics, including feather color in birds.

## Waxes Are Complex Lipids That Prevent Water Loss from Organisms

Many plants and animals produce lipids called waxes that are secreted onto their surface, such as the leaves of plants and the cuticles of insects. Although any wax may contain hundreds of different compounds, all waxes contain one or more hydrocarbons and long structures that resemble a fatty acid attached by its carboxyl group to another long hydrocarbon chain. Waxes are very nonpolar and therefore exclude water, providing a barrier to water loss. They also are used as structural elements, such as the beeswax that forms the honeycomb produced by honeybees.

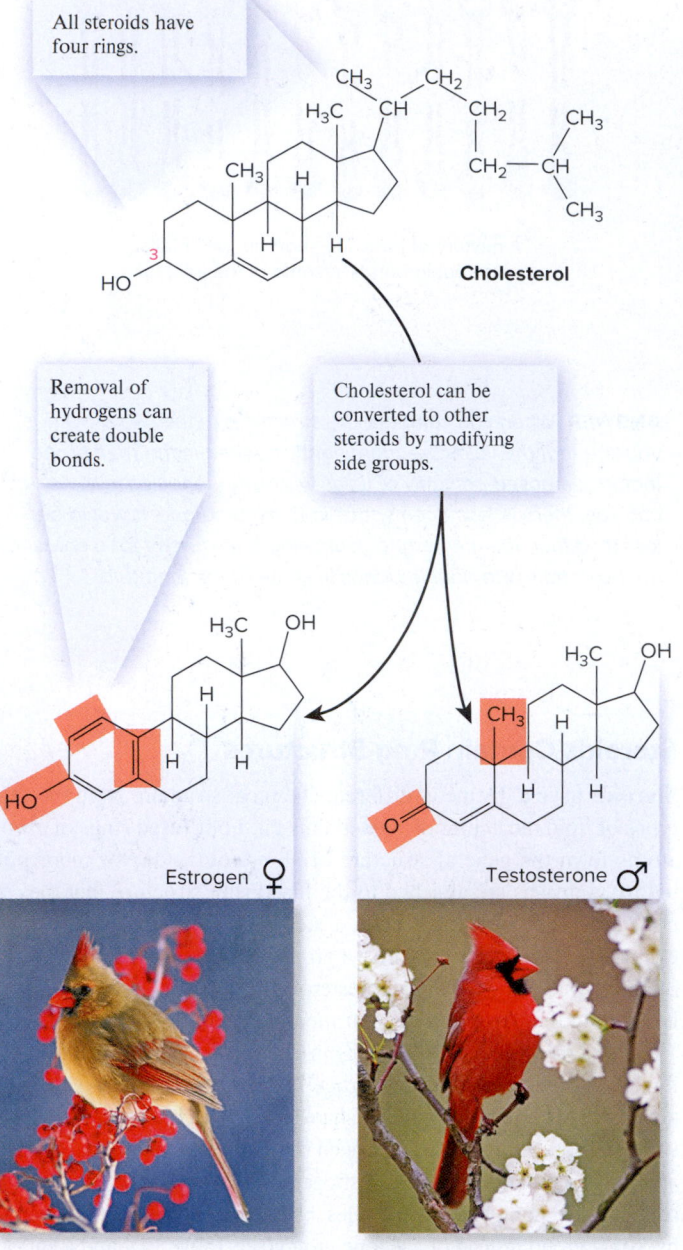

All steroids have four rings.

Removal of hydrogens can create double bonds.

Cholesterol can be converted to other steroids by modifying side groups.

Cholesterol

Estrogen ♀

Testosterone ♂

**Female cardinal**

**Male cardinal**

**Figure 3.12** **Structures of cholesterol and steroid hormones derived from cholesterol.** The structure of a steroid has four rings. Steroids include cholesterol and molecules derived from cholesterol, such as steroid hormones. These include the reproductive hormones estrogen and testosterone. (left, right): ©Adam Jones/Science Source

 **Core Concept: Structure and Function** The minor structural differences between estrogen and testosterone dramatically affect their biological functions. The differences between female and male cardinals are one example from the animal world of sex-dependent differences in form and function that are due to these two hormones.

## 3.6 Proteins

**Learning Outcomes:**

1. Give examples of the general functions that are carried out by different proteins.
2. Describe how amino acids are joined to form a polypeptide, and distinguish between a polypeptide and a protein.
3. Explain the four levels of protein structure.
4. Outline the factors that determine protein shape and function.
5. Define domain, as it relates to protein structure.

**Proteins** play critical roles in nearly all life processes (**Table 3.3**). The word protein comes from the Greek *proteios* (meaning of the first rank), which aptly describes their importance. Proteins account for about 50% of the organic material in a typical animal's body. In this section, we will survey the structure and function of proteins.

### Amino Acids Are the Building Blocks of Proteins

Proteins are composed of carbon, hydrogen, oxygen, nitrogen, and small amounts of other elements, notably sulfur. The monomers of proteins are **amino acids**, compounds with a structure in which a carbon atom, called the α-carbon, is linked to an amino group (—$NH_2$) and a carboxyl group (—COOH). The α-carbon also is linked to a hydrogen atom and a side chain, designated with the letter R. Proteins are polymers of amino acids.

When an amino acid is dissolved in water at neutral pH, the amino group accepts a hydrogen ion and is positively charged, whereas the

| Table 3.3 | Major Categories and Functions of Proteins | |
|---|---|---|
| **Category** | **Functions** | **Examples** |
| Proteins involved in gene expression and regulation | Make mRNA from a DNA template; synthesize polypeptides from mRNA; regulate genes | RNA polymerase catalyzes the synthesis of RNA using DNA as a template. |
| Motor proteins | Initiate movement | Myosin provides the contractile force of muscles. |
| Defense proteins | Protect organisms against disease | Antibodies help destroy bacteria or viruses. |
| Metabolic enzymes | Increase rates of chemical reactions | Hexokinase is an enzyme involved in sugar metabolism. |
| Cell-signaling proteins | Enable cells to communicate with each other and with the environment | Taste receptors in the tongue allow animals to taste molecules in food. |
| Structural proteins | Support and strengthen structures | Actin provides shape to the cytoplasm of plant and animal cells. Collagen gives strength to tendons. |
| Transporters | Promote movement of solutes across membranes | Glucose transporters move glucose from outside cells to inside cells, where it can be used for energy. |

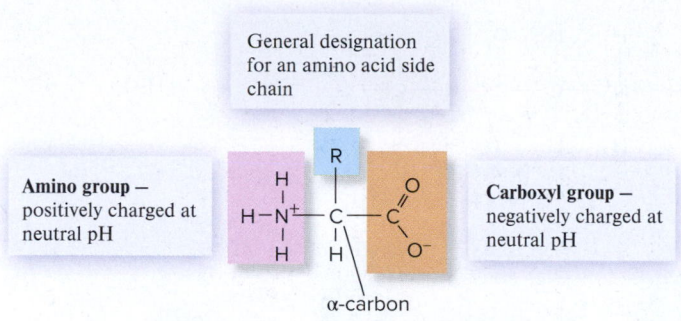

General designation for an amino acid side chain

**Amino group** — positively charged at neutral pH

**Carboxyl group** — negatively charged at neutral pH

α-carbon

carboxyl group loses a hydrogen ion and is negatively charged. The name amino acid was given to such molecules because they have an amino group and also a carboxyl group that acts as an acid.

All amino acids except glycine exist in more than one isomeric form, called the D and L forms, which are enantiomers. Only L-amino acids are found in proteins. D-amino acids are not found in most cells. An exception is in the cell walls of certain bacteria, where D-amino acids may play a protective role against molecules secreted by the host organism in which the bacteria live.

The 20 amino acids in proteins are distinguished by their side chains (**Figure 3.13**). The amino acids are categorized by whether

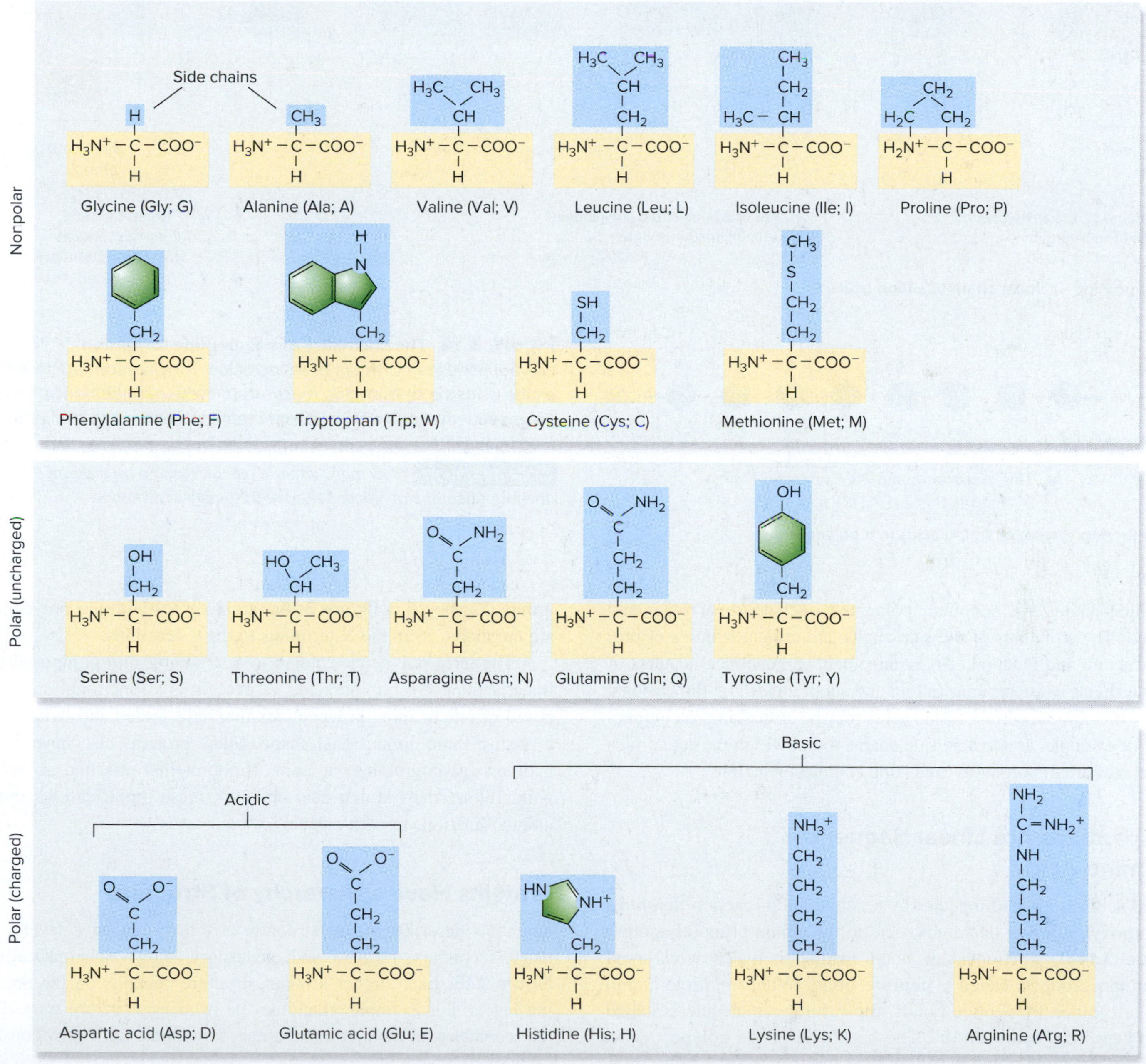

**Nonpolar**

Side chains

Glycine (Gly; G)   Alanine (Ala; A)   Valine (Val; V)   Leucine (Leu; L)   Isoleucine (Ile; I)   Proline (Pro; P)

Phenylalanine (Phe; F)   Tryptophan (Trp; W)   Cysteine (Cys; C)   Methionine (Met; M)

**Polar (uncharged)**

Serine (Ser; S)   Threonine (Thr; T)   Asparagine (Asn; N)   Glutamine (Gln; Q)   Tyrosine (Tyr; Y)

**Polar (charged)**

Acidic

Basic

Aspartic acid (Asp; D)   Glutamic acid (Glu; E)   Histidine (His; H)   Lysine (Lys; K)   Arginine (Arg; R)

**Figure 3.13  The 20 amino acids found in living organisms.**  Amino acids have different chemical properties (for example, nonpolar versus polar) due to their different side chains, which are highlighted in blue. These properties contribute to the differences in the three-dimensional shapes and chemical properties of proteins, which, in turn, influence proteins' biological functions. Note: Tyrosine has both polar and nonpolar characteristics and is listed in just one category for simplicity. The common three-letter and one-letter abbreviations for each amino acid are shown in parentheses.

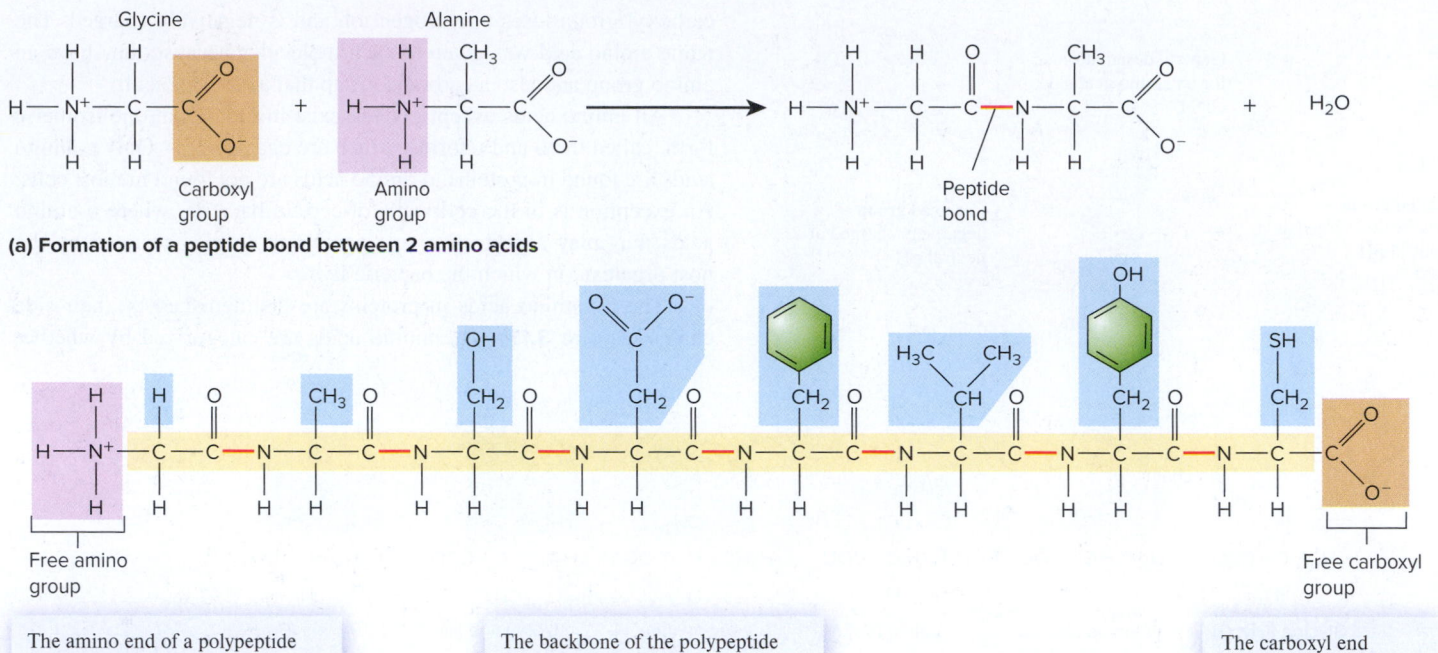

**(a) Formation of a peptide bond between 2 amino acids**

**(b) Polypeptide—a linear chain of amino acids**

The amino end of a polypeptide is called the N-terminus.

The backbone of the polypeptide is highlighted in yellow.

The carboxyl end of a polypeptide is called the C-terminus.

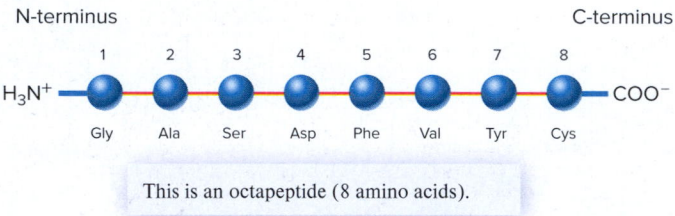

This is an octapeptide (8 amino acids).

**(c) Numbering system of amino acids in a polypeptide**

**Figure 3.14    The chemistry of polypeptide formation.** Polypeptides are polymers of amino acids. They are formed by linking amino acids via dehydration reactions to make peptide bonds. Every polypeptide has an amino end, or N-terminus, and a carboxyl end, or C-terminus.

**Concept Check:**  *How many water molecules would be produced during the formation of a polypeptide that is 72 amino acids long?*

their side chains are nonpolar, polar and uncharged, or polar and charged. The structures of the side chains are critical features of protein structure and function. The arrangement and chemical features of the side chains cause proteins to fold and adopt their three-dimensional shapes. In addition, certain amino acids may be critical in protein function. For example, amino acid side chains found within the active sites of enzymes are important in catalyzing chemical reactions.

## Polypeptides Are Linear Sequences of Amino Acids

Amino acids are joined together by a dehydration reaction that links the carboxyl group of one amino acid to the amino group of another (**Figure 3.14a**). The covalent bond formed between a carboxyl and amino group is called a **peptide bond**. When multiple amino acids are joined by peptide bonds, the resulting molecule is called a **polypeptide** (**Figure 3.14b**). The backbone of the polypeptide in Figure 3.14 is highlighted in yellow. The amino acid side chains project from the backbone. When two or more amino acids are linked together, one end of the resulting molecule has a free amino group. This is the amino end, or **N-terminus**. The other end of the polypeptide, called the carboxyl end, or **C-terminus**, has a free carboxyl

group. As shown in **Figure 3.14c**, amino acids within a polypeptide are numbered from the N-terminus to the C-terminus.

The term polypeptide refers to a structural unit composed of a linear sequence of amino acids. A **protein** is a functional unit composed of one or more polypeptides that have folded and twisted into a precise three-dimensional shape. Many proteins also have carbohydrates (glycoproteins) or lipids (lipoproteins) attached at various points along their amino acid chain(s); these modifications impart unique functions to such proteins.

## Proteins Have a Hierarchy of Structure

Scientists describe protein structure at four progressive levels: primary, secondary, tertiary, and quaternary, shown schematically in **Figure 3.15**. Each higher level of structure depends on the preceding levels. For example, changing the primary structure may affect the secondary, tertiary, and quaternary structures. Let's now consider each level separately.

*Primary Structure*    The **primary structure** (see Figure 3.15) of a protein is the amino acid sequence of its polypeptide(s). The primary structures of proteins are determined by genes. As we will explore in

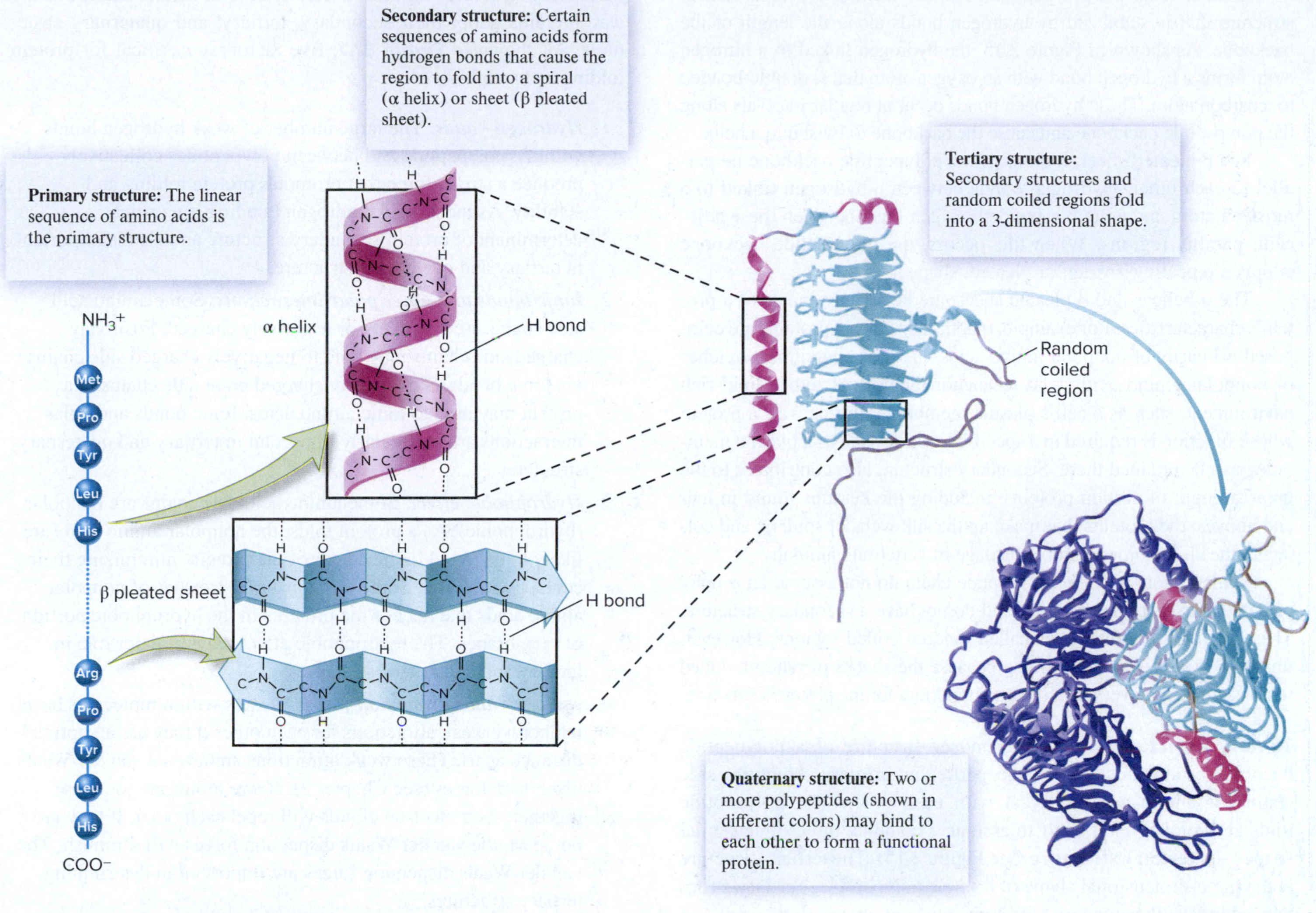

**Primary structure:** The linear sequence of amino acids is the primary structure.

**Secondary structure:** Certain sequences of amino acids form hydrogen bonds that cause the region to fold into a spiral (α helix) or sheet (β pleated sheet).

**Tertiary structure:** Secondary structures and random coiled regions fold into a 3-dimensional shape.

NH₃⁺

Met
Pro
Tyr
Leu
His

α helix

H bond

Random coiled region

β pleated sheet

H bond

Arg
Pro
Tyr
Leu
His

COO⁻

**Quaternary structure:** Two or more polypeptides (shown in different colors) may bind to each other to form a functional protein.

**Figure 3.15**  **The hierarchy of protein structure.**  The R groups are omitted for simplicity.

Chapter 12, genes carry the information for the production of polypeptides with specific amino acid sequences.

**Figure 3.16** shows the primary structure of ribonuclease, which functions as an enzyme to degrade ribonucleic acid (RNA) molecules after they are no longer required by a cell. As described later and in Unit III, RNA carries the information for protein synthesis. Ribonuclease is composed of a relatively short polypeptide consisting of 124 amino acids. An average polypeptide is about 300 to 500 amino acids in length, but some polypeptides in proteins are a few thousand amino acids long.

***Secondary Structure***  The amino acid sequence of a polypeptide, together with the laws of chemistry and physics, cause a protein to fold into a more compact structure. Amino acids can rotate around bonds within a polypeptide. Consequently, proteins are flexible and can fold into a number of shapes, just as a string of beads can be twisted into many configurations. Folding can be irregular, or certain regions can have a repeating folding pattern called **secondary structure**. The two basic types of a protein's secondary structure are the α helix and the β pleated sheet.

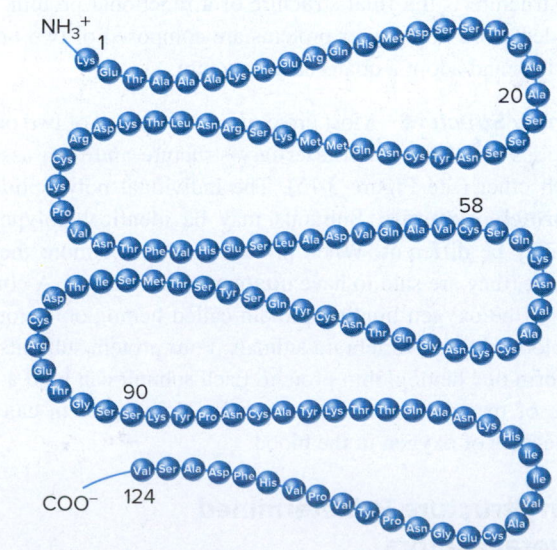

**Figure 3.16**  **The primary structure of ribonuclease.**  The example shown here is ribonuclease from cattle, which contains 124 amino acids.

In an α helix, the polypeptide backbone forms a repeating helical structure that is stabilized by hydrogen bonds along the length of the backbone. As shown in Figure 3.15, the hydrogen linked to a nitrogen atom forms a hydrogen bond with an oxygen atom that is double-bonded to a carbon atom. These hydrogen bonds occur at regular intervals along the polypeptide backbone and cause the backbone to twist into a helix.

In a β pleated sheet, regions of the polypeptide backbone lie parallel to each other. Hydrogen bonds between a hydrogen linked to a nitrogen atom and a double-bonded oxygen form between these adjacent, parallel regions. When this occurs, the polypeptide backbone adopts a repeating zigzag, or pleated, shape.

The α helices and β pleated sheets are key determinants of a protein's characteristics. For example, α helices in certain proteins are composed primarily of nonpolar amino acids. Proteins containing stretches of nonpolar amino acids tend to anchor themselves into a lipid-rich environment, such as a cell's plasma membrane. In this way, a protein whose function is required in a specific location such as a plasma membrane can be retained there. Secondary structure also contributes to the great strength of certain proteins, including the keratins found in hair and hooves; the proteins that make up the silk webs of spiders; and collagen, the chief component of cartilage in vertebrate animals.

Some regions along a polypeptide chain do not assume an α helix or β pleated sheet conformation and do not have a secondary structure. These regions are sometimes called random coiled regions. However, this term is somewhat misleading because the shapes of random coiled regions are usually very specific and important for the protein's function.

*Tertiary Structure*   As the secondary structure of a polypeptide becomes established due to the particular primary structure, side chains of amino acids interact with each other. The polypeptide folds and refolds upon itself to assume a complex three-dimensional shape—its **tertiary structure** (see Figure 3.15). The tertiary structure is the three-dimensional shape of a single polypeptide. Tertiary structure includes all secondary structures plus any interactions involving amino acid side chains. For some proteins, such as ribonuclease, the tertiary structure is the final structure of a functional protein. However, as described next, other proteins are composed of two or more polypeptides and adopt a quaternary structure.

*Quaternary Structure*   Most proteins are composed of two or more polypeptides that each adopt a tertiary structure and then assemble with each other (see Figure 3.15). The individual polypeptides are called **protein subunits**. Subunits may be identical polypeptides or they may be different. When proteins consist of more than one polypeptide, they are said to have **quaternary structure**. A common example is the oxygen-binding protein called hemoglobin, found in the red blood cells of vertebrate animals. Four protein subunits combine to form one hemoglobin protein. Each subunit can bind a single molecule of oxygen; therefore, each hemoglobin protein can carry four molecules of oxygen in the blood.

## Protein Structure Is Determined by Several Factors

The amino acid sequences of polypeptides distinguish the structure of one protein from another. As polypeptides are synthesized in a cell, they fold into secondary and tertiary structures, which assemble into quaternary structures for most proteins. Several factors determine the way proteins adopt their secondary, tertiary, and quaternary structures. As shown in **Figure 3.17**, five factors are critical for protein folding and stability:

1. *Hydrogen bonds.* The large number of weak hydrogen bonds within a polypeptide and between polypeptides collectively produce a strong force that promotes protein folding and stability. As mentioned, hydrogen bonding is a critical determinant of protein secondary structure and also is important in tertiary and quaternary structure.

2. *Ionic bonds and other polar interactions.* Some amino acid side chains are positively or negatively charged. Positively charged side chains may bind to negatively charged side chains via ionic bonds. Similarly, uncharged polar side chains in a protein may bind to ionic amino acids. Ionic bonds and polar interactions are particularly important in tertiary and quaternary structure.

3. *Hydrophobic effect.* Some amino acid side chains are nonpolar (hydrophobic). As a protein folds, the nonpolar amino acids are likely to be found in the center of the protein, minimizing their contact with water. Some proteins have stretches of nonpolar amino acids that anchor the proteins in the hydrophobic portion of membranes. The hydrophobic effect plays a major role in tertiary and quaternary structures.

4. *van der Waals dispersion forces.* Atoms within molecules have temporary weak attractions for each other if they are an optimal distance apart. These weak attractions are termed van der Waals dispersion forces (see Chapter 2). If two atoms are too close together, their electron clouds will repel each other. If they are far apart, the van der Waals dispersion forces will diminish. The van der Waals dispersion forces are important in determining tertiary structures.

5. *Disulfide bridges.* The side chain of the amino acid cysteine contains a sulfhydryl group (—SH), which can react with a sulfhydryl group in another cysteine side chain (see Figure 3.13). The result is a **disulfide bridge**, or disulfide bond, which links the two amino acid side chains together (—S—S—). Disulfide bridges are covalent bonds that can occur within a polypeptide or between different polypeptides. Though other forces are usually more important in protein folding, the covalent nature of disulfide bridges can help to stabilize the structure of a protein.

The first four factors just described are also important in the ability of different proteins to interact with each other. As discussed throughout Unit II and other parts of this textbook, many cellular processes involve steps in which two or more different proteins interact with each other. For such an interaction to occur, one protein must recognize and bind to the surface of the other. Such binding is usually very specific. The surface of one protein precisely fits into the surface of another (**Figure 3.18**). Such **protein-protein interactions** are critically important in allowing cellular processes to occur in a series of defined steps. In addition, protein-protein interactions are important in building complicated cellular structures that provide shape and organization to cells.

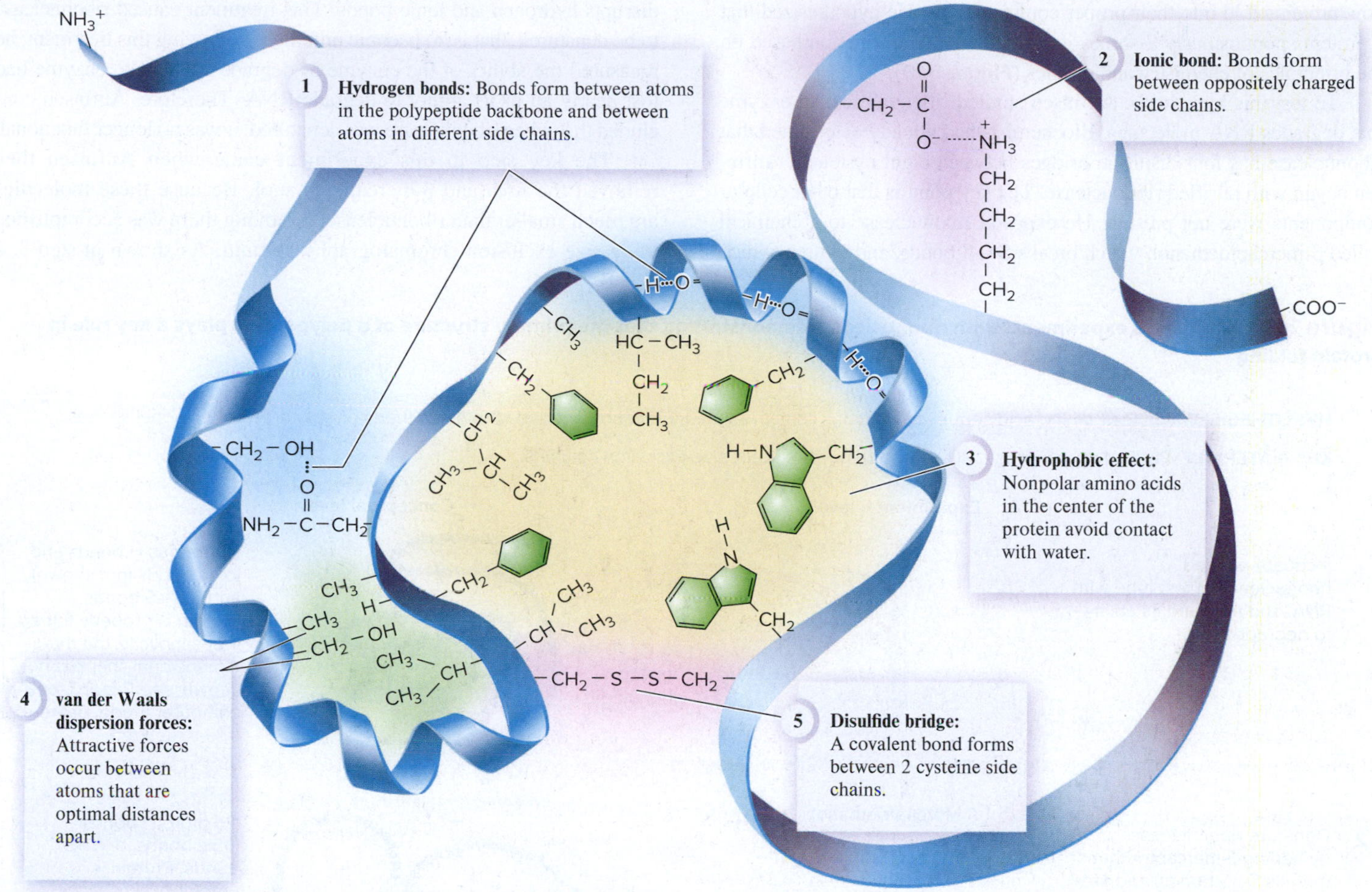

**1 Hydrogen bonds:** Bonds form between atoms in the polypeptide backbone and between atoms in different side chains.

**2 Ionic bond:** Bonds form between oppositely charged side chains.

**3 Hydrophobic effect:** Nonpolar amino acids in the center of the protein avoid contact with water.

**4 van der Waals dispersion forces:** Attractive forces occur between atoms that are optimal distances apart.

**5 Disulfide bridge:** A covalent bond forms between 2 cysteine side chains.

**Figure 3.17** **Factors that influence protein folding and stability.**

 **Core Concept: Structure and Function** This core concept of biology is apparent even at the molecular level. As seen in this figure, several distinct types of chemical interactions produce a protein with a complex shape. The three-dimensional shapes of different proteins determine their functions and their ability to interact with other cellular components.

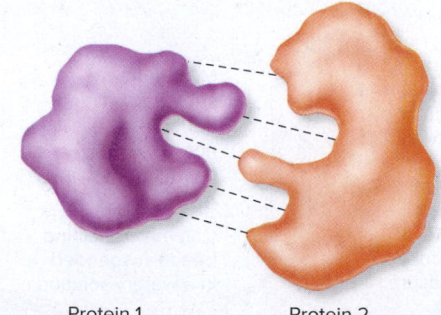

**Figure 3.18** **Protein-protein interaction.** Two different proteins may interact with each other through hydrogen bonding, ionic bonding, the hydrophobic effect, and van der Waals dispersion forces.

**Concept Check:** *If the primary structure of protein 1 in this figure was experimentally altered by the substitution of several incorrect amino acids, would protein 1 still be able to interact with protein 2?*

Protein 1        Protein 2

 **Core Skill: Process of Science**

# Feature Investigation | Anfinsen Showed That the Primary Structure of Ribonuclease Determines Its Three-Dimensional Structure

Prior to the 1960s, the mechanisms by which proteins assume their three-dimensional structures were not understood. Scientists hypothesized that the correct folding required cellular factors or that ribosomes, the sites where polypeptides are synthesized, somehow shaped proteins during their synthesis. American researcher Christian Anfinsen, however, postulated that the amino acid sequence is the primary factor that determines how proteins fold into their proper conformations. He hypothesized that a protein spontaneously assumes its most stable conformation based on the principles of chemistry and physics (**Figure 3.19**).

To test this hypothesis, Anfinsen studied ribonuclease, an enzyme that degrades RNA molecules. Biochemists had already determined that ribonuclease has four disulfide bridges between eight cysteines. Anfinsen began with purified ribonuclease. The key point is that other cellular components were not present. He exposed ribonuclease to a chemical called β-mercaptoethanol, which breaks S—S bonds, and to urea, which disrupts hydrogen and ionic bonds. This treatment caused ribonuclease to be denatured, that is, to become unfolded. Following this treatment, he measured the ability of the enzyme to degrade RNA. The enzyme had lost nearly all of its ability to degrade RNA. Therefore, Anfinsen concluded that when ribonuclease was denatured, it was no longer functional.

The key step in this experiment came when Anfinsen then removed the urea and β-mercaptoethanol. Because these molecules are much smaller than ribonuclease, removing them was accomplished via a size-exclusion chromatography column. As shown in step 3, a

**Figure 3.19** **Anfinsen's experiments with ribonuclease, demonstrating that the primary structure of a polypeptide plays a key role in protein folding.**

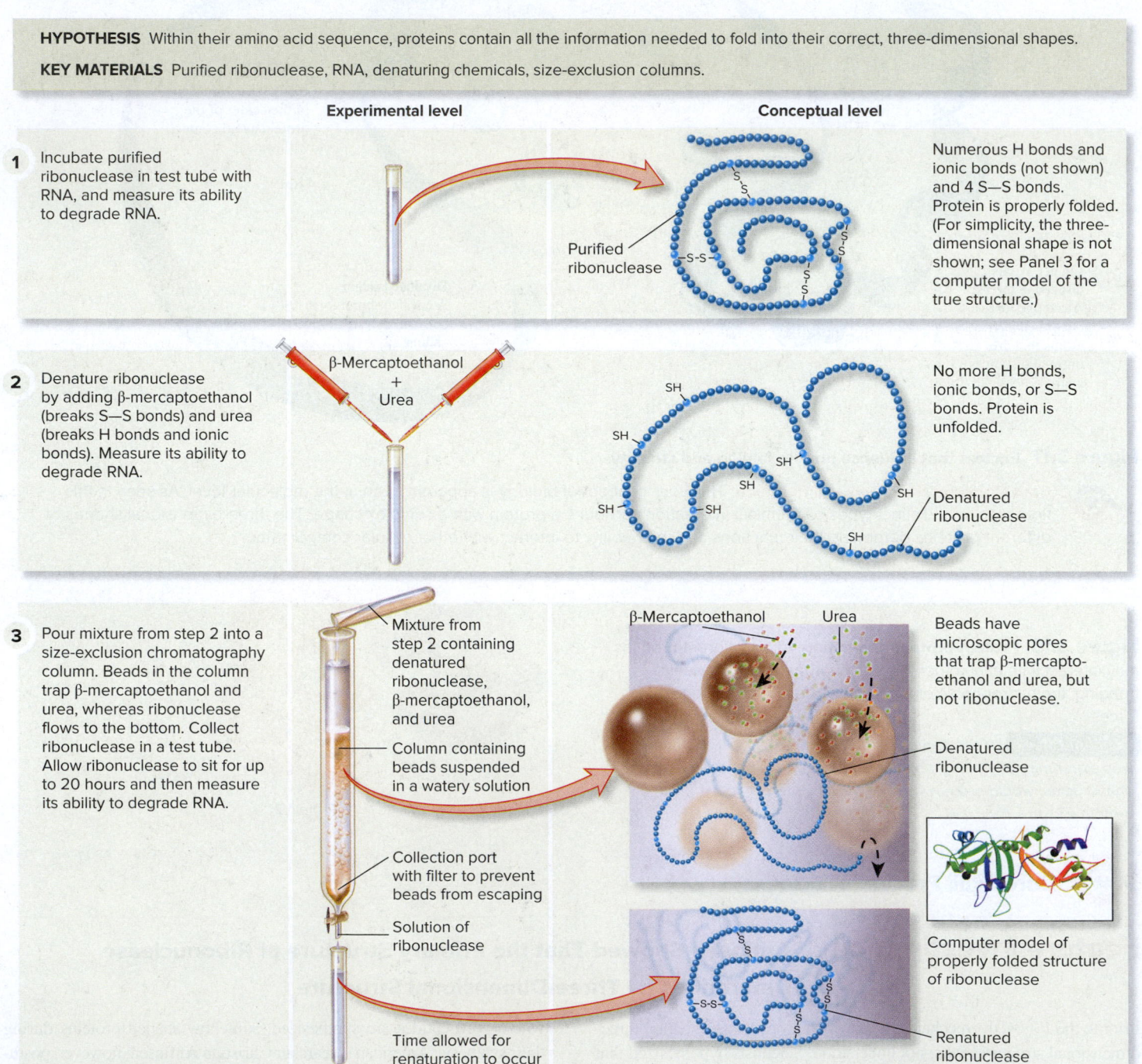

**4  THE DATA**

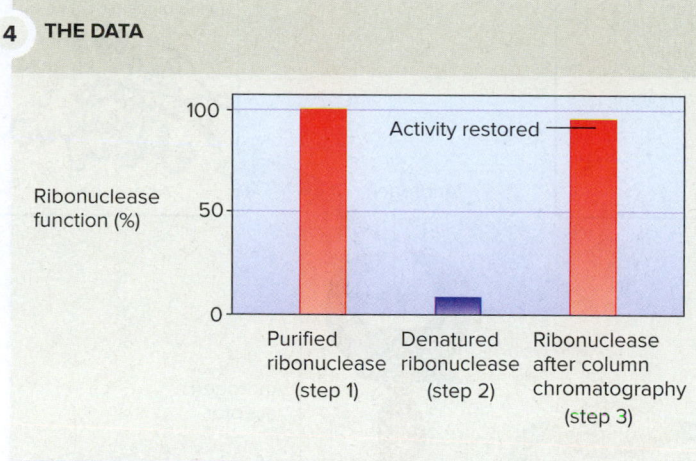

Ribonuclease function (%)

100 — Activity restored

50

0

| Purified ribonuclease (step 1) | Denatured ribonuclease (step 2) | Ribonuclease after column chromatography (step 3) |

**5  CONCLUSION**  Certain proteins, like ribonuclease, can spontaneously fold into their final, functional shapes without assistance from other cellular structures or factors. (However, as described in the text, this is not true of many other proteins.)

**6  SOURCE**  Haber, E., and Anfinsen, C.B. 1961. Regeneration of enzyme activity by air oxidation of reduced subtilisin-modified ribonuclease. *Journal of Biological Chemistry* 236: 422–424.

solution containing ribonuclease, β-mercaptoethanol, and urea was poured on top of a column of small beads and allowed to flow down the column to an open collection port at the bottom. The beads in the column had microscopic pores that trapped small molecules like urea and β-mercaptoethanol but allowed large proteins such as ribonuclease to pass down the length of the column and out the collection port.

Using this chromatography column, Anfinsen separated ribonuclease from β-mercaptoethanol and urea. He allowed the ribonuclease to sit for up to 20 hours and then retested its ability to degrade RNA. The result revolutionized our understanding of proteins. The activity of the ribonuclease was almost completely restored! Therefore, even in the complete absence of any cellular factors or ribosomes, this unfolded protein can refold into its correct, functional structure.

Researchers have since learned that ribonuclease's ability to refold into its functional structure does not occur with all proteins.

Some proteins require assistance from enzymes and other proteins to achieve their proper folding. Even so, Anfinsen's experiments provided compelling evidence that the primary structure of a polypeptide is a key determinant of a protein's tertiary structure, an observation that earned him the Nobel Prize in Chemistry in 1972.

### Experimental Questions

1. What hypothesis was Anfinsen testing?

2. Why did Anfinsen use urea and β-mercaptoethanol in his experiments?

3. **CoreSKILL »** Explain the result that was crucial to the discovery that the tertiary structure of ribonuclease may depend entirely on the primary structure.

 **Core Concept: Evolution**

## Proteins Contain Functional Domains

By comparing the structures of many different proteins, researchers have determined many of them have a modular design. This means that portions within these proteins, called **domains**, have distinct structures and functions. These domains have been duplicated during evolution, so the same kind of domain is found in many different proteins. A domain that is found in different proteins has the same three-dimensional tertiary structure in all of them and performs a characteristic function.

As an example, **Figure 3.20** shows two members of a family of related proteins, known as nuclear receptors, which function in the nucleus of animal cells by regulating how certain genes are turned on. These types of proteins are involved in animal development, reproduction, metabolism, and homeostasis. Nuclear receptors contain four or more domains within their structure:

- One domain found in nuclear receptors is a ligand-binding domain. These receptors are activated by a ligand, which is a molecule that binds to the receptor. For the two examples in

Figure 3.20, the ligands are the steroid hormones estrogen and testosterone (a type of androgen), and the nuclear receptors are called the estrogen receptor and the androgen receptor, respectively. Though the ligand-binding domains in these two nuclear receptors have similar structures, slight structural differences allow one to bind estrogen and the other to bind testosterone (see the insets in Figure 3.20).

- A second domain, called the DNA-binding domain, binds to DNA once the receptor is activated by its ligand. The DNA-binding domains of these two receptors are sufficiently similar to enable both of them to bind to DNA, but dissimilar enough that they bind to different genes. Therefore, estrogen and testosterone have different effects, because they regulate different sets of genes.

- Another domain, called a nuclear localization domain, facilitates the movement of the protein into the cell nucleus, where the DNA is located.

- Once the protein binds to DNA, a fourth domain, called the activation domain, activates the transcription of the target gene.

Overall, a nuclear receptor is a single protein with multiple domains, each with a unique function.

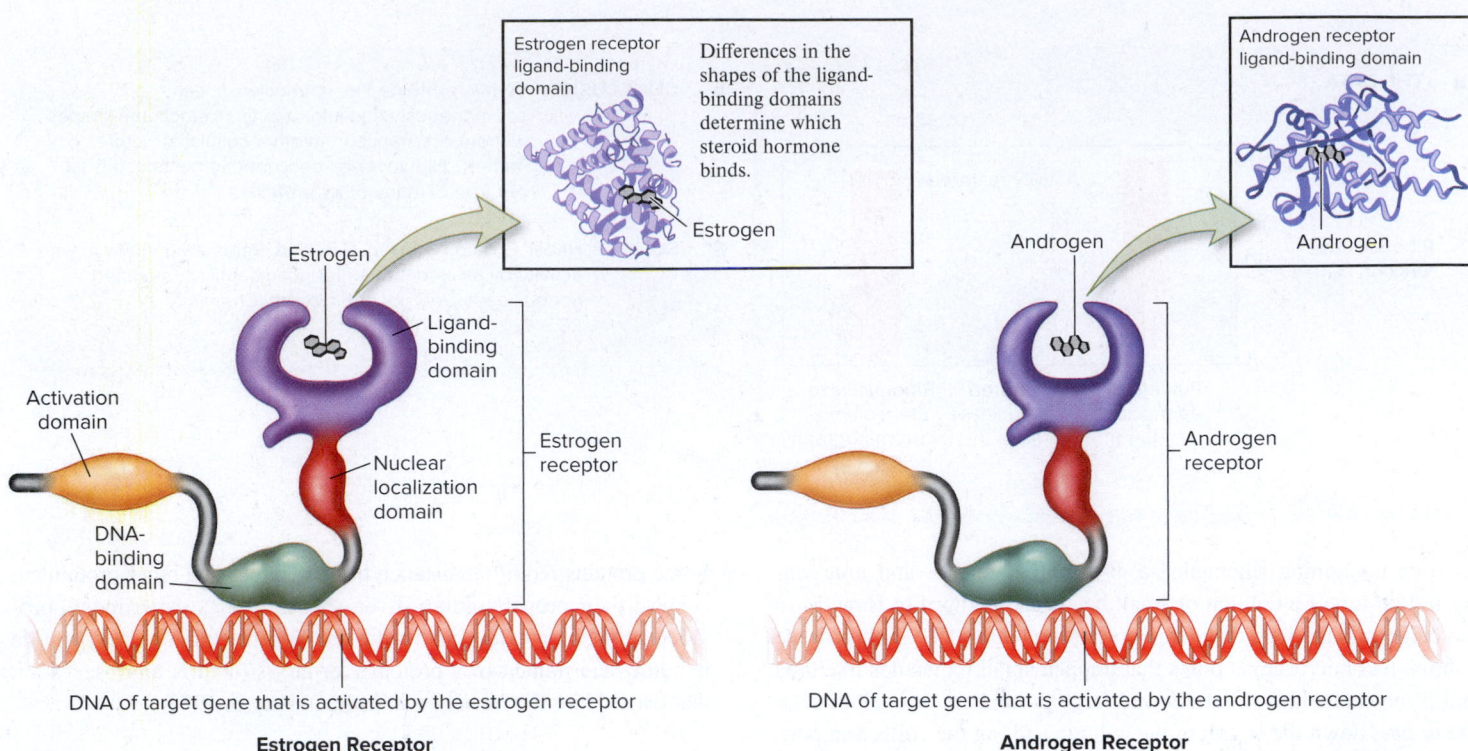

**Figure 3.20** **The domains in two related proteins called nuclear receptors.** These schematic drawings do not depict the actual three-dimensional structures of the estrogen and androgen receptors but rather emphasize that both receptors are composed of a series of different domains. The insets show the actual structures of the ligand-binding domains for both receptors. Note: The estrogen and androgen receptors are each composed of two identical polypeptides. For simplicity, only one polypeptide is shown here.

## 3.7 Nucleic Acids

### Learning Outcomes:

1. Describe the three components of a nucleotide.
2. Distinguish between the structures of DNA and RNA.
3. Explain how bases form hydrogen bonds with other bases in DNA and RNA.

**Nucleic acids** account for only about 2% of the weight of animals like humans, yet these molecules are extremely important because they are responsible for the storage, expression, and transmission of genetic information. The expression of genetic information in the form of specific proteins determines whether an organism is a human, a frog, an onion, or a bacterium. In this section, we will survey the general features of DNA and RNA, which are discussed in greater detail in Chapter 11.

### Nucleotides Are the Building Blocks of DNA and RNA

The two classes of nucleic acids are **deoxyribonucleic acid (DNA)** and **ribonucleic acid (RNA)**. DNA molecules store genetic information coded in the sequence of their building blocks. RNA molecules are involved in decoding this information into instructions for linking a specific sequence of amino acids to form a polypeptide. The monomers in DNA must be arranged in a precise way so that the correct code can be read.

Like other macromolecules, DNA and RNA are polymers consisting of linear sequences of repeating monomers. Each monomer, known as a **nucleotide**, has three components: (1) a phosphate group, (2) a pentose (five-carbon) sugar (either ribose or deoxyribose), and (3) a single or a double ring of carbon and nitrogen atoms known as a base (**Figure 3.21**). A nucleotide of DNA is called a deoxyribonucleotide; one of RNA is a ribonucleotide. The nucleotides in DNA contain the five-carbon sugar **deoxyribose**. Four different nucleotides are present in DNA, corresponding to the four different bases that can be linked to deoxyribose. The **purine** bases, **adenine (A)** and **guanine (G)**, have a fused double ring of carbon and nitrogen atoms, and the **pyrimidine** bases, **cytosine (C)** and **thymine (T)**, have a single-ring structure (look ahead to Figure 3.22).

### DNA Is Composed of Two Strands of Nucleotides

Nucleotides are covalently linked together to form strands of DNA. The phosphates and sugar molecules form the backbone of a DNA strand, with the bases projecting from the backbone. The carbon atoms of the sugar are numbered 1′ through 5′ (**Figure 3.22**). The phosphate groups link the 3′ carbon of one nucleotide to the 5′ carbon of the next. A DNA molecule consists of two strands of nucleotides coiled around each other to form a double helix (**Figure 3.23**). The two strands are held together by hydrogen bonds between a purine base in one strand and a pyrimidine base in the opposite strand. The ring structure of each base lies in a flat plane perpendicular to the sugar-phosphate backbone, somewhat like steps on a spiral staircase.

Only certain bases can pair with each other, due to the locations of the hydrogen-bonding groups in the four bases (see Figure 3.23). In

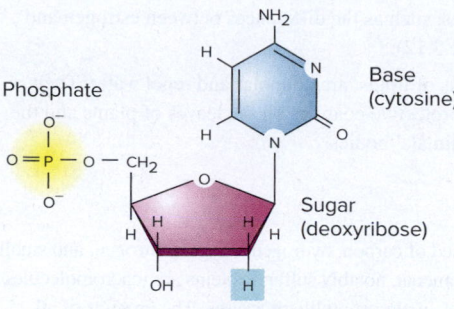

**Figure 3.21**
**Examples of two nucleotides found in RNA or DNA.**
A nucleotide has a phosphate group, a five-carbon sugar, and a base.

**Example of a deoxyribonucleotide**

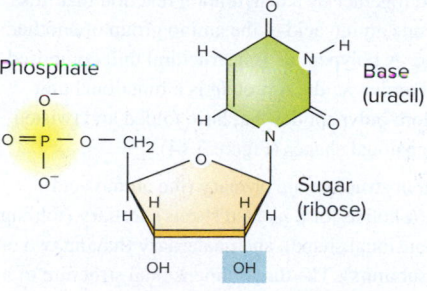

**Example of a ribonucleotide**

a DNA molecule, A on one strand is always paired with T on the opposite strand, and G is always paired with C. Two hydrogen bonds form between adenine and thymine (A-T pairing), whereas three hydrogen bonds are formed between guanine and cytosine (G-C pairing).

If we know the amount of one type of base in a DNA molecule, we can predict the relative amounts of each of the other three bases. For example, if a DNA molecule is composed of 20% A, it must also have 20% T. That leaves 60% of the bases that must be G and C combined. Because the amounts of G and C must be equal, this particular DNA molecule must contain 30% each of G and C. This specificity provides the mechanism for duplicating and transferring genetic information (see Chapter 11).

## RNA Strands Have a Similar Structure to DNA Strands

RNA structure differs in only a few respects from DNA structure. Like DNA, RNA consists of nucleotides covalently linked together. RNA usually consists of a single strand of nucleotides. In RNA, the sugar in each nucleotide is **ribose** instead of deoxyribose. Also, the pyrimidine base thymine found in DNA is replaced in RNA with the pyrimidine base **uracil (U)** (see Figure 3.21). The other three bases—adenine, guanine, and cytosine—are found in both DNA and RNA.

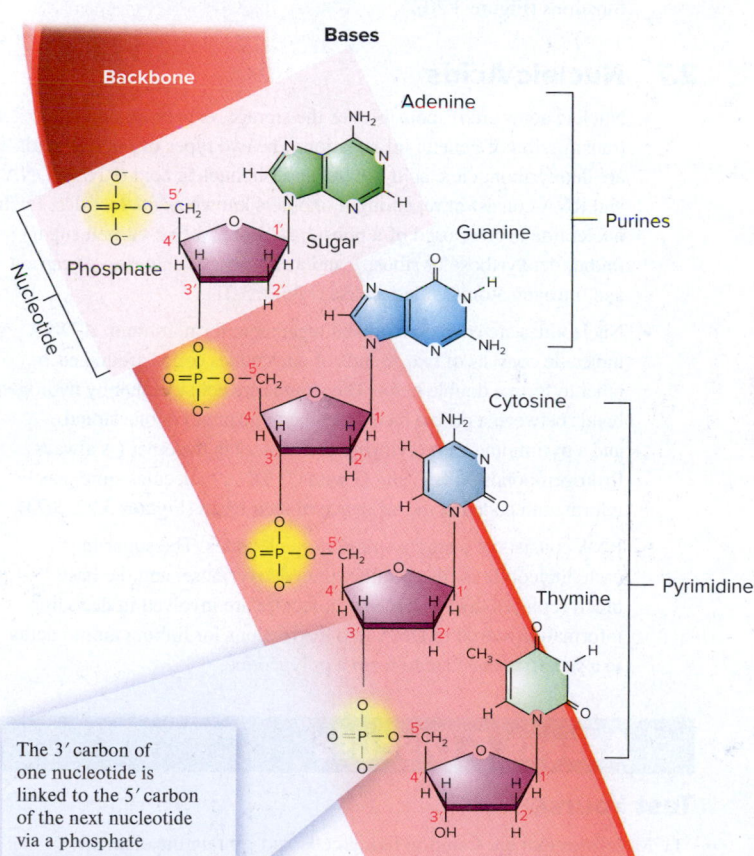

The 3′ carbon of one nucleotide is linked to the 5′ carbon of the next nucleotide via a phosphate group.

**Figure 3.22  Structure of a DNA strand.** Nucleotides are linked to each other to form a strand of DNA. The four bases found in DNA are shown. A strand of RNA is similar except the sugar is ribose, and uracil is substituted for thymine.

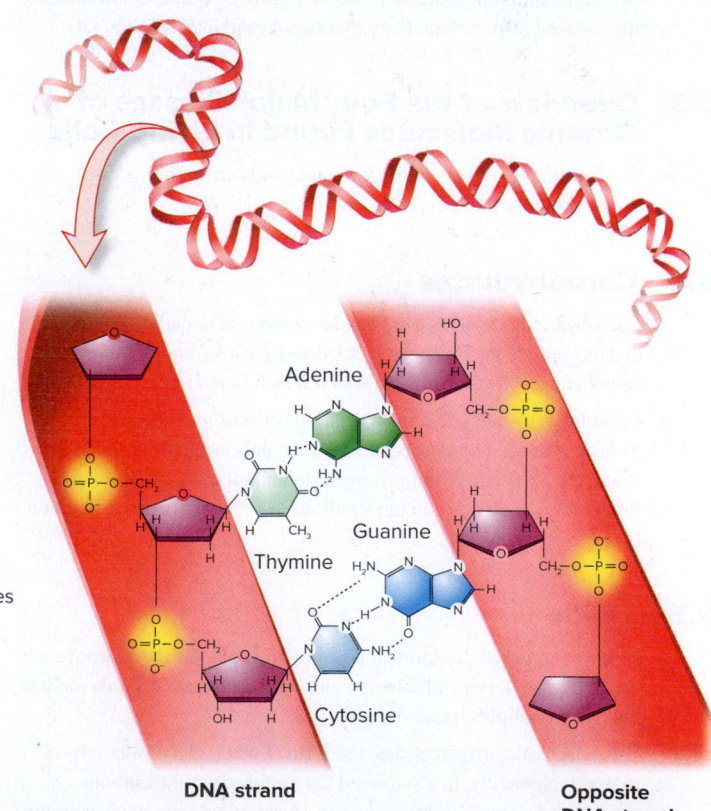

**DNA strand**

**Opposite DNA strand**

**Figure 3.23  The double-stranded structure of DNA.** DNA consists of two strands coiled around each other to form a double helix. The bases form hydrogen bonds (dashed lines) in which A pairs with T, and G pairs with C.

*Concept Check:  If the sequence of bases in one strand of a DNA double helix is known, can the base sequence of the opposite strand be predicted?*

# Summary of Key Concepts

## 3.1 The Carbon Atom

- Organic chemistry is the science of carbon-containing molecules, which are found in living organisms.

- A key property of the carbon atom is its ability to form four covalent bonds (polar or nonpolar) with other atoms. The combination of different elements and different types of bonds allows a vast number of organic compounds to be formed from relatively few chemical elements (Figures 3.1, 3.2).

- Carbon bonds are stable at the different temperatures associated with life.

- Organic compounds may contain functional groups with specific structures and chemical properties (Table 3.1).

- Carbon-containing molecules can exist as isomers, which have the same chemical formula but different structures and characteristics (Figure 3.3).

## 3.2 Formation of Organic Molecules and Macromolecules

- Organic molecules exist as monomers or polymers. Polymers are large macromolecules formed by dehydration reactions, in which individual monomers are attached to a growing polymer. Monomers are released from polymers by hydrolysis reactions (Figure 3.4).

## 3.3 Overview of the Four Major Classes of Organic Molecules Found in Living Cells

- The four major classes of organic molecules are carbohydrates, lipids, proteins, and nucleic acids (Table 3.2).

## 3.4 Carbohydrates

- Carbohydrates are organic molecules composed of carbon, hydrogen, and oxygen atoms. Cells can break down glucose, an important carbohydrate, releasing energy, which is then stored in the bonds of ATP.

- Carbohydrates include monosaccharides (the simplest sugars), disaccharides, and polysaccharides. The polysaccharides called starch (in plant cells) and glycogen (in animal cells) store energy. Some polysaccharides, notably cellulose, serve a structural function (Figures 3.5, 3.6, 3.7).

## 3.5 Lipids

- Lipids, composed predominantly of hydrogen and carbon atoms, are nonpolar and very insoluble in water. Major classes of lipids include fats, phospholipids, steroids, and waxes.

- Fats, also called triglycerides, are formed when glycerol bonds to three fatty acids. In a saturated fatty acid, all of the carbons are linked by single covalent bonds. Unsaturated fatty acids contain one or more C=C double bonds (Figures 3.8, 3.9, 3.10).

- Phospholipids are similar in structure to triglycerides, except that one glycerol is linked to a phosphate group instead of a fatty acid. Phospholipids contain both hydrophilic and hydrophobic regions, making them amphipathic (Figure 3.11).

- Steroids are constructed of four fused rings of carbon atoms. Small differences in steroid structure can lead to profoundly different biological properties, such as the differences between estrogen and testosterone (Figure 3.12).

- Waxes, another class of lipids, are nonpolar and repel water. They are often found as protective coatings on the leaves of plants and the outer surfaces of animals' bodies.

## 3.6 Proteins

- Proteins are composed of carbon, hydrogen, oxygen, nitrogen, and small amounts of other elements, notably sulfur. Proteins are macromolecules that play critical roles in almost all life processes. The proteins of all living organisms are composed of the same set of 20 amino acids, which contain 20 different side chains (Figure 3.13, Table 3.3).

- Amino acids are joined together by a dehydration reaction that links the carboxyl group of one amino acid to the amino group of another, forming a peptide bond. A polypeptide is a structural unit composed of a linear sequence of amino acids. A protein is a functional unit composed of one or more polypeptides that have folded and twisted into precise three-dimensional shapes (Figure 3.14).

- The four levels of protein structure are primary (the amino acid sequence), secondary (α helices or β pleated sheets), tertiary (folding to assume a three-dimensional shape), and quaternary (having two or more polypeptides as subunits). The three-dimensional structure of a protein determines its function (Figures 3.15, 3.16, 3.17, 3.18, 3.19).

- Proteins contain regions called domains that have particular functions (Figure 3.20).

## 3.7 Nucleic Acids

- Nucleic acids are responsible for the storage, expression, and transmission of genetic information. The two types of nucleic acids are deoxyribonucleic acid (DNA) and ribonucleic acid (RNA). DNA and RNA consist of repeating monomers known as nucleotides. Each nucleotide is composed of a phosphate group, a five-carbon sugar (either deoxyribose or ribose), and a single or double ring of carbon and nitrogen atoms called a base (Figure 3.21).

- Nucleotides are covalently linked together to form a strand. A DNA molecule consists of two strands of nucleotides coiled around each other to form a double helix. The strands are held together by hydrogen bonds between a purine base (adenine or guanine) in one strand and a pyrimidine base (cytosine or thymine) in the other (A always hydrogen-bonds with T, and G with C). DNA molecules store genetic information coded in the sequence of their bases (Figures 3.22, 3.23).

- RNA consists of a single strand of nucleotides. The sugar in each nucleotide is ribose rather than deoxyribose, and the base uracil replaces thymine. RNA molecules are involved in decoding information stored in DNA into instructions for linking amino acids in a specific sequence to form a polypeptide.

# Assess & Discuss

## Test Yourself

1. Molecules that are found in living cells and contain the element _____ are considered organic molecules.
   a. hydrogen
   b. carbon
   c. oxygen
   d. nitrogen
   e. calcium

2. The versatility of carbon that allows it to serve as the backbone for a variety of different molecules is due to
   a. the ability of carbon atoms to form four covalent bonds.
   b. the fact that carbon usually forms ionic bonds with many different atoms.
   c. the abundance of carbon in the environment.
   d. the ability of carbon to form covalent bonds with many different types of atoms.
   e. both a and d.

3. Which of the following type(s) of bonds are nonpolar?
   a. C—C
   b. C—O
   c. C—H
   d. both a and b
   e. both a and c

4. Which of the following molecules are isomers of each other?

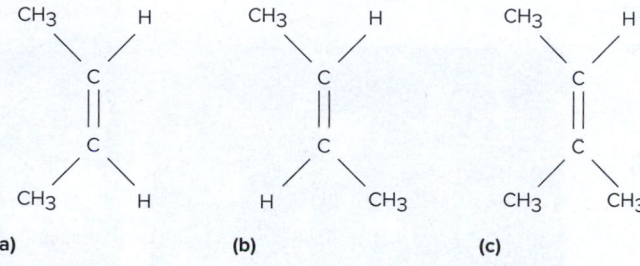

(a)    (b)    (c)

   a. a and b
   b. a and c
   c. b and c
   d. a, b, and c
   e. None of them are isomers of each other.

5. A polysaccharide that is commonly found in animal cells and stores energy is
   a. glucose.
   b. sucrose.
   c. glycogen.
   d. starch.
   e. cellulose.

6. In contrast to other fatty acids, essential fatty acids
   a. are always saturated fats.
   b. cannot be synthesized by a living organism and are necessary for survival.
   c. can act as building blocks for large, more complex macromolecules.
   d. are the simplest form of lipids found in plant cells.
   e. are structural components of plasma membranes.

7. The structures of three molecules are shown below. Which of them is (are) amphipathic?

(a)

(b)

(c)

   a. a only
   b. b only
   c. c only
   d. b and c
   e. a and c

8. The monomers of proteins are _____, and these are linked by polar covalent bonds commonly referred to as _____.
   a. nucleotides, peptide bonds
   b. amino acids, ester bonds
   c. hydroxyl groups, ester bonds
   d. amino acids, peptide bonds
   e. monosaccharides, glycosidic linkages

9. A _____ is a portion of protein with a particular structure and function.
   a. peptide bond
   b. domain
   c. phospholipid
   d. wax
   e. monosaccharide

10. A DNA molecule contains 30% G. What percentage of its bases are A?
   a. 30%
   b. 20%
   c. 15%
   d. 10%
   e. none of the above

## Conceptual Questions

1. Explain the similarities and differences between molecules that are isomers.

2. What is the difference between a saturated and an unsaturated fatty acid? How do the structural differences contribute to differences in their properties?

3. **Core Concept: Structure and Function**  A core concept of biology is that *structure determines function*. What does this mean for organic molecules such as carbohydrates, lipids, and proteins?

## Collaborative Questions

1. Discuss the differences between the various types of carbohydrates.

2. Discuss some of the roles that proteins play in living organisms.

# UNIT II
# CELL

Cell biology is the study of life at the cellular level. Although cells are the simplest units of life, biologists have come to realize that they are wonderfully complex and interesting, providing information about all living things. In this unit, Chapter 4 begins with a description of the evolutionary origin of cells and then provides an overview of cell structure and function. In Chapter 5, we will examine the structure and synthesis of cell membranes and the transport of substances in and out of the cell.

Chapters 6, 7, and 8 are largely devoted to metabolism, the sum of the chemical reactions in a cell or organism. Chapter 6 explores the topic of energy and considers how enzymes facilitate chemical reactions in a cell. Chapter 7 examines the pathways for carbohydrate breakdown and their production of an energy intermediate called ATP. In Chapter 8, we will explore the process of photosynthesis, in which the energy of sunlight drives the synthesis of carbohydrates.

Finally, Chapters 9 and 10 consider the ways that cells interact with their environment and with each other. In Chapter 9, we will examine how cells respond to signals, either those that come directly from their environment or those that are made by other cells. In Chapter 10, we will explore how cells interact with each other to produce a multicellular organism.

 **The following Core Concepts and Core Skills will be emphasized in this unit:**

- **Evolution:** *At the beginning of this unit, we will consider the evolutionary origin of cells, and also the origin of organelles, such as mitochondria and chloroplasts.*
- **Systems:** *As you will learn, a cell is a system with many interacting parts.*
- **Energy and Matter:** *Chapters 6, 7, and 8 examine how cells utilize and store energy contained within organic molecules such as glucose.*
- **Information:** *This unit provides many examples in which a cell's genome (its genetic makeup) results in a proteome (the proteins it makes) that largely determines cell structure and function.*
- **Structure and Function:** *Throughout this unit, we will see many examples in which the structure of proteins determines their cellular functions.*
- **Process of Science:** *Every chapter in this unit has a Feature Investigation that describes a pivotal experiment that provided insights into the workings of cells.*
- **Modeling:** *Every chapter has a Modeling Challenging that refines this important skill.*

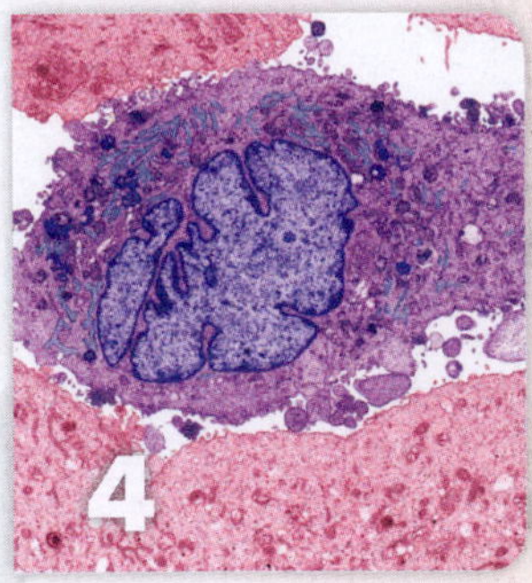

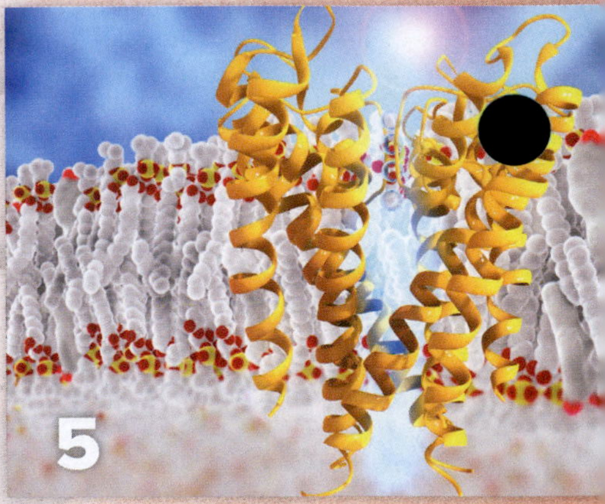

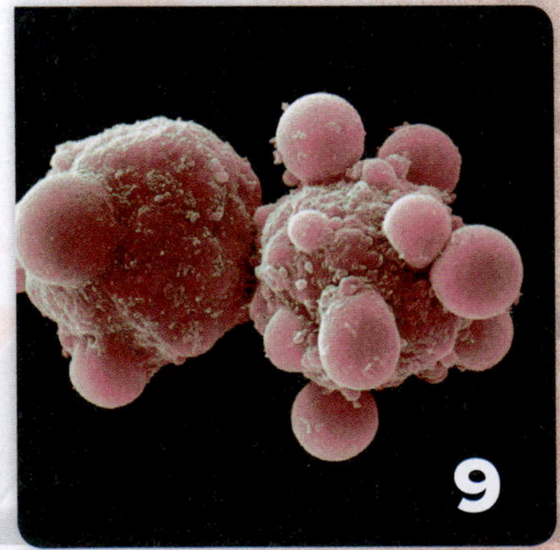

# Evolutionary Origin of Cells and Their General Features

# 4

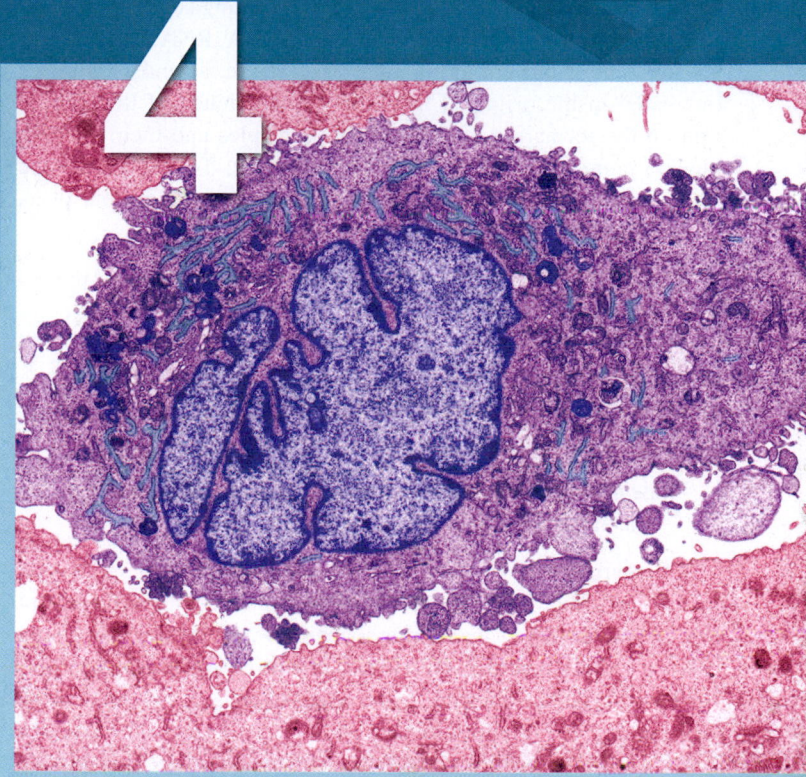

**A cell from the lung of a person with lung cancer.** The cell in this micrograph was viewed by a technique called transmission electron microscopy, which is described in this chapter. The micrograph was artificially colored using a computer to enhance the visualization of certain cell structures. ©Steve Gschmeissner/Science Source

**E**mily had a persistent cough ever since she started smoking cigarettes in college. However, at age 35, it seemed to be getting worse, and she was alarmed by the occasional pain in her chest. When she began to lose weight and became easily fatigued, Emily decided to see a doctor. The diagnosis was lung cancer. Despite aggressive treatment with chemotherapy and radiation therapy, she succumbed to lung cancer 14 months after the initial diagnosis. Emily was 36.

Topics such as cancer are within the field of **cell biology**—the study of individual cells and their interactions with each other. Researchers in this field want to understand the basic features of cells and apply their knowledge in the treatment of diseases such as cystic fibrosis, sickle cell disease, and lung cancer.

The idea that organisms are composed of cells originated in the mid-1800s. German botanist Matthias Schleiden studied plant material under the microscope and was struck by the presence of many similar-looking compartments, each of which contained a dark area. Today we call those compartments cells and the dark area the nucleus. In 1838, Schleiden speculated that cells are living entities and that plants are aggregates of cells arranged according to definite laws.

Schleiden was a good friend of the German physiologist Theodor Schwann. Over dinner one evening, their conversation turned to the nuclei of plant cells, and Schwann remembered having seen similar structures in animal tissue. Schwann conducted additional studies that showed that animal tissue contains large numbers of nuclei, which are located in cell-like compartments and occur at regular intervals. In 1839, Schwann extended Schleiden's hypothesis to animals. About two decades later, German biologist Rudolf Virchow proposed that *omnis cellula e cellula*, or "every cell originates from another cell." This idea arose from his research, which showed that diseased cells divide to produce more diseased cells.

The **cell theory**, which is credited to both Schleiden and Schwann with contributions from Virchow, has three parts.

1. All living organisms are composed of one or more cells.
2. Cells are the smallest units of life.
3. New cells come only from pre-existing cells by cell division.

Most cells are so small they cannot be seen with the unaided eye. However, as cell biologists have begun to unravel cell structure and function at the molecular level, the cell has emerged as a unit of incredible complexity and adaptability. In this chapter, we will begin our exploration of cells with a description of hypotheses of how living cells arose on Earth. We will then consider the general features of cell structure and function. Later chapters in this unit will explore certain aspects of cell biology in greater detail.

## 4.1 Origin of Living Cells on Earth

**Learning Outcomes:**

1. Outline the four overlapping stages that are hypothesized to have led to the origin of living cells.
2. List various hypotheses about how complex organic molecules formed.
3. Explain the concept of an RNA world, and describe how it could have evolved into a DNA/RNA/protein world.

Living cells are complex collections of molecules and macromolecules. DNA stores genetic information, RNA acts as an intermediary in the process of protein synthesis and plays other important roles, and proteins form the foundation for the structure and activities of living cells. Life as we know it requires this interplay between DNA, RNA, and proteins for its existence and perpetuation. On modern Earth, every living cell is made from a pre-existing cell.

But how did life get started? Because DNA, RNA, and proteins are the central players in the enterprise of life, scientists who are interested in the origin of life have focused much of their attention on the formation of these macromolecules and their building blocks, namely, nucleotides and amino acids. To understand the origin of cells, we can view the process as occurring in four overlapping stages:

*Stage 1:* Nucleotides and amino acids were produced prior to the existence of cells.

*Stage 2:* Nucleotides became polymerized to form RNA and/or DNA, and amino acids became polymerized to form proteins.

*Stage 3:* Polymers became enclosed in membranes.

*Stage 4:* Polymers enclosed in membranes acquired properties that are associated with living cells.

Researchers have followed a variety of experimental approaches to determine how life may have begun, including the synthesis of organic molecules in the laboratory without the presence of living cells or cellular material. This work has led researchers to propose a variety of hypotheses regarding the origin of cells. In this section, we will examine the origin of life at each of these stages and consider a few scientific viewpoints that wrestle with the question "How did life on Earth begin?"

## Stage 1: Organic Molecules Formed Prior to the Existence of Cells

Let's begin our inquiry into the first stage of the origin of life by considering how nucleotides and amino acids may have been made prior to the existence of living cells. In the 1920s, the Russian biochemist Alexander Oparin and the Scottish biologist J. B. S. Haldane independently proposed that organic molecules, such as nucleotides and amino acids, arose spontaneously under the conditions that occurred on early Earth. According to this hypothesis, the spontaneous appearance of organic molecules produced what they called a "primordial soup," which eventually gave rise to living cells.

The conditions on early Earth, which were much different from today, may have been more conducive to the spontaneous formation of organic molecules. Current hypotheses suggest that organic molecules, and eventually macromolecules, formed spontaneously. The formation of such molecules is termed prebiotic (before life) or abiotic (without life) synthesis. These organic molecules slowly accumulated because there was little free oxygen gas, so they were not spontaneously oxidized, and there were as yet no living organisms, so they were also not metabolized. The slow accumulation of these molecules in the early oceans over a long period of time formed what is now called the **prebiotic soup**. The formation of this medium was a key event that preceded the origin of life.

Though most scientists agree that life originated from the assemblage of nonliving matter on early Earth, the mechanism of how and where these molecules originated is widely debated. Many intriguing hypotheses have been proposed, which are not mutually exclusive. A few of the more widely debated ideas are the reducing atmosphere hypothesis, the extraterrestrial hypothesis, and the deep-sea vent hypothesis.

***Reducing Atmosphere Hypothesis*** Based largely on geological data, many scientists in the 1950s proposed that the atmosphere on early Earth was rich in water vapor ($H_2O$), hydrogen gas ($H_2$), methane ($CH_4$), and ammonia ($NH_3$). These components, along with a lack of atmospheric oxygen ($O_2$), produced a reducing atmosphere because methane and ammonia readily give up electrons to other molecules, thereby reducing them. Such oxidation-reduction reactions, or redox reactions, are required for the formation of complex organic molecules from simple inorganic molecules.

In 1953, American chemist Stanley Miller, a student in the laboratory of physical chemist Harold Urey, was the first scientist to use experimentation to test whether the prebiotic synthesis of organic molecules is possible. His experimental apparatus was intended to simulate the conditions on early Earth that were postulated in the 1950s (**Figure 4.1**). Water vapor from a flask of boiling water rose into another chamber containing hydrogen gas ($H_2$), methane ($CH_4$), and ammonia

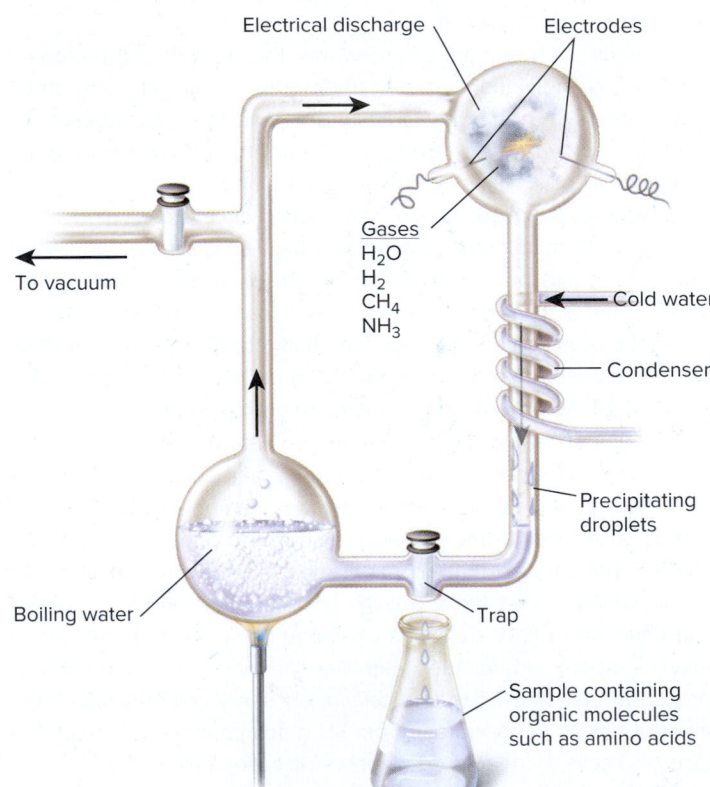

**Figure 4.1** **Testing the reducing atmosphere hypothesis for the origin of life—the Miller and Urey experiment.**

 **Core Skill: Process of Science** By conducting experiments, researchers were able to demonstrate the feasibility of the synthesis of organic molecules prior to the emergence of living cells.

(NH$_3$). Miller inserted two electrodes that sent electrical discharges into the chamber to simulate lightning bolts. A condenser jacket cooled some of the gases from the chamber, causing droplets to form that fell into a trap. He then took samples from this trap for chemical analysis. In his first experiments, he observed the formation of hydrogen cyanide (HCN) and formaldehyde (CH$_2$O). Such molecules are precursors of more complex organic molecules. These precursors also combined to make larger molecules such as the amino acid glycine. At the end of 1 week of operation, 10–15% of the carbon had been incorporated into organic compounds. Later experiments by Miller and others demonstrated the formation of sugars, a few types of amino acids, lipids, and bases found in nucleic acids (for example, adenine).

In a study published in 2011, researchers analyzed samples that Miller had preserved from a 1958 experiment in which he used a mixture of CH$_4$, NH$_3$, hydrogen sulfide (H$_2$S), and carbon dioxide (CO$_2$). For unknown reasons, Miller had not analyzed what products were made in this experiment. When these preserved samples were analyzed using modern technology, they were found to contain 23 different amino acids and 4 amines (another type of organic molecule), more organic compounds than seen in Miller's classic experiments.

Why were these studies important? The work of Miller and Urey was the first attempt to apply scientific experimentation to our quest to understand the origin of life. Their pioneering strategy showed that the prebiotic synthesis of organic molecules is possible, although it could not prove that such synthesis did in fact occur. In spite of the importance of these studies, critics of the so-called reducing atmosphere hypothesis have argued that Miller and Urey were wrong about the composition of early Earth's environment. More recently, many scientists have suggested that the atmosphere on early Earth was not reducing, but instead was a neutral gaseous mixture consisting mostly of carbon monoxide (CO), carbon dioxide (CO$_2$), nitrogen (N$_2$), and H$_2$O. These newer ideas are derived from studies of volcanic gas, which has much more CO$_2$ and N$_2$ than CH$_4$ and NH$_3$, and from the observation that ultraviolet (UV) radiation destroys CH$_4$ and NH$_3$, so these molecules would have been short-lived on early Earth, which had high levels of UV radiation. Nevertheless, since the experiments of Miller and Urey, many newer investigations have shown that organic molecules can be made under a variety of conditions. For example, organic molecules can be made prebiotically from a neutral gaseous mixture composed primarily of CO, CO$_2$, N$_2$, and H$_2$O.

***Extraterrestrial Hypothesis*** Many scientists have argued that sufficient organic molecules may have been present in the materials from asteroids and comets that reached the surface of early Earth in the form of meteorites. A significant proportion of meteorites belong to a class known as carbonaceous chondrites. Such meteorites may contain a substantial amount of organic carbon, including amino acids and nucleic acid bases. Based on this observation, some scientists have postulated that such meteorites could have transported a significant amount of organic molecules to early Earth.

Opponents of this hypothesis argue that most of this material would have been destroyed by the intense heating that accompanies the passage of large bodies through the atmosphere and their subsequent collision with the surface of the Earth. Though some organic molecules are known to reach the Earth via such meteorites, the degree to which heat would have destroyed those organic molecules remains a matter of controversy.

***Deep-Sea Vent Hypothesis*** In 1988, German lawyer and organic chemist Günter Wächtershäuser proposed that key organic molecules may have originated in deep-sea vents, which are cracks in the Earth's surface where superheated water rich in metal ions and hydrogen sulfide (H$_2$S) mixes abruptly with cold seawater. These vents release hot gaseous substances from the interior of the Earth at temperatures in excess of 300°C (572°F). Supporters of this hypothesis propose that biologically important molecules may have been formed in the temperature gradient between the extremely hot vent water and the cold water that surrounds the vent (**Figure 4.2a**).

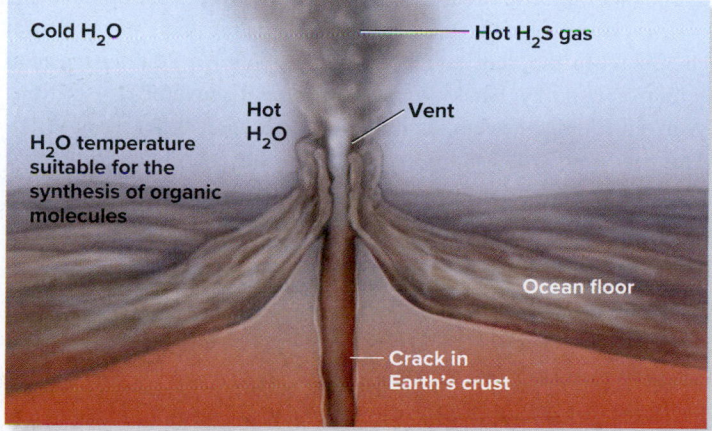

**(a) Deep-sea vent hypothesis**

**(b) A deep-sea vent community**

**Figure 4.2   The deep-sea vent hypothesis for the origin of life.**
**(a)** Deep-sea vents are cracks in the Earth's surface that release hot gases such as hydrogen sulfide (H$_2$S). This heats the water near the vent and results in a gradient between the very hot water adjacent to the vent and the cold water farther from the vent. The synthesis of organic molecules can occur in this gradient. **(b)** Photograph of a biological community near a deep-sea vent, which includes giant tube worms and crabs. b: ©CSSF/Neptune Canada

*Concept Check:*  *What properties of deep-sea vents made them suitable for the prebiotic synthesis of molecules?*

Experimentally, the temperatures within this gradient are known to be suitable for the synthesis of molecules that form components of biological molecules. For example, the reaction between iron and $H_2S$ yields pyrites and $H_2$ and has been shown to provide the energy necessary for the reduction of $N_2$ to $NH_3$. Nitrogen is an essential component of both nucleic acids and amino acids—the molecular building blocks of life. But $N_2$, which is found abundantly on Earth, is chemically inert, so it is unlikely to have given rise to life. The conversion of $N_2$ to $NH_3$ at deep-sea vents may have led to the production of amino acids and nucleic acids.

Interestingly, complex biological communities are found in the vicinity of modern deep-sea vents. Various types of fish, worms, clams, crabs, shrimp, and bacteria are found in significant abundance in those areas (**Figure 4.2b**). Unlike most other forms of life on our planet, these organisms receive their energy from chemicals in the vent and not from the Sun. In 2007, American scientist Timothy Kusky and colleagues discovered 1.4-billion-year-old fossils of deep-sea microbes near ancient deep-sea vents. This study provided more evidence that life may have originated on the bottom of the ocean. However, debate continues as to the primary way organic molecules were made prior to the existence of life on Earth.

## Stage 2: Organic Polymers May Have Formed on the Surface of Clay

The preceding three hypotheses provide reasonable mechanisms whereby small organic molecules could have accumulated on early Earth. Scientists hypothesize that the second stage in the origin of life was a period in which simple organic molecules polymerized to form more complex organic polymers such as DNA, RNA, or proteins. Most ideas regarding the origin of life assume that polymers with at least 30–60 monomers are needed to store enough information to make a viable genetic system. Because hydrolysis competes with polymerization, many scientists have speculated that the synthesis of polymers did not occur in a watery prebiotic soup, but instead took place on a solid surface or in evaporating tidal pools.

In 1951, Irish X-ray crystallographer John Bernal first suggested that the prebiotic synthesis of polymers took place on clay. In his book *The Physical Basis of Life*, he wrote that "clays, muds and inorganic crystals are powerful means to concentrate and polymerize organic molecules." Many clay minerals are known to bind organic molecules such as nucleotides and amino acids. Experimentally, many research groups have demonstrated the formation of nucleic acid polymers and polypeptides on the surface of clay, given the presence of monomer building blocks. During the prebiotic synthesis of RNA, the purine bases of the nucleotides could have interacted with the silicate surfaces of the clay. Cations, such as $Mg^{2+}$, bound the nucleotides to the negative surfaces of the clay, thereby positioning the nucleotides in a way that promoted bond formation between the phosphate of one nucleotide and the ribose sugar of an adjacent nucleotide. In this way, polymers such as RNA may have been formed.

Though the formation of polymers on clay remains a reasonable hypothesis, studies by American chemist Luke Leman and his colleagues English chemist Leslie Orgel and Iranian-American chemist M. Reza Ghadiri indicate that polymers can also form in aqueous solutions, contrary to the prevalent view. Their work in 2004 showed that carbonyl sulfide, a simple gas present in volcanic gases and deep-sea vent emissions, can bring about the formation of peptides from amino acids under mild conditions in water. These results indicate that the synthesis of polymers could have taken place in the prebiotic soup.

## Stage 3: Cell-like Structures May Have Originated When Polymers Were Enclosed by a Boundary

The third stage in the origin of living cells is hypothesized to be the formation of a boundary that separated the polymers such as RNA from the environment. The term **protobiont** is used to describe an aggregate of prebiotically produced molecules and macromolecules that acquired a boundary, such as a lipid bilayer, that allowed it to maintain an internal chemical environment distinct from that of its surroundings. What characteristics make protobionts possible precursors of living cells? Scientists envision the existence of four key features:

1. A boundary, such as a membrane, separated the internal contents of the protobiont from the external environment.
2. Polymers inside the protobiont contained information.
3. Polymers inside the protobiont had catalytic functions.
4. The protobionts eventually developed the capability of self-replication.

Protobionts were not capable of precise self-reproduction like living cells, but could divide to increase in number. Such protobionts are thought to have exhibited basic metabolic pathways in which the structures of organic molecules were changed. In particular, the polymers inside protobionts must have gained the catalytic ability to link organic building blocks to produce new polymers. This would have been a critical step in the process that eventually provided protobionts with the ability to self-replicate. According to this scenario, metabolic pathways became more complex, and the ability of protobionts to self-replicate became more refined over time. Eventually, these structures exhibited the characteristics that we attribute to living cells. As described next, researchers have hypothesized that protobionts may have existed as coacervates or liposomes.

***Coacervates*** In 1924, Alexander Oparin hypothesized that living cells evolved from **coacervates**, droplets that form spontaneously from the association of charged polymers such as proteins, carbohydrates, or nucleic acids surrounded by water. Their name derives from the Latin *coacervare*, meaning to assemble together or cluster. Coacervates measure 1–100 µm (micrometers) across, are surrounded by a tight skin of water molecules, and possess osmotic properties (**Figure 4.3a**). The skin of water allows the selective absorption of simple molecules from the surrounding medium.

Enzymes trapped within coacervates can perform primitive metabolic functions. For example, researchers have made coacervates containing the enzyme glycogen phosphorylase. When

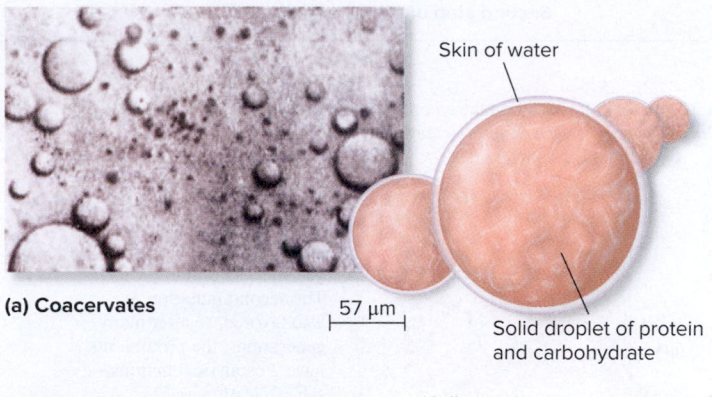

(a) Coacervates    |—— 57 µm ——|

Skin of water

Solid droplet of protein
and carbohydrate

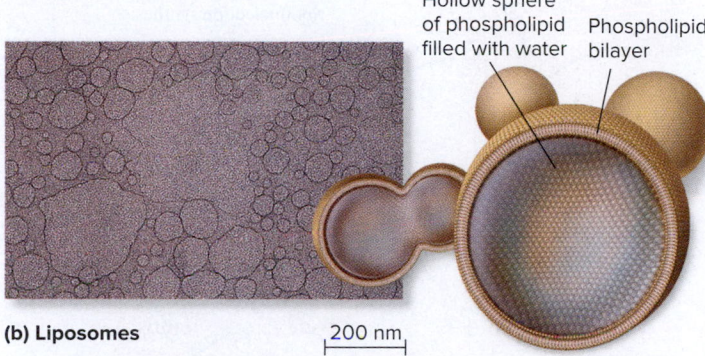

Hollow sphere
of phospholipid    Phospholipid
filled with water    bilayer

(b) Liposomes    |—— 200 nm ——|

**Figure 4.3  Possible structures of protobionts.**  Primitive cell-like structures such as coacervates or liposomes could have given rise to living cells. **(a)** A micrograph and illustration of coacervates, which are droplets of charged polymers surrounded by a skin of water molecules. **(b)** An electron micrograph and illustration of liposomes. Each liposome is made of a phospholipid bilayer surrounding an aqueous compartment. a: Source: A. I. Oparin. From *The Origin of Life*, New York: Dover, 1952; b: ©Mary Kraft

**Concept Check:**  *Which protobiont seems most similar to today's cells? Explain.*

  **Core Skill: Connections**  Look back at Figure 3.11. What is the physical/chemical reason why phospholipids tend to form a bilayer?

glucose-1-phosphate was made available to these coacervates, it was taken up into them, and starch was produced. The starch merged with the wall of the coacervates, and they increased in size and eventually divided into two. When the enzyme amylase was included, the starch was broken down to maltose, which was released from the coacervates.

***Liposomes***  As a second possibility, protobionts may have resembled **liposomes**—vesicles surrounded by a phospholipid bilayer (**Figure 4.3b**). When certain types of lipids are dissolved in water, they spontaneously form liposomes. As discussed in Chapter 5, phospholipid bilayers are selectively permeable (look ahead to Figure 5.10), and some liposomes can even store energy in the form of an electrical gradient. Such liposomes can discharge this energy in a neuron-like fashion, showing rudimentary signs of excitability, which is characteristic of living cells.

In 2003, Danish chemist Martin Hanczyc, American chemist Shelly Fujikawa, and Canadian-American biologist Jack Szostak showed that clay can catalyze the formation of liposomes that grow and divide, a primitive form of self-replication. Furthermore, if RNA was on the surface of the clay, the researchers discovered that liposomes that enclosed RNA were formed. These experiments are compelling because they showed that the formation of membrane vesicles containing RNA molecules is a plausible explanation for the emergence of cell-like structures based on simple physical and chemical properties.

## Stage 4: Cellular Characteristics May Have Evolved via Chemical Selection, Beginning with an RNA World

The majority of scientists favor RNA as the first macromolecule that was found in protobionts. Unlike other polymers, RNA exhibits three key functions:

1. RNA has the ability to store information in its nucleotide base sequence.
2. Due to base pairing, its nucleotide sequence has the capacity for self-replication.
3. RNA can perform a variety of catalytic functions. The results of many experiments have shown that some RNA molecules can function as **ribozymes**—RNA molecules that catalyze chemical reactions.

By comparison, DNA and proteins are not as versatile as RNA. DNA has very limited catalytic activity, and proteins are not known to undergo self-replication. RNA can perform functions that are characteristic of proteins and, at the same time, can serve as genetic material with replicative and informational functions.

How did the RNA molecules that were first made prebiotically evolve into more complex molecules that developed cell-like characteristics? Researchers propose that a process called chemical selection was responsible. **Chemical selection** occurs when a chemical within a mixture has special properties or advantages that cause it to increase in number relative to other chemicals in the mixture. (As we will discuss in Chapter 23, natural selection is a similar process except that it promotes change in a population of living organisms over time due to survival and reproductive advantages.) Chemical selection results in **chemical evolution**—a population of molecules changes over time to become a new population with a different chemical composition.

Scientists speculate that initially the special properties that enabled certain RNA molecules to undergo chemical selection were their ability to self-replicate and to perform other catalytic functions. As a way to understand the concept of chemical selection, let's consider a hypothetical scenario showing two steps of this process. Step 1 of **Figure 4.4** shows a group of protobionts that contain RNA molecules that were made prebiotically. RNA molecules inside these protobionts could be used as templates for the prebiotic synthesis of complementary RNA molecules. Such a process of self-replication, however, would be very slow because it would not be catalyzed by enzymes in the protobiont. In a first step of chemical selection, the sequence of one of the RNA molecules has undergone

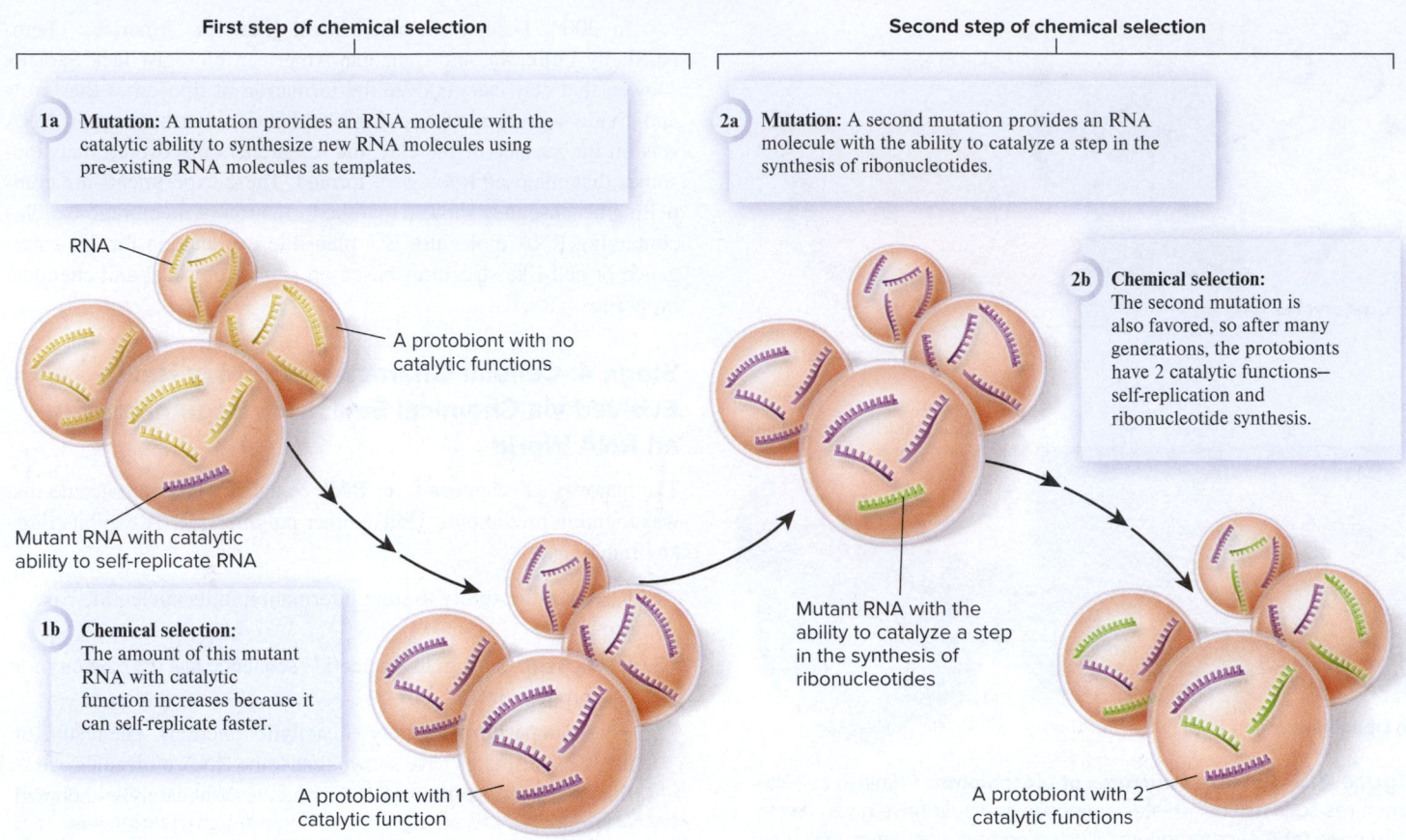

**First step of chemical selection**

1a **Mutation:** A mutation provides an RNA molecule with the catalytic ability to synthesize new RNA molecules using pre-existing RNA molecules as templates.

RNA

A protobiont with no catalytic functions

Mutant RNA with catalytic ability to self-replicate RNA

1b **Chemical selection:** The amount of this mutant RNA with catalytic function increases because it can self-replicate faster.

A protobiont with 1 catalytic function

**Second step of chemical selection**

2a **Mutation:** A second mutation provides an RNA molecule with the ability to catalyze a step in the synthesis of ribonucleotides.

2b **Chemical selection:** The second mutation is also favored, so after many generations, the protobionts have 2 catalytic functions—self-replication and ribonucleotide synthesis.

Mutant RNA with the ability to catalyze a step in the synthesis of ribonucleotides

A protobiont with 2 catalytic functions

**Figure 4.4  A hypothetical scenario illustrating the process of chemical selection.**  This figure shows a two-step scenario. In the first step, RNAs that can self-replicate are selected, and in the second step, RNAs with the ability to catalyze a step in ribonucleotide synthesis are selected.

**Concept Check:**  *What is meant by the term chemical selection?*

a mutation that gives it the catalytic ability to attach nucleotides together, using other RNA molecules as a template. This protobiont would have an advantage over others because it would be capable of faster self-replication of its RNA molecules. Over time, due to this enhanced rate of replication, protobionts carrying such RNA molecules would increase in number compared with the others. Eventually, the group of protobionts shown in the figure contains only this type of catalytic RNA.

In the second step of chemical selection (Figure 4.4, right side), a second mutation in an RNA molecule could produce the catalytic function that would help to promote the synthesis of ribonucleotides, the building blocks of RNA. For example, a hypothetical ribozyme may catalyze the attachment of a base to a ribose, thereby catalyzing one of the steps necessary for making a ribonucleotide. This protobiont would not rely solely on the prebiotic synthesis of ribonucleotides, which is a very slow process. Therefore, the protobiont having the ability to both self-replicate and synthesize ribonucleotides would have an advantage over a protobiont that could only self-replicate. Over time, the faster rate of self-replication and ribonucleotide synthesis would cause an increase in the numbers of protobionts with both functions.

The **RNA world** is a hypothetical period on early Earth when both the information needed for life and the catalytic activity of living cells were contained solely in RNA molecules. In this scenario, protobionts containing RNA exhibited the properties of life due to RNA genomes that were copied and maintained through the catalytic function of RNA molecules. Over time, scientists envision that mutations occurred in these RNA molecules, occasionally introducing new functional possibilities. Chemical selection would have eventually produced an increase in complexity in these protobionts, with RNA molecules accruing abilities such as the ability to link amino acids together into proteins and other catalytic functions.

But is an RNA world a plausible scenario? Remarkably, scientists have been able to perform experiments in the laboratory that can select for RNA molecules with a particular function. American biologists David Bartel and Jack Szostak conducted the first study of this type in 1993, in which they selected for RNA molecules with the catalytic ability to link nucleotides together. After 10 rounds of chemical selection, they obtained a collection of RNA molecules that had catalytic activity that was 3 million times higher than their original random collection of molecules!

Like the work of Miller and Urey, the study by Bartel and Szostak showed the feasibility of another phase of the prebiotic process that led to life. In this case, chemical selection resulted in chemical evolution. The results showed that chemical selection can change the functional characteristics of a group of RNA molecules over time by increasing the proportion of those molecules with enhanced function.

## The RNA World Was Superseded by the Modern DNA/RNA/Protein World

Assuming that an RNA world was the origin of life, researchers have asked, "Why and how did the RNA world evolve into the DNA/RNA/protein world we see today?" The RNA world may have been superseded by a DNA/RNA world or an RNA/protein world before the emergence of the modern DNA/RNA/protein world. Let's now consider the advantages of a DNA/RNA/protein world as opposed to the simpler RNA world and explore how this modern biological world might have come into being.

*Information Storage* RNA can store information in its base sequence. If so, why did DNA take over that function in modern cells? During the RNA world, RNA had to perform two roles: the storage of information and the catalysis of chemical reactions. Scientists have speculated that the incorporation of DNA into cells would have relieved RNA of its storage role, thereby allowing RNA to perform a variety of other functions. For example, if DNA stored the information for the synthesis of RNA molecules, such RNA molecules could bind cofactors, have modified bases, or bind peptides that might enhance their catalytic function. Cells with both DNA and RNA would have had an advantage over those with just RNA, and so they would have been selected. Another advantage of DNA is its stability. Compared with RNA, DNA strands are less likely to spontaneously break.

A second issue is how DNA came into being. Scientists have proposed that an ancestral RNA molecule had the ability to make DNA using RNA as a template. This function, known as reverse transcription, is described in Chapter 19 in the discussion of retroviruses. Interestingly, modern eukaryotic cells can use RNA as a template to make DNA. For example, an RNA sequence in the enzyme telomerase is used as a template to copy the ends of chromosomes, thus preventing progressive shortening of the chromosomes (look ahead to Figure 11.21).

*Metabolism and Other Cellular Functions* Now let's consider the origin of proteins. The emergence of proteins as catalysts may have been a great benefit to early cells. Due to the different chemical properties of the 20 amino acids, proteins have vastly greater catalytic ability than do RNA molecules, again providing a major advantage to cells that had both RNA and proteins. In modern cells, proteins have taken over most, but not all, catalytic functions. In addition, proteins can perform other important tasks. For example, cytoskeletal proteins carry out structural roles, and certain membrane proteins are responsible for the uptake of substances into living cells.

How would proteins have come into being in an RNA world? Chemical selection experiments have shown that RNA molecules can catalyze the formation of peptide bonds and even attach amino acids to primitive tRNA molecules. Similarly, modern protein synthesis still includes a central role for RNA in the synthesis of polypeptides. First, mRNA provides the information for a polypeptide sequence. Second, tRNA molecules act as adaptors for the formation of polypeptides. And finally, ribosomes containing rRNA provide a site for polypeptide synthesis. Furthermore, rRNA within ribosomes acts as a ribozyme to catalyze peptide bond formation. Taken together, the analysis of translation in modern cells is consistent with an evolutionary history in which RNA molecules were instrumental in the emergence and formation of proteins.

## 4.2 Microscopy

### Learning Outcomes:

1. **CoreSKILL »** Explain the three key parameters in microscopy: resolution, contrast, and magnification.
2. **CoreSKILL »** Compare and contrast the different types of light and electron microscopes and their uses.

Before we examine the general features of modern cells, we will first consider the **microscope**, which is a magnification tool that enables researchers to visualize the structure and inner workings of cells. A **micrograph** is an image taken with the aid of a microscope. The first compound microscope—a microscope with more than one lens—was first constructed in 1595 by Zacharias Jansen of Holland. In 1665, an English biologist, Robert Hooke, studied cork under a primitive compound microscope he had made. He actually observed cell walls because cork cells are dead and have lost their internal components. Hooke coined the word cell, derived from the Latin word *cellula*, meaning small compartment, to describe the structures he observed.

Three important parameters in microscopy are resolution, contrast, and magnification.

- **Resolution**, a measure of the clarity of an image, is the ability to observe two adjacent objects as distinct from one another. For example, a microscope with good resolution enables a researcher to distinguish as separate objects two adjacent chromosomes, which would appear as a single, blurry object under a microscope with poor resolution.

- **Contrast** refers to relative differences in the lightness, darkness, or color between adjacent regions in a sample. The ability to visualize a particular cell structure may depend on how different it looks from adjacent structures. Staining the cellular structure of interest with a dye can make viewing much easier. The application of stains, which selectively label individual components of the cell, greatly improves contrast. However, staining should not be confused with colorization. Many of the micrographs shown in this textbook are colorized, or artificially colored, to emphasize certain cellular structures, such as different parts of a cell (see the chapter opener, for example). In colorization, particular colors are added to micrographs with the aid of a computer.

- **Magnification** is the ratio between the size of an image produced by a microscope and the object's actual size. For example, if the image size is 100 times larger than the object's actual size, the magnification is designated 100×. Depending on the quality of the lens and the illumination source, every microscope has an optimal range of magnification before objects appear too blurry to be readily observed.

Microscopes are categorized into two groups based on the source of illumination. A **light microscope** utilizes light for illumination, whereas an **electron microscope** uses a beam of electrons for illumination. Very good light microscopes resolve structures that are as close as 0.2 μm (micron, or micrometer) from each other.

Resolution is improved when the illumination source has a shorter wavelength. A major advance in microscopy occurred in 1931 when Max Knoll and Ernst Ruska invented the first electron microscope. Because the wavelength of an electron beam is much shorter than visible light, the resolution of an electron microscope is far better than that of any light microscope. The resolution limit of an electron microscope is typically around 2 nm (nanometers), which is about 100 times better than a light microscope. **Figure 4.5** shows the ranges of resolving powers of the electron microscope, light microscope, and unaided eye and compares them with the sizes of various chemical and biological structures.

Over the past several decades, technological advances have made light microscopy a powerful research tool. Improvements in lens technology, microscope organization, sample preparation,

sample illumination, and computerized image processing have enabled researchers to invent different types of light microscopes, each with its own advantages and disadvantages (**Figure 4.6**).

Similarly, improvements in electron microscopy occurred during the 1930s and 1940s, and by the 1950s, the electron microscope was playing a major role in advancing our understanding of cell structure. Two general types of electron microscopy have been developed: transmission electron microscopy and scanning electron microscopy. In **transmission electron microscopy (TEM)**, a beam of electrons is transmitted through a biological sample. To provide contrast, the sample is stained with a heavy metal, which binds to certain cellular structures such as membranes. The sample is then adhered to a copper grid and placed in a transmission electron microscope. When the beam of electrons strikes the sample, some of them hit the heavy metal and are scattered, while those that pass through without being scattered are focused to form an image on a photographic plate or screen (**Figure 4.7a**). The metal-stained regions of the sample that scatter electrons appear as darker areas, because of reduced electron penetration of those regions. TEM provides a cross-sectional view of a cell and its organelles and gives the best resolution compared with other forms of microscopy. However, such microscopes are expensive and cannot be used to view living cells.

**Scanning electron microscopy (SEM)** is used to view the surface of a biological sample. The sample is coated with a thin layer of heavy metal, such as gold or palladium, and then is exposed to an electron beam that scans its surface. Secondary electrons are emitted from the sample, which are detected and create an image of its three-dimensional surface (**Figure 4.7b**).

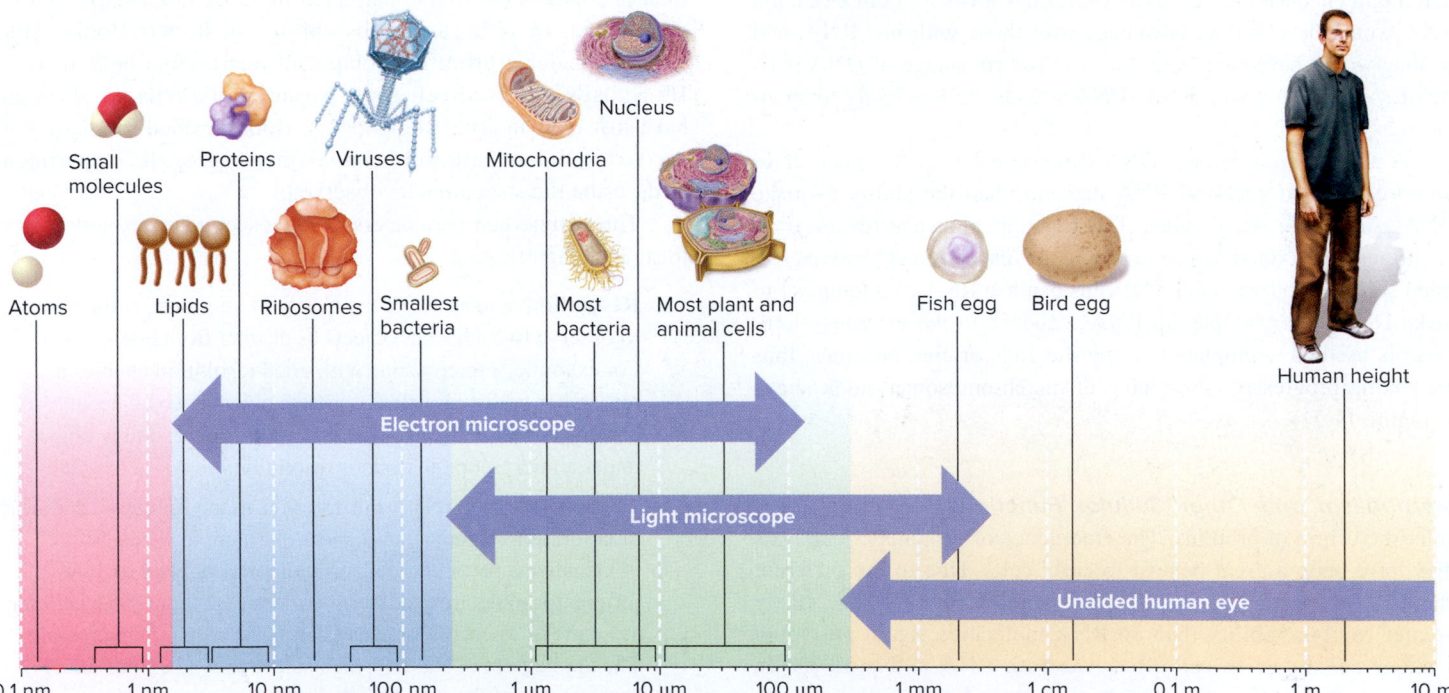

**Figure 4.5** **A comparison of the sizes of various chemical and biological structures with the resolving powers of the unaided eye, light microscope, and electron microscope.** The scale at the bottom is logarithmic to accommodate the wide range of sizes in this drawing.

`Concept Check:` *Which type of microscope would you use to observe a virus?*

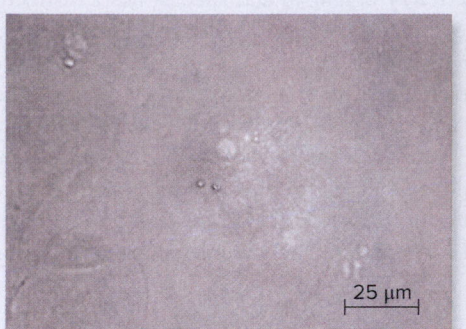

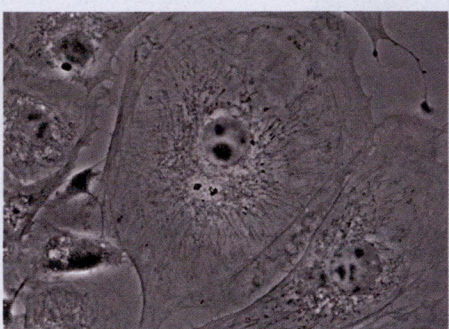

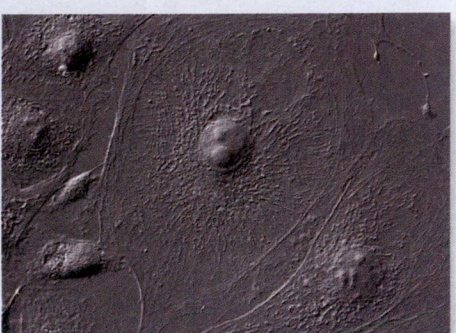

**Standard light microscopy (bright field, unstained sample).**
Light is passed directly through a sample, and the light is focused using glass lenses. Simple, inexpensive, and easy to use but offers little contrast with unstained samples.

**Phase contrast microscopy.**
As an alternative to staining, this microscope controls the path of light and amplifies differences in the phase of light transmitted or reflected by a sample. The dense structures appear darker than the background, thereby improving the contrast in different parts of the specimen. Can be used to view living, unstained cells.

**Differential interference contrast (Nomarski) microscopy.**
Similar to a phase contrast microscope in that it uses optical modifications to improve contrast in unstained specimens. Can be used to visualize the internal structures of cells and is commonly used to view whole cells or large cell structures such as nuclei.

**(a) Three different methods of light microscopy on the same unstained sample**

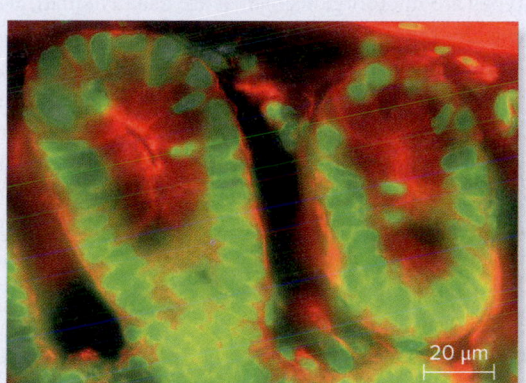

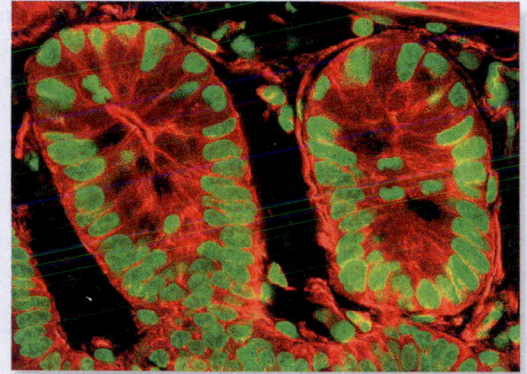

**Standard (wide-field) fluorescence microscopy.**
Fluorescent molecules specifically label a particular type of cellular protein or organelle. A fluorescent molecule absorbs light at a particular wavelength and emits light at a longer wavelength. This microscope has filters that illuminate the sample with the wavelength of light that a fluorescent molecule absorbs, and then only the light that is emitted by the fluorescent molecules is allowed to reach the observer. To detect their cellular location, researchers often label specific cellular proteins using fluorescent antibodies that bind specifically to a particular protein.

**Confocal fluorescence microscopy.**
Uses lasers that illuminate various points in the sample. These points are processed by a computer to give a very sharp focal plane. In this example, this microscope technique is used in conjunction with fluorescence microscopy to view fluorescent molecules within a cell.

**(b) Two different methods of fluorescence microscopy on the same sample**

**Figure 4.6**  **Examples of light microscopy.**  **(a)** These micrographs compare the use of three types of light microscopy to view the same unstained sample of endothelial cells that line the interior surface of arteries in the lungs. **(b)** These two micrographs compare standard (wide-field) fluorescence microscopy with confocal fluorescence microscopy. The sample is a section through a mouse intestine, showing two villi, projections from the small intestine that are described in Chapter 46. In this sample, the nuclei are stained green, and the actin filaments (discussed later in this chapter) are stained red. a-b: Courtesy of Molecular Expressions

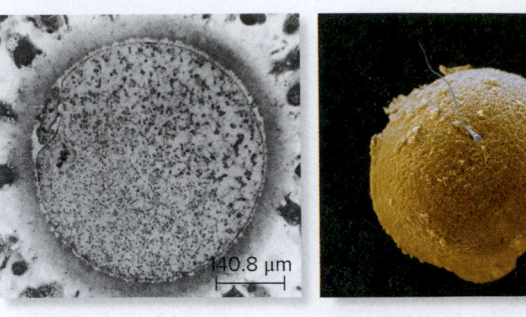

(a) Transmission electron
micrograph (TEM)

(b) Scanning electron
micrograph (SEM)

**Figure 4.7** **A comparison of transmission and scanning electron microscopy.** **(a)** Section through a developing human egg cell, observed by TEM, shortly before it was released from an ovary. **(b)** An egg cell, with an attached sperm, was coated with heavy metal and observed via SEM. This SEM is colorized. a: ©Don W. Fawcett/Science Source; b: ©Eye of Science/Science Source

**Concept Check:** *What is the primary advantage of SEM?*

# **4.3** Overview of Cell Structure and Function

### Learning Outcomes:

1. Compare and contrast the general features of prokaryotic and eukaryotic cells.
2. Explain how the proteome underlies the structure and function of a cell.
3. **CoreSKILL »** Analyze how cell size and shape affect the surface area/volume ratio.

Cell structure and function are primarily determined by four factors: (1) matter, (2) energy, (3) organization, and (4) information. In Chapters 2 and 3, we considered the first factor. The matter found in living organisms is composed of atoms, molecules, and macromolecules. Each type of cell synthesizes a unique set of molecules and macromolecules that contribute to cell structure and function. We will discuss the second factor, energy, throughout this unit, particularly in Chapters 6 through 8. Energy is needed to produce molecules and macromolecules and to carry out many cellular functions.

The third phenomenon that underlies a cell's structure and function is organization. A cell is not a haphazard bag of components. The molecules and macromolecules that constitute a cell are found at specific sites. For example, if we compare muscle cells from two different humans, or two muscle cells within the same individual, we would see striking similarities in their overall structures. All living cells have the ability to build and maintain their internal organization. Proteins often bind to each other in much the same way that toy building blocks snap together. These types of **protein-protein interactions** create intricate cell structures and also facilitate processes in which proteins interact in a consistent series of steps.

The fourth critical factor underlying cell structure and function is information. This information consists of instructions found in the blueprint of life, namely, the genetic material (DNA), which is discussed in Unit III. Every organism and species has a distinctive **genome**, the entire complement of its genetic material. Likewise, each living cell has a copy of the genome. This genetic information is passed from cell to cell and from parent to offspring to yield new generations of cells and new generations of offspring. The **genes** within each species' genome contain the information to produce cellular proteins, which are largely responsible for determining cell structure and function. In this section, we will explore the general features of cells and examine how the genome contributes to those features.

## Prokaryotic Cells Have a Simple Structure

Based on cell structure, all forms of life can be placed into two categories called prokaryotes and eukaryotes. We will first consider **prokaryotic cells**, which have a relatively simple structure. The term comes from the Greek *pro* and *karyon*, meaning before a kernel—a reference to the kernel-like appearance of what would later be named the cell nucleus. Prokaryotic cells lack a membrane-enclosed nucleus.

From an evolutionary perspective, the two categories of organisms that are composed of prokaryotic cells are **bacteria** and **archaea**. Both types are microorganisms that are usually small, with cell sizes that typically range between 1 micrometer (μm) and 10 μm in diameter. Bacteria are abundant throughout the world, being found in soil, water, and even our digestive tracts. Most bacterial species are not harmful to humans, and they play vital roles in ecology. However, some species are pathogenic—they cause disease. Examples of pathogenic bacteria include *Vibrio cholerae*, the source of cholera, and *Bacillus anthracis*, which causes anthrax. Archaea are also widely found throughout the world, though they are less common than bacteria and often occupy extreme environments such as hot springs and deep-sea vents.

**Figure 4.8** shows a typical bacterial cell. The **plasma membrane**, which is a double layer of phospholipids and embedded proteins, forms an important barrier between the interior of the cell and its external environment. The cytoplasm is the region of the cell contained within the plasma membrane. Certain features in the bacterial cytoplasm are visible via microscopy. These include the **nucleoid** (not to be confused with the eukaryotic nucleus), where the genetic material is located. The nucleoid is not a membrane-bound compartment. **Ribosomes**, which are involved in polypeptide synthesis, are also found in the cytoplasm.

Some bacterial structures are located outside the plasma membrane. Nearly all species of bacteria and archaea have a relatively rigid **cell wall** that supports and protects the plasma membrane and cytoplasm. The cell-wall composition varies widely among prokaryotic cells but commonly contains peptides and carbohydrates. It is relatively porous, allowing most nutrients in the environment to reach the plasma membrane. Many bacteria also secrete a **glycocalyx**, an outer viscous covering surrounding the bacterium. The glycocalyx traps water and helps protect the bacterium from drying out. Certain strains of bacteria that invade animals' bodies produce a very thick, gelatinous glycocalyx called a **capsule** that may help them avoid being destroyed by an animal's immune (defense) system or may aid in the attachment to cell surfaces. Finally, many prokaryotic cells

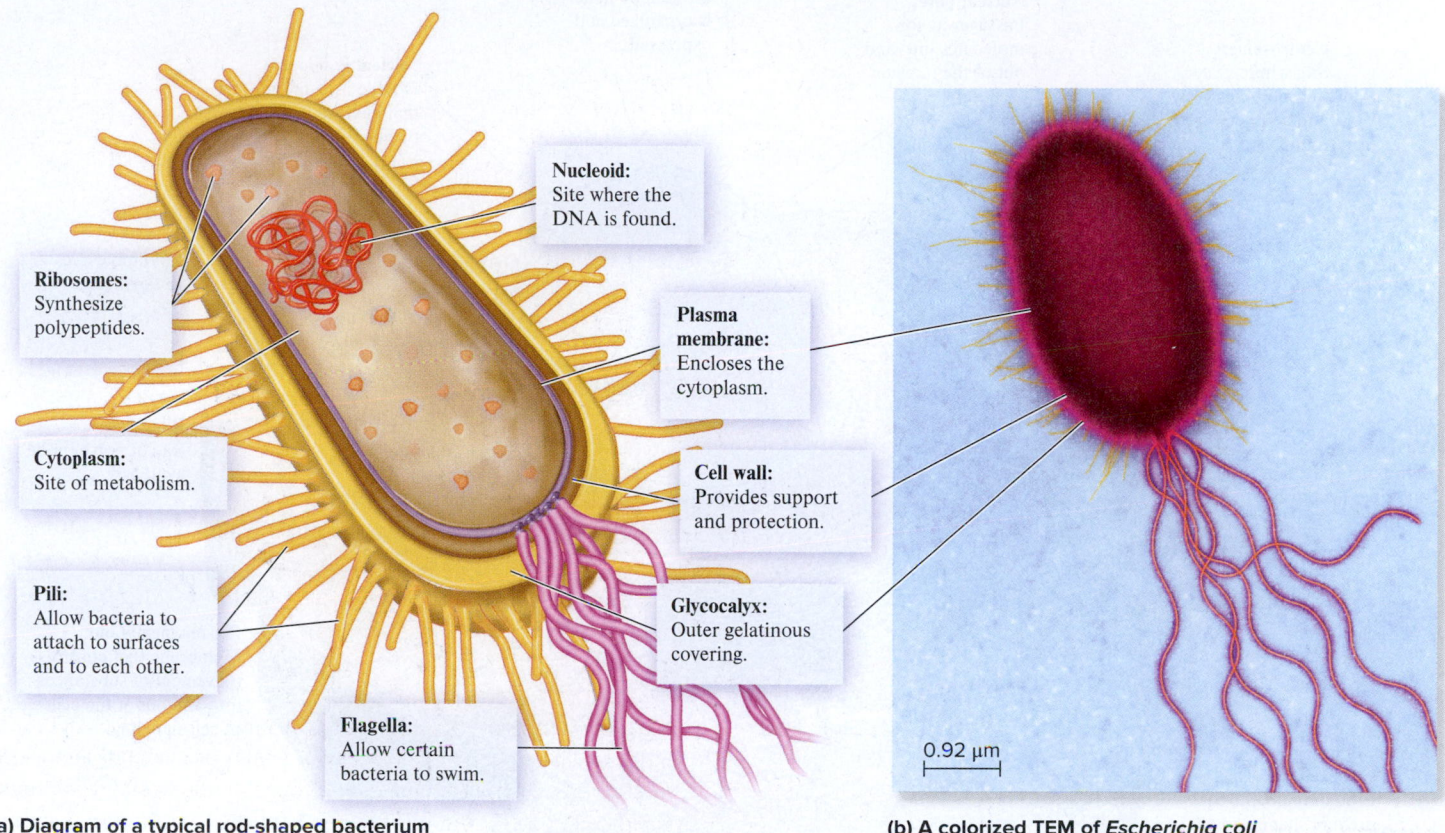

**Nucleoid:** Site where the DNA is found.

**Ribosomes:** Synthesize polypeptides.

**Plasma membrane:** Encloses the cytoplasm.

**Cytoplasm:** Site of metabolism.

**Cell wall:** Provides support and protection.

**Pili:** Allow bacteria to attach to surfaces and to each other.

**Glycocalyx:** Outer gelatinous covering.

**Flagella:** Allow certain bacteria to swim.

0.92 μm

(a) Diagram of a typical rod-shaped bacterium

(b) A colorized TEM of *Escherichia coli*

**Figure 4.8** **Structure of a typical bacterial cell. Prokaryotic cells, which include bacteria and archaea, lack internal compartmentalization.**
b: ©Dennis Kunkel Microscopy, Inc./Phototake

have appendages such as pili and flagella. **Pili** allow cells to attach to surfaces and to each other. **Flagella** provide prokaryotic cells with the ability to move, also called motility.

## Eukaryotic Cells Are Compartmentalized by Internal Membranes to Create Organelles

Aside from bacteria and archaea, all other species are **eukaryotes** (from the Greek, meaning true nucleus), which include protists, fungi, plants, and animals. Paramecia and algae are types of protists; yeasts and molds are types of fungi. **Figure 4.9** illustrates the general structure of a typical animal cell. Eukaryotic cells possess a true nucleus, where most of the DNA is housed. A nucleus is a type of **organelle**—a membrane-bound compartment with its own unique structure and function. In contrast to prokaryotic cells, eukaryotic cells exhibit extensive **compartmentalization**, which means they have many membrane-bound organelles that separate the cell into different regions. Cellular compartmentalization allows a cell to carry out specialized chemical reactions in different places.

Some general features of cell organization, such as a nucleus, are found in nearly all eukaryotic cells. However, the shape, size, and organization of cells vary considerably among different species and even among different cell types of the same species. For example, micrographs of a human skin cell and a human neuron (a nervous system cell) show that, although these cells contain the same types of organelles, their overall morphologies are quite different (**Figure 4.10**).

Plant cells possess a collection of organelles similar to those found in animal cells (**Figure 4.11**). Additional structures found in plant cells but not animal cells include chloroplasts, a central vacuole, and a cell wall.

## Droplet Organelles Are a Category of Organelles Whose Boundary Is Due to Phase Separation

Most of the organelles that are shown in Figures 4.9 and 4.11 are surrounded by a single or double membrane. Recently, however, researchers have discovered that cells can also become compartmentalized by a second mechanism called **liquid-liquid phase separation** in which aggregated solutes, such as proteins and RNA molecules, separate from the bulk solvent and form a droplet. The droplet has a spherical shape with a measurable surface tension and viscosity. Molecules can diffuse within the droplet and occasionally leave it and pass into the surrounding liquid phase. Cell biologists are beginning to recognize these droplets as organelles, hence the name, **droplet organelle**. An example is the nucleolus, the site for rRNA processing and the assembly of ribosomal subunits.

The internal environment of droplet organelles is thought to serve two purposes. First, molecules are brought close together and can assemble into complexes. For example, ribosomal subunits assemble within the nucleolus. Second, the environment within the droplet is chemically different from the surrounding medium, which may affect events such as RNA folding.

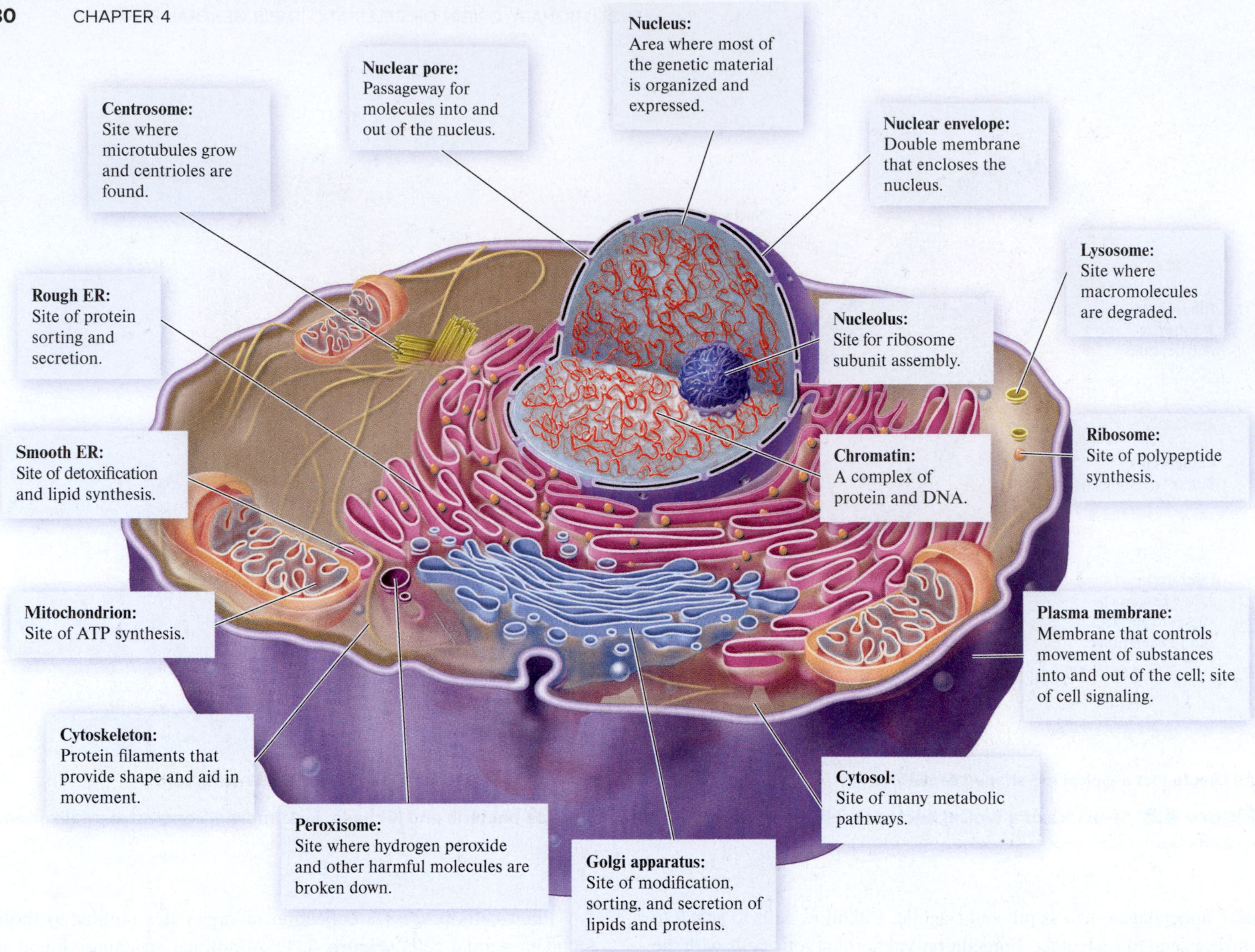

**Centrosome:**
Site where
microtubules grow
and centrioles are
found.

**Nuclear pore:**
Passageway for
molecules into and
out of the nucleus.

**Nucleus:**
Area where most of
the genetic material
is organized and
expressed.

**Nuclear envelope:**
Double membrane
that encloses the
nucleus.

**Rough ER:**
Site of protein
sorting and
secretion.

**Lysosome:**
Site where
macromolecules
are degraded.

**Nucleolus:**
Site for ribosome
subunit assembly.

**Smooth ER:**
Site of detoxification
and lipid synthesis.

**Chromatin:**
A complex of
protein and DNA.

**Ribosome:**
Site of polypeptide
synthesis.

**Mitochondrion:**
Site of ATP synthesis.

**Plasma membrane:**
Membrane that controls
movement of substances
into and out of the cell; site
of cell signaling.

**Cytoskeleton:**
Protein filaments that
provide shape and aid in
movement.

**Cytosol:**
Site of many metabolic
pathways.

**Peroxisome:**
Site where hydrogen peroxide
and other harmful molecules are
broken down.

**Golgi apparatus:**
Site of modification,
sorting, and secretion of
lipids and proteins.

**Figure 4.9** General structure of an animal cell.

**Core Concept: Systems** A cell, such as the animal cell illustrated here, is the smallest unit of life. As this diagram shows, it is composed of many interacting parts.

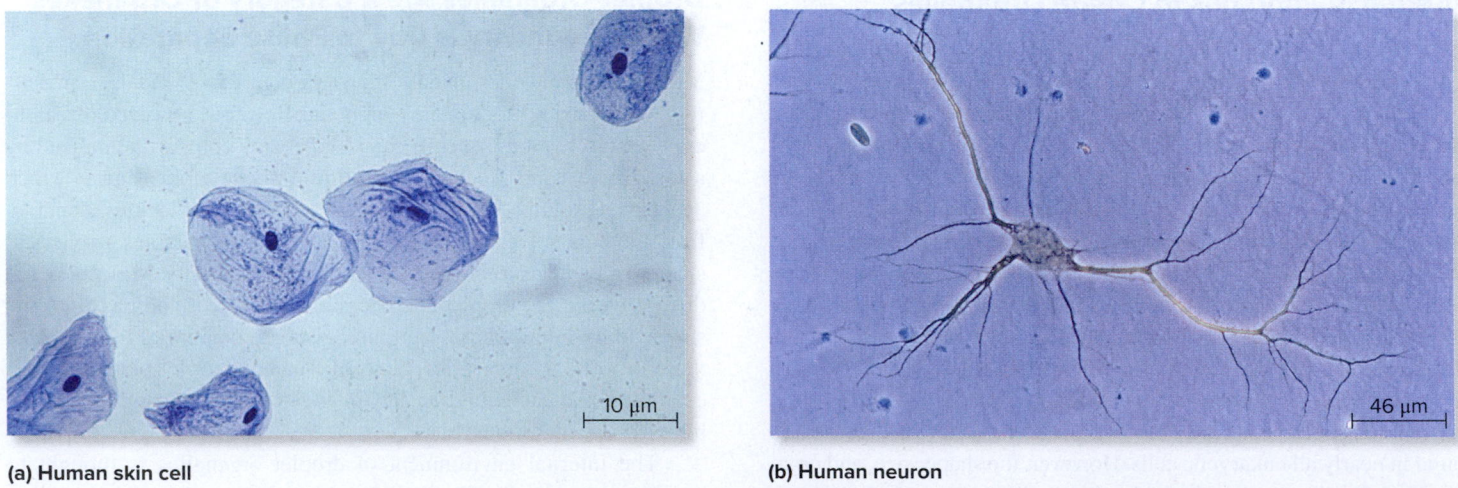

(a) **Human skin cell**                10 μm

(b) **Human neuron**                46 μm

**Figure 4.10** **Variation in morphology of eukaryotic cells.** Light micrographs of **(a)** a human skin cell and **(b)** a human neuron (a cell of the nervous system). Although these cells have the same genome and the same types of organelles, their general morphologies are quite different.

a: ©Ed Reschke/Getty Images; b: ©Eye of Science/Science Source

**Core Skill: Connections** Look ahead to Figure 14.21. How does alternative splicing affect protein structure and function?

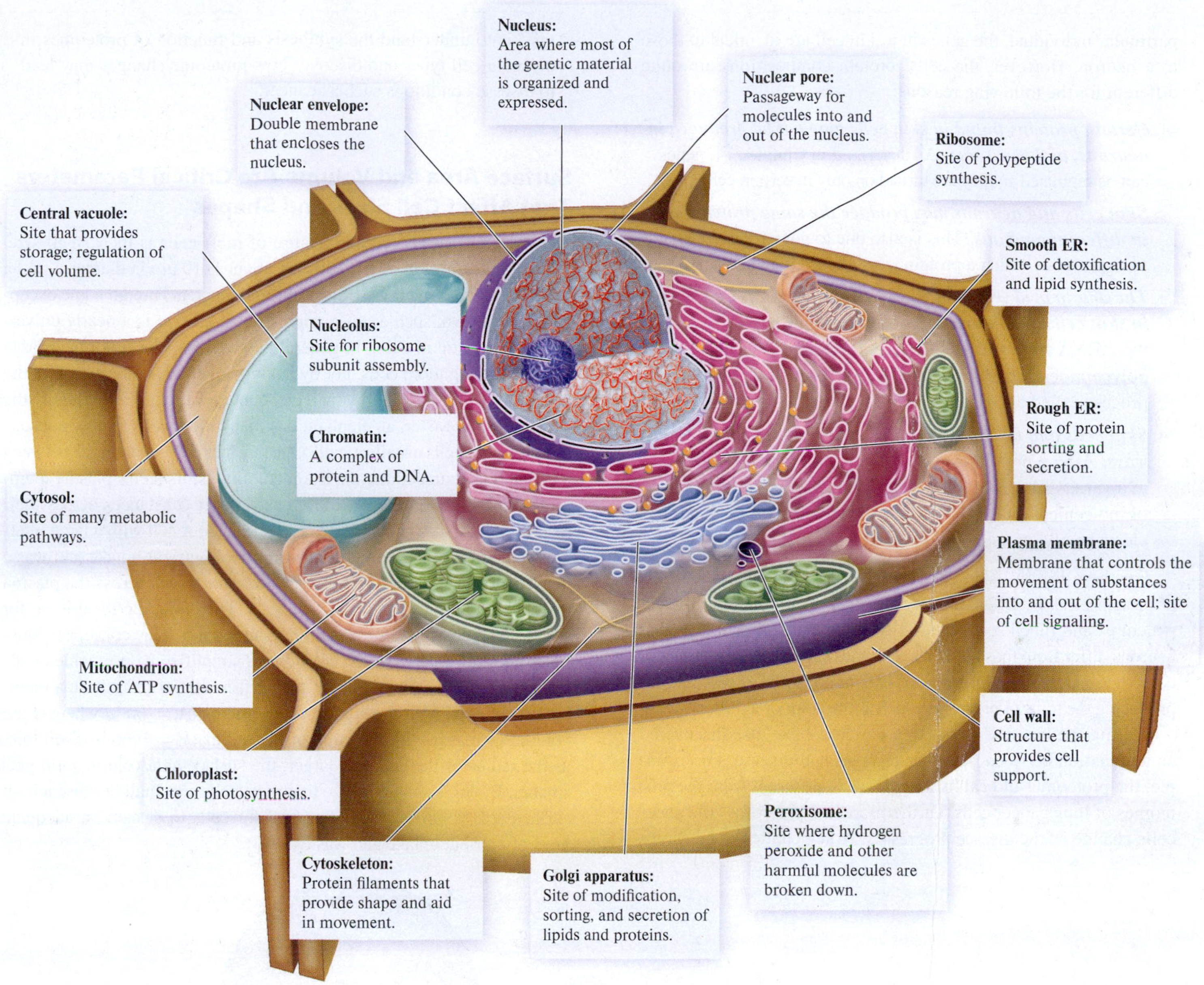

**Nucleus:**
Area where most of the genetic material is organized and expressed.

**Nuclear pore:**
Passageway for molecules into and out of the nucleus.

**Nuclear envelope:**
Double membrane that encloses the nucleus.

**Ribosome:**
Site of polypeptide synthesis.

**Central vacuole:**
Site that provides storage; regulation of cell volume.

**Smooth ER:**
Site of detoxification and lipid synthesis.

**Nucleolus:**
Site for ribosome subunit assembly.

**Rough ER:**
Site of protein sorting and secretion.

**Chromatin:**
A complex of protein and DNA.

**Cytosol:**
Site of many metabolic pathways.

**Plasma membrane:**
Membrane that controls the movement of substances into and out of the cell; site of cell signaling.

**Mitochondrion:**
Site of ATP synthesis.

**Cell wall:**
Structure that provides cell support.

**Chloroplast:**
Site of photosynthesis.

**Peroxisome:**
Site where hydrogen peroxide and other harmful molecules are broken down.

**Cytoskeleton:**
Protein filaments that provide shape and aid in movement.

**Golgi apparatus:**
Site of modification, sorting, and secretion of lipids and proteins.

**Figure 4.11** **General structure of a plant cell.** Plant cells lack lysosomes and centrioles. Unlike animal cells, plant cells have an outer cell wall; a large central vacuole that functions in storage and the regulation of cell volume; and chloroplasts, which carry out photosynthesis.

 *Concept Check:* *What are the functions of the cell structures and organelles that are found in animal cells but not plant cells or found in plant cells but not animal cells?*

## 👁 Core Concepts: Information, Structure and Function

### The Characteristics of a Cell Are Largely Determined by the Proteins It Makes

Many organisms, such as animals and plants, are multicellular, meaning that a single organism is composed of many cells. However, the cells of most multicellular organisms are not all identical. For example, your body contains skin cells, neurons, muscle cells, and many other types. An intriguing question, therefore, is how does a single organism produce different types of cells?

To answer this question, we need to consider the distinction between a cell's genome and its proteome. Recall that the genome consists of all of an organism's genetic material, namely its DNA, which contains many different genes. Most genes encode the production of polypeptides, which assemble into functional proteins. The **proteome** is defined as the complete set of proteins that a cell is currently making or an organism can make. The set of proteins that is made by a given cell type is largely responsible for determining the structure and function of that cell.

The set of proteins made in one cell type is not the same as that made in a different cell type. As an example, let's consider human skin cells and neurons—two cell types that have dramatically different organization and structure (look back at Figure 4.10). In any particular individual, the genes in a skin cell are identical to those in a neuron. However, the cells' protein compositions are quite different for the following reasons:

1. *Certain proteins found in skin cells may not be produced in neurons, and vice versa.* As described in Chapter 14, genes can be regulated so they are turned on only in certain cell types.

2. *Skin cells and neurons may produce the same protein but in different amounts.* This is also due to gene regulation and to the rates at which a protein is synthesized and degraded.

3. *The amino acid sequences of particular proteins can vary in skin cells and neurons.* As discussed in Chapter 14, the mRNA from a single gene can produce two or more polypeptides with different amino acid sequences via a process called alternative splicing.

4. *Skin cells and neurons may alter their proteins in different ways.* After a protein is made, its structure may be changed in a variety of ways. These include the covalent attachment of molecules, such as phosphate and carbohydrates, and the cleavage of a protein to a smaller size.

For these reasons, skin cells and neurons produce different sets of proteins, that is, different proteomes, and therefore have different structures and functions. Likewise, the proteomes of skin cells and neurons differ from those of other cell types such as muscle and liver cells. Ultimately, the proteomes of cells are largely responsible for producing the traits of organisms, such as the color of a person's eyes.

During the last few decades, researchers have also discovered an association between proteome changes and disease. For example, the proteomes of healthy lung cells are different from the proteomes of lung cancer cells. Furthermore, the proteomes of cancer cells change as the disease progresses. A key challenge for biolo-gists is to understand the synthesis and function of proteomes in different cell types and discover how proteome changes may lead to disease conditions such as cancer.

## Surface Area and Volume Are Critical Parameters That Affect Cell Sizes and Shapes

As we have seen, a common feature of most cells is their small size. For example, most bacterial cells are about 1–10 μm in diameter, and a typical eukaryotic cell is 10–100 μm in diameter. Though some exceptions are known, such as an ostrich egg, small size is a nearly universal characteristic of cells. In general, large organisms attain their large sizes by having more cells, not by having larger cells. For example, the various types of cells found in an elephant and a mouse are roughly the same sizes. However, an elephant has many more cells than a mouse.

Why are cells usually small? One key factor is the interface between a cell and its extracellular environment, that is, the cell's plasma membrane. For cells to survive, they must import substances across their plasma membranes and export waste products. A cell with a large internal volume will require a greater amount of nutrient uptake and waste export. However, the rate of transport of substances across the plasma membrane is limited by its surface area. Therefore, a critical issue for sustaining a cell is the surface area/volume ratio. This concept is illustrated in **Figure 4.12**, which considers a simplified case in which cells are spherical. As cells get larger, the surface area of their plasma membrane increases with the square of the radius ($A = 4\pi r^2$), whereas the volume increases with the cube of the radius ($V = 4/3\pi r^3$). Therefore, as the radius of the cell gets larger, the surface area/volume ratio gets smaller. Biologists hypothesize that most cells are small because a high surface area/volume ratio is required for cells to sustain an adequate level of nutrient uptake and waste export.

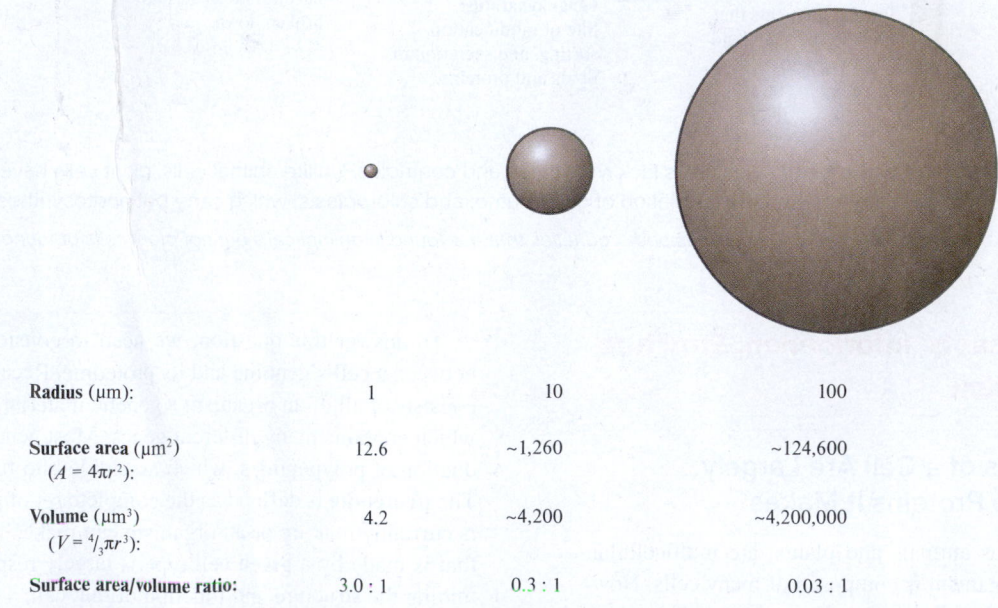

| Radius (μm): | 1 | 10 | 100 |
|---|---|---|---|
| Surface area (μm²) ($A = 4\pi r^2$): | 12.6 | ~1,260 | ~124,600 |
| Volume (μm³) ($V = 4/3\pi r^3$): | 4.2 | ~4,200 | ~4,200,000 |
| Surface area/volume ratio: | 3.0 : 1 | 0.3 : 1 | 0.03 : 1 |

**Figure 4.12   Relationship between cell size and the surface area/volume ratio.** As cells get larger, the surface area/volume ratio gets smaller. Note: The three spheres shown here are not drawn precisely to scale.

**Core Skill: Connections**  Look ahead to Figure 41.8a. How does the surface area/volume ratio relate to the shapes of structures involved with gas exchange?

## Core Skill: Quantitative Reasoning

**BIO TIPS**        **THE QUESTION** *One way for cells to partially overcome the limitation imposed by the surface area/volume ratio is to be elongated and have an irregularly shaped surface. For example, look back at Figure 4.10. The skin cell is roughly spherical, whereas the neuron is very elongated and has an irregularly shaped surface. If a skin cell and a neuron cell had the same internal volume, the neuron would have a much higher surface area/volume ratio.*

*Let's suppose that one cell is spherical with a radius of 21 μm. Another cell is cylindrical, with a smaller radius of 3 μm but a much greater length of 1,372 μm. Which of these cells has the greater volume? Which has the greater surface area/volume ratio? Note: For a cylinder, the volume and surface area are calculated as follows:*

$$\text{Volume} = \pi r^2 h$$

$$\text{Surface area} = 2\pi r^2 + 2\pi rh$$

*In these equations, h (height) corresponds to cell length.*

**T**OPIC **What topic in biology does this question address?** The topic is cell volume and the surface area/volume ratio. More specifically, the question asks you to compare these parameters for two cells with different shapes.

**I**NFORMATION **What information do you know based on the question and your understanding of the topic?** In the question, you are given the dimensions of two different cells and equations to calculate the volume and surface area of a cylindrical cell. From your understanding of the topic, you may remember the equations for finding the volume and surface area of a spherical cell (look back at Figure 4.12).

**P**ROBLEM-SOLVING **S**TRATEGY **Make a drawing. Make a calculation.** To solve this problem, you need to use the equations for calculating cell volume and surface area. It may be helpful to begin with a drawing that depicts the variables in these equations, as shown below:

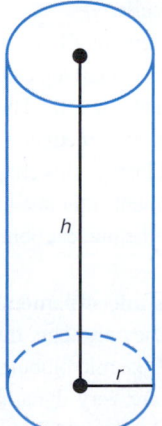
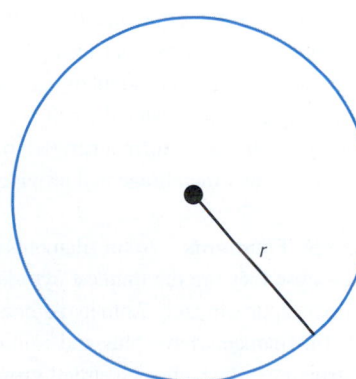

Let's begin by calculating the volume for both cell types. For the cylinder, we have

Volume = $\pi r^2 h$

Volume = $3.14(3 \text{ μm})^2 (1372 \text{ μm}) = 38{,}773 \text{ μm}^3$

For the sphere, we have

Volume = $^4/_3 \pi r^3$

Volume = $^4/_3 (3.14)(21 \text{μm})^3 = 38{,}773 \text{ μm}^3$

Now let's calculate the surface area for both cell types. For the cylinder,

Surface area = $2\pi r^2 + 2\pi rh$

Surface area = $2(3.14)(3 \text{ μm})^2 + 2(3.14)(3 \text{ μm})(1372 \text{ μm})$

$= 25{,}905 \text{ μm}^2$

For the sphere,

Surface area = $4\pi r^2$

Surface area = $4(3.14)(21 \text{ μm})^2 = 5{,}539 \text{ μm}^2$

**ANSWER** *Though they have very different shapes, the two cell types have the same volume. The surface area/volume ratio for the cylindrical cell is 25,905/38,773 = 0.67. By comparison, the surface area/volume ratio for the spherical cell is 5,539/38,773 = 0.14. If we divide 0.67 by 0.14, we find that the cylindrical cell's surface area/volume ratio is about 4.8 times greater than that of the spherical cell.*

## 4.4 The Cytosol

**Learning Outcomes:**

1. Identify the location of the cytosol in a eukaryotic cell, and list its general functions.
2. Describe the three types of protein filaments that make up the cytoskeleton.
3. Explain how motor proteins interact with microtubules or actin filaments to promote cellular movements.

In Section 4.3, we focused on the general features of prokaryotic and eukaryotic cells. In the rest of this chapter, we will survey the various compartments of eukaryotic cells with an emphasis on structure and function. **Figure 4.13** highlights various regions in an animal cell and a plant cell. We will start with the **cytosol** (shown in yellow), the region of a eukaryotic cell that is outside the organelles but inside the plasma membrane. The other regions of the cell, which we will examine later in this chapter, include the interior of the nucleus (blue), the endomembrane system (purple and pink), and the semiautonomous organelles (orange and green). As in prokaryotic cells, the term cytoplasm refers to the region enclosed by the plasma membrane, which includes the cytosol and the organelles.

### Synthesis and Breakdown of Molecules Occur in the Cytosol

**Metabolism** is defined as the sum of the chemical reactions by which cells produce the materials and utilize the energy necessary to sustain life. Although many steps of metabolism also occur in cell organelles, the cytosol is a central coordinating region for many metabolic activities of eukaryotic cells. Metabolism often involves a series of steps called a metabolic pathway. Each step in a metabolic pathway is catalyzed by a specific **enzyme**—a protein that accelerates the rate of a chemical

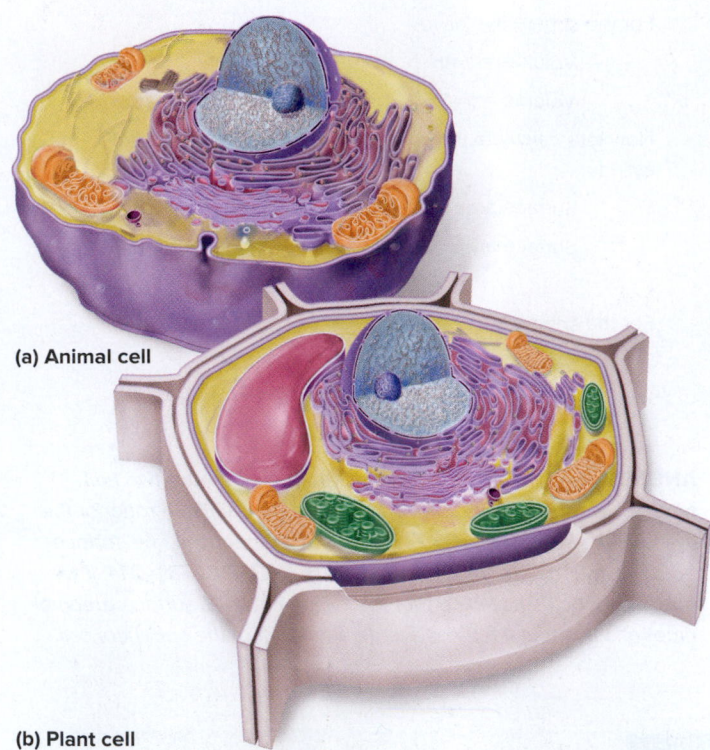

(a) Animal cell

(b) Plant cell

**Figure 4.13** **Compartments within (a) animal and (b) plant cells.** The cytosol, which is outside the organelles but inside the plasma membrane, is shown in yellow. The membranes of the endomembrane system are shown in purple, and the fluid-filled interiors are pink. The peroxisome is dark purple. The interior of the nucleus is blue. Semiautonomous organelles are shown in orange (mitochondria) and green (chloroplasts).

reaction. In Chapters 6 and 7, we will examine enzymes and consider a few metabolic pathways that occur in the cytosol and cell organelles.

Some metabolic pathways involve the breakdown of a molecule into smaller components, a process termed **catabolism**. Such pathways are needed by the cell to utilize energy and also to generate molecules that provide the building blocks to construct macromolecules. Conversely, other pathways are involved in **anabolism**, the synthesis of molecules and macromolecules. For example, polysaccharides are made by linking sugar molecules. To make proteins, amino acids are covalently connected to form a polypeptide, using the information within an mRNA (see Chapter 12). Translation of that information occurs on ribosomes, which are found in various locations in the cell. Some ribosomes may float freely in the cytosol, others are attached to the outer membrane of the nuclear envelope and endoplasmic reticulum membrane, and still others are found within the mitochondria or chloroplasts.

## The Cytoskeleton Provides Cell Shape, Organization, and Movement

The **cytoskeleton** is a network of three different types of protein filaments: **microtubules**, **intermediate filaments**, and **actin filaments** (**Table 4.1**). Each type is constructed from many protein monomers. The cytoskeleton is a striking example of protein-protein interactions. The cytoskeleton is found primarily in the cytosol and also in the nucleus along the inner nuclear membrane. Let's first consider the structure of

cytoskeletal filaments and their roles in the construction and organization of cells. Later, we will examine how they are involved in cell movement.

**Microtubules** Microtubules are long, hollow, cylindrical structures about 25 nm in diameter composed of protein subunits called α- and β-tubulin. The assembly of tubulin to form a microtubule results in a structure with a plus end and a minus end (see Table 4.1). Microtubules grow only at the plus end, but can shorten at either the plus or minus end. A single microtubule can oscillate between growing and shortening phases, a phenomenon termed **dynamic instability**. This phenomenon is important in many cellular activities, including the sorting of chromosomes during cell division.

The sites where microtubules form within a cell vary among different types of organisms. Nondividing animal cells contain a single structure near their nucleus called the **centrosome**, also called a microtubule-organizing center (see Table 4.1). Within the centrosome are the **centrioles**, a conspicuous pair of structures arranged perpendicularly to each other. In animal cells, microtubule growth typically starts at the centrosome in such a way that the minus end is anchored there. In contrast, most plant cells and many protists lack centrosomes and centrioles. Microtubules are created at many sites that are scattered throughout a plant cell. In plants, the nuclear membrane appears to function as a microtubule-organizing center.

Microtubules are important for cell shape and organization. Organelles such as the Golgi apparatus are attached to microtubules. In addition, microtubules are involved in the organization and movement of chromosomes during mitosis and in the orientation of cells during cell division, events we will examine in Chapter 16.

**Intermediate Filaments** Intermediate filaments are another class of cytoskeletal filaments found in the cells of many but not all animal species. Their name is derived from the observation that they are intermediate in diameter between actin filaments and microtubules. Intermediate filament proteins bind to each other in a staggered array to form a twisted, ropelike structure with a diameter of approximately 10 nm (see Table 4.1). They function as tension-bearing fibers that help maintain cell shape and rigidity. The lengths of intermediate filaments tend to be relatively permanent. By comparison, microtubules and actin filaments readily lengthen and shorten in cells.

Several types of proteins assemble into intermediate filaments. Keratins form intermediate filaments in skin, intestinal, and kidney cells, where they are important for cell shape and mechanical strength. They are also a major constituent of hair and nails. In addition, intermediate filaments are found inside the cell nucleus. As discussed later in this chapter, nuclear lamins form a network of intermediate filaments that line the inner nuclear membrane and provide anchor points for the nuclear pores.

**Actin Filaments** Actin filaments are also known as **microfilaments**, because they are the thinnest cytoskeletal filaments. They are long, thin fibers approximately 7 nm in diameter (see Table 4.1). Like microtubules, actin filaments have plus and minus ends, and they are very dynamic structures in which each strand grows at the plus end by the addition of actin monomers. This assembly process produces a fiber composed of two strands of actin monomers that spiral around each other.

Despite their thinness, actin filaments play a key role in cell shape and strength. Although actin filaments are dispersed throughout the cytosol,

| Table 4.1 | Types of Cytoskeletal Filaments Found in Eukaryotic Cells | | |
|---|---|---|---|
| Characteristic | Microtubules | Intermediate filaments | Actin filaments |
| Diameter | 25 nm | 10 nm | 7 nm |
| Structure | Hollow tubule | Twisted filament | Spiral filament |

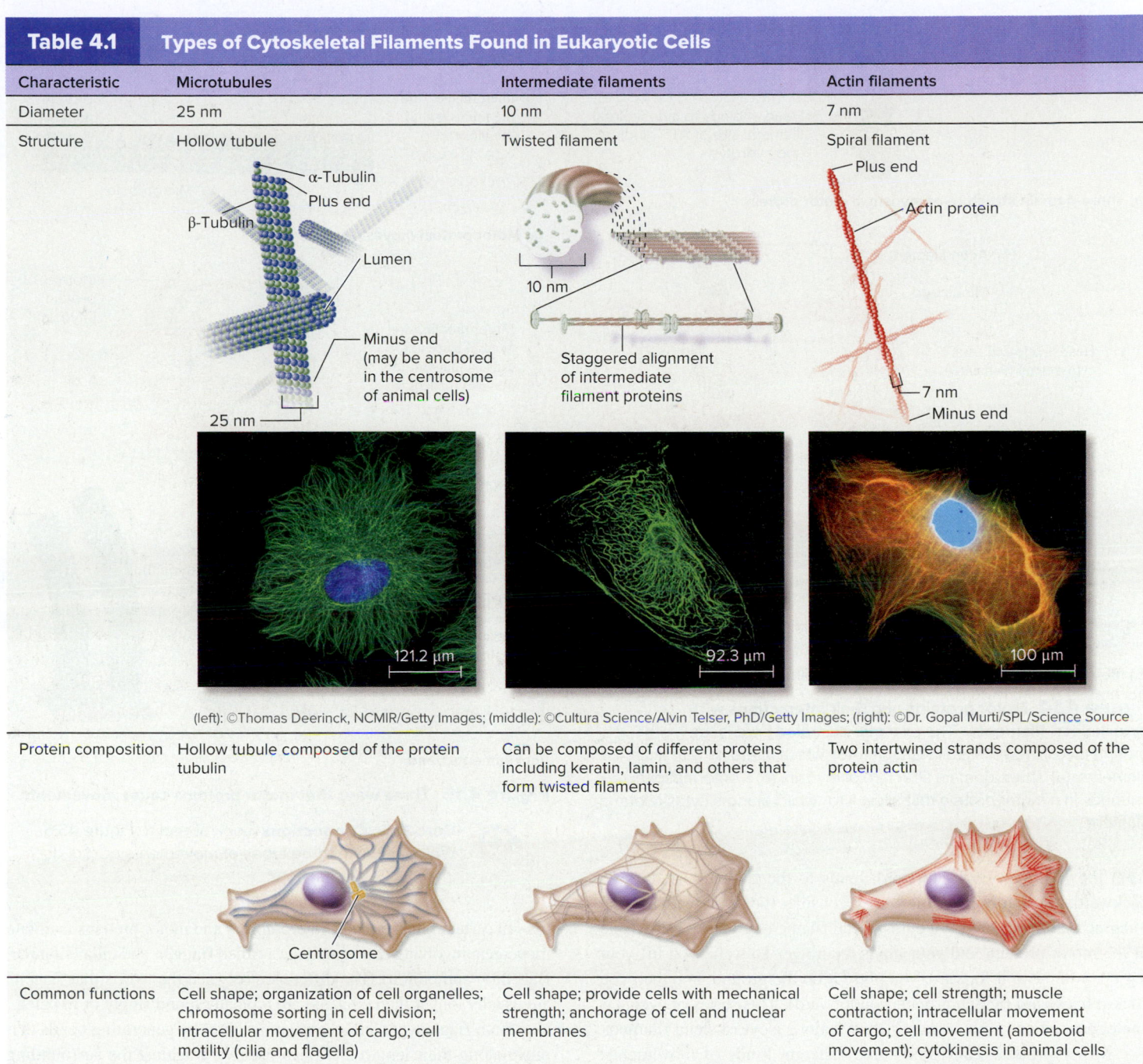

(left): ©Thomas Deerinck, NCMIR/Getty Images; (middle): ©Cultura Science/Alvin Telser, PhD/Getty Images; (right): ©Dr. Gopal Murti/SPL/Science Source

| | | | |
|---|---|---|---|
| Protein composition | Hollow tubule composed of the protein tubulin | Can be composed of different proteins including keratin, lamin, and others that form twisted filaments | Two intertwined strands composed of the protein actin |
| Common functions | Cell shape; organization of cell organelles; chromosome sorting in cell division; intracellular movement of cargo; cell motility (cilia and flagella) | Cell shape; provide cells with mechanical strength; anchorage of cell and nuclear membranes | Cell shape; cell strength; muscle contraction; intracellular movement of cargo; cell movement (amoeboid movement); cytokinesis in animal cells |

they tend to be highly concentrated near the plasma membrane. In many types of cells, actin filaments support the plasma membrane and provide shape and strength to the cell. The sides of actin filaments are often anchored to other proteins near the plasma membrane, which explains why actin filaments are typically found there. The plus ends grow toward the plasma membrane and play a key role in cell shape and movement.

## Motor Proteins Interact with Cytoskeletal Filaments to Promote Movements

**Motor proteins** are a category of proteins that use ATP as a source of energy to promote various types of movements. As shown in Figure 4.14a, a motor protein consists of three domains: head, hinge, and tail. Together, the hinge and tail make up a structure called the lever arm. The head is the site where ATP binds and is hydrolyzed to adenosine diphosphate (ADP) and inorganic phosphate ($P_i$). ATP binding and hydrolysis cause a bend in the hinge, which results in movement. The tail region is attached to other proteins or to other kinds of cellular molecules.

To implement movement, the head region of a motor protein interacts with a cytoskeletal filament, such as an actin filament (Figure 4.14b). When ATP binds and is hydrolyzed, the motor protein interacts with the filament in a series of steps. The head of the motor protein is initially attached to a filament. To move forward, the head detaches

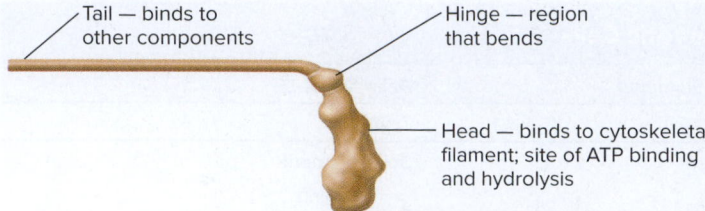

**(a) Three-domain structure of myosin, a motor protein**

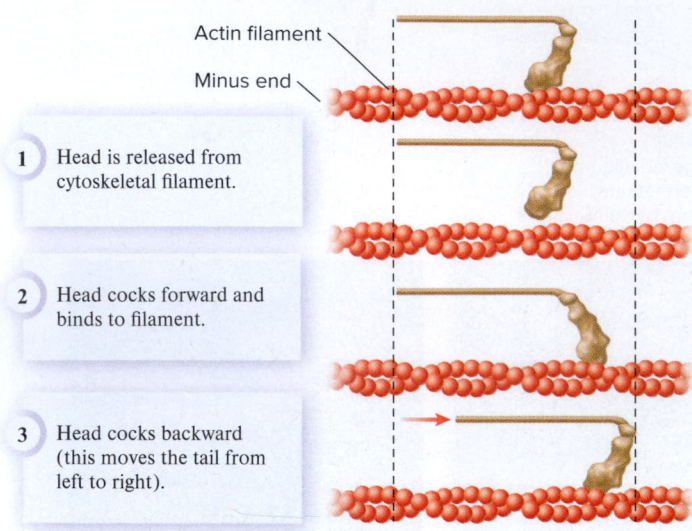

1. Head is released from cytoskeletal filament.

2. Head cocks forward and binds to filament.

3. Head cocks backward (this moves the tail from left to right).

**(b) Movement of a motor protein along a cytoskeletal filament**

**Figure 4.14  Motor proteins and their interactions with cytoskeletal filaments.** The example illustrated here is the motor protein myosin (discussed in Chapter 45), which interacts with actin filaments. **(a)** Three-domain structure of myosin. **(b)** Conformational changes in a motor protein that allow it to "walk" along a cytoskeletal filament.

from the filament, cocks forward, binds to the filament, and cocks backward. To picture how this works, consider the act of walking and imagine that the ground is a cytoskeletal filament, your leg is the head of the motor protein, and your hip is the hinge. To walk, you lift your leg up, you move it forward, you place it on the ground, and then you cock it backward (which propels you forward). This series of events is analogous to how a motor protein moves along a cytoskeletal filament.

Motor proteins can cause three different kinds of movements: movement of cargo via the motor protein, movement of the filament, or bending of the filament.

- In the example shown in **Figure 4.15a**, the tail region of a motor protein called kinesin is attached to a cargo, and the motor protein moves the cargo from one location to another.

- Alternatively, a motor protein called myosin can remain in place and cause the filament to move (**Figure 4.15b**). This occurs during muscle contraction, which is described in Chapter 45.

- A third possibility is that both the motor protein and filament are restricted in their movement due to the presence of linking proteins. In this case, when motor proteins called dynein attempt to walk toward the minus end, they exert a force that causes the microtubules to bend (**Figure 4.15c**).

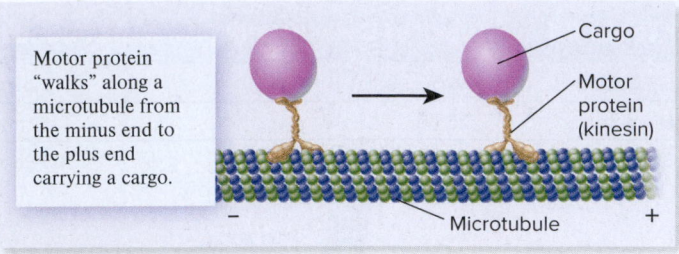

**(a) Motor protein moves**

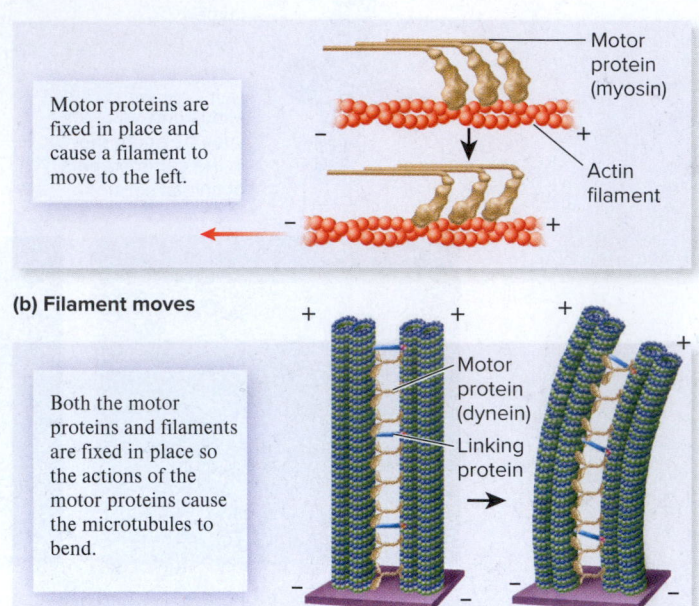

**(b) Filament moves**

**(c) Filaments bend**

**Figure 4.15  Three ways that motor proteins cause movements.**

 **Core Skill: Connections** Look ahead to Figure 45.6. Which of these three types of movements occurs during muscle contraction?

In certain kinds of cells, microtubules and motor proteins facilitate movement involving cell appendages called **flagella** and **cilia** (singular, flagellum and cilium). The difference between the two is that flagella are usually longer than cilia and are typically found singly or in pairs.

Both flagella and cilia cause movement by generating bends that move along their length and push backward against the surrounding fluid. The flagellum of a sperm cell generates bends alternatively in each direction, which begin at the head and move (propagate) toward the tip of the flagellum (**Figure 4.16a**). Alternatively, a pair of flagella may move in a synchronized manner to pull a microorganism through the water (think of a human swimmer doing the breaststroke). Certain unicellular algae swim in this manner (**Figure 4.16b**). By comparison, cilia are typically shorter than flagella and tend to cover all or part of the surface of a cell. Protists such as paramecia may have hundreds of adjacent cilia that beat in a coordinated fashion to propel the organism through the water (**Figure 4.16c**).

Despite their differences in length, flagella and cilia have the same internal structure called the **axoneme**. The axoneme contains microtubules, the motor protein dynein, and linking proteins (**Figure 4.17**). In the cilia and flagella of most eukaryotic

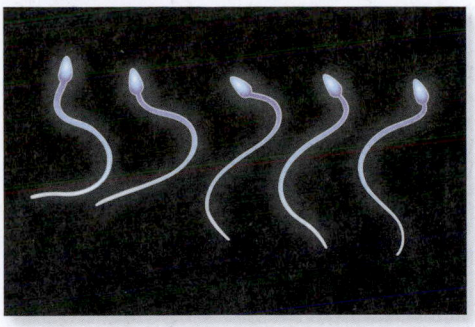

**(a) Drawing of a sperm moving its flagellum**

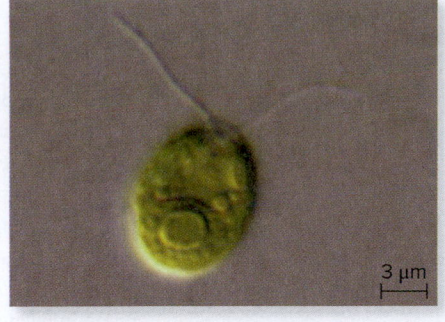

3 μm

**(b) *Chlamydomonas* with 2 flagella**

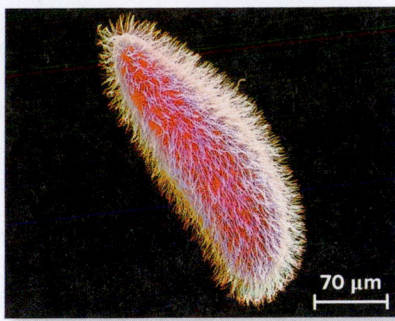

70 μm

**(c) *Paramecium* with many cilia**

**Figure 4.16   Cellular movements due to the actions of flagella and cilia.   (a)** Spermatozoa (singular, spermatozoon) are sperm cells that are motile. They swim by producing repeated bends of a single, long flagellum, which move along its length. **(b)** The swimming of *Chlamydomonas reinhardtii*, a unicellular green algae, also involves a bending motion beginning at the base of flagella, but the motion is precisely coordinated between two flagella. This results in swimming behavior that resembles a human doing the breaststroke. **(c)** Ciliated protozoa such as this paramecium swim by coordinated beating of many shorter cilia. b: Courtesy of Dr. Barbara Surek, Culture Collection of Algae at the University of Cologne (CCAC); c: ©SPL/Science Source

**Concept Check:**   *Describe the type of movements that occur between the motor proteins and microtubules when flagella or cilia bend.*

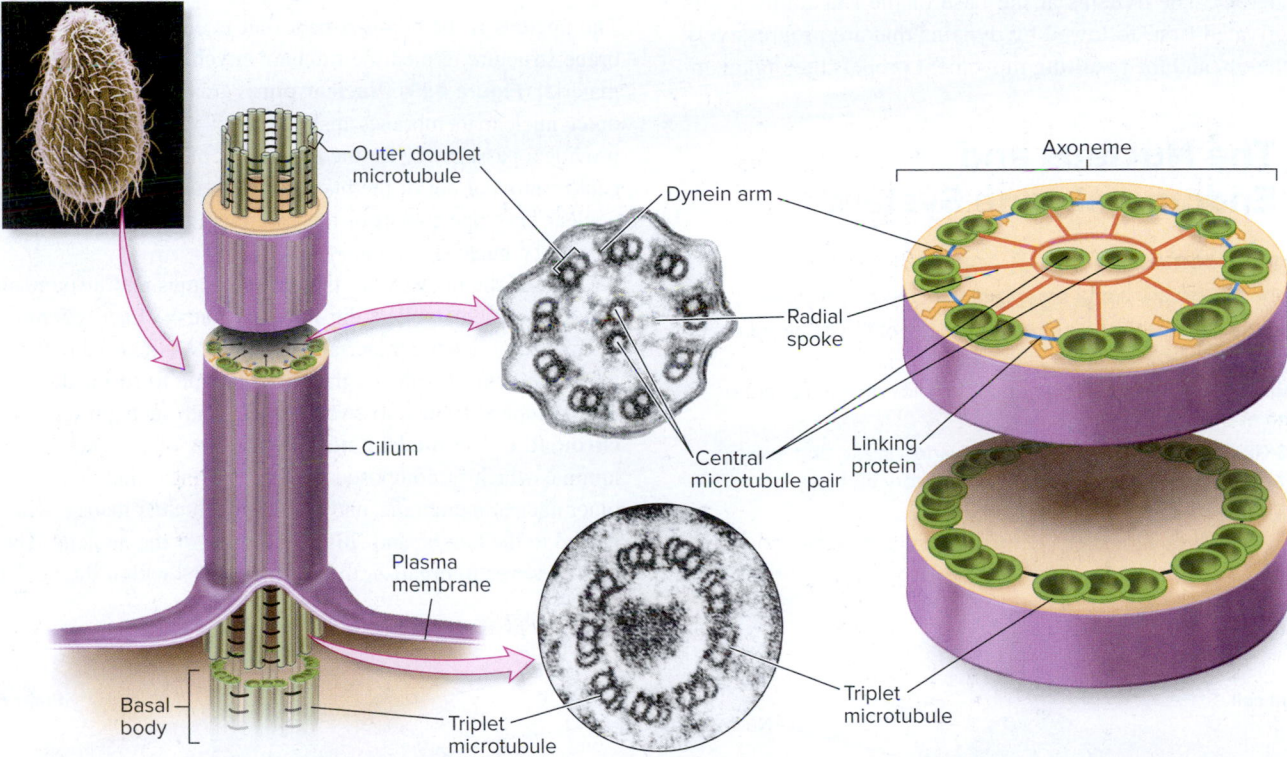

**Figure 4.17   Structure of a eukaryotic cilium or flagellum.**   The structure of a cilium of a protist, *Tetrahymena thermophila* (see inset), consists of a 9 + 2 arrangement of nine outer doublet microtubules and two central microtubules. This structure is anchored to the basal body, which has nine triplet microtubules, in which three microtubules are fused together. Note: The structure of the basal body is very similar to that of centrioles in animal cells. (top left): ©Aaron J. Bell/Science Source; (top middle, bottom middle): ©Dr. William Dentler/University of Kansas

 **Core Skill: Modeling**   The goal of this modeling challenge is to make a model of the circular structure formed by SAS-6 proteins based on information regarding which sites (A, B, and C) bind to other SAS-6 proteins or to doublet microtubules.

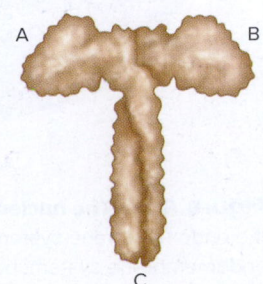

**Modeling Challenge:** A key protein that is a component of the radial spokes in an axoneme determines the nine-fold symmetry of the doublet microtubules. This protein is called SAS-6. A schematic drawing of its structure is shown to the right, with three sites labeled A, B, and C. Several SAS-6 proteins bind to each other to form a circular structure. During this process, the A site in one SAS-6 protein binds to the B site in another one. The C site binds to a doublet microtubule. Draw a model of the circular structure formed by SAS-6 proteins and indicate how many proteins make up this structure.

organisms, the microtubules form an arrangement called a 9 + 2 array. The outer nine are doublet microtubules, which are composed of a partial microtubule attached to a complete microtubule. Each of the two central microtubules is a single microtubule. Radial spokes project from the outer doublet microtubules toward the central pair. The microtubules in flagella and cilia emanate from **basal bodies**, which are anchored to the cytoplasmic side of the plasma membrane. At the basal body, the microtubules form a triplet structure. Much like the centrosome of animal cells, the basal bodies provide a site for microtubules to grow.

The movement of both flagella and cilia involves the propagation of a bend, which begins at the base of the structure and proceeds toward the tip (look back at Figure 4.16a). The bending occurs because dynein is activated to walk toward the minus end of the microtubules. However, the microtubules and dynein are not free to move relative to each other because of linking proteins. Therefore, instead of freely walking along the microtubules, the dyneins exert a force that bends the microtubules (look back at Figure 4.15c). The dyneins at the base of the flagellum or cilium are activated first, followed by dyneins that are progressively closer to the tip, and the resulting movement propels the organism.

## 4.5 The Nucleus and Endomembrane System

### Learning Outcomes:

1. Describe the structure and organization of the cell nucleus.
2. Outline the structures and general functions of the components of the endomembrane system.
3. Distinguish between the rough endoplasmic reticulum and the smooth endoplasmic reticulum.
4. **CoreSKILL »** Analyze the results of Palade's study, and explain how they indicate the existence of a secretory pathway in eukaryotic cells.
5. List three important functions of the plasma membrane.

In Chapter 2, we learned that the nucleus of an atom contains protons and neutrons. In cell biology, the term **nucleus** has a different meaning. It is an organelle found in eukaryotic cells that contains most of the cell's genetic material. A small amount of genetic material is also found outside the nucleus, in mitochondria and chloroplasts.

The membranes that enclose the nucleus are part of a larger network of membranes called the **endomembrane system**. This system includes not only the nuclear envelope, which encloses the nucleus, but also the endoplasmic reticulum, Golgi apparatus, lysosomes, vacuoles, and peroxisomes. The prefix endo- (from the Greek, meaning inside) originally referred only to these organelles and internal membranes. However, we now know that the plasma membrane is also part of this integrated membrane system (**Figure 4.18**). In this section, we will examine the nucleus and survey the structures and functions of the organelles and membranes of the endomembrane system.

### The Eukaryotic Nucleus Contains Chromosomes

The nucleus is the compartment that is enclosed by a double-membrane structure termed the **nuclear envelope** and houses the genetic material (**Figure 4.19**). **Nuclear pores** are formed where the inner and outer nuclear membranes make contact with each other. The pores provide a passageway for the movement of molecules and macromolecules into and out of the nucleus. Although cell biologists view the nuclear envelope as part of the endomembrane system, the materials within the nucleus are not.

Inside the nucleus are the chromosomes and a filamentous network of proteins called the nuclear matrix. Each **chromosome** is composed of genetic material, namely DNA, and many types of proteins that help to compact the chromosome to fit inside the nucleus. The complex formed between DNA and such proteins is termed **chromatin**. The **nuclear matrix** consists of two parts: the nuclear lamina, which is composed of intermediate filaments that line the inner nuclear membrane, and an internal nuclear matrix, which is connected to the lamina and fills the interior of the nucleus. The nuclear matrix serves to organize the chromosomes within the nucleus. Each

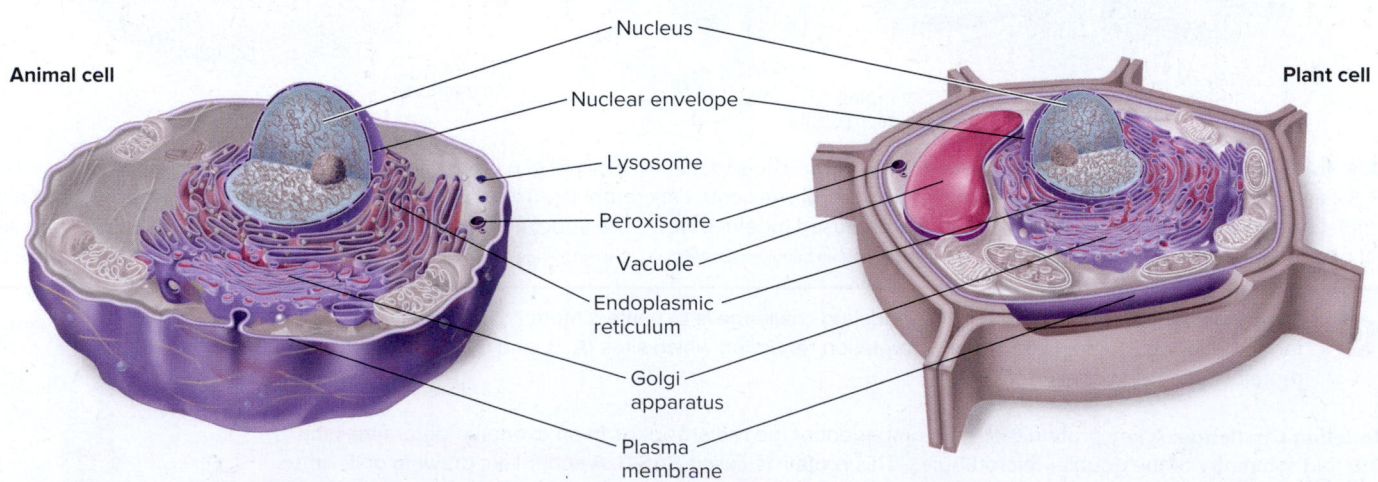

**Figure 4.18   The nucleus and endomembrane system.** This figure highlights the internal compartment of the nucleus (blue), the membranes of the endomembrane system (purple), and the fluid-filled interiors of the endomembrane system (pink). The nuclear envelope is part of the endomembrane system, but the interior of the nucleus is not.

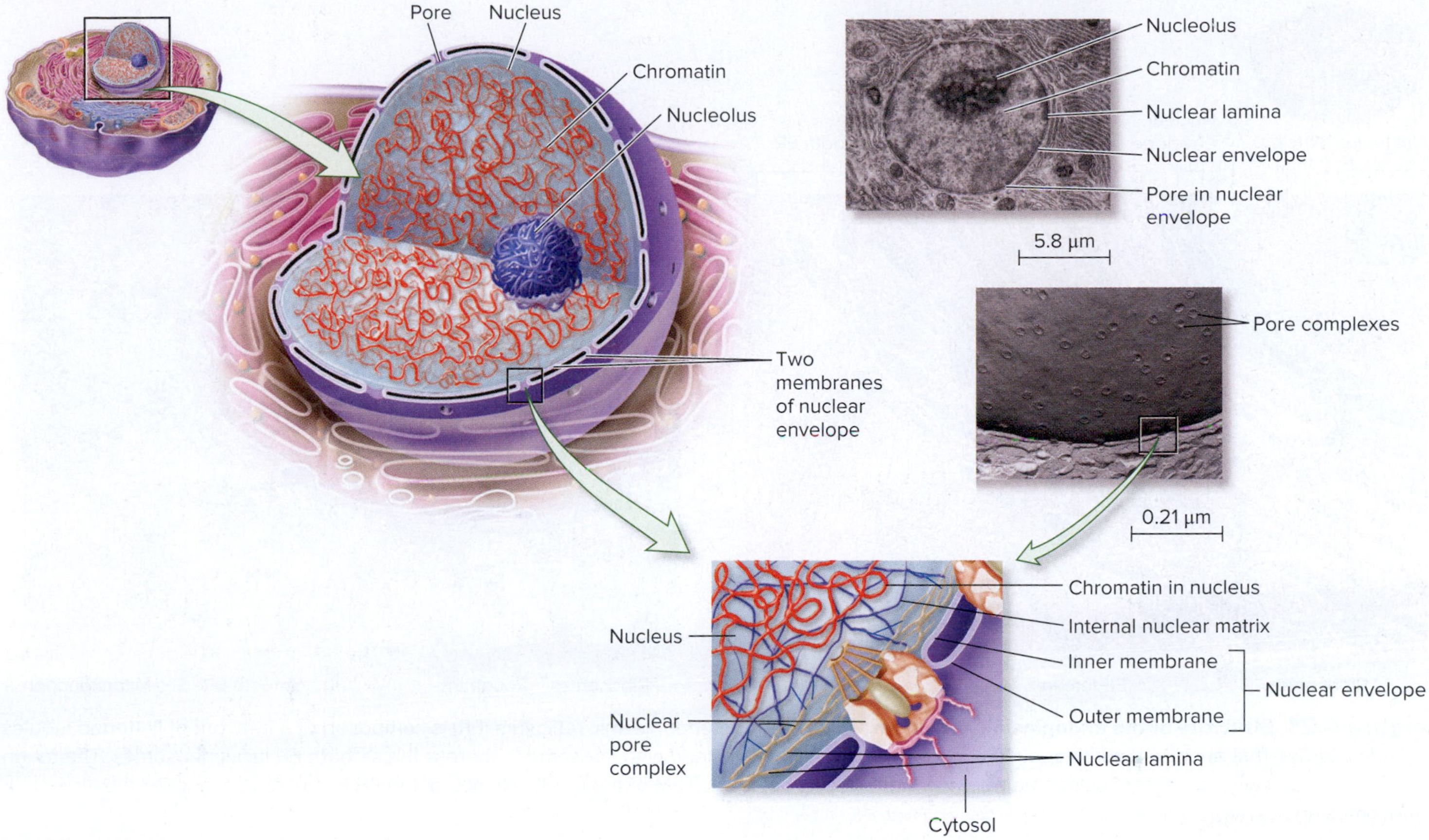

Pore    Nucleus

Chromatin

Nucleolus

Nucleolus

Chromatin

Nuclear lamina

Nuclear envelope

Pore in nuclear envelope

5.8 μm

Two membranes of nuclear envelope

Pore complexes

0.21 μm

Chromatin in nucleus

Internal nuclear matrix

Inner membrane

Nucleus

Nuclear envelope

Outer membrane

Nuclear pore complex

Nuclear lamina

Cytosol

**Figure 4.19** **The nucleus and nuclear envelope.** The nuclear envelope is composed of an inner membrane and an outer membrane that come into contact at the nuclear pores. The inner nuclear membrane is lined with lamin proteins to form the nuclear lamina. The interior of the nucleus contains chromatin, which is attached to the nuclear matrix, and a nucleolus, where ribosome subunits are assembled. (top right, middle right): ©Don W. Fawcett/Science Source

**Concept Check:** *What is the function of the nuclear matrix?*

chromosome is located in a distinct **chromosome territory**, which is visible when cells are exposed to dyes that label specific types of chromosomes (**Figure 4.20**).

The primary function of the nucleus is the protection, organization, replication, and expression of the genetic material. These topics are discussed in Unit III. Another important function is the assembly of ribosome subunits—cellular structures involved in producing polypeptides during the process of translation (look ahead to Table 12.3). The assembly of ribosome subunits occurs in the **nucleolus** (plural, nucleoli), a droplet organelle in the nucleus of nondividing cells. A ribosome is composed of two subunits: one small and one large. Each subunit contains one or more RNA molecules and several types of proteins. Most of the RNA molecules that are components of ribosomes are made in the vicinity of the nucleolus. By comparison, the ribosomal proteins are produced in the cytosol and then imported into the nucleus through the nuclear pores. The ribosomal proteins and RNA molecules then enter the nucleolus and are assembled into ribosomal subunits. Finally, the subunits exit the nucleolus and then move through the nuclear pores into the cytosol, where they carry out polypeptide synthesis.

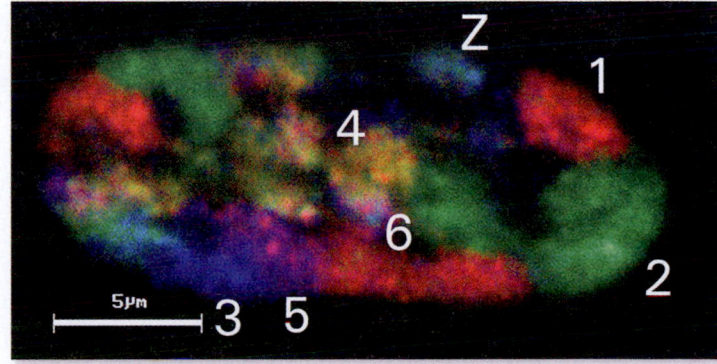

**Figure 4.20** **Chromosome territories in the cell nucleus.** Chromosomes from a chicken were labeled with chromosome-specific probes. Seven types of chicken chromosomes are stained with a different dye. Each chromosome occupies its own distinct, nonoverlapping territory within the cell nucleus. Courtesy of Felix A. Habermann

 **Core Skill: Connections** Look ahead to Figure 16.8. What happens to chromosome territories during cell division?

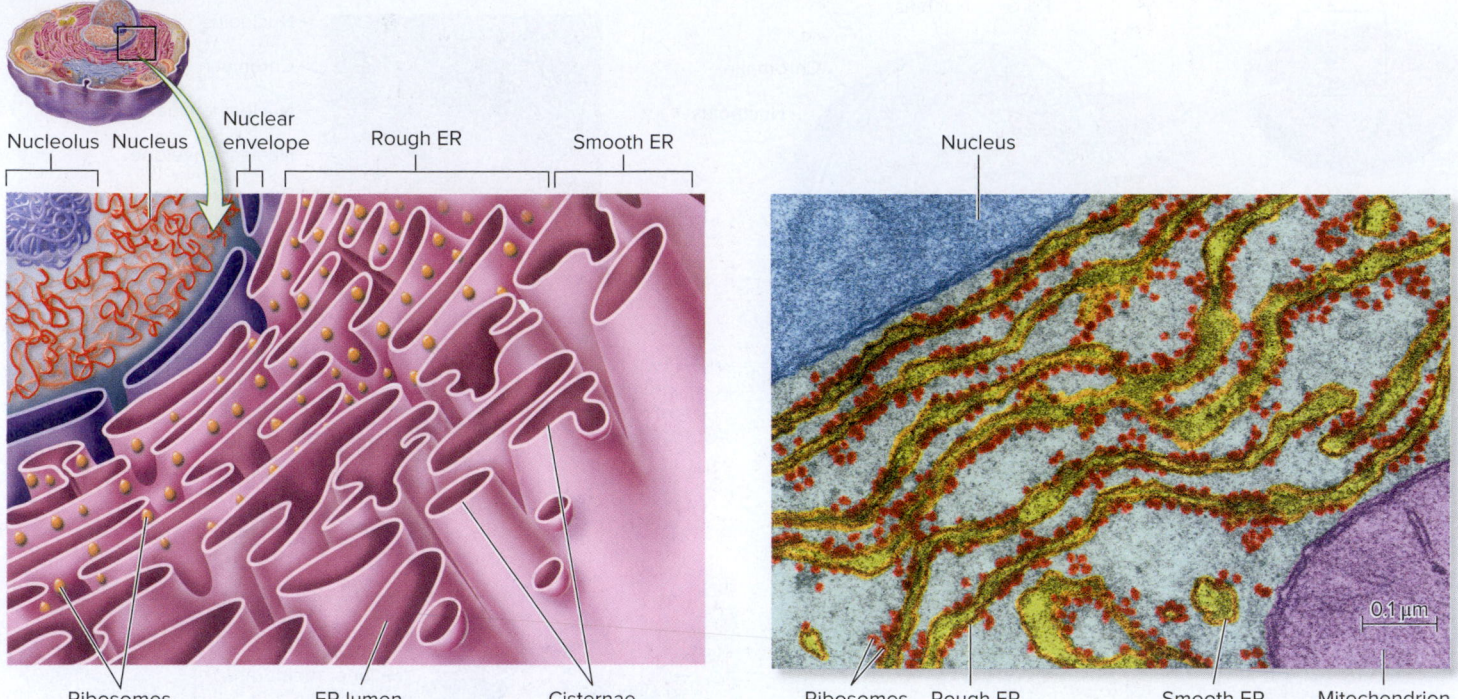

**Figure 4.21** **Structure of the endoplasmic reticulum.** (Left side) The endoplasmic reticulum (ER) is composed of a network of flattened tubules called cisternae that enclose a continuous ER lumen. The rough ER is studded with ribosomes, whereas the smooth ER lacks ribosomes. The rough ER is continuous with the outer nuclear membrane. (Right side) A colorized TEM of the ER. The lumen of the ER is colored yellow and the ribosomes are red. (right): ©Dennis Kunkel Microscopy, Inc./Phototake

## The Endoplasmic Reticulum Initiates the Sorting of Some Proteins and Carries Out Metabolic Functions

The **endoplasmic reticulum (ER)** is a network of membranes that form flattened, fluid-filled tubules, or **cisternae** (**Figure 4.21**). The terms endoplasmic (Greek, for in the cytoplasm) and reticulum (Latin, for little net) refer to the location and shape of this organelle when viewed under a microscope. The term **lumen** refers to the internal space of an organelle. The ER membrane encloses a single compartment called the **ER lumen**. There are two distinct, but continuous, types of ER: rough ER and smooth ER.

**Rough ER**   The outer surface of the **rough endoplasmic reticulum (rough ER)** is studded with ribosomes, giving it a bumpy appearance. Rough ER plays a key role in the sorting of proteins that are destined for the ER, Golgi apparatus, lysosomes, vacuoles, plasma membrane, or extracellular environment. Proteins are packaged into **membrane vesicles**—small spheres enclosed by a membrane—and moved from one location in the endomembrane system to another. This sorting process is described in Section 4.7. In conjunction with protein sorting, a second function of the rough ER is the insertion of certain newly made proteins into the ER membrane. A third important function of the rough ER is the attachment of carbohydrates to proteins and lipids. This process is called **glycosylation**. The topics of membrane protein insertion and protein glycosylation will be discussed in Chapter 5.

**Smooth ER**   The **smooth endoplasmic reticulum (smooth ER)**, which lacks ribosomes, functions in diverse metabolic processes. The extensive network of smooth ER membranes provides a large surface area for enzymes that play important metabolic roles. In liver cells, enzymes in the smooth ER detoxify many potentially harmful organic molecules, including barbiturate drugs and ethanol. These enzymes convert hydrophobic toxic molecules into more hydrophilic molecules, which are easily excreted from the body. Chronic alcohol consumption, as in alcoholics, leads to a greater amount of smooth ER in liver cells, which increases the rate of alcohol breakdown. This explains why people who consume alcohol regularly must ingest more alcohol to experience its effects.

The smooth ER of liver cells also plays a role in carbohydrate metabolism. The liver cells of animals store energy in the form of glycogen, which is a polymer of glucose. Glycogen granules sit very close to the smooth ER membrane. When chemical energy is needed, enzymes are activated that break down the glycogen to glucose-6-phosphate. Then, an enzyme in the smooth ER called glucose-6-phosphatase removes the phosphate group, and glucose is exported from the liver cell into the bloodstream.

Another important function of the smooth ER in all eukaryotes is the accumulation of calcium ions ($Ca^{2+}$). The smooth ER contains calcium pumps that transport $Ca^{2+}$ into the ER lumen. The regulated release of $Ca^{2+}$ into the cytosol is involved in many vital cellular processes, including muscle contraction in animals.

Finally, enzymes in the smooth ER are critical in the synthesis and modification of lipids. For example, the smooth ER is the primary site for the synthesis of phospholipids, which are the main lipid component of eukaryotic cell membranes. This topic is discussed in Chapter 5. In addition, enzymes in the smooth ER are necessary for certain modifications of the lipid cholesterol to produce steroid hormones such as estrogen and testosterone.

## The Golgi Apparatus Directs the Processing, Sorting, and Secretion of Cellular Molecules

The **Golgi apparatus** (also called the Golgi body, Golgi complex, or simply Golgi) was discovered by the Italian microscopist Camillo Golgi in 1898. It consists of a stack of flattened membranes, with each flattened membrane enclosing a single compartment. The Golgi compartments are named according to their orientation in the cell. The *cis* Golgi is near the ER membrane, the *trans* Golgi is closest to the plasma membrane, and the medial Golgi is found in the middle.

Two models have been proposed to explain how materials move through the Golgi apparatus:

- *Vesicular transport model.* Materials are transported between the Golgi cisternae via membrane vesicles that bud from one compartment in the Golgi (for example, the *cis* Golgi) and fuse with another compartment (for example, the medial Golgi).

- *Cisternal maturation model.* Vesicles from the ER fuse to form a cisterna at the *cis* face; the cisterna that was previously at the *cis* face becomes a medial cisterna. This addition of a cisterna moves the other medial cisternae toward the *trans* face.

A cisterna at the *trans* face is lost as a result of the export of vesicles from its surface.

Further research is needed to determine the validity of these models.

The Golgi apparatus performs three overlapping functions: (1) processing, (2) protein sorting, and (3) secretion. We will discuss protein sorting in Section 4.7. Enzymes in the Golgi apparatus process, or modify, certain proteins and lipids. As mentioned earlier, carbohydrates can be attached to proteins and lipids in the endoplasmic reticulum. Glycosylation continues in the Golgi. For this to occur, a protein or lipid is transported via vesicles from the ER to the *cis* Golgi. Most glycosylation occurs in the medial Golgi.

A second type of processing event is **proteolysis**, whereby enzymes called **proteases** make cuts in polypeptides. For example, the hormone insulin is first made as a large precursor molecule termed proinsulin. In the Golgi apparatus, proinsulin is packaged with proteases into vesicles. The proteases cut out a portion of the proinsulin to create a smaller insulin polypeptide that is a functional hormone. This happens just prior to secretion, which is described next.

The Golgi apparatus packages different types of materials (cargo) into **secretory vesicles** that fuse with the plasma membrane, thereby releasing their contents outside the cell. Proteins destined for secretion are synthesized into the ER, travel to the Golgi, and then are transported by vesicles to the plasma membrane. The vesicles then fuse with the plasma membrane, and the proteins are secreted to the outside of the cell. The entire route is called the **secretory pathway** (**Figure 4.22**). In addition to secretory vesicles, the Golgi also produces vesicles that travel to other parts of the cell, such as the lysosomes.

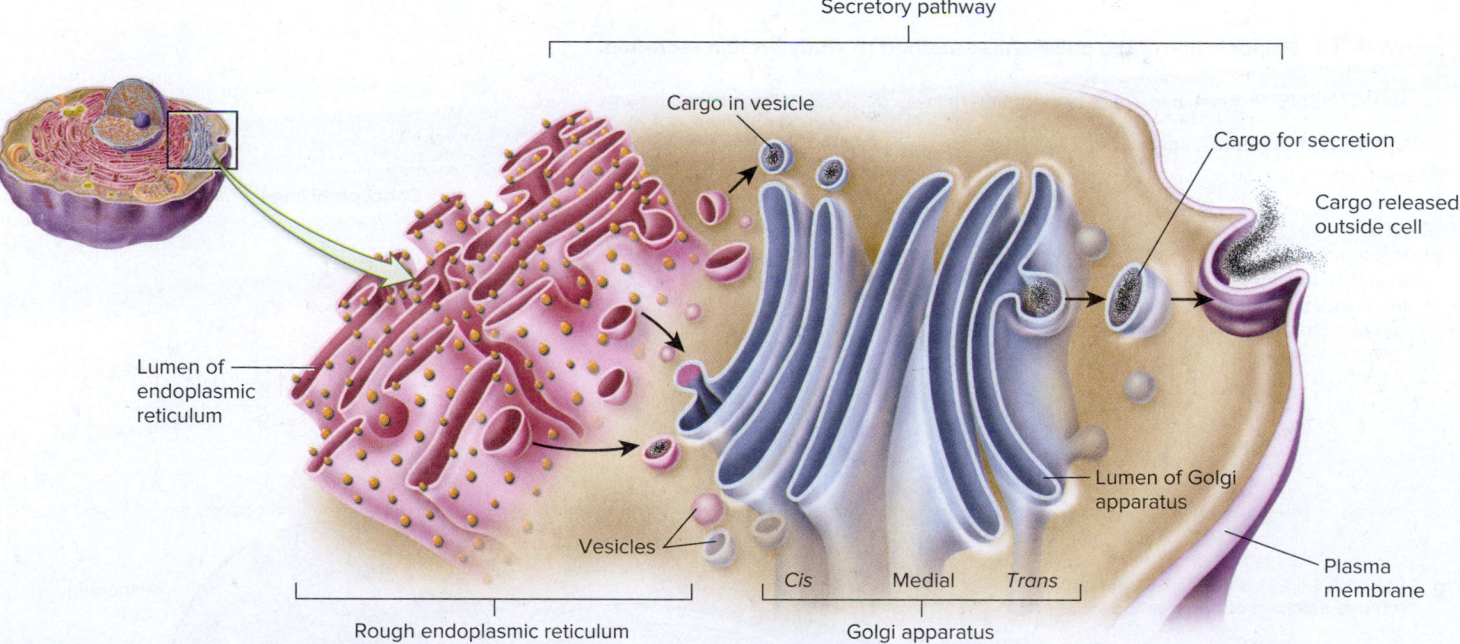

**Figure 4.22  The Golgi apparatus and secretory pathway.** The Golgi is composed of stacks of membranes that enclose distinct compartments. Transport to and from the Golgi compartments occurs via membrane vesicles. Vesicles bud from the ER and go to the Golgi, and vesicles from the Golgi fuse with the plasma membrane to release cargo to the outside. The pathway from the ER to the Golgi to the plasma membrane is termed the secretory pathway.

*Concept Check:* If we consider the Golgi apparatus as three compartments (cis, medial, and trans), in what order does a protein travel through them before being secreted?

## Core Skill: Process of Science

# Feature Investigation | Palade Discovered That Proteins Destined for Secretion Move Sequentially Through Organelles of the Endomembrane System

As we have seen, a key function of the endomembrane system is protein secretion. The identification of the secretory pathway came from studies of George Palade and his colleagues in the 1960s. He hypothesized that proteins follow an intracellular pathway to be secreted. Palade's team conducted pulse-chase experiments, in which the researchers administered a pulse of radioactive amino acids to cells so they made radioactive proteins. A few minutes later, the cells were given a large amount of nonradioactive amino acids. This step is called a "chase" because it chases away the ability of the cells to make any more radioactive proteins. In this way, radioactive proteins were produced only briefly. Because they were labeled with radioactivity, the fate of these proteins could be monitored over time. The goal of a pulse-chase experiment is to determine where the radioactive proteins are produced and the pathway they take as they travel through a cell.

Palade chose to study the cells of the pancreas. This organ secretes enzymes and protein hormones that play a role in digestion and metabolism. Therefore, these cells were chosen because their primary activity is protein secretion. To study the pathway for protein secretion, Palade and colleagues injected a radioactive version of the amino acid leucine into the bloodstream of male guinea

pigs. The radiolabeled leucine traveled in the bloodstream and was quickly taken up by cells of the body, including those in the pancreas. Three minutes later, the researchers injected nonradiolabeled leucine (**Figure 4.23**). At various times after the second injection, samples of pancreatic cells were removed from the animals. The cells were then prepared for transmission electron microscopy (TEM). The sample was stained with osmium tetroxide, a heavy metal compound that became bound to membranes and showed the locations of the cell organelles. In addition, the sample was coated with a radiation-sensitive emulsion containing silver. When radiation was emitted from the radiolabeled proteins, it interacted with the emulsion in a way that caused the precipitation of silver, which became tightly bound to the sample. In this way, the precipitated silver marked the location of the radiolabeled proteins. Unprecipitated silver in the emulsion was later washed away. Because silver atoms are electron-dense (allowing few electrons to pass), they produce dark spots in a TEM. Therefore, dark spots revealed the locations of radiolabeled proteins.

The schematic drawings shown as the data indicate the path of the proteins as they moved through the secretory pathway. Very dark objects, namely radiolabeled proteins, were first observed in the

**Figure 4.23** Palade's use of the pulse-chase method to study protein secretion.

**HYPOTHESIS** Proteins that are to be secreted follow a particular intracellular pathway.

**KEY MATERIALS** Male guinea pigs.

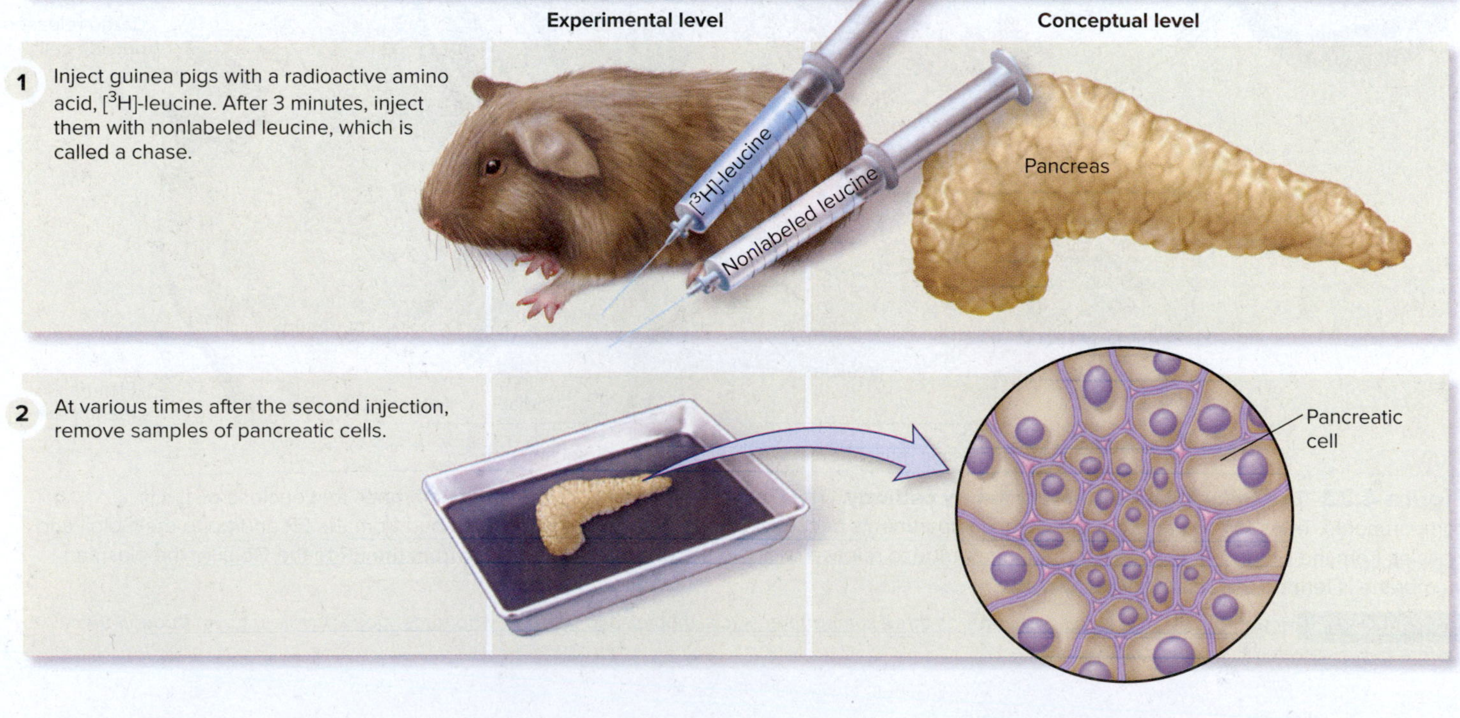

Experimental level      Conceptual level

1  Inject guinea pigs with a radioactive amino acid, [³H]-leucine. After 3 minutes, inject them with nonlabeled leucine, which is called a chase.

[³H]-leucine

Nonlabeled leucine

Pancreas

2  At various times after the second injection, remove samples of pancreatic cells.

Pancreatic cell

**3** Stain the sample with osmium tetroxide, which is a heavy metal that binds to membranes.

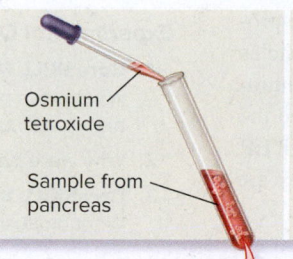

Osmium tetroxide

Sample from pancreas

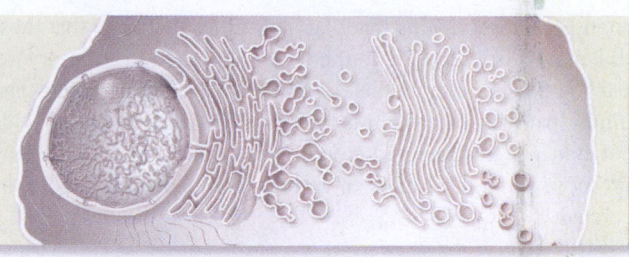

**4** Cut thin sections of the samples, and place a thin layer of radiation-sensitive emulsion over the sample. Allow time for radioactive emission from radiolabeled proteins to precipitate silver atoms in the emulsion. Wash away unprecipitated silver atoms.

Thin section

Add radiation-sensitive emulsion

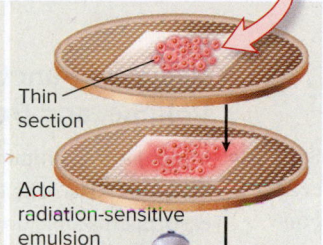

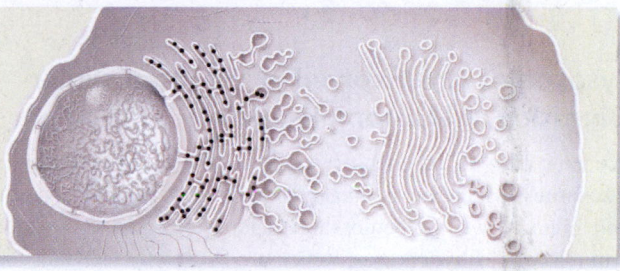

**5** Observe the sample under a transmission electron microscope.

**6** **THE DATA**

Schematic drawings of transmission electron micrographs

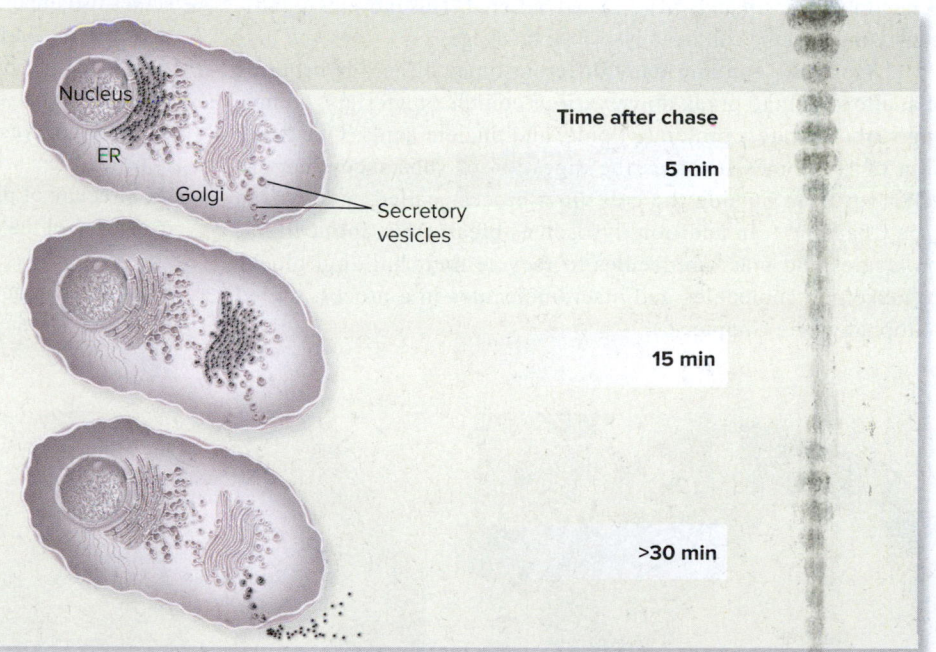

Nucleus

ER

Golgi

Secretory vesicles

Time after chase

**5 min**

**15 min**

**>30 min**

**7** **CONCLUSION** To be secreted, proteins move from the ER to the Golgi to secretory vesicles and then to the plasma membrane, where they are released to the outside of the cell.

**8** **SOURCE** Caro, L.G., and Palade, G.E. 1964. Protein synthesis, storage, and discharge in the pancreatic exocrine cell. An autoradiographic study. *Journal of Cell Biology* 20: 473–495.

rough ER. Observations made at later times indicated that these proteins moved from the ER to the Golgi, and then to secretory vesicles near the plasma membrane. In this way, Palade followed the intracellular pathway of protein movement. His experiments provided the first evidence that secreted proteins are synthesized into the rough ER and move through a series of cellular compartments before they are secreted.

### Experimental Questions

1. **CoreSKILL »** Explain the procedure of a pulse-chase experiment. What is the pulse, and what is the chase? What was the purpose of this approach?

2. Why were pancreatic cells used for this investigation?

3. **CoreSKILL »** Analyze the results of the experiment of Figure 4.23. What did the researchers conclude?

## Lysosomes Are Involved in the Intracellular Digestion of Macromolecules

We now turn to another organelle of the endomembrane system, **lysosomes**, which are small organelles that are found in animal cells and break down molecules and macromolecules. Lysosomes contain many **acid hydrolases**, which are hydrolytic enzymes that use a molecule of water to break a covalent bond. As described in Chapter 3 (refer back to Figure 3.4b), this type of chemical reaction is called hydrolysis:

$$R_1—R_2 + H_2O \xrightarrow{\text{Acid hydrolase}} R_1—OH + R_2—H$$

The acid hydrolases in a lysosome function optimally at an acidic pH. The fluid-filled interior of a lysosome has a pH of approximately 4.8. If a lysosomal membrane breaks, releasing acid hydrolases into the cytosol, the enzymes are not very active because the cytosolic pH is neutral (approximately pH 7.2) and buffered. This prevents significant damage to the cell from lysosome breakage.

Lysosomes contain many different types of acid hydrolases that allow them to break down various complex materials, including carbohydrates, proteins, lipids, and nucleic acids. One function of lysosomes involves the digestion of substances that are taken up from outside the cell via a process called endocytosis (see Chapter 5). In addition, lysosomes break down intracellular molecules and macromolecules to recycle their building blocks to make new molecules and macromolecules in a process called autophagy (see Chapter 6).

## Vacuoles Function in Storage, Regulation of Cell Volume, and Degradation

**Vacuoles** are prominent organelles in plant cells, fungal cells, and certain protists. The term vacuole (Latin, for empty space) came from early microscopic observations of these compartments. We now know that vacuoles are not empty but instead contain fluid and sometimes even solid substances. Most vacuoles are made from the fusion of many smaller membrane vesicles. Vacuoles in animal cells tend to be smaller than those in plants and are more commonly used to temporarily store materials or transport substances. Such vacuoles are sometimes called storage vesicles.

The functions of vacuoles are extremely varied, and they differ among cell types and even with environmental conditions. The best way to appreciate vacuole function is to consider a few examples. Mature plant cells usually have a large **central vacuole** that occupies 80% or more of the cell volume (**Figure 4.24a**). The central vacuole serves two important purposes. First, it stores a large amount of water, enzymes, and inorganic ions such as calcium. It also stores other materials including proteins and pigments. Second, it performs a space-filling function. The central vacuole exerts a pressure on the cell wall, called turgor pressure. If a plant becomes dehydrated and this pressure is lost, a plant will wilt. Turgor pressure is important in maintaining the structure of plant cells and the plant itself, and it helps to drive the expansion of the cell wall, which is necessary for growth.

Certain species of protists use vacuoles to maintain cell volume. Freshwater organisms such as the alga *Chlamydomonas reinhardtii* have small, water-filled **contractile vacuoles** that expand as water

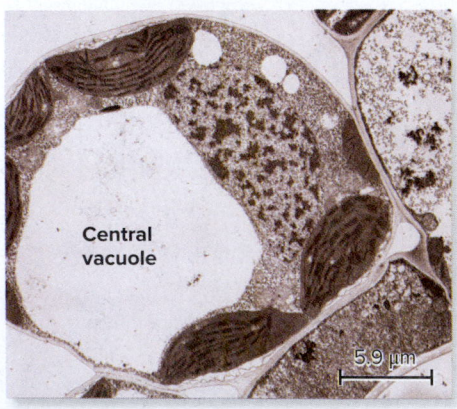

**(a) Central vacuole in a plant cell**

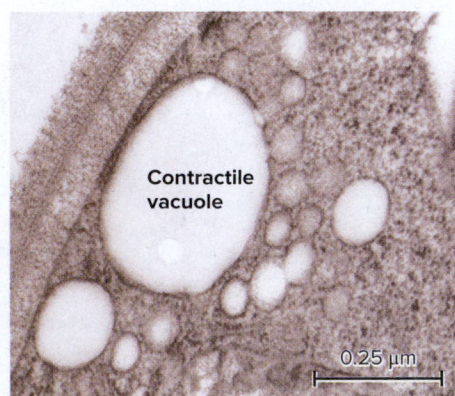

**(b) Contractile vacuoles in an algal cell**

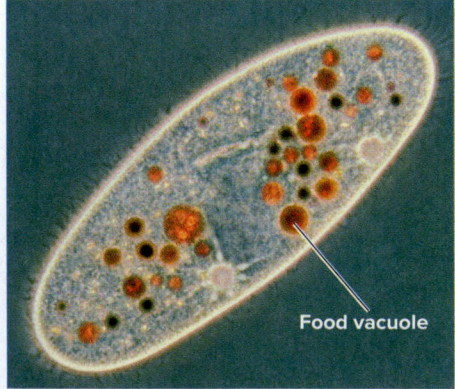

**(c) Food vacuoles in a paramecium**

**Figure 4.24** **Examples of vacuoles.** These are TEMs. Part (c) is colorized. a: ©Biophoto Associates/Science Source; b: Courtesy of Dr. Peter Luykx, Biology, University of Miami; c: ©Dr. David Patterson/Science Source

enters the cell (Figure 4.24b). Once they reach a certain size, the vacuoles fuse with the plasma membrane, expelling their contents to the exterior of the cell (look ahead to Figure 5.14). This mechanism is necessary to remove excess water that continually enters the cell by diffusion across the plasma membrane.

Another function of vacuoles is degradation. Some protists engulf their food into large food vacuoles in the process of phagocytosis (Figure 4.24c). Food vacuoles contain hydrolytic enzymes that break down macromolecules within food. Macrophages, a type of cell found in animals' immune systems, engulf bacterial cells into phagocytic vacuoles, which then fuse with lysosomes, where the bacteria are destroyed.

## Peroxisomes Catalyze Detoxifying Reactions

**Peroxisomes**, discovered by Christian de Duve in 1965, are small organelles found in all eukaryotic cells. Peroxisomes consist of a single membrane that encloses a fluid-filled lumen. A typical eukaryotic cell contains several hundred of them.

Peroxisomes catalyze a variety of chemical reactions, including some reactions that break down organic molecules and others that are biosynthetic. In mammals, large numbers of peroxisomes are found in liver cells, where toxic molecules accumulate and are broken down. A common byproduct of the breakdown of toxins is hydrogen peroxide, $H_2O_2$:

$$RH_2 + O_2 \longrightarrow R + H_2O_2$$
$$\text{(toxin)}$$

Hydrogen peroxide has the potential to damage cellular components. In the presence of metals such as iron ($Fe^{2+}$), which are found naturally in living cells, $H_2O_2$ is broken down to form a hydroxide ion ($OH^-$) and a molecule called a hydroxide free radical ($\cdot OH$):

$$Fe^{2+} + H_2O_2 \rightarrow Fe^{3+} + OH^- + \cdot OH \text{ (hydroxide free radical)}$$

The $\cdot OH$ is highly reactive and can damage proteins, lipids, and DNA. Therefore, it is beneficial for cells to break down $H_2O_2$ in an alternative manner that does not form $\cdot OH$. Peroxisomes contain an enzyme called **catalase** that breaks down hydrogen peroxide to make water and oxygen gas (hence the name peroxisome):

$$2 H_2O_2 \xrightarrow{\text{Catalase}} 2 H_2O + O_2$$

Aside from detoxification, peroxisomes can play a role in the metabolism of fats and amino acids. For example, plant seeds contain specialized organelles called **glyoxysomes**, which are similar to peroxisomes. Seeds often store fats instead of carbohydrates. Because fats have higher energy per unit mass, seeds that store fats are smaller and less heavy than seeds that store carbohydrates would be. Glyoxysomes contain enzymes that are needed to convert fats to sugars. These enzymes become active when a seed germinates and the seedling begins to grow.

A general model for peroxisome formation is shown in **Figure 4.25**, though the details may differ among animal, plant, and fungal cells. To initiate peroxisome formation, vesicles bud from the ER membrane and form a premature peroxisome. Following the import of additional proteins, the premature peroxisome becomes a mature peroxisome. Once the mature peroxisome has formed, it may then divide to further increase the number of peroxisomes in the cell.

## The Plasma Membrane Is the Interface Between a Cell and Its Environment

The cytoplasm of eukaryotic cells is surrounded by a plasma membrane, which is part of the endomembrane system and provides a boundary between a cell and the extracellular environment. Proteins

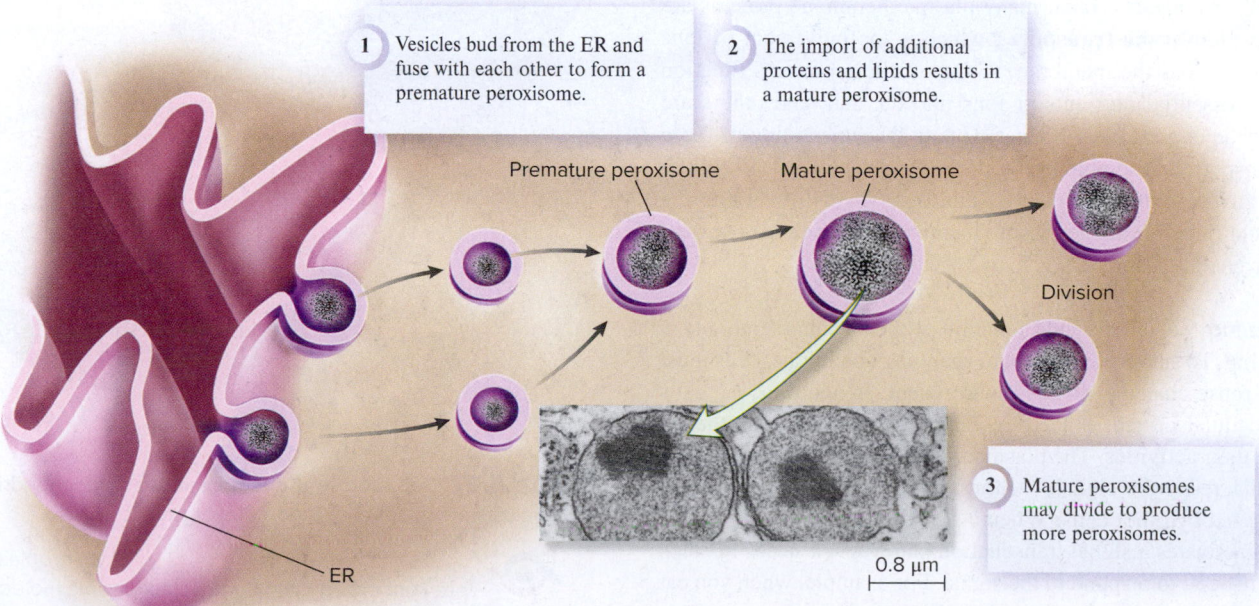

1 Vesicles bud from the ER and fuse with each other to form a premature peroxisome.

2 The import of additional proteins and lipids results in a mature peroxisome.

Premature peroxisome

Mature peroxisome

Division

ER

3 Mature peroxisomes may divide to produce more peroxisomes.

0.8 μm

**Figure 4.25  Formation of peroxisomes.**  The inset is a TEM of mature peroxisomes. (inset): ©Don W. Fawcett/Science Source

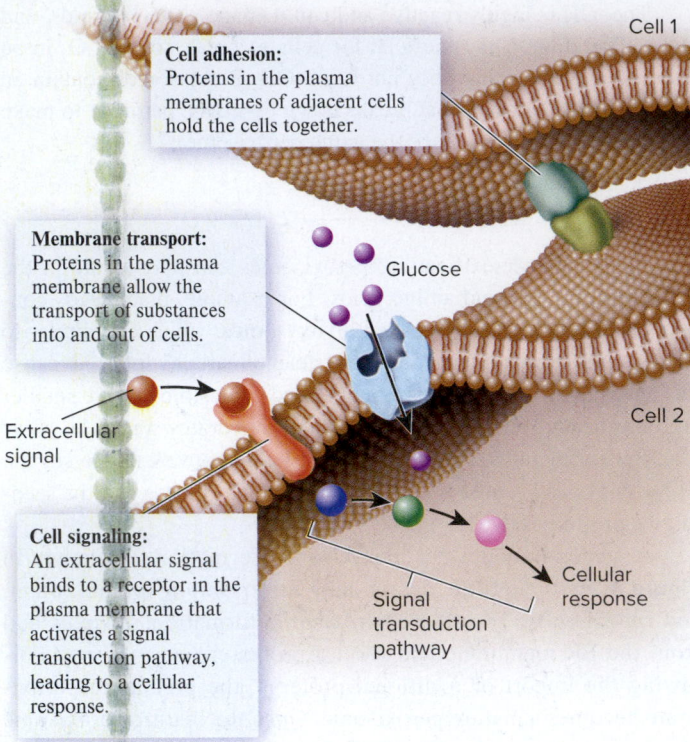

**Cell adhesion:** Proteins in the plasma membranes of adjacent cells hold the cells together.

Cell 1

**Membrane transport:** Proteins in the plasma membrane allow the transport of substances into and out of cells.

Glucose

Extracellular signal

Cell 2

**Cell signaling:** An extracellular signal binds to a receptor in the plasma membrane that activates a signal transduction pathway, leading to a cellular response.

Signal transduction pathway

Cellular response

**Figure 4.26  Major functions of the plasma membrane.** Three important roles are membrane transport, cell signaling, and cell adhesion.

*Concept Check:* *Which of these three functions do you think is the most important for cellular metabolism?*

in the plasma membrane perform many important functions that affect the activities inside the cell (**Figure 4.26**).

**Membrane Transport**   First, many plasma membrane proteins are involved in **membrane transport** , which is the movement of ions or molecules across the membrane. Some of these proteins function to transport essential nutrients or ions into the cell, and others are involved in the export of substances. Due to the functioning of these protein transporters, the plasma membrane is selectively permeable; it allows only certain substances in and out. We will examine the structure and function of the plasma membrane, as well as a variety of transporters, in Chapter 5.

**Cell Signaling**   A second vital function of the plasma membrane is **cell signaling**. To survive and adapt to changing conditions, cells must be able to sense changes in their environment. In addition, the cells of a multicellular organism need to communicate with each other to coordinate their activities. The plasma membrane of all cells contains receptors that recognize signals—either environmental agents or molecules secreted by other cells. When a signaling molecule binds to a receptor, it activates a signal transduction pathway—a series of steps that cause the cell to respond to the signal. For example, when you eat a meal, the hormone insulin is secreted into your bloodstream. This hormone binds to receptors in the plasma membranes of your cells, which results in a cellular response that allows your cells to increase

their uptake of certain molecules found in food, such as glucose. We will explore the details of cell signaling in Chapter 9.

**Cell Adhesion**   A third important role of the plasma membrane in animal cells is **cell adhesion**. Protein-protein interactions among proteins in the plasma membranes of adjacent cells promote cell-to-cell adhesion. This phenomenon is critical for animal cells to properly interact to form a multicellular organism and for cells to recognize each other. The structures and functions of proteins involved in cell adhesion will be examined in Chapter 10.

## 4.6  Semiautonomous Organelles

**Learning Outcomes:**

1. Outline the structures and general functions of mitochondria and chloroplasts.

2. **CoreSKILL »** Evaluate the evidence for the endosymbiosis theory.

We now turn to those organelles in eukaryotic cells that are considered semiautonomous: mitochondria and chloroplasts. These organelles grow and divide, but they are not completely autonomous because they depend on other parts of the cell for their internal components (**Figure 4.27**). For example, most of the proteins found in mitochondria are imported from the cytosol. In this section, we will survey the structures and functions of the semiautonomous organelles in eukaryotic cells and consider

**Animal cell**

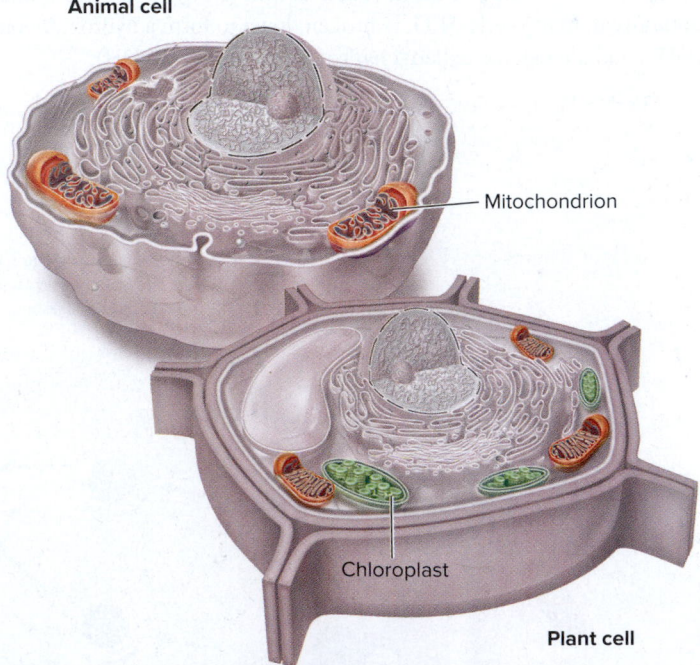

Mitochondrion

Chloroplast

**Plant cell**

**Figure 4.27  Semiautonomous organelles.** Mitochondria and chloroplasts are the semiautonomous organelles.

 **Core Concept: Energy and Matter** Chloroplasts capture light energy and synthesize organic molecules. Mitochondria break down organic molecules and make ATP that is used as an energy source to drive many different cellular processes.

their evolutionary origins. In Chapters 7 and 8, we will explore the functions of mitochondria and chloroplasts in greater depth.

## Mitochondria Supply Cells with Most of Their ATP

**Mitochondrion** (plural, mitochondria) literally means thread granule, which is what mitochondria look like under a light microscope—either threadlike or granular-shaped. They are similar in size to bacteria. A typical cell may contain a few hundred to a few thousand mitochondria. Cells with particularly heavy energy demands, such as muscle cells, have more mitochondria than other cells. Research has shown that regular exercise increases the number and size of mitochondria in human muscle cells to meet the expanded demand for energy.

A mitochondrion has an outer membrane and an inner membrane separated by a region called the intermembrane space (**Figure 4.28**). The inner membrane is highly invaginated (folded) to form projections called **cristae**. The cristae greatly increase the surface area of the inner membrane, which is the site where ATP is made. The compartment enclosed by the inner membrane is the **mitochondrial matrix**.

The primary role of mitochondria is to make ATP. Even though mitochondria produce most of a cell's ATP, mitochondria do not create energy. Rather, their primary function is to convert chemical energy that is stored within the covalent bonds of organic molecules into a form that can be readily used by cells. Covalent bonds in sugars, fats, and amino acids store a large amount of energy. The breakdown of these molecules into simpler molecules releases energy that is used

to make ATP. Many proteins in living cells use ATP as a source of energy to carry out their functions, such as muscle contraction, the uptake of nutrients, cell division, and many other cellular processes.

Mitochondria perform other functions as well. They are involved in the synthesis, modification, and breakdown of several types of cellular molecules. For example, the synthesis of certain hormones requires enzymes that are found in mitochondria. Another interesting role of mitochondria is to generate heat in specialized fat cells known as brown fat cells. Groups of brown fat cells serve as heating pads that help to revive hibernating animals and protect sensitive areas of young animals from the cold.

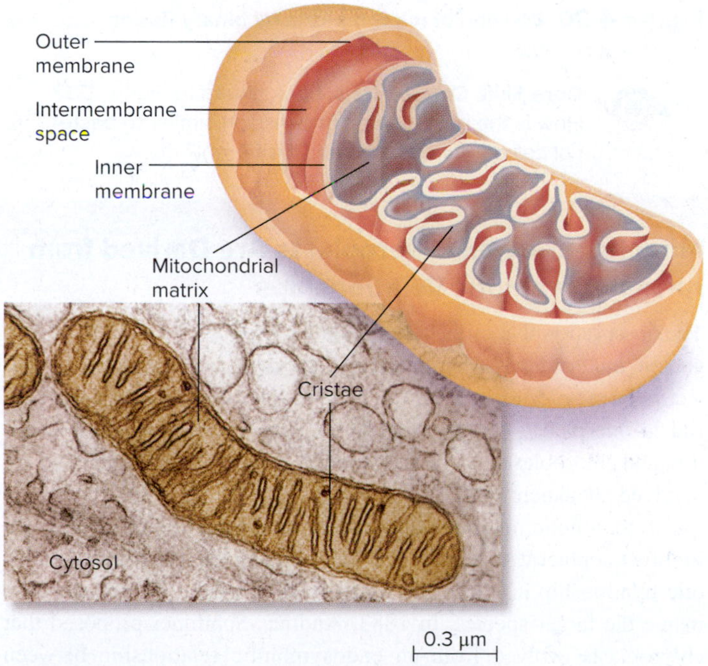

## Chloroplasts Carry Out Photosynthesis

**Chloroplasts** are semiautonomous organelles that capture light energy and use some of that energy to synthesize organic molecules such as glucose. This process, called **photosynthesis**, is described in Chapter 8. Chloroplasts are found in nearly all species of plants and algae. **Figure 4.29** shows the structure of a typical chloroplast. Like a mitochondrion, a chloroplast contains an outer and inner membrane. An intermembrane space lies between these two membranes. A third membrane, the **thylakoid membrane**, forms many flattened, fluid-filled tubules that enclose a single, convoluted compartment called the thylakoid lumen. These tubules stack on top of each other to form a structure called a **granum** (plural, grana). The **stroma** is the compartment of the chloroplast that is enclosed by the inner membrane but outside the thylakoid membrane.

Chloroplasts are a specialized version of plant organelles that are more generally known as **plastids**. All plastids are derived from unspecialized **proplastids**. The various types of plastids are

Outer membrane
Intermembrane space
Inner membrane
Mitochondrial matrix
Cristae
Cytosol

0.3 μm

**Figure 4.28  Structure of a mitochondrion.**  This figure emphasizes the membrane organization of a mitochondrion, which has an outer and inner membrane. The invaginations of the inner membrane are called cristae. The mitochondrial matrix lies inside the inner membrane. The micrograph is a colorized TEM. ©Don W. Fawcett/Science Source

**Concept Check:**  *What is the advantage of the mitochondrion's highly invaginated inner membrane?*

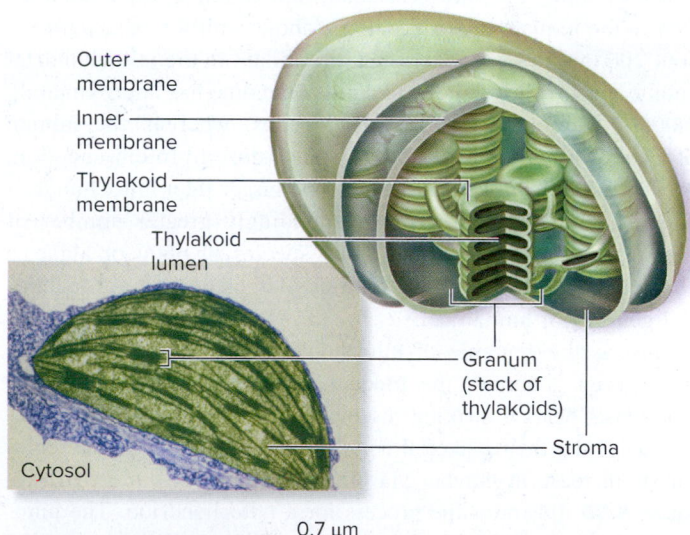

Outer membrane
Inner membrane
Thylakoid membrane
Thylakoid lumen
Granum (stack of thylakoids)
Stroma
Cytosol

0.7 μm

**Figure 4.29  Structure of a chloroplast.**  Like a mitochondrion, a chloroplast is enclosed in a double membrane. In addition, it has an internal thylakoid membrane that forms flattened tubular compartments. These compartments stack on each other to form grana. The stroma is located inside the inner membrane but outside the thylakoid membrane. This micrograph is a colorized TEM. ©Dr. Jeremy Burgess/Science Source

distinguished by their synthetic abilities and the types of pigments they contain. Chloroplasts, which carry out photosynthesis, contain the green pigment chlorophyll. The abundant number of chloroplasts in the leaves of plants gives them their green color. Chromoplasts, a second type of plastid, function in synthesizing and storing the yellow, orange, and red pigments known as carotenoids. Chromoplasts give many fruits and flowers their colors. In autumn, the chromoplasts also give many leaves their yellow, orange, and red colors. A third type of plastid, leucoplasts, typically lacks pigment molecules. An amyloplast is a leucoplast that synthesizes and stores starch. Amyloplasts are common in underground plant structures such as roots and tubers.

## Mitochondria and Chloroplasts Contain Their Own Genetic Material and Divide by Binary Fission

To fully appreciate the structure and organization of mitochondria and chloroplasts, we also need to briefly examine their genetic properties. In 1951, Yasutane Chiba exposed plant cells to Feulgen stain, a DNA-specific dye, and discovered that the chloroplasts became stained. Based on this observation, he was the first to suggest that chloroplasts contain their own DNA. Researchers in the 1970s and 1980s isolated DNA from both chloroplasts and mitochondria. These studies revealed that the DNA of these organelles resembled smaller versions of bacterial chromosomes.

The chromosomes found in mitochondria and chloroplasts are referred to as the **mitochondrial genome** and **chloroplast genome**, respectively, whereas the chromosomes found in the nucleus of a eukaryotic cell constitute the **nuclear genome**. Like bacterial genomes, the genomes of most mitochondria and chloroplasts are composed of a single circular chromosome. Compared with the nuclear genome, they are very small. For example, the amount of DNA in the human nuclear genome (about 3 billion base pairs) is about 200,000 times greater than the amount in the mitochondrial genome. In terms of genes, the human genome has approximately 22,000 different protein-encoding genes, whereas the human mitochondrial genome has only about a dozen protein-encoding genes. Chloroplast genomes tend to be larger than mitochondrial genomes, and they have a correspondingly greater number of genes. Depending on the particular species of plant or algae, a chloroplast genome is about 10 times larger than the mitochondrial genome of human cells.

Just as the genomes of mitochondria and chloroplasts resemble bacterial genomes, the production of new mitochondria and chloroplasts bears a striking resemblance to the division of bacterial cells. Like their bacterial counterparts, mitochondria and chloroplasts increase in number via **binary fission**, or splitting in two. **Figure 4.30** illustrates the process for a mitochondrion. The mitochondrial chromosome, which is found in a region called the nucleoid, is duplicated, and the organelle divides into two separate organelles. Mitochondrial and chloroplast divisions are needed to maintain a full complement of these organelles when cell growth occurs following cell division. In addition, environmental conditions may influence the sizes and numbers of these organelles. For example, when plants are exposed to more sunlight, the number of chloroplasts in leaf cells increases.

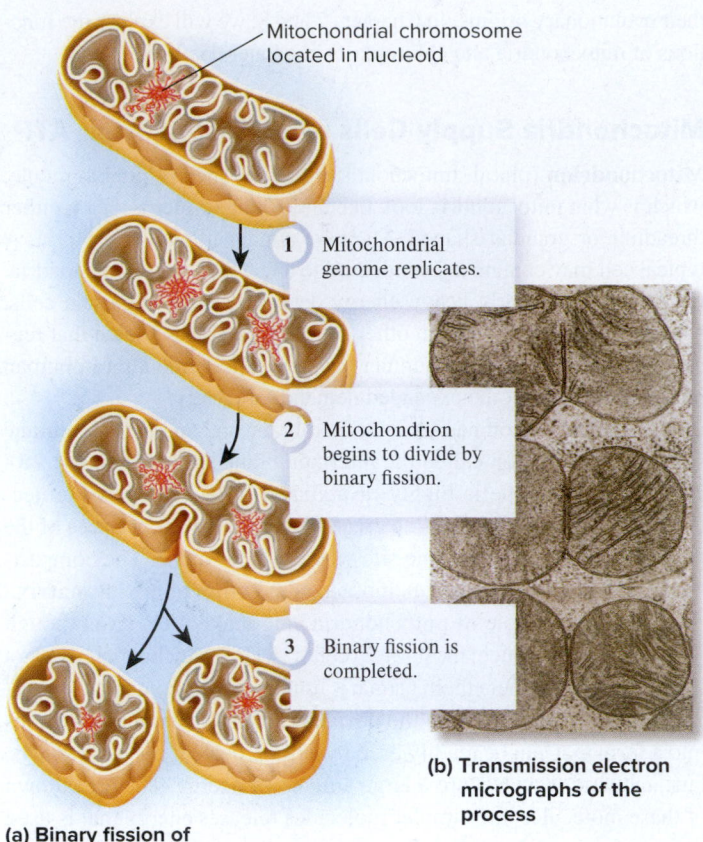

Mitochondrial chromosome located in nucleoid

1. Mitochondrial genome replicates.

2. Mitochondrion begins to divide by binary fission.

3. Binary fission is completed.

**(a) Binary fission of mitochondria**

**(b) Transmission electron micrographs of the process**

**Figure 4.30**  **Division of mitochondria by binary fission.** b: ©Don W. Fawcett/Science Source

 **Core Skill: Connections**  Look ahead to Figure 19.13. How is the process of binary fission similar to bacterial cell division, and how is it different?

## Mitochondria and Chloroplasts Are Derived from Ancient Symbiotic Relationships

The observation that mitochondria and chloroplasts contain their own genetic material may seem puzzling. Perhaps you might think that it would be simpler for a eukaryotic cell to have all of its genetic material in one place—the nucleus. The distinct genomes of mitochondria and chloroplasts can be traced to their evolutionary origin, which involved an ancient symbiotic association.

A symbiotic relationship occurs when two different species live in direct contact with each other. **Endosymbiosis** describes a symbiotic relationship in which the smaller species—the symbiont—lives inside the larger species. In 1883, Andreas Schimper proposed that chloroplasts evolved from an endosymbiotic relationship between cyanobacteria (a bacterium capable of photosynthesis) and eukaryotic cells. In 1922, Ivan Wallin also hypothesized an endosymbiotic origin for mitochondria.

In spite of these interesting proposals, the question of whether endosymbiosis gave rise to mitochondria and chloroplasts was largely ignored until the discovery that these organelles contain their own genetic material. In 1970, the idea of endosymbiosis as the origin of mitochondria and chloroplasts was revived by Lynn Margulis in her

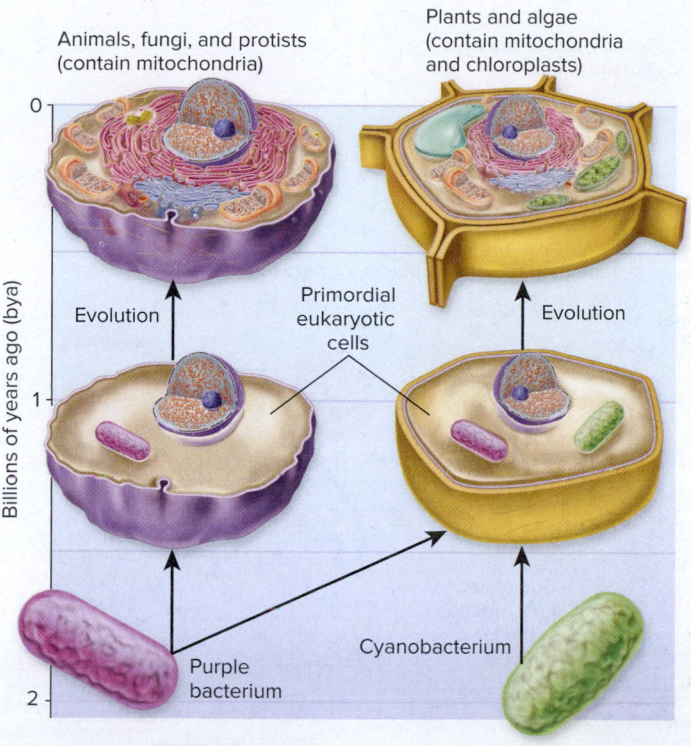

**(a) Mitochondria originated from endosymbiotic purple bacteria.**

**(b) Chloroplasts originated from endosymbiotic cyanobacteria.**

**Figure 4.31  A simplified view of the endosymbiosis theory.**

**Core Concept: Evolution (a)** According to the endosymbiosis theory, modern mitochondria were derived from purple bacteria, also called α-proteobacteria. Over the course of evolution, their characteristics evolved into those found in mitochondria today. **(b)** A similar phenomenon occurred for chloroplasts, which were derived from cyanobacteria (blue-green bacteria), which are capable of photosynthesis.

book *Origin of Eukaryotic Cells*. During the 1970s and 1980s, the advent of molecular genetic techniques allowed researchers to analyze genes from mitochondria, chloroplasts, bacteria, and eukaryotic nuclear genomes. Researchers discovered that genes in mitochondria and chloroplasts are very similar to bacterial genes. Likewise, mitochondria and chloroplasts are strikingly similar in size and shape to certain bacterial species. These observations provided strong support for the **endosymbiosis theory**, which proposes that mitochondria and chloroplasts originated from bacteria that took up residence within primordial eukaryotic cells (**Figure 4.31**). Over the next 2 billion years, the characteristics of these intracellular bacterial cells gradually changed to those of mitochondria or chloroplasts. The origin of eukaryotic cells is discussed in more detail in Chapter 26.

Symbiosis occurs because the relationship is beneficial to one or both species. According to the endosymbiosis theory, such a relationship provided eukaryotic cells with useful cellular characteristics. Chloroplasts, which were derived from cyanobacteria, have the ability to carry out photosynthesis. This benefits plant cells by giving them the ability to use the energy from sunlight. By comparison,

mitochondria are thought to have been derived from a different type of bacteria known as purple bacteria, or α-proteobacteria. In this case, the endosymbiotic relationship enabled eukaryotic cells to synthesize greater amounts of ATP. How the relationship would have been beneficial to a cyanobacterium or purple bacterium is less clear, though the cytosol of a eukaryotic cell may have provided a stable environment with an adequate supply of nutrients.

During the evolution of eukaryotic species, many genes that were originally found in the genomes of the primordial purple bacteria and cyanobacteria have been transferred from the organelles to the nucleus. This has occurred many times throughout evolution, so modern mitochondria and chloroplasts have lost most of the genes that still exist in present-day purple bacteria and cyanobacteria. Some researchers speculate that the movement of genes into the nucleus makes it easier for the cell to control the structure, function, and division of mitochondria and chloroplasts. In modern cells, hundreds of different proteins that make up these organelles are encoded by genes that have been transferred to the nucleus. These proteins are made in the cytosol and then taken up into mitochondria or chloroplasts. We will discuss this topic next.

## 4.7  Protein Sorting to Organelles

**Learning Outcomes:**

1. List the categories of proteins that are sorted cotranslationally and post-translationally.
2. Describe the steps that occur during the cotranslational sorting of proteins to the endoplasmic reticulum.
3. Outline the steps of post-translational sorting of proteins to mitochondria.

As we have seen, eukaryotic cells contain a variety of membrane-bound organelles. Each protein that a cell makes usually functions within one cellular compartment or is secreted from the cell. How does each protein reach its appropriate destination? For example, how does a mitochondrial protein get sent to the mitochondrion rather than to a different organelle such as a lysosome? In eukaryotes, most proteins contain short stretches of amino acid sequences that direct them to their correct cellular location. These sequences are called **sorting signals**, or **traffic signals**. Each sorting signal is recognized by specific cellular components that facilitate the proper movement of the protein carrying that signal to its correct location.

The synthesis of most eukaryotic proteins begins on ribosomes in the cytosol, using messenger RNA (mRNA) that contains the information for polypeptide synthesis (**Figure 4.32**). The cytosol provides amino acids, which are used as building blocks to make the proteins during translation. Cytosolic proteins lack any sorting signal, so they remain there. By comparison, the synthesis of proteins destined for the ER, Golgi, lysosomes, vacuoles, or secretory vesicles begins in the cytosol and then halts temporarily until the ribosome has become bound to the ER membrane. After this occurs, translation resumes and the polypeptide is synthesized into the ER. Proteins that are destined for the ER, Golgi, lysosomes, vacuoles, plasma membrane, or secretion are first directed to the ER. This is called **cotranslational sorting** because the first step in the sorting process begins while translation is occurring. In contrast, the uptake of most proteins into the nucleus,

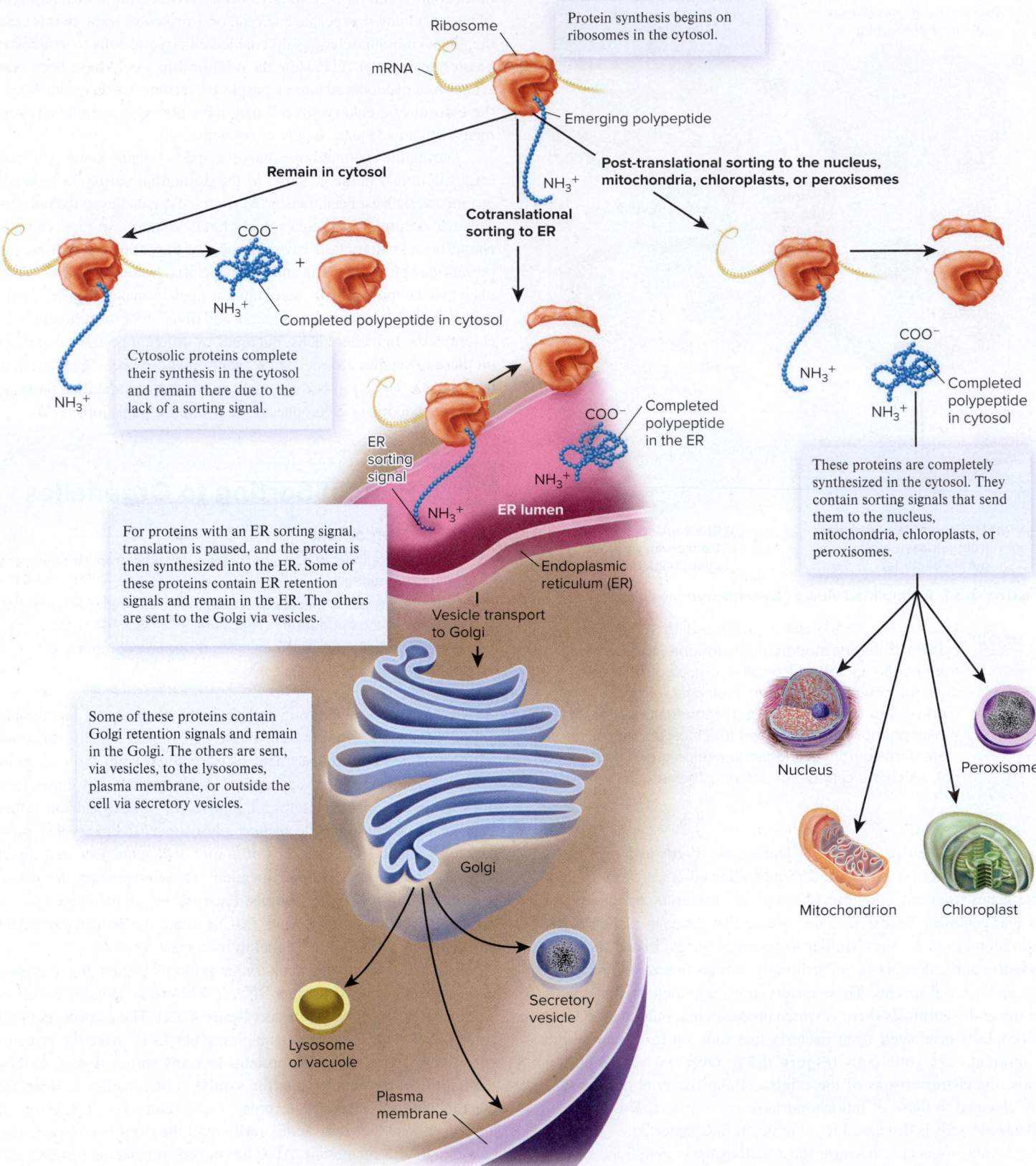

Protein synthesis begins on ribosomes in the cytosol.

Ribosome

mRNA

Emerging polypeptide

**Remain in cytosol**

**Post-translational sorting to the nucleus, mitochondria, chloroplasts, or peroxisomes**

**Cotranslational sorting to ER**

COO⁻

NH₃⁺

+

NH₃⁺

Completed polypeptide in cytosol

NH₃⁺

Cytosolic proteins complete their synthesis in the cytosol and remain there due to the lack of a sorting signal.

Completed polypeptide in the ER

COO⁻

NH₃⁺

NH₃⁺

Completed polypeptide in cytosol

ER sorting signal

NH₃⁺

**ER lumen**

These proteins are completely synthesized in the cytosol. They contain sorting signals that send them to the nucleus, mitochondria, chloroplasts, or peroxisomes.

For proteins with an ER sorting signal, translation is paused, and the protein is then synthesized into the ER. Some of these proteins contain ER retention signals and remain in the ER. The others are sent to the Golgi via vesicles.

Endoplasmic reticulum (ER)

Vesicle transport to Golgi

Some of these proteins contain Golgi retention signals and remain in the Golgi. The others are sent, via vesicles, to the lysosomes, plasma membrane, or outside the cell via secretory vesicles.

Nucleus

Peroxisome

Golgi

Mitochondrion

Chloroplast

Secretory vesicle

Lysosome or vacuole

Plasma membrane

**Figure 4.32  Three pathways for protein sorting in a eukaryotic cell.** Proteins either remain in the cytosol, are sorted to the ER (cotranslational sorting), or are sorted after they are completely synthesized (post-translational sorting).

mitochondria, chloroplasts, and peroxisomes occurs after the protein is completely made (that is, completely translated) in the cytosol. This is called **post-translational sorting** because it does not happen until translation is finished. In this section, we will consider how cells carry out cotranslational and post-translational sorting.

## The Cotranslational Sorting of Some Proteins Occurs at the Endoplasmic Reticulum Membrane

The concept of sorting signals in proteins was first proposed by Günter Blobel in the 1970s. Blobel and colleagues discovered a sorting signal in proteins that sends them to the ER membrane, which is the first step in cotranslational sorting (**Figure 4.33**, also see Figure 4.32). To be directed to the rough ER membrane, a polypeptide must contain a sorting signal called an **ER signal sequence**, which is a sequence of about 6–12 amino acids that are predominantly hydrophobic and usually located near the N-terminus. As the ribosome is making the polypeptide in the cytosol, the ER signal sequence emerges from the ribosome and is recognized by a protein-RNA complex called **signal recognition particle (SRP)**. SRP has two functions. First, it recognizes the ER signal sequence and pauses translation. Second, SRP binds to an SRP receptor in the ER membrane, which docks the ribosome over a channel. At this stage, SRP is released and translation resumes. The growing polypeptide is threaded through the channel to cross the ER membrane. If the protein is not a membrane protein, it will be released into the lumen of the ER. In most cases, the ER signal sequence is removed by an enzyme, signal peptidase. In 1999, Blobel won the Nobel Prize in Physiology or Medicine for his discovery of sorting signals in proteins. The process shown in Figure 4.33 illustrates another important role of protein-protein interactions—a series of interactions causes the steps of a process to occur in a specific order.

Some proteins are meant to function in the ER. Such proteins contain ER retention signals in addition to the ER signal sequence. Alternatively, other proteins that are destined for the Golgi, lysosomes, vacuoles, plasma membrane, or secretion leave the ER and are transported to their correct location. This transport process occurs via vesicles that are formed from one compartment and then move through the cytosol and fuse with another compartment. Vesicles from the ER may go to the Golgi, and then vesicles from the Golgi may go to the lysosomes, vacuoles, or plasma membrane. Sorting signals within proteins' amino acid sequences are responsible for directing them to the correct location.

## Proteins Are Sorted Post-translationally to the Nucleus, Peroxisomes, Mitochondria, and Chloroplasts

The organization and function of the nucleus, peroxisomes, mitochondria, and chloroplasts depend on the uptake of proteins from the cytosol. Most of these proteins are synthesized in the cytosol and then taken up into their respective organelles. For example, most proteins involved in ATP synthesis are made in the cytosol and taken up into mitochondria after they have been completely synthesized. For this to occur, a protein must have the appropriate sorting signal as part of its amino acid sequence.

As one example of post-translational sorting, let's consider how a protein is directed to the mitochondrial matrix. Such a protein has a short amino acid sequence at the N-terminus called a matrix-targeting sequence. As shown in **Figure 4.34**, the process of protein import into the matrix involves a series of intricate protein-protein interactions. A protein destined for the mitochondrial matrix is first made in the cytosol, where proteins called **chaperones** keep it in an unfolded state. A

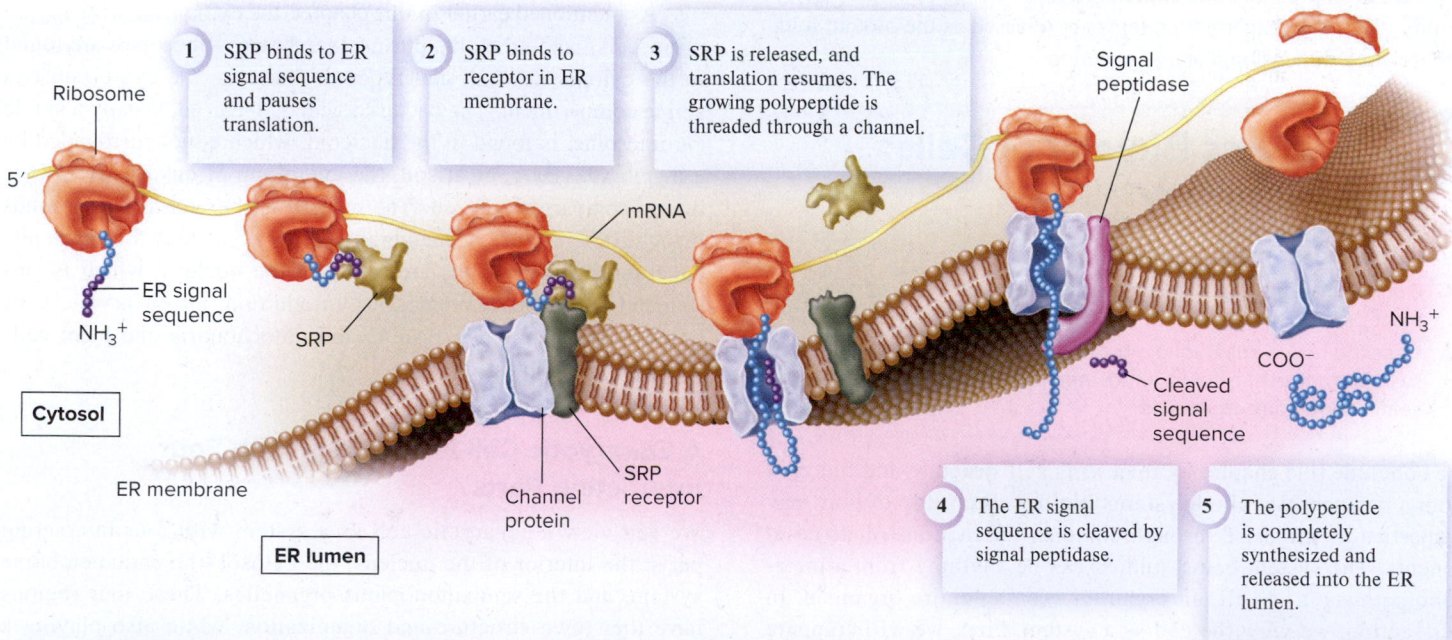

**Figure 4.33** First step in cotranslational sorting: sending proteins to the ER.

*Concept Check:* *What prevents a protein destined for the ER from being completely synthesized in the cytosol?*

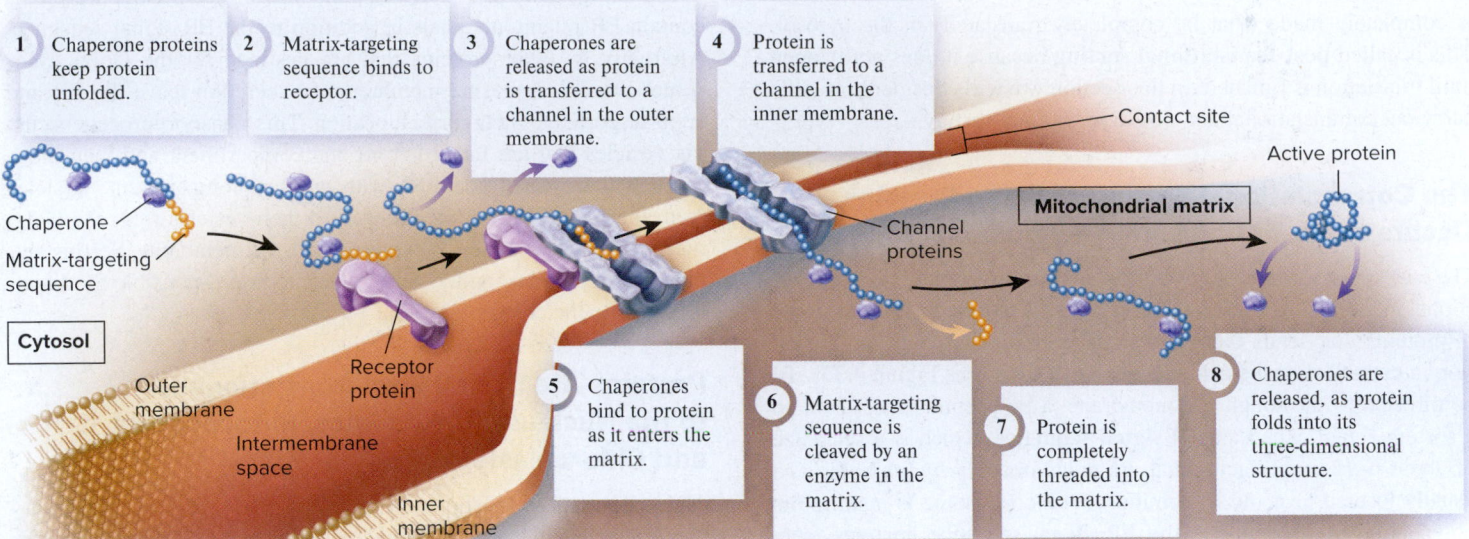

**Figure 4.34** Post-translational sorting of a protein to the mitochondrial matrix.

*Concept Check:* *What do you think would happen if chaperone proteins did not bind to a protein before it was imported into the mitochondrial matrix?*

receptor protein in the outer mitochondrial membrane recognizes the matrix-targeting sequence. The protein is released from the chaperone as it is transferred to a channel in the outer mitochondrial membrane. Because it is in an unfolded state, the mitochondrial protein can be threaded through this channel, and then through another channel in the inner mitochondrial membrane. These channels lie close to each other at contact sites between the outer and inner membranes. As the protein emerges in the matrix, other chaperone proteins already in the matrix continue to keep it unfolded. Eventually, the matrix-targeting sequence is cleaved, and the entire protein is threaded into the matrix. At this stage, the chaperone proteins are released as the protein folds into its three-dimensional active structure.

# 4.8 Systems Biology of Cells: A Summary

## Learning Outcomes:

1. Outline the differences in complexity among bacteria, animal, and plant cells.
2. Describe how a eukaryotic cell can be viewed as four interacting systems: nucleus, cytosol, endomembrane system, and semiautonomous organelles.

We conclude this chapter by reviewing cell structure and function from a perspective called **systems biology**, the study of how new properties of life arise through complex interactions of its components. The system being studied can be anything from a metabolic pathway to a cell, an organ, or even an entire organism. In this section, we view the cell as a system. First, we will compare prokaryotic and eukaryotic cells as systems, and then examine the four interconnected parts that make up the system that is the eukaryotic cell.

## Bacterial Cells Are Relatively Simple Systems Compared to Eukaryotic Cells

Bacterial cells are relatively small and lack the extensive internal compartmentalization characteristic of eukaryotic cells (**Table 4.2**). On the outside, bacterial cells are surrounded by a cell wall, and many species have flagella. Animal cells lack a cell wall, and only certain cell types have flagella or cilia. Like bacteria, plant cells also have cell walls but the chemical composition of these walls is different from that of bacterial cells. Plant cells rarely have flagella.

As mentioned earlier in this chapter, the cytoplasm is the region of the cell enclosed by the plasma membrane. Ribosomes are found in the cytoplasm of all cell types. In bacteria, the cytoplasm is a single compartment. The bacterial genetic material, usually a single chromosome, is found in the nucleoid, which is not surrounded by a membrane. By comparison, the cytoplasm of eukaryotic cells is highly compartmentalized. The cytosol is the area that surrounds many different types of membrane-bound organelles. For example, eukaryotic chromosomes are found in the nucleus, which is surrounded by a double membrane. In addition, all eukaryotic cells have an endomembrane system, and mitochondria and plant cells also have chloroplasts.

## A Eukaryotic Cell Is a System with Four Interacting Parts

We can view a eukaryotic cell as a system with four interacting parts: the interior of the nucleus, the cytosol, the endomembrane system, and the semiautonomous organelles. These four regions have their own structure and organization, while also playing a role in the structure and organization of the entire cell. The structures and functions of these four interacting parts are described in **Figure 4.35**.

| Table 4.2 | A Comparison of Cell Complexity Among Bacterial, Animal, and Plant Cells | | |
|-----------|---------|---------|---------|
| **Structures** | **Bacteria** | **Animal cells** | **Plant cells** |
| *Extracellular structures* | | | |
| Cell wall* | Present | Absent | Present |
| Flagella/cilia | Flagella sometimes present | Cilia or flagella present on certain cell types | Rarely present† |
| Plasma membrane | Present | Present | Present |
| *Interior structures* | | | |
| Cytoplasm | Usually a single compartment inside the plasma membrane | Composed of membrane-bound organelles that are surrounded by the cytosol | Composed of membrane-bound organelles that are surrounded by the cytosol |
| Ribosomes | Present | Present | Present |
| Chromosomes | Typically one circular chromosome per nucleoid; a nucleoid is not a membrane-bound compartment. | Multiple linear chromosomes in the nucleus, which is surrounded by a double membrane. Mitochondria also have chromosomes. | Multiple linear chromosomes in the nucleus, which is surrounded by a double membrane. Mitochondria and chloroplasts also have chromosomes. |
| Endomembrane system | Absent | Present | Present |
| Mitochondria | Absent | Present | Present |
| Chloroplasts | Absent | Absent | Present |

*Note that the biochemical composition of bacterial cell walls is very different from plant cell walls.
†Some plant species produce sperm cells with flagella, but flowering plants produce sperm within pollen grains that lack flagella.

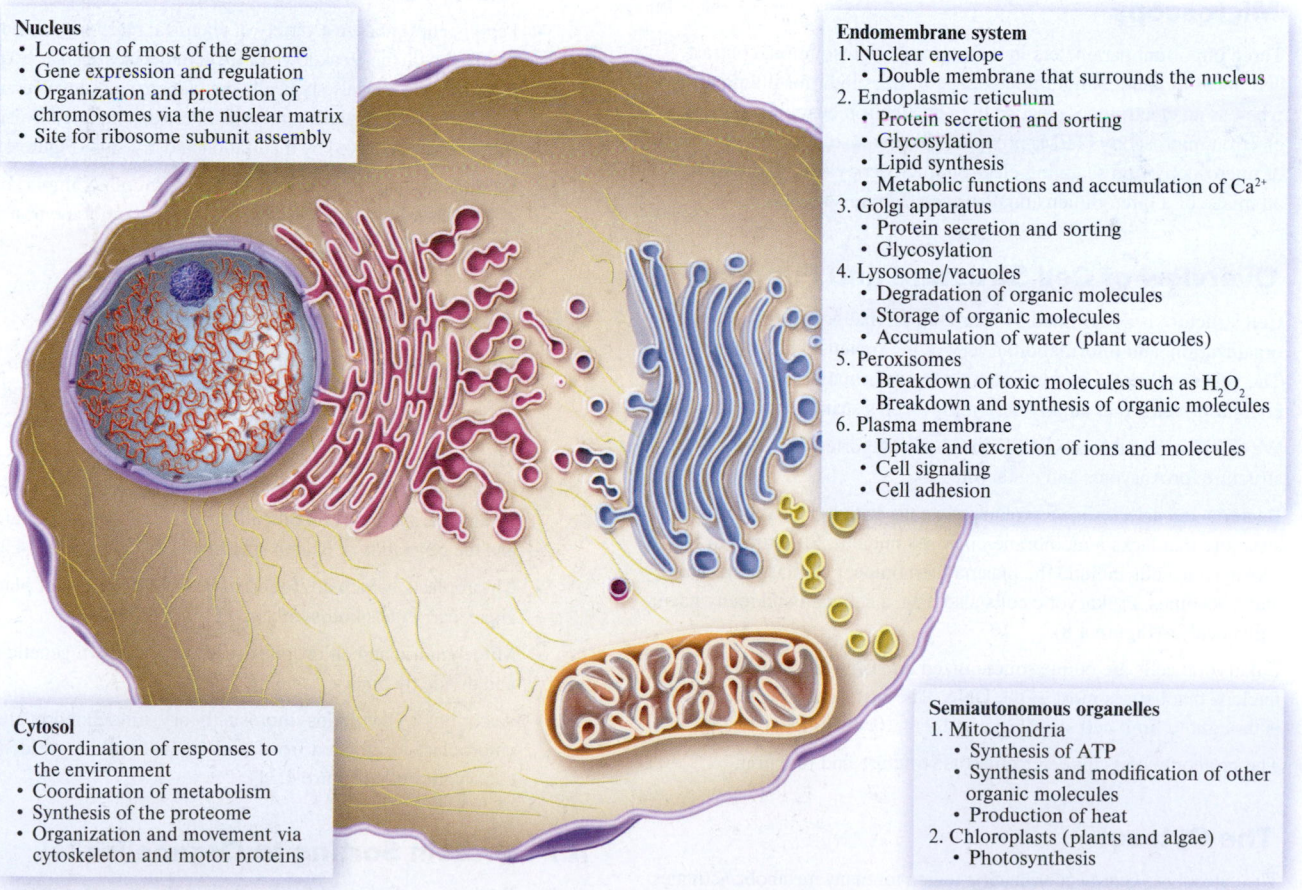

**Nucleus**
- Location of most of the genome
- Gene expression and regulation
- Organization and protection of chromosomes via the nuclear matrix
- Site for ribosome subunit assembly

**Endomembrane system**
1. Nuclear envelope
   - Double membrane that surrounds the nucleus
2. Endoplasmic reticulum
   - Protein secretion and sorting
   - Glycosylation
   - Lipid synthesis
   - Metabolic functions and accumulation of $Ca^{2+}$
3. Golgi apparatus
   - Protein secretion and sorting
   - Glycosylation
4. Lysosome/vacuoles
   - Degradation of organic molecules
   - Storage of organic molecules
   - Accumulation of water (plant vacuoles)
5. Peroxisomes
   - Breakdown of toxic molecules such as $H_2O_2$
   - Breakdown and synthesis of organic molecules
6. Plasma membrane
   - Uptake and excretion of ions and molecules
   - Cell signaling
   - Cell adhesion

**Cytosol**
- Coordination of responses to the environment
- Coordination of metabolism
- Synthesis of the proteome
- Organization and movement via cytoskeleton and motor proteins

**Semiautonomous organelles**
1. Mitochondria
   - Synthesis of ATP
   - Synthesis and modification of other organic molecules
   - Production of heat
2. Chloroplasts (plants and algae)
   - Photosynthesis

**Figure 4.35** **The four interacting parts of eukaryotic cells: nucleus, cytosol, endomembrane system, and semiautonomous organelles.** An animal cell is illustrated here.

 **Core Concept: Systems** A eukaryotic cells is a system with four interacting parts.

# Summary of Key Concepts

## 4.1   Origin of Living Cells on Earth

- Life on Earth is hypothesized to have arisen in four overlapping stages. The first stage involved the synthesis of organic molecules to form a prebiotic soup. Possible scenarios as to how this occurred are the reducing atmosphere, extraterrestrial, and deep-sea vent hypotheses (Figures 4.1, 4.2).

- The second stage was the formation of polymers from simple organic molecules. This may have occurred on the surface of clay.

- The third stage involved the emergence of protobionts, which were aggregates of polymers with a boundary that separated them from the external environment (Figure 4.3).

- In the fourth stage, polymers enclosed in membranes acquired properties of cells, such as self-replication and catalytic functions (Figure 4.4).

- During the hypothetical period called the RNA world, the first living cells used RNA for both information storage and catalytic functions.

- Bartel and Szostak demonstrated experimentally that chemical selection for RNA molecules that can catalyze covalent bond formation is possible.

- The RNA world was eventually superseded by the modern DNA/RNA/protein world.

## 4.2   Microscopy

- Three important parameters in microscopy are resolution, contrast, and magnification. A light microscope utilizes light for illumination, whereas an electron microscope uses an electron beam. Transmission electron microscopy (TEM) provides the best resolution of any form of microscopy, and scanning electron microscopy (SEM) produces an image of a three-dimensional surface (Figures 4.5, 4.6, 4.7).

## 4.3   Overview of Cell Structure and Function

- Cell structure is determined by four factors: matter, energy, organization, and information. Every living organism has a genome. The genes within the genome contain the information to produce the cellular proteins that largely determine a cell's structure and function.

- We can classify all forms of life into two categories based on cell structure: prokaryotes and eukaryotes.

- Bacteria and archaea have prokaryotic cells with a relatively simple structure that lacks a membrane-enclosed nucleus. Structures in prokaryotic cells include the plasma membrane, cytoplasm, nucleoid, and ribosomes. Prokaryotic cells also have a cell wall and many have a glycocalyx (Figure 4.8).

- Eukaryotic cells are compartmentalized into organelles and contain a nucleus that houses most of the DNA. The surface area/volume ratio is thought to limit cell size (Figures 4.9, 4.10, 4.11, 4.12).

- The proteome of a cell determines its structure and function.

## 4.4   The Cytosol

- The cytosol is a central coordinating region for many metabolic activities of eukaryotic cells, including polypeptide synthesis (Figure 4.13).

- The cytoskeleton is a network of three different types of protein filaments: microtubules, intermediate filaments, and actin filaments. Microtubules are important for cell shape, organization, and

movement. Intermediate filaments help maintain cell shape, rigidity, and strength. Actin filaments support the plasma membrane and play a key role in cell strength, shape, and movement (Table 4.1, Figures 4.14, 4.15, 4.16, 4.17).

## 4.5   The Nucleus and Endomembrane System

- The primary function of the nucleus is the organization and expression of the genetic material. A second important function is the assembly of ribosomal subunits in the nucleolus (Figures 4.18, 4.19, 4.20).

- The endomembrane system includes the nuclear envelope, endoplasmic reticulum (ER), Golgi apparatus, lysosomes, vacuoles, peroxisomes, and plasma membrane. The rough endoplasmic reticulum (rough ER) plays a key role in the initial sorting of proteins. The smooth endoplasmic reticulum (smooth ER) functions in metabolic processes such as detoxification, carbohydrate metabolism, accumulation of calcium ions, and synthesis and modification of lipids. The Golgi apparatus performs three overlapping functions: processing, protein sorting, and secretion. Lysosomes degrade macromolecules and help digest substances taken up from outside the cell (endocytosis) and inside the cell (autophagy) (Figures 4.21, 4.22).

- Palade's pulse-chase experiments demonstrated that secreted proteins move sequentially through the ER and Golgi apparatus (Figure 4.23).

- Types of vacuoles include central vacuoles, contractile vacuoles, and food or phagocytic vacuoles (Figure 4.24).

- Peroxisomes catalyze a variety of chemical reactions, including those involved with the breakdown of toxic molecules such as hydrogen peroxide, and they also typically contain enzymes involved in the metabolism of fats and amino acids. Peroxisomes are formed by budding from the ER, followed by maturation and division (Figure 4.25).

- Proteins in the plasma membrane perform many important roles that affect activities inside the cell, including membrane transport, cell signaling, and cell adhesion (Figure 4.26).

## 4.6   Semiautonomous Organelles

- Mitochondria and chloroplasts are considered semiautonomous organelles because they grow and divide, but still depend on other parts of the cell for their internal components (Figure 4.27).

- Mitochondria produce most of a cell's ATP, which is utilized by many proteins to carry out their functions. Other mitochondrial functions include the synthesis, modification, and breakdown of cellular molecules and the generation of heat in specialized fat cells (Figure 4.28).

- Chloroplasts, which are found in nearly all species of plants and algae, carry out photosynthesis (Figure 4.29).

- Mitochondria and chloroplasts contain their own genetic material and divide by binary fission (Figure 4.30).

- According to the endosymbiosis theory, mitochondria and chloroplasts originated from bacteria that took up residence in early eukaryotic cells (Figure 4.31).

## 4.7   Protein Sorting to Organelles

- Proteins synthesized in eukaryotic cells are sorted to their correct cellular destination (Figure 4.32).

- The cotranslational sorting of proteins to the ER, Golgi, lysosomes, vacuoles, plasma membrane, and secretory vesicles begins in the cytosol, while translation is occurring, and involves sorting signals and vesicle transport (Figure 4.33).

- Most proteins destined for the nucleus, mitochondria, chloroplasts, and peroxisomes are synthesized in the cytosol and taken up after synthesis is complete; this is called post-translational sorting (Figure 4.34).

## 4.8   Systems Biology of Cells: A Summary

- Systems biology is the study of how new properties of life arise by complex interactions of its components. In systems biology, the cell is viewed in terms of its structural and functional connections, rather than its individual molecular components.

- Prokaryotic and eukaryotic cells differ in their levels of organization. In eukaryotic cells, four parts—nucleus, cytosol, endomembrane system, and semiautonomous organelles—work together to produce dynamic organization (Table 4.2, Figure 4.35).

## Assess & Discuss

### Test Yourself

1. The cell theory states that
   a. all living things are composed of cells.
   b. cells are the smallest units of living organisms.
   c. new cells come from pre-existing cells by cell division.
   d. all of the above.
   e. only a and b are true.

2. For a microscope, resolution refers to
   a. the ratio between the size of the image produced by the microscope and the actual size of the object.
   b. the degree to which a particular structure looks different from other structures around it.
   c. how well a structure takes up certain dyes.
   d. the ability to observe two adjacent objects as being distinct from each other.
   e. the degree to which the image is magnified.

3. A spherical cell has a radius of 34 μm. What is its surface area/volume ratio?
   a. 0.088
   b. 0.12
   c. 11.3
   d. 55.7
   e. 127

4. If a motor protein was held in place and a cytoskeletal filament was free to move, what type of motion would occur when the motor protein was active?
   a. The motor protein would "walk" along the filament.
   b. The filament would move.
   c. The filament would bend.
   d. all of the above
   e. Only b and c would happen.

5. Each of the following is part of the endomembrane system except
   a. the nuclear envelope.
   b. the endoplasmic reticulum.
   c. the Golgi apparatus.
   d. lysosomes.
   e. mitochondria.

6. Vesicle transport occurs between the ER and the Golgi in both directions. Let's suppose a researcher exposed some cells to a drug that inhibited vesicle transport from the Golgi to the ER but did not affect vesicle transport from the ER to the Golgi. If you observed the cells microscopically after the drug was added, what would you expect to see happen over the course of 1 hour?
   a. The ER would get smaller, and the Golgi would get larger.
   b. The ER would get larger, and the Golgi would get smaller.
   c. The ER and Golgi would stay the same size.
   d. Both the ER and Golgi would get larger.
   e. Both the ER and Golgi would get smaller.

7. Functions of the smooth endoplasmic reticulum include
   a. detoxification of harmful organic molecules.
   b. metabolism of carbohydrates.
   c. protein sorting.
   d. all of the above.
   e. a and b only.

8. The central vacuole in many plant cells is important for
   a. storage.
   b. photosynthesis.
   c. structural support.
   d. all of the above.
   e. a and c only.

9. Let's suppose an abnormal protein contains three sorting signals: an ER signal sequence, an ER retention sequence, and a mitochondrial matrix-targeting sequence. The ER retention sequence is supposed to keep a protein within the ER. Where would you expect this abnormal protein to go? Note: Think carefully about the timing of events in protein sorting and which events occur cotranslationally and which occur post-translationally.
   a. It would go to the ER.
   b. It would go the mitochondria.
   c. It would go to both the ER and mitochondria equally.
   d. It would remain in the cytosol.
   e. It would be secreted.

10. Which of the following observations would *not* be considered evidence for the endosymbiosis theory?
    a. Mitochondria and chloroplasts have genomes that resemble smaller versions of bacterial genomes.
    b. Mitochondria, chloroplasts, and bacteria all divide by binary fission.
    c. Mitochondria, chloroplasts, and bacteria all have ribosomes.
    d. Mitochondria, chloroplasts, and bacteria all have similar sizes and shapes.
    e. All of the above are considered evidence for the theory.

### Conceptual Questions

1. What are the four stages that led to the origin of living cells?

2. Explain how motor proteins and cytoskeletal filaments interact to promote three different types of movements: movement of a cargo, movement of a filament, and bending of a filament.

3. **Core Concept: Structure and Function** A core concept of biology is that *structure determines function*. Explain how the invaginations of the inner mitochondrial membrane are related to mitochondrial function.

### Collaborative Questions

1. Discuss the roles of the genome and proteome in determining cell structure and function.

2. Discuss and draw the structural relationship between the nucleus, the rough endoplasmic reticulum, and the Golgi apparatus.

# Membrane Structure, Synthesis, and Transport

# 5

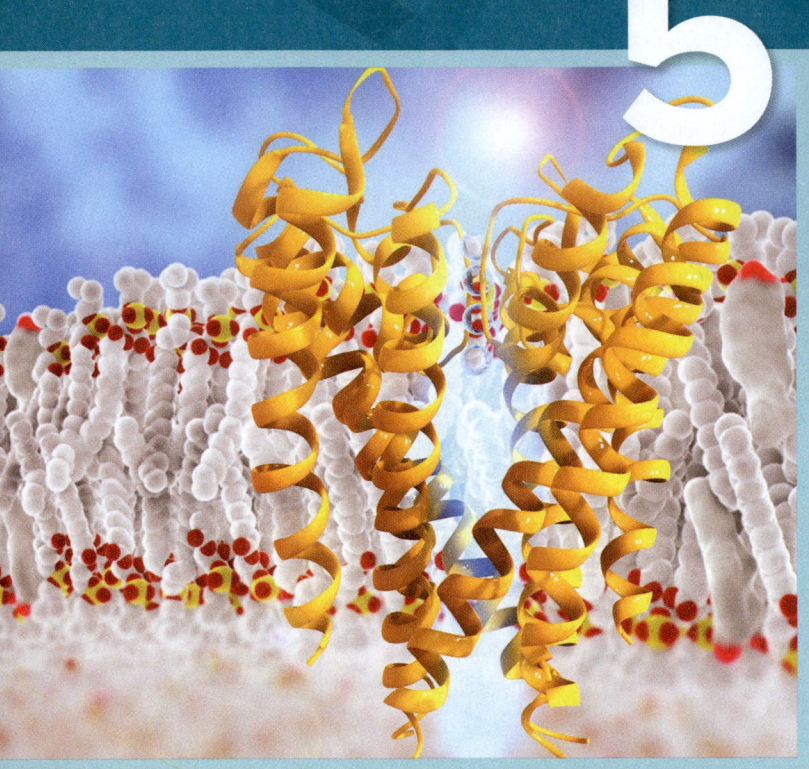

This is a model of a membrane in which a protein (shown in yellow) is embedded in a bilayer of lipids. The protein functions as a channel that allows ions to cross the membrane.

©Ramón Andrade, 3Dciencia/Science Source

**W**hen he was 28, Andrew began to develop a combination of symptoms that included fatigue, joint pain, abdominal pain, and a loss of sex drive. His doctor conducted some tests and discovered that Andrew had abnormally high levels of iron in his body. Iron is a mineral found in many foods. Andrew was diagnosed with a genetic disease called hemochromatosis, which caused him to absorb more iron than he needed. This was due to an overactive protein involved in the transport of iron across the membranes of intestinal cells and into the bloodstream. Unfortunately, when the human body takes up too much iron, it is stored in body tissues, especially the liver, heart, pancreas, and joints. The extra iron can damage a person's organs.

In Andrew's case, the disease was caught relatively early, and treatment—which includes a modification in diet along with medication that inhibits the absorption of iron—prevented more severe symptoms. Without treatment, however, hemochromatosis can cause organ failure. Later signs and symptoms include skin discoloration, arthritis, liver disease, diabetes mellitus, and heart failure.

The disease hemochromatosis illustrates the importance of membranes in regulating the traffic of ions and molecules into and out of cells.

All cells have a **plasma membrane** that encloses the cytoplasm, and eukaryotic cells have internal membranes that surround organelles (see Chapter 4). Both types are also called **biological membranes**. The plasma membrane separates the internal contents of a cell from its external environment. With such a role, you might imagine that the plasma membrane would be thick and rigid. Remarkably, the opposite is true. All biological membranes, including the plasma membrane, are thin (typically 5–10 nm) and somewhat fluid. It would take 5,000–10,000 of these membranes stacked on top of each other to equal the thickness of a piece of paper! Despite their thinness, however, membranes are impressively dynamic structures that effectively maintain the separation between a cell and its surroundings and also provide interfaces where many vital cellular activities occur (**Table 5.1**).

In this chapter, we will begin by considering the components that provide the structure and fluid properties of membranes and then explore how they are synthesized. Finally, we will examine one of a membrane's primary functions—membrane transport. Biological membranes regulate the traffic of substances into and out of the cell and its organelles. As you will learn, this occurs via transport proteins and via exocytosis and endocytosis.

| Table 5.1 | Important Functions of Biological Membranes |
|---|---|
| **Function** | |
| Selective uptake and export of ions and molecules | |
| Cell compartmentalization | |
| Protein sorting | |
| Anchoring of the cytoskeleton | |
| Production of energy intermediates such as ATP and NADPH | |
| Cell signaling | |
| Cell and nuclear division | |
| Adhesion of cells to each other and to the extracellular matrix | |

## 5.1  Membrane Structure

### Learning Outcomes:

1. Describe the fluid-mosaic model of membrane structure.
2. Identify the three different types of membrane proteins.

The two primary components of membranes are phospholipids, which form the basic matrix of a membrane, and proteins, which are embedded in the membrane or loosely attached to its surface. A third component is carbohydrates, which may be attached to membrane lipids and proteins. In this section, we will examine the organization of these components to form a biological membrane and their importance in the overall function of membranes.

### Biological Membranes Are a Mosaic of Lipids, Proteins, and Carbohydrates

**Figure 5.1** shows the biochemical organization of a membrane, which is similar in composition among all living organisms. The framework of the membrane is the **phospholipid bilayer**, which consists of two layers of phospholipids. Recall from Chapter 3 that phospholipids are **amphipathic** molecules. They have hydrophobic (water-fearing) or nonpolar tails, and also a hydrophilic (water-loving) or polar head. The nonpolar tails of the lipids are found in the interior of the membrane, and the polar heads are on the surface. Biological membranes also contain proteins, and most membranes have carbohydrates attached to lipids and proteins. Overall, the membrane is considered a mosaic of lipid, protein, and carbohydrate molecules.

The membrane structure illustrated in Figure 5.1 is referred to as the **fluid-mosaic model**, originally proposed by S. Jonathan Singer and Garth Nicolson in 1972. As discussed later, the membrane exhibits properties that resemble a fluid because lipids and proteins can move relative to each other within the membrane.

Half of a phospholipid bilayer is termed a **leaflet**. Each leaflet faces a different region. For example, the plasma membrane contains a cytosolic leaflet and an extracellular leaflet (see Figure 5.1). With regard to lipid composition, the two leaflets of membranes are asymmetrical. Certain types of lipids may be more abundant in one leaflet compared to the other. A striking asymmetry occurs with glycolipids—lipids with a carbohydrate attached. These are found primarily in the extracellular leaflet of the plasma membrane. The carbohydrate portion of a glycolipid protrudes into the extracellular medium.

### Proteins Associate with Membranes in Three Different Ways

Although the phospholipid bilayer forms the basic foundation of cellular membranes, the protein component carries out many key functions. Some of these functions were considered in Chapter 4. For example, we saw how membrane proteins in the smooth ER membrane function as enzymes that break down glycogen. Later in this chapter, we will explore how membrane proteins are involved in transporting ions and molecules across membranes. In other chapters, we will examine how membrane proteins are responsible for other functions, including ATP synthesis (Chapter 7), photosynthesis (Chapter 8), cell signaling (Chapter 9), and cell-to-cell adhesion (Chapter 10).

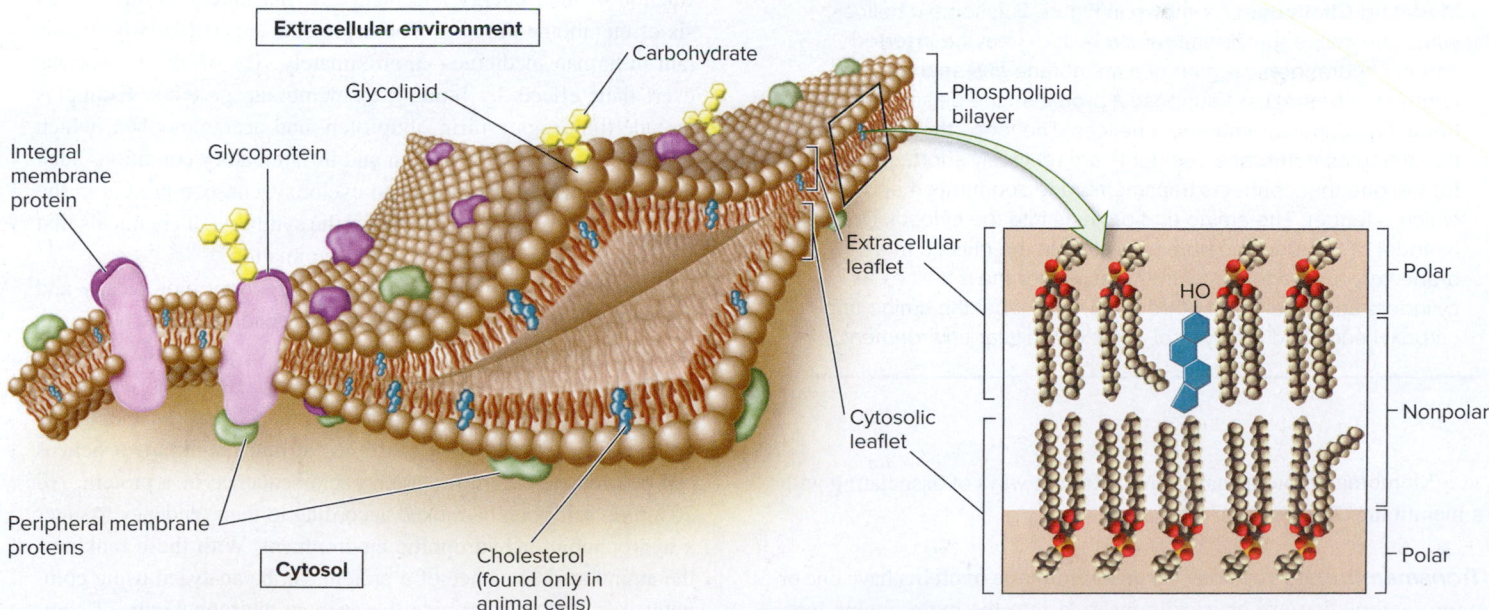

**Figure 5.1** **Fluid-mosaic model of membrane structure.** The membrane shown here is a plasma membrane of a eukaryotic cell, which separates the extracellular environment from the cytosol. The basic framework of a membrane is a phospholipid bilayer, which may also contain other lipids such as cholesterol. Integral membrane proteins have regions that span the membrane. Peripheral membrane proteins are noncovalently attached to integral membrane proteins or to lipids. Proteins and lipids that have covalently bound carbohydrates are called glycoproteins and glycolipids, respectively. The inset shows nine phospholipids and one cholesterol molecule in a bilayer, and it emphasizes the polar and nonpolar regions of the two leaflets. Note: A portion of the bilayer is artificially pealed apart so you can more easily see the two leaflets.

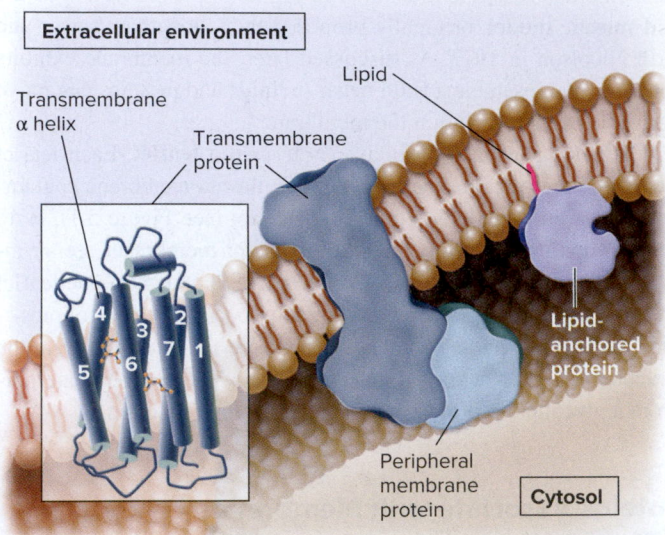

Extracellular environment

Transmembrane α helix

Transmembrane protein

Lipid

Lipid-anchored protein

Peripheral membrane protein

Cytosol

**Figure 5.2** **Types of membrane proteins.** Integral membrane proteins are of two types: transmembrane proteins and lipid-anchored proteins. Peripheral membrane proteins are noncovalently bound to the hydrophilic regions of integral membrane proteins or to the polar head groups of lipids. Inset: The protein bacteriorhodopsin contains seven transmembrane segments, depicted as cylinders, each having an α-helical structure. Bacteriorhodopsin is found in halophilic (salt-loving) archaea.

 **Core Skill: Modeling** The goal of this modeling challenge is to propose a model for a transmembrane protein.

**Modeling Challenge:** As shown in Figure 5.2, some α helices, which are called transmembrane α helices, may be inserted into the hydrophobic region of a membrane and span the entire membrane. Let's suppose a protein in the plasma membrane has 5 transmembrane α helices. The loops that connect these 5 transmembrane segments are relatively short, except for the one that connects transmembrane segments 4 and 5, which is longer. The amino end projects into the cytosol. Draw a model of this transmembrane protein in the plasma membrane. In your model, draw the transmembrane α helices as cylinders and label them 1 through 5. Also label the amino and carboxyl ends and the cytosol and extracellular environment.

Membrane proteins have three different ways of associating with a membrane (**Figure 5.2**).

***Transmembrane Proteins*** **Transmembrane proteins** have one or more regions that are physically inserted into the hydrophobic interior of the phospholipid bilayer. These regions, the transmembrane segments, are stretches of nonpolar amino acids that span or traverse the membrane from one leaflet to the other. In most transmembrane proteins, each transmembrane segment is folded into an α helix. Such a segment is stable in a membrane because the nonpolar amino acids interact favorably with the nonpolar lipid tails.

***Lipid-Anchored Proteins*** A **lipid-anchored protein** associates with a membrane because it has a lipid molecule that is covalently attached to an amino acid side chain within the protein. The lipid tails are inserted into the hydrophobic portion of the membrane and thereby keep the protein firmly attached to the membrane. Both lipid-anchored proteins and transmembrane proteins are considered to be **integral membrane proteins** because they cannot be released from the membrane unless the membrane is dissolved with an organic solvent or detergent. In other words, they cannot be removed without disrupting the integrity of the membrane.

***Peripheral Membrane Proteins*** **Peripheral membrane proteins** associate with membranes in a third way. They do not interact with the hydrophobic interior of the phospholipid bilayer. Instead, they are noncovalently bound to regions of integral membrane proteins that project out from the membrane (see Figure 5.2), or they are bound to the polar head groups of phospholipids. Peripheral membrane proteins are typically attached to the membrane by hydrogen and/or ionic bonds.

 **Core Concept: Information**

## Approximately 20–30% of All Genes Encode Transmembrane Proteins

Membrane proteins participate in some of the most important cellular processes, including transport, energy transduction, cell signaling, secretion, cell recognition, metabolism, and cell-to-cell contact. Research studies have revealed that cells devote a sizable fraction of their energy and metabolic machinery to the synthesis of membrane proteins. These proteins are particularly important in human medicine—approximately 70% of all medications exert their effects by binding to membrane proteins. Examples include the drugs aspirin, ibuprofen, and acetaminophen, which are widely used to relieve pain and inflammatory conditions such as arthritis. These drugs bind to cyclooxygenase, a protein in the ER membrane that is necessary for the synthesis of chemicals that play a role in pain sensation and inflammation.

Because membrane proteins are so important biologically and medically, researchers have analyzed the genomes of many species and asked the question "What percentage of genes encode transmembrane proteins?" To answer this question, they have developed tools to predict the likelihood that a gene encodes a transmembrane protein. For example, the occurrence of transmembrane α helices can be predicted from the amino acid sequence of a protein. All 20 amino acids can be ranked according to their tendency to enter a hydrophobic or hydrophilic environment. With these rankings, the amino acid sequence of a protein can be analyzed using computer software to determine the average hydrophobicity of short amino acid sequences within the protein. A stretch of 18–20 amino acids in an α helix is long enough to span the membrane. If such a stretch contains a high percentage of hydrophobic amino acids, it is predicted to be a transmembrane α helix. However, such computer predictions must eventually be verified by experimentation.

| Table 5.2 | Estimated Percentage of Genes That Encode Transmembrane Proteins* |
|---|---|
| Organism | Percentage of protein-encoding genes that encode transmembrane proteins |
| **Archaea** | |
| *Archaeoglobus fulgidus* | 24.2 |
| *Methanococcus jannaschii* | 20.4 |
| *Pyrococcus horikoshii* | 29.9 |
| **Bacteria** | |
| *Escherichia coli* | 29.9 |
| *Bacillus subtilis* | 29.2 |
| *Haemophilus influenzae* | 25.3 |
| **Eukaryotes** | |
| *Homo sapiens* | 29.7 |
| *Drosophila melanogaster* | 24.9 |
| *Arabidopsis thaliana* | 30.5 |
| *Saccharomyces cerevisiae* | 28.2 |

* Source: Stevens, A. J., and Arkin, T. I. 2000. Do More Complex Organisms Have a Greater Proportion of Membrane Proteins in Their Genomes? *Proteins* 39: 417–420.

Using computer analysis, many research groups have attempted to calculate the percentage of genes that encode transmembrane proteins in various species. **Table 5.2** shows the results of one such study. The estimated percentage of transmembrane proteins is substantial: 20–30% of all genes may encode transmembrane proteins. This trend is found throughout all domains of life, including archaea, bacteria, and eukaryotes. For example, about 30% of human genes encode transmembrane proteins. With a genome size of about 22,000 different protein-encoding genes, the total number of human genes that encode transmembrane proteins is estimated to be 6,600. The functions of many of the proteins have yet to be determined. Identifying their functions will help researchers gain a better understanding of human biology. Likewise, medical researchers and pharmaceutical companies are interested in the identification of new transmembrane proteins that could be targets for effective new medications.

# 5.2 Fluidity of Membranes

## Learning Outcomes:

1. Describe the fluidity of membranes.
2. **CoreSKILL »** Predict how changes in lipid composition affect membrane fluidity.
3. **CoreSKILL »** Analyze the results of experiments that showed the lateral diffusion of membrane proteins.

Let's now turn our attention to the dynamic properties of membranes. Although a membrane provides a critical interface between a cell or an organelle and its environment, it is not a solid, rigid structure. Rather, biological membranes exhibit properties of **fluidity**, which means that individual molecules remain in close association yet have the ability to readily move within the membrane. In this section, we will examine the fluid properties of biological membranes.

## Membranes Are Semifluid

Though membranes are often described as fluid, it is more appropriate to say they are **semifluid**, because the movement of membrane components occurs only in two dimensions. In a fluid substance, molecules can move in three dimensions. By comparison, most phospholipids can rotate freely around their long axes and move laterally within the membrane leaflet (**Figure 5.3a**). This type of motion is considered two-dimensional, which means it occurs within the plane of the membrane. Because rotational and lateral movements keep the lipid tails within the hydrophobic interior, such movements are energetically favorable. At 37°C, a typical lipid molecule exchanges places with its neighbors about $10^7$ times per second, and it can move several micrometers per second. At this rate, a lipid can traverse the length of a bacterial cell (approximately 1 μm) in only 1 second and the length of a typical animal cell in 10 to 20 seconds.

In contrast to rotational and lateral movements, the flip-flop of lipids from one leaflet to the opposite leaflet does not occur spontaneously. Flip-flop is energetically unfavorable because the polar head of a phospholipid would have to travel through the hydrophobic interior of the membrane. How are lipids moved from one leaflet to the other? The transport of lipids between leaflets is due to the action of the enzyme flippase, which requires energy input in the form of ATP (**Figure 5.3b**).

Although most lipids diffuse rotationally and laterally within the plane of the lipid bilayer, researchers have discovered that certain types of lipids in animal cells tend to strongly associate with each other to form structures called lipid rafts. As the word raft suggests, a **lipid raft** is a group of lipids that float together as a unit within a larger sea of lipids. Lipid rafts have a lipid composition that differs from the surrounding membrane. For example, they usually have a high amount of cholesterol. In addition, lipid rafts may contain unique sets of lipid-anchored proteins and transmembrane proteins. The functional importance of lipid rafts is the subject of a large amount of current research. Lipid rafts may play an important role in endocytosis (discussed later in this chapter) and cell signaling.

## Lipid Composition Affects Membrane Fluidity

The biochemical properties of phospholipids affect the fluidity of the phospholipid bilayer.

***Length of Phospholipid Tails*** One key factor that affects membrane fluidity is the length of the lipid tails, which range from 14 to 24 carbon atoms, with 16 to 18 carbons being the most common. Shorter tails are less likely to interact with each other, which makes the membrane more fluid.

***Double Bonds in Phospholipid Tails*** A second important factor is the presence of double bonds in the lipid tails. When a double bond is present, the lipid is said to be **unsaturated** with respect to the number of hydrogens that are bound to the carbon atoms (refer back to Figure 3.9). A double bond creates a kink in a lipid tail (see inset to Figure 5.1), making it more difficult for neighboring tails to interact and making the bilayer more fluid. As described in Chapter 3, unsaturated lipids tend to be more liquid than saturated lipids, which often form solids at room temperature (refer back to Figure 3.10).

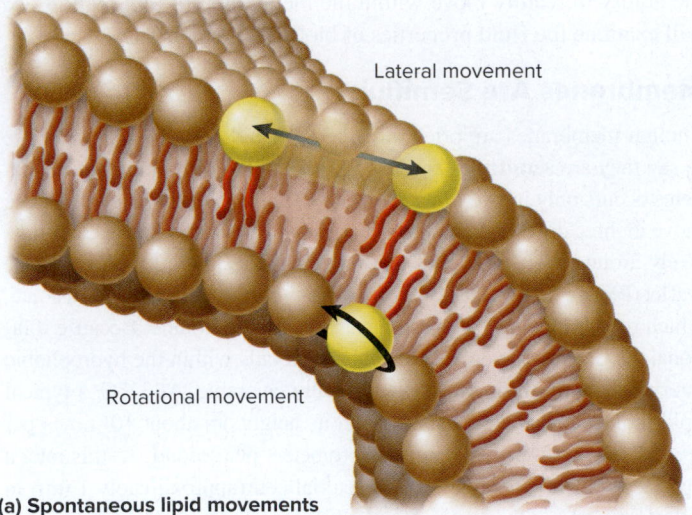

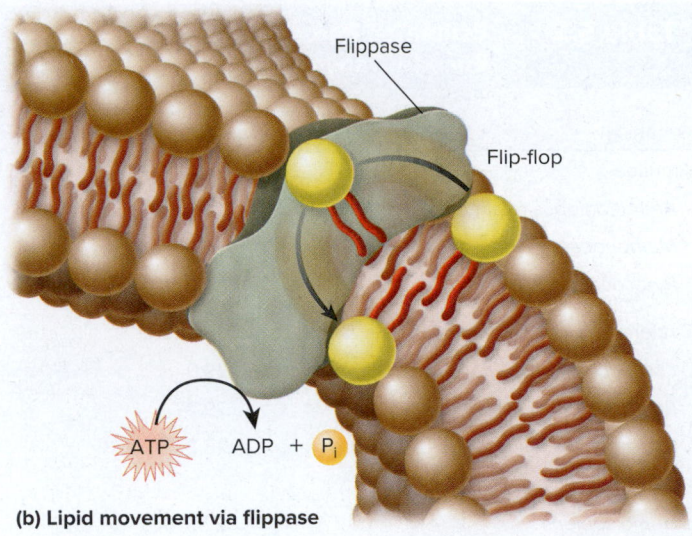

**(a) Spontaneous lipid movements**

**(b) Lipid movement via flippase**

**Figure 5.3  Semifluidity of the lipid bilayer. (a)** Spontaneous movements in the bilayer. Lipids can rotate (that is, move 360°) and move laterally (for example, from left to right in the plane of the bilayer). **(b)** Flip-flop does not happen spontaneously, because the polar head group would have to pass through the hydrophobic region of the bilayer. Instead, the enzyme flippase uses ATP to flip phospholipids from one leaflet to the other.

*Cholesterol*   A third factor affecting fluidity is the presence of cholesterol, a short and rigid molecule produced by animal cells (see inset to Figure 5.1). Plant cell membranes contain phytosterols that resemble cholesterol in their chemical structure. Cholesterol tends to stabilize membranes; its effects depend on temperature. At higher temperatures, such as those observed in mammals that maintain a constant body temperature, cholesterol makes the membrane less fluid. At lower temperatures, such as icy water, cholesterol has the opposite effect. It makes the membrane more fluid and prevents it from freezing.

An optimal level of bilayer fluidity is essential for normal cell function, growth, and division. If a membrane is too fluid, which may occur at higher temperatures, it can become leaky. However, if a membrane becomes too solid, which may occur at lower temperatures, the functioning of membrane proteins will be inhibited. How can organisms cope with changes in temperature? The cells of many species adapt to changes in temperature by altering the lipid composition of their membranes. For example, when the water temperature drops, the cells of certain fish will incorporate more cholesterol into their membranes, making the membrane more fluid. If a plant cell is exposed to high temperatures for many hours or days, it will alter the lipid composition of its cell membrane to have longer lipid tails and fewer double bonds, which will make the membrane less fluid.

## Many Transmembrane Proteins Can Rotate and Move Laterally, but Some Are Restricted in Their Movement

Like lipids, many transmembrane proteins may rotate and move laterally throughout the plane of a membrane. Because transmembrane proteins are larger than lipids, they move within the membrane at a much slower rate. Flip-flop of transmembrane proteins does not occur, because the proteins also contain hydrophilic regions that project out from the phospholipid bilayer, and it would be energetically unfavorable for the hydrophilic regions of membrane proteins to pass through the hydrophobic portion of the phospholipid bilayer.

In 1970, Larry Frye and Michael Edidin conducted an experiment that verified the lateral movement of transmembrane proteins (**Figure 5.4**). Mouse and human cells were mixed together and

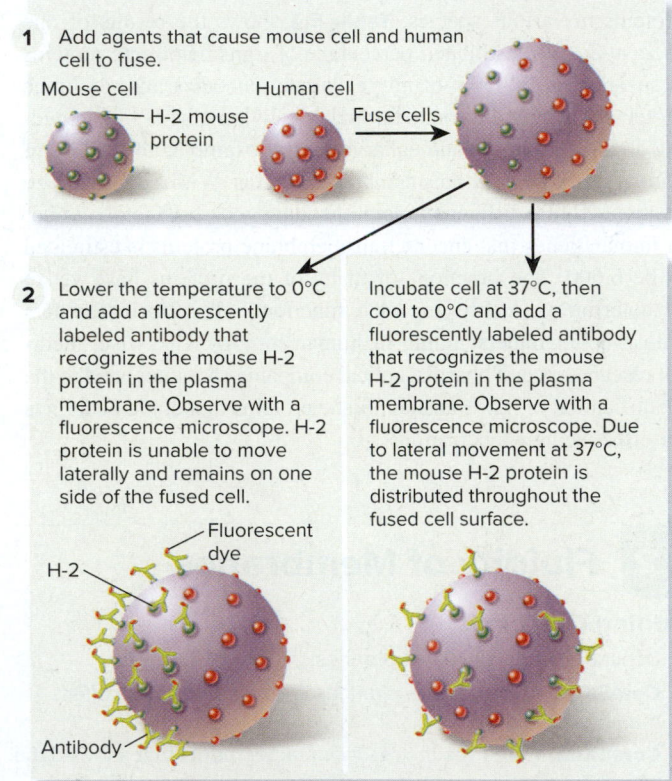

**Figure 5.4  A method to measure the lateral movement of membrane proteins.**

 **Core Skill: Process of Science** This experiment verified that membrane proteins can diffuse laterally within the plane of the lipid bilayer.

exposed to agents that caused them to fuse with each other to produce mouse-human cell hybrids. Some cells were cooled to 0°C, while others were incubated at 37°C before being cooled. Both sets of cells were then exposed to fluorescently labeled antibodies that became specifically bound to a mouse transmembrane protein called H-2. The fluorescent label was observed with a fluorescence microscope. If the cells were maintained at 0°C, a temperature that greatly inhibits lateral movement, the fluorescence was seen on only one side of the fused cell. However, if the cells were incubated for several hours at 37°C and then cooled to 0°C, the fluorescence was distributed throughout the plasma membrane of the fused cell. This occurred because the higher temperature allowed the lateral movement of the H-2 protein throughout the fused cell.

Unlike the example shown in Figure 5.4, not all transmembrane proteins are capable of rotational and lateral movement. Depending on the cell type, 10–70% of membrane proteins may be restricted in their movement. Transmembrane proteins may be bound to components of the cytoskeleton, which restricts the proteins from moving (**Figure 5.5**), or they may be attached to molecules that are outside the cell, such as the interconnected network of proteins that forms the extracellular matrix of animal cells (see Chapter 10).

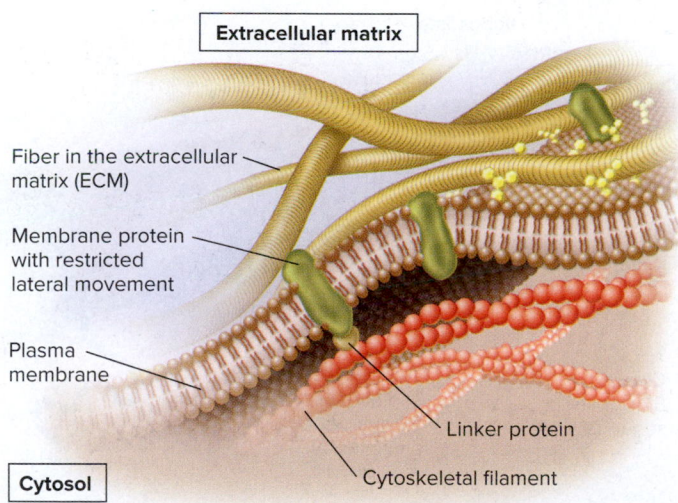

**Figure 5.5 Attachment of transmembrane proteins to the cytoskeleton and extracellular matrix of an animal cell.** Some transmembrane proteins have regions that extend into the cytosol and are anchored to large cytoskeletal filaments via linker proteins. Being bound to these large filaments restricts the movement of these proteins. Similarly, some transmembrane proteins are bound to large, immobile fibers in the extracellular matrix, which restricts their movement.

 **Core Skill: Connections** Look ahead to Figure 10.8. Discuss how transmembrane proteins are important in the binding of cells to each other and the binding of cells to the extracellular matrix.

## 5.3 Synthesis of Membrane Components in Eukaryotic Cells

### Learning Outcomes:

1. Outline the synthesis of lipids at the ER membrane.
2. Explain how transmembrane proteins are inserted into the ER membrane.
3. Describe the process of glycosylation, and explain its functional consequences.

As we have seen, membranes are composed of lipids, proteins, and carbohydrates. Most of the membrane components of eukaryotic cells are made at the endoplasmic reticulum (ER). In this section, we will begin by considering how phospholipids are synthesized at the ER membrane. We will then examine the process by which transmembrane proteins are inserted into the ER membrane and explore how carbohydrates are attached to some proteins.

### Lipid Synthesis Occurs at the ER Membrane

In eukaryotic cells, the cytosol and endomembrane system work together to synthesize most lipids. This process occurs at the cytosolic leaflet of the smooth ER membrane. **Figure 5.6** shows a simplified pathway for the synthesis of phospholipids. The building blocks of a phospholipid are two fatty acids, each with a long tail, one glycerol molecule, one phosphate, and a polar head group. These building blocks are made via enzymes in the cytosol, or they are taken into cells from food. To begin the process of phospholipid synthesis, the fatty acids are activated by attachment to an organic molecule called coenzyme A (CoA). This activation promotes the bonding of the two fatty acids to a glycerol-phosphate molecule, and the resulting molecule is inserted into the cytosolic leaflet of the ER membrane. The phosphate is removed from glycerol, and then a polar molecule already linked to phosphate is attached to glycerol. In the example shown in Figure 5.6, the polar head group contains choline, but many other types of head groups are possible. Phospholipids are initially inserted into the cytosolic leaflet. Flippases in the ER membrane transfer about half of the newly made phospholipids to the other leaflet so similar amounts of lipids are found in both leaflets.

The lipids made in the ER membrane are transferred to other membranes in the cell by a variety of mechanisms. Phospholipids in the ER can diffuse laterally to the nuclear envelope. In addition, lipids are transported via vesicles to the Golgi, lysosomes, vacuoles, or plasma membrane. A third mode of lipid transfer involves **lipid exchange proteins**, which extract a lipid from one membrane, diffuse through the cell, and insert the lipid into another membrane. Such transfer can occur between any two membranes, even between the endomembrane system and semiautonomous organelles. For example, lipid exchange proteins transfer lipids between the ER and mitochondria. In addition, chloroplasts and mitochondria synthesize certain types of lipids that are transferred from these organelles to other cellular membranes via lipid exchange proteins.

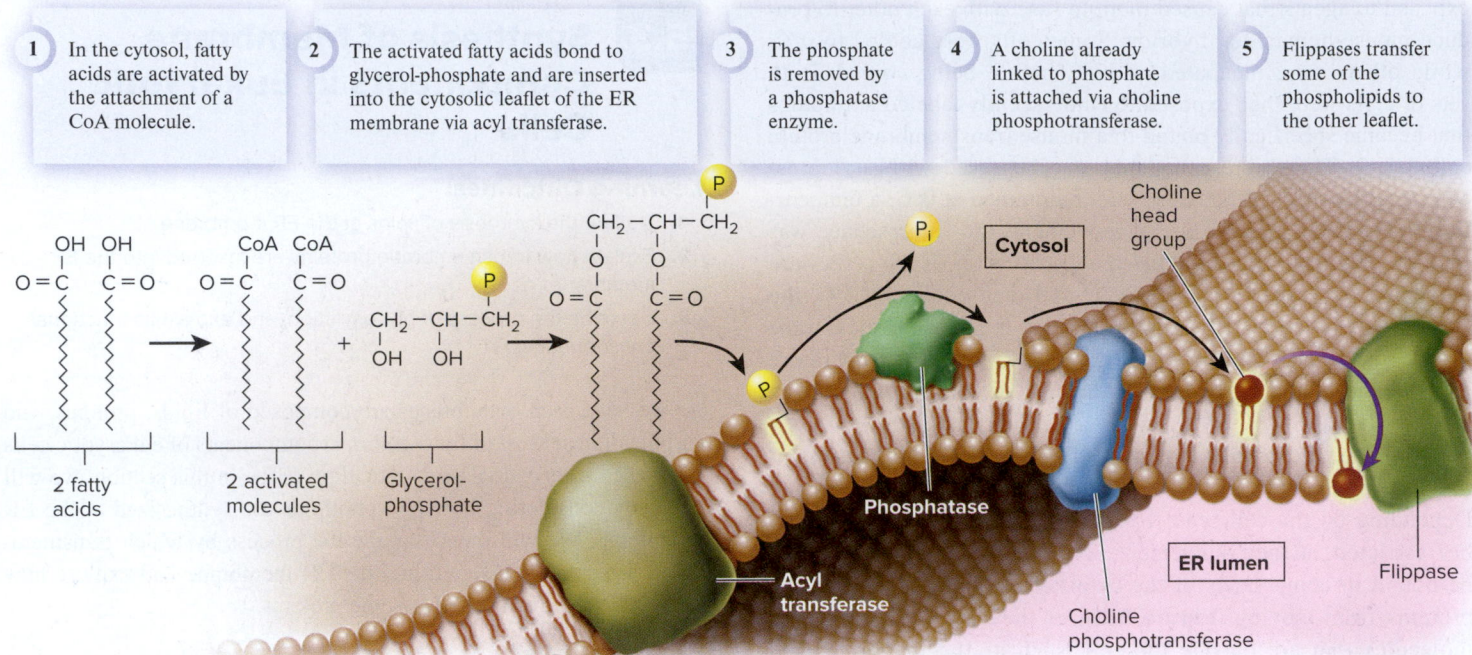

**1** In the cytosol, fatty acids are activated by the attachment of a CoA molecule.

**2** The activated fatty acids bond to glycerol-phosphate and are inserted into the cytosolic leaflet of the ER membrane via acyl transferase.

**3** The phosphate is removed by a phosphatase enzyme.

**4** A choline already linked to phosphate is attached via choline phosphotransferase.

**5** Flippases transfer some of the phospholipids to the other leaflet.

**Figure 5.6** A simplified pathway for the synthesis of membrane phospholipids at the ER membrane. Note: Phosphate is abbreviated P when it is attached to an organic molecule and $P_i$ when it is unattached. The subscript i refers to the inorganic form of phosphate.

**Concept Check:** *How are phospholipids transferred to the leaflet of the ER membrane that faces the ER lumen?*

## Most Transmembrane Proteins Are First Inserted into the ER Membrane

In Chapter 4, we considered how eukaryotic proteins contain sorting signals that direct them to their proper destination (look back at Figure 4.32). With the exception of proteins destined for semiautonomous organelles, most transmembrane proteins contain an ER signal sequence that directs them to the ER membrane. If a polypeptide also contains a stretch of 20 amino acids that are mostly hydrophobic and form an α helix, this region will become a transmembrane segment. In the example shown in **Figure 5.7**, the polypeptide contains one such sequence. After the ER signal sequence is removed by signal peptidase (refer back to Figure 4.33), a membrane protein with a single transmembrane segment is the result. Other polypeptides may contain more than one transmembrane segment. Each time a polypeptide sequence contains a region of 20 amino acids that are mostly hydrophobic and form an α helix, an additional transmembrane segment is synthesized into the membrane. From the ER, membrane proteins can be transferred via vesicles to other regions of the cell, such as the Golgi, lysosomes, vacuoles, or plasma membrane.

## The Attachment of Carbohydrates to Proteins Occurs in the ER and Golgi Apparatus

**Glycosylation** refers to the process of covalently attaching a carbohydrate to a lipid or protein. When a carbohydrate is attached to a lipid, a **glycolipid** is created, whereas attachment of a carbohydrate to a protein produces a **glycoprotein**.

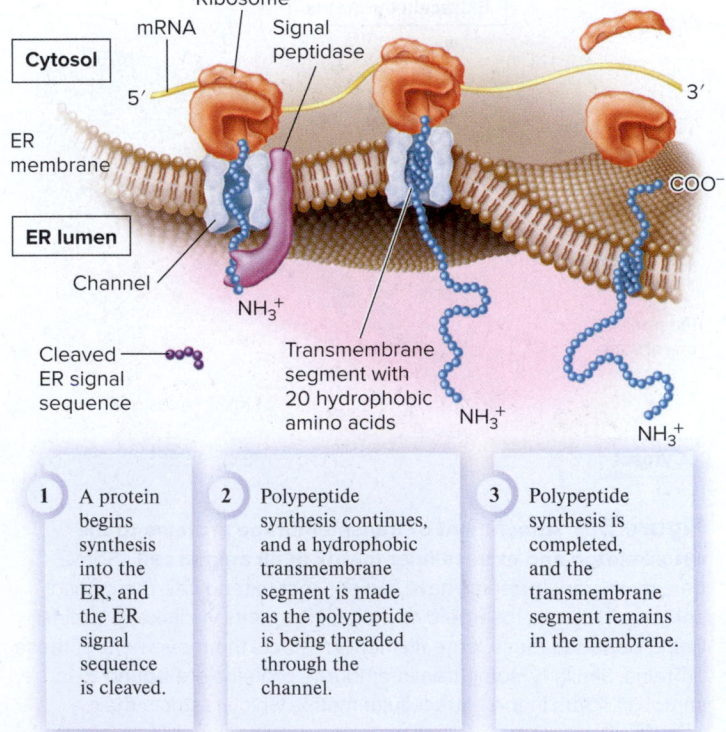

**1** A protein begins synthesis into the ER, and the ER signal sequence is cleaved.

**2** Polypeptide synthesis continues, and a hydrophobic transmembrane segment is made as the polypeptide is being threaded through the channel.

**3** Polypeptide synthesis is completed, and the transmembrane segment remains in the membrane.

**Figure 5.7** Insertion of membrane proteins into the ER membrane.

**Concept Check:** *What structural feature of a polypeptide causes a region of it to form a transmembrane segment?*

What is the function of glycosylation? Though the roles of carbohydrates in cell structure and function are not entirely understood, some functional consequences of glycosylation have emerged. Glycolipids and glycoproteins often play a role in cell surface recognition. When glycolipids and glycoproteins are found in the plasma membrane, the carbohydrate portion is located in the extracellular region. During embryonic development in animals, significant cell movement occurs. Layers of cells slide over each other to create body structures such as the spinal cord and internal organs. The proper migration of individual cells and cell layers relies on the recognition of cell types via the carbohydrates on their cell surfaces.

Carbohydrates often have a protective effect. The carbohydrate-rich zone on the surface of certain animal cells shields the cell from mechanical and physical damage. Similarly, the carbohydrate portion of glycosylated proteins protects them from the harsh conditions of the extracellular environment and degradation by extracellular proteases, which are enzymes that digest proteins.

Two forms of protein glycosylation occur in eukaryotes: N-linked and O-linked. N-linked glycosylation, which also occurs in archaea, involves the attachment of a carbohydrate to the amino acid asparagine in a polypeptide. It is called N-linked because the carbohydrate is attached to a nitrogen atom of the asparagine side chain. For this to occur, a group of 14 sugar molecules, called a carbohydrate tree, is first built onto a lipid found in the ER membrane (**Figure 5.8**). An enzyme in the ER, oligosaccharide transferase, transfers the carbohydrate tree from the lipid

to an asparagine in the polypeptide. N-linked glycosylation commonly occurs on membrane proteins that are transported to the cell surface.

The second form of glycosylation, O-linked glycosylation, occurs only in the Golgi apparatus. This form involves the addition of a string of sugars to the oxygen atom of a serine or threonine side chain in a polypeptide. In animals, O-linked glycosylation is important for the production of proteoglycans, which are highly glycosylated proteins that are secreted from cells and help to organize the extracellular matrix that surrounds cells. Proteoglycans are also a component of mucus, a slimy material that coats many cell surfaces and is secreted into fluids such as saliva. High concentrations of carbohydrates give mucus its slimy texture.

## 5.4 Overview of Membrane Transport

**Learning Outcomes:**

1. Compare and contrast simple diffusion, facilitated diffusion, passive transport, and active transport.
2. Describe the process of osmosis, and explain how it affects cell structure.
3. **CoreSKILL »** Predict the direction of water movement in response to solute gradients.

We now turn to one of the key functions of membranes, **membrane transport**—the movement of ions and molecules across biological membranes. All cells contain a plasma membrane that exhibits **selective permeability**, allowing the passage of some ions and molecules but not others. Essential molecules such as glucose and amino acids enter the cell, metabolic intermediates remain in the cell, and waste products exit. The selective permeability of the plasma membrane allows the cell to maintain a favorable internal environment.

Substances can move directly across a membrane in three general ways (**Figure 5.9**).

- **Simple diffusion** occurs when a substance moves across a membrane from an area of high concentration to one of lower concentration by passing directly through the phospholipid bilayer. This diffusion is an example of **passive transport**—the movement of a substance across a membrane from an area of high concentration to one of lower concentration, which does not require an input of energy.

- A second mechanism of passive transport is **facilitated diffusion**, in which a transport protein provides a passageway for a substance to cross a membrane from an area of higher concentration to one of lower concentration.

- A third mode of membrane transport, called **active transport**, moves a substance from an area of low concentration to one of high concentration with the aid of a transport protein. Active transport requires an input of energy from a source such as ATP.

In this section, we will begin with a discussion of how the phospholipid bilayer presents a barrier to the simple diffusion of ions and polar molecules across membranes. We will then consider the concept of gradients across membranes and how such gradients affect the movement of water.

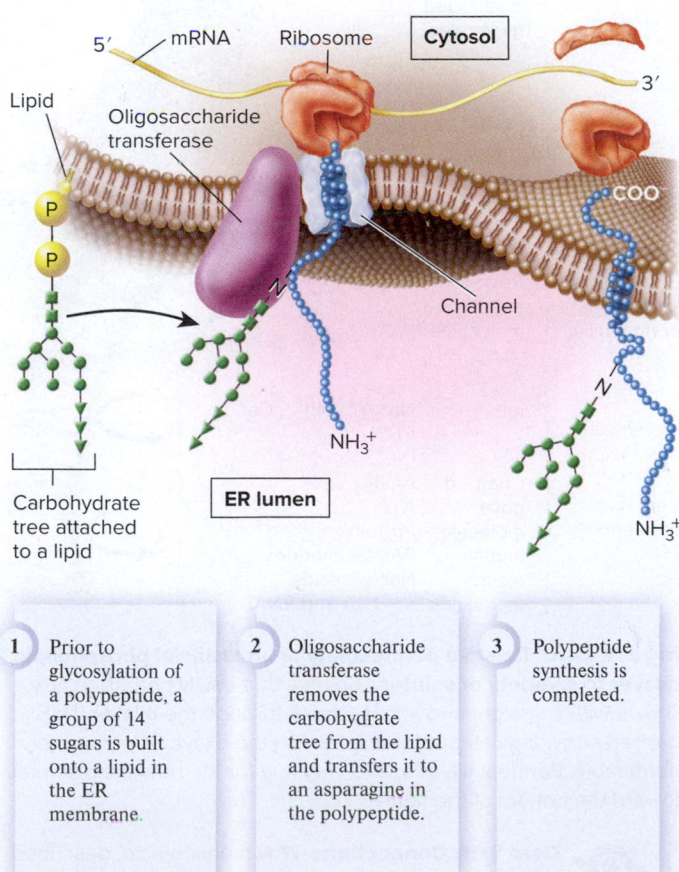

1. Prior to glycosylation of a polypeptide, a group of 14 sugars is built onto a lipid in the ER membrane.

2. Oligosaccharide transferase removes the carbohydrate tree from the lipid and transfers it to an asparagine in the polypeptide.

3. Polypeptide synthesis is completed.

**Figure 5.8** N-linked glycosylation in the ER.

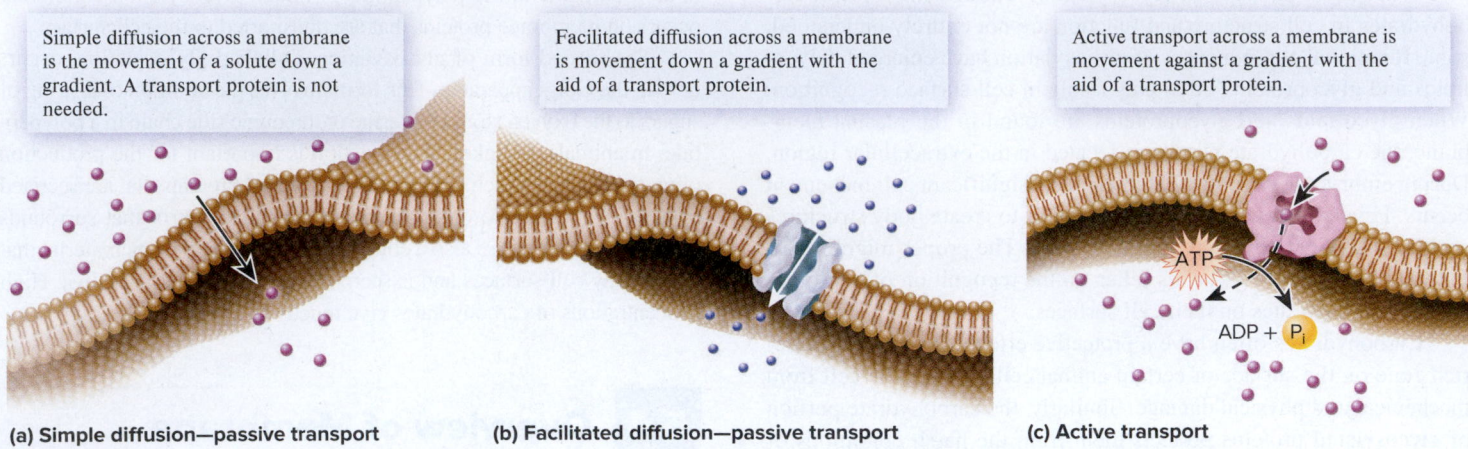

Simple diffusion across a membrane is the movement of a solute down a gradient. A transport protein is not needed.

Facilitated diffusion across a membrane is movement down a gradient with the aid of a transport protein.

Active transport across a membrane is movement against a gradient with the aid of a transport protein.

ATP

ADP + P$_i$

(a) Simple diffusion—passive transport

(b) Facilitated diffusion—passive transport

(c) Active transport

**Figure 5.9**  Three general types of membrane transport.

## The Phospholipid Bilayer Is a Barrier to the Simple Diffusion of Hydrophilic Solutes

Because of their hydrophobic interiors, phospholipid bilayers are a barrier to the simple diffusion of ions and polar (hydrophilic) molecules. Such ions and molecules are called **solutes**; they are dissolved in water, which is a **solvent**. Four factors affect the ability of solutes to pass through a phospholipid bilayer.

- *Size.* Small solutes cross bilayers faster than larger ones.
- *Polarity.* Nonpolar solutes cross bilayers faster than polar ones.
- *Charge.* Noncharged solutes cross bilayers faster than charged ones.
- *Concentration.* The rate of movement of a solute across a membrane will be higher when its concentration is higher.

**Figure 5.10** compares the relative permeabilities of an artificial phospholipid bilayer to various solutes. This artificial bilayer does not contain any proteins or carbohydrates. Gases and a few small, uncharged molecules can readily cross the bilayer by simple diffusion. However, the permeability of the bilayer to ions and larger polar molecules, such as sugars, is relatively low, and the permeability to macromolecules, such as proteins and polysaccharides, is even lower.

## Cells Maintain Gradients Across Their Membranes

A hallmark of living cells is their ability to maintain a relatively constant internal environment that is distinctively different from their external environment. Solute gradients are formed across the plasma membrane and across internal membranes. When we speak of a **transmembrane gradient**, or **concentration gradient**, we mean that the concentration of a solute is higher on one side of a membrane than on the other. Transmembrane gradients of solutes are a universal feature of all living cells. For example, immediately after you eat a meal containing carbohydrates, a higher concentration of glucose is found outside your cells than inside; this is an example of a chemical gradient (**Figure 5.11a**).

Gradients involving ions have two components—electrical and chemical. An **electrochemical gradient** is a dual gradient with both electrical and chemical components (**Figure 5.11b**). It occurs with solutes that have a net positive or negative charge. For example, let's

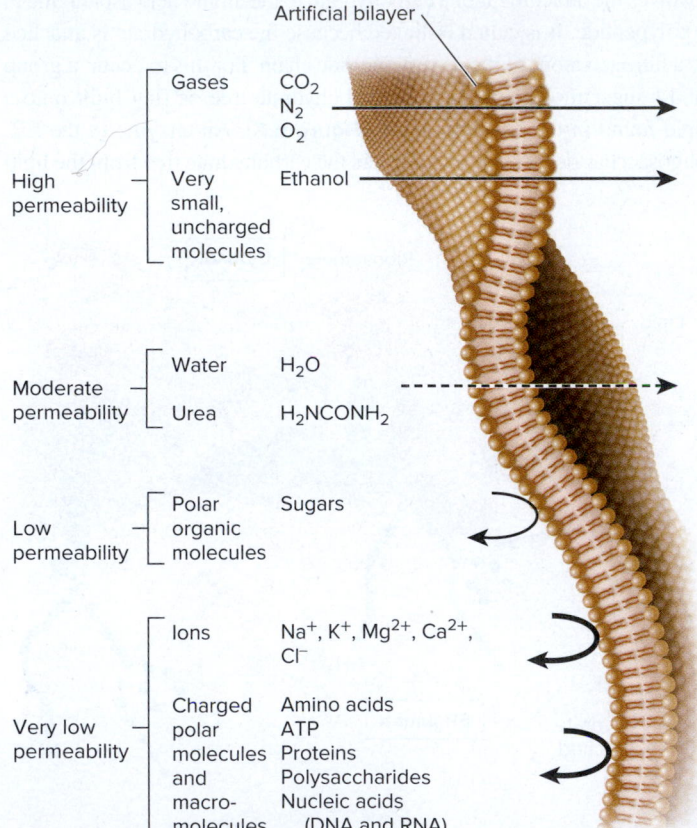

Artificial bilayer

| | | |
|---|---|---|
| High permeability | Gases | $CO_2$ $N_2$ $O_2$ |
| | Very small, uncharged molecules | Ethanol |
| Moderate permeability | Water | $H_2O$ |
| | Urea | $H_2NCONH_2$ |
| Low permeability | Polar organic molecules | Sugars |
| Very low permeability | Ions | $Na^+$, $K^+$, $Mg^{2+}$, $Ca^{2+}$, $Cl^-$ |
| | Charged polar molecules and macro-molecules | Amino acids ATP Proteins Polysaccharides Nucleic acids (DNA and RNA) |

**Figure 5.10**  Relative permeability of an artificial phospholipid bilayer to a variety of solutes.  Solutes that easily penetrate are shown with a straight arrow that passes through the bilayer. The dashed arrow indicates solutes for which the bilayer is moderately permeable. Permeability is low to very low for the remaining solutes, toward the bottom of the figure.

 **Core Skill: Connections**  Which amino acid, described in Chapter 3 (see Figure 3.13), would you expect to be more likely to cross an artificial phospholipid bilayer, leucine or lysine?

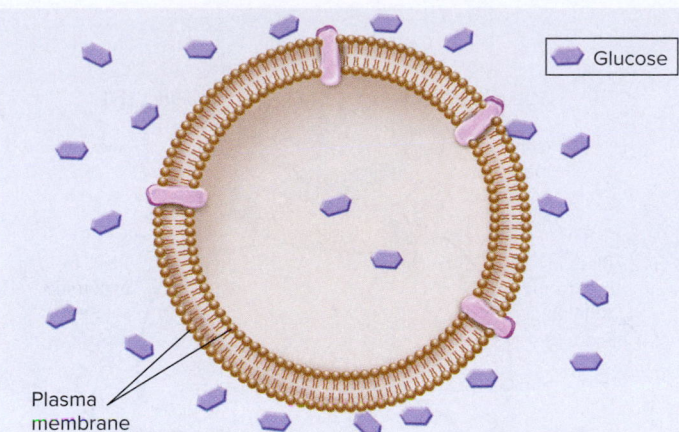

**(a) Chemical gradient for glucose—a higher glucose concentration outside the cell**

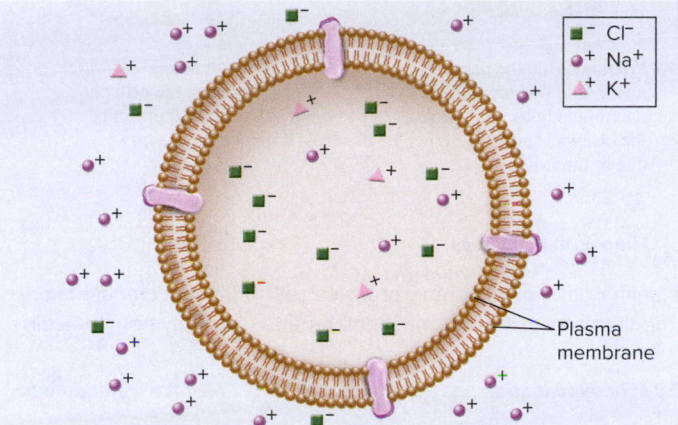

**(b) Electrochemical gradient for Na⁺—more positive charges outside the cell and a higher Na⁺ concentration outside the cell**

**Figure 5.11** Gradients across cell membranes.

 **Core Skill: Connections** Look ahead to Figure 42.9. What types of ion gradients are important for the conduction of action potentials across the plasma membrane of a neuron?

consider a gradient involving $Na^+$. An electrical gradient can exist in which the amount of net positive charge outside a cell is greater than inside. In Figure 5.11b, an electrical gradient is due to differences in the amounts of different types of ions—sodium, potassium, and chloride ($Na^+$, $K^+$, and $Cl^-$)—on the two sides of the plasma membrane. At the same time, a chemical gradient—a difference in $Na^+$ concentration across the membrane—also exists in which the concentration of $Na^+$ outside is greater than inside. The $Na^+$ electrochemical gradient is composed of both an electrical gradient due to charge differences across the membrane and a chemical gradient for $Na^+$.

One way to view the transport of solutes across membranes is to consider how the transport process affects the pre-existing transmembrane gradients. Passive transport tends to dissipate a pre-existing gradient. Such a process is energetically favorable and does not require an input of energy. As noted earlier, passive transport can occur in two ways, via simple diffusion or facilitated diffusion (see Figure 5.9a,b).

By comparison, active transport produces a chemical or electrochemical gradient. The formation of a gradient requires an input of energy.

## Osmosis Is the Movement of Water Across a Membrane to Balance Solute Concentrations

Let's now turn our attention to how gradients affect the movement of water across membranes. When the concentrations of solutes on both sides of the plasma membrane are equal, the two concentrations are said to be **isotonic** (**Figure 5.12a**). However, we have also seen that transmembrane gradients commonly exist across membranes. When the concentration of solutes outside the cell is higher, the outside is said to be **hypertonic** relative to the inside of the cell (**Figure 5.12b**). Alternatively, the outside of the cell could be **hypotonic**—have a lower concentration of solutes than the inside (**Figure 5.12c**).

If solutes cannot readily move across the membrane, water will do so and tend to balance the solute concentrations. In this process, called **osmosis**, water moves across a membrane from the hypotonic compartment (with a lower solute concentration) into the hypertonic compartment (with a higher solute concentration). Animal cells, which are not surrounded by a rigid cell wall, must maintain a balance between the extracellular and intracellular solute concentrations; the two solutions need to be isotonic. Animal cells contain a variety of transport proteins that sense changes in cell volume and allow the necessary movements of solutes across the membrane to prevent

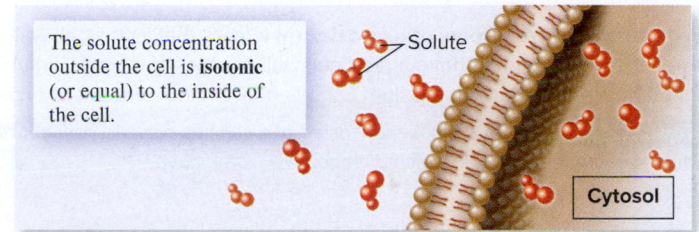

**(a) Outside isotonic**

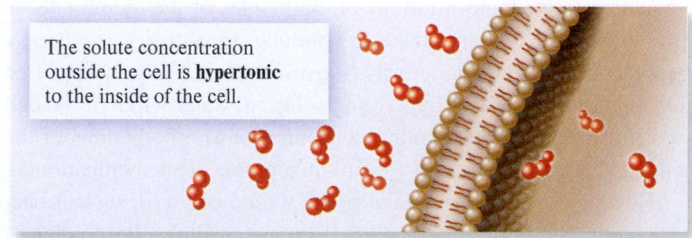

**(b) Outside hypertonic**

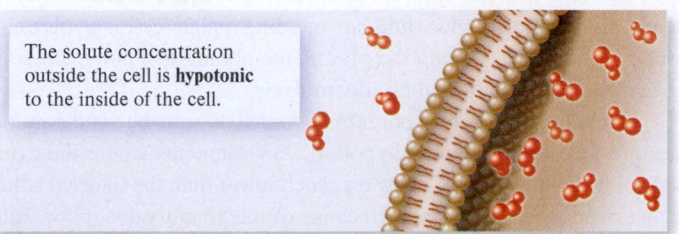

**(c) Outside hypotonic**

**Figure 5.12** Relative solute concentrations outside and inside cells.

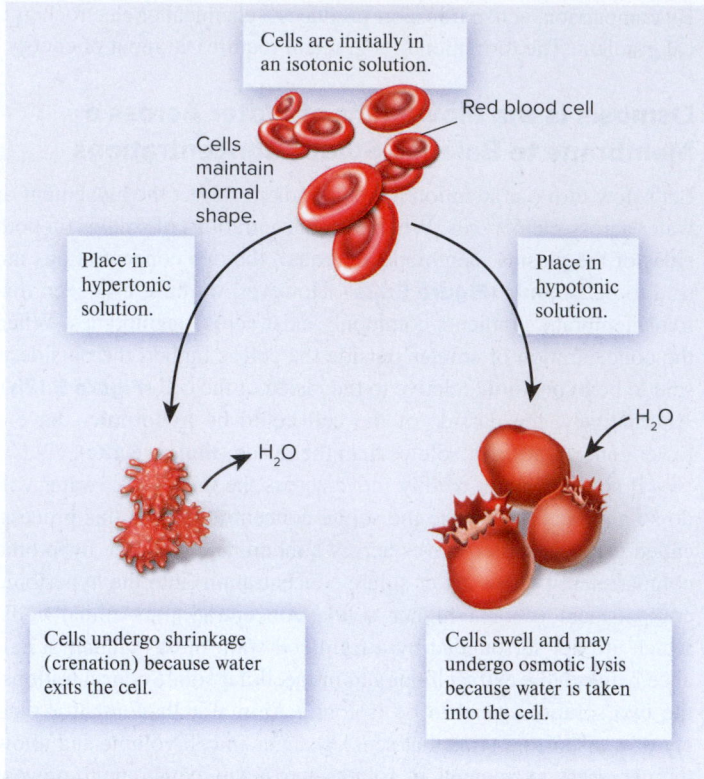

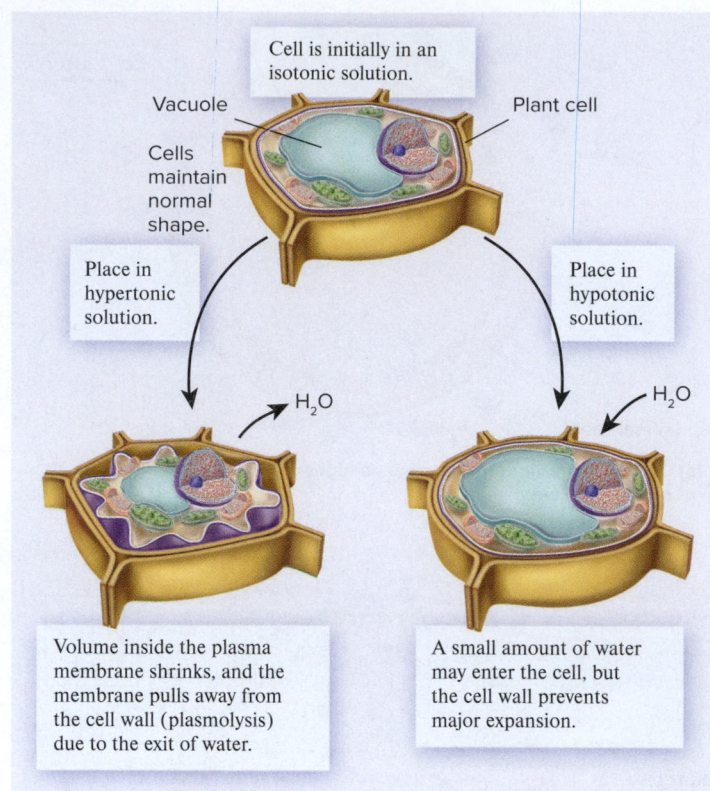

**(a) Osmosis in animal cells**

**(b) Osmosis in plant cells**

**Figure 5.13    The effects of osmosis.  (a)** In cells that lack a cell wall, such as animal cells, osmosis may promote cell shrinkage (crenation) or swelling. **(b)** In cells that have a rigid cell wall, such as plant cells, a hypertonic medium causes the plasma membrane to pull away from the cell wall, whereas a hypotonic medium causes only a minor amount of expansion.

**Concept Check:**   *Let's suppose the inside of a cell has a solute concentration of 0.3 M and the outside has a concentration of 0.2 M. If the membrane is impermeable to solutes, in which direction will water move?*

osmotic changes and maintain normal cell shape. However, if an animal cell is placed in a hypotonic solution, water will enter the cell to equalize solute concentrations on both sides of the membrane. In extreme cases, a cell may take up so much water that it ruptures, a phenomenon called osmotic lysis (**Figure 5.13a**). Alternatively, if an animal cell is placed in a hypertonic solution, water will exit the cell via osmosis and equalize solute concentrations on both sides of the membrane, causing the cell to shrink in a process called crenation.

How does osmosis affect cells with a rigid cell wall, such as bacteria, fungi, algae, and plant cells? If the extracellular fluid is hypotonic, a plant cell will take up a small amount of water, but the cell wall prevents osmotic lysis from occurring (**Figure 5.13b**). Alternatively, if the extracellular fluid surrounding a plant cell is hypertonic, water will exit the cell and the plasma membrane will pull away from the cell wall, a process called **plasmolysis**.

Some freshwater microorganisms, such as amoebae and paramecia, are found in extremely hypotonic environments where the external solute concentration is always much lower than the internal solute concentration in the cytosol. Because of the great tendency for water to move into these cells by osmosis, such organisms contain one or more contractile vacuoles to prevent osmotic lysis. A contractile vacuole takes up water from the cytosol and periodically discharges it by fusing with the plasma membrane (**Figure 5.14**).

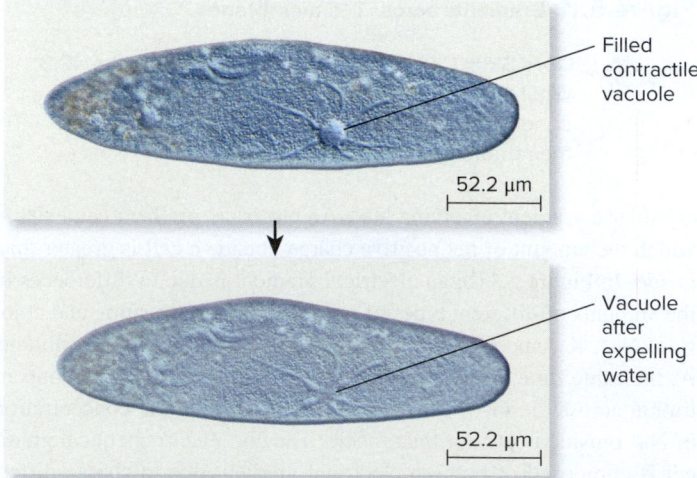

**Figure 5.14    The contractile vacuole in *Paramecium caudatum.*** In the upper photo, a contractile vacuole is filled with water from radiating canals that collect fluid from the cytosol. The lower photo shows the cell after the contractile vacuole has fused with the plasma membrane (which would be above the plane of this page) and released the water from the cell. (photos): ©Michael Abbey/ Science Source

## 5.5 Transport Proteins

### Learning Outcomes:

1. Outline the functional differences between channels and transporters.
2. Compare and contrast uniporters, symporters, and antiporters.
3. **CoreSKILL »** Analyze the results of Agre, and explain how they indicated the presence of a water channel.
4. Explain the difference between primary active transport and secondary active transport.
5. Describe the structure and function of pumps.

Because the phospholipid bilayer is a physical barrier to the movement of most polar molecules and ions across membranes, cells can separate their internal contents from the external environment. However, this barrier also poses a potential problem because cells must take up nutrients from the environment and export waste products. How do cells resolve this dilemma? Over the course of millions of years, species have evolved a multitude of **transport proteins**— transmembrane proteins that provide passageways for the movement of ions and hydrophilic molecules across the phospholipid bilayer. Transport proteins play a central role in the selective permeability of biological membranes. In this section, we will examine the two categories of transport proteins—channels and transporters—and see how they move solutes across the membrane.

### Channels Provide Open Passageways for Solute Movement

A **channel** is a transmembrane protein that forms an open passageway for the facilitated diffusion of ions or molecules across the membrane (**Figure 5.15**). Solutes move directly through a channel to get to the

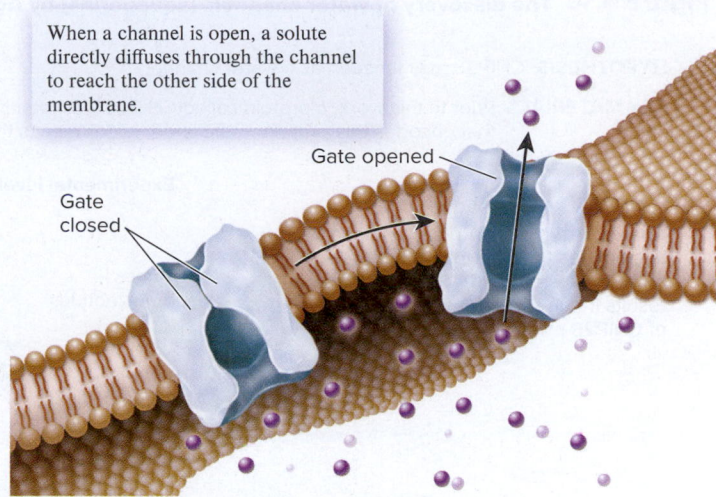

When a channel is open, a solute directly diffuses through the channel to reach the other side of the membrane.

Gate opened

Gate closed

**Figure 5.15**  **Mechanism of solute transport via a channel.**

*Concept Check:*  *What is the purpose of gating?*

other side. When a channel is open, the transmembrane movement of solutes can be extremely rapid, up to 100 million ions or molecules per second!

Most channels are **gated**, which means they open to allow the diffusion of solutes and close to prohibit diffusion. The phenomenon of gating allows cells to regulate the movement of solutes. For example, gating may involve the direct binding of a molecule to the channel protein itself. These gated channels are controlled by the noncovalent binding of small molecules—called ligands—such as hormones or neurotransmitters. The ligands are often important in the transmission of signals between neurons and muscle cells or between two neurons.

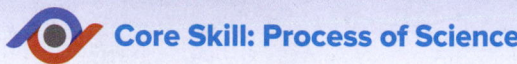

## Core Skill: Process of Science

## Feature Investigation | Agre Discovered That Osmosis Occurs More Quickly in Cells with a Channel That Allows the Facilitated Diffusion of Water

As discussed earlier in this chapter, osmosis is the movement of water to balance solute concentrations. Water can cross biological membranes slowly by simple diffusion through the phospholipid bilayer. However, in the 1980s, researchers discovered that certain cell types allow water to move across the plasma membrane at a much faster rate than would occur with simple diffusion alone. For example, water moves very quickly across the membrane of red blood cells, which causes them to shrink and swell in response to changes in extracellular solute concentrations. Likewise, bladder and kidney cells, which play a key role in regulating water balance in the bodies of vertebrates, allow the rapid movement of water across their membranes. Based on these observations, researchers speculated that certain cell types might have channels in their plasma membranes that enable the rapid movement of water.

One approach to characterizing a new protein is to first identify a protein based on its relative abundance in a particular cell type and then attempt to determine the protein's function. This rationale was applied to the discovery of proteins that allow the rapid movement of water across membranes. Peter Agre and his colleagues first identified a protein that was abundant in red blood cells and kidney cells but not found in high amounts in many other cell types. Though they initially did not know the function of the protein, its physical structure was similar to other proteins that were already known to function as channels. They named this protein CHIP28, which stands for channel-forming integral membrane protein with a molecular mass of 28,000 daltons. During the course of their studies, they also identified and isolated the gene that encodes CHIP28.

In 1992, Agre and his colleagues conducted experiments to determine if CHIP28 functions in the transport of water across membranes (**Figure 5.16**). Because they already had isolated the gene that encodes CHIP28, they could make many copies of

**Figure 5.16** **The discovery of water channels (aquaporins) by Agre.** (4): Courtesy Dr. Peter Agre

**HYPOTHESIS** CHIP28 may function as a water channel.

**KEY MATERIALS** Prior to this work, a protein called CHIP28 was identified that is abundant in red blood cells and kidney cells. The gene that encodes this protein was cloned, which means that many copies of the gene were made in a test tube.

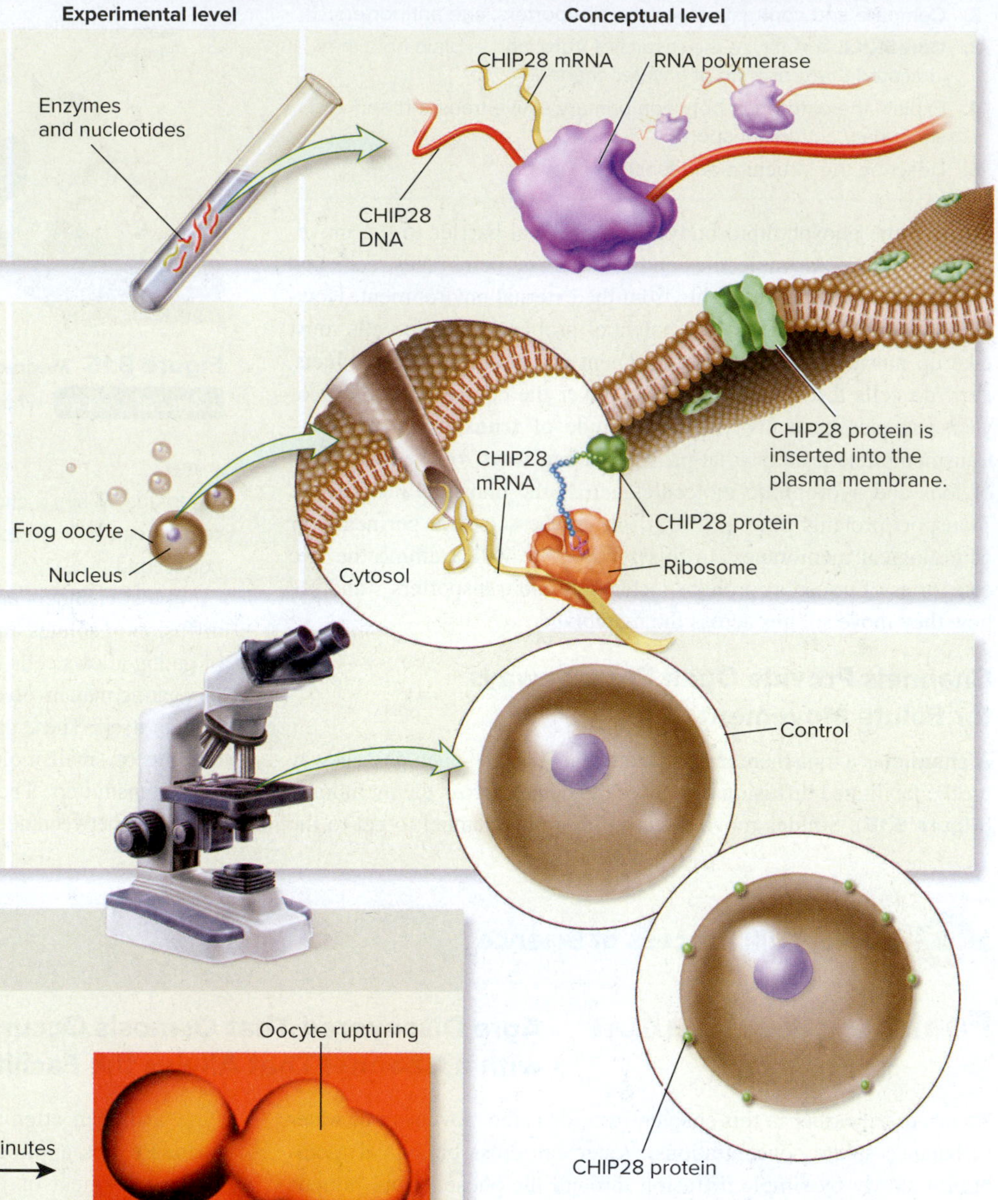

Experimental level  Conceptual level

1 Add an enzyme (RNA polymerase) and nucleotides to a test tube that contains many copies of the CHIP28 gene. This results in the synthesis of many copies of CHIP28 mRNA.

Enzymes and nucleotides

CHIP28 mRNA  RNA polymerase

CHIP28 DNA

2 Inject the CHIP28 mRNA into frog eggs (oocytes). Wait several hours to allow time for the mRNA to be translated into CHIP28 protein at the ER membrane and then moved via vesicles to the plasma membrane.

CHIP28 protein is inserted into the plasma membrane.

CHIP28 mRNA

Frog oocyte

Nucleus

Cytosol

CHIP28 protein

Ribosome

3 Place oocytes into a hypotonic medium and observe under a light microscope. As a control, also place oocytes that have not been injected with CHIP28 mRNA into a hypotonic medium and observe by microscopy.

Control

CHIP28 protein

4 **THE DATA**

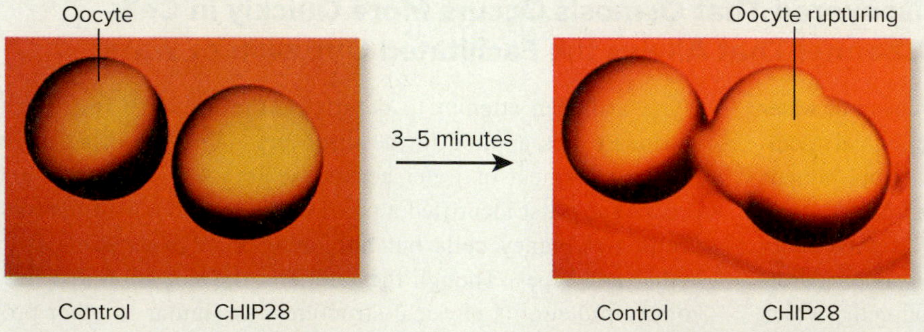

Oocyte

Oocyte rupturing

3–5 minutes

Control    CHIP28

Control    CHIP28

5 **CONCLUSION** The CHIP28 protein, now called aquaporin, allows the rapid movement of water across the membrane.

6 **SOURCE** Preston, G. M., Carroll, T. P., Guggino, W. B., and Agre, P. "Appearance of water channels in Xenopus oocytes expressing red cell CHIP28 protein." *Science*. 1992.

this gene in a test tube (in vitro) using gene cloning techniques (see Chapter 21). Starting with many copies of the gene in vitro, they added an enzyme to transcribe the gene into mRNA that encodes the CHIP28 protein. This mRNA was then injected into frog oocytes, chosen because these oocytes are large, easy to inject, and lack pre-existing proteins in their plasma membranes that allow the rapid movement of water. Following injection, the mRNA was translated into CHIP28 proteins that were inserted into the plasma membrane of the oocytes. After sufficient time had been allowed for this to occur, the oocytes were placed in a hypotonic medium. As a control, oocytes that had not been injected with CHIP28 mRNA were also exposed to a hypotonic medium.

As you can see in the data, a striking difference was observed between oocytes that expressed CHIP28 versus the control oocytes. Within minutes, oocytes that contained the CHIP28 protein were seen to swell due to the rapid uptake of water. Three to five minutes after being placed in a hypotonic medium, they actually ruptured! By comparison, the control oocytes did not swell as rapidly, and

they did not rupture even after 1 hour. Taken together, these results are consistent with the hypothesis that CHIP28 functions as a channel that allows the facilitated diffusion of water across the membrane. Many subsequent studies confirmed this observation. Later, CHIP28 was renamed **aquaporin** to indicate its newly identified function of allowing water to diffuse through a channel in the membrane. In 2003, Agre was awarded the Nobel Prize in Chemistry for this work.

### Experimental Questions

1. What observations about particular cell types in the human body led to the experimental strategy of Figure 5.16?

2. What were the characteristics of CHIP28 that made Agre and associates speculate that it may transport water? In your own words, briefly explain how they tested the hypothesis that CHIP28 has this function.

3. **CoreSKILL »** Explain how the results of the experiment of Figure 5.16 support the proposed hypothesis.

## Transporters Bind Their Solutes and Undergo Conformational Changes

Let's now turn our attention to a second category of transport proteins known as **transporters**.[*] These transmembrane proteins bind one or more solutes in a hydrophilic pocket and undergo a conformational change that switches the exposure of the pocket from one side of the membrane to the other side (**Figure 5.17**). For example, in 1995, American biologist Robert Brooker and colleagues proposed that a transporter called lactose permease, which is found in the bacterium *E. coli*, has a hydrophilic pocket that binds lactose. They further proposed that the two halves of the transporter protein come together at an interface that moves in such a way that the lactose-binding site alternates between an outwardly accessible pocket and an inwardly accessible pocket, as shown in Figure 5.17. This idea was later confirmed by studies that determined the structure of the lactose permease and related transporters.

Transporters provide the principal pathway for the cellular uptake of organic molecules, such as sugars, amino acids, and nucleotides. In animals, they also allow cells to take up certain hormones and neurotransmitters. In addition, many transporters play a key role in export. Waste products of cellular metabolism must be released from cells before they reach toxic levels. For example, a transporter removes lactic acid, a by-product of muscle cells during exercise. Other transporters, which are involved with ion transport, play an important role in regulating internal pH and controlling cell volume. Transporters tend to be much slower than channels. Their rate of transport is typically 100 to 1,000 ions or molecules per second.

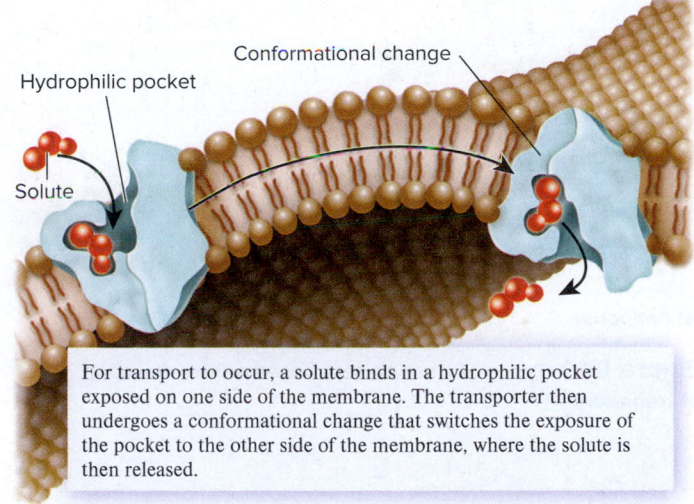

Conformational change

Hydrophilic pocket

Solute

For transport to occur, a solute binds in a hydrophilic pocket exposed on one side of the membrane. The transporter then undergoes a conformational change that switches the exposure of the pocket to the other side of the membrane, where the solute is then released.

**Figure 5.17** Mechanism of transport by a transporter, also called a carrier.

**Core Concept: Structure and Function** Two structural features—a hydrophilic pocket and the ability to switch back and forth between two conformations—allow transporters to move ions and molecules across the membrane.

Transporters are named according to the number of solutes they bind and the direction in which they transport those solutes (**Figure 5.18**). **Uniporters** bind a single ion or molecule and transport it across the membrane. **Symporters** bind two or more ions or molecules and transport them in the same direction. **Antiporters** bind two or more ions or molecules and transport them in opposite directions.

---
[*] Transporters are also called carriers. However, this term is misleading because transporters do not physically carry the solutes across the membrane.

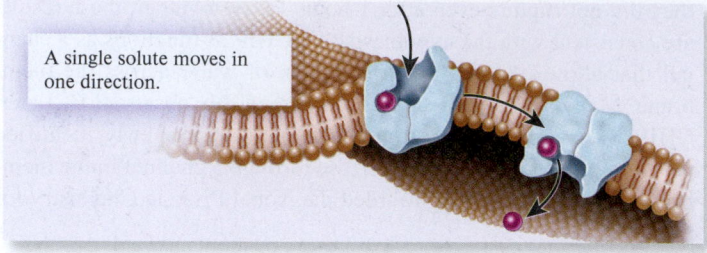

A single solute moves in one direction.

**(a) Uniporter**

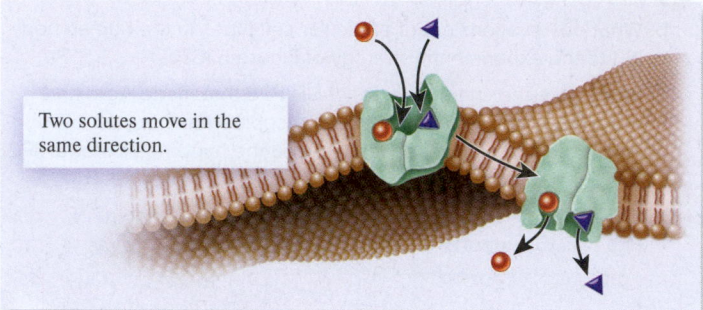

Two solutes move in the same direction.

**(b) Symporter**

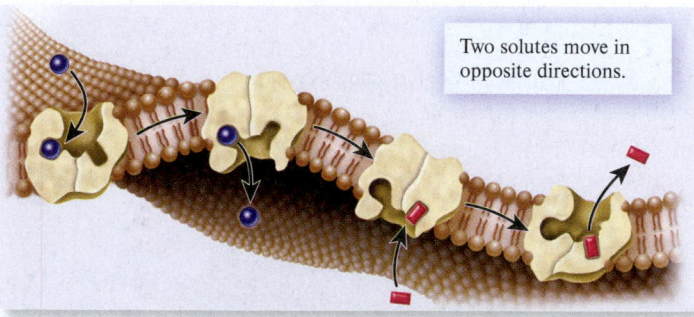

Two solutes move in opposite directions.

**(c) Antiporter**

**Figure 5.18** **Types of transporters based on the direction of transport.**

**PROBLEM-SOLVING STRATEGY** *Make a drawing. Relate structure and function.* One strategy to solve this problem is to make a drawing that compares the structures of channels and transporters and shows their abilities to transport solutes across membranes.

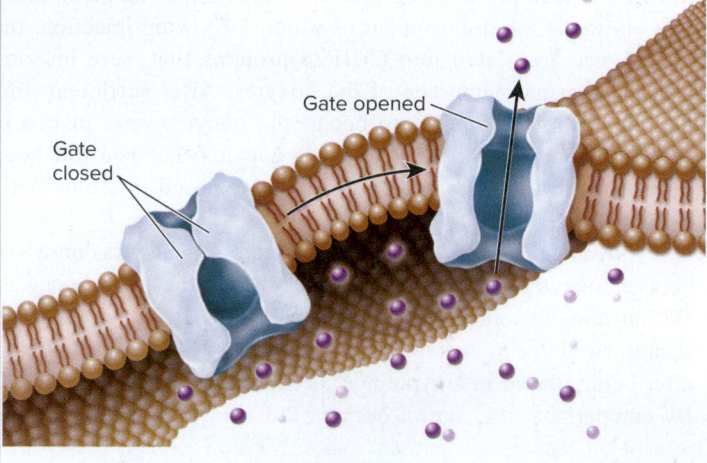

Gate closed

Gate opened

**Channel**

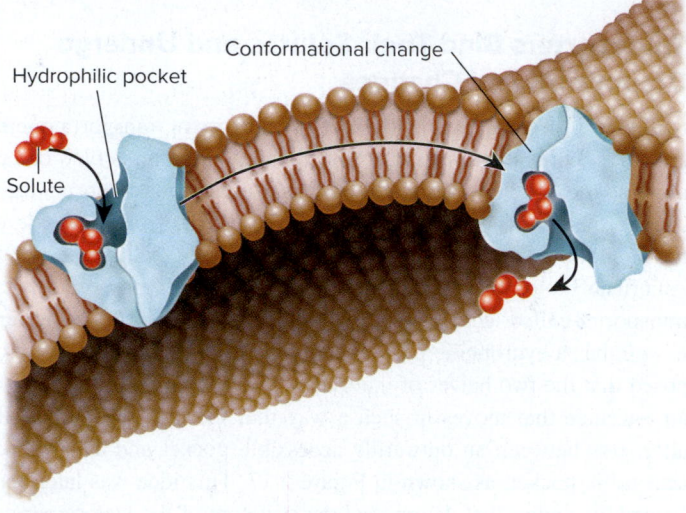

Conformational change

Hydrophilic pocket

Solute

**Transporter**

**BIO TIPS**    **THE QUESTION** *Channels and transporters allow the passage of solutes across membranes. However, at the molecular level, they work in fundamentally different ways. Explain how each type can transport solutes across a membrane.*

**TOPIC** *What topic in biology does this question address?* The topic is membrane transport. More specifically, the question asks you to compare the mechanisms of transport used by channels and transporters.

**INFORMATION** *What information do you know based on the question and your understanding of the topic?* From the question, you know that channels and transporters allow solutes to move across membranes. From your understanding of the topic, you may remember that channels and transporters are transmembrane proteins with different structures.

**ANSWER** *When its gate is open, a channel provides a direct passageway for the movement of a solute across a membrane. A transporter does not provide a direct passageway for such movement. Instead, a solute must first enter a hydrophilic pocket on one side of the membrane. The transporter then undergoes a conformational change that exposes the pocket on the other side of the membrane, where the solute is released.*

## Active Transport Is the Movement of Solutes Against a Gradient

As described earlier, active transport is the movement of a solute across a membrane against its concentration gradient—that is, from a region of lower concentration to one of higher concentration. Active

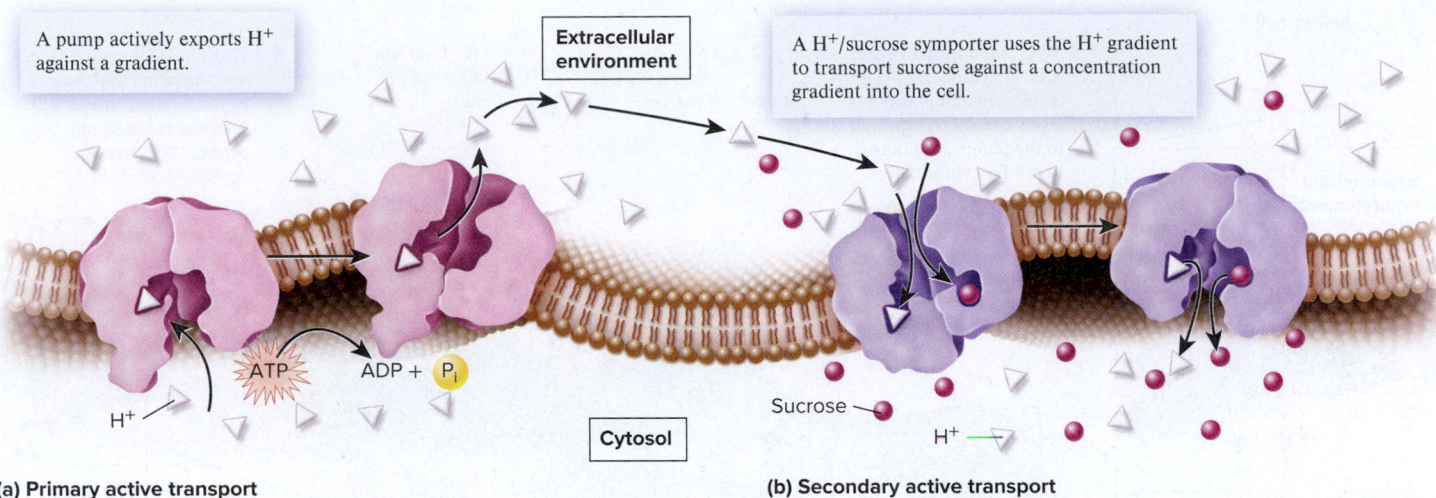

A pump actively exports H$^+$ against a gradient.

**Extracellular environment**

A H$^+$/sucrose symporter uses the H$^+$ gradient to transport sucrose against a concentration gradient into the cell.

ATP    ADP + P$_i$

H$^+$

**Cytosol**

Sucrose

H$^+$

**(a) Primary active transport**

**(b) Secondary active transport**

**Figure 5.19  Types of active transport.** **(a)** During primary active transport, a pump directly uses energy, in this case from ATP, to transport a solute against a concentration gradient. The pump shown here uses ATP to establish an H$^+$ electrochemical gradient. **(b)** Secondary active transport via a symporter involves the use of this gradient to drive the active transport of a solute, such as sucrose.

transport is energetically unfavorable and requires an input of energy. **Primary active transport** involves the functioning of a **pump**—a type of transporter that directly uses energy to transport a solute against a concentration gradient. **Figure 5.19a** shows a pump that uses ATP to transport H$^+$ against a gradient. Such a pump can establish an H$^+$ electrochemical gradient across a membrane.

**Secondary active transport** is a process in which a pre-existing gradient drives the active transport of another solute. For example, an H$^+$/sucrose symporter uses an H$^+$ electrochemical gradient, established by a pump, to move sucrose against its concentration gradient (**Figure 5.19b**). In this case, only sucrose is actively transported. Hydrogen ions move down their electrochemical gradient. H$^+$/solute symporters are more common in bacteria, fungi, algae, and plant cells, because H$^+$ pumps are found in their plasma membranes. In animal cells, a pump that exports Na$^+$ maintains a Na$^+$ gradient across the plasma membrane. Na$^+$/solute symporters are prevalent in animal cells.

Symporters enable cells to actively import nutrients against a gradient. These proteins use the energy stored in the electrochemical gradient of H$^+$ or Na$^+$ to power the uphill movement of organic solutes such as sugars, amino acids, and other necessary molecules. Therefore, with symporters in their plasma membrane, cells can scavenge nutrients from the extracellular environment and accumulate them to high levels within the cytoplasm.

## ATP-Driven Ion Pumps Generate Ion Electrochemical Gradients

The phenomenon of active transport was discovered in the 1940s based on the study of the transport of sodium ions (Na$^+$) and potassium ions (K$^+$). In animal cells, the concentration of Na$^+$ is lower inside the cell than outside, whereas the concentration of K$^+$ is higher inside the cell than outside. After analyzing the movement of these ions across the plasma membranes of muscle cells, neurons, and red blood cells, researchers determined that the export of Na$^+$ is coupled to the import of K$^+$. In the late 1950s, Danish biochemist Jens Skou

proposed that a single transporter is responsible for this phenomenon. He was the first to describe an ATP-driven ion pump, which was later named Na$^+$/K$^+$-ATPase. This pump actively transports Na$^+$ and K$^+$ against their gradients by using the energy from ATP hydrolysis. The plasma membrane of a typical animal cell contains thousands of Na$^+$/K$^+$-ATPase pumps that maintain large concentration gradients in which the concentration of Na$^+$ is higher outside the cell and the concentration of K$^+$ is higher inside the cell.

Let's take a closer look at the Na$^+$/K$^+$-ATPase that Skou discovered. Every time one ATP is hydrolyzed, the Na$^+$/K$^+$-ATPase functions as an antiporter that pumps three Na$^+$ into the extracellular environment and two K$^+$ into the cytosol (**Figure 5.20a**). Because one cycle of pumping results in the net export of one positive charge, the Na$^+$/K$^+$-ATPase also produces an electrical gradient across the membrane. For this reason, it is called an **electrogenic pump**, because it generates an electrical gradient.

By studying the interactions of Na$^+$, K$^+$, and ATP with the Na$^+$/K$^+$-ATPase pump, researchers have pieced together a molecular road map of the steps that direct the pumping of ions across the membrane (**Figure 5.20b**). The Na$^+$/K$^+$-ATPase alternates between two conformations, designated E1 and E2. In E1, the ion-binding sites are accessible from the cytosol—Na$^+$ binds tightly to this conformation, whereas K$^+$ has a low affinity. In E2, the ion-binding sites are accessible from the extracellular environment—Na$^+$ has a low affinity, and K$^+$ binds tightly.

To examine the pumping mechanism of Na$^+$/K$^+$-ATPase, let's begin with the E1 conformation. Three Na$^+$ bind to the Na$^+$/K$^+$-ATPase from the cytosol (Figure 5.20b). When this occurs, ATP is hydrolyzed to ADP and phosphate. Temporarily, the phosphate is covalently bound to the pump, an event called phosphorylation. The pump then switches to the E2 conformation. The three Na$^+$ are released into the extracellular environment, because they have a lower affinity for the E2 conformation. In this conformation, two K$^+$ bind from the outside. The binding of two K$^+$ causes the release of phosphate, which, in turn, causes a switch to E1. Because the E1 conformation has a low affinity

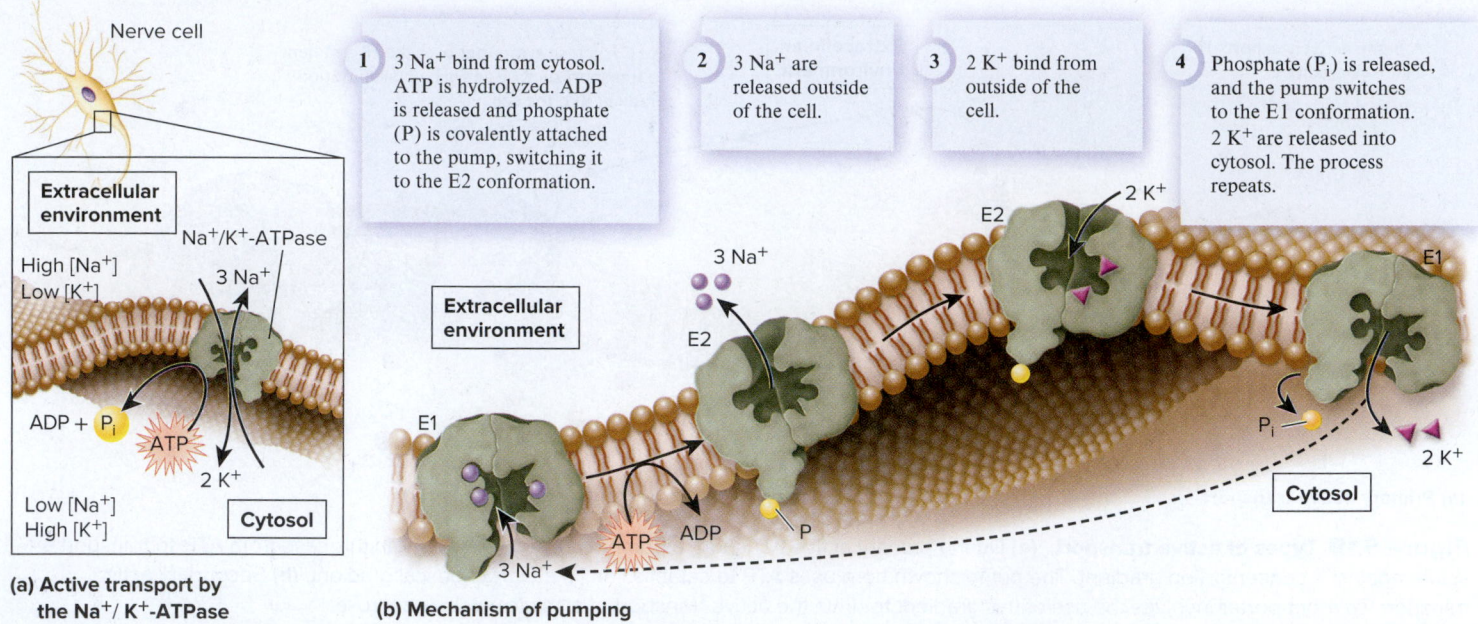

Nerve cell

**①** 3 Na⁺ bind from cytosol. ATP is hydrolyzed. ADP is released and phosphate (P) is covalently attached to the pump, switching it to the E2 conformation.

**②** 3 Na⁺ are released outside of the cell.

**③** 2 K⁺ bind from outside of the cell.

**④** Phosphate (Pᵢ) is released, and the pump switches to the E1 conformation. 2 K⁺ are released into cytosol. The process repeats.

**Extracellular environment**

Na⁺/K⁺-ATPase

High [Na⁺]
Low [K⁺]

3 Na⁺

ADP + Pᵢ

ATP

2 K⁺

Low [Na⁺]
High [K⁺]

**Cytosol**

**(a) Active transport by the Na⁺/K⁺-ATPase**

2 K⁺

E2

**Extracellular environment**

3 Na⁺

E2

E1

E1

ATP    ADP    P

3 Na⁺

Pᵢ

2 K⁺

**Cytosol**

**(b) Mechanism of pumping**

**Figure 5.20** **Structure and function of the Na⁺/K⁺-ATPase.** **(a)** Active transport by Na⁺/K⁺-ATPase. Each time this protein hydrolyzes one ATP molecule, it pumps out three Na⁺ and pumps in two K⁺. **(b)** Pumping mechanism. This figure illustrates the protein conformational changes between E1 and E2. As this occurs, ATP is hydrolyzed to ADP and phosphate. During the process, phosphate is covalently attached to the protein but is released after two K⁺ bind.

 **Core Concept: Energy and Matter** The Na⁺/K⁺-ATPase uses energy to establish Na⁺ and K⁺ gradients across the plasma membrane of animal cells.

for K⁺, the two K⁺ are released into the cytosol. The Na⁺/K⁺-ATPase is now ready for another round of pumping.

Na⁺/K⁺-ATPase is a critical ion pump in animal cells because it maintains Na⁺ and K⁺ gradients across the plasma membrane. Many other types of ion pumps are also found in the plasma membrane and in the membranes of organelles. Ion pumps play the primary role in the formation and maintenance of ion electrochemical gradients that drive many important cellular processes (**Table 5.3**). ATP is commonly the source of energy that drives ion pumps, and cells typically use a

substantial portion of their ATP to keep these pumps working. For example, neurons use up to 70% of their ATP to operate ion pumps!

## 5.6 Exocytosis and Endocytosis

**Learning Outcome:**

**1.** Describe the steps in exocytosis and endocytosis.

We have seen that most small substances are transported via transmembrane proteins such as channels and transporters, which provide passageways for the movement of ions and molecules directly across the membrane. Eukaryotic cells have two other mechanisms, exocytosis and endocytosis, for transporting larger molecules such as proteins and polysaccharides, and even very large particles. Both mechanisms involve the packaging of the transported substance, sometimes called the cargo, into a membrane vesicle or vacuole. **Table 5.4** describes some examples.

### Exocytosis

During **exocytosis**, material inside the cell is packaged into vesicles and then excreted into the extracellular environment (**Figure 5.21**). These vesicles are usually derived from the Golgi apparatus. As a vesicle forms, a specific cargo is loaded into the interior. The budding process involves the formation of a protein coat around the emerging vesicle. The assembly of the proteins to make the coat on the surface of the Golgi membrane causes the bud to form. Eventually, the bud separates from the membrane to form a vesicle. After the vesicle is

| Table 5.3 | Important Functions of Ion Electrochemical Gradients |
|---|---|
| **Function** | **Description** |
| Transport of ions and molecules | Symporters and antiporters use H⁺ and Na⁺ gradients to take up nutrients and export waste products (see Figure 5.19). |
| Production of energy intermediates | In the mitochondrion and chloroplast, H⁺ gradients are used to synthesize ATP. |
| Osmotic regulation | Animal cells control their internal volume by regulating ion gradients between the cytosol and extracellular fluid. |
| Neuronal signaling | Na⁺ and K⁺ gradients are involved in conducting action potentials, the signals transmitted by neurons. |
| Muscle contraction | Ca²⁺ gradients regulate the ability of muscle fibers to contract. |
| Bacterial swimming | H⁺ gradients drive the rotation of bacterial flagella. |

| Table 5.4 | Examples of Exocytosis and Endocytosis |
| --- | --- |
| **Exocytosis** | **Description** |
| Hormones | Certain hormones, such as insulin, are composed of polypeptides. To exert its effect, insulin is secreted via exocytosis into the bloodstream from beta cells of the pancreas. |
| Digestive enzymes | Digestive enzymes that function in the lumen of the small intestine are secreted via exocytosis from exocrine cells of the pancreas. |
| **Endocytosis** | **Description** |
| Uptake of vital nutrients | Many important nutrients are highly insoluble in the blood. Therefore, they are bound to proteins in the blood and then taken into cells via endocytosis. Examples include the uptake of lipids (bound to low-density lipoprotein) and iron (bound to transferrin protein). |
| Root nodules | Nitrogen-fixing root nodules found in certain species of plants, such as legumes, are formed by the endocytosis of bacteria. After being taken up, the bacterial cells are contained within a membrane-enclosed compartment in the nitrogen-fixing tissue of root nodules. |
| Immune system | Cells of the immune system, known as macrophages, engulf and destroy bacteria via phagocytosis. |

released, the coat is shed. Finally, the vesicle fuses with the plasma membrane and releases the cargo into the extracellular environment.

## Endocytosis

During **endocytosis**, the plasma membrane invaginates, or folds inward, to form a vesicle that brings substances into the cell. Three types of endocytosis are receptor-mediated endocytosis, pinocytosis, and phagocytosis.

*Receptor-Mediated Endocytosis*  In **receptor-mediated endocytosis**, a receptor in the plasma membrane is specific for a given cargo

(**Figure 5.22**). Cargo molecules binding to their specific receptors stimulate many receptors to aggregate, and then coat proteins bind to the membrane. The protein coat causes the membrane to invaginate and form a vesicle. Once it is released into the cell, the vesicle sheds its coat. In most cases, the vesicle fuses with an internal organelle, such as a lysosome, and the receptor releases its cargo. Then, the cargo may be directly released into the cytosol, or it may be digested into simpler building blocks before release.

*Pinocytosis*  Other specialized types of endocytosis occur in certain cells. **Pinocytosis** (from the Greek, meaning cell-drinking) involves the formation of membrane vesicles from the plasma membrane as a

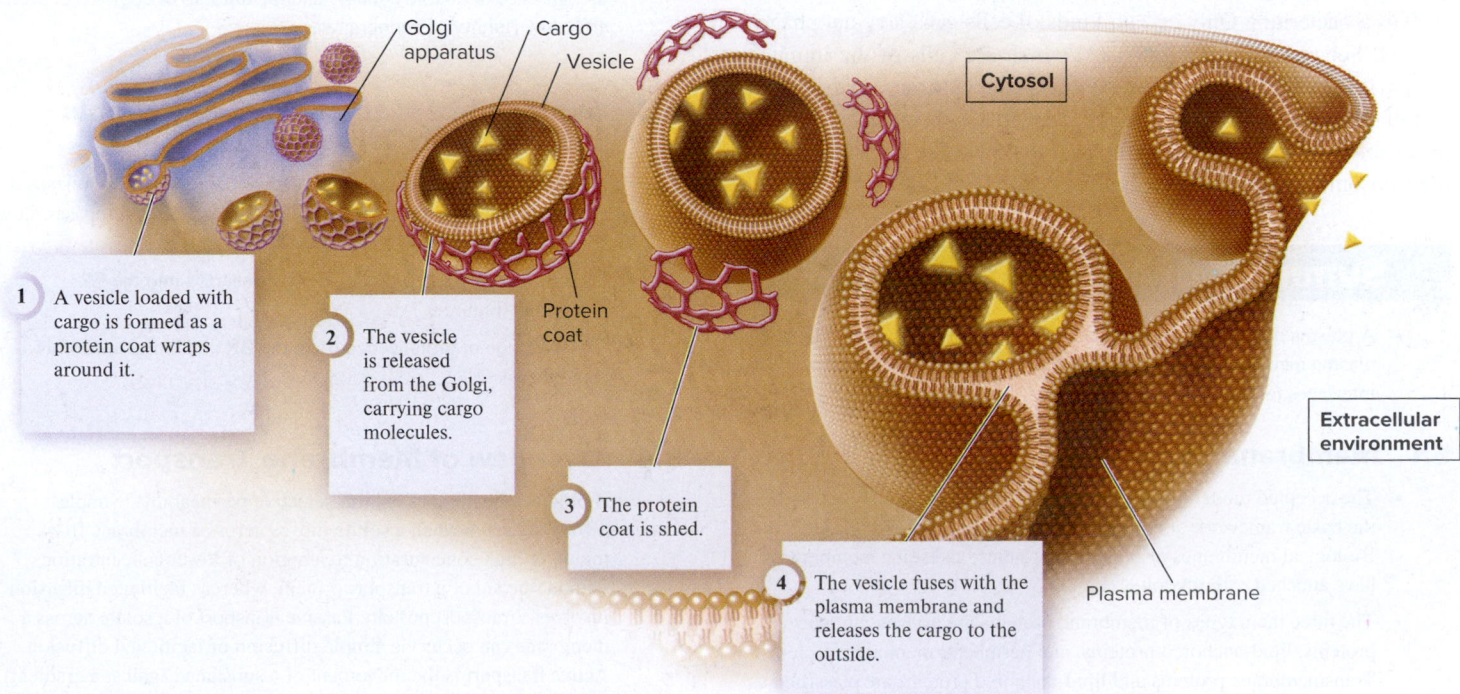

**1** A vesicle loaded with cargo is formed as a protein coat wraps around it.

**2** The vesicle is released from the Golgi, carrying cargo molecules.

**3** The protein coat is shed.

**4** The vesicle fuses with the plasma membrane and releases the cargo to the outside.

Golgi apparatus

Cargo

Vesicle

Cytosol

Protein coat

Extracellular environment

Plasma membrane

**Figure 5.21** **Exocytosis.**

**Concept Check:** *What is the function of the protein coat?*

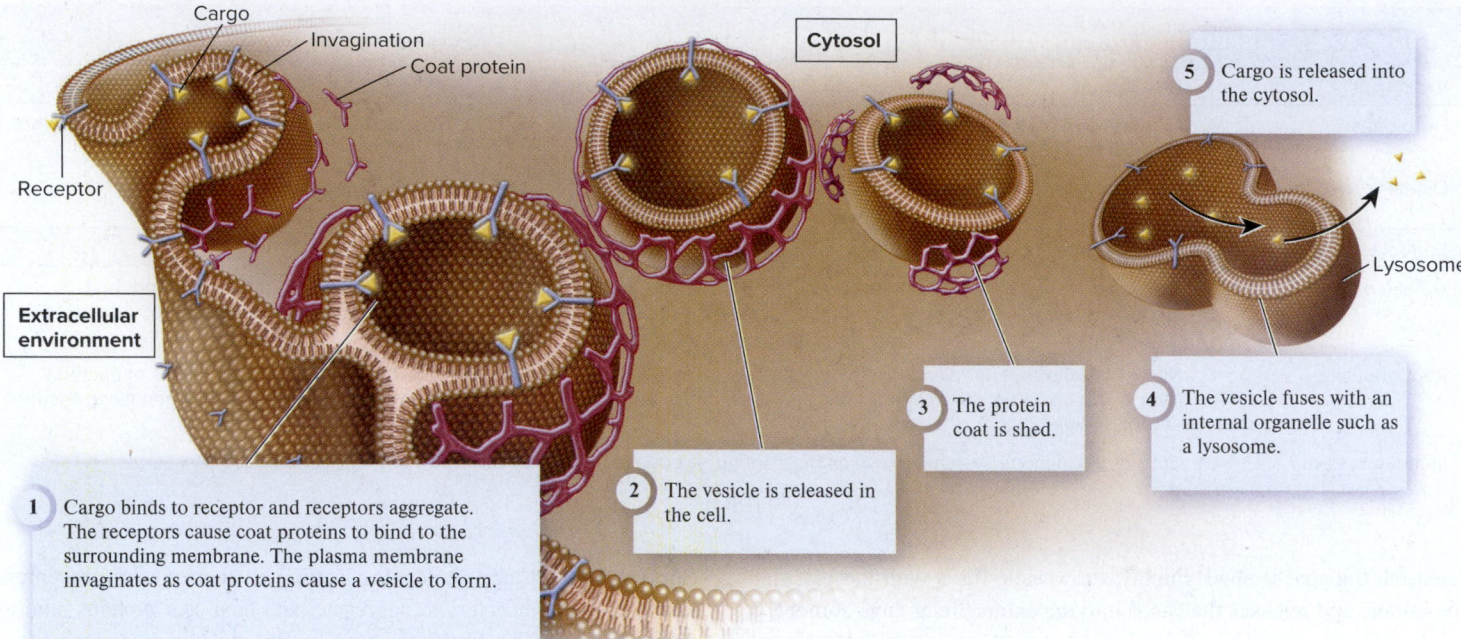

**Figure 5.22** Receptor-mediated endocytosis.

way for cells to internalize the extracellular fluid. This allows cells to sample the extracellular solutes. Pinocytosis is particularly important in cells that are actively involved in nutrient absorption, such as cells that line the intestine in animals.

*Phagocytosis*   Phagocytosis (from the Greek, meaning cell-eating) involves the formation of an enormous membrane vesicle called a phagosome, or phagocytic vacuole, which engulfs a large particle such as a bacterium. Only certain kinds of cells can carry out phagocytosis. For example, macrophages, which are cells of the immune system in mammals, kill bacteria via phagocytosis. Macrophages engulf bacterial cells into phagosomes. Once inside the cell, the phagosome fuses with a lysosome, and the digestive enzymes within the lysosome destroy the bacterium.

## Summary of Key Concepts

- A plasma membrane separates a cell from its surroundings. The plasma membrane and the membranes of organelles provide interfaces for carrying out vital cellular activities (Table 5.1).

### 5.1   Membrane Structure

- The accepted model of membranes is the fluid-mosaic model, and the basic framework of a membrane is the phospholipid bilayer. Biological membranes also contain proteins, and some membranes have attached carbohydrates (Figure 5.1).
- The three main types of membrane proteins are transmembrane proteins, lipid-anchored proteins, and peripheral membrane proteins. Transmembrane proteins and lipid-anchored proteins are classified as integral membrane proteins. Researchers are working to identify new membrane proteins and their functions because these proteins are important biologically and medically (Figure 5.2, Table 5.2).

### 5.2   Fluidity of Membranes

- Membrane fluidity is essential for normal cell function, growth, and division. Lipids and many proteins can move rotationally and laterally, but the flip-flop of lipids from one leaflet to the opposite does not occur spontaneously. Some membrane proteins are restricted in their movements (Figures 5.3, 5.4, 5.5).
- The chemical properties of phospholipids—such as tail length and the presence of double bonds—and the amount of cholesterol present affect the fluidity of membranes.

### 5.3   Synthesis of Membrane Components in Eukaryotic Cells

- In eukaryotic cells, most membrane phospholipids are synthesized at the cytosolic leaflet of the smooth ER membrane. Flippases move some phospholipids to the other leaflet (Figure 5.6).
- Most transmembrane proteins are first inserted into the ER membrane (Figure 5.7).
- Glycosylation of proteins occurs in the ER and Golgi apparatus (Figure 5.8).

### 5.4   Overview of Membrane Transport

- Biological membranes exhibit selective permeability. Simple diffusion occurs when a solute moves across a membrane from a region of high concentration to a region of lower concentration without the aid of a transport protein, whereas facilitated diffusion involves a transport protein. Passive transport of a solute across a membrane can occur via simple diffusion or facilitated diffusion. Active transport is the movement of a substance against a gradient (Figure 5.9).
- The phospholipid bilayer is relatively impermeable to many hydrophilic substances (Figure 5.10).

- Living cells maintain an internal environment that is separated from their external environment. Transmembrane gradients are established across the plasma membrane and across internal membranes (Figure 5.11).

- In the process of osmosis, water moves through a membrane from an area of lower concentration of solute (hypotonic) into an area of higher concentration of solute (hypertonic). Solutions with identical concentrations are isotonic. Some microorganisms have contractile vacuoles to eliminate excess water (Figures 5.12, 5.13, 5.14).

## 5.5   Transport Proteins

- Two classes of transport proteins are channels and transporters.

- Channels provide open passageways for the facilitated diffusion of solutes across the membrane. One example is aquaporin, which allows the movement of water. Most channels are gated, which allows cells to regulate the movement of solutes (Figures 5.15, 5.16).

- Transporters, which tend to function at a slower rate than channels, bind their solutes in a hydrophilic pocket and undergo a conformational change that switches the exposure of the pocket to the other side of the membrane. They can be uniporters, symporters, or antiporters (Figures 5.17, 5.18).

- Primary active transport involves pumps that directly use energy to generate a solute gradient. Secondary active transport uses a pre-existing gradient (Figure 5.19).

- $Na^+/K^+$-ATPase is an electrogenic pump that uses energy from ATP to transport ions across the membrane. Ion electrochemical gradients perform several important cellular functions (Figure 5.20, Table 5.3).

## 5.6   Exocytosis and Endocytosis

- In eukaryotes, exocytosis and endocytosis are used to transport large molecules and particles. Exocytosis is a process in which material inside the cell is packaged into vesicles and excreted into the extracellular environment. During endocytosis, the plasma membrane folds inward to form a vesicle that brings substances into the cell. Receptor-mediated endocytosis, pinocytosis, and phagocytosis are types of endocytosis (Figures 5.21, 5.22, Table 5.4).

## Assess & Discuss

### Test Yourself

1. Which of the following statements best describes the chemical composition of biological membranes?
   a. Biological membranes are bilayers of proteins with associated lipids and carbohydrates.
   b. Biological membranes are composed of two layers—one layer of phospholipids and one layer of proteins.
   c. Biological membranes are bilayers of phospholipids with associated proteins and carbohydrates.
   d. Biological membranes are composed of equal numbers of phospholipids, proteins, and carbohydrates.
   e. Biological membranes are composed of lipids with proteins attached to the outer surface.

2. Which of the following events can never be energetically favorable in a biological membrane and therefore will not occur spontaneously?
   a. the rotation of phospholipids
   b. the lateral movement of phospholipids
   c. the flip-flop of phospholipids to the opposite leaflet
   d. the rotation of membrane proteins
   e. the lateral movement of membrane proteins

3. Let's suppose an insect, which doesn't maintain a constant body temperature, was exposed to a shift in temperature from 60°F to 80°F. Which of the following types of membrane changes would be the most beneficial in helping the insect cope with the temperature shift?
   a. increase the number of double bonds in the lipid tails of phospholipids
   b. increase the length of the lipid tails of phospholipids
   c. decrease the amount of cholesterol in the membrane
   d. decrease the amount of carbohydrate attached to membrane proteins
   e. decrease the amount of carbohydrate attached to phospholipids

4. Carbohydrates of the plasma membrane
   a. are bonded to a protein or lipid.
   b. are located on the outer surface of the plasma membrane.
   c. can function as cell markers for recognition by other cells.
   d. All of the above are true of the carbohydrates.
   e. Only a and c are true.

5. A transmembrane protein in the plasma membrane is glycosylated at two sites in the polypeptide sequence. Where in this protein would you expect these two sites to be?
   a. in transmembrane segments
   b. in hydrophilic regions that project into the extracellular environment
   c. in hydrophilic regions that project into the cytosol
   d. could be anywhere
   e. b and c only

6. The tendency for $Na^+$ to move into the cell can be due to
   a. the higher numbers of $Na^+$ outside the cell, resulting in a chemical concentration gradient.
   b. the net negative charge inside the cell attracting the positively charged $Na^+$.
   c. the attractive force of $K^+$ inside the cell pulling $Na^+$ into the cell.
   d. all of the above.
   e. a and b only.

7. Let's suppose the solute concentration inside the cells of a plant is 0.3 M and the concentration outside is 0.2 M. If we assume that the solute does not readily cross the membrane, which of the following statements best describes what will happen?
   a. The plant cells will lose water, and the plasma membrane will push against the cell wall.
   b. The plant cells will lose water, and the plasma membrane will pull away from the cell wall (plasmolysis).
   c. The plant cells will take up a lot of water and undergo osmotic lysis.
   d. The plant cells will take up a little water, and the plasma membrane will push against the cell wall.
   e. both a and b.

8. What features of a biological membrane are major contributors to its selective permeability?
   a. phospholipid bilayer
   b. transport proteins
   c. glycolipids on the outer surface of the membrane
   d. peripheral membrane proteins
   e. both a and b

9. What is the name given to the process in which solutes are moved across a membrane against their concentration gradient?
   a. simple diffusion
   b. facilitated diffusion
   c. osmosis
   d. passive diffusion
   e. active transport

10. Large particles or large volumes of fluid can be brought into the cell by
    a. facilitated diffusion.
    b. active transport.
    c. endocytosis.
    d. exocytosis.
    e. all of the above.

## Conceptual Questions

1. With your textbook closed, draw and describe the fluid-mosaic model of membrane structure.

2. Describe two different ways that integral membrane proteins associate with a membrane. How do peripheral membrane proteins associate with a membrane?

3. **Core Concept: Energy and Matter**  Discuss how the lipid bilayer, channels, and transporters influence the ability of cells to control the amounts of solutes they contain.

## Collaborative Questions

1. Proteins in the plasma membrane are often the target of medicines. Discuss why you think this is the case. How would you determine experimentally that a specific membrane protein was the target of a drug?

2. With regard to bringing solutes into the cell across the plasma membrane, discuss the advantages and disadvantages of simple diffusion, facilitated diffusion, active transport, and endocytosis.

# An Introduction to Energy, Enzymes, and Metabolism

# 6

**Common drugs that act as enzyme inhibitors.** Drugs such as aspirin and ibuprofen exert their effects by inhibiting an enzyme that speeds up a cellular chemical reaction. ©McGraw-Hill Education/Aaron Roeth, photographer

**H**ave you ever taken aspirin or ibuprofen to relieve a headache or reduce a fever? Do you know how these medications work? If you answered "no" to the second question, you're not alone. Over 2,000 years ago, humans began treating pain with powder from the bark and leaves of the willow tree, which contains a compound called salicylic acid. Modern aspirin is composed of a derivative of salicylic acid called acetylsalicylic acid, which is gentler to the stomach. Only recently, however, have we learned how these drugs work. Aspirin and ibuprofen are examples of drugs that inhibit a specific enzyme called cyclooxygenase. This enzyme is needed to synthesize molecules called prostaglandins, which play a role in inflammation, pain, and fever. Aspirin and ibuprofen exert their effects by inhibiting cyclooxygenase, thereby blocking the production of prostaglandins and in turn relieving pain and fever.

Enzymes are proteins that act as critical catalysts to speed up thousands of different reactions in cells. As discussed in Chapter 2, a **chemical reaction** is a process in which one or more substances are changed into other substances. Such reactions may involve molecules attaching to each other to form larger molecules, molecules breaking apart to form two or more smaller molecules, rearrangements of atoms within molecules, or the transfer of electrons from one atom to another. Every living cell continuously performs thousands of such chemical reactions to sustain life. **Metabolism** is the sum total of all chemical reactions that occur within an organism. Metabolism also refers to a specific set of chemical reactions occurring at the cellular level. For example, biologists may speak of sugar metabolism or fat metabolism. Most types of metabolism involve the breakdown or synthesis of organic molecules. Cells maintain their structure by using organic molecules. Such molecules provide the building blocks for constructing cells, and the chemical bonds within organic molecules store energy that is used to drive cellular processes.

In this chapter, we begin with a general discussion of energy and chemical reactions. We will examine what factors control the direction of a chemical reaction and what determines its rate, paying particular attention to the role of enzymes. We then consider metabolism at the cellular level, examining how chemical reactions are often coordinated with each other in metabolic pathways. We will also explore the variety of ways in which metabolic pathways are regulated and how organic molecules are recycled.

## 6.1 Energy and Chemical Reactions

**Learning Outcomes:**

1. Define energy, and distinguish between potential and kinetic energy.
2. State the first and second laws of thermodynamics, and discuss how they relate to living things.
3. Explain how the change in free energy determines the direction of a chemical reaction and how chemical reactions eventually reach a state of equilibrium.
4. Distinguish between exergonic and endergonic reactions in terms of the energy of the reactants and products and the free energy change.
5. Describe how cells use the energy released by the hydrolysis of ATP to drive endergonic reactions.

Two general factors govern the fate of a given chemical reaction in a living cell—its direction and rate. To illustrate this point, let's consider this generalized chemical reaction:

$$a\text{A} + b\text{B} \rightleftharpoons c\text{C} + d\text{D}$$

where A and B are the reactants, C and D are the products, and $a$, $b$, $c$, and $d$ are the number of moles of reactants and products. This reaction is reversible, which means that A + B could be converted to C + D, or C + D could be converted to A + B. The direction of the reaction, whether C + D are made (the forward direction) or A + B are made (the reverse direction), depends on energy and on the concentrations of A, B, C, and D. In this section, we will begin by examining the interplay of energy and reactants' concentrations as they govern the direction of a chemical reaction. You will learn that cells use energy intermediate molecules, such as ATP, to drive chemical reactions in a desired direction.

## Energy Exists in Different Forms

To understand why a chemical reaction occurs, we first need to consider **energy**, which is the ability to promote change or do work. Physicists often consider energy in two general forms: kinetic energy and potential energy (**Figure 6.1**). **Kinetic energy** is energy associated with movement, such as the movement of a baseball bat from one location to another. By comparison, **potential energy** is the energy that a substance or object possesses due to its structure or location. An electron in an atom has potential energy based on its position relative to other electrons and the positively charged nucleus. As you may recall from Chapter 2, electrons occupy orbitals of different shapes and sizes, which are found within electron shells, or energy levels (refer back to Figure 2.4). An electron in an outer shell has a higher amount of potential energy than one in an inner shell. If an electron drops to a lower shell, some of its potential energy is converted to kinetic energy.

The energy that is stored in atoms and in the bonds between atoms is called **chemical potential energy** (or simply, chemical energy). This energy can be released during chemical reactions. Organic molecules, such as glucose, store a great deal of potential energy. As discussed in Chapter 7, the breakdown of glucose releases energy that is harnessed to make molecules such as ATP that are energy

**(a) Kinetic energy**  **(b) Potential energy**

**Figure 6.1** **Examples of energy. (a)** Kinetic energy, such as that of a swinging bat, is energy associated with motion. **(b)** Potential energy is stored energy, as in a bow that is ready to shoot an arrow. a: ©moodboard/Corbis; b: ©amanaimages/Corbis

intermediates. **Table 6.1** summarizes chemical potential energy and other forms of energy that are common in biological systems.

An important phenomenon in biology is the ability of energy to be converted from one form to another. The study of energy interconversions is called **thermodynamics**. Let's consider two laws of thermodynamics that govern energy interconversions:

1. *The first law of thermodynamics.* The first law of thermodynamics, also called the law of conservation of energy, states that energy cannot be created or destroyed. However, energy can be transferred from one place to another and can be transformed from one type to another (as when, for example, chemical energy is transformed into heat).

2. *The second law of thermodynamics.* The second law states that any energy transfer or transformation from one form to another increases the degree of disorder of a system, called **entropy** (**Figure 6.2**). Entropy is a measure of the randomness of molecules in a system. When a physical system becomes more disordered, the entropy increases. As the energy becomes more evenly distributed, that energy is less able to promote change or do work. When energy is converted from one form to another, some energy may become unusable by living organisms. For example, a chemical reaction may release unusable heat.

| Table 6.1 | Types of Energy That Are Important in Biology | |
| --- | --- | --- |
| **Energy type** | **Description** | **Biological example** |
| Light | Light is a form of electromagnetic radiation that is visible to the eye. The energy of light is packaged in photons. | During photosynthesis, light energy is captured by pigments in chloroplasts (described in Chapter 8). Ultimately, this energy is used to produce organic molecules. |
| Heat | Heat is the transfer of kinetic energy from one object to another or from an energy source to an object. In biology, heat is often viewed as kinetic energy that can be transferred due to a difference in temperature between two objects or locations. | Many organisms, including humans, maintain their bodies at a constant temperature. This is achieved, in part, by chemical reactions that generate heat. |
| Mechanical | Mechanical energy is the energy possessed by an object due to its motion or its position relative to other objects. | In animals, mechanical energy is associated with movement due to muscle contraction, such as walking. |
| Chemical potential | Chemical potential energy is potential energy stored in the electrons of molecules. When bonds are broken and rearranged, energy may be released. | The covalent bonds in organic molecules, such as glucose and ATP, store large amounts of energy. When bonds are broken in larger molecules to form smaller molecules, the energy that is released can be used to drive cellular processes. |
| Electrical/ion gradient | The movement of charge or the separation of charges can provide energy. Also, a difference in ion concentration across a membrane constitutes an electrochemical gradient, which is a source of potential energy. | During a stage of cellular respiration called oxidative phosphorylation (described in Chapter 7), an $H^+$ gradient provides the energy to drive ATP synthesis. |

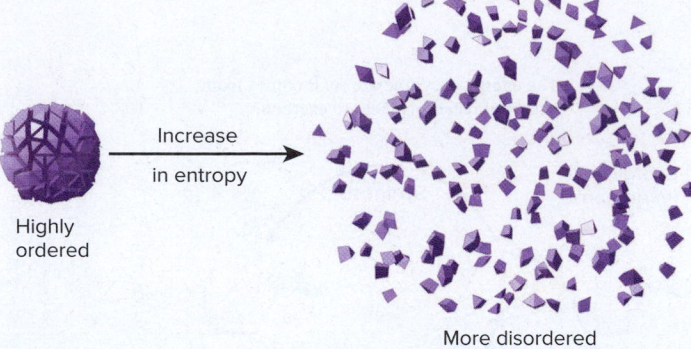

**Figure 6.2 Entropy, a measure of the disorder of a system.** An increase in entropy means an increase in disorder.

**Concept Check:** *Which do you think has more entropy, a NaCl crystal at the bottom of a beaker of water or the solution that would be formed after the Na⁺ and Cl⁻ ions forming the crystal have dissolved in the water?*

Next, we will see how the two laws of thermodynamics place limits on the ways that living cells use energy for their own needs.

## The Change in Free Energy Determines the Direction of a Chemical Reaction

Energy is required for many cellular processes, including chemical reactions, cellular movements such as those occurring in muscle contraction, and the maintenance of cell organization. To understand how organisms use energy, we need to distinguish between the energy that can be used to promote change or do work (usable energy) and the energy that cannot (unusable energy).

$$\text{Total energy} = \text{Usable energy} + \text{Unusable energy}$$

Why is some energy unusable? The main culprit is entropy. As stated by the second law of thermodynamics, energy transfers or transformations involve an increase in entropy, a degree of disorder that cannot be harnessed in a useful way. The total energy of a system is termed **enthalpy (H)**, and the usable energy—the amount of energy that is available and can be used to promote change or do work—is called the **free energy (G)**. The use of the letter G is in recognition of American physicist J. Willard Gibbs, who proposed the concept of free energy in 1878. The unusable energy is the system's entropy (S). Gibbs proposed that these three components of a system's energy are related to each other in the following way:

$$H = G + TS$$

where T is the absolute temperature in kelvins (K). Because our focus is on free energy, we can rearrange this equation as

$$G = H - TS$$

A critical issue in biology is whether a process does or does not occur spontaneously. For example, will glucose be broken down into carbon dioxide and water? Another way of framing this question is to ask: "Is the breakdown of glucose a spontaneous, or favorable, reaction?" A spontaneous reaction or process is one that occurs without being driven by an input of energy. However, a spontaneous reaction does not necessarily proceed quickly. In some cases, the rate of a spontaneous reaction can be quite slow. For example, the breakdown of sugar is a spontaneous reaction, but the rate at which sugar in a sugar bowl breaks down into $CO_2$ and $H_2O$ is very slow.

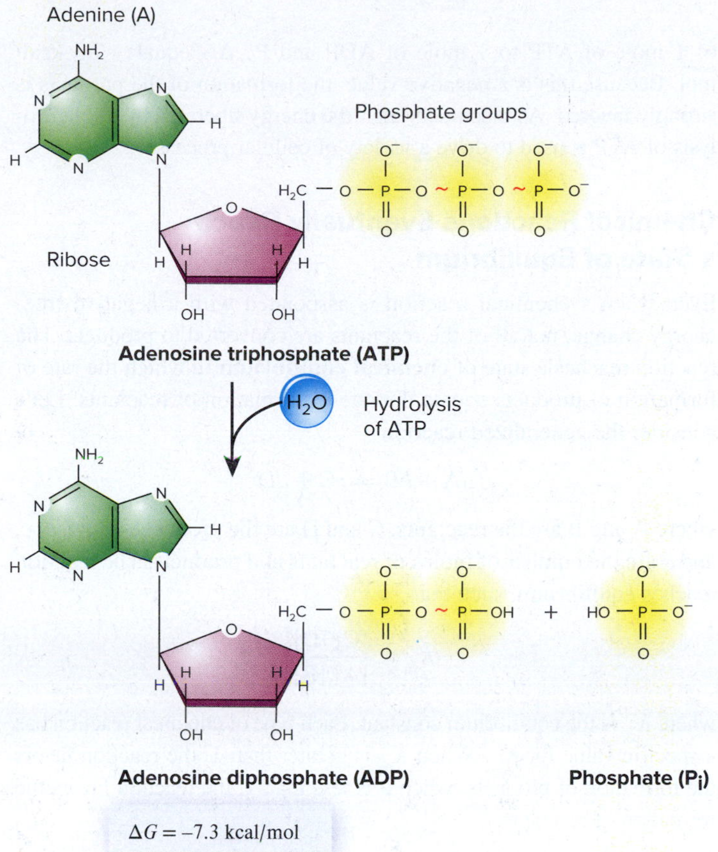

Adenine (A)

Phosphate groups

**Adenosine triphosphate (ATP)**

$H_2O$  Hydrolysis of ATP

**Adenosine diphosphate (ADP)**       **Phosphate (Pᵢ)**

$$\Delta G = -7.3 \text{ kcal/mol}$$

**Figure 6.3 The hydrolysis of ATP to ADP and Pᵢ.** As shown in this figure, ATP has a net charge of −4, while ADP and Pᵢ are shown with net charges of −2 each. When these compounds are shown in chemical reactions with other molecules, the net charges are also indicated. Otherwise, these compounds are simply designated ATP, ADP, and Pᵢ. At neutral pH, $ADP^{2-}$ dissociates to $ADP^{3-}$ and $H^+$.

**Core Concept: Energy and Matter** The hydrolysis of ATP is an exergonic reaction that is used to drive many cellular processes, such as chemical reactions.

The key way to evaluate if a chemical reaction is spontaneous is to determine the free-energy change that occurs as a result of the reaction:

$$\Delta G = \Delta H - T \Delta S$$

where the symbol $\Delta$ (the Greek letter delta) indicates a change, such as before and after a chemical reaction. If a chemical reaction has a negative free-energy change ($\Delta G < 0$), the products have less free energy than the reactants, and, therefore, free energy is released during product formation. Such a reaction is said to be **exergonic**. Exergonic reactions are spontaneous. Alternatively, if a reaction has a positive free-energy change ($\Delta G > 0$), requiring the addition of free energy, it is termed **endergonic**. An endergonic reaction is not a spontaneous reaction.

If $\Delta G$ for a chemical reaction is negative, the reaction favors the conversion of reactants to products, whereas a reaction with a positive $\Delta G$ favors the formation of reactants. Chemists have determined free-energy changes for a variety of chemical reactions, which allows them to predict their direction. As an example, let's consider **adenosine triphosphate (ATP)**, which is a molecule that is a common energy source for all cells. ATP is broken down to adenosine diphosphate (ADP) and inorganic phosphate ($HPO_4^{2-}$, abbreviated Pᵢ). Because water is used to remove a phosphate group, chemists refer to this reaction as the hydrolysis of ATP (**Figure 6.3**). For the conversion of 1 mole of ATP to 1 mole of ADP and $P_i$, $\Delta G$ equals $-7.3$ kcal/mol. Because this is a negative value, the formation of the products is strongly favored. As discussed later, the energy liberated by the hydrolysis of ATP is used to drive a variety of cellular processes.

## Chemical Reactions Eventually Reach a State of Equilibrium

Even when a chemical reaction is associated with a negative free-energy change, not all of the reactants are converted to products. The reaction reaches a state of **chemical equilibrium** in which the rate of formation of products equals the rate of formation of reactants. Let's consider the generalized reaction

$$a\text{A} + b\text{B} \rightleftharpoons c\text{C} + d\text{D}$$

where A and B are the reactants, C and D are the products, and $a$, $b$, $c$, and $d$ are the number of moles of reactants and products. The reaction reaches equilibrium, such that

$$K_{eq} = \frac{[\text{C}]^c[\text{D}]^d}{[\text{A}]^a[\text{B}]^b}$$

where $K_{eq}$ is the equilibrium constant. Each type of chemical reaction has a specific value for $K_{eq}$. When $K_{eq}$ is greater than 1, the reaction favors the formation of products; when it is less than 1, the reaction favors the formation of reactants.

## Cells Use ATP to Drive Endergonic Reactions

Many biological processes require the addition of free energy; that is, they are endergonic and do not occur spontaneously. How do cells overcome this problem? One strategy is to couple exergonic reactions with endergonic reactions. If an exergonic reaction is coupled with an endergonic reaction, the endergonic reaction will proceed spontaneously if the net free-energy change for both processes combined is negative. For example, consider the following reactions:

$$\text{Glucose} + \text{Phosphate}^{2-} \rightarrow \text{Glucose-6-phosphate}^{2-} + \text{H}_2\text{O}$$
$$\Delta G = +3.3 \text{ kcal/mol}$$

$$\text{ATP}^{4-} + \text{H}_2\text{O} \rightarrow \text{ADP}^{2-} + \text{P}_i^{2-}$$
$$\Delta G = -7.3 \text{ kcal/mol}$$

Coupled reaction:

$$\text{Glucose} + \text{ATP}^{4-} \rightarrow \text{Glucose-6-phosphate}^{2-} + \text{ADP}^{2-}$$
$$\Delta G = -4.0 \text{ kcal/mol}$$

The first reaction, in which phosphate is covalently attached to glucose, is endergonic, and by itself is not spontaneous. The second reaction, the hydrolysis of ATP, is exergonic. If the two reactions are coupled, however, the combined net free-energy change for both is negative ($\Delta G = -4.0$ kcal/mol), and the coupled reaction is exergonic. In the coupled reaction, a phosphate is directly transferred from ATP to glucose, in a process called **phosphorylation**. This coupled reaction proceeds spontaneously because the net free-energy change is negative. Exergonic reactions, such as the breakdown of ATP, are commonly coupled to chemical reactions and other cellular processes that would otherwise be endergonic.

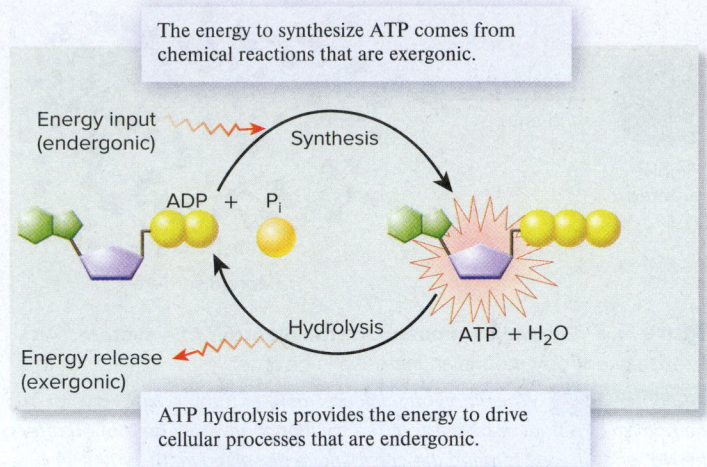

The energy to synthesize ATP comes from chemical reactions that are exergonic.

Energy input (endergonic)

Synthesis

ADP + $P_i$

Hydrolysis

ATP + $H_2O$

Energy release (exergonic)

ATP hydrolysis provides the energy to drive cellular processes that are endergonic.

**Figure 6.4** **The ATP cycle.** Living cells continuously recycle ATP. The energy released from the breakdown of food molecules into smaller molecules is used to synthesize ATP from ADP and $P_i$. The hydrolysis of ATP to ADP and $P_i$ is used to drive many different endergonic reactions and processes that occur in cells.

**Concept Check:** *If a large amount of ADP was broken down in a cell, how would this affect the ATP cycle?*

In humans, a typical cell uses millions of ATP molecules per second to drive endergonic processes. At the same time, the breakdown of food molecules to form smaller molecules (an exergonic reaction) releases energy that allows cells to make more ATP from the phosphorylation of ADP (an endergonic reaction). The recycling of ATP from ADP and phosphate occurs at a remarkable pace. An average person hydrolyzes about 100 pounds of ATP per day, yet at any given time we do not have 100 pounds of ATP in our bodies. For this to happen, each molecule of ATP undergoes about 10,000 cycles of hydrolysis and regeneration during an ordinary day (**Figure 6.4**).

### Core Concepts: Information, Energy and Matter

### Genomes Encode Many Proteins That Use ATP as a Source of Energy

Over the past several decades, researchers have studied the functions of many proteins and discovered numerous examples in which a protein uses the hydrolysis of ATP to drive a chemical reaction or other type of cellular process (**Table 6.2**). By studying the structures of proteins that use ATP in this way, biochemists have discovered that particular amino acid sequences within proteins function as ATP-binding sites. This information has allowed researchers to predict whether a newly discovered protein uses ATP or not. When an entire genome sequence of a species has been determined, the genes that encode proteins can be analyzed to find out if the encoded proteins have ATP-binding sites in their amino acid sequences. Using this approach, researchers have been able to analyze **proteomes**—all of the proteins that a given cell

| Table 6.2 | Examples of Proteins That Use ATP for Energy |
|---|---|
| **Type** | **Description** |
| Metabolic enzymes | Many enzymes use ATP to catalyze endergonic reactions. For example, hexokinase uses ATP to attach phosphate to glucose, producing glucose-6-phosphate. |
| Transporters | Ion pumps, such as $Na^+/K^+$-ATPase, use ATP to pump ions against a gradient (see Chapter 5). |
| Motor proteins | Motor proteins, such as myosin, use ATP to facilitate cellular movement, as in muscle contraction (see Chapter 45). |
| Chaperones | Chaperones are proteins that use ATP to aid in the folding and unfolding of cellular proteins (see Chapter 4). |
| DNA-modifying enzymes | Many proteins, such as helicases and topoisomerases, use ATP to modify the conformation of DNA (see Chapter 11). |
| Aminoacyl-tRNA synthetases | These synthetases are enzymes that use ATP to attach amino acids to tRNAs (transfer RNAs; see Chapter 12). |
| Protein kinases | Protein kinases are regulatory proteins that use ATP to attach a phosphate to a protein, thereby phosphorylating the protein and affecting its function (see Chapter 9). |

or organism makes—and estimate the percentage of proteins that are able to bind ATP. This approach has been applied to the proteomes of bacteria, archaea, and eukaryotes.

On average, over 20% of all proteins bind ATP. However, this number is likely an underestimate because all of the types of ATP-binding sites in proteins may not have been identified. In humans, whose genome has an estimated size of 22,000 different protein-encoding genes, a minimum of 4,400 of those genes encode proteins that use ATP. From these numbers, we can see the enormous importance of ATP as a source of energy for living cells.

## 6.2 Enzymes and Ribozymes

### Learning Outcomes:

1. Explain how enzymes increase the rates of chemical reactions by lowering the activation energy.
2. Describe how enzymes bind their substrates with high specificity and undergo induced fit.
3. **CoreSKILL »** Analyze the velocity of chemical reactions, and evaluate the effects of competitive and noncompetitive inhibitors.
4. Explain how additional factors, such as nonprotein molecules or ions, temperature, and pH, influence enzyme activity.
5. Identify the unique feature of ribozymes.

For most chemical reactions in cells to proceed at a rapid pace, a catalyst is needed. A **catalyst** is an agent that speeds up the rate of a chemical reaction without being permanently changed or consumed by it. In living cells, the most common catalysts are **enzymes**, which are proteins. The term was coined in 1876 by a German physiologist,

Wilhelm Kühne, who discovered trypsin, an enzyme in pancreatic juice that is needed for the digestion of food proteins. In this section, we will explore how enzymes increase the rates of chemical reactions. Interestingly, some biological catalysts are RNA molecules called ribozymes. We will examine a few examples in which RNA molecules carry out catalytic functions.

### Enzymes Increase the Rates of Chemical Reactions

If a chemical reaction has a negative free-energy change, the reaction will be spontaneous; it will tend to proceed in the direction of reactants to products. Although thermodynamics governs the direction of an energy transformation, it does not determine the rate of a chemical reaction. For example, the breakdown of the molecules in gasoline to smaller molecules is an exergonic reaction. Even so, we could place gasoline and oxygen in a container and nothing much would happen (provided the container wasn't near a flame). If we came back several days later, we would expect to see the gasoline still sitting there. Perhaps if we came back in a few million years, the gasoline would have been broken down. On a timescale of months or a few years, however, the chemical reaction would proceed very slowly.

In living cells, the rates of enzyme-catalyzed reactions typically occur millions of times faster than the corresponding uncatalyzed reactions. A dramatic example involves the enzyme catalase, which catalyzes the breakdown of hydrogen peroxide ($H_2O_2$) into water and oxygen. Catalase speeds up this reaction so that it occurs $10^{15}$-fold faster than the uncatalyzed reaction!

Why are catalysts necessary to speed up a chemical reaction? Chemical reactions between molecules involve bond breaking and bond forming. When a covalent bond is broken or formed, this process initially involves the straining or stretching of one or more bonds in the starting molecule(s) and/or the positioning of two molecules so that they interact with each other properly. Enzymes help to facilitate these kinds of events.

As an example, let's consider the reaction in which ATP is used to phosphorylate glucose:

$$\text{Glucose} + \text{ATP}^{4-} \rightarrow \text{Glucose-6-phosphate}^{2-} + \text{ADP}^{2-}$$

For a reaction to occur between glucose and ATP, the molecules must collide in the correct orientation and possess enough energy so the chemical bonds can be changed. As glucose and ATP get close together, the electrons in the outer shells of their atoms repel each other. To overcome this repulsion, an initial input of energy, called the **activation energy**, is required (**Figure 6.5**). Activation energy ($E_A$) allows the molecules to get close enough to cause a rearrangement of bonds. With the input of activation energy, glucose and ATP can achieve a **transition state** in which the original bonds have stretched to their limit. Once the reactants have reached the transition state, the chemical reaction can readily proceed to the formation of products, which in this case are glucose-6-phosphate and ADP.

The activation energy required to achieve the transition state is a barrier to the formation of products. This barrier is the reason why the rate of many chemical reactions is very slow. Enzymes lower the activation energy to a point where a small amount of available heat can push the reactants to the transition state.

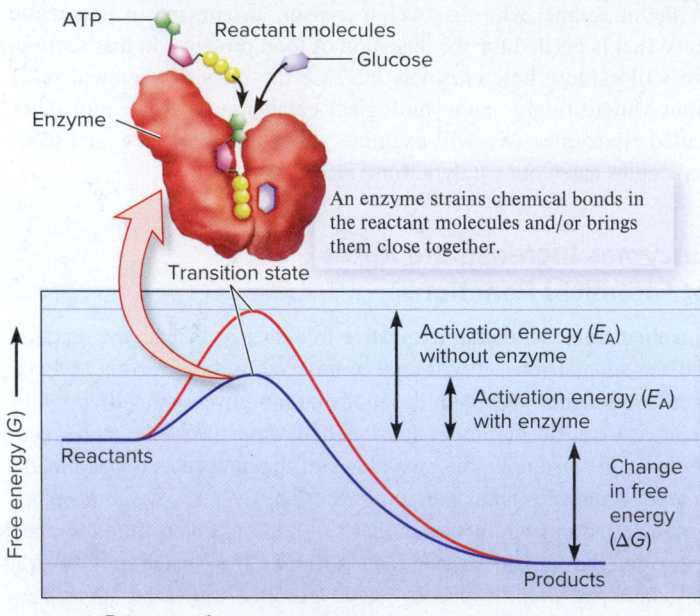

**Figure 6.5** **Activation energy of a chemical reaction.** This figure depicts an exergonic reaction. The activation energy ($E_A$) is needed for molecules to achieve a transition state. One way that enzymes lower the activation energy is by straining chemical bonds in the reactants so less energy is required to attain the transition state. A second way is by binding two reactants so they are close to each other and in a favorable orientation.

> **Concept Check:** *How does lowering the activation energy affect the rate of a chemical reaction? How does it affect the direction?*

How do enzymes lower the activation energy barrier of chemical reactions? Let's consider two common ways that enzymes exert their effects.

- Enzymes are proteins that bind relatively small reactants. When reactant molecules are bound to an enzyme, their bonds can be strained, thereby making it easier for them to achieve the transition state (see Figure 6.5).

- In addition, when a chemical reaction involves two or more reactants, the enzyme provides a site where the reactants are positioned very close to each other in an orientation that facilitates the formation of new covalent bonds. This favorable orientation also lowers the necessary activation energy for a chemical reaction.

## Enzymes Recognize Their Substrates with High Specificity and Undergo Conformational Changes

Thus far, we have considered how enzymes lower the activation energy of a chemical reaction, and thereby increase its rate. Let's consider some other features of enzymes that enable them to serve as effective catalysts in chemical reactions. The **active site** is the location in an enzyme where the chemical reaction takes place. The **substrates** for an enzyme are the reactant molecules that bind to an enzyme at the active site and participate in the chemical reaction. For example, hexokinase is an enzyme whose substrates are glucose and ATP (**Figure 6.6**). The binding between enzyme and substrate produces an **enzyme-substrate complex**.

A key feature of nearly all enzymes is their ability to bind their substrates with a high degree of specificity. For example, hexokinase recognizes glucose but does not recognize other similar sugars, such as fructose and galactose, very well. In 1894, a German chemist Emil Fischer proposed that the recognition of a substrate by an enzyme resembles the interaction between a lock and key: Only the correctly shaped key (the substrate) will fit into the keyhole (active site) of the lock (the enzyme). Further research

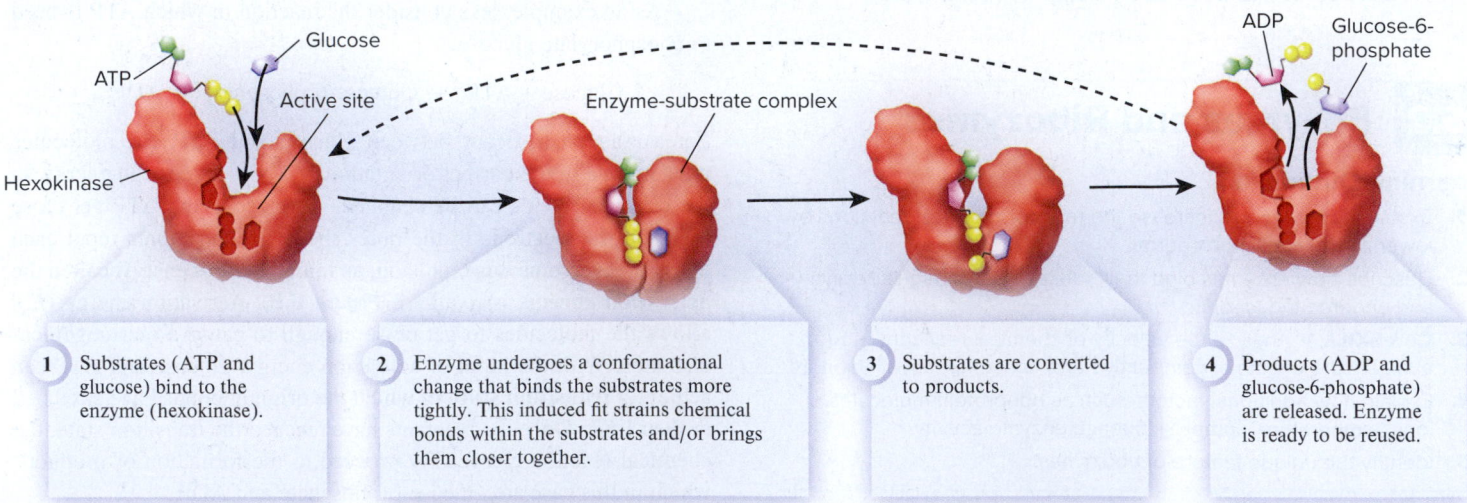

1. Substrates (ATP and glucose) bind to the enzyme (hexokinase).

2. Enzyme undergoes a conformational change that binds the substrates more tightly. This induced fit strains chemical bonds within the substrates and/or brings them closer together.

3. Substrates are converted to products.

4. Products (ADP and glucose-6-phosphate) are released. Enzyme is ready to be reused.

**Figure 6.6** **The steps of an enzyme-catalyzed reaction.** The example shown here involves the enzyme hexokinase, which binds glucose and ATP. The products are glucose-6-phosphate and ADP, which are released from the enzyme.

 **Core Concept: Structure and Function** A key function of enzymes is their ability to bind their substrates with high specificity. This specificity is due to the structure of the enzyme's active site.

revealed that the interaction between an enzyme and its substrates also involves movements or conformational changes in the enzyme itself. As shown in step 2 in Figure 6.6, these conformational changes cause the substrates to bind more tightly to the enzyme, a phenomenon called **induced fit**, which was proposed by American biochemist Daniel Koshland in 1958. Only after induced fit takes place does the enzyme catalyze the conversion of reactants to products. Induced fit is a key phenomenon that lowers the activation energy.

## Enzyme Function Is Influenced by the Substrate Concentration and by Inhibitors

The degree of attraction between an enzyme and its substrate(s) is called the **affinity** of the enzyme for its substrate(s). Some enzymes have very high affinity for their substrates, which means they readily recognize them. Such enzymes bind their substrates even when the substrate concentration is relatively low. Other enzymes have lower affinity for their substrates; the enzyme-substrate complex is likely to form only when the substrate concentration is higher.

Let's consider how biologists analyze the relationship between substrate concentration and enzyme function. In the experiment of **Figure 6.7a**, tubes labeled A, B, C, and D each contained 1 µg of enzyme, but they varied in the amount of substrate that was added. This enzyme recognizes a single substrate and converts it to a product. The samples were incubated for 60 seconds, and then the amount of product in each tube was measured. The velocity, or rate, of the chemical reaction is expressed as the amount of product produced per second. As we see in Figure 6.7a, the velocity increases as the substrate concentration increases, but eventually reaches a plateau. Why does the plateau occur? At high substrate concentrations, nearly all of the active sites of the enzyme are occupied with substrate, so further increasing the substrate concentration has a negligible effect. At this point, the enzyme is saturated with substrate, and the velocity of the chemical reaction is near its maximal rate, called its $V_{max}$.

Figure 6.7a also helps us understand the relationship between substrate concentration and velocity. The $K_M$ is the substrate concentration at which the velocity is half its maximal value. The $K_M$ is also called the Michaelis constant in honor of German biochemist Leonor Michaelis, who carried out pioneering work with Canadian biochemist Maud Menten on the study of enzymes. The $K_M$ is a measure of the substrate concentration required for a chemical reaction to occur. An enzyme with a high $K_M$ requires a higher substrate concentration to achieve a particular reaction velocity compared to an enzyme with a lower $K_M$.

For an enzyme-catalyzed reaction, we can view the formation of product as occurring in two steps: (1) binding or release of substrate and (2) formation of product:

$$E + S \rightleftharpoons ES \rightarrow E + P$$

where E is the enzyme, S is the substrate, ES is the enzyme-substrate complex, and P is the product.

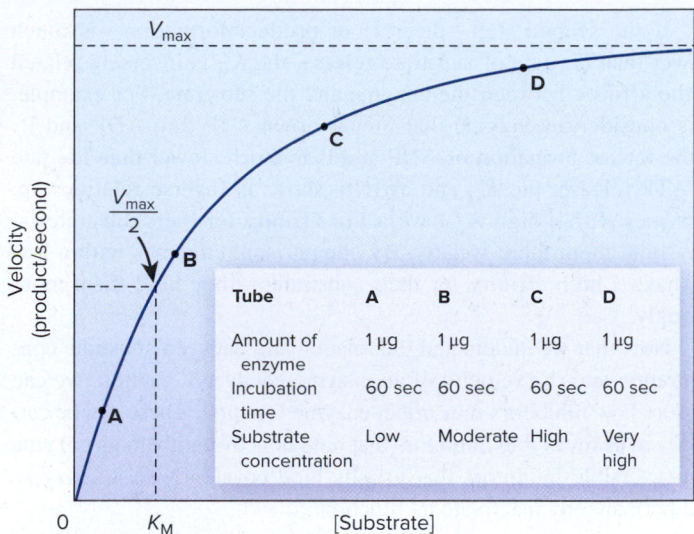

**(a) Reaction velocity in the absence of inhibitors**

| Tube | A | B | C | D |
|---|---|---|---|---|
| Amount of enzyme | 1 µg | 1 µg | 1 µg | 1 µg |
| Incubation time | 60 sec | 60 sec | 60 sec | 60 sec |
| Substrate concentration | Low | Moderate | High | Very high |

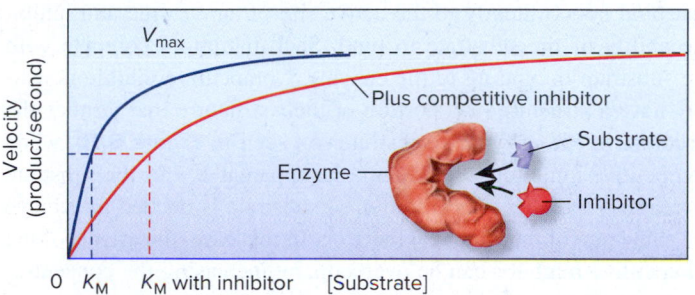

**(b) Competitive inhibition**

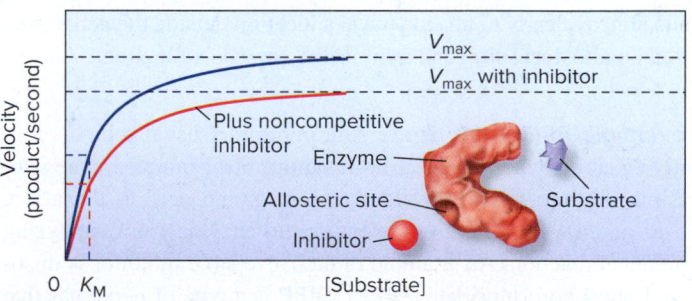

**(c) Noncompetitive inhibition**

**Figure 6.7** **The relationship between velocity and substrate concentration in an enzyme-catalyzed reaction, and the effects of inhibitors.** **(a)** In the absence of an inhibitor, the maximal velocity ($V_{max}$) of an enzyme-catalyzed reaction is achieved when the substrate concentration is high enough to be saturating. The $K_M$ value for an enzyme is the substrate concentration at which the velocity of the reaction is half the maximal velocity. **(b)** A competitive inhibitor binds to the active site of an enzyme and raises the $K_M$. **(c)** A noncompetitive inhibitor binds to an allosteric site outside the active site and lowers the $V_{max}$.

*Concept Check:* *Enzyme A has a $K_M$ of 0.1 mM, whereas enzyme B has a $K_M$ of 1.0 mM. The reactions the two enzymes catalyze both have the same $V_{max}$. If the substrate concentration was 0.5 mM, which reaction—the one catalyzed by enzyme A or the one catalyzed by enzyme B—would have the higher velocity?*

If the second step—the rate of product formation—is much slower than the rate of substrate release, the $K_M$ is inversely related to the affinity between the enzyme and the substrate. For example, let's consider an enzyme that breaks down ATP into ADP and $P_i$. If the rate of formation of ADP and $P_i$ is much slower than the rate of ATP release, the $K_M$ and affinity show an inverse relationship. Enzymes with a high $K_M$ have a low affinity for their substrates—they bind them more weakly. By comparison, enzymes with a low $K_M$ have a high affinity for their substrates—they bind them more strongly.

Now that we understand the relationship between substrate concentration and the velocity of an enzyme-catalyzed reaction, we can explore how inhibitors may affect enzyme function. These can be categorized as reversible inhibitors that bind noncovalently to an enzyme or irreversible inhibitors that usually bind covalently to an enzyme and permanently inactivate its function.

*Reversible Inhibitors*  Cells often use reversible inhibitors to modulate enzyme function. **Competitive inhibitors** are molecules that bind noncovalently to the active site of an enzyme and inhibit the ability of the substrate to bind. Such inhibitors compete with the substrate in binding to the enzyme. Competitive inhibitors usually have a structure or a portion of their structure that mimics the structure of the enzyme's substrate. As seen in **Figure 6.7b**, when competitive inhibitors are present, the apparent $K_M$ for the substrate increases—a higher concentration of substrate is needed to achieve the same rate of the chemical reaction. In this case, the effects of the competitive inhibitor can be overcome by increasing the concentration of the substrate.

By comparison, **Figure 6.7c** illustrates the effects of a **noncompetitive inhibitor**. This type of inhibitor lowers the $V_{max}$ for the reaction without affecting the $K_M$. A noncompetitive inhibitor binds noncovalently to an enzyme at a location outside the active site, called an **allosteric site**, and inhibits the enzyme's function.

*Irreversible Inhibitors*  Irreversible inhibitors usually bind covalently to an enzyme to inhibit its function. For example, some irreversible inhibitors bind covalently to an amino acid at the active site of an enzyme, thereby preventing the enzyme from catalyzing a chemical reaction. An example of an irreversible inhibitor is diisopropyl phosphorofluoridate (DIFP). DIFP is a type of nerve gas that was developed as a chemical weapon. This molecule covalently reacts with the enzyme acetylcholinesterase, which is important for the proper functioning of neurons.

Irreversible inhibition is not a common way for cells to control enzyme function. Why do cells usually control enzymes via reversible inhibitors? The answer is that a reversible inhibitor allows an enzyme to be used again, when the inhibitor concentration becomes lower. Being able to reuse an enzyme is energy-efficient. In contrast, irreversible inhibitors permanently inactivate an enzyme, thereby preventing its further use.

## Additional Factors Influence Enzyme Function

Enzymes, which are proteins, sometimes require nonprotein molecules or ions to carry out their functions.

- **Prosthetic groups** are small molecules that are permanently attached to the surface of an enzyme and aid in enzyme function.

- **Cofactors** are usually inorganic ions, such as $Fe^{3+}$ or $Zn^{2+}$, that temporarily bind to the surface of an enzyme and promote a chemical reaction.

- Some enzymes use **coenzymes**, organic molecules that temporarily bind to an enzyme and participate in the chemical reaction that the enzyme catalyzes, but are left unchanged when the reaction is completed.

The ability of enzymes to increase the rate of a chemical reaction is also affected by their environment. In particular, the temperature, pH, and ionic conditions play an important role in the proper functioning of enzymes. Most enzymes function maximally in a narrow range of temperature and pH. For example, many human enzymes work best at 37°C (98.6°F), which is normal body temperature. If the temperature is several degrees above or below this optimal temperature due to infection or environmental causes, the function of many enzymes is greatly inhibited (**Figure 6.8**). Very high temperatures may denature a protein, causing it to unfold and lose its three-dimensional shape, thereby inhibiting its function.

Enzyme function is also sensitive to pH. Certain enzymes in the stomach function best at the acidic pH found in this organ. For example, pepsin is a protease—an enzyme that digests proteins into peptides—that is released into the stomach. The optimal pH for pepsin function is around pH 2.0, which is extremely acidic. By comparison, many cytosolic enzymes function optimally at a more neutral pH, such as pH 7.2, which is the pH normally found in the cytosol of human cells. If the pH was significantly above or below this value, function would be decreased for cytosolic enzymes.

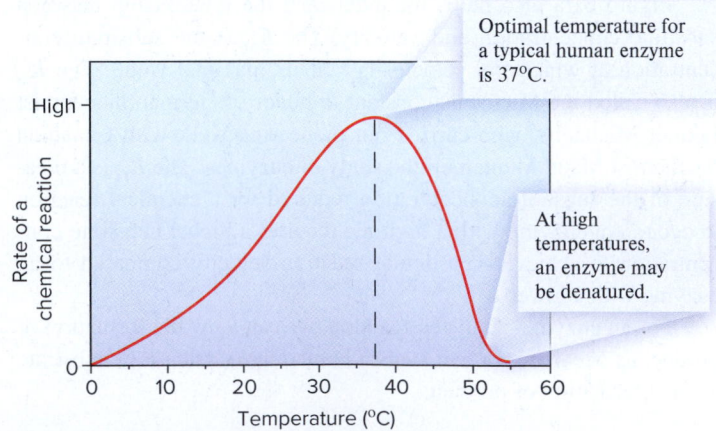

**Figure 6.8  Effects of temperature on a typical human enzyme.** Most enzymes function optimally within a narrow range of temperature. Many human enzymes function best at 37°C, which is normal body temperature.

## Feature Investigation | The Discovery of Ribozymes by Sidney Altman Revealed That RNA Molecules May Also Function as Catalysts

Until the 1980s, scientists thought that all biological catalysts are proteins. An avenue of study that dramatically changed this view came from the analysis of ribonuclease P (RNase P), a catalyst involved in the processing of tRNA molecules—a type of molecule required for protein synthesis. tRNA molecules are synthesized as longer precursor molecules called ptRNAs, which have 5′ and 3′ ends. (The 5′ and 3′ directionality of RNA molecules is described in Chapter 12.) RNase P breaks a covalent bond at a specific site in a ptRNA, which releases a fragment at the 5′ end and makes the precursor molecule shorter (**Figure 6.9**).

Sidney Altman and his colleagues became interested in the processing of tRNA molecules and turned their attention to RNase P in *E. coli*. During the course of their studies, they purified this enzyme and, to their surprise, discovered it has two subunits—one is an RNA molecule that contains 377 nucleotides, and the other is a small protein with a mass of 14 kDa. In the 1980s, the finding that a catalyst had an RNA subunit was very unexpected. Even so, a second property of RNase P would prove even more exciting.

Altman and colleagues were able to purify RNase P and study its properties in vitro. Cecilia Guerrier-Takada in Altman's laboratory determined that magnesium ion (Mg²⁺) has a stimulatory effect on RNase P function. In the experiment described in **Figure 6.10**, the effects of Mg²⁺ were studied in greater detail. The researchers analyzed the effects of low (10 mM MgCl₂) and high (100 mM MgCl₂) magnesium concentrations on the processing of a ptRNA. At low or high magnesium concentrations,

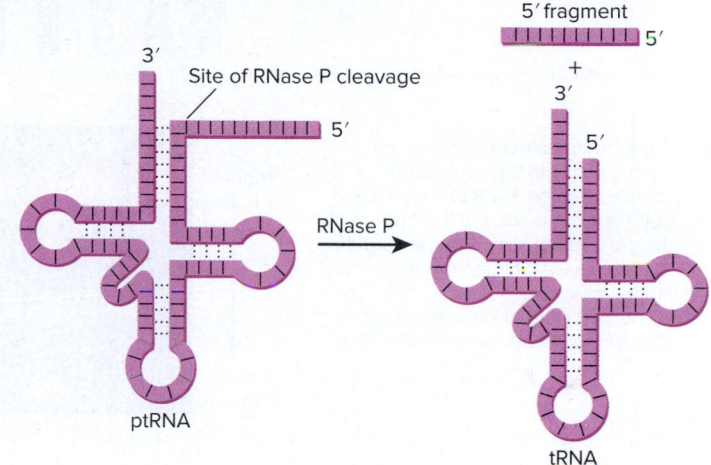

**Figure 6.9  The function of RNase P.** A specific bond in a precursor tRNA (ptRNA) is cleaved by RNase P, which releases a small fragment at the 5′ end. This results in the formation of a mature tRNA.

 **Core Skill: Connections** Look ahead to Figure 12.20. How do you think translation would be affected if RNase P did not function properly?

**Figure 6.10  The discovery that the RNA subunit of RNase P is a catalyst.**

**HYPOTHESIS** The catalytic function of RNase P is carried out by its RNA subunit or by its protein subunit.

**KEY MATERIALS** Purified precursor tRNA (ptRNA) and purified RNA and protein subunits of RNase P from *E. coli*.

|  | Experimental level | Conceptual level |
|---|---|---|
| 1 | Into each of five tubes, add ptRNA. | 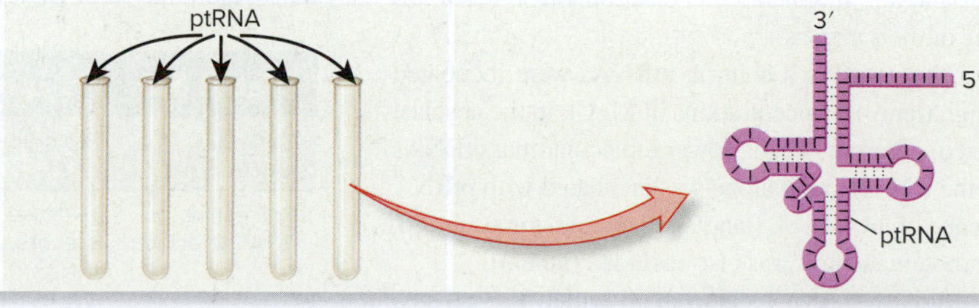 ptRNA / ptRNA |
| 2 | In tubes 1–3, add a low concentration of MgCl₂; in tubes 4 and 5, add a high MgCl₂ concentration. | MgCl₂ — Low MgCl₂ (10 mM)  High MgCl₂ (100 mM) |

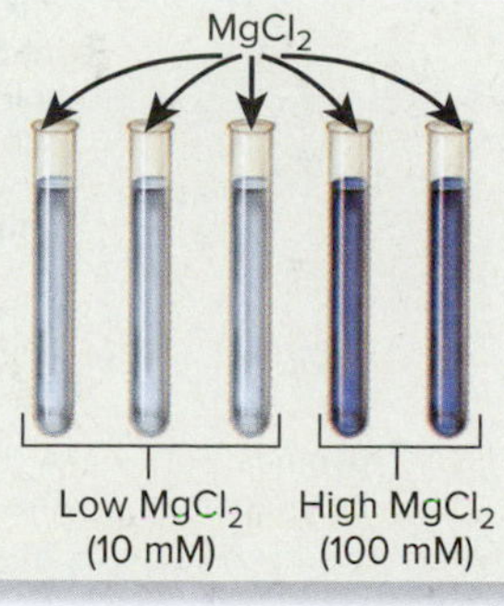

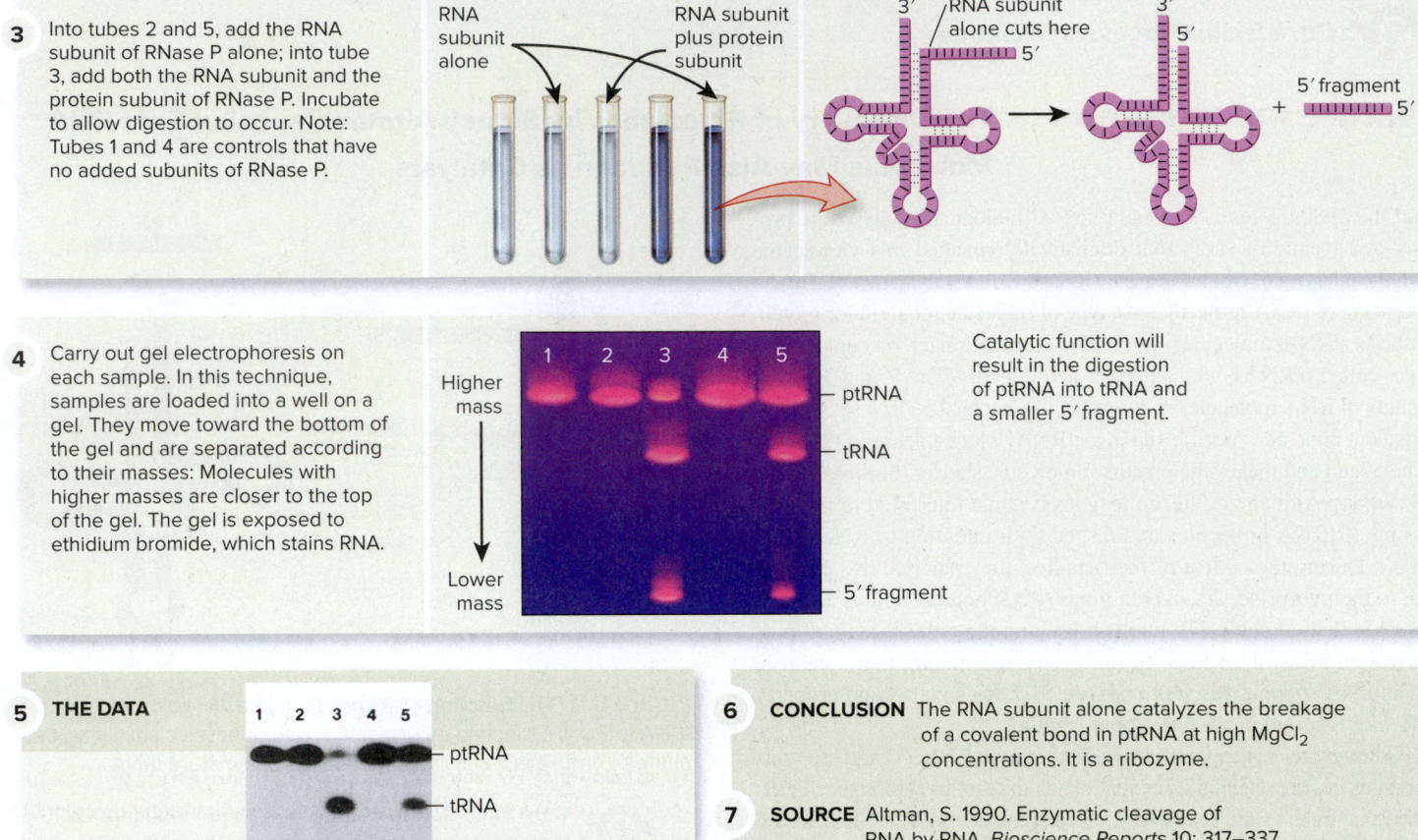

**3** Into tubes 2 and 5, add the RNA subunit of RNase P alone; into tube 3, add both the RNA subunit and the protein subunit of RNase P. Incubate to allow digestion to occur. Note: Tubes 1 and 4 are controls that have no added subunits of RNase P.

RNA subunit alone

RNA subunit plus protein subunit

RNA subunit alone cuts here

5′ fragment

**4** Carry out gel electrophoresis on each sample. In this technique, samples are loaded into a well on a gel. They move toward the bottom of the gel and are separated according to their masses: Molecules with higher masses are closer to the top of the gel. The gel is exposed to ethidium bromide, which stains RNA.

Higher mass

Lower mass

ptRNA

tRNA

5′ fragment

Catalytic function will result in the digestion of ptRNA into tRNA and a smaller 5′ fragment.

**5** THE DATA

ptRNA

tRNA

5′ fragment

**6** CONCLUSION The RNA subunit alone catalyzes the breakage of a covalent bond in ptRNA at high $MgCl_2$ concentrations. It is a ribozyme.

**7** SOURCE Altman, S. 1990. Enzymatic cleavage of RNA by RNA. *Bioscience Reports* 10: 317–337.

(4-5): Altman, S. 1990. Enzymatic Cleavage of RNA by RNA. *Bioscience Reports*, 10:317–337, Fig. 7. ©The Nobel Foundation

ptRNAs were incubated without RNase P (as a control), with the RNA subunit alone, or with intact RNase P (RNA subunit and protein subunit). Following incubation, the researchers performed gel electrophoresis on the samples to determine if the ptRNAs had been cleaved into two pieces—the tRNA and a 5′ fragment. (Gel electrophoresis separates molecules on the basis of their masses.)

Let's now look at the data. As a control, ptRNAs were incubated with low (lane 1) or high (lane 4) concentrations of $MgCl_2$ in the absence of RNase P. As expected, no processing to lower molecular mass tRNAs was observed. When the RNA subunit alone was incubated with ptRNA molecules in the presence of low $MgCl_2$ (lane 2), no processing occurred, but it did occur if the protein subunit was also included (lane 3).

The surprising result is shown in lane 5, in which the ptRNA was incubated with the RNA subunit alone in the presence of a high concentration of $MgCl_2$. The RNA subunit by itself was sufficient to cleave the ptRNA to a smaller tRNA and a 5′ fragment! Presumably, the high $MgCl_2$ concentration helps to keep the RNA subunit in a conformation that is catalytically active. Alternatively, the protein subunit plays a similar role in a living cell.

Subsequent work confirmed these observations and showed that the RNA subunit of RNase P is a true catalyst—it accelerates the rate of a chemical reaction and is not permanently altered by it. Around the same time, Thomas Cech and colleagues determined that a different RNA molecule found in the protist *Tetrahymena thermophila* also has

catalytic activity. The term **ribozyme** is now used to describe an RNA molecule that catalyzes a chemical reaction. In 1989, Altman and Cech received the Nobel Prize in Chemistry for their discovery of ribozymes.

Since the pioneering work of Altman and Cech, researchers have discovered that ribozymes play key catalytic roles in cells (**Table 6.3**).

| Table 6.3 | Types of Ribozymes |
|---|---|
| **General function** | **Biological examples** |
| Processing of RNA molecules | 1. RNase P: As described earlier, RNase P cleaves precursor tRNA molecules (ptRNAs) to form mature tRNAs. |
| | 2. Spliceosomal RNA: As described in Chapter 12, eukaryotic pre-mRNAs often have regions called introns. These introns are later removed by a spliceosome that is composed of RNA and protein subunits. The RNA within the spliceosome is believed to function as a ribozyme that removes the introns from pre-mRNA. |
| | 3. Certain introns found in mitochondrial, chloroplast, and bacterial RNAs are removed by a self-splicing mechanism. |
| Synthesis of polypeptides | A ribosome has an RNA component that catalyzes the formation of covalent bonds between adjacent amino acids during polypeptide synthesis. |

They are primarily involved in the processing of RNA molecules from precursor to mature forms. In addition, a ribozyme in ribosomes catalyzes the formation of covalent bonds between adjacent amino acids during polypeptide synthesis.

### Experimental Questions

1. Briefly explain why it was necessary to purify the individual subunits of RNase P to show that it is a ribozyme.

2. **CoreSKILL** » Explain why the researchers conducted experiments in which they measured the formation of mature tRNAs without adding the protein subunit or without adding the RNA subunit.

3. **CoreSKILL** » Analyze the results of Altman and colleagues, and explain how they indicated that RNase P is a ribozyme. How does the concentration of $Mg^{2+}$ affect the function of the RNA subunit in RNase P?

---

## 6.3 Overview of Metabolism

### Learning Outcomes:

1. Explain the concept of a metabolic pathway, and distinguish between catabolic and anabolic reactions.

2. Describe how catabolic reactions are used to generate building blocks to make larger molecules and to produce energy intermediates.

3. Define redox reaction.

4. Compare and contrast three ways that metabolic pathways are regulated.

In the previous sections, we examined the underlying factors that govern individual chemical reactions and explored the properties of enzymes and ribozymes. In living cells, chemical reactions are coordinated with each other and often occur in a series of steps called a **metabolic pathway**, with each step catalyzed by a specific enzyme (**Figure 6.11**). These pathways are categorized according to whether the reactions lead to the breakdown or synthesis of substances. **Catabolic reactions** result in the breakdown of larger molecules into smaller ones. Such reactions are often exergonic. By comparison, **anabolic reactions** involve the synthesis of larger molecules from smaller precursor molecules. These reactions usually are endergonic and, in living cells, must be coupled to an exergonic reaction. In this section, we will survey the general features of catabolic and anabolic reactions and explore the ways in which metabolic pathways are controlled.

### Catabolic Reactions Recycle Organic Building Blocks and Produce Energy Intermediates Such as ATP

Catabolic reactions result in the breakdown of larger molecules into smaller ones. Such catabolic reactions have two uses.

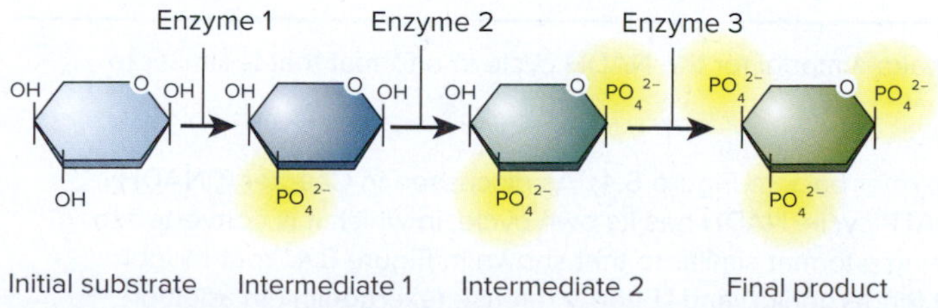

*Recycling of Organic Building Blocks*   One reason to break down macromolecules is to recycle their organic molecules, which are used as building blocks to construct new molecules and macromolecules. For example, polypeptides, which make up proteins, are composed of a linear sequence of amino acids. When a protein is improperly folded or is no longer needed by a cell, the peptide bonds between the amino acids in the protein are broken by enzymes called proteases. This generates amino acids that can be used in the construction of new proteins.

$$\text{Protein} \xrightarrow{\text{Proteases}} \rightarrow \rightarrow \rightarrow \rightarrow \rightarrow \rightarrow \rightarrow \rightarrow \rightarrow \text{Many individual amino acids}$$

We will consider the mechanisms of recycling in Section 6.4.

*Breakdown of Organic Molecules to Obtain Energy*   A second reason to break down macromolecules into smaller organic molecules is to obtain energy that is used to drive endergonic processes in the cell. Covalent bonds store a large amount of energy. However, when cells break covalent bonds in organic molecules such as glucose, they do not directly use the energy released in this process. Instead, the released energy is stored in **energy intermediates**, molecules such as ATP, which are directly used to drive endergonic reactions in cells.

As an example, let's consider the breakdown of glucose into two molecules of pyruvate. As discussed in Chapter 7, the breakdown of glucose to pyruvate involves a catabolic pathway called glycolysis. Some of the energy released during the breakage of covalent bonds in glucose is harnessed to synthesize ATP. Glycolysis involves a series of steps in which covalent bonds are broken and rearranged. This process produces molecules that readily donate a phosphate group to ADP, thereby producing ATP. For example, phosphoenolpyruvate has a phosphate group attached to pyruvate. Due to the arrangement of bonds in phosphoenolpyruvate, this phosphate bond is unstable and easily broken. Therefore, the phosphate can be readily transferred from phosphoenolpyruvate to ADP:

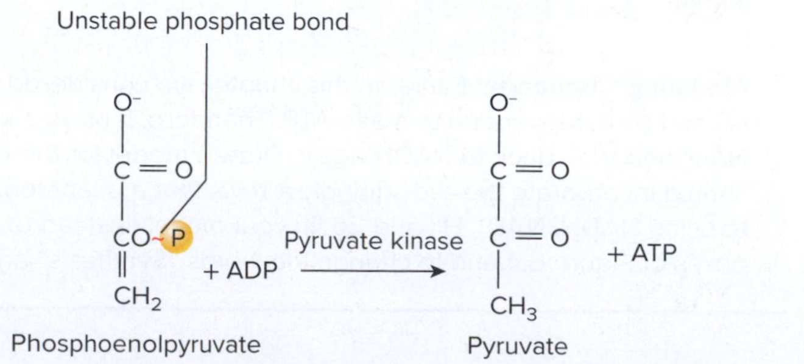

**Figure 6.11  A metabolic pathway.** In this metabolic pathway, a series of different enzymes catalyze the attachment of phosphate groups at various positions on a sugar molecule, beginning with a starting substrate and ending with a final product.

This is an exergonic reaction ($\Delta G = -7.5$ kcal/mol) and therefore favors the formation of products. In this step of glycolysis, the breakdown of an organic molecule, namely phosphoenolpyruvate, results in the formation of pyruvate and the synthesis of an energy intermediate, a molecule of ATP, which can then be used by a cell to drive an endergonic reaction. This way of synthesizing ATP, termed **substrate-level phosphorylation**, occurs when an enzyme directly transfers a phosphate from an organic molecule to ADP, thereby making ATP.

Another way to make ATP is via **chemiosmosis**. In this process, energy stored in an ion electrochemical gradient is used to make ATP from ADP and $P_i$. We will consider this other mechanism in Chapter 7.

## Redox Reactions Involve the Transfer of Electrons

During the breakdown of small organic molecules, **oxidation**—the removal of one or more electrons from an atom or molecule—may occur. This process is called oxidation because oxygen is frequently involved in chemical reactions that remove electrons from other atoms or molecules. By comparison, **reduction** is the addition of one

or more electrons to an atom or molecule. Reduction is so named because the addition of a negatively charged electron reduces the net charge of an atom or molecule.

Electrons do not exist freely in solution. When an atom or molecule is oxidized, the electron that is removed must be transferred to another atom or molecule, which becomes reduced. This type of reaction is termed a **redox reaction**, which is short for a <u>red</u>uction-<u>ox</u>idation reaction. An electron may be transferred from molecule A to molecule B as shown in the following generalized equation:

$$A e^- + B \rightarrow A + B e^-$$
$$\text{(oxidized) (reduced)}$$

As shown on the right side of this reaction, A has been oxidized (that is, had an electron removed), and B has been reduced (that is, had an electron added). In general, a substance that has been oxidized has less energy, whereas a substance that has been reduced has more energy.

During the oxidation of organic molecules such as glucose, the electrons that are removed may be used to produce energy intermediates such as NADH (**Figure 6.12**). In this process, an organic molecule

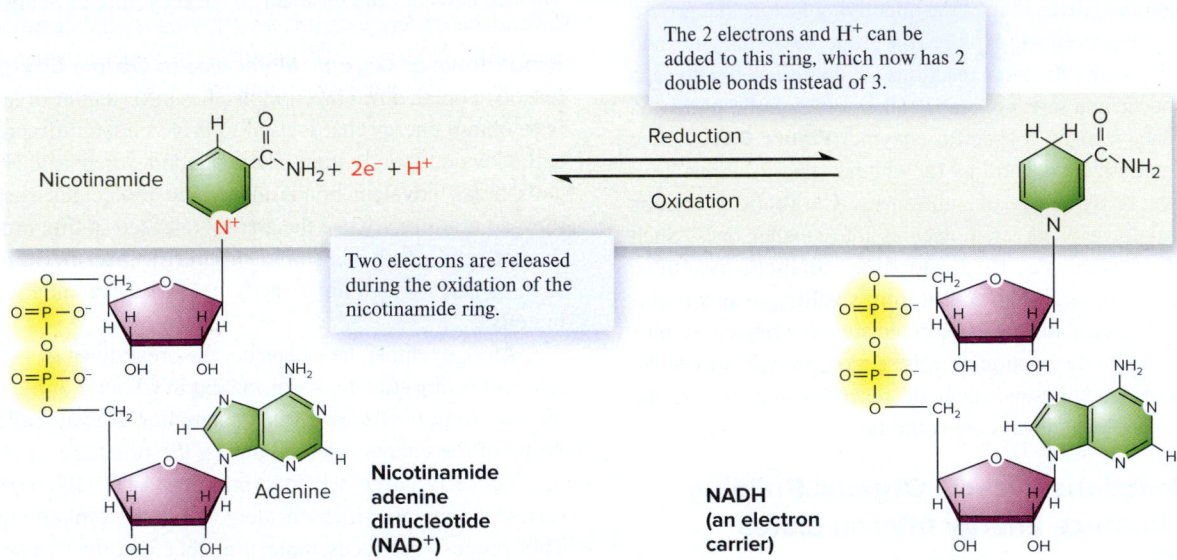

**Figure 6.12   The reduction of NAD⁺ to produce NADH.**  NAD⁺ is composed of two nucleotides, one with an adenine base and one with a nicotinamide base. The oxidation of organic molecules releases electrons that bind to NAD⁺ (and along with a hydrogen ion) result in the formation of NADH. The two electrons and H⁺ are incorporated into the nicotinamide ring. Note: The actual net charges of NAD⁺ and NADH are −1 and −2, respectively. They are designated NAD⁺ and NADH to emphasize the net charge of the nicotinamide ring, which is involved in reduction-oxidation reactions.

**Core Skill: Modeling**  The goal of this modeling challenge is to make a model for the NADH cycle in a format that is similar to Figure 6.4.

**Modeling Challenge:** Earlier in this chapter, we considered the ATP cycle (refer back to Figure 6.4). As discussed in Chapter 7, NADH is used by mitochondria to make ATP. Therefore, it plays a key role in the ATP cycle. NADH has its own cycle, in which it is converted to NAD⁺ and then back to NADH again. Draw a model for the NADH cycle using a format similar to that shown in Figure 6.4. Your model should incorporate the red squiggly arrows that are labeled "Energy input (endergonic)" and "Energy release (exergonic)." In addition to using NADH, NAD⁺, H⁺, and 2e⁻ in your model (instead of ATP, etc.), you will need to change the sentences in the top and bottom text boxes of Figure 6.4 and to change the words "Synthesis" and "Hydrolysis."

is oxidized, and **NAD+** (**nicotinamide adenine dinucleotide**) is reduced to NADH. Cells use NADH in two common ways. First, as we will see in Chapter 7, the oxidation of NADH is a highly exergonic reaction that can be used to make ATP. Second, NADH can donate electrons to other organic molecules and thereby energize them. Such energized molecules can more readily form covalent bonds. Therefore, as described next, NADH is often needed in reactions that involve the synthesis of larger molecules through the formation of covalent bonds between smaller molecules.

## Anabolic Reactions Require an Input of Energy to Make Larger Molecules

Anabolic reactions are also called **biosynthetic reactions**, because they are necessary to make larger molecules and macromolecules. We will examine the synthesis of macromolecules in several chapters of this textbook. For example, RNA and protein biosynthesis are described in Chapter 12. Cells also need to synthesize small organic molecules, such as amino acids and fats, if they are not readily available from food sources. Such molecules are made by the formation of covalent linkages between precursor molecules. For example, glutamate (an amino acid) is made by covalently linking α-ketoglutarate (a product of sugar metabolism) and ammonium ($NH_4^+$).

α-Ketoglutarate                    Glutamate

An energy intermediate, a molecule of NADH, is needed to drive this reaction forward.

## Metabolic Pathways Are Regulated in Three General Ways

The regulation of metabolic pathways is important for a variety of reasons. Catabolic pathways are regulated so organic molecules are broken down only when they are no longer needed or when the cell requires energy. During anabolic reactions, regulation ensures that a cell synthesizes molecules only when they are needed. The regulation of catabolic and anabolic pathways occurs at the genetic, cellular, and biochemical levels.

***Gene Regulation***    Enzymes are protein molecules that are encoded by genes. One way that cells control metabolic pathways is via gene regulation. For example, if a bacterial cell is not exposed to a particular sugar in its environment, it will turn off the genes that encode the enzymes that are needed to break down

that sugar. Then, if the sugar becomes available, the genes are switched back on. Chapter 14 examines the steps of gene regulation in detail.

***Cellular Regulation***    Metabolism is also coordinated at the cellular level. Cells integrate signals from their environment and adjust their metabolic pathways to adapt to those signals. As discussed in Chapter 9, cell-signaling pathways often lead to the activation of protein kinases—enzymes that covalently attach a phosphate group to a target protein. For example, when people are frightened, they secrete a hormone called epinephrine into their bloodstream. This hormone binds to the surface of muscle cells and stimulates an intracellular pathway that leads to the phosphorylation of specific enzymes involved in carbohydrate metabolism. These activated enzymes promote the breakdown of carbohydrates, an event that supplies the frightened individual with more energy. Epinephrine is sometimes called the fight-or-flight hormone because the added energy prepares an individual to either stay and fight or run away quickly. After a person no longer feels frightened, hormone levels drop, and other enzymes called phosphatases remove the phosphate groups from enzymes, thereby restoring the original level of carbohydrate metabolism.

***Biochemical Regulation***    A third and very prominent way that metabolic pathways are controlled is at the biochemical level. In this case, the noncovalent binding of a molecule to an enzyme directly regulates the enzyme's function. As discussed earlier, one form of biochemical regulation involves the binding of molecules called competitive or noncompetitive inhibitors (see Figure 6.7). An example of noncompetitive inhibition is a type of regulation called **feedback inhibition**, in which the product of a metabolic pathway inhibits an enzyme that acts early in the pathway, thus preventing the overaccumulation of the product (**Figure 6.13**).

Many metabolic pathways use feedback inhibition as a form of biochemical regulation. In such cases, the inhibited enzyme has two binding sites. One site is the active site, where the reactants are converted to products. In addition, enzymes controlled by feedback inhibition also have an allosteric site, where a molecule can bind noncovalently and affect the enzyme's function. The binding of a molecule to an allosteric site causes a conformational change in the enzyme that inhibits its catalytic function. Allosteric sites are often found in the enzymes that catalyze the early steps in a metabolic pathway. Such allosteric sites typically bind molecules that are the products of the metabolic pathway. When the products bind to these sites, they inhibit the function of these enzymes, thereby preventing the formation of too much product. As described earlier, in Figure 6.7c, this phenomenon is also called noncompetitive inhibition.

***Regulation of the Rate-Limiting Step***    Cellular regulation and biochemical regulation are important ways to control chemical reactions in a cell. For a metabolic pathway composed of several enzyme-catalyzed reactions, which enzyme should be controlled? In many cases, a metabolic pathway has a **rate-limiting step**, which is the slowest step in the pathway. If the rate-limiting step is inhibited or enhanced, such changes will have the greatest influence on the

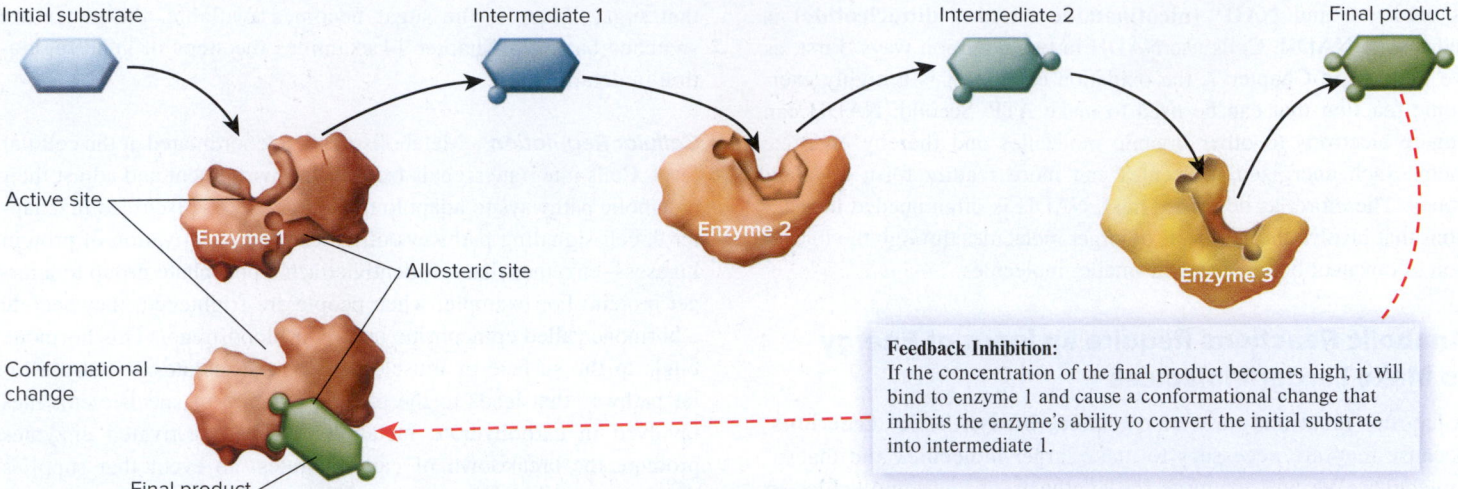

**Figure 6.13 Feedback inhibition.** In this process, the final product of a metabolic pathway inhibits an enzyme that functions early in the pathway, thereby preventing the overaccumulation of the product.

**Core Skill: Connections** Look ahead to Figure 7.3, which describes a metabolic pathway called glycolysis. Feedback inhibition occurs during this process in that high levels of ATP inhibit phosphofructokinase, an enzyme that catalyzes the conversion of fructose-6-phosphate and ATP to fructose-1,6-bisphosphate and ADP. How is this beneficial to the cell?

formation of the product of the metabolic pathway. Rather than affecting all of the enzymes in a metabolic pathway, cellular or biochemical regulation is often directed at the enzyme that catalyzes the rate-limiting step. This is an efficient and rapid way to control the amount of product of a pathway.

 **BIO TIPS**

**THE QUESTION** *The enzyme called 3-phosphoglycerate dehydrogenase catalyzes the following chemical reaction:*

3-phospho-D-glycerate + NAD⁺ ⇌
3-phosphonooxypyruvate + NADH + H⁺

*This reaction is the rate-limiting step in a metabolic pathway that synthesizes serine, which is an amino acid. Serine inhibits 3-phosphoglycerate dehydrogenase by binding to an allosteric site on the enzyme, thereby preventing the overaccumulation of serine in a cell. This is an example of feedback inhibition. Researchers have identified a mutant version of 3-phosphoglycerate dehydrogenase that does not exhibit feedback inhibition. Cells that make the mutant enzyme tend to overaccumulate serine. Make a drawing that depicts how the mutant enzyme is different from the normal one. Your drawing should include binding sites for 3-phospho-D-glycerate, NAD⁺, and serine.*

**T** **OPIC** *What topic in biology does this question address?* The topic is enzymes and feedback inhibition. More specifically, the question is about a mutant version of 3-phosphoglycerate dehydrogenase that does not exhibit feedback inhibition.

**I** **NFORMATION** *What information do you know based on the question and your understanding of the topic?* From the question, you know that 3-phosphoglycerate dehydrogenase catalyzes the rate-limiting step in a metabolic pathway that

produces serine. Serine causes feedback inhibition of the normal version of the enzyme but does not inhibit a mutant version. From your understanding of the topic, you may recall that feedback inhibition occurs via an allosteric site.

**P** **ROBLEM-SOLVING** **S** **TRATEGY** *Compare and contrast. Make a drawing.* To solve this problem, you could begin by comparing the properties of the normal and mutant enzyme. The mutant enzyme does not exhibit feedback inhibition. However, because cells harboring the mutant version of the enzyme overaccumulate serine, you know that the catalytic properties of the enzyme must be functioning normally. In other words, the active site is functional. When making the drawing, you need to remember that the enzyme has two sites: an active site and an allosteric site.

**ANSWER** *The mutation, designated by an X in the enzyme on the right, is an alteration in the structure of the allosteric site that prevents serine from binding there.*

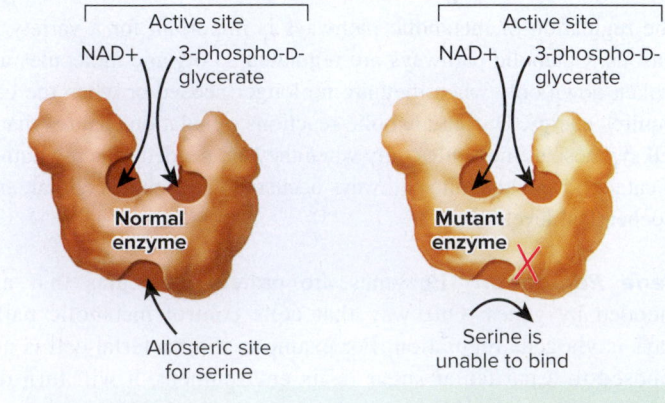

# 6.4 Recycling of Organic Molecules

## Learning Outcomes:

1. Explain the relationship between the recycling of organic molecules and cellular efficiency.
2. Outline how the building blocks of proteins are recycled.
3. Describe how the components of cellular organelles are recycled via autophagy.

As mentioned earlier in this chapter, an important feature of metabolism is the recycling of organic molecules, such as amino acids, which are the building blocks of proteins. Except for DNA, which is stably maintained and inherited from cell to cell, other large molecules such as RNA, proteins, lipids, and polysaccharides typically exist for a relatively short period of time. Biologists often speak of the **half-life** of molecules, which is the time it takes for 50% of a specific type of molecule in a cell to be broken down and recycled. For example, mRNA molecules in bacterial cells have an average half-life of about 5 minutes, whereas mRNAs in eukaryotic cells tend to exist for longer periods of time, on the order of 30 minutes to 24 hours or even several days.

Why is recycling important? To compete effectively in their native environments, all living organisms must efficiently use and recycle the organic molecules that are needed as building blocks to construct larger molecules and macromolecules. Otherwise, they would waste a great deal of energy making such building blocks from smaller molecules. For example, organisms conserve an enormous amount of energy by reusing the amino acids that are needed to construct proteins. In this section, we will explore how amino acids are recycled and consider a mechanism for recycling all of the materials found in an organelle.

## Proteins in Eukaryotes and Archaea Are Broken Down in the Proteasome

Cells continually degrade proteins that are faulty or no longer needed. To be degraded, proteins are recognized by **proteases**—enzymes that cleave the bonds between adjacent amino acids. The primary pathway for protein degradation in archaea and eukaryotic cells occurs via a protein complex called a **proteasome**. The core of the proteasome consists of four stacked rings, each composed of seven protein subunits (**Figure 6.14a**). The proteasomes of eukaryotic cells also contain caps at each end that control the entry of proteins into the proteasome.

**Figure 6.14b** describes the steps of protein degradation via eukaryotic proteasomes. A string of small proteins called **ubiquitins** is covalently attached to the target protein. This event directs the target protein to a proteasome cap, which has binding sites for ubiquitins. The cap also has enzymes that unfold the protein and inject it into the internal cavity of the proteasome core. The ubiquitins are removed during entry and released to the cytosol for reuse. Inside the proteasome, proteases degrade the target protein into small peptides and amino acids. The process is completed when the peptides and

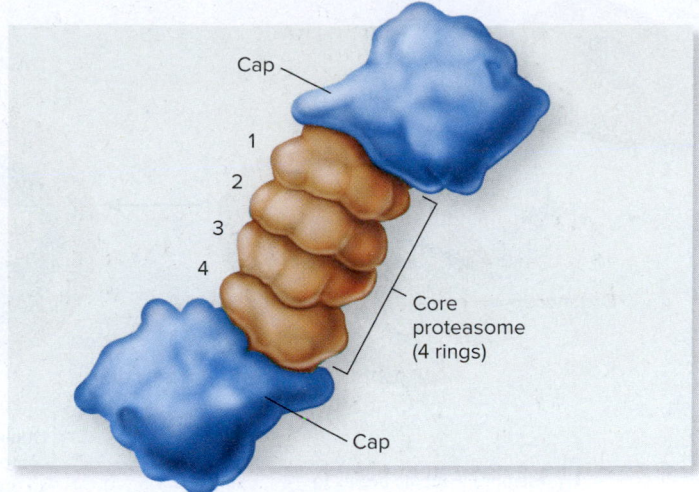

**(a) Structure of the eukaryotic proteasome**

Cap
1
2
3
4
Core proteasome (4 rings)
Cap

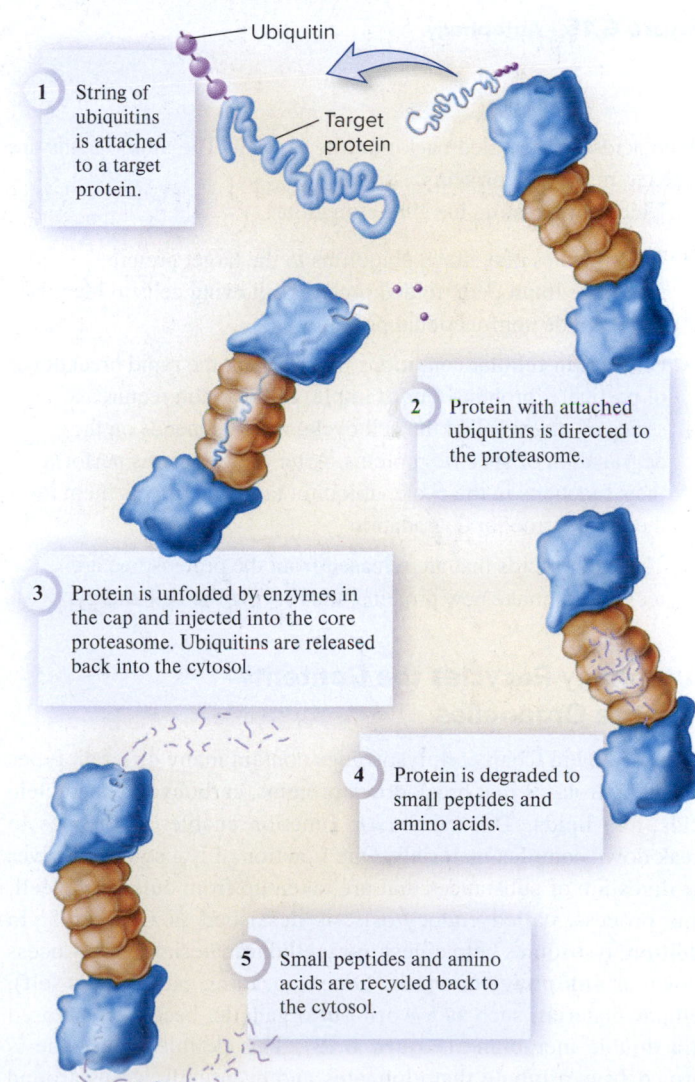

Ubiquitin

1. String of ubiquitins is attached to a target protein.

Target protein

2. Protein with attached ubiquitins is directed to the proteasome.

3. Protein is unfolded by enzymes in the cap and injected into the core proteasome. Ubiquitins are released back into the cytosol.

4. Protein is degraded to small peptides and amino acids.

5. Small peptides and amino acids are recycled back to the cytosol.

**(b) Steps of protein degradation in eukaryotic cells**

**Figure 6.14** **Protein degradation via the proteasome.**

*Concept Check* *What are advantages of protein degradation?*

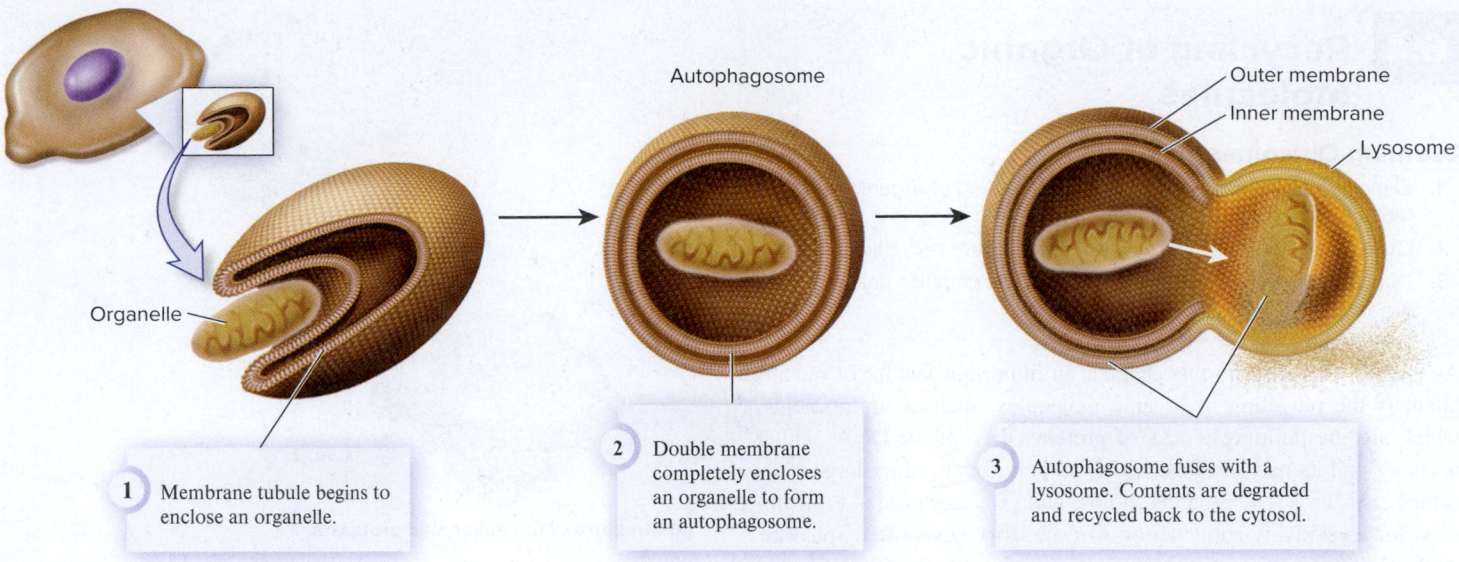

**Autophagosome**

**Outer membrane**
**Inner membrane**
**Lysosome**

**Organelle**

1 | Membrane tubule begins to enclose an organelle.

2 | Double membrane completely encloses an organelle to form an autophagosome.

3 | Autophagosome fuses with a lysosome. Contents are degraded and recycled back to the cytosol.

**Figure 6.15**   **Autophagy.**

amino acids are recycled back into the cytosol. The amino acids are reused to make new proteins.

Ubiquitin targeting has three functions.

- The enzymes that attach ubiquitins to the target protein recognize improperly folded proteins, allowing cells to identify and degrade nonfunctional proteins.

- Changes in cellular conditions may warrant the rapid breakdown of particular proteins. For example, cell division requires a series of stages called the cell cycle, which depends on the degradation of specific proteins. After these proteins perform their functions in the cycle, ubiquitin targeting directs them to the proteasome for degradation.

- The amino acids that are released from the proteosome are recycled to make new proteins, thus saving the cell energy.

## Autophagy Recycles the Contents of Entire Organelles

As described in Chapter 4, lysosomes contain many different types of acid hydrolases that break down proteins, carbohydrates, nucleic acids, and lipids. This enzymatic function enables lysosomes to break down complex materials. One function of lysosomes involves the digestion of substances that are taken up from outside the cell. This process, called endocytosis, is described in Chapter 5. In addition, lysosomes help digest intracellular materials. In a process known as **autophagy** (from the Greek, meaning eating one's self), cellular material, such as a worn-out organelle, becomes enclosed in a double membrane (**Figure 6.15**). This double membrane is formed from a tubule that elongates and eventually wraps around the organelle to form an **autophagosome**. The autophagosome then fuses with one or more lysosomes, and the material inside the autophagosome is digested. The small molecules released from this digestion are recycled back into the cytosol.

## Summary of Key Concepts

### 6.1   Energy and Chemical Reactions

- The fate of a chemical reaction is determined by its direction and rate.

- Energy, the ability to promote change or do work, exists in many forms. According to the first law of thermodynamics, energy cannot be created or destroyed, but it can be converted from one form to another. The second law of thermodynamics states that energy interconversions involve an increase in entropy (Figures 6.1, 6.2, Table 6.1).

- Free energy is the amount of available energy that can be used to promote change or do work. Spontaneous or exergonic reactions, which release free energy, have a negative free-energy change, whereas endergonic reactions have a positive free-energy change (Figure 6.3).

- Chemical reactions proceed until they reach a state of chemical equilibrium, where the rate of formation of products equals the rate of formation of reactants.

- Exergonic reactions, such as the hydrolysis of ATP, are commonly coupled to cellular processes that would otherwise be endergonic. Cells continuously synthesize ATP from ADP and $P_i$ and then hydrolyze it to drive endergonic reactions (Figure 6.4).

- Estimates from genome analysis indicate that over 20% of all proteins bind ATP (Table 6.2).

### 6.2   Enzymes and Ribozymes

- Enzymes are proteins that speed up the rate of a chemical reaction by lowering the activation energy ($E_A$) needed to achieve a transition state (Figure 6.5).

- Enzymes recognize reactant molecules, also called substrates, with high specificity. Conformational changes in an enzyme cause its substrate to bind more tightly to it, a phenomenon called induced fit (Figure 6.6).

- Each enzyme-catalyzed reaction has a maximal velocity ($V_{max}$). The $K_M$ value for an enzyme is the substrate concentration at which the velocity of the reaction is half of the maximal value. Competitive inhibitors raise the $K_M$ for the substrate, whereas noncompetitive inhibitors lower the $V_{max}$ (Figure 6.7).

- Enzyme function may be affected by a variety of other factors, including prosthetic groups, cofactors, coenzymes, temperature, and pH (Figure 6.8).

- Altman and colleagues discovered that the RNA subunit within RNase P is a ribozyme, an RNA molecule that catalyzes a chemical reaction. Other ribozymes play key roles in the cell (Figures 6.9, 6.10, Table 6.3).

## 6.3  Overview of Metabolism

- Metabolism is the sum of the chemical reactions in a living organism. Metabolic pathways consist of coordinated chemical reactions that occur in steps and are catalyzed by specific enzymes (Figure 6.11).

- Catabolic reactions involve the breakdown of larger molecules into smaller ones. These reactions recycle organic molecules that are used as building blocks to make new molecules. The organic molecules are also broken down to make energy intermediates such as ATP.

- Some chemical reactions are redox reactions, in which electrons are transferred from one molecule to another. These reactions can be used to make energy intermediates such as NADH (Figure 6.12).

- Anabolic reactions require an input of energy to synthesize larger molecules and macromolecules.

- Metabolic pathways are controlled by gene regulation, cellular regulation, and biochemical regulation. An example of biochemical regulation is feedback inhibition. The enzyme that catalyzes the rate-limiting step in a pathway is often the target of cellular or biochemical regulation (Figure 6.13).

## 6.4  Recycling of Organic Molecules

- Recycling of organic molecules saves a great deal of energy for living organisms.

- Proteins in the cells of eukaryotes and archaea are degraded by proteasomes (Figure 6.14).

- Lysosomes digest intracellular material through the process of autophagy (Figure 6.15).

## Assess & Discuss

### Test Yourself

1. Reactions that release free energy are
   a. exergonic.
   b. spontaneous.
   c. endergonic.
   d. endothermic.
   e. both a and b.

2. Enzymes speed up reactions by
   a. providing chemical energy to fuel a reaction.
   b. lowering the activation energy necessary to initiate a reaction.

   c. causing an endergonic reaction to become an exergonic reaction.
   d. substituting for one of the reactants necessary for a reaction.
   e. none of the above.

3. For the idealized reaction $aA + bB \rightleftharpoons cC + dD$, suppose that the equilibrium constant, $K_{eq}$, is 0.01. If the starting concentrations for A, B, C, and D are 1 M each, what would you predict based on the value of $K_{eq}$?
   a. The forward reaction is favored.
   b. The reverse reaction is favored.
   c. The forward reaction is fast.
   d. The reverse reaction is fast.
   e. both b and d.

4. Researchers analyzed a cell extract—a mixture of molecules isolated from a certain type of cell—and studied a chemical reaction in which a carbohydrate was broken down into smaller molecules. When they added a protease to the cell extract, they discovered that the protease greatly inhibited the rate of the reaction. Based on this observation, you could conclude that the reaction is
   a. exergonic.
   b. endergonic.
   c. catalyzed by an enzyme.
   d. catalyzed by a ribozyme.
   e. Both b and c are true of this reaction.

5. In biological systems, ATP functions by
   a. providing the energy to drive endergonic reactions.
   b. acting as an enzyme and lowering the activation energy of certain reactions.
   c. adjusting the pH of intracellular solutions to maintain optimal conditions for enzyme activity.
   d. regulating the speed at which endergonic reactions proceed.
   e. interacting with enzymes as a cofactor to stimulate chemical reactions.

6. In a chemical reaction, NADH is converted to $NAD^+ + H^+$. We say that NADH has been
   a. reduced.
   b. phosphorylated.
   c. oxidized.
   d. decarboxylated.
   e. methylated.

7. Scientists identify proteins that use ATP as an energy source by
   a. determining whether a protein functions in anabolic or catabolic reactions.
   b. determining if a protein has a known ATP-binding site.
   c. predicting the free energy necessary for a protein to function.
   d. determining if a protein has an ATP synthase subunit.
   e. all of the above.

8. For a particular chemical reaction, an inhibitor raises the $K_M$ but does not affect the $V_{max}$. This inhibitor
   a. is a competitive inhibitor.
   b. is a noncompetitive inhibitor.
   c. binds to the active site of the enzyme.
   d. binds to an allosteric site of the enzyme.
   e. is a competitive inhibitor and binds to the active site of the enzyme.

9. Which of the following is (are) key benefits of catabolic reactions?
   a. recycling of organic building blocks
   b. breakdown of organic molecules to obtain energy
   c. synthesis of important polymers, such as polypeptides
   d. all of the above
   e. a and b only

10. Autophagy provides a way for cells to
    a. degrade entire organelles and recycle their components.
    b. control the level of ATP.
    c. engulf bacterial cells.
    d. export unwanted organelles out of the cell.
    e. inhibit the first enzyme in a metabolic pathway.

## Conceptual Questions

1. With regard to rate and direction, discuss the differences between endergonic and exergonic reactions.

2. Describe the mechanism and purpose of feedback inhibition in a metabolic pathway.

3. **Core Concept: Energy and Matter** A core concept of biology is that *living organisms use energy*. Explain why the recycling of amino acids and nucleotides is energy-efficient.

## Collaborative Questions

1. Living cells are highly ordered units, yet the entropy of the universe is increasing. Discuss how life can maintain its order in spite of the second law of thermodynamics. Are we defying this law?

2. What is the advantage of using ATP as a common energy source; that is, how is using just ATP better than using a bunch of different food molecules? For example, instead of just having $Na^+/K^+$-ATPase in a cell, why not have many different ion pumps, each driven by a different food molecule, like $Na^+/K^+$-glucosase (a pump that uses glucose), $Na^+/K^+$-sucrase (a pump that uses sucrose), $Na^+/K^+$-fatty acidase (a pump that uses fatty acids), and so on?

# Cellular Respiration and Fermentation

# 7

**Physical endurance.** Conditioned athletes, like these marathon runners, metabolize organic molecules such as glucose very efficiently. ©Michael Dwyer/AP Images

Carmen became inspired while watching the 2008 Summer Olympics and set a personal goal to run a marathon—a distance of 42.2 kilometers, or 26.2 miles. Although she was active in volleyball and downhill skiing in high school, she had never attempted distance running. At first, running an entire mile was pure torture. She was out of breath, overheated, and unhappy, to say the least. However, she became committed to endurance training and within a few weeks discovered that running a mile was a "piece of cake." Two years later, Carmen participated in her first marathon and finished with a time of 4 hours and 11 minutes—not bad for someone who had previously struggled to run a single mile!

How had Carmen's training allowed her to achieve this goal? Perhaps the biggest factor is that the training altered the metabolism in her leg muscles. For example, the network of small blood vessels supplying oxygen to her leg muscles became more extensive, providing more efficient delivery of oxygen and removal of wastes. Second, her muscle cells developed more mitochondria. Recall from Chapter 4 that the primary role of mitochondria is to make ATP, which cells use as a source of energy. With these changes, Carmen's leg muscles were better able to break down organic molecules in her food and utilize them to make ATP.

The cells in Carmen's leg muscles had become more efficient at **cellular respiration**, which comprises the metabolic reactions that a cell uses to get energy from food molecules and release waste products. When we eat food, we use much of that food for energy. People often speak of "burning calories." Although metabolism does generate some heat, the chemical reactions that take place in the cells of living organisms are uniquely different from those that occur, say, in a fire. When wood is burned, the reaction produces enormous amounts of heat in a short period of time—the reaction lacks control. In contrast, the metabolism that occurs in living cells is extremely controlled. The food molecules from which we harvest energy give up that energy in a very restrained manner rather than all at once, as in a fire. An underlying theme of metabolism is the remarkable control that cells possess when they coordinate chemical reactions. A key emphasis of this chapter is how cells use the energy stored within the chemical bonds of organic molecules.

We will begin by surveying a group of chemical reactions that accomplish the breakdown of the main carbohydrate cells use as an energy source, namely, the sugar glucose. As you will learn, cells carry out an intricate series of reactions so that glucose can be "burned" in a very controlled fashion when oxygen is available. We will then examine how cells use organic molecules in the absence of oxygen via processes known as anaerobic respiration and fermentation.

## 7.1 Overview of Cellular Respiration

**Learning Outcome:**

1. List and briefly describe the four metabolic pathways that are needed to break down glucose to $CO_2$ and $H_2O$.

Cellular respiration is a process by which living cells obtain energy from organic molecules and release waste products. A primary aim of cellular respiration is to make adenosine triphosphate, or ATP.

When oxygen ($O_2$) is used, this process is termed **aerobic respiration**. During aerobic respiration, $O_2$ is consumed, and carbon dioxide ($CO_2$) is released via the oxidation of organic molecules. When we breathe, we inhale the oxygen needed for aerobic respiration and exhale $CO_2$, a by-product of the process. For this reason, the term respiration has a second meaning, which is the act of breathing.

$$C_6H_{12}O_6 + 6\,O_2 \rightarrow 6\,CO_2 + 6\,H_2O$$

(Glucose)

$$\Delta G = -685\,\text{kcal/mol}$$

We will focus on the breakdown of glucose in a eukaryotic cell in the presence of oxygen. Certain covalent bonds within glucose store a large amount of chemical potential energy. When glucose is broken down via oxidation, ultimately to $CO_2$ and water,

a tremendous amount of free energy is released ($-685$ kcal/mol). Some of the energy is lost as heat, but much of it is used to make three energy intermediates: ATP, NADH, and $FADH_2$. This process involves four metabolic pathways: (1) glycolysis, (2) the breakdown of pyruvate, (3) the citric acid cycle, and (4) oxidative phosphorylation (**Figure 7.1**):

1. ***Glycolysis.*** In glycolysis, glucose (a compound with six carbon atoms) is broken down to two pyruvate molecules (with three carbons each), producing a net energy yield of two ATP molecules and two NADH molecules. The two ATP molecules are synthesized via **substrate-level phosphorylation**, which occurs when an enzyme directly transfers a phosphate from an organic molecule to ADP. In eukaryotes, glycolysis occurs in the cytosol.

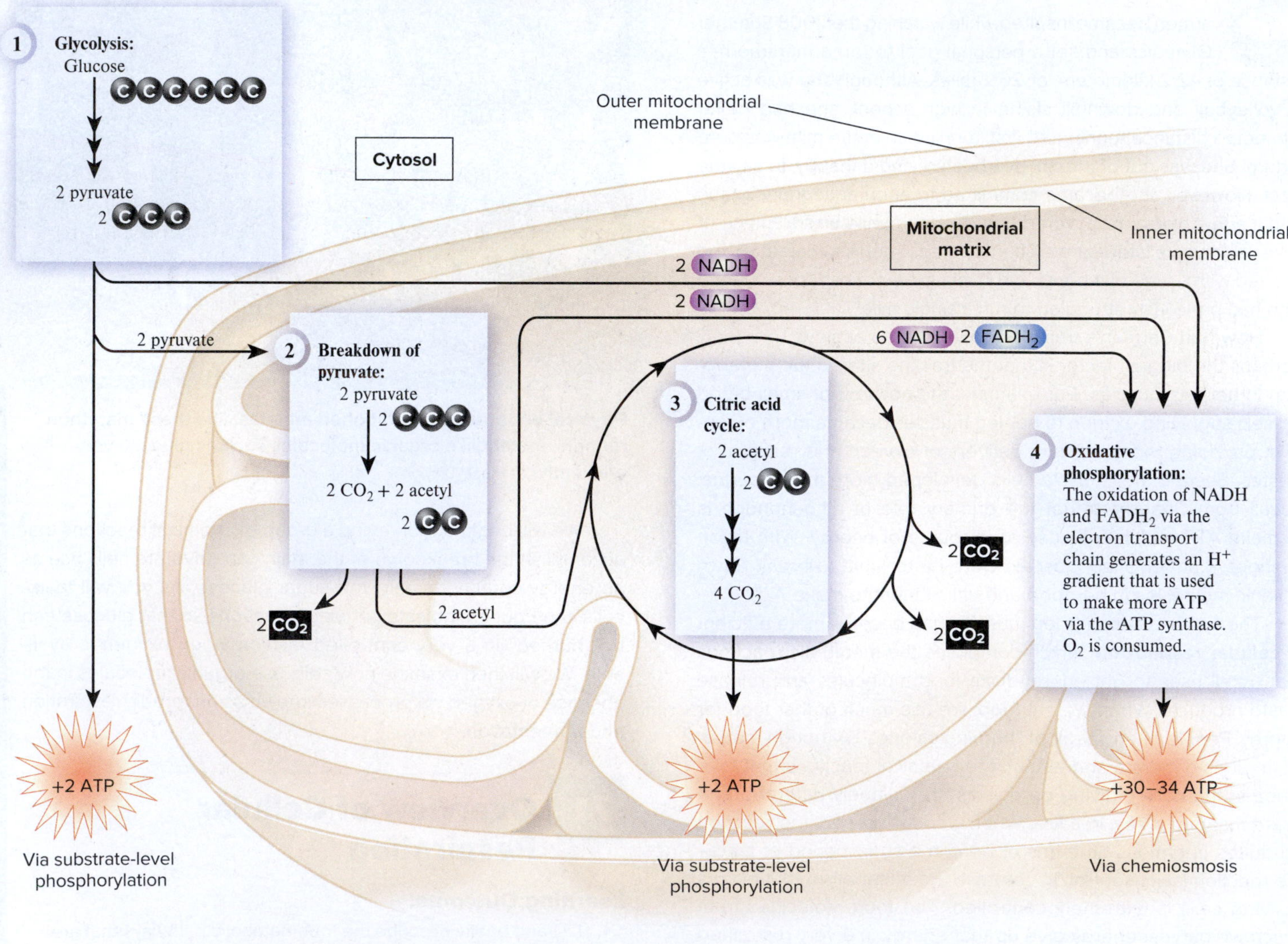

**Figure 7.1** **An overview of cellular respiration.** The 30–34 ATP molecules produced via chemiosmosis represent the maximum number possible. As described later in this chapter, mitochondria may use NADH, $FADH_2$, and the $H^+$ electrochemical gradient for purposes other than ATP synthesis.

 **Core Concept: Energy and Matter** Molecules such as glucose store a large amount of energy. The breakdown of glucose is used to make energy intermediates, such as ATP molecules, which drive many types of cellular processes.

2. ***Breakdown of pyruvate.*** The two pyruvate molecules enter the mitochondrial matrix, where each one is broken down to an acetyl group (with two carbons each) and one $CO_2$ molecule. For each pyruvate broken down via oxidation, one NADH molecule is made by the reduction of $NAD^+$.

3. ***Citric acid cycle.*** Each acetyl group is incorporated into an organic molecule, which is later oxidized to liberate two $CO_2$ molecules. One ATP, three NADH, and one $FADH_2$ are made in this process. Because there are two acetyl groups (one from each pyruvate), the total yield is four $CO_2$, two ATP via substrate-level phosphorylation, six NADH, and two $FADH_2$. This process occurs in the mitochondrial matrix.

4. ***Oxidative phosphorylation.*** The NADH and $FADH_2$ made in the three previous stages contain high-energy electrons that can be readily transferred in a redox reaction to other molecules. Once removed from NADH or $FADH_2$, these high-energy electrons release some energy, and through an electron transport chain, that energy is harnessed to produce an $H^+$ electrochemical gradient. In **chemiosmosis**, energy stored in the $H^+$ electrochemical gradient is used to synthesize ATP from ADP and $P_i$. The overall process of electron transport and ATP synthesis is called oxidative phosphorylation because NADH or $FADH_2$ has been oxidized and ADP has become phosphorylated to make ATP. Approximately 30–34 ATP molecules can be made via oxidative phosphorylation.

In eukaryotes, oxidation phosphorylation occurs along the **cristae**, which are projections formed by the invagination of the inner mitochondrial membrane. The cristae greatly increase the surface area of the inner membrane and thereby increase the amount of ATP that can be made. In bacteria and archaea, oxidative phosphorylation occurs along the plasma membrane.

## 7.2 Glycolysis

### Learning Outcomes:

1. Outline the three phases of glycolysis, and identify the net products.
2. Describe the series of enzymatic reactions that constitute glycolysis.
3. **CoreSKILL »** Explain the underlying basis for the use of positron-emission tomography to detect cancer.

Thus far, we have examined the general features of the four metabolic pathways that are involved in the breakdown of glucose. We will now turn our attention to a more detailed understanding of these pathways for glucose metabolism, beginning with glycolysis.

### Glycolysis Is a Metabolic Pathway That Breaks Down Glucose to Pyruvate

**Glycolysis** (from the Greek *glykos*, meaning sweet, and *lysis*, meaning splitting) involves the breakdown of glucose, a simple sugar, into two molecules of a compound called pyruvate. This process can occur in the presence of oxygen, that is, under aerobic conditions, and it can also occur in the absence of oxygen. During the 1930s, the efforts of several German biochemists, including Gustav Embden, Otto Meyerhof, and Jacob Parnas, established that glycolysis involves 10 steps, each one catalyzed by a different enzyme. The elucidation of these steps was a major achievement in the field of **biochemistry**—the study of the chemistry of living organisms. Researchers have since discovered that glycolysis is the common pathway for glucose breakdown in bacteria, archaea, and eukaryotes. Remarkably, the steps of glycolysis are virtually identical in nearly all living species, suggesting that glycolysis arose very early in the evolution of life on our planet.

The 10 steps of glycolysis can be grouped into three phases (**Figure 7.2**).

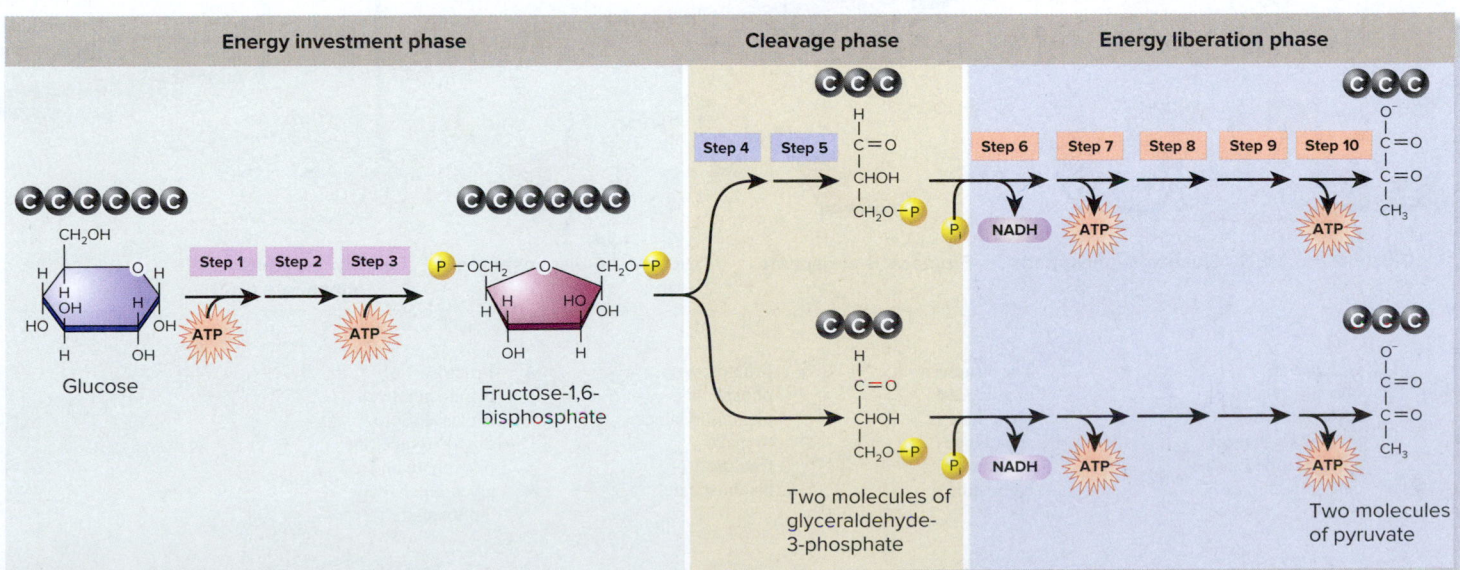

**Figure 7.2  Overview of glycolysis.**

**Core Skill: Connections** Look ahead to Table 45.1. With regard to oxygen needs, what advantage do glycolytic muscle fibers provide?

- The first phase (steps 1–3) involves an energy investment. Two ATP molecules are hydrolyzed, and the phosphates from those ATP molecules are attached to glucose, which is converted to fructose-1,6-bisphosphate. The energy investment phase raises the free energy of glucose, thereby allowing later reactions to be exergonic.

- The cleavage phase (steps 4–5) breaks this six-carbon molecule into two molecules of glyceraldehyde-3-phosphate.

- The energy liberation phase (steps 6–10) produces four ATP, two NADH, and two molecules of pyruvate. Because two molecules of ATP are used in the energy investment phase, the net yield is two molecules of ATP.

**Figure 7.3** describes the details of the 10 reactions of glycolysis. The net reaction of glycolysis is as follows:

$$C_6H_{12}O_6 \ + \ 2\,NAD^+ \ + \ 2\,ADP^{2-} \ + \ 2\,P_i^{2-} \ \rightarrow$$
Glucose

$$2\,CH_3(C{=}O)COO^- \ + \ 2\,H^+ \ + \ 2\,NADH \ + \ 2\,ATP^{4-} \ + \ 2\,H_2O$$
Pyruvate

**Regulation of Glycolysis** How do cells control glycolysis? The rate of glycolysis is regulated by the availability of substrates, such as glucose, and by feedback inhibition. A key control point involves the enzyme phosphofructokinase, which catalyzes the third step in glycolysis, the step believed to be the slowest, or rate-limiting, step. When a cell has a sufficient amount of ATP, feedback inhibition occurs. At high concentrations, ATP binds to an allosteric site in phosphofructokinase, causing a conformational change that renders the enzyme functionally inactive. This prevents the further breakdown of glucose and thereby inhibits the overproduction of ATP.

**BIO:TIPS**

**THE QUESTION** *During the process of glycolysis, glucose is broken down into two pyruvate molecules. As shown in Figure 7.3, this metabolic pathway consists of 10 consecutive chemical reactions. Describe the three major phases of glycolysis.*

**T** **OPIC** *What topic in biology does this question address?* The topic is glycolysis. More specifically, the question asks you to describe the three major phases of this process.

**I** **NFORMATION** *What information do you know based on the question and your understanding of the topic?* In the question, you are reminded that glycolysis consists of 10 consecutive chemical reactions. From your understanding of the topic, you may remember that different types of chemical reactions are occurring.

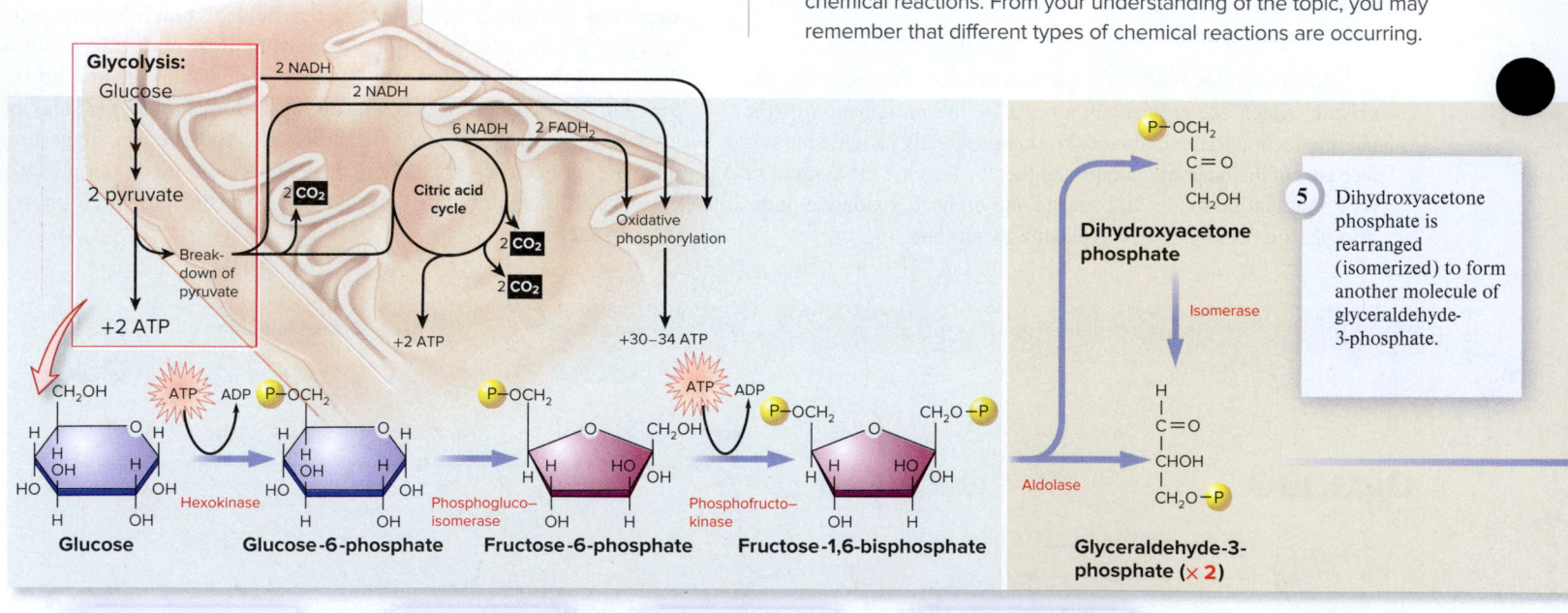

| 1 | Glucose is phosphorylated by ATP. Glucose-6-phosphate is more easily trapped in the cell than glucose. | 2 | The structure of glucose-6-phosphate is rearranged to fructose-6-phosphate. | 3 | Fructose-6-phosphate is phosphorylated to make fructose-1,6-bisphosphate. | 4 | Fructose-1,6-bisphosphate is cleaved into dihydroxyacetone phosphate and glyceraldehyde-3-phosphate. |

**Figure 7.3 A detailed look at the steps of glycolysis.** The pathway begins with a six-carbon molecule (glucose) that is broken down into two molecules that contain three carbons each. The notation **x2** in the figure indicates that two of these three-carbon molecules are produced from each glucose molecule.

**Concept Check:** *Which organic molecules donate a phosphate group to ADP during substrate-level phosphorylation?*

**PROBLEM-SOLVING STRATEGY** *Sort out the steps in a complicated process.* To solve this problem, it may be helpful to examine the process in a step-by-step manner to identify the key events.

**ANSWER** *First phase: During steps 1–3 of glycolysis, ATP is used to phosphorylate two different sites in the glucose molecule. This stage is called the energy investment phase because ATP is used to fuel the process. The energy investment phase prepares the glucose molecule for the next two phases.*

*Second phase: During steps 4 and 5, glucose is cleaved into two three-carbon molecules, and then one of those is isomerized to glyceraldehyde-3-phosphate. This phase is called the cleavage phase because a six-carbon molecule is split (cleaved) into two three-carbon molecules.*

*Third phase: During steps 6–10, ATP and NADH are made, molecules that are energy intermediates. ATP is made by substrate-level phosphorylation, in which a phosphate is removed from 1,3-bisphosphoglycerate or phosphoenolpyruvate and directly transferred to ADP. NADH is made when glyceraldehyde-3-phosphate is oxidized. This last phase is called the energy liberation phase because energy that was stored in organic molecules was released (liberated) and used to make energy intermediates (ATP and NADH).*

## Core Concept: Information

### The Overexpression of Certain Genes Causes Cancer Cells to Exhibit High Levels of Glycolysis

In 1931, the German physiologist Otto Warburg discovered that certain cancer cells preferentially use glycolysis for ATP production, in contrast to healthy cells, which mainly generate ATP from oxidative phosphorylation. This phenomenon, termed the Warburg effect, is very common among different types of tumors. The Warburg effect is the basis for the detection of cancer via a procedure called positron-emission tomography (PET, see Figure 2.6). In this technique, patients are injected with a radioactive glucose analogue called [$^{18}$F]-fluorodeoxyglucose (FDG). FDG is taken up by cells that use high amounts of glucose, such as cancer cells. The scanner detects regions of the body that accumulate high amounts of FDG, which are visualized as bright spots on the PET scan.

**Figure 7.4** shows a PET scan of a patient with lung cancer. The bright regions that the arrows point at are tumors that show abnormally high levels of glycolysis. The tumors show up so well because the genome found in cancer cells exhibits an increased expression of genes that encode enzymes involved with glycolysis. Research has shown that the enzymes of glycolysis are over-expressed in approximately 80% of all types of cancer, including

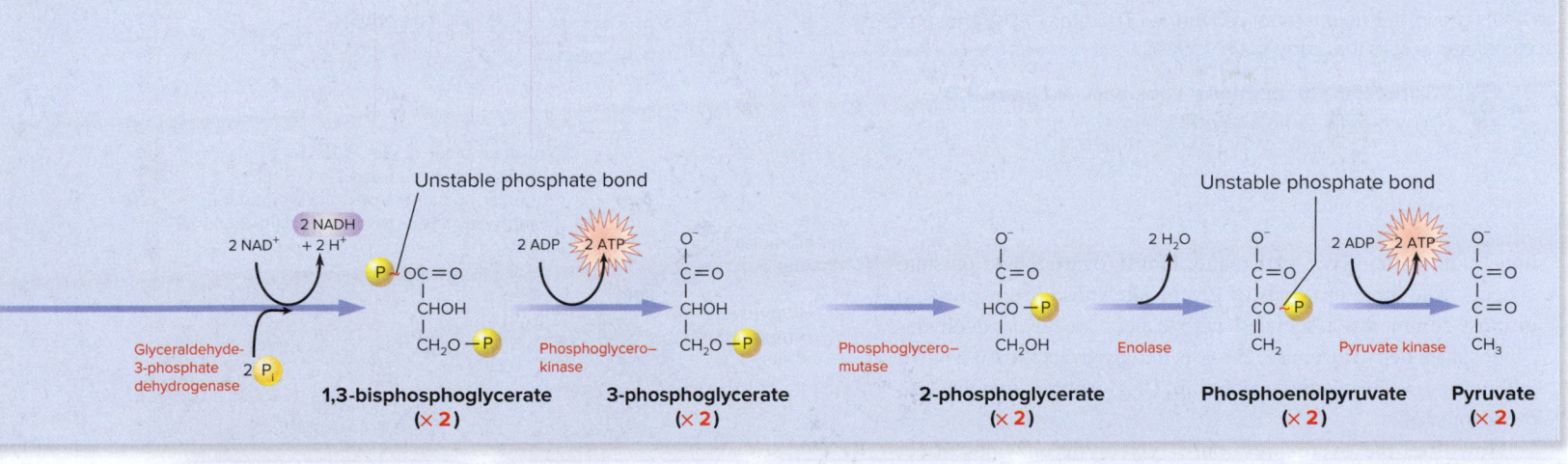

| 6 | Glyceraldehyde-3-phosphate is oxidized to 1,3-bisphosphoglycerate. NADH is produced. In 1,3-bisphosphoglycerate, the phosphate group in the upper left is destabilized, meaning that the bond will break in a highly exergonic reaction. | 7 | A phosphate is removed from 1,3-bisphosphoglycerate to form 3-phosphoglycerate. The removed phosphate is transferred to ADP to make ATP via substrate-level phosphorylation. | 8 | The phosphate group in 3-phosphoglycerate is moved to a new location, creating 2-phosphoglycerate. | 9 | A water molecule is removed from 2-phosphoglycerate to form phosphoenol-pyruvate. In phosphoenol-pyruvate, the phosphate group is destabilized, meaning that the bond will break in a highly exergonic reaction. | 10 | A phosphate is removed from phosphoenolpyruvate to form pyruvate. The removed phosphate is transferred to ADP to make ATP via substrate-level phosphorylation. |

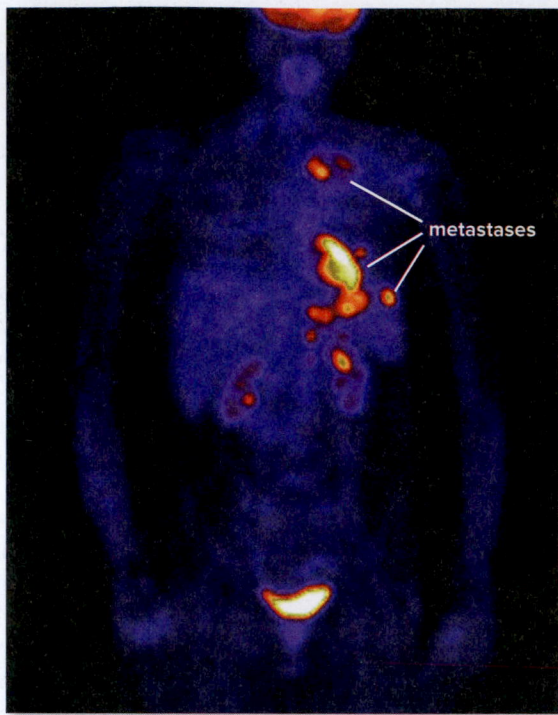

**Figure 7.4  A PET scan of a patient with lung cancer.** The bright regions in the lungs are tumors (indicated by the arrows). The brain, which is not cancerous in this patient, appears bright because it performs high levels of glucose metabolism. Also, the kidneys and bladder appear bright because they filter and accumulate FDG. (Note: FDG is taken up by cells and converted to FDG-phosphate by hexokinase, the first enzyme in glycolysis. However, because FDG lacks an —OH group, it is not metabolized further. Therefore, FDG-phosphate accumulates in cells that carry out glycolysis.) ©Steven Needell/Science Source

👁 **Core Skill: Connections**  Look back at **Figure 2.6**. Why is FDG radiolabeled?

lung, skin, colon, liver, pancreatic, breast, ovarian, and prostate cancers. The three enzymes of glycolysis whose overexpression is most commonly associated with cancer are glyceraldehyde-3-phosphate dehydrogenase, enolase, and pyruvate kinase (shown in Figure 7.3). In many cancers, all 10 glycolytic enzymes are overexpressed!

How does the overexpression of glycolytic enzymes affect tumor growth? While the genetic changes associated with tumor growth are complex, researchers have speculated that an increase in glycolysis favors the growth as a result of changes in oxygen levels. As a tumor grows, the internal regions of the tumor tend to become hypoxic, or deficient in oxygen. The hypoxic state inside a tumor may contribute to the overexpression of glycolytic genes and lead to a higher level of glycolytic enzymes within the cancer cells. This favors glycolysis as a means of making ATP, because glycolysis does not require oxygen. Making ATP via glycolysis is an advantage to cancer cells, because such cells would have trouble making ATP via

oxidative phosphorylation, which requires oxygen. Based on these findings, some current research is aimed at discovering drugs that inhibit glycolysis in cancer cells as a way to prevent their growth.

## 7.3  Breakdown of Pyruvate

**Learning Outcome:**

1. Describe how pyruvate is broken down and acetyl CoA is made.

In eukaryotes, glycolysis produces pyruvate in the cytosol, which is then transported into the mitochondria. Once in the mitochondrial matrix, pyruvate molecules are broken down (oxidized) by an enzyme complex called pyruvate dehydrogenase (**Figure 7.5**). A molecule of $CO_2$ is removed from pyruvate, and the remaining acetyl group is attached to an organic molecule called coenzyme A (CoA) to produce acetyl CoA. (In chemical equations, CoA is depicted as CoA—SH to emphasize how the —SH group participates in the chemical reaction.) During this process, two high-energy electrons are removed from pyruvate and transferred to NAD$^+$, together with H$^+$, to produce a molecule of NADH. For each pyruvate, the net reaction is as follows:

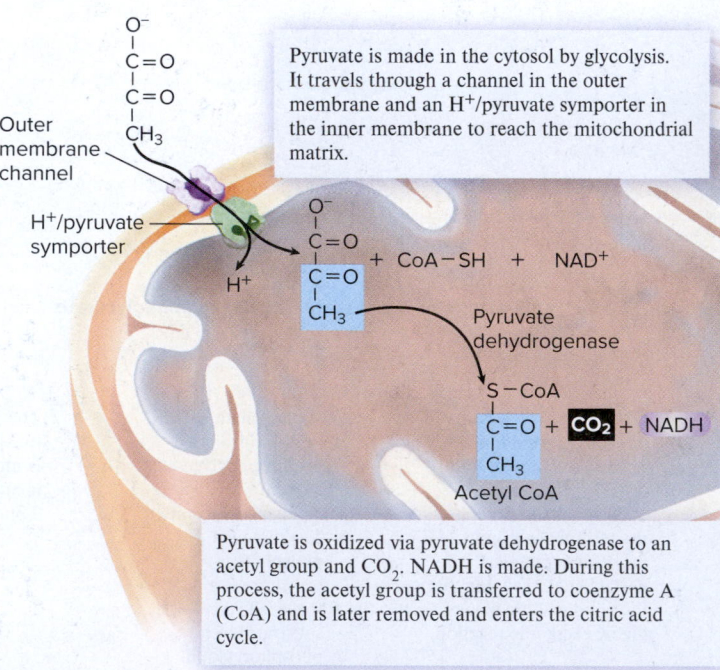

**Figure 7.5  Breakdown of pyruvate and the attachment of an acetyl group to CoA.**

The acetyl group is attached to CoA via a covalent bond to a sulfur atom. The hydrolysis of this bond releases a large amount of free energy, making it possible for the acetyl group to be transferred to other organic molecules. As described next, the acetyl group is removed from CoA and enters the citric acid cycle.

## 7.4 Citric Acid Cycle

### Learning Outcomes:

1. Explain the concept of a metabolic cycle.
2. Describe how an acetyl group enters the citric acid cycle, and list the net products of the cycle.

The third stage of glucose metabolism introduces a new concept, that of a **metabolic cycle**. During a metabolic cycle, particular molecules enter the cycle while others leave. The process is cyclical because it involves a series of organic molecules that are regenerated with each turn of the cycle. The idea of a metabolic cycle was first proposed in the early 1930s by German biochemist Hans Krebs. While studying carbohydrate metabolism in England, he analyzed cell extracts from pigeon muscle and determined that citric acid and other organic molecules participated in a cycle that resulted in the breakdown of carbohydrates to carbon dioxide. This cycle is called the **citric acid cycle**, or the Krebs cycle, in honor of Krebs, who was awarded the Nobel Prize in Physiology or Medicine in 1953.

An overview of the citric acid cycle is shown in **Figure 7.6**. In the first step of the cycle, the acetyl group (with two carbons) is removed from acetyl CoA and attached to oxaloacetate (with four carbons) to form citrate (with six carbons), also called citric acid. Then, in a series of several steps, two $CO_2$ molecules are released. As this occurs, three molecules of NADH, one molecule of $FADH_2$, and one molecule of guanosine triphosphate (GTP) are made. The GTP, which is made via substrate-level phosphorylation, is used to make ATP. After a total of eight steps, oxaloacetate is regenerated, so the cycle can begin again, provided acetyl CoA is available. **Figure 7.7** shows a more detailed view of the citric acid cycle. For each acetyl group attached to CoA, the net reaction of the citric acid cycle is as follows:

$$\text{Acetyl-CoA} + 2\,H_2O + 3\,NAD^+ + FAD + GDP^{2-} + Pi^{2-} \rightarrow$$

$$\text{CoA—SH} + 2\,CO_2 + 3\,NADH + FADH_2 + GTP^{4-} + 3\,H^+$$

***Regulation of the Citric Acid Cycle*** How is the citric acid cycle controlled? The rate of the cycle is largely regulated by the availability of substrates, such as acetyl-CoA and $NAD^+$, and by feedback

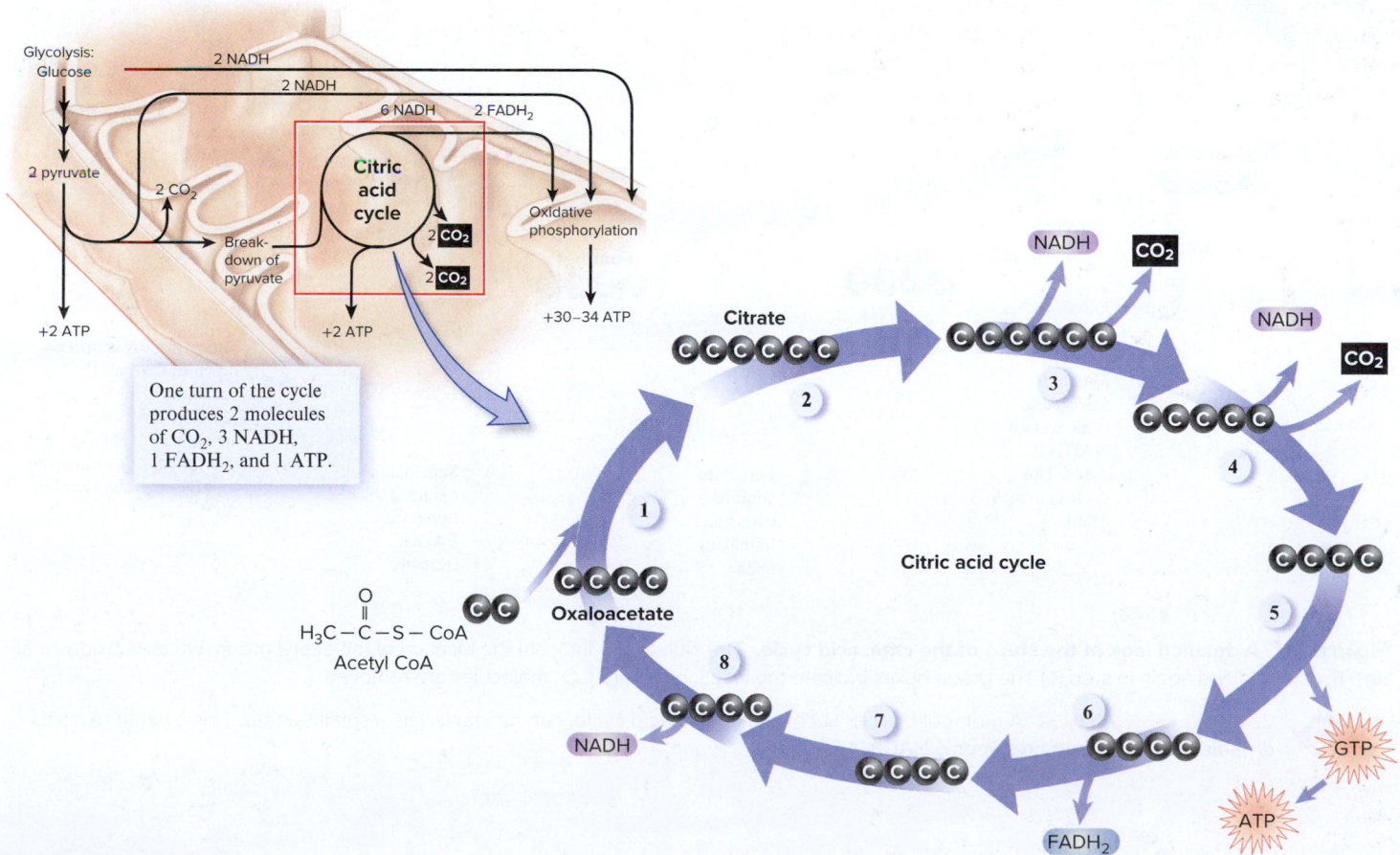

**Figure 7.6** **Overview of the citric acid cycle.**

*Concept Check:* *What are the main products of the citric acid cycle?*

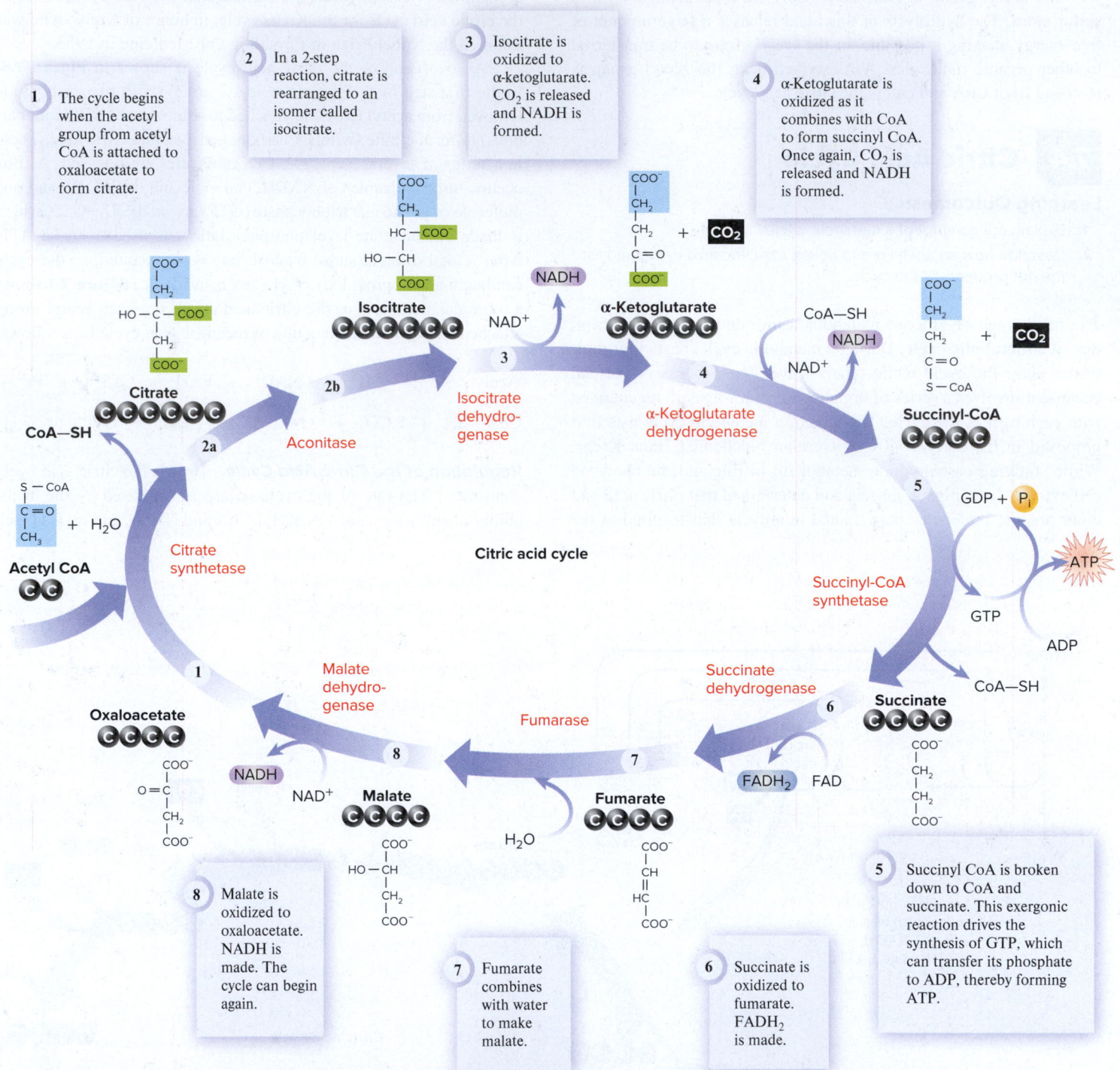

**1** The cycle begins when the acetyl group from acetyl CoA is attached to oxaloacetate to form citrate.

**2** In a 2-step reaction, citrate is rearranged to an isomer called isocitrate.

**3** Isocitrate is oxidized to α-ketoglutarate. $CO_2$ is released and NADH is formed.

**4** α-Ketoglutarate is oxidized as it combines with CoA to form succinyl CoA. Once again, $CO_2$ is released and NADH is formed.

**Citric acid cycle**

Citrate synthetase

Aconitase

Isocitrate dehydrogenase

α-Ketoglutarate dehydrogenase

Succinyl-CoA synthetase

Succinate dehydrogenase

Fumarase

Malate dehydrogenase

**5** Succinyl CoA is broken down to CoA and succinate. This exergonic reaction drives the synthesis of GTP, which can transfer its phosphate to ADP, thereby forming ATP.

**6** Succinate is oxidized to fumarate. $FADH_2$ is made.

**7** Fumarate combines with water to make malate.

**8** Malate is oxidized to oxaloacetate. NADH is made. The cycle can begin again.

**Figure 7.7  A detailed look at the steps of the citric acid cycle.** The blue boxes indicate the location of the acetyl group, which is oxidized at step 6. (It is oxidized again in step 8.) The green boxes indicate the locations where $CO_2$ molecules are removed.

**Core Concept: Systems** A metabolic cycle, such as the citric acid cycle, can be viewed as a small system. This system oxidizes organic molecules and produces 3 NADH, 1 $FADH_2$, 1 ATP, and 2 $CO_2$.

inhibition. The three steps in the cycle that are highly exergonic are those catalyzed by citrate synthase, isocitrate dehydrogenase, and α-ketoglutarate dehydrogenase (see Figure 7.7). Each of these steps is rate-limiting under certain circumstances, and the way that each enzyme is regulated varies among different species. Let's consider an example. In mammals, NADH and ATP act as feedback inhibitors of isocitrate dehydrogenase, whereas NAD+ and ADP act as activators. In this way, the citric acid cycle is inhibited when NADH and ATP levels are high, but it is stimulated when NAD+ and ADP levels are high.

## 7.5 Overview of Oxidative Phosphorylation

### Learning Outcomes:

1. Describe how the electron transport chain produces an H+ electrochemical gradient.
2. Explain how ATP synthase utilizes the H+ electrochemical gradient to synthesize ATP.

During the first three stages of glucose metabolism, the oxidation of glucose yields 6 molecules of $CO_2$, 4 molecules of ATP, 10 molecules of NADH, and 2 molecules of $FADH_2$. Let's now consider how high-energy electrons are removed from NADH and $FADH_2$ to produce more ATP. This process is called **oxidative phosphorylation**. As mentioned earlier, the term refers to the observation that electrons are removed from NADH and $FADH_2$, that is, these molecules are oxidized, and ATP is made by the phosphorylation of ADP. In this section, we will examine how the oxidative process involves the electron transport chain, whereas the phosphorylation of ADP occurs via ATP synthase.

### The Electron Transport Chain Establishes an Electrochemical Gradient

The **electron transport chain (ETC)** consists of a group of protein complexes and small organic molecules embedded in the inner mitochondrial membrane. These components are referred to as an electron transport chain because electrons are passed from one component to the next in a series of redox reactions (**Figure 7.8**). Most members of the ETC are protein complexes (designated I–IV in the figure) that have prosthetic groups, which are small molecules permanently attached to the surface of proteins that aid in their function. For example, cytochrome oxidase contains two prosthetic groups, each with an iron atom. The iron in each prosthetic group can readily accept and release an electron. One member of the ETC, ubiquinone (Q), is not a protein. Rather, ubiquinone is a small organic molecule that can accept and release an electron.

The red line in Figure 7.8 shows the path of electron flow. The electrons, which are originally found in NADH or $FADH_2$, are transferred to components of the ETC. The electron path is

a series of redox reactions in which electrons are transferred to components with increasingly higher electronegativity. At the end of the chain is oxygen, which is the most electronegative component and the final electron acceptor. The ETC is also called the **respiratory chain** because the oxygen we breathe is used in this process.

NADH and $FADH_2$ donate their electrons at different points in the ETC. Two high-energy electrons from NADH are first transferred one at a time to NADH dehydrogenase (complex I). They are then transferred to ubiquinone (Q), cytochrome $b\text{-}c_1$ (complex III), cytochrome $c$, and cytochrome oxidase (complex IV). The final electron acceptor is $O_2$. By comparison, $FADH_2$ transfers electrons to succinate reductase (complex II), then to ubiquinone, and the rest of the chain.

As shown in Figure 7.8, some of the energy that is released during the movement of electrons is used to pump H+ across the inner mitochondrial membrane into the intermembrane space. This active transport establishes a large **H+ electrochemical gradient**, in which the concentration of H+ is higher outside of the mitochondrial matrix than inside and an excess of positive charge exists outside the matrix.

Chemicals that inhibit the flow of electrons along the ETC have lethal effects. For example, one component of the ETC, cytochrome oxidase (complex IV), is inhibited by cyanide. The deadly effects of cyanide ingestion occur because the ETC is shut down, preventing cells from making enough ATP for survival.

### ATP Synthase Makes ATP via Chemiosmosis

The second event of oxidative phosphorylation is the synthesis of ATP by an enzyme called **ATP synthase**. The H+ electrochemical gradient across the inner mitochondrial membrane is a source of potential energy. How is this energy used? The passive flow of H+ back into the matrix is an exergonic process. The lipid bilayer is relatively impermeable to H+. However, H+ can pass through the membrane-embedded portion of ATP synthase. This enzyme harnesses some of the free energy that is released as the H+ ions flow through its membrane-embedded region to synthesize ATP from ADP and $P_i$ (see bottom of Figure 7.8). This is an example of an energy conversion: Energy in the form of an H+ gradient is converted to chemical potential energy in ATP. The synthesis of ATP that occurs as a result of pushing H+ across a membrane is called chemiosmosis (from the Greek *osmos*, meaning to push). The theory behind it was proposed by Peter Mitchell, a British biochemist who was awarded the Nobel Prize in Chemistry in 1978.

***Regulation of Oxidative Phosphorylation*** How is oxidative phosphorylation controlled? This process is regulated by a variety of factors, including the availability of ETC substrates, such as NADH and $O_2$, and by the ATP/ADP ratio. When ATP levels are high, ATP binds to a subunit of cytochrome oxidase (complex IV), thereby inhibiting the ETC and oxidative phosphorylation. By comparison, when ADP levels are high, oxidative phosphorylation is

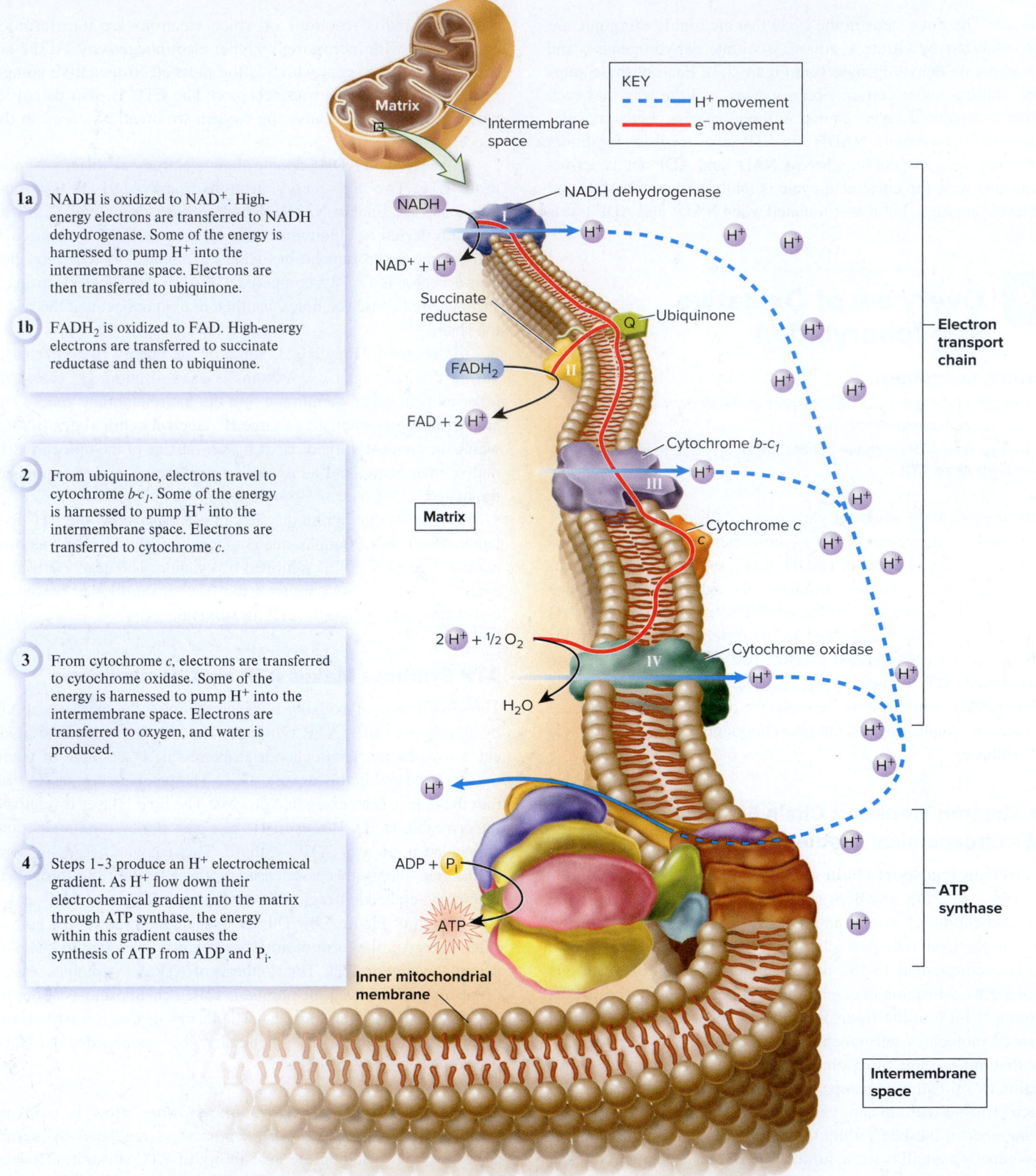

**KEY**
- ---- $H^+$ movement
- —— $e^-$ movement

Matrix

Intermembrane space

**1a** NADH is oxidized to $NAD^+$. High-energy electrons are transferred to NADH dehydrogenase. Some of the energy is harnessed to pump $H^+$ into the intermembrane space. Electrons are then transferred to ubiquinone.

**1b** $FADH_2$ is oxidized to FAD. High-energy electrons are transferred to succinate reductase and then to ubiquinone.

**2** From ubiquinone, electrons travel to cytochrome $b$-$c_1$. Some of the energy is harnessed to pump $H^+$ into the intermembrane space. Electrons are transferred to cytochrome $c$.

**3** From cytochrome $c$, electrons are transferred to cytochrome oxidase. Some of the energy is harnessed to pump $H^+$ into the intermembrane space. Electrons are transferred to oxygen, and water is produced.

**4** Steps 1–3 produce an $H^+$ electrochemical gradient. As $H^+$ flow down their electrochemical gradient into the matrix through ATP synthase, the energy within this gradient causes the synthesis of ATP from ADP and $P_i$.

NADH

NADH dehydrogenase

$NAD^+ + H^+$

Succinate reductase

Ubiquinone

$FADH_2$

$FAD + 2 H^+$

Cytochrome $b$-$c_1$

**Matrix**

Cytochrome $c$

$2 H^+ + \frac{1}{2} O_2$

Cytochrome oxidase

$H_2O$

$H^+$

ADP + $P_i$

ATP

**Inner mitochondrial membrane**

**Electron transport chain**

**ATP synthase**

**Intermembrane space**

**Figure 7.8 Oxidative phosphorylation.** This process consists of two distinct events: the electron transport chain (ETC) and ATP synthesis. The ETC oxidizes, or removes electrons from, NADH or $FADH_2$ and pumps $H^+$ across the inner mitochondrial membrane. In chemiosmosis, ATP synthase uses the energy in this $H^+$ electrochemical gradient to phosphorylate ADP, thereby synthesizing ATP. In this figure, an oxygen atom is represented as $\frac{1}{2} O_2$ to emphasize that the ETC reduces oxygen when it is in its molecular ($O_2$) form.

*Concept Check:* *Explain the meaning of the name cytochrome oxidase.*

stimulated for two reasons: (1) ADP stimulates cytochrome oxidase, and (2) ADP is a substrate that is used (with $P_i$) to make ATP.

## NADH Oxidation Makes a Large Proportion of a Cell's ATP

For each molecule of NADH that is oxidized and each molecule of ATP that is made, the two chemical reactions of oxidative phosphorylation can be represented as follows:

$$NADH + H^+ + \tfrac{1}{2}O_2 \rightarrow NAD^+ + H_2O$$

$$ADP^{2-} + P_i^{2-} \rightarrow ATP^{4-} + H_2O$$

When we add up the maximal amount of ATP that can be made by oxidative phosphorylation, most researchers agree it is in the range of 30–34 ATP molecules for each glucose molecule that is broken down to $CO_2$ and $H_2O$. However, that maximal amount of ATP is rarely achieved, for two reasons.

- First, although 10 NADH and 2 $FADH_2$ are available to make the $H^+$ electrochemical gradient across the inner mitochondrial membrane, a cell uses some of these molecules for anabolic pathways. For example, NADH is used in the synthesis of organic molecules such as glycerol (a component of phospholipids).

- Second, the mitochondrion may use some of the energy in the $H^+$ electrochemical gradient for other purposes. For example, the gradient is used for the uptake of pyruvate into the matrix via an $H^+$/pyruvate symporter (see Figure 7.5).

Therefore, the actual amount of ATP synthesized is usually a little less than the maximum of 30 to 34 molecules. Even so, when we compare the amount of ATP that is made by glycolysis (2), the citric acid cycle (2), and oxidative phosphorylation (30–34), we see that oxidative phosphorylation provides a cell with a much greater capacity to make ATP.

## Free-Energy Changes Drive Oxidative Phosphorylation and Other Stages of Glucose Breakdown

Thus far, we have considered (1) glycolysis, (2) the breakdown of pyruvate, (3) the citric acid cycle, and (4) oxidative phosphorylation. All four of these stages are ultimately driven by the oxidation of glucose, which is a highly exergonic process that releases free energy. However, the energy is not released in one big blast, as in an explosion, but rather in small step-wise increments. Releasing the energy in small increments allows cells to couple the breakdown of glucose with useful chemical processes. For example, as we saw earlier in this chapter, the breakdown of glucose to pyruvate is coupled to the synthesis of ATP. **Figure 7.9** shows how free energy is released as electrons move along the electron transport chain. At particular points along the ETC, some of the energy is used to pump $H^+$ across the inner mitochondrial membrane and establish an $H^+$ electrochemical gradient. This gradient is then used to power ATP synthesis.

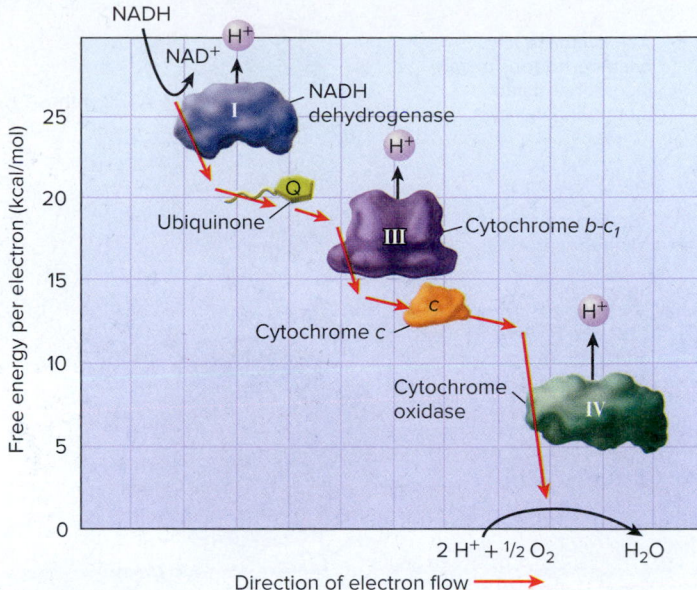

**Figure 7.9** **The relationship between free energy and electron movement along the electron transport chain.** As electrons are transferred from one site to another along the electron transport chain, they release energy. Some of this energy is harnessed to pump $H^+$ across the inner mitochondrial membrane. The total energy released by a single electron is approximately −25 kcal/mol.

## 7.6  A Closer Look at ATP Synthase

**Learning Outcomes:**

1. **CoreSKILL »** Analyze the results of an experiment that verified that ATP synthase uses an $H^+$ electrochemical gradient to make ATP.
2. Describe the structure of ATP synthase.
3. Explain how a series of three conformational changes enables ATP synthase to make ATP.
4. **CoreSKILL »** Analyze the results of an experiment that showed that ATP synthase is a rotary machine.

The structure and function of ATP synthase are particularly intriguing and have received much attention over the past few decades. In this section, we will consider experiments that were aimed at elucidating this enzyme's function and explore, in greater depth, how it is able to synthesize ATP.

## Experiments with Purified Proteins in Membrane Vesicles Verified Chemiosmosis

To show experimentally that ATP synthase makes ATP using an $H^+$ electrochemical gradient, researchers needed to purify the enzyme and study its function in vitro. In 1974, Efraim Racker and Walther Stoeckenius purified ATP synthase and another protein called bacteriorhodopsin, which is found in certain species of archaea. Previous research had shown that bacteriorhodopsin is a light-driven $H^+$ pump. Racker and Stoeckenius took both purified proteins and experimentally inserted them into membrane vesicles, a process called reconstitution (**Figure 7.10**). ATP synthase was oriented so its ATP-synthesizing region was on the outside of the

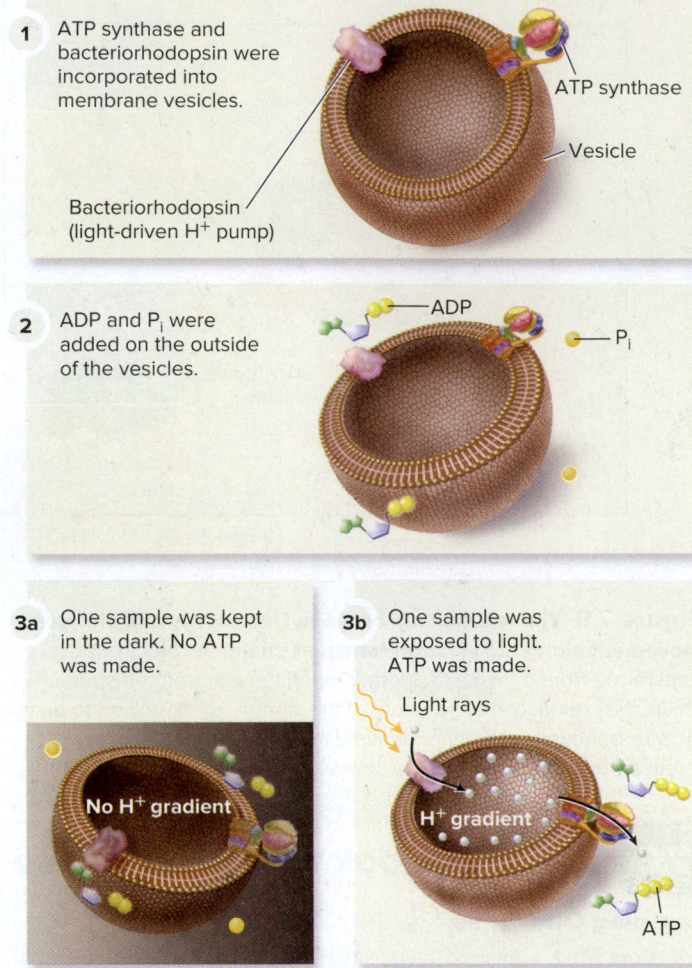

1. ATP synthase and bacteriorhodopsin were incorporated into membrane vesicles.

ATP synthase

Vesicle

Bacteriorhodopsin (light-driven H⁺ pump)

2. ADP and $P_i$ were added on the outside of the vesicles.

ADP

$P_i$

3a. One sample was kept in the dark. No ATP was made.

No H⁺ gradient

3b. One sample was exposed to light. ATP was made.

Light rays

H⁺ gradient

ATP

**Figure 7.10   The Racker and Stoeckenius experiment.** In this experiment, bacteriorhodopsin pumped H⁺ into vesicles, and the resulting H⁺ electrochemical gradient was sufficient to drive ATP synthesis via ATP synthase.

**Concept Check:** *Is the functioning of the electron transport chain always needed to make ATP via ATP synthase?*

vesicles. Bacteriorhodopsin was oriented so it would pump H⁺ into the vesicles. The researchers added ADP and $P_i$ on the outside of the vesicles. In the dark, no ATP was made. However, when they shone light on the vesicles, a substantial amount of ATP was synthesized. Because bacteriorhodopsin was already known to be a light-driven H⁺ pump, these results convinced the researchers that ATP synthase uses an H⁺ electrochemical gradient as an energy source to make ATP.

## ATP Synthase Is a Rotary Machine That Makes ATP as It Spins

ATP synthase is a rotary machine (**Figure 7.11**). It spins! The region embedded in the membrane is composed of three types of subunits called *a*, *b*, and *c*. Approximately 10–14 *c* subunits form a ring in the membrane. One *a* subunit is bound to this ring, and

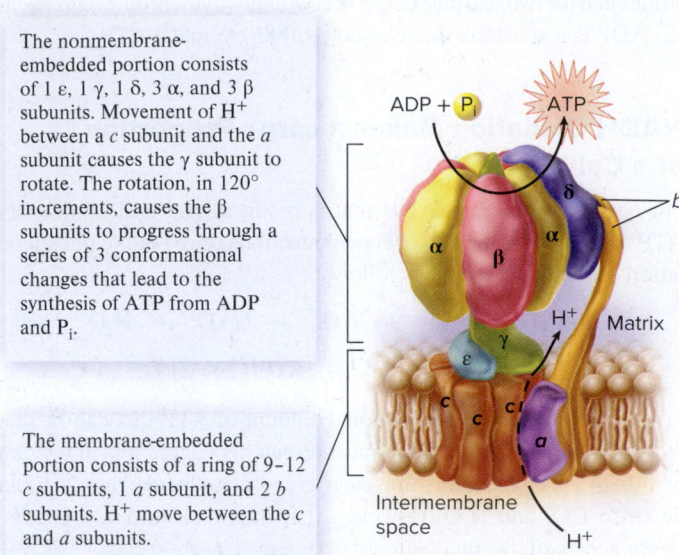

The nonmembrane-embedded portion consists of 1 ε, 1 γ, 1 δ, 3 α, and 3 β subunits. Movement of H⁺ between a *c* subunit and the *a* subunit causes the γ subunit to rotate. The rotation, in 120° increments, causes the β subunits to progress through a series of 3 conformational changes that lead to the synthesis of ATP from ADP and $P_i$.

ADP + $P_i$

ATP

δ

α   β   α

H⁺   Matrix

γ

ε

The membrane-embedded portion consists of a ring of 9–12 *c* subunits, 1 *a* subunit, and 2 *b* subunits. H⁺ move between the *c* and *a* subunits.

c   c   c

b

a

Intermembrane space

H⁺

**Figure 7.11   The subunit structure and function of ATP synthase.**

two *b* subunits are attached to the *a* subunit and protrude from the membrane. The nonmembrane-embedded subunits are designated with Greek letters. One ε and one γ subunit bind to the ring of *c* subunits. The γ subunit forms a long stalk that pokes into the center of another ring of three α and three β subunits. Each β subunit contains a catalytic site where ATP is made. Finally, the δ subunit forms a connection between the ring of α and β subunits and the two *b* subunits.

When hydrogen ions pass through a narrow channel at the contact site between a *c* subunit and the *a* subunit, a conformational change causes the γ subunit to turn clockwise (when viewed from the intermembrane space). Each time the γ subunit turns 120°, it changes its contacts with the three β subunits, which, in turn, causes the β subunits to change their conformations. How do these conformational changes promote ATP synthesis? The answer is that the conformational changes occur in a way that favors ATP synthesis and release. As shown in **Figure 7.12**, the conformational changes in the β subunits happen in the following order:

- Conformation 1: ADP and $P_i$ bind with good affinity.
- Conformation 2: ADP and $P_i$ bind very tightly, which strains chemical bonds so that ATP is made.
- Conformation 3: ATP binds very weakly and is released.

Each time the γ subunit turns 120°, it causes a β subunit to change to the next conformation. After conformation 3, a 120° turn by the γ subunit returns a β subunit back to conformation 1, and the cycle of ATP synthesis can begin again. Because ATP synthase has three β subunits, each subunit is in a different conformation at any given time.

American biochemist Paul Boyer proposed the concept of a rotary machine in the late 1970s. In his model, the three β subunits alternate between three conformations, as described previously. Boyer's

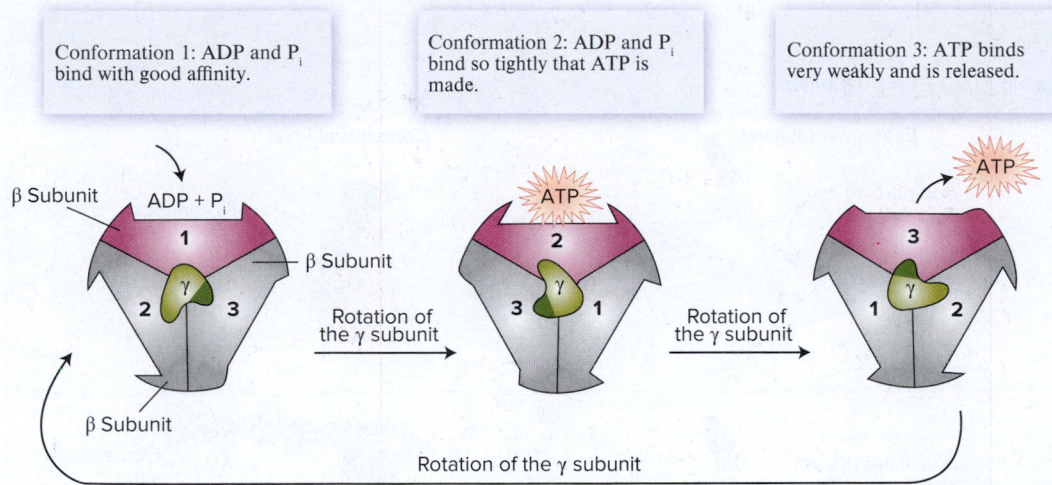

Conformation 1: ADP and P$_i$ bind with good affinity.

Conformation 2: ADP and P$_i$ bind so tightly that ATP is made.

Conformation 3: ATP binds very weakly and is released.

β Subunit

ADP + P$_i$

β Subunit

β Subunit

Rotation of the γ subunit

Rotation of the γ subunit

Rotation of the γ subunit

**Figure 7.12 Conformational changes that result in ATP synthesis.** For simplicity, the α subunits are not shown. This drawing emphasizes the conformational changes in the β subunit shown at the top. The other two β subunits also make ATP. All three β subunits alternate between three conformational states due to their interactions with the γ subunit.

original idea was met with great skepticism, because the concept that part of an enzyme could spin was very novel, to say the least. In 1994, British biochemist John Walker and his colleagues determined the three-dimensional structure of the nonmembrane-embedded portion of the ATP synthase. The structure revealed that each of the three β subunits had a different conformation—one with ADP bound, one with ATP bound, and one without any nucleotide bound. This result supported Boyer's model. In 1997, Boyer and Walker shared the Nobel Prize in Chemistry for their work on ATP synthase. As described next in the Feature Investigation, other researchers subsequently visualized the rotation of the γ subunit.

**Core Skill: Modeling** The goal of this modeling challenge is to predict how a mutation in the β subunit of ATP synthase would affect ATP synthesis.

**Modeling Challenge:** Let's suppose a researcher has identified a mutation in the β subunit that only affects conformation 3. A model that depicts the shape of the mutant β subunit is shown to the right. Look very carefully at the shape of the mutant subunit in conformation 3 and compare it to the normal β subunit in that conformation, as shown in Figure 7.12. Predict how this mutant subunit would affect ATP synthesis.

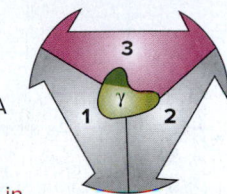

Mutant β subunit

## Core Skill: Process of Science

# Feature Investigation | Yoshida and Kinosita Demonstrated That the γ Subunit of ATP Synthase Spins

In 1997, Japanese biochemist Masasuke Yoshida, biophysicist Kazuhiko Kinosita, and colleagues set out to experimentally visualize the rotary nature of ATP synthase (Figure 7.13). The membrane-embedded region of ATP synthase can be separated from the rest of the protein by treating mitochondrial membranes with a high concentration of salt, releasing the portion of the protein containing one γ, three α, and three β subunits. The researchers adhered the γα$_3$β$_3$ complex to a glass slide so that the γ subunit was protruding upward. Because the γ subunit is too small to be seen with a light microscope, the rotation of this subunit cannot be visualized directly. To overcome this problem, the researchers attached a long, fluorescently labeled actin filament to the γ subunit via linker proteins. The fluorescently labeled actin filament is very long compared to the γ subunit and can be readily seen with a fluorescence microscope.

Because the membrane-embedded portion of the protein was missing, you may be wondering how the researchers could get the γ subunit to rotate. The answer is that they added ATP. Although the normal function of ATP synthase is to make ATP, it can also hydrolyze ATP. In other words, ATP synthase can run backward. As shown in the data

in **Figure 7.13**, when the researchers added ATP, they observed that the fluorescently labeled actin filament rotated in a counterclockwise direction, which is opposite to the direction that the γ subunit rotates when ATP is synthesized. Actin filaments were observed to rotate for more than 100 revolutions in the presence of ATP. These results convinced the scientific community that ATP synthase is a rotary machine.

### Experimental Questions

1. **CoreSKILL »** The components of ATP synthase are too small to be visualized by light microscopy. For the experiment of Figure 7.13, how did the researchers observe the movement of ATP synthase?

2. **CoreSKILL »** In the experiment of Figure 7.13, what observation indicated to the researchers that ATP synthase is a rotary machine? What was the control of this experiment? What did it indicate?

3. **CoreSKILL »** Were the rotations seen by the researchers in the data of Figure 7.13 in the same direction as they are expected to occur in mitochondria during ATP synthesis? Why or why not?

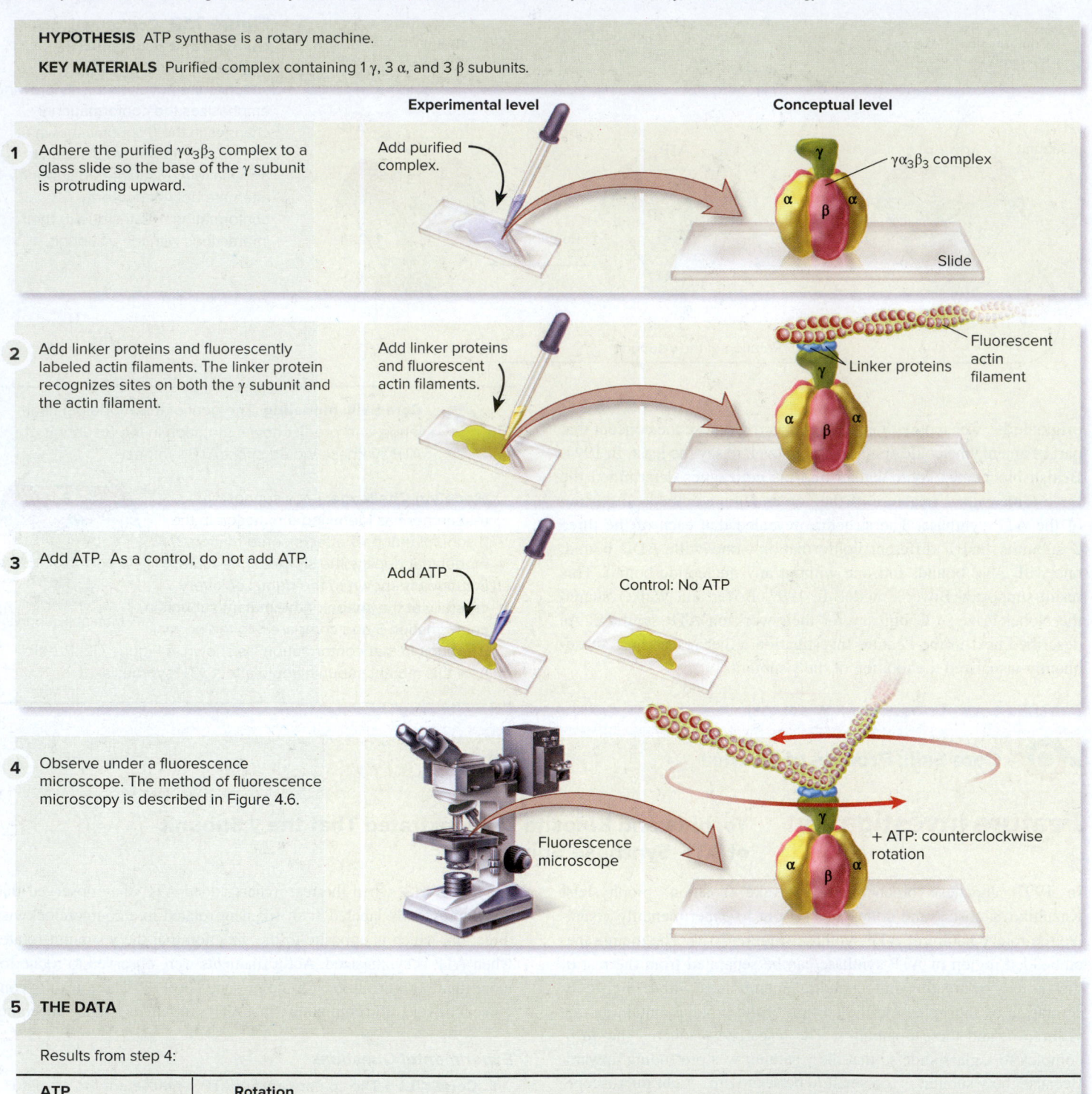

**Figure 7.13** **Yoshida and Kinosita provide evidence that ATP synthase is a rotary machine.** (5): From Noji, H., Yoshida, M. 2001. The Rotary Machine in the Cell, ATP Synthase. *Journal of Biological Chemistry 276: 1665–1668.* ©2001 The American Society for Biochemistry and Molecular Biology

**HYPOTHESIS** ATP synthase is a rotary machine.

**KEY MATERIALS** Purified complex containing 1 γ, 3 α, and 3 β subunits.

Experimental level

Conceptual level

1. Adhere the purified $\gamma\alpha_3\beta_3$ complex to a glass slide so the base of the γ subunit is protruding upward.

Add purified complex.

γ
$\gamma\alpha_3\beta_3$ complex
α
β
α
Slide

2. Add linker proteins and fluorescently labeled actin filaments. The linker protein recognizes sites on both the γ subunit and the actin filament.

Add linker proteins and fluorescent actin filaments.

Fluorescent actin filament
Linker proteins
γ
α
β
α

3. Add ATP. As a control, do not add ATP.

Add ATP

Control: No ATP

4. Observe under a fluorescence microscope. The method of fluorescence microscopy is described in Figure 4.6.

Fluorescence microscope

+ ATP: counterclockwise rotation
γ
α
β
α

5. **THE DATA**

Results from step 4:

| ATP | Rotation |
|---|---|
| No ATP added | No rotation observed. |
| ATP added | Rotation was observed as shown below. This is a time-lapse view of the rotation in action. |

| | |
|---|---|
| Row 1 | |
| Row 2 | |

**6    CONCLUSION** The γ subunit rotates counterclockwise when ATP is hydrolyzed. It would be expected to rotate clockwise when ATP is synthesized.

**7    SOURCE** Noji, H., Yoshida, M. 2001. The rotary machine in the cell, ATP synthase. *Journal of Biological Chemistry* 276: 1665–1668.

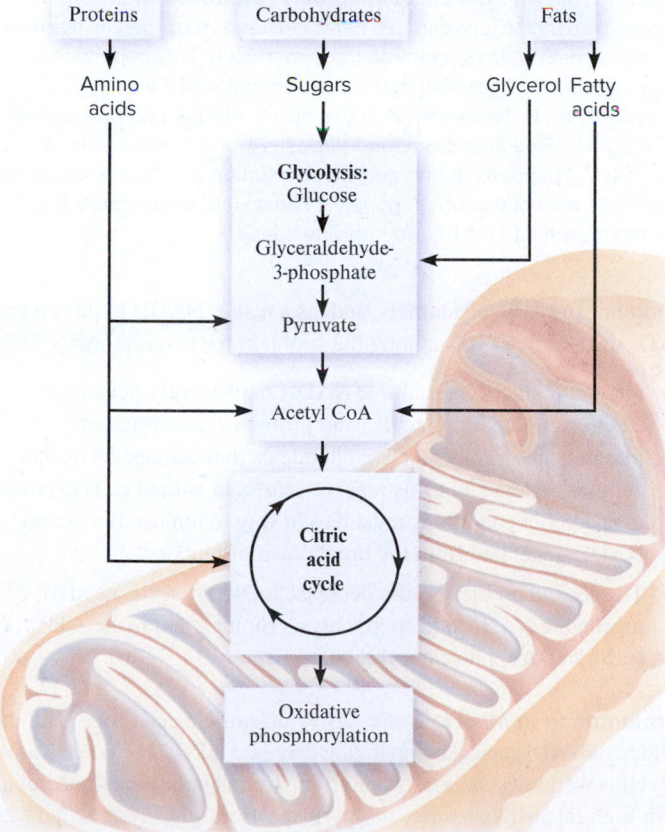

**Figure 7.14 Integration of carbohydrate, protein, and fat metabolism.** Breakdown products of proteins and fats are used as fuel for cellular respiration, entering the same pathways used to break down carbohydrates. ©Ernie Friedlander/Cole Group/Getty Images

*Concept Check:* *What advantage does integrating protein, carbohydrate, and fat metabolism have for cells?*

## 7.7 Connections Among Carbohydrate, Protein, and Fat Metabolism

**Learning Outcome:**

1. Explain how carbohydrate, protein, and fat metabolism are interconnected.

When you eat a meal, it usually contains not only carbohydrates (including glucose), but also proteins and fats. These molecules are broken down by some of the same enzymes involved with glucose metabolism. The use of the same pathways for the breakdown of sugars, amino acids, and fats makes cellular metabolism more efficient because the same enzymes are used for the breakdown of different starting molecules.

As shown in **Figure 7.14**, proteins and fats can enter glycolysis or the citric acid cycle at different points.

- Proteins are first acted on by enzymes, either in digestive juices or within cells, that cleave the bonds connecting individual amino acids. Because the 20 amino acids differ in their side chains, amino acids and their breakdown products can enter at different points in the pathway. Breakdown products of some amino acids can enter at later steps of glycolysis, or an acetyl group can be removed from certain amino acids and become attached to CoA and then enter the citric acid cycle (see Figure 7.14). Other amino acids are modified and enter the citric acid cycle.

- Fats are typically broken down to glycerol and fatty acids. Glycerol can be modified to glyceraldehyde-3-phosphate and enter glycolysis. Lipid tails can have two carbon acetyl units removed, which bind to CoA and enter the citric acid cycle.

## 7.8 Anaerobic Respiration and Fermentation

**Learning Outcomes:**

1. Describe how certain microorganisms make ATP using a final electron acceptor in the electron transport chain that is not oxygen.

2. Explain how muscle and yeast cells use fermentation to synthesize ATP under anaerobic conditions.

Thus far, we have surveyed catabolic pathways that result in the complete breakdown of glucose in the presence of oxygen. Cells also commonly metabolize organic molecules in the absence of oxygen. The term **anaerobic** is used to describe an environment that lacks oxygen. Many bacteria and archaea and some fungi exist in anaerobic environments but still have to oxidize organic molecules to obtain sufficient amounts of energy. Examples include microbes living in your intestinal tract and those living deep in the soil. Similarly, when a person exercises strenuously, the rate of oxygen consumption by muscle cells may greatly exceed the rate of oxygen delivery—particularly at the start of the strenuous exercise. Under these conditions, muscle cells become anaerobic and must obtain sufficient energy in the absence of oxygen to maintain their level of activity.

Two different strategies may be used by cells to metabolize organic molecules in the absence of oxygen. One mechanism is to use a substance other than $O_2$ as the final electron acceptor of the electron transport chain, a process called **anaerobic respiration**. A second approach is to produce ATP via substrate-level phosphorylation only, without any net oxidation of organic molecules, a process called **fermentation**. In this section, we will consider examples of both strategies.

## Some Microorganisms Carry Out Anaerobic Respiration

At the end of the ETC, as shown earlier in Figure 7.8, cytochrome oxidase recognizes $O_2$ and catalyzes its reduction to $H_2O$. The final electron acceptor of the chain is $O_2$. Many species of bacteria that live under anaerobic conditions have evolved enzymes that function similarly to cytochrome oxidase but recognize molecules other than $O_2$ and use them as the final electron acceptor.

For example, under anaerobic conditions *Escherichia coli*, a bacterial species found in your intestinal tract, produces an enzyme called nitrate reductase. This enzyme uses nitrate ($NO_3^-$) as the final electron acceptor of the electron transport chain. **Figure 7.15** shows a simplified ETC in *E. coli* in which nitrate is the final electron acceptor. In *E. coli* and other bacterial species, the ETC is in the plasma membrane that surrounds the cytoplasm. Electrons travel from NADH to NADH dehydrogenase to ubiquinone to cytochrome *b* and then to nitrate reductase. At the end of the chain, $NO_3^-$ is converted to nitrite ($NO_2^-$). This process generates an $H^+$ electrochemical gradient in three ways. First, NADH dehydrogenase pumps $H^+$ out of the cytoplasm. Second, ubiquinone picks up $H^+$ in the cytoplasm and carries it to the other side of the membrane. Third, the reduction of nitrate to nitrite consumes $H^+$ in the cytoplasm. The generation of an $H^+$ electrochemical gradient via these three processes allows *E. coli* cells to make ATP via chemiosmosis under anaerobic conditions.

## Fermentation Is the Breakdown of Organic Molecules Without Net Oxidation

Many organisms, including animals and yeast, use only $O_2$ as the final electron acceptor of their ETCs. When confronted with anaerobic conditions, these organisms must have a different way of producing sufficient ATP. One strategy is to make ATP via glycolysis, which can occur under either anaerobic or aerobic conditions. Under anaerobic conditions, cells do not use the citric acid cycle or the ETC, but make ATP only via glycolysis.

A key issue is that glycolysis requires $NAD^+$ and generates NADH. Under aerobic conditions, NADH is oxidized to $NAD^+$ to make more ATP. However, this cannot occur under anaerobic

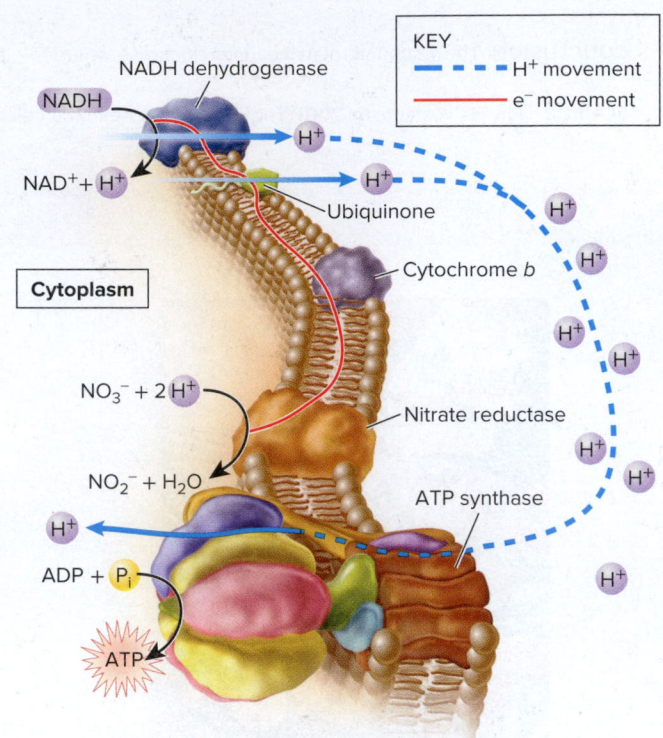

**Figure 7.15** **An example of anaerobic respiration in *E. coli*.** When oxygen is absent, *E. coli* can use nitrate instead of oxygen as the final electron acceptor of the electron transport chain. This generates an $H^+$ electrochemical gradient that is used to make ATP via chemiosmosis. Note: As shown in this figure, ubiquinone picks up $H^+$ on one side of the membrane and deposits it on the other side. A similar event happens during aerobic respiration in mitochondria (see Figure 7.8), except that ubiquinone transfers $H^+$ to cytochrome $b$-$c_1$, which pumps it into the intermembrane space.

conditions in yeast and animals, and, as a result, NADH builds up and $NAD^+$ decreases. This is a potential problem for two reasons:

- First, at high concentrations, NADH haphazardly donates its electrons to other molecules and promotes the formation of free radicals, highly reactive chemicals that damage DNA and cellular proteins. For this reason, yeast and animal cells exposed to anaerobic conditions must have a way to remove the excess NADH generated from the breakdown of glucose.

- The second problem is the decrease in $NAD^+$. Cells need to regenerate $NAD^+$ to keep glycolysis running and make ATP via substrate-level phosphorylation.

*Fermentation in Muscle Cells* How do muscle cells cope with the buildup of NADH and accompanying decrease in $NAD^+$? When a muscle cell is working strenuously and its environment becomes anaerobic, as in high-intensity exercise, the pyruvate from glycolysis is reduced to make lactate. (The uncharged, or protonated, form is called lactic acid.) The electrons to reduce pyruvate are derived from NADH, which is oxidized to $NAD^+$ (**Figure 7.16a**). Therefore, this process decreases NADH and reduces its potentially harmful effects. It also increases the level of $NAD^+$, thereby allowing glycolysis to continue. The lactate is secreted from muscle cells. Once sufficient oxygen is restored, the

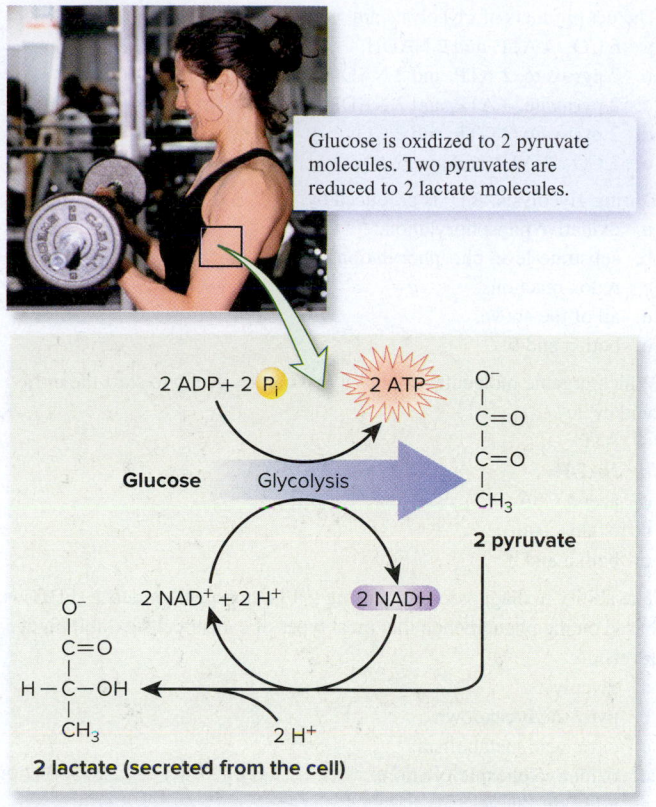

Glucose is oxidized to 2 pyruvate molecules. Two pyruvates are reduced to 2 lactate molecules.

**(a) Production of lactic acid**

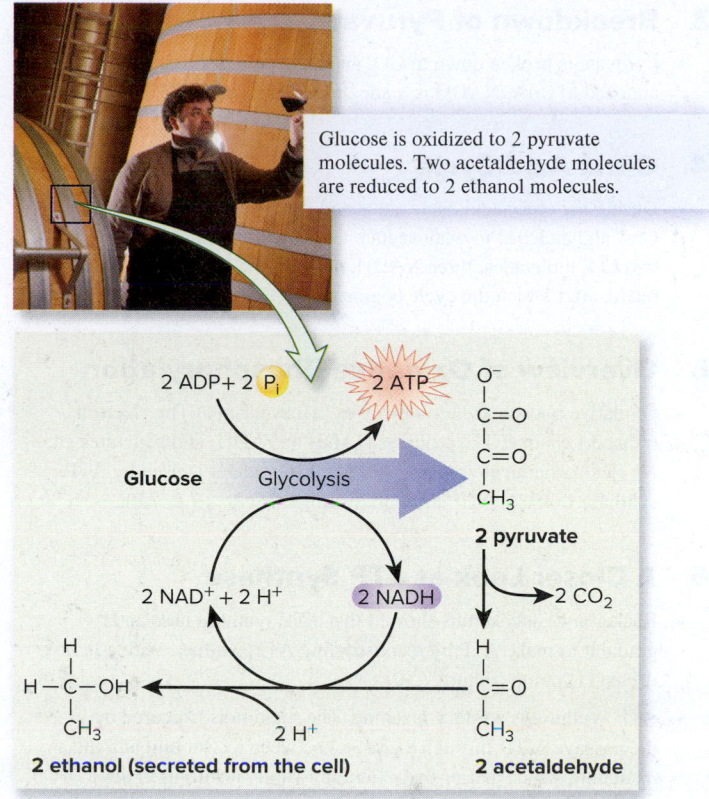

Glucose is oxidized to 2 pyruvate molecules. Two acetaldehyde molecules are reduced to 2 ethanol molecules.

**(b) Production of ethanol**

**Figure 7.16   Examples of fermentation.**  In these examples, NADH is produced by the oxidation of an organic molecule, and then the NADH is converted back to NAD$^+$ when it donates electrons to a different organic molecule such as pyruvate **(a)** or acetaldehyde **(b)**. a: ©Homer W Sykes/ Alamy Stock Photo; b: ©FreeProd/Alamy Stock Photo

 **Core Skill: Science and Society**  Fermentation by microorganisms is used in wine making, beer brewing, and bread making.

lactate produced during strenuous exercise can be taken up by cells, converted back to pyruvate, and used for energy, or this lactate may be used by the liver and other tissues to make glucose.

***Fermentation in Yeast Cells***  Yeast cells cope with anaerobic conditions differently. During wine making, a yeast cell metabolizes sugar under anaerobic conditions. The pyruvate is broken down to $CO_2$ and a two-carbon molecule called acetaldehyde. The acetaldehyde is then reduced by NADH to make ethanol, while NADH is oxidized to NAD$^+$ (**Figure 7.16b**). Similar to lactate production in muscle cells, this process decreases NADH and increases NAD$^+$, thereby preventing the harmful effects of NADH and allowing glycolysis to continue.

The term fermentation is used to describe the breakdown of organic molecules to harness energy without any net oxidation (that is, without any removal of electrons). The pathways for breaking down glucose to lactate or ethanol are examples of fermentation. Although electrons are removed from an organic molecule such as glucose to make pyruvate and NADH, the electrons are donated back to an organic molecule in the production of lactate or ethanol. Therefore, there is no net removal of electrons from an organic molecule. Compared with oxidative phosphorylation, fermentation produces far less ATP, for two reasons. First, glucose is not oxidized completely to $CO_2$ and $H_2O$. Second, the NADH

made during glycolysis cannot be used to make more ATP. Overall, the complete breakdown of glucose in the presence of oxygen yields 34–38 ATP molecules. By comparison, the anaerobic breakdown of glucose to lactate or ethanol yields only 2 ATP molecules.

## Summary of Key Concepts

### 7.1   Overview of Cellular Respiration

- Cells obtain energy via cellular respiration, which involves the breakdown of organic molecules and the export of waste products.

- The breakdown of glucose occurs in four stages: glycolysis, pyruvate breakdown, citric acid cycle, and oxidative phosphorylation (Figure 7.1).

### 7.2   Glycolysis

- During glycolysis, which occurs in the cytosol, glucose is split into two molecules of pyruvate, with a net yield of two ATP and two NADH. The ATP is made by substrate-level phosphorylation (Figures 7.2, 7.3).

- Cancer cells exhibit high levels of glycolysis, which enables the detection of tumors via a procedure called positron-emission tomography (PET) (Figure 7.4).

## 7.3 Breakdown of Pyruvate

- Pyruvate is broken down to $CO_2$ and an acetyl group that becomes attached to CoA. NADH is made during this process (Figure 7.5).

## 7.4 Citric Acid Cycle

- During the citric acid cycle, an acetyl group is removed from acetyl CoA and attached to oxaloacetate to make citrate. In a series of steps, two $CO_2$ molecules, three NADH, one $FADH_2$, and one ATP are made, after which the cycle begins again (Figures 7.6, 7.7).

## 7.5 Overview of Oxidative Phosphorylation

- Oxidative phosphorylation involves two events: (1) The electron transport chain (ETC) oxidizes NADH or $FADH_2$ and generates an $H^+$ electrochemical gradient, and (2) this gradient is used by ATP synthase to make ATP via chemiosmosis (Figures 7.8, 7.9).

## 7.6 A Closer Look at ATP Synthase

- Racker and Stoeckenius showed that ATP synthase uses an $H^+$ gradient to make ATP by reconstituting ATP synthase with a light-driven $H^+$ pump (Figure 7.10).

- ATP synthase is a rotary machine. The rotation is triggered by the passage of $H^+$ through a channel between a $c$ subunit and the $a$ subunit, which causes the γ subunit to spin, resulting in three conformational changes in the β subunits that promote ATP synthesis (Figures 7.11, 7.12).

- Yoshida and Kinosita demonstrated rotation of the γ subunit of ATP synthase by attaching a fluorescently labeled actin filament and observing its movement during the hydrolysis of ATP (Figure 7.13).

## 7.7 Connections Among Carbohydrate, Protein, and Fat Metabolism

- Proteins and fats can enter into glycolysis or the citric acid cycle at different points (Figure 7.14).

## 7.8 Anaerobic Respiration and Fermentation

- Anaerobic respiration occurs in the absence of oxygen. Certain microorganisms carry out anaerobic respiration by using as the final electron acceptor of the ETC a substance other than oxygen, such as nitrate (Figure 7.15).

- During fermentation, organic molecules are broken down without any net oxidation (that is, without any net removal of electrons). Examples include lactic acid production in muscle cells and ethanol production in yeast cells (Figure 7.16).

## Assess & Discuss

### Test Yourself

1. Which of the following pathways occurs in the cytosol?
   a. glycolysis
   b. breakdown of pyruvate to an acetyl group
   c. citric acid cycle
   d. oxidative phosphorylation
   e. all of the above

2. The net products of glycolysis are
   a. 6 $CO_2$, 4 ATP, and 2 NADH.
   b. 2 pyruvate, 2 ATP, and 2 NADH.
   c. 2 pyruvate, 4 ATP, and 2 NADH.
   d. 2 pyruvate, 2 GTP, and 2 $CO_2$.
   e. 2 $CO_2$, 2 ATP, and glucose.

3. During glycolysis, ATP is produced by
   a. oxidative phosphorylation.
   b. substrate-level phosphorylation.
   c. redox reactions.
   d. all of the above.
   e. both a and b.

4. Which organic molecule supplies a two-carbon group to start the citric acid cycle?
   a. ATP
   b. NADH
   c. acetyl CoA
   d. oxaloacetate
   e. both a and b

5. The ability to diagnose tumors using [$^{18}$F]-fluorodeoxyglucose (FDG) is based on the phenomenon that most types of cancer cells exhibit higher levels of
   a. glycolysis.
   b. pyruvate breakdown.
   c. citric acid metabolism.
   d. oxidative phosphorylation.
   e. all of the above.

6. In the experiment of Racker and Stoeckenius, bacteriorhodopsin was oriented in such a way that it pumped $H^+$ into a vesicle. Each vesicle actually contained many molecules of bacteriorhodopsin. How would the results of the experiment have been affected if 50% of the bacteriorhodopsin molecules pumped $H^+$ into the vesicle and 50% pumped $H^+$ out of the vesicles?
   a. The same amount of ATP would be made in the presence of light, and no ATP would be made in the dark.
   b. More ATP would be made in the presence of light, and no ATP would be made in the dark.
   c. No ATP would be made in the presence of light, and no ATP would be made in the dark.
   d. No ATP would be made in the presence of light, but some ATP would be made in the dark.
   e. Some ATP would be made in the presence of light, and some ATP would be made in the dark.

7. Certain drugs, which are called ionophores, cause the mitochondrial membrane to be highly permeable to $H^+$. How would such drugs affect oxidative phosphorylation?
   a. Movement of electrons down the ETC would be inhibited.
   b. ATP synthesis would be inhibited.
   c. ATP synthesis would be unaffected.
   d. ATP synthesis would be stimulated.
   e. Both a and b would occur.

8. The source of energy that *directly* drives the synthesis of ATP during oxidative phosphorylation is the
   a. oxidation of NADH.
   b. oxidation of glucose.
   c. oxidation of pyruvate.
   d. $H^+$ electrochemical gradient.
   e. reduction of $O_2$.

9. Compared with oxidative phosphorylation in mitochondria, anaerobic respiration in bacteria differs in that
   a. more ATP is made.
   b. ATP is made only via substrate-level phosphorylation.
   c. $O_2$ is converted to $H_2O_2$ rather than $H_2O$.
   d. something other than $O_2$ acts as a final electron acceptor of the ETC.
   e. both b and d occur.

10. When conditions in a muscle become anaerobic during strenuous exercise, why is it necessary to convert pyruvate to lactate?
    a. to decrease $NAD^+$ and increase NADH
    b. to decrease NADH and increase $NAD^+$
    c. to increase NADH and increase $NAD^+$
    d. to decrease NADH and decrease $NAD^+$
    e. to keep oxidative phosphorylation running

## Conceptual Questions

1. The electron transport chain is so named because electrons are transported from one component to another. Describe the purpose of the ETC.

2. What causes the rotation of the γ subunit of ATP synthase? How does this rotation promote ATP synthesis?

3. **Core Concept: Energy and Matter** How is glucose breakdown regulated to avoid the overproduction of ATP and NADH? What would be some potentially harmful consequences if glucose metabolism was not regulated properly?

## Collaborative Questions

1. Discuss the advantages and disadvantages of aerobic respiration, anaerobic respiration, and fermentation.

2. Read more about PET scans in other sources. Which types of cancers are most easily detected by this procedure, and which types are not readily detected? Is the ability to detect cancer via a PET scan related to the level of oxygen within a tumor?

# Photosynthesis

**8**

**A tropical rain forest in the Amazon.** Plant life in tropical rain forests carries out a large amount of the world's photosynthesis and supplies the atmosphere with a sizable fraction of its oxygen.

©Travelpix Ltd/Getty Images

**T**ake a deep breath. Nearly all of the oxygen in every breath you take is made by Earth's abundant plants, algae, and cyanobacteria. More than 20% of the world's oxygen is produced in the Amazon rain forest in South America alone (see the chapter opening photo). Biologists are alarmed about the rate at which such forests are being destroyed by human activities such as logging, mining, and oil extraction. Rain forests once covered 14% of the Earth's land surface, but they now occupy less than 6%. At their current rate of destruction, rain forests may be nearly eliminated in less than 40 years. Such a development may lower the level of oxygen in the atmosphere and thereby have a harmful effect on living organisms on a global scale.

In rain forests and across all of the Earth, the most visible color on land is green. The green color of plants is due to a pigment called chlorophyll. This pigment provides the starting point for the process of **photosynthesis**, in which the energy from light is captured and used to synthesize glucose and other organic molecules. Nearly all living organisms ultimately rely on photosynthesis for their nourishment, either directly or indirectly. Photosynthesis is also responsible for producing the oxygen that makes up a large portion of the Earth's atmosphere. Therefore, all aerobic organisms rely on photosynthesis for cellular respiration.

We begin this chapter with an overview of photosynthesis as it occurs in plants and algae. We will then explore the two stages of photosynthesis in more detail. In the first stage, called the **light reactions**, light energy is absorbed by chlorophyll and converted to chemical energy in the form of two energy intermediates: ATP and NADPH. During the second stage, known as the **Calvin cycle**, ATP and NADPH are used to drive the synthesis of carbohydrates. We will conclude with a consideration of the variations in photosynthesis that occur in plants existing in hot and dry conditions.

## 8.1 Overview of Photosynthesis

**Learning Outcomes:**

1. Write the general equations that represent the process of photosynthesis.
2. Explain how photosynthesis powers the biosphere.
3. Describe the general structure of chloroplasts.
4. Compare and contrast the two phases of photosynthesis: the light reactions and the Calvin cycle.

In the mid-1600s, a Flemish physician, Jan Baptista Van Helmont, conducted an experiment in which he transplanted the shoot of a young willow tree into a bucket of soil and allowed it to grow for 5 years. After this time, the willow tree had added 164 pounds to its original weight, but the soil had lost only 2 ounces. Van Helmont correctly concluded that the willow tree did not get most of its nutrients from the soil. He also hypothesized that the mass of the tree came from the water he had added over the 5 years. This hypothesis was partially correct, but we now know that $CO_2$ from the air is also a major contributor to the growth and mass of plants.

In the 1770s, Jan Ingenhousz, a Dutch physician, immersed green plants under water and discovered that they released bubbles of oxygen. Ingenhousz determined that sunlight was necessary for oxygen production. During this same period, Jean Senebier, a Swiss botanist, found that $CO_2$ is required for plant growth. With this accumulating information, Julius von Mayer, a German physicist, proposed in 1845 that plants convert light energy from the Sun into chemical energy.

For the next several decades, plant biologists studied photosynthesis in plants, algae, and species of bacteria that are capable of photosynthesis. They discovered that some photosynthetic bacteria use hydrogen sulfide ($H_2S$) instead of water ($H_2O$) for photosynthesis, and these organisms release sulfur instead of oxygen. In the 1930s, based on this information, Dutch-American microbiologist Cornelis van Niel proposed a general equation for photosynthesis that applies to plants, algae, and photosynthetic bacteria:

$$CO_2 + 2 H_2A + \text{Light energy} \rightarrow CH_2O + A_2 + H_2O$$

where A is oxygen (O) or sulfur (S) and $CH_2O$ is the general formula for a carbohydrate. This is a redox reaction in which $CO_2$ is reduced and $H_2A$ is oxidized.

In plants and algae, A is oxygen and $A_2$ is a molecule of oxygen that is designated $O_2$. Therefore, this equation becomes

$$CO_2 + 2 H_2O + \text{Light energy} \rightarrow CH_2O + O_2 + H_2O$$

When the carbohydrate produced is glucose ($C_6H_{12}O_6$), we multiply each side of the equation by 6 to obtain:

$$6\,CO_2 + 12\,H_2O + \text{Light energy} \rightarrow C_6H_{12}O_6 + 6\,O_2 + 6\,H_2O$$
$$\Delta G = +685 \text{ kcal/mol}$$

In this redox reaction, $CO_2$ is reduced during the formation of glucose, and $H_2O$ is oxidized during the formation of $O_2$. Notice that the free-energy change required for the production of 1 mole of glucose from carbon dioxide and water is a whopping +685 kcal/mol! As we learned in Chapter 6, an endergonic reaction is driven forward by being coupled with an exergonic process that releases free energy. In this case, the energy from sunlight ultimately drives the synthesis of glucose.

In this section, we will survey the general features of photosynthesis as it occurs in plants and algae. The sections that follow will examine the various steps in this process.

## Photosynthesis Powers the Biosphere

The term **biosphere** describes the regions on the surface of the Earth and in the atmosphere where living organisms exist. Organisms can be categorized as heterotrophs and autotrophs. **Heterotrophs** must consume food—organic molecules from their environment—to sustain life. Most species of bacteria and protists, as well as all species of fungi and animals, are heterotrophs. By comparison, **autotrophs** sustain themselves by producing organic molecules from inorganic sources such as $CO_2$ and $H_2O$. **Photoautotrophs** are autotrophs that use light as a source of energy to make organic molecules. These include plants, algae, and some bacterial species such as cyanobacteria.

Life in the biosphere is largely driven by the photosynthetic power of plants, algae, and cyanobacteria. The existence of most species relies on a key energy cycle that involves the interplay between organic molecules (such as glucose) and inorganic molecules, namely, $O_2$, $CO_2$, and $H_2O$ (**Figure 8.1**). Photoautotrophs make a large proportion of the Earth's organic molecules via photosynthesis, using light

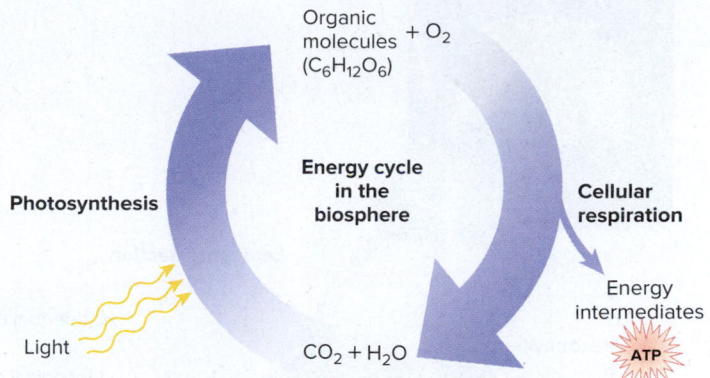

**Figure 8.1 An important energy cycle between photosynthesis and cellular respiration.** Photosynthesis is a process in which light, $CO_2$, and $H_2O$ are used to produce $O_2$ and organic molecules. The organic molecules are broken down to $CO_2$ and $H_2O$ via cellular respiration to supply energy in the form of ATP; $O_2$ is reduced to $H_2O$.

 **Core Skill: Modeling** The goal of this modeling challenge is to increase the complexity of the model shown in Figure 8.1 by adding organisms that carry out photosynthesis and those that carry out cellular respiration.

**Modeling Challenge:** The figure shows a simplified model for an energy cycle in the biosphere. Increase the complexity of the model in the following ways: On the left side, add drawings of three different broad categories of organisms that carry out photosynthesis. On the right, add drawings of three or more categories that carry out cellular respiration.

energy, $CO_2$, and $H_2O$. During this process, they also produce $O_2$. To supply their energy needs, both photoautotrophs and heterotrophs metabolize organic molecules via cellular respiration. As described in Chapter 7, cellular respiration generates $CO_2$ and $H_2O$ and is used to make ATP. The $CO_2$ is released into the atmosphere and can be reused by photoautotrophs to make more organic molecules such as glucose. In this way, an energy cycle between photosynthesis and cellular respiration sustains life on our planet.

## In Plants and Algae, Photosynthesis Occurs in the Chloroplasts

**Chloroplasts** are organelles found in plant and algal cells that carry out photosynthesis. These organelles contain large quantities of **chlorophyll**, which is a pigment that gives plants their green color. All green parts of a plant contain chloroplasts and can perform photosynthesis, although the majority of photosynthesis in most species of plants occurs in the leaves (**Figure 8.2**). The tissue in the internal part of the leaf, called the **mesophyll**, contains cells with chloroplasts. For photosynthesis to occur, the mesophyll cells must receive light, and also obtain water and carbon dioxide. The water is taken up by the roots of the plant and is transported to the leaves by small veins. Carbon dioxide gas enters the leaf, and oxygen exits, via pores called stomata (singular, stoma or stomate; from the Greek, meaning mouth).

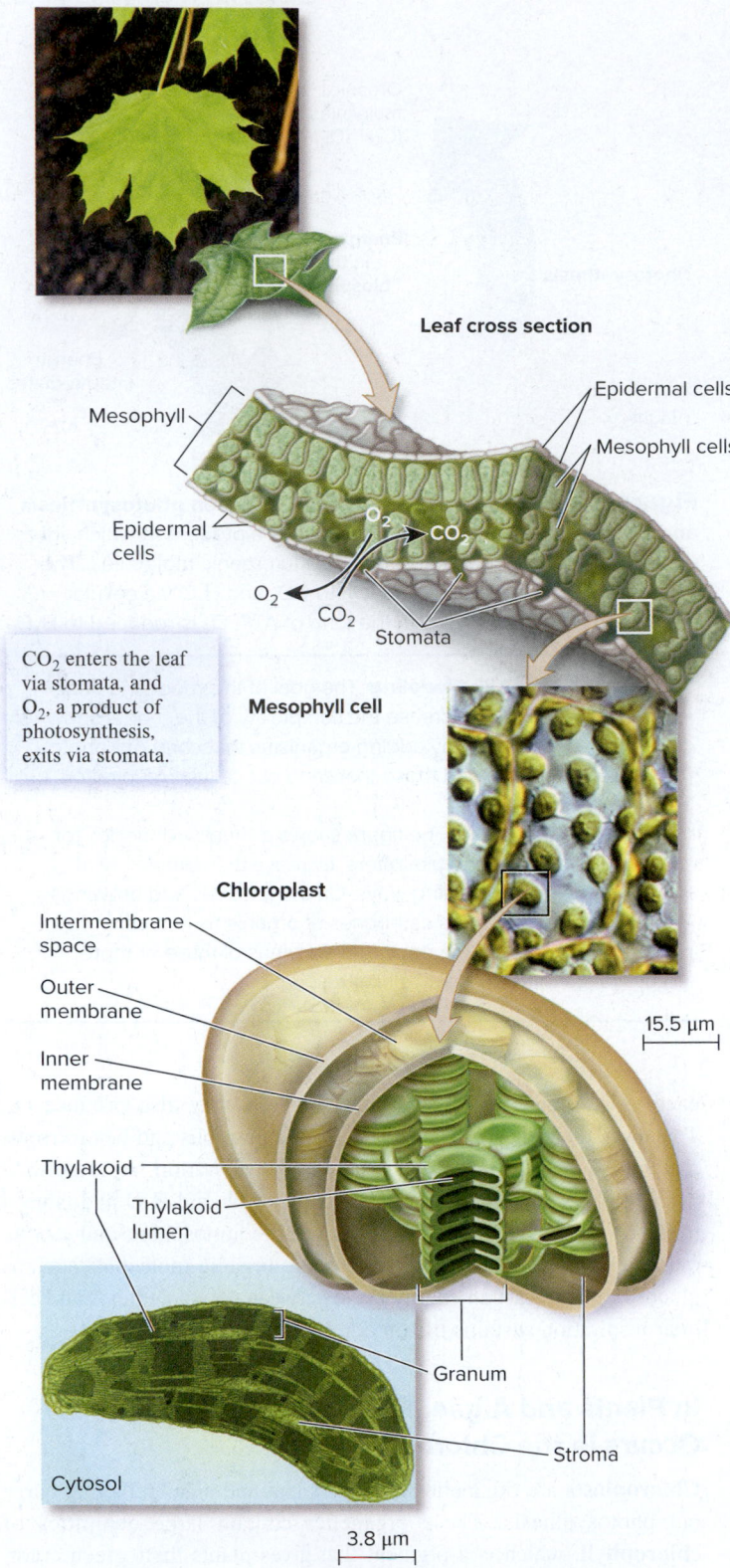

Like a mitochondrion, a chloroplast contains an outer and an inner membrane, with an intermembrane space lying between the two. A third membrane, called the **thylakoid membrane**, contains pigment molecules, including chlorophyll. The thylakoid membrane forms many flattened, fluid-filled tubules called **thylakoids**, each of which encloses a single compartment known as the **thylakoid lumen**. Thylakoids stack on top of each other to form a structure called a **granum** (plural, grana). The **stroma** is the fluid-filled region of the chloroplast between the thylakoid membrane and the inner membrane (see Figure 8.2).

## Photosynthesis Occurs in Two Stages: Light Reactions and the Calvin Cycle

How does photosynthesis take place? As mentioned, the process of photosynthesis occurs in two stages called the light reactions and the Calvin cycle. The term photosynthesis is derived from the association between these two stages: Photo refers to the light reactions that capture the energy from sunlight needed for the synthesis of carbohydrates that occurs in the Calvin cycle. The light reactions take place at the thylakoid membrane, and the Calvin cycle occurs in the stroma (**Figure 8.3**).

The light reactions involve an amazing series of energy conversions, starting with light energy and ending with chemical energy that is stored in the form of covalent bonds. The light reactions produce three chemical products: ATP, NADPH, and $O_2$. ATP and NADPH are energy intermediates that provide the needed

**Figure 8.2 Leaf organization.** Leaves are composed of layers of cells. The epidermal cells are on the outer surface, both top and bottom, with mesophyll cells sandwiched in the middle. The mesophyll cells contain chloroplasts and are the primary sites of photosynthesis in most plants. (1): ©McGraw-Hill Education/Mark Dierker, photographer; (2): ©Biophoto Associates/SPL/Science Source; (3): ©Omikron/Science Source

**Core Skill: Connections** Look ahead to Figure 39.17. How many guard cells make up a stoma (plural, stomata)?

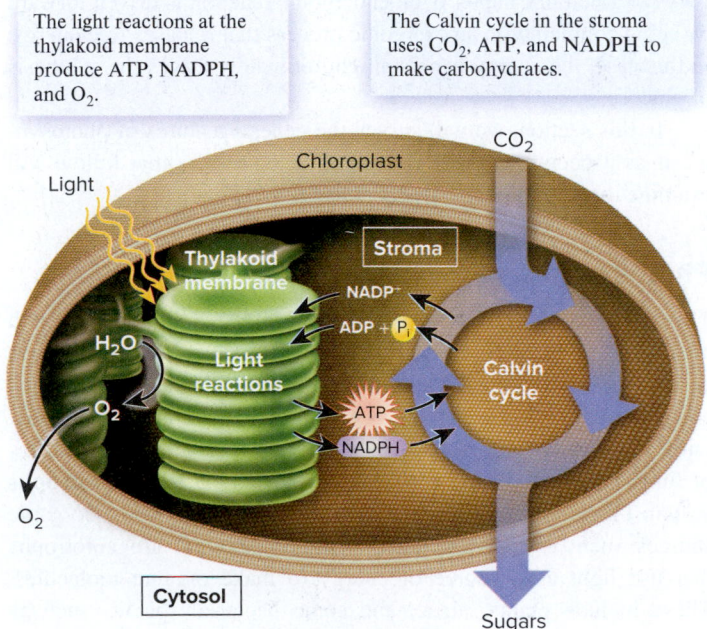

**Figure 8.3 An overview of the two stages of photosynthesis: light reactions and Calvin cycle.** The light reactions, through which ATP, NADPH, and $O_2$ are made, occur at the thylakoid membrane. The Calvin cycle, in which enzymes use ATP and NADPH to incorporate $CO_2$ into carbohydrates, occurs in the stroma.

**Concept Check:** *Can the Calvin cycle occur in the dark?*

energy and electrons to drive the Calvin cycle. Like NADH, **NADPH (nicotinamide adenine dinucleotide phosphate)** is an electron carrier that can accept two electrons. Its structure differs from NADH by the presence of an additional phosphate group. The structure of NADH is described in Chapter 6 (see Figure 6.12).

## 8.2 Reactions That Harness Light Energy

### Learning Outcomes:

1. Describe the general properties of light.
2. Explain how pigments absorb light energy, and list the types of pigments found in plants and green algae.
3. Outline the steps by which photosystems II and I capture light energy and produce $O_2$, ATP, and NADPH.
4. Describe the process of cyclic photophosphorylation, which produces only ATP.

According to the first law of thermodynamics, discussed in Chapter 6, energy cannot be created or destroyed, but it can be transferred from one place to another and transformed from one form to another. During photosynthesis, energy in the form of light is transferred from the Sun, some 92 million miles away, to a pigment molecule in a photosynthetic organism such as a plant. What follows is an interesting series of energy transformations in which light energy is transformed into electrochemical energy and then into energy stored within chemical bonds.

In this section, we will explore this series of transformations, collectively called the light reactions of photosynthesis. We begin by examining the properties of light and then consider the features of chloroplasts that allow them to capture light energy. The remainder of this section focuses on how the light reactions of photosynthesis generate three important products: ATP, NADPH, and $O_2$.

### Light Energy Is a Form of Electromagnetic Radiation

Light is essential to support life on Earth. Light is a type of electromagnetic radiation, so named because it consists of energy in the form of electric and magnetic fields. Electromagnetic radiation travels as waves caused by the oscillation of the electric and magnetic fields. The **wavelength** is the distance between the peaks in a wave pattern. The **electromagnetic spectrum** encompasses all possible wavelengths of electromagnetic radiation, from relatively short wavelengths (gamma rays) to much longer wavelengths (radio waves) (**Figure 8.4**). Visible light is the range of wavelengths detected by the human eye, commonly between 380 and 740 nm. As discussed later, visible light provides the energy to drive photosynthesis.

Physicists have also discovered that light has properties that are characteristic of particles. Albert Einstein formulated the photon theory of light, in which he proposed that light is composed of discrete particles called **photons**—massless particles traveling in a wavelike pattern and moving at the speed of light (about 300 million m/sec). Each photon contains a specific amount of energy. An important

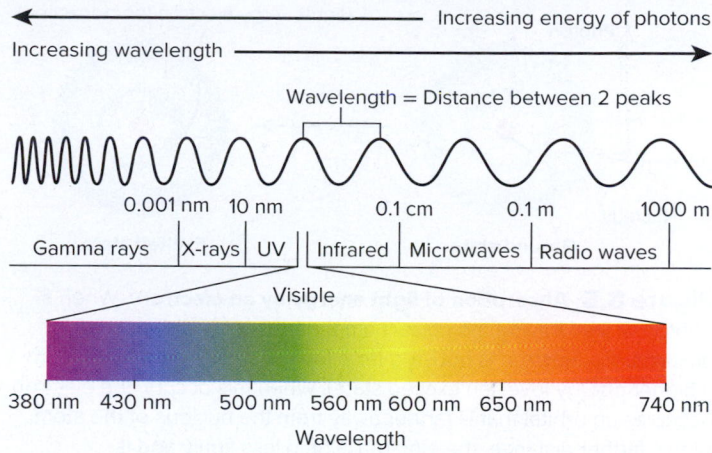

**Figure 8.4** **The electromagnetic spectrum.** The bottom portion of this figure emphasizes visible light—the wavelengths of electromagnetic radiation visible to the human eye. Light in the visible portion of the electromagnetic spectrum drives photosynthesis.

*Concept Check:* *Which has higher energy, gamma rays or radio waves?*

difference between the various types of electromagnetic radiation, shown in Figure 8.4, is the amount of energy of the photons. Shorter wavelength radiation carries more energy per unit of time than longer wavelength radiation. For example, the photons of gamma rays carry more energy than those of radio waves.

The Sun radiates the entire spectrum of electromagnetic radiation, but the atmosphere prevents much of this radiation from reaching the Earth's surface. For example, the ozone layer forms a thin shield in the upper atmosphere, protecting life on Earth from much of the Sun's ultraviolet (UV) radiation. Even so, a substantial amount of electromagnetic radiation does reach the Earth's surface. The effect of light on living organisms is critically dependent on the energy of the photons that reach them. The photons in gamma rays, X-rays, and UV radiation have very high energy. When molecules in cells absorb such energy, the effects can be devastating. Such radiation can cause mutations in DNA and even lead to cancer. By comparison, the energy of photons in visible light is much less intense. Molecules can absorb this energy in a way that does not cause damage. Next, we will consider how molecules in living cells absorb the energy within visible light.

### Pigments Absorb Light Energy

When light strikes an object, one of three things happens. First, light may simply pass through the object. Second, the object may change the path of light toward a different direction. A third possibility is that the object may absorb the light. The term **pigment** is used to describe a molecule that can absorb light energy. When light strikes a pigment, some of the wavelengths of light energy are absorbed, while others are reflected. For example, leaves look green to us because they reflect light energy with wavelengths in the green region of the visible spectrum. Various pigments in the leaves absorb the energy of other wavelengths. At the extremes of color reflection are white and black. A white object reflects nearly all of

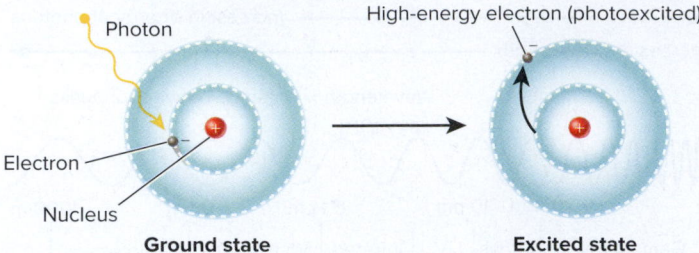

**Ground state**     **Excited state**

**Figure 8.5  Absorption of light energy by an electron.** When a photon of light having the correct amount of energy strikes an electron, the electron is boosted from the ground (unexcited) state to a higher energy level (an excited state). When this occurs, the electron occupies an orbital that is farther away from the nucleus of the atom. At this farther distance, the electron is held less firmly and is considered unstable.

**Concept Check:** *Describe the three events that can enable a photoexcited electron to become more stable.*

the visible light energy falling on it, whereas a black object absorbs nearly all of the light energy. This is why it is coolest to wear white clothes on a sunny, hot day.

What do we mean when we say that light energy is absorbed? Light energy in the visible spectrum can be absorbed by an atom when it boosts an electron to a higher energy level (**Figure 8.5**). The location in which an electron is found is called its orbital. Electrons in different orbitals possess different amounts of energy. For an electron to absorb light energy and be boosted to an orbital with a higher energy, it must overcome the difference in energy between the orbital it is in and the orbital to which it is going. For this to happen, an electron must absorb a photon that contains precisely that amount of energy. Different pigment molecules contain a variety of electrons that can be shifted to different energy levels. The wavelength of light that a pigment absorbs depends on the amount of energy needed to boost an electron to a higher orbital.

After an electron absorbs energy, it is said to be in an excited state. Usually, this is an unstable condition. To become stable again, one of four things can happen.

- To become stable, an excited electron may drop back down to a lower energy level and release heat. For example, on a sunny day, the sidewalk heats up because it absorbs light energy that is released as heat.

- Alternatively, an electron can become stable by releasing energy in the form of light. Certain organisms, such as jellyfish, possess molecules that make them glow. This glowing is due to the release of light when electrons drop down to lower energy levels, a phenomenon called fluorescence.

- An excited electron can transfer its extra energy to an electron in a nearby molecule, a process called **resonance energy transfer**.

- Rather than releasing energy or transferring it to another molecule, an excited electron can be removed from the molecule in which it is unstable and transferred to another molecule where it is stable. When this occurs, the energy in the electron is said to be captured, because the electron does not readily drop down to a lower energy level and release heat or light.

## Plants Contain Different Types of Photosynthetic Pigments

In plants, different pigment molecules absorb the light energy used to drive photosynthesis. Two types of chlorophyll pigments, termed **chlorophyll *a*** and **chlorophyll *b***, are found in green plants and green algae. Their structure was determined in the 1930s by German chemist Hans Fischer (**Figure 8.6a**). In the chloroplast, both chlorophylls *a* and *b* are bound to integral membrane proteins in the thylakoid membrane.

The chlorophylls contain a porphyrin ring and a phytol tail. A magnesium ion ($Mg^{2+}$) is bound to the porphyrin ring. An electron in the porphyrin ring is able to hop from one atom in the ring to another. Because this electron isn't restricted to a single atom, it is called a delocalized electron. The delocalized electron can absorb light energy. The phytol tail in chlorophyll is a long hydrocarbon chain that is hydrophobic. Its function is to anchor the pigment to the surface of hydrophobic proteins within the thylakoid membrane of chloroplasts.

**Carotenoids** are another type of pigment found in chloroplasts (**Figure 8.6b**). These pigments impart a color that ranges from yellow to orange to red. Carotenoids are often the major pigments in flowers and fruits. In leaves, the more abundant chlorophylls usually mask the colors of carotenoids. In temperate climates where the leaves change colors, the quantity of chlorophyll in the leaf declines during autumn. The carotenoids become readily visible and produce the yellows, oranges, and reds of autumn foliage.

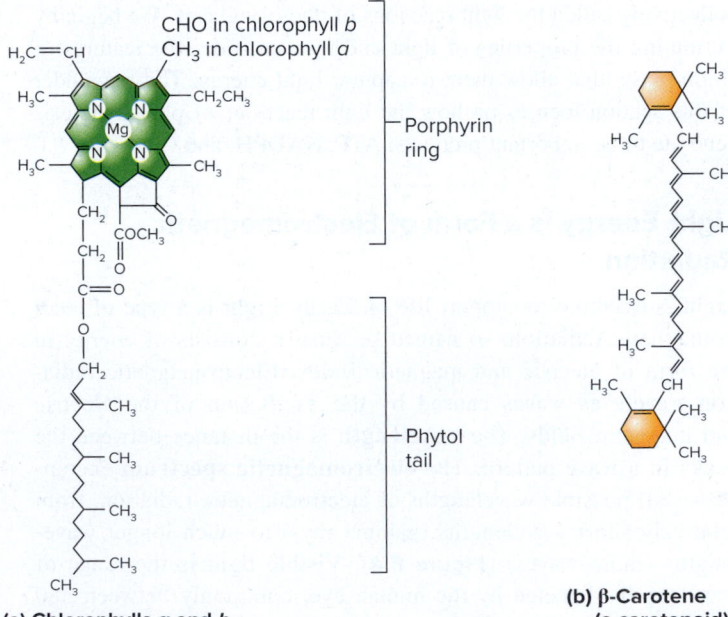

**(a) Chlorophylls *a* and *b***

**(b) β-Carotene (a carotenoid)**

**Figure 8.6  Structures of pigment molecules.** **(a)** The structure of chlorophylls *a* and *b*. As indicated, chlorophylls *a* and *b* differ only at a single site, at which chlorophyll *a* has a —CH₃ group and chlorophyll *b* has a —CHO group. **(b)** The structure of β-carotene, an example of a carotenoid. The green- and orange-shaded areas of the structures in parts (a) and (b) are regions where a delocalized electron can hop from one atom to another.

An **absorption spectrum** is a graph that plots a pigment's light absorption as a function of the light's wavelength. Each of the photosynthetic pigments shown in **Figure 8.7a** absorbs light in different regions of the visible spectrum. The absorption spectra of chlorophylls *a* and *b* are slightly different, though both chlorophylls absorb light most strongly in the red and violet parts of the visible spectrum and absorb green light poorly. Green light is reflected, which is why leaves appear green during the growing season. Carotenoids absorb light in the blue and blue-green regions of the visible spectrum, reflecting yellow and red.

Why do plants have different pigments? Having different pigments allows plants to absorb light at many different wavelengths. In this way, plants are more efficient at capturing the energy in sunlight. This phenomenon is highlighted in an **action spectrum**, which plots the rate of photosynthesis as a function of wavelength (**Figure 8.7b**). The highest rates of photosynthesis in green plants correlate with the wavelengths that are strongly absorbed by the chlorophylls and carotenoids. Photosynthesis is poor in the green region of the spectrum, because these pigments do not readily absorb this wavelength of light.

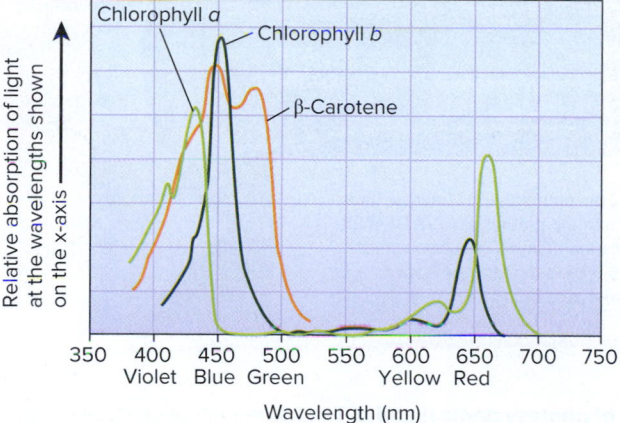

**(a) Absorption spectra**

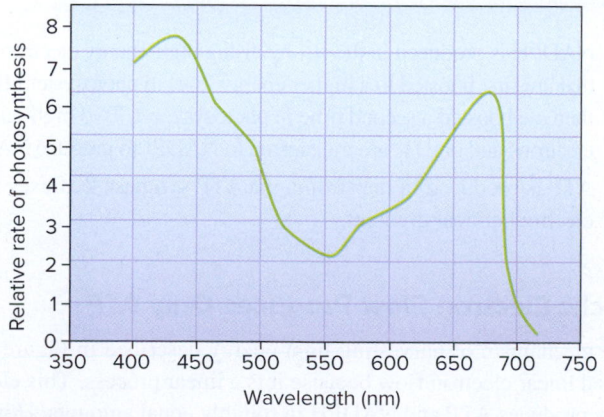

**(b) Action spectrum**

**Figure 8.7** **Properties of pigment function: absorption and action spectra.** **(a)** These absorption spectra show the absorption of light by chlorophyll *a*, chlorophyll *b*, and β-carotene. **(b)** An action spectrum of photosynthesis depicting the relative rate of photosynthesis in green plants at different wavelengths of light.

**Concept Check:** *What is the advantage of having different pigment molecules?*

## Photosystems II and I Work Together to Produce ATP and NADPH via Linear Electron Flow

A key feature of photosynthesis is the ability of pigments to absorb light energy and transfer it to other molecules that can hold the energy in a stable fashion and ultimately produce energy-intermediate molecules that can do cellular work. Let's now consider how chloroplasts capture light energy. The thylakoid membranes of the chloroplast contain two distinct complexes of proteins and pigment molecules called **photosystem I (PSI)** and **photosystem II (PSII)** (**Figure 8.8**). Photosystem I was discovered before photosystem II, but photosystem II is the initial step in photosynthesis. Working together, these two systems enable chloroplasts to capture light energy and synthesize ATP, NADPH, and $O_2$.

***Events within Photosystem II*** As described in step 1a of Figure 8.8, light excites electrons in pigment molecules, such as chlorophylls, which are located in a region of PSII called a light-harvesting complex. Rather than releasing their energy in the form of heat, the excited electrons begin to follow a path shown by the red arrow. Initially, the excited electrons move sequentially from a pigment molecule called P680 in PSII to other electron carriers called pheophytin (Pp), $Q_A$, and $Q_B$. PSII also oxidizes water, which generates $O_2$ and adds $H^+$ into the thylakoid lumen (see step 1b of Figure 8.8). The electrons released from oxidized water molecules replenish the electrons that are removed from P680.

***Electron Transport Chain*** Via $Q_B$, the electrons exit PSII and enter an electron transport chain (ETC)—a series of electron carriers—located in the thylakoid membrane (see Figure 8.8, step 1a). This ETC functions similarly to the one found in mitochondria. From $Q_B$, an electron goes to a cytochrome complex; then to plastocyanin (Pc), a small protein; and then to photosystem I. Along its journey from photosystem II to photosystem I, the electron releases some of its energy at particular steps and is transferred to the next component that has a higher electronegativity. The energy released is harnessed to pump $H^+$ into the thylakoid lumen.

***Photosystem I and NADPH Synthesis*** A key role of photosystem I is to make NADPH (see Figure 8.8, steps 2 and 3). When light strikes the light-harvesting complex of photosystem I, this energy is also transferred to a reaction center, where a high-energy electron is removed from a pigment molecule, designated P700, and transferred to a primary electron acceptor. A protein called ferredoxin (Fd) can accept two high-energy electrons, one at a time, from the primary electron acceptor. Fd then transfers the two electrons to the enzyme NADP+ reductase. This enzyme transfers the two electrons to $NADP^+$, which also accepts an $H^+$ to produce NADPH. The formation of NADPH results in fewer $H^+$ in the stroma. The combined action of photosystem II and photosystem I is termed **linear electron flow** because the electrons move linearly from PSII to PSI and ultimately reduce $NADP^+$ to NADPH.

A key difference between PSII and PSI lies in the source of the electrons received by their respective pigment molecules. An oxidized pigment in PSII called P680 receives an electron from water. By comparison, an oxidized pigment in PSI called P700 receives an electron from the protein Pc. Therefore, PSI does not need to split water to reduce this pigment and does not generate oxygen.

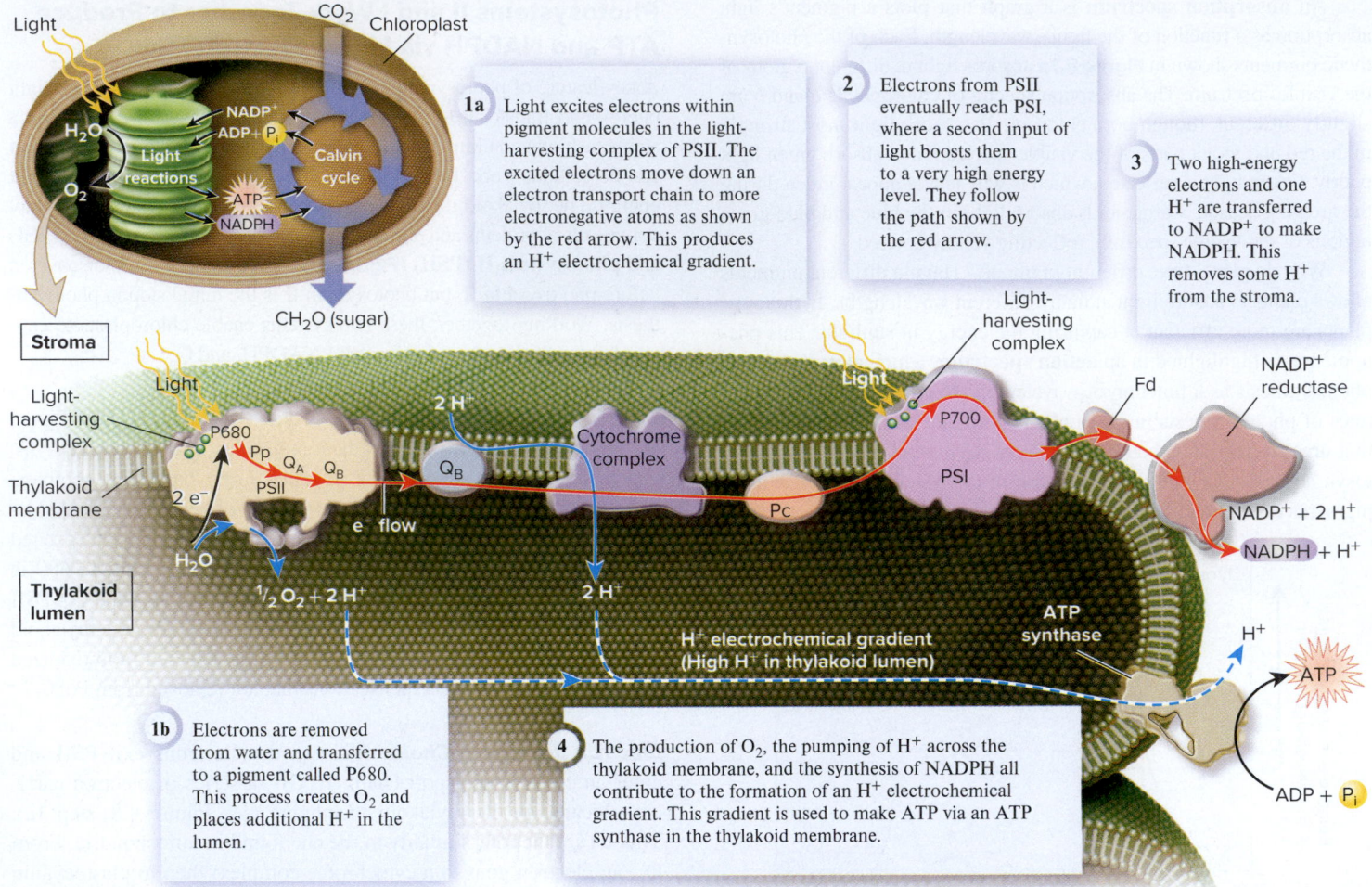

**Figure 8.8  The synthesis of ATP, NADPH, and O₂ by the concerted actions of photosystems II and I.** The movement of electrons from photosystem II to photosystem I to NADPH is called linear electron flow.

Concept Check:  *Are ATP, NADPH, and O₂ produced in the stroma or in the thylakoid lumen?*

***ATP Synthesis***   The synthesis of ATP in chloroplasts is achieved by a chemiosmotic mechanism called **photophosphorylation**, which is similar to the oxidative phosphorylation used to make ATP in mitochondria. In chloroplasts, ATP synthesis is driven by the flow of $H^+$ from the thylakoid lumen into the stroma via ATP synthase (Figure 8.8, step 4). The light reactions produce an $H^+$ electrochemical gradient in which more $H^+$ is in the thylakoid lumen and less in the stroma. The gradient is generated in three ways:

1. The splitting of water places $H^+$ in the thylakoid lumen.
2. The movement of high-energy electrons along the ETC from photosystem II to photosystem I pumps $H^+$ into the thylakoid lumen.
3. The formation of NADPH consumes $H^+$ in the stroma.

***Products of Photosynthesis***   In summary, the steps of the light reactions of photosynthesis produce three chemical products: $O_2$, NADPH, and ATP:

1. $O_2$ is produced in the thylakoid lumen by the oxidation of water by photosystem II. Two electrons are removed from water, producing 2 $H^+$ and $\frac{1}{2} O_2$. The two electrons are transferred to P680 molecules.

2. NADPH is produced in the stroma using high-energy electrons that are first boosted to a higher energy level in photosystem II and then are boosted a second time in photosystem I. Two high-energy electrons and one $H^+$ are transferred to $NADP^+$ to produce NADPH.

3. ATP is produced in the stroma via ATP synthase that uses an $H^+$ electrochemical gradient.

## Cyclic Electron Flow Produces Only ATP

The mechanism of harvesting light energy described in Figure 8.8 is called linear electron flow because it is a linear process. This electron flow produces ATP and NADPH in roughly equal amounts. However, as we will see later, the Calvin cycle uses more ATP than NADPH. How can plant cells avoid making too much NADPH and not enough ATP? In 1959, Daniel Arnon discovered a pattern of electron flow that is cyclic and generates only ATP (**Figure 8.9**). Arnon termed the process **cyclic photophosphorylation** because (1) the path of electrons is cyclic, (2) light energizes the electrons, and (3) ATP is made via the phosphorylation of ADP. Due to the path of electrons, the mechanism is also called **cyclic electron flow**.

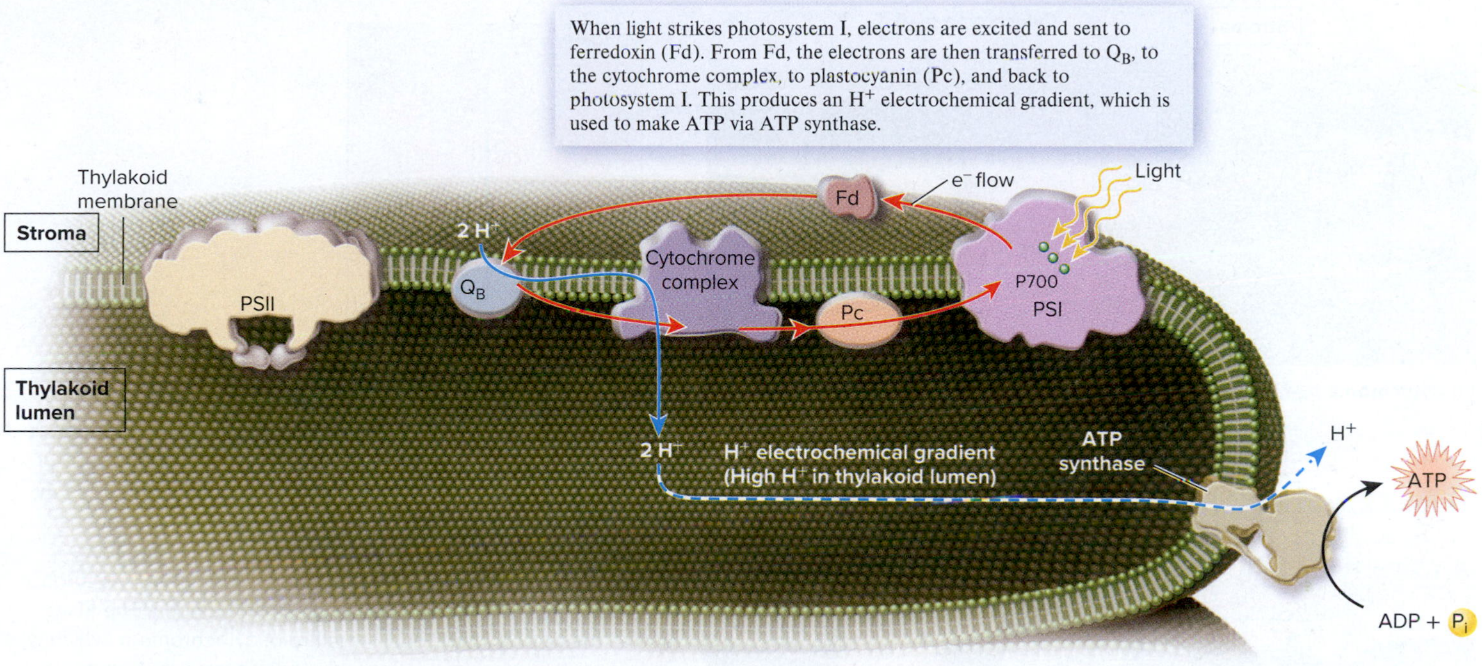

When light strikes photosystem I, electrons are excited and sent to ferredoxin (Fd). From Fd, the electrons are then transferred to $Q_B$, to the cytochrome complex, to plastocyanin (Pc), and back to photosystem I. This produces an $H^+$ electrochemical gradient, which is used to make ATP via ATP synthase.

Thylakoid membrane

Stroma

$e^-$ flow

Light

Fd

2 H⁺

$Q_B$

Cytochrome complex

P700
PSI

PSII

Pc

Thylakoid lumen

2 H⁺

$H^+$ electrochemical gradient (High $H^+$ in thylakoid lumen)

ATP synthase

$H^+$

ATP

ADP + Pᵢ

**Figure 8.9   Cyclic photophosphorylation.**  In this process, electrons follow a cyclic path that is powered by photosystem I (PSI). This contributes to the formation of an $H^+$ electrochemical gradient, which is then used by ATP synthase to make ATP.

*Concept Check:*  *Why does cyclic photophosphorylation provide an advantage to a plant over using only linear electron flow?*

When light strikes photosystem I, high-energy electrons are sent to the primary electron acceptor and then to ferredoxin (Fd). The key difference in cyclic photophosphorylation is that the high-energy electrons are transferred from Fd to $Q_B$. From $Q_B$, the electrons then go to the cytochrome complex, then to plastocyanin ($P_c$), and back to photosystem I. As the electrons travel along this cyclic route, they release energy, and some of this energy is used to transport $H^+$ into the thylakoid lumen. The resulting $H^+$ gradient drives the synthesis of ATP via ATP synthase.

Cyclic photophosphorylation is favored when the level of $NADP^+$ is low and NADPH is high. Under these conditions, there is sufficient NADPH to run the Calvin cycle, which is described later. Alternatively, when $NADP^+$ is high and NADPH is low, linear electron flow is favored, so more NADPH can be made. Cyclic photophosphorylation is also favored when ATP levels are low.

## Core Concepts: Evolution, Structure and Function

## The Cytochrome Complexes of Mitochondria and Chloroplasts Contain Evolutionarily Related Proteins

A recurring theme in cell biology is that evolution has resulted in groups of genes that encode proteins that play similar but specialized roles in cells—an example of descent with modification. When two or more genes are similar because they are derived from the same ancestral gene, they are called **homologous genes**. As discussed in Chapter 22, homologous genes encode proteins that have similar amino acid sequences and often perform similar functions.

A comparison of the electron transport chains of mitochondria and chloroplasts reveals homologous genes. In particular, let's consider the cytochrome complex found in the thylakoid membrane of plants and algae, called cytochrome $b_6$-$f$ (**Figure 8.10a**), and the complex cytochrome $b$-$c_1$, which is found in the ETC of mitochondria (**Figure 8.10b**; also refer back to Figure 7.8). Both of these cytochrome complexes are composed of several proteins. One of the proteins is called cytochrome $b_6$ in cytochrome $b_6$-$f$ and cytochrome $b$ in cytochrome $b$-$c_1$.

By analyzing the sequences of the genes that encode these proteins, researchers discovered that cytochrome $b_6$ and cytochrome $b$ are homologous proteins. These proteins carry out similar functions: Both of them accept electrons from a quinone ($Q_B$, or ubiquinone), and both donate an electron to another protein within their respective complexes (cytochrome $f$ or cytochrome $c_1$). Likewise, both proteins function as $H^+$ pumps that capture some of the energy that is released from electrons to transport $H^+$ across the membrane. In this way, evolution has produced a family of cytochrome $b$ proteins that play similar but specialized roles.

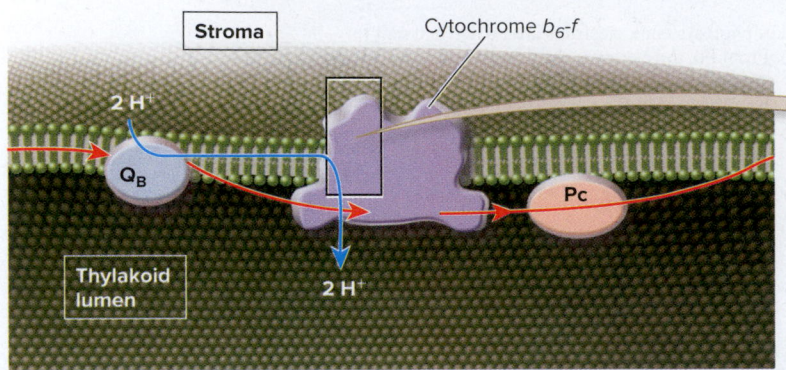

**Stroma**

Cytochrome $b_6$-f

2 H⁺

Q_B

**Thylakoid lumen**

2 H⁺

Pc

(a) Cytochrome $b_6$-f in the chloroplast

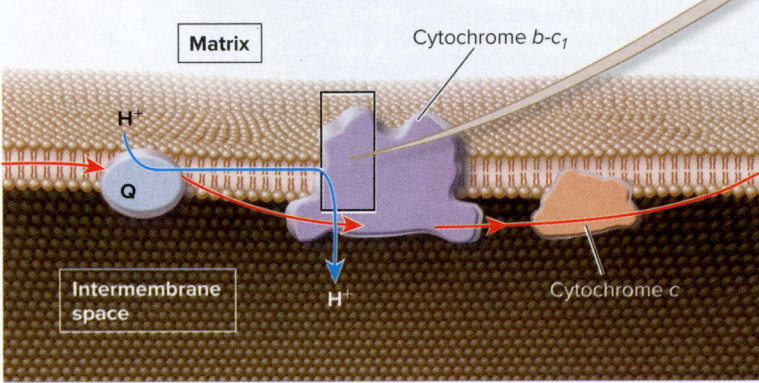

**Matrix**

Cytochrome b-c₁

H⁺

Q

**Intermembrane space**

H⁺

Cytochrome c

(b) Cytochrome b-c₁ in the mitochondrion

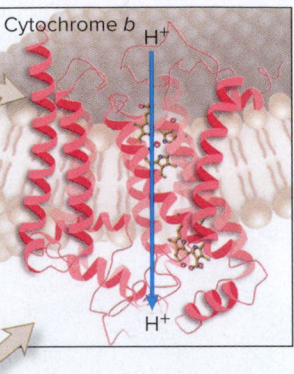

Cytochrome b

H⁺

H⁺

**Figure 8.10** **Homologous proteins in the electron transport chains of chloroplasts and mitochondria.** **(a)** Cytochrome $b_6$-f is a complex of proteins involved in electron and H⁺ transport in chloroplasts, and **(b)** cytochrome b-c₁ is a complex of proteins involved in electron and H⁺ transport in mitochondria. These complexes contain homologous proteins designated cytochrome $b_6$ in chloroplasts and cytochrome b in mitochondria. The inset shows the three-dimensional structure of cytochrome b, which was determined by X-ray crystallography. It is an integral membrane protein with several transmembrane helices and two heme groups, which are prosthetic groups involved in electron transfer. The structure of cytochrome $b_6$ has also been determined and found to be very similar.

**Concept Check:** *Explain why the three-dimensional structures of cytochrome b and cytochrome $b_6$ are very similar.*

## 8.3 Molecular Features of Photosystems

### Learning Outcomes:

1. Explain how PSII absorbs and captures light energy and how it produces $O_2$.
2. Diagram the variation in the energy of an electron as it moves from PSII to PSI to NADP⁺.

The previous section provided an overview of how chloroplasts absorb light energy and produce ATP, NADPH, and $O_2$. As you have learned, two photosystems—PSI and PSII—play critical roles in two aspects of photosynthesis. First, both PSI and PSII absorb light energy and capture that energy in the form of excited electrons. Second, PSII oxidizes water, thereby producing $O_2$. In this section, we will take a closer look at how these events occur at the molecular level.

### Photosystem II Captures Light Energy and Produces $O_2$

PSI and PSII have two main components: a light-harvesting complex and a reaction center. **Figure 8.11** shows how these components function in PSII.

***Absorption of Energy by the Light-Harvesting Complex and Its Transfer to P680 via Resonance Energy Transfer*** In 1932, American biologist Robert Emerson and an undergraduate student,

William Arnold, originally discovered the **light-harvesting complex** in the thylakoid membrane. It is composed of several dozen pigment molecules that are anchored to transmembrane proteins. The role of the complex is to directly absorb photons of light. When a pigment molecule absorbs a photon, an electron is boosted to a higher energy level. As shown in Figure 8.11, the energy (not the electron itself) is transferred to adjacent pigment molecules by a process called resonance energy transfer. The energy may be transferred among multiple pigment molecules until it is eventually transferred to a special pigment molecule designated P680, which is located within the reaction center of PSII. The P680 pigment is so named because it can directly absorb light at a wavelength of 680 nm. However, P680 is more commonly excited by resonance energy transfer from a chlorophyll pigment in the light-harvesting complex. In either case, when an electron in P680 is excited, the molecule is designated P680*. The light-harvesting complex is also called the antenna complex because it acts like an antenna that absorbs energy from light and funnels that energy to P680 in the reaction center.

***Rapid Transfer of a High-Energy Electron from P680* to the Primary Electron Acceptor*** A high-energy (photoexcited) electron in a pigment molecule is very unstable. It may abruptly release its energy by giving off heat or light. Unlike the pigments in the light-harvesting complex that undergo resonance energy transfer, P680* can actually release its high-energy electron and become P680⁺.

$$P680^* \rightarrow P680^+ + e^-$$

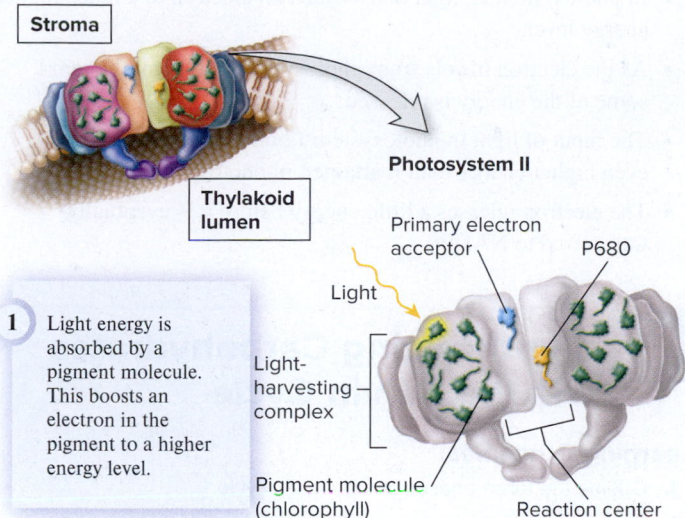

**Stroma**

**Photosystem II**

**Thylakoid lumen**

Primary electron acceptor

P680

Light

1 Light energy is absorbed by a pigment molecule. This boosts an electron in the pigment to a higher energy level.

Light-harvesting complex

Pigment molecule (chlorophyll)

Reaction center

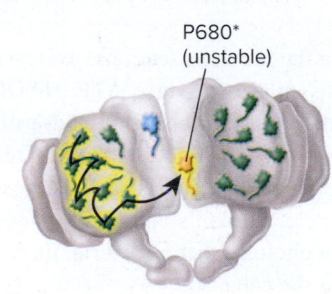

P680* (unstable)

2 Energy is transferred among pigment molecules via resonance energy transfer until it reaches P680, converting it to P680*.

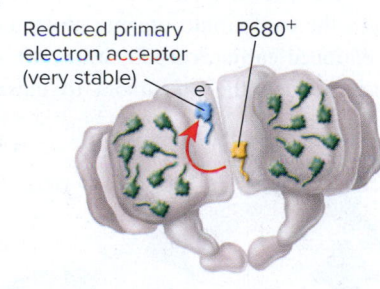

Reduced primary electron acceptor (very stable)

P680+

e−

3 The high-energy electron on P680* is transferred to the primary electron acceptor (pheophytin), where it is very stable. P680* becomes P680+.

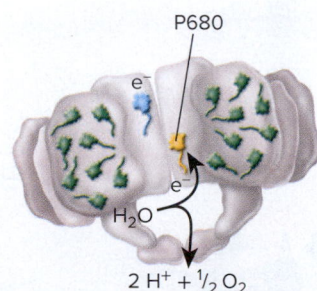

P680

e−

4 A low-energy electron from water is transferred to P680+ to convert it to P680. O2 is produced.

e−

H2O

2 H+ + 1/2 O2

**Figure 8.11  A closer look at how photosystem II harvests light energy and oxidizes water.** Note: Two electrons are released during the oxidation of water, but they are transferred one at a time to P680+.

**Core Concept: Energy and Matter**  The pigments in PSII absorb light energy that is captured in a stable form when an excited electron is transferred to the primary electron acceptor (pheophytin).

The role of the reaction center is to quickly remove the high-energy electron from P680* and transfer it to another molecule, where the electron is stable. This molecule is called the **primary electron acceptor** (see Figure 8.11). The transfer of the electron from P680* to the primary electron acceptor is remarkably fast. It occurs in less than a few picoseconds! (One picosecond equals one-trillionth of a second, or $10^{-12}$ sec.) Because this occurs so quickly, the excited electron does not have much time to release its energy in the form of heat or light.

When the primary electron acceptor (pheophytin) has received this high-energy electron, the light energy has been captured and can be used to perform cellular work. As discussed earlier, the work it performs is to synthesize the energy intermediates ATP and NADPH (look back at Figure 8.8).

***Transfer of a Low-Energy Electron from Water to P680$^+$***    Let's now consider what happens to P680$^+$, which has given up its high-energy electron. After P680$^+$ is formed, the electron that has been removed must be replaced so that P680 can function again. Therefore, another role of the reaction center is to replace the electron that is removed when P680* becomes P680$^+$. This missing electron of P680$^+$ is replaced with a low-energy electron from water (see Figure 8.11).

$$H_2O \rightarrow 2H^+ + {}^1\!/_2\, O_2 + 2e^-$$

$$2\, P680^+ + 2e^- \rightarrow 2\, P680$$

(from water)

The oxidation of water results in the formation of oxygen gas ($O_2$), which is used by many organisms for cellular respiration. Photosystem II is the only known protein complex that can oxidize water, resulting in the release of $O_2$ into the atmosphere.

**BIO TIPS**

**THE QUESTION**  *Describe the roles of the light-harvesting complex, P680, and the primary electron acceptor during the absorption of light energy by photosystem II (PSII). At which step is the light energy captured?*

**T** **OPIC**  *What topic in biology does this question address?*  The topic is the absorption of light by PSII. More specifically, the question asks you to describe the various roles played by the light-harvesting complex, P680, and the primary electron acceptor.

**I** **NFORMATION**  *What information do you know based on the question and your understanding of the topic?*  In the question, you are reminded that PSII has a light-harvesting complex, P680, and a primary electron acceptor. From your understanding of the topic, you may remember that light absorption begins at the light-harvesting complex, which funnels energy to P680. The energy transforms P680 to P680*, which then transfers a high-energy electron to the primary electron acceptor.

**P** **ROBLEM-SOLVING S** **TRATEGY**  *Sort out the steps in a complicated process. Compare and contrast.*  To begin to solve this problem, it may be helpful to review the steps of the light-absorption process shown in Figure 8.11. As you do so, compare and contrast the roles of the light-harvesting complex, P680, and the primary electron acceptor.

**ANSWER** *Light-harvesting complex: The role of the light-harvesting complex is to absorb light. Because it is composed of many pigment molecules (chlorophylls and β-carotene), the light-harvesting complex is the most likely place for visible light to be absorbed. When a pigment molecule absorbs light, an electron is boosted to a higher energy level, and that energy is transferred via resonance energy transfer to P680.*

*P680: The role of P680 is to provide a link between the light-harvesting complex and the primary electron acceptor. Although P680 can directly absorb light, P680 is far more likely to gain energy from the light-harvesting complex, which converts it to P680\*. The high-energy electron of P680\* is then transferred to the primary electron acceptor.*

*Primary electron acceptor: The role of the primary electron acceptor is to capture the light energy. The high-energy electron of P680\* is unstable. However, when it is transferred to the primary electron acceptor, the electron becomes stable, meaning that it will not drop down to a lower energy level.*

### Electrons Vary in Energy as They Move from Photosystem II to Photosystem I to NADP⁺

In 1960, Robin Hill and Fay Bendall proposed that the light reactions of photosynthesis involve two photoactivation events. According to their model, known as the **Z scheme**, an electron proceeds through a series of energy changes during photosynthesis (**Figure 8.12**). The Z refers to the zigzag shape of this energy curve. Based on our modern understanding of photosynthesis, we now know that these events involve increases and decreases in the energy of an electron as it moves linearly from photosystem II through photosystem I to NADP⁺.

- An electron on a nonexcited pigment molecule in photosystem II has the lowest energy.

- In photosystem II, light boosts such an electron to a much higher energy level.

- As the electron travels from photosystem II to photosystem I, some of the energy is released.

- The input of light in photosystem I boosts the electron to an even higher energy than it attained in photosystem II.

- The electron releases a little energy before it is eventually transferred to NADP⁺.

## 8.4 Synthesizing Carbohydrates via the Calvin Cycle

**Learning Outcomes:**

1. Outline the three phases of the Calvin cycle.
2. **CoreSKILL »** Analyze the results of Calvin and Benson, and explain how they identified the components of the Calvin cycle.

In the previous sections, we learned how the light reactions of photosynthesis produce ATP, NADPH, and O₂. We will now turn our attention to the second phase of photosynthesis, the Calvin cycle, in which ATP and NADPH are used to make carbohydrates. The Calvin cycle consists of a series of steps that occur in a metabolic cycle. In plants and algae, it occurs in the stroma of chloroplasts. In photosynthetic bacteria, the Calvin cycle occurs in the cytoplasm of the cell.

The Calvin cycle takes CO₂ from the atmosphere and incorporates the carbon into organic molecules, namely, carbohydrates. As mentioned earlier, carbohydrates are critical for two reasons. First, they provide the precursors to make the organic molecules and

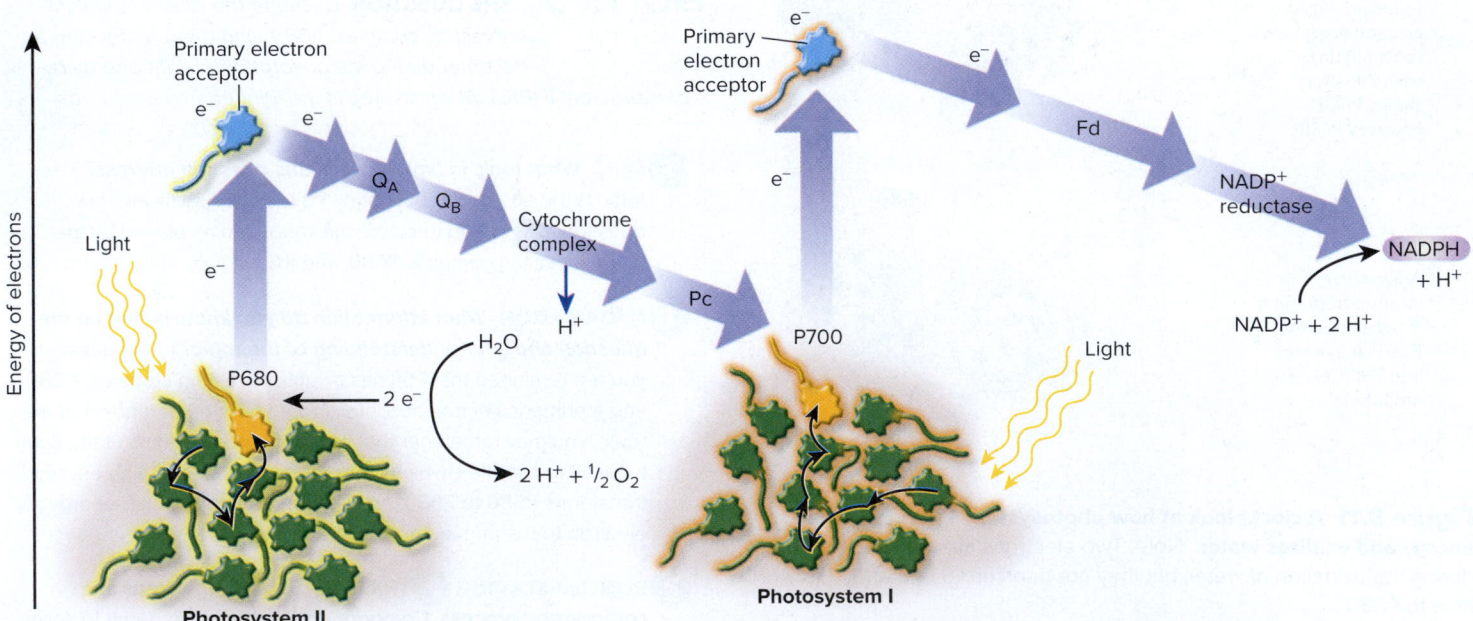

**Figure 8.12  The Z scheme, showing the energy of an electron moving from photosystem II to NADP⁺.**  The oxidation of water releases two electrons that travel one at a time from photosystem II to NADP⁺. As seen here, the input of light boosts the energy of the electron twice. At the end of the pathway, two electrons are used to make NADPH.

**Concept Check:**  *During its journey from photosystem II to NADP⁺, at what point does an electron have the highest amount of energy?*

macromolecules of nearly all living cells. The second key reason is the storage of energy. The Calvin cycle produces carbohydrates, which store energy. These carbohydrates are accumulated inside plant cells. When a plant is in the dark and not carrying out photosynthesis, the stored carbohydrates are used as a source of energy. Similarly, when an animal consumes a plant, it uses the carbohydrates as an energy source.

In this section, we will examine the three phases of the Calvin cycle. We will also explore the experimental approach of Melvin Calvin and his colleagues that enabled them to elucidate the steps of this cycle.

## The Calvin Cycle Incorporates $CO_2$ into a Carbohydrate

The Calvin cycle, also called the Calvin-Benson cycle, was determined by chemists Melvin Calvin and Andrew Adam Benson and their colleagues in the 1940s and 1950s. This cycle requires a massive input of energy. For every 6 carbon dioxide molecules that are incorporated into a carbohydrate such as glucose ($C_6H_{12}O_6$), 18 ATP molecules are hydrolyzed and 12 NADPH molecules are oxidized:

$$6\ CO_2 + 12\ H_2O \rightarrow C_6H_{12}O_6 + 6\ O_2 + 6\ H_2O$$
$$18\ ATP + 18\ H_2O \rightarrow 18\ ADP + 18\ P_i$$
$$12\ NADPH \rightarrow 12\ NADP^+ + 12\ H^+ + 24\ e^-$$

Although biologists commonly describe glucose as a product of photosynthesis, glucose is not directly made by the Calvin cycle. Instead, molecules of glyceraldehyde-3-phosphate, which are products of the Calvin cycle, are used as starting materials for the synthesis of glucose and other molecules, including sucrose. After glucose molecules are made, they may be linked together to form a polymer of glucose called starch, which is stored in the chloroplast for later use. Alternatively, the disaccharide sucrose may be made and transported out of the leaf to other parts of the plant.

The Calvin cycle can be divided into three phases: carbon fixation, reduction and carbohydrate production, and regeneration of ribulose bisphosphate (RuBP) (**Figure 8.13**).

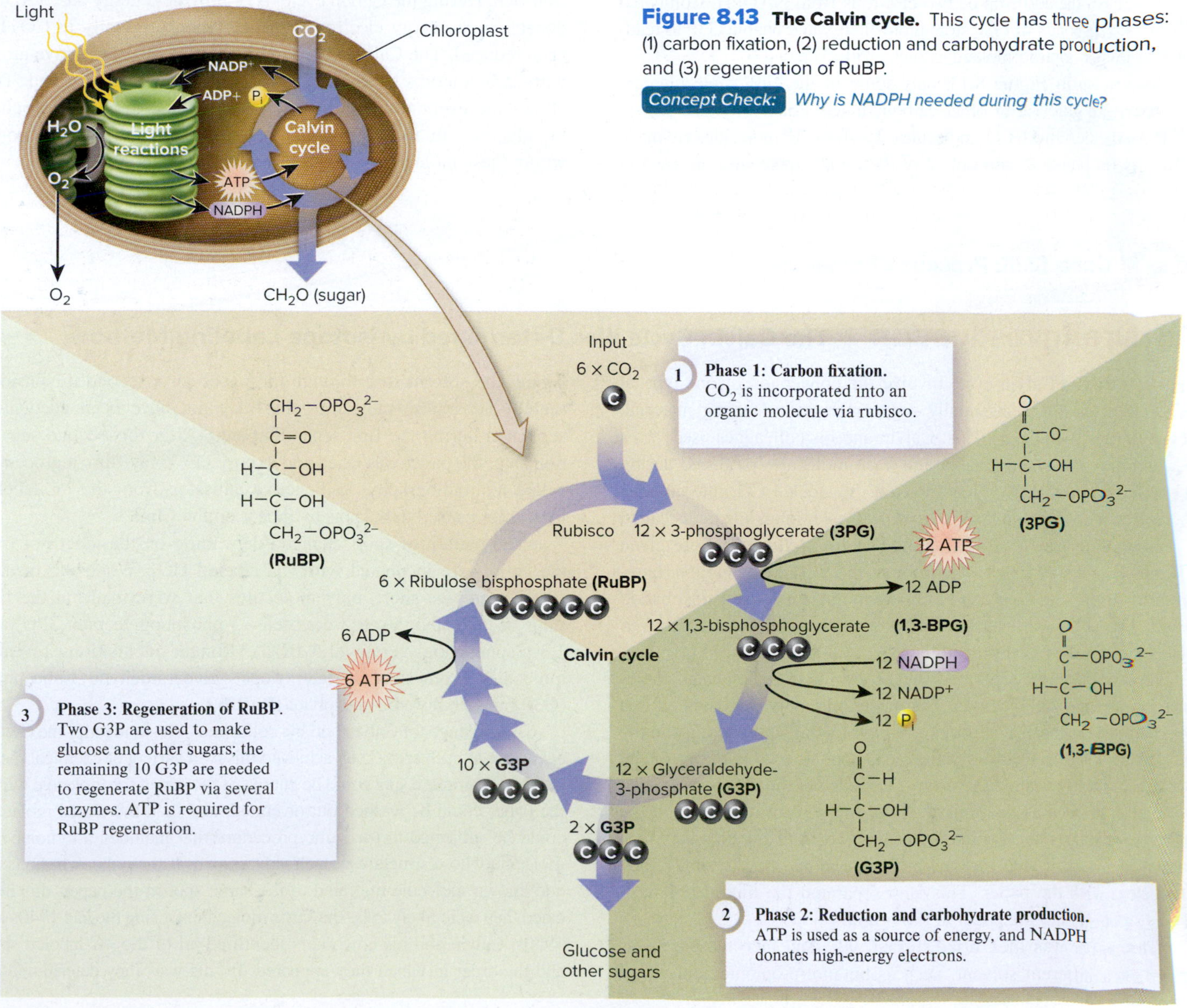

**Figure 8.13** **The Calvin cycle.** This cycle has three phases: (1) carbon fixation, (2) reduction and carbohydrate production, and (3) regeneration of RuBP.

**Concept Check:** *Why is NADPH needed during this cycle?*

***Carbon Fixation (Phase 1)***    During **carbon fixation**, $CO_2$ is incorporated into RuBP, a five-carbon sugar. The term fixation means that the carbon has been removed from the atmosphere and incorporated into an organic molecule that is not a gas. More specifically, the product of this reaction in phase 1 is a six-carbon intermediate that immediately splits in half to form two molecules of 3-phosphoglycerate (3PG). The enzyme that catalyzes this step is named RuBP carboxylase/oxygenase, or **rubisco**. It is the most abundant protein in chloroplasts and perhaps the most abundant protein on Earth! This observation underscores the massive amount of carbon fixation that happens in the biosphere.

***Reduction and Carbohydrate Production (Phase 2)***    In the second phase of the Calvin cycle, ATP is used to convert 3PG to 1,3-bisphosphoglycerate (1,3-BPG). Next, electrons from NADPH reduce 1,3-BPG to glyceraldehyde-3-phosphate (G3P). G3P is a carbohydrate with three carbon atoms. The key difference between 3PG and G3P is that 3PG has a C—O bond, whereas the analogous carbon in G3P has a C—H bond (see Figure 8.13). The C—H bond results when the G3P molecule is reduced by the addition of two electrons from NADPH. Compared with 3PG, the bonds in G3P store more energy and enable G3P to readily form larger organic molecules such as glucose.

As shown in Figure 8.13, only some of the G3P molecules are used to make glucose or other carbohydrates. Phase 1 begins with 6 RuBP molecules and 6 $CO_2$ molecules. Twelve G3P molecules are made at the end of phase 2, and only 2 of these G3P molecules are used in

carbohydrate production. As described next, the other 10 G3P molecules are needed to keep the Calvin cycle turning by regenerating RuBP.

***Regeneration of RuBP (Phase 3)***    In the last phase of the Calvin cycle, a series of enzymatic steps converts the 10 G3P molecules into 6 RuBP molecules, using 6 molecules of ATP. After the RuBP molecules are regenerated, they serve as acceptors for $CO_2$, thereby allowing the cycle to continue.

As we have just seen, the Calvin cycle begins by using carbon from an inorganic source, that is, $CO_2$, and ends with organic molecules that will be used by the plant to make other molecules. You may be wondering why $CO_2$ molecules cannot be directly linked to form these larger molecules. The answer lies in the number of electrons that are around the carbon atoms. In $CO_2$, the carbon atom is considered electron poor. Oxygen is a very electronegative atom that monopolizes the electrons it shares with other atoms. In a covalent bond between carbon and oxygen, the shared electrons are closer to the oxygen atom.

By comparison, in an organic molecule, the carbon atom is electron-rich. During the Calvin cycle, ATP provides energy and NADPH donates high-energy electrons, so the carbon originally in $CO_2$ has been reduced. The Calvin cycle combines less electronegative atoms with carbon atoms so that C—H and C—C bonds are formed. This allows the eventual synthesis of larger organic molecules, including glucose, amino acids, and so on. In addition, the covalent bonds within these molecules store large amounts of energy.

 **Core Skill: Process of Science**

# Feature Investigation | The Calvin Cycle Was Determined by Isotope-Labeling Methods

The steps in the Calvin cycle involve the conversion of one type of molecule to another, eventually regenerating the starting material, RuBP. In the 1940s and 1950s, Calvin and his colleagues used $^{14}C$, a radioisotope of carbon, to label and trace molecules produced during the cycle (**Figure 8.14**). They injected $^{14}C$-labeled $CO_2$ into cultures of the green alga *Chlorella pyrenoidosa* grown in an apparatus called a "lollipop" (because of its shape). The *Chlorella* cells were given different lengths of time to incorporate the $^{14}C$-labeled carbon, ranging from fractions of a second to many minutes. After this incubation period, the cells were abruptly placed into a solution of alcohol to inhibit enzymatic reactions and thereby stop the cycle.

The researchers separated the newly made radiolabeled molecules by a variety of methods. The most commonly used method was two-dimensional paper chromatography. In this approach, a sample containing radiolabeled molecules was spotted onto a corner of the paper at a location called the origin. The edge of the paper was placed in a solvent, such as phenol-water. As the solvent rose through the paper, so did the radiolabeled molecules. The rate at which they rose depended on their structures, which determined how strongly they interacted with the paper. This step separated the mixture of molecules spotted onto the paper at the origin.

The paper was then dried, turned 90°, and then the edge was placed in a different solvent, such as butanol-propionic acid-water.

Again, the solvent rose through the paper (in a second dimension), thereby separating molecules that may not have been adequately separated during the first separation step. After this second separation step, the paper was dried and exposed to X-ray film, a procedure called autoradiography. Radioactive emission from the $^{14}C$-labeled molecules caused dark spots to appear on the film.

The pattern of spots changed depending on the length of time the cells were incubated with $^{14}C$-labeled $CO_2$. When the incubation period was short, only molecules that were made in the first steps of the Calvin cycle were seen—3-phosphoglycerate (3PG) and 1,3-bisphosphoglycerate (1,3-BPG). Longer incubations revealed molecules synthesized in later steps—glyceraldehyde-3-phosphate (G3P) and ribulose bisphosphate (RuBP).

A challenge for Calvin and his colleagues was to identify the chemical nature of each spot. They achieved this by a variety of chemical methods. For example, a spot could be cut out of the paper, the molecule within the paper could be washed out or eluted, and then the eluted molecule could be subjected to the same procedure that included a radiolabeled molecule whose structure was already known. If the unknown molecule and known molecule migrated to the same spot in the paper, this indicated they were likely to be the same molecule. During the late 1940s and 1950s, Calvin and his coworkers identified all of the $^{14}C$-labeled spots and the order in which they appeared. In this way, they determined the

**Figure 8.14** **The determination of the Calvin cycle using $^{14}$C-labeled $CO_2$ and paper chromatography.** (6): Calvin, M. 1961. The path of carbon in photosynthesis, *Nobel Lecture* pp. 618–644, Fig. 4. ©The Nobel Foundation

**GOAL** The incorporation of $CO_2$ into carbohydrate involves a biosynthetic pathway. The aim of this experiment was to identify the steps.

**KEY MATERIALS** The green alga *Chlorella pyrenoidosa* and $^{14}$C-labeled $CO_2$.

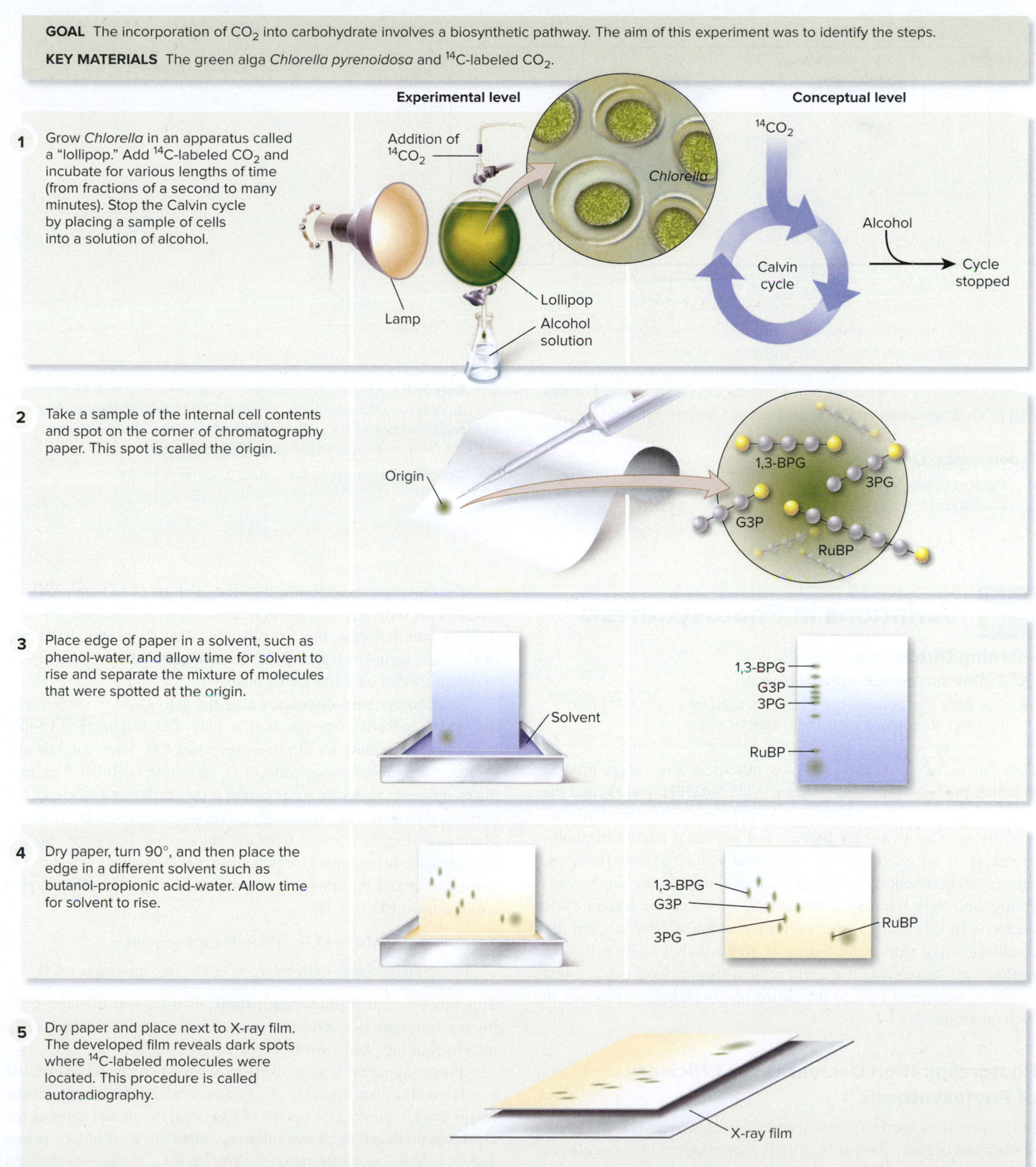

**Experimental level**                **Conceptual level**

1  Grow *Chlorella* in an apparatus called a "lollipop." Add $^{14}$C-labeled $CO_2$ and incubate for various lengths of time (from fractions of a second to many minutes). Stop the Calvin cycle by placing a sample of cells into a solution of alcohol.

Addition of $^{14}$CO$_2$ — Chlorella — $^{14}$CO$_2$ — Alcohol — Calvin cycle — Cycle stopped

Lamp — Lollipop — Alcohol solution

2  Take a sample of the internal cell contents and spot on the corner of chromatography paper. This spot is called the origin.

Origin — 1,3-BPG — 3PG — G3P — RuBP

3  Place edge of paper in a solvent, such as phenol-water, and allow time for solvent to rise and separate the mixture of molecules that were spotted at the origin.

Solvent — 1,3-BPG — G3P — 3PG — RuBP

4  Dry paper, turn 90°, and then place the edge in a different solvent such as butanol-propionic acid-water. Allow time for solvent to rise.

1,3-BPG — G3P — 3PG — RuBP

5  Dry paper and place next to X-ray film. The developed film reveals dark spots where $^{14}$C-labeled molecules were located. This procedure is called autoradiography.

X-ray film

**6**  **THE DATA***

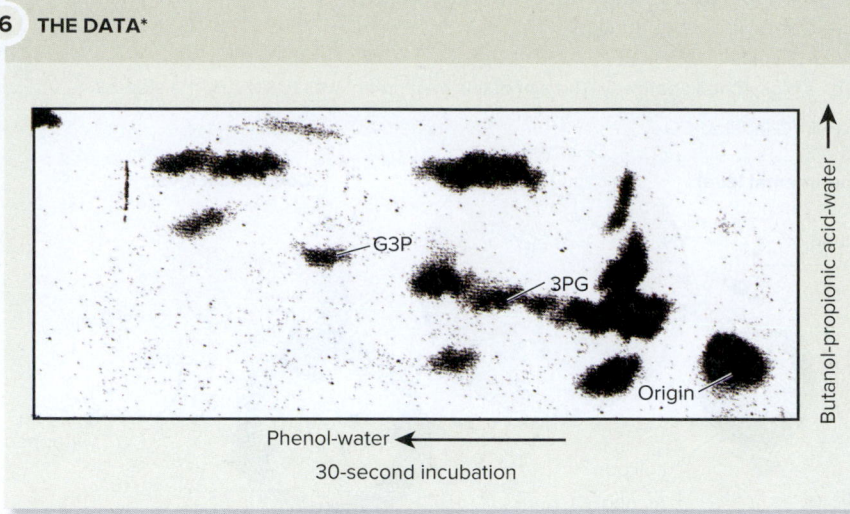

*An autoradiograph from one of Calvin's experiments.

**7**  **CONCLUSION**  The identification of the molecules in each spot elucidated the steps of the Calvin cycle.

**8**  **SOURCE**  Calvin, M. 1961. The path of carbon in photosynthesis, *Nobel Lecture*, 618–644.

series of reactions of what we now know as the Calvin cycle. For this work, Calvin was awarded the Nobel Prize in Chemistry in 1961.

***Experimental Questions***

1. What was the purpose of the study conducted by Calvin and his colleagues?

2. **CoreSKILL »** In Calvin's experiment shown in Figure 8.14, why did the researchers use [14]C-labeled $CO_2$? Why did they examine samples taken after several different time periods? How were the different molecules in the samples identified?

3. **CoreSKILL »** Interpret the results of Calvin's study.

## 8.5  Variations in Photosynthesis

**Learning Outcomes:**

1. Explain the concept of photorespiration.
2. Compare and contrast the strategies used by $C_4$ and CAM plants to avoid photorespiration and conserve water.

Thus far, we have considered photosynthesis as a two-stage process in which the light reactions produce ATP, NADPH, and $O_2$ and the Calvin cycle uses the ATP and NADPH for the synthesis of carbohydrates. This two-stage process is a universal feature of photosynthesis in all green plants, algae, and cyanobacteria. However, certain environmental conditions such as temperature, water availability, and light intensity alter the way in which the Calvin cycle operates. In this section, we begin by examining how hot and dry conditions may reduce the output of photosynthesis. We will then explore two adaptations that certain plant species have evolved that conserve water and help to maximize photosynthetic efficiency in such environments.

### Photorespiration Decreases the Efficiency of Photosynthesis

In the previous section, we learned that rubisco adds a $CO_2$ molecule to an organic molecule, RuBP, to produce two molecules of 3-phosphoglycerate (3PG):

$$RuBP + CO_2 \rightarrow 2\ 3PG$$

For most species of plants, the incorporation of $CO_2$ into 3PG via RuBP is the only way for carbon fixation to occur. Because 3PG is a three-carbon molecule, these plants are called $C_3$ **plants**. Examples of $C_3$ plants include wheat and oak trees (**Figure 8.15**). About 90% of the plant species on Earth are $C_3$ plants.

Researchers have discovered that the active site of rubisco can also add $O_2$ to RuBP, although its affinity for $CO_2$ is more than 10-fold better than its affinity for $O_2$. Even so, when $CO_2$ levels are low and $O_2$ levels are high, rubisco adds an $O_2$ molecule to RuBP. This produces only one molecule of 3PG and a two-carbon molecule called phosphoglycolate. The phosphoglycolate is then dephosphorylated to glycolate, which is released from the chloroplast. In a series of several steps, the two-carbon glycolate molecule is eventually oxidized in peroxisomes and mitochondria to produce an organic molecule plus a molecule of $CO_2$:

$$RuBP + O_2 \rightarrow 3PG + Phosphoglycolate$$

$$Phosphoglycolate \rightarrow Glycolate \rightarrow \rightarrow Organic\ molecule + CO_2$$

This process, called **photorespiration**, uses $O_2$ and liberates $CO_2$. Photorespiration is considered wasteful because it releases $CO_2$, thereby limiting plant growth.

Photorespiration is more likely to occur when plants are exposed to a hot and dry environment. To conserve water, the stomata of the leaves close, inhibiting the uptake of $CO_2$ from the air and trapping the $O_2$ that is produced by photosynthesis. When the level of $CO_2$ is low and $O_2$ is high, photorespiration is favored. If $C_3$ plants are subjected to hot and dry environmental conditions, as much as 25–50% of their photosynthetic work is reversed by the process of photorespiration.

**(a) Wheat plants**

**(b) Oak leaves**

**Figure 8.15** **Examples of C$_3$ plants.** The structures of **(a)** wheat and **(b)** oak leaves are similar to that shown in Figure 8.2. a: ©David Noton Photography/Alamy Stock Photo; b: ©McGraw-Hill Education/Vicki Copeland, photographer

Why do plants carry out photorespiration? The answer is not entirely clear. One possibility is that photorespiration may have a protective advantage. On hot and dry days when the stomata are closed, $CO_2$ levels within the leaves fall, and $O_2$ levels rise. Under these conditions, highly toxic oxygen-containing molecules such as free radicals may be produced that could damage the plant. Therefore, plant biologists have hypothesized that the role of photorespiration may be to protect the plant against the harmful effects of such toxic molecules by consuming $O_2$ and releasing $CO_2$.

## C$_4$ Plants Have Evolved a Mechanism to Minimize Photorespiration

Certain species of plants have developed a way to minimize photorespiration. In the early 1960s, Hugo Kortschak discovered that the first product of carbon fixation in sugarcane is not 3GP but instead is a molecule with four carbon atoms. Species such as sugarcane are called **C$_4$ plants** because of this four-carbon molecule. Later, Marshall Hatch and Roger Slack confirmed this result and identified the molecule as oxaloacetate. For this reason, the pathway is sometimes called the Hatch-Slack pathway.

Some C$_4$ plants have a unique leaf anatomy that allows them to avoid photorespiration (**Figure 8.16**). An interior layer in the leaves of many C$_4$ plants has a two-cell organization composed of mesophyll cells and bundle-sheath cells. $CO_2$ from the atmosphere enters the mesophyll cells via stomata. Once inside, the enzyme PEP carboxylase attaches $CO_2$ to phosphoenolpyruvate (PEP), a three-carbon molecule, to produce oxaloacetate, a four-carbon molecule. PEP carboxylase does not recognize $O_2$. Therefore, unlike rubisco, PEP carboxylase does not promote photorespiration when $CO_2$ is low and $O_2$ is high. Instead, PEP carboxylase continues to fix $CO_2$.

As shown in Figure 8.16, oxaloacetate is converted to the four-carbon molecule malate, which is transported into the bundle-sheath cell. Malate is then broken down into pyruvate and $CO_2$. The pyruvate returns to the mesophyll cell, where it is converted to PEP via ATP, and the cycle in the mesophyll cell can begin again. The $CO_2$ enters the Calvin cycle in the chloroplasts of the bundle-sheath cells. Because the mesophyll cell supplies the bundle-sheath cell with a steady supply of $CO_2$, the concentration of $CO_2$ remains high in the

bundle-sheath cell. Also, the mesophyll cells shield the bundle-sheath cells from high levels of $O_2$. This strategy minimizes photorespiration, which requires low $CO_2$ and high $O_2$ levels to proceed.

Which is better—being a C$_3$ or a C$_4$ plant? The answer is that it depends on the environment. In warm and dry climates, C$_4$ plants have an advantage. During the day, they can keep their stomata partially closed to conserve water. Furthermore, they minimize photorespiration. Examples of C$_4$ plants are sugarcane, crabgrass, and corn. In cooler climates, C$_3$ plants have the edge because they use less energy to fix $CO_2$. The process of carbon fixation that occurs in C$_4$ plants uses ATP to regenerate PEP from pyruvate (see Figure 8.16), and C$_3$ plants do not have to expend that ATP.

## CAM Plants Are C$_4$ Plants That Take Up $CO_2$ at Night

We have just learned that certain C$_4$ plants prevent photorespiration by providing $CO_2$ to the bundle-sheath cells, where the Calvin cycle occurs. This mechanism spatially separates the processes of carbon fixation and the Calvin cycle. Another strategy followed by other C$_4$ plants, called **CAM plants**, separates these processes in time. CAM stands for crassulacean acid metabolism, because the process was first studied in members of the plant family Crassulaceae. Most CAM plants are water-storing succulents such as cacti, bromeliads (including pineapple), and sedums. To avoid water loss, CAM plants keep their stomata closed during the day and open them at night, when it is cooler and the relative humidity is higher.

How, then, do CAM plants carry out photosynthesis? **Figure 8.17** compares CAM plants with the other type of C$_4$ plants we considered in Figure 8.16. Photosynthesis in CAM plants occurs entirely within mesophyll cells, but the synthesis of a C$_4$ molecule and the Calvin cycle occur at different times. During the night when temperatures are cooler, the stomata of CAM plants open, thereby allowing the entry of $CO_2$ into mesophyll cells. $CO_2$ is joined with PEP to form the four-carbon molecule oxaloacetate. This is then converted to malate, which accumulates during the night in the central vacuoles of the cells. In the morning, the stomata close to conserve moisture. The accumulated malate in the mesophyll cells leaves the vacuole and is broken down to release $CO_2$, which then drives the Calvin cycle during the daytime.

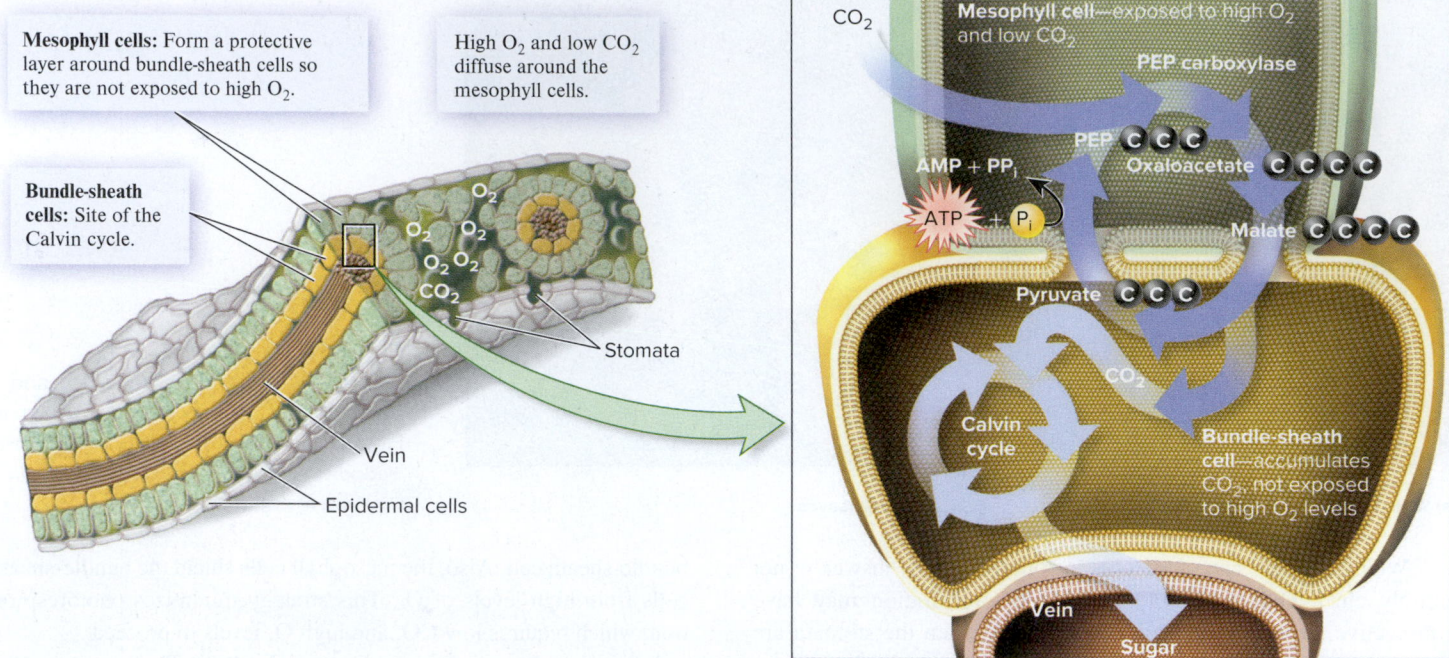

**Figure 8.16** **Leaf structure and its relationship to the C₄ cycle.** C₄ plants have mesophyll cells, which initially take up $CO_2$, and bundle-sheath cells, where much of the carbohydrate synthesis occurs. Compare this leaf structure with the structure of C₃ leaves shown in Figure 8.2.

**Concept Check:** *How does the cellular arrangement in C₄ plants minimize photorespiration?*

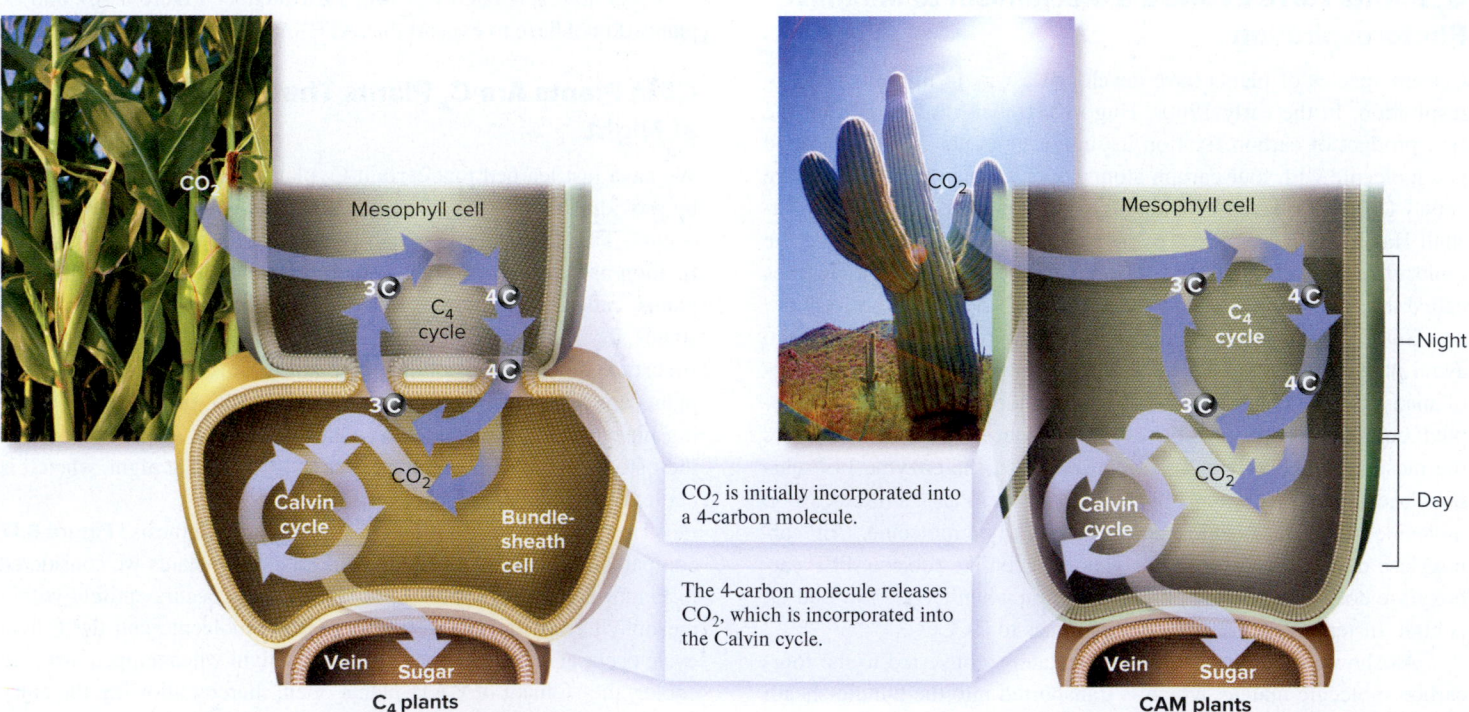

$CO_2$ is initially incorporated into a 4-carbon molecule.

The 4-carbon molecule releases $CO_2$, which is incorporated into the Calvin cycle.

**Figure 8.17** **A comparison of C₄ and CAM plants.** The name C₄ designates those plants in which the first organic product of carbon fixation is a four-carbon molecule. Using this definition, CAM plants are a type of C₄ plant. CAM plants, however, do not separate the functions of making a four-carbon molecule and the Calvin cycle into different types of cells. Instead, they make a four-carbon molecule at night and break down that molecule during the day so the $CO_2$ can be used in the Calvin cycle. (left): ©Wesley Hitt/Getty Images; (right): ©John Foxx/Getty Images

**Concept Check:** *What are the advantages for C₃, C₄, and CAM plants?*

## Summary of Key Concepts

### 8.1   Overview of Photosynthesis

- Photosynthesis is the process by which plants, algae, and photosynthetic bacteria capture light energy that is used to synthesize carbohydrates.

- During photosynthesis, carbon dioxide, water, and energy are used to make carbohydrates and oxygen.

- Heterotrophs must obtain organic molecules in their food, whereas autotrophs make organic molecules from inorganic sources. Photoautotrophs use the energy from light to make organic molecules.

- An energy cycle occurs in the biosphere in which photosynthesis uses light, $CO_2$, and $H_2O$ to make organic molecules, and the organic molecules are broken back down to $CO_2$ and $H_2O$ via cellular respiration to supply energy in the form of ATP (Figure 8.1).

- In plants and algae, photosynthesis occurs within chloroplasts, organelles with an outer membrane, inner membrane, and thylakoid membrane. The stroma is the fluid-filled region between the thylakoid membrane and inner membrane. In plants, the leaves are the major site of photosynthesis (Figure 8.2).

- The light reactions of photosynthesis capture light energy to make ATP, NADPH, and $O_2$. These reactions occur at the thylakoid membrane. Carbohydrate synthesis via the Calvin cycle uses ATP and NADPH from the light reactions and happens in the stroma (Figure 8.3).

### 8.2   Reactions That Harness Light Energy

- Light is a form of electromagnetic radiation that travels in waves and is composed of photons with discrete amounts of energy (Figure 8.4).

- Electrons can absorb light energy and be boosted to a higher energy level—an excited state (Figure 8.5).

- Photosynthetic pigments include chlorophylls *a* and *b* and carotenoids. These pigments absorb light energy in the visible spectrum to drive photosynthesis (Figures 8.6, 8.7).

- During linear electron flow, electrons from photosystem II (PSII) follow a pathway along an electron transport chain (ETC) in the thylakoid membrane. This pathway generates an $H^+$ gradient that is used to make ATP. In addition, light energy striking photosystem I (PSI) boosts electrons to a very high energy level that allows the synthesis of NADPH (Figure 8.8).

- During cyclic photophosphorylation, electrons are activated in PSI and flow through the ETC back to PSI. This cyclic electron flow produces an $H^+$ gradient that is used to make ATP (Figure 8.9).

- Cytochrome $b_6$ in chloroplasts and cytochrome $b$ in mitochondria are homologous proteins involved in electron transport and $H^+$ pumping (Figure 8.10).

### 8.3   Molecular Features of Photosystems

- In the light-harvesting complex of PSII, pigment molecules absorb light energy that is transferred to the reaction center via resonance energy transfer. A high-energy electron from P680* is transferred to a primary electron acceptor. An electron from water then replaces the electron lost by P680* (Figure 8.11).

- The Z scheme proposes that an electron absorbs light energy twice, at both PSII and PSI, losing some of that energy as it flows along the ETC in the thylakoid membrane (Figure 8.12).

### 8.4   Synthesizing Carbohydrates via the Calvin Cycle

- The Calvin cycle is composed of three phases: carbon fixation, reduction and carbohydrate production, and regeneration of ribulose bisphosphate (RuBP). In this cycle, ATP is used as a source of energy, and NADPH is used as a source of high-energy electrons to incorporate $CO_2$ into a carbohydrate (Figure 8.13).

- Calvin and Benson determined the steps in the Calvin cycle by isotope-labeling methods in which the products of the Calvin cycle were separated by paper chromatography (Figure 8.14).

### 8.5   Variations in Photosynthesis

- $C_3$ plants incorporate $CO_2$ into RuBP to make 3PG, a three-carbon molecule (Figure 8.15).

- Photorespiration occurs when the level of $O_2$ is high and $CO_2$ is low, which happens under hot and dry conditions. During this process, some $O_2$ is used and $CO_2$ is liberated. Photorespiration is inefficient because it reverses the incorporation of $CO_2$ into an organic molecule.

- Some $C_4$ plants avoid photorespiration by first incorporating $CO_2$, via PEP carboxylase, into a four-carbon molecule, which is pumped from mesophyll cells into bundle-sheath cells. This maintains a high concentration of $CO_2$ in the bundle-sheath cells, where the Calvin cycle occurs. The high $CO_2$ concentration minimizes photorespiration (Figure 8.16).

- CAM plants, a type of $C_4$ plant, prevent photorespiration by fixing $CO_2$ into a four-carbon molecule at night and then running the Calvin cycle during the day with their stomata closed to reduce water loss (Figure 8.17).

## Assess & Discuss

### Test Yourself

1. The water necessary for photosynthesis
   a. is split into $H_2$ and $O_2$.
   b. is directly involved in the synthesis of carbohydrates.
   c. provides the electrons to replace those lost in photosystem II.
   d. provides the $H^+$ needed to synthesize G3P.
   e. does none of the above.

2. In PSII, P680 differs from the pigment molecules of the light-harvesting complex in that it
   a. is a carotenoid.
   b. absorbs light energy and transfers that energy to other molecules via resonance energy transfer.
   c. transfers an excited electron to the primary electron acceptor.
   d. transfer an excited electron to $O_2$.
   e. acts like ATP synthase to produce ATP.

3. The cyclic electron flow that occurs via photosystem I produces
   a. NADPH.
   b. oxygen.
   c. ATP.
   d. all of the above.
   e. a and c only.

4. During linear electron flow, the high-energy electron from P680*
    a. eventually moves to NADP$^+$.
    b. becomes incorporated in water molecules.
    c. is pumped into the thylakoid space to drive ATP production.
    d. provides the energy necessary to split water molecules.
    e. falls back to the low-energy state in photosystem II.

5. During the first phase of the Calvin cycle, carbon dioxide is incorporated into ribulose bisphosphate (RuBP) by
    a. oxaloacetate.
    b. rubisco.
    c. RuBP.
    d. quinone.
    e. G3P.

6. The NADPH produced during the light reactions is necessary for
    a. the carbon fixation phase, which incorporates carbon dioxide into an organic molecule during the Calvin cycle.
    b. the reduction phase, which produces carbohydrates in the Calvin cycle.
    c. the regeneration of RuBP of the Calvin cycle.
    d. all of the above.
    e. a and b only.

7. The majority of the G3P produced during the reduction and carbohydrate production phase is used in making
    a. glucose.
    b. ATP.
    c. RuBP to continue the cycle.
    d. rubisco.
    e. all of the above.

8. Photorespiration
    a. is the process in which plants use sunlight to make ATP.
    b. is an inefficient way that plants can produce organic molecules by using $O_2$ and releasing $CO_2$.
    c. is a process that plants use to convert light energy to NADPH.
    d. occurs in the thylakoid lumen.
    e. is the normal process of carbohydrate production in cool, moist environments.

9. Photorespiration is avoided by $C_4$ plants because
    a. these plants separate the formation of a four-carbon molecule from the rest of the Calvin cycle in different cells.
    b. these plants carry out only anaerobic respiration.
    c. the enzyme PEP carboxylase functions to maintain high $CO_2$ concentrations in the bundle-sheath cells.
    d. all of the above.
    e. a and c only.

10. Plants commonly found in hot and dry environments that carry out carbon fixation at night are
    a. oak trees.
    b. $C_3$ plants.
    c. CAM plants.
    d. all of the above.
    e. a and b only.

## Conceptual Questions

1. What are the two stages of photosynthesis? What are the key products of each stage?

2. What is the function of NADPH in the Calvin cycle?

3.  **Core Concept: Energy and Matter**  At the level of the biosphere, what is the role of photosynthesis in the utilization of energy by living organisms?

## Collaborative Questions

1. Discuss the advantages and disadvantages of being a heterotroph or a photoautotroph.

2. Biotechnologists are trying to genetically modify $C_3$ plants to convert them to $C_4$ or CAM plants. Why would this be useful? What genes might you introduce into $C_3$ plants to convert them to $C_4$ or CAM plants?

# Cell Communication

# 9

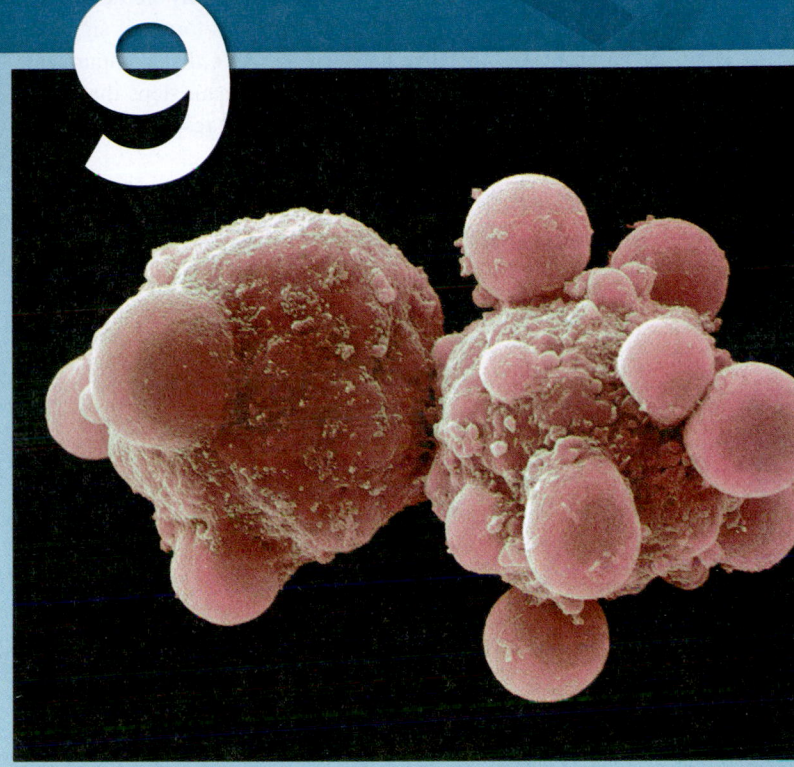

Over 2 billion cells will die in your body during the next hour. In an adult human body, approximately 50–70 billion cells die each day due to programmed cell death—the process in which a cell breaks apart into small fragments (see the chapter opening photo). In a year, your body produces and purposely destroys a mass of cells that is equal to its total weight! Though this may seem like a scary process, it's actually keeping you healthy. Programmed cell death, also called apoptosis, ensures that your body maintains a proper number of cells. It also eliminates cells that are worn out or potentially harmful, such as cancer cells. Programmed cell death can occur via signals that intentionally cause particular cells to die, or it can result from a failure of proper cell communication. It may also happen when environmental agents cause damage to a cell. Programmed cell death is one example of a response that involves **cell communication**—the process by which cells can detect, interpret, and respond to signals in their environment. A **signal** is an agent that can influence the properties of cells.

In this chapter, we will examine how cells detect environmental signals and also how they produce signals that enable them to communicate with other cells. Communication at the cellular level involves not only receiving and sending signals but also their interpretation. For this to occur, a signal must be recognized by a cellular protein called a **receptor**. When a signal and a receptor interact, the receptor changes shape, or conformation, thereby changing the way the receptor interacts with cellular factors. These interactions eventually lead to some type of response in the cell. We begin the chapter with the general features of cell communication, and then discuss the main ways in which cells receive, process, and respond to signals sent by other cells. As you will learn, cell communication involves an amazing diversity of signaling molecules and cellular proteins that are devoted to this process. We conclude by looking at the role of cell communication in apoptosis, in which a cell becomes programmed to die.

**Programmed cell death.** The two cells shown here are breaking apart because signaling molecules initiated a pathway that programmed their death. ©David McCarthy/SPL/Science Source

## 9.1 General Features of Cell Communication

**Learning Outcomes:**

1. Explain the two general reasons for cell signaling: responding to environmental changes and cell-to-cell communication.
2. Compare and contrast the five ways that cells communicate with each other based on the distance between them.
3. Outline the three-stage process of cell signaling.

All living cells, including those of bacteria, archaea, protists, fungi, plants, and animals, must engage in cell communication to survive. Cell communication, also known as cell signaling, involves both incoming and outgoing signals. For example, on a sunny day, cells can sense their exposure to ultraviolet (UV) light—a physical signal—and respond accordingly. In humans, UV light acts as an incoming signal to promote the synthesis of melanin, a protective pigment that helps to prevent the harmful effects of UV radiation. In addition, cells produce outgoing signals that influence the behavior of neighboring cells. Plant cells, for example, produce hormones that influence the pattern of cell elongation so the plant grows toward light. Cells of all living organisms both respond to incoming signals and produce outgoing signals. Cell communication is a two-way street.

In this section, we begin by considering why cells need to respond to signals. We will then examine various forms of signaling that are based on the distance between the cells that communicate with each other. Finally, we will examine the main steps that occur when a cell is exposed to a signal and produces a response to it.

## Cells Detect and Respond to Signals from Their Environment and from Other Cells

Before getting into the details of cell communication, let's take a general look at why cell communication is necessary.

*Responding to Changes in the Environment* The first reason for cell communication is that cells need to respond to a changing environment. Changes in the environment are a persistent feature of life, and living cells are continually faced with alterations in temperature and availability of nutrients, water, and light. A cell may even be exposed to a toxic chemical in its environment. Being able to respond to change at the cellular level is called a **cellular response**.

As an example, let's consider the response of a yeast cell to glucose in its environment (**Figure 9.1**). Some of the glucose acts as a signaling molecule that binds to a receptor and causes a cellular response. In this case, the cell responds by increasing the number of glucose transporters needed to take glucose into the cell and also by increasing the number of metabolic enzymes required to utilize glucose once it is inside. The cellular response allows the cell to use glucose efficiently.

*Cell-to-Cell Communication* A second reason for cell communication is the need for cells to share information with each other—a

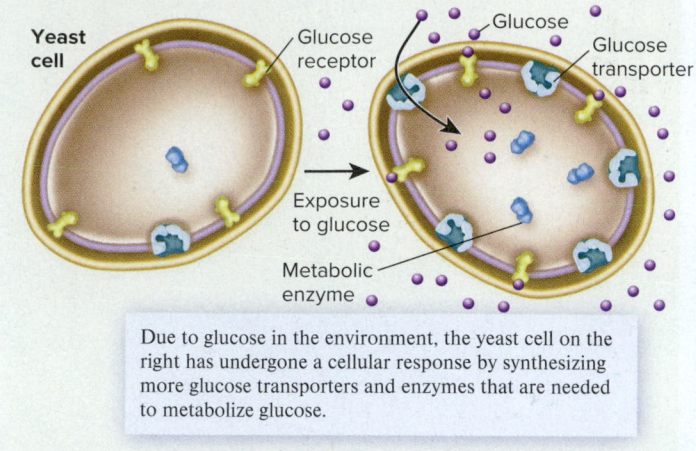

Due to glucose in the environment, the yeast cell on the right has undergone a cellular response by synthesizing more glucose transporters and enzymes that are needed to metabolize glucose.

**Figure 9.1 Response of a yeast cell to glucose.** When glucose is absent from the extracellular environment, the cell is not well prepared to take up and metabolize this sugar. However, when glucose is present, some of that glucose binds to receptors in the membrane, which leads to changes in the amounts and properties of intracellular and membrane proteins so the cell can readily use glucose.

*Concept Check:* *What is the signaling molecule in this example?*

type of cell communication called **cell-to-cell communication**. In one of the earliest experiments demonstrating cell-to-cell communication, Charles Darwin and his son Francis Darwin studied phototropism, the phenomenon in which plants grow toward light (**Figure 9.2**). The Darwins observed that the actual bending occurs in a zone below the growing shoot tip. They concluded that a signal must be transmitted from the growing tip to lower parts of the shoot. Later research revealed that the signal is a molecule called auxin, which is transmitted from cell to cell. A higher amount of auxin accumulates on the nonilluminated side of the shoot and promotes cell elongation on that side of the shoot only, thereby causing the shoot to bend toward the light source.

Cells in the growing shoot tip sense light and send a signal (auxin) to cells on the nonilluminated side of the shoot.

Growing shoot tip of plant

Phototropism

Cells located below the growing tip receive this signal and elongate, thereby causing a bend in the shoot. In this way, the tip grows toward the light.

**Figure 9.2 Phototropism in plants.** This process involves cell-to-cell communication that leads to a shoot bending toward light just beneath its actively growing tip. (inset): ©Cordelia Molloy/SPL/Science Source

 **Core Skill: Connections** Look ahead to Figure 37.5. How does light affect the distribution of auxin produced by a plant's growing shoot tip?

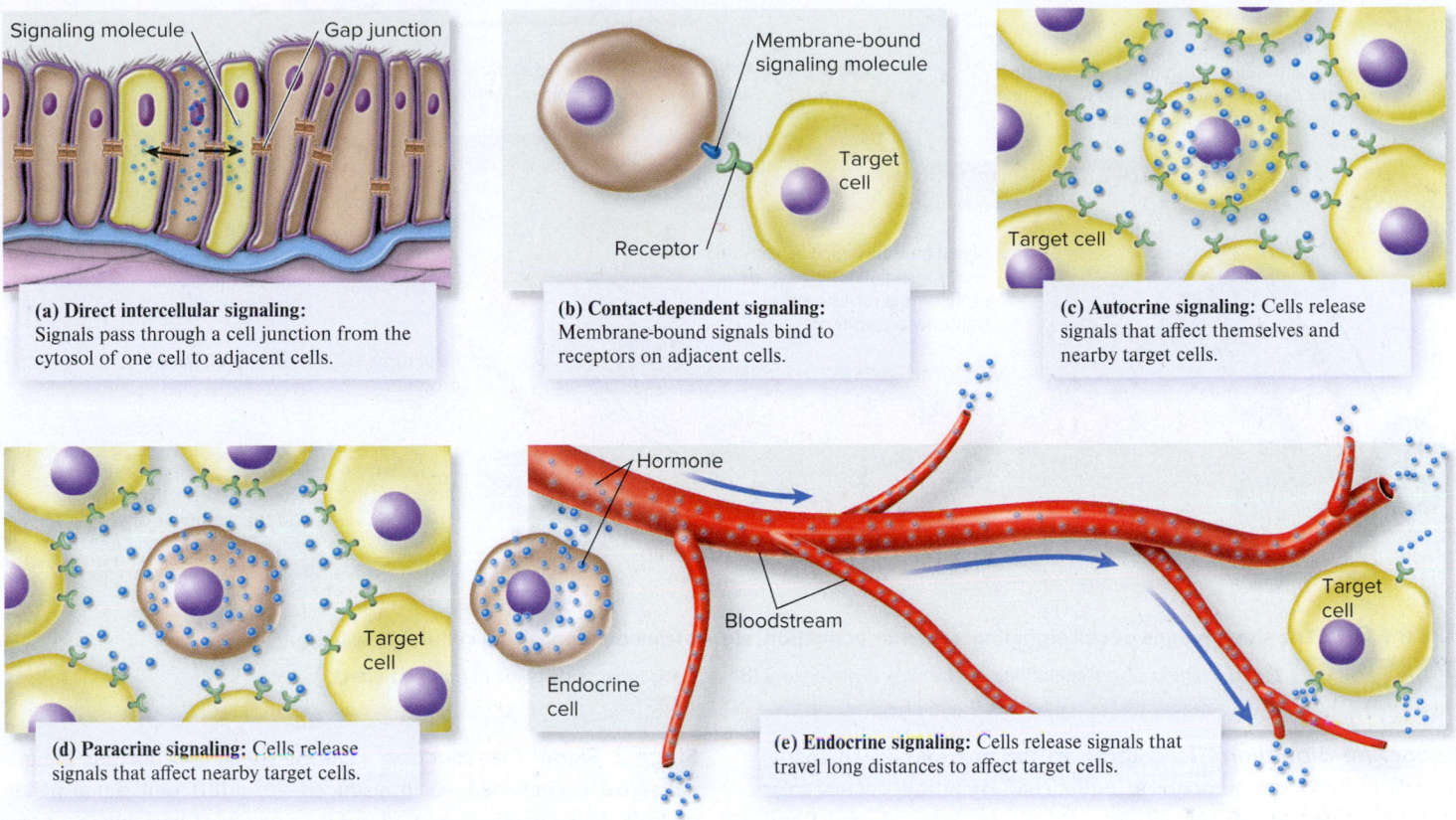

**Figure 9.3** **Types of cell-to-cell communication based on the distance between cells.**

*Concept Check:* *Which type of signal, paracrine or endocrine, is likely to exist for a longer period of time? Explain why this longer existence is necessary.*

## Cell-to-Cell Communication Can Occur Between Adjacent Cells and Between Cells That Are Long Distances Apart

Organisms have a variety of different mechanisms to achieve cell-to-cell communication. The mode of communication depends, in part, on the distance between the cells that need to communicate with each other. Let's first examine the various ways in which signals are transferred between cells. Later in this chapter, we will learn how such signals elicit a cellular response.

One way to categorize cell signaling is by the manner in which the signal is transmitted from one cell to another. Signals are relayed between cells in five common ways, all of which involve a cell that produces a signal and a target cell that receives the signal (**Figure 9.3**).

*Direct Intercellular Signaling*    In a multicellular organism, cells adjacent to each other may have contacts, called cell junctions, that enable them to pass ions, signaling molecules, and other materials between the cytosol of one cell and the cytosol of the other (Figure 9.3a). For example, cardiac muscle cells, which cause your heart to beat, have intercellular connections called gap junctions that allow the passage of ions needed for the coordinated contraction of these cells. We will examine how gap junctions work in Chapter 10.

*Contact-Dependent Signaling*    Not all signaling molecules diffuse from one cell to another. Some molecules are bound to the surface of

a cell and provide a signal to other cells that make contact with the surface of that cell (Figure 9.3b). In the case of contact-dependent signaling, one cell has a membrane-bound signaling molecule that is recognized by a receptor on the surface of another cell. This type of cell-to-cell communication occurs, for example, when portions of neurons (nerve cells) grow and make contact with other neurons. This is important for the formation of the proper connections between neurons.

*Autocrine Signaling*    In autocrine signaling, a cell secretes signaling molecules that bind to receptors on its own cell surface and on the surfaces of neighboring cells of the same cell type, stimulating a response (Figure 9.3c). What is the purpose of autocrine signaling? It is often important for groups of cells to sense cell density. When cell density is high, the concentration of autocrine signals is also high. In some cases, such signals inhibit further cell growth, thereby limiting cell density.

*Paracrine Signaling*    In paracrine signaling, a specific cell secretes a signaling molecule that does not affect that cell but instead influences the behavior of target cells in close proximity (Figure 9.3d). Paracrine signaling is typically of short duration. Usually, the signal is broken down too quickly to be carried to other parts of the body and affect distant cells. A specialized form of paracrine signaling occurs in the nervous systems of animals. Neurotransmitters—molecules made in neurons that transmit a signal to an adjacent cell—are released at the end of a neuron and traverse a narrow space called the synapse (see Chapter 42). The neurotransmitter then binds to a receptor in a target cell.

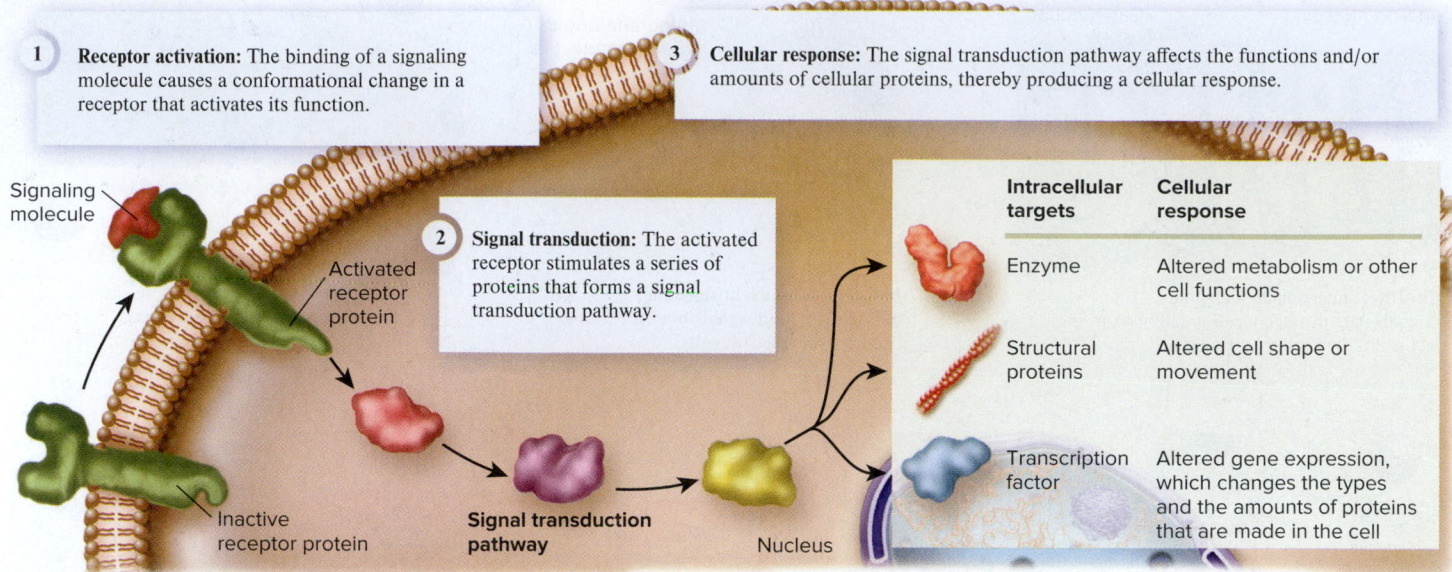

**Figure 9.4** The three stages of cell signaling: receptor activation, signal transduction, and a cellular response.

**Concept Check:** *Explain why a signal transduction pathway is necessary for most signaling molecules to have an effect.*

***Endocrine Signaling*** In contrast to the previous mechanisms of cell-to-cell communication, endocrine signaling occurs over relatively long distances (Figure 9.3e). In both animals and plants, molecules involved in long-distance signaling are called **hormones**. They usually last longer than signaling molecules involved in autocrine and paracrine signaling. In mammals, endocrine signaling involves the secretion of hormones into the bloodstream, which may affect virtually all cells of the body, including those that are far from the cells that secrete the signaling molecules. In flowering plants, hormones move through the plant vascular system and also move through adjacent cells. Some hormones are even gases that diffuse into the air. Ethylene, a gas given off by plants, plays a variety of roles, such as accelerating the ripening of fruit.

## Cells Usually Respond to Signals via a Three-Stage Process

Up to this point, we have seen how signals influence the behavior of cells in close proximity or at long distances, interacting with receptors to elicit a cellular response. What events occur when a cell receives a signal? In most cases, the binding of a signaling molecule to a receptor causes the receptor to activate a signal transduction pathway, which then leads to a cellular response. **Figure 9.4** diagrams the three common stages of cell signaling: receptor activation, signal transduction, and a cellular response.

***Stage 1: Receptor Activation*** In the initial stage, a signaling molecule binds to a receptor of the target cell, causing a conformational change in the receptor that activates its function. In most cases, the activated receptor initiates a response by causing changes in a series of proteins that collectively forms a signal transduction pathway, as described next.

***Stage 2: Signal Transduction*** During signal transduction, the initial signal is converted—or transduced—to a different signal inside the cell. This process is carried out by a group of proteins that form a **signal transduction pathway**. These proteins undergo a series of changes that may result in the production of an intracellular signaling molecule. However, some receptors are intracellular and do not activate a signal transduction pathway. As discussed later, certain types of intracellular receptors directly cause a cellular response.

***Stage 3: Cellular Response*** Cells respond to signals in several different ways. Figure 9.4 shows three common categories of proteins that are controlled by cell signaling: enzymes, structural proteins, and transcription factors.

1. Many signaling molecules exert their effects by altering the activity of one or more enzymes. For example, certain hormones provide a signal that the body needs energy. These hormones activate enzymes that are required for the breakdown of molecules such as carbohydrates.

2. Cells also respond to signals by altering the functions of structural proteins in the cell. For example, when animal cells move during embryonic development or when an amoeba moves toward food, signals play a role in the rearrangement of actin filaments, which are components of the cytoskeleton. The coordination of signaling and changes in the cytoskeleton enables a cell to move in the correct direction.

3. Signaling molecules may also affect the function of transcription factors—proteins that regulate the transcription of genes. Some transcription factors activate gene expression. For example, when cells are exposed to sex hormones, transcription factors activate genes that change the properties of cells, which can lead to changes in the sexual characteristics of entire organisms. As discussed in Section 51.3, estrogens and androgens are

responsible for the development of secondary sex characteristics in humans, including breast development in females and beard growth in males.

### Learning Outcomes:

1. **CoreSKILL »** Calculate the affinity, measured as a dissociation constant, that a receptor has for its signaling molecule, or ligand.
2. Explain how a signaling molecule activates a receptor.
3. Identify three general types of cell surface receptors.
4. Describe intracellular receptors, using estrogen receptors as an example.

In this section, we will take a closer look at receptors and their interactions with signaling molecules. We will compare receptors based on whether they are located on the cell surface or inside the cell. In this chapter, our focus will be on receptors that respond to chemical signaling molecules. Other receptors discussed in Units VI and VII respond to mechanical motion (mechanoreceptors), temperature changes (thermoreceptors), and light (photoreceptors).

### Signaling Molecules Bind to Receptors

The ability of cells to respond to a signal usually requires precise recognition between a signal and its receptor. In many cases, the signal is a molecule, such as a steroid or a protein, that binds to the receptor. A signaling molecule binds to a receptor in much the same way that a substrate binds to the active site of an enzyme, as described in Chapter 6. The signaling molecule, which is called a **ligand**, binds noncovalently to the receptor with a high degree of specificity. The binding occurs when the ligand and receptor happen to collide in the correct orientation with enough energy to form a **ligand•receptor complex**.

$$[\text{Ligand}] + [\text{Receptor}] \underset{k_{\text{off}}}{\overset{k_{\text{on}}}{\rightleftharpoons}} [\text{Ligand•Receptor complex}]$$

Square brackets, [ ], indicate concentration. The value $k_{\text{on}}$ is the rate at which binding occurs. After a complex forms between the ligand and its receptor, the noncovalent interaction between ligand and receptor remains stable for a finite period of time. The term $k_{\text{off}}$ is the rate at which the ligand·receptor complex falls apart or dissociates.

In general, the binding and dissociation of a ligand and its receptor occur relatively rapidly, and therefore an equilibrium is reached when the rate of formation of new ligand·receptor complexes equals the rate at which existing ligand•receptor complexes dissociate:

$$k_{\text{on}}[\text{Ligand}][\text{Receptor}] = k_{\text{off}}[\text{Ligand•Receptor complex}]$$

Rearranging the equation gives

$$\frac{[\text{Ligand}][\text{Receptor}]}{[\text{Ligand•Receptor complex}]} = \frac{k_{\text{off}}}{k_{\text{on}}} = K_{\text{d}}$$

$K_{\text{d}}$ is called the **dissociation constant** between a ligand and its receptor. The $K_{\text{d}}$ value is inversely related to the affinity between the ligand and receptor. A low $K_{\text{d}}$ value indicates that a receptor has a high affinity for its ligand.

Let's look carefully at the left side of this equation and consider what it means. At a ligand concentration at which half of the receptors are bound to a ligand, the concentration of the ligand•receptor complex equals the concentration of receptor that doesn't have ligand bound. At this ligand concentration, [Receptor] and [Ligand•Receptor complex] cancel out of the equation because they are equal. Therefore, at a ligand concentration at which half of the receptors have bound ligand:

$$K_{\text{d}} = [\text{Ligand}]$$

When the ligand concentration is above the $K_{\text{d}}$ value, most of the receptors are likely to have ligand bound to them. In contrast, if the ligand concentration is substantially below the $K_{\text{d}}$ value, most receptors will not be bound by their ligand. The $K_{\text{d}}$ values for many different ligands and their receptors have been experimentally determined. How is this information useful? It allows researchers to predict when a signaling molecule is likely to cause a cellular response. If the concentration of a signaling molecule is far below the $K_{\text{d}}$ value, a cellular response is not likely because relatively few receptors will form a complex with a signaling molecule.

### Receptors Undergo Conformational Changes

Unlike enzymes, which convert their substrates into products, receptors do not usually alter the structure of their ligands. Instead, the ligands alter the structure of their receptors, causing a conformational change (**Figure 9.5**). In this case, the binding of the ligand to its receptor changes the receptor in a way that activates its ability to initiate a cellular response.

Because the binding of a ligand to its receptor is a reversible process, the ligand and receptor will dissociate. Once the ligand is released, the receptor is no longer activated.

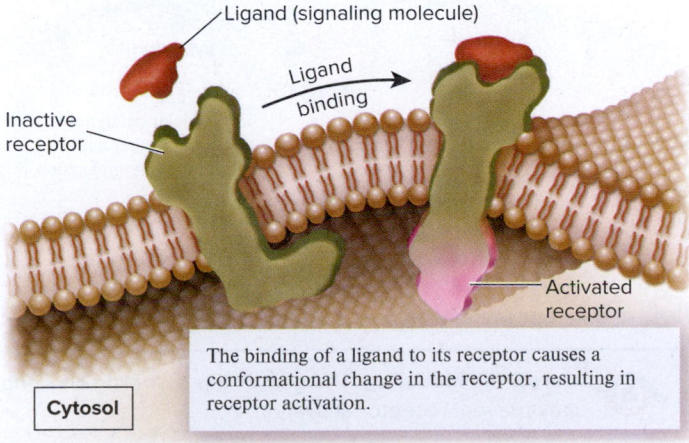

The binding of a ligand to its receptor causes a conformational change in the receptor, resulting in receptor activation.

**Figure 9.5** **Receptor activation.**

**Core Skill: Connections** Look back at Figure 6.6. How is the binding of a ligand to its receptor similar to the binding of a substrate to an enzyme? How are these processes different?

## Cells Contain a Variety of Cell Surface Receptors That Respond to Extracellular Signals

Most signaling molecules are either small hydrophilic molecules or large molecules that do not readily pass through the plasma membrane of cells. Such extracellular signaling molecules bind to **cell surface receptors**—receptors found in the plasma membrane. A typical cell usually contains dozens or even hundreds of different cell surface receptors that enable the cell to respond to different kinds of extracellular signaling molecules. By analyzing the functions of cell surface receptors from many different organisms, researchers have determined that most fall into one of three categories: enzyme-linked receptors, G-protein-coupled receptors, and ligand-gated ion channels, which are described next.

***Enzyme-Linked Receptors*** Receptors known as **enzyme-linked receptors** are found in all living species. Many human hormones bind to this type of receptor. For example, when insulin binds to an enzyme-linked receptor in muscle cells, it enhances the ability of those cells to use glucose. Enzyme-linked receptors typically have two important domains: an extracellular domain, which binds to a signaling molecule, and an intracellular domain, which has a catalytic function (**Figure 9.6a**). When a signaling molecule binds to the extracellular domain, a conformational change is transmitted through the membrane-embedded portion of the protein and affects the conformation of the intracellular catalytic domain. In most cases, this conformational change causes the intracellular catalytic domain to become functionally active.

Most types of enzyme-linked receptors function as **protein kinases**, enzymes that transfer a phosphate group from ATP to specific amino acids in a protein (**Figure 9.6b**). For example, tyrosine kinases attach phosphate to the amino acid tyrosine, whereas serine/threonine kinases attach phosphate to the amino acids serine and threonine. In the example shown in Figure 9.6b, the catalytic domain of the receptor remains inactive when no signaling molecule is present. However, when a signal binds to the extracellular domain, the catalytic domain is activated. Under these conditions, the receptor may phosphorylate itself, or it may phosphorylate intracellular proteins. The attachment of a negatively charged phosphate changes the structure of a protein and thereby alters its function. Later in this chapter, we will explore how this event leads to a cellular response, such as the activation of enzymes that affect cell function.

***G-Protein-Coupled Receptors*** Receptors called **G-protein-coupled receptors (GPCRs)** are found in the cells of all eukaryotic species and are particularly common in animals. GPCRs typically contain seven transmembrane segments that wind back and forth through the plasma membrane. The receptors interact with intracellular proteins called **G proteins**, which are so named because of their ability to bind guanosine triphosphate (GTP) and guanosine diphosphate (GDP). GTP is similar in structure to ATP except it has guanine as a base instead of adenine. In the 1970s, the existence of G proteins was first proposed by Martin Rodbell and colleagues, who found that GTP is needed for certain hormone receptors to cause an intracellular response. Later,

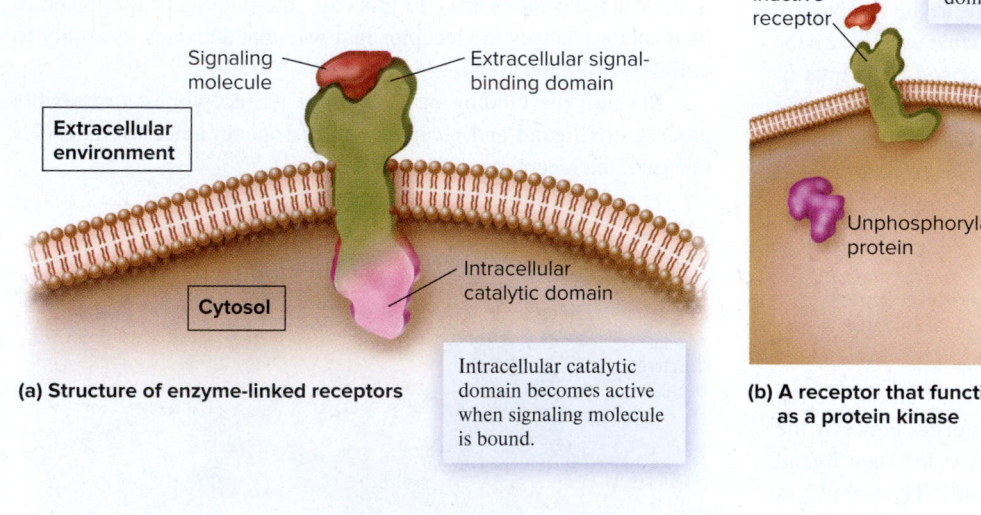

Signaling molecule — Extracellular signal-binding domain

**Extracellular environment**

**Cytosol**

Intracellular catalytic domain

Intracellular catalytic domain becomes active when signaling molecule is bound.

**(a) Structure of enzyme-linked receptors**

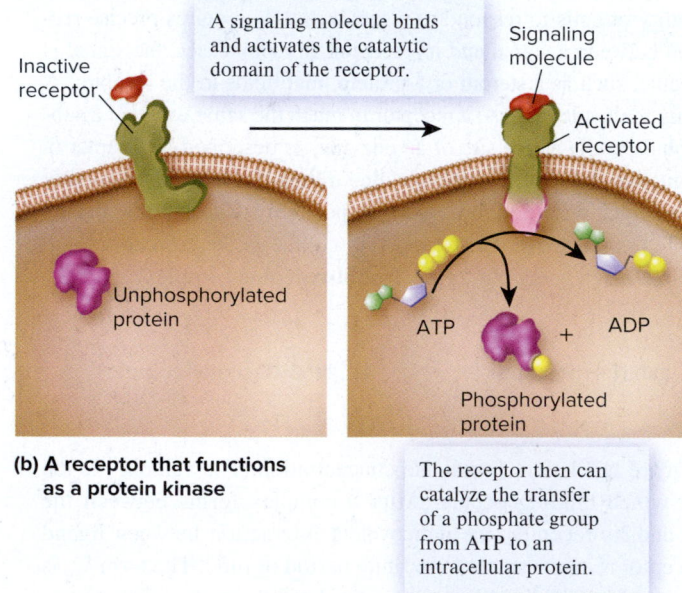

Inactive receptor

A signaling molecule binds and activates the catalytic domain of the receptor.

Signaling molecule

Activated receptor

Unphosphorylated protein

ATP    +    ADP

Phosphorylated protein

**(b) A receptor that functions as a protein kinase**

The receptor then can catalyze the transfer of a phosphate group from ATP to an intracellular protein.

**Figure 9.6** Enzyme-linked receptors.

**Core Skill: Modeling** The goal of this modeling challenge is to predict the possible locations where an amino acid substitution may prevent receptor activation.

**Modeling Challenge:** Figure 9.6 is a general model that shows how the binding of a ligand to an enzyme-linked receptor results in receptor activation. Let's suppose that researchers have identified a mutant version of this type of receptor in which the ligand can still bind to the receptor correctly, but the receptor is not activated. In other words, the binding of the ligand does not cause the receptor to phosphorylate intracellular proteins. The mutation changes just one amino acid in the receptor protein by substituting a glycine (found in the normal protein) to a glutamic acid. On the model shown in part (a) of the figure, put two X's in places where you think the glutamic acid might be found in the mutant receptor, and place a Y where you think it would not be found. Briefly explain your chosen locations.

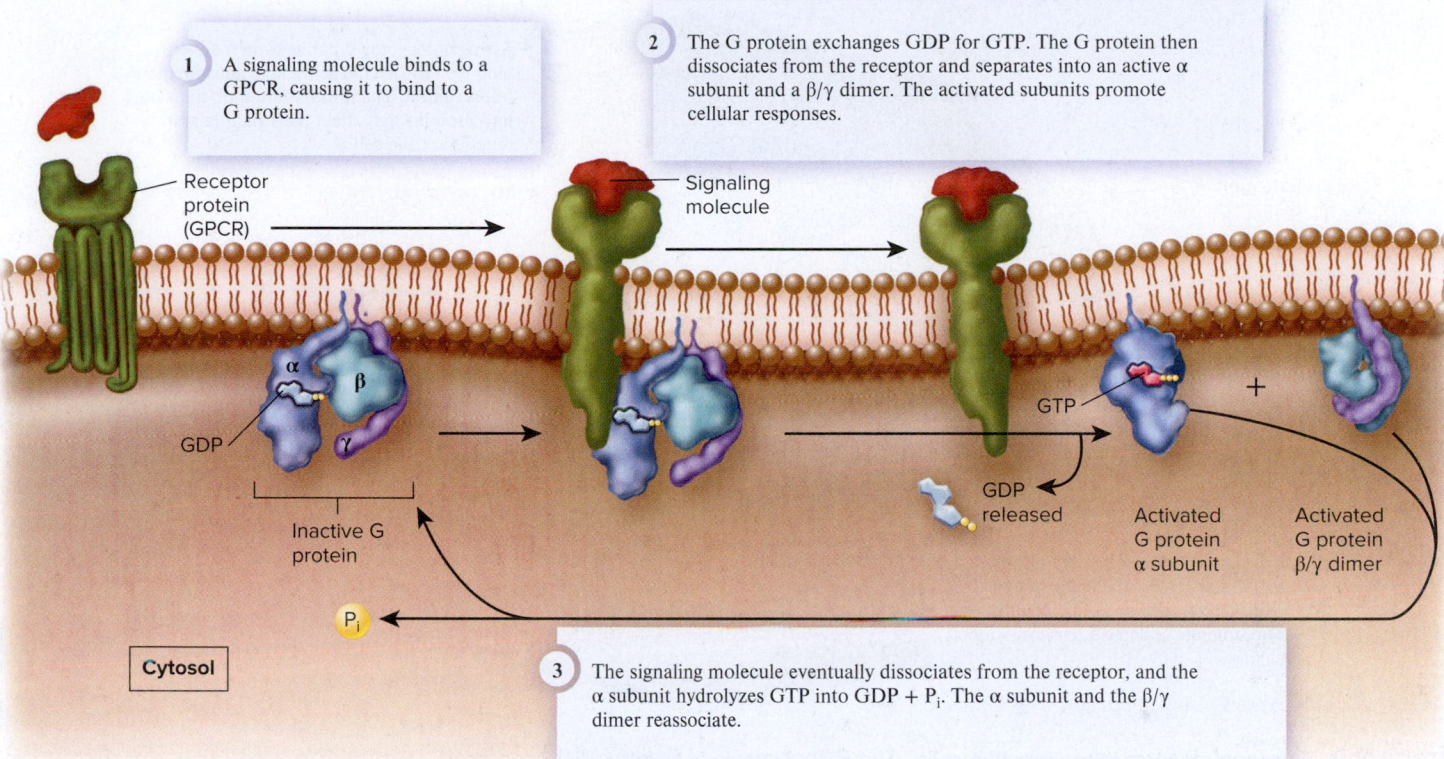

**1** A signaling molecule binds to a GPCR, causing it to bind to a G protein.

**2** The G protein exchanges GDP for GTP. The G protein then dissociates from the receptor and separates into an active α subunit and a β/γ dimer. The activated subunits promote cellular responses.

Receptor protein (GPCR)

Signaling molecule

GTP

GDP

α

β

γ

Inactive G protein

GDP released

Activated G protein α subunit

Activated G protein β/γ dimer

+

P_i

Cytosol

**3** The signaling molecule eventually dissociates from the receptor, and the α subunit hydrolyzes GTP into GDP + P_i. The α subunit and the β/γ dimer reassociate.

**Figure 9.7** **The activation of G-protein-coupled receptors (GPCRs) and G proteins.** Note: All three receptors shown in this figure are the same receptor, but the one on the left is drawn with greater detail to emphasize that it has seven transmembrane segments.

*Concept Check:* *What has to happen before the α and β/γ subunits of the G protein can reassociate with each other?*

Alfred Gilman and coworkers used genetic and biochemical techniques to identify and purify a G protein. In 1994, Rodbell and Gilman won the Nobel Prize in Physiology or Medicine for their pioneering work.

**Figure 9.7** shows how a GPCR and a G protein interact. At the cell surface, a signaling molecule binds to a GPCR, causing a conformational change that activates the receptor, enabling it to bind to a G protein. The G protein, which is a lipid-anchored protein, releases GDP and binds GTP instead. The binding of GTP changes the conformation of the G protein, causing it to dissociate into an α subunit and a β/γ dimer. Later in this chapter, we will examine how the α subunit interacts with other proteins in a signal transduction pathway to elicit a cellular response. The β/γ dimer also plays a role in signal transduction. For example, it can regulate the function of ion channels in the plasma membrane.

When a signaling molecule and a GPCR dissociate, the GPCR is no longer activated, and the cellular response is reversed. For the G protein to return to the inactive state, the α subunit first hydrolyzes its bound GTP to GDP and P_i. After this occurs, the α and β/γ subunits reassociate with each other to form an inactive G protein.

***Ligand-Gated Ion Channels*** As described in Chapter 5, ion channels are proteins that allow the diffusion of ions across cell membranes. **Ligand-gated ion channels** are a third type of cell surface receptor found in the plasma membranes of animal, plant, and fungal cells. When signaling molecules (ligands) bind to this type of receptor, the ion channel opens and allows the flow of ions through the membrane, changing the concentration of the ions in the cell (**Figure 9.8**).

In animals, ligand-gated ion channels are important in the transmission of signals between neurons and muscle cells and between two

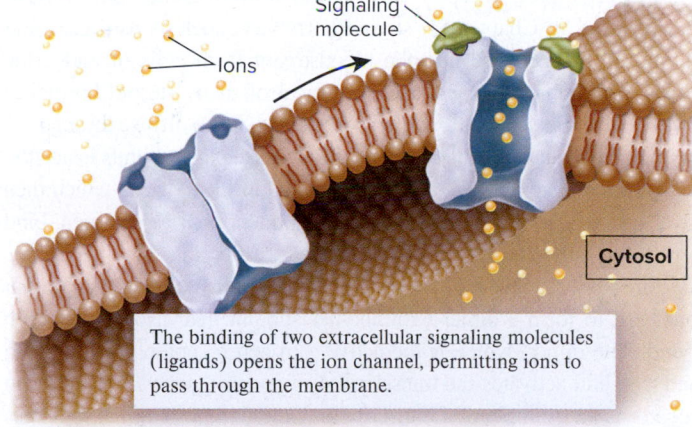

Signaling molecule

Ions

Cytosol

The binding of two extracellular signaling molecules (ligands) opens the ion channel, permitting ions to pass through the membrane.

**Figure 9.8** **The function of a ligand-gated ion channel.**

neurons. In addition, ligand-gated ion channels in the plasma membrane allow the influx of $Ca^{2+}$ into the cytosol. Changes in the cytosolic concentration of $Ca^{2+}$ often play a role in signal transduction.

## Cells Also Have Intracellular Receptors Activated by Signaling Molecules That Pass Through the Plasma Membrane

Although most receptors for signaling molecules are located in the plasma membrane, some are found inside the cell. In these cases, an extracellular signaling molecule must diffuse through the plasma membrane to gain access to its receptor.

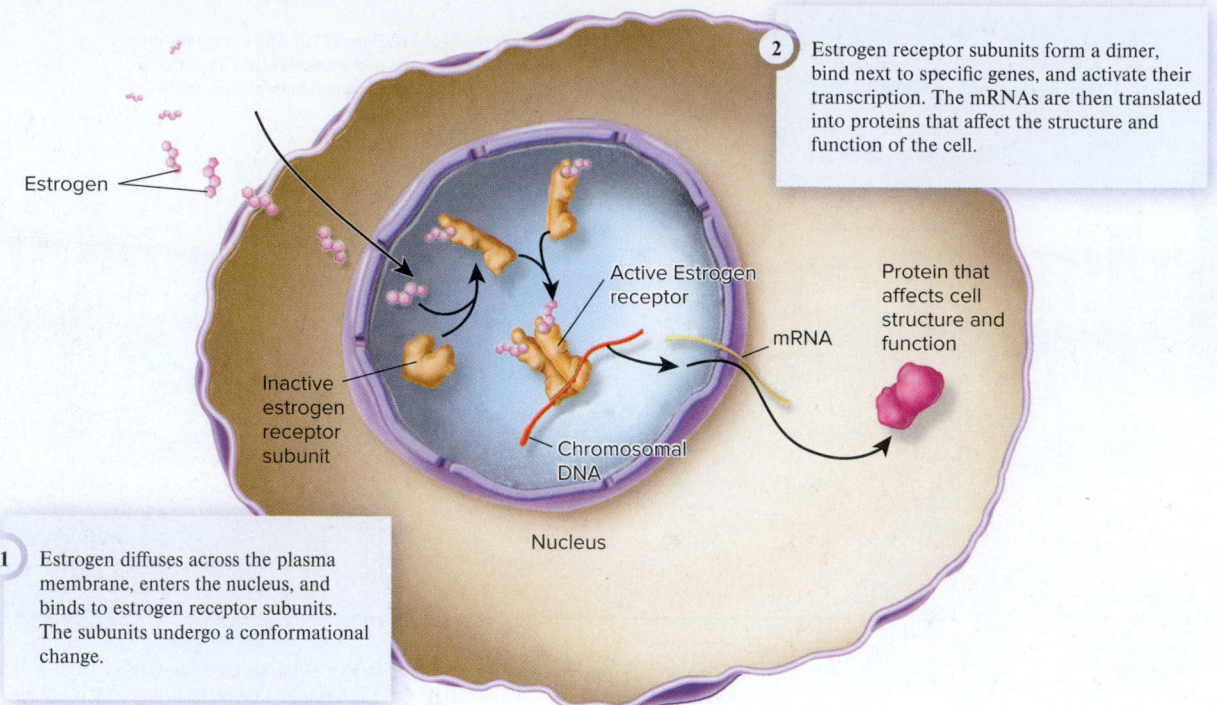

**2** Estrogen receptor subunits form a dimer, bind next to specific genes, and activate their transcription. The mRNAs are then translated into proteins that affect the structure and function of the cell.

Estrogen

Active Estrogen receptor

Inactive estrogen receptor subunit

Chromosomal DNA

mRNA

Protein that affects cell structure and function

Nucleus

**1** Estrogen diffuses across the plasma membrane, enters the nucleus, and binds to estrogen receptor subunits. The subunits undergo a conformational change.

**Figure 9.9   Estrogen receptor in mammalian cells.** This is an example of an intracellular receptor.

 **Core Concept: Structure and Function**   The structure of the estrogen receptor, which is a dimer, has two important sites: the estrogen-binding site and the DNA-binding site. When estrogen binds to its receptor, a conformational change occurs that allows the DNA-binding site to function. The estrogen receptor then binds to the DNA and activates the transcription of specific genes.

In vertebrates, receptors for steroid hormones are intracellular. As discussed in Chapter 51, steroid hormones, such as estrogens and androgens, are secreted into the bloodstream from cells of endocrine glands. The behavior of estrogen is typical of many steroid hormones (**Figure 9.9**). Because estrogen is hydrophobic, it can diffuse through the plasma membrane of a target cell and bind to receptor subunits inside the cell. Some steroids bind to receptor subunits in the cytosol, which then travel into the nucleus. Other steroid hormones, such as estrogen, bind to receptor subunits already in the nucleus. After this binding occurs, the estrogen receptor subunit undergoes a conformational change that enables it to form a dimer with another subunit that also has estrogen bound. The dimer, which is the active estrogen receptor, then binds to the DNA and activates the transcription of specific genes. The estrogen receptor is an example of a transcription factor—a protein that regulates the transcription of genes. The expression of specific genes changes cell structure and function in a way that results in a cellular response.

## 9.3   Signal Transduction and the Cellular Response

### Learning Outcomes:

**1.** For signaling molecules that bind to receptor tyrosine kinases or G-protein-coupled receptors, describe the signal transduction pathways and how those pathways lead to a cellular response.

**2.** Relate the function of second messengers to signal transduction pathways.

**3.** List examples of second messengers, and explain how they exert their effects.

We now turn our attention to the intracellular events that enable a cell to respond to a signaling molecule that binds to a cell surface receptor: signal transduction and a cellular response. In most cases, the binding of a signaling molecule to its receptor stimulates a signal transduction pathway. We will begin by examining a pathway that is controlled by an enzyme-linked receptor, and then consider G-protein-coupled receptors.

### Receptor Tyrosine Kinases Activate Signal Transduction Pathways Involving a Protein Kinase Cascade That Alters Gene Transcription

**Receptor tyrosine kinases** are a category of enzyme-linked receptors that are found in all animals and also in choanoflagellates, which are the protists that are most closely related to animals (see Chapter 28). However, they are not found in bacteria, archaea, or other eukaryotic species. (Bacteria do have receptor histidine kinases, and all eukaryotes have receptor serine/threonine kinases.) The human genome contains about 60 different genes that encode receptor tyrosine kinases that recognize various types of signaling molecules such as hormones.

**Figure 9.10** shows a simplified signal transduction pathway for epidermal growth factor (EGF). A **growth factor** is a signaling molecule that promotes cell division. Multicellular organisms, such as plants and animals, produce a variety of different growth factors to coordinate cell division throughout the body. In vertebrate animals,

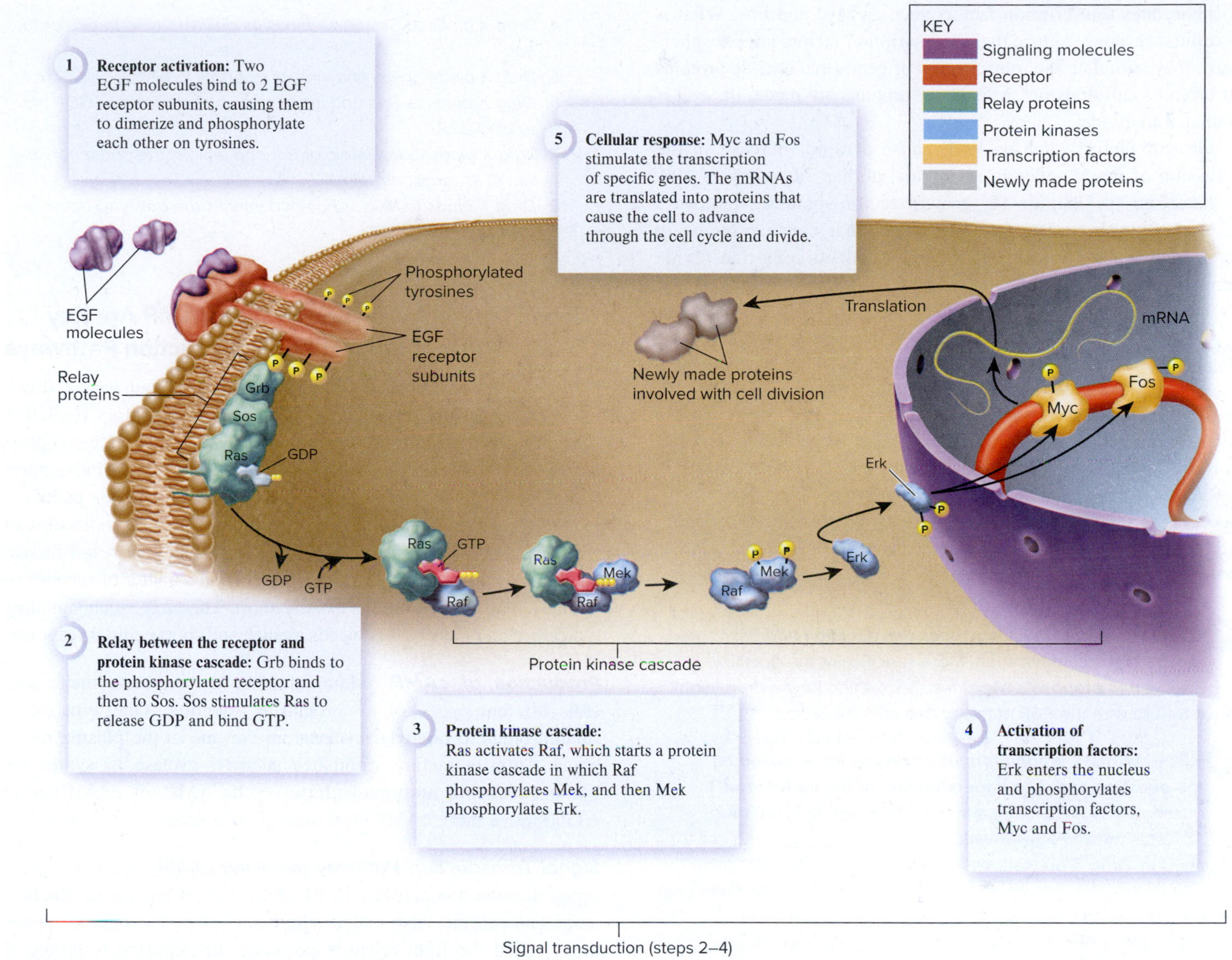

**1** **Receptor activation:** Two EGF molecules bind to 2 EGF receptor subunits, causing them to dimerize and phosphorylate each other on tyrosines.

**5** **Cellular response:** Myc and Fos stimulate the transcription of specific genes. The mRNAs are translated into proteins that cause the cell to advance through the cell cycle and divide.

KEY

Signaling molecules
Receptor
Relay proteins
Protein kinases
Transcription factors
Newly made proteins

Phosphorylated tyrosines

EGF molecules

EGF receptor subunits

Relay proteins

Grb

Sos

Ras

GDP

GDP    GTP

Translation

mRNA

Newly made proteins involved with cell division

Erk

Fos

Myc

Ras    GTP

Ras    Mek

Mek    Erk

Raf

Raf

Raf

Protein kinase cascade

**2** **Relay between the receptor and protein kinase cascade:** Grb binds to the phosphorylated receptor and then to Sos. Sos stimulates Ras to release GDP and bind GTP.

**3** **Protein kinase cascade:** Ras activates Raf, which starts a protein kinase cascade in which Raf phosphorylates Mek, and then Mek phosphorylates Erk.

**4** **Activation of transcription factors:** Erk enters the nucleus and phosphorylates transcription factors, Myc and Fos.

Signal transduction (steps 2–4)

**Figure 9.10** **The epidermal growth factor (EGF) pathway that promotes cell division.**

 **Core Skill: Connections** Look ahead to Figures 15.11 and, in particular, 15.12. Certain mutations alter the structure of the Ras protein so it does not hydrolyze GTP. Such mutations cause cancer. Explain why.

EGF is secreted from endocrine cells, travels through the bloodstream, and binds to a receptor tyrosine kinase, which is located on target cells and called the EGF receptor. EGF is responsible for stimulating epidermal cells, such as skin cells, to divide. Following receptor activation, the three general parts of the signal transduction pathway are (1) relay proteins activate a protein kinase cascade; (2) the protein kinase cascade phosphorylates intracellular proteins such as transcription factors; and (3) the phosphorylated transcription factors stimulate gene transcription. Next, we will consider the details of this pathway.

**EGF Receptor Activation**     For receptor activation to occur, two EGF receptor subunits each bind to a molecule of EGF. The binding of EGF causes the subunits to dimerize and phosphorylate each other on tyrosines within the receptors, which is why they are named receptor tyrosine kinases. Once the EGF receptor is activated, the signal transduction pathway starts.

**Relay Proteins**     The phosphorylated form of the EGF receptor is first recognized by a relay protein of the signal transduction pathway called Grb. This interaction changes the conformation of Grb, causing it to bind another relay protein in the signal transduction pathway termed Sos, causing it to undergo a conformational change. This activation of Sos causes a third relay protein called Ras to release GDP and bind GTP. The GTP form of Ras is the active form.

**Protein Kinase Cascade**     The function of the relay proteins is to activate a **protein kinase cascade**. This cascade involves the sequential activation of three protein kinases. Activated Ras binds to Raf, the first protein kinase in the cascade. Raf then phosphorylates Mek, which becomes active and, in turn, phosphorylates Erk.

**Activation of Transcription Factors and the Cellular Response**
The phosphorylated form of Erk enters the nucleus and

phosphorylates transcription factors such as Myc and Fos. What is the cellular response? Once these transcription factors are phosphorylated, they stimulate the transcription of genes that encode proteins that promote cell division. After these proteins are made, the cell is stimulated to divide.

Growth factors such as EGF cause a rapid increase in the expression of many genes in mammals, perhaps as many as 100. As discussed in Chapter 15, growth factor signaling pathways are often involved in cancer. Mutations that cause proteins in these pathways to become hyperactive result in cells that divide uncontrollably!

## BIO TIPS

**THE QUESTION** *One of the genes that is activated by the EGF signaling pathway is a gene called HSF1, which encodes a protein that is thought to be important for regulating cell division. Let's suppose that researchers have identified a drug that prevents EGF from activating the HSF1 gene. In the laboratory, this drug seems to prevent the growth of certain types of cancer cells. Propose a hypothesis for how this drug exerts its effect. In other words, which protein in the cell might drug X be binding to, and how does drug X affect that protein's function?*

**T** **OPIC** *What topic in biology does this question address?* The topic is cell communication. More specifically, the question asks you to propose a hypothesis explaining how a drug might interfere with the EGF pathway and prevent cancer.

**I** **NFORMATION** *What information do you know based on the question and your understanding of the topic?* From the question, you have learned that the EGF signaling pathway activates the *HSF1* gene, which plays a role in regulating cell division. Drug X prevents EGF from turning on the *HSF1* gene and inhibits the growth of certain kinds of cancer cells. From your understanding of the topic, you may remember that the EGF pathway involves a series of steps, beginning with the binding of EGF to its receptor.

**P** **ROBLEM-SOLVING** **S** **STRATEGY** *Sort out the steps in a complicated process. Propose a hypothesis.* One strategy to begin to solve this problem is to analyze the steps in the EGF pathway (see Figure 9.10) and identify the proteins involved. Any of these proteins could potentially be the target of drug X. Propose a hypothesis for how drug X could bind to one of these proteins and alter its function in a way that would prevent the expression of the *HSF1* gene and thus prevent cancer cells from dividing.

**ANSWER** *For drug X to exert its effect, it must be inhibiting one of the steps of the EGF pathway. Here are some possible hypotheses for how drug X works:*

1. *Drug X binds to the EGF receptor and inhibits the ability of EGF to bind to the receptor.*
2. *Drug X binds to the EGF receptor and inhibits its ability to phosphorylate itself.*
3. *Drug X binds to Grb and inhibits its ability to bind to the EGF receptor or to Sos.*
4. *Drug X binds to Sos and inhibits its ability to bind to Grb or to Ras.*
5. *Drug X binds to Ras and inhibits its ability to bind to Sos or Raf.*
6. *Drug X binds to Ras and inhibits its ability to release GDP or to bind GTP.*
7. *Drug X binds to Raf, Mek, or Erk and inhibits the phosphorylation of its target protein.*
8. *Drug X binds to Myc or Fos and inhibits the ability to activate a gene.*

## Second Messengers Such as Cyclic AMP Are Key Components of Many Signal Transduction Pathways

Let's now turn to examples of signal transduction pathways and cellular responses that involve G-protein-coupled receptors (GPCRs). Extracellular signaling molecules that bind to cell surface receptors are sometimes referred to as first messengers. After first messengers bind to receptors such as GPCRs, many signal transduction pathways lead to the production of **second messengers**—small molecules or ions that relay signals inside the cell. The signals that result in second messenger production often act quickly, in a matter of seconds or minutes, but their duration is usually short. Therefore, such signaling typically occurs when a cell needs a quick and short cellular response.

***Production of cAMP*** Mammalian and plant cells make several different types of G protein α subunits. One type of α subunit binds to **adenylyl cyclase**, an enzyme in the plasma membrane. This interaction stimulates adenylyl cyclase to synthesize **cyclic adenosine monophosphate (cyclic AMP, or cAMP)** from ATP (**Figure 9.11**). cAMP is an example of a second messenger.

***Signal Transduction Pathway Involving cAMP*** Let's explore a signal transduction pathway in which the GPCR recognizes the hormone epinephrine (also called adrenaline). This hormone is sometimes called the fight-or-flight hormone. Epinephrine is produced when an individual is confronted with a stressful situation and helps the individual deal with a perceived threat or danger.

First, epinephrine binds to its receptor and activates a G protein (**Figure 9.12**). The α subunit then activates adenylyl cyclase, which catalyzes the production of cAMP from ATP. One effect of cAMP is to activate protein kinase A (PKA), which is composed of four subunits: two catalytic subunits that phosphorylate specific cellular proteins, and two regulatory subunits that inhibit the catalytic subunits when they are bound to each other. cAMP binds to the regulatory subunits of PKA. The binding of cAMP separates the regulatory and catalytic subunits, which allows each catalytic subunit to be active.

***Cellular Response via PKA*** How does PKA activation lead to a cellular response? The catalytic subunit of PKA phosphorylates specific cellular proteins such as enzymes, structural proteins, and transcription factors. The phosphorylation of enzymes and structural proteins influences the structure and function of the cell. Likewise, the phosphorylation of transcription factors leads to the synthesis of new proteins that affect cell structure and function.

As a specific example of a cellular response, **Figure 9.13** shows how a skeletal muscle cell responds to elevated levels of epinephrine.

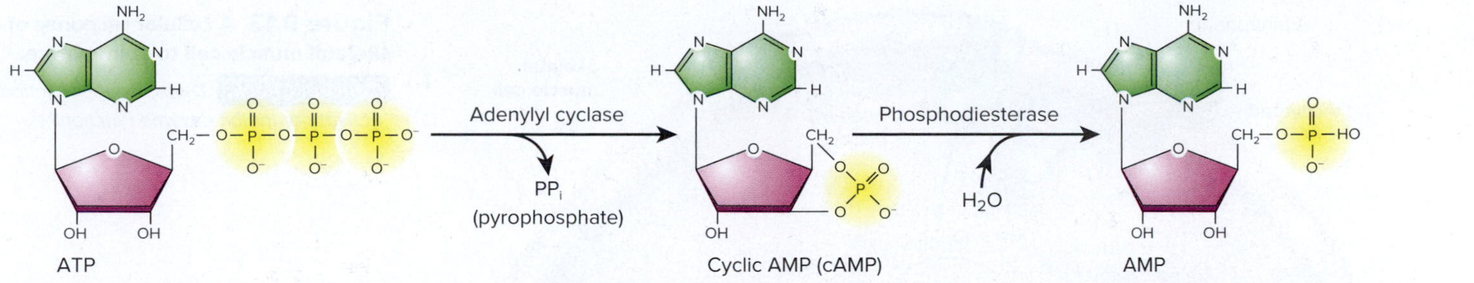

The synthesis and breakdown of cyclic AMP. Cyclic AMP (cAMP) is a second messenger formed from ATP by adenylyl cyclase, an enzyme in the plasma membrane. cAMP is inactivated by the action of an enzyme called phosphodiesterase, which converts cAMP to AMP.

**Figure 9.11** **The synthesis and breakdown of cyclic AMP.** Cyclic AMP (cAMP) is a second messenger formed from ATP by adenylyl cyclase, an enzyme in the plasma membrane. cAMP is inactivated by the action of an enzyme called phosphodiesterase, which converts cAMP to AMP.

When PKA becomes active, it phosphorylates two enzymes—phosphorylase kinase and glycogen synthase. Both of these enzymes are involved with the metabolism of glycogen, which is a polymer of glucose used to store energy.

- When phosphorylase kinase is phosphorylated, it becomes activated. The function of phosphorylase kinase is to phosphorylate another enzyme in the cell called glycogen phosphorylase, which then becomes activated. This enzyme causes glycogen breakdown by phosphorylating glucose units at the ends of a glycogen polymer, which releases individual glucose-phosphate molecules from glycogen:

$$\text{Glycogen}_n + P_i \xrightarrow{\begin{array}{c}\text{Glycogen}\\\text{phosphorylase}\end{array}} \text{Glycogen}_{n-1} + \text{Glucose-phosphate}$$

where $n$ is the number of glucose units in the glycogen polymer.

- When PKA phosphorylates glycogen synthase, the function of this enzyme is inhibited rather than activated (see Figure 9.13). The function of glycogen synthase is to make glycogen. Therefore, the effect of cAMP is to prevent glycogen synthesis.

Taken together, the effects of epinephrine in skeletal muscle cells are to stimulate glycogen breakdown and inhibit glycogen synthesis. This provides these cells with more glucose molecules, which they can use for the energy needed for muscle contraction. In this way, the individual is better prepared to fight or flee.

***Reversal of the Cellular Response***    As mentioned, signaling that involves second messengers is typically of short duration. When the signaling molecule is no longer produced and its concentration falls, a larger percentage of the receptors are not bound by their ligands. When a ligand dissociates from a GPCR, the GPCR becomes deactivated. Intracellularly, the α subunit hydrolyzes its GTP to GDP, and the α subunit and β/γ dimer reassociate to form an inactive G protein

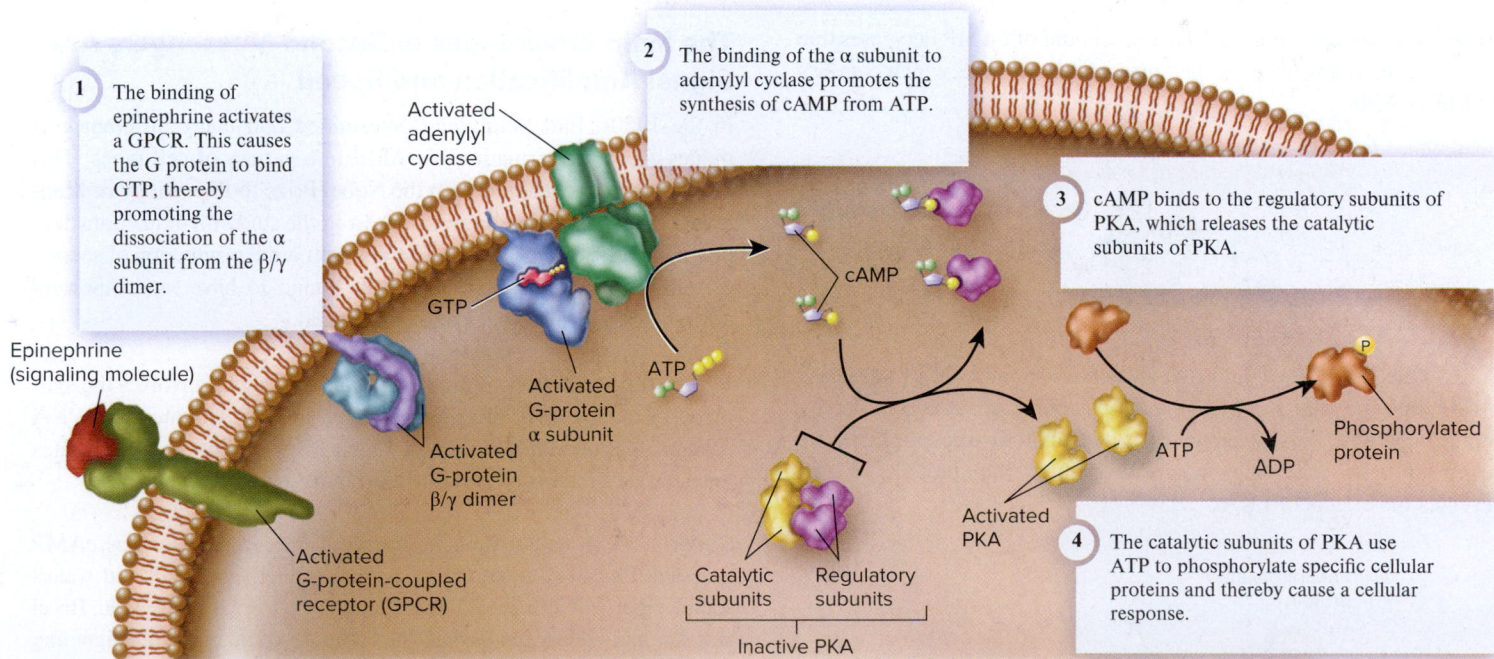

**Figure 9.12** **A signal transduction pathway involving cAMP.** The pathway leading to the formation of cAMP and subsequent activation of protein kinase A (PKA), which is mediated by a G-protein-coupled receptor (GPCR).

*Concept Check:* *In this figure, where does the signal transduction pathway begin and end, and what is the cellular response?*

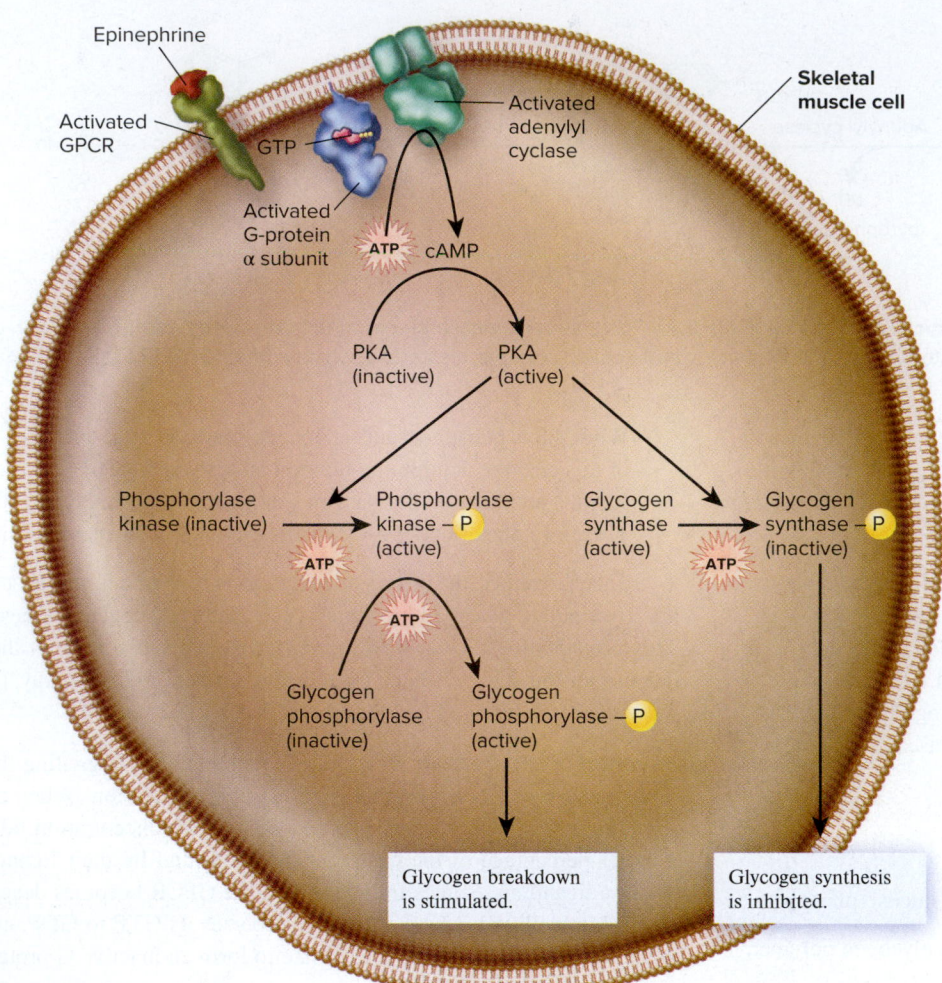

Epinephrine

Activated GPCR

GTP

Activated G-protein α subunit

ATP    cAMP

Activated adenylyl cyclase

Skeletal muscle cell

PKA (inactive)

PKA (active)

Phosphorylase kinase (inactive)    Phosphorylase kinase – P (active)    ATP

Glycogen synthase (active)    Glycogen synthase – P (inactive)    ATP

Glycogen phosphorylase (inactive)    Glycogen phosphorylase – P (active)    ATP

Glycogen breakdown is stimulated.

Glycogen synthesis is inhibited.

**Figure 9.13**  A cellular response of a skeletal muscle cell to epinephrine.

*Concept Check:*  *Does phosphorylation activate or inhibit enzyme function?*

---

(refer back to step 3, Figure 9.7). The amount of cAMP decreases due to the action of an enzyme called **phosphodiesterase**, which converts cAMP to AMP:

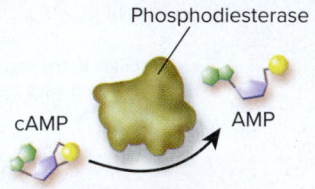

Phosphodiesterase

cAMP        AMP

As the cAMP level falls, the regulatory subunits of PKA release cAMP, and the regulatory and catalytic subunits reassociate, thereby inhibiting PKA. Finally, enzymes called **protein phosphatases** are responsible for removing phosphate groups from proteins, which reverses the effects of PKA:

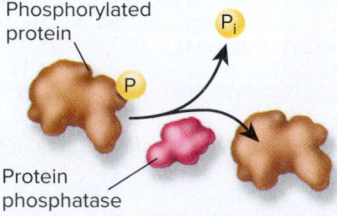

Phosphorylated protein

P

P_i

Protein phosphatase

## The Main Advantages of Second Messengers Are Signal Amplification and Speed

In the 1950s, Earl Sutherland determined that many different hormones cause the formation of cAMP in a variety of cell types. This observation, for which he won the Nobel Prize in Physiology or Medicine in 1971, stimulated great interest in the study of signal transduction pathways. Since Sutherland's discovery, the production of second messengers such as cAMP has been found to have two important advantages: signal amplification and speed.

***Signal Amplification***   Amplification of the signal involves the synthesis of many cAMP molecules, which, in turn, activate many PKA proteins (**Figure 9.14**). Likewise, each PKA protein phosphorylates many target proteins in the cell to promote a cellular response.

***Speed***   A second advantage of second messengers such as cAMP is speed. Because second messengers are relatively small and water-soluble, they can diffuse rapidly through the cytosol. For example, Brian Bacskai and colleagues studied the response of neurons to a signaling molecule called serotonin, which is a neurotransmitter that binds to a GPCR. In humans, low serotonin is believed to play a role in depression, anxiety, and other behavioral disorders. To monitor cAMP levels, neurons grown in a laboratory were injected with a fluorescent protein that

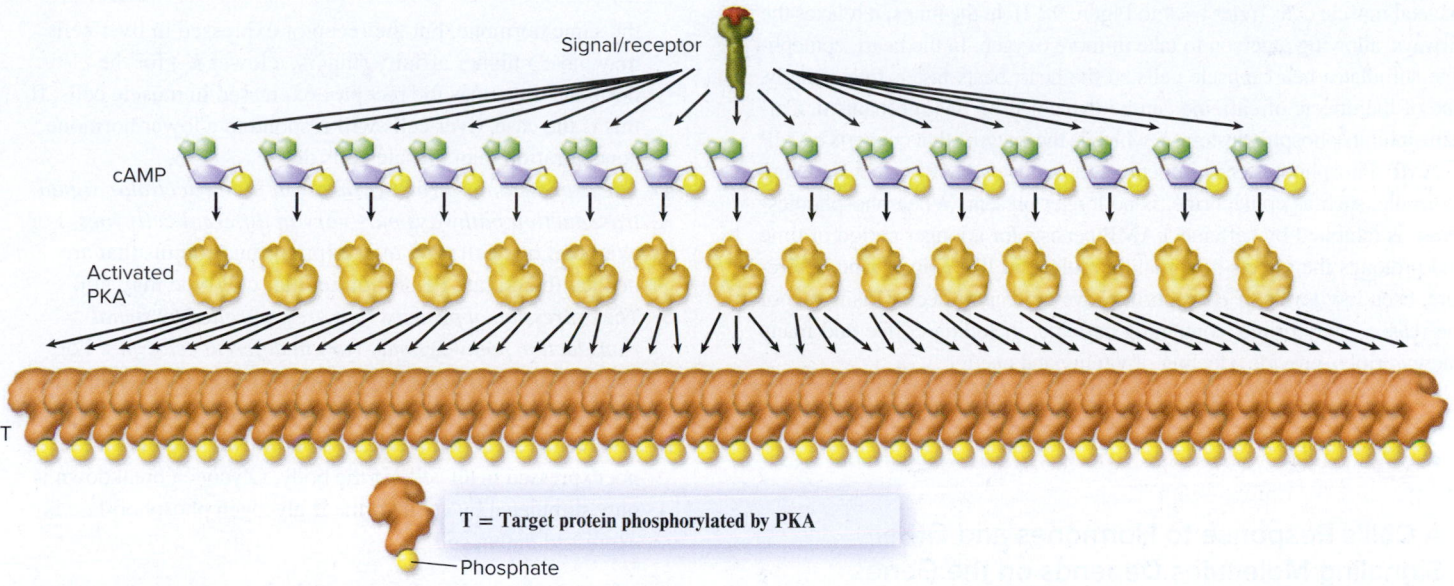

**Figure 9.14 Signal amplification.** An advantage of a signal transduction pathway is the amplification of a signal. In this case, a single signaling molecule leads to the phosphorylation of many, perhaps hundreds or thousands of, target proteins (designated T).

**Concept Check:** *In the case of signaling pathways involving hormones, why is signal amplification an advantage?*

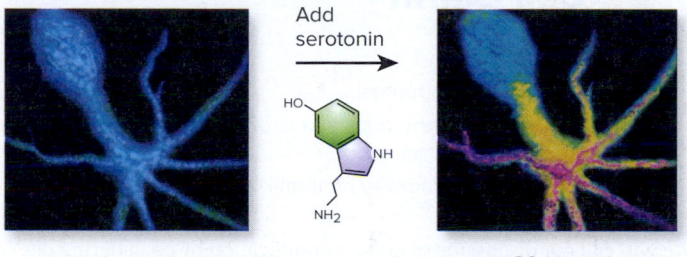

**Figure 9.15 The rapid speed of cAMP production.** The schematic drawing on the left shows a neuron prior to its exposure to serotonin, a signaling molecule; the drawing on the right shows the same cell 20 seconds after exposure. Blue indicates a low level of cAMP, yellow is an intermediate level, and purple is a high level.

changes its fluorescence when cAMP is made. As shown schematically in the drawing on the right in **Figure 9.15**, such cells made a substantial amount of cAMP within 20 seconds after the addition of serotonin.

# 9.4 Hormonal Signaling in Multicellular Organisms

**Learning Outcomes:**

1. Explain how the cellular response to a particular hormone can vary among different cell types.
2. Describe how a cell's response to a hormone depends on the genes it expresses.

Thus far, we have considered how signaling molecules bind to particular types of receptors, thereby activating a signal transduction pathway that leads to a cellular response. In this section, we will consider how hormones in multicellular organisms exert a variety of responses. As you will learn, the type of cellular response that is caused by a given hormone depends on the type of cell. Each cell type responds to a particular hormone in its own unique way. The variation in a cellular response is determined by the types of proteins, such as receptors and signal transduction proteins, that each cell type makes, which is determined by the genes expressed in that type of cell.

## The Cellular Response to a Given Hormone Varies Among Different Cell Types

As we have seen, signaling molecules usually exert their effects on cells via signal transduction pathways that control the functions and/ or synthesis of specific proteins. In multicellular organisms, one of the amazing effects of hormones is their ability to coordinate cellular activities. One example is epinephrine, which is secreted from endocrine cells. As mentioned earlier, epinephrine is also called the fight-or-flight hormone because it quickly prepares the body for strenuous physical activity in response to a perceived danger. Epinephrine is also secreted into the bloodstream when a person is exercising.

Epinephrine has different effects throughout the body (**Table 9.1**). We have already discussed how it promotes the breakdown of glycogen in

| Table 9.1 | Effects of Epinephrine in Humans |
|---|---|
| **Organ/Tissue** | **Effect** |
| Eye | Dilates pupils |
| Salivary glands | Inhibits the production of saliva |
| Skeletal muscle | Stimulates cells to break down glycogen and release glucose |
| Skin | Constricts blood vessels; stimulates sweating |
| Lungs | Relaxes airways so more oxygen is taken in |
| Heart | Increases the rate of beating |

skeletal muscle cells (refer back to Figure 9.13). In the lungs, it relaxes the airways, allowing a person to take in more oxygen. In the heart, epinephrine stimulates heart muscle cells so the heart beats faster. Interestingly, one of the effects of caffeine can be explained by this mechanism. Caffeine inhibits phosphodiesterase, which is the enzyme that converts cAMP to AMP. Phosphodiesterase functions to remove cAMP once a signaling molecule, such as epinephrine, is no longer present. When phosphodiesterase is inhibited by caffeine, cAMP persists for a longer period of time and prolongs the effects of signaling molecules like epinephrine. Therefore, even low levels of epinephrine have a greater effect. This is one of the reasons why drinks containing caffeine, including coffee and many energy drinks, provide a feeling of vitality and energy.

## Core Concept: Information

### A Cell's Response to Hormones and Other Signaling Molecules Depends on the Genes It Expresses

As Table 9.1 shows, the hormone epinephrine produces diverse responses throughout the body. How do we explain the observation that various cell types respond so differently to the same hormone? As a multicellular organism develops from a fertilized egg, the cells of the body become differentiated into particular types, such as heart and lung cells. The mechanisms that underlie this differentiation process are described in Chapter 20. Although different cell types, such as heart and lung cells, contain the same set of genes—the same genome—those genes are not expressed in the same pattern in all cells. Certain genes that are turned off in heart cells are turned on in lung cells, whereas some genes that are turned on in heart cells are turned off in lung cells. This phenomenon, which is called **differential gene regulation**, causes each cell type to have its own distinct proteome. The set of proteins made in any given cell type is critical to a cell's ability to respond to signaling molecules. The following are examples of how differential gene regulation affects the cellular response:

1. *A cell may or may not express a receptor for a particular signaling molecule*. For example, not all cells of the human body express a receptor for epinephrine. Cells without such a receptor are not affected when epinephrine is released into the bloodstream.

2. *Different cell types have different cell surface receptors that recognize the same signaling molecule*. In humans, for example, a signaling molecule called acetylcholine has two different types of receptors. One acetylcholine receptor is a ligand-gated ion channel that is expressed in skeletal muscle cells. Another acetylcholine receptor is a G-protein-coupled receptor (GPCR) that is expressed in heart muscle cells. Because of this, acetylcholine activates different signal transduction pathways in skeletal and heart muscle cells. Therefore, these cells respond differently to acetylcholine.

3. *Two (or more) receptors may work the same way in different cell types but have different affinities for the same signaling molecule*. For example, two different GPCRs may recognize the same hormone, but the receptor expressed in liver cells may have a higher affinity (that is, a lower $K_d$) for the hormone than does the receptor expressed in muscle cells. If this is the case, liver cells will respond to a lower hormone concentration than muscle cells do.

4. *The expression of proteins involved in intracellular signal transduction pathways may vary in different cell types*. For example, one cell type may express the proteins that are needed to activate PKA, but another cell type may not.

5. *The expression of proteins that are controlled by signal transduction pathways may vary in different cell types*. For example, the presence of epinephrine in skeletal muscle cells leads to the activation of glycogen phosphorylase, an enzyme involved in glycogen breakdown. However, this enzyme is not expressed in all cells of the body. Glycogen breakdown is only stimulated by epinephrine if glycogen phosphorylase is expressed in that cell.

## 9.5 Apoptosis: Programmed Cell Death

**Learning Outcomes:**

1. Define and describe apoptosis.
2. **CoreSKILL »** Analyze the results of experiments indicating that certain hormones control apoptosis.
3. Outline the extrinsic pathway of apoptosis.

We will end our discussion of cell communication by considering one of the most dramatic responses that eukaryotic cells exhibit—**apoptosis**, or programmed cell death. During this process, a cell orchestrates its own destruction! The cell first shrinks and becomes rounder due to the internal destruction of its nucleus and cytoskeleton (**Figure 9.16**). The plasma membrane then forms irregular extensions that eventually become blebs—small cell fragments that break away from the cell as it destroys itself (also look back at the chapter opening photo).

Cell biologists have discovered that apoptosis plays many important roles.

- During embryonic development in animals, it is needed to sculpt the tissues and organs. For example, the fingers on a human hand are initially webbed, but become separated during embryonic development when the cells between the fingers undergo apoptosis (see Figure 20.4).

- Apoptosis is also necessary in adult organisms to maintain the proper numbers of cells in tissues and organs.

- Programmed cell death also eliminates cells that have become worn out or infected by viruses, or have the potential to cause cancer.

During the past few decades, clinical research has revealed that many human diseases are associated with irregularities in apoptosis. **Table 9.2** describes a few examples. In this section, we will examine the pioneering work that led to the discovery of apoptosis and explore its molecular mechanism.

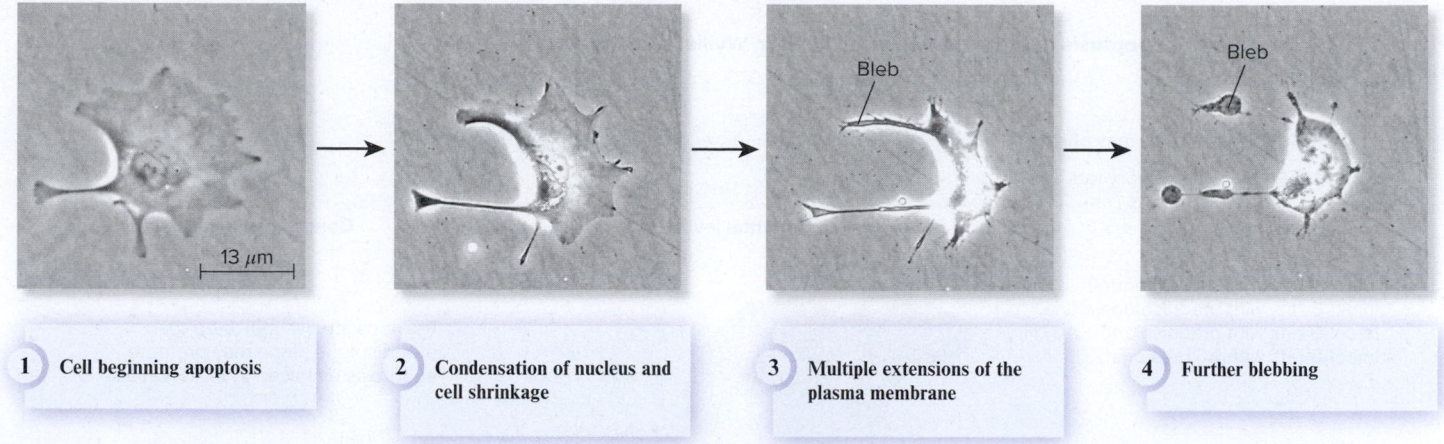

| ① | Cell beginning apoptosis | ② | Condensation of nucleus and cell shrinkage | ③ | Multiple extensions of the plasma membrane | ④ | Further blebbing |

**Figure 9.16**  **Stages of apoptosis.** (1–4): ©Prof. Guy Whitley/Reproductive and Cardiovascular Disease Research Group at St. George's University of London

| Table 9.2 | Relationship Between Certain Diseases and Abnormal Levels of Apoptosis |
|---|---|
| **Disease** | **Description/Examples** |
| *Diminished levels of apoptosis* | |
| Cancer | Cancer cells proliferate in an uncontrolled manner. In some forms of cancer, a decrease in the normal rate of apoptosis contributes to the faster proliferation rate. Examples include particular types of prostate and ovarian cancers. |
| *Elevated levels of apoptosis* | |
| Viral diseases | Certain viral diseases are associated with elevated levels of apoptosis. For example, infection by human immunodeficiency virus (HIV) results in an increased rate of apoptosis of helper T cells. |
| Neurodegenerative diseases | Some neurodegenerative diseases occur because specific neurons undergo an unusually high rate of apoptosis. An example is Parkinson's disease, which arises from a loss of dopaminergic neurons. |

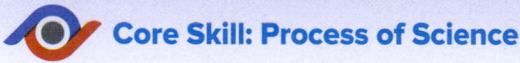 **Core Skill: Process of Science**

## Feature Investigation | Kerr, Wyllie, and Currie Found That Hormones May Control Apoptosis

How was apoptosis discovered? One line of evidence involved the microscopic examination of tissues in mammals. In the 1960s, British pathologist John Kerr microscopically examined liver tissue that was deprived of oxygen. He observed that, within hours of oxygen deprivation, some cells underwent a process that involved cell shrinkage. Around this time, similar results had been noted by other researchers, such as Scottish pathologists Andrew Wyllie and Alastair Currie, who had studied cell death in the adrenal glands. In 1973, Kerr, Wyllie, and Currie joined forces to study this process further.

Prior to that collaboration, other researchers had already established that certain hormones affect the growth of the adrenal glands, which sit atop the kidneys. Adrenocorticotropic hormone (ACTH) was known to increase the number of cells in the adrenal cortex, which is the outer layer of the adrenal glands. By contrast, the drug prednisolone was shown to suppress the synthesis of ACTH and cause a decrease in the number of cells in the cortex. In the experiment described in **Figure 9.17**, Kerr, Wyllie, and Currie wanted to understand how ACTH and prednisolone exert their effects. They subjected rats to four types of treatments. The control rats were injected with saline (salt water). Other rats were injected with prednisolone alone, prednisolone plus ACTH, or ACTH alone. After 2 days, samples of adrenal cortex were obtained from the rats and observed by light microscopy. Even in control samples,

the researchers occasionally observed cell death via apoptosis (see the micrograph in step 4). However, in prednisolone-treated rats, the cells in the adrenal cortex were found to undergo a dramatically higher rate of apoptosis. Multiple cells undergoing apoptosis were found in 9 out of every 10 samples observed under the light microscope. Such a high level of apoptosis was not observed in control samples or in samples obtained from rats treated with both prednisolone and ACTH or with ACTH alone. Therefore, ACTH appears to prevent apoptosis.

The results of Kerr, Wyllie, and Currie are important for two reasons. First, their results indicated that tissues decrease their cell number via a mechanism that involves cell shrinkage and eventually blebbing. Second, they showed that cell death followed a program that, in this case, was induced by the presence of prednisolone (which decreases ACTH). They coined the term apoptosis to describe this process.

As you may know, prednisone is an anti-inflammatory and immunosuppressive drug that is used to treat a wide variety of disorders, including asthma and rheumatoid arthritis. When taken into the body, it is converted to prednisolone by the liver. In recent years, prednisone has been used in conjunction with other therapies to treat certain forms of cancer, such as leukemia, which is cancer of white blood cells. Prednisone is thought to exert its effect by promoting apoptosis in the cancer cells.

**Figure 9.17  Discovery of apoptosis in the adrenal cortex by Kerr, Wyllie, and Currie.** (4): ©Dr. Thomas Caceci, Virginia-Maryland Regional College of Veterinary Medicine

**HYPOTHESIS**  Hormones may affect cell number in the adrenal gland by controlling the rate of apoptosis.

**KEY MATERIALS**  Laboratory rats, prednisolone, and ACTH.

| | Experimental level | Conceptual level |
|---|---|---|

**1**  Inject 5 rats with saline (control).
Inject 5 rats with prednisolone alone.
Inject 5 rats with prednisolone + ACTH.
Inject 5 rats with ACTH alone.

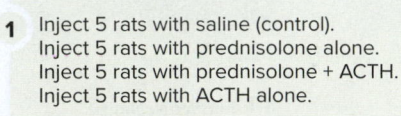

Previous studies indicated that prednisolone alone may promote apoptosis by lowering ACTH levels.

**2**  After 2 days, obtain samples of adrenal tissue from all 20 rats.

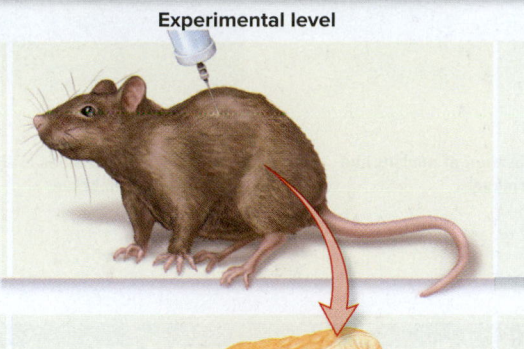
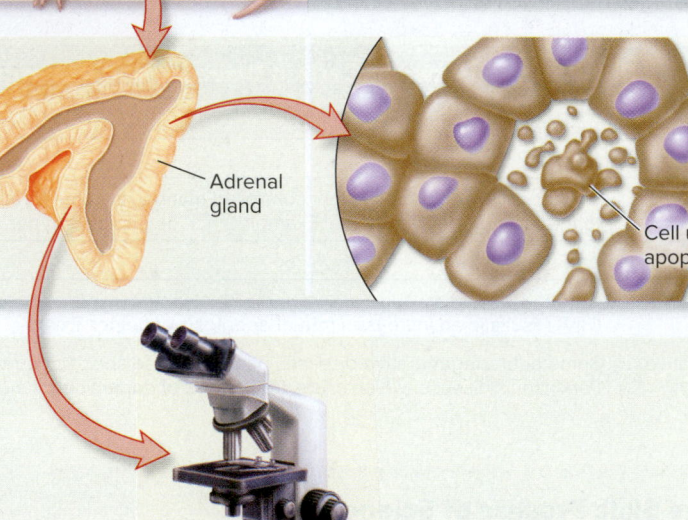

Adrenal gland

Cell undergoing apoptosis

**3**  Observe the samples via light microscopy, described in Chapter 4.

**4**  THE DATA

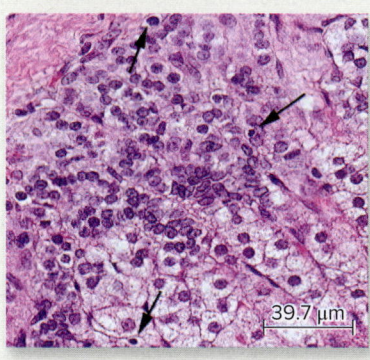

39.7 μm

Micrograph of adrenal tissue showing occasional cells undergoing apoptosis (see arrows)

| Treatment | Number of animals | Glands with enhanced apoptosis*/ Total number of animals |
|---|---|---|
| Saline | 5 | 0/10 |
| Prednisolone | 5 | 9/10 |
| Prednisolone + ACTH | 5 | 0/10 |
| ACTH | 5 | 0/10 |

*Samples from two adrenal glands were removed from each animal. Enhanced apoptosis means that cells undergoing apoptosis were observed in every sample under the light microscope.

**5**  CONCLUSION  Prednisolone alone, which lowers ACTH levels, causes some cells to undergo apoptosis. During this process, the cells shrink and form blebs as they kill themselves. Apoptosis is controlled by hormones.

**6**  SOURCE  Wyllie, A. H., Kerr, J. F. R., Macaskill, I. A. M., and Currie, A. R. 1973. Adrenocortical cell deletion: the role of ACTH. *Journal of Pathology* 111: 85–94.

*Experimental Questions*

1. **CoreSKILL »** In the experiment of Figure 9.17, explain the effects on apoptosis in the control rats (injected with saline) versus those injected with prednisolone alone, predinisolone + ACTH, or ACTH alone.

2. Prednisolone inhibits the production of ACTH in rats. Do you think it inhibited the ability of rats to make their own ACTH when they were injected with both prednisolone and ACTH? Explain.

3. **CoreSKILL »** Of the four groups of rats—control, prednisolone alone, prednisolone + ACTH, and ACTH alone—which would you expect to have the lowest level of apoptosis? Explain.

## Signal Transduction Pathways Lead to Apoptosis

Apoptosis involves the activation of cell-signaling pathways. One pathway, called the **extrinsic pathway**, begins with the activation of **death receptors** on the cell surface. When death receptors bind to extracellular signaling molecules, a pathway is stimulated that leads to apoptosis. **Figure 9.18** shows a simplified pathway for this process. In this example, the signaling molecule is a protein composed of three identical subunits—a trimeric protein. Such trimeric signaling molecules are typically produced by cells of the immune system that recognize abnormal cells and target them for destruction. For example, when a cell is infected with a virus, cells of the immune system may

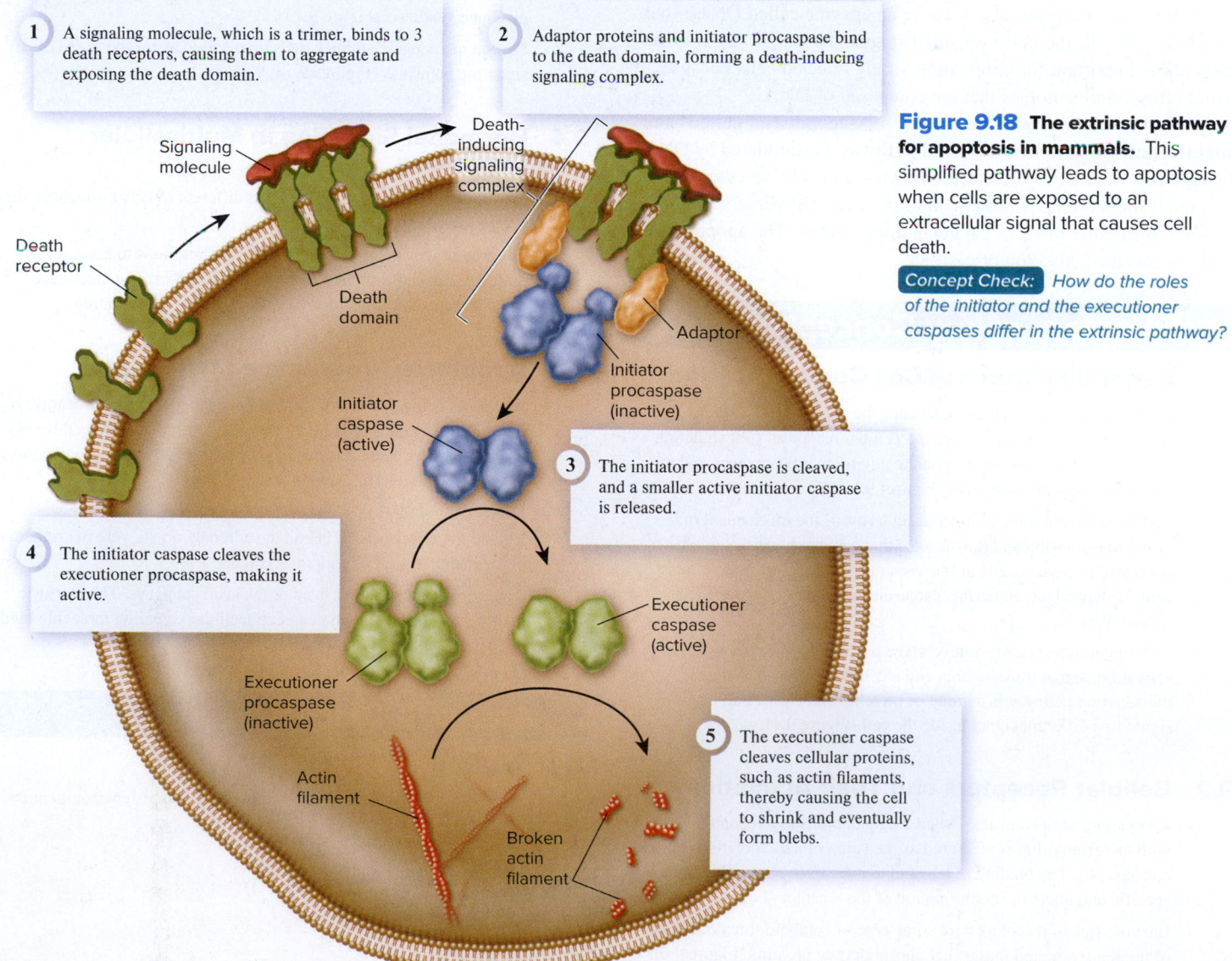

1 A signaling molecule, which is a trimer, binds to 3 death receptors, causing them to aggregate and exposing the death domain.

2 Adaptor proteins and initiator procaspase bind to the death domain, forming a death-inducing signaling complex.

3 The initiator procaspase is cleaved, and a smaller active initiator caspase is released.

4 The initiator caspase cleaves the executioner procaspase, making it active.

5 The executioner caspase cleaves cellular proteins, such as actin filaments, thereby causing the cell to shrink and eventually form blebs.

**Figure 9.18 The extrinsic pathway for apoptosis in mammals.** This simplified pathway leads to apoptosis when cells are exposed to an extracellular signal that causes cell death.

**Concept Check:** *How do the roles of the initiator and the executioner caspases differ in the extrinsic pathway?*

Signaling molecule

Death-inducing signaling complex

Death receptor

Death domain

Adaptor

Initiator procaspase (inactive)

Initiator caspase (active)

Executioner procaspase (inactive)

Executioner caspase (active)

Actin filament

Broken actin filament

target the infected cell for apoptosis. The signaling molecule binds to three death receptors, which causes them to aggregate into a trimer. This results in a conformational change that exposes a domain on the death receptors called the death domain. Once the death domain is exposed, it binds to adaptors, which then bind to an initiator procaspase. The complex between the death receptors, adaptors, and initiator procaspase is called the death-inducing signaling complex (DISC).

Once the initiator procaspase, which is inactive, is part of the death-inducing signaling complex, it is converted by proteolytic cleavage to an initiator caspase, which is active. An active **caspase** functions as a protease—an enzyme that digests other proteins. After it is activated, the initiator caspase is then released from the DISC. This caspase is called an initiator caspase because it initiates the activation of many other caspases in the cell. These other caspases are called executioner, or effector, caspases because they are directly responsible for digesting intracellular proteins and causing the cell to die. The executioner caspases digest a variety of intracellular proteins, including the proteins that constitute the cytoskeleton and nuclear lamina as well as proteins involved with DNA replication and repair. In this way, the executioner caspases cause the cellular changes shown in Figure 9.16. The caspases also activate an enzyme called DNase that chops the DNA in the cell into small fragments. This event may be particularly important for eliminating virally infected cells because it also destroys viral genomes that are composed of DNA.

Alternatively, another pathway of apoptosis, called the **intrinsic pathway** or **mitochondrial pathway**, is stimulated by DNA damage that could cause cancer. Mitochondria release cytochrome *c* (a small mitochondrial protein) into the cytosol, where it forms a complex with other proteins called an **apoptosome**. The apoptosome then initiates the activation of caspases.

## Summary of Key Concepts

### 9.1 General Features of Cell Communication

- A signal is an agent that can influence the properties of cells. A signal binds to a receptor to elicit a cellular response. Cell signaling enables cells to sense and respond to environmental changes and to communicate with each other (Figures 9.1, 9.2).

- Cell-to-cell communication varies in terms of the mechanism of signal transmission and the distance that a signal travels. Signals are relayed between cells in five common ways: direct intercellular, contact-dependent, autocrine, paracrine, and endocrine signaling (Figure 9.3).

- Cell signaling is usually a three-stage process involving receptor activation, signal transduction, and a cellular response. A signal transduction pathway is a group of proteins that convert an initial signal to a different signal inside the cell (Figure 9.4).

### 9.2 Cellular Receptors and Their Activation

- A signaling molecule, also called a ligand, binds to a receptor with an affinity that is measured as the value of a dissociation constant, $K_d$. The binding of a ligand to a receptor is usually very specific and alters the conformation of the receptor (Figure 9.5).

- Enzyme-linked receptors have some type of catalytic function. Many of them are protein kinases that phosphorylate proteins (Figure 9.6).

- G-protein-coupled receptors (GPCRs) interact with G proteins to initiate a cellular response (Figure 9.7).

- Ligand-gated ion channels are receptors that allow the flow of ions across the plasma membrane (Figure 9.8).

- Although most receptors involved in cell signaling are found on the cell surface, some receptors, such as the estrogen receptor, are intracellular receptors (Figure 9.9).

### 9.3 Signal Transduction and the Cellular Response

- Signaling pathways influence whether or not a cell divides. An example is the pathway that is stimulated by epidermal growth factor, which binds to a receptor tyrosine kinase (Figure 9.10).

- Second messengers, such as cAMP, play a key role in signal transduction pathways, such as those that occur via GPCRs. These pathways are reversible once the signal is degraded (Figures 9.11, 9.12).

- An example of a pathway that uses cAMP is found in skeletal muscle cells responding to elevated levels of epinephrine, the fight-or-flight hormone. Epinephrine enhances the function of enzymes that increase glycogen breakdown and inhibits enzymes that cause glycogen synthesis (Figure 9.13).

- Second messengers amplify the signal and increase the speed of signaling pathways (Figures 9.14, 9.15).

### 9.4 Hormonal Signaling in Multicellular Organisms

- Hormones such as epinephrine exert different effects throughout the body (Table 9.1).

- The way in which any particular cell type responds to a signaling molecule depends on the set of proteins it makes. The amounts of these proteins are controlled by differential gene regulation.

### 9.5 Apoptosis: Programmed Cell Death

- Apoptosis is the process of programmed cell death in which the nucleus and cytoskeleton break down and eventually the cell breaks apart into blebs. Irregularities in apoptosis are associated with some diseases (Figure 9.16, Table 9.2).

- Microscopy studies of Kerr, Wyllie, and Currie, in which they studied the effects of ACTH on the adrenal cortex, were instrumental in the identification of apoptosis (Figure 9.17).

- Apoptosis occurs via extrinsic or intrinsic pathways. The extrinsic pathway is stimulated when an extracellular signaling molecule binds to death receptors (Figure 9.18).

## Assess & Discuss

### Test Yourself

1. An agent that allows a cell to respond to changes in its environment is termed
   a. a cell surface receptor.
   b. an intracellular receptor.
   c. a structural protein.
   d. a signal.
   e. apoptosis.

2. When a cell secretes a signaling molecule that binds to receptors on neighboring cells as well as the cell itself, this is called _____ signaling.
   a. direct intercellular
   b. contact-dependent
   c. autocrine
   d. paracrine
   e. endocrine

3. Which of the following does *not* describe a typical cellular response to signaling molecules?
   a. activation of enzymes within the cell
   b. change in the function of structural proteins, which determine cell shape
   c. alteration of levels of certain proteins in the cell by changing the level of gene expression
   d. change in a gene sequence that encodes a particular protein
   e. all of the above are examples of cellular responses.

4. A receptor has a $K_d$ for its ligand of 50 nM. This receptor
   a. has a higher affinity for its ligand than does a receptor with a $K_d$ of 100 nM.
   b. has a higher affinity for its ligand than does a receptor with a $K_d$ of 10 nM.
   c. is mostly bound by its ligand when the ligand concentration is 100 nM.
   d. must be an intracellular receptor.
   e. both a and c are true of this ligand.

5. _____ binds to receptors inside cells.
   a. Estrogen
   b. Epinephrine
   c. Epidermal growth factor
   d. all of the above bind to such receptors.
   e. none of the above binds to such receptors.

6. The relay protein Ras is part of the EGF pathway that promotes cell division (see Figure 9.10). The active form of Ras has GTP bound to it, whereas the inactive form has GDP. GTP is hydrolyzed to GDP and $P_i$ to switch Ras from the active to the inactive form. Researchers have discovered that certain forms of cancer involve mutations in the gene that encodes the Ras protein. Which of the following types of mutations would you expect to promote cell division and thereby lead to cancer?
   a. a mutation that prevents the synthesis of Ras
   b. a mutation that causes Ras to bind GDP more tightly
   c. a mutation that prevents the GTP bound to Ras from being hydrolyzed
   d. a mutation that prevents Ras from binding to Raf
   e. both b and c

7. The benefit of second messengers in signal transduction pathways is
   a. an increase in the speed of a cellular response.
   b. duplication of the ligands in the system.
   c. amplification of the signal.
   d. all of the above.
   e. a and c only.

8. All cells of a multicellular organism may not respond in the same way to a particular ligand (signaling molecule) that binds to a cell surface receptor. The difference in response may be due to
   a. the type of receptor for the ligand that the cell expresses.
   b. the affinity of the ligand for the receptor in a given cell type.
   c. the type of signal transduction pathways that the cell expresses.
   d. the type of target proteins that the cell expresses.
   e. all of the above.

9. Apoptosis is the process of
   a. cell migration.
   b. cell signaling.
   c. signal transduction.
   d. signal amplification.
   e. programmed cell death.

10. Which statement best describes the extrinsic pathway for apoptosis?
    a. Caspases recognize an environmental signal and expose their death domain.
    b. Death receptors recognize an environmental signal, which then leads to the activation of caspases.
    c. Initiator caspases digest the nuclear lamina and cytoskeleton.
    d. Executioner caspases are part of the death-inducing signaling complex (DISC).
    e. All of the above are true of the extrinsic pathway.

## Conceptual Questions

1. What are the two general reasons that cell communication is essential?

2. What are the three stages of cell signaling? What stage does *not* occur when the estrogen receptor is activated?

3. **Core Concept: Systems** Discuss how cell signaling helps organisms to interact with their environment.

## Collaborative Questions

1. Discuss and compare several different types of cell-to-cell communication. What are some advantages and disadvantages of each type?

2. How does differential gene regulation enable various cell types to respond differently to the same signaling molecule? Why is this useful to multicellular organisms?

# Multicellularity

# 10

**The General Sherman in Sequoia National Park, a striking example of the size that multicellular organisms can reach.** This tree is thought to be the largest organism (by mass) in the world. ©Altrendo Panoramic/Getty Images

**W**hat is the largest living organism on Earth? The size of an organism can be defined by its volume, mass, height, length, or the area it occupies. A giant fungus (*Armillaria ostoyae*), growing in the soil in Malheur National Forest in Oregon, spans 8.9 km², or 2,200 acres, which makes it the largest known organism by area. Most of this organism lies below ground, so it is not visible from the surface. In the Mediterranean Sea, marine biologists discovered a giant aquatic plant (*Posidonia oceanica*) whose length is 8 km, or 4.3 miles, making it the world's longest known organism. With regard to mass, the largest organism is probably a tree named the General Sherman, which is 83.8 m tall (275 feet), nearly the length of a football field (see the chapter opening photo). This giant sequoia tree (*Sequoiadendron giganteum*) is estimated to weigh nearly 2 million kg (over 2,000 tons)—equivalent to a herd of 400 elephants!

An organism composed of more than one cell is said to be **multicellular**. The preceding examples illustrate the amazing sizes that certain multicellular organisms have attained.

As we will discuss in Chapter 26, multicellular organisms came into being approximately 1 billion years ago. Some species of protists are multicellular, as are most species of fungi. In this chapter, we will focus on plants and animals, which are always multicellular species.

The main benefit of multicellularity arises from the division of labor between different types of cells in an organism. For example, the intestinal cells of animals and the root cells of plants have become specialized for nutrient uptake. Other types of cells in a multicellular organism perform different roles, such as reproduction. In animals, most of the cells of the body—somatic cells—are devoted to the growth, development, and survival of the organism, whereas specialized cells—gametes—function in sexual reproduction.

Multicellular species usually have much larger genomes than unicellular species. The increase in genome size is associated with an increase in proteome size—multicellular organisms produce a larger array of proteins than do unicellular species. The additional proteins play a role in three general phenomena.

- First, in a multicellular organism, cell communication is vital for the proper organization and functioning of cells. Many more proteins involved in cell communication are made in multicellular species.
- Second, both the arrangement of cells within the organism and the attachment of cells to each other require a greater variety of proteins in multicellular species than in unicellular species.
- Finally, additional proteins play a role in cell specialization because proteins that are needed for the structure and function of one cell type may not be needed in a different cell type, and vice versa. Likewise, additional proteins are needed to regulate the expression of genes so that all proteins are expressed in the proper cell types.

In this chapter, we will consider the cellular characteristics that are specific to multicellular organisms. We begin by exploring the material that is produced by animal and plant cells to form an extracellular matrix or cell wall, respectively. This material plays many important roles in the structure, organization, and functioning of cells within multicellular organisms. We will then turn our attention to cell junctions, specialized structures that enable cells to make physical contact with one another. Cells within multicellular organisms form junctions that help the cells function in a cohesive and well-organized way. Finally, we will examine the organization and function of tissues, groups of cells that have a similar structure and function. In this chapter, we will survey the general features of tissues from a cellular perspective. Units VI and VII will explore the characteristics of particular plant and animal tissues in greater detail.

# 10.1 Extracellular Matrix and Cell Walls

**Learning Outcomes:**

1. Explain the functional roles of the extracellular matrix in animals.
2. Outline the major structural components of the ECM of animals.
3. Describe the structure and function of plant cell walls.

Organisms are not composed solely of cells. A large portion of an animal or plant consists of a network of material that is secreted from cells and forms a complex meshwork outside of cells. In animals, this is called the **extracellular matrix (ECM)**, whereas plant cells are surrounded by a **cell wall**. The ECM and cell walls are a major component of certain parts of animals and plants, respectively. For example, bones and cartilage in animals are composed largely of ECM, and the woody portions of plants are composed mostly of cell walls. Although the cells within wood eventually die, the cell walls they have produced provide a rigid structure that supports the plant for years or even centuries.

In this section, we begin by examining the structure and functions of the ECM in animals, focusing on the major ECM components: proteins and polysaccharides. We will then explore the cell wall that surrounds plant cells.

## The Extracellular Matrix in Animals Supports and Organizes Cells and Plays a Role in Cell Signaling

Unlike the cells of bacteria, archaea, fungi, and plants, the cells of animals are not surrounded by a rigid cell wall that provides structure and support. However, animal cells secrete materials that form an ECM that provides support and helps to organize cells. Certain animal cells are completely embedded within an extensive ECM, whereas other cells may adhere to the ECM on only one side. **Figure 10.1** illustrates the general features of the ECM and its relationship to cells. The major macromolecules of the ECM are proteins and polysaccharides. The most abundant proteins are those that form large fibers. The polysaccharides attract water and give the ECM a gel-like character.

As we will see, the ECM found in animals performs many important functions, including strength, structural support, organization, and cell signaling.

- *Strength*. The ECM is the "tough stuff" of animals' bodies. The strength of the ECM in the skin of mammals prevents tearing.

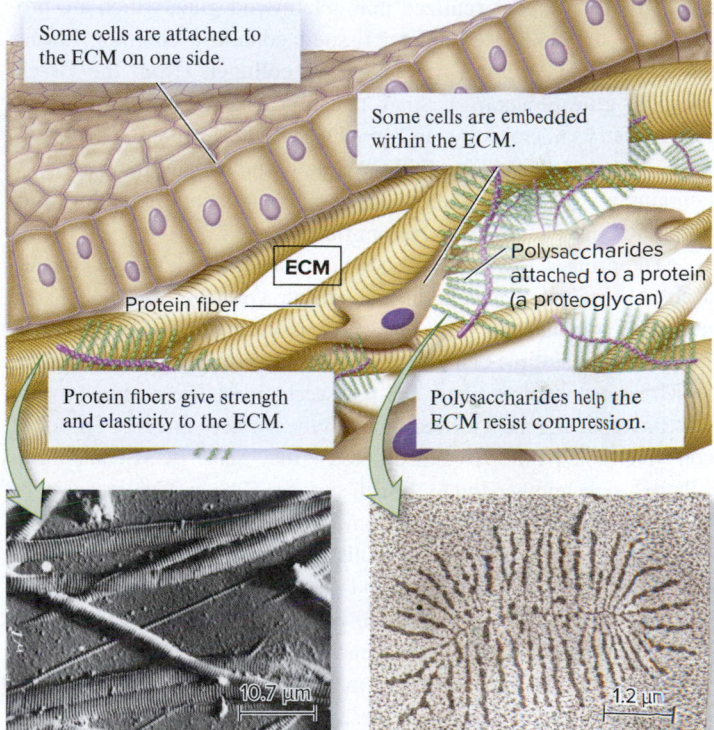

**Figure 10.1** **The extracellular matrix (ECM) of animal cells.** The micrograph (SEM) at the bottom left shows collagen fibers, a type of protein fiber found in the ECM. The micrograph (TEM) at the bottom right shows a proteoglycan, which consists of polysaccharides attached to a protein. (left): ©Biophoto Associates/Science Source; (right): Courtesy of Dr. Joseph Buckwalter/University of Iowa

*Concept Check:* What are the four functions of the ECM in animals?

The ECM found in cartilage resists compression and provides protection to the joints. Similarly, the ECM protects the soft parts of the body, such as the internal organs.

- *Structural support*. The bones of many animals are composed primarily of ECM. Skeletons not only provide structural support but also facilitate movement via the functioning of attached muscles.

- *Organization*. The attachment of cells to the ECM plays a key role in the proper arrangement of cells throughout the body. In addition, the ECM binds many body parts together, such as tendons to bones.

• *Cell signaling*. A less obvious role of the ECM is cell signaling. One way that cells in multicellular organisms sense their environment is via changes in the ECM.

Let's now consider the synthesis and structure of ECM components found in animals.

## Adhesive and Structural Proteins Are Major Components of the ECM of Animals

In the 1850s, German biologist Rudolf Virchow suggested that all extracellular materials are made and secreted by cells. Around the same time, biologists realized that gelatin and glue, which are produced by the boiling of animal tissues, contain a common fibrous substance. This substance was named **collagen** (from the Greek, meaning glue-producing). Since that time, experimental techniques in chemistry, microscopy, and biophysics have enabled scientists to probe the structure of the ECM. We now understand that the ECM contains a mixture of several different components, including proteins such as collagen, which form fibers.

The proteins found in the ECM are grouped into adhesive proteins, such as fibronectin and laminin, and structural proteins, such as collagen and elastin (**Table 10.1**). How do adhesive proteins work? Fibronectin and laminin have multiple binding sites that bind to other components in the ECM, such as protein fibers and polysaccharides. These same proteins also have binding sites for receptors on the surfaces of cells. Therefore, adhesive proteins are so named because they make ECM components adhere to one another and to the cell surface. They provide organization to the ECM and facilitate the attachment of cells to the ECM.

Structural proteins, such as collagen and elastin, form large fibers that give the ECM its strength and elasticity. A key function of collagen is to impart tensile strength, which is a measure of how much stretching force a material can bear without tearing apart. Collagen provides high tensile strength to many parts of an animal's body. It is the main protein found in bones, cartilage, tendons, skin, and the lining of blood vessels and internal organs. In the bodies of mammals, more than 25% of the total protein mass consists of collagen, much more than any other protein. Approximately 75% of the protein in mammalian skin is composed of collagen. Leather is largely a pickled and tanned form of collagen.

As described in Chapter 4 (see Figure 4.32), proteins, such as collagen, that are secreted from eukaryotic cells are first directed from the cytosol to the endoplasmic reticulum (ER), then to the Golgi apparatus, and subsequently are secreted from the cell via vesicles that fuse with the plasma membrane. **Figure 10.2** depicts the synthesis and assembly of collagen. Individual procollagen polypeptides (called α chains) are synthesized into the lumen of the ER. Three procollagen polypeptides then associate with each other to form a procollagen triple helix. The amino acid sequences at both ends of the polypeptides, termed extension sequences, promote the formation of procollagen and prevent the formation of a larger fiber. After procollagen is secreted from the cell, extracellular enzymes remove the extension sequences. Once this occurs, the protein, now called collagen, can form larger structures. Collagen proteins assemble in a staggered way to form relatively thin collagen fibrils, which then align and produce large collagen fibers. The many layers of these proteins give collagen fibers their great tensile strength.

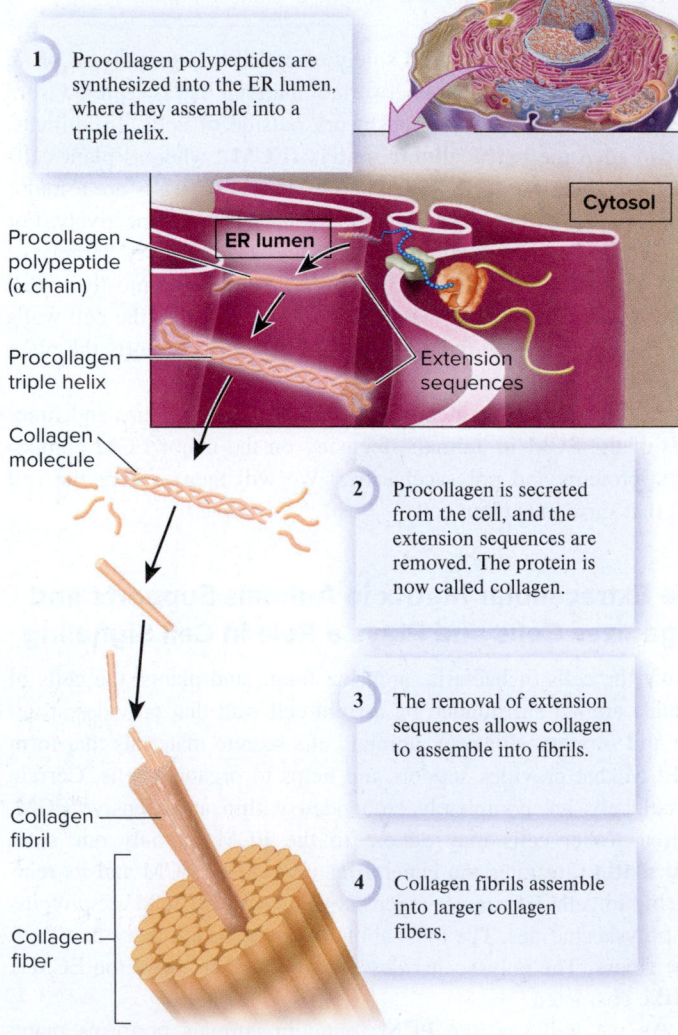

1  Procollagen polypeptides are synthesized into the ER lumen, where they assemble into a triple helix.

Procollagen polypeptide (α chain)

**ER lumen**    **Cytosol**

Procollagen triple helix

Extension sequences

Collagen molecule

2  Procollagen is secreted from the cell, and the extension sequences are removed. The protein is now called collagen.

3  The removal of extension sequences allows collagen to assemble into fibrils.

Collagen fibril

Collagen fiber

4  Collagen fibrils assemble into larger collagen fibers.

**Figure 10.2  Formation of collagen fibers.** Collagen is one type of structural protein found in the ECM of animal cells.

*Concept Check:* *What prevents large collagen fibers from forming intracellularly?*

| Table 10.1 | Proteins in the ECM of Animals | |
|---|---|---|
| **General type** | **Example** | **Function** |
| **Adhesive** | Fibronectin | Connects cells to the ECM and helps to organize components in the ECM. |
| | Laminin | Connects cells to the ECM and helps to organize components in the ECM. |
| **Structural** | Collagen | Forms large fibers and interconnected fibrous networks in the ECM. Provides tensile strength. |
| | Elastin | Forms elastic fibers in the ECM that can stretch and recoil. |

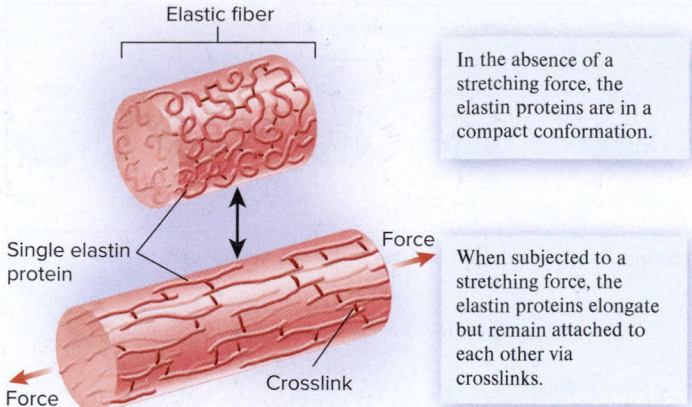

In the absence of a stretching force, the elastin proteins are in a compact conformation.

When subjected to a stretching force, the elastin proteins elongate but remain attached to each other via crosslinks.

**Figure 10.3** **Structure and function of elastic fibers.** Elastic fibers are made of elastin, one type of structural protein found in the ECM surrounding animal cells.

**Concept Check:** *Suppose you started with an unstretched elastic fiber and treated it with a chemical that breaks the crosslinks between adjacent elastin proteins. What would happen when the fiber was stretched?*

In addition to tensile strength, elasticity is needed in regions of the body such as the lungs and blood vessels, which regularly expand and return to their original shape. In these places, the ECM contains elastic fibers composed primarily of the protein **elastin** (**Figure 10.3**). Elastin proteins form many covalent crosslinks to make a fiber with remarkable elastic properties. In the absence of a stretching force, each protein tends to adopt a compact conformation. When subjected to a stretching force, however, the compact proteins become more linear, with the covalent crosslinks holding the fiber together. When the stretching force stops, the proteins naturally return to their compact conformation. In this way, an elastic fiber behaves much like a rubber band, stretching under tension and snapping back when the tension is released.

**THE QUESTION** *Two structural proteins found in the ECM of animals are collagen and elastin. How are the structures of these proteins related to their functions?*

**T OPIC** *What topic in biology does this question address?* The topic is structural proteins in the ECM. More specifically, the question asks you to relate the structures and functions of collagen and elastin.

**I NFORMATION** *What information do you know based on the question and your understanding of the topic?* In the question, you are reminded that collagen and elastin are structural proteins found in the ECM of animals. From your understanding of the topic, you may remember that collagen is composed of long, relatively thick fibers, and its role is to provide tensile strength. Elastin is a more compact protein that forms crosslinked elastic fibers, which provide elasticity.

**P ROBLEM-SOLVING S TRATEGY** *Relate structure and function.* Take a closer look at the structures of these proteins, and consider how the structures determine the proteins' functions.

**ANSWER** *Collagen: A collagen fiber is composed of many smaller fibrils. Three procollagen polypeptides associate with each other to form a protein with a triple helix structure. These collagen proteins then assemble in a staggered way to form relatively thin collagen fibrils. The fibrils, in turn, align with each other and produce larger collagen fibers. The many layers of fibrils give collagen fibers their tensile strength.*

*Elastin: Elastin has a very different structure from collagen. It is a fairly compact protein that forms elastic fibers with many covalent crosslinks between the proteins. In the absence of a stretching force, the elastin proteins remain in the compact conformation. However, when subjected to a stretching force, they become more linear. The covalent crosslinks keep the proteins within the elastic fiber from coming apart. When the stretching force ends, the proteins naturally return to their compact conformation.*

## Core Concepts: Evolution, Structure and Function

### Collagens Are a Family of Proteins That Give the ECM of Animals a Variety of Properties

Researchers have determined that animals make many different types of collagen fibers. These are designated as type I, type II, and so on. At least 27 different types of collagens have been identified in humans. To make different types of collagens, the human genome, as well as the genomes of other animals, has many different genes that encode procollagen polypeptides. Some inherited human diseases are caused by mutations in genes that encode collagen proteins. For example, Ehlers-Danlos syndrome is caused by mutations in one of several different collagen genes. Characteristic symptoms are very stretchable skin and hyperflexible joints.

Why are different collagens made? Each of the many different types of collagen polypeptides has a similar yet distinctive amino acid sequence that affects the structure of not only individual collagen proteins but also the resulting collagen fibers. For example, the amino acid sequence may cause the α chains within each collagen protein to bind to each other very tightly, thereby creating rigid proteins that form a relatively stiff fiber. Such collagen fibers are found in bone and cartilage.

The amino acid sequence of the α chains also influences the interactions between the collagen proteins within a fiber. For example, the amino acid sequences of certain α chains promote a looser interaction that produces a more bendable or thinner fiber. More flexible collagen fibers support the lining of your lungs and intestines. In addition, domains within the collagen polypeptide affect the spatial arrangement of collagen proteins. The collagen shown in Figure 10.2 forms fibers in which collagen proteins align themselves in parallel arrays. However, not all collagen proteins form long fibers. For example, type IV collagen proteins interact with each other in a meshwork pattern. This meshwork acts as a filter around capillaries.

Gene regulation controls which types of collagens are made throughout the body and in what amounts they are made. Of the

| Table 10.2 | Examples of Collagen Types in Humans | |
|---|---|---|
| **Type** | **Sites of synthesis*** | **Structure and function** |
| I | Tendons, ligaments, bones, and skin | Forms a relatively rigid and thick fiber. Very abundant, provides most of the tensile strength to the ECM. |
| II | Cartilage, discs between vertebrae | Forms a fairly rigid and thick fiber but is more flexible than type I. Permits smooth movements of joints. |
| III | Arteries, skin, internal organs, and around muscles | Forms thin fibers, often arranged in a meshwork pattern. Allows for greater elasticity in tissues. |
| IV | Skin, intestine, and kidneys; also found around capillaries | Does not form long fibers. Instead, the proteins are arranged in a meshwork pattern that provides organization and support to cell layers. Functions as a filter around capillaries. |

*The sites of synthesis indicate where a large amount of the collagen type is made.

27 types of collagens identified in humans, **Table 10.2** considers types I to IV, each of which varies as to where it is primarily synthesized and its structure and function. In skin cells, for example, the genes that encode the polypeptides that make up collagen types I, III, and IV are turned on, but the synthesis of type II collagen is minimal.

The regulation of collagen synthesis has received a great deal of attention due to the phenomenon of wrinkling. As we age, the amount of collagen that is synthesized in our skin significantly decreases. The underlying network of collagen fibers, which provides scaffolding for the surface of our skin, loosens and unravels. This is one factor that causes the skin of older people to sink, sag, and form wrinkles. Various therapeutic and cosmetic agents have been developed to prevent or reverse the appearance of wrinkles, most with limited benefits. For example, many face and skin creams contain collagen as an ingredient. Another approach is collagen injections, in which small amounts of collagen (from cows) are injected into areas where the body's collagen has weakened, filling the depressions to the level of the surrounding skin. Because collagen is naturally broken down in the skin, the injections are not permanent and last only about 3 to 6 months.

## Animal Cells Also Secrete Polysaccharides into the ECM

Polysaccharides are the second major component of the ECM of animals. As discussed in Chapter 3, polysaccharides are polymers of many simple sugars. Among vertebrates, the most abundant types of polysaccharides in the ECM are **glycosaminoglycans (GAGs)**. These macromolecules are long, unbranched polysaccharides containing a repeating disaccharide unit (**Figure 10.4a**). GAGs are highly negatively charged molecules that tend to attract positively charged ions and water. The majority of GAGs in the ECM are linked to core proteins, forming **proteoglycans** (**Figure 10.4b**).

Providing resistance to compression is the primary function of GAGs and proteoglycans. Once secreted from cells, these

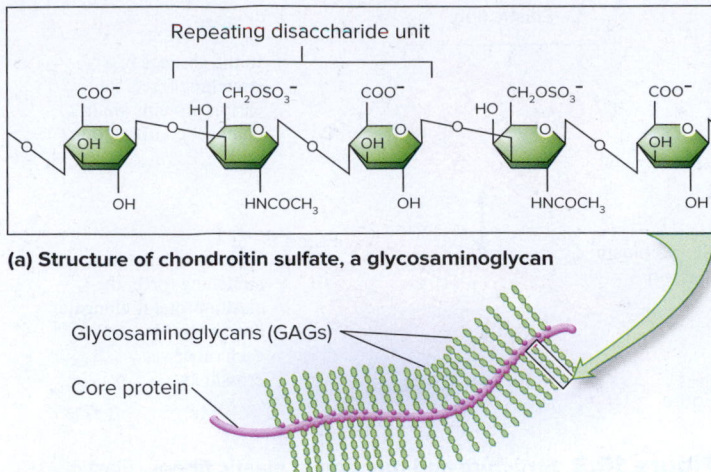

**(a) Structure of chondroitin sulfate, a glycosaminoglycan**

**(b) General structure of a proteoglycan**

**Figure 10.4 Structures of glycosaminoglycans and proteoglycans.** These macromolecules are found in the ECM, which is located outside of animal cells. **(a)** Glycosaminoglycans (GAGs) are composed of repeating disaccharide units. They range in length from several dozen to 25,000 disaccharide units. The GAG shown here is chondroitin sulfate, which is a component of cartilage. **(b)** Proteoglycans are composed of a long, linear core protein with many GAGs attached. Note that each GAG is typically 80 disaccharide units long but only a short chain of sugars is shown in this illustration.

**Concept Check:** *What structural feature of GAGs gives the ECM a gel-like character?*

macromolecules form a gel-like component in the ECM. How is this gel-like property important? Due to its high water content, the ECM is difficult to compress and thereby serves to protect cells. GAGs and proteoglycans are found abundantly in regions of the body that are subjected to harsh mechanical forces, such as the joints of the human body. Two examples of GAGs are chondroitin sulfate, which is a major component of cartilage, and hyaluronic acid, which is found in the skin, eyes, and joint fluid. Purified hyaluronic acid is also used to treat wrinkles and give skin fullness.

Among many invertebrates, an important ECM component is **chitin**, a nitrogen-containing polysaccharide. Chitin forms the hard protective outer covering (called an exoskeleton) of insects, such as crickets and grasshoppers, and crustaceans, such as lobsters and shrimp. As these animals grow, they periodically shed this rigid outer layer and secrete a new, larger one—a process called molting (look ahead to Figure 33.13).

## The Cell Wall of Plants Provides Strength and Resistance to Compression

Let's now turn our attention to the cell walls of plants. Plant cells are surrounded by a cell wall, a protective layer that forms outside of the plasma membrane. Like animal cells, the cells of plants are surrounded by material that provides tensile strength and resistance to compression. The cell walls of plants, however, are usually thicker, stronger, and more rigid than the ECM found in animals. Plant cell walls provide rigidity for mechanical support and also play a role in the maintenance of cell shape and the direction of cell growth.

As described in Chapter 5, the cell wall also prevents expansion when water enters the cell, thereby preventing osmotic lysis.

The main macromolecule of the plant cell wall is **cellulose**, a polysaccharide made of repeating molecules of glucose attached end to end. These glucose polymers associate with each other via hydrogen bonding to form microfibrils that provide great tensile strength (**Figure 10.5**).

Cellulose was discovered in 1838 by French chemist Anselme Payen, who was the first scientist to attempt to separate wood into its component parts. After treating different types of wood with nitric acid, Payen obtained a fibrous substance that was also found in cotton and other plants. His chemical analysis revealed that the fibers were made of the carbohydrate glucose. Payen called this substance cellulose (from the Latin, meaning consisting of cells). Cellulose is probably the single most abundant organic molecule on Earth. Wood consists mostly of cellulose, and cotton and paper are almost pure cellulose.

## Plant Cell Walls Consist of Primary and Secondary Walls

The cell walls of plants are composed of a primary cell wall and a secondary cell wall (**Figure 10.6**). These walls are named based on the timing of their synthesis—the primary cell wall is made before the secondary cell wall.

***Primary Cell Wall***   During cell division, the **primary cell wall** develops between two newly formed daughter cells. It is usually very flexible and allows the new cells to increase in size. The main constituent of the primary cell wall is cellulose.

In addition to cellulose, other components found in the primary cell wall include hemicellulose, glycans, and pectins (see Figure 10.6).

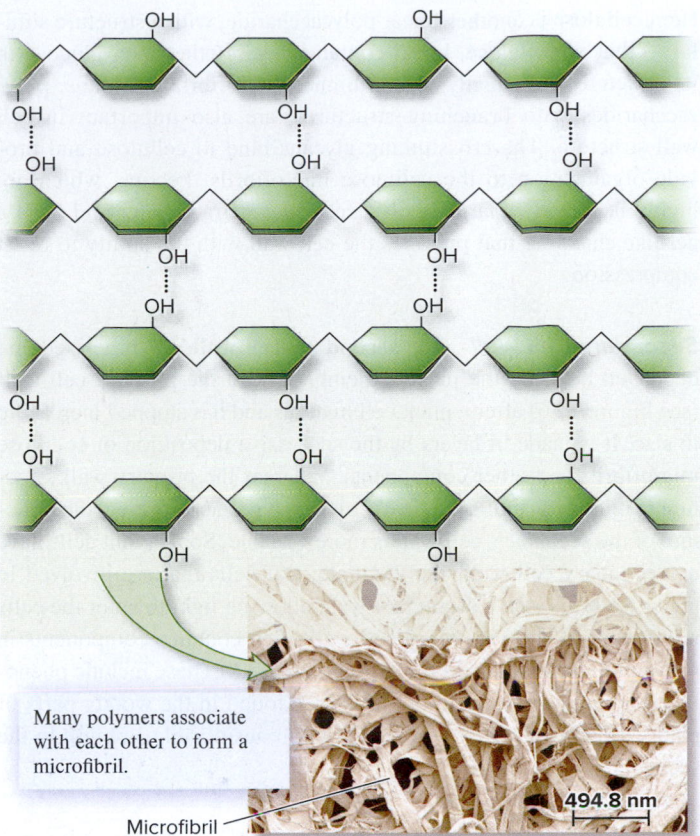

Many polymers associate with each other to form a microfibril.

494.8 nm

Microfibril

**Figure 10.5   Structure of cellulose, the main macromolecule of the plant cell wall.** Cellulose is made of repeating glucose units linked end to end that hydrogen-bond to each other to form microfibrils (SEM). ©SciMAT/Science Source

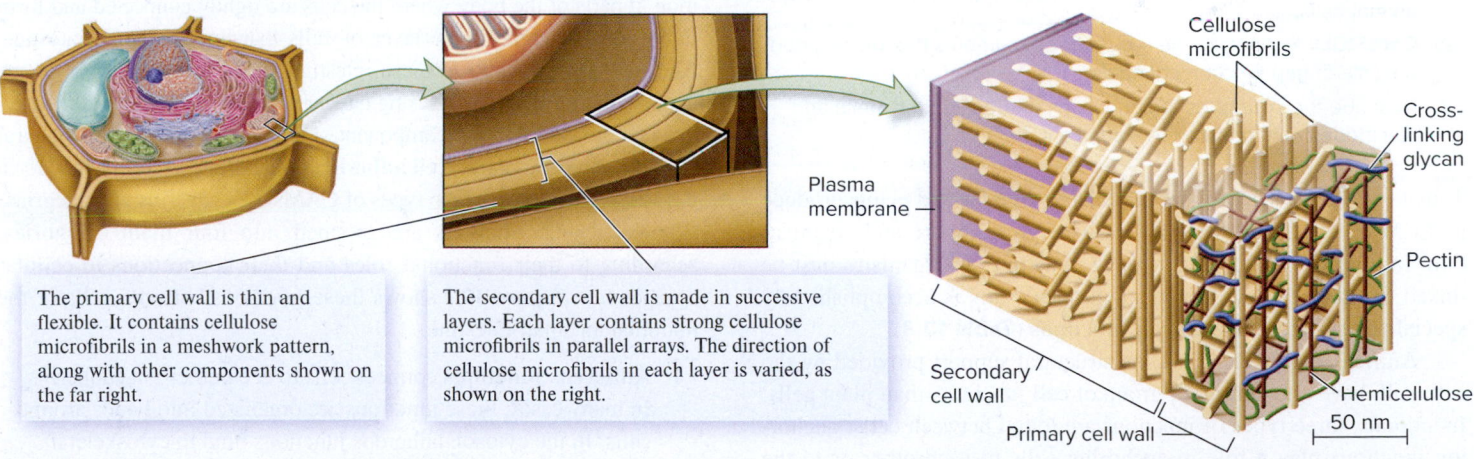

The primary cell wall is thin and flexible. It contains cellulose microfibrils in a meshwork pattern, along with other components shown on the far right.

The secondary cell wall is made in successive layers. Each layer contains strong cellulose microfibrils in parallel arrays. The direction of cellulose microfibrils in each layer is varied, as shown on the right.

Plasma membrane

Cellulose microfibrils

Cross-linking glycan

Pectin

Secondary cell wall

Hemicellulose

Primary cell wall

50 nm

**Figure 10.6   Structure of the cell wall of plant cells.** The primary cell wall is relatively thin and flexible. It contains cellulose (tan), hemicellulose (red), crosslinking glycans (blue), and pectin (green). The secondary cell wall, which is produced only by certain plant cells, is made after the primary cell wall and is synthesized in successive layers.

**Core Skill: Modeling**   The goal of this modeling challenge is to draw layers of a plant's secondary cell wall in colors that reflect the timing of their synthesis.

**Modeling Challenge:** After making its primary cell wall, a particular type of plant cell makes its secondary cell wall in three successive layers. Draw a model that is similar to the model shown in the middle of Figure 10.6, but don't show any components in the cytoplasm. The colors of your model should be as follows: primary cell wall, blue; first-made layer of the secondary cell wall, yellow; second-made layer of the secondary cell wall, green; and third-made layer of the secondary cell wall, black.

Hemicellulose is another linear polysaccharide, with a structure similar to that of cellulose, but it contains sugars other than glucose in its structure and usually forms thinner microfibrils. Glycans, polysaccharides with branching structures, are also important in cell wall structure. The crosslinking glycans bind to cellulose and provide organization to the cellulose microfibrils. Pectins, which are highly negatively charged polysaccharides, attract water and have a gel-like character that provides the cell wall with the ability to resist compression.

*Secondary Cell Wall*  The **secondary cell wall** is synthesized and deposited between the plasma membrane and the primary cell wall (see Figure 10.6) after a plant cell matures and has stopped increasing in size. It is made in layers by the successive deposition of cellulose microfibrils and other components. Whereas the primary wall structure is relatively similar in nearly all cell types and species, the structure of the secondary cell wall is more variable. Some plant cells have no secondary cell wall. For example, leaf cells that are involved in photosynthesis lack a secondary wall, allowing light to enter the cells more readily. The secondary cell wall often contains components in addition to those found in the primary cell wall. These include phenolic compounds called lignins, which are found in the woody parts of plants. Lignins are very hard and impart considerable strength to the secondary wall structure.

## 10.2  Cell Junctions

### Learning Outcomes:

1. Compare and contrast the structures and functions of anchoring junctions, tight junctions, and gap junctions found between animal cells.

2. **CoreSKILL »** Analyze the results of experiments that determined the size of gap junction channels.

3. Describe the structures and functions of the middle lamella and plasmodesmata that connect adjacent plant cells.

Thus far, we have learned that the cells of animals and plants produce an ECM or a cell wall that provides strength, support, and organization. In a multicellular organism, cells within the organism must be linked to each other. In animals and plants, this is accomplished by specialized structures called **cell junctions** (**Table 10.3**).

Animal cells, which lack the structural support provided by the cell wall, have a more varied group of cell junctions than plant cells. In animals, three types of junctions are found between cells: anchoring junctions play a role in anchoring cells to each other or to the ECM; tight junctions seal cells together to prevent small molecules from leaking across a layer of cells; and gap junctions allow the passage of materials between adjacent cells.

In plants, cellular organization is somewhat different because plant cells are surrounded by a rigid cell wall. Plant cells are connected to each other by a component called the middle lamella, which cements their cell walls together. They also have junctions termed plasmodesmata that allow the passage of materials between adjacent cells. In this section, we will examine these various types of junctions found between the cells of animals and plants.

| Table 10.3 | Common Types of Cell Junctions |
|---|---|
| **Type** | **Description** |
| *Animals* | |
| Anchoring junctions | Cell junctions that hold adjacent cells together or attach cells to the ECM. Anchoring junctions are mechanically strong. |
| Tight junctions | Junctions between adjacent cells in a layer that prevent the leakage of material between cells. |
| Gap junctions | A cluster of channels that permit the direct exchange of ions and small molecules between the cytosols of adjacent cells. |
| *Plants* | |
| Middle lamella | A polysaccharide layer that cements together the cell walls of adjacent cells. |
| Plasmodesmata | Passageways between the cell walls of adjacent cells that can be opened or closed. When open, they permit the direct diffusion of ions and molecules between the cytosols of the adjacent cells. |

## Anchoring Junctions Link Animal Cells to Each Other and to the ECM

Electron microscopy allows researchers to explore the types of junctions that occur between adjacent cells and between cells and the ECM. In the 1960s, Marilyn Farquhar, George Palade, and colleagues conducted several studies showing that various types of cell junctions connect cells to each other. Collectively called **anchoring junctions**, these junctions attach cells to each other and to the ECM. Anchoring junctions are common in parts of the body where the cells are tightly connected and form linings. An example is the layer of cells that line the small intestine. Anchoring junctions keep these intestinal cells tightly adhered to one another, thereby forming a strong barrier between the lumen of the intestine and the blood. A key component of anchoring junctions are integral membrane proteins called **cell adhesion molecules (CAMs)**, which form the actual connections. Two types of CAMs are cadherins and integrins.

Anchoring junctions are grouped into four main categories, according to their functional roles and their connections to cellular components. **Figure 10.7** shows these junctions between cells of the mammalian small intestine.

1. **Adherens junctions** connect cells to each other via cadherins. In many cases, these junctions are organized into bands around cells. In the cytosol, adherens junctions bind to cytoskeletal filaments called actin filaments.

2. **Desmosomes** also connect cells to each other via cadherins. They are spotlike points of intercellular contact that rivet cells together. Desmosomes are connected to cytoskeletal filaments called intermediate filaments.

3. **Hemidesmosomes** connect cells to the extracellular matrix via integrins. Like desmosomes, they interact with intermediate filaments.

4. **Focal adhesions** also connect cells to the ECM via integrins. In the cytosol, focal adhesions bind to actin filaments.

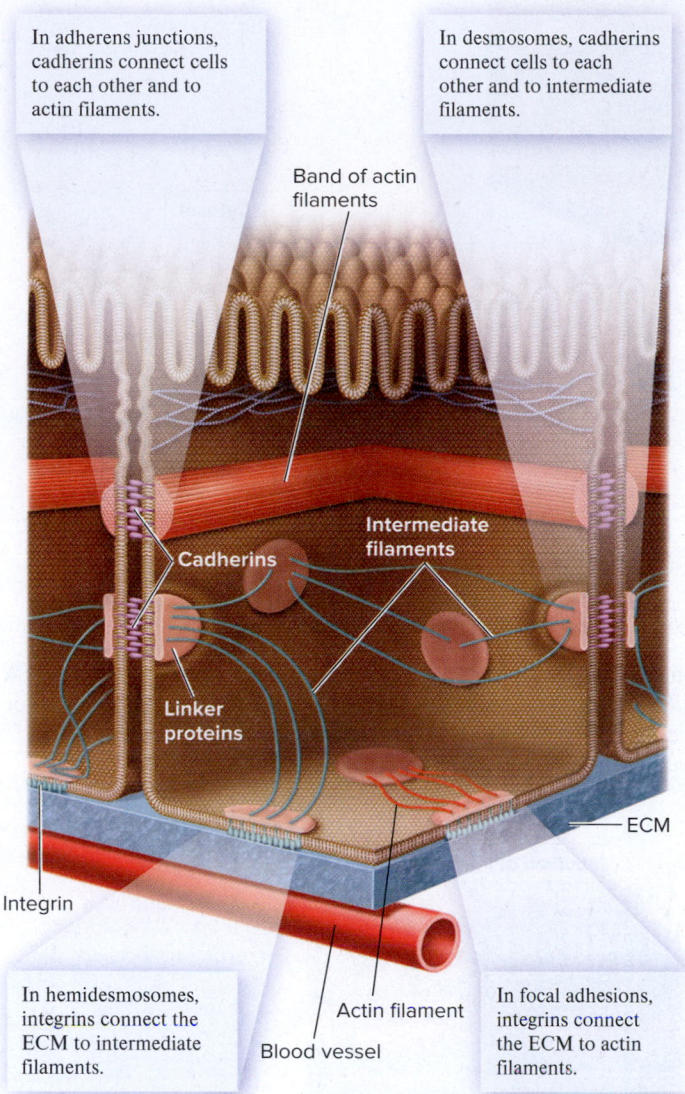

In adherens junctions, cadherins connect cells to each other and to actin filaments.

In desmosomes, cadherins connect cells to each other and to intermediate filaments.

Band of actin filaments

Intermediate filaments

Cadherins

Linker proteins

ECM

Integrin

In hemidesmosomes, integrins connect the ECM to intermediate filaments.

Actin filament

Blood vessel

In focal adhesions, integrins connect the ECM to actin filaments.

**Figure 10.7 Types of anchoring junctions.** This figure shows the four types of anchoring junctions in three adjacent intestinal cells. The tops of these cells face the lumen of the intestine, whereas the bottoms are adjacent to the ECM and a blood vessel.

**Concept Check:** *Which anchoring junctions are cell-to-cell junctions and which are cell-to-ECM junctions?*

## Cell Adhesion Molecules (CAMs) Form Links Between Cells and to the ECM

Let's now consider the molecular components of anchoring junctions.

**Cadherins** As shown in Figure 10.7, **cadherins** are CAMs that create cell-to-cell junctions. The extracellular domains of two cadherin proteins, each in adjacent cells, bind to each other to promote cell-to-cell adhesion (**Figure 10.8a**). This binding requires the presence of calcium ions ($Ca^{2+}$), which change the conformation of the cadherin protein such that cadherins in adjacent cells bind to each other. (This calcium dependence gives cadherin its name—$Ca^{2+}$-dependent adhering molecule.) On the interior of

the cell, linker proteins connect cadherins to actin or intermediate filaments of the cytoskeleton. This promotes a more stable interaction between two cells because their strong cytoskeletons are connected to each other.

The genomes of vertebrates and invertebrates contain multiple cadherin genes, which encode slightly different cadherin proteins. The expression of cadherins in particular cell types allows cells to recognize each other. Dimer formation follows a homophilic, or like-to-like, binding mechanism. To understand the concept of homophilic binding, let's consider an example. One type of cadherin is called E-cadherin, and another is N-cadherin. E-cadherin in one cell binds to E-cadherin in an adjacent cell to form a homodimer. However, E-cadherin in one cell does not bind to N-cadherin in an adjacent cell to form a heterodimer. Similarly, N-cadherin binds to N-cadherin but not to E-cadherin in an adjacent cell. Why is such homophilic binding important? By expressing only certain types of cadherins, each cell binds only to other cells that express the same cadherin types. This phenomenon plays a key role in the proper arrangement of cells throughout the body, particularly during embryonic development.

**Integrins** Another type of CAM is a group of proteins called **integrins**, which form connections between cells and the ECM. Integrins do not require $Ca^{2+}$ to function. Each integrin protein is composed of two nonidentical subunits. In the example shown in **Figure 10.8b**, an integrin is bound to fibronectin, an adhesive protein in the ECM that binds to other ECM components such as collagen fibers. Like cadherins, integrins also bind to actin or intermediate filaments in the cytosol of the cell, via linker proteins, to promote a strong association between the cytoskeleton and the ECM. Thus, integrins have an extracellular domain for the binding of ECM components and an intracellular domain for the binding of cytosolic proteins.

When CAMs were first discovered, researchers imagined that cadherins and integrins played only a mechanical role. In other words, their functions were described as holding cells together or to the ECM. More recently, however, experiments have shown that cadherins and integrins are important in cell communication. The formation or breaking of cell-to-cell and cell-to-ECM anchoring junctions affects signal transduction pathways within the cell. Similarly, intracellular signal transduction pathways affect cadherins and integrins in ways that alter intercellular junctions and the binding of cells to ECM components.

Abnormalities in CAMs such as integrins are associated with the ability of cancer cells to metastasize, that is, to move to other parts of the body. CAMs are critical for keeping cells in their correct locations. When these adhesion molecules become defective due to cancer-causing mutations, cells lose their proper connections with the ECM and adjacent cells and may move to other parts of the body.

## Tight Junctions Prevent the Leakage of Materials Across Animal Cell Layers

In animals, **tight junctions** are a second type of junction, one that forms a tight seal between adjacent cells, thereby preventing material from leaking between the cells. As an example, let's consider the

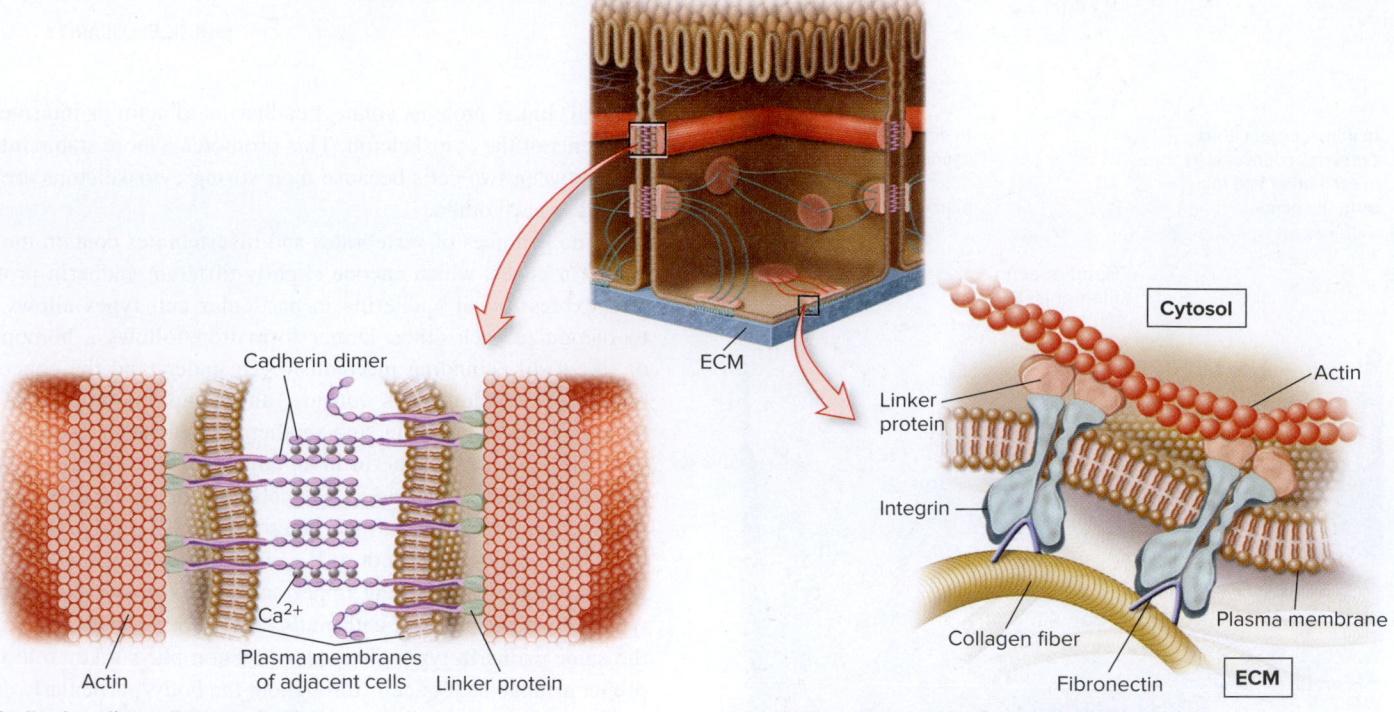

Cadherin dimer

ECM

Cytosol

Actin

Linker protein

Integrin

Plasma membrane

Ca²⁺

Collagen fiber

Fibronectin

ECM

Actin

Plasma membranes of adjacent cells

Linker protein

**(a) Cadherins—link cells to each other**

**(b) Integrins—link cells to the extracellular matrix**

**Figure 10.8** **Types of cell adhesion molecules (CAMs).** Cadherins and integrins are CAMs that form connections in anchoring junctions. **(a)** A cadherin in one cell binds to a cadherin of an identical type in an adjacent cell. This binding requires Ca²⁺. In the cytosol, cadherins bind to actin or intermediate filaments of the cytoskeleton via linker proteins. **(b)** Integrins link cells to the ECM and form intracellular connections to actin or intermediate filaments. Each integrin protein is a heterodimer, composed of two nonidentical subunits.

intestine. The cells that line the intestine form a sheet that is one cell thick. One side of each cell faces the intestinal lumen, and the other faces the ECM and a blood vessel (**Figure 10.9**). Tight junctions between these cells prevent the leakage of materials from the lumen of the intestine into the blood, and vice versa.

Tight junctions are made by integral membrane proteins, called occludin and claudin, that form interlaced strands in the plasma membrane (see inset in Figure 10.9). These strands of proteins, located in adjacent cells, bind to each other, thereby forming a tight seal between cells. Tight junctions are not mechanically strong like anchoring junctions, because they do not have strong connections with the cytoskeleton. Therefore, adjacent cells that have tight junctions also have anchoring junctions to hold them in place.

Tight junctions perform several important roles. Let's consider a few examples.

- Tight junctions between intestinal cells prevent leakage of materials between the lumen of the intestine and the blood.

- Tight junctions help maintain the polarity of intestinal cells by preventing the lateral diffusion of integral membrane proteins between the apical side (which faces the lumen of the intestine) and the basolateral side (which faces a blood vessel). For example, proteins involved with receptor-mediated endocytosis are restricted to the apical side, and proteins involved with exocytosis are located at the basolateral side. Thus, intestinal cells are able to take up nutrients from the intestinal lumen and export them into the bloodstream, a phenomenon called **transepithelial transport**.

- Tight junctions prevent microbes from entering the body. In mammals, the skin on the exterior of the body and the lining of

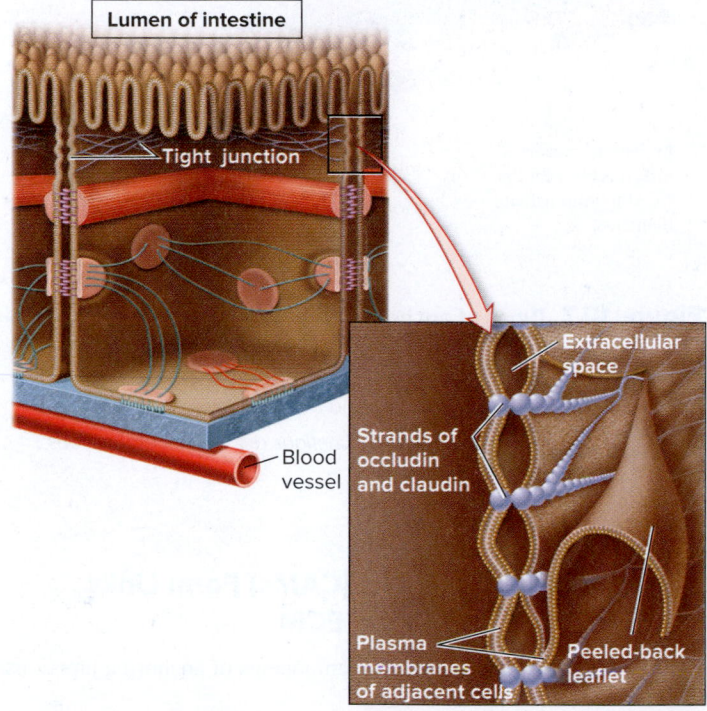

Lumen of intestine

Tight junction

Blood vessel

Extracellular space

Strands of occludin and claudin

Plasma membranes of adjacent cells

Peeled-back leaflet

**Figure 10.9** **Tight junctions between adjacent intestinal cells.** In this example, tight junctions form a seal between cells of the intestinal lining. The inset shows the interconnected network of occludin and claudin that forms the tight junction.

 **Core Skill: Connections** Look ahead to Figure 46.8. What problems might arise if tight junctions did not connect the cells that line your small intestine?

the digestive tract are formed from interconnected cells that have tight junctions. Some pathogenic microorganisms, such as those that cause certain forms of diarrhea, are able to cause infection by disrupting tight junctions.

The amazing ability of tight junctions to prevent the leakage of material across cell layers has been demonstrated by dye-injection studies. In 1972, Daniel Friend and Norton Gilula injected lanthanum into the bloodstream of a rat. Lanthanum is an electron-dense element that can be visualized using electron microscopy. A few minutes later, a sample of a cell layer in the digestive tract was removed and observed under an electron microscope. As seen in the micrograph in **Figure 10.10**, lanthanum diffused into the region between the cells that faces the blood, but it could not move past the tight junction to the side of the cell layer facing the lumen of the digestive tract.

## Gap Junctions Between Animal Cells Provide Passageways for Intercellular Transport

A third type of junction found between animal cells is called a **gap junction**, because a small gap occurs between the plasma membranes of cells connected by these junctions (**Figure 10.11**). Gap junctions are abundant in tissues and organs where the cells need to communicate with each other. For example, cardiac muscle cells, which cause your heart to beat, are interconnected by many gap junctions. Because gap junctions allow the passage of ions, electrical changes in one cardiac muscle cell are easily transmitted to an adjacent cell that is connected via gap junctions. These connections are needed for the coordinated contraction of cardiac muscle cells.

In vertebrates, gap junctions are composed of an integral membrane protein called connexin. Invertebrates have a structurally similar protein called innexin. Six connexin proteins in one vertebrate cell form a channel called a **connexon**. A connexon in one cell aligns with a connexon in an adjacent cell to form an intercellular channel (see the middle drawing in Figure 10.11). The term gap junction refers to

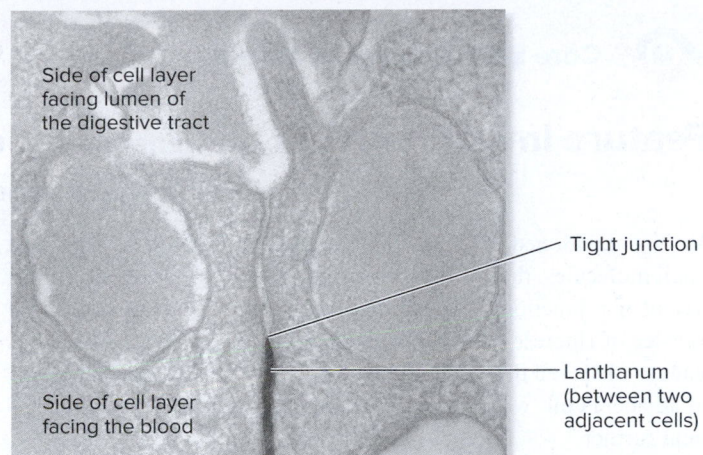

Side of cell layer facing lumen of the digestive tract

Tight junction

Side of cell layer facing the blood

Lanthanum (between two adjacent cells)

**Figure 10.10  An experiment demonstrating the function of a tight junction.** When lanthanum was injected into the bloodstream of a rat, it diffused between the cells in the region up to a tight junction but could not diffuse past the junction to the other side of the cell layer. ©Dr. Daniel Friend

**Concept Check:** *What results would you expect if a rat was fed lanthanum and then a sample of a cell layer in the digestive tract was observed under an electron microscope?*

a cluster of many connexons that are close to each other in the plasma membrane and form many intercellular channels.

Gap junction channels allow the passage of ions and small molecules, including amino acids, sugars, and signaling molecules such as $Ca^{2+}$ and cAMP, between cells. In this way, gap junctions allow adjacent cells to share metabolites and directly signal each other. However, gap junction channels are too small to allow the passage of RNA, proteins, or polysaccharides. Therefore, cells that communicate via gap junctions still maintain their own distinctive sets of macromolecules.

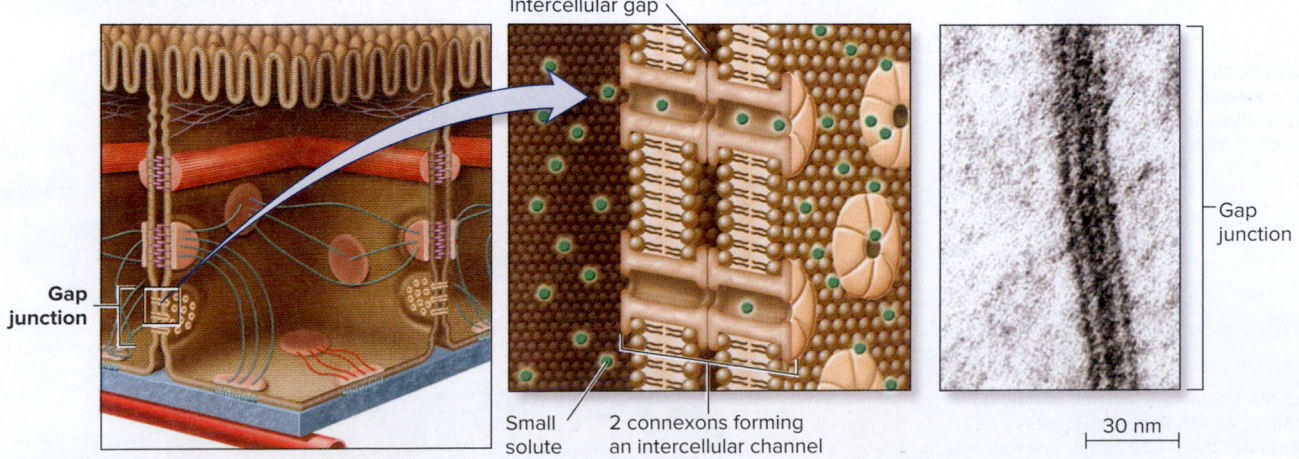

Intercellular gap

Gap junction

Gap junction

Small solute

2 connexons forming an intercellular channel

Gap junction

30 nm

**Figure 10.11  Gap junctions between adjacent cells.** Gap junctions form intercellular channels that allow the passage of small solutes with masses less than 1,000 Da. One connexon consists of six proteins called connexins. Two connexons align to form an intercellular channel. The micrograph shows a gap junction, which is composed of many connexons, between intestinal cells. (right): Courtesy Dr. Dan Goodenough/Harvard Medical School

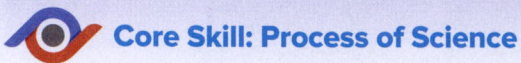

## Core Skill: Process of Science

# Feature Investigation | Loewenstein and Colleagues Followed the Transfer of Fluorescent Dyes to Determine the Size of Gap-Junction Channels

As just mentioned, gap junctions allow the passage of ions and small molecules, those with a mass up to about 1,000 Da. This property of gap junctions was determined in experiments involving the transfer of fluorescent dyes. In 1964, Werner Loewenstein and colleagues observed that a fluorescent dye could move from one cell to an adjacent cell, which prompted them to investigate this phenomenon further.

In the experiment shown in **Figure 10.12**, Loewenstein and colleagues grew rat liver cells in the laboratory, where they formed a single layer. The adjacent cells formed gap junctions. Single cells were injected with various dyes composed of fluorescently labeled amino acids or peptide molecules with different masses, and then the cell layers were observed via fluorescence microscopy. As the data in Figure 10.12 show, dyes with a molecular mass up to 901 Da passed from cell to cell. Dyes with a larger mass, however, did not move

intercellularly. Loewenstein and other researchers subsequently investigated dye transfer in other cell types and species. Though some variation is found among different cell types and species, the researchers generally observed that molecules with a mass greater than 1,000 Da do not pass through gap junctions.

---

### Experimental Questions

1. What was the purpose of the study conducted by Loewenstein and colleagues?

2. **CoreSKILL »** Explain the experimental procedure used by Loewenstein and colleagues to determine the sizes of molecules that can pass through gap-junction channels.

3. **CoreSKILL »** What do the results of the experiment in Figure 10.12 indicate about the size of gap-junction channels?

---

**Figure 10.12** Use of fluorescent molecules by Loewenstein and colleagues to determine the size of gap-junction channels.

**HYPOTHESIS** Gap-junction channels allow the passage of ions and molecules, but there is a limit to how large the molecules can be.

**KEY MATERIALS** Rat liver cells grown in the laboratory, a collection of fluorescent dyes.

| Experimental level | Conceptual level |
|---|---|

1. Grow rat liver cells in a laboratory on solid growth medium until they become a single layer. At this point, adjacent cells have formed gap junctions.

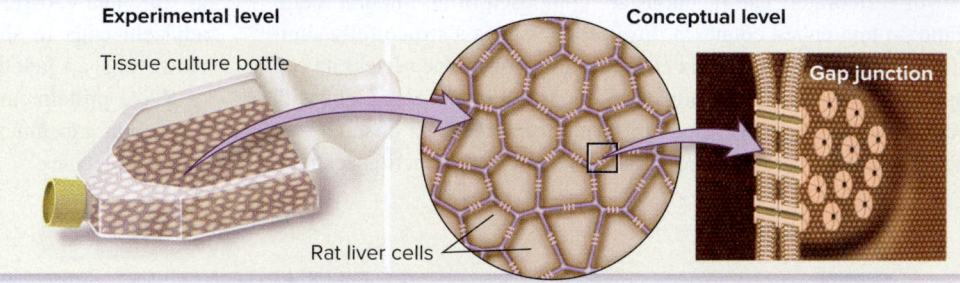

2. Inject 1 cell in the layer with fluorescently labeled amino acids or peptides. Note: Several dyes with different molecular masses were tested.

3. Incubate for various lengths of time (for example, 40–45 minutes). Observe cell layer under the fluorescence microscope to determine if the dye has moved to adjacent cells.

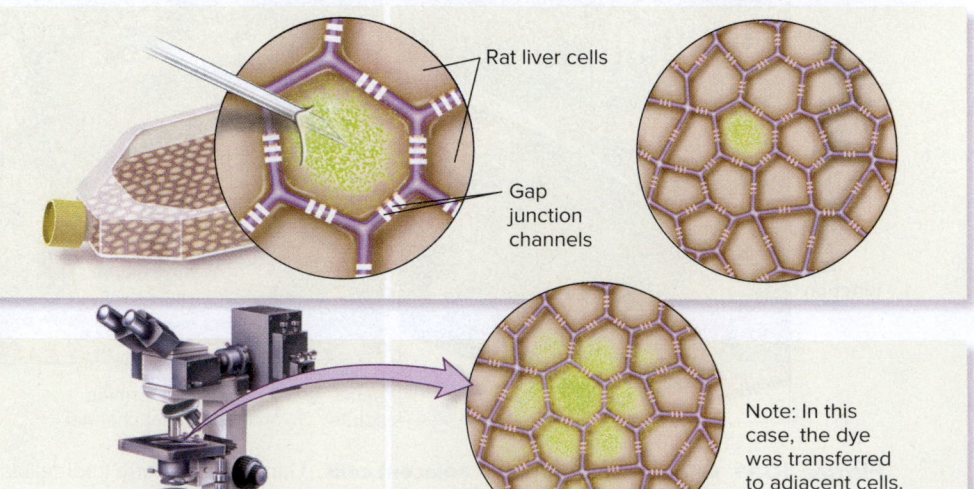

Note: In this case, the dye was transferred to adjacent cells.

**4    THE DATA**

| Mass of dye (in daltons) | Transfer to adjacent cells* | Mass of dye | Transfer to adjacent cells* |
|---|---|---|---|
| 376 | ++++ | 851** | – |
| 464 | ++++ | 901 | +++ |
| 536 | +++ | 946 | – |
| 559 | ++++ | 1004 | – |
| 665 | + | 1158 | – |
| 688 | ++++ | 1678 | – |
| 817 | +++ | 1830 | – |

*The number of pluses indicates the relative speed of transfer. Four pluses denotes fast transfer, whereas one plus is slow transfer. A minus indicates that transfer between cells did not occur. **In some cases, molecules with less mass did not pass between cells compared with molecules with a higher mass. This may be due to differences in their structures (for example, charges) that influence whether or not they can easily penetrate the channel.

**5    CONCLUSION** Gap junctions allow the intercellular movement of molecules that have a mass of approximately 900 Da or less.

**6    SOURCE** Flagg-Newton, J., Simpson, II, and Loewenstein, W. R. 1973. Permeability of the cell-to-cell membrane channels in mammalian cell junctions. *Science* 205: 404–407.

## The Middle Lamella Cements Adjacent Plant Cell Walls Together

In animals, cell-to-cell contact via anchoring junctions, tight junctions, and gap junctions involves interactions between membrane proteins in adjacent cells. In plants, cell junctions are biochemically different. Rather than using membrane proteins to form cell-to-cell connections, plant cells make an additional component called the **middle lamella** (plural, lamellae), which is found between most adjacent plant cells (**Figure 10.13**). When plant cells are dividing, the middle lamella is the first layer formed. The primary cell wall is then made. The middle lamella is rich in pectins, negatively charged polysaccharides that are also found in the primary cell wall (see Figure 10.6). Pectins attract water and thus produce a hydrated gel. $Ca^{2+}$ and $Mg^{2+}$ interact with the negative charges in the pectins and cement the cell walls of adjacent cells together.

The process of fruit ripening illustrates the importance of pectins in holding plant cells together. An unripened fruit, such as a green tomato, is very firm because the rigid cell walls of adjacent cells are firmly attached to each other. During ripening, the cells secrete a group of enzymes called pectinases, which digest pectins in the middle lamella as well as those in the primary cell wall. As this process continues, the attachments between cells are broken, and the cell walls become less rigid. For this reason, a red ripe tomato is much less firm than an unripe tomato.

## Plasmodesmata Are Channels Connecting the Cytoplasm of Adjacent Plant Cells

In 1879, Eduard Tangl, a Russian botanist, observed intercellular connections in the seeds of the strychnine tree and hypothesized that the cytoplasm of adjacent cells is connected by ducts in the cell walls.

He was the first to propose that direct cell-to-cell communication integrates the functioning of plant cells. The ducts or intercellular channels that Tangl observed are now known as **plasmodesmata** (singular, plasmodesma).

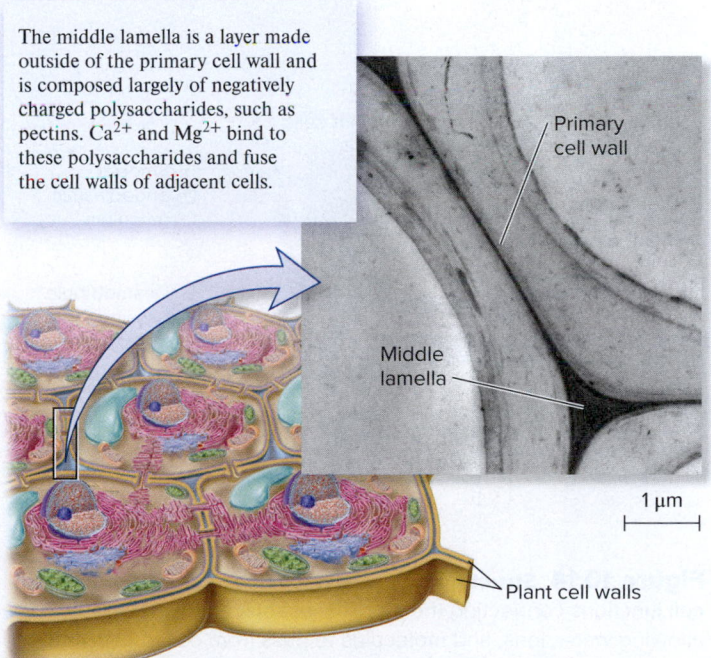

The middle lamella is a layer made outside of the primary cell wall and is composed largely of negatively charged polysaccharides, such as pectins. $Ca^{2+}$ and $Mg^{2+}$ bind to these polysaccharides and fuse the cell walls of adjacent cells.

Primary cell wall

Middle lamella

1 µm

Plant cell walls

**Figure 10.13** **Plant cell-to-cell connections consist of middle lamellae.** ©Purbasha Sarkar

*Concept Check:* *How are middle lamellae similar to the anchoring junctions and desmosomes found between animal cells? How are they different?*

Plasmodesmata are functionally similar to gap junctions in animal cells because they are open pores that allow the passage of ions and molecules between the cytosols of adjacent plant cells. However, the structure of the plasmodesmata is quite different from that of gap junctions. As shown in **Figure 10.14**, the plasma membrane of one cell is continuous with the plasma membrane of the adjacent cell, which forms a pore that permits the diffusion of molecules from the cytosol of one cell to the cytosol of the other. In addition to a cytosolic connection, plasmodesmata also have a central tubule, called a desmotubule, connecting the smooth ER membranes of adjacent cells.

Plant cells can alter the diameter of the channel formed by plasmodesmata. The channel can occur in the closed, open, and dilated states. In the open state, plasmodesmata allow the passage of ions and small molecules, such as sugars and cAMP. In this state, plasmodesmata

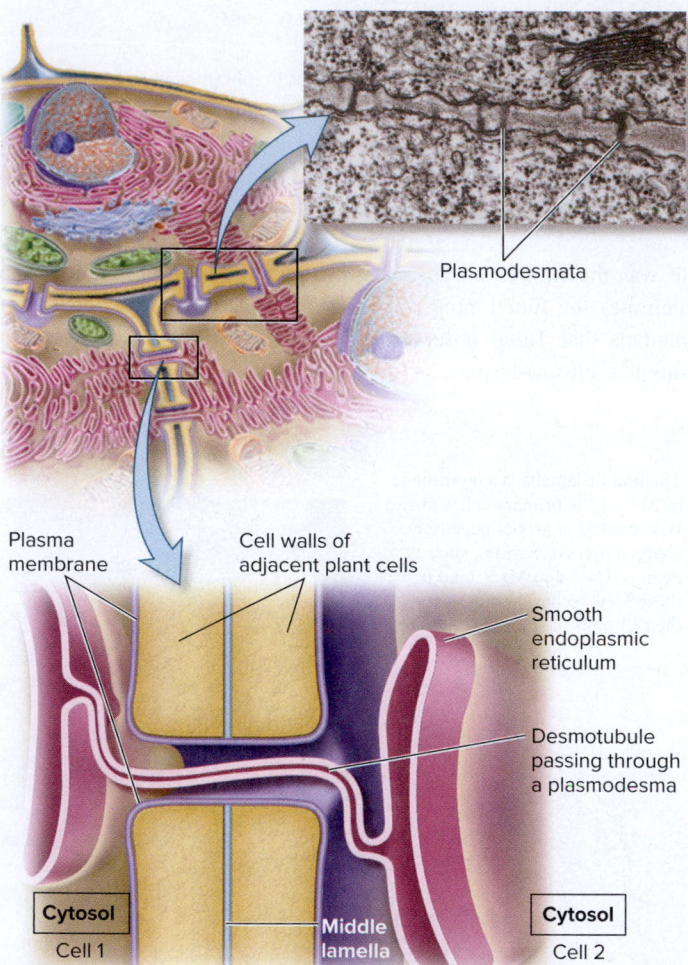

Plasmodesmata

Plasma membrane

Cell walls of adjacent plant cells

Smooth endoplasmic reticulum

Desmotubule passing through a plasmodesma

Cytosol — Cell 1

Middle lamella

Cytosol — Cell 2

**Figure 10.14** **Structure of a plasmodesma.** Plasmodesmata are cell junctions connecting the cytosols of adjacent plant cells, allowing water, ions, and molecules to pass from cell to cell. At these pores, the plasma membrane of one cell is continuous with the plasma membrane of an adjacent cell. In addition, the smooth ER from one cell is connected to that of the adjacent cell via a desmotubule. ©Biophoto Associates/Science Source

 **Core Skill: Connections** Look ahead to Figure 39.8. How do plasmodesmata play a role in the movement of nutrients through a plant root?

play a similar role to gap junctions between animal cells. Plasmodesmata tend to close when a large pressure difference occurs between adjacent cells. Why does this happen? One reason is related to cell damage. When a plant is wounded, damaged cells lose their turgor pressure. (Turgor pressure is described in Chapter 39, look ahead to Figure 39.4.) The closure of plasmodesmata between adjacent cells helps to prevent the loss of water and nutrients from the wound site.

Unlike gap junctions between animal cells, plasmodesmata can dilate to also allow the passage of macromolecules and even viruses between adjacent plant cells. Though the mechanism of dilation is not well understood, the wider opening of plasmodesmata is important for the passage of proteins and mRNA during plant development. It also provides a key mechanism whereby viruses can move from cell to cell.

## 10.3 Tissues

### Learning Outcomes:

1. List the six basic cell processes that produce tissues and organs.
2. Outline the structures and functions of the four types of animal tissues: epithelial, connective, nervous, and muscle tissues.
3. Summarize the structures and functions of the three types of plant tissues: dermal, ground, and vascular tissues.

A **tissue** is a part of an animal or plant consisting of a group of cells having a similar structure and function. In this section, we will view tissues from the perspective of cell biology. Animals and plants contain many different types of cells. Humans, for example, have over 200 different cell types, each with a specific structure and function. Even so, these cells can be grouped into a few general categories. For example, muscle cells found in your heart (cardiac muscle cells), in your biceps (skeletal muscle cells), and around your arteries (smooth muscle cells) look somewhat different under the microscope and have unique roles in the body. Yet due to structural and functional similarities, all three types are categorized as muscle tissue. In this section, we begin by surveying the basic processes that cells undergo to make tissues. Then, we will examine the main categories of animal and plant tissues.

### Six Different Cellular Processes Produce Tissues and Organs

A multicellular organism, such as a plant or animal, contains many cells. For example, an adult human has somewhere between 10 and 100 trillion cells in her or his body. Cells are organized into tissues, and tissues are organized into organs. An **organ** is a collection of two or more tissues that performs a specific function or set of functions. The heart is an organ found in the bodies of complex animals, and a leaf is an organ found in plants. We will examine the structures and functions of organs in Units VI and VII.

How are tissues and organs formed? To form tissues and organs, cells undergo six different processes that influence their morphology, arrangement, and number: division, growth, differentiation, migration, apoptosis, and formation of connections.

1. ***Division.*** As discussed in Chapter 16, eukaryotic cells advance through a cell cycle that leads to cell division.

2. *Growth.* Following cell division, cells take up nutrients and usually expand in volume. Cell division and growth are the primary mechanisms for increasing the size of tissues, organs, and organisms.

3. *Differentiation.* Due to gene regulation, cells differentiate into specialized types of cells. Cell differentiation is described in Chapter 20.

4. *Migration.* During embryonic development in animals, cells migrate to their appropriate positions within the body. Also, adults have cells that can move into regions that have become damaged. Cell migration does not occur during plant development.

5. *Apoptosis.* Programmed cell death, also known as apoptosis (discussed in Chapter 9), is necessary to produce certain morphological features of the body. For example, during development in mammals, the formation of individual fingers and toes requires the removal, by apoptosis, of the skin cells between them.

6. *Formation of connections.* In the first section of this chapter, we learned that cells produce an extracellular matrix or cell wall that provides strength and support. In animals, the ECM serves to organize cells within tissues and organs. In plants, the connections and structures of cell walls are largely responsible for the shapes of plant tissues. Different types of cell junctions in both animal and plant cells enable cells to maintain physical contact and communicate with one another.

## Animals Are Composed of Epithelial, Connective, Nervous, and Muscle Tissues

The body of an animal contains four general types of tissue—epithelial, connective, nervous, and muscle—that serve very different purposes (**Figure 10.15**).

*Epithelial Tissue*  **Epithelial tissue** is composed of cells that are joined together via tight junctions and form continuous sheets. (Epithelial cells are shown in Figure 10.9.) Epithelial tissue covers or forms the lining of all internal and external body surfaces. For example, epithelial tissue lines organs such as the lungs and digestive tract. In addition, epithelial tissue forms the outer layer of the skin, a protective surface that shields the body from the outside environment.

*Connective Tissue*  Most **connective tissue** provides support to the body and/or helps to connect different tissues to each other. Connective tissue is rich in ECM. Examples of connective tissue include cartilage, tendons, bone, fat tissue, and the inner layers of the skin. Blood is also considered a form of connective tissue because it provides liquid connections to various regions of the body.

**Figure 10.16** shows a micrograph of cartilage, a connective tissue found in joints such as your knees. The cells that synthesize cartilage, known as chondrocytes, actually represent a small proportion of the total volume of cartilage. As shown in Figure 10.16, the chondrocytes are found in small cavities within the cartilage called lacunae (singular, lacuna). In some types of cartilage, the chondrocytes represent only 1–2% of the total volume of the tissue! Chondrocytes are the only cells found in cartilage. They are solely responsible for the synthesis of protein fibers, such as collagen, as well as the glycosaminoglycans and proteoglycans that are found in cartilage.

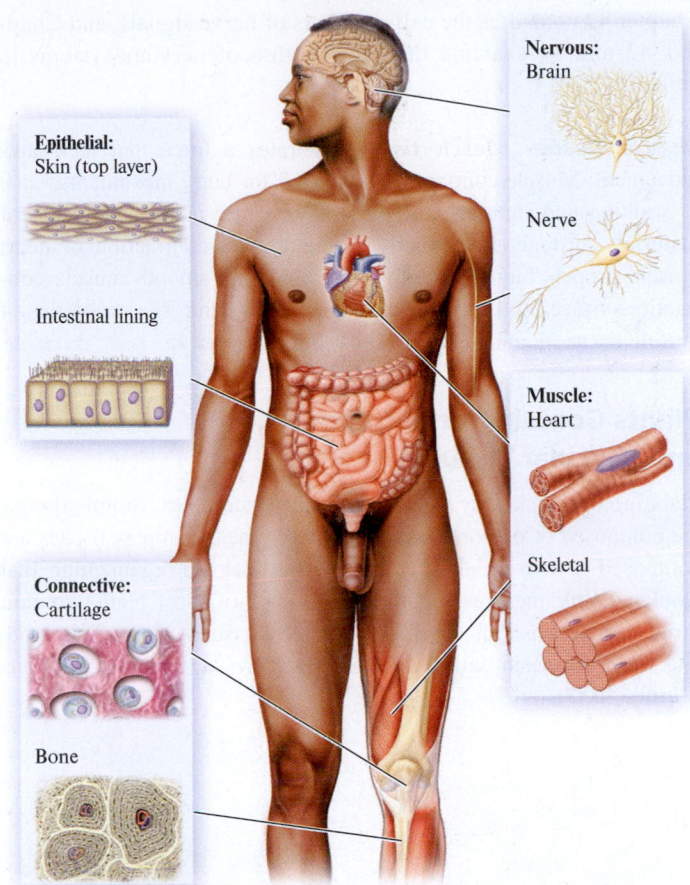

**Figure 10.15**  **Examples of the four general types of tissues—epithelial, connective, nervous, and muscle—found in animals.**

**Concept Check:**  *Which of the four general types of tissues has the most extensive ECM?*

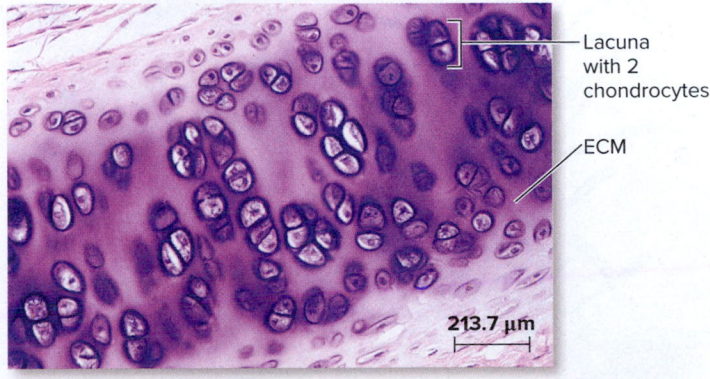

**Figure 10.16**  **An example of connective tissue in animals that is rich in extracellular matrix.** This micrograph of cartilage shows chondrocytes in the ECM. The chondrocytes, which are responsible for making the components of cartilage, are found in cavities called lacunae. ©Victor P. Eroschenko

*Nervous Tissue*  **Nervous tissue** receives, generates, and conducts electrical signals throughout the body. In vertebrates, these electrical signals are integrated by nervous tissue in the brain and transmitted down the spinal cord to the rest of the body.

Chapter 42 considers the cellular basis of nerve signals, and Chapters 43 and 44 examine the organization of nervous systems in animals.

**Muscle Tissue**    **Muscle tissue** generates a force that facilitates movement. Muscle contraction is needed for body movements, such as walking and running, and also plays a role in the movement of materials throughout the body. For example, contraction of heart muscle propels blood through your body, and smooth muscle contractions move food through the digestive system. The properties of muscle tissue in animals are examined in Chapter 45.

## Plants Contain Dermal, Ground, and Vascular Tissues

Plant biologists classify tissues as simple or complex. Simple tissues are composed of one or possibly two cell types. Complex tissues are composed of two or more cell types but lack an organization that would qualify them as organs. The bodies of most plants contain three general types of simple or complex tissues—dermal, ground, and vascular—each with a different structure suited to its functions (**Figure 10.17**).

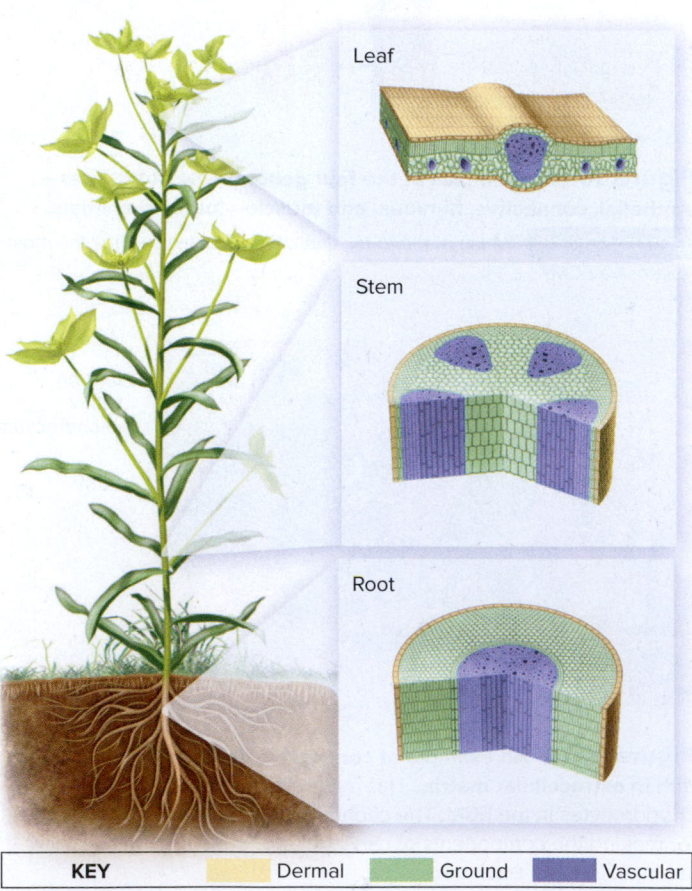

**KEY**  ▢ Dermal  ▢ Ground  ▢ Vascular

**Figure 10.17** **Locations of the three general types of tissues—dermal, ground, and vascular—found in plants.**

*Concept Check:* *Which of these three types of plant tissues is found on the surfaces of leaves, stems, and roots?*

**Dermal Tissue**    **Dermal tissue** is a complex tissue that forms a covering on various parts of the plant. The term **epidermis** refers to the newly made dermal tissue on the surfaces of leaves, stems, and roots. Plant epidermal cells have a thick primary cell wall and are tightly interlocked by their middle lamellae. As a consequence, these cells are held closely together, much like epithelial cell layers in animals.

The epidermal cells of leaves usually secrete a waxy cuticle to prevent water loss. In addition, leaf epidermis often has hairs, or trichomes, which are specialized types of epidermal cells. Trichomes have diverse functions, including the secretion of oils and leaf protection. In leaves, epidermal cells called guard cells form pores known as stomata, which permit gas exchange. The function of the root epidermis is the absorption of water and nutrients. The root epidermis does not have a waxy cuticle because such a cuticle would inhibit water and nutrient absorption.

**Ground Tissue**    Most of a plant's body is made of **ground tissue**, which has a variety of functions, including photosynthesis, storage of carbohydrates, and support. Ground tissue is subdivided into three types of simple tissues: parenchyma, collenchyma, and sclerenchyma. Let's look briefly at each of these types of ground tissue (also see Figure 36.7).

1. Parenchyma is very active metabolically. The mesophyll, the central part of the leaf that carries out the bulk of photosynthesis, is composed of parenchyma. Parenchyma also functions in the storage of carbohydrates. The cells of parenchyma usually lack a secondary cell wall.

2. Collenchyma provides structural support to the plant body, particularly to growing regions such as the periphery of the stems and leaves. Cells in collenchyma tend to have thick, secondary cell walls but do not contain much lignin. Therefore, they provide support but are also able to stretch.

3. Sclerenchyma also provides structural support to the plant body, particularly to those parts that are no longer growing, such as the dense, woody parts of stems. The secondary cell walls of sclerenchyma cells tend to have large amounts of lignin, which provides rigid support. In many cases, sclerenchyma cells are dead at maturity, but their cell walls continue to provide structural support during the life of the plant.

**Vascular Tissue**    Some types of modern plants, such as mosses, are nonvascular plants that lack conducting vessels. These plants tend to be small and live in damp, shady places. Most plants living today, however, are vascular plants. In these species, which include ferns and seed plants, the **vascular tissue** is a complex tissue composed of cells that are interconnected and form conducting vessels for water and nutrients. As described in greater detail in Chapter 39, the two types of vascular tissue are called xylem and phloem. The xylem transports water and mineral ions from the root to the rest of the plant, and the phloem distributes the products of photosynthesis and a variety of other nutrients throughout the plant.

# Summary of Key Concepts

## 10.1 Extracellular Matrix and Cell Walls

- The extracellular matrix (ECM) is a network of material that forms a complex meshwork outside of animal cells; the cell wall is a similar component of plant cells.

- Proteins and polysaccharides are the major constituents of the ECM in animals. These materials are involved in strength, structural support, organization, and cell signaling (Figure 10.1).

- Adhesive proteins, such as fibronectin and laminin, help adhere cells to the ECM. Structural proteins include collagen, which forms fibers and fibrous networks that provide tensile strength, and elastin, which forms elastic fibers that stretch and recoil (Table 10.1, Figures 10.2, 10.3).

- Animals make many different types of collagen fibers, and gene regulation controls the locations in the body where they are made (Table 10.2).

- Glycosaminoglycans (GAGs) are polysaccharides of repeating disaccharide units that give a gel-like character to the ECM of animals. Proteoglycans consist of a long, linear core protein with many GAGs attached (Figure 10.4).

- Plant cells are surrounded by a cell wall composed largely of cellulose. The primary cell wall is made first and tends to be thin and flexible. The secondary cell wall is made after the primary cell wall and is often thick and rigid (Figures 10.5, 10.6).

## 10.2 Cell Junctions

- The three common types of cell junctions found in animals are anchoring, tight, and gap junctions (Table 10.3).

- Key components of anchoring junctions are cell adhesion molecules (CAMs), which bind cells to each other or to the ECM. The four types of anchoring junctions are adherens junctions, desmosomes, hemidesmosomes, and focal adhesions (Figure 10.7).

- Cadherins and integrins are two types of CAMs. Cadherins link cells to each other, whereas integrins link cells to the ECM. In the cytosol, CAMs bind to actin or intermediate filaments (Figure 10.8).

- Tight junctions between cells, composed of occludin and claudin, prevent the leakage of materials across a layer of cells (Figures 10.9, 10.10).

- Gap junctions consist of many channels called connexons, which permit the direct passage of ions and small molecules between adjacent cells (Figure 10.11).

- Loewenstein and colleagues showed that gap junctions permit the passage of substances with a molecular mass of less than about 1,000 Da (Figure 10.12).

- The cell walls of adjacent plant cells are cemented together via middle lamellae, which are rich in pectins—negatively charged polysaccharides (Figure 10.13).

- The plasma membranes and endoplasmic reticula of adjacent plant cells are connected via plasmodesmata that allow the passage of water, ions, and molecules between the cytosols of adjacent cells (Figure 10.14).

## 10.3 Tissues

- Cells are organized into tissues, and tissues are organized into organs. A tissue is a group of cells that have a similar structure and function, and an organ is composed of two or more tissues that carry out a particular function or set of functions.

- Six processes—cell division, cell growth, differentiation, migration, apoptosis, and formation of cell-to-cell connections—produce tissues and organs.

- The four general kinds of tissues found in animals are epithelial, connective, nervous, and muscle tissues (Figures 10.15, 10.16).

- The three general kinds of tissues found in plants are dermal, ground, and vascular tissues (Figure 10.17).

# Assess & Discuss

## Test Yourself

1. The function of the extracellular matrix (ECM) in animals is
   a. to provide strength.
   b. to provide structural support.
   c. to organize cells and other body parts.
   d. cell signaling.
   e. all of the above.

2. The protein found in the ECM of animals that provides strength and resistance to tearing when stretched is
   a. elastin.
   b. cellulose.
   c. collagen.
   d. laminin.
   e. fibronectin.

3. The polysaccharide that forms the hard outer covering of many invertebrates is
   a. collagen.
   b. chitin.
   c. chondroitin sulfate.
   d. pectin.
   e. cellulose.

4. The extension sequence found in procollagen polypeptides
   a. causes procollagen to be synthesized into the ER lumen.
   b. causes procollagen to form a triple helix.
   c. prevents procollagen from forming large collagen fibers.
   d. causes procollagen to be secreted from the cell.
   e. both b and c.

5. The dilated state of plasmodesmata allows the passage of
   a. water.
   b. ions.
   c. small molecules.
   d. macromolecules and viruses.
   e. all of the above.

6. The gap junctions of animal cells differ from the plasmodesmata of plant cells in that
   a. gap junctions serve as communicating junctions and plasmodesmata serve as anchoring junctions.
   b. gap junctions prevent extracellular material from moving between adjacent cells but plasmodesmata do not.
   c. gap junctions allow for direct exchange of cellular material between cells but plasmodesmata cannot allow the same type of exchange.
   d. gap junctions are formed by specialized proteins that form channels through the membranes of adjacent cells and plasmodesmata are formed by connecting the plasma membranes of adjacent cells.
   e. all of the above are correct.

7. Which of the following is (are) involved in the process of tissue and organ formation in multicellular organisms?
   a. cell division
   b. cell growth
   c. cell differentiation
   d. cell connections
   e. all of the above

8. The tissue type common to animals that functions in the conduction of electrical signals is
   a. epithelial.
   b. dermal.
   c. muscle.
   d. nervous.
   e. ground.

9. A type of tissue that is rich in ECM or has cells with a thick cell wall is
   a. dermal tissue in plants.
   b. ground tissue in plants.
   c. nervous tissue in animals.
   d. connective tissue in animals.
   e. both b and d.

10. Which of the following is *not* a correct statement comparing plant tissues and animal tissues?
   a. Nervous tissue of animals plays the same role as vascular tissue in plants.
   b. The dermal tissue of plants is similar to epithelial tissue of animals in that both provide a covering for the organism.
   c. The epithelial tissue of animals and the dermal tissue of plants have special characteristics that limit the movement of material between cell layers.
   d. The ground tissue of plants and the connective tissue of animals provide structural support for the organism.
   e. The ground tissue of plants and the connective tissue of animals have large amounts of extracellular material (that is, thick cell walls in plants and lots of ECM in animals).

## Conceptual Questions

1. What are key differences between the primary cell wall and the secondary cell wall of plant cells?

2. What are similarities and differences in the structures and functions of cadherins and integrins, proteins found in animal cells?

3. **Core Concept: Systems** We can view the body of a multicellular organism, such as a plant or animal, as a system of interconnected cells. Discuss how cell junctions play a key role in forming this system.

## Collaborative Questions

1. Discuss the similarities and differences between the ECM of animals and the cell walls of plants.

2. Cell junctions in animals are important in preventing cancer cells from metastasizing—moving to other parts of the body. Certain drugs bind to CAMs and influence their structure and function. Some of these drugs may help to prevent the spread of cancer cells. What would you hypothesize to be the mechanism by which such drugs work? What might be some harmful side effects?

# UNIT III
# GENETICS

**Genetics** is the branch of biology that deals with **inheritance**—the transmission of characteristics from parent to offspring. We begin this unit by examining the structure of the genetic material, namely DNA, at the molecular and cellular levels. We will explore the structure and replication of DNA and see how it is packaged into chromosomes (Chapter 11). We will then consider how segments of DNA are organized into units called genes, and how those genes are expressed at the molecular level to produce mRNA, proteins, and noncoding RNAs (Chapters 12 and 13). In Chapter 14, we will consider how the expression of genes is regulated. We will also examine how mutations alter the properties of genes and even lead to diseases such as cancer (Chapter 15).

In Chapter 16, we turn our attention to the mechanisms by which genes are transmitted from parent to offspring, beginning with a discussion of how chromosomes are sorted and transmitted during cell division. Chapters 17 and 18 explore the relationships between the transmission of genes and the outcome of an offspring's traits. We will look at genetic patterns called Mendelian inheritance and more complex patterns that could not have been predicted from Mendel's work.

The remaining chapters of this unit explore additional topics that are of interest to biologists. In Chapter 19, we will examine some of the unique genetic properties of bacteria and viruses. Chapter 20 considers the central role genes play in the development of animals and plants from a fertilized egg to an adult. We end this unit by exploring genetic technologies that are used by researchers, clinicians, and biotechnologists to unlock the mysteries of genes and provide tools and applications that benefit humans (Chapter 21).

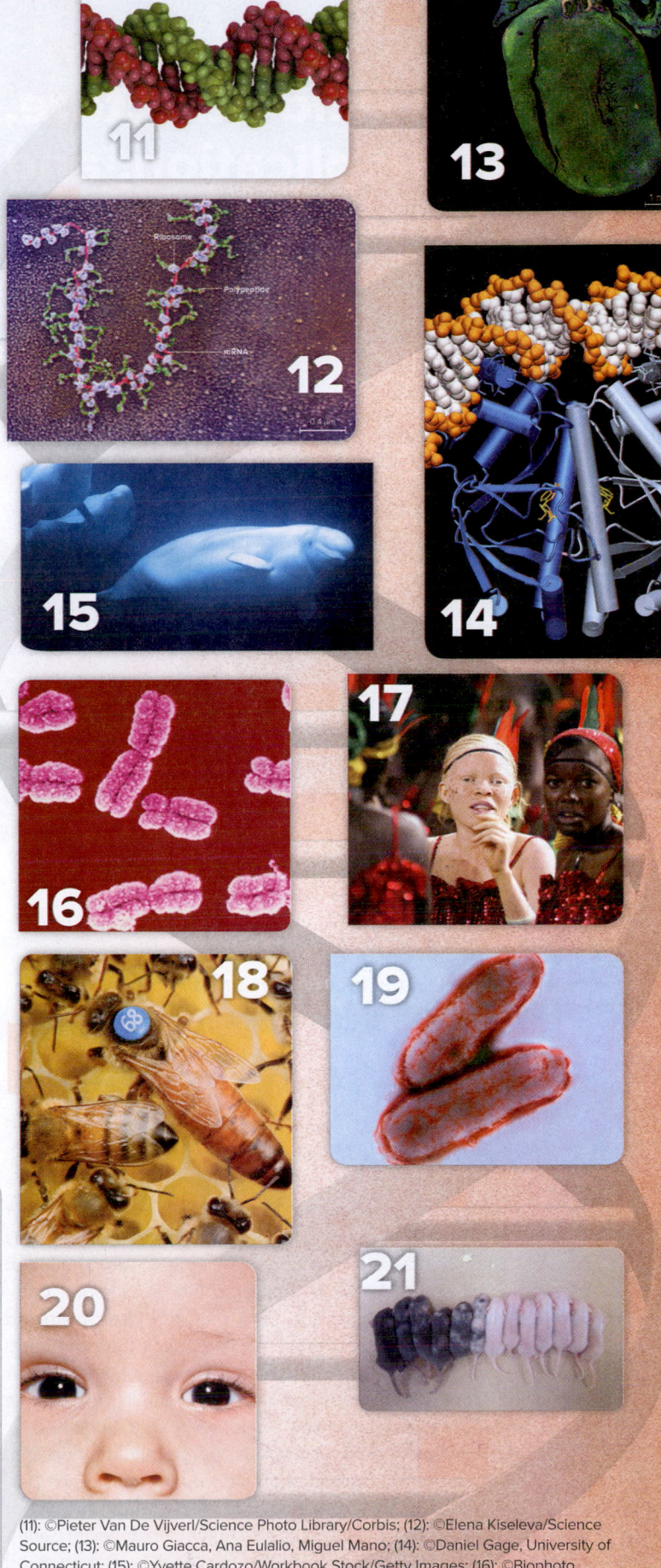

The following **Core Concepts** and **Core Skills** will be emphasized in this unit:

- *Information:* Throughout this unit, we will see how the genetic material carries the information to sustain life.
- *Structure and Function:* In Chapters 11 through 15, we will examine how the structures of DNA, RNA, genes, and chromosomes underlie their functions.
- *Quantitative Reasoning:* In Chapters 17 and 18, we will consider methods used to predict the outcome of genetic crosses.
- *Science and Society:* In Chapter 21, we will examine genetic technologies that have many applications in our society.
- *Process of Science:* Every chapter in this unit has a Feature Investigation that describes a pivotal experiment that provided insights into our understanding of genetics.

# Nucleic Acid Structure, DNA Replication, and Chromosome Structure

## 11

**A molecular model for the structure of a DNA double helix.**
©Pieter Van De Vijverl/Science Photo Library/Corbis

characteristics of unicellular and multicellular organisms. The past several decades have seen exciting advances in techniques and approaches for investigating and even altering the genetic material. These advances have greatly expanded our understanding of molecular genetics, and the techniques are widely used in related disciplines, including biochemistry, cell biology, and microbiology. Likewise, genetic techniques have many important applications in biotechnology and are used in the field of criminal justice, especially in forensics, to provide evidence of guilt or innocence.

To a large extent, our understanding of genetics comes from our knowledge of the molecular structure of DNA. In this chapter, we begin by considering some classic experiments that provided evidence that DNA is the genetic material. We will then survey the molecular features of DNA, which will allow us to appreciate how DNA can store information and be accurately copied. We will also consider the components of ribonucleic acid (RNA), which show striking similarities to those of DNA. Lastly, we will examine the molecular composition of chromosomes, where the DNA is found.

**O**n October 17, 2001, Mario K. was set free after serving 16 years in prison. He had been convicted of a sexual assault and murder that occurred in 1985. The charges were dropped because investigators discovered that another person, Edwin M., had actually committed the crime. How was Edwin M. identified as the real murderer? In 2001, he committed another crime, and his DNA was entered into a computer database. Edwin's DNA matched the DNA that had been collected from the victim in 1985, and other evidence was then gathered indicating that Edwin M. was the true murderer. Like Mario K., over 200 other inmates have been exonerated when DNA tests have shown that a different person was responsible for the crime.

**Deoxyribonucleic acid**, or **DNA**, is the genetic material that provides the blueprint to produce an individual's traits. Each person's DNA is distinct and unique. Even identical twins show minor differences in their DNA sequences. We begin our survey of genetics by examining DNA at the molecular level. Once we understand how DNA works at this level, it becomes easier to see how DNA functions to control the properties of cells and ultimately the

## 11.1 Biochemical Identification of the Genetic Material

**Learning Outcomes:**

1. List the four key criteria that the genetic material must fulfill.
2. **CoreSKILL »** Analyze the results of experiments that identified DNA as the genetic material.

DNA carries the genetic instructions for the traits of living organisms. In the case of multicellular organisms such as plants and animals, the information stored in the genetic material enables a fertilized egg to develop into an embryo and eventually into an adult organism. In addition, the genetic material allows organisms to survive in their native environments. For example, an individual's DNA provides the blueprint to produce enzymes that are needed to metabolize nutrients in food. To fulfill its role, the genetic material must meet the following key criteria:

1. *Information.* The genetic material must contain the information necessary to construct an entire organism.

2. *Replication.* The genetic material must be accurately copied, a process known as **replication**.

3. *Transmission.* After it is replicated, the genetic material can be passed from parent to offspring. It also must be passed from cell to cell during the process of cell division.

4. *Variation.* Differences in the genetic material must account for the known variation within each species and among different species.

How was the genetic material discovered? The quest to identify the genetic material began in the late 19th century, when a few scientists postulated that living organisms possess a blueprint that has a biochemical basis. In 1883, German biologist August Weismann and his Swiss colleague Karl Nägeli championed the idea that a chemical substance exists within living cells that is responsible for the transmission of traits from parents to offspring. During the next 30 years, experimentation along these lines centered on the behavior of **chromosomes**, the cellular structures that we now know contain the genetic material. The term chromosome is from the Greek words *chromo* and *soma*, meaning colored body, which refers to the observation of early microscopists that chromosomes are easily stained by colored dyes. By studying the transmission patterns of chromosomes from cell to cell and from parent to offspring, researchers were convinced that chromosomes carry the determinants that control the outcome of traits.

Ironically, the study of chromosomes initially misled researchers regarding the biochemical identity of the genetic material. Chromosomes contain two classes of macromolecules: proteins and DNA. Scientists of that era viewed proteins as being more biochemically complex because they are made from 20 different amino acids. Furthermore, biochemists already knew that proteins perform an amazingly wide range of functions, and complexity seemed an important prerequisite for the blueprint of an organism. By comparison, DNA seemed less complex, because it contains only four types of repeating units, called nucleotides, which will be described later in this chapter. In addition, the functional role of DNA in the nucleus had not been extensively investigated prior to the 1920s. Therefore, from the 1920s to the 1940s, most scientists were expecting research studies to reveal that proteins are the genetic material. Contrary to this expectation, however, several different experiments revealed that DNA carries out this critical role. In this section, we will examine one early line of study that involved research in microbiology.

## Griffith's Bacterial Transformation Experiments Indicated the Existence of a Genetic Material

In the late 1920s, an English microbiologist, Frederick Griffith, studied a type of bacterium known then as pneumococci and now classified as *Streptococcus pneumoniae*. Some strains of *S. pneumoniae* secrete a polysaccharide capsule, but other strains do not. When streaked on petri plates containing solid growth media, capsule-secreting strains have a smooth colony morphology. Those strains unable to secrete a capsule have a colony morphology that looks rough. In mammals, smooth strains of *S. pneumoniae* may cause pneumonia and other symptoms. In mice, such infections are usually fatal.

As shown in **Figure 11.1**, Griffith injected live and/or heat-killed bacteria into mice and then observed whether or not the bacteria caused them to die. He investigated the effects of two strains of *S. pneumoniae*: type S for smooth and type R for rough.

1. When injected into a live mouse, the type S strain killed the mouse (Figure 11.1, step 1). The capsule made by type S strains prevents the mouse's immune system from killing the bacterial cells. Following the death of the mouse, many type S bacteria were found in the mouse's blood.

2. When type R bacteria were injected into a mouse, the mouse survived, and after several days, living bacteria were not found in the live mouse's blood (Figure 11.1, step 2).

3. Griffith also heat-killed the type S bacteria and then injected them into a mouse. As expected, the mouse survived (Figure 11.1, step 3).

4. A surprising result occurred when Griffith mixed live type R bacteria with heat-killed type S bacteria and then injected them into a mouse—the mouse died (Figure 11.1, step 4). The blood from the dead mouse contained living type S bacteria! How did Griffith explain these results? He postulated that a substance from dead type S bacteria transformed the type R bacteria into

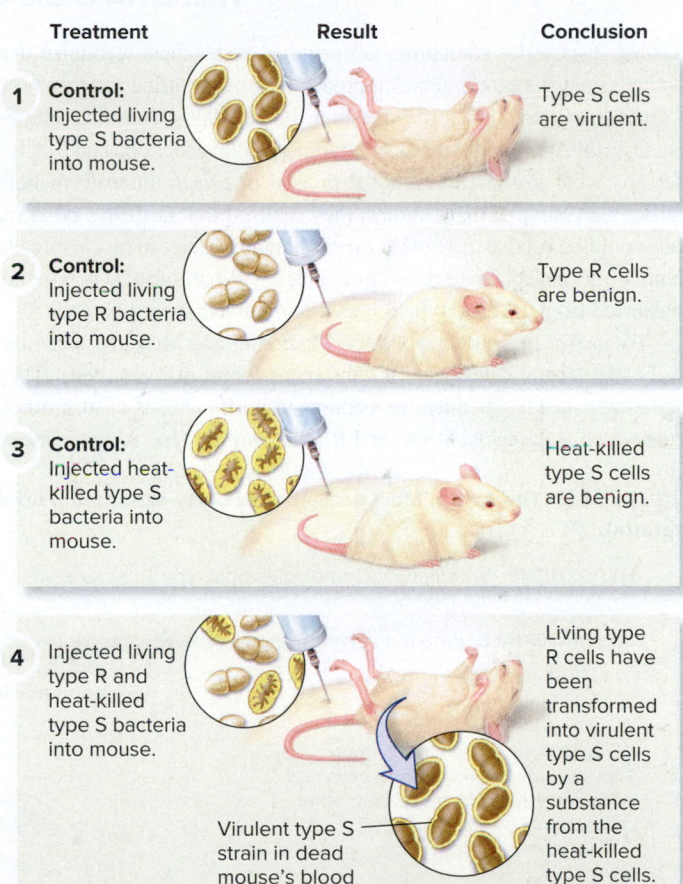

**Figure 11.1** **Griffith's experiments showing that genetic material can be transferred from one bacterium to another.** Note: To determine if a mouse's blood contained live bacteria, a sample of blood was also applied to solid growth media. (This part of the procedure is not shown.) For steps 1 and 4, smooth bacterial colonies were observed. For step 2, no bacterial colonies were observed because the type R cells were killed by the immune system of the mouse.

 **Core Skill: Connections** Look ahead to Figure 19.17. How does bacterial transformation play a role in the transfer of genes, such as antibiotic resistance genes, from one bacterial species to another?

type S bacteria. Griffith called this process **transformation**, and he termed the unidentified material responsible for this phenomenon the "transformation principle."

Let's consider what these observations mean with regard to the four criteria of the genetic material: information, replication, transmission, and variation. According to Griffith's results, the transformed bacteria had acquired the information (criterion 1) to make a capsule from the heat-killed cells. For the transformed bacteria to proliferate and thereby kill the mouse, the substance conferring the ability to make a capsule must be replicated (criterion 2) and then transmitted

(criterion 3) from mother to daughter cells during cell division. Finally, Griffith already knew that variation (criterion 4) existed in the ability of his strains to produce a capsule (S strain) or not produce a capsule (R strain). Taken together, these observations are consistent with the idea that the formation of a capsule is governed by genetic material. In the experiment of Figure 11.1, step 4 indicated that some genetic material from the heat-killed type S bacteria had been transferred to the living type R bacteria and provided those bacteria with a new trait. At the time of his studies, however, Griffith could not determine the biochemical composition of the transforming substance.

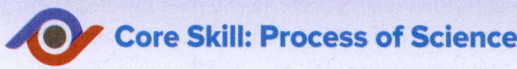

## Core Skill: Process of Science

## Feature Investigation | Avery, MacLeod, and McCarty Used Purification Methods to Reveal That DNA Is the Genetic Material

Exciting discoveries sometimes occur when researchers recognize that another scientist's experimental approach may be modified and then used to dig deeper into a scientific question. In the 1940s, American physician Oswald Avery and American biologists Colin MacLeod and Maclyn McCarty were also interested in the process of bacterial transformation. During the course of their studies, they realized that Griffith's observations could be used as part of an experimental strategy to biochemically identify the genetic material. They asked, "What substance is being transferred from the dead type S bacteria to the live type R bacteria?"

To answer this question, Avery, MacLeod, and McCarty needed to purify the general categories of substances found in living cells. They used established biochemical procedures to purify classes of macromolecules, such as proteins, DNA, and RNA, from the type S streptococcal

strain. Initially, they discovered that only the purified DNA could convert type R bacteria into type S. To further verify that DNA is the genetic material, they performed the investigation outlined in **Figure 11.2**. They purified DNA from the type S bacteria and mixed it with type R bacteria. After allowing time for DNA uptake into the type R bacteria, they added an antibody that aggregated any nontransformed type R bacteria, which were then removed by centrifugation. The remaining bacteria were placed on solid growth media within petri plates and incubated overnight to allow the division and growth of cells to form visible bacterial colonies.

As a control, no DNA extract was added, and no type S bacterial colonies were observed on the petri plates (see plate A in step 6). When the researchers mixed their S strain DNA extract with type R bacteria, some of the bacteria were converted to type S bacteria (see

**Figure 11.2**  The Avery, MacLeod, and McCarty experiments that identified DNA as Griffith's transformation principle—the genetic material.

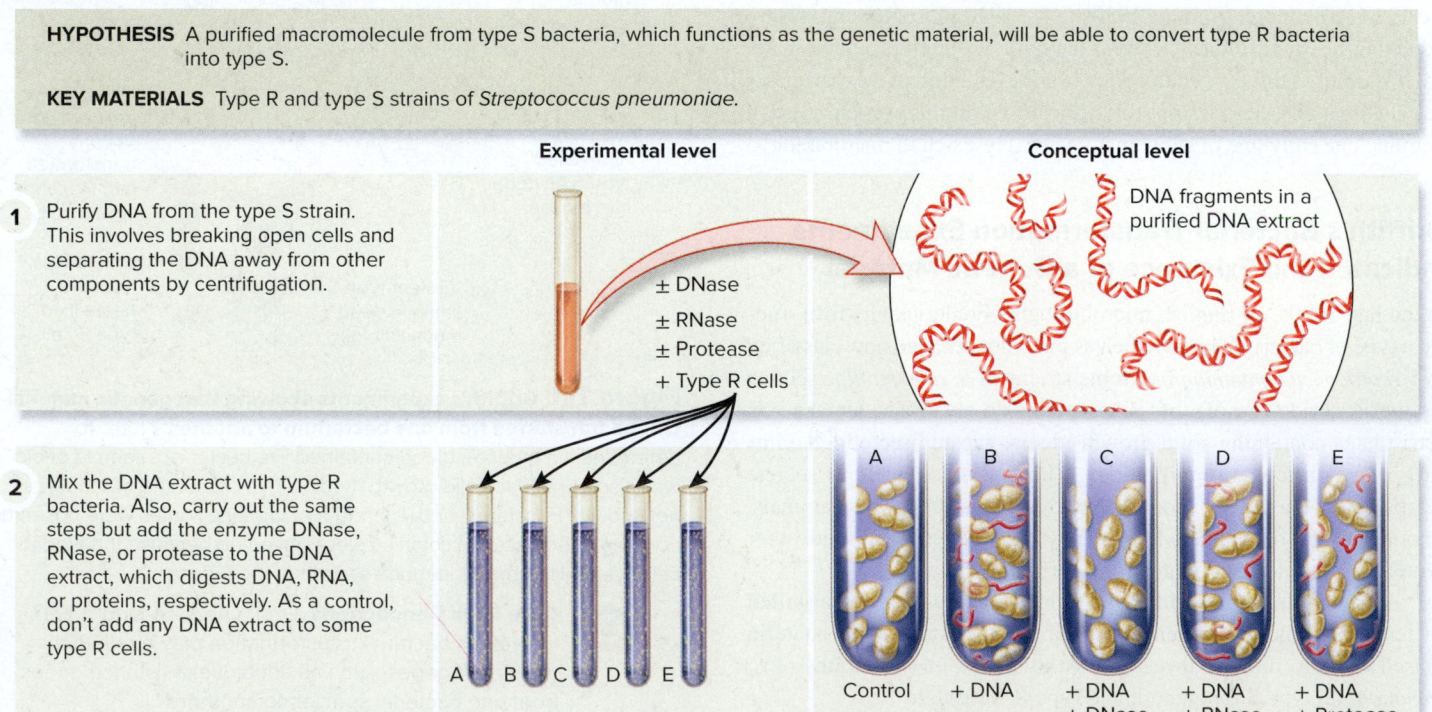

**HYPOTHESIS** A purified macromolecule from type S bacteria, which functions as the genetic material, will be able to convert type R bacteria into type S.

**KEY MATERIALS** Type R and type S strains of *Streptococcus pneumoniae*.

Experimental level    Conceptual level

1 Purify DNA from the type S strain. This involves breaking open cells and separating the DNA away from other components by centrifugation.

± DNase
± RNase
± Protease
+ Type R cells

DNA fragments in a purified DNA extract

2 Mix the DNA extract with type R bacteria. Also, carry out the same steps but add the enzyme DNase, RNase, or protease to the DNA extract, which digests DNA, RNA, or proteins, respectively. As a control, don't add any DNA extract to some type R cells.

A B C D E

A    B    C    D    E
Control    + DNA    + DNA + DNase    + DNA + RNase    + DNA + Protease

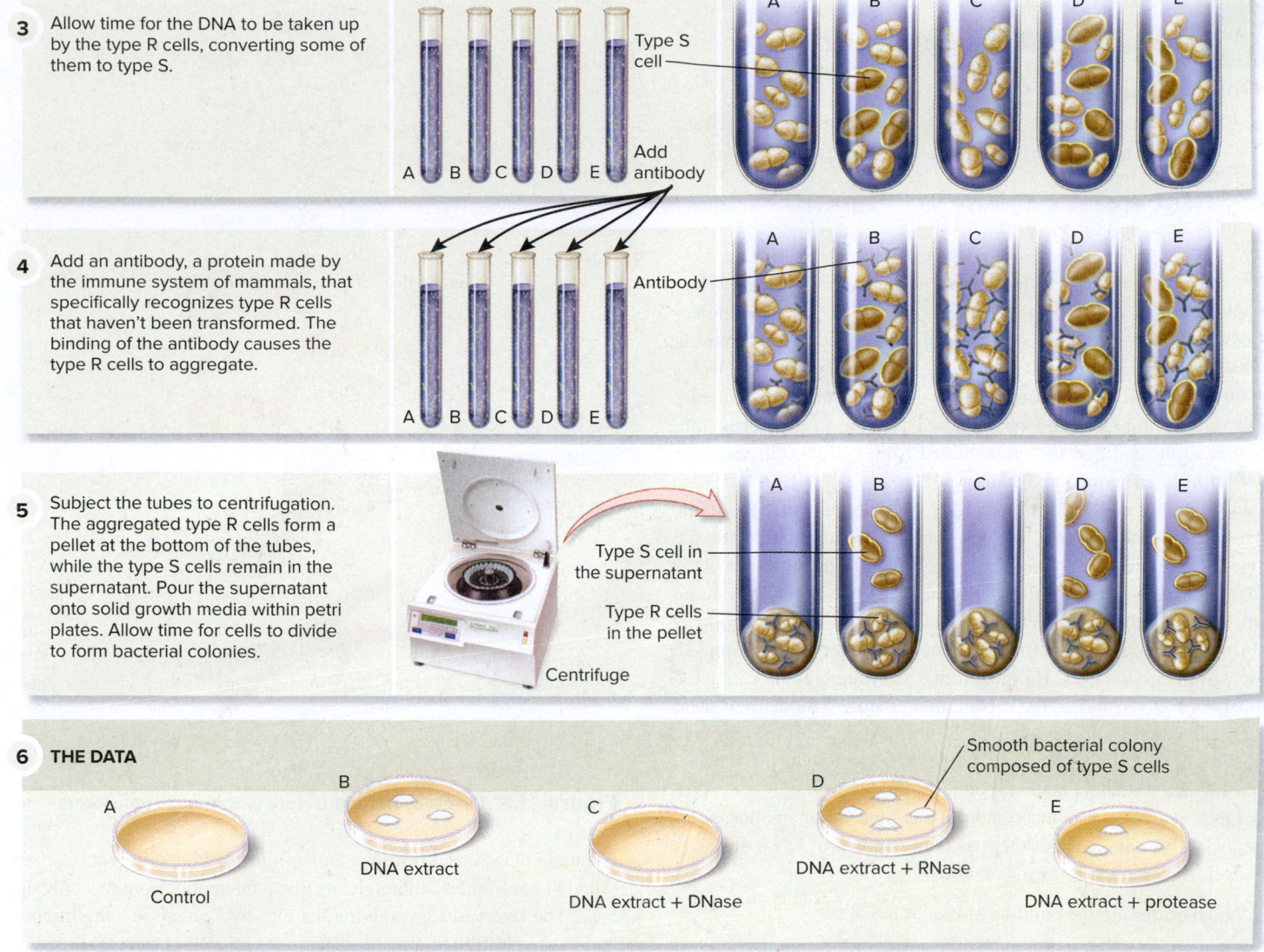

**3** Allow time for the DNA to be taken up by the type R cells, converting some of them to type S.

Type S cell

Add antibody

**4** Add an antibody, a protein made by the immune system of mammals, that specifically recognizes type R cells that haven't been transformed. The binding of the antibody causes the type R cells to aggregate.

Antibody

**5** Subject the tubes to centrifugation. The aggregated type R cells form a pellet at the bottom of the tubes, while the type S cells remain in the supernatant. Pour the supernatant onto solid growth media within petri plates. Allow time for cells to divide to form bacterial colonies.

Type S cell in the supernatant

Type R cells in the pellet

Centrifuge

**6** **THE DATA**

Smooth bacterial colony composed of type S cells

A

Control

B

DNA extract

C

DNA extract + DNase

D

DNA extract + RNase

E

DNA extract + protease

**7** **CONCLUSION** DNA is responsible for transforming type R cells into type S cells.

**8** **SOURCE** Avery, O.T., MacLeod, C.M., and McCarty, M. 1944. Studies on the Chemical Nature of the Substance Inducing Transformation of Pneumococcal Types. *Journal of Experimental Medicine* 79: 137–158.

plate B in step 6 of Figure 11.2). This result was consistent with the idea that DNA is the genetic material.

Even so, a careful biochemist could argue that the DNA extract might not have been 100% pure. For this reason, the researchers realized that a small amount of contaminating material in the DNA extract could actually be the genetic material. The most likely contaminating substances in this case would be RNA or protein.

To address this possibility, Avery, MacLeod, and McCarty treated the DNA extract with an enzyme that digests either DNA (called **DNase**), RNA (**RNase**), or protein (**protease**) (see step 2). When the DNA extracts were treated with RNase or protease, the type R bacteria were still converted into type S bacteria, indicating that contaminating RNA or protein in the extract was not acting as the genetic material (see step 6, plates D and E). Moreover, when

the extract was treated with DNase, it lost the ability to convert type R bacteria into type S bacteria (see plate C). Taken together, these results were consistent with the idea that DNA is the genetic material.

**Experimental Questions**

1. **CoreSKILL »** Avery, MacLeod, and McCarty worked with two strains of *Streptococcus pneumoniae* to determine the biochemical identity of the genetic material. Explain the characteristics of the *S. pneumoniae* strains that made them particularly well suited for the researchers' experiment.

2. What is a DNA extract?

3. **CoreSKILL »** In the experiment of Avery, MacLeod, and Mc̸
what was the purpose of using protease, RNase, and D̸
only the DNA extract caused transformation?

# 11.2 Nucleic Acid Structure

**Learning Outcomes:**

1. Outline the structural features of DNA at five levels of complexity.
2. Describe the structures of nucleotides, a DNA strand, and the DNA double helix.
3. **CoreSKILL »** Discuss and interpret the work of Franklin; Chargaff; and Watson and Crick.

A core concept in biology is that structure determines function. When biologists want to understand the function of a material at the molecular and cellular level, they focus some of their efforts on determining its biochemical structure. In this regard, an understanding of DNA's structure has proven to be particularly exciting because the structure makes it easier for us to understand how DNA can store information, how it is replicated and then transmitted from cell to cell, and how variation in its structure can occur.

DNA and its molecular cousin, RNA, are known as **nucleic acids**, polymers consisting of nucleotides, which are responsible for the storage, expression, and transmission of genetic information. This term is derived from the discovery of DNA by Swiss physician Friedrich Miescher in 1869. He identified a novel phosphorus-containing substance from the nuclei of white blood cells found in waste surgical bandages. He named this substance nuclein. As the structure of DNA and RNA became better understood, they were found to be acidic molecules, which means they release hydrogen ions ($H^+$) in solution and have a net negative charge at neutral pH. Thus, the name nucleic acid was coined.

DNA is a very large macromolecule composed of smaller building blocks. We can consider the structural features of DNA at different levels of complexity (**Figure 11.3**):

1. **Nucleotides** are the building blocks of DNA.
2. A **strand** of DNA is formed by the covalent linkage of nucleotides in a linear manner.
3. Two strands of DNA hydrogen-bond with each other to form a **double helix**. In a DNA double helix, two DNA strands are twisted together to form a structure that resembles a spiral staircase.
4. In living cells, DNA is associated with an array of different proteins to form chromosomes. The association of proteins with DNA organizes the long double helix into a compact structure.
5. A **genome** is the complete complement of an organism's genetic material. For example, the genome of most bacteria is a single circular chromosome, whereas eukaryotic cells have DNA in their nucleus, mitochondria, and chloroplasts.

The first three levels of complexity will be the focus of this section. Level 4 will be discussed in Section 11.5, and level 5 is examined in Chapter 21.

## Nucleotides Contain a Phosphate, a Sugar, and a Base

A nucleotide has three components: a phosphate group, a pentose (five-carbon) sugar, and a nitrogen-containing base (**Figure 11.4**). The nucleotides in DNA and RNA contain different sugars. Deoxyribose

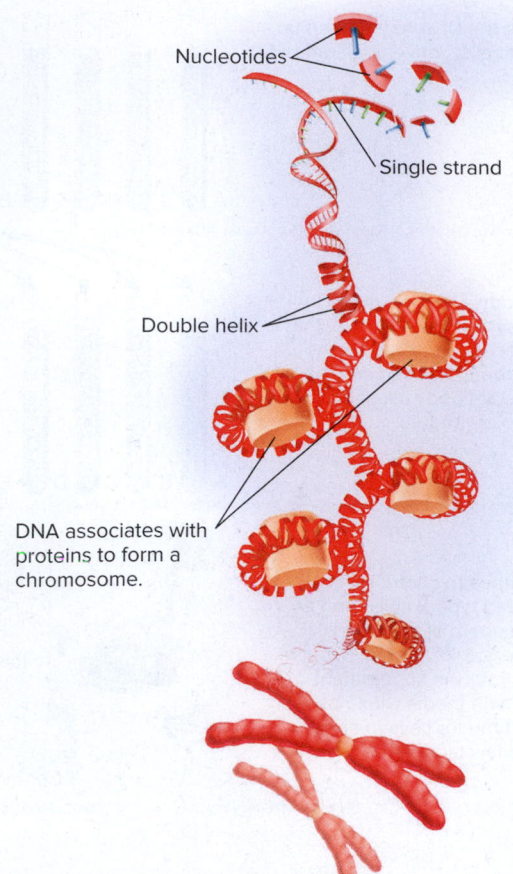

Nucleotides

Single strand

Double helix

DNA associates with proteins to form a chromosome.

**Figure 11.3  Levels of DNA structure within a chromosome.**

is found in DNA, and ribose is found in RNA. Five different bases are found in nucleotides, although any given nucleotide contains only one base. The five bases are subdivided into two categories, the **purines** and the **pyrimidines**, due to differences in their structures (see Figure 11.4). The purine bases, **adenine (A)** and **guanine (G)**, have a double-ring structure, whereas the pyrimidine bases, **thymine (T)**, **cytosine (C)**, and **uracil (U)**, have a single-ring structure. Adenine, guanine, and cytosine are found in both DNA and RNA. Thymine is found only in DNA, whereas uracil is found only in RNA.

A conventional numbering system describes the locations of carbon and nitrogen atoms in the sugars and bases (**Figure 11.5**). The prime symbol ($'$) is used to distinguish the numbering of carbons in the sugar. The atoms in the ring structures of the bases are not given the prime designation. The sugar carbons are designated $1'$ (read as "one prime"), $2'$, $3'$, $4'$, and $5'$, with the carbon atoms numbered in a clockwise direction starting with the carbon atom to the right of the ring oxygen atom. The fifth carbon is outside the ring. A base is attached to the $1'$ carbon atom, and a phosphate group is attached at the $5'$ position. Compared with ribose (see Figure 11.4), deoxyribose lacks a single oxygen atom at the $2'$ position; the prefix deoxy- (meaning without oxygen) refers to this missing atom.

## A Strand Is a Linear Linkage of Nucleotides with Directionality

The next level of nucleic acid structure is the formation of a strand of DNA or RNA in which nucleotides are covalently attached to each other

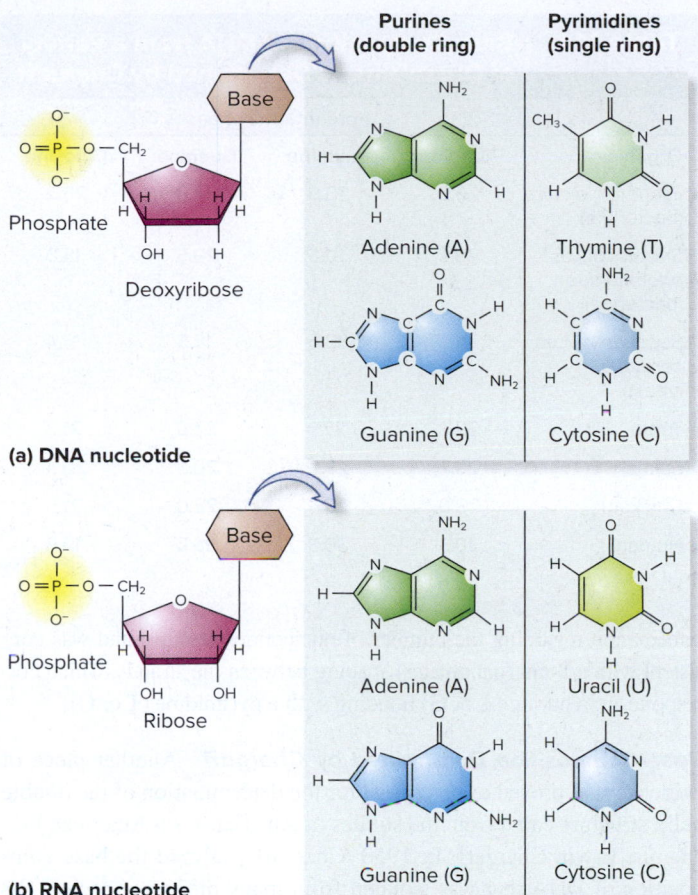

**(a) DNA nucleotide**

**(b) RNA nucleotide**

**Figure 11.4  Nucleotides and their components.** For simplicity, the carbon atoms in the ring structures are shown only for guanine and cytosine in part (a).

**Concept Check:** *Which pyrimidine(s) is (are) found in both DNA and RNA?*

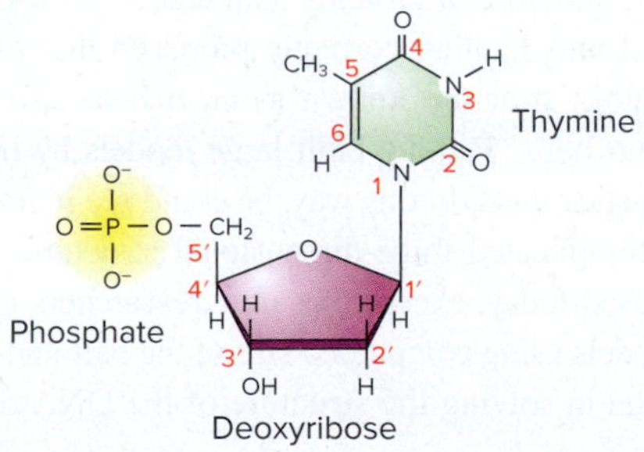

**Figure 11.5  Conventional numbering in a DNA nucleotide.** The carbons in the sugar are given a prime designation, whereas those in the base are not.

**Concept Check:** *What is the numbering designation of the carbon atom to which the phosphate is attached?*

in a linear fashion. **Figure 11.6** depicts a short strand of DNA with four nucleotides. The linkage is a phosphoester bond (a covalent bond between phosphorus and oxygen) involving a sugar molecule in one nucleotide and a phosphate group in the next nucleotide. Another way of viewing this linkage is to notice that a phosphate group connects two sugar molecules. From this perspective, the linkage in DNA and RNA strands is called a **phosphodiester linkage**, which has two phosphoester bonds.

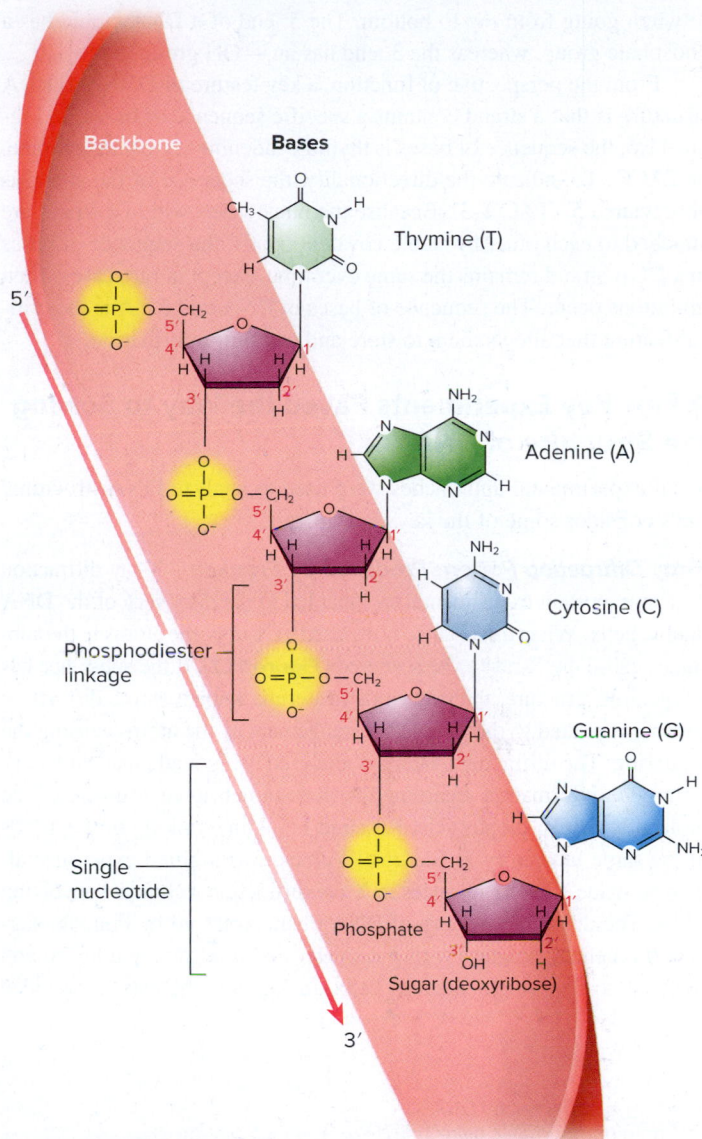

**Figure 11.6  The structure of a DNA strand.** Nucleotides are covalently bonded to each other in a linear manner. Notice the directionality of the strand and that it carries a particular sequence of bases. An RNA strand has a very similar structure, except the sugar is ribose rather than deoxyribose, and uracil is substituted for thymine.

**Core Concept: Information**  The covalent linkage of a sequence of bases allows DNA to store information.

The phosphates and sugar molecules form the **backbone** of a DNA or RNA strand, and the bases project from the backbone. The backbone is negatively charged due to the negative charges of the phosphate groups.

An important structural feature of a DNA strand is the orientation of the nucleotides. Each phosphate in a phosphodiester linkage is covalently bonded to the 5′ carbon in one nucleotide and to the 3′ carbon in the other. In a strand, all sugar molecules are oriented in the same direction. For example, in the strand shown in Figure 11.6, all of the 5′ carbons in every sugar molecule are above the 3′ carbons. A strand has a **directionality** based on the orientation of the sugar molecules within that strand. In Figure 11.6, the direction of the strand is said to be 5′ to

3′ when going from top to bottom. The 5′ end of a DNA strand has a phosphate group, whereas the 3′ end has an —OH group.

From the perspective of function, a key feature of DNA and RNA structure is that a strand contains a specific sequence of bases. In Figure 11.6, the sequence of bases is thymine–adenine–cytosine–guanine, or TACG. To indicate the directionality, the sequence of the strand is abbreviated 5′–TACG–3′. Because the nucleotides within a strand are attached to each other by stable covalent bonds, the sequence of bases in a DNA strand remains the same over time, except in rare cases when mutations occur. The sequence of bases in DNA and RNA is the critical feature that allows them to store and transmit information.

## A Few Key Experiments Paved the Way to Solving the Structure of DNA

What experimental approaches were used to analyze DNA structure? Let's consider some of the key experiments.

**X-ray Diffraction Pattern Produced by Franklin** X-ray diffraction was an important experimental tool that led to the discovery of the DNA double helix. When a substance is exposed to X-rays, the atoms in the substance cause the X-rays to be scattered (**Figure 11.7**). If the substance has a repeating structure, the pattern of scattering, known as the diffraction pattern, is related to the structural arrangement of the atoms causing the scattering. The diffraction pattern is analyzed using mathematical theory to provide information regarding the three-dimensional structure of the molecule. British biophysicist Rosalind Franklin, working in the 1950s in the same laboratory as Maurice Wilkins, was a gifted experimentalist who made marked advances in X-ray diffraction techniques involving DNA. The diffraction pattern of DNA fibers produced by Franklin suggested a helical structure with a diameter that is relatively uniform and too wide to be a single-stranded helix. In addition, the pattern provided

| Table 11.1 | Base Composition of DNA from a Variety of Organisms as Determined by Chargaff | | | |
|---|---|---|---|---|
| | Percentages of bases (%) | | | |
| **Organism** | **Adenine** | **Thymine** | **Guanine** | **Cytosine** |
| *Escherichia coli* (bacterium) | 26.0 | 23.9 | 24.9 | 25.2 |
| *Streptococcus pneumoniae* (bacterium) | 29.8 | 31.6 | 20.5 | 18.0 |
| *Saccharomyces cerevisiae* (yeast) | 31.7 | 32.6 | 18.3 | 17.4 |
| Turtle | 28.7 | 27.9 | 22.0 | 21.3 |
| Salmon | 29.7 | 29.1 | 20.8 | 20.4 |
| Chicken | 28.0 | 28.4 | 22.0 | 21.6 |
| Human | 30.3 | 30.3 | 19.5 | 19.9 |

information regarding the number of nucleotides per turn and was consistent with a 2-nm (nanometer) spacing between the strands, which corresponds to a purine (A or G) bonding with a pyrimidine (T or C).

**Base Composition Determined by Chargaff** Another piece of evidence that proved to be critical for the determination of the double helix structure came from the studies of Austrian-born American biochemist Erwin Chargaff. In 1950, Chargaff analyzed the base composition of DNA that was isolated from many different species. His experiments consistently showed that the amount of adenine in each sample was similar to the amount of thymine, and the amount of cytosine was similar to the amount of guanine (**Table 11.1**).

**Model Building by Pauling** In the early 1950s, more information was known about the structure of proteins than that of nucleic acids. American biochemist Linus Pauling correctly proposed that some regions of proteins fold into a structure known as an α helix. To determine the structure of the α helix, Pauling built large models by linking together simple ball-and-stick units. In this way, he could see if atoms fit together properly in a complicated three-dimensional structure. This approach is still widely used today, except that now researchers construct three-dimensional models using computers. Use of the ball-and-stick approach was instrumental in solving the structure of the DNA double helix.

## Watson and Crick Deduced the Double Helix Structure of DNA

Thus far, we have considered the experimental studies that led to the determination of the DNA double helix. American biologist James Watson and English biologist Francis Crick, working together at Cambridge University, assumed that nucleotides are linked together in a linear fashion and that the chemical linkage between two nucleotides is always the same. In collaboration with Wilkins, they then set out to build ball-and-stick models that incorporated all of the known experimental observations.

Modeling of chemical structures involves trial and error. Watson and Crick initially considered several incorrect models. One model was a double helix in which the bases were on the outside of the helix. In another model, each base formed hydrogen bonds with the identical base in the

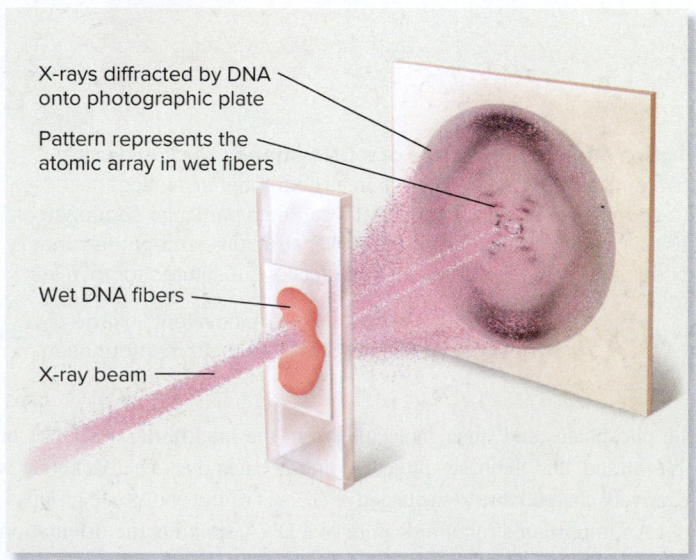

**Figure 11.7 Franklin's X-ray diffraction of DNA fibers.** The exposure of DNA wet fibers to X-rays causes the X-rays to be scattered and the pattern of scattering is related to the position of the atoms in the DNA fibers.

X-rays diffracted by DNA onto photographic plate
Pattern represents the atomic array in wet fibers
Wet DNA fibers
X-ray beam

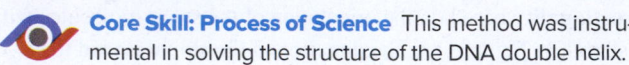

**Core Skill: Process of Science** This method was instrumental in solving the structure of the DNA double helix.

opposite strand (A to A, T to T, G to G, and C to C). However, model-building revealed that purine-purine pairs were too wide and pyrimidine-pyrimidine pairs were too narrow to fit the uniform diameter of DNA revealed from Franklin's work. Eventually, they realized that the hydrogen bonding of adenine to thymine was structurally similar to that of guanine to cytosine. In both cases, a purine base (A or G) bonds with a pyrimidine base (T or C). With an interaction between A and T and between G and C, the ball-and-stick models showed that the two strands would form a double helix structure in which all atoms would fit together properly.

Watson and Crick proposed the structure of DNA, which was published in the journal *Nature* in 1953. In 1962, Watson, Crick, and Wilkins were awarded the Nobel Prize in Physiology or Medicine.

Unfortunately, Rosalind Franklin had died before this time, and the Nobel Prize is awarded only to living recipients.

## DNA Has a Repeating, Antiparallel Helical Structure Formed by the Complementary Base Pairing of Nucleotides

The structure that Watson and Crick proposed is a double-stranded, helical structure with the sugar-phosphate backbone on the outside and the bases on the inside (**Figure 11.8a**). This structure is stabilized by hydrogen bonding between the bases in opposite strands to form **base pairs**. A distinguishing feature of base pairing is its specificity. An adenine (A) base in one strand forms two hydrogen bonds with a

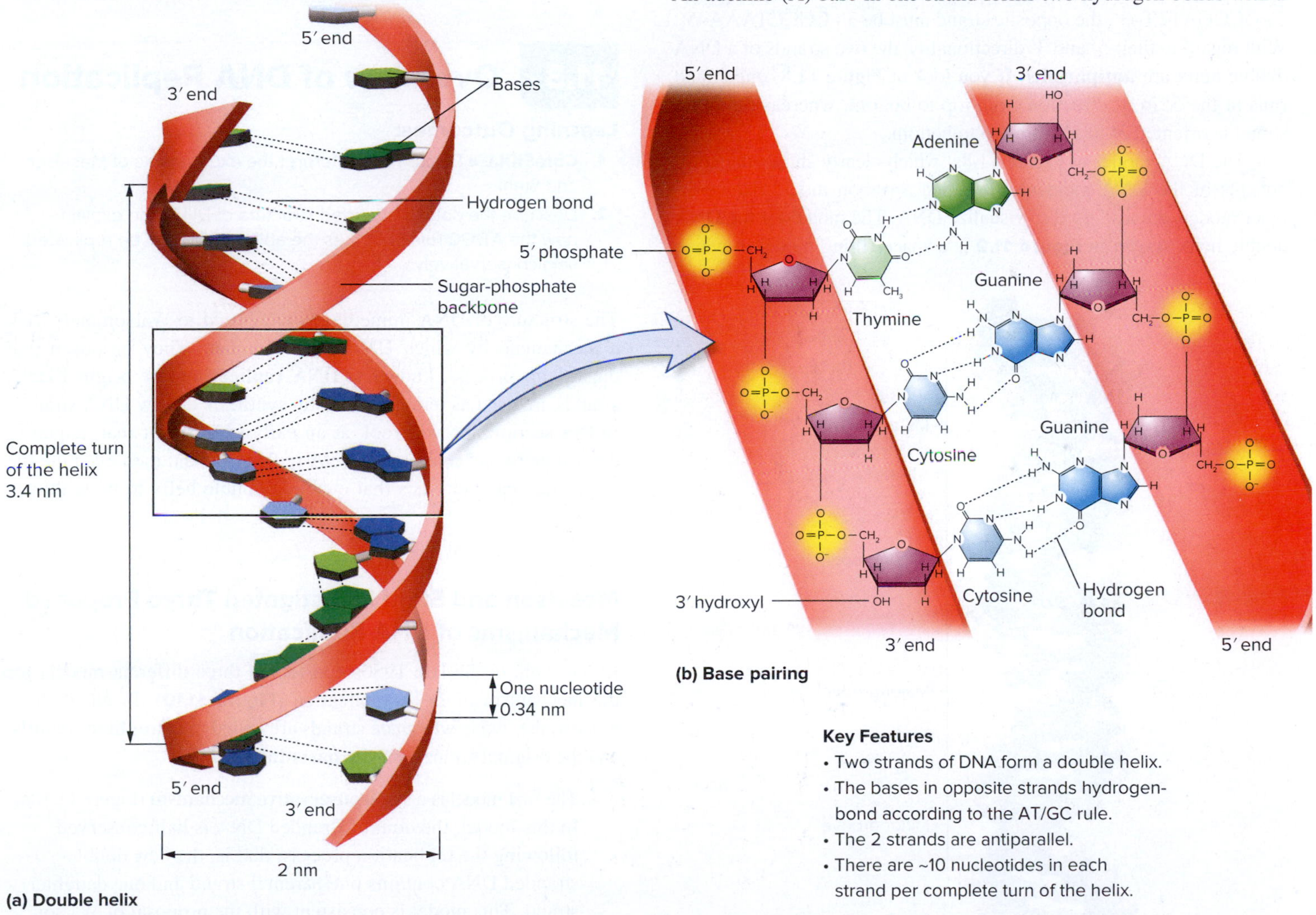

**Key Features**
- Two strands of DNA form a double helix.
- The bases in opposite strands hydrogen-bond according to the AT/GC rule.
- The 2 strands are antiparallel.
- There are ~10 nucleotides in each strand per complete turn of the helix.

**Figure 11.8** **Structure of the DNA double helix.** As seen in part **(a)**, DNA is a helix composed of two antiparallel strands. Part **(b)** shows the AT/GC base pairing that holds the strands together via hydrogen bonds.

 **Core Skill: Modeling** The goal of this modeling challenge is to predict the hydrogen-bonding relationship between O⁶-MeG and cytosine.

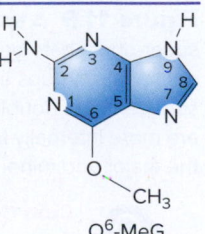

**Modeling Challenge:** As discussed in Chapter 15, certain chemicals, such as nitrogen mustard and ethyl methanesulfonate, can modify the structures of DNA bases. For example, a methyl group (–CH₃) can be attached to the oxygen atom on guanine, thereby creating 6-O-methylguanine (O⁶-MeG), as shown to the right. When O⁶-MeG is included in a DNA strand, it can form only two hydrogen bonds with cytosine instead of three. Draw a model for the base pairing between O⁶-MeG and cytosine.

thymine (T) base in the opposite strand, or a guanine (G) base forms three hydrogen bonds with a cytosine (C) base (**Figure 11.8b**). This **AT/GC rule** is consistent with Chargaff's observation that DNA contains approximately equal amounts of A and T, and equal amounts of G and C. According to the AT/GC rule, purines (A and G) always bond with pyrimidines (T and C) (recall that purines have a double-ring structure, whereas pyrimidines have single rings). This keeps the width of the double helix relatively constant. One complete turn of the double helix is 3.4 nm in length and comprises about 10 base pairs.

Due to the AT/GC rule, the base sequences of two DNA strands are **complementary** to each other. That is, you can predict the sequence in one DNA strand if you know the sequence in the opposite strand. For example, if one DNA strand has the sequence 5′–GCGGATTT–3′, the opposite strand must be 3′–CGCCTAAA–5′. With regard to their 5′ and 3′ directionality, the two strands of a DNA double helix are **antiparallel**. If you look at Figure 11.8, one strand runs in the 5′ to 3′ direction from top to bottom, whereas the other strand is oriented 3′ to 5′ from top to bottom.

The DNA model in Figure 11.8a, which clearly shows the components of the DNA molecule, is called a ribbon model. However, other models are also used to visualize DNA. The model for the DNA double helix shown in **Figure 11.9** is a space-filling model in which

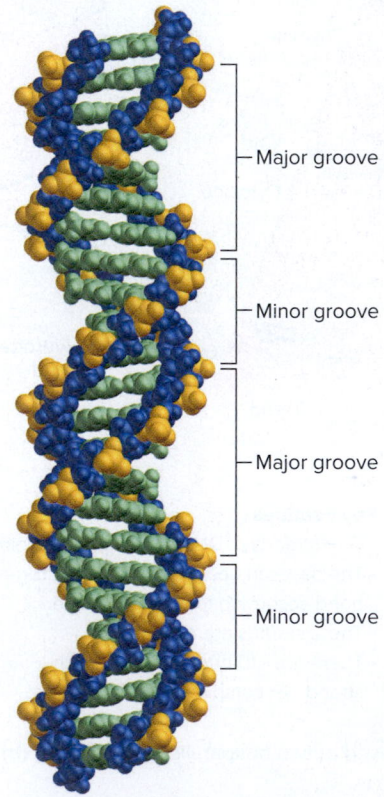

- Major groove
- Minor groove
- Major groove
- Minor groove

**Figure 11.9    A space-filling model of the DNA double helix.** In the sugar-phosphate backbone, sugar molecules are shown in blue, and phosphate groups are yellow. The backbone is on the outermost surface of the double helix. The atoms of the bases, shown in green, are more internally located within the double-stranded structure. Notice the major and minor grooves that are formed by this arrangement.

 **Core Concept: Structure and Function** The major groove provides a binding site for proteins that control the expression of genes.

the atoms are depicted as spheres. Why is this model useful? This type of structural model emphasizes the surface of DNA. As you can see in this model, the sugar-phosphate backbone is on the outermost surface of the double helix; the backbone has the most direct contact with water in the surroundings. The atoms of the bases are more internally located within the double-stranded structure. The indentations where the atoms of the bases make contact with the surrounding water are termed grooves. Two grooves, called the **major groove** and the **minor groove**, spiral around the double helix. As discussed in later chapters, the major groove provides a location where a protein can bind to a particular sequence of bases and affect the expression of a gene (for example, look ahead to Figure 14.10).

## 11.3    Overview of DNA Replication

**Learning Outcomes:**

1. **CoreSKILL »** Discuss and interpret the experiments of Meselson and Stahl.
2. Describe the double-stranded structure of DNA, and explain how the AT/GC rule underlies the ability of DNA to be replicated semiconservatively.

The structure of DNA immediately suggested to Watson and Crick a mechanism by which DNA can be copied. They proposed that during this process, known as **DNA replication**, the original DNA strands are used as templates for the synthesis of new DNA strands. In this section, we will look at an early experiment that helped to determine the mechanism of DNA replication and then examine the structural characteristics that enable a double helix to be faithfully copied.

### Meselson and Stahl Investigated Three Proposed Mechanisms of DNA Replication

Researchers in the late 1950s considered three different models for the mechanism of DNA replication (**Figure 11.10**). In all of these models, the two newly made strands are called the **daughter strands**, and the original strands are the **parental strands**.

- The first model is a **semiconservative mechanism** (Figure 11.10a). In this model, the double-stranded DNA is half conserved following the replication process; that is, the new double-stranded DNA contains one parental strand and one daughter strand. This model is consistent with the proposal of Watson and Crick.

- According to a second model, called a **conservative mechanism**, both parental strands of DNA remain together following DNA replication (Figure 11.10b). The original arrangement of parental strands is completely conserved, and the two newly made daughter strands are also together following replication.

- A third possibility, called a **dispersive mechanism**, proposed that segments of parental DNA and newly made daughter DNA are interspersed in both strands following the replication process (Figure 11.10c).

| Original double helix | First round of replication | Second round of replication |
| --- | --- | --- |

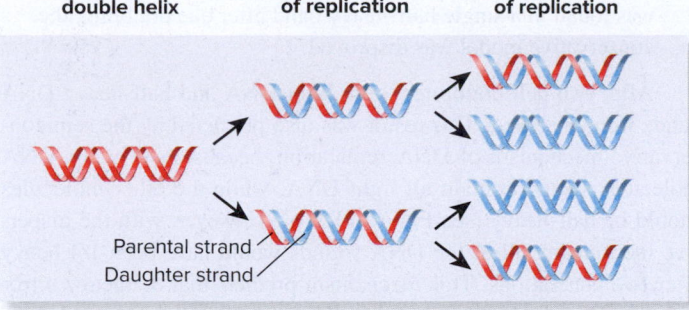

Parental strand
Daughter strand

**(a) Semiconservative mechanism. DNA replication produces DNA molecules with 1 parental strand and 1 newly made daughter strand.**

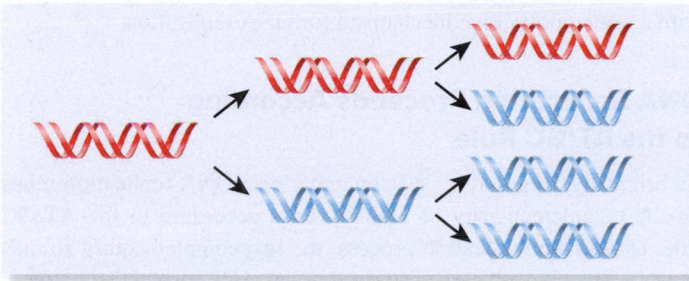

**(b) Conservative mechanism. DNA replication produces 1 double helix with both parental strands and the other with 2 new daughter strands.**

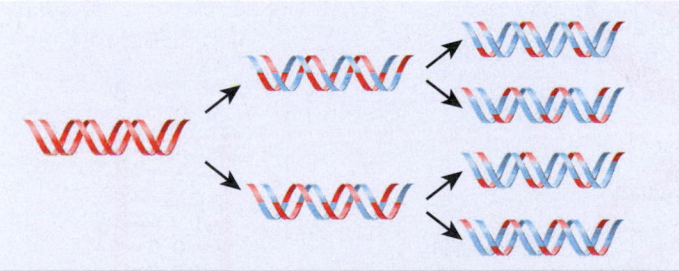

**(c) Dispersive mechanism. DNA replication produces DNA strands in which segments of new DNA are interspersed with the parental DNA.**

**Figure 11.10   Three proposed mechanisms for DNA replication.** The strands of the original (parental) double helix are shown in red. Two rounds of replication are illustrated with the daughter strands, which are shown in blue.

In 1958, American biologists Matthew Meselson and Franklin Stahl devised an experimental approach to distinguish among these three mechanisms. An important feature of their research was the use of isotope labeling. Nitrogen, which is found in DNA, occurs in a common light ($^{14}$N) form and a rare heavy ($^{15}$N) form. Meselson and Stahl studied DNA replication in the bacterium *Escherichia coli*.

1. They grew *E. coli* cells for many generations in a medium that contained only the $^{15}$N form of nitrogen (**Figure 11.11**). This produced a population of bacterial cells in which all of the DNA was $^{15}$N-labeled.

2. Then they switched the bacteria to a medium that contained only $^{14}$N as the nitrogen source. The cells were allowed to divide,

**1**  Grow bacteria in $^{15}$N media.

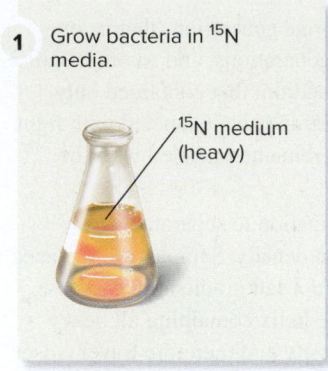

$^{15}$N medium (heavy)

**2**  Transfer to $^{14}$N media and continue growth for <1.0, 1.0, 2.0, or 3.0 generations.

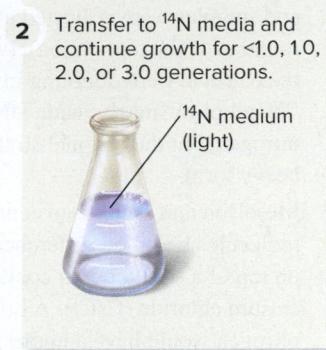

$^{14}$N medium (light)

**3**  Isolate DNA after each generation. Transfer DNA to CsCl gradient, and centrifuge.

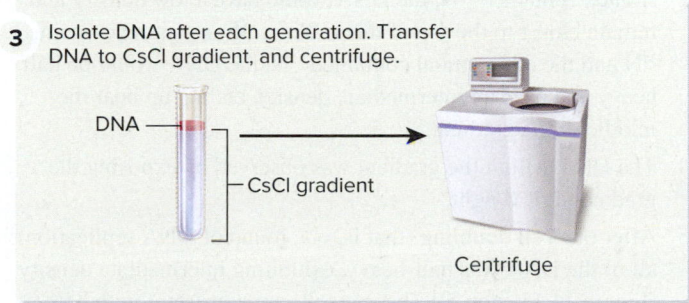

DNA

CsCl gradient

Centrifuge

**4**  Observe DNA under UV light.

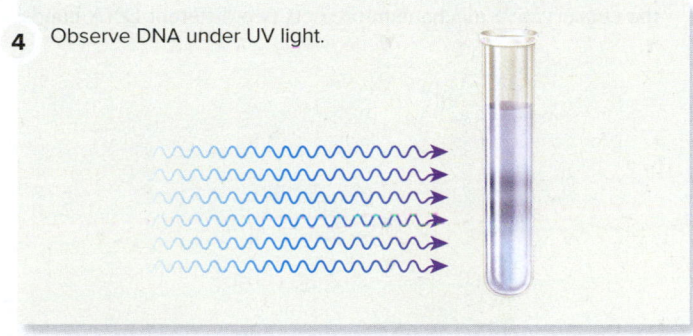

**5**  **THE DATA**

Approximate generations after transfer to $^{14}$N medium.

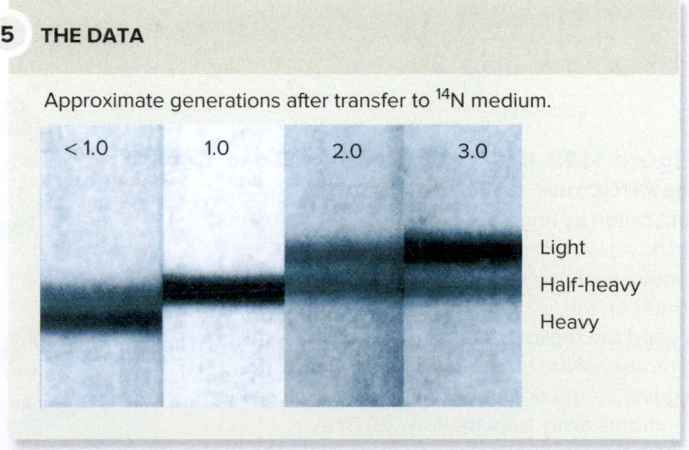

< 1.0    1.0    2.0    3.0

Light
Half-heavy
Heavy

**Figure 11.11   The Meselson and Stahl experiment showing that DNA replication is semiconservative.** (5): ©Meselson, M., and Stahl, F. 1958. The replication of DNA in *Escherichia coli. PNAS* 44(7): 671–682, Fig. 4a

*Concept Check:*  *If this experiment was conducted for four rounds of DNA replication (that is, four generations), what would be the expected fractions of light DNA and half-heavy DNA according to the semiconservative model?*

and samples were collected after one generation (that is, one round of DNA replication), two generations, and so on. Because the bacteria were doubling in a medium that contained only $^{14}N$, all of the newly made DNA strands were labeled with light nitrogen, but the original strands remained labeled with the heavy form.

3. Meselson and Stahl used centrifugation to separate DNA molecules based on differences in density. Samples were placed on top of a solution that contained a salt gradient, in this case, cesium chloride (CsCl). A double helix containing all heavy nitrogen would have a higher density and therefore travel closer to the bottom of the gradient. By comparison, if both DNA strands contained $^{14}N$, the DNA would have a low density and remain closer to the top of the gradient. If one strand contained $^{14}N$ and the other strand contained $^{15}N$, the DNA would be half-heavy and have an intermediate density, ending up near the middle of the gradient.

4. The DNA within the gradient was observed by exposing the gradient to UV light.

5. After one cell doubling (that is, one round of DNA replication), all of the DNA was half-heavy, exhibiting intermediate density (Figure 11.11, step 5). These results are consistent with both the semiconservative and dispersive mechanisms. In contrast, the conservative mechanism predicts two different DNA bands:

one of high density and one of low density. Because the DNA was found in a single half-heavy band after one doubling, the conservative model was disproved.

After two cell doublings, both light DNA and half-heavy DNA bands were observed. This result was also predicted by the semiconservative mechanism of DNA replication, because half of the DNA molecules should contain all light DNA, while the other molecules should be half-heavy (see Figure 11.10a). However, with the dispersive mechanism, all of the DNA strands would have been 1/4 heavy after two generations. This mechanism predicts that the heavy nitrogen would be evenly dispersed among four double helices, each strand containing 1/4 heavy nitrogen and 3/4 light nitrogen (see Figure 11.10c). This prediction did not agree with the data. Taken together, the results of the Meselson and Stahl experiment are consistent only with a semiconservative mechanism for DNA replication.

## DNA Replication Proceeds According to the AT/GC Rule

As originally proposed by Watson and Crick, DNA replication relies on the complementarity of DNA strands according to the AT/GC rule. During the replication process, the two complementary strands of DNA separate and serve as **template strands**, also called parental strands, for the synthesis of daughter strands of DNA (**Figure 11.12a**).

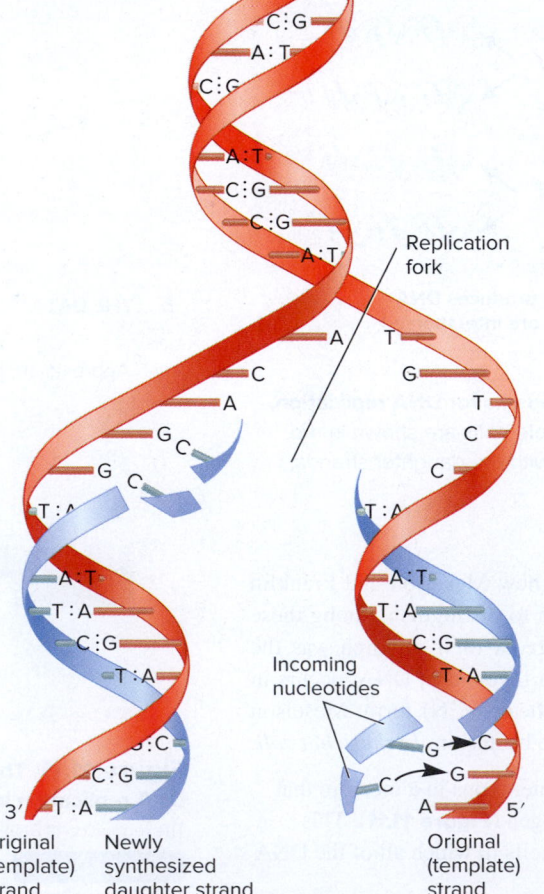

**Figure 11.12 DNA replication according to the AT/GC rule. (a)** The mechanism of DNA replication as originally proposed by Watson and Crick. As we will see in Section 11.4, the synthesis of one newly made strand (the leading strand on the left side) occurs in the direction toward the replication fork, whereas the synthesis of the other newly made strand (the lagging strand on the right side) occurs in small segments away from the fork. **(b)** DNA replication produces two copies of DNA with the same sequence as the original DNA molecule.

 **Core Concept: Structure and Function** A double-stranded structure whose base sequence obeys the AT/GC rule underlies the function of DNA replication.

Original (template) strand    Newly synthesized daughter strand

Original (template) strand

Replication fork

Incoming nucleotides

**(a) The mechanism of DNA replication**

**(b) The products of replication**

After the double helix has separated, individual nucleotides have access to the template strands in a region called the replication fork. First, individual nucleotides hydrogen-bond to the template strands according to the AT/GC rule: An adenine (A) base in one strand bonds with a thymine (T) base in the opposite strand, or a guanine (G) base bonds with a cytosine (C). Next, a covalent bond is formed between the phosphate of one nucleotide and the sugar of the previous nucleotide. The end result is that two double helices are made that have the same base sequence as the original DNA molecule (**Figure 11.12b**). This is a critical feature of DNA replication, because it enables the replicated DNA molecules to retain the same information (that is, the same base sequence) as the original molecule. In this way, DNA has the remarkable ability to direct its own duplication.

## 11.4 Molecular Mechanism of DNA Replication

### Learning Outcomes:

1. Describe how the synthesis of new DNA strands begins at an origin of replication.
2. List the functions of helicase, topoisomerase, single-strand binding protein, primase, and DNA polymerase at the replication fork.
3. Outline the key differences in the synthesis of the leading and lagging strands.

4. List three reasons why DNA replication is very accurate.
5. Explain how DNA replication occurs at telomeres in eukaryotic chromosomes.

Thus far, we have examined the general mechanism of DNA replication, known as semiconservative replication, and considered how DNA synthesis obeys the AT/GC rule. In this section, we will explore the details of DNA replication as it occurs inside living cells. As you will learn, several different proteins are needed to initiate DNA replication and allow it to proceed quickly and accurately.

### DNA Replication Begins at an Origin of Replication

Where does DNA replication begin? An **origin of replication** is a site within a chromosome that serves as a starting point for DNA replication. At the origin, the two DNA strands unwind (**Figure 11.13a**). DNA replication proceeds outward from two **replication forks**, a process termed **bidirectional replication**. The number of origins of replication varies among different organisms. In bacteria, which typically have a small circular chromosome, a single origin of replication is found. Bidirectional replication starts at the origin of replication and proceeds until the new strands meet on the opposite side of the chromosome (**Figure 11.13b**). Eukaryotes have larger chromosomes that are linear. They have multiple origins of replication so the DNA can be replicated in a reasonable length of time. The newly made

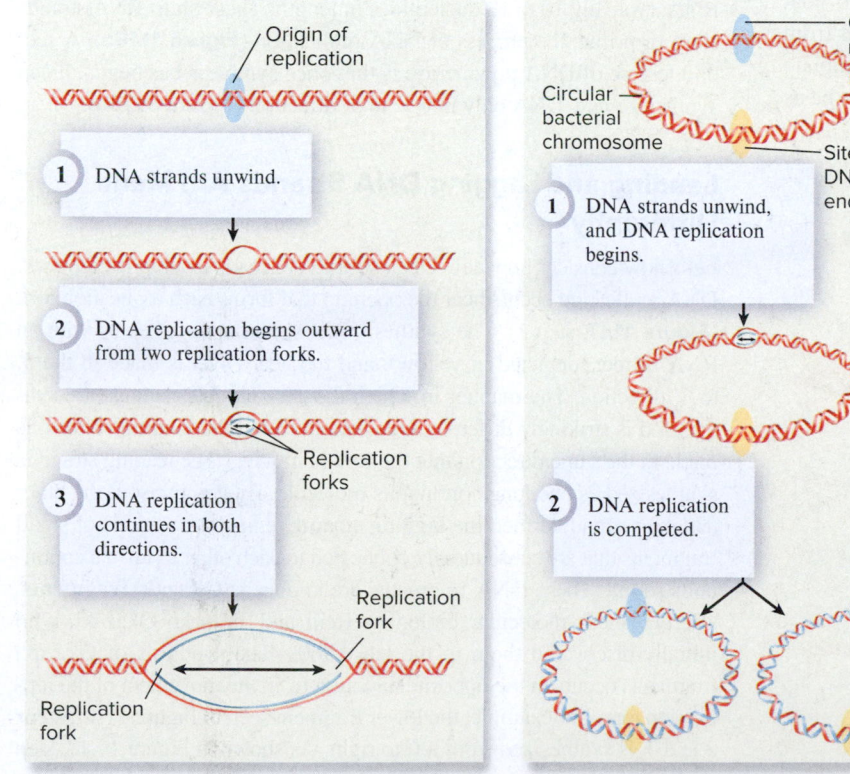

(a) Bidirectional replication

(b) Single origin of replication in bacteria

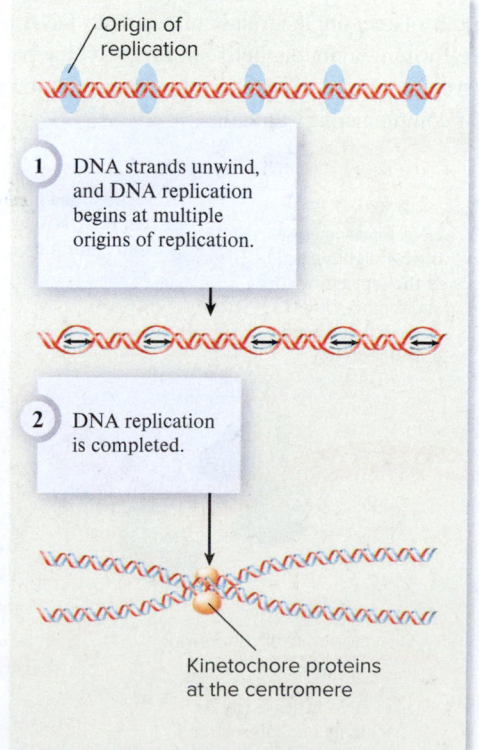

(c) Multiple origins of replication in eukaryotes

**Figure 11.13  The bidirectional replication of DNA.** **(a)** DNA replication proceeds in both directions from an origin of replication. **(b)** Bacterial chromosomes have a single origin of replication, whereas **(c)** eukaryotes have multiple origins. Following DNA replication in eukaryotes, the two copies remain attached to each other at the centromere via kinetochore proteins.

strands from each origin eventually make contact with each other to complete the replication process (**Figure 11.13c**).

## DNA Replication Requires the Action of Several Different Proteins

Thus far, we have considered how DNA replication occurs outward from an origin of replication in a region called a DNA replication fork. In all living species, a set of several different proteins is involved in this process. An understanding of the functions of these proteins is critical to explaining the replication process at the molecular level.

***Helicase, Topoisomerase, and Single-Strand Binding Proteins: Formation and Movement of the Replication Fork*** To act as a template for DNA replication, the strands of a double helix must separate, and the resulting fork must move. As mentioned, an origin of replication serves as a site where this separation initially occurs. The strand separation at each fork then moves outward from the origin via the action of an enzyme called **DNA helicase**. At each fork, DNA helicase binds to one of the DNA strands and travels in the 5′ to 3′ direction toward the fork (**Figure 11.14**). It uses energy from ATP to break hydrogen bonds between base pairs. This separates the DNA strands and keeps the fork moving forward. The action of DNA helicase can cause knots (called supercoils) to form just ahead of the replication fork. These knots are removed by another enzyme called **DNA topoisomerase**.

After the two template DNA strands have separated, they must remain that way until the complementary daughter strands have been made. The function of **single-strand binding proteins** is to coat both of the single strands of template DNA and prevent them from re-forming a double helix. In this way, the bases within the template strands are kept exposed so they can act as templates for the synthesis of complementary strands.

***DNA Polymerase and Primase: Synthesis of DNA Strands*** The enzyme **DNA polymerase** is responsible for covalently linking nucleotides together to form DNA strands. American biochemist Arthur Kornberg discovered this enzyme in the 1950s. The structure of DNA polymerase resembles a human hand with the DNA threaded through it (**Figure 11.15a**). As DNA polymerase slides along the DNA, individual nucleotides with three phosphate groups, called **deoxynucleoside triphosphates**, hydrogen-bond to the exposed bases in the template strand according to the AT/GC rule. At the catalytic site, DNA polymerase breaks a bond between the first and second phosphate and then attaches the resulting nucleotide with one phosphate group (a deoxynucleoside monophosphate) to the 3′ end of a growing strand via a phosphoester bond. The breakage of the covalent bond releases pyrophosphate; this is an exergonic reaction that provides the energy to covalently connect adjacent nucleotides (**Figure 11.15b**). The pyrophosphate is broken down to two phosphates.

The rate of DNA synthesis is truly remarkable. In bacteria, DNA polymerase synthesizes DNA at a rate of 500 nucleotides per second, whereas eukaryotic species make DNA at a rate of about 50 nucleotides per second.

DNA polymerase has two additional enzymatic features that affect how DNA strands are made. First, if a DNA or RNA strand is already attached to a template strand, DNA polymerase can elongate such a pre-existing strand by making DNA. However, DNA polymerase is unable to begin DNA synthesis on a bare template strand. A different enzyme called **DNA primase** is required if the template strand is bare. DNA primase makes a complementary primer that is actually a short segment of RNA, typically 10 to 12 nucleotides in length. These short RNA strands start, or prime, the process of DNA replication (**Figure 11.16a**). A second feature of DNA polymerase is that once synthesis has begun, it can synthesize new DNA only in a 5′ to 3′ direction (**Figure 11.16b**).

## Leading and Lagging DNA Strands Are Made Differently

Let's now consider how new DNA strands are made at a replication fork. DNA replication occurs near the opening that forms each replication fork (**Figure 11.17**, step 1). The synthesis of a strand always begins with an RNA primer (depicted in yellow), and the new DNA is made in the 5′ to 3′ direction. The manner in which the two daughter strands are synthesized is strikingly different. One strand, called the **leading strand**, is made in the same direction that the fork is moving. The leading strand is synthesized as one long continuous molecule. By comparison, the other daughter strand, termed the **lagging strand**, is made as a series of small fragments that are subsequently connected to each other to form a continuous strand. These DNA fragments are known as **Okazaki fragments**, after Japanese molecular biologists Reiji and Tsuneko Okazaki, who initially discovered them in the late 1960s. The synthesis of Okazaki fragments occurs in the opposite direction from the movement of the replication fork. For example, the lower fragment seen in Figure 11.17, steps 2 and 3, is synthesized from left to right. As shown in Figure 11.17, step 4, the RNA primer is eventually removed, and adjacent Okazaki fragments are connected to each other to form a continuous strand of DNA.

**Figure 11.18** shows how the leading and lagging strands are made during bidirectional DNA replication from a single origin of replication. The fork moving to the left uses the top template strand to make

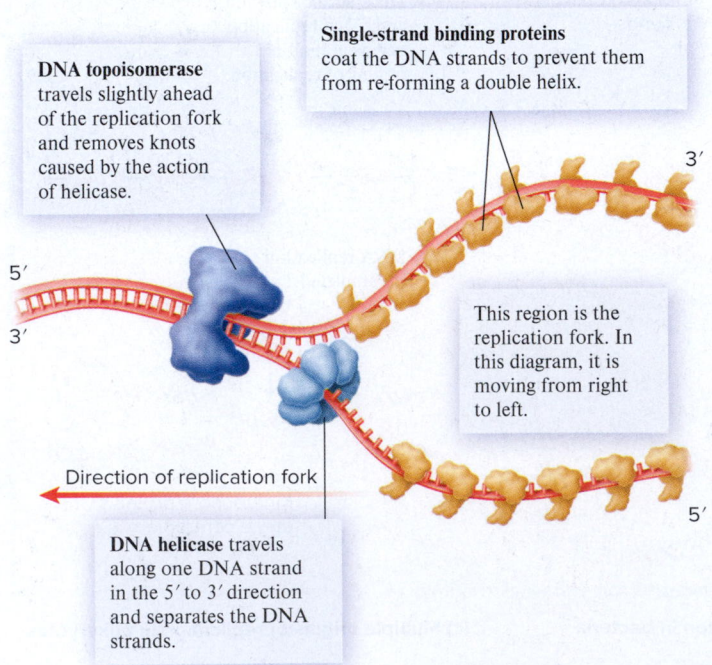

**DNA topoisomerase** travels slightly ahead of the replication fork and removes knots caused by the action of helicase.

**Single-strand binding proteins** coat the DNA strands to prevent them from re-forming a double helix.

This region is the replication fork. In this diagram, it is moving from right to left.

Direction of replication fork

**DNA helicase** travels along one DNA strand in the 5′ to 3′ direction and separates the DNA strands.

**Figure 11.14** Proteins that facilitate the formation and movement of a replication fork.

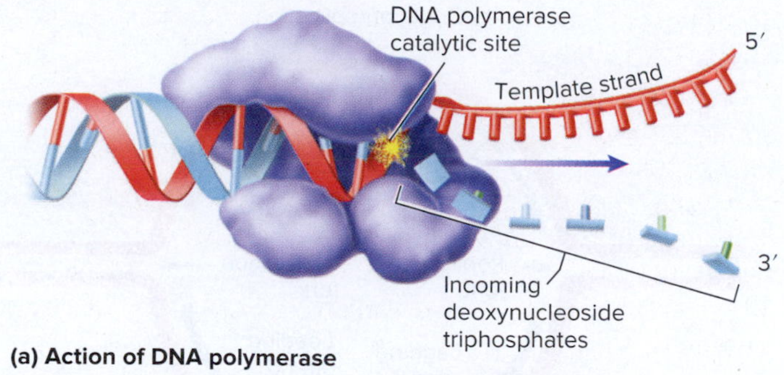

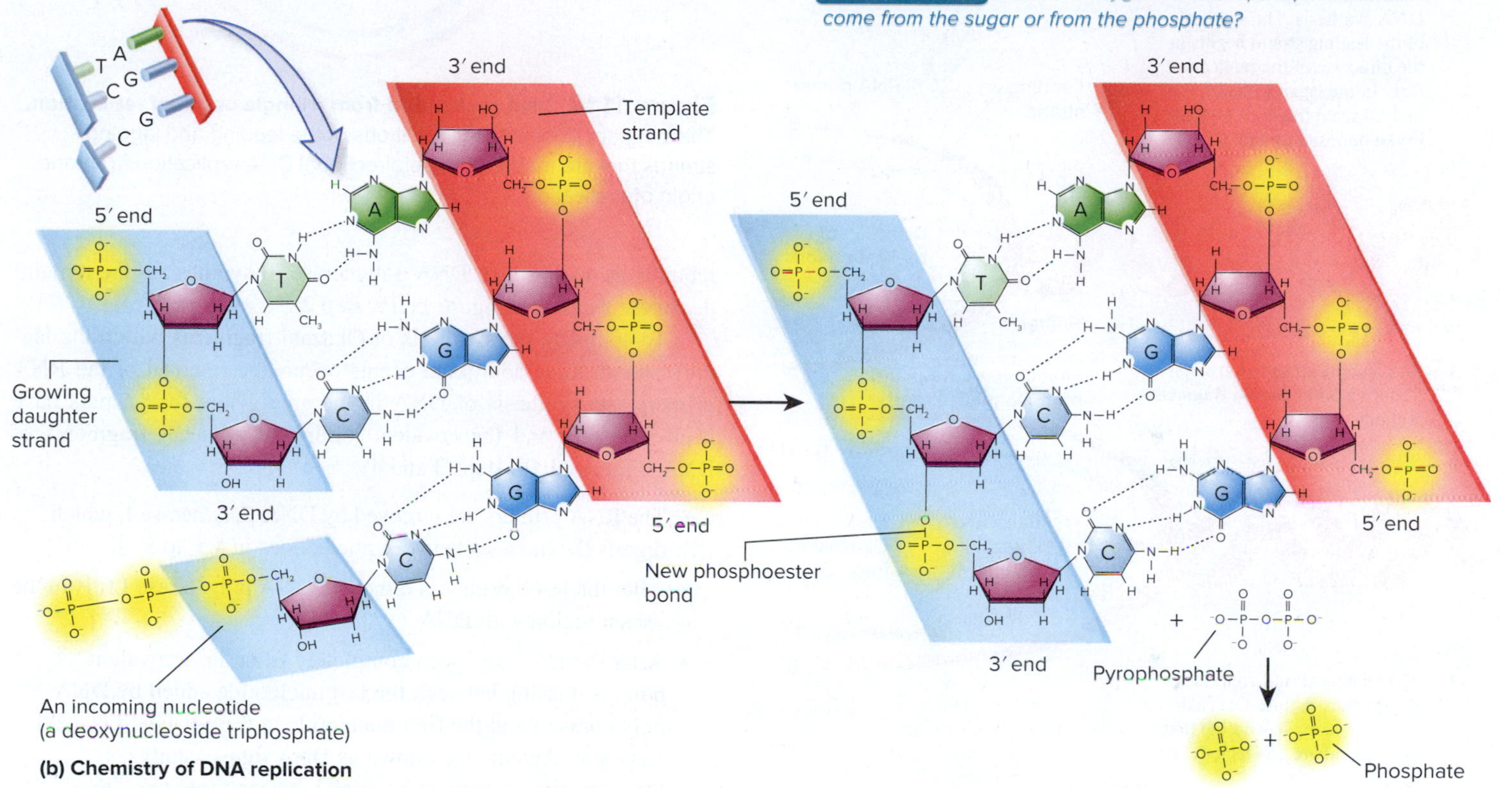

**Figure 11.15** **Enzymatic synthesis of DNA.** **(a)** Incoming deoxynucleoside triphosphates first hydrogen-bond to the template strand according to the AT/GC rule. DNA polymerase recognizes these deoxynucleoside triphosphates and attaches a deoxynucleoside monophosphate to the 3′ end of a growing strand. **(b)** DNA polymerase breaks the bond between the first and second phosphate in a deoxynucleoside triphosphate, causing the release of pyrophosphate. This provides the energy to form a covalent bond between the resulting deoxynucleoside monophosphate and the previous nucleotide in the growing strand. The pyrophosphate is broken down to two phosphates.

**Concept Check:** *Does the oxygen in a new phosphoester bond come from the sugar or from the phosphate?*

**(a) Action of DNA polymerase**

**(b) Chemistry of DNA replication**

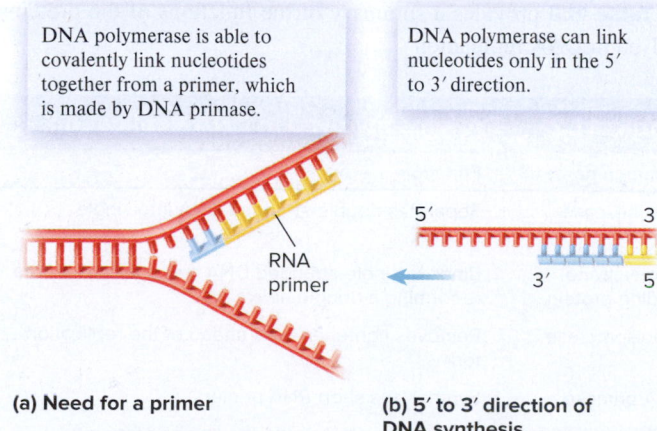

**(a) Need for a primer**

DNA polymerase is able to covalently link nucleotides together from a primer, which is made by DNA primase.

DNA polymerase can link nucleotides only in the 5′ to 3′ direction.

**(b) 5′ to 3′ direction of DNA synthesis**

**Figure 11.16** **Enzymatic features of DNA polymerase.** **(a)** DNA polymerase needs a primer to begin DNA synthesis, and **(b)** it can synthesize DNA only in the 5′ to 3′ direction.

the leading strand and the bottom template strand to make the lagging strand. In contrast, the fork moving to the right uses the top strand to make the lagging strand and bottom strand to make the leading strand.

The synthesis of DNA shown in Figures 11.17 and 11.18 emphasizes the synthesis of new DNA strands. **Figure 11.19** also includes the proteins involved in the synthesis of the leading and lagging strands in *E. coli*. In this bacterium, two different DNA polymerases, called DNA polymerase I and III, are primarily responsible for DNA replication. In the leading strand, DNA primase makes one RNA primer at the origin, and then DNA polymerase III attaches nucleotides in a 5′ to 3′ direction as it slides toward the opening of the replication fork. DNA polymerase III has a subunit called the clamp protein that allows the enzyme to slide along the template strand without falling off, a characteristic called **processivity**.

In the lagging strand, DNA is also synthesized in a 5′ to 3′ direction, but this synthesis occurs in the direction away from the replication fork. In the lagging strand, short segments of DNA are made discontinuously as a series of Okazaki fragments, each of which

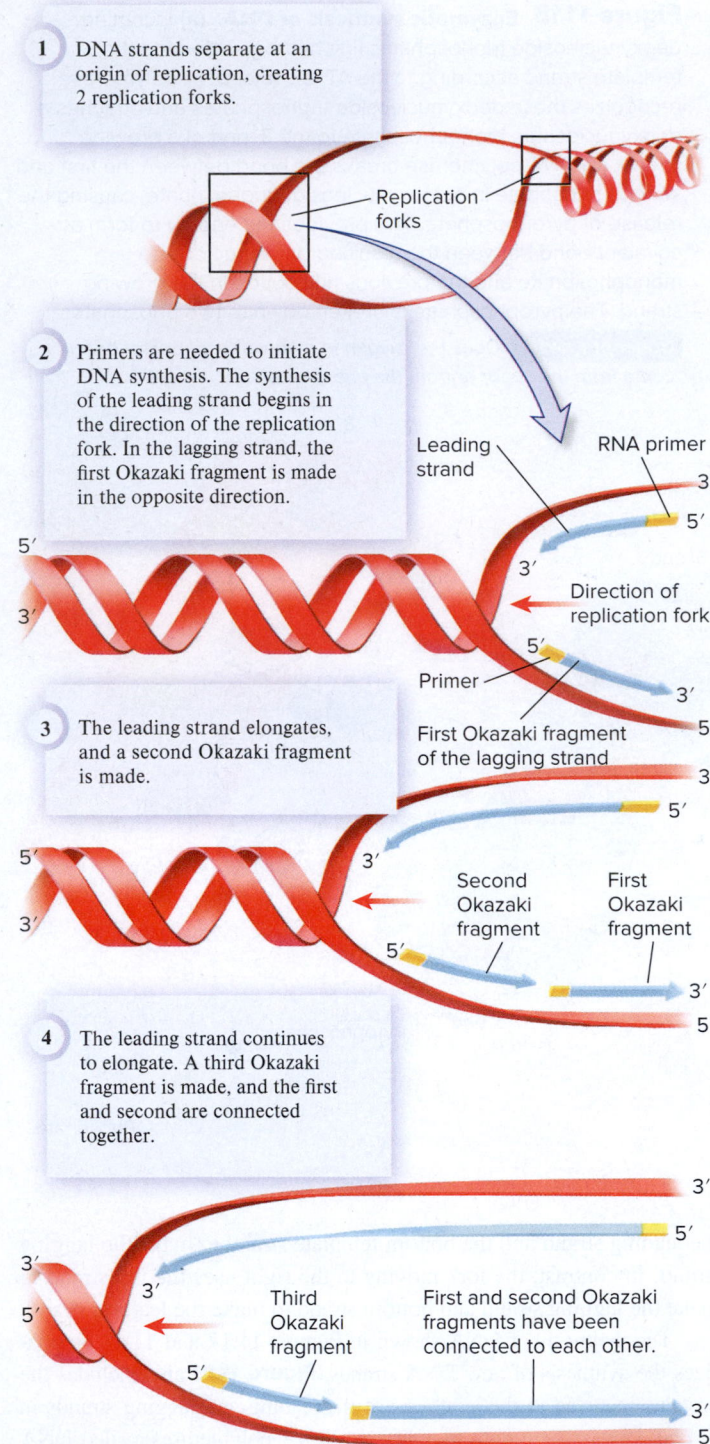

1. DNA strands separate at an origin of replication, creating 2 replication forks.

Replication forks

2. Primers are needed to initiate DNA synthesis. The synthesis of the leading strand begins in the direction of the replication fork. In the lagging strand, the first Okazaki fragment is made in the opposite direction.

Leading strand

RNA primer

Direction of replication fork

Primer

First Okazaki fragment of the lagging strand

3. The leading strand elongates, and a second Okazaki fragment is made.

Second Okazaki fragment

First Okazaki fragment

4. The leading strand continues to elongate. A third Okazaki fragment is made, and the first and second are connected together.

Third Okazaki fragment

First and second Okazaki fragments have been connected to each other.

**Figure 11.17** **Synthesis of new DNA strands.** The separation of DNA at the origin of replication produces two replication forks that move in opposite directions. New DNA strands are made near the opening of each fork. The leading strand is made continuously in the same direction the fork is moving. The lagging strand is made as small pieces in the opposite direction. These small pieces are then connected to each other to form a continuous lagging strand.

**Concept Check:** *Which strand, the leading or lagging strand, is made discontinuously in the direction opposite to the movement of the replication fork?*

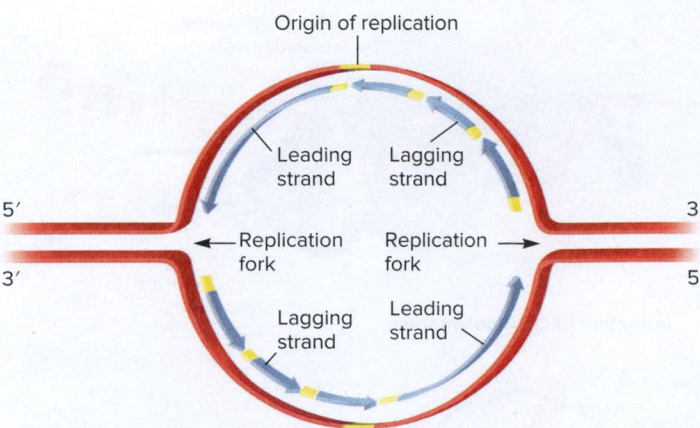

Origin of replication

Leading strand

Lagging strand

Replication fork

Replication fork

Lagging strand

Leading strand

**Figure 11.18** **DNA replication from a single origin of replication.** This diagram illustrates the locations of the leading and lagging strands that are made during bidirectional DNA replication from one origin of replication.

requires its own primer. DNA polymerase III synthesizes the remainder of the fragment (Figure 11.19, step 2).

To complete the synthesis of Okazaki fragments within the lagging strand, three additional events occur: the removal of the RNA primers, the synthesis of DNA in the area where the primers have been removed, and the covalent joining of adjacent fragments of DNA (Figure 11.19, steps 3 and 4).

- The RNA primers are removed by DNA polymerase I, which digests the linkages between nucleotides in a 5′ to 3′ direction.
- After the RNA primer is removed, DNA polymerase I fills in the vacant region with DNA.
- After the DNA has been completely filled in, a covalent bond is missing between the last nucleotide added by DNA polymerase I and the first nucleotide in the adjacent Okazaki fragment. An enzyme known as **DNA ligase** catalyzes the formation of a covalent bond between these two DNA fragments to complete the replication process in the lagging strand (Figure 11.19, step 4).

**Table 11.2** provides a summary of the functions of the proteins involved in DNA replication.

| Table 11.2 | Proteins Involved in DNA Replication |
|---|---|
| **Common name** | **Function** |
| DNA helicase | Separates double-stranded DNA into single strands |
| Single-strand binding protein | Binds to single-stranded DNA and prevents it from re-forming a double helix |
| Topoisomerase | Removes tightened coils ahead of the replication fork |
| DNA primase | Synthesizes short RNA primers |
| DNA polymerase | Synthesizes DNA in the leading and lagging strands, removes RNA primers, and fills in gaps |
| DNA ligase | Covalently attaches adjacent Okazaki fragments in the lagging strand |

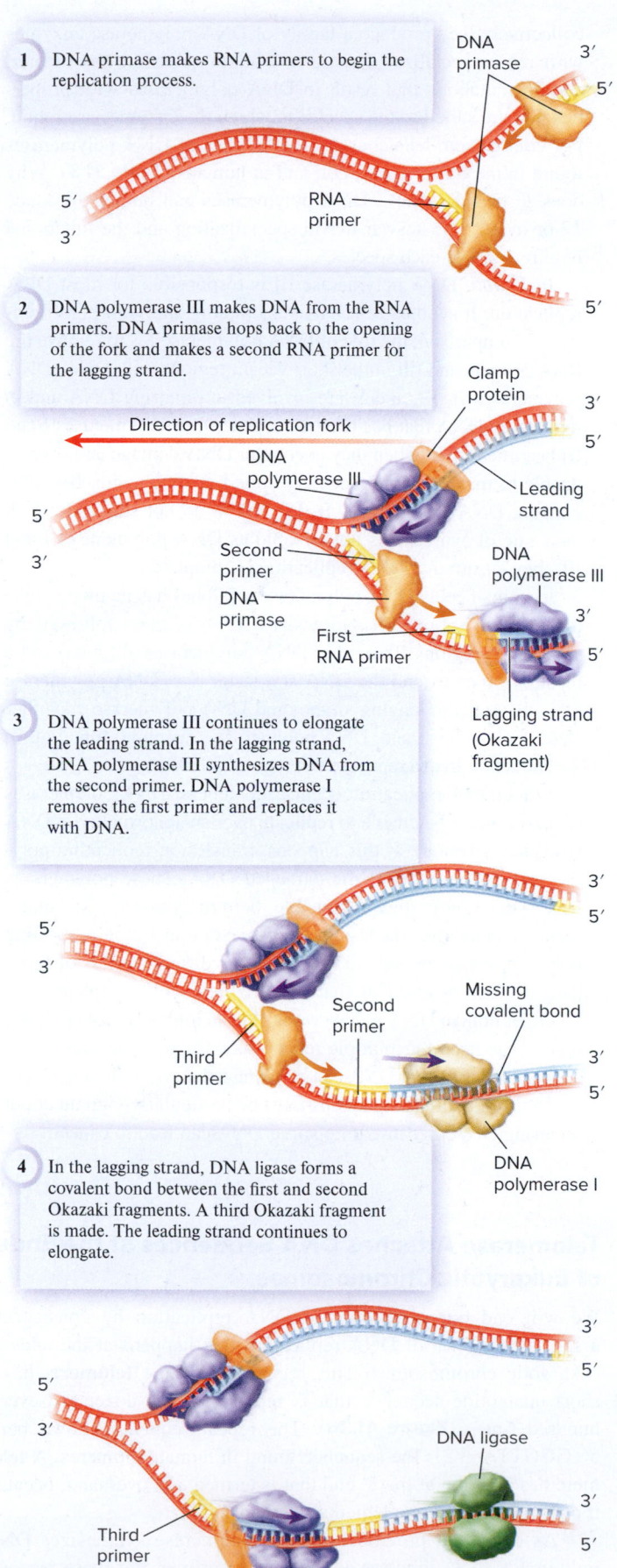

**1** DNA primase makes RNA primers to begin the replication process.

DNA primase

RNA primer

3'
5'
5'
3'

**2** DNA polymerase III makes DNA from the RNA primers. DNA primase hops back to the opening of the fork and makes a second RNA primer for the lagging strand.

Direction of replication fork

DNA polymerase III

Clamp protein

Leading strand

Second primer

DNA primase

First RNA primer

DNA polymerase III

Lagging strand (Okazaki fragment)

5'
3'
3'
5'
3'
5'

**3** DNA polymerase III continues to elongate the leading strand. In the lagging strand, DNA polymerase III synthesizes DNA from the second primer. DNA polymerase I removes the first primer and replaces it with DNA.

Missing covalent bond

Second primer

Third primer

DNA polymerase I

5'
3'
3'
5'

**4** In the lagging strand, DNA ligase forms a covalent bond between the first and second Okazaki fragments. A third Okazaki fragment is made. The leading strand continues to elongate.

DNA ligase

Third primer

5'
3'
3'
5'

**Figure 11.19** Proteins involved with the synthesis of the leading and lagging strands in *E. coli.*

*Concept Check:* *Briefly describe the movement of primase in the lagging strand in this figure. In which direction does it move when it is making a primer, from left to right or right to left? Describe how it must move after it is done making a primer and has to start making the next primer at a new location. Does it have to hop from left to right or from right to left?*

## BIO TIPS

**THE QUESTION** *The drawing below shows the synthesis of the lagging strand of a DNA molecule. The RNA primers are yellow, and the newly made DNA is blue. The top (red) strand is the template strand.*

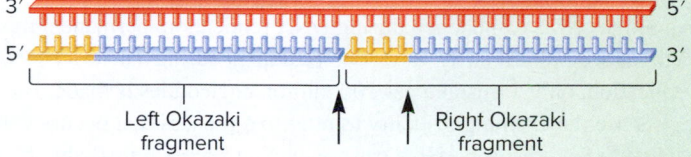

3'                                                        5'
5'                                                        3'

Left Okazaki fragment          Right Okazaki fragment

*After DNA polymerase I removes the right RNA primer and fills in the gap with DNA, where will DNA ligase act? See the arrows on either side of the right RNA primer. Is the ligase needed at the left arrow, the right arrow, or both?*

**T** **OPIC** *What topic in biology does this question address?* The topic is DNA replication. More specifically, the question asks you to determine where DNA ligase needs to act during replication of the lagging strand.

**I** **NFORMATION** *What information do you know based on the question and your understanding of the topic?* In the question, you are given the arrangement of two adjacent Okazaki fragments, with arrows pointing to two possible places where DNA ligase may be needed. From your understanding of the topic, you may remember that in *E. coli*, DNA polymerase I removes the RNA primer and fills in the region with DNA; then DNA ligase connects the adjacent Okazaki fragments.

**P** **ROBLEM-SOLVING** **S** **TRATEGY** *Sort out the steps in a complicated process. Make a drawing.* To solve this problem, it may be helpful to list the steps of the process. Also, you should consider how DNA polymerase I and DNA ligase function.

1. First, DNA polymerase I removes the right RNA primer.

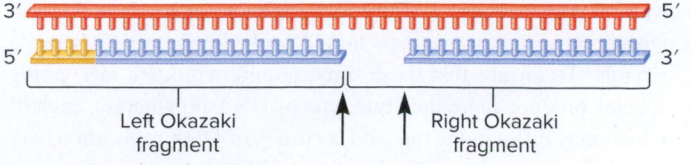

3'                                                        5'
5'                                                        3'

Left Okazaki fragment          Right Okazaki fragment

2. Starting at the 3' end of the left Okazaki fragment, DNA polymerase I then extends the DNA until it fills in the region where the RNA primer was removed. DNA ligase is not needed at the left arrow.

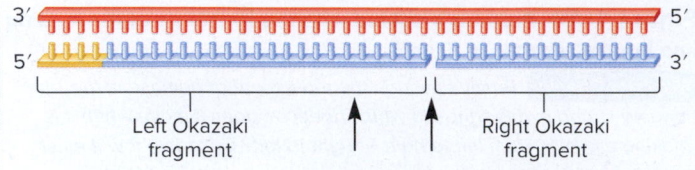

3. As shown above, when DNA polymerase I reaches the (blue) DNA of the right Okazaki fragment, a gap exists between the 3′ end of the left Okazaki fragment and the 5′ end of the DNA of the right Okazaki fragment.

**ANSWER** *DNA ligase is needed at the right arrow only.*

## DNA Replication Is Very Accurate

Although errors can happen during DNA replication, permanent mistakes are extraordinarily rare. For example, during bacterial DNA replication, only 1 mistake per 100 million nucleotides is made. Biologists use the term high fidelity to refer to a process that occurs with relatively few mistakes. How can we explain such a remarkably high fidelity for DNA replication?

- First, hydrogen bonding between A and T or between G and C is more stable than hydrogen bonding between mismatched pairs of bases.

- Second, the active site of DNA polymerase is unlikely to catalyze bond formation between adjacent nucleotides if a mismatched base pair is formed.

- Third, DNA polymerase can identify a mismatched nucleotide and remove it from the daughter strand. This event, called **proofreading**, occurs when DNA polymerase detects a mismatch and then reverses its direction and digests the linkages between nucleotides at the end of a newly made strand in the 3′ to 5′ direction. Once it passes the mismatched base and removes it, DNA polymerase then changes direction again and continues to synthesize DNA in the 5′ to 3′ direction.

## Core Concepts: Evolution, Structure and Function

## DNA Polymerases Are a Family of Enzymes with Specialized Functions

Three important properties of DNA replication are speed, fidelity, and completeness. DNA replication must proceed quickly and with great accuracy, and gaps should not be left in the newly made strands. To ensure that these three requirements are met, living species produce more than one type of DNA polymerase, each of which may differ in the rate and accuracy of DNA replication and/or the ability to prevent the formation of DNA gaps.

The genomes of living species have multiple DNA polymerase genes, which were produced by random gene duplication events. During evolution, mutations have altered each gene so that

collectively they produce a family of DNA polymerase enzymes with more specialized functions. Natural selection has favored certain mutations that result in DNA polymerases with properties that are suited to the species in which the enzymes are found. For comparison, let's consider the families of DNA polymerases found in the bacterium *E. coli* and in humans (**Table 11.3**). Why does *E. coli* produce 5 DNA polymerases and humans produce 12 or more? The answer lies in specialization and the functional requirements of each species.

In *E. coli*, DNA polymerase III is responsible for most DNA replication. It synthesizes DNA very rapidly and with high fidelity. By comparison, the role of DNA polymerase I is to remove the RNA primers and fill in the short vacant regions with DNA. DNA polymerases II, IV, and V are involved in repairing DNA and in replicating DNA that has been damaged. DNA polymerases I and III become stalled when they encounter DNA damage and may be unable to make a complementary strand at such a site. By comparison, DNA polymerases II, IV, and V do not stall. Although their rate of synthesis is not as rapid as DNA polymerases I and III, they ensure that DNA replication is complete.

In human cells, DNA polymerase α (alpha) has its own "built-in" primase subunit. It synthesizes RNA primers followed by short DNA regions. Two other DNA polymerases, δ (delta) and ε (epsilon), then extend the DNA at a faster rate. DNA polymerase δ synthesizes the lagging strand, and DNA polymerase ε synthesizes the leading strand. DNA polymerase γ (gamma) functions in the mitochondria to replicate mitochondrial DNA.

When DNA replication occurs, the general DNA polymerases (α, δ, or γ) may be unable to replicate over an abnormality in DNA structure (a lesion). If this happens, translesion-replicating polymerases are attracted to the damaged DNA. These polymerases have special properties that enable them to synthesize a complementary strand over the lesion. Each type of translesion-replicating polymerase may be able to replicate over different kinds of DNA damage, thereby ensuring that DNA replication is complete.

Other human DNA polymerases play an important role in DNA repair. The need for multiple repair enzymes arises because there are various ways that DNA can be damaged, as described in Chapter 15. Multicellular organisms must be particularly vigilant about repairing DNA, because unrepaired DNA can lead to cancer.

## Telomerase Attaches DNA Sequences at the Ends of Eukaryotic Chromosomes

We will end our discussion of DNA replication by considering a specialized form of DNA replication that happens at the ends of eukaryotic chromosomes. This region, called the **telomere**, has a short nucleotide sequence that is repeated a few dozen to several hundred times (**Figure 11.20**). The repeat sequence shown here, 5′–GGGTTA–3′, is the sequence found in human telomeres. A telomere has a region at the 3′ end that is termed a 3′ overhang, because it does not have a complementary strand.

As discussed previously, DNA polymerase synthesizes DNA only in a 5′ to 3′ direction and requires a primer. For these reasons,

| Table 11.3 | DNA Polymerases in *E. coli* and Humans |
|---|---|
| Polymerase types* | Functions |
| *E. coli* | |
| III | Replicates most of the DNA during cell division |
| I | Removes RNA primers and fills in the gaps |
| II, IV, and V | Repairs damaged DNA and replicates over DNA abnormalities |
| *Humans* | |
| α (alpha) | Makes RNA primers and synthesizes short DNA strands |
| δ (delta), ε (epsilon) | Displaces DNA polymerase α and then replicates DNA at a rapid rate |
| γ (gamma) | Replicates the mitochondrial DNA |
| η (eta), κ (kappa), ι (iota), ζ (zeta) | Replicates over DNA abnormalities |
| α, β (beta), δ, ε, σ (sigma), λ (lambda), μ (mu), φ (phi), θ (theta) | Repairs DNA or has other functions |

*Certain DNA polymerases have more than one function.

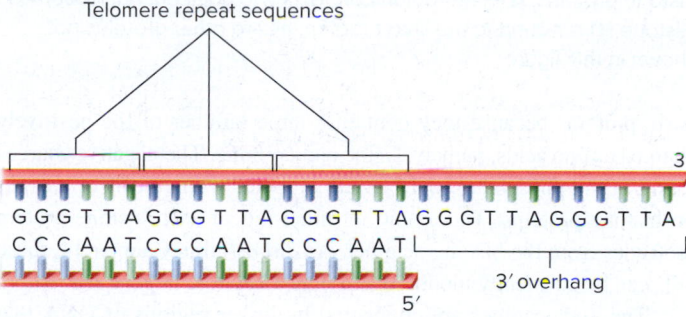

**Figure 11.20** **Telomere sequences at the end of a human chromosome.** The telomere sequence shown here is found in humans and other mammals. The length of the telomere and the 3′ overhang varies among different species and cell types.

DNA polymerase cannot copy the tip of a DNA strand with a 3′ end. Therefore, if this replication problem was not overcome, a linear chromosome would become progressively shorter with each round of DNA replication. In 1984, American molecular biologist Carol Greider and Australian-born American molecular biologist Elizabeth Blackburn discovered an enzyme called **telomerase** that prevents chromosome shortening by attaching many copies of a DNA repeat sequence to the ends of chromosomes (**Figure 11.21**). Telomerase contains both protein and RNA. The RNA part of telomerase has a sequence that is complementary to the DNA repeat sequence. This allows telomerase to bind to the 3′ overhang region of the telomere. Following binding, the RNA sequence beyond the binding site functions as a template, allowing telomerase to synthesize a 6-nucleotide sequence at the end of the DNA strand. The enzyme then moves to the new end of this DNA strand and attaches another 6 nucleotides to the end. This occurs many times, thereby greatly lengthening the 3′ end of the DNA in the telomere. This lengthening

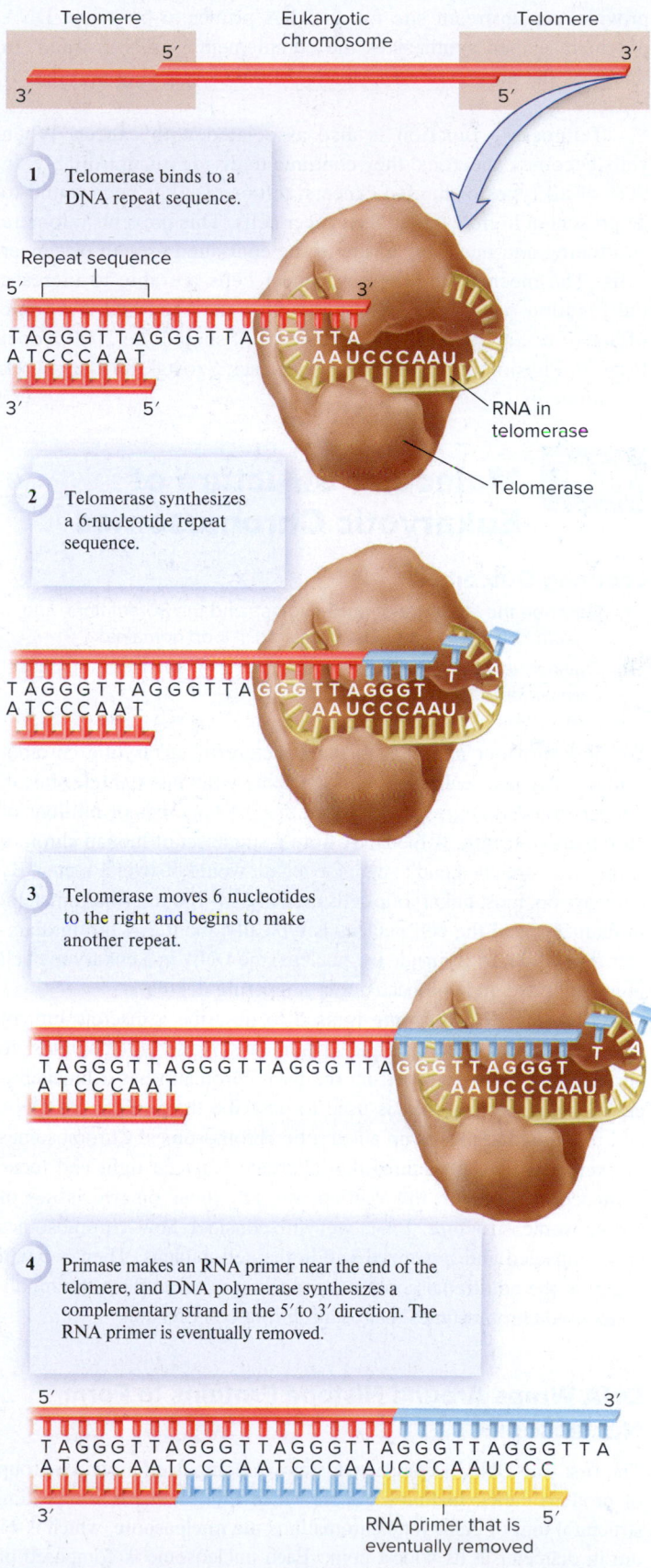

**Figure 11.21** **Mechanism of DNA replication by telomerase.**

*Concept Check:* *What does telomerase use as a template to make the DNA repeat sequence?*

provides an upstream site for an RNA primer to be made. DNA polymerase then synthesizes the complementary DNA strand. In this way, the progressive shortening of eukaryotic chromosomes is prevented.

Telomerase function is also associated with cancer. When cells become cancerous, they continue to divide uncontrollably. In 90% of all types of human cancers, telomerase has been found to be present at high levels in the cancer cells. This prevents telomere shortening and may play a role in the continued growth of cancer cells. The mechanism whereby cancer cells are able to increase the function of telomerase is not well understood and is a topic of active research. Greider and Blackburn shared the 2009 Nobel Prize in Physiology or Medicine with Jack Szostak for their work on telomeres.

# 11.5 Molecular Structure of Eukaryotic Chromosomes

**Learning Outcomes:**

1. Describe the structure of nucleosomes and the 30-nm fiber, and explain how the 30-nm fiber forms radial loop domains.
2. Outline the various levels of compaction that lead to a metaphase chromosome.

We now turn our attention to the structure of eukaryotic chromosomes. A typical eukaryotic chromosome contains a single, linear, double-stranded DNA molecule that may be hundreds of millions of base pairs in length. If the DNA from a single set of human chromosomes was stretched end to end, the length would be over 1 meter! By comparison, most eukaryotic cells are only 10–100 μm (micrometers) in diameter, and the cell nucleus is typically about 2–4 μm in diameter. Therefore, to fit inside the nucleus, the DNA in a eukaryotic cell must be folded and compacted to a staggering degree.

The term **chromosome** is used to describe a discrete unit of genetic material. For example, a human somatic cell contains 46 chromosomes. By comparison, the term chromatin has a biochemical meaning. **Chromatin** is used to describe the complex of DNA and proteins that makes up eukaryotic chromosomes. Chromosomes are very dynamic structures that alternate between tight and loose compaction states. In this section, we will focus on two issues of chromosome structure. First, we will consider how chromosomes are compacted and organized within the cell nucleus. Then, we will examine the additional compaction necessary to produce the highly condensed chromosomes that occur during cell division.

## DNA Wraps Around Histone Proteins to Form Nucleosomes

The first way DNA is compacted is by wrapping itself around a group of proteins called **histones**. As shown in **Figure 11.22**, a repeating structural unit of eukaryotic chromatin is the **nucleosome**, which is 11 nm in diameter at its widest point. Each nucleosome is composed of 146 or 147 bp (base pairs) of DNA wrapped around an octamer of histone proteins. An octamer contains two molecules each of four types of histone proteins: H2A, H2B, H3, and H4. Histone proteins are very

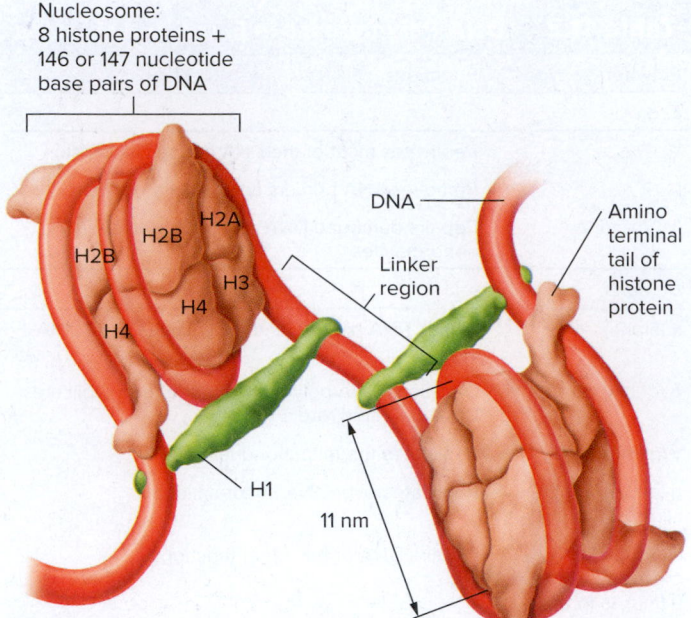

**Figure 11.22** **Structure of a nucleosome.** A nucleosome is composed of double-stranded DNA wrapped around an octamer of histone proteins. A linker region connects two adjacent nucleosomes. Histone H1 is bound to the linker region, as are other proteins not shown in this figure.

basic proteins because they contain a large number of the positively charged amino acids, namely, lysine and arginine. The negative charges found in the phosphates of DNA are attracted to the positive charges on histone proteins. The amino terminal tail of each histone protein protrudes from the histone octamer. As discussed in Chapter 14, these tails can be covalently modified and play a key role in gene regulation.

The nucleosomes are connected by linker regions of DNA that vary in length from 20 to 100 bp, depending on the species and cell type. A particular histone named H1 is bound to the linker region, as are other types of proteins. The overall structure of connected nucleosomes resembles beads on a string. This structure shortens the length of the DNA molecule about sevenfold.

## Nucleosomes Form a 30-nm Fiber

Nucleosome units are organized into a more compact structure that is 30 nm in diameter, known as the **30-nm fiber** (**Figure 11.23a**). Histone H1 and other proteins are important in the formation of the 30-nm fiber, which shortens the nucleosome structure another sevenfold. The structure of the 30-nm fiber has proven difficult to determine because the conformation of the DNA may be substantially altered when extracted from living cells. A model for the 30-nm fiber was proposed by Rachel Horowitz-Scherer and Christopher Woodcock in the 1990s (**Figure 11.23b**). According to their model, linker regions in the 30-nm structure are variably bent and twisted, with little direct contact between nucleosomes. The 30-nm fiber forms an asymmetric, three-dimensional zigzag of nucleosomes. At this level of compaction, the overall picture of chromatin that emerges is an irregular, fluctuating structure with stable nucleosome units connected by bendable linker regions.

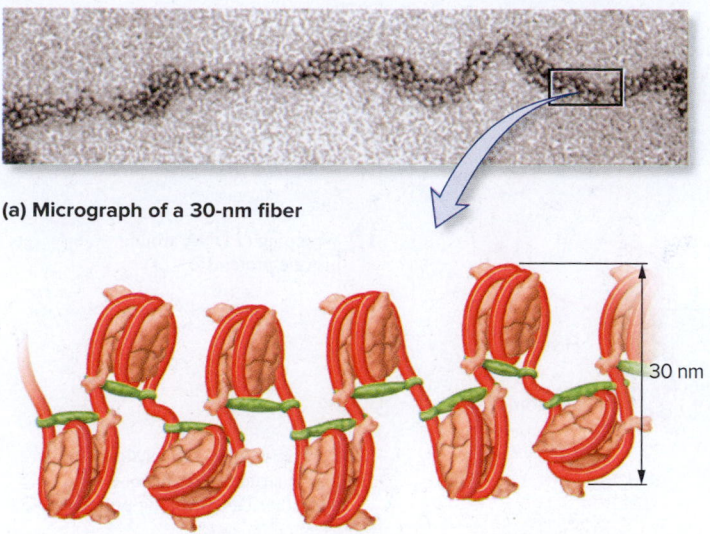

**(a) Micrograph of a 30-nm fiber**

30 nm

**(b) Three-dimensional zigzag model**

**Figure 11.23** **The 30-nm fiber.** **(a)** A photomicrograph of the 30-nm fiber. **(b)** In this three-dimensional zigzag model, the linker DNA forms a bendable structure with little contact between adjacent nucleosomes. a: ©Dr. Barbara A. Hamkalo

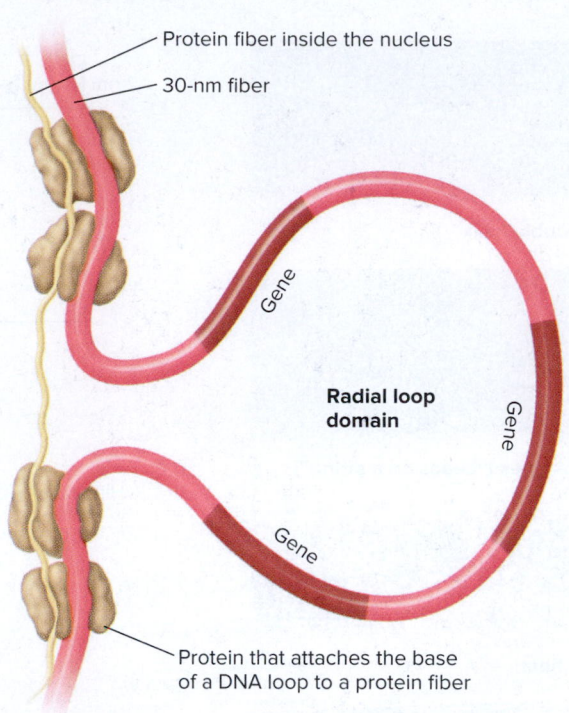

Protein fiber inside the nucleus

30-nm fiber

Gene

Gene

**Radial loop domain**

Gene

Protein that attaches the base of a DNA loop to a protein fiber

**Figure 11.24** **Attachment of the 30-nm fiber to a protein fiber to form a radial loop domain.**

**Concept Check:** *What holds the bases of the radial loop domains in place?*

## Chromatin Loops Are Anchored to the Nuclear Matrix

Thus far, we have examined two mechanisms that compact eukaryotic DNA: the formation of nucleosomes and their arrangement into a 30-nm fiber. Taken together, these two events shorten the folded DNA about 49-fold. A third level of compaction involves interactions between the 30-nm fibers and a filamentous network of proteins in the nucleus called the **nuclear matrix**. This matrix consists of the **nuclear lamina**, which is composed of protein fibers that line the inner nuclear membrane (refer back to Figure 4.19), and an internal nuclear matrix that is connected to the lamina and fills the interior of the nucleus. The internal nuclear matrix is an intricate network of irregular protein fibers plus many other proteins that bind to these fibers. The nuclear matrix aids in compaction by binding to the 30-nm fiber to form **radial loop domains**. These loops, often containing 25,000–200,000 bp, are anchored to the nuclear matrix (**Figure 11.24**).

How are chromosomes organized within the cell nucleus? In nondividing cells, each chromosome occupies its own discrete region in the cell nucleus that usually does not overlap with the territory of adjacent chromosomes (refer back to Figure 4.20). In other words, different chromosomes are not substantially intertwined with each other even when they are in a noncompacted condition.

The compaction level of chromosomes in the cell nucleus is not completely uniform. This variability can be seen with a light microscope and was first observed by German cytologist Emil Heitz in 1928. He coined the term **heterochromatin** to describe the highly compacted regions of chromosomes. The less condensed regions are known as **euchromatin**, which is the form of chromatin in which the 30-nm fiber forms radial loop domains. In heterochromatin, these radial loop domains are compacted even further. In nondividing cells, most chromosomal regions are euchromatic, and these are the regions where genes are usually located. By comparison, the centromeric and telomeric regions are often heterochromatic and may not contain genes.

## During Cell Division, Chromosomes Undergo Maximum Compaction

When cells prepare to divide, the chromosomes become even more compacted or condensed. Each chromosome becomes entirely compacted into heterochromatin. This aids in their proper alignment during metaphase, which is a stage of eukaryotic cell division described in Chapter 16. **Figure 11.25** illustrates the levels of compaction that contribute to the formation of a metaphase chromosome. DNA in the nucleus is always compacted by forming nucleosomes and condensing into a 30-nm fiber (Figure 11.25a, b, c). In euchromatin, the 30-nm fibers are arranged in radial loop domains that are relatively loose, meaning that a fair amount of space is between the 30-nm fibers (Figure 11.25d). The average width of such loops is about 300 nm.

By comparison, heterochromatin involves a much tighter packing of the loops, so little space is left between the 30-nm fibers (Figure 11.25e). When cells prepare to divide, all of the euchromatin becomes highly compacted. The compaction of euchromatin greatly shortens the chromosomes. For a metaphase chromosome that contains two copies of the DNA (Figure 11.25f), the width averages about 1,400 nm, but the length of a metaphase chromosome is much shorter than the same chromosome in the nucleus of a nondividing cell.

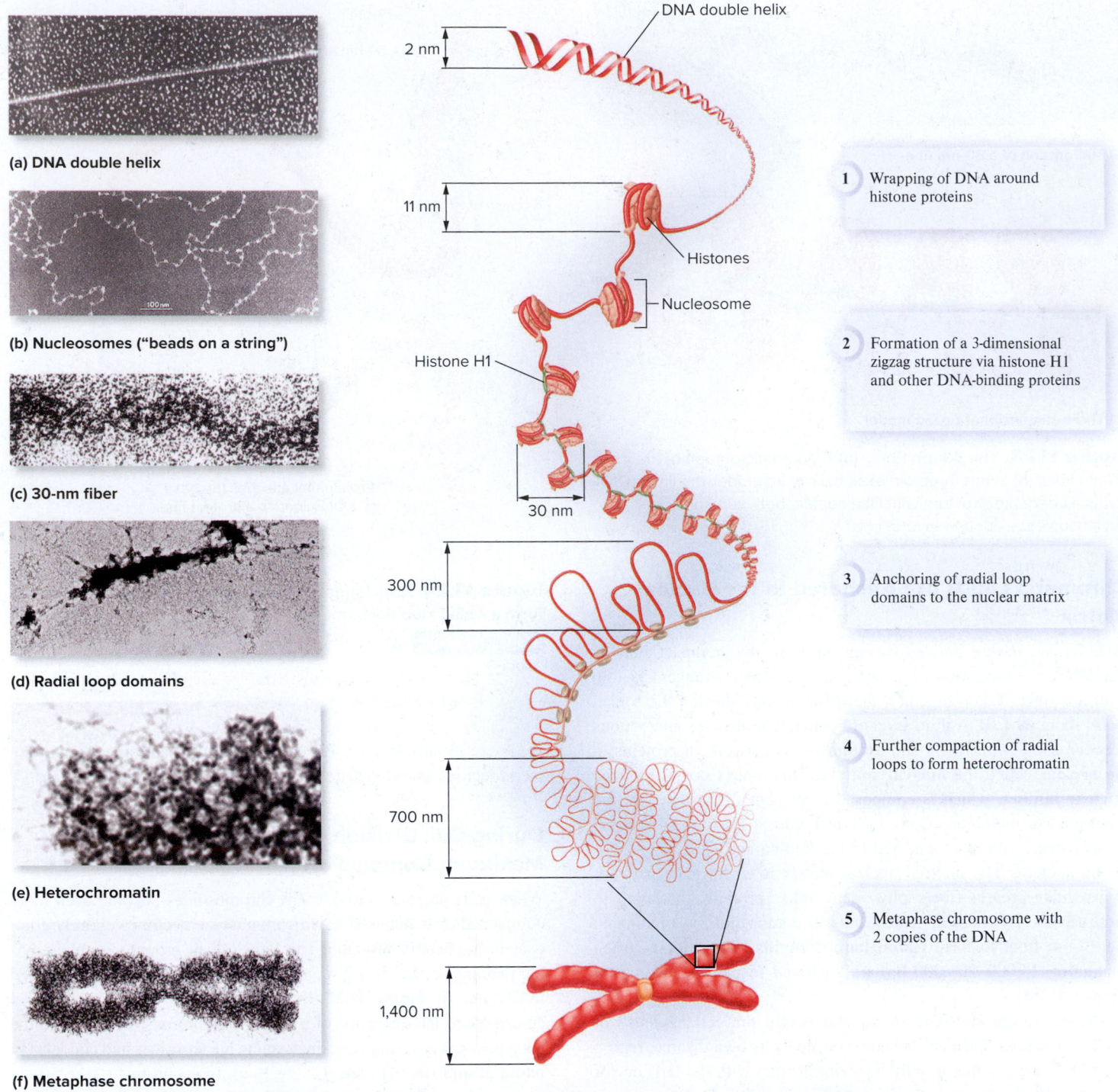

**(a) DNA double helix**

**(b) Nucleosomes ("beads on a string")**

**(c) 30-nm fiber**

**(d) Radial loop domains**

**(e) Heterochromatin**

**(f) Metaphase chromosome**

DNA double helix

2 nm

11 nm

Histones

Nucleosome

Histone H1

30 nm

300 nm

700 nm

1,400 nm

**1** Wrapping of DNA around histone proteins

**2** Formation of a 3-dimensional zigzag structure via histone H1 and other DNA-binding proteins

**3** Anchoring of radial loop domains to the nuclear matrix

**4** Further compaction of radial loops to form heterochromatin

**5** Metaphase chromosome with 2 copies of the DNA

**Figure 11.25** **The steps in eukaryotic chromosomal compaction leading to the metaphase chromosome.** a: ©Science Source; b: ©Dr. Barbara A. Hamkalo; c: Courtesy Dr. Jerome B. Rattner, Cell Biology and Anatomy, University of Calgary; d: ©Dr. James Paulson, Ph.D.; e-f: ©Peter Engelhardt/Department of Virology, Haartman Institute

**Core Skill: Connections** Look ahead to Figure 16.8. Why do you think it is necessary for the chromosomes to become highly compacted in preparation for cell division?

## Summary of Key Concepts

### 11.1 Biochemical Identification of the Genetic Material

- The genetic material contains the information that produces the traits of organisms. It is accurately replicated and transmitted from cell to cell and parent to offspring. The genetic material has differences that explain the variation among different organisms.

- Griffith's work with type S and type R bacteria indicated the existence of a genetic material, which he called the transformation principle (Figure 11.1).

- Avery, MacLeod, and McCarty used biochemical methods to show that DNA is the genetic material (Figure 11.2).

### 11.2 Nucleic Acid Structure

- DNA is composed of nucleotides, which are covalently linked to form DNA strands. Two DNA strands are held together by hydrogen bonds between the bases to form a double helix. DNA associates with various proteins to form a chromosome (Figure 11.3).

- Nucleotides are composed of a phosphate, a sugar, and a nitrogen-containing base. The sugar can be deoxyribose (DNA) or ribose (RNA). The purine bases are adenine and guanine, and the pyrimidine bases are thymine (DNA only), cytosine, and uracil (RNA only). The atoms in a nucleotide are numbered in a conventional way (Figures 11.4, 11.5).

- In a strand of DNA (or RNA), the sugars are connected by covalent bonds in a 5′ to 3′ direction (Figure 11.6).

- The X-ray diffraction data of Franklin, the biochemical data of Chargaff, and the ball-and-stick modeling of Pauling helped reveal the structure of DNA (Figure 11.7, Table 11.1).

- Watson and Crick determined that DNA is a double helix in which the DNA strands are complementary, with base sequences that are antiparallel and obey the AT/GC rule (Figures 11.8, 11.9).

### 11.3 Overview of DNA Replication

- Meselson and Stahl used isotope labeling methods to show that DNA is replicated by a semiconservative mechanism in which the products are two DNA molecules, each with one parental strand and one daughter strand (Figures 11.10, 11.11).

- During DNA replication, the parental double-stranded DNA separates and each strand serves as a template for the synthesis of daughter strands. The bases within nucleotides hydrogen-bond to each other according to the AT/GC rule: An adenine (A) base in one strand bonds with a thymine (T) base in the opposite strand, or a guanine (G) base bonds with a cytosine (C) base. The result of DNA replication is two double helices with the same base sequence as the original DNA (Figure 11.12).

### 11.4 Molecular Mechanism of DNA Replication

- DNA synthesis occurs bidirectionally from an origin of replication. The synthesis of new DNA strands happens near each replication fork (Figure 11.13).

- DNA helicase separates DNA strands, single-strand binding proteins keep them separated, and DNA topoisomerase alleviates supercoiling ahead of the fork (Figure 11.14).

- Deoxynucleoside triphosphates bind to the template strands according to the AT/GC rule. DNA polymerase recognizes these deoxynucleoside triphosphates and attaches a deoxynucleoside monophosphate to the 3′ end of a growing strand (Figure 11.15).

- DNA polymerase requires a primer and synthesizes new DNA strands only in the 5′ to 3′ direction (Figure 11.16).

- The leading strand is made continuously, in the same direction that the fork is moving. The lagging strand is made in the opposite direction as short Okazaki fragments that are synthesized and connected together (Figure 11.17).

- The two replication forks from a single origin of replication use opposite DNA strands to synthesize the leading and lagging strands (Figure 11.18).

- DNA primase makes one RNA primer in the leading strand and multiple RNA primers in the lagging strand. In *E. coli*, DNA polymerase III extends these primers with DNA, and DNA polymerase I removes the primers when they are no longer needed and fills in with DNA. DNA ligase connects adjacent Okazaki fragments in the lagging strand (Figure 11.19, Table 11.2).

- DNA replication is very accurate because (1) hydrogen bonding that follows the AT/CG rule is more stable; (2) DNA polymerase is unlikely to catalyze bond formation if a mismatched base pair is formed; and (3) DNA polymerase carries out proofreading.

- Living organisms have several different types of DNA polymerases with specialized functions (Table 11.3).

- The ends of linear, eukaryotic chromosomes have telomeres composed of repeat sequences. Telomerase binds to the telomere repeat sequence and synthesizes a 6-nucleotide repeat. This happens many times in a row to lengthen one DNA strand of the telomere. DNA primase, DNA polymerase, and DNA ligase are needed to synthesize the complementary DNA strand (Figures 11.20, 11.21).

### 11.5 Molecular Structure of Eukaryotic Chromosomes

- Chromosomes are structures in living cells that carry the genetic material. Chromatin is the name given to the complex of DNA and proteins that makes up eukaryotic chromosomes.

- In eukaryotic chromosomes, the DNA is wrapped around histone proteins to form nucleosomes. Nucleosomes are further compacted into 30-nm fibers. The linker regions are variably twisted and bent into a zigzag pattern (Figures 11.22, 11.23).

- A third level of compaction of eukaryotic chromosomes involves the formation of radial loop domains in which the 30-nm fibers are anchored to a network of proteins called the nuclear matrix. This level of compaction is called euchromatin. In heterochromatin, the loops are even more closely packed together (Figure 11.24).

- During cell division, chromosomes become entirely heterochromatic (Figure 11.25).

## Assess & Discuss

### Test Yourself

1. Why did researchers initially believe that the genetic material was composed of proteins?
   a. Proteins are more biochemically complex than DNA.
   b. Proteins are found only in the nucleus, but DNA is found in many areas of the cell.
   c. Proteins are much larger molecules and can store more information than DNA.
   d. all of the above
   e. both a and c

2. Which component is always different when comparing a nucleotide in a DNA strand to one in an RNA strand?
   a. phosphate group
   b. pentose sugar
   c. nitrogenous base
   d. both b and c
   e. a, b, and c

3. Which of the following equations is accurate concerning DNA base composition?
   a. %A + %T = %G + %C
   b. %A = %G
   c. %A = %G = %T = %C
   d. %A + %G = %T + %C

4. If the sequence of a segment of DNA in one strand is 5′–CGCAACTAC–3′, what is the sequence of the corresponding segment in the opposite strand?
   a. 5′–GCGTTGATG–3′
   b. 3′–ATACCAGCA–5′
   c. 5′–ATACCAGCA–3′
   d. 3′–GCGTTGATG–5′

5. Which of the following statements about the process of DNA replication is correct?
   a. New DNA molecules are composed of two completely new strands.
   b. New DNA molecules are composed of one strand from the old molecule and one new strand.
   c. New DNA molecules are composed of strands that are a mixture of sections from the old molecule and sections that are new.
   d. None of the above statements is correct.

6. Meselson and Stahl were able to demonstrate semiconservative replication in *E. coli* by
   a. using radioactive isotopes of phosphorus to label the original strand and visually determining the relationship of original and new DNA strands.
   b. using different enzymes to eliminate old strands from DNA.
   c. using isotopes of nitrogen to label the DNA and determining the relationship of original and new DNA strands by density differences of the new DNA molecules.
   d. labeling viral DNA before it was incorporated into a bacterial cell and visually determining the location of the DNA after centrifugation.

7. During replication of a DNA molecule, the daughter strands are not produced in exactly the same manner. One strand, the leading strand, is made toward the replication fork, while the lagging strand is made in fragments in the opposite direction. This difference in the synthesis of the two strands is the result of which of the following?
   a. DNA polymerase is not fast enough to make two leading strands of DNA.
   b. The two template strands are antiparallel, and DNA polymerase makes DNA only in the 5′ to 3′ direction.
   c. The lagging strand is the result of DNA breakage due to UV light.
   d. The cell does not contain enough nucleotides to make two complete strands.

8. In living cells, chromosomes consist of
   a. DNA and RNA.
   b. DNA only.
   c. RNA and proteins.
   d. DNA and proteins.
   e. RNA only.

9. A nucleosome is
   a. a dark-staining body composed of RNA and proteins found in the nucleus.
   b. a protein that helps organize the structure of chromosomes.
   c. another word for a chromosome.
   d. a structure composed of DNA wrapped around eight histones.
   e. the short arm of a chromosome.

10. The conversion of euchromatin into heterochromatin involves
    a. the formation of more nucleosomes.
    b. the formation of less nucleosomes.
    c. a greater compaction of loop domains.
    d. a lesser compaction of loop domains.
    e. both a and c.

### Conceptual Questions

1. What are the four key criteria that the genetic material must fulfill? What was Griffith's contribution to the study of DNA, and why was it so important?

2. A double-stranded DNA molecule contains 560 nucleotides. How many complete turns occur in this double helix?

3. **Core Concept: Structure and Function** Discuss how the structure of DNA underlies different aspects of its function.

### Collaborative Questions

1. **CoreSKILL »** A trait that some bacterial strains exhibit is resistance to being killed by antibiotics. For example, certain strains of bacteria are resistant to the drug tetracycline, whereas other strains are sensitive to this antibiotic. Describe an experiment you would carry out to demonstrate that tetracycline resistance is an inherited trait carried in the DNA of the resistant strain.

2. **CoreSKILL »** How might you provide evidence that DNA is the genetic material in mice?

# Gene Expression at the Molecular Level I: Production of mRNA and Proteins

# 12

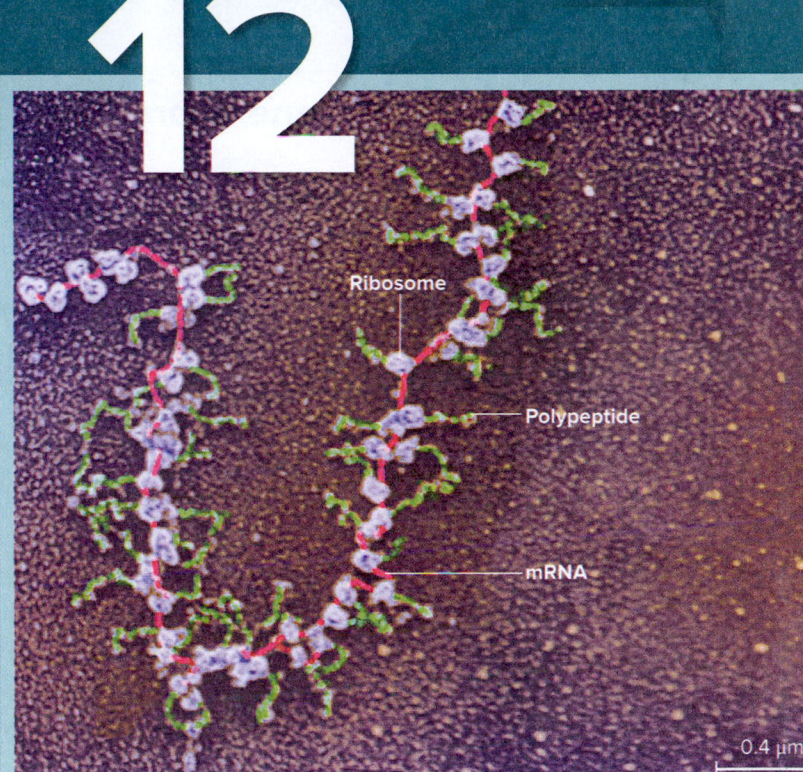

**An electron micrograph of many ribosomes that are translating an mRNA into polypeptides.** This colorized electron micrograph shows ribosomes attached to a bacterial mRNA and synthesizing polypeptides. Ribosomes are blue, mRNA is red, and emerging polypeptides are green. The complex of one mRNA and many ribosomes is called a polysome. ©Elena Kiseleva/Science Source

Mina, age 21, works part-time in an ice-cream shop and particularly enjoys the double-dark chocolate and chocolate fudge brownie flavors on her breaks. She exercises little and spends most of her time studying or watching television. Mina is effortlessly thin. She never worries about what or how much she eats. By comparison, her close friend, Rezzy, has struggled with her weight as long as she can remember. Compared with Mina, she feels like she must constantly deprive herself of food just to maintain her current weight—a weight she would describe as 30 pounds too much.

How do we explain the differences between Mina and Rezzy? Two fundamental factors are involved. Our weight is strongly influenced by the environment, especially our diet, as well as by social and behavioral factors. The amount and types of food we eat are correlated with weight gain. However, there is little doubt that our weight is also influenced by variation in our genes. *Obesity, the condition of having too much body fat, runs in families. The degree of obesity is often similar between genetically identical twins who have been raised apart. Why has genetic variation resulted in some genes that cause certain people to gain weight? A popular hypothesis is that some people have inherited "thrifty genes" as hand-me-downs from their ancestors, who periodically faced famines and food scarcity. Such thrifty genes would be advantageous in allowing people to store body fat more easily and to use food resources more efficiently when times are lean. The negative side is that when food is abundant, unwanted weight gain, and associated diseases such as diabetes and heart disease, can constitute a serious health problem.

Why is knowing about our genes important? Let's consider this question with regard to obesity. Researchers have identified several key genes that influence a person's predisposition to becoming obese. Dozens more are likely to play a minor role. By identifying those genes and studying the proteins specified by them, researchers gain a better understanding of how genetic variation causes certain people to gain weight more easily than others. In addition, this knowledge has led to the development of drugs that are used to combat obesity.

*People are generally considered obese when their body mass index, a measurement obtained by dividing a person's weight by the square of the person's height, is over 30 kg/m².

We can broadly define a gene as a unit of heredity. Geneticists view gene function at different biological levels. In Chapter 17, we will examine how genes affect the traits, or characteristics, of individuals. For example, we will consider how the transmission of genes from parents to offspring affects the color of the offspring's eyes and the likelihood that the offspring will be color blind. In this chapter, we will begin to explore how genes work at the molecular level. You will learn how DNA sequences are organized to form genes and how those genes are used as a template to make RNA copies, ultimately leading to the synthesis of a functional protein. The term **gene expression** can refer to gene function either at the level of traits or at the molecular level. In reality, the two phenomena are intricately woven together. The expression of genes at the molecular level affects the structure and function of cells, which, in turn, determine the traits that an organism expresses.

We begin this chapter by considering how researchers came to realize that most genes store the information to make proteins. We then explore the steps of gene expression as they occur at the molecular level. These steps include the use of a gene as a template to make an RNA molecule, the modifications of the RNA into a functional molecule (in eukaryotes), and the use of RNA to direct the formation of a protein. This chapter focuses on the expression of genes that encode polypeptides, which are called **protein-encoding genes**. The next chapter focuses on some of the interesting functions of genes that produce **non-coding RNAs (ncRNAs)**, which are RNAs that do not encode polypeptides.

## 12.1 Overview of Gene Expression

### Learning Outcomes:

1. **CoreSKILL »** Analyze the results of the experiments of Garrod and of Beadle and Tatum.
2. Outline the general steps of gene expression at the molecular level, which together constitute the central dogma.
3. Explain how proteins are largely responsible for determining an organism's characteristics.

Even before DNA was identified as the genetic material, scientists had asked, "How does the functioning of genes produce the traits of living organisms?" At the molecular level, a similar question can be asked: "How do genes affect the composition and/or function of molecules found within living cells?" An approach that was successful in answering these questions involved the study of **mutations**, which are changes in the genetic material that can be inherited. Mutations may affect the genetic blueprint by altering gene function. For this reason, research that focused on the effects of mutations proved instrumental in determining the molecular function of genes.

In this section, we will consider two early experiments in which researchers studied the effects of mutations in humans and in a bread mold. Both studies led to the conclusion that the role of some genes is to carry the information to produce enzymes, which are a type of protein. Then we will examine the general features of gene expression at the molecular level.

### The Study of Inborn Errors of Metabolism Suggested That Some Genes Carry the Information to Make Enzymes

In 1908, Archibald Garrod, a British physician, proposed a relationship between genes and the production of enzymes. Prior to his work, biochemists had studied many metabolic pathways that consist of a series of conversions of one molecule to another, each step catalyzed by an enzyme. **Figure 12.1** illustrates part of the metabolic pathway for the breakdown of phenylalanine, an amino acid commonly found in human diets. The enzyme phenylalanine hydroxylase catalyzes the conversion of phenylalanine to tyrosine, another amino acid. A different enzyme, tyrosine aminotransferase, converts tyrosine into the next molecule in the pathway, called *p*-hydroxyphenylpyruvic acid. In each case, a specific enzyme catalyzes a single chemical reaction.

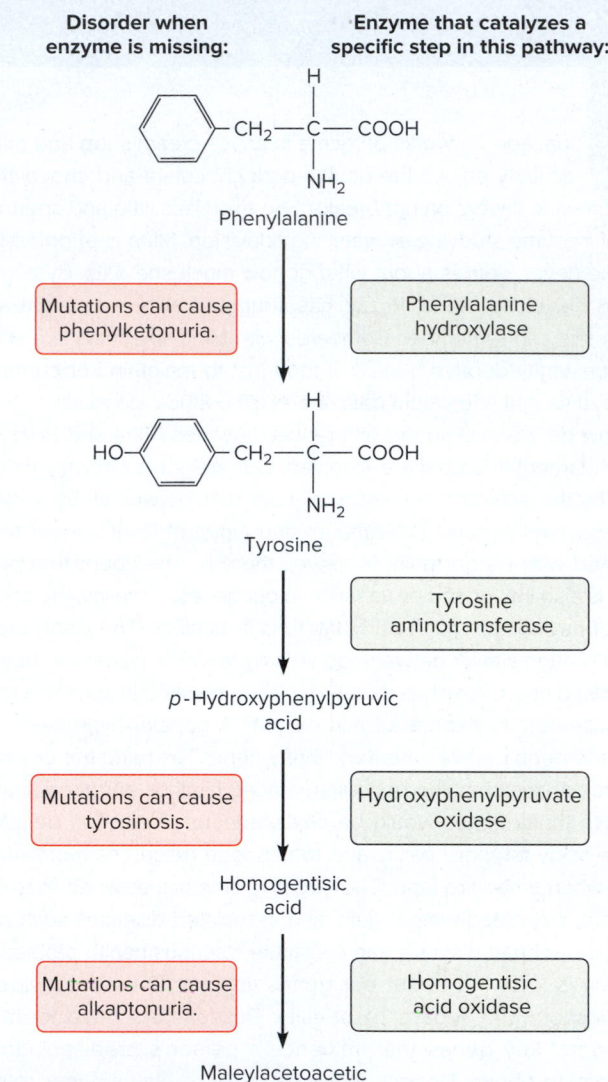

**Figure 12.1 The metabolic pathway that breaks down phenylalanine and its relationship to certain genetic diseases.** Each step in the pathway is catalyzed by a different enzyme, shown in the boxes on the right. If one of the enzymes is not functioning, the product of the previous step builds up, causing the disorders named in the boxes on the left.

**Concept Check:** *What disease would result if a person inherited two defective copies of the gene that encodes phenylalanine hydroxylase?*

Much of Garrod's early work centered on the inherited disease alkaptonuria, in which the patient's body accumulates abnormal levels of homogentisic acid (also called alkapton). This compound, which is bluish black, results in discoloration of the skin and cartilage and causes the urine to appear black. Garrod hypothesized that the accumulation of homogentisic acid in these patients is due to a defect in an enzyme, namely, homogentisic acid oxidase (see Figure 12.1). Furthermore, he already knew that alkaptonuria is an inherited condition that follows a recessive pattern of inheritance. As discussed in Chapter 17, if a disease is recessive, an individual with the disease has inherited the mutant (defective) gene that causes it from both parents.

How did Garrod explain these observations? In 1908, he proposed a relationship between the inheritance of a mutant gene and a defect in metabolism. In the case of alkaptonuria, if an individual inherited the mutant gene from both parents, she or he would not produce any normal enzyme and would be unable to metabolize homogentisic acid. Garrod described alkaptonuria as an **inborn error of metabolism**. An inborn error refers to a mutation in a gene that is inherited from one or both parents. At the turn of the last century, this was a particularly insightful idea because the structure and function of the genetic material were completely unknown.

## Beadle and Tatum Proposed the One-Gene/One-Enzyme Hypothesis

In early 1940s, American geneticists George Beadle and Edward Tatum became aware of Garrod's work and were interested in the relationship between genes and enzymes. They focused their studies on *Neurospora crassa*, a common bread mold. *Neurospora* is easily grown in the laboratory and has only a few nutritional requirements: a carbon source (namely, sugar), inorganic salts, and one vitamin known as biotin. *Neurospora* has many different enzymes that synthesize the molecules, such as amino acids and vitamins, which are essential for growth.

Like Garrod, Beadle and Tatum hypothesized that genes carry the information to make specific enzymes. They reasoned that a mutation, that is, a change in a gene, might cause a defect in an enzyme required for the synthesis of an essential molecule, such as an amino acid. A mutant *Neurospora* strain (one that carries such a mutation) would be unable to grow unless the amino acid was supplemented in the growth medium. Strains without a mutation are called wild-type strains. One line of study involved the amino acid arginine. At the time of Beadle and Tatum's work in the early 1940s, the pathway leading to arginine synthesis was known to involve certain precursor molecules, including ornithine and citrulline. A simplified pathway for arginine synthesis is shown in **Figure 12.2a**. Each step is catalyzed by a different enzyme.

Beadle and Tatum exposed *Neurospora* cells to X-rays, which caused mutations to occur, and studied the resulting cells. By plating the cells on growth media with or without arginine, they were able to identify several different mutant strains that required arginine for growth. They hypothesized that each mutant strain might be blocked at only a single step in the consecutive series of reactions that lead to arginine synthesis. To test this hypothesis, the mutant strains were examined for their ability to grow in the presence of ornithine, citrulline, or arginine (**Figure 12.2b**). The wild-type strain could grow on minimal growth media that did not contain ornithine, citrulline, or arginine. Based on their growth properties, the mutant strains that had been originally identified as requiring arginine for growth could be placed into three groups, designated 1, 2, and 3:

- Group 1 mutants were missing enzyme 1, needed for the conversion of a precursor molecule into ornithine. They could grow only if ornithine, citrulline, or arginine was added to the growth medium.

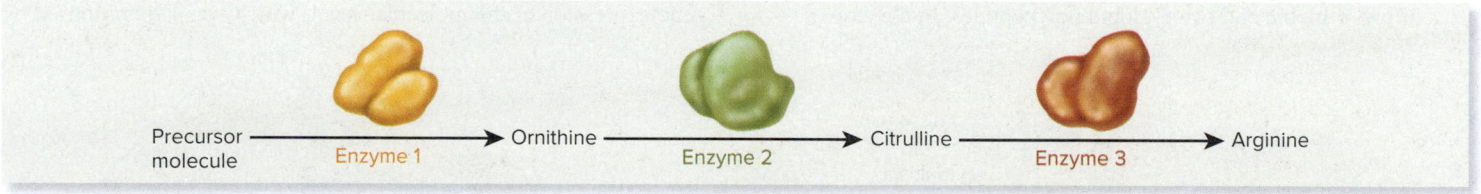

**(a) Simplified pathway for arginine synthesis**

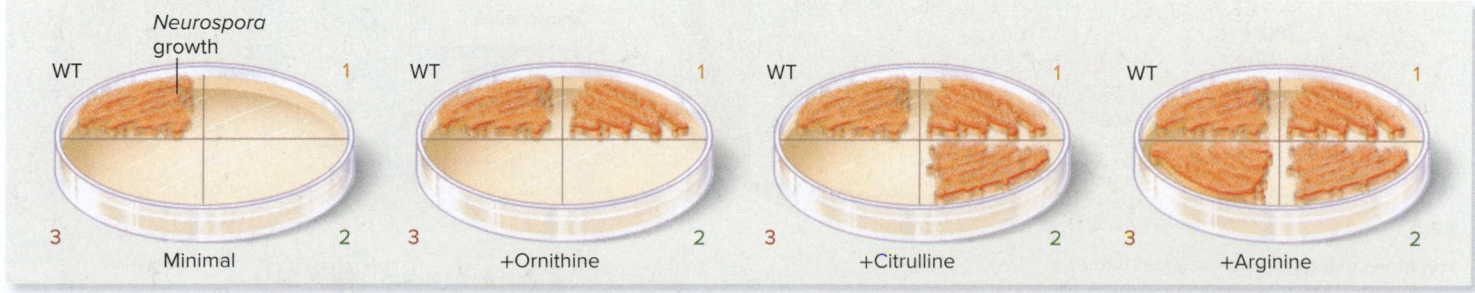

**(b) Growth of strains on minimal and supplemented growth media**

**Figure 12.2   An experiment that supported Beadle and Tatum's one-gene/one-enzyme hypothesis.** (a) This simplified metabolic pathway shows three enzymes that are required for arginine synthesis. (b) Growth of wild-type (WT) and mutant *Neurospora* strains (groups 1, 2, and 3) on minimal growth medium or in the presence of ornithine, citrulline, or arginine.

**Concept Check:** *What enzyme function was missing in group 2 mutants in this experiment?*

- Group 2 mutants were missing the second enzyme in this pathway that is needed for the conversion of ornithine into citrulline. The group 2 mutants would not grow if only ornithine was added, but could grow if citrulline or arginine was added.

- Group 3 mutants were missing the enzyme needed for the conversion of citrulline into arginine. These mutants could grow only if arginine was added.

How were these results interpreted? The researchers were able to order the functions of the genes involved in arginine synthesis in the following way:

$$\text{Precursor} \xrightarrow{\text{Group 1}} \text{Ornithine} \xrightarrow{\text{Group 2}} \text{Citrulline} \xrightarrow{\text{Group 3}} \text{Arginine}$$

From these results and earlier studies, Beadle and Tatum concluded that a single gene controlled the synthesis of a single enzyme. This was referred to as the **one-gene/one-enzyme hypothesis**. Beadle and Tatum received the 1958 Nobel Prize in Physiology or Medicine for their work on the role of genes in metabolism.

In later decades, their hypothesis was modified in four ways.

1. The information to make all proteins is contained within genes, and many proteins do not function as enzymes.

2. Some proteins are composed of two or more different polypeptides. The term **polypeptide** refers to a linear sequence of amino acids; it denotes structure. Most genes carry the information to make a particular polypeptide. By comparison, the term **protein** denotes function. Some proteins are composed of one polypeptide. In such cases, a single gene does contain the information to make a single protein. In other cases, however, a functional protein is composed of two or more different polypeptides. An example is hemoglobin—the protein that carries oxygen in red blood cells—which is composed of two α-globin and two β-globin polypeptides. In the case of

hemoglobin, the expression of two genes (that is, the α-globin and β-globin genes) is needed to produce a functional protein.

3. As described in Chapter 14, some mRNAs (messenger RNAs) are spliced in alternative ways so they produce two or more polypeptides. This allows a single gene to encode more than one polypeptide.

4. A fourth modification to the one-gene/one-enzyme hypothesis is that some genes produce non-coding RNAs that do not specify the amino acid sequence of a polypeptide. This topic is discussed in Chapter 13.

Because of these additional complexities, the one-gene/one-enzyme hypothesis was modified and expanded as the functions of genes became better understood.

## Molecular Gene Expression Involves the Processes of Transcription and Translation

Let's now examine the general steps of gene expression at the molecular level. The first step, known as **transcription**, produces an RNA copy of a gene, also called an RNA transcript (**Figure 12.3**). The term transcription literally means the act of making a copy. Most genes, which are termed protein-encoding genes, produce an RNA molecule that contains the information to specify a polypeptide with a particular amino acid sequence. This type of RNA is called **messenger RNA** (abbreviated **mRNA**), because its function is to carry information from the DNA to cellular components called ribosomes. As discussed later, ribosomes play a key role in the synthesis of polypeptides. The process of synthesizing a specific polypeptide on a ribosome is called **translation**. The term translation is used because a base sequence in an mRNA is "translated" into an amino acid sequence of a polypeptide.

Together, the transcription of DNA into mRNA and the translation of mRNA into a polypeptide constitute the **central dogma** of gene expression at the molecular level, which was first proposed by

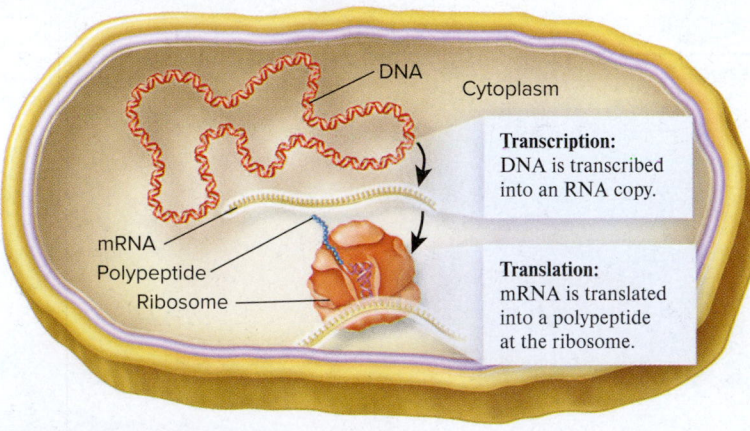

**(a) Molecular gene expression in bacteria**

**Figure 12.3** **The central dogma of gene expression at the molecular level.** **(a)** In bacteria, transcription and translation occur in the cytoplasm. **(b)** In eukaryotes, transcription and RNA modification occur in the nucleus, whereas translation takes place in the cytosol.

*Concept Check:* *What is the direction of flow of genetic information?*

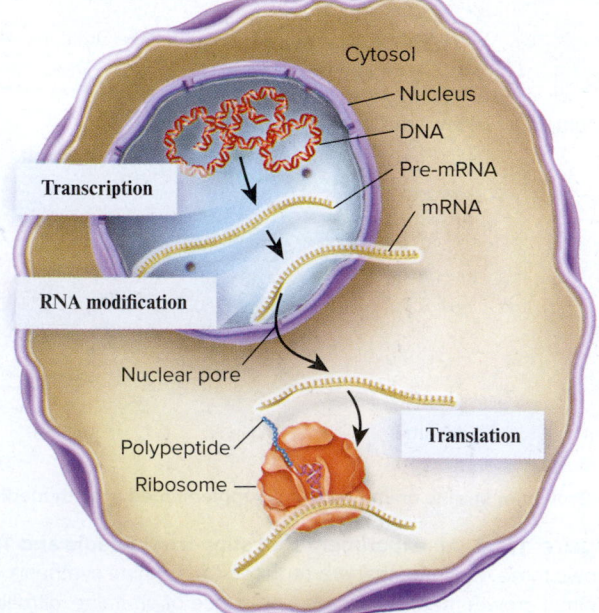

**(b) Molecular gene expression in eukaryotes**

Francis Crick in 1958 (see Figure 12.3). The central dogma applies equally to bacteria, archaea, and eukaryotes. However, in eukaryotes, an additional step occurs between transcription and translation. During **RNA modification**, which is described later in this chapter, the RNA transcript, termed **pre-mRNA**, is modified in ways that make it a functionally active mRNA (Figure 12.3b).

Another difference between bacteria and eukaryotes is the cellular location of transcription and translation. In bacteria, both events occur in the same location, namely, the cytoplasm. In eukaryotes, transcription occurs in the nucleus. The mRNA then exits the nucleus through a nuclear pore, and translation occurs in the cytosol.

Though the direction of information flow, that is, from DNA to RNA to protein, is the most common pathway, exceptions do occur. For example, certain viruses use RNA as a template to synthesize DNA. Such viruses are described in Chapter 19.

### The Protein Products of Genes Largely Determine an Organism's Characteristics

The genes that constitute the genetic material provide a blueprint for the characteristics of every organism. They contain the information necessary to produce an organism and allow it to interact appropriately with its environment. Each protein-encoding gene stores the information for the production of a polypeptide, which then becomes a unit within a functional protein. The activities of proteins largely determine the structure and function of cells. Furthermore, the characteristics of an organism are rooted in the activities of cellular proteins.

A key purpose of the genetic material is to encode the production of proteins in the correct cell, at the proper time, and in suitable amounts. This is an intricate task, because living cells make thousands of different kinds of proteins. Genetic analyses have shown that a typical bacterium can make a few thousand different proteins, and estimates for eukaryotes range from several thousand in simpler eukaryotes to tens of thousands in more complex eukaryotes like humans.

## 12.2 Transcription

**Learning Outcomes:**

1. Give a molecular definition of the term gene.
2. Outline the three stages of transcription and the role of RNA polymerase in this process.
3. Explain how genes within the same chromosome vary in their direction of transcription.
4. Compare and contrast transcription in bacteria and eukaryotes.

DNA is an information storage unit. For genes to be expressed, the information in them must be accessed at the molecular level. Rather than accessing the information directly, however, a working copy of the DNA, composed of RNA, is made. This occurs by the process of transcription, in which a DNA sequence is copied into an RNA sequence. Importantly, transcription does not permanently alter the structure of DNA. Therefore, the same DNA can continue to store information even after an RNA copy has been made. In this section, we will examine the steps necessary for genes to act as transcriptional units. We will also consider some differences in these steps between bacteria and eukaryotes.

### At the Molecular Level, a Gene Is Transcribed and Produces a Functional Product

What is a gene? At the molecular level, a **gene** is defined in the following way:

*At the molecular level, a gene is defined as an organized unit of base sequences that enables a segment of DNA to be transcribed into RNA and ultimately results in the formation of a functional product.*

When a protein-encoding gene is transcribed, an mRNA is made that specifies the amino acid sequence of a polypeptide. After it is made, the polypeptide becomes a functional product; one or more polypeptides form a functional protein. The mRNA is an intermediary in polypeptide synthesis. Among all species, most genes are protein-encoding genes. However, for some genes, the functional product is the RNA itself. The RNA from such a gene, which is called a non-coding RNA, is never translated. Two important examples of non-coding RNAs are transfer RNA and ribosomal RNA. **Transfer RNA (tRNA)** translates the language of mRNA into that of amino acids. **Ribosomal RNA (rRNA)** forms part of ribosomes, which provide the site where translation occurs. We'll learn more about these two types of RNA later in this chapter.

A gene is composed of specific base sequences organized in a way that allows the DNA to be transcribed into RNA. **Figure 12.4** shows the general organization of sequences in a protein-encoding gene. The **promoter** is a sequence of DNA that controls when and where transcription will begin. By comparison, the **terminator** specifies the end of transcription. Therefore, transcription occurs between these two boundaries. As shown in Figure 12.4, the DNA is transcribed into mRNA

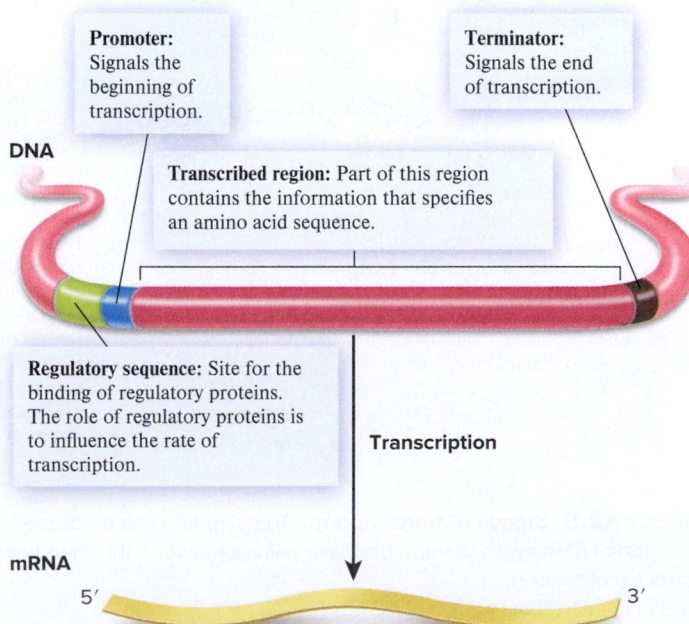

**Promoter:** Signals the beginning of transcription.

**Terminator:** Signals the end of transcription.

**DNA**

**Transcribed region:** Part of this region contains the information that specifies an amino acid sequence.

**Regulatory sequence:** Site for the binding of regulatory proteins. The role of regulatory proteins is to influence the rate of transcription.

**Transcription**

**mRNA**

5'                                                    3'

**Figure 12.4** **A protein-encoding gene as a transcriptional unit.**

 **Core Concept: Information** Transcription is the first step in accessing the information that is stored in DNA.

from the end of the promoter through the coding sequence to the terminator. Within this transcribed region is the information that will specify the amino acid sequence of a polypeptide when the mRNA is translated.

Other DNA sequences are involved in the regulation of transcription. These **regulatory sequences** function as sites for the binding of regulatory proteins, which are discussed in Chapter 14. When a regulatory protein binds to a regulatory sequence, the rate of transcription is affected. Some regulatory proteins enhance the rate of transcription, whereas others inhibit it.

## During Transcription, RNA Polymerase Uses a DNA Template to Make RNA

Transcription occurs in three stages, called initiation, elongation, and termination, during which various proteins interact with DNA sequences (**Figure 12.5**).

**Initiation**   The stage called **initiation** is a recognition step. In bacteria such as *E. coli*, a protein called **sigma factor** binds to

**RNA polymerase**, the enzyme that synthesizes strands of RNA. Sigma factor recognizes the base sequence of a promoter and binds there. An example of a promoter sequence is described in the legend to Figure 12.5. The role of sigma factor is to cause RNA polymerase to bind to the promoter. The initiation stage is completed when the DNA strands are separated near the promoter to form an **open complex** that is approximately 10–15 bp long.

**Elongation**   During **elongation**, RNA polymerase synthesizes the RNA transcript. For this to occur, sigma factor is released and RNA polymerase slides along the DNA in a way that maintains an open complex as it goes. The DNA strand that is used as a template for RNA synthesis is called the **template strand**. For protein-encoding genes, the opposite DNA strand is called the **coding strand**. The coding strand has the same sequence of bases as the resulting mRNA, except that the RNA has uracil instead of the thymine found in the DNA. The coding strand is so named because, like mRNA, it carries the information that codes for a polypeptide.

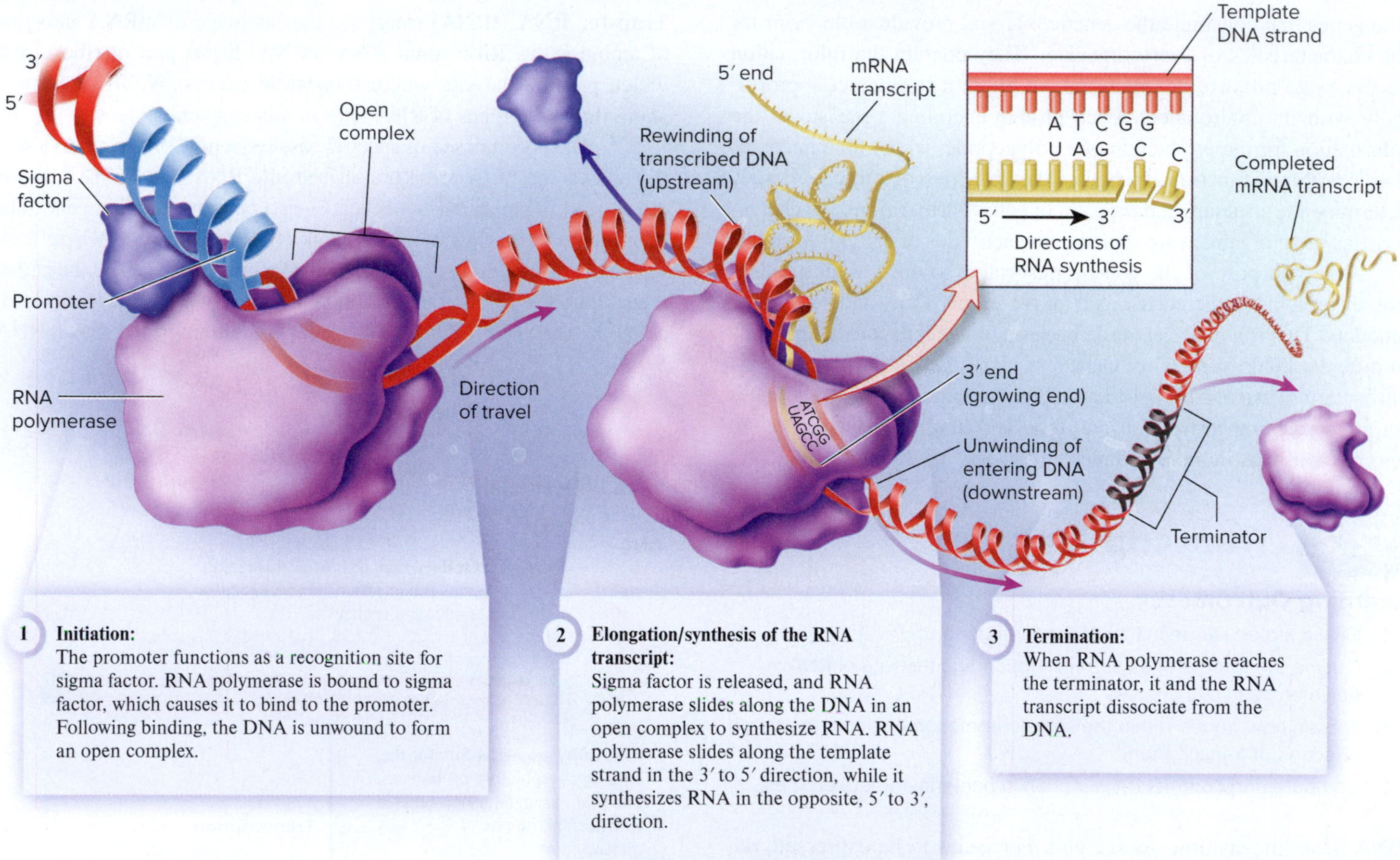

**1 Initiation:**
The promoter functions as a recognition site for sigma factor. RNA polymerase is bound to sigma factor, which causes it to bind to the promoter. Following binding, the DNA is unwound to form an open complex.

**2 Elongation/synthesis of the RNA transcript:**
Sigma factor is released, and RNA polymerase slides along the DNA in an open complex to synthesize RNA. RNA polymerase slides along the template strand in the 3′ to 5′ direction, while it synthesizes RNA in the opposite, 5′ to 3′, direction.

**3 Termination:**
When RNA polymerase reaches the terminator, it and the RNA transcript dissociate from the DNA.

**Figure 12.5   Stages of transcription.**   Transcription can be divided into initiation, elongation, and termination. The inset emphasizes the direction of RNA synthesis and the base pairing between the DNA template strand and RNA. An example of a promoter sequence in the DNA of *E. coli* is as follows:

5′-TTGACATGATAGAAGCACTCTACTATATT-3′
3′-AACTGTACTATCTTCGTGAGATGATATAA-5′

This region is 29 bp long, and it immediately precedes the site where transcription begins. The bases that are specifically recognized by sigma factor are shown in red. The sequences of promoters for different genes are fairly diverse, particularly in eukaryotic species.

 **Core Skill: Connections**   Look back at the role of DNA polymerase shown in Figure 11.15. What are similarities and differences between the function of DNA polymerase and that of RNA polymerase?

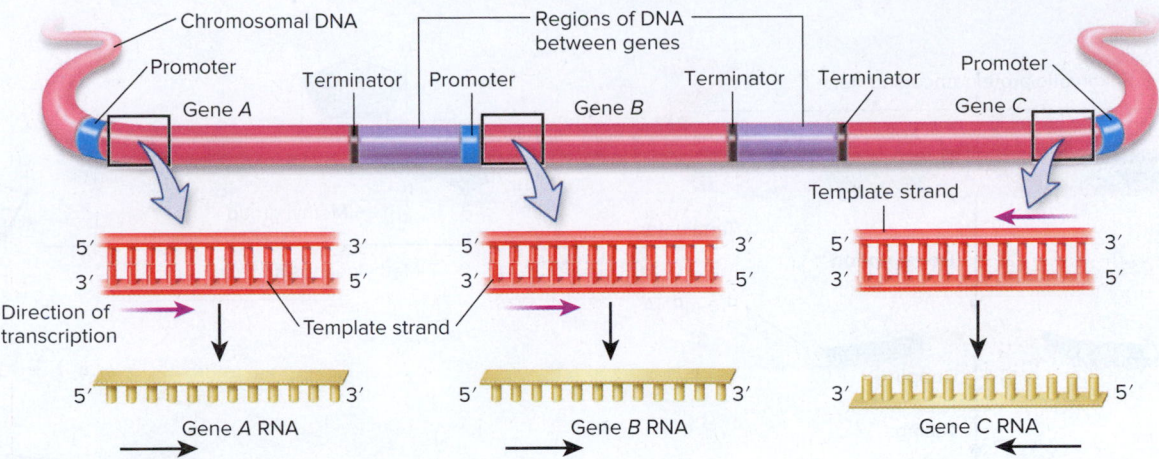

**Figure 12.6** **The transcription of three different genes found in the same chromosome.** RNA polymerase synthesizes each RNA transcript in a 5′ to 3′ direction, sliding along a DNA template strand in a 3′ to 5′ direction. However, the strand used as the template can vary from gene to gene. For example, for transcribing genes A and B, the bottom strand is used as the template strand, while the top strand is used for transcribing gene C.

During the elongation stage of transcription, nucleotides bind to the template strand and are covalently connected in the 5′ to 3′ direction (see inset of step 2, Figure 12.5). The complementarity rule used in this process is similar to the AT/GC rule of DNA replication, except that uracil (U) in RNA substitutes for thymine (T) in DNA. For example, a DNA template with the sequence 3′–TACAATGTAGCC–5′ will be transcribed into an RNA sequence reading 5′–AUGUUACAUCGG–3′. In bacteria, the rate of RNA synthesis is about 40 nucleotides per second! Behind the open complex, the DNA rewinds back into a double helix.

*Termination* Eventually, RNA polymerase reaches a terminator, which causes it and the newly made RNA transcript to dissociate from the DNA. This event constitutes the **termination** of transcription.

When multiple genes within a chromosome are transcribed, the DNA strand that is used as the template strand varies among the genes. **Figure 12.6** shows three genes adjacent to each other within a chromosome. Genes A and B are transcribed from left to right, using the bottom DNA strand as the template strand. By comparison, gene C is transcribed from right to left, using the top DNA strand as a template strand. In all three cases, however, the synthesis of the RNA transcript begins at a promoter and always occurs in a 5′ to 3′ direction. The template strand is read in the 3′ to 5′ direction.

## Transcription in Eukaryotes Involves More Proteins

The basic features of transcription are similar among all organisms. The genes of all species have promoters, and the transcription process occurs in the stages of initiation, elongation, and termination. However, the transcription of eukaryotic genes tends to involve a greater complexity of protein components than does the transcription of bacterial genes. For example, three forms of RNA polymerase, designated I, II, and III, are found in eukaryotes. RNA polymerase II is responsible for transcribing the mRNA from eukaryotic protein-encoding genes, whereas RNA polymerases I and III transcribe genes that specify non-coding RNAs, such as tRNAs and rRNAs. By comparison, bacteria have a single type of RNA polymerase that

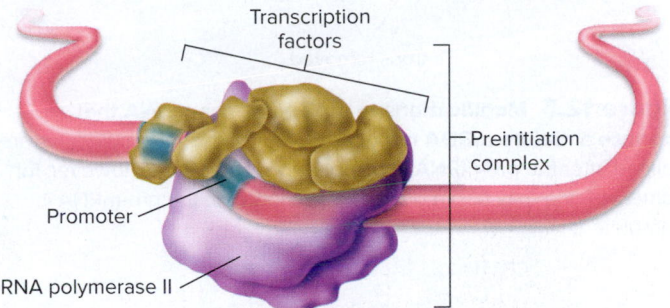

**Figure 12.7** **The preinitiation complex.** Transcription factors and RNA polymerase II assemble into the preinitiation complex at the promoter in eukaryotic protein-encoding genes.

transcribes all genes, though many bacterial species have more than one type of sigma factor that can recognize different promoters.

The initiation stage of transcription in eukaryotes is also more complex. Recall that in bacteria such as *E. coli*, sigma factor recognizes the promoter of a gene. By comparison, RNA polymerase II of eukaryotes always requires five transcription factors to initiate transcription. **Transcription factors** are proteins that influence the ability of RNA polymerase to transcribe genes. The binding of RNA polymerase II to the promoter is an assembly process in which RNA polymerase II and the five transcription factors form a **preinitiation complex** (**Figure 12.7**). The complex then unwinds the DNA to initiate transcription.

## 12.3  RNA Modification in Eukaryotes

**Learning Outcomes:**

1. Describe the addition of the 5′ cap and 3′ poly A tail to eukaryotic mRNA.
2. Explain the process of splicing that produces mature eukaryotic mRNA.

As noted previously, eukaryotic mRNA transcripts undergo modifications to produce functional mRNAs. Transcription initially produces

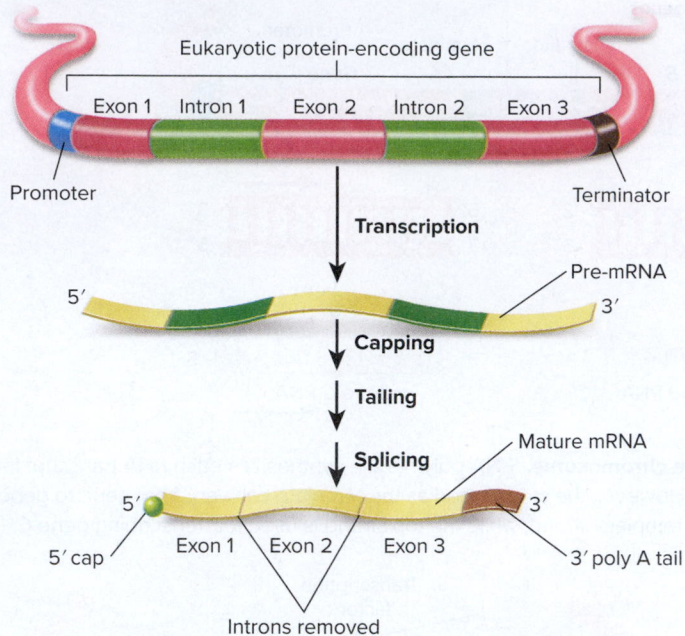

**Figure 12.8** **Modifications to eukaryotic pre-mRNA that produce a mature mRNA molecule.** Note: Most RNA molecules are spliced after the pre-mRNA is completely synthesized. However, for some, splicing may begin before transcription of the pre-mRNA is completed.

a longer RNA, called a **pre-mRNA**, which undergoes certain modifications before it exits the nucleus. The final product is called a **mature mRNA**, or simply an mRNA. As shown in **Figure 12.8**, three different modifications are common. Two of them, called capping and tailing, are modifications to the ends of the pre-mRNA. As described later, these add a 5′ cap and a polyA tail. A third modification, called splicing, involves the removal of internal segments called introns. The segments that are retained in a mature mRNA are exons. After all of the RNA modifications have been completed, the mRNA leaves the nucleus and enters the cytosol, where translation occurs. In this section, we will examine the molecular mechanisms that account for these RNA modifications and consider why they are functionally important.

## The Ends of Eukaryotic Pre-mRNAs Are Modified by the Addition of a 5′ Cap and a 3′ Poly A Tail

Mature mRNAs of eukaryotes have a modified form of guanine covalently attached at the 5′ end, an event known as **capping** (**Figure 12.9a**). Capping occurs while a pre-mRNA is being made by RNA polymerase, usually when the transcript is only 20 to 25 nucleotides in length. What are the functions of the cap?

- The 7-methylguanosine structure, called a **5′ cap**, is recognized by cap-binding proteins, which are needed for the proper exit of mRNAs from the nucleus.
- After an mRNA is in the cytosol, the cap structure helps to prevent its degradation.

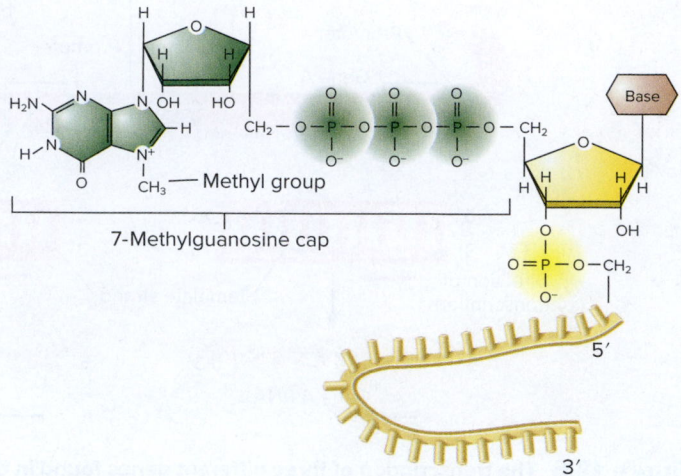

**(a) Cap structure at the 5′ end of eukaryotic mRNA**

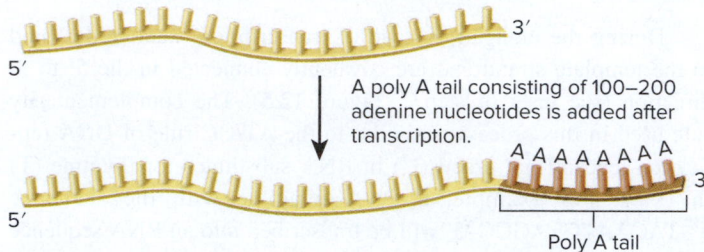

**(b) A poly A tail at the 3′ end of eukaryotic mRNA**

**Figure 12.9** **Modifications that alter the ends of mRNA in eukaryotic cells.** **(a)** A 7-methylguanosine cap is attached to the 5′ end of a pre-mRNA. This is a guanine base modified by the attachment of a methyl group. The linkage between the cap and the mRNA is a 5′ to 5′ linkage. **(b)** A poly A tail is added to the 3′ end of a pre-mRNA.

**Concept Check:** *Do the ends of protein-encoding genes have a poly T region that acts as a template for the synthesis of a poly A tail in mRNA? Explain.*

- The cap structure is recognized by cap-binding proteins that enable the mRNA to bind to a ribosome for translation.

At the 3′ end, most mature eukaryotic mRNAs have a string of adenine nucleotides, typically 100 to 200 nucleotides in length, referred to as a **poly A tail** (**Figure 12.9b**). The poly A tail is not encoded in the gene sequence. Instead, the tail is added enzymatically after a pre-mRNA has been completely transcribed.

- A long poly A tail aids in the export of mRNA from the nucleus.
- It also stabilizes a eukaryotic mRNA so it can exist for a longer period of time in the cytosol.

Interestingly, new research has shown that some bacterial mRNAs also have poly A tails attached to them. However, the poly A tail has an opposite effect in bacteria, where it causes the mRNA to be rapidly degraded.

## Splicing Involves the Removal of Introns and the Linkage of Exons

In the late 1970s, when the experimental tools became available to study eukaryotic genes at the molecular level, the scientific community was astonished by the discovery that the coding sequences within many eukaryotic protein-encoding genes are separated by DNA sequences that are transcribed but not translated into protein. These intervening sequences that are not translated are called **introns**, whereas sequences contained in the mature mRNA are termed **exons**. Exons are considered to be expressed regions, because they contain the coding sequence for a polypeptide. In contrast, introns are intervening regions that are not expressed, because they are removed from the pre-mRNA. To become a mature mRNA, the pre-mRNA transcribed from eukaryotic genes that contain introns must undergo a third RNA modification known as **RNA splicing**, or simply **splicing** (see Figure 12.8). During this process, introns are removed and the remaining exons are connected to each other.

Introns are found in many but not all eukaryotic genes. Splicing is less frequent among unicellular eukaryotic species, such as yeast, but is a widespread phenomenon among more complex eukaryotes. In animals and flowering plants, most protein-encoding genes have one or more introns. For example, an average human gene has about nine introns. The sizes of introns vary from a few dozen nucleotides to over 100,000! A few bacterial genes have been found to have introns, but they are rare among bacterial and archaeal species.

Introns are precisely removed from eukaryotic pre-mRNA by a large complex called a **spliceosome** that is composed of several different snRNPs (pronounced "snurps"); each snRNP contains small nuclear RNA and a set of proteins. An intron in pre-mRNA is defined by a particular sequence within the intron termed the branch site and by two intron-exon boundaries, called the 5′ splice site and the 3′ splice site (**Figure 12.10**). Particular snRNPs bind to specific sequences at these three locations. This binding causes the intron to loop outward, which brings the two exons close together. The 5′ splice site is then cut, and the 5′ end of the intron becomes covalently attached to the branch site. In the final step, the 3′ splice site is cut, and the two exons are covalently attached to each other. The intron is released and eventually degraded.

In some cases, the function of the spliceosome is regulated so the splicing of a given mRNA can occur in two or more ways. This phenomenon, called **alternative splicing**, allows a single gene to encode two or more polypeptides with differences in their amino acid sequences. As described in Chapter 14, alternative splicing allows complex eukaryotic species to use the same gene to make different proteins at different stages of development or in different cell types. This increases the size of the proteome while minimizing the size of the genome.

Although primarily found in mRNAs, introns occasionally occur in rRNA and tRNA molecules of certain species. These introns, however, are not removed by the action of a spliceosome. Instead, such rRNAs and tRNAs are **self-splicing**, which means the RNA itself can catalyze the removal of its own intron(s). Portions of the RNA act like an enzyme to cleave the covalent bonds at the intron-exon boundaries and connect the exons together. An RNA molecule that catalyzes a chemical reaction is termed a **ribozyme**.

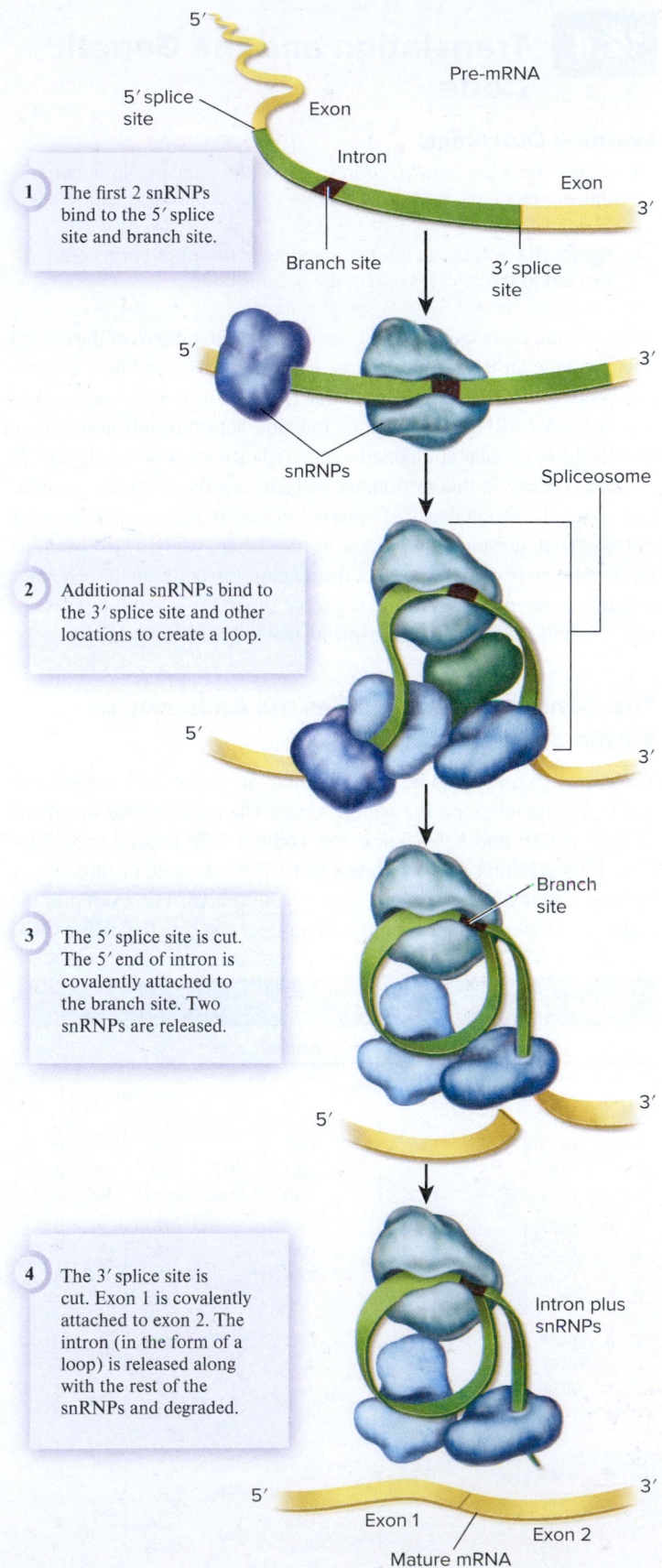

**1** The first 2 snRNPs bind to the 5′ splice site and branch site.

**2** Additional snRNPs bind to the 3′ splice site and other locations to create a loop.

**3** The 5′ splice site is cut. The 5′ end of intron is covalently attached to the branch site. Two snRNPs are released.

**4** The 3′ splice site is cut. Exon 1 is covalently attached to exon 2. The intron (in the form of a loop) is released along with the rest of the snRNPs and degraded.

**Figure 12.10** **The splicing of a eukaryotic pre-mRNA by a spliceosome.**

## 12.4 Translation and the Genetic Code

### Learning Outcomes:

1. Explain how the genetic code specifies the relationship between the sequence of codons in mRNA and the amino acid sequence of a polypeptide.
2. **CoreSKILL »** Analyze the experiments of Nirenberg and Leder that led to the deciphering of the genetic code.

In the two previous sections, we considered the first stages of the central dogma—how an RNA transcript is made from DNA and how eukaryotes process that transcript. Recall that this type of RNA is called messenger RNA (mRNA), because its function is to transmit information from DNA to cellular components called ribosomes, where polypeptide synthesis occurs. In this section, we will consider the next stage, translation, which is the synthesis of polypeptides using information from the mRNA. To understand the process of translation, we will first examine the **genetic code**, which specifies the relationship between the sequence of bases in the mRNA and the sequence of amino acids in a polypeptide. We will also explore experiments that helped to "crack" the code.

### The Genetic Code Specifies the Amino Acids Within a Polypeptide

The ability of mRNA to be translated into an amino acid sequence of a polypeptide relies on the genetic code. The code is read in groups of three nucleotide bases known as **codons**. The genetic code consists of 64 different codons (**Table 12.1**). The sequence of three bases in most codons specifies a particular amino acid. For example, the codon CCC specifies the amino acid proline, whereas the codon GGC

encodes the amino acid glycine. From the analysis of many different species, including bacteria, archaea, protists, fungi, plants, and animals, researchers have found that the genetic code is nearly universal. Only a few rare exceptions to the genetic code have been discovered.

Why are there 64 codons, as shown in Table 12.1? Because amino acids are found in 20 different types, at least 20 different codons are needed so each amino acid can be specified by a codon. With four types of bases in mRNA (U, C, A, and G), a genetic code containing two bases in a codon would not be sufficient, because only $4^2$, or 16, different codons would be possible. A three-base system can specify $4^3$, or 64, different codons, which is far more than the number of amino acids. The genetic code is said to be **degenerate** because more than one codon can specify the same amino acid (see Table 12.1). For example, the codons GGU, GGC, GGA, and GGG all code for the amino acid glycine. In most instances, the third base in the codon is the degenerate, or variable, base.

### During Translation, mRNA Is Used to Make a Polypeptide with a Specific Amino Acid Sequence

Let's look at the organization of a bacterial mRNA to see how translation occurs (**Figure 12.11**). A ribosomal-binding site is located near the 5′ end of the mRNA. The **start codon**, which is AUG, is the site where translation begins. AUG specifies the amino acid methionine. The start codon is only a few nucleotides from the ribosomal-binding site. Beyond

| Table 12.1 | The Genetic Code* | | | | |
|---|---|---|---|---|---|

**Second position**

| First Position | | U | C | A | G | Third Position |
|---|---|---|---|---|---|---|
| **U** | | UUU ⎱ Phe<br>UUC ⎰<br>UUA ⎱ Leu<br>UUG ⎰ | UCU ⎱<br>UCC ⎮ Ser<br>UCA ⎮<br>UCG ⎰ | UAU ⎱ Tyr<br>UAC ⎰<br>UAA Stop<br>UAG Stop | UGU ⎱ Cys<br>UGC ⎰<br>UGA Stop<br>UGG Trp | U<br>C<br>A<br>G |
| **C** | | CUU ⎱<br>CUC ⎮ Leu<br>CUA ⎮<br>CUG ⎰ | CCU ⎱<br>CCC ⎮ Pro<br>CCA ⎮<br>CCG ⎰ | CAU ⎱ His<br>CAC ⎰<br>CAA ⎱ Gln<br>CAG ⎰ | CGU ⎱<br>CGC ⎮ Arg<br>CGA ⎮<br>CGG ⎰ | U<br>C<br>A<br>G |
| **A** | | AUU ⎱<br>AUC ⎮ Ile<br>AUA ⎰<br>AUG Met/<br>start | ACU ⎱<br>ACC ⎮ Thr<br>ACA ⎮<br>ACG ⎰ | AAU ⎱ Asn<br>AAC ⎰<br>AAA ⎱ Lys<br>AAG ⎰ | AGU ⎱ Ser<br>AGC ⎰<br>AGA ⎱ Arg<br>AGG ⎰ | U<br>C<br>A<br>G |
| **G** | | GUU ⎱<br>GUC ⎮ Val<br>GUA ⎮<br>GUG ⎰ | GCU ⎱<br>GCC ⎮ Ala<br>GCA ⎮<br>GCG ⎰ | GAU ⎱ Asp<br>GAC ⎰<br>GAA ⎱ Glu<br>GAG ⎰ | GGU ⎱<br>GGC ⎮ Gly<br>GGA ⎮<br>GGG ⎰ | U<br>C<br>A<br>G |

*The sequences of bases are in the 5' to 3' direction (from left to right) in the mRNA. Exceptions to the genetic code are sporadically found among various species. For example, AUA encodes methionine in yeast and mammalian mitochondria. The three-letter abbreviations for the amino acids are given in Chapter 3 (see Figure 3.13).

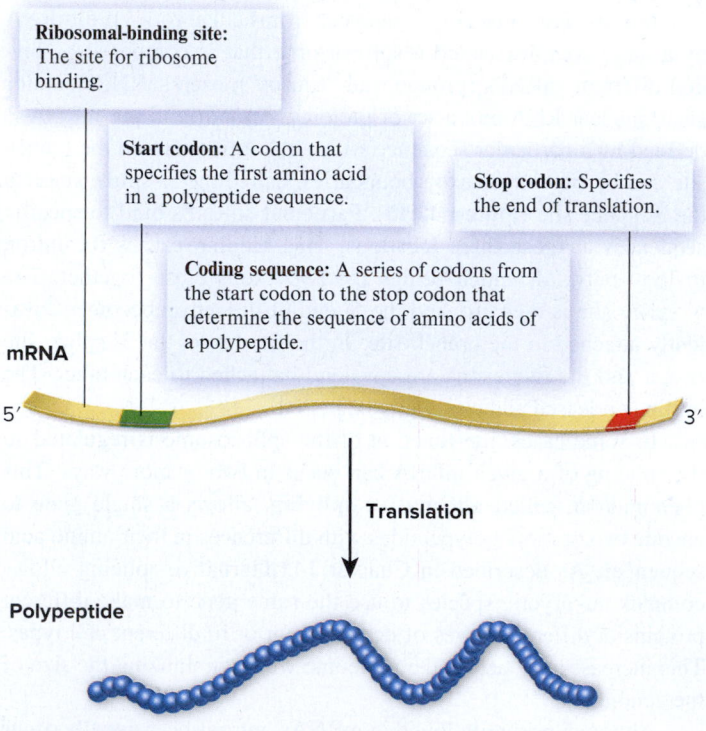

**Figure 12.11** **The organization of a bacterial mRNA as a translational unit.** The string of blue balls represents a sequence of amino acids in a polypeptide. During and following translation, a sequence of amino acids folds into a more compact structure as described in Chapter 3 (see Figure 3.15).

**Concept Check:** *If a mutation eliminated the start codon from a protein-encoding gene, how would the mutation affect transcription, and how would it affect translation?*

this, a large portion of an mRNA functions as a **coding sequence**—a region that specifies the linear amino acid sequence of a polypeptide. A typical polypeptide is a few hundred amino acids in length. The coding sequence consists of a series of codons. Finally, one of three **stop codons** signals the end of translation. These codons, also known as **termination codons**, are UAA, UAG, and UGA.

The start codon also defines the **reading frame** of an mRNA, which refers to the order in which codons are read during translation. Beginning at the start codon, each adjacent codon is read as a group of three bases, also called a **triplet**, in the 5′ to 3′ direction. For example, look at the following two mRNA sequences and their corresponding amino acid sequences.

|  | Ribosomal-binding site | Start codon |
|---|---|---|

mRNA 5′–<u>AUAAGGAGG</u>UUACG(<u>AUG</u>)(CAG)(CAG)(GGC)(UUU)(ACC)–3′
**Polypeptide**                       Met - Gln - Gln - Gly - Phe - Thr

|  | Ribosomal-binding site | Start codon |
|---|---|---|

mRNA 5′–<u>AUAAGGAGG</u>UUACG(<u>AUG</u>)(**U**CA)(GCA)(GGG)(CUU)(UAC)C–3′
**Polypeptide**                       Met - Ser - Ala - Gly - Leu - Tyr

The first sequence shows how the mRNA codons would be correctly translated into amino acids. In the second sequence, an additional U has been added to the sequence after the start codon. This shifts the reading frame, thereby changing the codons as they occur in the 5′ to 3′ direction. The polypeptide produced from this series of codons has a very different sequence of amino acids. From this comparison, we can also see that the reading frame is not overlapping, which means that each base functions within a single codon.

## DNA Stores Information, Whereas mRNA and tRNA Access That Information to Make Polypeptides

The relationships among the DNA sequence of a gene, the mRNA transcribed from the gene, and the polypeptide sequence are shown schematically in **Figure 12.12**. Recall that the template strand is used to make mRNA. The resulting mRNA strand corresponds to the coding strand of DNA, except that U in the mRNA substitutes for T in the DNA. The 5′ end of the mRNA contains an <u>un</u>translated region (5′ UTR) as does the 3′ end (3′ UTR). The middle portion contains a series of codons that specify the amino acid sequence of a polypeptide.

To translate a nucleotide sequence of mRNA into an amino acid sequence, recognition occurs between mRNA and transfer RNA (tRNA) molecules. Transfer RNA, which is described in

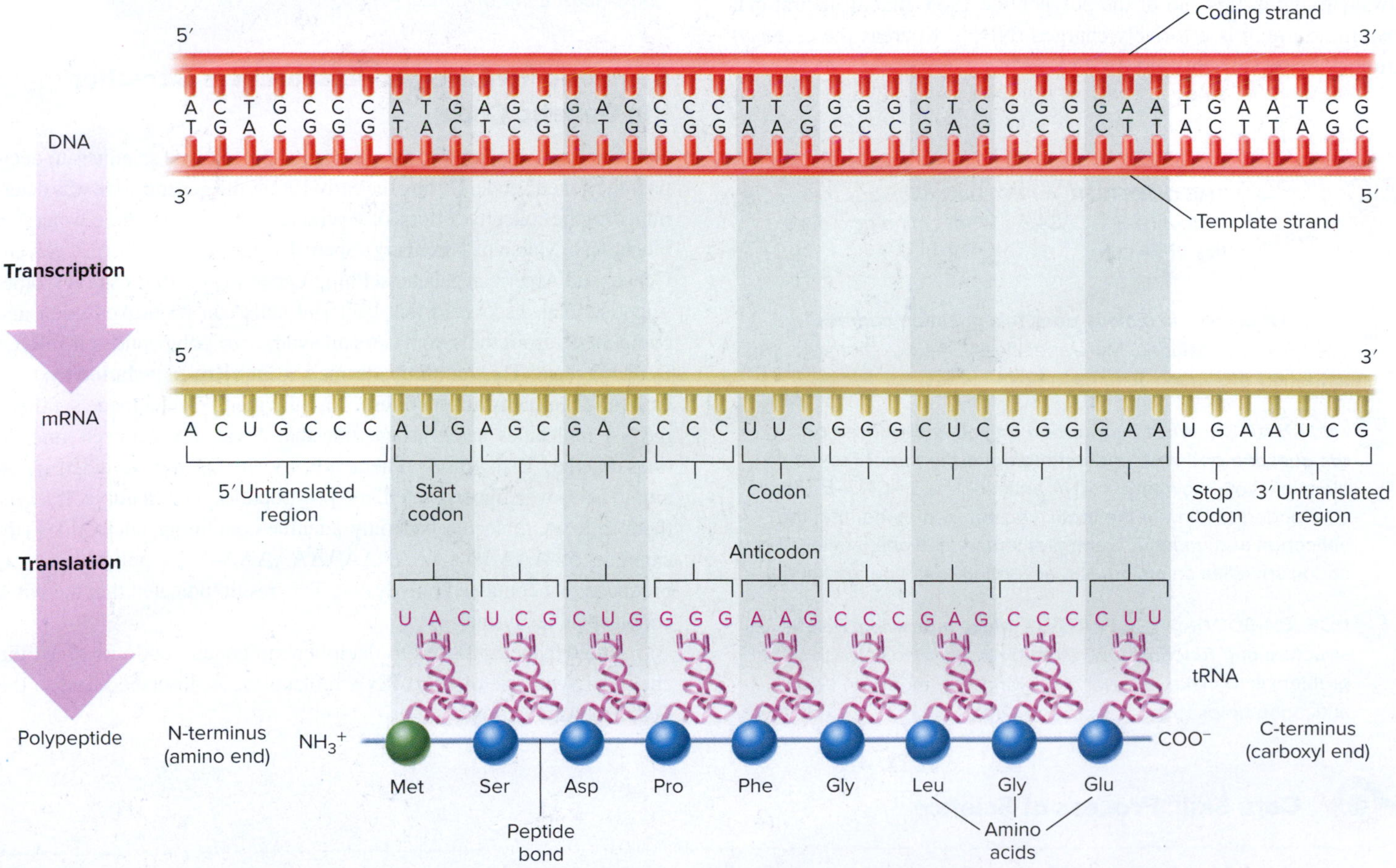

**Figure 12.12** **Relationships among the coding sequence of a gene, the codon sequence of an mRNA, the anticodons of tRNA, and the amino acid sequence of a polypeptide.** Note: The tRNAs are detached from the polypeptide as it being synthesized. Also, this gene does not contain any introns.

 **Core Concept: Information** Figure 12.12 illustrates how DNA stores the information to make a polypeptide with a particular amino acid sequence. Messenger RNA is a temporary copy of that information that is directly used to make a polypeptide.

Section 12.5, functions as the "translator" or intermediary between an mRNA codon and an amino acid. The **anticodon** is a three-base sequence in a tRNA molecule that is complementary to a codon in mRNA. Due to this complementarity, the anticodon in a tRNA and a codon in an mRNA bind to each other. Furthermore, the anticodon in a tRNA corresponds to the amino acid that it carries. For example, if the anticodon in a tRNA is 3′–AAG–5′, it is complementary to a 5′–UUC–3′ codon. According to the genetic code, a UUC codon specifies phenylalanine (Phe). Therefore, a tRNA with a 3′–AAG–5′ anticodon must carry phenylalanine. As another example, a tRNA with a 3′–GGG–5′ anticodon is complementary to a 5′–CCC–3′ codon, which specifies proline. This tRNA must carry proline (Pro).

As seen at the bottom of Figure 12.12, the direction of polypeptide synthesis parallels the 5′ to 3′ orientation of mRNA. The first amino acid is said to be at the amino end, or **N-terminus**, of the polypeptide. The term N-terminus refers to the presence of a nitrogen atom (N) at this end, whereas amino end indicates the presence of an amino group ($NH_2$). **Peptide bonds** connect the amino acids together. These covalent bonds form between the carboxyl group (COOH) of the previous amino acid and the amino group of the next amino acid. The last amino acid in a completed polypeptide does not have another amino acid attached to its carboxyl group. This last amino acid is said to be located at the carboxyl end, or **C-terminus**. A carboxyl group is always found at this end of the polypeptide. Note that at neutral pH, the amino group is positively charged ($NH_3^+$), whereas the carboxyl group is negatively charged ($COO^-$).

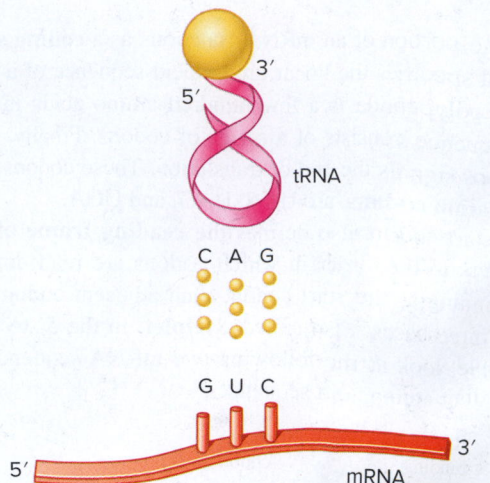

The basis of anticodon/codon recognition is that the anticodon and codon are complementary and antiparallel. The function of the codon is to specify a particular amino acid according to the genetic code.

**ANSWER** *The anticodon is 3′–CAG–5′, so it is complementary to a codon with the sequence 5′–GUC–3′. According to the genetic code, this codon specifies the amino acid valine. Therefore, this tRNA must carry a valine.*

## Synthetic RNA Helped Researchers Decipher the Genetic Code

Now let's look at some early experiments that allowed scientists to decipher the genetic code. During the early 1960s, the genetic code was determined by the collective efforts of several researchers, including American biochemist Marshall Nirenberg, Spanish-American biochemist Severo Ochoa, and American geneticist Philip Leder. Prior to their studies, other scientists had discovered that bacterial cells can be broken open and components from the cytoplasm can synthesize polypeptides if mRNA is also present. This mixture is termed an **in vitro translation system**, or a **cell-free translation system**. Nirenberg and Ochoa made synthetic mRNA molecules using an enzyme that covalently connects nucleotides together. Using this synthetic mRNA, they then determined which amino acids were incorporated into polypeptides. For example, if a synthetic mRNA molecule had only adenine-containing nucleotides (for example, 5′–AAAAAAAAAAAAAAAAAAAA–3′), a polypeptide was produced that contained only lysine. This result indicated that the AAA codon specifies lysine.

Another method used to decipher the genetic code involved the chemical synthesis of short RNA molecules, as described next in the Feature Investigation.

**BIO TIPS**

**THE QUESTION** *A tRNA anticodon has the sequence 3′–CAG–5′. What amino acid does this tRNA carry?*

**T OPIC** *What topic in biology does this question address?* The topic is translation. More specifically, the question asks you to identify the amino acid that a tRNA carries.

**I NFORMATION** *What information do you know based on the question and your understanding of the topic?* From the question, you know that a tRNA anticodon is 3′–CAG–5′. From your understanding of the topic, you may remember that the anticodon and codon are complementary and antiparallel. The codon specifies an amino acid according to the genetic code.

**P ROBLEM-SOLVING S TRATEGY** *Make a drawing. Relate structure and function.* One strategy to begin to solve this problem is to make a drawing showing how the given tRNA anticodon binds to a codon in an mRNA.

 **Core Skill: Process of Science**

## Feature Investigation | Nirenberg and Leder Found That RNA Triplets Can Promote the Binding of tRNA to Ribosomes

In 1964, Nirenberg and Leder discovered that RNA molecules containing three nucleotides (that is, a triplet) can cause a tRNA molecule to bind to a ribosome. In other words, an RNA triplet acts like a codon within an mRNA molecule. To establish the relationship between triplet sequences and specific amino acids, Nirenberg and Leder made triplets with specific base sequences (**Figure 12.13**). For example, in one experiment they studied 5′–CCC–3′ triplets. This particular triplet was added to 20 different tubes. To each tube, they next added an in vitro translation system, which contained ribosomes and tRNAs that already had amino acids attached to them. However, each translation system had only one type of radiolabeled amino acid. One translation system had only proline that was radiolabeled, a second translation system had only serine that was radiolabeled, and so on.

As shown in step 2 of Figure 12.13, the triplets became bound to the ribosomes just like mRNAs bind to ribosomes. The tRNA with an anticodon that was complementary to the added triplet then bound to the triplet, which was already bound to the ribosome. For example, when the triplet was 5′–CCC–3′, a tRNA with a 3′–GGG–5′ anticodon bound to the triplet/ribosome complex. This tRNA carries proline.

To determine which tRNA had bound, the contents from each tube were poured through a filter that trapped the large ribosomes but did not trap tRNAs that were not bound to ribosomes (see step 3). If the tRNA carrying the radiolabeled amino acid was bound to the triplet/ribosome complex, radioactivity would be trapped on the filter. Using a scintillation counter, the researchers determined the amount of radioactivity on each filter. Because only one amino acid was radiolabeled in each in vitro translation system, they could determine which triplet corresponded to which amino acid. In the example shown here, CCC corresponds to proline. Therefore, the in vitro translation system containing radiolabeled proline showed a large amount of radioactivity on the filter. As seen in the data, by studying triplets with different sequences, Nirenberg and Leder identified many codons of the genetic code.

### Experimental Questions

1. Briefly explain how a triplet mimics the role of an mRNA molecule. How was this observation useful in the study done by Nirenberg and Leder?

3. **CoreSKILL »** What was the benefit of using radiolabeled amino acids in the Nirenberg and Leder experiment?

3. **CoreSKILL »** Predict the results that Nirenberg and Leder would have found for the following triplets: AUG, UAA, UAG, and UGA.

**Figure 12.13** Nirenberg and Leder's use of triplet binding method to decipher the genetic code.

**HYPOTHESIS** An RNA triplet can bind to a ribosome and promote the binding of the tRNA that carries the amino acid that the RNA triplet specifies.

**KEY MATERIALS** The researchers made 20 in vitro translation systems, which included ribosomes, tRNAs, and 20 amino acids. The 20 translation systems differed with regard to which amino acid was radiolabeled. For example, in 1 translation system, radiolabeled glycine was added, and the other 19 amino acids were unlabeled. In another system, radiolabeled proline was added, and the other 19 amino acids were unlabeled. The in vitro translation systems also contained the enzymes that attach amino acids to tRNAs.

| | Experimental level | Conceptual level |

1 Mix together RNA triplets of a specific sequence and 20 in vitro translation systems. In the example shown here, the triplet is 5′–CCC–3′. Each translation system contained a different radiolabeled amino acid. (Note: Only 3 tubes are shown here.)

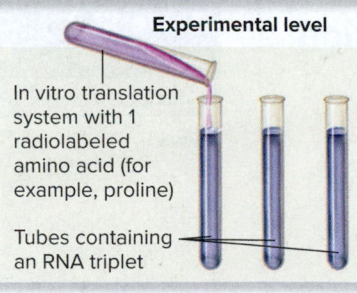 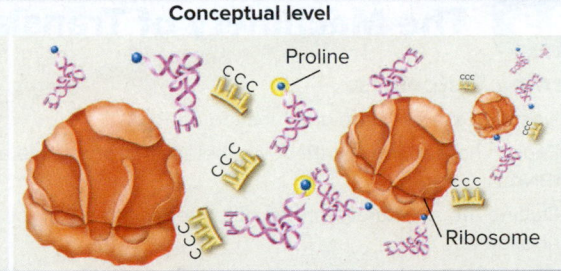

In vitro translation system with 1 radiolabeled amino acid (for example, proline)

Tubes containing an RNA triplet

Proline

Ribosome

2 Allow time for the RNA triplet to bind to the ribosome and for the appropriate tRNA to bind to the RNA triplet.

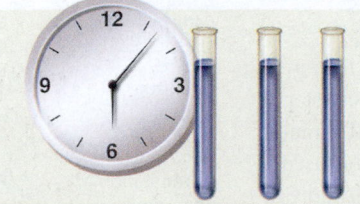 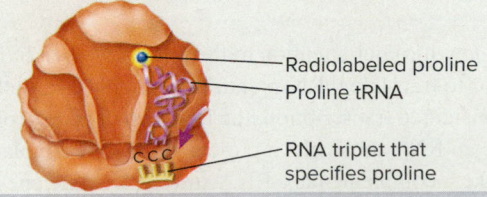

Radiolabeled proline
Proline tRNA

RNA triplet that specifies proline

3 Pour each mixture through a filter that allows the passage of unbound tRNA but does not allow the passage of ribosomes.

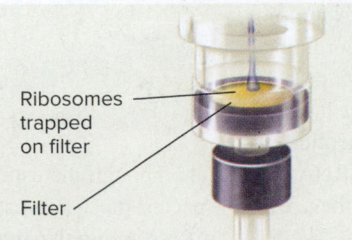 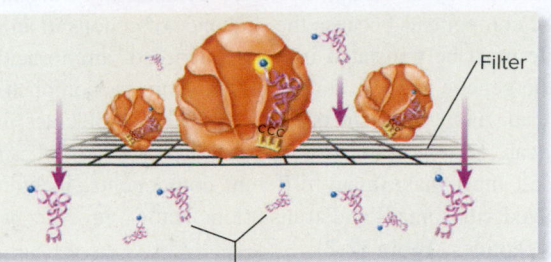

Ribosomes trapped on filter

Filter

Filter

tRNAs not bound to a ribosome

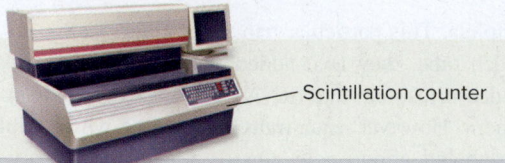

**4**  Count radioactivity on the filter.

Scintillation counter

**5**  **THE DATA**

| Triplet | Radiolabeled amino acid trapped on the filter | Triplet | Radiolabeled amino acid trapped on the filter |
|---------|----------------------------------------------|---------|----------------------------------------------|
| 5′ – AAA – 3′ | Lysine | 5′ – GAC – 3′ | Aspartic acid |
| 5′ – ACA – 3′ | Threonine | 5′ – GCC – 3′ | Alanine |
| 5′ – ACC – 3′ | Threonine | 5′ – GGU – 3′ | Glycine |
| 5′ – AGA – 3′ | Arginine | 5′ – GGC – 3′ | Glycine |
| 5′ – AUA – 3′ | Isoleucine | 5′ – GUU – 3′ | Valine |
| 5′ – AUU – 3′ | Isoleucine | 5′ – UAU – 3′ | Tyrosine |
| 5′ – CCC – 3′ | Proline | 5′ – UGU – 3′ | Cysteine |
| 5′ – CGC – 3′ | Arginine | 5′ – UUG – 3′ | Leucine |
| 5′ – GAA – 3′ | Glutamic acid | | |

**6**  **CONCLUSION**  This method enabled the researchers to identify many of the codons of the genetic code.

**7**  **SOURCE**  Leder, P., and Nirenberg, M. W. 1964. RNA Codewords and Protein Synthesis, III. On the nucleotide sequence of a cysteine and a leucine RNA codeword. *Proceedings of the National Academy of Sciences* 52: 1521–1529.

## 12.5  The Machinery of Translation

### Learning Outcomes:

1. Describe the structure and function of tRNA.
2. Explain how aminoacyl-tRNA synthetases attach amino acids to tRNAs.
3. Outline the structural features of bacterial and eukaryotic ribosomes.
4. **CoreSKILL »** Analyze how ribosomal RNA (rRNA) is used to evaluate evolutionary relationships among different species.

Let's now turn our attention to the components in living cells that are needed to translate mRNAs into polypeptides. Earlier in this chapter, we considered transcription, the first step in gene expression. To transcribe an RNA molecule, a pre-existing DNA template strand is used to make a complementary RNA strand. A single enzyme, RNA polymerase, catalyzes this reaction. By comparison, translation requires more components because the sequence of codons in an mRNA molecule must be translated into a sequence of amino acids according to the genetic code. A single protein cannot accomplish such a task. Instead, many different proteins and RNA molecules interact in an intricate series of steps to achieve the synthesis of a polypeptide. A cell must make many different components, including mRNAs, tRNAs, ribosomes, and translation factors, in order to synthesize polypeptides (**Table 12.2**).

| Table 12.2 | Components of the Translation Machinery |
|------------|------------------------------------------|
| **Component** | **Function** |
| mRNA | Contains the information for the amino acid sequence of a polypeptide according to the genetic code. |
| tRNA | A molecule with two functional sites: one site, termed the anticodon, binds to a codon in mRNA, and a second site is where an appropriate amino acid is attached. |
| Ribosome | Composed of many proteins and rRNA molecules, the ribosome provides a location where mRNA and tRNA molecules can properly interact with each other. The ribosome catalyzes the formation of covalent bonds between adjacent amino acids to make a polypeptide. |
| Translation factors | Proteins needed for the three stages of translation. Initiation factors are required for the assembly of mRNA, the first tRNA, and ribosomal subunits. Elongation factors are needed to synthesize the polypeptide. Release factors are needed to recognize the stop codon and disassemble the translation machinery. Several translation factors use GTP as an energy source to carry out their functions. |

Though the estimates vary from cell to cell and from species to species, most cells use a substantial amount of their energy to translate mRNA into polypeptides. In *E. coli*, for example, approximately 90%

of the cellular energy is used for this process. This value underscores the complexity and importance of translation in living organisms. In this section, we will focus on the components of the translation machinery. The last section of the chapter will describe the stages of translation as they occur in living cells.

## Transfer RNAs Share Common Structural Features

To understand how tRNAs function as carriers of the correct amino acids during translation, researchers have examined their structural characteristics. The tRNAs of bacteria, archaea, and eukaryotes share common features. As originally proposed in 1965 by American biochemist Robert Holley, the two-dimensional structure of a tRNA resembles a cloverleaf. The structure has three stem-loops and a fourth stem with a 3′ single-stranded region (**Figure 12.14a**). The stem in a stem-loop is a region where the RNA is double-stranded due to complementary base pairing via hydrogen bonding, whereas the loop is a region without base pairing. The anticodon is located in the loop of the middle stem-loop region. The 3′ single-stranded region is the amino acid attachment site. The three-dimensional structure of tRNA molecules involves additional folding of the secondary structure (**Figure 12.14b**).

The cells of every organism make many different tRNA molecules, each encoded by a different gene. A tRNA is named according to the amino acid it carries. For example, tRNA^Ser carries a serine. Because the genetic code contains six different serine codons, as shown in Table 12.1, a cell produces more than one type of tRNA^Ser.

## Aminoacyl-tRNA Synthetases Charge tRNAs by Attaching an Appropriate Amino Acid

To perform its role during translation, a tRNA must have the appropriate amino acid attached to its 3′ end. The enzymes that catalyze the attachment of amino acids to tRNA molecules are known as **aminoacyl-tRNA synthetases**. Cells make 20 distinct types of aminoacyl-tRNA synthetase enzymes, with each type recognizing just one of the 20 different amino acids. Each aminoacyl-tRNA synthetase is named for the specific amino acid it attaches to tRNA. For example, alanyl-tRNA synthetase recognizes alanine and attaches this amino acid to all tRNAs with alanine anticodons.

Aminoacyl-tRNA synthetases catalyze chemical reactions involving an amino acid, a tRNA molecule, and ATP (**Figure 12.15**).

1. First, a specific amino acid and ATP bind to the enzyme.
2. Next, the amino acid is activated by the covalent attachment of adenosine monophosphate (AMP), and pyrophosphate is released.
3. In a third step, the activated amino acid is covalently attached to the 3′ end of a tRNA molecule, and AMP is released.
4. Finally, the tRNA with its attached amino acid, called a **charged tRNA**, or an **aminoacyl tRNA**, is released from the enzyme.

The ability of each aminoacyl-tRNA synthetase to recognize an appropriate tRNA has been called the second genetic code. A precise recognition process is necessary to maintain the fidelity of genetic information. If the wrong amino acid was attached to a tRNA, the amino acid sequence of the translated polypeptide would be incorrect. However, aminoacyl-tRNA synthetases are amazingly accurate enzymes. The wrong amino acid is attached to a tRNA less than

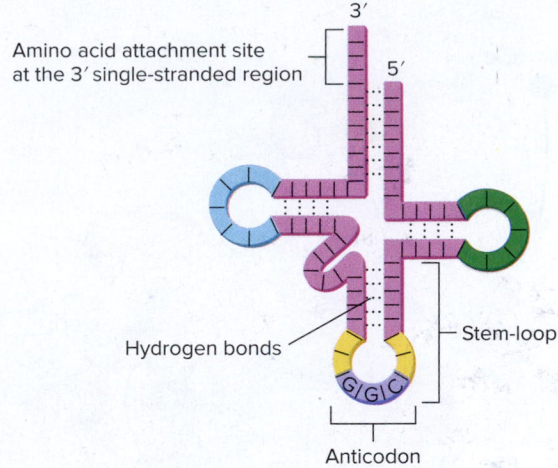

(a) **Two-dimensional structure of tRNA**

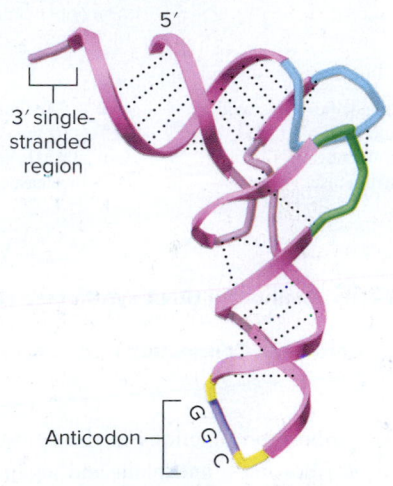

(b) **Three-dimensional structure of tRNA**

**Figure 12.14** **Structure of tRNA.** **(a)** The two-dimensional or secondary structure of tRNA resembles a cloverleaf, with the anticodon in the loop of the middle stem-loop structure. The 3′ single-stranded region is where an amino acid can attach. **(b)** The actual three-dimensional structure folds in on itself.

**Core Concept: Structure and Function** The structure of tRNA has two functional sites: a 3′ single-stranded region where an amino acid is attached and an anticodon that binds to a codon on mRNA.

once in 100,000 times! The anticodon region of the tRNA is usually important for recognition by the correct aminoacyl-tRNA synthetase. In addition, the base sequences in other regions may facilitate binding to an aminoacyl-tRNA synthetase.

## Ribosomes Are Composed of rRNA and Proteins

Let's now turn our attention to the **ribosome**, the site where translation takes place. The ribosome is often described as a molecular machine. Bacterial cells have one type of ribosome, which translates all mRNAs in the cytoplasm. Because eukaryotic cells are compartmentalized into cellular organelles, biochemically distinct ribosomes are found in different cellular compartments. The most abundant type

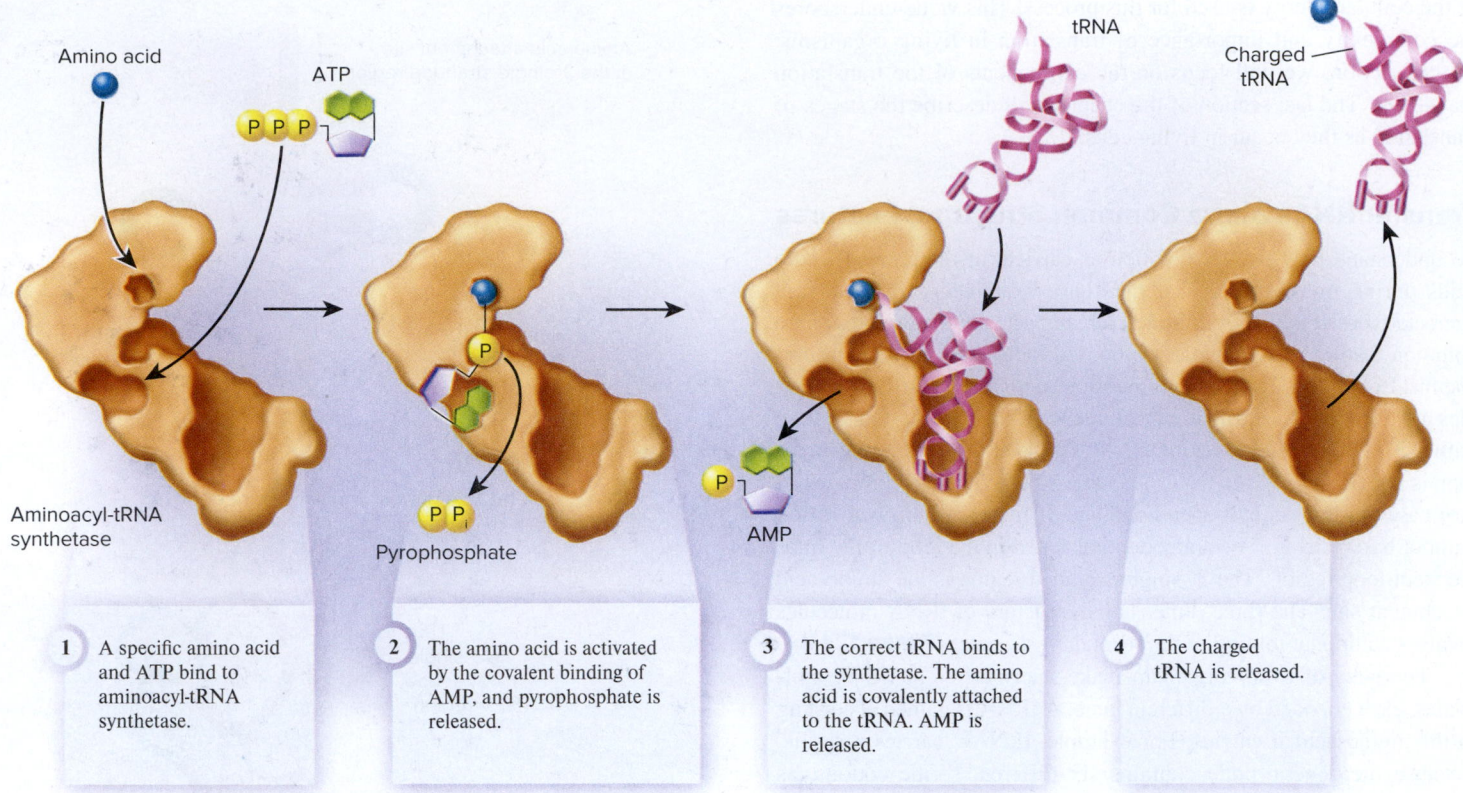

| 1 | A specific amino acid and ATP bind to aminoacyl-tRNA synthetase. | 2 | The amino acid is activated by the covalent binding of AMP, and pyrophosphate is released. | 3 | The correct tRNA binds to the synthetase. The amino acid is covalently attached to the tRNA. AMP is released. | 4 | The charged tRNA is released. |

**Figure 12.15** **Aminoacyl-tRNA synthetase charging a tRNA.**

 **Core Skill: Connections** Look back at Figure 6.3, which describes the hydrolysis of ATP. Why is ATP needed to charge a tRNA?

of eukaryotic ribosome functions in the cytosol. In addition, mitochondria have ribosomes, and plant and algal cells have ribosomes in their chloroplasts. Unless otherwise noted, the term eukaryotic ribosome refers to ribosomes in the cytosol, not to those found in organelles.

A ribosome is a large complex composed of structures called the large and small subunits. The term subunit is perhaps misleading, because each ribosomal subunit is itself assembled from many different proteins and one or more RNA molecules. In the bacterium *E. coli*, the small ribosomal subunit is called 30S, and the large subunit is 50S (**Table 12.3**). The designations 30S and 50S refer to the rate at which these subunits sediment when subjected to a centrifugal force. This rate is described as a sedimentation coefficient in Svedberg units (S) in honor of Swedish chemist Theodor Svedberg, who invented the ultracentrifuge. The 30S subunit is formed from the assembly of 21 different ribosomal proteins and one 16S rRNA molecule. The 50S subunit contains 34 different proteins and two different rRNA molecules, called 5S and 23S. Together, the 30S and 50S subunits form a 70S ribosome. (Svedberg units don't add up linearly, because the sedimentation coefficient is a function of both size and shape.) In bacteria, ribosomal proteins and rRNA molecules are synthesized in the cytoplasm, and the ribosomal subunits are assembled there as well.

Eukaryotic ribosomes consist of subunits that are slightly larger than their bacterial counterparts (Table 12.3). In eukaryotes, 40S and 60S subunits combine to form an 80S ribosome. The 40S subunit is composed of 33 proteins and an 18S rRNA, and the 60S subunit has

| Table 12.3 | Composition of Bacterial and Eukaryotic Ribosomes | | |
|---|---|---|---|
| | Small subunit | Large subunit | Assembled ribosome |
| **Bacterial** | | | |
| Sedimentation coefficient | 30S | 50S | 70S |
| Number of proteins | 21 | 34 | 55 |
| rRNA | 16S rRNA | 5S rRNA, 23S rRNA | 16S rRNA, 5S rRNA, 23S rRNA |
| **Eukaryotic** | | | |
| Sedimentation coefficient | 40S | 60S | 80S |
| Number of proteins | 33 | 49 | 82 |
| rRNA | 18S rRNA | 5S rRNA, 5.8S rRNA, 28S rRNA | 18S rRNA, 5S rRNA, 5.8S rRNA, 28S rRNA |

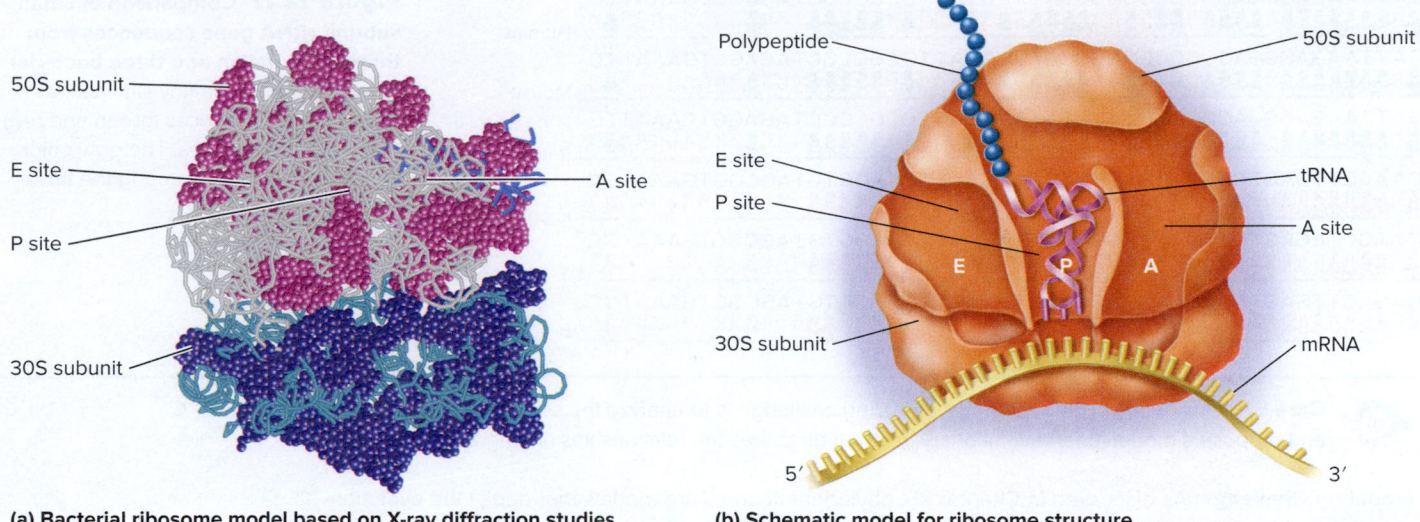

(a) Bacterial ribosome model based on X-ray diffraction studies

(b) Schematic model for ribosome structure

**Figure 12.16** **Ribosome structure.** **(a)** A model for the structure of a bacterial ribosome based on X-ray diffraction studies, showing the large and small subunits and the major binding sites. The rRNA is shown in gray (large subunit) and turquoise (small subunit), whereas the ribosomal proteins are magenta (large subunit) and dark blue (small subunit). **(b)** A schematic model emphasizing functional sites in the ribosome, and showing bound mRNA and tRNA with an attached polypeptide.

 **Core Concept: Structure and Function**   The structure of a ribosome has three discrete sites, called the E, P, and A sites, which carry out different functions in polypeptide synthesis, described in Section 12.6.

49 proteins and 5S, 5.8S, and 28S rRNAs. The synthesis of eukaryotic rRNA occurs in the nucleolus, a droplet organelle in the nucleus that is specialized for that purpose. The ribosomal proteins are made in the cytosol and imported into the nucleus. The rRNAs and ribosomal proteins are then assembled within the nucleolus to make the 40S and 60S subunits. The 40S and 60S subunits are exported into the cytosol, where they associate to form an 80S ribosome during translation.

### Components of Ribosomal Subunits Form Functional Sites for Translation

To understand the structure and function of the ribosome at the molecular level, researchers have determined the locations and functional roles of individual ribosomal proteins and rRNAs. In recent years, a few research groups have succeeded in purifying ribosomes and causing them to crystallize in a test tube. When researchers use X-ray diffraction to study the crystallized ribosomes, they gain detailed information about ribosome structure. **Figure 12.16a** shows a model of a bacterial ribosome.

During bacterial translation, the mRNA lies on the surface of the 30S subunit, within a space between the 30S and 50S subunits (**Figure 12.16b**). As a polypeptide is synthesized, it exits through a hole within the 50S subunit. Ribosomes contain discrete sites where tRNAs bind and the polypeptide is synthesized. In 1964, James Watson proposed a two-site model for tRNA binding to the ribosome. These sites are known as the **peptidyl site (P site)** and **aminoacyl site (A site)**. In 1981, German geneticists Knud Nierhaus and Hans-Jörg Rheinberger expanded this to a three-site model (Figure 12.16b). The third site is known as the exit site (E site). In Section 12.6, we will examine the roles of these sites in the synthesis of a polypeptide.

 **Core Concept: Evolution**

### Comparisons of Small Subunit rRNAs Among Different Species Provide a Basis for Establishing Evolutionary Relationships

Translation is a fundamental process that is vital for the existence of all living species. Research indicates that the components needed for translation arose very early in the evolution of life on our planet in ancestors that gave rise to all known living species. For this reason, all organisms have translational components that are evolutionarily related to each other. For example, the rRNA found in the small subunit of ribosomes is similar in all forms of life, though it is slightly larger in eukaryotic species (18S) than in bacterial species (16S). The gene for the small subunit rRNA (SSU rRNA) is found in the genomes of all organisms.

How is this observation useful? One way that geneticists explore evolutionary relationships is to compare the sequences of evolutionarily related genes. At the molecular level, gene evolution involves changes in DNA sequences. After two different species have diverged from each other during evolution, the genes of each species can accumulate changes, or mutations, that alter the sequences of those genes. After many generations, evolutionarily related species contain genes that are similar but not identical to each other, because each species accumulates different mutations. In general, if a very long time has elapsed since two species diverged evolutionarily, their genes tend to be quite different. In contrast, if two species diverged relatively recently on an evolutionary time scale, their genes tend to be more similar.

GATTAAGAGGGGACGGCCGGGGGCATTCGTATTGCGCCGCTAGAGGTGAAATTC
Human

GATTAAGAGGGGACGGCCGGGGGCATTCGTATTGCGCCGCTAGAGGTGAAATTC
Mouse

GATTAAGAGGGACGGCCGGGGGCATTCGTATTGCGCCGCTAGAGGTGAAATTC
Rat

CAAGCTTGAGTCTCGTAGAGGGGGGTAGAATTCCAGGTGTAGCGGTGAAATGC
E. coli

CAAGCTAGAGTCTCGTAGAGGGGGGTAGAATTCCAGGTGTAGCGGTGAAATGC
S. marcescens

GAGACTTGAGTACAGAAGAAGAGAGTGGAATTCCACGTGTAGCGGTGAAATGC
B. subtilis

**Figure 12.17** **Comparison of small subunit rRNA gene sequences from three mammalian and three bacterial species.** Note the many similarities (yellow) and differences (green and red) among the sequences. The gray color indicates differences among the three bacterial species.

 **Core Skill: Modeling** The goal of this modeling challenge is to analyze the sequences in Figure 12.17 and propose a model for an evolutionary tree that describes the relationships among these six species.

**Modeling Challenge:** As discussed in Chapter 25, phylogenetic trees are models that depict the evolutionary relationships among different species. Those species that are more closely related are closer together on a tree. The model to the right shows a tree with the tips of the branches labeled A through E. Relying only on the sequences shown in Figure 12.17, replace the 6 letters on this tree with the names of the six species. Note: More than one model is possible.

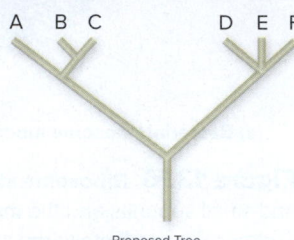
Proposed Tree

**Figure 12.17** compares a portion of the sequence of the SSU rRNA gene from three mammalian and three bacterial species. The colors highlight different types of comparisons. The bases shaded in yellow are identical in five or six species. Sequences of bases that are identical or very similar in different species are said to be **evolutionarily conserved**. Presumably, these sequences were found in the primordial gene that gave rise to modern species. Perhaps because these sequences may have some critical function, they have not changed over evolutionary time. Those sequences shaded in green are identical in all three mammals, but differ from the sequences in one or more bacterial species. Actually, if you scan the mammalian species, you may notice that all three have identical sequences in this region. The sequences shaded in red are identical or very similar in the bacterial species, but differ from the mammalian SSU rRNA genes. The sequences from *Escherichia coli* and *Serratia marcescens* are more similar to each other than the sequence from *Bacillus subtilis* is to either of them. This observation suggests that *E. coli* and *S. marcescens* are more closely related evolutionarily than either of them is to *B. subtilis*.

## 12.6 The Stages of Translation

### Learning Outcomes:

1. Describe the three stages of translation.
2. Summarize the similarities and differences between translation in bacteria and in eukaryotes.
3. Explain how some antibiotics inhibit the growth of bacteria by interfering with translation.

Like transcription, the process of translation occurs in three stages called initiation, elongation, and termination. **Figure 12.18** provides an overview of the process. During initiation, an mRNA, the first tRNA,

and the ribosomal subunits assemble into a complex. Next, in the elongation stage, the ribosome moves in the 5′ to 3′ direction from the start codon in the mRNA toward the stop codon, synthesizing a polypeptide according to the sequence of codons in the mRNA. Finally, the process is terminated when the ribosome reaches a stop codon and the complex disassembles, releasing the completed polypeptide. In this section, we will examine the steps in this process as they occur in living cells.

### Translation Is Initiated with the Assembly of mRNA, tRNA, and the Ribosomal Subunits

During **initiation**, a complex is formed between an mRNA molecule, the first tRNA, and the ribosomal subunits. In all species, the assembly of this complex requires the help of proteins called **initiation factors** that facilitate the interactions between these components (see Table 12.2). The assembly also requires an input of energy. Guanosine triphosphate (GTP) is hydrolyzed by certain initiation factors to provide the necessary energy.

When they are not involved in translation, the small and large ribosomal subunits exist separately.

1. To begin assembly of the complex in bacteria, mRNA binds to the small ribosomal subunit (**Figure 12.19**). The binding of mRNA to this subunit is facilitated by a short ribosomal-binding site near the 5′ end of the mRNA. The ribosomal binding site is a sequence of bases that is complementary to a portion of the 16S rRNA within the small ribosomal subunit. The mRNA becomes bound to the ribosome because the ribosomal-binding site and rRNA hydrogen-bond to each other by base-pairing. The start codon is usually just a few nucleotides downstream (that is, toward the 3′ end) from the ribosomal-binding site.

2. A specific tRNA, which functions as the initiator tRNA, recognizes the start codon in mRNA (AUG) and binds to it. In eukaryotes, this tRNA carries a methionine, whereas in bacteria it carries a methionine that has been modified by the attachment of a formyl group.

**Figure 12.18** **An overview of the stages of translation.**

**Core Concept: Energy and Matter** The process of translation uses a sizable amount of a cell's energy and results in the synthesis of a cell's proteins.

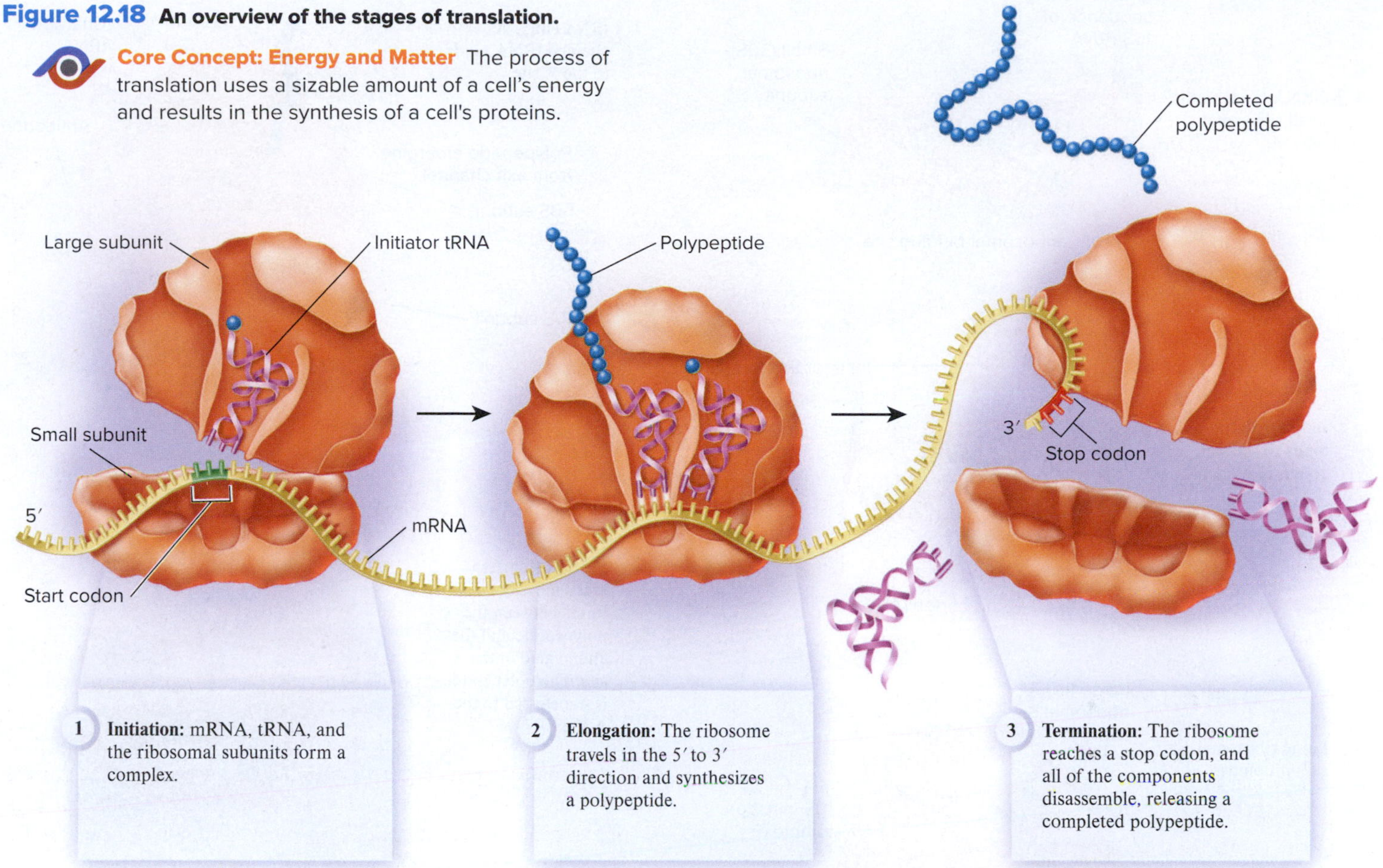

| 1 | **Initiation:** mRNA, tRNA, and the ribosomal subunits form a complex. | 2 | **Elongation:** The ribosome travels in the 5′ to 3′ direction and synthesizes a polypeptide. | 3 | **Termination:** The ribosome reaches a stop codon, and all of the components disassemble, releasing a completed polypeptide. |

3. To complete the initiation stage, the large ribosomal subunit associates with the small subunit. At the end of this stage, the initiator tRNA is located in the P site of the ribosome.

In eukaryotic species, the initiation phase of translation differs in two ways from the process in bacteria.

- First, instead of an RNA sequence that functions as a ribosomal-binding site, eukaryotic mRNAs have a 7-methylguanosine cap (5′ cap) at their 5′ end. This 5′ cap is recognized by cap-binding proteins that promote the binding of the mRNA to the small ribosomal subunit.

- Also, in bacteria, the start codon is very close to a ribosomal-binding site, but the location of start codons in eukaryotes is more variable.

In 1978, American biochemist Marilyn Kozak proposed that the small ribosomal subunit identifies a start codon by beginning at the 5′ end and then scanning along the mRNA in the 3′ direction in search of an AUG sequence. In many, but not all, cases, the first AUG codon is used as a start codon. By analyzing the sequences of many eukaryotic mRNAs, Kozak and her colleagues discovered that the sequence around an AUG codon is important for it to be recognized as a start codon. The sequence for optimal start codon recognition is shown here:

| Upstream of start codon | Start codon | Downstream coding region |
|---|---|---|

...G C C (A or G) C C **A U G** G...............

Aside from the AUG codon itself, a guanine just past the start codon and the sequence of six bases directly upstream from the start codon are important for start codon selection. If the first AUG codon is within a site that deviates markedly from this optimal sequence, the small subunit may skip this codon and instead use another AUG codon farther downstream. Once the small subunit selects a start codon, an initiator tRNA binds to the start codon, and then the large ribosomal subunit associates with the small subunit to complete the assembly process.

## Polypeptide Synthesis Occurs During the Elongation Stage

As its name suggests, the stage of translation called **elongation** involves the covalent bonding of amino acids to each other, one at a time, to produce a polypeptide. Even though this process involves several different components, translation occurs at a remarkable rate. Under normal cellular conditions, the translation machinery can elongate a polypeptide at a rate of 15 to 18 amino acids per second in bacteria and 6 amino acids per second in eukaryotes!

*tRNA Entry* To elongate a polypeptide by one amino acid, a tRNA brings a new amino acid to the ribosome, where it is attached to the end of a growing polypeptide. In step 1 of **Figure 12.20**, translation has already proceeded to a point where a short polypeptide is attached to the tRNA located in the P site of the ribosome. This is called peptidyl tRNA. In the first step of elongation, a charged tRNA

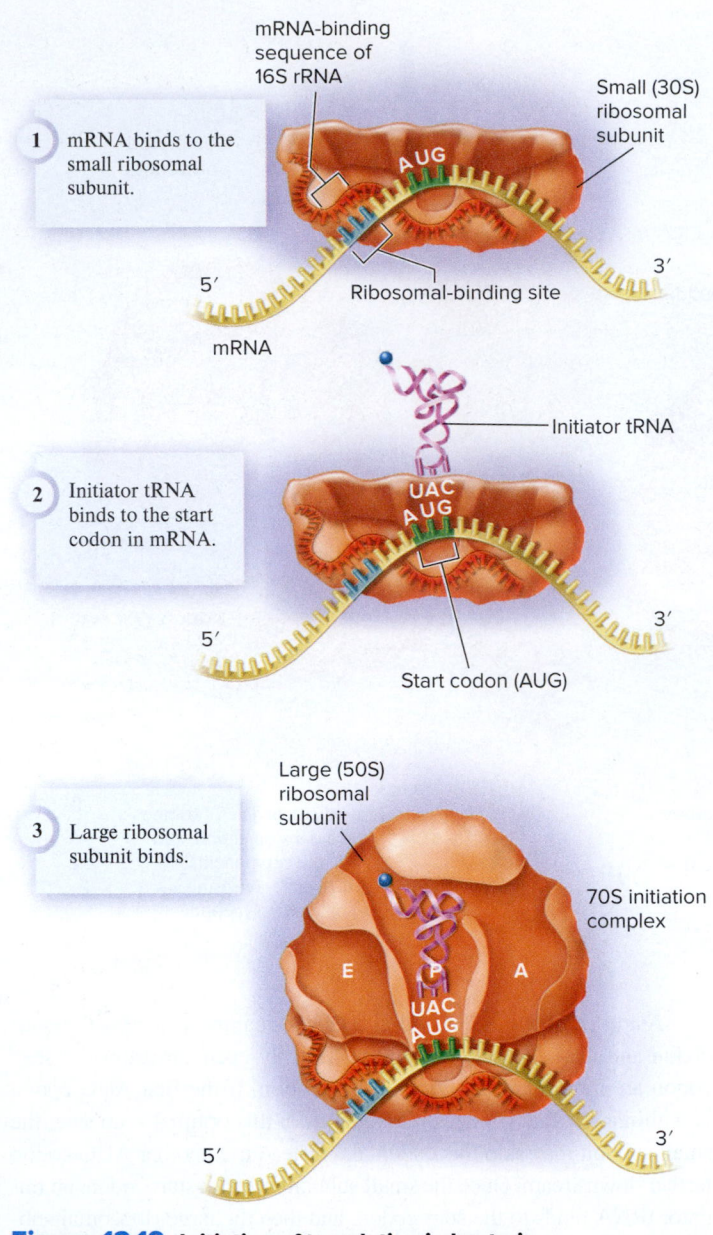

**Figure 12.19** **Initiation of translation in bacteria.**

Concept Check: *What promotes the binding between the mRNA and the small ribosomal subunit?*

carrying a single amino acid binds to the A site. This binding occurs because the anticodon in the tRNA is complementary to the codon in the mRNA. The hydrolysis of GTP by proteins that function as **elongation factors** provides the energy for the binding of the tRNA to the A site (see Table 12.2). At this step in elongation, a peptidyl tRNA is in the P site and a charged tRNA (an aminoacyl tRNA) is in the A site. This is how the P and A sites came to be named.

*Peptidyl Transfer Reaction*   In the second step, a peptide bond is formed between the amino acid at the A site and the growing polypeptide, thereby lengthening the polypeptide by one amino acid. As this occurs, the polypeptide is removed from the tRNA in the P site and transferred to the amino acid at the A site, an event termed a **peptidyl transfer reaction**. This reaction is catalyzed by a region of the 50S subunit known as the peptidyltransferase center, which is composed of several proteins and rRNA. In 2000, American biochemist Thomas Steitz,

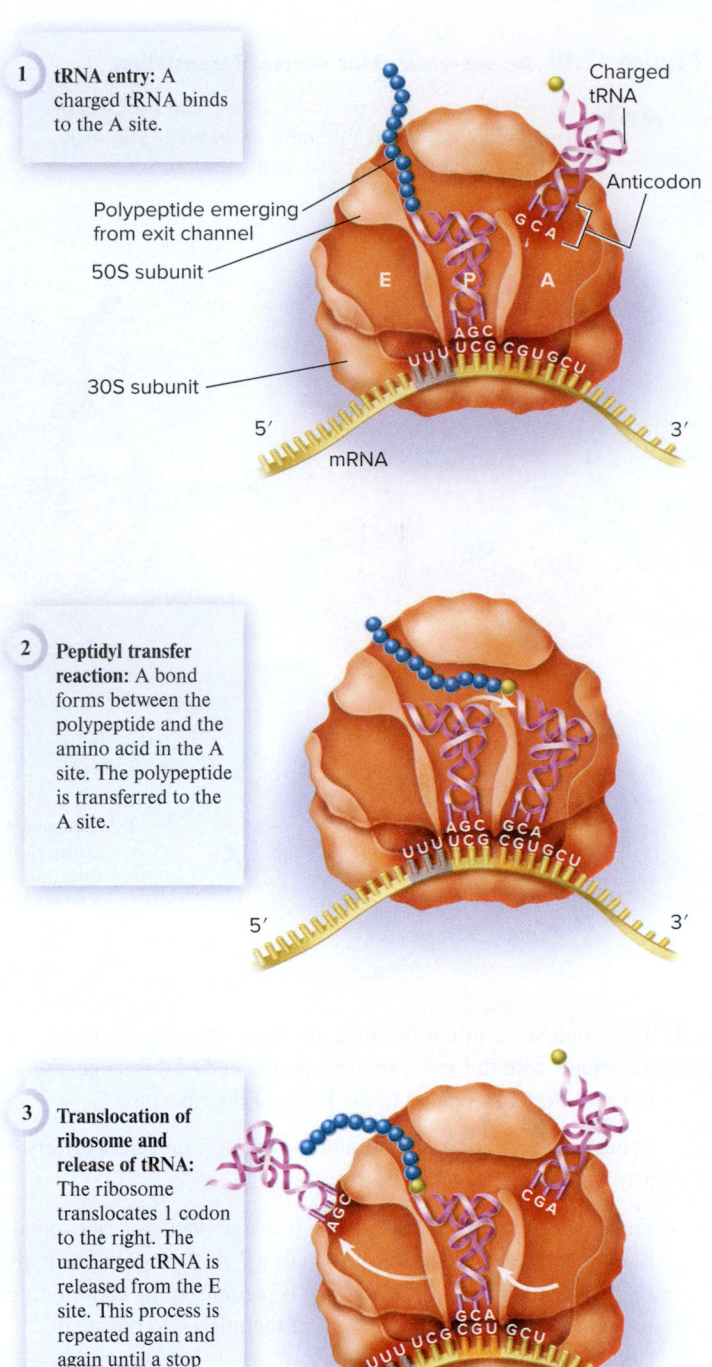

**Figure 12.20** **Elongation stage of translation in bacteria.** In this drawing, the amino acids that were already part of the polypeptide are shown in blue. The amino acids attached to incoming tRNAs are shown in yellow.

American biophysicist Peter Moore, and their colleagues proposed that the rRNA is responsible for catalyzing the peptide bond formation between adjacent amino acids. It is now accepted that the ribosome is a ribozyme.

*Translocation of the Ribosome and Release of tRNA*   After the peptidyl transfer reaction is complete, the third step involves the movement or translocation of the ribosome toward the 3′ end of the mRNA by exactly one codon. This shifts the tRNAs in the P and A

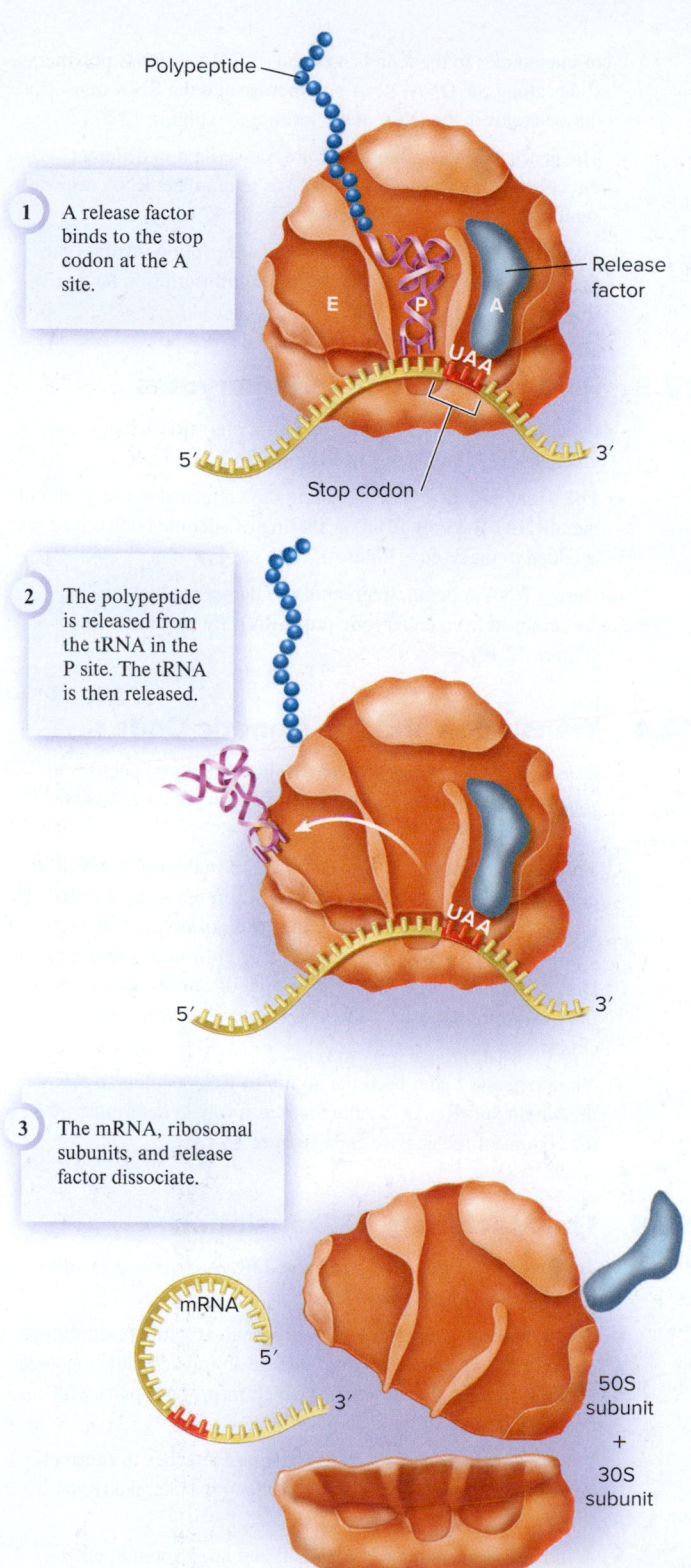

1  A release factor binds to the stop codon at the A site.

Release factor

E  P  A

UAA

5′                    3′

Stop codon

2  The polypeptide is released from the tRNA in the P site. The tRNA is then released.

UAA

5′                    3′

3  The mRNA, ribosomal subunits, and release factor dissociate.

mRNA
5′
3′

50S subunit
+
30S subunit

**Figure 12.21  Termination of translation in bacteria.**

Polypeptide

Figure 12.20 shows a single ribosome in the act of translating an mRNA. In living cells, it is common for multiple ribosomes to be gliding along the same mRNA and synthesizing polypeptides. The complex of a single mRNA and multiple ribosomes is called a **polysome** (see the chapter opening micrograph).

## Termination Occurs When a Stop Codon Is Reached in the mRNA

Elongation continues until a stop codon moves into the A site of a ribosome. The three stop codons, UAA, UAG, and UGA, are recognized by a protein known as a **release factor**. The three-dimensional structure of a release factor protein mimics the structure of tRNAs, which allows it to fit into the A site.

**Figure 12.21** illustrates the **termination** of translation.

1. In step 1 of this figure, a release factor binds to the stop codon at the A site. The completed polypeptide is attached to a tRNA in the P site.

2. In step 2, the bond between the polypeptide and the tRNA is hydrolyzed, causing the polypeptide and tRNA to be released from the ribosome.

3. In step 3, the mRNA, ribosomal subunits, and release factor dissociate.

The termination stage of translation is similar in bacteria and eukaryotes except that bacteria have two different termination factors that recognize stop codons (RF1 and RF2), whereas eukaryotes have only one (eRF). **Table 12.4** compares some of the key differences between bacterial and eukaryotic translation.

## Antibiotics That Inhibit Bacterial Translation Are Used to Treat Bacterial Infections

Many diseases that affect humans and domesticated animals are caused by pathogenic bacteria. An **antibiotic** is any substance produced by a

sites to the E and P sites, respectively. The uncharged tRNA exits the E site. Notice that the next codon in the mRNA (GCU in Figure 12.20) is now exposed at the unoccupied A site. At this point, a charged tRNA can enter the A site, and the same series of steps will add the next amino acid to the polypeptide.

| Table 12.4 | Comparison of Bacterial and Eukaryotic Translation | |
|---|---|---|
| | **Bacterial** | **Eukaryotic** |
| Cellular location | Cytoplasm | Cytosol* |
| Ribosome composition | 70S ribosomes: | 80S ribosomes: |
| | 30S subunit: 21 proteins + 1 rRNA | 40S subunit: 33 proteins + 1 rRNA |
| | 50S subunit: 34 proteins + 2 rRNAs | 60S subunit: 49 proteins + 3 rRNAs |
| Initiator tRNA | tRNA^Formyl-methionine | tRNA^Methionine |
| Initial binding of mRNA | Requires a ribosomal-binding site | Requires a 7-methylguanosine cap |
| Selection of a start codon | Just downstream from the ribosomal-binding site | According to Kozak's sequences |
| Termination factors | Two factors: RF1 and RF2 | One factor: eRF |

*The components for eukaryotic translation described in this table refer to those that are used in the cytosol. Different types of ribosomes are used for translation inside mitochondria and chloroplasts.

| Table 12.5 | Mechanisms of Inhibition of Bacterial Translation by Selected Antibiotics |
| --- | --- |
| **Antibiotic** | **Description** |
| Chloramphenicol | Blocks elongation by acting as a competitive inhibitor of the peptidyltransferase complex. |
| Erythromycin | Binds to the 23S rRNA and blocks elongation by interfering with the translocation step. |
| Puromycin | Binds to the A site and causes premature release of the polypeptide. This early termination of translation results in polypeptides that are shorter than normal and usually nonfunctional. |
| Tetracycline | Blocks elongation by inhibiting the binding of aminoacyl tRNAs to the ribosome. |
| Streptomycin | Interferes with normal pairing between aminoacyl tRNAs and codons. This causes misreading of the code and thereby produces abnormal proteins. |

microorganism that inhibits the growth of other microorganisms, such as pathogenic bacteria. Most antibiotics are small organic molecules, with masses of less than 2,000 Da. In some cases, antibiotics exert their effect because they inhibit or interfere with bacterial translation. Because the components of translation differ somewhat between bacteria and eukaryotes, some antibiotics inhibit bacterial translation without affecting eukaryotic translation. Therefore, they can be used to treat bacterial infections in humans, pets, and livestock. **Table 12.5** describes a few of these antibiotics.

# Summary of Key Concepts

## 12.1 Overview of Gene Expression

- Based on his studies of inborn errors of metabolism, Garrod hypothesized that certain genetic diseases are caused by a defect in a gene encoding an enzyme (Figure 12.1).
- Based on their study of the nutritional requirements of a bread mold, Beadle and Tatum proposed the one-gene/one-enzyme hypothesis, in which a single gene controls the synthesis of a single enzyme (Figure 12.2).
- A polypeptide is a unit of structure. A protein, composed of one or more polypeptides, is a unit of function.
- At the molecular level, the central dogma states that most genes are transcribed into mRNA, and then the mRNA is translated into a polypeptide. Eukaryotes modify their RNA transcripts to make them functional (Figure 12.3).
- The molecular expression of genes is an underlying factor that determines an organism's characteristics.

## 12.2 Transcription

- A site in a gene called a promoter specifies where transcription begins. A terminator specifies where transcription will end (Figure 12.4).
- In bacteria, the initiation of transcription begins when sigma factor binds to RNA polymerase and to a promoter. During elongation, synthesis of an RNA transcript occurs via base pairing

of nucleotides to the template strand of DNA as RNA polymerase slides along the DNA. RNA polymerase and the RNA transcript dissociate from the DNA at the terminator (Figure 12.5).

- The genes along a chromosome are transcribed in different directions using either DNA strand as a template. RNA is always synthesized in a 5′ to 3′ direction (Figure 12.6).
- In eukaryotes, the initiation stage of transcription involves the assembly of RNA polymerase with five transcription factors (Figure 12.7).

## 12.3 RNA Modification in Eukaryotes

- In eukaryotes, transcription produces a pre-mRNA that is capped, given a poly A tail, and spliced (Figure 12.8).
- The 5′ cap is a methylated guanine base attached at the 5′ end of the mRNA. The poly A tail is a string of adenine nucleotides that is added to the 3′ end (Figure 12.9).
- During RNA splicing, intervening sequences called introns are removed from eukaryotic pre-mRNA by a spliceosome (Figure 12.10).

## 12.4 Translation and the Genetic Code

- Based on the genetic code, each of the 64 codons specifies a start codon (methionine), other amino acids, or a stop codon (Table 12.1, Figure 12.11).
- The template strand of DNA is used to make mRNA that contains a series of codons. Recognition between mRNA and many tRNA molecules determines the amino acid sequence of a polypeptide. A polypeptide has a directionality in which the first amino acid is at the N-terminus, or amino end, whereas the last amino acid is at the C-terminus, or carboxyl end (Figure 12.12).
- Nirenberg and Leder used the ability of RNA triplets to promote the binding of tRNAs to ribosomes as a way to determine many of the codons of the genetic code (Figure 12.13).

## 12.5 The Machinery of Translation

- Translation requires mRNA, charged tRNAs, ribosomes, and many translation factors (Table 12.2).
- tRNA molecules have a two-dimensional structure resembling a cloverleaf. Two important sites are the amino acid attachment site at the 3′ end and the anticodon, which forms base pairs with a codon in mRNA (Figure 12.14).
- The enzyme aminoacyl-tRNA synthetase attaches the correct amino acid to a tRNA molecule, producing a charged tRNA (Figure 12.15).
- A ribosome is composed of a small and large subunit, each consisting of rRNA molecules and many proteins. Bacterial and eukaryotic ribosomes differ in their molecular composition (Table 12.3).
- Ribosomes have three sites, termed the A, P, and E sites, which are locations for the binding and release of tRNA molecules (Figure 12.16).
- The gene that encodes the small subunit rRNA (SSU rRNA) has been used to determine evolutionary relationships among different species (Figure 12.17).

## 12.6   The Stages of Translation

- Translation occurs in three stages called initiation, elongation, and termination (Figure 12.18).

- During the initiation stage of translation, an mRNA assembles with the ribosomal subunits and an initiator tRNA molecule, which carries methionine, the first amino acid (Figure 12.19).

- During elongation, amino acids are added one at a time to a growing polypeptide (Figure 12.20).

- Termination of translation occurs when the binding of a release factor to a stop codon causes the release of the completed polypeptide from the tRNA and the disassembly of the mRNA, ribosomal subunits, and the release factor (Figure 12.21).

- Though translation in bacteria and eukaryotes is strikingly similar, some key differences have been observed (Table 12.4).

- Some antibiotics inhibit bacterial growth by interfering with translation (Table 12.5).

## Assess & Discuss

### Test Yourself

1. Which of the following best represents the central dogma of gene expression?
   a. During transcription, DNA codes for polypeptides.
   b. During transcription, DNA codes for mRNA, which codes for polypeptides during translation.
   c. During translation, DNA codes for mRNA, which codes for polypeptides during transcription.
   d. none of the above

2. A mutation prevents a gene from being transcribed into an mRNA. The mutation most likely disrupts
   a. the promoter.
   b. the terminator.
   c. the start codon.
   d. the stop codon.
   e. both a and c.

3. The functional product of a protein-encoding gene is
   a. tRNA.
   b. mRNA.
   c. rRNA.
   d. a polypeptide.
   e. a, b, and c.

4. Which of the following is *not* a property of the genetic code?
   a. It specifies the amino acids within a polypeptide.
   b. It is composed of codons, which are a specific sequences of three bases.
   c. It has a start codon, which specifies the starting point for polypeptide synthesis.
   d. It has stop codons, which specify the end of polypeptide synthesis.
   e. It determines the rate of transcription.

5. If a eukaryotic mRNA failed to have a cap attached to its 5′ end, what would the negative consequence(s) be?
   a. The mRNA would not properly exit the nucleus.
   b. The mRNA would not properly bind to a ribosome.
   c. The mRNA would not receive a poly A tail.
   d. The mRNA would not use the correct start codon.
   e. both a and b

6. The small subunit of a ribosome is composed of
   a. a protein.
   b. an rRNA molecule.
   c. many proteins.
   d. many rRNA molecules.
   e. many proteins and one rRNA molecule.

7. The part of a tRNA that is complementary to a codon in an mRNA is the
   a. acceptor stem.
   b. codon.
   c. peptidyl site.
   d. anticodon.
   e. adaptor loop.

8. During the initiation of translation, the first codon, _____, enters the _____ and associates with the initiator tRNA.
   a. UAG, A site
   b. AUG, A site
   c. UAG, P site
   d. AUG, P site
   e. AUG, E site

9. The movement of the polypeptide from the tRNA in the P site to the tRNA in the A site is referred to as
   a. peptide bonding.
   b. aminoacyl binding.
   c. translation.
   d. the peptidyl transfer reaction.
   e. initiation.

10. During which stage of translation does the synthesis of a polypeptide occur?
   a. initiation
   b. elongation
   c. termination
   d. splicing

### Conceptual Questions

1. Briefly explain how studying the pathway that leads to arginine synthesis allowed Beadle and Tatum to conclude that one gene sometimes encodes one enzyme.

2. What is the function of an aminoacyl-tRNA synthetase?

3. **Core Concept: Information** The genetic material provides a blueprint for producing the characteristics of living organisms. Explain how the information within the blueprint is accessed at the molecular level.

### Collaborative Questions

1. Why do you think some complexes, such as spliceosomes and ribosomes, have both protein and RNA components?

2. Discuss and make a list of the similarities and differences in the events that occur during the initiation, elongation, and termination stages of transcription and translation.

# Gene Expression at the Molecular Level II: Non-coding RNAs

# 13

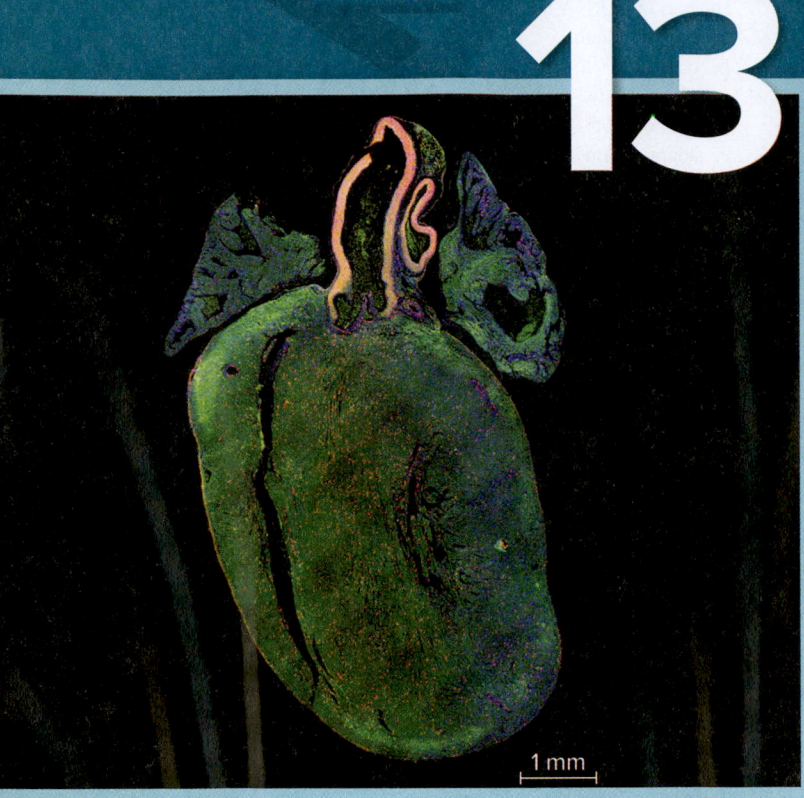

1 mm

**Non-coding RNAs and heart repair.** The muscles of the mammalian heart have poor regenerating abilities. Researchers have identified several non-coding RNAs that stimulate cardiac muscle regeneration. This heart is from a mouse that was treated with such a non-coding RNA, and it showed a significant increase in proliferating cells. This discovery may lead to new therapies to help heart attack victims regenerate new cardiac muscle. ©Mauro Giacca, Ana Eulalio, Miguel Mano

**P**eople with a rare genetic disorder called cartilage-hair hypoplasia (CHH) have short stature, underdeveloped hair, and short limbs with malformations in the cartilage. In addition, they have a higher predisposition to develop lymphomas and other cancers, and they may lack a normal immune response. Though CHH was first identified by American geneticist Victor McKusick in 1965, the underlying genetic cause remained a mystery for 36 years. In 2001, the function of a single mutant gene was linked to this disease. The gene specifies an RNA molecule that does not encode a protein. Rather, the RNA made from this gene is the RNA component of RNase MRP, which is an RNA-protein complex involved in the modification of some ribosomal and mitochondrial RNAs. This was the first case in which geneticists determined that a human genetic disease was due to a mutation in a nuclear gene that is not a protein-encoding

gene. By comparison, the first human genetic disease involving a protein-encoding gene was discovered nearly a century earlier! In 1909, Archibald Garrod proposed that patients with a disease called alkaptonuria, which is characterized by bluish-black discoloration of the cartilage and skin, is due to a mutation in a gene that encodes the protein homogentisic acid oxidase.

Why such a big time gap in our understanding of protein-encoding genes versus other types of genes? It's all about tools. The experimental tools to study the structure and function of proteins and to identify protein-encoding genes have been around for a long time. In contrast, the tools to study RNA structure and function and to identify genes that do not specify proteins are much more recent and are under rapid development. We are witnessing a revolution in molecular biology that is uncovering an unprecedented number of functions for RNA molecules.

In Chapter 12, we focused our attention on gene expression at the molecular level. The emphasis was on protein-encoding genes, which are transcribed into mRNA. During translation, the information within mRNAs is used to make polypeptides, which then assemble into functional proteins. The human genome has about 22,000 protein-encoding genes. In contrast, other genes are transcribed into **non-coding RNAs (ncRNAs),** which are RNA molecules that do not encode polypeptides. As this chapter will reveal, ncRNAs perform a very diverse set of functions. In humans, the number of genes that specify ncRNAs is still difficult to measure and a matter of controversy. Estimates range from several thousand to tens of thousands.

In the past, educators have tended to emphasize proteins and DNA in the teaching of biology at the molecular level. For example, in the cell unit (Chapters 4 through 10), we saw many examples of how proteins affect cell structure and function, and the genetics unit (Chapters 11 through 21) largely focuses on the structure and function of DNA. Although DNA, RNA, and proteins are key molecular players in living cells, a historical bias has existed against RNA. With a few exceptions, the educational exploration of RNA has been limited to its role in making proteins (see Chapter 12).

The purpose of this chapter is to lessen the bias against RNA. New molecular tools have enabled researchers to discover that ncRNAs perform a spectacular array of cellular functions in bacteria, archaea, protists, fungi, plants, and animals. ncRNAs play important roles in a variety of processes, including DNA replication, chromatin modification, transcription, translation, and genome defense. In most cell types, ncRNAs are more abundant than mRNAs. For example, in a typical human cell, only about 20% of transcription involves the production of mRNAs, whereas 80% is associated with making ncRNAs! This observation underscores the importance of RNA in the enterprise of life, and indicates why it deserves greater recognition and deeper study. Furthermore, abnormalities in ncRNAs are associated with a wide range of human diseases, including CHH, cancer, neurological disorders, and cardiovascular diseases. Many ncRNAs are also critical to the growth of plants, including the crop plants that are so essential to human survival.

In this chapter, we will begin with an overview of the general properties of ncRNAs, and then examine specific examples of the functions they perform. We will end the chapter by considering the role of ncRNAs in different human diseases and in plant health.

## 13.1 Overview of Non-coding RNAs

**Learning Outcomes:**

1. Describe the ability of ncRNAs to bind to other molecules and macromolecules.
2. Outline the general functions of ncRNAs.
3. Define ribozyme.
4. List several examples of ncRNAs, and describe their functions.

The study of ncRNAs is a rapidly expanding field, and researchers speculate that many ncRNAs have yet to be discovered. Also, due to the relative youth of this field, not all researchers agree on the names of certain ncRNAs or their primary functions. Even so, some broad themes are beginning to emerge. In this section, we will survey the general features of ncRNAs, and in later sections, we will discuss specific examples in greater detail.

### ncRNAs Can Bind to Different Types of Molecules

The ability of ncRNAs to carry out an amazing array of functions is largely related to their ability to bind to different types of molecules. **Figure 13.1a** shows four common types of molecules that are recognized by ncRNAs. Some ncRNAs bind to DNA or another RNA through complementary base pairing. This allows ncRNAs to affect processes such as DNA replication, transcription, and translation. In addition, ncRNAs can bind to proteins or small molecules.

As described in Chapter 12, RNA molecules, such as tRNAs, can form stem-loop structures (refer back to Figure 12.14). Similar structures in other ncRNAs may bind to pockets on the surface of proteins, or multiple stem-loops may form a binding site for a small molecule. In some cases, a single ncRNA may contain multiple binding sites. This allows an ncRNA to facilitate the formation of a large structure composed of multiple molecules, such as an ncRNA and three different proteins, as shown in **Figure 13.1b**.

### ncRNAs Can Perform a Diverse Set of Functions

In recent decades, researchers have uncovered many examples in which ncRNAs play a critical role in different biological processes. Let's first consider how ncRNAs work in a general way. The common functions of ncRNAs are the following.

**Scaffold** Some ncRNAs contain binding sites for multiple components, such as a group of different proteins. Much like the beams in a building, an ncRNA can act as a scaffold for the formation of a complex, as in Figure 13.1b.

**Guide** Some ncRNAs guide a molecule to a specific location in a cell. For example, an ncRNA may bind to a protein and guide it to a target site in the DNA that is part of a particular gene (**Figure 13.2**). This function also relies on the ncRNA having multiple binding sites: one for the protein and another for the target site in the DNA.

**Alteration of Protein Function or Stability** When it binds to a protein, an ncRNA can alter that protein's structure, which in turn can have a variety of effects. The binding of an ncRNA may affect

- the ability of the protein to act as a catalyst;
- the ability of the protein to bind to other molecules, such as proteins, DNA, or RNA;
- the stability of the protein.

**Ribozyme** Another interesting feature of some ncRNAs is that they function as **ribozymes**, which are RNA molecules with catalytic function. For example, in Chapter 12, we saw how peptidyltransferase, which is a component of the large ribosomal subunit, catalyzes peptide bond formation during translation (refer back to Figure 12.20). An rRNA within peptidyltransferase plays the key role in this catalysis. In other words, a part of the ribosome is a ribozyme.

**Blocker** An ncRNA may physically prevent or block a cellular process from happening. For example, in bacteria, an antisense RNA is a type of ncRNA that is complementary to an mRNA. When an antisense RNA binds to an mRNA, it blocks the ability of a ribosome to bind to the mRNA, thereby inhibiting translation.

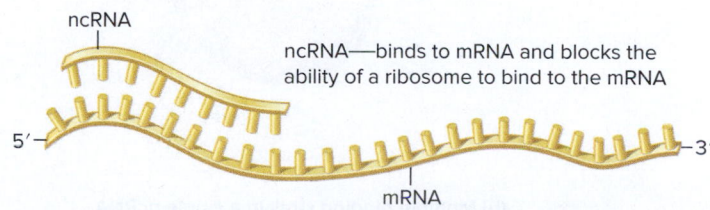

ncRNA

ncRNA—binds to mRNA and blocks the ability of a ribosome to bind to the mRNA

5′—

—3′

mRNA

**Decoy** Some ncRNAs recognize other ncRNAs and sequester them, thereby preventing them from working. For example, a decoy ncRNA may bind to a different ncRNA called a microRNA (miRNA), which is described later in this chapter (**Figure 13.3**). The function of

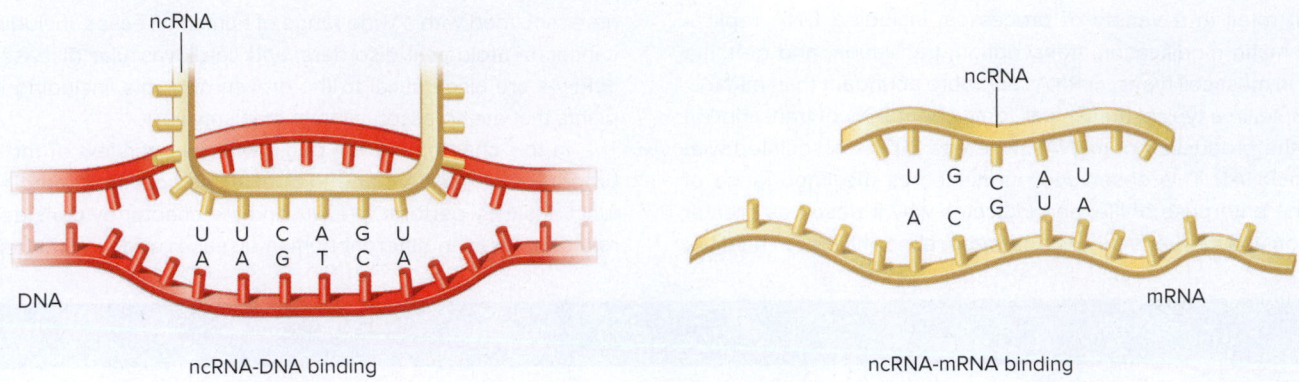

ncRNA-DNA binding

ncRNA-mRNA binding

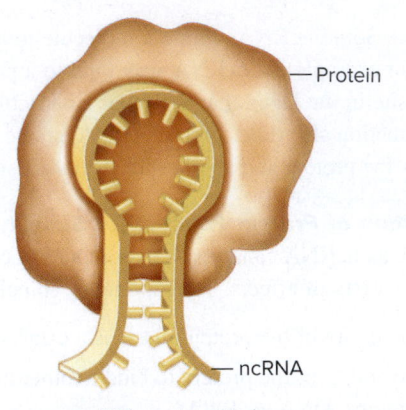

ncRNA-protein binding

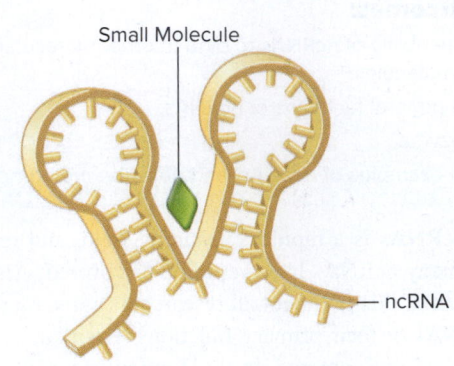

ncRNA-Small molecule binding

**(a) Common binding interactions between ncRNAs and other molecules**

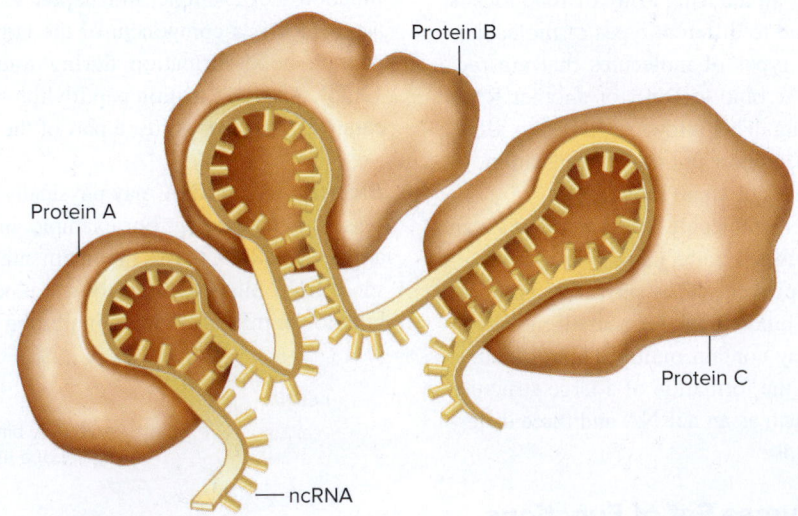

**(b) Multiple binding sites in a single ncRNA**

**Figure 13.1** **Ability of ncRNAs to bind to other molecules.** **(a)** ncRNA molecules can bind to DNA, mRNA, proteins, and small molecules. **(b)** Some ncRNAs have multiple binding sites for different molecules, such as proteins.

*Concept Check:* *Which of these binding interactions might be inhibited by the formation of a stem-loop within the ncRNA?*

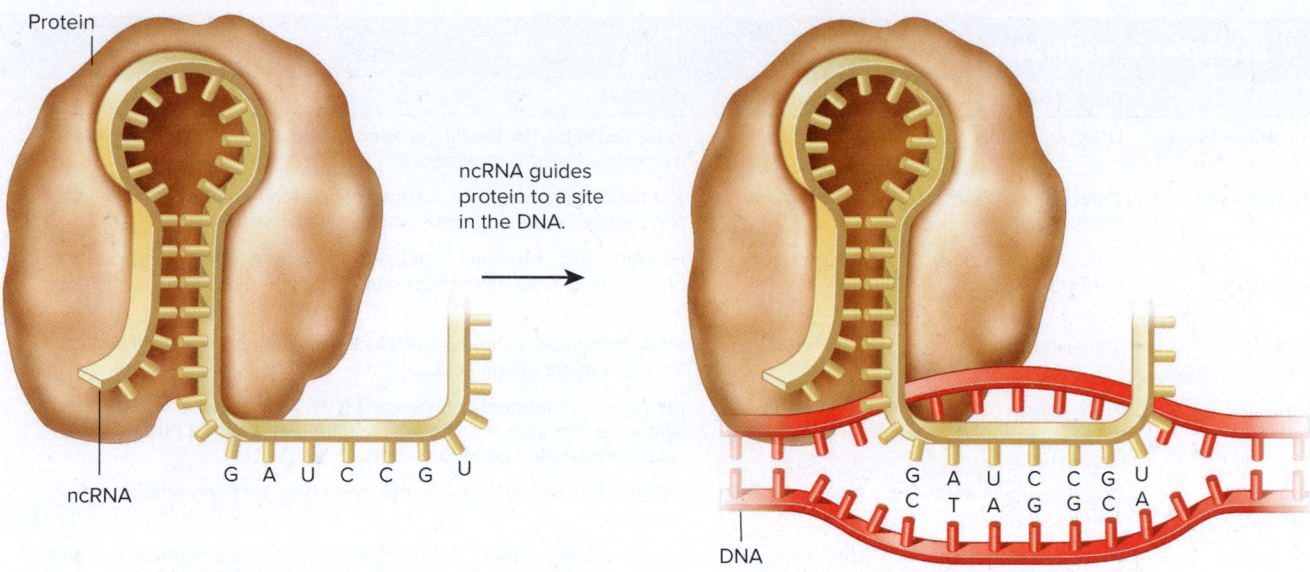

**Figure 13.2** **Ability of an ncRNA to function as a guide.**

 **Core Concept: Structure and Function** This guide RNA has two key structural features: One part binds to a protein and another part binds to a specific sequence of DNA. These two structural features allow the ncRNA to carry out its function of guiding the protein to the DNA.

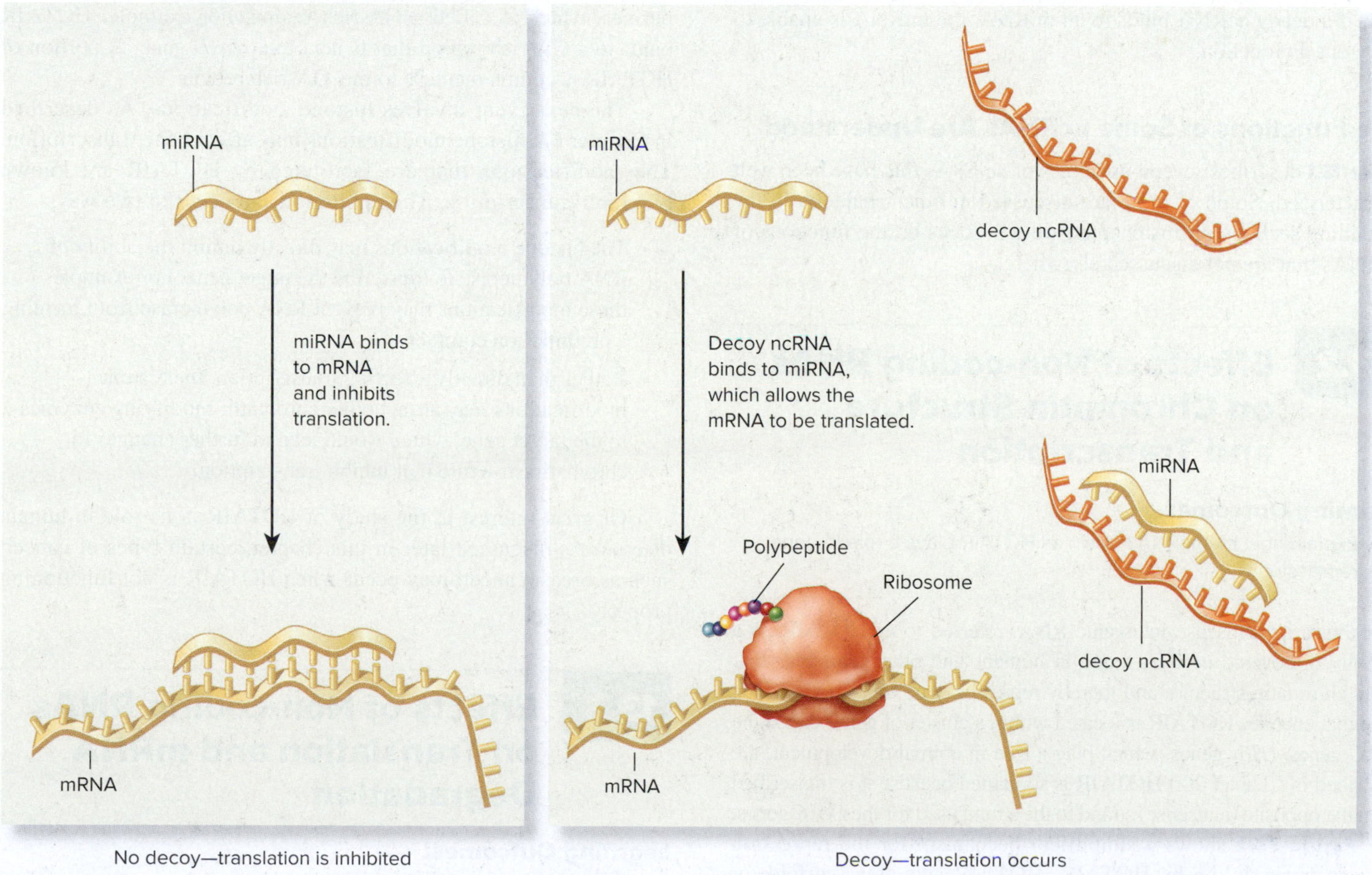

**Figure 13.3** **Ability of an ncRNA to function as a decoy.**

| Table 13.1 | Examples of ncRNAs | | |
|---|---|---|---|
| **Type of ncRNA** | **Plays a role in** | **Discussed in** | **Description** |
| Telomerase RNA component (TERC) | DNA replication | Chapter 11 | TERC facilitates the binding of telomerase to the telomere and acts as a template for DNA replication. |
| X inactive specific transcript (Xist RNA) | Chromatin structure, transcription | Chapter 18 | Xist RNA coats one of the X chromosomes in female mammals and plays a role in its compaction and resulting inactivation. |
| *Hox* transcript antisenseintergenic RNA (HOTAIR) | Chromatin structure, transcription | This chapter | HOTAIR alters chromatin structure and thereby represses transcription by guiding histone-modifying complexes to target genes. |
| Transfer RNA (tRNA) | Translation | Chapter 12 | tRNA molecules recognize mRNA codons during translation and carry the appropriate amino acid. |
| Ribosomal RNA (rRNA) | Translation | Chapter 12 | rRNAs are components of ribosomes, which are the site of polypeptide synthesis. Ribosomes contain an ncRNA that acts as a ribozyme by catalyzing peptide bond formation. |
| microRNA (miRNA), small-interfering RNA (siRNA) | Translation and RNA degradation | This chapter | miRNAs and siRNAs regulate the expression and degradation of mRNAs. |
| RNA component of signal recognition particle (SRP RNA) | Protein sorting and secretion | This chapter | In bacteria, SRP directs some polypeptides to the plasma membrane. In eukaryotes, it directs polypeptides to the endoplasmic reticulum. |
| CRISPR RNA (crRNA) | Genome defense | This chapter | crRNA, found in bacteria and archaea, guides an endonuclease to foreign DNA, such as the DNA of a bacteriophage. |

an miRNA is to inhibit the translation of a particular mRNA. However, if a decoy ncRNA binds to an miRNA, the miRNA is unable to carry out its function.

## The Functions of Some ncRNAs Are Understood

**Table 13.1** describes several examples of ncRNAs that have been well characterized. Some of these are discussed in other chapters. In the remaining sections of this chapter, we will focus on the functions of ncRNAs that are not discussed elsewhere.

## 13.2 Effects of Non-coding RNAs on Chromatin Structure and Transcription

### Learning Outcome:

1. Explain how the ncRNA known as HOTAIR plays a role in gene repression.

*Hox* transcript antisense intergenic RNA, referred to as HOTAIR, is a recently discovered ncRNA found in humans and other mammals that alters chromatin structure and thereby represses gene transcription. The gene that encodes HOTAIR is located within a cluster of genes called the *HoxC* genes. (*Hox* genes, which play a role in animal development, are described in Chapter 20.) HOTAIR is so named because it is transcribed from the opposite (antisense) strand to the strand used for the *HoxC* genes.

**Figure 13.4** shows a simplified mechanism for the repression of gene transcription by HOTAIR. HOTAIR acts as a scaffold for the binding of two protein complexes that covalently modify histone proteins. One of these complexes binds to the 5′ end of HOTAIR, and the other binds to the 3′ end. HOTAIR then guides these complexes to

a target gene by binding to a region near the gene that contains many purines, which is called a GA-rich region. For example, HOTAIR binds to a GA-rich region that is next to a *HoxD* gene. A portion of HOTAIR is complementary to this GA-rich region.

The next event involves histone modifications. As described in Chapter 14, histone modifications may affect gene transcription. The modifications that are facilitated by HOTAIR are known to inhibit transcription. This inhibition can occur in two ways:

- The histone modifications may directly inhibit the ability of RNA polymerase to transcribe the target gene. For example, these modifications may prevent RNA polymerase from forming a preinitiation complex.
- Rather than directly affecting transcription, the histone modifications may attract other chromatin-modifying enzymes to the target gene, which would lead to further changes in chromatin structure that inhibit transcription.

Of great interest in the study of HOTAIR is its role in human disease. As discussed later in this chapter, certain types of cancer, such as breast cancer, may occur when HOTAIR is not functioning properly.

## 13.3 Effects of Non-coding RNAs on Translation and mRNA Degradation

### Learning Outcomes:

1. **CoreSKILL »** Analyze experimental evidence that double-stranded RNA is more potent in silencing mRNA than antisense RNA is.
2. Outline the steps of RNA interference.

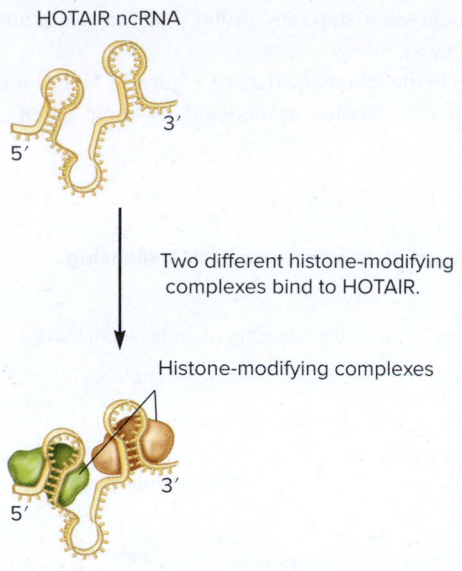

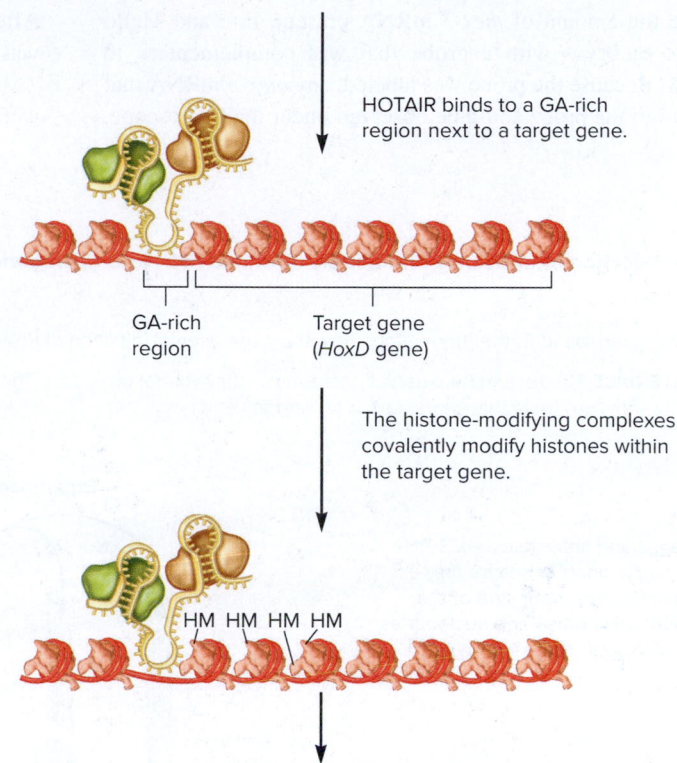

HOTAIR ncRNA

Two different histone-modifying complexes bind to HOTAIR.

Histone-modifying complexes

HOTAIR binds to a GA-rich region next to a target gene.

GA-rich region

Target gene (*HoxD* gene)

The histone-modifying complexes covalently modify histones within the target gene.

HM HM HM HM

**Figure 13.4** **Simplified mechanism for inhibition of gene transcription by HOTAIR.** This is just one proposed role of HOTAIR. This ncRNA is known to interact with other proteins as well. Note: The abbreviation HM stands for histone modification.

**Concept Check:** *Explain why HOTAIR binds to the target gene. Why doesn't it bind next to every gene?*

These histone modifications may directly inhibit transcription, or they may lead to further changes in chromatin structure that inhibit transcription.

In the previous section, we considered how an ncRNA can affect the process of transcription. In recent years, researchers have discovered that ncRNAs often exert their effects on RNA molecules that are already made. In this section, we will consider how ncRNAs can affect the ability of mRNAs to be translated or degraded, as well as the ability of rRNAs to be covalently modified.

---

### Core Skill: Process of Science

## Feature Investigation | Fire and Mello Showed That Double-Stranded RNA Is More Potent Than Antisense RNA in Silencing mRNA

Specific mRNAs can be targeted for translational inhibition or degradation by a mechanism involving double-stranded RNA. This mechanism was discovered during research involving plants and the nematode worm *Caenorhabditis elegans*. The study described here involved an examination of gene expression in *C. elegans*.

American biologists Andrew Fire, Craig Mello, and colleagues used *C. elegans* as their experimental organism, because it is relatively easy to inject with RNA and the expression of many of its genes had been established. In 1998, Fire and Mello investigated the effects of injected RNA on the expression of specific mRNAs. In the investigation described in **Figure 13.5**, we will focus on one of their experiments involving an mRNA encoded by a gene called *mex-3*. This mRNA had already been shown to be made in high amounts in early embryos of *C. elegans*.

Prior to this work, the *mex-3* gene had been identified and inserted into a plasmid. The process of inserting genes into plasmids is described in Chapter 21 (look ahead to Figure 21.2). Let's first look at the upper plasmid shown in the "Conceptual level" column for step 1. When RNA polymerase, nucleotides, and this plasmid were mixed together in a test tube, *mex-3* mRNA was made, which is called the sense strand. In living cells, the sense strand is used to make the mex-3 protein. Fire and Mello also switched the location of the promoter so that it was at the other end of the gene and signaled transcription of the opposite strand, which is called the antisense strand (see the second plasmid in the "Conceptual level" column). The sense and antisense strands are complementary to each other.

Next, they injected these RNAs into the gonads of *C. elegans*. Into some worms, they injected antisense RNA alone. Alternatively, they mixed sense and antisense RNA, which formed double-stranded RNA, and injected this double-stranded RNA into the gonads of other worms. They also used uninjected worms as controls. After injection, the RNA was taken up by eggs, which later developed into embryos.

To determine the amount of *mex-3* mRNA present, Fire and Mello incubated the embryos with a probe that was complementary to *mex-3* mRNA. Because the probe was labeled, any *mex-3* mRNA that became bound to the probe could be observed under the microscope.

After this incubation step, any probe that was not bound to mRNA was washed away.

As seen in the schematic data of Figure 13.5, the control embryos were very darkly labeled as denoted by their green color. These

**Figure 13.5** Injection of antisense and double-stranded RNA into *C. elegans* to compare their effects on mRNA silencing.

**GOAL** The goal was to further understand how the experimental injection of RNA was responsible for the silencing of particular mRNAs.

**KEY MATERIALS** The researchers used *C. elegans* as their model organism. They also had the cloned *mex-3* gene, which had been previously shown to be highly expressed in *C. elegans* embryos.

| Experimental level | Conceptual level |
|---|---|

1. Make sense and antisense *mex-3* RNA in vitro using cloned genes for *mex-3* with promoters on either side of the gene. RNA polymerase and nucleotides are added to synthesize the RNAs.

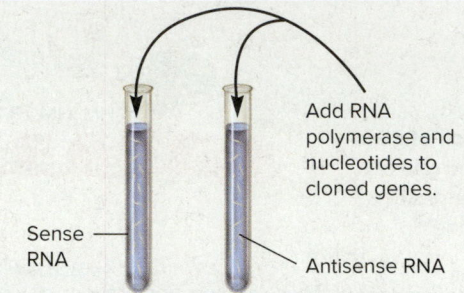

Sense RNA

Add RNA polymerase and nucleotides to cloned genes.

Antisense RNA

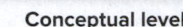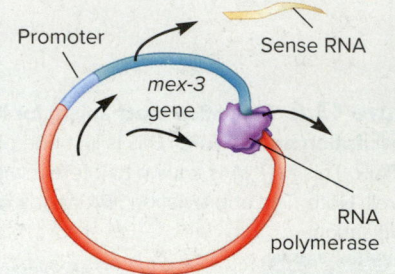

Promoter    Sense RNA

*mex-3* gene

RNA polymerase

2. Inject either *mex-3* antisense RNA or a mixture of *mex-3* sense and antisense RNA into the gonads of *C. elegans*. This RNA is taken up by the eggs and early embryos. As a control, do not inject any RNA.

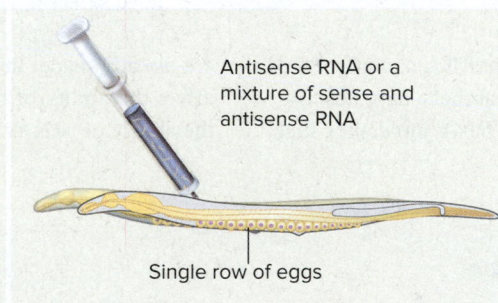

Antisense RNA or a mixture of sense and antisense RNA

Single row of eggs

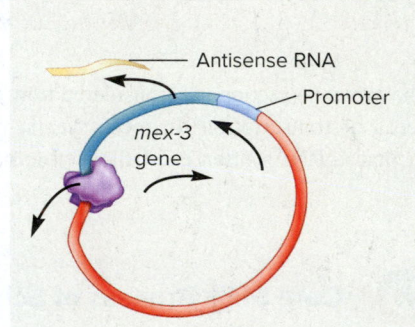

Antisense RNA

Promoter

*mex-3* gene

3. Incubate and then subject early embryos to in situ hybridization. In this method, a labeled probe is added that is complementary to *mex-3* mRNA. If cells express *mex-3*, the mRNA in the cells will bind to the probe and become labeled. After incubation with a labeled probe, the cells are washed to remove unbound probe.

Add labeled probe

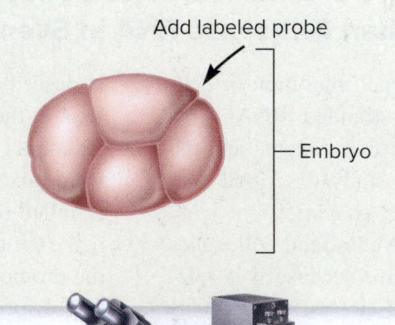

Embryo

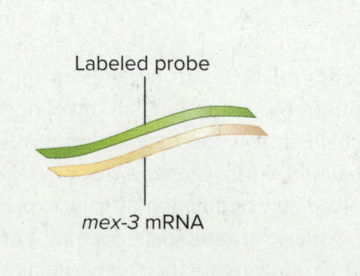

Labeled probe

*mex-3* mRNA

4. Observe embryos under the microscope.

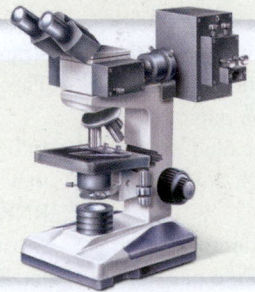

**5  THE DATA**

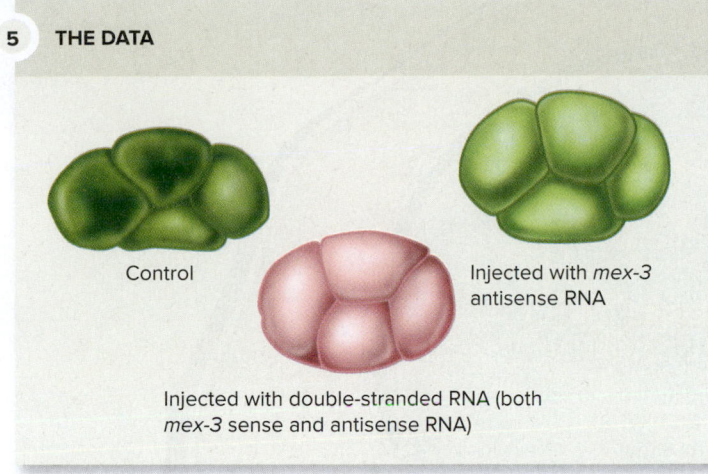

Control

Injected with *mex-3* antisense RNA

Injected with double-stranded RNA (both *mex-3* sense and antisense RNA)

**6  CONCLUSION**  Double-stranded RNA is more potent at inhibiting *mex-3* mRNA than antisense RNA alone.

**7  SOURCE**  Fire, A., Xu, S., Montgomery, M.K., et al. 1998. Potent and Specific Genetic Interference by Double-Stranded RNA in *Caenorhabditis elegans*. *Nature* 391: 806–811.

results indicated that the control embryos contained a high amount of *mex-3* mRNA, which was known from previous research. In the embryos that had received antisense RNA, *mex-3* mRNA levels were decreased, but detectable, as shown by faint labeling. Remarkably, in embryos that had received double-stranded RNA, no *mex-3* mRNA was detected! These results indicated that double-stranded RNA is more potent at silencing mRNA than is antisense RNA. In this case, the double-stranded RNA caused the *mex-3* mRNA to be degraded. Fire and Mello used the term **RNA interference (RNAi)** to describe the phenomenon in which double-stranded RNA causes the silencing of mRNA. This surprising observation led researchers to investigate

the underlying molecular mechanism that accounts for this phenomenon, as described next.

**Experimental Questions**

1. In this experiment, does the *mex-3* mRNA correspond to the sense strand or antisense strand?

2. **CoreSKILL »** Explain how the sense and antisense *mex-3* RNAs were made.

3. **CoreSKILL »** According to the data, which material was the most effective at causing the degradation of *mex-3* mRNA?

## RNA Interference Is Mediated by MicroRNAs or Small-Interfering RNAs via the RNA-Induced Silencing Complex

RNA interference is found in most eukaryotic species, including animals and plants. It can arise from two sources: microRNAs and small-interfering RNAs. **MicroRNAs (miRNAs)** are ncRNAs that are transcribed from endogenous eukaryotic genes—genes that are normally found in the genome. They play key roles in regulating gene expression, particularly during embryonic development in animals and plants. Most commonly, a single type of miRNA inhibits the translation of several different mRNAs. An miRNA and an mRNA bind to each other because they have base sequences that are partially complementary. In humans, over 2,000 genes encode miRNAs. Researchers estimate that 60% of human protein-encoding genes are regulated by miRNAs.

By comparison, **small-interfering RNAs (siRNAs)** are ncRNAs that usually originate from sources that are exogenous, which means they are not normally made by cells. The siRNAs can come from viruses that infect a cell, or they might be synthesized by researchers to study gene function experimentally, as in Figure 13.5. In most cases, siRNAs are a perfect match to a single type of mRNA. The functioning of siRNAs is thought to play a key role in preventing certain types of viral infections. In addition, siRNAs have become important experimental tools in molecular biology.

How do miRNAs and siRNAs cause the silencing of specific mRNAs? **Figure 13.6** shows how an miRNA or an siRNA leads to

RNA interference. The miRNA is first synthesized as a pri-miRNA (for primary-miRNA) in the nucleus. Due to complementary base pairing, the pri-miRNA folds into a hairpin structure (also called a stem-loop) with long, single-stranded 5′ and 3′ ends. The pri-miRNA is cleaved at both ends to form a pre-miRNA (for precursor-miRNA, not to be confused with pri-miRNA). The pre-miRNA is then exported from the nucleus.

As shown in Figure 13.6, siRNAs do not go through the processing events that occur in the nucleus. Instead, pre-siRNAs may be derived from viral RNAs, or they may be made by researchers and taken up by cells. For example, in the work of Fire and Mello described in Figure 13.5, the double-stranded *mex-3* RNA is an example of a pre-siRNA. The pre-siRNA is formed from two complementary RNA molecules that base-pair with each other.

In the cytosol, both pre-miRNAs and pre-siRNAs are cut by an endonuclease called dicer (see Figure 13.6). This releases a double-stranded RNA molecule that is typically 20–25 bp long. This double-stranded RNA associates with proteins to form a complex called the **RNA-induced silencing complex (RISC)**. One of the RNA strands is degraded. The remaining single-stranded miRNA or siRNA is complementary to specific mRNAs that will be silenced. The miRNA or siRNA acts as a guide that causes RISC to recognize and bind to such mRNA molecules.

After RISC binds to an mRNA, one of two things may happen:

- RISC may inhibit translation without degrading the mRNA. This is more common for miRNAs, which often are only partially complementary to their target mRNAs.

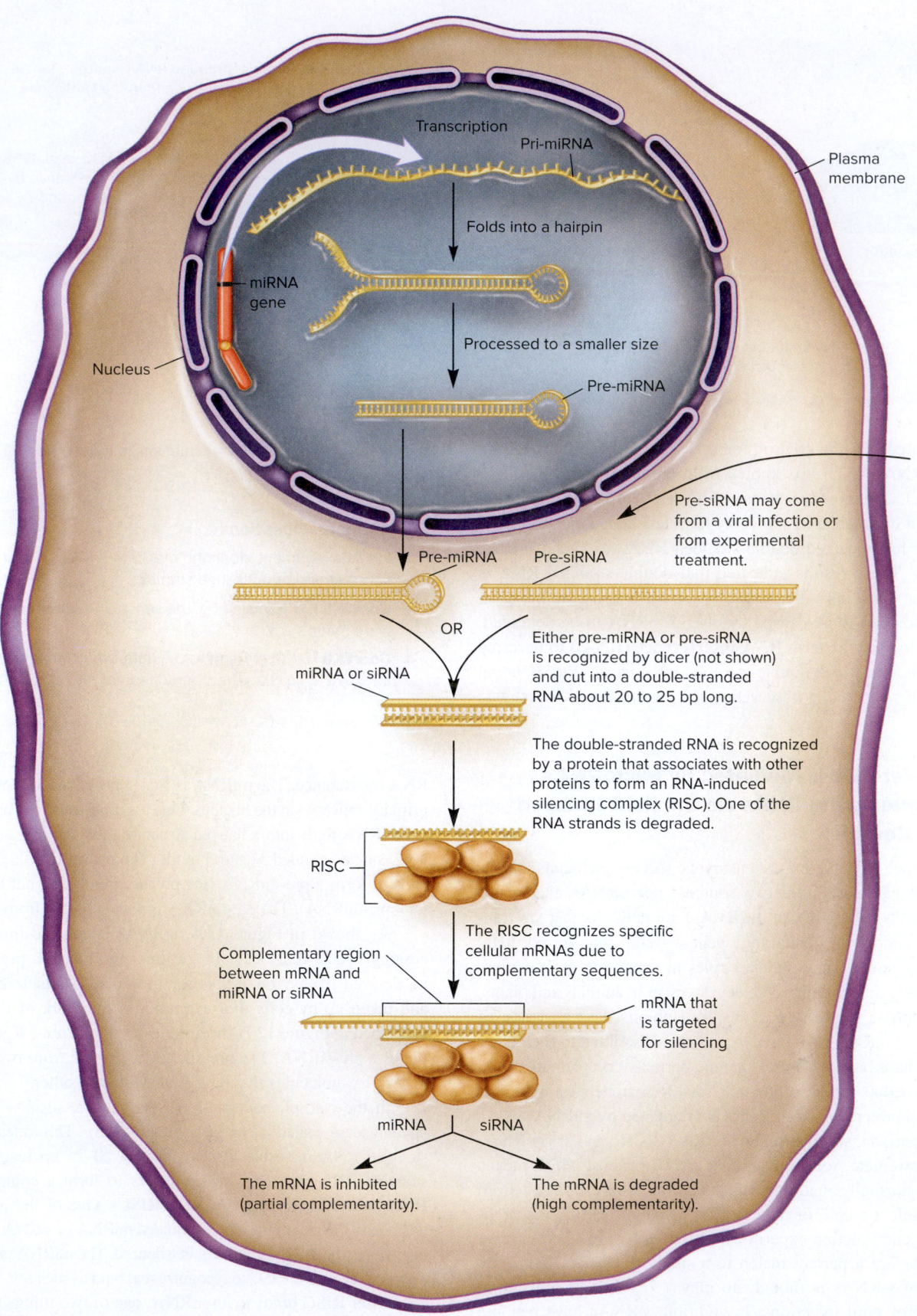

**Figure 13.6**  Mechanism of RNA interference.

*Concept Check:*  *Explain why RISC binds to a specific mRNA. What type of bonding occurs?*

- RISC may direct the degradation of the mRNA. One of the proteins in RISC can cleave the mRNA. This usually occurs for siRNAs that typically are a perfect match (or highly complementary) to their target mRNA.

These two effects are termed RNA interference because the miRNA or siRNA interferes with the proper expression of an mRNA. In 2006, Fire and Mello received the Nobel Prize in Physiology or Medicine for their discovery of this mechanism.

RNA interference is believed to have at least two benefits:

- This mechanism represents an important form of regulation. When genes encoding pri-miRNAs are turned on, the production of miRNAs silences the expression of specific mRNAs.

- RNA interference provides a defense against viruses. This mechanism is widely used by plants to prevent viral infections.

## 13.4 Non-coding RNAs and Protein Sorting

### Learning Outcome:

1. Describe the function of SRP, and explain the roles of SRP RNA with regard to its function.

To carry out their functions, proteins need to be directed to particular locations (refer back to Figure 4.32). For example, some proteins function extracellularly and need to be secreted from the cell. For such proteins to be secreted, they are first sorted to the plasma membrane in bacteria and archaea, or to the endoplasmic reticulum (ER) membrane in eukaryotic cells (refer back to Figure 4.33). This process is facilitated by a protein-RNA complex called **signal recognition particle (SRP).** In bacteria, SRP is composed of one ncRNA and one protein. In eukaryotes, SRP is composed of one ncRNA and six different proteins.

**Figure 13.7** takes a closer look at how SRP works in eukaryotes. To be directed to the ER membrane, a polypeptide must contain a sorting signal called an **ER signal sequence**, which is a sequence of about 6–12 amino acids that are predominantly hydrophobic and usually located near the N-terminus. As the ribosome is making the polypeptide in the cytosol, the ER signal sequence emerges from the ribosome and is recognized by a protein in SRP. The binding of SRP to the polypeptide pauses translation.

SRP then binds to an SRP receptor in the ER membrane, which docks the ribosome over a channel. For this binding to occur, proteins within SRP and the SRP receptor must also be bound by GTP. Next, these GTP-binding proteins hydrolyze their GTP, which causes the release of SRP from the SRP receptor and the polypeptide. Once SRP is released, translation resumes and the growing polypeptide is threaded through a channel to cross the ER membrane. In the case of a secreted protein, the newly made polypeptide then travels through the Golgi apparatus and then to the plasma membrane, where it is released outside of the cell.

Researchers have identified at least two key roles for SRP RNA:

1. SRP RNA provides a scaffold for the binding of SRP proteins.

2. After SRP binds to the SRP receptor in the ER membrane, the SRP RNA stimulates proteins within both SRP and the SRP

receptor to hydrolyze their GTP. In other words, SRP RNA alters the structures of these proteins to enhance their GTPase activities. This stimulation is essential for the release of SRP.

## 13.5 Non-coding RNAs and Genome Defense

### Learning Outcome:

1. Explain how the CRISPR-Cas system defends bacteria against bacteriophages.

Much like the immune system found in vertebrates, a system called the **CRISPR-Cas system** provides some species of bacteria and archaea with a means of defense against foreign invaders. CRISPR-Cas systems are an effective defense against bacteriophages, which are viruses that infect bacteria (discussed in Chapter 19), and transposons, which are small segments of DNA that can be inserted into the chromosomes of all species (discussed in Chapter 21). ncRNAs play a key role in CRISPR-Cas systems. About half of all bacterial species and most archaeal species have such a system. Three general types are known, designated type I, II, and III. In this section, we will focus on the type II CRISPR-Cas system and its role in providing bacteria with defense against bacteriophages.

### The CRISPR-Cas System Provides Bacteria with Defense Against Bacteriophages

In 1993, Spanish microbiologist Francisco Mojica and colleagues were the first to recognize that different species of bacteria and archaea have a site in their chromosome, now called the CRISPR locus, that contains a series of repeated sequences. In 2005, by analyzing the DNA sequences of the CRISPR locus in a variety of bacterial species, Mojica, Spanish geneticist Giles Vergnaud, and Russian microbiologist Alexander Bolotin independently proposed that this locus provides protection against bacteriophage infection. This hypothesis was based on the observation that the CRISPR locus contains segments that are derived from bacteriophage DNA. The hypothesis was confirmed in 2007 by French microbiologist Philippe Horvath and colleagues, who showed experimentally that the CRISPR-Cas locus provides defense against bacteriophage infection.

**Figure 13.8a** shows a common organization of the CRISPR-Cas system (also called the CRISPR locus), which has five genes: *tracr, Cas9, Cas1, Cas2,* and *Crispr*. A key feature of the *Crispr* gene is a group of clustered, regularly interspaced, short, palindromic repeats—hence the name CRISPR. The repeats within the *Crispr* gene are interspersed by short, unique sequences, which are called spacers. The CRISPR-Cas type II system also includes a gene that encodes an ncRNA called tracrRNA and a few protein-encoding CRISPR-associated genes (*Cas* genes), which are usually adjacent to the *Crispr* gene. These genes are needed to mediate the defense against bacteriophages.

The CRISPR-Cas system is considered an adaptive defense system because a bacterial cell must first be exposed to an agent, such as a bacteriophage, to elicit a response from the system. As shown in **Figure 13.8b–d**, the defense mechanism occurs in three phases.

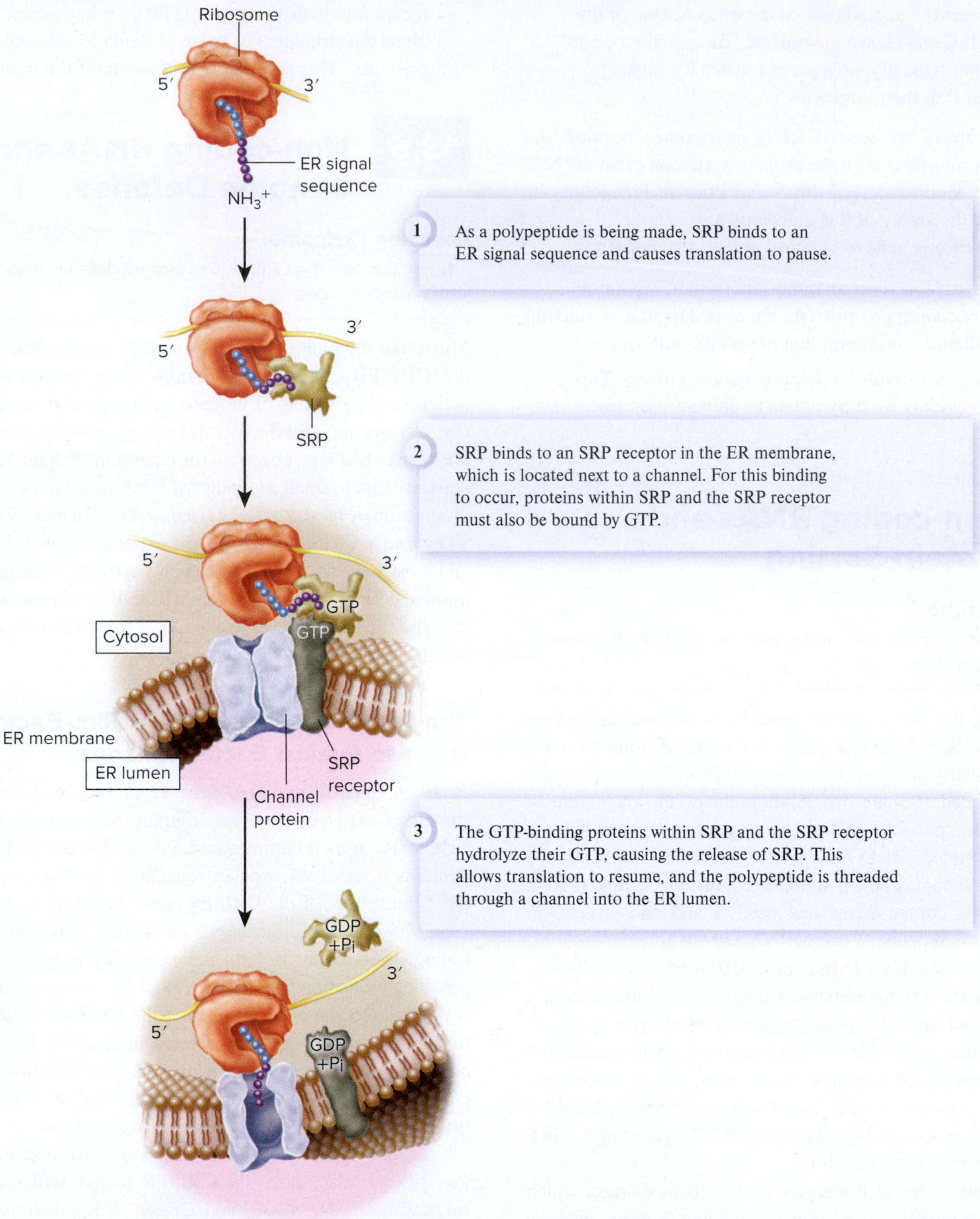

Ribosome

5′ 3′

ER signal
sequence

NH₃⁺

**1** As a polypeptide is being made, SRP binds to an
ER signal sequence and causes translation to pause.

5′ 3′

SRP

**2** SRP binds to an SRP receptor in the ER membrane,
which is located next to a channel. For this binding
to occur, proteins within SRP and the SRP receptor
must also be bound by GTP.

5′ 3′

GTP
GTP

Cytosol

ER membrane

ER lumen

Channel
protein

SRP
receptor

**3** The GTP-binding proteins within SRP and the SRP receptor
hydrolyze their GTP, causing the release of SRP. This
allows translation to resume, and the polypeptide is threaded
through a channel into the ER lumen.

GDP
+Pi

3′

5′

GDP
+Pi

**Figure 13.7 Directing of polypeptides to the endoplasmic reticulum membrane via SRP.** In eukaryotes, several categories of proteins are first directed to the ER via SRP. These include proteins that are secreted from the cells as well as proteins that are destined to stay in the ER, Golgi, lysosomes, or vacuoles.

 **Core Skill: Connections** Refer back to Figure 4.32. Which types of proteins need SRP to reach their proper location, and which do not?

*Adaptation* The process of adaptation (also called spacer acquisition) occurs after a bacterial cell has been exposed to a bacteriophage. The proteins encoded by the *Cas1* and *Cas2* genes form a complex that recognizes the bacteriophage DNA as being foreign and cleaves it into small pieces. As shown in

Figure 13.8b, a piece of bacteriophage DNA, usually between 20 and 50 bp in length, is inserted into the *Crispr* gene. The mechanism of insertion is not entirely understood. The newly inserted piece of bacteriophage DNA is called a spacer because it acts as a space between adjacent repeats. The different spacers found in

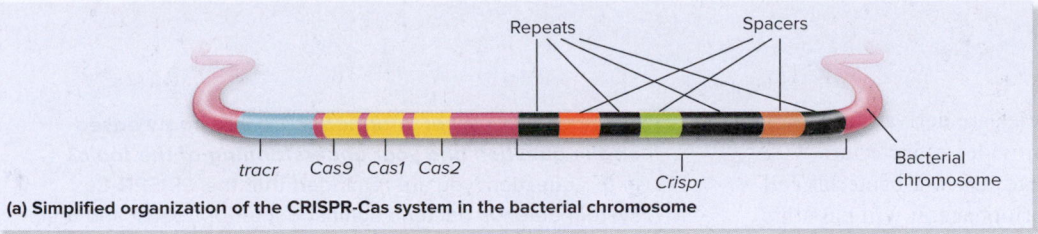

**(a) Simplified organization of the CRISPR-Cas system in the bacterial chromosome**

Repeats    Spacers

tracr   Cas9  Cas1  Cas2        Crispr        Bacterial chromosome

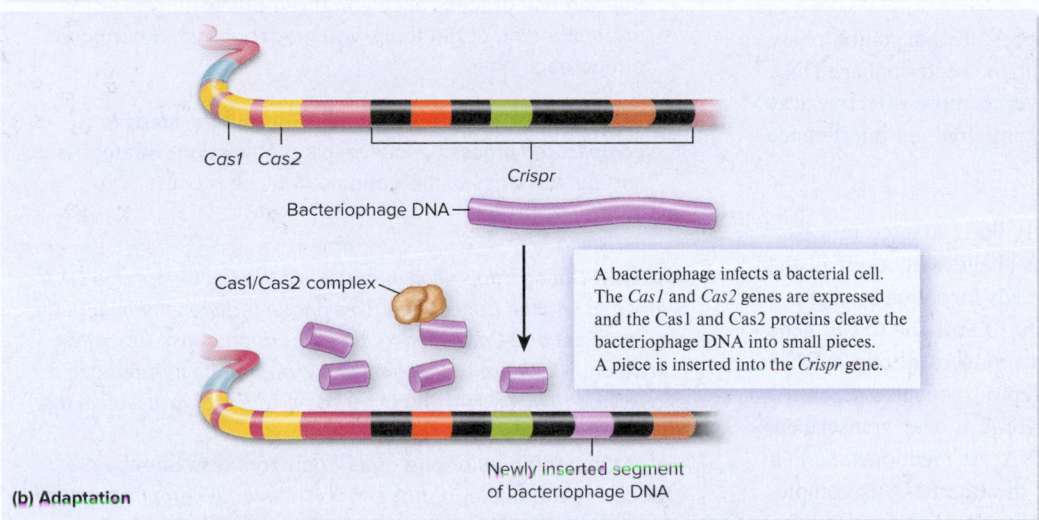

Cas1  Cas2        Crispr

Bacteriophage DNA

Cas1/Cas2 complex

A bacteriophage infects a bacterial cell. The *Cas1* and *Cas2* genes are expressed and the Cas1 and Cas2 proteins cleave the bacteriophage DNA into small pieces. A piece is inserted into the *Crispr* gene.

Newly inserted segment of bacteriophage DNA

**(b) Adaptation**

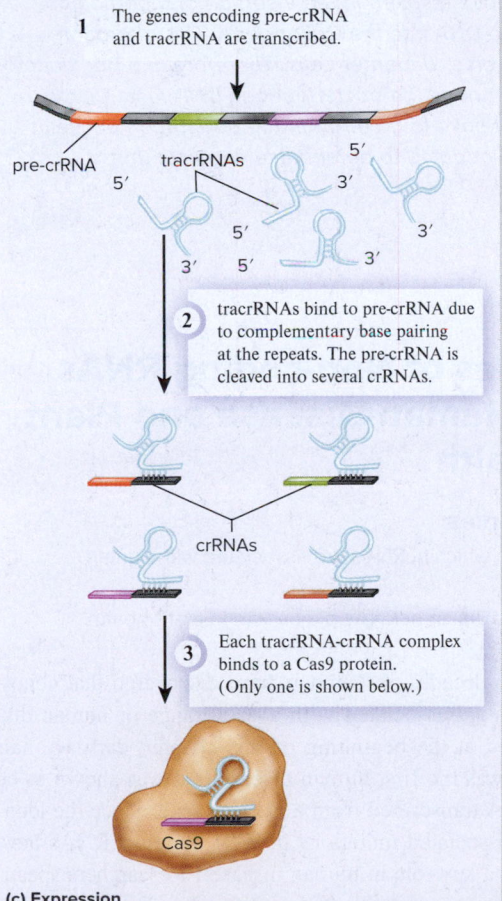

**1** The genes encoding pre-crRNA and tracrRNA are transcribed.

pre-crRNA    tracrRNAs    5′
5′                          3′
        5′          3′
    3′  5′      3′

**2** tracrRNAs bind to pre-crRNA due to complementary base pairing at the repeats. The pre-crRNA is cleaved into several crRNAs.

crRNAs

**3** Each tracrRNA-crRNA complex binds to a Cas9 protein. (Only one is shown below.)

Cas9

**(c) Expression**

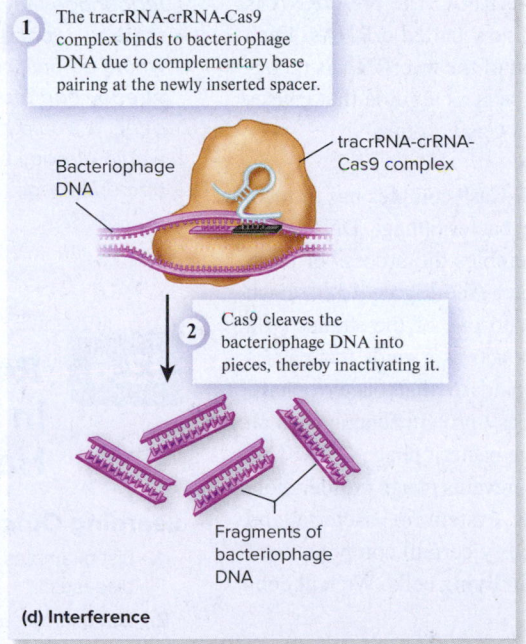

**1** The tracrRNA-crRNA-Cas9 complex binds to bacteriophage DNA due to complementary base pairing at the newly inserted spacer.

Bacteriophage DNA

tracrRNA-crRNA-Cas9 complex

**2** Cas9 cleaves the bacteriophage DNA into pieces, thereby inactivating it.

Fragments of bacteriophage DNA

**(d) Interference**

**Figure 13.8  The CRISPR-Cas system of genome defense in bacteria.** The system shown here is a type II system, which is found in the chromosome of certain bacterial species but not in archaea. **(a)** Organization of the CRISPR-Cas system in a bacterial chromosome. This drawing shows a typical organization, but different species have variations. **(b–d)** A simplified mechanism of the CRISPR-Cas system. The defense occurs in three phases, called adaptation, expression, and interference.

 **Core Skill: Modeling**  The goal of this modeling challenge is to increase the complexity of the model shown in Figure 13.8d by including sequences in the crRNA and bacteriophage DNA that bind to each other.

**Modeling Challenge:** For simplicity, the structure of the crRNAs in Figure 13.8 is shown shorter than it really is. Let's suppose that the part of a crRNA that recognizes the bacteriophage DNA is 20 nucleotides long with the following base sequence: 3'-GAUCCCAUACGGCU-AAUCAA-5'. (Note: This is only part of the sequence of the crRNA; it does not include the part that binds to tracrRNA.) Suppose this crRNA is within a tracrRNA-crRNA-Cas9 complex, and draw a model that shows the binding of the complex to a bacteriophage. The general features of the model should be similar to those in Figure 13.8d, except you should include the sequences of the crRNA and the bacteriophage DNA that bind to each other. Label the 5' and 3' ends of the tracrRNA, crRNA and bacteriophage DNA.

the *Crispr* gene of modern bacterial species are derived from past bacteriophage infections. Each spacer provides a bacterium with defense against a particular bacteriophage. Once a bacterial cell has become adapted to a particular bacteriophage, it will pass this trait on to its daughter cells.

By cleaving the bacteriophage into pieces, the adaptation phase can protect a bacterial cell, because it cuts up the bacteriophage DNA and thereby inactivates the phage. However, a more effective way of destroying phages is provided by the expression and interference phases of this system.

**Expression**    If a bacterial cell has already been adapted to a bacteriophage, a subsequent infection by that phage will result in the expression phase, in which the system gets ready for action by expressing the *Crispr*, *tracr*, and *Cas9* genes (Figure 13.8c). The *Crispr* gene is transcribed from a single promoter and produces a long ncRNA called pre-crRNA, which contains several repeat sequences separated by spacers. The gene encoding the tracrRNA is also transcribed, which produces many molecules of tracrRNA. As mentioned earlier, tracrRNA is also an ncRNA. A region of the tracrRNA is complementary to the repeat sequences of the pre-crRNA. Several molecules of tracrRNA base-pair with the pre-crRNA. The pre-crRNA is then cleaved into many small molecules, now called crRNAs. Each crRNA is attached to a tracrRNA. A region of the tracrRNA is recognized by the Cas9 protein. The tracrRNA acts as a guide that causes the tracrRNA-crRNA complex to bind to a Cas9 protein.

**Interference**    After the tracrRNA-crRNA-Cas9 complex has formed, the bacterial cell is ready to destroy the bacteriophage DNA. This phase is called interference because it resembles the process of RNA interference described earlier in this chapter (see Figure 13.6). Each spacer within a crRNA is complementary to one of the strands of a bacteriophage DNA. Therefore, the crRNA acts as a guide that causes the tracrRNA-crRNA-Cas9 complex to bind to that bacteriophage DNA (Figure 13.8d). After binding, the Cas9 protein functions as an endonuclease that breaks both strands in the bacteriophage DNA. This cleavage inactivates the phage and thereby prevents phage proliferation.

Since discovering the CRISPR-Cas system in bacteria and archaea, researchers have been able to modify certain components of this system and use them to mutate genes in living cells. We will consider this technology in Chapter 21.

BIO TIPS    **THE QUESTION** *With regard to the CRISPR-Cas system that defends bacteria against bacteriophages, what happens during the adaptation, expression, and interference phases? When a bacterium is exposed to a particular bacteriophage, is the adaptation phase always necessary?*

**T**OPIC    *What topic in biology does this question address?* The topic is the CRISPR-Cas system that provides bacteria with defense against bacteriophages. More specifically, the question asks you to sort out what happens during each phase of the genome defense process and decide whether or not the first phase is always needed.

**I**NFORMATION    *What information do you know based on the question and your understanding of the topic?* In the question, you are reminded that the CRISPR-Cas system defends bacteria against bacteriophages, and that the defense process occurs in three phases. From your understanding of the topic, you may recall what happens during each phase.

**P**ROBLEM-SOLVING **S**TRATEGY    *Sort out the steps in a complicated process.* To solve this problem, one strategy is to sort out the steps of this genome defense process.

**ANSWER** *During adaptation, a portion of the bacteriophage DNA is inserted into the Crispr gene. This phase requires the help of the proteins Cas1 and Cas2. During the expression phase, tracrRNA, crRNA, and Cas9 are produced. Finally, during the interference phase, tracrRNA, crRNA, and Cas9 come together and cleave the bacteriophage DNA, thereby inactivating it.*

*If the bacterium or one of its ancestors was already exposed to the bacteriophage that is currently infecting it, the adaptation phase is not necessary. Prior exposure to the bacteriophage may have resulted in the insertion of a portion of the bacteriophage DNA into the Crispr gene. This alteration would be passed on to daughter cells. Therefore, if a bacterium already had a portion of the bacteriophage DNA in its Crispr gene, it would not have to go through the adaptation phase; it would already be adapted to defend itself against that bacteriophage.*

## 13.6    Roles of Non-coding RNAs in Human Disease and Plant Health

**Learning Outcomes:**

1. List examples in which ncRNAs are associated with human diseases.
2. List examples in which ncRNAs play a role in plant health.

During the past two decades, researchers have discovered that abnormalities in ncRNAs are associated with a wide range of human diseases. As mentioned at the beginning of this chapter, cartilage-hair hypoplasia (CHH) was the first human disease that was shown to be caused by an ncRNA transcribed from a nuclear gene. Since the identification of CHH-associated mutations in 2001, many ncRNAs have been shown to play a key role in human diseases. Researchers speculate that we are still seeing only the "tip of the iceberg" with regard to identifying the roles of ncRNAs in human pathology. Likewise, the impact of ncRNAs on plant health is only beginning to be appreciated but has exciting potential in the field of agriculture. In this section, we will focus on the roles of ncRNAs in the development of cancer, neurological disorders, and cardiovascular diseases, as well as their effects on plant health.

| Table 13.2 | Examples of ncRNAs Associated with Human Diseases |
|---|---|
| **Type of ncRNA** | **Disease(s)*** |
| A group of miRNAs called the miR-200 family | Several types of cancer, including bladder cancer, melanoma, stomach cancer, and colorectal cancer |
| HOTAIR | Several types of cancer, including breast cancer, lung cancer, and colorectal cancer |
| Many miRNAs | Alzheimer disease |
| Many miRNAs | Multiple sclerosis |
| An miRNA called miR-1 | Heart arrhythmias |
| Several different miRNAs, including miR10a, miR145, and miR143 | Formation of arterial plaques |

*The diseases listed in this table show an association with abnormal levels of ncRNAs. In many cases, it is not yet clear if the disease symptoms are caused, in part, by the abnormal levels of ncRNAs or if the abnormal levels are a consequence of the disease symptoms.

## ncRNAs Play a Role in Many Forms of Cancer and Other Human Diseases

As we have seen throughout this chapter, ncRNAs play important roles in chromatin modification, gene transcription, mRNA translation, and protein function. When certain ncRNAs are expressed abnormally, that is, at too high or too low a level, disease conditions are known to occur. Such abnormal expression levels can be caused by mutations in specific genes or by epigenetic changes, described in Chapter 18, that alter the expression of genes that encode ncRNAs. Several examples of human diseases associated with the abnormal expression of ncRNAs are listed in **Table 13.2**.

**ncRNAs and Cancer** The topic of cancer is described in Chapter 15 (look ahead to Section 15.4). The roles of ncRNAs in cancer have been most thoroughly studied with respect to miRNAs. In nearly all forms of human cancer, levels of expression of particular miRNAs differ between normal and cancer cells. In some cases, the genes that encode the miRNAs behave as tumor-suppressor genes, because a lower level of expression of particular miRNAs allows tumor growth. In other cases, the genes that encode certain miRNAs act as oncogenes; their overexpression promotes cancer.

A well-studied example of the role of miRNAs in cancer involves a group of several different miRNAs called the miR-200 family. The miR-200 family plays an essential role in tumor suppression by inhibiting metastasis—the process by which cancer cells can spread through the bloodstream to other parts of the body. Low levels of expression of miR-200 members have been associated with many types of cancer, including bladder cancer, melanoma, stomach cancer, and colorectal cancer.

Though they have been less well studied, other ncRNAs are also associated with particular types of human cancers. HOTAIR, which was discussed in Section 13.2, is an ncRNA that is highly expressed in a variety of cancers, including breast cancer, lung cancer, and colorectal cancer. When overexpressed, the gene that encodes HOTAIR behaves as an oncogene. High levels of HOTAIR expression in primary breast tumors are a significant predictor of metastasis and death. HOTAIR is known to interact with a variety of cellular components, but the mechanism by which it promotes cancer is not well understood.

**ncRNAs and Neurological Disorders** Many miRNAs are essential for the proper development and functioning of the nervous system. Approximately 70% of all miRNAs are expressed in the brain, and many of them are specific to neurons. miRNAs are involved in neuron growth and the overall development of the nervous system. Abnormal levels of expression of miRNAs have been associated with nearly all neurological disorders in which these ncRNAs have been investigated! Table 13.2 describes some examples in which the expression of miRNAs has been altered and associated with neurological disorders. For example, in Alzheimer disease, abnormally expressed miRNAs are thought to be involved in down-regulating the expression of the enzyme β-secretase, which leads to the overproduction of certain β-amyloid peptides—a key feature of the disease. miRNAs are also known to control the inflammatory process that leads to the development of multiple sclerosis.

**ncRNAs and Cardiovascular Diseases** Abnormalities in miRNA levels have been linked to several cardiovascular diseases. A particular miRNA, called miR-1, is associated with the development of arrhythmias—irregularities in the rate or rhythm of the heartbeat. This miRNA regulates the expression of genes that encode ion channel proteins, which are important for proper signaling between cardiac muscle cells. Other miRNAs appear to play a role in vascular disease. The formation of arterial plaques is associated with abnormal expression levels of several miRNAs, including miR10a, miR145, and miR143.

## ncRNAs Are Essential to Plant Health

In parallel to the growing knowledge of the role of ncRNAs in human diseases, plant biologists are discovering that abnormalities in ncRNAs play many essential roles that contribute to the health of plants. This realization is likely to have great impact in the field

| Table 13.3 | Importance of ncRNAs in Plant Health |
|---|---|
| **Type of ncRNA\*** | **Normal role in plant structure and function** |
| Several miRNAs, including miR156, miR157, and miR159 | Control the time of year when flowering occurs |
| An ncRNA called COOLAIR | Promotes vernalization, the process in which certain plants will only flower after being exposed to cold winter temperatures |
| Two miRNAs called miR167 and miR397 | Play a role in seed development |
| An miRNA called miR402 | Affects the rate of seed germination and seedling growth under stress conditions |
| An miRNA called miR824 | Plays a role in the development of stomata |
| An ncRNA called IPS1 | Affects the ability of plants to cope with phosphate starvation |

*Most of the examples listed are miRNAs, which are short ncRNAs. COOLAIR and IPS1 are longer ncRNAs.

of agriculture as we develop methods to change the expression of ncRNAs in order to modify the characteristics of agriculturally important plants. **Table 13.3** describes several ncRNAs that are known to play key roles in plant health.

## Summary of Key Concepts

### 13.1 Overview of Non-coding RNAs

- Non-coding RNAs (ncRNAs) are RNA molecules that do not encode polypeptides.
- ncRNAs bind to different types of molecules, including DNA, other RNAs, proteins, and small molecules (Figure 13.1).
- An ncRNA can provide a scaffold, act as a guide, alter protein function or stability, function as a ribozyme, function as a blocker, and/or act as a decoy (Figures 13.2, 13.3).
- ncRNAs play a role in DNA replication, chromatin structure, transcription, translation, RNA degradation, RNA modification, protein sorting and secretion, and genome defense (Table 13.1).

### 13.2 Effects of Non-coding RNAs on Chromatin Structure and Transcription

- HOTAIR is an ncRNA found in humans and other mammals that regulates transcription by forming a scaffold that binds to two protein complexes and guides them to particular genes. The protein complexes covalently modify histones, and these modifications inhibit transcription of the target genes (Figure 13.4).

### 13.3 Effects of Non-coding RNAs on Translation and mRNA Degradation

- Fire and Mello showed that double-stranded RNA is more potent at silencing mRNA than is antisense RNA (Figure 13.5).
- RNA interference is a mechanism of mRNA silencing in which miRNA or siRNA becomes part of an RNA-induced silencing complex (RISC) that inhibits the translation of a specific mRNA or causes its degradation, respectively (Figure 13.6).

### 13.4 Non-coding RNAs and Protein Sorting

- Signal recognition particle (SRP), which is composed of one or more proteins and an ncRNA, plays a role in directing proteins to the plasma membrane of prokaryotic cells and to the ER membrane of eukaryotic cells (Figure 13.7).

### 13.5 Non-coding RNAs and Genome Defense

- The CRISPR-Cas system in bacteria and archaea provides defense against bacteriophages and transposons. The defense occurs in three phases: adaptation, expression, and interference (Figure 13.8).

### 13.6 Role of Non-coding RNAs in Human Disease and Plant Health

- Abnormalities in the expression of ncRNAs have been associated with many human diseases, including cancer, neurological disorders, and cardiovascular diseases (Table 13.2).
- The proper level of expression of ncRNAs is also important for plant health (Table 13.3).

## Assess & Discuss

### Test Yourself

1. Which of the following types of molecules could bind to an ncRNA through base pairing?
   a. DNA
   b. RNA
   c. protein
   d. small molecule
   e. both a and b

2. Which of the following is *not* a general function of an ncRNA?
   a. encoding a polypeptide
   b. acting as a ribozyme
   c. acting as a guide
   d. acting as a scaffold
   e. acting as a decoy

3. ncRNAs play an important role in
   a. DNA replication.
   b. chromatin structure and transcription.
   c. translation and RNA degradation.

  d.  genome defense.
  e.  all of the above.

4.  HOTAIR causes certain genes to be repressed by facilitating
  a.  the binding of a repressor protein.
  b.  the release of an activator protein.
  c.  the covalent modification of histones.
  d.  the removal of nucleosomes.
  e.  both a and c.

5.  One of the roles of the RNA component of signal recognition particle
    (SRP) is to stimulate certain proteins to hydrolyze GTP. If this function
    of SRP RNA did not work properly, what would you expect to happen?
  a.  SRP would not bind to the ER signal sequence of a polypeptide.
  b.  SRP would not cause translation to pause.
  c.  SRP would not bind to an SRP receptor in the ER membrane.
  d.  SRP would not be released from the ER membrane.
  e.  both a and b

6.  During RNA interference, what binds to an mRNA to inhibit translation?
  a.  a pri-miRNA
  b.  a pre-miRNA or pre-siRNA
  c.  a double-stranded miRNA or double-stranded siRNA
  d.  a single-stranded miRNA or single-stranded siRNA
  e.  dicer

7.  With regard to miRNAs and siRNAs, which of the following statements
    is (are) correct?
  a.  miRNAs are transcribed from endogenous genes.
  b.  miRNAs are usually a perfect match to an mRNA.
  c.  siRNAs are transcribed from endogenous genes.
  d.  siRNAs cause mRNA degradation.
  e.  both a and d

8.  Cas1 and Cas2 proteins play a role during which of the following phases
    of genome defense?
  a.  adaptation
  b.  expression
  c.  interference
  d.  both adaptation and expression
  e.  both expression and interference

9.  Which of the following components bind to tracrRNA?
  a.  crRNA and Cas1 protein
  b.  crRNA and Cas2 protein
  c.  crRNA and Cas9 protein
  d.  crRNA only
  e.  Cas1 and Cas2 proteins

10.  Abnormalities in the expression of ncRNAs are associated with
  a.  many forms of cancer.
  b.  neurological disorders.
  c.  cardiovascular diseases.
  d.  all of the above.
  e.  only a and b.

## Conceptual Questions

1.  An ncRNA may have one or more of the following functions: scaffold,
    guide, alterer of protein function or stability, ribozyme, blocker,
    and decoy. Which of those functions are exhibited by the following
    examples: HOTAIR, SRP RNA, miRNA, and crRNA? Note: A single
    ncRNA may have more than one function.

2.  What is RNA interference (RNAi)? Explain how the double-stranded
    RNA is processed during RNAi and how it leads to the silencing of a
    complementary mRNA.

3.  **Core Concept: Structure and Function**  Explain how the
    structure of HOTAIR allows it to carry out its function.

## Collaborative Questions

1.  Review the concept of an RNA world described in Section 4.1. Discuss
    which ncRNAs described in Table 13.1 may have arisen during the RNA
    world, and which probably arose after the modern DNA/RNA/protein
    world came into being.

2.  Go to the PubMed website and search for "non-coding RNA and
    disease." Scan through the journal articles you retrieve, and make a list
    of the roles that ncRNAs may play in human diseases.

# Gene Expression at the Molecular Level III: Gene Regulation

# 14

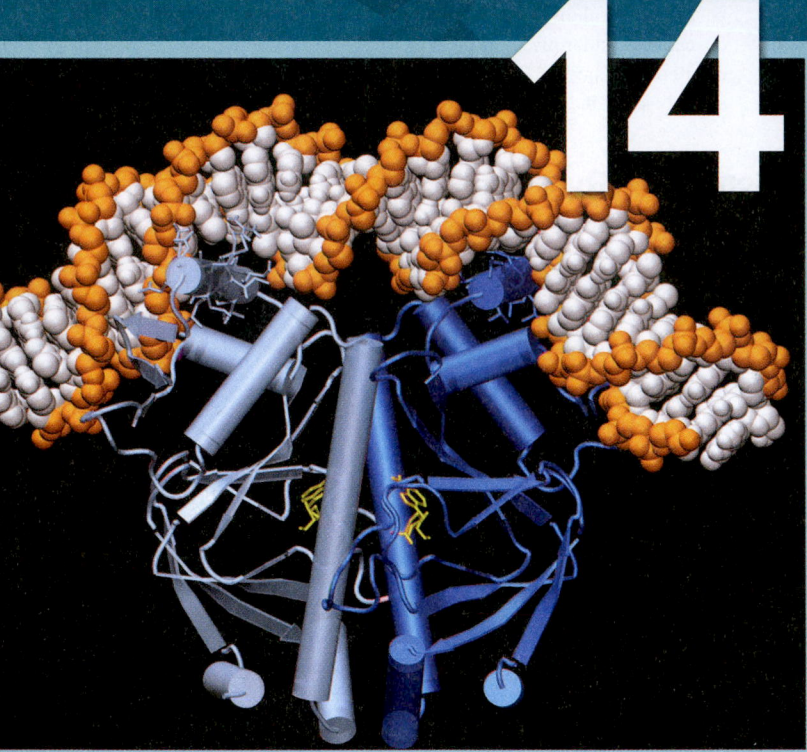

**A model for a protein that binds to DNA and regulates genes.** The catabolite activator protein (CAP), shown in dark and light blue, is binding to the DNA double helix, shown in orange and white. CAP, shown again in Figure 14.10, activates gene transcription. ©Daniel Gage, University of Connecticut

produced at appropriate times and in specific amounts. The term **gene regulation** refers to the ability of cells to control the expression of their genes. By comparison, some genes have relatively constant levels of expression in all conditions over time. These are called **constitutive genes**. In most cases, constitutive genes encode proteins that are constantly required for the survival of an organism, such as certain metabolic enzymes.

The importance of gene regulation is underscored by the number of genes devoted to this process in an organism. For example, in *Arabidopsis thaliana*, a plant that is studied by many plant geneticists, over 5% of the genome is involved with regulating gene transcription. This species has more than 1,500 different genes that encode proteins that regulate the transcription of other genes.

In this chapter, we will begin with an overview that emphasizes the benefits of gene regulation and the general mechanisms that achieve such regulation in bacteria and in eukaryotes. The following sections will describe how bacteria regulate gene expression in the face of environmental change and the more complex nature of gene regulation in eukaryotes.

**E**milio took a weight-lifting class in college and was surprised by the results. Within a few weeks, he was able to lift substantially more weight. He was inspired by his progress and continued lifting weights after the semester-long course ended. A year later, he was not only much stronger, but he could see physical changes in his body. Certain muscles, such as the biceps and triceps in his upper arms, were noticeably larger. How can we explain the increase in mass of Emilio's muscles? Unknowingly, when he was lifting weights, Emilio was affecting the regulation of his genes. Certain genes in his muscle cells were being "turned on" during his workouts, which then led to the synthesis of proteins that increased the mass of Emilio's muscles.

At the molecular level, **gene expression** is the process by which the information within a gene is made into a functional product, such as an RNA molecule or a protein. Most genes in all species are regulated so that the proteins they specify are

## 14.1   Overview of Gene Regulation

### Learning Outcomes:

1. Discuss the various ways that organisms benefit from gene regulation.
2. Identify where gene regulation can occur during the process of gene expression in bacteria and eukaryotes.

How do living organisms benefit from gene regulation? One benefit is that gene regulation conserves energy. Proteins that are encoded by genes are produced only when needed. In multicellular organisms, gene regulation also ensures that genes are expressed in the appropriate cell types and at the correct stage of development. In this section, we will examine a few examples that illustrate the important consequences of gene regulation. We will also survey the major points in the gene expression process at which genes are regulated in bacterial and eukaryotic cells.

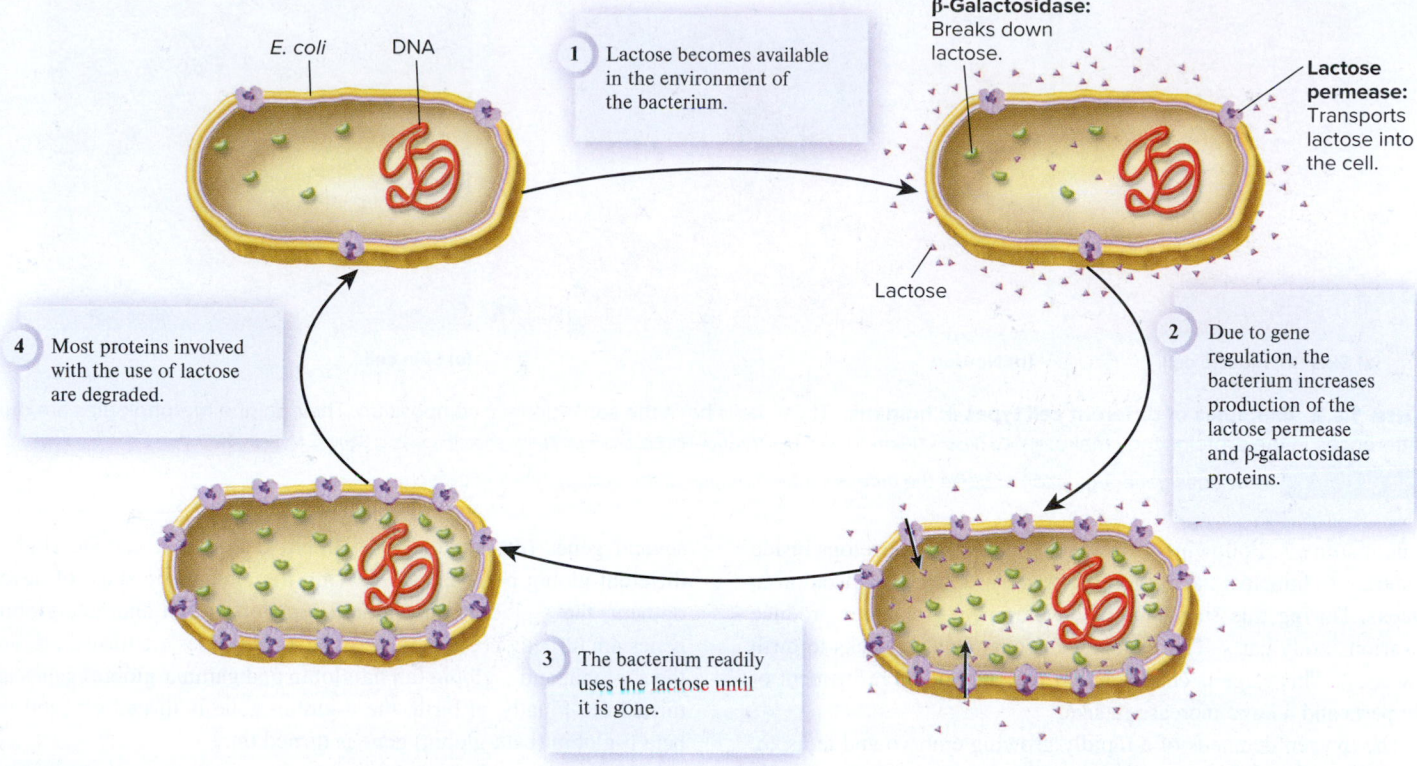

**β-Galactosidase:** Breaks down lactose.

**1** Lactose becomes available in the environment of the bacterium.

**Lactose permease:** Transports lactose into the cell.

*E. coli*    DNA

Lactose

**4** Most proteins involved with the use of lactose are degraded.

**2** Due to gene regulation, the bacterium increases production of the lactose permease and β-galactosidase proteins.

**3** The bacterium readily uses the lactose until it is gone.

**Figure 14.1** **Gene regulation of lactose utilization in *E. coli*.**

 **Core Concept: Energy and Matter** Gene regulation provides a way for organisms to avoid wasting energy. Their cells make proteins only when the proteins are needed.

## Bacteria Regulate Genes in Response to Changes in Their Environment

The bacterium *Escherichia coli* can use many types of sugars as food sources, thereby increasing its chances of survival. With regard to gene regulation, we will focus on how it uses lactose, which is a sugar found in milk. The genome of *E. coli* carries genes that code for proteins that enable the bacterium to take up lactose from the environment and metabolize it.

**Figure 14.1** illustrates the effects of lactose on the regulation of those genes. In order to utilize lactose, an *E. coli* cell requires a transporter, called lactose permease, that facilitates the uptake of lactose into the cell, and an enzyme, called β-galactosidase, that catalyzes the breakdown of lactose. When lactose is not present in the environment, an *E. coli* cell makes very little of these proteins. However, when lactose becomes available, the bacterium produces many more of these proteins, enabling it to readily use lactose from its environment. Eventually, all of the lactose in the environment will be used up. At this point, the genes encoding these proteins will be shut off, and most of the proteins will be degraded. Overall, gene regulation conserves energy because it ensures that the proteins needed for lactose utilization are made only when lactose is present in the environment.

## Eukaryotic Gene Regulation Produces Different Cell Types in a Single Organism

One of the most amazing examples of gene regulation is the phenomenon of **cell differentiation**, the process by which cells become specialized into particular types. In humans, for example, cells may differentiate into muscle cells, neurons, skin cells, or other types. **Figure 14.2** shows micrographs of three types of cells found in humans. As the images show, their morphologies are strikingly different. Likewise, their functions within the body are also quite different. Muscle cells are important in body movements, neurons function in cell signaling, and skin cells form a protective outer surface to the body.

Gene regulation is responsible for producing different types of cells within a multicellular organism. The three cell types shown in Figure 14.2 contain the same **genome**, meaning they carry the same set of genes. However, their **proteomes**—the sets of proteins they make—are quite different. Certain proteins are found in particular cell types but not in others. Alternatively, a protein may be present in all three cell types, but the relative amounts of the protein may be different. The amount of a given protein depends on many factors, including how strongly the corresponding gene is turned on and how much protein is synthesized from mRNA. Gene regulation plays a major role in determining the proteome of each cell type.

## Eukaryotic Gene Regulation Enables Multicellular Organisms to Proceed Through Developmental Stages

In multicellular organisms that progress through developmental stages, certain genes are expressed at particular stages of development but not others. Let's consider an example of such gene regulation in mammals. Early stages of development occur in the uterus of

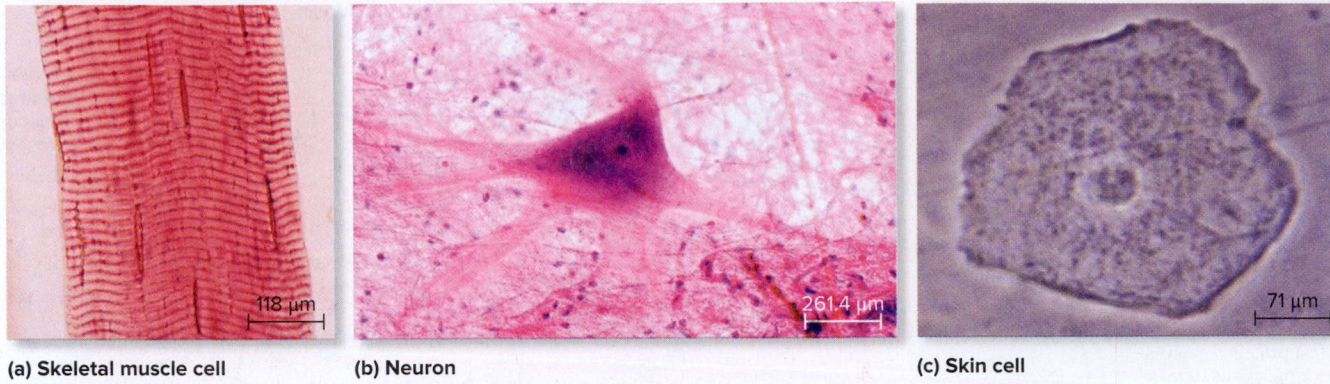

| (a) Skeletal muscle cell | (b) Neuron | (c) Skin cell |

**Figure 14.2** **Examples of different cell types in humans.** These cells have the same genetic composition. Their unique morphologies are due to differences in the proteins they make. a: ©Ed Reschke/Getty Images; b: ©McGraw-Hill Education/Al Telser, photographer; c: Source: M. Rein/CDC

*Concept Check:* *How does gene regulation underlie the different morphologies of the various types of cells?*

female mammals. Following fertilization, an embryo develops inside the uterus. In humans, the embryonic stage lasts from fertilization to 8 weeks. During this stage, major developmental changes produce the various body parts. The fetal stage occurs from 8 weeks to birth (41 weeks). This stage is characterized by a continued refinement of body parts and a large increase in size.

The oxygen demands of a rapidly growing embryo and fetus are quite different from the needs of the mother. Gene regulation plays a vital role in ensuring that an embryo and fetus get the proper amount of oxygen. Hemoglobin is a protein that delivers oxygen to the cells of a mammal's body. A hemoglobin protein is composed of four globin polypeptides, two encoded by one globin gene and two encoded by another globin gene (**Figure 14.3**). The genomes of mammals carry several genes (designated with Greek letters) that encode slightly different globin polypeptides. During the embryonic stage of development, the ε-globin and ζ-globin (epsilon-globin and zeta-globin) genes are turned on. At the fetal stage, these genes are turned off, and the α-globin and γ-globin (alpha-globin and gamma-globin) genes are turned on. Finally, at birth, the γ-globin gene is turned off, and the beta β-globin (beta-globin) gene is turned on.

How do the embryo and fetus acquire oxygen from their mother's bloodstream? The hemoglobin produced during the embryonic and fetal stages has a higher binding affinity for oxygen than does the hemoglobin produced after birth. Therefore, the embryo and fetus can remove oxygen from the mother's bloodstream and use that oxygen for their own needs. This occurs across the placenta, where the mother's bloodstream is adjacent to the bloodstream of the embryo or fetus. In this way, gene regulation enables mammals to develop inside the mother's body, even though the embryo and fetus are not breathing on their own. Gene regulation ensures that the correct hemoglobin protein is produced at the right time in development. We'll discuss how gene expression controls the process of development in greater detail in Chapter 20.

## Gene Regulation Occurs at Different Points in the Process from DNA to Protein

Thus far, we have learned that gene regulation has a dramatic influence on the ability of organisms to respond to environmental changes, produce different types of cells, and progress through developmental stages. For protein-encoding genes, the regulation of gene expression can occur at any of the steps in the process that produces a functional protein.

In bacteria, gene regulation most commonly occurs at the level of transcription, which means that bacteria regulate how much mRNA is made from genes (**Figure 14.4a**). When geneticists say a gene is "turned off," they mean that very little or no mRNA is made from that gene, whereas a gene that is "turned on" is transcribed into mRNA. Because transcription is the first step in gene expression, transcriptional regulation is a particularly efficient way to regulate genes because cells avoid wasting energy when the product of the gene is not needed. A second way for bacteria to regulate gene expression is to control the ability of an mRNA to be translated into a protein. This form of gene regulation is less common in bacteria. Last, gene

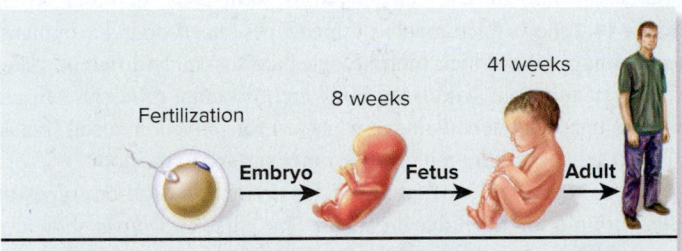

|  | Embryo | Fetus | Adult |
|---|---|---|---|
| **Hemoglobin protein** | 2 ζ-globins 2 ε-globins | 2 α-globins 2 γ-globins | 2 α-globins 2 β-globins |
| **Oxygen affinity** | Highest | High | Moderate |
| **Gene expression** | | | |
| α-globin gene | Off | On | On |
| β-globin gene | Off | Off | On |
| γ-globin gene | Off | On | Off |
| ζ-globin gene | On | Off | Off |
| ε-globin gene | On | Off | Off |

**Figure 14.3** **Regulation of human globin genes at different stages of development.**

 **Core Concept: Information** Gene regulation is an important process that allows organisms to properly access the information within their genomes and proceed through developmental stages.

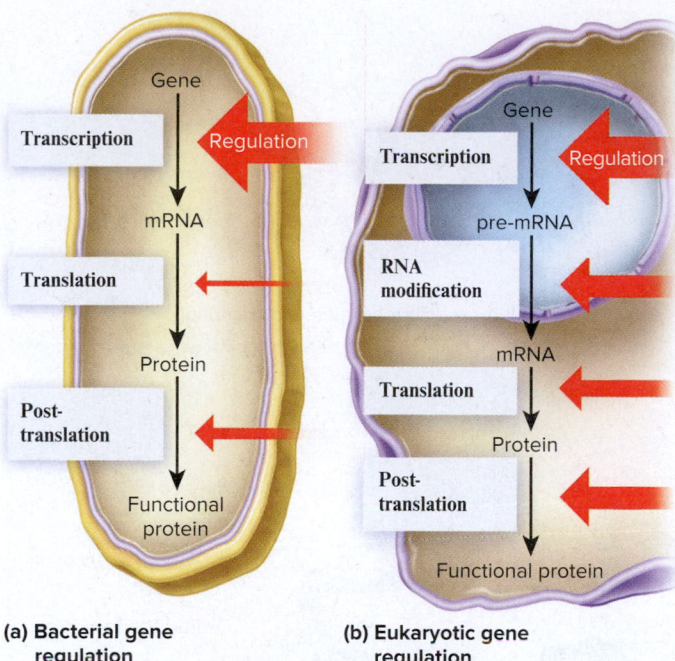

Gene

Transcription

Regulation

mRNA

Translation

Protein

Post-translation

Functional protein

**(a) Bacterial gene regulation**

Gene

Transcription

Regulation

pre-mRNA

RNA modification

mRNA

Translation

Protein

Post-translation

Functional protein

**(b) Eukaryotic gene regulation**

**Figure 14.4  Overview of gene regulation in (a) bacteria and (b) eukaryotes.** The relative widths of the red arrows indicate the prominence with which gene regulation is used to control the production of functional proteins.

expression can be regulated at the post-translational level in which a protein's function is controlled after it is synthesized via translation.

In eukaryotes, gene regulation occurs at many levels, including transcription, RNA modification, translation, and after translation of a protein is completed (**Figure 14.4b**). As it is for their bacterial counterparts, transcriptional regulation is a prominent form of gene regulation for eukaryotes. As discussed later in this chapter, eukaryotic genes are transcriptionally regulated in several different ways, some of which do not occur in bacteria. Regulation of RNA modification and of the rate of translation of mRNAs is also common.

Like bacterial proteins, eukaryotic proteins can be regulated in a variety of ways, including cellular regulation and biochemical regulation (such as feedback inhibition). These various types of regulation are best understood within the context of cell biology, so they were primarily discussed in Unit II (look back at Chapter 6, especially Figure 6.13).

## 14.2  Regulation of Transcription in Bacteria

### Learning Outcomes:

1. Explain how regulatory transcription factors and small effector molecules are involved in the regulation of transcription.
2. Describe the organization of the *lac* operon and how it is under negative and positive control.
3. **CoreSKILL »** Analyze the results of the experiments of Jacob, Monod, and Pardee.
4. Describe how the *trp* operon is under negative control.

As we have seen, when a bacterium is exposed to a particular nutrient in its environment, such as a sugar, the genes are expressed that encode proteins needed for the uptake and metabolism of that sugar. In addition, bacteria have genes that encode enzymes that synthesize molecules such as particular amino acids. For these genes, the control of expression often occurs at the level of transcription. In this section, we will examine the underlying molecular mechanisms that bring about transcriptional regulation in bacteria.

## Transcriptional Regulation Involves Regulatory Transcription Factors and Small Effector Molecules

In most cases, regulation of transcription involves the actions of **regulatory transcription factors**—proteins that bind to **regulatory sequences**, usually in the DNA in the vicinity of a promoter, and affect the rate of transcription of one or more nearby genes. These transcription factors either decrease or increase the rate of transcription of a gene. **Repressors** are regulatory transcription factors that bind to the DNA and decrease the rate of transcription. This is a form of regulation called **negative control**. **Activators** bind to the DNA and increase the rate of transcription, a form of regulation termed **positive control** (**Figure 14.5a**).

In conjunction with regulatory transcription factors, molecules called **small effector molecules** often play a critical role in transcriptional regulation. A small effector molecule exerts its effects by binding to a regulatory transcription factor and causing a conformational change in the protein. In many cases, the effect of the conformational change determines whether or not the protein can bind to the DNA. **Figure 14.5b** illustrates an example involving a repressor. When the small effector molecule is not present in the cytoplasm, the repressor binds to the DNA and inhibits transcription. However, when the small effector molecule is present in the cytoplasm, it will bind to the repressor and cause a conformational change that inhibits the ability of the protein to bind to the DNA. Transcription can occur because the repressor is not able to bind to the DNA. Repressors and activators that respond to small effector molecules have two functional regions called domains. One domain is a site where the protein binds to the DNA, whereas the other is the binding site for the small effector molecule.

## The *lac* Operon Contains Genes That Encode Proteins Involved in Lactose Metabolism

In bacteria, a set of two or more genes may be under the transcriptional control of a single promoter. This arrangement is known as an **operon**. The group of genes are transcribed as a single unit, resulting in the production of a **polycistronic mRNA**, an mRNA that encodes more than one protein. What advantage does this arrangement provide? An operon allows a bacterium to coordinately regulate a group of genes that encode proteins whose functions are used in a common pathway.

The genome of *E. coli* carries an operon, called the **lac operon**, that contains the genes for the proteins that allow the bacterium to

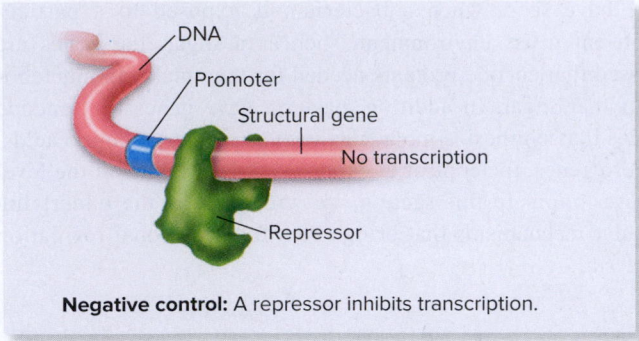

**Negative control:** A repressor inhibits transcription.

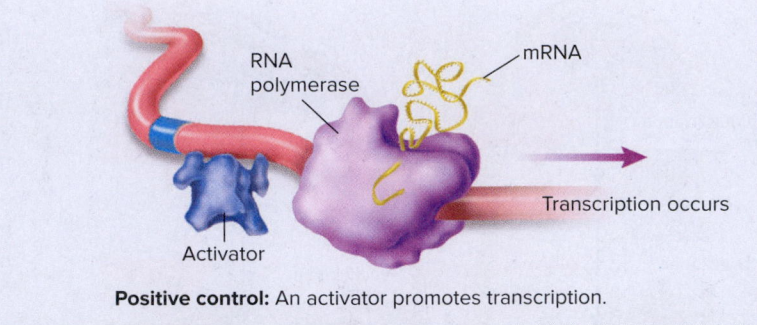

**Positive control:** An activator promotes transcription.

**(a) Actions of regulatory transcription factors**

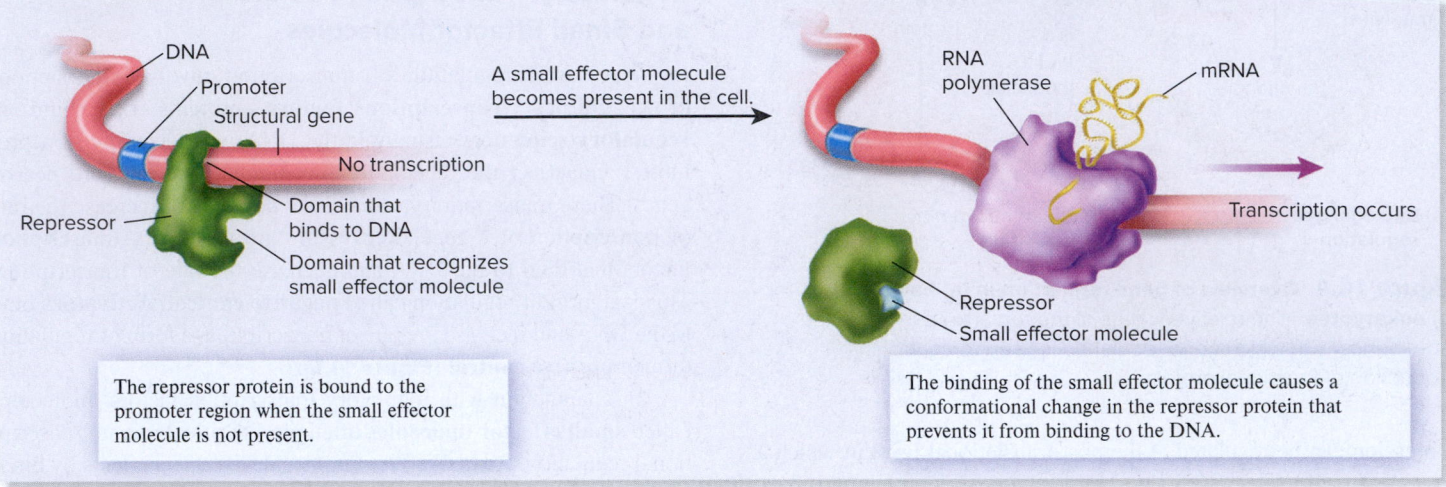

A small effector molecule becomes present in the cell.

The repressor protein is bound to the promoter region when the small effector molecule is not present.

The binding of the small effector molecule causes a conformational change in the repressor protein that prevents it from binding to the DNA.

**(b) Action of a small effector molecule on a repressor**

**Figure 14.5   Actions of regulatory transcription factors and small effector molecules.  (a)** Regulatory transcription factors are proteins that exert negative or positive control. **(b)** One way that a small effector molecule may exert its effects is by preventing a repressor protein from binding to the DNA.

metabolize lactose (refer back to Figure 14.1). **Figure 14.6a** shows the organization of this operon as it is found in the *E. coli* chromosome, as well as the polycistronic mRNA that is transcribed from it. The *lac* operon contains a promoter, *lacP*, that is involved in the transcription of three protein-encoding genes: *lacZ*, *lacY*, and *lacA*.

- The *lacZ* gene encodes β-galactosidase, which is an enzyme that breaks down lactose (**Figure 14.6b**). As a side reaction, β-galactosidase also converts a small percentage of lactose into allolactose, a structurally similar sugar, or lactose analogue. As described later, allolactose is important in the regulation of the *lac* operon.

- The *lacY* gene encodes lactose permease, which is a membrane protein required for the transport of lactose into the cytoplasm of the bacterium.

- The *lacA* gene encodes galactoside transacetylase, which covalently modifies lactose and lactose analogues by attaching an acetyl group (—COCH₃). The attachment of acetyl groups to nonmetabolizable lactose analogues prevents their toxic buildup in the cytoplasm.

Near the *lac* promoter are two regulatory sequences designated the operator and the CAP site (see Figure 14.6a). The **operator**

(*lacO*) is a regulatory sequence in the DNA. The sequence of bases at the operator provides a binding site for a repressor protein. The **CAP site** is a regulatory sequence recognized by an activator protein.

Adjacent to the *lac* operon is the *lacI* gene, which encodes the **lac repressor**. This repressor protein is important for the regulation of the *lac* operon. The *lacI* gene, which is constitutively expressed at a fairly low level, has its own promoter called the *i* promoter. The *lacI* gene is not considered a part of the *lac* operon. Let's now take a look at how the *lac* operon is regulated by the lac repressor.

## The *lac* Operon Is Under Negative Control by a Repressor Protein

In the late 1950s, the first researchers to investigate gene regulation were French biologists François Jacob and Jacques Monod at the Pasteur Institute in Paris, France. Their focus on gene regulation stemmed from an interest in the phenomenon known as enzyme adaptation, which had been identified early in the 20th century. Enzyme adaptation occurs when a particular enzyme appears within a living cell only after the cell has been exposed to the substrate for that enzyme. Jacob and Monod studied lactose metabolism in *E. coli* to investigate this phenomenon. When they exposed bacteria to lactose, the levels of lactose-utilizing enzymes in the cells increased by

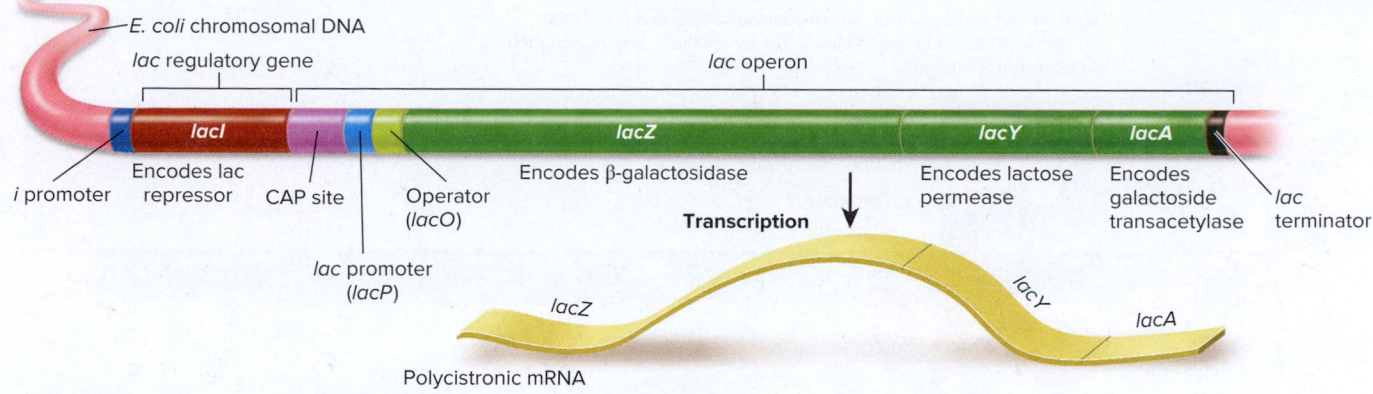

**(a) Organization of DNA sequences in the *lac* region of the *E. coli* chromosome**

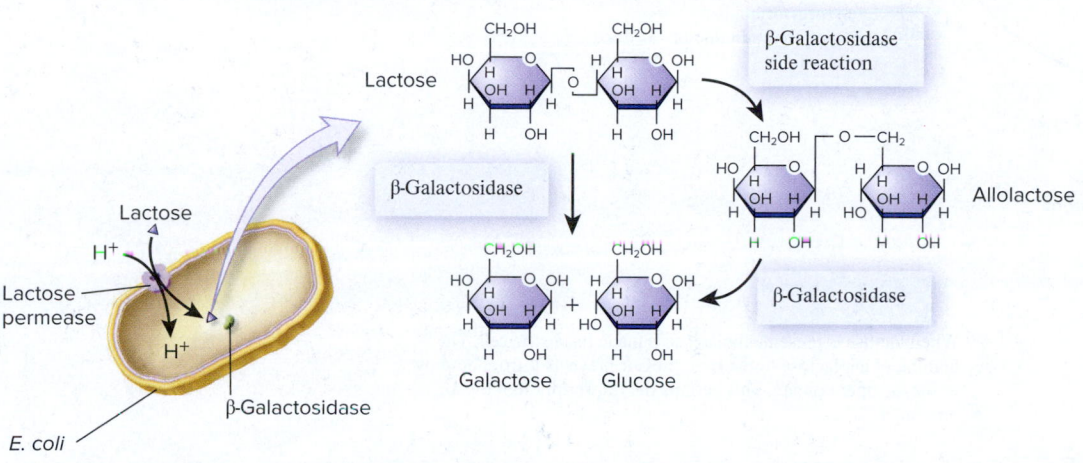

**(b) Functions of lactose permease and β-galactosidase**

**Figure 14.6    The *lac* operon.    (a)** This diagram depicts a region of the *E. coli* chromosome that contains the *lacI* gene and the adjacent *lac* operon, as well as the polycistronic mRNA transcribed from the operon. The mRNA is translated into three proteins: β-galactosidase, lactose permease, and galactoside transacetylase. **(b)** Lactose permease cotransports H⁺ with lactose. Bacteria maintain an H⁺ gradient across their cytoplasmic membrane that drives the active transport of lactose into the cytoplasm. β-Galactosidase cleaves lactose into galactose and glucose. As a side reaction, it can also convert lactose into allolactose.

**Concept Check:**    *Which genes are under the control of the lac promoter?*

1,000- to 10,000-fold. After lactose was removed, the synthesis of the enzymes abruptly stopped.

The first mechanism of regulation that Jacob and Monod discovered involved the lac repressor, which binds to the sequence of bases found at the *lac* operator site. Once bound, the lac repressor prevents RNA polymerase from transcribing the *lacZ*, *lacY*, and *lacA* genes (**Figure 14.7a**). RNA polymerase can bind to the promoter when the lac repressor is bound to the operator site, but cannot move past the operator to transcribe the *lacZ*, *lacY*, and *lacA* genes.

Whether or not the lac repressor binds to the operator site depends on allolactose, the previously mentioned side product of the β-galactosidase enzyme (see Figure 14.6b). How does allolactose control the lac repressor? Allolactose is an example of a small effector molecule. The lac repressor protein contains four identical subunits, each of which recognizes a single allolactose molecule. When four allolactose molecules bind to the lac repressor, a conformational change occurs that prevents the repressor from binding to the operator. Under these conditions, RNA polymerase is free to transcribe the operon (**Figure 14.7b**).

The regulation of the *lac* operon enables *E. coli* to conserve energy because lactose-utilizing proteins are made only when lactose is present in the environment. Allolactose is an **inducer**, a small effector molecule that increases the rate of transcription, and the *lac* operon is said to be an **inducible operon**. When the bacterium is not exposed to lactose, no allolactose is available to bind to the lac repressor. Therefore, the lac repressor binds to the operator site and inhibits transcription. In reality, the repressor does not completely inhibit transcription, so very small amounts of β-galactosidase, lactose permease, and galactoside transacetylase are made. However, the levels are far too low for the bacterium to readily use lactose. When the bacterium is exposed to lactose, a small amount can be transported into the cytoplasm via lactose permease, and β-galactosidase converts some of it to allolactose (see Figure 14.6b). The cytoplasmic level of allolactose gradually rises until allolactose binds to the lac repressor, which induces the *lac* operon and promotes a high rate of transcription of the *lacZ*, *lacY*, and *lacA* genes. Translation of the encoded polypeptides produces the proteins needed for lactose uptake and metabolism, as described previously in Figure 14.1.

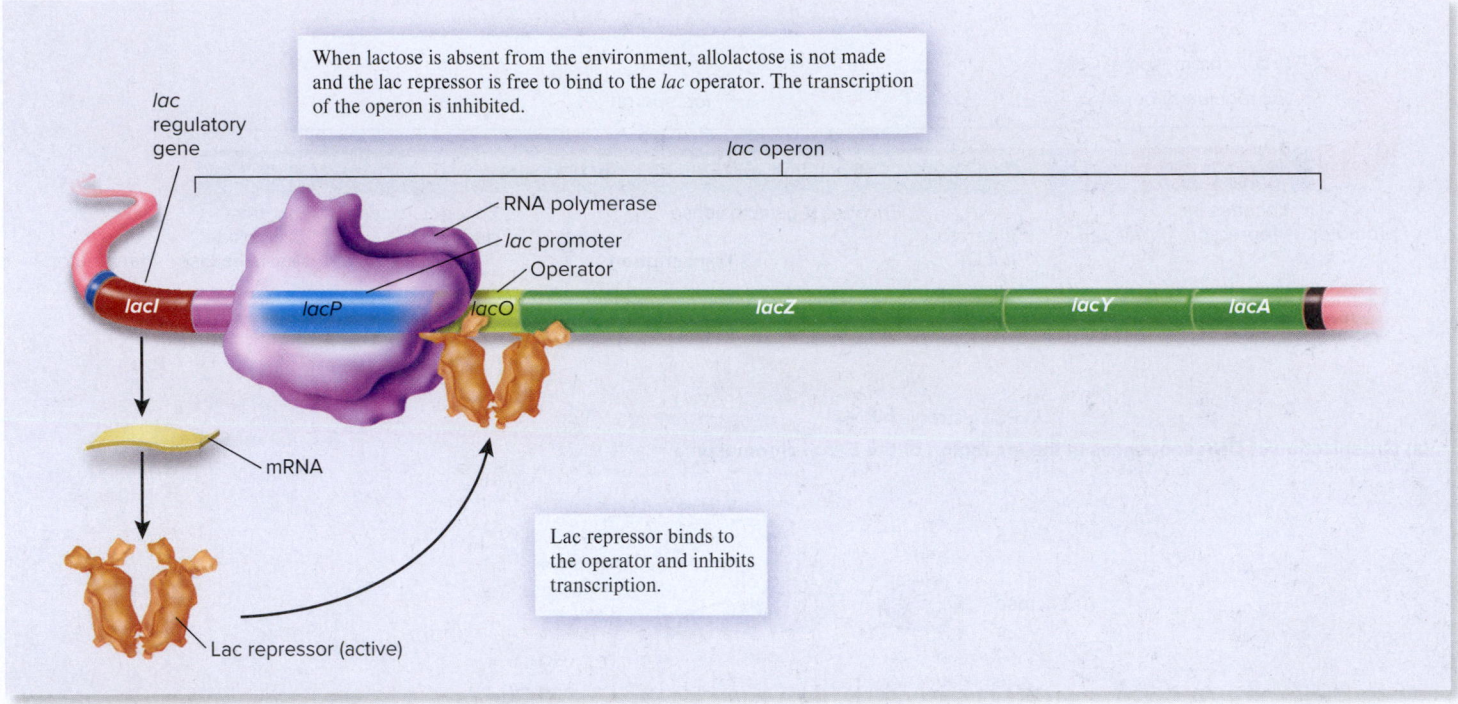

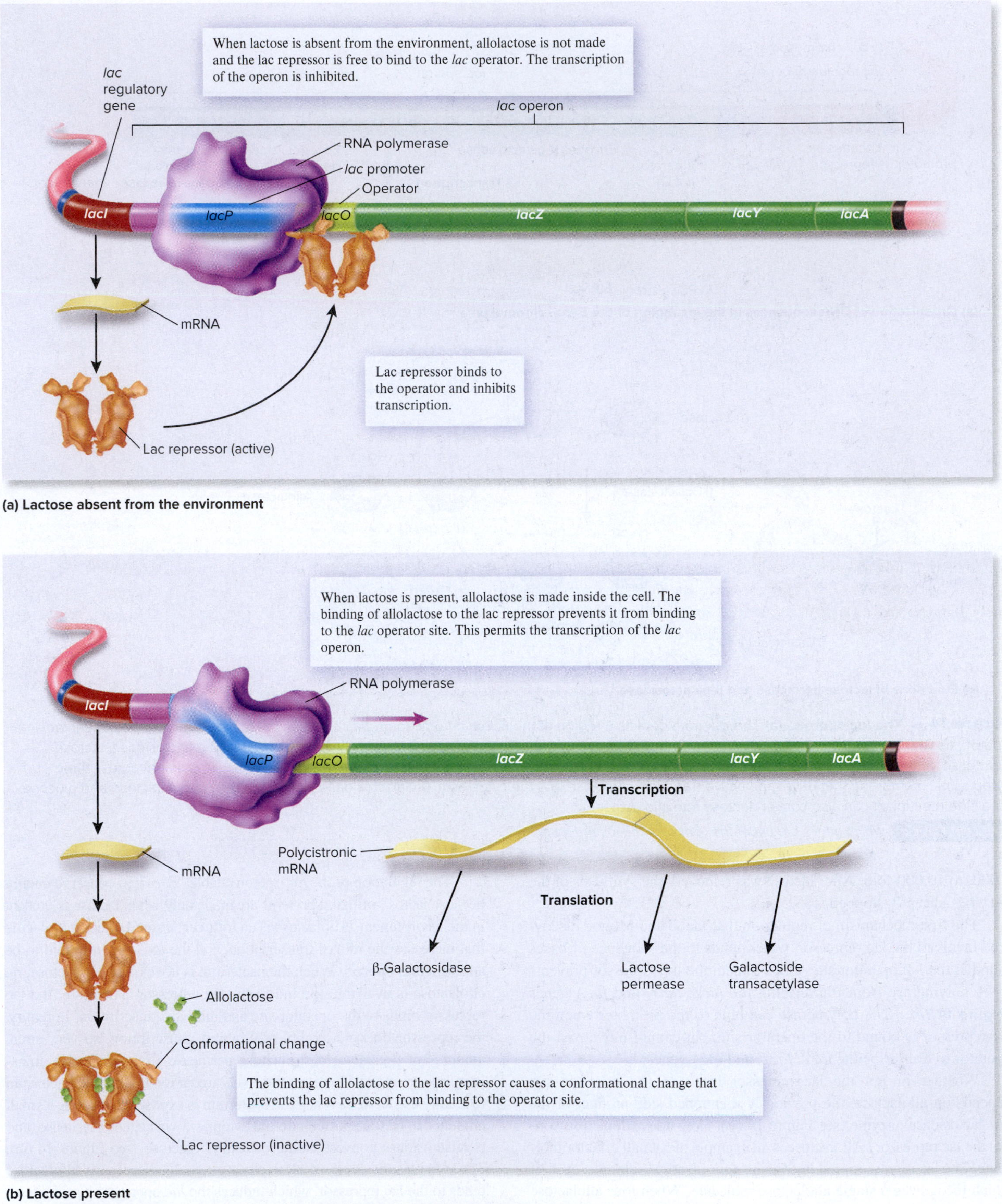

When lactose is absent from the environment, allolactose is not made and the lac repressor is free to bind to the *lac* operator. The transcription of the operon is inhibited.

*lac* regulatory gene

*lac* operon

RNA polymerase

*lac* promoter

Operator

lacI

lacP

lacO

lacZ

lacY

lacA

mRNA

Lac repressor binds to the operator and inhibits transcription.

Lac repressor (active)

**(a) Lactose absent from the environment**

When lactose is present, allolactose is made inside the cell. The binding of allolactose to the lac repressor prevents it from binding to the *lac* operator site. This permits the transcription of the *lac* operon.

RNA polymerase

lacI

lacP

lacO

lacZ

lacY

lacA

Transcription

mRNA

Polycistronic mRNA

Translation

β-Galactosidase

Lactose permease

Galactoside transacetylase

Allolactose

Conformational change

The binding of allolactose to the lac repressor causes a conformational change that prevents the lac repressor from binding to the operator site.

Lac repressor (inactive)

**(b) Lactose present**

**Figure 14.7** **Negative control of an inducible set of genes: function of the lac repressor in regulating the *lac* operon.**

**Concept Check:** *With regard to regulatory proteins and small effector molecules, explain the meaning of negative control and inducible.*

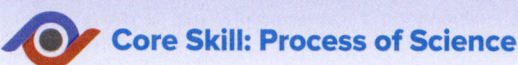 

# Feature Investigation | Jacob, Monod, and Pardee Studied a Constitutive Mutant to Determine the Function of the Lac Repressor

Thus far, we have learned that the lac repressor binds to the *lac* operator site to exert its effects. Let's now take a look at experiments that helped researchers determine the function of the lac repressor. Our understanding of *lac* operon regulation came from studies involving *E. coli* strains that showed abnormalities in the process. In the 1950s, French biologist François Jacob, French biochemist Jacques Monod, and their colleague, American biochemist Arthur Pardee, had identified a few rare mutant bacteria that expressed the genes of the *lac* operon constitutively, meaning that the *lacZ*, *lacY*, and *lacA* genes were expressed even in the absence of lactose in the environment. The researchers discovered that some mutations that caused this abnormality had occurred in the *lacI* region of the DNA. Such strains were termed *lacI⁻* (*lacI* minus) to indicate that the *lacI* region was not functioning properly. Normal, or wild-type, *lacI* strains of *E. coli* are called *lacI⁺* (*lacI* plus).

The researchers initially hypothesized that the *lacI* gene encodes an enzyme that degrades an internal inducer of the *lac* operon. The *lacI⁻* mutation was thought to inhibit this enzyme, thereby allowing the internal inducer to be synthesized continuously. In this way, the *lacI⁻* mutation made it unnecessary for cells to be exposed to lactose for induction. However, over the course of this study and other studies, the researchers eventually arrived at the hypothesis that the *lacI* gene encodes a repressor protein, which proved to be correct (**Figure 14.8**). A mutation in the *lacI* gene that eliminates the synthesis of a functional lac repressor prevents the lac repressor protein from inhibiting transcription. At the time of Jacob, Monod, and Pardee's work, however, the function of the lac repressor was not yet known.

To understand the nature of the *lacI⁻* mutation, Jacob, Monod, and Pardee applied a genetic approach. Although the transfer of DNA from one bacterial cell to another is described in Chapter 19, let's briefly examine this process in order to understand this experiment. Bacteria sometimes exchange circular segments of DNA known as F factors. Some F factors also carry genes that were originally found within the bacterial chromosome. These types of F factors are called F′ factors (F prime factors). A bacterial cell that contains an F′ factor is called a **merozygote**. The study of merozygotes was instrumental in allowing Jacob, Monod, and Pardee to elucidate the function of the *lacI* gene.

As shown in **Figure 14.9**, these researchers studied the *lac* operon in a bacterial strain carrying a *lacI⁻* mutation that caused constitutive expression of the *lac* operon. In addition, a F′ factor was transferred to the mutant strain, thereby producing a merozygote that also carried a normal *lac* operon and a normal *lacI⁺* gene on this F′ factor. The merozygote contained both *lacI⁺* and *lacI⁻* genes. The constitutive mutant and corresponding merozygote were grown separately in liquid media and then divided into two tubes each. In half of the tubes, the cells were incubated with lactose to determine if lactose was needed to induce the expression of the operon. In the other tubes, lactose was omitted. To monitor the expression of the *lac* operon, the cells were broken open and then tested for the amount of β-galactosidase they released by measuring the ability of any β-galactosidase present to convert a colorless compound into a yellow product.

The data table in Figure 14.9 summarizes the effects of this constitutive mutation and its analysis in a merozygote. As Jacob, Monod, and Pardee already knew, the *lacI⁻* mutant strain expressed the *lac* operon constitutively, in both the presence and the absence of lactose. However, when a normal *lac* operon and *lacI⁺* gene on an F′ factor were introduced into a cell harboring the mutant *lacI⁻* gene on the chromosome, the normal *lacI⁺* gene could regulate both operons. In the absence of lactose, both operons were shut off. How did Jacob, Monod, and Pardee explain these results? They concluded that a single *lacI⁺* gene on the F′ factor produces enough repressor protein to bind to both operator sites. Furthermore, this protein is diffusible—can spread through the cytoplasm—and binds to *lac* operons that are on the F′ factor and on the bacterial chromosome. Taken together, the data indicated that the normal *lacI* gene encodes a diffusible protein that represses the *lac* operon.

The interactions between regulatory proteins and DNA sequences illustrated in this experiment led to the definition of three genetic terms. A **cis-acting element** is a DNA segment that must be adjacent to the gene(s) that it regulates. The *lac* operator site is an example of a *cis*-acting element. A **trans-effect** is a form of gene regulation that can occur even though two DNA segments are not physically adjacent. The action of the lac repressor on the *lac* operon is a *trans*-effect. A **cis-effect** is mediated by a *cis*-acting element that binds regulatory proteins, whereas a *trans*-effect is mediated by genes that encode diffusible regulatory proteins.

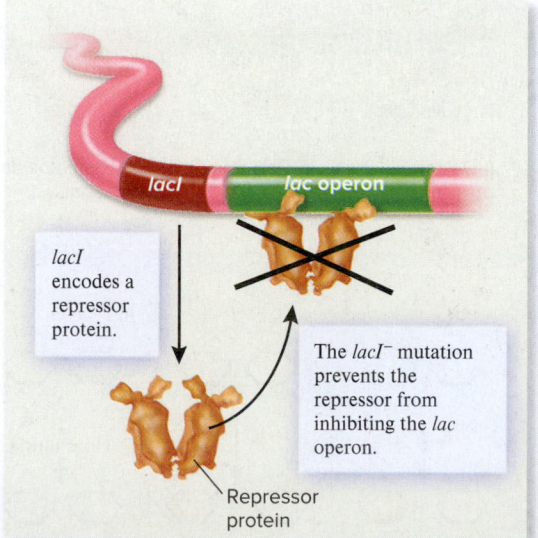

lacI
encodes a repressor protein.

*lacI* encodes a repressor protein.

The *lacI⁻* mutation prevents the repressor from inhibiting the *lac* operon.

Repressor protein

**Figure 14.8** **A hypothesis for the function of the *lacI* gene.**

**Figure 14.9** The experiment performed by Jacob, Monod, and Pardee to study a constitutive *lacI⁻* mutant.

**HYPOTHESIS** The *lacI⁻* mutation inhibits the lac repressor and thereby allows the constitutive expression of the *lac* operon. Note: This correct hypothesis actually arose from the results of this study and other studies.

**KEY MATERIALS** A constitutive *lacI⁻* mutant strain was already characterized. An F′ factor carrying a normal *lacI⁺* gene and *lac* operon was introduced into this strain to produce a merozygote strain. Note: POZ⁺Y⁺A⁺ refers to a normal *lac* operon.

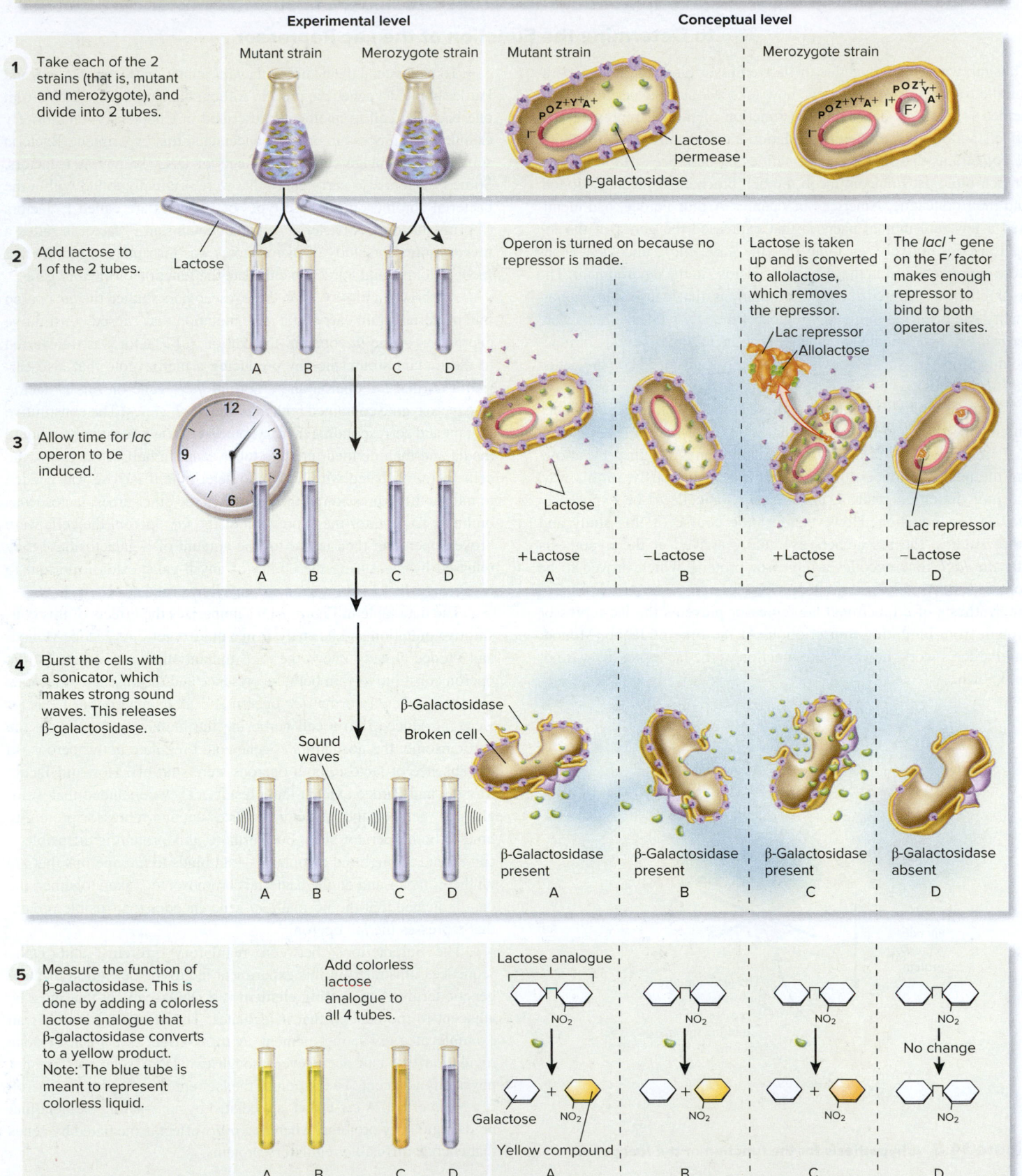

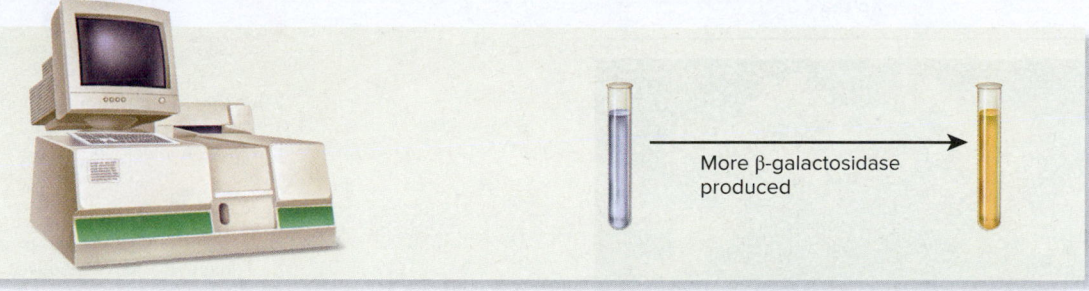

**6** The amount of yellow color is measured with a spectrophotometer; the deeper the yellow color, the more β-galactosidase was produced.

More β-galactosidase produced

---

**7   THE DATA**

Results from step 6:

|  | Expression of the *lac* operon With lactose | Without lactose |
|---|---|---|
| Mutant strain | 100% | 100% |
| Merozygote strain | 220% | <1% |

**8   CONCLUSION** The *lacI* gene encodes a diffusible repressor protein.

**9   SOURCE** Jacob, F., and Monod, J. 1961. Genetic regulatory mechanisms in the synthesis of proteins. *Journal of Molecular Biology* 3: 318–356.

---

*Experimental Questions*

1. What were the key observations made by Jacob, Monod, and Pardee that led to the development of their hypothesis regarding the *lacI* gene and the regulation of the *lac* operon?

2. **CoreSKILL »** What was the eventual hypothesis proposed by the researchers to explain the function of the *lacI* gene and the regulation of the *lac* operon?

3. **CoreSKILL »** How did Jacob, Monod, and Pardee test the hypothesis? What were the results of the experiment? How do these results support the idea that the *lacI* gene produces a repressor protein?

---

## The *lac* Operon Is Also Under Positive Control by an Activator Protein

In addition to being under negative control by a repressor protein, the *lac* operon is also positively regulated by an activator called the **catabolite activator protein (CAP)**. CAP is controlled by a small effector molecule, **cyclic AMP (cAMP)**, which is produced from ATP via an enzyme known as adenylyl cyclase. Gene regulation involving CAP and cAMP is an example of positive control (**Figure 14.10**). When cAMP binds to CAP, the cAMP-CAP complex binds to the CAP site near the *lac* promoter. This causes a bend in the DNA that enhances the ability of RNA polymerase to bind to the promoter. In this way, the rate of transcription is increased.

The key functional role of CAP is to allow *E. coli* to choose between different sugars as an energy source. In a process known as **catabolite repression**, the presence of a preferred energy source inhibits the use of other energy sources. In this case, transcription of the *lac* operon is inhibited by the presence of glucose, which is a catabolite (it is broken down—catabolized—inside the cell). This gene regulation allows *E. coli* to preferentially use glucose instead of other sugars, such as lactose. How does this occur? Glucose inhibits the production of cAMP, thereby preventing the binding of CAP to the DNA. In this way, glucose blocks the activation of the *lac* operon by inhibiting transcription. Though it may seem puzzling, the term catabolite repression was coined before the action of the cAMP-CAP complex was understood at the molecular level. Historically, the primary observation of researchers was that glucose (a catabolite) inhibited (repressed) lactose metabolism. Further experimentation revealed that CAP is actually an activator protein.

**Figure 14.11** considers the four possible environmental conditions that an *E. coli* bacterium might experience with regard to these two sugars.

- When both lactose and glucose levels are high (Figure 14.11a), CAP does not bind to the CAP site, which inhibits transcription. However, a low level of transcription does occur. Under these conditions, the bacterium primarily uses glucose rather than lactose. Why is this a benefit to the bacterium? The bacterium conserves energy by using one type of sugar at a time.

- If the lactose level is high and the glucose level is low (Figure 14.11b), the transcription rate of the *lac* operon is very high because CAP is bound to the CAP site and the lac repressor is not bound to the operator site. Under these conditions, the bacterium metabolizes lactose.

- When the lactose level is low, the lac repressor prevents transcription of the *lac* operon, whether the glucose level is high or low (Figure 14.11c,d).

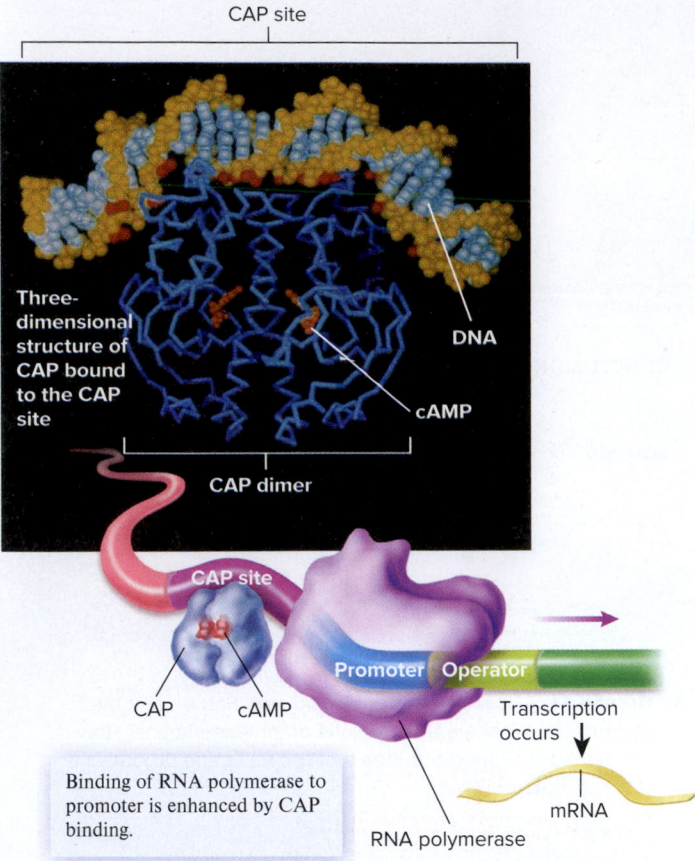

CAP site

Three-dimensional structure of CAP bound to the CAP site

DNA

cAMP

CAP dimer

CAP site

CAP    cAMP

Promoter    Operator

Transcription occurs

mRNA

RNA polymerase

Binding of RNA polymerase to promoter is enhanced by CAP binding.

**Figure 14.10** **Positive control of the *lac* operon by the catabolite activator protein (CAP).** When cAMP is bound to CAP, CAP binds to the DNA and causes it to bend. This bend facilitates the binding of RNA polymerase. ©Thomas Steitz, Howard Hughes Medical Institution, Yale University

**Core Skill: Connections** Look back at Figure 9.12. What is the function of cAMP in eukaryotic cells?

**BIO TIPS**    **THE QUESTION** *Let's suppose you have isolated a mutant strain of E. coli in which the lac operon is constitutively expressed. In other words, the operon is turned on in the presence or the absence of lactose. One possibility is that the mutation is blocking transcription of the lacI gene, thereby preventing the synthesis of lac repressor. A second possibility is that the mutation altered the sequence of the lac operator site in a way that prevents the lac repressor protein from binding there. How could you distinguish between these two possibilities?*

**T OPIC** *What topic in biology does this question address?* The topic is gene regulation. More specifically, the question asks how you could determine the way a mutation is affecting the expression of the *lac* operon.

**I NFORMATION** *What information do you know based on the question and your understanding of the topic?* From the question, you know that the mutation is either inhibiting the expression of *lacI* or has altered *lacO* in a way that prevents the binding of lac repressor. From your understanding of the topic, you may

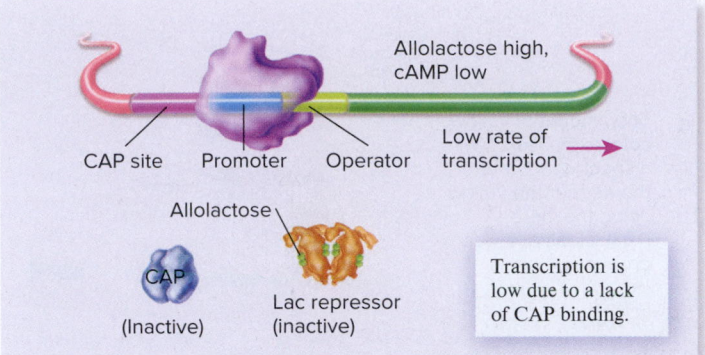

CAP site    Promoter    Operator

Allolactose high, cAMP low

Low rate of transcription

Allolactose

CAP (Inactive)

Lac repressor (inactive)

Transcription is low due to a lack of CAP binding.

**(a) Lactose high, glucose high**

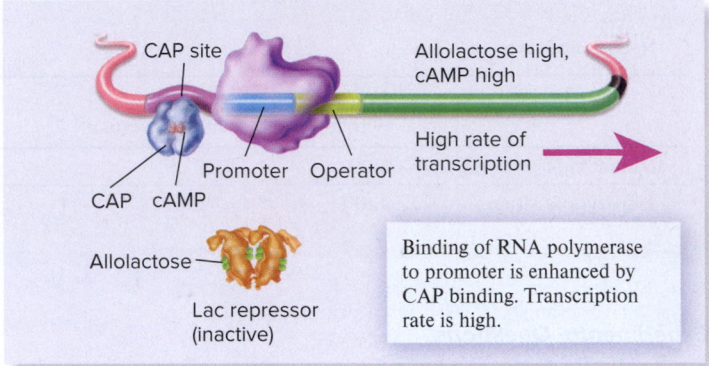

CAP site

Allolactose high, cAMP high

Promoter    Operator

High rate of transcription

CAP    cAMP

Allolactose

Lac repressor (inactive)

Binding of RNA polymerase to promoter is enhanced by CAP binding. Transcription rate is high.

**(b) Lactose high, glucose low**

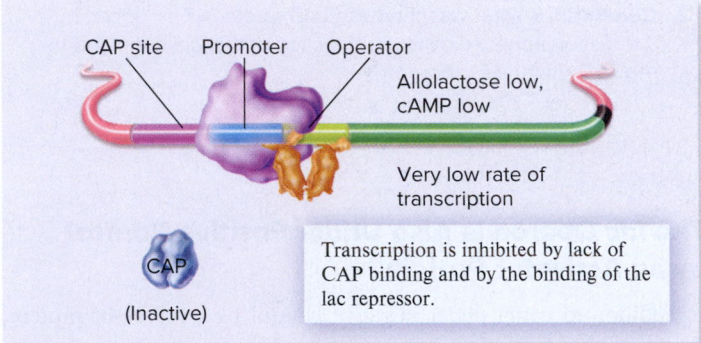

CAP site    Promoter    Operator

Allolactose low, cAMP low

Very low rate of transcription

CAP (Inactive)

Transcription is inhibited by lack of CAP binding and by the binding of the lac repressor.

**(c) Lactose low, glucose high**

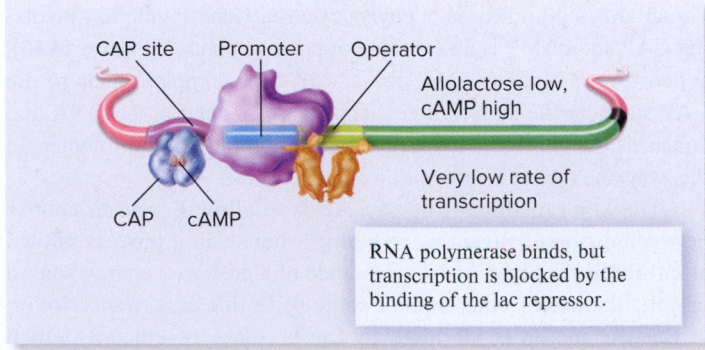

CAP site    Promoter    Operator

Allolactose low, cAMP high

Very low rate of transcription

CAP    cAMP

RNA polymerase binds, but transcription is blocked by the binding of the lac repressor.

**(d) Lactose low, glucose low**

**Figure 14.11** **Effects of lactose and glucose on the expression of the *lac* operon.**

**Concept Check:** *What are the advantages of having both an activator and a repressor protein?*

remember that the lac repressor exhibits a *trans*-effect because it is a diffusible protein, whereas *lacO* is a *cis*-acting element.

**P**ROBLEM-SOLVING **S**TRATEGY *Design an experiment.* *Predict the outcome.* One strategy to solve this problem is to design an experiment that can distinguish between a mutation that results in a *cis*-effect versus one that produces a *trans*-effect. The use of a merozygote is one way to accomplish that goal.

**ANSWER**

*Key materials: The constitutive strain of E. coli and a merozygote that carries a normal lac operon and a normal lacI gene on an F′ factor (see Figure 14.9).*

*Procedure:*

*1. Place each strain into separate tubes with or without lactose.*

*2. Allow induction to occur.*

*3. Burst the cells with a sonicator.*

*4. Add the lactose analogue, and measure yellow color production.*

*Expected results: If the mutation is in lacI, the repressor encoded on the F′ factor will inhibit the expression of the lac operon on the chromosome and the one on the F′ factor. There will be very little yellow color in the absence of lactose in the tube with the merozygote. (This was the result obtained in Figure 14.9.) Alternatively, if the mutation is in lacO, the lac operon on the chromosome will still be turned on even in the absence of lactose. The merozygote will produce a strong yellow color in the absence of lactose.*

## The *trp* Operon Is Under Negative Control by a Repressor Protein

So far in this section, we have examined the regulation of the *lac* operon. Let's now consider an example of an operon that encodes enzymes involved in biosynthesis rather than breakdown. Our example is the **trp operon** of *E. coli*, which encodes enzymes that are required to make the amino acid tryptophan, a building block of proteins. More specifically, the *trpE*, *trpD*, *trpC*, *trpB*, and *trpA* genes encode enzymes that are involved in a pathway that leads to tryptophan synthesis.

The *trp* operon is regulated by a repressor protein that is encoded by the *trpR* gene. The binding of the repressor to the *trp* operator site inhibits transcription. The ability of the trp repressor to bind to the *trp* operator is controlled by tryptophan, which is the product of the metabolic pathway controlled by the enzymes that are encoded by the operon.

- When the tryptophan level within the cell is very low, the trp repressor cannot bind to the operator site. Under these conditions, RNA polymerase readily transcribes the operon (**Figure 14.12a**). In this way, the cell expresses the genes that encode enzymes that result in the synthesis of tryptophan, which is in short supply.

- When the tryptophan level within the cell is high, tryptophan turns off the *trp* operon. Tryptophan acts as a small effector molecule, or **corepressor**, by binding to the trp repressor protein. This causes a conformational change in the repressor that allows it to bind to the *trp* operator site, inhibiting

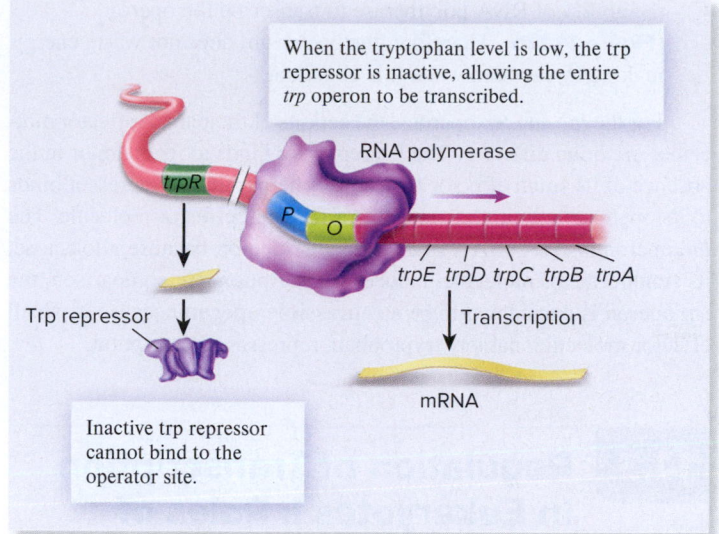

When the tryptophan level is low, the trp repressor is inactive, allowing the entire *trp* operon to be transcribed.

RNA polymerase

Trp repressor

Inactive trp repressor cannot bind to the operator site.

mRNA

**(a) Low tryptophan**

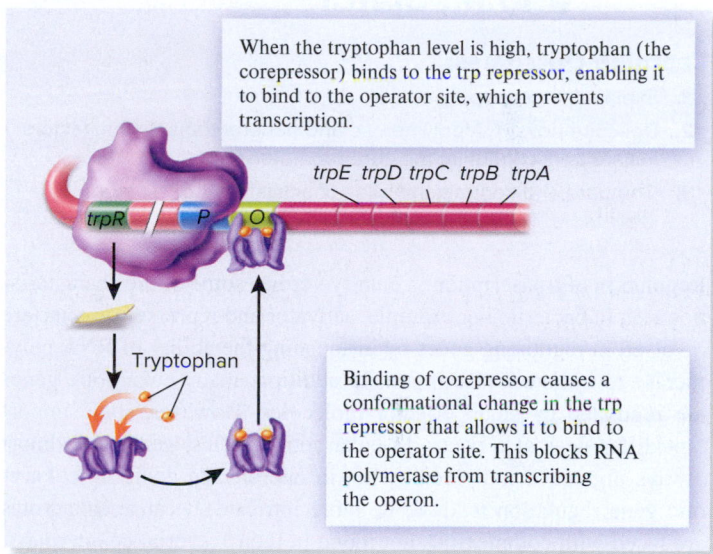

When the tryptophan level is high, tryptophan (the corepressor) binds to the trp repressor, enabling it to bind to the operator site, which prevents transcription.

Tryptophan

Binding of corepressor causes a conformational change in the trp repressor that allows it to bind to the operator site. This blocks RNA polymerase from transcribing the operon.

**(b) High tryptophan**

**Figure 14.12** **Negative control of a repressible set of genes: function of the trp repressor and corepressor (tryptophan) in regulating the *trp* operon.**

 **Core Skill: Modeling** The goal of this modeling challenge is to propose a model for the structure of a mutant trp repressor that cannot bind to tryptophan and to predict the mutant's effect on the expression of the *trp* operon.

**Modeling Challenge:** Let's suppose that researchers identified a mutant form of the trp repressor that cannot bind to tryptophan. Draw a model for the structure of the repressor in the presence of tryptophan and in its absence. Based on your model, would you predict that the *trp* operon would be repressed in the presence of tryptophan?

the ability of RNA polymerase to transcribe the operon (**Figure 14.12b**). Therefore, the bacterium does not waste energy making tryptophan when it is abundant.

For the *lac* and *trp* operons, the actions of their small effector molecules are quite different. The lac repressor binds to its operator in the absence of its small effector molecule, whereas the trp repressor binds to its operator only in the presence of its small effector molecule. The *lac* operon is categorized as an inducible operon because allolactose, its small effector molecule, induces transcription. By comparison, the *trp* operon is considered to be a **repressible operon** because its small effector molecule, namely tryptophan, represses transcription.

## 14.3 Regulation of Transcription in Eukaryotes I: Roles of Transcription Factors and Mediator

**Learning Outcomes:**

1. Explain the concept of combinatorial control.
2. Describe how RNA polymerase and general transcription factors initiate transcription at the core promoter.
3. Compare and contrast the roles of activators, coactivators, repressors, TFIID, and mediator in gene regulation.

Regulation of transcription in eukaryotes has some of the characteristics seen in bacteria. For example, activator and repressor proteins are involved in regulating genes by influencing the ability of RNA polymerase to initiate transcription. In addition, many eukaryotic genes are regulated by small effector molecules. However, some important differences also occur. In eukaryotic species, genes are almost always organized individually, not in operons. In addition, eukaryotic gene regulation tends to be more intricate, because eukaryotes are faced with complexities not found in their bacterial counterparts. For example, eukaryotes have more complicated cell structures that contain many more proteins and a variety of cell organelles. Many eukaryotes, such as animals and plants, are multicellular and contain different cell types. As discussed earlier in this chapter, animal cells may differentiate into neurons, muscle cells, and skin cells, and so on. Furthermore, animals and plants progress through developmental stages that require changes in gene expression.

By studying transcriptional regulation, researchers have discovered that most eukaryotic genes, particularly those found in multicellular species, are regulated by many factors. This phenomenon is called **combinatorial control** because the combination of many factors determines the expression of any given gene. At the level of transcription, the following factors contribute to combinatorial control:

1. One or more activators may stimulate the ability of RNA polymerase to initiate transcription.
2. One or more repressors may inhibit the ability of RNA polymerase to initiate transcription.
3. The function of activators and repressors may be modulated in several ways, which include the binding of small effector

molecules, protein-protein interactions, and covalent modifications.

4. Activators are necessary to alter chromatin structure in the region where a gene is located, thereby making it easier for the gene to be recognized and transcribed by RNA polymerase.
5. DNA methylation usually inhibits transcription, either by preventing the binding of an activator or by recruiting proteins that inhibit transcription.

All five of these factors may contribute to the regulation of a single gene, or possibly only three or four will play a role. In most cases, transcriptional regulation is aimed at controlling the initiation of transcription at the promoter. In this section and the following section, we will survey these basic types of gene regulation in eukaryotic species.

### Eukaryotic Protein-Encoding Genes Have a Core Promoter and Regulatory Elements

To understand gene regulation in eukaryotes, we first need to consider the DNA sequences that are needed to initiate transcription. For eukaryotic protein-encoding genes, three features are common among most promoters: **regulatory elements**, a **TATA box**, and a **transcriptional start site** (**Figure 14.13**).

The TATA box and transcriptional start site form the **core promoter**. The transcriptional start site is the place in the DNA where transcription actually begins. The TATA box, which is a 5′–TATAAA–3′ sequence, is usually about 25 bp upstream from a transcriptional start site. The TATA box determines the precise starting point for transcription. If it is missing from the core promoter, transcription may start at a variety of different locations. The core promoter, acting alone, results in a low level of transcription that is termed **basal transcription**.

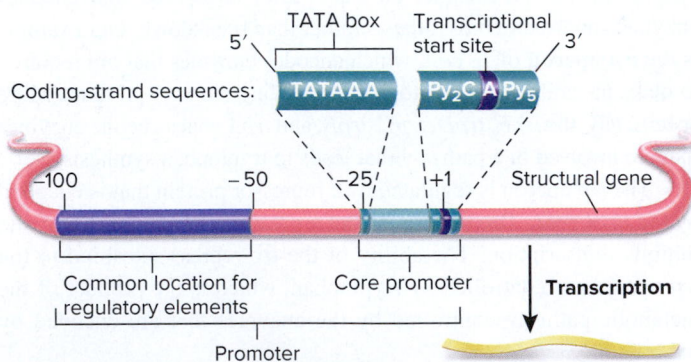

**Figure 14.13** **A common organization of sequences for the promoter of a eukaryotic protein-encoding gene.** The core promoter has a TATA box and a transcriptional start site. The TATA box sequence is 5′–TATAAA–3′. However, not all protein-encoding genes in eukaryotes have a TATA box. The A highlighted in dark blue is the transcriptional start site. This A marks the site of the first adenine in the RNA transcript. The sequence that flanks the A of the transcriptional start site is two pyrimidines, then a cytosine (C), then five pyrimidines. Py stands for a pyrimidine—cytosine or thymine. Regulatory elements, such as enhancers and silencers, are usually found upstream from the core promoter.

Regulatory elements (or regulatory sequences) are DNA segments that regulate eukaryotic genes. As described later, regulatory elements are recognized by regulatory transcription factors that control the ability of RNA polymerase to initiate transcription at the core promoter. Some regulatory elements, known as **enhancers**, play a role in the ability of RNA polymerase to begin transcription, thereby enhancing the rate of transcription. When enhancers are not functioning, most eukaryotic genes have very low levels of transcription. Other regulatory elements, known as **silencers**, prevent transcription of a given gene when its expression is not needed. When these sequences function, the rate of transcription is decreased.

A common location for regulatory elements is the region that is 50–100 bp upstream from the transcriptional start site (see Figure 14.13). However, the locations of regulatory elements vary greatly among different eukaryotic genes. Regulatory elements can be quite distant from the promoter, even 100,000 bp away, yet exert strong effects on the ability of RNA polymerase to initiate transcription at the core promoter! Regulatory elements were first discovered by Japanese molecular biologist Susumu Tonegawa and coworkers in the 1980s. While studying genes that play a role in immunity, they identified a region that was far away from the core promoter but was needed for high levels of transcription to take place.

## RNA Polymerase II, General Transcription Factors, and Mediator Are Needed to Transcribe Eukaryotic Protein-Encoding Genes

As discussed in Chapter 12, three forms of RNA polymerases, designated I, II, and III, are found in eukaryotes. RNA polymerase II transcribes protein-encoding genes. By studying transcription in a variety of eukaryotic species, researchers have identified three types of proteins that play a role in initiating transcription at the core promoter of protein-encoding genes. These are RNA polymerase II, five different proteins called **general transcription factors (GTFs)**, and a large protein complex called mediator.

RNA polymerase II and GTFs must come together at the TATA box of the core promoter so transcription can be initiated. A series of interactions that occurs between these proteins enables RNA polymerase II to bind to the DNA. The completed assembly of RNA polymerase II and GTFs at the TATA box is known as the **preinitiation complex** (**Figure 14.14**).

Another component needed for transcription in eukaryotes is the protein complex called mediator. **Mediator** is composed of many proteins that bind to each other to form an elliptically shaped complex that partially wraps around RNA polymerase II and the GTFs. Mediator derives its name from the observation that it mediates interactions between the preinitiation complex and regulatory transcription factors such as activators or repressors that bind to enhancers or silencers. The function of mediator is to control the rate at which RNA polymerase begins to transcribe RNA at the transcriptional start site.

## Activators and Repressors May Influence the Function of GTFs or Mediator

In eukaryotes, regulatory transcription factors called activators and repressors bind to enhancers or silencers, respectively, and regulate

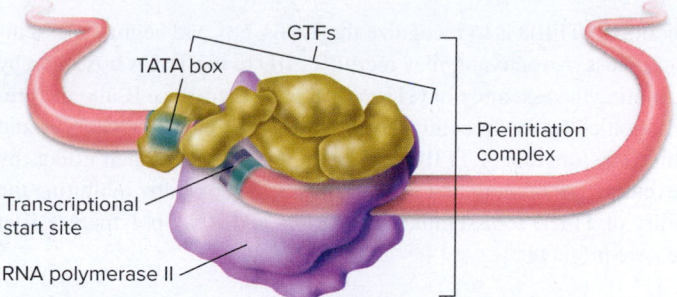

**Figure 14.14  The preinitiation complex.** General transcription factors (GTFs) and RNA polymerase II assemble into the preinitiation complex at the core promoter of eukaryotic protein-encoding genes.

the rate of transcription of genes. Activators and repressors commonly regulate the function of RNA polymerase II by binding to GTFs or mediator.

***Affecting the Function of GTFs***   As shown in **Figure 14.15**, some activators bind to an enhancer and then influence the function of GTFs. For example, an activator may improve the ability of a GTF called transcription factor II D (TFIID) to initiate transcription. The

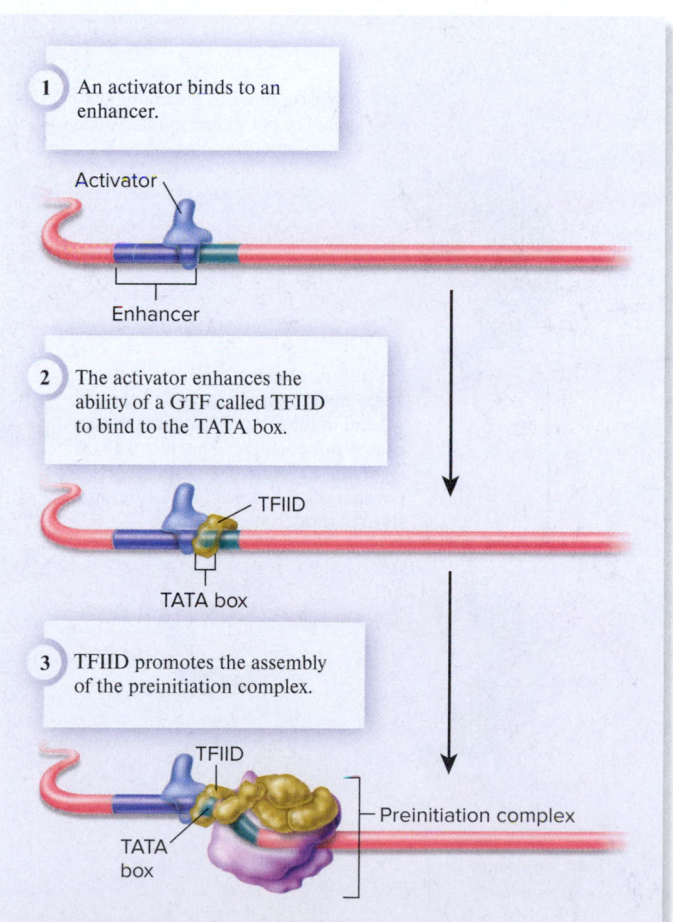

1  An activator binds to an enhancer.

2  The activator enhances the ability of a GTF called TFIID to bind to the TATA box.

3  TFIID promotes the assembly of the preinitiation complex.

**Figure 14.15  Effect of an activator via TFIID, a general transcription factor.**

function of TFIID is to recognize the TATA box and begin the assembly process. An activator may recruit TFIID to the TATA box, thereby promoting the assembly of GTFs and RNA polymerase II into the pre-initiation complex. In contrast, repressors may bind to a silencer and inhibit the function of TFIID. Certain repressors exert their effects by preventing the binding of TFIID to the TATA box or by inhibiting the ability of TFIID to assemble other GTFs and RNA polymerase II at the core promoter.

**Affecting the Function of Mediator**  In addition to affecting GTFs, a second way that regulatory transcription factors control RNA polymerase II is via mediator (**Figure 14.16**). In this example, an activator also interacts with a **coactivator**—a protein that increases the rate of transcription but does not directly bind to the DNA itself. The activator-coactivator complex stimulates the function of mediator, thereby causing RNA polymerase II to proceed to the elongation phase of transcription more quickly. Alternatively, repressors have the opposite effect to those seen in Figure 14.16. When a repressor inhibits mediator, RNA polymerase II cannot progress to the elongation stage.

A third way that regulatory transcription factors influence transcription is by recruiting proteins that affect chromatin structure in the promoter region, as described in the next section.

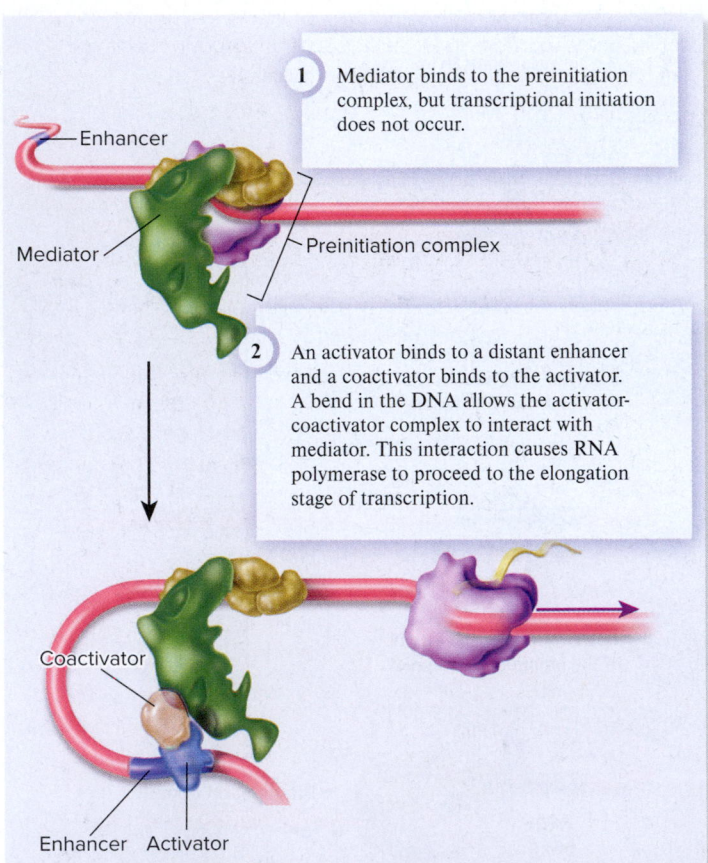

1  Mediator binds to the preinitiation complex, but transcriptional initiation does not occur.

Enhancer

Mediator    Preinitiation complex

2  An activator binds to a distant enhancer and a coactivator binds to the activator. A bend in the DNA allows the activator-coactivator complex to interact with mediator. This interaction causes RNA polymerase to proceed to the elongation stage of transcription.

Coactivator

Enhancer  Activator

**Figure 14.16**  **Effect of an activator via mediator.**

**Concept Check:**  *When an activator interacts with mediator, how does this affect the function of RNA polymerase?*

## 14.4  Regulation of Transcription in Eukaryotes II: Changes in Chromatin Structure and DNA Methylation

**Learning Outcomes:**

1. Describe the flanking of eukaryotic genes by nucleosome-free regions, and explain how nucleosomes are altered during gene transcription.
2. Explain how DNA methylation affects transcription.

In eukaryotes, DNA is associated with proteins to form a structure called **chromatin**—the complex of DNA and proteins that makes up eukaryotic chromosomes (refer back to Figures 11.22 through 11.25). How does the structure of chromatin affect gene transcription? Recall from Chapter 11 that nucleosomes are composed of DNA wrapped around an octamer of histone proteins. Depending on the locations and arrangements of nucleosomes, a region of chromatin containing a gene may be in a **closed conformation**, and transcription may be difficult or impossible. Transcription requires changes in chromatin structure that allow transcription factors to gain access to and bind to the DNA in the promoter region. Such chromatin, said to be in an **open conformation**, is accessible to GTFs and RNA polymerase II, so transcription can take place. In this section, we will examine how chromatin is converted from a closed to an open conformation. We will also explore how **DNA methylation**—the attachment of methyl groups to cytosine bases—affects chromatin conformation and gene expression.

### Transcription Is Controlled by Changes in Chromatin Structure

In recent years, geneticists have been trying to identify the steps that promote the interconversion between the closed and open conformations of chromatin. One way to change chromatin structure is through **ATP-dependent chromatin-remodeling complexes**, which are complexes of proteins that alter chromatin structure. Such complexes use energy from ATP hydrolysis to drive a change in the locations and/or compositions of nucleosomes, thereby making the DNA more or less amenable to transcription. Therefore, chromatin remodeling is important for both the activation and repression of transcription.

How do ATP-dependent chromatin-remodeling complexes change chromatin structure? Three effects are possible.

- One effect is that these complexes may bind to chromatin and change the locations of nucleosomes (**Figure 14.17a**). This may involve a shift of the relative positions of a few nucleosomes or a change in the relative spacing of nucleosomes over a long stretch of DNA.

- A second effect is that remodeling complexes may evict histone octamers from the DNA, thereby creating gaps where nucleosomes are not found (**Figure 14.17b**).

- A third possibility is that chromatin-remodeling complexes may change the composition of nucleosomes by removing

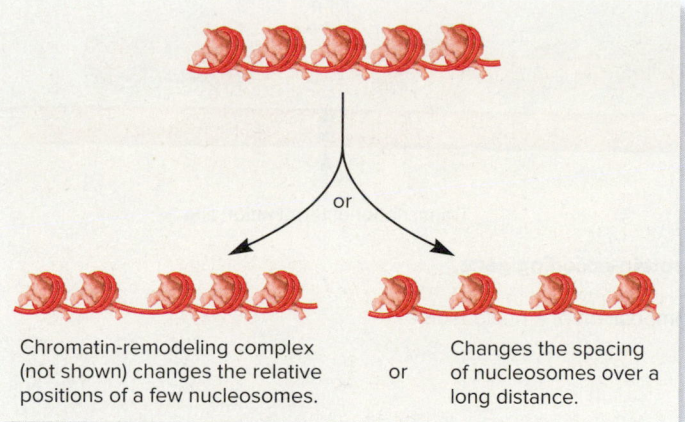

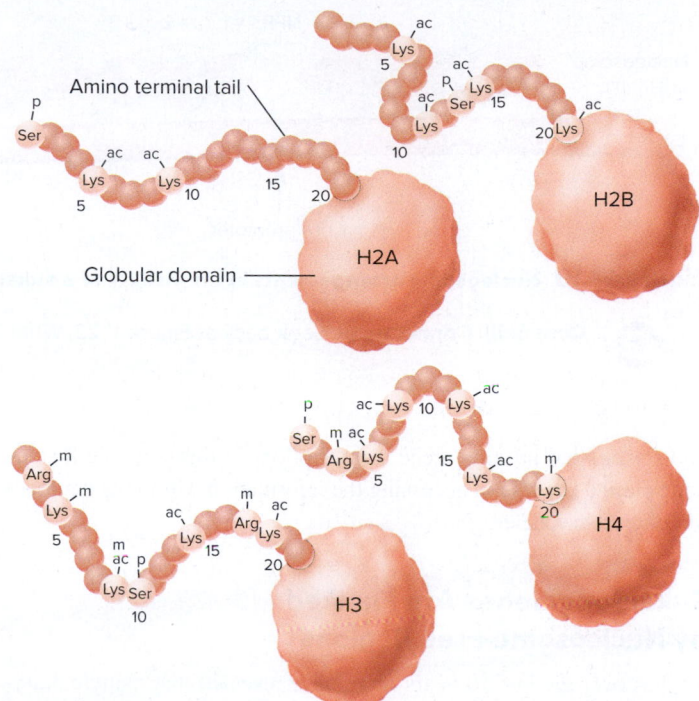

**(a) Change in nucleosome position**

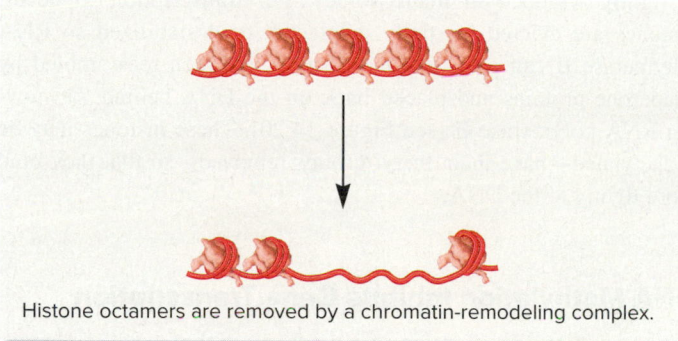

**(b) Histone eviction**

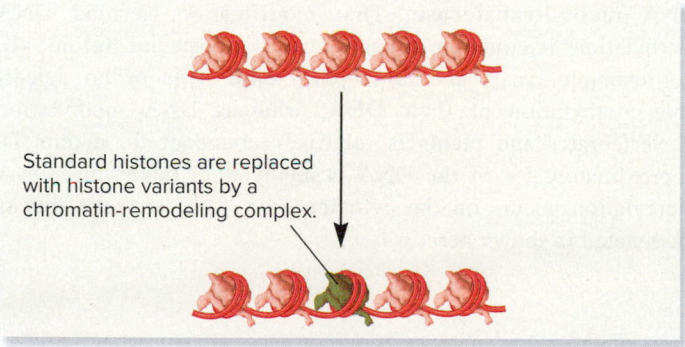

**(c) Replacement with histone variants**

**Figure 14.17** **ATP-dependent chromatin remodeling.** Chromatin-remodeling complexes may **(a)** change the locations of nucleosomes, **(b)** remove histones from the DNA, or **(c)** replace standard histones with histone variants. The chromatin-remodeling complex, which is a complex of proteins, is not shown in this figure.

standard histone proteins from an octamer and replacing them with histone variants (**Figure 14.17c**). A **histone variant** is a histone protein that has a slightly different amino acid sequence than that of the standard histone proteins described in Chapter 11. Some histone variants promote gene transcription, whereas others inhibit it.

**Figure 14.18** **Examples of covalent modifications of the amino terminal tails of histone proteins.** The amino acids are numbered from the N-terminus, or amino end. The modifications shown here are labeled m for methylation, p for phosphorylation, and ac for acetylation. Many more modifications can be made to the amino terminal tails. These modifications are reversible.

**Concept Check:** *What are the two opposing effects that histone modifications may have on transcription?*

## Histone Modifications Affect Gene Transcription

In recent years, researchers have learned that the amino terminal tails of histone proteins are subject to several types of covalent modifications. For example, an enzyme called **histone acetyltransferase** attaches acetyl groups (—$COCH_3$) to the amino terminal tails of histone proteins. When acetylated, histone proteins do not bind as tightly to the DNA, which aids in transcription. Over 50 different enzymes that selectively modify amino terminal tails have been found in mammals. **Figure 14.18** shows how the amino terminal tails of histone proteins H2A, H2B, H3, and H4 can be modified by the attachment of acetyl, methyl, and phosphate groups.

What are the effects of covalent modifications of histones? First, modifications may directly influence interactions between DNA and histone proteins, and between adjacent nucleosomes. As mentioned, the acetylation of histones loosens their binding to DNA and aids in transcription. Second, histone modifications provide binding sites that are recognized by other proteins. According to the **histone code hypothesis**, proposed by American biologists Brian Strahl and David Allis in 2000, the pattern of histone modification is recognized by proteins much like a language or code. One pattern of histone modification may attract proteins that inhibit transcription. Alternatively, a different combination of histone modifications may attract proteins, such as ATP-dependent chromatin-remodeling

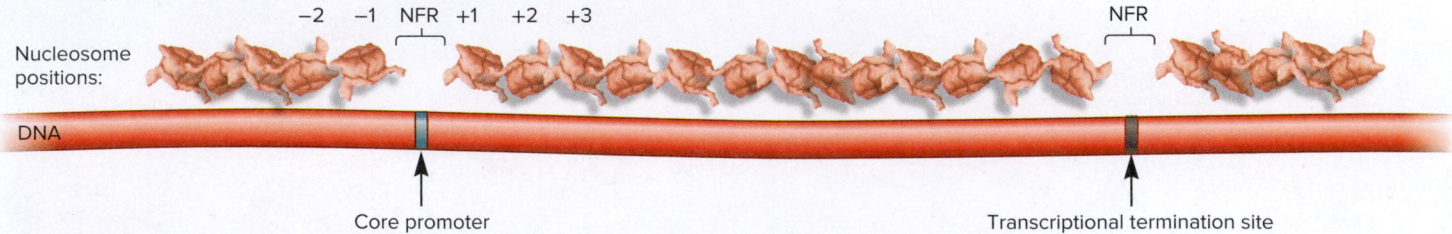

**Figure 14.19** **Nucleosome arrangements in the vicinity of a eukaryotic protein-encoding gene.**

**Core Skill: Connections** Look back at Figure 11.22. What is the composition of a nucleosome?

complexes, that promote gene transcription. In this way, the histone code plays a key role in accessing the information within the genomes of eukaryotic species.

## Eukaryotic Genes Are Flanked by Nucleosome-Free Regions

Studies over the last 10 years or so have revealed that many eukaryotic genes show a common pattern of nucleosome organization (**Figure 14.19**). For active genes or those genes that can be activated, the core promoter is found at a **nucleosome-free region (NFR)**, which is a site in the chromatin that is missing nucleosomes. The NFR is typically 150 bp in length. Although the NFR may be required for transcription, it is not, by itself, sufficient for gene activation. At any given time in the life of a eukaryotic cell, many genes that contain an NFR are not being actively transcribed. The NFR is flanked by two nucleosomes that are termed the –1 and +1 nucleosomes. These nucleosomes often contain histone variants that promote transcription. The end of many eukaryotic genes is followed by another NFR. This arrangement at the end of genes may be important for transcriptional termination.

## Transcriptional Activation Involves Changes in Nucleosome Locations and Changes in Histones

A key role of certain activators is to recruit ATP-dependent chromatin-remodeling complexes and histone-modifying enzymes to the promoter region of eukaryotic genes. Though the order of recruitment may differ among specific activators, the recruiting appears to be critical for transcriptional initiation and elongation. In the scenario shown in **Figure 14.20**, an activator binds to an enhancer in the NFR. The activator then recruits chromatin-remodeling complexes and histone-modifying enzymes to this region. A chromatin-remodeling complex may shift nucleosomes or temporarily evict nucleosomes from the promoter region. Nucleosomes containing certain histone variants are thought to be more easily removed from the DNA than those containing the standard histones. Histone-modifying enzymes, such as histone acetyltransferase, covalently modify histone proteins and may affect nucleosome contact with the DNA. The actions of chromatin-remodeling complexes and histone-modifying enzymes facilitate the binding of general transcription factors and RNA polymerase II to the core promoter, thereby allowing the formation of a preinitiation complex (see Figure 14.20, step 2).

Further changes in chromatin structure are necessary for elongation to occur. RNA polymerase II cannot transcribe DNA that is tightly wrapped in nucleosomes. For transcription to occur, histones are evicted, partially displaced, or destabilized so RNA polymerase II can pass. Evicted histones are then reassembled by chaperone proteins and placed back on the DNA behind the moving RNA polymerase II (see Figure 14.20). These histones may be deacetylated—have their acetyl groups removed—so that they bind more tightly to the DNA.

## DNA Methylation Inhibits Gene Transcription

Let's now turn our attention to a mechanism that usually silences gene expression. DNA structure can be modified by the covalent attachment of methyl groups (—$CH_3$) by an enzyme called **DNA methyltransferase**. This modification, termed DNA methylation, is common in some eukaryotic species but not all. For example, yeast and *Drosophila* have little or no detectable methylation of their DNA, whereas DNA methylation in vertebrates and plants is relatively abundant. In mammals, approximately 5% of the DNA is methylated. Eukaryotic DNA methylation occurs on the cytosine base. The sequence that is methylated is shown here:

$$
\begin{array}{c}
CH_3 \\
| \\
5'—CG—3' \\
3'—GC—5' \\
| \\
CH_3
\end{array}
$$

DNA methylation usually inhibits the transcription of eukaryotic genes, particularly when it occurs in the vicinity of the promoter. In vertebrates and flowering plants, many genes contain sequences called **CpG islands** near their promoters. CpG refers to the bases cytosine (C) and guanine (G) in DNA whose nucleotides are connected by a phosphodiester linkage. A CpG island is a cluster of CpG sites. Unmethylated CpG islands are usually correlated with active genes, whereas repressed genes contain methylated CpG islands. In this way, DNA methylation may play an important role in the silencing of particular genes.

How does DNA methylation inhibit transcription? This can occur in two general ways. First, methylation of CpG islands may prevent an activator from binding to an enhancer element, thus inhibiting

Many genes are flanked by nucleosome-free regions (NFR) and well-positioned nucleosomes.

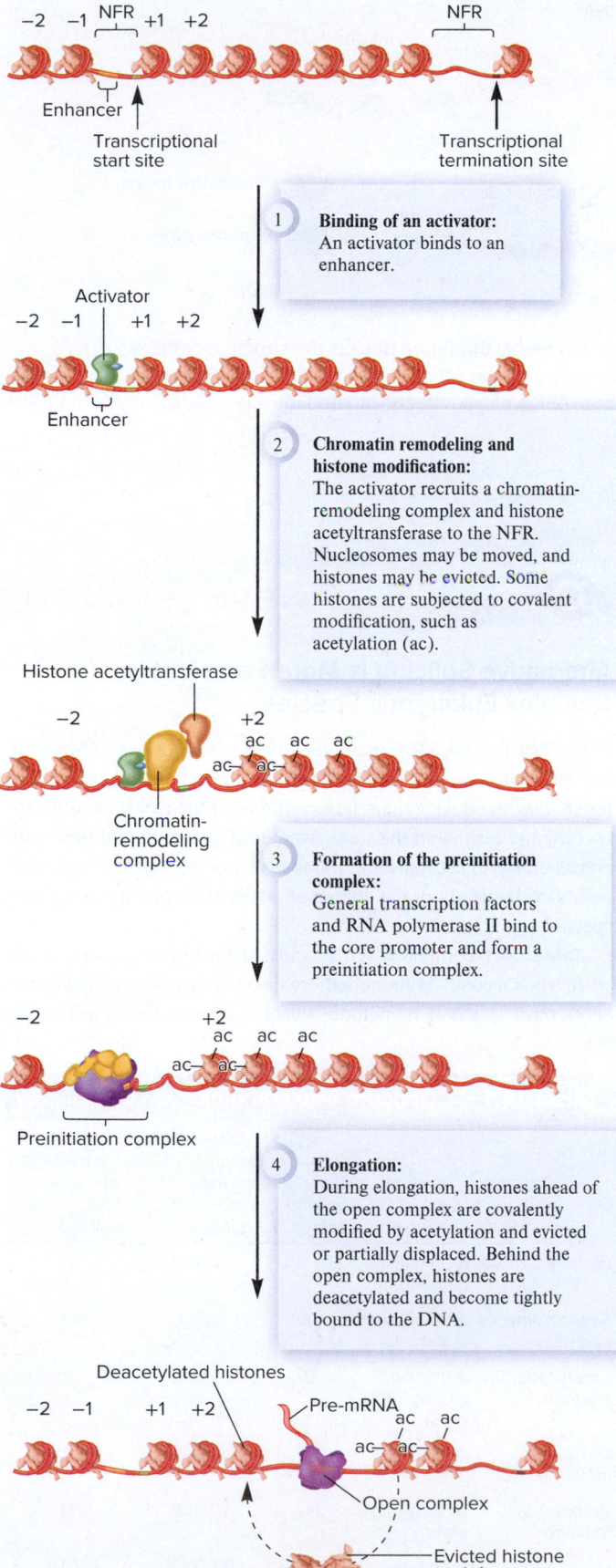

1. **Binding of an activator:**
   An activator binds to an enhancer.

2. **Chromatin remodeling and histone modification:**
   The activator recruits a chromatin-remodeling complex and histone acetyltransferase to the NFR. Nucleosomes may be moved, and histones may be evicted. Some histones are subjected to covalent modification, such as acetylation (ac).

3. **Formation of the preinitiation complex:**
   General transcription factors and RNA polymerase II bind to the core promoter and form a preinitiation complex.

4. **Elongation:**
   During elongation, histones ahead of the open complex are covalently modified by acetylation and evicted or partially displaced. Behind the open complex, histones are deacetylated and become tightly bound to the DNA.

**Figure 14.20** **A simplified model for the transcriptional activation of a eukaryotic protein-encoding gene.**

the initiation of transcription. A second way that methylation inhibits transcription is by altering chromatin structure. Proteins known as **methyl-CpG-binding proteins** bind methylated sequences. Once bound to the DNA, the methyl-CpG-binding protein recruits to the site other proteins that inhibit transcription.

<table>
<tr><td>**14.5**</td><td>**Regulation of RNA Modification and Translation in Eukaryotes**</td></tr>
</table>

**Learning Outcomes:**

1. Outline the process of alternative splicing, and explain how it increases protein diversity.
2. Explain how RNA-binding proteins regulate the translation of specific mRNAs, using the regulation of iron absorption in mammals as an example.

In the preceding sections of this chapter, we focused on gene regulation at the level of transcription in bacteria and eukaryotes. Eukaryotic gene expression is also commonly regulated at the levels of RNA modification and translation. These added levels of regulation provide important benefits to eukaryotic species. First, by regulating RNA modification, eukaryotes can produce more than one mRNA transcript from a single gene. This allows a gene to encode two or more polypeptides, thereby increasing the complexity of eukaryotic proteomes. A second issue is timing. Regulation of transcription in eukaryotes takes a fair amount of time before its effects are observed at the cellular level. During transcription (1) the chromatin must be converted to an open conformation, (2) the gene must be transcribed, (3) the RNA must be modified and exported from the nucleus, and (4) the protein must be made via translation. All four steps take time, on the order of several minutes. One way to achieve faster regulation is to control steps that occur after an RNA transcript is made. In eukaryotes, regulation of translation provides a faster way to regulate the levels of gene products, namely, proteins.

During the past few decades, many critical advances have been made in our knowledge of the regulation of RNA modification and translation. Even so, molecular geneticists are still finding new forms of regulation, making this an exciting area of modern research. In Chapter 13, we considered RNA interference (RNAi), which is a mechanism for regulating translation that involves the use of noncoding RNAs. In this section, we will examine two other mechanisms of RNA regulation: (1) alternative splicing and (2) translational regulation via RNA-binding proteins.

### Alternative Splicing of Pre-mRNAs Increases Protein Diversity

In eukaryotes, a pre-mRNA transcript is modified before it becomes a mature mRNA (refer back to Figure 12.8). When a pre-mRNA has multiple introns and exons, splicing may occur in more than one way, resulting in the production of two or more different polypeptides. Such **alternative splicing** is a form of gene regulation that allows an organism to use the same gene to make different proteins at different

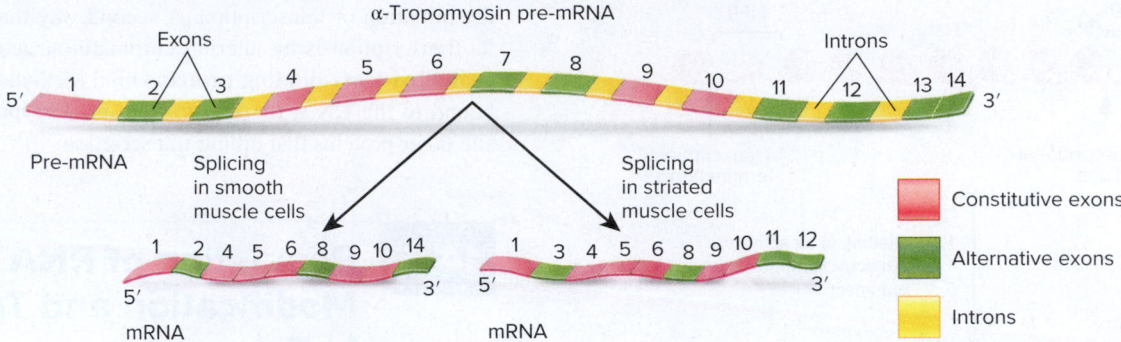

Figure 14.21 **Alternative splicing of the rat α-tropomyosin pre-mRNA.** The top part of this figure depicts the structure of the rat α-tropomyosin pre-mRNA. Exons are red or green, and introns are yellow. The lower part of the figure shows the final mRNA products in smooth and striated muscle cells after alternative splicing. Note: Exon 8 is found in the final mRNA of smooth and striated muscle cells, but not in the mRNA of some other cell types.

*Concept Check:* *What is the biological advantage of alternative splicing?*

stages of development, in different cell types, and/or in response to a change in the environmental conditions. Alternative splicing is an important form of gene regulation in complex eukaryotes such as animals and plants. An advantage of alternative splicing is that two or more different polypeptides can be derived from a single gene, thereby increasing the size of the proteome while minimizing the size of the genome.

Let's consider an example of alternative splicing for a pre-mRNA that encodes a protein known as α-tropomyosin, which functions in the regulation of cell contraction in animals. It is located along the thin filaments found in smooth muscle cells, such as those in the uterus and small intestine, and in striated muscle cells that are found in cardiac and skeletal muscle. α-Tropomyosin is also synthesized in many types of nonmuscle cells but in lower amounts. Within a multicellular organism, different types of cells must regulate their contractibility in subtly different ways. One way this may be accomplished is by the production of different forms of α-tropomyosin.

**Figure 14.21** shows the intron-exon structure of the rat α-tropomyosin pre-mRNA and two alternative ways that the pre-mRNA can be spliced. The pre-mRNA contains 14 exons, 6 of which are constitutive exons (shown in red), which are always found in the mature mRNA from all cell types. Presumably, constitutive exons encode polypeptide segments of the α-tropomyosin protein that are necessary for its general structure and function. By comparison, alternative exons (shown in green) are not always found in the mRNA after splicing has occurred. The polypeptide sequences encoded by alternative exons may subtly change the function of α-tropomyosin to meet the needs of the cell type in which it is found. For example, Figure 14.21 shows the predominant splicing products found in smooth muscle cells and striated muscle cells. Exon 2 encodes a segment of the α-tropomyosin protein that alters its function to make it suitable for smooth muscle cells. By comparison, the α-tropomyosin mRNA found in striated muscle cells does not include exon 2. Instead, this mRNA contains exon 3, leading to the production of an α-tropomyosin more suitable for that cell type.

 **Core Concepts: Evolution, Information**

## Alternative Splicing Is More Prevalent in Complex Eukaryotic Species

In the past few decades, many technical advances have improved our ability to analyze the genomes and proteomes of many different species. Researchers have sequenced the DNA from many species and estimated the total number of genes. In addition, scientists can also estimate the number of polypeptides if information is available about the degree of alternative splicing in a given species.

**Table 14.1** compares six species: a bacterium (*Escherichia coli*), a eukaryotic single-celled organism (yeast—*Saccharomyces cerevisiae*), a small nematode worm (*Caenorhabditis elegans*),

| Table 14.1 | Genome Size and Biological Complexity | | | |
|---|---|---|---|---|
| Species | Level of complexity | Genome size (million bp) | Approximate number of protein-encoding genes | Percentage of genes alternatively spliced |
| *Escherichia coli* | A unicellular bacterium | 4.2 | 4,300 | 0 |
| *Saccharomyces cerevisiae* | A unicellular eukaryote | 12 | 6,300 | <1 |
| *Caenorhabditis elegans* | A tiny worm (about 1,000 cells) | 97 | 20,500 | 2 |
| *Drosophila melanogaster* | An insect | 137 | 15,600 | 7 |
| *Arabidopsis thaliana* | A flowering plant | 142 | 27,000 | 11 |
| *Homo sapiens* | A complex mammal | 3,000 | 22,000 | 70 |

a fruit fly (*Drosophila melanogaster*), a flowering plant (*Arabidopsis thaliana*), and a human (*Homo sapiens*). One general trend is that less complex organisms tend to have fewer genes. For example, unicellular organisms have only a few thousand genes, whereas multicellular species have tens of thousands. However, the trend is by no means a linear one. If we compare *C. elegans* and *D. melanogaster*, the fruit fly actually has fewer genes, even though it is more complex morphologically.

A second trend you can see in Table 14.1 concerns alternative splicing. This phenomenon does not occur in bacteria and is rare in *S. cerevisiae*. The frequency of alternative splicing increases from worms to flies to humans. For example, the level of alternative splicing is 10-fold higher in humans than in *Drosophila*. This trend can partially explain the increase in complexity among these species. Even though humans have only about 22,000 different protein-encoding genes, their cells make well over 100,000 different polypeptides because most genes are alternatively spliced in multiple ways. This increases the level of information contained within the human genome.

## The Prevention of Iron Toxicity in Mammals Involves the Regulation of Translation

In Chapter 13, we considered how microRNAs (miRNAs) and small-interfering RNAs (siRNAs) can silence mRNAs via RNA interference (RNAi). Another way to regulate mRNAs involves RNA-binding proteins that directly affect the initiation of translation. The regulation of iron absorption in mammals provides a well-studied example. Although iron is a vital cofactor for many cellular enzymes, it is toxic at high levels. To prevent toxicity, mammalian cells synthesize a protein called ferritin, which forms a hollow, spherical complex that stores excess iron.

The mRNA that encodes ferritin is controlled by an RNA-binding protein known as the **iron regulatory protein (IRP)**. When the iron level in the cytosol is low and more ferritin is not needed, IRP binds to a regulatory element within the ferritin mRNA known as the **iron regulatory element (IRE)**. The IRE is located between the 5′ cap, where the ribosome binds, and the start codon where translation begins. Due to base pairing, it forms a stem-loop structure. The binding of IRP to the IRE inhibits translation of the ferritin mRNA (**Figure 14.22a**). However, when iron is abundant in the cytosol, the iron binds directly to IRP, which changes its conformation and prevents it from binding to the IRE. Under these conditions, the ferritin mRNA is translated to make more ferritin protein (**Figure 14.22b**).

Why is translational regulation of ferritin mRNA an advantage over transcriptional regulation of the ferritin gene? This mechanism of translational control allows cells to rapidly respond to changes in their environment. When cells are confronted with high levels of iron, they can quickly make more ferritin protein to prevent the toxic buildup of iron. This mechanism is faster than transcriptional regulation, which would require the activation of the ferritin gene and the transcription of ferritin mRNA prior to the synthesis of more ferritin protein.

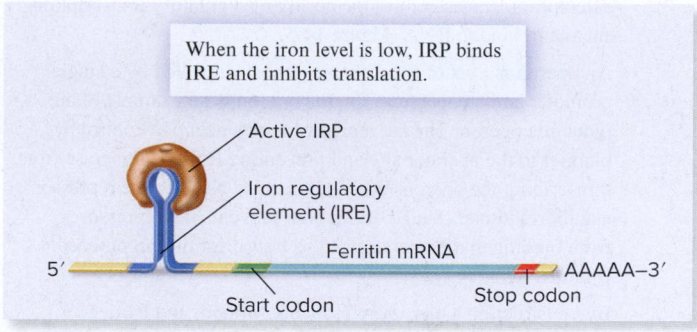

When the iron level is low, IRP binds IRE and inhibits translation.

**(a) Low iron level**

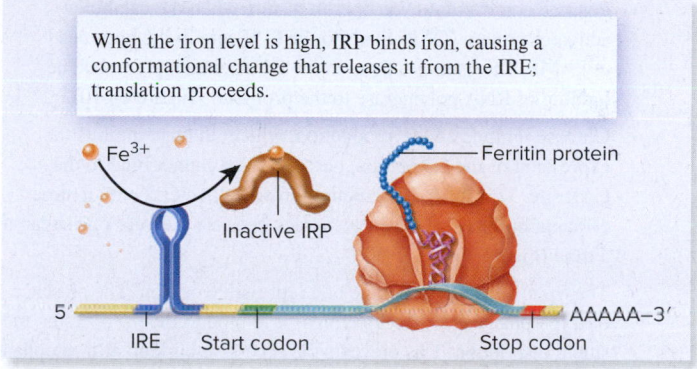

When the iron level is high, IRP binds iron, causing a conformational change that releases it from the IRE; translation proceeds.

**(b) High iron level**

**Figure 14.22** **Translational regulation of ferritin mRNA by the iron regulatory protein (IRP).**

**Concept Check:** *Poisoning may occur if a young child finds a bottle of vitamins, such as those that taste like candy, and eats a large number of them. One of the toxic effects involves the ingestion of too much iron. How does the IRP protect people from the toxic effects of too much iron?*

## Summary of Key Concepts

### 14.1 Overview of Gene Regulation

- Most genes are regulated so that the level of gene expression can vary under different conditions. By comparison, constitutive genes are expressed at constant levels.

- Gene regulation ensures that gene products are made only when they are needed. An example is the synthesis of the gene products needed for lactose utilization in bacteria (Figure 14.1).

- In eukaryotes, gene regulation leads to the production of different cell types, such as neurons, muscle cells, and skin cells, within an organism (Figure 14.2).

- In eukaryotes, gene regulation also enables gene products to be produced at different developmental stages (Figure 14.3).

- All organisms regulate gene expression at a variety of levels, including transcription, translation, and post-translation. Eukaryotes also regulate RNA modification (Figure 14.4).

### 14.2 Regulation of Transcription in Bacteria

- Repressors and activators are regulatory transcription factors that bind to the DNA and affect the transcription of genes. Small effector molecules control the ability of regulatory transcription factors to bind to DNA (Figure 14.5).

- An operon is a set of two or more genes controlled by a single promoter and an operator. The *lac* operon is an example of an inducible operon. The lac repressor exerts negative control by binding to the operator site and preventing RNA polymerase from transcribing the operon. When allolactose binds to the repressor, a conformational change occurs that prevents the repressor from binding to the operator site so transcription can proceed (Figures 14.6, 14.7).

- By constructing a merozygote, Jacob, Monod, and Pardee determined that the *lacI* gene encodes a diffusible protein that represses the *lac* operon (Figures 14.8, 14.9).

- Positive control of the *lac* operon occurs when the catabolite activator protein (CAP) binds to the CAP site in the presence of cAMP. This causes a bend in the DNA, which promotes the binding of RNA polymerase to the promoter (Figure 14.10).

- Glucose inhibits cAMP production, which, in turn, inhibits the expression of the *lac* operon, because CAP cannot bind to the CAP site. This form of regulation provides bacteria with a more efficient utilization of their resources because they use one sugar at a time (Figure 14.11).

- The *trp* operon is an example of a repressible operon. The presence of tryptophan causes the trp repressor to bind to the *trp* operator and stop transcription. This prevents the excessive buildup of tryptophan in the cell, which would be a waste of energy (Figure 14.12).

## 14.3    Regulation of Transcription in Eukaryotes I: Roles of Transcription Factors and Mediator

- Eukaryotic genes exhibit combinatorial control, meaning that many factors control the expression of a single gene.

- Eukaryotic promoters consist of a core promoter (containing a TATA box and transcriptional start site) and regulatory elements, such as enhancers or silencers, that regulate the rate of transcription (Figure 14.13).

- General transcription factors (GTFs) are needed for RNA polymerase II to bind to the core promoter, forming a preinitiation complex (Figure 14.14).

- Activators and repressors regulate RNA polymerase II by affecting the function of TFIID (a GTF) or mediator, a protein complex that wraps around RNA polymerase II and the GTFs (Figures 14.15, 14.16).

## 14.4    Regulation of Transcription in Eukaryotes II: Changes in Chromatin Structure and DNA Methylation

- ATP-dependent chromatin-remodeling complexes change the positions and compositions of nucleosomes (Figure 14.17).

- The pattern of covalent modification of the amino terminal tails of histone proteins, also called the histone code, can inhibit or promote transcription (Figure 14.18).

- Eukaryotic genes are usually flanked by nucleosome-free regions (Figure 14.19).

- For eukaryotic protein-encoding genes, a preinitiation complex forms at a nucleosome-free region. During elongation,

nucleosomes are displaced ahead of RNA polymerase and re-form after RNA polymerase has passed (Figure 14.20).

- DNA methylation, which occurs at CpG islands near promoters, usually inhibits transcription by (1) preventing the binding of activator proteins or (2) promoting the binding of proteins that inhibit transcription.

## 14.5    Regulation of RNA Modification and Translation in Eukaryotes

- In alternative splicing, a single type of pre-mRNA can be spliced in more than one way, producing polypeptides with somewhat different sequences. This is a common way for complex eukaryotes to increase the size of their proteomes (Figure 14.21, Table 14.1).

- RNA-binding proteins can regulate the translation of specific mRNAs. An example is the regulation of iron absorption, in which the iron regulatory protein (IRP) regulates the translation of ferritin mRNA (Figure 14.22).

---

## Assess & Discuss

### Test Yourself

1. Genes that are expressed at all times at relatively constant levels are known as _____ genes.
   a. inducible
   b. repressible
   c. positive
   d. constitutive
   e. negative

2. Which of the following is *not* a level at which gene regulation occurs in bacteria?
   a. transcription
   b. RNA modification
   c. translation
   d. post-translation
   e. All of the above are levels at which bacteria are able to regulate gene expression.

3. Transcription factors that bind to DNA and stimulate transcription are
   a. repressors.
   b. small effector molecules.
   c. activators.
   d. promoters.
   e. operators.

4. In bacteria, the unit of DNA that contains multiple genes under the control of a single promoter is called _____. The mRNA produced from this unit is referred to as _____ mRNA.
   a. an operator, a polycistronic
   b. a template, a protein-encoding
   c. an operon, a polycistronic
   d. an operon, a monocistronic
   e. a template, a monocistronic

5. For the *lac* operon, what would be the expected effects of a mutation in the operator site that prevented the binding of the repressor protein?
   a. The operon would always be turned on.
   b. The operon would always be turned off.
   c. The operon would always be turned on, except when glucose is present.
   d. The operon would be turned on only in the presence of lactose.

e. The operon would be turned on only in the presence of lactose and the absence of glucose.

6. The presence of _____ in a bacterium's environment prevents CAP from binding to the DNA, resulting in _____ in transcription of the *lac* operon.
   a. lactose, an increase
   b. glucose, an increase
   c. cAMP, a decrease
   d. glucose, a decrease
   e. lactose, a decrease

7. The *trp* operon is considered _____ operon because the protein-encoding genes necessary for tryptophan synthesis are not expressed when the level of tryptophan in the cell is high.
   a. an inducible
   b. a positive
   c. a repressible
   d. a negative
   e. both c and d

8. Regulatory elements that function to increase transcription levels in eukaryotes are called
   a. promoters.
   b. silencers.
   c. enhancers.
   d. transcriptional start sites.
   e. activators.

9. The iron regulatory protein (IRP) binds to the iron regulatoty element (IRE) when iron levels are _____ and _____ translation of ferritin mRNA.
   a. high, stimulates
   b. high, inhibits
   c. low, stimulates
   d. low, inhibits
   e. both a and d

10. _____ refers to the process that allows a single type of pre-mRNA to give rise to multiple types of mRNAs due to different patterns of intron and exon removal.
    a. Spliceosomes
    b. Variable expression
    c. Alternative splicing
    d. Polycistronic mRNA
    e. Induced silencing

## Conceptual Questions

1. What is the difference between inducible and repressible operons? Give an example of each.

2. Transcriptional regulation often involves a regulatory protein that binds to a segment of DNA and a small effector molecule that binds to the regulatory protein. Does each of the following terms apply to a regulatory protein, a segment of DNA, or a small effector molecule?
   a. repressor
   b. inducer
   c. operator
   d. corepressor
   e. activator

3.  **Core Concept: Information** Explain the importance of gene regulation as a mechanism for properly accessing the information within genes.

## Collaborative Questions

1. Discuss the advantages and disadvantages of genetic regulation at the different levels shown in Figure 14.4.

2. Discuss the advantages and disadvantages of combinatorial control of eukaryotic genes.

# Mutation, DNA Repair, and Cancer

# 15

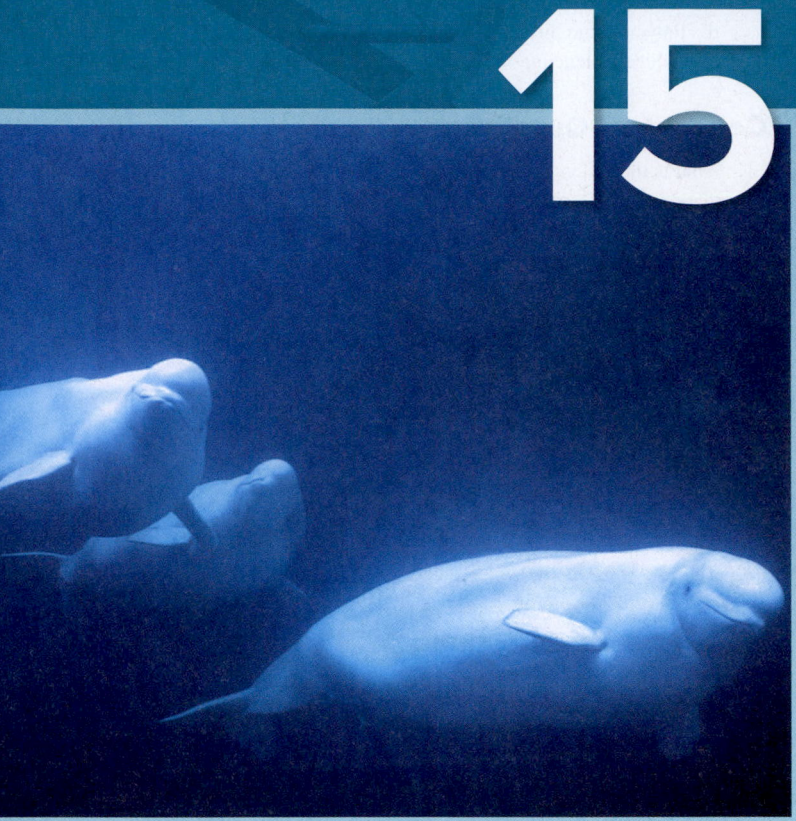

**During the past two decades, over 25% of the beluga whales in Canada's St. Lawrence Seaway have died of cancer.** Biologists speculate that these deaths are caused by cancer-causing pollutants, such as polycyclic aromatic hydrocarbons (PAHs).

©Yvette Cardozo/Workbook Stock/Getty Images

**A**t a summer camp, the children enjoy ice cream, horseback riding, hay rides, swimming, and learning about the habits of owls. Not such an unusual camp, you might be thinking. However, what makes Camp Sundown unique is that the outdoor fun begins at dusk and runs all night. The children at this camp have inherited a disorder called xeroderma pigmentosum (XP), which makes them highly sensitive to sunlight. Their skin will blister or freckle on minimum exposure to sunlight. Of greater concern, however, is skin cancer. Persons with XP may have a 1,000-fold greater risk of developing skin cancer, though such a risk is greatly decreased if exposure to sunlight is minimized.

What explains the symptoms of XP? Individuals with this condition are highly susceptible to **mutation,** which is defined as a heritable change in the genetic material. When a mutation occurs, the order of nucleotide bases in a DNA molecule, its base sequence, is changed permanently, an alteration that can be passed from mother to daughter cells during cell division. Mutations that lead to cancer cause particular genes to be expressed in an abnormal way.

For example, a mutation could affect the transcription of a gene, or it could alter the functional properties of the polypeptide that is specified by a gene. In addition to mutations, changes in chromatin structure can also affect gene expression and contribute to cancer.

Should we be afraid of mutations? Yes and no. On the positive side, mutations are essential to the long-term continuity of life. Mutations provide the foundation for evolutionary change. They supply the variation that enables species to evolve and become better adapted to their environments. On the negative side, however, new mutations are more likely to be harmful than beneficial to the individual. The genes within modern species are the products of billions of years of evolution and have evolved to work properly. Random mutations are more likely to disrupt genes rather than enhance their function. As we will see in this chapter, mutations can cause cancer. In addition, many inherited disorders, such as XP and cystic fibrosis, are caused by gene mutations. For these and many other reasons, understanding the molecular nature of mutations is a compelling area of research.

All species have evolved several ways to repair damaged DNA. Such DNA repair systems reverse DNA damage before a permanent mutation can occur. DNA repair systems are vital to the survival of all organisms. If these systems did not exist, mutations would be so prevalent that few species, if any, would survive. In this chapter, we will examine how these DNA repair systems operate. But first, let's explore the consequences and causes of mutations.

## 15.1 Consequences of Mutations

**Learning Outcomes:**

1. List several ways that mutations can alter the amino acid sequence of a polypeptide.
2. Outline how mutations in protein-encoding genes may affect the amino acid sequence of a polypeptide.
3. Explain how mutations that occur outside of the coding sequence can affect the expression of a gene.
4. Compare and contrast the effects of mutations in somatic cells versus germ-line cells.

How do mutations affect traits? To answer this question at the molecular level, we must understand how changes in the DNA sequence of a gene ultimately affect gene function. Most of our understanding of mutations has come from the study of experimental organisms, such as bacteria and *Drosophila*. Researchers can expose these organisms to agents that cause mutations and then study the consequences of the mutations that arise. In addition, because these organisms have a short generation time, researchers can investigate the effects of mutations when they are passed from cell to cell and from parent to offspring.

The structure and amount of genetic material can be altered in a variety of ways. For example, the structure and number of chromosomes can change. We will examine these types of genetic changes in Chapter 16. In this section, we will focus our attention on gene mutations, which are relatively small changes in the sequence of bases in a particular gene. We will also consider how the timing of new mutations during an organism's development has important consequences.

## Gene Mutations Alter the DNA Sequence of a Gene

Mutations cause two basic types of changes to a gene: (1) the base sequence within a gene can be changed; and (2) one or more base pairs can be added to or removed from a gene. A **point mutation** affects only a single base pair within the DNA. For example, the DNA sequence shown here has been altered by a **base substitution** in which a T (in the top strand) has been replaced by a G and the corresponding A in the bottom strand is replaced with a C:

5′–CCCGCTAGATA–3′          5′–CCCGCGAGATA–3′
3′–GGGCGATCTAT–5′    →    3′–GGGCGCTCTAT–5′

A point mutation could also involve the addition or deletion of a single base pair to a DNA sequence. For example, in the following sequence, a single base pair (A-T) has been added to the DNA:

5′–GGCGCTAGATC—3′          5′–GGCAGCTAGATC–3′
3′–CCGCGATCTAG—5′    →    3′–CCGTCGATCTAG–5′

Though point mutations may seem like small changes to a DNA sequence, they can have important consequences when genes are expressed, as we will see next.

## Gene Mutations May Affect the Amino Acid Sequence of a Polypeptide

If a mutation occurs within the coding region of a protein-encoding gene, the mutation may alter that sequence in a variety of ways. **Table 15.1** considers the potential effects of point mutations.

*Silent Mutations*   **Silent mutations** do not alter the amino acid sequence of the polypeptide, even though the nucleotide sequence has changed. As discussed in Chapter 12, the genetic code is degenerate; that is, more than one codon can specify the same amino acid. Silent mutations occur in the third base of many codons without changing the type of amino acid that is encoded.

*Missense Mutations*   A **missense mutation** is a base substitution that changes a single amino acid in a polypeptide sequence.

| Table 15.1 | Consequences of Point Mutations Within the Coding Sequence of a Protein-Encoding Gene | |
|---|---|---|
| **Mutation in the DNA** | **Effect on polypeptide** | **Example*** |
| None | None | |
| Base substitution | **Silent**—causes no change | |
| Base substitution | **Missense**—changes one amino acid in the polypeptide | |
| Base substitution | **Nonsense**—changes a normal codon to a stop codon | |
| Addition of a single base | **Frameshift**—produces a different amino acid sequence | |

*DNA sequence in the coding strand. This sequence is the same as the mRNA sequence except that RNA contains uracil (U) instead of thymine (T).

A missense mutation may not alter protein function because it changes only a single amino acid within a polypeptide that is typically hundreds of amino acids in length. A missense mutation that substitutes an amino acid with a chemistry similar to the original amino acid is less likely to alter protein function. For example, a missense mutation that substitutes a glutamic acid for an aspartic acid may not alter protein function because both amino acids are negatively charged and have similar side chain structures.

Alternatively, some missense mutations have a dramatic effect on protein function. A striking example occurs in the human disease known as **sickle cell disease**. This disease involves a missense mutation in the β-globin gene, which encodes one of the polypeptide subunits that make up hemoglobin, the oxygen-carrying protein in red blood cells. In the most common form of this disease, a missense mutation alters the polypeptide sequence such that the sixth amino acid is changed from a glutamic acid to a valine (**Figure 15.1**). Because glutamic acid is hydrophilic but valine is hydrophobic, this single amino acid substitution alters the structure and function of the hemoglobin protein. The mutant hemoglobin subunits tend to stick to one another when the oxygen concentration is low. The aggregated proteins form fiber-like structures within red blood cells, which causes the cells to lose their normal disc-shaped morphology and become sickle-shaped. It is amazing that a single amino acid substitution could have such a profound effect on the structure of cells.

*Nonsense Mutations*   A **nonsense mutation** involves a change from a normal codon to a stop, or termination, codon. This causes translation to be terminated earlier than expected, producing a truncated polypeptide (see Table 15.1). Compared with a normal polypeptide, a shorter polypeptide is much less likely to function properly.

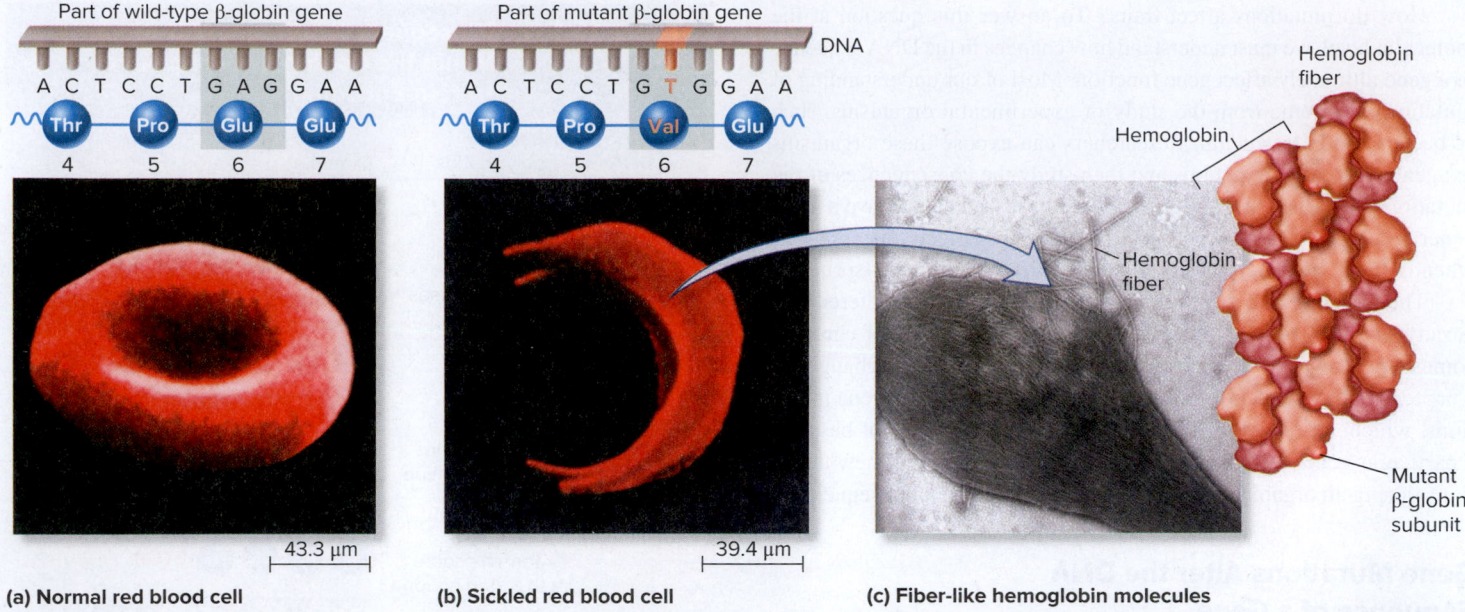

**Figure 15.1** **A missense mutation that causes red blood cells to sickle in sickle cell disease.** Scanning electron micrographs of **(a)** a normal red blood cell and **(b)** a sickled red blood cell. As shown above the micrographs, a missense mutation in the β-globin gene (which codes for a subunit of hemoglobin) changes the sixth amino acid in the β-globin polypeptide from glutamic acid (Glu) to valine (Val). **(c)** This micrograph shows how this alteration to the structure of β-globin causes the formation of abnormal fiber-like structures. In normal red blood cells, hemoglobin proteins do not form fibers. a: ©Mary Martin/Science Source; b: ©Science Source; c: Courtesy of Thomas Wellems and Robert Josephs. Electron Microscopy and Image Processing Laboratory, University of Chicago

**Concept Check:**  *Based on the fiber-like structures seen in part (c) of this figure, what aspect of hemoglobin structure does a glutamic acid at the sixth position in normal β-globin prevent? Speculate as to how the charge of this amino acid may play a role.*

*Frameshift Mutations*   Finally, a **frameshift mutation** involves the addition or deletion of a number of nucleotides that is not a multiple of three. For example, a frameshift mutation could involve the addition or deletion of one, two, four, or five nucleotides. Because the codons are read in multiples of three, these types of insertions or deletions shift the reading frame so a completely different amino acid sequence occurs downstream from the mutation (see Table 15.1). Such a large change in polypeptide structure is likely to inhibit protein function.

Changes in protein function may affect the ability of an organism to survive and to reproduce. Except for silent mutations, new mutations are more likely to produce polypeptides that have reduced rather than enhanced function. However, mutations can occasionally produce a polypeptide that has an enhanced function. Such mutations may change in frequency in a population over the course of many generations due to natural selection. This topic is discussed in Chapter 23.

## Gene Mutations That Occur Outside of Coding Sequences Can Influence Gene Expression

Thus far, we have focused our attention on mutations in the coding regions of protein-encoding genes. In Chapters 12 and 14, we explored the role of DNA sequences in gene expression. A mutation can occur within a noncoding DNA sequence and affect gene expression (Table 15.2). For example, a mutation may alter the sequence within the promoter of a gene, thereby affecting the rate of transcription. A mutation that improves the ability of RNA polymerase to bind

to the promoter may enhance transcription, whereas other mutations may inhibit transcription.

Mutations in regulatory elements or operator sites can alter the regulation of gene transcription. For example, in Chapter 14, we considered the roles of regulatory elements such as the *lac* operator site in *E. coli*, which is recognized by the lac repressor protein (refer back to Figure 14.7). Mutations in the *lac* operator site can disrupt the proper regulation of the *lac* operon. An operator mutation may change the DNA sequence so the lac repressor protein does not bind to it. This mutation would cause the operon to be constitutively expressed.

| Table 15.2 | Effects of Mutations Outside of the Coding Sequence of a Gene |
|---|---|
| **Sequence** | **Effect of mutation** |
| Promoter | May increase or decrease the rate of transcription |
| Transcriptional regulatory element/operator site | May alter the regulation of transcription |
| Splice sites | May alter the ability of pre-mRNA to be properly spliced |
| Translational regulatory element | May alter the ability of mRNA to be translationally regulated |
| Intergenic region | Not as likely to have an effect on gene expression |

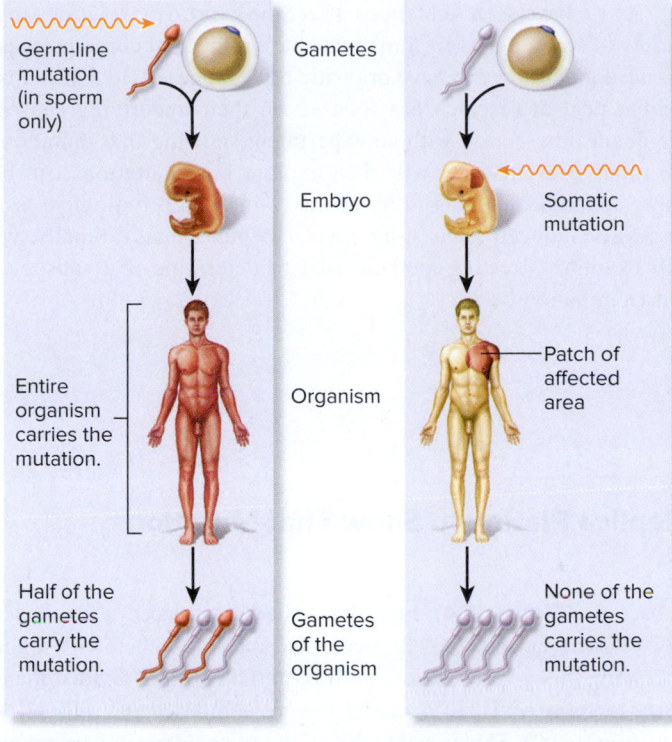

(a) **Germ-line mutation**          (b) **Somatic cell mutation**

**Figure 15.2**  **The effects of germ-line versus somatic cell mutations.** The red color indicates which cells carry the mutation. **(a)** In this example, a mutation occurs in a gamete. This germ-line mutation will be passed to every cell of the body. Because humans have two copies of most genes, a germ-line mutation in one of those two copies is transmitted to only half of the gametes. **(b)** Somatic mutations affect a limited area of the body and are not transmitted to offspring.

**Core Concept: Information**  As a multicellular organism grows and develops, a germ-line mutation is transmitted to all cells of the body, whereas a somatic mutation is found only in a particular region.

## Mutations Can Occur in Germ-Line or Somatic Cells

Let's now consider how the timing of a mutation may have important consequences for its potential effects. Multicellular organisms typically begin their lives as a single fertilized egg cell that divides many times to produce all the cells of an adult organism. A mutation can occur in any cell of the body, either very early in life, such as in a gamete (egg or sperm) or a fertilized egg, or later in life, such as in the embryonic or adult stages. The number and location of cells with a mutation are critical both to the severity of the genetic effect and to the ability of the mutation to be passed on to offspring.

Geneticists classify the cells of animals into two types: germ-line cells and somatic cells. The term **germ line** refers to cells that give rise to gametes, such as egg and sperm cells. A germ-line mutation can occur directly in an egg or sperm cell, or it can occur in a precursor cell that produces the gamete. If a mutant human

gamete participates in fertilization, all the cells of the resulting offspring will contain the mutation, as indicated by the red color in **Figure 15.2a**. Likewise, when such an individual produces gametes, the mutation may be transmitted to future generations of offspring. Because humans carry two copies of most genes, a new mutation in a single gene has a 50% chance of being transmitted from parent to offspring.

The **somatic cells** constitute all cells of the body except for the germ line. Examples include skin cells and muscle cells. Mutations can also occur within somatic cells at early or late stages of development. What are the consequences of a mutation that happens during the embryonic stage? As shown in **Figure 15.2b**, a mutation occurred within a single embryonic cell. This single somatic cell was the precursor for many cells of the adult. Therefore, in the adult, a patch of tissue contains cells that carry the mutation. The size of any patch depends on the timing of a new mutation. In general, the earlier a mutation occurs during development, the larger the patch. An individual with somatic regions that are genetically different from each other is called a **mosaic**.

**Figure 15.3** illustrates a child who had a somatic mutation during an early stage of development. In this case, the child has a streak of white hair while the rest of his hair is black. Presumably, a single mutation happened in an embryonic cell that ultimately gave rise to the patch that produced the white hair.

Although a change in hair color is not a harmful consequence, mutations during early stages of life can be quite harmful, especially if they disrupt essential developmental processes. Even though it is sensible to avoid environmental agents that cause mutations at any stage of life, the possibility of somatic mutations is a compelling reason to avoid such agents during the early stages of life such as embryonic and fetal development, infancy, and early childhood.

**Figure 15.3**  **Example of a somatic mutation.** This child has a streak of white hair. This is due to a somatic mutation in a single cell during embryonic development. This cell continued to divide to produce a streak of white hair. ©Otero/GTphoto

**Concept Check:**  *Can this child with a streak of white hair transmit this trait to his future offspring?*

## 15.2  Causes of Mutations

### Learning Outcomes:

1. **CoreSKILL** » Analyze the replica plating experiments of the Lederbergs.
2. Describe the difference between spontaneous and induced mutations.
3. **CoreSKILL** » Analyze the results of an Ames test for determining if a substance is a mutagen.

As we have seen, mutations affect the expression of genes in a variety of ways, and their timing can have important consequences. Because mutations can have dramatic effects on individuals' traits, a great deal of research has focused on their underlying causes. We begin this section with an experiment showing that mutations are random events. We will then explore how mutations can be either spontaneous (caused by mistakes in natural biological processes) or induced (caused by environmental agents). Finally, we will examine a testing method used to determine if a substance causes mutations.

 **Core Skill: Process of Science**

## Feature Investigation | The Lederbergs Used Replica Plating to Show That Mutations Are Random Events

Prior to understanding the causes of mutations at the molecular level, scientists considered the following question: Are mutations that affect the traits of an individual caused by pre-existing circumstances, or are they random events that may happen in any gene of any individual? In the 19th century, French naturalist Jean-Baptiste Lamarck proposed that physiological events (such as use or disuse) determine whether traits are passed along to offspring. For example, his hypothesis suggested that an individual who practiced and became adept at a physical activity, such as the long jump, would pass that quality on to his or her offspring. Alternatively, geneticists in the early 20th century suggested that genetic variation occurs as a matter of chance. According to this view, those individuals whose genes happen to contain beneficial mutations are more likely to survive and pass those genes to their offspring.

These opposing views were tested in bacterial studies in the 1940s and 1950s. One such study, by American microbiologists Joshua and Esther Lederberg, focused on the occurrence of mutations in bacteria (**Figure 15.4**). First, the Lederbergs placed a few dozen *E. coli* bacteria onto growth media and incubated them overnight. Following this growth period, each bacterial cell had divided many times to form a visible bacterial colony composed of millions of cells (see step 2). This is called the master plate. Next, in a technique known as **replica plating,** a sterile piece of velvet cloth was lightly touched to the master plate to pick up bacterial cells from each colony on the master plate. The Lederbergs then transferred this replica to two secondary plates containing an agent that selected for the growth of bacterial cells with a particular mutation.

In the example shown in Figure 15.4, the secondary plates contained T1 bacteriophages, which are viruses that infect bacteria and cause them to lyse. On these plates, only those rare cells that had acquired a mutation conferring resistance to T1, termed *ton*[r], could grow. All other cells were lysed by the proliferation of bacteriophages in the bacteria. Therefore, only a few colonies were observed on the secondary plates. Strikingly, these colonies occupied the same

**Figure 15.4**  The experiment performed by the Lederbergs showing that mutations are random events.

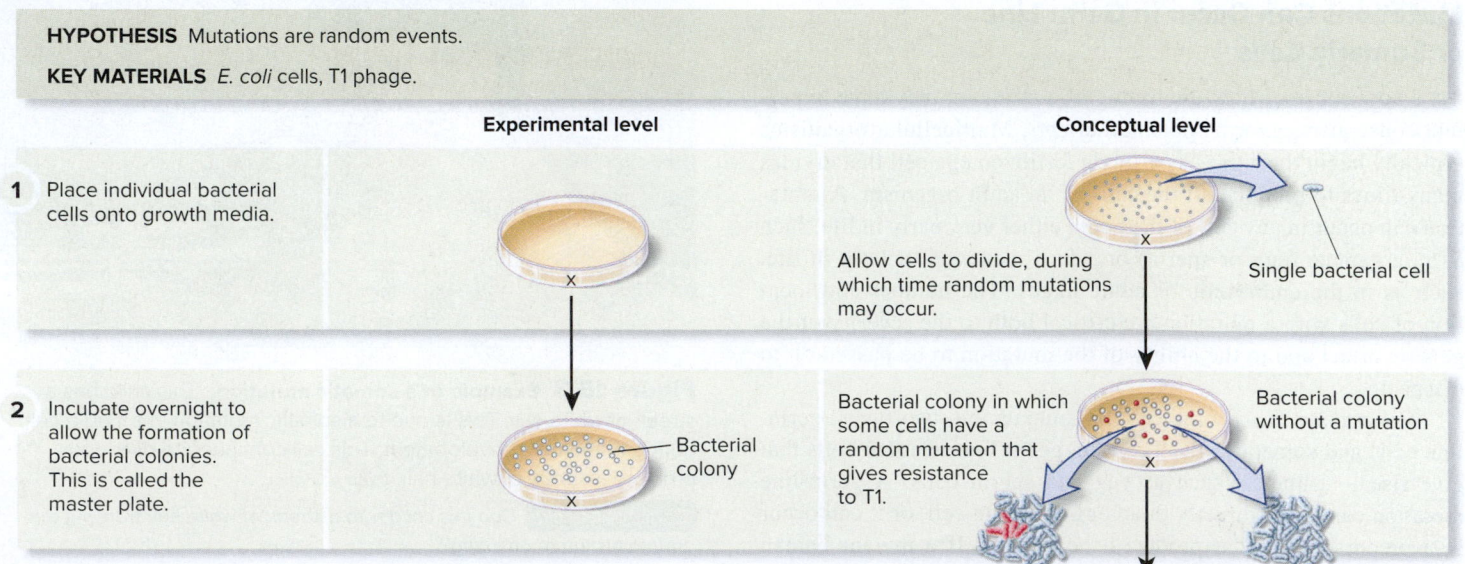

**HYPOTHESIS**  Mutations are random events.

**KEY MATERIALS**  *E. coli* cells, T1 phage.

Experimental level | Conceptual level

1  Place individual bacterial cells onto growth media.

Allow cells to divide, during which time random mutations may occur.

Single bacterial cell

2  Incubate overnight to allow the formation of bacterial colonies. This is called the master plate.

Bacterial colony

Bacterial colony in which some cells have a random mutation that gives resistance to T1.

Bacterial colony without a mutation

**3**  Press a velvet cloth (wrapped over a cylinder) onto the master plate, and then lift gently to obtain a replica of each bacterial colony. Press the replica onto 2 secondary plates that contain T1 bacteriophage. Incubate overnight to allow bacterial growth.

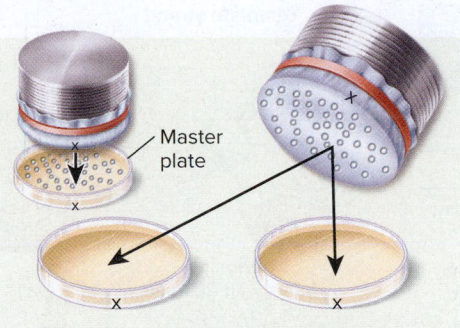

Master plate

Secondary plates containing T1 phage

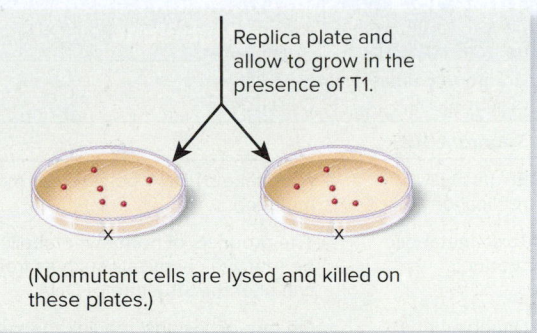

Replica plate and allow to grow in the presence of T1.

(Nonmutant cells are lysed and killed on these plates.)

**4**  **THE DATA**

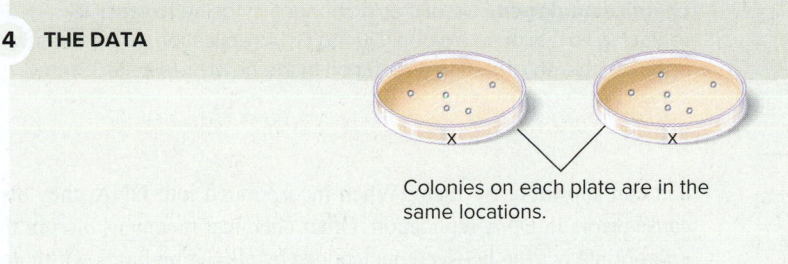

Colonies on each plate are in the same locations.

**5**  **CONCLUSION**  Mutations are random events. In this case, the mutations occurred on the master plate prior to exposure to T1 bacteriophage.

**6**  **SOURCE**  Lederberg, J., and Lederberg, E. M. 1952. Replica Plating and Indirect Selection of Bacterial Mutants. *Journal of Bacteriology* 63: 399–406.

locations on each plate. How did the Lederbergs interpret these results? The data indicated that the *ton*<sup>r</sup> mutations occurred randomly while the bacterial cells were forming colonies on the nonselective master plate. The presence of T1 bacteriophages in the secondary plates did not cause the mutations to develop. Rather, the T1 bacteriophages simply selected for the growth of *ton*<sup>r</sup> mutants that were already in the population. These results supported the idea that mutations are random events.

**Experimental Questions**

1. Explain the opposing views of mutation prior to the Lederbergs' study.
2. **CoreSKILL »** What hypothesis was being tested by the Lederbergs?
3. **CoreSKILL »** How did the results of the Lederbergs support or falsify the hypothesis?

## Mutations May Be Spontaneous or Induced

Biologists categorize the causes of mutation as spontaneous or induced (**Table 15.3**). **Spontaneous mutations** result from abnormalities in biological processes. Spontaneous mutations reflect the observation that biology isn't perfect. Enzymes, for example, can function abnormally. In Chapter 11, we learned that DNA polymerase can make mistakes during DNA replication by putting the wrong base in a newly synthesized daughter strand. Though such errors are rare, due to the proofreading function of DNA polymerase, they do occur. In addition, normal metabolic processes within the cell may produce toxic chemicals such as free radicals that can react directly with the DNA and alter its structure. Finally, the structures of nucleotides are not absolutely stable. On occasion, the structure of a base may spontaneously change, and such a change may cause a mutation if it occurs immediately prior to DNA replication.

The rates of spontaneous mutations vary from species to species and from gene to gene. Larger genes are usually more likely to incur a mutation than are smaller ones. A common rate of spontaneous mutation among various species is approximately 1 mutation for every 1 million genes per cell division, which equals 1 in $10^6$, or simply $10^{-6}$.

This is the expected rate of spontaneous mutation, which creates the variation that is the raw material of evolution.

**Induced mutations** are caused by environmental agents that enter the cell and alter the structure of DNA. They cause the mutation rate to be higher than the spontaneous mutation rate. Agents that cause mutation are called **mutagens**. Mutagenic agents can be categorized as **chemical** or **physical mutagens** (**Table 15.4**). We will consider their effects next.

## Mutagens Alter DNA Structure in Different Ways

Researchers have discovered that an enormous array of agents act as mutagens. We often hear in the news media that we should avoid these agents in our foods and living environments. We even use products such as sunscreens that help us avoid the mutagenic effects of ultraviolet (UV) light from the Sun. The public is often concerned about mutagens for two important reasons. First, mutagenic agents are usually involved in the development of human cancers. Second, because new mutations may be deleterious, people want to avoid mutagens to prevent mutations that may have harmful effects in their future offspring.

## Table 15.3   Some Common Causes of Gene Mutations

| Common causes of mutations | Description |
|---|---|
| **Spontaneous** | |
| Errors in DNA replication | A mistake by DNA polymerase may cause a point mutation. |
| Toxic metabolic products | The products of normal metabolic processes may be reactive chemicals such as free radicals that can alter the structure of DNA. |
| Changes in nucleotide structure | On rare occasions, the linkage between a purine and deoxyribose can spontaneously break. Changes in base structure (isomerization) may cause mispairing during DNA replication. |
| Transposons | As discussed in Chapter 21, transposons are small segments of DNA that can insert at various sites in the genome. If they insert into a gene, they may inactivate the gene. |
| **Induced** | |
| Chemical agents | Chemical substances, such as benzo(a)pyrene, a chemical found in cigarette smoke, may cause changes in the structure of DNA. |
| Physical agents | Physical agents such as UV (ultraviolet) light and X-rays can damage DNA. |

## Table 15.4   Examples of Mutagens

| Mutagen | Effect(s) on DNA structure |
|---|---|
| **Chemical** | |
| Nitrous acid | Deaminates bases |
| 5-Bromouracil | Acts as a base analogue |
| 2-Aminopurine | Acts as a base analogue |
| Nitrogen mustard | Alkylates bases |
| Ethyl methanesulfonate (EMS) | Alkylates bases |
| Benzo(a)pyrene | Inserts between bases in the DNA double helix and causes additions or deletions |
| **Physical** | |
| X-rays | Causes base deletions, single nicks in DNA strands, crosslinking, and chromosomal breaks |
| UV light | Promotes pyrimidine dimer formation, which involves covalent bonds between adjacent pyrimidines (C and T) |

**Chemical Mutagens** How do mutagens affect DNA structure? Some chemical mutagens act by covalently modifying the structure of nucleotides. For example, nitrous acid ($HNO_2$) deaminates bases by replacing amino groups with keto groups (replacing —$NH_2$ with =O). This can change cytosine to uracil. When this altered DNA replicates, the modified base does not pair with the appropriate base in the newly made strand. In this case, uracil pairs with adenine (**Figure 15.5**).

Similarly, 5-bromouracil and 2-aminopurine, which are called base analogues, have structures that are similar to particular bases in DNA

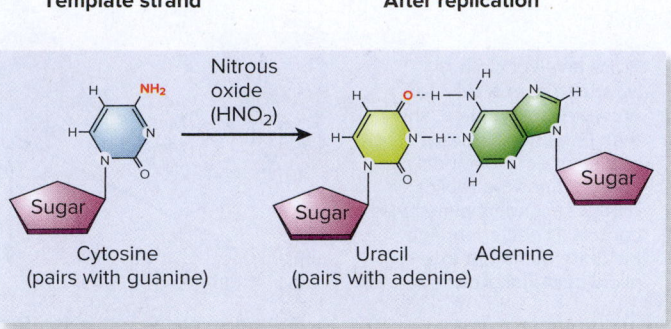

**Template strand**      **After replication**

Cytosine (pairs with guanine) → Uracil (pairs with adenine)   Adenine

**Figure 15.5 Deamination and mispairing of modified bases by a chemical mutagen.** Nitrous acid changes cytosine to uracil by replacing $NH_2$ with an oxygen. During DNA replication, uracil pairs with adenine, thereby creating a mutation in the newly replicated strand.

and can substitute for them. When incorporated into DNA, they also cause errors in DNA replication. Other chemical mutagens disrupt the appropriate pairing between nucleotides by alkylating bases within the DNA. During alkylation, methyl or ethyl groups are covalently attached to the bases. Examples of alkylating agents include nitrogen mustards (used as a chemical weapon during World War I) and ethyl methanesulfonate (EMS), which is used as a mutagen in laboratory experiments.

Some chemical mutagens exert their effects by interfering with DNA replication. For example, benzo(a)pyrene, which is found in automobile exhaust, cigarette smoke, and charbroiled food, is metabolized to a compound (benzopyrene diol epoxide) that inserts between the bases of the double helix, thereby distorting the helical structure. When DNA containing such a mutagen is replicated, single-nucleotide additions and deletions may be incorporated into the newly made strands.

**Physical Mutagens** DNA molecules are also sensitive to physical agents such as radiation. In particular, radiation of short wavelength and high energy, known as ionizing radiation, is known to alter DNA structure. Ionizing radiation includes X-rays and gamma rays. This type of radiation can penetrate deeply into biological materials, where it creates free radicals. These molecules can alter the structure of DNA in a variety of ways. Exposure to high doses of ionizing radiation can cause base deletions, breaks in one DNA strand, or even a break in both DNA strands.

Nonionizing radiation, such as UV light, contains less energy, and so it penetrates only the surface of biological materials, such as the skin. Nevertheless, UV light is known to cause mutations. For example, UV light can cause the formation of a **thymine dimer**, which is a site where two adjacent thymine bases become covalently crosslinked to each other (**Figure 15.6**).

Thymine dimers are typically repaired before or during DNA replication. However, if such repair fails to occur, a thymine dimer may cause a mutation when that DNA strand is replicated. When DNA polymerase attempts to replicate over a thymine dimer, proper base pairing does not occur between the template strand and the incoming nucleotides. This mispairing can cause gaps in the newly made strand or the incorporation of incorrect bases. Plants, in particular, must have effective ways to prevent UV damage because they are exposed to sunlight throughout the day.

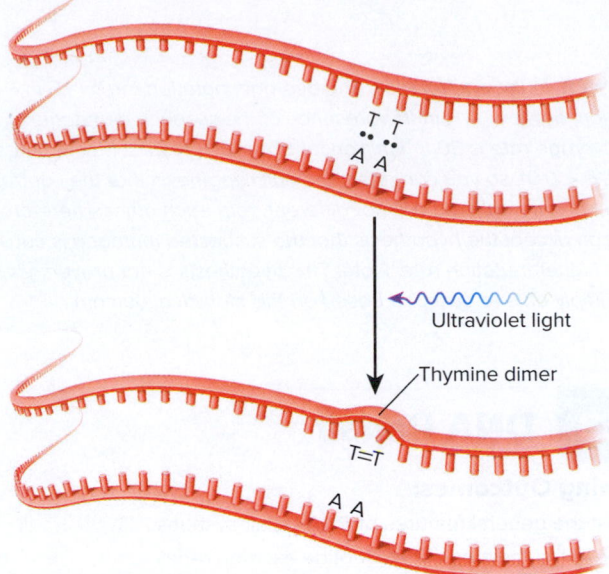

**Figure 15.6  Formation and structure of a thymine dimer.**

**Concept Check:**  *Why is a thymine dimer harmful?*

## Testing Methods Determine If an Agent Is a Mutagen

Because mutagens are harmful, researchers have developed testing methods to evaluate the ability of a substance to cause mutation. One commonly used test is the **Ames test**, which was developed by American biochemist Bruce Ames in the 1970s. This test uses a strain of a bacterium, *Salmonella typhimurium*, that cannot synthesize the amino acid histidine. This strain contains a point mutation within a gene that encodes an enzyme required for histidine biosynthesis. The mutation renders the enzyme inactive. The bacteria cannot grow unless histidine has been added to the growth medium. However, a second mutation may correct the first mutation, thereby restoring the ability to synthesize histidine. The Ames test monitors the rate at which this second mutation occurs and thereby indicates whether an agent increases the mutation rate above the spontaneous rate.

**Figure 15.7** outlines the steps in the Ames test. The suspected mutagen is mixed with a rat liver extract and the strain of *S. typhimurium* that cannot synthesize histidine. Because some potential mutagens may require activation by cellular enzymes, the rat liver extract provides a mixture of enzymes that may cause such activation. This step improves the ability to identify agents that cause mutations in mammals. As a control, bacteria that have not been exposed to the mutagen are also tested. After an incubation period in which mutations may occur, a large number of bacteria are plated on a growth medium that does not contain histidine. The *S. typhimurium* strain is not expected to grow on these plates. However, if a mutation has occurred that allows a cell to synthesize histidine, the bacterium harboring this second mutation will proliferate during an overnight incubation period to form a visible bacterial colony.

To estimate the mutation rate, the colonies that grow in the absence of histidine are counted and compared with the total number of bacterial cells that were originally placed on the plate for both the suspected-mutagen sample and the control. The control condition is a measure of the spontaneous mutation rate, whereas

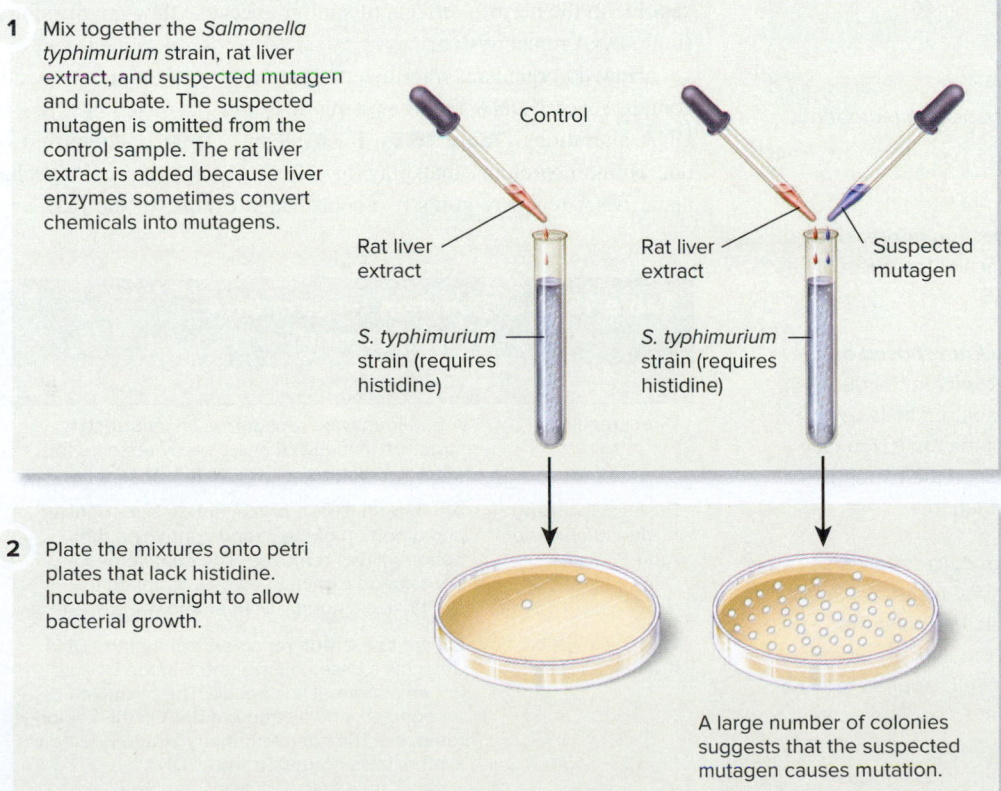

1  Mix together the *Salmonella typhimurium* strain, rat liver extract, and suspected mutagen and incubate. The suspected mutagen is omitted from the control sample. The rat liver extract is added because liver enzymes sometimes convert chemicals into mutagens.

Control

Rat liver extract

*S. typhimurium* strain (requires histidine)

Rat liver extract

*S. typhimurium* strain (requires histidine)

Suspected mutagen

2  Plate the mixtures onto petri plates that lack histidine. Incubate overnight to allow bacterial growth.

A large number of colonies suggests that the suspected mutagen causes mutation.

**Figure 15.7  The Ames test for mutagenicity.** In this example, 2 million bacterial cells were placed on plates lacking histidine. Two colonies arose from the control sample, whereas 44 arose from the sample exposed to a suspected mutagen.

**Core Skill: Science and Society** Biologists have developed many methods, including the Ames test, for determining if a substance is a mutagen. The results of these tests have prevented the use of many different chemicals in the production of food and also resulted in warning labels on products such as cigarettes.

the other sample measures the rate of mutation in the presence of the suspected mutagen. As an example, let's suppose that 2 million bacteria were plated from both the control and the suspected-mutagen tubes. In the control experiment, 2 bacterial colonies were observed. The spontaneous mutation rate is calculated by dividing 2 (the number of mutants) by 2 million (the number of original cells). This equals 1 in 1 million, or $1 \times 10^{-6}$. By comparison, 44 colonies arose from the suspected-mutagen sample (see Figure 15.7). In this case, the mutation rate would be 44 divided by 2 million, which equals $2.2 \times 10^{-5}$. The mutation rate in the presence of the mutagen is over 20 times higher than the spontaneous mutation rate.

How do we judge if an agent is a mutagen? Researchers compare the mutation rate in the presence and absence of the suspected mutagen. The experimental procedure shown in Figure 15.7 is conducted several times. If statistics reveal that the mutation rate in the suspected-mutagen sample is significantly higher than in the control sample, they may tentatively conclude that the agent is a mutagen. Interestingly, many studies have used the Ames test to compare the urine from cigarette smokers with that from nonsmokers. This research has shown that urine from smokers contains much higher levels of mutagens.

## BIO TIPS

**THE QUESTION** *Let's suppose a researcher studied the effects of a suspected mutagen, mutagen X, using the protocol described in Figure 15.7. The following data were obtained after placing 2 million cells on each plate:*

| | Number of colonies | |
| Trial | Control (no mutagen) | With mutagen X |
| --- | --- | --- |
| 1 | 3 | 62 |
| 2 | 2 | 77 |
| 3 | 5 | 46 |
| 4 | 2 | 55 |

*Calculate the average mutation rate in the presence and absence of mutagen X. Conduct a t-test to determine if suspected mutagen X is significantly affecting the mutation rate.*

**T**OPIC *What topic in biology does this question address?* The topic is identifying a mutagen. More specifically, the question is about analyzing results from the Ames test.

**I**NFORMATION *What information do you know based on the question and your understanding of the topic?* In the question, you are given data regarding the outcome of four trials using the Ames test. From your understanding of the topic, you may remember that a higher number of colonies on the experimental plate may indicate that a substance is a mutagen.

**P**ROBLEM-SOLVING **S**TRATEGY *Make a calculation. Use statistics.* To begin to solve this problem, you first need to calculate the average mutation rates. To do this, take the average of the four trials and then divide the average number of mutant colonies by the total number of cells applied to each plate (in this case, 2 million). You also need to conduct a t-test to determine if the control and experimental data are significantly different. A description of a t-test can be found in various statistics textbooks.

**ANSWER** *In the control trials, the average mutation rate is 1.5 in 1 million, or $1.5 \times 10^{-6}$. In the presence of the suspected mutagen, the average rate is 30 in 1 million, or $30 \times 10^{-6}$. From a t-test of these data, $P < 0.01$, so you can reject the null hypothesis that the control and experimental data are not different from each other. Therefore, you can accept the hypothesis that the suspected mutagen is causing a higher mutation rate. Note: This hypothesis is not proven; you are simply able to accept it based on this statistical outcome.*

## 15.3    DNA Repair

**Learning Outcomes:**

1. List the general features of DNA repair systems.
2. Describe the steps of nucleotide excision repair.
3. Explain the connection between a defect in DNA repair and the inherited human disease xeroderma pigmentosum.

In the previous sections, we considered the consequences and causes of mutations. As we have seen, mutations are random events that often have negative consequences. To minimize mutations, all living organisms have the ability to repair changes that occur in the structure of DNA. For example, in Chapter 11, we considered how DNA polymerase has a proofreading function that helps to prevent mutations from arising during DNA replication. In this section, we will examine DNA repair systems that can detect abnormalities in DNA structure and repair them. The importance of these systems becomes evident when they are missing. For example, as discussed at the beginning of this chapter, persons with xeroderma pigmentosum are highly susceptible to the harmful effects of sunlight because they are missing a single DNA repair system.

How do organisms minimize the occurrence of mutations? Cells contain several DNA repair systems that can fix different types of DNA alterations (**Table 15.5**). Each repair system is composed of one or more proteins that play specific roles in the repair mechanism. DNA repair requires two coordinated events. In the first step,

| Table 15.5 | Common Types of DNA Repair Systems* |
| --- | --- |
| **System** | **Description** |
| Direct repair | A repair enzyme recognizes an incorrect structure in the DNA and directly restores the correct structure. |
| Base excision and nucleotide excision repair | An abnormal base or nucleotide is recognized, and a portion of the strand containing the abnormality is removed. The complementary DNA strand is then used as a template to synthesize a normal DNA strand. |
| Methyl-directed mismatch repair | Similar to excision repair except that the DNA defect is a base pair mismatch in the DNA, not an abnormal nucleotide. The mismatch is recognized, and a strand of DNA in this region is removed. The complementary strand is used to synthesize a normal strand of DNA. |

*Other types of repair systems exist; these are common examples.

one or more proteins in the repair system detect an irregularity in DNA structure. In the second step, the abnormality is repaired. In some cases, the change in DNA structure can be directly repaired. For example, DNA may be modified by the attachment of an alkyl group, such as —$CH_2CH_3$, to a base. In **direct repair**, an enzyme removes this alkyl group, thereby restoring the structure of the original base. More commonly, however, the altered DNA is removed, and a new segment of DNA is synthesized. In this section, we will examine nucleotide excision repair as an example of how such systems operate. This system, which is found in all species, is an important mechanism of DNA repair.

## Nucleotide Excision Repair Removes Segments of Damaged DNA

In **nucleotide excision repair** (**NER**), a region encompassing several nucleotides in the damaged strand is removed from the DNA, and the intact undamaged strand is used as a template for the resynthesis of a normal complementary strand. NER can fix many different types of DNA damage, including UV-induced damage, chemically modified bases, missing bases, and various types of crosslinks (such as thymine dimers). The system is found in all species, although its molecular mechanism is best understood in bacteria.

In *E. coli*, the NER system is composed of four key proteins: UvrA, UvrB, UvrC, and UvrD. They are named Uvr because they are involved in ultraviolet light repair of thymine dimers, although these proteins are also important in repairing chemically damaged DNA. In addition, DNA polymerase and DNA ligase are required to complete the repair process.

How does the NER system work?

1. Two UvrA proteins and one UvrB protein form a complex that tracks along the DNA (**Figure 15.8**). Damaged DNA will have a distorted double helix, which is sensed by the UvrA-UvrB complex.

2. When the complex identifies a damaged site, the two UvrA proteins are released, and UvrC binds to UvrB at the site.

3. The UvrC protein makes incisions in one DNA strand on both sides of the damaged site.

4. After this incision process, UvrC is released. UvrD binds to UvrB. UvrD then begins to separate the DNA strands, and UvrB is released. The action of UvrD unravels the DNA, which removes a short DNA strand that contains the damaged region. UvrD is released.

5. After the damaged DNA strand is removed, a gap is left in the double helix. DNA polymerase fills in the gap using the undamaged strand as a template. Finally, DNA ligase makes the final covalent connection between the newly made DNA and the original DNA strand.

## Human Genetic Diseases Occur When a Component of the NER System Is Missing

Thus far, we have considered the NER system in *E. coli*. In humans, NER systems were discovered by the analysis of genetic diseases that affect DNA repair. These include xeroderma pigmentosum (XP),

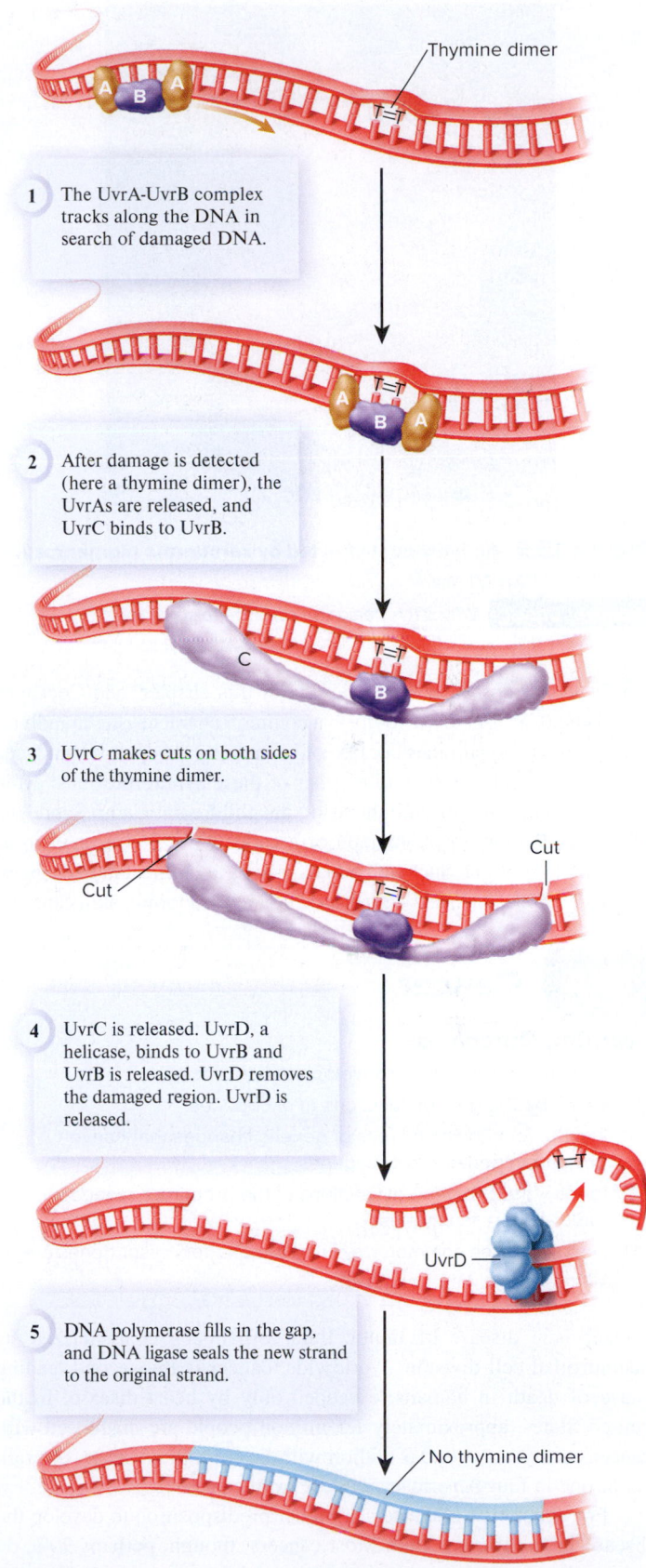

1. The UvrA-UvrB complex tracks along the DNA in search of damaged DNA.

2. After damage is detected (here a thymine dimer), the UvrAs are released, and UvrC binds to UvrB.

3. UvrC makes cuts on both sides of the thymine dimer.

4. UvrC is released. UvrD, a helicase, binds to UvrB and UvrB is released. UvrD removes the damaged region. UvrD is released.

5. DNA polymerase fills in the gap, and DNA ligase seals the new strand to the original strand.

**Figure 15.8** **Nucleotide excision repair in *E. coli*.**

*Concept Check:* *Which components of the NER system are responsible for removing the damaged DNA?*

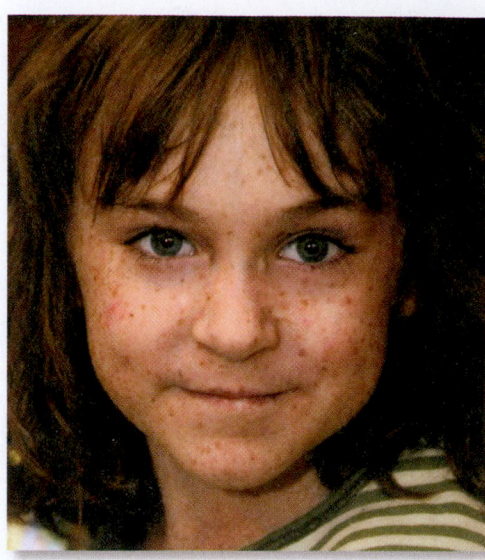

**Figure 15.9** **An individual affected by xeroderma pigmentosum.**
©Barcroft Media/Getty Images

**Concept Check:** *Why is this person so sensitive to sunlight?*

which was discussed at the beginning of this chapter, and Cockayne syndrome (CS). Photosensitivity is a common characteristic in individuals with these syndromes because of an inability to repair UV-induced lesions. Therefore, people with either of these syndromes must avoid prolonged exposure to sunlight, as do the children at Camp Sundown. **Figure 15.9** shows a photograph of a child with XP who had some exposure to sunlight. Such individuals may have pigmentation changes, precancerous lesions, and a predisposition to developing skin cancer.

## 15.4 Cancer

**Learning Outcomes:**

1. Outline the steps in the development of cancer.
2. Describe the general functions of oncogenes.
3. List the four common types of genetic changes that convert proto-oncogenes into oncogenes.
4. Identify the two general functions of the proteins encoded by tumor-suppressor genes.
5. Describe three common ways that tumor-suppressor genes are silenced.

Cancer is a disease of multicellular organisms characterized by uncontrolled cell division. Worldwide, cancer is the second leading cause of death in humans, exceeded only by heart disease. In the United States, approximately 1.5 million people are diagnosed with cancer each year; over 0.5 million will die from the disease. Overall, about one in four Americans will die from cancer.

For about 10% of cancers, a higher predisposition to develop the disease is an inherited trait. Most cancers, though, perhaps 90%, do not involve genetic changes that are passed from parent to offspring. Rather, cancer is usually an acquired condition that typically occurs later in life. At least 80% of all human cancers are related to exposure to **carcinogens**, agents that increase the likelihood of developing

cancer. Most carcinogens, such as UV light and certain chemicals in cigarette smoke, are mutagens that promote genetic changes in somatic cells. These genetic changes can alter gene expression in a way that ultimately affects cell division, leading to cancer. In this section, we will explore such genetic abnormalities.

How does cancer occur? In most cases, the development of cancer is a multistep process (**Figure 15.10**). Cancers originate from a single cell. This single cell and its lineage of daughter cells undergo a series of mutations and other genetic changes that cause the cells to grow abnormally. At an early stage, the cells form a **tumor**, which is an abnormal overgrowth of cells. For most types of cancer, growth begins as a precancerous mass, or a **benign tumor**. Such tumors do not invade adjacent tissues and do not spread throughout the body. This may be followed by additional genetic changes that cause some cells in the tumor to lose their normal growth regulation and it becomes a **malignant tumor**. At this stage, the individual has cancer. Cancerous tumors invade adjacent healthy tissues, and cancer cells may spread through the bloodstream or surrounding body fluids, a process called **metastasis**. If left untreated, malignant cells will cause the death of the organism.

Over the past few decades, researchers have identified many genes that promote cancer when they are mutant. By comparing the function of each mutant gene with the corresponding nonmutant gene found in healthy cells, these cancer-promoting genes have been placed into two categories.

- In some cases, a mutation causes a gene to be overactive— have an abnormally high level of expression. This overactivity contributes to the uncontrolled cell growth that is observed in cancer cells. This type of mutant gene is called an **oncogene**.

- Alternatively, when a **tumor-suppressor gene** is normal (that is, not mutant), it encodes a protein that helps to prevent cancer. However, when a mutation eliminates its function, cancer may occur.

Thus, the two categories of cancer-causing genes are based on the effects of mutations. Oncogenes are the result of mutations that cause overactivity, whereas cancer-causing mutations in tumor-suppressor genes are due to a loss of activity. In this section, we will begin with a discussion of oncogenes and then consider tumor-suppressor genes.

### Oncogenes May Result from Mutations That Cause the Overactivity of Proteins Involved with Cell Division

Over the past four decades, researchers have identified many oncogenes. A large number of oncogenes encode proteins that function in signal transduction pathways involved in cell growth. Cell division is regulated, in part, by growth factors. A growth factor binds to a receptor, which results in receptor activation (**Figure 15.11**). This stimulates an intracellular signal transduction pathway that activates transcription factors. In this way, the transcription of specific genes is activated in response to a growth factor. After they are made, the gene products promote cell division.

Eukaryotic species produce many different growth factors that play a role in cell division. Likewise, cells have several different types of signal transduction pathways, which are composed of proteins that respond to growth factors and promote cell division. Mutations in the

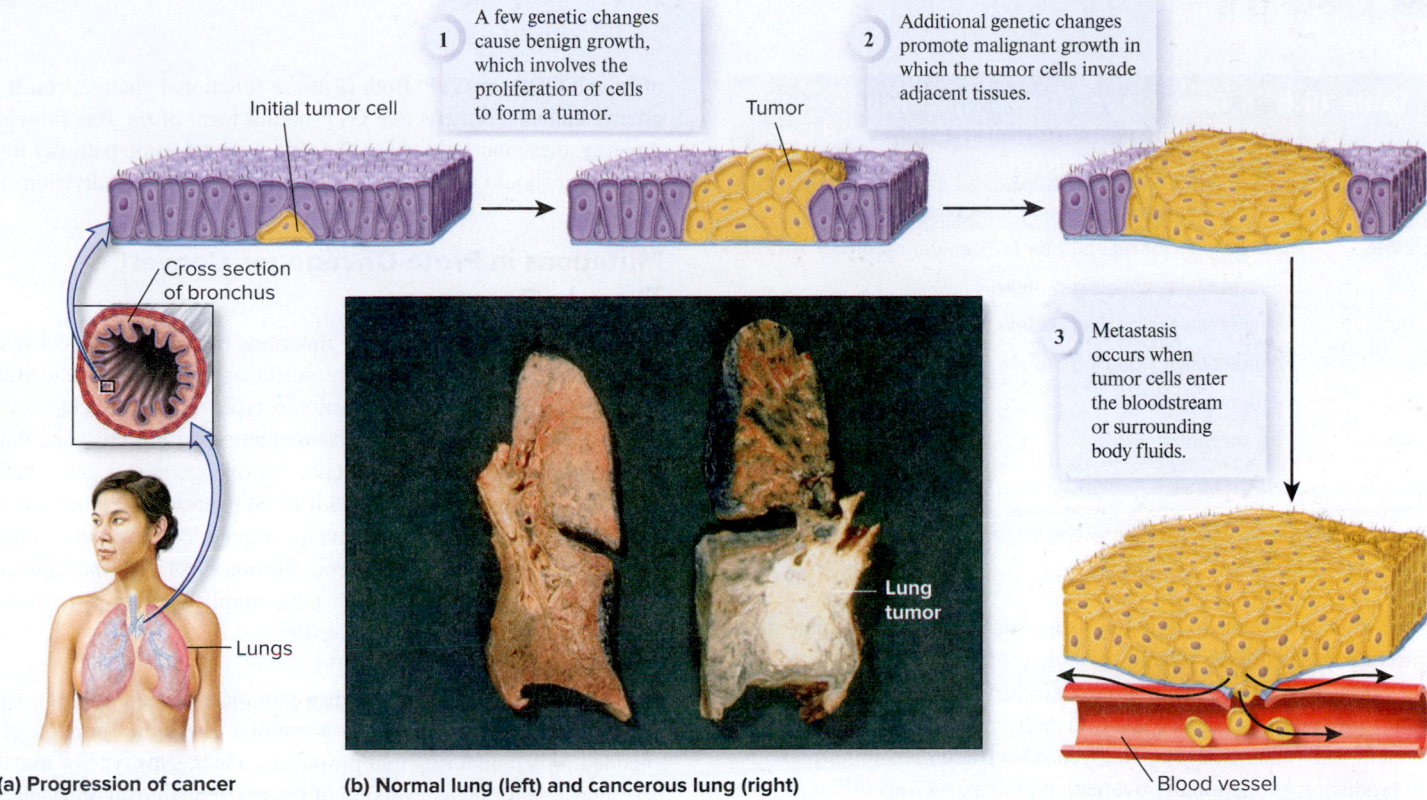

1 A few genetic changes cause benign growth, which involves the proliferation of cells to form a tumor.

2 Additional genetic changes promote malignant growth in which the tumor cells invade adjacent tissues.

Initial tumor cell

Tumor

Cross section of bronchus

3 Metastasis occurs when tumor cells enter the bloodstream or surrounding body fluids.

Lungs

Lung tumor

Blood vessel

(a) Progression of cancer

(b) Normal lung (left) and cancerous lung (right)

**Figure 15.10  Cancer: its typical progression and effects.** **(a)** In a healthy individual, a few mutations convert a normal cell into a tumor cell. This cell divides to produce a benign tumor. Additional mutations and other changes in the tumor cells may occur, leading to a malignant tumor. At a later stage in malignancy, the tumor cells invade surrounding tissues, and some malignant cells may metastasize by traveling through the bloodstream to other parts of the body. **(b)** On the left of the photo is a human lung that was obtained from a healthy nonsmoker. The lung shown on the right has been ravaged by lung cancer. This lung was taken from a person who was a heavy smoker. b: ©St. Bartholomew's Hospital/Science Source

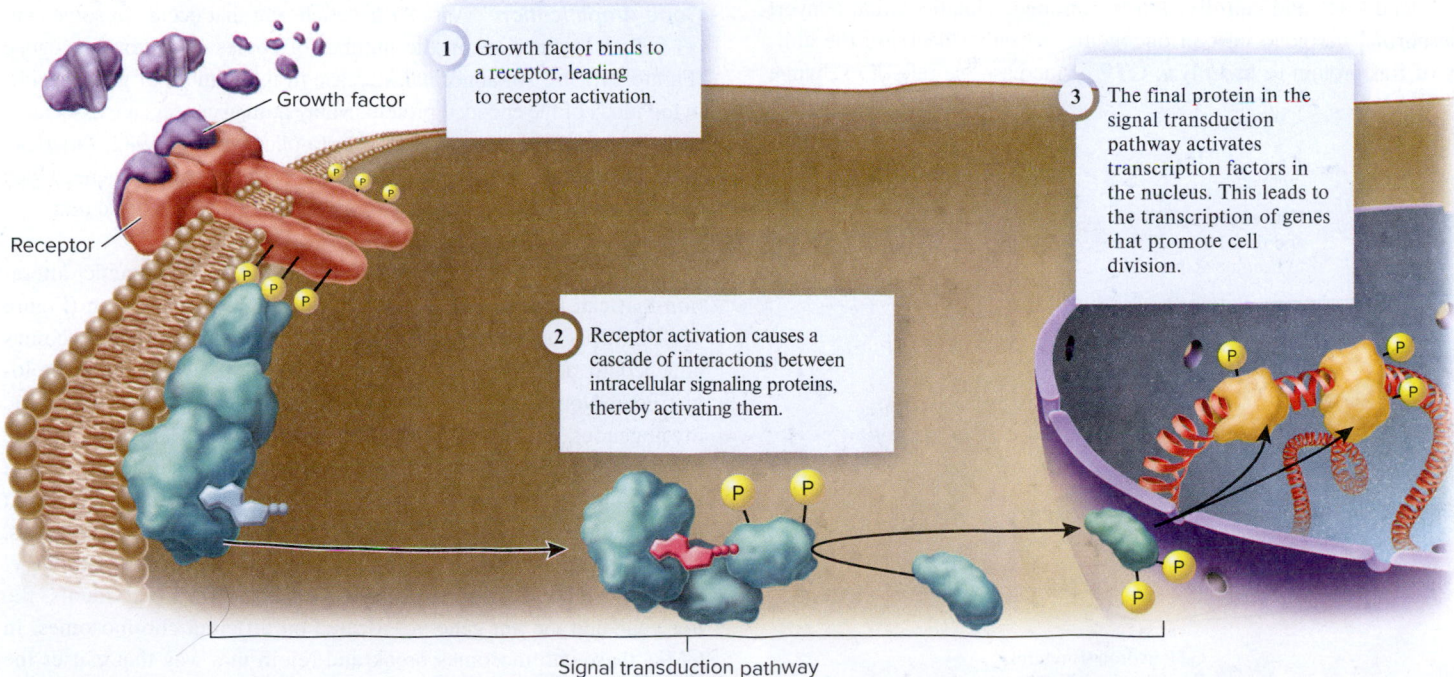

1 Growth factor binds to a receptor, leading to receptor activation.

Growth factor

3 The final protein in the signal transduction pathway activates transcription factors in the nucleus. This leads to the transcription of genes that promote cell division.

Receptor

2 Receptor activation causes a cascade of interactions between intracellular signaling proteins, thereby activating them.

Signal transduction pathway

**Figure 15.11  General features of a signal transduction pathway involving a growth factor that promotes cell division.** A detailed description of this pathway is found in Chapter 9 (look back at Figure 9.10).

**Concept Check:** *How does the presence of a growth factor ultimately affect the function of a cell?*

 **Core Skill: Connections** Look back at Figure 9.10. Could drugs that inhibit protein kinases be used to combat cancer? Explain.

| Table 15.6 | Examples of Genes That Encode Signal Transduction Proteins and Can Become Oncogenes |
|------------|------------------------------------------------------------------------------------|
| **Gene*** | **Cellular function of encoded protein** |
| *erbB* | Growth factor receptor for EGF (epidermal growth factor) |
| *ras* | Intracellular signaling protein |
| *raf* | Intracellular signaling protein |
| *src* | Intracellular signaling protein |
| *fos* | Transcription factor |
| *jun* | Transcription factor |

*The genes described in this table are found in humans as well as other vertebrate species. Most of the genes have been given three-letter names that are abbreviations for the type of cancer the oncogene causes or the type of virus in which the gene was first identified.

genes that encode these signal transduction proteins can change them into oncogenes (**Table 15.6**).

How does an oncogene promote cancer? In some cases, an oncogene may keep a signal transduction pathway for cell division in a permanent "on" state. One way oncogenes keep cell division turned on is by producing a functionally overactive protein. As a specific example, let's consider how a mutation alters an intracellular signaling protein called Ras (refer back to Figure 9.10). The Ras protein is a GTPase that hydrolyzes GTP to GDP + P$_i$ (**Figure 15.12**). When a signal transduction pathway is activated, the Ras protein releases GDP and binds GTP. When GTP is bound, the activated Ras protein promotes cell division. The Ras protein returns to its inactive state by hydrolyzing its bound GTP, and cell division is inhibited. Mutations that convert the normal *ras* gene into an oncogenic *ras* either decrease the ability of Ras protein to hydrolyze GTP or increase the rate of exchange

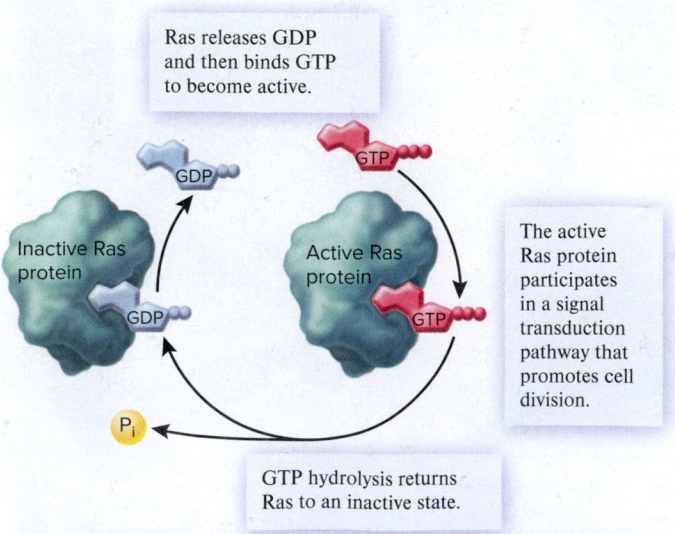

Ras releases GDP and then binds GTP to become active.

Inactive Ras protein

Active Ras protein

The active Ras protein participates in a signal transduction pathway that promotes cell division.

GTP hydrolysis returns Ras to an inactive state.

**Figure 15.12 The function of Ras, a protein that is part of signal transduction pathways.** When GTP is bound, the activated Ras protein promotes cell division. When GTP is hydrolyzed to GDP and P$_i$, Ras is inactivated, and cell division is inhibited.

of bound GDP for GTP. Both of these functional changes result in a greater amount of the active GTP-bound form of the Ras protein. In this way, these mutations keep the signal transduction pathway turned on when it should not be, resulting in uncontrolled cell division.

## Mutations in Proto-Oncogenes Convert Them to Oncogenes

Thus far, we have examined the functions of proteins that cause cancer when they become overactive, resulting in uncontrolled cell division. Let's now consider the common types of genetic changes that create such oncogenes. A **proto-oncogene** is a normal gene that, if mutated, can become an oncogene. An oncogene is a gene that has been altered in a way that causes it to be overexpressed or expressed in the wrong cell type. Several types of genetic changes may convert a proto-oncogene into an oncogene. **Figure 15.13** describes four common types: missense mutations, gene amplifications, chromosomal translocations, and retroviral insertions.

***Missense Mutation*** A missense mutation (Figure 15.13a), which changes a single amino acid in a protein, alters the function of the encoded protein in a way that promotes cancer. This type of mutation is responsible for the conversion of the *ras* gene into an oncogene. An example is a mutation in the *ras* gene that changes a specific glycine to a valine in the Ras protein. This mutation decreases the ability of the Ras protein to hydrolyze GTP, which promotes cell division (see Figure 15.12). Experimentally, chemical mutagens have been shown to cause this missense mutation, thereby leading to cancer.

***Gene Amplification*** Another genetic event that occurs in some cancer cells is an increase in the number of copies of a proto-oncogene (Figure 15.13b). An abnormal increase in the number of genes results in too much of the encoded protein. Many human cancers are associated with the amplification of particular proto-oncogenes. In 1982, American molecular biologist Mark Groudine discovered that the *myc* gene, which encodes a transcription factor, was amplified in a human leukemia.

***Chromosomal Translocation*** A third type of genetic alteration that can lead to cancer is a chromosomal translocation (Figure 15.13c). This occurs when one segment of a chromosome becomes attached to a different chromosome. In 1960, American pathologist Peter Nowell discovered that a form of leukemia called chronic myelogenous leukemia (CML)—a type of cancer involving white blood cells—was correlated with the presence of a shortened version of a human chromosome. This shortened chromosome is the result of a chromosome translocation in which two different chromosomes, chromosomes 9 and 22, exchange pieces. This activates a proto-oncogene, *abl*, in an unusual way (**Figure 15.14**). In healthy individuals, the *bcr* gene and the *abl* gene are located on different chromosomes. In CML, these chromosomes break and rejoin in a way that causes the promoter and the first part of *bcr* to fuse with part of *abl*. This fused gene acts as an oncogene and encodes a fusion protein whose functional overactivity leads to leukemia.

***Retroviral Insertion*** Certain types of viruses convert proto-oncogenes into oncogenes during the viral replication cycle (see

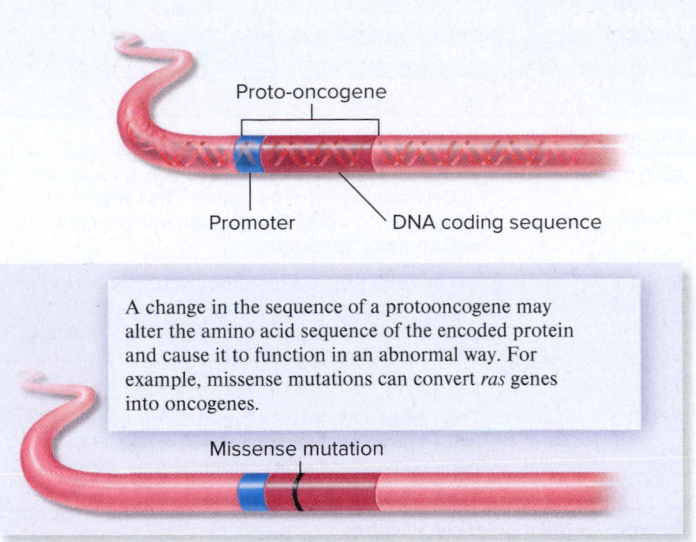

A change in the sequence of a protooncogene may alter the amino acid sequence of the encoded protein and cause it to function in an abnormal way. For example, missense mutations can convert *ras* genes into oncogenes.

**(a) Missense mutation**

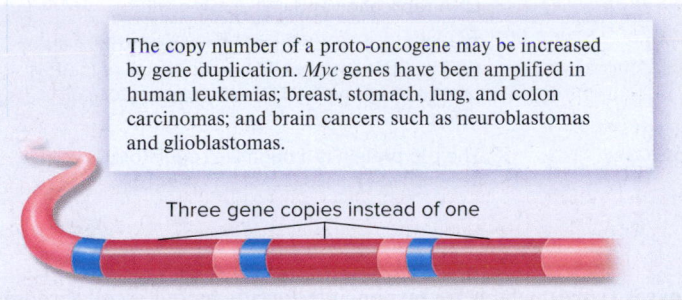

The copy number of a proto-oncogene may be increased by gene duplication. *Myc* genes have been amplified in human leukemias; breast, stomach, lung, and colon carcinomas; and brain cancers such as neuroblastomas and glioblastomas.

**(b) Gene amplification**

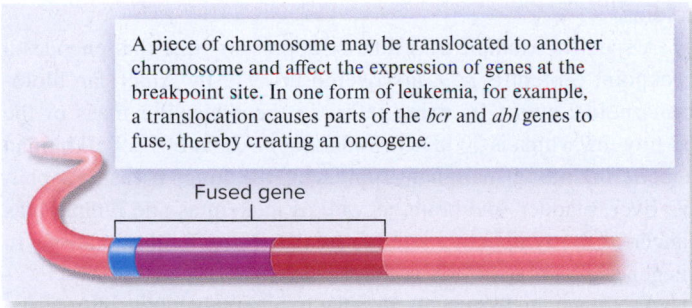

A piece of chromosome may be translocated to another chromosome and affect the expression of genes at the breakpoint site. In one form of leukemia, for example, a translocation causes parts of the *bcr* and *abl* genes to fuse, thereby creating an oncogene.

**(c) Chromosomal translocation**

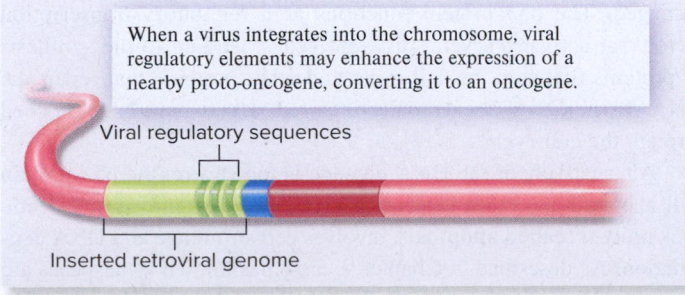

When a virus integrates into the chromosome, viral regulatory elements may enhance the expression of a nearby proto-oncogene, converting it to an oncogene.

**(d) Retroviral insertion**

**Figure 15.13** **Common genetic changes that convert proto-oncogenes to oncogenes.**

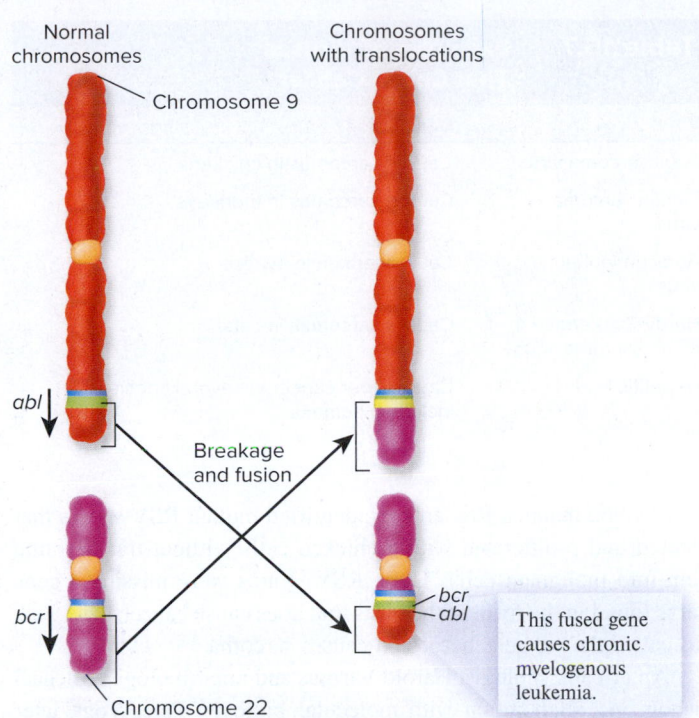

**Figure 15.14** **The formation of a fused gene found in people with certain forms of leukemia.** The fusion of the *bcr* and *abl* genes creates a fused gene that encodes a fusion protein, leading to leukemia. The blue regions are the promoters for the *bcr* and *abl* genes.

**Concept Check:** *The bcr gene is normally expressed in white blood cells. Explain how this observation is related to the type of cancer that the translocation between chromosomes 9 and 22 causes.*

Figure 15.13d). Retroviruses insert their DNA into the chromosomal DNA of the host cell. The viral genome contains promoter and regulatory elements that cause a high level of expression of viral genes. On occasion, the viral DNA may insert into a host chromosome in such a way that a viral promoter and regulatory elements are next to a proto-oncogene. This may result in the overexpression of the proto-oncogene, thereby promoting cancer. This is one way for a virus to cause cancer. Alternatively, a virus may cause cancer because it carries an oncogene in its viral genome. This phenomenon is described next.

## Some Types of Cancer Are Caused by Viruses

The majority of cancers are caused by mutagens or other changes that alter the structure and expression of genes that are found in somatic cells. A few viruses, however, are known to cause cancer in plants and animals, including humans (**Table 15.7**).

In 1911, the first cancer-causing virus to be discovered was isolated from chicken sarcomas by American pathologist Peyton Rous. A **sarcoma** is a tumor of connective tissue such as bone or cartilage. The virus was named the Rous sarcoma virus (RSV). In the 1970s, research involving RSV led to the identification of a viral gene that acts as an oncogene. Researchers investigated RSV by using it to infect chicken cells grown in the laboratory. This infection causes the chicken cells to grow like cancer cells, continuously and in an

| Table 15.7 | Examples of Viruses That Cause Cancer |
|---|---|
| **Virus** | **Description** |
| Rous sarcoma virus | Causes sarcomas in chickens |
| Simian sarcoma virus | Causes sarcomas in monkeys |
| Abelson leukemia virus | Causes leukemia in mice |
| Hardy-Zuckerman 4 feline sarcoma virus | Causes sarcomas in cats |
| Hepatitis B | Causes liver cancer in several species, including humans |

| Table 15.8 | Functions of Selected Tumor-Suppressor Genes |
|---|---|
| **Gene** | |
| ***Maintenance of genome integrity*** | |
| *p53* | p53 is a transcription factor that acts as a sensor of DNA damage. It can promote DNA repair, prevent the progression through the cell cycle, and promote apoptosis. |
| *BRCA-1 BRCA-2* | BRCA-1 and BRCA-2 proteins are both involved in the cellular defense against DNA damage. They play a role in sensing DNA damage and facilitate DNA repair. These genes are mutant in persons with certain inherited forms of breast cancer. |
| *XPD* | This represents one of several different genes whose products function in DNA repair. These genes are defective in patients with xeroderma pigmentosum. |
| ***Negative regulation of cell division*** | |
| *Rb* | The Rb protein is a negative regulator that represses the transcription of genes required for DNA replication and cell division. |
| *NF1* | The NF1 protein stimulates Ras to hydrolyze its GTP to GDP. Loss of NF1 function causes the Ras protein to be overactive, which promotes cell division. |
| *p16* | The p16 protein is a negative regulator of cyclin-dependent kinases (cdks). |

uncontrolled manner. Researchers identified mutant RSV strains that infected and proliferated within chicken cells without transforming them into malignant cells. These RSV strains were missing a gene that is found in the form of the virus that does cause cancer. This gene was called the *src* gene because it causes sarcoma.

Americans, biologist Harold Varmus and microbiologist Michael Bishop, in collaboration with molecular biologist Peter Vogt, later discovered that normal (nonviral-infected) chicken cells also contain a copy of the *src* gene in their chromosomes. This gene is a proto-oncogene. When it is incorporated into a viral genome, it is overexpressed because it is transcribed from a very active viral promoter. This overexpression ultimately produces too much of the Src protein in infected cells and promotes uncontrolled cell division.

## Tumor-Suppressor Genes Prevent Mutation or Cell Proliferation

Thus far, we have examined one category of genes that promote cancer, namely oncogenes. We now turn our attention to the second category, those called tumor-suppressor genes. The functioning of a normal (nonmutant) tumor-suppressor gene prevents cancerous growth. The proteins encoded by tumor-suppressor genes usually have one of two functions: maintenance of genome integrity or negative regulation of cell division (**Table 15.8**).

***Maintenance of Genome Integrity***   Some tumor-suppressor genes encode proteins that maintain the integrity of the genome by monitoring and/or repairing genome alterations. The proteins encoded by these genes are vital for the prevention of abnormalities such as gene mutations, DNA breaks, and improperly segregated chromosomes. Therefore, when these proteins are functioning properly, they minimize the chance that a cancer-causing mutation will occur. In some cases, the proteins encoded by tumor-suppressor genes prevent a cell from progressing through the cell cycle if an abnormality is detected. These are termed **checkpoint proteins** because their role is to check the integrity of the genome and prevent a cell from progressing past a certain point in the cell cycle. Checkpoint proteins are not always required to regulate normal, healthy cell division, but they can stop cell division if an abnormality is detected.

How do checkpoint proteins stop the cell cycle? One way is by controlling proteins called cyclins and cyclin-dependent

kinases (cdks), which are responsible for advancing a cell through the four phases of the cell cycle (see Chapter 16). The formation of activated cyclin/cdk complexes can be stopped by checkpoint proteins.

A specific example of a tumor-suppressor gene that encodes a checkpoint protein is *p53*, discovered in 1979 by American biologist Arnold Levine. Its name refers to the molecular mass of the p53 protein, which is 53 kDa (kilodaltons). About 50% of all human cancers, including malignant tumors of the lung, breast, esophagus, liver, bladder, and brain, as well as leukemias and lymphomas (cancer of the lymphatic system), are associated with mutations in this gene.

As shown in **Figure 15.15**, p53 is a protein that controls the ability of cells to advance from the $G_1$ stage of the cell cycle to the S phase. The expression of the *p53* gene is induced when DNA is damaged. The p53 protein functions as a regulatory transcription factor that activates several different genes, leading to the synthesis of proteins that stop the cell cycle and other proteins that repair the DNA. If the DNA is eventually repaired, a cell may later proceed through the cell cycle.

Alternatively, if the DNA damage is too severe, the p53 protein will also activate other genes that promote programmed cell death. This process, called **apoptosis**, involves cell shrinkage and DNA degradation. As described in Chapter 9, enzymes known as caspases are activated during apoptosis (refer back to Figure 9.18). They function as proteases that are sometimes called the "executioners" of the cell. Caspases digest selected cellular proteins such as microfilaments, which are components of the cytoskeleton. The destruction of the

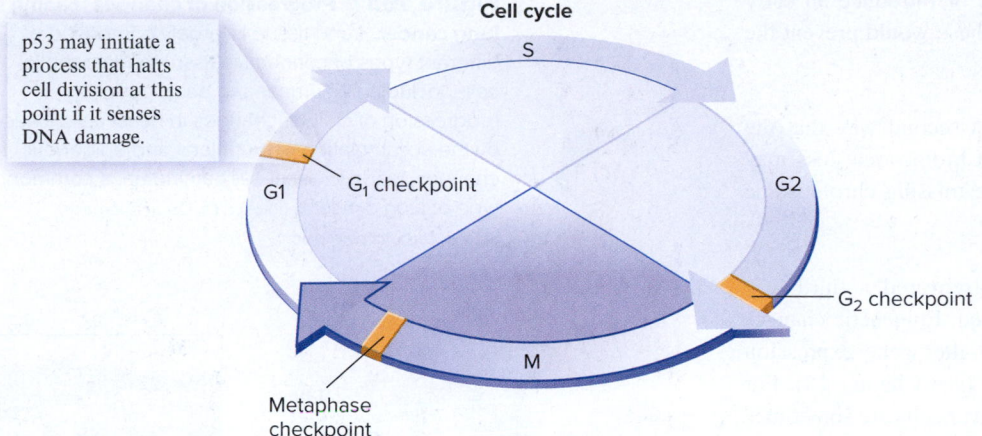

p53 may initiate a process that halts cell division at this point if it senses DNA damage.

Cell cycle

S

G1    G₁ checkpoint    G2    G₂ checkpoint

M

Metaphase checkpoint

**Figure 15.15** **The cell cycle and checkpoints.** As discussed in Chapter 16, eukaryotic cells advance through a cell cycle composed of G₁, S, G₂, and M phases (look ahead to Figure 16.2). The yellow bars indicate common checkpoints where the cell cycle is stopped if genetic abnormalities are detected. The p53 protein stops a cell at the G₁ checkpoint if it senses DNA damage.

*Concept Check:* *Why is it an advantage for an organism to have checkpoints where the cell cycle can be stopped?*

microfilaments causes the cell to break into small vesicles that are eventually phagocytized by cells of the immune system. It is beneficial for a multicellular organism to kill an occasional cell with cancer-causing potential.

**Negative Regulation of Cell Division** A second category of tumor-suppressor genes encodes proteins that are negative regulators or inhibitors of cell division. These proteins must function properly to halt cell division. If their function is lost, cell division is abnormally accelerated.

An example of such a tumor-suppressor gene is the *Rb* gene. It was the first tumor-suppressor gene to be identified in humans, from studies of patients with a disease called retinoblastoma, a cancerous tumor that occurs in the retina of the eye. The Rb protein negatively controls a regulatory transcription factor called E2F that activates genes required for cell cycle progression from G₁ to S phase. The binding of the Rb protein to E2F inhibits its activity and prevents cell division (**Figure 15.16**). When a normal cell is supposed to divide, cyclins bind to cyclin-dependent kinases (cdks). This binding activates the kinases, which catalyze the transfer of a phosphate to the Rb protein. The phosphorylated form of the Rb protein is released from E2F, thereby allowing E2F to activate genes needed to advance through the cell cycle. When both copies of Rb are defective due to mutations, the E2F protein is always active. This explains why uncontrolled cell division occurs in retinoblastoma.

## Gene Mutations, Chromosome Loss, and Changes in Chromatin Structure Can Inhibit the Expression of Tumor-Suppressor Genes

Cancer biologists want to understand how tumor-suppressor genes are inactivated, because this knowledge may ultimately help them to prevent or combat cancer. How are tumor-suppressor genes silenced? The function of tumor-suppressor genes is lost in three common ways.

**Inactivation of Tumor-Suppressor Genes via Mutation** First, a mutation can occur within a tumor-suppressor gene to inactivate its function. For example, a mutation could abolish the function

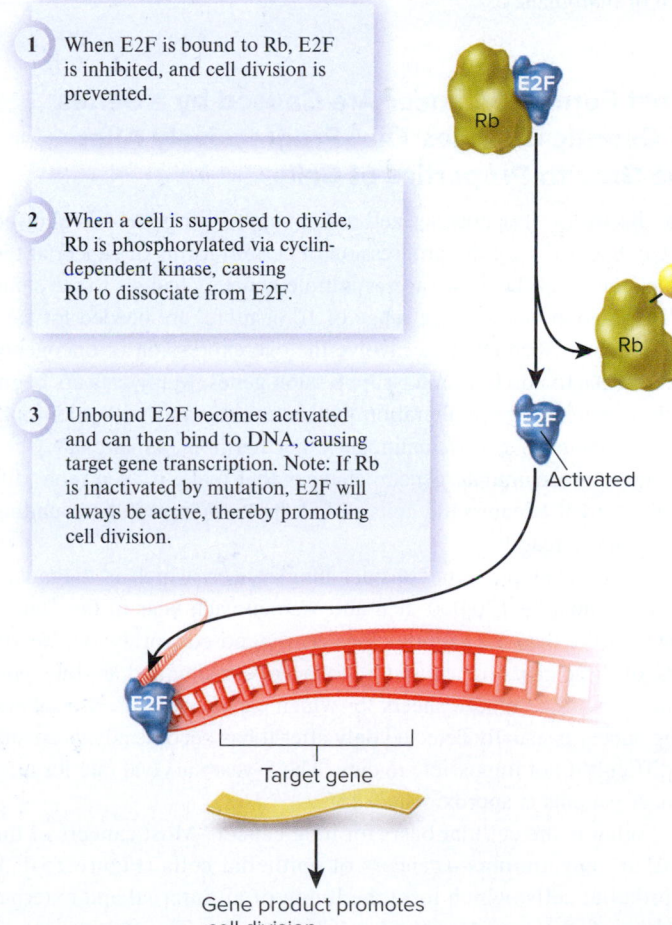

1  When E2F is bound to Rb, E2F is inhibited, and cell division is prevented.

2  When a cell is supposed to divide, Rb is phosphorylated via cyclin-dependent kinase, causing Rb to dissociate from E2F.

3  Unbound E2F becomes activated and can then bind to DNA, causing target gene transcription. Note: If Rb is inactivated by mutation, E2F will always be active, thereby promoting cell division.

Activated

Target gene

Gene product promotes cell division.

**Figure 15.16** **Function of the Rb protein.** The Rb protein inhibits the function of E2F, which turns on genes that cause a cell to divide. When cells are supposed to divide, Rb is phosphorylated by cyclin-dependent kinase, which allows E2F to function.

*Concept Check:* *Would cancer occur if both copies of the Rb gene and both copies of the E2F gene were rendered inactive due to mutations?*

of the promoter for a tumor-suppressor gene or introduce an early stop codon in its coding sequence. Either of these would prevent the expression of a functional protein.

**Chromosome Loss**    Chromosome loss is a second way that the function of a tumor-suppressor gene is lost. Chromosome loss may contribute to the progression of cancer if the missing chromosome carries one or more tumor-suppressor genes.

**Epigenetic Changes**    Researchers have discovered a third way that tumor-suppressor genes may be inactivated. Epigenetic changes involve changes in chromatin structure that alter gene expression without altering the base sequence of DNA (see Chapter 18). For example, tumor-suppressor genes found in cancer cells are sometimes abnormally methylated. As discussed in Chapter 14, transcription is inhibited when CpG islands near a promoter are methylated. Such DNA methylation near the promoters of tumor-suppressor genes has been found in many types of tumors, suggesting that this form of gene inactivation plays an important role in the formation and/or progression of malignancy.

## Most Forms of Cancer Are Caused by a Series of Genetic Changes That Progressively Alter the Growth Properties of Cells

The discovery of oncogenes and tumor-suppressor genes has allowed researchers to study the progression of certain forms of cancer at the molecular level. In most cases, multiple genetic changes to the same cell lineage, perhaps in the range of 10 or more, are needed for cancer to occur. Such changes involve the overexpression of oncogenes and the inactivation of tumor-suppression genes. Many cancers begin with a benign genetic alteration that, over time and with additional genetic changes, leads to malignancy. Furthermore, a malignancy can continue to accumulate genetic changes that make it even more difficult to treat because the cells divide faster or invade surrounding tissues more readily.

As an example, lets consider lung cancer, which is diagnosed in approximately 170,000 men and women each year in the United States. More than 1.2 million cases are diagnosed worldwide. Nearly 90% of these cases are caused by tobacco smoking and are thus preventable. Unlike other cancers for which early diagnosis is possible, lung cancer is usually detected only after it has become advanced and is difficult if not impossible to cure. The 5-year survival rate for lung cancer patients is approximately 15%.

What is the cellular basis for lung cancer? Most cancers of the lung are **carcinomas**—cancers of epithelial cells (Figure 15.17). (Epithelial cells, which form the lining of all internal and external body surfaces, are described in Chapter 10.) The top images in **Figure 15.17** show the normal epithelium found in a healthy lung. The rest of the figure shows the progression of a carcinoma that is due to mutations in basal cells, a type of epithelial cell. Keep in mind that cancer occurs due to the accumulation of mutations in a cell lineage, beginning with an initial mutant cell that then divides multiple times to produce a population of many daughter cells (see Figure 15.10). As mutations accumulate in a lineage of basal cells, the number of these cells increases dramatically. This

**Figure 15.17  Progression of changes leading to lung cancer.**  Lung tissue is largely composed of different types of connective tissue and epithelial cells, including columnar and basal cells. A progression of cellular changes in basal cells, caused by the accumulation of mutations and epigenetic changes, leads to basal cell carcinoma, a common type of lung cancer. (photos): ©Dr. Oscar Auerbach, reproduced with permission

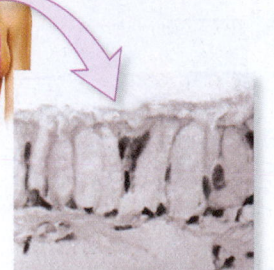
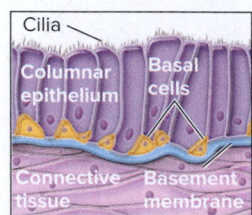

Normal lung epithelium

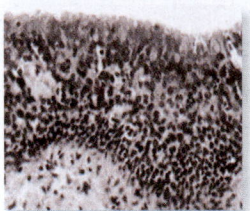
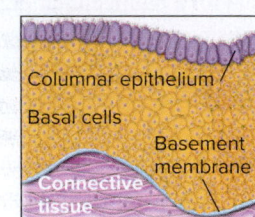

Hyperplasia

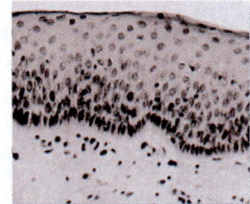
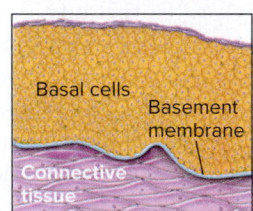

Loss of ciliated cells

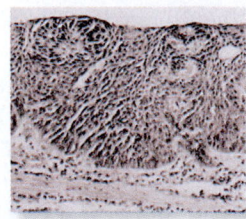
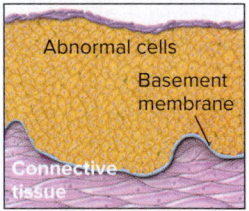

Dysplasia (initially precancerous, then cancerous)

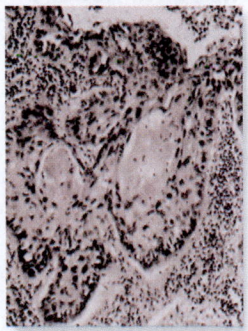
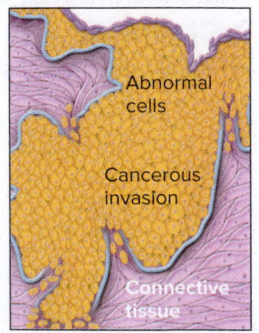

Invasive cancerous cells that can metastasize

causes a thickening of the epithelium, a condition called hyperplasia. The proliferation of the basal cells causes the loss of the ciliated, columnar epithelial cells that normally line the airways and help remove mucus and its trapped particles from the lungs. As additional mutations accumulate in this cell lineage, the basal cells develop more abnormal morphologies, a condition known as dysplasia. In the early stages of dysplasia, the abnormal basal cells are precancerous. If the source of chronic irritation (usually cigarette smoke) is eliminated, the abnormal cells are likely to disappear. Alternatively, if smoking continues, these abnormal cells may accumulate additional genetic changes and lose the ability to stop dividing. Such cells have become cancerous—the person has basal cell carcinoma.

The basement membrane is a sheetlike layer of extracellular matrix that provides a barrier between the lung cells and the bloodstream. If the cancer cells have not yet penetrated the basement membrane, they will not have metastasized, that is, spread into the blood and to other parts of the body. If the entire tumor is removed at this stage, the patient should be cured. The lower images in Figure 15.17 show a tumor that has broken through the basement membrane. The metastasis of these cells to other parts of the body will likely kill the patient, usually within a year of diagnosis.

The cellular changes that lead to lung cancer are correlated with genetic changes. These include the occurrence of mutations that create oncogenes and inhibit tumor-suppressor genes. The order of mutations is not absolute. It takes time for multiple changes to accumulate, so cancer is usually a disease of older people. Reducing your exposure to mutagens such as cigarette smoke throughout your lifetime helps minimize the risk of mutations to your genes that could promote cancer.

 **Core Concept: Evolution**

## Mutations in Approximately 300 Human Genes May Promote Cancer

Researchers have identified a large number of genes that are mutated in cancer cells. Though not all of these mutant genes have been directly shown to affect the growth rate of cells, such mutations are likely to be found in tumors because they provide some type of growth advantage for the cell population from which the cancer developed. For example, certain mutations may affect the functions of proteins that enable cells to metastasize to neighboring locations. These mutations may not affect growth rate, but they provide a growth advantage in that cancer cells are not limited to growing in a particular location. They can migrate to new locations.

How many genes can contribute to cancer when they become mutant? Researchers have estimated that about 300 different genes may play a role in the development of human cancer. With an approximate human genome size of 22,000 genes, this observation indicates that over 1% of our genes have the potential to promote cancer if their expression is altered by a mutation or an epigenetic change.

# Summary of Key Concepts

## 15.1 Consequences of Mutations

- A mutation is a heritable change in the genetic material; gene mutations are relatively small changes in the base sequence of a gene.

- Point mutations affect a single base pair and can alter the coding sequence of genes in several ways. These mutations include silent, missense, nonsense, and frameshift mutations (Table 15.1).

- Sickle cell disease is caused by a missense mutation that results in a single amino acid substitution in β-globin (Figure 15.1).

- Gene mutations also alter gene function by changing DNA sequences that are not within the coding region (Table 15.2).

- Germ-line mutations affect gametes, whereas mutations in somatic cells affect only a part of the body and cannot be passed to offspring (Figures 15.2, 15.3).

## 15.2 Causes of Mutations

- The Lederbergs used replica plating to show that mutations are random events (Figure 15.4).

- Spontaneous mutations are the result of abnormalities in biological processes. Induced mutations are caused by agents in the environment that alter DNA structure (Table 15.3).

- Mutagens are chemical or physical agents that lead to mutations in DNA (Table 15.4, Figures 15.5, 15.6).

- The Ames test is a method of testing whether an agent is a mutagen (Figure 15.7).

## 15.3 DNA Repair

- DNA repair systems consist of proteins that sense DNA damage and repair it before a mutation occurs (Table 15.5).

- In nucleotide excision repair (NER), proteins recognize various types of DNA damage, such as thymine dimers. A region in the damaged strand is excised, and a new strand is synthesized, using the intact strand as a template (Figure 15.8).

- Certain inherited diseases in humans, such as xeroderma pigmentosum (XP), are due to defects in the NER system (Figure 15.9).

## 15.4 Cancer

- Cancer is due to the accumulation of mutations and epigenetic changes in a lineage of cells that leads to uncontrolled cell growth (Figure 15.10).

- Mutations in proto-oncogenes that result in overactivity produce cancer-causing genes called oncogenes.

- Oncogenes often encode proteins involved in signal transduction pathways that promote cell division (Figures 15.11, 15.12, Table 15.6).

- Four common types of genetic changes, namely, missense mutations, gene amplifications, chromosomal translocations, and retroviral insertions, can change proto-oncogenes into oncogenes (Figures 15.13, 15.14).

- Some types of cancer are caused by viruses (Table 15.7).

- The normal function of tumor-suppressor genes is to prevent cancer. Mutations of these genes may affect this function and thus promote cancer. Tumor-suppressor genes often encode proteins that maintain the integrity of the genome or function as negative regulators of cell division (Table 15.8).

- Checkpoint proteins such as p53 monitor the integrity of the genome and prevent the cell from progressing through the cell cycle if abnormalities are detected (Figure 15.15).

- The Rb protein is an inhibitor of the cell cycle, because it negatively controls E2F, a transcription factor that promotes cell division (Figure 15.16).

- Tumor-suppressor genes can be inactivated by gene mutations, chromosome loss, and epigenetic changes such as DNA methylation.

- Most forms of cancer, such as lung cancer, involve multiple genetic changes that lead to malignancy (Figure 15.17).

- Over 300 human genes, or over 1% of the total, are known to be associated with cancer when they become mutant.

# Assess & Discuss

## Test Yourself

1. A mutation removes a single base pair within the coding sequence of a gene and inactivates the protein encoded by the gene. Such a mutation is
   a. a silent mutation.
   b. a missense mutation.
   c. a nonsense mutation.
   d. a frameshift mutation.
   e. both b and c.

2. Some point mutations lead to an mRNA that produces a shorter polypeptide. This type of mutation is known as a _____ mutation.
   a. neutral
   b. silent
   c. missense
   d. nonsense
   e. chromosomal

3. In which location is a mutation least likely to affect gene function?
   a. promoter
   b. coding region
   c. splice junction
   d. intergenic region
   e. regulatory site

4. Mutagens can cause mutations by
   a. chemically altering DNA nucleotides.
   b. disrupting DNA replication.
   c. altering the genetic code of an organism.
   d. doing all of the above.
   e. doing a and b only.

5. The mutagenic effect of UV light is
   a. the alteration of cytosine bases to adenine bases.
   b. the formation of adenine dimers that interfere with genetic expression.
   c. the breaking of the sugar-phosphate backbone of the DNA molecule.
   d. the formation of thymine dimers that disrupt DNA replication.
   e. the deletion of thymine bases along the DNA molecule.

6. The Ames test
   a. provides a way to determine if any type of cell has experienced a mutation.
   b. provides a way to determine if an agent is a mutagen.
   c. allows researchers to experimentally disrupt gene activity by causing a mutation in a specific gene.
   d. provides a way to repair mutations in bacterial cells.
   e. does all of the above.

7. Xeroderma pigmentosum
   a. is a genetic disorder that results in uncontrolled cell growth.
   b. is a genetic disorder in which the NER system is not fully functional.
   c. is a genetic disorder that results in the loss of pigment in certain patches of skin.
   d. results from the lack of DNA polymerase proofreading.
   e. both b and d are true of this disorder.

8. If a mutation eliminated the function of UvrC, which aspect of the nucleotide excision repair system would not work?
   a. sensing a damaged DNA site
   b. endonuclease cleavage of the damaged strand
   c. removal of the damaged strand
   d. synthesis of a new strand, using the undamaged strand as a template
   e. none of the above

9. Cancer cells are said to be metastatic when they
   a. begin to divide uncontrollably.
   b. invade healthy tissue.
   c. migrate to other parts of the body.
   d. cause mutations in other healthy cells.
   e. do all of the above.

10. Oncogenes can be produced by
    a. missense mutations.
    b. gene amplification.
    c. chromosomal translocation.
    d. retroviral insertion.
    e. all of the above.

## Conceptual Questions

1. Is a random mutation more likely to be beneficial or harmful? Explain your answer.

2. Distinguish between spontaneous and induced mutations. Which are more harmful? Which are avoidable?

3. **Core Concept: Information** Explain how mutations may cause alterations to the genetic material that are detrimental for reproduction and sustaining life.

## Collaborative Questions

1. Discuss the pros and cons of mutation.

2. A large amount of research is aimed at studying mutation. However, there is not an infinite supply of research funding. Where would you put your money for mutation research?
   a. testing of potential mutagens
   b. investigating molecular effects of mutagens
   c. investigating DNA repair mechanisms
   d. some other area

# The Eukaryotic Cell Cycle, Mitosis, and Meiosis

## 16

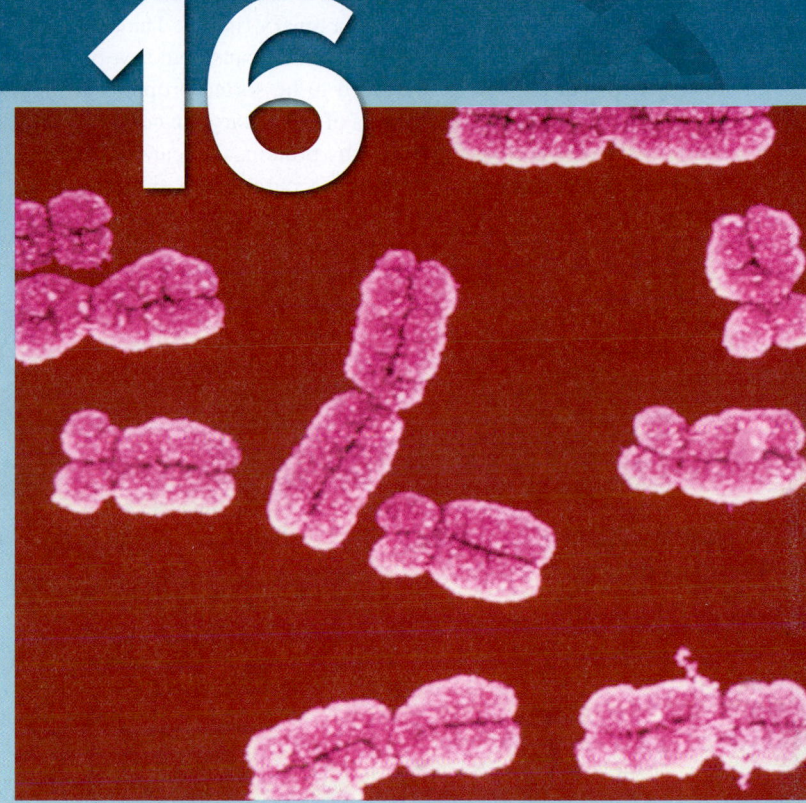

**A scanning electron micrograph of human chromosomes.** These highly compacted chromosomes were found in a dividing cell.
©Biophoto Associates/Science Source

O ver 10,000,000,000,000! Researchers estimate that the adult human body contains somewhere between 10 trillion and 50 trillion cells. It is an almost incomprehensible number. Even more amazing is the accuracy of the process that produces these cells. After a human sperm and egg unite, the fertilized egg goes through a long series of cell divisions to produce an adult with over 10 trillion cells. Let's suppose you randomly removed a cell from your arm and compared it with a cell from your foot. If you examined the chromosomes in both cells under the microscope, they would look identical. The DNA sequences along those chromosomes would also be the same, barring rare mutations. Similar comparisons could be made among the trillions of cells in your body. When you consider how many cell divisions are needed to produce an adult human, the precision of cell division is truly remarkable.

What accounts for this high level of accuracy? As we will examine in this chapter, **cell division**, the reproduction of cells, is a highly regulated process that distributes and monitors the integrity of the genetic material. The eukaryotic cell cycle is a series of phases needed for cell division. The cells of eukaryotic species follow one of two different sorting processes so that new daughter cells receive the correct number and types of chromosomes. The first sorting process we will explore, called mitosis, ensures that two daughter cells receive the same amount of genetic material as the mother cell that produced them. The second sorting process we will consider, called meiosis, is needed for sexual reproduction. In meiosis, cells that have two sets of chromosomes produce daughter cells with a single set of chromosomes. Lastly, we will look at variation in the structure and number of chromosomes. As you will see, certain mechanisms that alter chromosome structure and number have important consequences for the organisms that carry them.

## 16.1 The Eukaryotic Cell Cycle

**Learning Outcomes:**

1. **CoreSKILL »** Describe the features of chromosomes, and explain how sets of chromosomes are examined microscopically.
2. Outline the phases of the eukaryotic cell cycle.
3. Explain how cyclins and cdks work together to advance a cell through the eukaryotic cell cycle.

Life is a continuum in which new living cells are formed by the division of pre-existing cells. The Latin axiom *Omnis cellula e cellula*, meaning "Every cell originates from another cell," was first proposed in 1858 by Rudolf Virchow, a German biologist. From an evolutionary perspective, cell division has a very ancient origin. All living organisms, from unicellular bacteria to multicellular plants and animals, have been produced by a series of repeated rounds of cell growth and division extending back to the beginnings of life nearly 4 billion years ago.

A **cell cycle** is a series of events that leads to cell division. In all species, it is a highly regulated process, to ensure that cell division occurs at the appropriate time. As discussed in Chapter 19, bacterial cells produce more cells via binary fission. The cell cycle in eukaryotes is more complex, in part, because eukaryotic cells have sets of chromosomes that need to be sorted properly. In this section, we will examine the phases of the eukaryotic cell cycle and see how the cell cycle is controlled by proteins that carefully monitor the division process to ensure its accuracy. But first, we need to consider some general features of chromosomes in eukaryotic species.

## Chromosomes Are Inherited in Sets and Occur in Homologous Pairs

To understand the chromosomal composition of cells and the behavior of chromosomes during cell division, scientists use microscopes to observe cells and chromosomes. **Cytogenetics** is the field of genetics that involves the microscopic examination of chromosomes. When a cell prepares to divide, the chromosomes become more tightly compacted, a process that decreases their apparent length and increases their diameter. A consequence of this compaction is that distinctive shapes and numbers of chromosomes become visible under a light microscope.

*Microscopic Examination of Chromosomes* **Figure 16.1** shows the general procedure for preparing and viewing chromosomes from a eukaryotic cell. In this example, the cells were obtained from a sample of human blood. Specifically, the chromosomes within leukocytes

**1** A sample of blood is collected and treated with drugs that stimulate cell division. The sample is then subjected to centrifugation.

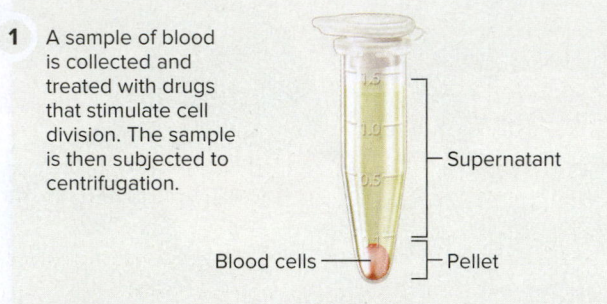

Supernatant

Blood cells — Pellet

**4** The slide is viewed by a light microscope equipped with a camera; the sample is seen on a computer screen. The chromosomes can be photographed and arranged electronically on the screen.

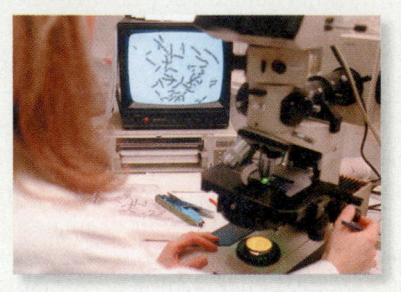

**2** The supernatant is discarded, and the cell pellet is suspended in a hypotonic solution. This causes the cells to swell and the chromosomes to spread out from each other.

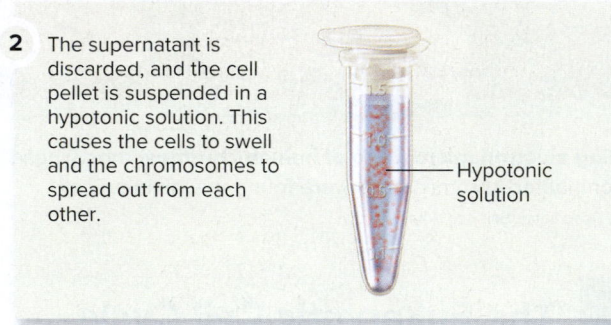

Hypotonic solution

**5**

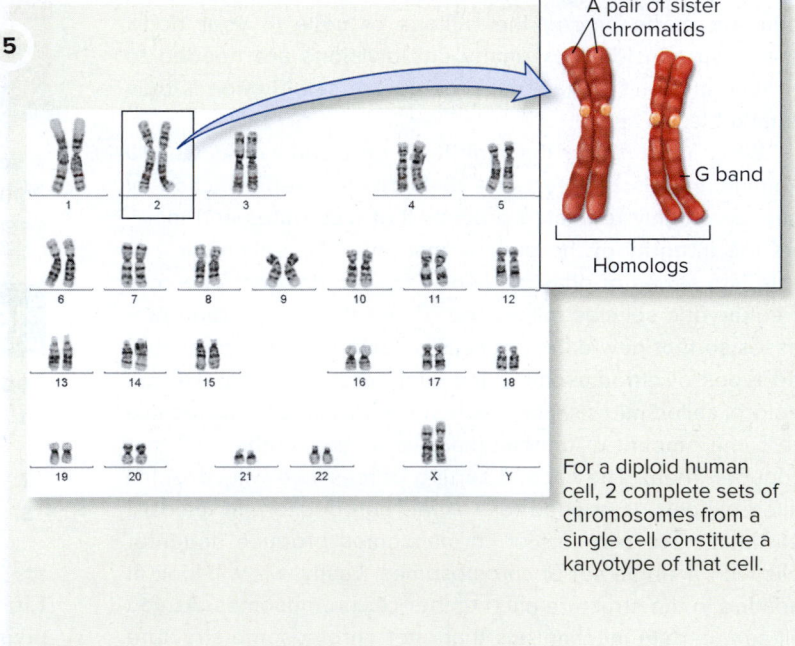

A pair of sister chromatids

G band

Homologs

For a diploid human cell, 2 complete sets of chromosomes from a single cell constitute a karyotype of that cell.

**3** The sample is subjected to centrifugation a second time to concentrate the cells. The cells are suspended in a fixative, stained, and placed on a slide.

Fix    Stain

Blood cells

**Figure 16.1** **The procedure for making a karyotype.** In this example, the chromosomes were treated with Giemsa stain, and the resulting bands are called G bands. (4): ©Burger/Science Source; (5): Courtesy of the Genomic Centre for Cancer Research and Diagnosis, CancerCare Manitoba, University of Manitoba, Winnipeg, Manitoba, Canada

**Concept Check:** *Researchers usually treat cells with drugs that stimulate them to divide before beginning the procedure for making a karyotype. Why is this treatment useful?*

(white blood cells) were examined. A sample of the blood cells was treated with drugs that stimulated them to divide. The actively dividing cells were centrifuged to concentrate them into a pellet, which was then mixed with a hypotonic solution that caused the cells to swell. The expansion of the cells caused the chromosomes to spread out from each other, making it easier to see each individual chromosome.

Next, the cells were concentrated by a second centrifugation and treated with a fixative, which chemically fixed them in place so the chromosomes could no longer move around. The cells were then exposed to a chemical dye, such as Giemsa stain, that binds to the chromosomes and stains them. This gives chromosomes a distinctive banding pattern that greatly enhances their contrast and ability to be uniquely identified; in this case, the bands are called G bands.

The cells were then placed on a slide and viewed with a light microscope. In a cytogenetics laboratory, microscopes are equipped with an electronic camera to photograph the chromosomes. On a computer screen, the images of the chromosomes are organized in a standard way, usually from largest to smallest. This type of photographic representation of chromosomes, such as the photo in step 5 of Figure 16.1, is called a **karyotype**. A karyotype reveals the number, size, and form of chromosomes found within an actively dividing cell. It should also be noted that the chromosomes viewed in actively dividing cells have already replicated. The two copies are still joined to each other and referred to as a pair of **sister chromatids** (see inset to Figure 16.1).

*Sets of Chromosomes* What type of information is learned from a karyotype? By studying the karyotypes of many species, scientists have discovered that eukaryotic chromosomes occur in sets. Each set is composed of several different types of chromosomes. For example, one set of human chromosomes contains 23 different types of chromosomes (see Figure 16.1). By convention, the chromosomes are numbered according to size, with the largest chromosomes having the smallest numbers. For example, human chromosomes 1, 2, and 3 are relatively large, whereas 21 and 22 are the two smallest. This numbering system does not apply to the **sex chromosomes**, which determine the sex of the individual. Sex chromosomes in humans are designated with the letters X and Y; females are XX and males are XY. The chromosomes that are not sex chromosomes are called **autosomes**. Humans have 22 different types of autosomes.

A second feature of many eukaryotic species is that most cells contain two sets of chromosomes. The karyotype shown in Figure 16.1 contains two sets of chromosomes, with 23 different chromosomes in each set. Therefore, this human cell contains a total of 46 chromosomes. Each cell has two sets because the individual inherited one set from the father and one set from the mother. When the cells of an organism carry two sets of chromosomes, that organism is said to be **diploid**. Geneticists use the letter $n$ to represent a set of chromosomes. Diploid organisms are referred to as $2n$, because they have two sets of chromosomes. For example, humans are $2n$, where $n = 23$. Most human cells are diploid. An exception is the **gametes**, the sperm and egg cells. Gametes are **haploid**, or $1n$, which means they contain one set of chromosomes.

*Homologous Pairs of Chromosomes* When an organism is diploid, the members of a pair of chromosomes are called **homologs** (see inset to Figure 16.1). The term **homology** refers to any similarity that is due to common ancestry. Pairs of homologous chromosomes are evolutionarily derived from the same chromosome. However, homologous chromosomes are not usually identical because over many generations they have accumulated some genetic changes that make them distinct.

How similar are homologous chromosomes to each other? Each of the two chromosomes in a homologous pair is nearly identical in size and contains a very similar composition of genetic material. A particular gene found on one copy of a chromosome is usually found on the homolog. However, because one homolog is received from each parent, the two homologs may vary in the way that a gene affects an organism's traits. As an example, let's consider a gene in humans called *OCA2*, which plays a major role in determining eye color. The *OCA2* gene is found on chromosome 15. One copy of chromosome 15 might carry the form of this eye color gene that confers brown eyes, whereas the gene on the homolog could confer blue eyes. The topic of how genes affect an organism's traits will be considered in Chapter 17.

The DNA sequences on homologous chromosomes are very similar. In most cases, the sequence of bases on one homolog differs by less than 1% from the sequence on the other homolog. For example, the DNA sequence of chromosome 1 that you inherited from your mother is likely to be more than 99% identical to the DNA sequence of chromosome 1 that you inherited from your father. Nevertheless, keep in mind that the sequences are not identical. The slight differences in DNA sequence provide important variation in gene function. Again, if we use the eye color gene *OCA2* as an example, a minor difference in DNA sequence distinguishes two forms of the gene, brown versus blue.

The striking similarity between homologous chromosomes does not apply to the sex chromosomes (for example, X and Y). These chromosomes differ in size and genetic composition. Certain genes found on the X chromosome are not found on the Y chromosome, and vice versa. The X and Y chromosomes are not considered homologous chromosomes, although they do have short regions of homology.

## The Cell Cycle Is a Series of Phases That Lead to Cell Division

Eukaryotic cells that are destined to divide advance through the cell cycle, a series of changes that involves growth, replication, and division, and ultimately produces new cells. **Figure 16.2** provides an overview of the cell cycle. In this diagram, the mother cell has three pairs of chromosomes, for a total of six individual chromosomes. Such a cell is diploid ($2n$) and contains three chromosomes per set ($n = 3$). The paternal set is shown in blue, and the homologous maternal set is shown in red.

The phases of the cell cycle are $G_1$ (first gap), **S** (synthesis of DNA, the genetic material), $G_2$ (second gap), and **M phase** (mitosis and cytokinesis). The $G_1$ and $G_2$ phases were originally described as gap phases to indicate the periods between DNA synthesis and mitosis. In actively dividing cells, the $G_1$, S, and $G_2$ phases are collectively known as **interphase**. During interphase, the cell grows and copies its chromosomes in preparation for cell division. Alternatively, a cell may exit the cell cycle and remain for long periods of time in a phase called $G_0$ (G zero). The $G_0$ phase is an alternative to proceeding through $G_1$. A cell in the $G_0$ phase has postponed division or, in the case of terminally differentiated cells (such as muscle cells in an adult animal), will never divide again. $G_0$ is a nondividing phase.

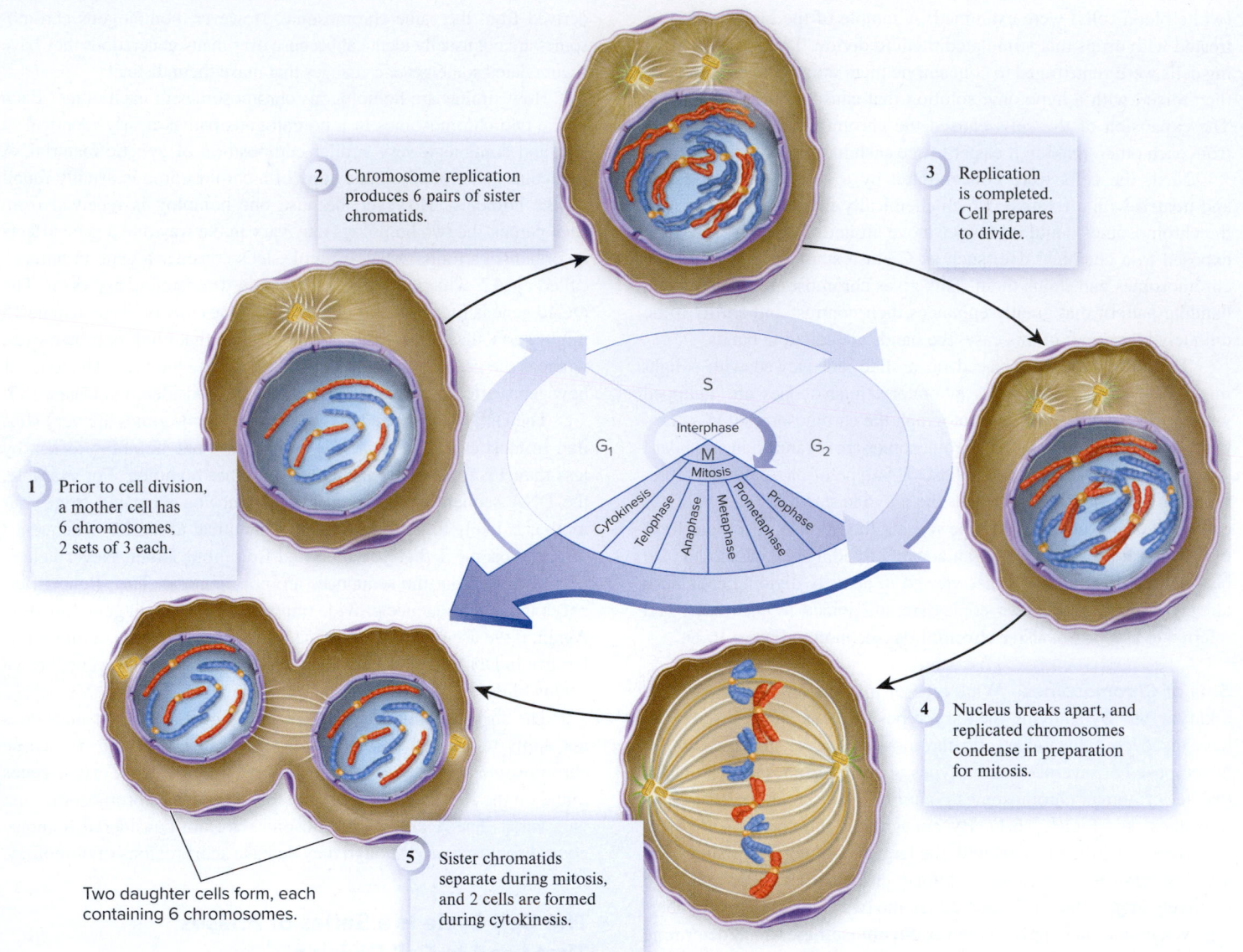

**2** Chromosome replication produces 6 pairs of sister chromatids.

**3** Replication is completed. Cell prepares to divide.

**1** Prior to cell division, a mother cell has 6 chromosomes, 2 sets of 3 each.

**4** Nucleus breaks apart, and replicated chromosomes condense in preparation for mitosis.

**5** Sister chromatids separate during mitosis, and 2 cells are formed during cytokinesis.

Two daughter cells form, each containing 6 chromosomes.

S

Interphase

$G_1$

M

$G_2$

Mitosis

Cytokinesis  Telophase  Anaphase  Metaphase  Prometaphase  Prophase

**Figure 16.2** **The eukaryotic cell cycle.** Dividing cells advance through a series of phases denoted $G_1$, S, $G_2$, and M. This diagram shows the advancement of a cell through the cell cycle to produce two daughter cells. The original diploid cell had three pairs of chromosomes, for a total of six individual chromosomes. During S phase, these replicate to yield 12 chromatids. After mitosis is complete, the two daughter cells each contain six individual chromosomes. The width of the phases shown in this figure is not meant to reflect their actual length. $G_1$ is typically the longest phase of the cell cycle, whereas M phase is relatively short.

**Concept Check:** *Which phases make up interphase?*

**$G_1$ Phase**    The $G_1$ phase is a period in a cell's life when it may become committed to divide. Depending on the environmental conditions and the presence of signaling molecules, a cell in the $G_1$ phase may accumulate molecular changes that cause it to advance through the rest of the cell cycle. Cell growth typically occurs during the $G_1$ phase.

**S Phase**    During the S phase, each chromosome is replicated to form a pair of sister chromatids (see Figure 16.1). When S phase is completed, a cell has twice as many chromatids as the number of chromosomes in the $G_1$ phase. For example, a human cell in $G_1$ phase has 46 distinct chromosomes, whereas the same cell in $G_2$ phase will have 46 pairs of sister chromatids, for a total of 92 chromatids.

**$G_2$ Phase**    During the $G_2$ phase, a cell synthesizes the proteins necessary for chromosome sorting and cell division. Some cell growth may occur.

**M Phase**    The first part of M phase is **mitosis**. The purpose of mitosis is to divide one cell nucleus into two nuclei, distributing the duplicated chromosomes so that each daughter cell receives the same complement of chromosomes. As noted previously, a human cell in $G_2$ phase has 92 chromatids, which are found in 46 pairs. During mitosis, these pairs of chromatids are separated and sorted so that each daughter cell receives 46 chromosomes. In most cases, mitosis is followed by **cytokinesis**, which is the division of the cytoplasm to produce two distinct daughter cells.

The length of the cell cycle varies considerably among different cell types, ranging from several minutes in quickly growing embryos to several months in slow-growing adult cells. For fast-dividing mammalian cells in adults, such as skin cells, the length of the cycle is often in the range of 10 to 24 hours. The various phases within the cell cycle also vary in length. $G_1$ is often the longest and also the most variable phase, and M is the shortest. For a cell that divides in 24 hours, the following lengths of time for each phase are typical:

- $G_1$ phase: 11 hours
- S phase: 8 hours
- $G_2$ phase: 4 hours
- M phase: 1 hour

What factors determine whether or not a cell will divide? First, cell division is controlled by external factors, such as environmental conditions and signaling molecules. The effects of growth factors on cell division are discussed in Chapter 9 (refer back to Figure 9.10). Second, internal factors affect cell division. These include cell cycle control molecules and checkpoints, as we will discuss next.

## The Cell Cycle Is Controlled by Checkpoint Proteins

The advancement through the cell cycle is a process that is highly regulated to ensure that the genome remains intact and that the conditions are appropriate for a cell to divide. As discussed in Chapter 15, this regulation is necessary to minimize the occurrence of mutations, which could have harmful effects and potentially lead to cancer. Proteins called **cyclins** and **cyclin-dependent kinases (cdks)** are responsible for advancing a cell through the phases of the cell cycle. Cyclins

are so named because their amount varies throughout the cell cycle. To be active, the cyclin-dependent kinases controlling the cell cycle must bind to (are dependent on) cyclins. The numbers of different types of cyclins and cdks vary from species to species.

**Figure 16.3** provides a simplified description of how cyclins and cdks work together to advance a cell through $G_1$ and mitosis. During $G_1$, the amount of a particular cyclin termed $G_1$ cyclin increases in response to sufficient nutrients and growth factors. The $G_1$ cyclin binds to a cdk to form an activated $G_1$ cyclin/cdk complex. Once activated, cdk functions as a protein kinase that phosphorylates other proteins needed to advance the cell to the next phase in the cell cycle. For example, certain proteins involved with DNA synthesis are phosphorylated and activated, thereby allowing the cell to replicate its DNA in S phase. After the cell passes into the S phase, $G_1$ cyclin is degraded. Similar events advance the cell through other phases of the cell cycle. A different cyclin, called mitotic cyclin, accumulates late in $G_2$. It binds to a cdk to form an activated mitotic cyclin/cdk complex. This complex phosphorylates proteins that are needed to advance the cell into M phase.

Three critical regulatory points called **checkpoints** are found in the cell cycle of eukaryotic cells (see Figure 16.3). At these checkpoints, a variety of proteins, referred to as checkpoint proteins, act as sensors to determine if a cell is in the proper condition to divide. At the $G_1$ checkpoint, also called the **restriction point**, the checkpoint proteins determine if conditions are favorable for cell division. In addition, $G_1$ checkpoint proteins sense if the DNA has incurred damage. What happens if DNA damage is detected? The checkpoint proteins prevent the formation of active cyclin/cdk complexes, thereby stopping the advancement of the cell cycle.

A second checkpoint exists in $G_2$. At this checkpoint, proteins also check the DNA for damage and ensure that all of the DNA has been replicated. In addition, the $G_2$ checkpoint proteins monitor the

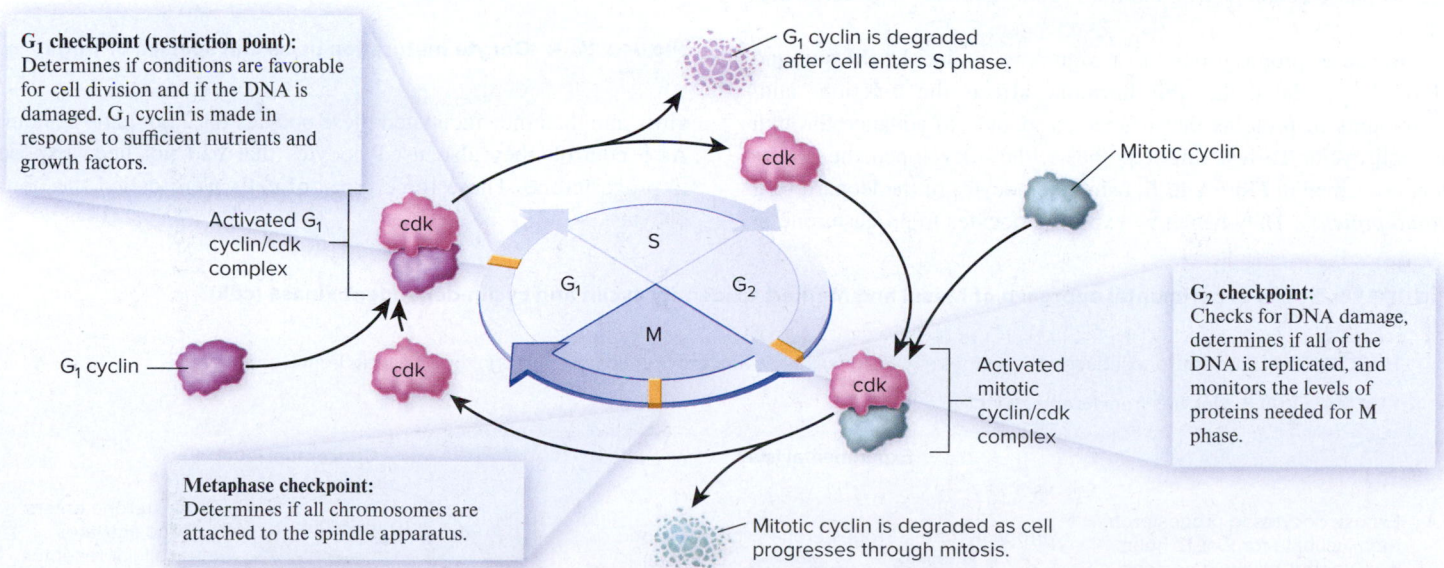

**G₁ checkpoint (restriction point):** Determines if conditions are favorable for cell division and if the DNA is damaged. $G_1$ cyclin is made in response to sufficient nutrients and growth factors.

$G_1$ cyclin is degraded after cell enters S phase.

Activated $G_1$ cyclin/cdk complex

cdk

Mitotic cyclin

$G_1$ cyclin

cdk

S

$G_1$

$G_2$

M

**G₂ checkpoint:** Checks for DNA damage, determines if all of the DNA is replicated, and monitors the levels of proteins needed for M phase.

Activated mitotic cyclin/cdk complex

cdk

**Metaphase checkpoint:** Determines if all chromosomes are attached to the spindle apparatus.

Mitotic cyclin is degraded as cell progresses through mitosis.

**Figure 16.3 Checkpoints in the cell cycle.** This is a general diagram of the eukaryotic cell cycle. Advancement through the cell cycle requires the formation of activated cyclin/cdk complexes. Cells make different types of cyclin proteins, which are typically degraded after the cell has advanced to the next phase. The formation of activated cyclin/cdk complexes is regulated by checkpoint proteins.

 **Core Skill: Connections** Look back at Figure 15.15. How do checkpoint proteins prevent cancer?

levels of the proteins that are needed to advance through M phase. A third checkpoint, called the metaphase checkpoint, has proteins that monitor the integrity of the spindle apparatus. As we will see later, the spindle apparatus is involved in chromosome sorting. Metaphase is a step in mitosis during which all of the chromosomes should be attached to the spindle apparatus. If a chromosome is not correctly attached, the metaphase checkpoint proteins will stop the cell cycle. This checkpoint prevents cells from incorrectly sorting their chromosomes during division.

Checkpoint proteins delay the cell cycle until problems are fixed or prevent cell division when problems cannot be fixed. A primary aim of checkpoint proteins is to prevent the division of a cell that has incurred DNA damage or harbors abnormalities in chromosome number. As discussed in Chapter 15, when the functions of checkpoint genes are lost due to mutation, the likelihood increases that undesirable genetic changes will occur that can cause additional mutations and cancerous growth.

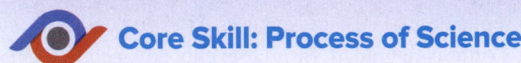

## Core Skill: Process of Science

# Feature Investigation | Masui and Markert's Study of Oocyte Maturation Led to the Identification of Cyclins and Cyclin-Dependent Kinases

During the 1960s and 1970s, researchers were intensely searching for the factors that promote cell division. In 1971, Japanese zoologist Yoshio Masui and American biologist Clement Markert developed a way to test whether a substance causes a cell to advance from one phase of the cell cycle to the next. They chose to study frog oocytes— cells that mature into egg cells. At the time of their work, researchers had already determined that frog oocytes naturally become dormant in the $G_2$ phase of the cell cycle for up to 8 months (**Figure 16.4**). During mating season, female frogs produce a hormone called progesterone. After progesterone enters an oocyte and binds to intracellular receptors, the oocyte advances from $G_2$ to the beginning of M phase, where the chromosomes condense and become visible under the microscope. This phenomenon is called maturation. When a sperm fertilizes the egg, M phase is completed, and the zygote continues to undergo cellular divisions.

Because progesterone is a signaling molecule, Masui and Markert speculated that this hormone affects the functions and/or amounts of proteins that trigger the oocyte to advance through the cell cycle. To test this hypothesis, they developed the procedure described in **Figure 16.5**, using the oocytes of the leopard frog (*Rana pipiens*). They began by exposing oocytes to progesterone in

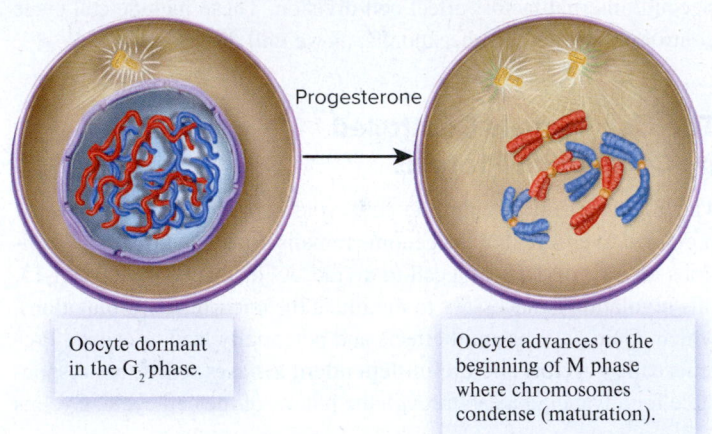

Oocyte dormant in the $G_2$ phase.

Oocyte advances to the beginning of M phase where chromosomes condense (maturation).

**Figure 16.4** Oocyte maturation in certain species of frogs.

vitro, and then they incubated these oocytes for 2 hours or 12 hours. As a control, they also used oocytes that had not been exposed to progesterone. These three types of cells were called the donor oocytes.

**Figure 16.5** The experimental approach of Masui and Markert to identify cyclin and cyclin-dependent kinase (cdk).

**HYPOTHESIS** Progesterone induces the synthesis of factor(s) that advance(s) frog oocytes through the cell cycle from $G_2$ to M phase.

**KEY MATERIALS** Oocytes from *Rana pipiens*.

Experimental level

Conceptual level

1 Expose oocytes to progesterone, then incubate for 2 or 12 hours. As a control, also use oocytes that have not been exposed to progesterone. All 3 types are donor oocytes.

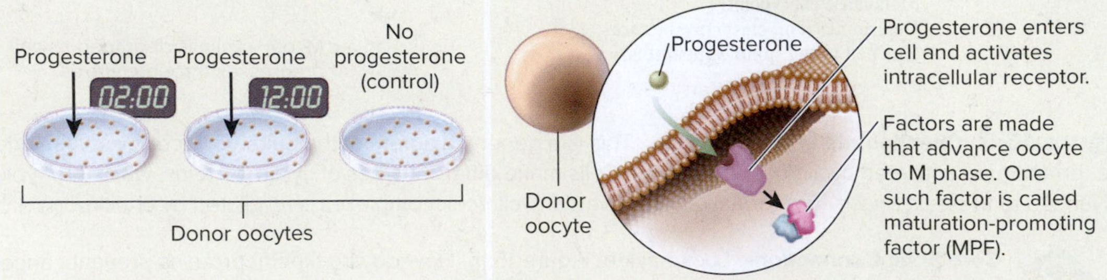

Progesterone 02:00   Progesterone 12:00   No progesterone (control)

Donor oocytes

Donor oocyte

Progesterone

Progesterone enters cell and activates intracellular receptor.

Factors are made that advance oocyte to M phase. One such factor is called maturation-promoting factor (MPF).

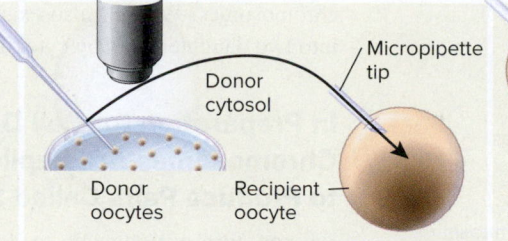
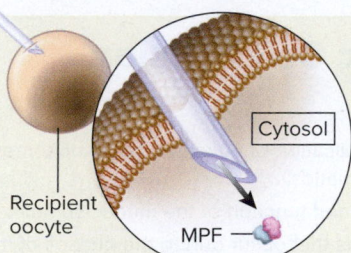

**2** Using a micropipette, transfer some cytosol from the 3 types of donor oocytes to recipient oocytes that have not been exposed to progesterone.

Donor cytosol

Micropipette tip

Donor oocytes

Recipient oocyte

Recipient oocyte

Cytosol

MPF

Recipient oocyte received MPF from donor oocyte if donor oocyte was incubated for 12 hours with progesterone.

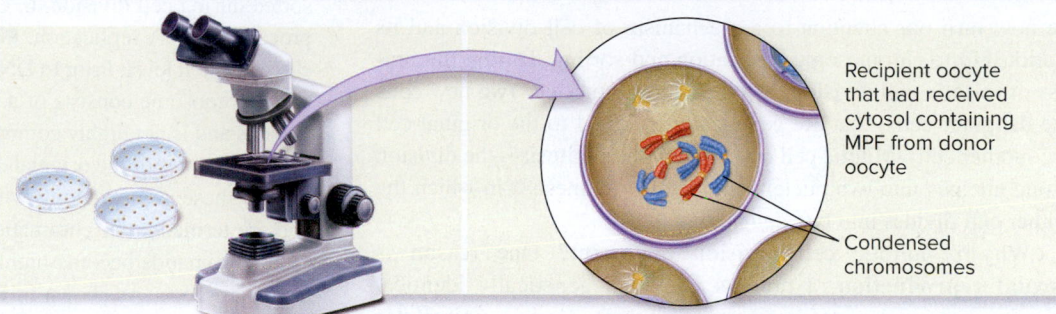

**3** Incubate for several hours, and observe the recipient oocytes under the microscope to determine if the recipient oocytes advance to M phase. Advancement to M phase can be determined by the condensation of the chromosomes.

Recipient oocyte that had received cytosol containing MPF from donor oocyte

Condensed chromosomes

**4    THE DATA**

| Donor oocytes | Recipient oocytes proceeded to M phase? |
|---|---|
| Control, no progesterone exposure | No |
| Progesterone exposure, incubation for 2 hours | No |
| Progesterone exposure, incubation for 12 hours | Yes |

**5    CONCLUSION**  Exposure of oocytes to progesterone for 12 hours results in the synthesis of factor(s) that advance(s) frog oocytes through the cell cycle from $G_2$ to M phase.

**6    SOURCE**  Masui, Y., and Markert, C. L. 1971. Cytoplasmic Control of Nuclear Behavior During Meiotic Maturation of Frog Oocytes. *Journal of Experimental Zoology* 177: 129–145.

Next, Masui and Markert used a micropipette to transfer a small amount of cytosol from the three types of donor oocytes to recipient oocytes that had not been exposed to progesterone. As seen in the data, the recipient oocytes that had been injected with cytosol from the control donor oocytes or from oocytes that had been incubated with progesterone for only 2 hours did not advance to M phase. However, cytosol from donor oocytes that had been incubated with progesterone for 12 hours caused the recipient oocytes to advance to M phase. Masui and Markert concluded that a cytosolic factor, which required more than 2 hours to be synthesized after progesterone treatment, had been transferred to the recipient oocytes and induced maturation. The factor that caused the oocytes to advance (or mature) from $G_2$ to M phase was originally called the maturation-promoting factor (MPF).

After MPF was discovered in frogs, it was found in all eukaryotic species that researchers studied. MPF is important in the division of all types of cells, not just oocytes. It took another 17 years before Manfred Lohka, Marianne Hayes, and James Maller were

able to purify the components that make up MPF. This was a difficult undertaking because these components are found in very small amounts in the cytosol and are easily degraded during purification procedures. We now know that MPF is a complex made of a mitotic cyclin and a cyclin-dependent kinase (cdk), as shown in Figure 16.3.

### Experimental Questions

1. At the time of Masui and Markert's study, summarized in Figure 16.5, what was known about the effects of progesterone on oocytes?

2. **CoreSKILL »** What hypothesis did Masui and Markert propose to explain the function of progesterone? Describe the procedure used to test the hypothesis.

3. **CoreSKILL »** How did the researchers explain the difference between the results with 2-hour-exposed donor oocytes versus 12-hour-exposed donor oocytes?

## 16.2  Mitotic Cell Division

**Learning Outcomes:**

1. Describe how the replication of eukaryotic chromosomes produces sister chromatids.
2. Explain the structure and function of the mitotic spindle.
3. Outline the key events that occur during the phases of mitosis.

We now turn our attention to a mechanism of cell division and its relationship to chromosome replication and sorting. During the process of **mitotic cell division**, a cell divides to produce two new cells (the daughter cells) that are genetically identical to the original cell (the mother cell). Mitotic cell division involves mitosis—the division of one nucleus into two nuclei—and then cytokinesis—in which the mother cell divides into two daughter cells.

Why is mitotic cell division important? One reason is **asexual reproduction**, a process in which genetically identical offspring are produced from a single parent. Certain unicellular eukaryotic organisms, such as baker's yeast (*Saccharomyces cerevisiae*) and the amoeba, increase their numbers in this manner. A second important reason for mitotic cell division is the production and maintenance of multicellularity. Organisms such as plants, animals, and most fungi are derived from a single cell that subsequently undergoes repeated cell divisions to become a multicellular organism.

In this section, we will explore the process of mitotic cell division, which requires the replication, organization, and sorting of

chromosomes. We will also examine how a single cell is separated into two daughter cells by cytokinesis.

### In Preparation for Cell Division, Eukaryotic Chromosomes Are Replicated and Compacted to Produce Pairs Called Sister Chromatids

We now turn our attention to how chromosomes are replicated and sorted during cell division. In Chapter 11, we examined the molecular process of DNA replication. **Figure 16.6** describes the process at the chromosomal level. Prior to DNA replication, the DNA of each eukaryotic chromosome consists of a linear double helix that is found in the nucleus and is not highly compacted. When the DNA is replicated, two identical copies of the original double helix are produced. As discussed earlier, these copies, along with associated proteins, lie side-by-side and are termed sister chromatids. When a cell prepares to divide, the sister chromatids become highly compacted and readily visible under the microscope. As shown in Figure 16.6b, the two sister chromatids are tightly associated at a region called the **centromere**. A protein called cohesin holds the sister chromatids together. In addition, the centromere serves as an attachment site for a group of proteins that form the **kinetochore**, a structure necessary for sorting the chromosomes.

### The Mitotic Spindle Organizes and Sorts Chromosomes During Cell Division

What structure is responsible for organizing and sorting the chromosomes during cell division? The answer is the **mitotic spindle**

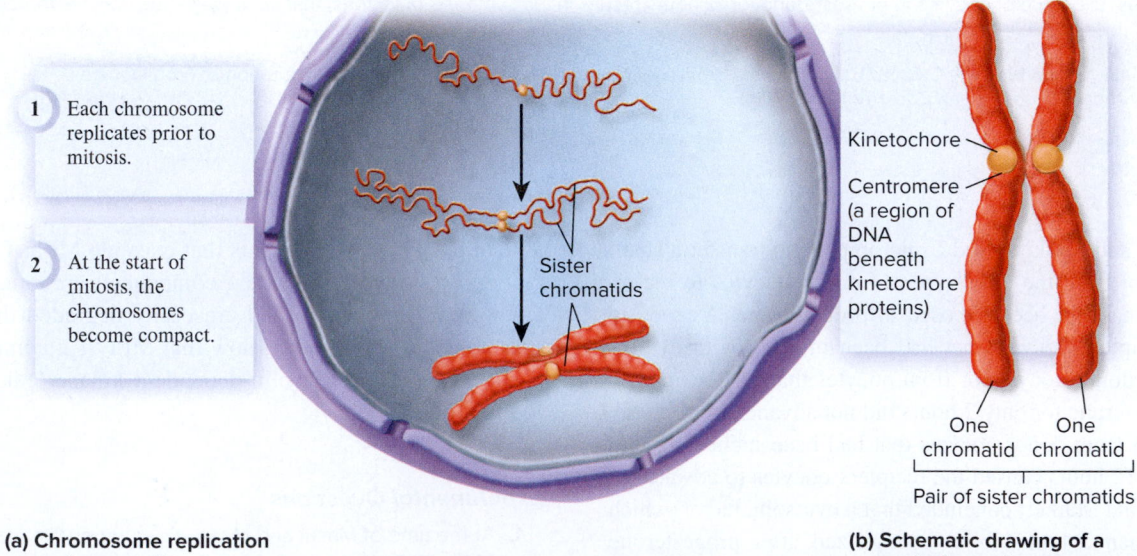

① Each chromosome replicates prior to mitosis.

② At the start of mitosis, the chromosomes become compact.

Sister chromatids

Kinetochore

Centromere (a region of DNA beneath kinetochore proteins)

One chromatid    One chromatid

Pair of sister chromatids

(a) Chromosome replication and compaction

(b) Schematic drawing of a metaphase chromosome

**Figure 16.6  Replication and compaction of chromosomes into pairs of sister chromatids.** **(a)** Chromosomal replication produces a pair of sister chromatids. While the chromosomes are elongated, they are replicated to produce two copies that are connected and lie parallel to each other. This is a pair of sister chromatids. Later, when the cell is preparing to divide, the sister chromatids condense into more compact structures that are easily seen with a light microscope. **(b)** A schematic drawing of a metaphase chromosome. This structure has two chromatids that lie side-by-side. The two chromatids are held together by cohesin proteins (not shown in this drawing). The kinetochore is a group of proteins that are attached to the centromere and play a key role during chromosome sorting.

**Core Concept: Information**  The process of mitosis ensures that each daughter cell receives a complete copy of the genetic material.

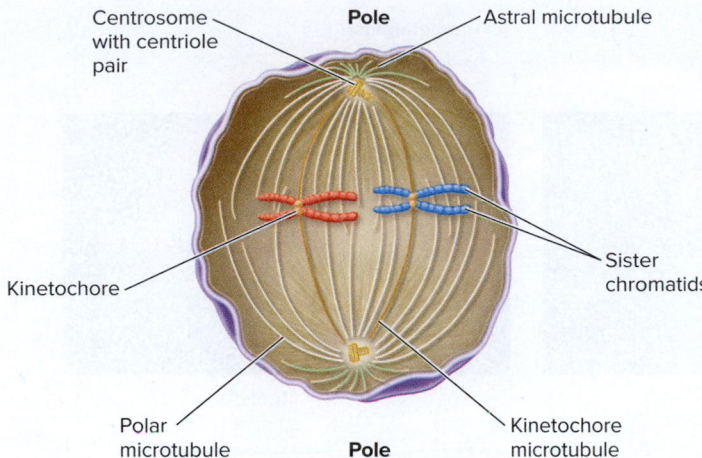

**Figure 16.7** **The structure of the mitotic spindle.** The mitotic spindle in animal cells is formed by the centrosomes, which produce three types of microtubules. The astral microtubules emanate away from the region between the poles. The polar microtubules project into the region between the two poles. The kinetochore microtubules are attached to the kinetochores of sister chromatids. Note: For simplicity, this diagram shows only one pair of homologous chromosomes. Eukaryotic species typically have multiple chromosomes per set.

(**Figure 16.7**). It is composed of microtubules—protein fibers that are components of the cytoskeleton (refer back to Table 4.1). In animal cells, microtubule growth and organization start at two **centrosomes**, structures that are also referred to as microtubule-organizing centers (MTOCs). A single centrosome duplicates during interphase. When the cell enters mitosis, each centrosome defines a **pole** of the spindle apparatus, one within each of the future daughter cells. The centrosome in animal cells has a pair of **centrioles**. Each one is composed of nine sets of triplet microtubules. However, centrioles are not found in many other eukaryotic species, such as plants, and are not required for spindle formation.

Each centrosome organizes the construction of the microtubules by rapidly polymerizing tubulin proteins. The three types of spindle microtubules are termed astral, polar, and kinetochore microtubules (see Figure 16.7).

## The Transmission of Chromosomes Requires a Sorting Process Known as Mitosis

Mitosis is the sorting process for dividing one cell nucleus into two nuclei (**Figure 16.8**). The duplicated chromosomes are distributed so that each daughter cell receives the same complement of chromosomes. Mitosis was first observed microscopically in the 1870s by a German biologist, Walther Flemming, who coined the term (from the Greek *mitos*, meaning thread). He studied the large, transparent skin cells of salamander larvae as they were dividing and noticed that chromosomes are constructed of "threads" that are doubled in appearance along their length. These double threads divided and moved apart, one going to each of the two daughter nuclei. By this

mechanism, Flemming pointed out, the two daughter cells receive an identical group of threads, the same as the number of threads in the mother cell.

Figure 16.8 depicts the process of mitosis in an animal cell, though the process is quite similar in a plant cell. Mitosis occurs as a continuum of phases known as prophase, prometaphase, metaphase, anaphase, and telophase. In the simplified diagrams shown along the bottom of Figure 16.8, the original cell contains six chromosomes. One set of chromosomes is depicted in red, whereas the homologous set is blue. The different colors are intended to distinguish maternal and paternal chromosomes.

*Interphase*  Prior to mitosis, the cells are in interphase, which consists of the $G_1$, S, and $G_2$ phases of the cell cycle. The chromosomes have replicated in S phase and are decondensed and found in the nucleus (Figure 16.8a). The nucleolus, which is the site where the components of ribosomes assemble into ribosomal subunits, is visible during interphase.

*Prophase*  At the start of mitosis, in **prophase**, the chromosomes have already replicated to produce 12 chromatids, joined as six pairs of sister chromatids that have condensed into highly compacted structures readily visible by light microscopy (Figure 16.8b). As prophase proceeds, the nuclear envelope begins to dissociate into small vesicles. The nucleolus is no longer visible.

*Prometaphase*  During **prometaphase**, the nuclear envelope completely fragments into small vesicles, and the mitotic spindle is fully formed (Figure 16.8c). As prometaphase advances, the centrosomes move apart and demarcate the two poles. Once the nuclear envelope has dissociated, the spindle fibers can interact with the sister chromatids. How do the sister chromatids become attached to the spindle apparatus? Initially, microtubules are rapidly formed and can be seen under a microscope growing out from the two poles. As it grows, if a microtubule happens to make contact with a kinetochore, it is said to be captured and remains firmly attached to the kinetochore. Alternatively, if a microtubule does not collide with a kinetochore, the microtubule eventually depolymerizes and retracts to the centrosome. This random process is how sister chromatids become attached to kinetochore microtubules. As the end of prometaphase nears, the two kinetochores on each pair of sister chromatids are attached to kinetochore microtubules from opposite poles. As these events are occurring, the sister chromatids are seen under the microscope to undergo jerky movements as they are tugged, back and forth, between the two poles by the kinetochore microtubules.

*Metaphase*  Eventually, the pairs of sister chromatids are aligned in a single row along the **metaphase plate**, a plane halfway between the poles of the spindle. When this alignment is complete, the cell is in **metaphase** of mitosis (Figure 16.8d). The chromatids can then be equally distributed into two daughter cells.

*Anaphase*  During **anaphase**, the connections between the pairs of sister chromatids are broken (Figure 16.8e). Each chromatid, now an

**(a) Interphase**    **(b) Prophase**    **(c) Prometaphase**

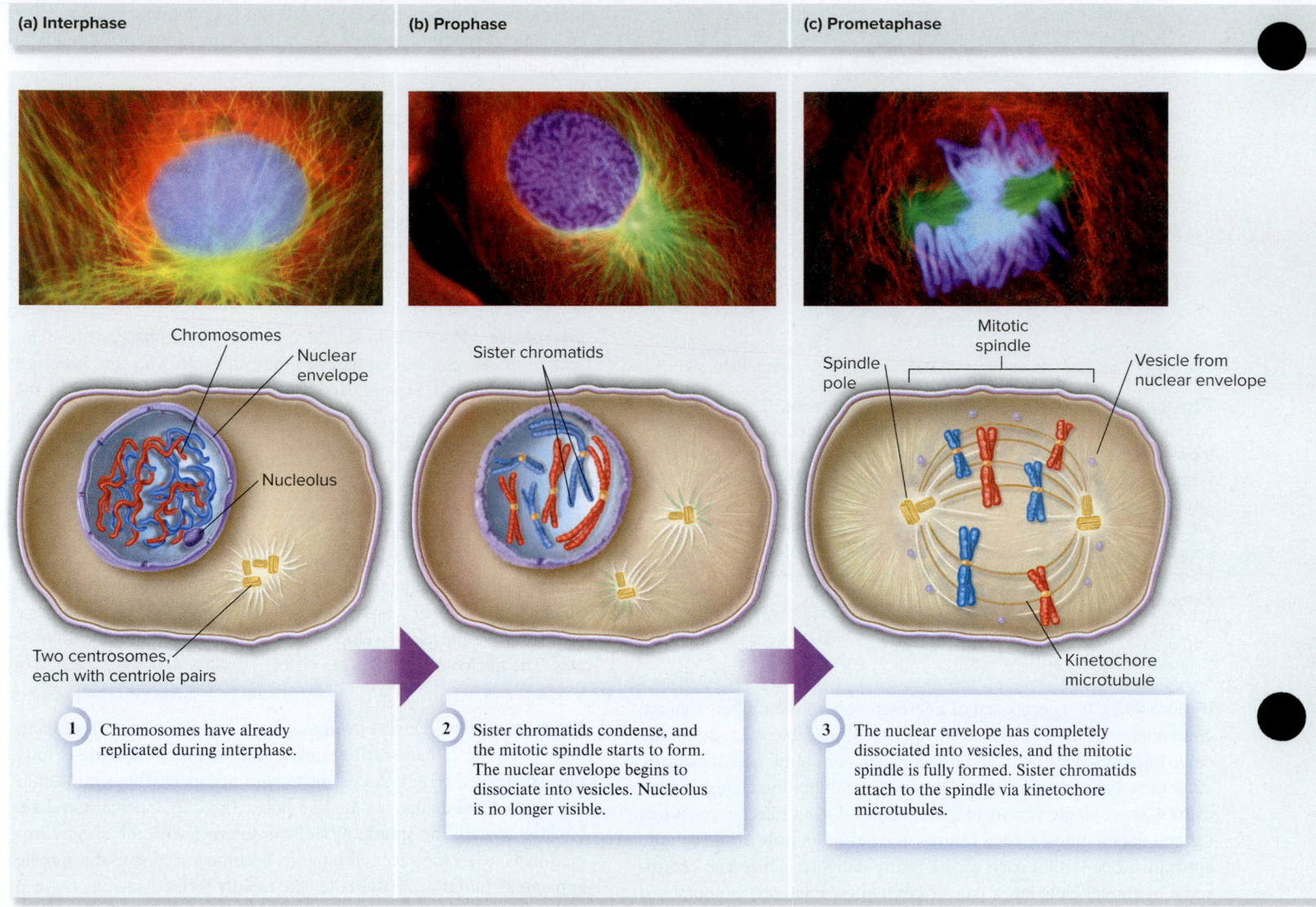

Chromosomes
Nuclear envelope
Sister chromatids
Mitotic spindle
Spindle pole
Vesicle from nuclear envelope
Nucleolus
Two centrosomes, each with centriole pairs
Kinetochore microtubule

1 │ Chromosomes have already replicated during interphase.

2 │ Sister chromatids condense, and the mitotic spindle starts to form. The nuclear envelope begins to dissociate into vesicles. Nucleolus is no longer visible.

3 │ The nuclear envelope has completely dissociated into vesicles, and the mitotic spindle is fully formed. Sister chromatids attach to the spindle via kinetochore microtubules.

**Figure 16.8** **The process of mitosis in an animal cell.** The top panels are fluorescence micrographs of a newt cell advancing through mitosis. The drawings below emphasize the sorting of the chromosomes, in which the diploid mother cell had six chromosomes (three in each set). At the start of mitosis, these have already replicated into 12 chromatids. The final result is two daughter cells, each containing six chromosomes.

a–f: ©Photographs by Dr. Conly L. Rieder, East Greenbush, New York, 12061

**Concept Check:**   *What are the functions of the three types of microtubules?*

individual chromosome, is linked to only one of the two poles by one or more kinetochore microtubules. As anaphase proceeds, the kinetochore microtubules shorten, pulling the chromosomes toward the pole to which they are attached. In addition, the two poles move farther away from each other. This occurs because the overlapping polar microtubules lengthen and push against each other, thereby pushing the poles farther apart.

**Telophase**   During **telophase**, the chromosomes have reached their respective poles and decondense. The nuclear envelope now re-forms to produce two separate nuclei. In Figure 16.8f, two nuclei that contain six chromosomes each are being produced.

**Cytokinesis**   In most cases, mitosis is quickly followed by cytokinesis, in which the two nuclei are segregated into separate daughter cells. Whereas the phases of mitosis are similar between plant and animal cells, the process of cytokinesis is quite different.

- In animal cells, cytokinesis involves the formation of a **cleavage furrow**, which constricts like a drawstring to separate the cells (**Figure 16.9a**).

- In plants, vesicles from the Golgi apparatus move along microtubules to the center of the cell and coalesce to form a **cell plate** (**Figure 16.9b**), which then forms a cell wall between the two daughter cells.

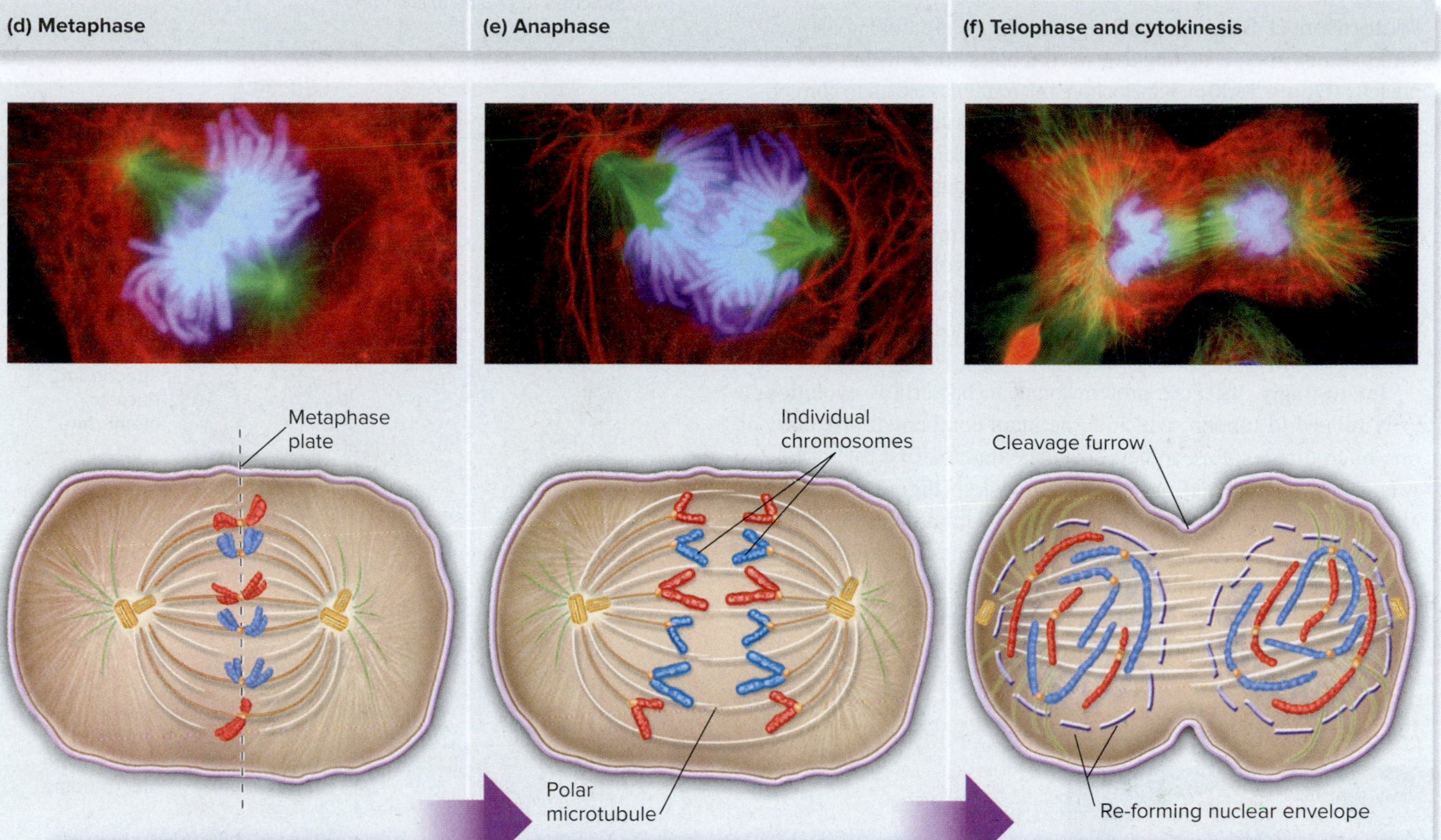

**(d) Metaphase**

**(e) Anaphase**

**(f) Telophase and cytokinesis**

Metaphase plate

Individual chromosomes

Cleavage furrow

Polar microtubule

Re-forming nuclear envelope

**4** Sister chromatids align along the metaphase plate.

**5** Sister chromatids separate, and individual chromosomes move toward the poles as kinetochore microtubules shorten. Polar microtubules lengthen and push the poles apart.

**6** Chromosomes decondense, and the nuclear envelope re-forms. Cytokinesis separates the mother cell into two daughter cells, and it begins with a cleavage furrow in animal cells.

What are the results of mitosis and cytokinesis? These processes ultimately produce two daughter cells with the same number of chromosomes as the mother cell. Barring rare mutations, the two daughter cells are genetically identical to each other and to the mother cell from which they were derived. The critical consequence of this sorting process is ensuring genetic consistency from one cell to the next. The development of multicellularity relies on the repeated process of mitosis and cytokinesis.

 **Core Concept: Evolution**

### Mitosis in Eukaryotes Evolved from the Binary Fission That Occurs in Prokaryotic Cells

The process of mitosis allows eukaryotic cells to properly sort their chromosomes during cell division. By comparing cell division among prokaryotic cells, simple eukaryotes, and more com-

plex eukaryotes, biologists have pieced together a progression of how mitosis may have evolved.

**Binary Fission in Bacterial Cells** As described in Chapter 19 (look ahead to Figure 19.13), bacterial cells divide by a relatively simple process known as binary fission (**Figure 16.10a**). After chromosome replication, each copy of the bacterial chromosome becomes anchored to the plasma membrane. Proteins called FtsZ form a ring at the site where the mother cell will divide into two daughter cells.

**Dinoflagellates** In protists known as dinoflagellates, nuclear division is much simpler than in animal and plant cells (**Figure 16.10b**). After chromosome replication, the chromosomes become attached to the nuclear envelope. The nuclear envelope does not break apart. Microtubules, which are described in Chapter 4 (see Table 4.1), are formed in the cytosol and pass through tunnels in the nuclear envelope. The nucleus then divides by a process that resembles binary fission.

**Diatoms and Some Yeasts** In diatoms (a type of protist) and some yeasts (a type of fungus), microtubules form within the cell nucleus (**Figure 16.10c**). Kinetochore microtubules attach to chromosomes and facilitate their sorting, and other microtubules promote the separation of the nucleus into two separate nuclei. As in dinoflagellates, the nuclear envelope does not break apart during this process.

**Complex Eukaryotes** As we have seen, mitosis in complex eukaryotes, such as animals and plants, involves the breaking apart of the nuclear envelope and the formation of the spindle apparatus (**Figure 16.10d**). After the chromosomes are sorted, the nuclear envelope then re-forms.

Interestingly, the FtsZ protein found in bacteria is evolutionarily related to tubulin, which is the main component of eukaryotic microtubules. Researchers speculate that the first role of FtsZ was to form a ring and promote cell division (Figure 16.10a). In eukaryotic cells, the homologous protein, tubulin, forms linear microtubules and has acquired additional roles in cell division. These include the division of the cell nucleus and the sorting of chromosomes. The relationship of tubulin to FtsZ is an example of descent with modification.

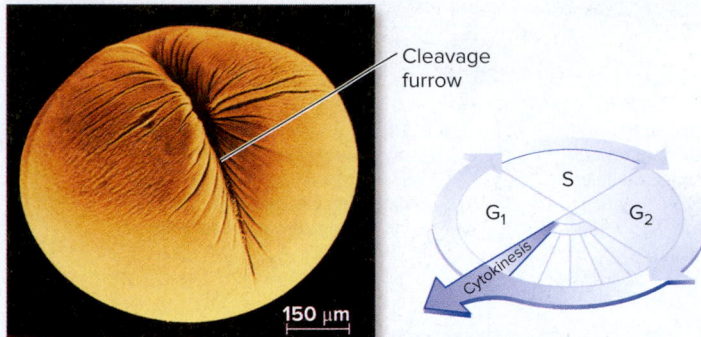

(a) Cleavage of an animal cell

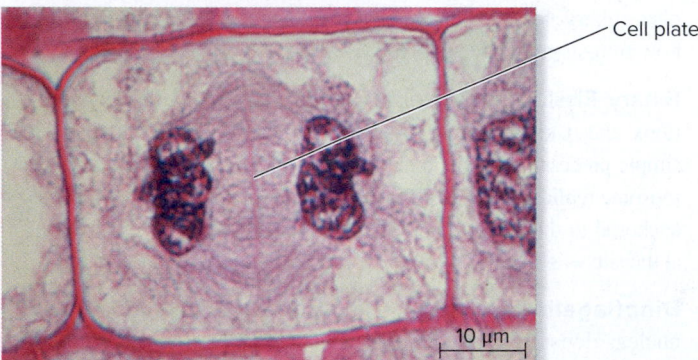

(b) Formation of a cell plate in a plant cell

**Figure 16.9** **Micrographs showing cytokinesis in animal and plant cells.** a: ©Don W. Fawcett/Science Source; b: ©Carolina Biological Supply Company/Phototake

*Concept Check:* *What are the similarities and differences between animal and plant cells with regard to cytokinesis?*

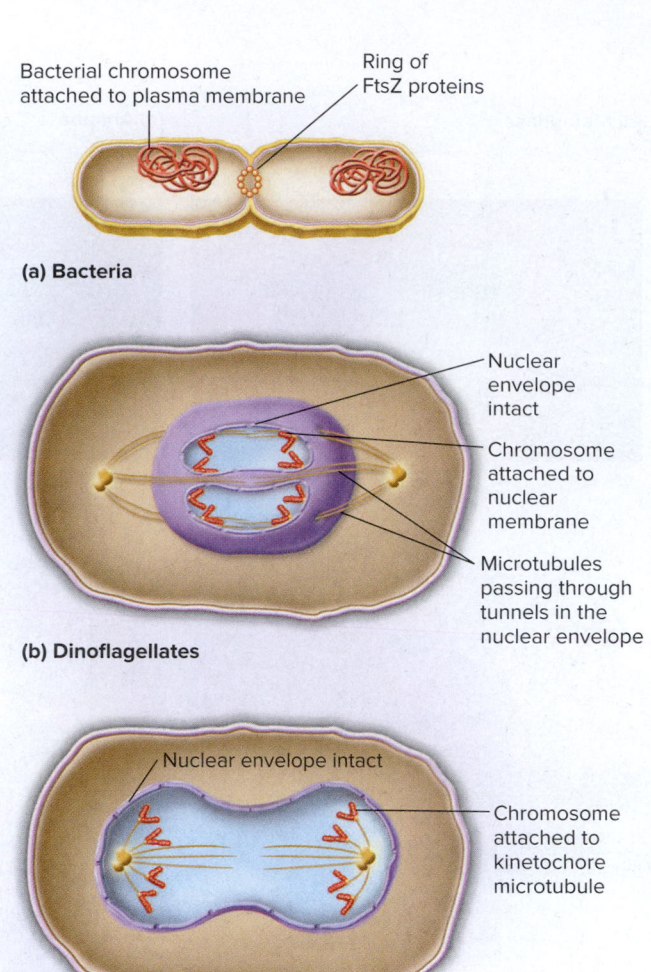

(a) Bacteria

(b) Dinoflagellates

(c) Diatoms and some yeasts

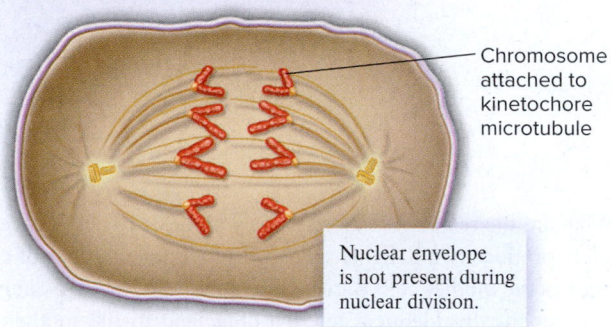

Nuclear envelope is not present during nuclear division.

(d) Most eukaryotes

**Figure 16.10** **A comparison of cell division among bacteria, simple eukaryotes, and more complex eukaryotes.**

## 16.3 Meiosis

**Learning Outcomes:**

1. Describe the processes of synapsis and crossing over.
2. Outline the key events that occur during the phases of meiosis.
3. Compare and contrast mitosis and meiosis, focusing on key steps that account for the different outcomes of these two processes.

As discussed earlier, a diploid cell contains two homologous sets of chromosomes, whereas a haploid cell contains a single set. For example, a diploid human cell contains 46 chromosomes, but a human gamete—sperm or egg cell—is a haploid cell that contains only 23 chromosomes. **Meiosis** is the process by which haploid cells are produced from a cell that was originally diploid. The term meiosis, which means to make smaller, refers to the fewer chromosomes found in cells that have undergone this process. For haploid cells to be produced, the chromosomes must be correctly sorted and distributed in a way that reduces the chromosome number to half its original diploid value. In the case of human gametes, for example, each gamete must receive one chromosome from each of the 23 pairs. For this to happen, two rounds of divisions are necessary, termed meiosis I and meiosis II (**Figure 16.11**). When

a cell begins meiosis, it contains chromosomes that are found in homologous pairs. When meiosis is completed, a single diploid cell with homologous pairs of chromosomes has produced four haploid cells.

In this section, we will examine the cellular events of meiosis that reduce the chromosome number from diploid to haploid. In the following section, we will consider how this process plays a role in the sexual reproduction of animals, plants, fungi, and protists.

## Bivalent Formation and Crossing Over Occur at the Beginning of Meiosis

Like mitosis, meiosis begins after a cell has progressed through the $G_1$, S, and $G_2$ phases of the cell cycle. However, two key events occur at the beginning of meiosis that do not occur in mitosis. First, homologous pairs of sister chromatids associate with each other, lying side by side to form a **bivalent**, also called a tetrad (**Figure 16.12**). The process of forming a bivalent is termed **synapsis**. In most eukaryotic species, a protein structure called the synaptonemal complex connects homologous chromosomes during a portion of meiosis. However, the synaptonemal complex is not required for the pairing of homologous chromosomes because some species of fungi completely lack such a complex, yet their chromosomes associate with each other correctly.

The second event that occurs at the beginning of meiosis, but not usually during mitosis, is **crossing over**, which involves a physical exchange between chromosome segments of the bivalent (Figure 16.12). As discussed in Chapter 18, crossing over increases the genetic variation of sexually reproducing species. After crossing over occurs, the arms of the chromosomes tend to separate but remain adhered at a crossover site. This connection is called a chiasma (plural, chiasmata), because the connected chromosomal arms resemble the Greek letter chi, χ. The number of crossovers is carefully controlled by cells and depends on the size of the chromosome and the species. The range of crossovers for eukaryotic chromosomes is typically one or two to a couple dozen. During the formation of sperm in humans, for example, an average chromosome undergoes slightly more than two crossovers, whereas chromosomes in certain plant species may undergo 20 or more.

## Meiosis I Separates Homologous Chromosomes

Now that you have an understanding of bivalent formation and crossing over, we are ready to consider the phases of meiosis (**Figure 16.13**). These simplified diagrams depict a diploid cell (2n) that contains a total of six chromosomes (as in the diagram of mitosis in Figure 16.8). Prior to meiosis, the chromosomes are replicated in S phase to produce pairs of sister chromatids. This single replication event is then followed by sequential divisions called meiosis I and II. Like mitosis, each of these processes is a continuous series of stages called prophase, prometaphase, metaphase, anaphase, and telophase. The sorting that occurs during **meiosis I** separates homologous chromosomes from each other (Figure 16.13a–e).

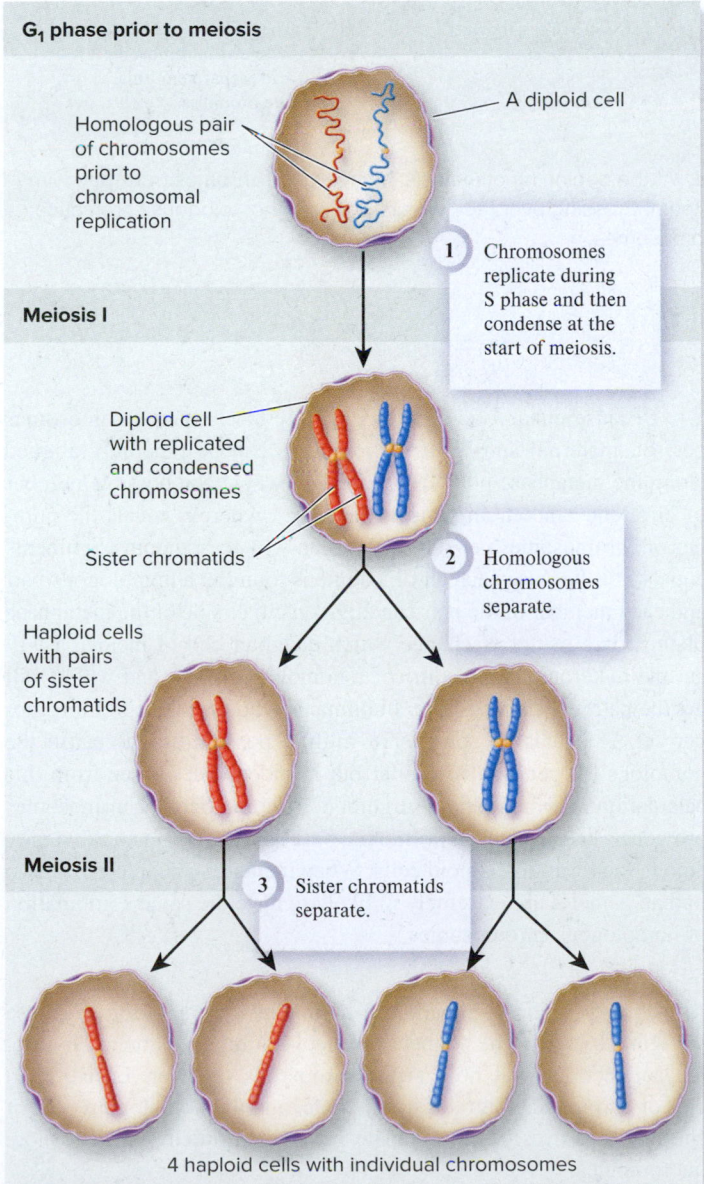

**Figure 16.11  How the process of meiosis reduces chromosome number.** This simplified diagram emphasizes the reduction in chromosome number as a diploid cell divides by meiosis to produce four haploid cells.

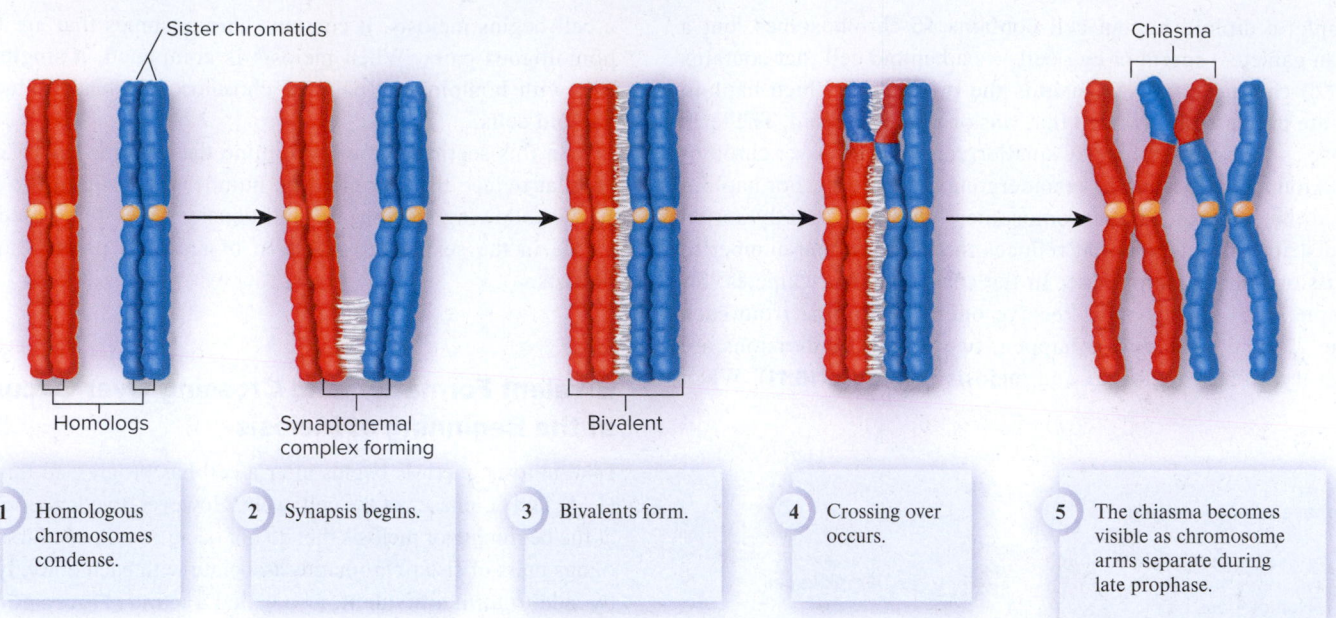

Sister chromatids

Chiasma

Homologs | Synaptonemal complex forming | Bivalent

1. Homologous chromosomes condense.
2. Synapsis begins.
3. Bivalents form.
4. Crossing over occurs.
5. The chiasma becomes visible as chromosome arms separate during late prophase.

**Figure 16.12** **Formation of a bivalent and crossing over during meiosis I.** At the beginning of meiosis, homologous chromosomes pair with each other to form a bivalent, usually with a synaptonemal complex between them. Crossing over then occurs between homologous chromatids within the bivalent. During this process, homologs exchange segments of chromosomes.

**Prophase I** During prophase I, the replicated chromosomes condense, the homologous chromosomes form bivalents, and crossing over occurs. The nuclear envelope then starts to fragment into small vesicles.

**Prometaphase I** In prometaphase I, the nuclear envelope is completely broken apart into vesicles, and the spindle apparatus is entirely formed. The sister chromatids become attached to kinetochore microtubules. However, a key difference exists between mitosis and meiosis I. In mitosis, each pair of sister chromatids is attached to both poles (see Figure 16.8c). In meiosis I, each pair of sister chromatids is attached to just one pole via kinetochore microtubules (Figure 16.13b).

**Metaphase I** At metaphase I, the bivalents are organized along the metaphase plate. Notice how this pattern of alignment is strikingly different from that observed during mitosis (see Figure 16.8d). In particular, the sister chromatids are aligned in a double row rather than a single row (as in mitosis). Furthermore, the arrangement of sister chromatids within this double row is random with regard to the (red and blue) homologs. (Remember that these different colors indicate maternal and paternal chromosomes.) In Figure 16.13c, one of the red homologs is to the left of the metaphase plate, and the other two are to the right, whereas two of the blue homologs are to the left of the metaphase plate and the other one is to the right. In other cells, homologs could be arranged differently along the metaphase plate (for example, three blues to the left and none to the right, or none to the left and three to the right).

Because eukaryotic species typically have many chromosomes per set, maternal and paternal homologs can be randomly aligned along the metaphase plate in a variety of ways. The possible number of different, random alignments equals $2^n$, where $n$ equals the number of chromosomes per set. The reason why the random alignments equals $2^n$ is because each chromosome is found in a homologous pair and each member of the pair can align on either side of the metaphase plate. It is a matter of chance which daughter cell of meiosis I will get the maternal chromosome of a homologous pair, and which will get the paternal chromosome. In humans, who have 23 chromosomes per set, $2^n$ equals $2^{23}$, or over 8 million possibilities. Because the homologs are genetically similar but not identical, we see from this calculation that the random alignment of homologous chromosomes provides a mechanism to promote a vast amount of genetic diversity among the resulting haploid cells. When meiosis is complete, any two human gametes are extremely unlikely to have the same combination of homologous chromosomes.

**Anaphase I** The segregation of homologs occurs during anaphase I (Figure 16.13d). The connections between bivalents break, but not the connections that hold sister chromatids together. Each joined pair of chromatids migrates to one pole, and the homologous pair of chromatids moves to the opposite pole, both pulled by kinetochore microtubules.

**Telophase I** At telophase I, the sister chromatids have reached their respective poles and then decondense. The nuclear envelope now reforms to produce two separate nuclei.

If we consider the end result of meiosis I, we see that two nuclei are produced, each with three pairs of sister chromatids; this is called a reduction division. The original diploid cell had its chromosomes in homologous pairs, whereas the two cells produced as a result of meiosis I and cytokinesis are considered haploid—they do not have pairs of homologous chromosomes.

## Meiosis II Separates Sister Chromatids

Meiosis I is followed by cytokinesis and then **meiosis II** (see Figure 16.13f–j). DNA replication does not occur between meiosis I and meiosis II. The sorting events of meiosis II are similar to those of mitosis, but the starting point is different. For a diploid cell with six chromosomes, mitosis begins with 12 chromatids that are joined as six pairs of sister chromatids (see Figure 16.8). By comparison, the two cells that begin meiosis II each have six chromatids that are joined as three pairs of sister chromatids. Otherwise, the steps that occur during prophase, prometaphase, metaphase, anaphase, and telophase of meiosis II are analogous to a mitotic division. Sister chromatids are separated during anaphase II.

## Mitosis and Meiosis Differ in a Few Key Steps

How are the outcomes of mitosis and meiosis different? Mitosis produces two diploid daughter cells that are genetically identical. In our example shown in Figure 16.8, the starting cell had six chromosomes (three homologous pairs of chromosomes), and both daughter cells received copies of the same six chromosomes. By comparison, meiosis reduces the number of sets of chromosomes. In the example shown in Figure 16.13, the starting cell also had six chromosomes, whereas the resulting four daughter cells have only three chromosomes. However, the daughter cells do not contain a random mix of three chromosomes. Each haploid daughter cell contains one complete set of chromosomes, whereas the original diploid mother cell had two complete sets.

How do we explain the different outcomes of mitosis and meiosis? **Table 16.1** emphasizes the differences between certain key steps in mitosis and meiosis that account for the different outcomes of these two processes. DNA replication occurs prior to mitosis and meiosis I, but not between meiosis I and II. During prophase of meiosis I, the homologs synapse to form bivalents. This explains why crossing over occurs commonly during meiosis, but rarely during mitosis. During prometaphase of mitosis and meiosis II, pairs of sister chromatids are attached to both poles. In contrast, during meiosis I, each pair of sister chromatids (within a bivalent) is attached to a single pole. Bivalents align along the metaphase plate during metaphase of meiosis I, whereas sister chromatids align along the metaphase plate during metaphase of mitosis and meiosis II. At anaphase of meiosis I, the homologous chromosomes separate, but the sister chromatids remain together. In contrast, sister chromatid separation occurs during anaphase of mitosis and meiosis II. Taken together, the steps of mitosis produce two diploid cells that are genetically identical, whereas the steps of meiosis involve two sequential cell divisions that produce four haploid cells that may not be genetically identical.

 **THE QUESTION** *A diploid cell has 12 chromosomes, or 6 pairs. In the following diagram, in what phase of mitosis, meiosis I or meiosis II, is this cell?*

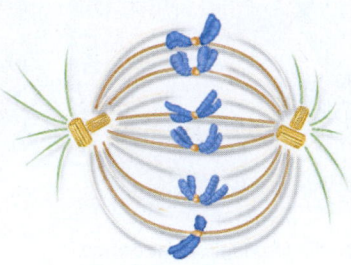

**T** OPIC *What topic in biology does this question address?* The topic is cell division. More specifically, the question is asking you to be able to look at a drawing and discern which phase of cell division a particular cell is in.

**I** NFORMATION *What information do you know based on the question and your understanding of the topic?* In the question, you are given a diagram of a cell at a particular phase of the cell cycle. This cell is derived from a mother cell with 6 pairs of chromosomes. From your understanding of the topic, you may remember the various phases of mitosis, meiosis I, and meiosis II, which are described in Figures 16.8 and 16.13. If so, you may initially realize that the cell is in metaphase.

**P** ROBLEM-SOLVING **S** TRATEGY *Sort out the steps in a complicated process.* To solve this problem, you may need to describe the steps, starting with a mother cell that has 6 pairs of chromosomes. Keep in mind that a mother cell with 6 pairs of chromosomes has 12 chromosomes during $G_1$, which then replicate to form 12 pairs of sister chromatids during S phase. Therefore, at the beginning of M phase, this mother cell will have 12 pairs of sister chromatids. During mitosis, the 12 pairs of sister chromatids will align at metaphase. During meiosis I, 6 bivalents will align along the metaphase plate in the mother cell. During meiosis II, 6 pairs of sister chromatids will align along the metaphase plate in the two cells.

**ANSWER** *The cell is in metaphase of meiosis II. You can tell because the chromosomes are lined up in a single row along the metaphase plate, and the cell has only 6 pairs of sister chromatids. If it were mitosis, the cell would have 12 pairs of sister chromatids. If it were in meiosis I, bivalents would be aligned along the metaphase plate.*

## Meiosis I

### (a) Prophase I

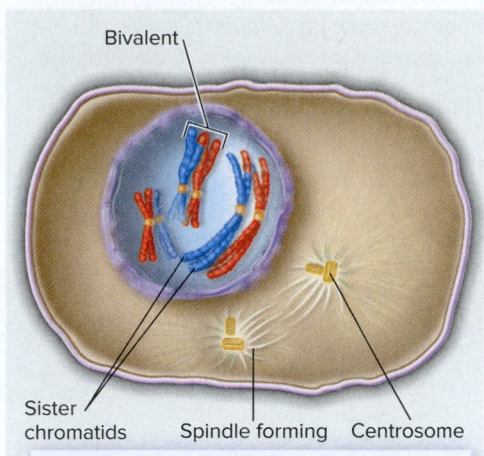

Bivalent

Sister chromatids  Spindle forming  Centrosome

**1** Homologous chromosomes synapse to form bivalents, and crossing over occurs. Chromosomes condense, and the nuclear envelope begins to dissociate into vesicles.

### (b) Prometaphase I

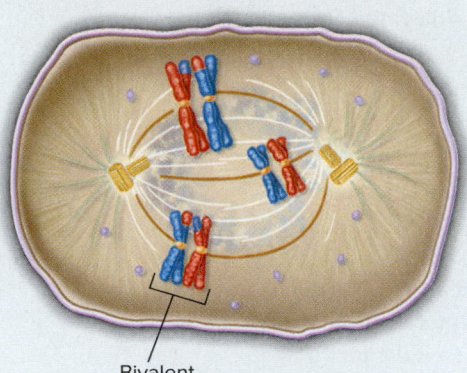

Bivalent

**2** The nuclear envelope completely dissociates into vesicles, and bivalents become attached to kinetochore microtubules.

### (c) Metaphase I

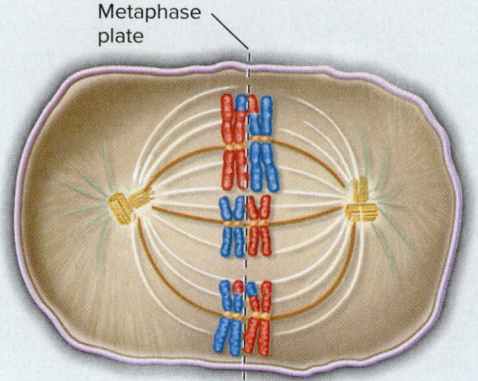

Metaphase plate

**3** Bivalents randomly align along the metaphase plate. Each pair of sister chromatids is attached to one pole.

## Meiosis II

### (f) Prophase II

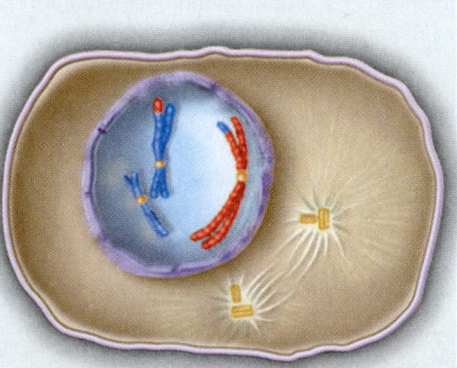

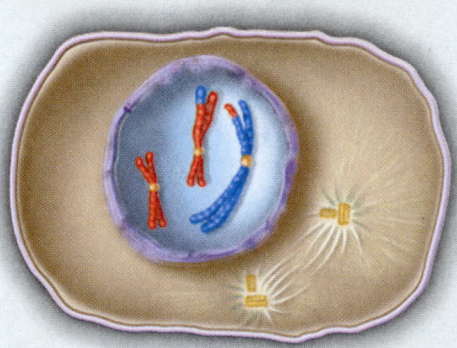

**6** Sister chromatids condense, and the spindle starts to form. The nuclear envelope begins to dissociate into vesicles.

### (g) Prometaphase II

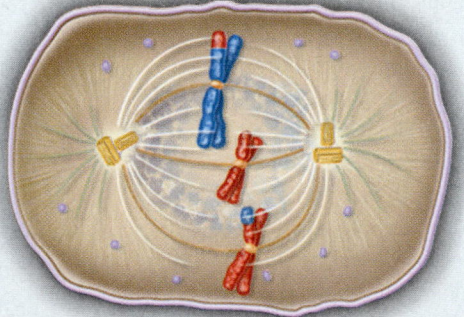

**7** The nuclear envelope completely dissociates into vesicles. Sister chromatids attach to the spindle via kinetochore microtubules.

### (h) Metaphase II

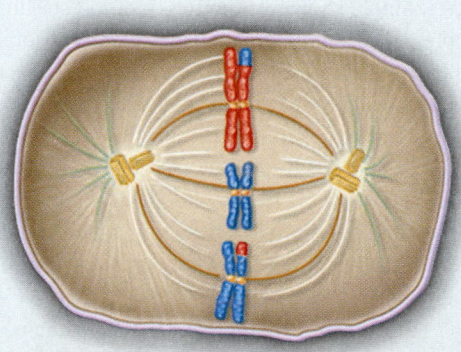

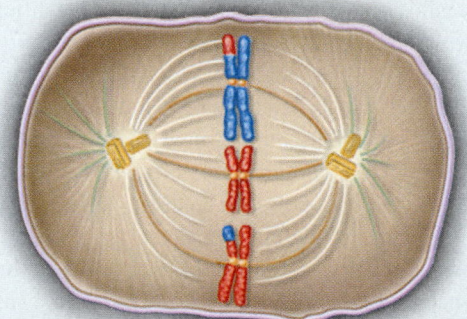

**8** Sister chromatids align along the metaphase plate. Each pair of sister chromatids is attached to both poles.

**(d) Anaphase I**

**(e) Telophase I and cytokinesis**

Cleavage furrow

**4**  Homologous chromosomes separate and pairs of sister chromatids move toward opposite poles.

**5**  The chromosomes decondense, and the nuclear envelope re-forms. The 2 daughter cells are separated by a cleavage furrow.

**(i) Anaphase II**

**(j) Telophase II and cytokinesis**

Four haploid cells

**9**  Sister chromatids separate, and individual chromosomes move toward the poles as kinetochore microtubules shorten. Polar microtubules lengthen and push the poles apart.

**10**  Chromosomes decondense, and the nuclear envelope re-forms. Cleavage furrows separate the 2 cells into 4 cells.

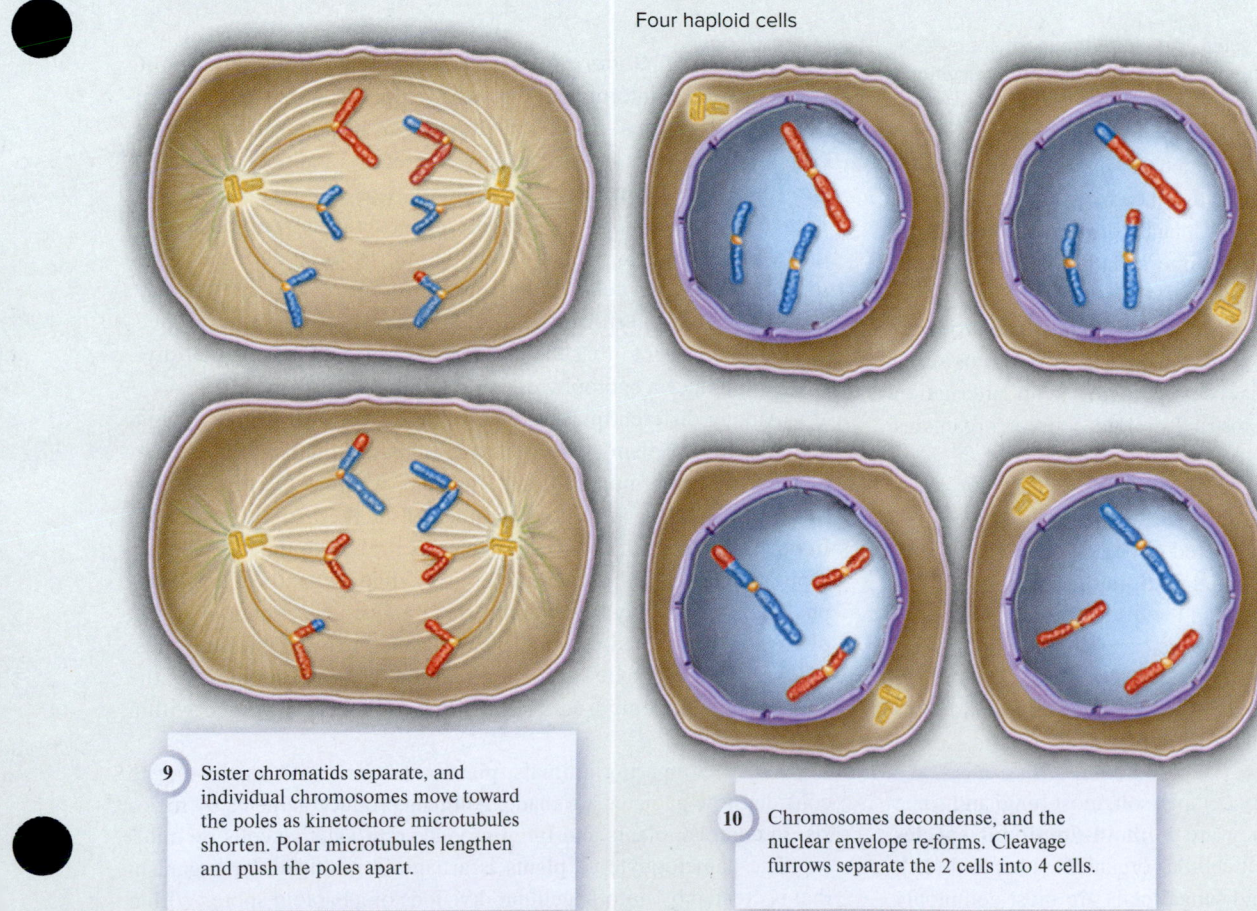

**Figure 16.13**  **The phases of meiosis in an animal cell.**

**Core Skill: Modeling**
The goal of this modeling challenge is to predict the outcome of meiosis if one pair of chromosomes does not separate properly during meiosis II.

**Modeling Challenge:** In Figure 16.13, the starting cell in meiosis I has 3 homologous pairs of chromosomes that differ in length: short, medium, and long. Let's suppose that during meiosis I, the segregation of the long chromosomes into the two daughter cells occurs abnormally, and both of the long chromosomes go into the same daughter cell. Draw a model showing the chromosomal composition of the four daughter cells at the end of meiosis II. For each pair of homologs, draw one red and the other blue, as in the figure. You do not need to include crossovers in your model.

## Table 16.1    A Comparison of Mitosis, Meiosis I, and Meiosis II

| Event | Mitosis | Meiosis I | Meiosis II |
|---|---|---|---|
| DNA replication: | Occurs prior to mitosis | Occurs prior to meiosis I | Does not occur between meiosis I and II |
| Synapsis during prophase: | No | Yes, bivalents are formed. | No |
| Crossing over during prophase: | Rarely | Commonly | Rarely |
| Attachment to poles at prometaphase: | A pair of sister chromatids is attached to kinetochore microtubules from both poles. | A pair of sister chromatids is attached to kinetochore microtubules from just one pole. | A pair of sister chromatids is attached to kinetochore microtubules from both poles. |
| Alignment along the metaphase plate: | Sister chromatids align. | Bivalents align. | Sister chromatids align. |
| Type of separation at anaphase: | Sister chromatids separate. A single chromatid, now called a chromosome, moves to each pole. | Homologous chromosomes separate. A pair of sister chromatids moves to each pole. | Sister chromatids separate. A single chromatid, now called a chromosome, moves to each pole. |
| End result when the mother cell is diploid: | Two daughter cells that are diploid | — | Four daughter cells that are haploid |

## 16.4  Sexual Reproduction

### Learning Outcome:

1. Distinguish between the life cycles of diploid-dominant species, haploid-dominant species, and species that exhibit an alternation of generations.

**Sexual reproduction** is a process in which two haploid gametes unite in a fertilization event to form a diploid cell called a zygote. For multicellular species such as animals and plants, the zygote then grows and divides by mitotic cell divisions into a multicellular organism with many diploid cells.

For any given species, the sequence of events that produces another generation of organisms is known as a **life cycle**. For sexually reproducing organisms, this usually involves an alternation between haploid cells or organisms and diploid cells or organisms (**Figure 16.14**).

***Diploid-Dominant Species***    Most species of animals are diploid, and their haploid gametes are considered to be a specialized type of cell. For this reason, animals are viewed as **diploid-dominant species** (Figure 16.14a). Certain diploid cells in the testes or ovaries undergo meiosis to produce haploid sperm or eggs, respectively. During fertilization, sperm and egg unite to form a diploid zygote, which then undergoes repeated mitotic cell divisions to produce a diploid multicellular organism.

***Haploid-Dominant Species***    By comparison, most fungi and some protists are just the opposite; they are **haploid-dominant species** (Figure 16.14b). In fungi, the multicellular organism is haploid (1n); only the zygote is diploid. Haploid fungal cells are most commonly produced by mitosis. During sexual reproduction, haploid cells unite to form a diploid zygote, which then immediately proceeds through meiosis to produce four haploid cells called spores. Each spore goes through mitotic cellular divisions to produce a haploid multicellular organism.

***Alternation of Generations***    Plants and some algae have life cycles that are intermediate between diploid and haploid dominance. Such species exhibit an **alternation of generations** (Figure 16.14c). The species alternate between diploid multicellular organisms called **sporophytes**, and haploid multicellular organisms called **gametophytes**. Meiosis in certain cells within the sporophyte produces haploid spores, which divide by mitosis to produce the gametophyte. Particular cells within the gametophyte differentiate into haploid gametes. Fertilization occurs between two gametes, producing a diploid zygote that then undergoes repeated mitotic cell divisions to produce a sporophyte.

Among different plant species, the relative sizes of the haploid and diploid organisms vary greatly. In mosses, the haploid gametophyte is a visible multicellular organism, whereas the diploid sporophyte is smaller and remains attached to the haploid organism. In other plants, such as ferns (Figure 16.14c), both the diploid sporophyte and haploid gametophyte grow independently. The sporophyte is considerably larger and is the organism we commonly think of as a fern. In seed-bearing plants, such as roses and oak trees, the diploid sporophyte is the large multicellular plant, whereas the gametophyte is composed of only a few cells and is formed within the sporophyte.

When comparing animals, plants, and fungi, it's interesting to consider how gametes are made. Animals produce gametes by meiosis. In contrast, plants and fungi produce reproductive cells by mitosis. The gametophyte of plants is a haploid multicellular organism that is created by mitotic cellular divisions of a haploid spore. Within the multicellular gametophyte, certain cells become specialized as gametes.

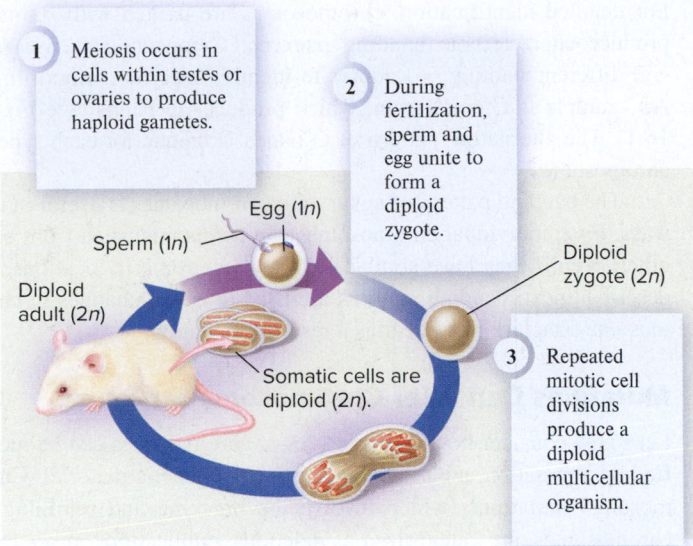

**(a) Animal life cycle—diploid dominant**

1. Meiosis occurs in cells within testes or ovaries to produce haploid gametes.
2. During fertilization, sperm and egg unite to form a diploid zygote.
3. Repeated mitotic cell divisions produce a diploid multicellular organism.

Sperm (1n)
Egg (1n)
Diploid adult (2n)
Somatic cells are diploid (2n).
Diploid zygote (2n)

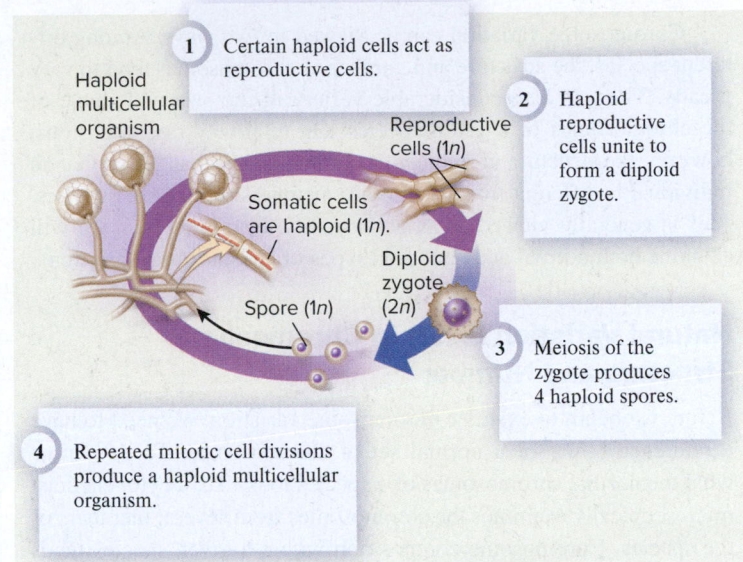

**(b) Fungal life cycle—haploid dominant**

1. Certain haploid cells act as reproductive cells.
2. Haploid reproductive cells unite to form a diploid zygote.
3. Meiosis of the zygote produces 4 haploid spores.
4. Repeated mitotic cell divisions produce a haploid multicellular organism.

Haploid multicellular organism
Reproductive cells (1n)
Somatic cells are haploid (1n).
Diploid zygote (2n)
Spore (1n)

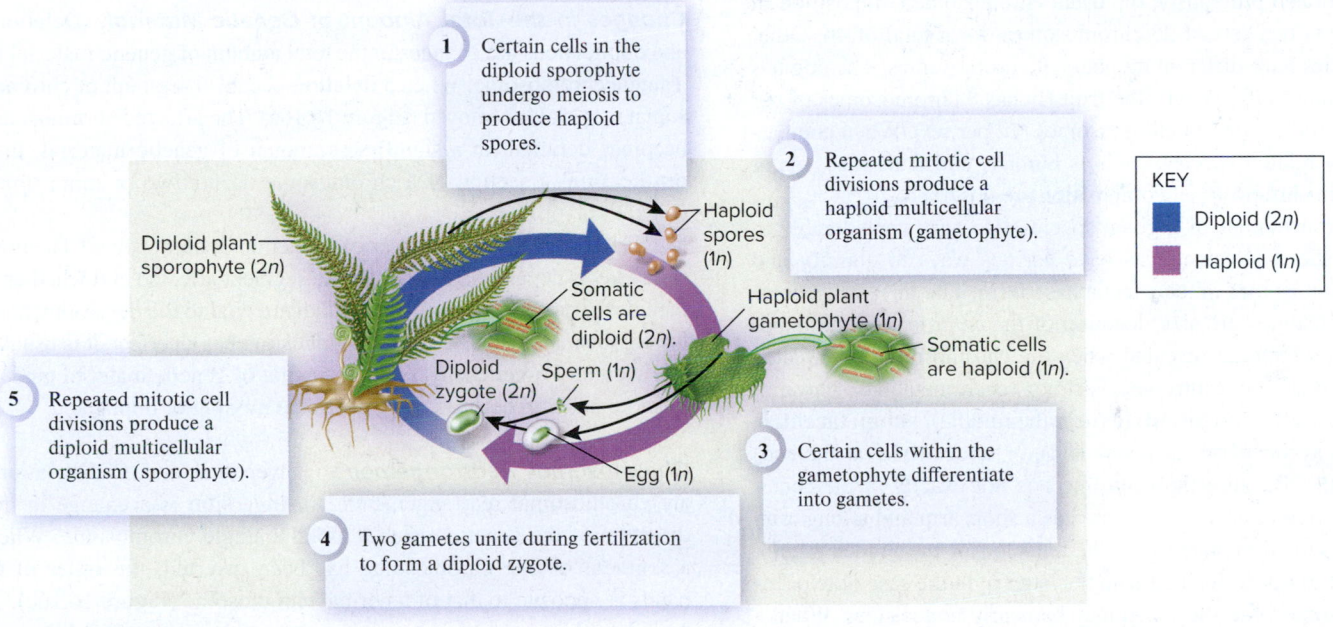

**(c) Plant life cycle—alternation of generations**

1. Certain cells in the diploid sporophyte undergo meiosis to produce haploid spores.
2. Repeated mitotic cell divisions produce a haploid multicellular organism (gametophyte).
3. Certain cells within the gametophyte differentiate into gametes.
4. Two gametes unite during fertilization to form a diploid zygote.
5. Repeated mitotic cell divisions produce a diploid multicellular organism (sporophyte).

Diploid plant sporophyte (2n)
Haploid spores (1n)
Somatic cells are diploid (2n).
Diploid zygote (2n)
Sperm (1n)
Haploid plant gametophyte (1n)
Somatic cells are haploid (1n).
Egg (1n)

**KEY**
■ Diploid (2n)
■ Haploid (1n)

**Figure 16.14** A comparison of three types of life cycles for sexually reproducing organisms.

*Concept Check:* *What is the main purpose of meiosis in animals? What is the main purpose of mitosis in animals?*

## 16.5 Variation in Chromosome Structure and Number

**Learning Outcomes:**

1. Describe how chromosomes can vary in size, centromere location, and number.
2. Identify the four ways that the structure of a chromosome can be changed via mutation.
3. Compare and contrast changes in the number of sets of chromosomes and changes in the number of individual chromosomes.
4. Give examples of how changes in chromosome number affect the characteristics of animals and plants.

In the previous sections of this chapter, we examined two important features of chromosomes: They occur in sets, and two sorting processes determine the number of sets of chromosomes following cell division. In this section, we will examine how the structures and numbers of chromosomes may vary between different species and within the same species.

Why is the study of chromosomal variation important? First, geneticists have discovered that variations in chromosome structure and number can have major effects on the characteristics of an organism. We now know that several human genetic diseases are caused by such changes. In addition, changes in chromosome structure and number have been an important factor in the evolution of new species, which is a topic we will consider in Chapter 24.

Chromosome variation can be viewed in two ways. Among different species, the structure and number of chromosomes tend to vary greatly. There is also considerable variety in the size and shape of the chromosomes of a given species. On relatively rare occasions, however, the structure or number of chromosomes changes so that an individual is different from most other members of the same species. This is generally viewed as an abnormality. In this section, we will examine both normal and abnormal types of chromosome variation.

## Natural Variation Exists in Chromosome Structure and Number

Before we begin to examine chromosome variation, we need to have a reference point for a normal set of chromosomes. To determine what the normal chromosomes of a species look like, a cytogeneticist microscopically examines the chromosomes from several members of the species. Chromosome composition within a given species tends to remain relatively constant. In most cases, individuals of the same species have the same number and types of chromosomes. For example, as mentioned previously, the usual chromosome composition of human cells is two sets of 23 chromosomes, for a total of 46. Other diploid species have different numbers of chromosomes. The dog has 78 chromosomes (39 per set), the fruit fly has 8 chromosomes (4 per set), and the tomato has 24 chromosomes (12 per set). When comparing distantly related species, such as humans and fruit flies, major differences in chromosomal composition are observed.

The chromosomes of a given species also vary considerably in size and shape. Cytogeneticists have various ways to classify and identify chromosomes in their metaphase form. The three most commonly used features are size, location of the centromere, and banding patterns, which are revealed when the chromosomes are treated with stains. Based on centromere location, each metaphase chromosome is classified as **metacentric** (near the middle), **submetacentric** (off center), **acrocentric** (near one end), or **telocentric** (at the end) (**Figure 16.15**). Because the centromere is not exactly in the center of a chromosome, each chromosome has a short arm and a long arm. The short arm is designated with the letter p (for the French *petite*), and the long arm is designated with the letter q. In the case of telocentric chromosomes, the short arm may be nearly nonexistent. When a karyotype is prepared (see Figure 16.1), the chromosomes are aligned with the short arms on top and the long arms on the bottom.

Because different chromosomes often have similar sizes and centromeric locations, cytogeneticists must use additional methods to accurately identify each type of chromosome within a karyotype.

For detailed identification, chromosomes are treated with stains to produce characteristic banding patterns. Cytogeneticists use several different staining procedures to identify specific chromosomes. An example is Giemsa stain, which produces G bands (see Figure 16.1). The alternating pattern of G bands is unique for each type of chromosome.

The banding pattern of eukaryotic chromosomes is useful in two ways. First, individual chromosomes can be distinguished from each other, even if they have similar sizes and centromeric locations. As described next, banding patterns are used to detect changes in chromosome structure that occur as a result of mutation.

## Mutations Can Alter Chromosome Structure

Let's now consider how the structures of chromosomes can be modified by a mutation, a heritable change in the genetic material. Chromosomal mutations, which involve the breaking and rejoining of chromosomes, are categorized as deletions, duplications, inversions, and translocations (**Figure 16.16**).

***Changes in the Total Amount of Genetic Material*** Deletions and duplications are changes in the total amount of genetic material in a single chromosome. When a **deletion** occurs, a segment of chromosomal material is removed (Figure 16.16a). The affected chromosome becomes deficient in a significant amount of genetic material. In a **duplication**, a section of a chromosome occurs two or more times (Figure 16.16b).

What are the consequences of a deletion or duplication? The possible effects depend on the size of the segment affected and whether it includes genes or portions of genes that are vital to the development of the organism. When a deletion or duplication has an effect, it is usually detrimental. Larger changes in the amount of genetic material tend to be more harmful because more genes are missing or duplicated.

***Chromosomal Rearrangements*** Inversions and translocations are chromosomal rearrangements. An **inversion** is a change in the direction of the genetic material along a single chromosome. When a segment of one chromosome has been inverted, the order of G bands is opposite to that of a normal chromosome (Figure 16.16c). A **translocation** occurs when one segment of a chromosome becomes attached to a different chromosome. In a **simple translocation**, a single piece of chromosome is attached to another chromosome (Figure 16.16d). In a **reciprocal translocation**, two different types of chromosomes exchange pieces, thereby producing two abnormal chromosomes carrying translocations (Figure 16.16e).

## Variation Occurs in the Number of Chromosome Sets and the Number of Individual Chromosomes

Variations in chromosome number can be categorized in two ways: variation in the number of sets of chromosomes and variation in the number of particular chromosomes within a set. The suffix -ploid or -ploidy refers to a complete set of chromosomes. Organisms that are **euploid** (the prefix eu- means true) have chromosomes that occur in one or more complete sets. For example, in a species that is diploid, a euploid organism would have two sets of chromosomes in its somatic cells. In *Drosophila melanogaster*, for example, a normal individual

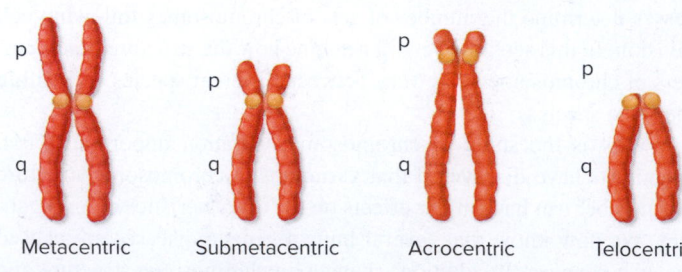

p — q (Metacentric)   p — q (Submetacentric)   p — q (Acrocentric)   p — q (Telocentric)

**Figure 16.15**  **A comparison of centromeric locations among metaphase chromosomes.**

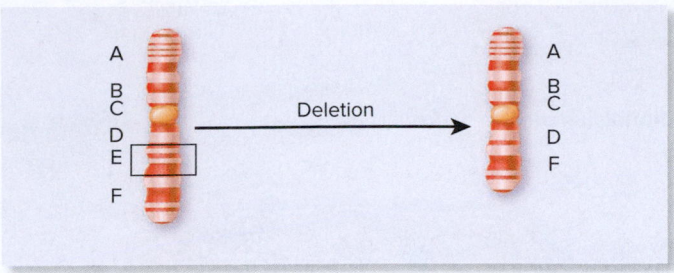

(a) Deletion: Removes a segment of chromosome.

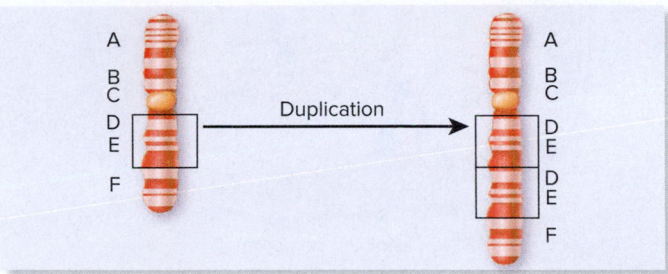

(b) Duplication: Doubles a particular region.

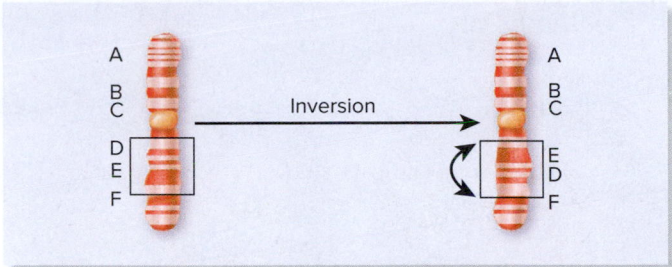

(c) Inversion: Flips a region to the opposite orientation.

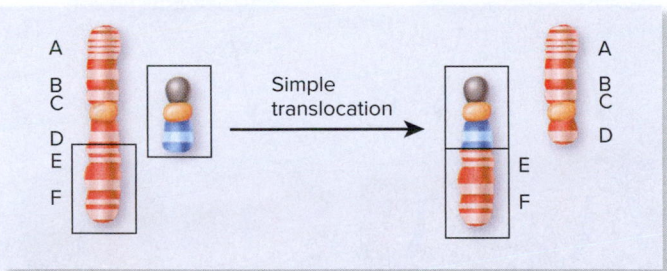

(d) Simple translocation: Moves a segment of 1 chromosome to another chromosome.

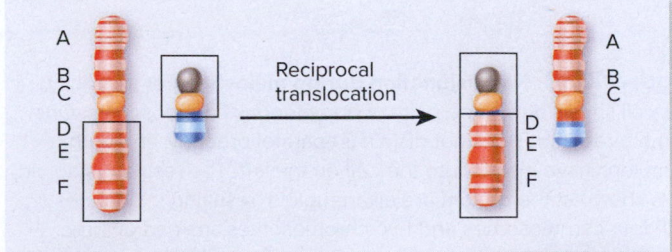

(e) Reciprocal translocation: Exchanges pieces between 2 different chromosomes.

**Figure 16.16** **Types of changes in chromosome structure.** The letters alongside the chromosomes are placed there as frames of reference.

*Concept Check:* *Which types of changes in chromosome structure do not affect the total amount of genetic material?*

has eight chromosomes. The species is diploid, having two sets of four chromosomes each (**Figure 16.17a**).

***Variation in the Number of Sets of Chromosomes*** Organisms can vary in the number of sets of chromosomes they have. For example, on rare occasions, an abnormal fruit fly can be produced with 12 chromosomes, that is, having three sets of 4 chromosomes each (**Figure 16.17b**). Organisms with three or more sets of chromosomes are called **polyploid**. A diploid organism is referred to as $2n$, a **triploid** organism as $3n$, a **tetraploid** organism as $4n$, and so forth. All such organisms are euploid because they have complete sets of chromosomes.

***Aneuploidy*** A second way that chromosome number can vary is called **aneuploidy**. This refers to an alteration in the number of a particular chromosome, so the total number of chromosomes is not an exact multiple of a set. For example, an abnormal fruit fly could have nine chromosomes instead of eight because it had three copies

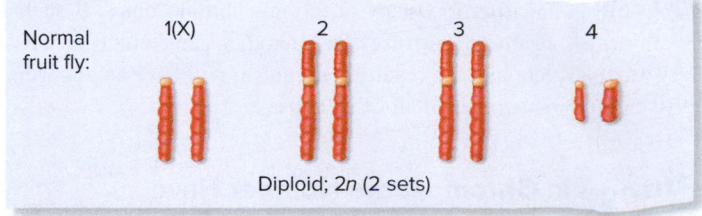

(a) Normal fruit fly chromosome composition

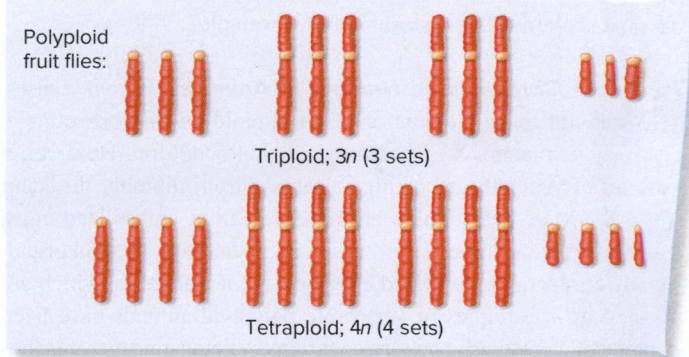

(b) Polyploidy

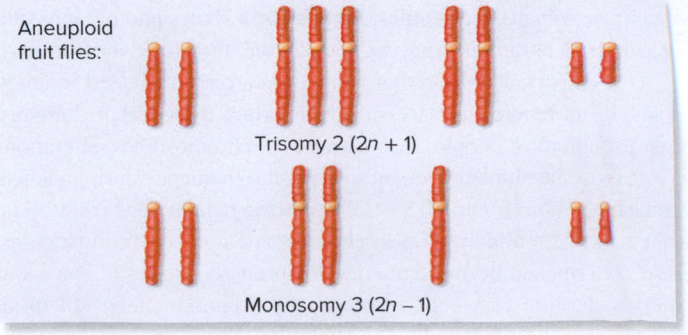

(c) Aneuploidy

**Figure 16.17** **Types of variation in chromosome number. (a)** The normal diploid number of chromosomes in a female *Drosophila*. The X chromosome is also called chromosome 1. Examples of chromosomes of **(b)** polyploid flies and **(c)** aneuploid flies.

of chromosome 2 instead of the normal two copies (**Figure 16.17c**). Instead of being perfectly diploid, a trisomic animal is $2n + 1$. Such an animal is said to have trisomy 2. By comparison, a fruit fly could be lacking a single chromosome, such as chromosome 3, and have a total of seven chromosomes ($2n - 1$). This animal is said to be monosomic and is described as having monosomy 3.

Variations in chromosome number are fairly widespread and have a significant effect on the characteristics of plants and animals. For these reasons, researchers want to understand the mechanisms that cause these variations. In some cases, a change in chromosome number is the result of the abnormal sorting of chromosomes during cell division. The term **nondisjunction** refers to an event in which the chromosomes do not separate properly during cell division. Nondisjunction can occur during meiosis I or meiosis II and produces haploid cells that have too many or too few chromosomes. **Figure 16.18** illustrates the consequences of nondisjunction during meiosis I. In this case, one pair of homologs moved into the cell on the left instead of separating from each other. This results in the production of aneuploid cells, with either too many or too few chromosomes. If such a cell becomes a gamete that fuses with another gamete during fertilization, the zygote and the resulting organism will have an abnormal number of chromosomes in all of its cells.

## Changes in Chromosome Number Have Important Consequences

How do changes in chromosome number affect the characteristics of animals and plants? Let's consider a few examples.

*Changes in Chromosome Number in Animals* In many cases, animals do not tolerate deviations from diploidy well. For example, polyploidy in mammals is generally a lethal condition. However, a few cases of naturally occurring variations from diploidy do occur in animals. Male bees, which are produced from unfertilized eggs, contain a single set of chromosomes and are therefore haploid organisms. By comparison, fertilized eggs become female bees, which are diploid. A few examples of vertebrate polyploid animals have been discovered. Interestingly, on rare occasions, animals that are morphologically very similar to each other can be found as a diploid species as well as a separate polyploid species. This situation occurs among certain amphibians and reptiles. **Figure 16.19** shows photographs of a diploid and a tetraploid frog. As you can see, they look very similar.

One important reason that geneticists are so interested in aneuploidy is its relationship to certain inherited disorders in humans. Even though most people are born with 46 chromosomes, alterations in chromosome number occur at a surprising frequency during gamete formation. About 5% to 10% of all fertilized human eggs result in an embryo with an abnormality in chromosome number. In most cases, these abnormal embryos do not develop properly and result in a spontaneous abortion very early in pregnancy. Approximately 50% of all spontaneous abortions are due to alterations in chromosome number.

In some cases, an abnormality in chromosome number produces an offspring that can survive. Several human disorders are the result of abnormalities in chromosome number. The most common are trisomies of chromosomes 21, 18, or 13 and abnormalities in the number of the sex chromosomes (**Table 16.2**). These syndromes are most likely

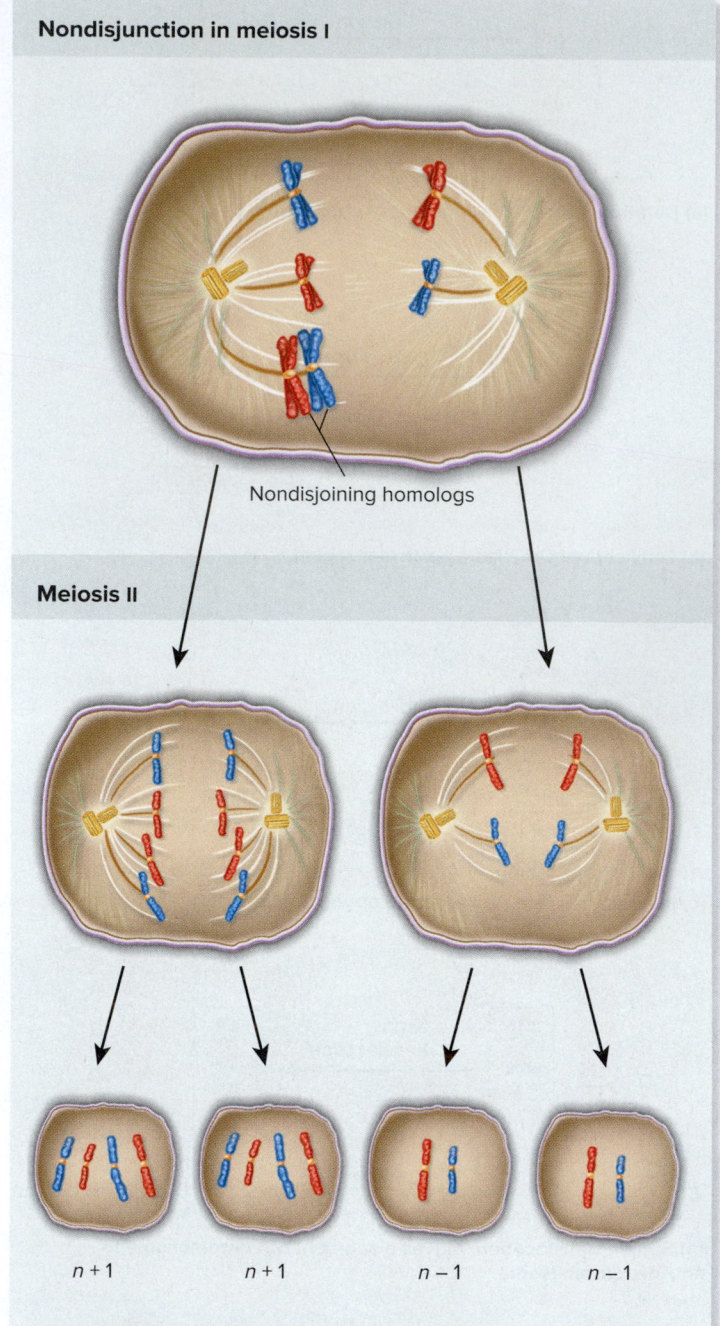

**Figure 16.18 Nondisjunction during meiosis I.** For simplicity, this cell shows only three pairs of homologous chromosomes. One of the three pairs does not disjoin (separate) properly, and both homologs have moved into the cell on the left. The resulting haploid cells shown at the bottom are all aneuploid, resulting in gametes with four chromosomes and two chromosomes, instead of three.

due to nondisjunction. For example, Turner syndrome (XO) may occur when a gamete that is lacking a sex chromosome due to nondisjunction has fused with a gamete carrying an X chromosome. By comparison, triple X syndrome (XXX) occurs when a gamete carrying two X chromosomes fuses with a gamete carrying a single X chromosome.

Most of the known trisomies involve chromosomes that are relatively small, so they carry fewer genes. Trisomies of the other human

**(a)** *Hyla chrysoscelis* (diploid)

**(b)** *Hyla versicolor* (tetraploid)

**Figure 16.19** **Differences in chromosome number in two closely related frog species.** The frog in **(a)** is diploid, whereas the frog in **(b)** is tetraploid. These frogs are in the act of performing their mating calls, which is why the skin under their mouths is distended, forming a large bubble. a–b: ©A. B. Sheldon

| Table 16.2 | Aneuploid Conditions in Humans | | |
|---|---|---|---|
| Condition | Frequency (per number of live births) | Syndrome | Characteristics |
| *Autosomal* | | | |
| Trisomy 21 | 1/800 | Down | Mental impairment, abnormal pattern of palm creases, slanted eyes, flattened face, short stature |
| Trisomy 18 | 1/6,000 | Edward | Mental and physical impairment, facial abnormalities, extreme muscle tone, early death |
| Trisomy 13 | 1/15,000 | Patau | Mental and physical impairment, wide variety of defects in organs, large triangular nose, early death |
| *Sex chromosomal* | | | |
| XXY | 1/1,000 (males) | Klinefelter | Sexual immaturity (no sperm), breast swelling (males) |
| XYY | 1/1,000 (males) | Jacobs | Tall |
| XXX | 1/1,500 (females) | Triple X | Tall and thin, menstrual irregularity |
| XO | 1/5,000 (females) | Turner | Short stature, webbed neck, sexually undeveloped |

chromosomes and most monosomies are presumed to be lethal and have been found in spontaneously aborted embryos and fetuses.

Human abnormalities in chromosome number are influenced by the age of the parents. Older parents are more likely to produce children with abnormalities in chromosome number, possibly because meiotic nondisjunction is more likely to occur in older cells. **Down syndrome**, which was first described by the English physician John Langdon Down in 1866, provides an example. This disorder is caused by the inheritance of three copies of chromosome 21 (see Table 16.2). The incidence of Down syndrome rises with the age of either parent. In males, however, the rise occurs relatively late in life, usually past the age when most men have children. By comparison, the likelihood of having a child with Down syndrome rises dramatically during the later reproductive ages of women.

***Changes in Chromosome Number in Plants*** In contrast to animals, plants commonly exhibit polyploidy, which is important in agriculture. In many instances, polyploid strains of plants display characteristics that are helpful to humans. They are often larger in size and more robust. These traits are clearly advantageous in the production of food. For example, the species of wheat that we use to make bread, *Triticum aestivum*, is a hexaploid (with six sets of chromosomes) that arose from the union of diploid genomes from three closely related species (**Figure 16.20**). During the course of its domestication, a

**Figure 16.20** **Example of a polyploid plant** Cultivated wheat, *Triticum aestivum*, is a hexaploid. It was derived from three different diploid species of grasses that originally were found in the Middle East and were cultivated by ancient farmers in that region. Modern varieties of wheat have been produced from this hexaploid species. ©irin-k/age fotostock

wild species became a tetraploid, and then a second diploid species interbred with the tetraploid to produce a hexaploid. Plant polyploids tend to exhibit a greater adaptability, which allows them to withstand harsher environmental conditions. Polyploid ornamental plants commonly produce larger flowers than their diploid counterparts.

Although polyploidy is often beneficial in plants, aneuploidy usually has detrimental consequences for the characteristics of an organism of any eukaryotic species. Why is aneuploidy usually detrimental? To answer this question, we need to consider the relationship between gene expression and chromosome number. For many, but not all genes, the level of gene expression is correlated with the number of genes per cell. For example, if a gene is carried on a chromosome that is present in a cell as three copies instead of two, approximately 150% of the normal amount of gene product is usually made. Alternatively, if only one copy of that gene is present due to a missing chromosome, only 50% of the gene product is typically made. For some genes, producing too much or too little of the gene product may not have adverse effects. However, for other genes, over- or underexpression may interfere with the proper functioning of cells.

## Summary of Key Concepts

### 16.1 The Eukaryotic Cell Cycle

- Cytogeneticists examine cells microscopically to determine their chromosome composition. A micrograph that shows the alignment of chromosomes from a given cell is called a karyotype. Eukaryotic chromosomes are inherited in sets. A diploid cell has two sets of chromosomes. The members of each pair are called homologs (Figure 16.1).

- The eukaryotic cell cycle consists of four phases called $G_1$ (first gap), S (synthesis of DNA), $G_2$ (second gap), and M (mitosis and cytokinesis). The $G_1$, S, and $G_2$ phases are collectively known as interphase (Figure 16.2).

- An interaction between cyclin and cyclin-dependent kinase (cdk) is necessary for cells to advance through the cell cycle. Checkpoint proteins sense the environmental conditions and the integrity of the genome and control whether or not the cell advances through the cell cycle (Figure 16.3).

- Masui and Markert studied the maturation of frog oocytes to identify a substance necessary for the oocytes to progress through the cell cycle. This substance was initially called maturation-promoting factor (MPF) and was later identified as a complex of mitotic cyclin and cyclin-dependent kinase (Figures 16.4, 16.5).

### 16.2 Mitotic Cell Division

- In the process of mitotic cell division, a cell divides to produce two new cells (the daughter cells) that are genetically identical to the original cell.

- During S phase, eukaryotic chromosomes are replicated to produce a pair of identical sister chromatids that remain attached to each other (Figure 16.6).

- The mitotic spindle is a network of microtubules that plays a central role in chromosome sorting during cell division (Figure 16.7).

- Mitosis occurs in five phases called prophase, prometaphase, metaphase, anaphase, and telophase. During prophase, the chromosomes condense, and the nuclear envelope begins to dissociate. The spindle apparatus is completely formed by the end of prometaphase. At metaphase, the chromosomes are aligned in a single row along the metaphase plate of the spindle. During anaphase, the sister chromatids separate from each other and move to opposite poles; the poles themselves also move farther apart. During telophase, the chromosomes decondense, and the nuclear envelope re-forms (Figure 16.8).

- Cytokinesis, which occurs after mitosis, is the division of the cytoplasm to produce two distinct daughter cells. In animal cells, cytokinesis involves the formation of a cleavage furrow. In plant cells, two separate cells are produced by the formation of a cell plate (Figure 16.9).

- The analysis of cell division in prokaryotic cells and in simple and complex eukaryotes has revealed an evolutionary progression in which the protein FtsZ plays a role in bacterial cell division and microtubules gain new functions in eukaryotes, such as sorting chromosomes and promoting nuclear division (Figure 16.10).

### 16.3 Meiosis

- The process of meiosis begins with a diploid cell and produces four haploid cells with one set of chromosomes each (Figure 16.11).

- During prophase of meiosis, homologous pairs of sister chromosomes synapse, and crossing over occurs. After crossing over, chiasmata—the sites where crossing over occurs—become visible (Figure 16.12).

- Meiosis consists of two divisions—meiosis I and II—each composed of prophase, prometaphase, metaphase, anaphase, and telophase. During meiosis I, the homologs are separated into two different cells, and during meiosis II, the sister chromatids are separated into four different cells (Figure 16.13, Table 16.1).

### 16.4 Sexual Reproduction

- Animals are diploid-dominant species, whereas most fungi and some protists are haploid-dominant species. Plants alternate between diploid and haploid forms (Figure 16.14).

### 16.5 Variation in Chromosome Structure and Number

- Chromosomes are classified as metacentric, submetacentric, acrocentric, and telocentric, based on their centromere location. Each type of chromosome can be uniquely identified by its banding pattern after staining (Figure 16.15).

- Deletions, duplications, inversions, and translocations are different ways in which mutations alter chromosome structure (Figure 16.16).

- A euploid organism has chromosomes that occur in complete sets. A polyploid organism has three or more sets of chromosomes. An organism that has one too many (trisomy) or one too few (monosomy) chromosomes is termed aneuploid. Aneuploidy can be caused by nondisjunction, an event in which the chromosomes do not separate properly during cell division (Figures 16.17, 16.18).

- Aneuploidy in humans is responsible for several types of inherited disorders, including Down syndrome (Table 16.2).

- Polyploid animals are relatively rare, but polyploid plants are common and tend to be larger and more robust than their diploid counterparts (Figures 16.19, 16.20).

## Assess & Discuss

### Test Yourself

1. In which phase of the cell cycle are chromosomes replicated?
   a. $G_1$ phase
   b. S phase
   c. M phase
   d. $G_2$ phase
   e. none of the above

2. If two chromosomes are homologous, they
   a. look similar under the microscope.
   b. have very similar DNA sequences.
   c. carry the same types of genes.
   d. may carry different versions of the same gene.
   e. are all of the above.

3. Checkpoints during the cell cycle are important because they
   a. allow the organellar activity to catch up to cellular demands.
   b. ensure the integrity of the cell's DNA.
   c. allow the cell to generate sufficient ATP for cellular division.
   d. are the only time DNA replication can occur.
   e. do all of the above.

4. Which of the following is a reason for mitotic cell division?
   a. asexual reproduction
   b. gamete formation in animals
   c. multicellularity
   d. all of the above
   e. both a and c

5. A replicated chromosome is composed of
   a. two homologous chromosomes held together at the centromere.
   b. four sister chromatids held together at the centromere.
   c. two sister chromatids held together at the centromere.
   d. four homologous chromosomes held together at the centromere.
   e. one chromosome with a centromere.

6. Which of the following is *not* an event of anaphase of mitosis?
   a. The nuclear envelope breaks apart.
   b. Sister chromatids separate.
   c. Kinetochore microtubules shorten, pulling the chromosomes to the poles.
   d. Polar microtubules push against each other, moving the poles farther apart.
   e. All of the above occur during anaphase.

7. A student is looking at cells under the microscope. The cells are from an organism that has a diploid chromosome number of 14. In one particular case, the cell has seven replicated chromosomes (sister chromatids) aligned at the metaphase plate of the cell. Which of the following statements accurately describes this particular cell?
   a. The cell is in metaphase of mitosis.
   b. The cell is in metaphase of meiosis I.
   c. The cell is in metaphase of meiosis II.
   d. All of the above are correct.
   e. Both b and c are correct.

8. Which of the following statements accurately describes a difference between mitosis and meiosis?
   a. Mitosis may produce diploid cells, whereas meiosis produces haploid cells.
   b. Homologous chromosomes synapse during meiosis but do not synapse during mitosis.
   c. Crossing over commonly occurs during meiosis, but it does not commonly occur during mitosis.
   d. All of the above are correct.
   e. Both a and c are correct.

9. During crossing over in meiosis I,
   a. homologous chromosomes are not altered.
   b. homologous chromosomes exchange genetic material.
   c. chromosomal damage occurs.
   d. genetic information is lost.
   e. cytokinesis occurs.

10. Aneuploidy may be the result of
   a. duplication of a region of a chromosome.
   b. inversion of a region of a chromosome.
   c. nondisjunction during meiosis.
   d. interspecies breeding.
   e. all of the above.

### Conceptual Questions

1. Distinguish between homologous chromosomes and sister chromatids.

2. The *OCA2* gene, which influences eye color in humans, is found on chromosome 15. How many copies of this gene are found in the karyotype in the inset in Figure 16.1? Is it one, two, or four?

3. **Core Concept: Information** Explain why mitosis is a key process for passing genetic information to new cells.

### Collaborative Questions

1. Why is it necessary for chromosomes to condense during mitosis and meiosis? What do you think might happen if chromosomes did not condense?

2. A diploid eukaryotic cell has 10 chromosomes (5 per set). As a group, take turns having one student draw the cell as it would look during a phase of mitosis, meiosis I, or meiosis II; then have the other students guess which phase it is.

# Mendelian Patterns of Inheritance

# 17

**An African girl with albinism.** This condition results in very light skin and hair color. ©Radu Sigheti/Reuters

**N**tombi knew she looked different as long as she can remember. Born in Nigeria in 1997, she has accepted her appearance, though she still finds the occasional stare from strangers to be disturbing. Ntombi has albinism, a condition characterized by a total or partial lack of pigmentation of the skin, hair, and eyes. As a result, she has very fair skin, blond hair, and blue eyes.* In contrast, her parents and three brothers have dark skin, black hair, and brown eyes, as do most of her relatives and most of the people in the city where she lives. Ntombi is very close to her aunt, who also has albinism.

Cases like Ntombi's have intrigued people for many centuries. How do we explain the traits that are found in people, plants, and other organisms? Can we predict what types of offspring two parents will produce? To answer such questions, researchers have studied the characteristics among related individuals and tried to make some sense of the data. Their goal is to understand

inheritance—the acquisition of traits by their transmission from parent to offspring.

The first systematic attempt to understand inheritance was carried out by German plant breeder Joseph Kolreuter between 1761 and 1766. In crosses between two strains of tobacco plants, Kolreuter found that the offspring were usually intermediate in appearance between the two parents. He concluded that parents make equal genetic contributions to their offspring and that their genetic material blends together as it is passed to the next generation. This interpretation was consistent with the concept known as blending inheritance, which was incorrect but widely accepted at that time. In the late 1700s, Jean-Baptiste Lamarck, a French naturalist, hypothesized that physiological events (such as use or disuse) could modify traits and such modified traits would be inherited by offspring. For example, an individual who became adept at archery would pass that skill to his or her offspring. Overall, the prevailing view prior to the 1800s was that hereditary traits were rather malleable and could change and blend over the course of one or two generations.

In contrast, microscopic observations of chromosome transmission during mitosis and meiosis in the second half of the 19th century provided compelling evidence for **particulate inheritance**— the idea that the determinants of hereditary traits are transmitted in discrete units, or particles, from one generation to the next. Remarkably, this idea was first put forward in the 1860s by a researcher who knew nothing about chromosomes. Gregor Mendel used statistical analysis of carefully designed plant breeding experiments to arrive at the concept of a gene, which is broadly defined as a unit of heredity. Forty years later, through the convergence of Mendel's work and that of cell biologists, this concept became the foundation of the modern science of genetics.

In this chapter, we will consider inheritance patterns known as Mendelian inheritance. Although these patterns can vary, they all obey Mendel's laws of inheritance, which we will examine in Section 17.1. As summarized in **Table 17.1**, our emphasis will be on two aspects of inheritance. First, we will explore how the various types of inheritance patterns produce different outcomes in a genetic cross. Second, we will examine the underlying molecular mechanisms that explain these different outcomes. In Chapter 18, we will explore some inheritance patterns that violate Mendel's laws.

---

*In contrast to popular belief, most people with albinism have blue eyes, not pink eyes. This is particularly the case among Africans with albinism.

| Table 17.1 | Different Types of Mendelian Inheritance Patterns and Their Molecular Basis |
|---|---|
| **Type** | **Description** |
| Simple Mendelian inheritance | **Inheritance pattern:** Pattern of traits is determined by a pair of alleles that display a dominant/recessive relationship and are located on an autosome. The presence of the dominant allele masks the presence of the recessive allele. |
| | **Molecular basis:** In many cases, the recessive allele is nonfunctional. Though a heterozygote may produce 50% of the functional protein compared with a dominant homozygote, this is sufficient to produce the dominant trait. |
| X-linked inheritance | **Inheritance pattern:** Pattern of traits is determined by genes that display a dominant/recessive relationship and are located on the X chromosome. In mammals and fruit flies, males are hemizygous for X-linked genes. In these species, X-linked recessive traits occur more frequently in males than in females. |
| | **Molecular basis:** In a female with one recessive X-linked allele (a heterozygote), the protein encoded by the dominant allele is sufficient to produce the dominant trait. A male with a recessive X-linked allele does not have a dominant allele and does not make any of the functional protein. |
| Incomplete dominance | **Inheritance pattern:** Pattern that occurs when the heterozygote has a phenotype intermediate to the phenotypes of the homozygotes, as when a cross between red-flowered and white-flowered plants produces pink-flowered offspring. |
| | **Molecular basis:** Fifty percent of the protein encoded by the functional (wild-type) allele results in an intermediate phenotype. |
| Codominance | **Inheritance pattern:** Pattern that occurs when the heterozygote expresses both alleles simultaneously. For example, a human carrying the A and B alleles for the ABO antigens of red blood cells produces both the A and the B antigens (has an AB blood type). |
| | **Molecular basis:** The codominant alleles encode proteins that function somewhat differently from each other. In a heterozygote, the function of each protein affects the phenotype uniquely. |
| Epistasis | **Inheritance pattern:** A type of gene interaction in which the alleles of one gene mask the effects of an allele of another gene. |
| | **Molecular basis:** Two different genes are needed to produce a given phenotype. Loss of function of one of the genes alters the phenotype. |
| Continuous variation | **Inheritance pattern:** A pattern in which the offspring display a continuous range of phenotypes. |
| | **Molecular basis:** This pattern is produced by the additive interactions of several genes, along with environmental influences. |

## 17.1 Mendel's Laws of Inheritance

**Learning Outcomes:**

1. List the advantages of using the garden pea to study inheritance.
2. Describe the difference between dominant and recessive traits.
3. Distinguish between genotype and phenotype.
4. **CoreSKILL »** Predict the outcome of genetic crosses using a Punnett square.
5. State Mendel's law of segregation and law of independent assortment.

Gregor Johann Mendel (**Figure 17.1**) grew up on a small farm in northern Moravia, then a part of the Austrian Empire and now in the Czech Republic. At the age of 21, he entered the Augustinian monastery of St. Thomas in Brno and was ordained a priest in 1847. Mendel then worked for a short time as a substitute teacher, but to continue teaching he needed a license. Surprisingly, he failed the licensing exam due to poor answers in physics and natural history, so he enrolled at the University of Vienna to expand his knowledge in these two areas. Mendel's training in physics and mathematics taught him to perceive the world as an orderly place, governed by natural laws that could be stated as simple mathematical relationships.

In 1856, Mendel began his historic studies on pea plants. For 8 years, he analyzed thousands of pea plants that he grew on a small plot in his monastery garden. In 1866, he published the results of his work in

**Figure 17.1 Gregor Johann Mendel.** ©SPL/Science Source

a paper entitled "Experiments on Plant Hybrids." This paper was largely ignored by scientists at that time, partly because of its title. Also, Mendel was clearly ahead of his time. During this period, biology had not yet become a quantitative, experimental science. In addition, the behavior of chromosomes during mitosis and meiosis, which provides a framework for understanding inheritance patterns, had yet to be studied. Prior to his death in 1884, Mendel reflected, "My scientific work has brought me a great deal of satisfaction and I am convinced it will be appreciated before long by the whole world." Sixteen years later, in 1900, Mendel's work was independently rediscovered by three biologists with an interest in plant genetics: Hugo de Vries of Holland, Carl Correns of Germany, and Erich von Tschermak of Austria. Within a few years, the influence of Mendel's landmark studies was felt around the world.

In this section, we will examine Mendel's experiments and see how they led to the formulation of basic genetic principles known as Mendel's laws. We will discover that these principles apply not only to the pea plants Mendel studied, but also to a wide variety of sexually reproducing organisms, including humans.

## Mendel Chose the Garden Pea to Study Inheritance

When two individuals of the same species with different characteristics are bred or crossed to each other, the process is called **hybridization**, and the offspring are referred to as hybrids. For example, a hybridization experiment could involve breeding a purple-flowered plant to a white-flowered plant. Mendel was particularly intrigued by the consistency with which offspring of such crosses showed characteristics of one or the other parent in successive generations. His intellectual foundation in physics and the natural sciences led him to consider that this regularity might be rooted in natural laws that could be expressed mathematically. To uncover these laws, he carried out quantitative experiments in which he carefully analyzed the numbers of offspring carrying specific traits.

Mendel chose the garden pea, *Pisum sativum*, to investigate the natural laws that govern inheritance. Why did he choose this species? Several properties of the garden pea were particularly advantageous for studying inheritance. First, it was available in many varieties that differed in characteristics, such as the appearance of seeds, pods, flowers, and stems. Such general features of an organism are called **characters**. **Figure 17.2** illustrates the seven characters that Mendel eventually chose to follow in his breeding experiments. Each of these characters was found in two discrete variants. For example, one character he followed was height, which had the variants known as tall and dwarf. Another was seed color, which had the variants yellow and green. A **trait** is an identifiable characteristic of an organism. The term trait usually refers to a variant for a character.* For example, seed color is a character, and green and yellow seed colors are traits.

A second advantageous property of garden peas is they are normally self-fertilizing. In **self-fertilization**, a female gamete is fertilized by a male gamete from the same plant. Like many flowering plants, peas have male and female sex organs in the same flower (**Figure 17.3**). Male gametes (sperm cells) are produced within pollen grains, which are formed in structures called stamens. Female gametes (egg cells) are produced in structures called ovules, which form within an organ called an ovary. For fertilization to occur, a pollen grain must land on the receptacle called a stigma, enabling a sperm to migrate to an ovule and fuse with an egg cell. In peas, the stamens and the ovaries are enclosed by a modified petal, an arrangement that greatly favors self-fertilization. Self-fertilization makes it easy to produce plants that breed true for a given trait, meaning the trait does not vary from generation to generation. For example, if a pea plant with yellow seeds breeds true for seed color, all of the plants that grow from these seeds will also produce yellow seeds. A strain that continues

---

*Geneticists may also use the term trait to refer to a character.

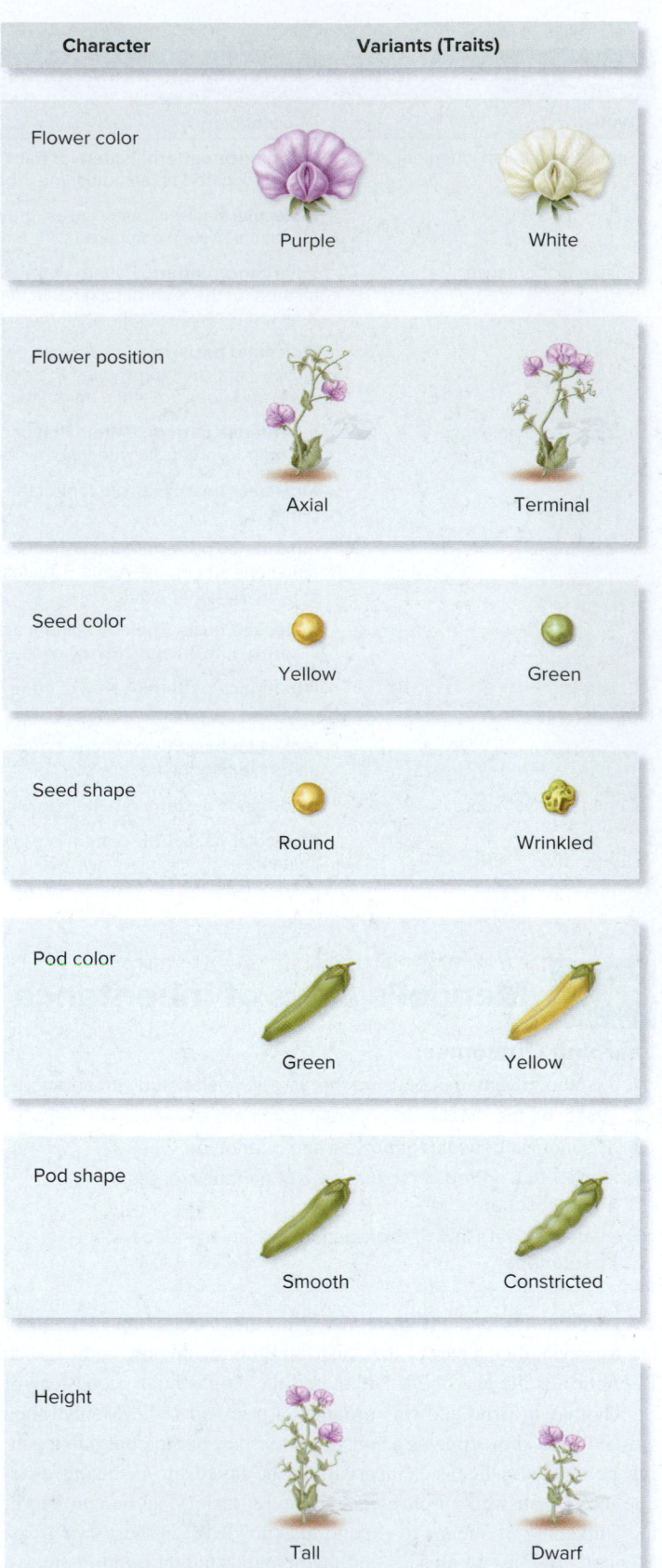

| Character | Variants (Traits) | |
|---|---|---|
| Flower color | Purple | White |
| Flower position | Axial | Terminal |
| Seed color | Yellow | Green |
| Seed shape | Round | Wrinkled |
| Pod color | Green | Yellow |
| Pod shape | Smooth | Constricted |
| Height | Tall | Dwarf |

**Figure 17.2** **The seven characters that Mendel studied.**

 **Core Concept: Information** The traits that Mendel studied in pea plants are governed by the genetic material of this species.

**Figure 17.3  Flower structure in pea plants.** The pea flower produces both male and female gametes. Sperm form in the pollen produced within the stamens; egg cells form in ovules within the ovary. A modified petal encloses the stamens and stigma, encouraging self-fertilization. ©Nigel Cattlin/Science Source

to exhibit the same trait after several generations of self-fertilization is called a **true-breeding line**. Prior to conducting the studies described in this chapter, Mendel had already established that the seven characters he chose to study were true-breeding in the strains of pea plants he had obtained.

A third reason for using garden peas in hybridization experiments is the ease of making crosses: The flowers are fairly large and easy to manipulate. In some cases, Mendel wanted his pea plants to self-fertilize, but in others, he wanted to cross plants that differed with respect to some character, a process called hybridization, or **cross-fertilization**. In garden peas, cross-fertilization requires placing pollen from one plant onto the stigma of a flower on a different plant (**Figure 17.4**). Mendel would pry open an immature flower

and remove the stamens before they produced pollen, so the flower could not self-fertilize. He then used a paintbrush to transfer pollen from another plant to the stigma of the flower that had its stamens removed. In this way, Mendel was able to cross-fertilize any two of his true-breeding pea plants and obtain any type of hybrid he wanted.

## By Following the Inheritance Pattern of Single Traits, Mendel's Work Revealed the Law of Segregation

Mendel began his investigations by studying the inheritance patterns of pea plants that differed in a single character. A cross in which an experimenter follows the variants of only one character is called a **single-factor cross**. As an example, we will consider a single-factor cross in which Mendel followed the tall and dwarf variants for height (**Figure 17.5**). The left side of Figure 17.5a shows his experimental approach. The true-breeding parents are termed the **P generation** (parental generation), and their offspring constitute the $F_1$ **generation** (first filial generation, from the Latin *filius*, meaning son). When the true-breeding parents differ in a single character, their $F_1$ offspring are called single-trait hybrids, or **monohybrids**. When Mendel crossed true-breeding tall and dwarf plants, he observed that all plants of the $F_1$ generation were tall.

Next, Mendel followed the transmission of this character for a second generation. To do so, he allowed the $F_1$ monohybrids to self-fertilize, producing a generation called the $F_2$ **generation** (second filial generation). The dwarf trait reappeared in the $F_2$ offspring: Three-fourths of the plants were tall and one-fourth were dwarf. Mendel obtained similar results for each of the seven characters he studied, as shown in the data of Figure 17.5b. A quantitative analysis of his data allowed Mendel to postulate three important ideas about the properties of traits and their transmission from parents to offspring:

1. Traits may exist in two forms, dominant and recessive.
2. An individual carries two genes for a given character, and genes have variant forms (now called **alleles**).
3. The two alleles of a gene separate during the process that gives rise to haploid cells and gametes, so each sperm and egg receives only one allele.

***Dominant and Recessive Traits***    Perhaps the most surprising outcome of Mendel's work was that the data argued strongly against the prevailing notion of blending inheritance. In each of the seven cases, the $F_1$ generation displayed a trait distinctly like one of the two parents rather than an intermediate trait. Using genetic terms that Mendel originated, we describe the alternative traits as dominant and recessive. The term **dominant** describes the displayed trait, whereas the term **recessive** describes a trait that is masked by the presence of a dominant trait. Tall stems and purple flowers are examples of dominant traits; dwarf stems and white flowers are examples of recessive traits. In this case, we say that tall is dominant over dwarf, and purple is dominant over white.

***Genes and Alleles***    Mendel's results were consistent with particulate inheritance, in which the determinants of traits are inherited as unchanging, discrete units. In all seven cases, the recessive trait reappeared in the $F_2$ generation: Most $F_2$ plants displayed the dominant trait, whereas a smaller proportion showed the recessive trait. This

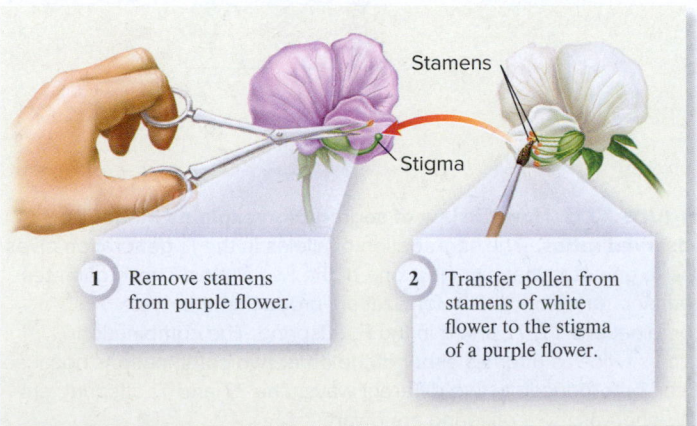

**Figure 17.4  A procedure for cross-fertilizing pea plants.**

*Concept Check:*  *Why are the stamens removed from the purple flower in this cross-fertilization procedure?*

1    Remove stamens from purple flower.

2    Transfer pollen from stamens of white flower to the stigma of a purple flower.

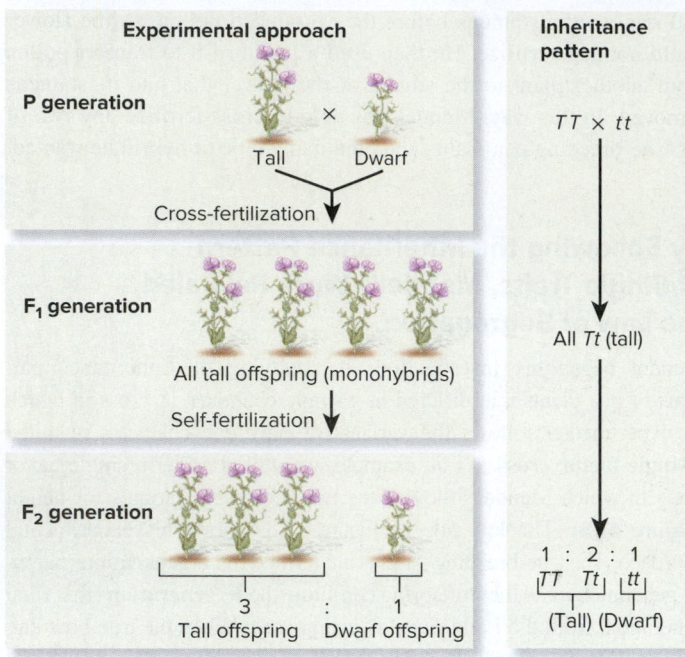

**(a) Mendel's protocol for making single-factor crosses**

**THE DATA**

| P cross | $F_1$ generation | $F_2$ generation | Ratio |
|---|---|---|---|
| Purple × white flowers | All purple | 705 purple, 224 white | 3.15:1 |
| Axial × terminal flowers | All axial | 651 axial, 207 terminal | 3.14:1 |
| Yellow × green seeds | All yellow | 6,022 yellow, 2,001 green | 3.01:1 |
| Round × wrinkled seeds | All round | 5,474 round, 1,850 wrinkled | 2.96:1 |
| Green × yellow pods | All green | 428 green, 152 yellow | 2.82:1 |
| Smooth × constricted pods | All smooth | 882 smooth, 299 constricted | 2.95:1 |
| Tall × dwarf stem | All tall | 787 tall, 277 dwarf | 2.84:1 |
| **Total** | **All dominant** | **14,949 dominant, 5,010 recessive** | **2.98:1** |

**(b) Mendel's observed data for all 7 traits**

**Figure 17.5  Mendel's analyses of single-factor crosses.**
*Concept Check:*  *Why do offspring of the $F_1$ generation exhibit only one variant of each character?*

observation led Mendel to conclude that the genetic determinants of traits are "unit factors" that are passed intact from generation to generation. These unit factors are what we now call **genes** (from the Greek *genos*, meaning birth), a term coined by the Danish botanist Wilhelm Johannsen in 1909. Mendel postulated that every individual carries two genes for a given character and that the gene for each character in his pea plants exists in two variant forms, which we now call alleles.

For example, the gene controlling height in Mendel's pea plants occurs in two variants, called the tall allele and the dwarf allele. The right side of Figure 17.5a shows Mendel's conclusions, using genetic symbols (italic letters) that were adopted later. The letters *T* and *t* represent the alleles of the gene for plant height. By convention, the uppercase letter represents the dominant allele (in this case, tall), and the same letter in lowercase represents the recessive allele (dwarf).

***Segregation of Alleles***   When Mendel compared the numbers of $F_2$ offspring exhibiting dominant and recessive traits, he noticed a recurring pattern. Although some experimental variation occurred, he always observed a 3:1 ratio between the dominant and the recessive trait (Figure 17.5b). How did Mendel interpret this ratio? He concluded that each $F_1$ plant carried two versions (alleles) of a gene affecting height (or another character) and that the two alleles carried by such an $F_1$ plant separate, or segregate, from each other during the process that gives rise to gametes. Therefore, each sperm or egg carried only one allele. The diagram in **Figure 17.6** shows that

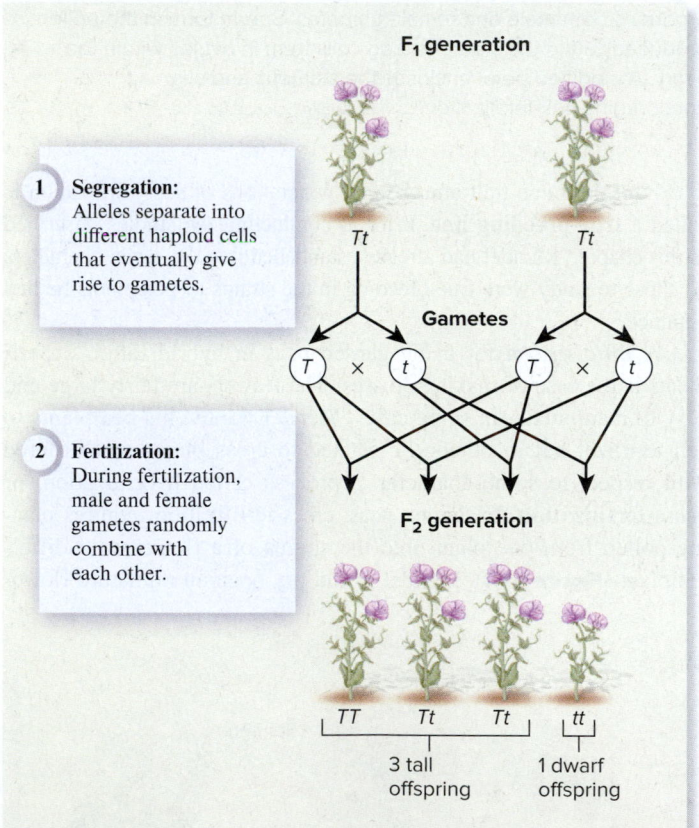

**Figure 17.6  How the law of segregation explains Mendel's observed ratios.**  The segregation of alleles in the $F_1$ generation gives rise to gametes that carry just one of the two alleles. These gametes combine randomly during fertilization, producing the allele combinations *TT*, *Tt*, and *tt* in the $F_2$ offspring. The combination *Tt* occurs twice as often as either of the other two combinations because it can be produced in two different ways. The *TT* and *Tt* offspring are tall, whereas the *tt* offspring are dwarf.

*Concept Check:*  *What is the ratio of the T allele to the t allele in the $F_2$ generation? Does this ratio differ from the 3:1 phenotype ratio? If so, explain why.*

the segregation of the $F_1$ alleles should result in equal numbers of gametes carrying the dominant allele ($T$) and the recessive allele ($t$). If these gametes combine with one another randomly at fertilization, as shown in the figure, this would account for the 3:1 ratio of the $F_2$ generation. Note that a $Tt$ individual can be produced by two different combinations of alleles—the $T$ allele can come from the male gamete and the $t$ allele from the female gamete, or vice versa. This accounts for the observation that $Tt$ offspring are produced twice as often as either $TT$ or $tt$. The results of the study of traits in pea plants gave rise to **Mendel's law of segregation**, which can be stated as follows:

> *The two alleles of a gene separate (segregate) from each other during the process that gives rise to gametes, so every gamete receives only one allele.*

## Genotype Describes an Organism's Genetic Makeup, Whereas Phenotype Describes Its Characteristics

To continue our discussion of Mendel's results, we need to introduce a few more genetic terms. The term **genotype** refers to the genetic composition of an individual. In the example shown in Figure 17.5a, $TT$ and $tt$ are the genotypes of the P generation, and $Tt$ is the genotype of the $F_1$ generation. In the P generation, both parents are true-breeding plants, which means that each has identical copies of the allele of the gene for height. An individual with two identical alleles of a gene is said to be **homozygous** with respect to that gene. In the specific cross we are considering, the tall plant ($TT$) is homozygous for $T$, and the dwarf plant ($tt$) is homozygous for $t$. In contrast, a **heterozygous** individual carries two different alleles of a gene. Plants of the $F_1$ generation are heterozygous, with the genotype $Tt$, because every individual carries one copy of the tall allele ($T$) and one copy of the dwarf allele ($t$). The $F_2$ generation includes both homozygous individuals (homozygotes) and heterozygous individuals (heterozygotes).

The term **phenotype** refers to the characteristics of an organism that are the result of the expression of its genes. In the example in Figure 17.5a, one of the parent plants is phenotypically tall, and the other is phenotypically dwarf. Although the $F_1$ offspring are heterozygous ($Tt$), they are phenotypically tall because each of them has a copy of the dominant tall allele ($T$). In contrast, the $F_2$ plants display both phenotypes in a ratio of 3:1.

## A Punnett Square Is Used to Predict the Outcome of Crosses

A common way to predict the outcome of simple genetic crosses is to make a **Punnett square**, a method originally proposed by the British geneticist Reginald Punnett. To construct a Punnett square, you must know the genotypes of the parents. What follows is a step-by-step description of the Punnett-square approach, using a cross of heterozygous tall plants.

> **Step 1.** *Write down the genotypes of both parents.* In this example, a heterozygous tall plant is crossed to another heterozygous tall plant. The plant providing the pollen is considered the male parent and the plant providing the eggs, the female parent. (In self-pollination, a single individual produces both types of gametes.)

Male parent: $Tt$
Female parent: $Tt$

**Step 2.** *Write down the possible gametes that each parent can make.* Remember the law of segregation tells us that a gamete contains only one copy of each allele.

Male gametes: $T$ or $t$
Female gametes: $T$ or $t$

**Step 3.** *Create an empty Punnett square.* The number of columns equals the number of male gametes, and the number of rows equals the number of female gametes. Our example has two rows and two columns. Place the male gametes across the top of the Punnett square and the female gametes along the side.

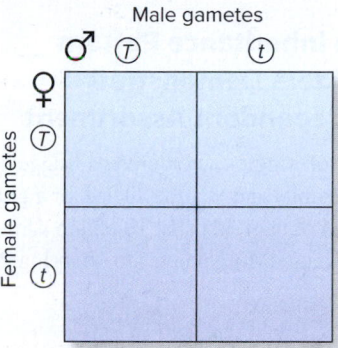

**Step 4.** *Fill in the possible genotypes of the offspring by combining the alleles of the gametes in the empty boxes.*

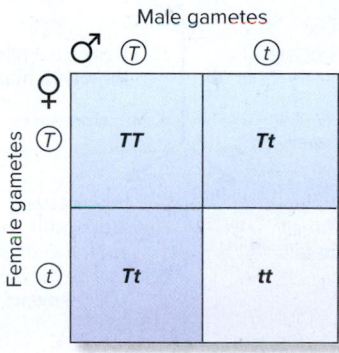

**Step 5.** *Determine the relative proportions of genotypes and phenotypes of the offspring.* The genotypes are obtained directly from the Punnett square. In this example, the genotype ratio is 1 $TT$ : 2 $Tt$ : 1 $tt$. To determine the phenotypes, you must know which allele is dominant. For plant height, $T$ (tall) is dominant to $t$ (dwarf). The genotypes $TT$ and $Tt$ are tall, whereas the genotype $tt$ is dwarf. Therefore, our Punnett square shows us that the phenotype ratio is expected to be 3 tall : 1 dwarf.

## A Testcross Is Used to Determine an Individual's Genotype

When a character has two variants, one of which is dominant over the other, we know that an individual with a recessive phenotype is homozygous for the recessive allele. A dwarf pea plant, for example, must have the genotype *tt*. But an individual with a dominant phenotype may be either homozygous or heterozygous—a tall pea plant may have the genotype *TT* or *Tt*. How can we distinguish between these two possibilities? Mendel devised a method called a **testcross** to address this question. In a testcross, the researcher crosses the individual of interest to a homozygous recessive individual and observes the phenotypes of the offspring.

**Figure 17.7** shows how this procedure can be used to determine the genotype of a tall pea plant. If the testcross produces some dwarf offspring, as shown in the Punnett square on the right side, these offspring must have two copies of the recessive allele, one inherited from each parent. Therefore, the tall parent must be a heterozygote, with the genotype *Tt*. Alternatively, if all of the offspring are tall, as shown in the Punnett square on the left, the tall parent is likely to be a homozygote, with the genotype *TT*.

## Analyzing the Inheritance Pattern of Two Characters Demonstrated the Law of Independent Assortment

Mendel's analysis of single-factor crosses suggested that traits are inherited as discrete units and that the alleles for a given gene segregate during the formation of haploid cells. To obtain additional insights into how genes are transmitted from parents to offspring, Mendel conducted

crosses in which he simultaneously followed the inheritance of two different characters. A cross of this type is called a **two-factor cross**. We will examine a two-factor cross in which Mendel simultaneously followed the inheritance of seed color and seed shape (**Figure 17.8**). He began by crossing strains of pea plants that bred true for both characters. The plants of one strain had yellow, round seeds, and plants of the other strain had green, wrinkled seeds. He then allowed the $F_1$ offspring to self-fertilize and observed the phenotypes of the $F_2$ generation.

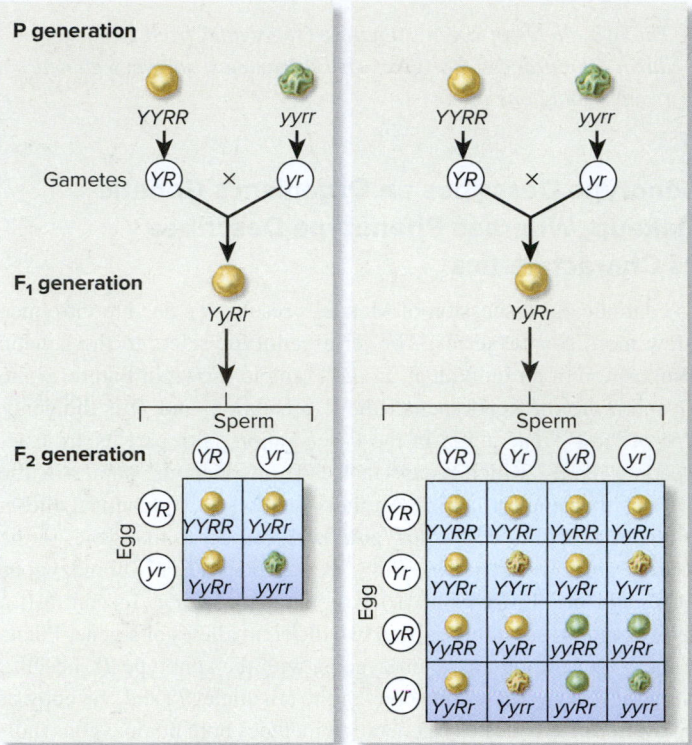

**(a) Hypothesis: linked assortment**

**(b) Hypothesis: independent assortment**

| P cross | $F_1$ generation | $F_2$ generation |
|---|---|---|
| Yellow, round seeds × Green, wrinkled seeds | Yellow, round seeds | 315 yellow, round seeds |
| | | 101 yellow, wrinkled seeds |
| | | 108 green, round seeds |
| | | 32 green, wrinkled seeds |

**(c) The data observed by Mendel**

**Figure 17.8  Two hypotheses for the assortment of two different genes.** In a cross between two true-breeding pea plants, one with yellow, round seeds and one with green, wrinkled seeds, all of the $F_1$ offspring have yellow, round seeds. When the $F_1$ offspring self-fertilize, the two hypotheses predict different ratios of phenotypes in the $F_2$ generation. **(a)** Linked assortment, in which parental alleles stay associated with each other, or **(b)** independent assortment, in which each allele assorts independently. **(c)** Mendel's data supported the independent assortment hypothesis.

*Concept Check:* What ratio of phenotypes would have occurred in the $F_2$ generation if the linked assortment hypothesis had been correct?

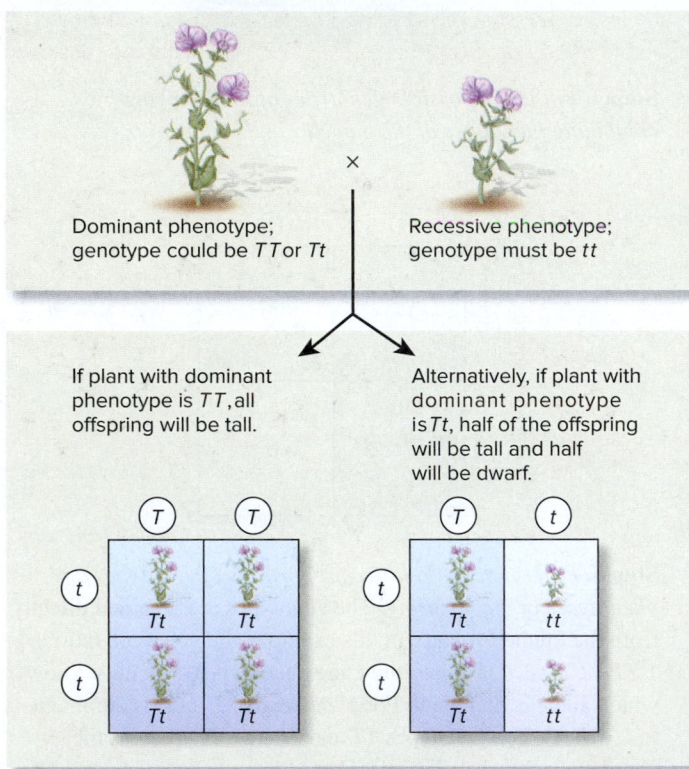

**Figure 17.7  A testcross.** The purpose of this experiment is to determine if an organism with the dominant phenotype, in this case a tall pea plant, is a homozygote (*TT*) or a heterozygote (*Tt*).

*Concept Check:* Let's suppose you had a pea plant with purple flowers and unknown genotype and conducted a testcross to determine its genotype. You obtained 41 offspring plants: 20 with white flowers and 21 with purple flowers. What was the genotype of the original purple-flowered plant?

What are the possible patterns of inheritance for two characters? One possibility is that the two genes are linked in some way, so variants that occur together in the parents are always inherited as a unit. In our example, the allele for yellow seeds ($Y$) would always be inherited with the allele for round seeds ($R$), and the alleles for green seeds ($y$) would always be inherited with the allele for wrinkled seeds ($r$), as shown in Figure 17.8a. A second possibility is that the two genes are independent of one another, so their alleles are randomly distributed into gametes (Figure 17.8b). By following the transmission pattern of two characters simultaneously, Mendel could determine whether the genes that determine seed shape and seed color assort (are distributed) together as a unit or independently of each other.

What experimental results could Mendel predict for each of these two models? The two homozygous plants of the P generation can produce only two kinds of gametes, $YR$ and $yr$, so in either case the $F_1$ offspring would be heterozygous for both genes; that is, they would have the genotype $YyRr$. Because Mendel knew from his earlier experiments that yellow was dominant over green and round over wrinkled, he could predict that all the $F_1$ plants would have yellow, round seeds. In contrast, as shown in Figure 17.8, the ratios he obtained in the $F_2$ generation would depend on whether the alleles of both genes assort together or independently.

If the two genes are linked, as in Figure 17.8a, the $F_1$ plants could produce gametes that are only $YR$ or $yr$. These gametes would combine to produce offspring with the genotypes $YYRR$ (yellow, round), $YyRr$ (yellow, round), and $yyrr$ (green, wrinkled). The ratio of phenotypes would be 3 yellow, round to 1 green, wrinkled. Every $F_2$ plant would be phenotypically like one P-generation plant or the other. None would display a new combination of the parental traits. However, if the alleles assorted independently, the $F_2$ generation would have a wider range of genotypes and phenotypes, as shown by the Punnett square in Figure 17.8b. In this case, each $F_1$ parent produces four kinds of gametes—$YR$, $Yr$, $yR$, and $yr$—instead of two, so the square is constructed with four rows on each side and shows 16 possible genotypes. The $F_2$ generation includes plants with yellow, round seeds; yellow, wrinkled seeds; green, round seeds; and green, wrinkled seeds, in a ratio of 9:3:3:1.

The actual results of this two-factor cross are shown in Figure 17.8c. Crossing the true-breeding parents produced **dihybrid** offspring—offspring that are hybrids with respect to both traits. These $F_1$ dihybrids all had yellow, round seeds, confirming that yellow and round are dominant traits. This result was consistent with either hypothesis. However, the data for the $F_2$ generation were consistent only with the independent assortment hypothesis. The $F_2$ offspring showed four different phenotypes in a ratio that was reasonably close to 9:3:3:1.

In his original studies, Mendel reported that he had obtained similar results for every pair of characters he analyzed. This work gave rise to **Mendel's law of independent assortment**, which can be stated as follows:

*The alleles of different genes assort independently of each other during the process that gives rise to gametes.*

Independent assortment means that a specific allele for one gene may be found in a gamete regardless of which allele for a different gene is found in the same gamete. In our example, the yellow and green alleles assort independently of the round and wrinkled alleles.

The union of gametes from $F_1$ plants carrying these alleles produces the $F_2$ genotype and phenotype ratios shown in Figure 17.8b.

As we will see in Chapter 18, not all two-factor crosses exhibit independent assortment. In some cases, the alleles of two genes that are physically located near each other on the same chromosome do not assort independently.

## 17.2 The Chromosome Theory of Inheritance

**Learning Outcomes:**

1. Outline the principles of the chromosome theory of inheritance.
2. Relate the behavior of chromosomes during meiosis to Mendel's laws of inheritance.

Mendel's studies with pea plants eventually led to the concept of a gene, which is the foundation for our understanding of inheritance. However, at the time of Mendel's work, the physical nature and location of genes were a complete mystery. The idea that inheritance has a physical basis was not even addressed until 1883, when German biologist August Weismann and Swiss botanist Karl Nägeli championed the idea that a substance in living cells is responsible for the transmission of hereditary traits. This idea challenged other researchers to identify the genetic material. Several scientists, including German biologists Eduard Strasburger and Walther Flemming, observed dividing cells under the microscope and suggested that the chromosomes are the carriers of the genetic material. As we now know, the genetic material is the DNA within chromosomes.

In the early 1900s, the idea that chromosomes carry the genetic material gained increasing support as researchers continued to study the processes of mitosis, meiosis, and fertilization. It became increasingly clear that the characteristics of organisms are rooted in the continuity of cells during the life of an organism and from one generation to the next. Several scientists noted striking parallels between the segregation and assortment of traits noted by Mendel and the behavior of chromosomes during meiosis. Among these scientists were German biologist Theodor Boveri and American biologist Walter Sutton, who independently proposed the chromosome theory of inheritance. According to this theory, the inheritance patterns of traits can be explained by the transmission of chromosomes during meiosis and fertilization.

A modern version of the **chromosome theory of inheritance** consists of a few fundamental principles:

1. Chromosomes contain DNA, which is the genetic material. Genes are found within the chromosomes.
2. Chromosomes are replicated and passed from parent to offspring. They are also passed from cell to cell during the development of a multicellular organism.
3. The nucleus of a diploid cell contains two sets of chromosomes, which are found in homologous pairs. The maternal and paternal sets of homologous chromosomes are functionally equivalent; each set carries a full complement of genes.
4. At meiosis, one member of each chromosome pair segregates into one daughter nucleus, and its homolog segregates into the other daughter nucleus. During the formation of haploid

cells, the members of different chromosome pairs segregate independently of each other.

5. Gametes are haploid cells that combine to form a diploid cell during fertilization, with each gamete transmitting one set of chromosomes to the offspring.

In this section, we will relate the chromosome theory of inheritance to Mendel's laws of inheritance.

## Mendel's Law of Segregation Is Explained by the Segregation of Homologous Chromosomes During Meiosis

Now that you have an understanding of the basic tenets of the chromosome theory of inheritance, let's relate these ideas to Mendel's laws of inheritance. To do so, it will be helpful to introduce another genetic term. The physical location of a gene on a chromosome is called the gene's **locus** (plural, loci). As shown in **Figure 17.9**, each member of a homologous chromosome pair carries an allele of the same gene at the same locus. The individual in this example is heterozygous (*Tt*), so each homolog has a different allele.

How can we relate the chromosome theory of inheritance to Mendel's law of segregation? **Figure 17.10** follows a pair of homologous chromosomes through the events of meiosis. This example involves a pea plant, heterozygous for height, *Tt*. The top of Figure 17.10 shows the two homologous chromosomes prior to DNA replication. When a cell prepares to divide, the homologs replicate to produce pairs of sister chromatids. Each chromatid carries a copy of the allele found on the original homolog, either *T* or *t*. During meiosis I, the homologs, each consisting of two sister chromatids, pair up and then segregate into two daughter cells. One of these cells has two copies of the *T* allele, and the other has two copies of the *t* allele. The sister chromatids separate during meiosis II, which produces four haploid cells. The end result of meiosis is that each haploid cell has a copy of just one of the two original homologs. Two of the cells have a chromosome carrying the *T* allele, and the other two have a chromosome carrying the *t* allele at the same locus.

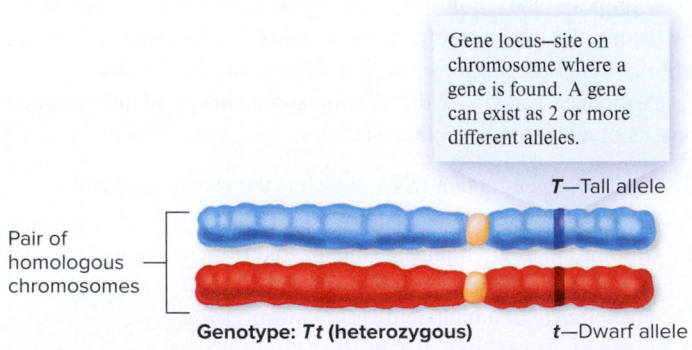

Gene locus—site on chromosome where a gene is found. A gene can exist as 2 or more different alleles.

*T*—Tall allele

Pair of homologous chromosomes

Genotype: *Tt* (heterozygous)          *t*—Dwarf allele

**Figure 17.9** **A gene locus.** The locus (location) of a gene is the same for each member of a homologous pair, whether the individual is homozygous or heterozygous for that gene. This individual is heterozygous (*Tt*) for a gene for plant height.

**Core Skill: Connections** Look back at Section 16.4. Explain the relationship between sexual reproduction and homologous chromosomes.

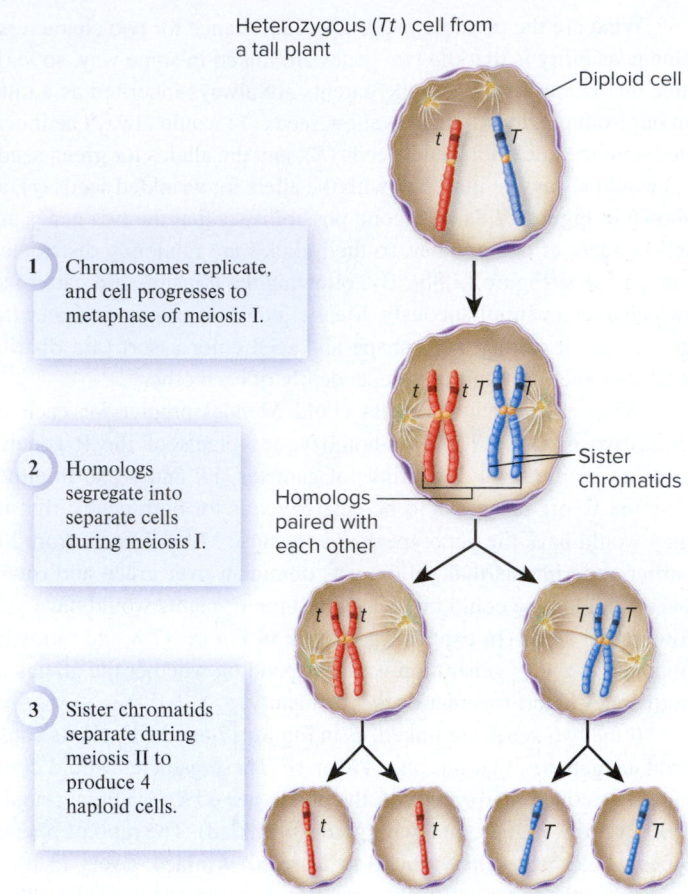

Heterozygous (*Tt*) cell from a tall plant

Diploid cell

1 Chromosomes replicate, and cell progresses to metaphase of meiosis I.

2 Homologs segregate into separate cells during meiosis I.

Sister chromatids

Homologs paired with each other

3 Sister chromatids separate during meiosis II to produce 4 haploid cells.

Four haploid cells

**Figure 17.10** **The chromosomal basis of allele segregation.** This example shows a pair of homologous chromosomes in a cell of a pea plant. The blue chromosome was inherited from the male parent, and the red chromosome was inherited from the female parent. This individual is heterozygous (*Tt*) for a height gene. The two homologs segregate from each other during meiosis, leading to segregation of the tall allele (*T*) and the dwarf allele (*t*) into different haploid cells. Note: For simplicity, this diagram shows a single pair of homologous chromosomes, though eukaryotic cells typically have several different pairs of homologous chromosomes.

**Core Skill: Connections** When we say that alleles segregate, what does the word segregate mean? How is this related to meiosis, shown in Figure 16.13?

If the haploid cells shown at the bottom of Figure 17.10 give rise to gametes that combine randomly during fertilization, they produce diploid offspring with the genotype and phenotype ratios shown earlier in Figure 17.6.

## Mendel's Law of Independent Assortment Is Explained by the Independent Alignment of Different Chromosomes During Meiosis

How can we relate the chromosome theory of inheritance to Mendel's law of independent assortment? **Figure 17.11** shows the alignment and segregation of two pairs of chromosomes in a pea plant. One pair carries the gene for seed color: The yellow allele (*Y*) is on one chromosome, and the green allele (*y*) is on its homolog. The other

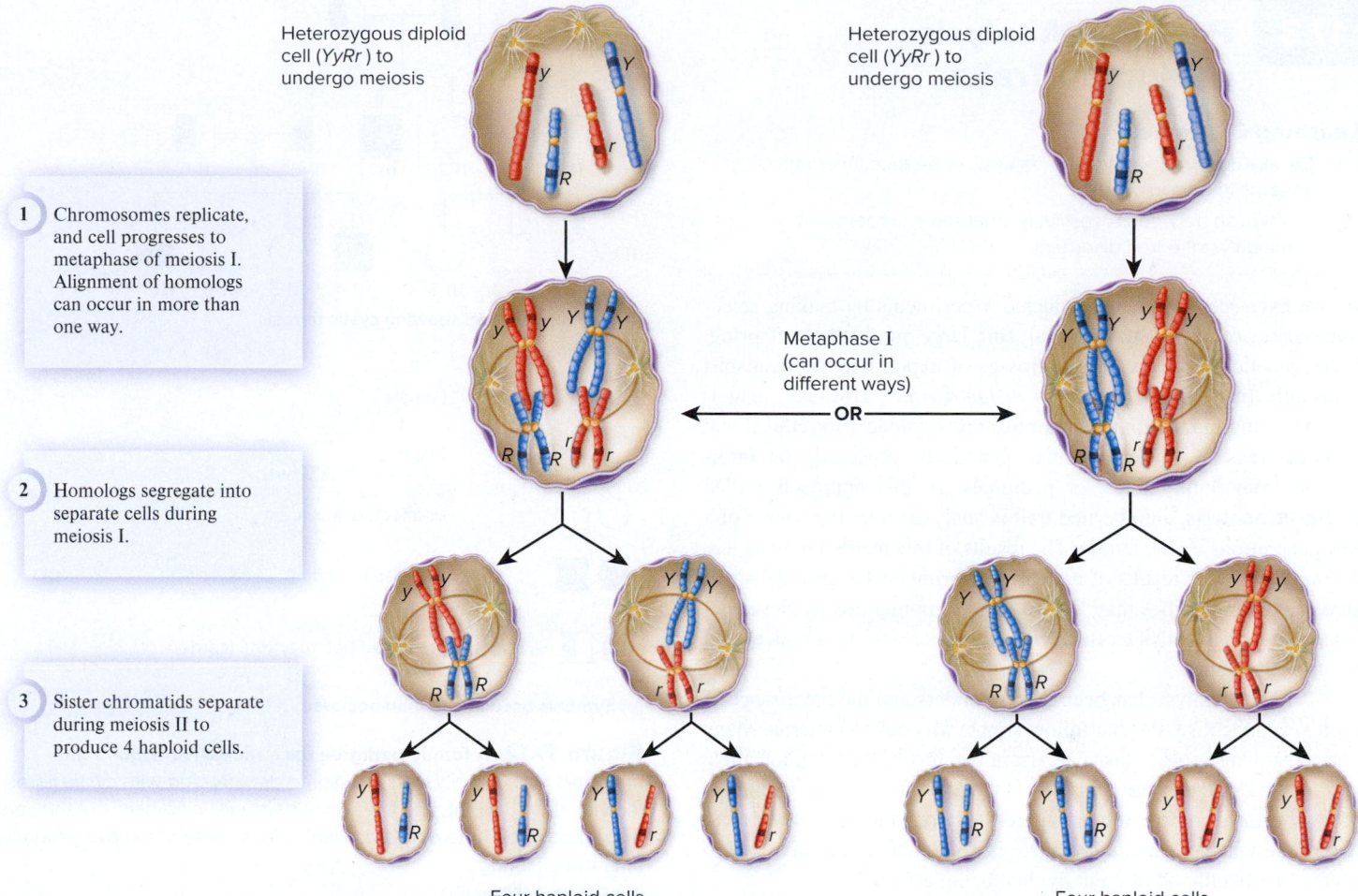

**Figure 17.11** **The chromosomal basis of independent assortment.** The alleles for seed color (*Y or y*) and seed shape (*R or r*) in peas are on different chromosomes. During metaphase of meiosis I, different arrangements of the two chromosome pairs lead to different combinations of the alleles in the resulting haploid cells. On the left, the chromosome carrying the recessive y allele has segregated with the chromosome carrying the dominant *R* allele. On the right, the two chromosomes carrying the dominant alleles (*Y and R*) have segregated together. Note: For simplicity, this diagram shows only two pairs of homologous chromosomes, though eukaryotic cells typically have several different pairs of homologous chromosomes.

> **Concept Check:** *Let's suppose that a cell is heterozygous for three different genes (Aa, Bb, and Cc) and that each gene is on a different chromosome. How many different ways can the three pairs of homologous chromosomes align themselves during metaphase I, and how many different types of gametes can be produced?*

pair of chromosomes carries the gene for seed shape: One member of the pair has the round allele (*R*), whereas its homolog carries the wrinkled allele (*r*). Therefore, this individual is heterozygous for both genes, with the genotype *YyRr*.

When meiosis begins, the DNA in each chromosome has already replicated, producing two sister chromatids. At metaphase I of meiosis, the two pairs of chromosomes randomly align themselves along the metaphase plate. This alignment can occur in two equally probable ways, shown on the two sides of the figure. On the left, the chromosome carrying the y allele is aligned on the same side of the metaphase plate as the chromosome carrying the *R* allele; *Y* is aligned with *r*. On the right, the opposite has occurred: *Y* is aligned with *R*, and *y* is with *r*. In each case, the chromosomes that aligned on the same side of the metaphase plate segregate into the same daughter cell. In this way, the random alignment of chromosome pairs during meiosis I

leads to the independent assortment of alleles found on different chromosomes. For two genes found on different chromosomes, each with two variant alleles, meiosis produces four allele combinations in equal numbers (*yR, Yr, YR,* and *yr*), as seen at the bottom of the figure.

If a *YyRr* (dihybrid) plant undergoes self-fertilization, any two gametes can combine randomly during fertilization. Because four kinds of gametes are made, $4^2$, or 16, possible allele combinations are possible in the offspring. These genotypes, in turn, produce four phenotypes in a 9:3:3:1 ratio, as seen earlier in Figure 17.8. This ratio is the expected outcome when a heterozygote for two genes on different chromosomes undergoes self-fertilization.

But what if two different genes are located on the same chromosome? In this case, the transmission pattern may not conform to the law of independent assortment. We will discuss this phenomenon, known as linkage, in Chapter 18.

## 17.3 Pedigree Analysis of Human Traits

**Learning Outcomes:**

1. **CoreSKILL »** Apply pedigree analysis to deduce inheritance patterns in humans.
2. Distinguish between recessively inherited disorders and dominantly inherited disorders.

As we have seen, Mendel conducted experiments by making selective crosses of pea plants and analyzing large numbers of offspring. Later geneticists also relied on crosses of experimental organisms, especially fruit flies (*Drosophila melanogaster*). However, geneticists studying human traits cannot use this approach, for ethical and practical reasons. Instead, human geneticists must rely on information from family trees, or pedigrees. In this approach, called **pedigree analysis**, an inherited trait is analyzed over the course of a few generations in one family. The results of this method may be less definitive than the results of breeding experiments because the small size of human families may lead to large sampling errors. Nevertheless, a pedigree analysis often provides important clues concerning human inheritance.

Pedigree analysis has been used to understand the inheritance of human genetic diseases that follow simple Mendelian patterns. Many genes that play a role in disease exist in two forms: the common allele and a rare allele that has arisen by mutation. The disease symptoms are associated with the mutant allele. Pedigree analysis allows us to determine whether the mutant allele is dominant or recessive and to predict the likelihood of an individual being affected.

Let's consider a recessive condition to illustrate pedigree analysis. The pedigree in **Figure 17.12** concerns a human genetic disease known as cystic fibrosis (CF), which involves a mutation in a gene that encodes the cystic fibrosis transmembrane regulator (the *CFTR* gene, also see Figure 1.15). Approximately 3% of Americans of European descent are heterozygous carriers of the recessive (disease-causing) *CFTR* allele. Individuals who are homozygous for this allele exhibit the disease symptoms, which include abnormalities of the lungs, pancreas, intestine, and sweat glands. A human pedigree, like the one in Figure 17.12, shows the oldest generation (designated by the Roman numeral I) at the top, with later generations (II and III) below it. A male (represented by a square) and a female (represented by a circle) who produce offspring are connected by a horizontal line; a vertical line connects parents with their offspring. Siblings (brothers and sisters) are placed on downward projections from a single horizontal line, from left to right in the order of their birth. For example, individuals I-1 and I-2 are the parents of individuals II-2, II-3, and II-4, who are all siblings. Individuals affected by the disease, such as individual II-3, are depicted by filled symbols.

Why does this pedigree indicate a recessive pattern of inheritance for CF? The answer is that two unaffected individuals can produce an affected offspring. Such individuals are presumed to be heterozygotes (designated by a half-filled symbol). However, the same unaffected parents can also produce unaffected offspring (depicted by an unfilled symbol), because an individual must inherit two copies of the mutant allele to exhibit the disease. A recessive

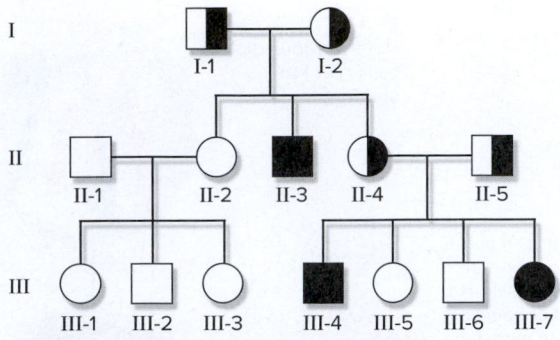

**(a) Human pedigree showing cystic fibrosis**

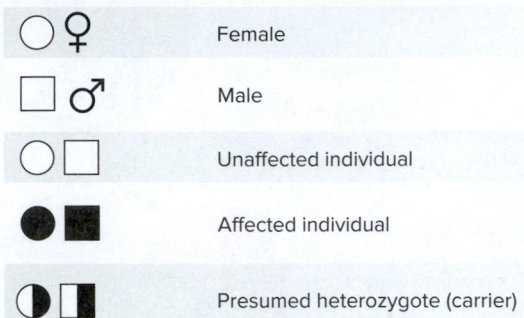

| | |
|---|---|
| ○ ♀ | Female |
| □ ♂ | Male |
| ○ □ | Unaffected individual |
| ● ■ | Affected individual |
| ◐ ◧ | Presumed heterozygote (carrier) |

**(b) Symbols used in a human pedigree**

**Figure 17.12 A family pedigree for a recessive trait.** Some members of the family in this pedigree are affected with cystic fibrosis. Individuals I-1, I-2, II-4, and II-5 do not have cystic fibrosis, but they are presumed to be heterozygotes (carriers) because they have produced affected offspring.

*Concept Check:* *Let's suppose a genetic disease is caused by a mutant allele. If two affected parents produce an unaffected offspring, can the mutant allele be recessive?*

mode of inheritance is also indicated by the observation that all of the offspring of two affected individuals are affected themselves. However, for genetic diseases that limit survival or fertility, there are rarely if ever cases where two affected individuals produce offspring.

Although many of the alleles causing human genetic diseases are recessive, some are known to be dominant. Let's consider Huntington disease, a condition that causes the degeneration of brain cells involved in emotions, intellect, and movement. The symptoms of Huntington disease, which usually begin to appear when people are 30 to 50 years old, include uncontrollable jerking movements of the limbs, trunk, and face; progressive loss of mental abilities; and the development of psychiatric problems. If you examine the pedigree shown in **Figure 17.13**, you will see that every affected individual has one affected parent. This pattern is characteristic of most dominant disorders. However, affected parents do not always produce affected offspring. For example, II-6 is a heterozygote that has passed the nondisease-causing allele to his offspring, thereby producing unaffected offspring (III-3 and III-4).

Most human genes are found on the paired chromosomes known as **autosomes**, which are the same in both sexes. Mendelian inheritance patterns involving these autosomal genes are described as autosomal inheritance patterns. Huntington disease is an example of a

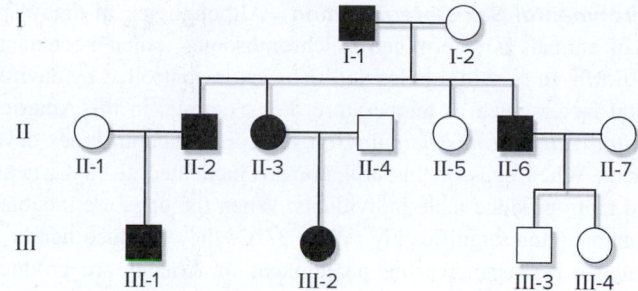

**Figure 17.13** **A family pedigree for a dominant trait.** Huntington disease is caused by a dominant allele.

*Concept Check:* *What observation in a pedigree suggests a dominant pattern of inheritance?*

trait with an autosomal dominant inheritance pattern, whereas cystic fibrosis displays an autosomal recessive pattern. However, some human genes are located on sex chromosomes, which are different in males and females. These genes have their own characteristic inheritance patterns, which we will consider next.

## 17.4 Sex Chromosomes and X-Linked Inheritance Patterns

**Learning Outcomes:**

1. Describe different systems of sex determination in animals and plants.
2. **CoreSKILL »** Predict the outcome of crosses when genes are located on sex chromosomes.
3. Explain why X-linked recessive traits are more likely to occur in males.

Earlier in this chapter, we discussed Mendel's experiments that established the basis for understanding how traits are transmitted from parents to offspring. We also examined the chromosome theory of inheritance, which provided a framework for explaining Mendel's observations. Mendelian patterns of gene transmission are observed for most genes located on autosomes in a wide variety of eukaryotic species.

We will now turn our attention to genes located on **sex chromosomes**. This term refers to a distinctive pair of chromosomes that are different in males and females and that determine the sex of the individuals. Sex chromosomes are found in many but not all species with two sexes. The study of sex chromosomes proved pivotal in confirming the chromosome theory of inheritance. The distinctive transmission patterns of genes on sex chromosomes helped early geneticists show that particular genes are located on particular chromosomes. Later, other researchers became interested in these genes because some of them were found to cause inherited disorders in humans.

In this section, we will consider a few mechanisms by which sex chromosomes in various species determine an individual's sex. We

will then explore the inheritance patterns of genes on sex chromosomes and see that recessive alleles are expressed more frequently in males than in females. Last, we will examine some of the early research involving sex chromosomes that provided convincing evidence for the chromosome theory of inheritance.

### In Many Species, Sex Differences Are Due to the Presence of Sex Chromosomes

Some early evidence supporting the chromosome theory of inheritance involved a consideration of sex determination. In 1901, American biologist C. E. McClung suggested that the inheritance of particular chromosomes is responsible for determining sex in fruit flies. Following McClung's initial observations, several mechanisms of sex determination were found in different species of animals. Some examples are described in **Figure 17.14**. All of these mechanisms involve chromosomal differences between the sexes, and most involve a difference in a single pair of sex chromosomes.

***X-Y System*** In the X-Y system of sex determination, which operates in mammals, the somatic cells of males have one X and one Y chromosome, whereas female somatic cells contain two X chromosomes (Figure 17.14a). For example, the 46 chromosomes carried by human cells consist of 22 pairs of autosomes and one pair of sex chromosomes (either XY or XX). Which chromosome, the X or Y, determines sex? In mammals, the presence of the Y chromosome causes maleness. This is known from the analysis of rare individuals who carry chromosomal abnormalities. For example, mistakes that occasionally occur during meiosis may produce an individual who carries two X chromosomes and one Y chromosome. Such an individual develops into a male. A gene called the *SRY* gene located on the Y chromosome of mammals plays a key role in the developmental pathway that leads to maleness.

***X-O System*** The X-O system operates in many insects (Figure 17.14b). Unlike the X-Y system in mammals, the presence of the Y chromosome in the X-O system does not determine maleness. Females in this system have a pair of sex chromosomes and are designated XX. In some insect species that follow the X-O system, the male has only one sex chromosome, the X. In other X-O insect species, such as *Drosophila melanogaster*, the male has both an X chromosome and a Y chromosome. In all cases, an insect's sex is determined by the ratio between its X chromosomes and its sets of autosomes. If a fly has one X chromosome and is diploid for the autosomes (2n), this ratio is 1/2, or 0.5. This fly will become a male whether or not it receives a Y chromosome. On the other hand, if a diploid fly receives two X chromosomes, the ratio is 2/2, or 1.0, and the fly becomes a female.

***Z-W System*** Thus far, we have considered examples where females have two similar copies of a sex chromosome, the X. However, in other animal species, such as birds and some fish, the male carries two similar chromosomes (Figure 17.14c). This is called the Z-W system to distinguish it from the X-Y system found in mammals. The male is ZZ, and the female is ZW.

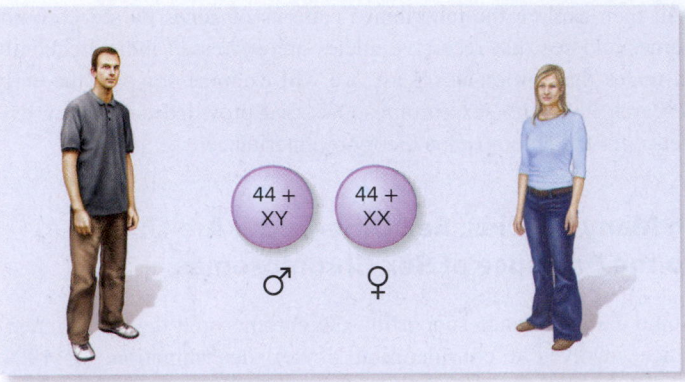

(a) The X–Y system in mammals

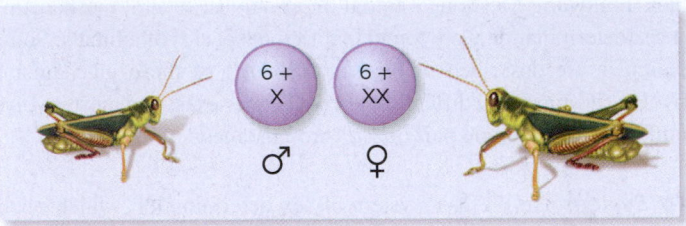

(b) The X–O system in certain insects

(c) The Z–W system in birds

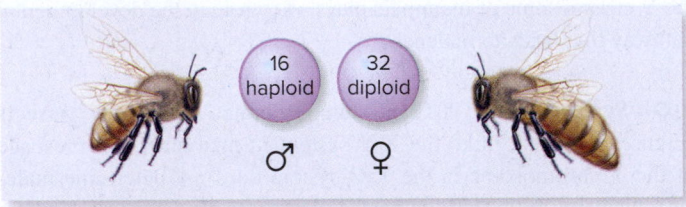

(d) The haplodiploid system in bees

**Figure 17.14  Different mechanisms of sex determination in animals.** The numbers shown in the circles indicate the numbers of autosomes.

Concept Check:  *If a person is born with only one X chromosome and no Y chromosome, would you expect that person to be a male or a female? Explain your answer.*

**Haplodiploid System**  Not all chromosomal mechanisms of sex determination involve a special pair of sex chromosomes. An interesting mechanism known as the haplodiploid system is found in bees (Figure 17.14d). The male bee, or drone, is produced from an unfertilized haploid egg. Therefore, male bees are haploid individuals. Females, both worker bees and queen bees, are produced from fertilized eggs and are diploid.

**Environmental Sex Determination**  Although sex in many species of animals is determined by chromosomes, other mechanisms are known. In certain reptiles and fish, sex is controlled by environmental factors such as temperature. For example, in the American alligator (*Alligator mississippiensis*), temperature controls sex development. When eggs of this alligator are incubated at 33°C, nearly all of them produce male individuals. When the eggs are incubated at a temperature significantly below 33°C, they produce nearly all females, whereas increasing percentages of females are produced above 34°C.

**Sex Determination in Plants**  Most species of flowering plants, including pea plants, have a single type of diploid plant, or sporophyte, that makes both male and female gametophytes. However, the sporophytes of some species have two sexually distinct types of individuals, one with flowers that produce male gametophytes, and the other with flowers that produce female gametophytes. Examples include hollies, willows, poplars, and date palms. Sex chromosomes, designated X and Y, are responsible for sex determination in many such species. The male plant is XY, whereas the female plant is XX. However, in some plant species with separate sexes, microscopic examination of the chromosomes does not reveal distinct types of sex chromosomes.

## In Humans, Recessive X-Linked Traits Are More Likely to Occur in Males

In humans, the X chromosome is rather large and carries over 1,000 genes, whereas the Y chromosome is quite small and has less than 100 genes. Therefore, many genes are found on the X chromosome but not on the Y; these are known as **X-linked genes**. By comparison, fewer genes are known to be Y-linked, meaning they are found on the Y chromosome but not on the X. **Sex-linked genes** are found on one sex chromosome but not on the other. Because fewer genes are found on the Y chromosome, the term usually refers to X-linked genes. In mammals, a male cannot be described as being homozygous or heterozygous for an X-linked gene, because these terms apply to genes that are present in two copies. Instead, the term **hemizygous** is used to describe an individual with only one copy of a particular gene. A male mammal is said to be hemizygous for an X-linked gene.

Many recessive X-linked alleles cause diseases in humans, and these diseases occur more frequently in males than in females. An example is the X-linked recessive disorder called classical hemophilia (hemophilia A). In individuals with hemophilia, blood does not clot properly, and a minor cut may bleed for a long time. Common accidental injuries that are minor in most people pose a threat of severe internal or external bleeding for hemophiliacs. Hemophilia A is caused by a recessive X-linked allele that encodes a defective form of a clotting protein. If a mother is a heterozygous carrier of hemophilia A, each of her children has a 50% chance of inheriting the recessive allele. The following Punnett square shows a cross between an unaffected father and a heterozygous mother. $X^H$ designates an X chromosome carrying the dominant functional allele, and $X^{h-A}$ is the X chromosome that carries the recessive nonfunctional allele for hemophilia A.

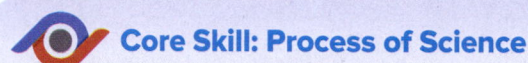

Male gametes

♂ $X^H$   Y

Female gametes

♀ $X^H$

| $X^H X^H$ Unaffected female | $X^H Y$ Unaffected male |
| $X^H X^{h-A}$ Carrier female | $X^{h-A} Y$ Male with hemophilia |

Although each child has a 50% chance of inheriting the hemophilia allele from the mother, only 1/2 of the sons will exhibit the disorder. Because a son inherits only one X chromosome, a son who inherits the recessive (disease-causing) allele from his mother will have hemophilia. However, a daughter inherits an X chromosome from both her mother and her father. In this example, a daughter who inherits the recessive allele from her mother also inherits a dominant allele from her father. This daughter will not have hemophilia, but if she passes the recessive allele to a son, he will have hemophilia.

---

### Core Skill: Process of Science

## Feature Investigation | Morgan's Experiments Showed a Correlation Between a Genetic Trait and the Inheritance of a Sex Chromosome in *Drosophila*

The distinctive inheritance pattern of X-linked alleles provides a way of demonstrating that a specific gene is on an X chromosome. An X-linked gene was the first gene to be located on a specific chromosome. In 1910, American geneticist Thomas Hunt Morgan began work on a project in which he reared large populations of fruit flies, *Drosophila melanogaster*, in the dark to determine if their eyes would atrophy from disuse and disappear in future generations. Even after many consecutive generations, the flies showed no noticeable changes. After 2 years of looking at many flies, Morgan happened to discover a male fly with white eyes rather than red,

which is the common (wild-type) color. The white-eye trait must have arisen from a new mutation that converted a red-eye allele into a white-eye allele.

To study the inheritance of the white-eye trait, Morgan followed an approach similar to Mendel's in which he made crosses and quantitatively analyzed their outcome. In the experiment described in **Figure 17.15**, Morgan crossed his white-eyed male to a red-eyed female. All of the $F_1$ offspring had red eyes, indicating that red is dominant to white. The $F_1$ offspring were then mated to each other to obtain an $F_2$ generation. As seen in the data table, this cross

**Figure 17.15** **Morgan's crosses of red-eyed and white-eyed *Drosophila*.**

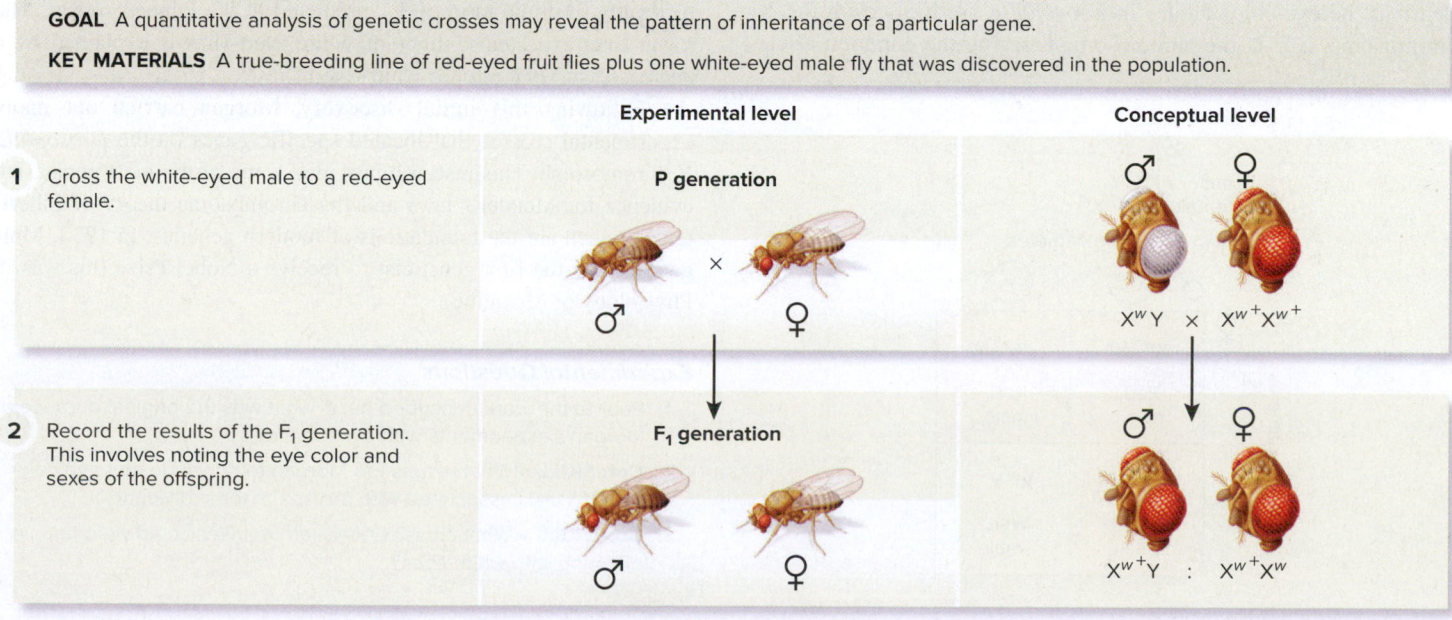

**GOAL** A quantitative analysis of genetic crosses may reveal the pattern of inheritance of a particular gene.

**KEY MATERIALS** A true-breeding line of red-eyed fruit flies plus one white-eyed male fly that was discovered in the population.

| | Experimental level | Conceptual level |
| **1** Cross the white-eyed male to a red-eyed female. | **P generation** ♂ × ♀ | ♂ ♀ $X^w Y$ × $X^{w+} X^{w+}$ |
| **2** Record the results of the $F_1$ generation. This involves noting the eye color and sexes of the offspring. | **$F_1$ generation** ♂ ♀ | ♂ ♀ $X^{w+} Y$ : $X^{w+} X^w$ |

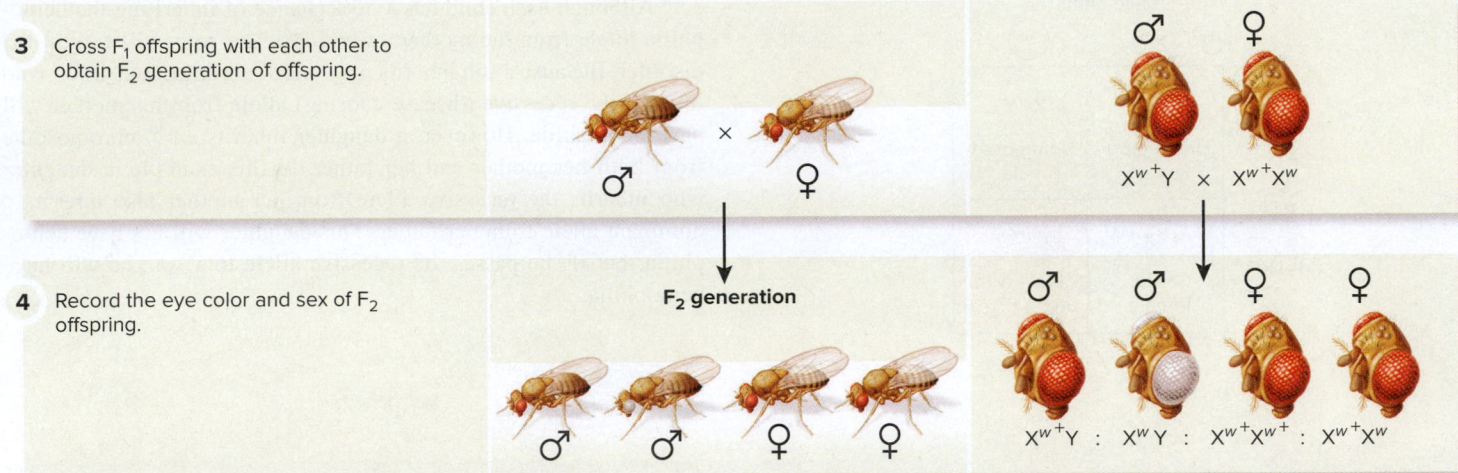

**3** Cross F₁ offspring with each other to obtain F₂ generation of offspring.

$X^{w^+}Y \times X^{w^+}X^w$

**4** Record the eye color and sex of F₂ offspring.

**F₂ generation**

$X^{w^+}Y : X^wY : X^{w^+}X^{w^+} : X^{w^+}X^w$

**5 THE DATA**

| Cross | Results | |
|---|---|---|
| Original white-eyed male to a red-eyed female | F₁ generation | All red-eyed flies |
| F₁ males to F₁ females | F₂ generation | 1,011 red-eyed males<br>782 white-eyed males<br>2,459 red-eyed females<br>0 white-eyed females |

**6 CONCLUSION** The data are consistent with an inheritance pattern in which an eye-color gene is located on the X chromosome.

**7 SOURCE** Morgan, T. H. 1910. Sex limited inheritance in *Drosophila*. *Science* 32: 120–122.

---

produced 1,011 red-eyed males, 782 white-eyed males, and 2,459 red-eyed females. Surprisingly, no white-eyed females were observed in the F₂ generation.

How did Morgan interpret these results? The results suggested a connection between the alleles for eye color and the sex of the offspring. As shown in the conceptual column of Figure 17.15 and in the following Punnett square, his data were consistent with the idea that the eye-color alleles in *Drosophila* are located on the X chromosome. $X^{w^+}$ is the chromosome carrying the common allele

for red eyes, and $X^w$ is the chromosome with the mutant allele for white eyes.

The Punnett square predicts that the F₂ generation will not have any white-eyed females. This prediction was confirmed by Morgan's experimental data. However, it should also be pointed out that the experimental ratio of red eyes to white eyes in the F₂ generation is (2,459 + 1,011) : 782, which equals 4.4 : 1. This ratio deviates significantly from the ratio of 3:1 predicted in the Punnett square. The lower than expected number of white-eyed flies is explained by a decreased survival rate for white-eyed flies.

Following this initial discovery, Morgan carried out many experimental crosses that located specific genes on the *Drosophila* X chromosome. This research provided some of the most persuasive evidence for Mendel's laws and the chromosome theory of inheritance, which are the foundations of modern genetics. In 1933, Morgan became the first geneticist to receive a Nobel Prize (his was in Physiology or Medicine).

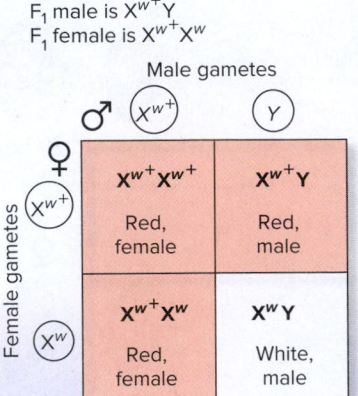

F₁ male is $X^{w^+}Y$
F₁ female is $X^{w^+}X^w$

Male gametes

♂ $X^{w^+}$    $Y$

Female gametes

♀ $X^{w^+}$
| $X^{w^+}X^{w^+}$<br>Red, female | $X^{w^+}Y$<br>Red, male |
$X^w$
| $X^{w^+}X^w$<br>Red, female | $X^wY$<br>White, male |

### Experimental Questions

1. Prior to the work described here, what was the original purpose of Morgan's experiments with *Drosophila*?

2. **CoreSKILL »** What results led Morgan to conclude that eye color in fruit flies is associated with the sex of the individual?

3. **CoreSKILL »** What crosses between fruit flies could yield female offspring with white eyes?

# Variations in Inheritance Patterns and Their Molecular Basis

**Learning Outcomes:**

1. Relate dominant and recessive traits to protein function.
2. Define pleiotropy, and explain why it occurs.
3. **CoreSKILL »** Predict the outcomes of crosses that exhibit incomplete dominance and codominance.
4. Discuss how the environment plays a critical role in determining the expression of traits.

The term **Mendelian inheritance** describes the inheritance patterns of genes that segregate and assort independently. In the first section of this chapter, we considered the inheritance pattern of traits affected by a single gene that is found in two variants, one of which is dominant over the other. This pattern is called **simple Mendelian inheritance**, because the phenotype ratios in the offspring clearly demonstrate Mendel's laws. We will begin this section by discussing the molecular basis of dominant and recessive traits and see how the molecular expression of a gene can have widespread effects on an organism's phenotype. In addition, we will examine the inheritance patterns of genes that segregate and assort independently but do not display a simple dominant/recessive relationship. The transmission of these genes from parents to offspring does not usually produce the ratios of phenotypes we would expect on the basis of Mendel's observations. This does not mean that Mendel was wrong. Rather, the inheritance patterns of many traits are different from the simple patterns he chose to study. In this section, we will explore these variations in Mendelian inheritance.

## Protein Function Explains the Phenomenon of Dominance

As described at the beginning of this chapter, Mendel studied seven characters that were found in two variants each (see Figure 17.2). The dominant variants are caused by the common alleles for these traits in pea plants. For any given gene, geneticists refer to a prevalent allele in a population as a **wild-type allele**. In most cases, a wild-type allele encodes a protein that is made in the proper amount and functions properly. By comparison, alleles that have been altered by mutation are called **mutant alleles**; these tend to be rare in natural populations. In the case of Mendel's seven characters in pea plants, the recessive alleles are due to rare mutations.

How do we explain why one allele is dominant and another allele is recessive? By studying genes and their gene products at the molecular level, researchers have discovered that a recessive allele is often defective in its ability to express a functional protein. In other words, mutations that produce recessive alleles are likely to decrease or eliminate the synthesis or functional activity of a protein. These are called loss-of-function alleles. To understand why many loss-of-function alleles are recessive, we need to take a quantitative look at protein function.

In a simple dominant/recessive relationship, the recessive allele does not affect the phenotype of the heterozygote. In this type of relationship, a single copy of the dominant (wild-type) allele is sufficient to mask the effects of the recessive allele. How do we explain the dominant phenotype of the heterozygote? **Figure 17.16** considers the example of flower color in a pea plant. The gene encodes an enzyme (protein P) that is needed to convert a colorless molecule into a purple pigment. The *P* allele is dominant because one *P* allele encodes enough of the functional protein—50% of the amount found in a *PP* homozygote—to provide a purple phenotype. Therefore, the *PP* homozygote and the *Pp* heterozygote both make enough of the purple pigment to yield purple flowers. The *pp* homozygote cannot make any of the functional protein required for pigment synthesis, so its flowers are white.

This explanation—that 50% of the functional protein is enough—is true for many dominant alleles. In such cases, the homozygote with two dominant alleles is making much more of the protein than necessary, so if the amount is reduced to 50%, as it is in the heterozygote, the individual still has plenty of this protein to accomplish whatever cellular function it performs. In other cases, however, an allele may be dominant because the heterozygote actually produces more than 50% of the functional protein. This increased production is due to the phenomenon of gene regulation. The dominant allele is up-regulated in the heterozygote to compensate for the lack of function of the recessive allele.

| Genotype | *PP* | *Pp* | *pp* |
|---|---|---|---|
| Amount of functional protein P produced | 100% | 50% | 0% |
| Phenotype | Purple | Purple | White |

Only 50% of the functional protein is needed to produce the purple phenotype

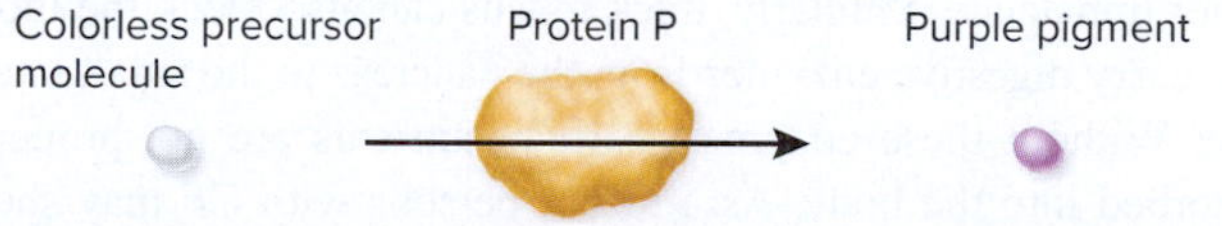

Colorless precursor molecule        Protein P        Purple pigment

**Figure 17.16  How genes give rise to traits during simple Mendelian inheritance.** In the heterozygote, the amount of protein encoded by a single dominant allele is sufficient to produce the dominant phenotype. In this example, the gene encodes an enzyme that is needed to produce a purple pigment. A plant with one or two copies of the dominant allele makes enough pigment to produce purple flowers. In a *pp* homozygote, the complete lack of the functional protein (enzyme) results in white flowers.

**Core Skill: Quantitative Reasoning** In a simple dominant/recessive relationship, even though the heterozygote may produce less of a functional protein compared to the homozygote that has two copies of the dominant allele, the amount made by the heterozygote is sufficient to yield the dominant phenotype.

---

### Core Concept: Systems

## The Expression of a Single Gene Often Has Multiple Effects on Phenotype

By studying mutations in humans and model organisms, researchers have discovered that genes usually exhibit **pleiotropy**, which means that a mutation in a single gene can have multiple effects on an individual's phenotype. Pleiotropy occurs for several reasons, including the following:

1. The expression of a single gene can affect cell function in more than one way. For example, a defect in a microtubule protein may affect cell division and cell movement.

2. A gene may be expressed in different cell types in a multicellular organism.

3. A gene may be expressed at different stages of development.

In this genetics unit, we tend to discuss genes as they affect a single trait. This educational approach allows us to appreciate how genes function and how they are transmitted from parents to offspring. However, in all or nearly all cases, the expression of a gene is pleiotropic with regard to the characteristics of an organism. The expression of any given gene influences the expression of many other genes in the genome, and vice versa. Pleiotropy is revealed when researchers study the effects of gene mutations.

As an example of a pleiotropic effect, let's consider cystic fibrosis (CF), which we discussed earlier as an example of a recessive human disorder (see Figure 17.12). In the late 1980s, the gene for CF was identified. The gene encodes a protein called the cystic fibrosis transmembrane regulator (CFTR), which regulates ion balance by allowing the transport of chloride ions ($Cl^-$) across cell membranes. The mutation that causes CF diminishes the function of this $Cl^-$ transporter, affecting several parts of the body in different ways. Because the movement of $Cl^-$ affects water transport across membranes, the most severe symptom of CF is the production of thick mucus in the lungs, which occurs because of a water imbalance. Similarly, thick mucus can also block the tubes that carry digestive enzymes from the pancreas to the small intestine. Without these enzymes, certain nutrients are not properly absorbed into the body. As a result, persons with CF may show poor weight gain. Another effect is seen in the sweat glands. The $Cl^-$ transporter has the function of recycling salt out of the glands and back into the skin before it can be lost to the outside world. Persons with CF have excessively salty sweat due to their inability to recycle salt back into their skin cells. A common test for CF is the measurement of salt on the skin. Taken together, we can see that a defect in CFTR has multiple effects throughout the body.

## Incomplete Dominance Results in an Intermediate Phenotype

We will now turn our attention to examples in which the alleles for a given gene do not show a simple dominant/recessive relationship. In some cases, a heterozygote that carries two different alleles exhibits a phenotype that is intermediate between those of the corresponding homozygous individuals. This phenomenon is known as **incomplete dominance**.

In 1905, Carl Correns discovered this pattern of inheritance for alleles affecting flower color in the four-o'clock plant (*Mirabilis jalapa*). **Figure 17.17** shows a cross between two four-o'clock plants: a red-flowered homozygote and a white-flowered homozygote. The allele for red flower color is designated $C^R$, and the white allele is $C^W$. These alleles are designated with superscripts rather than upper- and lowercase letters because neither allele is dominant. The offspring of this cross have pink flowers—they are $C^R C^W$ heterozygotes with an intermediate phenotype. If these $F_1$ offspring are allowed to self-fertilize, the $F_2$ generation has 1/4 red-flowered plants, 1/2 pink-flowered plants, and 1/4 white-flowered plants. This is a 1:2:1 phenotype ratio rather than the 3:1 ratio observed for simple Mendelian inheritance. What is the molecular explanation for this ratio? In this

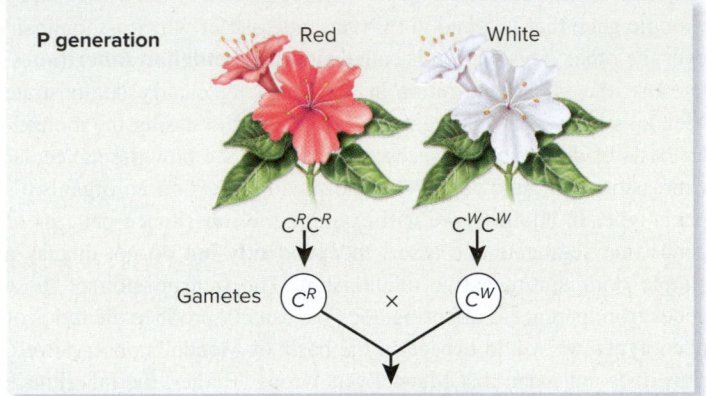

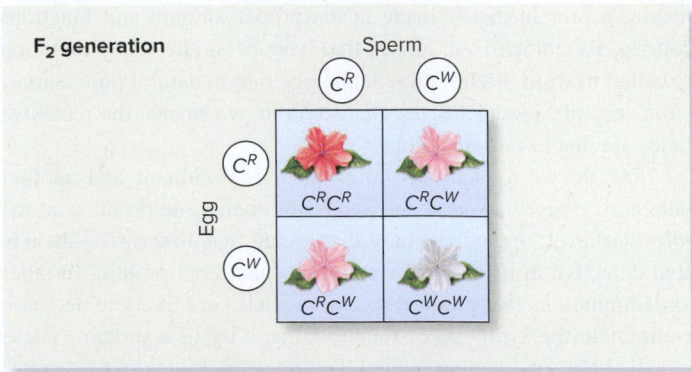

**Figure 17.17  Incomplete dominance in the four-o'clock plant.** When red-flowered and white-flowered homozygotes ($C^R C^R$ and $C^W C^W$) are crossed, the resulting heterozygote ($C^R C^W$) has an intermediate phenotype of pink flowers.

case, the red allele encodes a functional protein needed to produce a red pigment, whereas the white allele is a mutant allele that is nonfunctional. In the $C^R C^W$ heterozygote, 50% of the protein encoded by the $C^R$ allele is not sufficient to produce the red-flower phenotype, but it does provide enough pigment to give pink flowers.

## ABO Blood Types Provide an Example of Multiple Alleles and Codominance

Although diploid individuals have only two copies of most genes, the majority of genes have three or more variants in natural populations. We describe such genes as occurring in **multiple alleles**. Particular phenotypes depend on which two alleles each individual inherits. ABO blood types in humans are an example of phenotypes produced by multiple alleles.

As shown in **Table 17.2**, human red blood cells have structures on their plasma membrane known as surface antigens, which are constructed from several sugar molecules that are connected to form a carbohydrate tree. The carbohydrate tree is attached to a lipid or membrane protein to form a glycolipid or a glycoprotein, respectively. As noted in Chapter 5, glycolipids and glycoproteins often play a role in cell surface recognition.

Antigens are substances (in this case, carbohydrates) that may be recognized as foreign material when introduced into the body of an animal. Let's consider two types of surface antigens, known as A and B, which may be found on red blood cells. The synthesis of these antigens is determined by enzymes that are encoded by a gene that exists in three alleles, designated $I^A$, $I^B$, and $i$, respectively. The $i$ allele is recessive to both $I^A$ and $I^B$. A person who is $ii$ homozygous does not produce surface antigen A or B and has blood type O. The red blood cells of an $I^A I^A$ homozygous or $I^A i$ heterozygous

individual have surface antigen A (blood type A). Similarly, a homozygous $I^B I^B$ or heterozygous $I^B i$ individual produces surface antigen B (blood type B). A person who is $I^A I^B$ heterozygous makes both antigens, A and B, on every red blood cell (blood type AB). The phenomenon in which a single individual expresses two alleles is called **codominance**.

What is the molecular explanation for codominance? Biochemists have analyzed the carbohydrate tree produced in people of differing blood types. The differences are shown schematically in Table 17.2. In type O, the carbohydrate tree is smaller than in type A or type B because a sugar has not been attached to a specific site on the tree. People with blood type O have a loss-of-function mutation in the gene that encodes the enzyme that attaches a sugar at this site. This enzyme, called a glycosyl transferase, is inactive in type O individuals. In contrast, the type A and type B antigens have sugars attached to this site, but each of them has a different sugar. This difference occurs because the enzymes encoded by the $I^A$ allele and the $I^B$ allele have slightly different active sites. As a result, the enzyme encoded by the $I^A$ allele attaches a sugar called *N*-acetylgalactosamine to the carbohydrate tree, whereas the enzyme encoded by the $I^B$ allele attaches galactose. *N*-Acetylgalactosamine is represented by an orange hexagon in Table 17.2, and galactose by a green triangle.

## The Environment Plays a Vital Role in the Making of a Phenotype

In this chapter, we have been mainly concerned with the effects of genes on phenotypes. In addition, phenotypes are shaped by an organism's environment. An organism cannot exist without its genes or without an environment in which to live. Both are indispensable for life. An organism's genotype provides the plan to create a phenotype,

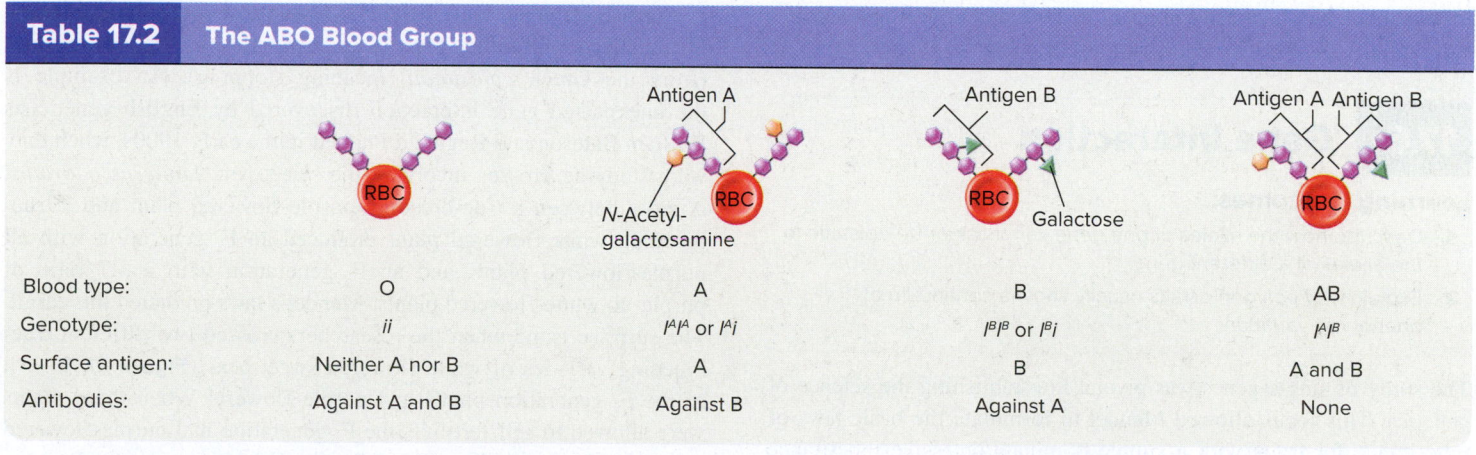

| Table 17.2 | The ABO Blood Group | | | |
|---|---|---|---|---|
| **Blood type:** | O | A | B | AB |
| **Genotype:** | $ii$ | $I^A I^A$ or $I^A i$ | $I^B I^B$ or $I^B i$ | $I^A I^B$ |
| **Surface antigen:** | Neither A nor B | A | B | A and B |
| **Antibodies:** | Against A and B | Against B | Against A | None |

**Core Skill: Modeling** The goal of this modeling challenge is to create a pair of models that depict the key difference between the forms of glycosyl transferase encoded by the $I^A$ and $I^B$ alleles.

**Modeling Challenge:** Glycosyl transferase is an enzyme that recognizes the structure of a carbohydrate tree on the surface of a cell and attaches an additional sugar to that tree. Based on the schematic drawings of carbohydrate trees in Table 17.2, draw a pair of models that highlight the key difference between the glycosyl transferase encoded by the $I^A$ allele and that encoded by the $I^B$ allele. Before drawing your models, refer back to Figure 6.6, which shows a model for a different enzyme, called hexokinase, and use that style for your models for glycosyl transferase.

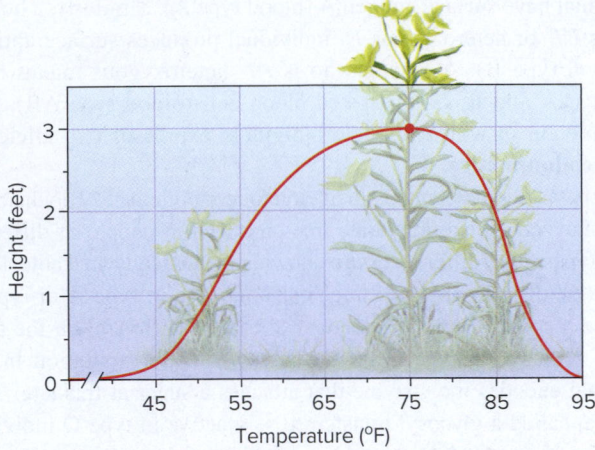

**Figure 17.18 The norm of reaction.** The norm of reaction is the range of phenotypes that a population of organisms with a particular genotype exhibit under different environmental conditions. In this example, genetically identical plants were grown at different temperatures in a greenhouse and then measured for height.

**Concept Check:** *Could you study the norm of reaction in a wild population of squirrels?*

and the environment provides nutrients and energy so that plan can be executed.

The **norm of reaction** is the phenotype range that individuals with a particular genotype exhibit under differing environmental conditions. To evaluate the norm of reaction, researchers study members of true-breeding strains that have the same genotypes and subject them to different environmental conditions. For example, **Figure 17.18** shows the norm of reaction for genetically identical plants raised at different temperatures. As shown in the figure, these plants attain a maximal height when raised at 75°F. At 50°F and 85°F, the plants are substantially shorter. Growth cannot occur below 40°F or above 95°F.

## 17.6 Gene Interaction

### Learning Outcomes:

1. Describe how the alleles of one gene can mask or be epistatic to the alleles of a different gene.
2. Explain why polygenic traits usually show a continuum of phenotype variation.

The study of single genes was pivotal in establishing the science of genetics. This focus allowed Mendel to formulate the basic laws of inheritance for traits with a simple dominant/recessive inheritance pattern. Likewise, this approach helped later researchers understand inheritance patterns involving incomplete dominance and codominance, as well as traits that are influenced by an individual's sex. However, all or nearly all traits are influenced by many genes. For example, in both plants and animals, height is affected by genes that encode proteins involved in the production of growth hormones, cell division, the uptake of nutrients, metabolism, and many other functions. Variation in any of the genes involved in these processes is likely to influence an individual's height.

If height is controlled by many genes, how was Mendel able to study the effects of a single gene that produced tall or dwarf pea plants? The answer lies in the genotypes of his strains. Although many genes affect the height of pea plants, Mendel chose true-breeding strains that differed with regard to only one of those genes. As a hypothetical example, let's suppose that pea plants have 10 genes affecting height, which we will call *K, L, M, N, O, P, Q, R, S,* and *T*. The genotypes of two hypothetical strains of pea plants may be:

Tall strain:   *KK LL MM NN OO PP QQ RR SS TT*
Dwarf strain:   *KK LL MM NN OO PP QQ RR SS tt*

In this example, the tall and dwarf strains differ at only a single gene. One strain is *TT* and the other is *tt*, and this accounts for the difference in their height. If we make crosses of tall and dwarf plants, the genotypes of the $F_2$ offspring will differ with regard to only one gene; the other nine genes will be identical in all of them. This approach allows a researcher to study the effects of a single gene even though many genes may affect a single character.

In this section, we will examine situations in which a single character is controlled by two or more different genes, each of which has two or more alleles. This phenomenon is called **gene interaction**. As you will see, allelic variation of two or more genes may affect the outcome of traits in different ways. First, we will look at a gene interaction in which an allele of one gene prevents the phenotypic expression of an allele of a different gene. Then we will discuss an interaction in which multiple genes have additive effects on a single character. These additive effects, together with environmental influences, account for the continuous phenotypic variation that we see in most traits.

### In an Epistatic Gene Interaction, the Alleles of One Gene Mask the Phenotypic Effects of a Different Gene

In some gene interactions, the alleles of one gene mask the expression of the alleles of another gene. This phenomenon is called **epistasis** (from the Greek *ephistanai*, meaning stopping). An example is the unexpected gene interaction discovered by English geneticists William Bateson and Reginald Punnett in the early 1900s, when they were studying crosses involving the sweet pea, *Lathyrus odoratus*. A cross between a true-breeding purple-flowered plant and a true-breeding white-flowered plant produced an $F_1$ generation with all purple-flowered plants and an $F_2$ generation with a 3:1 ratio of purple- to white-flowered plants. Mendel's laws predicted this result. The surprise came when the researchers crossed two different true-breeding varieties of white-flowered sweet peas (**Figure 17.19**). All of the $F_1$ generation plants had purple flowers! When these plants were allowed to self-fertilize, the $F_2$ generation had purple-flowered and white-flowered plants in a 9:7 ratio. From these results, Bateson and Punnett deduced that two different genes were involved. To have purple flowers, a plant must have one or two dominant alleles for each of these genes. The relationships among the alleles are as follows:

*C* (one allele for purple) is dominant to *c* (white)
*P* (an allele of a different gene for purple) is dominant to *p* (white)
*cc* masks *P*, or *pp* masks *C*, resulting in white flowers in either case

A plant that is homozygous for either *c* or *p* has white flowers even if it has a dominant purple-producing allele for the other gene.

How do we explain these results at the molecular and cellular level? Epistatic interactions often arise because two or more different proteins are involved in a single cellular function. For example, two or more proteins may be part of a metabolic pathway leading to the formation of a single product. This is the case for the formation of a purple pigment in the sweet pea strains we have been discussing:

Colorless precursor $\xrightarrow{\text{Enzyme C}}$ Colorless intermediate $\xrightarrow{\text{Enzyme P}}$ **Purple pigment**

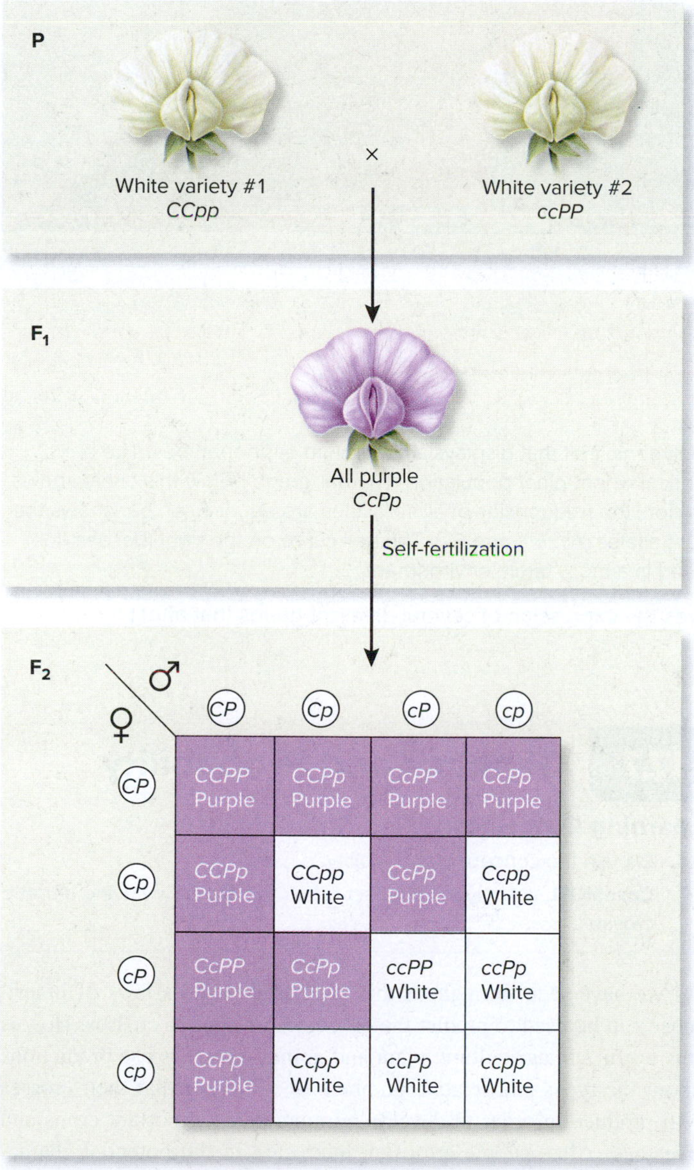

**Figure 17.19  Epistasis in the sweet pea.**  The color of the sweet pea flower is controlled by two genes, each with a dominant and a recessive allele. Each of the dominant alleles (*C* and *P*) encodes an enzyme required for the synthesis of purple pigment. A plant that is homozygous recessive for either gene (*cc* or *pp*) cannot synthesize the pigment and has white flowers.

*Concept Check:*  *In a Ccpp individual, which functional enzyme is missing? Is it the enzyme encoded by the C or P gene?*

In this example, a colorless precursor molecule must be acted on by two different enzymes to produce the purple pigment. Gene *C* encodes a functional protein called enzyme C, which converts the colorless precursor into a colorless intermediate. The recessive *c* allele results in a lack of production of enzyme C in the *cc* homozygote. Gene *P* encodes the functional enzyme P, which converts the colorless intermediate into the purple pigment. Like the *c* allele, the *p* allele results in an inability to produce a functional enzyme. A plant homozygous for either of the recessive alleles does not make any functional enzyme C or enzyme P. When either of these enzymes is missing, the plant cannot make the purple pigment and has white flowers. Note that the results observed in Figure 17.19 do not conflict with Mendel's laws of segregation or independent assortment. Mendel investigated the effects of only a single gene on a given character. The 9:7 ratio is due to a gene interaction in which two genes affect a single character.

## Polygenic Inheritance and Environmental Influences Produce Continuous Phenotypic Variation

Until now, we have discussed the inheritance of characters with clearly defined phenotypic variants, such as red or white eyes in fruit flies and round or wrinkled seeds in garden peas. These are known as **discrete traits**, because the phenotypes do not overlap. For most traits, however, the phenotypes cannot be sorted into discrete categories. Traits that show continuous variation over a range of phenotypes are called **quantitative traits**. In humans, quantitative traits include height, weight, skin color, metabolic rate, and heart size. In the case of domestic animals and plant crops, many of the traits that people consider desirable are quantitative in nature, such as the number of eggs a chicken lays, the amount of milk a cow produces, and the number of apples on an apple tree. Consequently, much of our modern understanding of quantitative traits comes from agricultural research.

Quantitative traits are usually **polygenic**, which means that multiple genes contribute to the outcome of the trait. For many polygenic traits, genes contribute to the phenotype in an additive way. Also, environmental factors often have a major effect on quantitative traits. For example, an animal's diet affects its weight, and the amounts of rain and sunlight that fall on an apple tree affect how many apples it produces.

Because quantitative traits are polygenic and greatly influenced by environmental conditions, the phenotypes among different individuals may vary substantially in any given population. As an example, let's consider grain pigmentation in wheat. In certain strains of wheat, this character is influenced by three genes that interact in an additive way. Let's call them genes *A*, *B*, and *C*. Each gene may exist as a red allele, designated $A^R$, $B^R$, or $C^R$, or a white allele, designated $A^W$, $B^W$, or $C^W$, respectively. The red alleles encode enzymes that cause the synthesis of grain pigment, whereas the white alleles are the result of loss-of-function mutations. The color of wheat grains varies along a continuum, ranging from white to dark red.

**Figure 17.20** considers a hypothetical case in which wheat plants that were heterozygous for all three genes produced a large population of offspring. The bar graph shows the genotypes of the offspring, grouped according to the total number of red and white alleles. As shown by the shading in the figure, red pigmentation increases as the number of red alleles increases. Offspring that have all white alleles

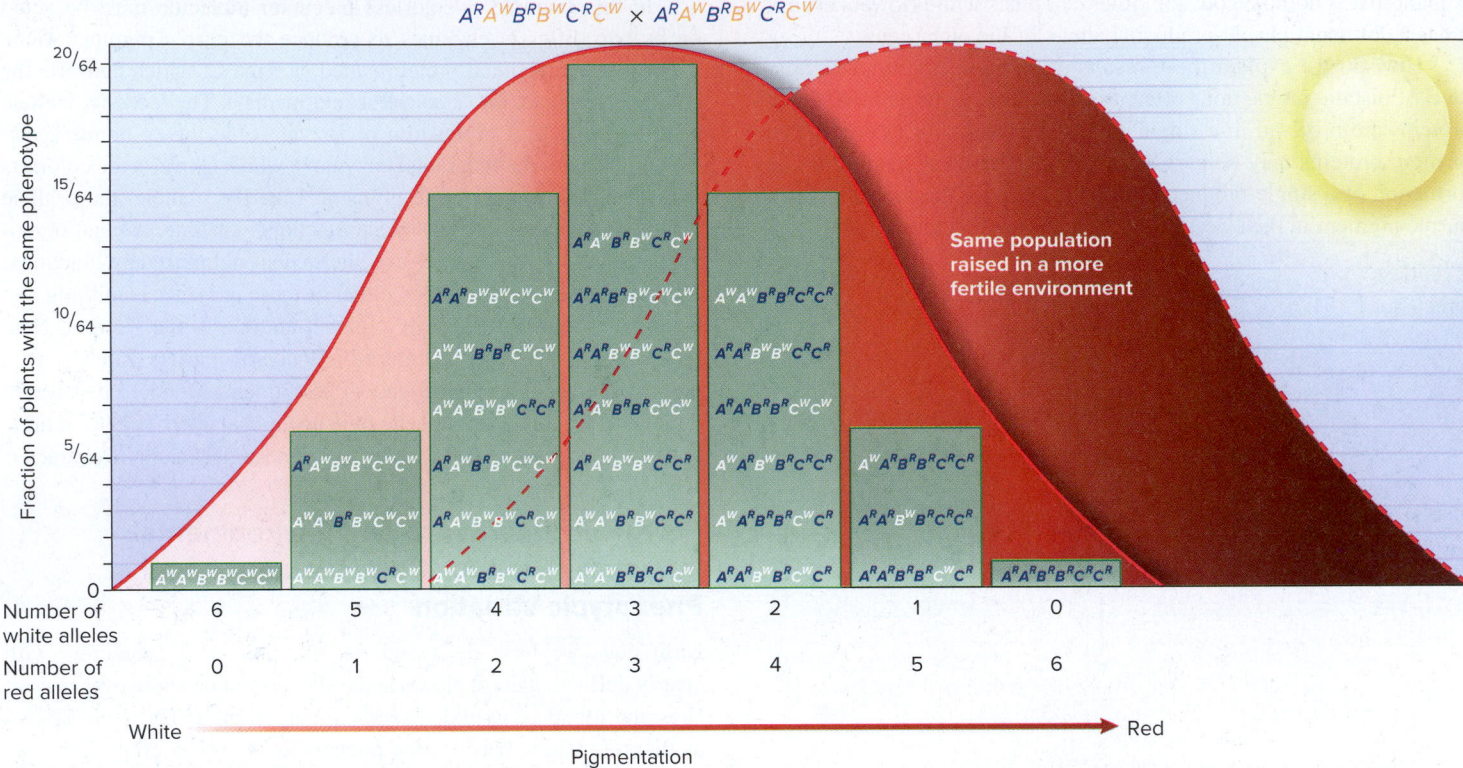

$A^R A^W B^R B^W C^R C^W \times A^R A^W B^R B^W C^R C^W$

**Figure 17.20  Continuous variation in a polygenic trait.** Grain color is a polygenic trait that displays a continuum of phenotypes. The bell curve on the left (solid line) shows the range of grain pigmentation in a hypothetical wheat-plant population. The bar graph below this curve shows the additive effects of three genes that affect pigment production in this population; the frequencies of all six alleles are equal. Each bar shows the fraction of plants with a particular number of dark alleles ($A^R$, $B^R$, and $C^R$) and light alleles ($A^W$, $B^W$, and $C^W$). The bell curve on the right (dashed line) represents the expected range of phenotypes if the same population were raised in a more fertile environment.

 **Core Concept: Information**  The genetic blueprint usually involves the expression of several different genes that affect the same trait. In many cases, this results in continuous variation.

or have all red alleles—that is, those that are homozygous for all three genes—are fewer in number than those with some combination of light and dark alleles. As indicated by the bell-shaped curve above the bar graph, the phenotypes of the offspring fall along a continuum. This continuous phenotypic variation, which is typical of quantitative traits, is produced by genotypic differences together with environmental effects.

A second bell-shaped curve (dashed) depicts the expected phenotypic range if the same population of offspring had been raised in a more fertile environment (i.e., richer soil and more sunlight), which increases pigment production. This curve illustrates how the environment can also have a significant influence on the range of phenotypes.

In our discussion of genetics, we tend to focus on discrete traits because this makes it easier to relate a specific genotype with a phenotype. Identifying such a clear relationship is usually not possible for continuous traits. For example, as depicted in the middle bar of Figure 17.20, seven different genotypes can produce plants with a medium amount of pigmentation. It is important to emphasize that the majority of traits in all organisms are continuous, not discrete. Most traits are influenced by multiple genes, and the environment has an important influence on the phenotypic outcome.

## 17.7  Genetics and Probability

**Learning Outcomes:**

1. Explain the concept of probability.
2. **CoreSKILL »** Apply the product rule to problems involving genetic crosses.

As we have seen throughout this chapter, Mendel's laws of inheritance can be used to predict the outcomes of genetic crosses. How is this useful? In agriculture, plant and animal breeders use predictions about the types and relative numbers of offspring that their crosses will produce in order to develop commercially important crops and livestock. Also, people are often interested in the potential characteristics of their future children. This has particular importance to individuals who may carry alleles that cause inherited diseases. Of course, no one can see into the future and definitively predict what will happen. Nevertheless, genetic counselors can help couples predict the likelihood of having an affected child. This probability is one factor that may influence a couple's decision about whether to have children.

Earlier in this chapter, we considered how a Punnett square can be used to predict the outcome of simple genetic crosses. In addition to Punnett squares, we can apply the tools of mathematics and probability to solve more complex genetic problems. In this section, we will examine a couple of ways to calculate the outcomes of genetic crosses using these tools.

## Genetic Predictions Are Based on the Rules of Probability

The chance that an event will have a particular outcome is called the **probability** of that outcome. The probability of a given outcome depends on the number of possible outcomes. For example, if you draw a card at random from a 52-card deck, the probability that you will get the jack of diamonds is 1 in 52, because 52 outcomes are possible. In contrast, only two outcomes are possible when you flip a coin, so the probability is one in two (1/2, or 0.5, or 50%) that the heads side will be showing when the coin lands. The general formula for the probability ($P$) that an event will have a specific outcome is

$$P = \frac{\text{Number of times an event occurs}}{\text{Total number of possible outcomes}}$$

For a single coin toss, the chance of getting heads is

$$P_{\text{heads}} = \frac{1 \text{ heads}}{(1 \text{ heads} + 1 \text{ tails})} = \frac{1}{2}$$

Earlier in this chapter, we used Punnett squares to predict the fractions of offspring with a given genotype or phenotype. In a cross between two pea plants that were heterozygous for the height gene ($Tt$), our Punnett square predicted that one-fourth of the offspring would be dwarf. We can make the same prediction by using a probability calculation.

$$P_{\text{dwarf}} = \frac{1 \text{ } tt}{(1 \text{ } TT + 2 \text{ } Tt + 1 \text{ } tt)} = \frac{1}{4} = 0.25, \text{ or } 25\%$$

A probability calculation allows us to predict the likelihood that a future event will have a specific outcome. However, the accuracy of this prediction depends to a great extent on the number of events we observe—in other words, on the size of our sample. For example, if we toss a coin six times, the calculation we just presented for $P_{\text{heads}}$ suggests we should get heads three times and tails three times. However, each coin toss is an independent event, meaning that every time we toss the coin there is an equal chance that it will come up heads or tails, regardless of the outcome of the previous toss. With only six tosses, we would not be too surprised if we got four heads and two tails instead of the expected three heads and three tails. The deviation between the observed and expected outcomes due to random chance is called the **random sampling error**. With a small sample, the random sampling error may cause the observed data to be quite different from the expected outcome. By comparison, if we flipped a coin 1,000 times, the percentage of heads would be fairly close to the predicted 50%. With a larger sample, we expect the sampling error to be smaller.

## The Product Rule Is Used to Predict the Outcome of Independent Events

Punnett squares allow us to predict the likelihood that a genetic cross will produce an offspring with a particular genotype or phenotype. To predict the likelihood of producing multiple offspring with particular genotypes or phenotypes, we can use the **product rule**, which states:

*The probability that two or more independent events will occur is equal to the product of their individual probabilities.*

As we have already discussed, events are independent if the outcome of one event does not affect the outcome of another. In our coin-toss example, each toss is an independent event—if one toss comes up heads, another toss still has an equal chance of coming up either heads or tails. If we toss a coin twice, what is the probability that we will get heads both times? The product rule says that it is equal to the probability of getting heads on the first toss (1/2) times the probability of getting heads on the second toss (1/2), or one in four ($1/2 \times 1/2 = 1/4$).

To see how the product rule can be applied to a genetics problem, let's consider a rare recessive human trait known as congenital analgesia. (Congenital refers to a condition present at birth; analgesia means insensitivity to pain.) People with this trait can distinguish between sensations such as sharp and dull, or hot and cold, but they do not perceive extremes of sensation as painful. The first known case of congenital analgesia, described in 1932, was a man who made his living entertaining the public as a "human pincushion." For a couple, each heterozygous for the recessive allele causing congenital analgesia, we can ask, "What is the probability that their first three offspring will have the disorder?" To answer this question, we must first determine the probability of a single offspring having the disorder. By using a Punnett square, we would find that the probability of an individual offspring being homozygous recessive is 1/4. Thus, each of this couple's children has a one in four chance of having congenital analgesia.

We can now use the product rule to calculate the probability of this couple having three affected offspring in a row. The phenotypes of the first, second, and third offspring are independent events; that is, the phenotype of the first offspring does not affect the phenotype of the second or third offspring. The product rule tells us that the probability of all three children having the disorder is

$$\frac{1}{4} \times \frac{1}{4} \times \frac{1}{4} = \frac{1}{64} = 0.016, \text{ or } 1.6\%$$

The probability of the first three offspring having the disorder is 0.016, or 1.6%. In other words, we can say that this couple's chance of having three children in a row with congenital analgesia is very small—only 1.6 out of 100.

The product rule can also be used to predict the outcome of a cross involving two or more genes. Let's suppose a pea plant with the genotype $TtYy$ was crossed with a plant with the genotype $Ttyy$. We could ask, "What is the probability that an offspring will have the genotype $ttYy$?" If the two genes independently assort, the probability of inheriting alleles for one gene is independent of the probability for other gene. Therefore, we can separately calculate the probability of the desired outcome for each gene. By constructing two small Punnett squares, we can determine the probability of genotypes for each gene individually, as shown in the following Punnett squares.

Cross: *TtYy × Ttyy*

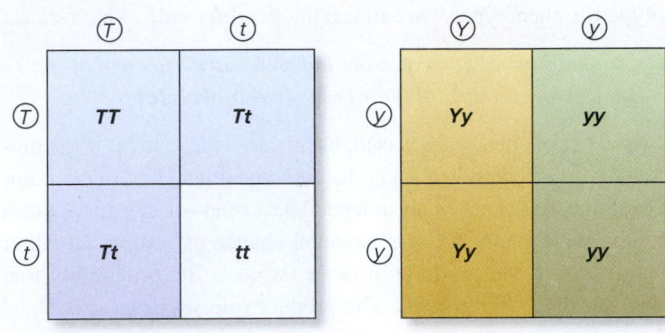

The probability that an offspring will be *tt* is 1/4, or 0.25.
The probability that an offspring will be *Yy* is 1/2, or 0.5.

We can now use the product rule to determine the probability that an offspring will be *ttYy*:

$$P = (0.25)(0.5) = 0.125, \text{ or } 12.5\%$$

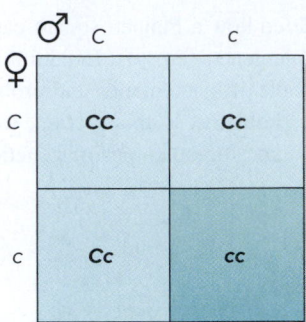

*C* = dominant allele
*c* = recessive, disease-causing allele

To calculate the probability, you need to consider two things. First, you need to know the probability of having an unaffected offspring. This probability can be deduced from the Punnett square. Once this is known, you can use the product rule to calculate the likelihood of having two unaffected offspring in a row, because these are independent outcomes.

**ANSWER** *The genotype ratio for the offspring is 1 CC : 2 Cc : 1 cc. The ratio of the phenotypes is 3 unaffected with CF : 1 affected with CF.*

*The probability of these parents having a single unaffected offspring is*

$$P_{unaffected} = \frac{3}{(3 + 1)} = \frac{3}{4}$$

*To obtain the probability that they will have two unaffected offspring in a row, you need to apply the product rule.*

$$\frac{3}{4} \times \frac{3}{4} = \frac{9}{16} = 0.56, \text{ or } 56\%$$

*The chance that their next two children will not have CF is 56%.*

## Core Skill: Quantitative Reasoning

**BIO TIPS** · **THE QUESTION** *As described earlier in this chapter, the human disease known as cystic fibrosis (CF) is inherited via a recessive allele. Two individuals, who do not have disease symptoms, have a first child who has CF. What is the probability that their next two children will not have the disease?*

**T** **OPIC** *What topic in biology does this question address?*
The topic is Mendelian inheritance. More specifically, the question is about a single-factor cross involving cystic fibrosis.

**I** **NFORMATION** *What information do you know based on the question and your understanding of the topic?*
From the question, you know that both parents are unaffected, but they produced an affected offspring, who must be homozygous for the recessive allele. Therefore, both parents must be heterozygotes. If *C* is the dominant allele and *c* is the recessive, disease-causing allele, the genotype of each parent must be *Cc*. From your understanding of the topic, you may remember that alleles segregate during the process that yields gametes and parents pass one allele to their offspring, when gametes combine at fertilization.

**P** **ROBLEM-SOLVING** **S** **TRATEGY** *Predict the outcome. Make a calculation.* One strategy to predict the outcome is to use a Punnett square, as shown next.

## Summary of Key Concepts

- Mendelian inheritance patterns obey Mendel's laws (Table 17.1).

### 17.1 Mendel's Laws of Inheritance

- Mendel studied seven characters of garden peas that existed in two variants each (Figures 17.1, 17.2).
- Mendel allowed his pea plants to self-fertilize, or he carried out cross-fertilization, also known as hybridization (Figures 17.3, 17.4).
- Mendel determined that certain traits exist in two forms, dominant and recessive. An individual carries two genes for a given character, and genes have variant forms, which are called alleles.
- By following the inheritance pattern of a single character (using a single-factor cross) for two generations, Mendel proposed the law of segregation, which states that two alleles segregate during the process that gives rise to gametes, so every gamete receives only one allele (Figures 17.5, 17.6).
- The genotype is the genetic makeup of an organism. Phenotype is a description of the traits that an organism displays.
- A Punnett square is constructed to predict the outcome of crosses.

- A testcross is conducted to determine if an individual displaying a dominant trait is a homozygote or a heterozygote (Figure 17.7).

- By conducting a two-factor cross, Mendel determined the law of independent assortment, which states that the alleles of different genes assort independently of each other during the process that gives rise to gametes. In a two-factor cross, this yields a 9:3:3:1 ratio in the $F_2$ generation (Figure 17.8).

## 17.2  The Chromosome Theory of Inheritance

- The chromosome theory of inheritance explains how the behavior of chromosomes during meiosis accounts for Mendel's laws of inheritance. Each gene is located at a particular site, or locus, on a chromosome (Figures 17.9, 17.10, 17.11).

## 17.3  Pedigree Analysis of Human Traits

- The inheritance patterns in humans are determined from pedigree analysis (Figures 17.12, 17.13).

## 17.4  Sex Chromosomes and X-Linked Inheritance Patterns

- Most species of animals and some species of plants have separate male and female sexes. In many species, sex is determined by differences in sex chromosomes (Figure 17.14).

- X-linked genes are found on the X chromosome but not the Y chromosome. Recessive X-linked alleles in humans can cause disorders such as hemophilia, which are more likely to occur in males.

- Morgan's experiments showed that an eye-color gene in *Drosophila* is located on the X chromosome (Figure 17.15).

## 17.5  Variations in Inheritance Patterns and Their Molecular Basis

- Several inheritance patterns have been discovered that obey Mendel's laws but yield differing ratios of offspring compared with Mendel's crosses.

- Recessive inheritance is often due to a loss-of-function mutation. In many simple dominant/recessive relationships, the heterozygote has a dominant phenotype because 50% of the functional protein is sufficient to produce that phenotype (Figure 17.16).

- The effects of a mutant gene are often pleiotropic, meaning the gene affects several different aspects of bodily structure and function.

- Incomplete dominance occurs when a heterozygote has a phenotype that is intermediate between either homozygote. This occurs because 50% of the functional protein is not enough to produce the same phenotype as a homozygote (Figure 17.17).

- ABO blood types are produced by the expression of a gene that exists in multiple alleles in humans. The $I^A$ and $I^B$ alleles show codominance, a phenomenon in which both alleles are expressed in a heterozygous individual (Table 17.2).

- Genes and the environment interact to determine an individual's phenotype. The norm of reaction is the phenotype range that individuals with the same genotype exhibit under different environmental conditions (Figure 17.18).

## 17.6  Gene Interaction

- Epistasis is a gene interaction that occurs when the alleles of one gene mask the effects of the alleles of a different gene (Figure 17.19).

- Quantitative traits such as height and weight are polygenic, which means that they are determined by multiple genes. Often, the alleles of such genes contribute in an additive way to the phenotype and are greatly affected by the environment. This interaction produces continuous variation in the trait, resulting in a phenotype range that can be graphed as a bell-shaped curve (Figure 17.20).

## 17.7  Genetics and Probability

- Probability is the likelihood that an event will occur in the future. Random sampling error is the deviation between observed and expected values.

- The product rule states that the probability of two or more independent events occurring is equal to the product of their individual probabilities.

## Assess & Discuss

### Test Yourself

1. Based on Mendel's experiments, what is the expected phenotype ratio in the $F_2$ generation from a single-factor cross?
   a. 1:2:1
   b. 2:1
   c. 3:1
   d. 9:3:3:1
   e. 4:1

2. During which phase of nuclear division does the phenomenon described in Mendel's law of segregation occur?
   a. mitosis
   b. meiosis I
   c. meiosis II
   d. all of the above
   e. b and c only

3. An individual that has two different alleles of a particular gene is said to be
   a. dihybrid.              d. heterozygous.
   b. recessive.            e. hemizygous.
   c. homozygous.

4. Which of Mendel's laws cannot be observed in a single-factor cross?
   a. segregation
   b. dominance/recessiveness
   c. independent assortment
   d. codominance
   e. All of the above can be observed in a single-factor cross.

5. During a_____ , an individual with the dominant phenotype and an unknown genotype is crossed with a_____ individual to determine the unknown genotype.
   a. single-factor cross, homozygous recessive
   b. two-factor cross, heterozygous
   c. testcross, homozygous dominant
   d. single-factor cross, homozygous dominant
   e. testcross, homozygous recessive

6. A woman is heterozygous for an X-linked disorder, hemophilia A. If she has a child with a man without hemophilia A, what is the probability that the child will be a male with hemophilia A? (Note: The child could be a male or female.)
   a. 100%          d. 25%
   b. 75%           e. 0%
   c. 50%

7. A gene that affects more than one trait is said to be
   a. dominant.          d. pleiotropic.
   b. wild-type.         e. heterozygous.
   c. dihybrid.

8. A hypothetical flowering plant species produces blue, light blue, and white flowers. To determine the inheritance pattern, the following crosses were conducted with the results indicated:
   blue × blue → all blue
   white × white → all white
   blue × white → all light blue
   What type of inheritance pattern does this represent?
   a. simple Mendelian     d. incomplete dominance
   b. X-linked             e. pleiotropy
   c. codominance

9. Genes located on a sex chromosome are said to be
   a. X-linked.          d. sex-linked.
   b. dominant.          e. autosomal.
   c. hemizygous.

10. A man and woman are both heterozygous for the recessive allele that causes cystic fibrosis. What is the probability that their first two offspring will have the disorder?
    a. 1              d. 1/32
    b. 1/4            e. 0
    c. 1/16

## Conceptual Questions

1. Describe one observation in a human pedigree that rules out a recessive pattern of inheritance. Describe an observation that rules out a dominant pattern.

2. A cross is made between individuals having the following genotypes: *AaBbCCDd* and *AabbCcdd*. What is the probability that an offspring will be *AAbbCCDd*? Hint: Don't waste your time making a really large Punnett square. Make four small Punnett squares instead and use the product rule.

3. **Core Concept: Systems** We can view life as a complex system in which organisms interact with their surrounding environments. Discuss how the environment plays a key role in determining the outcome of an individual's traits.

## Collaborative Questions

1. Discuss the principles of the chromosome theory of inheritance. Which principles do you think were deduced via light microscopy, and which were deduced from crosses? What modern techniques could be used to support the chromosome theory of inheritance?

2. When examining a human pedigree, what patterns do you look for to distinguish between X-linked recessive inheritance and autosomal recessive inheritance? How would you distinguish X-linked dominant inheritance from autosomal dominant inheritance from an analysis of a human pedigree?

# Epigenetics, Linkage, and Extranuclear Inheritance

# 18

Female honeybees that are fed royal jelly throughout their entire larval stage and into adulthood develop into queen bees. The larger queen bee is shown with a blue disk labeled 68. By comparison, those larvae that do not receive this diet become smaller worker bees. These differences in development are caused by epigenetic modifications. ©Andia/Alamy Stock Photo

**F**emale honeybees (*Apis mellifera*) are of two types: queen bees and worker bees (see the chapter opening photo). Queens are larger, live for years, and produce up to 2,000 eggs each day. By comparison, the smaller worker bees are sterile, typically live only for weeks, and engage in specialized types of work, such as constructing and cleaning comb cells, nurturing larvae, guarding the hive entrance, and foraging for pollen and nectar. The striking differences between queen and worker bees are caused by differences in their diets. Certain worker bees, called nurse bees, produce a nutritive substance called royal jelly from glands in their mouths. All female larvae are initially fed royal jelly, but those that are bathed in royal jelly throughout their larval development and feed on it into adulthood become queens. Alternatively, female larvae that are weaned at an early stage of development and switched to a diet of pollen and nectar become worker bees.

The differences between queen and worker bees are not due to differences in the alleles they carry. We cannot predict the phenotype of female bees based on their genotype. Likewise, we cannot carry out crosses and predict the ratio between queen and worker bees based on Mendel's laws. This observation does not mean Mendel was wrong, however. A large number of genes do obey his laws and follow the types of inheritance patterns that we examined in Chapter 17. However, many genes, including those that control female development in honeybees, do not.

How do we distinguish genes that follow a Mendelian pattern from genes that do not? Mendelian inheritance patterns follow three general rules:

1. Except in the case of rare mutations, genes are passed unaltered from cell to cell, and from generation to generation.
2. The genes obey Mendel's law of segregation.
3. For crosses involving two or more genes, the genes obey Mendel's law of independent assortment.

For Mendelian patterns of inheritance, we must also consider other factors, such as the dominant/recessive relationship of alleles and whether or not the gene is on the X chromosome. Once these additional factors are understood, we can predict the phenotypes of offspring from their genotypes and predict the outcome of crosses based on Mendel's laws. Most genes in eukaryotic species follow a Mendelian pattern of inheritance.

In this chapter, we will examine additional types of inheritance patterns that deviate from a Mendelian pattern because one of the three rules is broken. In Sections 18.1–18.4, we will explore epigenetics, which breaks rule number 1. In epigenetics, genes that are passed to offspring are altered, either in the gametes that produce the offspring or during embryonic development of the offspring. This alteration does not change the DNA sequence of a gene but, instead, alters the structure of chromatin and thereby affects the gene's expression. We will then examine inheritance patterns that arise because some genetic material is not located in the cell nucleus. Certain organelles, such as mitochondria and chloroplasts, contain their own genetic material. The inheritance patterns of genes within mitochondria and chloroplasts do not obey rule number 2; they do not follow the law of segregation. Finally, we will consider inheritance patterns that violate rule number 3, which is Mendel's law of independent assortment. As you will learn, some genes do not independently assort because they are located on the same chromosome.

**Learning Outcomes:**

1. Define epigenetics and epigenetic inheritance.
2. Outline the types of molecular changes that underlie epigenetic effects on gene expression.

The term epigenetics was first coined by British biologist Conrad Waddington in 1941. The prefix *epi-,* which means over, suggests that some types of changes in gene expression are at a level that goes beyond changes in DNA sequences. How do geneticists distinguish an epigenetic effect from other types of gene regulation, such as those described in Chapter 14 (refer back to Section 14.4)? An epigenetic effect begins with an initial event that causes a change in gene expression. For example, DNA methylation may inhibit transcription. However, for this to be an epigenetic effect, the change must be passed from cell to cell and must not involve a change in the sequence of DNA.

Thus, a key feature of an epigenetic effect is the long-term maintenance of a change in gene expression. As an example, let's consider muscle cells in humans. Some genes in the human genome should be expressed in muscle cells, such as the genes that encode the proteins called actin and myosin, which are required for muscle contraction (see Chapter 45). In embryonic muscle cells, these genes are subjected to epigenetic changes that promote their expression through adulthood. Alternatively, in other cell types, these same genes are inhibited by epigenetic changes such as DNA methylation. As the embryo grows and eventually becomes an adult, these epigenetic changes are passed from cell to cell so that adult muscle cells express actin and myosin genes at very high levels, whereas many other cell types do not.

Some epigenetic changes, such as those involving the expression or inhibition of actin and myosin genes, are relatively permanent during the life of an individual. Alternatively, other epigenetic changes may be reversible during the life of an individual or may be reversible from one generation to the next. For example, many species of flowering plants undergo **vernalization**, which is the process in which certain species of plants require an exposure to cold temperatures in order to flower. After vernalization, plants do not necessarily initiate flowering, but they acquire the ability to do so. Most commonly, plants are vernalized by exposure to cold winter temperatures and then flower the following spring or summer. Vernalization involves epigenetic changes to genes that play a role in flowering. These epigenetic changes are induced by cold winter temperatures, and they are maintained during the flowering season. However, the epigenetic changes are reversed after the flowering season is over.

Although researchers are still debating the proper definition, one way to define epigenetics is the following.

> **Epigenetics** *is the study of mechanisms that lead to changes in gene expression that can be passed from cell to cell and are reversible, but do not involve a change in the sequence of DNA. This type of change may also be called an* **epimutation**—*a heritable change in gene expression that does not alter the sequence of DNA.*

In multicellular species that reproduce via gametes (that is, sperm and egg cells), some types of epigenetic changes are passed from parent to offspring, a phenomenon called **epigenetic inheritance**. In Section 18.2, we will consider an example of epigenetic inheritance called genomic imprinting. However, not all epigenetic changes are inherited from parents. For example, a person may be exposed to an environmental agent in cigarette smoke that causes an epigenetic change in a lung cell that is subsequently transmitted from cell to cell and promotes lung cancer. Such a change would not be transmitted to offspring.

The molecular mechanisms of epigenetics are the subject of a large amount of recent research. The most common types of molecular changes that underlie epigenetic effects on gene expression are DNA methylation, chromatin remodeling, covalent histone modification, and the localization of histone variants (**Table 18.1**). These types of changes are also involved in transient (nonepigenetic) gene regulation that is not transmitted from cell to cell. The details of these mechanisms were

| Table 18.1 | Chromatin Modifications That May Have an Epigenetic Effect on Gene Expression |
|---|---|
| DNA methylation | Methyl groups may be attached to cytosine bases in DNA. When this occurs near promoters, transcription is usually repressed. |
| Chromatin remodeling | Nucleosomes may be moved to new locations or evicted. When such changes occur in the vicinity of promoters, the level of transcription may be altered. Also, larger-scale changes in chromatin structure may occur, such as those that happen during X-chromosome inactivation in female mammals, discussed later in this chapter. |
| Covalent histone modification | Specific amino acid side chains in the amino terminal tails of histones can be covalently modified. For example, they can be acetylated or phosphorylated. Such modifications may repress or activate transcription. |
| Localization of histone variants | Histone variants may become localized to specific locations, such as near the promoters of genes, and affect transcription. |

examined in Chapter 14 (refer back to Section 14.4). In some cases, epigenetic changes repress the transcription of a given gene; in other cases, they activate gene transcription. In Sections 18.2–18.4, we will consider examples in which these types of changes have an epigenetic effect on gene expression and thereby affect an individual's phenotype.

## 18.2 Epigenetics I: Genomic Imprinting

### Learning Outcomes:

1. **CoreSKILL »** Predict the outcome of crosses for imprinted genes.
2. Explain the molecular basis of genomic imprinting.

The term imprinting implies a type of marking process that has a memory. For example, newly hatched birds identify marks on their parents, which allows them to distinguish their parents from other individuals. The term **genomic imprinting** refers to an analogous situation in which a segment of DNA is marked, and that mark is retained and recognized throughout the life of the organism inheriting the marked DNA. However, depending on the gene, it may be marked by females during egg formation or by males during sperm formation, but not both. This marking process, which involves epigenetic modifications, affects whether or not the gene is expressed in the offspring. Depending on how a particular gene is marked by one of the parents, the offspring expresses either the maternal or the paternal allele, but not both. Imprinted genes do not follow a Mendelian pattern of inheritance because imprinting causes the offspring to distinguish between maternal and paternal alleles. From an epigenetic

perspective, the cells in the offspring are able to "remember" an event that occurred during gamete formation in their parents.

Genomic imprinting, which was discovered in the early 1980s, occurs in numerous species, including insects, plants, and mammals. The number of human genes that are imprinted is still a matter of controversy, but most estimates are in the range of a few hundred to over 1,000. In this section, we will consider an example of genomic imprinting that occurs in mammals.

### For Imprinted Genes, the Gene from Only One Parent Is Expressed

One of the first imprinted genes to be identified is a gene called *Igf2*, found in mice and other mammals. This gene encodes a growth hormone called insulin-like growth factor 2 (Igf2), which is needed for proper growth. If a functional copy of this gene is not expressed, a mouse will be dwarf. The *Igf2* gene is known to be located on an autosome, not on a sex chromosome. Because mice are diploid, they have two copies of this gene, one from each parent.

Researchers have discovered mutations in the *Igf2* gene that block the function of the Igf2 hormone. Such a mutant allele is designated *Igf2⁻*. When a homozygous *Igf2 Igf2* mouse and a homozygous *Igf2⁻Igf2⁻* mouse are crossed, the results are surprising (**Figure 18.1**). If the female parent is homozygous for the mutant allele and the male parent is homozygous for the functional allele (left side), all of the offspring grow to a normal size. By contrast, if the female is homozygous for the functional allele and the male is homozygous for the mutant allele (right side), all of the offspring are dwarf. The reason this result is so surprising is that

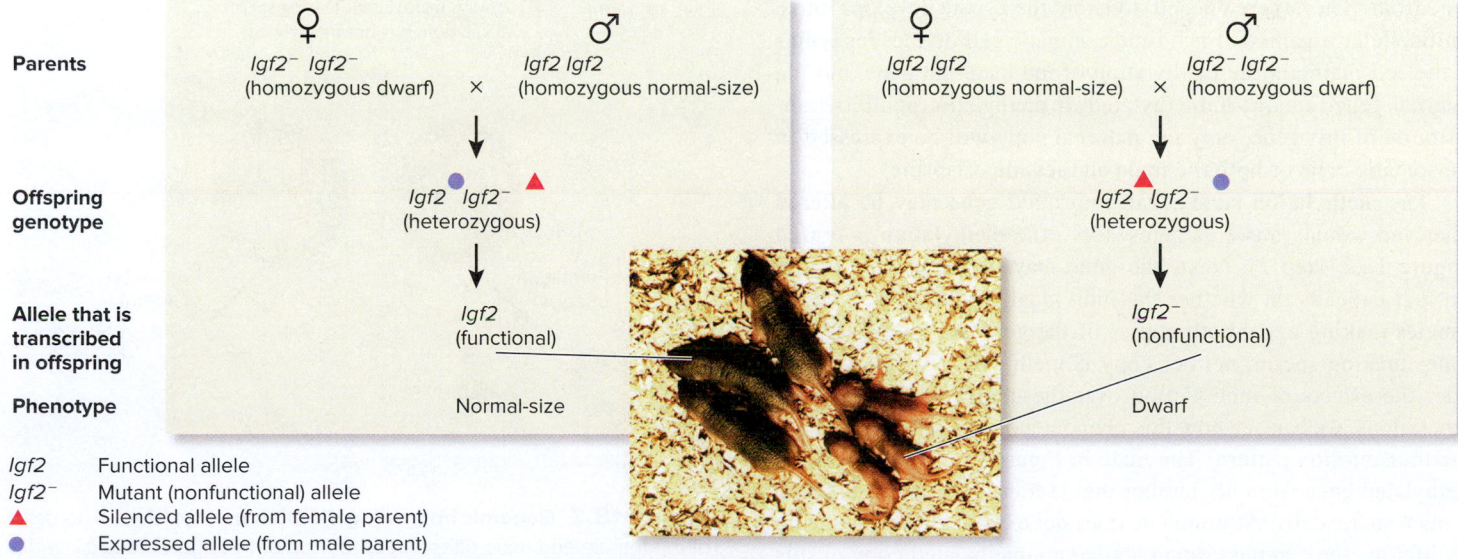

| | | |
|---|---|---|
| **Parents** | ♀ *Igf2⁻ Igf2⁻* (homozygous dwarf) × ♂ *Igf2 Igf2* (homozygous normal-size) | ♀ *Igf2 Igf2* (homozygous normal-size) × ♂ *Igf2⁻ Igf2⁻* (homozygous dwarf) |
| **Offspring genotype** | *Igf2 Igf2⁻* (heterozygous) | *Igf2 Igf2⁻* (heterozygous) |
| **Allele that is transcribed in offspring** | *Igf2* (functional) | *Igf2⁻* (nonfunctional) |
| **Phenotype** | Normal-size | Dwarf |

*Igf2*    Functional allele
*Igf2⁻*   Mutant (nonfunctional) allele
▲        Silenced allele (from female parent)
●        Expressed allele (from male parent)

**Figure 18.1   An example of genomic imprinting in the mouse.** In the cross on the left, a homozygous female carrying the mutant allele *Igf2⁻* was crossed to a homozygous male with the functional *Igf2* allele. Offspring are normal size, because the paternal allele is expressed. In the cross on the right, a homozygous *Igf2 Igf2* female was crossed to a homozygous male carrying the mutant allele. In this case, offspring are dwarf because the paternal allele is defective and the maternal allele is not expressed. The photograph shows normal-size (left) and dwarf littermates (right) from a cross between a wild-type female (*Igf 2 Igf 2*) and a heterozygous male carrying a loss-of-function allele (*Igf 2 Igf 2⁻*). The loss-of-function allele was made using methods described in Chapter 21. Courtesy of Dr. Argiris Efstratiadis

**Concept Check:** *If you crossed an Igf2 Igf2⁻ male mouse to an Igf2 Igf2⁻ female mouse, what would be the expected results?*

the normal-size and dwarf offspring have the same genotype (*Igf2 Igf2⁻*) but different phenotypes! In mice, the *Igf2* gene is imprinted in such a way that only the paternal allele is expressed, which means it is transcribed into mRNA. The maternal allele is not transcribed. The newborn mice shown on the left side of the photograph of Figure 18.1 are normal size, because they express a functional paternal allele. By contrast, the mice on the right side are dwarf because they express a paternal allele that is defective and results in a nonfunctional hormone.

## Transcription of an Imprinted Gene Depends on Methylation

Why is the maternal gene encoding Igf2 not transcribed into mRNA? To answer this question, we need to consider the regulation of gene transcription in eukaryotes. As discussed in Chapter 14, DNA methylation, which is the attachment of methyl (—CH₃) groups to bases of DNA, can alter gene transcription. Researchers have discovered that DNA methylation is the marking process that occurs during the imprinting of certain genes, including the *Igf2* gene. For most genes, DNA methylation silences gene expression by inhibiting the initiation of transcription or by causing the chromatin in a region to become more compact. By contrast, for a few imprinted genes, DNA methylation may enhance gene expression by attracting activator proteins to the promoter or by preventing the binding of repressor proteins.

**Figure 18.2** shows the imprinting process in which a maternal gene is methylated. The left side of the figure follows the marking process during the life of a female individual; the right side follows the same process in a male. Both individuals received a methylated gene from their mother and a nonmethylated copy of the same gene from their father. Via cell division, the zygote develops into a multicellular organism. Each time a somatic cell divides, enzymes in the cell maintain the methylation of the maternal gene, but the paternal gene remains unmethylated. If methylation inhibits transcription of this gene, only the paternal copy will be expressed in the somatic cells of both the male and female offspring.

The methylation state of an imprinted gene may be altered when individuals make gametes. First, the methylation is erased (Figure 18.2, step 2). Next, the gene may be methylated again, but that depends on whether the individual is female or male. In females making eggs, both copies of the gene are methylated; in males making sperm, neither copy is methylated. When we consider the effects of methylation over the course of two or more generations, we can see how this phenomenon results in an epigenetic transmission pattern. The male in Figure 18.2 has inherited a methylated gene from his mother that is transcriptionally silenced in his somatic cells. Although he does not express this gene during his lifetime, he can pass on an active, nonmethylated copy of this exact same gene to his offspring.

Genomic imprinting is a recently discovered phenomenon that has been shown to occur for a many genes in mammals. For some genes, such as *Igf 2*, the maternal allele is silenced, but for other genes, the paternal allele is silenced. Although several hypotheses have been advanced, biologists are still trying to identify possible advantages that this curious marking process may confer.

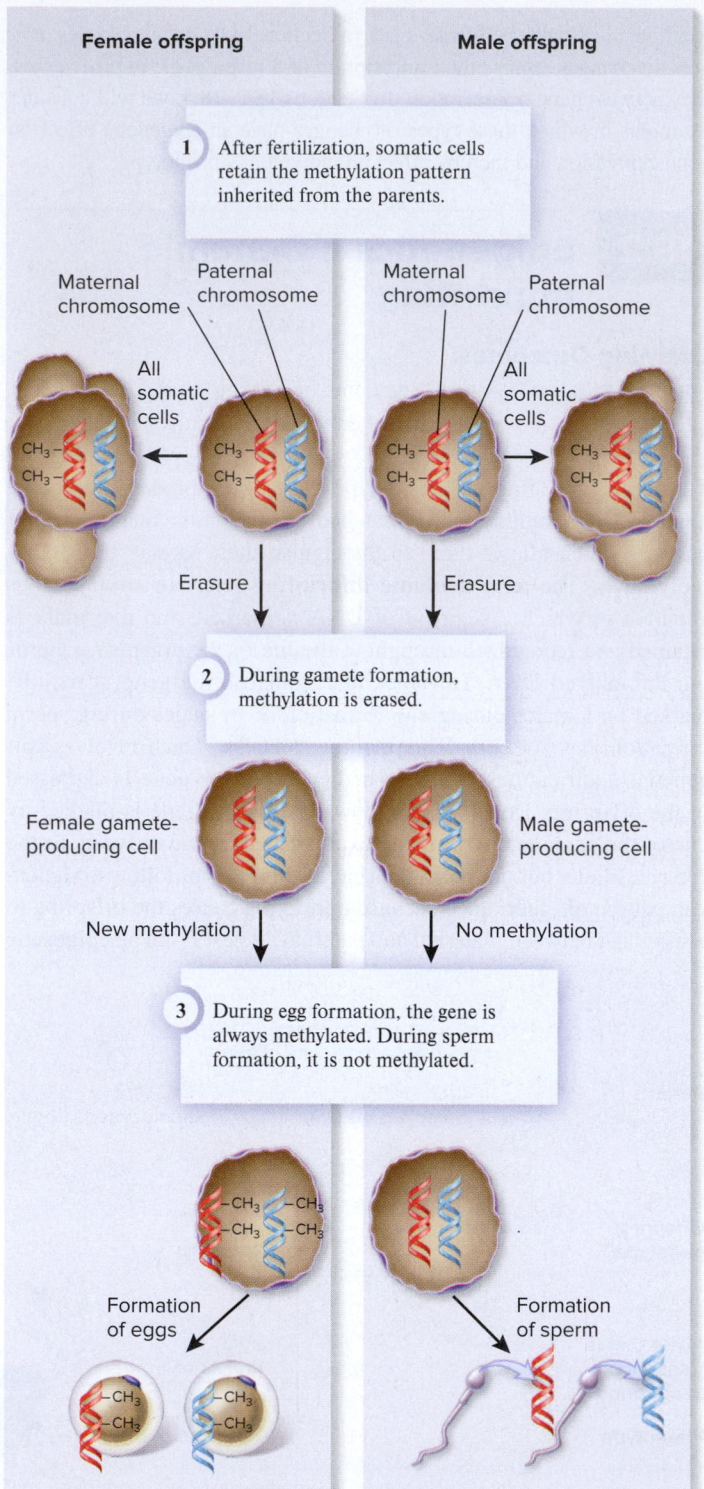

**Figure 18.2 Genomic imprinting via DNA methylation.** The cells of the female and male offspring at the top of this figure have a methylated gene inherited from the mother and a nonmethylated version of the same gene inherited from the father. This pattern of methylation is maintained in the somatic cells of both female and male offspring. The methylation is erased during gamete formation, but in females, the gene is methylated again at a later stage in the formation of eggs. Therefore, females always transmit a methylated, transcriptionally silent copy of this gene, whereas males transmit a nonmethylated, transcriptionally active copy.

## 18.3 Epigenetics II: X-Chromosome Inactivation

**Learning Outcomes:**

1. Explain how X-chromosome inactivation may affect the phenotype of female mammals.
2. Describe the process of X-chromosome inactivation at the cellular level.

As we have seen, genomic imprinting involves epigenetic changes that occur during gamete formation and are passed from parent to offspring. Other types of epigenetic changes begin later, such as during embryonic development, and then persist into adulthood. For example, as mentioned in Section 18.1, epigenetic changes may silence actin and myosin genes in some embryonic cells, and these changes are passed from cell to cell so that the same genes are silenced in nonmuscle cells of the adult.

In this section, we will consider an example of epigenetic silencing that begins during embryonic development in female mammals. As discussed in Chapter 17, female mammals carry two X chromosomes in their cells, whereas males carry one X and one Y. During embryonic development in female mammals, one of the X chromosomes undergoes an epigenetic change called **X-chromosome inactivation (XCI)**. This process causes that X chromosome to become highly compacted, which silences the genes that it carries. In this section, we will examine how XCI may affect a female mammal's phenotype and how the process occurs at the cellular and molecular levels.

### In Female Mammals, One X Chromosome Is Inactivated in Each Somatic Cell

In 1961, British geneticist Mary Lyon proposed the epigenetic phenomenon of XCI. Its discovery was based on two lines of evidence. The first came from microscopic studies of mammalian cells. In 1949, Canadian physicians Murray Barr and Ewart Bertram identified a highly condensed structure in the cells of female cats that was not found in the cells of male cats. This structure was named a **Barr body** after one of its discoverers (**Figure 18.3**). In 1960, Asian-American geneticist Susumu Ohno correctly proposed that a Barr body is a highly condensed X chromosome. Lyon's second line of evidence was the inheritance pattern of variegated coat colors in certain female mammals. A classic case is the calico cat, which has randomly distributed patches of black and orange fur (look ahead to the bottom of **Figure 18.4**).

How do we explain this patchwork phenotype? According to Lyon's hypothesis, the calico pattern is due to the permanent inactivation of one X chromosome in each cell that forms a patch of the cat's skin, as shown in Figure 18.4. The gene involved is an X-linked gene that occurs as an orange allele, $X^O$, and a black allele, $X^B$. A female cat heterozygous for this gene will be calico. (The cat's white underside is due to a dominant allele of a different autosomal gene.) At an early stage of embryonic development, one of the two X chromosomes is randomly inactivated in each of the cat's somatic cells, including those that will give rise to the hair-producing skin cells. As the embryo grows and matures, the pattern of XCI is maintained during

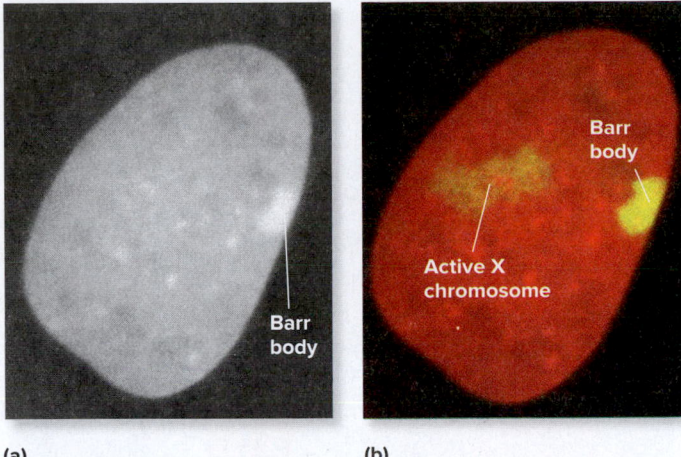

(a)                              (b)

**Figure 18.3** **X-chromosome inactivation in female mammals.** **(a)** A Barr body is seen on the periphery of a human nucleus (during interphase) after the nucleus has been stained with a DNA-specific dye. Because it is compact, the Barr body is the most brightly stained feature. **(b)** The same nucleus was labeled using a yellow fluorescent probe that recognizes the X chromosome. The Barr body is more compact than the active X chromosome, which is to the left of the Barr body. a–b: Courtesy of I. Solovei, University of Munich (LMU)

*Concept Check:* *How is the Barr body different from the other X chromosome in this cell?*

subsequent cell divisions. Skin cells derived from a single embryonic cell in which the $X^B$-carrying chromosome has been inactivated produce a patch of orange fur, because they express only the $X^O$ allele that is carried on the active chromosome. Alternatively, a group of skin cells in which the chromosome carrying $X^O$ has been inactivated express only the $X^B$ allele, producing a patch of black fur. If female mammals are heterozygous for X-linked genes, approximately half of their somatic cells express one allele, whereas the rest of their somatic cells express the other allele. The result is an animal with randomly distributed patches of black and orange fur. These heterozygotes are called **mosaics** because they have somatic regions that are composed of two types of cells.

For many X-linked traits in humans, females who are heterozygous for recessive X-linked alleles usually show the dominant trait because the expression of the dominant allele in 50% of their cells is sufficient to produce the dominant phenotype. For example, let's consider the recessive X-linked form of hemophilia that was discussed in Chapter 17 (hemophilia A). This type of hemophilia is caused by a defect in a gene that encodes a blood-clotting protein, called factor VIII, which is made by cells in the liver and secreted into the bloodstream. In a heterozygous female, approximately half of her liver cells make and secrete this clotting factor, which is sufficient to prevent hemophilia. Therefore, she exhibits the dominant trait of proper blood clotting.

On rare occasions, a female who is heterozygous may show mild or even severe disease symptoms. How is this possible? X-chromosome inactivation in humans occurs when an embryo is 10 days old. At this stage, the liver contains only about a dozen cells. In most females who are heterozygous for the dominant and recessive (hemophilia-causing) alleles, roughly half of their liver cells express the dominant

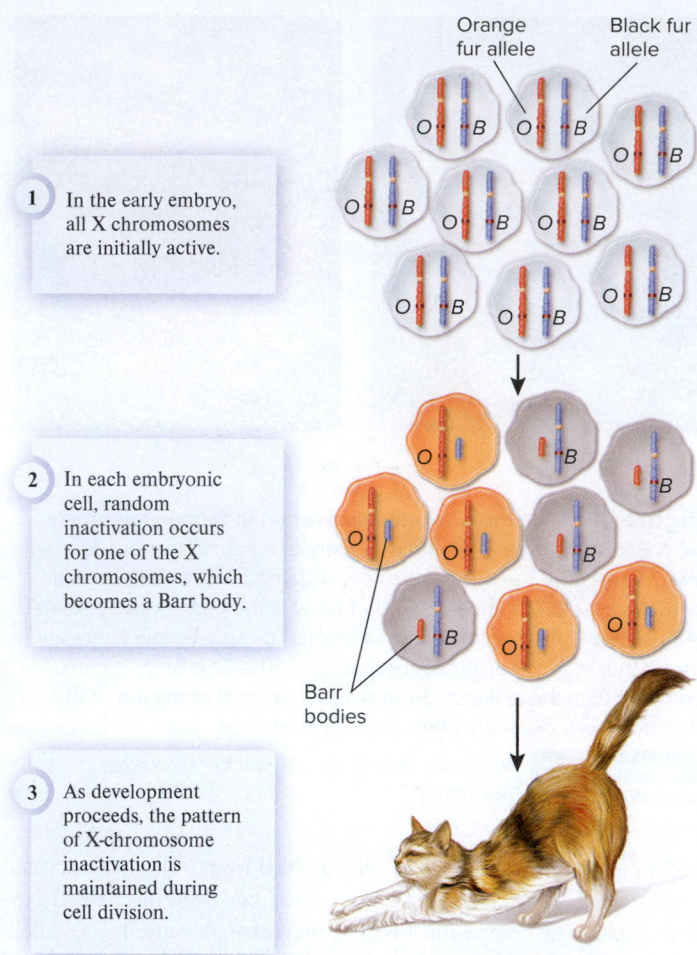

Orange fur allele    Black fur allele

1  In the early embryo, all X chromosomes are initially active.

2  In each embryonic cell, random inactivation occurs for one of the X chromosomes, which becomes a Barr body.

Barr bodies

3  As development proceeds, the pattern of X-chromosome inactivation is maintained during cell division.

**Figure 18.4 X-chromosome inactivation (XCI) in a calico cat.** The calico pattern is due to random XCI in a female that is heterozygous for an X-linked gene with black ($X^B$) and orange ($X^O$) alleles. The cells at the top of this figure represent a small mass of cells making up the very early embryo. In these cells, both X chromosomes are active. At an early stage of embryonic development, one X chromosome is randomly inactivated in each cell. The initial inactivation pattern is an epigenetic change that is maintained in the descendants of each cell as the embryo matures into an adult. The pattern of orange and black fur in the adult cat reflects the pattern of XCI in the embryo.

 **Core Skill: Modeling** The goal of this modeling challenge is to create a model that shows the pattern of X-chromosome inactivation (XCI) in a clump of cells after XCI has occurred.

**Modeling Challenge:** Step 2 of Figure 18.4 shows a model of early embryonic cells after XCI has occurred. Let's suppose you could remove the (orange) embryonic cell on the top left and the (gray) embryonic cell on the top right from this embryo. You then place each of these cells on solid growth media and allow them to grow and divide to form a small clump of cells. Draw a model that depicts the pattern of XCI in these two clumps of cells after 2 cell divisions.

allele. However, on rare occasions, all or most of the dozen embryonic liver cells may inactivate the X chromosome carrying the dominant functional allele. Following growth and development, such a female will have a very low level of factor VIII and as a result will show symptoms of hemophilia.

Why does X-chromosome inactivation occur? Researchers have proposed that XCI achieves **dosage compensation**, an equalization of the expression of X-linked genes in male and female mammals. The inactivation of one X chromosome in the female reduces the number of expressed copies (doses) of X-linked genes from two to one, the same as expressed in males. As a result, the expression of X-linked genes in females and males is roughly equal.

## The X Chromosome Has an X Inactivation Center That Controls Compaction into a Barr Body

After Lyon's hypothesis was confirmed, researchers became interested in the cellular and molecular control of X-chromosome inactivation. The cells of humans and other mammals have the ability to count their X chromosomes and allow only one of them to remain active. Additional X chromosomes are converted to Barr bodies. In females, two X chromosomes are counted and one is inactivated. In males, one X chromosome is counted and none inactivated.

On rare occasions, people are born with abnormalities in the number of their sex chromosomes. In the disorders known as Turner syndrome, triple X syndrome, and Klinefelter syndrome, the cells inactivate the number of X chromosomes necessary to leave a single active chromosome (**Table 18.2**). For example, in triple X syndrome, in which an extra X chromosome is found in each cell, two X chromosomes are converted to Barr bodies. In spite of X-chromosome inactivation, people with these three syndromes do exhibit some phenotypic abnormalities. The symptoms associated with these disorders may be due to effects that occur prior to XCI or may arise because not all of the genes on the Barr body are completely silenced.

Although the mechanism of inactivation is not entirely understood at the molecular level, a short region on the X chromosome called the **X inactivation center (Xic)** is known to play a critical role. Finnish-born American geneticist Eeva Therman and German-born American geneticist Klaus Patau determined that XCI is

| Table 18.2 | Relationship Between X-Chromosome Inactivation and the Number of X Chromosomes | | |
|---|---|---|---|
| Phenotype | Chromosome composition | Number of Barr bodies | Number of active X chromosomes |
| Female | XX | 1 | 1 |
| Male | XY | 0 | 1 |
| Turner syndrome (female) | XO | 0 | 1 |
| Triple X syndrome (female) | XXX | 2 | 1 |
| Klinefelter syndrome (male) | XXY | 1 | 1 |

accomplished by counting the number of Xics and inactivating all X chromosomes except for one. In cells with two X chromosomes, if one of them is missing its Xic due to a chromosome mutation, neither X chromosome will be inactivated, because only one Xic is counted. Having two active X chromosomes is a lethal condition for a human female embryo.

The expression of a specific gene within the Xic is required for compaction of the X chromosome into a Barr body. This gene, discovered in 1991, is named *Xist* (for **X** **i**nactive **s**pecific **t**ranscript). The *Xist* gene product is a long RNA molecule that does not encode a protein. Instead, the role of *Xist* RNA is to coat one of the two X chromosomes during the process of X-chromosome inactivation. After coating, proteins associate with the *Xist* RNA and promote compaction of the chromosome into a Barr body. The *Xist* gene on the Barr body continues to be expressed after other genes on this chromosome have been silenced. The expression of the *Xist* gene also maintains a chromosome as a Barr body during cell division. Whenever a somatic cell divides in a female mammal, the Barr body is replicated to produce two Barr bodies.

## 18.4 Epigenetics III: Effects of Environmental Agents

### Learning Outcomes:

1. Explain how chemicals in the diet may affect an individual's phenotype.
2. List examples of chemicals that cause epigenetic changes that may contribute to cancer.

One of the most active fields in genetics is the study of how certain environmental agents cause epigenetic changes and thereby affect gene expression. Two areas that have received a great deal of attention are the effects of diet and the potential effects of toxic agents, such as carcinogens—cancer-causing agents. In this section, we will consider examples in which environmental agents promote epigenetic changes that affect an individual's phenotype or cause a disease such as cancer.

### Chemicals in an Individual's Diet May Cause Epigenetic Changes That Affect Phenotype

At the beginning of this chapter, we considered how chemicals in royal jelly are responsible for producing queen bees (see the chapter opening photo). Another striking example of how chemicals in the diet can promote epigenetic changes is illustrated by studies of the *Agouti* gene (designated *A*) found in mice. This gene encodes the Agouti signaling peptide that controls the deposition of yellow pigment in developing hairs. In mice that are homozygous for a functional allele, *AA,* the expression of this gene promotes the synthesis of pheomelanin, a yellow pigment. During the growth of a hair, melanocytes (pigment-producing cells) within a hair follicle initially make eumelanin, which is black. The transient expression of the *Agouti* gene causes the melanocytes to make pheomelanin. The melanocytes then revert to making black pigment. The result is a band of yellow pigment sandwiched between layers of black

pigment, which gives a brown color. The yellow pigment is not synthesized near the tip of the hair, so the hair of *AA* mice is brown with black tips.

Researchers have identified several mutations that affect the expression of the *Agouti* gene. For example, mice that are homozygous for a loss-of-function allele, *aa,* have black fur because pheomelanin is not made. Alternatively, a gain-of-function mutation that causes the *Agouti* gene to be overexpressed results in a mouse with yellow fur. One such mutation is designated $A^{vy}$. (*A* refers to Agouti, *v* refers to viable, and *y* refers to yellow. The letter *v* is used because other mutations in the *Agouti* gene are not viable.) By characterizing the $A^{vy}$ allele at the molecular level, researchers determined that it is due to the insertion of a new promoter next to the normal promoter of the *Agouti* gene. This new promoter is very active, which causes the overexpression of the *Agouti* gene.

An intriguing observation of mice carrying the $A^{vy}$ allele is that they exhibit a wide phenotypic variation, ranging from yellow to mottled to pseudo-agouti (**Figure 18.5a**). Why should mice with the same genotype show such a wide range of phenotypic variation? Although the answer is not entirely understood, researchers have speculated that the new promoter in mice carrying the $A^{vy}$ allele is very sensitive to epigenetic changes. In particular, this promoter may be more likely to be modified by DNA methylation, which would inhibit its function. Furthermore, a variety of environmental factors may cause this epigenetic change to occur. The sensitivity of this promoter to epigenetic modifications, together with variation in environmental factors, may explain the phenotypic variation seen in these mice.

A key factor that may affect epigenetic changes is diet. With regard to the $A^{vy}$ allele, the exposure of pregnant female mice to different diets affects the phenotypes of the resulting offspring. In 2003, American geneticists Robert Waterland and Randy Jirtle conducted a study in which they investigated the effects of certain dietary supplements. Their goal was to determine if nutrients that are known to inhibit DNA methylation would alter the expression of the *Agouti* gene and thereby affect coat color. A variety of dietary factors can inhibit the enzyme DNA methyltransferase, which methylates DNA. These include folic acid, vitamin $B_{12}$, betaine, and choline chloride.

Waterland and Jirtle began with female mice carrying the $A^{vy}$ allele and divided them into a control group (which was fed a standard diet) and an experimental group (which was fed a diet supplemented with folic acid, vitamin $B_{12}$, betaine, and choline chloride). Both groups were fed their respective diets before and during pregnancy and up to the stage of weaning. Offspring that inherited the $A^{vy}$ allele were then analyzed with regard to their coat color and levels of DNA methylation.

As expected, a range of coat colors was observed among the offspring (**Figure 18.5b**). However, the offspring of females that had been fed a supplemented diet tended to have darker coats. For example, over 25% of the offspring with heavily mottled coats had mothers that were fed a supplemented diet (blue bars), whereas less than 10% had mothers that were given a standard diet (red bars).

The coat colors of the offspring were correlated with the degree of methylation that occurred at the new promoter—offspring

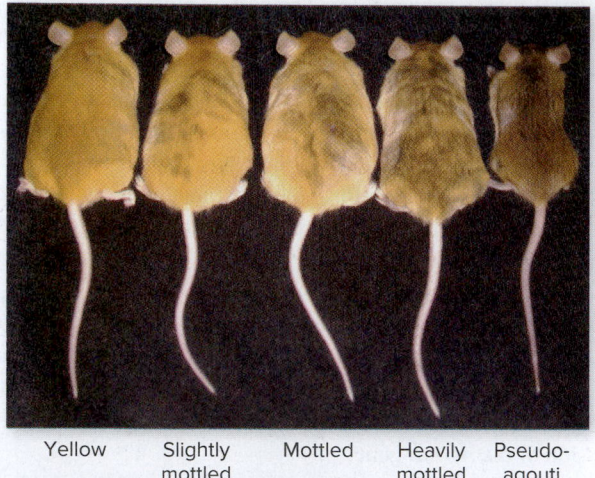

(a) Range in coat-color phenotypes in $A^{vy}a$ mice due to epigenetic changes

Yellow  Slightly mottled  Mottled  Heavily mottled  Pseudo-agouti

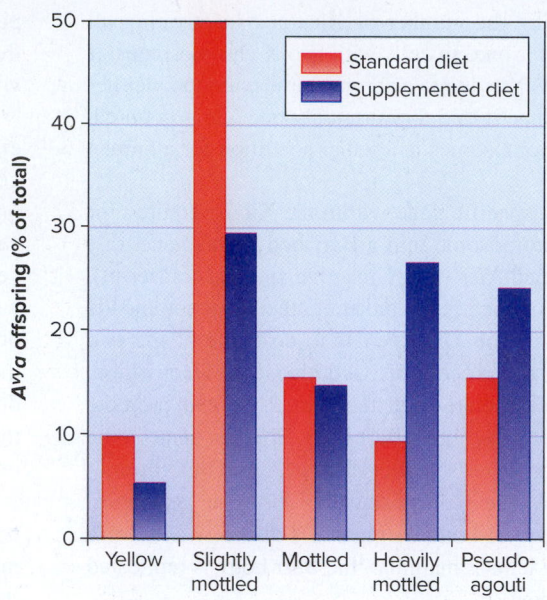

(b) Effect of diet on coat color

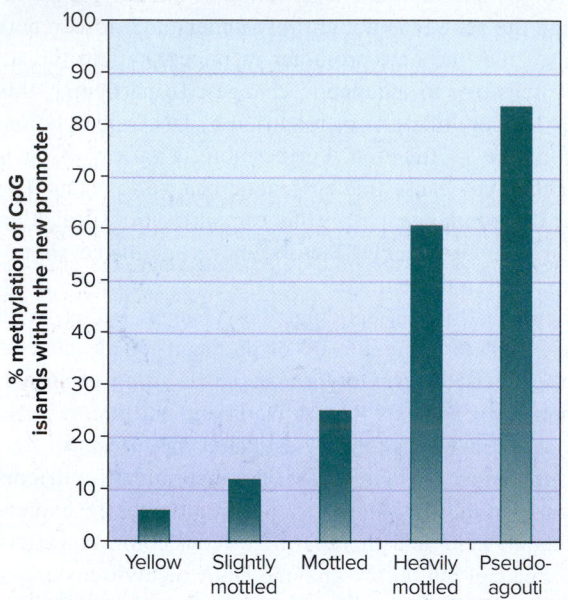

(c) Level of DNA methylation of CpG islands within the new promoter among mice with different coat colors

**Figure 18.5 Dietary effects on coat color in mice.** (a) Mice carrying the $A^{vy}$ allele exhibit a range of phenotypes. The mice shown here are heterozygotes, $A^{vy}a$; they carry the $A^{vy}$ allele and a loss-of-function allele, $a$. (b) Effects of diet supplementation on coat color. Red bars represent offspring from females given a standard diet, and blue bars represent offspring from females given a supplemented diet. (c) DNA methylation patterns among mice with different coat colors. a–c: Source: Waterland, R. A., and Jirtle, R. L. 2003. Transposable Elements: Targets for Early Nutritional Effects on Epigenetic Gene Regulation, *Molecular and Cellular Biology* 23: 5293–5300

with darker coat color had greater levels of DNA methylation (**Figure 18.5c**). How do we explain these results? In the mice that are more yellow, the new promoter has undergone less DNA methylation. Therefore, the promoter remains active, leading to

the transcription of the *Agouti* gene and the overproduction of yellow pigment. By contrast, the new promoter in the darker mice has undergone more methylation. Such methylation inhibits the overexpression of the *Agouti* gene and thereby prevents the overproduction of yellow pigment, resulting in darker fur.

## Environmental Agents May Cause Epigenetic Changes That Are Associated with Human Diseases, Such as Cancer

One of the most active fields in genetics involves the study of epigenetic changes that contribute to human diseases. We have probably seen only "the tip of the iceberg" with respect to our understanding of this topic. Many medical studies have identified correlations between epigenetic changes and particular diseases. For example, some research studies have compared one variable, such as the level of DNA methylation of a specific gene, to a second variable, such as the severity of a disease. If a high level of DNA methylation is associated with an increase in disease severity, this is a positive correlation. Researchers analyze the data to decide if such a correlation is statistically significant. When a statistically significant correlation is obtained, how do we interpret its meaning? Such a result suggests a true **association**—changes in the two variables follow a pattern. For example, in a positive correlation, when one variable increases, the other variable also increases.

However, an association does not necessarily imply a cause-and-effect relationship. When considering the role of epigenetic changes and human disease, an association can arise in three common ways:

- The epigenetic changes directly contribute to the disease symptoms. There is a cause-and-effect relationship.

- Conversely, the disease symptoms may arise first, and then they cause subsequent epigenetic changes to happen. This is also a cause-and-effect relationship, but in the opposite direction.

- The association is indirect because a third factor is involved. For example, a toxic agent in the environment may cause a disease and also cause particular types of epigenetic changes even though those epigenetic changes do not contribute to the disease.

Correlations identify associations between two variables. We should use caution, however, because the correlation by itself cannot prove that the association is due to cause and effect. Even so, research studies that identify associations are very useful because they provide the rationale to carry out further research to determine if a cause-and-effect relationship exists.

Researchers have identified many examples in which epigenetic changes are associated with a particular disease. These include Alzheimer disease, cardiovascular diseases, diabetes, multiple sclerosis, and asthma. For these diseases, further research is needed to determine if these epigenetic changes are directly contributing to the disease symptoms. The role of epigenetics in disease has been most extensively studied with regard to cancer. **Table 18.3** describes several examples in which an environmental factor is associated with a particular type of cancer.

In some of the examples listed in Table 18.3, scientific evidence indicates that the association is causative. For example, certain agents in tobacco smoke have been shown to cause epigenetic changes that underlie lung cancer. As described in Chapter 15, cancer-causing genes are placed into two categories: oncogenes, which are overexpressed in cancer, and tumor-suppressor genes, which are inhibited. In cases where research has shown that an environmental agent results in an epigenetic effect that contributes to cancer, it is more common for that agent to inhibit tumor-suppressor genes. However, in some cases, it may activate oncogenes. Alternatively, some of the agents listed in

Table 18.3 show an association with particular cancers, but researchers are still trying to determine if the epigenetic changes caused by these agents actually promote the changes that result in cancer.

## 18.5 Extranuclear Inheritance: Organelle Genomes

**Learning Outcomes:**

1. Describe the general features of mitochondrial and chloroplast genomes.
2. **CoreSKILL »** Predict the outcome of crosses that exhibit maternal inheritance.
3. List human diseases associated with mutations in mitochondrial genes.

In this section, we will explore inheritance patterns that violate the law of segregation. As described in Chapter 17, the segregation of genes is explained by the pairing and segregation of homologous chromosomes during meiosis. However, some genes are not found on the chromosomes in the cell nucleus, and these genes do not segregate in the same way. The transmission of genes located outside the cell nucleus is called **extranuclear inheritance**. Two important types of extranuclear inheritance patterns involve genes found in chloroplasts and mitochondria. Extranuclear inheritance is also called cytoplasmic inheritance because these organelles are in the cytoplasm of the cell. In this section, we will examine the transmission patterns observed for genes found in the chloroplast and mitochondrial genomes and consider how mutations in these genes may affect an individual's traits.

 **Core Concepts: Evolution, Information**

### Chloroplast and Mitochondrial Genomes Are Relatively Small, but Contain Genes That Encode Important Proteins

As discussed in Chapter 4, mitochondria and chloroplasts are found in eukaryotic cells because of an ancient endosymbiotic relationship. They contain their own genetic material, called the mitochondrial genome and chloroplast genome, respectively (**Figure 18.6**). Mitochondrial and chloroplast genomes are composed of a single, circular DNA molecule. The mitochondrial genome of many mammalian species has been analyzed and usually contains a total of 37 genes. Twenty-four genes encode tRNAs and rRNAs, which are needed for translation inside the mitochondrion, and 13 genes encode proteins that are involved in oxidative phosphorylation. As discussed in Chapter 7, the primary function of the mitochondrion is the synthesis of ATP via oxidative phosphorylation. Among different species of flowering plants, chloroplast genomes typically contain between 100 and 200 genes. Many of these genes encode proteins that are vital to the process of photosynthesis, which is discussed in Chapter 8.

| Table 18.3 | Environmental Agents That Are Associated with Cancer and Are Known to Cause Epigenetic Changes | |
|---|---|---|
| **Environmental agent** | **Occurrence** | **Associations with particular cancers** |
| Polycyclic aromatic hydrocarbons | Tobacco smoke, automobile exhaust, charbroiled food | Lung, breast, stomach, and skin cancer |
| Benzene | Tobacco smoke, automobile exhaust | Leukemia, lymphoma, multiple myeloma |
| Endocrine disruptors (such as diethylstilbestrol) | Insecticides, fungicides, herbicides, some types of plastic | Breast, prostate, and thyroid cancer |
| Cadmium | Tobacco products, production of batteries | Lung and breast cancer |
| Nickel | Occupational exposure in mining, welding, and electroplating and in the manufacturing of jewelry, stainless steel, and batteries | Lung and nasal cancer |
| Arsenic | Lead alloy, feed additive in agriculture, insecticides | Skin, bladder, kidney, and liver cancer |

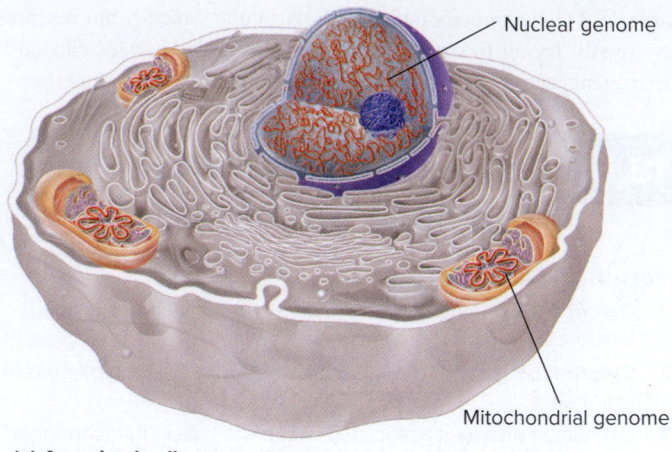

Nuclear genome

Mitochondrial genome

**(a) An animal cell**

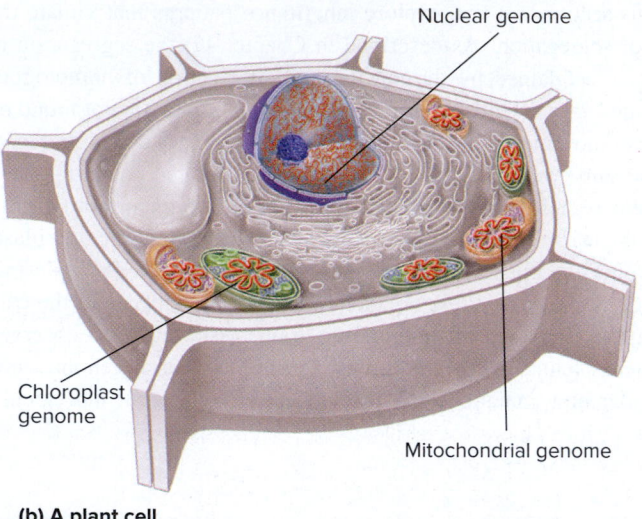

Nuclear genome

Chloroplast genome

Mitochondrial genome

**(b) A plant cell**

**Figure 18.6   The locations of genetic material in animal and plant cells.** The chromosomes in the cell nucleus are collectively known as the nuclear genome. Mitochondria and chloroplasts have small circular chromosomes called the mitochondrial and chloroplast genomes, respectively.

 **Core Skill: Connections**   Look back at Figure 4.31. What is the evolutionary origin of mitochondria and chloroplasts in eukaryotic cells?

## Chloroplast Genomes Are Often Maternally Inherited

One of the first experiments showing an extranuclear inheritance pattern was carried out by German botanist Carl Correns in 1909. Correns discovered that leaf pigmentation in the four-o'clock plant (*Mirabilis jalapa*) follows a pattern of inheritance that does not obey Mendel's law of segregation. Four-o'clock leaves may be green, white, or variegated. Correns observed that the pigmentation of the offspring depended solely on the pigmentation of the female parent, a phenomenon called **maternal inheritance** (**Figure 18.7**). In cross 1, if the female parent providing the eggs had white leaves and the male parent providing the pollen had green leaves, all of the offspring had white leaves like the

Correns's crosses

**Cross 1**

×

♀                    ♂

All white offspring

×

Reciprocal cross of cross 1

♀                    ♂

All green offspring

**Cross 2**

×

♀                    ♂

Green, white, or variegated offspring

×

Reciprocal cross of cross 2

♀                    ♂

All green offspring

**Figure 18.7   Maternal inheritance in the four-o'clock plant.** In four-o'clocks, the egg contains all of the proplastids, which develop into chloroplasts, that are inherited by the offspring. The phenotype of the offspring is determined by the maternal parent. The green phenotype is due to the presence of normal chloroplasts. The white phenotype is due to chloroplasts with a mutant allele that greatly reduces green pigment production. The variegated phenotype is due to a mixture of normal and mutant chloroplasts.

**Concept Check:**   *In this example, where is the gene located that causes the green color of four-o'clock leaves? How is this gene transmitted from parent to offspring?*

female parent. Correns also conducted a **reciprocal cross**—a cross in which the sexes and phenotypes are reversed compared to another cross. In the reciprocal cross of cross 1, the female parent had green leaves and the male parent had white leaves. Again, the offspring exhibited a phenotype like the female parent. Similarly, if the female was green, so were all of the offspring. In cross 2, the offspring of a variegated female parent could be green, white, or variegated.

What accounts for maternal inheritance? At the time, Correns did not understand that chloroplasts contain genetic material. Subsequent research identified DNA present in chloroplasts as responsible for the unusual inheritance pattern observed. We now know that the pigmentation of four-o'clock leaves can be explained by the occurrence of genetically different types of chloroplasts in the leaf cells. As discussed in Chapter 8, chloroplasts are the site of photosynthesis, and their green color is due to the presence of the pigment called chlorophyll. Certain genes required for chlorophyll synthesis are found within the chloroplast DNA. For four-o'clock plants, the green phenotype is due to the presence of chloroplasts that have functional genes and synthesize the usual quantity of chlorophyll. The white phenotype is caused by a mutation in a gene within the chloroplast DNA that prevents the synthesis of most of the chlorophyll. (Enough chlorophyll is made for the plant to survive.) The variegated phenotype occurs in leaves that have a mixture of the two types of chloroplasts.

Leaf pigmentation follows a maternal inheritance pattern because the chloroplasts in four-o'clocks are transmitted only through the cytoplasm of the egg (**Figure 18.8**). In most species of plants, the egg cell provides most of the zygote's cytoplasm, whereas the much smaller male gamete often provides little more than a nucleus. Therefore, chloroplasts are most often inherited via the egg. Recall from Chapter 4 that chloroplasts are derived from proplastids. In four-o'clocks, the egg cell contains several proplastids that are inherited by the offspring. The sperm cell does not contribute any proplastids. For this reason, the phenotype of a four-o'clock plant reflects the types of proplastids it inherits from the maternal parent.

- If the maternal parent transmits only normal proplastids, all offspring will have green leaves (**Figure 18.8a**).
- Alternatively, if the maternal parent transmits only mutant proplastids, all offspring will have white leaves (**Figure 18.8b**).
- Because an egg cell contains several proplastids, an offspring from a variegated maternal parent may inherit only normal proplastids, only mutant proplastids, or a mixture of normal and mutant proplastids. Consequently, the offspring of a variegated maternal parent can be green, white, or variegated individuals (**Figure 18.8c**).

How do we explain the variegated phenotype at the cellular level? This phenotype is due to events that occur after fertilization. As a zygote containing both types of proplastids grows via cellular division to produce a multicellular plant, some cells may receive mostly protoplastids that develop into normal chloroplasts. Further division of these cells gives rise to a patch of green tissue. Alternatively, as a matter of chance, other cells may receive all or mostly mutant protoplastids that develop into chloroplasts that are defective in chlorophyll synthesis. The result is a patch of white tissue.

In seed-bearing plants, maternal inheritance of chloroplasts is the most common transmission pattern. However, certain species

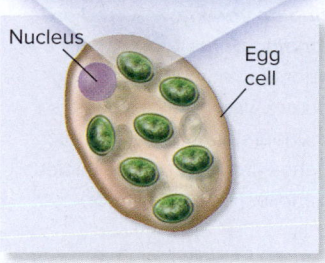

Normal proplastid will produce chloroplasts with a normal amount of green pigment.

Mutant proplastid will produce chloroplasts with very little pigment.

Nucleus        Egg cell

**(a) Egg cell from a maternal parent with green leaves**

**(b) Egg cell from a maternal parent with white leaves**

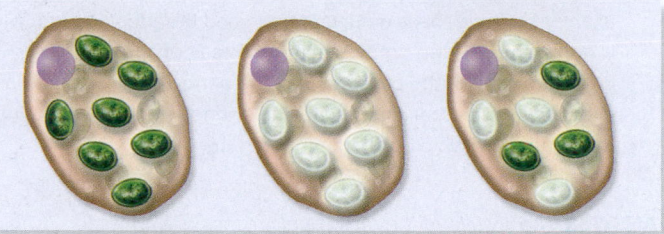

**(c) Possible egg cells from a maternal parent with variegated leaves**

**Figure 18.8   Plastid compositions of egg cells from green, white, and variegated four-o'clock plants.** In this drawing of four-o'clock egg cells, normal proplastids are represented as green and mutant proplastids as white. (Note: This drawing is schematic. Proplastids do not differentiate into chloroplasts in egg cells, and they are not actually green.) **(a)** A green plant produces eggs carrying normal proplastids. **(b)** A white plant produces eggs carrying mutant proplastids. **(c)** A variegated plant produces eggs that may contain either or both types of proplastids.

exhibit a pattern called **biparental inheritance**, in which both the pollen and the egg contribute chloroplasts to the offspring. Others exhibit **paternal inheritance**, in which only the pollen contributes these organelles. For example, most types of pine trees show paternal inheritance of chloroplasts.

**BIO TIPS**  **THE QUESTION**  *One strain of periwinkle plants has green leaves, and another strain has variegated leaves. You do not know if the difference in the phenotypes is due to a mutation in a nuclear gene or in an organellar gene (one found in the chloroplasts or mitochondria). Using reciprocal crosses of the two strains, the following results were obtained:*

| Variegated plant pollinated by green plant | Green plant pollinated by variegated plant |
| --- | --- |
| ↓ | ↓ |
| Offspring with green, white, or variegated leaves | All offspring with green leaves |

*Explain this pattern of inheritance.*

**T**OPIC  **What topic in biology does this question address?** The topic is inheritance. More specifically, the question is about distinguishing nuclear and extranuclear (organellar) inheritance patterns.

**I**NFORMATION  **What information do you know based on the question and your understanding of the topic?** From the question, you know there are green-leaved and variegated-leaved strains of periwinkles. You are also given the results of reciprocal crosses. From your understanding of the topic, you may remember that some genes are found in the nucleus, whereas others are found in organelles, chloroplasts or mitochondria.

**P**ROBLEM-SOLVING **S**TRATEGY  **Predict the outcome. Compare and contrast.** Crosses of these two strains will yield different results depending on the mode of inheritance. For example, if the gene is a nuclear gene and the green-leaved allele is dominant, you can predict that all of the F$_1$ offspring will be green-leaved. On the other hand, if the gene is in the chloroplasts and follows maternal inheritance, the phenotype of the offspring will depend on which plant contributed the egg.

**ANSWER** *The results of the reciprocal crosses are consistent with maternal inheritance, because the phenotypes of the offspring correlate with inheriting the gene from the plant contributing the egg cells. Since this gene affects the synthesis of chlorophyll, it is likely to be found in the chloroplasts.*

## Mitochondrial Genomes Are Maternally Inherited in Humans and Most Other Species

Mitochondria are found in nearly all eukaryotic species. As with the transmission of chloroplasts in plants, maternal inheritance is the most common pattern of mitochondrial transmission in eukaryotes, although some species do exhibit biparental or paternal inheritance.

In humans, mitochondria are maternally inherited. Researchers have discovered that mutations in human mitochondrial genes cause a variety of rare diseases (**Table 18.4**). These are usually chronic degenerative disorders that affect organs, such as the brain, eyes, heart, muscle, kidneys, and endocrine glands, whose cells require high levels of ATP. For example, Leber's hereditary optic neuropathy (LHON) affects the optic nerve and leads to the progressive loss of vision in one or both eyes. LHON is caused by point mutations in several different mitochondrial genes.

| Table 18.4 | Examples of Human Mitochondrial Disease | |
|---|---|
| **Disease** | **Causes and symptoms** |
| Leber's hereditary optic neuropathy (LHON) | A mutation in one of several mitochondrial genes that encode electron transport proteins. The main symptom is loss of vision. |
| Neurogenic muscle weakness | A mutation in a mitochondrial gene that encodes a subunit of mitochondrial ATP synthase, which is required for ATP synthesis. Symptoms involve abnormalities in the nervous system that affect the muscles and eyes. |
| Maternal myopathy and cardiomyopathy | A mutation in a mitochondrial gene that encodes a tRNA for leucine. The primary symptoms involve muscle abnormalities, most notably in the heart. |
| Myoclonic epilepsy and ragged-red muscle fibers | A mutation in a mitochondrial gene that encodes a tRNA for lysine. Symptoms include epilepsy, dementia, blindness, deafness, and heart and kidney malfunctions. |

## 18.6 Genes on the Same Chromosome: Linkage and Recombination

### Learning Outcomes:

1. Describe how linkage violates the law of independent assortment.
2. Explain how experimental crosses can demonstrate linkage.
3. **CoreSKILL »** Calculate the distance between genes that are linked on the same chromosome.

In Chapter 17, we saw that the independent assortment of alleles is due to the random alignment of homologous chromosomes during meiosis (refer back to Figure 17.11). But what happens when the alleles of different genes are on the same chromosome and do not independently assort? A typical chromosome contains many hundreds or even a few thousand different genes. When two genes are close together on the same chromosome, they tend to be transmitted as a unit, a phenomenon known as **linkage**. A group of genes that usually stay together during meiosis is called a **linkage group**, and the genes in the group are said to be linked. In a two-factor cross, linked genes that are close together on the same chromosome do not follow the law of independent assortment.

In this section, we will begin by examining the first experimental cross that demonstrated linkage. This pattern was subsequently explained by Thomas Hunt Morgan, who proposed that different genes located close to each other on the same chromosome tend to be inherited together. We will also see how crossing over between such genes provided the first method of mapping genes along chromosomes.

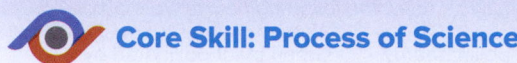
**Core Skill: Process of Science**

## Feature Investigation | Bateson and Punnett's Cross of Sweet Peas Showed That Genes Do Not Always Assort Independently

The first study showing linkage between two different genes was a cross of sweet peas carried out by William Bateson and Reginald Punnett in 1911. A surprising result occurred when they conducted a two-factor cross involving flower color and pollen shape (**Figure 18.9**). One of the parent plants had purple flowers (*PP*) and long pollen (*LL*); the other had red flowers (*pp*) and round pollen (*ll*).

**Figure 18.9** **A cross of sweet peas showing that independent assortment does not always occur.**

**HYPOTHESIS** The alleles of different genes assort independently of each other.

**KEY MATERIALS** True-breeding sweet pea strains that differ with regard to flower color and pollen shape.

| | Experimental level | Conceptual level |
|---|---|---|
| **1** Cross a plant with purple flowers and long pollen to a plant with red flowers and round pollen. | Purple flowers, long pollen × Red flowers, round pollen | $PPLL \times ppll$ |
| **2** Observe the phenotypes of the $F_1$ offspring. | Purple flowers, long pollen | $PpLl$ |
| **3** Allow the $F_1$ offspring to self-fertilize. | Purple flowers, long pollen × Purple flowers, long pollen | Meiosis $PL$ and $pl$ gametes — more frequent $Pl$ and $pL$ gametes — less frequent |
| **4** Observe the phenotypes of the $F_2$ offspring. | Purple flowers, long pollen 15.6 : Purple flowers, round pollen 1.0 : Red flowers, long pollen 1.4 : Red flowers, round pollen 4.5 | Fertilization $F_2$ offspring having phenotypes of purple flowers with long pollen or red flowers with round pollen occurred more frequently than expected from Mendel's law of independent assortment. |

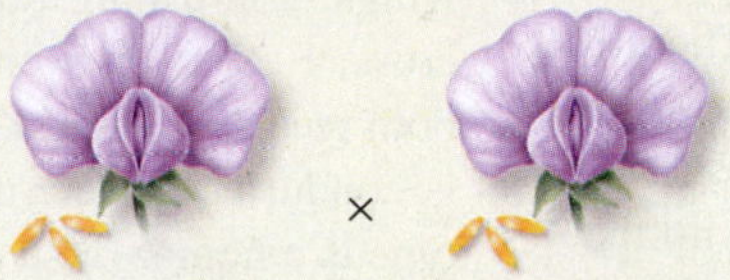

**5** **THE DATA**

| Phenotypes of $F_2$ offspring | Observed number | Observed ratio | Expected number | Expected ratio |
|---|---|---|---|---|
| Purple flowers, long pollen | 296 | 15.6 | 240 | 9 |
| Purple flowers, round pollen | 19 | 1.0 | 80 | 3 |
| Red flowers, long pollen | 27 | 1.4 | 80 | 3 |
| Red flowers, round pollen | 85 | 4.5 | 27 | 1 |

**6** **CONCLUSION** The data are not consistent with the law of independent assortment.

**7** **SOURCE** Bateson, W., and Punnett, R. C. 1911. On the inter-relations of genetic factors. *Proceedings of the Royal Society of London, Series B* 84: 3–8.

As Bateson and Punnett expected, the $F_1$ plants all had purple flowers and long pollen (*PpLl*). The unexpected result came in the $F_2$ generation.

Although the $F_2$ offspring displayed the four phenotypes predicted by Mendel's laws, the observed numbers of offspring did not conform to the predicted 9:3:3:1 ratio (refer back to Figure 17.8). Rather, as seen in the data in Figure 18.9, the $F_2$ generation had a much higher proportion of the two phenotypes found in the parental generation: purple flowers with long pollen, and red flowers with round pollen. How did Bateson and Punnett explain these results? They suggested that the transmission of flower color and pollen shape was somehow coupled, so these traits did not always assort

independently. Although the law of independent assortment applies to many other genes, in this example, the hypothesis of independent assortment was rejected.

### Experimental Questions

1. What hypothesis were Bateson and Punnett testing when conducting the crosses of sweet peas?

2. **CoreSKILL »** What were the expected results of Bateson and Punnett's cross of $F_1$ plants?

3. **CoreSKILL »** How did the observed results differ from the expected results?

## Linkage and Crossing Over Produce Parental and Recombinant Types

Bateson and Punnett realized their results did not conform to Mendel's law of independent assortment. However, they did not know why the genes were not assorting independently. A few years later, Thomas Hunt Morgan obtained similar ratios in crosses of fruit flies while studying the transmission pattern of genes in *Drosophila*. Like Bateson and Punnett, Morgan observed many more $F_2$ offspring with the combination of traits found in the parental generation than predicted on the basis of independent assortment. To explain his data, Morgan proposed three ideas:

1. When different genes are located on the same chromosome, the traits determined by those genes are more likely to be inherited together. This violates the law of independent assortment.

2. Due to crossing over during meiosis, homologous chromosomes can exchange pieces of chromosomes and create new combinations of alleles.

3. The likelihood of a crossover occurring in the region between two genes depends on the distance between the two genes. Crossovers between homologous chromosomes are much more likely to occur between two genes farther apart along a chromosome compared to two genes that are closer together.

To illustrate the first two ideas, **Figure 18.10** considers a series of crosses involving two genes linked on the same chromosome in *Drosophila*. The two genes are located on an autosome, not on a sex chromosome. The P generation cross is between flies that are homozygous for alleles that affect body color and wing shape. The female is homozygous for the dominant wild-type alleles that produce gray body color ($b^+b^+$) and straight wings ($c^+c^+$); the male is homozygous for recessive mutant alleles that produce black body color (*bb*) and curved wings (*cc*). The symbols for the genes are based on the names of the mutant phenotypes; the dominant wild-type allele is indicated by a superscript plus sign ($^+$). The chromosomes next to the flies in Figure 18.10 show the arrangement of these alleles. If the two genes are on the same chromosome, we know the arrangement of alleles in the P generation flies because these flies are homozygous for both genes ($b^+b^+c^+c^+$ for one parent and *bbcc* for the other parent). In the

P generation female, on the left, $b^+$ and $c^+$ are linked, whereas $b$ and $c$ are linked in the male, on the right.

Let's now look at the outcome of the crosses in Figure 18.10. As expected, the $F_1$ offspring ($b^+bc^+c$) all had gray bodies and straight wings, confirming that these are the dominant traits. In the next cross, $F_1$ females were mated to males that were homozygous for both recessive alleles (*bbcc*). Recall from Chapter 17 that a **testcross** is conducted to determine if an individual with a dominant phenotype is a homozygote or a heterozygote. However, in the crosses we are discussing here, the purpose of the testcross is to determine if the genes for body color and wing shape are linked. If the genes are on different chromosomes and assort independently, this testcross will produce $F_2$ offspring with the four possible phenotypes in a 1:1:1:1 ratio. The observed numbers clearly conflict with this prediction. The two most abundant phenotypes are those with the combinations of characteristics in the P generation: gray bodies and straight wings or black bodies and curved wings. These offspring are termed **nonrecombinants**. The smaller number of offspring that have a combination of traits not found in the parental generation—gray bodies and curved wings or black bodies and straight wings—are called **recombinants**.

How do we explain the occurrence of recombinants when genes are linked on the same chromosome? As shown beside the flies of the $F_2$ generation in Figure 18.10, each recombinant individual has a chromosome that is the product of a crossover. The crossover occurred during the process of egg formation in the $F_1$ female fly. As shown below, four different egg cells are possible:

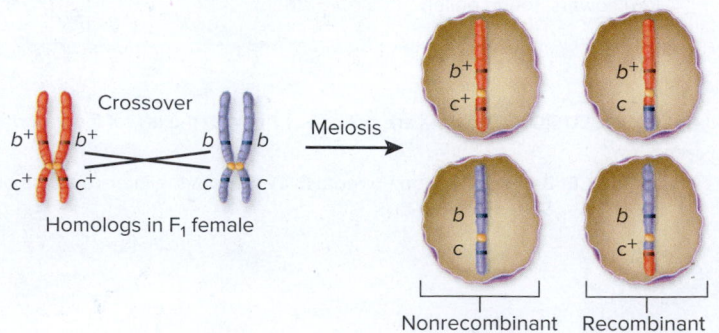

Homologs in $F_1$ female

Nonrecombinant chromosomes    Recombinant chromosomes

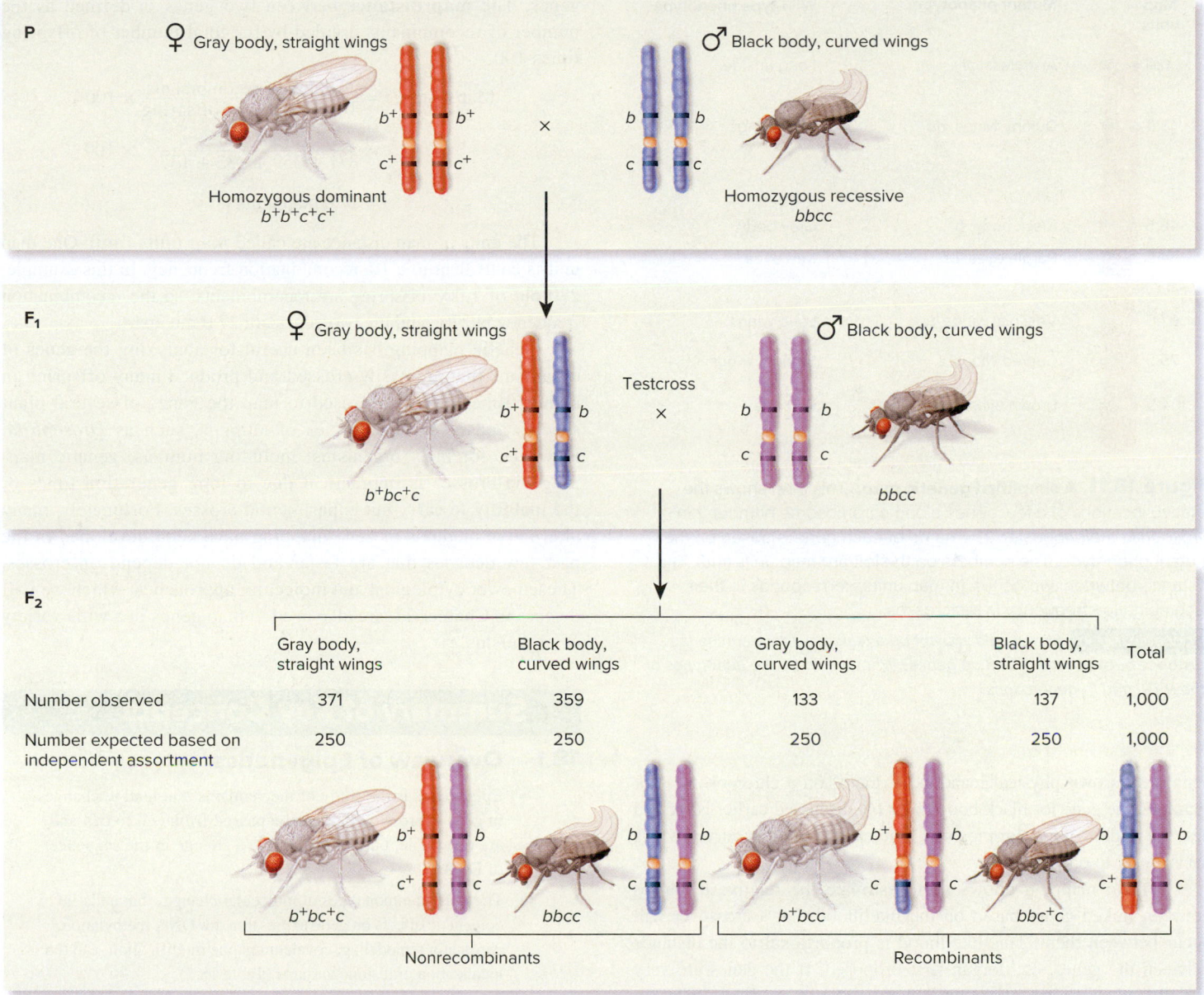

**Figure 18.10  Linkage and recombination of alleles.** An experimenter crossed $b^+b^+c^+c^+$ and $bbcc$ flies to produce $F_1$ heterozygotes. $F_1$ females were then testcrossed to $bbcc$ males. $F_2$ nonrecombinants have phenotypes that are the same as the parental (P) generation, whereas recombinants have a new combination of traits that is different from the P generation. The large number of nonrecombinant phenotypes in the $F_2$ generation suggests that the two genes are linked on the same chromosome. $F_2$ recombinant phenotypes occur because the alleles can be rearranged by crossing over. Note: The $b^+$ and $c^+$ alleles are dominant, and the $b$ and $c$ alleles are recessive.

*Concept Check:*  *In which fly or flies did crossing over occur to produce the recombinant offspring of the $F_2$ generation?*

Due to crossing over, two of the four egg cells produced by meiosis have recombinant chromosomes. What happens when eggs containing such chromosomes are fertilized in the testcross? Each of the male fly's sperm cells carries a chromosome with the two recessive alleles. If the egg contains the recombinant chromosome carrying the $b^+$ and $c$ alleles, the testcross will produce an $F_2$ offspring with a gray body and curved wings. If the egg contains the recombinant chromosome carrying the $b$ and $c^+$ alleles, $F_2$ offspring will have a black body and straight wings. Therefore, crossing over in the $F_1$ female can explain the occurrence of both types of $F_2$ recombinant offspring.

Morgan's third idea regarding linkage was that the frequency of crossing over between linked genes depends on the distance between them. This suggested a method for determining the relative positions of genes on a chromosome, as we will discuss next.

## Recombination Frequencies Provide a Method for Mapping Genes Along Chromosomes

The study of the arrangement of genes in a species' genome is called **genetic mapping**. As depicted in **Figure 18.11**, the linear order of genes along a chromosome is shown in a chart known as a **genetic map**. Each

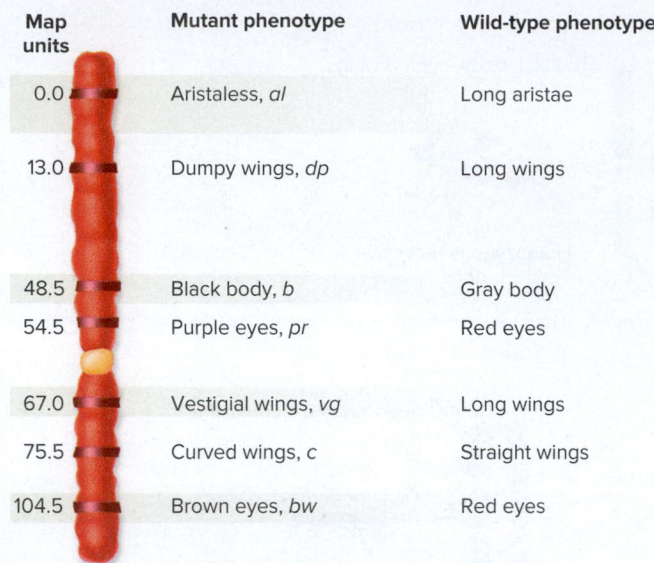

| Map units | Mutant phenotype | Wild-type phenotype |
|---|---|---|
| 0.0 | Aristaless, *al* | Long aristae |
| 13.0 | Dumpy wings, *dp* | Long wings |
| 48.5 | Black body, *b* | Gray body |
| 54.5 | Purple eyes, *pr* | Red eyes |
| 67.0 | Vestigial wings, *vg* | Long wings |
| 75.5 | Curved wings, *c* | Straight wings |
| 104.5 | Brown eyes, *bw* | Red eyes |

**Figure 18.11  A simplified genetic map.** This map shows the relative locations of a few genes along chromosome number 2 in *Drosophila melanogaster*. The name of each gene is based on the mutant phenotype. The numbers on the left are map units (mu). The distance between two genes, in map units, corresponds to their recombination frequency in testcrosses.

**Concept Check:** *How would you set up a testcross to determine the distance between the al and dp genes? What would be the genotypes of the P, $F_1$, and $F_2$ generations?*

gene has its own physical location, or locus, on a chromosome. For example, the gene for black body color (*b*) discussed earlier is located near the middle of the chromosome, whereas the gene for curved wings (*c*) is closer to one end.

Genetic mapping allows us to estimate the relative distances between linked genes based on the likelihood that a crossover will occur between them. This likelihood is proportional to the distance between the genes, as Morgan first proposed. If the genes are very close together, a crossover is unlikely to occur in the region between them. However, if the genes are very far apart, a crossover is more likely to occur in the region between the genes and thereby recombine the alleles. Therefore, in a testcross involving two genes on the same chromosome, the percentage of recombinant offspring is correlated with the distance between the genes. If a two-factor testcross produces many recombinants, the experimenter concludes that the two genes are far apart. If very few recombinants are observed, the two genes must be close together.

To find the distance between two genes, the experimenter must determine the frequency of crossing over between them, called their **recombination frequency**. This is accomplished by conducting a testcross. As an example, let's refer back to the *Drosophila* testcross described in Figure 18.10. As we discussed, the genes for body color and wing shape are on the same chromosome. The recombinants are the result of crossing over during egg formation in the $F_1$ female. We can use the data from the testcross shown in Figure 18.10 to estimate the distance between these two

genes. The **map distance** between two genes is defined as the number of recombinants divided by the total number of offspring times 100.

$$\text{Map distance} = \frac{\text{Number of recombinants}}{\text{Total number of offspring}} \times 100$$

$$= \frac{133 + 137}{371 + 359 + 133 + 137} \times 100$$

$$= 27.0 \text{ map units}$$

The units of map distance are called **map units (mu)**. One map unit is equivalent to a 1% recombination frequency. In this example, 270 out of 1,000 offspring are recombinants, so the recombination frequency is 27%, and the two genes are 27.0 mu apart.

Genetic mapping has been useful for analyzing the genes of organisms that are easily crossed and produce many offspring in a short time. It has been used to map the genes of several plant species and of certain species of animals, such as *Drosophila*. However, for most organisms, including humans, genetic mapping via crosses is impractical due to long generation times or the inability to carry out experimental crosses. Fortunately, many alternative methods of gene mapping have been developed in the past few decades that are faster and do not depend on crosses. These newer cytological and molecular approaches, which we will discuss in Chapter 21, are also used to map genes in a wide variety of organisms.

## Summary of Key Concepts

### 18.1  Overview of Epigenetics

- Epigenetics is the study of mechanisms that lead to changes in gene expression that can be passed from cell to cell and are reversible, but do not involve a change in the sequence of DNA.

- The most common types of molecular changes that underlie epigenetic effects on gene expression are DNA methylation, chromatin remodeling, covalent histone modification, and the localization of histone variants (Table 18.1).

### 18.2  Epigenetics I: Genomic Imprinting

- As a result of genomic imprinting, offspring express either a maternal or paternal allele, depending on how a particular gene is marked, or imprinted (Figure 18.1).

- During gamete formation, DNA methylation of an allele from one parent is a mechanism to achieve imprinting (Figure 18.2).

### 18.3  Epigenetics II: X-Chromosome Inactivation

- If a female is heterozygous for an X-linked gene, this can lead to a mosaic phenotype, with some of the somatic cells expressing one allele and some expressing the other (Figures 18.3, 18.4).

- XCI in female mammals occurs when one X chromosome in every somatic cell is randomly inactivated. The counting of X inactivation centers (Xics) causes somatic cells to have only one active X chromosome (Table 18.2)

## 18.4  Epigenetics III: Effects of Environmental Agents

- Dietary factors during early stages of development cause epigenetic changes that affect phenotype (Figure 18.5).

- Environmental agents have been shown to cause epigenetic changes that are associated with human diseases such as cancer (Table 18.3).

## 18.5  Extranuclear Inheritance: Organelle Genomes

- Mitochondria and chloroplasts carry a small number of genes. The inheritance of such genes is called extranuclear inheritance (Figure 18.6).

- Chloroplasts in the four-o'clock plant are transmitted via the egg, a pattern called maternal inheritance (Figures 18.7, 18.8).

- Several human diseases are known to be caused by mutations in mitochondrial genes, which follow a maternal inheritance pattern (Table 18.4).

## 18.6  Genes on the Same Chromosome: Linkage and Recombination

- When two different genes are on the same chromosome, they are said to be linked. Linked genes tend to be inherited as a unit, unless crossing over separates them (Figures 18.9, 18.10).

- Genetic mapping allows us to determine the order of genes along a chromosome and the relative distances between them, based on the frequency of crossing over observed in testcrosses (Figure 18.11).

## Assess & Discuss

### Test Yourself

1. Which of the following is an example of an epigenetic change that alters gene expression?
   a. chromatin remodeling
   b. covalent histone modification
   c. localization of histone variants
   d. DNA methylation
   e. all of the above

2. In mice, the allele of the *Igf2* gene that is inherited from the mother is never expressed in her offspring. This happens because the *Igf2* gene from the mother
   a. always undergoes a mutation that inactivates its function.
   b. is deleted during oogenesis.
   c. is deleted during embryonic development.
   d. is not transcribed in the somatic cells of her offspring.
   e. is affected by all of the above.

3. A female mouse that is *Igf2 Igf2⁻* is crossed to a male that is also *Igf2 Igf2⁻*. The expected outcome for the phenotypes of the offspring from this cross is
   a. all normal size.
   b. all dwarf.
   c. 1 normal size : 1 dwarf.
   d. 3 normal size : 1 dwarf.
   e. 1 normal size : 3 dwarf.

4. The marking process for genomic imprinting initially occurs during
   a. gametogenesis.
   b. fertilization.
   c. embryonic development.
   d. adulthood.
   e. both b and c.

5. According to Lyon's hypothesis, the patchwork pattern on a calico cat can be explained by which of the following?
   a. One of the X chromosomes is converted to a Barr body in somatic cells of female mammals.
   b. One of the X chromosomes is converted to a Barr body in all cells of female mammals.
   c. Both of the X chromosomes are converted to Barr bodies in somatic cells of female mammals.
   d. Both of the X chromosomes are converted to Barr bodies in all cells of female mammals.
   e. One of the X chromosomes is lost in the somatic cells of female mammals.

6. A female mouse that is homozygous for the $A^{vy}$ allele is mated to a male that is homozygous for a loss-of-function *(a)* allele. How would you expect the diet of this female during pregnancy to affect her offspring?
   a. Dietary agents that promote a greater level of DNA methylation would produce offspring with more yellow fur.
   b. Dietary agents that promote a lower level of DNA methylation would produce offspring with more yellow fur.
   c. Dietary agents that promote a greater level of DNA methylation would produce offspring with darker fur.
   d. Dietary agents that promote a lower level of DNA methylation would produce offspring with darker fur.
   e. Both b and c are correct.

7. Environmental agents may cause epigenetic changes that alter the expression of specific genes. Which of the following epigenetic changes would be expected to promote cancer?
   a. a change that resulted in the overexpression of a tumor-suppressor gene
   b. a change that resulted in the inhibition of a tumor-suppressor gene
   c. a change that resulted in the overexpression of an oncogene
   d. a change that resulted in the inhibition of an oncogene
   e. Both b and c are correct.

8. In many organisms, organelles, such as the mitochondria, are transmitted only by the egg. This phenomenon is known as
   a. biparental inheritance.
   b. paternal inheritance.
   c. X-linked inheritance.
   d. maternal inheritance.
   e. both c and d.

9. Based on the ideas proposed by Morgan, which of the following statements concerning linkage is *false?*
   a. Traits determined by genes located close together on the same chromosome are likely to be inherited together.
   b. Crossing over between homologous chromosomes can create new allele combinations.
   c. A crossover is more likely to occur in a region between two genes that are close together than in a region between two genes that are farther apart.
   d. The probability of crossing over depends on the distance between the genes.
   e. Genes that tend to be transmitted together are physically located on the same chromosome.

10. Extranuclear inheritance occurs because
    a. certain genes are found on the X chromosome.
    b. chromosomes in the nucleus may be transferred to the cytoplasm.
    c. some cellular organelles contain DNA.
    d. the nuclear membrane breaks apart during cell division.
    e. both a and c.

## Conceptual Questions

1. Define epigenetics. Are all epigenetic changes passed from parent to offspring? Explain.

2. What is a Barr body? How is its structure different from that of other chromosomes in the cell? How does the structure of a Barr body affect the level of X-linked gene expression?

3.  **Core Concept: Information** A core concept of biology is that the genetic material provides a blueprint for reproduction. Explain how epigenetics affects that blueprint.

## Collaborative Questions

1. Go to the PubMed website and search the words epigenetic and cancer. Scan through the journal articles you retrieve, and make a list of environmental agents that may cause epigenetic changes that contribute to cancer.

2. Mendel studied seven traits in garden pea plants, and this species happens to have seven different chromosomes. It has been pointed out that Mendel was very lucky not to have conducted crosses involving two traits governed by genes that are closely linked on the same chromosome, because the results would have confounded his theory of independent assortment. It has even been suggested that Mendel may not have published data involving traits that were linked! An article by Stig Blixt 1975. (Why Didn't Gregor Mendel Find Linkage? *Nature* 256: 206, 1975) considers this issue. Look up this article, and discuss why Mendel did not find linkage.

# Genetics of Viruses and Bacteria

# 19

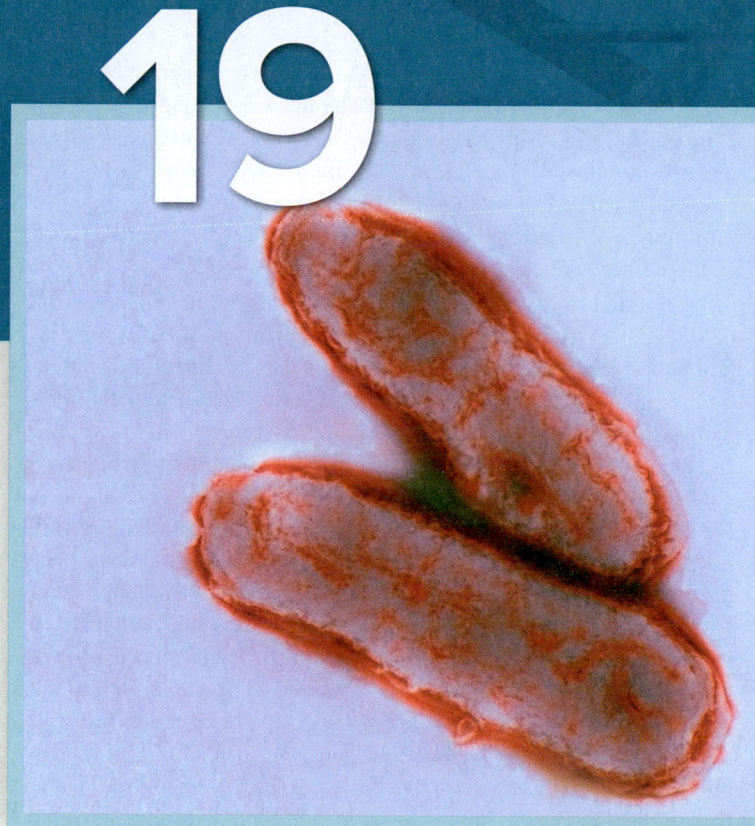

**A colorized micrograph of** *Haemophilus influenzae*, **type b. This bacterium is a common cause of meningitis—a serious infection of the fluid in the spinal cord and the fluid that surrounds the brain.** ©CAMR/A. Barry Dowsett/Science Source

**W**hile studying for his calculus test, Jason was having trouble concentrating due to a severe headache and fever. He thought he must be coming down with a cold. Though he had taken some aspirin, it didn't seem to be working. As he was eating some potato chips, one dropped in his lap. When he tried to look down to see where the chip had fallen, he realized that his neck was extremely stiff; he could barely move his head to look downward. Also, the brightness of his desk light seemed freakishly painful to his eyes. Over the course of that evening, Jason became confused and lethargic, and his roommate urged him to see a doctor. Fortunately, Jason took his advice and went to the college clinic. The diagnosis was bacterial meningitis—an inflammation of the protective membranes that cover the brain and spinal cord, collectively called the meninges. Although a relatively rare disease, bacterial meningitis is up to six times more common among people living in close quarters, such as college dormitories. Because Jason sought help early enough, his disease was successfully treated with antibiotics. Had he not gotten help, the disease could have progressed to the point of causing severe brain damage and even death.

Jason's story highlights a primary reason why biologists are so interested in viruses and bacteria. Many of them are **pathogens**— agents that cause disease symptoms in their hosts. Infectious diseases caused by viruses and bacteria are a leading cause of human suffering and death, accounting for one-quarter to one-third of deaths worldwide. The spread of infectious diseases results from human behavior, and in recent times, it has been accelerated by changes in land-use patterns, increased trade and travel, and the inappropriate use of antibiotic drugs. Although the incidence of fatal infectious diseases in the U.S. is low compared to the worldwide average, an alarming increase in more deadly strains of viruses and bacteria has occurred over the past few decades. Since 1980, the number of deaths in the U.S. due to infectious diseases has approximately doubled.

In this chapter, we turn our attention to the genetic analyses of viruses and bacteria. We will begin by examining viruses and other nonliving particles that infect living cells. All organisms are thought to be susceptible to infection by one or more types of viruses, which use the host's cellular machinery to replicate. Once a cell is infected, the genetic material of a virus orchestrates a series of events that ultimately leads to the production of new virus particles. We will consider the biological complexity of viruses and explore viral reproductive cycles. We will also examine some of the simplest and smallest infectious agents, called viroids and prions.

In the remaining sections of this chapter, we will examine the bacterial genome and the methods used in its investigation. Like their eukaryotic counterparts, bacteria have genetic differences that affect their cellular traits, and techniques of modern microbiology make many of these differences, such as sensitivity to antibiotics and differences in nutritional requirements, easy to study. Although bacteria reproduce asexually by cell division, their genetic variety is enhanced by the phenomenon called gene transfer, in which genes are passed from one bacterial cell to another. Like sexual reproduction in eukaryotes, gene transfer enhances the genetic diversity observed among bacterial species. In this chapter, we will explore three interesting ways that bacteria transfer genetic material.

# 19.1    General Properties of Viruses

## Learning Outcome:

1. Compare and contrast types of viruses with regard to their host range, structure, and genomes.

Viruses are nonliving particles with nucleic acid genomes. Why are viruses considered nonliving? The answer is that they do not exhibit key properties associated with living organisms. Viruses are not composed of cells, and by themselves, they do not use energy or carry out metabolism, maintain homeostasis, or even reproduce. A virus or its genetic material must be taken up by a living cell to replicate.

The first virus to be discovered was tobacco mosaic virus (TMV). This virus infects many species of plants and causes mosaic-like patterns in which normal-colored patches are interspersed with light green or yellowish patches on the leaves (**Figure 19.1**). TMV damages leaves, flowers, and fruit but almost never kills the plant. In 1883, German chemist Adolf Mayer determined that this disease could be spread by spraying the sap from one plant onto another. By subjecting this sap to filtration, Russian scientist Dmitri Ivanovski demonstrated that the disease-causing agent was not a bacterium. Sap that had been passed through filters with pores small enough to prevent the passage of bacterial cells was still able to spread the disease. At first, some researchers suggested the agent was a chemical toxin. However, Dutch botanist Martinus Beijerinck ruled out this possibility by showing that sap continued to transmit the disease after many plant generations. A toxin would have been diluted after many generations, but Beijerinck's results indicated the disease agent was multiplying in the plant. Around the same time, animal viruses were discovered in connection with a disease of cattle called foot-and-mouth disease. In 1900, the first human virus, the virus that causes yellow fever, was identified.

**Figure 19.1** **A plant infected with tobacco mosaic virus.** ©Norm Thomas/Science Source

Since these early studies, microbiologists, geneticists, and molecular biologists have taken a great interest in the structure, genetic composition, and replication of viruses. In this section, we will discuss their general properties.

## Viruses Are Remarkably Varied, Despite Their Simple Structure

A **virus** is a small infectious particle that consists of nucleic acid enclosed in a protein coat. Researchers have identified and studied over 4,000 different types of viruses. Although all viruses share some similarities, such as small size and the reliance on a living cell for replication, they vary greatly in their characteristics, including their host range, structure, and genome composition. Some of the major differences are described next, and characteristics of selected viruses are shown in **Table 19.1**.

| Table 19.1 | Hosts and Characteristics of Selected Viruses | | | | | |
|---|---|---|---|---|---|---|
| Virus or group of viruses | Host | Effect on host | | Nucleic acid* | Genome size (kb)† | Number of genes† |
| Phage λ | E. coli | Can exist harmlessly in the host cell or cause lysis | | dsDNA | 48.5 | 36 |
| Phage T4 | E. coli | Causes lysis | | dsDNA | 169 | 288 |
| Tobacco mosaic virus (TMV) | Many plants | Causes mottling and necrosis of leaves and other plant parts | | ssRNA | 6.4 | 6 |
| Baculoviruses | Insects | Most baculoviruses are species specific; they usually kill the insect | | dsDNA | 133.9 | 154 |
| Influenza virus | Mammals | Causes classical "flu," with fever, cough, sore throat, and headache | | ssRNA | 13.5 | 11 |
| Epstein-Barr virus | Humans | Causes mononucleosis, with fever, sore throat, and fatigue | | dsDNA | 172 | 80 |
| Adenovirus | Humans | Causes respiratory symptoms and diarrhea | | dsDNA | 34 | 35 |
| Herpes simplex type II | Humans | Causes blistering sores around the genital region | | dsDNA | 158.4 | 77 |
| HIV (type I) | Humans | Causes AIDS, an immunodeficiency syndrome eventually leading to death | | ssRNA | 9.7 | 9 |

*The abbreviations ss and ds refer to single-stranded and double-stranded, respectively.
†Several of the viruses listed in this table are found in different strains that vary with regard to genome size and number of genes. The numbers reported in this table are typical values. The abbreviation kb means kilobase, which equals 1,000 bases.

**Differences in Host Range**   A cell that is infected by a virus is called a **host cell**, and a species that can be infected by a specific virus is called a host species for that virus. Viruses differ greatly in their **host range**—the number of species and cell types they can infect. Table 19.1 lists a few examples of viruses with widely different ranges of host species. Tobacco mosaic virus, which we discussed earlier, has a broad host range. TMV is known to infect over 150 different species of plants. By comparison, other viruses have a narrow host range, with some infecting only a single species. Furthermore, a virus may infect only a specific cell type in a host species. **Figure 19.2** shows some viruses that infect particular human cells and cause disease.

**Structural Differences**   Viruses cannot be resolved by even the best light microscope. Although the existence of viruses was postulated in the 1890s, viruses were not observed until the 1930s when the electron microscope was invented. Viruses range in size from 20 to 400 nm in diameter (1 nm [nanometer] = $10^{-9}$ meter).

For comparison, a typical bacterium is 1,000 nm in diameter, and the diameter of most eukaryotic cells is 10 to 1,000 times that of a bacterium. Adenoviruses, which cause infections of the respiratory and gastrointestinal tracts, have an average diameter of 75 nm. Over 50 million adenoviruses could fit into an average-sized human cell.

What are the common structural features of all viruses? As shown in **Figure 19.3**, all viruses have a protein coat called a **capsid**, which encloses a genome consisting of one or more molecules of nucleic acid (DNA or RNA). Capsids are composed of one or several different protein subunits called capsomers. Capsids have a variety of shapes, including helical and polyhedral. Figure 19.3a shows the structure of TMV, which has a helical capsid made of identical capsomers. Figure 19.3b shows an adenovirus, which has a polyhedral capsid. Protein fibers with a terminal knob are located at the corners of the polyhedral capsid. Many viruses that infect animal cells, such as the influenza virus shown in Figure 19.3c, have a **viral envelope** enclosing the capsid. The envelope consists of a lipid bilayer that is derived from the plasma membrane of the host cell and is embedded with virally encoded spike glycoproteins.

In addition to encasing and protecting the genetic material, the capsid and envelope enable viruses to infect their hosts. Many viruses have protein fibers with a knob (Figure 19.3b) or spike glycoproteins (Figure 19.3c) that help them bind to the surface of a host cell. Viruses that infect bacteria, called **bacteriophages**, or simply **phages**, may have more complex protein coats, with accessory structures used for anchoring the virus to a host cell and injecting the viral nucleic acid (Figure 19.3d). As discussed later, the tail fibers of such bacteriophages are needed to attach the virus to the bacterial cell wall.

**Genome Differences**   The genetic material in a virus is called a **viral genome**. The composition of viral genomes varies markedly among different types of viruses, as suggested by the examples in Table 19.1. The nucleic acid of some viruses is DNA, whereas in others it is RNA. These are referred to as DNA viruses and RNA viruses, respectively. It is striking that some viruses use RNA for their genome, whereas all living organisms use DNA. In some viruses, the nucleic acid is single stranded, whereas in others, it is double stranded. The genome can be linear or circular, depending on the type of virus. Some kinds of viruses have more than one copy of the genome.

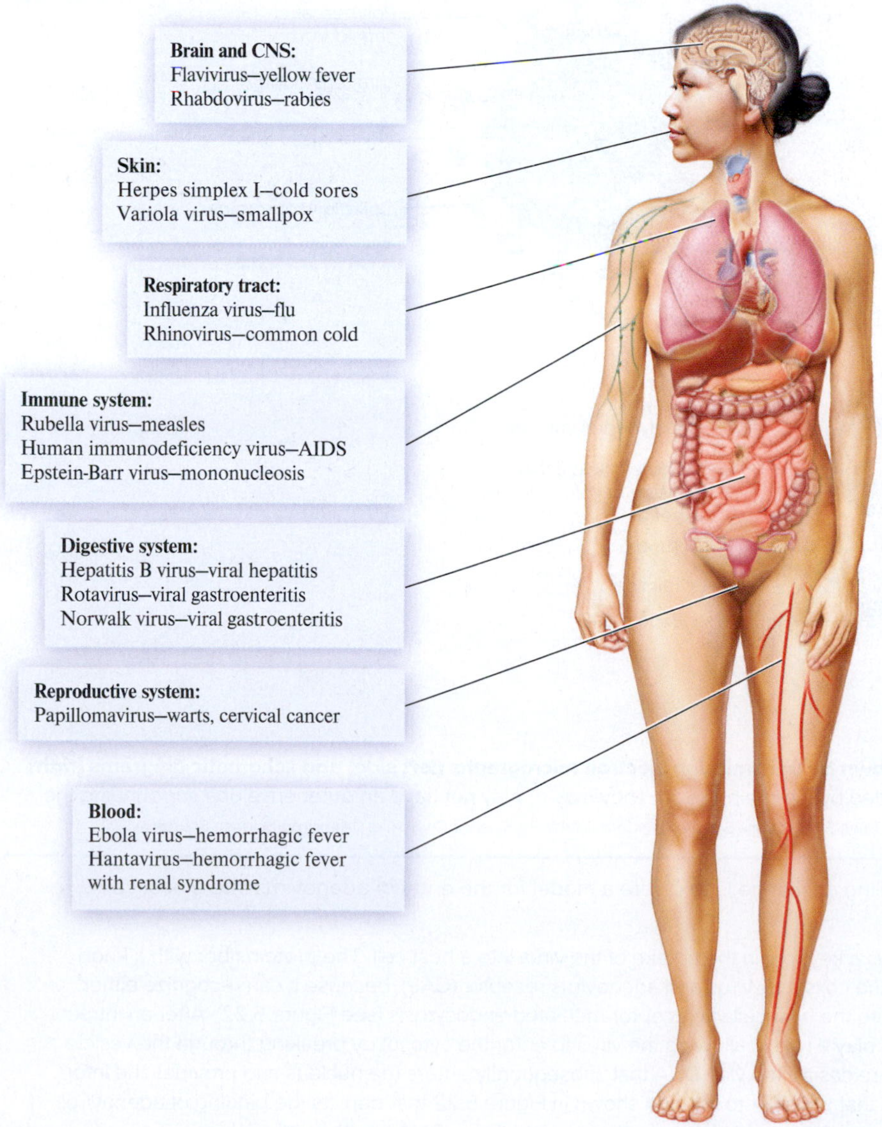

**Brain and CNS:**
Flavivirus—yellow fever
Rhabdovirus—rabies

**Skin:**
Herpes simplex I—cold sores
Variola virus—smallpox

**Respiratory tract:**
Influenza virus—flu
Rhinovirus—common cold

**Immune system:**
Rubella virus—measles
Human immunodeficiency virus—AIDS
Epstein-Barr virus—mononucleosis

**Digestive system:**
Hepatitis B virus—viral hepatitis
Rotavirus—viral gastroenteritis
Norwalk virus—viral gastroenteritis

**Reproductive system:**
Papillomavirus—warts, cervical cancer

**Blood:**
Ebola virus—hemorrhagic fever
Hantavirus—hemorrhagic fever with renal syndrome

**Figure 19.2   Some viruses that cause human diseases.** Most viruses that cause disease in humans infect cells of specific tissues, as illustrated by the examples in this figure.

 **Core Concept: Science and Society** By studying viruses, biologists have developed vaccines and drugs to help prevent their spread.

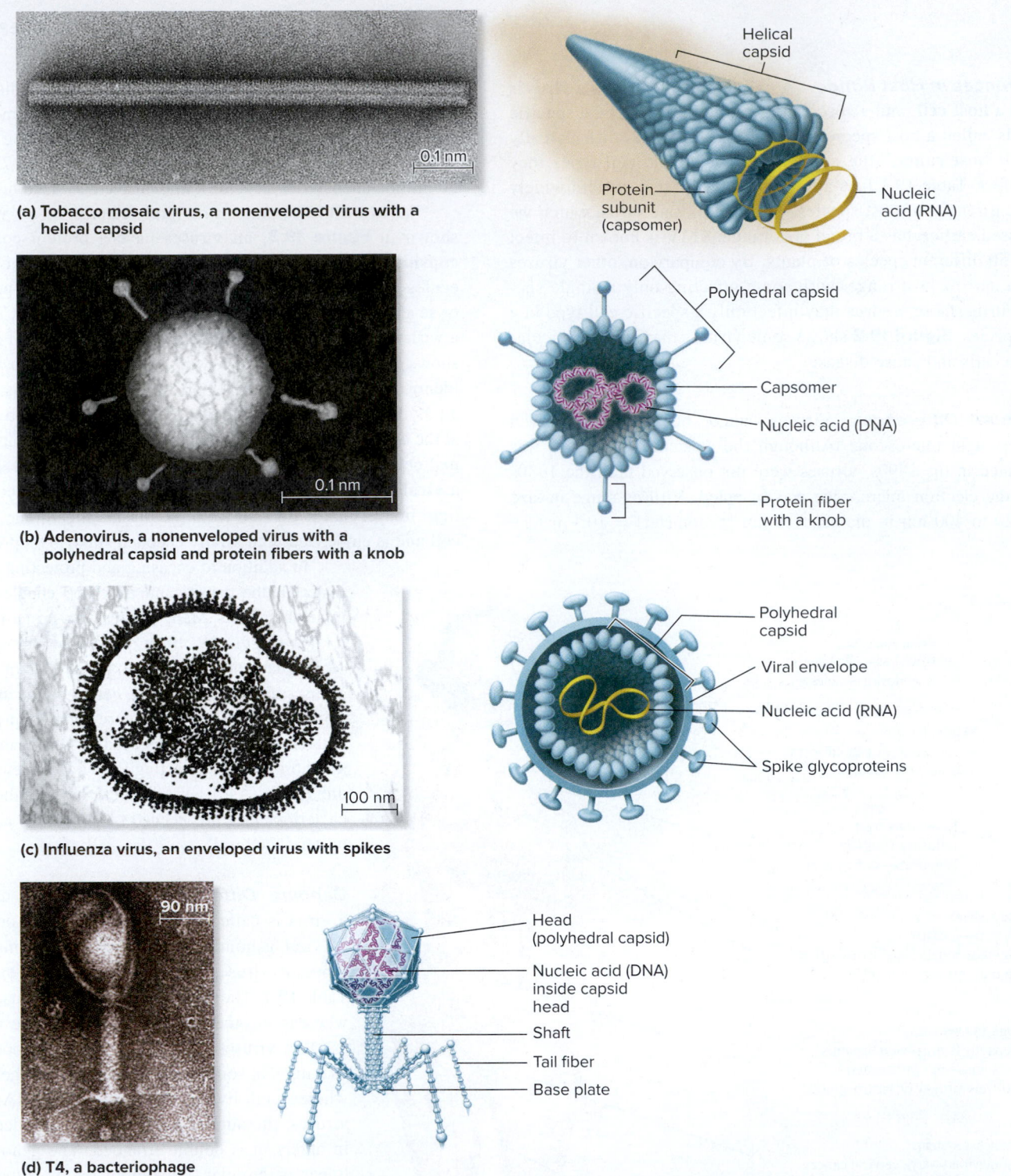

**(a) Tobacco mosaic virus, a nonenveloped virus with a helical capsid**

Helical capsid

Protein subunit (capsomer)

Nucleic acid (RNA)

**(b) Adenovirus, a nonenveloped virus with a polyhedral capsid and protein fibers with a knob**

Polyhedral capsid

Capsomer

Nucleic acid (DNA)

Protein fiber with a knob

**(c) Influenza virus, an enveloped virus with spikes**

Polyhedral capsid

Viral envelope

Nucleic acid (RNA)

Spike glycoproteins

**(d) T4, a bacteriophage**

Head (polyhedral capsid)

Nucleic acid (DNA) inside capsid head

Shaft

Tail fiber

Base plate

**Figure 19.3 Variations in the structure of viruses, as shown by transmission electron micrographs (left side) and schematic diagrams (right side).** All viruses contain nucleic acid (DNA or RNA) surrounded by a protein capsid. They may or may not have an outer envelope surrounding the capsid. a: ©Science Source; b: ©Dr. Linda M. Stannard, University of Cape Town/SPL/Science Source; c: ©Chris Bjornberg/Science Source; d: ©Omikron/Science Source

**Core Skill: Modeling** The goal of this modeling challenge is to create a model for the entry of adenovirus into a host cell.

**Modeling Challenge** The capsid of an adenovirus plays a key role in the uptake of the virus into a host cell. The protein fiber with a knob binds to a receptor on the surface of the host cell called the coxsackievirus and adenovirus receptor (CAR), because it can recognize either coxsackievirus or adenovirus. Adenovirus is then taken into the host cell via receptor-mediated endocytosis (see Figure 5.22). After an intracellular vesicle carrying the virus forms, capsid proteins also play a role in allowing the virus to enter the cytosol by breaking through the vesicle membrane. During this process, the capsid breaks apart, releasing the viral DNA that subsequently enters the nucleus and provides the information to make thousands of new viruses. Draw a model that is similar to the one shown in Figure 5.22 that depicts the binding of adenovirus to a host cell and its subsequent uptake into the cell. Your model should also include the entry of viral DNA into the cell nucleus.

Viral genomes also vary considerably in size, ranging from a few thousand to more than a hundred thousand nucleotides in length (see Table 19.1). For example, the genome of TMV is only 6,400 base pairs in length and contains only six genes. Other viruses, particularly those with a complex structure, such as phage T4, contain many more genes. These extra genes encode many different proteins that are involved in the formation of the elaborate structure shown in Figure 19.3d.

## 19.2 Viral Reproductive Cycles

**Learning Outcomes:**

1. List the steps in a viral reproductive cycle, and distinguish between the lysogenic and lytic cycles.

2. Describe how emerging viruses such as HIV arise and spread through a population.

3. Explain how an understanding of viral structure and reproduction can aid in the development of drugs to combat viruses.

4. Outline three hypotheses regarding the origin of viruses.

When a virus infects a host cell, the expression of viral genes leads to a series of steps, called a **viral reproductive cycle**, which results in the production of new viruses. The details of the steps differ among various types of viruses, and a given type of virus may follow alternative cycles. Even so, by studying hundreds of different viruses, researchers have determined that the viral reproductive cycle consists of five or six common steps. In this section, we will examine viral reproductive cycles in detail, paying particular attention to human immunodeficiency virus (HIV), which causes acquired immunodeficiency syndrome (AIDS) in humans.

### Viral Reproductive Cycles Consist of a Few Common Steps

To illustrate the general features of viral reproductive cycles, **Figure 19.4** considers these steps for two types of viruses. Figure 19.4a shows the cycle of phage λ (lambda), a bacteriophage with double-stranded DNA as its genome, and Figure 19.4b depicts the cycle of HIV, an enveloped animal virus containing single-stranded RNA. The descriptions that follow compare the reproductive cycles of these two very different viruses.

**Step 1: Attachment**    In the first step of a viral reproductive cycle, the virus attaches to the surface of a host cell. This attachment is usually specific for one or just a few types of cells because proteins in the virus recognize and bind to specific molecules on the cell surface. In the case of phage λ, the tail fibers bind to proteins in the outer bacterial cell membrane of *E. coli* cells. In the case of HIV, spike glycoproteins in the viral envelope bind to protein receptors in the plasma membrane of human white blood cells called helper T cells.

**Step 2: Entry**    After attachment, the viral genome enters the host cell. Attachment of phage λ stimulates a conformational change in its coat proteins, so the shaft contracts, and the phage injects its DNA into the bacterial cytoplasm. In contrast, the envelope of HIV fuses with the plasma membrane of the host cell, so both the capsid and its contents are released into the cytosol. Some of the HIV capsid proteins are then removed by host cell enzymes, a process called uncoating. This releases two copies of the viral RNA and molecules of two enzymes called reverse transcriptase and integrase into the cytosol. As discussed shortly, these enzymes are needed for step 3.

Once a viral genome has entered the cell, one or a few viral genes are expressed immediately due to the action of host cell enzymes and ribosomes. Expression of these early genes leads quickly to either step 3 or step 4 of the reproductive cycle, depending on the specific virus. The genome of some viruses, including both phage λ and HIV, can integrate into a chromosome of the host cell. For such viruses, the cycle may proceed from step 2 to step 3 as described next, delaying the production of new viruses. Alternatively, the cycle for phage λ may proceed directly from step 2 to step 4 and quickly lead to the production of new viruses.

**Step 3: Integration**    Viruses capable of integration carry a gene that encodes an enzyme called **integrase**, which cuts the host's chromosomal DNA and inserts the viral genome into the chromosome. In the case of phage λ, the double-stranded DNA that entered the cell can be directly integrated into the double-stranded DNA of the chromosome. Once integrated, the phage DNA in a bacterium is called a **prophage**. When a bacterial cell divides, the prophage DNA is copied and transmitted to daughter cells along with the bacterial chromosomal DNA. When a prophage has been integrated into a bacterial chromosome, this type of viral reproductive cycle is called the **lysogenic cycle**. As discussed later, new phages are not made during the lysogenic cycle, and the host cell is not destroyed. On occasion, a prophage can be excised from the bacterial chromosome and proceed to step 4.

How can an RNA virus integrate its genome into the host cell's DNA? For this to occur, the viral genome must be copied into DNA. HIV accomplishes this by means of a viral enzyme called **reverse transcriptase**, which is carried within the capsid and released into the host cell along with the viral RNA. Reverse transcriptase uses the viral RNA strand to make a complementary copy of DNA. The complementary DNA is then used as a template to make double-stranded viral DNA. This process is called reverse transcription because it is the reverse of the usual transcription process, in which a DNA strand is used to make a complementary strand of RNA. The viral double-stranded DNA enters the host cell nucleus and is inserted into a host chromosome via integrase. Like reverse transcriptase, integrase is carried within the HIV capsid and released into the host cell during uncoating. Once integrated, the viral DNA in a eukaryotic cell is called a **provirus**. Viruses that follow this mechanism are called **retroviruses**.

**Step 4: Synthesis of Viral Components**    The production of new viruses by a host cell involves the replication of the viral genome and the synthesis of viral proteins that make up the protein coat. A prophage must be excised as described in step 3 before the synthesis of new viral components can occur. An enzyme called excisionase is required for this process. Following excision, host cell enzymes make many copies of the phage DNA and transcribe the genes within these copies into mRNA. Host cell ribosomes translate this viral mRNA

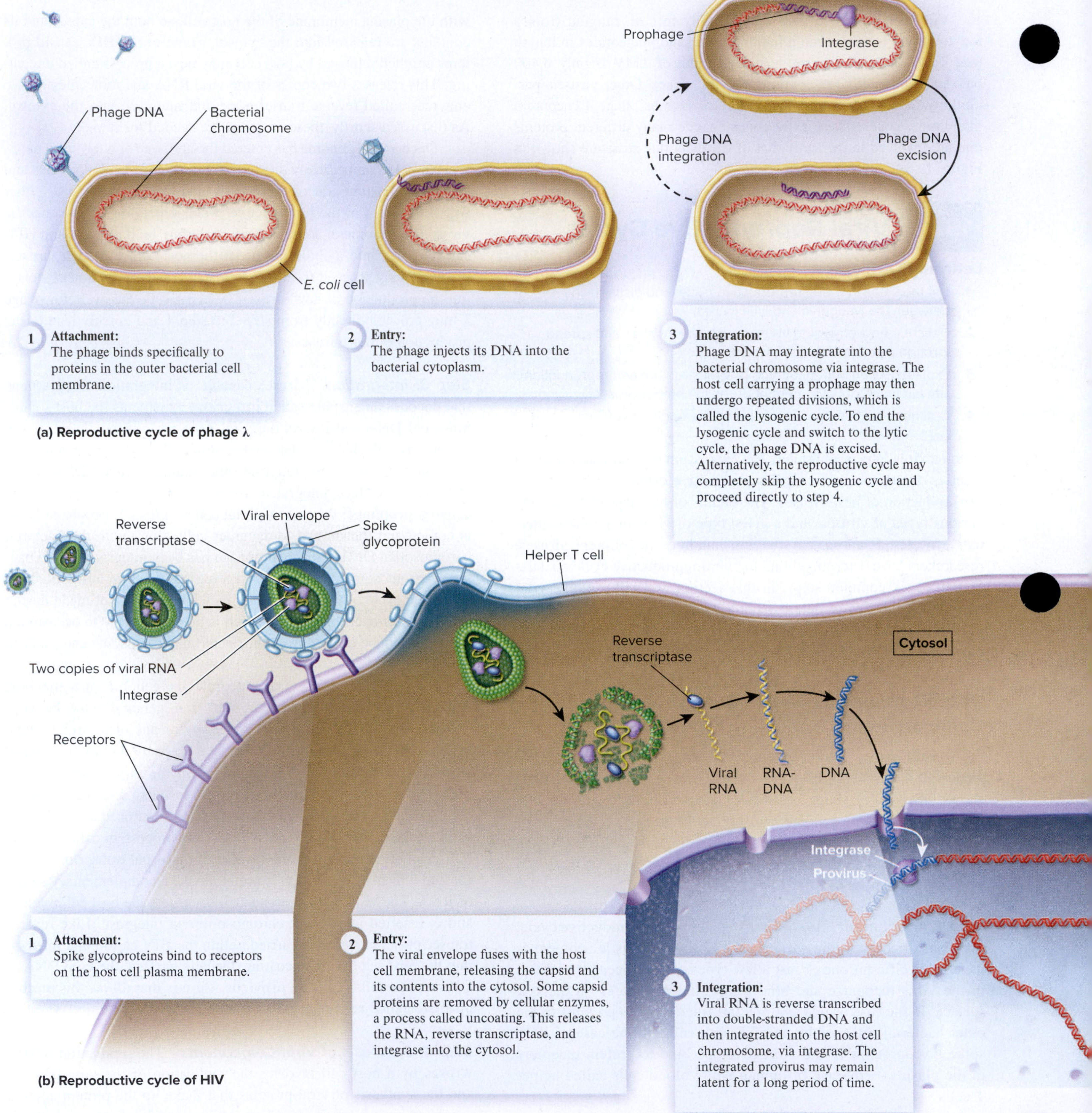

**Phage DNA**   **Bacterial chromosome**

**Prophage**   **Integrase**

**Phage DNA integration**   **Phage DNA excision**

*E. coli* cell

**1  Attachment:**
The phage binds specifically to proteins in the outer bacterial cell membrane.

**2  Entry:**
The phage injects its DNA into the bacterial cytoplasm.

**3  Integration:**
Phage DNA may integrate into the bacterial chromosome via integrase. The host cell carrying a prophage may then undergo repeated divisions, which is called the lysogenic cycle. To end the lysogenic cycle and switch to the lytic cycle, the phage DNA is excised. Alternatively, the reproductive cycle may completely skip the lysogenic cycle and proceed directly to step 4.

**(a) Reproductive cycle of phage λ**

**Reverse transcriptase**   **Viral envelope**   **Spike glycoprotein**

**Helper T cell**

**Two copies of viral RNA**

**Integrase**

**Reverse transcriptase**

**Cytosol**

**Receptors**

**Viral RNA**   **RNA-DNA**   **DNA**

**Integrase**
**Provirus**

**1  Attachment:**
Spike glycoproteins bind to receptors on the host cell plasma membrane.

**2  Entry:**
The viral envelope fuses with the host cell membrane, releasing the capsid and its contents into the cytosol. Some capsid proteins are removed by cellular enzymes, a process called uncoating. This releases the RNA, reverse transcriptase, and integrase into the cytosol.

**3  Integration:**
Viral RNA is reverse transcribed into double-stranded DNA and then integrated into the host cell chromosome, via integrase. The integrated provirus may remain latent for a long period of time.

**(b) Reproductive cycle of HIV**

**Figure 19.4  Comparison of the steps of two viral reproductive cycles.** (a) The reproductive cycle of phage λ, a bacteriophage with a double-stranded DNA genome. (b) The reproductive cycle of HIV, an enveloped animal virus with a single-stranded RNA genome.

 **Core Skill: Connections** Look back at Figure 5.21. How does the release of HIV resemble exocytosis?

**4**  **Synthesis of viral components:**
In the lytic cycle, phage DNA directs the synthesis of viral components. During this process, the phage DNA circularizes, and the host chromosomal DNA is degraded.

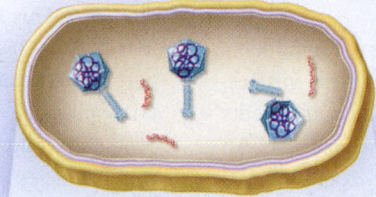

**5**  **Viral assembly:**
Phage components are assembled with the help of noncapsid proteins to make many new phages.

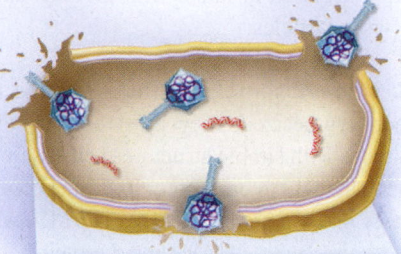

**6**  **Release:**
The viral enzyme called lysozyme causes cell lysis, and new phages are released from the broken cell.

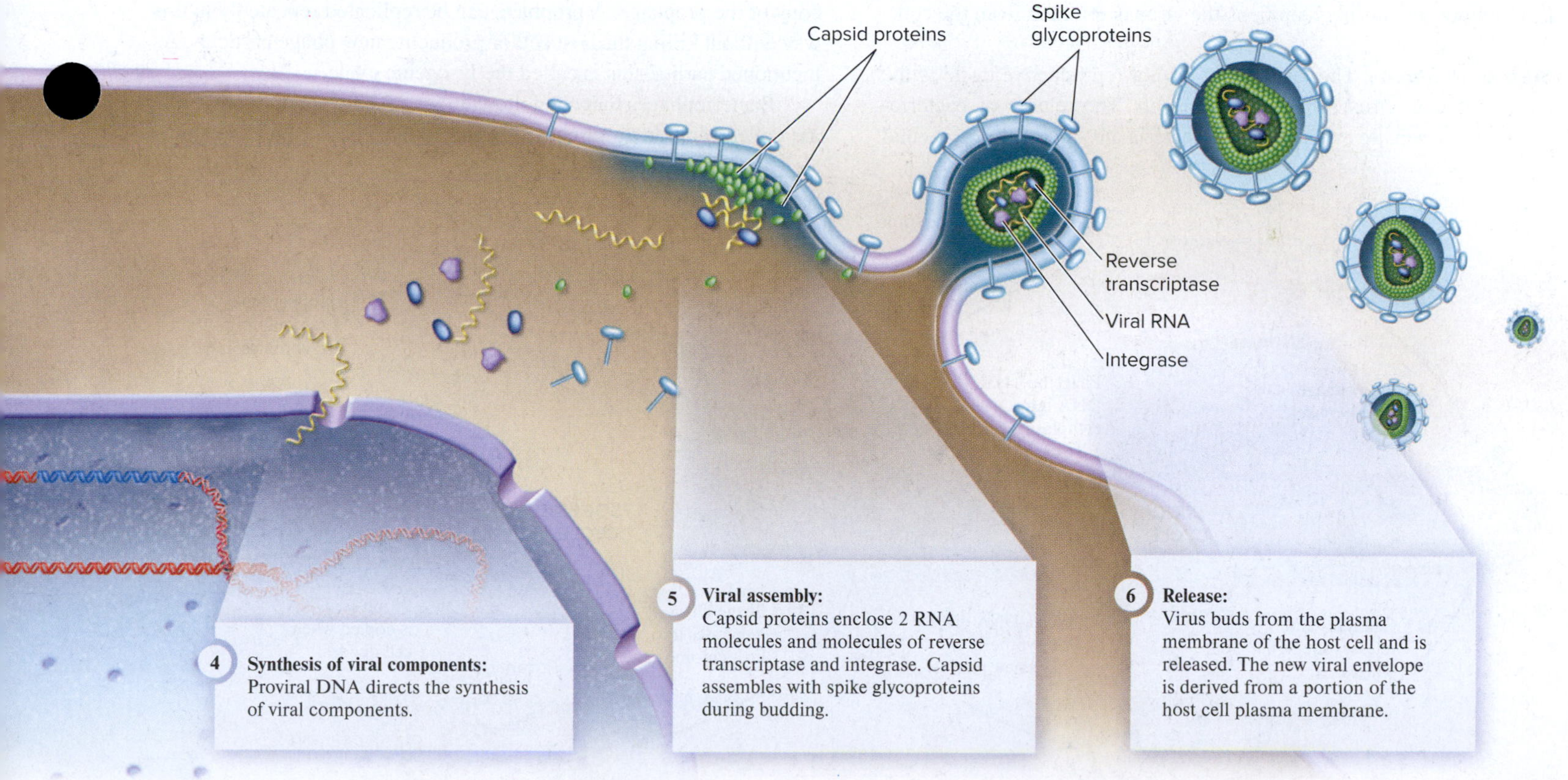

Capsid proteins

Spike glycoproteins

Reverse transcriptase

Viral RNA

Integrase

**4**  **Synthesis of viral components:**
Proviral DNA directs the synthesis of viral components.

**5**  **Viral assembly:**
Capsid proteins enclose 2 RNA molecules and molecules of reverse transcriptase and integrase. Capsid assembles with spike glycoproteins during budding.

**6**  **Release:**
Virus buds from the plasma membrane of the host cell and is released. The new viral envelope is derived from a portion of the host cell plasma membrane.

into viral proteins. The expression of phage genes also leads to the degradation of the host chromosomal DNA.

In the case of HIV, the provirus DNA is not excised from the host chromosome. Instead, it is transcribed in the nucleus to produce many copies of viral RNA. These viral RNA molecules enter the cytosol, where they are used to make viral proteins and serve as the genome for new viral particles.

### Step 5: Viral Assembly

After all of the necessary components have been synthesized, they must be assembled into new viruses. Some viruses with simple structures self-assemble—viral components spontaneously bind to each other to form a complete virus particle. An example of a self-assembling virus is TMV, which we examined earlier (see Figure 19.3a). TMV capsid proteins assemble around a TMV RNA molecule, which becomes trapped inside the hollow capsid.

Other viruses, including the two shown in Figure 19.4, do not self-assemble. The assembly of phage λ requires the help of noncapsid proteins not found in the completed phage particle. Some of these noncapsid proteins function as enzymes that modify capsid proteins, whereas others serve as scaffolding for the assembly of the capsid.

The assembly of HIV occurs in two stages. First, capsid proteins assemble around two molecules of viral RNA and molecules of reverse transcriptase and integrase. Next, the newly formed capsid acquires its outer envelope in a budding process. This second phase of assembly occurs during step 6, as the virus is released from the cell.

### Step 6: Release

The last step of a viral reproductive cycle is the release of new viruses from the host cell. The release of bacteriophages is a dramatic event. Because bacteria are surrounded by a rigid cell wall, the phages must burst, or lyse, their host cell to escape. After the phages have been assembled, a phage-encoded enzyme called lysozyme digests the bacterial cell wall, causing the cell to burst. Lysis releases many new phages into the environment, where they can infect other bacteria and begin the cycle again. Collectively, steps 1, 2, 4, 5, and 6 are called the **lytic cycle** because they lead to cell lysis.

The release of enveloped viruses from an animal cell is far less dramatic. This type of virus escapes by a mechanism called budding that does not lyse the cell. In the case of HIV, a newly assembled virus particle associates with a portion of the plasma membrane containing HIV spike glycoproteins. The membrane enfolds the viral capsid and eventually buds from the surface of the cell. This is how the virus acquires its envelope, which is a piece of host cell membrane studded with viral glycoproteins.

### Latency in Bacteriophages

As we saw in step 3, viruses can integrate their genomes into a host chromosome. In some cases, the prophage or provirus may remain inactive, or **latent**, for a long time. Most of the viral genes are silent during latency, and the viral reproductive cycle does not progress to step 4.

Latency in bacteriophages is also called lysogeny. When this occurs, both the prophage and its host cell are said to be lysogenic. When a lysogenic bacterium prepares to divide, it copies the prophage DNA along with its own DNA, so each daughter cell inherits a copy of the prophage. A prophage can be replicated repeatedly in this way without killing the host cell or producing new phage particles. As mentioned earlier, this is called the lysogenic cycle.

Bacteriophages that can follow either a lysogenic or a lytic cycle are called **temperate phages** (**Figure 19.5**). Phage λ is an example

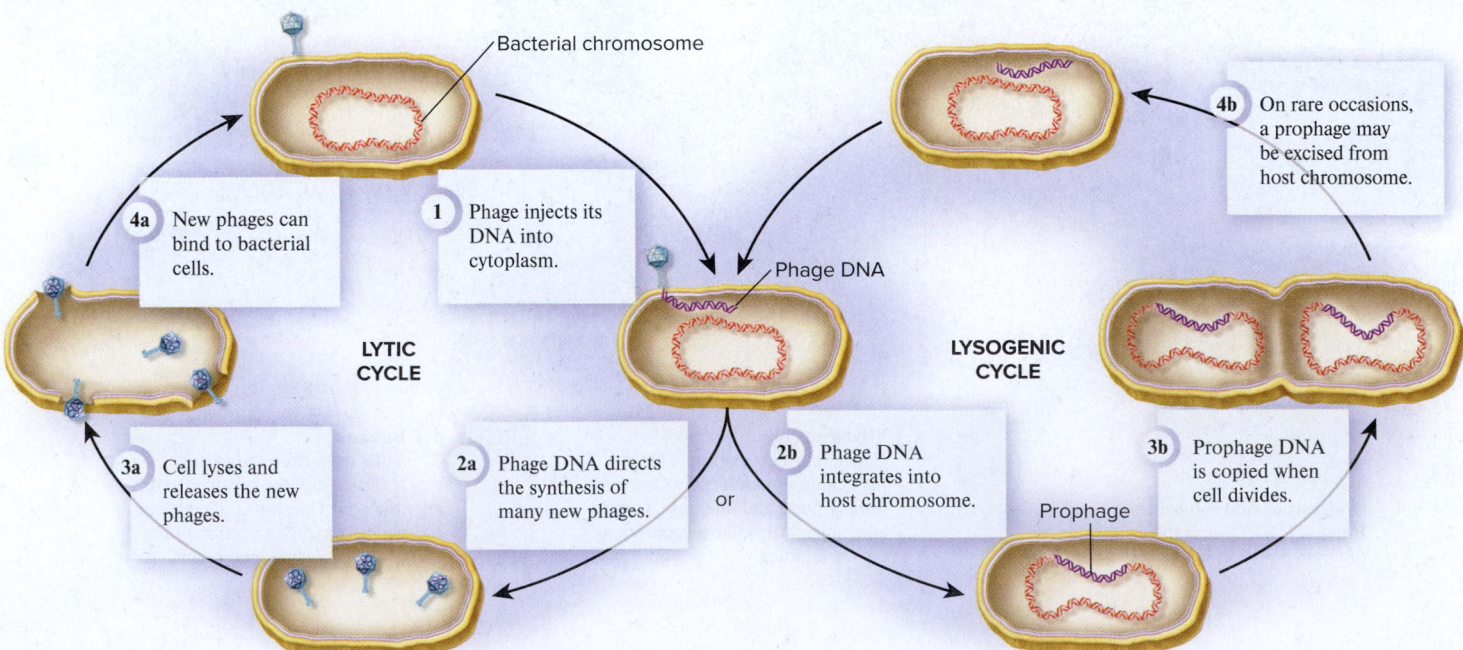

**Figure 19.5  Lytic and lysogenic cycles of bacteriophages.**  Some phages, such as phage λ, may follow either a lytic or a lysogenic reproductive cycle. During the lytic cycle, new phages are made, and the bacterial cell lyses. During the lysogenic cycle, the integrated phage DNA, or prophage, is replicated along with the DNA of the host cell. Environmental conditions influence how long the phage remains in the lysogenic cycle.

**Concept Check:**  *From the perspective of the bacteriophage, what are the primary advantages of the lytic and lysogenic cycles?*

of a temperate phage. Upon infection, it can either enter the lysogenic cycle or proceed directly to the lytic cycle. Other phages, called **virulent phages**, have only lytic cycles. The genome of a virulent phage is not capable of integrating into a host chromosome. Phage T4 is a virulent phage that infects *E. coli* (see Figure 19.3d). Unlike phage λ, which may coexist harmlessly with *E. coli*, T4 always lyses the host cell.

For temperate phages such as phage λ, environmental conditions influence whether or not viral DNA is integrated into a host chromosome and how long the virus remains in the lysogenic cycle. If nutrients are readily available, phage λ usually proceeds directly to the lytic cycle after its DNA enters the cell. Alternatively, if nutrients are in short supply, the lysogenic cycle is favored because sufficient material may not be available to make new viruses. If more nutrients become available later, the prophage may become activated. At this point, the viral reproductive cycle switches to the lytic cycle, and new viruses are made and released.

***Latency in Human Viruses*** Latency among human viruses can occur in two different ways. For HIV, latency occurs because the virus has integrated into the host genome and may remain dormant for a long time. In addition, the genome of some other viruses can exist as an **episome**—a genetic element that replicates independently of the chromosomal DNA but also can occasionally integrate into chromosomal DNA. Examples of viral genomes that exist as episomes include different types of herpesviruses that cause cold sores (usually herpes simplex type I), genital herpes (usually herpes simplex type II), and chickenpox (varicella-zoster). A person infected with a given type of herpesvirus may have periodic outbreaks of disease symptoms when the virus switches from the latent, episomal form to the active form that produces new virus particles.

As an example, let's consider the herpesvirus called varicella-zoster. The initial infection by this virus causes chickenpox, after which the virus may remain latent for many years as an episome. The disease called shingles occurs when varicella-zoster switches from the latent state and starts making new virus particles. Shingles begins as a painful rash that eventually erupts into blisters. The blisters follow the path of the neurons that carry the latent varicella-zoster virus. The blisters often form a ring around the back of the patient's body. The name shingles is derived from a Latin word meaning girdle, referring to the observation that the blisters girdle a part of the body.

## Emerging Viruses, Such as HIV and Zika Virus, May Spread Rapidly Through a Population

A key reason researchers have been interested in viral reproductive cycles is the ability of many viruses to cause diseases in humans and other hosts. **Emerging viruses** are ones that have arisen recently or have recently shown a greater probability of causing infection. Because the base sequences of many viruses are already known, researchers have determined that emerging viruses typically result from mutations in pre-existing viruses. Emerging viruses can lead to significant loss of human life and often cause public alarm. New strains of influenza virus arise fairly regularly due to mutations. An example is the strain H1N1, also called swine flu. In the U.S., despite attempts to minimize deaths by vaccination, over 30,000 people die annually from influenza.

Another example of an emerging virus is Zika virus, an enveloped virus with a genome composed of single-stranded RNA. Its name comes from the Zika Forest of Uganda, where the virus was first isolated in 1947. Zika virus is primarily spread by mosquitoes of the genus *Aedes*. The most common symptoms of a Zika infection are fever, rash, joint pain, and conjunctivitis. In most people, the illness is usually mild, but in rare cases, a Zika infection in an adult can cause a more serious illness called Guillain-Barré syndrome. In addition, Zika virus infection during pregnancy can result in serious brain abnormalities, including the condition microcephaly, in which the infant's head is smaller than normal. The Zika virus has spread globally from Africa into Asia, South America, and North America. Though estimates for infection rates vary, some epidemiologists predict that millions of people will become infected with the virus in the coming years.

During the past few decades, the most devastating example of an emerging virus is **human immunodeficiency virus (HIV)**, the causative agent of acquired immune deficiency syndrome (AIDS). HIV is primarily spread by sexual contact between infected and uninfected individuals, but it can also be spread by the transfusion of HIV-infected blood, by the sharing of needles among drug users, and from infected mother to unborn child. Since AIDS was first recognized in 1981, the total number of AIDS deaths has been nearly 40 million, making it one of the most deadly diseases in human history. More than 0.6 million of these deaths have occurred in the U.S. In 2016, over 30 million people were living with HIV; approximately 3 million of them were infected that year. In that same year, nearly 2 million died from AIDS. Worldwide, nearly 1 in every 100 adults between ages 15 and 49 is infected. In the U.S., about 50,000 new HIV infections occur each year, 70% of those infections in men and 30% in women.

The devastating effects of AIDS result from viral destruction of helper T cells, a type of white blood cell that plays an essential role in the immune system of mammals. **Figure 19.6** shows HIV virus particles invading a helper T cell. As described in Chapter 52, helper T cells interact with other cells of the immune system to facilitate the production of antibodies and other molecules that target and kill

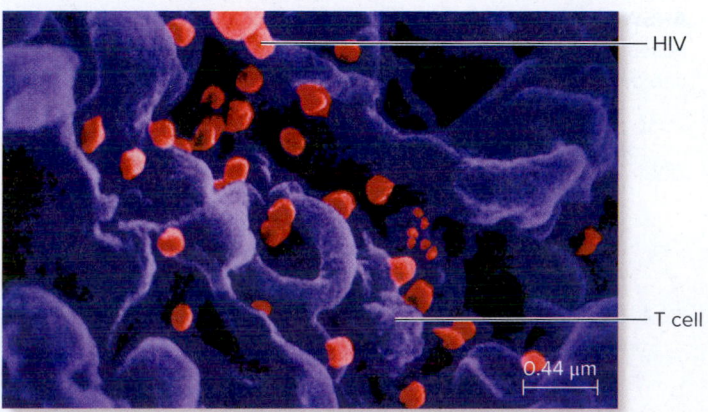

**Figure 19.6  Micrograph of HIV invading a human helper T cell.** This is a colorized scanning electron micrograph. The surface of the T cell is purple, and HIV particles are red. Source: Cynthia Goldsmith/CDC

foreign invaders of the body. When large numbers of helper T cells are destroyed by HIV, the function of the immune system is seriously compromised, and the individual becomes susceptible to infectious diseases called opportunistic infections that would not normally occur in a healthy person. For example, *Pneumocystis jiroveci*, a fungus that causes a type of pneumonia, is easily destroyed by a healthy immune system. However, in people with AIDS, *Pneumocystis jiroveci* pneumonia can be fatal.

An insidious feature of HIV replication, which is summarized in Figure 19.4b, is that reverse transcriptase, the enzyme that copies the RNA genome into DNA, lacks a proofreading function. In Chapter 11, we learned that DNA polymerase can identify and remove mismatched nucleotides in newly synthesized DNA. Because reverse transcriptase lacks this function, it makes more errors and thereby tends to create mutant strains of HIV. This undermines the ability of the body to combat HIV because mutant strains may be resistant to the body's defenses.

 **BIO TIPS**

**THE QUESTION** *Propose an experiment to explain how an emerging virus could arise.*

**T**OPIC **What topic in biology does this question address?** The topic is emerging viruses. More specifically, the question asks you to propose an experiment whose results could explain how they come into existence.

**I**NFORMATION **What information do you know based on the question and your understanding of the topic?** From your understanding of the topic, you may remember that emerging viruses typically arise via mutations in pre-existing viruses.

**P**ROBLEM-SOLVING **S**TRATEGY **Design an experiment. Compare and contrast.** In designing an experiment that may explain how an emerging virus could arise, you want to keep in mind that new forms of viruses typically result from the genetic alteration of pre-existing viruses. Another important point to keep in mind is that the base sequences of many viruses have already been determined.

**ANSWER** *The rationale behind the design of this experiment is based on the premise that emerging viruses arise from genetic alterations in pre-existing viruses. Here is a possible procedure:*

1. *Determine the base sequence of the emerging virus that you are interested in. (Note: The topic of DNA sequencing is described in Chapter 21.)*
2. *Compare its base sequence with those of other viruses. The expectation is that the emerging virus will be closely related to some other virus that is already known.*
3. *Analyze the differences in DNA sequences between the emerging virus and its closest relative. You may identify differences in particular genes in the emerging virus that may have altered the infectivity of the virus. For example, a mutation may have altered a viral protein in a way that allows the virus to bind more easily to host cells and enter them.*

## Drugs Have Been Developed to Combat the Proliferation of HIV

A compelling reason to understand the reproductive cycle of HIV and other disease-causing viruses is that such knowledge may be used to develop drugs that stop viral proliferation. For example, in the U.S., the highest rate of AIDS-related deaths was approximately 17 per year per 100,000 people in 1994 and 1995. The current rate is about 4 to 5 deaths per year per 100,000 people, owing in part to the use of new antiviral drugs. These drugs inhibit viral proliferation, though they cannot eliminate the virus from the body.

One approach to the design of antiviral treatments is to develop drugs that specifically bind to proteins encoded by the viral genome. For example, azidothymidine (AZT) mimics the structure of a normal nucleotide and binds to the enzyme reverse transcriptase. In this way, AZT inhibits reverse transcription, thereby inhibiting viral replication. Another way to combat HIV involves the use of antiviral drugs that inhibit proteases, enzymes that are needed during the assembly of the HIV capsid. Certain HIV proteases cut capsid proteins, which makes them smaller and able to assemble into a capsid structure. If the proteases do not function, the capsid will not assemble, and new HIV particles will not be made. Several drugs known as protease inhibitors have been developed that bind to HIV proteases and inhibit their function.

A major challenge in AIDS research is to discover drugs that inhibit viral proteins without also binding to host cell proteins and inhibiting normal cellular functions. A second challenge is to develop drugs to which mutant strains will not become resistant. As mentioned, HIV readily accumulates mutations during viral replication. A current strategy is to treat HIV patients with a "cocktail" of three or four HIV drugs, making it less likely that any mutant strain will overcome all of the inhibitory effects.

 **Core Concept: Evolution**

### Several Hypotheses Have Been Proposed to Explain the Origin of Viruses

Because viruses are such small particles, no fossil record is available to provide evidence about their evolution. Researchers must rely on analyses of modern viruses to develop hypotheses about their origin. Viral genomes follow the same rules of gene expression as the genomes of their host cells. Viral genes have promoter sequences similar to those of their host cells, and the translation of viral proteins relies on the genetic code. Viruses depend entirely on host cells for their proliferation. No known virus makes its own ribosomes or generates the energy it requires to make new viruses. Therefore, many biologists have argued that cells must have evolved before viruses.

How did viruses come into existence? A few hypotheses have been proposed.

***Progressive Evolution from Genetic Elements Within Cells*** A common hypothesis for the origin of viruses is that they evolved from macromolecules inside living cells. The precursors of the first viruses may have been RNA molecules or they may have been plasmids—small, circular DNA molecules that exist independently of chromosomal DNA. (Plasmids are described later in this chapter.) Biologists have hypothesized that such RNA or DNA molecules may have become more complex by acquiring genes that code for proteins that facilitate their own replication.

***Regressive Evolution from Cells***    Though many biologists favor the idea that viruses originated from primitive plasmids or other chromosomal elements, some have suggested they are an example of regressive evolution—the reduction of a trait or traits over time. This hypothesis proposes that viruses are degenerate cells that have retained the minimal genetic information essential for reproduction. For example, some viruses may have originated as small cells that infected larger cells. Over time, genes not required for their independent existence were lost.

***Parallel Evolution with Cells***    A new and interesting hypothesis is that viruses did not evolve from living cells but instead evolved in parallel with cellular organisms. As discussed in Chapter 4, the precursors of cellular DNA genomes may have been RNA molecules that could replicate independently of cells. An early stage of evolution, termed the RNA world, could have involved the parallel evolution of both viruses and cellular organisms.

## 19.3    Viroids and Prions

**Learning Outcomes:**

1. Distinguish between a viroid and a virus.
2. Describe the structure of prions, and explain how they cause disease.

Some nonliving infectious agents are simpler than viruses. Viroids are composed solely of RNA, and prions are composed solely of protein. In this section, we begin by examining viroids, infectious agents that cause diseases in plants. Next, we will discuss infectious proteins known as prions, which cause devastating neurological diseases in humans and other mammals.

### Viroids Are RNA Molecules That Infect Plant Cells

In 1971, Swiss-born American plant pathologist Theodor Diener discovered that the agent of potato spindle tuber disease is a small RNA molecule devoid of any protein. He coined the term **viroid** for this newly discovered infectious particle. Viroids are composed solely of a single-stranded circular RNA molecule that is a few hundred nucleotides in length.

Viroids infect plant cells, where they depend entirely on host enzymes for their replication. Some viroids are replicated in the host cell nucleus, whereas others replicate in a chloroplast. In contrast to viral genomes, the RNA genomes of viroids do not code for any proteins. How do viroids affect plant cells? The RNA of some viroids has ribozyme activity, and researchers have hypothesized that this activity may damage plants by interfering with the function of host cell molecules. However, the mechanism by which viroids induce disease is not well understood.

Since Diener's initial discovery, many more viroids have been characterized as the agents of diseases that affect many economically important plants, including potato, tomato, cucumber, orange, coconut, grape, avocado, peach, apple, pear, and plum. Some viroids have devastating effects, as illustrated by the case of the coconut cadang-cadang viroid, which has killed more than 20 million coconut trees in Southeast Asia and New Guinea (**Figure 19.7**). Other viroids produce less severe damage, causing necrosis on leaves, shortening of stems, bark cracking, and delays in foliation, flowering, and fruit ripening. A few viroids induce mild symptoms or no symptoms at all.

## Prions Are Infectious Proteins That Cause Neurodegenerative Diseases

Before we end our discussion of nonliving, infectious particles, let's consider an unusual infectious agent that causes a group of rare, fatal brain diseases affecting humans and other mammals. Until the 1980s, biologists thought that any infectious agent, whether living or

**Figure 19.7    Effects of a viroid.**    This palm tree (left foreground) in Papua, New Guinea, has been infected with the coconut cadang-cadang viroid and shows symptoms of stunting and yellowing.
©Photograph by J. W. Randles, at Albay Research Center, Philippines

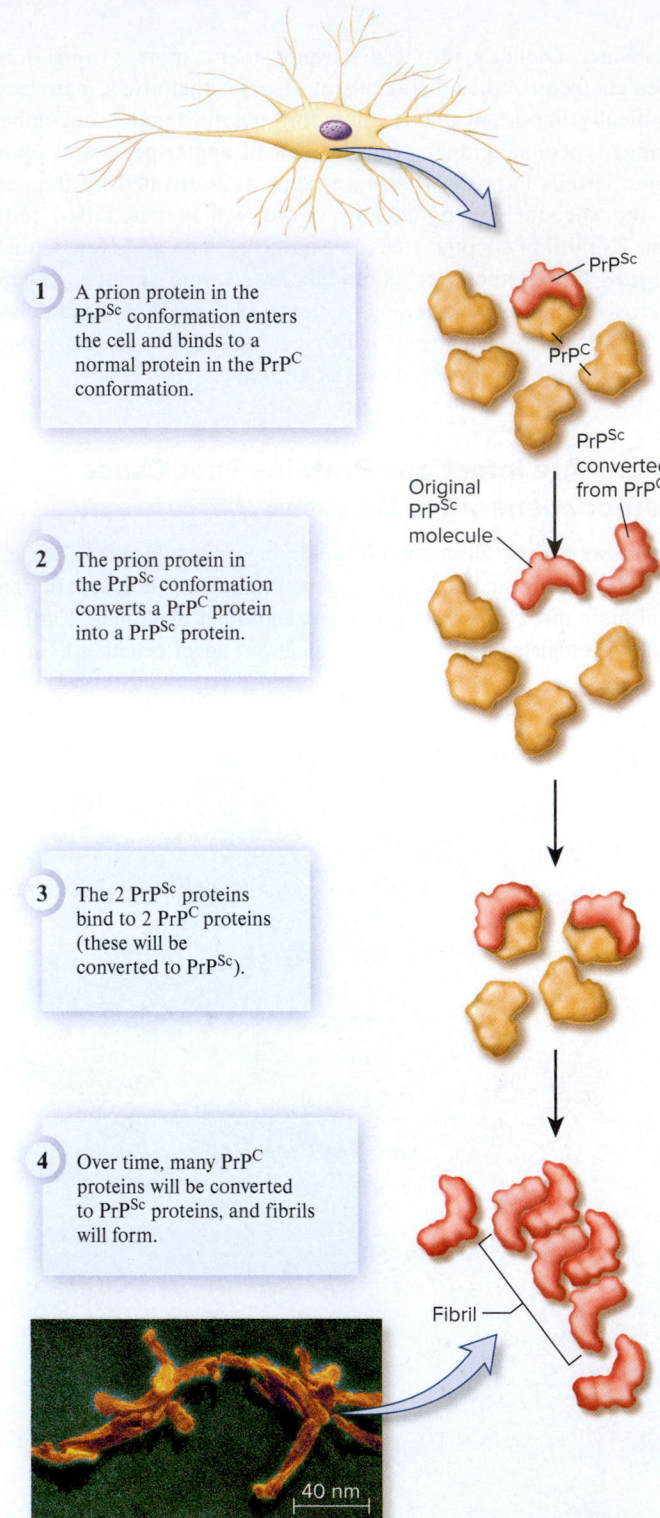

1. A prion protein in the PrP^Sc conformation enters the cell and binds to a normal protein in the PrP^C conformation.

2. The prion protein in the PrP^Sc conformation converts a PrP^C protein into a PrP^Sc protein.

3. The 2 PrP^Sc proteins bind to 2 PrP^C proteins (these will be converted to PrP^Sc).

4. Over time, many PrP^C proteins will be converted to PrP^Sc proteins, and fibrils will form.

PrP^Sc

PrP^C

PrP^Sc converted from PrP^C

Original PrP^Sc molecule

Fibril

40 nm

**Figure 19.8  A proposed molecular mechanism of prion diseases.** A healthy neuron contains only the PrP^C conformation of the protein. The abnormal PrP^Sc conformation catalyzes the conversion of PrP^C proteins into PrP^Sc proteins, thereby causing the symptoms of the prion disease. ©Eye of Science/Science Source

 **Core Concept: Structure and Function** The prion and normal protein conformations have different structures. The abnormal conformation of the prion has the functional ability to convert normal proteins to the abnormal conformation.

nonliving, must have genetic material. It seemed logical that genetic material is needed to store the information to produce new infectious particles.

In the 1960s, British researchers Tikvah Alper and John Stanley Griffith discovered that preparations from animals with certain neurodegenerative diseases remained infectious even after exposure to radiation that would destroy any DNA or RNA. They suggested that the infectious agent was a protein. In the early 1970s, American neurologist Stanley Prusiner, moved by the death of a patient from such a neurodegenerative disease, began to search for the causative agent. In 1982, Prusiner isolated a disease-causing particle composed entirely of protein, which he called a **prion**. The term was based on his characterization of the particle as a proteinaceous infectious agent. In 1997, Prusiner was awarded the Nobel Prize in Physiology or Medicine for his work on prions.

Prion diseases arise from the ability of the prion to induce abnormal folding in normal protein molecules (**Figure 19.8**). The prion exists in a disease-causing conformation designated PrP^Sc. The superscript Sc refers to scrapie, an example of a prion disease. A normal conformation of this same protein, which does not cause disease, is termed PrP^C. The superscript C stands for cellular. The normal protein is encoded by an individual's genome, and it is expressed at low levels in certain types of neurons.

How does someone contract a prion disease? A healthy person may become "infected" with the abnormal protein by eating meat of an animal with the disease. Unlike most other proteins in the diet, the prion escapes digestion in the stomach and small intestine and is absorbed into the bloodstream. After being taken up by neurons, the prion gradually converts the cell's normal proteins to the abnormal conformation. As a prion disease progresses, the PrP^Sc proteins are deposited as dense aggregates that form tough fibrils in the cells of the brain and peripheral nervous tissues, causing the disease symptoms. Some of the abnormal proteins are also secreted from infected cells, where they travel through the bloodstream. In this way, a prion disease spreads through the body like many viral diseases.

Prions cause several types of fatal neurodegenerative diseases affecting humans, livestock, and wildlife (**Table 19.2**). Prion diseases are termed transmissible spongiform encephalopathies (TSEs). The postmortem examination of the brains of affected individuals

| Table 19.2 | Examples of Neurodegenerative Diseases Caused by Infectious Prions |
|---|---|
| **Disease** | **Description** |
| Scrapie | A disease of sheep and pigs characterized by intense itching, causing the animals to scrape themselves against trees or other objects, followed by neurodegeneration. |
| Mad cow disease | Begins with changes in posture and temperament, followed by loss of coordination and neurodegeneration. |
| Chronic wasting disease | A disease of deer (genus *Odocoileus*) and Rocky Mountain elk (*Cervus elaphus*). A consistent symptom is weight loss over time. The disease is progressive and fatal. |

reveals a substantial destruction of brain tissue. The brain has a spongy appearance. Most prion diseases progress fairly slowly. Over the course of a few years, symptoms proceed from a loss of motor control to dementia, paralysis, wasting, and eventually death. These symptoms are correlated with an increase in the level of prions in the neurons of infected individuals. No current treatment can halt the progression of any of the TSEs. For this reason, great public alarm occurs when an outbreak of a TSE is reported. For example, in 2003, a report of a single cow in the U.S. with a TSE commonly known as mad cow disease prompted several countries to restrict the import of American beef.

## 19.4  Genetic Properties of Bacteria

### Learning Outcomes:

1. Outline the key features of a bacterial chromosome.
2. Explain the two processes that compact the bacterial chromosome.
3. Outline the structure and functions of plasmids.
4. Diagram the process of cell division in bacteria.

Many bacteria exist as unicellular organisms. However, some of them may remain associated with each other after cell division, forming pairs, chains, or clumps. Bacteria are widespread on Earth, and numerous species are known to cause various types of infectious diseases, such as bacterial meningitis, discussed at the beginning of this chapter (see the chapter opening photo). We begin this section by exploring the structure and replication of the bacterial genome and the organization of DNA sequences along a bacterial chromosome. We then examine how the chromosome is compacted to fit inside a bacterium and how it is transmitted during asexual reproduction.

### Bacteria Typically Have Circular Chromosomes That Carry a Few Thousand Genes

The genes of bacteria are within structures known as bacterial chromosomes.

- Although a bacterial cell usually has a single type of chromosome, it may have more than one copy of that chromosome. The number of copies depends on the bacterial species and on growth conditions, but a bacterium typically has one to four identical chromosomes.

- Each bacterial chromosome is tightly packed within a distinct **nucleoid** of the cell (**Figure 19.9**). Unlike the eukaryotic nucleus, the bacterial nucleoid is not a separate cellular compartment bounded by a membrane. The DNA in a nucleoid is in direct contact with the cytoplasm of the cell.

- Bacterial chromosomes contain molecules of double-stranded DNA along with many different proteins. They are usually circular and are typically a few million base pairs (bp) long. For example, the chromosome of *Escherichia coli* has approximately 4.6 million bp, and the *Haemophilus influenzae* chromosome has roughly 1.8 million bp.

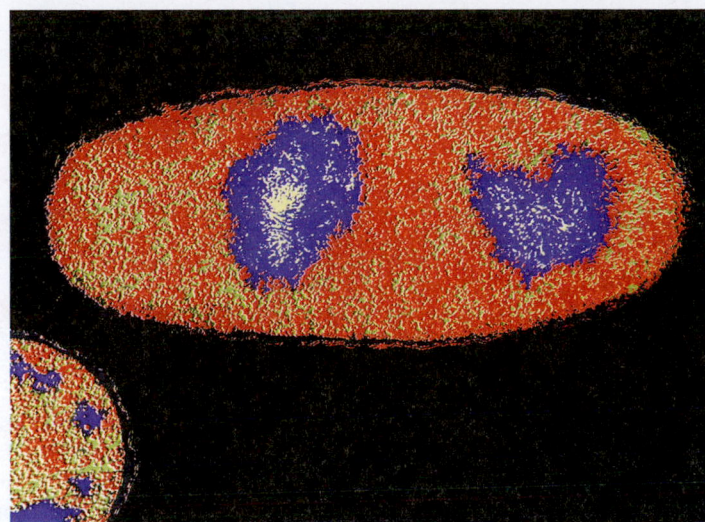

**Figure 19.9** **Nucleoids within the bacterium *Bacillus subtilis*.** In this light micrograph, the nucleoids are fluorescently labeled and seen as purple, oval-shaped areas within the bacterial cytoplasm. Two or more nucleoids are usually found within each cell. ©M. Wurtz/Biozentrum, University of Basel/Science Source

 **Core Skill: Connections** Look back at Figures 4.8 and 4.9. How is a nucleoid different from a nucleus found in a eukaryotic cell?

- A typical bacterial chromosome contains a few thousand genes that are found throughout the chromosome. Gene sequences, primarily those that encode proteins, account for the largest part of bacterial DNA.

- Other nucleotide sequences in the chromosome play a role in DNA replication, gene expression, and chromosome structure. One of these sequences is the origin of replication, which is a few hundred base pairs long. Bacterial chromosomes have a single origin of replication that functions as an initiation site for the assembly of several proteins that are required for DNA replication (refer back to Figure 11.13b).

### The Formation of Chromosomal Loops and DNA Supercoiling Make the Bacterial Chromosome Compact

Bacterial cells are much smaller than most eukaryotic cells. *E. coli* cells, for example, are approximately 1 μm wide and 2 μm long. To fit within a bacterial cell, the DNA of a typical bacterial chromosome must be compacted about 1,000-fold. How does this occur? The compaction of a bacterial chromosome, shown in **Figure 19.10**, occurs by two processes: the formation of loops and DNA supercoiling.

Unlike eukaryotic DNA, bacterial DNA is not wound around histone proteins to form nucleosomes. However, the binding of proteins to bacterial DNA is important in the formation of **loop domains**—chromosomal segments that are folded into loops. As seen in Figure 19.10, DNA-binding proteins anchor the bases of the loops in place. The number of loops varies according to the size of a bacterial chromosome and the species. The *E. coli* chromosome has 400 to 500 loop domains, each with about 10,000 bp. This looping

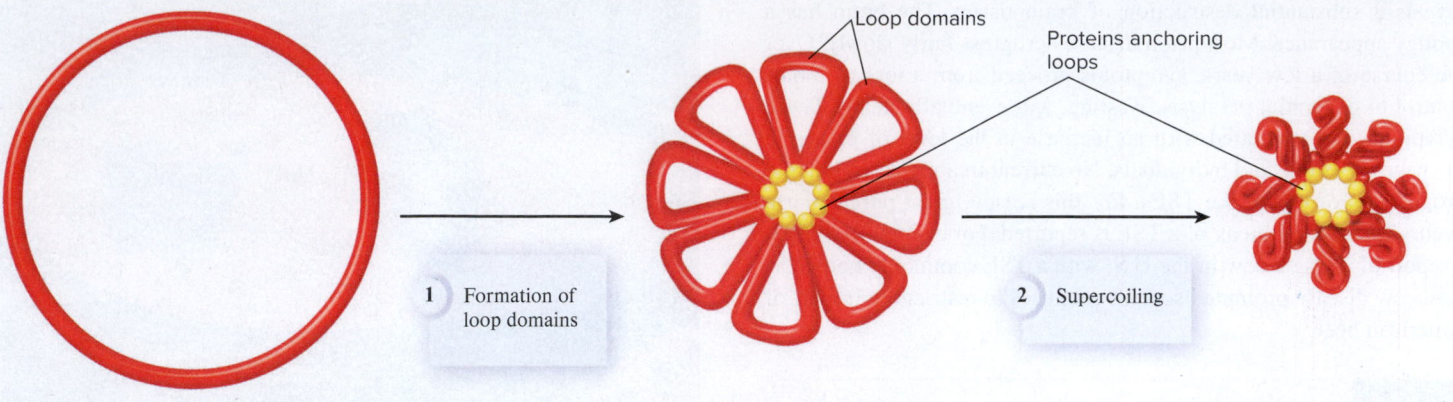

Loop domains

Proteins anchoring loops

1   Formation of loop domains

2   Supercoiling

Circular chromosomal DNA

Looped chromosomal DNA with associated proteins

Supercoiled and looped DNA

**Figure 19.10**  **The compaction of a bacterial chromosome.** To compact the large, circular chromosome, segments are organized into smaller loop domains by binding to proteins at the bases of the loops. These loops are made more compact by DNA supercoiling. Note: This is a simplified drawing; bacterial chromosomes typically have between 400 and 500 loop domains

*Concept Check:*  *Describe how the loop domains are held in place.*

compacts the circular chromosome about 10-fold. A similar process of loop-domain formation occurs in eukaryotic chromatin compaction, which is described in Chapter 11 (Figure 11.24).

**DNA supercoiling** is a second important compaction process for the bacterial chromosome. Because DNA is a long, thin molecule, twisting can dramatically change its conformation. This compaction process is similar to what happens to a rubber band if you twist it in one direction. Because the two strands of DNA already coil around each other, the formation of additional coils due to twisting is referred to as supercoiling. Bacterial enzymes called topoisomerases twist the DNA and control the degree of DNA supercoiling.

## Plasmids Are Small, Circular Pieces of Extrachromosomal DNA

In addition to chromosomal DNA, bacterial cells commonly contain **plasmids**, small, circular pieces of DNA that exist separately from the bacterial chromosome (**Figure 19.11**). Plasmids occur naturally in many strains of bacteria and in a few types of eukaryotic cells, such as yeast. The smallest plasmids consist of just a few thousand base pairs and carry only a gene or two. The largest are in the range of 100,000 to 500,000 bp and carry several dozen or even hundreds of genes. A plasmid has its own origin of replication that allows it to be replicated independently of the bacterial chromosome. The DNA sequence of the origin of replication influences how many copies of the plasmid are found within a cell. Some origins are said to be very strong because they result in many copies of the plasmid, perhaps as many as 100 per cell. Other origins of replication have sequences that are much weaker, so the number of copies is relatively low, such as one or two per cell.

Why do bacteria have plasmids? Certain genes within a plasmid usually provide some type of growth advantage to the cell or may aid in survival under certain conditions. By studying plasmids in many different species, researchers have discovered that most plasmids fall into a few different categories:

1. Resistance plasmids, also known as R factors, contain genes that confer resistance against antibiotics and other types of toxins.

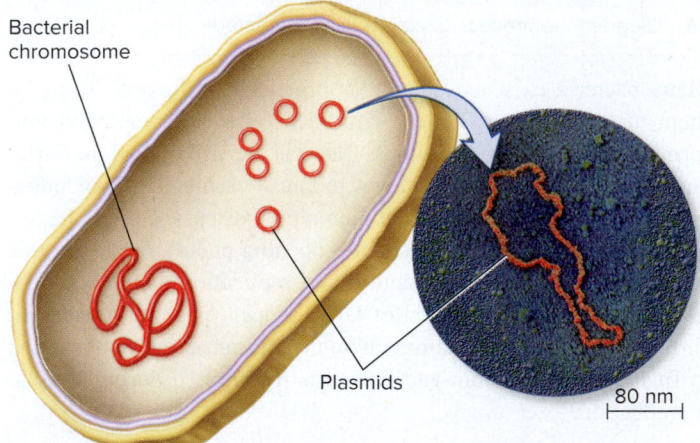

Bacterial chromosome

Plasmids

80 nm

**Figure 19.11**  **Plasmids in a bacterial cell.** Plasmids are small, circular DNA molecules that exist independently of the bacterial chromosome. (right): ©Stanley Cohen/Science Source

*Concept Check:*  *Describe the similarities and differences between a bacterial chromosome and a plasmid.*

2. Degradative plasmids carry genes that enable the bacterium to digest and utilize an unusual substance. For example, a degradative plasmid may carry genes that allow a bacterium to digest an organic solvent such as toluene.

3. Virulence plasmids carry genes that turn a bacterium into a pathogenic strain.

4. Fertility plasmids, also known as F factors, allow bacteria to transfer genes to each other, as described later in this chapter.

On occasion, a plasmid may integrate into the bacterial chromosome. Such plasmids, which can integrate or remain independent of the chromosome, are also termed episomes.

## Most Bacteria Reproduce by Binary Fission

Thus far, we have considered the genetic material of bacteria and the compaction of the bacterial chromosome to fit inside the cell. Let's now turn our attention to the process of cell division. The capacity of bacteria to divide is really quite astounding. The cells of some species, such as *E. coli*, can divide every 20–30 minutes. When placed on a solid growth medium in a petri dish, an *E. coli* cell and its daughter cells undergo repeated cellular divisions and form a clone of genetically identical cells called a **bacterial colony** (**Figure 19.12**). Starting with a single cell that is invisible to the unaided eye, a visible bacterial colony containing 10–100 million cells forms in less than a day!

Cell division of most bacterial species occurs by a process called **binary fission**, during which a cell divides into two daughter cells. **Figure 19.13** shows this process for a cell with a single chromosome.

Before it divides, the cell replicates its DNA. This produces two identical copies of the chromosome. Next, the cell's plasma membrane is drawn inward and deposits new cell-wall material, separating the two daughter cells. Each daughter cell receives one of the copies of the original chromosome. Therefore, except when a mutation occurs, each daughter cell contains an identical copy of the mother cell's genetic material.

Binary fission in most bacterial species requires proteins named FtsA and FtsZ, which are evolutionarily related to eukaryotic actin and tubulin proteins, respectively. With the aid of FtsA, the FtsZ proteins assemble into a ring at the site where a septum will be formed

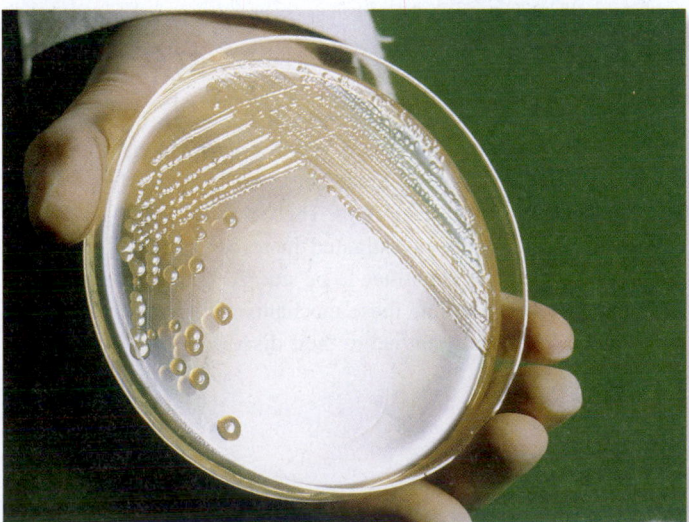

**Figure 19.12  Growth of a bacterial colony.** Through successive cell divisions, a single bacterial cell of *E. coli* forms a genetically identical group of cells called a bacterial colony.

©Dr. Jeremy Burgess/SPL/Science Source

*Concept Check:* *Suppose a bacterial strain divides every 30 minutes. If a single cell is placed on a plate, how many cells will be in the colony after 16 hours?*

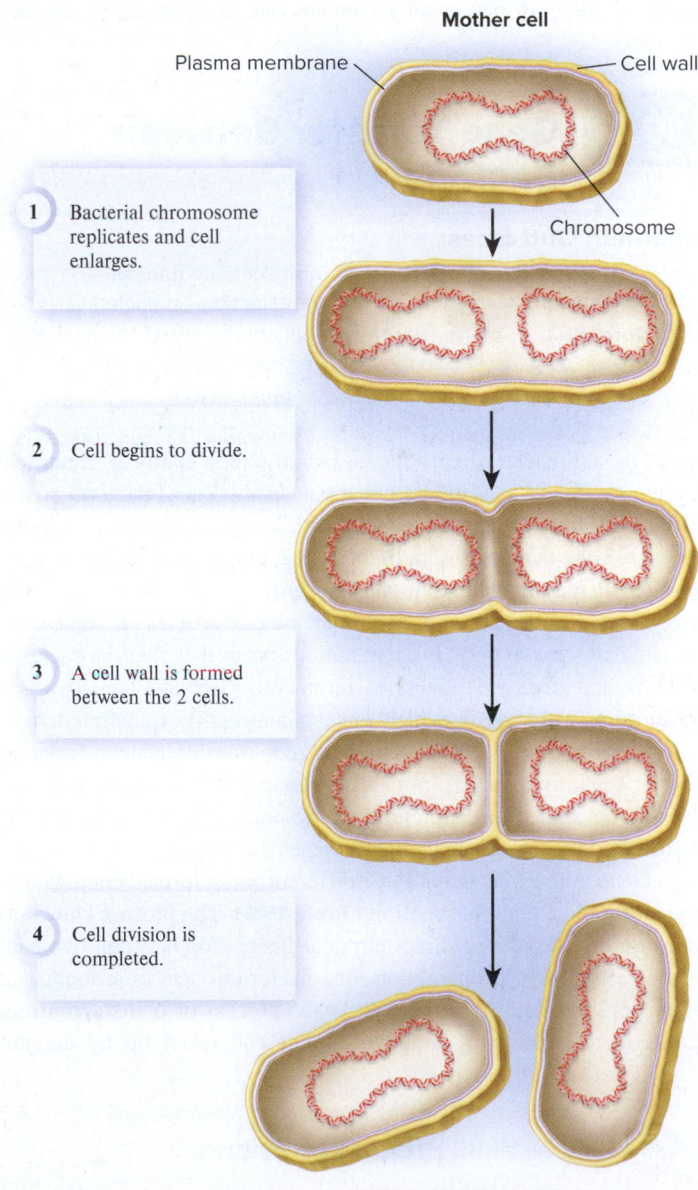

1  Bacterial chromosome replicates and cell enlarges.

2  Cell begins to divide.

3  A cell wall is formed between the 2 cells.

4  Cell division is completed.

**Figure 19.13  Bacterial cell division.** Bacteria reproduce by binary fission. Before a bacterium divides, the bacterial chromosome is replicated to produce two identical copies. These two copies segregate from each other, with one copy going to each daughter cell.

 **Core Concept: Information** After cell division, each daughter cell has a copy of the genetic material that was in the mother cell.

that separates the daughter cells. While the mechanism by which Fts proteins promote septum formation is not well understood, their role is thought to involve the binding of other proteins that are needed in the process. For example, FtsZ proteins attract enzymes that are involved with synthesizing peptidoglycan. Researchers have proposed that FtsZ proteins may guide the insertion of new cell wall by building smaller and smaller rings of peptidoglycan to eventually divide the cell into two daughter cells. Some archaea also use Fts proteins during cell division, but others utilize different mechanisms.

Plasmids replicate independently of the bacterial chromosome. During binary fission, the plasmids are distributed to the daughter cells so that each one usually receives one or more copies of each plasmid.

## 19.5 Gene Transfer Between Bacteria

### Learning Outcomes:

1. Compare and contrast the three forms of gene transfer—conjugation, transformation, and transduction—in bacteria.
2. **CoreSKILL »** Analyze the results of Lederberg and Tatum, and explain how they led to the discovery of conjugation.
3. Describe the process of horizontal gene transfer.

Even though bacteria reproduce asexually, they exhibit a great deal of genetic diversity. Within a given bacterial species, the term **strain** refers to a lineage that has genetic differences compared to another lineage. For example, one strain of *E. coli* may be resistant to an antibiotic, whereas another strain may be sensitive to the same antibiotic. How does genetic diversity arise in an asexual species? It comes primarily from two sources. First, mutations occur that alter the bacterial genome and affect the traits of bacterial cells. Second, diversity arises from **gene transfer**, in which genetic material is transferred from one bacterial cell to another. Through gene transfer, genetic variation that arises in one bacterium can be spread to other strains and even to other species. For example, an antibiotic-resistance gene may be transferred from a resistant strain to a sensitive strain.

Gene transfer occurs in three different ways, termed conjugation, transformation, and transduction (**Table 19.3**). The process known as **conjugation** involves a direct physical interaction between two bacterial cells. During conjugation, one bacterium acts as a donor and transfers DNA to a recipient cell. In the process of **transformation**, DNA is released into the environment and taken up by another

| Table 19.3 | Mechanisms of Gene Transfer Between Bacterial Cells |
|---|---|
| **Mechanism** | **Description** |
| **Conjugation:** 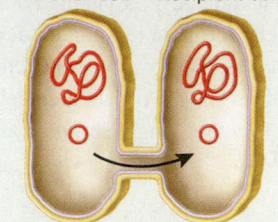 Donor cell    Recipient cell | Requires direct contact between a donor cell and a recipient cell. The donor cell transfers a strand of DNA to the recipient. In the example shown here, DNA from a plasmid is transferred to the recipient cell. The end result is that both donor and recipient cells have a plasmid. |
| **Transformation:** 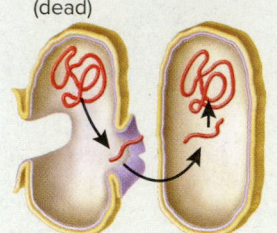 Donor cell    Recipient cell (dead) | A fragment of DNA from a donor cell is released into the environment. This may happen when a bacterial cell dies. This DNA fragment is taken up by a recipient cell, which incorporates the DNA into its chromosome. |
| **Transduction:** Donor cell    Recipient cell (infected by a bacteriophage) | When a bacteriophage infects a donor cell, it causes the bacterial chromosome of the donor cell to break up into fragments. A fragment of bacterial chromosomal DNA is incorporated into a newly made bacteriophage. The bacteriophage then transfers this fragment of DNA to a recipient cell. |

bacterial cell. **Transduction** occurs when a bacteriophage infects a bacterial cell and then a newly made bacteriophage transfers some of that cell's DNA to another bacterium. These three types of gene transfer have been extensively investigated in research laboratories, and their molecular pathways continue to be studied with great interest. In this section, we will examine these mechanisms in greater detail and consider the experiments that led to their discovery.

 **Core Skill: Process of Science**

## Feature Investigation | Lederberg and Tatum's Work with *E. coli* Demonstrated Gene Transfer Between Bacteria and Led to the Discovery of Conjugation

In 1946 and 1947, Joshua Lederberg and Edward Tatum carried out the first experiments that showed gene transfer from one bacterial cell to another (**Figure 19.14**). The researchers studied strains of *E. coli* that had different nutritional requirements for growth. They designated

one strain *met⁻bio⁻thr⁺pro⁺* because its growth required the amino acid methionine (met) and the vitamin biotin (bio) in the growth medium. This strain did not require the amino acids threonine (thr) or proline (pro) for growth. Another strain, designated *met⁺bio⁺thr⁻pro⁻*, had just

**Figure 19.14** Experiment of Lederberg and Tatum demonstrating gene transfer between *E. coli* cells.

**HYPOTHESIS** Genetic material can be transferred from one bacterial cell to another.

**KEY MATERIALS** Two bacterial strains, one that was *met⁻bio⁻thr⁺pro⁺* and the other that was *met⁺bio⁺thr⁻pro⁻*.

Experimental level | Conceptual level

1  In 3 separate tubes, add the *met⁻bio⁻thr⁺pro⁺* strain, the *met⁺bio⁺thr⁻pro⁻* strain, or a mixture of both strains. Incubate several hours.

*met⁻bio⁻thr⁺pro⁺*    *met⁺bio⁺thr⁻pro⁻*

2  Remove 10⁸ cells from each tube and spread onto plates that lack methionine, biotin, threonine, and proline.

10⁸    10⁸    10⁸

Nutrient agar plates lacking amino acids and biotin

Genetic material was transferred between the two strains.

3  Incubate overnight to allow growth of bacterial colonies.

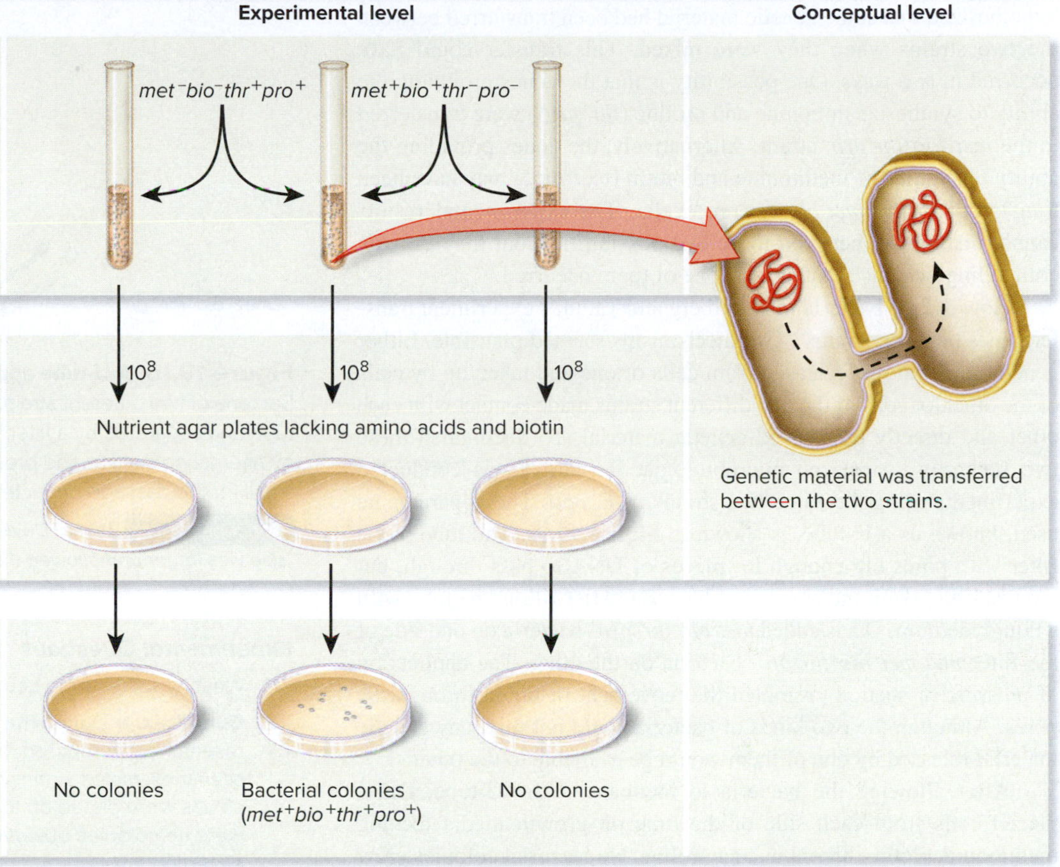

No colonies    Bacterial colonies (*met⁺bio⁺thr⁺pro⁺*)    No colonies

4  **THE DATA**

| Strain | Number of colonies after overnight growth |
|---|---|
| *met⁻bio⁻thr⁺pro⁺* | 0 |
| *met⁺bio⁺thr⁻pro⁻* | 0 |
| Both strains together | ~10 |

5  **CONCLUSION** Gene transfer has occurred from one bacterial cell to another.

6  **SOURCES** Lederberg, J., and Tatum, E. L. 1946. Novel genotypes in mixed cultures of biochemical mutants of bacteria, *Cold Spring Harbor Symposia on Quantitative Biology* 11: 113–114.

Tatum, E. L., and Lederberg, J. 1947. Genetic recombination in the bacterium *Escherichia coli. Journal of Bacteriology* 53: 673–684.

the opposite requirements. It needed threonine and proline, but not methionine or biotin. These differences in nutritional requirements correspond to allelic differences between the two strains. The *met⁻bio⁻thr⁺pro⁺* strain had defective genes encoding enzymes necessary for methionine and biotin synthesis, whereas the *met⁺bio⁺thr⁻pro⁻* strain had defective genes encoding the enzymes required to make threonine and proline.

Figure 19.14 compares the results of mixing the two *E. coli* strains with the results when they were not mixed. The tube on the left contained only *met⁻bio⁻thr⁺pro⁺* cells, and the tube on the right had only *met⁺bio⁺thr⁻pro⁻* cells. The middle tube contained a mixture

of the two kinds of cells. In each case, the researchers applied about 100 million (10⁸) cells to plates containing a growth medium lacking amino acids and the vitamin biotin. When the unmixed strains were applied to these plates, no colonies were observed to grow. This result was expected because the plates did not contain the methionine and biotin that the *met⁻bio⁻thr⁺pro⁺* cells needed for growth or the threonine and proline that the *met⁺bio⁺thr⁻pro⁻* cells required. The striking result occurred when the researchers plated 10⁸ cells from the tube containing the mixture of the two strains. In this case, approximately 10 cells multiplied and formed visible bacterial colonies on the plates. Because these cells multiplied without supplemental amino acids or vitamins, their genotype must have been *met⁺bio⁺thr⁺pro⁺*. Mutations cannot account for the occurrence of this new genotype because colonies were not observed on the other two plates, which had the same number of cells and also could have incurred mutations.

To explain the results of their experiment, Lederberg and Tatum hypothesized that some genetic material had been transferred between the two strains when they were mixed. This transfer could have occurred in two ways. One possibility is that the genes providing the ability to synthesize threonine and proline (*thr⁺pro⁺*) were transferred to the *met⁺bio⁺thr⁻pro⁻* strain. Alternatively, the genes providing the ability to synthesize methionine and biotin (*met⁺bio⁺*) may have been transferred to the *met⁻bio⁻thr⁺pro⁺* cells. The experimental results cannot distinguish between these two possibilities, but they provide compelling evidence that at least one of them occurred.

How did the bacteria in Lederberg and Tatum's experiment transfer genes between strains? Two mechanisms seemed plausible. Either genetic material was released from cells of one and taken up by cells of the other, or cells of the two different strains made contact with each other and directly transferred genetic material. To distinguish these two scenarios, American microbiologist Bernard Davis conducted experiments using the same two strains of *E. coli*. The apparatus he used, known as a U-tube, is shown in **Figure 19.15**. The tube had a filter with pores big enough for pieces of DNA to pass through, but too small to permit the passage of bacteria. After filling the tube with a liquid medium, Davis added *met⁻bio⁻thr⁺pro⁺* bacteria on one side of the filter and *met⁺bio⁺thr⁻pro⁻* bacteria on the other. The application of pressure or suction promoted the movement of liquid through the pores. Although the two kinds of bacteria could not mix, any genetic material released by one of them would be available to the other.

After allowing the bacteria to incubate in the U-tube, Davis placed cells from each side of the tube on growth media lacking methionine, biotin, threonine, and proline. No bacterial colonies grew on the media. How did Davis interpret these results? He proposed that without physical contact, the two *E. coli* strains could not transfer genetic material from one cell to the other. The conceptual level of Figure 19.14, step 1, shows the physical connection that explains Lederberg and Tatum's results. Conjugation is the process of gene transfer that requires direct cell-to-cell contact. It has been subsequently observed in other species of bacteria. Many, but not all, species of bacteria can conjugate.

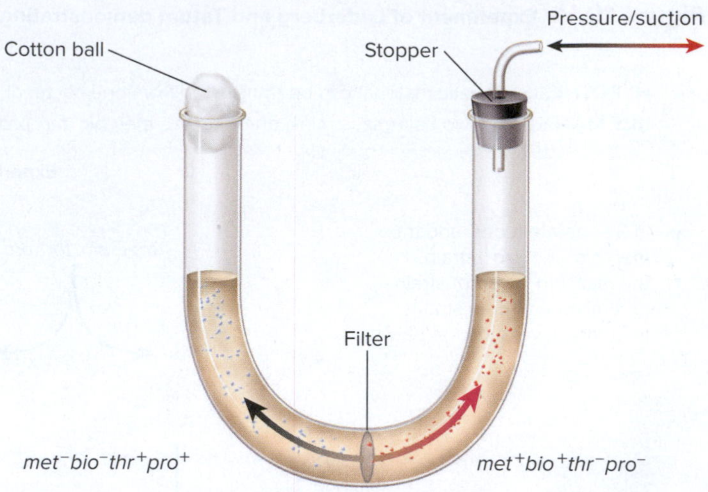

**Figure 19.15  A U-tube apparatus like the one used by Davis.** Bacteria of two different strains were suspended in the liquid in the tube and separated by a filter. The liquid was forced through the filter by alternating suction and pressure. The pores in the filter were too small for the passage of bacteria, but they allowed the passage of DNA.

**Concept Check:** *Would Davis's results have been different if the pore size was larger and allowed the passage of bacterial cells?*

---

***Experimental Questions***

1. What hypothesis did Lederberg and Tatum test?

2. **CoreSKILL »** During the Lederberg and Tatum experiment, the researchers compared the growth of mutant strains under two scenarios: mixed strains or unmixed strains. When the unmixed strains were plated on the experimental growth medium, why were no colonies observed to grow? When the mixed strains were plated on the experimental growth medium, 10 colonies were observed. What was the significance of these colonies?

3. **CoreSKILL »** The gene transfer seen in the Lederberg and Tatum experiment could have occurred in one of two ways: by bacteria taking up DNA released into the environment or by contact between two bacterial cells that allowed for direct transfer. Davis conducted an experiment to determine the correct process. Explain how his results indicated the correct gene transfer process.

---

## During Conjugation, DNA Is Transferred from a Donor Cell to a Recipient Cell

In the early 1950s, American microbiologists Joshua and Esther Lederberg, Irish physician William Hayes, and Italian geneticist Luca Cavalli-Sforza independently discovered that only certain bacterial strains can donate genetic material during conjugation. For example, about 5% of *E. coli* strains found in nature act as donor strains. Further research showed that a strain that is incapable of acting as a donor can acquire this ability after being mixed with a donor strain. Hayes correctly proposed that donor strains contain a type of plasmid called a fertility factor, or **F factor**, that can be transferred to recipient strains. Also, other donor *E. coli* strains were later identified that

transfer portions of the bacterial chromosome at high frequencies. After a segment of the chromosome is transferred, it then inserts, or recombines, into the chromosome of the recipient cell. Such donor strains were named *Hfr* (for *H*igh *f*requency of *r*ecombination). In our discussion, we will focus on donor strains that carry F factors.

The micrograph in **Figure 19.16a** shows two conjugating *E. coli* cells. The cell on the left is designated *F⁺*, meaning that it has an F factor. This donor cell is transferring genetic material to the recipient cell on the right, which lacks an F factor and is designated *F⁻*. F factors carry several genes that are required for conjugation and also may carry genes that confer a growth advantage for the bacterium.

**Figure 19.16b** describes the events that occur during conjugation in *E. coli*. **Sex pili** (singular, pilus) are made by *F⁺* cells that bind

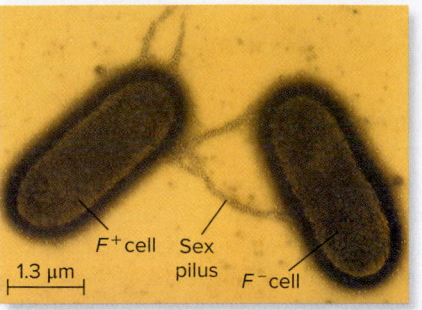

$F^+$cell    Sex pilus    $F^-$cell

1.3 μm

**(a) Micrograph of conjugating cells**

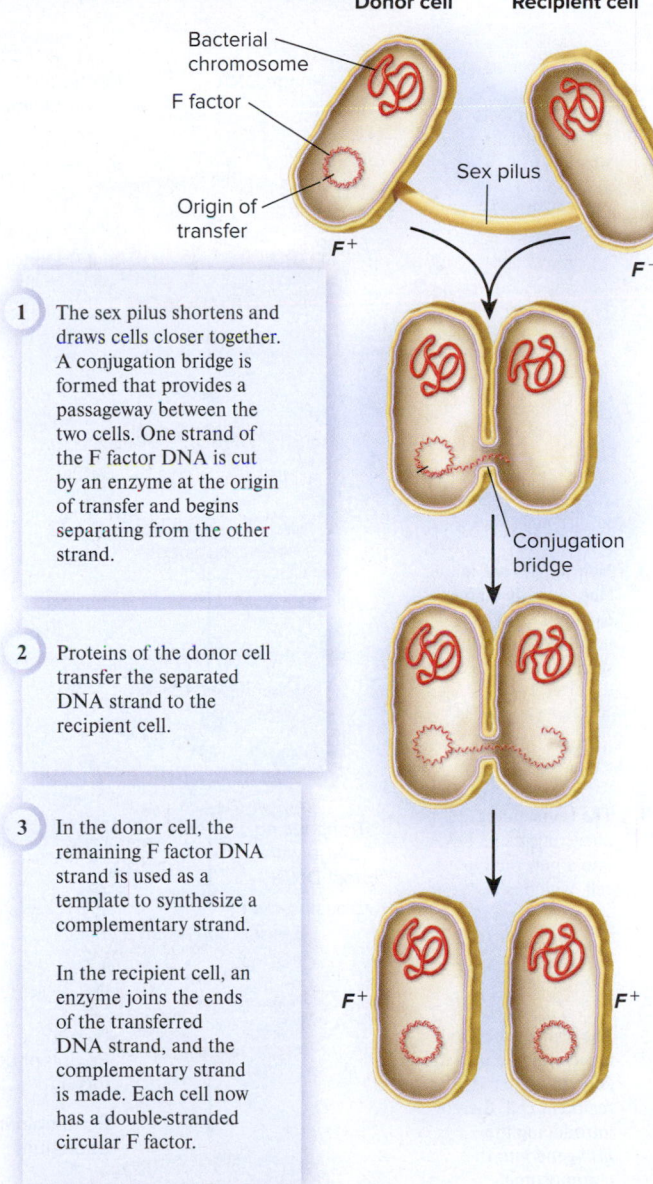

Donor cell    Recipient cell

Bacterial chromosome

F factor

Sex pilus

Origin of transfer

$F^+$    $F^-$

1  The sex pilus shortens and draws cells closer together. A conjugation bridge is formed that provides a passageway between the two cells. One strand of the F factor DNA is cut by an enzyme at the origin of transfer and begins separating from the other strand.

Conjugation bridge

2  Proteins of the donor cell transfer the separated DNA strand to the recipient cell.

3  In the donor cell, the remaining F factor DNA strand is used as a template to synthesize a complementary strand.

In the recipient cell, an enzyme joins the ends of the transferred DNA strand, and the complementary strand is made. Each cell now has a double-stranded circular F factor.

$F^+$    $F^+$

**(b) Transfer of an F factor**

**Figure 19.16  Bacterial conjugation.** **(a)** A micrograph of two *E. coli* cells that are conjugating. The cell on the left, designated $F^+$, is the donor; the cell on the right, designated $F^-$, is the recipient. The two cells make contact via sex pili made by the $F^+$ cell. **(b)** The transfer of an F factor during conjugation. At the end of conjugation, both the donor cell and the recipient cell are $F^+$. a: ©Dr. L. Caro/SPL/Science Source

*Concept Check:* *If a donor cell has only one F factor, explain how the donor and recipient cell both contain one F factor following the transfer of an F factor during conjugation.*

specifically to $F^-$ cells. They are so named because conjugation has been called bacterial mating. However, this term is a bit misleading because the process does not involve equal genetic contributions from two gametes and it does not produce offspring. Donor strains have genes responsible for the formation of sex pili. In $F^+$ strains, the genes are located on the F factor. In *E. coli* and some other species, $F^+$ cells make very long pili that attempt to make contact with nearby $F^-$ cells. Once contact is made, the pili shorten, drawing the donor and recipient cells closer together.

After the pili have shortened, contact between donor and recipient cell stimulates the donor cell to begin the transfer process. First, a conjugation bridge is formed that provides a direct passageway for DNA transfer. One strand of F factor DNA is cut at the origin of transfer and then travels through the conjugation bridge into the recipient cell. The other strand remains in the donor cell, and the complementary strand is synthesized, thereby restoring the F factor DNA to its original double-stranded condition. In the recipient cell, the two ends of the newly acquired F factor DNA strand are joined to form a circular molecule, and its complementary strand is synthesized to produce a double-stranded F factor. If conjugation is successful, the end result is that the recipient cell has acquired an F factor, converting it from an $F^-$ to an $F^+$ cell. The genetic composition of the donor strain has not been changed.

## In Transformation, Bacteria Take Up DNA from the Environment

In contrast to conjugation, the process of gene transfer known as bacterial transformation does not require direct contact between bacterial cells. Frederick Griffith first discovered this process in 1928 while working with strains of *Streptococcus pneumoniae*. We discussed early experiments involving transformation in Chapter 11 (refer back to Figures 11.1 and 11.2).

How does a bacterial cell become transformed? First, it imports a strand of DNA from the environment. This DNA strand, which is typically derived from a dead bacterial cell, may then insert or recombine into the bacterial chromosome. The live bacterium then carries genes from the dead bacterium—the live bacterium has been transformed.

Not all bacterial strains have the ability to take up DNA. Those that do have this ability are described as naturally **competent**, and they have genes that encode proteins called competence factors. Competence factors facilitate the binding of DNA fragments to the bacterial cell surface, the uptake of DNA into the cytoplasm, and the incorporation of the imported DNA into the bacterial chromosome. Temperature, ionic conditions, and the availability of nutrients affect whether or not a bacterium will be competent to take up genetic material.

In recent years, biologists have unraveled some of the steps that occur when competent bacterial cells are transformed by taking up genetic material from the environment. In the example shown in **Figure 19.17**, the DNA released from a dead bacterium carries a gene, $tet^R$, that confers resistance to the antibiotic tetracycline. First, a large fragment of the DNA binds to a cell surface receptor on the outside of a bacterial cell that is sensitive to tetracycline. Enzymes secreted by the bacterium cut this large fragment into fragments small enough to enter the cell. In our example, one of the two strands of a fragment of DNA containing the $tet^R$ gene is degraded. The other strand enters the bacterial cytoplasm via a DNA uptake system that transports the DNA across the

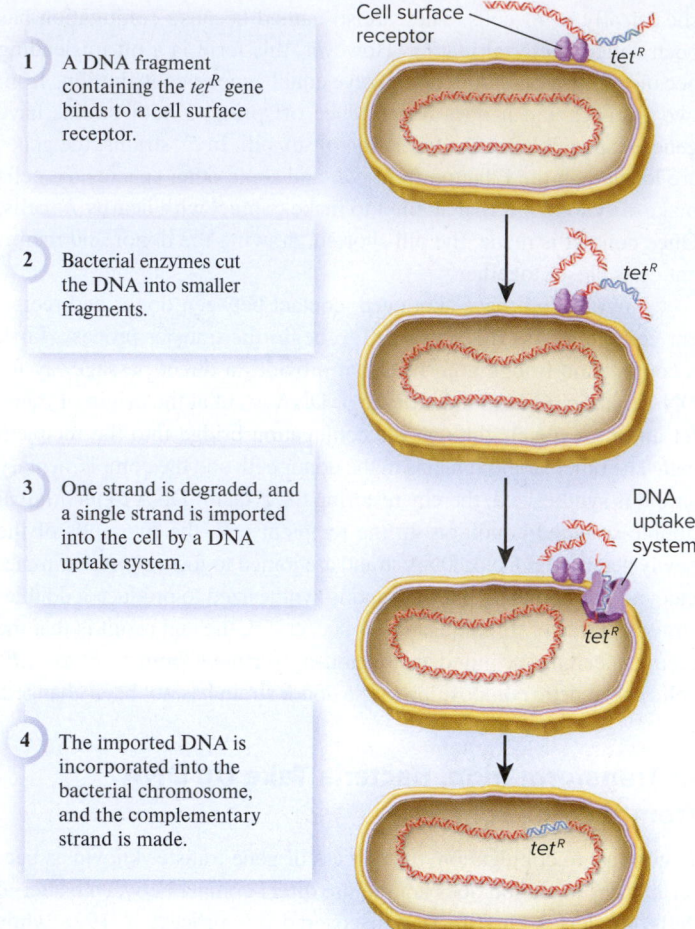

1  A DNA fragment containing the *tet*$^R$ gene binds to a cell surface receptor.

Cell surface receptor

*tet*$^R$

2  Bacterial enzymes cut the DNA into smaller fragments.

*tet*$^R$

3  One strand is degraded, and a single strand is imported into the cell by a DNA uptake system.

DNA uptake system

*tet*$^R$

4  The imported DNA is incorporated into the bacterial chromosome, and the complementary strand is made.

*tet*$^R$

Transformed cell that is resistant to the antibiotic tetracycline

**Figure 19.17  Bacterial transformation.** This process has transformed a bacterium that was sensitive to the antibiotic tetracycline into one that is resistant to this antibiotic.

 **Core Skill: Connections** Look back at Figures 11.1 and 11.2. How did the phenomenon of transformation allow researchers to demonstrate that DNA is the genetic material?

plasma membrane. Finally, the imported DNA strand is incorporated into the bacterial chromosome, and the complementary strand is synthesized. Following transformation, the recipient cell has been transformed from a tetracycline-sensitive cell to a tetracycline-resistant cell.

## In Transduction, Bacteriophages Transfer Genetic Material from One Bacterium to Another

A third mechanism of gene transfer is transduction, in which bacteriophages transfer bacterial genes from one bacterium to another. As discussed earlier in this chapter, a bacteriophage (or simply phage) is a virus that uses the cellular machinery of a bacterium for its own replication. The new viral particles made in this way usually contain only viral genes. On rare occasions, however, a phage may pick up a piece of DNA from the bacterial chromosome. When a phage carrying a

segment of bacterial DNA infects another bacterium, it transfers this segment into the chromosome of its new bacterial host.

Transduction is actually an error in a phage lytic cycle, as shown in **Figure 19.18**. In this example, a phage called P1 infects an *E. coli* cell that has a gene (*his*$^+$) for histidine synthesis. Phage P1 causes the host cell chromosome to be degraded into small pieces. New phage DNA and proteins are synthesized. When new phages are assembled, coat proteins may accidentally enclose a piece of host DNA that carries the *his*$^+$ gene, creating a phage that carries the gene. In the

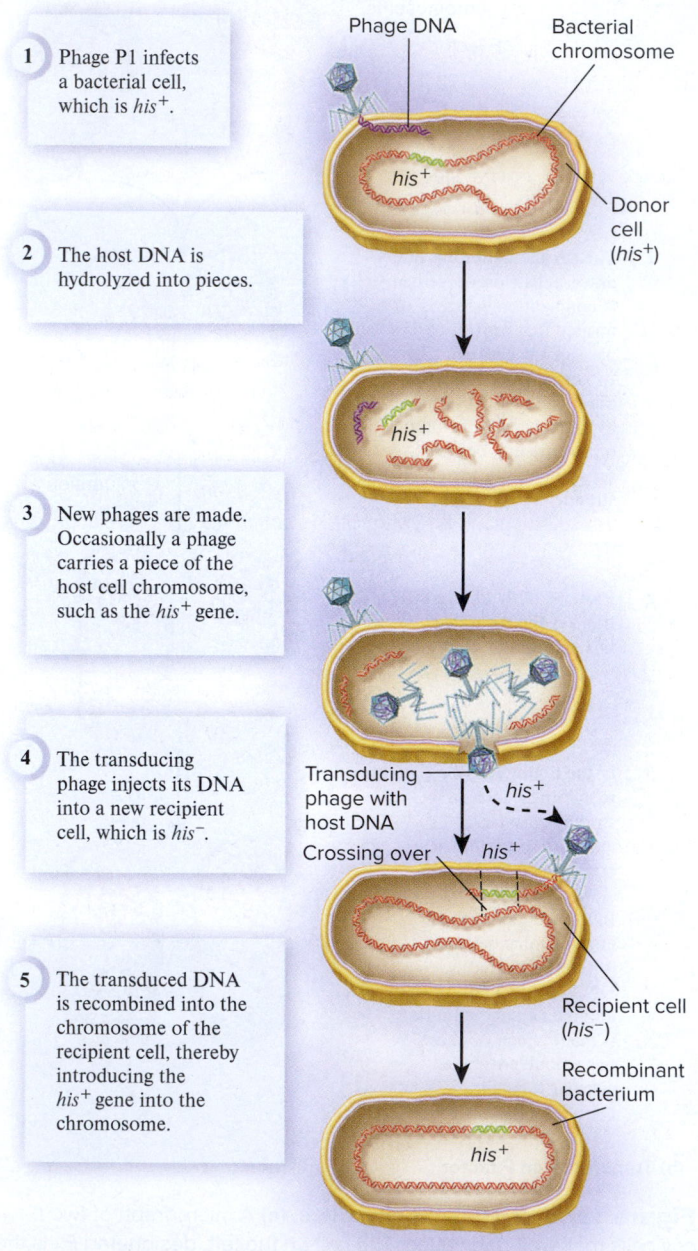

1  Phage P1 infects a bacterial cell, which is *his*$^+$.

Phage DNA

Bacterial chromosome

*his*$^+$

Donor cell (*his*$^+$)

2  The host DNA is hydrolyzed into pieces.

*his*$^+$

3  New phages are made. Occasionally a phage carries a piece of the host cell chromosome, such as the *his*$^+$ gene.

4  The transducing phage injects its DNA into a new recipient cell, which is *his*$^-$.

Transducing phage with host DNA

*his*$^+$

Crossing over

*his*$^+$

5  The transduced DNA is recombined into the chromosome of the recipient cell, thereby introducing the *his*$^+$ gene into the chromosome.

Recipient cell (*his*$^-$)

Recombinant bacterium

*his*$^+$

The recombinant bacterium has a genotype (*his*$^+$) that is different from the original recipient bacterial cell (*his*$^-$).

**Figure 19.18  Bacterial transduction by phage P1.**

*Concept Check:* *Is transduction a normal part of the phage life cycle? Explain.*

example shown in Figure 19.18, this transducing phage is released and binds to an *E. coli* cell that lacks the *his⁺* gene. It inserts the bacterial DNA fragment into the recipient cell, which then integrates this fragment into its own chromosome. In this case, gene transfer by transduction converts a *his⁻* strain of *E. coli* to a *his⁺* strain.

 **Core Concept: Evolution**

## Horizontal Gene Transfer Can Occur Within a Species or Between Different Species

The term **horizontal gene transfer** refers to a process in which an organism incorporates genetic material from another organism without being the offspring of that organism. Conjugation, transformation, and transduction are examples of horizontal gene transfer. In contrast, vertical gene transfer occurs when genes are passed from one generation to the next—from parents to offspring and from mother cells to daughter cells.

Conjugation, transformation, and transduction occasionally occur between cells of different bacterial species. In recent years, analyses of bacterial genomes have shown that a sizable percentage of bacterial genes are derived from horizontal gene transfer. For example, roughly 17% of the genes of *E. coli* and of *Salmonella typhimurium* have been acquired from other species by horizontal gene transfer during the past 100 million years. Many of these acquired genes affect traits that give cells a selective advantage, including genes that confer antibiotic resistance, the ability to degrade toxic compounds, and the ability to withstand extreme environments. Some horizontally transferred genes confer pathogenicity, turning a harmless bacterial strain into one that can cause disease. Geneticists have suggested that horizontal gene transfer has played a major role in the evolution of different bacterial species. In many cases, the acquisition of new genes allows a bacterium to survive in a new type of environment and can eventually lead to the formation of a new species.

A second reason why horizontal gene transfer is important is its medical relevance. Let's consider the topic of antibiotic resistance. Antibiotics are widely prescribed to treat bacterial infections in humans. They are also used in agriculture to control bacterial diseases. Unfortunately, the widespread use of antibiotics has greatly increased the prevalence of antibiotic-resistant strains of bacteria, strains that have a selective advantage over those that are susceptible to antibiotics. Resistant strains carry genes that counteract the action of antibiotics in various ways. A resistance gene may encode a protein that breaks down the antibiotic, pumps it out of the cell, or prevents it from inhibiting cellular processes.

The term **acquired antibiotic resistance** refers to the common phenomenon in which a previously susceptible strain becomes resistant to a specific antibiotic. This change may result from genetic alterations in the genome of the susceptible strain, but it is often due to the horizontal transfer of resistance genes from a resistant strain. As often mentioned in the news media, antibiotic resistance has increased dramatically worldwide over the past few decades, with resistant strains reported in almost all pathogenic strains of bacteria. As an example, some *Staphylococcus aureus* strains have developed resistance to methicillin and all penicillins. Evidence suggests that these so-called methicillin-resistant strains of *Staphlococcus aureus* (MRSA) acquired the methicillin-resistance gene by horizontal gene transfer, possibly from a strain of *Enterococcus faecalis*. MRSA strains cause skin infections that are more difficult to treat than ordinary "staph" infections caused by nonresistant strains of *S. aureus*.

## Summary of Key Concepts

### 19.1  General Properties of Viruses

- Viruses are nonliving particles that do not exhibit all of the properties associated with living organisms. A virus or its genetic material must be taken up by a living cell to replicate. Tobacco mosaic virus (TMV) was the first virus to be discovered (Figure 19.1).

- Viruses vary with regard to their host range, structure, and genome composition (Table 19.1, Figures 19.2, 19.3).

### 19.2  Viral Reproductive Cycles

- The viral reproductive cycle consists of a series of steps: attachment, entry, integration, synthesis of components, assembly, and release (Figure 19.4).

- Some bacteriophages can follow two different reproductive cycles: the lytic cycle and the lysogenic cycle (Figure 19.5).

- Emerging viruses, such as human immunodeficiency virus (HIV) and Zika virus, can cause significant loss of human life. HIV, the causative agent of the disease AIDS, is a retrovirus whose reproductive cycle involves the integration of the viral genome into a chromosome in the host cell (Figure 19.6).

- Drugs to combat viral proliferation are often developed to specifically inhibit viral proteins.

### 19.3  Viroids and Prions

- Viroids are RNA molecules that infect plant cells (Figure 19.7).

- Prions are infectious proteins that induce abnormal folding in normal proteins. They cause several fatal neurodegenerative diseases in humans (Figure 19.8, Table 19.2).

### 19.4  Genetic Properties of Bacteria

- Bacteria typically have a single type of circular chromosome found in the nucleoid of the cell. The chromosome contains many genes and one origin of replication (Figure 19.9).

- The bacterial chromosome is made more compact by the formation of loop domains and by DNA supercoiling (Figure 19.10).

- Plasmids are small, circular DNA molecules that exist independently of the bacterial chromosome (Figure 19.11).

- When placed on solid growth media, a single bacterial cell will divide many times to produce a colony composed of many cells (Figure 19.12).

- Bacterial cells reproduce by asexual reproduction in a process called binary fission, during which a cell divides to form two daughter cells (Figure 19.13).

## 19.5 Gene Transfer Between Bacteria

- Three common modes of gene transfer among bacteria are conjugation, transformation, and transduction (Table 19.3).

- Lederberg and Tatum's work demonstrated the transfer of bacterial genes between different strains of *E. coli* by conjugation. Davis showed that direct contact was needed for this type of gene transfer (Figures 19.14, 19.15).

- During conjugation, a strand of DNA from an F factor is transferred from a donor to a recipient cell by physical contact (Figure 19.16).

- Transformation is the process in which a segment of DNA from the environment is taken up by a naturally competent bacterial cell and incorporated into the bacterial chromosome (Figure 19.17).

- Bacterial transduction is a form of gene transfer in which a bacteriophage transfers a segment of bacterial chromosomal DNA from one bacterial cell to another (Figure 19.18).

- Horizontal gene transfer is a process in which an organism incorporates genetic material from another organism without being the offspring of that organism. It has played an important role in the evolution of bacterial species.

## Assess & Discuss

### Test Yourself

1. The _____ is the protein coat of a virus that surrounds the genetic material.
   a. host
   b. prion
   c. capsid
   d. viroid
   e. capsule

2. The characteristics of viral genomes show many variations. Which of the following does *not* describe a typical characteristic of viral genomes?
   a. The genetic material may be DNA or RNA.
   b. The nucleic acid may be single stranded or double stranded.
   c. The genome may carry just a few genes or several dozen.
   d. The number of copies of the genome may vary.
   e. All of the above describe typical variation in viral genomes.

3. During viral infection, attachment is usually specific to a particular cell type because
   a. the virus is attracted to the appropriate host cells by proteins secreted into the extracellular fluid.
   b. the virus recognizes and binds to specific molecules in the cytoplasm of the host cell.
   c. the virus recognizes and binds to specific molecules on the surface of the host cell.
   d. the host cell produces channel proteins that provide passageways for viruses to enter the cytoplasm.
   e. the virus releases specific proteins that make holes in the membrane large enough for the virus to enter.

4. HIV, a retrovirus, has a high mutation rate because
   a. the DNA of its genome is less stable than other viral genomes.
   b. the viral enzyme reverse transcriptase has a high likelihood of making replication errors.
   c. the viral genome is altered every time it is incorporated into the host genome.
   d. antibodies produced by the host cell mutate the viral genome when infection occurs.
   e. All of the above are true.

5. A _____ is an infectious agent composed solely of RNA, whereas a _____ is an infectious agent composed solely of protein.
   a. retrovirus, bacteriophage
   b. viroid, virus
   c. prion, virus
   d. retrovirus, prion
   e. viroid, prion

6. Genetic diversity is maintained in bacterial populations by all of the following except
   a. binary fission.
   b. mutation.
   c. transformation.
   d. transduction.
   e. conjugation.

7. Bacterial cells divide by a process known as
   a. mitosis.
   b. cytokinesis.
   c. meiosis.
   d. binary fission.
   e. glycolysis.

8. Gene transfer in which a bacterial cell takes up bacterial DNA from the environment is called
   a. conjugation.
   b. binary fission.
   c. asexual reproduction.
   d. transformation.
   e. transduction.

9. A bacterial cell can donate DNA during conjugation when it
   a. produces competence factors.
   b. contains an F factor.
   c. is virulent.
   d. has been infected by a bacteriophage.
   e. All of the above allow a bacterial cell to donate DNA during conjugation.

10. A bacterial species that becomes resistant to a certain antibiotic may have acquired the resistance gene from another bacterial species. The phenomenon of acquiring genes from another organism without being the offspring of that organism is known as
    a. hybridization.
    b. integration.
    c. horizontal gene transfer.
    d. vertical gene transfer.
    e. competence.

### Conceptual Questions

1. How are viruses similar to living cells, and how are they different?

2. Describe the three mechanisms of gene transfer in bacteria.

3. **Core Concept: Information** Discuss how horizontal gene transfer alters the genetic material.

### Collaborative Questions

1. Discuss the possible origin of viruses. Which idea(s) do you think is (are) the most likely?

2. Conjugation is sometimes called "bacterial mating." Discuss how conjugation is similar to sexual reproduction in eukaryotes and how it is different.

# Developmental Genetics

# 20

**A child with aniridia.** The *Pax6* gene plays an important role in eye development. A mutation in this gene causes aniridia, which causes the iris of the eye to be greatly reduced or absent.
©Medical-on-Line/Alamy Stock Photo

T ake a close look at the child in the chapter opening photo. Do you notice anything unusual? Though it may not be immediately apparent, this child has a disorder called aniridia, in which the iris in each eye does not develop properly. The iris is the part of the eye, usually blue, green, or brown, that regulates the amount of light entering the eye. In aniridia, some or all of the iris is missing, giving the appearance of very large pupils. People with aniridia cannot adjust the amount of light entering their eyes, which results in a decreased quality of vision and leads to eye diseases such as glaucoma and cataracts. In addition, other structures within the eye, such as the retina and optic nerve, may not develop correctly. What is the underlying cause of aniridia? It is due to a mutation in a gene called *Pax6*, which is responsible for the development of the eye. People with aniridia are heterozygotes, having one functional copy of the *Pax6* gene and one mutant copy; it is a dominant disorder. The proper development of the eye requires two functional copies of the *Pax6* gene. This disorder illustrates how genes play a key role in the development of our bodies.

In biology, the term **development** refers to a series of changes in the state of a cell, tissue, organ, or organism. Development is the underlying process that gives rise to the structures and functions of living organisms. The structure or form of an organism is called its **morphology**. As we have learned throughout this textbook, an important core concept of biology is that structure (morphology) determines function.

How do developmental changes occur? Since the 1940s, the genetic makeup of an organism has emerged as the fundamental factor behind development. The science of **developmental genetics** is concerned with understanding how gene expression controls the process of development.

In this chapter, we will examine how the sequential actions of genes provide a program for the development of an organism from a fertilized egg to an adult. Utilizing a few organisms, such as the fruit fly, a nematode worm, the mouse, and the plant *Arabidopsis thaliana*, scientists are working to identify and characterize the genes required to run developmental programs and are exploring how proteins encoded by these genes control the course of development. In this chapter, we begin with an overview that emphasizes the general principles of development. We will then examine specific examples of development in animals and plants, focusing on the role of genes in embryonic development. Chapters 40 and 51 also consider plant and animal development, respectively, with an emphasis on structure and function.

## 20.1 General Themes in Development

**Learning Outcomes:**

1. List several invertebrate, vertebrate, and plant model organisms, and describe the reasons why biologists study them.
2. Outline the process of pattern formation in plants and animals.
3. Describe the four ways that cells respond to positional information.
4. Explain how morphogens and cell-to-cell contacts convey positional information.
5. Outline four phases of pattern formation in animals, and describe how they are controlled by transcription factors.

Animals and plants begin to develop when a sperm and an egg unite to produce a **zygote**, a diploid cell that divides and develops into a multicellular **embryo**, and eventually into an adult organism. During the early stages of development, cells divide and begin to arrange themselves into ordered units. As this occurs, each cell also becomes **determined**, which means it is committed to become a particular cell type, such as a muscle or intestinal cell. The commitment to become a specific cell type occurs long before a cell differentiates. During the process of **cell differentiation**, a cell's morphology and function have changed, usually permanently, into a highly specialized cell type. In an adult, each cell type plays its own particular role. In animals, for example, muscle cells enable an organism to move, and intestinal cells facilitate the absorption of nutrients. This division of labor among various cells of an organism works collectively to promote its survival.

The genomes of living organisms contain a set of genes that constitute a program of development. In unicellular species, the program controls the structure and function of the cell. In multicellular species such as animals and plants, the program not only controls cell morphology but also determines the arrangement of cells. In this section, we will examine some of the general concepts associated with the development of multicellular species.

## Developmental Biologists Use Model Organisms to Study Development

The development of even a simple multicellular organism involves many types of changes in form and function. For this reason, the research community has focused its efforts on only a few **model organisms**—organisms studied by many different researchers

so that they can compare their results and determine scientific principles that apply more broadly to other species.

With regard to animal development, the model organisms that have been the most extensively investigated are two invertebrate species: the fruit fly (*Drosophila melanogaster*) and the nematode worm (*Caenorhabditis elegans*) (**Figure 20.1a,b**). Why have these two organisms been chosen as models to investigate development? *Drosophila* has been studied for a variety of reasons. First, researchers have exposed this organism to mutagens and identified many mutant strains with altered developmental pathways. Second, in all of its life stages, *Drosophila* has distinct morphological features and is large enough to easily identify the effects of mutations. *C. elegans* is used by developmental geneticists because of its simplicity. The adult organism is a small transparent worm about 1 mm in length and composed of only about 1,000 somatic cells. Starting with a fertilized egg, the pattern of cell division and the fate of each cell within the embryo are completely known. This pattern is essentially identical from one worm to another, which allows researchers to predict the fate of cells in this organism.

Embryologists have also studied the morphological features of development in vertebrate species. Historically, amphibians and birds have been studied extensively, because their eggs are rather large and easy to manipulate. From a morphological point of view, the developmental stages of the chicken (*Gallus gallus*) and the African clawed frog (*Xenopus laevis*) have been described in great detail. More recently, a few vertebrate species have been the subject of genetic studies of development. These include the house mouse (*Mus musculus*) and the small aquarium zebrafish (*Danio rerio*) (**Figure 20.1c,d**).

In the study of plant development, the model organism for genetic analysis is a small flowering plant known as thale cress (*Arabidopsis thaliana*), which is typically called *Arabidopsis* by researchers

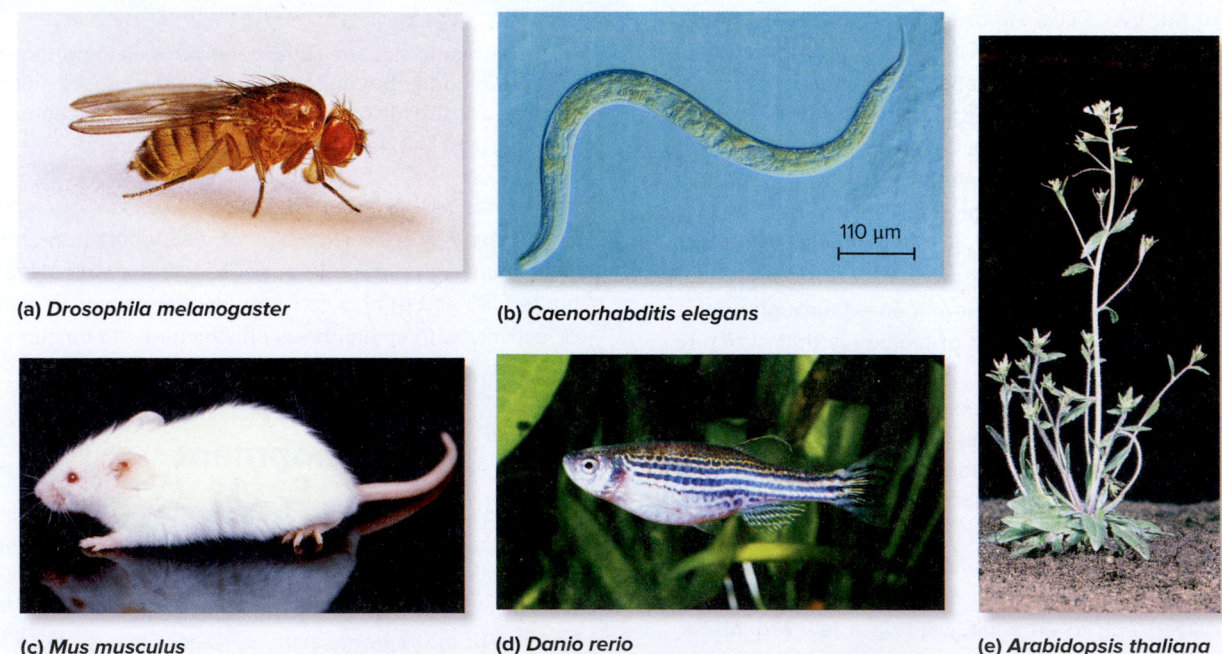

**(a)** *Drosophila melanogaster*

**(b)** *Caenorhabditis elegans*

110 µm

**(c)** *Mus musculus*

**(d)** *Danio rerio*

**(e)** *Arabidopsis thaliana*

**Figure 20.1**  **Model organisms used to study developmental genetics.** a: ©Herman Eisenbeiss/Science Source; b: ©Sinclair Stammers/Science Source; c: ©J–M. Labat/Science Source; d: ©Mark Smith/Science Source; e: ©Nigel Cattlin/Science Source

 **Core Skill: Process of Science** Biologists often focus their attention on model organisms with the expectation that the results will apply more broadly to many species.

(**Figure 20.1e**). *Arabidopsis* is an annual (a plant that lives out its entire life cycle during a single growing season) belonging to the wild mustard family. It occurs naturally throughout temperate regions of the world. *Arabidopsis* has a short generation time of about 6 weeks and a small genome size of $12 \times 10^7$ bp, which is similar to the genome sizes of *Drosophila* and *C. elegans*. A flowering *Arabidopsis* plant is small enough to be grown in the laboratory and produces a large number of seeds.

## Both Animals and Plants Develop by Pattern Formation

Development in animals and plants produces a body plan, or pattern. At the cellular level, the body pattern is due to the arrangement of cells and their differentiation. The process, called **pattern formation**, gives rise to the formation of a body with a particular morphology. Pattern formation in most animals is organized along three axes: the **dorsoventral axis**, the **anteroposterior axis**, and the **left-right axis** (**Figure 20.2a**). In addition, many animal bodies are segmented into separate sections containing specific body parts such as wings or legs.

Pattern formation in plants is quite different, being organized along a **root-shoot axis** and in a **radial pattern**, in which

the cells found in roots and shoots form concentric rings of tissues (**Figure 20.2b**). The root-shoot axis is determined at the first division of the fertilized egg. As we'll see later, the identification of mutant alleles that disrupt development has permitted great insight into the genes controlling pattern formation.

## Pattern Formation Depends on Positional Information

For an organism to develop the correct morphological features or pattern, each cell of the body must become the appropriate cell type based on its position relative to other cells. How does this occur? At appropriate times during development, cells receive **positional information**—information regarding a cell's location relative to other cells of the body. Later in this chapter, we will examine how the expression of genes at the correct times provides this information.

A cell may respond to positional information in one of four ways: cell division, cell migration, cell differentiation, and cell death (**Figure 20.3**).

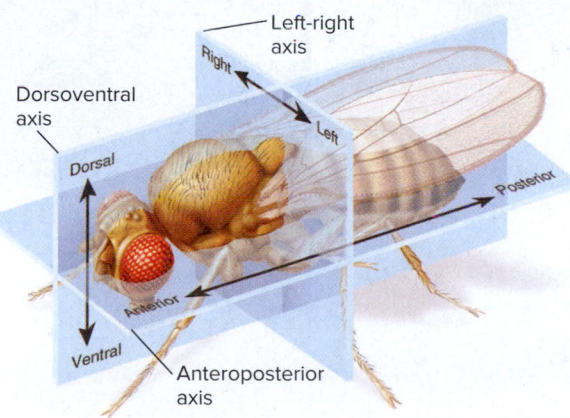

**(a) Body plan found in many animals**

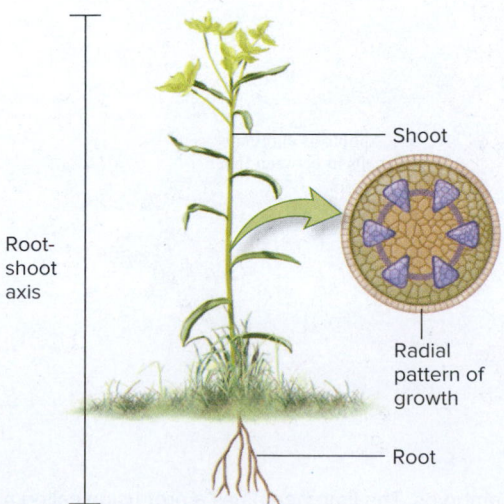

**(b) Body plan found in many seed-bearing plants**

**Figure 20.2**  **Body plan axes in animals and plants.**

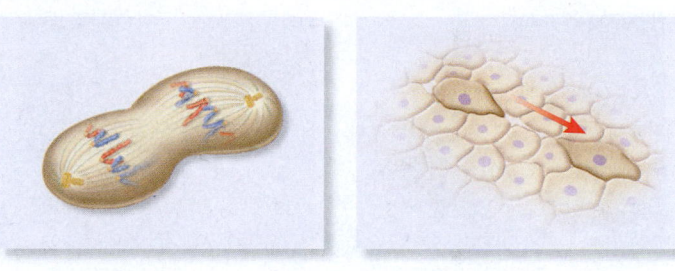

**(a) Cell division**          **(b) Cell migration**

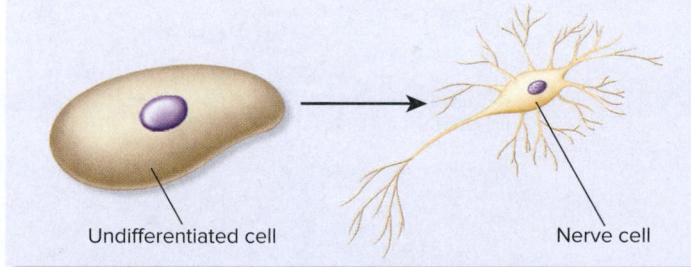

**(c) Cell differentiation**

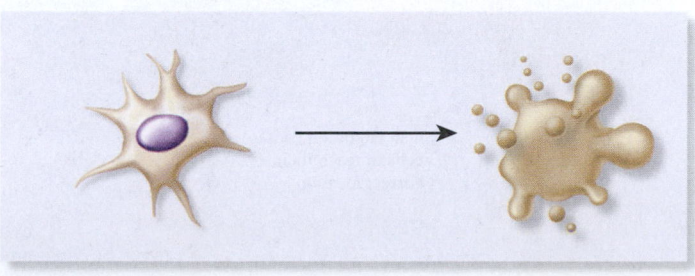

**(d) Cell death (apoptosis)**

**Figure 20.3**  **Four types of cellular responses to positional information in animals.**

**Concept Check:**  *Which of these four responses do you expect to be more prevalent in the early stages of development, and which would become more prevalent in later stages?*

- Positional information may stimulate a cell to divide.

- Positional information in animals may cause the migration of a cell or group of cells in a particular direction from one region of the embryo to another. (Cell migration does not occur during development in plants.)

- Positional information may cause a cell to differentiate into a specific cell type such as a neuron.

- Positional information may promote **apoptosis**, or programmed cell death, which is described in Chapter 9. Apoptosis plays a key role in sculpting the bodies of animals. In vascular plants, certain cells undergo programmed cell death to form tracheids, specialized cells that function in water transport.

As an example of how the coordination of these four processes is required for pattern formation, **Figure 20.4** shows the embryonic development of a human limb, which has an arm and hand.

- Cell division with accompanying cell growth increases the size of the limb.

- Cell migration is also important for limb development. For example, embryonic cells that eventually form muscles in the arm and hand must migrate from outside the limb to reach their correct location within the limb.

- As development proceeds, cell differentiation produces the various tissues that will eventually be found in the fully developed limb. Some cells become neurons, others muscle cells, and still others become epidermal cells, forming the outer layer of skin.

- Finally, apoptosis is important in the formation of fingers. If apoptosis did not occur, a human hand would have webbed fingers.

## Morphogens and Cell-to-Cell Contacts Convey Positional Information

How does positional information lead to the development of a body plan? Though the details of pattern formation vary widely among different species, two main mechanisms are commonly used to

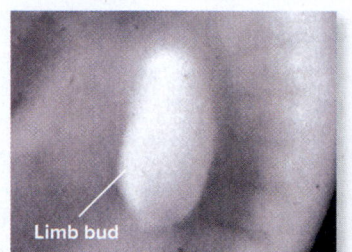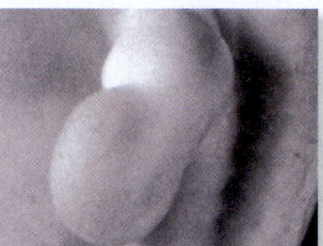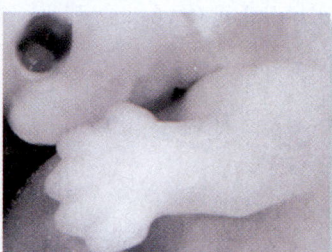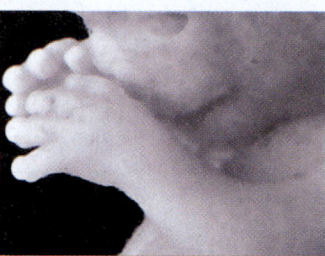

Limb bud

**(a) Limb development in a human embryo**

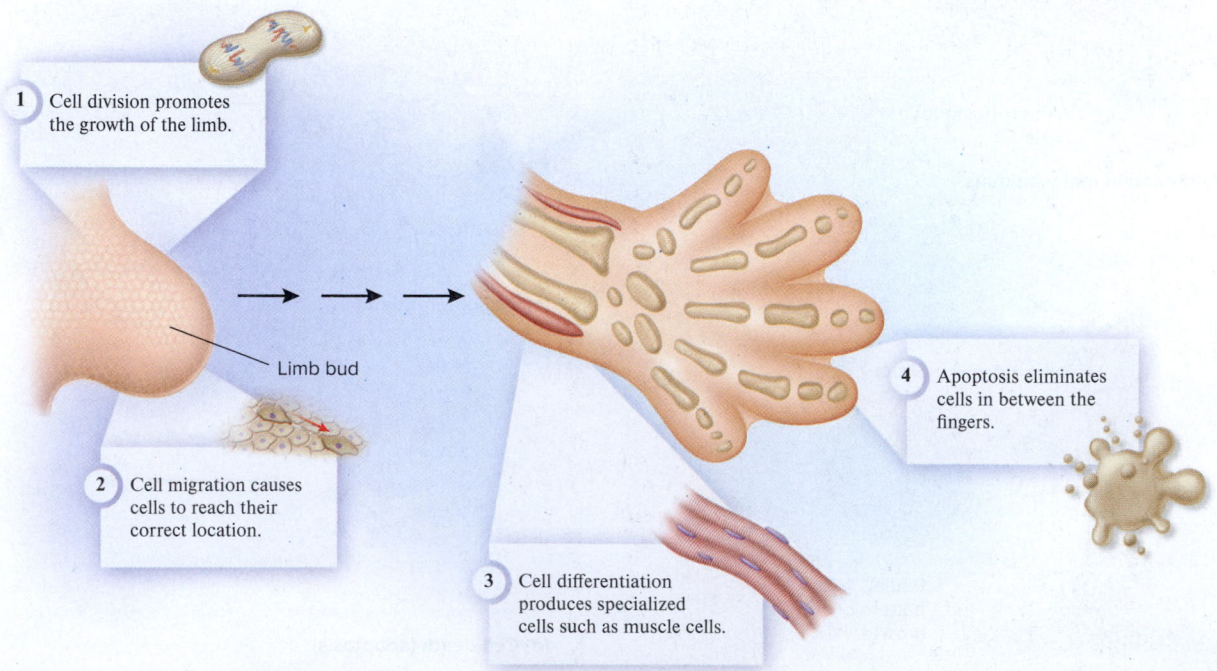

1 Cell division promotes the growth of the limb.

Limb bud

2 Cell migration causes cells to reach their correct location.

3 Cell differentiation produces specialized cells such as muscle cells.

4 Apoptosis eliminates cells in between the fingers.

**(b) Four cellular processes that promote limb formation**

**Figure 20.4  Limb development in humans.  (a)** Photographs of limb development in human embryos. The limb begins as a protrusion called a limb bud that eventually forms an arm and hand. **(b)** The development of a human hand from an embryonic limb bud. a: Courtesy of the National Museum of Health and Medicine, Washington, D.C.

**Concept Check:** *How would human finger formation be affected if apoptosis did not occur?*

communicate positional information. One of these mechanisms involves molecules called morphogens. **Morphogens** impart positional information and promote developmental changes at the cellular level. Many morphogens are proteins, but they can also be small signaling molecules. A morphogen influences the fate of a cell by promoting cell division, cell migration, cell differentiation, or apoptosis. A key feature of morphogens is that they act in a concentration-dependent manner. At a high concentration, a morphogen restricts a cell into a particular developmental pathway, whereas at a lower concentration, it does not. There is often a critical **threshold concentration** above which the morphogen exerts its effects.

Morphogens typically are distributed asymmetrically along a concentration gradient. Morphogen gradients may be established in the **oocyte**, a cell that matures into an egg cell (**Figure 20.5a**). In addition, a morphogen gradient can be established in the embryo by secretion and diffusion (**Figure 20.5b**). A certain cell or group of cells may synthesize and secrete a morphogen at a specific stage of development.

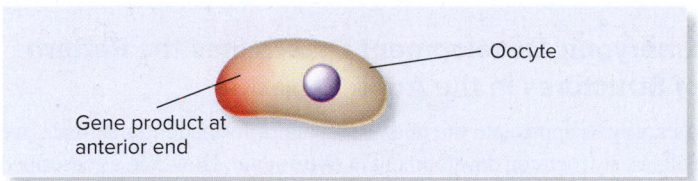

**(a) Asymmetric distribution of morphogens in the oocyte**

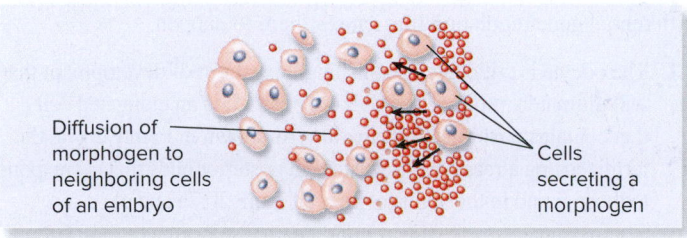

**(b) Induction: asymmetric synthesis and extracellular distribution of a morphogen**

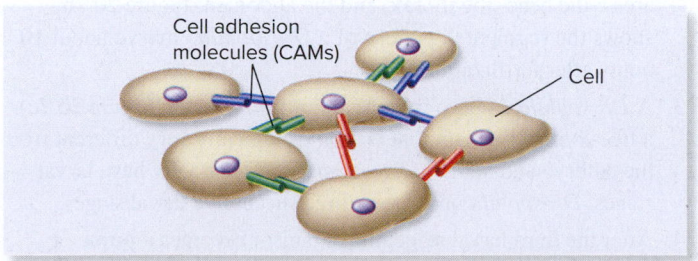

**(c) Cell adhesion: Cell-to-cell contact conveys positional information.**

**Figure 20.5** **Molecular mechanisms that convey positional information.** Morphogen gradients may be established in the **(a)** oocyte or **(b)** embryo. **(c)** Positional information may also be conveyed by cell-to-cell contact. Different colors indicate different CAMs on the surfaces of the cells.

 **Core Skill: Connections** Look back at Chapter 9. Discuss the role of cell surface receptors in responding to positional information.

After secretion, the morphogen diffuses to neighboring cells, as in Figure 20.5b, or it may be transported to cells that are distant from the cells that secrete it. The morphogen may then influence the fate of cells exposed to it. The process by which a cell or group of cells governs the developmental fate of other cells is known as **induction**.

Another mechanism to convey positional information is **cell adhesion** (**Figure 20.5c**). Each animal cell makes its own collection of surface receptors that enable it to adhere to other cells and to the extracellular matrix (ECM). Such receptors, known as **cell adhesion molecules (CAMs)**, are described in Chapter 10 (refer back to Figure 10.8). The positioning of a cell within a multicellular organism is strongly influenced by the combination of contacts it makes with other cells and with the ECM.

The phenomenon of cell adhesion and its role in multicellular development was first recognized by American biologist Henry V. Wilson in 1907. He took multicellular sponges and passed them through a sieve, dissociating them into individual cells. Remarkably, the cells actively migrated until they adhered to one another to form a new sponge, complete with the chambers and canals that characterize a sponge's internal structure! When sponge cells from different species were mixed, they sorted themselves properly, adhering only to cells of the same species. Overall, these results indicate that cells possess specific CAMs, which are critical in cell-to-cell recognition. Cell adhesion plays an important role in governing the position that an animal cell will adopt during development.

## Pattern Formation Occurs in Phases That Are Controlled by Transcription Factors

The formation of a body, in both animals and plants, occurs in a series of overlapping organizational phases. As an overview of this process, let's consider four general phases of pattern formation in an animal (**Figure 20.6**). This example involves human development, but research suggests that pattern formation in all complex animals follows a similar plan.

1. The first phase organizes the body along major axes. The anteroposterior axis is organized from head to tail, the dorsoventral axis is organized from back (dorsal) to front/abdomen (ventral), and the left-right axis is organized from side to side (refer back to Figure 20.2a).

2. During the second phase, the body becomes organized into smaller regions, a process called **segmentation**. In insects, these regions form well-defined segments. In mammals, some segmentation of the body is apparent during embryonic development, but defined boundaries are lost as the embryo proceeds to the fetal and adult stages.

3. In the third phase, the cells within the segments organize themselves in ways that will produce particular body parts.

4. Finally, during the fourth phase, the cells change their morphologies and become differentiated. This final phase of development produces an organism with many types of tissues, organs, and other body parts with specialized functions.

It should be noted that the four phases of development are overlapping. For example, cell differentiation begins to occur as the cells are adopting their correct locations.

**Hierarchy of transcription factors**

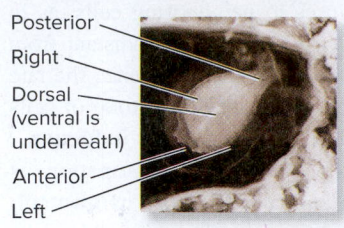

Posterior
Right
Dorsal
(ventral is
underneath)
Anterior
Left

**1  Phase 1:**
Transcription factors determine the formation of the body axes and control the expression of transcription factors of phase 2.

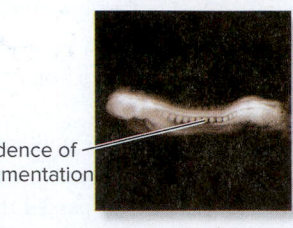

Evidence of segmentation

**2  Phase 2:**
Transcription factors cause the embryo to become subdivided into regions that have properties of individual segments. They also control transcription factors of phase 3.

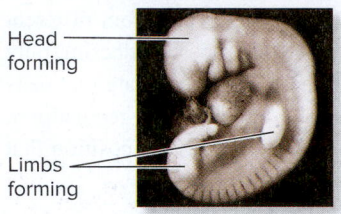

Head forming

Limbs forming

**3  Phase 3:**
Transcription factors cause each segment and groups of segments to develop specific characteristics. They also control transcription factors of phase 4.

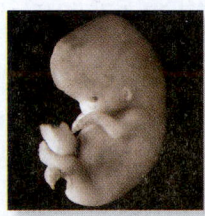

**4  Phase 4:**
Transcription factors cause cells to differentiate into specific cell types such as skin, nerve, and muscle cells.

**Figure 20.6  Pattern formation in a human embryo.** The ideas in this scenario are based largely on analogies between pattern formation in *Drosophila* and mammals. Many of the transcription factors that control the early phases of pattern formation in mammals have yet to be identified. (1, 3, 4): Courtesy of the National Museum of Health and Medicine, Washington, D.C.; (2): ©Congenital Anomaly Research Center of Kyoto University

 **Core Concept: Information** Pattern formation in animals occurs in four phases controlled by a hierarchy of transcription factors that cause some genes to be expressed and others to be repressed.

How does genetics underlie the phases of animal development? Geneticists have discovered a parallel between the expression of specific transcription factors, which are described in Chapters 12 and 14, and the four major phases of animal development. As noted in Figure 20.6, a hierarchy of transcription factors controls whether or not certain genes are expressed at a specific phase of development in a particular cell type, a phenomenon called **differential gene regulation**. Many morphogens, particularly those that act during early phases of development, function as transcription factors.

**Learning Outcomes:**

1.  **CoreSKILL »** Explain how the analysis of mutants has been an important tool in our understanding of development in animals.
2.  Distinguish the functions of maternal effect genes, segmentation genes, and homeotic genes in animal development.

In this section, we will begin by examining the general stages of *Drosophila* development and then focus our attention on its embryonic stage. During this stage, the overall body plan is determined. We will see how the differential expression of particular genes within the embryo controls pattern formation. Although the roles of genes in the organization of mammalian embryos are not as well understood as they are in *Drosophila*, the analysis of the genomes of mammals and many other species has revealed many interesting parallels in the developmental program of all animals.

## Embryonic Development Determines the Pattern of Structures in the Adult

As a way to appreciate the phases of pattern formation in animals, we will largely focus on development in *Drosophila*. However, as described in Chapter 51, animal development is quite varied among different species. **Figure 20.7** illustrates a simplified sequence of events in *Drosophila* development. Let's examine these steps before we consider the differential gene regulation that causes them to happen.

1.  The oocyte is critical to establishing the pattern of development that will ultimately produce an adult organism. It is an elongated cell that contains positional information. As shown in Figure 20.7a, the fertilized egg already has anterior and posterior ends that correspond to those found in the adult (compare Figure 20.7a and e).

2.  A key process in the embryonic development of *Drosophila* is the formation of a segmented body pattern. The embryo is subdivided into visible segments grouped into three general areas: the head, the thorax, and the abdomen. Figure 20.7b shows the segmented pattern of a *Drosophila* embryo about 10 hours after fertilization.

3.  A *Drosophila* embryo then develops into a **larva** (Figure 20.7c), a free-living organism that is morphologically very different from the embryo and adult. Many animal species do not have larval stages. *Drosophila* undergoes three successive larval stages.

4.  After the third larval stage, the organism becomes a **pupa** (Figure 20.7d), a transitional stage between the larva and the adult.

5.  Through a process known as **metamorphosis**, the pupa transforms into a mature adult that emerges from the pupal case (Figure 20.7e). Each segment in the adult has its own characteristic structures. For example, the wings are on a thoracic segment.

From beginning to end, this developmental process takes about 10 days.

### (a) Fertilized oocyte (0 hours)

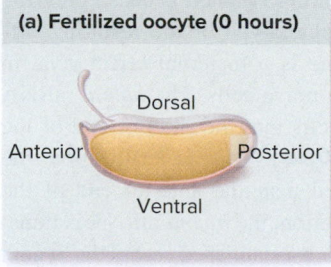

Dorsal
Anterior · Posterior
Ventral

### (b) Embryo (10 hours)

Segments

### (c) Newly hatched larva (24 hours)

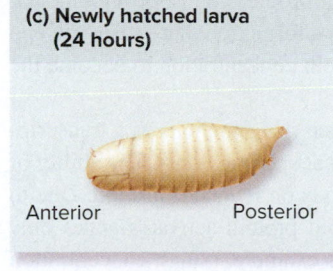

Anterior · Posterior

### (d) Pupa (5 days)

### (e) Adult (10 days)

**Figure 20.7** Developmental stages of the fruit fly, *Drosophila*.

## Core Skill: Quantitative Reasoning

**BIO TIPS** — **THE QUESTION** *One approach to studying development is to isolate mutants that cause abnormal development. This approach helps researchers identify the genes that regulate normal development. However, one problem with this approach is that mutations in genes that control the early stages of development are often lethal because they prevent the embryo from developing properly. To circumvent this problem, developmental biologists may try to isolate temperature-sensitive developmental mutations, or ts alleles. If an embryo carries a ts allele, it will develop correctly if incubated at the permissive temperature (for example, 25°C) but will fail to develop if incubated at the nonpermissive temperature (for example, 30°C). In most cases, ts alleles are mutations that slightly alter the amino acid sequence of a protein, causing a change in its structure that prevents it from working properly at the nonpermissive temperature. Such mutations are useful because they can provide insight regarding the stage of development at which the protein encoded by the gene is necessary. Researchers can take groups of embryos that carry the same ts allele and expose them to the permissive and nonpermissive temperatures at different stages of development. In the experiment described next, embryos carrying a ts allele were divided*

*into five groups and exposed to the permissive or nonpermissive temperature at different times after fertilization. This experiment yielded the following results:*

| Time after fertilization (hours) | Group | | | | |
|---|---|---|---|---|---|
| | 1 | 2 | 3 | 4 | 5 |
| 0–1 | 30°C | 25°C | 25°C | 25°C | 25°C |
| 1–2 | 25°C | 30°C | 25°C | 25°C | 25°C |
| 2–3 | 25°C | 25°C | 30°C | 25°C | 25°C |
| 3–4 | 25°C | 25°C | 25°C | 30°C | 25°C |
| 4–5 | 25°C | 25°C | 25°C | 25°C | 30°C |
| Survival | Yes | Yes | Yes | No | Yes |

*Explain these results.*

**T OPIC** *What topic in biology does this question address?* The topic is development. More specifically, the question asks you to analyze the effects of a temperature-sensitive mutation that affects development.

**I NFORMATION** *What information do you know based on the question and your understanding of the topic?* From the question, you have learned that developmental biologists can isolate ts alleles that are lethal only at nonpermissive temperatures. You are also given data about the survival of embryos in an experiment in which the temperature was shifted to a nonpermissive temperature at different times after fertilization.

**P ROBLEM-SOLVING S TRATEGY** *Interpret data. Compare and contrast.* To solve this problem, you need to interpret data that relate embryo survival to the timing of exposure to the nonpermissive temperature. You want to compare and contrast: When do the embryos survive, and when don't they survive?

**ANSWER** *The embryos do not survive if they are subjected to the nonpermissive temperature between 3 and 4 hours after fertilization, but they do survive if subjected to the nonpermissive temperature at other times during development. These results indicate that the protein encoded by the ts allele plays a crucial role in development at 3–4 hours after fertilization.*

## Phase 1 Pattern Formation: Maternal Effect Genes Promote the Formation of the Body Axes

The first phase in *Drosophila* pattern formation is the establishment of the body axes, which occurs before the embryo becomes segmented. The morphogens necessary to establish these axes are distributed prior to fertilization. In most invertebrates and some vertebrates, certain morphogens, which are important in early developmental stages, are deposited asymmetrically within the egg as it develops (refer back to Figure 20.5a). Later, after the egg has been fertilized and development begins, these morphogens initiate developmental programs that govern the formation of the body axes of the embryo.

As an example of one morphogen that plays a role in axis formation, let's consider the product of a gene in *Drosophila* called *bicoid*. Its name is derived from the observation that a mutation

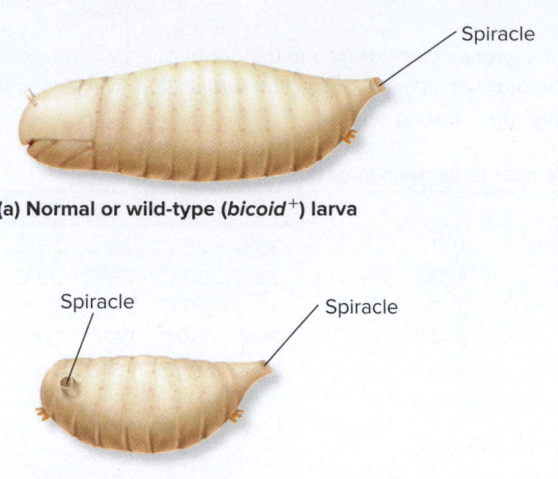

**(a) Normal or wild-type (*bicoid⁺*) larva**

Spiracle

Spiracle                Spiracle

**(b) Mutant (*bicoid⁻*) larva**

**Figure 20.8   The bicoid mutation in *Drosophila*. (a)** A normal larva from a *bicoid⁺* mother. **(b)** An abnormal larva from a *bicoid⁻* mother, in which both ends of the larva develop posterior structures. For example, both ends develop a spiracle, a small pore that normally is found only at the posterior end.

**Concept Check:** *What would you expect to be the phenotype of a larva in which the bicoid gene product accumulated in both the anterior region and the posterior region of the oocyte?*

that inactivates the gene results in a larva with two posterior ends (**Figure 20.8**). It lacks a head and a thorax! During normal oocyte development, the *bicoid* gene product accumulates in the anterior region of the oocyte. This gene product later acts as a morphogen to cause the development of the anterior end of the embryo.

How does the *bicoid* gene product accumulate in the anterior region of the oocyte? The answer involves specialized nurse cells that are found next to the oocyte, which matures in a follicle within the ovary of a female fly. Nurse cells supply oocytes with the products

from **maternal effect genes**. They are so named because only the mother's gene product affects the phenotype of the resulting offspring. For example, the *bicoid* gene is a maternal effect gene in *Drosophila* that is transcribed in the nurse cells. The *bicoid* mRNA is then transported from the nurse cells into the anterior end of the oocyte and trapped there (**Figure 20.9a**). Prior to fertilization, the *bicoid* mRNA is highly concentrated near the anterior end of the oocyte (**Figure 20.9b**). After fertilization, the *bicoid* mRNA is translated, and a gradient of Bicoid protein is established across the zygote (**Figure 20.9c**). This gradient starts a progression of developmental events that will provide the positional information that causes the end of the zygote with a high Bicoid protein concentration to become the anterior region of the embryo.

The Bicoid protein is a morphogen that functions as a transcription factor to activate particular genes at specific times. The ability of Bicoid to activate a given gene depends on its concentration. Due to its asymmetric distribution, the Bicoid protein activates genes only in certain regions of the embryo. For example, a high concentration of Bicoid stimulates the expression of a gene called *hunchback* (that also encodes a transcription factor) in the anterior half of the embryo, but its concentration is too low in the posterior half to activate the *hunchback* gene. The ability of Bicoid to activate genes in certain regions but not others plays a role in the second phase of pattern formation—segmentation.

## The Study of *Drosophila* Mutants Has Identified Genes That Control the Development of Segments

As shown in Figure 20.7b, the second phase of pattern formation is the development of segments. The *Drosophila* embryo is subdivided into 15 segments: three head segments, three thoracic segments, and nine abdominal segments (**Figure 20.10**). Each segment of the embryo gives rise to unique morphological features in the adult. For example, the second thoracic segment (T2) produces a pair of legs and a pair of wings.

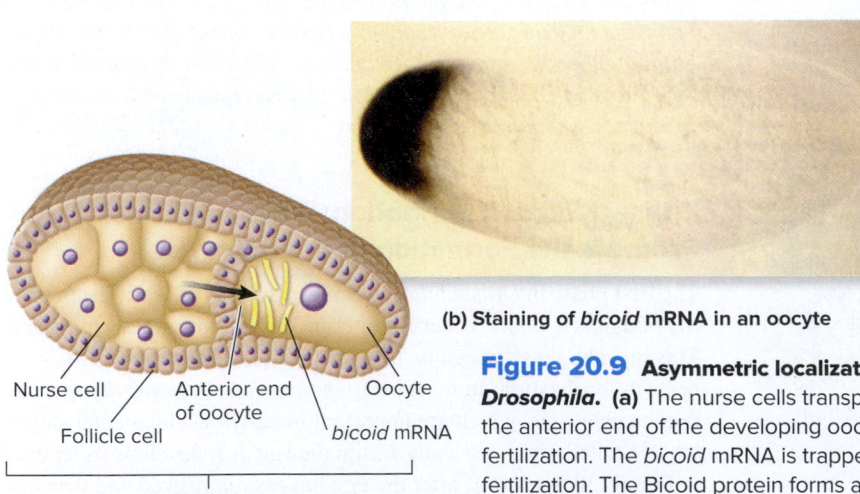

**(a) Transport of maternal effect gene products (*bicoid* mRNA) into the oocyte**

Nurse cell

Follicle cell

Anterior end of oocyte

*bicoid* mRNA

Oocyte

Follicle

**(b) Staining of *bicoid* mRNA in an oocyte**

**(c) Staining of Bicoid protein in an early embryo**

**Figure 20.9   Asymmetric localization of gene products during egg development in *Drosophila*. (a)** The nurse cells transport maternal effect gene products such as *bicoid* mRNA into the anterior end of the developing oocyte. **(b)** Staining of *bicoid* mRNA in an oocyte prior to fertilization. The *bicoid* mRNA is trapped at the anterior region. **(c)** Staining of Bicoid protein after fertilization. The Bicoid protein forms a gradient, with its highest concentration near the anterior end. b: Christiane Nüsslein Volhard 1991. Determination of the embryonic axes of *Drosophila*, *Development*, Supplement 1, pp. 1–10, Fig. 5A, ©The Company of Biologists Limited 1991; c: Christiane Nüsslein Volhard 1991. Determination of the embryonic axes of *Drosophila*, *Development*, Supplement 1, pp. 1–10, Fig. 5B, ©The Company of Biologists Limited 1991

**Concept Check:** *What is the function of the Bicoid protein? After fertilization, in which part of the resulting zygote will its function be highest?*

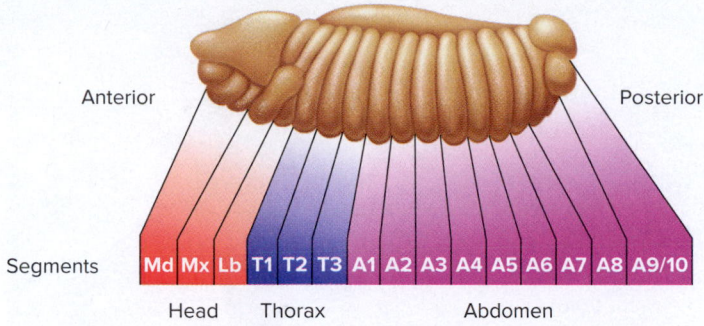

**Figure 20.10   The organization of segments in the *Drosophila* embryo.** The *Drosophila* embryo is subdivided into 15 segments, each of which gives rise to unique structures in the adult.

In the 1970s, German biologist Christiane Nüsslein-Volhard and American developmental biologist Eric Wieschaus undertook a systematic search for *Drosophila* mutants with disrupted development. They focused their search on **segmentation genes**, genes that control the segmentation pattern of the *Drosophila* embryo and larva. Based on the characteristics of abnormal larva, they identified three classes of segmentation genes: gap genes, pair-rule genes, and

segment-polarity genes. When a mutation inactivates a **gap gene**, several adjacent segments are missing in the larva—a gap occurs. A defect in a **pair-rule gene** causes alternating segments or parts of segments to be absent. Finally, mutations of **segment-polarity genes** cause portions of segments to be missing and cause adjacent regions to become mirror images of each other. The role of these segmentation genes during normal *Drosophila* development is described next.

## Phase 2 Pattern Formation: Segmentation Genes Act Sequentially to Divide the *Drosophila* Embryo into Segments

The study of segmentation genes has revealed how segments are formed. To make a segment, particular genes act sequentially to govern the fate of a given region of the body. A simplified scheme of gene expression that leads to a segmented pattern in the *Drosophila* embryo is shown in **Figure 20.11**. Many more genes are actually involved in this process.

In general, the products of maternal effect genes such as *bicoid*, which promote the formation of body axes, activate gap genes. This activation is seen as broad bands of gap proteins in the embryo (Figure 20.11, step 2). Next, products from the gap genes and maternal effect

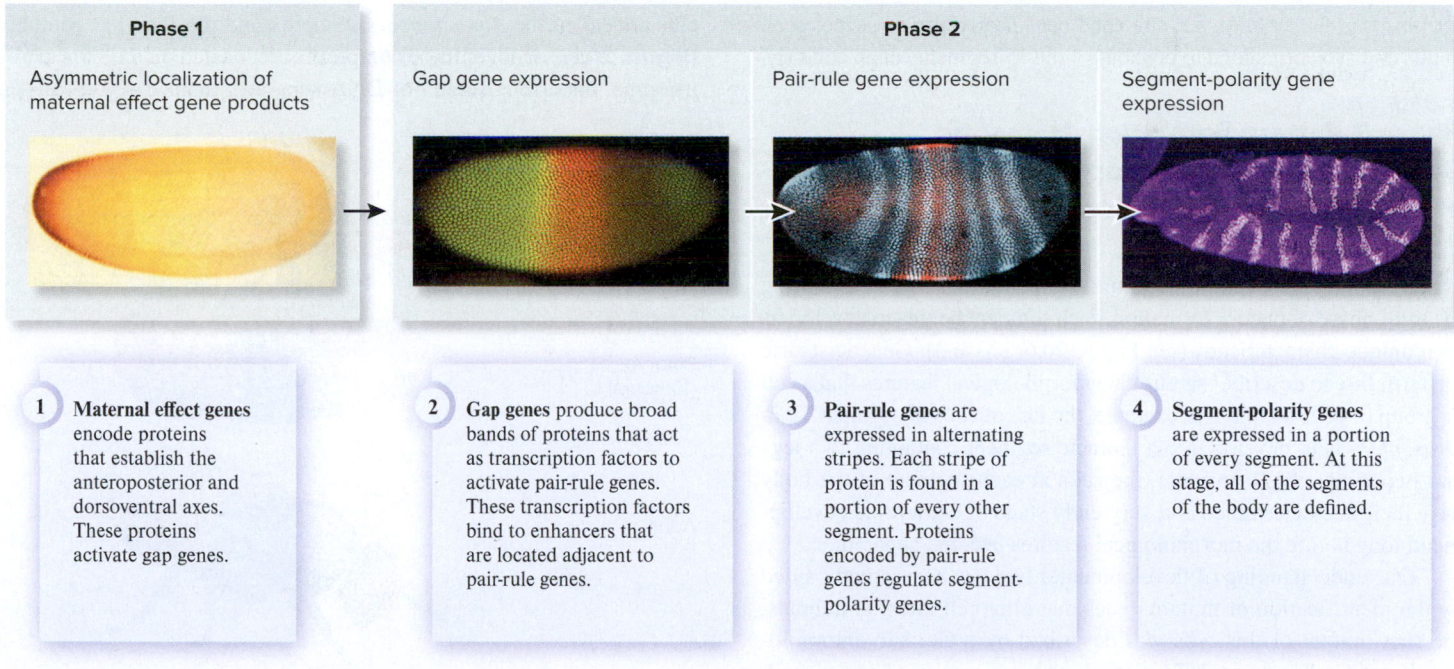

**Figure 20.11   Overview of segmentation in *Drosophila*.** The micrographs depict the progression of *Drosophila* development during the first few hours following fertilization. The micrographs also show the expression of protein products of a maternal effect gene (step 1) and segmentation genes (steps 2–4). In step 1, the protein is stained brown and is found in the left side of the early embryo, which is the anterior end. In step 2, one protein encoded by a gap gene is stained in green and another is stained in red. The yellow region is where the two different gap proteins overlap. In step 3, a protein encoded by a pair-rule gene is stained in light blue. In step 4, a protein encoded by a segment-polarity gene is stained pink (the rest of the embryo is purple). When comparing steps 3 and 4, note that the embryo has undergone a 180° turn, folding back on itself. (1): Christiane Nüsslein Volhard 1991. Determination of the embryonic axes of *Drosophila, Development*, Supplement 1, pp. 1–10, Fig. 5B, ©The Company of Biologists Limited 1991; (2–4): ©Jim Langeland, Steve Paddock and Sean Carroll/University of Wisconsin-Madison

 **Core Concept: Systems**   New properties of life emerge from complex interactions. Body segmentation is an emergent property that has arisen as the result of complex interactions among segmentation gene products.

**Figure 20.12   The bithorax mutation in *Drosophila*.** (a) A normal fly has two wings on the second thoracic segment, and two halteres on the third thoracic segment. (b) This fly carries mutations in a complex of genes called the *bithorax* complex. In this fly, the third thoracic segment has the same characteristics as the second thoracic segment, resulting in four wings instead of two. a–b: Courtesy of the Archives, California Institute of Technology

**(a) Normal fly with two wings**

**(b) Mutant fly with four wings**

genes function as transcription factors to activate the pair-rule genes in alternating stripes in the embryo (Figure 20.11, step 3). Once the pair-rule genes are activated, their gene products then regulate the segment-polarity genes. As you follow the progression from maternal effect genes to segment-polarity genes, notice that a body pattern is emerging in the embryo that matches the segmentation pattern found in the larva and adult animal. As you can see in step 4 of Figure 20.11, the 15 locations where a segment-polarity gene is expressed correspond to portions of segments in the adult fly. To appreciate this phenomenon, notice that the embryo at this stage is curled up and folded back on itself. If you imagine that the embryo was stretched out linearly, the 15 stripes seen in this embryo correspond to portions of the 15 segments of an adult fly.

## Phase 3 Pattern Formation: Homeotic Genes Control the Development of Segment Characteristics

Thus far, we have considered how the *Drosophila* embryo becomes organized along axes and then into a segmented body pattern. During the third phase of pattern formation, each segment begins to develop its own unique characteristics (see Figure 20.6, phase 3). Geneticists use the term **fate** to describe the ultimate morphological features that a cell or group of cells adopts. For example, the fate of cells in segment T2 in *Drosophila* is to develop into a thoracic segment containing two legs and two wings. In *Drosophila*, the cells in each segment of the body have their fate determined at a very early stage of embryonic development, long before the morphological features become apparent.

Our understanding of developmental fate has been greatly aided by the identification of mutant genes that alter cell fates. In animals, the first mutant of this type was described by a German entomologist Ernst G. Kraatz in 1876. He observed a sawfly (*Cimbex axillaris*) in which part of an antenna was replaced with a leg. During the late 19th century, an English zoologist William Bateson collected many of these types of observations and published them in a book. He coined the term homeotic to describe changes in which one body part is replaced by another. These abnormalities are caused by mutant alleles of **homeotic genes**—genes that specify the fate of particular segments or regions of an animal's body.

As an example, **Figure 20.12** shows a normal fly and one with mutations in a complex of homeotic genes called the *bithorax* complex. In a normal fly, two wings are found on the second thoracic segment, and two halteres, which together function as a balancing

organ that resembles a pair of miniature wings, are found on the third thoracic segment. In this mutant fly, the third thoracic segment has the characteristics of the second, so the fly has no halteres and four wings. The term *bithorax* refers to the duplicated characteristics of the second thoracic segment. Edward Lewis, an American pioneer in the genetic study of development, became interested in the bithorax phenotype and began investigating it in 1946. He discovered that the mutant chromosomal region contains a complex of genes that play a role in the third phase of development.

*Drosophila* has eight homeotic genes that are found in two clusters called the *Antennapedia* complex and the *bithorax* complex (**Figure 20.13**). Both of these complexes are located on the same chromosome, but a long stretch of DNA separates them. As you can see

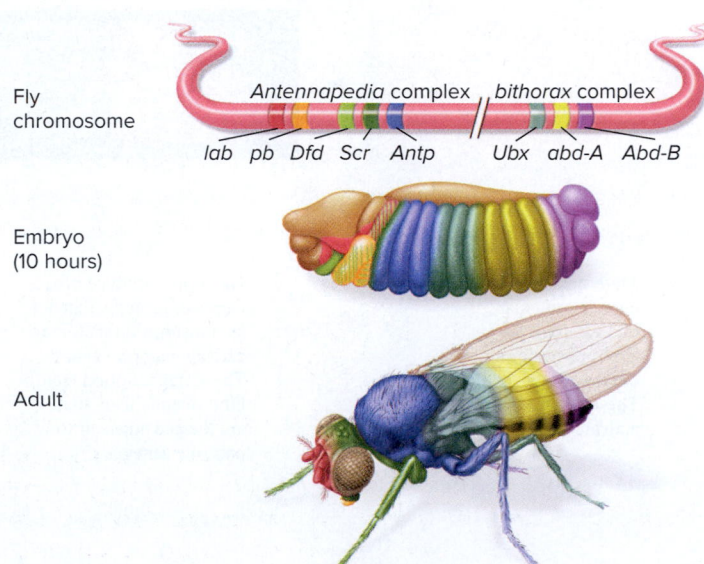

**Figure 20.13   Expression pattern of homeotic genes in *Drosophila*.** The order of homeotic genes, *labial (lab), proboscipedia (pb), Deformed (Dfd ), Sex combs reduced (Scr ), Antennapedia (Antp), Ultrabithorax (Ubx), abdominal A (abd-A),* and *Abdominal B (Abd-B),* correlates with their spatial order of expression in the embryo. (Note: The capitalization of the gene names is not consistent because it is based on the identification of mutations. If the first mutation isolated for a particular homeotic gene was recessive, the gene name is lowercase. If the first mutation discovered was dominant, the gene name begins with a capital letter.)

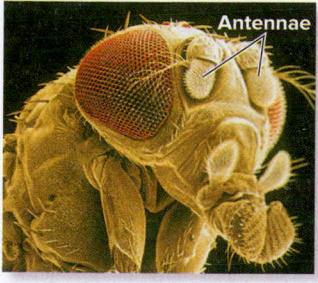

**(a) Normal fly**

**(b) Mutant fly**

**Figure 20.14** **The Antennapedia mutation in *Drosophila*.** **(a)** A normal fly with antennae. **(b)** This fly has a mutation in which the *Antp* gene is expressed in the embryonic segment that normally gives rise to antennae. The abnormal expression of *Antp* causes this region to have legs rather than antennae. a: ©Juergen Berger/Science Source; b: ©F. R. Turner, Indiana University

**Concept Check:**  *What phenotype would you expect if the Antp gene was expressed where the abd-A gene was supposed to be expressed?*

in Figure 20.13, the order of homeotic genes along the chromosome correlates with their expression along the anteroposterior axis of the body. This phenomenon is called the **colinearity rule**. For example, *lab* (for labial) is expressed in the anterior segment and governs the formation of mouth structures. The *Antp* (for *Antennapedia*) gene is expressed in the thoracic region during embryonic development and controls the formation of thoracic structures such as legs.

As we have seen in Figure 20.12, the role of homeotic genes in determining the identity of particular segments has been revealed by mutations that alter their function. As a second example, a mutation in the *Antp* gene has been identified in which the gene is incorrectly expressed in an anterior segment (**Figure 20.14**). A fly with this mutation has the bizarre trait in which it develops legs where antennae are normally found!

How do homeotic genes work at the molecular level? Homeotic genes encode homeotic proteins that function as transcription factors. The coding sequence of homeotic genes contains a 180-bp sequence known as a **homeobox** (**Figure 20.15a**). This sequence was first discovered in the *Antp* and *Ubx* genes, and it has since been found in many *Drosophila* homeotic genes. The homeobox is also found in other genes affecting pattern formation. The homeobox encodes a region of the homeotic protein called a **homeodomain**, which can bind to DNA (**Figure 20.15b**). The arrangement of α helices in the homeodomain promotes the binding of the protein to the DNA.

The primary function of homeotic proteins is to activate the transcription of specific genes that promote developmental changes in the animal. The homeodomain protein binds to DNA sequences called enhancers, which are described in Chapter 14. The enhancers recognized by homeotic proteins are found in the vicinity of specific genes that control development. Most homeotic proteins also contain a transcriptional activation domain (see Figure 20.15b). After the homeodomain binds to an enhancer, the transcriptional activation domain of the homeotic protein activates RNA polymerase to begin transcription. Some homeotic proteins also function as repressors of certain genes.

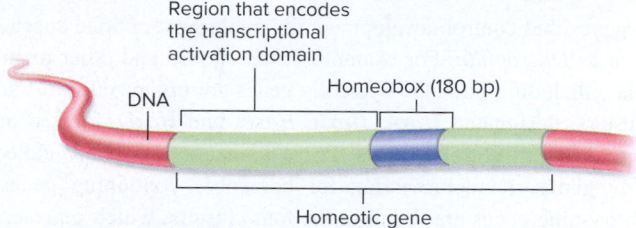

**(a) Homeotic gene containing homeobox**

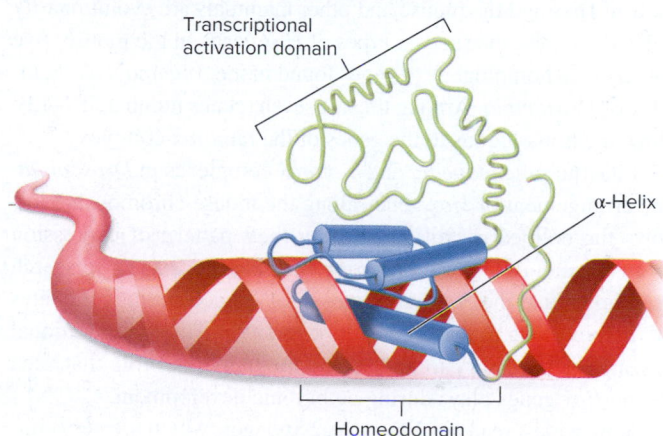

**(b) Homeotic protein binding to the DNA via its homeodomain**

**Figure 20.15** **Molecular features of homeotic genes and proteins.** **(a)** A homeotic gene (shown mostly in green) contains a 180-bp sequence called the homeobox (shown in blue). **(b)** Homeotic genes encode homeotic proteins that function as transcription factors. The homeotic protein contains two key domains called the homeodomain and the transcriptional activation domain. The homeodomain binds to the DNA at a regulatory site such as an enhancer. After this binding occurs, the transcriptional activation domain activates RNA polymerase to begin transcription.

 **Core Skill: Modeling**  The goal of this modeling challenge is to make a model that shows how a homeotic protein can interact with mediator.

**Modeling Challenge:** A gene, which we will call gene *X*, has an enhancer that is activated by a homeotic protein. The homeodomain binds to the enhancer, as shown in Figure 20.15b. The enhancer is a long distance from the core promoter of gene *X*. After the homeotic protein binds to the enhancer, it interacts with mediator (refer back to Figure 14.16) to stimulate the transcription of gene *X*. Draw a model that shows how this homeotic protein interacts with mediator.

 **Core Concept: Evolution**

## A Homologous Group of Homeotic Genes Is Found in Nearly All Animals

**Homologous genes** are evolutionarily derived from the same ancestral gene and have similar DNA sequences. Researchers have found that homeotic genes in vertebrate species are homologous to genes that control development in simpler invertebrate species such as *Drosophila*. For example, in the mouse and other mammals, including humans, homeotic genes are organized into four clusters, designated *HoxA*, *HoxB*, *HoxC*, and *HoxD*, located on four different chromosomes. These homeotic genes are called **Hox genes**, an abbreviation for *homeobox*-containing genes. Thirty-nine genes are found in the four clusters, which represent 13 different gene types. As shown in **Figure 20.16**, the *Hox* genes in fruit flies and the mouse and other mammals are evolutionarily related. Among the first six types of *Hox* genes in the mouse, five of them are homologous to genes found in the *Antennapedia* complex of *Drosophila*. Among the last seven (genes numbered 7–13), three are homologous to the genes of the *bithorax* complex.

Like the *Antennapedia* and *bithorax* complexes in *Drosophila*, the arrangement of *Hox* genes along the mouse chromosome follows the colinearity rule, reflecting their pattern of expression from the anterior to the posterior end (**Figure 20.17**). Research has shown that the *Hox* genes play a role in determining the fates of regions along the anteroposterior axis. Nevertheless, additional research is necessary to understand the individual role that each of the *Hox* genes plays during embryonic development.

How widespread are *Hox* genes? Sponges, which are very simple animals with no true tissues, do not have *Hox* genes, though they have an evolutionarily related gene called an *NK-like* gene. Complex species of animals, including invertebrates such as fruit flies and vertebrates such as mice, have *Hox* genes, but in different numbers. The role of *Hox* genes in the evolution of animal body plans is discussed in Chapter 24 (look ahead to Figure 24.15).

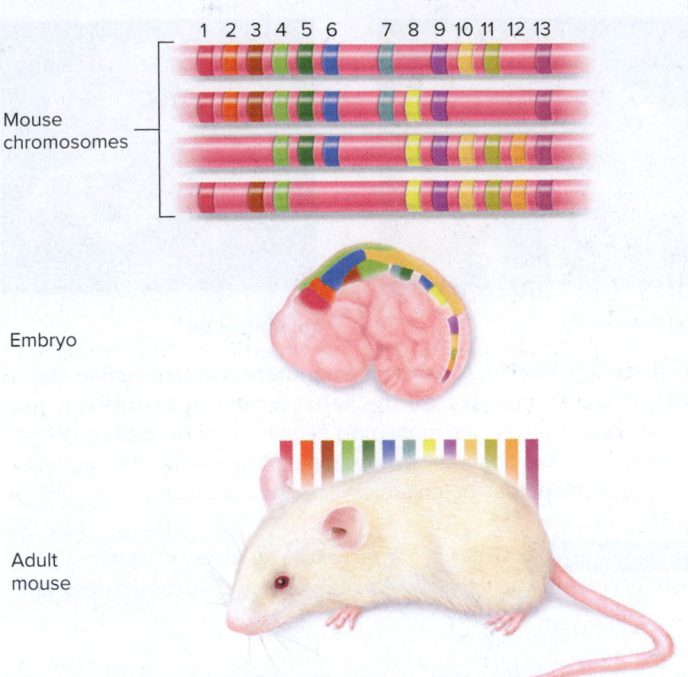

**Figure 20.17** **Expression pattern of *Hox* genes in the mouse.** This schematic illustration shows *Hox* gene expression in the embryo and in the corresponding regions in the adult. The order of *Hox* gene expression, from anterior to posterior, parallels the order of genes along four different chromosomes.

> **Core Skill: Connections** Look ahead to Figure 24.15. How is the number of *Hox* genes related to the complexity of animal bodies?

## 20.3 Development in Animals II: Cell Differentiation

**Learning Outcomes:**

1. Define the general properties of stem cells.
2. Describe how transcription factors control cell differentiation.
3. **CoreSKILL »** Analyze the results of experiments that indicated that master transcription factors control cell differentiation.

In this section, we will focus on phase 4 of animal development, which is cell differentiation. This process is better understood in mammals than in *Drosophila* because researchers have been studying mammalian cells in the laboratory for many decades. To explore cell differentiation, we will consider mammals as our primary example.

### Stem Cells Can Divide and Differentiate into Specialized Cell Types

In the previous section, we focused our attention on patterns of gene expression that occur during the early stages of development. These genes control the basic body plan of the organism. During the fourth phase of pattern formation, the emphasis shifts to cell differentiation (see Figure 20.6, phase 4).

Although invertebrates have been instrumental in extending our understanding of pattern formation in animals, cell differentiation has been studied more extensively in mammals. One reason is because

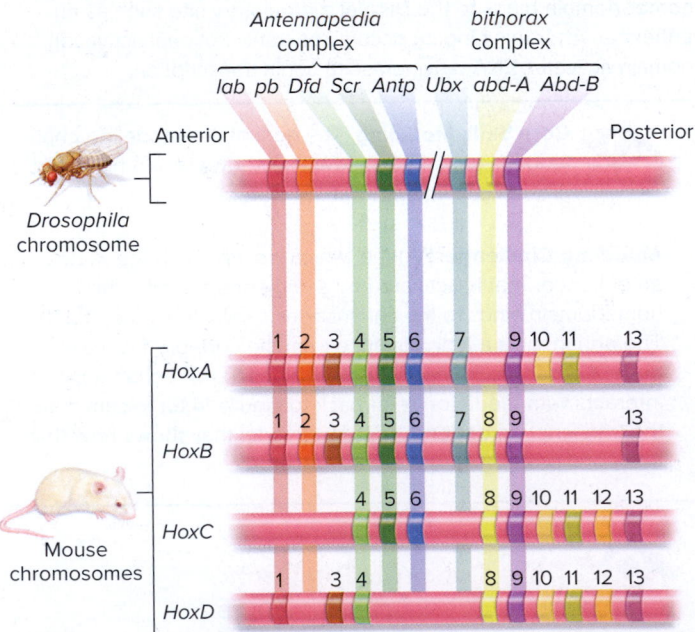

**Figure 20.16** **A comparison of homeotic genes in *Drosophila* and the mouse.** The mouse and other mammals have four gene complexes, *HoxA* through *HoxD*, that are homologous to certain homeotic genes found in *Drosophila*. Thirteen different types of homeotic genes are found in the mouse, although each *Hox* complex does not contain all 13 genes. In this drawing, homologous genes are aligned in columns. For example, *lab* is the homolog to *HoxA-1*, *HoxB-1*, and *HoxD-1*.

researchers have been able to grow mammalian cells in the laboratory for many decades. The availability of laboratory-grown cells makes it much easier to analyze the process of cell differentiation.

By studying mammalian cells in the laboratory, geneticists have determined that the morphological differences between two different types of differentiated cells, such as muscle cells and neurons, arise through gene regulation. Though muscle cells and neurons in a given organism contain the same set of genes, they regulate the expression of their genes in very different ways. Certain genes that are transcriptionally active in muscle cells are inactive in neurons, and vice versa. Therefore, muscle cells and neurons express different proteins that affect the characteristics of the respective cells in distinct ways. In this manner, differential gene regulation underlies cell differentiation.

***General Properties of Stem Cells***  To understand the process of cell differentiation in a multicellular organism, we need to consider the special properties of **stem cells**, undifferentiated cells that divide and supply the cells that constitute the bodies of all animals and plants. Stem cells have two common characteristics. First, they have the capacity to divide, and second, their daughter cells can differentiate into one or more specialized cell types. The two daughter cells that are produced from the division of a stem cell can have different fates (**Figure 20.18**). One of the cells may remain an undifferentiated stem cell, and the other daughter cell can differentiate into a specialized cell type. With this asymmetric pattern of division and differentiation, stem cells continue dividing throughout life and generate a population of specialized cells. For example, in mammals, this mechanism is needed to replenish cells that have a finite life span, such as skin cells and red blood cells.

***Stem Cells During Development***  In mammals, stem cells are commonly categorized according to their developmental stage and their ability to differentiate (**Figure 20.19**). The ultimate stem cell

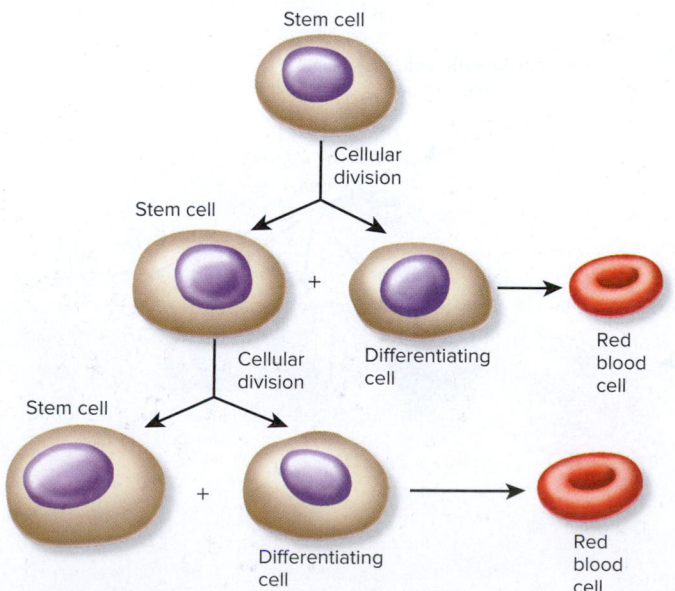

**Figure 20.18  Growth pattern of stem cells.**  When a stem cell divides, one of the two daughter cells may remain a stem cell, while the other cell can differentiate into a specialized cell type, such as the red blood cells shown here.

*Concept Check:*  *What are the two common characteristics of stem cells?*

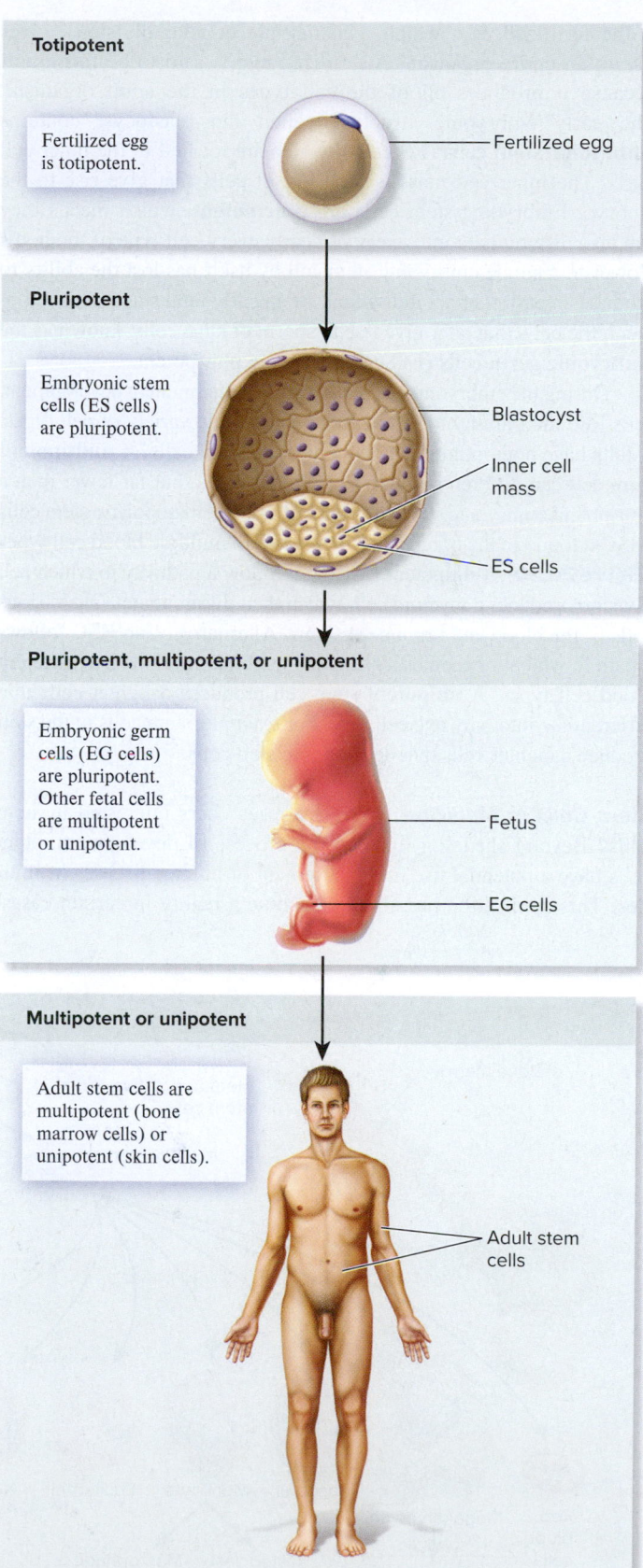

**Figure 20.19**  **Occurrence of stem cells at different stages of mammalian development.**

is the fertilized egg, which, via multiple cellular divisions, gives rise to an entire organism. A fertilized egg is said to be **totipotent** because it produces all of the cell types in the adult organism. The early embryonic structure called the blastocyst contains **embryonic stem cells (ES cells)**, which are located in the inner cell mass. The inner cell mass is a cluster of cells that give rise to the embryo. Embryonic stem cells are **pluripotent**, which means they can also differentiate into every or nearly every cell type of the body. However, a single embryonic stem cell by itself has lost the ability to produce an entire, intact individual. At an early fetal stage of development, the cells that later give rise to sperm or eggs cells, known as the **embryonic germ cells (EG cells)**, also are pluripotent.

During the embryonic and fetal stages of mammalian development, cells lose their ability to differentiate into a wide variety of cell types. Adults have both multipotent and unipotent stem cells. A **multipotent** stem cell can differentiate into several cell types, but far fewer than a pluripotent embryonic stem cell. For example, hematopoietic stem cells (HSCs) found in the bone marrow give rise to multiple blood cell types (**Figure 20.20**). Multipotent HSCs can follow a pathway in which cell division produces a myeloid cell, which then differentiates into various cells of the blood and immune systems. Alternatively, an HSC follows a path in which it becomes a lymphoid cell that develops into different blood cell types. A **unipotent** stem cell produces daughter cells that differentiate into only one cell type. For example, stem cells in the skin produce daughter cells that develop into skin cells.

***Stem Cells in Medicine*** Why are researchers interested in stem cells? Beyond shedding light on the process of development, stem cells have a potential use in the treatment of human diseases or injuries. This application has already become a reality in certain cases.

| Table 20.1 | Some Potential Uses of Stem Cells to Treat Diseases |
|---|---|
| **Cell/tissue type** | **Disease treatment** |
| Nerve | Implantation of cells into the brain to treat Parkinson disease; treatment of spinal cord injuries |
| Skin | Treatment of burns and skin disorders |
| Cardiac | Repair of heart damage associated with heart attacks |
| Cartilage | Repair of joints damaged by injury or arthritis |
| Bone | Repair or replacement of damaged bone |
| Liver | Repair or replacement of liver tissue damaged by injury or disease |
| Skeletal muscle | Repair or replacement of damaged muscle |

For example, bone marrow transplants are used to treat patients with certain forms of cancer, such as leukemia. When bone marrow from a healthy person is injected into the body of a patient whose immune system has been wiped out via radiation, the stem cells within the transplanted marrow have the ability to proliferate and differentiate into various types of blood cells within the body of the patient.

Renewed interest in the use of stem cells in the potential treatment of many other diseases has been fostered by studies in 1998 in which researchers obtained ES cells from blastocysts and EG cells from aborted fetuses and successfully propagated them in the laboratory. Because ES and EG cells are pluripotent, they could potentially be used to treat a wide variety of diseases associated with cell and tissue damage (**Table 20.1**). Much progress has been made in testing

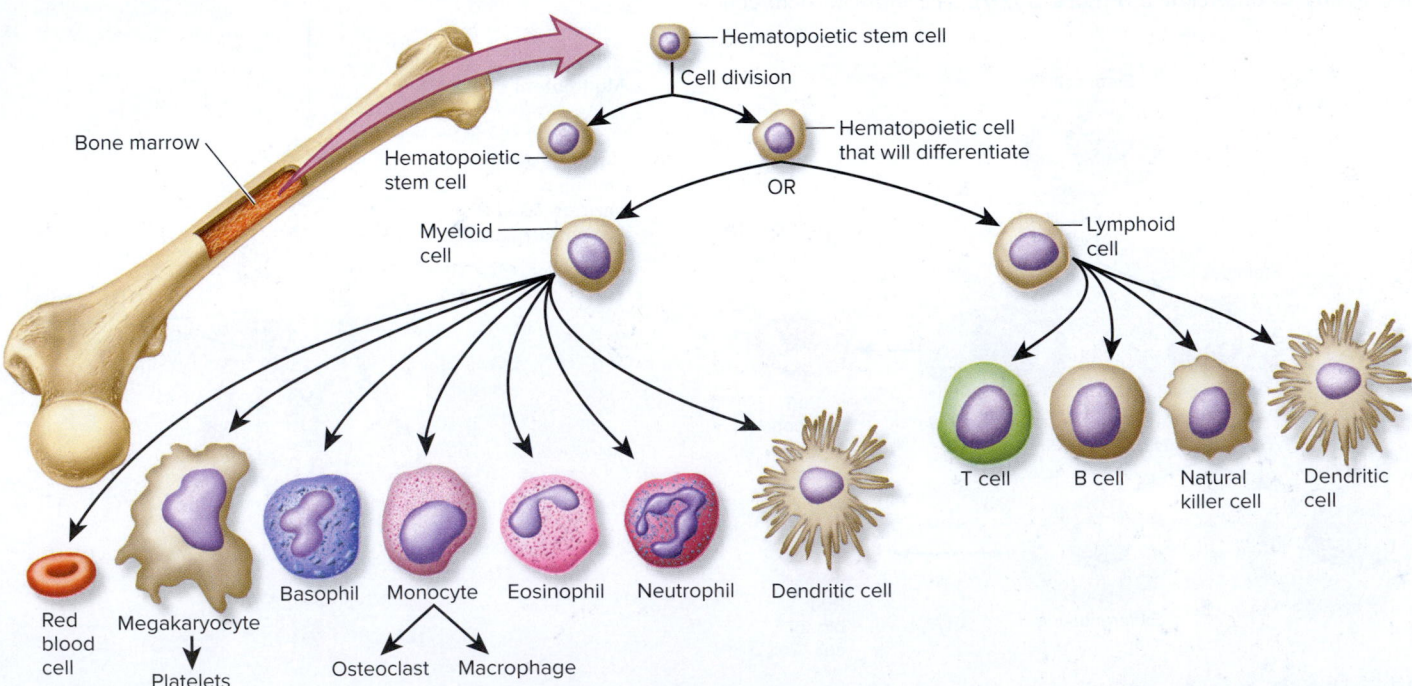

**Figure 20.20 Fates of hematopoietic stem cells (HSCs).** HSCs can follow two pathways: one in which cell division produces a myeloid cell and one that produces a lymphoid cell. Each develops into different blood cell types.

*Concept Check:* *Are hematopoietic stem cells totipotent, pluripotent, multipotent, or unipotent?*

the use of stem cells in animal models. However, more research is needed before the use of stem cells to treat such diseases in humans is realized.

From an ethical perspective, the primary issue that raises debate is the source of stem cells for research and potential treatments. Most ES cells have been derived from human embryos that were produced from in vitro fertilization, a method of assisted conception in which fertilization occurs outside of the mother's body and a limited number of the resulting embryos are transferred to the uterus. Most EG cells are obtained from aborted fetuses, either those that spontaneously aborted or those in which the decision to abort was not related to donating the fetal tissue to research. Some feel that it is morally wrong to use such tissue in research and/or the treatment of disease. Furthermore, some people fear this technology could lead to intentional abortions for the sole purpose of obtaining fetal tissues for transplantation. Others feel the embryos and fetuses that have been the sources of ES and EG cells were not going to become living individuals, and therefore it is beneficial to study these cells and to use them in a positive way to treat human diseases and injury. It is not clear whether these two opposing viewpoints can reach a common ground.

If stem cells could be obtained from adult cells and induced to become pluripotent cells in the laboratory, an ethical dilemma may be avoided, because most people do not have moral objections to current procedures that use adult cells such as bone marrow transplantation. In 2006, work by Japanese physician Shinya Yamanaka and colleagues showed that adult mouse fibroblasts (a type of connective tissue cell) could become pluripotent by the introduction of four different genes that encode transcription factors. In 2007, Yamanaka's laboratory and two other research groups were able to show that such induced pluripotent stem cells can differentiate into all cell types when injected into mouse blastocysts and grown into baby mice. These results indicate that adult cells can be reprogrammed to become embryonic stem cells.

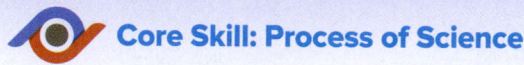

## Core Skill: Process of Science

## Feature Investigation | Davis, Weintraub, and Lassar Identified Genes That Promote Muscle Cell Differentiation

A key question regarding stem cells is "What causes a stem cell to differentiate into a particular cell type?" Researchers have discovered that certain proteins function as master transcription factors that cause cells to differentiate into specific types of cells. The investigation described here was one of the first studies to reveal this phenomenon.

In 1987, Robert Davis, Harold Weintraub, and Andrew Lassar conducted a study to identify genes that promote skeletal muscle cell differentiation. The initial strategy for their experiments was to identify genes that are expressed only in differentiating skeletal muscle cells, not in nonmuscle cells. Though methods of gene cloning are described in Chapter 21, we will briefly consider these scientists' cloning methods in order to understand their approach. The researchers began with two different laboratory cell lines that could differentiate into muscle cells. From these two cell lines, they cloned and identified about 10,000 different genes that were transcribed into mRNA. Next, they compared the expressed genes in these two muscle cell lines with genes that were expressed in a nonmuscle cell line. Their comparison revealed 26 genes that were expressed only in the two muscle cell lines but not in the nonmuscle cell line. To narrow their search further, they compared these 26 genes with other nonmuscle cell lines they had available. Among the 26, only 3 of them, which the researchers termed *MyoA*, *MyoD*, and *MyoH*, were expressed exclusively in the two muscle cell lines.

In the experiment shown in **Figure 20.21**, the scientists' goal was to determine if any of these three genes could cause nonmuscle cells to differentiate into muscle cells. Using techniques described in Chapter 21, the coding sequence of each gene was placed next to an active promoter that caused a high level of transcription, and then the genes were introduced into fibroblasts, which are a type of cell that normally differentiates into osteoblasts (bone cells), chondrocytes (cartilage cells), adipocytes (fat cells), and smooth muscle cells, but never differentiates into skeletal muscle cells in vivo. The cells were plated on growth media and allowed to grow for 3 to 5 days. When the cloned *MyoD* gene was introduced into fibroblast cells in a laboratory, the fibroblasts differentiated into skeletal muscle cells! These cells contained large amounts of myosin, which is a protein expressed in muscle cells. The other two cloned genes (*MyoA* and *MyoH*) did not cause muscle cell differentiation or promote myosin production.

Since this initial discovery, researchers have found that *MyoD* belongs to a small group of genes that initiate muscle cell development. These myogenic genes encode transcription factors. They are found in all vertebrates and have been identified in several invertebrates, such as *Drosophila* and *C. elegans*. In all cases, myogenic genes are activated during skeletal muscle cell development.

---

### Experimental Questions

1. What was the goal of the research conducted by Davis, Weintraub, and Lassar?

2. **CoreSKILL** » How did Davis, Weintraub, and Lassar's research identify the candidate genes for muscle cell differentiation?

3. **CoreSKILL** » Once the researchers identified the candidate genes for muscle cell differentiation, how did they test the effect of each gene on cell differentiation? What were the results of the study?

**Figure 20.21**  Davis, Weintraub, and Lassar study showing that promotion of skeletal muscle cell differentiation in fibroblasts is caused by the expression of *MyoD*.

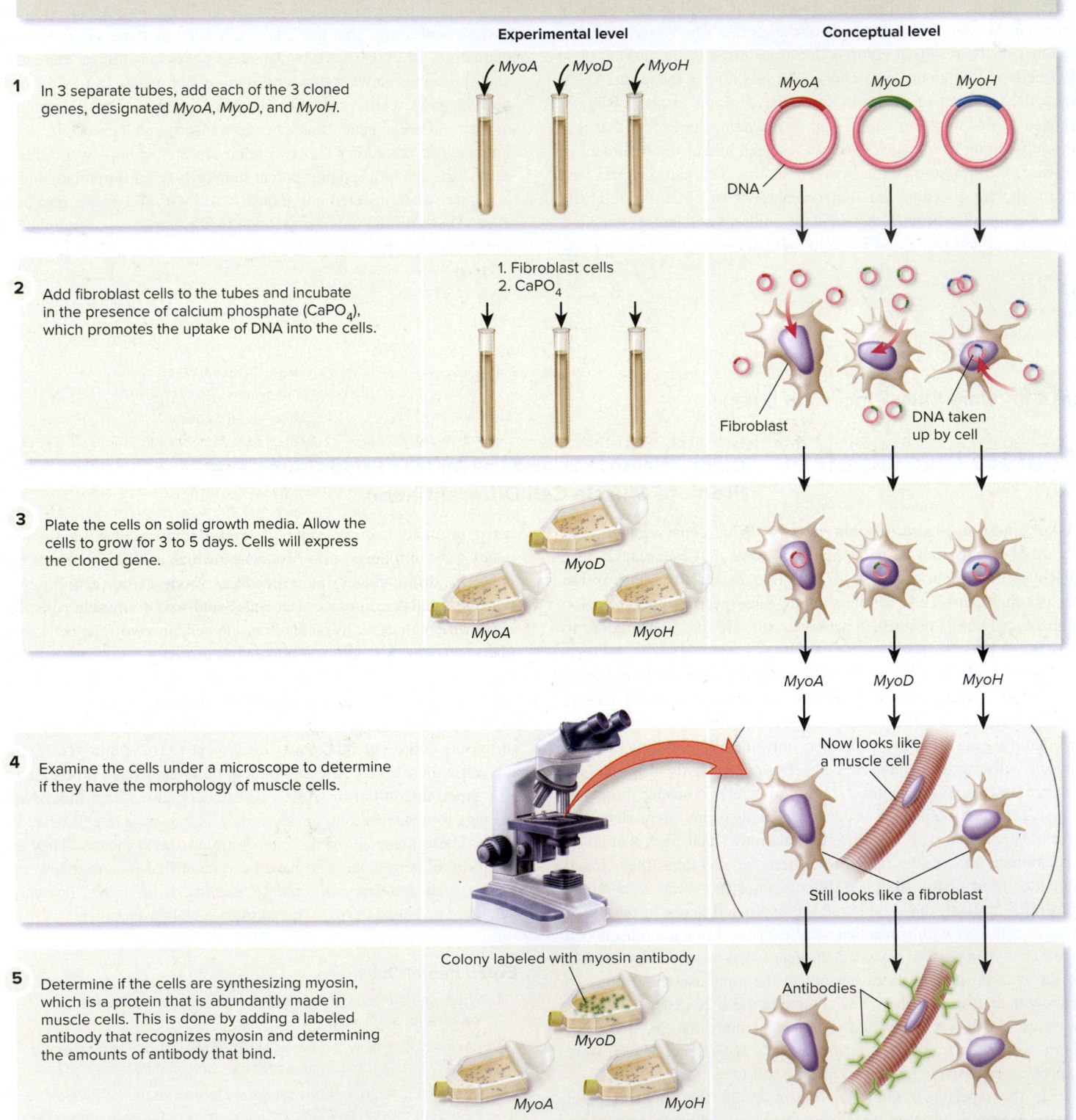

**HYPOTHESIS**  Muscle differentiation is induced by particular genes.

**KEY MATERIALS**  Three cloned genes had been identified that were expressed only in differentiating muscle cell lines. The researchers also had fibroblast cell lines, which do not normally differentiate into muscle cells.

Experimental level                    Conceptual level

1  In 3 separate tubes, add each of the 3 cloned genes, designated *MyoA*, *MyoD*, and *MyoH*.

*MyoA*    *MyoD*    *MyoH*

*MyoA*    *MyoD*    *MyoH*

DNA

2  Add fibroblast cells to the tubes and incubate in the presence of calcium phosphate ($CaPO_4$), which promotes the uptake of DNA into the cells.

1. Fibroblast cells
2. $CaPO_4$

Fibroblast          DNA taken up by cell

3  Plate the cells on solid growth media. Allow the cells to grow for 3 to 5 days. Cells will express the cloned gene.

*MyoD*

*MyoA*          *MyoH*

*MyoA*    *MyoD*    *MyoH*

4  Examine the cells under a microscope to determine if they have the morphology of muscle cells.

Now looks like a muscle cell

Still looks like a fibroblast

5  Determine if the cells are synthesizing myosin, which is a protein that is abundantly made in muscle cells. This is done by adding a labeled antibody that recognizes myosin and determining the amounts of antibody that bind.

Colony labeled with myosin antibody

*MyoD*

*MyoA*          *MyoH*

Antibodies

**6   THE DATA**

Results from step 4:

| DNA added | Microscopic morphology of cells |
|-----------|--------------------------------|
| MyoA | Fibroblasts |
| MyoD | Muscle cells |
| MyoH | Fibroblasts |

Results from step 5:

| DNA added | Colonies labeled with antibody that binds to myosin? |
|-----------|------------------------------------------------------|
| MyoA | No |
| MyoD | Yes |
| MyoH | No |

**7   CONCLUSION** The *MyoD* gene encodes a protein that causes cells to differentiate into skeletal muscle cells.

**8   SOURCE** Davis, R. L., Weintraub, H., and Lassar, A. B. 1987. Expression of a single transfected cDNA converts fibroblasts to myoblasts. *Cell* 51: 987–1000.

## 20.4   Development in Plants

### Learning Outcomes:

1. Outline the stages of pattern formation in plants.
2. **CoreSKILL »** Explain the ABC model for flower development.

Because all eukaryotic organisms share an evolutionary history, animals and plants have many common features, including the types of events that occur during development. However, the general morphology of plants is quite different from that of animals. Plant morphology exhibits two key features (see Figure 20.2b). The first is the root-shoot axis. Most plant growth occurs via cell division near the tips of the shoots and the bottoms of the roots. Second, this growth occurs in a well-defined radial pattern, which means that growth in the stems and roots occurs in concentric rings of tissues (**Figure 20.22**).

At the cellular level too, plant development shows some differences from animal development. For example, cell migration does not occur during plant development. In addition, the development of a plant does not rely on morphogens that are deposited in the oocyte, as in many animals. In plants, an entirely new individual can be regenerated from many types of somatic cells—cells that do not give rise to gametes. Such somatic cells of plants are totipotent.

In spite of these apparent differences, the underlying molecular mechanisms of pattern formation in plants share striking similarities with those in animals. Like animals, plants use the mechanism of differential gene regulation to coordinate the development of a body plan. Like their animal counterparts, plants have a developmental program that relies on transcription factors to determine when gene products are made and in what quantity. In this section, we will consider pattern formation in plants and examine how transcription factors play a key role in plant development.

### Plant Development Occurs from Meristems That Are Formed in the Embryo

How does pattern formation occur in plants? **Figure 20.23** illustrates a common order of events in the embryonic development of flowering plants such as *Arabidopsis*. After fertilization, the first cellular division is asymmetrical and produces a smaller apical cell and a larger basal cell (Figure 20.23a). In 2009, Danish geneticist Martin Bayer and colleagues conducted experiments indicating that the sperm carries mRNA molecules that are critical for this asymmetric cell division. The apical cell gives rise to most of the embryo and later develops into the shoot of the plant. In *Arabidopsis*, the basal cell gives rise to the root, along with a structure called the suspensor, which channels nutrients from the parent plant to the young embryo (Figure 20.23b).

At the heart stage, which is composed of only about 100 cells, the basic organization of the plant has been established (Figure 20.23c). Plants have organized groups of actively dividing stem cells called **meristems**. As discussed earlier, stem cells retain the ability both to divide and to differentiate into multiple cell types. The meristem produces offshoots of proliferating and differentiating cells.

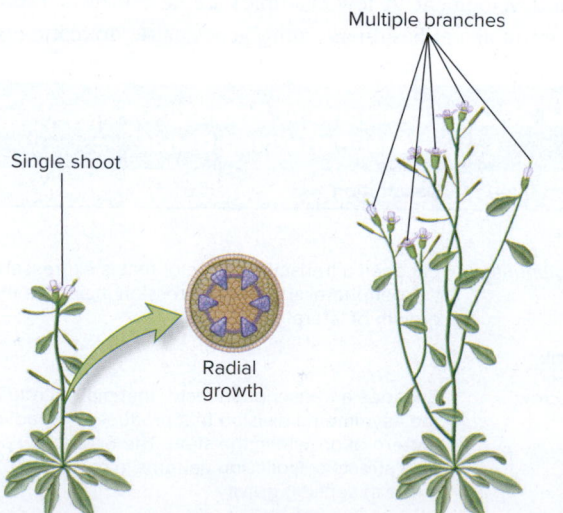

**Figure 20.22   Pattern of shoot growth in plants.** Early in development, as shown here in *Arabidopsis*, a single shoot promotes the formation of early leaves on the plant. Later, buds will form from this main shoot and grow into branches.

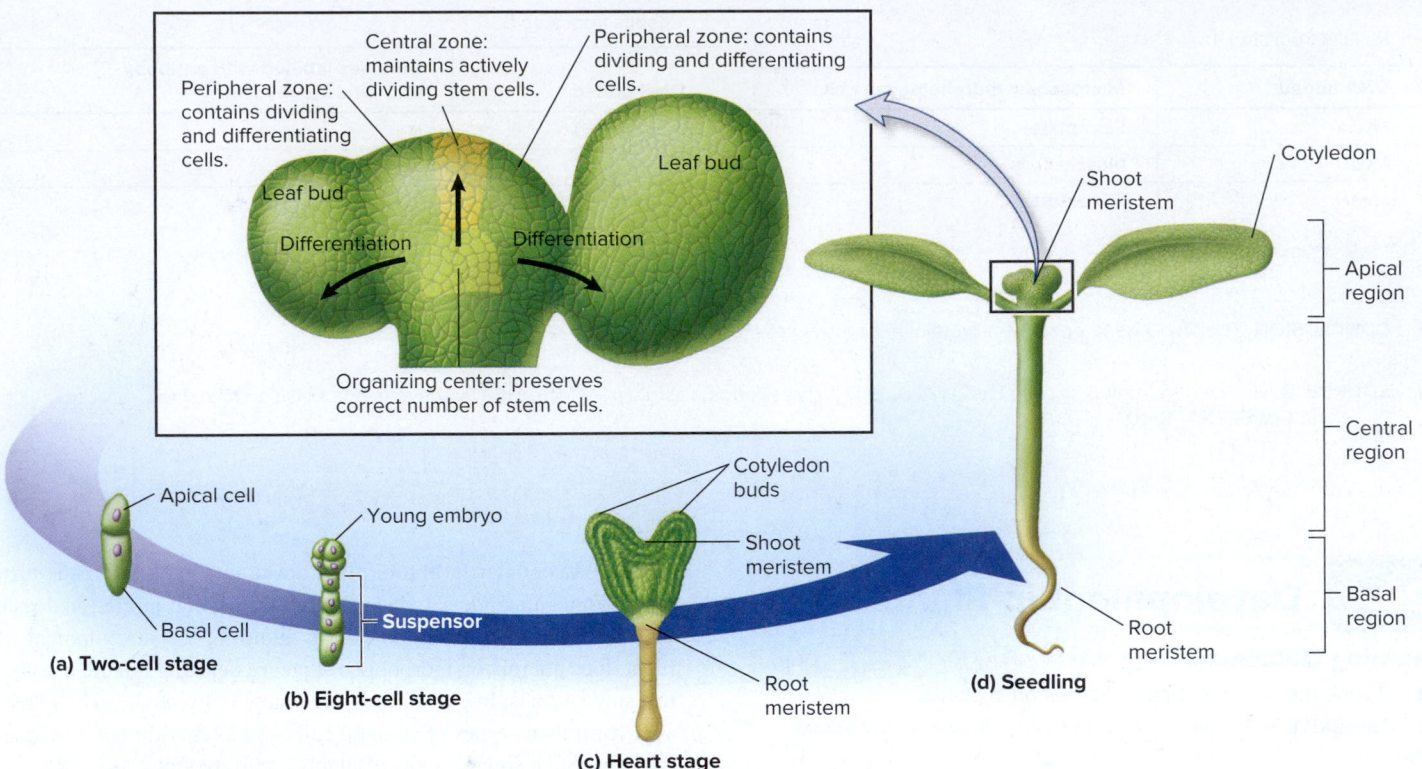

**Figure 20.23** **Developmental steps in the formation of a plant embryo.** **(a)** The two-cell stage consists of the apical cell and basal cell. **(b)** The eight-cell stage consists of a young embryo and a suspensor. The suspensor channels nutrients to the young embryo from the parent plant. **(c)** At the heart stage, all of the plant tissues have begun to form. The shoot meristem is located between the future cotyledons, and the root meristem is on the opposite side. **(d)** A seedling showing apical, central, and basal regions. The inset shows the organization of the shoot meristem. Note: The steps shown in parts (a), (b), and (c) occur during seed formation, when the embryo is enclosed within a seed.

*Concept Check:* *Where are stem cells found in a growing plant?*

The **root apical meristem** gives rise only to the root, whereas the **shoot apical meristem** produces all aerial parts of the plant, which include the stem as well as lateral structures such as branches, leaves, and flowers.

The heart stage then progresses to the formation of a seedling that has two cotyledons, which are embryonic leaves that store nutrients for the developing embryo and seedling. In the seedling shown in Figure 20.23d, you can see three main regions. The **apical region** produces the leaves and flowers of the plant. The **central region** creates the stem. Finally, the **basal region** produces the roots. Each of these three regions develops differently, as indicated by their unique cell division patterns and distinct morphologies.

As seen in the inset to Figure 20.23d, the shoot meristem is organized into three areas: the organizing center, the central zone, and the peripheral zone. The **organizing center** ensures the proper organization of the meristem and preserves the correct number of actively dividing stem cells. The **central zone** is an area where undifferentiated stem cells are always maintained. The **peripheral zone** contains dividing cells that eventually differentiate into plant structures. For example, the peripheral zone may form a bud that will produce a leaf or flower.

By analyzing mutations that disrupt the developmental process, researchers have discovered that the apical, central, and basal regions of a growing plant express different sets of genes. A category of genes

called **apical-basal-patterning genes** are important in early stages of plant development. A few examples are described in **Table 20.2**. Mutations in apical-basal-patterning genes cause dramatic effects in

| Table 20.2 | Examples of *Arabidopsis* Apical-Basal-Patterning Genes |
|---|---|
| **Region:** *Gene* | **Description** |
| **Apical:** | |
| *Aintegumenta* | Encodes a transcription factor that is expressed in the peripheral zone. Its expression maintains the growth of lateral buds. |
| **Central:** | |
| *Scarecrow* | Encodes a transcription factor that plays a role in the asymmetric division that produces the radial pattern of growth in the stem. The Scarecrow protein also affects cell division patterns in roots and plays a role in sensing gravity. |
| **Basal:** | |
| *Monopterous* | Encodes a transcription factor. When the *Monopterous* gene is defective, the plant embryo cannot initiate the formation of root structures, although root structures can be formed postembryonically. This gene seems to be required for organizing root formation in the embryo. |

one of these three regions. For example, the *Aintegumenta* gene is necessary for apical development. When it is defective, the growth of lateral buds is defective.

## Plant Homeotic Genes Control Flower Development

Although William Bateson coined the term homeotic to describe mutations in animals in which one body part is replaced by another, the first known homeotic mutations were described in plants. Naturalists in ancient Greece and Rome, for example, recorded their observations of double flowers in which stamens were replaced by petals. In current research, geneticists are studying these types of mutations to better understand developmental pathways in plants. Many homeotic mutations affecting flower development have been identified in *Arabidopsis* and also in the snapdragon (*Antirrhinum majus*).

A normal *Arabidopsis* flower is composed of four concentric whorls of structures (**Figure 20.24a**). The first, outer whorl has four **sepals**, which protect the flower bud before it opens. The second whorl is composed of four **petals**, and the third whorl has six **stamens**, structures that make male gametophytes, pollen. Finally, the fourth, innermost whorl contains two carpels that are fused together. The **carpels** produce, enclose, and nurture the female gametophytes.

After analyzing the effects of many different homeotic mutations in *Arabidopsis*, British plant biologist Enrico Coen and his American colleague, plant geneticist, Elliot Meyerowitz, proposed the **ABC model** for flower development in 1991. In this model, three classes of genes, called *A*, *B*, and *C*, govern the formation of sepals, petals, stamens, and carpels. More recently, a fourth class, called the *E* genes, was found to be required for this process. All four types of genes encode transcription factors that control flower development in *Arabidopsis* (Figure 20.24a). In whorl 1, gene *A* product is made. This promotes sepal formation. In whorl 2, *A*, *B*, and *E* gene products are made, which promotes petal formation. In whorl 3, the expression of genes *B*, *C*, and *E* causes stamens to be made. Finally, in whorl 4, the products of *C* and *E* genes promote carpel formation.

What happens in certain homeotic mutants that undergo transformations of particular whorls? According to the original ABC model, genes *A* and *C* repress each other's expression, and gene *B* functions independently. In a mutant defective in gene *A* expression, gene *C* is also expressed in whorls 1 and 2. This produces a carpel-stamen-stamen-carpel arrangement in which the sepals have been transformed into carpels and the petals into stamens (**Figure 20.24b**). When gene *B* is defective, a flower cannot make petals or stamens. Therefore, a gene *B* defect yields a flower with a sepal-sepal-carpel-carpel arrangement. When gene *C* is defective, gene *A* is expressed in all four whorls. This results in a sepal-petal-petal-sepal pattern. If the expression of *E* genes is defective, the flower consists entirely of sepals.

Working together, the genes shown in Figure 20.24 promote a pattern of development that leads to sepal, petal, stamen, or carpel structures. But what happens if genes *A*, *B*, and *C* are all defective? This produces a flower composed entirely of leaves (**Figure 20.24c**). These results indicate that the leaf structure is the default pathway

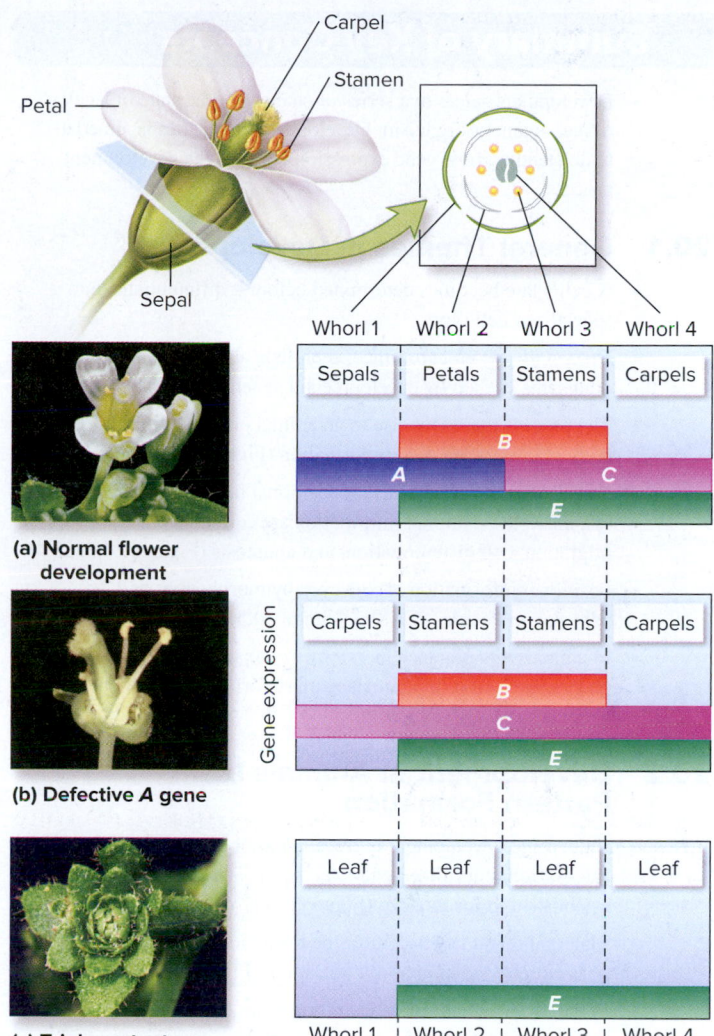

(a) Normal flower development

(b) Defective *A* gene

(c) Triple mutant

**Figure 20.24  Normal and mutant homeotic gene action in *Arabidopsis*.** **(a)** A normal flower is composed of four concentric whorls of structures: sepals, petals, stamens, and carpels. To the right is the ABC model of homeotic gene action that has been revised to include *E* genes. **(b)** A homeotic mutant defective in gene *A* in which the sepals have been transformed into carpels and the petals have been transformed into stamens. **(c)** A triple mutant defective in the *A*, *B*, and *C* genes, producing a flower with all leaves. a: ©Darwin Dale/ Science Source; b–c: ©John Bowman

**Concept Check:** *What flower pattern would you expect if the B gene was expressed in whorls 2, 3, and 4?*

and that the *A*, *B*, and *C* genes cause development to deviate from a leaf structure in order to make something else. In this regard, the sepals, petals, stamens, and carpels of flowers can be viewed as modified leaves.

Like the *Drosophila* homeotic genes, plant homeotic genes are part of a hierarchy of gene regulation. Most plant homeotic genes belong to a family of genes called MADS-box genes, which encode transcription factor proteins that contain a DNA-binding domain called a MADS domain. The *Arabidopsis* homeotic genes do not contain a sequence similar to the homeobox found in animal homeotic genes.

# Summary of Key Concepts

- Development refers to a series of changes in the state of a cell, tissue, organ, or organism. Developmental genetics is aimed at understanding how gene expression controls the development process.

## 20.1 General Themes in Development

- A cell's fate becomes determined before it differentiates into a specialized cell type.

- *Drosophila*, *C. elegans*, mice, zebrafish, and *Arabidopsis* are model organisms studied by developmental geneticists (Figure 20.1).

- The process that gives rise to an animal or plant with a particular body plan is called pattern formation (Figure 20.2).

- Pattern formation depends on positional information. Four responses to positional information are cell division, cell migration, cell differentiation, and apoptosis (Figures 20.3, 20.4).

- Positional information is conveyed by morphogens and cell adhesion molecules (CAMs) (Figure 20.5).

- In animals and plants, a hierarchy of transcription factors controls pattern formation. In animals, pattern formation occurs in four general phases (Figure 20.6).

## 20.2 Development in Animals I: Pattern Formation

- Embryonic development in *Drosophila* occurs in a sequence of stages, from a fertilized egg to an adult. The basic body plan is established in the embryo (Figure 20.7).

- Maternal effect genes control the formation of body axes, the first phase in *Drosophila* pattern formation (Figures 20.8, 20.9).

- In the second phase of pattern formation, the sequential expression of three categories of segmentation genes divides the embryo into segments. Mutations that alter *Drosophila* development have allowed scientists to understand the normal process (Figures 20.10, 20.11).

- During the third phase of pattern formation, each segment begins to develop its own unique characteristics. Homeotic genes control the development of a particular segment or group of segments (Figures 20.12, 20.13, 20.14, 20.15).

- Invertebrates and vertebrates have a homologous set of homeotic genes. These genes are called *Hox* genes (Figures 20.16, 20.17).

## 20.3 Development in Animals II: Cell Differentiation

- In the fourth phase of pattern formation, stem cells divide and differentiate into specialized cell types (Figure 20.18).

- Stem cells are categorized according to their developmental stage and their ability to differentiate. In mammals, a fertilized egg is totipotent; certain embryonic and fetal cells are pluripotent; and stem cells in the adult are multipotent or unipotent (Figures 20.19, 20.20).

- Stem cells have the potential to be used to treat a variety of human diseases (Table 20.1).

- The differentiation of cell types within certain tissues or organs is controlled by master transcription factors. An example is a transcription factor encoded by *MyoD*, a gene that initiates skeletal muscle cell development (Figure 20.21).

## 20.4 Development in Plants

- Plants grow along a root-shoot axis and in a well-defined radial pattern. Cell migration does not occur in plants (Figure 20.22).

- Plant stem cells within meristems promote the development of plant structures such as roots, stems, leaves, and flowers (Figure 20.23).

- The apical, central, and basal regions of the growing plant express different apical-basal-patterning genes (Table 20.2).

- According to the ABC model, four classes of homeotic genes in plants, *A*, *B*, *C*, and *E*, control flower development (Figure 20.24).

# Assess & Discuss

## Test Yourself

1. The process whereby a cell's morphology and function change is called
   a. determination.
   b. cell fate.
   c. differentiation.
   d. genetic engineering.
   e. both a and c.

2. Pattern formation in plants occurs along the _____ axis.
   a. dorsoventral
   b. anteroposterior
   c. left-right
   d. root-shoot
   e. all of the above are correct.

3. Positional information is important in determining the fate of a cell in a multicellular organism. Animal cells may respond to positional information by
   a. dividing.
   b. migrating.
   c. differentiating.
   d. undergoing apoptosis.
   e. doing any of the above.

4. Morphogens are
   a. molecules that disrupt normal development.
   b. molecules that convey positional information and promote changes in development.
   c. mutagenic agents that cause apoptosis.
   d. receptors that allow cells to adhere to the extracellular matrix.
   e. both a and c.

5. What group of proteins play a key role in controlling the program of developmental changes?
   a. motor proteins
   b. transporters
   c. transcription factors
   d. restriction endonucleases
   e. cyclins

6. Arrange the following phases of pattern formation in animals in the correct sequence:
   1. Tissues, organs, and other body structures in each segment are formed.
   2. Axes of the entire animal are determined.
   3. Cells become differentiated.
   4. The entire animal is divided into segments.
   a. 2, 3, 4, 1
   b. 1, 2, 4, 3
   c. 2, 4, 3, 1
   d. 3, 2, 4, 1
   e. 2, 4, 1, 3

7. Homeotic genes in *Drosophila*
   a. determine the structural and functional characteristics of different segments of the developing fly.
   b. encode motor proteins that transport morphogens throughout the embryo.
   c. are dispersed randomly throughout the genome.
   d. are expressed in similar levels in all parts of the developing embryo.
   e. Both a and c are correct.

8. Which of the following genes do *not* play a role in the process whereby segments are formed in the fruit fly embryo?
   a. *MyoD*
   b. gap genes
   c. pair-rule genes
   d. segment-polarity genes
   e. All of the above play a role in segmentation.

9. An embryonic stem cell that can give rise to any type of cell of an adult organism but cannot produce an entire, intact individual is called
   a. totipotent.
   b. pluripotent.
   c. multipotent.
   d. unipotent.
   e. antipotent.

10. During plant development, the leaves and the flowers of the plant are derived from
    a. the central region.
    b. the basal region.
    c. the suspensor.
    d. the apical region.
    e. both a and d.

## Conceptual Questions

1. If you observed fruit flies with each of the following developmental abnormalities, would you guess that a mutation had occurred in a segmentation gene or a homeotic gene? Explain your guess.

   a. Three abdominal segments are missing.
   b. One abdominal segment has legs.

2. The *MyoD* gene in mammals plays a role in muscle cell differentiation. The *Hox* genes are homeotic genes that play a role in the differentiation of particular regions of the body. Compare and contrast the functions of these genes.

3. **Core Concept: Information** Discuss how maternal effect genes, segmentation genes, and homeotic genes control the process of development in animals.

## Collaborative Questions

1. It seems that developmental genetics boils down to a complex network of gene regulation. Starting with maternal effect genes and ending with master transcription factors, draw or describe this network for the development of *Drosophila*. How many genes do you think are necessary to specify a complete developmental network for the fruit fly? How many genes do you think are needed for a network to specify one segment?

2. Is it possible for a phenotypically normal female fly to be homozygous for a loss-of-function allele in the *bicoid* gene? What would be the phenotype of the offspring that such a fly would produce if it was mated to a male that was homozygous for the normal *bicoid* allele?

# Genetic Technologies and Genomics

# 21

Japanese researchers used an exciting new genetic approach, called CRISPR-Cas technology, to silence a gene in mice that encodes tyrosinase, an enzyme needed for pigment production. In the seven albino newborn mice on the left, both copies of this gene have been silenced. In the ones with a mottled appearance, partial silencing was achieved. The gene was not silenced in the black newborn on the right. ©Fumihiro Sugiyama

**J**on was born with hemophilia A, which is a blood clotting disorder. As discussed in Chapter 17, hemophilia is inherited as an X-linked recessive trait. Jon's blood is unable to clot properly because he is missing a protein, called factor VIII, that is needed in a pathway required for normal blood clotting. Fortunately, Jon can take injections of purified factor VIII and thereby minimize the harmful effects of hemophilia. You might be surprised to learn that purified factor VIII is not obtained from human blood. Instead, it's made by cells grown in the laboratory. These cells have been genetically modified to synthesize factor VIII in large amounts. This process is just one example of how researchers have been able to apply **recombinant DNA technology**—the use of laboratory techniques to bring together fragments of DNA from multiple sources—to benefit humans.

In the early 1970s, the first successes in making recombinant DNA molecules were accomplished independently by two groups at Stanford University: David Jackson, Robert Symons, and Paul Berg; and Peter Lobban and A. Dale Kaiser. Both groups were able to isolate and purify pieces of DNA in a test tube and then covalently link two or more DNA fragments. Once inside a host cell, the recombinant molecules were replicated to produce many identical copies. The process of making multiple copies of a particular gene is known as **gene cloning**. In the first section of this chapter, we will explore recombinant DNA technology and gene cloning, techniques that have enabled geneticists to probe the relationships between gene sequences and phenotypic consequences.

In the second section, we will consider the topic of **genomics**—the molecular analysis of the entire genome of a species. In recent years, molecular techniques have progressed to the point where researchers can study the structure and function of many genes as large, integrated networks and can make changes to the genome in order to silence a particular gene (see the chapter opening photo) or alter its DNA sequence in a specific way.

In the remaining sections of this chapter, we will consider the genome characteristics of bacteria, archaea, and eukaryotes. In many species, such as multicellular plants and animals, the genome shows a high abundance of repetitive sequences—sequences that are repeated multiple times in the same genome. We will examine how such repetitive sequences increase in number.

## 21.1 Gene Cloning

**Learning Outcomes:**

1. Outline the steps of gene cloning using vectors.
2. Distinguish between a genomic library and a cDNA library.
3. Explain how gel electrophoresis is used to separate DNA fragments.
4. Describe the steps of gene cloning using polymerase chain reaction (PCR).

As already mentioned, the term gene cloning refers to the process of making many copies of a particular gene. Why is gene cloning useful? **Figure 21.1** provides an overview of the steps and goals of

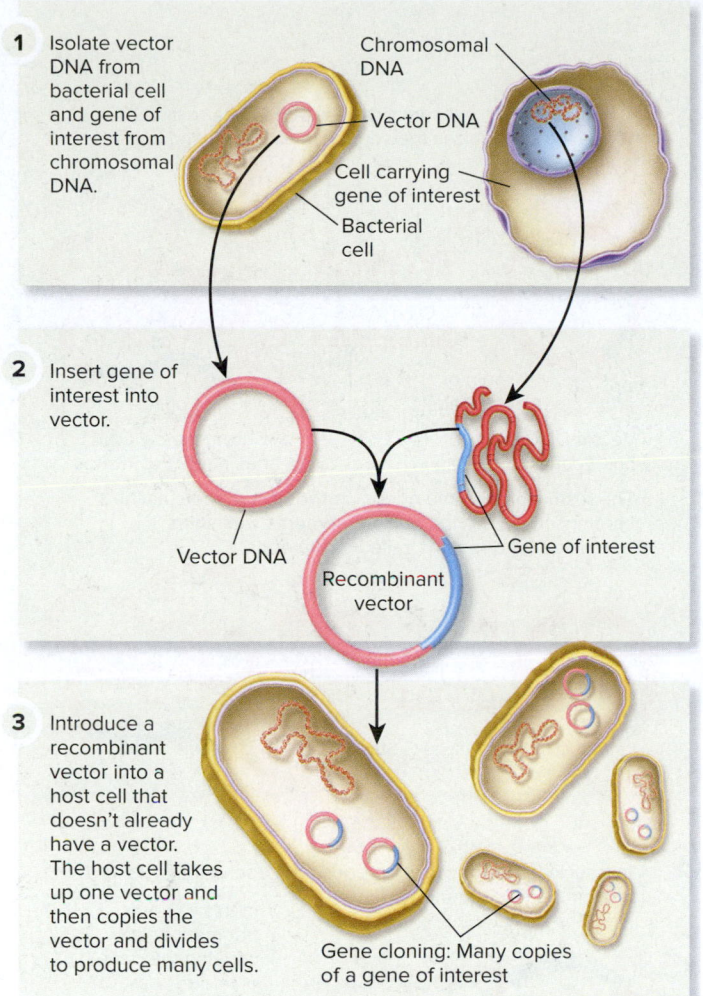

**1** Isolate vector DNA from bacterial cell and gene of interest from chromosomal DNA.

Chromosomal DNA

Vector DNA

Cell carrying gene of interest

Bacterial cell

**2** Insert gene of interest into vector.

Vector DNA

Gene of interest

Recombinant vector

**3** Introduce a recombinant vector into a host cell that doesn't already have a vector. The host cell takes up one vector and then copies the vector and divides to produce many cells.

Gene cloning: Many copies of a gene of interest

**Gene cloning is done to achieve one of two main goals:**

| Producing large amounts of DNA of a specific gene | Expressing the cloned gene to produce the encoded protein |
|---|---|
| *Examples* <br> • Cloned genes provide enough DNA for DNA sequencing. The sequence of a gene can help us understand how a gene works and identify mutations that cause diseases. <br><br> • Cloned DNA can be used as a probe to identify the same gene or similar genes in other organisms. | *Examples* <br> • Large amounts of the protein can be purified to study its structure and function. <br><br> • Cloned genes can be introduced into bacteria or livestock to make pharmaceutical products such as insulin. <br><br> • Cloned genes can be introduced into plants and animals to alter their traits. <br><br> • Cloned genes can be used to treat diseases—a clinical approach called gene therapy. |

**Figure 21.1 Overview of gene cloning.** The process of gene cloning is used to produce large amounts of a gene or its protein product.

 **Core Skill: Process of Science** The technique of gene cloning allows researchers to study genes and gene products in greater detail.

gene cloning. The process is usually done with one of two goals in mind. One is that a researcher or clinician wants many copies of the gene, perhaps to study the DNA directly or to use the DNA as a tool. For example, geneticists may want to determine the sequence of a gene from a person with a disease to see if the gene carries a mutation. Alternatively, the goal may be to obtain a large amount of the gene product, such as a specific protein. For example, biochemists use gene cloning to obtain large amounts of proteins to study their structure and function. In recent years, gene cloning has provided the foundation for critical technical advances in a variety of disciplines, including molecular biology, genetics, cell biology, biochemistry, and medicine. In this section, we will examine the procedures that are used in gene cloning.

### Step 1: Vector DNA and Chromosomal DNA Are the Starting Materials for Gene Cloning

One way to carry out gene cloning uses a type of DNA known as a **vector** (from the Latin, for "one who carries") (see Figure 21.1). Vector DNA acts as a carrier of the DNA segment that is to be cloned. In cloning experiments, a vector may carry a small segment of chromosomal DNA, perhaps only a single gene. By comparison, a chromosome carries up to a few thousand genes. When a vector is introduced into a living cell, it can replicate, and so the DNA that it carries is also replicated. This produces many identical copies of the inserted gene.

The vectors commonly used in gene cloning experiments were originally derived from two natural sources: plasmids or viruses.

- **Plasmids** are small, circular pieces of DNA that are found naturally in many strains of bacteria and exist independently of the bacterial chromosome. Commercially available plasmids have been genetically engineered for effective use in cloning experiments. They contain unique sites into which geneticists can easily insert pieces of chromosomal DNA.

- **Viral vectors** are derived from viruses, which can infect living cells and propagate themselves by taking control of the host cell's metabolic machinery. When a chromosomal gene is inserted into a viral vector, the gene is replicated whenever the viral DNA is replicated. Therefore, viruses can be used as vectors to carry other pieces of DNA.

The second material necessary for cloning a gene is the gene itself, which we will call the gene of interest. The source of the gene is the chromosomal DNA that carries the gene. The preparation of chromosomal DNA involves breaking open cells and extracting and purifying the DNA using biochemical separation techniques such as chromatography and centrifugation.

### Step 2: Cutting Chromosomal and Vector DNA into Pieces and Linking Them Together Produces Recombinant Vectors

The second step in a gene cloning experiment is the insertion of the gene of interest into the vector (see Figure 21.1, step 2). How is this accomplished? DNA molecules are cut and pasted into vectors to produce recombinant vectors. To cut DNA, researchers use enzymes known as **restriction enzymes**. These enzymes, which were

| Table 21.1 | Examples of Restriction Enzymes Used in Gene Cloning | |
|---|---|---|
| Restriction enzyme* | Bacterial source | Sequence recognized† |
| EcoRI | *Escherichia coli* (strain RY13) | ↓<br>5′–GAATTC–3′<br>3′–CTTAAG–5′<br>↑ |
| SacI | *Streptomyces achromogenes* | ↓<br>5′–GAGCTC–3′<br>3′–CTCGAG–5′<br>↑ |

*Restriction enzymes are named according to the species in which they are found. The first three letters are italicized because they indicate the genus and species names. Because a species may produce more than one restriction enzyme, the enzymes are designated I, II, III, and so on, to indicate the order in which they were discovered in a given species.
†The arrows show the locations in the two DNA strands where the restriction enzymes cleave the DNA backbone.

discovered by Swiss geneticist Werner Arber and American microbiologists Hamilton Smith and Daniel Nathans in the 1960s and 1970s, are made naturally by many different species of bacteria. Restriction enzymes protect bacterial cells from invasion by viruses by degrading the viral DNA into small fragments. Several hundred different restriction enzymes from various bacterial species have been identified and are commercially available to molecular biologists.

The restriction enzymes used in cloning experiments bind to a specific base sequence and then cleave the DNA backbone at two defined locations, one in each strand. Most restriction enzymes recognize sequences that are palindromic, which means a sequence is identical when read in the opposite directions on the two strands (**Table 21.1**). For example, the sequence recognized by the restriction enzyme *Eco*RI is 5′–GAATTC–3′ in the top strand shown in Table 21.1. Read in the opposite direction in the bottom (complementary) strand, this sequence is also 5′–GAATTC–3′. Certain restriction enzymes are useful in cloning because they digest DNA into fragments with single-stranded ends (termed "sticky" ends) that hydrogen-bond to other DNA fragments that are cut with the same enzyme and thus have complementary sequences.

**Figure 21.2** shows the action of a restriction enzyme (*Eco*RI) and the insertion of a gene into a vector. This vector, which is a plasmid, carries the *amp^R* and *lacZ* genes, whose useful functions will be discussed later. *Eco*RI binds to specific sequences in both the vector and chromosomal DNA. It then cleaves the DNA backbones, producing DNA fragments with sticky ends (Figure 21.2, step 1). The sticky ends of a piece of chromosomal DNA and the vector DNA can hydrogen-bond with each other (Figure 21.2, step 2), a process called annealing. However, this interaction is not stable, because it involves only a few hydrogen bonds between complementary bases. To establish a permanent connection between two DNA fragments, the sugar-phosphate backbones of the DNA strands must be covalently linked, or ligated. This linkage is catalyzed by DNA ligase (Figure 21.2, step 3). Recall from Chapter 11 that DNA ligase is an enzyme that catalyzes the formation of a covalent bond between two adjacent DNA fragments.

In some cases, the two ends of the vector simply ligate back together, restoring it to its original circular structure; this forms

**1** Cut vector and chromosomal DNA with *Eco*RI, a restriction enzyme that recognizes the sequence **GAATTC** and cuts at the arrows. **CTTAAG**

Vector DNA has one *Eco*RI site.

Chromosomal DNA has many *Eco*RI sites.

The restriction enzyme opens up the vector and cuts the chromosomal DNA into many fragments with short single-stranded regions called sticky ends.

**2** Allow sticky ends to hydrogen-bond with each other due to complementary sequences.

In this example, a fragment of DNA carrying the gene of interest has hydrogen-bonded to the vector. Four gaps are found where covalent bonds in the DNA backbone are missing.

**3** Add DNA ligase to close the gaps by catalyzing the formation of covalent bonds in the DNA backbone.

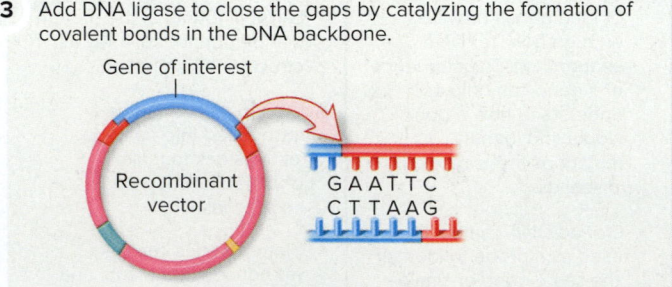

**Figure 21.2 Step 2 of gene cloning: the actions of a restriction enzyme and DNA ligase to produce a recombinant vector.**

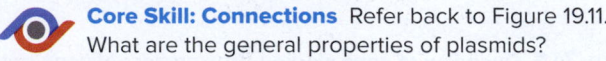

**Core Skill: Connections** Refer back to Figure 19.11. What are the general properties of plasmids?

a recircularized vector. In other cases, a fragment of chromosomal DNA may become ligated to both ends of the vector. When this happens, a segment of chromosomal DNA has been inserted into the vector. The result is a vector containing a piece of chromosomal DNA, which is called a **recombinant vector**. Such a vector is ready to be cloned. A recombinant vector may contain the gene of interest, or it may contain a different piece of chromosomal DNA.

## Step 3: Putting Recombinant Vectors into Host Cells and Allowing Those Cells to Propagate Achieves Gene Cloning

The third step in gene cloning is the actual cloning of the gene of interest. In this step, the goal is for the recombinant vector carrying the desired gene to be taken up by bacterial cells that have been treated with agents that render them permeable to DNA molecules (**Figure 21.3**). The process, called transformation, is described in Chapter 19 (refer back to Figure 19.17). Bacterial cells with the ability to take up DNA are called **competent**. After the bacteria are combined with a mixture of recircularized vectors and recombinant vectors, some bacterial cells take up a single vector, although most cells fail to take up any vector. The bacteria are then inoculated onto petri plates containing a bacterial growth medium and ampicillin.

In the experiment shown in Figure 21.3, the bacterial cells were originally sensitive to ampicillin. The vector, which is a plasmid, carries an antibiotic-resistance gene, called the $amp^R$ gene. What is the purpose of this gene in a cloning experiment? Such a gene is called a **selectable marker** because the presence of the antibiotic selects for the growth of cells expressing the $amp^R$ gene. The $amp^R$ gene encodes an enzyme known as β-lactamase that degrades the antibiotic ampicillin, which normally kills bacteria. Bacteria that have not taken up a plasmid are killed by the antibiotic. In contrast, any bacterium that has taken up a plasmid carrying the $amp^R$ gene grows and divides many times to form a visible bacterial colony containing tens of millions of cells. Because each cell in a single colony is derived from the same original cell that took up a single plasmid, all cells within a colony contain the same type of plasmid DNA.

The experimenter also needs a way to distinguish bacterial colonies that contain cells with a recombinant vector from those containing cells with a recircularized vector. In a recombinant vector, a piece of chromosomal DNA has been inserted into a region of the vector that contains the *lacZ* gene, which encodes the enzyme β-galactosidase. The insertion of chromosomal DNA into the vector disrupts the *lacZ* gene, thereby preventing the synthesis of β-galactosidase. By comparison, a recircularized vector has a functional *lacZ* gene. The functionality of *lacZ* can be determined by adding to the growth medium a colorless compound, X-Gal, which is cleaved by β-galactosidase into a blue dye. Bacteria grown in the presence of X-Gal form blue colonies if they produce a functional β-galactosidase enzyme, and white colonies if they do not. In this experiment, therefore, bacterial colonies containing recircularized vectors form blue colonies, whereas colonies containing recombinant vectors carrying a segment of chromosomal DNA are white.

After a bacterial cell has taken up a recombinant vector, two subsequent events lead to the production of many copies of that vector. First, when the vector has a highly active origin of replication, the bacterial host cell produces many copies of the recombinant vector per cell. Second, the bacterial cells divide approximately every

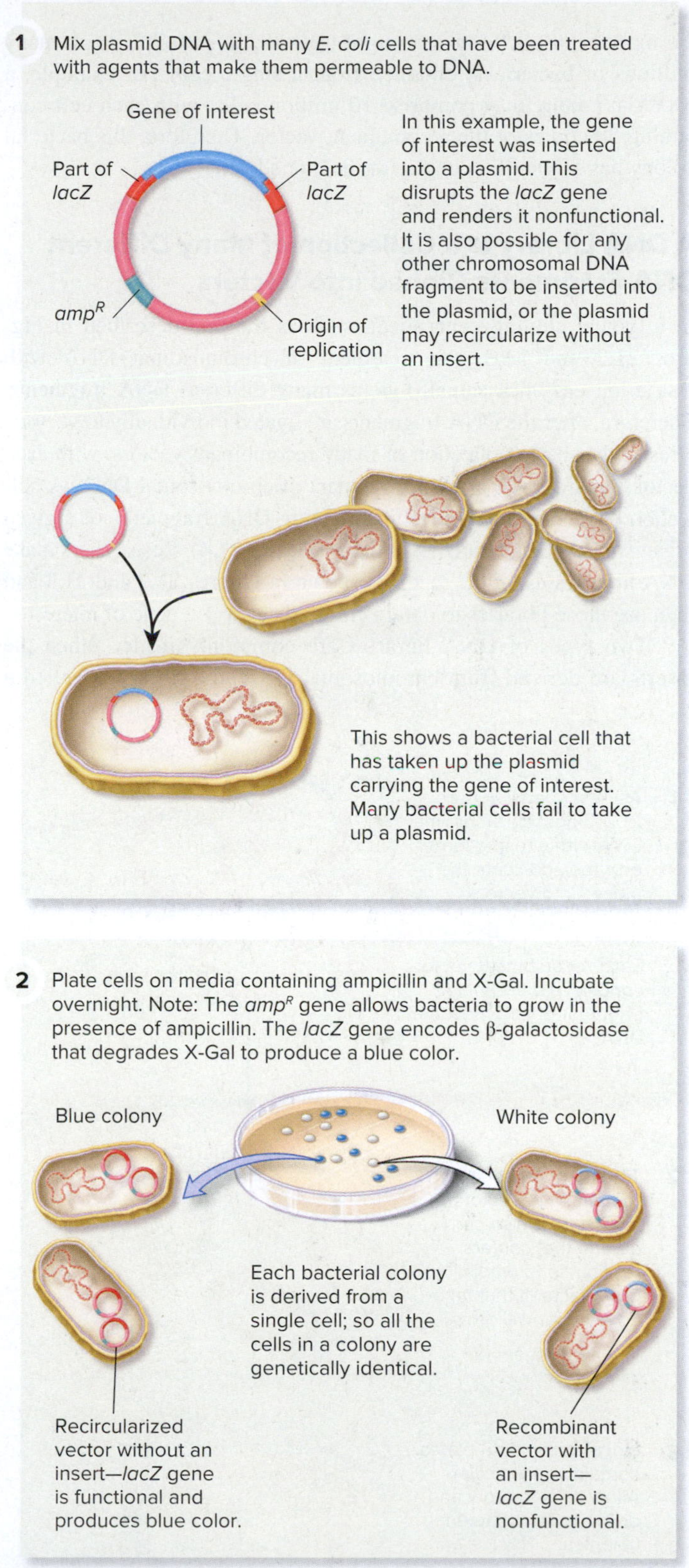

**1** Mix plasmid DNA with many *E. coli* cells that have been treated with agents that make them permeable to DNA.

In this example, the gene of interest was inserted into a plasmid. This disrupts the *lacZ* gene and renders it nonfunctional. It is also possible for any other chromosomal DNA fragment to be inserted into the plasmid, or the plasmid may recircularize without an insert.

This shows a bacterial cell that has taken up the plasmid carrying the gene of interest. Many bacterial cells fail to take up a plasmid.

**2** Plate cells on media containing ampicillin and X-Gal. Incubate overnight. Note: The $amp^R$ gene allows bacteria to grow in the presence of ampicillin. The *lacZ* gene encodes β-galactosidase that degrades X-Gal to produce a blue color.

Each bacterial colony is derived from a single cell; so all the cells in a colony are genetically identical.

Recircularized vector without an insert—*lacZ* gene is functional and produces blue color.

Recombinant vector with an insert—*lacZ* gene is nonfunctional.

**Figure 21.3** **Step 3 of gene cloning: introduction of a recombinant vector into a host cell.** For cloning to occur, a recombinant vector is introduced into a host cell, which copies the vector and divides to produce many cells. This produces many copies of the gene of interest.

*Concept Check:* *In this cloning experiment, what is the purpose of having the lacZ gene in the vector?*

20 minutes. Following overnight growth, a population of many millions of bacteria is obtained from a single cell. For example, a bacterial colony may comprise 10 million cells, with each cell containing 50 copies of the recombinant vector. Therefore, this bacterial colony has 500 million copies of the cloned gene!

## A DNA Library Is a Collection of Many Different DNA Fragments Cloned into Vectors

In a typical cloning experiment, such as the one described in Figures 21.2 and 21.3, the treatment of chromosomal DNA with restriction enzymes actually yields many different DNA fragments. Therefore, after the DNA fragments are ligated individually to vectors, a researcher has a collection of many recombinant vectors, with each vector containing a particular fragment of chromosomal DNA. A collection of recombinant vectors containing DNA fragments of a given organism is known as a **DNA library** (**Figure 21.4**). Researchers make DNA libraries using the methods shown in Figures 21.2 and 21.3 and then use those libraries to obtain clones that carry a gene of interest.

Two types of DNA libraries are commonly made. When the inserts are derived from chromosomal DNA, the library is called a **genomic library**. Alternatively, researchers can isolate mRNA and use the enzyme reverse transcriptase, which is described in Chapter 19, to make DNA molecules using mRNA as a starting material. Such DNA is called **complementary DNA**, or **cDNA**. A **cDNA library** is a collection of recombinant vectors that have inserts derived from cDNA. From a research perspective, an important advantage of cDNA is that it lacks introns—intervening sequences that are not

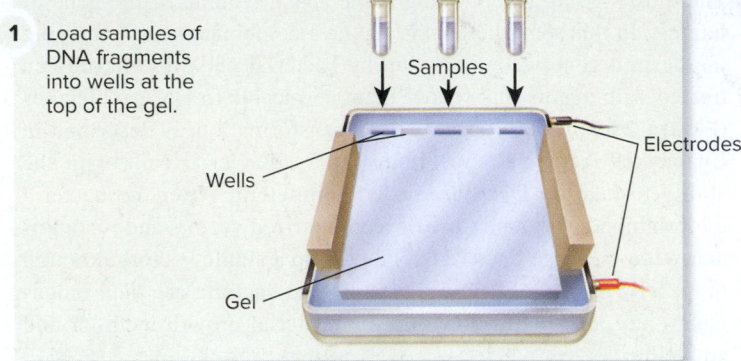

**1** Load samples of DNA fragments into wells at the top of the gel.

Samples

Electrodes

Wells

Gel

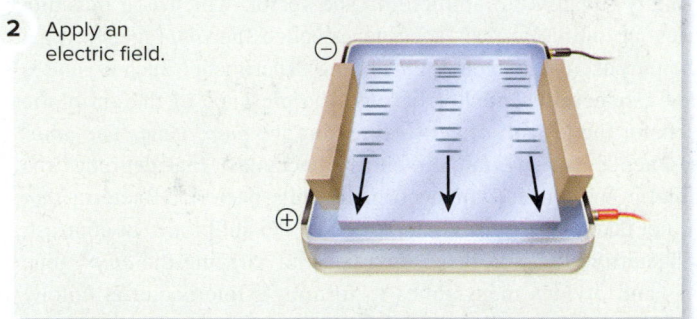

**2** Apply an electric field.

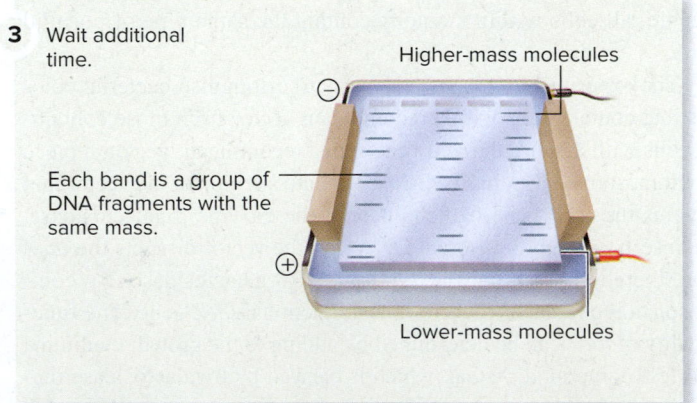

**3** Wait additional time.

Higher-mass molecules

Each band is a group of DNA fragments with the same mass.

Lower-mass molecules

**Figure 21.5** **Separation of molecules by gel electrophoresis.** In this example, samples containing many fragments of DNA are loaded into wells at the top of the gel and then subjected to an electric current that causes the fragments to move toward the positively charged electrode at the bottom of the gel. This separates the fragments according to their masses, with the smaller DNA fragments near the bottom of the gel.

*Concept Check:* *One DNA fragment contains 600 bp and another has 1,300 bp. Following electrophoresis, which will be closer to the bottom of the gel?*

**1** As described in Figure 21.2, digest chromosomal DNA with a restriction enzyme and ligate the pieces into vectors.

Each recombinant vector contains a different fragment of chromosomal DNA.

Vector

**2** Transform bacteria with recombinant vectors. The vectors also carry a gene that confers resistance to ampicillin. Only bacteria that take up a vector will grow.

**3** Inoculate on petri plates containing ampicillin. Allow cells to grow and divide to form bacterial colonies.

Each bacterial colony contains millions of cells that were derived from a single transformed cell.

**Figure 21.4** **A DNA library.** Each colony in a DNA library contains a vector with a different piece of chromosomal DNA.

translated into proteins. Because introns can be quite large, it is much simpler for researchers to insert cDNAs into vectors rather than chromosomal DNA segments if they want to focus their attention on the coding sequence of a gene. In addition, because bacteria do not splice out introns, using cDNAs provides an advantage if researchers want to express the gene of interest in bacteria.

## Gel Electrophoresis Separates Macromolecules, Such as DNA Fragments

**Gel electrophoresis** is a technique for separating macromolecules, such as DNA and proteins, as they migrate through a gel. This method is often used to evaluate the results of a cloning experiment. For example, gel electrophoresis is used to determine the sizes of DNA fragments that have been inserted into recombinant vectors.

Gel electrophoresis can separate molecules based on their charge, size/length, and mass. In the example shown in **Figure 21.5**, gel electrophoresis is used to separate different fragments of chromosomal DNA based on their masses. The flat slab of semisolid gel has depressions at the top called wells, where samples are added. Electrodes are located at each end of the gel. An electric current is applied to the gel, which causes charged molecules, either proteins or nucleic acids, to migrate from the top of the gel toward the bottom—a process called electrophoresis. DNA is negatively charged and moves toward the positively charged electrode, which is at the bottom in this figure. As gel electrophoresis occurs, the DNA fragments are separated into distinct bands within the gel. Smaller DNA fragments move more quickly through the gel than larger ones in a given amount of time and therefore are located closer to the bottom of the gel than the larger ones. The fragments in each band can then be stained with a dye for identification.

## Polymerase Chain Reaction (PCR) Is Also Used to Make Many Copies of DNA

As we have seen, one method of cloning involves an approach in which the gene of interest is inserted into a vector, introduced into a host cell, and then propagated. Another cloning technique, in which DNA is copied without the aid of vectors and host cells, is a process called **polymerase chain reaction (PCR)**, which was developed by American biochemist Kary Mullis in 1985 (**Figure 21.6**). The goal of PCR is to make many copies of DNA in a defined region, perhaps encompassing a gene or part of a gene. Several reagents are required for the synthesis of DNA. First, two different primers are needed that

**Figure 21.6  Polymerase chain reaction (PCR).** During each PCR cycle, the steps of denaturation, primer annealing, and primer extension take place. The net result of PCR is the synthesis of many copies of DNA in the region that is flanked by the two primers. To conduct this type of PCR experiment, researchers must have prior knowledge about the base sequence of the template DNA so they can make primers with base sequences that are complementary to the ends of the template DNA. Note: The temperatures shown in this figure are approximate and may vary depending on the primer sequence and length.

*Concept Check:* *Why do the primers used in PCR bind specifically to the primer-annealing sites?*

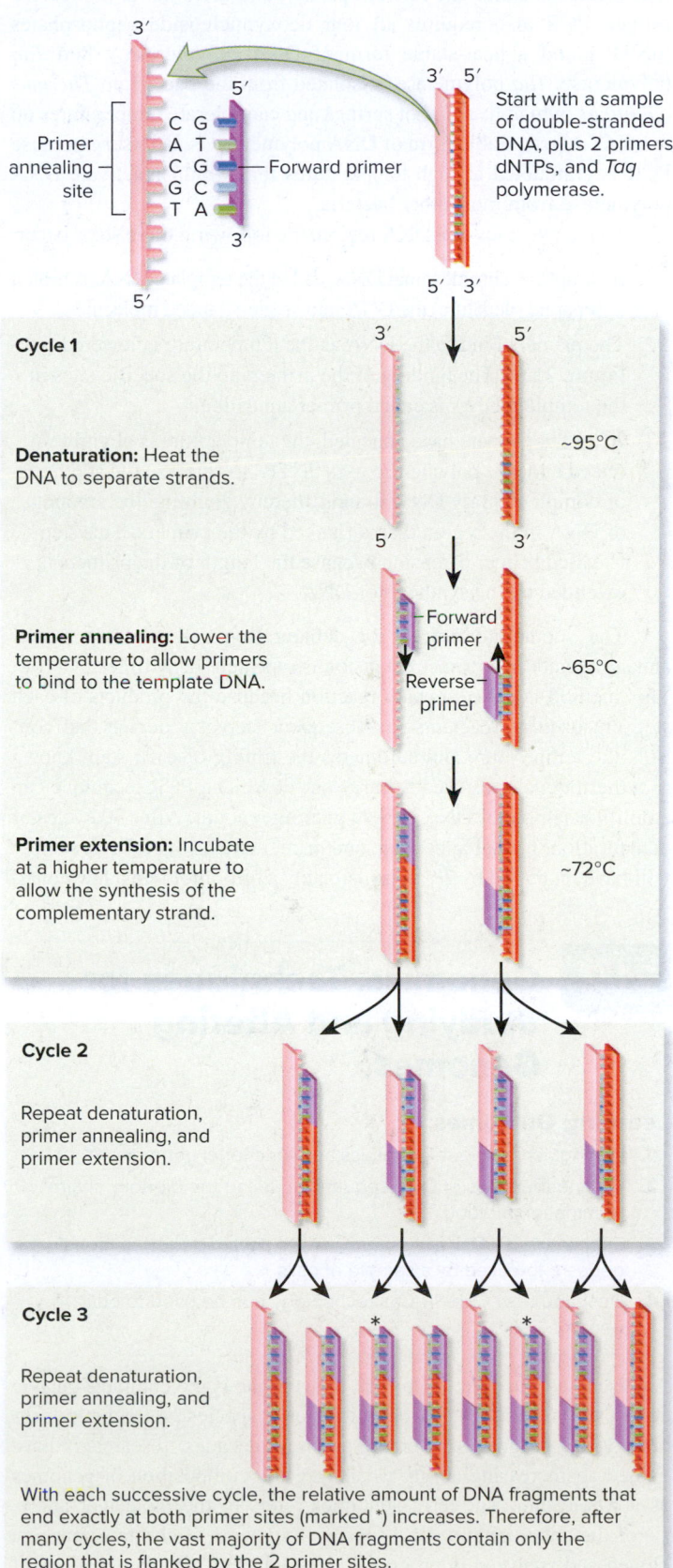

are complementary to sequences at each end of the DNA region to be amplified. These primers are usually about 20 nucleotides long. One primer is called the forward primer, and the other is the reverse primer. PCR also requires all four deoxynucleoside triphosphates (dNTPs) and a heat-stable form of DNA polymerase called *Taq* polymerase. *Taq* polymerase is isolated from the bacterium *Thermus aquaticus*, which lives in hot springs and can tolerate temperatures up to 95°C. A heat-stable form of DNA polymerase is necessary because PCR is conducted at high temperatures that would inactivate DNA polymerase from most other bacteria.

To make copies of a DNA region, the following three steps occur:

1. A sample of chromosomal DNA, called the template DNA, is heated to separate (denature) the DNA into single-stranded molecules.

2. The primers bind to the DNA as the temperature is lowered (see Figure 21.6). The binding of the primers to the specific sites in the template DNA is called primer annealing.

3. After the primers have annealed, the temperature is slightly raised and *Taq* polymerase uses dNTPs to catalyze the synthesis of complementary DNA strands, thereby doubling the amount of DNA in the region that is flanked by the primers. This step is called primer extension because the length of the primers is extended by the synthesis of DNA.

The sequential process of denaturation followed by primer annealing and then primer extension is repeated many times in a row. This method is called a chain reaction because the products of each step are used as reactants in subsequent steps. A device that controls the temperature and automates the timing of each step, known as a thermocycler, is used to carry out PCR. The PCR technique can amplify a sample of DNA by a staggering amount. After 30 cycles of denaturation, primer annealing, and primer extension, a DNA sample will have increased by $2^{30}$, approximately a billionfold, in a few hours!

## 21.2 Genomics: Techniques for Studying and Altering Genomes

### Learning Outcomes:

1. Distinguish between genomics and functional genomics.
2. Outline the steps of DNA sequencing using the dideoxy chain-termination method.
3. Explain what a DNA microarray is and how it is used to identify the genes expressed by a sample of cells.
4. Describe how CRISPR-Cas technology can be used to edit genes.

As discussed throughout Unit III, the genome is the complete genetic composition of a cell, an organism, or a species. As genetic technology has progressed over the past few decades, researchers have gained an increasing ability to analyze the composition of genomes as a whole unit. The term genomics refers to the molecular analysis of the entire genome of a species. Segments of chromosomes are cloned and analyzed in progressively smaller pieces, the locations of which are known on the intact chromosomes. This is the mapping

phase of genomics. The mapping of a genome ultimately progresses to the determination of the complete DNA sequence, which provides the most detailed description available of an organism's genome at the molecular level. By comparison, **functional genomics** studies the expression of a genome. For example, functional genomics can be used to analyze which genes are turned on or off in normal versus cancer cells. In this section, we will consider a few of the methods that are used in genomics and functional genomics.

### The Dideoxy Chain-Termination Method Is Used to Determine the Base Sequence of DNA

The term **DNA sequencing** refers to a procedure that is aimed at determining the base sequence of DNA. Scientists can learn a great deal about the function of a gene if its nucleotide sequence is known. For example, the investigation of genetic sequences has been vital in our understanding of the genetic basis of human diseases.

One type of DNA sequencing, developed in 1977 by English biochemist Frederick Sanger and colleagues, is known as the **dideoxy chain-termination method**, or more simply, **dideoxy sequencing**. Dideoxy sequencing is based on our knowledge of DNA replication. As described in Chapter 11, DNA polymerase connects adjacent deoxynucleoside triphosphates (dNTPs) by catalyzing a covalent linkage between the phosphate group at the 5′ position on an incoming nucleotide and the —OH group at the 3′ position on the growing strand. Chemists, however, can synthesize nucleotides, called dideoxynucleoside triphosphates (ddNTPs), that are missing the —OH group at the 3′ position (**Figure 21.7**). What happens if a ddNTP is incorporated during DNA replication? If a ddNTP is added to a growing DNA strand, the strand can no longer grow because the 3′—OH group, the site of attachment for the next nucleotide, is missing. This ending of DNA synthesis is called chain termination.

Before describing the steps of this DNA sequencing procedure, let's first consider the DNA segment that is analyzed in such an experiment. The segment of DNA to be sequenced, the target DNA, is obtained in large amounts by using the gene cloning techniques that were described earlier in this chapter. In **Figure 21.8a**, the target DNA has been inserted into a vector next to a primer-annealing site, the place where a primer will bind. The target DNA is initially double stranded, but Figure 21.8a shows the DNA after it has been denatured into a single strand by heat treatment.

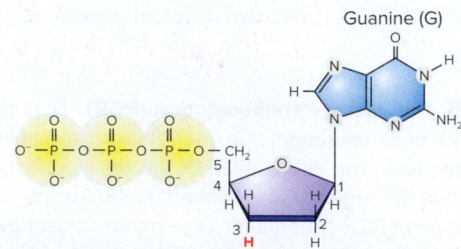

Guanine (G)

**Figure 21.7** **Structure of a dideoxynucleotide triphosphate.** This figure shows the structure of dideoxyguanosine triphosphate (ddGTP). It has a hydrogen, shown in red, instead of a hydroxyl group at the 3′ position. The prefix dideoxy- means that the sugar has two (di) missing (de) oxygens (oxy) compared with ribose, which has —OH groups at both the 2′ and 3′ positions.

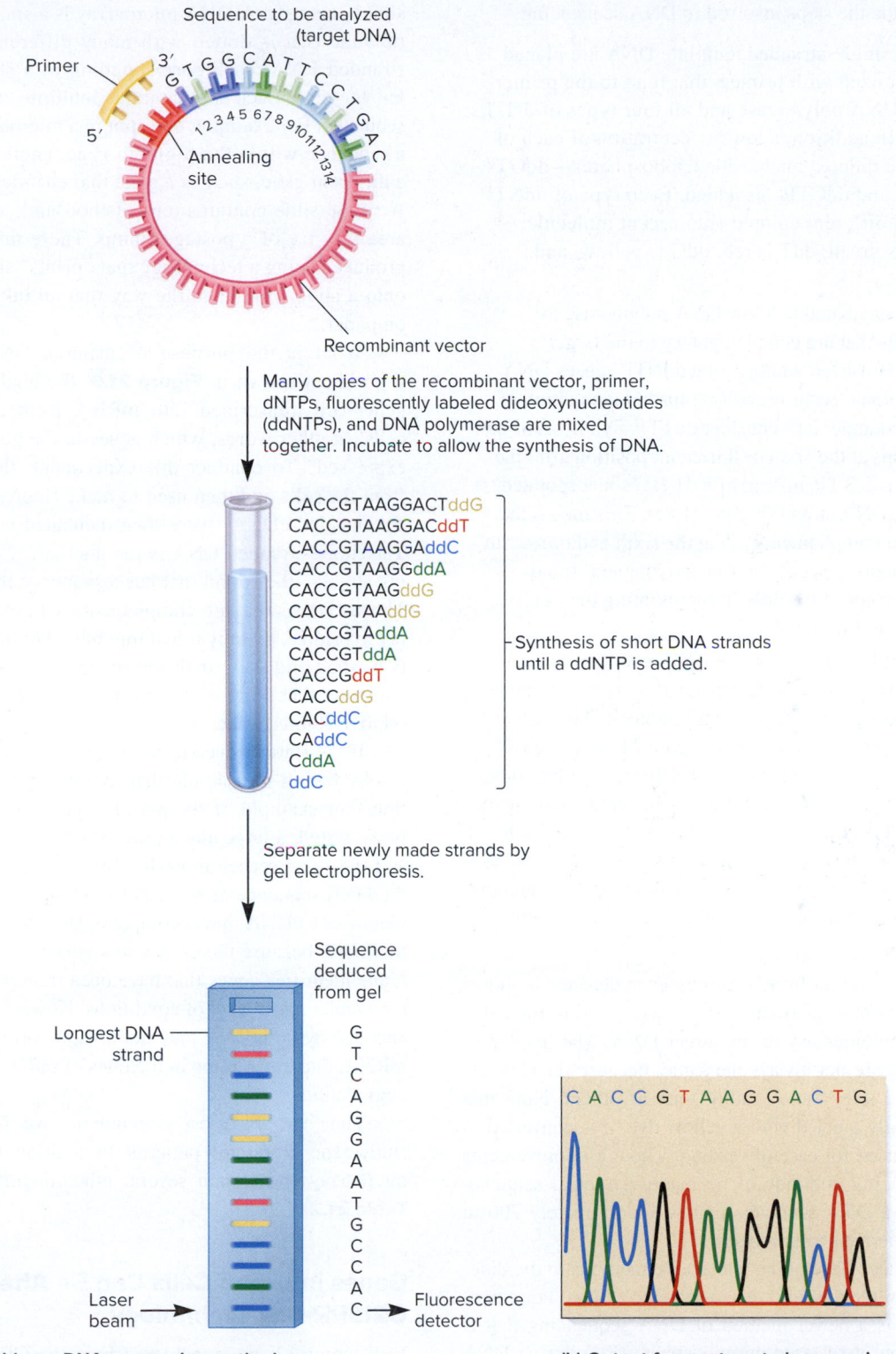

(a) Dideoxy DNA sequencing method

(b) Output from automated sequencing

**Figure 21.8** **Dideoxy sequencing of DNA.** **(a)** The procedure of dideoxy sequencing uses fluorescently labeled ddNTPs. **(b)** This method uses a fluorescence detector that measures the four kinds of ddNTPs as they emerge from the gel.

*Concept Check:* *What happens when a ddNTP is incorporated into a growing DNA strand?*

Let's now examine the steps involved in DNA sequencing.

1. Many copies of single-stranded template DNA are placed into a tube and mixed with primers that bind to the primer-annealing site. DNA polymerase and all four types of dNTPs are also added. In addition, a low concentration of each of the four possible dideoxynucleoside triphosphates—ddGTP, ddATP, ddTTP, and ddCTP—is added. Each type of ddNTP is tagged with a different colored fluorescent molecule; typically, ddA is green, ddT is red, ddG is yellow, and ddC is blue.

2. The tube is then incubated to allow DNA polymerase to synthesize strands that are complementary to the target DNA sequence. However, addition of a ddNTP causes DNA synthesis to terminate early, preventing further elongation of the strand. For example, let's consider ddTTP. Synthesis of new DNA strands stops at the sixth or thirteenth position after the annealing site if a ddTTP, instead of a dTTP, is incorporated into the growing DNA strand (Figure 21.8a). This means the target DNA has a complementary A at the sixth and thirteenth positions. Eventually, a set of fluorescently tagged strands results, with the color of the ddNTP representing the last nucleotide added to the strand.

3. After the samples have been incubated for several minutes, the newly made DNA strands are separated according to their lengths by subjecting them to gel electrophoresis. This can be done using a slab gel, as shown in Figure 21.8a, or more commonly by running them through a gel-filled capillary tube. The shorter strands move to the bottom of the gel more quickly than the longer ones. Electrophoresis is continued until each band emerges from the bottom of the gel, where a laser excites the fluorescent dye. A fluorescence detector records the amount of fluorescence emission at four wavelengths, corresponding to the four dyes.

An example of a printout from a fluorescence detector is shown in **Figure 21.8b**. The peaks of fluorescence correspond to the DNA sequence that is complementary to the target DNA. The heights of the fluorescent peaks are not always the same, because ddNTPs get incorporated at some sites more readily than at others. Note that although ddG is usually labeled with a yellow dye, it is converted to black ink on the printout for ease of reading. Though improvements in automated sequencing continue to be made, a typical sequencing run can provide a DNA sequence that is approximately 700 to 900 bases long, and perhaps even longer.

Researchers are also developing alternative methods to the dideoxy chain-termination method in order to sequence DNA. For example, pyrosequencing is a newer method of DNA sequencing that is based on the detection of released pyrophosphate ($PP_i$) during DNA synthesis.

## A Microarray Can Identify Which Genes Are Transcribed by a Cell

Let's now turn our attention to functional genomics. Researchers have developed a technology, called a **DNA microarray**, that is used to monitor the expression of thousands of genes simultaneously. A DNA microarray is a small silica, glass, or plastic slide that is dotted with many different sequences of single-stranded DNA, each corresponding to a short sequence within a known gene. Each spot contains multiple copies of a known DNA sequence. For example, one spot in a microarray may correspond to a sequence within the β-globin gene; another might correspond to a different gene, such as a gene that encodes a glucose transporter. A single slide contains tens of thousands of different spots in an area the size of a postage stamp. These microarrays are typically produced using a technology that "prints" spots of DNA sequences onto a slide, similar to the way that an inkjet printer deposits ink on paper.

What is the purpose of using a DNA microarray? In the experiment shown in **Figure 21.9**, the goal is to determine which genes are transcribed into mRNA from a particular sample of cells. In other words, which genes in the genome of these cells are expressed? To conduct this experiment, the mRNA was isolated from the cells and then used to make fluorescently labeled cDNAs. The labeled cDNAs were then incubated with a DNA microarray. The single-stranded DNA in the microarray corresponds to the coding strand—the strand that has a sequence that is similar to mRNA. Those cDNAs that are complementary to the DNAs in the microarray hybridize, thereby remaining bound to the microarray. The array is washed and then analyzed using a microscope equipped with a computer that scans all of the spots and generates an image of their relative fluorescence.

If the fluorescence intensity in a spot is high, a large amount of cDNA was in the sample that hybridized to the DNA at this location. For example, if the β-globin gene was expressed in the cells being tested, a large amount of cDNA for this gene would be made, and the fluorescence intensity for that spot would be high. Because the DNA sequence of each spot is already known, a fluorescent spot identifies a cDNA that is complementary to that DNA sequence. Furthermore, because the cDNA was generated from mRNA, this technique identifies genes that have been transcribed in a particular cell type under a given set of conditions. However, the amount of protein encoded by an mRNA may not always correlate with the amount of mRNA, due to variation in the rates of mRNA translation and protein degradation.

Thus far, the most common use of DNA microarrays is to study gene expression patterns. In addition, the technology of DNA microarrays has found several other important uses, described in **Table 21.2**.

## Genes in Living Cells Can Be Altered Using CRISPR-Cas Technology

In Chapter 13, we considered how the CRISPR-Cas system provides bacteria with a defense against bacteriophages. Researchers realized that the components of this system can be used to alter genes, an approach called **gene editing**. This particular method is named **CRISPR-Cas technology**. Researchers have made a modification to the natural system to make it efficient for gene editing. They create a single RNA in which tracrRNA and crRNA are linked to each other (**Figure 21.10a**). This is called the single guide RNA (sgRNA). The spacer region of the sgRNA is designed

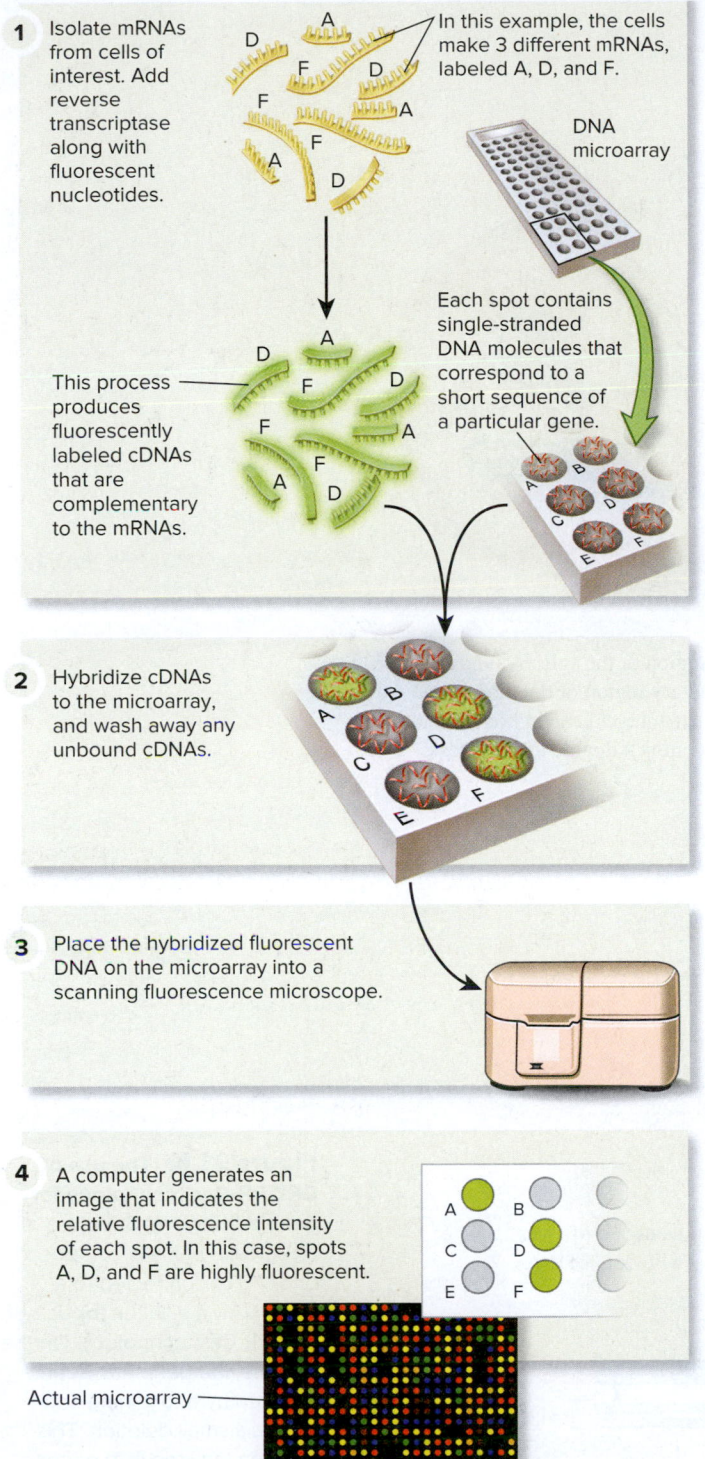

1  Isolate mRNAs from cells of interest. Add reverse transcriptase along with fluorescent nucleotides.

In this example, the cells make 3 different mRNAs, labeled A, D, and F.

DNA microarray

This process produces fluorescently labeled cDNAs that are complementary to the mRNAs.

Each spot contains single-stranded DNA molecules that correspond to a short sequence of a particular gene.

2  Hybridize cDNAs to the microarray, and wash away any unbound cDNAs.

3  Place the hybridized fluorescent DNA on the microarray into a scanning fluorescence microscope.

4  A computer generates an image that indicates the relative fluorescence intensity of each spot. In this case, spots A, D, and F are highly fluorescent.

Actual microarray

**Figure 21.9  Identifying transcribed genes within a DNA microarray.** In this simplified example, only three cDNAs specifically hybridize to spots on the microarray. Those genes were expressed in the cells from which the mRNA was isolated. In an actual experiment, the array typically contains hundreds or thousands of different cDNAs and tens of thousands of different spots.

*(bottom)* ©Alfred Pasieka/Science Source

**Concept Check:** *If a fluorescent spot appears on a microarray, what information does this provide regarding gene expression?*

| Table 21.2 | Applications of DNA Microarrays |
|---|---|
| **Application** | **Description** |
| Cell-specific gene expression | A comparison of microarray data using cDNAs derived from mRNAs of different cell types can identify genes that are expressed in a cell-specific manner. |
| Gene regulation | Because environmental conditions play an important role in gene regulation, a comparison of microarray data using cDNAs derived from mRNAs from cells exposed to two different environmental conditions may reveal genes that are induced under one set of conditions and repressed under another set. |
| Elucidation of metabolic pathways | Genes that encode proteins that participate in a common metabolic pathway are often expressed together and can be revealed from a microarray analysis. |
| Tumor profiling | Different types of cancer cells exhibit striking differences in their gene expression profiles, which can be revealed by a DNA microarray analysis. This approach is gaining use as a tool to classify tumors that are sometimes morphologically indistinguishable. |
| Genetic variation | A mutant allele may not hybridize to a spot on a microarray as well as a wild-type allele. Therefore, microarrays are gaining use as a tool for detecting genetic variation. This application has been used to identify disease-causing alleles in humans and to identify mutations that contribute to quantitative traits in plants and other species. |
| Microbial strain identification | Microarrays can distinguish between closely related bacterial species and subspecies. |

to be complementary to one of the DNA strands of a target gene that a researcher wants to edit. The sgRNA binds to Cas9 and guides it to the target gene. Cas9 then makes a double-strand break in this gene.

Following this break, two different DNA repair events are possible. If the break is repaired by a process called end joining, the gene may incur a small deletion (see the left side of **Figure 21.10b**). This deletion may inactivate the gene, particularly if it causes a frameshift mutation in the coding sequence. Alternatively, a researcher can add a double-stranded segment of DNA, called the donor DNA, that is homologous to the region where the break occurs (see the right side of Figure 21.10b). This homologous DNA is made synthetically and is designed to carry a particular mutation, such as a point mutation, that the researcher wants to introduce. A different DNA repair system swaps in the donor DNA by a double crossover. In this way, researchers can introduce a specific mutation into a gene.

The experiment described in Figure 21.10b can be performed on different cell types and even on whole organisms. By studying how a gene mutation affects cell structure and function or by examining the phenotypic effects of a mutation in a whole organism (see the chapter opening photo), researchers often gain a better understanding of how genes function.

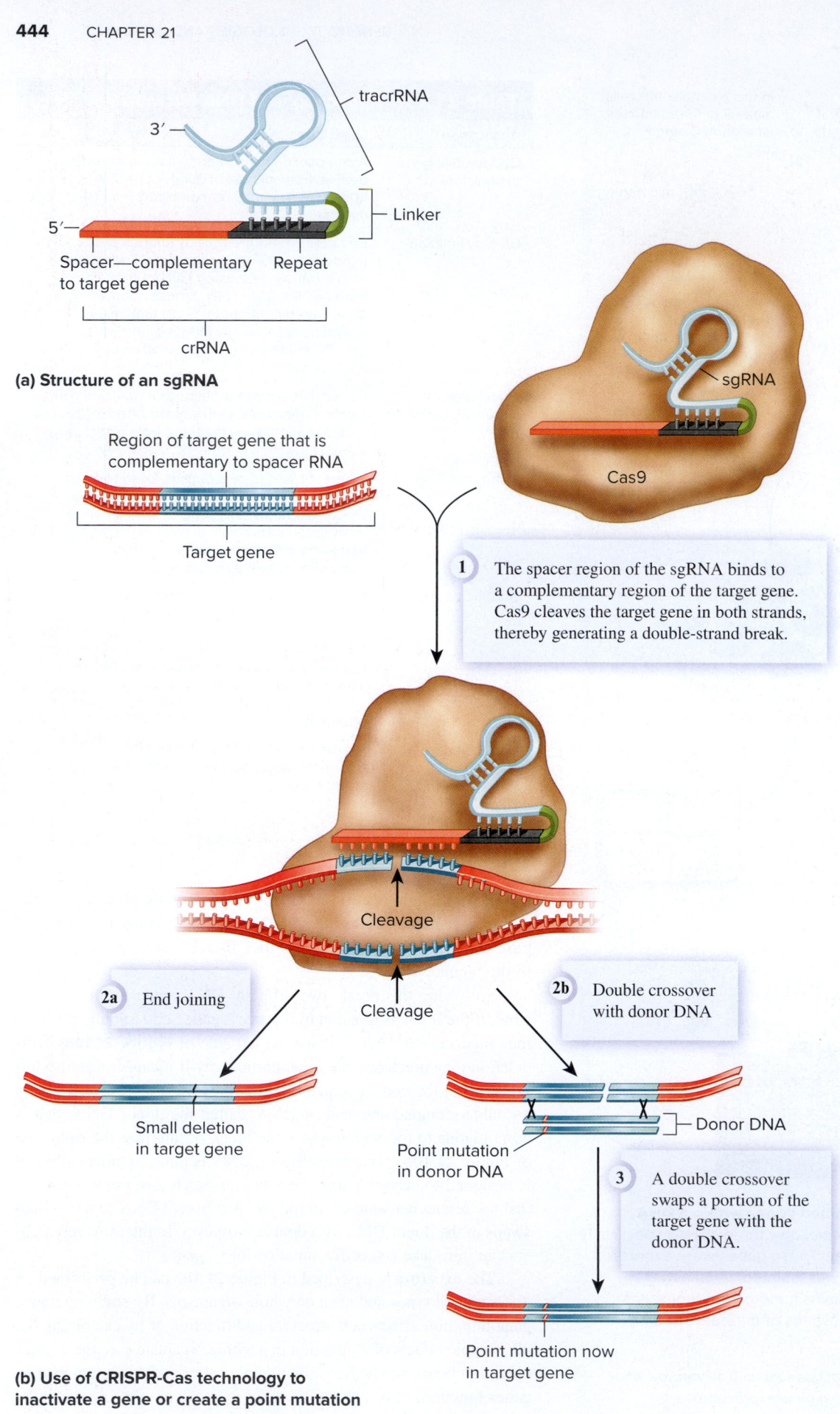

**(a) Structure of an sgRNA**

1. The spacer region of the sgRNA binds to a complementary region of the target gene. Cas9 cleaves the target gene in both strands, thereby generating a double-strand break.

**2a** End joining

**2b** Double crossover with donor DNA

Small deletion in target gene

Point mutation in donor DNA

Donor DNA

3. A double crossover swaps a portion of the target gene with the donor DNA.

Point mutation now in target gene

**(b) Use of CRISPR-Cas technology to inactivate a gene or create a point mutation**

**Figure 21.10** **The use of CRISPR-Cas technology for gene editing.** **(a)** Structure of an sgRNA, which is composed of a crRNA connected to a tracrRNA via a linker. **(b)** Use of CRISPR-Cas technology. On the left side, the gene has been repaired by end joining and suffers a small deletion. This deletion may result in gene inactivation. On the right side, a double crossover has occurred that swaps in the donor DNA. The target gene now carries a point mutation.

*Concept Check:* *How is the sgRNA different from certain components of the bacterial defense system described in Chapter 13 (refer back to Figure 13.8)?*

## Core Skills: Process of Science, Quantitative Reasoning

**BIO TIPS**

**THE QUESTION** *Gene editing is often used to explore the structure and function of proteins. For example, changes can be made to the coding sequence of a gene to determine how the resulting alterations in the amino acid sequence affect the function of a protein. Let's suppose that you are interested in the functional importance of a particular asparagine (an amino acid) within a protein you are studying. By gene editing, you make mutant proteins in which this asparagine has been changed to other amino acids. You then test the encoded mutant proteins for functionality. The results are as follows:*

|  | Functionality (%) |
|---|---|
| *Wild-type (nonmutant) protein* | *100* |
| *Mutant proteins in which the asparagine is changed to:* |  |
| *Leucine* | *7* |
| *Phenylalanine* | *3* |
| *Glutamine* | *98* |
| *Proline* | *4* |

*From these results, what would you conclude about the functional significance of this asparagine within the protein? Note: The structures of the amino acids are shown in Chapter 3 (refer back to Figure 3.13).*

**T**OPIC *What topic in biology does this question address?* The topic is gene editing. More specifically, the question is about the effects of gene editing on protein structure and function.

**I**NFORMATION *What information do you know based on the question and your understanding of the topic?* From the question, you know that gene editing can be used to study protein structure and function. You are also given data on the results of a gene editing experiment, and reminded where to find the structures of the amino acids in this textbook. From your understanding of the topic, you may recall that if an amino acid is important for a protein's structure and function, then changing that amino acid is likely to inhibit the protein's function. Conservative substitutions, which involve changes to amino acids with similar side chains, are more likely than nonconservative ones to retain functionality.

**P**ROBLEM-SOLVING **S**TRATEGY *Interpret data. Compare and contrast.* One strategy to solve this problem is to look at the results of this experiment and compare and contrast the level of functionality that the mutant proteins have compared to the wild-type protein.

**ANSWER** *These results suggest that the asparagine is important for this protein's function. When this asparagine is replaced with glutamine, which has a very similar structure, the protein retains most of its functionality. However, if it is replaced with other types of amino acids, most of the functionality is lost.*

## 21.3 Bacterial and Archaeal Genomes

**Learning Outcomes:**

1. List the key characteristics of bacterial and archaeal genomes.
2. Describe the method of shotgun DNA sequencing.

The past decade has seen remarkable advances in our overall understanding of the entire genomes of many species. As genetic technology has progressed, researchers have gained an increasing ability to analyze the composition of a genome as a whole unit. The complete DNA sequence is now known for many species, providing the most detailed description available of an organism's genome at the molecular level. In this section, we will survey the sizes and compositions of genomes in selected species of bacteria and archaea.

### Studying the Genomes of Bacteria and Archaea Has Important Applications

Why are researchers interested in the genomes of bacteria and archaea?

1. Bacteria cause many different diseases that affect humans as well as other animals and plants. Studying the genomes of bacteria reveals important clues about the process of infection, which may also help us find ways to combat bacterial infections.

2. The knowledge that is obtained by studying bacterial and archaeal genomes often applies to larger and more complex organisms.

3. A third reason is evolution. The origin of the first eukaryotic cell probably involved a union between an archaeal cell and a bacterial cell, as we will explore in Chapter 26. The study of bacterial and archaeal genomes helps us understand how all living species evolved.

4. Bacteria are often used as tools in research, as discussed in Section 21.1. A better understanding of their genomes can make them more effective tools.

### The Genomes of Bacteria and Archaea Typically Consist of a Circular Chromosome with a Few Thousand Genes

Geneticists have made great progress in the study of bacterial and archaeal genomes. The genomes of thousands of bacterial and archaeal species have been sequenced and analyzed. The chromosomes of bacteria and archaea are usually a few million base pairs in length. Genomic researchers refer to 1 million base pairs as 1 megabase pair, abbreviated 1 Mb. Most bacteria and archaea contain a single type of chromosome, though multiple copies may be present in a single cell. However, some bacteria are known to have more than one type of chromosome. For example, *Vibrio cholerae*, the bacterium that causes the diarrheal disease cholera, has two different chromosomes in each cell, one has 2.9 Mb and the other 1.1 Mb.

Bacterial and archaeal chromosomes are usually circular. For example, the two chromosomes in *V. cholerae* are circular, as is the single type of chromosome found in *E. coli*. However, linear chromosomes are found in some species, such as *Borrelia burgdorferi*,

| Table 21.3 | Examples of Bacterial and Archaeal Genomes That Have Been Sequenced* | | |
|---|---|---|---|
| **Species** | **Genome size (Mb)†** | **Number of genes‡** | **Description** |
| *Methanobacterium thermoautotrophicum* | 1.7 | 1,921 | An archaeal species that produces methane |
| *Haemophilus influenzae* | 1.8 | 1,753 | One of several different bacterial species that cause respiratory illness and meningitis |
| *Sulfolobus solfataricus* | 3.0 | 3,032 | An archaeal species that metabolizes sulfur-containing compounds |
| *Lactobacillus plantarum* | 3.3 | 3,052 | A type of lactic acid–producing bacterium used in the production of cheese and yogurt |
| *Mycobacterium tuberculosis* | 4.4 | 4,294 | The bacterium that causes the respiratory disease tuberculosis |
| *Escherichia coli* | 4.6 | 4,377 | A naturally occurring intestinal bacterium; certain strains cause human illness |
| *Bacillus anthracis* | 5.2 | 5,439 | The bacterium that causes the disease anthrax |

*Bacterial and archaeal species often exist in different strains that may differ slightly in their genome size and number of genes. The data are from common strains of the indicated species. The species shown in this table have only one type of chromosome.

†Mb equals 1 million base pairs, or a megabase pair.

‡The number of genes is an estimate based on the analysis of genome sequences.

the bacterium that causes Lyme disease, the most common tick-borne disease in the United States. Certain bacterial species even contain both linear and circular chromosomes. *Agrobacterium tumefaciens*, which infects plants and causes crown gall tumors, has one linear chromosome (2.1 Mb) and one circular chromosome (3.0 Mb).

**Table 21.3** compares the sequenced genomes from several bacterial and archaeal species. These genomes range in size from 1.7 to 5.2 Mb. The total number of genes is correlated with the total genome size. Roughly 1,000 genes are found for every megabase pair of DNA. Compared with eukaryotic genomes, bacterial and archaeal genomes are less complex. Their chromosomes lack centromeres and telomeres and have a single origin of replication. Also, chromosomes of bacteria and archaea have relatively little repetitive DNA, whereas repetitive sequences, which are discussed later in this chapter, are often abundant in eukaryotic genomes.

In addition to one or more chromosomes, bacteria may have plasmids, circular pieces of DNA that exist independently of the bacterial chromosome (refer back to Figure 19.11). Plasmids are typically small, ranging in length from a few thousand to tens of thousands of base pairs, though some can be quite large, with hundreds of thousands of base pairs. The various functions of plasmids were described in Chapter 19, and their use as vectors in gene cloning was discussed in Section 21.1.

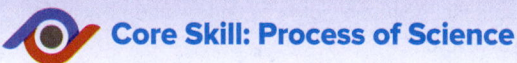

## Core Skill: Process of Science

## Feature Investigation | Venter, Smith, and Colleagues Sequenced the First Genome in 1995

The first genome to be entirely sequenced was that of the bacterium *Haemophilus influenzae*. This bacterium causes a variety of diseases in humans, including respiratory illnesses and bacterial meningitis. *H. influenzae* has a relatively small genome consisting of approximately 1.8 Mb of DNA in a single circular chromosome.

The most commonly used strategy for sequencing an entire genome is called **shotgun DNA sequencing**. In this approach, researchers use a DNA sequencing method, such as the dideoxy chain-termination method (see Figure 21.8), to randomly sequence many DNA fragments from the genome. As a matter of chance, some of the fragments are overlapping—the end of one fragment contains the same DNA region as the beginning of another fragment. Computers are used to align the overlapping regions and assemble the DNA fragments into a contiguous sequence identical to that found in the intact chromosome. This procedure is called shotgun DNA sequencing because the process generates many tiny pieces of DNA, reminding people of the tiny metal pellets of shot sprayed by a shotgun blast.

To obtain a complete sequence of a genome with the shotgun approach, how do researchers decide how many fragments

to sequence? We can calculate the probability that a base will not be sequenced (*P*) using this equation:

$$P = e^{-m}$$

where *e* is the base of the natural logarithm ($e = 2.72$), and *m* is the number of sequenced bases divided by the total genome size. For example, in the case of *H. influenzae*, with a genome size of 1.8 Mb, if researchers sequenced 9.0 Mb, $m = 5$ (that is, 9.0 Mb divided by 1.8 Mb):

$$P = e^{-m} = e^{-5} = 0.0067, \text{ or } 0.67\%.$$

This means that if researchers randomly sequence 9.0 Mb, which is five times the length of a single genome, they are likely to miss only 0.67% of the genome. With a genome size of 1.8 Mb, they would miss about 12,000 bp out of approximately 1.8 million. Such missed sequences are typically on small DNA fragments that, as a matter of random chance, did not happen to be sequenced. Though it is beyond the scope of this textbook, these missed sequences can be identified and sequenced using more advanced types of cloning methods.

In their discovery-based investigation, American biologists Craig Venter and Hamilton Smith and their colleagues used a shotgun DNA sequencing approach (**Figure 21.11**). The researchers isolated

**Figure 21.11** **Determination of the complete genome sequence of *Haemophilus influenzae*.**

**GOAL** The goal is to obtain the entire genome sequence of *Haemophilus influenzae*. This information will reveal its genome size and also which genes the organism has.

**KEY MATERIALS** A strain of *H. influenzae*.

|  | Experimental level | Conceptual level |
|---|---|---|
| **1** Purify DNA from a strain of *H. influenzae*. This involves breaking the cells open by adding phenol and chloroform. Most protein and lipid components go into the phenol-chloroform phase. DNA remains in the aqueous (water) phase. | DNA in aqueous (water) phase / Proteins and lipids in phenol-chloroform phase | *H. influenzae* chromosomal DNA |
| **2** Sonicate the DNA to break it into small fragments of about 2,000 bp in length. | Sound waves / DNA fragments in aqueous phase | Sound waves |
| **3** Clone the DNA fragments into vectors. The procedures for cloning are described in Section 21.1. This produces a DNA library. | Refer back to Figures 21.2 and 21.3. | Vector DNA / Piece of *H. influenzae* DNA / A DNA library |
| **4** Subject many clones to the procedure of dideoxy sequencing, described in Section 21.2. A total of 10.8 Mb was sequenced. | Refer back to Figure 21.8. | CCAGTCCCATGCCATGGCCCAGTCCC / Produces a large number of sequences with overlapping regions. |
| **5** Use computer methods to identify various types of genes in the genome. | | CCATGCCATGGCCCC... / Explores the genome sequence and identifies and characterizes genes. |

**6** **THE DATA**

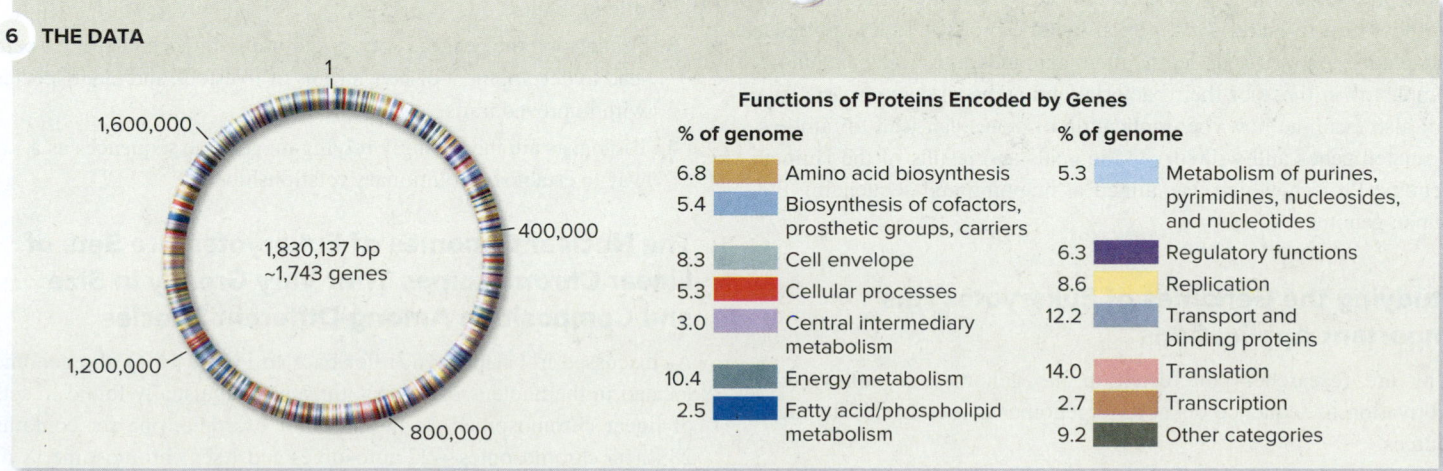

1,830,137 bp
~1,743 genes

**Functions of Proteins Encoded by Genes**

| % of genome | | | % of genome | |
|---|---|---|---|---|
| 6.8 | Amino acid biosynthesis | | 5.3 | Metabolism of purines, pyrimidines, nucleosides, and nucleotides |
| 5.4 | Biosynthesis of cofactors, prosthetic groups, carriers | | 6.3 | Regulatory functions |
| 8.3 | Cell envelope | | 8.6 | Replication |
| 5.3 | Cellular processes | | 12.2 | Transport and binding proteins |
| 3.0 | Central intermediary metabolism | | 14.0 | Translation |
| 10.4 | Energy metabolism | | 2.7 | Transcription |
| 2.5 | Fatty acid/phospholipid metabolism | | 9.2 | Other categories |

---

**7**    **CONCLUSION** *H. influenzae* has a genome size of 1.83 Mb with approximately 1,743 genes. The functions of many of those genes could be inferred by comparing them to genes in other species.

**8**    **SOURCE** Fleischmann, R. D. et al. 1995. Whole-genome random sequencing and assembly of *Haemophilius Influenzae* Rd, *Science* 269: 496–512.

---

chromosomal DNA from *H. influenzae* and used sound waves to break the DNA into small fragments of approximately 2,000 bp in length. These fragments were randomly inserted into vectors, allowing the DNA to be propagated in *E. coli*. Each *E. coli* clone carried a vector with a different piece of DNA from *H. influenzae*. The complete set of vectors, each containing a different fragment of DNA, is called a DNA library (refer back to Figure 21.4). The researchers then subjected many of these clones to the procedure of DNA sequencing. They sequenced a total of approximately 10.8 Mb of DNA.

The outcome of this genome-sequencing project was a very long DNA sequence. In 1995, Venter, Smith, and colleagues published the entire DNA sequence of *H. influenzae*. The researchers then analyzed the genome sequence using a computer to obtain information about the properties of the genome. Questions they asked included "How many genes does the genome contain, and what are the likely functions of those genes?" The data in Figure 21.11 summarize these researchers' results. The *H. influenzae* genome is composed of 1,830,137 bp

of DNA. The computer analysis predicted 1,743 genes. Based on the similarities of the sequences in *H. influenzae* to sequences of genes identified in other species, the researchers also predicted the functions of proteins encoded by nearly two-thirds of the genes. The diagram shown in the data of Figure 21.11 places proteins in various categories based on their predicted function. These results gave the first comprehensive "genome picture" of a living organism!

### Experimental Questions

1. What was the goal of the investigation conducted by Venter, Smith, and colleagues?

2. **CoreSKILL** » Let's suppose that researchers used the shotgun DNA sequencing approach and sequenced 20 Mb of DNA from a bacterium with a genome size of 4.1 Mb. What percentage of the genome would be left unsequenced?

3. **CoreSKILL** » What were the results of the study by Venter, Smith, and colleagues?

---

## 21.4   Eukaryotic Genomes

### Learning Outcomes:

1. Describe the key features of eukaryotic genomes.
2. Explain how gene duplications occur and lead to the formation of gene families.
3. List the goals and results of the Human Genome Project.

In the previous section, we examined bacterial and archaeal genomes. We now turn to eukaryotes, which include protists, fungi, animals, and plants. As you will learn, their genomes are larger and more complex than those of their bacterial and archaeal counterparts. We will also examine how the duplication of genes can lead to families of related genes and will survey the goals and results of the Human Genome Project, which was aimed at mapping and sequencing the human genome.

### Studying the Genomes of Eukaryotes Has Important Applications

Why are researchers interested in the genomes of eukaryotes? Motivation to sequence eukaryotic genomes comes from four main sources.

1. The availability of genome sequences makes it easier for researchers to identify and characterize the genes of model

organisms. This was the impetus for genome projects involving baker's yeast (*Saccharomyces cerevisiae*), the fruit fly (*Drosophila melanogaster*), a nematode worm (*Caenorhabditis elegans*), the flowering plant called thale cress (*Arabidopsis thaliana*), and the mouse (*Mus musculus*).

2. Studying eukaryotic genomes enables researchers to gather more information for identifying and treating human diseases. Researchers hope that knowing the DNA sequence of the human genome will help to identify genes in which mutation plays a role in disease.

3. Sequencing the genomes of agriculturally important species can lead to development of new strains of livestock and plant species with improved traits.

4. Biologists are increasingly relying on genome sequences as a way to establish evolutionary relationships.

### The Nuclear Genomes of Eukaryotes Are Sets of Linear Chromosomes That Vary Greatly in Size and Composition Among Different Species

As discussed in Chapter 16 (refer back to Figure 16.1), the genome located in the nucleus of eukaryotic species is usually found in sets of linear chromosomes. In humans, for example, one set contains 23 linear chromosomes—22 autosomes and 1 sex chromosome, X or Y. In addition, certain organelles in eukaryotic cells contain a small amount of DNA, including the mitochondrion, which plays a role in

## Table 21.4 Examples of Eukaryotic Nuclear Genomes That Have Been Sequenced

| Species | Nuclear Genome size (Mb)* | Number of protein-encoding genes | Description |
|---|---|---|---|
| *Saccharomyces cerevisiae* (baker's yeast) | 12.1 | ~6,600 | One of the simplest eukaryotic species, which has been extensively studied by researchers trying to understand eukaryotic molecular biology |
| *Caenorhabditis elegans* (a nematode worm) | 100 | ~20,000 | A model organism used to study animal development |
| *Drosophila melanogaster* (fruit fly) | 180 | ~15,000 | A model organism used to study many genetic phenomena, including development |
| *Arabidopsis thaliana* (thale cress) | 120 | ~26,000 | A model organism studied by plant biologists |
| *Oryza sativa* (rice) | 440 | ~40,000 | A cereal grain that has a relatively small genome and is very important worldwide as a food crop |
| *Mus musculus* (mouse) | 2,500 | ~21,000 | A model mammalian organism used to study genetics, cell biology, and development |
| *Homo sapiens* (humans) | 3,200 | ~22,000 | Our own genome, the sequencing of which will help in our understanding of inherited traits and aid in the identification and treatment of diseases |

*The genome size refers to the number of megabase (Mb) pairs in one set of chromosomes. For species with sex chromosomes, it includes both sex chromosomes.

ATP synthesis, and the chloroplast (found in plants and algae), which carries out photosynthesis. The genetic material in these organelles is referred to as the mitochondrial or the chloroplast genome to distinguish it from the nuclear genome, which is located in the cell nucleus. In this chapter, we will focus on the nuclear genome of eukaryotes.

***Sizes of Nuclear Genomes*** In the past decade or so, the DNA sequence of entire nuclear genomes has been determined for hundreds of eukaryotic species, including several dozen mammalian genomes. Examples are shown in **Table 21.4**.

***Relationship Between Genome Sizes and Repetitive Sequences*** Eukaryotic genomes are generally larger than bacterial and archaeal genomes, in terms of both the number of genes and genome size. The genomes of simpler eukaryotes, such as yeast, carry several thousand different protein-encoding genes, whereas the genomes of more complex eukaryotes contain tens of thousands (see Table 21.4). Note that the number of genes is not the same as genome size. When we speak of genome size, we mean the total amount of DNA, often measured in megabase pairs (1 million bp). The relative sizes of nuclear genomes vary dramatically among different eukaryotic species (**Figure 21.12a**). In general, increases in the amount of DNA are correlated with increases in cell complexity and body complexity. For example, yeasts have smaller genomes than animals.

However, major variations in genome sizes are often observed among species that are similar in form and function. For example, the total amount of DNA found within different species of amphibians varies over 100-fold. As another example, let's consider two closely related species of the plant called the globe thistle, *Echinops bannaticus* and *Echinops nanus* (**Figure 21.12b, c**). These species have similar numbers of chromosomes, but *E. bannaticus* has nearly double the amount of DNA that *E. nanus* has. What is the

explanation for the larger genome of *E. bannaticus*? The genome of *E. bannaticus* is not likely to contain twice as many genes. Rather, its genome composition includes many repetitive sequences, which are

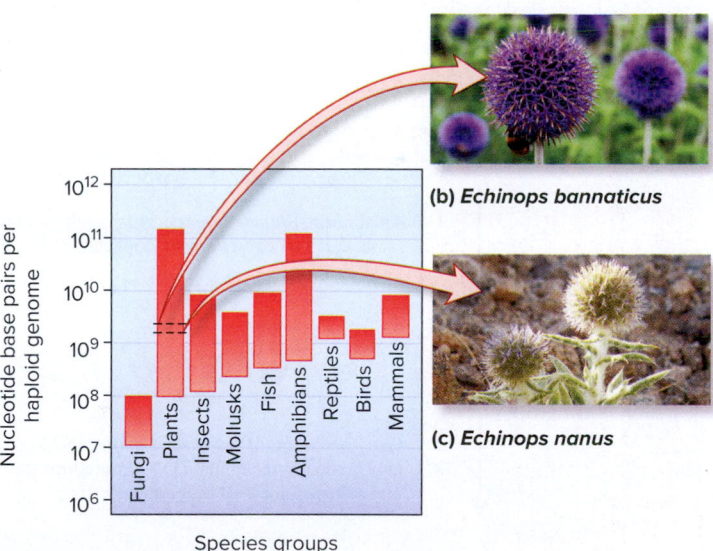

**(b)** *Echinops bannaticus*

**(c)** *Echinops nanus*

**(a)** Genome size

**Figure 21.12 Genome sizes among selected groups of eukaryotes. (a)** Genome sizes among various groups of eukaryotes are shown on a log scale. As an example for comparison, two closely related species of globe thistle are pictured. These species have similar characteristics, but *Echinops bannaticus* **(b)** has nearly double the amount of DNA that *E. nanus* **(c)** has, as a result of the accumulation of repetitive DNA sequences. b: ©The Picture Store/SPL/ Science Source; c: ©Photo by Michael Beckmann, Institute of Geobotany and Botanical Garden, Halle, Germany

***Concept Check:*** *What are two reasons that the groups of species shown in part (a) vary in their total amount of DNA?*

short DNA sequences that are present in many copies throughout the genome. Repetitive sequences are often abundant in eukaryotic species. We will examine the characteristics of such repetitive sequences in Section 21.5.

### Core Concept: Evolution

## Gene Duplications Provide Additional Material for Genome Evolution, Sometimes Leading to the Formation of Gene Families

Let's now turn our attention to gene duplications, a way of increasing the number of genes in a genome. Gene duplications are important because they provide raw material for the addition of more genes into a species' genome. Such duplications produce **homologous genes**, two or more genes that are derived from the same ancestral gene (**Figure 21.13a**). Over the course of many generations, each version of the gene accumulates different mutations, resulting in genes with similar but not identical DNA sequences.

How do gene duplications occur? One mechanism that produces gene duplications is a misaligned crossover (**Figure 21.13b**). In this example, two homologous chromosomes have paired with each other during meiosis, but the homologs are misaligned. A crossover produces one chromosome with a gene duplication, one with a gene deletion, and two normal chromosomes. Each of these chromosomes is segregated into different haploid cells. If a haploid cell carrying the chromosome with the gene duplication participates in fertilization with another gamete, an offspring with a gene duplication is produced. In this way, gene duplications can form and be transmitted to future generations.

During evolution, gene duplications can occur several times. Two or more homologous genes within a single species are also called paralogous genes, or **paralogs**. Multiple gene duplications followed by the accumulation of mutations in each paralog result in a **gene family**—a group of paralogs that carry out related functions. A well-studied example is the globin gene family found in animals. The globin genes encode polypeptides that are subunits of proteins that function in oxygen binding. Hemoglobin, which is made in red blood cells, carries oxygen throughout the body. In humans, the globin gene family is composed of 14 paralogs that were originally derived from a single ancestral globin gene (**Figure 21.14**). According to an evolutionary analysis, the ancestral globin gene duplicated between 500 and 600 mya. Since that time, additional duplication events and chromosomal rearrangements have occurred to produce the current number of 14 genes on three different human chromosomes. Four of these are pseudogenes—genes that have been produced by gene duplication but have accumulated mutations that make them nonfunctional, so they are not transcribed into RNA.

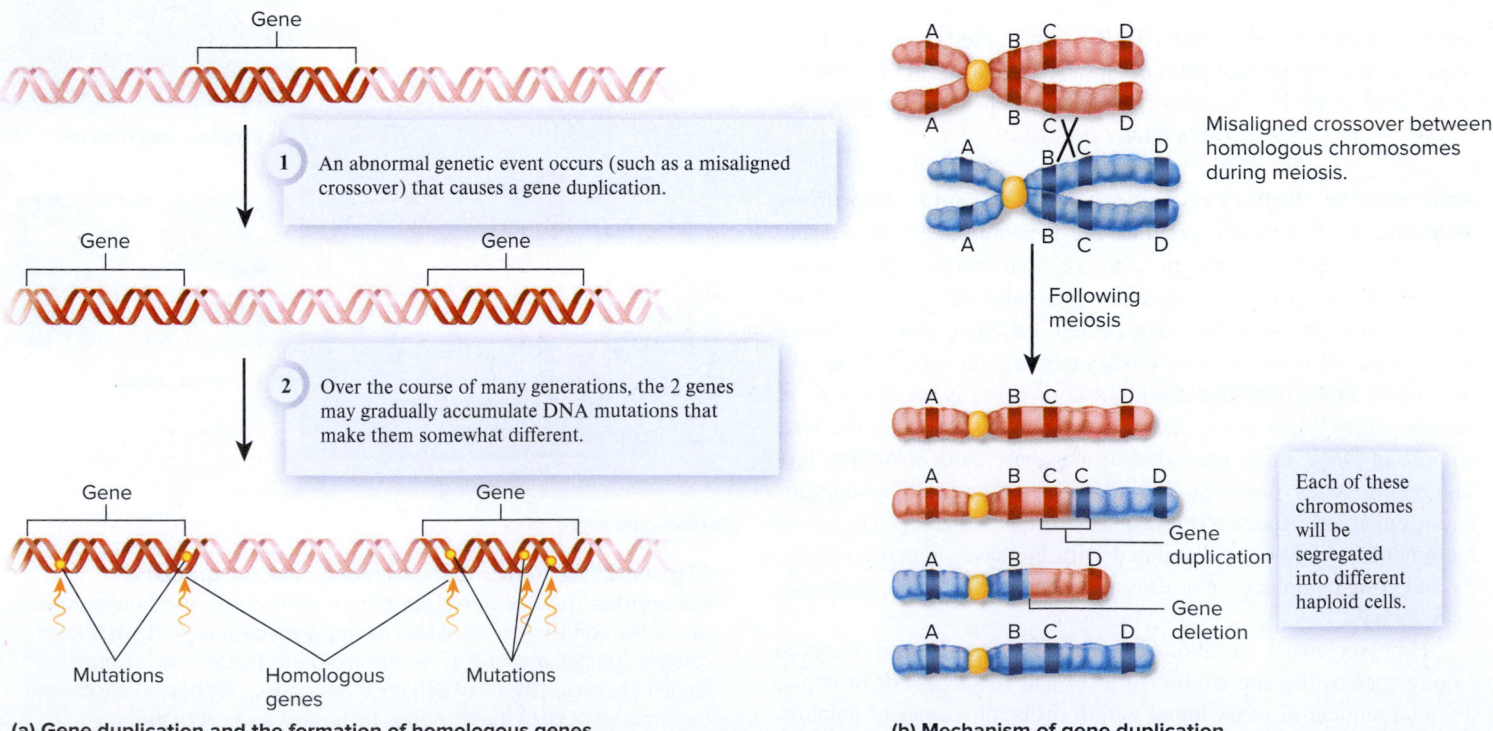

1   An abnormal genetic event occurs (such as a misaligned crossover) that causes a gene duplication.

2   Over the course of many generations, the 2 genes may gradually accumulate DNA mutations that make them somewhat different.

Gene

Mutations    Homologous genes    Mutations

**(a) Gene duplication and the formation of homologous genes**

Misaligned crossover between homologous chromosomes during meiosis.

Following meiosis

Gene duplication

Gene deletion

Each of these chromosomes will be segregated into different haploid cells.

**(b) Mechanism of gene duplication**

**Figure 21.13  Gene duplication and the evolution of homologous genes.** **(a)** A gene duplication produces two copies of the same gene. Over time, these copies accumulate different random mutations, which results in homologous genes with similar but not identical DNA sequences. **(b)** A mechanism of gene duplication. If two homologous chromosomes misalign during meiosis, a crossover may produce a chromosome with a gene duplication.

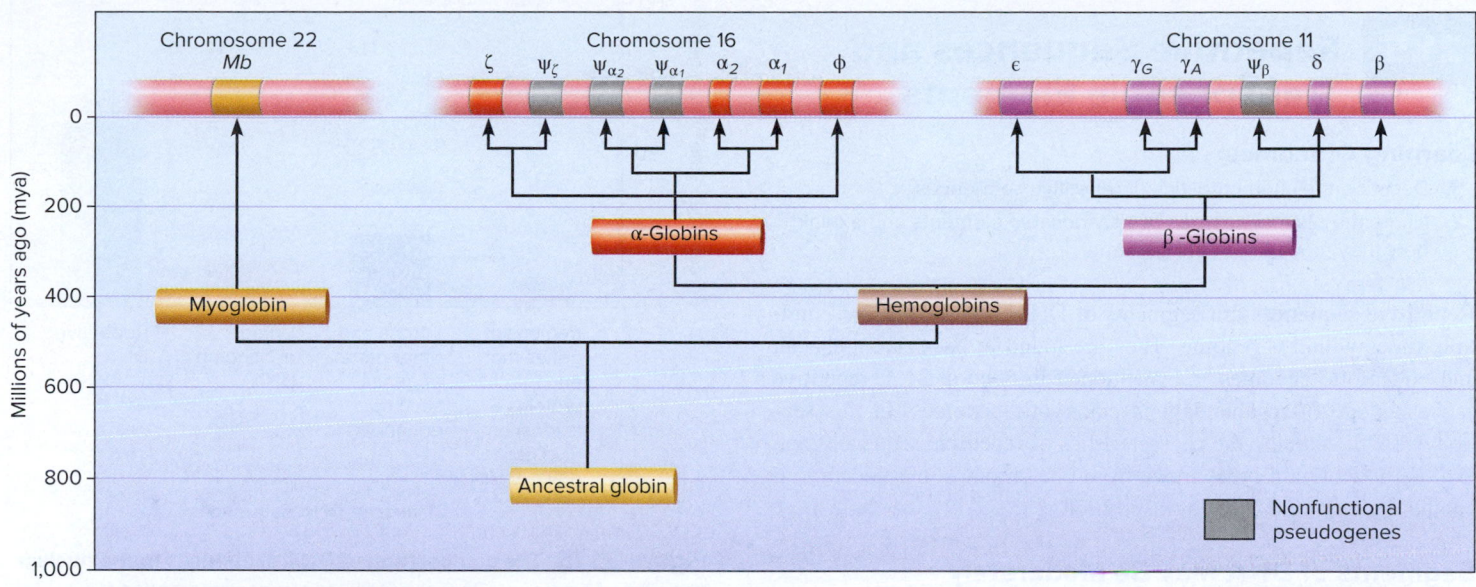

**Figure 21.14** **The evolution of the globin gene family in humans.** The globin gene family evolved from a single ancestral globin gene.

 **Core Skill: Connections** Refer back to Figure 14.3. How do the proteins encoded by different members of the globin gene family vary in their affinity for oxygen?

The accumulation of different mutations in the various family members has produced globins that are specialized in their function. For example, myoglobin binds and stores oxygen in muscle cells, whereas the hemoglobins bind and transport oxygen via red blood cells. Also, different globin genes are expressed during different stages of development. The zeta (ζ)-globin and epsilon (ε)-globin genes are expressed very early in embryonic life. During the second trimester of gestation, the alpha (α)-globin and gamma (γ)-globin genes are turned on. Following birth, the γ-globin genes are turned off, and the β-globin gene is turned on. These differences in the expression of the globin genes reflect the differences in the oxygen transport needs of humans during the embryonic, fetal, and postpartum stages of life (refer back to Figure 14.3).

## The Human Genome Project Has Stimulated Genomic Research

Before ending our discussion of genomes, let's consider the **Human Genome Project**, a research effort to identify and map all human genes. Scientists had been discussing how to undertake this project beginning in the mid-1980s. In 1988, the National Institutes of Health (NIH) in Bethesda, Maryland, established an Office of Human Genome Research, with James Watson as its first director. The Human Genome Project officially began on October 1, 1990, and was largely finished by the end of 2003. It was an international consortium that included research institutions in the U.S., U.K., France, Germany, Japan, and China. From its outset, the Human Genome Project had the following goals:

- *To obtain the DNA sequence of the entire human genome.* The first draft of a nearly completed DNA sequence was published in February 2001, and a second draft was published in 2003. The entire human genome is approximately 3.2 billion base pairs in length.

- *To identify all human genes.* This involved mapping the locations of genes throughout the entire genome.

- *To develop technology for the generation and management of human genome information.* Some of the efforts of the Human Genome Project have involved improvements in molecular genetic technology, such as gene cloning, DNA sequencing, and so forth. The Human Genome Project has also developed computer tools that allow scientists to easily access up-to-date information from the project and analytical tools to interpret genomic information.

- *To analyze the genomes of model organisms.* These include *E. coli, S. cerevisiae, D. melanogaster, C. elegans, A. thaliana,* and *M. musculus.*

- *To develop programs focused on understanding and addressing the ethical, legal, and social implications of the results obtained from the Human Genome Project.* The Human Genome Project raised many ethical issues regarding genetic information and genetic engineering. Who should have access to genetic information? Should employers, insurance companies, law enforcement agencies, and schools have access to our genetic makeup?

Some current and potential applications of the results from the Human Genome Project include the improved diagnosis and treatment of genetic diseases such as cystic fibrosis, Huntington disease, and Duchenne muscular dystrophy. The results may also enable researchers to identify the genetic basis of common disorders such as cancer, diabetes, and heart disease, which involve alterations in several genes.

# 21.5 | Repetitive Sequences and Transposable Elements

## Learning Outcomes:

1. Describe the characteristics of repetitive sequences.
2. Name the two major types of transposable elements and explain how they move about the genome.

**Repetitive sequences** are segments of DNA that are repeated multiple times within a genome. They are found in bacterial, archaeal, and eukaryotic genomes. As mentioned in Section 21.4, repetitive sequences are often abundant in eukaryotic genomes. In this section, we will examine the characteristics of repetitive sequences and explore how certain types move from one chromosomal location to another by a process called transposition.

## Segments of DNA May Be Moderately or Highly Repetitive

Repetitive sequences fall into two broad categories, moderately and highly repetitive. Sequences that are repeated a few hundred to several thousand times are called **moderately repetitive sequences**. In some cases, these sequences are multiple copies of the same gene. For example, the genes that encode ribosomal RNA (rRNA) are found in many copies. The cell needs a large amount of rRNA for its cellular ribosomes. This is accomplished by having and expressing multiple copies of the genes that encode rRNA. In addition, other types of functionally important sequences can be moderately repetitive. For example, multiple copies of origins of replication are found in eukaryotic chromosomes. Other moderately repetitive sequences may play a role in the regulation of gene transcription and translation.

**Highly repetitive sequences** are those that are repeated tens of thousands to millions of times throughout the genome. Each copy of a highly repetitive sequence is relatively short, ranging from a few to several hundred nucleotides in length.

Some highly repetitive sequences are clustered together in a tandem array, in which a very short nucleotide sequence is repeated many times in a row. In *Drosophila*, for example, 19% of the chromosomal DNA is highly repetitive DNA found in tandem arrays. An example is shown here:

In this particular tandem array, two related sequences, AATAT and AATATAT (in the top strand), are repeated multiple times. Highly repetitive sequences, which contain tandem arrays of short sequences, can be quite long, sometimes more than 1 million bp in length!

Other highly repetitive sequences are interspersed throughout the genome. A widely studied example is the *Alu* family of sequences found in humans and other primates. The *Alu* sequence is approximately 300 bp long. This sequence derives its name from the observation that it contains a site for cleavage by a restriction enzyme known as *Alu*I. It represents about 10% of the total human DNA and occurs (on average) approximately every 5,000–6,000 bases. Evolutionary studies suggest that the *Alu* sequence arose 65 million years ago (mya)

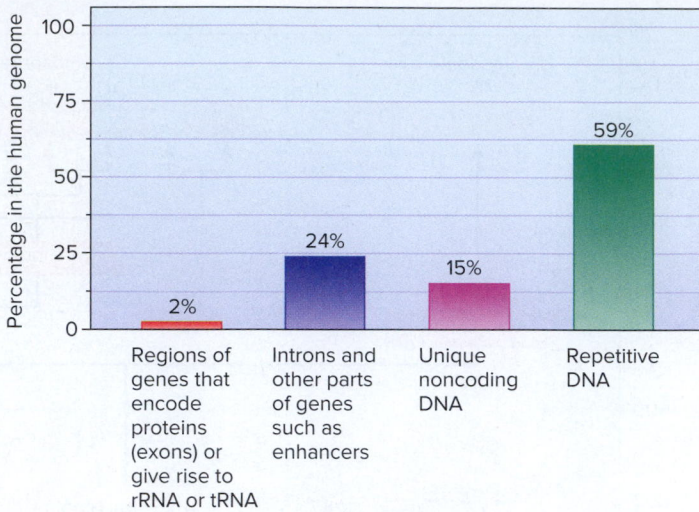

**Figure 21.15  The composition of DNA sequences in the nuclear genome of humans.** Only about 2% of our genome codes for proteins. Most of our genome is made up of repetitive sequences.

**Core Concept: Information**  Only a small percentage of the human genome is involved with encoding the proteins that are largely responsible for human traits.

from a section of a single ancestral gene known as the 7SL RNA gene. Remarkably, over the course of 65 million years, the *Alu* sequence has been copied and inserted into the human genome so often that it now occurs more than 1 million times! The mechanism for the proliferation of *Alu* sequences will be described later in this section.

**Figure 21.15** shows the composition of the classes of DNA sequences found in the nuclear genome of humans. Surprisingly, exons, the coding regions of protein-encoding genes, and the genes that give rise to rRNA and tRNA make up only about 2% of our genome! The other 98% is composed of noncoding sequences. Though we often think of genomes as being the repository of sequences that code for proteins, most eukaryotic genomes are largely composed of other types of sequences. Intron DNA comprises about 24% of the human genome, and unique noncoding DNA constitutes 15%. Repetitive DNA makes up 59% of the DNA in the genome.

Scientists once described noncoding sequences as "junk DNA" because it was believed to have no biological function. However, beginning in 2003, the Encyclopedia of DNA Elements (ENCODE) Project has involved more than 440 researchers in 32 laboratories. The overall goal of the ENCODE Project is to identify functional sequences in the human genome. In 2012, the researchers announced that they were able to assign function to approximately 80% of the human genome. Much of the previously described "junk DNA" appears to play a role in a complex network of gene regulation, a topic discussed in Chapter 14.

## Transposable Elements Move from One Chromosomal Location to Another

As we have seen, repetitive sequences can be moderately or highly repetitive, and they can be found in tandem arrays or interspersed throughout a genome. Some types of repetitive sequences, which are interspersed in genomes, are **transposable elements (TEs)**—DNA segments that can move throughout the genome. The *Alu* sequence discussed earlier is an

**(a) Barbara McClintock**

**(b) Speckled corn kernels caused by transposable elements**

**Figure 21.16** **Barbara McClintock, who discovered transposable elements (TEs).** As shown in part **(b)**, when a TE is found within a pigment gene in corn, its frequent movement disrupts the gene, causing the kernel color to be speckled. a: ©Topham/The Image Works; b: ©Kenneth Keifer/Getty Images

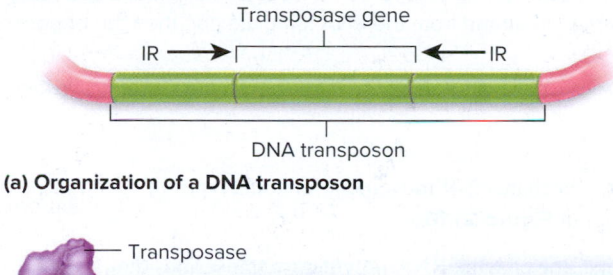

**(a) Organization of a DNA transposon**

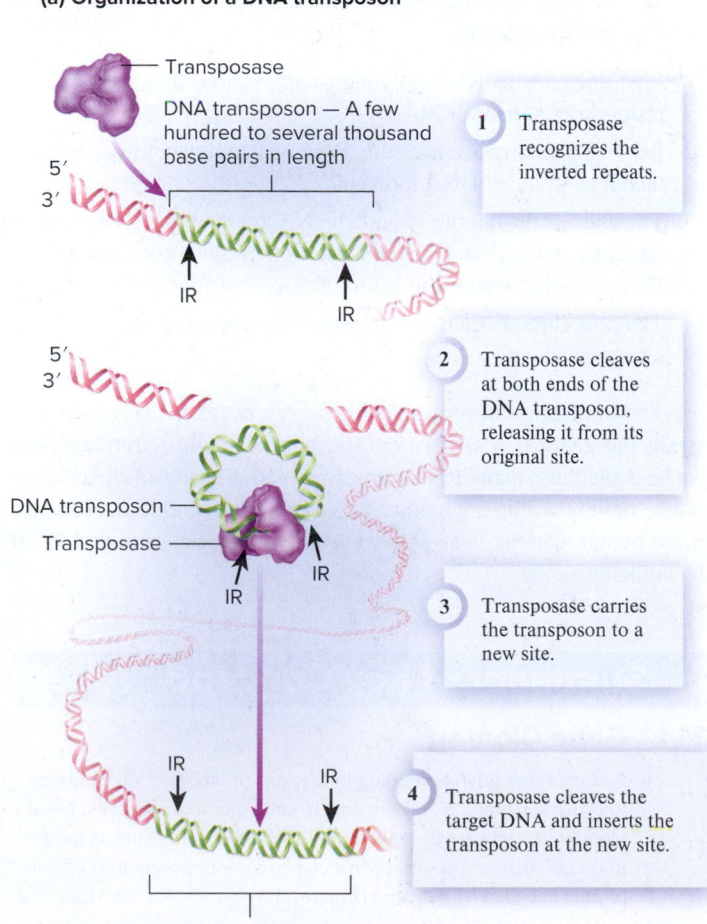

**(b) Cut-and-paste mechanism of transposition**

example of a TE. The process in which a TE moves to a new site in a genome is called **transposition**. TEs range from a few hundred to several thousand base pairs in length. They have sometimes been referred to as "jumping genes," because they are inherently mobile.

American cytogeneticist Barbara McClintock first identified TEs in the late 1940s from her studies with corn plants (**Figure 21.16**). She identified a segment of DNA that could move into and out of a gene that affected the color of corn kernels, producing a speckled appearance. Since that time, biologists have discovered many different types of TEs in nearly all species they have examined.

Though McClintock identified TEs in corn in the late 1940s, her work was met with great skepticism because many researchers had trouble believing that DNA segments could be mobile. The advent of molecular technology in the 1960s and 1970s allowed scientists to understand more about the characteristics of TEs that enable their movement. Most notably, research involving bacterial TEs eventually progressed to a molecular understanding of the transposition process. In 1983, more than 30 years after her initial discovery, McClintock was awarded the Nobel Prize in Physiology or Medicine.

Researchers have studied TEs from many species, including bacteria, archaea, and eukaryotes. They have discovered that TEs fall into two groups—DNA transposons and retrotransposons—based on different mechanisms of movement.

***DNA Transposons*** Transposable elements that move as DNA molecules are called **DNA transposons**. Both ends of DNA transposons have inverted repeats (IRs)—DNA sequences that are identical (or very similar) but run in opposite directions (**Figure 21.17a**), such as the following:

5′–CTGACTCTT–3′          5′–AAGAGTCAG–3′
                    and
3′–GACTGAGAA–5′          3′–TTCTCAGTC–5′

**Figure 21.17** **DNA transposons and their mechanism of transposition.** **(a)** DNA transposons contain inverted repeat (IR) sequences at each end and may contain a gene that encodes transposase in the middle. **(b)** Transposition occurs by a cut-and-paste mechanism.

**Core Skill: Modeling** The goal of this modeling challenge is to predict the outcome of transposition depending on the pattern of DNA replication.

**Modeling Challenge:** As shown below and to the right, a eukaryotic chromosome has a single DNA transposon at a site called site A. Let's suppose that it transposes to site B. Draw 2 models that show the location(s) of the transposon in this chromosome after transposition has occurred and after the chromosome has completely replicated. In your first model, assume that site A has replicated before transposition but site B has not. In your second model, assume that neither site has replicated prior to transposition. Note: Each of your two models will show a pair of sister chromatids. Refer back to Figure 11.13c for a description of chromosome replication in eukaryotes.

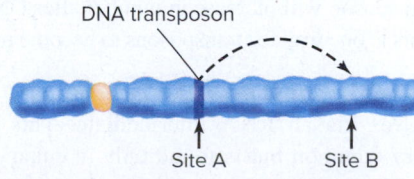

DNA transposon

Site A      Site B

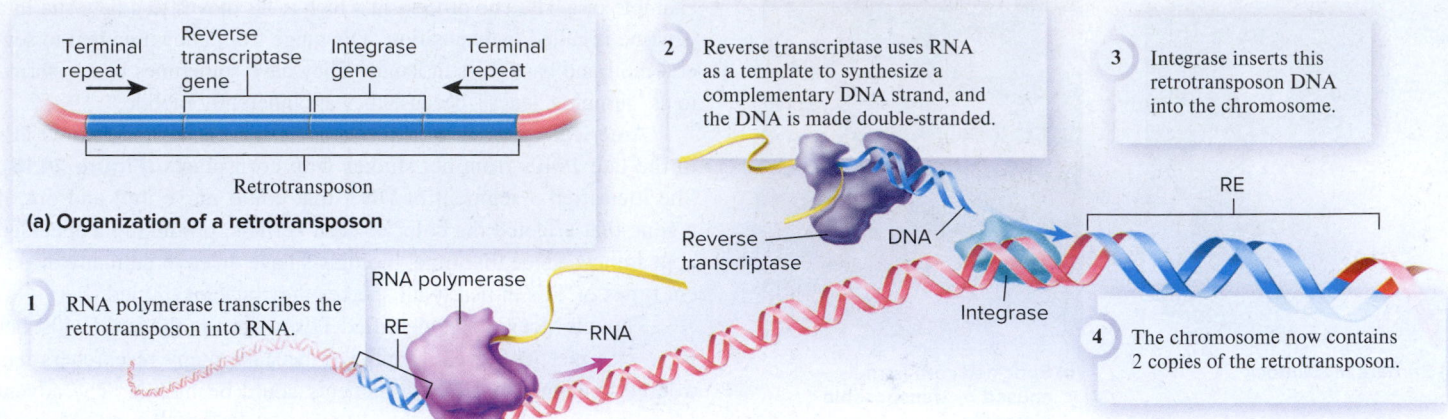

**(a) Organization of a retrotransposon**

Terminal repeat | Reverse transcriptase gene | Integrase gene | Terminal repeat

Retrotransposon

**1** RNA polymerase transcribes the retrotransposon into RNA.

RNA polymerase

RE

RNA

**2** Reverse transcriptase uses RNA as a template to synthesize a complementary DNA strand, and the DNA is made double-stranded.

Reverse transcriptase

DNA

**3** Integrase inserts this retrotransposon DNA into the chromosome.

RE

Integrase

**4** The chromosome now contains 2 copies of the retrotransposon.

**(b) Mechanism of movement of a retrotransposon**

**Figure 21.18 Retrotransposons and their mechanism of transposition.** Retrotransposons are found only in eukaryotic species. **(a)** Some retrotransposons contain terminal repeats and genes that encode the enzymes reverse transcriptase and integrase, which are needed in the transposition process. **(b)** The process that inserts a copy of a retrotransposon into a host chromosome. Note: In addition to using RNA as a template, some forms of reverse transcriptase can also use a DNA template to make a complementary DNA strand. As depicted here, these forms can make double-stranded DNA using a strand of RNA as a starting material. However, some forms of reverse transcriptase can make DNA only using an RNA template. In those cases, reverse transcriptase makes a complementary DNA strand from the RNA template and then the opposite DNA strand is made via a host-cell DNA polymerase.

**Concept Check:** *Based on their mechanism of movement, which type of TE do you think would proliferate more rapidly in a genome, DNA transposons (see Figure 21.17) or retrotransposons?*

Depending on the particular TE, inverted repeats range from 9 to 40 bp in length. In addition, DNA transposons may contain a central region that encodes **transposase**, an enzyme that facilitates transposition.

As shown in **Figure 21.17b**, transposition of DNA transposons occurs by a cut-and-paste mechanism.

1. Transposase first recognizes the inverted repeats (IR) in the transposon.
2. It then cleaves both ends of the DNA transposon and removes it from its original site.
3. Next, the transposase/transposon complex moves to a new location, where transposase cleaves the target DNA and inserts the transposon into the site.

Transposition may occur when a cell is in the process of DNA replication. If a TE is removed from a site that has already replicated and is inserted into a chromosomal site that has not yet replicated, the transposon will increase in number after DNA replication is complete. This is one way for transposons to become more prevalent in a genome.

***Retrotransposons*** Another category of transposable element moves via an RNA intermediate. This form of transposition is very common but is found only in eukaryotic species. These types of TEs are known as **retrotransposons**. The *Alu* sequence in the human genome is an example of a retrotransposon. Some retrotransposons contain genes that encode the enzymes reverse transcriptase and integrase, which are needed in the transposition process (**Figure 21.18a**). Recall from Chapter 19 that reverse transcriptase uses RNA as a template to synthesize a complementary copy of DNA. Retrotransposons may also contain repeated sequences called terminal repeats at each end that facilitate their recognition.

The mechanism of movement for some types of retrotransposons is shown in **Figure 21.18b**.

1. First, the enzyme RNA polymerase transcribes the retrotransposon into RNA.
2. Reverse transcriptase uses this RNA as a template to synthesize a double-stranded DNA molecule.
3. The ends of the double-stranded DNA are then recognized by integrase, which catalyzes the insertion of the retrotransposon DNA into the host chromosomal DNA.
4. The host chromosome now contains two copies of the retrotransposon.

The integration of retrotransposons can occur at many locations within the genome. Furthermore, because a single retrotransposon can be copied into many RNA transcripts, retrotransposons may accumulate rapidly within a genome. This explains how the *Alu* sequence in the human genome was able to proliferate and constitute 10% of the human genome.

## Summary of Key Concepts

### 21.1 Gene Cloning

- Recombinant DNA technology is the use of laboratory techniques to bring together DNA fragments from two or more sources. Gene cloning, the process of making multiple copies of a gene, is used to obtain many copies of a particular gene or large amounts of the protein encoded by the gene (Figure 21.1).

- In one method of gene cloning, both a vector and chromosomal DNA are cut with restriction enzymes. The DNA fragments hydrogen-bond to each other at their sticky ends, and the pieces are covalently linked together via DNA ligase, producing recombinant vectors. When a recombinant vector is introduced into a bacterial cell, the cell replicates the vector and divides to produce many cells (Figures 21.2, 21.3).

- A collection of recombinant vectors, each with a particular piece of chromosomal DNA, is introduced into bacterial cells to produce a DNA library. If the DNA molecules are derived from mRNA, this creates a cDNA (complementary DNA) library (Figure 21.4).

- Gel electrophoresis is used to separate macromolecules by applying an electric current that causes them to move through a gel matrix. Gel electrophoresis typically separates molecules according to their charge, size/length, and mass (Figure 21.5).

- Polymerase chain reaction (PCR) is an alternative technique for gene cloning that does not involve vectors and host cells. Primers that flank the region of DNA to be amplified are required (Figure 21.6).

## 21.2 Genomics: Techniques for Studying and Altering Genomes

- Genomics is the molecular analysis of an entire genome of a species; functional genomics studies the expression of a genome.

- Dideoxy sequencing uses ddNTPs to determine the base sequence of a segment of DNA (Figures 21.7, 21.8).

- A DNA microarray is a small silica, glass, or plastic slide dotted with different sequences of single-stranded DNA, each corresponding to a short sequence within a known gene. It is used to study gene expression patterns (Figure 21.9, Table 21.2).

- CRISPR-Cas technology can be used to edit genes (Figure 21.10).

## 21.3 Bacterial and Archaeal Genomes

- The genome is the complete genetic makeup of a cell, an organism, or a species.

- Bacterial and archaeal genomes are typically a single circular chromosome with a few million base pairs of DNA. Such genomes usually have a few thousand different genes. Bacteria often have plasmids in addition to one or more chromosomes (Table 21.3).

## 21.4 Eukaryotic Genomes

- The nuclear genomes of eukaryotic species are composed of sets of linear chromosomes with a total length of several million to billions of base pairs. They typically contain several thousand to tens of thousands of genes. Genome sizes vary greatly among eukaryotic species (Figure 21.12, Table 21.4).

- One way that the number of genes in a genome can increase is via gene duplication. Gene duplication may occur as a result of a misaligned crossover during meiosis. This is one mechanism that can produce a gene family, two or more homologous genes in a species that have related functions (Figures 21.13, 21.14).

- The Human Genome Project, an international effort to map and sequence the entire human genome, was completed by an international consortium in 2003.

## 21.5 Repetitive Sequences and Transposable Elements

- Repetitive sequences are segments of DNA found in multiple copies in a genome; they can be moderately or highly repetitive. Repetitive DNA can consist of multiple copies of the same gene, tandem arrays, or transposable elements (TEs). It is often abundant in eukaryotic genomes, such as the human genome (Figure 21.15).

- Transposable elements are segments of DNA that can move from one site to another within a genome through a process called transposition (Figure 21.16).

- Transposable elements fall into two groups that move by different molecular mechanisms. DNA transposons move by a cut-and-paste mechanism facilitated by the enzyme transposase. Retrotransposons move to new sites in the genome via RNA intermediates (Figures 21.17, 21.18).

## Assess & Discuss

### Test Yourself

1. Restriction enzymes used in most cloning experiments
   a. are used to cut DNA into pieces for gene cloning.
   b. are produced by bacteria cells to prevent viral infection.
   c. produce sticky ends on DNA fragments.
   d. All of the above are true of restriction enzymes.
   e. Only a and c are true of restriction enzymes.

2. DNA ligase is needed in a cloning experiment
   a. to promote hydrogen bonding between sticky ends.
   b. to covalently link the backbone of DNA strands.
   c. to digest the chromosomal DNA into small pieces.
   d. to do only a and b.
   e. to do a, b, and c.

3. Let's suppose you performed the steps of gene cloning described in Figures 21.2 and 21.3. Which experiment(s) would you conduct to confirm that a white colony really contained a recombinant vector with an insert?
   a. Pick a white bacterial colony and restreak it on plates containing X-Gal to confirm that the cells really form white colonies.
   b. Pick a white bacterial colony, isolate plasmid DNA, digest the plasmid DNA with a restriction enzyme, and then perform gel electrophoresis with the DNA.
   c. Pick a white bacterial colony and test it to see if β-galactosidase is functional within the bacterial cells.
   d. Pick a white bacterial colony and retest it on ampicillin-containing plates to double-check that the cells are really ampicillin resistant.
   e. Both c and d should be conducted.

4. Why is *Taq* polymerase used in PCR rather than other DNA polymerases?
   a. *Taq* polymerase is a synthetic enzyme that produces DNA strands at a faster rate than natural polymerases.
   b. *Taq* polymerase is a heat-stable form of DNA polymerase that can function after exposure to the high temperatures necessary for PCR.
   c. *Taq* polymerase is easier to isolate than other DNA polymerases.
   d. *Taq* polymerase is the DNA polymerase commonly produced by most eukaryotic cells.
   e. All of the above are correct.

5. Let's suppose you want to clone a gene that has never been analyzed before by DNA sequencing. Which of the following statements is the most accurate?
   a. Do PCR to clone the gene because it is much faster.
   b. Do PCR to clone the gene because it is very specific and gives a high yield.
   c. You can't do PCR because you can't make forward and reverse primers.
   d. Do cloning using a vector because it will give you a higher yield.
   e. Do cloning by insertion into a vector because it is easier than PCR.

6. In the CRISPR-Cas technology for editing genes, what is (are) the function(s) of sgRNA?
   a. to bind to the target gene
   b. to bind to Cas9
   c. to cause a double-strand break in the target gene
   d. all of the above
   e. both a and b

7. Which of the following is *not* an important reason for studying the genomes of bacteria and archaea?
   a. It may provide information that helps us understand how bacteria infect other organisms.
   b. It may provide a basic understanding of cellular processes that allows us to determine eukaryotic cellular function.
   c. It may provide the means of understanding evolutionary processes.
   d. It will reveal the approximate number of genes that an organism has in its genome.
   e. All of the above are important reasons.

8. The enzyme that helps short segments of DNA move from one chromosomal location to another is
   a. transposase.
   b. DNA polymerase.
   c. protease.
   d. restriction endonuclease.
   e. DNA ligase.

9. A gene family includes
   a. one specific gene found in several different species.
   b. all of the genes on the same chromosome.
   c. two or more homologous genes found within a single species.
   d. genes that code for structural proteins.
   e. both a and c.

10. Which of the following was *not* a goal of the Human Genome Project?
    a. identify all human genes
    b. sequence the entire human genome
    c. address the legal and ethical implications resulting from the project
    d. develop programs to manage the information gathered from the project
    e. be able to clone a human

## Conceptual Questions

1. Draw the structure of a dideoxyribonucleotide triphosphate, and explain how it causes chain termination.

2. Briefly describe whether or not each of the following can be appropriately described as a genome.
   a. the *E. coli* chromosome
   b. human chromosome 11
   c. a complete set of 10 chromosomes in corn
   d. a copy of the single-stranded RNA packaged into the human immunodeficiency virus (HIV)

3. **Core Concept: Information** Explain the role of the genome at the molecular, cellular, and organism levels.

## Collaborative Questions

1. Identify and discuss three important advances that have resulted from gene cloning.

2. Compare and contrast the characteristics of the genomes of bacteria, archaea, and eukaryotes.

# UNIT IV
# EVOLUTION

Evolution is a heritable change in one or more characteristics of a population from one generation to the next. This process not only alters the characteristics of populations, it also leads to the formation of new species.

We will begin this unit by considering the fundamental concepts of evolution, with an emphasis on natural selection (Chapter 22). We will examine observations of evolutionary change, which includes (1) the fossil record, (2) a comparison of the characteristics of modern species, and (3) an analysis of molecular data. Chapter 23 continues our discussion of evolution at the molecular level and focuses on how changes in allele and genotype frequencies from one generation to the next are driven by a variety of different factors. By comparison, Chapter 24 shifts the emphasis of evolution to the level of species. We will examine how species are identified and discuss the mechanisms by which new species arise via evolution. In Chapter 25, we will examine how biologists determine the evolutionary relationships among different species and produce diagrams called evolutionary trees to describe those relationships. Finally, in Chapter 26, we will examine a timeline for the evolution of species from 4 billion years ago to the present and consider the topic of human evolution.

**The following Core Concepts and Core Skills will be emphasized in this unit:**

- *Evolution:* This concept will be emphasized throughout the entire unit.
- *Information:* As discussed in Chapters 22 and 23, evolution involves changes in genes.
- *Systems:* Living organisms interact with their environment. As discussed in Chapters 22 through 24, natural selection is a process in which certain individuals have greater reproductive success. This success is often due to their ability to survive in a given environment.
- *Structure and Function:* Chapters 22 and 24 will also consider how structural features change during the evolution of new species. Such changes are related to changes in function.
- *Process of Science:* Every chapter has a Feature Investigation describing a pivotal experiment that provided insights into our understanding of evolution.
- *Quantitative Reasoning:* Chapter 23 focuses on changes in allele and genotype frequencies in populations.
- *Modeling:* Every chapter has a Modeling Challenging to help you refine this important skill.

# An Introduction to Evolution

# 22

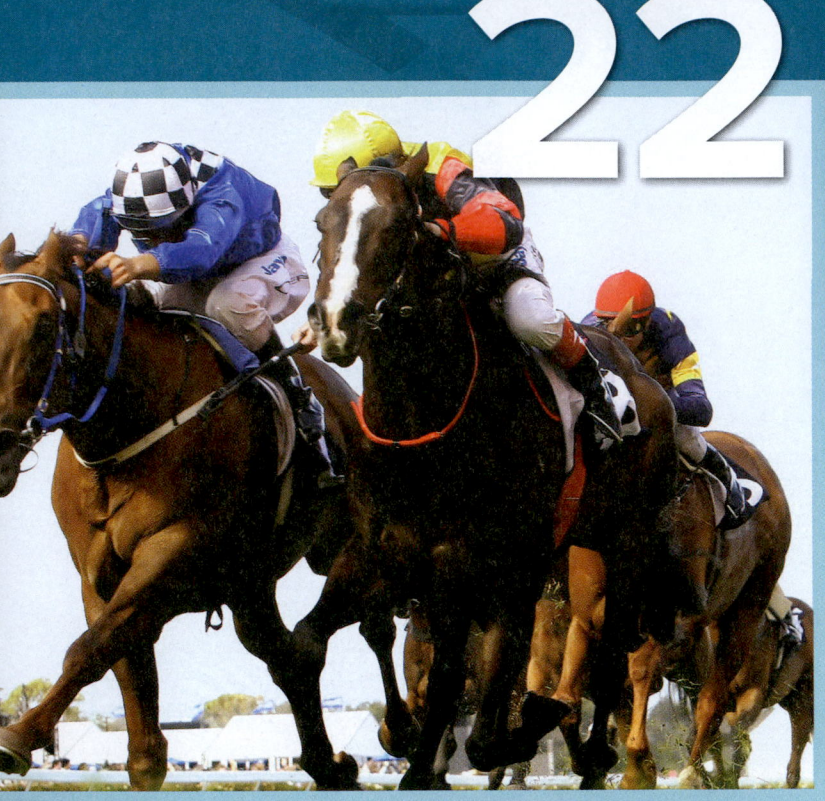

**Selective breeding.** The horses in this race have been bred for a particular trait: speed. Such a practice, called selective breeding, can dramatically change the traits of organisms over several generations. ©Mark Dadswell/Getty Images

Organic life beneath the shoreless waves
Was born and nurs'd in Ocean's pearly caves
First forms minute, unseen by spheric glass,
Move on the mud, or pierce the watery mass;
These, as successive generations bloom,
New powers acquire, and larger limbs assume;
Whence countless groups of vegetation spring,
And breathing realms of fin, and feet, and wing.

By Erasmus Darwin, grandfather of Charles Darwin.
Published posthumously in 1803.

The term **evolution** is used to describe a heritable change in one or more characteristics of a population from one generation to the next. Evolution can be viewed on different scales. **Microevolution** involves changes in a single gene or in allele frequencies in a population over time. **Macroevolution** refers to the formation of a new species or groups of related species.

For our discussion of evolution, let's begin with a working definition of a species. Biologists often define a **species** as a group of related organisms that share a distinctive form. Among species that reproduce sexually, such as plants and animals, members of the same species are capable of interbreeding in nature to produce viable and fertile offspring. The term **population** refers to all members of a species that live in the same area at the same time and have the opportunity to interbreed. As we will see in Chapter 24, some of the emphasis in the study of evolution is on understanding how populations change over the course of many generations to produce new species.

In the first section of this chapter, we will examine the development of evolutionary thought and some of the basic tenets of evolution, particularly those proposed by the British naturalist Charles Darwin in the mid-1800s. The theory of evolution has been refined over the past 170 years or so, but the fundamental principle of evolution remains unchanged and has provided a cornerstone for our understanding of biology. Ukrainian-born American geneticist Theodosius Dobzhansky, an influential evolutionary scientist of the 1900s, once said, "Nothing in biology makes sense except in the light of evolution." The extraordinarily diverse and often seemingly bizarre array of species on our planet can be explained within the context of evolution. As is the case with all scientific theories, evolution is called a theory because it is supported by a substantial body of evidence and because it explains a wide range of observations. The theory of evolution provides answers to many questions related to the diversity of life. In biology, theories such as this are viewed as scientific knowledge.

In the second section of this chapter, we will survey the extensive data that illustrate the processes by which evolution occurs. These data not only support the theory of evolution but also allow us to understand the interrelatedness of different species, whose similarities are often due to descent from a common ancestor. Much of the early evidence supporting evolution came from direct observations and comparisons of living and extinct species. More recently, advances in molecular genetics, particularly those related to DNA sequencing and genomics, have revolutionized the study of evolution. Scientists now have information that allows us to understand how evolution involves changes in the DNA sequences of a given species. These changes affect both a species' genes and the proteins they encode.

**Molecular evolution** refers to the process of evolution at the level of genes and proteins. Comparisons of gene or protein sequences in different organisms can reveal evolutionary relationships that cannot be seen in morphology. A major focus of this textbook is to provide an understanding of these changes. In the last section of this chapter, we consider some of the exciting new ways of exploring evolutionary change at the molecular level. In the following chapters of this unit, we will examine how such changes are acted upon by evolutionary factors in ways that alter the traits of a given species and may eventually lead to the formation of new species.

## 22.1 Overview of Evolution

**Learning Outcomes:**

1. Define evolution.
2. Describe the factors that led Darwin to the theory of evolution.
3. Explain the process of natural selection.

Undoubtedly, the question "Where did we come from?" has been asked and debated by people for thousands of years. Many of the early ideas regarding the existence of living organisms were strongly influenced by religion and philosophy. Some of these ideas suggested that all forms of life have remained the same since their creation. In the 1600s, however, scholars in Europe began a revolution that created the basis of empirical and scientific thought. **Empirical thought** relies on observation to form an idea or hypothesis rather than trying to understand life from a nonphysical or spiritual point of view. As described in this section, the shift toward empirical thought encouraged scholars to look for the basic rationale behind a given process or phenomenon. This perspective played a key role in the development of the theory of evolution.

### The Work of Several Scientists Set the Stage for Darwin's Ideas

In the mid- to late-1600s, the first scientist to carry out a thorough study of the living world was an English naturalist named John Ray, who developed an early classification system for plants and animals based on anatomy and physiology. He established the modern concept of a species, noting that organisms of one species do not interbreed with members of another, and used it as the basic unit of his classification system. Ray's ideas on classification were later expanded by the Swedish naturalist Carolus Linnaeus. How did their work contribute to the development of evolutionary theory? Neither Ray nor Linnaeus proposed that evolutionary change promotes the formation of new species. However, their systematic classification of plants and animals helped scholars of this period perceive the similarities and differences among living organisms.

Late in the 1700s, a small number of European scientists began to quietly suggest that life-forms are not fixed and unchanging. A French zoologist, Georges Buffon, actually proposed that populations of living things change through time. However, Buffon was careful to hide his views in a 44-volume series of books on natural history. Around the same time, a French naturalist named Jean-Baptiste Lamarck suggested an intimate relationship between variation and evolution. By examining fossils, he realized that some species had remained the same over the millennia and others had changed. Lamarck hypothesized that species change over the course of many generations by adapting to new environments. He believed that living things evolved in a continuously upward direction, from dead matter, through simple to more complex forms, toward "human perfection." According to Lamarck, organisms altered their behavior in response to environmental change. He thought that behavioral changes could modify traits and hypothesized that these modified traits were inherited by offspring. He called this idea the inheritance of acquired characteristics. For example, according to Lamarck's hypothesis, giraffes developed their elongated necks and front legs by feeding on the leaves at the top of trees. The exercise of stretching up to the leaves altered the neck and legs, and Lamarck presumed that these acquired characteristics were transmitted to offspring. However, further research has rejected Lamarck's idea that most acquired traits can be inherited. (Note: An acquired trait can sometimes be transmitted from parent to offspring via epigenetic changes, which are described in Chapter 18.) Lamarck's work was important in promoting the idea of evolutionary change.

Interestingly, Erasmus Darwin, the grandfather of Charles Darwin, was a contemporary of Buffon and Lamarck and an early advocate of evolutionary change. He was a physician, a plant biologist, and also a poet (see the poem at the beginning of the chapter). He was aware that modern species were different from similar types of fossilized organisms, and he noted how plant and animal breeders used breeding practices to change the traits of domesticated species (see the chapter opening photo). He knew that offspring inherited features from their parents and went so far as to say that life on Earth could have descended from a common ancestor.

### Darwin Suggested That Existing Species Are Derived from Pre-existing Species

Charles Darwin played a central role in developing the theory that existing species have evolved from pre-existing ones. Darwin's unique perspective and his ability to formulate evolutionary theory were shaped by several different fields of study, including ideas of his time about geology and population growth, as well as his own observations.

*Hypotheses about Geology*  Two main hypotheses about geological processes predominated in the early 19th century. Catastrophism was first proposed by French zoologist and paleontologist Georges Cuvier to explain the age of the Earth. Cuvier suggested that the Earth is just 6,000 years old and that only catastrophic events have changed its geological structure. This idea fit well with certain religious teachings. Alternatively, uniformitarianism, proposed by Scottish geologist James Hutton and popularized by another Scottish geologist, Charles Lyell, suggested that changes in the Earth are directly caused by recurring events. For example, they suggested that geological processes such as erosion existed in the past and happened at the same gradual rate as they do now. For such slow geological processes to eventually lead to substantial changes in the Earth's characteristics, a great deal of time is required. Hutton and Lyell were the first to propose that the age of the Earth is well beyond 6,000 years. The ideas of Hutton and Lyell helped to shape Darwin's view of the world.

**Population Growth** Darwin's thinking was also influenced by a paper published in 1798 called *An Essay on the Principle of Population* by Thomas Malthus, an English economist. Malthus asserted that the population size of humans can, at best, increase linearly due to increased land usage and improvements in agriculture, whereas our reproductive potential is exponential (for example, doubling with each generation). He argued that famine, war, and disease, especially among the poor, keep population growth within existing resources. The relevant message from Malthus's work was that not all members of any population will survive and reproduce.

**Voyage of the Beagle** Darwin's ideas, however, were most influenced by his own experiences and observations. His work as a young man aboard the HMS *Beagle*, a survey ship, lasted from 1831 to 1836 and involved a careful examination of many different species (**Figure 22.1**). The main mission of the *Beagle* was to map the coastline of southern South America and take oceanographic measurements. As the ship's naturalist, Darwin's job was to record information about the weather, geological features, plants, animals, fossils, rocks, minerals, and indigenous people.

Though Darwin made many interesting observations on his journey, he was particularly struck by the distinctive traits of island species. For example, Darwin observed several species of finches found on the Galápagos Islands, a group of volcanic islands 600 miles from the coast of Ecuador. Though it is often assumed that Darwin's personal observations of these finches directly inspired his theory of evolution, this is not the case. Initially, Darwin thought the birds were various species of blackbirds, grosbeaks, and finches. Later, however, the bird specimens from the islands were given to the British ornithologist John Gould, who identified them as several new finch species. Gould's observations helped Darwin in the later formulation of his theory.

As seen in **Table 22.1**, the finches differed widely in the size and shape of their beaks and in their feeding habits. Darwin clearly saw the similarities among these species, yet he noted the differences that provided them with specialized feeding strategies. We now know that these finches all evolved from a single species similar to the dull-colored grassquit finch (*Tiaris obscura*), commonly found along the Pacific Coast of South America. Once they arrived on the Galápagos Islands, the finches' ability to obtain particular types of food in their new habitat depended, in part, on the relative sizes and shapes of their beaks, which, in turn, influenced their abilities to survive and reproduce.

With an understanding of geology and population growth, and his observations from his voyage on the *Beagle*, Darwin had formulated his theory of evolution by the mid-1840s. He had also catalogued and described all of the species he had collected on his *Beagle* voyage except for one type of barnacle. Some have speculated that Darwin may have felt that he should establish himself as an expert on one species before making generalizations about all of them. Therefore, he spent several additional years studying barnacles. During this time, the geologist Charles Lyell, who had greatly influenced Darwin's thinking, strongly encouraged Darwin to publish his theory of evolution. In 1856, Darwin began to write a long book to explain his

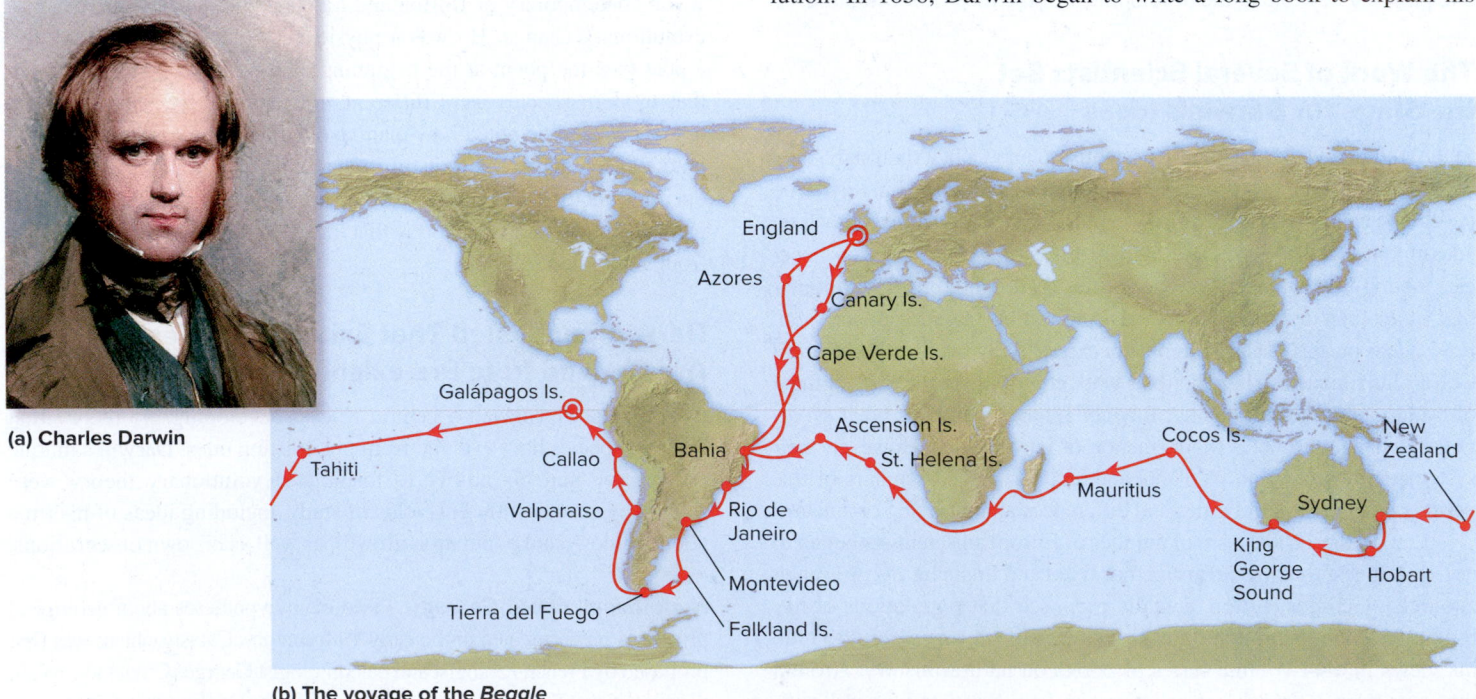

**(a) Charles Darwin**

**(b) The voyage of the Beagle**

**Figure 22.1** **Charles Darwin and the voyage of the *Beagle*, 1831–1836.** **(a)** A portrait of Charles Darwin (1809–1882) at age 31. **(b)** Darwin's voyage on the *Beagle*, which took almost 5 years to circumnavigate the world. a: ©GraphicaArtis/Archive Photos/Getty Images

| Table 22.1 | A Comparison of Beak Type and Diet Among the Galápagos Finches That Darwin Studied | |
|---|---|---|
| **Type of finch/diet** | **Species** | **Type of beak** |
| **Ground finches**<br><br>Ground finches have beaks shaped to crush various sizes of seeds; large beaks can crush large seeds, whereas smaller beaks are better for crushing small seeds. | Large ground finch (*Geospiza magnirostris*) | Crushing |
| | Medium ground finch (*G. fortis*) | |
| | Small ground finch (*G. fuliginosa*) | |
| | Sharp-billed ground finch (*G. difficilis*) | |
| **Vegetarian finch**<br><br>Vegetarian finches have crushing beaks to pull buds from branches. | Vegetarian finch (*Platyspiza crassirostris*) | Crushing |
| **Tree finches**<br><br>Tree finches have grasping beaks to pick insects from trees. Those with heavier beaks can also break apart wood in search of insects. | Large tree finch (*Camarhynchus psittacula*) | Grasping |
| | Medium tree finch (*Camarhynchus pauper*) | |
| | Small tree finch (*Camarhynchus parvulus*) | |
| **Tree and warbler finches**<br><br>These finches have probing beaks to search for insects in crevices and then pick them up. The woodpecker finch can also use a cactus spine for probing. | Mangrove finch (*Cactospiza heliobates*) | Probing |
| | Woodpecker finch (*Camarhynchus pallidus*) | |
| | Warbler finch (*Certhidea olivacea*) | |
| **Cactus finches**<br><br>Cactus finches have probing beaks to open cactus fruits and take out seeds. | Large cactus finch (*G. conirostris*) | Probing |
| | Cactus finch (*G. scandens*) | |

ideas. In 1858, however, Alfred Wallace, a British naturalist working in the East Indies, sent Darwin an unpublished manuscript to read prior to its publication. In it, Wallace proposed the same ideas concerning evolution. In response to this, Darwin decided to use some of his own writings on this subject, and two papers, one by Darwin and one by Wallace, were published in the *Proceedings of the Linnaean Society of London*. These papers were not widely recognized. A year later, however, Darwin finished his book *On the Origin of Species* (1859), which described his ideas in greater detail and included observational support. Although some of his ideas were incomplete because the genetic basis of traits was not understood at that time, Darwin's work remains a foundation of our understanding of biology.

## Natural Selection Changes Populations from Generation to Generation

Darwin hypothesized that existing life-forms on our planet result from the modification of pre-existing life-forms. He expressed this concept of evolution as "the theory of descent with modification through variation and natural selection." The term evolution refers to change. What factors bring about evolutionary change? According to Darwin's ideas, evolution occurs from generation to generation due to two interacting factors, genetic variation and natural selection:

1. Variation in traits may occur among individuals of a given species. The heritable traits are then passed from parents to offspring. The genetic basis for variation within a species was not understood at the time Darwin proposed his theory of evolution. We now know that such variation is due to different types of genetic changes such as random mutations in genes. Even though Darwin did not fully appreciate the genetic basis of variation, he and many other people before him observed that offspring resemble their parents more than they do unrelated individuals. Therefore, he assumed that some traits are passed from parent to offspring.

2. In each generation, many more offspring are usually produced than will survive and reproduce. Often, resources in the environment are limiting for an organism's survival. During the process of **natural selection**, individuals with heritable traits that make them better suited to their native environment tend to flourish and reproduce, whereas other individuals are less likely to survive and reproduce. As a result of natural selection, certain traits that favor reproductive success become more prevalent in a population over time.

As an example, we can consider a population of finches that migrates from the South American mainland to a distant island (**Figure 22.2**). Variation exists in the beak sizes among the migrating birds. Let's suppose the seeds produced on the distant island are larger than those produced on the mainland. Those birds with larger beaks would be better able to feed on these larger seeds and therefore would be more likely to survive and pass that trait to their offspring. What are the consequences of this selection process? In succeeding generations, the population tends to have a greater proportion of finches with larger beaks. Alternatively, if a trait happens to be detrimental to an individual's ability to survive and reproduce, natural selection is likely to eliminate this type of variation. For example, if a finch in the same environment had a small beak, this bird would be less likely to acquire enough food, which would decrease its ability

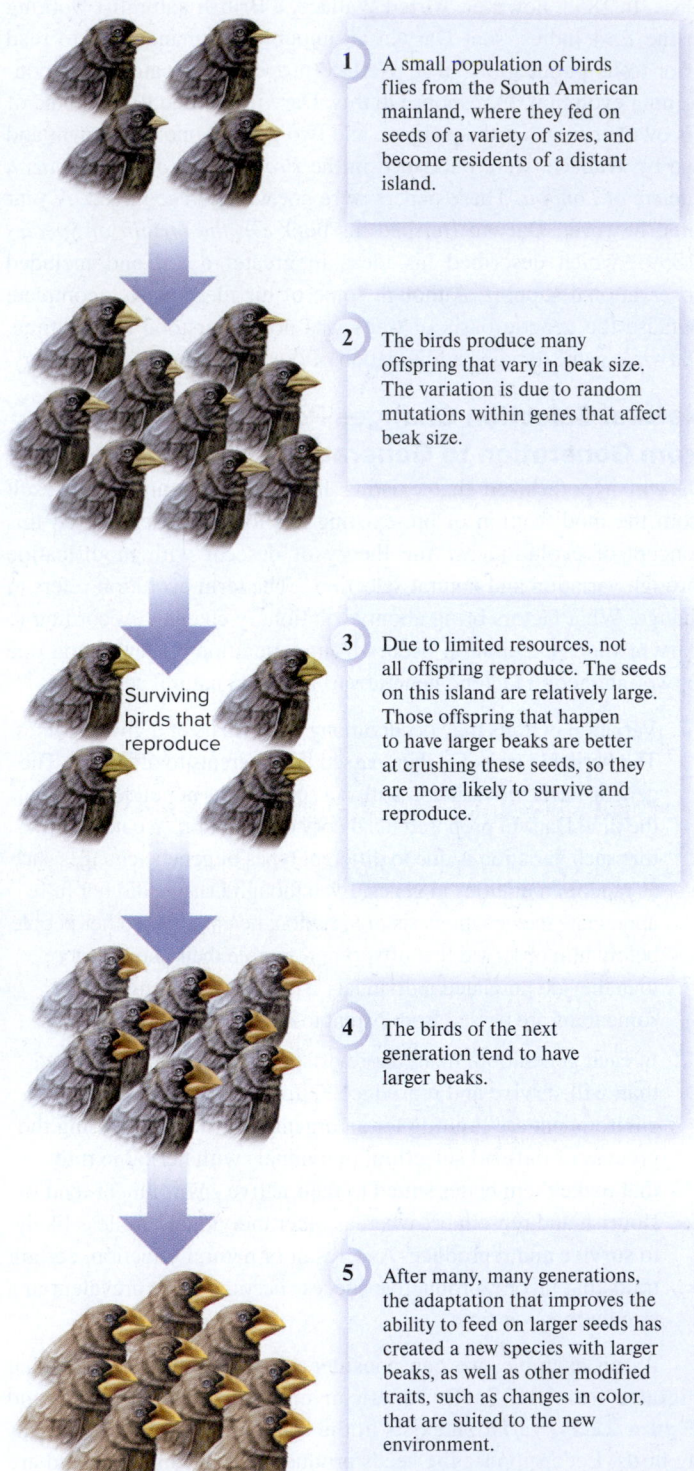

**Figure 22.2 Evolutionary adaptation to a new environment via natural selection.** The example shown here involves a species of finch adapting to a new environment on a distant island. According to Darwin's theory of evolution, the process of adaptation can lead to the formation of a new species with traits that are better suited to the new environment.

*Concept Check:* The phrase "an organism evolves" is incorrect. Explain why.

1. A small population of birds flies from the South American mainland, where they fed on seeds of a variety of sizes, and become residents of a distant island.

2. The birds produce many offspring that vary in beak size. The variation is due to random mutations within genes that affect beak size.

3. Due to limited resources, not all offspring reproduce. The seeds on this island are relatively large. Those offspring that happen to have larger beaks are better at crushing these seeds, so they are more likely to survive and reproduce.

Surviving birds that reproduce

4. The birds of the next generation tend to have larger beaks.

5. After many, many generations, the adaptation that improves the ability to feed on larger seeds has created a new species with larger beaks, as well as other modified traits, such as changes in color, that are suited to the new environment.

to survive and pass this trait to its offspring. Natural selection may ultimately result in a new species with a combination of multiple traits that are quite different from those of the original species, such as finches with larger beaks and changes in coloration. In other words, the newer species has evolved from a pre-existing one.

**BIO TIPS**

**THE QUESTION** *Antibiotics are medicines used to treat various types of bacterial infections. Examples include streptomycin, tetracycline, and amoxicillin. These drugs inhibit bacterial growth by interfering with processes such as the synthesis of the cell wall, bacterial translation, or vital metabolic pathways. When a strain of bacteria that was originally sensitive to an antibiotic is no longer sensitive, such a strain is said to be antibiotic resistant. Resistant to antibiotics can arise in two common ways:*

- *A new mutation in a bacterium may render the antibiotic ineffective. For example, suppose an antibiotic blocks bacterial growth by binding to a protein that is needed for cell wall synthesis and inhibiting its function. A mutation could occur in the bacterial gene that encodes this protein, altering the protein's structure in a way that prevents the antibiotic from binding.*

- *A bacterium may acquire a gene that confers antibiotic resistance via horizontal gene transfer, which is discussed later in this chapter. For example, a bacterium may acquire a gene that encodes an enzyme that breaks down the antibiotic, thereby rendering it ineffective.*

*The bacterium Staphylococcus aureus causes skin infections commonly called staph infections. One antibiotic that has been used to treat staph infections is methicillin. However, resistance to this antibiotic has become a widespread problem. The graph below shows the percentage of S. aureus strains that became resistant to methicillin over a 20-year period in the United States. As you may have heard, methicillin-resistant S. aureus strains are called MRSA (pronounced "mersa"). Explain these data. Is evolution occurring? If so, propose a hypothesis to explain why S. aureus populations are evolving with regard to antibiotic resistance.*

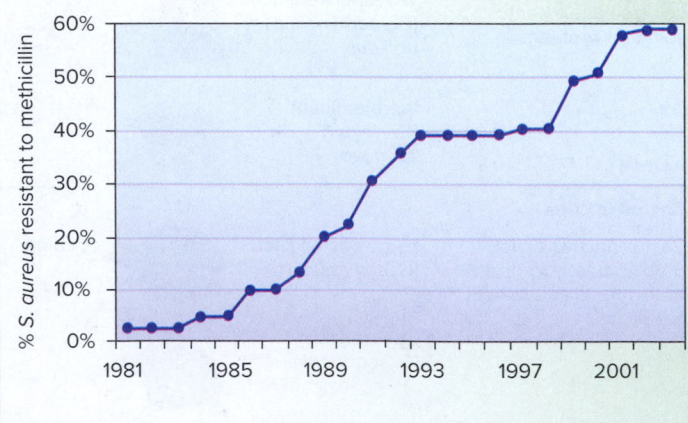

**T**OPIC  *What topic in biology does this question address?* The topic is antibiotic resistance. More specifically, the question asks you to analyze data regarding resistance to methicillin exhibited by *S. aureus*.

**I**NFORMATION  *What information do you know based on the question and your understanding of the topic?* From the question, you have learned what antibiotics are and how antibiotic resistance can arise. You are also given data regarding methicillin resistance in *S. aureus* over a 20-year period. From your understanding of the topic, you may remember that evolution involves heritable changes in a population that are passed from one generation to the next. Natural selection can facilitate evolution.

**P**ROBLEM-SOLVING **S**TRATEGY  *Interpret data. Propose a hypothesis.* To solve this problem, you can start by observing the trend in the data over time. As you can see in the graph, the level of methicillin resistance dramatically increased over a 20-year

period. The trait that is measured by these data is the ability of *S. aureus* to survive when exposed to methicillin. The population in 2001 had a much higher percentage of bacteria that survived and reproduced when exposed to methicillin compared to the population in 1981.

**ANSWER** *Evolution is occurring. The U.S. population of S. aureus changed over a 20-year period such that more of these bacteria are resistant to methicillin compared to the original population. One hypothesis to explain these data is natural selection. In the original population in 1981, relatively few bacteria were resistant. However, a very small percentage of them may have been resistant due to a new mutation or horizontal gene transfer. The use of methicillin to treat staph infections may have selected for the survival of these MRSA strains. Therefore, over the course of many bacterial generations, MRSA strains became more common.*

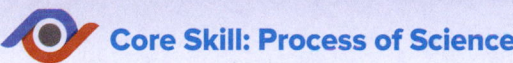

**Core Skill: Process of Science**

# Feature Investigation | The Grants Observed Natural Selection in Galápagos Finches

Since 1973, British evolutionary biologists Peter Grant, Rosemary Grant, and their colleagues have studied natural selection in finches found on the Galápagos Islands. For over 40 years, the Grants have focused much of their work on one of the Galápagos Islands known as Daphne Major (**Figure 22.3a**). This small island (0.34 km²) has a moderate degree of isolation (it is 8 km from the nearest island), an undisturbed habitat, and a resident population of *Geospiza fortis*, the medium ground finch (**Figure 22.3b**).

To study natural selection, the Grants observed various traits in finches over the course of many years. One trait they observed is beak size. The medium ground finch has a relatively small crushing beak, allowing it to more easily feed on small, tender seeds (see Table 22.1). The Grants quantified beak size among the medium ground finches of Daphne Major by carefully measuring beak depth—a measurement of the beak from top to bottom (**Figure 22.4**). The small size of the island made it possible for them to measure a large percentage of birds and their offspring. During the course of their studies, they compared the beak depths of parents and offspring by examining many broods over several years and found that the depth of the beak was transmitted from parents to offspring, regardless of environmental conditions, indicating that differences in beak depths are due to genetic differences in the population. In other words, they found that beak depth was a heritable trait.

By measuring many birds every year, the Grants were able to assemble a detailed portrait of natural selection in action. In the study shown in Figure 22.4, they measured beak depth from 1976 to 1978. In the wet year of 1976, the plants of Daphne Major produced an abundance of the small, tender seeds that these finches could easily eat. However, a severe drought occurred in 1977. During this year, the plants on Daphne Major tended to produce few of the smaller seeds, which the finches

**(a) Daphne Major**  **(b) Medium ground finch**

**Figure 22.3**  **The Grants' investigation of natural selection in finches. (a)** Daphne Major, one of the Galápagos Islands. **(b)** One of the medium ground finches (*Geospiza fortis*) that populate this island.
a: ©Worldwide Picture Library/Alamy Stock Photo; b: ©Ralph Lee Hopkins/Getty Images

 **Core Concept: Evolution**  This study was aimed at analyzing how beak size may change from one generation to the next.

rapidly consumed. Therefore, the finches resorted to eating larger, drier seeds, which are harder to crush. As a result, birds with larger beaks were more likely to survive and reproduce because they were better at breaking open the large seeds. As shown in the data, the average beak depth of birds in the population increased substantially, from 8.8 mm in predrought offspring to 9.8 mm in postdrought offspring.

How do we explain these results? According to evolutionary theory, birds with larger beaks were more likely to survive and pass this trait to their offspring. Overall, these results illustrate the

**Figure 22.4** **The Grants and natural selection of beak size in the medium ground finch.**

**HYPOTHESIS** Dry conditions produce larger seeds and may result in more birds with larger beaks in succeeding generations of *Geospiza fortis* due to natural selection.

**KEY MATERIALS** A population of *G. fortis* on the Galápagos Island called Daphne Major.

| | Experimental level | | Conceptual level |
|---|---|---|---|

**1** In 1976, measure beak depth in parents and offspring of the species *G. fortis*.

Capture birds and measure beak depth.

This is a way to measure a trait that may be subject to natural selection.

**2** Repeat the procedure on offspring that were born in 1978 and had reached mature size. A drought had occurred in 1977 that caused plants on the island to produce mostly large dry seeds and relatively few small seeds.

Capture birds and measure beak depth.

This is a way to measure a trait that may be subject to natural selection.

**3** **THE DATA**

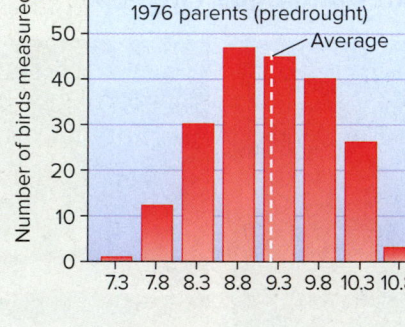
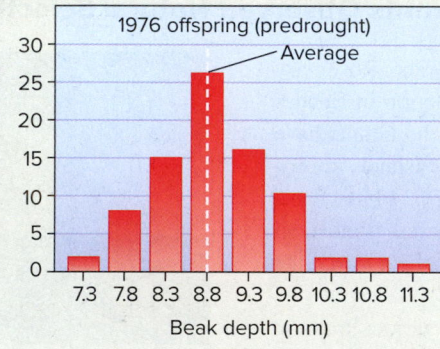
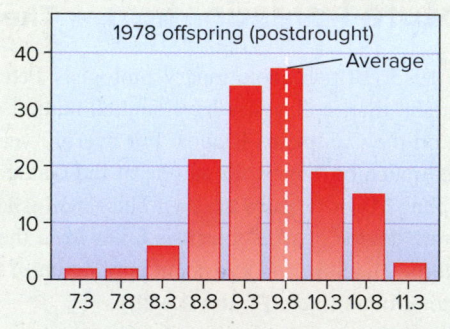

Beak depth (mm)

**4** **CONCLUSION** Because a drought produced larger seeds, birds with larger beaks were more likely to survive and reproduce. The process of natural selection produced postdrought offspring that had larger beaks compared to predrought offspring.

**5** **SOURCE** Grant, B. R., and Grant, P. R. 2003. What Darwin's finches can teach us about the evolutionary origin and regulation of biodiversity. *Bioscience* 53: 965–975.

power of natural selection to alter the features of a trait—in this case, beak depth—in a given population from one generation to the next.

### Experimental Questions

**1.** What features of Daphne Major made it a suitable field site for studying the effects of natural selection?

**2. CoreSKILL »** Why is beak depth in finches a good trait for a study of natural selection? What environmental conditions were important to allowing the Grants to collect information concerning natural selection?

**3. CoreSKILL »** Analyze the results of the Grants' study, and explain what they mean with regard to natural selection and evolution.

## 22.2 Evidence of Evolutionary Change

**Learning Outcomes:**

1. Summarize the different types of evidence for evolutionary change, including studies of natural selection, the fossil record, biogeography, convergent traits, selective breeding, and homologies.
2. Provide examples of three types of homologies.

Evidence that supports the theory of evolution has been gleaned from many sources (**Table 22.2**). As we have already seen, the Grants were able to observe changes in a finch population as a result of a drought. Historically, the first evidence of biological evolution came from studies of the fossil record, the distribution of related species on our planet, selective breeding experiments, and the comparison of similar anatomical features in different species. More recently, additional evidence that illustrates the process of evolution has been found at the molecular level. By comparing DNA sequences from many different species, evolutionary biologists have gained great insight into

| Table 22.2 | Evidence of Biological Evolution |
|---|---|
| **Type of evidence** | **Description** |
| Studies of natural selection | By following the characteristics of populations over time, researchers have observed how natural selection alters populations in response to environmental changes (see Figure 22.4). |
| Fossil record | When fossils are compared according to their age, from oldest to youngest, successive evolutionary change becomes apparent. |
| Biogeography | Unique species found on islands and other remote areas have arisen because the species in these locations have evolved in isolation from the rest of the world. |
| Convergent evolution | Two different species from different lineages sometimes become anatomically similar because they occupy similar environments. This indicates that natural selection results in adaptation to a given environment. |
| Selective breeding | The traits in domesticated species have been profoundly modified by selective breeding (also called artificial selection) in which breeders choose the parents that have desirable traits. |
| Homologies | |
| Anatomical | Homologous structures are structures that are anatomically similar to each other because they evolved from a structure in a common ancestor. In some cases, such structures have lost their original function and become vestigial. |
| Developmental | An analysis of embryonic development often reveals similar features that point to past evolutionary relationships. |
| Molecular | At the molecular level, certain characteristics are found in all living cells, suggesting that all living species are derived from an interrelated group of common ancestors. In addition, species that are closely related evolutionarily have DNA sequences that are more similar to each other than they are to the DNA sequences of distantly related organisms. |

the relationship between the evolution of species and the associated changes in the genetic material. In this section, we will survey the various types of evidence that show the process of evolutionary change.

### Fossils Show Successive Evolutionary Change

As discussed in Chapter 26, fossils are the preserved remains of past life. The fossil record has provided biologists with evidence of the history of life on Earth. Today, scientists have access to a far more extensive fossil record than was available to Darwin and other scientists of his time. Even though the fossil record is still incomplete, the many fossils that have been discovered provide detailed information regarding evolutionary change in a series of related organisms. When fossils are compared according to their age, from oldest to youngest, successive evolutionary change becomes apparent. Let's consider a couple of examples in which paleontologists have observed evolutionary change.

***Evolution of Terrestrial Tetrapods from Fish*** In 2005, fossils of *Tiktaalik roseae*, nicknamed fishapod, were discovered by paleontologists Ted Daeschler, Neil Shubin, and Farish Jenkins. The discovery of the fishapod illuminated one of several steps that led to the evolution of tetrapods, which are animals with four legs. *T. roseae* is called a **transitional form** because it displays an intermediate state between an ancestral form and the form of its descendants (**Figure 22.5**). In this case, the fishapod is a transitional form between fishes, which have fins for locomotion, and tetrapods, which are four-limbed animals. Unlike a true fish, *T. roseae* had a broad skull, a flexible neck, and eyes mounted on the top of its head like a crocodile. Its interlocking rib cage suggests it had primitive lungs. Perhaps the most surprising discovery was that its pectoral fins (those on the side of the body) revealed the beginnings of a primitive wrist and five finger-like bones. These appendages would have allowed *T. roseae* to support its body on shallow river bottoms and lift its head above the water to search for prey and perhaps even move out of the water for short periods. During the Devonian period (417–354 mya), this could have been an important advantage in the marshy floodplains of large rivers.

The early tetrapods gave rise to three main groups of modern tetrapods:

- Amphibians
- Birds and modern reptiles
- Mammals

Most tetrapods occupy a terrestrial environment. However, some, such as frogs and hippopotamuses, are semi-aquatic, spending a large amount of time in the water. Interestingly, tetrapod evolution has produced new species that are aquatic and no longer have hindlimbs. For example, whales evolved from terrestrial mammals, as described next.

***Evolution of Whales from Terrestrial Mammals*** Cetacea is an order of aquatic animals that includes whales, dolphins, and porpoises. The closest living relatives of cetaceans are the hippos. **Figure 22.6** presents a hypothesis for the evolution of whales based on the fossil record.

- The genus *Pakicetus* comprised the earliest known whales. However, these animals did not bear a striking resemblance to modern whales. They were wolflike meat eaters that spent some of their time in fresh water and ate fish. Their skull had a long shape like that of modern whales. The eyes were positioned

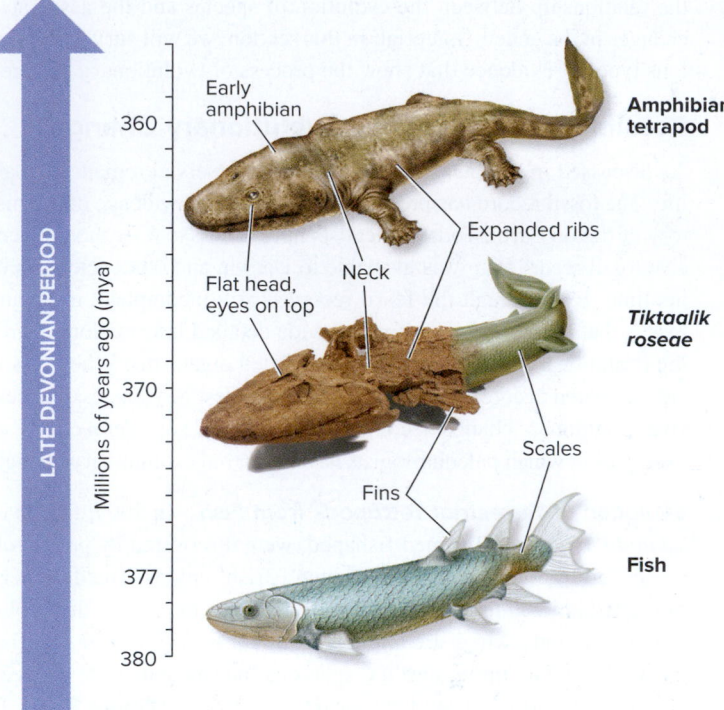

**Figure 22.5** **A transitional form in the tetrapod lineage.** This figure shows two early tetrapod ancestors, a Devonian fish and the transitional form *Tiktaalik roseae*, as well as one of their descendants, an early amphibian. An analysis of the fossils shows that *T. roseae*, also known as a fishapod, had both fish and amphibian characteristics, so it was likely able to survive brief periods out of the water.

(middle): ©Corbin17/Alamy Stock Photo

 **Core Skill: Modeling** The goal of this modeling challenge is to propose a model that describes a transitional form between dinosaurs and birds.

**Modeling Challenge:** Maniraptora includes a group of dinosaurs that is thought to have given rise to modern birds. As shown below, an example of a maniraptora is *Falcarius utahensis*. Propose a model that describes a transitional form between dinosaurs and birds. As in Figure 22.5, place the more recent species at the top and the earlier form at the bottom. In this case, copy and paste an image of a turkey at the top and *F. utahensis* at the bottom. In the middle, draw a model for a transitional form. Next to your model, list 4 key characteristics of the transitional form.

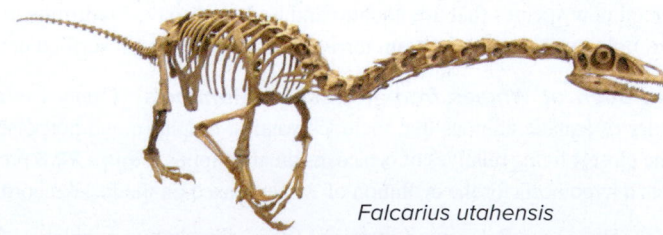

*Falcarius utahensis*

close together and high on the skull, which is characteristic of aquatic animals that peer out of the water. A particularly striking trait was a thick, bony wall around the middle ear, which is found in modern whales but not in other mammals.

- The genus *Ambulocetus* consisted of semi-aquatic whales of brackish (slightly salty) waters. They were roughly the size of a male sea lion and had short and powerful legs. The tail vertebrae were particularly large, suggesting that the tail was very muscular and possibly used for swimming. The eyes were more toward the sides but still high on the skull.

- The members of the genus *Remingtonocetus* were similar to those of *Ambulocetus* but with a longer snout and a fat pad in the jaw that aided in underwater hearing. They lived in saltwater habitats.

- In members of the genus *Rodhocetus*, the eyes were on the side of the head, and the nasal opening was beginning to shift away from the tip of the snout. The forelimbs had five fingers, and the hindlimbs had only four toes, suggesting the degeneration of the hindlimbs.

- The genus *Dorudon* was composed of whales that were completely aquatic animals. The nasal opening was shifted back toward the eyes to form a blowhole. The forelimbs became flippers, and the hindlimbs were very tiny. The tail was modified at the end to form a fluke.

- Odontoceti and Mysticeti are suborders of the order Cetacea, which includes many extinct species as well as all modern species of whales, dolphins, and porpoises. These animals show a complete loss of the hindlimbs in the adult. The nasal opening is the blowhole seen in modern whales. In odontocetes, echolocation is used for hunting. In mysticetes, baleen is used for filtering food.

Taken together, the changes observed in the fossil record of whales reveal a progression over the past 50 million years from a terrestrial tetrapod to aquatic animals that lack hindlimbs and have many adaptations that are beneficial in an aquatic environment.

## Biogeography Indicates That Species in a Given Area Have Evolved from Pre-existing Species

**Biogeography** is the study of the geographic distribution of extinct and living species. Patterns of past evolution are often found in the natural geographic distribution of related species. From such studies, scientists have discovered that isolated continents and island groups have evolved their own distinct plant and animal communities.

***Islands*** Islands, which are isolated from other landmasses, provide numerous examples in which geography has played a key role in the evolution of new species. Islands often have many species of plants and animals that are **endemic**, which means they are naturally found only in a particular location. Most endemic island species have closely related relatives on nearby islands or the mainland.

As an example, let's consider the island fox (*Urocyon littoralis*), which lives on the Channel Islands located off the coast near Santa Barbara in southern California (**Figure 22.7**). This type of fox is found nowhere else in the world. It weighs about 3–6 pounds and feeds largely on insects, mice, and fruits. The island fox evolved from the mainland gray fox (*Urocyon cinereoargenteus*), which is much larger, usually 7–11 pounds. During the last Ice Age, about 16,000–18,000 years ago, the Santa Barbara channel was frozen and narrow enough for ancestors of the mainland gray fox to cross over to the Channel Islands. When the Ice Age ended, the ice melted and sea levels rose, causing the foxes

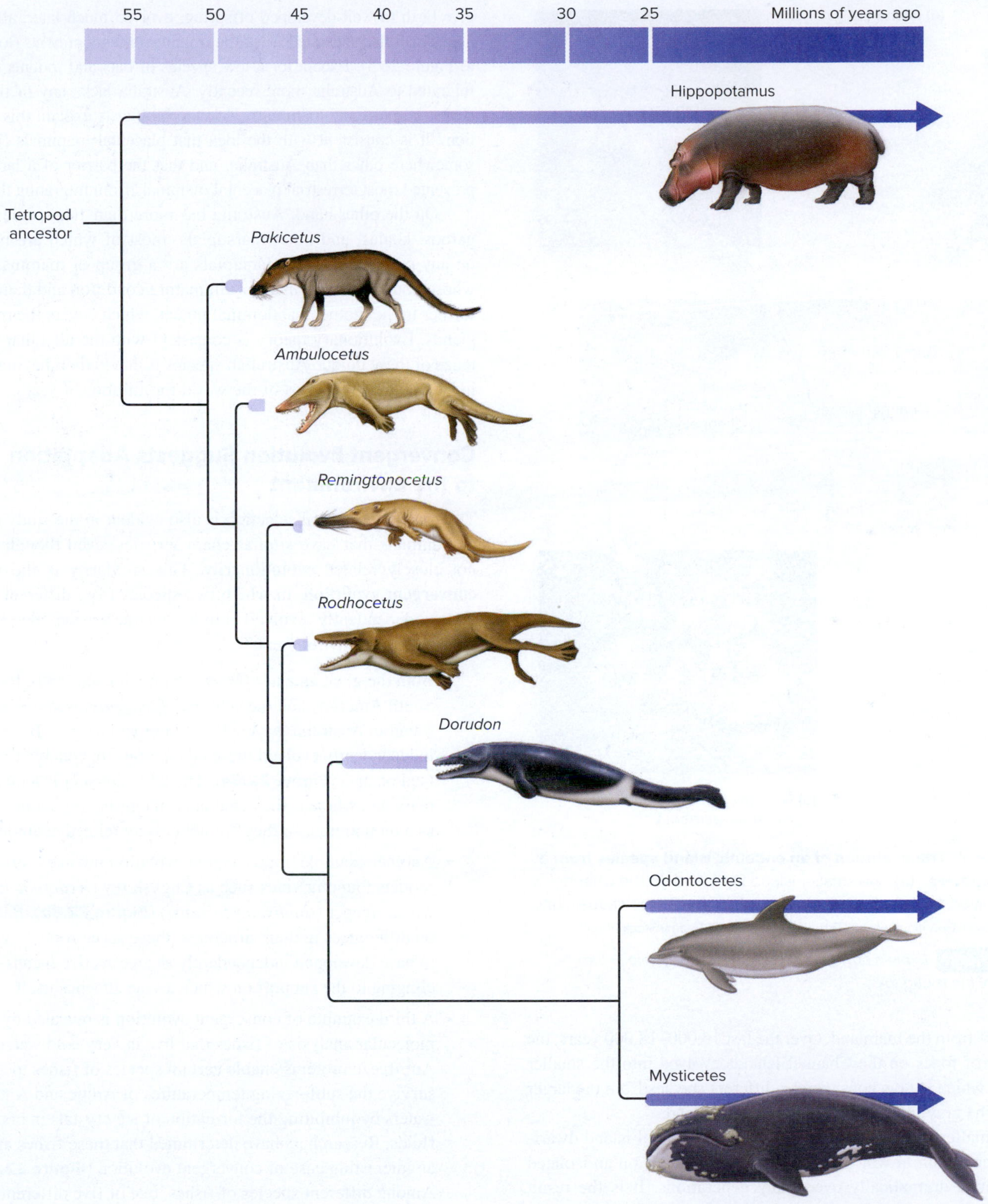

Millions of years ago

55  50  45  40  35  30  25

Hippopotamus

Tetropod
ancestor

*Pakicetus*

*Ambulocetus*

*Remingtonocetus*

*Rodhocetus*

*Dorudon*

Odontocetes

Mysticetes

**Figure 22.6 Evolution of whales.** *Pakicetus*, *Ambulocetus*, *Remingtonocetus*, *Rodhocetus*, and *Dorudon* are extinct genera of whales. Odontocetes and Mysticetes are suborders of the order Cetacea, which includes all modern species of whales, dolphins, and porpoises. This simplified representation of whale evolution is a type of diagram called a phylogenetic tree, which is explained in Chapter 25. Note: The genera described in this phylogenetic tree are not depicted as direct ancestors to modern whales, but they all shared common ancestors.

**Core Concept: Structure and Function** This diagram shows the morphological (structural) changes that occurred in the evolution of whales that made them better suited to an aquatic environment.

**(a) Island fox**
*(Urocyon littoralis)*

California

Santa Barbara

Channel Islands

Santa Barbara channel

San Miguel

Santa Rosa

Santa Cruz

Anacapa

**(b) Gray fox (*Urocyon cinereoargenteus*)**

**Figure 22.7**  **The evolution of an endemic island species from a mainland species. (a)** The smaller island fox found on the Channel Islands evolved from **(b)** the gray fox found on the California mainland.

a: ©Kevin Schafer/Getty Images; b: ©Prisma Bildagentur AG/Alamy Stock Photo

**Concept Check:**  *Explain how geography played a key role in the evolution of the island fox.*

to be cut off from the mainland. Over the last 16,000–18,000 years, the population of foxes on the Channel Islands evolved into the smaller island fox, which is now considered a different species from the larger gray fox. The gray fox is still found on the mainland.

The smaller size of the island fox is an example of island dwarfing, a phenomenon in which the size of large animals on an isolated island shrinks dramatically over many generations. It is the result of natural selection in which a smaller size provides a survival and reproductive advantage, probably because of the limited availability of food and other resources.

***Isolated Continents***  The evolution of major animal groups is also correlated with known changes in the distribution of landmasses on the Earth. The first mammals arose approximately 200 mya, when the area that is now Australia was still connected to the other continents. However, the first placental mammals, which have a long internal gestation and

give birth to well-developed offspring, evolved much later, after continental drift had separated Australia from the other continents (look ahead to Figure 26.5). Except for a few species of bats and rodents that have migrated to Australia more recently, Australia lacks any of the larger, terrestrial placental mammals. How do biologists explain this observation? It is consistent with the idea that placental mammals first arose somewhere other than Australia, and that the barrier of a large ocean prevented most terrestrial placental mammals from migrating there.

On the other hand, Australia has more than 100 species of kangaroos, koalas, and other marsupials, most of which are not found on any other continent. Marsupials are a group of mammal species whose young are born in a very immature condition and then develop further in the mother's abdominal pouch, which covers the mammary glands. Evolutionary theory is consistent with the idea that the existence of these unique Australian species is due to their having evolved in isolation from the rest of the world for millions of years.

## Convergent Evolution Suggests Adaptation to the Environment

The process of natural selection is also evident in the study of plants and animals that have similar characteristics, even though they are not closely related evolutionarily. This similarity is the result of **convergent evolution**, in which two species from different lineages have independently evolved similar characteristics because they occupy similar environments.

- Both the giant anteater (*Myrmecophaga tridactyla*), found in South America, and the echidna (*Tachyglossus aculeatus*), found in Australia, have a long snout and tongue. Both species independently evolved these adaptations that enable them to feed on ants (**Figure 22.8a**). The giant anteater is a placental mammal, whereas the echidna is an egg-laying mammal known as a monotreme, so they are not closely related evolutionarily.

- Another example of convergent evolution involves aerial rootlets found in vines such as English ivy (*Hedera helix*) and wintercreeper (*Euonymus fortunei*) (**Figure 22.8b**). Based on differences in their structures, these aerial rootlets appear to have developed independently as an effective means of clinging to the support on which a vine attaches itself.

- A third example of convergent evolution is revealed by the molecular analysis of fishes that live in very cold water. Antifreeze proteins enable certain species of fishes to survive the subfreezing temperatures of Arctic and Antarctic waters by inhibiting the formation of ice crystals in body fluids. Researchers have determined that these fishes are an interesting case of convergent evolution (**Figure 22.8c**). Among different species of fishes, one of five different genes has independently evolved to produce antifreeze proteins. For example, in the sea raven (*Hemitripterus americanus*), the antifreeze protein is rich in the amino acid cysteine, and the secondary structure of the protein is in a β pleated sheet conformation. In contrast, the antifreeze protein in the longhorn sculpin (*Trematomus nicolai*) is encoded by an entirely different gene. The antifreeze protein in this species is rich in the amino acid glutamine, and the secondary structure of the protein is largely composed of α helices.

(a) **The long snouts and tongues of the giant anteater (left) and the echidna (right) allow them to feed on ants.**

(b) **The aerial rootlets of English ivy (left) and wintercreeper (right) enable them to climb up supports.**

**Figure 22.8** **Examples of convergent evolution.** The species in each of the pairs shown in this figure are not closely related evolutionarily but occupy similar environments, suggesting that natural selection results in similar adaptations to a particular environment. a (left): ©Peter Schoen/Getty Images; a (right): ©Tom McHugh/Science Source; b (left): ©mm88/Getty Images; b (right): ©2003 Steve Baskauf/bioimages.vanderbilt.edu; c (left): ©Jonathan Bird/Getty Images; c (right): ©David Wrobel/SeaPics.com

*Concept Check:* *Can you think of another example in which two species that are not closely related have a similar adaptation?*

(c) **The sea raven (left) and the longhorn sculpin (right) have antifreeze proteins that enable them to survive in frigid waters.**

The similar characteristics in the examples shown in Figure 22.8—for example, the snouts of the anteater and the echidna—are called **analogous structures**, or convergent traits. They represent cases in which characteristics have arisen independently, two or more times, because different species have occupied similar types of environments on the Earth.

## Selective Breeding Is a Human-Driven Form of Selection

The term **selective breeding** refers to programs and procedures designed to modify traits in domesticated species. This practice, also called **artificial selection**, is related to natural selection. In forming his theory of evolution, Charles Darwin was influenced by his observations of selective breeding by pigeon breeders. The primary difference between natural and artificial selection is how the parents are

chosen. Natural selection occurs because of genetic variation in reproductive success. Organisms that are able to survive and reproduce are more likely to pass their genes to future generations. Environmental factors often determine which individuals will be successful parents. In artificial selection, the breeder chooses as parents those individuals with traits that are desirable from a human perspective.

The underlying phenomenon that makes selective breeding possible is genetic variation. Within a population, variation may exist in a trait of interest. For selective breeding to be successful, the underlying cause of the phenotypic variation is typically due to differences in **alleles**, variant forms of a particular gene, that determine the trait. The breeder chooses parents with desirable phenotypic characteristics. For centuries, humans have employed selective breeding to obtain domesticated species with interesting or agriculturally useful characteristics.

**(a) Bulldog**

**(b) Greyhound**

**(c) Dachshund**

**Figure 22.9** **Common breeds of dogs that have been obtained by selective breeding.** By selecting individuals carrying the alleles that influence traits desirable to humans, dog breeders have produced breeds with distinctive features. All the dogs in this figure carry the same kinds of genes (for example, genes that affect their size, shape, and fur color). However, the alleles for many of these genes are different among these dogs, thereby allowing dog breeders to select for or against them and produce breeds with strikingly different phenotypes. a: ©Willee Cole/Getty Images; b: ©Martin Rugner/age fotostock; c: ©Juniors Bildarchiv/age fotostock

**Artificial Selection in Dogs** The various breeds of dog are the result of selective breeding strategies (**Figure 22.9**). All dogs are members of the same species, *Canis lupus*, subspecies *familiaris*, so they can interbreed to produce offspring. Selective breeding can dramatically modify the traits in a species. When you compare certain breeds of dogs (for example, a greyhound and a dachshund), they hardly look like members of the same species! Recent work in 2007 by American geneticist Nathan Sutter and colleagues indicates that the size of dogs may be determined by alleles in the *Igf1* gene that encodes a growth hormone called insulin-like growth factor 1. A particular allele of this gene was found to be common to all small breeds of dogs and nearly absent from very large breeds, suggesting that this allele is a major contributor to body size in small breeds of dogs.

**Artificial Selection in Agricultural Species** Most of the food we eat—including products such as grains, fruits, vegetables, meat, milk, and juices—is also obtained from species that have been profoundly modified by selective breeding strategies. For example, certain characteristics in the wild mustard plant (*Brassica oleracea*) have been modified by selective breeding to produce several varieties of domesticated crops, including broccoli, Brussels sprouts, cabbage, and cauliflower (**Figure 22.10**). The wild mustard plant is native to Europe and Asia, and plant breeders began to modify its traits approximately 4,000 years ago. As seen here, certain traits in the domestic strains differ dramatically from those of the original wild species. These varieties are all members of the same species. They can interbreed to produce viable offspring. For example, in the grocery store you may have seen broccoflower, a vegetable produced from a cross between broccoli and cauliflower.

**Figure 22.10** **Crop plants developed by selective breeding of the wild mustard plant.** Although these six agricultural plants look quite different from each other, they carry many of the same alleles as the wild mustard plant. However, they differ from each other in alleles that affect the formation of stems, leaves, and flowers.

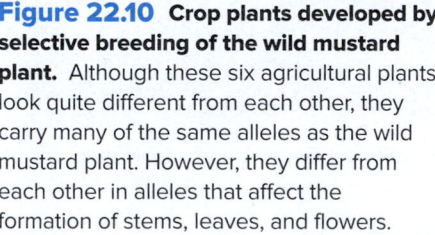

Wild mustard plant (*Brassica oleracea*)

| Strain | Kohlrabi | Kale | Broccoli | Brussels sprouts | Cabbage | Cauliflower |
|---|---|---|---|---|---|---|
| Modified trait | Stem | Leaves | Flower buds and stem | Lateral leaf buds | Terminal leaf bud | Flower buds |

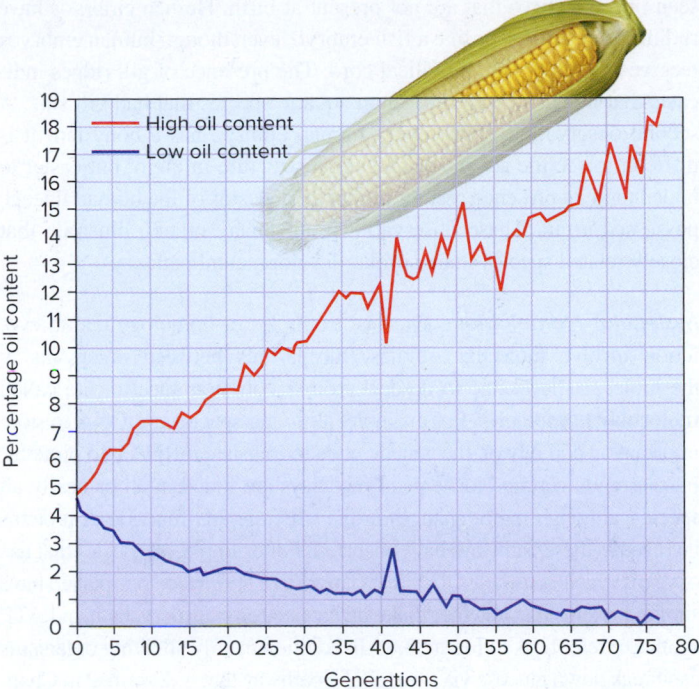

**Figure 22.11  Results of selective breeding for oil content in corn plants.** In this example, corn plants were selected for breeding based on high or low oil content of the kernels. Over the course of many generations, this artificial selection had a major influence on the amount of corn oil (an agriculturally important product) made by the two groups of plants.

**Concept Check:** *When comparing Figures 22.9, 22.10, and 22.11, what general effects of artificial selection do you observe?*

As another example, **Figure 22.11** shows the results of a selective breeding experiment on corn begun at the University of Illinois Agricultural Experiment Station in 1896, several years before the rediscovery of Mendel's laws. This study began with 163 ears of corn with an oil content ranging from 4% to 6%. In each of 80 succeeding generations, corn plants were divided into two separate groups. In one group, members with the highest oil content in the kernels were chosen as parents of the next generation. In the other group, members with the lowest oil content were chosen. After many generations, the oil content in the first group rose to over 18%. In the other group, it dropped to less than 1%. These results show that selective breeding can modify a trait in a very directed manner.

## A Comparison of Homologies Shows Evolution of Related Species from a Common Ancestor

Let's now consider other widespread observations of the process of evolution among living organisms. In biology, the term **homology** refers to a similarity that occurs due to descent from a common ancestor. Two species may have a similar trait because the trait was originally found in a common ancestor. As described next, such homologies may involve anatomical, developmental, or molecular features.

***Anatomical Homologies*** As noted by Theodosius Dobzhansky, many observations regarding the features of living organisms simply cannot be understood in any meaningful scientific way except as

a result of evolution. A comparison of vertebrate anatomy is a case in point. An examination of the limbs of modern vertebrate species reveals similarities that indicate the same set of bones has undergone evolutionary changes, becoming modified to perform different functions in different species. As seen in **Figure 22.12**, the forelimbs of vertebrates have a strikingly similar pattern of bone arrangements. These are termed **homologous structures**—structures that are similar to each other because they are derived from a structure in a common ancestor. The forearm has developed different functions among various vertebrates, including grasping, walking, flying, swimming, and climbing. The theory of evolution explains how these animals have descended from a common ancestor and how natural selection has resulted in modifications to the structure of the original set of bones in ways that ultimately allowed them to be used for several different functions.

Another result of evolution is the phenomenon of **vestigial structures**, anatomical features that have no current function but resemble structures of their presumed ancestors (**Table 22.3**). An interesting case is found in humans. People have a complete set of muscles for moving their ears, even though most people are unable to do so. By comparison, many modern mammals can move their ears, and presumably this was an important trait in a distant human ancestor.

Why would organisms have structures that are no longer useful? Within the context of evolutionary theory, vestigial structures

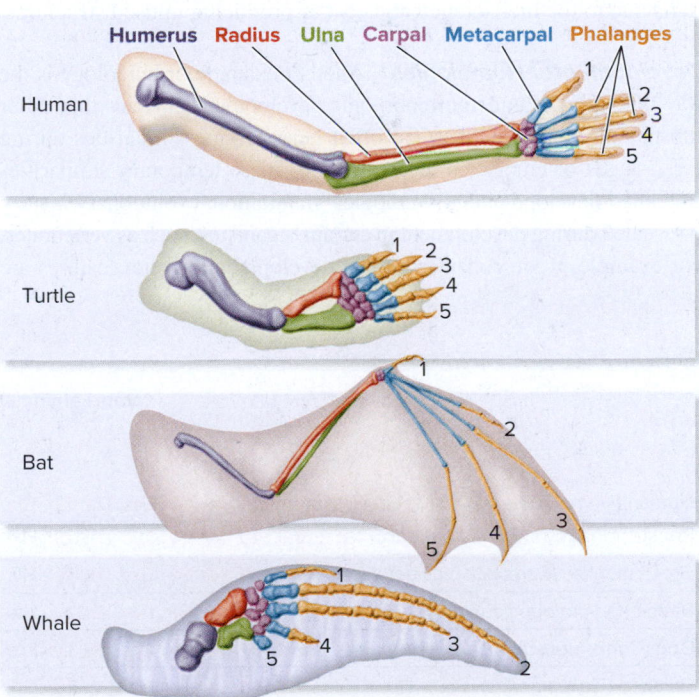

**Figure 22.12  An example of anatomical homology: Homologous structures are found in vertebrates.** The same set of bones is found in the human arm, turtle arm, bat wing, and whale flipper, although their relative sizes and shapes differ significantly. This homology suggests that all of these animals evolved from a common ancestor.

 **Core Concept: Structure and Function** These homologous sets of bones have evolved into somewhat different structures that provide functions that differ among humans, turtles, bats, and whales.

| Table 22.3 | Examples of Vestigial Structures in Animals |
|---|---|
| **Organism** | **Vestigial structure(s)** |
| Humans | Bony tail in embryo and muscles to wiggle ears in adult |
| Boa constrictors | Skeletal remnants of hip and hind leg bones |
| Whales | Skeletal remnants of a pelvis |
| Manatees | Fingernails on the flippers |
| Hornbills and cuckoos | Fibrous cords that were derived from the common carotid arteries. In certain families of birds, both of the common carotid arteries are nonfunctional, fibrous cords. Their vascular function has been assumed by other vessels. |

are evolutionary relics. Organisms having vestigial structures share a common ancestry with organisms in which the structure is functional. Natural selection maintains functional structures in a population of individuals. However, if a species changes its lifestyle so that the structure loses its purpose, the selection that would normally keep the structure in a functional condition is no longer present. When this occurs, the structure may degenerate over the course of many generations due to the accumulation of mutations that limit its size and shape. Natural selection may eventually eliminate such traits due to the inefficiency and cost of producing unused structures.

**Developmental Homologies**   Another example of homology is the way that animals undergo embryonic development. Species that differ substantially at the adult stage often bear striking similarities during early stages of embryonic development. These temporary similarities are called developmental homologies. In addition, evolutionary history is revealed during development in certain organisms, such as vertebrates. For example, if we consider human development, several features are seen in the embryo that are not present at birth. Human embryos have rudimentary gill ridges like a fish embryo, even though human embryos receive oxygen via the umbilical cord. The presence of gill ridges indicates that humans evolved from an aquatic species that had gill slits. A second observation is that every human embryo has a bony tail. It is difficult to see the advantage of such a structure in utero, but easier to understand its presence assuming that an ancestor of the human lineage possessed a tail. These observations, and many others, illustrate that closely related species share similar developmental pathways.

**Molecular Homologies**   Our last examples of homology due to evolution involve molecular studies. Similarities between organisms at the molecular level due to descent from a common ancestor are called **molecular homologies**. For example, all living species use DNA to store information and rely on the genetic code to translate mRNA into proteins. Furthermore, certain biochemical pathways are found in all or nearly all species, although minor changes in the structure and function of proteins involved in these pathways have occurred. For example, all species that use oxygen, which constitutes the great majority of species on our planet, have similar proteins that together make up an electron transport chain and ATP synthase (refer back to Figure 7.8). In addition, nearly all living organisms can break down glucose via a metabolic pathway that is described in Chapter 7. How do we explain these types of observations? Taken together, they indicate that such molecular phenomena arose very early in the origin of life and have been passed to all or nearly all modern forms.

Compelling molecular-level evidence indicating that modern life-forms are derived from an interrelated group of common ancestors is revealed by analyzing genetic sequences. The same type of gene is often found in diverse organisms. Furthermore, the degree of similarity between genetic sequences from different species reflects the evolutionary relatedness of those species. As an example, let's consider the *p53* gene, which encodes the p53 protein—a checkpoint protein of the cell cycle (refer back to Figure 15.15). **Figure 22.13** shows a short

| | Short amino acid sequence within the p53 protein | Percentages of amino acids in the whole p53 protein that are identical to human p53 |
|---|---|---|
| Human (*Homo sapiens*) | Val Pro Ser Gln Lys Thr Tyr Gln Gly Ser Tyr Gly Phe Arg Leu Gly Phe Leu His Ser Gly Thr | 100 |
| Rhesus monkey (*Macaca mulatta*) | Val Pro Ser Gln Lys Thr Tyr His Gly Ser Tyr Gly Phe Arg Leu Gly Phe Leu His Ser Gly Thr | 95 |
| Green monkey (*Cercopithecus aethiops*) | Val Pro Ser Gln Lys Thr Tyr His Gly Ser Tyr Gly Phe Arg Leu Gly Phe Leu His Ser Gly Thr | 95 |
| Rabbit (*Oryctolagus cuniculus*) | Val Pro Ser Gln Lys Thr Tyr His Gly Asn Tyr Gly Phe Arg Leu Gly Phe Leu His Ser Gly Thr | 86 |
| Dog (*Canis lupus familiaris*) | Val Pro Ser Pro Lys Thr Tyr Pro Gly Thr Tyr Gly Phe Arg Leu Gly Phe Leu His Ser Gly Thr | 80 |
| Chicken (*Gallus gallus*) | Val Pro Ser Thr Glu Asp Tyr Gly Gly Asp Phe Asp Phe Arg Val Gly Phe Val Glu Ala Gly Thr | 53 |
| Channel catfish (*Ictalurus punctatus*) | Val Pro Val Thr Ser Asp Tyr Pro Gly Leu Leu Asn Phe Thr Leu His Phe Gln Glu Ser Ser Gly | 48 |
| European flounder (*Platichthys flesus*) | Val Pro Val Val Thr Asp Tyr Pro Gly Glu Tyr Gly Phe Gln Leu Arg Phe Gln Lys Ser Gly Thr | 46 |
| Congo puffer fish (*Tetraodon miurus*) | Val Pro Val Thr Thr Asp Tyr Pro Gly Glu Tyr Gly Phe Lys Leu Arg Phe Gln Lys Ser Gly Thr | 41 |

**Figure 22.13**   **An example of genetic homology: a comparison of a short amino acid sequence within the p53 protein from nine different animals.**   This figure compares a short region of the p53 protein, a tumor suppressor that plays a role in preventing cancer. Amino acids are represented by three-letter abbreviations. The orange-colored amino acids in the sequences are identical to those in the human sequence. The numbers in the right column give the percentage of amino acids within the whole p53 protein of each species that is identical with the human p53 protein, which is 393 amino acids in length. For example, 95% of the amino acids, or 373 of 393, are identical between the p53 sequences found in humans and in rhesus monkeys.

amino acid sequence that makes up part of the p53 protein in a variety of species, including five mammals, one bird, and three fish. The top sequence is the human p53 sequence, and the right column gives the percentage of amino acids within each animal's entire sequence that are identical to those in the entire human sequence. Amino acids in the other species that are identical to those in humans are highlighted in orange. The sequences from the two monkeys are the most similar to those in humans, followed by the other two mammalian species (rabbit and dog). The three fish sequences are the least similar to the human sequence, but the fish sequences tend to be similar to each other.

Taken together, the data shown in Figure 22.13 illustrate two critical points about gene evolution. First, specific genes are found in a diverse array of species such as mammals, birds, and fishes. Second, the sequences of closely related species tend to be more similar to each other than they are to the sequences of distantly related species. The mechanisms underlying this second observation are discussed in the next section.

## 22.3 | The Molecular Processes That Underlie Evolution

### Learning Outcomes:

1. Explain how paralogs and orthologs are produced.
2. Distinguish between vertical evolution and horizontal gene transfer.

Historically, the study of evolution was based on comparing the anatomies of extinct and modern species to identify similarities between related species. However, the advent of molecular approaches for analyzing DNA sequences has revolutionized the field of evolutionary biology. Now researchers can analyze how changes in the genetic material are associated with changes in phenotype. In this section, we will examine some of the molecular changes in the genetic material that reveal evolutionary change.

### Homologous Genes Are Derived from a Common Ancestral Gene

Two or more genes derived from the same ancestral gene are called **homologous genes**. The analysis of homologous genes reveals evidence of evolutionary change at the molecular level. How do homologous genes arise? As an example, let's consider a gene that encodes a transport protein involved in the uptake of metal ions into the cells of two different species of bacteria (**Figure 22.14**). Homologous genes that are found in different species are termed **orthologs**. Millions of years ago, the two species of bacteria had a common ancestor. Over time, the common ancestor diverged into additional species, eventually evolving into *Escherichia coli*, *Clostridium acetylbutylicum*, and many other species. Since this divergence, the gene encoding the transport protein has accumulated mutations that altered its sequence, though the similarity between the *E. coli* and the *C. acetylbutylicum* genes remains striking. In this case, the two sequences are similar because they were derived from the same ancestral gene, but they are not identical due to the independent accumulation of different random mutations.

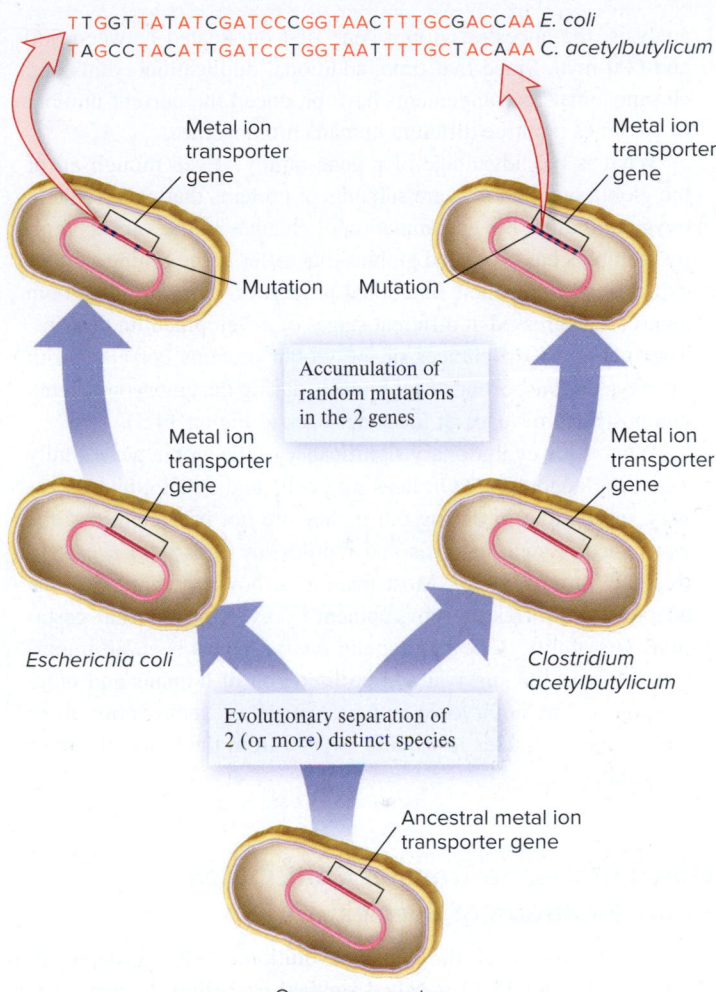

**Figure 22.14 The evolution of orthologs, homologous genes from different species.** After the two species diverged from each other, the genes accumulated random mutations that resulted in similar, but not identical, gene sequences called orthologs. These orthologs in *E. coli* and *C. acetylbutylicum* encode metal ion transporters. Only one of the two DNA strands is shown from each gene. Bases that are identical between the two genes are shown in orange.

**Concept Check:** *Why do these orthologs have similar gene sequences? Why aren't the sequences identical?*

 **Core Concept: Evolution**

### Gene Duplications Produce Gene Families

Evidence of evolutionary change is also found within a single species. Two or more homologous genes within a single species are termed **paralogs** of each other. Rare gene duplication events produce multiple copies of a gene and ultimately lead to the formation of a **gene family**—a set of paralogs within the genome of a single species. A well-studied example of a gene family is the globin gene family in humans, which is composed of 14 genes that are hypothesized to be derived from a single ancestral globin gene (refer back to Figure 21.14). According to an evolutionary analysis, the ancestral globin gene first duplicated between 500 and 600 mya. Since that time, additional duplication events and chromosomal rearrangements have produced the current number of 14 genes on three different human chromosomes.

What is the advantage of a gene family? Even though all of the globin polypeptides are subunits of proteins that play a role in oxygen binding, the accumulation of changes in the various family members has produced globins that differ in the timing of their expression and in their functional properties. The various globin genes are expressed at different stages of development in humans. The functional differences of the globin proteins correlate with the oxygen transport needs of humans during the embryonic, fetal, and postpartum stages of life (refer back to Figure 14.3).

What is the evolutionary significance of the globin gene family regarding adaptation? On land, egg cells and small embryos are very susceptible to drying out if they are not protected in some way. Species such as birds and reptiles lay eggs with a protective shell around them. Most mammals, however, have become adapted to a terrestrial environment by evolving internal gestation. The ability to develop young internally has been an important factor in the survival and proliferation of humans and other mammals. The embryonic and fetal forms of hemoglobin allow the embryo and fetus to capture oxygen from the bloodstream of the mother.

## Horizontal Gene Transfer Contributes to the Evolution of Species

At the molecular level, the type of evolutionary change depicted in Figures 22.13 and 22.14 is called **vertical evolution**. In these cases, new species arise from pre-existing species by the accumulation of genetic changes, such as gene mutations and gene duplications. Vertical evolution involves genetic changes in a series of ancestors that form a lineage. In addition to changing via vertical evolution, species accumulate genetic changes by **horizontal gene transfer**—a process in which an organism incorporates genetic material from another organism without being the offspring of that organism. Horizontal gene transfer can involve the exchange of genetic material between members of the same species or different species.

How does horizontal gene transfer occur? **Figure 22.15** illustrates one possible mechanism for horizontal gene transfer. In this example, a paramecium, which is a eukaryotic microorganism, has engulfed a bacterial cell. During the degradation of the bacterium in a phagocytic vesicle, a bacterial gene escapes to the nucleus of the eukaryotic cell, where it is inserted into one of the chromosomes. In this way, a gene has been transferred from a bacterial species to a eukaryotic species. By analyzing gene sequences among many different species, researchers have discovered that horizontal gene transfer is a common phenomenon. This process can occur from bacteria and archaea to eukaryotes, from eukaryotes to bacteria and archaea, between different species of bacteria and archaea, and between different species of eukaryotes. Therefore, our view of evolution should not only focus on one species evolving into one or more new species via the accumulation of random mutations. It also involves the horizontal transfer of genes among different species, enabling those species to acquire new traits that foster the evolutionary process.

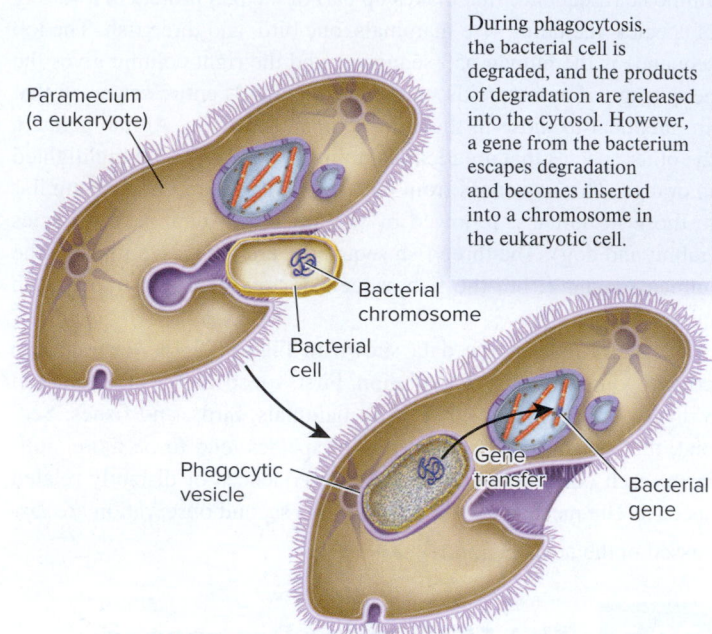

During phagocytosis, the bacterial cell is degraded, and the products of degradation are released into the cytosol. However, a gene from the bacterium escapes degradation and becomes inserted into a chromosome in the eukaryotic cell.

Paramecium (a eukaryote)

Bacterial chromosome

Bacterial cell

Phagocytic vesicle

Gene transfer

Bacterial gene

**Figure 22.15** **Horizontal gene transfer from a bacterium to a eukaryote.** In this example, a bacterium is engulfed by a paramecium (a ciliated protist), and a bacterial gene is transferred to one of the paramecium's chromosomes.

 **Core Skill: Connections** Look back at Table 19.3. What are three mechanisms of gene transfer that could result in horizontal gene transfer between two different bacterial species?

Horizontal gene transfer among bacterial species is relatively widespread. As discussed in Chapter 19, bacterial species carry out three natural mechanisms of gene transfer known as conjugation, transformation, and transduction. By analyzing the genomes of bacterial species, scientists have determined that many genes within a given bacterial genome are derived from horizontal gene transfer. Genome studies have suggested that as much as 20–30% of the variation in the genetic composition of modern bacterial species can be attributed to this process. The roles of the genes acquired by horizontal gene transfer are quite varied, though they commonly involve functions that are beneficial for survival and reproduction. These include genes that confer antibiotic resistance, the ability to degrade toxic compounds, and pathogenicity (the ability to cause disease).

## Evolution at the Genome Level Involves Changes in Chromosome Structure and Number

Thus far, we have considered several ways a species might acquire new genetic variation. These include mutations within pre-existing genes, gene duplications that produce gene families, and horizontal gene transfer. Evolution also involves changes in chromosome structure and number. Comparisons of chromosomes of closely related species have revealed that changes in chromosome structure and/or number are common.

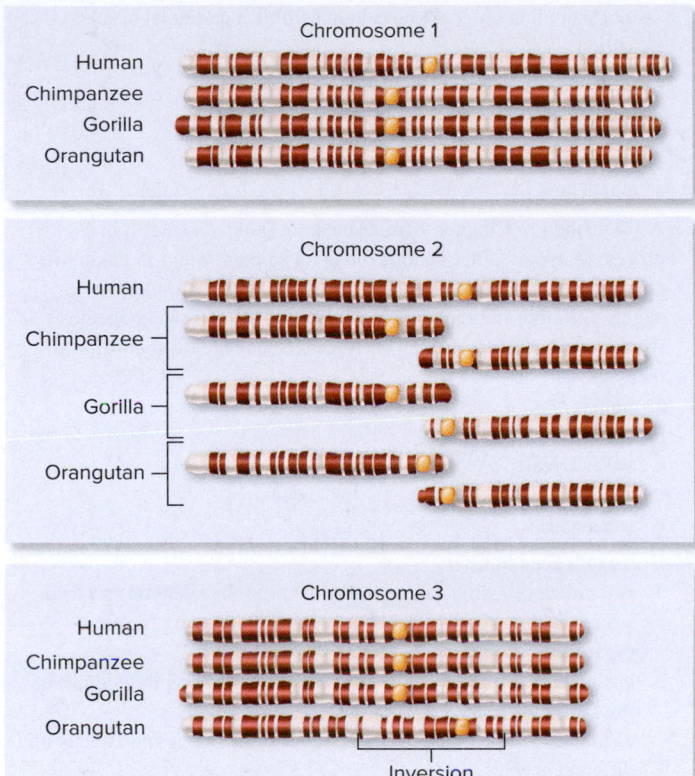

**Figure 22.16  Evolutionary changes in chromosome structure and number found in primates.**  This figure compares the three largest human chromosomes and the corresponding chromosomes in the chimpanzee, gorilla, and orangutan. It is a schematic drawing of Giemsa-stained chromosomes. The differences between these chromosomes illustrate the changes that have occurred during the evolution of these related primate species.

**Concept Check:**  *Describe two changes in chromosome structure that have occurred among these chromosomes.*

As an example, **Figure 22.16** compares the banding patterns of the three largest chromosomes in humans and the corresponding chromosomes in chimpanzees, gorillas, and orangutans. (Refer back to Figure 16.1 for an example of chromosome banding.) The banding patterns in the chromosomes are strikingly similar because these species are closely related evolutionarily. Chromosome 1 looks very similar in all species. Even so, you can see that some of the other chromosomes show interesting differences. Humans have one large chromosome 2, but this chromosome is divided into two separate chromosomes in the other three species. This explains why human cells have 23 pairs of chromosomes, whereas cells of chimpanzees, gorillas, and orangutans have 24. The fusion of the two smaller chromosomes during the development of the human lineage may have caused this difference in chromosome number. Another interesting change in chromosome structure is seen in chromosome 3. The banding patterns among humans, chimpanzees, and gorillas are very similar, but the orangutan has a large inversion that flips the order of the bands in the centromeric region. As discussed in Chapter 24, changes in chromosome structure and number may affect the ability of two organisms to breed with one another. Such changes have been important in the establishment of new species.

# Summary of Key Concepts

## 22.1  Overview of Evolution

- Evolution is a heritable change in one or more characteristics of a population from one generation to the next.

- Charles Darwin proposed the theory of evolution based on his understanding of geology and population growth and his observations of species in their natural settings. His voyage on the *Beagle*, during which he studied many species, including finches on the Galápagos Islands, was particularly influential in the development of his ideas (Figure 22.1, Table 22.1).

- Darwin expressed his theory of evolution as descent with modification through variation and natural selection. As a result of natural selection, genetic variation changes from generation to generation to produce populations of organisms with traits (adaptations) that favor greater reproductive success (Figure 22.2).

- The Grants' research on finches showed how differences in beak size (a heritable trait) were driven by natural selection (Figures 22.3, 22.4).

## 22.2  Evidence of Evolutionary Change

- Evidence of evolutionary change is found in studies of natural selection, the fossil record, biogeography, convergent evolution, selective breeding, and homologies (Table 22.2).

- Fossils provide evidence of evolutionary change in a series of related organisms. The fossil record often reveals transitional forms that link past ancestors to modern species (Figures 22.5, 22.6).

- Biogeography provides information on the geographic distribution of related species. When populations become isolated on islands or continents, they often evolve into new species (Figure 22.7).

- In convergent evolution, independent adaptations result in similar characteristics, called analogous structures, because different species occupy similar environments (Figure 22.8).

- Selective breeding, the selecting and breeding of organisms having desired traits, is a human-driven form of selection (Figures 22.9, 22.10, 22.11).

- Homologies are similarities that occur due to descent from a common ancestor. The set of bones in the forearms of vertebrates is an example of an anatomical homology. Homologies can also be seen during embryonic development and at the molecular level (Figures 22.12, 22.13).

- Vestigial structures, structures that were functional in an ancestor but no longer have a useful function in modern species, are evidence of evolutionary change (Table 22.3).

## 22.3  The Molecular Processes That Underlie Evolution

- Molecular evolution is the process of evolution at the level of genes and proteins. Molecular processes that underlie evolution include the formation of orthologs and paralogs, horizontal gene transfer, and changes in chromosome structure and number.

- Orthologs are homologous genes in different species that have accumulated random mutations over time (Figure 22.14).

- Paralogs are homologous genes in the same species that are produced by gene duplication events. Gene duplication can result in the formation of a gene family such as the globin gene family, which supported the evolutionary adaptation of internal gestation.

- Another mechanism that produces genetic variation is horizontal gene transfer, in which genetic material is transferred from one organism to another organism that is not its offspring. Such genetic changes are subject to natural selection (Figure 22.15).

- Molecular evolution can also involve changes in chromosome structure and number (Figure 22.16).

## Assess & Discuss

### Test Yourself

1. A change in one or more characteristics of a population that is heritable and occurs from one generation to the next is called
   a. natural selection.
   b. sexual selection.
   c. population genetics.
   d. evolution.
   e. inheritance of acquired characteristics.

2. Lamarck's vision of evolution differed from Darwin's in that Lamarck believed
   a. living things evolve in an upward direction.
   b. behavioral changes modify heritable traits.
   c. genetic differences among individuals in the population allow for evolution.
   d. a and b only.
   e. none of the above.

3. Which of the following scientists influenced Darwin's views on the nature of population growth?
   a. Cuvier
   b. Malthus
   c. Lyell
   d. Hutton
   e. Wallace

4. An evolutionary change in which a population of organisms changes its characteristics over many generations in ways that make it better suited to its environment is
   a. natural selection.
   b. an adaptation.
   c. an acquired characteristic.
   d. evolution.
   e. both a and c.

5. Vestigial structures are anatomical structures
   a. that have more than one function.
   b. that were functional in an ancestor but no longer have a useful function.
   c. that look similar in different species but have different functions.
   d. that have the same function in different species but have very different appearances.
   e. of the body wall.

6. Which of the following is an example of a developmental homology seen in human embryonic development and other vertebrate species that are not mammals?
   a. gill ridges
   b. umbilical cord
   c. tail
   d. both a and c
   e. all of the above

7. Two or more homologous genes found within a particular species are called
   a. homozygous.
   b. orthologs.
   c. paralogs.
   d. alleles.
   e. duplicates.

8. As described in Chapter 4 (refer back to Table 4.1), actin is a protein that is a component of the cytoskeleton found in eukaryotic cells. If you compared the sequence of an actin gene in bald eagles with that in other species, which of the following species would you expect to have an actin gene sequence most similar to the eagle's?
   a. human
   b. sea bass
   c. baker's yeast
   d. sparrow
   e. salamander

9. Horizontal gene transfer
   a. is a process in which an organism incorporates genetic material from another organism without being the offspring of that organism.
   b. can involve the exchange of genetic material among individuals of the same species.
   c. can involve the exchange of genetic material among individuals of different species.
   d. can be all of the above.
   e. can be a and b only.

10. Genetic variation can occur as a result of
    a. random mutations in genes.
    b. changes in chromosome structure and number.
    c. gene duplication.
    d. horizontal gene transfer.
    e. all of the above.

## Conceptual Questions

1. Evolution that results in adaptation is rooted in two phenomena: genetic variation and natural selection. In a very concise way (three sentences or less), describe how genetic variation and natural selection can bring about evolution.

2. What is convergent evolution? How does it support the theory of evolution?

3.  **Core Concept: Evolution** Explain how the homologous forelimbs of vertebrates indicate that populations evolve from one generation to the next.

## Collaborative Questions

1. The term natural selection is sometimes confused with the term evolution. Discuss the meanings of these two terms. Explain how the terms are different and how they are related to each other.

2. Make a list of the observations made by biologists that support the theory of evolution. Which of the observations on your list do you find the most convincing and the least convincing?

# Population Genetics

# 23

**Colorful African cichlids.** Color is a factor that influences the choice of mates in populations of cichlids. ©Dynamic Graphics Group/IT StockFree/Alamy Stock Photo

**K**imbareta, age 19, lives in the Democratic Republic of Congo (formerly Zaire) with his parents, one brother, and two sisters. Kimbareta has sickle cell disease, which causes some of his red blood cells to take on a crescent or sickled shape. The sickled cells may block the flow of blood through his vessels. This results in tissue and organ damage along with painful episodes involving his arms, legs, chest, and abdomen. In some people with this disease, stroke may even occur. Sickle cell disease follows a recessive pattern of inheritance. It is caused by a mutation in a gene that encodes β-globin, a subunit of hemoglobin that carries oxygen in the red blood cells.

Many different recessive diseases have been identified by geneticists. Most of them are very rare. However, in the village where Kimbareta lives, sickle cell disease is surprisingly common. Nearly 2% of the inhabitants have the disease—an incidence that is similar to other places in the country. How can we explain such a high occurrence of a serious inherited disease? If natural selection tends to eliminate detrimental genetic variation, as we saw in Chapter 22, why does the sickle cell allele persist in this population? As we will see later, biologists have discovered that the effect of the allele in heterozygotes is the underlying factor. Heterozygotes, who carry one copy of the sickle cell allele and one copy of the more common (non-disease-causing) allele, have an increased resistance to malaria.

**Population genetics** is the study of genes and genotypes in a population. A **population** is a group of individuals of the same species that occupy the same environment at the same time. For sexually reproducing species, members of a given population can interbreed with one another. The central issue in population genetics is genetic variation—its extent within populations, why it exists, how it is maintained, and how it changes over the course of many generations. Population genetics helps us understand the relationship between genetic variation and phenotypic variation.

Population genetics emerged as a branch of genetics in the 1920s and 1930s. Its mathematical foundations were developed by theoreticians who extended the principles of Darwin and Mendel by deriving equations to explain the occurrence of genotypes within populations. These foundations can be largely attributed to British evolutionary biologists J. B. S. Haldane and Ronald Fisher and American geneticist Sewall Wright. As we will see, several researchers who analyzed the genetic composition of natural and experimental populations provided support for the mathematical theories. More recently, population geneticists have used techniques to probe genetic variation at the molecular level. In addition, the staggering improvements in computer technology have aided population geneticists in the analysis of data and the testing of genetic hypotheses.

We will begin this chapter by exploring the extent of genetic variation that occurs in populations and how such variation may change. In many cases, genetic changes are associated with evolutionary adaptations, which are characteristics of a species that have evolved over a long period of time by the process of natural selection. In the second half of the chapter, we will examine various evolutionary mechanisms that promote genetic change in a population, including natural selection, genetic drift, migration, and nonrandom mating.

## 23.1 Genes in Populations

**Learning Outcomes:**

1. Define a gene pool.
2. Distinguish between allele frequency and genotype frequency.
3. **CoreSKILL »** Use the Hardy-Weinberg equation to calculate allele and genotype frequencies in a given population.
4. List the conditions that must be met for a population to be in Hardy-Weinberg equilibrium.
5. Describe the factors that cause microevolution to occur.

Population genetics is an extension of our understanding of Darwin's theory of natural selection, Mendel's laws of inheritance, and newer studies in molecular genetics. All of the alleles for every gene in a given population make up the **gene pool**. Each member of the population receives its genes from its parents. Individuals that reproduce contribute to the gene pool of the next generation. Population geneticists study the genetic variation within the gene pool and how such variation changes from one generation to the next. The emphasis is often on understanding the variation in alleles among members of a population. In this section, we will examine some of the general features of populations and gene pools.

### Populations Are Dynamic Units

Recall that a population is a group of individuals of the same species that occupy the same environment at the same time and can interbreed with one another. Certain species occupy a wide geographic range and are divided into discrete populations due to geographic isolation. For example, distinct populations of a given species may be located on different sides of a physical barrier, such as a mountain range.

Populations change from one generation to the next. How might populations become different? Populations may change in size and geographic location. As the size and location of a population change, the genetic composition generally changes as well. Some of the genetic changes involve adaptation, in which a population becomes better suited to its environment, making it more likely to survive and reproduce. For example, a population of mammals may move from a warmer to a colder geographic location. Over the course of many generations, natural selection may change the population such that animals whose fur is thicker and provides better insulation against the colder temperatures become more prevalent.

 **Core Concept: Evolution**

### Genes Are Usually Polymorphic

The term **polymorphism** (from the Greek, meaning many forms) refers to the presence of two or more variations for a given character within a population. **Figure 23.1** illustrates a striking example of polymorphism in the elder-flowered orchid (*Dactylorhiza sambucina*). Throughout the range of this species in Europe, both yellow- and red-flowered individuals are prevalent.

Polymorphism of a character is usually due to the existence of two or more alleles of a gene that influences the character.

**Figure 23.1** **An example of polymorphism: the two color variations found in the orchid *Dactylorhiza sambucina*.** ©Paul Harcourt Davies/SPL/Science Source

Geneticists also use the term polymorphism to describe the variation in the DNA sequence of genes. A gene that commonly exists as two or more alleles in a population is a **polymorphic gene**. To be considered polymorphic, a gene must exist in at least two alleles, and each allele must occur at a frequency that is greater than 1%. By comparison, a **monomorphic gene** exists predominantly as a single allele in a population. When 99% or more of the alleles of a given gene are identical in a population, the gene is considered to be monomorphic.

What types of molecular changes cause genes to be polymorphic? A polymorphism may involve various types of changes, such as a deletion of a significant region of the gene, a duplication of a region, or a change in a single nucleotide. This last type of variation is called a **single-nucleotide polymorphism (SNP)**. SNPs (or "snips") are the smallest type of genetic variation that can occur within a given gene and also the most common. For example, the sickle cell allele discussed at the beginning of the chapter involves a single-nucleotide change in the β-globin gene, which encodes a subunit of the oxygen-carrying protein called hemoglobin. The non-disease-causing allele and the sickle cell allele represent a SNP of the β-globin gene:

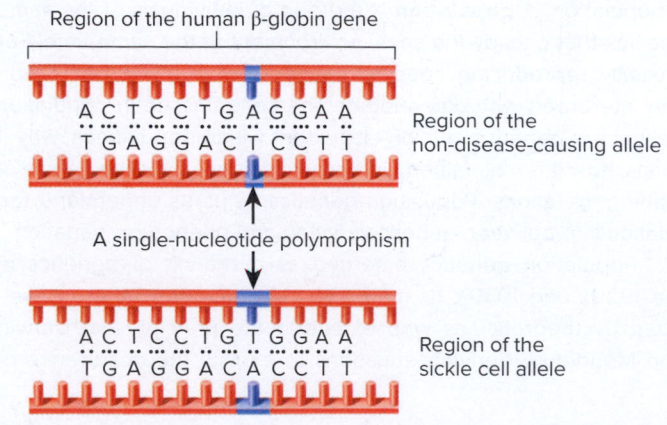

In this single-nucleotide polymorphism, the single nucleotide A (in the top strand) of the non-disease-causing allele changes to a T in the sickle cell allele.

SNPs represent 99% of all variation in human gene sequences that occurs among different people. In human populations, a gene that is 2,000–3,000 bp in length, on average, contains 10 different SNPs. Likewise, SNPs with a frequency of 1% or more are found very frequently among genes of nearly all species. Polymorphism is the norm for relatively large, healthy populations of nearly all species, as evidenced by the occurrence of SNPs within most genes.

Why do we care about SNPs? One reason is their importance in human health. By analyzing SNPs in human genes, researchers have determined that these small variations in DNA sequences can affect the function of the proteins encoded by the genes. These effects on the proteome, in turn, may influence how humans develop diseases, such as heart disease, diabetes, and sickle cell disease. Variations in SNPs in the human population are also associated with how people respond to viruses, drugs, and vaccines. The analysis of SNPs may be instrumental in the current and future development of **personalized medicine**—a medical practice in which information about a patient's genotype is used to tailor her or his medical care. For example, an analysis of a person's SNPs may be used to select between different types of medication or to customize the dosage. In addition, SNP analysis may reveal that a person has a high predisposition to develop a particular disease, such as heart disease. Such information may be used to initiate preventative measures to minimize the chances of developing the disease.

## Population Genetics Is Concerned with Allele and Genotype Frequencies

One approach to analyzing genetic variation in populations is to consider the frequency of specific alleles and genotypes in a quantitative way. Two fundamental calculations are central to population genetics: **allele frequency** and **genotype frequency**. Allele and genotype frequencies are defined as follows:

$$\text{Allele frequency} = \frac{\text{Number of copies of a specific allele in a population}}{\text{Total number of all alleles for that gene in the population}}$$

$$\text{Genotype frequency} = \frac{\text{Number of individuals with a particular genotype in a population}}{\text{Total number of individuals in the population}}$$

Although allele and genotype frequencies are related, make sure you clearly distinguish between them. As an example, let's consider a population of 100 four-o'clock plants (*Mirabilis jalapa*) with the following genotypes:

49 red-flowered plants with the genotype $C^R C^R$

42 pink-flowered plants with the genotype $C^R C^W$

9 white-flowered plants with the genotype $C^W C^W$

When calculating an allele frequency for a diploid species, remember that homozygous individuals have two copies of a given allele, whereas heterozygotes have only one. For example, in tallying the $C^W$ allele, each of the 42 heterozygotes has one copy of the $C^W$ allele, and each white-flowered plant has two copies. Therefore, the allele frequency for $C^W$ (the white color allele) equals

$$\text{Frequency of } C^W = \frac{(C^R C^W) + 2(C^W C^W)}{2(C^R C^R) + 2(C^R C^W) + 2(C^W C^W)}$$

$$\text{Frequency of } C^W = \frac{42 + (2)(9)}{(2)(49) + (2)(42) + (2)(9)}$$

$$= \frac{60}{200} = 0.3, \text{ or } 30\%$$

This result tells us that the allele frequency of $C^W$ is 0.3. In other words, 30% of the alleles for this gene in the population are the white color ($C^W$) allele.

Let's now calculate the genotype frequency of $C^W C^W$ homozygotes (white-flowered plants).

$$\text{Frequency of } C^W C^W = \frac{9}{49 + 42 + 9}$$

$$= \frac{9}{100} = 0.09, \text{ or } 9\%$$

We see that 9% of the individuals in this population have the white-flower genotype.

## The Hardy-Weinberg Equation Relates Allele and Genotype Frequencies in a Population

In 1908, Godfrey Harold Hardy, an English mathematician, and Wilhelm Weinberg, a German physician, independently derived a simple mathematical expression, now called the Hardy-Weinberg equation, that describes the relationship between allele and genotype frequencies when a population is not evolving. Let's examine the Hardy-Weinberg equation using the population of four-o'clock plants that we just considered. If the allele frequency of $C^R$ is denoted by the symbol $p$ and the allele frequency of $C^W$ by $q$, then

$$p + q = 1$$

For example, if $p = 0.7$, then $q$ must be 0.3. In other words, if the allele frequency of $C^R$ equals 70%, the remaining 30% of alleles must be $C^W$, because together they equal 100%.

For a gene that exists in two alleles, the **Hardy-Weinberg equation** states that

$$p^2 + 2pq + q^2 = 1$$

If we apply this equation to our flower color gene, then

$p^2$ = the genotype frequency of $C^R C^R$ homozygotes

$2pq$ = the genotype frequency of $C^R C^W$ heterozygotes

$q^2$ = the genotype frequency of $C^W C^W$ homozygotes

If $p = 0.7$ and $q = 0.3$, then

$$\text{Frequency of } C^R C^R = p^2 = (0.7)^2 = 0.49$$

$$\text{Frequency of } C^R C^W = 2pq = 2\,(0.7)\,(0.3) = 0.42$$

$$\text{Frequency of } C^W C^W = q^2 = (0.3)^2 = 0.09$$

In other words, if the allele frequency of $C^R$ is 70% and the allele frequency of $C^W$ is 30%, the expected genotype frequency of $C^R C^R$ is 49%, of $C^R C^W$ is 42%, and of $C^W C^W$ is 9%.

**Figure 23.2** uses a Punnett square to illustrate the relationship between allele frequencies and the way that gametes combine to produce genotypes. The validity of the Hardy-Weinberg equation rests on the assumption that two gametes combine randomly with each other to produce offspring. In a population, the frequency of a gamete carrying a particular allele is equal to the allele frequency in that population. For example, if the allele frequency of $C^R$ equals 0.7, the frequency of a gamete carrying the $C^R$ allele also equals 0.7. The probability of producing a $C^R C^R$ homozygote with red flowers is $0.7 \times 0.7 = 0.49$, or 49%. The probability of inheriting both $C^W$ alleles, which produces white flowers, is $0.3 \times 0.3 = 0.09$, or 9%. Two different gamete combinations produce heterozygotes with pink flowers. An offspring could inherit the $C^R$ allele from the pollen and $C^W$ from the egg, or $C^R$ from the egg and $C^W$ from the pollen. Therefore, the frequency of heterozygotes is $pq + pq$, which equals $2pq$. In our example, this is $2(0.7)(0.3) = 0.42$, or 42%. Note that the frequencies for all three genotypes total 100%.

The Hardy-Weinberg equation predicts that allele and genotype frequencies will remain the same, generation after generation, provided that a population is in equilibrium. To be in equilibrium, the population must not be affected by evolutionary mechanisms that can change allele and genotype frequencies. For this to occur, the following conditions must be met:

- No new mutations occur to alter allele frequencies.
- No natural selection occurs; that is, no survival or reproductive advantage exists for any of the genotypes.
- The population is so large that allele frequencies do not change due to random chance.
- No migration occurs between different populations, altering the allele frequencies.
- Random mating occurs; that is, the members of the population mate with each other without regard to their genotypes.

Why is Hardy-Weinberg equilibrium a useful concept? An equilibrium is a null hypothesis, which suggests that evolutionary change is not occurring. In reality, however, populations rarely achieve an equilibrium, though in large natural populations with little migration and negligible natural selection, Hardy-Weinberg equilibrium may be nearly approximated for certain genes. Sometimes, when researchers experimentally examine allele and genotype frequencies for one or more genes in a given species, they discover that the frequencies are not in Hardy-Weinberg equilibrium. In such cases, they assume that one or more of the conditions are being violated—in other words, mechanisms of evolutionary change are affecting the population. Conservation biologists and wildlife managers may wish to determine why such disequilibrium has occurred because it may affect the future survival of the species.

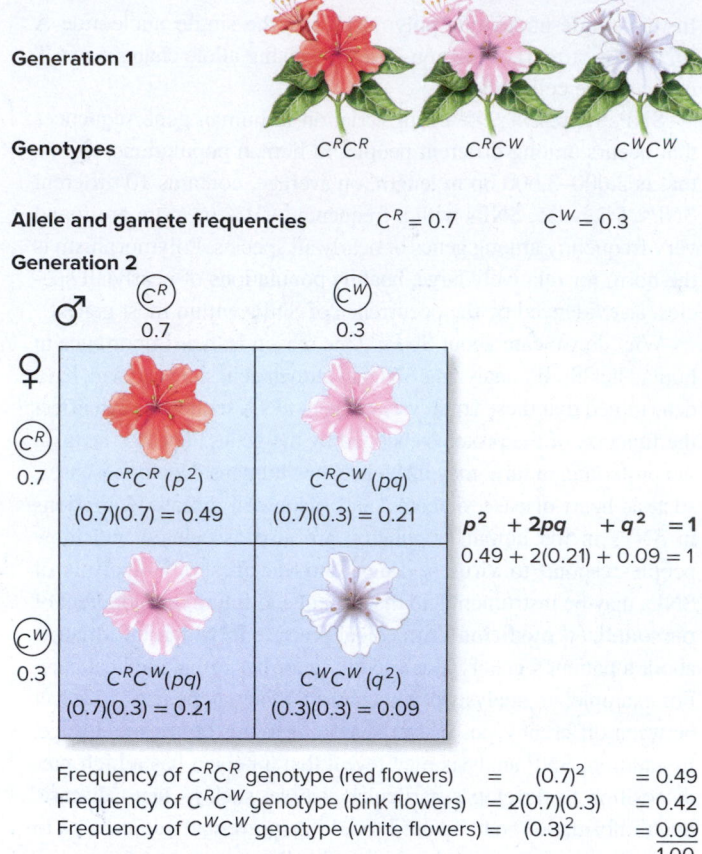

Generation 1

Genotypes          $C^R C^R$          $C^R C^W$          $C^W C^W$

Allele and gamete frequencies          $C^R = 0.7$          $C^W = 0.3$

Generation 2

$$p^2 + 2pq + q^2 = 1$$
$$0.49 + 2(0.21) + 0.09 = 1$$

| | $C^R$ (0.7) | $C^W$ (0.3) |
|---|---|---|
| $C^R$ (0.7) | $C^R C^R$ ($p^2$) $(0.7)(0.7) = 0.49$ | $C^R C^W$ ($pq$) $(0.7)(0.3) = 0.21$ |
| $C^W$ (0.3) | $C^R C^W$ ($pq$) $(0.7)(0.3) = 0.21$ | $C^W C^W$ ($q^2$) $(0.3)(0.3) = 0.09$ |

Frequency of $C^R C^R$ genotype (red flowers)     $=$     $(0.7)^2$     $= 0.49$
Frequency of $C^R C^W$ genotype (pink flowers)   $= 2(0.7)(0.3)$   $= 0.42$
Frequency of $C^W C^W$ genotype (white flowers) $=$   $(0.3)^2$   $= \underline{0.09}$
                                                                                    $1.00$

**Figure 23.2  Calculating allele and genotype frequencies with the Hardy-Weinberg equation.**  A population of four-o'clock plants has allele and gamete frequencies of 0.7 for the $C^R$ allele and 0.3 for the $C^W$ allele. Knowing the allele frequencies allows us to calculate the genotype frequencies in the population.

  **Core Skill: Modeling**  The goal of this modeling challenge is to make a mathematical model for a gene that exists in three alleles in a population that is in Hardy-Weinberg equilibrium.

**Modeling Challenge:** The Hardy-Weinberg equation is a mathematical model in which a gene exists as two alleles, designated $p$ and $q$. The equation is derived from the concept that alleles combine randomly at fertilization when diploid offspring are produced. If $p$ and $q$ are the only two alleles, $(p + q)^2 = 1$, which is the same as $p^2 + 2pq + q^2 = 1$. Let's suppose that a gene exists in a population in three alleles, designated A1, A2, and A3. A1 is represented by $p$, A2 is represented by $q$, and A3 is represented by $r$. Propose an equation that describes the relationship between allele and genotype frequencies, assuming that the population is in Hardy-Weinberg equilibrium. If $p = 0.2$, $q = 0.7$, and $r = 0.1$, what would be the genotype frequencies of A2A2 homozygotes and A2A3 heterozygotes?

**THE QUESTION** *In human populations, the phenotype frequency of the inability to taste the bitter substance phenylthiocarbamide (PTC) is approximately 0.3. This inability is due to a recessive allele. Assuming that there are only two alleles in a population (namely, tasters, T, and nontasters, t) and that the population is in Hardy-Weinberg equilibrium, calculate the frequencies of these two alleles.*

**T**OPIC *What topic in biology does this question address?* The topic is predicting the allele frequencies in a population. More specifically, the question is about predicting the frequency of alleles that affect the tasting of PTC.

**I**NFORMATION *What information do you know based on the question and your understanding of the topic?* From the question, you know the frequency of homozygotes who are nontasters. From your understanding of the topic, you may realize that you can use the Hardy-Weinberg equation to determine allele frequencies if you know the genotype frequencies.

**P**ROBLEM-SOLVING **S**TRATEGY *Make a calculation.* One strategy to solve this problem is to use the components of the Hardy-Weinberg equation to determine the allele frequencies. If $q$ represents the allele frequency of the recessive allele ($t$) that confers nontasting, then $q^2$ is the genotype frequency of homozygous nontasters:

$$q^2 = 0.3$$

We take the square root to determine $q$:

$$q = \sqrt{0.3}$$
$$q = 0.55$$

If $p$ represents the frequency of the taster allele ($T$), then

$$p = 1 - q$$
$$p = 1 - 0.55 = 0.45$$

**ANSWER** *The frequency of the nontaster allele is 0.55, or 55%, and that of the taster allele is 0.45, or 45%.*

## Microevolution Involves Changes in Allele Frequencies from One Generation to the Next

The term **microevolution** is used to describe changes in a population's gene pool, such as changes in allele frequencies, from generation to generation. What causes microevolution to happen? Such change is rooted in two related phenomena (**Table 23.1**). First, the introduction of new genetic variation into a population is one essential aspect of microevolution. New alleles of pre-existing genes arise by random mutation, and, as discussed in Chapters 21 and 22, new genes can be introduced into a population by gene duplication and horizontal gene transfer. Such mutations, albeit rare, provide a continuous source of new variation in populations. In 1926, Russian geneticist Sergei

| Table 23.1 | Factors That Govern Microevolution |
|---|---|
| **Sources of new genetic variation*** | |
| New mutations within genes that produce new alleles | Random mutations within pre-existing genes introduce new alleles into populations, but at a very low rate. New mutations may be neutral, deleterious, or beneficial. Because mutations are rare, the change from one generation to the next is very small. For alleles to rise to a significant percentage in a population, evolutionary mechanisms, such as natural selection, genetic drift, and migration, must operate on them. |
| Gene duplication[†] | Abnormal crossover events and transposable elements may increase the number of copies of a gene. Over time, the additional copies accumulate random mutations and constitute a gene family. |
| Horizontal gene transfer[‡] | A gene from one species may be introduced into another species. The transferred gene may be acted on by evolutionary mechanisms. |
| **Evolutionary mechanisms that alter the prevalence of a given allele or genotype** | |
| Natural selection | The process by which individuals that possess certain traits are more likely to survive and reproduce than individuals without those traits. Over the course of many generations, beneficial traits that are heritable become more common and detrimental traits become less common. |
| Genetic drift | A change in genetic variation from generation to generation due to random chance. Allele frequencies may change as a matter of chance from one generation to the next. Genetic drift has a greater influence in a small population. |
| Migration | Migration can occur between two populations that have different allele frequencies. The introduction of migrants into a recipient population may change the allele frequencies of that population. |
| Nonrandom mating | The phenomenon in which individuals select mates based on their genotypes or phenotypes. This alters the relative proportion of homozygotes and heterozygotes that is predicted by the Hardy-Weinberg equation, but it does not change allele frequencies. |

* These are examples that affect single genes. Other events, such as crossing over, independent assortment, and changes in chromosome structure and number, may alter the genetic variation among many genes.
[†] Described in Chapter 21. See Figures 21.13 and 21.14.
[‡] Described in Chapter 22. See Figure 22.15.

Chetverikov was the first to suggest that random mutations are the raw material for evolution. However, due to their low rate of occurrence, mutations by themselves do not play a major role in changing allele frequencies in a population over time. They do not significantly disrupt a Hardy-Weinberg equilibrium.

The second phenomenon that is required for evolution to occur is one or more mechanisms that alter the prevalence of a given allele or genotype in a population. These mechanisms are natural selection, genetic drift, migration, and nonrandom mating (see Table 23.1). Over the course of many generations, these mechanisms may promote widespread genetic changes in a population. In the remainder of this chapter, we will examine how natural selection, genetic drift, migration, and nonrandom mating affect the type of genetic variation that occurs when a gene exists as two alleles in a population.

## 23.2 Natural Selection

### Learning Outcomes:

1. Explain how natural selection can result in a population that is better adapted to its environment and more successful at reproduction.
2. **CoreSKILL** » Calculate the fitness values of given genotypes.
3. List and distinguish the four different types of natural selection.

Recall from Chapter 22 that **natural selection** is the process by which individuals with certain heritable traits tend to survive and reproduce at higher rates than those without those traits. As a result, favorable heritable traits become more common, while detrimental heritable traits become less common. Keep in mind that natural selection itself is not evolution. Rather it is a key mechanism that causes evolution to happen. Over time, natural selection results in **adaptations**—changes in populations of living organisms that increase their ability to survive and reproduce in a particular environment. In this section, we will examine various ways that natural selection produces such adaptations.

### Natural Selection Favors Individuals with Greater Reproductive Success

**Reproductive success** is the likelihood of an individual contributing fertile offspring to the next generation. Natural selection occurs because some individuals in a population have greater reproductive success compared to others. Those individuals having heritable traits that favor reproductive success are more likely to pass those traits to their offspring. Reproductive success is commonly attributed to two categories of traits:

1. Certain characteristics make organisms better adapted to their environment and therefore more likely to survive to reproductive age. Natural selection favors individuals with characteristics that provide a survival advantage.
2. Reproductive success may involve traits that are directly associated with reproduction, such as the abilities to find a mate and produce viable gametes and offspring. Traits that enhance the ability of individuals to reproduce, such as brightly colored plumage in male birds, are often subject to natural selection.

As discussed in Chapter 22, Charles Darwin and Alfred Wallace independently proposed the theory of evolution by natural selection. A modern description of the principles of natural selection can relate our knowledge of molecular genetics to the process of evolution:

1. Within a population, allelic variation arises from random mutations that cause differences in DNA sequences. A mutation that creates a new allele may alter the amino acid sequence of the encoded protein. This, in turn, may alter the function of the protein.
2. Some alleles encode proteins that enhance an individual's survival or reproductive capability over that of other members of the population. For example, an allele may produce a protein that is more efficient at a higher temperature, conferring on the individual a greater probability of survival in a hot climate.
3. Individuals with beneficial alleles are more likely to survive and contribute their alleles to the gene pool of the next generation.

4. Over the course of many generations, allele frequencies of many different genes may change through natural selection, thereby significantly altering the characteristics of a population. The net result of natural selection is a population that is better adapted to its environment and more successful at reproduction.

### Fitness Is a Quantitative Measure of Reproductive Success

As mentioned at the beginning of this chapter, Haldane, Fisher, and Wright developed mathematical relationships to explain the phenomenon of natural selection. To begin our discussion of natural selection, we need to consider the concept of **fitness**, which is the relative likelihood that one genotype will contribute to the gene pool of the next generation compared with other genotypes. Although this property often correlates with physical fitness, the two ideas should not be confused. Fitness is a quantitative measure of reproductive success. An extremely fertile individual may have a higher fitness than a less fertile individual that appears more physically fit.

To examine fitness, let's consider an example of a hypothetical gene existing in alleles $A$ and $a$. We can assign fitness values to each of the three possible genotypes according to their relative reproductive success. For example, let's suppose the average reproductive successes of the three genotypes are

> $AA$ produces 5 offspring
> $Aa$ produces 4 offspring
> $aa$ produces 1 offspring

By convention, the genotype with the highest reproductive success is given a fitness value of 1.0. Fitness values are denoted by the variable $w$. The fitness values of the other genotypes are assigned values relative to this 1.0 value.

> Fitness of $AA$: $w_{AA} = 1.0$
>
> Fitness of $Aa$: $w_{Aa} = 4/5 = 0.8$
>
> Fitness of $aa$: $w_{aa} = 1/5 = 0.2$

Variation in fitness occurs because certain genotypes result in individuals that have a greater reproductive success than individuals with other genotypes.

Likewise, the effects of natural selection can be viewed at the level of a population. The average reproductive success of members of a population is called the **mean fitness of the population**. Over many generations, as individuals with higher fitness values become more prevalent, natural selection also increases the mean fitness of the population. In this way, the process of natural selection results in a population of organisms that is well adapted to its native environment and more likely to be successful at reproduction.

### Natural Selection Follows Different Patterns

By studying species in their native environments, population geneticists have discovered that natural selection can occur in several ways. In most of the examples described next, natural selection leads to adaptations that make certain members of a species more likely to survive to reproductive age.

*Directional Selection* During **directional selection**, individuals at one extreme of a phenotypic range have greater reproductive success in a particular environment. Different phenomena may initiate the process of directional selection. A common reason for directional selection is that a population may be exposed to a prolonged change in its living environment. Under the new environmental conditions, the relative fitness values may change to favor one genotype, which will promote the elimination of other genotypes. As an example, let's suppose a population of finches on a mainland already has genetic variation that affects beak size (refer back to Figure 22.2). A small number of these birds migrate to an island where the seeds are generally larger than on the mainland. In this new environment, birds with larger beaks have a higher fitness because they are better able to crack open the larger seeds and thereby survive to reproduce. Over the course of many generations, directional selection will result in a population of finches carrying alleles that promote larger beak size.

Another way that directional selection may arise is that a new allele may be introduced into a population by mutation, and the new allele may confer a higher fitness in individuals that carry it (**Figure 23.3**). What are the long-term effects of such directional selection? If the homozygote carrying the favored allele has the highest fitness value, directional selection may cause this favored allele to eventually predominate in the population, perhaps even leading to a monomorphic gene.

*Stabilizing Selection* A type of natural selection called **stabilizing selection** favors the survival of individuals with intermediate phenotypes and selects against those with extreme phenotypes. Stabilizing selection tends to decrease genetic diversity. An example of stabilizing selection involves clutch size (number of eggs laid) of birds, which was first studied by British biologist David Lack in 1947. Under stabilizing selection, birds that lay too many or too few eggs per nest have lower fitness values than do those that lay an intermediate number. When a bird lays too many eggs, many offspring die due to inadequate parental care and food. In addition, the strain on the parents themselves may decrease their likelihood of survival and consequently their ability to produce more offspring. Having too few offspring, however, does not contribute many individuals to the next generation. Therefore, the most successful parents are those that produce an intermediate clutch size. In the 1980s, Swedish evolutionary biologist Lars Gustafsson and his colleagues examined the phenomenon of stabilizing selection in the collared flycatcher (*Ficedula albicollis*) on the Swedish island of

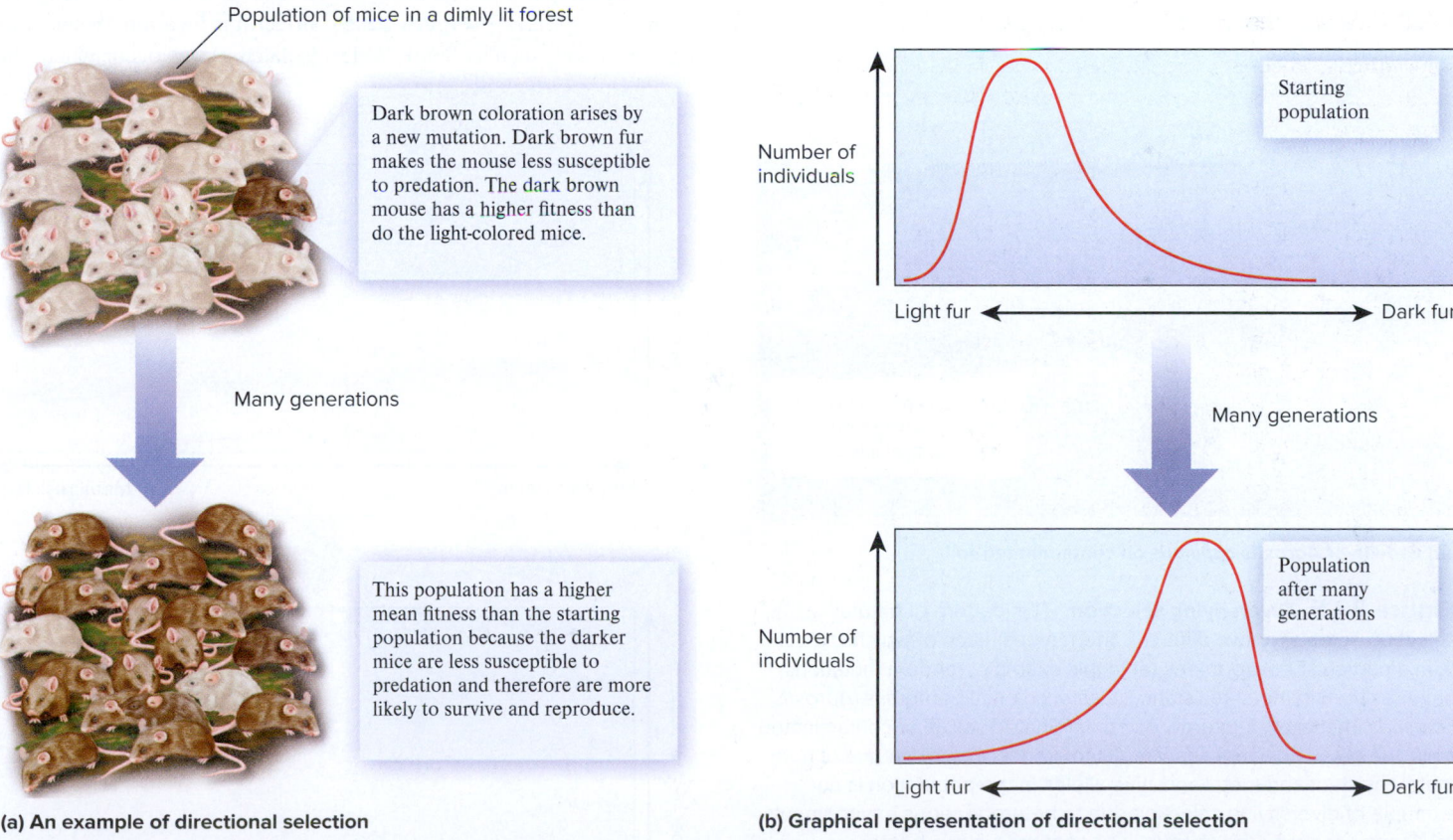

Population of mice in a dimly lit forest

Dark brown coloration arises by a new mutation. Dark brown fur makes the mouse less susceptible to predation. The dark brown mouse has a higher fitness than do the light-colored mice.

Many generations

This population has a higher mean fitness than the starting population because the darker mice are less susceptible to predation and therefore are more likely to survive and reproduce.

Number of individuals

Light fur ⟷ Dark fur

Starting population

Many generations

Number of individuals

Light fur ⟷ Dark fur

Population after many generations

**(a) An example of directional selection**

**(b) Graphical representation of directional selection**

**Figure 23.3** **Directional selection.** This pattern of natural selection selects for one extreme of a phenotype that confers the highest fitness in the population's environment. **(a)** In this example, a mutation causing darker fur arises in a population of mice. This new genotype confers higher fitness, because mice with darker fur can better evade predators and are more likely to survive and reproduce. Over many generations, directional selection favors the prevalence of individuals with darker fur. **(b)** These graphs show the change in fur color phenotypes before and after directional selection.

*Concept Check:* *Let's suppose the climate on an island abruptly changed so that the average temperature was 10°C higher. The climate change is permanent. How would directional selection affect the genetic diversity in a population of mice on the island (1) over the short run and (2) over the long run?*

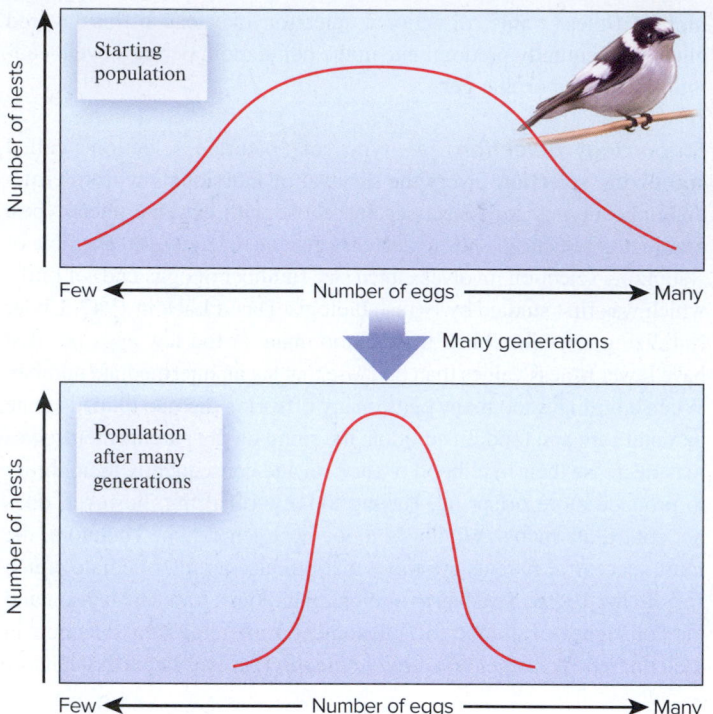

**Figure 23.4** **Stabilizing selection.** In this pattern of natural selection, the extremes of a phenotypic distribution are selected against. Those individuals with intermediate traits have the highest fitness. These graphs show the results of stabilizing selection on clutch size in a population of collared flycatchers (*Ficedula albicollis*). This process results in a population with less diversity and more uniform traits.

> *Concept Check:* *Why does stabilizing selection decrease genetic diversity?*

Gotland. They discovered that Lack's hypothesis concerning an optimal clutch size appears to be true for this species (**Figure 23.4**).

***Diversifying Selection*** **Diversifying selection** (also known as disruptive selection) favors the survival of two or more different genotypes that produce different phenotypes. In diversifying selection, the fitness values of a particular genotype are higher in one environment and lower in a different one, whereas the fitness values of the second genotype vary in an opposite manner. Diversifying selection is likely to occur in populations that occupy heterogeneous environments, so some members of the species are more likely to survive in each type of environmental condition.

An example of diversifying selection involves colonial bentgrass (*Agrostis capillaris*) (**Figure 23.5**). In certain locations where this grass is found, such as South Wales, isolated places occur where the

**(a) Growth of *Agrostis capillaris* on contaminated soil**

**Figure 23.5** **Diversifying selection.** This pattern of natural selection selects for two different phenotypes, each of which is most fit in a particular environment. **(a)** In this example, random mutations have resulted in metal-resistant alleles in colonial bentgrass (*Agrostis capillaris*) that allow it to grow on contaminated soil. In uncontaminated soils, the grass does not show metal tolerance. The existence of both metal-resistant and metal-sensitive alleles in the population is an example of diversifying selection due to heterogeneous environments. **(b)** Graphs showing the change in phenotypes in this bentgrass population before and after diversifying selection. a: Courtesy Mark MacNair, University of Exeter

 **Core Concept: Evolution** In this example, the frequencies of metal-resistant alleles become more prevalent when populations of *A. capillaris* are exposed to toxic metals in the soil.

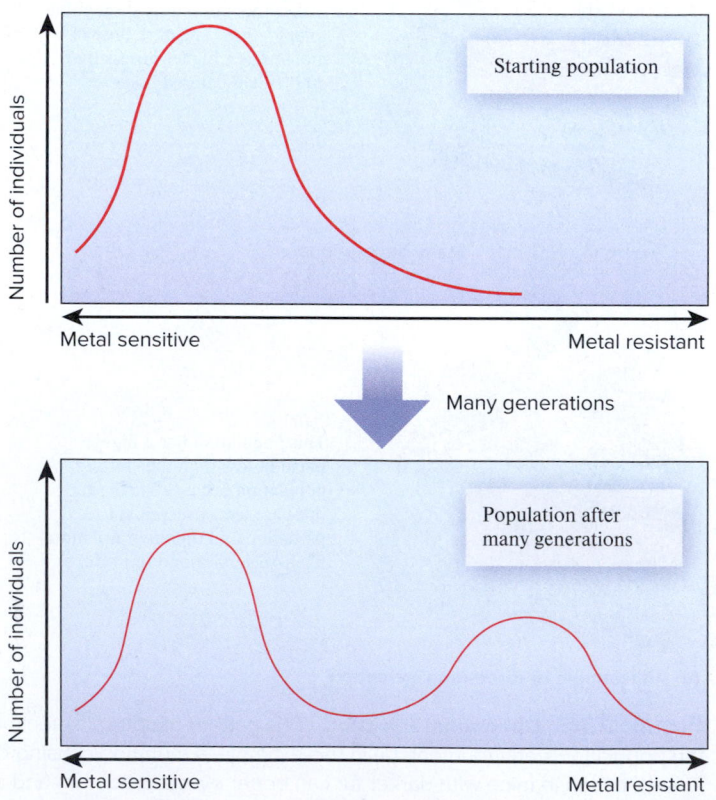

**(b) Graphical representation of disruptive selection**

soil is contaminated with high levels of heavy metals due to mining. The relatively recent metal contamination has selected for the proliferation of mutant strains of *A. capillaris* that are tolerant of the heavy metals (Figure 23.5a). Genetic changes enable these mutant strains to grow on contaminated soil but tend to inhibit their growth on normal, noncontaminated soil. These metal-resistant plants often grow on contaminated sites that are close to plants that grow on uncontaminated land and do not show metal tolerance.

*Balancing Selection*    Contrary to a popular misconception, natural selection does not always cause the elimination of "weaker" or less-fit alleles. **Balancing selection** is a type of natural selection that maintains genetic diversity in a population. Over many generations, balancing selection results in a **balanced polymorphism**, in which two or more alleles are kept in balance and therefore are maintained in a population over many generations.

How does balancing selection maintain polymorphism? Population geneticists have identified two common ways that balancing selection occurs. First, for genetic variation involving a single gene, balancing selection can favor the heterozygote over either corresponding homozygote. This phenomenon is called **heterozygote advantage**. Heterozygote advantage sometimes explains the persistence of alleles that are deleterious in a homozygous condition.

A classic example of heterozygote advantage involves the $H^S$ allele of the human β-globin gene. A homozygous $H^S H^S$ individual, such as Kimbareta, discussed at the beginning of the chapter, has sickle cell disease. This disease causes the red blood cells to form a sickle shape. Sickle-shaped cells deliver less oxygen to the body's tissues and can block the flow of blood through the vessels. The $H^S H^S$ homozygote has a lower fitness than a homozygote with two copies of the more common β-globin allele, $H^A H^A$. Heterozygotes, $H^A H^S$, do not typically show symptoms of the disease, but they have an increased resistance to malaria. Compared with $H^A H^A$ homozygotes, heterozygotes have the highest fitness because they have a 10–15% better chance of surviving if infected by the malarial parasite *Plasmodium falciparum*. Therefore, the $H^S$ allele is maintained in populations where malaria is prevalent, such as the Democratic Republic of Congo, even though the allele is detrimental in the homozygous state (**Figure 23.6**). This balanced polymorphism results in a higher mean fitness of the population. In areas where malaria is endemic, a population composed of all $H^A H^A$ individuals would have a lower mean fitness.

**Negative frequency-dependent selection** is a second way that natural selection produces balanced polymorphism. In this pattern of natural selection, the fitness of a genotype decreases when its frequency becomes higher. In other words, common individuals have a lower fitness, and rare individuals have a higher fitness. Therefore, common individuals are less likely to reproduce, whereas rare individuals are more likely to reproduce, thereby producing a balanced polymorphism in which no genotype becomes too rare or too common.

Negative frequency-dependent selection is thought to maintain polymorphisms among species that are preyed upon by predators. Research has shown that certain predators form a mental search image for their prey, which is usually based on the common type of prey in an area. A prey that exhibits a rare polymorphism that affects its appearance is less likely to be recognized by the predator. For example, a prey that is a different color from most other members of its

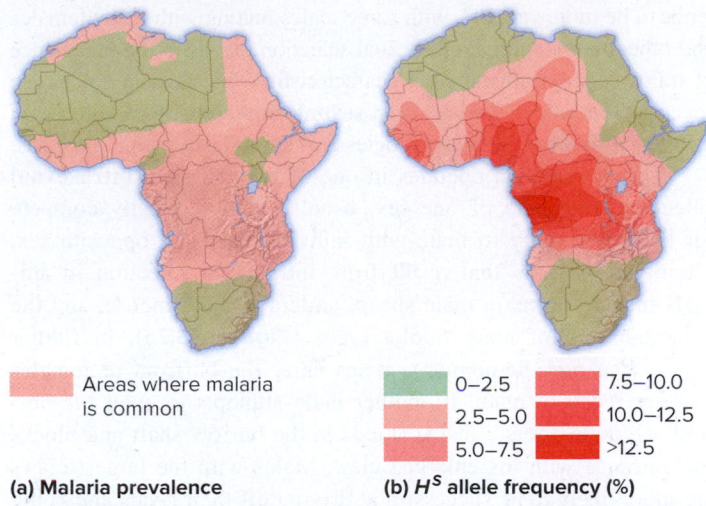

| | | |
|---|---|---|
| ![] Areas where malaria is common | ▨ 0–2.5 | ▨ 7.5–10.0 |
| | ▨ 2.5–5.0 | ▨ 10.0–12.5 |
| | ▨ 5.0–7.5 | ▨ >12.5 |

**(a) Malaria prevalence**          **(b) $H^S$ allele frequency (%)**

**Figure 23.6**  **Balancing selection and heterozygote advantage.** **(a)** The geographic prevalence of malaria in Africa. **(b)** The frequency of the $H^S$ allele of the β-globin gene in the same area. In the homozygous condition, the $H^S$ allele causes sickle cell disease. This allele is maintained in human populations in areas where malaria is prevalent, because the heterozygote ($H^A H^S$) has a higher fitness than either of the corresponding homozygotes ($H^A H^A$ or $H^S H^S$).

*Concept Check:* *If malaria was eradicated, what would you expect to happen to the frequencies of the $H^A$ and $H^S$ alleles over the long run?*

species may not be readily recognized by the predator. Such relatively rare organisms are subject to a lower rate of predation. This type of selection maintains polymorphism among certain prey populations.

## 23.3  Sexual Selection

**Learning Outcomes:**

1. Define sexual selection.
2. Distinguish between intrasexual and intersexual selection.
3. **CoreSKILL »** Analyze the results of Seehausen and van Alphen and explain how they relate to sexual selection.

Thus far, we have largely focused on examples of natural selection that produce adaptations for survival in particular environments. Now let's turn our attention to a form of natural selection called **sexual selection**, in which individuals with certain traits are more likely to engage in successful reproduction than other individuals. Darwin originally described sexual selection as "the advantage that certain individuals have over others of the same sex and species solely with respect to reproduction." In this section, we will explore how sexual selection alters traits that play a key role in reproduction.

### Sexual Selection Is a Type of Natural Selection Pertaining to Traits That Are Directly Involved with Reproduction

In many species of animals, sexual selection affects the characteristics of males more intensely than those of females. Unlike females, which tend to be fairly uniform in their reproductive success, male success tends to be more variable, with some males mating with many females and others not mating at all. Sexual selection results in the prevalence of traits, called secondary sex characteristics, that favor reproductive success. The process can result in **sexual dimorphism**—a significant difference between the morphologies of the two sexes within a species.

Sexual selection operates in one of two ways. In **intrasexual selection**, members of one sex, usually males, directly compete for the opportunity to mate with individuals of the opposite sex. Examples of traits that result from intrasexual selection in animals include horns in male sheep, antlers in male moose, and the enlarged claw of male fiddler crabs (**Figure 23.7a**). In fiddler crabs (*Uca paradussumieri*), males enter the burrows of females that are ready to mate. If another male attempts to enter the burrow, the male already inside stands in the burrow shaft and blocks the entrance with his enlarged claw. Males with the largest claws are more likely to be successful at driving off their rivals and being able to mate, and therefore more likely to pass on their genes to future generations.

In **intersexual selection**, also called mate choice, members of one sex, usually females, choose their mates from individuals of the other sex on the basis of certain desirable characteristics. This type of sexual selection often results in showy characteristics in males. **Figure 23.7b** shows a classic example of such a result—the Indian peafowl (*Pavo cristatus*), the national bird of India. Male peacocks have long and brightly colored tail feathers, which they fan out as a mating behavior. Female peahens select among males based on feather color and pattern as well as the physical prowess of the display.

A less obvious type of intersexual selection is cryptic female choice, in which the female reproductive system influences the relative success of sperm. For example, the female genital tract of certain animals selects for sperm that tend to be genetically unrelated to the female. Sperm from males closely related to the female, such as brothers or cousins, are less successful than are sperm from genetically unrelated males. The selection for sperm may occur over the journey through the reproductive tract. The egg itself may even have mechanisms to prevent fertilization by genetically related sperm. Cryptic female choice occurs in species in which females may mate with more than one male, such as many species of reptiles and ducks. A similar mechanism is found in many plant species in which pollen from genetically related plants, perhaps from the same flower, is unsuccessful at fertilization, whereas pollen from unrelated plants is successful. One possible advantage of cryptic female choice is that it inhibits inbreeding (described later in this chapter). At the population level, cryptic female choice may promote genetic diversity by favoring interbreeding among genetically unrelated individuals.

Sexual selection is sometimes a combination of both intrasexual and intersexual selection. During breeding season, male elk (*Cervus elaphus*) become aggressive and bugle loudly to challenge other male elk. Males spar with their antlers, which usually turns into a pushing match to determine which elk is stronger. Female elk then choose the strongest bulls as their mates.

Sexual selection can explain the existence of traits that could decrease an individual's chances of survival but increase the chances of reproducing. For example, the male guppy (*Poecilia reticulata*) is brightly colored compared with the female (**Figure 23.7c**).

**(a) Intrasexual selection**

**(b) Intersexual selection**

**(c) Sexual selection balanced by predation**

**Figure 23.7  Examples of the results of sexual selection, a type of natural selection.  (a)** An example of intrasexual selection. The enlarged claw of the male fiddler crab is used in direct male-to-male competition. In this photograph, a male inside a burrow is extending its claw out of the burrow to prevent another male from entering and mating with the female. **(b)** An example of intersexual selection. Female peahens choose male peacocks based on the males' colorful and long tail feathers and the robustness of their display. **(c)** Male guppies (on the right) are brightly colored to attract a female (on the left), but brightly colored males are less common where predation is high. Note: These photos also illustrate the concept of sexual dimorphism. a: ©Gerald Cubitt; b: ©Topham/The Image Works; c: ©Photo by Darren P. Croft, University of Exeter UK

**Concept Check:**  *Male birds of many species have loud and elaborate courtship songs. Is this trait likely to be involved in intersexual or intrasexual selection? Explain.*

In nature, females prefer brightly colored males. However, brightly colored males are more likely to be seen and eaten by predators. In places with few predators, the males tend to be brightly colored. In contrast, where predators are abundant, brightly colored males are less plentiful because they are subject to predation. In this case, the relative abundance of brightly and dully colored males depends on the balance between sexual selection, which favors bright coloring, and escape from predation, which favors dull coloring.

Many animals have secondary sexual characteristics, and evolutionary biologists generally agree that sexual selection is responsible for such traits. But why should males compete, and why should females be choosy? Researchers have proposed various hypotheses to explain the underlying mechanisms. One possible reason is related to the different roles that males and females play in the nurturing of offspring. In some animal species, the female is the primary caregiver, whereas the male plays a minor role. In such species, mating behavior may influence the fitness of both males and females. Males increase their fitness by mating with multiple females. This increases their likelihood of passing their genes on to the next generation. By comparison, females may produce relatively fewer offspring, and their reproductive success may not be limited by the number of available males. In these circumstances, females will have higher fitness if they choose males that are good defenders of their territory and have alleles that confer a survival advantage to their offspring.

One measure of alleles that confer higher fitness is age. Males that live to an older age are more likely to carry beneficial alleles. Many research studies involving female choice have shown that females tend to select traits that are more likely to be well developed in older males than in immature ones. For example, in certain species of birds, females tend to choose males with a larger repertoire of songs, which is more likely to occur in older males.

Sexual selection is governed by the same processes involved in the evolution of traits that are not directly related to sex. Sexual selection can occur by directional, stabilizing, diversifying, or balancing selection. For example, the evolution of the large and brightly colored tail of the male peacock is the result of directional selection.

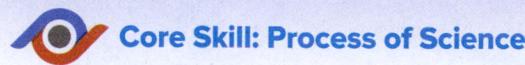

## Core Skill: Process of Science

## Feature Investigation | Seehausen and van Alphen Found That Male Coloration in African Cichlids Is Subject to Female Choice

In 1998, population geneticists Ole Seehausen and Jacques van Alphen investigated the possible role of sexual selection as it pertains to male coloration of two species of cichlid—a tropical freshwater fish popular among aquarium enthusiasts. The Cichlidae family is composed of more than 3,000 species that vary in body shape, coloration, behavior, and feeding habits, making it one of the largest and most diverse vertebrate families. By far, the greatest diversity of these fish occurs in Lake Victoria, Lake Malawi, and Lake Tanganyika in East Africa, where more than 1,800 species are found.

Cichlids have complex mating behavior, and females play an important role in choosing males with particular characteristics, such as color (see the chapter opening photo). In some locations, *Pundamilia pundamilia* and *P. nyererei* do not readily interbreed and behave like two distinct biological species, whereas in other places, they behave like a single interbreeding species with two color morphs. Males of both species have blackish underparts and blackish vertical bars on their sides (**Figure 23.8a**). *P. pundamilia* males are grayish white on top and on the sides, and they have a metallic blue and red dorsal fin—the uppermost fin. *P. nyererei* males are red-orange on top and yellow on their sides.

Seehausen and van Alphen hypothesized that females choose males for mates based, in part, on the males' coloration. The researchers took advantage of the observation that colors are obscured under orange monochromatic light. As seen in **Figure 23.8b**, males of both species look similar under these conditions. In this experiment, a female of one species was placed in an aquarium that contained one male of each species within an enclosure (**Figure 23.9**). The males

(a) **Males of two species in normal light**

(b) **Males of two species in artificial light**

**Figure 23.8**  **Male coloration in African cichlids. (a)** Two males (*Pundamilia pundamilia*, top, and *Pundamilia nyererei*, bottom) under normal illumination. **(b)** The same species under orange monochromatic light, which obscures their color differences.

a–b: ©Ole Seehausen

were within glass enclosures to prevent direct competition with each other, which would have likely affected female choice. The goal of the experiment was to determine which of the two males a female would prefer. Courtship between a male and female begins when a

**Figure 23.9**  **A study by Seehausen and van Alphen evaluating the effects of male coloration on female choice in African cichlids.**

**HYPOTHESIS** Female African cichlids choose mates based on the males' coloration.

**KEY MATERIALS** Two species of cichlid, *Pundamilia pundamilia* and *P. nyererei*, were chosen. The males differ with regard to their coloration. A total of 8 males and 8 females (4 males and 4 females from each species) were tested.

Experimental level     Conceptual level

1   Place 1 female and 2 males in an aquarium. Each male is within a separate glass enclosure. The enclosures contain 1 male from each species.

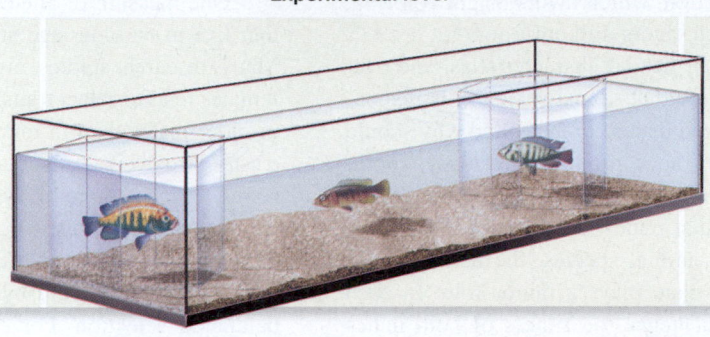

This is a method to evaluate sexual selection via female choice in 2 species of cichlid.

2   Observe potential courtship behavior for 1 hour. If a male exhibited lateral display (a courtship invitation) and then the female approached the enclosure that contained the male, this was scored as a positive encounter. This protocol was performed under normal light and under orange monochromatic light.

3   **THE DATA**

| Female | Male | Light condition | Percentage of positive encounters* |
|--------|------|-----------------|-----------------------------------|
| *P. pundamilia* | *P. pundamilia* | Normal | 16 |
| *P. pundamilia* | *P. nyererei* | Normal | 2 |
| *P. nyererei* | *P. nyererei* | Normal | 16 |
| *P. nyererei* | *P. pundamilia* | Normal | 5 |
| *P. pundamilia* | *P. pundamilia* | Monochromatic | 20 |
| *P. pundamilia* | *P. nyererei* | Monochromatic | 18 |
| *P. nyererei* | *P. nyererei* | Monochromatic | 13 |
| *P. nyererei* | *P. pundamilia* | Monochromatic | 18 |

*A positive encounter occurred when a male's lateral display was followed by the female approaching the male.

4   **CONCLUSION** Under normal light, where colors can be distinguished, *P. pundamilia* females prefer *P. pundamilia* males, and *P. nyererei* females prefer *P. nyererei* males.

5   **SOURCE** Seehausen, O., and van Alphen, J. J. M. 1998. The effect of male coloration on female mate choice in closely related Lake Victoria cichlids (Haplochromis nyererei complex). *Behavioral Ecology and Sociobiology* 42: 1–8.

male swims toward a female and exhibits a lateral display (that is, shows the side of his body to the female). If the female is interested, she will approach the male, and then the male will display a quivering motion. Such courtship behavior was examined under normal light and under orange monochromatic light.

As seen in the data, Seehausen and van Alphen found that the females' preference for males was dramatically different depending on the illumination conditions. Under normal light, *P. pundamilia* females preferred *P. pundamilia* males, and *P. nyererei* females preferred *P. nyererei* males. However, such mating preference was lost when colors were masked by artificial light. If the light conditions in their native habitats are similar to the normal light used in this experiment, female choice would be expected to separate cichlids into two populations, with *P. pundamilia* females mating with

*P. pundamilia* males and *P. nyererei* females mating with *P. nyererei* males. In this case, sexual selection appears to have followed a diversifying mechanism in which certain females prefer males with one color pattern, whereas other females prefer males with a different color pattern. A possible outcome of such sexual selection is that it can separate one large population into smaller populations that selectively breed with each other and eventually become distinct species. We will discuss the topic of species formation in more depth in Chapter 24.

**Experimental Questions**

1. What hypothesis was tested in the Seehausen and van Alphen experiment?

2. **CoreSKILL »** Describe the experimental design for this study, illustrated in Figure 23.9. What was the purpose of conducting the experiment under the two different light conditions?

3. **CoreSKILL »** Analyze the results of the experiment in Figure 23.9, and explain what they mean with regard to sexual selection.

## 23.4 Genetic Drift

**Learning Outcomes:**

1. Define genetic drift, and explain its effects on allele frequencies over time.
2. Compare and contrast the bottleneck and founder effects.
3. Explain how neutral mutations can spread through a population.

Thus far, we have focused on natural selection as a mechanism that can promote widespread genetic changes in a population. Let's now turn our attention to a second important way the gene pool of a population can change. In the 1930s, Sewall Wright played a large role in developing the concept of **genetic drift** (also called random genetic drift), which refers to changes in allele frequencies due to random chance. The term genetic drift is derived from the observation that allele frequencies may "drift" randomly from generation to generation as a matter of chance.

Changes in allele frequencies due to genetic drift happen regardless of the fitness of individuals that carry those alleles. For example, an individual with a high fitness value may, by chance, not encounter a member of the opposite sex. Likewise, random chance can influence which alleles happen to be carried in the gametes that fuse with each other in a successful fertilization. In this section, we will examine how genetic drift alters allele frequencies in populations.

### Genetic Drift Has a Greater Effect in Small Populations

What are the effects of genetic drift? Over the long run, genetic drift favors either the elimination (frequency of 0%) or the fixation (frequency of 100%) of an allele in a population. However, the number of generations it takes for an allele to be lost or fixed greatly depends on the population size. **Figure 23.10** illustrates the potential consequences of genetic drift in one large ($N = 1,000$) and two small ($N = 10$) populations ($N$ is the number of individuals that each population contains). This simulation involves the frequency of hypothetical $B$ and $b$ alleles of a gene for fur color in a population of mice—$B$ is the black allele, and $b$ is the white allele.

At the beginning of this hypothetical simulation, which ran for 50 generations, all three populations had identical allele frequencies: $B = 0.5$ and $b = 0.5$. In the small populations, the allele frequencies

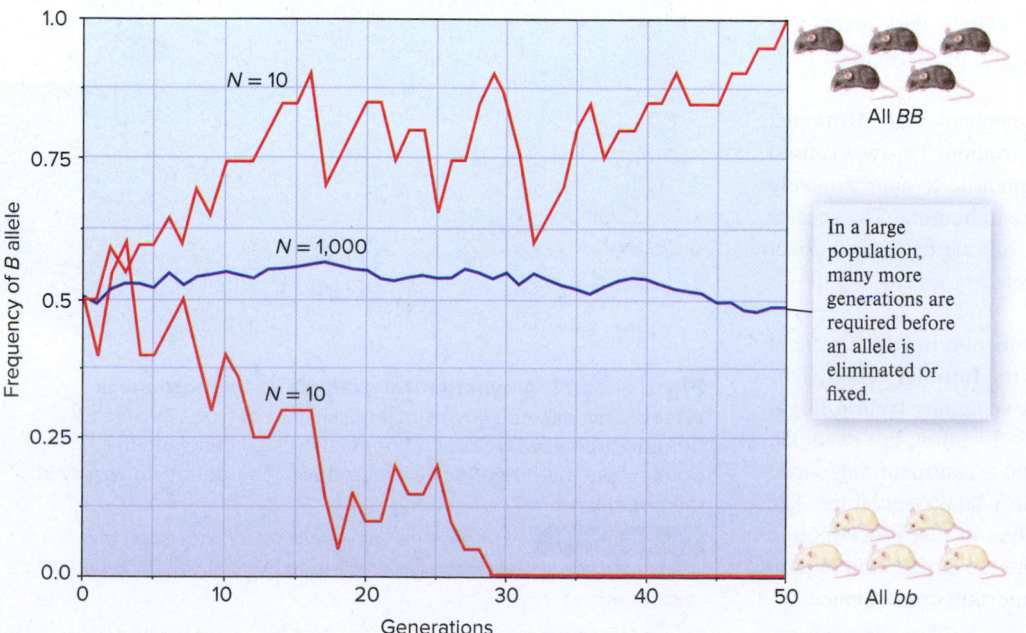

In a large population, many more generations are required before an allele is eliminated or fixed.

All *BB*

All *bb*

**Figure 23.10  Genetic drift and population size.** This graph shows three hypothetical simulations of genetic drift and their effects on small and large populations of black ($B$ allele) and white ($b$ allele) mice. In all cases, the starting allele frequencies are $B = 0.5$ and $b = 0.5$. The red lines illustrate two populations of mice in which $N = 10$; the blue line shows a population in which $N = 1,000$.

 **Core Skill: Quantitative Reasoning** After many generations, random fluctuations in allele frequencies can lead to dramatic differences in the genetic compositions of different populations.

fluctuated substantially from generation to generation. Eventually, in one population, the *b* allele was eliminated; in another, it was fixed at 100%. These small populations then consisted of only black mice or white mice, respectively. At this point, the gene became monomorphic, and allele frequencies could no longer fluctuate due to genetic drift.

By comparison, the frequencies of *B* and *b* in the large population fluctuated much less. As discussed in Chapter 17, the relative effect of random chance, termed random sampling error, is much smaller when the sample size is large. Nevertheless, genetic drift can eventually lead to allele loss or fixation even in large populations, but this will take many more generations to occur than it does in small populations.

In nature, genetic drift may rapidly alter allele frequencies when the size of a population dramatically decreases. Two examples of this phenomenon are the bottleneck effect and the founder effect, which are described next.

***Bottleneck Effect*** A population can be dramatically reduced in size by events such as earthquakes, floods, drought, and human destruction of habitat. These occurrences may eliminate most members of the population without regard to their genetic composition. The population is said to have passed through a bottleneck. The change in allele frequencies of the resulting population due to genetic drift is called the **bottleneck effect**. Some alleles may be over-represented whereas others may even be eliminated. Such changes may happen for two reasons. First, the surviving population often has allele frequencies that differ from those of the original population that was much larger. Second, as we saw in Figure 23.10, genetic drift acts more quickly to reduce genetic variation when the population size is small. Eventually, a population that has gone through a bottleneck may regain its original size. However, the new population is likely to have less genetic variation than the original one.

A hypothetical example of the bottleneck effect is shown with a population of frogs in **Figure 23.11**. In this example, a starting population of frogs is found in three phenotypes: yellow, dark green, and striped. Due to a bottleneck caused by a drought, the dark green variety is lost from the population.

As a real-life example, the northern elephant seal (*Mirounga angustirostris*) has lost much of its genetic variation. This was caused by a bottleneck in which the population decreased to approximately 20 to 30 surviving members in the 1890s due to hunting. The species has rebounded in numbers to over 100,000, but the bottleneck effect reduced its genetic variation to very low levels.

***Founder Effect*** Another common phenomenon in which genetic drift may rapidly alter allele frequencies is the **founder effect**. This occurs when a small group of individuals separates from a larger population and establishes a colony in a new location. For example, a few individuals from a large population on a continent may move to an island and become the founders of an island population. The founder effect differs from a bottleneck effect in that it occurs in a new location, although both effects are related to a reduction in population size. The founder effect has two important consequences.

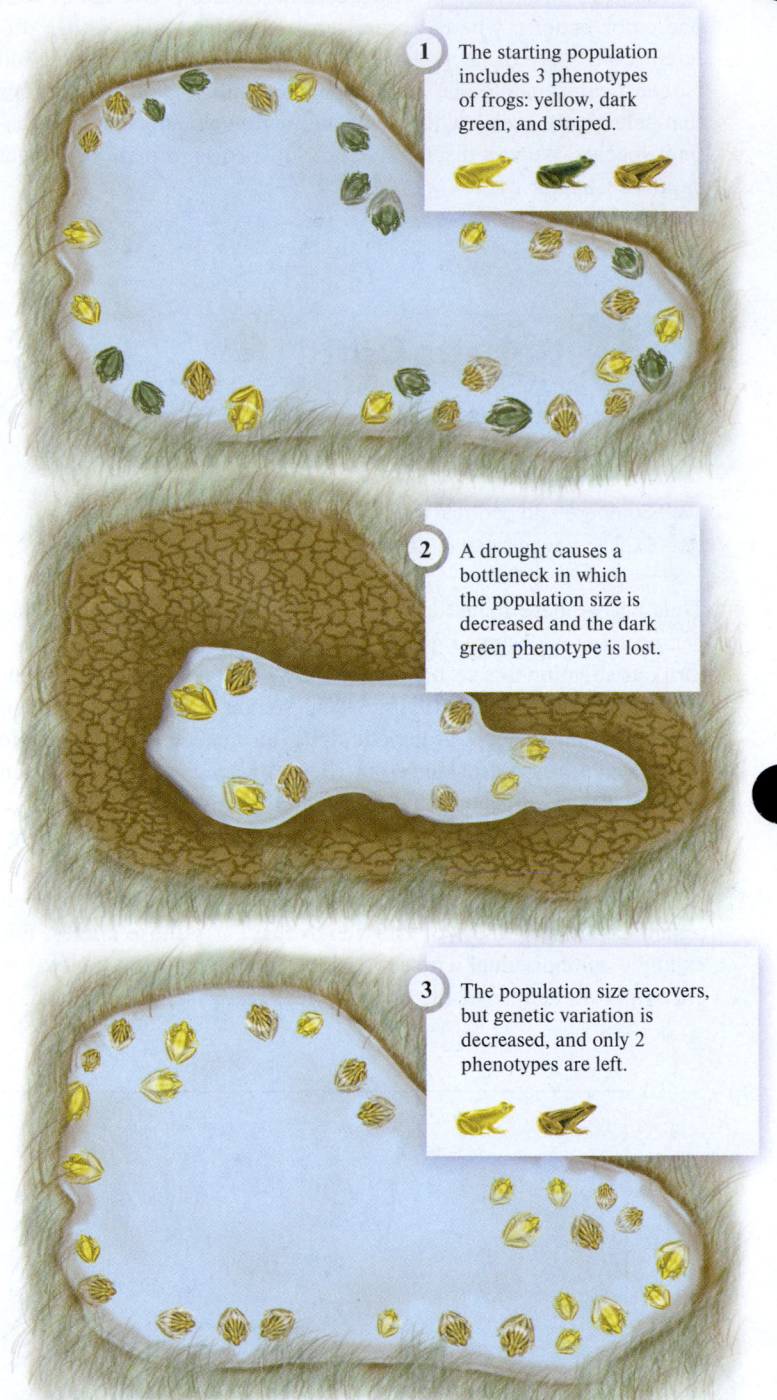

1. The starting population includes 3 phenotypes of frogs: yellow, dark green, and striped.

2. A drought causes a bottleneck in which the population size is decreased and the dark green phenotype is lost.

3. The population size recovers, but genetic variation is decreased, and only 2 phenotypes are left.

**Figure 23.11 A hypothetical example of the bottleneck effect.** This example involves a population of frogs in which a drought dramatically reduced population size, resulting in a bottleneck. The bottleneck effect reduced the genetic diversity in the population.

*Concept Check:* *How does the bottleneck effect undermine the efforts of conservation biologists who are trying to save species nearing extinction?*

- First, the founding population, which is relatively small, is expected to have less genetic variation than the larger original population from which it was derived.

- Second, as a matter of chance, the allele frequencies in the founding population may differ markedly from those of the original population.

Population geneticists have studied many examples in which isolated populations were founded via colonization by members of another population. For example, in the 1960s, American geneticist Victor McKusick studied allele frequencies in the Amish of Lancaster County, Pennsylvania. At that time, this group included about 8,000 people, descended from just three couples that immigrated to the U.S. in 1770. Among this population of 8,000, a genetic disease known as Ellis–van Creveld syndrome (a recessive form of dwarfism) was found at a frequency of 0.07, or 7%. By comparison, this disorder is extremely rare in other human populations, even the population from which the founding members had originated. Evidence suggests that the high frequency in the Lancaster County population can be traced to one couple, one of whom carried the mutated gene that causes the syndrome.

## Genetic Drift Plays an Important Role in Promoting Genetic Change

In 1968, Japanese evolutionary biologist Motoo Kimura proposed that much of the DNA sequence variation seen in genes in natural populations is the result of genetic drift rather than natural selection. Genetic drift is a random process that does not preferentially select for any particular allele—it can alter the frequencies of both beneficial and deleterious alleles. Much of the time, genetic drift promotes **neutral variation**—changes in genes and proteins that do not have an effect on reproductive success.

According to Kimura, most variation in DNA sequences is due to the accumulation of neutral mutations that have attained high frequencies in a population via genetic drift. For example, a new mutation within a gene that changes a glycine codon from GGG to GGC would not affect the amino acid sequence of the encoded protein. Both genotypes may be equal in fitness. However, such new mutations can spread throughout a population due to genetic drift (**Figure 23.12**). This phenomenon has been called **non-Darwinian evolution** and also "survival of the luckiest." Kimura agreed with Darwin that natural selection is responsible for adaptive changes in a species during evolution. The long snout of an anteater is the result of natural selection. His main idea is that much of the variation in DNA sequences is explained by neutral variation rather than adaptive variation.

The sequencing of genomes from many species is consistent with Kimura's proposal. When researchers examine changes of the coding sequence within protein-encoding genes, nucleotide substitutions are found to be more prevalent in the third base of a codon than in the first or second base. Mutations in the third base are often neutral; that is, they do not change the amino acid sequence of the protein (refer back to Table 12.1). In contrast, random mutations at the first or second base are more likely to be harmful than beneficial and tend to be eliminated from a population.

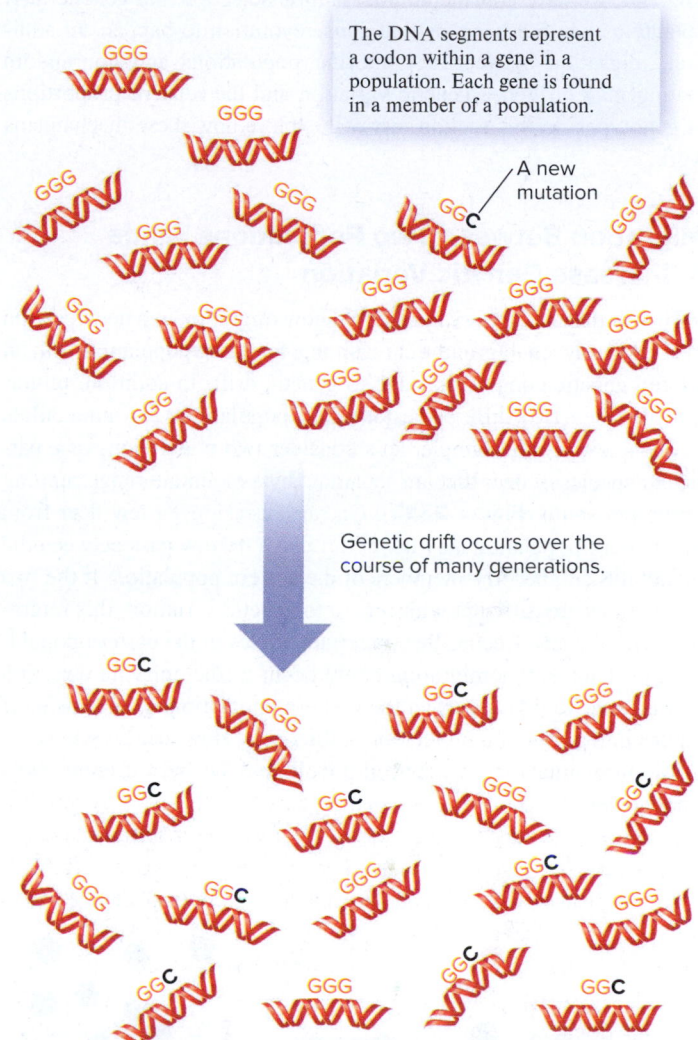

The DNA segments represent a codon within a gene in a population. Each gene is found in a member of a population.

A new mutation

Genetic drift occurs over the course of many generations.

**Figure 23.12  Non-Darwinian evolution in a population.**  In this example, a mutation within a gene changes a glycine codon from GGG to GGC, which does not affect the amino acid sequence of the encoded protein. Each gene shown represents a copy of the gene in a member of a population. Over the course of many generations, genetic drift may cause this altered allele to become prevalent in the population, perhaps even monomorphic.

 **Core Skill: Connections**  Look back at the genetic code described in Table 12.1. Describe three different genetic changes that you would expect to be neutral.

## 23.5  Migration and Nonrandom Mating

**Learning Outcomes:**

1. Describe how gene flow affects genetic variation in neighboring populations.
2. Define inbreeding, and explain how it may have detrimental consequences.

Thus far, we have considered how natural selection and genetic drift operate as key mechanisms that cause evolution to happen. In addition, migration between neighboring populations and nonrandom mating may influence genetic variation and the relative proportions of genotypes. In this section, we will explore how these mechanisms work.

## Migration Between Two Populations Tends to Increase Genetic Variation

Earlier in this chapter, we considered how migration to a new location by a relatively small group can result in a founding population with an altered genetic composition due to genetic drift. In addition, migration between two different established populations can alter allele frequencies. As an example, let's consider two populations of a particular species of deer that are separated by a mountain range running north and south (**Figure 23.13**). On rare occasions, a few deer from the western population may travel through a narrow pass between the mountains and become members of the eastern population. If the two populations are different with regard to genetic variation, this migration will alter the frequencies of certain alleles in the eastern population. Of course, this migration could occur in the opposite direction as well and would then affect the western population. This transfer of alleles into or out of a population, called **gene flow**, occurs whenever fertile individuals move between populations having different allele frequencies.

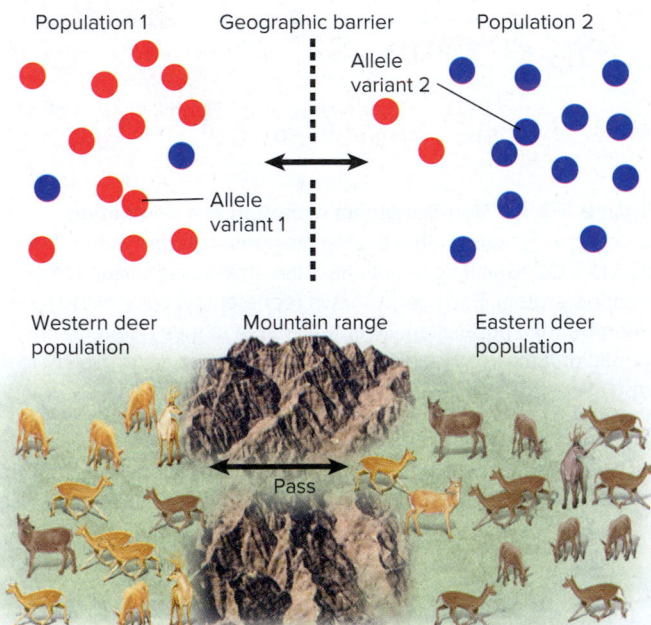

**Figure 23.13** **Migration and gene flow.** In this example, two populations of a deer species are separated by a mountain range. On rare occasions, a few deer from one population travel through a narrow pass and become members of the other population. If the two populations differ in regard to genetic variation, this migration will alter the frequencies of alleles in the populations.

**Concept Check:** *How does migration affect the genetic compositions of populations?*

What are the consequences of migration? First, migration tends to reduce differences in allele frequencies between neighboring populations. Population geneticists can evaluate the extent of migration between two populations by analyzing the similarities and differences between their allele frequencies. Populations that frequently mix their gene pools via migration tend to have similar allele frequencies, whereas the allele frequencies of isolated populations are more disparate, due to the effects of natural selection and genetic drift. Second, migration tends to increase genetic diversity within populations. As discussed earlier in this chapter, new mutations are relatively rare events. Therefore, a new mutation may arise in only one population, and migration may then introduce this new allele into a neighboring population.

## Nonrandom Mating Affects the Relative Proportion of Homozygotes and Heterozygotes in a Population

As mentioned earlier, one of the conditions required to establish Hardy-Weinberg equilibrium is random mating, which means that members of a population choose their mates irrespective of their genotypes or phenotypes. In many species, including human populations, this condition is violated. Such **nonrandom mating** takes different forms. Assortative mating occurs when individuals with similar phenotypes are more likely to mate. If the similar phenotypes are due to similar genotypes, assortative mating tends to increase the proportion of homozygotes and decrease the proportion of heterozygotes in the population. The opposite situation, where dissimilar phenotypes mate preferentially, causes heterozygosity to increase.

Another form of nonrandom mating involves the choice of mates based on their genetic history rather than their phenotypes. Individuals may choose a mate that is part of the same genetic lineage. The mating of two genetically related individuals, such as cousins, is called **inbreeding**. This sometimes occurs in human societies and is more likely to take place in nature when population size becomes very small.

In the absence of other evolutionary factors, nonrandom mating does not affect allele frequencies in a population. However, it will alter the balance of genotypes predicted by the Hardy-Weinberg equation. As an example, let's consider a human pedigree involving a mating between cousins (**Figure 23.14**). Individuals III-2 and III-3 are cousins and have produced the daughter labeled IV-1. She is said to be inbred, because her parents are genetically related. The parents of an inbred individual have one or more common ancestors. In the pedigree of Figure 23.14, I-2 is the grandfather of both III-2 and III-3.

Inbreeding increases the relative proportions of homozygotes and decreases the likelihood of heterozygotes in a population. Why does this happen? An inbred individual has a higher chance of being homozygous for any given gene because the same allele for that gene could be inherited twice from a common ancestor. For example, individual I-2 is a heterozygote, *Cc*. The *c* allele could pass from I-2 to II-2 to III-2 and finally to IV-1 (see red arrows in Figure 23.14). Likewise, the *c* allele could pass from I-2 to II-3 to III-3 and then to IV-1. Therefore, IV-1 has a chance of being homozygous because she inherited both copies of the *c* allele from a common ancestor to both of her parents. Inbreeding does not favor any particular allele—it does

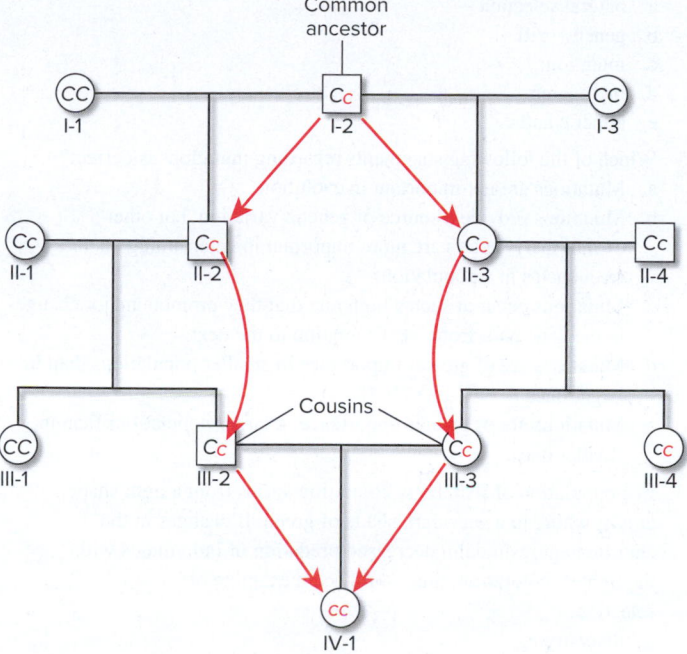

Common
ancestor

**Figure 23.14** **A human pedigree containing inbreeding.** The parents of individual IV-1 are genetically related (cousins), and, therefore, individual IV-1 is inbred. Inbreeding increases the likelihood that an individual will be homozygous for any given gene. The red arrows show how IV-1 could become homozygous by inheriting the same allele (c) that the common ancestor (I-2) passed to both of her parents.

not favor *c* over *C*—but it does increase the likelihood that an individual will be homozygous for any given gene.

Although inbreeding by itself does not affect allele frequencies, it may have negative consequences with regard to recessive alleles. Rare recessive alleles that are harmful in the homozygous condition are found in all natural populations. Such alleles do not usually pose a problem because heterozygotes carrying a rare recessive allele are also rare, making it very unlikely that two such heterozygotes will mate with each other. However, related individuals share some of their genes, including recessive alleles. Therefore, if inbreeding occurs, homozygous offspring are more likely to be produced. For example, rare recessive diseases in humans are more frequent when inbreeding occurs.

In natural populations, inbreeding lowers the mean fitness of the population if homozygous offspring have lower fitness values. This can be a serious problem as natural populations become smaller due to human destruction of habitat. As a population shrinks, inbreeding becomes more likely because individuals have fewer potential mates from which to choose. The inbreeding, in turn, produces homozygotes that are less fit, thereby decreasing the reproductive success of the population. This phenomenon is called **inbreeding depression**. Conservation biologists sometimes try to circumvent this problem by introducing individuals from one population into another. For example, the endangered Florida panther (*Puma concolor coryi*) suffers from inbreeding-related defects, which include poor sperm quality and quantity and morphological abnormalities. To alleviate these effects, panthers from Texas have been introduced into the Florida population of panthers.

## Summary of Key Concepts

### 23.1 Genes in Populations

- Population genetics is the study of genes and genotypes in a population. A population is a group of individuals of the same species that occupy the same environment at the same time and can interbreed if the species reproduces sexually. All of the alleles for every gene in a population constitute the population's gene pool.

- Polymorphism, which is very common in nearly all populations, refers to the presence of two or more variations of a character in a population. In contrast, a monomorphic gene exists as a single allele (comprising more than 99% of the alleles of the gene) in a population (Figure 23.1).

- Allele frequency is the number of copies of a specific allele divided by the total number of all alleles for that gene in a population. Genotype frequency is the number of individuals with a given genotype divided by the total number of individuals in a population.

- The Hardy-Weinberg equation ($p^2 + 2pq + q^2 = 1$) predicts that allele and genotype frequencies will remain in equilibrium if no new mutations arise, no natural selection occurs, the population size is very large, migration does not occur, and mating is random (Figure 23.2).

- Sources of new genetic variation include random gene mutations, gene duplications, and horizontal gene transfer. Natural selection, genetic drift, migration, and nonrandom mating may alter allele and genotype frequencies and cause a population to evolve (Table 23.1).

### 23.2 Natural Selection

- Natural selection is the process by which individuals with certain heritable traits that favor survival and reproduction tend to become more prevalent in a population. Fitness, the relative likelihood that a genotype will contribute to the gene pool of the next generation, is a measure of reproductive success.

- Directional selection is the process by which one extreme of a phenotypic distribution is favored (Figure 23.3).

- During stabilizing selection, individuals with an intermediate phenotype have greater reproductive success (Figure 23.4).

- Diversifying selection is the process by which two or more phenotypes are favored. An example is a population that occupies a heterogeneous environment (Figure 23.5).

- Balancing selection maintains balanced polymorphism in a population. Examples of the ways in which balancing selection occurs include heterozygote advantage and negative frequency-dependent selection (Figure 23.6).

### 23.3 Sexual Selection

- Sexual selection is a form of natural selection in which individuals with certain traits are more likely than others to engage in successful reproduction. In intrasexual selection, members of one sex compete for the opportunity to mate with individuals of the opposite sex. In intersexual selection, members of one sex choose their mates on the basis of certain desirable characteristics (Figure 23.7).

- Seehausen and van Alphen discovered that female cichlids' choice of mates is influenced by male coloration. This is an example of sexual selection (Figures 23.8, 23.9).

## 23.4 Genetic Drift

- Genetic drift involves changes in allele frequencies over time due to random chance. It occurs more rapidly in small populations and leads to either the elimination or the fixation of alleles (Figure 23.10).

- In the bottleneck effect, an environmental event dramatically reduces a population size, and the allele frequencies of the resulting population change due to genetic drift (Figure 23.11).

- The founder effect occurs when a small population moves to a new geographic location and genetic drift alters the genetic composition of that population.

- Kimura proposed that genetic drift promotes neutral variation, or the accumulation of changes in genes and proteins that do not affect reproductive success. Much of the genetic variation in DNA sequences in populations appears to be the result of genetic drift rather than natural selection (Figure 23.12).

## 23.5 Migration and Nonrandom Mating

- Gene flow occurs when individuals migrate between populations with different allele frequencies. It reduces differences in allele frequencies between populations and enhances genetic diversity (Figure 23.13).

- Inbreeding, a form of nonrandom mating in which genetically related individuals have offspring with each other, tends to increase the proportion of homozygotes relative to heterozygotes. When the resulting homozygotes have lower fitness, the phenomenon called inbreeding depression is the result (Figure 23.14).

## Assess & Discuss

### Test Yourself

1. Population geneticists are interested in the genetic variation in populations. The most common type of genetic change that causes polymorphism in a population is
   a. a deletion of a gene sequence.
   b. a duplication of a region of a gene.
   c. a rearrangement of a gene sequence.
   d. a single-nucleotide substitution.
   e. an inversion of a segment of a chromosome.

2. The Hardy-Weinberg equation characterizes the allele and genotype frequencies
   a. of a population that is experiencing selection for mating success.
   b. of a population that is extremely small.
   c. of a population that is very large and not evolving.
   d. of a community of species that is not evolving.
   e. of a community of species that is experiencing selection.

3. In the Hardy-Weinberg equation, what portion represents the frequency of individuals that do not exhibit a recessive disease but are carriers of a recessive allele?
   a. $q$
   b. $p^2$
   c. $2pq$
   d. $q^2$
   e. both b and d

4. By itself, which of the following is *not* likely to have a major influence on allele frequencies?
   a. natural selection
   b. genetic drift
   c. mutation
   d. inbreeding
   e. either c and d

5. Which of the following statements regarding mutations is correct?
   a. Mutations are not important in evolution.
   b. Mutations provide a source of genetic variation, but other evolutionary factors are more important in determining allele frequencies in a population.
   c. Mutations occur at such a high rate that they promote major changes in the gene pool from one generation to the next.
   d. Mutations are of greater importance in smaller populations than in larger ones.
   e. Mutations are of greater importance in larger populations than in smaller ones.

6. In a population of fish, body coloration varies from a light shade, almost white, to a very dark shade of green. If changes in the environment resulted in decreased predation of individuals with the lightest coloration, this would be an example of _____ selection.
   a. diversifying
   b. stabilizing
   c. directional
   d. sexual
   e. artificial

7. For the population of fish described in question 6, if the stream environment included some areas with a sandy, light-colored bottom and some with a lot of dark-colored vegetation, both the light- and dark-colored fish would have a selective advantage and increased survival in certain places. This scenario could explain the occurrence of
   a. genetic drift.
   b. diversifying selection.
   c. mutation.
   d. stabilizing selection.
   e. sexual selection.

8. The microevolutionary factor most sensitive to population size is
   a. mutation.
   b. migration.
   c. selection.
   d. genetic drift.
   e. all of the above.

9. Kimura's proposal regarding neutral variation differs from Darwinian evolution in that
   a. natural selection does not exist.
   b. most of the genetic variation in a population is due to mutations that do not affect reproductive success.
   c. neutral variation alters survival and reproductive success.
   d. neutral mutations are not affected by population size.
   e. it differs with respect to both b and c.

10. Populations that experience inbreeding may also experience
    a. a decrease in fitness due to an increased frequency of recessive genetic diseases.
    b. an increase in fitness due to increases in heterozygosity.
    c. very little genetic drift.
    d. no apparent change.
    e. increased mutation rates.

## Conceptual Questions

1. The percentage of individuals exhibiting a recessive disease in a population is 0.04, or 4%. Based on the Hardy-Weinberg equation, what percentage of individuals in this population are expected to be heterozygous carriers?

2. Compare and contrast the four patterns of natural selection that result in adaptation to a given environment and also describe sexual selection.

3. **Core Concept: Evolution**  Explain how genetic drift results in evolution.

## Collaborative Questions

1. Antibiotics are commonly used to combat bacterial and fungal infections. During the past several decades, however, antibiotic-resistant strains of microorganisms have become alarmingly prevalent. This has undermined the ability of physicians to treat many types of infectious disease. Discuss how the following processes that alter allele frequencies may have contributed to the emergence of antibiotic-resistant strains:

   a. random mutation
   b. genetic drift
   c. natural selection

2. Discuss the similarities and differences among directional, disruptive, balancing, and stabilizing selection.

# Origin of Species and Macroevolution

# 24

**Two different species of zebras.** Grevy's zebra (*Equus grevyi*) is shown on the left, and Grant's zebra (*Equus quagga boehmi*), which has fewer and thicker stripes, is shown on the right.
©Frederic B. Siskind

The origin of living organisms has been described by philosophers as the great "mystery of mysteries." Perhaps that is why so many different views have been put forth to explain the existence of living species. At the time of Aristotle (4th century B.C.E.), most people believed that some living organisms came into being by spontaneous generation—the idea that nonliving materials could give rise to living organisms. For example, some common beliefs were that worms and frogs could arise from mud, and mice could come from grain. By comparison, many religious teachings contend that species were divinely made and have remained the same since their creation. In contrast to these ideas, the work of Charles Darwin gave us the scientific theory of evolution by descent with modification. Darwin's work, and that of subsequent biologists, helps us to understand the diversity of life, and in particular,

it presents a scientific explanation for how new species can evolve from pre-existing species.

This chapter builds in an exciting way on the information that we have considered in previous chapters. In Chapter 22, we surveyed the tenets on which the theory of evolution is built. Next, Chapter 23 considered microevolution—evolution on the small scale of changes in allele frequencies in a population. In this chapter, we will explore evolution on a larger scale. **Macroevolution** refers to evolutionary changes that produce new species and groups of species.

To biologists, the concept of a **species** has come to mean a group of related organisms that share a distinctive set of attributes in nature. Members of the same species share an evolutionary history, which makes them more genetically similar to each other than they are to members of a different species. You may already have an intuitive sense of this concept. It is obvious that zebras and mice are different species. However, as we will see in the first section of this chapter, the distinction between different, closely related species is often blurred in natural environments. Two closely related species may look very similar, as the chapter opening photo illustrates. Species identification has several practical uses. For example, it allows biologists to plan for the preservation and conservation of endangered species. In addition, species identification is often important for a physician to correctly identify the microorganism that is causing a disease in a patient so the proper medication can be prescribed.

In this chapter, we will also focus on the mechanisms that promote the formation of new species, a phenomenon called **speciation**. Such macroevolution typically occurs by the accumulation of microevolutionary changes, those that occur in single genes (see Chapter 23). We will also consider how macroevolution can happen at a fast or slow pace and explore how variations in the genes that control development play a role in the evolution of new species.

## 24.1 | Identification of Species

### Learning Outcomes:

1. Outline the characteristics that biologists use to distinguish different species.
2. Describe different species concepts.
3. Compare and contrast prezygotic and postzygotic isolating mechanisms.

How many different species are on Earth? The number is astounding. A study done by American biologist E. O. Wilson and colleagues in 1990 estimated the known number of species at approximately 1.4 million. Currently, about 1.8 million species have been identified and catalogued. However, a vast number of species have yet to be classified. This is particularly true among bacteria and archaea, which are difficult to categorize into distinct species. Also, new invertebrate and even vertebrate species are still being found in the far reaches of pristine habitats. Common estimates of the total number of species range from 5 to 50 million!

When studying natural populations, evolutionary biologists are often confronted with situations in which some differences between two populations are apparent, but it is difficult to decide whether the two populations truly represent separate species. When two or more geographically restricted groups of the same species display one or more traits that are somewhat different but not enough to warrant their placement into different species, biologists sometimes classify such groups as **subspecies**. Similarly, many bacterial species are subdivided into **ecotypes**. Each ecotype is a genetically distinct population adapted to its local environment. In this section, we will consider the characteristics that biologists examine when deciding if two groups of organisms constitute different species.

### Each Species Is Established Using Characteristics and Histories That Distinguish It from Other Species

As mentioned, a species is a group of organisms that share a distinctive set of attributes in nature. In the case of sexually reproducing species, members of one species usually cannot successfully interbreed with members of other species. Members of the same species share an evolutionary history that is distinct from those of other species. Although this may seem like a reasonable way to characterize a given species, biologists would agree that distinguishing between species is a more difficult undertaking. What criteria do we use to distinguish species? How many differences must exist between two populations to classify them as different species? Such questions are often difficult to answer.

The characteristics that a biologist uses to identify a species depend, in large part, on the species in question. For example, the traits used to distinguish insect species are quite different from those used to identify different bacterial species. The relatively high level of horizontal gene transfer among bacteria presents special challenges in the grouping of bacterial species. Among bacteria, it is sometimes very difficult and perhaps arbitrary to divide closely related organisms into separate species.

The most commonly used characteristics for identifying species are morphological traits, the ability to interbreed, molecular features, ecological factors, and evolutionary relationships. A comparison of

these concepts will help you appreciate the various approaches that biologists use to identify the bewildering array of species on our planet.

***Morphological Traits***    One way to establish that a population constitutes a unique species is based on the physical characteristics of its members. Organisms are classified as the same species if their anatomical traits appear to be very similar. Likewise, microorganisms can be classified according to morphological traits at the cellular level. By comparing many different morphological traits, biologists may decide that certain populations constitute unique species.

Although an analysis of morphological traits is a common way for biologists to establish that a particular group constitutes a species, this approach has drawbacks. First, researchers may have difficulty deciding how many traits to consider. In addition, quantitative traits, such as size and weight, that vary in a continuous way among members of the same species are not easy to analyze. Another drawback is that the degree of dissimilarity that distinguishes different species may not show a simple relationship. The members of the same species sometimes look very different, and conversely, members of different species sometimes look remarkably similar to each other. For example, **Figure 24.1a** shows two different frogs of the species

**(a) Frogs of the same species**

**(b) Frogs of different species**

**Figure 24.1** **Difficulties of using morphological traits to identify species.** In some cases, members of the same species appear quite different. In other cases, members of different species look very similar. (a) Two frogs of the same species, the dyeing poison frog (*Dendrobates tinctorius*). (b) Two different species of frog, the northern leopard frog (*Rana pipiens*, left) and the southern leopard frog (*Rana utricularia*, right).
a (left): ©Mark Smith/Science Source; a (right): ©Pascal Goetgheluck/ardea.com; b (left): ©Ron Erwin/Getty Images; b (right): ©Robin Chittenden/Alamy Stock Photo

**Concept Check:** *Can you think of another example of two different species that look very similar?*

*Dendrobates tinctorius*, commonly called the dyeing poison frog. This species exists in many different-colored morphs, which are individuals of the same species that have noticeably dissimilar appearances. In contrast, **Figure 24.1b** shows two different species of frogs, the northern leopard frog (*Rana pipiens*) and the southern leopard frog (*Rana utricularia*), which look fairly similar.

**Reproductive Isolation**   Why would biologists describe two types of organisms, such as the northern leopard frog and southern leopard frog, as being different species if they are morphologically similar? One reason is that biologists have discovered that the two species of frogs are unable to breed with each other in nature. Therefore, a second way of identifying a species is by the ability of its members to interbreed. In the late 1920s, geneticist Theodosius Dobzhansky proposed that each species is reproductively isolated from other species. Such **reproductive isolation** prevents one species from successfully interbreeding with other species. In 1942, German evolutionary biologist Ernst Mayr expanded on the ideas of Dobzhansky to provide a definition of a species. According to Mayr, a key feature of sexually reproducing species is that, in nature, the members of one species have the potential to interbreed with one another to produce viable, fertile offspring but cannot successfully interbreed with members of other species. As discussed later in this section, reproductive isolation among species of plants and animals occurs by an amazing variety of different mechanisms.

Reproductive isolation has been used to distinguish many plant and animal species, especially those that look alike but do not interbreed. Even so, this criterion suffers from four main problems.

1. In nature, it may be difficult to determine if two populations are reproductively isolated, particularly if the populations have nonoverlapping geographic ranges.

2. Biologists have noted many cases in which two different species can interbreed in nature yet consistently maintain themselves as separate species. For example, different species of yucca plants, such as *Yucca pallida* and *Yucca constricta,* do interbreed in nature yet typically maintain populations with distinct characteristics. For this reason, they are viewed as distinct species.

3. Reproductive isolation does not apply to asexual species such as bacteria. Likewise, some species of plants and fungi reproduce only asexually.

4. Reproductive isolation cannot be applied to extinct species.

For these reasons, reproductive isolation has been primarily used to distinguish closely related species of modern animals and plants that reproduce sexually.

**Molecular Features**   Molecular features are now commonly used to determine if two different populations are different species. Evolutionary biologists often compare DNA sequences of genes, gene order along chromosomes, chromosome structure, and chromosome number in order to identify similarities and differences among different populations. For example, researchers may compare the DNA sequence of the *16S rRNA* gene between different bacterial populations as a way of determining if two populations represent different species. When the sequences are very similar, such populations would probably be judged to be the same species. However, it may be difficult to draw the line when separating groups into different species. How much difference must be present for species to be considered separate? Is a 2% difference in their genome sequences sufficient to warrant placement into two different species, or do we need a 5% difference?

**Ecological Factors**   A variety of factors related to an organism's habitat may be used to distinguish one species from another. For example, certain species of warblers are distinguished by the habitat in which they forage for food. Some species search the ground for food, others forage in bushes or small trees, and some species primarily forage in tall trees. Such habitat differences are used to distinguish different species that look morphologically similar.

Many bacterial species have been categorized as distinct based on ecological factors. Bacterial cells of the same species are likely to use the same resources (such as sugars and vitamins) and grow under the same conditions (such as particular temperature and pH ranges). However, a drawback of this approach is that different groups of bacteria sometimes display very similar growth characteristics, and even the same species may show great variation in the growth conditions it will tolerate.

**Evolutionary Relationships**   In Chapter 25, we will examine the methods that are used to produce evolutionary trees that describe the relationships between ancestral species and modern species. In some cases, such relationships are based on an analysis of the fossil record. For example, the fossil record was used to construct a tree that shows the ancestors that led to modern horse species. Alternatively, another way of establishing evolutionary relationships is by the analysis of DNA sequences. Researchers obtain samples of cells from different individuals and compare the genes within those cells to see how similar or different they are.

**BIO TIPS**

**THE QUESTION** *A biologist has discovered two populations of snakes that live on opposite sides of a large canyon. The snakes look very similar, but those on the western side of the canyon have a red spot on the top of their heads, whereas those on the eastern side have an orange spot. How would decide if these two populations represent one species or two species?*

**T OPIC** *What topic in biology does this question address?* The topic is species identification. More specifically, the question is about deciding if two different populations of snakes are the same or different species.

**I NFORMATION** *What information do you know based on the question and your understanding of the topic?* In the question, you are given information about two populations of snakes that are found on opposite sides of a canyon. From your understanding of the topic, you may remember that biologists analyze five different characteristics to identify species: (1) morphological traits, (2) reproductive isolation (the ability to interbreed), (3) molecular features, (4) ecological factors, and (5) evolutionary relationships.

**P**ROBLEM-SOLVING **S**TRATEGY *Design an experiment.*
*Interpret data.* One strategy for solving this problem is to design
an experiment to examine one of the five characteristics that
biologists use to identify species.

**ANSWER** *Here are two possible experimental designs:*

1. *Reproductive isolation. Move a few snakes from the western
   side to the eastern side, and vice versa. Observe whether or
   not the introduced snakes interbreed with the ones already
   there. If they do readily interbreed, you can probably conclude that they are not separate species. Alternatively, if they
   don't interbreed, you may conclude that the two populations
   are reproductively isolated and are different species.*

2. *Molecular features. Obtain samples of cells from several
   western and eastern snakes, and analyze their genetic material. If the gene sequences are very similar and the chromosomal composition is very similar, you may conclude that the
   two populations are the same species. Alternatively, if the
   gene sequences show significant differences and the chromosomes differ with regard to structure and/or number, you
   may conclude that the two populations are different species.*

## Species Concepts Emphasize Particular Features to Define and Distinguish Species

A **species concept** is a way of defining the concept of a species and/
or of providing an approach to distinguish one species from another.
Since 1942, over 20 different species concepts have been proposed by
a variety of evolutionary biologists.

**Biological Species Concept**    Ernst Mayr proposed one of the first
species concepts, called the **biological species concept**. According to
Mayr's concept, a species is a group of individuals whose members
have the potential to interbreed with one another in nature to produce
viable, fertile offspring but cannot successfully interbreed with members of other species. The biological species concept emphasizes reproductive isolation as the most important criterion for delimiting species.

**Evolutionary Lineage Concept**    Another example of a species concept is the **evolutionary lineage concept** proposed by American paleontologist George Gaylord Simpson in 1961. A **lineage** is a series of species
that forms a line of descent, with each new species the direct result of
speciation from an immediate ancestral species. According to Gaylord,
species should be defined based on their unique evolution of lineages.

**Ecological Species Concept**    A third example is the **ecological
species concept**, described by American evolutionary biologist Leigh
Van Valen in 1976. According to this viewpoint, each species occupies an ecological niche, which is the unique set of habitat resources
that a species requires, as well as its influence on the environment
and other species.

**General Lineage Concept**    Most evolutionary biologists would
agree that different methods are needed to distinguish the vast array of
species on Earth. Even so, some evolutionary biologists have questioned

whether it is valid to have many different species concepts. In 1998,
American zoologist Kevin de Queiroz suggested that there is only a
single general species concept, which concurs with Simpson's evolutionary lineage concept and includes all previous concepts. According
to de Queiroz's **general lineage concept**, each species is a population
of an independently evolving lineage. Each species has evolved from
a specific series of ancestors and, as a consequence, forms a group of
organisms with a particular set of characteristics. Multiple criteria are
used to determine if a population is part of an independent evolutionary lineage, and thus a species, which is distinct from others. Typically,
researchers use analyses of morphology, reproductive isolation, DNA
sequences, and ecology to determine if a population or group of populations is distinct from others. Because of its generality, the general
lineage concept has received significant support.

## Reproductive Isolating Mechanisms Help to Maintain the Distinctiveness of Each Species

Thus far in this section, we have considered various ways of differentiating species. From the discussion, you may have realized that
the identification of a species is not always a simple matter. The phenomenon of reproductive isolation has played a major role in the way
biologists study plant and animal species, partly because it identifies a possible mechanism for the process of forming new species.
For this reason, much research has been done to try to understand
**reproductive isolating mechanisms**, the mechanisms that prevent
interbreeding between different species.

Why do reproductive isolating mechanisms occur? Populations
do not intentionally erect these reproductive barriers. Rather, reproductive isolation is a consequence of genetic changes that occur
usually because a species becomes adapted to its own particular
environment. The view of evolutionary biologists is that reproductive isolation typically evolves as a by-product of genetic divergence.
Over time, as a species evolves its own unique characteristics, some
of those traits are likely to prevent breeding with other species.

Reproductive isolating mechanisms fall into two categories:
**prezygotic isolating mechanisms**, which prevent the formation of a
zygote, and **postzygotic isolating mechanisms**, which block the development of a viable and fertile individual after fertilization has taken
place. **Figure 24.2** summarizes some of the more common ways that
reproductive isolating mechanisms prevent reproduction between different species. When members of two different species interbreed and
produce offspring, such an offspring is called an **interspecies hybrid**.

**Prezygotic Isolating Mechanisms**    We will consider five types of
prezygotic isolating mechanisms.

**Habitat Isolation:**    One obvious way to prevent interbreeding is
for members of different species to never come in contact with each
other. This phenomenon, called habitat isolation, may involve a geographic barrier to interbreeding. For example, a large body of water
may separate two different plant species that live on nearby islands.

**Temporal Isolation:**    In temporal isolation, species happen to reproduce at different times of the day or year. In the northeastern U.S.,
for example, the two most abundant field crickets, *Gryllus veletis* and
*Gryllus pennsylvanicus* (spring and fall field crickets, respectively),

Species 1                              Species 2

**Prezygotic isolating mechanisms**

**Habitat isolation:** Species occupy different habitats, so they never come in contact with each other.

**Temporal isolation:** Species have different mating or flowering seasons or times of day or become sexually mature at different times of the year.

**Behavioral isolation:** Sexual attraction between males and females of different animal species is limited due to differences in behavior or physiology.

↓ Attempted mating

**Mechanical isolation:** Morphological features such as size and incompatible genitalia prevent 2 members of different species from interbreeding.

**Gametic isolation:** Gametic transfer takes place, but the gametes fail to unite with each other. This can occur because the male and female gametes fail to attract, because they are unable to fuse, or because the male gametes are inviable in the female reproductive tract of another species. In plants, the pollen of one species usually cannot generate a pollen tube to fertilize the egg cells of another species.

↓ Fertilization

**Postzygotic isolating mechanisms**

**Hybrid inviability:** The egg of one species is fertilized by the sperm from another species, but the fertilized egg fails to develop past the early embryonic stages.

**Hybrid sterility:** An interspecies hybrid survives, but it is sterile. For example, the mule, which is sterile, is produced from a cross between a male donkey (*Equus asinus*) and a female horse (*Equus ferus caballus*).

**Hybrid breakdown:** The $F_1$ interspecies hybrid is viable and fertile, but succeeding generations ($F_2$, and so on) become increasingly inviable. This is usually due to the formation of less-fit genotypes by genetic recombination.

↓ Interspecies hybrid

**Figure 24.2 Reproductive isolating mechanisms.** These mechanisms prevent successful breeding between different species. They can occur prior to fertilization (prezygotic) or after fertilization (postzygotic).

 **Core Skill: Connections** Look back at Figure 23.9. Is female choice an example of a prezygotic or postzygotic isolating mechanism?

**(a) Spring field cricket (*Gryllus veletis*)**   **(b) Fall field cricket (*Gryllus pennsylvanicus*)**

**Figure 24.3 Temporal isolation.** Interbreeding between these two species of crickets does not usually occur because *Gryllus veletis* matures in the spring, whereas *Gryllus pennsylvanicus* matures in the fall. a: ©C. Allan Morgan/Getty Images; b: ©Bryan E. Reynolds

**Concept Check:** *Is temporal isolation an example of a prezygotic or a postzygotic isolating mechanism?*

do not differ in song or habitat and are morphologically very similar (**Figure 24.3**). How do the two species maintain reproductive isolation? *G. veletis* matures in the spring, whereas *G. pennsylvanicus* matures in the fall. This minimizes interbreeding between the two species.

***Behavioral Isolation:*** In the case of animals, mating behavior and anatomy often play key roles in promoting reproductive isolation. An example of the third type of prezygotic isolation, behavioral isolation, separates the western meadowlark (*Sturnella neglecta*) and eastern meadowlark (*Sturnella magna*). Both species are nearly identical in shape, coloration, and habitat, and their ranges overlap in the central U.S. (**Figure 24.4**). For many years, they were thought to be the same species. When biologists discovered that the western meadowlark is a separate species, it was given the species name *S. neglecta* to reflect the long delay in its recognition. In the zone of overlap, very little interspecies mating takes place between western and eastern meadowlarks, largely due to differences in their songs. The song of the western meadowlark is a long series of flutelike gurgling notes that go down the scale. By comparison, the eastern meadowlark's song is a simple series of whistles, typically about four or five notes. These differences in songs enable meadowlarks to recognize potential mates as members of their own species.

***Mechanical Isolation:*** A fourth type of prezygotic isolation, called mechanical isolation, occurs when morphological features such as size or incompatible genitalia prevent two species from interbreeding. For example, male dragonflies use a pair of special appendages to grasp females during copulation. When a male tries to mate with a female of a different species, his grasping appendages do not fit her body shape.

***Gametic Isolation:*** A fifth type of prezygotic isolating mechanism occurs when two species attempt to interbreed, but the gametes fail to unite in a successful fertilization event. This phenomenon, called gametic isolation, is widespread among plant and animal species. In aquatic animals that release sperm and egg cells into the water, gametic isolation is important in preventing interspecies hybrids. For example, closely related species of sea urchins may release sperm and eggs into the water at the same time. Researchers have discovered that sea urchin sperm have a protein on their surface called bindin, which

(a) Western
meadowlark
(*Sturnella neglecta*)

**Western meadowlark**
**Eastern meadowlark**
**Zone of overlap**

(b) **Eastern**
**meadowlark**
(*Sturnella*
*magna*)

**Figure 24.4  Behavioral isolation. (a)** The western meadowlark (*Sturnella neglecta*) and **(b)** eastern meadowlark (*Sturnella magna*) are very similar in appearance. The red region in this map shows where the two species' ranges overlap. However, very little interspecies mating takes place due to differences in their songs. a: ©Rod Planck/Science Source; b: ©Ron Austing/Science Source

 **Core Concept: Evolution** For these two species of meadowlarks, one evolutionary change that took place is that their mating songs became different.

mediates sperm-egg attachment and membrane fusion. The structure of bindin is significantly different among different sea urchin species, thereby ensuring that fertilization occurs only between sperm and egg cells of the same species.

In flowering plants, gametic isolation is commonly associated with pollination. As discussed in Chapter 40, plant fertilization is initiated when a pollen grain lands on the stigma of a flower and sprouts a pollen tube that ultimately reaches an egg cell (look ahead to Figure 40.4). When pollen is released from a plant, it could be transferred to the stigma of many different plant species. In most cases, when a pollen grain lands on the stigma of a different species, it either fails to generate a pollen tube or the tube does not grow properly and thus does not reach the egg cell.

***Postzygotic Isolating Mechanisms*** Let's now turn to postzygotic mechanisms of reproductive isolation, of which there are three common types.

***Hybrid Inviability:*** The mechanism of hybrid inviability occurs when an egg of one species is fertilized by a sperm from another species, but the fertilized egg cannot develop past the early embryonic stages.

**Male donkey (*Equus asinus*)**        **Female horse (*Equus ferus*** **caballus*)**

**Mule**

**Figure 24.5  Hybrid sterility.** When a male donkey (*Equus asinus*) mates with a female horse (*Equus ferus caballus*), their offspring is a mule, which is usually sterile. (top left): ©Mark Boulton/Science Source; (top right): ©MyLoupe/UIG/Getty Images; (bottom): ©Stephen L. Saks/Science Source

*Concept Check:* *Is hybrid sterility an example of a prezygotic or a postzygotic isolating mechanism?*

***Hybrid Sterility:*** A second postzygotic isolating mechanism is hybrid sterility, in which an interspecies hybrid may be viable but sterile. A classic example of hybrid sterility is the mule, which is produced by a mating between a male donkey (*Equus asinus*) and a female horse (*Equus ferus caballus*) (**Figure 24.5**). All male mules and most female mules are sterile. Why are mules usually sterile? Two reasons explain the sterility. Because the horse has 32 chromosomes per set and a donkey has 31, a mule inherits 63 chromosomes (32 + 31). Due to the uneven number, all of the chromosomes cannot pair evenly. Also, the chromosomes of the horse and donkey have structural differences, which either prevent them from pairing correctly or lead to chromosomal abnormalities if crossing over occurs during meiosis. For these reasons, mules usually produce inviable gametes. Note that the mule has no species name because it is not considered a species due to this sterility.

***Hybrid Breakdown:*** Finally, interspecies hybrids may be viable and fertile, but the subsequent generation(s) may harbor genetic abnormalities that are detrimental. This third mechanism, called hybrid breakdown, can be caused by changes in chromosome structure. The chromosomes of closely related species may have structural differences from each other, such as inversions. In hybrids, a crossover may occur in the region that is inverted in one species but not the other. This will produce gametes with too little or too much genetic material. Such hybrids often have offspring with developmental abnormalities.

Postzygotic isolating mechanisms tend to be uncommon in nature compared with prezygotic mechanisms. Why are postzygotic mechanisms rare? One explanation is that they are more costly in terms of energy and resources used. For example, a female mammal uses a large amount of energy to produce an offspring that is sterile. Evolutionary biologists hypothesize that natural selection has favored prezygotic isolating mechanisms because they do not waste a lot of energy.

## 24.2 Mechanisms of Speciation

**Learning Outcomes:**

1. Describe how allopatric speciation can occur and how it can lead to adaptive radiation.
2. Outline three different mechanisms of sympatric speciation.

Speciation, the formation of a new species, is caused by genetic changes in a particular group that make it different from the species from which it was derived. As discussed in Chapter 23, mutations in genes can be acted on by natural selection and other evolutionary mechanisms to alter the genetic composition of a population. New species commonly evolve in this manner. In addition, interspecies breeding, changes in chromosome number, and horizontal gene transfer may also contribute to the formation of new species. In all of these cases, the underlying cause of speciation is the accumulation of genetic changes that ultimately promote enough differences that a population can be recognized as a unique species.

Even though genetic changes account for the phenotypic differences observed among living organisms, such changes do not fully explain the existence of many distinct species on our planet. Why does life often diversify into the more or less discrete populations that we recognize as species? Two main explanations have been proposed:

1. In some cases, speciation may occur due to abrupt events, such as changes in chromosome number, that cause reproductive isolation.
2. More commonly, species arise as a consequence of adaptation to different ecological niches. For sexually reproducing organisms, reproductive isolation is typically a by-product of that adaptation.

Depending on the species involved, one or both factors may play a dominant role in the formation of new species. In this section, we will consider how reproductive isolating mechanisms and adaptation to particular environments are critical aspects of the speciation process.

### Geographic and Habitat Isolation Can Promote Allopatric Speciation

**Cladogenesis** is the splitting or diverging of one species into two or more species. In the case of sexually reproducing organisms, the process of cladogenesis requires that gene flow becomes interrupted between two or more populations, limiting or eliminating reproduction between members of those populations. **Allopatric speciation** (from the Greek *allos*, meaning other, and the Latin *patria*, meaning homeland) is the most prevalent way for cladogenesis to occur. This form of speciation occurs when a population becomes isolated from other populations and evolves into one or more species. Typically, this isolation may involve a geographic barrier such as a large area of land or body of water.

In some cases, geographic separation may be caused by slow geological events that eventually produce quite large geographic barriers. For example, a mountain range may emerge and split one species that occupies the lowland regions, or a creeping glacier may divide

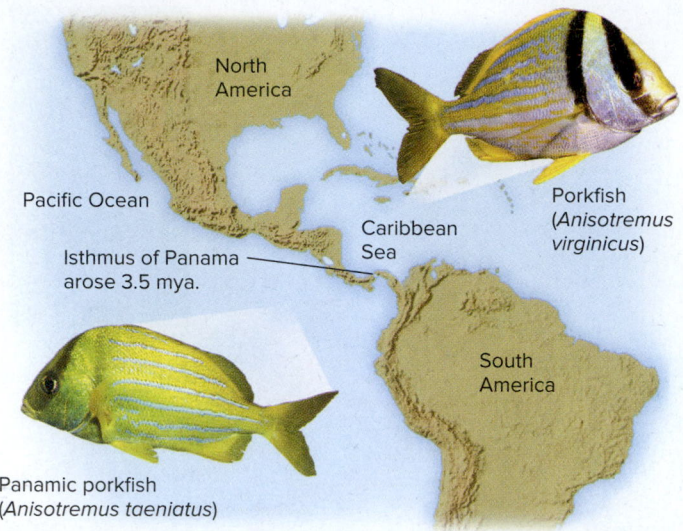

**Figure 24.6 Allopatric speciation.** An ancestral fish population was split into two by the formation of the Isthmus of Panama about 3.5 mya. Since that time, different genetic changes occurred in the two populations. These changes eventually led to the formation of different species: The Panamic porkfish (*Anisotremus taeniatus*) is found in the Pacific Ocean, and the porkfish (*Anisotremus virginicus*) is found in the Caribbean Sea. (left): ©Hal Beral/V&W/imagequestmarine.com; (right): ©Amar and Isabelle Guillen/Guillen Photography/Alamy Stock Photo

 **Core Concept: Evolution** These two species of fish look similar because they share a common ancestor that existed in the fairly recent past.

a population. **Figure 24.6** shows an interesting example in which geological separation promoted speciation. A fish called the Panamic porkfish (*Anisotremus taeniatus*) is found in the Pacific Ocean, whereas the porkfish (*Anisotremus virginicus*) is found in the Caribbean Sea. These two species were derived from an ancestral species whose population was split by the formation of the Isthmus of Panama about 3.5 mya. Before that geological event, the waters of the Pacific Ocean and Caribbean Sea mixed freely. Since the formation of the isthmus, the two populations have been geographically isolated and have evolved into distinct species.

Allopatric speciation can also occur when a small population moves to a new location that is geographically isolated from the main population. For example, a storm may force a small group of birds from a mainland to a distant island. In this case, migration between the island and the mainland population is an infrequent event. In a relatively short period of time, the small founding population on the island may evolve into a new species. How does speciation occur rapidly? Because the environment on the island may differ significantly from the mainland environment, natural selection may rapidly alter the genetic composition of the population, leading to adaptation to the new environment. In addition, as discussed in Chapter 23, a form of genetic drift known as the founder effect can have a larger influence in small founding populations.

The Hawaiian Islands are a showcase of allopatric speciation. The islands' extreme isolation coupled with their phenomenal array of ecological niches has enabled a small number of founding species to evolve into a vast assortment of different species. Biologists have investigated several examples of **adaptive radiation**, in which a single

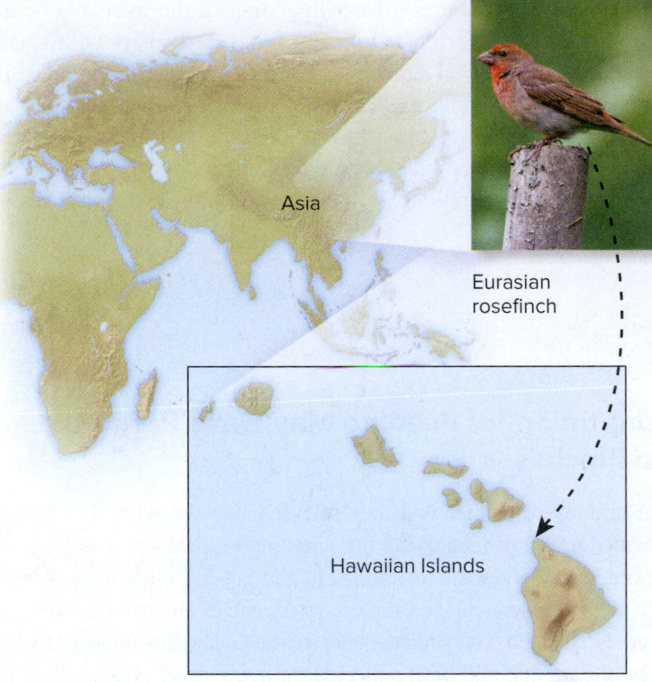

**(a) Migration of ancestor to the Hawaiian Islands**

**Figure 24.7** **Adaptive radiation.** **(a)** The honeycreepers' ancestor is believed to be related to a Eurasian rosefinch that arrived on the Hawaiian Islands approximately 3–7 mya. Since that time, at least 54 different species of honeycreepers (Drepanidinae) have evolved on the islands. **(b)** Adaptations to feeding have produced honeycreeper species with notable differences in beak morphology and other features. a: ©FLPA/Alamy Stock Photo; b (top left and center right): ©Jim Denny; b (top right, center left, bottom left and right): ©Jack Jeffrey Photography

 **Core Skill: Connections** Look back at Figure 23.5b. Discuss how diversifying selection played a role in the diversity of honeycreepers on the Hawaiian Islands.

ancestral species has evolved into a wide array of descendant species that differ in their habitat, form, or behavior. For example, approximately 1,000 species of *Drosophila* are found dispersed throughout the Hawaiian Islands. Evolutionary studies suggest that these evolved from a single colonization by one species of fruit fly! Natural selection resulted in changes in body form and function that produced the amazing diversity of *Drosophila* species that are now found on the islands.

As shown in **Figure 24.7**, an example of adaptive radiation is seen with a family of birds called honeycreepers (Drepanidinae). Researchers estimate that the honeycreepers' ancestor arrived in Hawaii 3–7 mya. This ancestor was a single species of finch, possibly a Eurasian rosefinch (genus *Carpodacus*) or, less likely, the North American house finch (*Carpodacus mexicanus*). At least 54 different species of honeycreepers, many of which are now extinct, evolved from this founding population to fill available niches in the islands' habitats. Natural selection resulted in the formation of many species with different feeding strategies. Seed eaters have stouter, stronger bills capable of cracking tough husks. Insect-eating honeycreepers have thin, warbler-like bills adapted for picking insects from foliage or strong, hooked bills to root out wood-boring insects. The curved bills of nectar-feeding honeycreepers enable them to extract nectar from the flowers of Hawaii's endemic plants.

**(b) Examples of Hawaiian honeycreepers**

Before ending our discussion of allopatric speciation, let's consider a common situation in which geographic separation is not complete. Areas where two populations can interbreed are known as **hybrid zones**. **Figure 24.8** shows a hybrid zone along a mountain

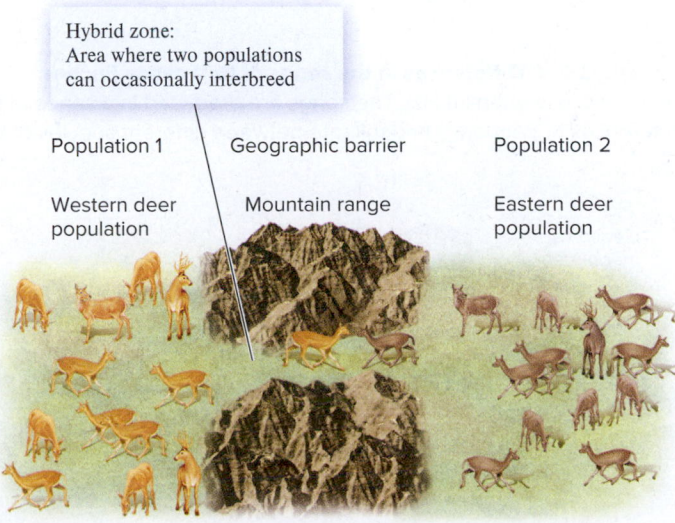

**Figure 24.8** **Hybrid zones.** Two populations of deer are separated by a mountain range. A hybrid zone exists in a mountain pass, where occasional interbreeding may occur.

pass that connects two deer populations. For speciation to occur, the amount of gene flow within hybrid zones must become very limited. How does this happen? As the two populations accumulate different genetic changes, the ability of individuals from different populations to mate with each other in the hybrid zone may decrease. For example, natural selection in the western deer population may favor an increase in body size that is not favored in the eastern population. Over time, as this size difference between members of the two populations becomes greater, breeding in the hybrid zone may decrease. Larger individuals may not interbreed easily with smaller ones due to mechanical isolation. In addition, larger individuals may prefer larger individuals as mates, and smaller individuals may also prefer each other. Once gene flow through the hybrid zone is greatly diminished, the two populations are reproductively isolated. Over the course of many generations, such populations may evolve into distinct species.

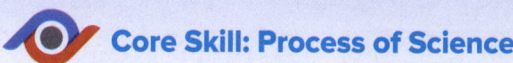

## Core Skill: Process of Science

## Feature Investigation | Podos Found That an Adaptation for Feeding May Have Promoted Reproductive Isolation in Finches

In 2001, American evolutionary biologist Jeffrey Podos analyzed the songs of Darwin's finches on the Galápagos Islands to determine how environmental adaptation may contribute to reproductive isolation. As in honeycreepers, the differences in beak sizes and shapes among the various species of finches are adaptations to different feeding strategies. Podos hypothesized that changes in beak morphology could also affect the songs that the birds produce, thereby having the potential to affect mate choice. The components of the vocal tract of birds, including the trachea, larynx, and beak, work collectively to produce a bird's song. Birds actively modify the shape of their vocal tracts during singing, and beak movements are normally very rapid and precise.

Podos focused on two aspects of a bird's song. The first feature is the frequency range, which is a measure of the minimum and maximum frequencies in a bird's song, measured in kilohertz (kHz). The second feature is the trill rate. A trill is a series of notes or group of notes repeated in succession. **Figure 24.9** shows a graphical depiction of the songs of Darwin's finches. As you can see, the song patterns of these finches are quite different from each other

To study the relationship between beak size and song in a quantitative way, Podos first captured male finches on Santa Cruz, one of the Galápagos Islands, and measured their beak sizes (**Figure 24.10**).

The birds were banded and then released into the wild. The banding provided a way of identifying the birds whose beaks had already been measured. After release, the songs of the banded birds were recorded on a tape recorder, and their range of frequencies and trill rate were analyzed. Podos then compared the data for the Galápagos finches to a large body of data that had been collected on many other bird species. This comparison was used to evaluate whether beak size, in this case, beak depth—the measurement of the beak from top to bottom, at its base—constrained the frequency range and/or the trill rate of the finches.

The results of this comparison are shown in the data of Figure 24.10. As seen here, the relative constraint on vocal performance became higher as the beak depth became larger. This means that birds with larger beaks had a narrower frequency range and/or a slower trill rate. Podos proposed that as jaws and beaks became adapted for strength to crack open larger, harder seeds, they became less able to perform the rapid movements associated with certain types of songs. In contrast, the finches with smaller beaks adapted to probe for insects or eat smaller seeds had less constraint on their vocal performance. From the perspective of evolution, the changes observed in song patterns for the Galápagos finches could have played an important role in promoting reproductive isolation, because song pattern is an important factor in mate selection

**Figure 24.9** **Differences in the songs of Galápagos finches.** These spectrograms depict the frequency of each bird's song over time, measured in kilohertz (kHz). The songs are produced in a series of trills that have a particular pattern and occur at regular intervals. Notice the differences in frequency and trill rate between different species of birds.

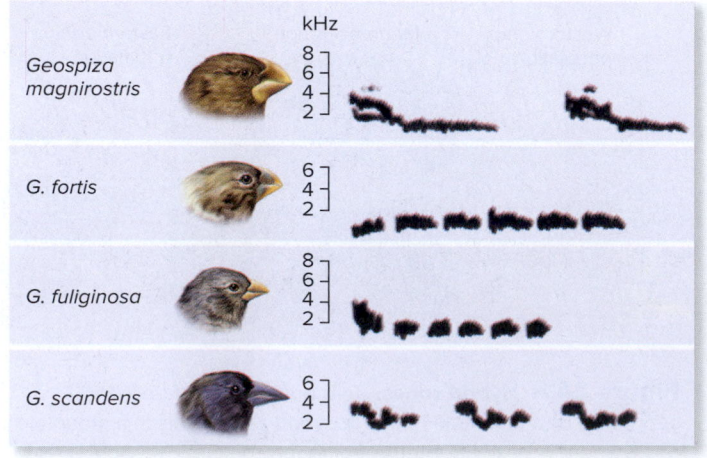

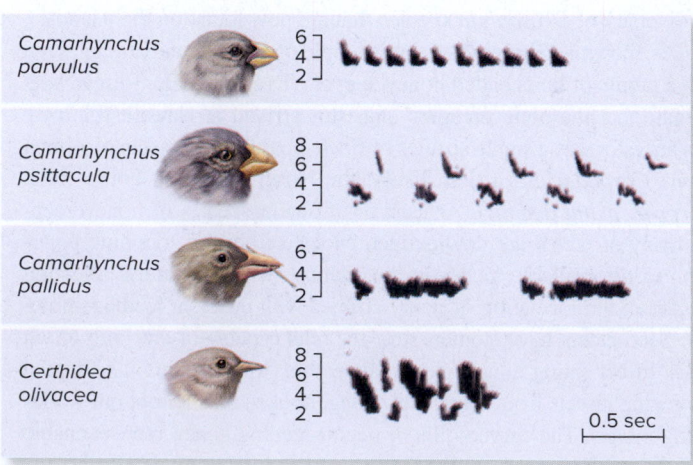

**Figure 24.10** **Study by Podos investigating the effects of beak depth on song among different species of Galápagos finches.**

**HYPOTHESIS** Changes in beak morphology that are an adaptation for feeding may also affect the songs of Galápagos finches and thereby lead to reproductive isolation between species.

**KEY MATERIALS** This study was conducted on finch populations of the Galápagos Island of Santa Cruz.

| | Experimental level | Conceptual level |
|---|---|---|
| 1 Capture male finches and measure their beak depth. Beak depth is measured at the base of the beak, from top to bottom. |  | This is a measurement of phenotypic variation in beak size. |
| 2 Band the birds and release them back into the wild. | 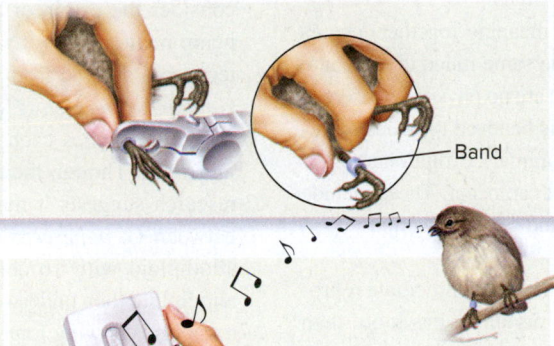 Band | Banding allows identification of birds with known beak depths. |
| 3 Record the bird's songs on a tape recorder. |  | This is a measurement of phenotypic variation in song. |
| 4 Analyze the songs with regard to frequency range and trill rate. | 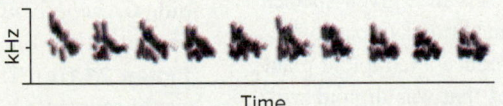 Time | The frequency range is the value between high and low frequencies. The trill rate is the number of repeats per unit time. |

**5 THE DATA**

The data for the Galápagos finches were compared to a large body of data that had been collected on many other bird species. The relative constraint on vocal performance is higher if a bird has a narrower frequency range and/or a slower trill rate. These constraints were analyzed with regard to each bird's beak depth.

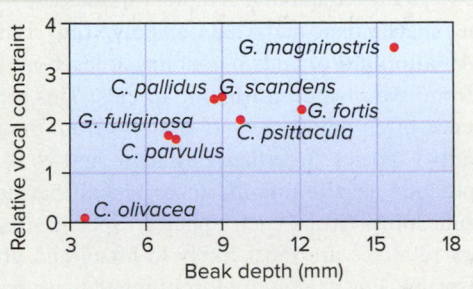

**6 CONCLUSION** Larger beak size, which is an adaptation to cracking open large, hard seeds, constrains vocal performance. This may affect mating song patterns and thereby promote reproductive isolation and, in turn, speciation.

**7 SOURCE** Podos, J. 2001. Correlated evolution of morphology and vocal signal structure in Darwin's finches. *Nature* 409: 185–188.

in birds. Therefore, a by-product of beak adaptation for feeding is that it also appears to have affected song pattern, possibly promoting reproductive isolation and eventually the formation of distinct species.

### Experimental Questions

1. What did Podos hypothesize regarding the effects of beak size on a bird's song? How could changes in beak size and shape lead to reproductive isolation among the finches?

2. **CoreSKILL** » How did Podos test the hypothesis that beak morphology caused changes in the birds' songs?

3. **CoreSKILL** » Analyze the results of Podos's study, and explain whether they support his original hypothesis. What is meant by the phrase "by-product of adaptation," and how does it apply to this particular study?

## Sympatric Speciation Occurs When Populations Are in Direct Contact

**Sympatric speciation** (from the Greek *sym*, meaning together) occurs when members of a species that are within the same range diverge into two or more different species even though there are no physical barriers to interbreeding. Although sympatric speciation is believed to be less common than allopatric speciation, particularly in animals, evolutionary biologists have discovered several ways in which it can occur. These include polyploidy, adaptation to local environments, and sexual selection.

*Polyploidy*    A type of genetic change that can cause immediate reproductive isolation is **polyploidy**, in which an organism has more than two sets of chromosomes. Plants tend to be more tolerant of changes in chromosome number than animals. For example, many crops and decorative species of plants are polyploid. How does polyploidy occur? One mechanism is complete nondisjunction of chromosomes, which increases the number of chromosome sets in a given species (autopolyploidy). Such changes can result in an abrupt sympatric speciation. For example, nondisjunction could produce a tetraploid plant with four sets of chromosomes from a species that was diploid with two sets. A cross between a tetraploid and a diploid produces a triploid offspring with three sets of chromosomes. Triploid offspring are usually sterile because an odd number of chromosomes cannot be evenly segregated during meiosis. This hybrid sterility causes reproductive isolation between the tetraploid and diploid species.

Another mechanism that leads to polyploidy is interspecies breeding. An **alloploid** organism contains at least one set of chromosomes from two or more different species. This term refers to the occurrence of chromosome sets (ploidy) from the genomes of different (allo-) species. Interbreeding between two different species may produce an allodiploid, an organism that has only one set of chromosomes from each species. Species that are close evolutionary relatives are most likely to breed and produce allodiploid offspring. For example, closely related species of grasses may interbreed to produce allodiploids. An organism containing two or more complete sets of chromosomes from two or more different species is called an allopolyploid. An allopolyploid can be the result of interspecies breeding between species that are already polyploid, or it can occur as a result of nondisjunction in an allodiploid organism. For example, complete nondisjunction in an allodiploid could produce an allotetraploid, which is an allopolyploid with two complete sets of chromosomes from two species for a total of four sets.

The formation of an allopolyploid can also abruptly lead to reproductive isolation, thereby promoting speciation. As an example, let's consider the origin of a natural species of plant called the common hemp nettle, *Galeopsis tetrahit*. This species is thought to be an allotetraploid derived from two diploid species, *Galeopsis pubescens* and *Galeopsis speciosa* (**Figure 24.11a**). These two diploid species have 16 chromosomes each ($2n = 16$), whereas *G. tetrahit* has 32 chromosomes. Though the origin of *G. tetrahit* is not completely certain, research suggests it may have originated from an interspecies cross between *G. pubescens* and *G. speciosa*, which initially produced an allodiploid with 16 chromosomes (one set from each species). The allodiploid then underwent complete nondisjunction to become an allotetraploid carrying four sets of chromosomes—two from each species.

How do these genetic changes cause reproductive isolation? The allotetraploid, *G. tetrahit*, is fertile, because all of its chromosomes occur in homologous pairs that can segregate evenly during meiosis. However, a cross between *G. tetrahit* and a diploid, *G. pubescens* or *G. speciosa*, produces an offspring that is monoploid for one chromosome set and diploid for the other set (**Figure 24.11b**). The chromosomes of the monoploid set cannot be evenly segregated during meiosis. These offspring are expected to be sterile, because they will produce gametes that have incomplete sets of chromosomes. This hybrid sterility causes the allotetraploid to be reproductively isolated from both diploid species. Therefore, this process could have led to the formation of a new species, *G. tetrahit*, by sympatric speciation.

Polyploidy is so frequent in plants that it is a major mechanism of their speciation. In ferns and flowering plants, about 40–70% of the species are polyploid. By comparison, polyploidy can occur in animals, but it is much less common. For example, less than 1% of reptiles and amphibians are polyploids derived from diploid ancestors. The reason why polyploidy is not usually tolerated in animals is not understood.

*Adaptation to Local Environments*    In some cases, populations that occupy different local environments, which are continuous with each other, may diverge into different species. An early example of this type of sympatric speciation was described by American biologists Jeffrey Feder, Guy Bush, and colleagues. They studied the North American apple maggot fly (*Rhagoletis pomenella*). This fly originally fed on native hawthorn trees. However, the introduction of apple trees approximately 200 years ago provided a new local environment for this species. The apple-feeding populations of this species develop more rapidly because apples mature more quickly than hawthorn fruit. The result is partial temporal isolation, which is an

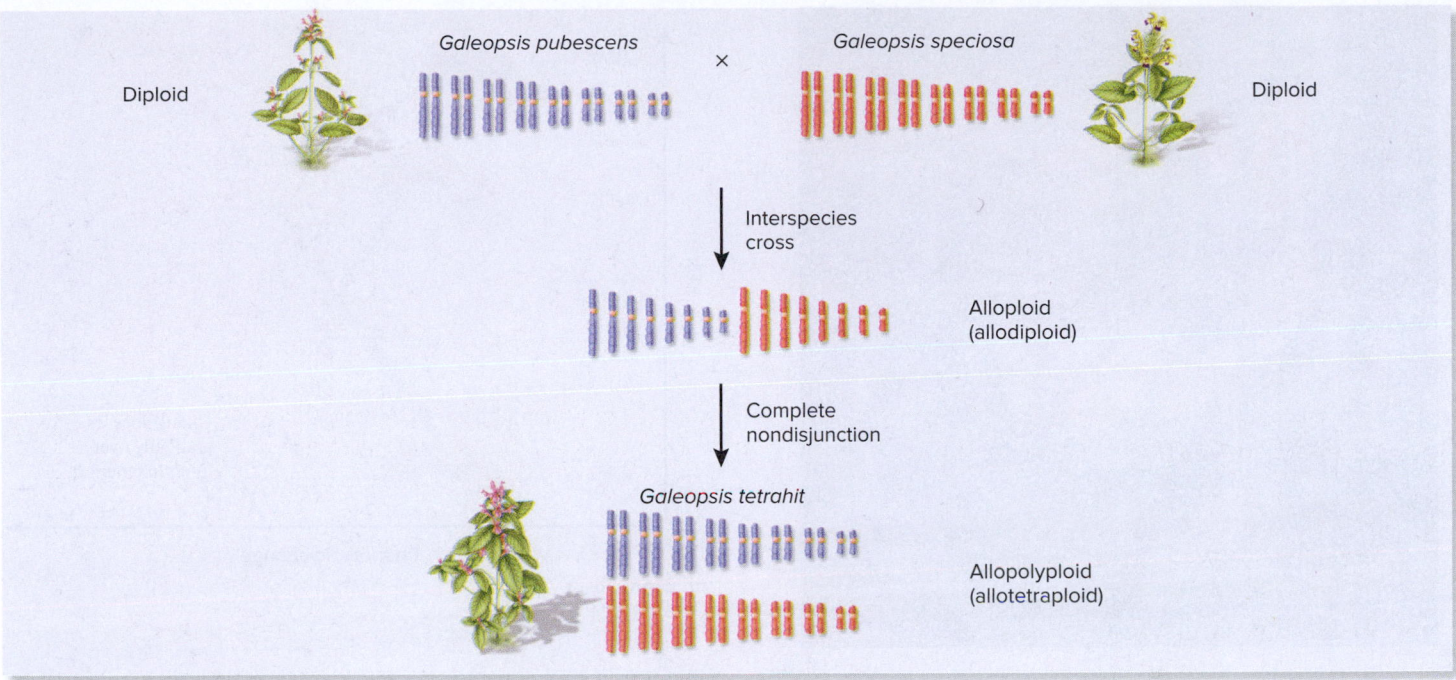

**(a) Possible formation of *G. tetrahit***

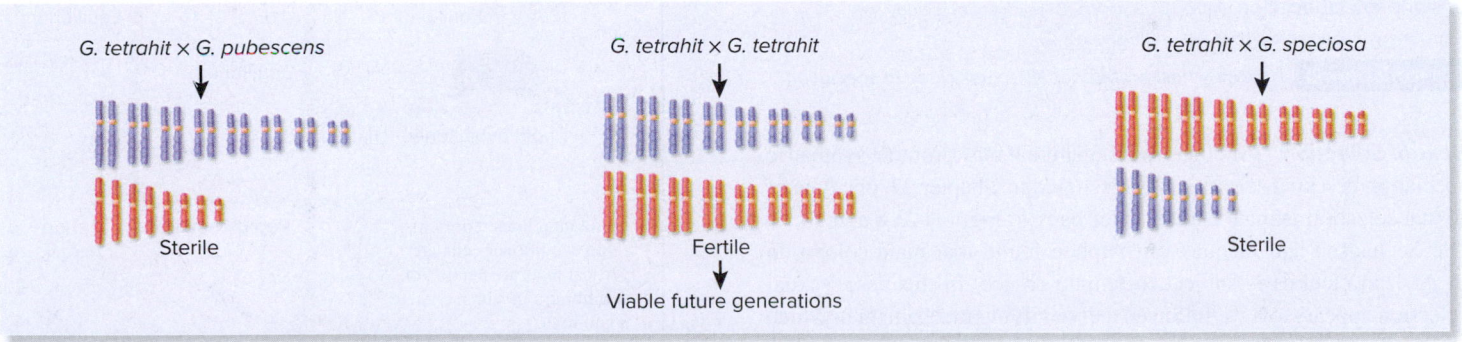

**(b) Outcome of breeding among *G. tetrahit*, *G. pubescens*, and *G. speciosa***

**Figure 24.11 Polyploidy and sympatric speciation.** **(a)** *Galeopsis tetrahit* may have arisen from an interspecies cross between *Galeopsis pubescens* and *Galeopsis speciosa*, which was followed by a subsequent nondisjunction event. **(b)** Polyploidy may have caused reproductive isolation between these three natural species of hemp nettle. If *G. tetrahit* is crossed with either of the other two species, the resulting offspring will be monoploid for one chromosome set and diploid for the other set, making them sterile. Therefore, *G. tetrahit* is reproductively isolated from the diploid species, making it a new species.

**Concept Check:** *Suppose that G. tetrahit was crossed to G. pubescens to produce an interspecies hybrid, as shown at the left side of part (b). If this interspecies hybrid was crossed to G. tetrahit, how many chromosomes do you think an offspring would have? The answer you give should be a range, not a single number.*

example of a prezygotic isolating mechanism. Although the two populations—those that feed on apple trees and those that feed on hawthorn trees—are considered subspecies, evolutionary biologists speculate they may eventually become distinct species due to reproductive isolation and the accumulation of independent mutations.

American entomologist Sara Via and colleagues have studied the beginnings of sympatric speciation in pea aphids (*Acyrthosiphon pisum*), a small, plant-eating insect. Pea aphids in the same geographic area can be found on both alfalfa (*Medicago sativa*) and red clover (*Trifolium pratenae*) (**Figure 24.12**). Although pea aphids on these two host plants look identical, they show significant genetic

differences and are highly ecologically specialized. Pea aphids that are found on alfalfa exhibit a lower fitness when transferred to red clover, whereas pea aphids found on red clover exhibit a lower fitness when transferred to alfalfa. The same traits involved in this host specialization cause these two groups of pea aphids to be substantially reproductively isolated. Taken together, the observations of the North American apple maggot fly, pea aphids, and other insect species suggest that diversifying selection (described in Chapter 23) occurs because some members within the same range evolve to feed on a different host. This may be an important mechanism of sympatric speciation among insects.

**Figure 24.12 Pea aphids, a possible example of sympatric speciation in progress.** Some pea aphids prefer alfalfa, whereas others prefer red clover. These two populations may be in the process of sympatric speciation. (left): ©Dr. Sara Via, Department of Biology and Department of Entomology, University of Maryland

**Concept Check:** *How may host preference eventually lead to speciation?*

***Sexual Selection*** Another mechanism that may promote sympatric speciation is sexual selection. As discussed in Chapter 23, one type of sexual selection is mate choice (refer back to Figures 23.8 and 23.9). Ole Seehausen and Jacques van Alphen found that male coloration in African cichlids is subject to female choice. In this case, sexual selection appears to have followed a diversifying mechanism in which certain females prefer males with one color pattern and other females prefer males with a different color pattern. A possible outcome of such sexual selection is that it can separate one large sympatric population into smaller populations that eventually become distinct species because they selectively breed among themselves.

## 24.3 The Pace of Speciation

### Learning Outcome:

1. Compare and contrast the concepts of gradualism and punctuated equilibrium.

Throughout the history of life on Earth, the rate of evolutionary change and speciation has not been constant. **Figure 24.13** illustrates two contrasting views concerning the rate of evolutionary change. These ideas are not mutually exclusive but represent two different ways to consider the tempo of evolution. The concept of **gradualism** suggests that each new species evolves continuously over long spans of time (Figure 24.13a). The principal idea is that large phenotypic differences that produce new species are due to the gradual accumulation of many small genetic changes.

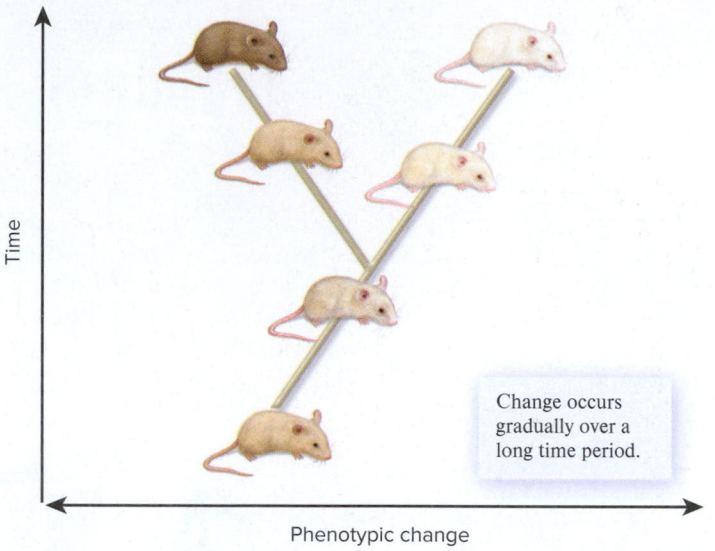

Change occurs gradually over a long time period.

**(a) Gradualism**

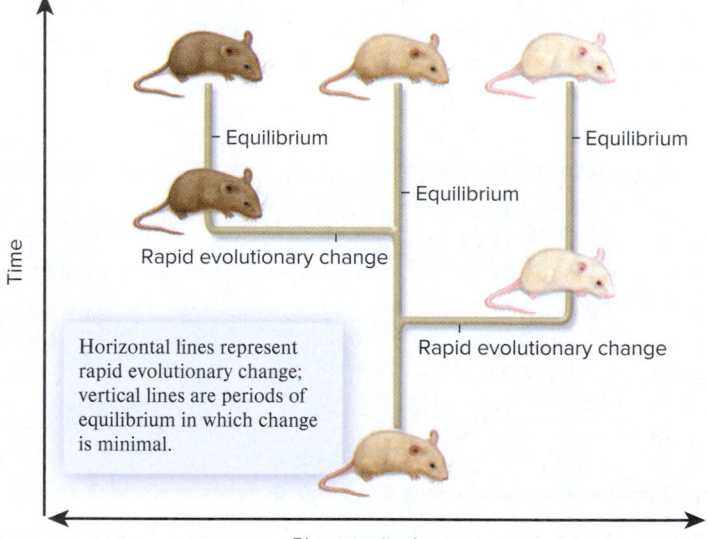

Equilibrium

Equilibrium

Equilibrium

Rapid evolutionary change

Rapid evolutionary change

Horizontal lines represent rapid evolutionary change; vertical lines are periods of equilibrium in which change is minimal.

**(b) Punctuated equilibrium**

**Figure 24.13 A comparison of gradualism and punctuated equilibrium.** **(a)** In gradualism, the phenotypic characteristics of a species gradually change due to the accumulation of small genetic changes. **(b)** In punctuated equilibrium, long periods of equilibrium in which species exist essentially unchanged are punctuated by relatively short periods of evolutionary change during which phenotypic characteristics may change rapidly.

 **Core Concept: Evolution** Gradualism and punctuated equilibrium are two different views regarding the pace of evolution.

By comparison, the concept of **punctuated equilibrium**, advocated in the 1970s by American paleontologist Niles Eldredge and American evolutionary biologist Stephen Jay Gould, suggests that the tempo of evolution is more sporadic (Figure 24.13b). According to this hypothesis, species exist relatively unchanged for many generations. During this equilibrium period, genetic changes are likely to accumulate, particularly neutral changes. However, genetic changes that significantly alter phenotype do not substantially change the overall composition of a population. These long periods of equilibria are punctuated by relatively short periods (that is, on a geological timescale) during which the frequencies of certain phenotypes in a population change substantially at a far more rapid rate.

A rapid rate of evolution could commonly occur via allopatric speciation, in which a small group migrates away from a larger population to a new environment in which different alleles provide better adaptation to the surroundings. By natural selection, the small population may rapidly evolve into a new species. In addition, events such as polyploidy may abruptly produce individuals with new phenotypic traits. On an evolutionary timescale, these types of events can be rather rapid, because a few genetic changes can have a major influence on phenotype.

In conjunction with genetic changes, species may also be subjected to sudden environmental shifts that quickly drive the gene pool in a particular direction via natural selection. For example, the climate may change or a new predator may infiltrate the geographic range of a species. Natural selection may lead to a rapid evolution of the gene pool by favoring those alleles that allow members of the population to survive the climatic change or to have phenotypic characteristics that allow them to avoid the predator.

Which viewpoint is correct, punctuated equilibrium or gradualism? Both have merit. The occurrence of punctuated equilibrium is often supported by the fossil record. New species seem to arise rather suddenly in a layer of rocks, persist relatively unchanged for a very long period of time, and then become extinct. In such cases, scientists hypothesize that the period during which a previous species evolved into a new species was so short that few, if any, of the transitional forms of the species were preserved as fossils. Even so, these rapid periods of change were probably followed by long periods that likely involved the additional accumulation of many small genetic changes, consistent with gradualism.

Finally, another issue associated with the speed of speciation is generation time. Species of large animals with long generation times tend to evolve much more slowly than do microbial species with short generations. Many new species of bacteria will come into existence during our lifetime, whereas new species of large animals tend to arise on a much longer timescale. This is an important consideration because bacteria have great environmental effects. They are decomposers of organic materials and pollutants in the environment, and they play a role in many diseases of plants and animals, including humans.

## 24.4 Evo-Devo: Evolutionary Developmental Biology

### Learning Outcomes:

1. Describe how the spatial expression of genes, such as *BMP4* and *Gremlin*, affects pattern formation.
2. Explain the relationship between the number of *Hox* genes and the body pattern of an animal species.

3. Outline how differences in the growth rates of body parts can change the characteristics of species.
4. Describe how the study of the *Pax6* gene suggests that the eyes of different animal species evolved from a common ancestor.

As described earlier in this chapter, the origin of new species involves genetic changes that lead to adaptations to environmental niches and/or to reproductive isolating mechanisms that prevent closely related species from interbreeding. These genetic changes result in morphological and physiological differences that distinguish one species from another. In recent years, many evolutionary biologists have begun to investigate how genetic variation produces species and groups of species with novel shapes and forms. The underlying reasons for such changes are often rooted in the developmental pathways that control an organism's morphology.

**Evolutionary developmental biology** (referred to as **evo-devo**) is an exciting and relatively new field of biology that compares the development of different organisms in an attempt to understand ancestral relationships between organisms and the mechanisms that bring about evolutionary change. During the past few decades, developmental geneticists have gained a better understanding of biological development at the molecular level. Much of this work has involved the discovery of genes that control development in model organisms. As the genomes of more organisms have been analyzed, researchers have become interested in the similarities and differences that occur between closely related and distantly related species. The field of evolutionary developmental biology arose out of this interest.

How do new morphological forms come into being? For example, how does a nonwebbed foot evolve into a webbed foot? How does a new organ, such as an eye, come into existence? As we will explore, such novelty arises through genetic changes, also called genetic innovations. Certain types of genetic innovations have been so advantageous they have resulted in groups of new species. For example, the innovation of wings resulted in the evolution of many different species of birds. In this section, we will see that proteins that control developmental changes, such as cell-signaling proteins and transcription factors, often play a key role in promoting the morphological changes that occur during evolution.

### The Spatial Expression of Genes That Affect Development Has a Dramatic Effect on Phenotype

In Chapter 20, we considered the role of genetics in the development of plants and animals. Genes that play a role in development influence cell division, cell migration, cell differentiation, and cell death. The interplay among these four processes produces an organism with a specific body pattern, a process called **pattern formation**. As you might imagine, genes that control development are very important to the phenotypes of individuals. They affect traits such as the shape of a bird's beak, the length of a giraffe's neck, and the size of a plant's flower. In recent years, the study of development has revealed that particular genes are key players in the evolution of many types of traits. Changes in such genes affect traits that can be acted on by natural selection. Furthermore, variation in the expression of these genes may be commonly involved in the acquisition of new traits that promote speciation.

As an example, let's compare pattern formation of a chicken's foot with that of a duck's foot. Two different patterns occur: a nonwebbed pattern in which the digits are not inconnected, and a webbed pattern in which the digits are connected by sheets of skin. Developmental biologists have discovered that the morphological differences between a nonwebbed and a webbed pattern are due to the differential expression of two different cell-signaling proteins called bone morphogenetic protein 4 (BMP4) and gremlin. The *BMP4* gene is expressed throughout the developing limb of both the chicken and duck; this is shown in **Figure 24.14a**, in which the BMP4 protein is stained purple. BMP4 causes cells to undergo apoptosis and die. The gremlin protein, which is stained brown in **Figure 24.14b**, inhibits the function of BMP4, thereby allowing cells to survive. In the developing chicken limb, the *Gremlin* gene is expressed throughout the limb, except in the regions between each digit. Therefore, in these regions, the cells die, and a chicken develops a nonwebbed foot (**Figure 24.14c**). By comparison, in the duck, *Gremlin* is expressed throughout the entire limb, including the interdigit regions, and the duck develops a webbed foot. Interestingly, researchers have been able to introduce gremlin protein into the interdigit regions of developing chicken limbs. This produces a chicken with webbed feet!

How are these observations related to evolution? During the evolution of birds, genetic variation arose such that some individuals expressed the *Gremlin* gene in the regions between each digit, but others did not. This variation determined whether or not a bird's feet were webbed. In terrestrial settings, having nonwebbed feet is an advantage because it enables the individual to hold onto perches, run along the ground, and snatch prey. Therefore, natural selection would favor nonwebbed feet in terrestrial environments. This process explains the occurrence of nonwebbed feet in chickens, hawks, crows, and many other terrestrial birds. In aquatic environments, however, webbed feet are an advantage because they act as paddles for swimming, so genetic variation that produced webbed feet in aquatic birds would have been acted on by natural selection. Over time, this gave rise to the webbed feet now found in a wide variety of aquatic birds, including ducks, geese, and penguins.

## The *Hox* Genes Have Been Important in the Evolution of a Variety of Body Patterns

The study of developmental genes has revealed interesting trends among large groups of species. *Hox* genes, which are discussed in Chapter 20, are found in nearly all animals, indicating that they originated very early in animal evolution. *Hox* genes are homeotic genes, which specify the fate of a particular segment or region of the body. Developmental biologists have hypothesized that variation in the *Hox* genes has spawned the formation of many new body patterns. As shown in **Figure 24.15**, the number and arrangement of *Hox* genes vary considerably among different types of animals. Sponges, the simplest of animals, have at least one gene that is homologous to *Hox* genes. Insects typically have nine or more *Hox* genes. In most cases, multiple *Hox* genes occur in a cluster, lying

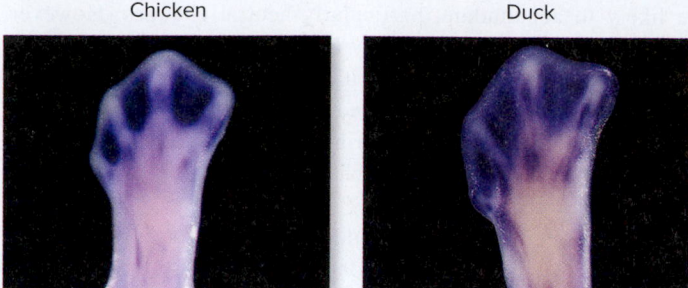

Chicken    Duck

**(a) BMP4 protein levels—similar expression in chicken and duck**

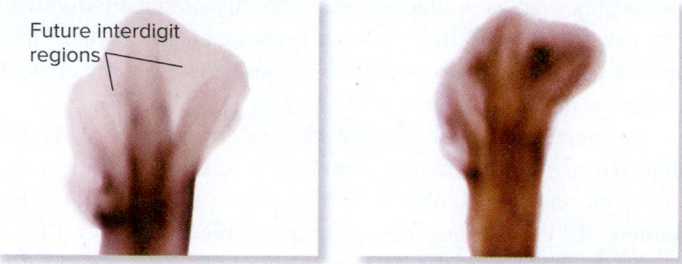

Future interdigit regions

**(b) Gremlin protein levels—not expressed in interdigit regions in chicken**

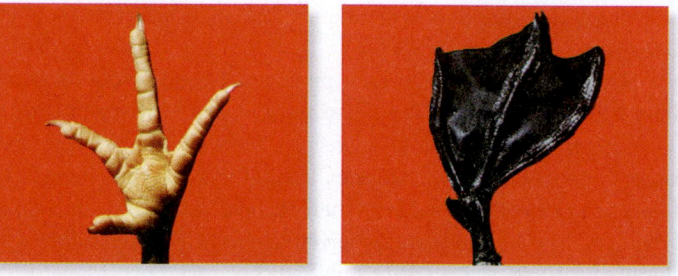

**(c) Comparison of a chicken foot and a duck foot**

**Figure 24.14** **The role of cell-signaling proteins in the pattern formation of birds' feet.** This figure shows how changes in gene expression can affect webbing between the toes. **(a)** Expression of the *BMP4* gene in the developing limbs. BMP4 protein is stained purple here and is expressed throughout the limb. **(b)** Expression of the *Gremlin* gene in the developing limbs. Gremlin protein is stained brown here. Note that *Gremlin* is not expressed in the interdigit regions of the chicken but is expressed in these regions of the duck. Gremlin inhibits the ability of BMP4 to cause programmed cell death. **(c)** Because BMP4 is not inhibited in the interdigit regions in the chicken, the cells in this region die, and the foot is not webbed. By comparison, inhibition of BMP4 in the interdigit regions in the duck results in a webbed foot.

a (left and right): Zou, H. and NISWANDER, L. 1996. Requirement for BMP signaling in interdigital apoptosis and scale formation. *Science* 272: 738–741. PMID: 8614838; b (left and right): Courtesy Ed Laufer; c (left): Courtesy of Dr. J. M. Hurle 1999. Originally published in *Development* 126(23):5515–22; c (right): Courtesy of Dr. J. M. Hurle 1999. Originally published in *Development* 126(23):5515–23

**Concept Check:** *What would you expect to happen to the pattern of the feet of ducks if the Gremlin gene was underexpressed?*

close to each other along a chromosome. In mammals, *Hox* gene clusters have been duplicated twice during the course of evolution to form four clusters, all slightly different, containing a total of 39 genes.

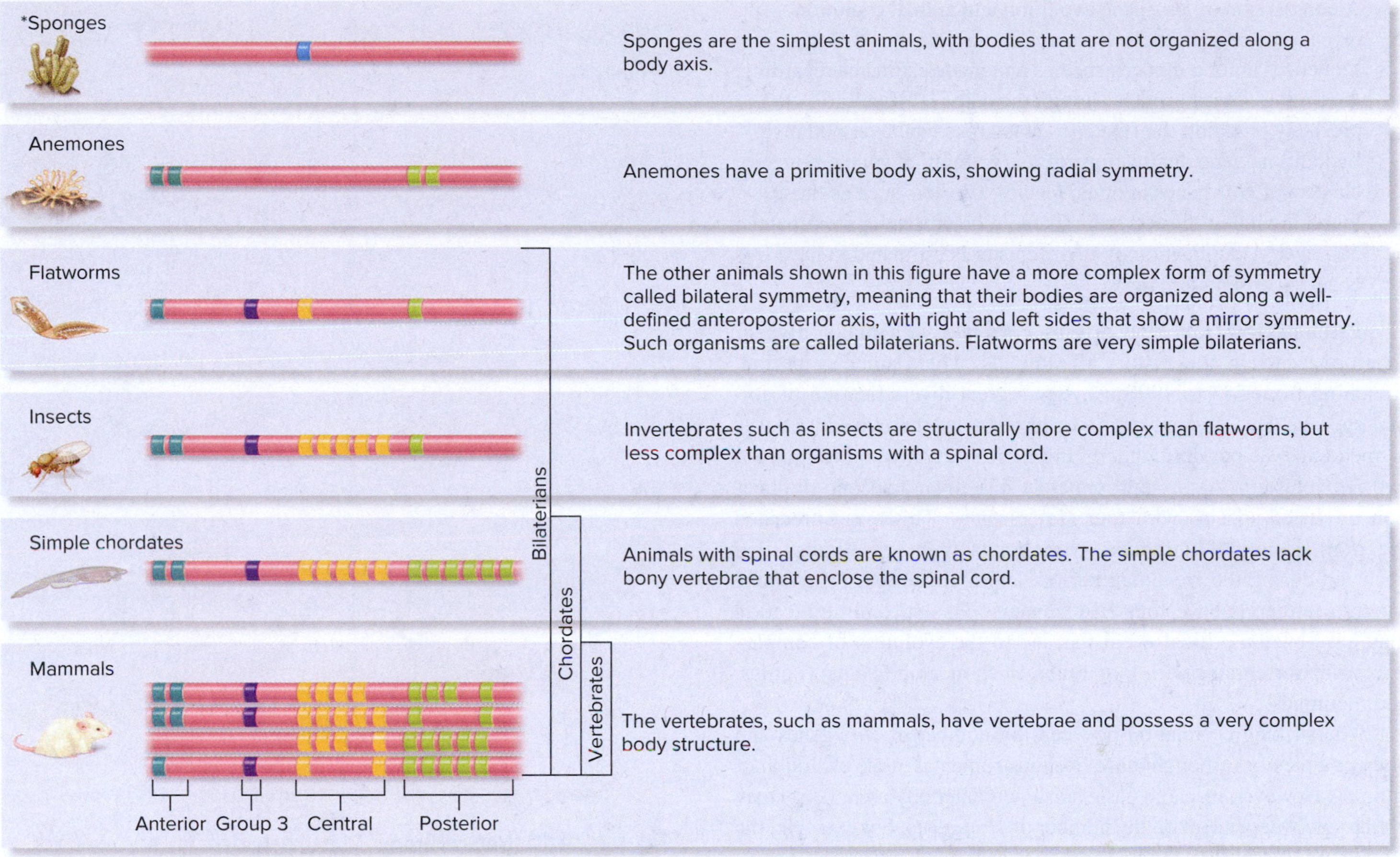

**\*Sponges**

Sponges are the simplest animals, with bodies that are not organized along a body axis.

**Anemones**

Anemones have a primitive body axis, showing radial symmetry.

**Flatworms**

The other animals shown in this figure have a more complex form of symmetry called bilateral symmetry, meaning that their bodies are organized along a well-defined anteroposterior axis, with right and left sides that show a mirror symmetry. Such organisms are called bilaterians. Flatworms are very simple bilaterians.

**Insects**

Invertebrates such as insects are structurally more complex than flatworms, but less complex than organisms with a spinal cord.

**Simple chordates**

Animals with spinal cords are known as chordates. The simple chordates lack bony vertebrae that enclose the spinal cord.

**Mammals**

The vertebrates, such as mammals, have vertebrae and possess a very complex body structure.

Bilaterians | Chordates | Vertebrates

Anterior   Group 3   Central   Posterior

\*Sponges are early diverging animals with no true tissues. They do not have true *Hox* genes, though they have an evolutionarily related gene called an *NK-like* gene.

**Figure 24.15** ***Hox* gene number and body complexity in different types of animals.**  Researchers speculate that the duplication of *Hox* genes and *Hox* gene clusters played a key role in the evolution of more complex body patterns in animals. A correlation is observed between increasing numbers of *Hox* genes and increasing complexity of body patterns. The *Hox* genes are divided into four groups, called anterior, group 3, central, and posterior, based on their relative similarities. Each group is represented with a different color in this figure.

 **Core Skill: Connections**  Look back at Figures 20.16 and 20.17. How is the expression of Hox genes related to segmentation and development along the anteroposterior axis?

Researchers propose that increases in the number of *Hox* genes have been instrumental in the evolution of many animal species whose body structures show greater complexity. To understand how, let's first consider *Hox* gene function. All *Hox* genes encode transcription factors that act as master control proteins for directing the formation of particular regions of the body. Each *Hox* gene controls a hierarchy of many regulatory genes that control the expression of genes encoding proteins that ultimately affect the morphology of the organism. The evolution of complex body patterns is associated with an increase not only in the number of regulatory genes—as evidenced by the increase in *Hox* gene complexity during evolution—but also in genes that encode proteins that directly affect an organism's form and function.

How would an increase in *Hox* genes enable more complex body patterns to evolve? Part of the answer lies in the spatial expression of the *Hox* genes. Different *Hox* genes are expressed in different regions of the body along the anteroposterior axis (refer back to **Figure 20.16**). Therefore, an increase in the number of *Hox* genes allows each of

these master control genes to become more specialized in the region that it controls. In fruit flies, one segment in the middle of the body can be controlled by a particular *Hox* gene and form wings and legs, whereas a segment in the head region can be controlled by a different *Hox* gene and develop antennae. Therefore, research suggests that one way for new, more complex body patterns to evolve is by increasing the number of *Hox* genes, thereby making it possible to form many specialized parts of the body that are organized along a body axis.

Three lines of evidence support the idea that increases in *Hox* gene number have been instrumental in the evolution and speciation of animals with more complex body patterns.

- As discussed in Chapter 20, *Hox* genes are known to control the fate of regions along the anteroposterior axis.

- As shown in Figure 24.15, a general trend is observed in which animals with a more complex body structure tend to have more *Hox* genes and *Hox* clusters in their genomes than do simpler animals.

- A comparison of *Hox* gene evolution and animal evolution reveals striking parallels. Researchers have analyzed *Hox* gene sequences among modern species and made estimates regarding the timing of past events. Though the date is difficult to precisely pinpoint, the first *Hox* gene arose well over 600 mya. In addition, gene duplications of this primordial gene produced clusters of *Hox* genes in other species. Clusters such as those found in modern insects were likely to be present approximately 600 mya. A duplication of a *Hox* cluster is estimated to have occurred around 520 mya.

Estimates of *Hox* gene origins correlate with major diversification events in the history of animals. The Cambrian period, stretching from 543 to 490 mya, saw a great diversification of animal species. This diversification occurred after the *Hox* cluster was formed and was possibly undergoing its first duplication to produce two *Hox* clusters. Also, approximately 420 mya, a second duplication produced species with four *Hox* clusters. This event preceded the proliferation of tetrapods—vertebrates with four limbs—that occurred during the Devonian period, approximately 417–354 mya. Modern tetrapods have four *Hox* clusters. This second duplication may have been a critical event that led to the evolution of complex terrestrial vertebrates with four limbs, such as amphibians, reptiles, and mammals.

The striking correlation between the number of *Hox* genes and body complexity is thought have been instrumental in the evolution of animals. However, research has also shown that body complexity may not be solely dependent on the number of *Hox* genes. For example, the number of *Hox* clusters in most tetrapods is four, whereas some fishes, which do not have more complex bodies than tetrapods, have seven or eight *Hox* clusters. In addition, researchers have discovered that specialized body structures can be formed by influencing the regulation of *Hox* genes and the other genes that are controlled by *Hox* genes. These findings indicate that changes in body complexity do not always have to be related to the total number of *Hox* genes or *Hox* clusters.

## Variation in Growth Rates Can Have a Dramatic Effect on Phenotype

Another way that genetic variation can influence morphology is by controlling the relative growth rates of different parts of the body during development. The term **heterochrony** refers to differences among species in the rate or timing of developmental events. The speeding up or slowing down of growth appears to be a common occurrence in evolution and leads to different species with striking morphological differences. With regard to the pace of evolution, such changes may rapidly lead to the formation of new species.

As an example, **Figure 24.16** compares the progressive growth of human and chimpanzee skulls. At the fetal stage, the sizes and shapes of the skulls look fairly similar. However, after this stage, the relative growth rates of certain regions become markedly different, thereby affecting the shape and size of the adult skull. In the chimpanzee, the jaw region grows faster, giving the adult chimpanzee a much larger and longer jaw. In the human, the jaw grows more slowly, and the region of the skull that surrounds the brain—the cranium—grows faster. The result is that adult humans have a smaller jaw but a larger cranium than adult chimpanzees.

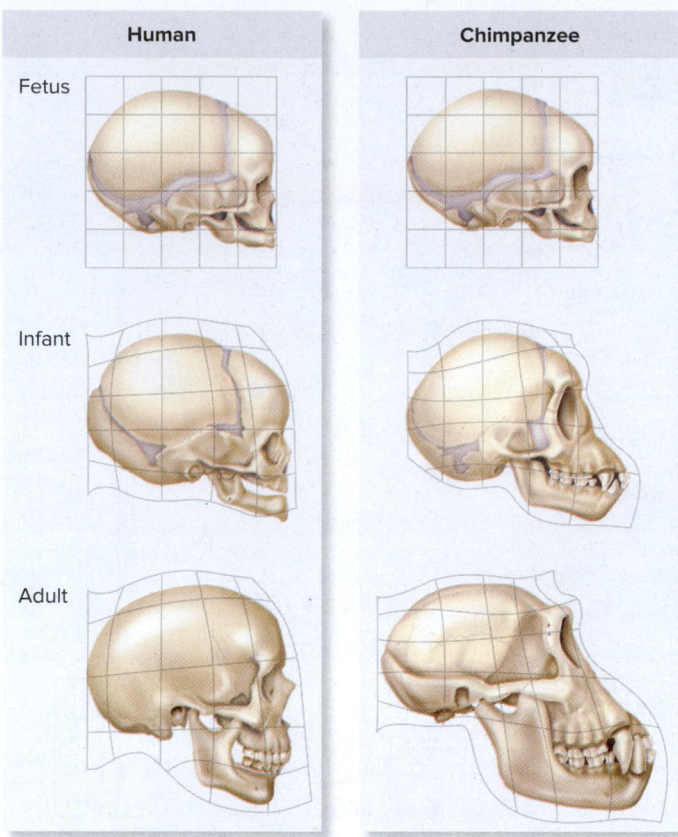

**Figure 24.16** **Heterochrony.** Due to heterochrony, one region of the body may grow faster than another during development in different species. For example, the skulls of adult chimpanzees and humans have different shapes even though their fetal skull shapes are quite similar.

 **Core Skill: Modeling** The goal of this modeling challenge is to make a series of models that show the differences in limb lengths among orangutans, chimpanzees, and humans.

**Modeling Challenge:** Search the Internet and look at photos of orangutans, chimpanzees, and humans. Even though these species look similar, one noticeable difference is the relative lengths of their limbs. Although the limbs in an early fetus look similar in all three species, the limbs in the adults show significant differences in their relative lengths. Draw models, similar to those in **Figure 24.16**, that show an early fetus, infant, and adult for all three species. Include an explanation of how heterochrony affects limb development.

 **Core Concept: Evolution**

## The Study of the *Pax6* Gene Indicates That Different Types of Eyes Evolved from One Simple Form

Thus far in this section, we have focused on the roles of particular genes as they influence the development of species with different body structures. Explaining how a complex organ comes into existence is another major challenge for evolutionary biologists. Although it is relatively easy to comprehend how a limb could undergo evolutionary modifications to become a wing, flipper, or arm, it is more difficult to understand how a body structure, such as a limb, comes into being in the first place. In his book *The Origin of Species*, Darwin addressed this question and admitted that the evolution and development of a complex organ such as the eye was difficult to understand.

As noted by Darwin, the eyes of vertebrate species are exceedingly complex, being able to adjust focus, let in different amounts of light, and detect a spectrum of colors. Darwin speculated that such complex eyes must have evolved from a simpler structure through the process of descent with modification. With amazing insight, he suggested that a very simple eye would be composed of two cell types, a photoreceptor cell and an adjacent pigment cell. The photoreceptor cell, which is a type of nerve cell, is able to absorb light and respond to it. The function of the pigment cell is to stop the light from reaching one side of the photoreceptor cell. This primitive, two-cell arrangement would allow an organism to sense both light and the direction from which the light comes.

A primitive eye would provide an additional way for an organism to sense its environment, possibly allowing it to avoid predators or locate food. Vision is nearly universal among animals, which indicates a strong selective advantage for eyesight. Over time, eyes could become more complex by enhancement of the ability to absorb different amounts and wavelengths of light and also by refinements in structures such as the addition of lenses that focus the incoming light.

Since the time of Darwin, many evolutionary biologists have wrestled with the question of eye evolution. From an anatomical point of view, researchers have discovered many different types of eyes. For example, the eyes of fruit flies, squid, and humans are quite different from each other. This observation led Austrian zoologist Luitfried von Salvini-Plawen and German evolutionary biologist Ernst Mayr to propose that eyes may have independently arisen multiple times during evolution. Based solely on morphology, such a hypothesis seemed reasonable and for many years was accepted by the scientific community.

The situation took a dramatic turn when geneticists began to study eye development. Researchers identified a master control gene, *Pax6*.* The protein encoded by the *Pax6* gene is a transcription factor that controls the expression of many other genes, including those involved in the development of the eye in both rodents and humans. In mice and rats, a mutation in the *Pax6* gene results in small eyes. A mutation in the human *Pax6* gene causes an eye disorder called aniridia, in which the iris and other structures of the eye do not develop properly. Similarly, *Drosophila* has a gene named *eyeless* that also causes a defect in eye development when mutant. *Eyeless* and *Pax6* are homologous genes; they are derived from the same ancestral gene.

In 1995, Swiss geneticist Walter Gehring and his colleagues were able to show experimentally that the expression of the

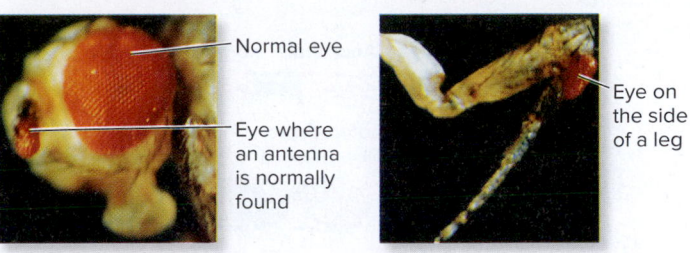

(a) **Abnormal expression of the *Drosophila eyeless* gene in the antenna region**

Normal eye

Eye where an antenna is normally found

(b) **Abnormal expression of the mouse *Pax6* gene in a fruit fly leg**

Eye on the side of a leg

**Figure 24.17** **Formation of additional eyes in *Drosophila* due to the abnormal expression of a master control gene for eye morphogenesis.** **(a)** When the *Drosophila eyeless* gene is expressed in the antenna region, eyes are formed where antennae should be located. **(b)** When the mouse *Pax6* gene is expressed in the leg region of *Drosophila*, a small eye is formed there. a–b: ©Prof. Walter J. Gehring, University of Basel

**Concept Check:** *What do you think would happen if the Drosophila eyeless gene was expressed at the tip of a mouse's tail?*

*eyeless* gene in parts of *Drosophila* where it is normally inactive could promote the formation of additional eyes. For example, using genetic engineering techniques, they were able to express the *eyeless* gene in the region where antennae should form. As seen in **Figure 24.17a**, this resulted in the formation of an eye where antennae are normally found! Remarkably, the expression of the mouse *Pax6* gene in *Drosophila* can also cause the formation of eyes in unusual places. For example, **Figure 24.17b** shows the formation of an eye on the leg of *Drosophila*.

Note that when the mouse *Pax6* gene switches on eye formation in *Drosophila*, the eye produced is a *Drosophila* eye, not a mouse eye. Why does this occur? It happens because the *Pax6* gene activates genes from the *Drosophila* genome. In *Drosophila*, the *Pax6* homolog called *eyeless* switches on a cascade involving several hundred genes required for eye morphogenesis.

Since the discovery of the *Pax6* and *eyeless* genes, homologs of this gene have been discovered in many different species. In all cases where it has been tested, the homologous gene is involved with eye development. Gehring and colleagues have hypothesized that the eyes of many different species have evolved from a common ancestral form consisting of, as proposed by Darwin, one photoreceptor cell and one pigment cell (**Figure 24.18**). As mentioned, such a very simple eye can accomplish a rudimentary form of vision by detecting light and its direction. Eyes such as these are still found in modern species, such as the larvae of certain types of mollusks. Over time, simple eyes evolved into more complex types of eyes by modifications that resulted in the addition of more types of cells, such as lens cells and nerve cells. Alternatively, other researchers propose that *Pax6* may control only certain features of eye development and that different types of eyes may have evolved independently. Future research will be needed to resolve this controversy.

---

*\*Pax is an abbreviation for paired box. The protein encoded by this gene contains a domain called a paired box.*

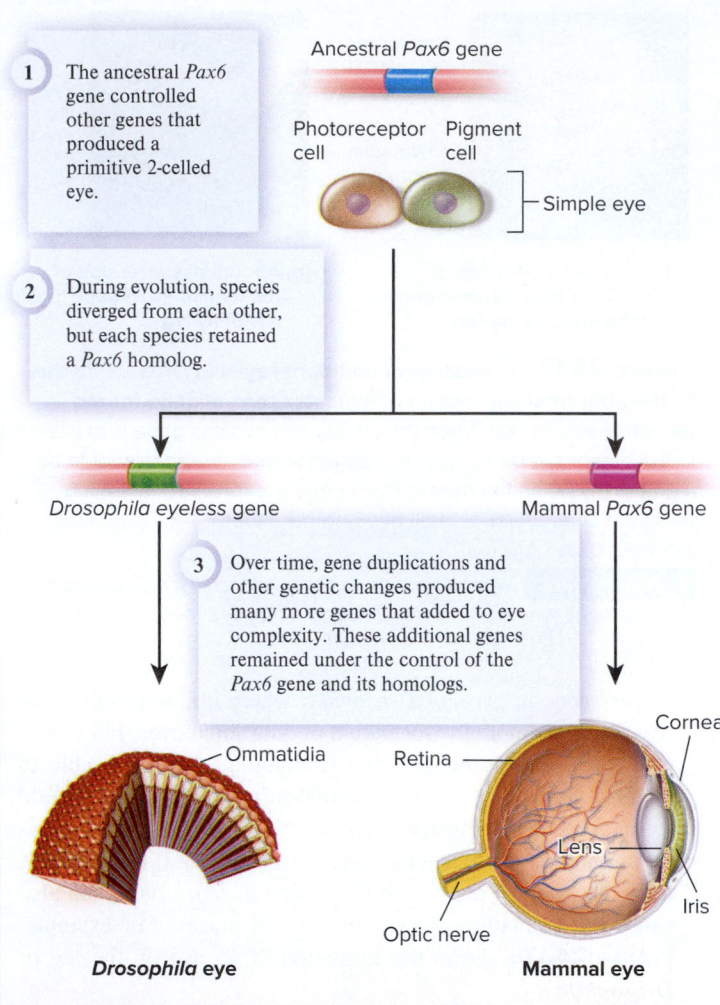

**1** The ancestral *Pax6* gene controlled other genes that produced a primitive 2-celled eye.

Ancestral *Pax6* gene

Photoreceptor cell    Pigment cell

Simple eye

**2** During evolution, species diverged from each other, but each species retained a *Pax6* homolog.

*Drosophila eyeless* gene

Mammal *Pax6* gene

**3** Over time, gene duplications and other genetic changes produced many more genes that added to eye complexity. These additional genes remained under the control of the *Pax6* gene and its homologs.

Ommatidia

*Drosophila* eye

Retina

Cornea

Lens

Iris

Optic nerve

**Mammal eye**

**Figure 24.18  Genetic control of eye evolution.** In this diagram, genetic changes, under the control of the ancestral Pax6 gene, led to the evolution of different types of eyes.

# Summary of Key Concepts

## 24.1  Identification of Species

- A species is a group of related organisms that shares a distinctive set of attributes in nature. Speciation is the process by which new species are formed. Macroevolution refers to evolutionary changes that produce new species and groups of species.

- Different characteristics, including morphological traits, reproductive isolation, molecular features, ecological factors, and evolutionary relationships, are used to identify species (Figure 24.1).

- Reproductive isolating mechanisms prevent two different species from breeding with each other (Figure 24.2).

- Prezygotic isolating mechanisms include habitat isolation, temporal isolation, behavioral isolation, mechanical isolation, and gametic isolation (Figures 24.3, 24.4).

- Postzygotic isolating mechanisms include hybrid inviability, hybrid sterility, and hybrid breakdown (Figure 24.5).

## 24.2  Mechanisms of Speciation

- Allopatric speciation occurs when a population becomes isolated from other populations and evolves into one or more new species. When speciation from a single ancestral species occurs multiple times, the process is called adaptive radiation. If two populations are incompletely separated, interbreeding may occur in hybrid zones (Figures 24.6, 24.7, 24.8).

- Podos hypothesized that changes in beak depth, associated with adaptation for feeding, promoted reproductive isolation by altering the song patterns of finches (Figures 24.9, 24.10).

- Sympatric speciation involves the formation of different species in populations that are not geographically isolated from one another. Polyploidy, adaptation to local environments, and sexual selection are mechanisms that promote sympatric speciation (Figures 24.11, 24.12).

## 24.3  The Pace of Speciation

- The pace of evolution may seem relatively constant or it may vary. Gradualism refers to steady evolution due to many small genetic changes, whereas punctuated equilibrium is a pattern of evolution in which new species arise more rapidly and then remain unchanged for long periods of time (Figure 24.13).

## 24.4  Evo-Devo: Evolutionary Developmental Biology

- Evolutionary developmental biology (evo-devo) compares the development of different species in order to understand ancestral relationships and the mechanisms that bring about evolutionary change. These changes often involve variation in the expression of cell-signaling proteins and transcription factors.

- The spatial expression of genes that affect development can affect phenotypes dramatically, as shown by the expression of the *BMP4* and *Gremlin* genes in birds, resulting in nonwebbed or webbed feet (Figure 24.14).

- An increase in the number of *Hox* genes played an important role in the evolution of more complex body patterns in animals (Figure 24.15).

- Heterochrony, which is a difference in the relative growth rates of body parts among different species, can have a major effect on phenotype (Figure 24.16).

- The *Pax6* gene and its homologs control eye development in many species of animals (Figures 24.17, 24.18).

# Assess & Discuss

## Test Yourself

1. Macroevolution refers to evolutionary changes that
   a  occur in multicellular organisms.
   b  produce new species and groups of species.
   c  occur over long periods of time.
   d  cause changes in allele frequencies.
   e  occur in large mammals.

2. The biological species concept classifies a species based on
   a  morphological characteristics.
   b  reproductive isolation.
   c  the niche the organism occupies in the environment.
   d  genetic relationships between an organism and its ancestors.
   e  both a and b.

3. Which of the following is considered an example of a postzygotic isolating mechanism?
   a incompatible genitalia
   b different mating seasons
   c incompatible gametes
   d mountain range separating two populations
   e failure of fertilized eggs to develop normally

4. Hybrid breakdown occurs when interspecies hybrids
   a do not develop past the early embryonic stages.
   b have a reduced life span.
   c are infertile.
   d are fertile but produce offspring with reduced viability and fertility.
   e produce offspring that express the traits of only one of the original species.

5. The evolution of one species into two or more species is called
   a gradualism.
   b punctuated equilibrium.
   c cladogenesis.
   d horizontal gene transfer.
   e microevolution.

6. The large number of honeycreeper species on the Hawaiian Islands is an example of
   a adaptive radiation.
   b genetic drift.
   c stabilizing selection.
   d horizontal gene transfer.
   e microevolution.

7. A major mechanism of speciation in plants but not in animals is
   a adaptation to new environments.
   b polyploidy.
   c hybrid breakdown.
   d genetic changes that alter the organism's niche.
   e both a and d.

8. The concept of punctuated equilibrium suggests that
   a the rate of evolution is constant, with short time periods of no evolutionary change.
   b evolution occurs gradually over time.
   c small genetic changes accumulate over time to allow for phenotypic change and speciation.
   d long periods of little evolutionary change are interrupted by short periods of major evolutionary change.
   e both b and c

9. Researchers suggest that an increase in the number of *Hox* genes
   a leads to reproductive isolation in all cases.
   b could explain the evolution of color vision.
   c allows for the evolution of more complex body patterns in animals.
   d results in the decrease in the number of body segments in insects.
   e does all of the above.

10. The observation that the mammalian *Pax6* gene and the *Drosophila eyeless* gene are homologous genes that promote the formation of different types of eyes suggests that
    a *Drosophila* eyes are more complex.
    b mammalian eyes are more complex.
    c eyes arose once during evolution.
    d eyes arose at least twice during evolution.
    e eye development is a simple process.

## Conceptual Questions

1. What is the key difference between prezygotic and postzygotic isolating mechanisms? Give an example of each type. Which type is more costly in terms of energy expenditure?

2. What are the key differences between gradualism and punctuated equilibrium? How are genetic changes related to these two models?

3. **Core Concept: Evolution** Describe one example in which genes that control development played an important role in the evolution of different species.

## Collaborative Questions

1. What is a species? Discuss how geographic isolation can lead to speciation, and explain how reproductive isolation plays a role.

2. Discuss the type of speciation (allopatric or sympatric) that is most likely to occur under each of the following conditions:

   a A pregnant female rat is transported by an ocean liner to a new continent.
   b A meadow containing several species of grasses is exposed to a pesticide that promotes nondisjunction.
   c In a very large lake containing several species of fishes, the water level gradually falls over the course of several years. Eventually, the large lake becomes subdivided into smaller lakes, some of which are connected by narrow streams.

# Taxonomy and Systematics

# 25

**The African forest elephant, *Loxodonta cyclotis*.** In 2001, biologists decided that this is a unique species of elephant.

©imageBROKER/Alamy Stock Photo

U ntil recently, biologists classified elephants into only two species—the African savanna elephant (*Loxodonta africana*) and the Asian elephant (*Elephas maximus*). However, by analyzing the DNA of African elephants, researchers have revised this classification, and proposed a third species, called the African forest elephant (*Loxodonta cyclotis*) (shown in the chapter opening photo). How was this new species identified? This surprising finding was made somewhat by accident in 2001. Elephants in Africa are being killed for their tusks at a high rate, despite the 1989 international ban on ivory sales. Scientists set up a DNA identification system to trace tusks to the region in Africa where the elephants were likely killed, information that could help law enforcement officials target poachers in those areas. By studying the DNA from captured tusks, researchers decided that Africa has two distinctly different *Loxodonta* species. The African forest elephant is found in the forests of central and western Africa. The African savanna elephant, which is larger and has longer tusks, lives on open, dry grasslands.

One consequence of this discovery is its effect on conservation efforts, which had previously been based on a single species of African elephants.

The rules for the classification of newly described species, such as the African forest elephant, are governed by the discipline of taxonomy (from the Greek *taxis*, meaning order, and *nomos*, meaning law). **Taxonomy** is the science of describing, naming, and classifying **extant** species, those that still exist today, as well as **extinct** species, those that have died out. Taxonomy results in the ordered division of species into groups based on similarities and dissimilarities in their characteristics. This task has been ongoing for over 300 years. British naturalist John Ray made the first attempt to broadly classify all known forms of life. Ray's ideas were later extended by Swedish naturalist Carolus Linnaeus in the mid-1700s, an achievement considered by some to be the official birth of taxonomy.

**Systematics** is the study of biological diversity and the evolutionary relationships among species, both extant and extinct. In the 1950s, German entomologist Willi Hennig began classifying species in a new way. Hennig proposed that evolutionary relationships should be inferred from features shared by descendants of a common ancestor. Since that time, biologists have applied systematics to the field of taxonomy. Researchers now try to place new species into taxonomic groups based on evolutionary relationships with other species. In addition, previously established taxonomic groups are revised as new data shed light on evolutionary relationships. Like any other scientific discipline, taxonomy should be viewed as a work in progress.

In this chapter, we will begin with a discussion of taxonomy and the concept of taxonomic groups. We will then examine how biologists use systematics to determine evolutionary relationships among species, looking in particular at how these relationships are portrayed in diagrams called phylogenetic trees. We will then explore how analyses of morphological data and molecular genetic data are used to understand the evolutionary history of life on Earth.

## 25.1 Taxonomy

### Learning Outcomes:

1. Identify the three domains of life.
2. Explain the hierarchy of groupings in taxonomy.
3. Describe how species are named using binomial nomenclature.

A hierarchy is a system of organization that involves successive levels. In biological taxonomy, every species is placed into several nested groups within a hierarchy. For example, a leopard and a fruit fly are both classified as animals, though they differ in many characteristics. By comparison, leopards and lions are placed together into a group with a smaller number of species called felines (formally named Felidae), which are predatory cats. The felines are a subset of the animal group, which has species that share many similar features. The species that are placed together into small taxonomic groups are likely to share many of the same characteristics. In this section, we will consider how biologists use a hierarchy to group similar species.

### Species Are Subdivided into Three Domains of Life

Modern taxonomy places species into progressively smaller hierarchical groups. Each group at any level is called a **taxon** (plural, taxa). The taxon called the **kingdom** was originally the highest and most inclusive. Linnaeus had classified all life into two kingdoms, plants and animals. In 1969, American ecologist Robert Whittaker proposed a five-kingdom system in which all life was classified into the kingdoms Monera, Protista, Fungi, Plantae, and Animalia. However, as biologists began to learn more about the evolutionary relationships among these groups, they found that this classification did not correctly reflect the relationships among them.

In the late 1970s, based on information in the sequences of genes, American biologist Carl Woese proposed the idea of creating a category called a **domain**. In the taxonomy hierarchy, a domain is above a kingdom. Under this system, all forms of life are grouped within three domains: **Bacteria**, **Archaea**, and **Eukarya** (**Figure 25.1**). The terms Bacteria and Archaea are capitalized when referring to the domains, but are not capitalized when referring to individual species. A single bacterial cell is called a bacterium, and a single archaeal cell is an archaeon.

The domain Eukarya formerly consisted of four kingdoms called **Protista**, **Fungi**, **Plantae**, and **Animalia**. However, researchers later discovered that protists do not constitute a separate kingdom but instead are a very broad collection of species. Taxonomists now place most eukaryotes into seven groups called supergroups. In the taxonomy of eukaryotes, a **supergroup** lies between a domain and a kingdom (see Figure 25.1). As discussed in Chapter 28, all seven supergroups contain a distinctive group of protists. In addition, the kingdoms Fungi and Animalia are within the supergroup Opisthokonta, because they are closely related to the protists in this supergroup. Kingdom Plantae is within the supergroup called Land Plants and Relatives. Plants are closely related to green algae, which are protists in this supergroup. **Table 25.1** compares a variety of molecular and cellular characteristics among the domains Bacteria, Archaea, and Eukarya.

### Every Species Is Placed into a Taxonomic Hierarchy

Why is it useful to categorize species into groups? The three domains of life contain millions of different species. Subdividing them into progressively smaller taxonomic groups makes it easier for biologists to appreciate the relationships among such a large number of species.

Below the domain and the supergroup is the kingdom, which is divided into **phyla** (singular, phylum). Each phylum is divided into

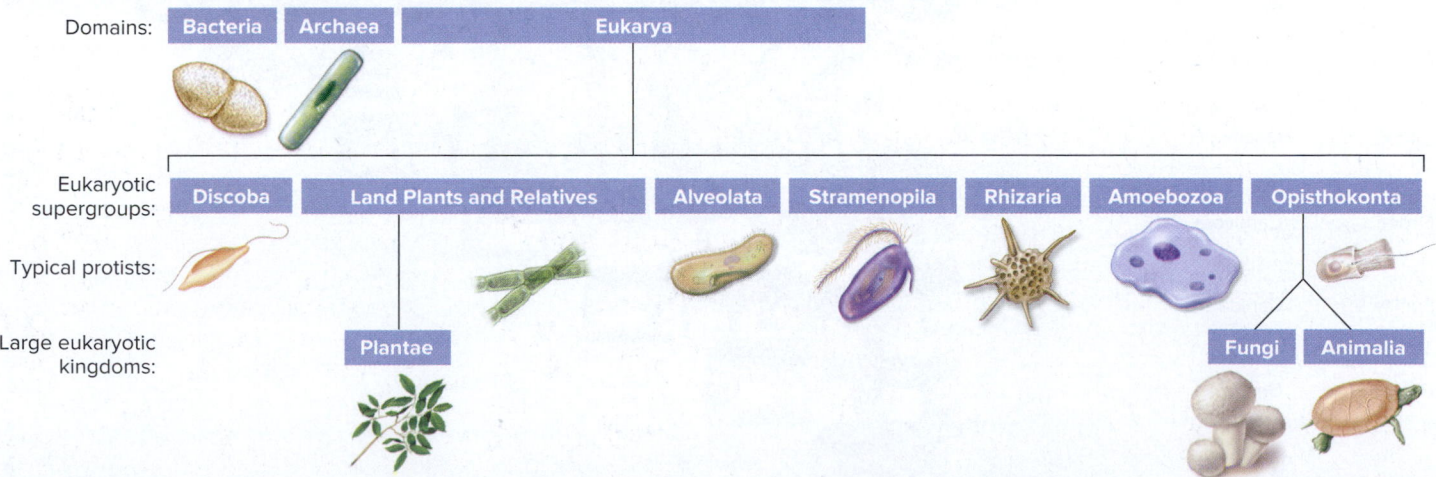

**Figure 25.1** **A classification system for living and extinct organisms.** All organisms are grouped into three domains: Bacteria, Archaea, and Eukarya. Eukaryotes are divided into seven supergroups. These seven supergroups encompass most eukaryotes. However, several small eukaryotic lineages are classified outside of the seven supergroups.

 **Core Skill: Connections** Look back at Figure 4.8. Which of the three domains contain(s) organisms with prokaryotic cells?

| Table 25.1 | Distinguishing Cellular and Molecular Features of Domains Bacteria, Archaea, and Eukarya* | | |
|---|---|---|---|
| Characteristic | Bacteria | Archaea | Eukarya |
| Chromosomes | Usually circular | Circular | Usually linear |
| Nucleosome structure | No | No | Yes |
| Chromosome segregation/cell division | Binary fission | Binary fission | Mitosis/meiosis |
| Introns in genes | Rarely | Rarely | Commonly |
| Ribosomes | 70S | 70S | 80S |
| Initiator tRNA | Formylmethionine | Methionine | Methionine |
| Operons | Yes | Yes | No |
| Capping of mRNA | No | No | Yes |
| RNA polymerases | One | Several | Three |
| Promoters of structural genes | −35 and −10 sequences | TATA box | TATA box |
| Cell compartmentalization | No | No | Yes |
| Membrane lipids | Ester-linked | Ether-linked | Ester-linked |

*The descriptions in this table represent the general features of most species in each domain. Some exceptions exist. For example, certain bacterial species have linear chromosomes, and operons occasionally are found in eukaryotes, such as the nematode worm *Caenorhabditis elegans*.

**classes**, then **orders**, **families**, **genera** (singular, genus), and **species**. As noted in Chapter 24, species may be divided into subspecies, often based on geographical distribution. Each of these taxa contains progressively fewer species that are more similar to each other than they are to the members of the taxa above them in the hierarchy. For example, the taxon Animalia, which is at the kingdom level, has a larger number of fairly diverse species than does the class Mammalia, which contains fewer species that are relatively similar to each other.

To further understand taxonomy, let's consider the classification of the gray wolf (*Canis lupus*) (**Figure 25.2**). The gray wolf is placed in the domain Eukarya, the supergroup Opisthokonta, and then within the kingdom Animalia. This kingdom contains all animals and has over 1 million species. Next, the gray wolf is classified in the phylum Chordata. The 50,000 species of animals in this phylum all have four common features at some stage of their development. These are a notochord (a cartilaginous rod that runs along the back of all chordates at some point in their life cycle), a tubular nerve or spinal cord located above the notochord, gill slits or arches, and a postanal tail. Examples of animals in the phylum Chordata include fishes, reptiles, and mammals.

The gray wolf is in the class Mammalia, which includes 5,513 species of mammals. Two distinguishing features of animals in this class are hair, which helps the body maintain a warm, constant body temperature, and mammary glands, which produce milk to nourish the young. There are 26 orders of mammals; the order that includes the gray wolf

| Taxonomic group | Gray wolf found in | Number of current species |
|---|---|---|
| Domain | Eukarya | ~4–10 million |
| Supergroup | Opisthokonta | >1 million |
| Kingdom | Animalia | >1 million |
| Phylum | Chordata | ~50,000 |
| Class | Mammalia | 5,513 |
| Order | Carnivora | 282 |
| Family | Canidae | 34 |
| Genus | *Canis* | 7 |
| Species | *lupus* | 1 |

**Figure 25.2** **A taxonomic classification of the gray wolf (*Canis lupus*).**

**Core Concept: Evolution** A goal of taxonomy is to relate the diversity of species to their evolutionary relationships. Note: The numbers in this figure will change as new species are discovered and some species become extinct.

*Concept Check:* *Which group is broader, a phylum or a family?*

is called Carnivora and has 282 species that are meat-eating animals with prominent canine teeth. The gray wolf is placed in the family Canidae, which is a relatively small family of 34 species, including different species of wolves, jackals, foxes, wild dogs, and the coyote and domestic dog. All species in the family Canidae are doglike animals. The smallest grouping that contains the gray wolf is the genus *Canis*, which has four species of jackals, the coyote, and two types of wolves. The species *Canis lupus* encompasses several subspecies, including the domestic dog (*Canis lupus familiaris*).

## Binomial Nomenclature Is Used to Name Species

As originally advocated by Linnaeus, **binomial nomenclature** is the standard format for naming species. The scientific name of every species has two names, its genus name and its unique specific epithet. For the gray wolf, the genus is *Canis* and the species epithet is *lupus*. The genus name is always capitalized, but the specific epithet is not. Both names are italicized. After the first mention, the genus name is abbreviated to a single letter. For example, we write that *Canis lupus* is the gray wolf, and in subsequent sentences, the species is referred to as *C. lupus*.

When naming a new species, genus names are always nouns or treated as nouns, whereas species epithets may be either nouns or adjectives. The names often have a Latin or Greek origin and refer to characteristics of the species or to features of its habitat. For example, the genus name of the newly discovered African forest elephant, *Loxodonta*, is from the Greek *loxo*, meaning slanting, and *odonta*, meaning tooth. The species epithet *cyclotis* refers to the observation that the ears of this species are rounder than those of *L. africana*.

The rules for naming animal species, such as *Canis lupus* and *Loxodonta africana*, were established by the International Commission on Zoological Nomenclature (ICZN). The ICZN provides and regulates a uniform system of nomenclature to ensure that every animal has a unique and universally accepted scientific name. Who is allowed to identify and name a new species? As long as ICZN rules are followed, new animal species can be named by anyone, not only scientists. The rules for naming plants are described in the International Code of Botanical Nomenclature (ICBN), and the naming of bacteria and archaea is overseen by the International Committee on Systematics of Prokaryotes (ICSP).

## 25.2 Phylogenetic Trees

### Learning Outcomes:

1. Define phylogeny, and explain how it is depicted in phylogenetic trees.
2. Explain the process of cladogenesis, the primary way that new species arise.
3. Describe how homology is used to construct phylogenetic trees.

Systematics is the study of biological diversity and evolutionary relationships. By studying the similarities and differences among species, biologists can construct a **phylogeny**, which is the evolutionary history of a species or group of species. To propose a phylogeny, biologists use the tools of systematics. For example, the classification of the gray wolf in Figure 25.2 is based on systematics. Therefore, one use of systematics is to place species into taxa and to understand the evolutionary relationships among different taxa.

In this section, we will consider the features of diagrams or trees that describe the evolutionary relationships among various species, both extant and extinct. As you will learn, such trees are usually based on morphological or genetic data.

## A Phylogenetic Tree Depicts Evolutionary Relationships Among Species

A **phylogenetic tree** is a diagram that describes the evolutionary relationships among various species, based on the information available to and gathered by systematists. Phylogenetic trees should be viewed as hypotheses that are proposed, tested, and later refined as additional data become available. Let's look at what information a phylogenetic tree contains and the form in which it is presented. **Figure 25.3** shows a hypothetical phylogenetic tree of the relationships among various flowering plant species, in which the species are labeled A through K. The vertical axis represents time, and the oldest species is at the bottom.

New species can be formed by **anagenesis**, in which a single species evolves into a different species. However, the primary way that new species arise is by **cladogenesis**, in which a species diverges into two or more species. The branch points in a phylogenetic tree, also called **nodes**, indicate times when cladogenesis has occurred. For example, approximately 12 mya, species A diverged into species A and species B. Figure 25.3 also shows anagenesis in which species C evolved into species G. The tips of branches represent species that became extinct in the past, such as species B and E, or living species, such as F, I, G, J, H, and K, which are at the top of the tree. Species A and D are also extinct but gave rise to species that are still in existence.

The branch points of a phylogenetic tree group species according to common ancestry. A **clade** consists of a common ancestral species and all of its descendant species. For example, the group highlighted in light green in Figure 25.3 is a clade derived from the common ancestral species labeled D. Likewise, the entire tree forms a clade, with species A as a common ancestor. Therefore, smaller and more recent clades are nested within larger clades that have older common ancestors.

## Systematics Constructs Taxonomic Groups Based on Evolutionary Relationships

A key goal of modern systematics is to create taxonomic groups that reflect evolutionary relationships. Systematics attempts to organize species into clades, so that each group includes an ancestral species and all of its descendants. A **monophyletic group** is a taxon that is a clade. Ideally, every taxon, whether it is a domain, supergroup, kingdom, phylum, class, order, family, or genus, should be a monophyletic group.

What is the relationship between a phylogenetic tree and taxonomy? The relationship depends on how far back we go to identify a common ancestor. For broader taxa, such as a kingdom, the common ancestor existed a very long time ago, on the order of hundreds of millions or even billions of years ago. For smaller taxa, such as a family or genus, the common ancestor occurred much more recently, on the order of millions or tens of millions of years ago. This concept is shown in a schematic way in **Figure 25.4**. This small, hypothetical kingdom is a clade that contains 64 living species. (Actual kingdoms

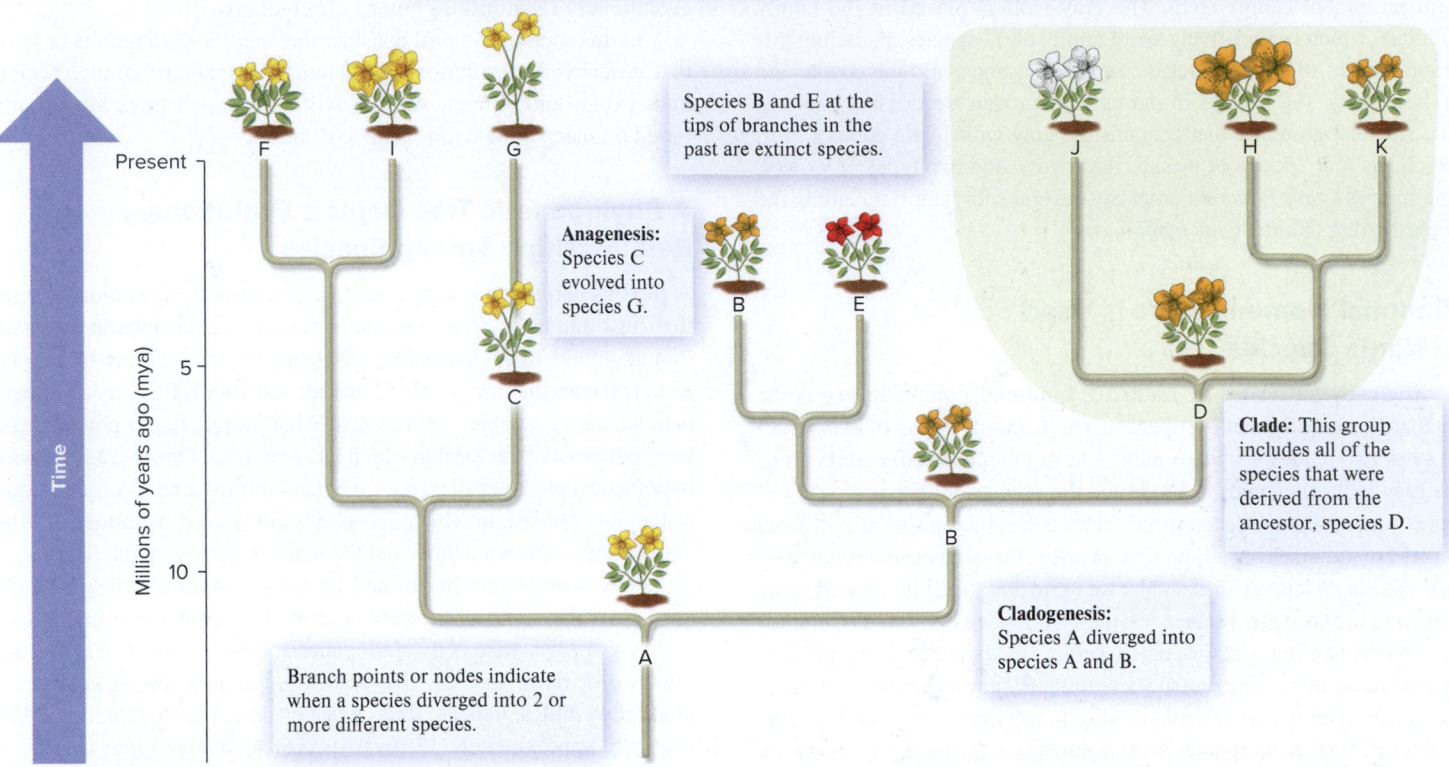

**Figure 25.3** **How to read a phylogenetic tree.** This hypothetical tree shows the proposed relationships between various flowering plant species. Species are placed into clades, groups of organisms containing an ancestral organism and all of its descendants. Note: Anagenesis is a possible way for a new species to arise, but as shown in this figure, cladogenesis is the primary mechanism.

**Concept Check:** *Can two different species have more than one common ancestor?*

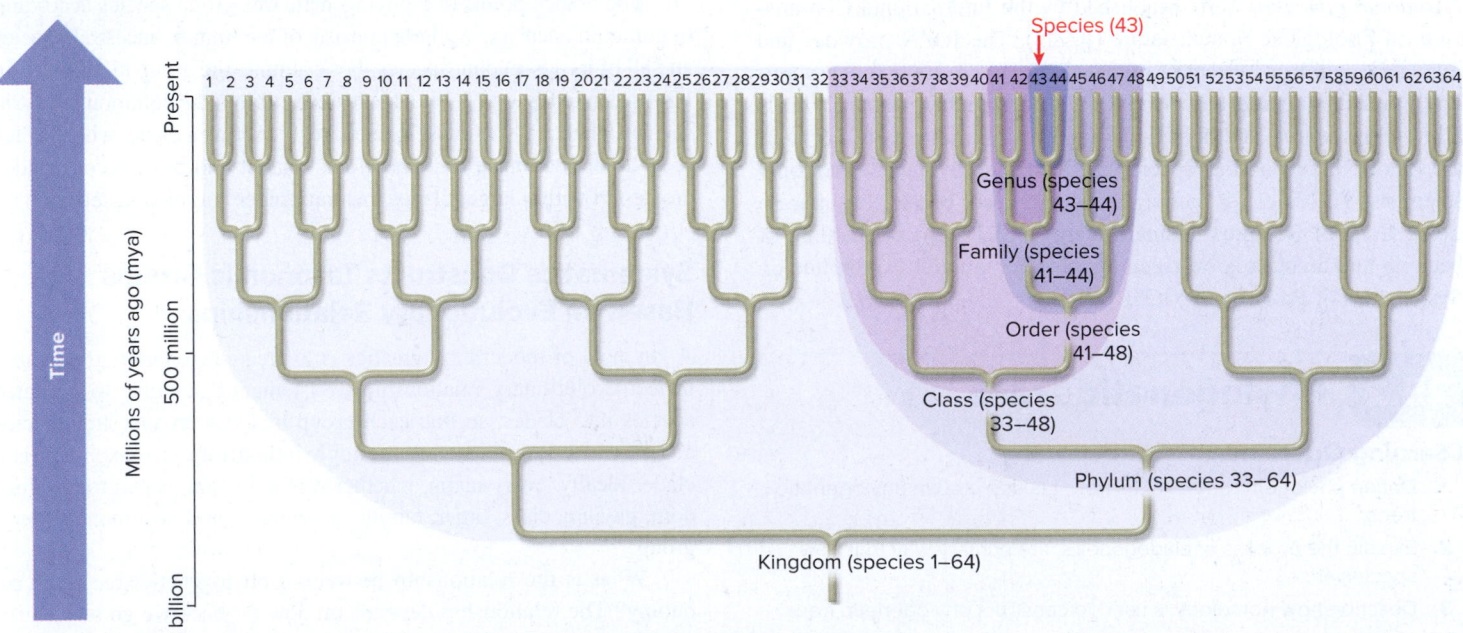

**Figure 25.4** **Schematic relationship between a phylogenetic tree and taxonomy, when taxonomy is correctly based on evolutionary relationships.** The shaded areas highlight the kingdom, phylum, class, order, family, and genus for species number 43. All of the taxa are clades. Broader taxa, such as phyla and classes, are derived from more ancient common ancestors. Smaller taxa, such as families and genera, are derived from more recent common ancestors. These smaller taxa are subsets of the broader taxa.

**Concept Check:** *Which taxon would have a more recent common ancestor, a phylum or an order?*

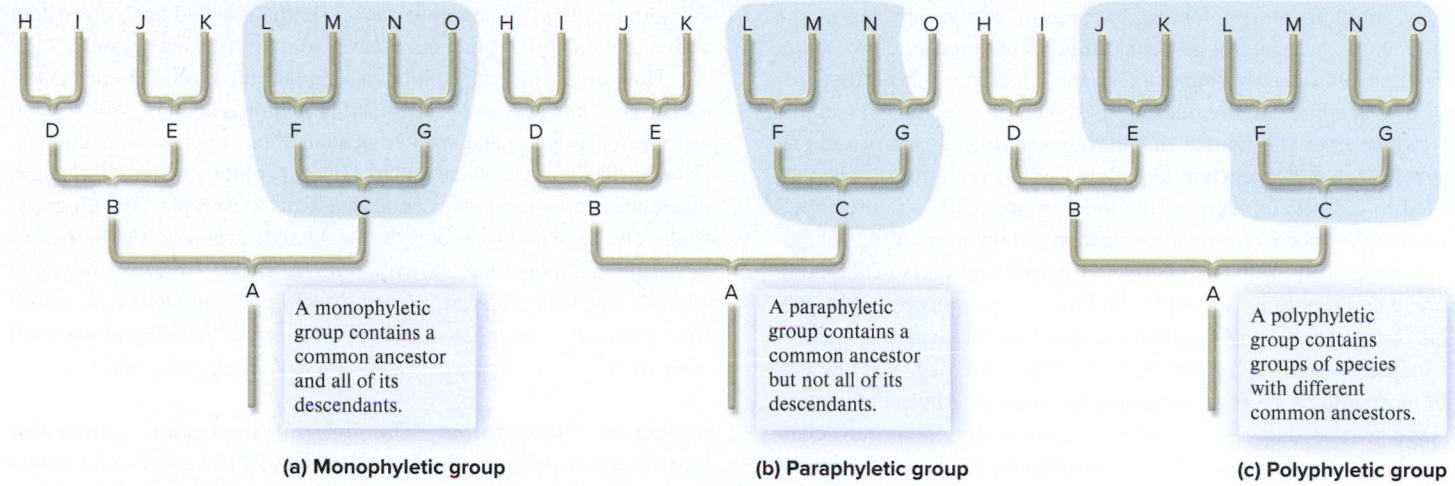

(a) **Monophyletic group** — A monophyletic group contains a common ancestor and all of its descendants.

(b) **Paraphyletic group** — A paraphyletic group contains a common ancestor but not all of its descendants.

(c) **Polyphyletic group** — A polyphyletic group contains groups of species with different common ancestors.

**Figure 25.5** **A comparison of monophyletic, paraphyletic, and polyphyletic taxonomic groups.**

are obviously larger and exceedingly more complex.) The diagram emphasizes the taxa that contain the species designated number 43. The common ancestor that gave rise to this kingdom existed approximately 1 billion years ago. Over time, more recent species arose that subsequently became the common ancestors to the phylum, class, order, family, and genus that contain species number 43.

How does research in systematics affect taxonomy? As researchers gather new information, they sometimes discover that some of the current taxonomic groups are not monophyletic. **Figure 25.5** compares a monophyletic group with taxonomic groups that are not. A **paraphyletic group** contains a common ancestor and some, but not all, of its descendants (Figure 25.5b). In contrast, a **polyphyletic group** consists of members of several evolutionary lines and does not include the most recent common ancestor of the included lineages (Figure 25.5c).

As scientists learn more about evolutionary relationships, taxonomic groups are reorganized to recognize only monophyletic groups. For example, traditional classification schemes once separated birds and reptiles into separate classes (**Figure 25.6a**). In this scheme, the reptile class (officially named Reptilia) contained orders that included turtles, lizards and snakes, and crocodiles, with birds constituting a different class. Research indicated that the reptile taxon was paraphyletic, because birds were excluded from the group. This group can be made monophyletic by including birds as a class within the reptile clade and elevating the other groups to a class status (**Figure 25.6b**).

## The Study of Systematics Is Usually Based on Morphological or Genetic Homology

As discussed in Chapter 22, the term **homology** refers to a similarity that occurs due to descent from a common ancestor. Such features are said to be homologous. For example, the arm of a human, the wing of a bat, and the flipper of a whale are homologous structures (refer back to Figure 22.12). Similarly, genes found in different species are homologous if they have been derived from the same ancestral gene (refer back to Figure 22.13).

In systematics, researchers identify homologous features that are shared by some species but not by others, which allows them to group

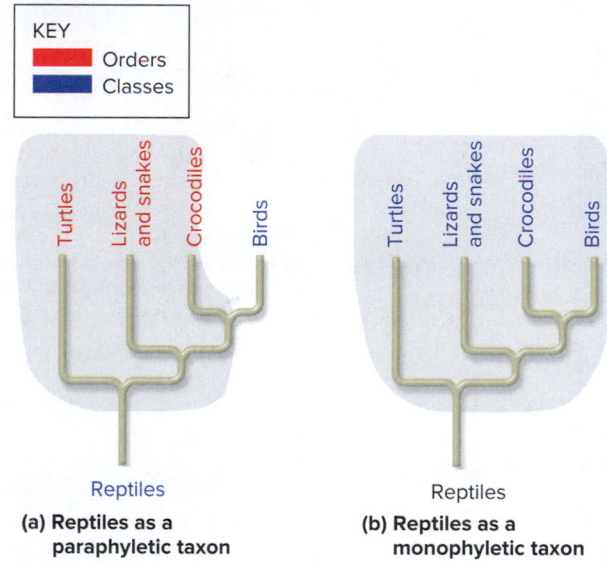

KEY
■ Orders
■ Classes

(a) **Reptiles as a paraphyletic taxon**

(b) **Reptiles as a monophyletic taxon**

**Figure 25.6** **An example of a taxon that is not monophyletic.** **(a)** The class of reptiles as a paraphyletic taxon. **(b)** The group can be made monophyletic if birds and the other orders are classified as classes within the reptile clade.

species based on shared similarities. Researchers usually study homology by examining morphological features or genetic data. In addition, the data they gather are viewed in light of geographic data. Many organisms do not migrate extremely long distances. Species that are closely related evolutionarily are relatively likely to inhabit neighboring or overlapping geographic regions, though many exceptions are known to occur.

***Morphological Analysis*** The first studies in systematics focused on morphological features of extinct and living species. Morphological traits continue to be widely used in systematic studies, particularly in those studies pertaining to extinct species and those involving groups that have not been extensively studied at the molecular level. To establish evolutionary relationships based on morphological homology, many traits have to be analyzed to identify similarities and differences.

By studying morphological features of extinct species in the fossil record, paleontologists can propose phylogenetic trees that chart the evolutionary lineages of species, including those that still exist. In this approach, the trees are based on morphological features that change over the course of many generations. As an example, **Figure 25.7** depicts a current hypothesis of the evolutionary changes that led to the development of the modern horse. This figure shows representative species from various genera. Many morphological features were used to propose this tree. Because hard parts of the body are more commonly preserved in the fossil record, this tree is largely based on the analysis of skeletal changes in foot structure, lengths and shapes of various leg bones, skull shape and size, and jaw and tooth morphology. Over an evolutionary timescale, the accumulation of many genetic changes has had a dramatic effect on species' characteristics. In the genera depicted in this figure, a variety of morphological changes occurred, such as an increase in size, a reduction in the number of toes, and modifications in the jaw and teeth consistent with a dietary shift from tender leaves to more fibrous grasses.

How do evolutionary biologists explain these changes in horses' traits? The changes can be attributed to natural selection, which acted on existing variation and resulted in adaptations to changes in climate. Over North America, where much of horse evolution occurred, changes in climate caused large areas of dense forests to be replaced with grasslands. The increase in size and changes in foot structure enabled horses to escape predators more easily and travel greater distances in search of food. The changes seen in horses' teeth are consistent with a shift from eating the tender leaves of bushes and trees to eating grasses and other more abrasive types of vegetation that require more chewing.

***Molecular Systematics*** The field of **molecular systematics** involves the analysis of genetic data, such as DNA sequences or amino acid sequences, to identify and study genetic homologies and propose

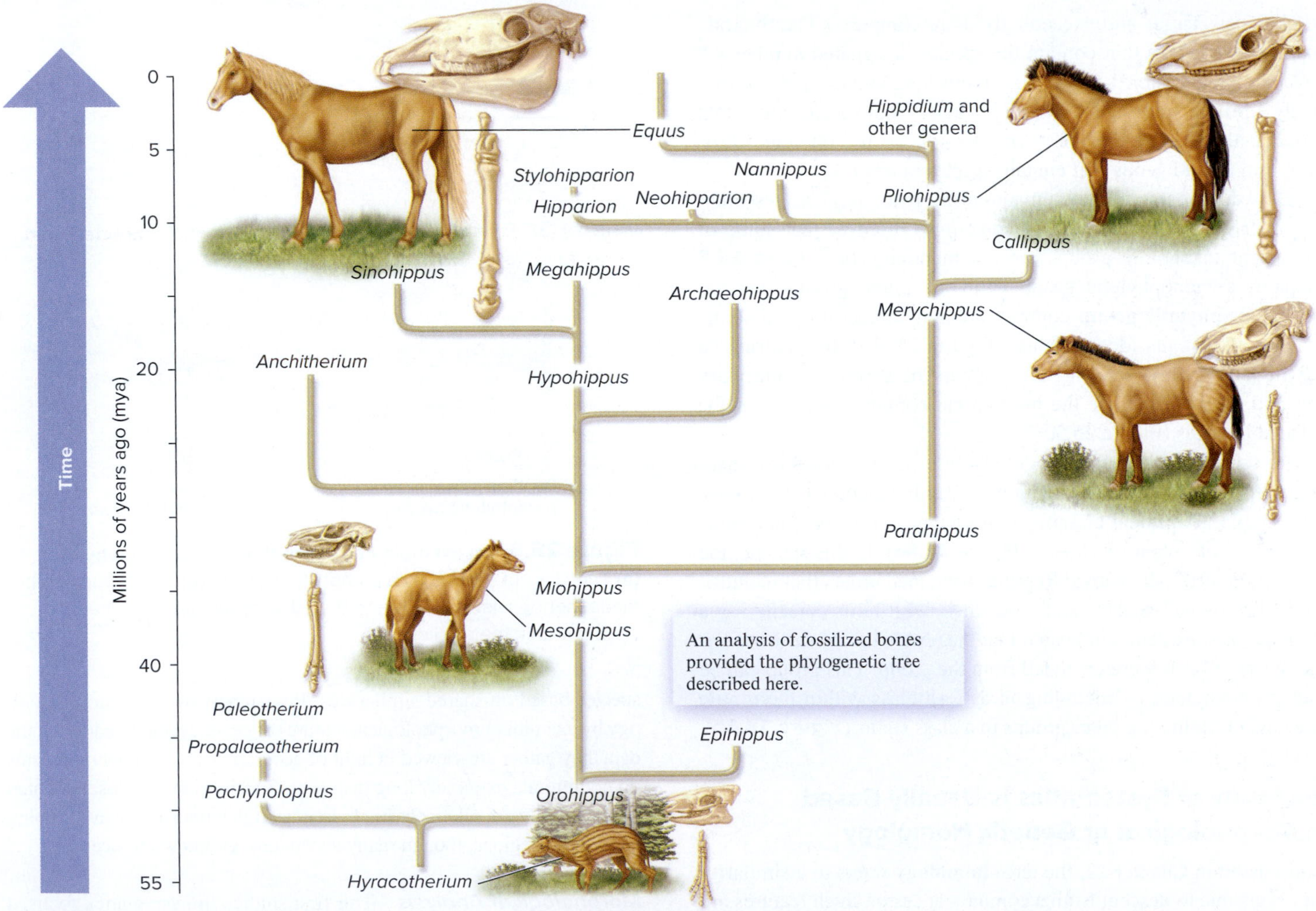

**Figure 25.7 Evolution of horse populations.** An analysis of morphological traits was used to produce this phylogenetic tree, which shows the evolutionary history that gave rise to the modern horse. As shown next to some of the genera, three important morphological changes were larger size, fewer toes, and a shift toward teeth suited for grazing.

 **Core Concept: Structure and Function** The changes in structural features during horse evolution are related to changes in functional needs. During the course of their evolution, horse populations shifted from feeding on leaves in forested regions to feeding on grasses in more open spaces.

phylogenetic trees. In 1963, Austrian biologist Emile Zuckerkandl and American chemist Linus Pauling were the first to suggest that molecular data could be used to establish evolutionary relationships. How can a comparison of genetic sequences help to establish evolutionary relationships? As discussed later in this chapter, DNA sequences change over the course of many generations due to the accumulation of mutations. Therefore, when comparing homologous sequences in different species, DNA sequences from closely related species are more similar to each other than they are to sequences from distantly related species.

## 25.3 Cladistics

### Learning Outcomes:

1. Distinguish between shared primitive characters and shared derived characters.

2. Outline the steps involved in using a cladistics approach to construct a phylogenetic tree, and explain how the principle of parsimony is used to choose among phylogenetic trees.

3. Describe how maximum likelihood is also used to discriminate among phylogenetic trees.

4. **CoreSKILL** » Explain how DNA can be analyzed to explore relationships among extant and extinct species.

**Cladistics** is the classification of species based on evolutionary relationships. A **cladistic approach** constructs phylogenetic trees by considering the possible pathways of evolutionary change that involve characteristics that are shared or not shared among various species. Such trees are known as **cladograms**. In this section, we will consider how biologists produce phylogenetic trees.

### Species Differ with Regard to Primitive and Derived Characters

A cladistic approach compares homologous features, also called **characters**, which may exist in two or more **character states**. For example, among different species, a front limb, which is a character, may exist in different character states such as a wing, an arm, or a flipper. The various character states are either shared or not shared by different species.

To understand the cladistic approach, let's take a look at a simplified phylogeny (**Figure 25.8**). We can place the species that currently exist into two groups: D and E, and F and G. The most recent common ancestor to D and E is B, whereas species C is the most recent common ancestor to F and G. With these ideas in mind, let's focus on the front limbs (flippers versus legs) and eyes.

A character that is shared by two or more different taxa and inherited from ancestors older than their last common ancestor is called a **shared primitive character**, or **symplesiomorphy**. Such characters are viewed as being older—ones that occurred earlier in evolution. With regard to species D, E, F, and G, having two eyes is a shared primitive character. It originated prior to species B and C.

By comparison, a **shared derived character**, or **synapomorphy**, is a character that is shared by two or more species or taxa and originated in their most recent common ancestor. With regard to species D and E, having two front flippers is a shared derived character that originated in species B, their most recent common ancestor (see Figure 25.8). Compared with shared primitive characters, shared derived characters are more

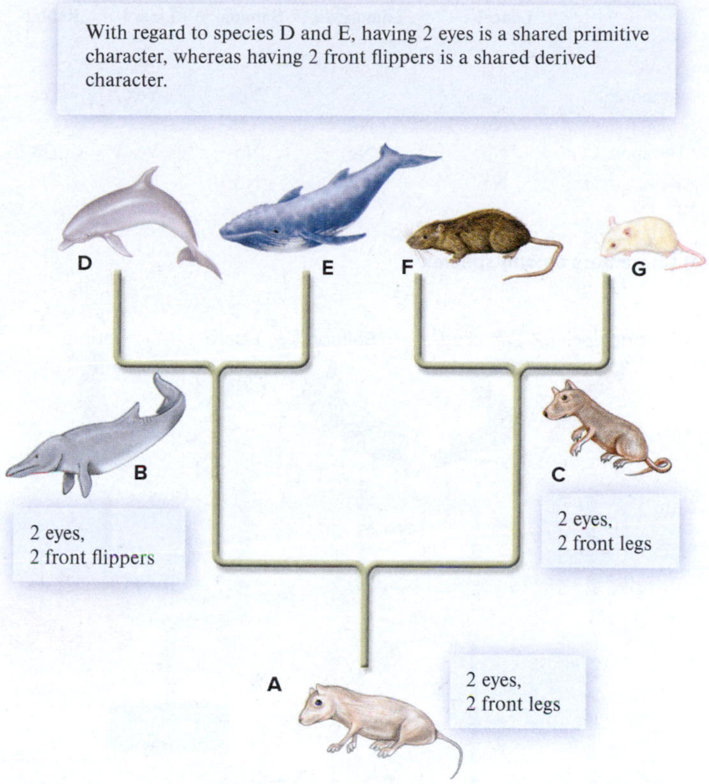

With regard to species D and E, having 2 eyes is a shared primitive character, whereas having 2 front flippers is a shared derived character.

**Figure 25.8**  **A comparison of shared primitive characters and shared derived characters.**

recent traits on an evolutionary timescale. For example, among mammals, only some species, such as whales and dolphins, have flippers. In this case, flippers were derived from the two front limbs of an ancestral species. The word derived indicates that evolution involves the modification of traits in pre-existing species. In other words, populations of organisms with new traits are derived from changes in pre-existing populations. The basis of the cladistic approach is to analyze many shared derived characters among groups of species to deduce the pathway that gave rise to those species.

The terms primitive and derived do not indicate the complexity of a character. For example, the flippers of a dolphin do not appear more complex than the front limbs of ancestral species A (see Figure 25.8), which were limbs with individual toes. Derived characters can be similar in complexity, less complex, or more complex than primitive characters.

### A Cladistic Approach Produces a Cladogram Based on Shared Derived Characters

To illustrate how shared derived characters are used to propose a phylogenetic tree, **Figure 25.9a** compares several characters among five species of animals. The proposed cladogram in **Figure 25.9b** is consistent with the distribution of shared derived characters among these species. A branch point is where two species differ in a character. The oldest common ancestor, which is now extinct, had a notochord and was an ancestor to all five species. Vertebrae are a shared derived character of the lamprey, salmon, lizard, and rabbit, but not the lancelet, which is an invertebrate. By comparison, a hinged jaw is a shared derived character of the salmon, lizard, and rabbit, but not of the lamprey or lancelet.

|  | Lancelet | Lamprey | Salmon | Lizard | Rabbit |
|---|---|---|---|---|---|
| Notochord | Yes | Yes | Yes | Yes | Yes |
| Vertebrae | No | Yes | Yes | Yes | Yes |
| Hinged jaw | No | No | Yes | Yes | Yes |
| Tetrapod | No | No | No | Yes | Yes |
| Mammary glands | No | No | No | No | Yes |

**(a) Characters among species**

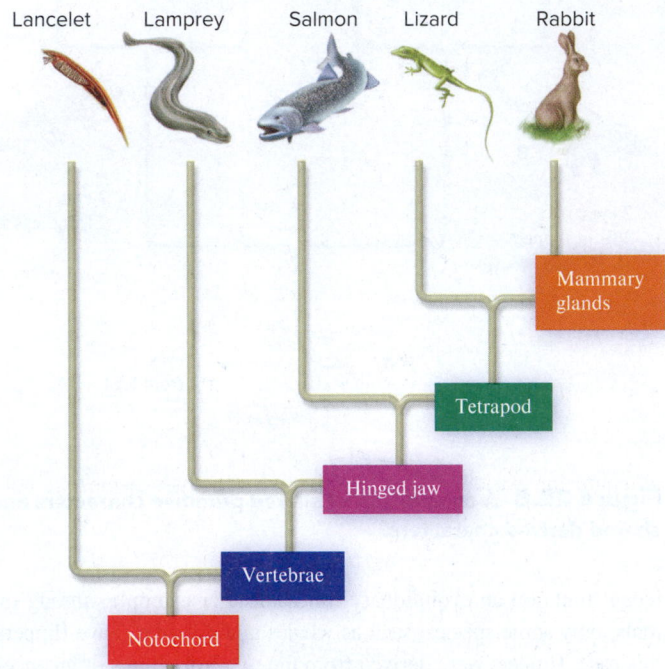

**(b) Cladogram based on morphological traits**

**Figure 25.9** **Using shared primitive characters and shared derived characters to propose a phylogenetic tree.** **(a)** A comparison of characters among these species. **(b)** This phylogenetic tree illustrates both shared primitive and shared derived characters in a cladogram of five animal species.

**Concept Check:** *What shared derived character is common to the salmon, lizard, and rabbit, but not the lamprey?*

In a cladogram, an **ingroup** is the group whose evolutionary relationships we wish to understand. By comparison, an **outgroup** is a species or group of species that is assumed to have diverged before the species in the ingroup. An outgroup lacks one or more shared derived characters that are found in the ingroup. A designated outgroup can be closely related or more distantly related to the ingroup. In the tree shown in Figure 25.9, if the salmon, lizard, and rabbit are an ingroup, the lamprey is an outgroup. The lamprey has a notochord and vertebrae but lacks a character shared by the ingroup, namely, a hinged jaw. Thus, for the ingroup, the notochord and vertebrae are shared primitive characters, whereas the hinged jaw is a shared derived character not found in the outgroup.

Likewise, the concept of shared derived characters can apply to molecular data, such as a gene sequence. Let's consider an example

to illustrate this idea. Our example involves molecular data obtained from seven different hypothetical plant species called A–G. In these species, a homologous region of DNA was sequenced as shown here:

12345678910

A: GATAGTACCC
B: GATAGTTCCC
C: GATAGTTCCG
D: GGTATTACCC
E: GGTATAACCC
F: GGTAGTACCA
G: GGTAGTACCC

The cladogram of **Figure 25.10** is a hypothesis of how these DNA sequences arose. A mutation that changes the sequence of nucleotides is comparable to a modification of a character. For example, let's designate species D as an outgroup and species A, B, C, F, and G as the ingroup. In this case, a G (guanine) at the fifth position is a shared derived character. The genetic sequence carrying this G is derived from an older primitive sequence.

Now that you have an understanding of shared primitive and derived characters, let's consider the steps a researcher would follow to propose a cladogram using a cladistic approach.

1. *Choose the species whose evolutionary relationships are of interest.* In a simple cladogram, such as those described in this chapter, individual species are compared with each other. In more complex cladograms, species may be grouped into larger taxa (for example, families) and compared with each other. If such grouping is done, the results are not reliable if the groups are not clades.

2. *Choose characters for comparing the species selected in step 1.* As mentioned, a character is a general feature of an organism and may come in different versions called character states. For example, a front limb is a character in mammals, which exists in different character states including wing, arm, and flipper.

3. *Determine the polarity of character states.* In other words, determine if a character state came first and is primitive or came later and is a derived character. This information may be available by examining the fossil record, for example, but is usually done by comparing the ingroup with the outgroup. For a character with two character states, an assumption is made that a character state shared by the outgroup and ingroup is primitive. A character state shared only by members of the ingroup is derived.

4. *Analyze cladograms based on the following principles:*

   - All species (or higher taxa) are placed on tips in a phylogenetic tree, not at branch points.

   - Each cladogram branch point should have a list of one or more shared derived characters that are common to all species above the branch point unless the character is later modified.

   - All shared derived characters appear together only once in a cladogram unless they arose independently during evolution more than once.

5. *Among many possible options, choose the cladogram that provides the simplest explanation for the data.* A common approach is to use a computer program that generates many

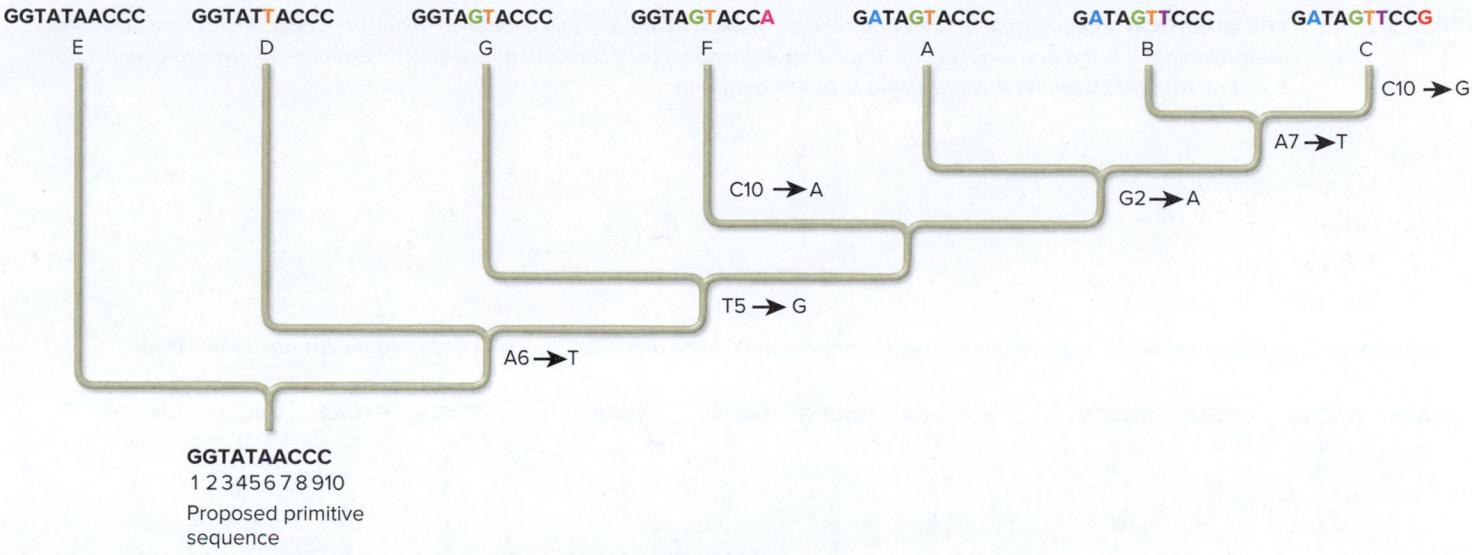

**Figure 25.10** **The use of shared derived characters applied to molecular data.** This phylogenetic tree illustrates a cladogram involving homologous gene sequences found in seven hypothetical plant species. Mutations that alter a primitive DNA sequence are shared among certain species but not others. Note: A, T, G, and C refer to nucleotide bases, and the numbers refer to the position of the base in the nucleotide sequences. For example, A6 refers to an adenine at the sixth position.

**Concept Check:** *What nucleotide change is a shared derived character for species A, B, and C, but not for species G?*

possible cladograms. Analyzing the data and choosing among the possibilities are key aspects of this process. As described later, different theoretical approaches, such as the principle of parsimony, can be used to choose among possible phylogenies.

6. ***Provide a root to the phylogenetic tree by choosing a noncontroversial outgroup.*** In this textbook, most phylogenetic trees are rooted, which means that a single node at the bottom of the tree corresponds to a common ancestor for all of the species or groups of species in the tree. A method for rooting trees is the use of a noncontroversial outgroup. Such an outgroup typically shares morphological traits and/or DNA sequence similarities with the members of the ingroup, to allow a comparison between the ingroup and outgroup. Even so, the outgroup must be noncontroversial in that it has enough distinctive differences with the ingroup to be considered a clear outgroup. For example, if the ingroup was a group of mammalian species, an outgroup could be a reptile species.

### The Principle of Parsimony Is Used to Choose from Among Possible Cladograms

One approach for choosing among possible cladograms is to assume that the best hypothesis is the one that requires the fewest number of evolutionary changes. This concept, called the **principle of parsimony**, states that the preferred hypothesis is the one that is the simplest for all the characters and their states. For example, if two species possess a tail, we would initially assume that a tail arose once during evolution and that both species have descended from a common ancestor with a tail. Such a hypothesis is simpler, and more likely to be correct, than assuming that tails arose

twice during evolution and that the tails in the two species are not due to descent from a common ancestor.

### Maximum Likelihood Is Also Used to Discriminate Among Possible Phylogenetic Trees

In addition to the principle of parsimony, evolutionary biologists also apply other approaches to the evaluation of phylogenetic trees. These methods involve the use of an evolutionary model—a set of assumptions about how evolution is likely to happen. For example, mutations affecting the third base in a codon are often neutral because they don't affect the amino acid sequence of the encoded protein and therefore don't affect the fitness of an organism. As discussed in Chapter 23, such neutral mutations are more likely to become prevalent in a population than are mutations in the first or second base. Therefore, one possible assumption of an evolutionary model is that neutral mutations are more likely than nonneutral mutations.

According to an approach called **maximum likelihood**, researchers may ask, "What is the probability that an evolutionary model and a proposed phylogenetic tree would give rise to observed molecular data?" To answer this question, they must devise rules about how DNA sequences change over time. For example, one rule may be that neutral mutations are more likely to occur than nonneutral mutations. A second rule might be that the rate of change of DNA sequences is relatively constant from one generation to the next in a particular lineage. With a set of probability rules, researchers can analyze different possible trees and predict the relative probabilities for each of them. The phylogenetic tree that gives the highest probability of producing the observed data is preferred to any trees that give lower probabilities.

 **THE QUESTION** *The principle of parsimony can be applied to the analysis of data on gene sequences. In this case, the most likely hypothesis is the one requiring the fewest base changes. Let's consider a hypothetical example involving molecular data from four taxa (A–D), where A is presumed to be the outgroup.*

    12345

A:  GTACA  (outgroup)

B:  GACAG

C:  GTCAA

D:  GACCG

*Given that B, C, and D comprise the ingroup, three possible phylogenetic trees that have the given base sequences are shown below.*

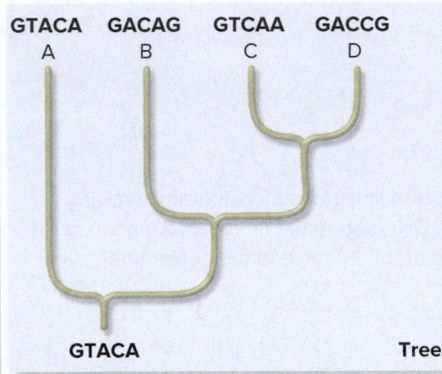
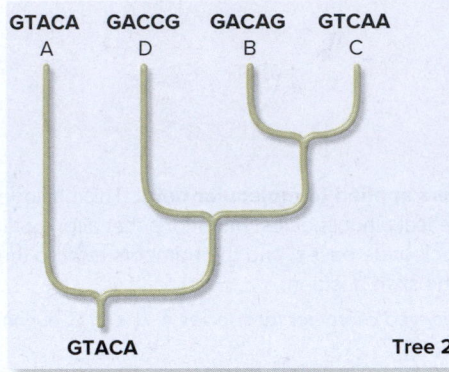
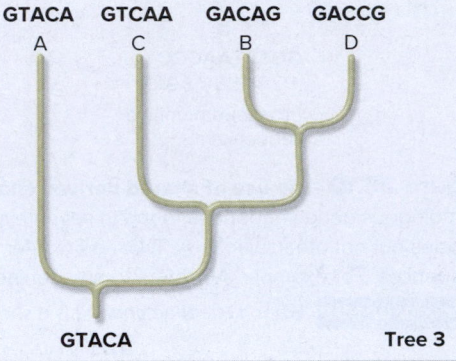

*Based on the principle of parsimony, which of these trees is the most likely to be correct?*

**T**OPIC  *What topic in biology does this question address?* The topic is phylogeny. More specifically, the question asks you to compare three phylogenetic trees and decide which one is most likely to be correct based on the principle of parsimony.

**I**NFORMATION  *What information do you know based on the question and your understanding of the topic?* In the question, you have learned that the principle of parsimony can be applied to the analysis of molecular data. You are given four base sequences and three possible phylogenetic trees. From your understanding of the topic, you may remember that the principle of parsimony states that the preferred hypothesis is the one that is the simplest for all the characters and their states.

**P**ROBLEM-SOLVING **S**TRATEGY  *Compare and contrast. Make a calculation.* One strategy for solving this problem is to compare the base sequence of the outgroup with those of the ingroup and identify the base changes. Then you can calculate which of the three possible phylogenetic trees proposes the fewest number of changes.

**ANSWER** *The diagrams below show the base changes in the three trees.*

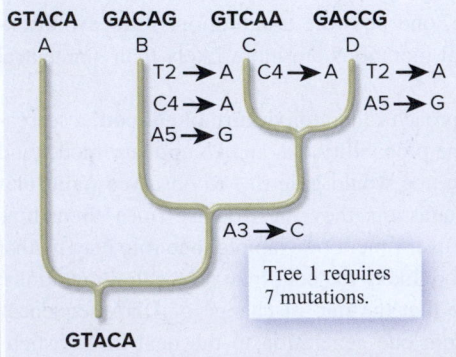
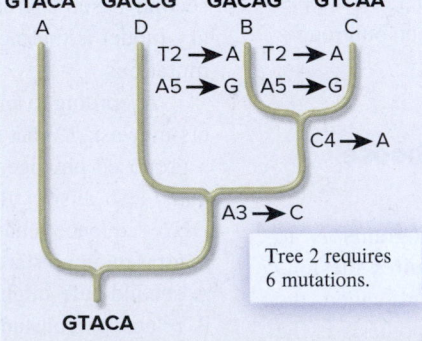
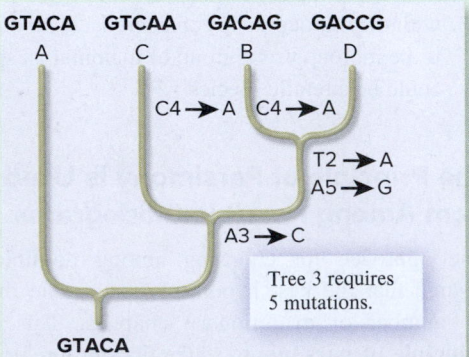

*Tree 1 requires seven mutations, tree 2 requires six, and tree 3 requires only five. Tree 3 requires the smallest number of mutations and is considered the most parsimonious, and therefore the most likely to be correct. (Note: In practice, researchers usually have multiple base sequences that are much longer than the ones shown here, so computer programs are used to find the most parsimonious tree.)*

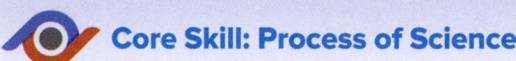

**Core Skill: Process of Science**

## Feature Investigation | Cooper and Colleagues Compared DNA Sequences from Extinct Flightless Birds and Existing Species to Propose a New Phylogenetic Tree

DNA sequencing is primarily used for studying relationships among existing species. However, in some cases, DNA can be obtained from extinct organisms. Starting with small tissue samples (usually bone, dried muscle, or preserved skin) from extinct species, scientists have discovered that it is occasionally possible to obtain DNA sequence information. This approach is called ancient DNA analysis, or molecular paleontology.

Since the mid-1980s, some researchers have become excited about the information derived from sequencing DNA of extinct specimens. Debate has centered on how long DNA can remain intact after an organism has died. Over time, the structure of DNA is degraded by hydrolysis and the loss of purines. Nevertheless, under certain conditions (such as cold temperature and low oxygen), DNA samples may be stable for as long as 50,000–100,000 years.

In recent years, this approach has been used to study evolutionary relationships between living and extinct species. As shown in **Figure 25.11**, Alan Cooper, Cécile Mourer-Chauviré, Geoffrey Chambers, Arndt von Haeseler, Allan Wilson, and Svante Pääbo

**Figure 25.11** DNA analysis of phylogenetic relationships among living and extinct flightless birds by Cooper and colleagues.

**GOAL** To gather molecular information to hypothesize about the evolutionary relationships among these species.

**KEY MATERIALS** Tissue samples from 4 extinct species of moas were obtained from museum specimens. Tissue samples were also obtained from 3 species of kiwis, 1 emu, 1 cassowary, 1 ostrich, and 2 species of rheas.

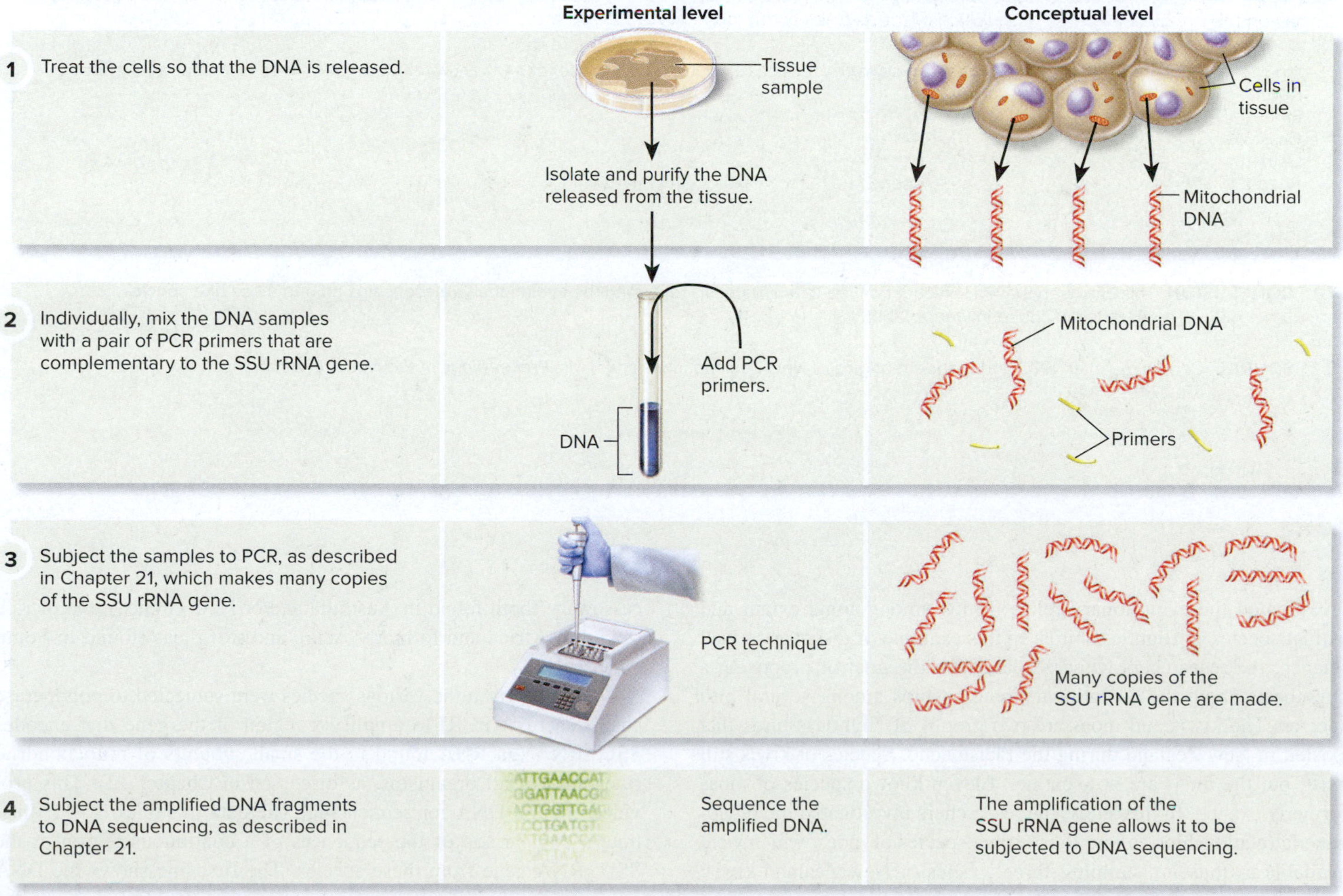

| | | Experimental level | Conceptual level |
|---|---|---|---|
| **1** | Treat the cells so that the DNA is released. | Tissue sample — Isolate and purify the DNA released from the tissue. | Cells in tissue — Mitochondrial DNA |
| **2** | Individually, mix the DNA samples with a pair of PCR primers that are complementary to the SSU rRNA gene. | DNA — Add PCR primers. | Mitochondrial DNA — Primers |
| **3** | Subject the samples to PCR, as described in Chapter 21, which makes many copies of the SSU rRNA gene. | PCR technique | Many copies of the SSU rRNA gene are made. |
| **4** | Subject the amplified DNA fragments to DNA sequencing, as described in Chapter 21. | Sequence the amplified DNA. | The amplification of the SSU rRNA gene allows it to be subjected to DNA sequencing. |

**5** Align the DNA sequences to each other, using computer techniques.

Align sequences, using computer programs.

Align sequences to compare the degree of similarity.

**6    THE DATA**

```
Moa 1      GCTTAGCCCTAAATCCAGATACTTACCCTACACAAGTATCCGCCCGAGAACTACGAGCACAAACGCTTAAAACTCTAAGGACTTGGCGGTGCCCCAAACCCA
Kiwi 1     ···········T·G·····GT···CT····C···············································T·····
Emu        ··········TT····C··T···CAG·C·····T················································T·····
Cassowary  ··········TT····CG·TA···CTG···············································T·····
Ostrich    ······T···AT·········C··CT··················································T·····
Rhea 1     ··········T···········C··CT··················································T·····

Moa 1      CCTAGAGGAGCCTGTTCTATAATCGATAATCCACGATACACCCGACCATCCCTCGCCCGT–GCAGCCTACATACCGCCGTCCCCAGCCCGCCT––AATGAAA
Kiwi 1     ·········································C·········A····T·T···AAC–A·····T·······G···T····AA····G
Emu        ·········································C·········A····T·T···AA–A·········G··········––····
Cassowary  ·········································C·········AG···T·T··AA·TA·········G········––·G··G
Ostrich    ·····································T···A···C··T··A––T·········G·······C––···G
Rhea 1     ·········································C·········T·T···A·–·········TA·G····A

Moa 1      G–AACAATAGCGAGCACAACAGCCCTCCCCCGCTAACAAGACAGGTCAAGGTATAGCATATGAGATGGAAGAAATGGGCTACATTTTCTAACATAGAACACC
Kiwi 1     ·–····C····A······TA·–··A···············C················A·····T·T
Emu        ·–······T···AC––TT···············G················T·T
Cassowary  ·–······T···AC––T···············G·················T·
Ostrich    ·–·········T···A––·············GAG············T·A
Rhea 1     ·–···C··AG··T·T··TA–––·············G··········TC·····A

Moa 1      C–––––––––––––ACGAAAGAGAAGGTGAAACCCTCCTCAAAAGGCGGATTTAGCAGTAAAATAGAACAAGAATGCCTATTTTAAGCCCGGCCCTGGGGC
Kiwi 1     –············A·GGT···T·–C···T·G··············C···T···GA·T·········–·T·····A····
Emu        –············AG·T·······T·AC·T···G··············C···T···GA·T········A–·T···T·A····
Cassowary  –············A·G·T·······T·A···T·G··············C········GA·T·······A––······A····
Ostrich    –············G·TA·····T·A·····G··············T···GA·T·········–T····T·A····
Rhea 1     –············G·····GGCA······–AC····CG··············G··G·TC····A···C·C······–·····A····
```

**7    CONCLUSION**  This discovery-based investigation led to a hypothesis regarding the evolutionary relationships among these bird species, which is described in Figure 25.12.

**8    SOURCE**  Cooper, A., et al. 1992. Independent origins of New Zealand moas and kiwis. *Proceedings of the National Academy of Sciences* 89: 8741–8744.

investigated the evolutionary relationships among some extant and extinct species of flightless birds. In this example of discovery-based science, the researchers gathered data with the goal of proposing a hypothesis about the evolutionary relationships among several bird species. The kiwis and moas are two groups of flightless birds that existed in New Zealand during the Pleistocene. Species of kiwis still exist, but the moas are now extinct. Eleven known species of moas formerly existed. In this study, the researchers investigated the phylogenetic relationships among four extinct species of moas, which were available as museum samples; three species of New Zealand kiwis; and living species of other flightless birds, including the emu and the cassowary (both found in Australia and/or New Guinea), the ostrich (found in Africa and formerly Asia), and two rheas (found in South America).

Samples from the various species were subjected to polymerase chain reaction (PCR) to amplify a region of the gene that encodes SSU rRNA (an RNA found in the small subunits of mitochondrial ribosomes of all organisms, as discussed in Chapter 12). This provided enough DNA for sequencing. The data in Figure 25.11 illustrate a comparison of the sequences of a continuous region of the SSU rRNA gene from these species. The first line shows the DNA sequence for one of the four extinct moa species. Below it are the sequences of several of the other species that were analyzed. When the other sequences are identical to the first sequence, a dot is placed in the corresponding position. When the sequences are different, the changed nucleotide base (A, T, G, or C) is placed there. In a few regions, the genes are different lengths. In these cases, a dash is placed to indicate missing nucleotides.

As you can see from the large number of dots, the gene sequences among these flightless birds are very similar, though some differences occur. If you look carefully at the data, you will notice that the sequence from the kiwi (a New Zealand species) is actually more similar to the sequence from the ostrich (an African species) than it is to that of the moa, which was once found in New Zealand. Likewise, the kiwi is more similar to the emu and cassowary (found in Australia and New Guinea) than to the moa. How were these results interpreted? The researchers concluded that the kiwis are more closely related to African and Australian flightless birds than they are to the moas. From these results, they concluded that New Zealand was colonized twice by ancestors of flightless birds. The researchers used a maximum likelihood analysis to propose a new phylogenetic tree that illustrates the revised relationships among these living and extinct species (**Figure 25.12**).

### Experimental Questions

1. What is molecular paleontology? What was the purpose of the study conducted by Cooper and colleagues?

2. What birds were examined in the Cooper study, and what are their geographic distributions? Why were the different species selected for this study?

3. **CoreSKILL »** What results did Cooper and colleagues obtain by comparing these DNA sequences? How did the results of this study affect the proposed phylogeny of flightless birds?

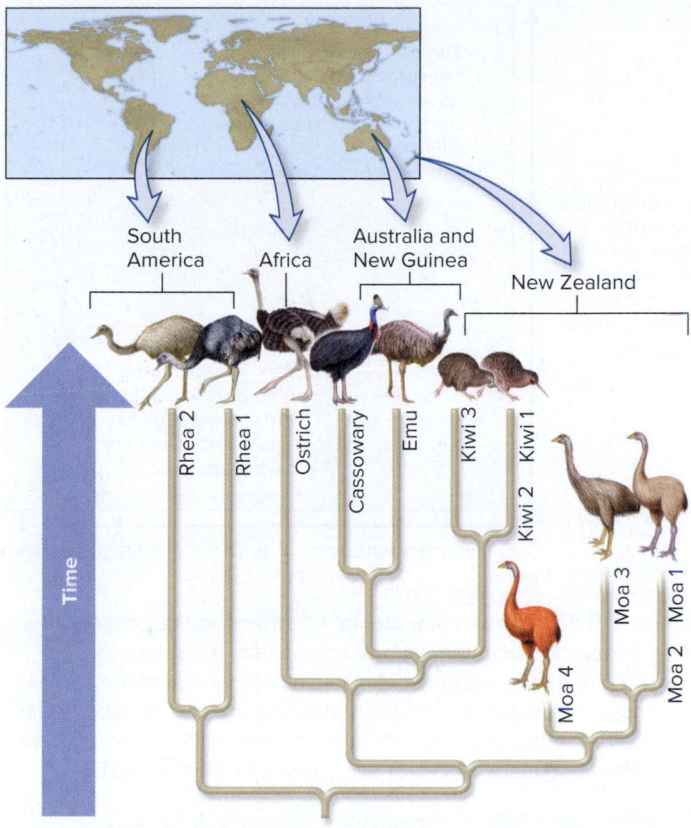

**Figure 25.12  A revised phylogenetic tree of flightless birds.** This tree is based on a comparison of DNA sequences from extinct and living flightless birds, as described in Figure 25.11.

**Concept Check:** *With regard to geography, why are the results of Cooper and his colleagues surprising?*

## 25.4  Molecular Clocks

### Learning Outcomes:

1. Explain how molecular clocks are used in the dating of evolutionary events.

2. **CoreSKILL »** Compare and contrast the use of different genes to produce phylogenetic trees.

As we have seen, researchers employ different methods to choose a phylogeny that describes the evolutionary relationships among various species. Researchers are interested not only in the most likely pathway of evolution (the branches of the trees), but also the timing of evolutionary change (the lengths of the branches). How can researchers determine when different species diverged from each other in the past? As shown earlier in Figure 25.7, the fossil record can sometimes help researchers apply a timescale to a phylogeny.

Another way to infer the timing of past events is by analyzing genetic sequences. The **neutral theory of evolution** proposes that most genetic variation that exists in populations is due to the accumulation of neutral mutations—changes in genes and proteins that are not acted on by natural selection. The reasoning behind this concept is that

favorable mutations are likely to be very rare and detrimental mutations are likely to be eliminated from a population by natural selection. A large body of evidence supports the idea that much of the genetic variation observed in living species is due to the accumulation of neutral mutations. From an evolutionary point of view, if neutral mutations occur at a relatively constant rate, they can serve as a **molecular clock** to measure evolutionary time. In this section, we will consider the concept of a molecular clock and its application in phylogenetic trees.

### The Timing of Evolutionary Change May Be Inferred from Molecular Clock Data

**Figure 25.13** illustrates the concept of a molecular clock. The graph's vertical axis measures the number of base differences in a homologous gene between different pairs of species. The horizontal axis plots the amount of time that has elapsed since each pair of species shared a common ancestor.

As an example, let's suppose a researcher compared a gene sequence that was 500 bp long. Between species A and species B, this sequence might differ at 10 places and be identical at 490 places. By comparison, the 500-bp sequence might differ at 20 places between species A and species C and be the same at 480 places. Such a result is consistent with

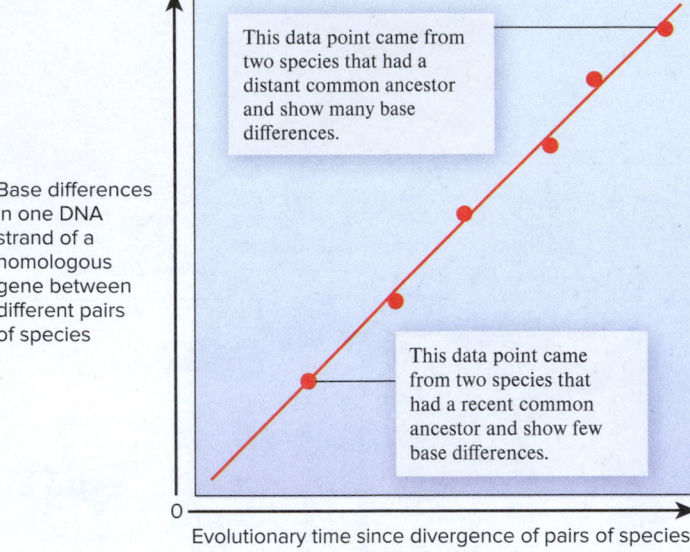

Base differences in one DNA strand of a homologous gene between different pairs of species

This data point came from two species that had a distant common ancestor and show many base differences.

This data point came from two species that had a recent common ancestor and show few base differences.

0

Evolutionary time since divergence of pairs of species (millions of years)

**Figure 25.13  A molecular clock.** According to the concept of a molecular clock, neutral mutations accumulate at a relatively constant rate over evolutionary time. In a comparison of the same homologous gene between pairs of different species, those species that diverged more recently tend to have fewer differences than do those whose common ancestor occurred in the distant past.

 **Core Skill: Connections**  Look back at Table 12.1, which shows the genetic code. Propose three changes to a codon sequence that you think would be neutral mutations.

the idea that species A and species B shared a more recent common ancestor than do species A and species C. The explanation for this phenomenon is that the gene sequences of various species accumulate independent mutations after they have diverged from each other. A longer period of time since their divergence allows for a greater accumulation of mutations, which makes their sequences of bases more different.

Figure 25.13 suggests a linear relationship between the number of base changes and the time of divergence. For example, a linear relationship predicts that a pair of species with, say, 20 base differences in a given gene sequence would have a common ancestor that lived roughly twice as long ago as a pair showing 10 nucleotide differences. Although actual data sometimes show a relatively linear relationship over a defined time period, evolutionary biologists do not think that molecular clocks are perfectly linear over very long periods of time. Several factors can contribute to nonlinearity of molecular clocks. These include differences in the generation times of the species being analyzed and variation in the mutation rates of genes between different species.

To obtain reliable data, researchers must calibrate their molecular clocks. How much time does it take to accumulate a certain percentage of base changes? To perform such a calibration, researchers must have information regarding the date when two species diverged from a common ancestor. Such information could come from the fossil record, for instance. The genetic differences between those species are then divided by the amount of time since their last common ancestor to calculate a rate of evolutionary change.

As an example of clock calibration, let's consider primates. The fossil evidence suggests that humans and chimpanzees diverged from a common ancestor approximately 6 mya. The percentage of base differences between the mitochondrial DNA of humans and chimpanzees is 12%. From these data, the molecular clock for changes in the sequence of bases in mitochondrial DNA of primates is calibrated at roughly 2% base changes per million years. However, molecular dating based on the use of a single fossil as a calibration point can lead to significant inaccuracies in the molecular clock. When possible, researchers advocate using multiple fossils in the calibration process.

## Different Genes Are Analyzed to Study Phylogeny and Evaluate the Timing of Evolutionary Change

For evolutionary comparisons, the DNA sequences of many genes have been obtained from a wide range of sources. Many different genes have been studied to propose phylogenetic trees and evaluate the timing of past events. For example, the SSU rRNA gene used by Cooper and colleagues in their research on flightless birds (see Figure 25.11) is commonly used in evolutionary studies. As noted in Chapter 12, the gene for SSU rRNA is found in the genomes of all living organisms. Therefore, its function must have been established at an early stage in the evolution of life on this planet, and its sequence has changed fairly slowly. Furthermore, SSU rRNA is a rather large molecule, so it contains a large amount of sequence information. This gene has been sequenced for thousands of different species (see Figure 12.17). Slowly changing genes such as the gene that encodes SSU rRNA are useful for evaluating distant evolutionary relationships, such as comparing higher taxa. For example, SSU rRNA data can be used to place eukaryotic species into their proper phyla or orders.

Other genes have changed more rapidly during evolution because of a greater tolerance of neutral mutations. For example, the mitochondrial genome and DNA sequences within introns can more easily incur neutral mutations (compared to the coding sequences of genes), and so their sequences change frequently during evolution. More rapidly changing DNA sequences have been used to study recent evolutionary relationships, particularly among eukaryotic species such as species of large animals that have long generation times and therefore tend to evolve more slowly. In these cases, slowly evolving genes may not be very useful for establishing evolutionary relationships because two closely related species may have identical or nearly identical DNA sequences for such genes.

**Figure 25.14** shows a simplified phylogeny of closely related species of primates. A molecular clock was used to give a timescale to this phylogenetic tree. The tree was proposed by comparing DNA sequence changes in the gene for cytochrome oxidase subunit II, one of several subunits of cytochrome oxidase, a protein in the mitochondrial inner membrane that is involved in cellular respiration. This gene tends to change fairly rapidly on an evolutionary timescale. The vertical scale of Figure 25.14 represents time, and the branch points labeled with letters represent common ancestors. Let's take a look at three branch points (labeled A, D, and E) and relate them to the accumulation of neutral mutations.

*Ancestor A:*  This ancestor diverged into two species that ultimately gave rise to siamangs and the other five species. Since this divergence,

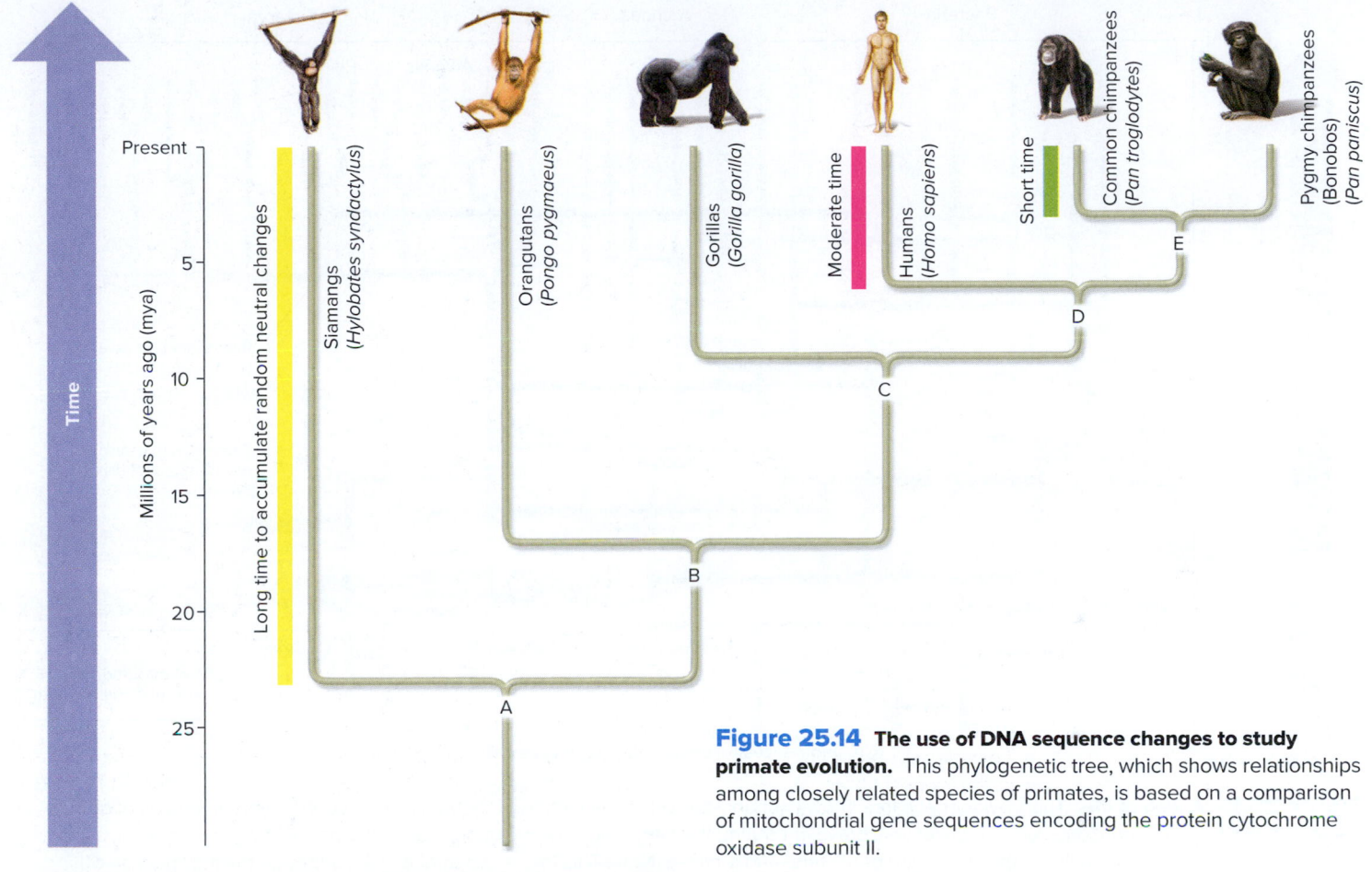

**Figure 25.14** **The use of DNA sequence changes to study primate evolution.** This phylogenetic tree, which shows relationships among closely related species of primates, is based on a comparison of mitochondrial gene sequences encoding the protein cytochrome oxidase subunit II.

**Core Skill: Modeling** The goal of this modeling challenge is to propose a phylogenetic tree based on molecular data.

**Modeling Challenge:** Figure 22.13 compares a short amino acid sequence within the p53 protein among nine species. Propose an evolutionary tree that describes the evolutionary relationships of these species. You should also consider the data in the right column of Figure 22.13, which shows the percentages of amino acids in the whole p53 protein that are identical to the human sequence. The top of your tree should show the nine species with their names (as in Figure 25.14). Above the name of each species, put the number amino acid differences that occur within the short amino acid sequence for each species compared with that of humans.

there has been a long time (approximately 23 million years) for the siamang genome to accumulate a relatively high number of random neutral changes that would be different from the random changes that have occurred in the genomes of the other five species (see the yellow bar in Figure 25.14). Therefore, the gene in the siamangs is fairly different from the genes in the other five species.

*Ancestor D:* This ancestor diverged into two species that eventually gave rise to humans and chimpanzees. This divergence occurred a moderate time ago, approximately 6 mya, as illustrated by the red bar. The differences in gene sequences between humans and chimpanzees are relatively moderate.

*Ancestor E:* This ancestor diverged into two species of chimpanzees. Since the divergence of species E into two species, approximately 3 mya, the time for the molecular clock to "tick" (that is, accumulate random mutations) is relatively short, as depicted by the green bar in

Figure 25.14. Therefore, the two existing species of chimpanzees have fewer differences in their gene sequences compared to other primates.

## 25.5 Horizontal Gene Transfer

### Learning Outcome:

**1.** Explain how horizontal gene transfer affects evolution and the relationships among different taxa.

Thus far, we have considered various ways to construct phylogenetic trees, which describe the relationships between ancestors and their descendents. The type of evolution depicted in the phylogenetic trees in previous sections, which involves changes in groups of species due to descent from a common ancestor, is sometimes called vertical evolution. Since the time of Darwin, vertical evolution has been the traditional way biologists have viewed the evolutionary process. However, over the past

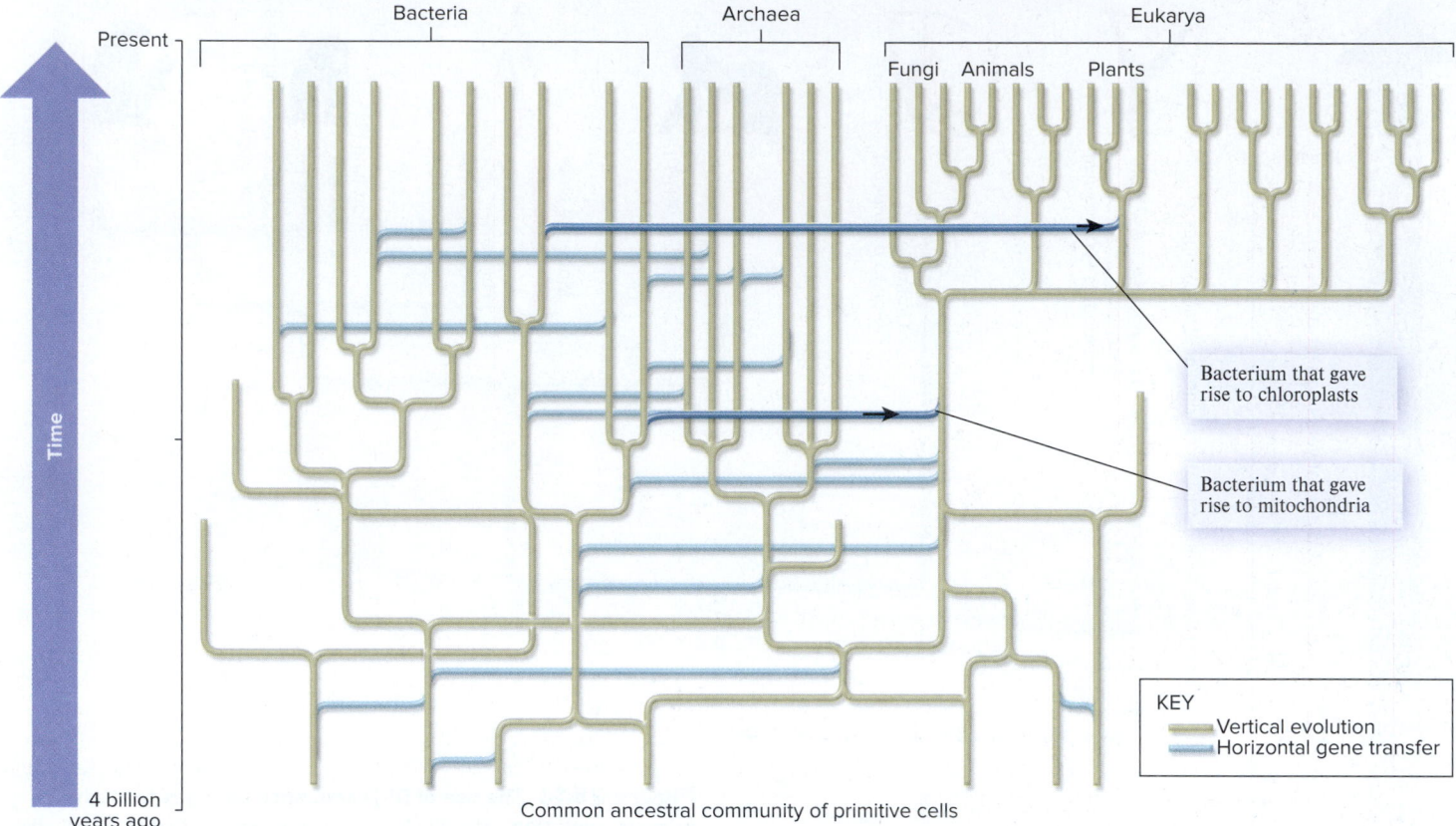

Present

Bacteria | Archaea | Eukarya

Fungi   Animals   Plods

Time

Bacterium that gave
rise to chloroplasts

Bacterium that gave
rise to mitochondria

KEY
— Vertical evolution
— Horizontal gene transfer

4 billion
years ago

Common ancestral community of primitive cells

**Figure 25.15  A web of life.** This phylogenetic tree shows not only the vertical evolution of life but also the contribution of horizontal gene transfer. In this scenario, horizontal gene transfer was prevalent during the early stages of evolution, when all organisms were unicellular, and continues to be a prominent factor in the speciation of bacteria and archaea. Note: This tree is schematic. Also, although the introduction of chloroplasts into the eukaryotic domain is shown as a single event, such events have occurred multiple times and by different mechanisms.

*Concept Check:*  *How does the phenomenon of horizontal gene transfer muddle the concept of monophyletic groups?*

couple of decades, researchers have come to realize that evolution is not so simple. In addition to vertical evolution, horizontal gene transfer has also played a significant role in the phylogeny of living species.

As discussed in Chapters 1 and 22, **horizontal gene transfer** (also called lateral gene transfer) is used to describe any process in which an organism incorporates genetic material from another organism without being the offspring of that organism. As discussed next, this phenomenon has reshaped the way biologists view the evolution of species.

### Core Concept: Evolution

### Due to Horizontal Gene Transfer, the "Tree of Life" Is Really a "Web of Life"

Horizontal gene transfer has played a major role in the evolution of many species. As discussed in Chapter 19, bacteria can transfer genes via conjugation, transformation, and transduction. Bacterial gene transfer can occur between strains of the same species or, occasionally, between cells of different bacterial species. The transferred genes may encode proteins that provide a survival

advantage, such as resistance to antibiotics or the ability to metabolize an organic molecule in the environment. Horizontal gene transfer is also fairly common among certain unicellular eukaryotes. However, its relative frequency and importance in the evolution of multicellular eukaryotes remains difficult to evaluate.

Scientists have debated the role of horizontal gene transfer in the earliest stages of evolution, prior to the divergence of the bacterial and archaeal domains. The traditional viewpoint was that the three domains of life—Bacteria, Archaea, and Eukarya—arose from a single type of prokaryotic (or pre-prokaryotic) cell called the universal ancestor. However, genomic research has suggested that horizontal gene transfer may have been particularly common during the early stages of evolution on Earth, when all species were unicellular. Horizontal gene transfer may have been so prevalent that the universal ancestor may have actually been an ancestral community of cell lineages that evolved as a whole. If that were the case, the tree of life cannot be traced back to a single ancestor.

**Figure 25.15** illustrates a schematic scenario for the evolution of life that includes the roles of both vertical evolution and horizontal gene transfer. This has been described as a "web of life" rather than a "tree of life." In this scenario, instead of a universal ancestor, a community of primitive cells frequently transferred genetic material in a horizontal fashion. Horizontal gene transfer was also prevalent during the early evolution of bacteria and archaea, and when eukaryotes first emerged as unicellular species. In living bacterial and archaeal species, it remains a prominent way to foster evolutionary change.

By comparison, the region of the diagram that contains most eukaryotic species has a more treelike structure. Researchers have speculated that multicellularity and sexual reproduction have presented barriers to horizontal gene transfer in most eukaryotes. For a gene to be transmitted to eukaryotic offspring, it would have to be transferred into a eukaryotic cell that is a gamete or a cell that gives rise to gametes. Horizontal gene transfer has become less common in eukaryotes, particularly among multicellular species, though it does occur occasionally.

## Summary of Key Concepts

### 25.1 Taxonomy

- Taxonomy is the field of biology concerned with describing, naming, and classifying extant and extinct species. Systematics is the study of biological diversity and classification of evolutionary relationships among species, both extant and extinct.

- Taxonomy places all living organisms into progressively smaller hierarchical groups called taxa (singular, taxon). The broadest groups are the three domains, called Bacteria, Archaea, and Eukarya, followed by supergroups, kingdoms, phyla, classes, orders, families, genera, and species (Figures 25.1, 25.2, Table 25.1).

- Binomial nomenclature is the standard format for naming species that provides each species with a genus name and a species epithet.

### 25.2 Phylogenetic Trees

- The evolutionary history of a species or group of species is its phylogeny. A phylogenetic tree is a diagram that describes the phylogeny of particular species and should be viewed as a hypothesis (Figure 25.3).

- A key goal of systematics is to construct taxa and phylogenetic trees based on evolutionary relationships. Smaller taxa, such as families and genera, are derived from more recent common ancestors than are broader taxa such as kingdoms and phyla (Figure 25.4).

- Ideally, all taxa should be monophyletic groups, consisting of the most recent common ancestor and all of its descendants, though previously established taxa sometimes turn out to be paraphyletic or polyphyletic groups (Figures 25.5, 25.6).

- Both morphological and genetic data are used to propose phylogenetic trees. Molecular systematics, which involves the analysis of genetic sequences, has led to major revisions in taxonomy (Figure 25.7).

### 25.3 Cladistics

- In the cladistic approach to creating a phylogenetic tree, also called a cladogram, species are grouped together according to shared derived characters (Figure 25.8).

- An ingroup is the group whose evolutionary relationships are of interest, whereas an outgroup is a species or group of species that lacks one or more shared derived characters (synapomorphies). A comparison of an ingroup and outgroup is used to determine which character states are derived and which are primitive (Figures 25.9, 25.10).

- The cladistic approach produces many possible cladograms. The most likely phylogenetic tree is chosen by a variety of methods, including analysis of fossils, the application of the principle of parsimony, and the approach of maximum likelihood.

- Cooper and colleagues analyzed DNA sequences from extinct and living flightless birds and proposed a new phylogenetic tree showing that New Zealand was colonized twice by ancestors of flightless birds (Figures 25.11, 25.12).

### 25.4 Molecular Clocks

- The neutral theory of evolution proposes that most genetic variation is due to the accumulation of neutral mutations. Assuming that neutral mutations occur at a relatively constant rate, genetic data can serve as a molecular clock to measure the timing of evolutionary changes (Figure 25.13).

- Slowly changing genes are useful for analyzing distant evolutionary relationships, whereas rapidly changing genes are used to analyze more recent evolutionary relationships, particularly among eukaryotic species that have long generation times and evolve more slowly (Figure 25.14).

### 25.5 Horizontal Gene Transfer

- Horizontal gene transfer is the phenomenon in which an organism incorporates genetic material from another organism without being the offspring of that organism. Due to horizontal gene transfer, the tree of life may more accurately be described as a web of life (Figure 25.15).

## Assess & Discuss

### Test Yourself

1. The study of biological diversity based on evolutionary relationships is
   a. paleontology.      d. phylogeny.
   b. evolution.          e. both a and b.
   c. systematics.

2. Which of the following is the correct order of the taxa used to classify organisms?
   a. kingdom, supergroup, domain, phylum, class, order, family, genus, species
   b. domain, supergroup, kingdom, class, phylum, order, family, genus, species
   c. domain, kingdom, supergroup, phylum, class, family, order, genus, species
   d. domain, supergroup, kingdom, phylum, class, order, family, genus, species
   e. supergroup, kingdom, domain, phylum, order, class, family, species, genus

3. Which type of taxon consists of organisms with the greatest similarity?
   a. kingdom      d. family
   b. class        e. genus
   c. order

4. Which of the following characteristics is (are) not shared by bacteria, archaea, and eukaryotes?
   a. DNA is the genetic material.
   b. Messenger RNA encodes the information to produce proteins.
   c. All cells are surrounded by a plasma membrane.
   d. The cytoplasm is compartmentalized into organelles.
   e. Both a and d are not shared by bacteria, archaea, and eukaryotes.

5. Which of the following occur at branch points, or nodes, in a phylogenetic tree?
   a. anagenesis            d. a and b
   b. cladogenesis          e. b and c
   c. horizontal gene transfer

6. The evolutionary history of a species is its
   a. systematics.          d. phylogeny.
   b. taxonomy.             e. embryology.
   c. evolution.

7. A taxon composed of all species derived from a common ancestor is referred to as
   a. a phylum.
   b. a monophyletic group or clade.
   c. a genus.
   d. an outgroup.
   e. all of the above.

8. A goal of modern taxonomy is to
   a. classify all organisms based on morphological similarities.
   b. classify all organisms into monophyletic groups.
   c. classify all organisms based solely on genetic similarities.
   d. determine the evolutionary relationships only between similar species.
   e. None of the above is a goal of modern taxonomy.

9. The concept that the preferred hypothesis is the one that is the simplest is
   a. phylogeny.            d. maximum likelihood.
   b. cladistics.           e. both b and d.
   c. the principle of parsimony.

10. Research indicates that horizontal gene transfer is less prevalent in eukaryotes because of
    a. the presence of organelles.
    b. multicellularity.
    c. sexual reproduction.
    d. all of the above.
    e. b and c only.

## Conceptual Questions

1. Explain how the names of species conform to binomial nomenclature. Give an example of a species' name.

2. What is a molecular clock? How is it useful in the construction of phylogenetic trees?

3. **Core Concept: Evolution** What are some advantages and potential pitfalls of using changes in morphology to construct phylogenetic trees?

## Collaborative Questions

1. Discuss how taxonomy is useful. Make a list of some practical applications that are derived from taxonomy.

2. Discuss how systematics is used to propose a phylogenetic tree and the rationale behind using the principle of parsimony to evaluate such a tree.

# History of Life on Earth and Human Evolution

# 26

**A fossil fish.** This 50-million-year-old fossil of a unicorn fish (*Naso rectifrons*) is an example of the many different kinds of organisms that have existed during the history of life on Earth.
©George Bernard/SPL/Science Source

**T**he amazing origin of the universe is difficult to comprehend. Astronomers think the universe began with a cosmic explosion called the Big Bang about 13.7 billion years ago (bya), after which the first clouds of the elements hydrogen and helium were formed. Over a long time period, gravitational forces collapsed these clouds to create stars that converted hydrogen and helium into heavier elements, including carbon, nitrogen, and oxygen, which are the atomic building blocks of life on Earth. These elements were returned to interstellar space by exploding stars called supernovas, which created clouds in which simple molecules such as water, carbon monoxide, and hydrocarbons formed. The clouds then collapsed to make a new generation of stars and solar systems.

Our solar system began about 4.6 bya after one or more local supernova explosions. According to one widely accepted scenario, hundreds of planetesimals (small celestial bodies like asteroids) occupied the region where Venus, Earth, and Mars are now found. The Earth, which is estimated to be 4.55 billion years old, grew from the aggregation of such planetesimals over a period of 100–200 million years. For the first half billion years or so after its formation, the Earth was too hot to allow liquid water to accumulate on its surface. By 4 bya, the Earth had cooled enough for the outer layers of the planet to solidify and for oceans to begin to form.

The period between 4.0 and 3.5 bya marked the emergence of life on our planet. The first forms of life that we know about produced well-preserved microscopic fossils, such as those found in western Australia. These fossils, estimated to be about 3.5 billion years old, resemble modern cyanobacteria, which are photosynthetic bacteria (**Figure 26.1**).

This chapter emphasizes when particular forms of life arose. We will begin by considering how researchers analyze and date fossils, which are the remains of past life-forms. We will then consider how fossils, such as the one shown in the chapter opening

photo, have provided biologists with evidence of the history of life on Earth from its earliest beginnings to the present day. The last section focuses on one of the most interesting stories of evolution, which is the lineage that gave rise to modern humans.

**(a) Fossil prokaryote**

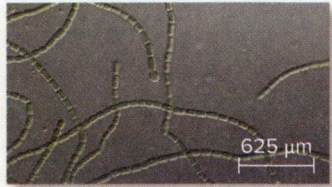

625 µm

**(b) Modern cyanobacteria**

**Figure 26.1  Earliest fossils and living cyanobacteria.  (a)** A fossilized prokaryote about 3.5 billion years old that is thought to be an early cyanobacterium. **(b)** A modern cyanobacterium, which has a similar morphology. Cyanobacterial cells are connected to each other to form chains, as shown here. a: ©J. W. Schopf; b: ©Michael Abbey/Science Source

## 26.1 The Fossil Record

**Learning Outcomes:**

1. Describe how fossils are formed.
2. **CoreSKILL »** Explain how radiometric dating is used to estimate the age of a fossil.
3. List several factors that affect the completeness of the fossil record.

We will begin this chapter by considering a process that has given us a window into the history of life over the past 3.5 billion years. **Fossils** are the preserved remains of past life on Earth. They can take many forms, including bones, shells, and leaves, and the impression of cells or other evidence, such as footprints or burrows. Scientists who study fossils are called **paleontologists** (from the Greek *palaios*, meaning ancient). Because our understanding of the history of life is derived primarily from the fossil record, it is important to appreciate how fossils are formed and dated and to understand why the fossil record cannot be viewed as complete.

### Fossils Are Formed Within Sedimentary Rock

How are fossils usually formed? Many of the rocks observed by paleontologists are sedimentary rocks that were formed from particles of older rocks broken apart by water or wind. These particles, in the form of gravel, sand, or mud, settle and bury living and dead organisms at the bottoms of rivers, lakes, and oceans. Over time, more particles pile up, and sediments at the bottom of the pile eventually become rock. Gravel particles form rock called conglomerate, sand becomes sandstone, and mud becomes shale. Most fossils are formed when organisms are buried quickly, and then during the process of sedimentary rock formation, their hard parts are gradually replaced over millions of years by minerals, producing a recognizable representation of the original organism (see, for example, the chapter opening photo).

The relative ages of fossils can sometimes be revealed by their locations in sedimentary rock formations. Because sedimentary rocks are formed by small particles settling in layers, the layers are piled one on top of the other. In a sequence of layered rock, the lower layers are usually older than the upper layers. Paleontologists often study changes in life-forms over time by studying the fossils in layers from bottom to top (**Figure 26.2**). The more ancient life-forms are found in the lower layers, and newer species are found in the upper layers. However, such an assumption can occasionally be misleading when geological processes such as folding have flipped the layers.

### The Analysis of Radioisotopes Is Used to Date Fossils

A common way to estimate the age of a fossil is by analyzing the decay of radioisotopes within the accompanying rock, a process called **radiometric dating**. As discussed in Chapter 2, many elements occur in multiple forms, called isotopes, that differ in

**Figure 26.2** **An example of layers of sedimentary rock that contain fossils.** ©Simon Fraser/SPL/Science Source

**Concept Check:** *Which rock layer in this photo is most likely to be the oldest?*

the number of neutrons their atoms contain. A radioisotope is an unstable isotope of an element that decays spontaneously, releasing radiation at a constant rate. The **half-life** is the length of time required for a radioisotope to decay to exactly one-half of its initial quantity. Each radioisotope has its own unique half-life (**Figure 26.3a**). Within a sample of rock, scientists can measure the amount of a given radioisotope as well as the amount of the decay product—the isotope that is produced when the original isotope decays. For dating geological materials, several types of isotope decay patterns are particularly useful: carbon to nitrogen, potassium to argon, rubidium to strontium, and uranium to lead (**Figure 26.3b**).

To determine the age of a rock using radiometric dating, paleontologists need to have a way to set the clock—extrapolate back to a starting point at which a rock did not have any amount of the decay product. Except for fossils less than 50,000 years old, in which carbon-14 ($^{14}C$) dating can be employed, fossil dating is not usually conducted on the fossil itself or on the sedimentary rock in which the fossil is found. Most commonly, igneous rock—rock formed through the cooling and solidification of lava—in the vicinity of the sedimentary rock is dated. Why is igneous rock chosen? One reason is that igneous rock derived from an ancient lava flow initially contains uranium-235 ($^{235}U$) but no lead-207 ($^{207}Pb$). The decay product of $^{235}U$ is $^{207}Pb$. By comparing the relative proportions of $^{235}U$ and $^{207}Pb$ in a sample, the age of igneous rock can be accurately determined.

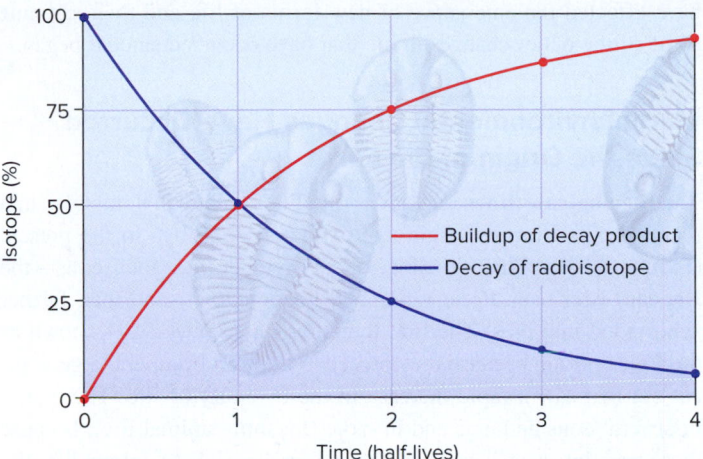

**(a) Decay of a radioisotope**

| Radioisotope | Decay product | Half-life (years) | Useful dating range (years) |
|---|---|---|---|
| Carbon-14 | Nitrogen-14 | 5,730 | 100–50,000 |
| Potassium-40 | Argon-40 | 1.3 billion | 100,000–4.5 billion |
| Rubidium-87 | Strontium-87 | 47 billion | 10 million–4.5 billion |
| Uranium-235 | Lead-207 | 710 million | 10 million–4.5 billion |
| Uranium-238 | Lead-206 | 4.5 billion | 10 million–4.5 billion |

**(b) Radioisotopes that are useful for geological dating**

**Figure 26.3  Radiometric dating of fossils.  (a)** A rock can be dated by measuring the relative amounts of a radioisotope and its decay product within the rock. **(b)** These five radioisotopes are particularly useful for the dating of fossils.

> **Concept Check:**  *If you suspected a fossil to be 50 million years old, which pair of radioisotope and decay product would you choose to analyze?*

 **BIO TIPS**

**THE QUESTION**  *The process of decay for a radio-isotope is represented by the following equation:*

$$N = N_0 e^{-(0.693t/T_{1/2})}$$

*where*
*N is the number of atoms of a radioisotope after a certain time,*
*$N_0$ is the number of atoms of the radioisotope that were originally present (prior to any decay),*
*e is the natural logarithm,*
*t is the time during which decay has occurred, and*
*$T_{1/2}$ is the half-life of the radioisotope.*

*A paleontologist discovered a fossil of a previously unidentified reptile. A sample of nearby igneous rock contained 0.11 mg of uranium-235 ($^{235}U$) and 0.035 mg of lead-207 ($^{207}Pb$). Estimate the age of this fossil.*

**T OPIC**  *What topic in biology does this question address?*
The topic is radiometric dating of fossils. More specifically, the question asks you to calculate an estimated age for a particular fossil.

**I NFORMATION**  *What information do you know based on the question and your understanding of the topic?*  In the question, you are given an equation that describes the decay process for a radioisotope. You are also given the relative amounts of uranium-235 and lead-207 in an igneous rock that was near the fossil of interest. From your understanding of the topic, you may recall that very old fossils are often dated by analyzing nearby igneous rock because such rock initially contains only uranium-235.

**P ROBLEM-SOLVING S TRATEGY**  *Make a calculation.*  You first need to calculate the number of atoms of $^{235}U$ and $^{207}Pb$ in the sample of igneous rock. The atomic masses of $^{235}U$ and $^{207}Pb$ are approximately 235 g/mol and 207 g/mol, respectively. For $^{235}U$, the number of atoms in the sample, which is $N$, is

$$N = \frac{0.11}{235} \times 6.022 \times 10^{23} = 2.82 \times 10^{20}$$

For $^{207}Pb$, the number of atoms in the sample is

$$\frac{0.035}{207} \times 6.022 \times 10^{23} = 1.02 \times 10^{20}$$

It is assumed that all of the $^{207}Pb$ is the decay product of $^{235}U$. Therefore, the original number of atoms of $^{235}U$, which is $N_0$, was

$$N_0 = (2.82 \times 10^{20}) + (1.02 \times 10^{20}) = 3.84 \times 10^{20}$$

The half-life ($T_{1/2}$) for $^{235}U$ is 710 million years (see Figure 26.3). You substitute these values for $N$, $N_0$, and $T_{1/2}$ into the given equation and solve for $t$.

**ANSWER**  *The fossil is approximately 316 million years old.*

## Several Factors Affect the Completeness of the Fossil Record

The fossil record should not be viewed as a complete and balanced representation of the species that existed in the past. Several factors affect the likelihood that extinct organisms have been preserved as fossils and will be identified by paleontologists (**Table 26.1**). First, certain organisms are more likely than others to become fossilized. Organisms with hard shells or bones tend to be over-represented in the fossil record. Factors such as anatomy, size, number, and the environment and time in which they lived also play important roles in determining the likelihood that organisms will be preserved in the fossil record. In addition, geological processes may favor the fossilization of certain types of organisms. Finally, unintentional biases arise that are related to the efforts of paleontologists. For example, scientific interests may favor searching for and analyzing certain species over others: Many paleontologists have been greatly interested in finding the remains of dinosaurs.

| Table 26.1 | Factors That Affect the Fossil Record |
|---|---|
| **Factor** | **Description** |
| Anatomy | Organisms with hard body parts, such as animals with a skeleton or thick shell, are more likely to be preserved than are organisms composed of soft tissues. |
| Size | The fossil remains of larger organisms are more likely to be found than those of smaller organisms. |
| Number | Species that existed in greater numbers or over a larger area are more likely to be preserved within the fossil record than those that existed in smaller numbers or in a smaller area. |
| Environment | Inland species are less likely to become fossilized than are those that lived in a marine environment or near the edge of water because sedimentary rock is more likely to be formed in or near water. |
| Time | Species that lived relatively recently or existed for a long time are more likely to be found as fossils than species that lived very long ago or for a relatively short time. |
| Geological processes | Due to the chemistry of fossilization, certain organisms are more likely to be preserved than are other organisms. |
| Paleontology | Certain types of fossils may be more interesting to paleontologists. In addition, a significant bias exists with regard to the locations where paleontologists search for fossils. For example, they tend to search in regions where other fossils have already been found. |

Although the fossil record is incomplete, it has provided a wealth of information regarding the history of the types of life that existed on Earth. The rest of this chapter will survey the emergence of life-forms from 3.5 bya to the present.

## 26.2 History of Life on Earth

**Learning Outcomes:**

1. List the types of environmental changes that have affected the history of life on Earth.
2. Describe the cell structure and energy utilization of the first living organisms that arose during the Archaean eon.
3. Explain how the origin of eukaryotic cells involved a union between bacterial and archaeal cells.
4. Describe the key features of multicellular organisms, which arose during the Proterozoic eon.
5. Outline the major events and changes in species diversity during the Paleozoic, Mesozoic, and Cenozoic eras.

In Chapter 4, we considered hypotheses concerning how the first cells came into existence. The first known fossils of single-celled organisms were preserved approximately 3.5 bya. In this section, we will begin with a brief description of the geological changes on Earth that

have affected the emergence of new forms of life and then examine some of the major changes in life that have occurred since it began.

### Many Environmental Changes Have Occurred Since the Origin of the Earth

The **geological timescale** is a timeline of Earth's history and major events from its origin approximately 4.55 bya to the present (**Figure 26.4**). This timeline is subdivided into four eons—the Hadean, Archaean, Proterozoic, and Phanerozoic—and then further subdivided into eras. The first three eons are collectively known as the Precambrian because they preceded the Cambrian era, a geological era that saw a rapid increase in the diversity of life. The names of several eons and eras end in -zoic (meaning animal life), because these time intervals have been defined on the basis of animal life. We will examine these time periods later in this chapter.

The changes that occurred in living organisms over the past 4 billion years are the result of two interactive processes. First, as discussed in the previous chapters, genetic changes in organisms can affect their characteristics. Such changes influence organisms' abilities to survive and reproduce in their native environment. Second, the environment on Earth has undergone dramatic changes that have profoundly influenced the types of organisms that have existed during different periods of time. In some cases, an environmental change has allowed new types of organisms to flourish. Alternatively, environmental changes have resulted in **extinction**—the complete loss of a species or group of species. Major types of environmental changes are described next.

*Temperature* During the first 2.5 billion years of its existence, the surface of the Earth gradually cooled. However, during the last 2 billion years, the Earth has undergone major fluctuations in temperature, producing Ice Ages that alternate with warmer periods. Furthermore, the temperature on Earth is not uniform, which produces environments where the temperatures are quite different, such as tropical rain forests and the arctic tundra.

*Atmosphere* The chemical composition of the gases surrounding the Earth has changed substantially over the past 4 billion years. One notable change involves the amount of oxygen. Prior to 2.4 bya, relatively little oxygen gas was in the atmosphere, but at that time, levels of oxygen in the form of $O_2$ began to rise significantly. The emergence of organisms that are capable of photosynthesis added oxygen to the atmosphere. Our current atmosphere contains about 21% $O_2$.

Increased levels of oxygen are thought to have a played a key role in various aspects of the history of life, including the following:

- The origin of many animal body plans coincided with a rise in atmospheric $O_2$.
- The conquest of land by arthropods (about 410 mya) and a second conquest by arthropods and vertebrates (about 350 mya) occurred during periods in which $O_2$ levels were high or increasing.
- Increases in animal body sizes are associated with higher $O_2$ levels.

Higher levels of $O_2$ could have contributed to these events because higher $O_2$ levels may enhance the ability of animals to carry out

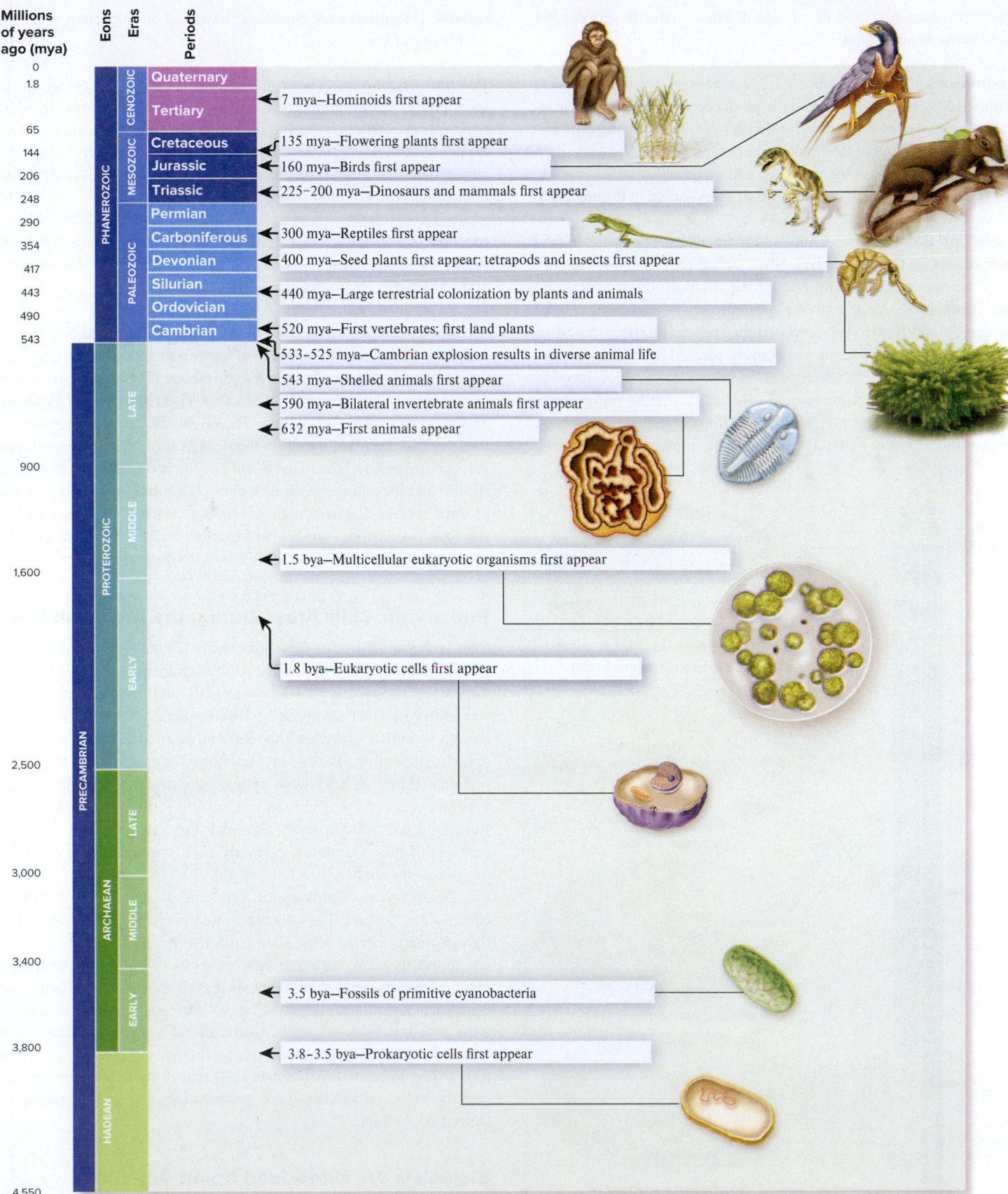

| Millions of years ago (mya) | Eons | Eras | Periods | |
|---|---|---|---|---|
| 0 | | | Quaternary | |
| 1.8 | | CENOZOIC | Tertiary | ← 7 mya—Hominoids first appear |
| 65 | | | Cretaceous | 135 mya—Flowering plants first appear |
| 144 | | MESOZOIC | Jurassic | ← 160 mya—Birds first appear |
| 206 | PHANEROZOIC | | Triassic | ← 225–200 mya—Dinosaurs and mammals first appear |
| 248 | | | Permian | |
| 290 | | | Carboniferous | ← 300 mya—Reptiles first appear |
| 354 | | PALEOZOIC | Devonian | ← 400 mya—Seed plants first appear; tetrapods and insects first appear |
| 417 | | | Silurian | ← 440 mya—Large terrestrial colonization by plants and animals |
| 443 | | | Ordovician | |
| 490 | | | Cambrian | ← 520 mya—First vertebrates; first land plants |
| 543 | | | | 533–525 mya—Cambrian explosion results in diverse animal life |
| | | LATE | | 543 mya—Shelled animals first appear |
| | | | | ← 590 mya—Bilateral invertebrate animals first appear |
| | | | | ← 632 mya—First animals appear |
| 900 | | | | |
| | PROTEROZOIC | MIDDLE | | |
| 1,600 | | | | ← 1.5 bya—Multicellular eukaryotic organisms first appear |
| | | EARLY | | |
| | PRECAMBRIAN | | | ← 1.8 bya—Eukaryotic cells first appear |
| 2,500 | | | | |
| | | LATE | | |
| 3,000 | | | | |
| 3,400 | ARCHAEAN | MIDDLE | | |
| | | EARLY | | ← 3.5 bya—Fossils of primitive cyanobacteria |
| 3,800 | | | | ← 3.8–3.5 bya—Prokaryotic cells first appear |
| | | HADEAN | | |
| 4,550 | | | | |

**Figure 26.4** The geological timescale and an overview of the history of life on Earth.

aerobic respiration. These events are discussed later in this chapter and in more detail in Unit V.

***Landmasses*** As the Earth cooled, landmasses formed that were surrounded by bodies of water. This produced two different environments: terrestrial and aquatic. Furthermore, over the course of billions of years, the major landmasses, known as the continents, shifted their positions, changed their shapes, and separated from each other. This phenomenon, called **continental drift**, is shown in **Figure 26.5**.

***Floods and Glaciations*** Catastrophic floods have periodically had major effects on the organisms in the flooded regions. Glaciers have moved across continents and altered the composition of species on those landmasses. As an extreme example, in 1992, American geobiologist Joseph Kirschvink proposed the snowball Earth hypothesis, which suggests that the Earth was entirely covered by ice during parts of the period from 790 to 630 mya. This hypothesis was developed to explain various types of geological evidence including sedimentary deposits of glacial origin that are found at tropical latitudes. Although the existence of a completely frozen Earth remains controversial, massive glaciations over our planet have had an important effect on the history of life.

***Volcanic Eruptions*** The eruptions of volcanoes harm organisms in the vicinity of an eruption, sometimes causing extinctions. In addition, volcanic eruptions in the oceans lead to the formation of new islands. Massive eruptions may also spew so much debris into the atmosphere that they affect global temperatures and limit solar radiation, which restricts photosynthetic production.

***Meteorite Impacts*** During its long history, the Earth has been struck by many meteorites. Large meteorites have significantly affected Earth's environment.

The effects of one or more of the changes described above have sometimes caused large numbers of species to go extinct at the same time. Such events are called **mass extinctions**. Five large mass extinctions occurred near the ends of the Ordovician, Devonian, Permian, Triassic, and Cretaceous periods. The boundaries between geological time periods are often based on the occurrences of mass extinctions. A recurring pattern seen in the history of life is the extinction of some species and the emergence of new ones. The rapid extinction of many modern species due to human activities is sometimes referred to as the sixth mass extinction. We will examine mass extinctions and the current biodiversity crisis in more detail in Chapter 60.

## Prokaryotic Cells Arose During the Archaean Eon

The Archaean (from the Greek, meaning ancient) was an eon when diverse microbial life flourished in the primordial oceans. As mentioned previously, the first known fossils of living cells were preserved in rocks that are about 3.5 billion years old (see Figure 26.1), though scientists postulate that cells arose many millions of years prior to this time. Based on the morphology of their fossilized remains, the first cells were prokaryotic. During the more than 1 billion years of the Archaean eon, all life-forms were prokaryotic. Because Earth's atmosphere had very little free $O_2$, the microorganisms of this eon almost certainly used only anaerobic (without oxygen) respiration.

Organisms with prokaryotic cells are divided into two groups: bacteria and archaea. Bacteria are more prevalent on modern Earth, though many species of archaea have also been identified. Archaea are found in many different environments, with some occupying extreme environments such as hot springs. Bacteria and archaea share fundamental similarities, indicating that they are derived from a common ancestor. Even so, certain differences suggest that these two types of prokaryotes diverged from each other quite early in the history of life. In particular, bacteria and archaea show some interesting differences in metabolism, lipid composition, and genetic pathways (refer back to Table 25.1).

## Biologists Are Undecided About Whether Heterotrophs or Autotrophs Came First

An important factor that greatly influenced the emergence of new species is the availability of energy. As we learned in Unit II, all

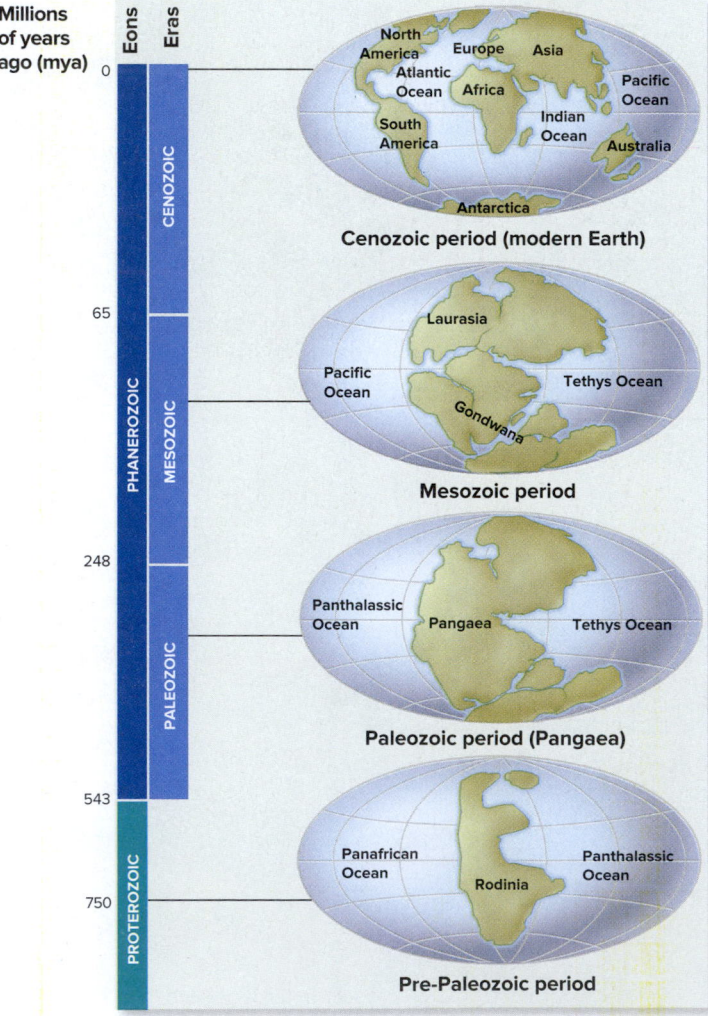

**Figure 26.5** **Continental drift.** The relative locations of the continents on Earth have changed dramatically over time.

organisms require energy to survive and reproduce. Organisms follow two different strategies to obtain energy.

- Some are **heterotrophs**, which means their energy is derived from the chemical bonds within the organic molecules they consume. Because the most common sources of organic molecules today are other organisms, heterotrophs typically consume other organisms or materials from other organisms.

- Alternatively, many organisms are **autotrophs**, which directly harness energy from either inorganic molecules or light. Among modern species, plants are an important example of autotrophs. Plants directly absorb light energy and use it (via photosynthesis) to synthesize organic molecules such as glucose. On modern Earth, heterotrophs ultimately rely on autotrophs for the production of food.

Were the first forms of life heterotrophs or autotrophs? The answer is not resolved. Some biologists have speculated that autotrophs, such as those living near deep-sea vents, may have arisen first. These organisms would have used chemicals that were made near the vents as an energy source to make organic molecules. Alternatively, many scientists have hypothesized that the first living cells were heterotrophs. They reason that it would have been simpler for the first primitive cells to use the organic molecules in the prebiotic soup as a source of energy.

If heterotrophs came first, why were cyanobacteria preserved in the earliest fossils, rather than heterotrophs? One possible reason is related to their manner of growth. Certain cyanobacteria promote the formation of a layered structure called a **stromatolite** (**Figure 26.6**). The aquatic environment where these cyanobacteria survive is rich in minerals such as calcium. The cyanobacteria grow in large mats that form layers. As they grow, they deplete the carbon dioxide ($CO_2$) in the surrounding water. This causes calcium carbonate in the water to gradually precipitate over the bacterial cells, calcifying the older cells in the lower layers and also trapping grains of sediment. Newer cells produce

a layer on top. Over time, many layers of calcified cells and sediment are formed, thereby producing a stromatolite. This process still occurs today in places such as Shark Bay in western Australia, which is renowned for the stromatolites along its beaches (Figure 26.6b).

The emergence and proliferation of ancient cyanobacteria had two critical consequences. First, the autotrophic nature of these bacteria enabled them to produce organic molecules from $CO_2$. This ability prevented the depletion of organic foodstuffs that would have been exhausted if only heterotrophs existed. Second, cyanobacteria produce $O_2$ as a waste product of photosynthesis. During the Archaean and Proterozoic eons, the activity of cyanobacteria led to the gradual rise in atmospheric $O_2$ noted earlier. The increase in $O_2$ spelled doom for many anaerobic species, which became restricted to a few anoxic (without oxygen) environments, such as deep within the soil. However, $O_2$ enabled the emergence of new bacterial and archaeal species that used aerobic (with oxygen) respiration (see Chapter 7). In addition, aerobic respiration is likely to have played a key role in the emergence and eventual explosion of eukaryotic life-forms, which typically have high energy demands. These eukaryotic life-forms are described next.

 **Core Concept: Evolution**

## The Origin of Eukaryotic Cells Involved a Union Between Bacterial and Archaeal Cells

Eukaryotic cells arose during the Proterozoic eon, which began 2.5 bya and ended 543 mya (see Figure 26.4). The manner in which the first eukaryotic cell originated is not entirely understood. In modern eukaryotic cells, genetic material is found in three distinct organelles. All eukaryotic cells contain DNA in the nucleus and mitochondria, and plant and algal cells also have DNA in their chloroplasts. To address the issue of the origin of

**(a) Fossil stromatolite**

**(b) Modern stromatolites**

**Figure 26.6   Fossil and modern stromatolites: evidence of autotrophic cyanobacteria.** Each stromatolite is a rocklike structure, typically 1 m in diameter. **(a)** Section of a fossilized stromatolite. The layers are mats of mineralized cyanobacteria, one layer on top of the other. The existence of fossil stromatolites provides evidence of early autotrophic organisms. **(b)** Modern stromatolites that have formed in western Australia.

a: ©Dirk Wiersma/SPL/Science Source; b: ©Horst Mahr/age fotostock

eukaryotic species, scientists have examined the DNA sequences found in these three organelles. They have concluded that the nuclear, mitochondrial, and chloroplast genomes appear to be derived from once-separate cells that came together.

***Nuclear Genome*** Both bacteria and archaea contributed substantially to the nuclear genome of eukaryotic cells. Eukaryotic nuclear genes encoding proteins involved in metabolic pathways and lipid biosynthesis appear to be derived from ancient bacteria, whereas genes involved with transcription and translation appear to be derived from an archaeal ancestor. To explain the origin of the nuclear genome, several hypotheses have been proposed. The most widely accepted involves an association between ancient bacteria and archaea, which is hypothesized to have been endosymbiotic. In an **endosymbiotic** relationship, a smaller organism (the endosymbiont) lives inside a larger organism (the host).

Researchers have suggested that an archaeal species evolved the ability to invaginate its plasma membrane, which could have two results (**Figure 26.7**). First, it could eventually lead to the formation of an extensive internal membrane system and enclosure of the genetic material in a nuclear envelope. Second, the ability to invaginate the plasma membrane provided a mechanism for taking up materials from the environment via endocytosis, which is described in Chapter 5. Along these lines, the closest modern relative to eukaryotes is thought to be a phylum of archaea called Lokiarchaeota, which carries many genes that are hypothesized to play a role in membrane remodeling. In the scenario described in Figure 26.7, an ancient archaeon engulfed a bacterium via endocytosis, maintaining the bacterium in its cytoplasm as an endosymbiont. Over time, some genes from the bacterium were transferred to the archaeal host cell, and the resulting genetic material eventually became the nuclear genome.

***Mitochondrial and Chloroplast Genomes*** As discussed in Chapter 4, the analyses of genes from mitochondria, chloroplasts, and bacteria are consistent with the endosymbiosis theory, which proposes that mitochondria and chloroplasts originated from bacteria that took up residence within a primordial eukaryotic cell (refer back to Figure 4.31). Mitochondria found in eukaryotic cells are likely derived from a bacterial species that resembled modern α-proteobacteria, a diverse group of bacteria that carry out oxidative phosphorylation to make ATP. One possibility is that an endosymbiotic event involving an ancestor of this bacterial species produced the first eukaryotic cell and that the mitochondrion is a remnant of that event. Alternatively, endosymbiosis may have produced the first eukaryotic cell, and then a subsequent endosymbiosis resulted in mitochondria (see Figure 26.7). DNA-sequencing data indicate that chloroplasts were derived from a separate endosymbiotic relationship between a primitive eukaryotic cell and a cyanobacterium. As discussed in Chapter 28, plastids, such as chloroplasts, have arisen on several independent occasions via primary, secondary, and tertiary endosymbiosis (see Figure 28.13).

Interestingly, an endosymbiotic relationship involving two different proteobacteria was reported in 2001. In mealybugs, bacteria

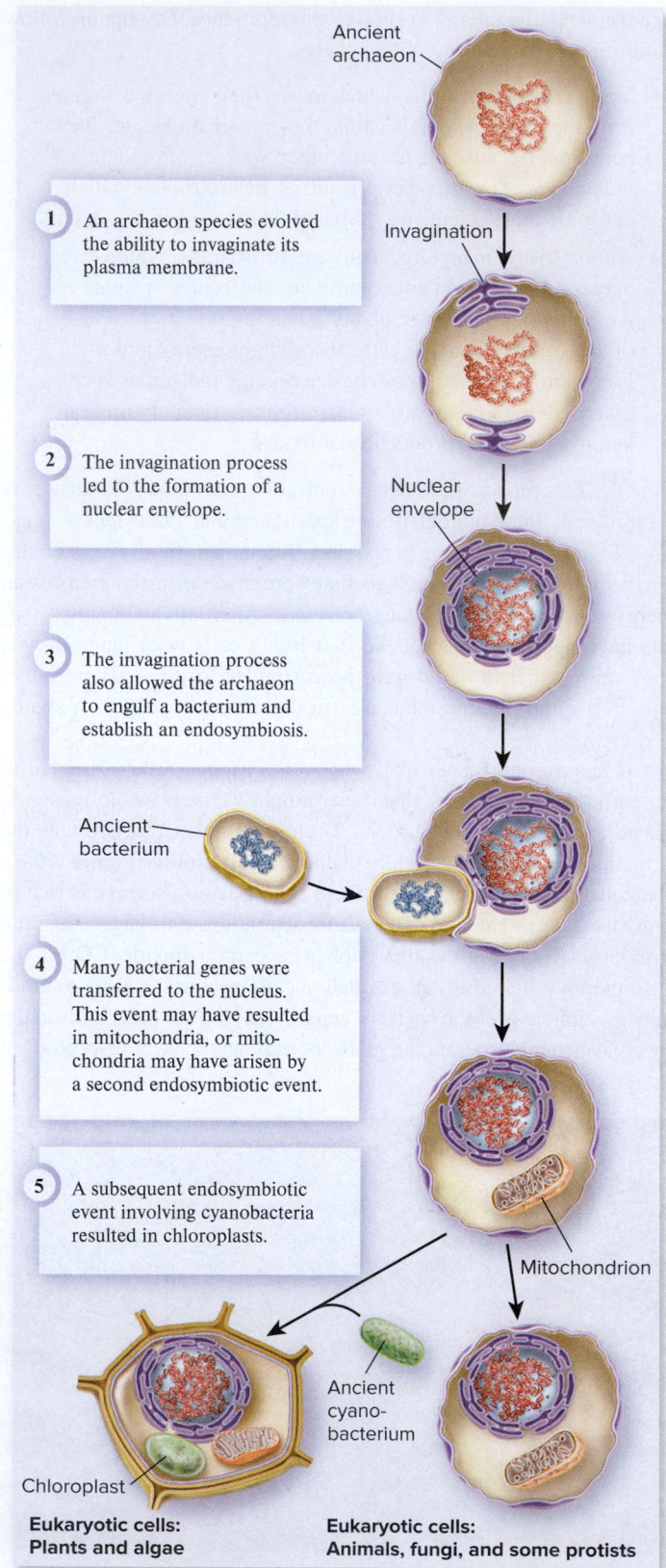

1. An archaeon species evolved the ability to invaginate its plasma membrane.

2. The invagination process led to the formation of a nuclear envelope.

3. The invagination process also allowed the archaeon to engulf a bacterium and establish an endosymbiosis.

4. Many bacterial genes were transferred to the nucleus. This event may have resulted in mitochondria, or mitochondria may have arisen by a second endosymbiotic event.

5. A subsequent endosymbiotic event involving cyanobacteria resulted in chloroplasts.

Ancient archaeon

Invagination

Nuclear envelope

Ancient bacterium

Mitochondrion

Ancient cyanobacterium

Chloroplast

**Eukaryotic cells: Plants and algae**

**Eukaryotic cells: Animals, fungi, and some protists**

**Figure 26.7** Possible endosymbiotic relationships that gave rise to the first eukaryotic cells.

 **Core Skill: Connections** Look back at Figure 5.22. Explain how endocytosis played a role in endosymbiosis.

survive within the cytoplasm of large host cells of a specialized organ called a bacteriome. Recent analysis has shown that different species of bacteria inside the host cells share their own endosymbiotic relationship. In particular, γ-proteobacteria live endosymbiotically inside β-proteobacteria. Such an observation demonstrates that an endosymbiotic relationship can occur between two prokaryotic cells.

## Multicellular Eukaryotes and the Earliest Animals Arose During the Proterozoic Eon

The first multicellular eukaryotes are thought to have emerged about 1.5 bya, in the middle of the Proterozoic eon. The oldest fossil evidence for multicellular eukaryotes was an organism that resembled modern red algae; this fossil was dated at approximately 1.2 billion years old.

Simple multicellular organisms are believed to have originated in one of two different ways. One possibility is that several individual cells found each other and aggregated to form a colony. Cellular slime molds, discussed in Chapter 28, are examples of modern organisms in which groups of single-celled organisms can come together to form a small multicellular organism. According to the fossil record, such organisms have remained very simple for hundreds of millions of years.

Alternatively, another way that multicellularity can occur is when a single cell divides and the resulting cells stick together. This pattern occurs in many simple multicellular organisms, such as algae and fungi, as well as in species with more complex body plans, such as plants and animals. Biologists cannot be certain whether the first multicellular eukaryotes arose by an aggregation process or by cell division and adhesion. However, the development of complex, multicellular organisms now occurs by cell division and adhesion.

An interesting example showing changes in the level of complexity from unicellular eukaryotes to more complex multicellular organisms is found among evolutionarily related species of volvocine green algae. These algae exist as unicellular species, as small clumps of cells of the same cell type, or as larger groups of cells with two distinct cell types. **Figure 26.8** compares four species of volvocine algae. *Chlamydomonas reinhardtii* is a unicellular alga (Figure 26.8a). It is called a biflagellate because each cell has two flagella. *Gonium pectorale* is a multicellular organism composed of 16 cells (Figure 26.8b). This simple multicellular organism is formed from a single cell by cell division and adhesion. All of the cells in this species are biflagellate. Other volvocine algae have evolved into larger and more complex organisms. *Pleodorina californica* has 64–128 cells (Figure 26.8c), and *Volvox aureus* has about 1,000–2,000 cells (Figure 26.8d). A feature of these more complex organisms is they have two cell types: somatic and reproductive cells. The somatic cells are biflagellate cells, but the reproductive cells are not. *V. aureus* has a higher percentage of somatic cells than *P. californica*.

Overall, an analysis of these four species of algae illustrates three important principles found among complex multicellular species:

1. Multicellular organisms arise from a single cell that divides to produce daughter cells that adhere to one another.

2. The daughter cells can follow different fates, thereby producing multicellular organisms with different cell types.

3. As organisms get larger, a greater percentage of their cells tend to be somatic cells. The somatic cells carry out the activities required for the survival of the multicellular organism, whereas the reproductive cells are specialized for the sole purpose of producing offspring.

Toward the end of the Proterozoic eon, multicellular animals emerged. The first animals were invertebrates—animals without a backbone. Most animals, except for organisms such as sponges and jellyfish, exhibit bilateral symmetry—a two-sided body plan with a right and left side that are mirror images. Because each side of the body has appendages such as legs, one advantage of bilateral symmetry is

Flagella

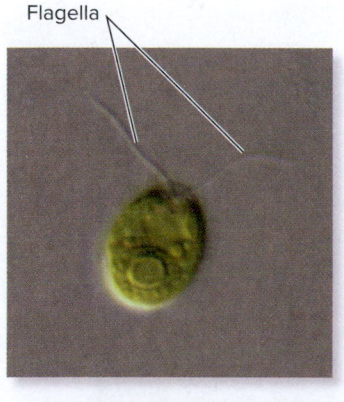  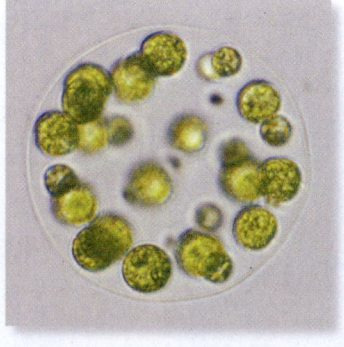 

| 3 μm | 10 μm | 30 μm | 100 μm |

**(a)** *Chlamydomonas reinhardtii*, a unicellular alga

**(b)** *Gonium pectorale*, composed of 16 identical cells

**(c)** *Pleodorina californica*, composed of 64–128 cells, has 2 cell types, somatic and reproductive

**(d)** *Volvox aureus*, composed of about 1,000–2,000 cells, has 2 cell types, somatic and reproductive

**Figure 26.8** **Variation in the level of multicellularity among volvocine algae.** a: Courtesy of Dr. Barbara Surek, Culture Colection of Algae at the University of Cologne (CCAC); b: ©William Bourland; c–d: ©Dr. Cristian A. Solari, Department of Ecology and Evolutionary Biology, University of Arizona

 **Core Concept: Systems** The formation of different cell types is an emergent property of multicellularity

that it facilitates locomotion. Bilateral animals also have anterior and posterior ends, with the mouth at the anterior end, as noted in Chapter 20. In southern China in 2004, Chinese paleontologist Jun-Yuan Chen, American paleobiologist David Bottjer, and their colleagues discovered a fossil, which they described as the earliest known ancestor of animals with bilateral symmetry. This minute creature, called *Vernanimalcula guizhouena*, has a shape like a flattened helmet and is barely visible to the naked eye (**Figure 26.9**). The fossil is approximately 580–600 million years old. However, the interpretation that *Vernanimalcula guizhouena* was the earliest ancestor of animals with bilateral symmetry remains controversial and is under active investigation.

## Phanerozoic Eon: The Paleozoic Era Saw the Diversification of Invertebrates and the Colonization of Land by Plants and Animals

The proliferation of multicellular eukaryotic life has been extensive during the Phanerozoic eon, which started 543 mya and extends to the present day. Phanerozoic means well-displayed life, referring to the abundance of fossils of plants and animals that have been identified from this eon. As shown in Figure 26.4, the Phanerozoic eon is subdivided into three eras: the Paleozoic, Mesozoic, and Cenozoic. Because they are relatively recent and we have many fossils from these eras, each of them is further subdivided into periods. We will consider each era with its associated conditions and prevalent forms of life separately.

The term Paleozoic means ancient animal life. The Paleozoic era covers approximately 300 million years, from 543 to 248 mya, and is subdivided into six periods: Cambrian, Ordovician, Silurian, Devonian, Carboniferous, and Permian. Periods are usually named after regions where rocks and fossils of that age were first discovered.

***Cambrian Period (543–490 mya)***    The climate in the Cambrian period was generally warm and wet, with no evidence of ice at the poles. During this time, the diversity of animal species increased

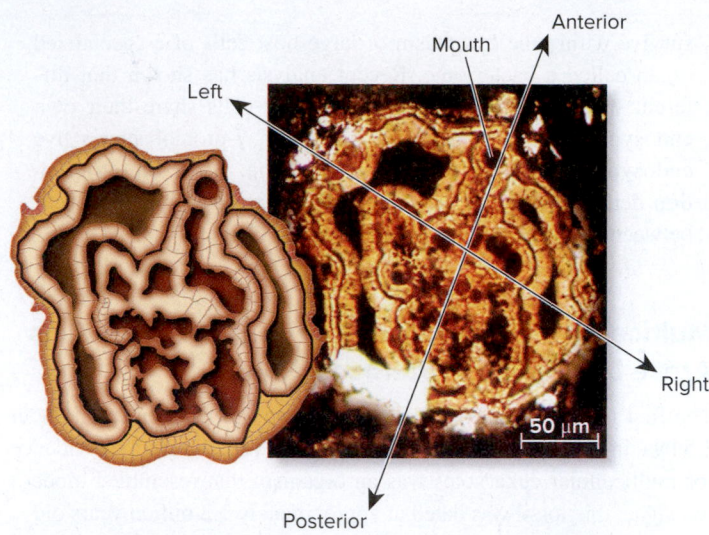

**Figure 26.9** **Fossil of an early invertebrate animal showing bilateral symmetry.** This fossil of an early animal, *Vernanimalcula guizhouena*, dates from 580 to 600 mya. Courtesy of Prof. Junyuan Chen.

**Concept Check:** *Name three other species that exhibit bilateral symmetry.*

rapidly, an event called the **Cambrian explosion**. However, recent evidence suggests that many types of animal groups present during the Cambrian period actually arose prior to this period.

Many fossils from the Cambrian period were found in the Canadian Rockies in a rock bed called the Burgess Shale, which was discovered by American paleontologist Charles Walcott in 1909. At this site, both soft- and hard-bodied (shelled) invertebrates were buried in an underwater mudslide and preserved in water that was so deep and oxygen-free that decomposition was minimal (**Figure 26.10a**). The excellent preservation of the softer tissues is what makes this deposit unique (**Figure 26.10b**).

**(a) The Burgess Shale**

**(b) A fossilized arthropod, *Marrella***

**Figure 26.10** **The Cambrian explosion and the Burgess Shale.** **(a)** This photograph shows the original site in the Canadian Rockies discovered by Charles Walcott. Since its discovery, this site has been made into a quarry for the collection of fossils. **(b)** A fossil of an extinct arthropod, Marrella, which was found at this site. a: ©L. Newman & A. Flowers/Science Source; b: ©O. Louis Mazzatenta/National Geographic/Getty Images

By the middle of the Cambrian period, all of the major types of invertebrates that exist today were present, plus many others that no longer exist. Over 100 major animal groups with significantly different body plans have been identified in the fossil record. Examples of groups that still exist include echinoderms (sea urchins and starfish), arthropods (insects, spiders, and crustaceans), mollusks (clams and snails), chordates (organisms with a dorsal nerve chord), and vertebrates (animals with backbones). Interestingly, although many new species of animals have arisen since this time, these later species have not shown a major reorganization of body plan, but instead exhibit variations on themes that were established during or prior to the Cambrian explosion.

Three possible causes of the Cambrian explosion are the following.

- Because it occurred shortly after marine animals evolved shells, some scientists have speculated that the changes observed in animal species may have allowed them to exploit new environments.

- Alternatively, others have suggested that the increase in diversity may be related to atmospheric oxygen levels. During this period, oxygen levels were increasing, and perhaps more complex body plans became possible only after the atmospheric oxygen surpassed a certain threshold. In addition, as atmospheric oxygen reached its present levels, an ozone ($O_3$) layer was produced that screens out harmful ultraviolet radiation, thereby allowing complex life to live in shallow water and eventually on land.

- Another possible contributor to the Cambrian explosion was an "evolutionary arms race" between interacting species. The ability of predators to capture prey and the ability of prey to avoid predators may have been major factors that resulted in a diversification of animals into many different species.

**Ordovician Period (490–443 mya)**    As in the Cambrian period, the climate of the early and middle parts of the Ordovician period was warm, and the atmosphere was moist. During this period, a diverse group of hard-shelled marine invertebrates, including trilobites and brachiopods, appeared in the fossil record (**Figure 26.11**). Marine communities consisted of invertebrates, algae, early jawless fishes (a type of early vertebrate), mollusks, and corals. Fossil evidence also suggests that early land plants and arthropods may have first invaded the land during this period.

Toward the end of the Ordovician period, the climate changed rather dramatically. Large glaciers formed, which drained the relatively shallow oceans, causing the water levels to drop. This resulted in a mass extinction in which as much as 60% of the existing marine invertebrates became extinct.

**Silurian Period (443–417 mya)**    In contrast to the dramatic climate changes observed during the Ordovician period, the climate during the Silurian period was relatively stable. The glaciers largely melted, which caused the ocean levels to rise. No new major types of invertebrate animals appeared during this period, but significant changes were observed among existing vertebrate and plant species. Many new types of fishes appeared in the fossil record. In addition, coral reefs made their first appearance during this period.

**(a) Trilobite**

**(b) Brachiopod**

**Figure 26.11    Shelled, invertebrate fossils of the Ordovician period.**  Trilobites existed for millions of years before becoming extinct about 250 mya. Many species of brachiopods exist today.
a: ©Francois Gohier/Science Source; b: ©kavring/Shutterstock

The Silurian marked a major colonization of land by terrestrial plants and animals. For this to occur, certain species evolved adaptations that prevented them from drying out, such as an external cuticle. Ancestral relatives of spiders and centipedes became prevalent. The earliest fossils of vascular plants, which have tissues that are specialized for the transport of water, sugar, and salts throughout the plant body, were observed in this period.

**Devonian Period (417–354 mya)**    In the Devonian period, generally dry conditions occurred across much of the northern landmasses. However, the southern landmasses were mostly covered by cool, temperate oceans.

The Devonian saw a major increase in the number of terrestrial species. At first, the vegetation consisted primarily of small plants, only a meter tall or less. Later, ferns, horsetails, and seed plants, such as gymnosperms, also emerged. By the end of the Devonian, the first trees and forests were formed. A major expansion of terrestrial animals also occurred. Insects first appeared in the fossil record, and other invertebrates became plentiful. In addition, the first tetrapods—vertebrates with four legs—are believed to have arisen in the

Devonian. Early tetrapods included amphibians, which lived on land but required water in which to lay their eggs.

In the oceans, many types of invertebrates flourished, including brachiopods, echinoderms, and corals. This period is sometimes called the Age of Fishes, as many new types of fishes emerged. During a period of approximately 20 million years near the end of the Devonian period, a prolonged series of extinctions eliminated many marine species. The cause of this mass extinction is not well understood.

**Carboniferous Period (354–290 mya)** The term Carboniferous refers to the deposits of coal, a sedimentary rock primarily composed of carbon, that were formed during this period. The Carboniferous period had the ideal conditions for the subsequent formation of coal. It was a cooler period, and much of the land was covered by forest swamps. Coal was formed over many millions of years from compressed layers of rotting vegetation.

Plants and animals further diversified during the Carboniferous period. Very large plants and trees became prevalent. For example, tree ferns such as *Psaronius* grew to a height of 15 m or more (**Figure 26.12**). The first flying insects emerged. Giant dragonflies with a wingspan of over 2 ft inhabited the forest swamps. Terrestrial vertebrates also became more diverse. Amphibians were very prevalent. One innovation that seemed particularly beneficial was the amniotic egg. In reptiles, the amniotic egg was covered with a leathery or hard shell, which prevented the desiccation of the embryo inside. This innovation was critical for the emergence of reptiles during this period.

**Permian Period (290–248 mya)** At the beginning of the Permian, continental drift had brought much of the Earth's total land together into a supercontinent known as Pangaea (see Figure 26.5). The interior regions of Pangaea were dry, with great seasonal fluctuations.

The forests of fernlike plants were replaced with gymnosperms. Species resembling modern conifers first appeared in the fossil record. Amphibians were prevalent, but reptiles became the dominant vertebrate species.

At the end of the Permian period, the largest known mass extinction in the history of life on Earth occurred; 90–95% of marine species and a large proportion of terrestrial species were eliminated. The cause of the Permian extinction is the subject of much research and controversy. One possibility is that glaciation destroyed the habitats of terrestrial species and lowered ocean levels, which would have caused greater competition among marine species. Another hypothesis is that enormous volcanic eruptions in Siberia produced large ash clouds that abruptly changed the climate on Earth.

## Phanerozoic Eon: The Mesozoic Era Saw the Rise and Fall of the Dinosaurs

The Permian extinction marks the division between the Paleozoic and Mesozoic eras. Mesozoic means middle animals. It was a time that saw great changes in animal and plant species. This era is sometimes called the Age of Dinosaurs, because those animals flourished during this time. The climate during the Mesozoic era was consistently hot, and terrestrial environments were relatively dry. Little if any ice was found at either pole. The Mesozoic is divided into three periods: the Triassic, Jurassic, and Cretaceous.

**Triassic Period (248–206 mya)** Reptiles were plentiful in this period, including new groups such as crocodiles and turtles. The first dinosaurs emerged during the middle of the Triassic. Dinosaurs were reptiles that shared certain anatomical features, including an erect posture. The first mammals also emerged, such as the small *Megazostrodon* (**Figure 26.13**). Gymnosperms were the dominant

*Psaronius*

**Figure 26.12** **A giant tree fern, *Psaronius*, from the Carboniferous period.** This genus became extinct during the Permian. The illustration is a re-creation based on fossil evidence. The inset shows a fossilized section of the trunk, also known as petrified wood. ©Natural History Museum, London/SPL/Science Source

**Figure 26.13** *Megazostrodon*, **the first known mammal of the Triassic period.** The illustration is a re-creation based on fossilized skeletons. The *Megazostrodon* was 10 to 12 cm long.

 **Core Skill: Connections** Look ahead to Table 35.1. What are the common characteristics of mammals?

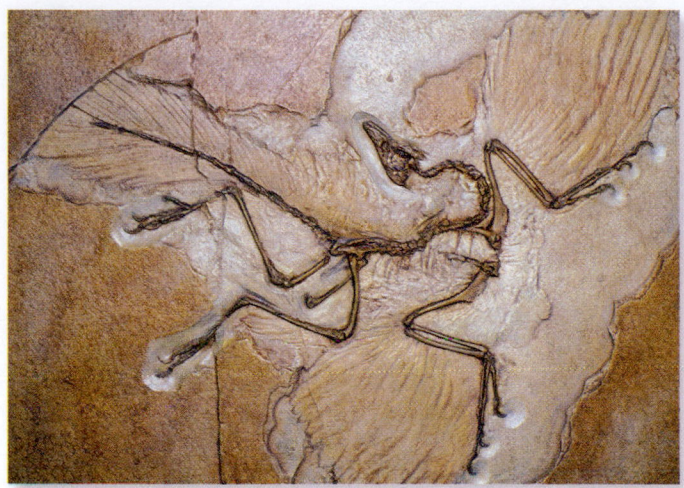

**Figure 26.14** A fossil of an early birdlike animal, *Archaeopteryx*, which emerged in the Jurassic period. ©Jason Edwards/Getty Images

land plant. Volcanic eruptions near the end of the Triassic are thought to have caused global warming, resulting in mass extinctions that eliminated many marine and terrestrial species.

***Jurassic Period (206–144 mya)***    Gymnosperms, such as conifers, continued to be the dominant vegetation in the Jurassic period. Mammals were not prevalent. Reptiles continued to be the dominant land vertebrate. Some dinosaurs attained enormous sizes, including the massive *Brachiosaurus*, which reached a length of 25 m (80 ft) and weighed up to 100 tons! Modern birds are descendants of a dinosaur lineage called theropod (meaning beast-footed) dinosaurs. *Tyrannosaurus rex* is one of the best known theropod dinosaurs. An early birdlike animal, *Archaeopteryx* (**Figure 26.14**), emerged in the Jurassic period. However, paleontologists are debating whether or not *Archaeopteryx* is a true ancestor of modern birds.

***Cretaceous Period (144–65 mya)***    On land, dinosaurs continued to be the dominant animals in the Cretaceous period. The earliest flowering plants, called angiosperms, which form seeds within a protective chamber, emerged and began to diversify.

The end of the Cretaceous witnessed another mass extinction, which brought an end to many previously successful groups of organisms. Except for the lineage that gave rise to birds, dinosaurs abruptly died out, as did many other species. As with the Permian extinction, the cause or causes of this mass extinction are still debated. One plausible hypothesis suggests that a large meteorite hit the region that is now the Yucatan Peninsula of Mexico, lifting massive amounts of debris into the air and thereby blocking the sunlight from reaching the Earth's surface. Such a dense haze could have cooled the Earth's surface by 11–15°C (20–30°F). Evidence also points to strong volcanic eruptions as a contributing factor for this mass extinction.

### Phanerozoic Eon: Mammals and Flowering Plants Diversified During the Cenozoic Era

The Cenozoic era spans the most recent 65 million years. It is divided into two periods: Tertiary and Quaternary. In many parts of the world,

tropical conditions were replaced by a colder, drier climate. During this time, mammals became the largest terrestrial animals, which is why the Cenozoic is sometimes called the Age of Mammals. However, the Cenozoic era also saw an amazing diversification of many types of organisms, including birds, fishes, insects, and flowering plants.

***Tertiary Period (65–1.8 mya)***    On land, the mammals that survived from the Cretaceous began to diversify rapidly during the early part of the Tertiary period. Angiosperms became the dominant land plant, and insects became important for their pollination. Fishes also diversified, and sharks became abundant.

Toward the end of the Tertiary period, about 7 mya, hominoids came into existence. **Hominoids** include humans, chimpanzees, gorillas, orangutans, and gibbons, plus all of their recent ancestors. The subset of hominoids called hominins includes modern humans, extinct human species (for example, of the *Homo* genus), and our immediate ancestors. In 2002, a fossil of the earliest known hominin, *Sahelanthropus tchadensis*, was discovered in Central Africa. This fossil was dated at between 6 and 7 million years old. Another early hominin genus, called *Australopithecus*, first emerged in Africa about 4 mya. Australopithecines walked upright and had a protruding jaw, prominent eyebrow ridges, and a small braincase.

***Quaternary Period (1.8 mya–present)***    Periodic Ice Ages have been prevalent during the last 1.8 million years, covering much of Europe and North America. This period has witnessed the widespread extinction of many species of mammals, particularly larger ones. Certain species of hominins became increasingly more like living humans. Fossils of *Homo habilis*, or handy man, so called because stone tools were found with the fossil remains, have been dated to close to the beginning of the Quaternary period. *Homo sapiens*—modern humans—first appeared about 200,000 years ago. The evolution of hominins is discussed in more detail in the next section.

## 26.3  Human Evolution

**Learning Outcomes:**

1. List the common characteristics of primates and describe their evolutionary relationships.
2. Explain how human species evolved from other primate species, and describe how they spread across the Earth.
3. Provide examples of how populations of *Homo sapiens* are still evolving.
4. Compare and contrast modern human variation at the phenotype and genotype levels.

Hardly a topic in biology has evoked more interest or public debate than human evolution. The question of "where did we come from" has been considered by people for thousands of years. In this section, we will tackle this question from an evolutionary perspective.

We begin with an overview of primate evolution, in which we explore how humans are evolutionarily related to their closest nonhuman relatives. We then take a closer look at the evolutionary events that gave rise to modern humans. As you will see, many extinct species of humans have existed, including the Denisovans, which were discovered as recently as 2010! Finally, we will turn our attention to the genetic variation found among modern humans and consider whether our species is still evolving.

## Primates Evolved from a Tree-Dwelling Species and Exhibit a Distinctive Set of Characteristics

Primates are primarily tree-dwelling species that evolved from a group of small, arboreal, insect-eating mammals about 85 mya, before dinosaurs went extinct. Primates have several defining characteristics, mostly relating to their tree-dwelling nature:

- *Binocular vision.* Primates have forward-facing eyes that are positioned close together on a flattened face. Jumping from branch to branch requires accurate judgment of distances. This is facilitated by binocular vision in which the field of vision for both eyes overlaps, producing a single image.

- *At least some digits with flat nails instead of claws.* This feature is believed to aid in the manipulation of objects.

- *Grasping hands.* All primates have grasping hands, a characteristic that enables them to hold onto branches. Monkeys, gibbons, orangutans, gorillas, chimpanzees, and humans also possess an opposable thumb, a thumb that can be placed opposite the fingers of the same hand, which gives them a precision grip and enables the manipulation of small objects. All of these primates except humans also have an opposable big toe.

- *Large brain.* Acute vision and other senses enhancing the ability to move quickly through the trees require the efficient processing of large amounts of information. As a result, primate brains are relatively large for their body sizes and are well developed.

- *Complex social behavior and well-developed parental care.* Compared to other mammals, primates have a tendency toward complex social behavior and relatively long parental care.

Some of these characteristics are found in other animals. For example, binocular vision occurs in owls and some other birds, grasping hands are found in raccoons, and relatively large brains occur in marine mammals. Primates are defined by possessing the whole suite of these characteristics.

The evolutionary relationships among primates are described in **Figure 26.15**. Taxonomists divide them into two broad groups: the strepsirrhini and the haplorrhini. The **strepsirrhini** are smaller species, such as bush babies, lemurs, and pottos. These are generally nocturnal and smaller-brained primates with eyes positioned a little more toward the side of their heads (**Figure 26.16a**). The strepsirrhini are named for their wet noses with no fur at the tip. The **haplorrhini** have dry noses and fully forward-facing eyes. This group consists of the tarsiers and the larger-brained and diurnal **anthropoidea**: monkeys and the **hominoidea** (gibbons, orangutans, gorillas, chimpanzees, and humans) (**Figure 26.16b** and **c**). A key feature of anthropoidea is an opposable thumb, which makes it easier to grasp and handle objects.

What differentiates monkeys from hominoids? Most monkeys have tails, but hominoids do not. In addition, hominoids

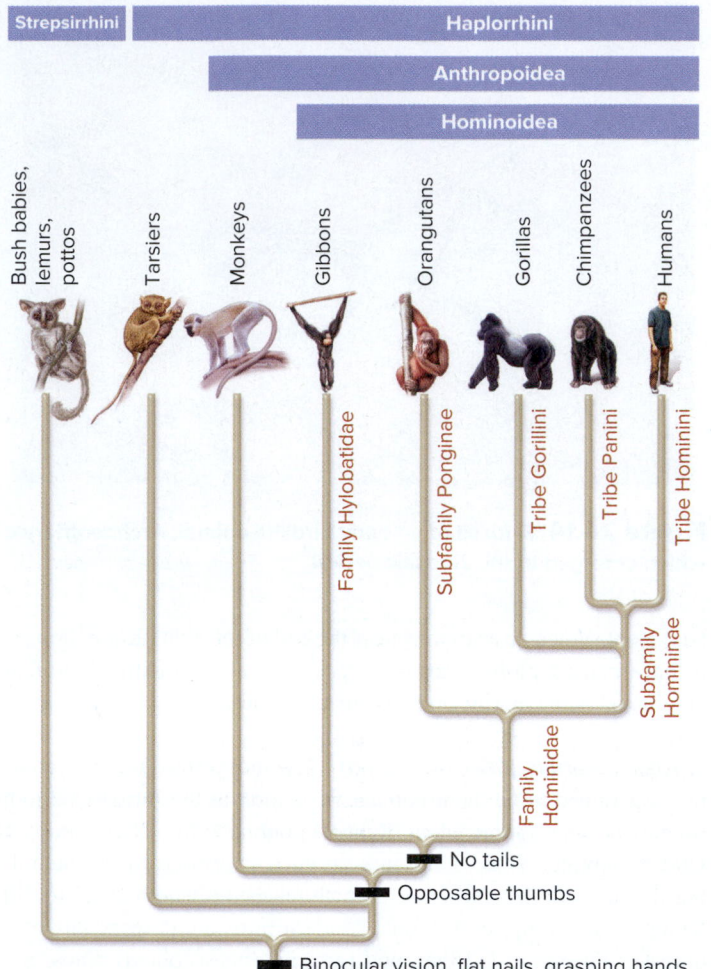

**Figure 26.15** **Evolutionary tree of the primates.**

have more mobile shoulder joints, broader rib cages, and a shorter spine. These features aid in brachiation, a swinging movement in trees. Hominoids also possess relatively long limbs and short legs and, with the exception of gibbons, are much larger than monkeys. The 20 species of hominoids are split into two groups: the lesser apes (family Hylobatidae), consisting of gibbons; and the greater apes (family Hominidae), consisting of orangutans, gorillas, chimpanzees, and humans (**Figure 26.17**). The lesser apes are strictly arboreal, whereas the greater apes often descend to the ground to feed.

Although humans are closely related to chimpanzees and gorillas, they did not evolve directly from them. Rather, all hominoid species shared a common ancestor. Recent molecular studies show that gorillas, chimpanzees, and humans are more closely related to one another than they are to gibbons and orangutans, so scientists have split the family Hominidae into groups, including the subfamily Ponginae (orangutans) and the subfamily Homininae (gorillas, chimpanzees, and humans and their ancestors) (see Figure 26.15). In turn, the Homininae are split into three tribes: the Gorillini (gorillas), the Panini (chimpanzees), and the Hominini (humans and their ancestors).

**(a) Strepsirrhini (lesser bush baby)**     **(b) Anthropoidea (capuchin monkey)**     **(c) Hominoidea (white-handed gibbon)**

**Figure 26.16    Primate classification.**  The primates are divided into two groups: **(a)** the strepsirrhini (smaller, nocturnal species such as this bush baby), and the haplorrhini (larger, diurnal species). Haplorrhini comprise **(b)** the monkeys and tarsiers, such as this capuchin monkey (*Cebus capucinus*), and **(c)** the hominoids, species such as this white-handed gibbon (*Hylobates lar*). a: ©David Haring/DUPC/Getty Images; b: ©Brand X Pictures/PunchStock/Getty Images; c: ©Katerina Novakova/catherinka/123RF

**Concept Check:**  *What are the defining features of primates?*

**(a) Gorilla (*Gorilla gorilla*)**     **(b) Chimpanzee (*Pan troglodytes*)**     **(c) Human (*Homo sapiens*)**

**Figure 26.17    Members of the family Hominidae.**  **(a)** Gorillas, the largest of the living primates, are ground-dwelling herbivores that inhabit the forests of Africa. **(b)** Chimpanzees are smaller, omnivorous primates that also live in Africa. The chimpanzees are close living relatives of modern humans. **(c)** Humans are members of the family Hominidae. The orangutan is also a member of this group. a: Source: Richard Ruggiero/USFWS; b: ©imageBROKER/Alamy Stock Photo; c: ©Tetra Images/Getty Images

## Comparing the Genomes of Humans and Chimpanzees

A male chimp called Clint, which lived at a primate research center in Atlanta provided the DNA used to sequence the chimp genome. In 2005, the Chimpanzee Sequencing and Analysis Consortium published an initial sequence of the chimpanzee genome. The draft sequence followed the 2003 publication of the human genome (see Chapter 21) and allowed scientists to make detailed comparisons between the two species. These comparisons revealed that the sequences of base pairs of the two genomes differ by only 1.23%, which is 10 times less than the difference between the mouse and rat genomes. Comparisons of human and chimpanzee proteomes also showed that 29% of all proteins are identical, with most others differing by one or two amino acid substitutions.

Many of the genetic differences between chimps and humans result from chromosome inversions and duplications. Over 1,500 inversions occur between the chimp and human genomes. Although many inversions are in the noncoding regions of the genome, the DNA in these regions may regulate the expression of the genes in the coding regions. Duplications and deletions are also common. For example, humans have lost a gene called *caspase-12,* which in other primates may protect against Alzheimer's disease.

Some interesting genetic differences were apparent between chimps and humans even before their entire genomes were sequenced. In 2002, Swedish molecular geneticist Svante Pääbo discovered differences between humans and chimps in a gene called *FOXP2,* which plays a role in speech development. Proteins encoded by this gene differ in just two amino acids of a 715-amino-acid sequence. Researchers propose that the mutations in this gene have been crucial for the development of human speech.

In 2006, a team led by American geneticist David Reich discovered that the human X chromosome diverged from the chimpanzee X chromosome about 1.2 million years more recently than the other chromosomes. This indicated to the researchers that the human and chimp lineages split apart, then began interbreeding before diverging again. If so, the interbreeding explains why many fossils appear to exhibit traits of both humans and chimps: Those primates may actually have been human-chimpanzee hybrids.

### Bipedalism Is a Distinguishing Feature of Humans

About 7 mya in Africa, a lineage that led to modern humans diverged from other primate lineages. The evolution of humans should not be viewed as a neat, stepwise progression from one species to another. Rather, human evolution, like the evolution of all species, can be visualized more like a tree, with one or two **hominin** species—members of the Hominini tribe—likely coexisting at the same point in time, with some eventually going extinct and some giving rise to other species (**Figure 26.18**).

The key characteristic differentiating hominins from other apes is that hominins walk on two feet; that is, they are **bipedal**. At about the time when hominins diverged from other ape lineages, the Earth's climate had cooled, and the forests of Africa gave way to grassy savannas. Bipedal locomotion and an upright stance may have been advantageous in allowing hominins to peer over the tall grasses of the savanna to see predators or prey.

Bipedalism is correlated with important anatomical changes in hominins. First, the opening of the skull where the spinal cord enters shifted forward, causing the spine to be more directly underneath the head. Second, the hominin pelvis became broader to support the additional weight. And third, the lower limbs, used for walking, became relatively larger than those in other apes. These are the types of anatomical changes paleontologists look for in the fossil record to determine whether fossil remains are hominins.

The earliest known hominin, *Sahelanthropus tchadensis,* was discovered in Central Africa in 2002. Fossils of this species are dated at 7 million years old. The evolutionary relationship between *S. tchadensis* and later hominin species is unclear. Another early group of hominins included several species of the genus *Australopithecus,* which first emerged in Africa about 4 mya. As shown in Figure 26.18, *Australopithecus afarensis* is generally regarded as the direct ancestor of most hominin species, but it could be a close relative of an unknown species that was the direct ancestor. From there, the evolution of different human species is still debated. It is generally agreed that two genera evolved from *Australopithecus:* the more robust *Paranthropus* and the genus *Homo.* The early stages of the evolution of *Homo* species, and their differentiation from at least two possible *Australopithecus* species, have not yet been determined with great certainty. However, the later divergence of various *Homo* species is better understood.

### *Australopithecus* and *Paranthropus* Are Early Human Genera

Let's consider some of the general features of the two early human genera, *Australopithecus* and *Paranthropus.* In 1924, the first fossil was found in South Africa for a member of the genus *Australopithecus* (from the Latin *austral,* meaning southern, and the Greek *pithecus,* meaning ape). Since then, hundreds of fossils of this group have been unearthed all over southern and eastern Africa, the areas where fossil deposits are best exposed to paleontologists. This group was widespread, with at least six species. In 1974, American paleontologist Donald Johanson discovered the skeleton of a female *A. afarensis* in the Afar region of Ethiopia and dubbed her Lucy. (The Beatles' song "Lucy in the Sky with Diamonds" was playing in the camp the night when Johanson was sorting the unearthed bones.) Over 40% of the skeleton had been preserved, enough to provide a good idea of the physical appearance of australopithecines. Compared with modern humans, they were relatively small, about 1–1.5 m in height and approximately 18 kg in weight (**Figure 26.19**). Females were much smaller than males, a characteristic known as sexual dimorphism. Examination of the bones revealed that *A. afarensis* walked on two legs. They had a facial structure and a brain size (about 500 cubic centimeters [$cm^3$]) similar to those of a chimp.

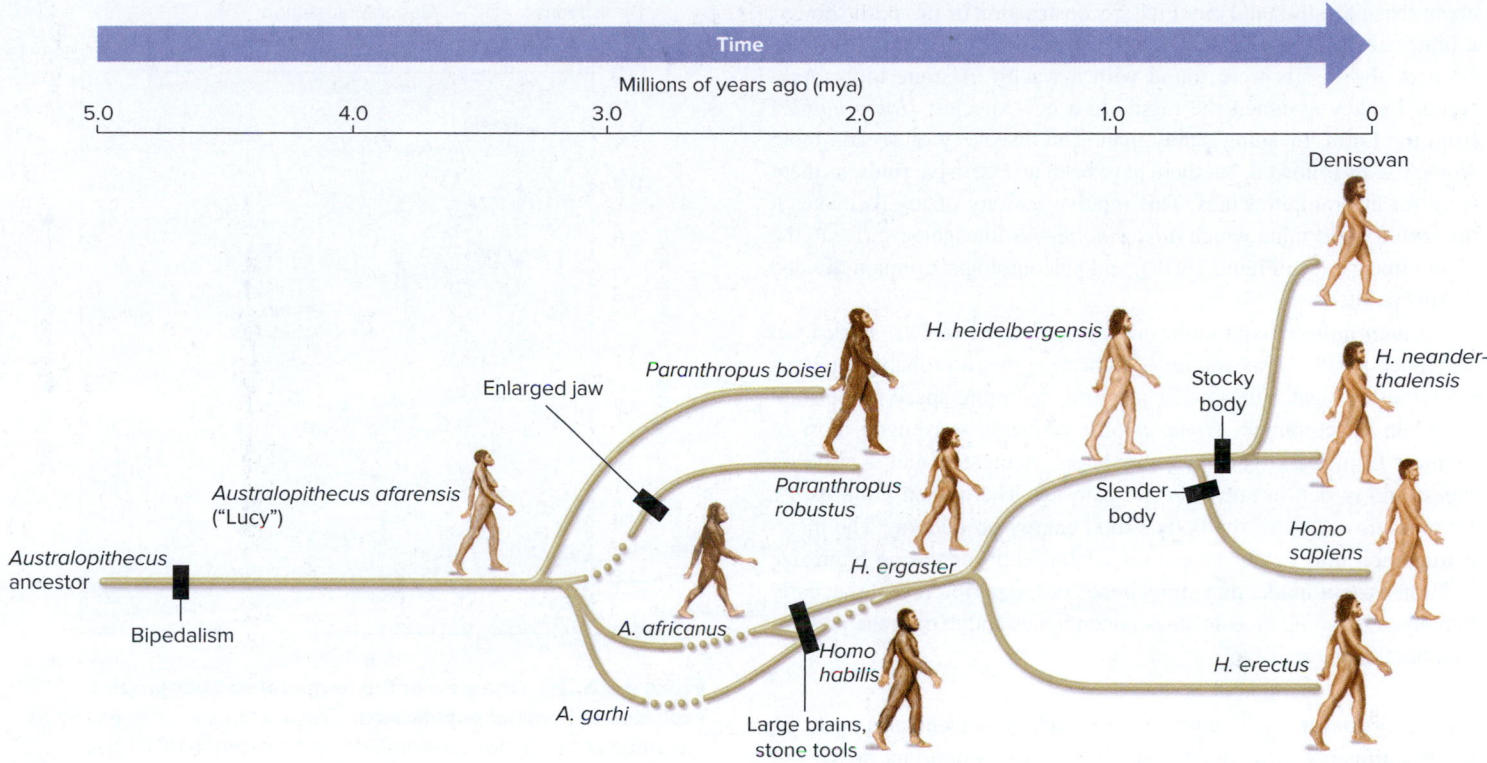

**Figure 26.18** **A possible scenario for human evolution.** In this human family tree (based on the work of Donald Johanson and others), several hominin species lived during the same time period as others, but only one lineage gave rise to modern humans (*Homo sapiens*).

In the 1930s, the remains of bigger-boned hominins were found. Two of these larger species, now considered to be a separate genus, *Paranthropus,* weighed about 40 kg and lived during the same time period as australopithecines and early *Homo* species. *Paranthropus* were vegetarians with enormous jaws used for grinding tough roots and tubers. Both *Paranthropus* species died out rather suddenly about 1.5–2.0 mya.

Although *Australopithecus africanus* was thought to have evolved slightly later than *A. afarensis,* its bones had been found much earlier than those of *A. afarensis.* In the 1920s, Australian anthropologist Raymond Dart described *A. africanus* from infant bones discovered in a cave in Taung, South Africa. The well-preserved skull was small but was well rounded, unlike the skulls of chimpanzees and gorillas. Also, the positioning of the head on the vertebral column indicated bipedalism. These observations suggested to Dart that he had found a transitional form between apes and humans. However, it would take another 20 years and the discovery of more fossils to convince the scientific world to support Dart's view. In 1996, remains of another species, *Australopithecus garhi,* were also found in the Afar region. The discoverers were surprised to find that the teeth had similarities with *Paranthropus boisei.* Garhi means surprise in the local Afar language. The proposal that both *A. garhi* and *A. africanus* are ancestors of modern humans has been the subject of much debate. They have been viewed as either dead-end cousins or the ancestors of the first members of the genus *Homo.*

## The Genus *Homo* Includes Modern Humans and Their Most Recent Relatives

Seven different species in the *Homo* genus are shown in Figure 26.18. Let's consider how they are evolutionarily related to each other.

*Homo habilis*  In the 1960s, British paleontologist Louis Leakey found hominin fossils estimated to be about 2 million years old in Olduvai Gorge, Tanzania. Two particularly interesting observations

**Figure 26.19** **A modern woman compared to an australopithecine.** Compared with modern humans, australopithecines, as illustrated by this reconstruction based on the famous fossil called Lucy, were much smaller and lighter.

about these fossils stand out. First, reconstruction of the skull showed a brain size of about 680 cm³, larger than that of *Australopithecus*. Second, the fossils were found with a wealth of stone tools. As a result, Leakey assigned the fossils to a new species, *Homo habilis,* from the Latin, meaning handy man. The discovery of several more *Homo* fossils followed, but there have been no extensive finds, as there were for australopithecines. This relative scarcity of fossils makes it difficult to determine which *Australopithecus* lineage gave rise to the *Homo* lineage (see Figure 26.17), and paleontologists remain divided on this point.

*Homo habilis* lived alongside *Paranthropus* in East Africa but had much smaller jaws and teeth, indicating that it probably ate large quantities of meat. The smaller jaw provided more space in the skull for brain development. *Homo habilis* probably scavenged most of its meat from the kills of large predators. A meatier diet is easier to digest and is rich in nutrients and calories. The human brain uses a lot of energy, 20% of the body's total energy production. The meat-eating habit thus helped propel the evolution of increasing brain size in humans. Cut marks on animal bones of the period reveal that early humans used stone tools to smash open bones and extract the protein-rich bone marrow.

**Homo ergaster** Although the evidence is not entirely clear, *H. habilis* probably gave rise to one of the most important species of *Homo: Homo ergaster.* This hominin species arose in Africa about 2 million years ago; it had a human-looking face and skull, with downward-facing nostrils. *H. ergaster* was also a tool user, and the tools, such as hand axes, were larger and more sophisticated than those associated with *H. habilis. H. ergaster* evolved in a period of global cooling and drying that reduced tropical forests even more and promoted savanna conditions. Hairlessness and the regulation of body temperature through sweating may also have evolved at this time as adaptations to the sunny environment. A leaner body shape was evident. We know this from so-called Turkana boy, a fossil teenage boy found in Kenya in 1984. Though only 13 years old, scientists predict he would have been about 185 cm (6 ft 1 in.) when adult, much the same height as the Masai tribesman that inhabit the area today. A dark skin probably protected *H. ergaster* from the Sun's rays. The pelvis had narrowed, promoting efficiencies in walking upright, and the size of the brain and hence the skull increased, which may have produced more difficulty in childbirth. Mothers had to push increasingly large-brained infants through a narrowed pelvis. Researchers think that as a result, the human gestation period was shortened. Earlier birth leads to prolonged care of human infants compared with that in other apes. Prolonged child care required well-nourished mothers, who would have benefited from the support of their male partner and other members of a social group. Some anthropologists have suggested this was the beginning of the family.

*H. ergaster* is thought to have given rise to many species, including *Homo erectus, Homo heidelbergensis, Homo neanderthalensis,* and *Homo sapiens.* A possible timeline and geographic locations for these species are shown in **Figure 26.20**. *H. ergaster* probably was the first type of human to leave Africa, as similar bones have been found in the Eurasian country of Georgia. This species is believed to be a direct ancestor of modern humans, with *Homo heidelbergensis* viewed as an intermediary step. Living at the same time as

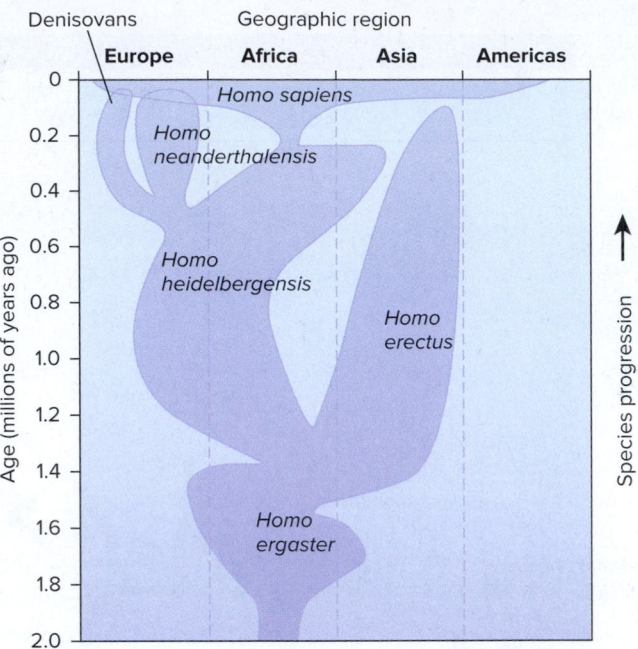

**Figure 26.20** **One view of the temporal and geographic evolution of hominid populations.** Though not shown in this figure, the range of the Denisovans and *Homo neanderthalensis* also extended into Asia.

*H. heidelbergensis* was another descendent of *H. ergaster, H. erectus,* though some researchers consider *H. ergaster* and *H. erectus* as the same species. We will treat them as separate species here.

**Homo erectus** *Homo erectus* was a large hominin, as large as a modern human but with heavier bones and a smaller brain capacity of between 750 and 1,225 cm³ (modern human brain size is about 1,350 cm³). Fossil evidence shows that *H. erectus* was a social species that used tools, hunted animals, and cooked over fires. The meat-eating habit may have sparked the migration of *H. erectus.* Carnivores tend to have larger ranges than similar-sized herbivores, because the food sources of carnivores (prey) are usually scarcer per unit area. *H. erectus* spread out of Africa soon after the species appeared, over a million years ago, and fossils have been found as far away as China and Indonesia. The first fossil was found by Dutch physician Eugene Dubois in 1891 on the Indonesian island of Java. Stone tools are rarely found in these Asian sites, suggesting *H. erectus* based their technology on other materials, such as bamboo, which was abundant at that time. Bamboo is strong yet lightweight and could have been used to make spears. These humans may even have used rafts to take to the seas. *H. erectus* went extinct about 100,000 years ago, for reasons that are unclear but may be related to the spread of *H. sapiens* into its range.

**Homo heidelbergensis** *Homo heidelbergensis* was similar in body form to modern humans. Large caches of this species' bones were found in Spain, at the bottom of a 14-m (45-ft) shaft known as La Sima de Los Huesos (the pit of bones). Similar remains were also found at Boxgrove in England. Shinbones recovered from Boxgrove suggest that males had heights around 180 cm (6 ft) and weighed 88 kg (196 pounds). Skulls were large, with brain volumes from 1,100 to

1,400 cm³, similar to those of modern humans. Animal bones from these sites showed cut marks from stone blades beneath tooth marks from carnivores. This indicated that humans were killing large prey before scavengers arrived. Horses, giant deer, and rhinoceroses were common prey, and successfully hunting them would have required much skill and cooperation. *H. heidelbergensis* gave rise to three species, *Homo neanderthalensis,* the Denisovans, and *Homo sapiens*. However, some scientists view these as three subspecies. In the discussion that follows, we will consider them to be three different species.

**Homo neanderthalensis**   *Homo neanderthalensis* was named for the Neander Valley of Germany, where the first fossils of this species were found. In the Pleistocene epoch, glaciers were locked in a cycle of advance and retreat, and the European landscape was often covered with snow. Over the course of many generations, the more slender body form of *H. heidelbergensis* evolved into a shorter, stockier build that was better equipped to conserve heat; we now call this type of human Neanderthal. Neanderthals also possessed a more massive skull and larger brain size than modern humans, about 1,450 cm³, perhaps associated with their bulk. Males were about 168 cm (5 ft 6 in.)

tall and would have been very strong by modern standards. They had a large face with a prominent bridge over the eyebrows, a large nose, and no chin (**Figure 26.21a**). They lived predominantly in Europe, with a range extending to the Middle East and Asia (**Figure 26.21b**). Their stocky and muscular physique was well suited to the rigors of cold climates and hunting prey. Paleontologists have found a high rate of head and neck injuries in Neanderthal bones, similar to that seen in present-day rodeo riders. This suggests that close encounters with large prey often resulted in blows that knocked the hunters off their feet. The hyoid bone, which holds the larynx (voice box) in place, was well developed, suggesting speech was used. The Neanderthals went extinct about 40,000 to 30,000 years ago.

**Denisovans**   In 2010, scientists sequenced DNA from a fossilized pinky finger of a young female found in Denisova Cave in Siberia, Russia, and found it to be genetically distinct from the DNA of *H. neanderthalensis* and *H. sapiens*. Because it is such a recent discovery, its species name is yet to be agreed upon. The common name given to this type of human is a Denisovan. Using carbon radioisotope dating, the fossil was estimated to be about 400,000 years old. Thus far, the remains of 4 Denisovans have been discovered.

**(a) An adult male Neanderthal**

**Figure 26.21   Neanderthals. (a)** Artist's rendition of a Neanderthal human. Neanderthals were shorter and stockier than modern humans, with larger elbow and ankle joints, shorter forearms, and a larger, broader rib cage. **(b)** The range of Neanderthals was confined to Europe and western Asia, with a northerly limit that corresponds to about 50° north, the southern limit of glaciation. Total population size may only have been 70,000 at its peak.

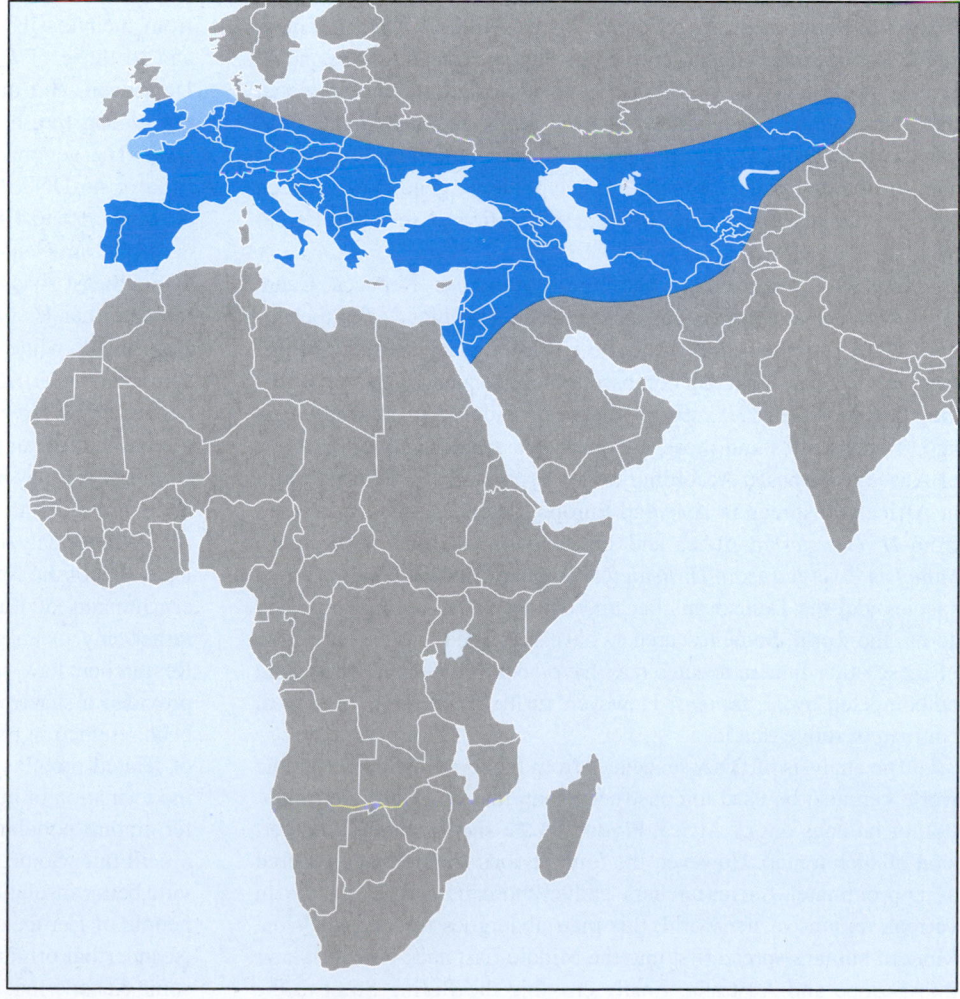

**(b) Geographic range of Neanderthals**

Scientists now hypothesize that Denisovans were a sister group to the Neanderthals, diverging from the Neanderthals about 400,000 years ago. It is difficult to reconstruct the physical traits of Denisovans, because only a finger bone, a toe bone, and a few molar teeth have been discovered thus far. Even so, the finger bone was unusually broad and robust, suggesting the Denisovans were similar in build to the Neanderthals. Like the Neanderthals, the Denisovans are thought to have gone extinct about 40,000 to 30,000 years ago.

**Homo sapiens**   *Homo sapiens* (from the Latin, meaning wise man) is our own species. *H. sapiens* is a slender, lighter-weight species with a slightly smaller brain capacity than that of the Neanderthals. Researchers hypothesize a variety of reasons why *H. sapiens* thrived while the Neanderthals disappeared; it may have been due to a more efficient body type with lower energy needs, increased longevity, and/or differences in social structure and cultural adaptations.

## Evidence Suggests That *Homo sapiens* Arose in Africa and Then Migrated to Other Parts of the World

Two models have been proposed to explain where the species of modern humans, *Homo sapiens,* arose. The first model, the out-of-Africa hypothesis, suggests that *H. sapiens* evolved in Africa from *H. heidelbergensis.* Some members of *H. sapiens* later migrated to other parts of the world, and gradually replaced species such as *H. erectus* and *H. neanderthalensis.* An alternative hypothesis, called the multiregional hypothesis, which is not widely accepted, suggests that human groups have evolved from *H. ergaster* populations in a number of different parts of the world. According to this hypothesis, gene flow between neighboring populations prevented the formation of several different *H. sapiens* species.

In 1987, American evolutionary biologists Rebecca Cann, Allan Wilson, and colleagues analyzed the sequences of mitochondrial DNA (mtDNA) that was collected from many different people from around the world. By comparing these sequences to each other, they concluded that *H. sapiens* arose in Africa about 200,000 years ago. These results and those of subsequent studies support the out-of-Africa hypothesis. According to this hypothesis, *H. ergaster* arose in Africa and spread to Asia and Europe. Later, *H. erectus* diverged from *H. ergaster* in Africa and spread into Asia, and *H. neanderthalensis* diverged from *H. heidelbergensis* in Europe. Both of these species and the Denisovans became extinct as *H. sapiens* migrated across the world. Some researchers have suggested that the extinction of these other human species may have occurred because they were outcompeted by *H. sapiens.* However, further research is needed to confirm or refute that idea.

The analysis of DNA sequences from modern humans across the world can also be used to construct a map that describes the migration of humans out of Africa. **Figure 26.22** shows a simplified version of such a map. However, the time periods should be considered as approximate. As researchers gather more data from people in various regions of the world, this map undergoes frequent revision. Modern humans spread first into the Middle East and Asia, then later into Europe and Australia, finally crossing the Bering Strait to the Americas.

Much remains to be resolved in our understanding of human evolution, and new data provide paleontologists with information to revise their hypotheses. For example, in 2004, the remains of a small human were discovered on the Indonesian island of Flores and were given the name *Homo florensiensis,* nicknamed hobbits by the media. Many species—for example, deer and elephants—develop into small forms in insular situations, so hobbit-sized humans seemed plausible. Since then, many researchers have suggested these people were modern humans who were suffering from a genetic disorder. Even modern humans on Flores are pygmies. Pathological dwarfism would have made these people even smaller. Only tools associated with *H. sapiens* have been found at the area where the bones occur, suggesting these individuals were indeed dwarf forms of modern humans.

A 2014 study showed that the small brain size of one of the *H. floresiensis* fossils was in the range predicted for an individual with Down syndrome, and it was suggested that this is evidence that *H. florensiensis* is an invalid species.

## Human Evolution Has Involved Interbreeding Among Closely Related Species

Because the remains of Neanderthals and Denisovans are less than 50,000 years old, researchers have been able to extract DNA from their fossils and compare their DNA sequences to each other and to those of *H. sapiens.* Detailed comparisons of Neanderthal, Denisovan, and modern human genomes have revealed interbreeding among the three species as *H. sapiens* spread into Europe and Asia. The genomes of modern humans of African descent contain little or no DNA that is derived from Neanderthals or Denisovans. However, 1% to 4% of the DNA from a person of European descent is derived from Neanderthals, and 4% to 6% of the DNA from a person of Southeast Asia descent is derived from Denisovans. These results indicate that *H. sapiens* ancestors interbred with Neanderthals and Denisovans while spreading across Europe and Asia. Interestingly, some people of African descent carry very small amounts of Neanderthal DNA. How is this possible? By analyzing the DNA sequences of certain African people, researchers speculate that some *H. sapiens* from Europe may have migrated back to Africa about 3,000 years ago and interbred with a few isolated African populations of *H. sapiens.*

When analyzing the human genome on a population level, at least 20% of the Neanderthal genome is found in the genome of modern humans of European descent. No one individual has all 20%; rather, any given person of European descent has about 1% to 4%. Researchers have speculated that certain Neanderthal genes may have provided a survival advantage, and that may explain why they have been retained in particular human populations. For example, a group of related proteins called keratins form filaments that play a role in the formation of human skin, hair, and nails. Such filaments may differ among populations that have evolved in a warm climate versus a cold one. Some alleles of keratin genes encode keratins that provide better insulation, a trait that is advantageous in cold climates. In people of European descent, genes that encode keratins are often of Neanderthal origin. This observation is consistent with the idea that some Neanderthal alleles may have helped European *H. sapiens* adapt to colder European environments.

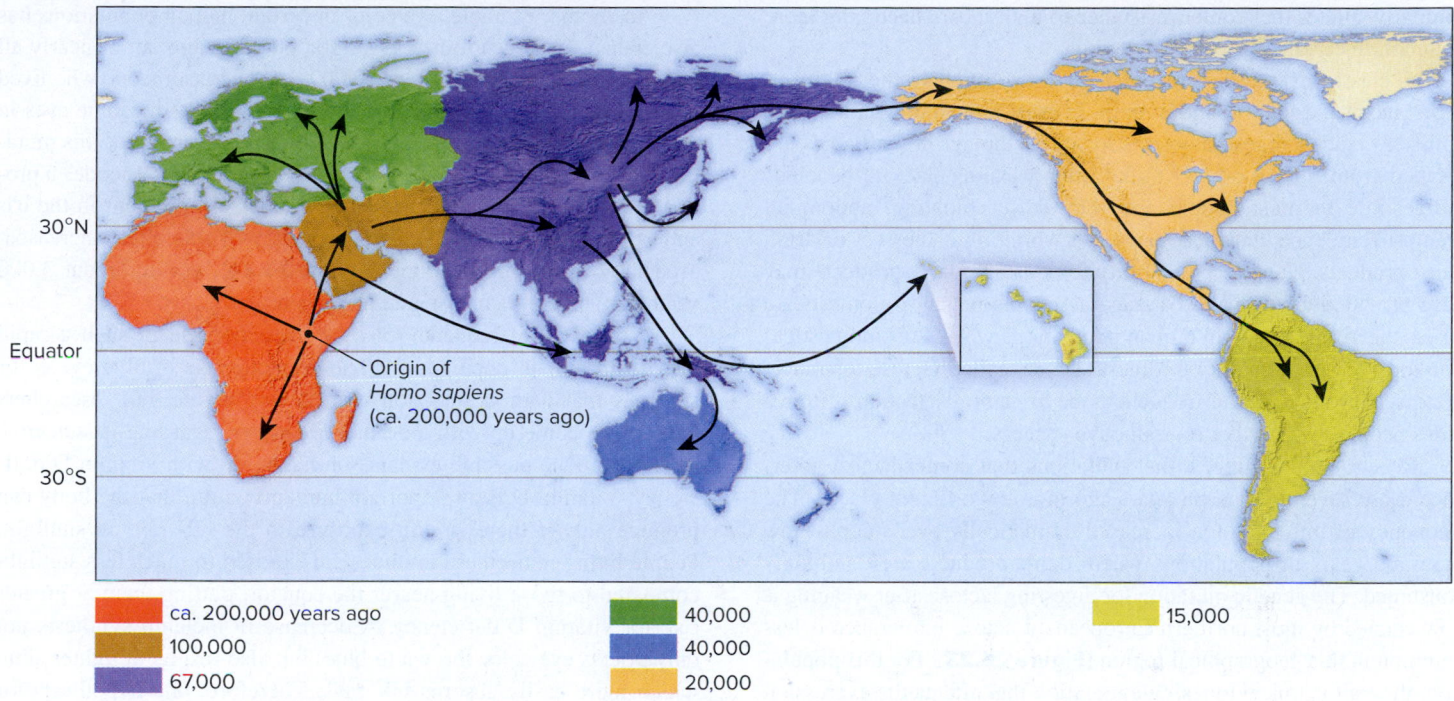

| | | |
|---|---|---|
| ca. 200,000 years ago | 40,000 | 15,000 |
| 100,000 | 40,000 | |
| 67,000 | 20,000 | |

**Figure 26.22  A simplified model for the origin and spread of *Homo sapiens* throughout the world.**  This map, based on differences of mtDNA throughout current members of the world's population, suggests *Homo sapiens* originated in East Africa. About 100,000 years ago, the species spread into the Middle East and from there to Europe, Asia, Australia, and the Americas.

**Core Skill: Modeling**  The goal of this modeling challenge is to revise the model shown in Figure 26.22 to account for recent data indicating that some modern Africans have a small amount of Neanderthal DNA.

**Modeling Challenge:** As discussed in this section, between 1% and 4% of the DNA of modern humans of European descent is derived from that of Neanderthals. Though most modern humans of African descent do not carry Neanderthal DNA, recent evidence has shown that some of them carry a very small amount. Assuming that the presence of this Neanderthal DNA is not due to recent interbreeding between modern humans of European and African descent, revise the model shown in Figure 26.22 to account for the observation.

On the downside, scientists also speculate that Neanderthals carried an allele of a gene involved with fatty acid uptake that helped them store fat better than *H. sapiens*. Such a gene was an advantage for a Neanderthal lifestyle in which hunter-gatherers gorged on prey and then went for days without eating. However, in modern humans, fat storage can put them at risk for developing various diseases, such as type 2 diabetes mellitus. In certain human populations, particularly native Americans and those of Latin American descent, this Neanderthal allele is fairly common. Some people are heterozygous, carrying one allele that is derived from the Neanderthal genome and one from the *H. sapiens* genome. Such individuals are 20% more likely to develop type 2 diabetes compared to people who are homozygous for the *H. sapiens* allele. Furthermore, people who are homozygous for the Neanderthal allele are 40% more likely to develop diabetes.

## Are Human Populations Still Evolving?

As discussed in Chapter 23, natural selection is the process by which individuals with greater reproductive success are more likely to pass

their genes to future generations. Natural selection results in evolution. (Other processes, such as genetic drift, can also promote evolutionary change.) One factor that often plays a role in natural selection is the environment. Various types of environmental factors such as temperature, predators, and food sources can affect reproductive success and thereby cause a population to evolve in a particular direction. For example, a prolonged decrease in temperature may favor the survival of mammals with thicker fur, a trait that will increase in frequency over the course of many generations.

Modern humans have great control over their environment. They live in dwellings where they can control the temperature, they can largely avoid predators, and they usually don't rely entirely on locally grown food for their survival. For such reasons, the impact of the natural environment on the evolution of human populations may have lessened compared to its impact on humans that lived long ago. Even so, human evolution via natural selection is still occurring. For example, we continue to evolve genetic resistance to infectious diseases. The bubonic plague of the 14th century killed about one-third of the Asian and European populations, yet many people survived and passed on alleles that confer greater resistance to this deadly disease.

Similarly, alleles that confer resistance to malaria are becoming more common in certain African populations.

A classic example of recent human evolution is the ability to digest lactose, a sugar found in milk. In most human populations, the ability to readily digest lactose is lost after the age when babies are weaned from their mother's milk. After weaning, lactose becomes indigestible to most people and they suffer bloating, abdominal cramps, flatulence, diarrhea, nausea, or vomiting if they eat or drink dairy products. However, the ability to consume dairy products may have provided a survival advantage as people began to domesticate cows, sheep, and goats. In human populations where such domestication took place, the ability to digest lactose, called lactose tolerance, is expected to increase in frequency due to natural selection if it provides people with greater reproductive success.

Recent studies suggest that mutations that confer lactose tolerance arose several thousand years ago in a few different places. The frequency of this mutation increased dramatically over the past few thousand years in populations where dairy products are commonly consumed. The genetic mutation for digesting lactose after weaning is now carried by most northern Europeans. Lactose intolerance is less common in this geographical region (**Figure 26.23**). For this population, the trait is linked to a single mutation that affects the expression of the lactase gene, which encodes an enzyme that is needed to digest lactose. The mutation prevents the gene from being turned off after weaning. Lactose tolerance has also been found in other populations, such as Africa and the Middle East, but these mutations occurred independently of the one that is usually found in lactose-tolerant people of European descent.

As another example, eye color in certain human populations has also changed in recent times. Over 10,000 years ago, all or nearly all humans had brown eyes. About 10,000 years ago, someone who lived near the Baltic Sea inherited a mutation that resulted in blue eyes in the homozygous state. In 2008, researchers discovered that this mutation decreases the expression of the *Oca2* gene, which encodes a protein that is needed for the production of melanin pigment in the iris and other parts of the body. The frequency of this allele increased, especially in northern Europe (**Figure 26.24**), and by about 3,000 years ago, blue eyes had spread across Europe.

Why did the frequency of blue eyes increase in such a rapid fashion in human populations? The dramatic rise in blue eye color suggests that natural selection was playing a role, but researchers have yet to come up with a definitive answer regarding its selective advantage. One possible explanation has to do with vitamin D deficiency. Vitamin D is an important human vitamin that the body can produce only if there is skin exposure to the UV rays in sunlight. People living in northern latitudes are exposed to much less sunlight compared to those living nearer the equator, putting them at greater risk for vitamin D deficiency. A decrease in melanin synthesis not only affects eye color (brown to blue) but also results in lighter skin, which more easily absorbs UV rays. Therefore, one hypothesis for the spread of blue eyes (and lighter skin) through the human population is that it may have enabled humans to avoid the harmful health effects of vitamin D deficiency, which includes weakness and bone abnormalities. In this scenario, natural selection acted on skin color, and the eye color phenotype increased due to its association with a lighter skin color.

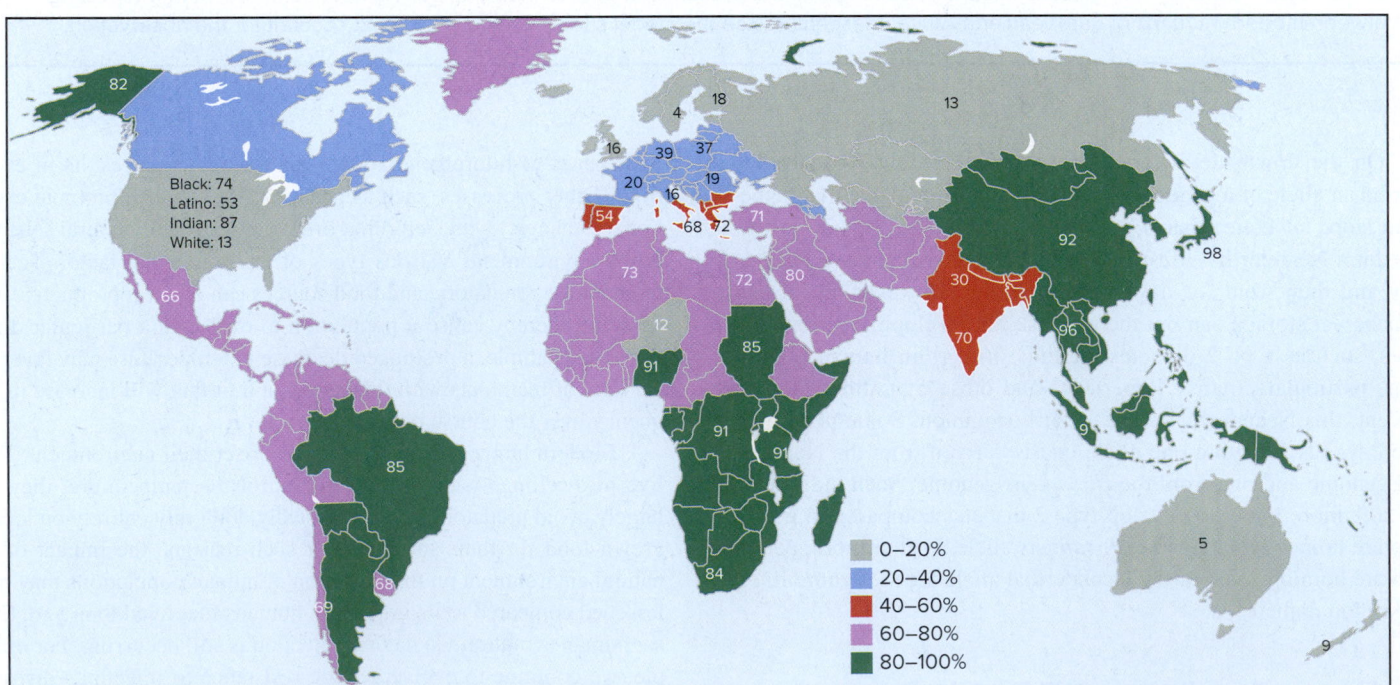

**Figure 26.23  The frequency of lactose intolerance in human populations.** This map emphasizes lactose intolerance, which is the inability to digest lactose after weaning. In contrast, most northern Europeans can digest the sugar lactose found in dairy products; they are lactose tolerant.

**(a) Blue eye color**

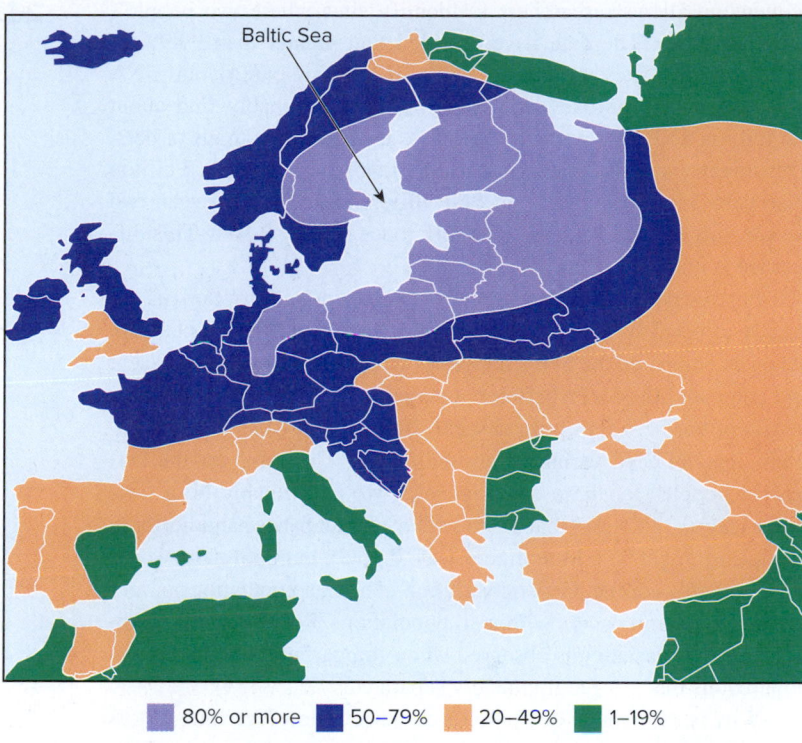

Baltic Sea

| 80% or more | 50–79% | 20–49% | 1–19% |

**(b) Percentage of people with blue eye color in Europe and surrounding regions**

**Figure 26.24   Blue eye color and its spread throughout Europe.   (a)** In humans, blue eye color is due to a lack of melanin in the iris. This mutation probably appeared only about 10,000 years ago. **(b)** Many people in western and northern Europe have blue eyes due to the spread of this single mutation that originated in someone living near the Baltic Sea. Note: The indicated percentages include blue eye color and other light eye colors, such as green. a: ©harpazo_hope/Getty Images

## Modern Humans Show Relatively Little Genetic Variation Yet Exhibit a Significant Amount of Phenotypic Variation

Looks can be deceiving. If you take a plane ride around the world and make stops in Japan, Nigeria, Norway, Brazil, and Australia, you might get the impression that the human species is genetically diverse (**Figure 26.25**). The physical differences among humans in many parts of the globe are striking. To assess the level of genetic variation, the **1000 Genomes Project** was launched in January 2008; it is an international research effort to establish the level of human genetic variation. By 2015, this project had sequenced the genomes of over 2,500 people from around the world and compared those sequences to each other. The results indicate that, genetically speaking, humans are very, very similar to each other. Our level of genetic variation is lower than most species of mammals whose genomes have been sequenced, and it is even lower than the variation within certain fruit fly species. This may seem surprising; if you compared two fruit flies of the same species to each other, they would probably look very similar!

Most human genetic variation is in the form of single nucleotide polymorphisms (SNPs), which are places where two humans' genomes differ at a single base pair (see Chapter 23). DNA

**(a) Japan**   **(b) Nigeria**   **(c) Norway**   **(d) Brazil**   **(e) Australia**

**Figure 26.25   Examples of human phenotypic variation in different parts of the world.**   The women seen here are from **(a)** Japan, **(b)** Nigeria, **(c)** Norway, **(d)** Brazil, and **(e)** Australia. They are genetically very similar even though they look somewhat different. a: ©William Perugini/123RF; b: ©Filipe Frazao/Shutterstock; c: ©Andrea Magugliani/Alamy Stock Photo; d: ©Daniel Ernst/Alamy Stock Photo; e: ©David Freund/Getty Images

sequencing allows researchers to identify sites where two people's genomes differ. The data have revealed that greater than 99.9% of those differing sites are due to SNPs. If you compared your DNA with that of any unrelated person, you would probably find about 4.5 million SNP sites where your SNP was different from his or hers. This might sound like a lot, but consider that your genome is 3 billion base pairs long. If 4.5 million SNP differences occur between two people, this amounts to a genome difference of only 0.15%. This difference is much less than 1%!

Furthermore, this very low value of genetic variation is not greatly affected by the pairs of people who are analyzed. Let's suppose you compared the SNPs between two people of Japanese descent and compared the SNPs between a person of Japanese descent and a person of northern European descent. Both comparisons would show a low level of SNP variation, around 0.15%. The value for the first pair of people (Japanese and Japanese) would probably be a little lower (say, 0.14%), and that for the other pair (Japanese and northern European) might be a little higher (say, 0.16%), but both pairs would be remarkably similar. Genetic variation in human populations is very low and most of it occurs within all populations. Relatively little additional genetic variation is observed when comparing individuals from populations that are geographically separated.

Why is the genetic diversity of our species low? Various factors are involved. First, *Homo sapiens* has not been around that long. Even though 200,000 years might seem like a long time, humans have a long generation time, and evolution occurs over the course of generations. Second, until recently, human populations have been relatively small. About 10,000 years ago, the human population size is estimated to have been about 5 million. Small populations tend to be less genetically diverse than larger ones. Although the human population has grown enormously over the past few centuries, this expansion is very recent on an evolutionary timescale.

How do we explain the disparity between a low level of genetic variation and a seemingly higher level of phenotypic diversity? The answer is not entirely understood, but it may be related to the traits that influence our perception of diversity. Some of the traits that are most visually obvious, such as eye and skin color, may be dramatically influenced by natural selection even though they involve changes in a relatively small number of genes. For example, we have already considered how blue eye color and light skin color spread rapidly throughout Europe, and this phenotypic change involved a mutation in a single gene. This observation underscores how our visual perception of diversity may be biased by traits not rooted in major genetic differences.

## Summary of Key Concepts

- Life began on Earth between 4.0 and 3.5 bya (Figure 26.1).

### 26.1  The Fossil Record

- Fossils, which are preserved remnants of past life-forms, are formed in sedimentary rock (Figure 26.2).
- Radiometric dating is one way of estimating the age of a fossil. Fossils provide an extensive record of the history of life on Earth, though the record is incomplete (Figure 26.3, Table 26.1).

### 26.2  History of Life on Earth

- The geological timescale, which is divided into four eons and many eras and periods, charts the major events that occurred during the history of life on Earth (Figure 26.4).
- Both the emergence of new species and mass extinctions are correlated with changes in temperature, amount of $O_2$ in the atmosphere, landmass locations, floods and glaciation, volcanic eruptions, and meteorite impacts (Figure 26.5).
- During the Archaean eon, bacteria and archaea arose. The proliferation of cyanobacteria led to a gradual rise in $O_2$ levels (Figure 26.6).
- Eukaryotic cells arose during the Proterozoic eon. This origin involved a union between bacterial and archaeal cells that is hypothesized to have been endosymbiotic. The origins of mitochondria and chloroplasts were also the result of endosymbiosis (Figure 26.7).
- Multicellular eukaryotes arose about 1.5 bya during the Proterozoic eon. Multicellularity now occurs via cell division and the adherence of the resulting cells to each other. A multicellular organism can have multiple cell types (Figure 26.8).
- The first animal showing bilateral symmetry emerged toward the end of the Proterozoic eon (Figure 26.9).
- The Phanerozoic eon is subdivided into the Paleozoic, Mesozoic, and Cenozoic eras. During the Paleozoic era, invertebrates greatly diversified, particularly during the Cambrian explosion, and the land became colonized by plants and animals. Terrestrial vertebrates, including tetrapods, became more diverse (Figures 26.10–26.12).
- Dinosaurs were prevalent during the Mesozoic era, particularly during the Jurassic period. Mammals and birds also emerged (Figures 26.13, 26.14).
- During the Cenozoic era, mammals diversified, and flowering plants became the dominant plant species. The first hominoids emerged approximately 7 mya. *Homo sapiens*, our species, first appeared about 200,000 years ago.

### 26.3  Human Evolution

- Many defining characteristics of primates relate to their tree-dwelling nature; these include grasping hands, large brain, nails instead of claws, and binocular vision (Figures 26.15–26.17).
- About 7 mya in Africa, a lineage that led to humans began to separate from other primate lineages. A key characteristic of hominins (extinct and modern humans) is bipedalism. Human evolution can be visualized like a tree, with a few hominin species coexisting at the same point in time, some eventually going extinct, and some giving rise to other species (Figures 26.18–26.21).
- Data from the sequencing of human mitochondrial DNA suggest that *H. sapiens* originated in East Africa. From there, *H. sapiens* spread to Asia and then to all other parts of the globe (Figure 26.22).
- Human evolution has involved interbreeding between closely related species, such as *H. sapiens* and the Neanderthals and the Denisovans.
- Evidence that human populations are still evolving includes traits such as lactose tolerance and blue eye color (Figures 26.23, 26.24).
- Modern humans show relatively little genetic variation yet exhibit a significant amount of phenotypic variation (Figure 26.25).

## Assess & Discuss

### Test Yourself

1. The movement of landmasses that has changed their positions, shapes, and association with other landmasses is called
   a. glaciation.
   b. Pangaea.
   c. continental drift.
   d. biogeography.
   e. geological scale.

2. Paleontologists estimate the dates of fossils using
   a. the layer of rock in which the fossils are found.
   b. analysis of radioisotopes found in nearby igneous rock.
   c. the complexity of the body plan of the organism.
   d. all of the above.
   e. a and b only.

3. The fossil record does not give us a complete picture of the history of life on Earth because
   a. not all past organisms have become fossilized.
   b. only organisms with hard skeletons can become fossilized.
   c. fossils of very small organisms have not been found.
   d. fossils of early organisms are located too deep in the crust of the Earth to be found.
   e. All of the above are true.

4. The endosymbiosis hypothesis explaining the evolution of eukaryotic cells is supported by
   a. DNA-sequencing analysis comparing bacterial genomes, mitochondrial genomes, and eukaryotic nuclear genomes.
   b. naturally occurring examples of endosymbiotic relationships between bacterial cells and eukaryotic cells.
   c. the presence of DNA in mitochondria and chloroplasts.
   d. all of the above.
   e. a and b only.

5. Which of the following evolutionary innovations was advantageous for survival in a terrestrial environment?
   a. the amniotic egg in animals
   b. the seed in plants
   c. the shell in marine invertebrates
   d. all of the above
   e. both a and b

6. Which of the following explanations of multicellularity in eukaryotes is seen in the development of complex, multicellular organisms today?
   a. endosymbiosis
   b. aggregation of cells to form a colony
   c. division of cells followed by cell adhesion of the resulting cells
   d. multiple cell types aggregating to form a complex organism
   e. None of the above phenomena are evident today.

7. The earliest fossils of vascular plants were formed during the _____ period.
   a. Ordovician
   b. Silurian
   c. Devonian
   d. Triassic
   e. Jurassic

8. The first mammal arose during the _____ period.
   a. Triassic
   b. Jurassic
   c. Cretaceous
   d. Tertiary
   e. Quaternary

9. The appearance of the first hominoids dates to the _____ period.
   a. Triassic
   b. Jurassic
   c. Cretaceous
   d. Tertiary
   e. Quaternary

10. Which of the following statements regarding modern humans, *H. sapiens,* is *false?*
    a. Some modern humans have a small amount of DNA that is derived from Neanderthals.
    b. Some modern humans have a small amount of DNA that is derived from Denisovans.
    c. *H. sapiens* probably arose in Africa.
    d. Modern humans are very genetically diverse compared to most other species.
    e. Modern humans are still evolving.

### Conceptual Questions

1. How are the ages of fossils determined? In your answer, you should discuss which types of rocks are analyzed and explain the concepts of radiometric dating and half-life.

2. How was the phenomenon of endosymbiosis important in the evolution of the first eukaryotic cells?

3. **Core Concept: Evolution** Describe two examples in which changes in the global climate affected the evolution of species.

### Collaborative Questions

1. Discuss the factors that have contributed to the dramatic changes in life-forms since the origin of life on Earth about 3.5 to 4 billion years ago.

2. Discuss how the human body has changed since the human lineage diverged from other primates about 7 million years ago.

# UNIT V
# DIVERSITY

**Biological diversity**, also called **biodiversity**, encompasses the variety of living things that exist now, as well as all the life-forms that lived in the past. Knowing about species that lived in the past and how they are related to modern microorganisms, plants, and animals aids our comprehension of evolution, which is the source of biological diversity. Knowing about the many different kinds of modern organisms also helps us understand how life-forms are structured in ways that allow them to function differently in nature (as discussed in Units VI and VII) and how species interact with each other and with their environments (described in Unit VIII).

This unit begins with bacteria and archaea, the oldest, simplest, and most numerous of Earth's life-forms, whose prominent ecological roles are described in Chapter 27. In Chapter 28, we survey the surprisingly diverse protists, which affect humans and other organisms in many important ways. The mysteries of the fungi, essential to the brewing and baking industries as well as to ecological stability, are revealed in Chapter 29. These first three chapters of the unit provide background essential to understanding microbiomes, systems of microbes that occur on and around us, explained in Chapter 30. Next, Chapter 31 explores the evolutionary origin of the first plants, a process that explains the features and functions of the seed plants that are vital sources of human food, fiber, and medicine, as described in Chapter 32. An overview of the diversity and evolutionary history of the animals in Chapter 33 provides the basis for exploring the simplest animals, the invertebrates, in Chapter 34. More complex vertebrate animals, including humans and their closest relatives, are the focus of Chapter 35.

 **The following Core Concepts and Core Skills will be emphasized in this unit:**

- *Structure and Function:* *Many of Earth's present and past species have bodies consisting of only one or a few cells, whereas other species display more complex bodies composed of many cells.*
- *Evolution:* *This unit provides many examples of the concept that all Earth species are related by an evolutionary history.*
- *Connections:* *Chapter 31 explains how ancient plants dramatically changed the composition of Earth's atmosphere, and how their modern descendants continue to influence today's atmosphere and climate.*
- *Process of Science:* *Every chapter has a Feature Investigation that illustrates how we understand ways in which diversity is important.*
- *Modeling:* *Every chapter has a Modeling Challenge to refine this important skill.*

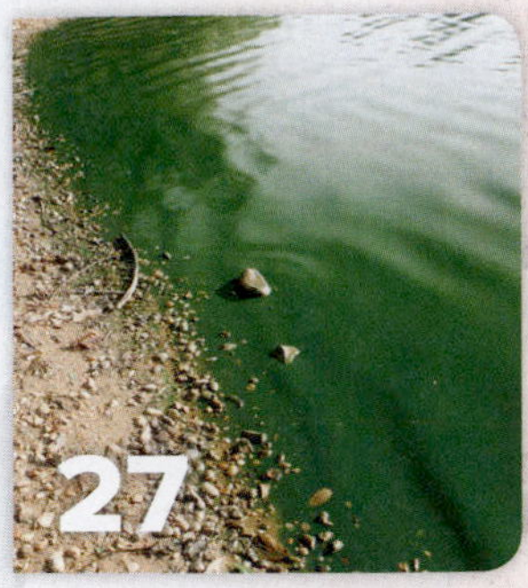

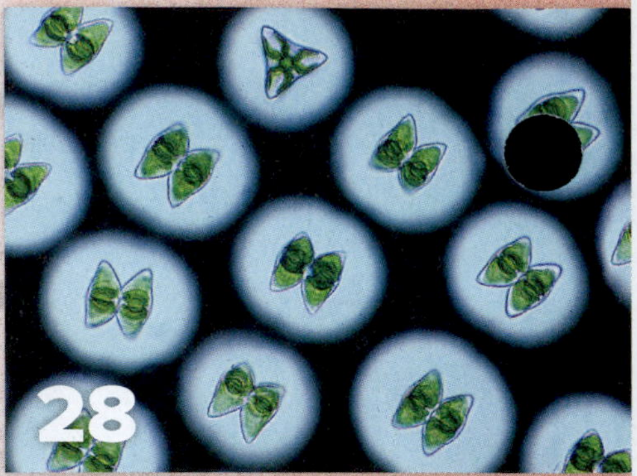

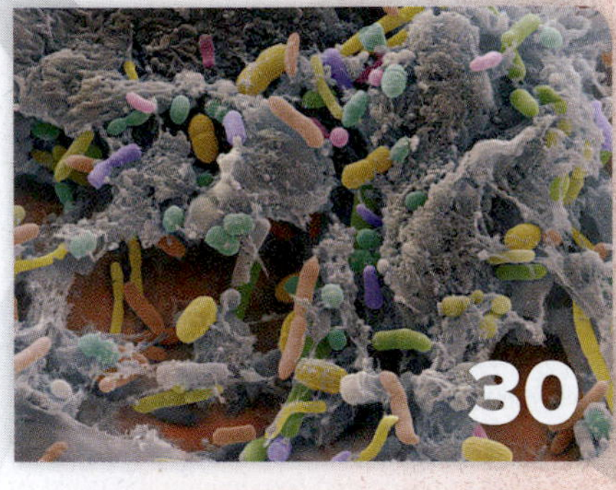

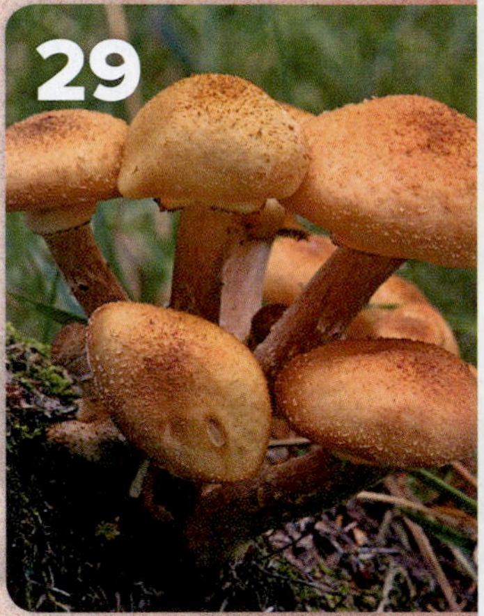

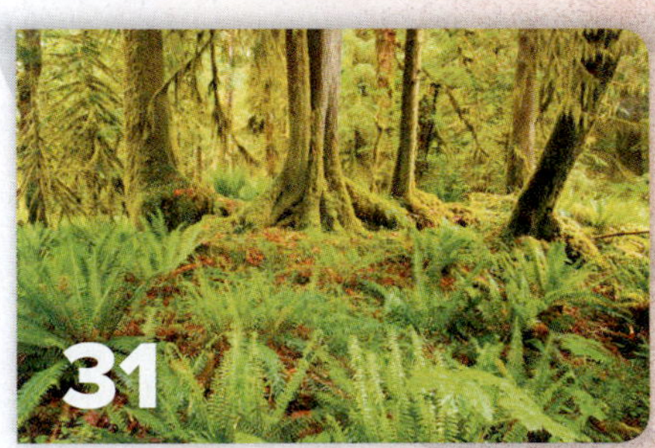

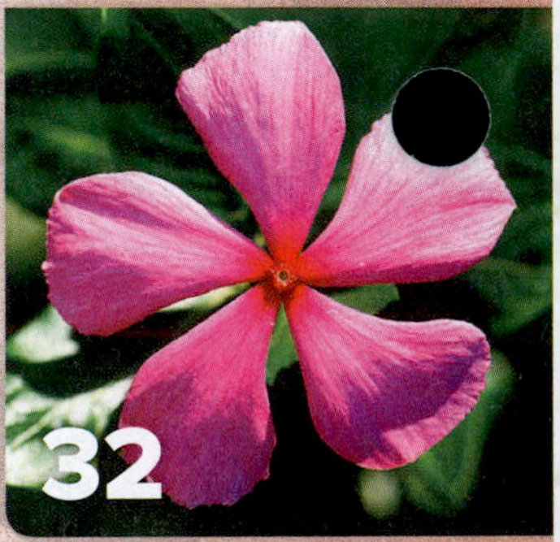

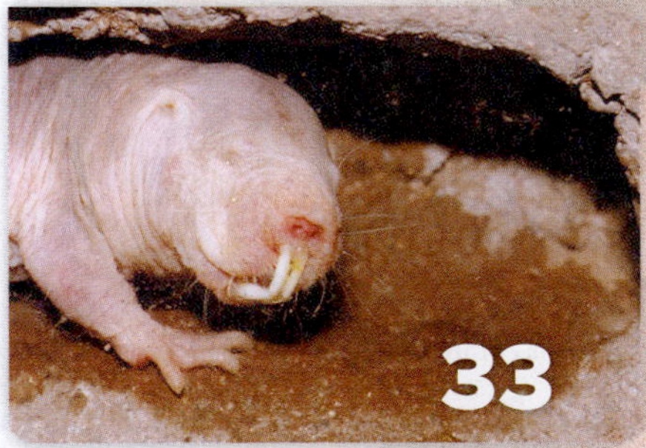

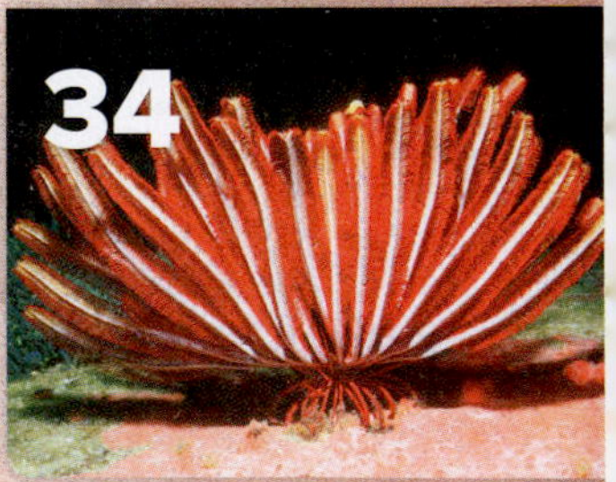

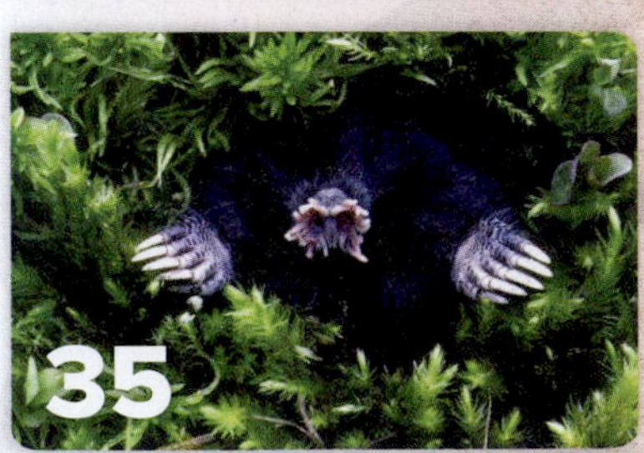

# Archaea and Bacteria

# 27

**Cyanobacterial bloom.** A visible growth of cyanobacteria, called a bloom, gives a blue-green coloration to this lake. ©Dr. Jeremy Burgess/ SPL/Science Source

**O**ne late-summer afternoon, a veterinarian was about to close the clinic at the end of a busy day when an emergency case, a very sick dog, arrived. The dog had collapsed after taking a lakeside walk with its owner. Responding to the vet's questions, the owner reported that the thirsty dog had consumed lake water thick with blue-green material. The vet deduced that the blue-green substance represented a large population of toxin-producing bacteria and that drinking the lake water had poisoned the dog—a conclusion that aided treatment.

Bacteria and archaea are examples of microorganisms, organisms so small they can usually be seen only with the use of a microscope. However, in phosphorus-rich bodies of water, some species of photosynthetic bacteria known as cyanobacteria grow rapidly into large, visible populations—known as blooms—that color the water blue-green or cyan (see the chapter opening photo). The individual cells release small amounts of toxins that help to keep small aquatic animals from eating them, but when blooms occur, toxins can rise to levels that poison humans, pets, livestock, and wildlife. Consequently, public health authorities often warn people that they should not swim in waters with visible blue-green blooms and should not allow pets and livestock to drink such water. People can prevent the formation of harmful cyanobacterial blooms by reducing the flow of phosphorus-rich fertilizers, manure, and sewage into bodies of water.

Despite the harmful effects of some species, cyanobacteria provide important benefits to humans and other organisms, such as producing atmospheric oxygen. Many cyanobacteria also have the ability to convert abundant but inert atmospheric nitrogen gas into ammonia, which algae and plants can use to synthesize amino acids and proteins. This process enriches nutrient-poor soils, particularly in wet paddy fields where rice is grown in many regions of the world, thereby helping to provide food for billions of people.

In this chapter, we will survey the diversity, structure, reproduction, metabolism, and ecology of archaea and bacteria. We

 **Core Skill: Science and Society**  Cyanobacterial blooms affect society by poisoning humans and organisms that people value. However, cyanobacteria can also have a positive impact by generating nitrogen-containing molecules that serve as fertilizer in soil and water.

will see how bacteria and archaea were important to the evolution of eukaryotic cells and other events in Earth's ancient history. The importance of some bacterial phyla to everyday human life illustrates the concept that relatively few bacterial species are harmful, and many benefit us. We will also learn how horizontal gene transfer increases the diversity of microbial genomes and how this process affects human societies. Our survey will reveal additional surprising ways in which these microorganisms affect the lives of humans and the world we inhabit.

## 27.1 Diversity and Evolution

**Learning Outcomes:**

1. **CoreSKILL »** Make a drawing that shows the evolutionary relationship among the domains Archaea, Bacteria, and Eukarya.

2. Explain how many species of archaea are able to grow in extreme habitats.

3. Discuss the medical, environmental, and evolutionary importance of cyanobacteria and proteobacteria.

4. List common mechanisms of horizontal gene transfer.

Life on Earth is classified into three domains. Members of the domain Eukarya—animals, plants, fungi, and protists—have cells with a eukaryotic structure. In contrast, **Archaea** (often referred to as simply archaea) and **Bacteria** (often referred to as simply bacteria) are domains of microorganisms whose cells have a prokaryotic structure. Archaeal and bacterial cells lack nuclei with porous envelopes and other cellular features typical of eukaryotes (see Chapter 4).

Although archaea and bacteria are sometimes collectively termed prokaryotes, such an aggregation is not a monophyletic group, but rather a paraphyletic group—one that does not include all of the descendants of a single common ancestor. That's because domain Archaea is more closely related to domain Eukarya than either is to domain Bacteria (**Figure 27.1**). Even so, archaea and bacteria display some common features in addition to a prokaryotic cell structure.

Archaea and bacteria include the smallest known cells and are the most abundant organisms on Earth. About half of Earth's total biomass consists of an estimated $10^{30}$ individual bacteria or archaea. Just a pinch of garden soil can contain 2 billion prokaryotic cells, and about a million occur in 1 mL of seawater. Archaea and bacteria

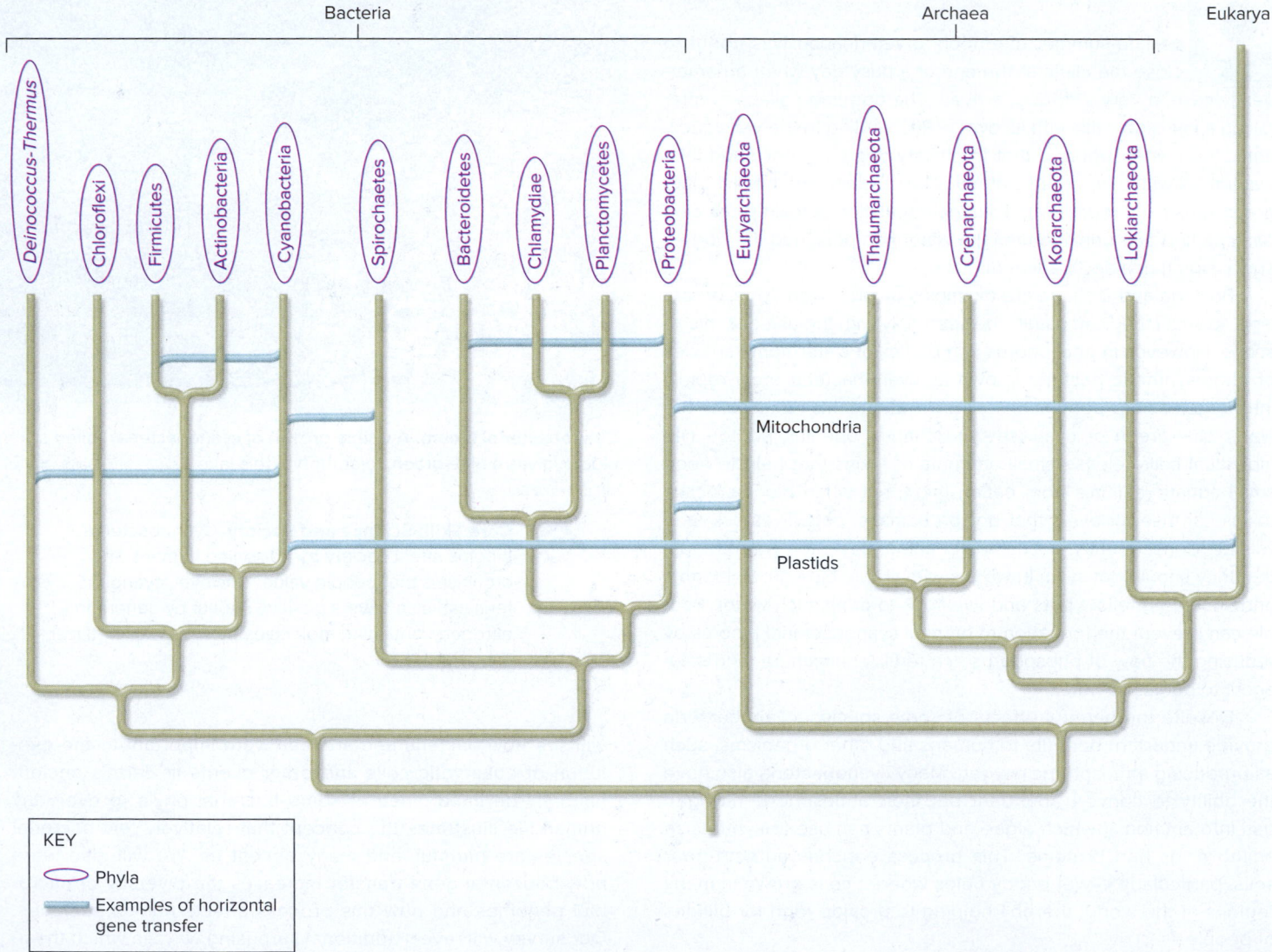

**Figure 27.1 Evolutionary relationships among selected phyla of Bacteria and Archaea to each other and to Eukarya.** Bacteria and Archaea are the two domains featuring prokaryotic cells. Eukarya is the domain consisting of organisms whose cells are eukaryotic. Archaea, particularly the phylum Lokiarchaeota and close relatives, is more closely related to Eukarya than is Bacteria. Each domain has diversified into multiple phyla (not all are shown here, for simplicity). Many cases of horizontal gene transfer among phyla and domains are known. Some of these are depicted with blue bars and include the acquisition of mitochondria and chloroplasts by eukaryotes.

live in nearly every conceivable habitat, including extremely hot or salty waters that support no other life, and they are also Earth's most ancient organisms, having originated more than 3 bya. Their long evolutionary history and varied habitats have resulted in extraordinary metabolic diversity.

Today, the many millions of species of archaea and bacteria collectively display more diverse metabolic processes than occur in any other group of organisms. Many of these metabolic processes are important on a global scale, influencing Earth's climate, atmosphere, soils, and water quality, as well as human health and technology. Most archaea and many bacteria have CRISPR-Cas systems that combat invading viruses and have proven useful in genetic engineering technology (see Chapter 21). In the past, microbiologists studied diversity by isolating archaea and bacteria from nature and growing cultures in the laboratory to observe variation in cell structure and metabolism. Today, biologists also use molecular techniques to assess diversity of archaea and bacteria and infer metabolic functions. Such molecular studies reveal that archaea and bacteria are vastly more diverse than was previously realized.

In this section, we will first survey the major kingdoms and phyla of the domains Archaea and Bacteria and then explore how horizontal gene transfer—the transfer of genes between different species—has influenced their evolution.

## Domain Archaea Was Ancestral to Domain Eukarya

Organisms classified in the domain Archaea, referred to as archaea, share a number of features with those classified in Eukarya,

suggesting common ancestry. For example, histone proteins are typically associated with the DNA of both archaea and eukaryotes, but they are absent from most bacteria. Archaea and eukaryotes share more than 30 ribosomal proteins that are not present in bacteria, and archaeal RNA polymerases are closely related to their eukaryotic counterparts. However, archaea possess distinctive membrane phospholipids, which are formed with ether bonds; in contrast, ester bonds characterize the membrane phospholipids of bacteria and eukaryotes (**Figure 27.2**). Ether-bonded membranes are resistant to damage by heat and other extreme conditions, which helps explain why many archaea are able to grow in extremely harsh environments. Also, archaea have isoprene chains instead of fatty-acid chains in their membranes. Another key difference is the biochemical composition of the cell wall. In most bacteria, the cell wall is composed of carbohydrates that are cross-linked by peptides, forming a substance called peptidoglycan, whereas the cell wall of archaea is usually a surface layer of proteins. Some bacterial species also have an outer envelope (membrane) composed of lipids and polysaccharides.

Though many archaea occur in soils and surface ocean waters in moderate conditions, diverse archaea occupy habitats with very high salt content, acidity, methane levels, or temperatures that would kill most bacteria and eukaryotes. Organisms that occur primarily in extreme habitats are known as **extremophiles**. One example is the methane producer *Methanopyrus*, which grows best at deep-sea thermal vent sites where the temperature is 98°C. At this temperature, the proteins of most organisms would denature, but those of *Methanopyrus* are resistant to such damage. *Methanopyrus* is so adapted to its extremely hot environment that it cannot grow when the temperature is less than 84°C. Such archaea are known as **hyperthermophiles**.

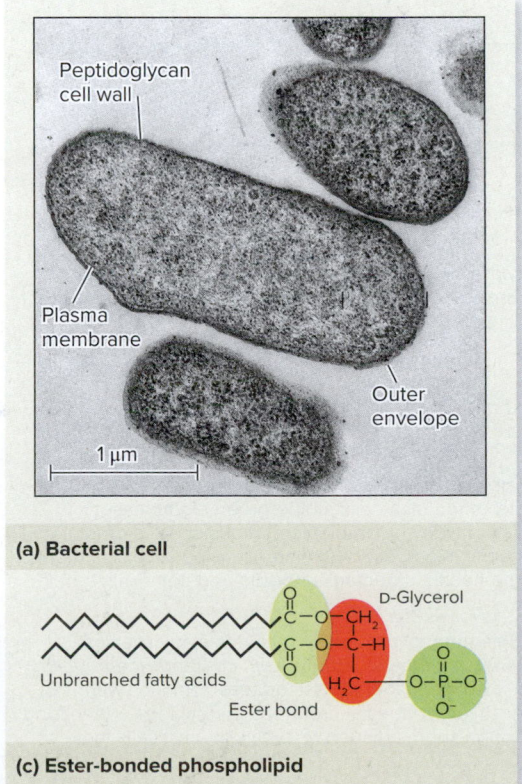

(a) Bacterial cell

(c) Ester-bonded phospholipid

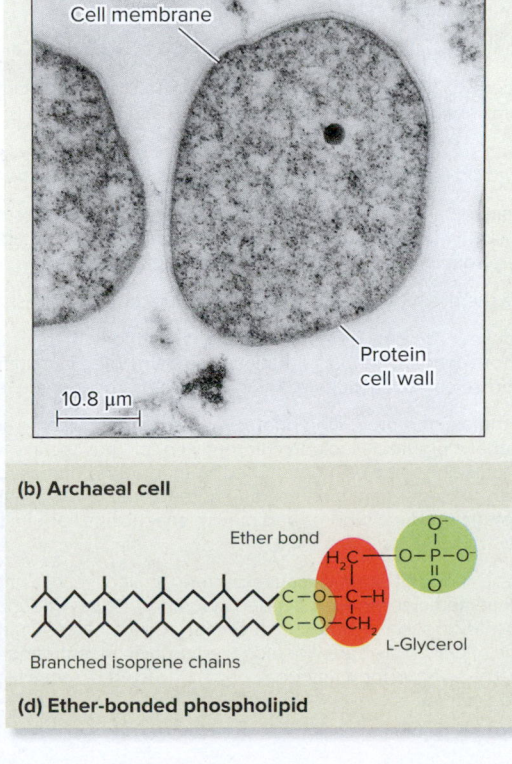

(b) Archaeal cell

(d) Ether-bonded phospholipid

**Figure 27.2  Bacteria and Archaea. (a)** Bacteria and **(b)** archaea both have prokaryotic cell structure, but **(c)** bacterial membrane lipids are formed with ester linkages, whereas **(d)** archaeal membrane lipids feature ether linkages, which are more stable under extreme environmental conditions. As shown in the transmission electron microscopic (TEM) images in (a) and (b), most bacteria feature a cell wall made of a material known as peptidoglycan and are often enclosed by an outer envelope, whereas archaea lack these features. Most archaea have outer coverings made of protein. a: ©Linda Graham; b: ©Eye of Science/Science Source

Some archaea prefer habitats having both high temperatures and extremely low pH. For example, the archaeal genus *Sulfolobus* was discovered in samples taken from sulfur hot springs having a pH of 3 or lower. Archaea help biologists better understand the origin of life, the origin of eukaryotes, how life on Earth has evolved in extreme environments, and what kinds of extraterrestrial life might exist.

The domain Archaea includes several phyla, including Lokiarchaeota, Korarchaeota, Thaumarchaeota, Crenarchaeota, and Euryarchaeota (see Figure 27.1). Lokiarchaeota and close relatives named for Norse deities are collectively known as the Asgard superphylum and are particularly closely related to eukaryotes (Eukarya). Members of Korarchaeota are primarily known from hot springs. Thaumarchaeota species are widespread in terrestrial and aquatic environments, and include archaea that oxidize ammonia, making them important in global nitrogen cycling. Crenarchaeota includes organisms that live in extremely hot or cold habitats and also some that are widespread in aquatic and terrestrial habitats. Early-diverging Euryarchaeota includes some hyperthermophiles, diverse methane producers, and extreme halophiles—species able to grow in higher than usual salt concentrations.

## Domain Bacteria Includes Cyanobacteria, Proteobacteria, and Many Other Phyla

Domain Bacteria is considerably more diverse than Archaea. Molecular studies suggest the existence of 50 or more bacterial phyla, though many are poorly characterized. Though some members of domain Bacteria live in extreme environments, most favor moderate conditions. Many bacteria form symbiotic associations with eukaryotes and are thus of concern in medicine and agriculture.

The characteristics of 10 prominent bacterial phyla are briefly summarized in **Table 27.1**. Bacterial phyla mentioned later in this chapter because of their medical, ecological, or evolutionary significance include Firmicutes, Bacteroidetes, Chlamydiae, Planctomycetes, Spirochaetes, Actinobacteria, Chloroflexi, and Deinococcus-Thermus. Because Cyanobacteria and Proteobacteria are particularly diverse and relevant to eukaryotic cell evolution, global ecology, and human affairs, we will consider them in greater detail next.

***Cyanobacteria*** The phylum Cyanobacteria contains photosynthetic bacteria that are abundant in fresh waters, oceans, and wetlands and on the surfaces of arid soils. Cyanobacteria are named for the typical blue-green (cyan) coloration of their cells. Blue-green pigmentation results from the presence of photosynthetic pigments called phycobilins that help chlorophyll absorb light energy. Cyanobacteria are the only bacteria known to generate oxygen as a product of photosynthesis. Ancient cyanobacteria produced Earth's first oxygen-rich atmosphere, which allowed the eventual rise of eukaryotes. The chloroplasts of eukaryotic algae and plants were derived from cyanobacteria.

Cyanobacteria display the greatest diversity in body type found among bacterial phyla (**Figure 27.3**). Some occur as single cells called unicells (Figure 27.3a); others form colonies of cells held

| Table 27.1 | Representative Bacterial Phyla |
| --- | --- |
| **Phyla** | **Characteristics** |
| Firmicutes | Diverse Gram-positive bacteria, some of which produce endospores. The disease-causing *Clostridium difficile* is an example. |
| Bacteroidetes | Includes representatives with diverse metabolic processes; some are common in the human intestinal tract, and others are primarily aquatic. |
| Chlamydiae | Notably tiny, obligate intracellular parasites. Some cause eye disease in newborns or sexually transmitted diseases. |
| Planctomycetes | Reproduce by budding rather than binary fission; cell walls lack peptidoglycan; cytoplasm may contain nucleus-like bodies, and endocytosis may occur. |
| Spirochaetes | Motile bacteria having distinctive corkscrew shapes, with flagella held close to the body. They include the pathogens *Treponema pallidum*, the agent of syphilis, and *Borrelia burgdorferi*, which causes Lyme disease. |
| Actinobacteria | Gram-positive bacteria producing branched filaments; many form spores. *Mycobacterium tuberculosis*, the agent of tuberculosis in humans, is an example. Actinobacteria are notable antibiotic producers; over 500 different antibiotics are known from this group. Some fix nitrogen in association with plants. |
| Chloroflexi | Known as the green nonsulfur bacteria; conduct photosynthesis without releasing oxygen (anoxygenic photosynthesis). |
| *Deinococcus-Thermus* | Extremophiles. The genus *Deinococcus* is known for high resistance to ionizing radiation, and the genus *Thermus* inhabits hot springs. *Thermus* aquaticus has been used in commercial production of *Taq* polymerase, an enzyme used in polymerase chain reaction (PCR), an important procedure in molecular biology laboratories. |
| Cyanobacteria | Includes the oxygen-producing photosynthetic bacteria (some are also capable of anoxygenic photosynthesis). Photosynthetic pigments include chlorophyll *a* and phycobilins, which often give cells a blue-green pigmentation. Occur as unicells, colonies, unbranched filaments, and branched filaments. Many of the filamentous species produce specialized cells: dormant akinetes and heterocytes in which nitrogen fixation occurs. In waters having excess nutrients, cyanobacteria produce blooms and may release toxins harmful to the health of humans and wild and domesticated animals. |
| Proteobacteria | A very large group of Gram-negative bacteria, collectively having high metabolic diversity. Includes many species important in medicine, agriculture, and industry such as *Agrobacterium tumifaciens*, *Escherichia coli*, and *Haemophilus influenzae*. *Myxococcus xanthus* is a Gram-negative bacterium that is able to glide across surfaces, forming swarms of thousands of cells. This behavior aids feeding by concentrating digestive enzymes secreted by the bacteria. When food is scarce, the swarms form tiny tree-shaped structures from which tough spores disperse. By this means, cells move to new, food-rich places. |

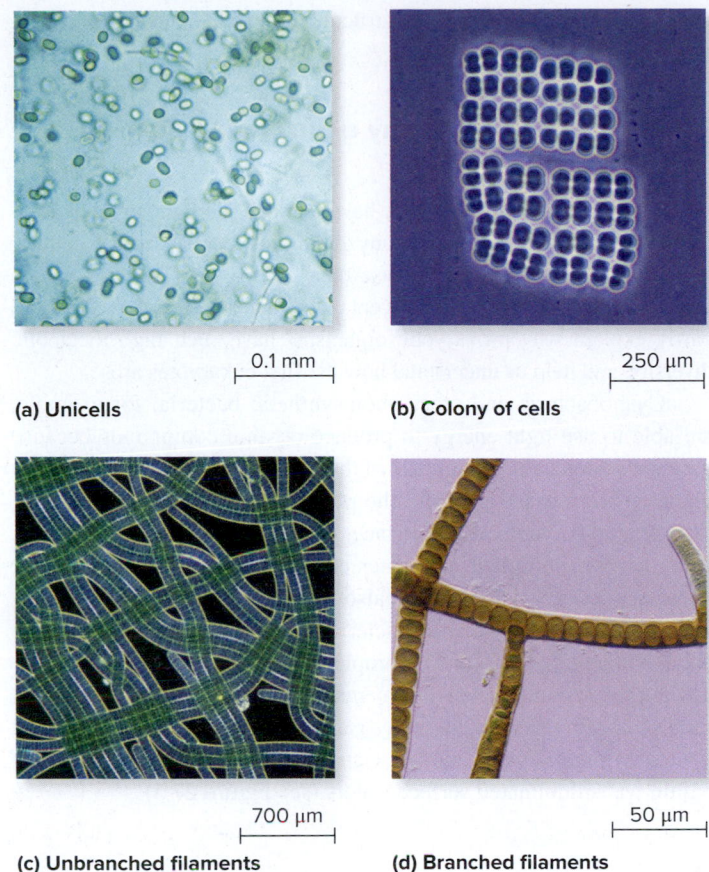

(a) Unicells    0.1 mm

(b) Colony of cells    250 μm

(c) Unbranched filaments    700 μm

(d) Branched filaments    50 μm

**Figure 27.3  Major body types found in the phylum Cyanobacteria. (a)** The genus *Chroococcus* occurs as unicells. **(b)** The genus *Merismopedia* forms a flat colony of cells held together by mucilage. **(c)** The genus *Oscillatoria* forms an unbranched filament. **(d)** The genus *Stigonema* forms a branched filament having a mucilage sheath; sunscreen compounds that protect the cells from damage by ultraviolet (UV) radiation cause the brown color of the sheath. a: ©Linda Graham; b: ©Michael Abbey/Science Source; c: ©Sinclair Stammers/SPL/Science Source; d: ©Lee W. Wilcox

together by a thick gluey substance called mucilage (Figure 27.3b), and many cyanobacteria form filaments in which cells are attached end-to-end (Figure 27.3c) or filaments that branch (Figure 27.3d). Some of the filamentous cyanobacteria display hallmarks of multicellularity: cellular attachment, specialized cells, intercellular chemical communication, and programmed cell death.

***Proteobacteria***  Though Proteobacteria share molecular and cell-wall features, this phylum displays amazing diversity of form and metabolism. Genera of this phylum are classified into five major subgroups: alpha (α), beta (β), gamma (γ), delta (δ), and epsilon (ε). As we saw in Chapter 26 (refer back to Figure 26.7), the ancestry of mitochondria can be traced to the α-proteobacteria, which also include several genera noted for mutually beneficial relationships with animals and plants. For example, *Rhizobium* and related genera of α-proteobacteria form nutritionally beneficial associations with the roots of legume plants such as beans and peas and are thus agriculturally important. Another α-proteobacterium, *Agrobacterium*

**Figure 27.4  *Agrobacterium tumifaciens* infection.** This proteobacterium causes cancer-like tumors to grow on plants (see the arrows). ©Linda Graham

*tumifaciens*, causes destructive cancer-like tumors called galls to develop on susceptible plants, including grapes and ornamental crops (**Figure 27.4**). *A. tumifaciens* induces gall formation by injecting DNA into plant cells. This property has allowed researchers to use the bacterium in the production of transgenic plants, which are plants that carry genes from another species.

The genus *Nitrosomonas*, soil inhabitants important in the global nitrogen cycle, represents the β-proteobacteria. *Neisseria gonorrhoeae*, the agent of the sexually transmitted disease gonorrhea, is one of the γ-proteobacteria. *Vibrio cholerae*, another γ-proteobacterium, causes cholera epidemics when drinking water becomes contaminated with animal waste during floods and other natural disasters. The γ-proteobacteria *Salmonella enterica* and *Escherichia coli* strain O157:H7 also cause human disease, so food and water are widely tested for their presence. The δ-proteobacteria include the colony-forming myxobacteria and predatory bdellovibrios, which drill through the cell walls of other bacteria in order to consume them. *Helicobacter pylori*, which causes stomach ulcers, belongs to the ε-proteobacteria.

## Horizontal Gene Transfer Influences the Diversity and Evolution of Archaea and Bacteria

As we have seen, Bacteria and Archaea are domains of life displaying an amazing level of diversity. One reason for this diversity is **horizontal gene transfer**, the process in which an organism receives genetic material from another organism without being the offspring of that organism. Horizontal gene transfer is common among archaea and bacteria, occurring most frequently between species that are closely related or that live in close proximity. Evolutionary change due to horizontal gene transfer contrasts with **vertical evolution**, in which gene transfer occurs from parent to progeny. During vertical evolution, genetic changes occur in a series of ancestors that form a lineage; species evolve from pre-existing species by the accumulation of mutations.

Horizontal gene transfer can result in large genetic changes that confer new metabolic capacities. For example, at least 17% of the genes present in the common human gut inhabitant *E. coli* were horizontally transferred from other bacteria. Studies of nearly 200 genomes have revealed that about 80% of prokaryotic genes have been involved in horizontal transfer at some point in their history. Genes also move among the bacterial, archaeal, and eukaryotic domains. For example, salt-tolerant (halophytic) archaea originated after an ancient horizontal transfer of more than a thousand genes from bacteria.

Horizontal gene transfer can occur between different bacterial species via transduction, transformation, and conjugation, as discussed in Chapter 19. In addition, horizontal gene transfer occurs by means of endosymbiosis, the process in which one species—the endosymbiont—lives in the body or cells of another species—the host. For example, certain γ-proteobacteria occupy the cells of β-proteobacterial hosts, which themselves live within insect cells. Such close proximity increases the odds that gene exchange will occur between distantly related species. During the process by which the mitochondria of eukaryotic cells originated from α-proteobacteria and plastids originated from cyanobacteria (see Chapter 26), so many bacterial genes were transferred to host nuclei that modern mitochondria and chloroplasts cannot reproduce outside the host cell.

## 27.2 Structure and Movement

### Learning Outcomes:

1. Discuss the structural adaptations that have increased the complexity of prokaryotic cells.
2. Explain how mucilage influences the behavior of bacterial cells.
3. **CoreSKILL** » Make a drawing that shows the structural differences between Gram-positive and Gram-negative bacterial cells, and predict how these features might influence disease treatments.
4. List the different means by which prokaryotic cells can move.

Bulbnose unicorn fish (*Naso tonganus*) living in Australian coastal ocean waters contain cigar-shaped bacterial symbionts (*Epulopiscium*) whose cells are more than 600 μm long, larger than most eukaryotic cells (whose largest dimension is between 10 and 100 μm). Spherical cells of the bacterial species *Thiomargarita namibiensis*, which lives in African coastal regions, likewise reach record-setting sizes, some being 800 μm in diameter and large enough to be seen without a microscope. However, most bacteria (and archaea) are much smaller: a few micrometers in diameter. Small cell size limits the amount of materials that can be stored within each cell but allows much faster cell division. When nutrients are sufficient, many bacteria can divide many times within a single day. This explains how bacteria can spoil food rapidly and why bacterial infections can spread quickly within the human body. Despite their generally small size, bacteria display

a high level of variation in cell structure and shape, surface and cell-wall features, and movement.

## Prokaryotic Cells Display a Surprising Degree of Complexity

Although bacteria, like archaea, have a much simpler cellular organization than do eukaryotes, many prokaryotic cells display cellular structural adaptations that increase their complexity. Features of this complexity illustrate the core concept that structure determines function, partly explain why prokaryotic organisms have such high metabolic diversity, and help us understand how the first eukaryotes arose.

Cyanobacteria and other photosynthetic bacteria, for example, are able to use light energy to produce organic compounds because their cells contain large numbers of thylakoids, flattened tubular membranes that grow inward from the plasma membrane (**Figure 27.5**). The extensive membrane surface of the thylakoids bears large amounts of chlorophyll and other components that are needed for photosynthesis. Thylakoids are also abundant in plant chloroplasts, which descended from cyanobacterial ancestors. Thylakoids enable photosynthetic bacteria and chloroplasts to take maximum advantage of light energy in their environments. Aquatic photosynthetic bacteria also commonly contain many gas vesicles. These protein-walled structures increase cell buoyancy and thus help the organisms float within well-illuminated surface waters (see Figure 27.5).

Thylakoids provide a greater surface area for chlorophyll and other molecules involved in photosynthesis.

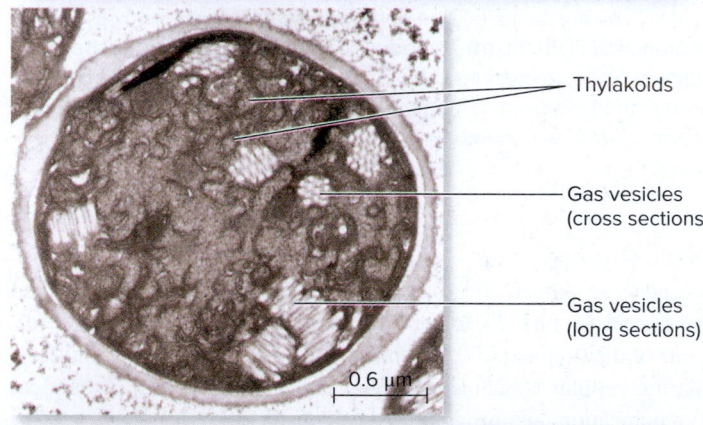

The gas vesicles buoy this photosynthetic organism to the lighted water surface, where it often forms conspicuous scums.

**Figure 27.5** **Photosynthetic thylakoid membranes and numerous gas vesicles in a cell of an aquatic cyanobacterium.**
©Norma Lang

**Core Concept: Structure and Function** Thylakoids, which contain chlorophyll and other components that are needed for photosynthesis, are the locations of the light-harvesting reactions. Gas vesicles allow photosynthetic cells to float in well-lit surface waters.

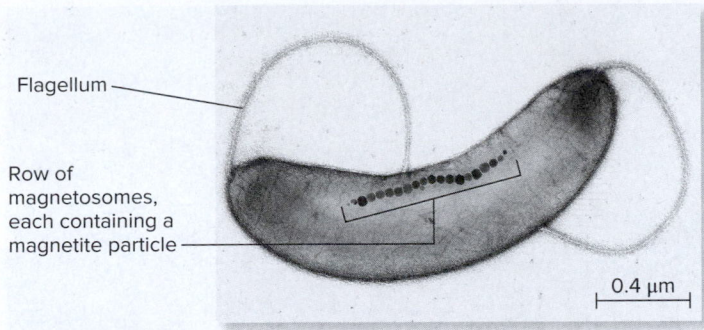

Flagellum

Row of
magnetosomes,
each containing a
magnetite particle

0.4 µm

**Figure 27.6** **Magnetosomes found in *Magnetospirillum magnetotacticum*.** An internal row of iron-rich magnetite crystals, each enclosed by a membrane derived from the plasma membrane, functions like a compass needle, allowing this bacterium to detect the Earth's magnetic field. This feature allows *M. magnetotacticum* to orient itself in space and thereby locate its preferred habitat, low-oxygen subsurface waters. These and other bacterial cells use flagella to move to more favorable locations. ©Dr. Richard P. Blakemore, University of New Hampshire

 **Core Skill: Connections** Look ahead to Section 44.4, which describes electromagnetic sensing by animals. What animals are like *M. magnetotacticum* in being able to sense and respond to magnetic fields?

In other bacteria, plasma membrane ingrowth has generated additional intriguing adaptations—magnetosomes and nucleus-like bodies—that are sometimes described as bacterial organelles. Magnetosomes are tiny crystals of an iron mineral known as magnetite, each surrounded by a membrane. These structures occur in the bacterium *Magnetospirillum* and related genera (**Figure 27.6**). In each of these bacteria, about 15 to 20 magnetosomes occur in a row, together acting as a compass needle that responds to the Earth's magnetic field. Magnetosomes help the bacteria to orient themselves in space and

thereby locate the submerged, low-oxygen habitats they prefer. Magnetosome development begins with invagination of the plasma membrane to form a row of spherical vesicles. If *Magnetospirillum* cells are grown in media having low iron levels, the vesicles remain empty. But if iron is available, a magnetite crystal forms within each vesicle. Fibrils of an actin-like protein keep the magnetosomes aligned in a row. (Recall from Chapter 4 that actin is a major cytoskeletal protein of eukaryotes.) Mutant bacteria lacking a functional form of this protein produce magnetosomes, but they do not remain aligned in a row. Instead, magnetosomes scatter around mutant cells, disrupting their ability to detect a magnetic field.

Plasma membrane invaginations produce nucleus-like bodies in *Gemmata obscuriglobus* and other members of the bacterial phylum Planctomycetes. In *G. obscuriglobus*, an envelope composed of a double membrane encloses all cellular DNA and some ribosomes. Although this bacterial envelope lacks the nuclear pores characteristic of the eukaryotic nuclear envelope, it likely plays a similar adaptive role in isolating DNA from other cellular influences. *G. obscuriglobus* and related bacterial species are also known to accomplish endocytosis by means of membrane coat proteins similar to those present in eukaryotic cells. The cellular diversity and surprising complexity of bacterial cell structure help us to understand not only how bacteria function in nature, but also how important features of eukaryotic cells first evolved.

## Prokaryotic Cells Vary in Shape

Although prokaryotic cells occur in multiple forms, they have five common shapes (**Figure 27.7**): spheres (**cocci**), elongate rods (**bacilli**), comma-shaped cells (**vibrios**), and spiral-shaped cells that are either flexible (**spirochaetes**) or rigid (**spirilli**; see Figure 27.6). Cytoskeletal proteins similar to those present in eukaryotic cells control these cell shapes. For example, helical strands of an actin-like protein are responsible for the rod shape of bacilli; if this protein is not produced, bacilli become spherical in shape. Cellular shape

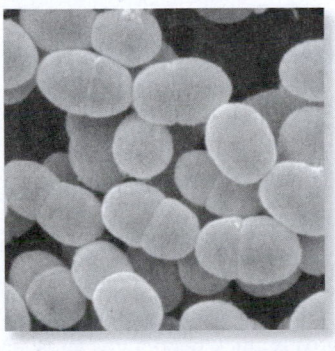

1 µm

**(a) Sphere-shaped cocci
(*Lactococcus lactis*)**

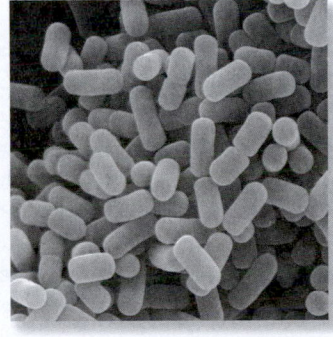

11.4 µm

**(b) Rod-shaped bacilli
(*Lactobacillus plantarum*)**

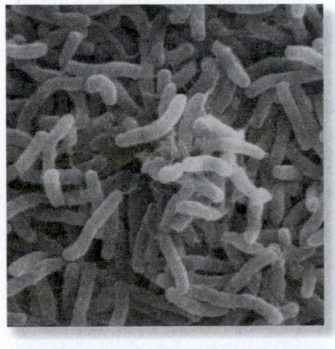

2.5 µm

**(c) Comma-shaped vibrios
(*Vibrio cholerae*)**

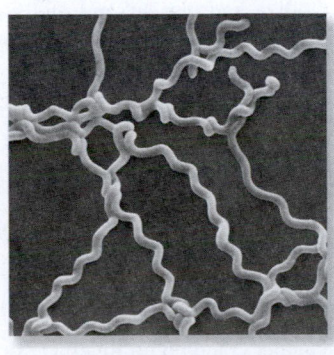

7.5 µm

**(d) Spiral-shaped spirochaetes
(*Leptospira* sp.)**

**Figure 27.7** **Major types of prokaryotic cell shapes.** These images are scanning electron micrographs. a: ©SciMAT/Science Source; b, d: ©Dennis Kunkel Microscopy, Inc./Phototake; c: ©Media for Medical/UIG/Getty Images

 **Core Concept: Systems** In the case of unicellular bacteria and archaea, a single cell constitutes an entire organism.

is an important component of bacterial function in nature. Cocci tend to have a greater surface area/volume ratio, which facilitates exchange of materials with the environment, but bacilli can often store more nutrients.

## Slimy Mucilage Often Coats Cellular Surfaces

Many bacteria exude a coat of slimy mucilage, sometimes called a glycocalyx, capsule, or extracellular polymeric substance (EPS). Mucilage, which varies in consistency and thickness, is composed of hydrated polysaccharides and proteins, as well as lipids and nucleic acids. A capsule helps some disease-causing bacteria evade the defense system of their host. You may recall that Frederick Griffith discovered the transfer of genetic material while experimenting with capsule-producing pathogenic strains and capsule-less nonpathogenic strains of the bacterium *Streptococcus pneumoniae* (refer back to Figure 11.1). The immune system cells of mice are able to destroy this bacterium only if it lacks a capsule.

Mucilage plays many additional roles: holding cells together closely enough for chemical communication and DNA exchange to occur, helping aquatic species to float in water, binding mineral nutrients, and repelling attack. Pigmented slime sheaths (see Figure 27.3d) coat some bacterial filaments, helping to prevent UV damage.

**Biofilms** are aggregations of microorganisms that secrete adhesive mucilage, thereby gluing themselves to surfaces. Formation of a biofilm helps microbes remain in favorable locations for growth; otherwise body or environmental fluids would wash them away. A mechanism known as **quorum sensing** fosters biofilm formation. During quorum sensing, individual microbes secrete small molecules having the potential to influence the behavior of nearby microbes. If enough individuals are present (a quorum), the concentration of signaling molecules builds to a level that causes collective behavior. In the case of biofilms, populations of microbes respond to chemical signals by moving to a common location and producing mucilage.

Biofilms are environmentally and medically important. From a human standpoint, biofilms have both beneficial and harmful consequences. In aquatic and terrestrial environments, biofilms help to stabilize and enrich sand and soil surfaces, and help form mineral deposits. Biofilms that form on the surfaces of animal tissues, however, can be harmful. Dental plaque is an example of a harmful biofilm (**Figure 27.8**); if allowed to remain, the bacterial community secretes acids that can damage tooth enamel. Biofilms may also develop in industrial pipelines, where the attached microbes can contribute to corrosion by secreting enzymes that chemically degrade metal surfaces.

## Prokaryotic Cells Vary in Cell-Wall Structure

Whether coated with mucilage or not, most prokaryotic cells possess a rigid cell wall outside the plasma membrane. Cell walls maintain cell shape and help protect against attack by viruses or predatory bacteria. Cell walls also help microbes avoid lysing in hypotonic conditions, when the solute concentration is higher inside the cell than outside. The structure and composition of bacterial cell walls are medically important.

Although some archaea lack cell walls, most possess a wall composed of protein. In contrast, the polymer known as **peptidoglycan**,

**Figure 27.8** **A biofilm composed of a community of microorganisms glued by mucilage to a surface.** This SEM shows a view of the top surface of dental plaque, consisting of several types of bacteria—falsely colored purple, green, and blue—attached by mucilage to a tooth surface lying beneath. ©Science Photo Library/Alamy Stock Photo

 **Core Skill: Modeling**  The goal of this modeling challenge is to make a model for the development of a biofilm of dental plaque. Such models are proving useful in finding new ways to reduce or prevent oral disease.

**Modeling Challenge:** Using the SEM of dental plaque shown in Figure 27.8, which illustrates the relative positions and abundances of three types of bacteria, draw a flow diagram showing several sequential stages that hypothetically model the process by which this biofilm might have developed. Your model should indicate which bacteria are most likely to have attached first (purple, green, or blue) and which most recently.

lacking in archaea, is an important component of most bacterial cell walls. Peptidoglycan is composed of carbohydrates that are cross-linked by peptides. Bacterial cell walls occur in two major forms that differ in thickness of the peptidoglycan layer, staining properties, and response to antibiotics. Bacteria having these chemically different walls are called Gram-positive or Gram-negative bacteria, after the staining process used to distinguish them (**Figure 27.9**). The stain is named for its inventor, Danish scientist Hans Christian Gram.

Gram-positive bacteria classified in the phyla Firmicutes and Actinobacteria have walls with a relatively thick peptidoglycan layer (**Figure 27.10a**). By contrast, the Gram-negative cell walls of Cyanobacteria, Proteobacteria, and other species have a thinner peptidoglycan layer and are enclosed by a thin, outer envelope whose outer leaflet is rich in **lipopolysaccharides**, which are lipids that have polysaccharides covalently attached to them (**Figure 27.10b**; see also Figure 27.2a). This outer envelope of Gram-negative bacteria contains a phospholipid bilayer that surrounds the outside of the cell wall, whereas the plasma membrane is found inside the cell wall.

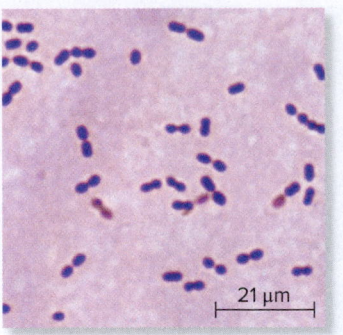

(a) Gram-positive bacteria

(b) Gram-negative bacteria

**Figure 27.9** **Gram-positive and Gram-negative bacteria.**
**(a)** *Streptococcus pneumoniae*, a member of the phylum Firmicutes, stains positive (purple) with the Gram stain. **(b)** *Escherichia coli*, a member of the Proteobacteria, stains negative (pink) when the Gram stain procedure is applied. a: ©CNRI/Science Source; b: ©Lee W. Wilcox

Peptidoglycan and lipopolysaccharides can affect disease symptoms, the composition of vaccines, and bacterial responses to antibiotics. For example, part of the peptidoglycan covering of the Gram-negative bacterial species *Bordetella pertussis* is responsible for the extensive tissue damage to the respiratory tract associated with whooping cough, and whooping cough vaccines are improved by including antibodies that reduce the ability of the lipopolysaccharide layer to attach to host cells.

The lipopolysaccharide-rich outer envelope of Gram-negative bacteria helps them to resist the entry of some antibiotics. However, this outer envelope also impedes the secretion of proteins from bacterial cells into the environment, a process that normally allows cells to communicate with each other, as in quorum sensing. Gram-negative bacteria have adapted to the presence of an outer envelope by evolving several types of protein systems that function in secretion. In Section 27.5, we will see how some of these secretion systems have been modified in ways that allow disease-causing bacteria to attack eukaryotic cells.

Distinguishing Gram-positive from Gram-negative bacteria is an important factor in choosing the best antibiotics for treating infectious diseases. For example, Gram-positive bacteria are typically more susceptible than Gram-negative bacteria to penicillin and related antibiotics because these antibiotics interfere with synthesis of peptidoglycan, which Gram-positive bacteria require in larger amounts. For this reason, penicillin and related antibiotics such as methicillin are widely used to treat infections caused by Gram-positive bacteria. However, it is of societal concern that some strains of bacteria have become resistant to some antibiotics, an example being methicillin-resistant *Staphylococcus aureus*, or MRSA.

## Bacteria and Archaea Display Diverse Types of Movements

Many bacteria and archaea have structures at the cell surface or within the cell that enable them to change position in their environment, an ability known as **motility**. Diverse motility adaptations allow microbes to respond to chemical signals emitted from other

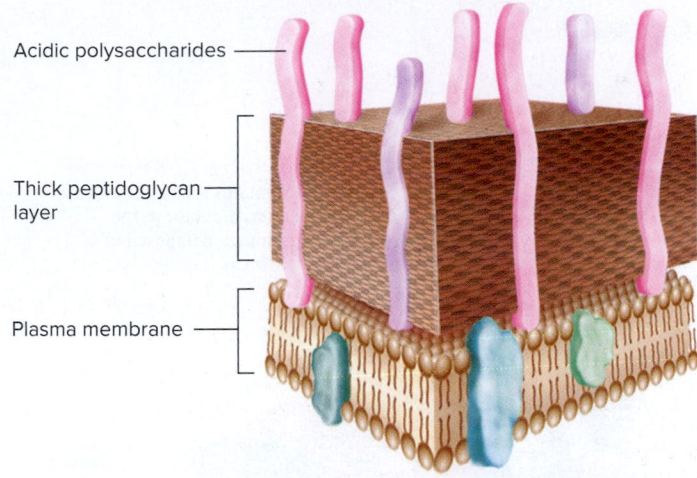

(a) Gram-positive: thick peptidoglycan layer, no outer envelope

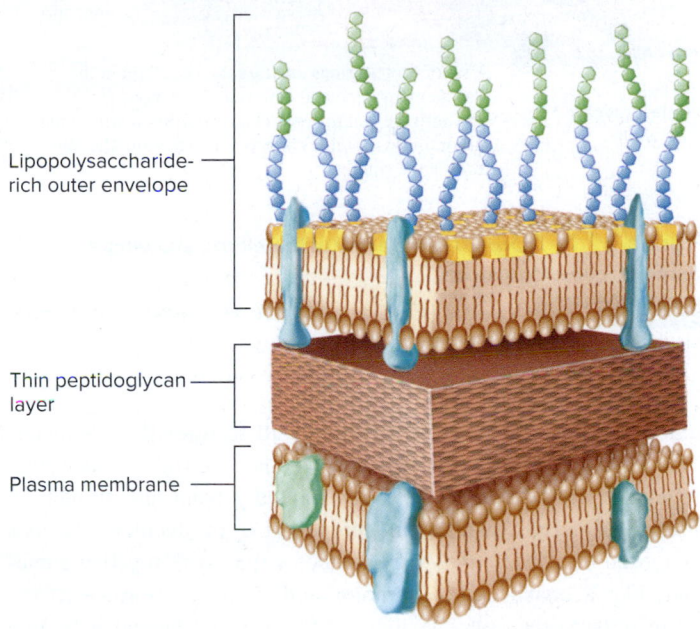

(b) Gram-negative: thinner peptidoglycan layer, with outer envelope

**Figure 27.10** **Cell-wall structures of Gram-positive and Gram-negative bacteria.** **(a)** The structure of the cell wall of Gram-positive bacteria. **(b)** The structure of the cell wall and lipopolysaccharide-rich envelope typical of Gram-negative bacteria.

cells during quorum sensing and mating, and to move to favorable conditions within gradients of light, gases, or nutrients. For example, we have already learned that gas vesicles help cyanobacteria float into well-illuminated waters close to the surface, where photosynthesis can occur (see Figure 27.5). In addition, prokaryotic cells may move by twitching, gliding, or swimming by means of flagella.

Bacterial **flagella** (singular, flagellum) differ from eukaryotic flagella in several ways. Although bacterial flagella are largely built of about 30 types of proteins, they lack a plasma membrane covering, an internal cytoskeleton of microtubules made of the protein

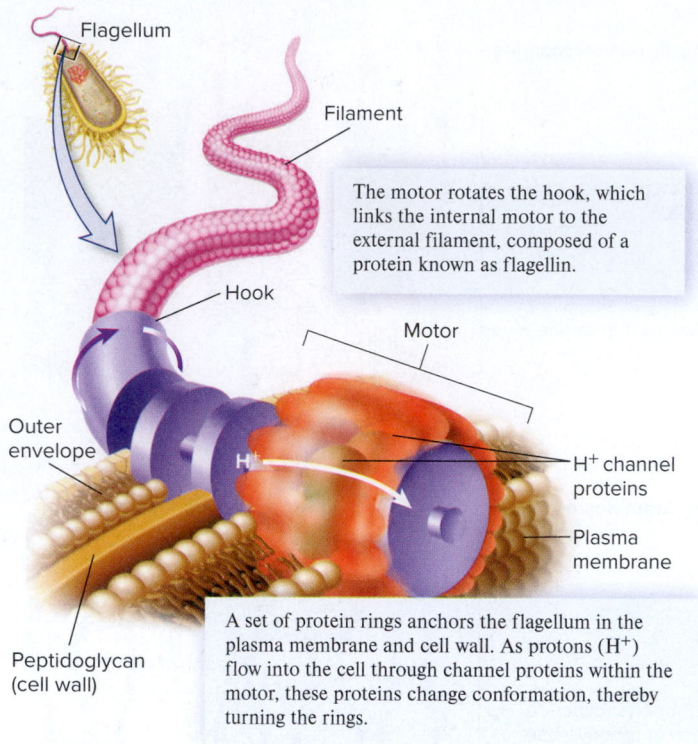

Flagellum

Filament

The motor rotates the hook, which links the internal motor to the external filament, composed of a protein known as flagellin.

Hook

Motor

Outer envelope

H⁺ channel proteins

Plasma membrane

Peptidoglycan (cell wall)

A set of protein rings anchors the flagellum in the plasma membrane and cell wall. As protons (H⁺) flow into the cell through channel proteins within the motor, these proteins change conformation, thereby turning the rings.

**Figure 27.11** Diagram of a bacterial flagellum, showing a filament, hook, and motor.

*Concept Check:* Does the filament move more like the arms of a human swimmer or the shaft of a boat propeller?

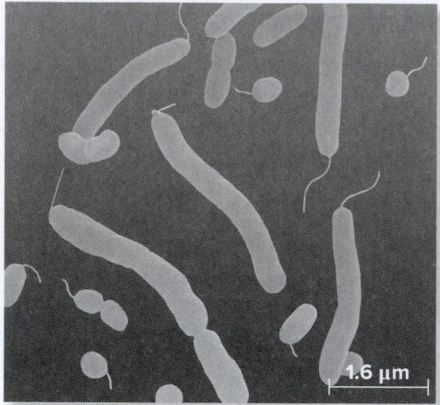

1.6 μm

**(a) Bacteria with a single short flagellum**

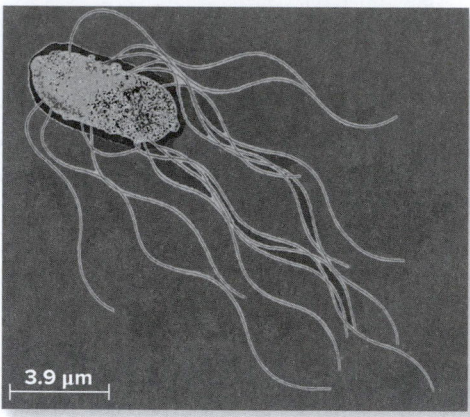

3.9 μm

**(b) Bacterium with multiple long flagella**

**Figure 27.12 Differences in the number and location of flagella.** Depending on the species, microbial cells can produce one or more flagella at the poles or numerous flagella around the periphery. **(a)** *Vibrio parahaemoliticus*, a bacterium that causes seafood poisoning, has a single short flagellum. **(b)** *Salmonella enterica*, another bacterium that causes food poisoning, has many flagella distributed around the cell periphery. a: ©Dennis Kunkel Microscopy, Inc./ Phototake; b: ©Dr. Linda Stannard, UCT/SPL/Science Source

 **Core Skill: Connections** Look ahead to Figure 29.8b. Like the bacteria shown here, what heterotrophic eukaryote moves to its food source in the human gut by means of flagella?

tubulin, and the motor protein dynein—all features that characterize eukaryotic flagella (look back to Chapter 4). Unlike eukaryotic flagella, bacterial flagella do not repeatedly bend and straighten. Instead, bacterial flagella spin, propelled by molecular machines composed of a filament, hook, and motor that work together somewhat like a boat's outboard motor and propeller (**Figure 27.11**). Lying outside the cell, the long, stiff, curved filament acts as a propeller. The hook links the filament with the motor that contains a set of protein rings at the cell surface. Hydrogen ions (protons), which have been pumped out of the cytoplasm, usually via an electron transport chain, diffuse back into the cell through channel proteins within the motor. This proton flow powers the turning of the hook and filament at rates of hundreds of revolutions per second. Archaeal flagella also rotate but are much thinner than bacterial flagella, are composed of different proteins, and are powered differently (by the hydrolysis of ATP).

Prokaryotic species differ in the number and location of flagella, which may occur singly or in tufts at one pole or may emerge from around the cell (**Figure 27.12**). Differences in flagellar number and location cause microorganisms to exhibit different modes or rates of swimming. Some bacterial species are known to swim at rates of more than 150 μm per second! By contrast, spirochaetes tend to move slowly. Their flagella are located outside the peptidoglycan cell wall but within the confines of an outer membrane that holds them close to the cell. Rotation of these flagella causes spirochaetes to display characteristic bending, flexing, and twirling

motions. This allows spirochaetes to move within the thick bodily fluids of their hosts.

Some prokaryotic species twitch or glide across surfaces, using threadlike cell surface structures known as **pili** (singular, pilus) (**Figure 27.13a**). *Myxococcus xanthus* cells, for example, move by alternately extending and retracting pili from one pole or the other. This process allows directional movement toward food materials. If nutrients are low, cells of these bacteria glide together to form tiny treelike colonies, which are part of a reproductive process **Figure 27.13b**. These and other motility adaptations help bacteria and archaea survive in their environments.

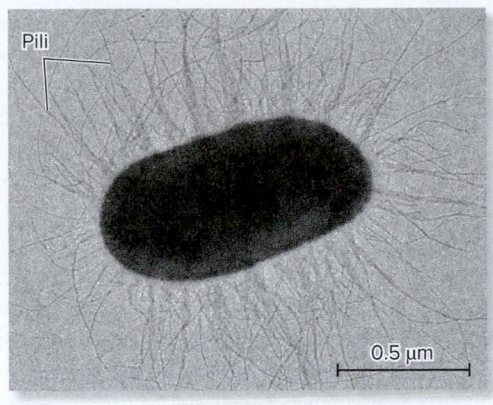

Pili

0.5 μm

(a)

(b)

**Figure 27.13** Pili that extend from prokaryotic cell surfaces may allow motility and sometimes foster the formation of complex colonies.
a: Courtesy Dr. Esther Bullitt; b: ©Yoav Levy/Medical Images.com

## 27.3 Reproduction

**Learning Outcomes:**

1. Explain how populations of prokaryotic organisms increase in number.
2. **CoreSKILL »** Describe how bacteria can be counted in medical and environmental samples.
3. Give examples of how some bacteria survive under stressful conditions.

Bacteria and archaea do not engage in the process of sexual reproduction used by eukaryotes, involving specialized gametes, gamete fusion, and meiosis. However, they can exchange some genes by conjugation, transformation, and transduction (described in Chapter 19). Bacteria and archaea usually reproduce asexually, generally by means of a type of cell division known as binary fission, but sometimes by forming small cells, known as buds, from one end. Both types of bacterial cell division increase the number of cells in populations. In addition, some bacteria produce tough cells that can withstand deleterious conditions for long periods in a dormant condition.

### Prokaryotic Cells Generally Divide by Binary Fission

The cells of most bacteria and archaea divide by splitting in two, a process known as **binary fission** (**Figure 27.14a**; refer back to Figure 19.13). When sufficient nutrients are available, an entire population of identical cells can be produced from a single parental cell by repeated binary fission. This growth process allows microbes to become very numerous in water, food, or animal tissues, potentially causing harm.

Binary fission is the basis of a widely used method for detecting and counting bacteria in food, water samples, or patients' body fluids. Microbiologists who study the spread of disease need to quantify bacterial cells in samples taken from the environment. Medical technicians often need to count bacteria in body fluid samples to assess the likelihood of infection. However, because bacterial cells are small and often unpigmented, they are difficult to count directly. One way that microbiologists count bacteria is to place a measured volume of sample into laboratory dishes filled with a semisolid nutrient medium. Bacteria in the sample undergo repeated binary fission to form colonies of cells visible to the unaided eye (**Figure 27.14b**). Because each colony represents a single cell that was present in the original sample,

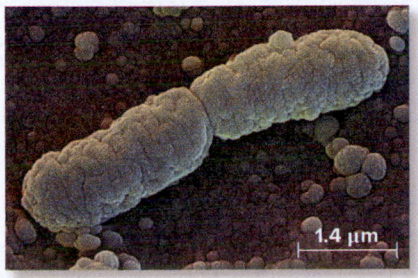

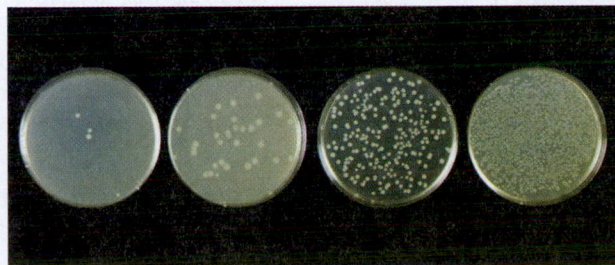

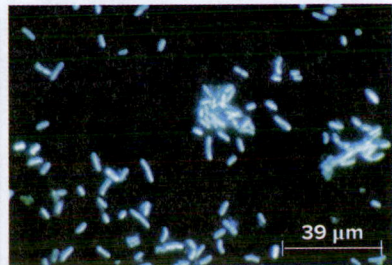

1.4 μm

39 μm

(a) Bacterium undergoing binary fission

(b) Colonies developed from single cells

(c) Bacteria stained with fluorescent DNA-binding dye

**Figure 27.14** **Binary fission and counting microbes.** **(a)** Division of a bacterial cell as viewed by scanning electron microscopy. **(b)** When samples are spread onto the surfaces of laboratory dishes containing nutrients, single cells of bacteria or archaea may divide repeatedly to form visible colonies, which can be easily counted. The number of colonies is an estimate of the number of culturable cells in the original sample. **(c)** If a fluorescence microscope is available, cells can be counted directly by applying a fluorescent stain that binds to their DNA. Each cell glows brightly when illuminated with UV light. a: ©David Scharf/Science Source; b: ©Linda Graham; c: ©Lee W. Wilcox

the number of colonies in the dish reflects the number of living bacteria in the original sample.

Another way to detect and count prokaryotic cells is to treat samples with a stain that binds to bacterial DNA, causing cells to glow brightly when illuminated with UV light. The glowing cells can be viewed and counted by the use of a fluorescence microscope (**Figure 27.14c**). The fluorescence method must be used when the microbes of interest cannot be cultivated in the laboratory. For many bacteria and archaea, the conditions needed to foster population growth in the laboratory are not known.

## Some Bacteria Survive Harsh Conditions as Akinetes or Endospores

Some bacteria produce thick-walled cells that are able to survive unfavorable conditions in a dormant state. These specialized cells develop when bacteria have experienced stressful conditions, such as low nutrients or unfavorable temperatures. Such dormant cells may be able to germinate into metabolically active cells when conditions improve again. For example, aquatic filamentous cyanobacteria often produce **akinetes**, thick-walled, food-filled cells, when winter approaches (**Figure 27.15a**). Akinetes are able to survive winter at the bottoms of lakes, and they produce new filaments in spring when they are carried by water currents to the brightly lit surface. Persistence of such akinetes explains how harmful cyanobacterial blooms can develop year after year in overly fertile lakes.

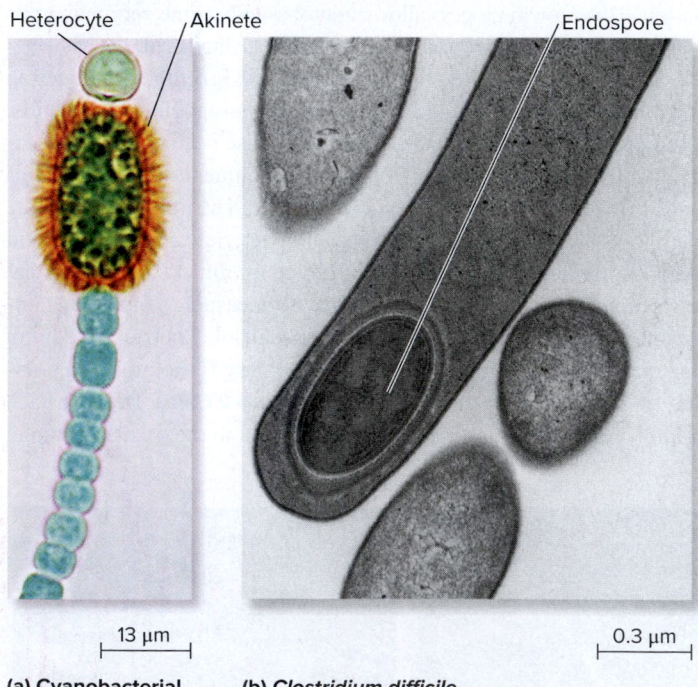

Heterocyte  Akinete  Endospore

13 μm    0.3 μm

(a) Cyanobacterial akinete

(b) *Clostridium difficile*

**Figure 27.15** **Specialized cells capable of dormancy.**
**(a)** Akinetes are thick-walled, food-filled cells produced by some cyanobacteria. Akinetes are able to resist stressful conditions and generate new populations when conditions improve. As discussed later, the heterocyte is a specialized cell in which nitrogen fixation occurs. **(b)** An endospore with a resistant wall develops within the cytoplasm of the pathogen *Clostridium difficile*. a: ©Lee W. Wilcox; b: ©Dr. Kari Lounatmaa/Science Source

*Concept Check:* *How do endospores affect the ability of some bacteria to cause disease?*

Endospores (**Figure 27.15b**) are produced inside bacterial cells by the enclosure of DNA and other materials within a tough coat, and then are released when the enclosing cell dies and breaks down. Endospores can remain alive, though in a dormant state, for long periods, then reactivate when conditions are suitable.

The ability to produce endospores allows some Gram-positive bacteria in the phylum Firmicutes to cause serious diseases. For example, *Bacillus anthracis* causes the disease anthrax and is thus a potential agent in bioterrorism and germ warfare. Most cases of human anthrax result when endospores of *B. anthracis* enter breaks in the skin, causing skin infections that are relatively easily cured by antibiotic treatment. But sometimes the endospores are inhaled or consumed in undercooked, contaminated meat, potentially causing more serious illness or death. *Clostridium botulinum* can contaminate improperly canned food that has not been heated to temperatures high enough to destroy its tough endospores. When the endospores germinate and bacterial cells grow in the food, they produce a deadly toxin, as well as $NH_3$ and $CO_2$ gases, which cause can lids to bulge. If humans consume the food, the toxin causes botulism, a severe type of food poisoning that can lead to respiratory and muscular paralysis. The botulism toxin is marketed commercially as Botox, which is injected into the skin, where it paralyzes facial muscles, thereby reducing the appearance of wrinkles. The toxin has also been used as a migraine treatment. *Clostridium tetani* produces a nerve toxin that causes lockjaw, also known as tetanus, when bacterial cells or endospores from soil enter wounds. The ability of the genera *Bacillus* and *Clostridium* to produce resistant endospores helps explain their widespread presence in nature and their effect on humans.

## 27.4 Nutrition and Metabolism

### Learning Outcomes:

1. List the major mechanisms of nutrition displayed by prokaryotic species.

2. **CoreSKILL »** Compare and contrast the effects of oxygen on the metabolism of different types of prokaryotic species; then predict the outcome when a gas gangrene patient is treated with oxygen.

3. Outline the process of biological nitrogen fixation, and explain why it is important and how oxygen interferes with this process.

All living cells require energy and a source of carbon to build their organic molecules. Bacteria and archaea use a wide variety of strategies to obtain energy and carbon for growth (**Table 27.2**). These microbes can be classified according to their energy source, carbon source, response to oxygen, and presence of specialized metabolic processes.

### Mechanisms of Nutrition and Responses to Oxygen

Cyanobacteria and some other prokaryotic species are **autotrophs** (from the Greek, meaning self-feeders), organisms that are able to produce all or most of their own organic molecules from inorganic sources. Autotrophs fall into two categories: photoautotrophs and chemoautotrophs. **Photoautotrophs**, including cyanobacteria, use light as a source of energy for the synthesis of organic molecules from $CO_2$ and $H_2O$ or $H_2S$. **Chemoautotrophs** use energy obtained by chemical modifications of inorganic compounds to synthesize organic compounds. Such chemical modifications include

| Table 27.2 | Major Types of Archaea and Bacteria Based on Energy and Carbon Source | | |
|---|---|---|---|
| **Type** | **Energy source** | **Carbon source** | **Example** |
| **Autotroph** | | | |
| Photoautotroph | Light | $CO_2$ | Cyanobacteria |
| Chemoautotroph | Inorganic compounds | $CO_2$ | *Sulfolobus* (Archaea) |
| **Heterotroph** | | | |
| Photoheterotroph | Light | Organic compounds | Chloroflexi (Bacteria) |
| Chemoheterotroph | Organic compounds | Organic compounds | Many |

nitrification (the conversion of ammonia to nitrate) and the oxidation of sulfur, iron, or hydrogen. For example, archaea of the genus *Sulfolobus* can oxidize certain sulfur-containing minerals.

**Heterotrophs** (from the Greek, meaning other feeders) are organisms that require at least one organic compound, and often more, from their environment. Some microorganisms, including bacteria in the phylum Chloroflexi, are **photoheterotrophs**, meaning that they are able to use light energy to generate ATP, but they must take in organic compounds from their environment as a source of carbon. **Chemoheterotrophs** must obtain organic molecules both for energy and as a carbon source.

Prokaryotic species differ in their need for and responses to oxygen. Like most eukaryotes (including humans), many prokaryotes are **obligate aerobes**, meaning that they require $O_2$ to survive. In contrast to obligate aerobes, obligate anaerobes, such as the Firmicutes genus *Clostridium*, are poisoned by $O_2$. People suffering from gas gangrene (caused by *Clostridium perfringens* and related species) are usually treated by placement in a chamber having a high oxygen content (called a hyperbaric chamber), which kills the organisms and deactivates the toxins. **Aerotolerant anaerobes** do not use $O_2$, but they are not poisoned by it either. These organisms obtain their energy by fermentation or anaerobic respiration, which are described in Chapter 7 (look back at Section 7.8). Anaerobic metabolic processes include denitrification (the conversion of nitrate into $N_2$ gas) and the reduction of manganese, iron, and sulfate, which are all important in the Earth's cycling of minerals.

**Facultative anaerobes** can use $O_2$ via aerobic respiration, obtain energy via anaerobic fermentation, or use inorganic chemical reactions to obtain energy—shifting between modes depending on environmental conditions. One fascinating example of a facultative anaerobe is the species *Thiomargarita namibiensis*, a large proteobacterium mentioned earlier in this chapter. This chemoheterotroph obtains its energy in two ways: by oxidizing sulfide with oxygen when oxygen is available or, when oxygen is low or unavailable, by oxidizing sulfide with nitrate. In either case, the cells convert sulfide to elemental sulfur, which is stored within the cells in large globules.

## Some Prokaryotic Species Play Important Roles as Nitrogen Fixers

Many cyanobacteria and some other prokaryotic organisms carry out a specialized metabolic process called biological **nitrogen fixation**. The removal of nitrogen from the gaseous phase is called fixation, and microbes that perform this process are known as nitrogen fixers.

Nitrogen fixation is an important component of the cycling of nitrogen on a global basis. During nitrogen fixation, the enzyme nitrogenase converts inert atmospheric nitrogen gas ($N_2$) into ammonia ($NH_3$). Plants and algae can use ammonia (though not $N_2$) to produce proteins and other essential nitrogen-containing molecules. As a result, many plants have developed close relationships with nitrogen fixers, which provide ammonia fertilizer to their plant partners. In addition to the aquatic photosynthetic cyanobacteria mentioned in the chapter opening, many types of heterotrophic soil bacteria also fix nitrogen. Examples include protobacteria of the genus *Rhizobium*, which live within the roots of legume plants (see Chapter 38).

Nitrogenase is inhibited by $O_2$, so most nitrogen fixers conduct nitrogen fixation only in low-oxygen conditions. Many cyanobacteria generate low-oxygen conditions in specialized cells known as heterocytes, allowing nitrogen fixation to occur in these cells (see Figure 27.15a). Heterocytes display adaptations that reduce nitrogenase exposure to oxygen. These include thick walls that reduce inward $O_2$ diffusion, increased occurrence of cellular reactions that consume oxygen, and down-regulation of the oxygen-producing components of photosynthesis. The latter adaptation, involving reduction in chlorophyll synthesis, explains why heterocytes are paler in color than neighboring photosynthetic cells.

## 27.5 Ecological Roles and Biotechnology Applications

**Learning Outcomes:**

1. Discuss the roles of bacteria and archaea in the carbon cycle.
2. List examples of bacteria-eukaryote symbiosis and of pathogenic microbes.
3. List ways in which bacteria contribute to industrial and biotechnology applications.

Bacteria and archaea play many key ecological roles. In addition to their roles in nitrogen fixation and other aspects of the global nitrogen cycle, bacteria and archaea produce or break down organic carbon, important in the global carbon cycle. Bacteria function as beneficial symbionts in plants and animals and as disease agents. In this section, we will focus on these ecological roles and also provide examples of ways that humans use the metabolic capabilities of bacteria in industry and biotechnology.

### Bacteria and Archaea Play Important Roles in Earth's Carbon Cycle

The Earth's carbon cycle is the sum of all the chemical changes that occur among compounds that contain carbon. (Look ahead to Chapter 59 for a detailed discussion of the carbon cycle.) One way that bacteria and archaea influence Earth's carbon cycle is by producing and consuming methane. Methane ($CH_4$)—the major component of natural gas—is a greenhouse gas more powerful than $CO_2$; $CH_4$ increases global warming over 20 times more per molecule than does $CO_2$. Therefore, atmospheric $CH_4$ has the potential to alter the Earth's climate, and in recent years the level of $CH_4$ has been increasing in Earth's atmosphere as the result of human activities. Several groups of anaerobic archaea known as **methanogens** convert $CO_2$, methyl groups, or acetate to $CH_4$ and release $CH_4$ from their cells into the atmosphere. Methanogens live in swampy wetlands, in deep-sea habitats, or in the digestive systems of animals, including cattle and humans. Marsh gas produced in wetlands is largely composed of $CH_4$, and large quantities of $CH_4$ produced long ago are trapped in deep-sea and subsurface Arctic deposits. Certain bacteria known as **methanotrophs** consume $CH_4$, thereby reducing its concentration in the atmosphere. In the absence of methanotrophs, Earth's atmosphere would be much richer in the greenhouse gas $CH_4$, which would substantially increase global temperatures.

Bacteria and archaea are also important in producing and degrading complex organic compounds. For example, cyanobacteria and other autotrophic bacteria are important **producers**. Such bacteria, together with algae and plants, remove $CO_2$ from the atmosphere and, via photosynthesis, synthesize the organic compounds that are used by themselves and other organisms for food. **Decomposers**, also known as saprobes, include heterotrophic microorganisms (as well as fungi and animals). These organisms break down dead organisms and organic matter, releasing minerals for uptake by living things. Astonishingly, many bacteria are able to break down antibiotics for use as a source of organic carbon, as discussed next.

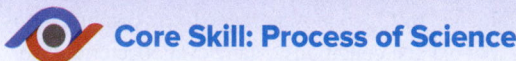

## Core Skill: Process of Science

## Feature Investigation | Dantas and Colleagues Found That Many Bacteria Can Break Down and Consume Antibiotics as a Sole Carbon Source

Many microorganisms naturally secrete antibiotics, which are organic compounds that inhibit the growth of other microorganisms. Antibiotics are evolutionary adaptations that allow bacteria and other microbes to avoid attack or reduce competition for resources. People have taken advantage of high antibiotic production by certain bacteria, particularly species of the phylum Actinomycetes, to make antibiotics commercially. Such antibiotics are used to treat bacterial infections in humans and domesticated species.

Due to the widespread production of antibiotics by bacteria, these organic compounds are commonly found in natural habitats. Many species of chemoheterotrophic bacteria have evolved the ability to metabolize antibiotics and use them as a source of carbon. In 2008, Gautam Dantas, George Church, and their colleagues reported this finding after experimentally testing their hypothesis that soil bacteria might be able to metabolize antibiotics (**Figure 27.16**). The investigators first cultivated bacteria from 11 different soils in the laboratory, finding a diverse

**Figure 27.16** **Diverse bacteria isolated from different soils are able to grow on many types of antibiotics.**

| | |
|---|---|
| **HYPOTHESIS** The soil bacterial community contains species that can take up and metabolize antibiotics. | |
| **KEY MATERIALS** Eleven diverse soil samples; 18 types of antibiotics. | |

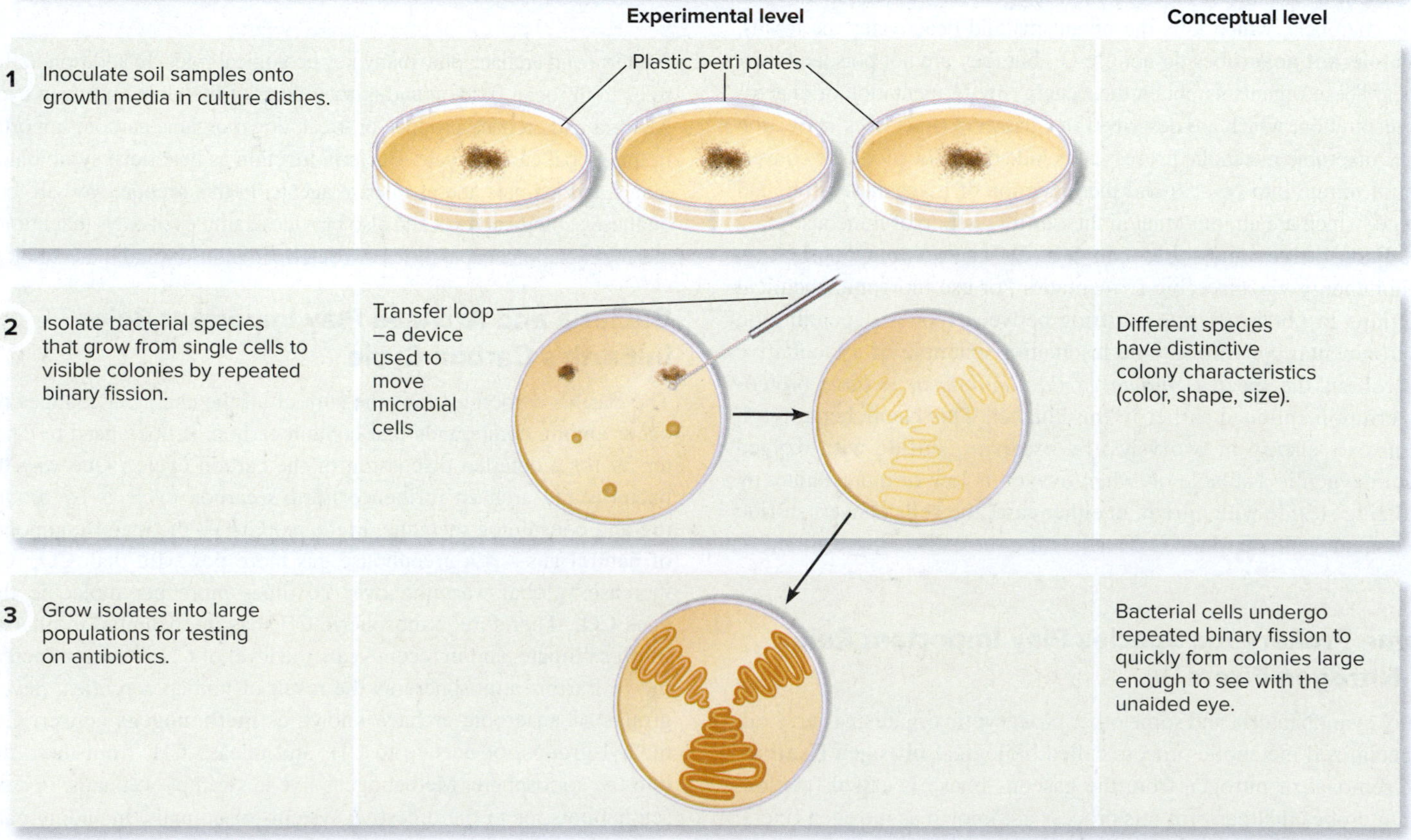

| | Experimental level | Conceptual level |
|---|---|---|
| **1** Inoculate soil samples onto growth media in culture dishes. | Plastic petri plates | |
| **2** Isolate bacterial species that grow from single cells to visible colonies by repeated binary fission. | Transfer loop —a device used to move microbial cells | Different species have distinctive colony characteristics (color, shape, size). |
| **3** Grow isolates into large populations for testing on antibiotics. | | Bacterial cells undergo repeated binary fission to quickly form colonies large enough to see with the unaided eye. |

**4** Inoculate each bacterial isolate onto replicate dishes containing a different antibiotic as the only food source.

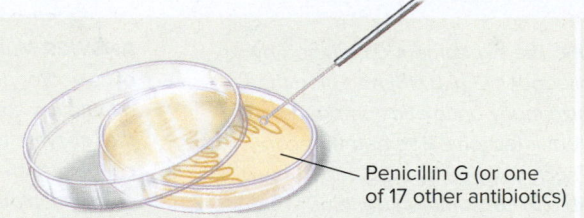

Penicillin G (or one of 17 other antibiotics)

Test the ability of each isolate to grow on a range of antibiotics.

**5** Allow time for bacterial population growth; compare growth among dishes.

Strong growth of forest soil bacterial isolate F1 on penicillin G food

Poor growth of urban soil bacterial isolate U3 on dicloxacillin food

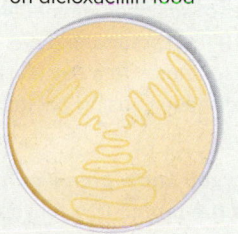

Compare isolate ability to grow on different antibiotics.

**6    THE DATA**

Most soils tested contained bacterial species that were able to use antibiotics of many types for food and thus were resistant to those antibiotics.

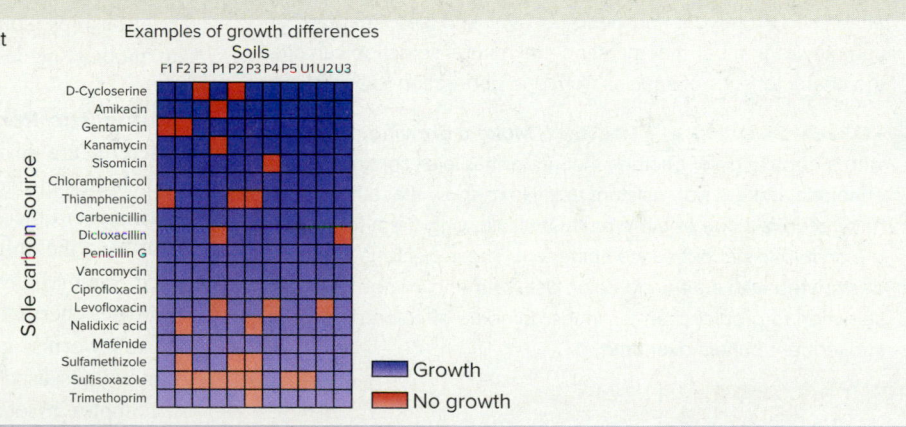

Examples of growth differences

**7    CONCLUSION** Natural soils contain bacteria that are able to utilize antibiotics produced naturally by other species as food. Soil bacteria are a previously unrecognized source of antibiotic resistance genes that can be transferred to other species.

**8    SOURCE** Dantas, G., Sommer, M. O. A., Oluwasegun, R. D., and Church, G. M. 1998. Bacteria subsisting on antibiotics. *Science* 320: 100–103.

collection of different species. Almost 90% of the cultured bacteria were Gram-negative Proteobacteria, some closely related to human pathogens, while 7% of the cultures were Gram-positive Actinomycetes.

These researchers then tested the ability of the bacteria cultured from different soils (isolates) to use various antibiotics as a sole carbon source for growth. The 18 antibiotics tested included penicillin and related compounds, as well as the widely prescribed ciprofloxacin (Cipro). Every antibiotic tested supported the growth of bacteria from soil. Importantly, each antibiotic-eating isolate was resistant to several antibiotics at concentrations used in the medical treatment of infections.

In today's society, the widespread use of antibiotics in medicine and agriculture is of concern because it is thought to foster increases in antibiotic resistance (refer back to Section 19.5). The experiment by

Dantas and associates revealed that natural evolutionary processes—the widespread development by diverse soil bacteria of metabolic processes to utilize many types of antibiotics as food—represent a previously unrecognized source of antibiotic resistance. The study also indicated that natural bacteria are a potential source of antibiotic-resistance genes that could be transferred to disease-causing bacteria

### Experimental Questions

1. What features of soil bacteria attracted the attention of the researchers?

2. **CoreSKILL »** What processes did the researchers use to test their hypothesis that soil bacteria might use antibiotics as a food source?

## BIO TIPS

**THE QUESTION** *The experiment conducted by Dantas and colleagues (Fig. 27.16) revealed that soil bacteria are able to metabolize many antibiotic compounds commonly used in medicine to treat infections. Bacteria that consume antibiotics are resistant to such compounds. Describe how a bacterial population that is not able to consume an antibiotic and is not resistant to that antibiotic evolves into one that is. Why is this evolutionary process important to society?*

**T**OPIC *What topic in biology does this question address?* The topic is the evolution of certain species of soil bacteria that are able to consume natural antibiotics as a food source and are also resistant to them. The question also relates to a societal problem, the increasing resistance of disease-causing bacteria to antibiotics used in medicine.

**I**NFORMATION *What information do you know based on the question and your understanding of the topic?* This section of the chapter summarizes the major ways in which bacteria acquire organic carbon and energy. Information provided in Figure 27.16 indicates that some soil bacteria can utilize carbon from antibiotics for growth and are also resistant to those same antibiotics. From your knowledge of evolution (see Chapter 22), you may remember that mutation and natural selection can alter the characteristics of species from one generation to the next.

**P**ROBLEM-SOLVING **S**TRATEGY *Make a drawing.* Begin with a population of bacteria that are not able to consume the antibiotic and are not resistant to it. Next, show that a few bacteria have acquired one or more mutations, allowing them to consume the antibiotic without succumbing to it. These bacteria can be designated with a different color. Use your knowledge of natural selection to predict changes in the numbers of colored bacteria in subsequent panels over time.

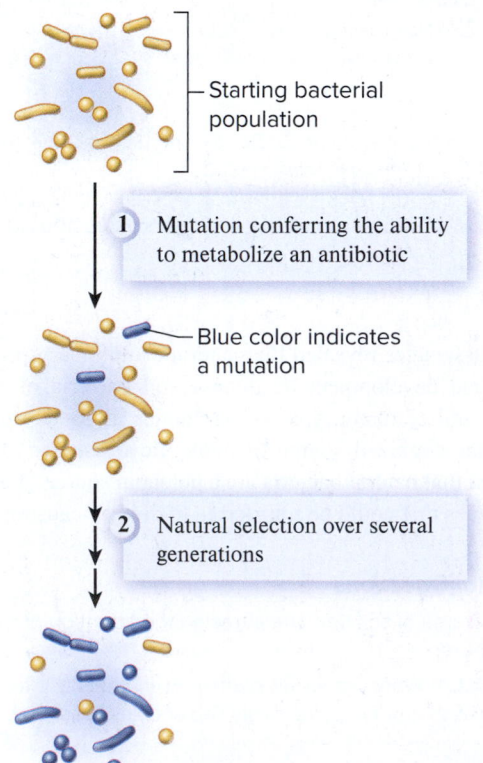

Starting bacterial population

1 Mutation conferring the ability to metabolize an antibiotic

Blue color indicates a mutation

2 Natural selection over several generations

**ANSWER** *Mutations arise that confer the ability to metabolize antibiotics and/or provide resistance to the same antibiotics. For example, a mutation could alter the structure of a metabolic enzyme, enabling that enzyme to recognize and break down the antibiotic. Bacteria carrying such a mutation may also be resistant to being killed by the same antibiotic because it would not accumulate in their cytoplasm. Over the course of many generations, such bacteria may increase in number due to natural selection. These populations of bacteria can quickly become resistant to antibiotics at levels that are used in medical treatments. The larger such populations are, the greater the chances they will cause infections for which antibiotic treatment may not work as expected.*

## Many Bacteria Live in Symbiotic Associations

Many bacterial species live in close associations with one or more other species, a phenomenon called **symbiosis** (from the Greek, meaning life together with). If symbiosis is beneficial to both partners, the interaction is known as **mutualism**. Many mutualistic bacteria live in associations of two or a few species that supply each other with essential nutrients. Alternatively, in other symbiotic relationships, some bacterial species may cause harm or even death to another species.

***Mutualistic Partnerships Between Bacteria and Eukaryotes*** Bacteria are involved in many mutually beneficial symbioses in which they provide aquatic or terrestrial eukaryotes with minerals or vitamins or other valuable services. Bioluminescent bacteria, bacteria that have the ability to produce and emit light (**Figure 27.17**), often form symbiotic relationships with squid and other marine animals. In deep-sea thermal vent communities, bacteria live within the tissues of tubeworms and mussels, supplying these animals with carbon compounds used as food. One terrestrial example of mutualism is a complex association involving four partners: ants, fungi that the ants cultivate for food, parasitic fungi that attack the food fungi, and Actinobacteria that produce antibiotics. The antibiotics control the growth of the parasitic fungi, preventing them from destroying the ants' fungal food supply. The ants rear the useful bacteria in cavities on their body surfaces; glands near these cavities supply the bacteria with nutrients.

***Pathogenic Microbes*** Microorganisms that cause disease in one or more types of host organism are known as **pathogens**. Cholera, leprosy, tetanus, pneumonia, whooping cough, diphtheria, Lyme disease, scarlet fever, rheumatic fever, typhoid fever, bacterial dysentery, and tooth decay are among the many examples of human diseases caused by bacterial pathogens. Bacteria also cause many plant diseases of importance in agriculture, including blights, soft rots, and wilts. How do microbiologists determine which bacteria cause these diseases? The pioneering research of the Nobel Prize–winning German physician Robert Koch provided the answer.

In the mid- to late 1800s, Koch established a series of four steps to determine whether a particular organism causes a specific disease.

1. The presence of the suspected pathogen must correlate with occurrence of symptoms.

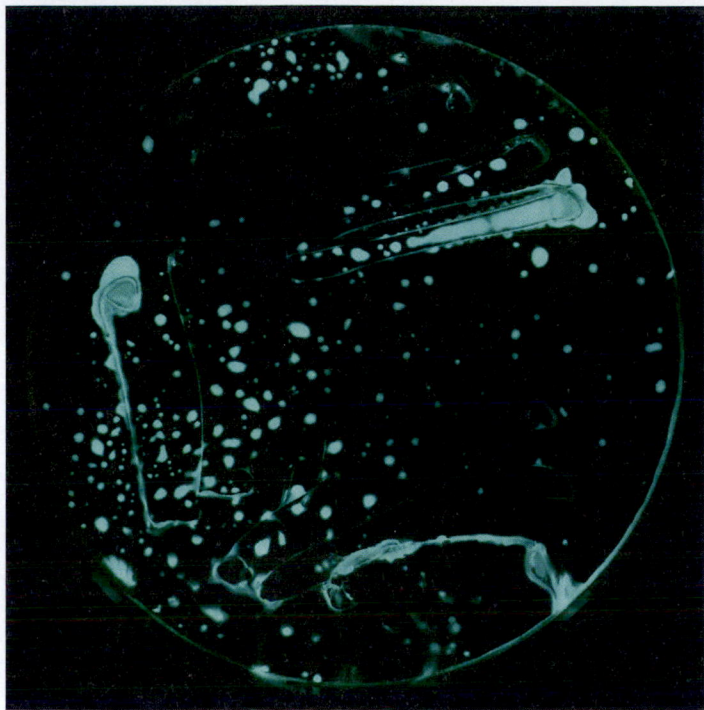

**Figure 27.17 Bioluminescent bacteria.** These blue-green colonies of *Vibrio fischeri* bacteria are growing on nutritive media in a culture plate. The colonies produce so much light that additional light was not needed to make this photo. ©Peter Durben.

2. The pathogen must be isolated from an infected host and grown in pure culture if possible.

3. Cells from the pure culture should cause disease when inoculated into a healthy host.

4. The same pathogen should be isolatable from the second infected host.

Using these steps, known as **Koch's postulates**, Koch discovered the bacterial causes of anthrax, cholera, and tuberculosis. Subsequent investigators have used Koch's postulates to establish the identities of bacterial species that cause other infectious diseases.

***How Pathogenic Bacteria Attack Cells*** An important aspect of symbiosis is the way in which bacteria interact with other species, such as human hosts. Understanding how disease-causing bacteria attack host cells aids in developing strategies for disease prevention and treatment. Many pathogenic bacteria attack cells by binding to the surface and injecting substances that help them utilize cellular components. During their evolution, some Gram-negative pathogenic bacteria developed needle-like systems, made of components also found in flagella, that inject proteins into animal or plant cells as part of the infection process. Such structures are known as injectisomes (**Figure 27.18a**). Examples of bacteria whose injectisomes allow them to attack human cells are *Yersinia pestis* (the agent of bubonic plague), *Salmonella enterica* (which causes the food poisoning called salmonellosis), and *Burkholderia pseudomallei* (the cause of melioidosis, a

deadly disease of growing concern in some parts of the world). More than 20 million people are infected by means of injectisomes every year. Several plant pathogenic bacteria also infect cells by means of injectisomes.

Some other Gram-negative bacterial pathogens synthesize a type IV secretion system, which functions as channel to deliver toxins or DNA into cells. Examples of such bacteria that cause human disease include *Helicobacter pylori*, *Legionella pneumophila*, and *Bordetella pertussis*. The plant pathogen *Agrobacterium tumifaciens* uses a type IV secretion system to

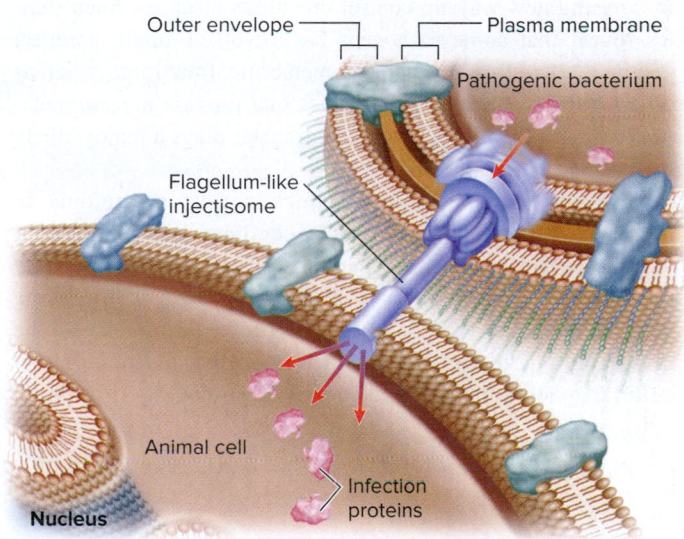

**(a) Injectisome**

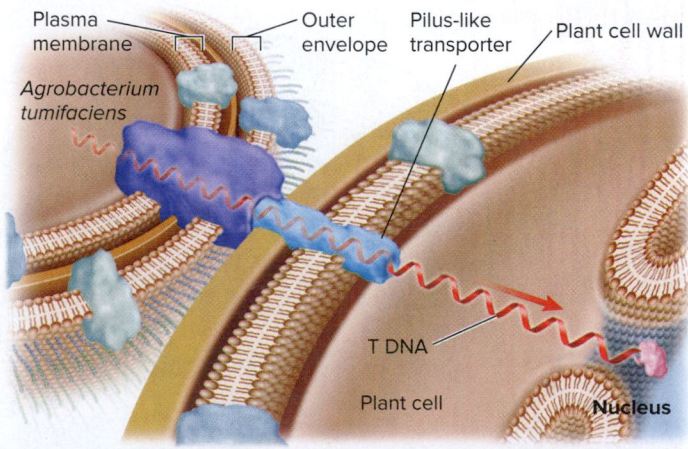

**(b) Type IV secretion system**

**Figure 27.18 Attack systems of pathogenic bacteria.** **(a)** The injectisome functions like a syringe to inject proteins into host cells, in this case an animal cell, thereby initiating the disease process. **(b)** A type IV secretion system forms a channel through which DNA can be transmitted from a pathogenic bacterium to a host cell, in this case from the bacterium *Agrobacterium tumifaciens* into a plant cell. Some of the components in this system are related to those found in pili, which are described in Chapter 19.

transfer DNA (T DNA) into plant cells (**Figure 27.18b**). The T DNA encodes an enzyme that affects normal plant growth, with the result that cancer-like tumors called galls develop (see Figure 27.4).

## Core Concept: Evolution

### The Evolution of Bacterial Pathogens

Genomic and proteomic studies have illuminated the evolution of bacterial pathogens, providing insights that are useful in devising new ways to control infectious diseases. Such studies reveal that some pathogens have evolved small, compact genomes encoding specialized metabolic functions, whereas others have acquired large genomes that provide diverse metabolic capabilities. Horizontal gene transfer plays a major role in increasing disease severity.

*Mycoplasma pneumoniae*, which causes pneumonia in humans, has one of the smallest genomes known to occur among self-replicating organisms. The tiny cells, only 0.3 μm in diameter, possess fewer than 700 protein-coding genes and make only 178 types of protein complexes. The bacterium is a model organism for investigating the minimal cellular machinery required for life. As one way of gaining insight into what forms a minimal proteome, the locations of five types of protein complexes have been mapped in these bacteria by means of specialized microscopic techniques (**Figure 27.19**).

By contrast, *Pseudomonas aeruginosa*—which causes respiratory disease in humans and other animals, and also infects plants—has a larger genome containing about 5,000 protein-encoding genes. Diverse strains share a common genome core but also have strain-specific genes acquired by horizontal transfer, which confer a wide variety of metabolic abilities. This genomic variation allows *P. aeruginosa* to survive in a wide range of environments. As an example of the species' wide metabolic capability, a strain of *P. aeruginosa* obtained from an infected human also possessed the genes and proteins necessary to degrade tough defensive resins produced by trees and use them as a food source!

*Escherichia coli* strain O157:H7, which causes deadly outbreaks of food-borne illness and is the leading cause of acute kidney failure in children, evolved from harmless strains by the step-wise gain via horizontal transfer of toxin genes. These genomic features enable this strain to use its flagellar tips to attach to host intestinal epithelium and then attack cells with an injectisome. Bacterial toxin produced in the intestine enters the host circulation system and reaches the kidneys, where the toxin inhibits protein synthesis, resulting in severe tissue damage.

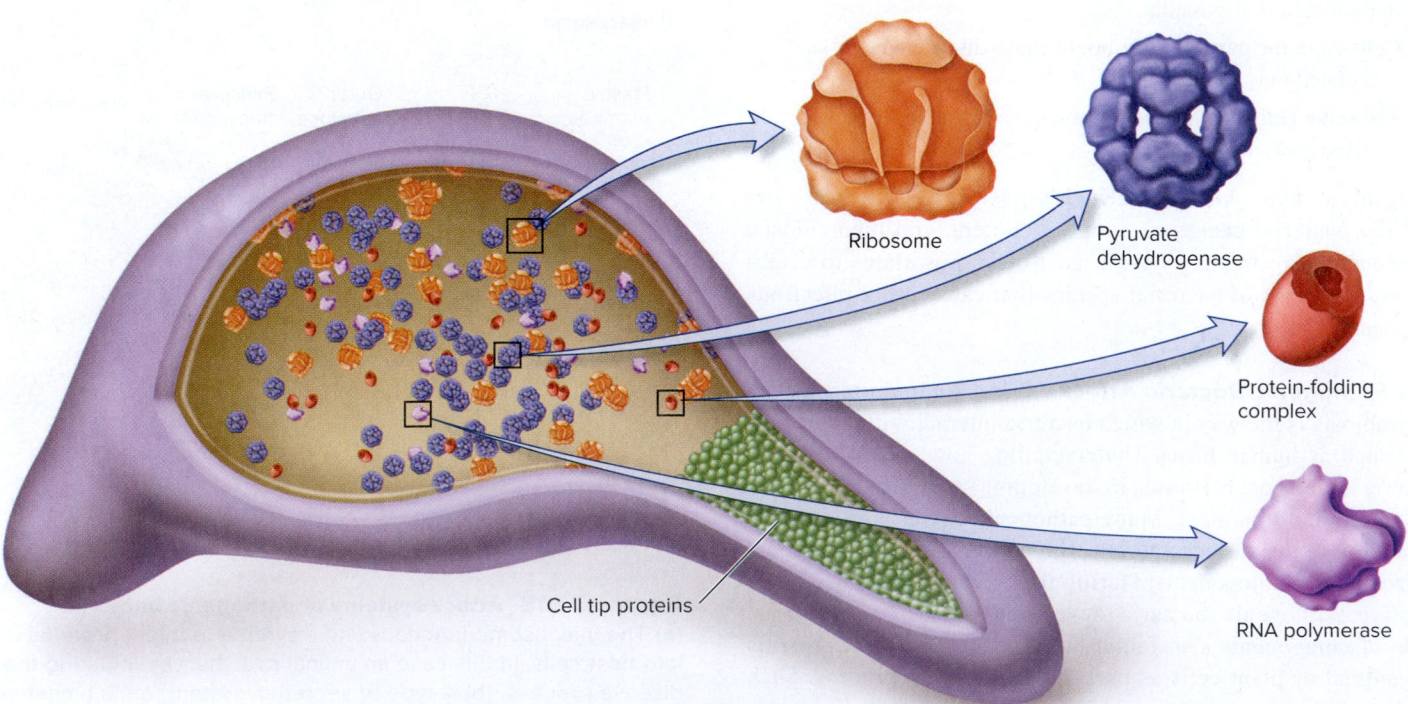

**Figure 27.19** **Locations of five protein complexes in the tiny pathogen *Mycoplasma pneumoniae*.** The cell tip is rich in proteins (colored green) that help these bacteria attach to host epithelial cells. Other mapped protein complexes are pyruvate dehydrogenase (involved in energy metabolism), ribosomes, RNA polymerase, and a protein-folding complex (colored red).

### Some Bacteria Are Useful in Industrial and Other Applications

Several industries have harnessed the metabolic capabilities of microbes obtained from nature. The food industry uses bacteria to produce chemical changes in food—for example, to make dairy products, including cheese and yogurt, from milk. Cheese makers add pure cultures of certain bacteria to milk. The bacteria consume milk sugar (lactose) and produce lactic acid, which aids in curdling the milk.

The chemical industry produces materials such as enzymes, vinegar, amino acids, vitamins, insulin, vaccines, antibiotics, and other useful pharmaceuticals by growing particular bacteria in giant vats. For example, the hot springs bacterial species *Thermus aquaticus* is a source of a form of DNA polymerase widely used in biology laboratories to amplify DNA in polymerase chain reaction (PCR). Industrially grown bacteria produce the antibiotics streptomycin, tetracycline, kanamycin, gentamycin, bacitracin, polymyxin-B, and neomycin.

The field of synthetic biology utilizes bacteria as chemical factories by genetically modifying bacterial genomes so that the bacteria produce particular useful compounds, such as pharmaceuticals and renewable biofuels. For example, biologists have modified the genome of the bacterium *Caldicellulosiruptor bescii* so it can transform switchgrass into ethanol. The ability of some microorganisms to break down organic compounds or precipitate metals makes them very useful in treating wastewater, industrial discharges, and harmful substances such as explosives, pesticides, and oil spills. This process, known as **bioremediation**, is used to reduce levels of harmful materials in the environment.

Agriculture employs several species of *Bacillus*, particularly *B. thuringiensis* (Bt), which produce toxins that kill the insects that ingest them but are harmless to many noninsect species. Tent caterpillars, potato beetles, gypsy moths, mosquitoes, and black flies are among the pests that can be controlled by the Bt toxin. For this reason, toxin-encoding genes from *B. thuringiensis* have been cloned and introduced into some crop plants, such as corn and cotton, to reduce conventional pesticide use and increase crop yields.

## Summary of Key Concepts

### 27.1 Diversity and Evolution

- The three domains of life are Bacteria, Archaea, and Eukarya (whose members are referred to as bacteria, archaea, and eukaryotes). Domain Archaea is more closely related to domain Eukarya than either is to domain Bacteria (Figure 27.1).

- Many representatives of the domain Archaea occur in extremely hot, salty, or acidic habitats. Ether-linked membrane phospholipids are among the features of archaea that enable their survival in extreme habitats (Figure 27.2).

- The domain Bacteria includes 50 or more phyla, including Cyanobacteria and Proteobacteria, which are particularly diverse and of great evolutionary and ecological importance (Table 27.1, Figures 27.3, 27.4).

- Widespread horizontal gene transfer has occurred among bacteria and archaea. Horizontal gene transfer by means of transduction, transformation, or conjugation allows microorganisms to evolve rapidly.

### 27.2 Structure and Movement

- Bacteria and archaea have prokaryotic cells that are smaller and simpler than eukaryotic cells. Structural adaptations that increased the complexity of some types of prokaryotic cells include thylakoids, magnetosomes, and nucleus-like bodies (Figures 27.5, 27.6).

- Common prokaryotic cell shapes are spherical cocci, rod-shaped bacilli, comma-shaped vibrios, and coiled spirochaetes and spirilli. Some cyanobacteria display features of multicellular organisms (Figure 27.7).

- Many microbes secrete a coating of slimy mucilage, which plays a role in disease resistance and in the development of biofilms. Biofilm development is influenced by quorum sensing, a mechanism in which group activity is coordinated by chemical communication (Figure 27.8).

- Most bacterial cell walls contain peptidoglycan, which is composed of carbohydrates cross-linked by peptides. Gram-positive bacterial cells have thick peptidoglycan walls, whereas Gram-negative cells have less peptidoglycan in their walls and are enclosed by an outer lipopolysaccharide envelope (Figures 27.9, 27.10).

- Motility enables microbes to change positions within their environment, which aids in locating favorable conditions for growth. Some bacteria have gas vesicles, which enable them to float, whereas others swim by means of flagella or twitch or glide by the action of pili (Figures 27.11, 27.12, 27.13).

### 27.3 Reproduction

- Populations of most bacteria and archaea enlarge by binary fission, a simple type of cell division that provides a means by which culturable microbes can be counted (Figure 27.14).

- Some bacteria are able to survive harsh conditions as dormant akinetes or endospores (Figure 27.15).

### 27.4 Nutrition and Metabolism

- Bacteria and archaea can be grouped according to mechanism of nutrition, response to oxygen, or presence of distinctive metabolic features. Major nutritional types are photoautotrophs, chemoautotrophs, photoheterotrophs, and chemoheterotrophs (Table 27.2).

- Obligate aerobes require oxygen, whereas obligate anaerobes are poisoned by oxygen. Aerotolerant anaerobes do not use oxygen but are not poisoned by it. Both obligate and aerotolerant anaerobes obtain their energy by anaerobic respiration. Facultative aerobes are able to live with or without oxygen by using different processes for obtaining energy.

- Nitrogen fixation is a distinctive metabolic process displayed only by certain microorganisms, and is a key component of the global nitrogen cycle.

### 27.5 Ecological Roles and Biotechnology Applications

- Bacteria and archaea play key roles in Earth's carbon cycle as producers, decomposers, beneficial symbionts, or pathogens. Methane-producing methanogens and methane-consuming methanotrophs influence the Earth's climate because methane is a powerful greenhouse gas.

- Some bacteria are able to consume antibiotics, a process linked to the evolution of antibiotic resistance (Figure 27.16).
- Bacteria may occur in symbiotic associations with other organisms. (Figure 27.17).
- Pathogenic bacteria obtain organic compounds from living host cells. Bacteria attack eukaryotic cells by means of injectisomes or type IV secretion systems (Figures 27.18).
- During their evolution, some pathogenic bacteria have reduced their genomes and proteomes, whereas others have acquired large genomes conferring diverse metabolic capacities; horizontal gene transfer is a major process by which disease severity increases (Figure 27.19).
- Many bacteria and archaea are useful in industrial and other applications; others are used to make food products or antibiotics or to clean up polluted environments.

## Assess & Discuss

### Test Yourself

1. Which of the following features is common to prokaryotic cells?
   a. a nucleus, featuring a nuclear envelope with pores
   b. mitochondria
   c. plasma membranes
   d. mitotic spindle
   e. none of the above

2. The bacterial phylum whose members typically produce oxygen gas as a product of photosynthesis is
   a. Proteobacteria.          d. all of the above.
   b. Cyanobacteria.           e. none of the above.
   c. the Gram-positive bacteria.

3. Gram-staining is a procedure that microbiologists use to
   a. determine if a bacterial strain is a pathogen.
   b. determine if a sample of bacteria sample can break down oil.
   c. infer the structure of the cell wall of bacteria and their response to antibiotics.
   d. count bacteria in medical or environmental samples.
   e. do all of the above.

4. Place the following steps in the correct order, according to Koch's postulates:
   I. Determine if pure cultures of bacteria cause disease symptoms when introduced to a healthy host.
   II. Determine if disease symptoms correlate with presence of a suspected pathogen.
   III. Isolate the suspected pathogen and grow it in pure culture, free of other possible pathogens.
   IV. Attempt to isolate pathogen from second-infected hosts.
   a. II, III, IV, I          d. II, III, I, IV
   b. II, IV, III, I          e. I, II, III, IV
   c. III, II, I, IV

5. What ecological role do cyanobacteria play?
   a. producers          d. parasites
   b. consumers          e. none of the above
   c. decomposers

6. Bacterial structures that pathogenic bacteria use in attacking host cells include
   a. injectisomes and IV secretion systems.
   b. magnetosomes.
   c. gas vesicles.
   d. thylakoids.
   e. none of the above.

7. The structures that enable some Gram-positive bacteria to remain dormant for extremely long periods of time are known as
   a. akinetes.                    d. lipopolysaccharide envelopes.
   b. endospores.                  e. pili.
   c. biofilms.

8. By means of what process do populations of bacteria or archaea increase their size?
   a. mitosis                      d. transduction
   b. meiosis                      e. none of the above
   c. conjugation

9. By what means do bacterial cells acquire new DNA?
   a. by conjugation, the mating of two cells of the same bacterial species
   b. by transduction, the injection of viral DNA into bacterial cells
   c. by transformation, the uptake of DNA from the environment
   d. by all of the above
   e. by none of the above

10. How do various types of bacteria move?
    a. by the use of flagella, composed of a filament, hook, and motor
    b. by means of pili, which help cells twitch or glide along a surface
    c. by using gas vesicles to regulate buoyancy in water bodies
    d. All of the above are used by bacteria for movement.
    e. None of the above are used by bacteria for movement.

### Conceptual Questions

1. Explain why many bacterial populations grow more rapidly than do populations of eukaryotes and how such population growth influences the rate of food spoilage or infection.

2. What processes contribute to antibiotic resistance?

3. **Core Concept: Systems** As we have seen in this chapter, living organisms interact with their environment. What organisms are responsible for the blue-green blooms that often occur in warm weather on lake surfaces? Think carefully; the answer is not just "cyanobacteria," as you might first guess.

### Collaborative Questions

1. How would you go about cataloging the phyla of bacteria and archaea that occur in a particular place?

2. How would you go about developing a bacterial product that could be used for remediation of a site contaminated with materials that are harmful to humans?

# Protists

# 28

P rotists are eukaryotes that live in moist habitats and are mostly microscopic in size. Despite their small size, protists have a greater influence on global ecology and human affairs than most people realize. For example, the photosynthetic protists known as algae generate at least half of the oxygen in the Earth's atmosphere and produce organic compounds that feed marine and freshwater animals. The oil that fuels our cars and industries is derived from pressure-cooked algae that accumulated on the ocean floor over millions of years. Today, algae are being engineered into systems for cleaning pollutants from water or air and for producing renewable biofuels.

Protists also include some parasites that cause serious human illnesses. For example, in 1993, the waterborne protist *Cryptosporidium parvum* sickened 400,000 people in Milwaukee, Wisconsin, costing $96 million in medical expenses and lost work time. Because this protist is exceptionally tolerant of the disinfectant chlorine, it is currently the major cause of diarrhea illness associated with aquatic recreational facilities such as swimming pools and waterparks. Species of the related genus *Plasmodium*, which is carried by mosquitoes in many warm regions of the world, cause the disease malaria. Every year, nearly 500 million people become ill with malaria, and more than 2 million die of this disease. As we will see, sequencing the genomes of these and other protist species has suggested new ways of battling such deadly pathogens.

In this chapter, we will survey protist diversity, including structural, nutritional, and ecological variations. We begin by exploring ways of informally labeling protists according to ecological role, habitat, and motility. We will then focus on the defining features, classification, and evolutionary importance of the major protist phyla. Next, the nutritional and defensive adaptations of protists are discussed, and we conclude by looking at the reproductive adaptations that allow protists to exploit and thrive in a variety of environments.

**Protists such as these green algal cells and their plant descendants produce much of the Earth's oxygen.** Each of the cells in this population is surrounded by a coating of protective mucilage. ©Photographs by H. Cantor-Lund reproduced with permission of the copyright holder Freshwater Biological Association and J. W. G. Lund.

## 28.1 An Introduction to Protists

**Learning Outcomes:**

1. List three features that define protists.
2. Label protists informally by ecological role, habitat, and type of motility.

The term protist comes from the Greek word *protos*, meaning first, reflecting the observation that protists were Earth's first eukaryotes. Protists are eukaryotes that are not classified in the plant, animal, or fungal kingdoms. Protists display two additional common characteristics: They are most abundant in moist habitats, and most of them are microscopic in size. Protists play diverse ecological roles, live in diverse habitats, and display diverse types of motility.

## Protists Play Diverse Ecological Roles

Protists occur in three major ecological types: algae, protozoa, and fungus-like protists. The term **algae** (singular, alga; from the Latin, meaning seaweeds) applies to protists that are generally photoautotrophic, meaning that most can produce organic compounds from inorganic sources by means of photosynthesis. In addition to organic compounds that can be used as food by heterotrophs—organisms that obtain their food from other organisms—photosynthetic algae produce oxygen, which is also needed by most heterotrophs. Thanks to their photosynthetic abilities, algae are increasingly important sources of renewable biofuels. Despite the general feature of photosynthesis, algae do not form a monophyletic group descended from a single common ancestor.

The term **protozoa** (from the Greek, meaning first life) is commonly used to describe diverse heterotrophic protists that feed by absorbing small organic molecules or by ingesting prey. For example, the protozoa known as ciliates consume smaller cells such as the single-celled photosynthetic algae known as diatoms (**Figure 28.1**). Like algae, protozoa do not form a monophyletic group.

Several types of heterotrophic **fungus-like protists** have bodies, nutritional mechanisms, or reproductive mechanisms similar to those of the true fungi. For example, fungus-like protists often have threadlike, filamentous bodies and absorb nutrients from their environment, as do the true fungi (see Chapter 29). However, fungus-like

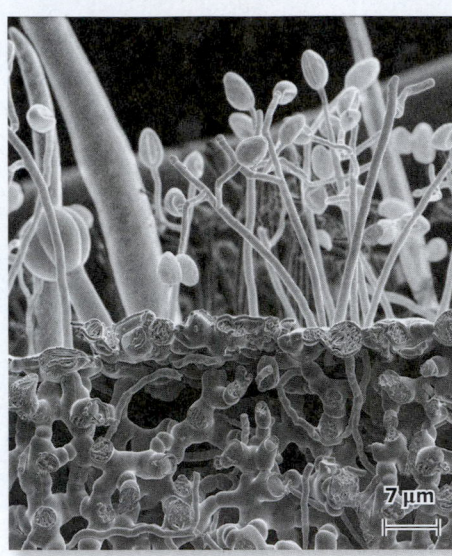

**Figure 28.2** A fungus-like protist, *Phytophthora infestans*. This organism causes the disease of potato plants known as late blight, or potato-blight. The image is an SEM of the protist growing on a host leaf. ©Andrew Syred/Science Source

protists are not actually related to fungi; their similar features represent cases of convergent evolution, in which species from different lineages have independently evolved similar characteristics (see Chapter 22). Water molds, some of which cause diseases of fish, and *Phytophthora infestans*, which causes diseases of many crops and wild plants, are examples of fungus-like protists (**Figure 28.2**). Various types of slime molds, some of which can be observed on decaying wood in forests, are also fungus-like, though not closely related to water molds and *Phytophthora*. These examples illustrate that the terms algae, protozoa, and fungus-like protists, although very useful in describing ecological roles, lack taxonomic or evolutionary meaning.

## Protists Live in Diverse Habitats

Although protists occupy nearly every type of moist habitat, they are particularly common and diverse in oceans, lakes, wetlands, and rivers. Even extreme aquatic environments such as Antarctic ice and acidic hot springs serve as habitats for some protists. In such places, protists may swim or float in open water or live attached to surfaces such as rocks or beach sand. These different habitats influence protists' structure and size.

Protists that swim or float in fresh or salt water are members of an informal aggregate of organisms known as plankton, which also includes bacteria, viruses, and small animals. The photosynthetic protists in plankton are called **phytoplankton** (plantlike plankton). Planktonic protists are necessarily quite small in size; otherwise they would readily sink to the bottom. Staying afloat is a particularly important characteristic of phytoplankton, which need light for photosynthesis. For this reason, planktonic protists occur primarily as single cells, colonies of cells held together with mucilage, or short filaments of cells linked end to end (**Figure 28.3a–c**).

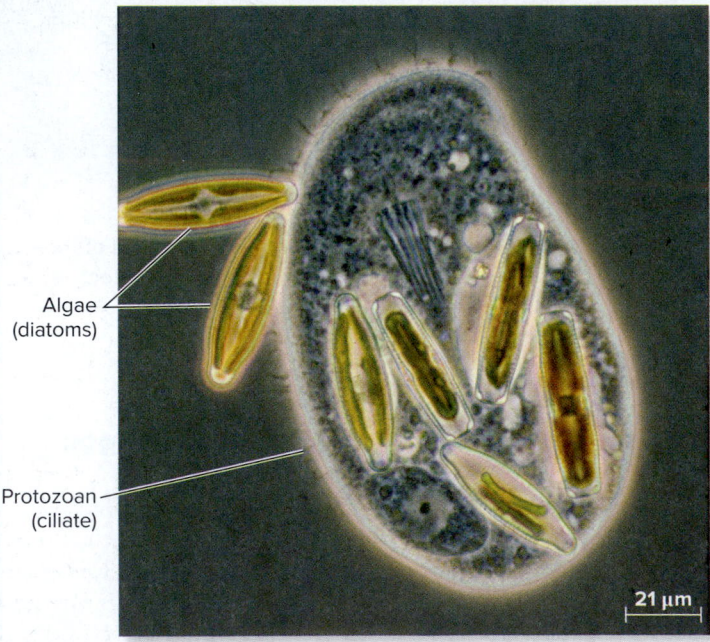

Algae (diatoms)

Protozoan (ciliate)

**Figure 28.1** A heterotrophic protozoan feeding on photosynthetic algae. The ciliate shown here has consumed several oil-rich, golden-pigmented, silica-walled algal cells known as diatoms. Diatom cells that have avoided capture glide nearby. ©Photographs by H. Cantor-Lund reproduced with permission of the copyright holder Freshwater Biological Association and J. W. G. Lund.

**Core Concept: Energy and Matter** The diatoms were ingested by the process of phagocytosis, and their organic components are digested by the ciliate as food.

**Planktonic protists**                                                      **Attached protists**

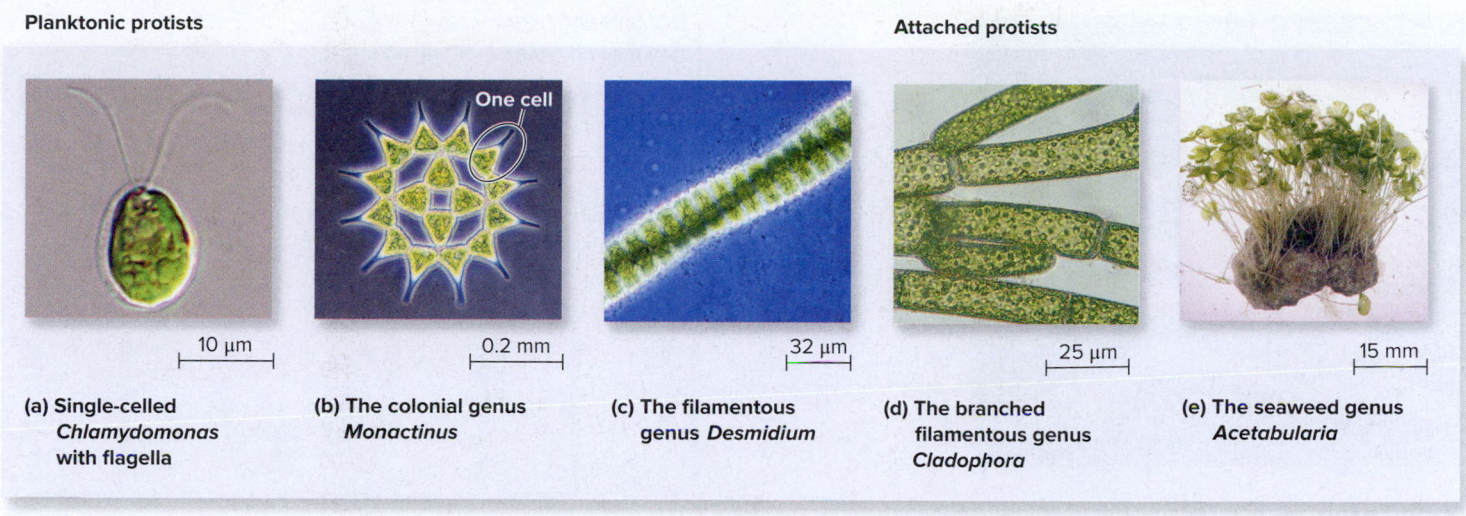

(a) **Single-celled** *Chlamydomonas* **with flagella**    (b) **The colonial genus** *Monactinus*    (c) **The filamentous genus** *Desmidium*    (d) **The branched filamentous genus** *Cladophora*    (e) **The seaweed genus** *Acetabularia*

**Figure 28.3**  **The diversity of algal body types reflects their habitats.** **(a)** The single-celled flagellate genus *Chlamydomonas* occurs in the phytoplankton of lakes. **(b)** The colonial genus *Monactinus* is composed of several cells arranged in a lacy star shape, which helps to keep this alga afloat in water and avoid being consumed by aquatic animals. **(c)** The filamentous genus *Desmidium* occurs as a twisted row of cells. **(d)** The branched filamentous genus *Cladophora* grows attached to nearshore surfaces and is large enough to see with the unaided eye. **(e)** The relatively large seaweed genus *Acetabularia* lives on rocks and coral rubble in shallow tropical oceans. The body of *Acetabularia* is a single very large cell.

a: ©Brian P. Piasecki; b: ©Roland Birke/Phototake; c–e: ©Linda Graham

Many protists live within **periphyton**—communities of microorganisms attached by mucilage to underwater surfaces such as rocks, sand, and plants. Because sinking is not a problem for attached protists, these often produce multicellular bodies, such as branched filaments (**Figure 28.3d**). Photosynthetic protists large enough to see with the unaided eye are known as **macroalgae**, or seaweeds. Although the bodies of some macroalgae are very large single cells (**Figure 28.3e**), most macroalgae are multicellular, often producing large and complex bodies. Macroalgae usually grow attached to underwater surfaces such as rocks, sand, docks, ship hulls, or offshore oil platforms. Seaweeds require sunlight and carbon dioxide for photosynthesis and growth, so most of them grow along coastal shorelines, fairly near the water's surface. Macroalgae serve as refuges for aquatic animals, generate large amounts of organic carbon that enters aquatic food chains, and play additional important ecological roles. Humans harvest some macroalgae for use as food or crop fertilizers or as sources of industrial chemicals to make diverse commercial products.

## Protists Display Diverse Types of Motility

Microscopic protists have evolved diverse ways to propel themselves in moist environments. Swimming by means of flagella or cilia, amoeboid movement, and gliding are major types of protist movements.

Many types of photosynthetic and heterotrophic protists are able to swim because they produce one or a few eukaryotic flagella—cellular extensions whose movement is based on interactions between microtubules and the motor protein dynein (refer back to Figure 4.17). Eukaryotic flagella rapidly bend and straighten, thereby pulling or pushing cells through the water. Protists that use flagella to move in water are commonly known as **flagellates** (Figure 28.3a). Flagellates are typically composed of one or only a few cells and are small—usually from 2 to 20 μm long—because flagellar motion is not powerful enough to keep larger bodies from sinking. Some flagellate protists are sedentary, living attached to underwater surfaces. These protists use flagella to collect bacteria and other small particles for food. Macroalgae and other immobile protists often produce small, flagellate reproductive cells that allow these protists to mate and disperse to new habitats.

An alternative type of protist motility relies on cilia, tiny hairlike extensions on the outsides of cells. Cilia are structurally similar to eukaryotic flagella but are shorter and more abundant (**Figure 28.4**). Protists that move by means of cilia are **ciliates**. Having many cilia allows ciliates to achieve larger sizes than flagellates yet still remain buoyant in water.

A third type of motility is amoeboid movement. This kind of motion involves extending cytoplasm into lobes, known as pseudopodia (from the Greek, meaning false feet). Once these pseudopodia move toward a food source or other stimulus, the rest of the cytoplasm flows after them, thereby changing the shape of the entire

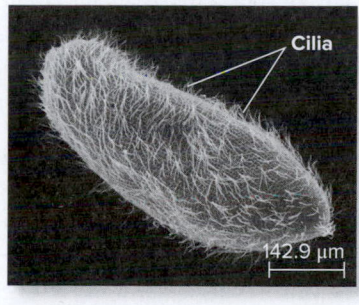

**Figure 28.4**  **SEM image of a member of the ciliate genus** *Paramecium*, **showing numerous cilia on the cell surface.** ©Dennis Kunkel Microscopy, Inc./Phototake

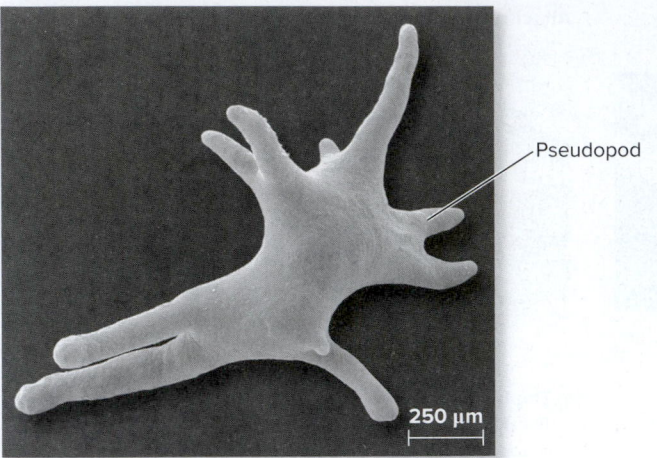

**Figure 28.5** SEM image of a member of the amoebozoan genus *Pelomyxa*, showing pseudopodia. ©Steve Gschmeissner/Science Source

 **Core Skill: Connections** Look ahead to Figure 34.3. What kind of mobile, amoeba-shaped cells carry materials within the bodies of the early-diverging animals known as sponges?

organism as it creeps along. Protist cells that move by pseudopodia are described as **amoebae** (**Figure 28.5**).

Finally, many diatoms, the malarial parasite genus *Plasmodium*, and some other protists glide along surfaces in a snail-like fashion by secreting protein or carbohydrate slime. With the exception of ciliates, the mode of motility does not correspond with the phylogenetic classification of protists, our next topic.

## 28.2 Evolution and Relationships

### Learning Outcomes:

1. Describe a distinctive structural characteristic for each of the seven eukaryotic supergroups.

2. List at least one species of each eukaryotic supergroup that is important to human life.

3. **CoreSKILL** » Draw a diagram showing how the process of endosymbiosis has affected eukaryotic diversity, and predict how this process could continue to affect biodiversity in the future.

At one time, protists were classified into a single kingdom. However, modern phylogenetic analyses based on comparisons of DNA sequences and cellular features reveal that protists do not form a monophyletic group. The relationships of some protists are uncertain or disputed, and new protist species are continually being discovered. As a result, concepts of protist evolution and relationships have been changing as new information becomes available.

Even so, molecular and cellular data reveal that many protist phyla can be classified within several eukaryotic **supergroups** that each display distinctive features (**Figure 28.6**). All of the eukaryotic supergroups include phyla of protists; some, in fact, contain only protist phyla. The supergroup Opisthokonta includes the multicellular animal and fungal kingdoms and related protists, whereas

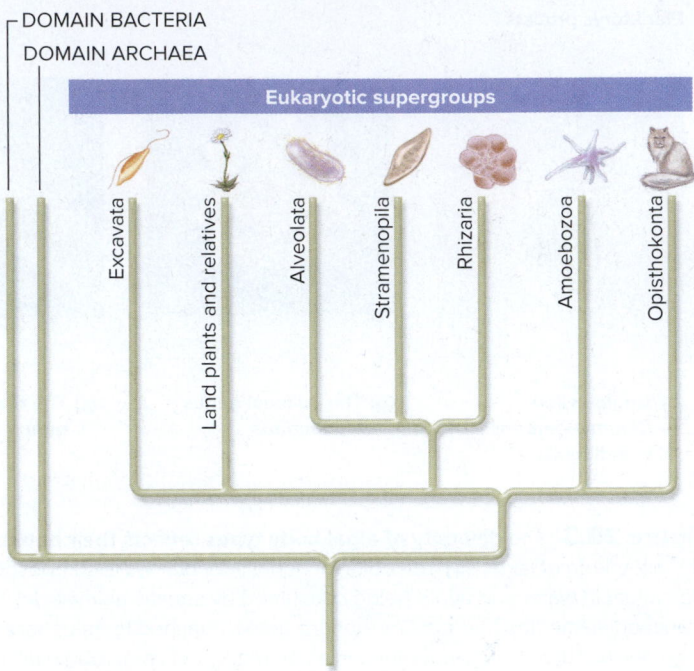

**Figure 28.6** A phylogenetic tree showing the major eukaryotic supergroups. Each of the eukaryotic supergroups shown here includes some protist phyla, and most supergroups consist only of protists. Additional eukaryotic lineages exist but are not shown in this streamlined diagram.

 **Core Concept: Evolution** All species (past and present) are related by an evolutionary history.

another supergroup includes the multicellular plant kingdom and the protists most closely related to it. The study of such protists helps to reveal how multicellularity originated in animals, fungi, and plants.

In this section, we survey the eukaryotic supergroups, focusing on the defining features and evolutionary importance of the major protist phyla. We will also examine ways in which protists are important ecologically or in human affairs.

### Cells of Many Protists Classified in Excavata have a Feeding Groove

The protist supergroup known as the Excavata originated very early among eukaryotes, so this supergroup is important in understanding the early evolution of eukaryotes.

Many of the Excavata feed by ingesting small particles of food in their aquatic habitats. Once food particles are collected within the feeding groove, they are then taken into cells by a type of endocytosis known as **phagocytosis** (from the Greek, meaning cellular eating; **Figure 28.7**). During phagocytosis, a vesicle of plasma membrane surrounds each food particle and pinches off within the cytoplasm. Enzymes within these food vesicles break the food particles down into small molecules that, upon their release into the cytosol, can be used for energy.

Phagocytosis is also the basis for an important evolutionary process known as **endosymbiosis**, a symbiotic association in which a

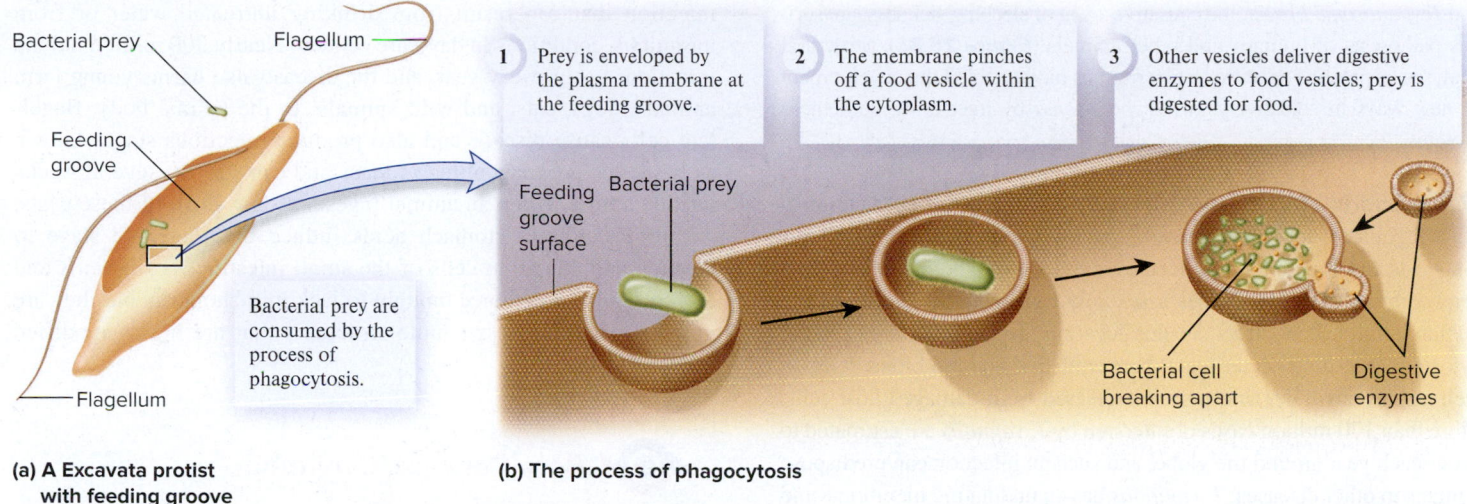

1  Prey is enveloped by the plasma membrane at the feeding groove.

2  The membrane pinches off a food vesicle within the cytoplasm.

3  Other vesicles deliver digestive enzymes to food vesicles; prey is digested for food.

Bacterial prey

Flagellum

Feeding groove

Bacterial prey are consumed by the process of phagocytosis.

Flagellum

Feeding groove surface

Bacterial prey

Bacterial cell breaking apart

Digestive enzymes

**(a) A Excavata protist with feeding groove**

**(b) The process of phagocytosis**

**Figure 28.7** **Feeding groove and phagocytosis displayed by many species of supergroup Excavata.** (a) Diagram of *Jakoba libera*, showing flagella emerging from the feeding groove. (b) Diagram of phagocytosis, the process by which food particles are consumed in a feeding groove.

**Concept Check:** *What happens to food particles after they enter a feeding groove?*

smaller species known as the endosymbiont lives within the body of a larger species known as the host. Phagocytosis provides a way for protist cells that function as hosts to take in prokaryotic or eukaryotic cells that function as endosymbionts. Such endosymbiotic cells confer valuable traits and are not digested. Endosymbiosis has played a particularly important role in protist evolution. For example, early in protist history, endosymbiotic bacterial cells gave rise to mitochondria, the organelles that are the major site of ATP synthesis in most eukaryotic cells (look back at Figure 4.31). Consequently, most protists possess mitochondria, though these may be highly modified in some species. Three protist groups classified within Excavata are described next.

***Euglenoids***   The members of Excavata known as euglenoids possess unique, interlocking, ribbon-like protein strips just beneath their plasma membranes (**Figure 28.8a**). These strips make the surfaces of some euglenoids so flexible that they can crawl through mud. Many euglenoids are colorless and heterotrophic, but *Euglena* and some other genera possess green plastids and are photosynthetic. Plastids are organelles in plant and algal cells that are distinguished by their synthetic abilities and that were acquired via endosymbiosis. Many euglenoids possess a light-sensing system that includes a conspicuous red structure known as an eyespot, or stigma, and light-detecting molecules located in a swollen region at the base of a flagellum. These structures enable green euglenoids to detect light environments that are optimal for photosynthesis. Most euglenoids produce conspicuous carbohydrate-storage particles that are held in the cytoplasm.

***Kinetoplastids***   The heterotrophic protists informally known as kinetoplastids (formally Kinetoplastea) are named for a large mass of DNA known as a kinetoplast that occurs in their single large mitochondrion (**Figure 28.8b**). These protists lack plastids, but they do possess an unusual modified peroxisome in which glycolysis takes place, known as a glycosome. (Recall that in most eukaryotes, glycolysis occurs in the cytosol.) Some kinetoplastids, including *Leishmania* (see Figure 28.8b), which causes an ulcerative skin disease and can result in organ damage,

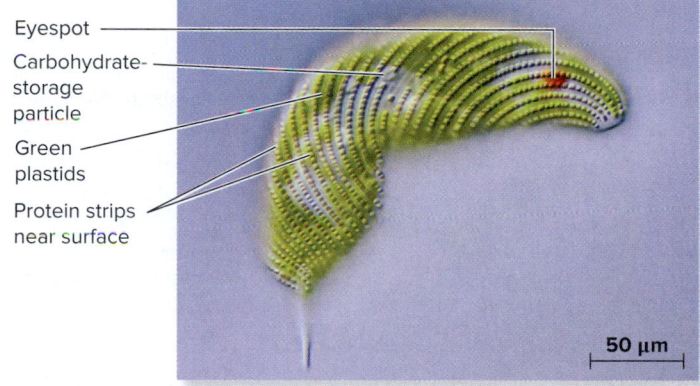

Eyespot

Carbohydrate-storage particle

Green plastids

Protein strips near surface

50 µm

**(a) *Euglena***

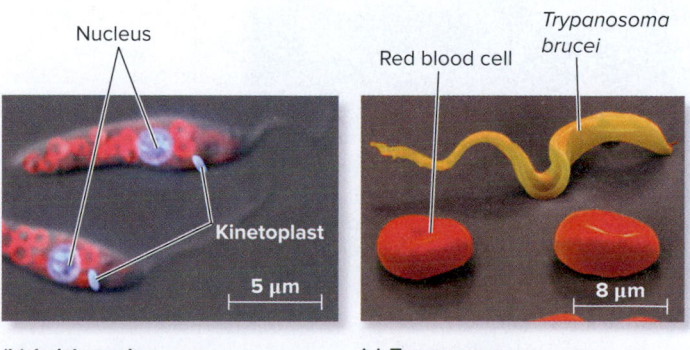

Nucleus

Kinetoplast

5 µm

**(b) *Leishmania***

Red blood cell

*Trypanosoma brucei*

8 µm

**(c) *Trypanosoma***

**Figure 28.8** **Representative euglenoids and kinetoplastids.** (a) *Euglena* has ribbon-like protein strips near its surface, internal green plastids, white carbohydrate-storage particles, and a red eyespot. (b) Fluorescence light micrograph of *Leishmania* showing the kinetoplast DNA mass typical of kinetoplastid mitochondria. (c) In this artificially colorized SEM, an undulating kinetoplastid (*Trypanosoma*) appears near disc-shaped red blood cells. a: ©Gerd Guenther/SPL/Science Source; b: ©Ross Waller, University of Cambridge, UK; c: ©Eye of Science/Science Source

and *Trypanosoma brucei*, the causative agent of sleeping sickness, are serious pathogens of humans and other animals (**Figure 28.8c**). Structural analyses of proteins specific to glycosomes have enabled the development of new ways to selectively kill *Trypanosoma* by interfering with these proteins.

**Metamonads**   Metamonads (formally Metamonada) are heterotrophic flagellates; some are parasitic species that attack the cells of animal hosts and absorb food molecules released from them. For example, *Trichomonas vaginalis* causes a sexually transmitted infection of the human genitourinary tract. In this location, *T. vaginalis* uses phagocytosis to consume bacteria and host epithelial and red blood cells, as well as carbohydrates and proteins released from damaged host cells. More than 170 million cases of infection by *T. vagnalis* are estimated to occur each year around the globe, and such an infection can predispose humans to other diseases. *T. vaginalis* has an undulating membrane and flagella that allow it to move over mucus-coated skin (**Figure 28.9a**).

*Giardia intestinalis* (previously known as *G. lamblia*), another parasitic protist, contains two active nuclei and produces eight flagella (**Figure 28.9b**). *G. intestinalis* causes giardiasis, an intestinal

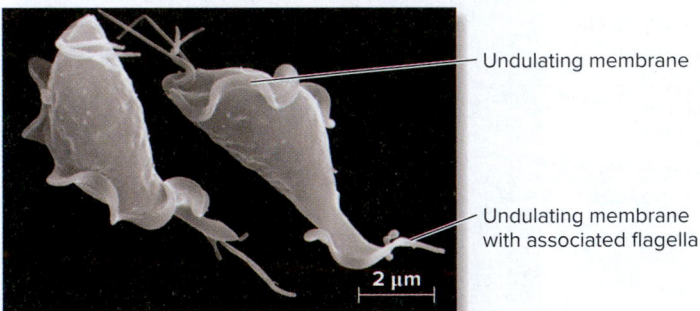

(a) *Trichomonas vaginalis*

Undulating membrane

Undulating membrane with associated flagella

2 μm

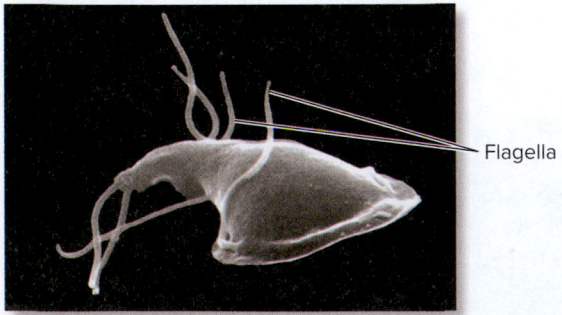

(b) *Giardia intestinalis*

Flagella

**Figure 28.9   Parasitic members of the supergroup Excavata.**
(a) *Trichomonas vaginalis*. (b) *Giardia intestinalis*. These specialized heterotrophic flagellates use flagella to disperse across the surfaces of moist host tissues; the flagellates then absorb nutrients from living cells. These images were made with a scanning electron microscope (SEM) that employs electrons rather than visible light, with the result that cellular structures do not appear in color. a: ©David M. Phillips/Science Source; b: Source: CDC/Dr. Stan Erlandsen, Dr. Dennis Feely

*Concept Check:*   *How do these two parasitic protists differ in how they are transmitted from one human host to another?*

infection that can result from drinking untreated water or from unsanitary conditions in day-care centers. Nearly 300 million human infections occur every year, and the disease also harms young farm animals, dogs, cats, and wild animals. In the animal body, flagellate cells cause disease and also produce infectious stages known as cysts that are transmitted in feces and can survive several weeks outside a host. When an animal ingests as few as 10 of these cysts, within 15 minutes stomach acids induce the flagellate stage to develop and adhere to cells of the small intestine. *T. vaginalis* and *G. intestinalis* were once thought to lack mitochondria, but they are now known to possess simple structures that are highly modified mitochondria.

### Core Concept: Evolution

## Genome Sequences Reveal the Different Evolutionary Pathways of *Trichomonas vaginalis* and *Giardia intestinalis*

In 2007, genome sequences were reported for *T. vaginalis* and *G. intestinalis*. A comparison of their genomes reveals similarities and differences in the evolution of their parasitic lifestyles. One common feature is that horizontal gene transfer from bacterial or archaeal donors has powerfully affected both genomes. About 100 *G. intestinalis* genes are likely to have been acquired via horizontal gene transfer. In *T. vaginalis*, more than 150 cases of likely horizontal gene transfer were identified, with most transferred genes encoding metabolic enzymes such as those involved in carbohydrate or protein metabolism. Another similarity between *T. vaginalis* and *G. intestinalis* revealed by comparative genomics is the apparent absence of the cytoskeletal protein myosin, which is present in most eukaryotic cells.

Despite these similarities, the genome sequences of *T. vaginalis* and *G. lamblia* reveal some dramatic differences. The *G. intestinalis* genome is quite compact, with only 11.7 megabase pairs (Mb), and the organism has relatively simple metabolic pathways and machinery for DNA replication, transcription, and RNA processing. In contrast, the *T. vaginalis* genome is a surprisingly large 160 Mb in size. *T. vaginalis* has a core set of about 60,000 protein-encoding genes, one of the greatest coding capacities known among eukaryotes. The additional genes provide an expanded capacity for biochemical degradation. Because most trichomonads inhabit animal intestines, the genomic data suggest that the large genome size of *T. vaginalis* is related to its ecological transition to a new habitat, the urogenital tract.

## Land Plants and Related Algae Share Similar Genetic Features

The supergroup that includes land plants also encompasses several protist phyla (**Figure 28.10**). The land plants, also known as the kingdom Plantae (described more fully in Chapters 31 and 32), evolved from green algal ancestors. Together, plants and some closely related

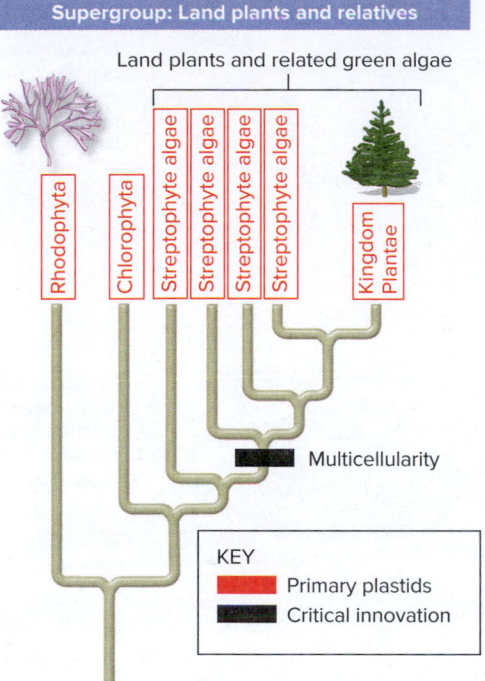

**Supergroup: Land plants and relatives**

Land plants and related green algae

Rhodophyta
Chlorophyta
Streptophyte algae
Streptophyte algae
Streptophyte algae
Streptophyte algae
Kingdom Plantae

Multicellularity

KEY
■ Primary plastids
■ Critical innovation

**Figure 28.10** **A phylogenetic tree of the supergroup that includes land plants and their close protist relatives.** Note that plant multicellularity first arose in closely related streptophyte algae. Many chlorophyte and red algae are macroalgae in which multicellularity arose independently.

green algae form the clade Streptophyta, informally known as streptophytes, whereas most green algae are classified in the phylum Chlorophyta. The red algae, classified in the phylum Rhodophyta, are also regarded as close relatives of green algae and land plants.

**Green Algae**  Diverse structural types of green algae (see Figure 28.3) occur in fresh water, the ocean, and on land or ice surfaces. Most of the green algae are photosynthetic, and their cells contain the same types of plastids and photosynthetic pigments that are present in land plants. Some green algae are responsible for harmful algal growths, but others are useful as food for aquatic animals, as model organisms, and as sources of renewable oil supplies. Many green algae possess flagella or the ability to produce them during the development of reproductive cells. Green algae are increasingly important in medicine because they produce channel rhodopsins, light-activated ion channels. Green algae use these channel proteins to detect and respond to light, and researchers are studying the proteins in an attempt to understand and possibly treat blindness in animals.

**Red Algae**  Most species of the protists known as red algae are multicellular marine macroalgae (**Figure 28.11**). The red appearance of these algae is caused by the presence of distinctive photosynthetic pigments that are absent from green algae and land plants. Red algae characteristically lack flagella—a feature that has strongly influenced the evolution of this group, resulting in unusually complex life cycles (illustrated in Figure 28.26b). These life cycles are important to humans because red algae are cultivated in ocean waters for the production of billions of dollars worth of food or industrial and scientific materials yearly. For

**(a)** *Calliarthron*          **(b)** *Chondrus crispus*

**Figure 28.11** **Representative red algae (Rhodophyta).** **(a)** The genus *Calliarthron* has cell walls that are impregnated with calcium carbonate. This stony, white material makes the red alga appear pink. **(b)** *Chondrus crispus* is an edible red seaweed. a: ©Lee W. Wilcox; b: ©Andrew J. Martinez/Science Source

example, the sushi wrappers called nori are composed of the sheetlike red algal genus *Porphyra*, which is grown in ocean farms. Carrageenan, agar, and agarose are complex polysaccharides that are extracted from red algae and are essential to the food industry and in biology laboratories for cultivating microorganisms and working with DNA.

***Primary Plastids and Primary Endosymbiosis***  The plastids of red algae resemble those of green algae and land plants (and differ from those of most other algae) in having an enclosing envelope composed of two membranes (**Figure 28.12**). Such plastids, known as **primary plastids**, are thought to have originated via a process known as **primary endosymbiosis** (**Figure 28.13**). During primary endosymbiosis, heterotrophic host cells captured cyanobacterial cells via phagocytosis but did not digest them. These endosymbiotic cyanobacteria provided host cells with photosynthetic capability and other useful biochemical pathways and eventually evolved into primary plastids (Figure 28.13a).

Endosymbiotic acquisitions of plastids and mitochondria resulted in massive horizontal gene transfer from the endosymbiont to the host nucleus. As a result of such gene transfer, many of the proteins needed by plastids and mitochondria are synthesized in the host cytosol and then targeted to these organelles. All cells of plants, green algae, and red algae contain one or more plastids, and most of these

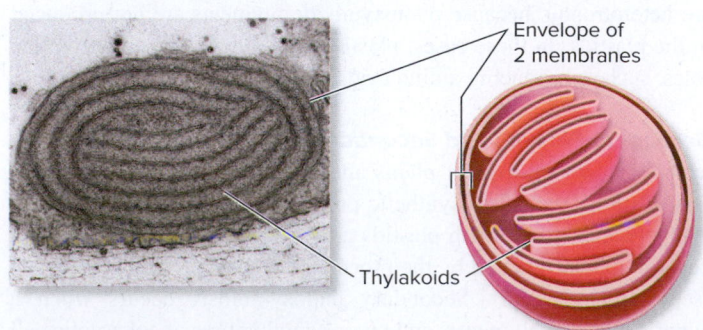

Envelope of 2 membranes

Thylakoids

**Figure 28.12** **A primary plastid, showing an envelope composed of two membranes.** The plastid shown here is red, but primary plastids can also be green or blue-green in color. ©Joe Scott, Department of Biology, College of William and Mary, Williamsburg, VA 23187

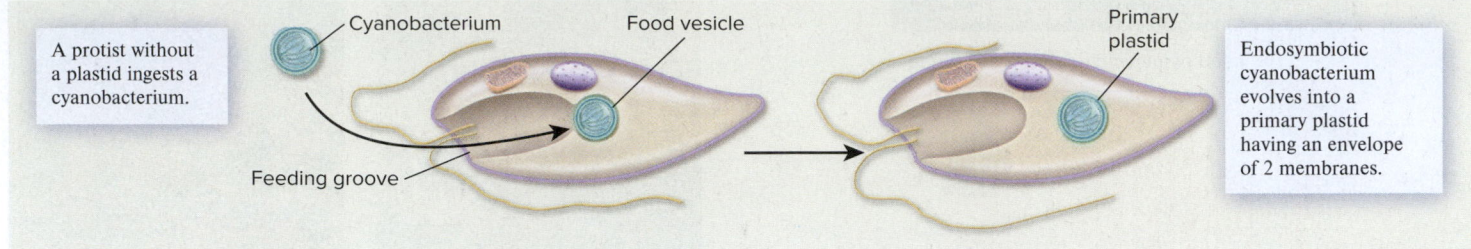

**(a) Primary endosymbiosis**

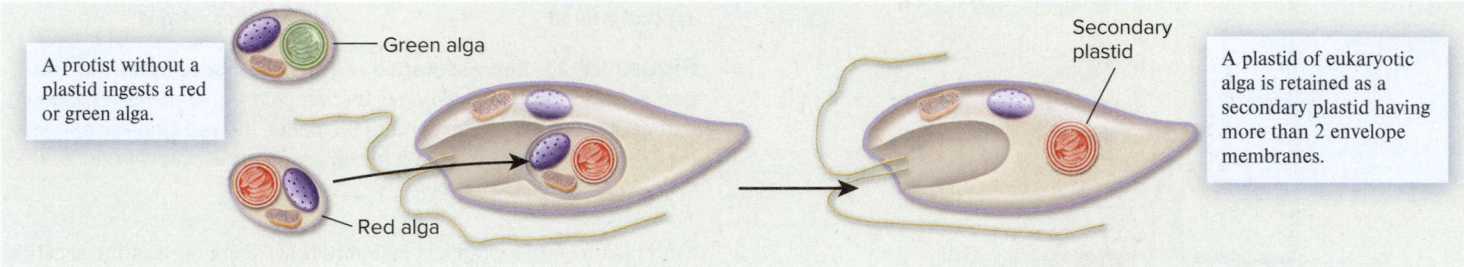

**(b) Secondary endosymbiosis**

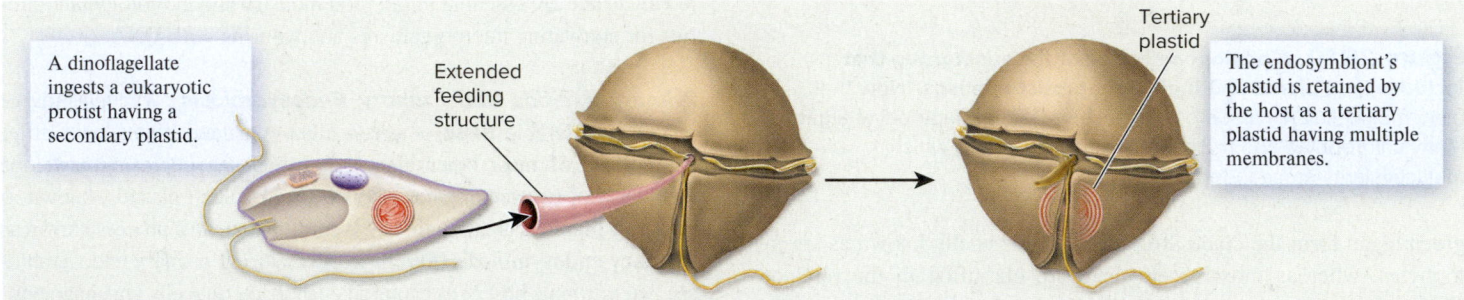

**(c) Tertiary endosymbiosis**

**Figure 28.13  Primary, secondary, and tertiary endosymbiosis.  (a)** Primary endosymbiosis involves the acquisition of a cyanobacterial endosymbiont by a host cell without a plastid. During the evolution of a primary plastid, the bacterial cell wall is lost, and most endosymbiont genes are transferred to the host nucleus. **(b)** Secondary endosymbiosis involves the acquisition by a host cell of a eukaryotic endosymbiont that contains one or more primary plastids. During the evolution of a secondary plastid, most components of the endosymbiont cell are lost, but a plastid is often retained within an envelope of endoplasmic reticulum. **(c)** Tertiary endosymbiosis involves the acquisition by a host cell of a eukaryotic endosymbiont that possesses secondary plastids.

organisms are photosynthetic. However, some species (or some of the cells within the multicellular bodies of photosynthetic species) are heterotrophic because photosynthetic pigments are not produced in the plastids. In these cases, plastids play other essential metabolic roles, such as producing amino acids and fatty acids.

***Secondary Plastids and Secondary Endosymbiosis***   In contrast to the primary plastids of plants and green and red algae, the plastids of most other photosynthetic protists are derived from a photosynthetic eukaryote. Such plastids are known as **secondary plastids** because they originated by the process of **secondary endosymbiosis** (see Figure 28.13b). Secondary endosymbiosis occurs when a eukaryotic host cell ingests and retains another type of eukaryotic cell that already has one or more primary plastids, such as a red or green alga. Such eukaryotic endosymbionts are often enclosed by the endoplasmic reticulum (ER), explaining why secondary plastids typically have envelopes of more than two membranes. Although most of the

endosymbiont's cellular components are digested over time, its plastids are retained, providing the host cell with photosynthetic capacity and other biochemical capabilities.

Cryptomonads (**Figure 28.14a**) and haptophytes are algal phyla that include single-celled flagellates whose plastids originated by secondary endosymbiosis involving the incorporation of plastids derived from a red alga. Occurring in marine and fresh waters, cryptomonads are excellent sources of the fatty-acid-rich food essential to aquatic animals. Haptophytes are primarily unicellular marine photosynthesizers; some have flagella and others do not. Some haptophytes are known as the coccolithophorids because they produce a covering of intricate white calcium carbonate discs known as coccoliths (**Figure 28.14b**). Coccolithophorids often form massive ocean growths that are visible from space and play important roles in Earth's climate by reflecting sunlight. In some places, coccoliths produced by huge populations of ancient coccolithophorids accumulated on the ocean floor, together with the calcium carbonate remains of other protists, for millions

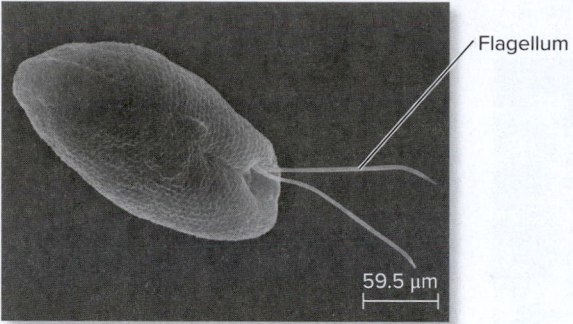

(a) A cryptomonad

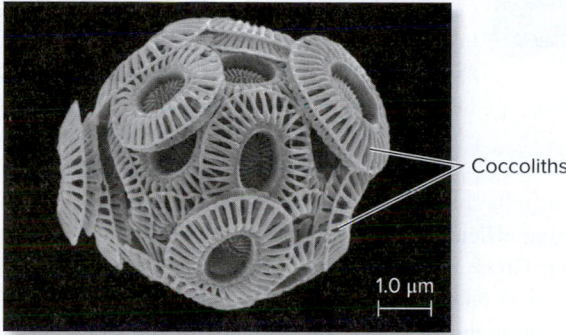

(b) A haptophyte coccolithophorid

(c) Fossil deposit containing
coccolithophorids

**Figure 28.14  Representative cryptomonads and haptophytes.**
**(a)** A cryptomonad flagellate. **(b)** A type of haptophyte known as a
coccolithophorid, covered with disc-shaped coccoliths made of calcium
carbonate. **(c)** Fossil carbonate remains of haptophyte algae and
protozoan protists known as foraminifera that were deposited over
millions of years formed the white cliffs of Dover in England. a: ©Dennis
Kunkel Microscopy, Inc./Phototake; b: ©Steve Gschmeissner/Science Source;
c: ©Stockbyte/Getty Images

of years. These deposits were later raised above sea level, forming
massive limestone formations or chalk cliffs such as those visible at
Dover, on the southern coast of England (**Figure 28.14c**).

## Membrane Sacs Lie at the Cell Periphery of Alveolata

The three supergroups Alveolata, Stramenopila, and Rhizaria seem to
be closely related, based on recent phylogenetic studies (**Figure 28.15**).
These photosynthetic protists have plastids acquired by secondary
endosymbiosis or, in some cases, by tertiary endosymbiosis.

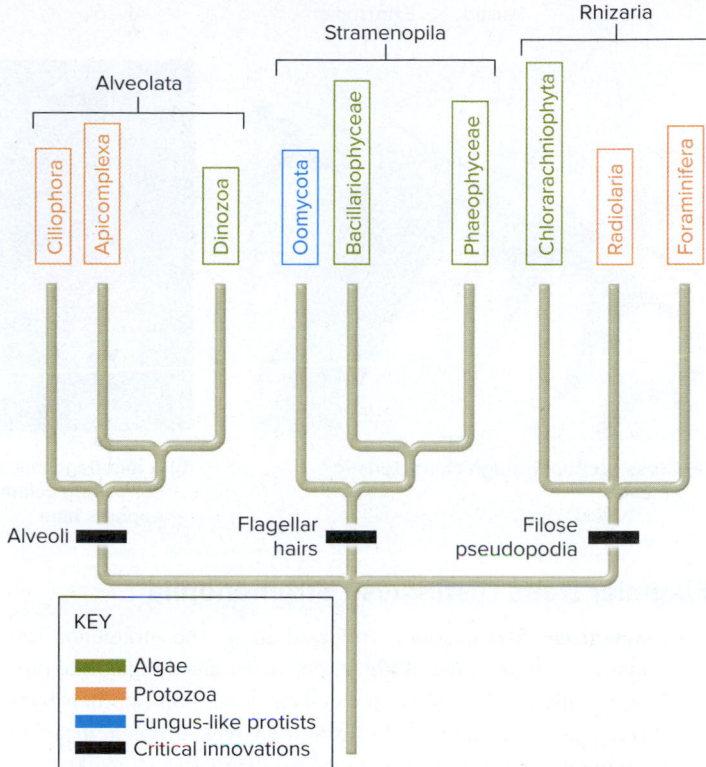

**Figure 28.15  A phylogenetic tree illustrating the close
relationship among the supergroups Alveolata, Stramenopila,
and Rhizaria.** Some stramenopiles, such as giant kelps, are
multicellular.

Turning first to Alveolata, we see that it includes three important
phyla: (1) the Ciliophora, or ciliates; (2) the Apicomplexa, a medically important group of parasites; and (3) the Dinozoa, known as
dinoflagellates. Apicomplexans include the malarial agent *Plasmodium* (see Section 28.4), the related protist *Cryptosporidium parvum*,
whose effects were noted in the chapter opening, and other serious
pathogens of humans and other animals. Dinoflagellates are recognized both for their mutualistic relationship with reef-building corals
(look ahead to Figure 54.26b) and for the harmful blooms (red tides)
that some species produce (see Section 28.3). The supergroup Alveolata is named for saclike membranous vesicles known as alveoli that
are present at the cell periphery in all of these phyla (**Figure 28.16a**).

The alveoli of some dinoflagellates seem empty, so the cell surface appears smooth. By contrast, the alveoli of many dinoflagellates contain plates of cellulose, which form an armor-like enclosure
(**Figure 28.16b**). These plates are often modified in ways that provide
an adaptive advantage, such as protection from predators or increased
ability to float.

About half of dinoflagellate species are heterotrophic, and half
possess photosynthetic plastids of diverse types that originated by
secondary or even tertiary endosymbiosis; therefore, these are known
as secondary or tertiary plastids. **Tertiary plastids** were obtained by
**tertiary endosymbiosis**—the acquisition by hosts of plastids from
cells that possessed secondary plastids (see Figure 28.13c). Species
having tertiary plastids have received genes by horizontal transfer
from diverse genomes.

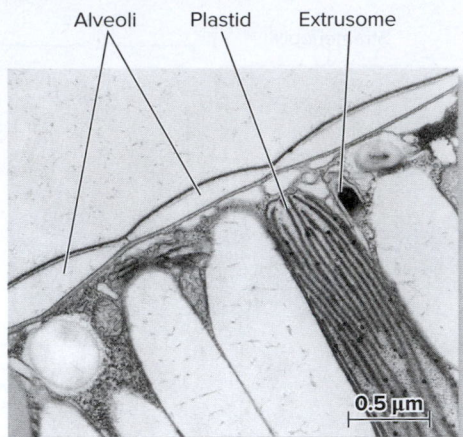

Alveoli    Plastid    Extrusome

0.5 µm

**(a)** Cross section through characteristic alveoli

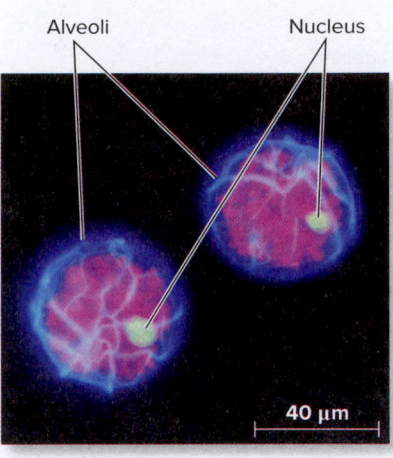

Alveoli                Nucleus

40 µm

**(b)** A dinoflagellate with alveoli containing cellulose that here appears blue

**Figure 28.16** **Dinoflagellates of the supergroup Alveolata and their characteristic alveoli.** **(a)** Sac-shaped membranous vesicles known as alveoli lie beneath the plasma membrane of a dinoflagellate, along with defensive projectiles, called extrusomes, that are ready for discharge. **(b)** Fluorescence microscopy reveals that alveoli of the dinoflagellate *Alexandrium catenella* contain cellulose plates, which glow blue when treated with a cellulose-binding dye. The nucleus appears green because DNA has bound a fluorescent dye, and chlorophyll self-fluoresces red. a: ©Lee W. Wilcox; b: Courtesy Joseph Wong and Alvin Kwok

## Flagellar Hairs Distinguish Stramenopila

The supergroup Stramenopila (referred to as the stramenopiles) encompasses a wide range of algae, protozoa, and fungus-like protists that usually produce flagellate cells at some point in their lives (see Figure 28.15). Stramenopila (from the Greek *stramen*, meaning straw, and *pila*, meaning hair) is named for distinctive strawlike hairs that occur on the surfaces of flagella of these protists (**Figure 28.17**).

These flagellar hairs function something like oars to greatly increase swimming efficiency. Stramenopiles are also known as heterokonts (from the Greek, meaning different flagella), because the two flagella often present on swimming cells have slightly different structures.

Heterotrophic stramenopiles include the fungus-like protist *Phytophthora infestans*, which causes the serious potato disease known as late blight. *P. infestans* is responsible for an estimated $7 billion in crop losses every year. Photosynthetic stramenopiles include diatoms (Bacillariophyceae), whose glasslike silicate cell walls are elaborately ornamented with pores, lines, and other intricate features (**Figure 28.18a**). Vast accumulations of the translucent walls of ancient diatoms, known as diatomite or diatomaceous earth, are mined for use in reflective paint and other industrial products. Recent genome sequencing projects have focused on the processes by which diatoms produce their detailed silicate structures; understanding

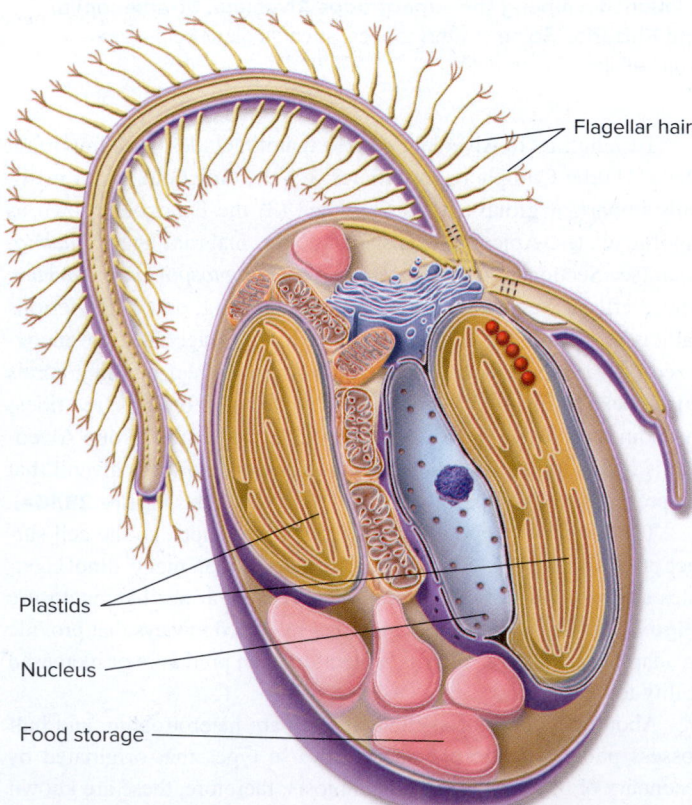

Flagellar hairs

Plastids

Nucleus

Food storage

**Figure 28.17** **A flagellate stramenopile cell, showing characteristic flagellar hairs.**

*Concept Check:* *How do the flagellar hairs aid stramenophile cell motion?*

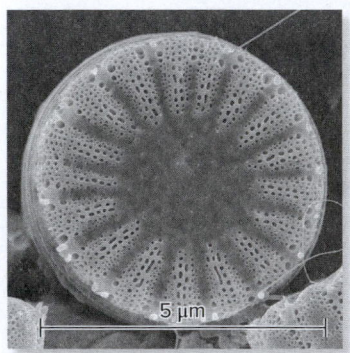

5 µm

**(a)** Diatom

**(b)** Kelp forest

**Figure 28.18** **Stramenopiles include diatoms and giant kelps.** **(a)** SEM of the silicate cell wall of the common diatom *Cyclotella meneghiniana*, showing elaborate ornamentation of the structure. The many pores in the colorless silicate wall lighten the cell, helping to keep it afloat in the water. **(b)** Forests of giant kelps occur along many ocean shores, providing habitat for diverse organisms. a: ©Linda Graham; b: ©Jeff Rotman/Science Source

*Concept Check:* *In what ways are kelp forests economically important?*

of these processes may prove useful in industrial microfabrication applications.

Diverse, photosynthetic brown algae (Phaeophyceae) are sources of industrial products such as the polysaccharide emulsifiers known as alginates. Brown algae acquired the ability to produce alginates by means of horizontal transfer of genes from bacteria. The brown algae known as giant kelps are ecologically important because they form extensive forests in cold and temperate coastal oceans (**Figure 28.18b**). Kelp forests are essential nurseries for fish and shellfish. The reproductive processes of diatoms and kelps are described in Section 28.4.

### Spiky Cytoplasmic Extensions Are Present on the Cells of Many Protists Classified in Rhizaria

Several groups of flagellates and amoebae that have thin, hairlike extensions of their cytoplasm—known as filose pseudopodia—are classified into the supergroup Rhizaria (from the Greek *rhiza*, meaning root) (see Figure 28.15). Rhizaria includes the phylum Chlorarachniophyta, whose spider-shaped cells possess secondary plastids obtained from endosymbiotic green algae. Other examples of Rhizaria are the Radiolaria (**Figure 28.19a**) and Foraminifera (**Figure 28.19b**)—two phyla of ocean plankton that produce exquisite mineral shells. Fossil shells of foraminiferans are widely used to infer past climatic conditions, because ratios of stable oxygen isotopes contained in the shells can be analyzed to reconstruct past water temperatures.

### Amoebozoa Includes Many Types of Amoebae with Pseudopodia

The supergroup Amoebozoa includes many types of amoebae that move by extension of pseudopodia (see Figure 28.5). Several types of protists known as slime molds are classified in this supergroup. One example, *Dictyostelium discoideum*, is widely used as a model organism for understanding movement, communication among cells, and development. In response to starvation, large numbers of *Dictyostelium* amoebae aggregate into a multicellular slug that

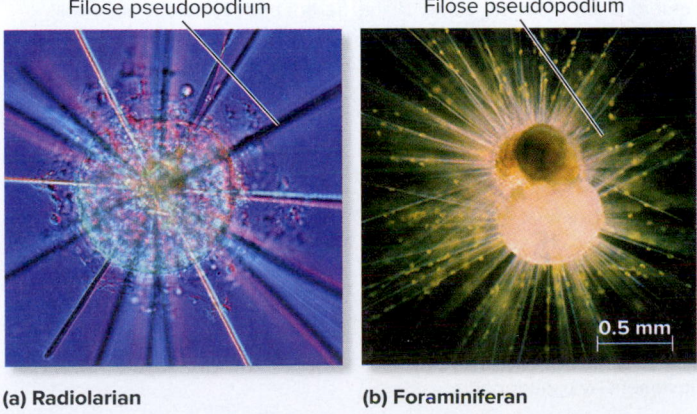

Filose pseudopodium    Filose pseudopodium

0.5 mm

**(a) Radiolarian**    **(b) Foraminiferan**

**Figure 28.19** **Representatives of supergroup Rhizaria. (a)** A radiolarian, *Acanthoplegma* spp., showing long filose pseudopodia. **(b)** A foraminiferan, showing calcium carbonate shell with long filose pseudopodia extending from pores in the shell. a: ©Claude Nuridsany & Marie Perennou/SPL/Science Source; b: ©O. Roger Anderson, Columbia University, Lamont-Doherty Earth Observatory

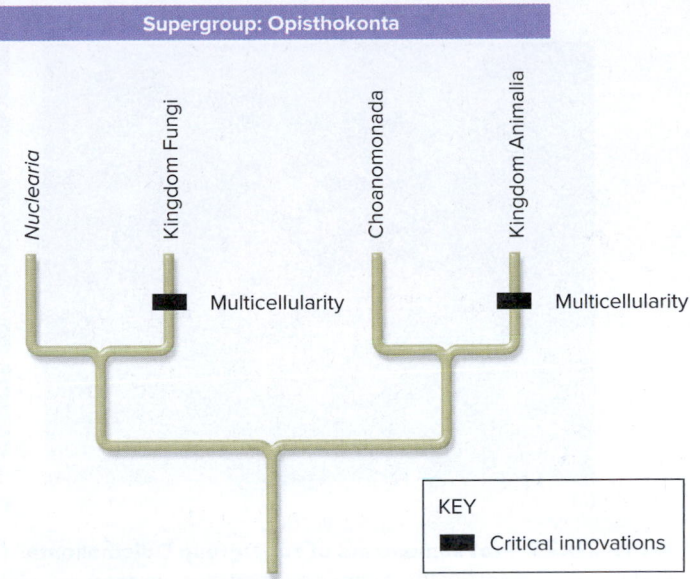

**Supergroup: Opisthokonta**

*Nuclearia*    Kingdom Fungi    Choanomonada    Kingdom Animalia

Multicellularity    Multicellularity

KEY
◼ Critical innovations

**Figure 28.20** **A phylogenetic tree of the supergroup Opisthokonta.** This supergroup includes protist phyla as well as the kingdoms Fungi and Animalia. Multicellularity arose independently in these kingdoms.

produces a cellulose-stalked structure containing many single-celled, asexual spores. In favorable conditions, these spores produce new amoebae, which feed on bacteria. A recent study revealed that some *Dictyostelium* clones carry favored bacterial food through these reproductive stages, showing a simple "farming" behavior.

### A Single Flagellum Occurs on Swimming Cells of Opisthokonta

The supergroup Opisthokonta includes the animal and fungal kingdoms and related protists (**Figure 28.20**). This supergroup is named for the presence of a single posterior flagellum on swimming cells. *Nuclearia* is a protist genus that feeds by phagocytosis and seems particularly closely related to the kingdom Fungi (**Figure 28.21a**). The more than 125 species of protists known as choanoflagellates (formally Choanomonada) are single-celled or colonial protists featuring a distinctive collar surrounding the single flagellum (**Figure 28.21b**). The collar is made of cytoplasmic extensions that filter bacterial food from water currents generated by flagellar motion.

Choanoflagellates are believed to represent the closest living relatives of animals. Evolutionary biologists interested in the origin of animals study choanoflagellates for molecular clues to this important event in our evolutionary history. Genomic studies have revealed genes that encode cell adhesion and extracellular matrix proteins. Such proteins help choanoflagellates attach to surfaces and were also essential to the evolution of multicellularity in animals. The choanoflagellate genome also encodes the p53 protein, a regulatory transcription factor that plays essential roles in the animal cell cycle, cancer, and reproduction (refer back to Figure 15.15).

The preceding survey of protist diversity, summarized in **Table 28.1**, illustrates the enormous evolutionary and ecological importance of protists. Next, we consider the diverse ways in which protists have become adapted to their environments.

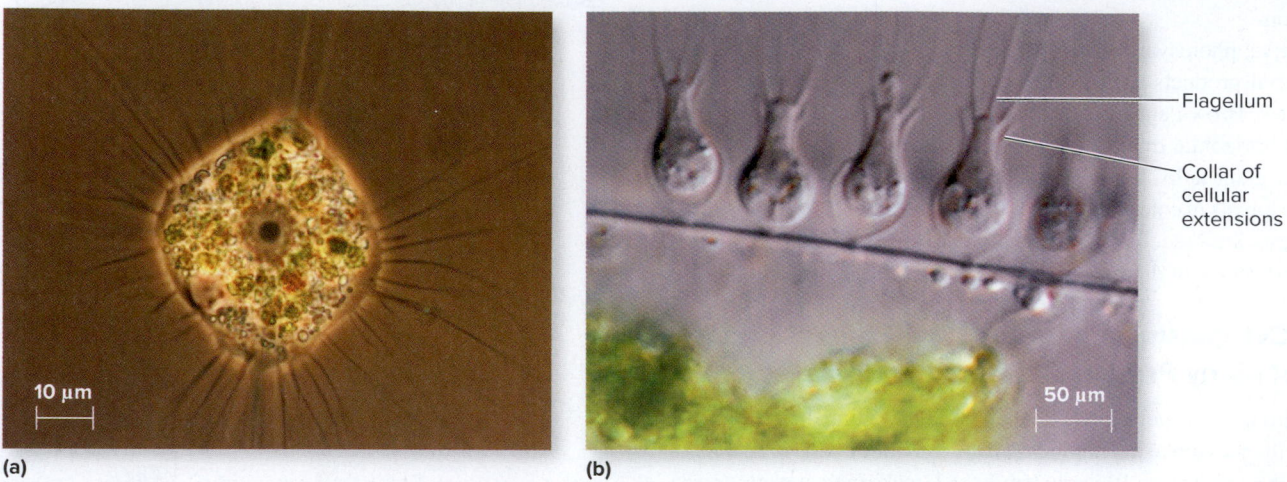

**Figure 28.21**  **Protist members of supergroup Opisthokonta.** **(a)** The amoeba *Nuclearia* represents protists most closely related to the kingdom Fungi. Some of the cytoplasmic particles are food materials that were ingested by phagocytosis. **(b)** Several choanoflagellate cells are shown on the surface of a green alga. Each cell has a distinctive collar of cellular extensions around the single flagellum, which is used to obtain food particles for ingestion by phagocytosis. a: ©Stephen Fairclough, King Lab, University of California at Berkeley; b: ©Lee W. Wilcox

*Concept Check:*  *What features of the ancient choanoflagellate ancestors of animals were important in the evolution of multicellularity, and what function do such features serve in modern choanoflagellates?*

| Table 28.1 | Eukaryotic Supergroups and Examples of Constituent Kingdoms, Phyla, Classes, or Species | |
|---|---|---|
| **Supergroup** | **KINGDOMS, Phyla, classes, or species** | **Distinguishing features** |
| Excavata | **Metamonada** (metamonads)<br>  *Giardia intestinalis*<br>  *Trichomonas vaginalis* | Excavata are unicellular flagellates that are often characterized by a feeding groove. Metamonads have modified mitochondria. |
| | **Kinetoplastea** (kinetoplastids)<br>  *Trypanosoma brucei* | Kinetoplastid mitochondria are characterized by a large mass of DNA, the kinetoplast. |
| | **Euglenida** (euglenoids) | Some euglenoids have secondary plastids derived from endosymbiotic green algae. |
| Land Plants and Relatives | **Rhodophyta** (red algae)<br>**Chlorophyta** (green algae)<br>**KINGDOM PLANTAE** and close green algal relatives | Land plants, green algae, and red algae have primary plastids derived from cyanobacteria; such plastids have two envelope membranes. |
| Alveolata | **Ciliophora** (ciliates)<br>**Apicomplexa** (apicomplexans)<br>  *Plasmodium falciparum*<br>  *Cryptosporidium parvum*<br>**Dinozoa** (dinoflagellates)<br>  *Alexandrium catenella* | Peripheral membrane sacs (alveoli); some ciliates harbor endosymbiotic algal cells or organelles; apicomplexans may possess nonphotosynthetic secondary plastids; some dinoflagellates have secondary plastids derived from red algae, some have secondary plastids derived from green algae, and some have tertiary plastids derived from diatoms, haptophytes, or cryptomonads. |
| Stramenopila | Bacillariophyceae (diatoms)<br>Phaeophyceae (brown algae)<br>*Phytophthora infestans* (fungus-like) | Strawlike flagellar hairs; secondary plastids (when present) derived from red algae; accessory pigment fucoxanthin is common in autotrophic forms. |
| Rhizaria | **Chlorarachniophyta**<br>**Radiolaria**<br>**Foraminifera** | Thin, cytoplasmic projections; secondary plastids (when present) derived from endosymbiotic green algae |
| Amoebozoa | *Entamoeba histolytica*<br>**Dictyostelia** (a slime mold phylum)<br>*Dictyostelium discoideum* | Amoeboid movement by pseudopodia |
| Opisthokonta | *Nuclearia* spp.<br>**KINGDOM FUNGI**<br>**Choanomonada** (choanoflagellates)<br>**KINGDOM ANIMALIA** | Swimming cells possess a single posterior flagellum. |

**Learning Outcomes:**

1. List four mechanisms of protist nutrition.
2. **CoreSKILL »** Predict how photosynthetic protists might respond to a persistent period of darkness.
3. Give examples of major types of protist defensive adaptations.

Protists play diverse and important ecological roles in all kinds of moist habitats. In this section, we will survey nutritional and defensive adaptations that occur widely among protists, that is, in more than one supergroup. Such adaptations help to explain protists' ecological roles.

### Protists Display Four Basic Mechanisms of Nutrition

Protists obtain nutrients by four basic mechanisms: phagotrophy, osmotrophy, photoautotrophy, and mixotrophy. Heterotrophic protists that feed by ingesting particles, or phagocytosis, are known as **phagotrophs** (see Figure 28.7). Protists that rely on osmotrophy—the uptake of small organic molecules across the cell membrane followed by their metabolism—are **osmotrophs**. Protists that feed on nonliving organic material function as decomposers, essential in breaking down wastes and releasing minerals for use by other organisms. Protists that feed on the living cells of other organisms are parasites that may cause disease in other organisms. *Trichomonas vaginalis, Giardia intestinalis,* and *Phytophthora infestans* are examples of pathogenic protists. Humans view such protists as pests when they harm us or our agricultural animals and crops, but pathogenic protists also play important roles in nature by controlling the population growth of other organisms.

Photosynthetic protists (algae) are **photoautotrophs,** organisms that can make their own organic nutrients from inorganic sources by using light energy. Because water absorbs much of the red component of sunlight, algae have evolved photosynthetic systems that compensate by capturing more of the blue-green light available underwater. For example, red algae produce the red pigment phycoerythrin, which absorbs blue-green light and transfers energy to chlorophyll *a* (see Figure 28.11). Likewise, blue-green light-absorbing fucoxanthin generates the golden and brown colors of other algae (see Figures 28.1, 28.18b). Carotene (the source of vitamin A) and lutein play similar light-absorbing roles in green algae and were inherited by their land plant descendants, today playing important roles in animal nutrition. Sunlight energy is captured in the bonds of polysaccharide and lipid molecules that function in food storage, explaining why algae of diverse types are good sources of food for aquatic animals and of renewable energy materials.

**Mixotrophs** are able to use photoautotrophy and phagotrophy or osmotrophy to obtain organic nutrients. The genus *Dinobryon* (**Figure 28.22**), consisting of photosynthetic stramenopiles that live in the phytoplankton of freshwater lakes, is a mixotrophic genus. These protists may switch back and forth between photoautotrophy and heterotrophy, depending on conditions in their environment. If sufficient light, carbon dioxide, and other minerals are available, *Dinobryon* cells produce their own organic food. If a shortage of

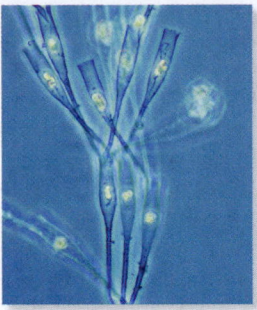

**Figure 28.22** **A mixotrophic protist.** The genus *Dinobryon* consists of colonial flagellates that occur in the phytoplankton of freshwater lakes. The photosynthetic cells have golden photosynthetic plastids and also capture and consume bacterial cells.
©Lee W. Wilcox

any of these resources limits photosynthesis or if organic food is especially abundant, *Dinobryon* cells can function as heterotrophs, consuming enormous numbers of bacteria. Mixotrophs thus have remarkable nutritional flexibility, explaining why many lineages of photosynthetic eukaryotes seem to have mixotrophic capability.

### Protists Defend Themselves in Diverse Ways

Protists use a wide variety of defensive adaptations to ward off attack. Major types of defenses are sharp projectiles that can be explosively shot from cells, light flashes, toxic compounds, and cell coverings.

Evolutionarily diverse protist cells contain structures known as extrusomes (extruded bodies) that are ejected when cells are disturbed, forming spear-like defenses (see Figure 28.16a). Some species of ocean dinoflagellates emit flashes of blue light when disturbed, explaining why ocean waters teeming with these protists display bioluminescence. The light flashes may deter herbivores by startling them, but when ingested, the dinoflagellates make the herbivores also glow, revealing them to hungry fishes. Light flashes benefit dinoflagellates by helping to reduce populations of herbivores that consume the algae.

Various protist species produce **toxins,** compounds that inhibit animal physiology and may function to deter small herbivores. Dinoflagellates are probably the most important protist toxin producers; they synthesize several types of toxins that affect humans and other animals. Why does this happen? Under natural conditions, small populations of dinoflagellates produce low amounts of toxin that do not harm large organisms. Dinoflagellate toxins become dangerous to humans when people contaminate natural waters with excess mineral nutrients such as nitrogen and phosphorus from untreated sewage, industrial discharges, or fertilizer that washes off agricultural fields. The excess nutrients fuel the development of harmful algal blooms, which then produce sufficient toxin to affect birds, aquatic mammals, fishes, and humans. Toxins can become concentrated in organisms. Humans who ingest shellfish that have accumulated dinoflagellate toxins can suffer poisoning.

The cell coverings produced by many protists also provide protection. Slimy mucilage (see the chapter opening photo) or spiny cell walls (see Figure 28.3b) provide protection from attack by herbivores or pathogens. Protective cell coverings made of polysaccharide polymers such as cellulose or minerals such as silica help to prevent osmotic damage and may enhance flotation in water. As we have seen, diatoms enclose themselves in glasslike silicate cell walls, haptophytes are covered with calcium carbonate scales, and the cells of brown and green algae secrete tough, cardboard-like cellulose walls. Cellulose-rich cell walls are also features of plant cells, but algal cellulose can be particularly resistant to chemical and microbial degradation, as described next.

## Core Skill: Process of Science

# Feature Investigation | Cook and Colleagues Demonstrated That Cellulose Helps Green Algae Avoid Chemical Degradation

The periphytic, branched green alga *Cladophora* (see Figure 28.3d) is common along marine and freshwater shorelines around the world, where it provides essential habitat for dense populations of diverse microorganisms on its extensive surfaces. Amazingly tough cellulose cell walls allow this alga to harbor extensive microbial diversity without readily decomposing. Acting in this host role, *Cladophora* provides such an important ecological service that it has been called an

ecological engineer, which is a species that strongly affects its habitat. *Cladophora* is known to be resistant to microbial decay, leading to the buildup of organic carbon in aquatic environments, but the basis of this resistance was unknown until recently. In 2013, American cell biologist Martha Cook and associates performed an experiment to examine the effects of extreme chemical treatment on the cell structure of *Cladophora* (**Figure 28.23**).

**Figure 28.23**  **Cook and colleagues showed that the green alga *Cladophora* produces tough cellulose walls that survive exposure to strong acids and high temperatures, suggesting the potential to leave fossil remains.** (left): ©Martha Cook; (middle): ©Linda Graham; (right): ©Nicholas Butterfield

> **GOAL** To determine the degree and chemical basis of the resistance to degradation of *Cladophora* and compare the results to ancient fossils of related algae.
>
> **KEY MATERIALS**  Samples of *Cladophora* collected from natural waters; concentrated acids

|  | **Experimental level** | **Conceptual level** |
|---|---|---|
| **1** Treat the algae with a standard mixture of acids, heated to boiling for 30 minutes, then centrifuge to sediment the algal remains and remove acid. | Holes in lid to release vapor — Acid mixture — Plastic tube — Green algae — Pipette tip — Acid to be removed — Organic algal remains | High-temperature acid treatment, known as acetolysis, would be expected to break down the cellulose microfibrils in land plants and other types of algae. |
| **2** Examine algal remains by means of scanning electron microscopy to determine their structure. | 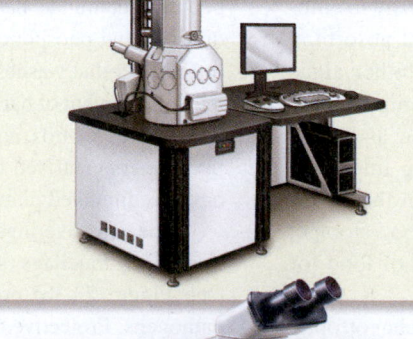 | Structure of the algal remains suggests that the biochemical composition includes fibrils made of celluose. Dimensions of microfibrils indicate the degree of chemical resistance: Microfibrils of greater width are known to have greater resistance to degradation. |
| **3** Examine algal remains using a light microscope equipped with crossed polarizers. Compare microscopic appearance of the remains to that of 750-million-year-old fossils. | 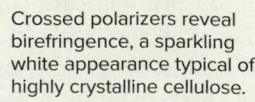 | Crossed polarizers reveal birefringence, a sparkling white appearance typical of highly crystalline cellulose. |

**4**  Compare the results of steps 3 and 4 to ancient fossils of related algae.

Compare modern *Cladophora* to ancient algal fossils.

*Cladophora* and the ancient algal fossils look similar. Biochemical materials that survive high-temperature acid treatment are also likely to have survived microbial degradation long enough to have formed organic fossils.

**5  THE DATA**

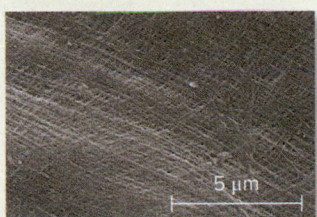 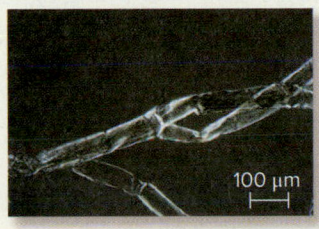 

SEM of *Cladophora* after acetolysis. Microfibrils of cellulose are 100 nanometers in diameter, thicker than those in land plant cell walls, which are only 3.5 nm thick.

Light microscope view of acetolyzed *Cladophora* using crossed-polarizers. The sparkling white appearance is typical of highly crystalline cellulose.

A 750-million-year-old fossil, *Proterocladus*, shows similar dimensions and distinctive branching pattern to modern *Cladophora*, even when the latter has been treated by acetolysis.

**6  CONCLUSION**  Tough cell-wall cellulose resists chemical degradation, allowing *Cladophora*-like algae to form as fossils and explaining modern ecological persistence in aquatic ecosystems.

**7  SOURCE**  Graham, L. E., Cook, M. E., et al. 2013. Resistance of filamentous chlorophycean, ulvophycean, and xanthophycean algae to acetolysis: Testing Proterozoic and Paleozoic microfossil attributions. *International Journal of Plant Sciences* 174: 947–957.

In the first step of the experiment, the investigators treated the algae with concentrated acids at a boiling temperature to mimic harsh chemical and microbial degradation processes. The second and third steps used two different microscopy methods to examine the algal remains and to determine their chemical makeup. In a final step, the investigators compared the algal remains to fossils of related algae from deposits about 750 million years old. As seen in the data, they found that the remains of modern *Cladophora* after acid treatment contained cellulose microfibrils that are thicker than those found in plants. The researchers inferred that the microbial resistance of modern *Cladophora* is derived from the relatively thick cellulose microfibrils present in its cell walls. In addition, the remains of *Cladophora* closely resembled the ancient fossils of related algae. These results are consistent with the idea that a cell wall that can withstand acetolysis may also have resisted microbial degradation long enough to allow the formation of fossils.

***Experimental Questions***

**1.** Why did the investigators collect the alga *Cladophora* from its natural habitat?

**2.** Why is boiling in concentrated acid a reasonable way to investigate the presence of biological materials that may resist decay long enough to form fossils?

**BIO TIPS**  **THE QUESTION** *The experiment by Cook and colleagues revealed that the relatively thick cellulose microfibrils found in the cell walls of* Cladophora *resist chemical degradation. How much thicker are these algal cellulose microfibrils than the cellulose microfibrils in less-resistant plant cell walls?*

**TOPIC** *What topic in biology does this question address?* The topic is protist defensive structures. More specifically, the question is about the biochemical makeup of the cellulose-rich cell walls of a common green alga.

**INFORMATION** *What information do you know based on the question and your understanding of the topic?* From Chapter 10, you may remember that plant cells are typically enclosed by a cell wall that is rich in cellulose, a biochemical material that provides strength and compression resistance. You learned that the strength of cellulose derives from extensive hydrogen bonding between adjacent chains. In this chapter, you learned that protist cells are enclosed by a variety of protective materials, including cellulose that differs in structure (for example, forms microfibrils that are thicker) and is consequently more resistant to degradation than is cellulose present in land plant cell walls. You also learned that the aquatic green alga *Cladophora* is known to host dense populations of hundreds of microbial species on its surfaces, suggesting that it has adapted to resist microbial attack.

**ⓅROBLEM-SOLVING ⓈTRATEGY** *Interpret data. Compare and contrast. Make a calculation.* The data in Figure 28.23 include an SEM image of *Cladophora* cell-wall cellulose microfibrils, with a scale bar enabling an estimation of their width. The data also provide information as to the width of land plant cellulose microfibrils. To calculate their relative widths, compare the two types of microfibrils.

**ANSWER** *The microfibrils of cellulose in* Cladophora *cell walls are 100 nm in diameter, whereas those of land plants are 3.5 nm wide. If you divide 100 nm by 3.5 nm, the* Cladophora *microfibrils are about 29 times thicker than those of land plants.*

*Note: This greater thickness reduces susceptibility to microbes, which start enzymatic attacks on cellulose at the microfibril surface. Thinner microfibrils have more exposed surface area where microbial attack can start. More extensive hydrogen bonding in the highly crystalline algal cellulose renders it less susceptible to chemical attack.*

## 28.4 Reproductive Adaptations

**Learning Outcomes:**

1. Briefly describe asexual reproduction of protists and three types of sexual life cycles observed in some protists.
2. **CoreSKILL** » Predict the number of chromosome sets (*n* or 2*n*) of an organism, given information about the type of cell that undergoes meiosis.
3. Give examples of how protist life cycles are important to humans.

Diverse reproductive adaptations allow protists to thrive in an amazing variety of environments, including the bodies of hosts in the cases of parasitic protists. These adaptations include specialized asexual reproductive cells, tough-walled dormant cells that allow protists to survive periods of environmental stress, and several types of sexual life cycles.

### Protist Populations Increase by Means of Asexual Reproduction

All protists are able to reproduce asexually by mitotic cell divisions of parental cells to produce progeny. When resources are plentiful, repeated mitotic divisions of single-celled protists generate large protist populations. Multicellular protists often generate specialized asexual cells that help disperse the organisms in their environment.

Many protists produce unicellular **cysts** as the result of asexual (and in some cases, sexual) reproduction (**Figure 28.24**). Cysts often have thick, protective walls and can remain dormant through periods of unfavorable climate or low food availability. Dinoflagellates commonly produce cysts that can be transported in the water of a ship's

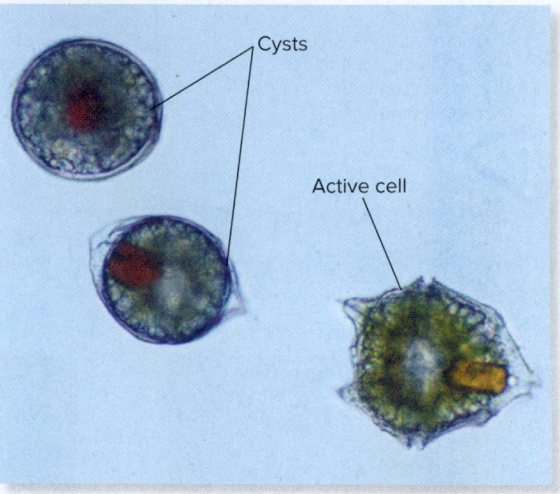

**Figure 28.24 Protistan cysts.** The round cells are dormant, tough-walled cysts of the dinoflagellate *Peridinium limbatum*. The pointed cell is an actively growing cell of the same species. As cysts develop, the outer cellulose plates present on actively growing cells are cast off. ©Linda Graham

**Concept Check:** *How are cysts involved in the spread of harmful algae and disease-causing parasitic protists?*

ballast from one port to another, a problem that has caused harmful dinoflagellate blooms to appear in harbors around the world. Ship captains can help to prevent such ecological disasters by heating ballast water before it is discharged from ships.

Many disease-causing protists spread from one host to another via cysts. As noted in the chapter opener, the alveolate pathogen *Cryptosporidium parvum* infects humans via waterborne cysts. The amoebozoan *Entamoeba histolytica* infects people who consume food or water that is contaminated with its cysts. Once inside the human digestive system, *E. histolytica* attacks intestinal cells, causing amoebic dysentery.

### Sexual Reproduction Provides Multiple Benefits to Protists

Eukaryotic sexual reproduction, featuring gametes, zygotes, and meiosis, first arose among protists. Sexual reproduction has not been observed in some protist phyla but is common in others. Sexual reproduction is generally adaptive because it produces diverse genotypes, thereby increasing the potential for faster evolutionary responses to environmental change. Many protists reap additional ecological benefits from sexual reproduction. Protists illustrate three major types of sexual life cycles that were introduced in Chapter 16: haploid dominant, alternation of generations, and diploid dominant. Ciliates and protistan parasites that exist in different life cycle stages in different host species display variations of these basic types.

***Haploid-Dominant Life Cycles*** Most unicellular protists that reproduce sexually display a haploid-dominant life cycle, meaning that most stages in the life cycle are haploid (**Figure 28.25**). In this type of life cycle, haploid cells may develop into gametes. Some protists produce nonmotile eggs and smaller flagellate sperm. However, many other protists have gametes that look similar to each other

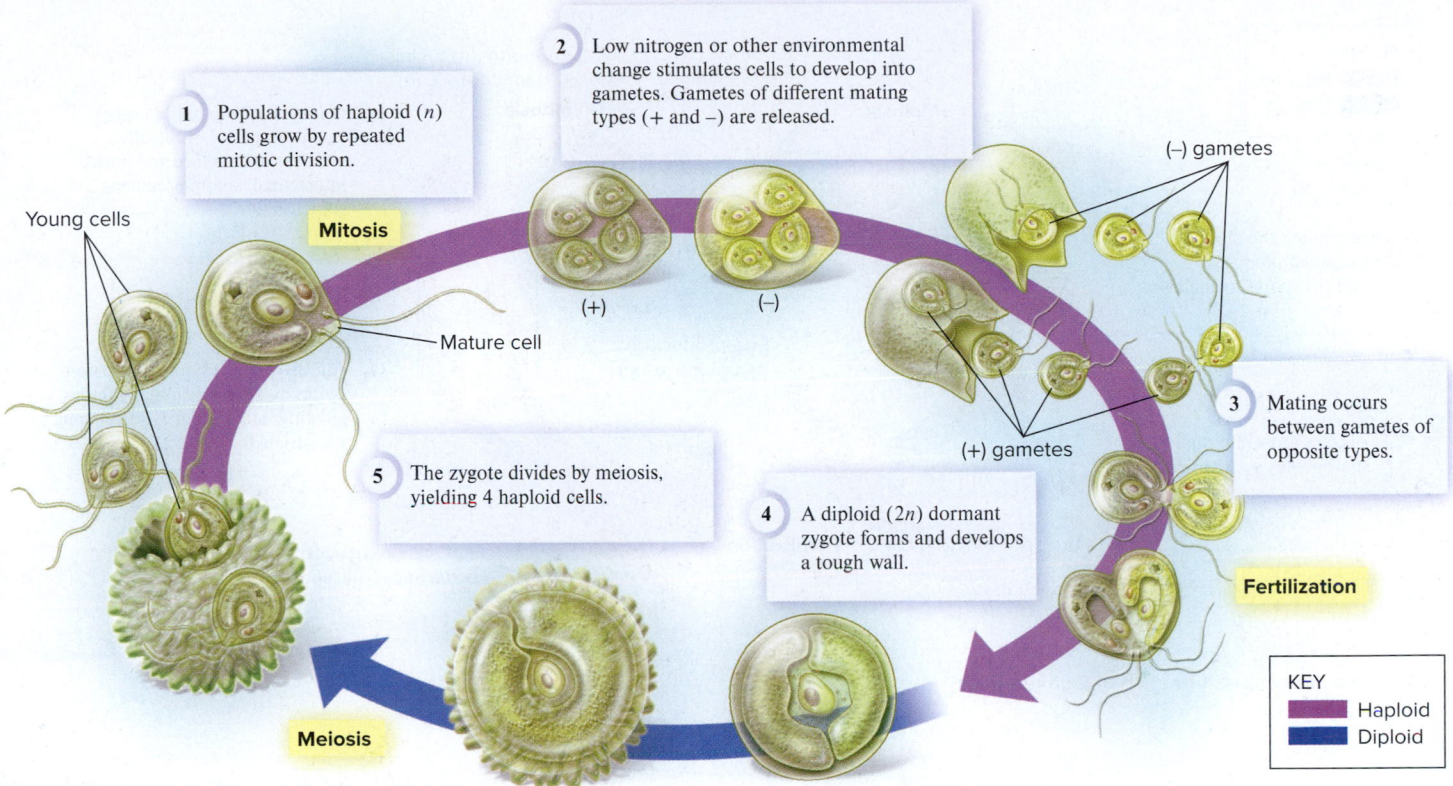

**Figure 28.25** **Haploid-dominant life cycle, illustrated by the unicellular flagellate genus *Chlamydomonas*.** In *Chlamydomonas*, most cells are haploid; only the zygote is diploid.

Labels within figure:

1  Populations of haploid (*n*) cells grow by repeated mitotic division.

2  Low nitrogen or other environmental change stimulates cells to develop into gametes. Gametes of different mating types (+ and −) are released.

3  Mating occurs between gametes of opposite types.

4  A diploid (2*n*) dormant zygote forms and develops a tough wall.

5  The zygote divides by meiosis, yielding 4 haploid cells.

Young cells

Mature cell

Mitosis

(+)    (−)

(−) gametes

(+) gametes

Fertilization

Meiosis

KEY
Haploid
Diploid

---

structurally but have distinctive biochemical features and hence are known as + and − mating types, as shown in Figure 28.25. Gametes fuse (mate) to produce thick-walled diploid zygotes, the only diploid stage in this type of life cycle. Such zygotes often have tough cell walls and can survive stressful conditions, much like cysts. When conditions permit, the zygote divides by meiosis to produce haploid cells that increase in number via mitotic cell divisions.

**Alternation of Generations**    Many multicellular green and brown seaweeds display a life cycle involving alternation of haploid and diploid generations (**Figure 28.26**). Giant kelps and some other protists having this kind of life cycle produce two types of multicellular organisms: a haploid gametophyte generation that produces gametes and a diploid sporophyte generation that produces spores by the process of meiosis (Figure 28.26a). Each of the two types of multicellular organisms can adapt to distinct habitats or seasonal conditions, allowing these protists to occupy more types of environments and for longer periods.

Many red seaweeds display a life cycle that involves the alternation of three distinct multicellular generations (Figure 28.26b). This unique type of sexual life cycle has evolved as compensation for the lack of flagella on red algal sperm. Because these sperm are unable to swim to eggs, fertilization occurs only when sperm carried by ocean currents happen to drift close to eggs. As a consequence, fertilization can be rare. Many red algae therefore amplify the products of a rare fertilization by copying the zygote genome into millions of spores produced by two successive sporophyte generations. Small

diploid sporophytes that are nourished by the parental gametophyte produce diploid spores that may each grow into a larger sporophyte. This larger sporophyte produces many haploid spores. Diverse economically valuable red algae possess this type of life cycle, an understanding of which is critical to growing seaweed crops.

**Diploid-Dominant Life Cycles**    Diatoms, which commonly occur as single cells that provide nutritious food for aquatic animals, represent protists known to display a diploid-dominant life cycle (**Figure 28.27**). In diploid-dominant life cycles, all cells except the gametes are diploid, and the gametes are produced by meiosis. Sexual reproduction in diatoms not only increases their genetic variability, but also has another major benefit related to cell size.

In many diatoms, one daughter cell arising from asexual reproduction, which involves mitosis, is smaller than the other, and it is also smaller than the parent cell (Figure 28.27a). This happens because diatom cell walls are composed of two overlapping halves, much like two-part round laboratory dishes having lids that overlap the bottoms. After each mitotic division, each daughter cell receives one-half of the parent cell wall. The daughter cell that inherits a larger, overlapping parental "lid" then produces a new "bottom" that fits inside. This daughter cell will be the same size as its parent. However, the daughter cell that inherits the parental "bottom" uses this as its lid and produces a new, even smaller "bottom." This cell will be smaller than its sibling or parent. Consequently, after many such mitotic divisions, the mean cell size of diatom populations declines over time. If diatom cells become too small, they cannot survive.

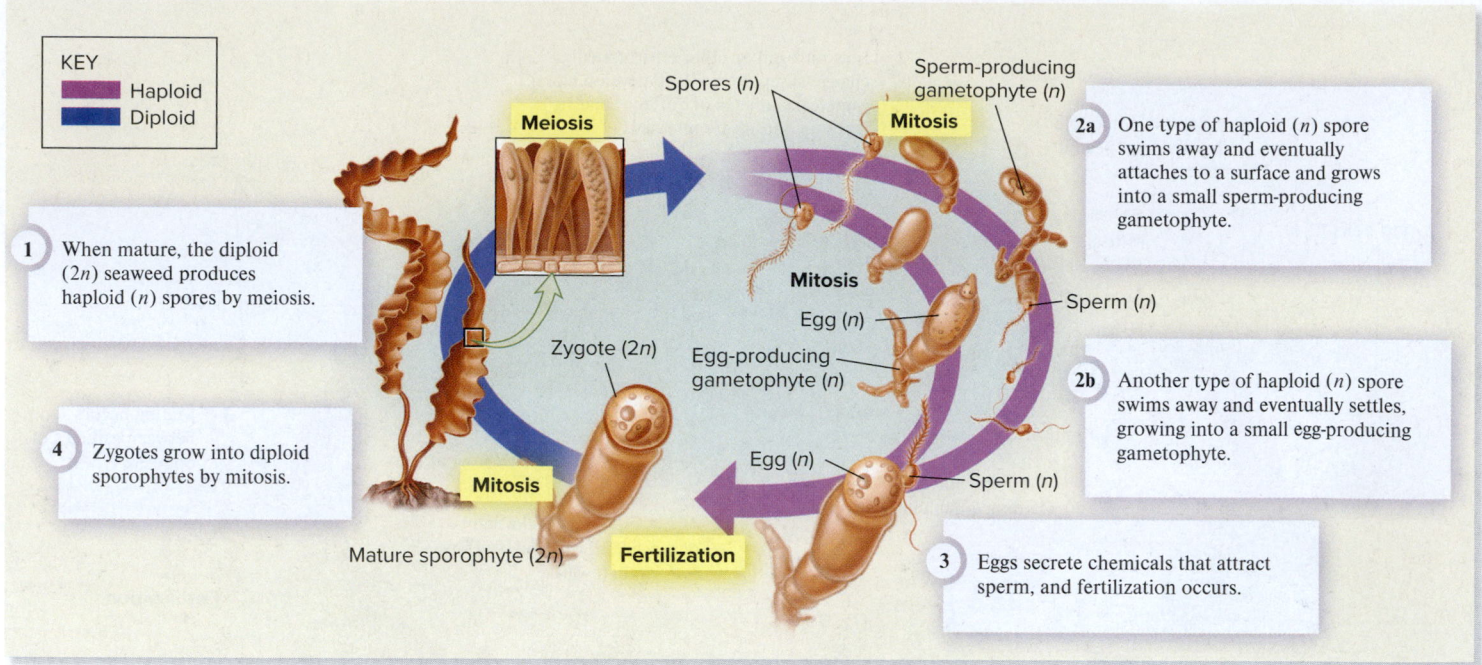

**KEY**
- Haploid
- Diploid

**Meiosis**

Spores (n)

**1** When mature, the diploid (2n) seaweed produces haploid (n) spores by meiosis.

Sperm-producing gametophyte (n)

**Mitosis**

**2a** One type of haploid (n) spore swims away and eventually attaches to a surface and grows into a small sperm-producing gametophyte.

**Mitosis**

Egg (n)

Sperm (n)

**4** Zygotes grow into diploid sporophytes by mitosis.

Zygote (2n)

Egg-producing gametophyte (n)

**2b** Another type of haploid (n) spore swims away and eventually settles, growing into a small egg-producing gametophyte.

**Mitosis**

Egg (n)

Sperm (n)

Mature sporophyte (2n)

**Fertilization**

**3** Eggs secrete chemicals that attract sperm, and fertilization occurs.

**(a) *Laminaria* life cycle—alternation of 2 generations**

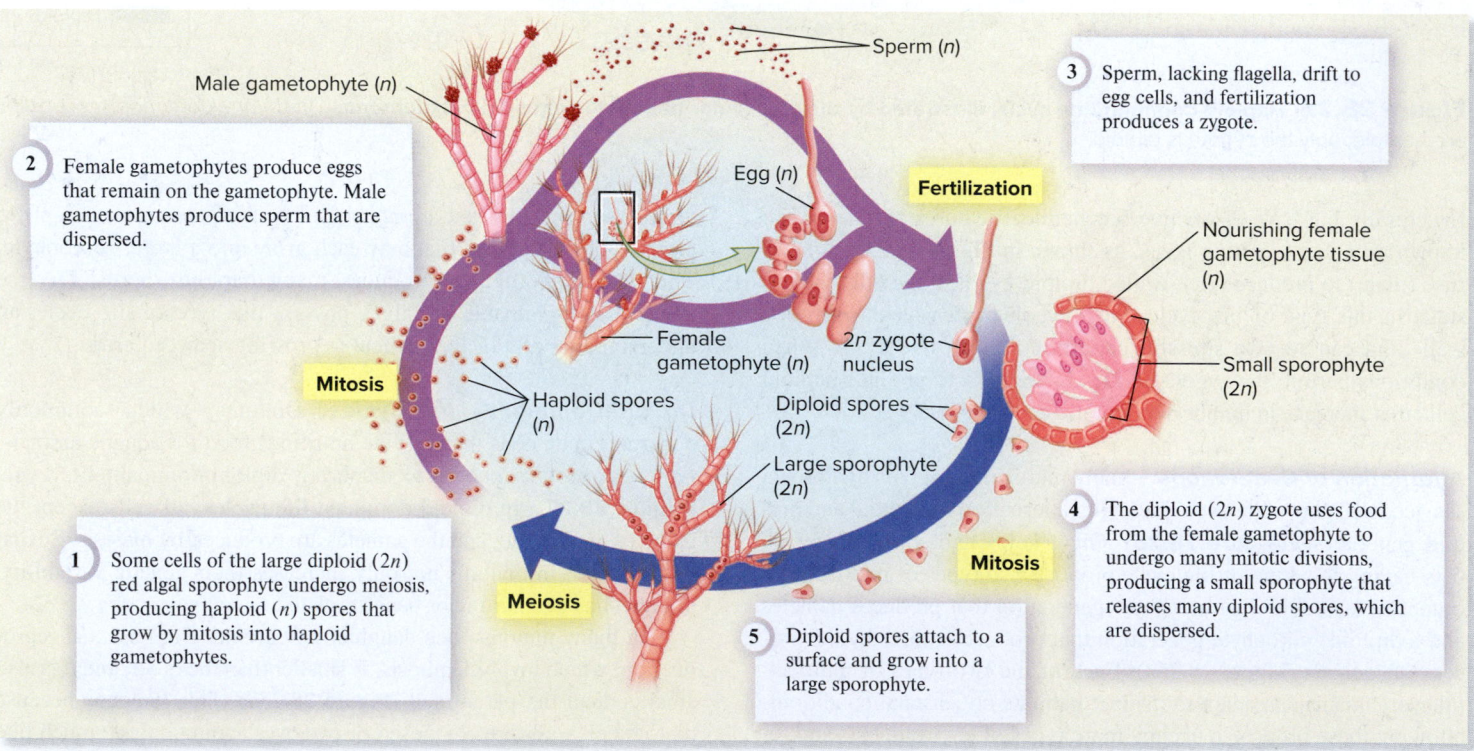

Sperm (n)

Male gametophyte (n)

**3** Sperm, lacking flagella, drift to egg cells, and fertilization produces a zygote.

**2** Female gametophytes produce eggs that remain on the gametophyte. Male gametophytes produce sperm that are dispersed.

Egg (n)

**Fertilization**

Nourishing female gametophyte tissue (n)

**Mitosis**

Female gametophyte (n)

2n zygote nucleus

Small sporophyte (2n)

Haploid spores (n)

Diploid spores (2n)

Large sporophyte (2n)

**4** The diploid (2n) zygote uses food from the female gametophyte to undergo many mitotic divisions, producing a small sporophyte that releases many diploid spores, which are dispersed.

**Meiosis**

**Mitosis**

**1** Some cells of the large diploid (2n) red algal sporophyte undergo meiosis, producing haploid (n) spores that grow by mitosis into haploid gametophytes.

**5** Diploid spores attach to a surface and grow into a large sporophyte.

**(b) *Polysiphonia* life cycle—alternation of 3 generations**

**Figure 28.26 Alternation of generations. (a)** Life cycle with two alternating generations, illustrated by the brown seaweed *Laminaria*. **(b)** Life cycle involving three alternating generations, illustrated by the common red seaweed *Polysiphonia*.

Sexual reproduction allows diatom species to attain maximal cell size. Diatom cells mate within a blanket of mucilage, with each partner undergoing meiotic divisions to produce gametes. The large, spherical diatom zygotes that result from fertilization (Figure 28.27b) later undergo a series of mitotic divisions to produce new diatom cells having the maximal size for the species. Maintaining size consistent with survival is important because diatoms provide a large proportion of the organic food at the base of marine and freshwater food chains.

***Ciliate Reproduction*** Ciliate protists reproduce asexually by mitosis and formation of cysts (**Figure 28.28a**). In addition, ciliates can reproduce

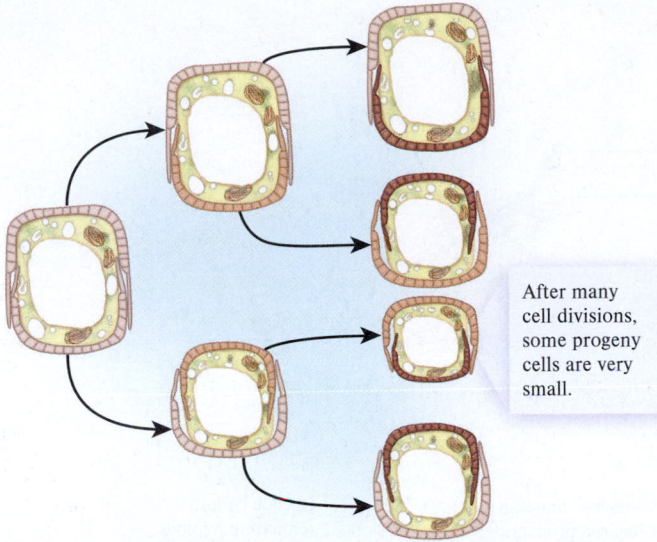

(a) Asexual reproduction in diatoms

After many cell divisions, some progeny cells are very small.

**Figure 28.27   Diploid-dominant life cycle, as illustrated by diatoms.** **(a)** Asexual reproduction in diatoms involves repeated mitotic division of cells having a two-piece cell wall—a top "lid" covering a "bottom." Because a new bottom cell-wall is always synthesized at the end of each mitotic division, asexual reproduction may eventually cause a decline in the mean cell size of a diatom population. **(b)** Small cell size may trigger sexual reproduction, which regenerates maximal cell size.

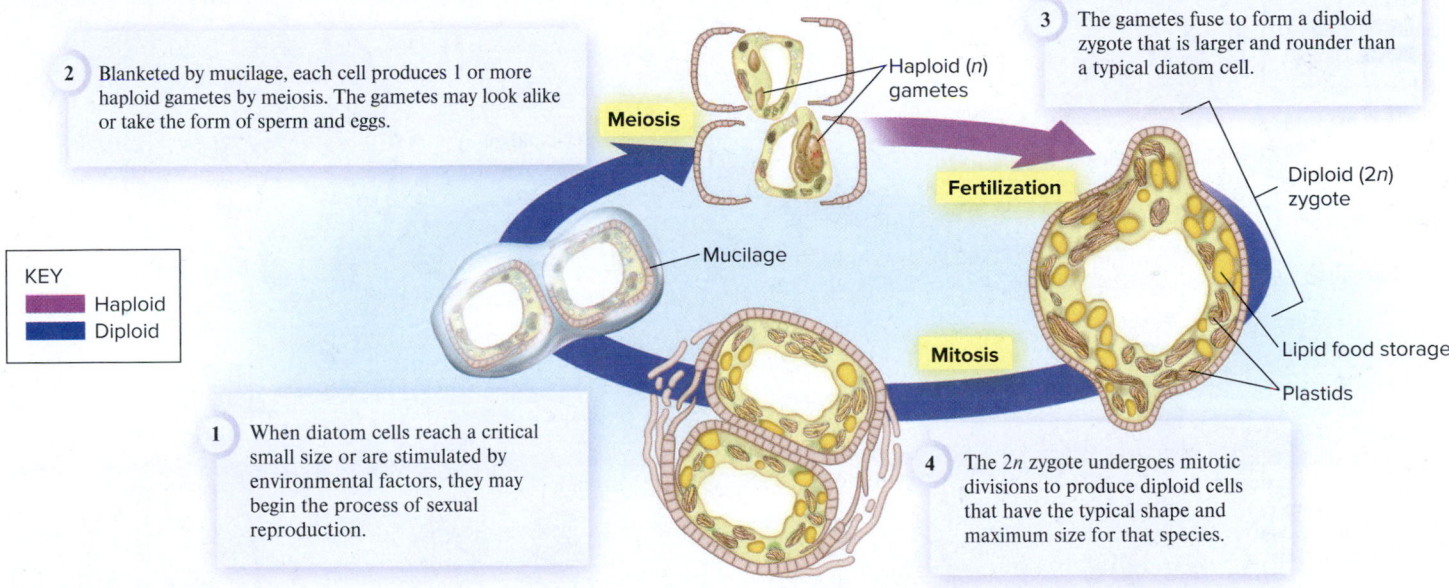

2   Blanketed by mucilage, each cell produces 1 or more haploid gametes by meiosis. The gametes may look alike or take the form of sperm and eggs.

Haploid ($n$) gametes

Meiosis

3   The gametes fuse to form a diploid zygote that is larger and rounder than a typical diatom cell.

Fertilization

Diploid ($2n$) zygote

Mucilage

KEY
- Haploid
- Diploid

Mitosis

Lipid food storage

Plastids

1   When diatom cells reach a critical small size or are stimulated by environmental factors, they may begin the process of sexual reproduction.

4   The $2n$ zygote undergoes mitotic divisions to produce diploid cells that have the typical shape and maximum size for that species.

(b) Sexual reproduction in diatoms

sexually by a process known as conjugation. Ciliates are unusual in having two types of nuclei: one or more smaller micronuclei and a single large macronucleus. Macronuclei, which contain many copies of the genome, serve as the source of information for cell function. Both macronuclei and micronuclei divide during asexual mitosis. The diploid micronuclei do not undergo gene expression during growth; instead, their role is to transmit the genome to the next generation during sexual reproduction via conjugation. Different species of ciliates vary as to the details of conjugation; the process for *Paramecium caudatum* is shown in **Figure 28.28b**.

## Parasitic Protists May Use Alternate Hosts for Different Life Stages

Many parasitic protists are notable for using more than one host organism, with different life stages occurring in each host. The malarial parasite genus *Plasmodium* is a prominent example. Several species of *Plasmodium* infect humans, some infect humans and/or other primates, and others infect rodents or birds. The malarial parasite's alternate host is the mosquito in the genus *Anopheles*. About 40% of humans live in tropical regions of the world where malaria occurs, and as noted earlier, millions of infections and human deaths result each year. Malaria is particularly deadly for young children. Insecticides can be used to control mosquito populations, mosquito nets help to reduce infection, and antimalarial drugs can be used to treat infections. However, malarial parasites can develop drug resistance, and experts are concerned that human cases may double in the next 20 years.

When an infected mosquito bites a human, *Plasmodium* enters the bloodstream in an asexual life stage known as a sporozoite (**Figure 28.29**). Upon reaching a victim's liver, sporozoites enter

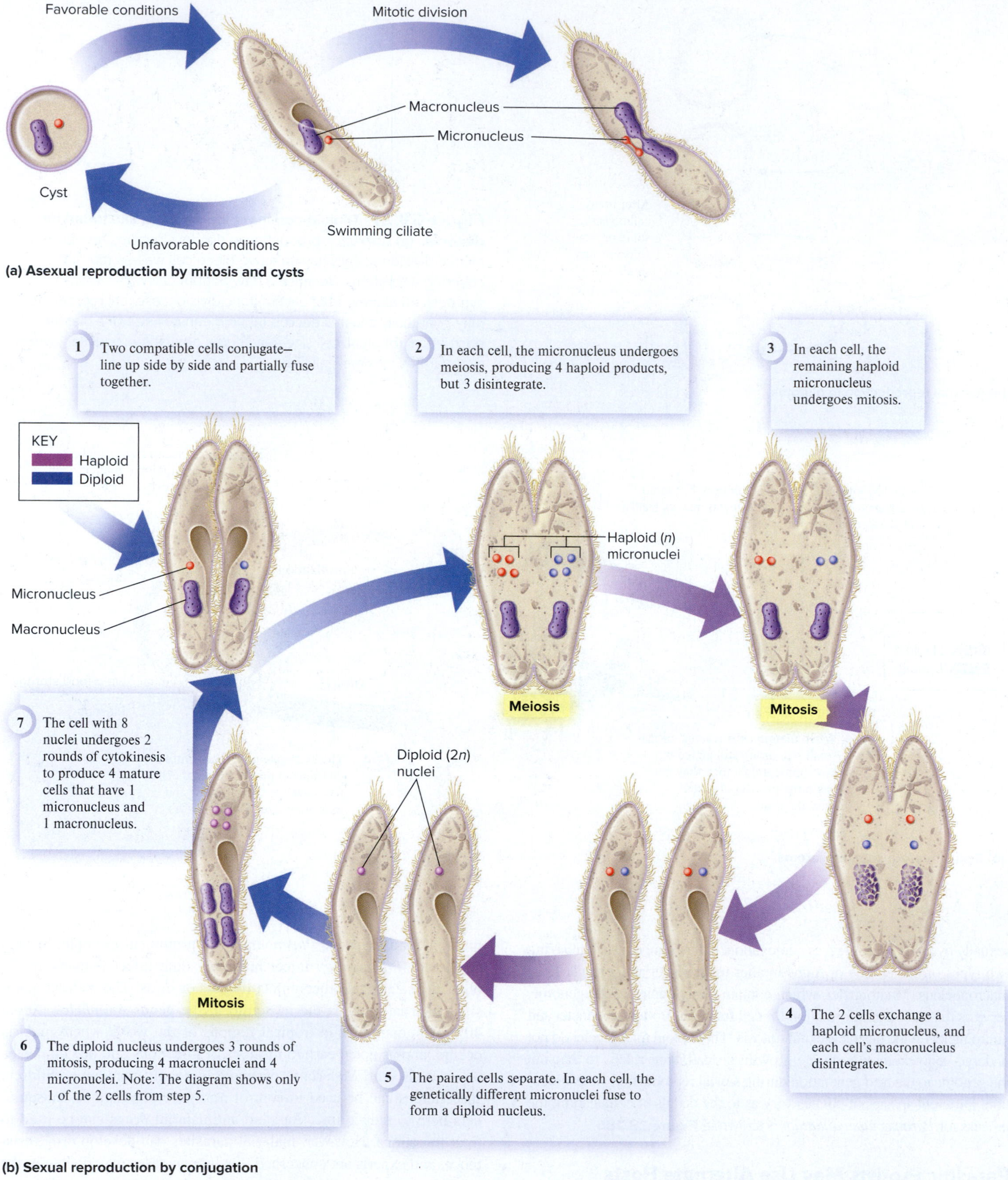

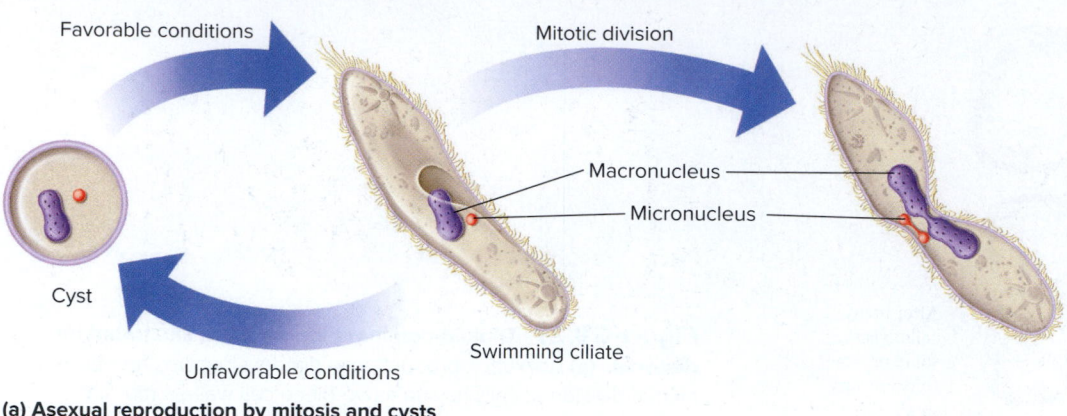

Favorable conditions

Mitotic division

Macronucleus

Micronucleus

Cyst

Unfavorable conditions

Swimming ciliate

**(a) Asexual reproduction by mitosis and cysts**

**1** Two compatible cells conjugate—line up side by side and partially fuse together.

**2** In each cell, the micronucleus undergoes meiosis, producing 4 haploid products, but 3 disintegrate.

**3** In each cell, the remaining haploid micronucleus undergoes mitosis.

KEY
- Haploid
- Diploid

Micronucleus

Macronucleus

Haploid (*n*) micronuclei

**Meiosis**

**Mitosis**

**7** The cell with 8 nuclei undergoes 2 rounds of cytokinesis to produce 4 mature cells that have 1 micronucleus and 1 macronucleus.

Diploid (2*n*) nuclei

**Mitosis**

**6** The diploid nucleus undergoes 3 rounds of mitosis, producing 4 macronuclei and 4 micronuclei. Note: The diagram shows only 1 of the 2 cells from step 5.

**5** The paired cells separate. In each cell, the genetically different micronuclei fuse to form a diploid nucleus.

**4** The 2 cells exchange a haploid micronucleus, and each cell's macronucleus disintegrates.

**(b) Sexual reproduction by conjugation**

**Figure 28.28 Ciliate reproduction. (a)** The asexual reproductive process in ciliates. **(b)** The sexual reproductive process, known as conjugation, of the ciliate *Paramecium caudatum*.

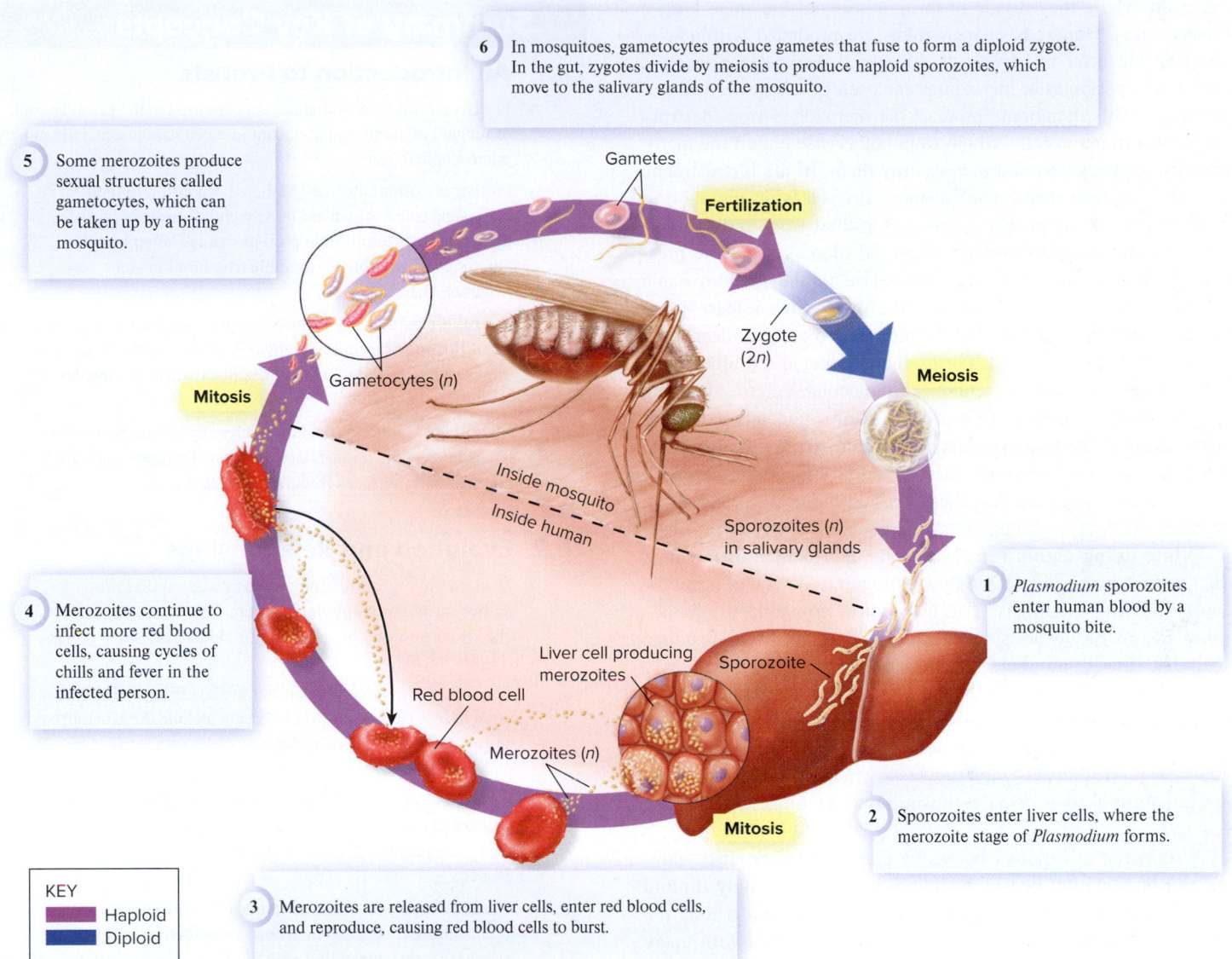

**6** In mosquitoes, gametocytes produce gametes that fuse to form a diploid zygote. In the gut, zygotes divide by meiosis to produce haploid sporozoites, which move to the salivary glands of the mosquito.

**5** Some merozoites produce sexual structures called gametocytes, which can be taken up by a biting mosquito.

Gametes

**Fertilization**

Zygote (2*n*)

**Meiosis**

**Mitosis**

Gametocytes (*n*)

Inside mosquito

Inside human

Sporozoites (*n*) in salivary glands

**1** *Plasmodium* sporozoites enter human blood by a mosquito bite.

**4** Merozoites continue to infect more red blood cells, causing cycles of chills and fever in the infected person.

Liver cell producing merozoites

Sporozoite

Red blood cell

Merozoites (*n*)

**Mitosis**

**2** Sporozoites enter liver cells, where the merozoite stage of *Plasmodium* forms.

KEY
- Haploid
- Diploid

**3** Merozoites are released from liver cells, enter red blood cells, and reproduce, causing red blood cells to burst.

**Figure 28.29** **Diagram of the life cycle of *Plasmodium falciparum*, a species that causes malaria in humans.** This life cycle requires two alternate hosts, humans and *Anopheles* mosquitoes. Parasite life stages known as sporozoites are transmitted by a mosquito bite, then infect the liver. The parasite life stages known as merozoites are produced in the liver, and infect red blood cells.

**Concept Check:** *In which of the hosts does sexual mating of P. falciparum gametes occur?*

**Core Skill: Modeling** The goal of this modeling challenge is to make a simplified model of the one shown in Figure 28.29, to determine if the latter represents a life cycle that is diploid dominant, haploid dominant, or an alternation of generations.

**Modeling Challenge:** Chapter 16 describes the three basic types of eukaryotic sexual life cycles (refer back to Figure 16.14). These cycles are modeled by circular diagrams that show life phases connected by the processes of fertilization, meiosis, and mitotic cell divisions. Animals display a diploid-dominant life cycle; fungi have a haploid-dominant life cycle, and the life cycle of plants involves an alternation of a haploid gametophyte with a diploid sporophyte. This chapter reveals that all of these basic life cycle types occur among protists. Use the detailed sexual life cycle shown in Figure 28.29 to identify the life cycle type (diploid dominant, haploid dominant, or alternation of generations) that is characteristic of *Plasmodium*. To make this decision, draw a very simple circular model representing the cycling of *Plasmodium* between haploid and diploid life phases and the occurrence of fertilization, meiosis and mitotic cell divisions. Then compare your simple model to the three models shown in Figure 16.14.

liver cells, where they divide to form an asexual life stage known as merozoites. Hundreds of merozoites are produced within liver cells (see the inset in Figure 28.29), which then release into the bloodstream packages of merozoites enclosed by a host-derived cell membrane. This membrane protects the merozoites from destruction by host immune cells, which would otherwise engulf the merozoites by phagocytosis and then destroy them. In the bloodstream, the protective host membranes disintegrate, releasing merozoites. The merozoites have protein complexes at their front ends, or apices, that allow them to invade human red blood cells. (The presence of these apical complexes gives rise to the phylum name Apicomplexa.) Within red blood cells, merozoites release more than 200 proteins, which enable the parasites to commandeer these cells, causing many changes. For example, infected red blood cells form surface knobs that function like molecular Velcro, attaching cells to capillary linings. This process allows infected red blood cells to avoid being transported to the spleen, where they would be destroyed. The attachment of infected red blood cells to capillary linings disrupts circulation in the brain and kidney, a process that can cause death of the animal host.

While living within red blood cells, merozoites form rings, which often can be visualized by staining and the use of a microscope, allowing diagnosis. The merozoites consume the hemoglobin in red blood cells, which gives them fuel to reproduce asexually. Large numbers of new merozoites synchronously break out of red blood cells at intervals of 48 or 72 hours. These merozoite reproduction cycles correspond to cycles of chills and fever that an infected person experiences. Some merozoites produce sexual structures—gametocytes—which, along with blood, are transmitted to a female mosquito as it bites an infected person.

Within the mosquito's body, the gametocytes produce gametes and fertilization occurs, yielding a zygote, the only diploid cell in *Plasmodium*'s life cycle. Within the mosquito gut, the zygote undergoes meiosis, generating structures filled with many sporozoites, the stage that can be transmitted to a new human host. Sporozoites move to the mosquito's salivary glands, where they remain until they are injected into a human host when the mosquito feeds.

In recent years, genomic information has added to our knowledge of these life stages, thereby helping medical scientists develop new ways to prevent or treat malaria. For example, about 10% of the nuclear-encoded proteins of *P. falciparum* are likely imported into a non-photosynthetic plastid known as an apicoplast, where they are needed for fatty-acid metabolism and other processes. *Plasmodium* and some other apicomplexan protists possess plastids because they are descended from algal ancestors that had photosynthetic plastids. Because plastids are not present in mammalian cells, enzymes in metabolic pathways in the apicoplast are possible targets for development of drugs that will kill the parasite without harming the host. Mammals also lack calcium-dependent protein kinases (CDPKs), enzymes that are essential to the release of parasite merozoites from red blood cells and also needed for the parasite's sexual development, offering another potential drug target.

## Summary of Key Concepts

### 28.1 An Introduction to Protists

- Protists are eukaryotes that are not classified in the plant, animal, or fungal kingdoms; are abundant in moist habitats; and are mostly microscopic in size.
- Protists are often informally labeled according to their ecological roles: Algae are mostly photosynthetic protists; protozoa are heterotrophic protists that are often mobile; and fungus-like protists resemble true fungi in some ways (Figures 28.1, 28.2).
- Protists are particularly diverse in aquatic habitats, occurring as small floating or swimming phytoplankton, attached members of a periphyton, and more complex macroalgae (seaweeds) (Figure 28.3).
- Microscopic protists propel themselves by means of flagella (flagellates), cilia (ciliates), or pseudopodia (amoebae) or by gliding across surfaces (Figures 28.4, 28.5).

### 28.2 Evolution and Relationships

- Modern phylogenetic analysis has revealed that protists do not form a monophyletic group; instead, many are classified into one of seven major eukaryotic supergroups (Figure 28.6).
- The supergroup Excavata consists of flagellate protists whose cells often have a feeding groove. Excavata include the kinetoplastids and euglenoids, some of which are photosynthetic (Figures 28.7, 28.8, 28.9).
- Land plants are related to green algae and red algae (having primary plastids). Cryptomonads and haptophytes display secondary plastids (Figures 28.10, 28.11, 28.12, 28.13, 28.14).
- The supergroup Alveolata includes the ciliates, apicomplexans, and dinoflagellates, all of whose cells feature saclike membrane vesicles called alveoli. Many dinoflagellates display secondary plastids, and some feature tertiary plastids (Figures 28.15, 28.16).
- The supergroup Stramenopila includes protists whose flagella have strawlike hairs that aid in swimming. Stramenopiles include diatoms, giant kelps, and other groups of algae, as well as some fungus-like protists (Figures 28.17, 28.18).
- The supergroup Rhizaria consists of flagellates and amoebae with thin hairlike extensions of cytoplasm called filose pseudopodia. Three prominent phyla are Chlorarachniophyta, with secondary green plastids; silicate-shelled Radiolaria; and Foraminifera, with calcium carbonate shells (Figure 28.19).
- The supergroup Amoebozoa is composed of many types of amoebae and includes slime molds such as *Dictyostelium discoideum*.
- The supergroup Opisthokonta includes organisms that are swimming cells having a single posterior flagellum. Opisthokonta includes the fungal and animal kingdoms and related protists (Figures 28.20, 28.21, Table 28.1).

## 28.3 Nutritional and Defensive Adaptations

- Protists display four mechanisms of nutrition: phagotrophs feed by ingesting particles; osmotrophs absorb small organic molecules; photoautotrophs make their own organic food by using light energy; and mixotrophs use both photoautotrophy and heterotrophy to obtain nutrients (Figure 28.22).

- Protists possess defensive adaptations such as sharp projectiles, light flashes, toxic compounds, and protective cell coverings. The green alga *Cladophora* is a protist that has cellulose-rich cell walls so tough that they help to explain ancient fossils, and it also plays an important modern ecological role (Figure 28.23).

## 28.4 Reproductive Adaptations

- Protist populations grow by means of asexual reproduction involving mitosis, and many persist through unfavorable conditions by producing tough-walled cysts (Figure 28.24).

- Sexual reproduction is also observed among protists. In the haploid-dominant life cycle, haploid cells develop into gametes, which fuse to produce diploid zygotes. These zygotes often have tough cell walls that enable them to survive unfavorable conditions (Figure 28.25).

- In protists displaying a life cycle called alternation of generations, a haploid generation produces gametes and a diploid generation produces spores. Each generation can adapt to different environments or conditions, allowing protists to occupy multiple habitats (Figure 28.26).

- In the diploid-dominant life cycle, all cells except gametes are diploid. Sexual reproduction in diatoms, which have this type of life cycle, increases genetic variability and allows the protists to maintain an adequate cell size (Figure 28.27).

- Ciliate protists display both asexual reproduction and sexual reproduction by conjugation (Figure 28.28).

- Parasitic protists may have life cycles involving alternate hosts for different life stages. One example is *Plasmodium*, which causes malaria in humans and infects some other animals; certain mosquitoes are the alternate hosts (Figure 28.29).

## Assess & Discuss

### Test Yourself

1. If you were studying the evolution of animal-specific cell-to-cell signaling systems, from which of the following would you choose representative species to observe?
   a. Rhodophyta
   b. Excavata
   c. Choanomonada
   d. Radiolaria
   e. Chlorophyta

2. If you were studying the origin of land plant traits, which of the following groups would you study?
   a. green algae
   b. radiolarians
   c. choanoflagellates
   d. diatoms
   e. ciliates

3. Which informal ecological group of protists includes photoautotrophs?
   a. protozoa
   b. algae
   c. fungus-like protists
   d. ciliates
   e. All of the above groups include photoautotrophs.

4. How would you recognize a primary plastid? It would
   a. have one envelope membrane.
   b. have two envelope membranes.
   c. have more than two envelope membranes.
   d. lack pigments.
   e. be golden brown in color.

5. What organisms have tertiary plastids?
   a. certain stramenopiles
   b. certain euglenoids
   c. certain cryptomonads
   d. certain opisthokonts
   e. certain dinoflagellates

6. What is surprising about mixotrophs?
   a. They have no plastids, but they occur in mixed communities with autotrophs.
   b. They have mixed heterotrophic and autotrophic mechanisms of nutrition.
   c. Their cells contain a mixture of red and green plastids.
   d. Their cells contain a mixture of haploid and diploid nuclei.
   e. They consume a mixed diet of algae.

7. What advantages do diatoms obtain from sexual reproduction?
   a. increased genetic variability
   b. increased ability of populations to respond to environmental change
   c. evolutionary potential
   d. regeneration of maximal cell size for the species
   e. all of the above

8. What are extrusomes?
   a. hairs on flagella
   b. membrane sacs beneath the cell surface
   c. tough-walled asexual cells
   d. spearlike defensive structures shot from cells under attack
   e. special types of survival cysts

9. How do pigments, such as phycoerythrin in red algae and fucoxanthin in brown algae, benefit these autotrophic protists?
   a. The pigments provide camouflage, so herbivores cannot see the algae.
   b. The pigments absorb blue-green underwater light and transfer the energy to chlorophyll a for use in photosynthesis.
   c. The pigments attract aquatic animals that carry gametes between seaweeds.
   d. The pigments absorb UV light that would harm the photosynthetic apparatus.
   e. All of the above are correct.

10. What are the alternate hosts of the malarial parasite *Plasmodium falciparum*?
    a. humans and ticks
    b. ticks and mosquitoes
    c. humans and *Anopheles* mosquitoes
    d. humans and all types of mosquitoes
    e. sporophytes and gametophytes

## Conceptual Questions

1. Explain why protists are classified into multiple supergroups, rather than as a single kingdom or phylum.

2. Why have molecular biologists sequenced the genomes of several parasitic protists?

3.  **Core Concept: Science and Society** Why are the cysts of protists important to epidemiologists, the biologists who study the spread of disease?

## Collaborative Questions

1. Imagine you are studying an insect species and you discover that the insects are dying of a disease that results in the production of cysts of the type that protists often generate. Thinking that the cysts might have been produced by a parasitic protist that could be used as an insect control agent, how would you go about identifying the disease-causing organism?

2. Imagine you are part of a marine biology team seeking to catalogue the organisms inhabiting a threatened coral reef. The team has found two new types of macroalgae (seaweeds), each of which occurs during a particular time of the year when the water temperature is at a certain point. You suspect that the two macroalgae might be different generations of the same species that have differing optimal temperature conditions. How would you go about testing your hypothesis?

# Fungi

# 29

**The aboveground reproductive parts of the fungus _Armillaria ostoyae._** Because of the large extent of its underground components, a member of this species may be the largest organism in the world. ©Brian Lightfoot/naturepl.com

**Y**ou might think that the largest organism in the world is a whale or perhaps a giant redwood tree. Amazingly, giant fungi would also be good candidates. For example, an individual of the fungus species _Armillaria ostoyae_ weighs hundreds of tons, is more than 2,000 years old, and spreads across 2,200 acres of Oregon forest soil! Scientists discovered the extent of this enormous fungus when they found identical DNA sequences in soil samples taken over this wide area. Other examples of such huge fungi have been found, and mycologists—scientists who study fungi—suspect that they may be fairly common, existing underfoot yet largely unseen.

Regardless of their size, fungi typically occur within soil or other materials, becoming conspicuous only when the reproductive portions such as mushrooms extend above the surface. Even though fungi can be inconspicuous, they play essential roles in the Earth's environment; are associated in diverse ways with other organisms, including humans; and have many applications in biotechnology. In this chapter, we will explore the distinctive features of fungal structure, growth, nutrition, reproduction, and diversity. In the process, we will see how fungi are connected to decomposition, forest growth, food production and food toxins, sick building syndrome, and other phenomena of great importance to humans.

## 29.1 Evolution and Distinctive Features of Fungi

**Learning Outcomes:**

1. **CoreSKILL »** Use information about groups of fungi to draw a diagram showing their evolutionary relationships.
2. Outline the distinctive features of fungi, including how they obtain food.
3. Discuss how fungal feeding is related to fungal growth.

The eukaryotes known as fungi are so distinct from other organisms that they are placed in their own kingdom, the kingdom Fungi (**Figure 29.1**). Together with certain closely related protists, the kingdom Fungi and the kingdom Animalia form a eukaryotic supergroup known as Opisthokonta (refer back to

Figure 28.6). The kingdom Fungi, also known as the true fungi, diverged from Animalia more than a billion years ago, during the Middle Proterozoic Era. Several types of slime molds, disease-causing oomycetes, and other fungus-like protists—though often studied with fungi—are not classified as true fungi (see Chapter 28).

The true fungi form a monophyletic group of an estimated 1.5 million species. Even greater diversity of this group is suggested by molecular evidence indicating the existence of many species that have not yet been cultivated or named. The earliest fungi probably originated in aquatic habitats, where they diverged from opisthokont protists closely related to the modern genus _Nuclearia_—an amoeba that feeds by ingesting algal and bacterial cells. Feeding on particles such as cells is a process known as phagocytosis (see Figure 28.7).

The earliest-diverging phylum of modern fungi, known as Cryptomycota, are unicellular and mostly occur in soil and water. Later-diverging fungi regularly produce a cell wall containing **chitin**, a tough polysaccharide polymer that contains nitrogen. Such a cell wall enables most fungi to resist the high osmotic pressure

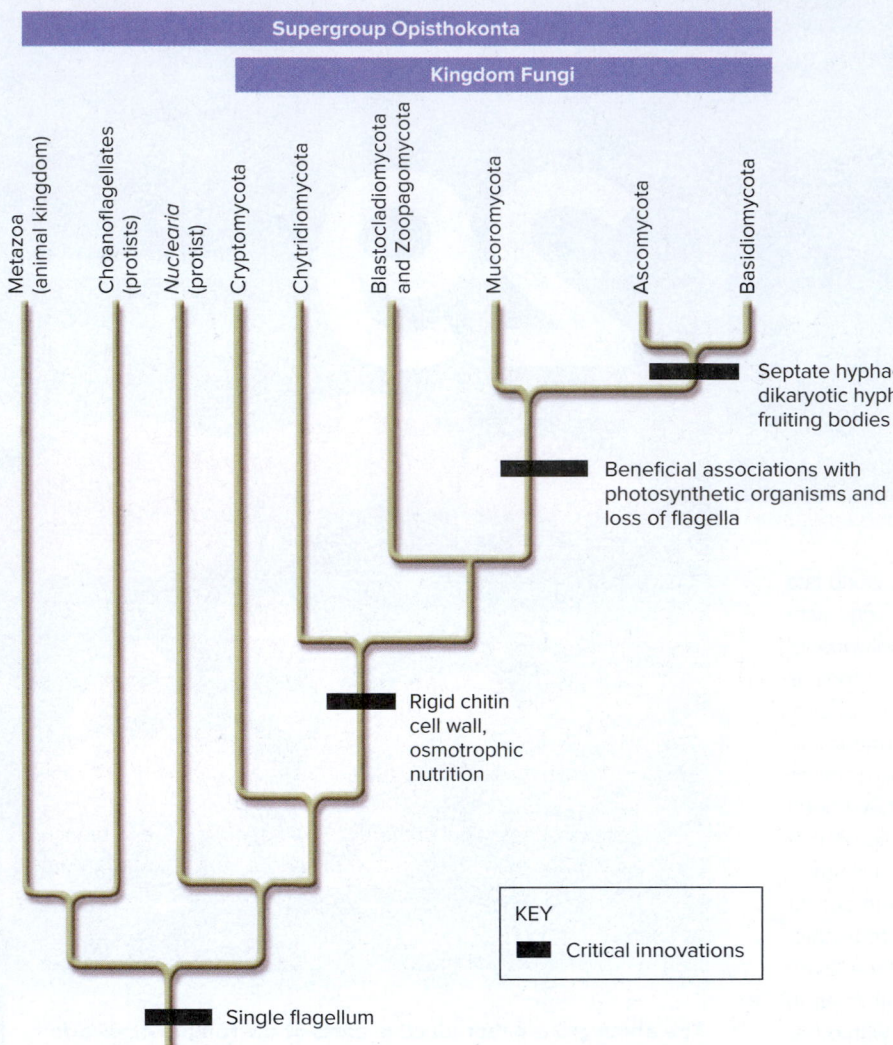

**Figure 29.1** **Evolutionary relationships of the fungi.** The kingdom Fungi arose from a protist ancestor similar to the modern genus *Nuclearia*. Several early-diverging phyla of fungi that commonly occupy moist environments are Cryptomycota, Chytridiomycota, and Blastocladiomycota. Later-diverging fungal phyla that mostly live on land include Mucoromycota, Ascomycota, and Basidiomycota.

that results when they feed by absorbing small organic molecules, a process known as **osmotrophy**. The evolution of a rigid cell wall accompanied a key evolutionary transition in fungal nutrition from phagocytosis to osmotrophy (see Figure 29.1).

Early-diverging phyla Cryptomycota, Blastocladiomycota, and Chytridiomycota, commonly live in aquatic habitats or moist soils, where they often reproduce by means of cells that swim using flagella. By contrast, later-diverging fungal phyla commonly occupy drier terrestrial environments and do not produce flagellate cells. Loss of flagella is regarded as an adaptation to life on land. Mucoromycota, Ascomycota (also known as **ascomycetes**, or sac fungi), and Basidiomycota (also known as **basidiomycetes**, or club fungi) are notable for displaying symbiotic associations with land organisms, particularly plants.

Because fungi are closely related to the animal kingdom, fungi and animals display some common features. For example, both are **heterotrophic**, meaning that they cannot produce their own food but must obtain it from the environment. Fungi use an amazing array of organic compounds as food, which is termed their **substrate**. The substrate can be soil, a rotting log, a piece of bread, a living tissue, or a wide array of other materials. Fungi are

also like animals in using **absorptive nutrition**. Both fungi and the cells of animal digestive systems secrete enzymes that break down complex organic materials and absorb the resulting small organic food molecules. In addition, both fungi and animals store surplus food as the carbohydrate glycogen in their cells. Despite these nutritional commonalities, fungal body structure, growth, and reproduction are distinct from those of animals and differ among fungal lineages. Because structure, growth, and reproductive differences are key to understanding fungal diversity, we will focus on these features before exploring fungal diversity in more detail in Section 29.3.

## Fungi Have a Unique Body Form

Most fungi have a distinctive body known as a **mycelium** (plural, mycelia), which is composed of individual microscopic, branched filaments known as **hyphae** (singular, hypha) (**Figure 29.2**). Even relatively large fungal structures such as fruiting bodies that function in reproduction are composed of hyphae. Hyphae and mycelia evolved even before fungi made the transition from aquatic to terrestrial habitats.

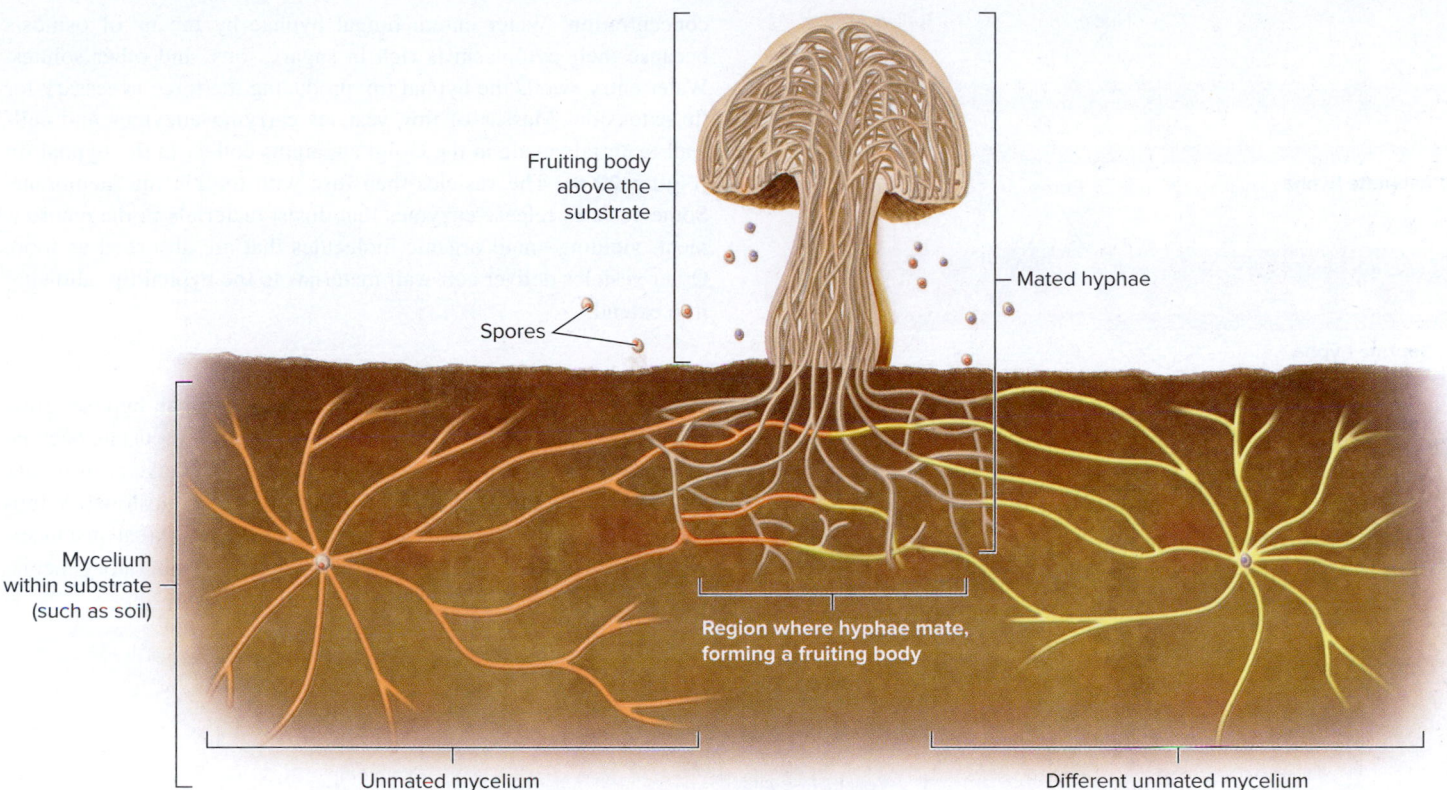

Fruiting body
above the
substrate

Spores

Mated hyphae

Mycelium
within substrate
(such as soil)

**Region where hyphae mate,
forming a fruiting body**

Unmated mycelium

Different unmated mycelium

**Figure 29.2  Fungal morphology.**  The greater part of a typical fungus consists of microscopic food-gathering hyphae that grow and branch from a central point to form a diffuse mycelium within a food substrate, such as soil. After a mating process occurs, mated hyphae may aggregate and grow out of the substrate, forming relatively large fruiting bodies that produce and disperse spores.

The hyphae of early-diverging fungi are not partitioned into smaller cells. Rather, these hyphae are **aseptate** and multinucleate (**Figure 29.3a**), a condition that results when nuclei repeatedly divide without intervening cytokinesis. Such aseptate hyphae are described as being coenocytic. By contrast, the hyphae of later-diverging fungi are subdivided into many small cells by cross walls known as **septa** (singular, septum) (**Figure 29.3b**). In such fungi, known as septate fungi, each round of nuclear division is followed by the formation of a septum that is perforated by a small pore. Septate hyphae appeared prior to the divergence of ascomycetes and basidiomycetes (see Figure 29.1).

Whether septate or aseptate, a fungal mycelium may be very extensive, as in the case of *Armillaria ostoyae* (see the chapter opening photo), but is often inconspicuous because the component hyphae are so tiny and spread out within the substrate. The diffuse form of the fungal mycelium makes sense because most hyphae function to absorb organic food from the substrate. By spreading out, hyphae can absorb food from a large volume of substrate. The absorbed food is used for mycelial growth and for reproduction.

Mushrooms are examples of fungal reproductive structures called **fruiting bodies** (see Figure 29.2). Fruiting bodies are composed of densely packed hyphae that have undergone a sexual mating process during which unmated hyphae of genetically different, but compatible, mycelia are attracted to each other and fuse. The resulting mated hyphae differ genetically and biochemically from unmated hyphae.

Researchers suspect that mating hyphae secrete signaling substances that cause many of them to cluster together and grow out of the substrate and into the air, where reproductive cells can be more easily dispersed. Amazingly diverse in form, color, and odor, mature fruiting bodies are specialized to produce and release reproductive cells known as **spores**. Produced by the process of meiosis and protected by tough cell walls, fungal spores are carried by wind or animals. When fungal spores settle in places where conditions are favorable for growth, they produce new mycelia. When the new mycelia undergo sexual reproduction, they produce new fruiting bodies.

## Fungi Have Distinctive Growth Processes

If you have ever watched bread or fruit become increasingly moldy over the course of several days, you have observed fungal growth. When a food source is plentiful, fungal mycelia can grow rapidly, adding as much as a kilometer of new hyphae per day. The mycelia grow at their edges as the fungal hyphae extend their tips through the undigested substrate. The narrow dimensions and extensive branching of hyphae provide a very high surface area for absorption of organic molecules, water, and minerals.

***Hyphal Tip Growth***  Cytoplasmic streaming and osmosis are important cellular processes in hyphal growth. Osmosis (see Chapter 5) is the diffusion of water through a membrane, from a solution with a lower solute concentration into a solution with a higher solute

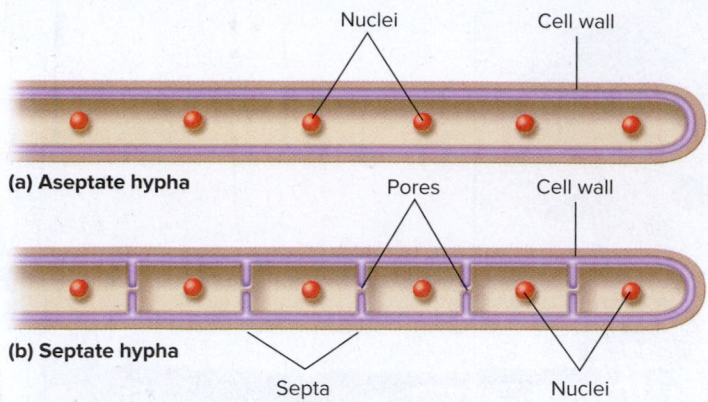

(a) Aseptate hypha

Nuclei    Cell wall

Pores    Cell wall

(b) Septate hypha

Septa    Nuclei

**Figure 29.3**  **Types of fungal hyphae.**

**Core Concept: structure and Function**  This figure compares the multinucleate hypha of an aseptate fungus in part (a) with a hypha of a septate fungus whose cells have a single nucleus in part (b).

concentration. Water enters fungal hyphae by means of osmosis because their cytoplasm is rich in sugars, ions, and other solutes. Water entry swells the hyphal tip, producing the force necessary for tip extension. Masses of tiny vesicles carrying enzymes and cell-wall materials made in the Golgi apparatus collect in the hyphal tip (**Figure 29.4**). The vesicles then fuse with the plasma membrane. Some vesicles release enzymes that digest materials in the environment, yielding small organic molecules that are absorbed as food. Other vesicles deliver cell-wall materials to the hyphal tip, allowing it to extend.

***Variations in Mycelium Growth Form***  Fungal hyphae grow rapidly through a substrate from areas where the food has become depleted to food-rich areas. In nature, mycelia may take an irregular shape, depending on the distribution of the food substrate. A fungal mycelium may extend into food-rich areas for great distances, as noted at the beginning of the chapter. In liquid laboratory media,

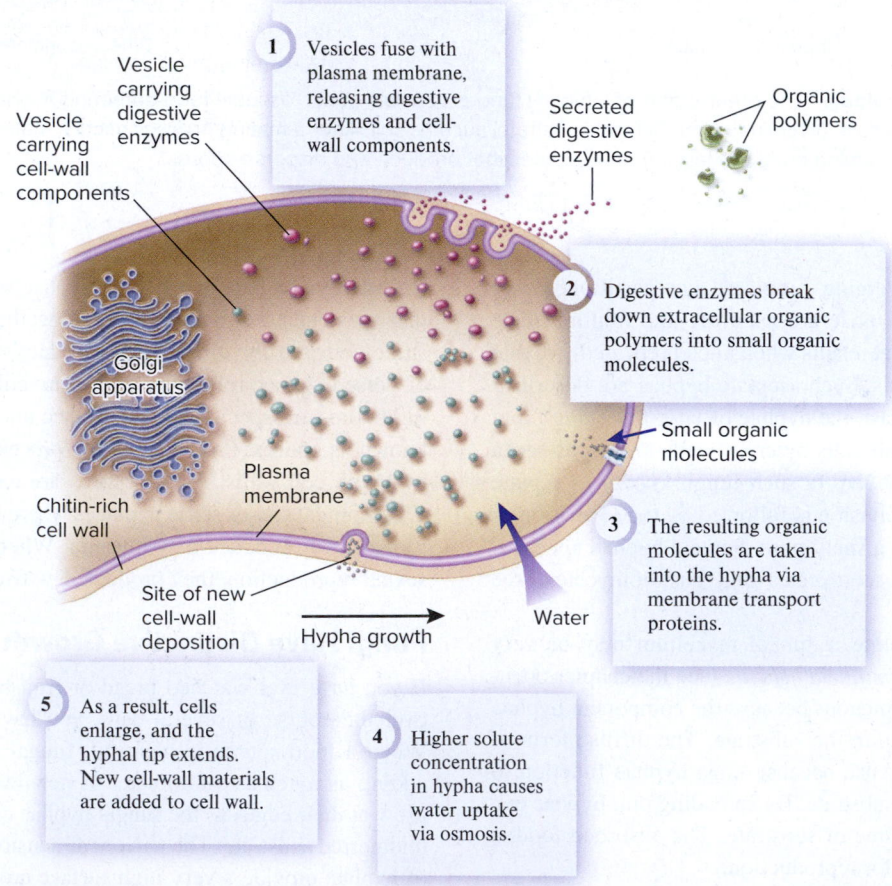

1  Vesicles fuse with plasma membrane, releasing digestive enzymes and cell-wall components.

2  Digestive enzymes break down extracellular organic polymers into small organic molecules.

3  The resulting organic molecules are taken into the hypha via membrane transport proteins.

4  Higher solute concentration in hypha causes water uptake via osmosis.

5  As a result, cells enlarge, and the hyphal tip extends. New cell-wall materials are added to cell wall.

Vesicle carrying cell-wall components

Vesicle carrying digestive enzymes

Secreted digestive enzymes

Organic polymers

Golgi apparatus

Small organic molecules

Chitin-rich cell wall

Plasma membrane

Site of new cell-wall deposition

Hypha growth

Water

**Mechanism of hyphal tip growth**

**Figure 29.4**  **Hyphal tip growth and absorptive nutrition.**  Diagram of a hyphal tip, with two types of vesicles, showing the steps of hyphal tip growth.

**Concept Check:**  *What do you think would happen to fungal hyphae that begin to grow into a substrate with a higher solute concentration? How might your answer be related to food preservation techniques such as drying or salting?*

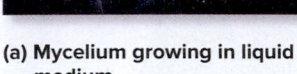

**(a) Mycelium growing in liquid medium**

**(b) Mycelium growing on flat, solid medium**

**Figure 29.5  Fungal shape shifting.  (a)** When a mycelium, such as that of this *Rhizoctonia solani,* is surrounded by food substrate in a liquid medium, it will grow into a spherical form. **(b)** When the food supply is limited to two dimensions, as shown by *Neotestudina rosatii* in a laboratory dish, the mycelium will form a disc. Likewise, distribution of the food substrate determines the mycelium shape in nature. a: ©Agriculture and Agri-Food Canada, Southern Crop Protection and Food Research Centre, London ON; b: Source: CDC

fungi will grow as a spherical mycelium that resembles a cotton ball floating in water (**Figure 29.5a**). When grown on the surface of solid media in flat laboratory dishes, the mycelium assumes a more two-dimensional form (**Figure 29.5b**).

## 29.2  Overview of Asexual and Sexual Reproduction in Fungi

### Learning Outcomes:

1. Give examples of fungal asexual reproduction.
2. Identify some of the distinctive sexual reproductive processes in fungi.
3. Explain why humans may safely consume some fungal fruiting bodies, whereas other fungal fruiting bodies are toxic to humans.

Many fungi reproduce either asexually or sexually by means of microscopic spores, each of which can grow into a new mature organism. Asexual reproduction is a natural cloning process; it produces genetically identical organisms. Production of asexual spores allows fungi that are well adapted to a particular environment to disperse to similar, favorable places. By comparison, sexual reproduction generates new allele combinations that may allow fungi to colonize new types of habitats.

### Fungi Reproduce Asexually by Dispersing Specialized Cells

Asexual reproduction is particularly important to fungi, allowing them to spread rapidly. To reproduce asexually, fungi do not need to find compatible mates or expend resources on fruiting-body formation and meiosis. More than 17,000 fungal species reproduce primarily or exclusively by asexual means. DNA-sequencing studies have revealed that many types of modern fungi that reproduce only asexually have evolved from ancestors that used both sexual and asexual reproduction.

Many fungi produce asexual spores known as **conidia** (from the Greek *konis*, meaning dust) at the tips of their hyphae (**Figure 29.6**). When they land on a favorable substrate, conidia germinate into a new mycelium that produces many more conidia. The green molds that form on citrus fruits are familiar examples of conidial fungi. A single fungus can produce as many as 40 million conidia per hour over a period of 2 days.

Because they can spread so rapidly, asexual fungi are responsible for costly fungal food spoilage, allergies, and diseases. Medically important fungi that reproduce primarily by asexual means include the athlete's foot fungus (*Epidermophyton floccosum*) and the infectious yeast *Candida albicans*. **Yeasts** are unicellular fungi of various lineages. Asexual reproduction in some yeasts occurs by budding (**Figure 29.7**).

### Fungi Have Distinctive Sexual Reproductive Processes

As is typical for eukaryotes, the fungal sexual reproductive cycle involves the union of gametes, the formation of zygotes, and the process of meiosis. In contrast to plants, whose life cycle is an alternation of haploid and diploid generations, and diploid-dominant animals, the fungal life cycle is typically haploid-dominant (look back at Figure 16.14). However, some early-diverging fungi are notable for a life cycle involving alternation of two generations, a haploid gametophyte and a diploid sporophyte. Some other aspects of fungal sexual reproduction are unique, including the function of hyphal branches as gametes and the development of fruiting bodies.

*Fungal Gametes and Mating*  Early-diverging fungi that live in the water produce flagellate sperm that swim to nonmotile eggs, like the gametes of animals and many protists and some plants. By contrast, the gametes of terrestrial fungi are cells of hyphal branches rather than distinguishable male and female gametes. Fungal mycelia occur in multiple mating types that differ biochemically. The compatibility of these mating types is controlled by particular genes. During fungal sexual reproduction, hyphal branches of different, but compatible mycelia are attracted to each other by secreted peptides, and when hyphae have grown sufficiently close, they fuse. This distinctive mating process represents an adaptation to terrestrial life.

**Figure 29.6**  **Asexual reproductive cells of fungi.** SEM of the asexual spores (conidia) of *Aspergillus versicolor,* which causes skin infections in burn victims and lung infections in AIDS patients. Each of these small cells is able to detach and grow into an individual that is genetically identical to the parent fungus and so is able to grow in similar conditions. ©Dennis Kunkel Microscopy, Inc./Phototake

69 µm

**Concept Check:**  *How might you try to protect a burn patient from infection by a conidial fungus?*

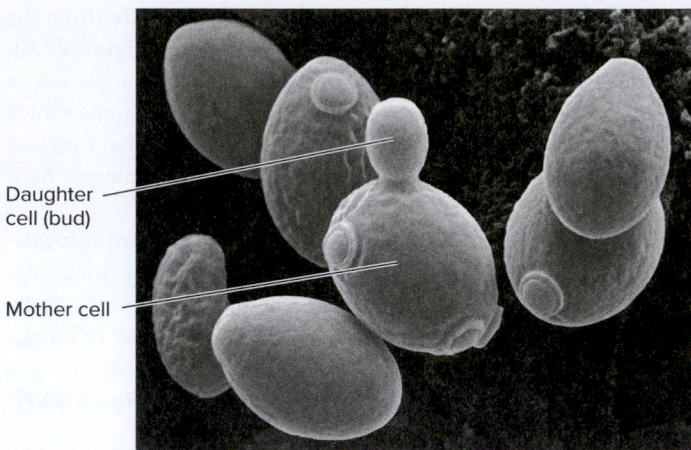

Daughter cell (bud)

Mother cell

**Figure 29.7** **The budding yeast *Saccharomyces cerevisiae*.** In budding, a small daughter cell is formed on the surface of a larger mother cell, eventually pinching off and forming a new cell. ©Medical-on-Line/Alamy Stock Photo

**Core Skill: Connections** Look back at Table 14.1, which shows the genome characteristics of some model organisms. How does the genome of *S. cerevisiae* compare with genomes of other model organisms?

***Fruiting Bodies*** Under appropriate environmental conditions, such as a seasonal change, a mated mycelium may produce a fleshy fruiting body, such as a mushroom. Fungal fruiting bodies typically emerge from the substrate and produce haploid spores (see Figure 29.2). Each spore acquires a tough chitin-rich wall that protects it from drying out and other stresses. Wind, rain, or animals disperse the mature spores, which grow into haploid mycelia. If a haploid mycelium encounters hyphae of an appropriate mating type, hyphal branches will fuse and start the sexual cycle over again.

Mycelium growth requires organic molecules, minerals, and water provided by the substrate, but in most cases, spores are more easily dispersed if released outside of the substrate. The structures of fruiting bodies vary in ways that reflect different adaptations that foster spore dispersal by wind, rain, or animals. For example, mature puffballs have delicate surfaces, and even a slight pressure on one causes the spores to puff out into wind currents (**Figure 29.8a**). Birds' nest fungi form characteristic egg-shaped spore clusters. Raindrops splash on these clusters and disperse the spores. The fruiting bodies of stinkhorn fungi smell and look like rotting meat (**Figure 29.8b**), which attracts carrion flies. The flies land on the fungi to investigate the potential meal and then fly away, in the process dispersing spores that stick to their bodies. The fruiting bodies of fungal truffles are unusual in being produced underground. Truffles have evolved a spore dispersal process that depends on animals that eat fungi. Mature truffles emit an odor that attracts wild pigs and dogs, which break up the fruiting structures while digging for them, thereby dispersing the spores (look ahead to Figure 29.19). Collectors use trained leashed pigs or dogs to locate valuable truffles from forests for the market.

Many fungal fruiting bodies such as truffles and morels are edible, and several species of edible fungi are cultivated for human consumption (**Figure 29.9**). However, many other fungi produce toxic substances that may deter animals from consuming them (**Figure 29.10**). For example, several fungi that attack stored grains,

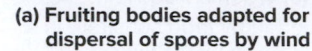

**(a) Fruiting bodies adapted for dispersal of spores by wind**

Spores in a sticky matrix

**(b) Fruiting body adapted for dispersal of spores by insects**

**Figure 29.8** **Fruiting body adaptations that foster spore dispersal. (a)** When disturbed by wind gusts or animal movements, spores puff from fruiting bodies of the puffball fungus (*Lycoperdon perlatum*). **(b)** The fruiting bodies of stinkhorn fungi, such as this *Phallus impudicus*, smell and look like dung or rotting meat. This attracts carrion flies, which come into contact with the sticky fungal spores, thereby dispersing them. a: ©Bob Gibbons/ardea.com; b: ©RF Company/Alamy Stock Photo.

fruits, and spices produce **aflatoxins** that cause liver cancer and are a major health concern worldwide. When people consume the forest mushroom *Amanita virosa*, known as the "destroying angel," they ingest a powerful toxin that may cause liver failure so severe that death may ensue unless a liver transplant is performed. Each year, many people in North America are poisoned when they consume similarly toxic mushrooms gathered in the wild. There is no reliable way for amateurs to distinguish poisonous from nontoxic fungi; it is essential to receive instruction from an expert before foraging for mushrooms in the woods. Therefore, many authorities recommend that it is better to search for mushrooms in the grocery store than in the wild.

Several types of fungal fruiting structures produce hallucinogenic or psychoactive substances. As in the case of fungal toxins, fungal hallucinogens may have evolved as herbivore deterrents, but humans have

**Figure 29.9** **Several types of edible fungi available in supermarkets.** ©Rob Casey/Alamy Stock Photo

**Figure 29.10** **Toxic fruiting body of _Amanita muscaria_.** Common in conifer forests, _A. muscaria_ is both toxic and hallucinogenic. Ancient people used this fungus to induce spiritual visions and to reduce fear during raids. This fungus produces a toxin, amanitin, which specifically inhibits RNA polymerase II in eukaryotes. ©MyLoupe/UIG CALC/ Universal Images Group/Getty Images

 **Core Skill: Connections** Look back at Figure 14.14, which illustrates the cellular role of RNA polymerase II in eukaryotes. What effect would the toxin amanitin have on human cells?

inadvertently experienced their effects. For example, _Claviceps purpurea_, which causes a disease of rye crops and other grasses known as ergot, produces a psychogenic compound related to LSD (lysergic acid diethylamide) (**Figure 29.11**). Some experts speculate that cases of hysteria, convulsions, infertility, and a burning sensation of the skin that occurred in Europe during the Middle Ages and that were attributed to witchcraft resulted from ergot-contaminated rye used in foods. Another example of a hallucinogenic fungus is the "magic mushroom" (_Psilocybe_), which is used in traditional rituals in some cultures. Like ergot, the magic mushroom produces a compound similar to LSD. Consuming hallucinogenic fungi is risky because the amount used to achieve psychoactive effects is dangerously close to a poisonous dose.

## 29.3 Diversity of Fungi

### Learning Outcome:

1. Outline the distinguishing features of the fungal phyla Mucoromycota, Ascomycota, and Basidiomycota.

As noted earlier, the kingdom Fungi is a monophyletic group that arose from a protist ancestor, diversifying first in aquatic habitats, then later in terrestrial environments (see Figure 29.1). In this section, we will

**Figure 29.11** **Ergot of rye.** The fungus _Claviceps purpurea_ infects rye and other grasses, producing hard masses of mycelia known as ergots in place of some of the grains (fruits). ©imageBROKER/Superstock

 **Core Skill: Science and Society** Ergots such as the one illustrated produce alkaloids related to LSD and thus cause psychotic delusions in humans. LSD also harms pets that may accidentally consume it and farm animals that eat infected grains, causing lameness among other symptoms.

begin with a brief description of the early-diverging fungi and then focus on the largely terrestrial fungal lineages listed in **Table 29.1**: Mucoromycota, Ascomycota, and Basidiomycota. We will survey the habitats and characteristics of these groups of fungi and examine their distinctive ecological, structural, growth, and reproductive features.

### Cryptomycota, Chytridiomycota, and Blastocladiomycota Occur in Moist Habitats

Single or few-celled Cryptomycota, Chytridiomycota, and Blastocladiomycota primarily live in moist locales, where they may reproduce using flagellate cells. Some classification schemes of

| Table 29.1 | Distinguishing Features of Later-Diverging Fungal Phyla | | | |
|---|---|---|---|---|
| **Phylum** | **Habitat** | **Ecological role** | **Reproduction** | **Examples cited in this chapter** |
| Mucoromycota | Terrestrial | Form mutually beneficial associations with plants | Distinctive multinucleate asexual spores or sexual zygospores | The genus _Glomus_, the genus _Rhizopus_ |
| Ascomycota | Mostly terrestrial | Decomposers; pathogens; many form lichens; some are plant symbionts | Asexual conidia; nonflagellate sexual spores (ascospores) in sacs (asci) on fruiting bodies (ascocarps) | _Aleuria aurantia, Venturia inaequalis, Saccharomyces cerevisiae, Tuber melanosporum_ |
| Basidiomycota | Terrestrial | Decomposers; many are plant symbionts; less commonly form lichens | Several types of asexual spores; nonflagellate sexual spores (basidiospores) on club-shaped basidia on fruiting bodies (basidiocarps) | _Coprinus disseminatus, Rhizoctonia solani, Armillaria mellea, Puccinia graminis, Ustilago maydis, Phanerochaete chrysosporium, Laccaria bicolor, Amanita muscaria, Phallus impudicus, Lycoperdon perlatum_ |

Cryptomycota include intracellular parasites known as microsporidia, which can be associated with animal disease. One example is *Nosema ceranae*, a single-celled organism that parasitizes honeybees, reproducing within host cells and then spreading to new cells and hosts by means of tough spore stages (**Figure 29.12**). However, recent evidence indicates that microsporidia may be protists rather than fungi.

Members of Chytridiomycota are informally known as chytrids. Some chytrids occur as single, spherical cells that may produce hyphae (**Figure 29.13**). Most chytrids are decomposers, but some are parasites of algal protists or cause diseases of plants or animals. For example, the chytrid *Batrachochytrium dendrobatidis* has been associated with declining frog populations in Australia and the Americas (see the chapter opening photo in Chapter 54).

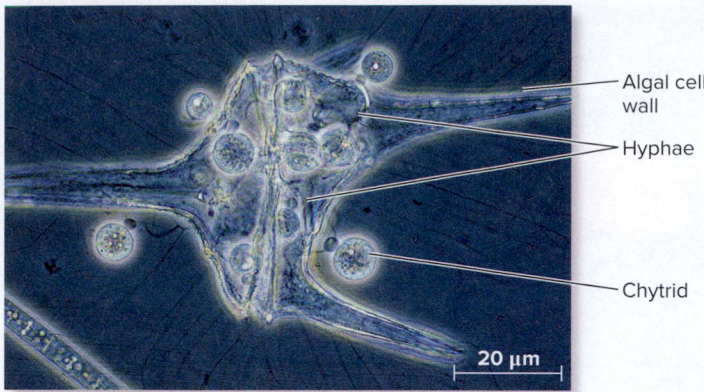

**Figure 29.13** **Chytrids growing on a freshwater algal protist.** The colorless chytrids produce hyphae that penetrate the cellulose cell walls of the alveolate protist *Ceratium hirundinella*, absorbing organic materials. Chytrids use these materials as food and produce spherical flagellate spores that swim away to attack other algal cells. ©Photographs by H. Canter-Lund reproduced with permission of the copyright holder Freshwater Biological Association and J.W.G.Lund.

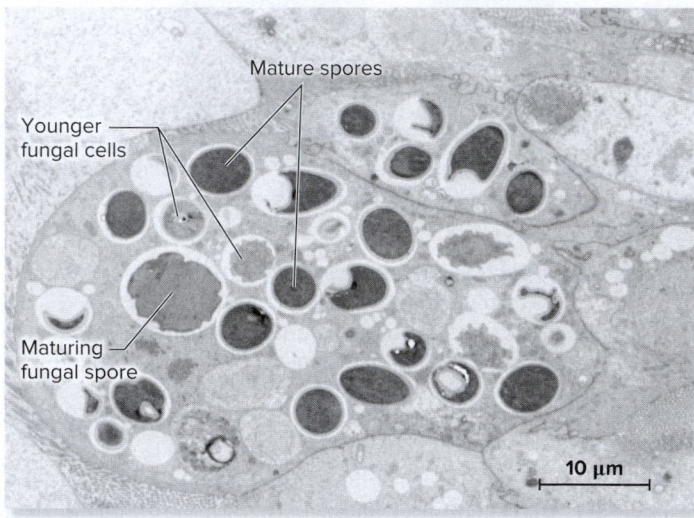

**Figure 29.12** **The microsporidian *Nosema ceranae*, linked with honeybee decline.** In this bee cell, the microsporidian occurs as relatively large round or oblong structures that are each surrounded by a membrane. This membrane separates microsporidian cells from the surrounding bee cell cytoplasm. A narrow white space can be observed between each microsporidian cell and surrounding membrane. The more lightly stained microsporidian cells showing cytoplasmic structure are relatively young products of cell division. As the microsporidian cells mature into spores, they become increasingly more dense and darkly stained. ©Dr. Raquel Martín Martinez and collaborators

 **Core Skill: Modeling** The goal of this modeling challenge is to make a diagram that models how a disease-causing microsporidian multiplies and develops inside an infected host cell.

**Modeling Challenge:** Figure 29.12 is a transmission electron microscopic image showing a honey bee gut cell that has been infected with the microsporidian parasite *N. ceranae*, sometimes classified within the phylum Cryptomycota. Microsporidians are named for their ability to use host resources to divide by mitosis and then form tough microscopic spores that can be transmitted to other cells and hosts. Use the image and information in the caption to draw a diagram that shows how the parasite reproduces and develops into spores.

## Mucoromycota Produce Distinctive Sexual or Asexual Spores

Mucoromycota includes the black bread mold *Rhizopus stolonifer,* which produces sexual spores in dark-pigmented enclosures known as **sporangia** (singular, sporangium) (**Figure 29.14a**). A sporangium is a structure that produces spores. The formation of large numbers of these dark sporangia is what makes moldy bread appear black. These asexual sporangia may each release up to 100,000 spores into the air! The great abundance of such spores means that mold can grow on bread rather easily unless the baker adds retardant chemicals.

*Rhizopus* can also reproduce sexually, forming distinctive zygospores (**Figures 29.14b, 29.15**). In black bread molds, zygospore production begins with the development of **gametangia** (from the Greek, meaning gamete-bearers). In these molds, gametangia are hyphal branches whose cytoplasm is isolated from the rest of the mycelium by cross walls. These gametangia enclose gametes that are basically a mass of cytoplasm containing several haploid nuclei.

When food supplies run low and if compatible mating strains are present, the gametangia of compatible mating types fuse, as do the gametes' cytoplasms. The resulting cell becomes a sporangium that contains many haploid nuclei. Eventually these haploid nuclei fuse in pairs, producing many diploid nuclei (zygote nuclei). For this reason, a zygomycete sporangium produced by sexual reproduction is called a zygosporangium. A single dark-pigmented, thick-walled, multinucleate **zygospore** matures within each zygosporangium.

The zygospore is capable of surviving stressful conditions, but when the environment is suitable, the diploid nuclei within the zygospore may undergo meiosis and germinate, dispersing many haploid spores. If the haploid spores land in a suitable place, they germinate to form aseptate hyphae that contain many haploid nuclei produced by mitosis. Most zygospore-producing fungi live

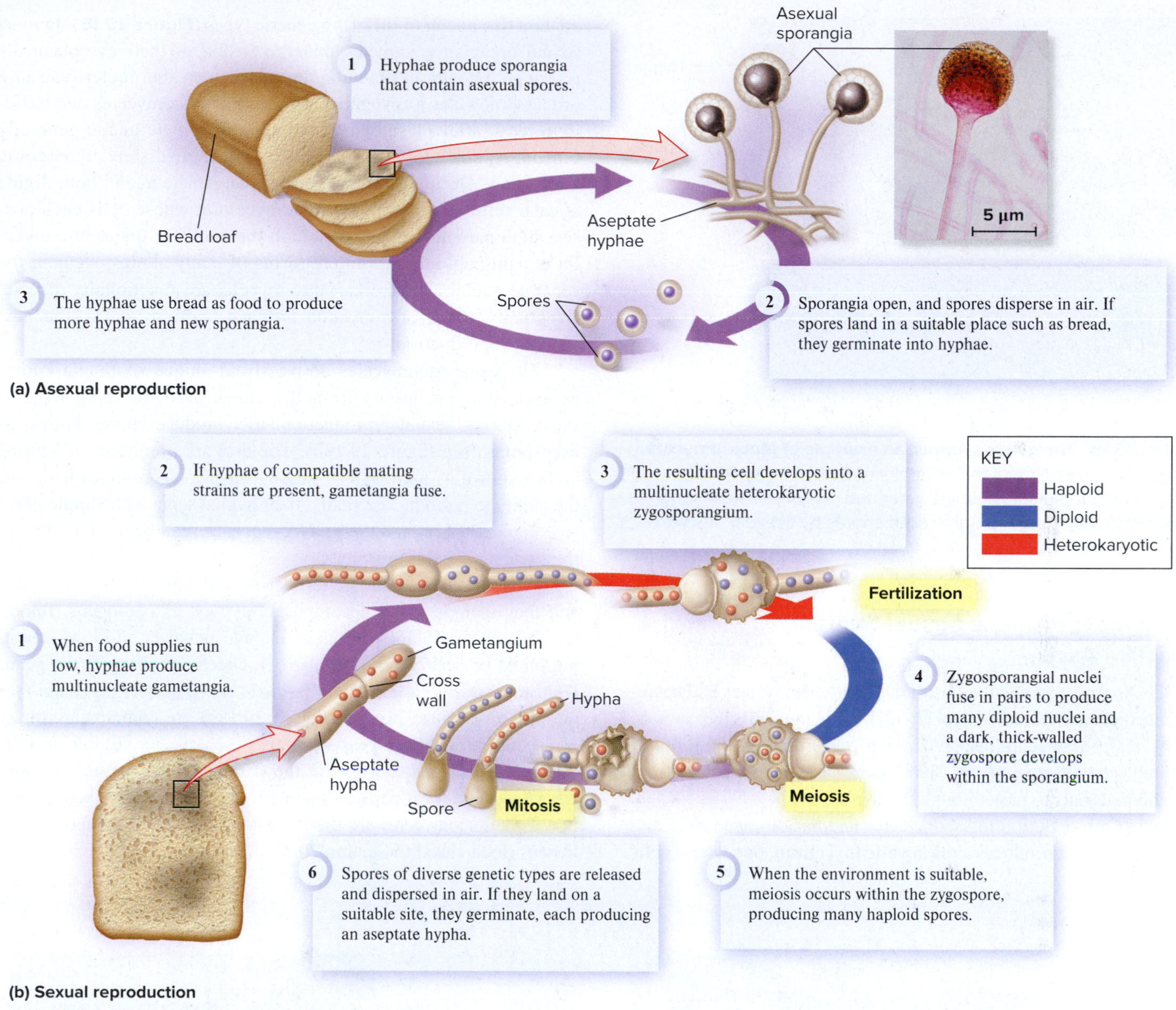

**(a) Asexual reproduction**

**1** Hyphae produce sporangia that contain asexual spores.

Asexual sporangia

Aseptate hyphae

5 μm

Bread loaf

Spores

**3** The hyphae use bread as food to produce more hyphae and new sporangia.

**2** Sporangia open, and spores disperse in air. If spores land in a suitable place such as bread, they germinate into hyphae.

**2** If hyphae of compatible mating strains are present, gametangia fuse.

**3** The resulting cell develops into a multinucleate heterokaryotic zygosporangium.

KEY
Haploid
Diploid
Heterokaryotic

**Fertilization**

**1** When food supplies run low, hyphae produce multinucleate gametangia.

Gametangium

Cross wall

Hypha

**4** Zygosporangial nuclei fuse in pairs to produce many diploid nuclei and a dark, thick-walled zygospore develops within the sporangium.

Aseptate hypha

Spore

**Mitosis**

**Meiosis**

**6** Spores of diverse genetic types are released and dispersed in air. If they land on a suitable site, they germinate, each producing an aseptate hypha.

**5** When the environment is suitable, meiosis occurs within the zygospore, producing many haploid spores.

**(b) Sexual reproduction**

**Figure 29.14** **The asexual and sexual life cycles of the black bread mold _Rhizopus stolonifer_.** (top right): ©Lee W. Wilcox

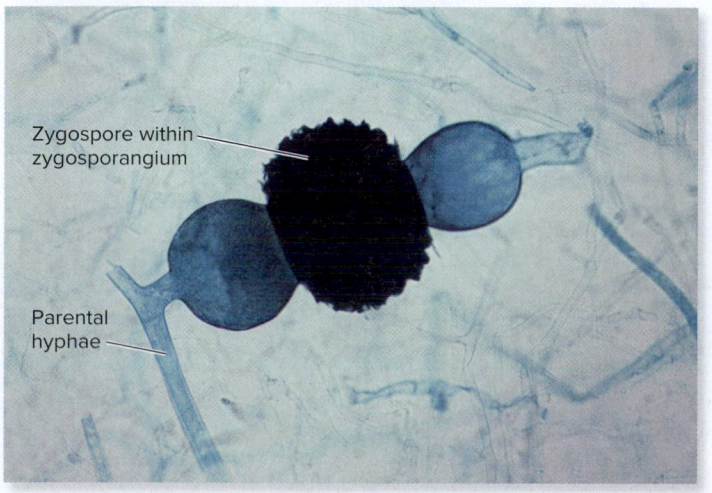

Zygospore within zygosporangium

Parental hyphae

**Figure 29.15** **Zygospores in zygosporangia.** These fungal structures arise from sexual reproduction. They have been stained with a green dye. ©Ed Reschke/Getty Images

on decaying materials in soil, but some are parasites of plants or animals.

Some Mucoromycota, in common with some Ascomycota and Basidiomycota, are notable for forming mutually beneficial partnerships with land plants, often associated with plant roots. Root-fungal partnerships, known as mycorrhizae (from the Greek, meaning fungus roots), are discussed in more detail in Chapter 30. Some mycorrhizal fungi classified in Mucoromycota reproduce only asexually by means of distinctive large, tough-walled spores that each contain many nuclei (**Figure 29.16**).

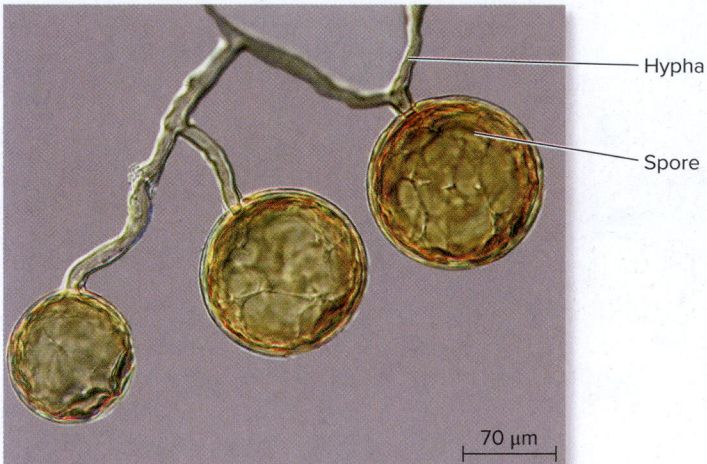

Hypha

Spore

70 μm

**Figure 29.16** **The genus *Glomus*, an example of Mucoromycota.** The hyphae of these fungi form symbioses with many types of plants, helping the plants acquire water and nutrients. These fungi produce distinctive large, tough-walled spores by asexual processes. ©Yolande Dalpé, Agriculture and Agri-Food Canada

## Ascomycota Produce Sexual Spores in Saclike Asci

Both the Ascomycota and the Basidiomycota (ascomycetes and basidiomycetes) are composed of hyphae subdivided into cells by septa. In ascomycetes, these septa display simpler pores than do the septa of basidiomycetes (**Figure 29.17**). Such pores allow cytoplasmic structures and materials to pass through the hyphae.

Sexual reproduction in ascomycetes and basidiomycetes is distinctive, because it produces a **dikaryotic mycelium**, one whose cells

contain two nuclei of differing genetic types (**Figure 29.18**). In most sexual organisms, gametes undergo fusion of their cytoplasms—a process known as **plasmogamy**—and then the nuclei fuse in a process known as **karyogamy**. However, in ascomycetes and basidiomycetes, after plasmogamy the haploid gamete nuclei generally remain separate for a time, rather than immediately undergoing karyogamy. During this time period, the gamete nuclei both divide at each cell division, producing a mycelium whose cells each possess both parental nuclei. Although the nuclei of dikaryotic mycelia remain haploid, alternative forms of many alleles occur in the separate nuclei. Thus, dikaryotic mycelia are functionally diploid. Eventually, dikaryotic mycelia produce fruiting bodies, the next stage of reproduction.

The name ascomycetes derives from unique sporangia known as **asci** (singular, ascus) from the Greek *asco*, meaning bags or sacs). During sexual reproduction, asci produce spores known as **ascospores** (see Figure 29.18b). The asci are produced on fruiting bodies known as **ascocarps**. Although many ascomycetes have lost the ability to reproduce sexually, their hyphal septa with simple pores (see Figure 29.17a) and their DNA sequences can be used to identify them as members of this phylum.

Ascomycetes occur in terrestrial and aquatic environments, and they include many decomposers as well as pathogens. Important ascomycete plant pathogens include powdery mildews, chestnut blight (*Cryphonectria parasitica*), Dutch elm disease (the genus *Ophiostoma*), and apple scab (*Venturia inaequalis*). Cup fungi (see the photo of an ascocarp in Figure 29.18b) are common examples of ascomycetes. Many yeasts are also ascomycetes. Edible truffles (**Figure 29.19**) and morels are the fruiting bodies of ascomycetes whose mycelia form partnerships with plant roots, described in Section 29.4. Ascomycetes are the most common fungal components of lichens (look ahead to Section 30.3).

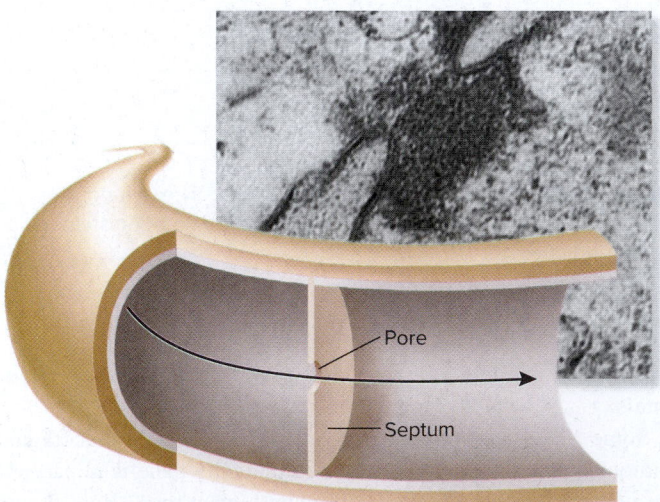

Pore

Septum

**(a) Simple pore—ascomycetes**

Septum

Endoplasmic reticulum (ER)

**(b) Complex pore—basidiomycetes**

**Figure 29.17** **Septal pores of ascomycetes and basidiomycetes.** **(a)** The septa of ascomycetes have simple pores at the centers. **(b)** More complex pores distinguish the septa of most types of basidiomycetes. a: Courtesy of William Whittingham, and Linda Graham; b: ©Charles Mims

**Figure 29.18  The asexual and sexual life cycles of ascomycete fungi.** During sexual reproduction, mating generates dikaryotic hyphae that may form a fruiting body. Nuclei in the dikaryotic surface cells of the fruiting body fuse to form zygotes that undergo meiosis to produce haploid spores. b (middle inset): ©Ed Reschke/Getty Images

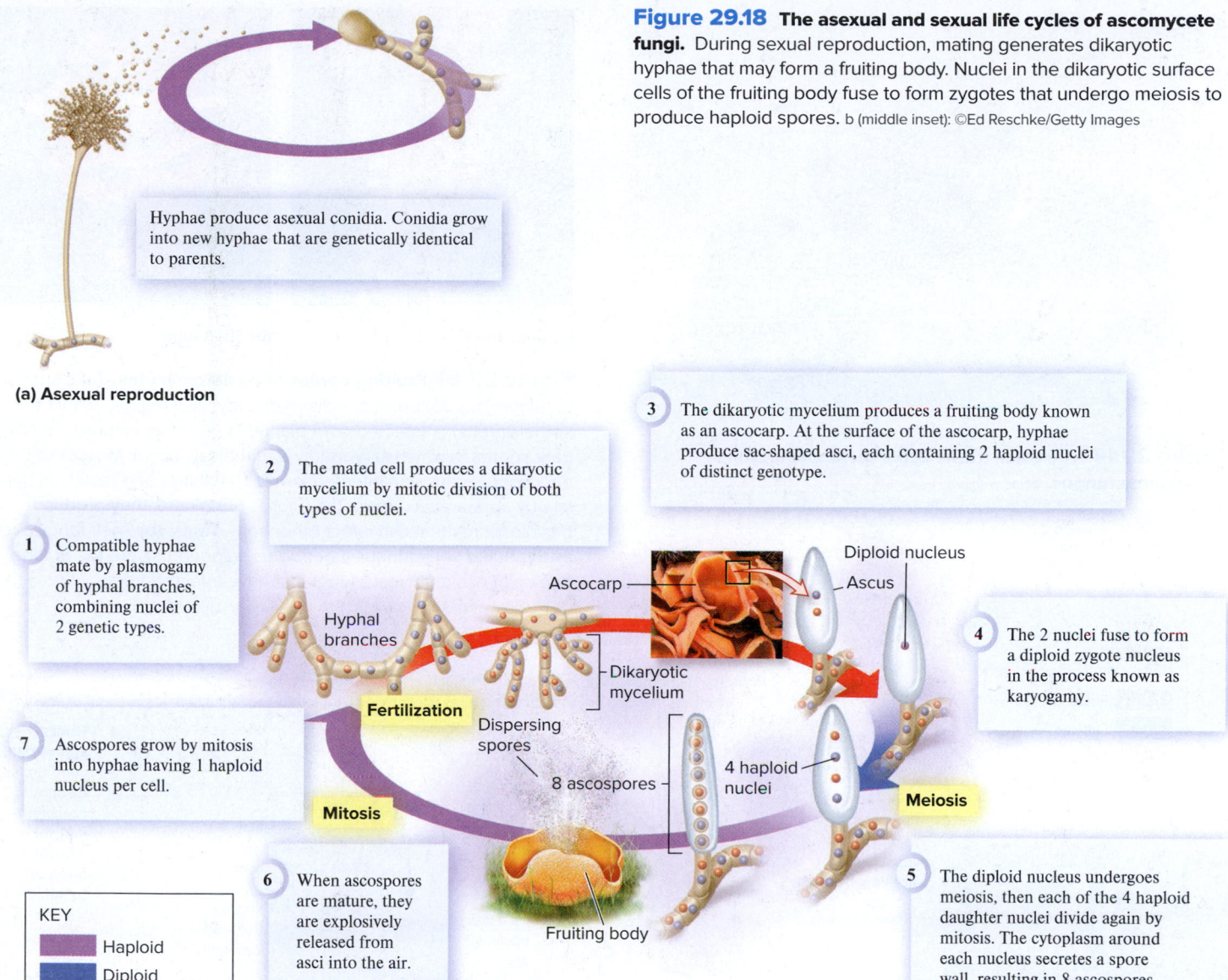

**(a) Asexual reproduction**

Hyphae produce asexual conidia. Conidia grow into new hyphae that are genetically identical to parents.

1 Compatible hyphae mate by plasmogamy of hyphal branches, combining nuclei of 2 genetic types.

2 The mated cell produces a dikaryotic mycelium by mitotic division of both types of nuclei.

3 The dikaryotic mycelium produces a fruiting body known as an ascocarp. At the surface of the ascocarp, hyphae produce sac-shaped asci, each containing 2 haploid nuclei of distinct genotype.

4 The 2 nuclei fuse to form a diploid zygote nucleus in the process known as karyogamy.

5 The diploid nucleus undergoes meiosis, then each of the 4 haploid daughter nuclei divide again by mitosis. The cytoplasm around each nucleus secretes a spore wall, resulting in 8 ascospores.

6 When ascospores are mature, they are explosively released from asci into the air.

7 Ascospores grow by mitosis into hyphae having 1 haploid nucleus per cell.

Hyphal branches

Fertilization

Dispersing spores

8 ascospores

Mitosis

Ascocarp

Dikaryotic mycelium

Diploid nucleus

Ascus

4 haploid nuclei

Meiosis

Fruiting body

**KEY**
- Haploid
- Diploid
- Dikaryotic

**(b) Sexual reproduction of the ascomycete *Aleuria aurantia***

## Basidiomycota Produce Diverse Fruiting Bodies

DNA-sequencing comparisons indicate that the Basidiomycota (or basidiomycetes) and the ascomycetes are the most recently diverged groups of fungi. The mated dikaryotic mycelia of basidiomycetes can live for hundreds of years and produce many fruiting bodies. The name given to the basidiomycetes derives from **basidia**, the club-shaped cells of fruiting bodies that produce sexual spores known as **basidiospores** (**Figure 29.20**). Basidia are typically located on the undersides of fruiting bodies, which are generally known as **basidiocarps**. Though some basidiomycetes have lost the property of sexual reproduction, they can be identified as members of this phylum by unique hyphal structures known as clamp connections that help distribute nuclei during cell division (see Figure 29.20). Basidiomycetes can also be identified by their distinctive septa having complex pores (see Figure 29.17b) and by DNA sequencing. Basidiomycetes reproduce asexually by various types of spores.

An estimated 30,000 modern basidiomycete species are known. Basidiomycetes are very important as decomposers and in symbiotic associations with plants, producing diverse basidiocarps commonly known as mushrooms, puffballs, stinkhorns, shelf fungi, rusts, and smuts (**Figure 29.21**). Basidiocarps are also shown in Figures 29.8, 29.9, and 29.10. The fairy rings of mushrooms that sometimes occur in open, grassy areas are ring- or arc-shaped arrays of basidiomycete fruiting bodies.

**Figure 29.19   The black truffle *Tuber melanosporum,* an ascomycete fungus.** ©Nacivet/Getty Images

**(a) Corn smut**          **(b) Shelf fungi**

**Figure 29.21   Fruiting bodies of basidiomycetes.  (a)** Corn smut (*Ustilago maydis*) produces dikaryotic mycelial masses within the kernels (fruits) of infected corn plants. These mycelia produce many dark spores in which karyogamy and meiosis occur. Masses of these dark spores cause an infected kernel to enlarge and results in the smutty appearance. When the spores germinate, they produce basidiospores that can infect other corn plants. **(b)** Shelf fungi, such as this sulfur shelf fungus (*Laetiporus sulphureus*), are the fruiting bodies of basidiomycete fungi that have infected trees. a: ©Scott Camazine/Alamy Stock Photo; b: ©Mark Turner/Botanica/Getty Imagess

**KEY**

| | |
|---|---|
| ▇ Haploid | |
| ▇ Diploid | |
| ▇ Dikaryotic | |

**3** Hyphal branches known as clamp connections bridge recently divided cells, ensuring that one of each nuclear type is regularly distributed to each daughter cell.

**2** The dikaryotic cell divides by mitosis to produce a dikaryotic mycelium, which can be very long-lived.

**4** Under appropriate conditions, dikaryotic mycelium may form a fruiting body, or basidiocarp.

Mitosis and cell growth in tip cell

Clamp connection forms

New septum forms

Hyphal branch carries 1 nucleus

Nuclear distribution complete

**1** Compatible hyphae mate by plasmogamy of hyphal branches, combining nuclei of 2 genetic types.

Gill of mushroom

**8** Basidiospores grow into mycelia, the cells of which each possess 1 haploid nucleus.

Basidium with haploid nuclei

Diploid nucleus

Basidiospore

Basidiospores

Basidium

**7** Basidia undergo meiosis to produce 4 haploid nuclei, which are incorporated into basidiospores that are dispersed.

**6** Nuclei in basidia fuse to form diploid nuclei.

**5** Dikaryotic basidia occur at the surfaces of gills (or pores of some mushrooms).

**Figure 29.20   The sexual life cycle of the basidiomycete fungus *Coprinus disseminatus.*** (left): ©Biophoto Associates/Science Source; (right): ©Dr. Jeremy Burgess/Science Source

# Fungal Ecology and Biotechnology

## Learning Outcomes:

1. List several ecological roles of fungi.
2. Give examples of fungal diseases of plants and animals, including humans.
3. List several uses of fungi in biochemistry, biological studies, and industrial processes.

Fungi play diverse important ecological roles in addition to previously mentioned beneficial associations with plant roots (mycorrhizae), which are discussed in detail in the next chapter. These additional roles, which include decomposition, predation, disease agent, and protection, allow fungi to be useful in technological applications.

## Many Fungi Play Ecological Roles as Decomposers and Some Fungi Are Predators

Decomposer fungi are essential components of the Earth's ecosystems. Together with bacteria, they decompose dead organisms and wastes, preventing the buildup of organic debris in ecosystems. For example, only certain bacteria and fungi can break down cellulose and lignin, the main components of wood. Decomposer fungi and bacteria are Earth's recycling engineers. They release $CO_2$ into the air and other minerals into the soil and water, making these essential nutrients available to plants and algae.

More than 200 species of predatory soil fungi use special adhesive or noose-like hyphae to trap tiny soil animals, such as nematodes, and absorb nutrients from their bodies (**Figure 29.22**). Such fungi help to control populations of nematodes, some of which attack plant roots. Other fungi obtain nutrients by attacking insects, and certain of these species have been used as biological control agents to kill black field crickets, red-legged earth mites, and other pests.

## Pathogenic Fungi Cause Plant and Animal Diseases

One of the most important ways in which fungi affect humans is by causing diseases of crop plants and animals. Five thousand fungal species are known to be plant pathogens that cause serious crop diseases. Plant pathogenic fungi typically produce specialized balloon-shaped cells known as haustoria, whose increased cell membrane surface area aids in the absorption of organic food from plant cells (**Figure 29.23**). Pathogenic fungi use the absorbed organic compounds to grow, attack more plant cells, and produce reproductive spores capable of infecting more plants.

Wheat rust is an example of a common crop disease caused by fungi (**Figure 29.24**). Rusts are named for the reddish spores that emerge from the surfaces of infected plants. Many types of plants can be attacked by rust fungi, but rusts are of particular concern when new strains attack crops. To control the spread of fungal diseases, agricultural experts work to identify effective fungicidal chemicals and develop resistant crop varieties. Agricultural customs inspectors closely monitor the entry of plants, soil, foods, and other materials that might harbor pathogenic fungi.

Some fungi cause disease in animals. For example, the ascomycete *Geomyces destructans* is associated with white nose syndrome in bats, which has killed more than 1 million hibernating bats in the U.S. Athlete's foot and ringworm are common human skin diseases caused by several types of fungi that are known as dermatophytes because they colonize the human epidermis. The ascomycete *Pneumocystis jiroveci* and the basidiomycete *Cryptococcus neoformans* are fungal pathogens that infect individuals with weakened immune systems, such those with AIDS, sometimes causing death.

**Dimorphic fungi** (from the Greek, meaning two forms) live as spore-producing hyphae in the soil but transform into pathogenic yeasts when mammals inhale their wind-dispersed spores

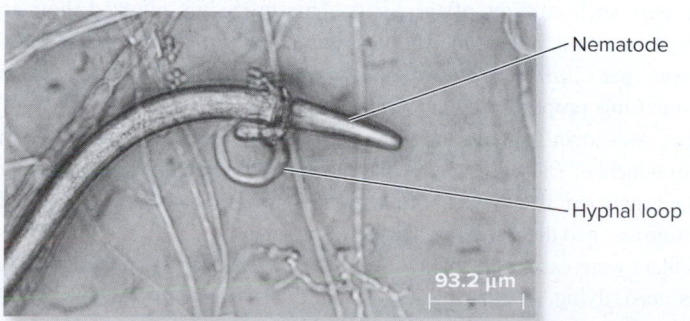

**Figure 29.22  A predatory fungus.** The fungus *Arthrobotrys anchonia* traps nematode worms in hyphal loops that suddenly swell in response to the animal's presence. Fungal hyphae then grow into the worm's body and digest it. ©Science Source

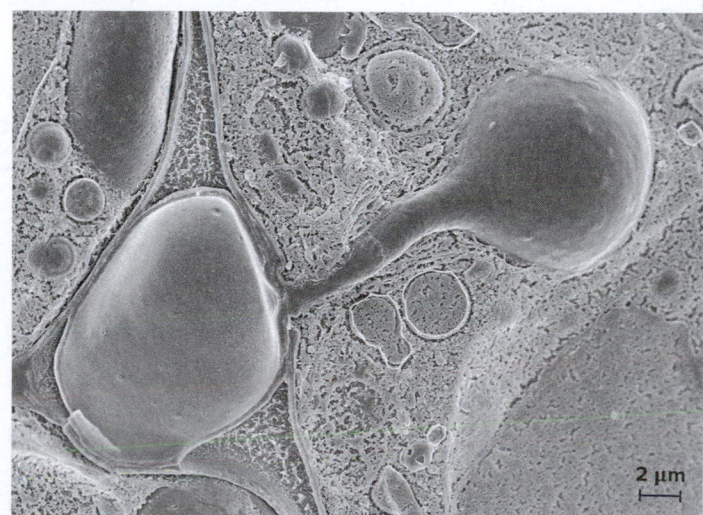

**Figure 29.23  Fungal haustoria.** Fungi that parasitize plants often produce specialized balloon-shaped cells called haustoria that absorb organic food from plant cells, as shown in this electron micrograph. ©Dr. Eric Kemen and Dr. Kurt W. Mendgen

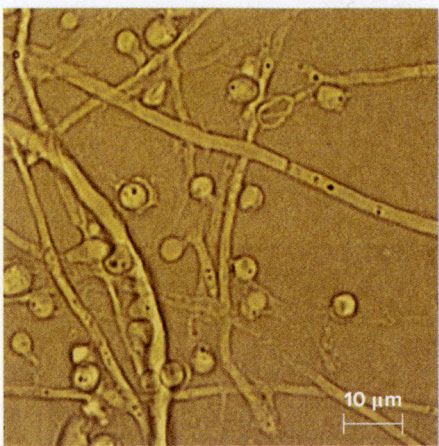

Wheat leaf tissue

*Puccinia graminis* spores

0.1 mm

**Figure 29.24 Wheat rust.** The plant pathogenic fungus *Puccinia graminis* grows within the tissues of wheat plants, using plant nutrients to produce rusty streaks of red spores that erupt at the stem and leaf surfaces where they can be dispersed. Red spore production is but one stage of a complex life cycle involving several types of spores. Rusts infect many other crops in addition to wheat, causing immense economic damage. (left): ©Nigel Cattlin/Science Source; (right): ©Herve Conge/ISM/Phototake

(**Figure 29.25**). Dimorphic fungi include the ascomycetes *Blastomyces dermatitidis*, which causes the disease blastomycosis; *Coccidioides immitis*, the cause of coccidiomycosis; and *Histoplasma capsulatum*, the agent responsible for histoplasmosis. These fungal diseases affect the lungs and may spread to other parts of the body, causing severe illness. The host's body temperature triggers the change from the hyphal form, which produces spores, to a yeast form. In an infected mammal, these pathogenic yeasts reproduce

Soil-dwelling hyphal phase

10 μm

**Figure 29.25 Dimorphic fungi.** The soil-dwelling hyphal stage reproduces by airborne spores. When a mammal inhales the spores, body heat causes the budding yeast phase to develop and attack host tissues. Courtesy Bruce Klein. Reprinted with permission

by forming buds that more effectively stick to lung cells, spread within lung tissue, and move to other organs.

## Some Fungi Play Protective Roles

Although some fungi cause disease, some have recently been discovered to have protective roles. For example, fungi known as endophytes commonly live within plant tissues, providing protection against pathogens and physical stresses such as heat. Some endophytic associations also involve viruses.

### Core Skill: Process of Science

## Feature Investigation | Márquez and Associates Discovered That a Three-Partner Symbiosis Allows Plants to Cope with Heat Stress

The endophytic fungus *Curvularia protuberata* commonly lives within aboveground tissues of the grass *Dichanthelium lanuginosum*. It can grow on very hot soils in areas of Yellowstone National Park. When the soil reaches 38°C, *D. lanuginosum* plants and *C. protuberata* fungi both die—unless they live together in a symbiosis. In the symbiotic association, the partners can survive temperatures near 65°C!

In 2007, a team of investigators led by Luis Márquez investigated the role of a virus in the symbiotic relationship between *D. lanuginosum* plants and *C. protuberata* fungi. Prior to the study described in **Figure 29.26**, these researchers discovered that *C. protuberata* may carry a virus, which they named *Curvularia* thermal tolerance virus (CthTV) to indicate its host and phenotype. In the laboratory, the investigators also noticed that some of their fungal cultures contained very little virus. They were able to use drying and freeze-thaw cycles

to cure such cultures of the virus. This procedure allowed them to experimentally determine the relative abilities of virus-infected and virus-free *C. protuberata* fungus to tolerate high temperatures and to confer this property to plant partners.

As shown in step 1 of Figure 29.26, the researchers began with containers of *D. lanuginosum* plants. One set of containers had neither the fungus nor the virus, another set had the virus-infected fungus, and a third set had the virus-free fungus. The plants in the three sets of containers were exposed to heat stress for 2 weeks, and then categorized as dead, dying, or healthy. As seen in the data, the researchers found that plants having virus-infected fungal endophytes tolerated high temperatures much better than plants that lacked fungal endophytes or possessed only virus-free fungal endophytes. In other experiments, the researchers determined that virus-infected fungi (but not virus-free fungi) could also protect a distantly related crop plant (tomato) from

**Figure 29.26** Márquez and associates discovered that a three-partner symbiosis allows plants to cope with heat stress.

**GOAL**  To determine if a virus is essential to the protective role of endophytic fungi to host plants under heat stress.

**KEY MATERIALS**  *Curvularia* thermal tolerance virus (CthTV), cultures of the endophytic fungus *Curvularia protuberata* infected with CthTV, *C. protuberata* cultures free of CthTV, and *Dichanthelium lanuginosum* plants.

| | Experimental level | Conceptual level |
|---|---|---|

**1**  Plant 25 replicate containers with *D. lanuginosum* lacking fungal symbionts (a) or having *C. protuberata* endophytes that either did (b) or did not (c) have virus.

(a) No fungus, no virus  (b) Fungus and virus  (c) Fungus, no virus

Compare the effects of virus on the ability of the fungus to confer heat stress protection.

**2**  Expose plants to heat stress treatment (up to 65°C) for 2 weeks in a greenhouse.

Keep environmental conditions constant to reduce experimental error.

**3**  Count the number of plants that were green (alive), yellow (dying), or brown (dead).

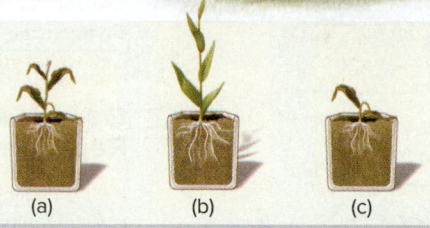

(a)    (b)    (c)

Assess plant survival in the presence or absence of fungus and/or virus.

**4  THE DATA**

Number of plants

KEY
Brown, dead
Yellow, dying
Green, healthy

(a) No fungus, no virus  (b) Fungus and virus  (c) Fungus, no virus

**5  CONCLUSION**  A virus enhances the protective role of endophytic fungi in this grass species. The next step will be to try to determine just how the virus changes the fungus so that the fungus is able to protect the plant from heat stress.

**6  SOURCE**  Marquez, L. M., et al. 2007. A virus in a fungus in a plant: three-way symbiosis required for thermal tolerance. *Science* 315: 513–515.

heat stress. These results add to accumulating evidence that multipartner symbioses are more common than previously realized and suggest that endophytic fungi may have useful agricultural applications.

---

**Experimental Questions**

**1.** Would you expect plants that grow on unusually hot soils to have endophytic fungi or not?

**2. CoreSKILL »** How did Márquez and associates demonstrate that a virus was important in the heat tolerance due to the symbiosis between *Dichanthelium lanuginosum* and *Curvularia protuberata*?

**3.** How might the results of the work by Márquez and associates be usefully applied in agriculture?

**BIO TIPS**

**THE QUESTION** *The data presented in step 4 of Figure 29.26 provide additional information about the relative effects of an endophytic fungus and a virus on heat tolerance in a species of plant. Compare the three bars in the graph and explain what these data mean.*

**T**OPIC *What topic in biology does this question address?* The topic is related to species diversity involved in protective interactions. More specifically, the question addresses the relative effects of fungal endophytes and viruses on plant heat tolerance.

**I**NFORMATION *What information do you know based on the question and your understanding of the topic?* Step 4 in Figure 29.26 presents experimental data in the form of a bar graph that shows the relative numbers of plants in each of three groups that were dead, dying, and healthy at the end of the heat stress experiment. The three groups of plants differed as to whether an endophytic fungus and/or a virus was present in their containers. You have learned in this chapter what endophytic fungi are and how they can affect plants.

**P**ROBLEM-SOLVING **S**TRATEGY *Compare and contrast. Interpret data.* Compare and contrast the bars in the graph, looking for differences in the responses of plants in each group.

**ANSWER** *First, compare the bars labeled (a) and (b). From this comparison, the presence of the virus-infected fungus is seen to confer heat resistance to the plant. However, from this comparison alone, you could infer that either the fungus or the virus or the combination of the two provided plants with protection from heat, but you could not discriminate among these possibilities. A comparison of the bars labeled (b) and (c) reveals that the fungus by itself does not confer thermal protection. Therefore, it is the virus-infected fungus that provides the plant with thermal protection.*

*Interestingly, a careful comparison of the bars labeled (a) and (c) reveals that the proportion of dead plants to dying plants in the (c) group (fungus present but no virus) is higher than in the (a) group (no fungus or virus). These results suggest that the fungus by itself might actually make the plant more sensitive to heat.*

## Fungi Have Many Applications in Biotechnology

In addition to the potential use of fungal endophytes to protect agricultural plants suggested by the experiments conducted by Márquez and associates, fungi have diverse additional technology applications. Enzymes extracted from fungi are widely used to break down tough plant materials for renewable bioenergy applications. A variety of industrial processes employ fungi to convert inexpensive organic compounds into valuable materials such as the citric acid used in the soft-drink industry, glycerol, antibiotics such as penicillin, and cyclosporine, a drug widely used to prevent rejection of organ transplants. In the food industry, fungi are used to produce the distinctive flavors of blue cheese and other cheeses. Other fungi secrete enzymes that are used in the manufacture of protein-rich tempeh and other food products from soybeans.

The brewing and winemaking industries find yeasts essential, and the baking industry depends on the yeast *Saccharomyces cerevisiae* (see Figure 29.7) for bread production. *S. cerevisiae* is also widely used as a model organism for fundamental biological studies. Yeasts are useful in the laboratory because they have short life cycles, they are easy and safe for lab workers to maintain, and their genomes show striking similarities to those of humans. About 31% of yeast proteins have human homologs, and nearly 50% of human genes that have been implicated in heritable diseases have homologs in yeasts.

## Summary of Key Concepts

### 29.1 Evolution and Distinctive Features of Fungi

- Fungi form a monophyletic group of heterotrophs that, together with the animal kingdom and certain protists, form the supergroup Opisthokonta (Figure 29.1).

- Fungal cells typically possess cell walls rich in the polysaccharide chitin. Fungal bodies, known as mycelia, are composed of microscopic branched filaments known as hyphae. Early-diverging fungi have aseptate hyphae that are not subdivided into cells. The hyphae of later-diverging fungi are subdivided into cells by cross walls known as septa (Figures 29.2, 29.3).

- Fungal hyphae feed and grow at their tips (Figure 29.4).

- Mycelial shape depends on the distribution of nutrients in the environment, which determines the direction in which cell division and hyphal growth will occur (Figure 29.5).

### 29.2 Overview of Asexual and Sexual Reproduction in Fungi

- Fungi spread rapidly by means of spores produced by asexual or sexual reproduction.

- Asexual reproduction does not involve mating or meiosis, and it occurs by means of asexual spores called conidia or by budding (Figures 29.6, 29.7).

- Fungi display a haploid-dominant sexual life cycle. During sexual reproduction of terrestrial fungi, hyphal branches (gametes) fuse with those of a different mycelium of compatible mating type. Mated hyphae form fruiting bodies in which haploid spores are produced by meiosis. Dispersed spores germinate to produce haploid fungal mycelia.

- Fungi produce diverse types of fruiting bodies that foster spore dispersal by wind, water, or animals. Although many fungal fruiting bodies are edible, many others produce defensive toxins or hallucinogens (Figures 29.8, 29.9, 29.10, 29.11).

### 29.3 Diversity of Fungi

- The fungi include several early-diverging lineages and the later-diverging phyla Mucoromycota, Ascomycota (ascomycetes), and Basidiomycota (basidiomycetes) (Table 29.1).

- Cryptomycota, Chytridiomycota, and Blastocladiomycota are among the simplest and earliest-diverging fungi. They commonly occur in aquatic habitats and moist soil, where they produce flagellate reproductive cells. Some are parasites of protists, animals, or plants (Figures 29.12, 29.13).

- Mucoromycota includes fungi that reproduce asexually or sexually by distinctive spores (Figures 29.14, 29.15, 29.16).

- Ascomycetes produce sexual ascospores in saclike asci located at the surfaces of fruiting bodies known as ascocarps. The septa of hyphae have simple pores (Figures 29.17, 29.18, 29.19).

- Basidiomycetes produce sexual basidiospores on club-shaped basidia located on the surfaces of fruiting bodies known as basidiocarps. Such fruiting bodies take a wide variety of forms, including mushrooms, puffballs, stinkhorns, shelf fungi, rusts, and smuts. The hyphae display complex septal pores and clamp connections. Mating commonly generates a long-lived dikaryotic mycelium that can produce many fruiting bodies (Figures 29.20, 29.21).

## 29.4  Fungal Ecology and Biotechnology

- Fungi play important roles in nature as decomposers, predators, and pathogens and in beneficial associations with other organisms. Pathogenic fungi cause plant and animal diseases (Figures 29.22, 29.23, 29.24, 29.25).

- Endophytic fungi live within the tissues of plants, providing protective services (Figure 29.26).

- Fungi are useful in the chemical, food-processing, waste-treatment, and renewable biofuel industries. The yeast *Saccharomyces cerevisiae* is a model organism and also important to the brewing and baking industries.

# Assess & Discuss

## Test Yourself

1. Fungal cells differ from animal cells in that fungal cells
   a. lack ribosomes, though these are present in animal cells.
   b. lack mitochondria, though these occur in animal cells.
   c. have chitin-rich cell walls, whereas animal cells lack cell walls.
   d. lack cell walls, whereas animal cells possess cell walls.
   e. None of the above is true.

2. Conidia are
   a. cells produced by some fungi as the result of sexual reproduction.
   b. fungal asexual reproductive cells produced by the process of mitosis.
   c. structures that occur in septal pores.
   d. the unspecialized gametes of fungi.
   e. none of the above.

3. What are mycelia?
   a. the bodies of fungi, composed of hyphae
   b. fungi that attack plant roots, causing disease
   c. fungal hyphae that are massed together into stringlike structures
   d. fungi that produce harmful toxins
   e. protists that are closely related to fungi

4. Where could you find diploid nuclei in an ascomycete or basidiomycete fungus?
   a. in spores
   b. in cells at the surfaces of fruiting bodies
   c. in conidia
   d. in zygospores
   e. in all of the above

5. Which fungi are examples of hallucinogen producers?
   a. *Claviceps* and *Psilocybe*
   b. *Epidermophyton* and *Candida*
   c. *Pneumocystis jiroveci* and *Histoplasma capsulatum*
   d. *Saccharomyces cerevisiae* and *Phanerochaete chrysosporium*
   e. *Cryphoenectria parasitica* and *Ventura inaequalis*

6. What role do fungal endophytes play in nature?
   a. They are decomposers.
   b. They are human pathogens that cause skin diseases.
   c. They are plant pathogens that cause serious crop diseases.
   d. They live within the tissues of plants, helping to protect them from herbivores, pathogens, and heat stress.
   e. All of the above are correct.

7. What determines whether a mycelium is flat or spherical?
   a. sunlight
   b. the nature of the substrate
   c. the amount of carbon dioxide in the air
   d. the amount of phosphorus available
   e. all of the above

8. Among fungi, nutrition is
   a. photosynthetic.
   b. mixotrophic.
   c. absorptive.
   d. all of the above.
   e. none of the above.

9. How can ascomycetes be distinguished from basidiomycetes?
   a. Ascomycete hyphae have simple pores in their septa, whereas basidiomycete hyphae display complex septal pores.
   b. Ascomycetes produce sexual spores in sacs, whereas basidiomycetes produce sexual spores on the surfaces of club-shaped structures.
   c. Ascomycetes lack clamp connections, whereas basidiomycetes display clamp connections.
   d. Ascomycetes fruiting bodies include cup structures and morels, whereas basidiomycete fruiting bodies take different forms that include shelf fungi on tree trunks.
   e. All of the above are correct.

10. Which of the following groups of organisms is most closely related to the kingdom Fungi?
    a. the animal kingdom
    b. the green algae
    c. the land plants
    d. the bacteria
    e. the archaea

## Conceptual Questions

1. List three ways in which fungi are like animals and two ways in which fungi resemble plants.

2. Explain why some fungi produce toxic or hallucinogenic compounds.

3. **Core Concept: Systems** Describe three ways in which fungi affect their environments.

## Collaborative Questions

1. Thinking about the natural habitats closest to you, where can you find fungi, and what roles do these fungi play?

2. Imagine that you are helping to restore the natural grassland vegetation in a region that has recently been used to grow crops. In what ways might you consider using fungi to aid in this restoration?

# Microbiomes: Microbial Systems On and Around Us

# 30

In this SEM image of a human fecal sample, the different colors indicate a diversity of bacteria present in the gut microbiome.

©Eye of Science/Science Source

Ideally, all of the world's children would have enough good food to start healthy lives. Unfortunately, malnutrition has left nearly 180 million children around the world stunted in their growth. Public health experts have tried to use dietary supplements to restore normal growth, a strategy that does not always work. New research on microbes living in the bodies of the stunted children has revealed why. These children retain an infantile set of gut microbes, in contrast to children whose more mature collection of gut microbes (illustrated by the chapter opening image) stimulates growth hormones. This understanding may help people to devise new ways to improve the health of millions of children.

Chapters 27–29 introduced the diverse groups of microorganisms—archaea, bacteria, protists, and fungi—that influence our health and environments in many ways. This chapter builds on that foundation by describing how these diverse microbes occur together in **microbiomes**, assemblages of microbes that are associated with a particular environment, such as the human body. Microbiologists are learning that oceans, ice, fresh waters, soils, and the bodies of organisms other than humans have distinctive microbiomes that influence nature and human life.

Because microbes are generally small, microbiomes are often inconspicuous. New molecular approaches described in this chapter have enabled biologists to study microbiome compositions and functions. Using these new methods, medical scientists are identifying new ways in which people can improve their health. Agricultural scientists are discovering new strategies for engineering crop microbiomes to promote plant health. Exploring the microbiomes of other organisms and environments has revealed previously unrecognized global ecological effects. In this chapter, we will learn why the study of microbiomes has become an important and fast-growing area of modern biology.

## 30.1 Microbiomes: Diversity of Microbes and Functions

**Learning Outcomes:**

1. Define microbiome.
2. List some reasons why microbiomes are considered to be complex systems.
3. Discuss how biologists use ribosomal RNA gene sequences and whole metagenomic sequencing to catalog the diverse types of microbes in a microbiome.
4. Explain how biologists identify microbiome functions.

As noted in the chapter opener, a microbiome is a particular assemblage of microbes (including their genes) that is associated with a particular environment. Microorganisms commonly present in microbiomes include archaea, bacteria, viruses, protists, and fungi (see Chapters 27–29), and sometimes microscopic animals. Visualizing such microbes requires the use of microscopes, such as scanning or transmission electron microscopes.

Microbiomes can be associated with physical biomes such as water, ice, or soil, or with living hosts, such as animals, plants, fungi, and algae. Bacteria dominate the microbiomes of humans and other animals, though certain protists and fungi may also be present (**Figure 30.1**). Microbes are found in many different places in the human body. The microbiomes of plants commonly harbor many types of fungi, and algal microbiomes often include many species of bacteria and protists (**Figure 30.2**). For example, the surface of *Cladophora* is covered with a biofilm of diverse bacteria (Figure 30.2b). A biofilm is a group of microbes that use mucilage to stick to each other and to surfaces.

## Microbiomes Are Complex Biological Systems

Microbiomes are complex systems, in part, because they contain many different microbial species that interact with each other in complicated ways. Microbiome studies often reveal new types of microbes that have not been studied in the lab and, so, have not even been formally named. These diverse life forms carry out many types of metabolism that influence their environments and other members of the same microbiome, but such ecological interactions are not fully understood. Microbes communicate with each other by means of chemical or electrical signals that biologists are just beginning to explore. Particular microbes seem to serve as network hubs, receiving information from the environment and transmitting information to the broader microbial community. In these ways, microbiomes resemble culturally diverse human groups whose social networks and responses to outside influences are complex. Identifying what species occur together, how microbes affect each other and their environments, and the effects of environmental change are major goals of microbiome research.

Determining the species compositions, functions, and responses of microbiomes are challenging because microbes are so small and difficult to distinguish. For example, different bacterial species often have similar body structure, such as single cells only

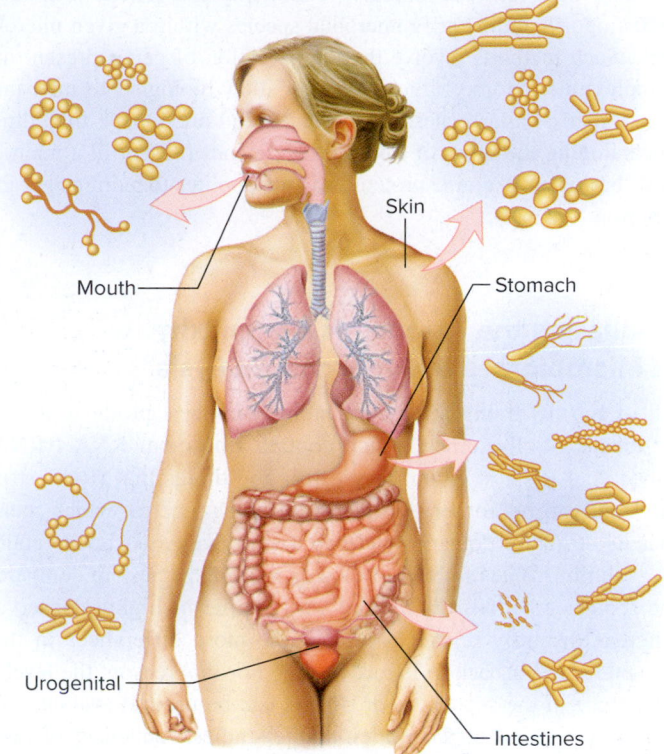

**Figure 30.1** **The human microbiome includes diverse bacteria, but also some protists and fungi.**

a few micrometers in diameter (see Figure 30.2b). Likewise, the bodies of millions of fungal species are composed of thin hyphae that often look alike, even with the use of a microscope (see Chapter 29).

(a)

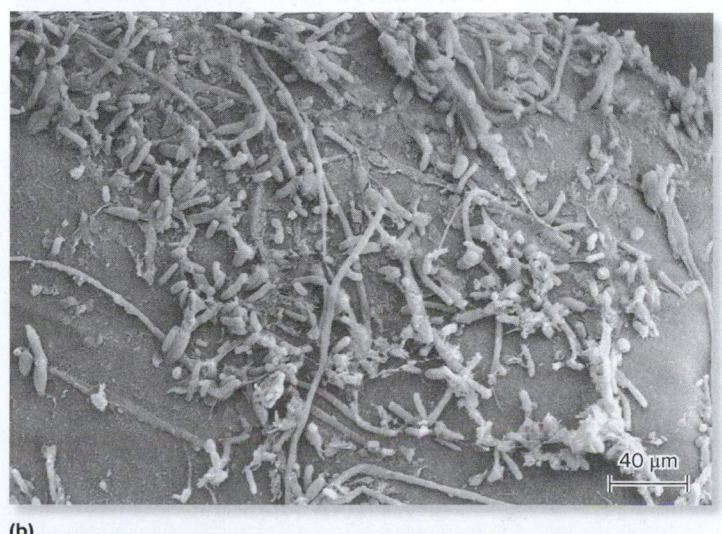

(b)

**Figure 30.2** **An algal microbiome that includes diverse bacterial and protist species. (a)** The green alga *Cladophora* provides living space, organic food, and oxygen to hundreds of species of bacteria, protists, fungi, and microscopic animals, visible as clouds of white, golden, and brown particles on the algal surface. **(b)** This SEM reveals that the algal surface is covered with a biofilm of structurally diverse bacteria. These bacteria have important ecological functions. a: ©Lee W. Wilcox; b: ©Linda Graham

For these reasons, biologists commonly use genetic techniques to distinguish and identify microbial species within a given microbiome. Such methods involve the identification of genes present in a complex microbiome, which may also allow biologists to infer their functions if similar genes have been studied previously. In addition to cataloging the types of microbes and genes present in a microbiome, biologists examine proteins and metabolites to gain insight into function.

## Evaluating the Taxonomic Complexity of Microbiomes by Amplicon Analysis

All of Earth's living things, including microbes, produce proteins by means of ribosomes, which contain ribosomal RNA (rRNA). Ribosomal RNA is so important to living things that rRNA structure tends to be conserved among different species, which means that its structure does not undergo major changes over the course of evolution. When rRNAs of closely related species are compared, the sequence of bases is usually very similar or highly conserved, whereas distantly related species show more differences in their sequences (refer back to Figure 12.17). For this reason, the level of difference in the sequences of rRNAs can be used to evaluate evolutionary relationships. In microbiome studies, sequences of genes that encode rRNAs are commonly used to identify and classify microbes, even if the microbiome includes thousands of microbial species. Other types of genes may also be amplified for evaluating microbiome complexity.

Biologists usually begin a microbiome study by obtaining a sample from a living organism or a physical environment and then extracting the DNA (**Figure 30.3**). As seen in step 3a, polymerase chain reaction (PCR) can be used to copy a particular region of an rDNA gene. This process yields many copies of that region from many different species that are in the sample. These copied rDNA regions are known as **amplicons**. Amplifying the DNA is generally required to generate sufficient DNA for sequencing (described in Chapter 21).

The amplicons are then subjected to DNA sequencing, which yields the base sequence of each gene that was amplified in the original sample. Biologists use computers to compare the DNA sequences of each amplicon to reference sequences in databases. These reference sequences come from microbes, the names and metabolic functions of which are already known. This relatively inexpensive way to examine the microbial diversity in a microbiome is called **amplicon analysis**.

In amplicon analysis, researchers often use PCR primers that amplify a region of genes that encode 16S, 23S, 18S, and/or 28S rRNA (refer back to Table 12.3). 16S rRNA and 23S rRNA occur in the small and large ribosomal subunits of bacteria and archaea; 18S rRNA and 28S rRNA occur in the corresponding ribosomal subunits of eukaryotes. As an example, **Figure 30.4** shows the evolutionary relationships for members of a particular microbiome that plays a role in mineral formation. This phylogenetic tree was constructed by comparing the genes encoding 16S rRNA. As you can see, the microbiome composition is complex, containing representatives of many bacterial phyla and one archaeal phylum.

### Core Skill: Quantitative Reasoning

**BIO TIPS**

**THE QUESTION** *Which prokaryotic phylum dominates the microbiome shown in Figure 30.4? Note: In this figure, the sizes of circles indicate the relative abundances of sequences, which are related to organism abundances. For reference, Figure 27.1 and Table 27.1 list prokaryotic phyla that occur in Figure 30.4.*

**T OPIC** *What topic in biology does this question address?* The topic is microbiomes. More specifically, the question focuses on the composition of a particular microbiome and identification of the prokaryotic phylum that is the most prevalent.

**I NFORMATION** *What information do you know based on the question and your understanding of the topic?* In the question, you are referred to Figure 30.4, which is a phylogenetic tree that provides information about the prokaryotic genera, families, orders, classes, and phyla present in one microbiome. You are also reminded that Figure 27.1 and Table 27.1 list prokaryotic phyla that occur in Figure 30.4.

**P ROBLEM-SOLVING S TRATEGY** *Make a calculation. Interpret data.* The common bacterial phyla Bacteroidetes, Verrucomicrobia, Chloroflexi, Cyanobacteria, Firmicutes, Planctomycetes, Proteobacteria, and Spirochaetes and the archaeal phylum Euryarchaeota, listed in Figure 27.1 and Table 27.1, are represented in this microbiome. Count the number of representatives of each phylum, giving greater weight to larger circles. Use these calculations to make a pie chart that groups bacteria and archaea into different phyla.

**ANSWER** *Of the nearly 60 taxa listed along the right edge of the phylogentic tree, more than half are classified in the phylum Proteobacteria, and the two most abundant genera inferred from sequence abundances (Rhodoferax and Rheinheimera) belong to this phylum. Your pie chart should show the phylum Proteobacteria making up its largest sector, indicating that this prokaryotic phylum dominates this particular microbiome.*

## Evaluating Taxonomic and Functional Complexity of Microbiomes by Whole Metagenomic Sequencing

An alternative method for characterizing microbiome diversity is to obtain base sequences of all the DNA present in a sample, a process known as **whole metagenomic sequencing (WMS)** (see Figure 30.3, step 3b). A **metagenome** is defined as the genomes of all the organisms present in a sample. WMS is carried out using an approach known as **shotgun DNA sequencing**, in which DNA fragments from a genome are randomly sequenced (refer back to Figure 21.11). This approach does not have to focus on the sequencing of one particular gene, such as the gene encoding 16S rRNA. Instead, many fragments of DNA are randomly sequenced, and then biologists use computers to identify places where the ends of DNA fragments have the same DNA

1  Obtain a sample to analyze.

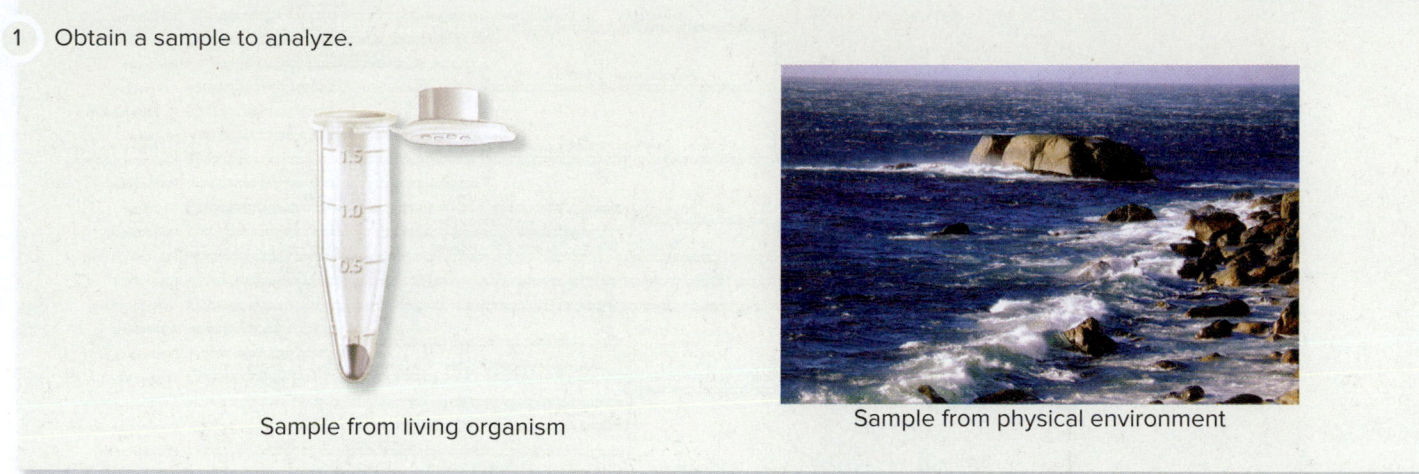

Sample from living organism

Sample from physical environment

2  Extract the DNA from the sample. This yields a collection of many DNA fragments from many different microbial species.

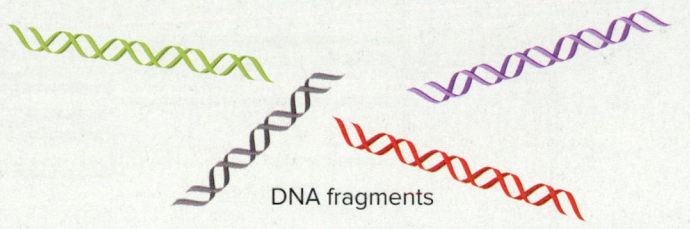

DNA fragments

3  **3a Amplicon Analysis:** Use primers that recognize a region of an rDNA gene and make many copies of that region using polymerase chain reaction (PCR). Subject the amplified segments of the *rDNA* gene to DNA sequencing.

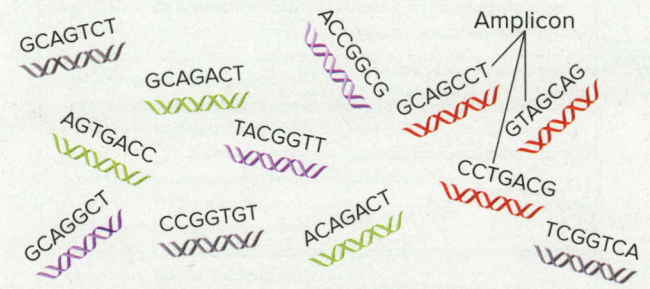

Sequences of rDNA genes that were in the sample.

**3b Whole metagenome sequencing:** Subject all of the DNA in the sample to shotgun DNA sequencing.

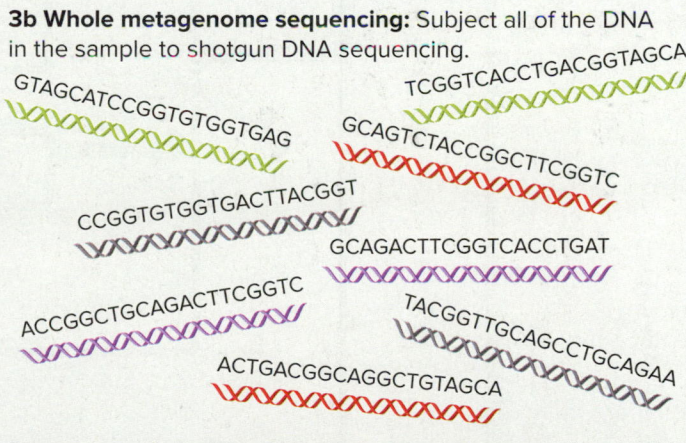

Sequences of all of the DNA fragments that were in the sample.

4  Compare the DNA sequences obtained in step 3 to DNA sequences in a database. The database sequences are already known to come from particular microbial species. This allows researchers to match the sequences obtained in step 3 to known sequences. For example, if a DNA sequence obtained in step 3a matches an rDNA sequence from the bacterium *E. coli*, this result indicates that *E. coli* was in the microbiome of that sample.

**Figure 30.3** **Characterization of microbiome taxonomic complexity via amplicon analysis or whole metagenome sequencing.**
*(right)* ©Goodshoot/Alamy Stock Photo

 **Core Concept: Information** The information contained in DNA sequences can be used to gain information about taxonomic and functional diversity in microbiomes.

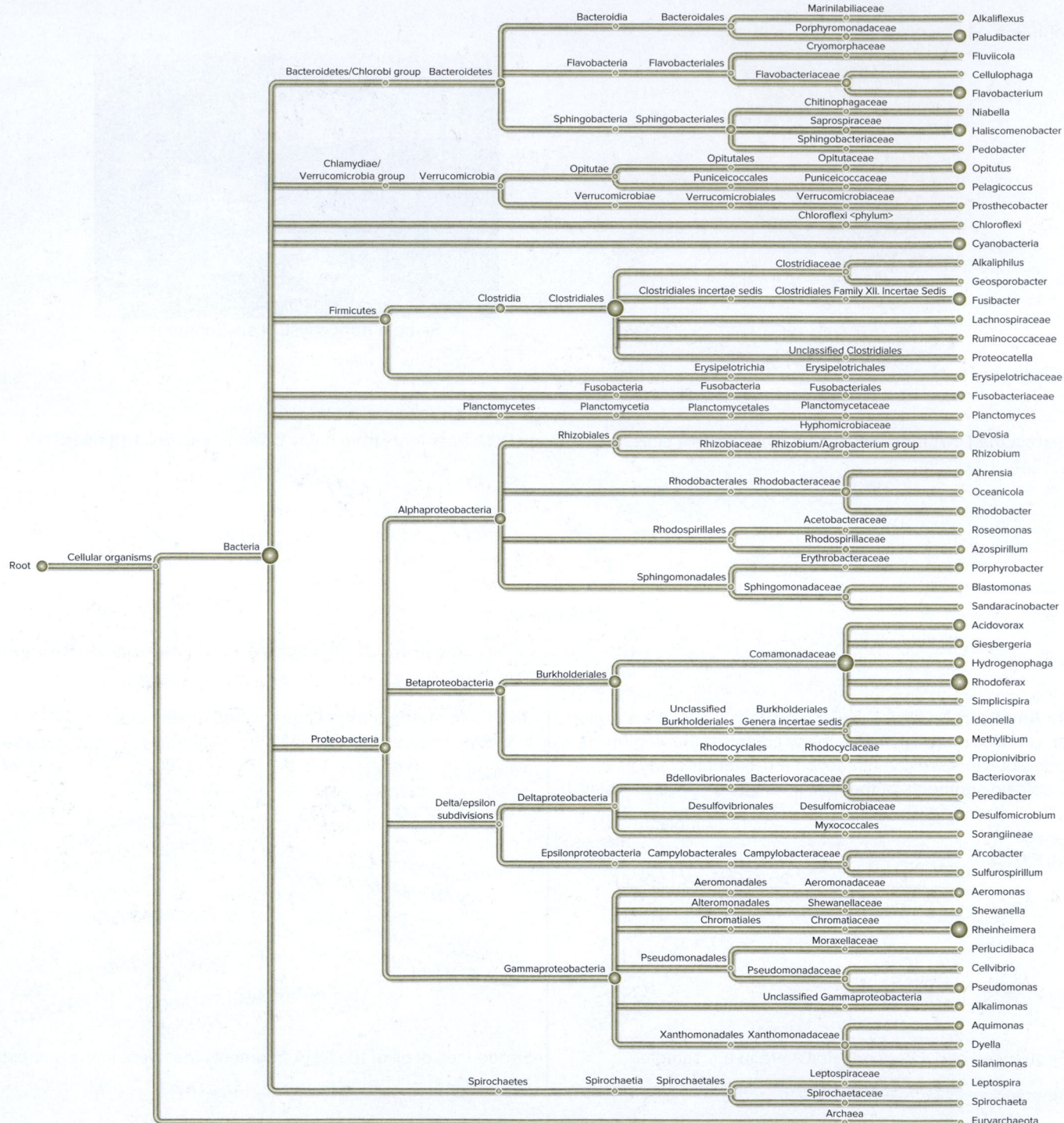

**Figure 30.4 Example of the use of gene sequences encoding 16S rRNA to detect bacteria and archaea present in the microbiome of a photosynthetic organism and to produce a phylogenetic tree.** This phylogenetic tree shows how genera are clustered into phyla and intermediate taxonomic groups. The sizes of circles indicate relative sequence abundances, which are approximately related to organism abundances. Graham, L. E., Knack, J. J., Piotrowski, M. J., Wilcox, L. W., Cook, M. E., Wellman, C. H., Taylor, W., Lewis, L. A., and Arancibia-Avila, P. 2014. Lacustrine Nostoc (Nostocales) and associated microbiome generate a new type of modern clotted microbialite. *Journal of Phycology* 50: 280–291. doi: 10.1111/jpy.12152. This work is licensed under a Creative Commons Attribution 3.0 License.

sequences (**Figure 30.5**). These overlapping regions allow researchers to align the DNA fragments into longer sequences known as contiguous sequences, or **contigs**. What is the advantage of constructing contigs? One advantage is that longer contigs provide greater amounts of information needed for more detailed classification.

If metagenomic sequences are relatively long, or if a microbiome contains relatively few microbes, it may be possible to use computer methods to assemble contigs into whole microbial genomes. A number of microbial species were discovered in this way and even today are known only from their genomic sequence. Some experts consider

Overlapping region

```
TTACGGTACCAGTTACAAATTCCAGACCTAGTACC
AATGCCATGGTCAATGTTTAAGGTCTGGATCATGG
                         GACCTAGTACCGGACTTATTCGATCCCCAATTTTGCAT
                         CTGGATCATGGCCTGAATAAGCTAGGGGTTAAAACGTA
```

**Figure 30.5** **A comparison of two DNA fragments that contain an overlapping region.** A contig consists of a series of DNA fragments that contain overlapping regions.

that one goal of WMS should be to assemble entire genome sequences of microbes, a process known as **genome-centric metagenomics**.

If sufficient DNA has been analyzed, WMS and computer methods can be used to identify both prokaryotic and eukaryotic species in a microbiome at the same time. By contrast, amplicon analyses typically focus on the amplification of a particular gene from a selected group of species. For this reason, many experts consider that the term microbiome should be limited to microbial communities characterized by WMS. The term **microbiota** is commonly used to describe collections of microbial life cataloged by amplicon analysis.

## Microbiome Functions Can Be Inferred by Identifying Protein-Encoding Genes

In the analysis of microbiomes by WMS, another goal is to find and classify protein-encoding genes, providing a deeper view of microbial function. To gain more information about microbiome function, biologists may look for particular protein-coding genes that indicate specialized microbial functions. For example, biologists have used metagenomic sequencing of DNA from natural microbiomes to find bacterial and archaeal genes that encode many previously undiscovered proteins involved in CRISPR-Cas systems. These proteins, which serve as microbial immune systems, have become important tools in modern genetics (see Chapter 21). Three additional examples of important microbiome functions that are inferred from protein-coding genes are described next.

***Nitrogen Fixation*** One important microbial function is nitrogen fixation, the process in which atmospheric nitrogen gas is reduced to form ammonia, which is useful as fertilizer. Only certain prokaryotic species have the natural ability to accomplish nitrogen fixation (see Chapter 27). Plants and other photosynthetic organisms commonly require ammonia or another source of fixed nitrogen to make amino acids, chlorophyll, and other essential molecules. Consequently, algal and plant microbiomes often include nitrogen-fixing prokaryotes. To obtain evidence for microbial species that are able to fix nitrogen, plant scientists may identify gene sequences known to encode enzymes essential for nitrogen fixation. One such gene is *nifH*, an indicator of nitrogen fixation. Such genes are known as **marker genes**, because they "mark" the occurrence of a particular function—in this case, nitrogen fixation.

***Methane Oxidation*** Additional marker genes encode subunits of the enzyme methane monooxygenase (MMO). This enzyme uses oxygen gas to oxidize the greenhouse gas methane, which plays an important role in global carbon cycles and climate warming. Lakes and wetlands are sources of methane, so it is not surprising that methane-oxidizing bacteria are commonly found in these places, often in association

with oxygen-producing algae and plants. For example, peat mosses that dominate vast wetland areas, known as peatlands, play an important role in global carbon cycling by hosting methane-oxidizing bacteria. Peat moss leaves display both oxygen-producing green photosynthetic cells and larger non-green cells that have undergone programmed cell death and whose cell walls are perforated by large pores. Methane-oxidizing bacteria, many other types of bacteria, and diverse protists enter through the pores (**Figure 30.6**), and many of these microbes use mucilage to attach to inner cell wall surfaces (**Figure 30.7**). Peat moss microbiomes commonly contain MMO marker genes, indicating that a microbiome function is to oxidize methane.

***Metabolites*** Some microbes produce very specific compounds that are not produced by most species of microbes. These compounds are called metabolites, because they are the products of metabolic pathways. Examples include vitamins, toxins, and antibiotics. In many cases, previous research has identified the enzymes that are needed to produce a particular metabolite. When analysis by WMS identifies the genes that encode these enzymes, this result indicates that one function of the microbiome is to produce that metabolite. Microbiomes are potential sources of new antibiotic compounds and other metabolites of industrial importance.

## The Analysis of mRNAs, Proteins, and Metabolites Provides Additional Information About Microbiome Function

Catalogs of microbial species and genes obtained by amplicon analysis or WMS don't reveal which genes were actually being transcribed and which transcripts were being translated at the time a sample was collected. Large data collections known as metatranscriptomes, metaproteomes, and meta-metabolomes can help provide the missing information.

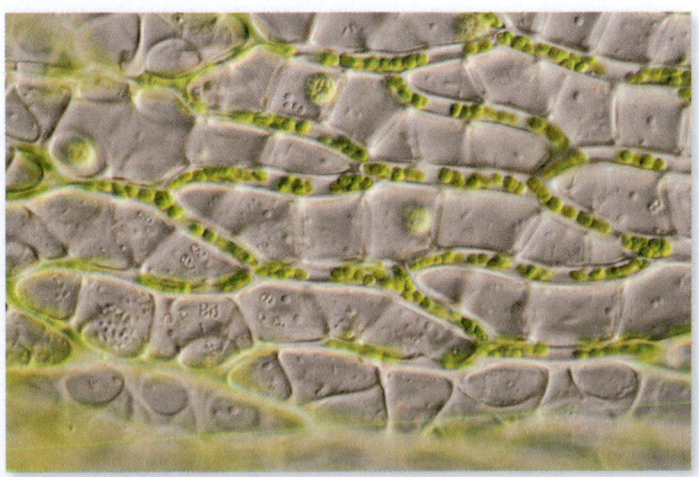

**Figure 30.6** **Peat moss leaf harboring microbes within specialized cells having wall pores.** Peat moss leaves feature narrow living cells having green chloroplasts and larger, non-green, water-filled cells having cell wall pores. Diverse prokaryotic and eukaryotic microbes enter through the pores and live within the larger cells. ©Lee W. Wilcox

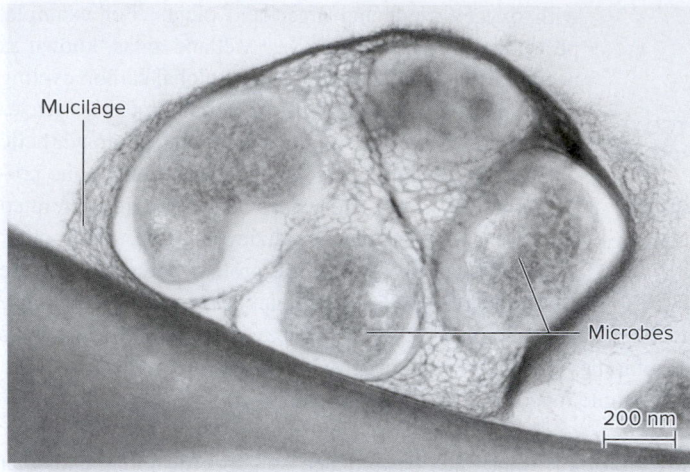

Mucilage

Microbes

200 nm

**Figure 30.7** TEM showing a biofilm of microbes attached by mucilage to the inner cell wall surfaces of peat moss leaf cells.
©Linda Graham

*Metatranscriptome* To learn which mRNAs were present in a microbiome at the time of sampling, biologists analyze transcriptomes. A **transcriptome** is a collection of all the mRNA sequences produced by a single organism under defined conditions. A **metatranscriptome** is a collection of all the mRNA sequences present in an environmental sample, that is, all of the mRNAs produced by all of the organisms sampled from a particular place at a particular time.

*Metaproteome* Biologists sometimes use the number of mRNA sequences of a particular type to infer abundance or activity level of a translated protein. However, mRNA abundance is influenced by the extent to which microbes were actively growing when they were collected, the lifetime of a particular transcript, and how often that transcript is translated. A proteome analysis can provide more direct information about what proteins are present in a particular microbiome. A proteome analysis, accomplished by chromatographic and spectroscopy methods, reveals which proteins are present in a particular sample. A **metaproteome** is all the proteins produced by all the members of a microbiome.

*Meta-metabolome* Because proteins, even if present, might not be functionally active, researchers may analyze the products of metabolism. **Metabolomes** are collections of information about the types and abundances of molecules, such as sugars and fatty acids, produced by metabolism in a single organism. A **meta-metabolome** provides similar information for an entire microbiome.

## 30.2 Microbiomes of Physical Systems

### Learning Outcomes:

1. List some types of microbiomes that are found within physical systems.
2. Explain how microbiome analyses can help monitor environments for microbial activities that affect human health.

Having discussed microbiomes as complex systems and examined the methods that biologists use to study microbiome composition and function, we have a foundation for surveying the diversity of Earth's microbiomes. We have seen that some microbiomes are found within **physical systems** such as oceans, ice, fresh waters, and soils, and others are associated with living organisms known as hosts. In this section, we focus on microbiomes within physical systems and the societal concerns related to them.

### Microbiomes Are Abundant in the World's Oceans and in Its Ice

Although you might imagine that animals such as fish and whales dominate Earth's oceans, in fact, microbes are far more numerous. Oceans occupy 71% of Earth's surface and have a volume of 1.37 billion cubic kilometers. The concentration of microbes in ocean water is typically $10^4$–$10^6$ microbial cells per mL. Therefore, the number of microbes in 1 liter of seawater reaches into the billions. That's a lot of microbes!

Collectively, ocean microbes represent an immense amount of genetic and functional diversity that influences the entire planet. For example, photosynthetic cyanobacteria and algae produce about half of the organic carbon and oxygen formed on Earth each year. Other ocean microbes play essential roles in degrading organic molecules and recycling dissolved minerals, a process essential to ocean productivity. The cyanobacterial genus *Synechococcus* and its phages, together with stramenopile, alveolate, and rhizarian protists (see Chapters 27–28), are key to the movement of organic molecules into deep-ocean waters. Retention of carbon in the deep oceans for long periods affects global climate and is the mechanism by which extensive oil and methane (fossil-fuel) deposits form in undersea locations. Biologists have recently used gene-sequencing techniques (described in Section 30.1) to catalog viruses, prokaryotic species, and small eukaryotes from 68 ocean locations worldwide. By also monitoring the physical features of these places, they have discovered that water temperature is a major factor influencing the compositions of ocean microbiomes, raising questions about the impact of global climate change.

Earth's icy environments—collectively known as the **cryosphere**—likewise contain a surprisingly large number of microbes, an estimated $10^{25}$–$10^{28}$ cells. The sampling of microbes from sea ice and glaciers is challenging, so adventurous biologists use ice-breaking ships, helicopters, planes, tractors, drilling rigs, and remotely operated vehicles to access polar oceans, sea ice sheets, snowfields, and glaciers (**Figure 30.8a**). Biologists are intrigued by the possibility that Earth's cold microbiomes might be similar to life on bodies such as Mars, Jupiter's moon Europa, and Saturn's moon Enceladus.

Microbiome studies of cold habitats have revealed surprisingly diverse types of microbiota that colorize otherwise white environments. Beneath floating sheets of sea ice live conspicuous growths of brown diatoms that dangle into the cold ocean. These photosynthetic algae supply organic molecules and oxygen to heterotrophic bacteria; ciliate, flagellate, and foraminiferan protists (see Chapter 28); and small animals. Algal cells on the surfaces of glaciers can be so abundant that they color the ice green, red,

**(a) Researchers sampling the cryosphere.** Darker areas of ice indicate growth of algae and other microbes.

**(b) A glacier colored by algae in ice microbiomes**

**(c) Blood Falls in Antarctica colored red by iron released by bacteria in ice microbiomes**

**Figure 30.8 Microbiomes within and on ice. (a)** Researchers obtaining samples growing on and within ice for microbiome analysis. **(b)** The color of the ice seen here is due to the presence of algae. **(c)** Blood Falls in Antarctica. The red color is from iron that is released by bacteria, which are part of a microbiome within the ice. a: Courtesy of Cody S. Sheik; b: ©Jason Edwards/National Geographic Creative; c: Source: Peter Rejcek, NSF

yellow, purple, or gray, in this way influencing the amount of light energy that glaciers reflect into space (**Figure 30.8b**). Blood Falls in Antarctica gets its dramatic red color from dissolved iron released from subsurface minerals by iron-metabolizing bacteria living in the cold darkness (**Figure 30.8c**).

## Microbiomes Affect the Quality of Fresh Water and Soil

Freshwater and soil microbiomes are also important to many human concerns, including drinking water safety and agricultural production. Marker genes are commonly analyzed to detect infectious or toxic organisms in the water used for drinking and recreation. Experts are particularly concerned about the effects of global warming on freshwater and soil cyanobacteria, because these organisms grow more abundantly in warmer temperatures, particularly where humans have polluted environments with excess minerals.

Some abundant cyanobacteria produce persistent and potent toxins that harm people and wildlife. For example, the cyanobacterial genus *Microcystis* (**Figure 30.9**) produces more than 100 different chemical forms of the toxin **microcystin**, which binds to and inhibits eukaryotic phosphatases, cellular enzymes that remove

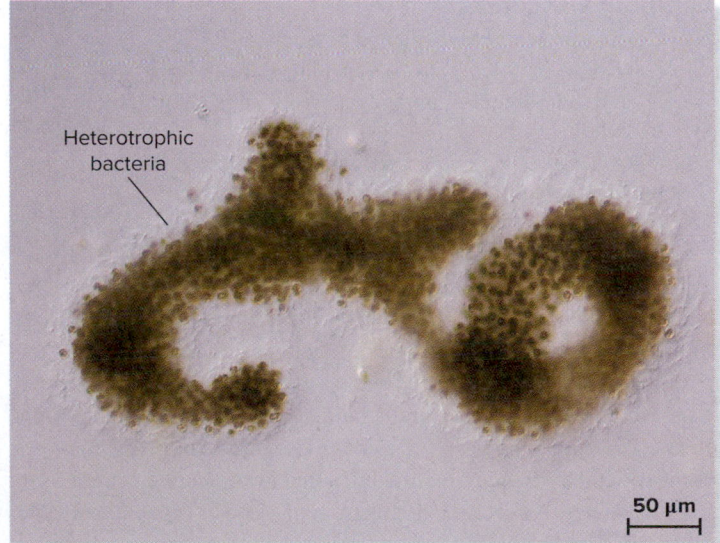

Heterotrophic bacteria

50 μm

**Figure 30.9** **The colony-forming cyanobacterial genus** *Microcystis.* This cyanobacterial colony has cells that look dark because they contain many light-refracting gas vesicles that aid in flotation. In addition to being an important microbial component of a physical (freshwater) microbiome, *Microcystis* hosts other microbes, such as smaller colorless heterotrophic bacteria. *Microcystis* and associated bacteria produce toxins harmful to human health. ©Lee W. Wilcox

 **Core Skill: Science and Society** Microbiome characterization is important in evaluating the safety of water used by humans for drinking and recreation.

phosphate from proteins and organic compounds. The inability to remove phosphate interferes with many cellular processes, including cell signaling (see Chapter 9). By this mechanism, microcystin can cause severe liver damage and hemorrhage in mammals, and chronic exposure to microcystin is associated with high rates of human liver cancer. Humans and other animals become exposed to microcystin when large numbers of cyanobacteria exist in water used for drinking or swimming, a situation that is a growing public health concern. Amplicon analyses of rRNA sequences are increasingly used to monitor aquatic microbiomes for cyanobacterial species likely to produce toxins. Because the genes encoding all enzymes required for the biosynthesis of microcystin and other cyanobacterial toxins are known, metagenomic data can be analyzed for the presence of these genes to assess the potential for toxin production.

Soil microbiomes, which include cyanobacteria and other types of microbes, are key to our ability to grow crops, because various soil microbes can foster or harm plant health (see Chapters 27 and 29). A single gram of soil contains as many as 50,000 bacterial species and fungi, many of which affect plants in some way. Soil microbiomes are important for understanding how terrestrial plants acquire microbiomes, described in Section 30.3.

## 30.3 Host-Associated Microbiomes

**Learning Outcomes:**
1. Define the terms holobiont and hologenome.
2. Describe a few examples in which animal and plant hosts acquire microbiomes, and how microbiomes change during evolution.
3. Make a list of the benefits that are derived from the associations between microbiomes and their hosts.
4. **CoreSKILL** » Analyze the data that host microbiomes play a role in human health.

Many people have heard news media stories about the human microbiome, which includes thousands of different microbial species that live in various locations within and on the surface of the human body (see Figure 30.1). Consequently, humans are said to be microbiome **hosts**. The human body serves as an environment for many microbes, some of which provide us with nutritional or protective benefits. Our microbes, in turn, receive benefits from us, such as organic molecules that serve as microbial food. As noted in previous sections, the bodies of other animals, plants, fungi, protists, and even some prokaryotes support complex microbiomes. In this section, we will learn how hosts acquire microbiomes, how microbiomes evolve, and how the functions of microbiomes are important for their fungal, plant, and animal hosts.

### Hosts Acquire Microbiomes in Different Ways

The combination of a host organism and its microbiome is known as a **holobiont**. The host and microbiome genomes together are called the **hologenome**. Microbiomes contribute many more genes to the hologenome than their hosts do. For example, the human genome contains about 22,000 protein-encoding genes, whereas the human microbiome is estimated to have millions of such genes! Having a functionally useful microbiome aids the survival of the young and thereby increases fitness. Different types of hosts acquire microbiomes in various ways.

- Certain insects coat the casings of their eggs with bacteria; when the young hatch, they become inoculated with beneficial microbes.
- Newborn bees get their microbiomes from sibling worker bees.
- Termites use specific behaviors to transfer among themselves microbes they need to break down plant materials into food.
- Mammals, including humans, transmit important microbes as the young transit the birth canal.
- Human intestinal bacteria, which often prefer low-oxygen environments, can be transmitted from one person to another as tough-walled spores that tolerate air exposure.
- Plant seedlings acquire their microbiomes from the surrounding soil and air and use inherited mechanisms, such as the secretion of particular organic compounds, to attract beneficial microbes.

These examples indicate that both host genetics and the environment play a role in microbiome establishment. The idea that genetics plays a role in microbiome composition is further supported by evidence that compares the microbiomes among different hosts. For example, closely related tropical plants have microbiomes more similar to each other than to the microbiomes of more distantly related plant species. Likewise, strong similarities are observed between the gut microbiomes of African apes (chimpanzees, bonobos, and gorillas) and their close relative, humans, as described next.

### Microbiomes Change During the Evolution of Their Hosts

Researchers are interested in the relationship between the evolution of hosts and the changes in the microbiomes that the hosts support. In the case of humans, evolutionary studies indicate that some bacteria typical of the modern human gut microbiome descended from microbial ancestors that diversified in human and African ape hosts over millions of years. Changes in gut microbiomes may be related to changes in their hosts' diet. After humans diverged from African apes, the ape diet remained plant-rich, whereas the human diet became increasingly animal-rich. These dietary differences affected the environment of gut microbiomes (**Figure 30.10**), which were exposed to different types of nutrients.

Compared to the gut microbes of African apes, some human-associated gut microbes have flourished under an animal-rich diet, whereas others were reduced. For example, humans have more gut microbes from the phylum Bacteroidetes, such as the genus *Bacteroides,* a feature associated with diets rich in animal fats and proteins. By contrast to apes, humans have lower numbers of the archaeon genus *Methanobrevibacter*, which degrades complex plant

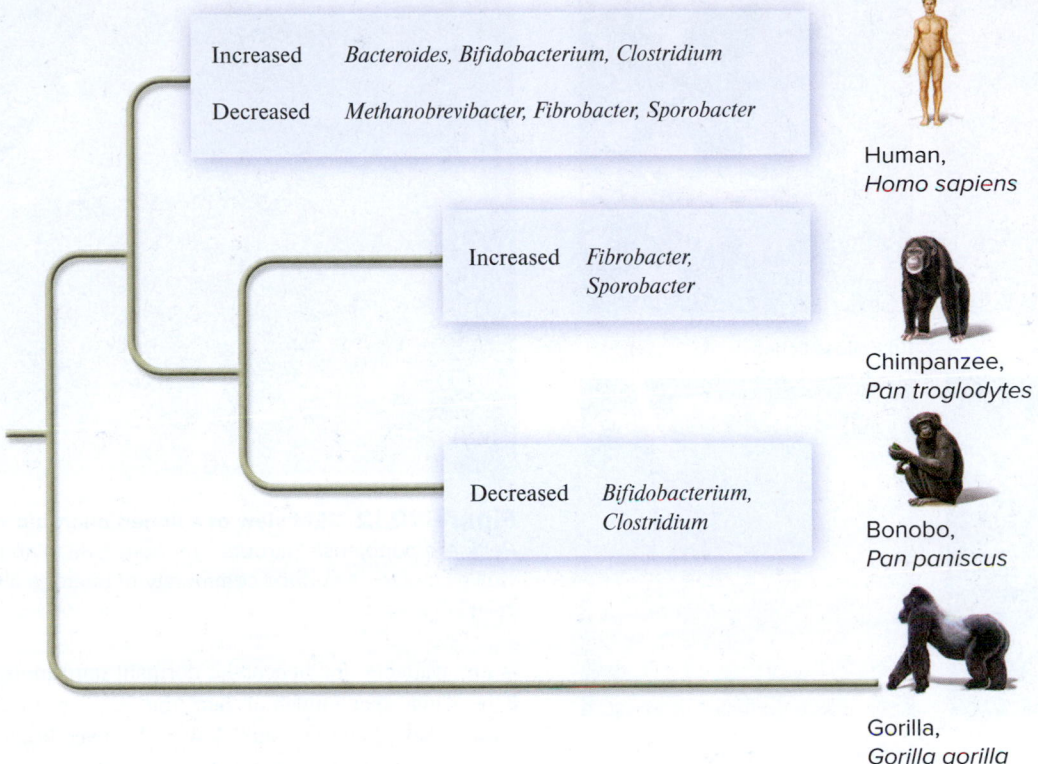

**Figure 30.10** **Changes in gut microbiomes during the evolution of African apes and humans.** The changes seen in humans, chimpanzees, and bonobos are relative to the gut microbiome of gorillas, which diverged earlier. Source: Moeller, A. H. et al. 2014. Rapid changes in the gut microbiome during human evolution. *Proceedings of the National Academy of Sciences of the United States of America* 111: 16431–16435.

 **Core Concept: Evolution** Comparing the microbiomes of evolutionarily related hosts illustrates the concept that host organisms and associated microbiota evolve together over long time periods.

polysaccharides to methane. Humans also have lower numbers of *Fibrobacter*, a bacterial genus common in ape microbiomes, where it helps to break down the plant foods that apes consume. In captivity, primates whose diets are decreased in plant content tend to lose some native gut microbes and gain microbes more common to human guts. Overall, these observations indicate that microbiome composition is influenced by host environment, genetics, and evolutionary history.

## Lichens Are Partnerships Between a Fungal Host and Many Microbial Species

We now turn our attention to microbiomes that are associated with fungi. **Lichens** are complex mutualistic associations between particular fungi and many other microbes, including photosynthetic green algae or cyanobacteria. Researchers estimate that 20% of known fungal species occur in approximately 18,500 kinds of lichen. The association between fungi and other microbes results in a distinctive body form that attaches to surfaces in different ways:

- Crustose lichens are flat and adhere tightly to an underlying surface (**Figure 30.11a**).

- Foliose lichens are flattened and leaflike (**Figure 30.11b**).
- Fruticose lichens grow upright (**Figure 30.11c**) or hang down from tree branches.

In the past, lichens were regarded as relatively simple associations between one fungal species and a single green algal or cyanobacterial species (or sometimes both). According to this simple model, the fungus obtained essential organic carbon and fixed nitrogen from the algal or cyanobacterial partner. The photosynthetic partner received inorganic nutrients and water from the spongy fungal body, and benefited from fungal compounds that protect against intense sunlight and predation by other organisms. However, WMS studies have recently revealed that lichens are complex microbiomes that include hundreds of bacterial species and multiple types of algae and fungal species, in addition to the most abundant fungus, which is considered the host. This new concept of a lichen as a fungal host with a complex microbiome offers new insight into lichen diversity and function in nature.

In a lichen, the diverse microbiota occur in distinct locations. The photosynthetic green algae or cyanobacteria typically occupy a distinct layer close to the lichen's surface, hyphae of the host fungus make up most of the body (**Figure 30.11d**), and other microbes primarily occur on the surface (**Figure 30.12**). Lichen body structure

**(a) Crustose lichen**

**(b) Foliose lichen**

**(c) Fruticose lichen**

**(d) Microscopic view of a cross section of a lichen**

**Figure 30.11** **Lichen structure.** **(a)** An orange-colored crustose lichen grows tightly pressed to the substrate. **(b)** The flattened, leaf-shaped genus *Umbilicaria* is a common foliose lichen. **(c)** The highly branched genus *Cladonia* is a common fruticose lichen. **(d)** A handmade thin slice of *Umbilicaria* viewed with a light microscope reveals that the photosynthetic algae occur in a thin upper layer. Fungal hyphae make up the rest of the lichen. a: ©Perry Mastrovito/Corbis/Getty Images; b, d: ©Lee W. Wilcox; c: ©Ed Reschke/Getty Images

 **Core Skill: Modeling** The goal of this modeling challenge is to propose a model that describes the location of algae within a fruticose lichen.

**Modeling Challenge:** Parts (a) through (c) of Figure 30.11 illustrate three major structural types of lichens: crustose, foliose, and fruticose. Figure 30.11d shows a representative light microscope view of a thin slice of a foliose lichen, which reveals that the green algal cells are located near the upper surface of the flat lichen body. In this location, the photosynthetic cells of the algae are best able to absorb sunlight. Use this information to sketch a structural model of the likely distribution of algal cells in the body of a fruticose lichen. Your model should be a circular cross section through one of the branchlike segments of the lichen. Label the fungal and algal layers.

differs dramatically from that of the main fungal species grown separately, indicating that microbiome components influence lichen form.

Most lichens occur in terrestrial environments, but some are found in aquatic locales. They often grow on rocks, buildings, tombstones, tree bark, soil, or other surfaces that easily become dry. When water

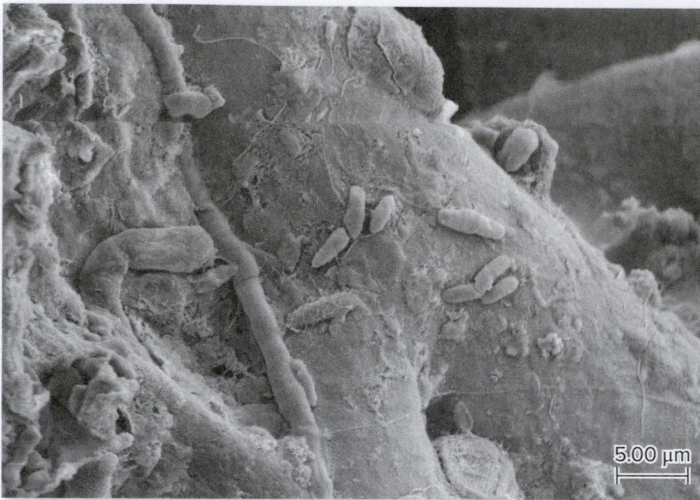

**Figure 30.12** **SEM view of a lichen microbiome.** The lichen *Peltigera ponojensis* harbors cyanobacteria in an internal layer as well as a diverse surface community of bacteria and protists. ©Linda Graham

is not available, the lichens lie dormant until moisture returns. Thus, lichens may spend much of their time in an inactive state, and for this reason, they often grow very slowly. However, because they can persist for long periods, lichens can be very old; some individuals are estimated to be more than 4,500 years old. Lichens grow in some of the most extreme, forbidding sites on Earth—deserts, mountaintops, and the Arctic and Antarctic—places where most plants cannot survive.

Lichens provide important ecological services. They are a food source for reindeer and other animals. Though unpalatable, lichens are not toxic to humans and some have served as survival foods in times of shortages. Soil building is another important lichen function. Lichen acids help to break up the surfaces of rocks, beginning the process of soil development. Lichens that include nitrogen-fixing cyanobacteria are known to increase environmental fertility. One study showed that such lichens released 20% of the nitrogen they fixed into the environment, where it is available for uptake by plants. Recent studies of lichen microbiomes have revealed that bacterial components play previously unknown roles in degrading complex organic materials, and that lichens serve as habitat for diverse protists.

## Plant Microbiomes Are Associated with Leaves, Stems, and Roots

WMS studies have been conducted on the microbiomes of modern bryophytes, which are nonvascular plants such as peat mosses (see Figures 30.6 and 30.7). The results revealed that bryophytes serve as hosts to diverse microbiomes, including microorganisms that provide important services such as nitrogen fixation. Because bryophytes are early-diverging land plants (see Chapter 31), this discovery of bryophyte microbiomes suggests that plants have hosted beneficial microbiomes throughout their evolutionary history.

Compared to bryophytes, vascular plants have a more complex body, which includes aboveground leaves and stems and subterranean roots. As described next, microbiomes are found on and within leaves, stems, and roots.

***Aboveground Plant Microbiomes***    Leaf surfaces commonly host as many as $10^7$ bacterial cells per square centimeter, and microbes also occur within leaf and stem tissues. Globally, leaf microbe numbers are estimated at $10^{26}$. Leaves of tropical forest trees may host more than 400 types of bacteria, and those of temperate forest trees likewise support diverse microbes whose diversity correlates with photosynthetic production.

***Subterranean Root Microbiomes***    Legumes and some other plants are known to form partnerships with soil bacteria that provide fixed nitrogen, greatly aiding plant growth, a process described more fully in Chapter 38. Plants foster specific partnerships with particular nitrogen-fixing bacteria by secreting peptides that cause the death of less valuable bacteria. Certain fungi are also important components of most plant root microbiomes. Plants acquire beneficial bacterial and fungal partners from the diverse assortment of microbes in the soil close to plant root surfaces, a region called the **rhizosphere**.

Associations between the hyphae of certain fungi and the roots of most seed plants are known as **mycorrhizae** (from the Greek, meaning fungus roots). Similar associations also occur between fungi and bryophytes, which lack roots, suggesting that fungi have aided plant success on land from the beginning. Modern fungus-root associations are very important in nature and agriculture; more than 80% of terrestrial plants form mycorrhizae. Plants that have mycorrhizal partners receive an increased supply of water and mineral nutrients, primarily phosphate, copper, and zinc. They do so because an extensive fungal mycelium is able to absorb minerals from a much larger volume of soil than can roots alone (**Figure 30.13**). Added together, the branches of a fungal mycelium in 1 m³ of soil can reach 20,000 km in total length. Experiments have shown that mycorrhizae greatly enhance the growth of the plants they are associated with compared with plants lacking fungal partners. In return, plants provide fungi with organic food molecules, sometimes contributing as much as 20% of their photosynthetic products.

The two most common types of mycorrhizae are endomycorrhizae, which occur within root tissue, and ectomycorrhizae, which coat roots. **Endomycorrhizae** (from the Greek *endo*, meaning inside) are partnerships between plants and fungi in which the fungal hyphae penetrate the spaces between root cell walls and plasma membranes and grow along the outer surface of the plasma membrane. In such spaces, endomycorrhizal fungi often form highly branched, bushy arbuscules (from the word arbor, referring to the tree-like shape of these structures). As the arbuscules develop, the root plasma membrane also expands. Consequently, the arbuscules and the root plasma membranes surrounding them have a very high surface area that facilitates rapid and efficient exchange of materials: Minerals flow from fungal hyphae to root cells, while organic food molecules move from root cells to hyphae. These fungus-root associations are known as **arbuscular mycorrhizae**, abbreviated **AM** (**Figure 30.14**). Fungi are associated in this way with apple and peach trees, coffee shrubs, and many herbaceous plants, including legumes, grasses, tomatoes, and strawberries.

**Ectomycorrhizae** (from the Greek *ecto*, meaning outside) are partnerships between temperate forest trees and soil fungi, particularly basidiomycetes. The fungi that engage in such associations are known as ectomycorrhizal fungi (**Figure 30.15a**). The hyphae of ectomycorrhizal fungi coat tree-root surfaces (**Figure 30.15b**) and grow into the spaces between root cells but do not penetrate the cell

Seedling root

Mycorrhizal hyphae

**Figure 30.13    Tree seedling with mycorrhizal fungi.** Hyphae of a mycorrhizal fungus extend farther into the soil than do the seedling's roots, helping the plant to obtain water and mineral nutrients. ©Dr. D. P. Donelley and Prof. J. R. Leake, University of Sheffield, Department of Animal & Plant Sciences

 **Core Skill: Connections**    Figure 30.13 illustrates the close connection between microbial science, soil science, and forestry science, which studies trees.

membrane; they occupy the spaces between the cell walls of adjacent cells (**Figure 30.15c**). Some species of oak, beech, pine, and spruce trees will not grow unless their ectomycorrhizal partners are also present. Mycorrhizae are thus essential to the success of commercial nursery tree production and reforestation projects.

## Animal Microbiomes Serve Many Useful Functions

Animal microbiomes are commonly dominated by bacteria, but may also include viruses, archaea, fungi, protists, and microscopic animals. These species affect animal health and may play important environmental roles or have medical applications. Termites, for example, are globally important in recycling plant biomass, a function enhanced by the hundreds of gut bacterial species these insects harbor. Tunicates, an early-diverging lineage of chordates (see Chapter 35), occur worldwide as colonies of animals that filter microbial food from seawater. The guts of tunicates contain complex microbiomes, including beneficial cyanobacteria that synthesize sterols useful to the hosts and other bacteria that produce defensive molecules.

Microbiomes are likewise important to mammals, such as the brown bear, *Ursus arctos*. Brown bears are notable for their seasonal lifestyle, which involves fat accumulation in summer

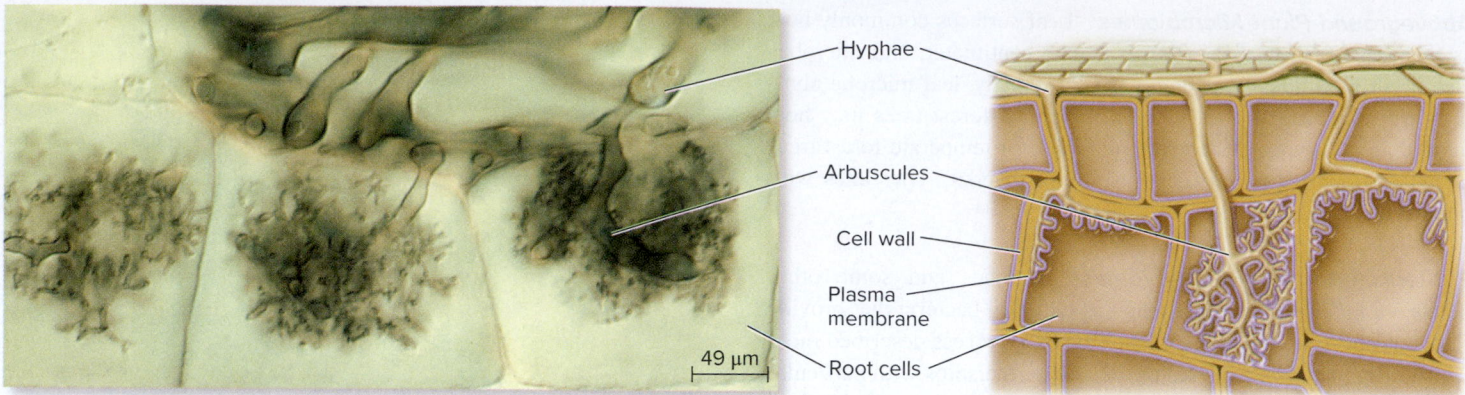

(a) Micrograph of arbuscular mycorrhizae

(b) Hyphae growing between cell walls and plasma membranes

**Figure 30.14** **Endomycorrhizae.** **(a)** Light micrograph showing black-stained endomycorrhizal fungi within the roots of the forest herb *Asarum canadensis*. The fungal hyphae penetrate plant root cell walls, then branch into the space between root cell walls and plasma membranes. **(b)** Diagram showing the position of highly branched arbuscules. Hyphal branches or arbuscules are found on the surface of the plasma membrane, which becomes highly invaginated. The result is that both hyphae and plant membranes have very large surface areas.

a: ©Mark Brundrett

 **Core Concept: Structure and Function** The highly branched structure of intracellular portions of endomycorrhizal fungi is key to the ability of the fungus to take up sufficient organic nutrients and efficiently deliver minerals to the host plant.

and reliance on that fat during hibernation in winter. Amplicon analysis of rRNAs (see Figure 30.3) reveals that the microbiomes in bear guts differ in winter versus summer. Winter microbiomes are less diverse and include higher numbers of bacteria from the phylum Bacteroidetes, which are associated with the breakdown of lipids and proteins. Bacteria linked to diets rich in plant fiber are also lower in winter than in summer, when plants are part of a bear's diet.

The most intensively studied animal microbiome is that of humans, the subject of several large scientific projects involving many biologists. The Human Microbiome Project, for example, characterized the microbes of 18 body sites on 300 healthy U.S. adults,

finding that humans have distinctive microbiomes in the gut, vagina, urogenital tract, mouth, nose, skin, and teeth. This and other studies have shown that human bodies host 100 trillion microbes that make up 1–3% of our body weight! The microbiomes associated with humans affect our health in many ways.

- The microbiomes of teeth form biofilms, known as plaque, that are detrimental to dental health.

- Up to 240 bacterial genera can be associated with human skin alone, performing beneficial functions. Some species within skin microbiomes break down dead skin, and others help to prevent infections or transform skin oil into natural moisturizer.

(a) Ectomycorrhizal fruiting body

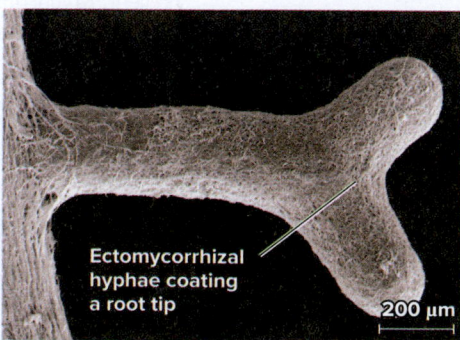

(b) SEM of ectomycorrhizal hyphae

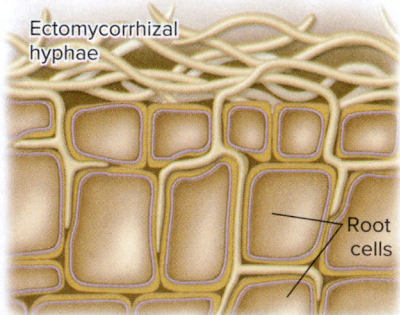

(c) Hyphae invading intercellular spaces

**Figure 30.15** **Ectomycorrhizae.** **(a)** The fruiting structure of the common forest fungus *Laccaria bicolor*. This is an ectomycorrhizal fungus that is associated with tree roots. **(b)** SEM showing ectomycorrhizal fungal hyphae of *L. bicolor* covering the surfaces of young *Pinus resinosa* root tips. **(c)** Diagram showing that the hyphae of ectomycorrhizal fungi do not penetrate root cell walls but grow within intercellular spaces. By doing this, fungal hyphae are able to obtain organic food molecules produced by plant photosynthesis. a: ©Jacques Landry, Mycoquebec.org; b: Courtesy of Larry Peterson and Hugues Massicotte

**Concept Check:** *What benefits do plants obtain from the association with fungi?*

- Microbes of the digestive system, particularly the gut, are very important in early life. *Bifidobacterium longum* subspecies *infantis*, for example, is the most prevalent microbe in the gut of healthy infants. This bacterium has genes encoding proteins that bind, import, and metabolize milk polysaccharides into short fatty acids such as acetate. Some of these fatty acids serve as food for the infant's colon cells, aid the immune system, and reduce gut pH, which deters some disease microbes. The best foods for *B. longum* are polysaccharides that are abundant in human milk but rare or absent in that of other animals. Breastfeeding thereby fosters the growth of these beneficial microbes.

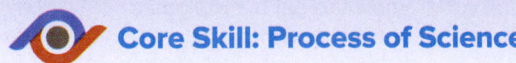 **Core Skill: Process of Science**

## Feature Investigation | Blanton, Gordon, and Associates Found That Gut Microbiomes Affect the Growth of Malnourished Children

In a study published in 2016, Laura Blanton, Jeffrey Gordon, and colleagues described the effects of differences in gut microbiomes on the growth of malnourished children. These researchers began their work by analyzing DNA sequences that encode 16S rDNA to determine how microbiome bacteria change during the first 32 months of life in malnourished children from the same locale (**Figure 30.16**). Some of these malnourished children appeared healthy based on their growth, but others were stunted in their growth and were underweight. The results revealed that the healthy children had microbiomes that changed over time, so that

**Figure 30.16** **Impact of the gut microbiome on growth.**

**HYPOTHESIS** Microbiomes from children who were stunted in their growth due to malnourishment will impair the growth of mice.

**KEY MATERIALS** Fecal samples from healthy and stunted children, germ-free mice which are mice that have been raised in an aseptic environment and do not have any microbes in or on their bodies.

| | Experimental level | Conceptual level |
|---|---|---|
| 1 Obtain fecal samples from many healthy and stunted children of the same locale from birth to 32 months of age. | 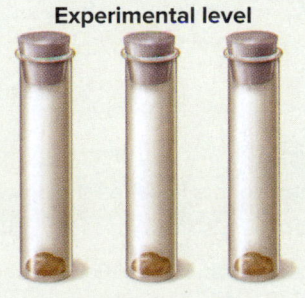 | 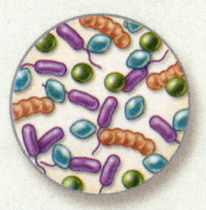 |
| 2 Perform an amplicon analysis on the fecal samples from birth to 32 months old. | See Figure 24.3. | This method reveals which prokaryote species are in the gut microbiome and how they change over the course of 32 months. |
| 3 Using a tube, introduce fecal samples from healthy and stunted children into the gut of 5-week-old germ-free mice. | 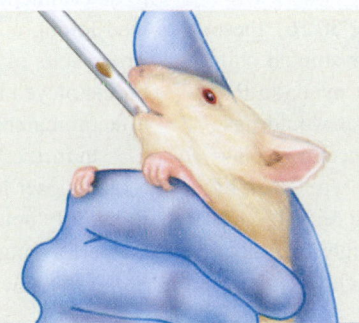 | Gut microbiome from children 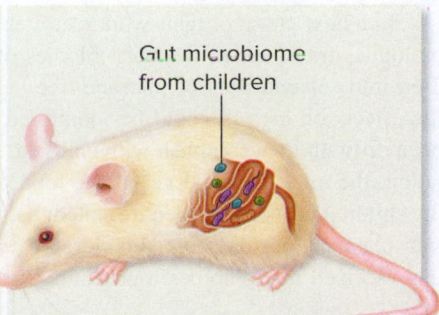 |

| 4 | Give the mice a diet that corresponds to a diet that the children eat. |  | The two different types of gut microbiomes may affect the ability of the mice to gain weight. |

| 5 | Monitor the weight of the mice over a 5-week period. | Received fecal material from a healthy donor | Received fecal material from a stunted donor | Weight gain is a measure of how the gut microbiomes affect growth. |

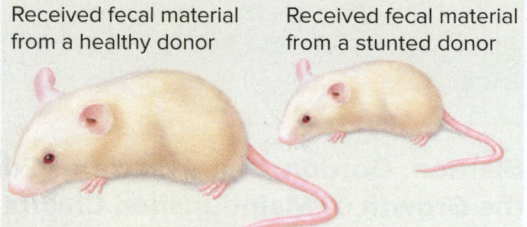

## 6  THE DATA

**From step 2:**

Healthy children: — From birth to 32 months → Microbiomes changed over 32 months

Stunted children: — From birth to 32 months → Microbiomes stayed similar over 32 months

**From step 6:**

- Mice that received fecal transplant from healthy children
- Mice that received fecal transplant from stunted children

Y-axis: % Initial weight (90, 100, 110, 120, 130)
X-axis: Days after the fecal transplant was established in the mouse (0, 10, 20, 30)

## 7  CONCLUSION  The gut microbiomes from stunted donors impaired the growth of mice.

## 8  SOURCE  Blanton, Laura B., et al. 2016. Gut Bacteria That Prevent Growth Impairments Transmitted by Microbiota from Malnourished Children. *Science* 351: 830–837.

older children had a different, more mature microbiome than did infants. By contrast, children who were stunted in growth retained an immature type of microbiome (see the left side of the data in Figure 30.16).

In a next phase of their work (see step 3 of Figure 30.16), these biologists transplanted the microbiomes of healthy or stunted children into separate sets of microbe-free ("germ-free") mice, so that the effects on growth could be monitored in ways that would have been difficult to accomplish with children. After this fecal transplantation, the mice were fed germ-free food of the same type eaten by the children. This experiment revealed that mice having microbiomes transplanted from the stunted children gained less weight than mice that received microbiomes transplanted from healthy children, even

though the mice consumed the same amount of food (see the right side of data in Figure 30.16). Though not shown in Figure 30.16, these researchers also used magnetic resonance imaging technology to determine lean and fat body mass, as well as micro-CT (computerized tomography) to evaluate the femur bone structure of the mice. The results of these imaging studies also showed that the gut microbiome played an important role in proper growth, but the basis of the effect remained unclear.

In further work that is not shown in Figure 30.16, the researchers took a closer look at the composition of the transferred microbiomes to identify potential microbes that were responsible for better growth. By performing an amplicon analysis, they determined that the microbiomes of mice that had received fecal samples from healthy donors tended to be dominated by a few bacterial species—*Ruminococcus gnavus*, *Clostridium symbiosum*, and a few others. To determine if any of these species were responsible for greater growth, researchers grew cultures of these bacterial species in the laboratory and then introduced them into mice that had received fecal samples from stunted children. The result was that the stunted mice grew larger! The researchers then characterized the gut microbiomes from these mice and determined that they were dominated by *Ruminococcus gnavus* and *Clostridium symbiosum*. Metabolomic studies revealed that these two bacterial species had the effect of decreasing amino acid oxidation and increasing amino acid incorporation into proteins

that contributed to increases in body mass. In this way, the microbes were inferred to help malnourished children better use amino acids for growth. Together, these experiments revealed a causal relationship between microbiome composition and mouse growth, rather than merely a correlation.

### Experimental Questions

1. Why did the investigators feed experimental mice germ-free food?

2. **CoreSKILL** » Explain how the microbiomes from healthy or stunted children affected the growth of mice.

## 30.4 Engineering Animal and Plant Microbiomes

### Learning Outcomes:

1. Explain why engineering animal and plant microbiomes may be useful.
2. Describe how biologists produce synthetic microbiomes.
3. Describe how artificial selection can be used to identify the most beneficial microbiomes for plants and animals.

Because microbiomes affect the health of animals and plants, researchers are investigating the potential of altering microbiomes to benefit humans, domesticated animals, and crop plants. Manipulating the composition of a microbiome to improve host characteristics is known as microbiome engineering. Much animal microbiome engineering research is done on mice or other laboratory animals, but the results may eventually be applied to humans.

### Microbiomes Can be Manipulated

As we saw in Figure 30.16, the gut microbiome can have an important impact on health. For this reason, fecal transplants are performed to introduce the entire gut microbiome of a healthy animal host into the gut of an unhealthy host. However, because thousands of microbial species and many complex interactions may be involved, the biological basis for success or failure of this treatment can be difficult to determine. Alternatively, biologists perform experiments to determine more precisely which particular microbiome members are consistently associated with host health. As shown by the studies of Blanton, Gordon, and colleagues, described in the previous section, comparing the microbiomes of healthy hosts with those of unhealthy hosts helps to reveal the key beneficial species.

Additional information about healthy human microbiomes is gained by studying humans in nonindustrial societies. For example, the microbiomes of isolated Amerindians (indigenous peoples of the Americas), who have had no previous contact with Westerners, have the highest known diversity of gut and skin bacteria and microbiome functions. Compared to humans living in industrialized societies, isolated Amerindians have lower diversity of bacterial species from the phylum Bacteroidetes and higher diversity of other important microbiome species. Such differences indicate that humans living in

industrialized countries have undergone shifts in their microbiomes that are implicated in gastrointestinal disorders, obesity, and autoimmune disease. For example, compared to people in isolated populations, people in industrialized countries consume less dietary fiber, composed of complex plant carbohydrates. Studies of mice show that the chronic lack of dietary fiber reduces gut microbiome diversity.

Researchers use such information to assemble **synthetic microbiomes** by mixing cultures of beneficial microbial species. The effects on host health are determined by implanting these synthetic microbiomes into germ-free hosts, also known as axenic or gnotobiotic hosts, such as mice, rats, or guinea pigs. The microbe combinations associated with the most positive effects on host health may be considered for use as treatments in humans.

### Probiotic Treatments Add Beneficial Microbes

A **probiotic treatment** involves the introduction of one or more microbial strains into the microbiome of a host organism. For example, the bacterial genus *Lactobacillus* (phylum Firmicutes) has been used as a probiotic to treat vaginal conditions in humans. Probiotic treatments may also help in the fight against AIDS, a disease caused by HIV (see Chapter 19). Studies of female AIDS patients revealed that decreases in vaginal *Lactobacillus* can be accompanied by increases in other bacteria that are normally rare. Some of these bacteria cause inflammation that enhances susceptibility to HIV infection, and others break down HIV drugs administered in vaginal gels. These results help to explain why anti-HIV drugs sometimes work less well in females than in males, and suggest that manipulating the vaginal microbiome may help in the fight against AIDS.

Recent studies in mice have shown that newborns lacking particular gut bacteria from the phylum Firmicutes are more susceptible to disease bacteria than newborns who have these Firmicutes species. Experimental addition of certain Firmicutes bacteria to the microbiota of newborn mice originally lacking them protected the animals from infection. Human newborns have less gut microbial diversity and are more susceptible to bacterial infections than adults whose gut microbiota include Firmicutes bacteria. These results suggest that probiotic treatments may offer the potential to increase protection from gut pathogens in early life.

Plants can also benefit from probiotic treatments. For example, strawberry plants that grew in a particular soil for many years without being attacked by harmful fungi were found to benefit from the presence in the soil of a fungus that produces an antifungal antibiotic. This observation suggests that the inoculation of soil with this fungus or the antibiotic could be beneficial to strawberry growers.

## Artificial Selection Can Be Used to Engineer Microbiomes

Yet another microbiome engineering approach relies on artificial selection, which is a human-controlled form of natural selection (see Chapter 23). In this process, biologists engineer bacterial communities by artificial selection of favorable microbiomes. As an example, the process of engineering plant microbiomes by artificial selection begins by planting seedlings bearing different microbiomes into sterilized soil. At the end of a period of growth, plants are assessed for an agriculturally important trait, such as height or time of flowering. Those soils that resulted in plants showing the highest levels of the desirable trait are presumed to contain microbes that make up the most beneficial microbiomes. The soil microbiomes associated with plants having favorable phenotypes are selected to incorporate into soils for the next generation of plants. In this way, scientists can use easily observed plant characteristics to select the most beneficial microbiomes (**Figure 30.17**). Amplicon analysis or WMS is then used to determine the composition of the winning microbiome.

Similar artificial selection strategies have been used to engineer the microbiomes of animals, such as bees (**Figure 30.18**). Like those of humans, bee gut microbiomes are dominated by bacteria that prefer low oxygen environments. Bee gut bacteria are well-adapted to consume sugars, which are abundant in flower nectar, a component of bees' diet. Phylogenetic analyses have revealed that, like the microbiomes of primates, bee microbiota have been co-evolving with their hosts over millions of years.

To begin the artificial selection process, bees having different gut microbiomes can be evaluated for health by examining particular phenotypic traits. Newborn bees lacking microbiomes can then be exposed to the microbiomes of the bees judged to have the best health. By repeating this process, the healthiest microbiomes can be inferred from indicators of bee health. Bees are important to human society because they aid plant reproduction in natural and agricultural systems and produce honey and wax. Bees around the world suffer from microbial infections (see Chapter 29), so microbiome engineering might be a valuable way to help improve bee survival.

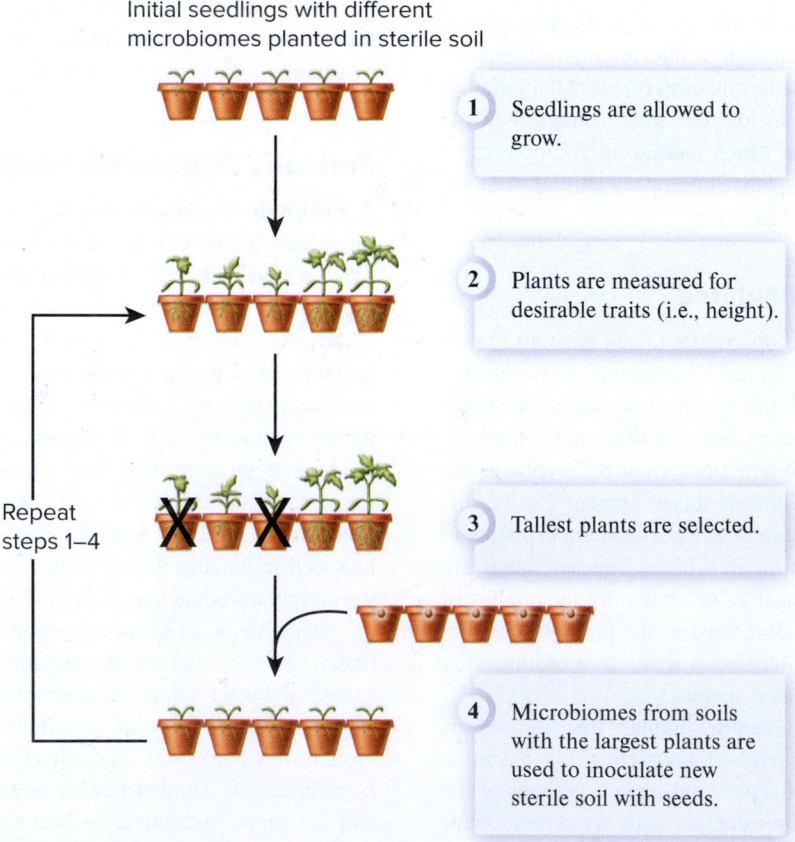

Initial seedlings with different microbiomes planted in sterile soil

1. Seedlings are allowed to grow.

2. Plants are measured for desirable traits (i.e., height).

3. Tallest plants are selected.

Repeat steps 1–4

4. Microbiomes from soils with the largest plants are used to inoculate new sterile soil with seeds.

**Figure 30.17 Artificial selection as a tool to engineer plant microbiomes.** Source: Mueller, U. G., and Sachs, J. L. 2015. Engineering Microbiomes to Improve Plant and Animal Health. *Trends in Microbiology* 23: 606–617.

Inital bee-microbiome associations before selection

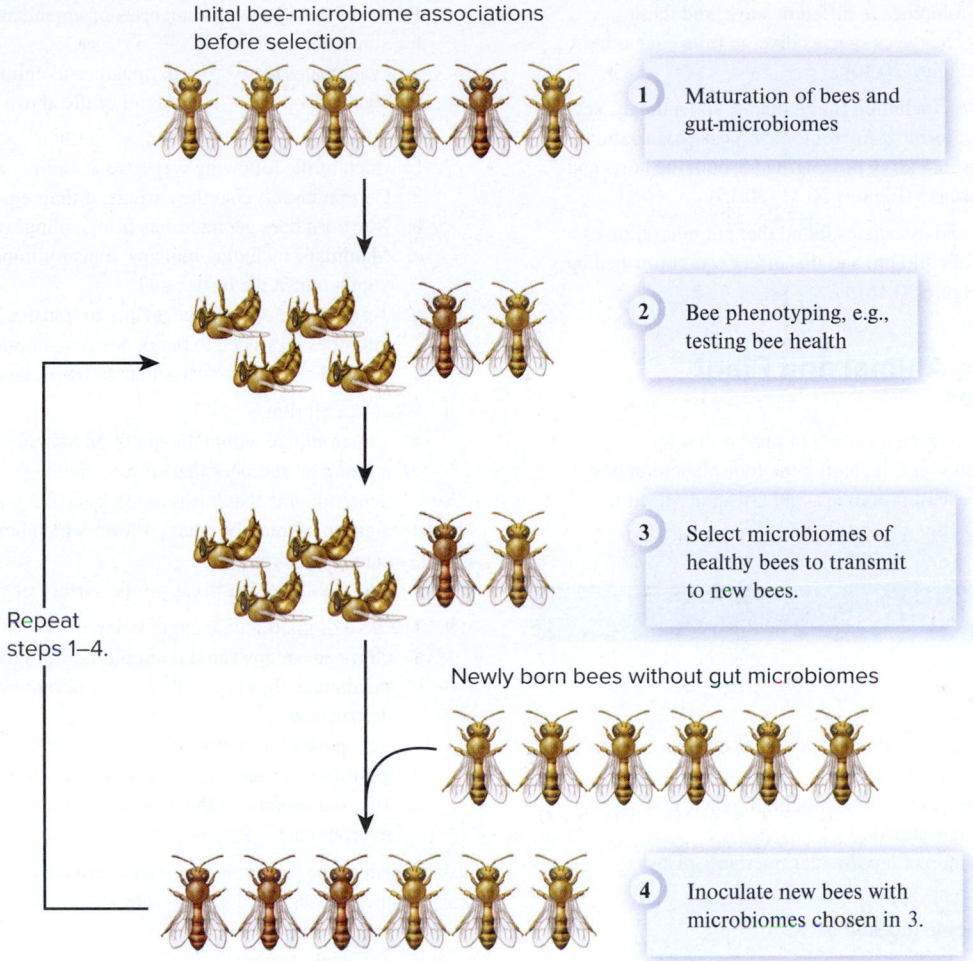

1   Maturation of bees and gut-microbiomes

2   Bee phenotyping, e.g., testing bee health

3   Select microbiomes of healthy bees to transmit to new bees.

Repeat steps 1–4.

Newly born bees without gut microbiomes

4   Inoculate new bees with microbiomes chosen in 3.

**Figure 30.18** **Artificial selection as a tool to engineer animal microbiomes.** Bee microbiomes provide an example of this method.
Source: Mueller, U. G., and Sachs, J. L. 2015. Engineering Microbiomes to Improve Plant and Animal Health. *Trends in Microbiology* 23: 606–617.

## Summary of Key Concepts

### 30.1   Microbiomes: Diversity of Microbes and Functions

- Microbiomes are complex biological systems that are difficult to characterize using a microscope alone (Figures 30.1, 30.2).

- The members of microbiomes are usually identified using genetic techniques such as amplicon analysis or whole metagenomic sequencing (WMS) (Figures 30.3–30.5).

- The analysis of protein-encoding genes can reveal important functions of microbiomes (Figure 30.6, Figure 30.7).

- Additional information regarding microbiome function can be obtained by characterizing the metatranscriptome, metaproteome, and meta-metabolome, which are all the types of mRNAs, proteins, and metabolic products, respectively, that are present in a microbiome.

### 30.2   Microbiomes of Physical Systems

- Seawater is a physical system containing varied microbiomes, which include important photosynthetic microbes such as cyanobacteria and algae.

- Biologists explore the microbiomes found within ice. These microbiomes can affect the color of ice and its ability to absorb light (Figure 30.8).

- Microbiomes are found in freshwater habitats and may produce human health concerns, such as toxins produced by *Microcystis* bacteria (Figure 30.9).

- The soil is another physical system containing microbiomes, which are known to have a great impact on the growth of crops.

### 30.3   Host-Associated Microbiomes

- The combination of a host organism and its microbiome is known as a holobiont. The host and microbiome genomes together form the hologenome.

- Hosts acquire microbiomes in different ways, and these microbiomes may change as species diverge from each other during evolution (Figure 30.10).

- A variety of species, including fungi, plants, and animals, act as hosts for microbes, forming microbiomes. These associations result in an astonishing array of benefits for both the hosts and their microbial partners (Figures 30.11–30.15).

- Blanton, Gordon, and associates found that gut microbiomes affect the growth of children, and this effect was confirmed by studies in mice (Figure 30.16).

## 30.4 Engineering Animal and Plant Microbiomes

- Researchers are exploring a variety of approaches for engineering microbiomes, including the transplantation of gut microbiomes, the use of probiotics, and artificial selection (Figures 30.17, 30.18).

## Assess & Discuss

### Test Yourself

1. A microbiome is
   a. an interaction between two different species of microorganisms.
   b. an environment that is microscopic.
   c. a particular assemblage of microbes (including their genes) that occurs in a defined environment.
   d. the entire genetic makeup of a particular microorganism.
   e. both a and d.

2. An example of a microbiome function is
   a. nitrogen fixation.
   b. methane oxidation.
   c. the production of particular metabolites.
   d. All of the above are microbiome functions.
   e. Only b and c are microbiome functions.

3. In what order should the four DNA fragments below be placed to form a contig? DNA regions with the same sequence are shown in the same color (red, green, or blue).

   Fragment 1: _____
   Fragment 2: _____
   Fragment 3: _____
   Fragment 4: _____

   a. 1, 2, 3, 4          d. 3, 2, 1, 4
   b. 2, 3, 4, 1          e. b or d
   c. 1, 3, 2, 4

4. Which of the following is a microbiome of a physical system?
   a. a microbiome on the surface of a leaf
   b. a microbiome in the human gut
   c. a microbiome in a soil sample
   d. a microbiome in a sample of human saliva
   e. all of the above

5. The combination of a host organism and its microbiome is known as
   a. a microbiome.
   b. a holobiont.
   c. a metagenome.
   d. a metabolome.
   e. both a and c.

6. Which of the following categories of organisms can function as a host for a microbiome?
   a. cyanobacteria          d. plants and animals
   b. algae                  e. all of the above
   c. fungi

7. In which of the following ways can an animal acquire a microbiome?
   a. Certain insects coat the casings of their eggs with bacteria.
   b. Newborn bees get microbes from sibling worker bees.
   c. Mammals, including humans, transmit important microbes as the young transit the birth canal.
   d. Termites use specific behaviors to transfer among themselves microbes they need to break down plant materials into food.
   e. All of the above are ways that animals can acquire microbiomes.

8. What is a biofilm?
   a. a microbiome within the gut of an animal
   b. a group of microbes that secrete mucilage and stick together
   c. a microbiome that forms an opaque film on ice
   d. a group of microbes that perform a metabolic function the host cannot perform
   e. a microbiome that floats on the surface of seawater

9. The goal of microbiome engineering is to
   a. eliminate an unwanted microbiome from the host.
   b. manipulate the composition of a microbiome to cause its self-destruction.
   c. manipulate the composition of a microbiome to benefit the host.
   d. identify all of the microbial species within a microbiome.
   e. alter the genome of the host so that it can support a different microbiome.

10. Which of the following is *not* an approach for microbiome engineering?
    a. transplantation of gut microbiomes
    b. probiotics
    c. artificial selection
    d. All of the above are approaches for microbiome engineering.
    e. Only a and c are not approaches for microbiome engineering.

### Conceptual Questions

1. Describe two examples of microbiomes that are found within a physical system. Explain how such microbiomes can affect Earth's environment on a global scale.

2. Pick one host-associated microbiome. Describe how the host benefits from the microbiome and how the microbes within the microbiome benefit from their association with the host.

3.  **Core Concept: Systems** Give one example of how a microbiome of a plant can influence the plant's ability to interact with the environment.

### Collaborative Questions

1. Microbiomes occur in places that you may have not imagined. Starting with the term microbiome or microbiota, search the literature using a search engine such as Pubmed, Google Scholar, or BioOne, and identify microbiomes that you never knew existed. Discuss your findings with fellow students.

2. This chapter described how the microbiomes of African apes and humans have changed since their evolutionary divergence. Pick your own example of a small group of closely related species and hypothesize how the microbiomes of these species may have changed since they diverged from each other. Explain your reasoning.

# Plants and the Conquest of Land

# 31

**A temperate rain forest containing diverse plant phyla in Olympic National Park in Washington state.** ©Craig Tuttle/Corbis/ Getty Images

**W**hen thinking about plants, people envision lush green lawns, shady street trees, garden flowers, or leafy fields of valuable crops. On a broader scale, they might imagine lush rain forests (like that shown in the chapter opening photo), vast grassy plains, or tough desert vegetation. Shopping in the produce section of the local grocery store may remind us that plant photosynthesis is the basic source of our food. Just breathing crisp fresh air might bring to mind the role of plants as oxygen producers—the ultimate air fresheners. Do you start your day with a "wake-up" cup of coffee, tea, or hot chocolate? Then you may appreciate the plants that produce these and many other materials we use in daily life: medicines, cotton, linen, wood, bamboo, cork, and paper.

In addition to their importance to humans and modern ecosystems, plants have played dramatic roles in the Earth's past. Throughout their evolutionary history, they have influenced Earth's atmospheric chemistry, climate, and soils. Plants have also affected the evolution of many other groups of organisms, including humans. In this chapter, we will survey the diversity of modern plant phyla and their distinctive features. This chapter also explains how early plants adapted to land and how plants have continued to adapt to changing terrestrial environments. During this process, we will gain insight into descent with modification and the core concept that all life is related by an evolutionary history.

## 31.1 Ancestry and Diversity of Modern Plants

**Learning Outcomes:**

1. List key derived features that land plants share with their closest algal relatives.

2. Name several characteristics unique to land plants.

3. Compare and contrast the features of vascular and nonvascular plants.

4. **CoreSKILL »** Explain how cuticles, stomata, and internal conduction systems work together in vascular plants to maintain stable water content.

Several hundred thousand modern species are formally classified into the kingdom Plantae, referred to informally as the plants or land plants (**Figure 31.1**). **Plants** are multicellular eukaryotic organisms that primarily live on land and are composed of cells that are surrounded by a cell wall that contains cellulose. Plant cells also contain plastids, such as chloroplasts. Most species of plants carry out photosynthesis, which is described in Chapter 8.

Molecular and other evidence indicates that the plant kingdom evolved from green algal ancestors whose modern representatives primarily occupy freshwater habitats. Modern plants and their closest green algal relatives are together known as **streptophytes**. Figure 31.1 shows the evolutionary relationships among green algae and modern land plants. Compared to the algal counterparts, plants display traits that foster survival in terrestrial conditions, which are drier, sunnier, hotter, colder, and less physically supportive than aquatic habitats.

In this section, we will examine the modern algae that are most closely related to plants and survey the diverse phyla of living land plants. These comparisons reveal how plants gradually acquired diverse structural, biochemical, and reproductive adaptations that fostered survival on land.

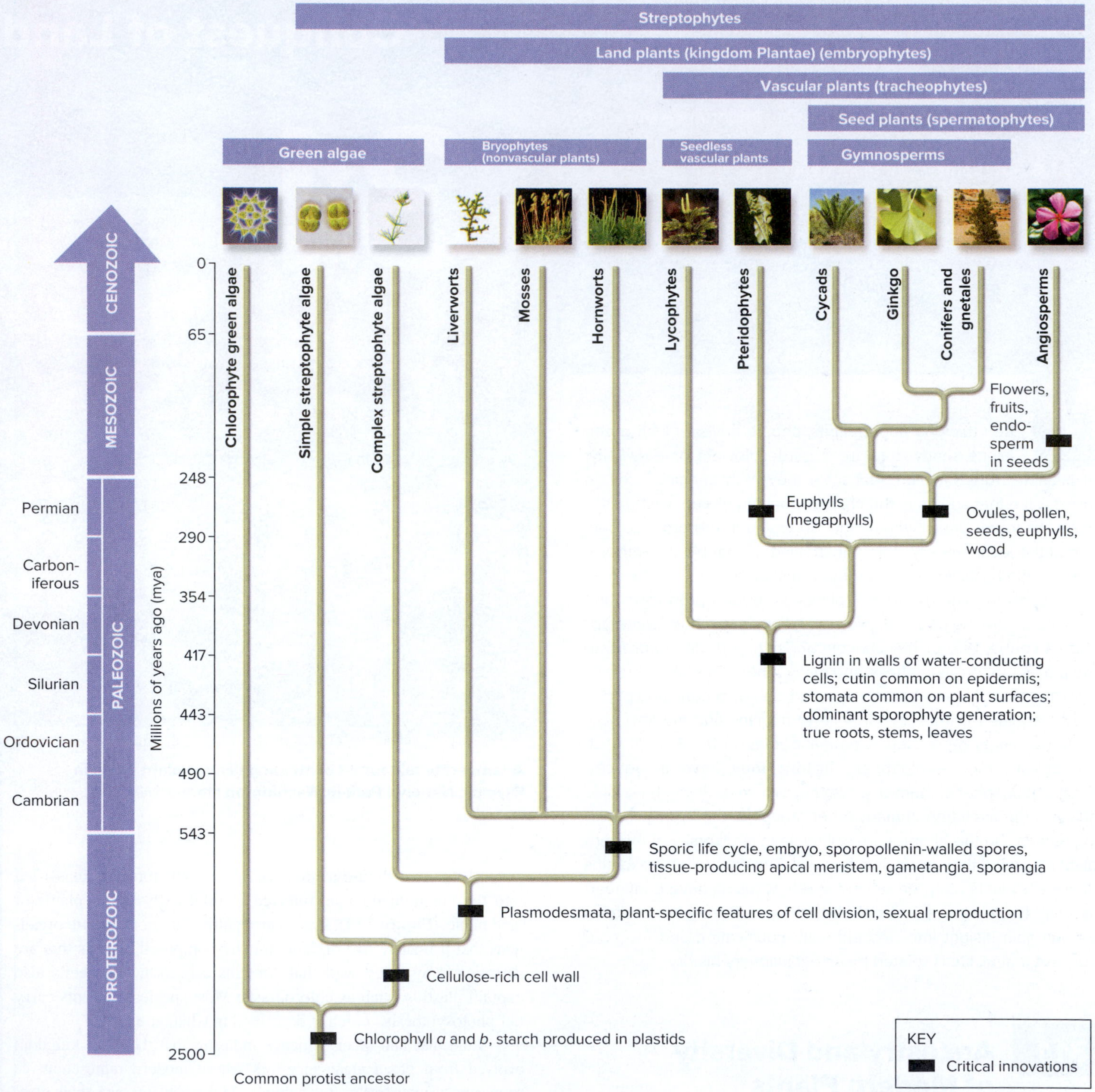

**Figure 31.1 Evolutionary relationships among green algae and modern plant phyla.** The blue bars in the arrow on the left side show maximal evolutionary divergence times indicated by molecular clock and some fossil evidence, suggesting when clades may first have arisen. This is a simplified diagram; fewer branches of streptophyte algae are shown here than actually exist. (inset 1): ©Roland Birke/Phototake; (inset 2): ©the CAUP image database, http://botany.natur.cuni.cz/algo/database; (inset 3–6): ©Lee W. Wilcox; (inset 7): ©Ed Reschke/Getty Images; (inset 8): ©Patrick Johns/Corbis/VCG/Getty Images; (inset 9): ©Philippe Psaila/Science Source; (inset 10): ©Fancy Photography/Veer/Getty Images; (inset 11): ©imageBROKER/Alamy Stock Photo; (inset 12): ©Gallo Images/Corbis/Getty Images

**Core Concept: Evolution** Land plants evolved from green algae and gradually acquired diverse structural, biochemical, and reproductive adaptations, allowing them to better survive in terrestrial habitats.

## Modern Green Algae Are Closely Related to the Ancestors of Land Plants

Molecular, biochemical, and structural data indicate that the kingdom Plantae originated from a photosynthetic protist ancestor that, if present today, would be classified among the **streptophyte algae** (**Figure 31.2**). The streptophyte algae are related to other green algae, but have more features in common with land plants. The more complex, later-diverging streptophyte algae display several **critical innovations**—derived features shared with land plants that fostered plant success on land. Examples of these shared features are a distinctive type of cytokinesis,

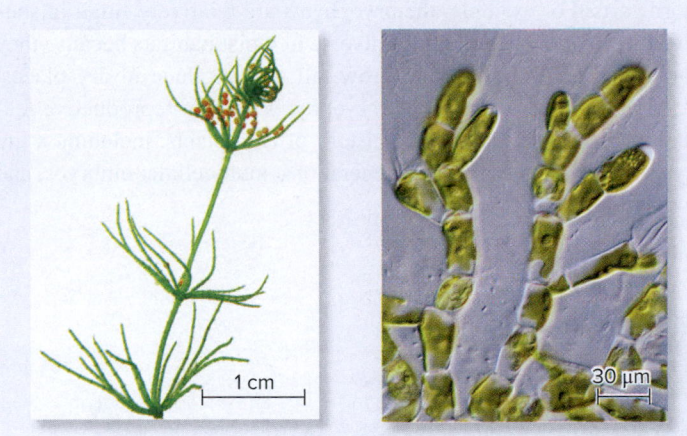

**(a)** Complex streptophyte algae: *Chara zeylanica* (left) and *Coleochaete pulvinata* (right)

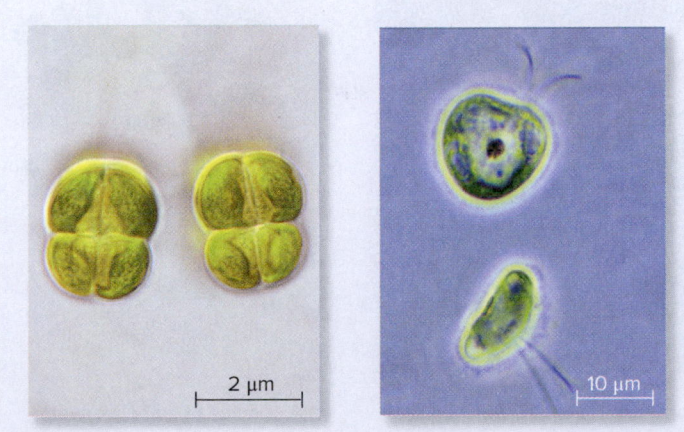

**(b)** Simple streptophyte algae: *Chlorokybus atmophyticus* (left) and *Mesostigma viridae* (right)

**Figure 31.2** **Streptophyte green algal relatives of the land plants.** Streptophyte green algae occur **(a)** as more complex branched filaments or **(b)** as small colonies (left side) or single cells (right side). Streptophyte algae share cellular, biochemical, and molecular features with land plants. a (left, right): ©Lee W. Wilcox; b (left): ©the CAUP image database, http://botany.natur.cuni.cz/algo/database; b (right): ©Lee W. Wilcox

intercellular connections known as plasmodesmata (see Chapter 10), and sexual reproduction (see Figure 31.1). For this reason, complex streptophyte algae are good sources of information about the ancestors of land plants.

## Distinctive Features of the Land Plants

Land plants can be distinguished from their close algae relatives by several features that represent early adaptations to the land habitat. For example, the bodies of all land plants are primarily composed of three-dimensional tissues, defined as close associations of cells of the same type. Tissues provide land plants with an increased ability to avoid water loss at their surfaces. Water loss is decreased in land plants because bodies composed of tissues have lower surface area/volume ratios than do branched filaments or simpler algal bodies. Land plant tissues arise from one or more actively dividing cells that occur at growing tips. Such localized regions of cell division are known as **apical meristems**. The tissue-producing apical meristems of land plants produce relatively thick, robust bodies able to withstand drought and mechanical stress and produce tissues and organs with specialized functions.

The land plants also have distinctive reproductive features.

- Land plants feature a life cycle involving alternation of generations. **Alternation of generations** means that two types of multicellular bodies alternate in time (refer back to Figure 16.14c). The diploid ($2n$) **sporophyte** generation produces spores by meiosis, and the haploid ($n$) **gametophyte** generation produces gametes by mitosis. By contrast, streptophyte algae feature a haploid-dominant life cycle (refer back to Figure 16.14b).

- During land plant sexual reproduction, a diploid zygote divides by mitosis to form a multicellular sporophyte embryo. A key feature is that the sporophyte embryo is nourished by maternal tissues. Although maternal cells of streptophyte algae may nourish zygotes, the algal zygotes do not develop into multicellular embryos.

- A mature land plant sporophyte undergoes meiosis to produce tough-walled non-flagellate reproductive cells known as **spores** that survive dispersal through dry air. Streptophyte algae differ in that spores produced by meiosis are adapted for dispersal in water; they possess flagella and lack protective walls.

We will take a closer look at the reproductive features of land plants in Sections 31.3–31.5, and also in Chapter 32.

How can we know about past events such as the origin and diversification of land plants? Some information comes from comparing molecular and other features of modern plants. For example, the genome sequence of the moss *Physcomitrella patens*, first reported in 2007, reveals the presence of genes that aid heat and drought tolerance, which are especially useful in the terrestrial habitat. Plant fossils, the preserved remains of plants that lived in earlier times, provide further information (**Figure 31.3**). The distinctive plant organic materials sporopollenin, cutin, and lignin (discussed later in this chapter) do not readily decay and thereby foster plant

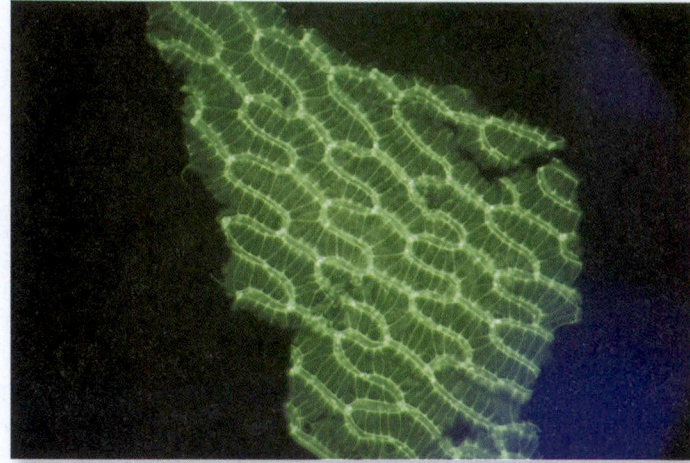

**Figure 31.3 Fossil moss leaf fragment** Fluorescence microscopy indicates the presence of decay-resistant lignin-like cell-wall materials that have fostered structural preservation over hundreds of millions of years of burial. The distinctive pattern of these cell-wall remains suggests a relationship to modern peat mosses, which are common worldwide and have important ecological roles, as discussed in Section 31.2. ©Linda Graham

fossilization. The study of fossils and the molecular, structural, and functional features of modern plants has revealed an amazing story—how plants gradually acquired adaptations, allowing them to conquer the land.

## Modern Land Plants Can Be Classified into Nine Phyla

Fossils reveal that a number of plant phyla that once lived are now extinct. In this textbook, nine phyla of living land plants are described (also see Figure 31.1):

- **liverworts**, formally known as Hepatophyta
- **mosses**, formally Bryophyta
- **hornworts**, Anthocerophyta
- **lycophytes**, Lycopodiophyta
- **pteridophytes**, Pteridophyta
- **cycads**, Cycadophyta
- **ginkgos**, Ginkgophyta
- **conifers**, Coniferophyta
- **flowering plants**, also known as **angiosperms**, Anthophyta

Although the order in which early land plant phyla diverged is not completely clear, fossil and molecular evidence indicates that relatively small and simple bryophytes—represented by modern liverworts, mosses, and hornworts—arose before the first vascular plants (see Figure 31.1). **Vascular plants** are distinguished by internal water and nutrient-conducting (vascular) tissues that also provide structural support, allowing these plants to become larger and more complex than are bryophytes. Among the modern vascular plants, lycophytes diverged earliest, pteridophytes arose next, and then seed plants. The

flowering plants dominate the modern world. A survey of these modern plant phyla reveals how land plants became increasingly better adapted to life on land.

## Liverworts, Mosses, and Hornworts Are the Simplest Land Plants

Liverworts, mosses, and hornworts are Earth's simplest land plants (**Figures 31.4**, **31.5**, and **31.6**), and each group is monophyletic. There are about 6,500 species of modern liverworts, 12,000 or more species of mosses, and about 100 species of hornworts. Collectively, liverworts, mosses, and hornworts are known informally as the **bryophytes** (from the Greek *bryon*, meaning moss, and *phyton*, meaning plant). This collective term reflects common structural, reproductive, and ecological features of liverworts, mosses, and hornworts. For example, the bryophytes are relatively small in stature and are most common and diverse in moist habitats because they lack traits allowing them to grow tall or reproduce in dry places. As described in Section 31.3, bryophytes display reproductive features that evolved early in the history of land plants, including a life cycle involving alternation of generations, multicellular embryos, and tough-walled spores.

**(a) The common liverwort,** *Marchantia polymorpha*

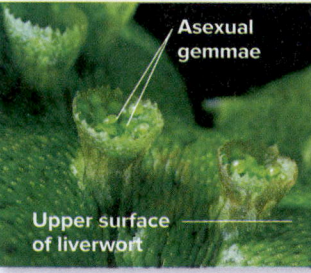

Asexual gemmae

Upper surface of liverwort

**(b) Close-up of liverwort structures**

**(c) A species of leafy liverwort**

**Figure 31.4 Liverworts. (a)** The common liverwort, *Marchantia polymorpha*, has a flat green body that produces raised, umbrella-shaped structures bearing sexually produced sporophytes on the undersides. Mature sporophytes generate spores, then release them into the air. **(b)** A close-up of *M. polymorpha* showing surface cups that contain multicellular, Frisbee-shaped asexual structures known as gemmae, which are dispersed by wind and grow into new liverworts. **(c)** A species of liverwort that has leaflike structures and so is known as a leafy liverwort. a: ©Dr. Jeremy Burgess/SPL/Science Source; b–c: ©Lee W. Wilcox

**Figure 31.5  Mosses.** The common moss genus *Mnium* has a leafy green gametophyte (multicellular body that generates gametes) and an unbranched, dependent sporophyte that bears a spore-producing sporangium at its tip. Inset: This SEM shows the tip of a moss sporangium. The sporangia often have teeth separated by spaces, so spores are sprinkled into the wind and dispersed over time, rather than being released all at once. (top): ©Eye of Science/Science Source; (bottom): ©Lee W. Wilcox

**Concept Check:** *Why might it be advantageous for a moss sporangium to release spores gradually?*

**Figure 31.6  Hornworts.** Gametophytes of hornworts grow close to the ground, whereas sporophytes generally grow up into the air. Mature hornwort sporophytes open at the top, dispersing spores into the wind. ©Lee W. Wilcox

## Lycophytes and Pteridophytes Are Vascular Plants That Do Not Produce Seeds

As mentioned, vascular plants produce internal water and nutrient-conducting (vascular) tissues that also provide structural support. Fossils and molecular comparisons indicate that the first vascular plants appeared later than the earliest bryophytes and that several early vascular plant lineages existed in the past but have become extinct. Molecular data indicate that the lycophytes are the oldest phylum of living vascular plants and that pteridophytes are the next oldest (see Figure 31.1). Hundreds of millions of years ago, lycophytes were more diverse than at present and included tall trees that contributed importantly to coal deposits. Now, only about 1,000 relatively small species exist (**Figure 31.7**). Pteridophytes have diversified more recently, and there are about 12,000 species of modern pteridophytes, including horsetails, whisk ferns, and other ferns (**Figure 31.8**).

Because the lycophytes and pteridophytes diverged prior to the origin of seeds, they are informally known as seedless vascular plants. Together, lycophytes, pteridophytes, and seed-producing plants are known as the **tracheophytes**. The latter term takes its name from **tracheids**, a type of specialized cell that conducts water and minerals and provides structural support. Tracheids and other cells involved in the conduction of materials within plants form vascular tissues that are described more fully in Chapters 36 and 39. Vascular tissues occur in stems, roots, and leaves, which are the organs of the vascular plant body.

***Stems, Roots, and Leaves*  Stems** of vascular plants are branching structures that contain vascular tissue and produce leaves and sporangia. Stems contain the specialized conducting tissues known as **phloem** and **xylem**, the latter of which contains tracheids. As described in Chapters 36 and 39, such conducting tissues enable vascular plants to move organic compounds, water, and minerals throughout the plant body. The xylem also provides structural support,

**Figure 31.7  An example of a lycophyte (*Lycopodium obscurum*).** The sporophyte stems bear many tiny leaves. The spore-producing sporangia generally occur in club-shaped clusters. For this reason, lycophytes are often referred to as club mosses or spike mosses, though they are not true mosses. The gametophytes of lycophytes are small structures that often occur underground. ©Lee W. Wilcox

**Figure 31.8**  **Pteridophyte diversity.**
**(a)** The leafless, rootless green stems of the
whisk fern (*Psilotum nudum*) branch by
forking and bear many clusters of yellow
sporangia that disperse spores via wind.
The gametophyte of this plant is a tiny pale
structure that lives underground in a
symbiotic partnership with fungi. **(b)** The
giant horsetail (*Equisetum telmateia*)
displays branches in whorls around the
green stems. The leaves of horsetail plants
are tiny, light brown structures that encircle
branches at intervals. Horsetail plants
produce spores in cone-shaped structures,
and the wind-dispersed spores grow into
small green gametophytes. **(c)** The early-
diverging fern *Botrychium lunaria*, showing
a green photosynthetic leaf with leaflets and
a modified leaf that bears many round
sporangia. **(d)** A later-diverging fern
showing leaves having leaflets, and young
leaves that are in the process of unrolling
from the bases to the tips. The stem of this
fern grows parallel to the ground and thus is
not shown. Most ferns produce spores in
sporangia on the undersides of leaves.

a: ©Lee W. Wilcox; b: ©José Julián Rico Cerdá/Alamy
Stock Photo; c: ©Patrick Johns/Corbis/VCG/Getty
Images; d: ©Lee W. Wilcox

**(a)  A whisk fern**

**(b)  The giant horsetail**

**(c)  An early-diverging fern**

**(d)  A later-diverging fern**

allowing vascular plants to grow taller than nonvascular plants. This
support function relies on the presence of a compression- and decay-
resistant waterproof material known as **lignin**, which occurs in the cell
walls of tracheids and some other types of plant cells. Most vascular
plants also produce **roots**, which are organs specialized for uptake of
water and minerals from the soil, and **leaves**, flattened plant organs
that emerge from stems and generally have a photosynthetic function.

Lycophyte roots and leaves differ from those of pteridophytes and
seed plants. For example, lycophyte roots fork at their tips, whereas
roots of pteridophytes branch from the inside like the roots of seed
plants (look ahead to Figure 36.24). Lycophyte leaves are relatively
small and possess only one unbranched vein, whereas pteridophyte

leaves are larger and have branched veins, as do those of seed plants
(compare Figures 31.7 and 31.8d).

***Adaptations That Foster Stable Internal Water Content***   The
stems, roots, and leaves of lycophytes and pteridophytes illustrate
adaptations acquired by early vascular plants that help to maintain
stable internal water content, also known as moisture homeostasis.
In relatively dry habitats, lycophytes, pteridophytes, and other vascular plants are able to grow to larger sizes and remain metabolically
active for longer periods than bryophytes. In addition to vascular tissue, two other adaptations that aid in moisture retention are a waxy
cuticle and stomata.

- A protective waxy **cuticle** is present on most surfaces of vascular plant sporophytes (**Figure 31.9a**). The plant cuticle contains a polyester polymer known as cutin, which helps to prevent attack by pathogens, as well as wax, which helps to prevent desiccation (drying out).

- The surface tissue of vascular plant stems and leaves contains many **stomata** (singular, stoma or stomate)—specialized surface cells associated with pores that are able to open and close (**Figure 31.9b**). Stomata first arose in bryophytes, whose mature

sporophytes must dry in order to discharge spores; stomatal pores may aid this drying process. By contrast, stomata occurring on the leaf and stem surfaces of vascular plants take in the carbon dioxide needed for photosynthesis and release oxygen to the air, while conserving plant water content. When the environment is moist, the pores open, allowing gas exchange to occur. When the environment is very dry, the pores close, thereby reducing water loss.

## Gymnosperms and Angiosperms Are the Modern Seed Plants

Collectively, the modern and extinct phyla of seed-producing plants are known as **spermatophytes** (the prefix sperm-, from the Greek, means seed) (see Figure 31.1). We will examine them more fully in Chapter 32. The modern seed plant phyla commonly known as cycads, ginkgos, conifers, and gnetophytes are collectively referred to as gymnosperms (**Figure 31.10** shows an example). **Gymnosperms** reproduce using both spores and seeds, as do the flowering plants, the angiosperms (**Figure 31.11**). For this reason, gymnosperms and angiosperms are known informally as the **seed plants**. **Seeds** are complex structures having specialized tissues that protectively enclose embryos and contain stores of carbohydrate, lipid, and protein. Embryos use such food stores to grow and develop into seedlings. The ability to produce seeds helps to free plants from reproductive limitations experienced by the seedless plants, explaining why seed plants are dominant today.

Though gymnosperms produce seeds, they lack several features unique to the flowering plants. The term gymnosperm comes from the Greek, meaning naked seeds, reflecting the observation that gymnosperm seeds are not enclosed within fruits. Despite their lack of flowers, fruits, and seed endosperm, the modern gymnosperms are diverse and abundant in many places (see Chapter 32).

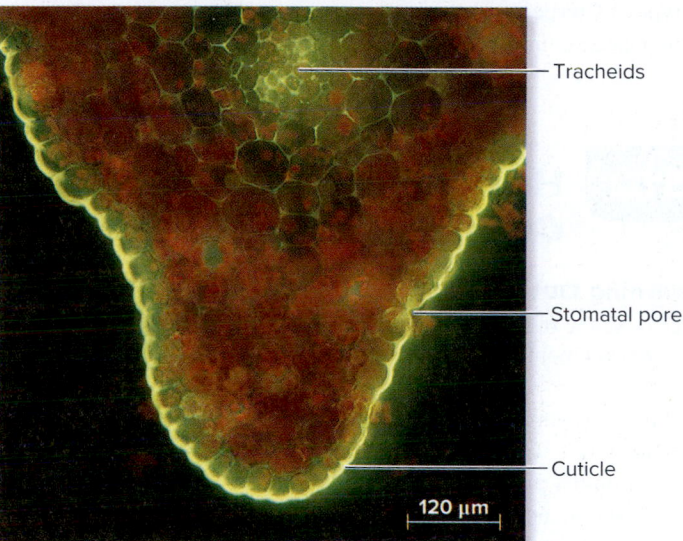

**(a) Stem showing tracheophyte adaptations**

Tracheids

Stomatal pore

Cuticle

120 μm

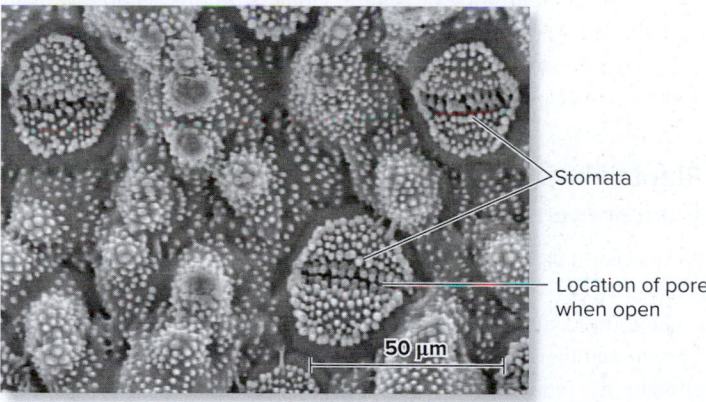

Stomata

Location of pore when open

50 μm

**(b) Close-up of stomata**

**Figure 31.9** **Tracheophyte adaptations for transporting and conserving water.** **(a)** A cross section through a stem of the leafless pteridophyte *Psilotum nudum*. When viewed with fluorescence microscopy and illuminated with violet light, an internal core of xylem tracheids glows yellow, as does the surface cuticle. Stomatal pores can be seen at indentations of the cuticle. **(b)** SEM view of tracheophyte surface pores associated with specialized cells— together known as stomata—allow for gas exchange between plant and atmosphere. a: ©Linda Graham; b: ©Martha Cook

 **Core Concept: Structure and Function** The structures illustrated in this figure help to explain how vascular land plants maintain moisture homeostasis.

**Figure 31.10** **An example of a gymnosperm, the pine (*Pinus*).** Gymnosperms produce and disperse seeds from cones, but do not produce flowers or fruits. ©Stephen P. Parker/Science Source

**Figure 31.11** **An example of an angiosperm, the bleeding heart plant (genus *Dicentra*).** The pink structures are flowers, which are a distinguishing feature of angiosperms. Flowers often develop into seed-containing fruits, another unique feature of angiosperm plants.
©imageBROKER/Alamy Stock Photo

The angiosperms are seed plants distinguished by the presence of flowers, fruits, and a specialized seed tissue known as endosperm. A **flower** is a short stem bearing reproductive organs that are specialized in ways that enhance seed production (see Figure 31.11). **Fruits** are structures that develop from flowers, enclose seeds, and foster seed dispersal in the environment. The term angiosperm comes from the Greek, meaning enclosed seeds, reflecting the observation that the flowering plants produce seeds within fruits. **Endosperm** is a nutritive seed tissue that increases the efficiency with which food is stored and used in the seeds of flowering plants (explained further in Section 31.5).

 **Core Concepts: Evolution, Information**

### Comparison of Plant Genomes Reveals Genetic Changes That Occurred During Plant Evolution

The first complete genome sequence for a seedless vascular plant was reported for the lycophyte *Selaginella moellendorffii* in 2011. This advance allowed plant evolutionary biologists to compare the new sequence with previously sequenced plant genomes, with the goal of identifying genes associated with major evolutionary transitions in plants.

Such genome comparisons revealed that 6,820 gene families, representing a basic set of embryophyte genes, are present in all land plants. The majority of gene families that are involved in flowering plant development were also observed in the lycophyte *Selaginella* and the bryophyte *Physcomitrella*, a moss. A comparison of the genomes of *Selaginella* and *Physcomitrella* indicated that more than 500 genes were gained and nearly 90 genes were lost during the transition from nonvascular plants (bryophytes) to vascular plants (lycophytes). By contrast, 1,350 more genes were gained in the evolution of traits specific to angiosperms. These

data indicate that the transition from bryophyte to lycophyte was less genetically complex than was the transition from lycophyte to angiosperm.

Another interesting finding involves terpenes, which are secondary metabolites that plants produce as chemical defenses against pathogens and herbivores. Both seedless and seed plants carry genes encoding proteins that are required to synthesize terpenes. However, genomic analysis revealed that seedless plants rely on genes also present in microbes, but seed plants use different terpene-synthesis genes. The diversification of new types of terpene-synthesis genes is an example of an increase in genetic complexity that accompanied the rise of seed plants.

## 31.2 How Land Plants Have Changed the Earth

**Learning Outcome:**

1. Describe examples of how plants have altered the Earth's physical environment and also affected other life on Earth.

A billion years ago, Earth's terrestrial surface was comparatively devoid of life. Green or brown crusts of cyanobacteria most likely grew in moist places, but there would have been very little soil, no plants, and no animal life. The origin of the first land plants was key to development of the first substantial soils, the rise to modern levels of atmospheric oxygen, the evolution of modern plant communities, and the colonization of land by animals. In this section, we will explore how plants have transformed Earth's physical environment, especially the atmosphere, and how the rise of seed plants has greatly impacted the evolution of animals.

### Plants Have Transformed the Earth's Physical Environment

As discussed in Chapter 8, the process of photosynthesis uses carbon dioxide ($CO_2$) from the atmosphere to produce carbon-containing organic molecules. A by-product of photosynthesis is oxygen ($O_2$). Throughout their evolutionary history, plants have influenced Earth's climate by reducing the concentration of atmospheric $CO_2$ and increasing the concentration of $O_2$, and they continue to do so. Ecologists are concerned that the negative impact of humans on plant communities may also affect $CO_2$ and $O_2$ levels.

***Ecological Effects of Ancient and Modern Bryophytes*** Several types of decay-resistant materials evolved in early seedless plants, such as sporopollenin, cutin, and lignin (discussed in Section 31.1). When the plants died, some of these organic molecules were not completely degraded, but instead were buried in sediments that eventually transformed into rock. Such fossil carbon can accumulate and remain buried for very long time periods with the consequence of lowering the level of atmospheric $CO_2$. As discussed in Chapter 59, $CO_2$ is a greenhouse gas that causes global temperatures to rise. Therefore, the accumulation of fossil carbon is expected to lower global temperatures.

Modern bryophytes also play important roles by storing $CO_2$ as decay-resistant organic compounds, suggesting that ancient relatives likewise played this role (see Figure 31.3). Plants of the modern peat moss genus *Sphagnum*, for example, contain so much decay-resistant body mass that in many places, dead moss accumulated over long time periods has formed deep peat deposits. By storing very large amounts of organic carbon for a long time, diverse species of *Sphagnum* moss help to keep Earth's climate steady. Under cooler than normal conditions, the mosses grow more slowly and thus absorb less $CO_2$, allowing atmospheric $CO_2$ to rise a bit, warming the climate a little. As the climate warms, the mosses grow faster and take up more $CO_2$, storing it in peat deposits. Such a reduction in atmospheric $CO_2$ returns the climate to slightly cooler conditions. Peat mosses also harbor bacteria that consume methane, another powerful greenhouse gas. In these ways, ancient and modern mosses have helped to moderate the world's climate. Experts are concerned that large regions currently dominated by these helpful mosses may be harmed by land use changes, peat harvesting for commercial use, and climate change.

Atmospheric $O_2$ has been an important factor affecting the evolution of species on Earth. Eukaryotic species tend to have high demands for $O_2$ because they use it to obtain energy via cellular respiration. Photosynthetic bacteria were the earliest organisms to produce $O_2$, and later in evolution, algae also contributed to atmospheric $O_2$. Even so, recent modeling studies have provided strong evidence that by 420–400 mya (early Devonian period) the activities of early seedless plants had raised oxygen levels in Earth's atmosphere to modern levels. These higher levels may have been key to the evolution of large, mobile, animals, which have a particularly high demand for oxygen.

***Ecological Effects of Ancient Vascular Plants***    Fossils tell us that extensive forests dominated by tree-sized lycophytes, pteridophytes, and early seed plants occurred in widespread swampy regions during the warm, moist Carboniferous period (354–290 mya) (**Figure 31.12**). As dead plants fell into the water, low oxygen levels there inhibited microbes that would have caused the plant matter to decay. The dead plants were then buried in sediments that later formed coal. Much of today's coal is derived from the abundant remains of ancient plants, explaining why the Carboniferous period is commonly known as the Coal Age. During that period, plants converted huge amounts of atmospheric $CO_2$ into decay-resistant organic materials. Long-term burial of these materials, compressed into coal, together with chemical interactions between soil and the roots of vascular plants, dramatically changed Earth's atmosphere and climate. The removal of large amounts of the greenhouse gas $CO_2$ from the atmosphere by plants had a cooling effect on the climate, which also became drier because cold air holds less moisture than warm air.

Mathematical models of ancient atmospheric chemistry, supported by measurements of natural carbon isotopes, led American paleoclimatologist Robert A. Berner to propose that the Carboniferous proliferation of vascular plants was correlated with a dramatic decrease in atmospheric $CO_2$, which reached the lowest known levels about 290 mya (**Figure 31.13**). During this period of very low $CO_2$, atmospheric $O_2$ levels rose to the highest known levels, because less $O_2$ was being used to convert organic carbon into $CO_2$. High atmospheric $O_2$ content may explain the occurrence during the Carboniferous period of giant dragonflies and other huge insects.

The great decline in $CO_2$ level ultimately caused cool, dry conditions to prevail in the late Carboniferous and early Permian periods. As a result of this relatively abrupt global climate change, many of the tall seedless lycophytes and pteridophytes that had dominated earlier Carboniferous forests became extinct, as did the giant dragonflies.

Giant dragonfly

Giant horsetail (pteridophyte)    Giant lycophyte

**Figure 31.12**  **Reconstruction of a Carboniferous period (Coal Age) forest.** This ancient forest was dominated by tree-sized lycophytes and pteridophytes, which later contributed to the formation of large coal deposits. ©Lee W. Wilcox

*Concept Check:*  *Why did giant dragonflies exist during the Carboniferous period, but not now?*

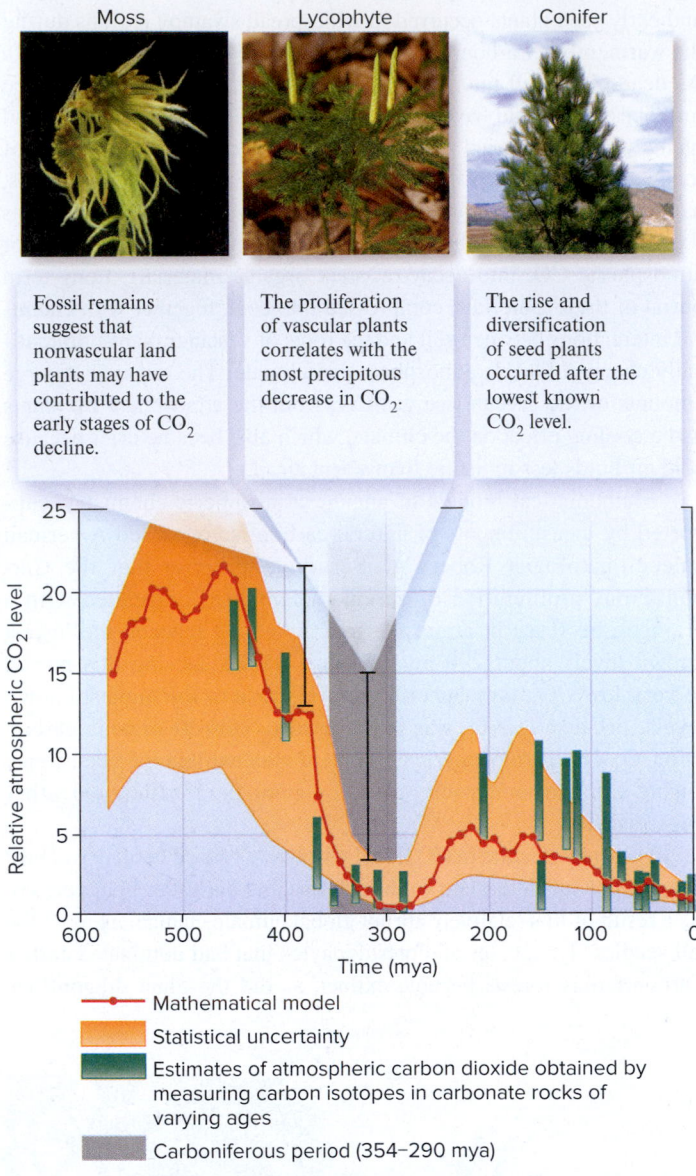

Moss

Lycophyte

Conifer

Fossil remains suggest that nonvascular land plants may have contributed to the early stages of $CO_2$ decline.

The proliferation of vascular plants correlates with the most precipitous decrease in $CO_2$.

The rise and diversification of seed plants occurred after the lowest known $CO_2$ level.

- — Mathematical model
- Statistical uncertainty
- Estimates of atmospheric carbon dioxide obtained by measuring carbon isotopes in carbonate rocks of varying ages
- Carboniferous period (354–290 mya)

**Figure 31.13  Changes in Earth's atmospheric carbon dioxide levels over geological time.** Geological evidence indicates that carbon dioxide levels in Earth's atmosphere were once higher than they are now, but that the rise of land plants caused atmospheric $CO_2$ to reach the lowest known level about 300 mya. (left): ©Linda Graham; (middle): ©Darlyne A. Murawski/Getty Images; (right): ©David R. Frazier/The Image Works

## Plant Evolution Has Greatly Impacted the Survival and Evolution of Animals

Diverse phyla of gymnosperms dominated Earth's vegetation through the Mesozoic era (248–65 mya), which is sometimes called the Age of Dinosaurs. In addition, fossils provide evidence that early

mammals and flowering plants existed in the Mesozoic. Gymnosperms and early angiosperms were probably sources of food for early mammals as well as for herbivorous dinosaurs. For example, fossils of an aquatic angiosperm named *Cobbania corrugata* have been found with bones of the dinosaur *Ornithomimus* in Dinosaur Park, Alberta, Canada. This dinosaur may have fed on the plant when alive (**Figure 31.14**).

One fateful day about 65 mya, disaster struck from the sky, causing a dramatic change in the types of plants and animals that dominated terrestrial ecosystems. That day, at least one large meteorite crashed into the Earth near the present-day Yucatán Peninsula in Mexico. This collision is known as the Cretaceous-Paleogene event (also sometimes referred to as the K/T event). The impact, together with substantial volcanic activity that also occurred at this time, is thought to have produced huge amounts of ash, smoke, and haze that dimmed the Sun's light long enough to kill many of the world's plants. Many types of plants, including *Cobbania*, became extinct, though others survived and their descendants persist to the present time. With a severely reduced food supply, most dinosaurs were also doomed, the exceptions being their descendants, the birds. The demise of the dinosaurs left room for birds and mammals to adapt to many kinds of terrestrial habitats formerly inhabited by dinosaurs.

After the Cretaceous-Paleogene event, ferns dominated long enough to leave huge numbers of fossil spores, and then surviving groups of flowering plants began to diversify into the space opened up by the extinction of earlier plants. The rise of angiosperms fostered the diversification of beetles (see Chapter 34) and other insects that associate with modern plants.

Cooler, drier conditions favored extensive diversification of the first seed plants, the gymnosperms. Compared with seedless plants, seed plants were better at reproducing in cooler, drier habitats (as we will see in Section 31.5). As a result, seed plants came to dominate Earth's terrestrial communities, as they still do.

**Figure 31.14  Early angiosperms as sources of food for large herbivorous dinosaurs of the Mesozoic era.** In this artist's habitat reconstruction from fossils, the extinct angiosperm *Cobbania corrugata* is shown growing in wetlands that were also inhabited by large dinosaurs such as *Ornithomimus*, whose head is illustrated here. ©Marjorie C. Leggitt

## 31.3 Evolution of Reproductive Features in Land Plants

**Learning Outcomes:**

1. Compare and contrast a haploid-dominant life cycle and alternation of generations.
2. List critical innovations in bryophytes that result in reproductive features different from those of streptophyte algae.
3. Discuss additional reproductive changes that occurred during the evolution of vascular plants.

In Section 31.1, we considered the diversity of land plants and some critical innovations associated with life in a terrestrial environment. In this section, we will take a closer look at the evolution of reproductive adaptations. We will compare and contrast such adaptions among green algae, bryophytes, and seedless vascular plants. In Chapter 32, we will examine some additional reproductive adaptions that have arisen in the seed plants—gymnosperms and angiosperms.

### A Comparison of Algal and Bryophyte Reproduction Highlights an Early Plant Adaptation to Life on Land: Alternation of Generations

Because bryophytes diverged early in the evolutionary history of land plants (see Figure 31.1), they serve as models of the earliest terrestrial plants. A comparison of the life cycle of aquatic streptophyte algae with that of bryophytes reveals the adaptive value on land of bryophyte reproductive features.

Streptophyte algae display a haploid-dominant life cycle in which the only diploid cell is the zygote, whose meiotic division produces relatively few spores (**Figure 31.15a**). By contrast, land plant zygotes do not undergo meiosis. Instead, they undergo mitosis to form a multicellular sporophyte in which many cells can undergo meiosis and thereby produce a large number of spores (**Figure 31.15b**). Producing more spores not only aids dispersal but also increases the genetic diversity of progeny. This life cycle difference allows bryophytes and other land plants to increase the number of spores generated per sexual cycle, an important advantage in terrestrial habitats.

Like related algae (see Figure 31.15a) and modern seedless plants, early land plants likely produced flagellate sperm that needed liquid water to swim to eggs to accomplish fertilization (see Figure 31.15b). On dry land, the number of successful fertilizations and resulting zygotes can be limited by lack of sufficient water. By producing multicellular sporophytes to greatly increase the number of spores resulting from each fertilization, land plants have overcome this problem. Genomic comparisons indicate that transcription factors encoded by *KNOX* genes were key to the origin of plant sporophytes. These transcription factors suppress gametophyte development, thereby allowing multicellular sporophytes to develop from zygotes.

### Bryophyte Reproduction Illustrates Other Terrestrial Adaptations

In addition to alternation of generations, other key features aided in terrestrial reproduction (**Figure 31.16**). As shown in step 3, the gametophytes of bryophytes and many other land plants produce

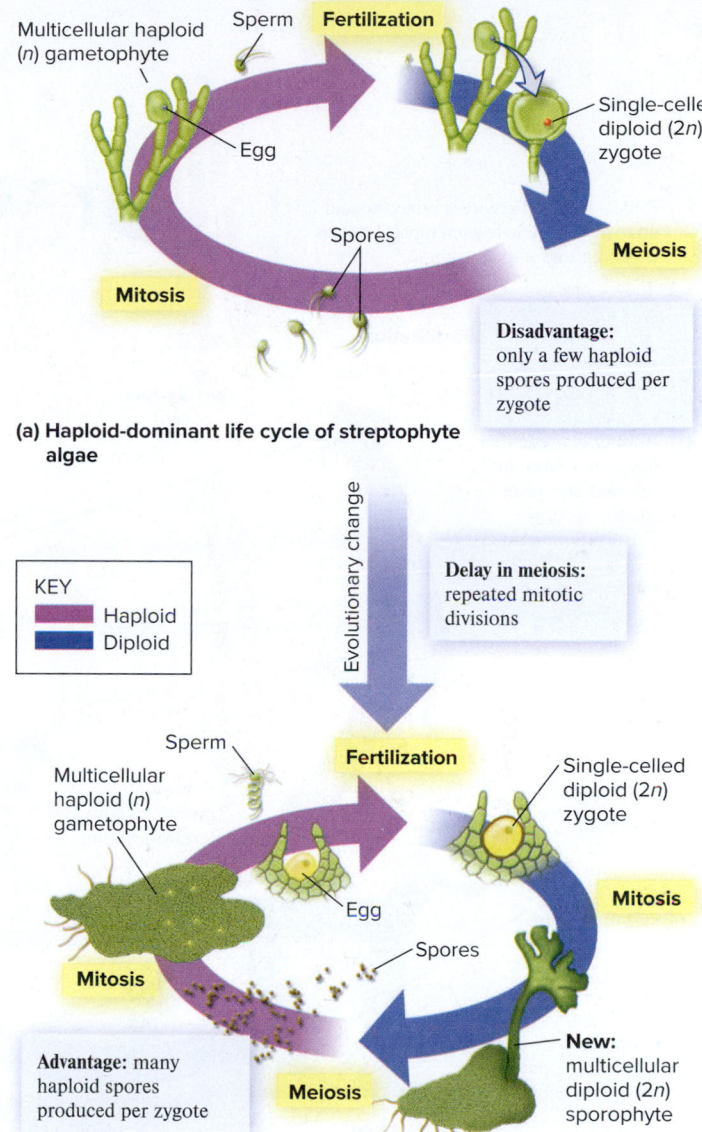

(a) **Haploid-dominant life cycle of streptophyte algae**

**KEY**
- Haploid
- Diploid

(b) **Alternation of generation of early plants**

**Figure 31.15** Evolutionary transition from **(a)** the life cycle of primarily aquatic streptophyte algae to **(b)** the derived life cycle of primarily terrestrial bryophytes.

gametes in structures known as **gametangia** (from the Greek, meaning gamete containers). Certain cells of gametangia develop into gametes, and other cells form an outer protective jacket of tissue. This jacket protects the delicate gametes from drying out and from microbial attack while they develop. Flask-shaped gametangia that each enclose a single egg cell are known as **archegonia** (singular, archegonium). Spherical or elongate gametangia that each produce many sperm are known as **antheridia** (singular, antheridium). When bryophyte sperm are mature and moist conditions exist, antheridia open and release sperm into films of water. Under the influence of sex-attractant molecules secreted from archegonia, the sperm swim toward the eggs, twisting their way

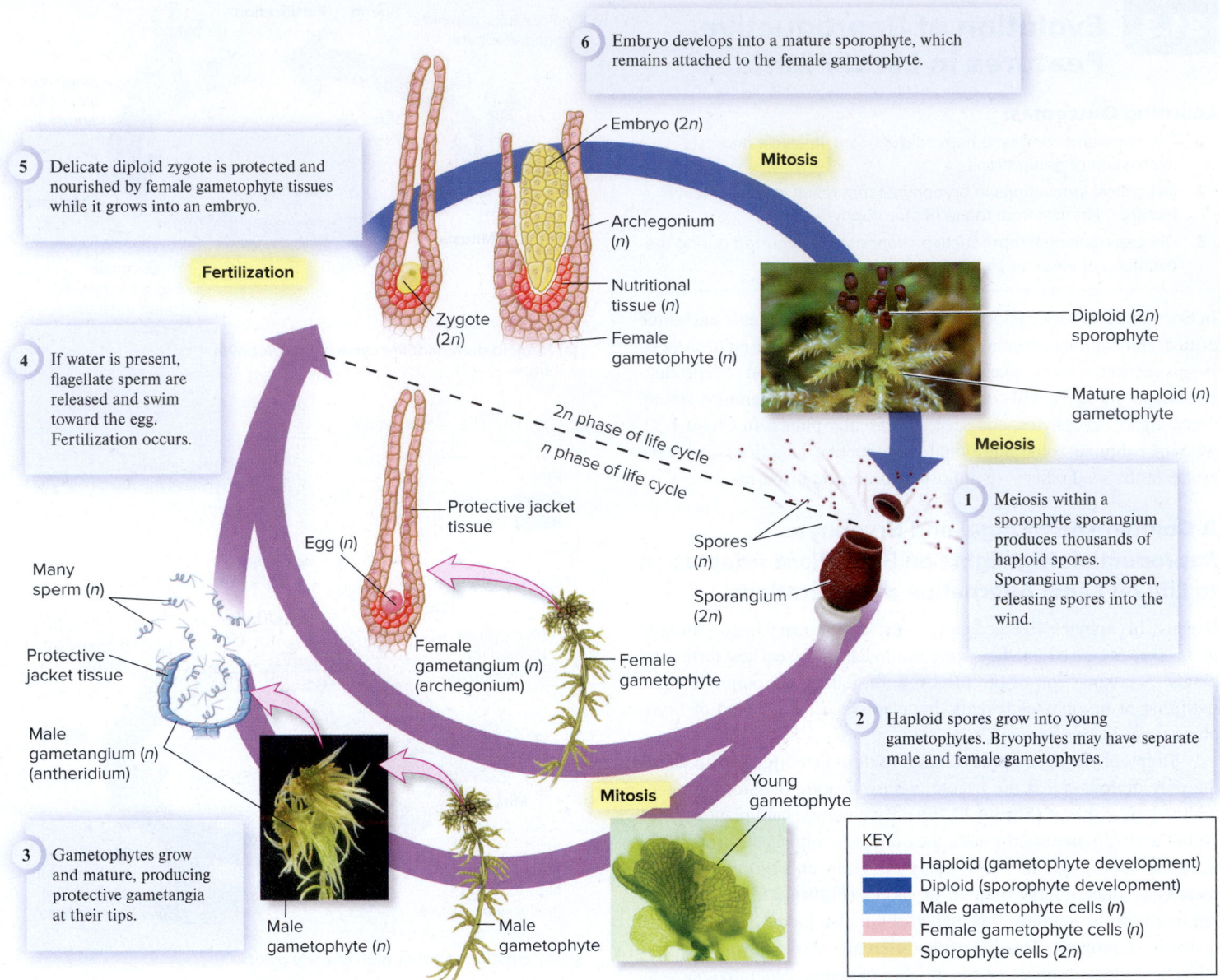

6 Embryo develops into a mature sporophyte, which remains attached to the female gametophyte.

Embryo (2n)

**Mitosis**

5 Delicate diploid zygote is protected and nourished by female gametophyte tissues while it grows into an embryo.

Archegonium (n)

**Fertilization**

Nutritional tissue (n)

Zygote (2n)

Female gametophyte (n)

Diploid (2n) sporophyte

4 If water is present, flagellate sperm are released and swim toward the egg. Fertilization occurs.

Mature haploid (n) gametophyte

2n phase of life cycle

n phase of life cycle

**Meiosis**

Protective jacket tissue

Egg (n)

Spores (n)

1 Meiosis within a sporophyte sporangium produces thousands of haploid spores. Sporangium pops open, releasing spores into the wind.

Many sperm (n)

Sporangium (2n)

Female gametangium (n) (archegonium)

Female gametophyte

Protective jacket tissue

2 Haploid spores grow into young gametophytes. Bryophytes may have separate male and female gametophytes.

Male gametangium (n) (antheridium)

**Mitosis**

Young gametophyte

3 Gametophytes grow and mature, producing protective gametangia at their tips.

Male gametophyte (n)

Male gametophyte

**KEY**
- Haploid (gametophyte development)
- Diploid (sporophyte development)
- Male gametophyte cells (n)
- Female gametophyte cells (n)
- Sporophyte cells (2n)

**Figure 31.16** **The life cycle of the early-diverging moss genus *Sphagnum*.** The life cycle of this bryophyte illustrates reproductive adaptations that likely helped early plants reproduce on land. Among modern bryophytes, *Sphagnum* is the single most abundant and ecologically important genus. (top right inset): ©Larry West/Science Source; (bottom right inset, bottom left inset): ©Linda Graham

down the tubular neck of the archegonium. The sperm then fertilize egg cells to form diploid zygotes, which grow into embryos (see steps 4 and 5, Figure 31.16).

A key reproductive advantage of the plant life cycle is that embryos remain enclosed by gametophyte tissues that provide protection and food. This critical innovation, known as **matrotrophy** (from the Latin, meaning mother, and the Greek, meaning food) gives zygotes a good start while they grow into embryos. Because all groups of land plants possess matrotrophic embryos, they are known as **embryophytes** (see Figure 31.1).

Plant reproductive advantages also involve the formation of spores. To produce haploid spores, meiosis occurs in cells that are

within enclosures known as **sporangia** (from the Greek, meaning spore containers) (see step 1, Figure 31.16). The cells of such enclosures are surrounded by tough cell walls that protect developing spores from harmful UV radiation and microbial attack. Bryophyte sporangia open in specialized ways that foster dispersal of mature spores into the air, allowing spore transport by wind (see Figures 31.5–31.8). Another important adaptation involves the spores themselves. The cell walls of mature plant spores contain a tough material, known as **sporopollenin**, that helps to prevent cellular damage during transport in air. If spores reach habitats favorable for growth, their cell walls crack open, and new gametophytes develop by mitotic divisions.

## The Evolution of Vascular Plants Coincided with an Increase in the Relative Size of the Mature Sporophyte and Its Independence

As we have seen, bryophytes illustrate a number of valuable reproductive features that appeared early in plant evolution and were inherited by vascular plant descendants: alternation of generations, gametangia, embryos, matrotrophy, sporangia, and sporopollenin-enclosed spores. In bryophytes, the gametophyte is relatively large and the mature sporophyte is usually small and depends on the gametophyte for its nutrition. The evolution of vascular plants involved a shift in the relative sizes of gametophytes and sporophytes (**Figure 31.17**). The gametophyte generation has become smaller in later-diverging phyla, whereas the sporophyte generation has become larger and more complex.

Although sporophyte embryos of vascular plants are dependent upon supportive gametophytes, these embryos eventually become independent, free-living organisms. Vascular-plant sporophytes can branch, continue to grow, and produce sporangia on lateral branches, often for many years (see Figure 31.17b,c). These features allow vascular plants to produce more progeny than bryophytes, whose sporophytes remain small, never become independent of parental gametophytes, are unable to branch, and have short lifetimes (see Figure 31.17a). In vascular plants the diploid sporophyte generation is the dominant generation, meaning that it is larger, more complex, and longer-lived than the gametophyte. The evolutionary shift toward sporophyte dominance explains why vascular plants are more prevalent in most modern terrestrial habitats.

*Life Cycle of Lycophytes and Pteridophytes*    The seedless vascular plants, which include lycophytes and pteridophytes, exhibit a life cycle in which the sporophyte is the larger, more conspicuous

organism whereas the gametophyte is smaller. **Figure 31.18** describes the life cycle of a fern, which is a pteridophyte. The organism that we associate with the name fern is the sporophyte. Stem branching allows adult plants to produce many leaves, and roots supply stems with large amounts of soil water and minerals. Lycophytes and pteridophytes use such resources to produce large numbers of sporangia. You might have seen clusters of sporangia as dark brown dots or lines on the undersides of fern leaves. As seen in steps 2 and 3 of Figure 31.18, the sporangia produce many spores that are released and dispersed by the wind.

When a spore lands in a favorable location, it will grow by mitotic divisions to form a gametophyte. Both lycophyte and pteridophyte gametophytes are small, delicate, and easily harmed by exposure to heat and drought. This explains why the gametophytes of lycophytes and pteridophytes are restricted to moist places, often existing underground, and why they have short lifetimes. The gametophyte of most ferns is thumbnail-sized. It doesn't have roots, stems or leaves, but it does have rhizoids that anchor it to the soil.

The gametophyte produces eggs in female gametangia (archegonia) and sperm in male gametangia (antheridia). Lycophyte and pteridophyte sperm are flagellate, a feature inherited from algal and bryophyte ancestors. When released from antheridia into water films, sperm must swim to eggs in archegonia (Figure 31.18, step 6). For this reason, lycophyte and pteridophyte reproduction is inhibited by dry conditions, as is also the case for bryophytes. However, when fertilization occurs, lycophytes and pteridophytes can produce many spores, because the spore-producing sporophyte generation grows to a much larger size than do bryophyte sporophytes.

After fertilization, lycophyte and pteridophyte embryos, like those of all land plants, are initially nourished by maternal gametophytes. As they mature, vascular plant sporophytes eventually become independent by producing leaves and roots able to harvest resources needed for photosynthesis (Figure 31.18, step 8).

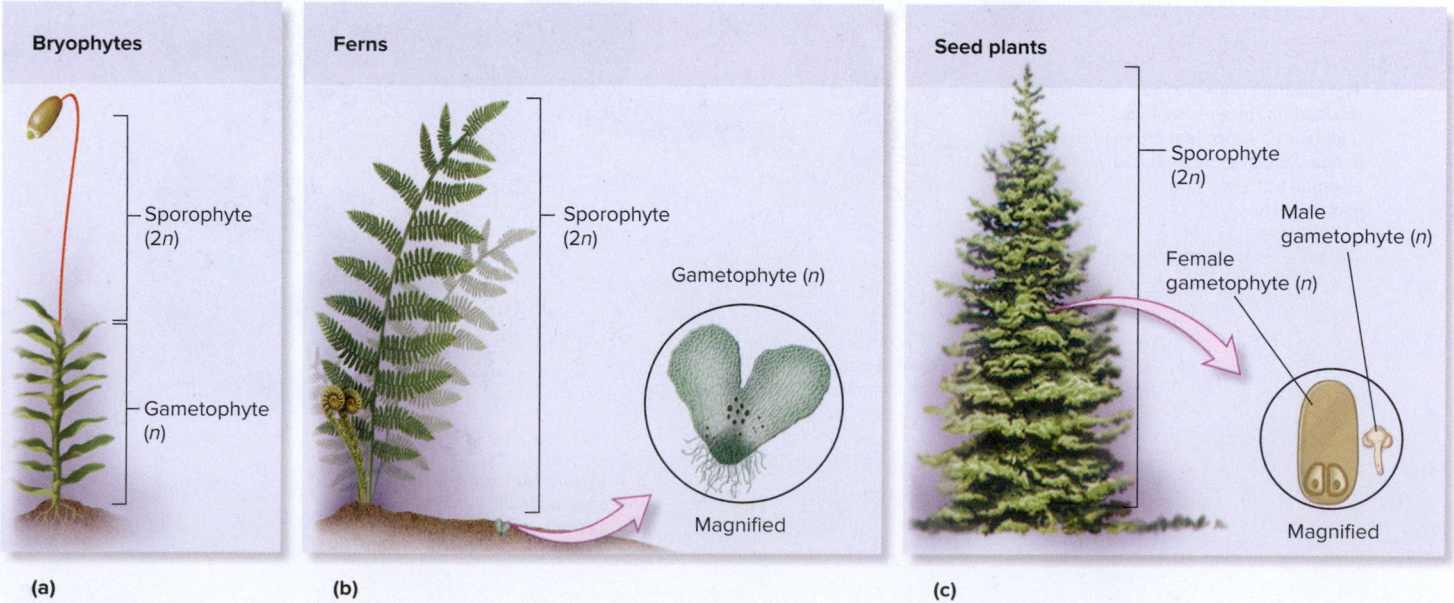

**Figure 31.17**    Relative sizes of the sporophyte and gametophyte generations of bryophytes, ferns, and seed plants.

*Concept Check:*    *What is the advantage to ferns and seed plants of having larger sporophytes relative to those of bryophytes?*

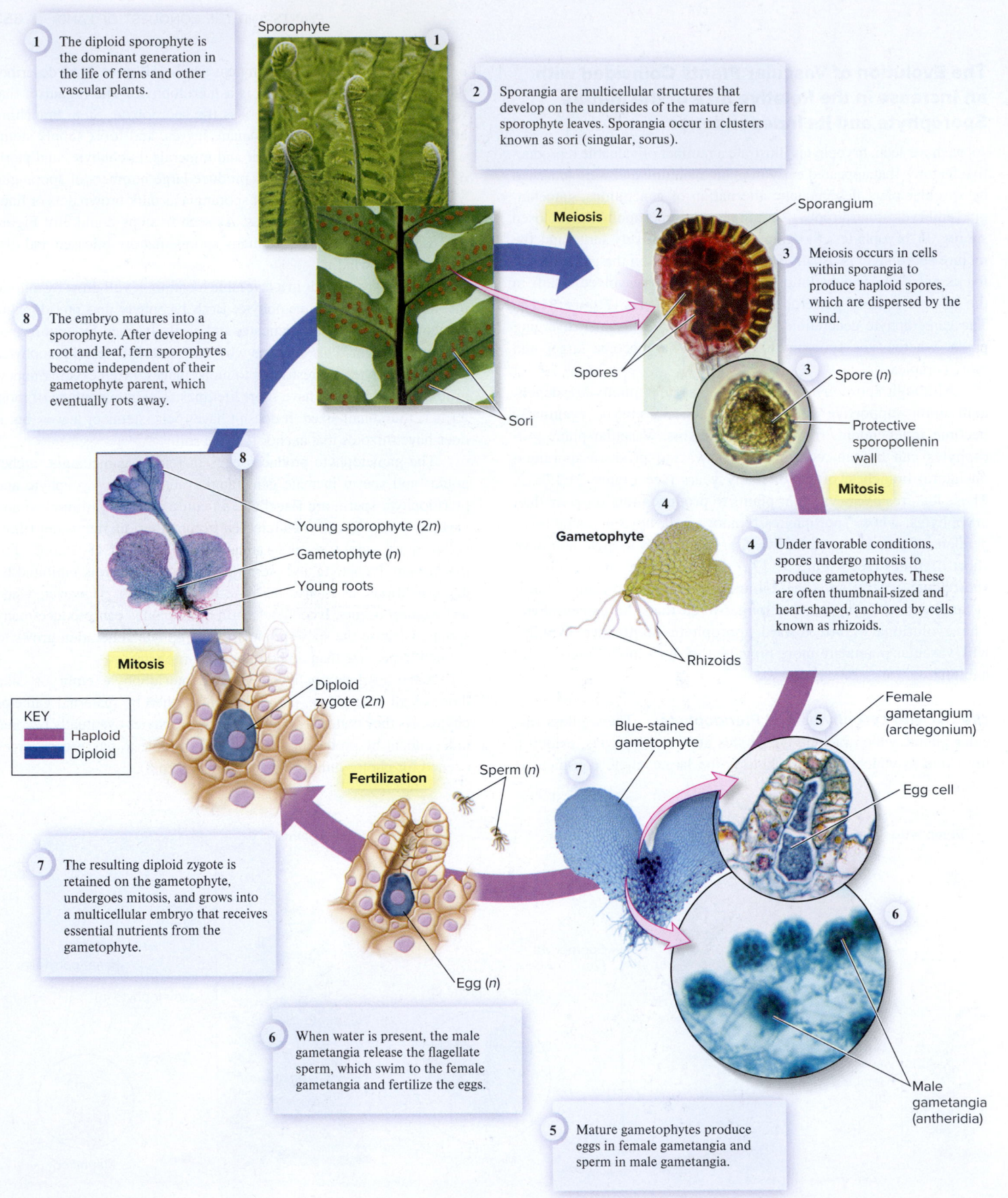

**1** The diploid sporophyte is the dominant generation in the life of ferns and other vascular plants.

Sporophyte

**2** Sporangia are multicellular structures that develop on the undersides of the mature fern sporophyte leaves. Sporangia occur in clusters known as sori (singular, sorus).

**Meiosis**

Sporangium

**3** Meiosis occurs in cells within sporangia to produce haploid spores, which are dispersed by the wind.

Spores

**8** The embryo matures into a sporophyte. After developing a root and leaf, fern sporophytes become independent of their gametophyte parent, which eventually rots away.

Sori

Spore (n)

Protective sporopollenin wall

**Mitosis**

Young sporophyte (2n)
Gametophyte (n)
Young roots

**Gametophyte**

**4** Under favorable conditions, spores undergo mitosis to produce gametophytes. These are often thumbnail-sized and heart-shaped, anchored by cells known as rhizoids.

Rhizoids

**Mitosis**

KEY
Haploid
Diploid

Diploid zygote (2n)

Female gametangium (archegonium)

Blue-stained gametophyte

Sperm (n)

Egg cell

**Fertilization**

**7** The resulting diploid zygote is retained on the gametophyte, undergoes mitosis, and grows into a multicellular embryo that receives essential nutrients from the gametophyte.

Egg (n)

**6** When water is present, the male gametangia release the flagellate sperm, which swim to the female gametangia and fertilize the eggs.

Male gametangia (antheridia)

**5** Mature gametophytes produce eggs in female gametangia and sperm in male gametangia.

**Figure 31.18** **The life cycle of a typical fern.** The fern life cycle is often used to illustrate plant alternation of generations because both sporophyte and gametophyte are large enough for people to see with the unaided eye. (foreground inset): ©Carolina Biological Supply Company/Phototake; (inset 1): ©Ernst Kucklich/Getty Images; (inset 2–3): ©Linda Graham; (inset 4–8): ©Lee W. Wilcox

# 31.4 Evolutionary Importance of the Plant Embryo

**Learning Outcomes:**

1. Describe how a plant embryo benefits from the maternal gametophyte.
2. **CoreSKILL »** Analyze the results of Browning and Gunning, and explain the role of placental transfer tissues in the movement of nutrients from mother plant to embryo.

The embryo was one of the first critical innovations acquired by land plants (see Figure 31.1). Recall that plant embryos are young sporophytes that develop from zygotes and are enclosed by maternal tissues that provide nutrients and protection. The presence of an embryo is critical to plant reproduction in terrestrial environments. Drought, heat, UV light, and microbial attack could kill delicate plant egg cells, zygotes, and embryos if these were not protected and nourished by enclosing maternal tissues. The first embryo-producing plants diversified into hundreds of thousands of modern species, as well as many species that have become extinct. A closer look at embryos reveals why their origin and evolution are so important.

## Plant Embryos Grow Protected Within the Maternal Plant Body

A plant embryo has several characteristic features, some of which were previously described. First, plant embryos are multicellular and diploid. Plant embryos develop by repeated mitotic divisions from a single-celled zygote (see Figure 31.15b). In addition, plant eggs are fertilized while still enclosed by the maternal plant body, and embryos begin their development within the protective confines of maternal tissues (see Figure 31.16). Plant biologists say that plants retain their zygotes and embryos. Third, plant embryo development depends on organic and mineral materials supplied by the mother plant, in the process known as matrotrophy. Nutritive tissues composed of specialized **placental transfer tissues** aid in the transfer of nutrients from mother to embryo. A closer look at placental transfer tissues will reveal their valuable role.

Placental transfer tissues function similarly to the placenta present in most mammals, which fosters nutrient movement from the mother's bloodstream to the developing fetus. Plant placental transfer

tissues often occur in haploid gametophyte tissues that lie closest to embryos and in the diploid tissues of young embryos themselves. Such transfer tissues contain cells that are specialized in ways that promote the movement of solutes from gametophyte to embryo. For example, the cells of placental transport tissues display complex arrays of finger-like cell-wall ingrowths (**Figure 31.19**). Because the plant plasma membrane lines the plant cell wall, the ingrowths vastly increase the surface area of the plasma membrane. This increase allows for more abundant membrane transport proteins, which move solutes into and out of cells. With more transport proteins present, materials can move at a faster rate from one cell to another. (Similar finger-like structures in animal intestines and placenta likewise foster nutrient flow by increasing cellular surface area.) Dissolved sugars, amino acids, and minerals first move from maternal plant cells into the intercellular space between maternal tissues and the embryo. Then, transport proteins in the membranes of nearby embryo cells efficiently import the nutrients into the embryo. As described next in the Feature Investigation, the role of placental transfer tissue was revealed in experiments involving the use of radiolabeled carbon dioxide ($CO_2$).

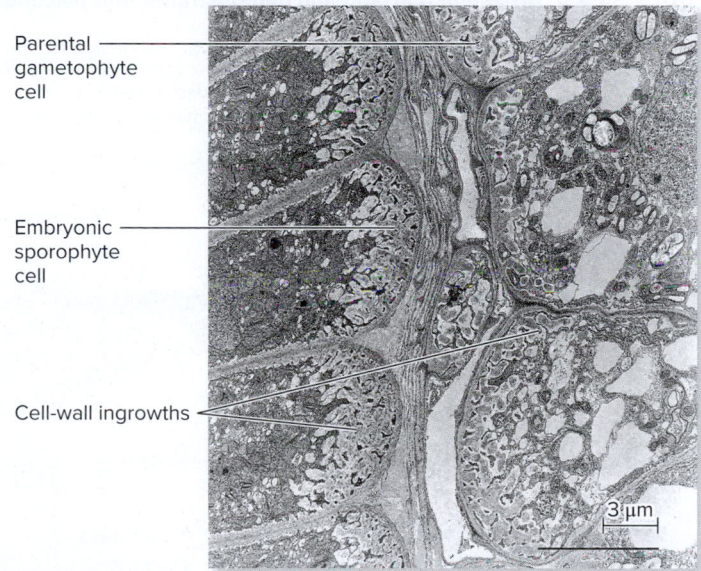

**Figure 31.19** **Placental transfer tissue from a plant in the liverwort genus _Monoclea_.** Similar structures occur at the gametophyte-sporophyte junction in all other land plant phyla. Courtesy Prof. Roberto Ligrone. Fig. 6 in Ligrone et al., _Protoplasma_ (1982), 154: 414–425

Parental gametophyte cell

Embryonic sporophyte cell

Cell-wall ingrowths

3 μm

---

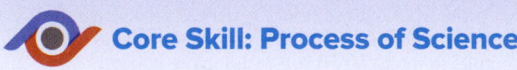 **Core Skill: Process of Science**

## Feature Investigation | Browning and Gunning Demonstrated That Placental Transfer Tissues Facilitate the Movement of Organic Molecules from Gametophytes to Sporophytes

In the 1970s, plant cell biologists Adrian Browning and Brian Gunning explored the function of placental transfer tissues. Using a simple moss as their experimental organism, they investigated the rate at which radiolabeled carbon moves through placental transfer tissues from green gametophytes into young sporophytes. Recall that embryos are very young,

few-celled sporophytes and that in mosses and other bryophytes, all stages of sporophyte development are nutritionally dependent on gametophyte tissues. Browning and Gunning investigated nutrient flow into young sporophytes because these slightly older and larger developmental stages were easier to manipulate in the laboratory than were tiny embryos.

In a first step, the investigators grew many gametophytes of the moss *Funaria hygrometrica* in a greenhouse until young sporophytes developed as the result of sexual reproduction (**Figure 31.20**). In a second step, they placed black glass tubing over young sporophytes as a shade to prevent photosynthesis, enclosed the whole plants—gametophytes and their attached sporophytes—within transparent jars, and supplied the plants with radiolabeled $^{14}CO_2$ for 15 minutes, an experimental procedure known as a pulse. Because the moss gametophytes were not shaded, their photosynthetic cells were able to incorporate some of the carbon from $^{14}CO_2$ into organic compounds, such as sugars and amino acids, thereby making these compounds radiolabeled. Shading prevented the young sporophytes, which possess some photosynthetic tissue, from using the $^{14}CO_2$ to produce organic compounds.

In a third step, the researchers added an excess amount of nonradioactive $CO_2$ to prevent the gametophytes from taking up more $^{14}CO_2$, an experimental procedure known as a chase. This process stopped the radiolabeling of photosynthetic products because the vast majority of $CO_2$ taken up was now unlabeled. (Experiments such as this one are known as pulse-chase experiments.) In a final step, Browning and Gunning plucked young sporophytes of different sizes (ages) from the gametophytes and measured the amount of $^{14}C$ present in the separated gametophyte and sporophyte tissues at various times following the chase.

$CO_2$ is rapidly incorporated into organic molecules via photosynthesis (refer back to Figure 8.13). Therefore, the researchers assumed that any transfer of radiolabeled carbon between gametophytes and sporophytes would be radiolabeled carbon within organic molecules. From their data, Browning and Gunning were able to calculate the relative amount of organic carbon that had moved from the moss gametophytes to their sporophytes. At the beginning of the chase, a group of gametophytes contained 228 units of radiolabeled carbon (see data set I in Figure 31.20). Eight hours after the chase, 51 units had been transferred to the young sporophytes. In other words, about 22% of the organic carbon produced by gametophyte photosynthesis was transferred to the young sporophytes during an 8-hour chase period. In addition, by comparing the amount of radioactive carbon accumulated by sporophytes of differing sizes, they

**Figure 31.20** Browning and Gunning demonstrated that placental transfer tissues increase plant reproductive success.

**HYPOTHESES** 1. Placental transfer tissues allow organic nutrients to flow from plant gametophytes to sporophytes faster than such nutrients move through plant tissues lacking transfer cells.
2. The rate of organic nutrient transfer into larger sporophytes is faster than into smaller sporophytes.

**KEY MATERIALS** Moss *Funaria hygrometrica*, $^{14}CO_2$ (radiolabeled carbon dioxide)

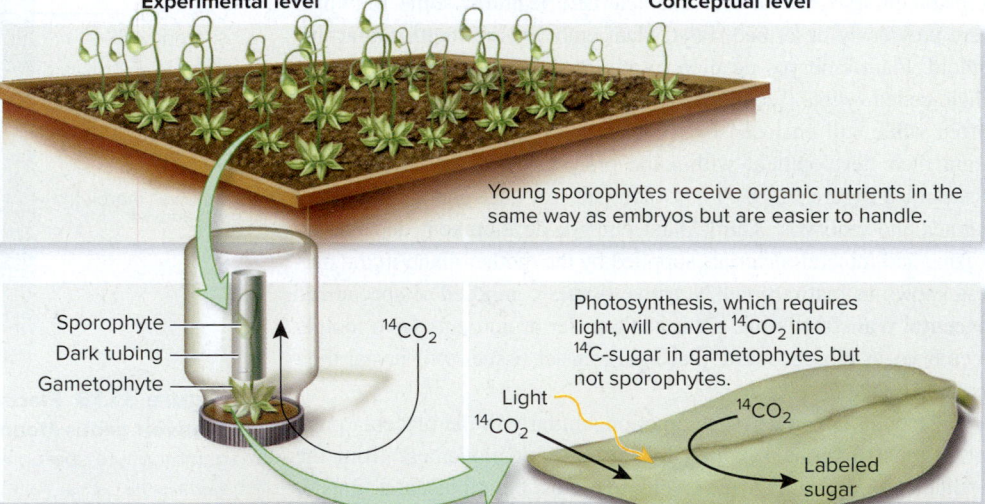

**Experimental level**      **Conceptual level**

1. Grow moss gametophytes until young sporophytes develop from embryos, and measure sporophyte size.

Young sporophytes receive organic nutrients in the same way as embryos but are easier to handle.

2. Shade young sporophytes from light with blackened glass tubing, and enclose whole plant in clear glass jar. Expose plants to $^{14}CO_2$ for 15 minutes. This is called a pulse.

Sporophyte
Dark tubing
Gametophyte
$^{14}CO_2$

Photosynthesis, which requires light, will convert $^{14}CO_2$ into $^{14}C$-sugar in gametophytes but not sporophytes.

Light
$^{14}CO_2$
$^{14}CO_2$
Labeled sugar

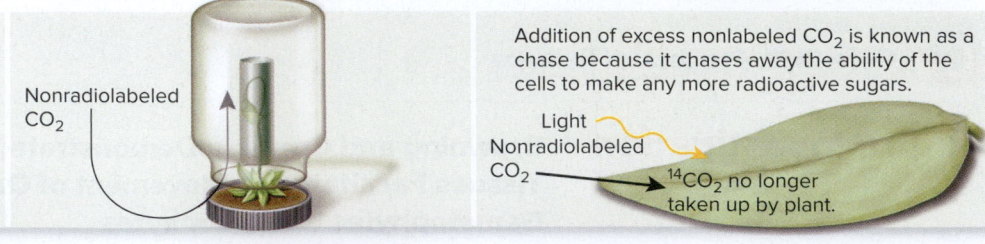

3. Expose plants to a large amount of nonradioactive $CO_2$. This is called a chase. Incubate up to 8 hours.

Nonradiolabeled $CO_2$

Addition of excess nonlabeled $CO_2$ is known as a chase because it chases away the ability of the cells to make any more radioactive sugars.

Light
Nonradiolabeled $CO_2$
$^{14}CO_2$ no longer taken up by plant.

4. Pluck young sporophytes of differing sizes from gametophytes. Assay $^{14}C$ in both sporophytes and gametophytes using a scintillation counter. This was done immediately following the chase, and at 2 or 8 hours after the chase.

Scintillation counter

Determine how much organic carbon flowed into sporophytes during each chase time.

**5  THE DATA I**

Organic carbon transfer from gametophyte to sporophyte:

| Mean $^{14}$C content of 5 gametophytes at 0 chase time | Mean $^{14}$C lost from gametophytes after 8-hour chase | Mean $^{14}$C gained by sporophytes after 8-hour chase |
|---|---|---|
| 228 units | 145 units | 51 units |

**6  THE DATA II**

Sporophyte size effect:

| Sporophyte size | Mean $^{14}$C content of 8 sporophytes after 2-hour chase |
|---|---|
| 5–7 mm | 8.47 ± 4.29 units |
| 11–13 mm | 9.93 ± 3.94 units |
| 23–25 mm | 24.97 ± 5.30 units |

**7  CONCLUSION**  Organic carbon moves from photosynthetic gametophytes into developing sporophytes, facilitated by placental transfer tissues. Larger sporophytes absorb more organic carbon than smaller ones.

**8  SOURCES**  Browning, A.J., and Gunning, B.E.S. 1979. Structure and function of transfer cells in the sporophyte haustorium of *Funaria hygrometrica*. III. Translocation of assimilate into the attached sporophyte and along the sets of attached and excised sporophytes. *Journal of Experimental Botany* 30: 1265–1273.

also learned that larger sporophytes absorbed $^{14}$C about three times faster than smaller ones (see data set II in Figure 31.20).

Browning and Gunning also calculated the rate of nutrient transfer from gametophyte to sporophyte and compared this rate with the rate (determined in other studies) at which organic carbon moves between several other plant tissues that lack specialized transfer cells. They discovered that organic carbon moved from moss gametophytes to young sporophytes nine times faster than organic carbon moves between these other plant tissues. These investigators inferred that the increased rate of nutrient movement could be attributed to placental transfer cell structure, namely, that cell-wall ingrowths enhance plasma membrane surface area.

Taken together, these data are consistent with the hypothesis that placental transfer tissues increase plant reproductive success by providing embryos and growing sporophytes with more nutrients than they would otherwise receive. Supplied with the greater amounts of nutrients, sporophytes are able to grow larger than they otherwise would, and eventually they produce more progeny spores.

### Experimental Questions

1. What were the goals of the Browning and Gunning investigation?

2. **CoreSKILL »** How did Browning and Gunning prevent photosynthesis from occurring in moss sporophytes during the experiment (shown in Figure 31.20), and why did they do this?

3. **CoreSKILL »** How did the measurements Browning and Gunning made after adding an excess amount of unlabeled $CO_2$ lead them to their conclusions?

## Core Skill: Quantitative Reasoning

**BIO TIPS**  **THE QUESTION**  *The experiment by Browning and Gunning revealed that placental transfer tissues are important for the movement of organic nutrients from moss gametophytes to young sporophytes. Organic nutrients are produced in photosynthetic tissues and transferred to nonphotosynthetic tissues. But diffusion from one generation to the next occurs too slowly, which explains why placental tissues evolved in plants (and placental mammals). According to the data of Browning and Gunning, how much more organic carbon do the largest sporophytes take up compared to the smallest ones? Based on the reproductive role of the sporophyte, propose a hypothesis that could explain this difference.*

**T OPIC**  *What topic in biology does this question address?*  The topic is the function of plant placental transfer tissues in moving organic carbon into embryonic sporophytes. More specifically, the question concerns the transfer of organic carbon into moss sporophytes of different sizes, which represent stages of developmental maturity.

**I NFORMATION**  *What information do you know based on the question and your understanding of the topic?*  In the question, you are reminded of the experiment by Browning and Gunning that revealed the importance of placental transfer tissues in supplying nutrients to young sporophytes and referred to their data. From your understanding of cell biology, you may recall that all cells require organic compounds for energy and to synthesize larger macromolecules. Photosynthetic cells are able to generate organic compounds from inorganic compounds in the presence of light. Thus, photosynthetic tissues are a potential source of organic compounds for plant tissues whose primary function is not photosynthesis, such as sporangia, which carry out spore production.

**P ROBLEM-SOLVING S TRATEGY**  *Make a calculation. Propose a hypothesis.*  Data set II in Figure 31.20 shows the amounts of $^{14}$C taken up by moss sporophytes of differing sizes. From these data, calculate the relative difference between the smallest and largest sporophytes. Consider the reproductive role of the moss sporophyte, and propose one or more reasons why the transfer of nutrients would be higher in the larger sporophytes.

**ANSWER** *If you compare the data for the smallest sporophytes (8.47 units) and the largest ones (24.97 units), you see that the largest have a mean content of radiolabeled carbon that is about three times that of the smallest. From this difference, you might infer that the rate of nutrient metabolism becomes three times higher as the sporophyte reaches maturity and produces spores that are ready for dispersal. One hypothesis to explain this difference is that the larger sporophytes are using nutrients at a faster rate because they are producing the tough spore coating containing sporopollenin, which would be an additional metabolic demand.*

## 31.5   The Origin and Evolutionary Importance of Leaves and Seeds

### Learning Outcomes:

1. Describe how the leaves of ferns and seed plants likely evolved from branched-stem systems.
2. Discuss how seeds develop from fertilized ovules.
3. Name several advantages that seeds provide.

Like embryos, leaves and seeds are critical innovations that increased plant fitness and fostered diversification. Unlike the plant embryo, which likely originated just once at the birth of the plant kingdom, leaves and seeds may have independently evolved several times during plant evolutionary history. Comparative studies of diverse types of leaves and seeds in fossil and living plants suggest how these critical innovations originated.

### The Large Leaves of Ferns Evolved from Branched-Stem Systems

Leaves are the solar panels of the plant world. Their flat structure provides a large surface area that effectively captures sunlight for photosynthesis. Among the vascular plants, lycophytes produce the simplest and most ancient type of leaves. Modern lycophytes have tiny leaves, known as **lycophylls** (also known as microphylls), which typically have only a single unbranched vein (**Figure 31.21a**). Some experts think that these small leaves may have evolved from sporangia.

By contrast to lycophytes, ferns and seed plants have leaves with extensively branched veins. Such leaves are known as **euphylls** (from the Greek, meaning true leaves) (**Figure 31.21b**). The branched veins of euphylls are able to supply relatively large areas of photosynthetic tissue with water and minerals. For this reason, euphylls are typically much larger than lycophylls, explaining why euphylls are also known as megaphylls. Euphylls provide considerable photosynthetic advantage to ferns and seed plants, because they provide more surface area for solar energy absorption than do small leaves. The evolution of relatively large leaves allowed plants to more effectively accomplish photosynthesis, enabling them to grow larger and produce more progeny.

The study of fern fossils indicates that euphylls likely arose from leafless, cylindrical, branched-stem systems by a series of steps (**Figure 31.21c**). First, one branch assumed the role of the main axis, while the other was reduced in size and became flattened in one plane, and then the spaces between the branches of this flattened system became filled with photosynthetic tissue. Such a process explains why euphylls have branched vascular systems; individual veins apparently originated from the separate branches of an ancestral branched stem. Plant evolutionary biologists suspect that euphylls arose several times by means of similar, parallel processes, and that leaves of ferns and seed plants are not homologous structures.

### Seeds Develop from the Interaction of Ovules and Pollen

The seed plants dominate modern ecosystems, suggesting that seeds offer reproductive advantages. Seed plants are also the plants with the greatest importance to humans, as described in Chapter 32. For these reasons, plant biologists are interested in understanding why seeds are so advantageous and how they evolved. To consider these questions, we must first take a closer look at seed structure and development.

As mentioned earlier, plants produce spores by meiosis within sporangia, and seed plants are no exception. However, seed plants produce two distinct types of spores in two types of sporangia, a trait known as **heterospory**, meaning different spores. Microsporangia produce small **microspores** that give rise to male gametophytes, which develop into pollen grains. Megasporangia produce larger **megaspores** that give rise to female gametophytes, which develop and produce eggs while enclosed by protective megaspore walls. The enclosed female gametophytes are not photosynthetic, so they need help in feeding the embryos that develop from fertilized eggs. Female gametophytes get this help by remaining attached to the previous sporophyte generation, which provides gametophytes with the nutrients needed for embryo development.

Plants produce seeds by reproductive structures known as ovules and pollen, which are unique to seed plants. An **ovule** is a sporangium that typically contains only a single spore that develops into a very small egg-producing gametophyte; the entire structure is enclosed by leaflike structures known as **integuments** (**Figure 31.22a**). You can think of an ovule as being like a nesting doll with four increasingly smaller dolls inside. The smallest doll corresponds to an egg cell; intermediate-sized doll represents the gametophyte, spore wall, and megasporangium; and the largest doll represents the integuments. Fertilization converts such layered ovules into seeds. In seed plants, the sperm needed for fertilization are supplied by **pollen**, tiny male gametophytes enclosed by protective sporopollenin-containing microspore walls.

Embryos and seeds develop as the result of fertilization, which cannot occur until after **pollination**, the process by which pollen comes into contact with ovules. Pollination typically occurs by means of wind or animal transport (see Chapter 32). Fertilization occurs in seed plants when a male gametophyte extends a slender pollen tube that carries two sperm toward an egg. The pollen tube enters through an opening in the integument called the micropyle and releases the

(a) Lycophyll (small leaf)

Single unbranched leaf vein

(b) Euphyll (large leaf)

Branched vascular system

1. Fern ancestors initially had a branched-stem system.

2. One branch began to dominate the stem system.

3. The branch system flattened into a single plane.

4. Photosynthetic tissue filled in the spaces between the branches of a system.

Euphyll

(c) Euphyll evolution process in pteridophytes

**Figure 31.21 Lycophylls and euphylls. (a)** Most lycophylls possess only a single unbranched leaf vein with limited conduction capacity, explaining why lycophylls are generally quite small. **(b)** Euphylls possess branched vascular systems with greater conduction capacity, explaining why many euphylls are relatively large. **(c)** Fossil evidence suggests how pteridophyte euphylls might have evolved from ancestors with branched-stem systems.

 **Core Skill: Modeling** The goal of this modeling challenge is to propose a visual model that compares the density of leaf veins between ferns and another plant group.

**Modeling Challenge:** Figure 31.21 provides a model of the process by which fern leaves having branched vascular systems (euphylls) are hypothesized to have evolved from ancestors with branching stems. Imagine that the leaves of some other plant group evolved similarly, but from stem systems that were twice as highly branched. In other words, when flattened, the stem system ancestral to this other plant group had twice as many branches per unit area. Assuming that the branch density of ancestors is directly related to vein density of leaves in descendant plants, draw a pair of models that compare the leaves of ferns and this other plant group, emphasizing the vein density in each type of leaf. How does the venation differ in the two leaf models?

sperm. The fertilized egg becomes an embryo, and the ovule's integument develops into a protective, often hard and tough **seed coat** (**Figure 31.22b** and **c**).

Gymnosperm seeds contain female gametophyte tissue that has accumulated large amounts of proteins, lipids, and carbohydrates prior to fertilization. These nutrients feed both embryo development and seed germination. Angiosperm seeds also contain this useful food supply, but most angiosperm ovules do not store food materials before fertilization. Instead, angiosperm seeds generally store food only after fertilization occurs, ensuring that the food is not wasted if an embryo does not form. How is this accomplished? The answer is a process known as **double fertilization**. This process produces

both a zygote and a food storage tissue known as endosperm, a feature unique to angiosperms. One of the two sperm delivered by each pollen tube fuses with the egg, producing a diploid zygote, as you might expect. The other sperm fuses with different gametophyte nuclei to form an unusual cell that has more than the diploid number of chromosomes; this cell continues to divide and generates the endosperm food tissue. Endosperm will be discussed in more detail in Chapter 40.

Seeds allow plant embryos access to food supplied by the previous sporophyte generation, an option not available to seedless plants. The layered structure of ovules explains why seeds are also layered, with a protective seed coat enclosing the embryo and

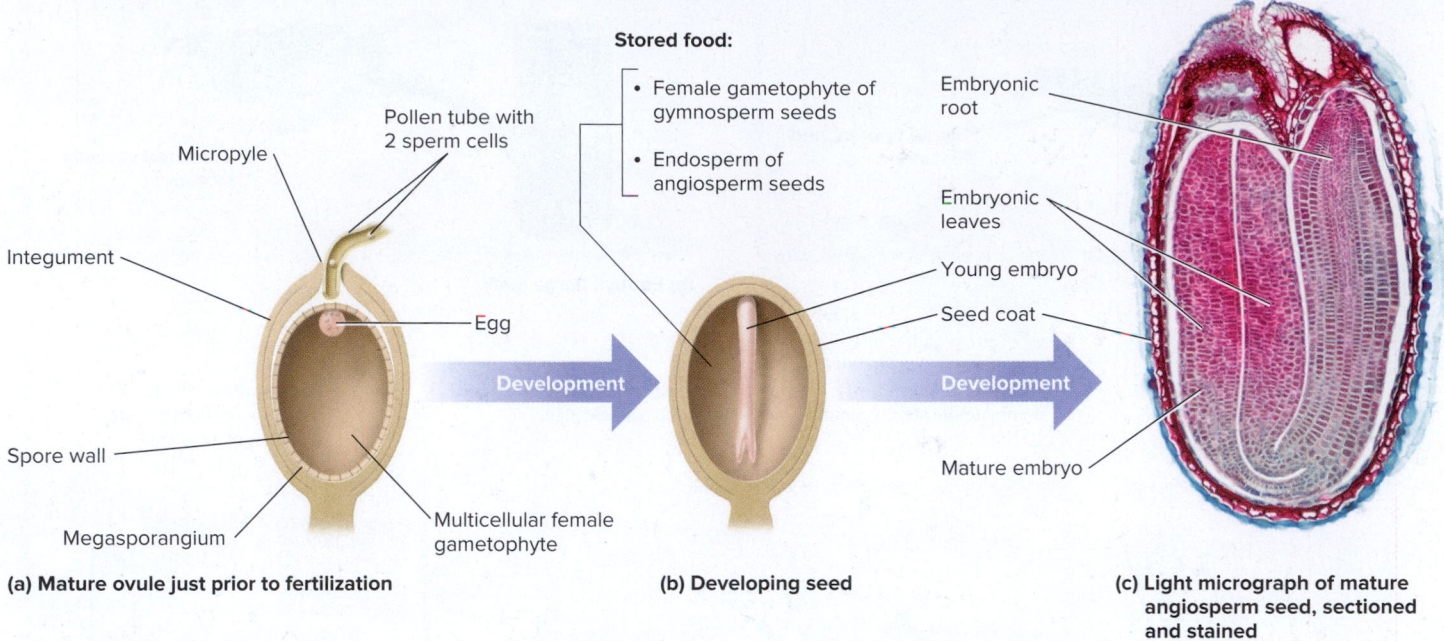

**Figure 31.22** **Structure of an ovule developing into a seed.** c: ©Lee W. Wilcox

*Concept Check:* *Can you hypothesize why the mature angiosperm seed does not show obvious endosperm tissue?*

stored food. As described next, these seed features improve the chances of embryo and seedling survival, thereby increasing seed plant fitness.

## Seeds Confer Important Reproductive Advantages

Seeds provide plants with numerous reproductive advantages.

- First, many seeds are able to remain dormant in the soil for long periods, until conditions become favorable for germination and seedling growth. Furthermore, seed coats are often adapted in ways that improve dispersal in diverse habitats. For example, many plants produce winged seeds that are effectively dispersed by wind. Other plants produce seeds with fleshy coverings that attract animals, which consume the seeds, digest the fleshy covering, and eliminate the bare seeds at some distance from the originating plants.

- Another advantage of seeds is that they can store considerable amounts of food, which supports embryo growth and helps plant seedlings grow large enough to compete for light, water, and minerals. This is especially important for seeds that must germinate in shady forests.

- Finally, the sperm of most seed plants can reach eggs without having to swim through water, because pollen tubes deliver sperm directly to ovules. Consequently, seed plant fertilization is not typically limited by lack of water, in contrast to fertilization of seedless plants. Consequently, seed plants are better able to reproduce in arid and seasonally dry habitats.

For these reasons, seeds are considered to be a key adaptation to reproduction in a land habitat.

## Ovule and Seed Evolution Illustrates Descent with Modification

As we have seen, seed plants reproduce using both spores and seeds, but seed plants have not replaced spores with seeds. Seed plants still produce spores. Ovules and seeds have evolved from spore-producing structures by descent with modification. Recall that this evolutionary principle involves changes in pre-existing structures and processes. Fossils provide some clues about ovule and seed evolution, and other information can be obtained by comparing reproduction in living lycophytes and pteridophytes.

Most modern lycophytes and pteridophytes release one type of spore that develops into one type of gametophyte. Such plants are considered to be homosporous, and their gametophytes live independently and produce both male and female gametangia (see Figure 31.18). However, some lycophytes and pteridophytes produce and release two distinct kinds of spores: relatively small microspores and larger megaspores, which grow into male and female gametophytes, respectively. As mentioned previously, production of two kinds of spores is called heterospory. As shown in steps 2a and 2b of **Figure 31.23**, an early step in the evolution of seed plants may have been a switch from homospory to heterospory.

What are advantages of heterospory? One advantage is that it mandates cross-fertilization. The eggs and sperm that fuse are derived from different gametophytes, which are likely to be genetically different. Cross-fertilization increases the potential for genetic variation. As described in Chapter 23, such variation may enhance the survival and reproduction of individuals with favorable phenotypes and result in evolution from one

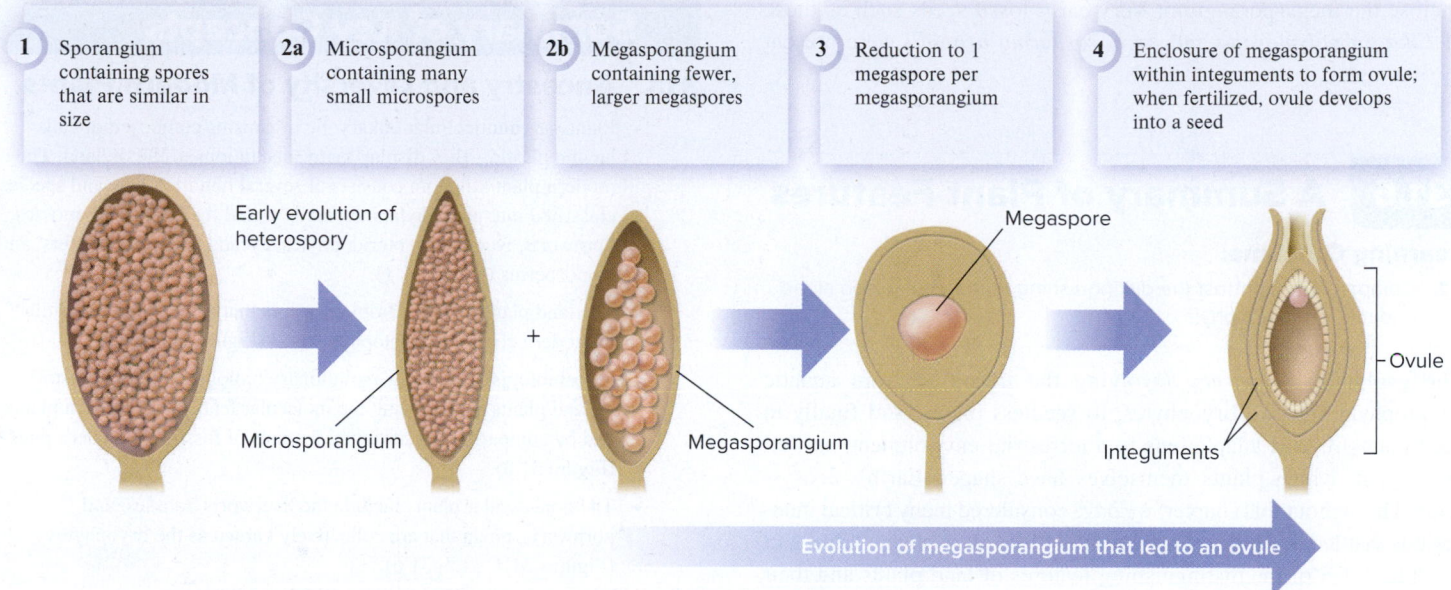

| 1 Sporangium containing spores that are similar in size | 2a Microsporangium containing many small microspores | 2b Megasporangium containing fewer, larger megaspores | 3 Reduction to 1 megaspore per megasporangium | 4 Enclosure of megasporangium within integuments to form ovule; when fertilized, ovule develops into a seed |

Early evolution of heterospory

Megaspore

+

Microsporangium    Megasporangium    Integuments    Ovule

Evolution of megasporangium that led to an ovule

**Figure 31.23** **Hypothetical stages in the evolution of seeds.** The parallel evolution of heterospory and endosporic gametophytes in some lycophytes and pteridophytes as well as in the seed plants suggests that these features were acquired early in the evolution of seeds. Later-occurring events in the evolution of seeds included reduction of the number of megaspores to one per megasporangium and enclosure of the megasporangium by protective integuments.

 **Core Skill: Connections** Look ahead to Figure 35.14, which illustrates the amniotic egg produced by many terrestrial animals. How is the plant seed like the amniotic egg?

generation to the next. A second advantage is that the gametophytes produced by heterosporous plants grow within the confines of microspore and megaspore walls and therefore are known as **endosporic gametophytes**. Endosporic gametophytes receive protection from environmental damage from the surrounding spore walls. Plant evolutionary biologists infer that heterospory and endosporic gametophytes were features of seed plant ancestors and constitute early steps toward seed evolution.

Whereas seedless plants produce multiple spores per sporangium, another key step in seed evolution may have been the production of only one megaspore per sporangium (see step 3 in Figure 31.23). Having a single megaspore allowed plants to channel more nutrients into each megaspore, thereby enabling megaspores to store more food. Following fertilization, this increased food confers an advantage by providing greater nutritive support to developing sporophytes.

A final step in seed evolution might have been the retention of megasporangia on parental sporophytes by the development of integuments (see step 4 in Figure 31.23). As mentioned earlier, this adaptation would allow food materials to flow from mature photosynthetic sporophytes to their dependent gametophytes and young embryos. Integuments also help ovules to receive pollen.

Fossils provide information about when and how the process of ovule and seed evolution first occurred. Fossil reproductive structures of an extinct Devonian plant named *Runcaria heinzelinii* may represent a precursor to an ovule or seed (see **Figure 31.24**). These fossil structures had a lacy integument that did not completely

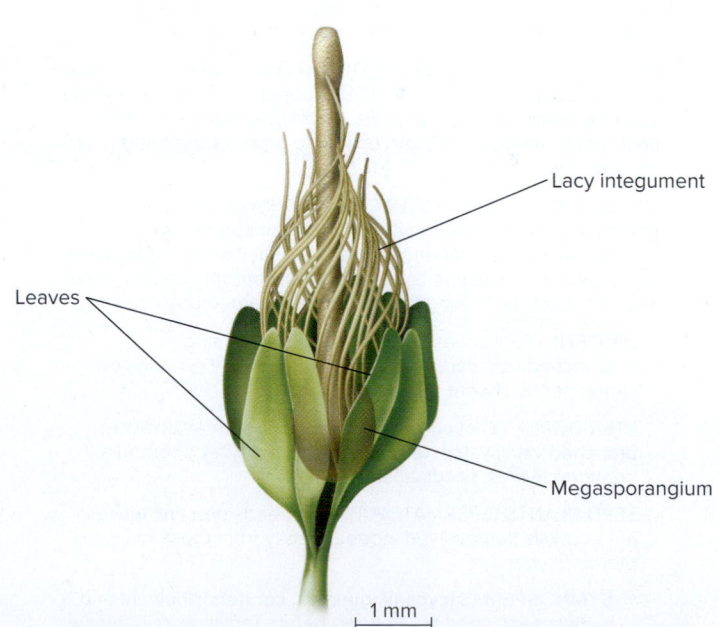

Lacy integument

Leaves

Megasporangium

1 mm

**Figure 31.24** **Reconstruction of reproductive parts of the fossil** *Runcaria heinzelinii*, **a plant with a probable precursor to an ovule or seed.**

**Concept Check:** *Based on your knowledge of integument function in modern seed plants, can you hypothesize a function for the lacy integument of R. heinzelinii?*

enclose the megasporangium. Very early fossil seeds such as those of *Elkinsia polymorpha* and *Archaeosperma arnoldii* were present by 365 mya.

## 31.6 A Summary of Plant Features

### Learning Outcome:

**1.** Compare and contrast the distinguishing features of green algae and modern plant phyla.

The evolutionary journey involving the transition from aquatic streptophyte algae to bryophytes, to seedless plants, and finally to seed plants reveals adaptations to a terrestrial environment, as well as ways in which plants themselves have shaped Earth's ecosystems. Throughout this chapter, we have considered many critical innovations that led to the development of modern plant phyla. **Table 31.1** provides a list of the distinguishing features of land plants and their algal relatives.

| Table 31.1 | Distinguishing Features of Modern Streptophyte Algae and Land Plants |
|---|---|

STREPTOPHYTES

Streptophyte algae: Primarily aquatic habitat; haploid-dominant life cycle; sporangia absent; sporophytes absent

EMBRYOPHYTES: Primarily terrestrial habitat; life cycle consisting of alternation of two multicellular generations (diploid sporophyte and haploid gametophyte); multicellular embryos nutritionally dependent on maternal gametophyte for at least some time during development; spore-producing sporangia; gamete-producing gametangia; spores with sporopollenin-containing walls

Nonvascular plants (bryophytes) (**liverworts**, **mosses**, **hornworts**): Dominant gametophyte generation; supportive, lignin-containing vascular tissue absent; true roots, stems, and leaves absent; sporophytes unbranched and unable to grow independently of gametophytes

VASCULAR PLANTS (TRACHEOPHYTES) (lycophytes, pteridophytes, spermatophytes): Dominant sporophyte generation; lignified water-conducting tissue (xylem); specialized organic food-conducting tissue (phloem); sporophytes branched and eventually becoming independent of gametophytes

LYCOPHYTES: Leaves generally small with a single, unbranched vein (lycophylls); sporangia borne on sides of stems; seeds absent

PTERIDOPHYTES: Leaves relatively large with extensively branched vein system (euphylls or megaphylls); sporangia borne on leaves; seeds absent

SEED PLANTS (SPERMATOPHYTES): Seeds present; leaves are euphylls that evolved independently from those of pteridophytes

GYMNOSPERMS (**cycads**, **ginkgos**, **conifers**): Flowers and fruits absent; seed food stored before fertilization in female gametophyte; endosperm absent

ANGIOSPERMS (flowering plants): Flowers and fruit present; seed food stored after fertilization in endosperm tissue

Key: **Phyla**; LARGER MONOPHYLETIC CLADES (FORMAL SYNONYMS). All other classification terms are not clades.

## Summary of Key Concepts

### 31.1 Ancestry and Diversity of Modern Plants

- Plants are multicellular eukaryotic organisms composed of cells having plastids; they display many adaptations to life on land. The modern plant kingdom consists of several hundred thousand species classified into nine phyla, informally called the liverworts, mosses, hornworts, lycophytes, pteridophytes, cycads, ginkgos, conifers, and angiosperms (Figure 31.1).

- The land plants evolved from ancestors that were probably similar to modern complex streptophyte algae (Figure 31.2).

- Paleobiologists and plant evolutionary biologists infer the history of land plants by analyzing the molecular features of modern plants and by comparing the structural features of fossil and modern plants (Figure 31.3).

- The nonvascular plants include the liverworts, mosses, and hornworts, phyla that are collectively known as the bryophytes (Figures 31.4, 31.5, 31.6).

- Lycophytes, pteridophytes, and other vascular plants generally possess stems, roots, and leaves having conductive vascular tissues composed of phloem and xylem, in addition to cuticle, and stomata. These features promote stable body water content (Figures 31.7, 31.8, 31.9).

- Cycads, ginkgos, conifers, and gnetophytes are collectively known as gymnosperms. Gymnosperms produce seeds, but not flowers and fruits. Angiosperms, the flowering plants, produce seeds, flowers, fruits, and seed endosperm (Figures 31.10, 31.11).

### 31.2 How Land Plants Have Changed the Earth

- Ancient seedless plants and later-emerging vascular plants transformed Earth's ecology by altering atmospheric chemistry and climate (Figures 31.12, 31.13, 31.14).

- The Cretaceous-Paleogene event, a probable meteorite collision with Earth that occurred 65 mya, helped cause the extinction of previously dominant dinosaurs and many types of gymnosperms, leaving space into which angiosperms, insects, birds, and mammals diversified.

### 31.3 Evolution of Reproductive Features in Land Plants

- Bryophytes illustrate early-evolved features of land plants, which include a life cycle featuring alternation of generations, involving embryos that develop within protective, nourishing gametophyte tissues (Figure 31.15).

- Bryophytes differ from other plants in having a dominant gametophyte generation and a dependent, nonbranching, short-lived sporophyte generation (Figure 31.16).

- The evolution of vascular plants involved a shift in the relative sizes of gametophytes and sporophytes, with the sporophyte becoming the dominant generation (Figure 31.17).

- The fern life cycle includes the dominant sporophyte characteristic of vascular plants (Figure 31.18).

### 31.4 Evolutionary Importance of the Plant Embryo

- The origin of the plant embryo was a critical innovation that fostered diversification of the land plants. Like placental mammal mothers, plant female gametophytes provide embryos with nutrients through specialized placental tissues (Figure 31.19).

- In a classic experiment, Browning and Gunning demonstrated that placental transfer tissues were responsible for an enhanced rate of nutrient flow from plant gametophytes to embryos (Figure 31.20).

## 31.5 The Origin and Evolutionary Importance of Leaves and Seeds

- Leaves are specialized photosynthetic organs that evolved more than once during plant evolutionary history. The lycophylls of lycophytes are relatively small leaves having a single unbranched vein. The larger leaves of ferns and seed plants, known as euphylls, have extensively branched vascular systems. Fossils indicate that fern euphylls evolved from branched-stem systems (Figure 31.21).

- Seeds develop from ovules, integument-enclosed sporangia that typically contain only a single spore that develops into an egg-producing gametophyte. Pollen produces thin cellular tubes that deliver sperm to eggs produced by female gametophytes. Following pollination and fertilization, ovules develop into seeds. Mature seeds contain stored food and an embryonic sporophyte that develops from the zygote (Figure 31.22).

- Seeds confer many reproductive advantages, including dormancy through unfavorable conditions, greater protection for embryos from mechanical and pathogen damage, seed coat modifications that enhance seed dispersal, and reduction of plant dependence on water for fertilization (Figures 31.23, 31.24).

## 31.6 A Summary of Plant Features

- The distinctive traits of modern streptophyte algae and the different phyla of land plants indicate the occurrence of descent with modification (Table 31.1).

## Assess & Discuss

### Test Yourself

1. The simplest and most ancient phylum of modern land plants is probably
   a. the pteridophytes.
   b. the cycads.
   c. the liverworts, mosses, or hornworts.
   d. the angiosperms.
   e. none of the above.

2. An important feature of land plants that originated during the diversification of streptophyte algae is
   a. the sporophyte.
   b. spores, which are dispersed in air and coated with sporopollenin.
   c. tracheids.
   d. plasmodesmata.
   e. fruits.

3. A seedless plant phylum that is included in the informal group known as bryophytes is
   a. liverworts.
   b. hornworts.
   c. mosses.
   d. All of the above phyla are included in the bryophytes.
   e. None of the above is included in the bryophytes.

4. Plants possess a life cycle that involves alternation of two multicellular generations: the gametophyte and
   a. the lycophyte.          d. the lignophyte.
   b. the bryophyte.          e. the sporophyte.
   c. the pteridophyte.

5. The seed plants are also known as
   a. bryophytes.
   b. spermatophytes.
   c. pteridophytes.
   d. lycophytes.
   e. none of the above.

6. A waxy cuticle is an adaptation that
   a. helps to prevent water loss from tracheophyte stem and leaf surfaces.
   b. helps to prevent water loss from streptophyte algae.
   c. helps to prevent water loss from spores.
   d. aids in water transport within the bodies of vascular plants.
   e. does all of the above.

7. Plant photosynthesis transformed a very large amount of atmospheric carbon dioxide into decay-resistant organic compounds, thereby contributing to a dramatic decrease in atmospheric carbon dioxide levels during the geological period known as the
   a. Cambrian.          d. Permian.
   b. Ordovician.        e. Pleistocene.
   c. Carboniferous.

8. Which of these plant phyla is likely to have the largest leaves?
   a. liverworts
   b. hornworts
   c. mosses
   d. lycophytes
   e. pteridophytes

9. Fern euphylls, also known as megaphylls, probably evolved from
   a. the leaves of mosses.
   b. lycophylls.
   c. branched-stem systems.
   d. modified roots.
   e. none of the above.

10. A seed develops from
    a. a spore.
    b. a fertilized ovule.
    c. a microsporangium covered by integuments.
    d. endosperm.
    e. none of the above.

### Conceptual Questions

1. List several common traits that lead evolutionary biologists to infer that land plants evolved from ancestors related to modern streptophyte algae.

2. Why have bryophytes such as mosses been able to diversify into so many species even though they have relatively small, dependent sporophytes?

3. **Core Concept: Structure and Function** Explain how several structural features help vascular plants maintain stable internal water content.

### Collaborative Questions

1. Discuss at least one difference in the environmental conditions experienced by early land plants and ancestral streptophyte algae.

2. Identify and describe as many plant adaptations to land as you can.

# The Evolution and Diversity of Modern Gymnosperms and Angiosperms

# 32

The Madagascar periwinkle (*Catharanthus roseus*), one of the many seed plants on which humans depend. ©Gallo Images/Corbis/Getty Images

**T**he seed plants—gymnosperms and angiosperms—are particularly important in our everyday lives because they are the sources of many products, including wood, paper, beverages, food, cosmetics, and medicines. Leukemia, for example, is effectively treated with vincristine, a drug extracted from the beautiful flowering plant known as the Madagascar periwinkle (*Catharanthus roseus*), pictured in the chapter opening photo. Vinblastine—another extract from *C. roseus*—is used to treat lymphatic cancers. Taxol, a compound used in the treatment of breast and ovarian cancers, was first discovered in extracts of the bark of the Pacific yew tree, a gymnosperm called *Taxus brevifolia*. Vincristine, vinblastine, taxol, and many other plant-derived medicines are examples of plant secondary metabolites, which are distinct from the products of primary metabolism (carbohydrates, lipids, proteins, and nucleic acids). Secondary metabolites play essential roles in protecting plants from disease-causing organisms and plant-eating animals, and they also aid plant growth and reproduction. Though all plants produce secondary metabolites, these natural products are exceptionally diverse in gymnosperms and angiosperms.

In this chapter, we will explore the many important roles that the hundreds of thousands of modern seed plants play in the lives of humans and modern ecosystems. This chapter builds on the introduction to seeds and seed plants provided in Chapter 31. We begin by focusing on the diversity of modern lineages of gymnosperms and angiosperms. Coevolutionary interactions among angiosperms and animals are presented as major factors influencing the diversification of these groups. This chapter concludes by considering human influences on seed plant evolution and the importance of seed plants in modern agriculture.

## 32.1 Overview of Seed Plant Diversity

**Learning Outcomes:**

1. Describe the evolutionary relationships among seedless vascular plants and seed plants.
2. List the critical innovations that occurred during the evolution of seed plants.

The seed plants—gymnosperms and angiosperms—evolved from the seedless vascular plants, which were described in Chapter 31. Fossils indicate that gymnosperms originated from now extinct seedless plants known as progymnosperms, some of which were woody, representing the first trees. Gymnosperms then diversified into multiple lineages, some of which became extinct. However, a few gymnosperm phyla, including the conifers, have persisted to the modern day. Angiosperms arose from an unknown gymnosperm lineage, thereby inheriting the capacity to produce wood and other seed plant features. **Figure 32.1** shows our current understanding of the evolutionary relationships among seedless vascular plants and modern seed plants, which include three modern phyla of gymnosperms and the flowering plants (angiosperms). **Table 32.1** provides a summary of the critical innovations of all modern seed plants, conifers, and angiosperms.

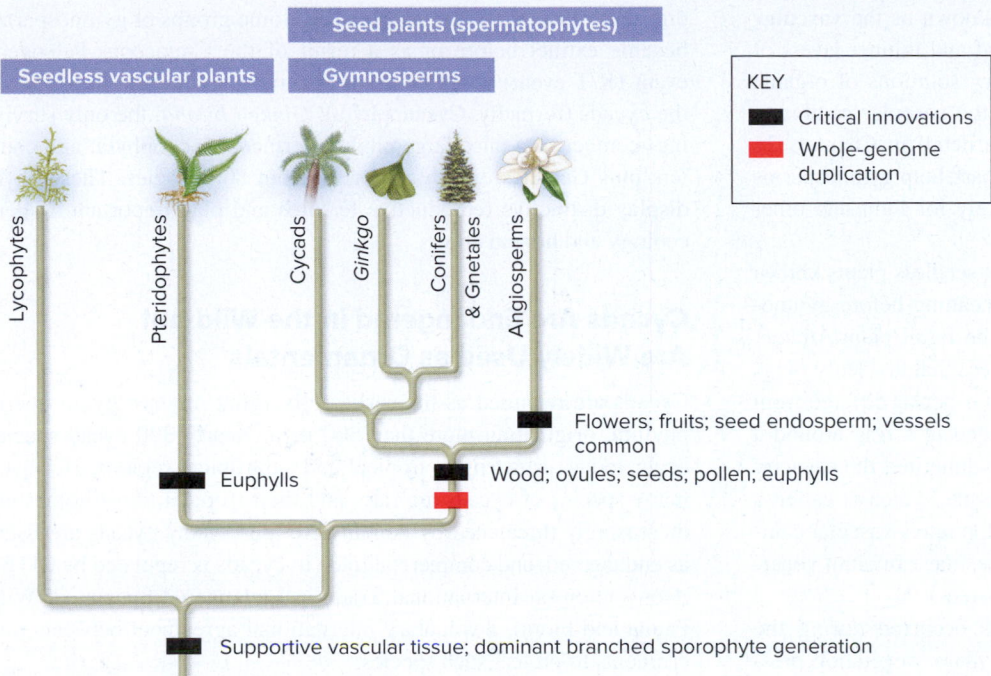

**Figure 32.1** A phylogenetic tree for modern seedless vascular plants and seed plants.

| Table 32.1 | Critical Innovations of Some Seed Plant Groups | |
|---|---|---|
| **Plant group** | **Innovation** | **Advantages** |
| All seed plants | Vascular cambium that makes vascular tissue and also makes wood and inner bark in woody plants | Seed plants have the potential to grow tall and produce many branches and reproductive structures. |
| | Pollen, ovules, seeds | Pollen allows seed plants to disperse male gametophytes. Ovules provide protection and nutrition to female gametophytes and developing embryos. Seeds allow seed plants to reproduce in arid or shady habitats. |
| Conifers | Tracheid torus | Fosters water flow in arid or cold conditions |
| | Scales or needle-shaped leaves | Retard water loss from leaf surfaces |
| | Conical shape | Sheds snow, preventing damage |
| | Resin | Protects against pathogens and herbivores |
| Angiosperms | Flowers | Foster pollen dispersal, ovule protection, pollination, and seed production |
| | Fruits | Foster seed dispersal |
| | Endosperm | Efficiently provides food to embryo of developing seed |
| | Vessels | Relatively wide diameter fosters water flow |
| | Many secondary compounds | Provide flower colors and fragrances and protect against herbivores |

## 32.2 The Evolution and Diversity of Modern Gymnosperms

**Learning Outcomes:**

1. Describe how gymnosperms diversified.
2. Identify three gymnosperm phyla, and describe their importance to humans.

**Gymnosperms** are defined as plants that produce seeds that are exposed rather than enclosed in fruits, as is the case for angiosperms. The word gymnosperm comes from the Greek *gymnos*, meaning naked (referring to the unclothed state of ancient athletes), and *sperma*, meaning seed. Most modern gymnosperms are woody plants that occur as shrubs or trees. Conifers, which are widely harvested for wood and produce other valuable materials, are familiar examples. In this section, we will examine the evolution and key features of the gymnosperms.

### Modern Gymnosperms Arose from Woody Ancestors

Modern gymnosperms include the famous giant sequoias (*Sequoiadendron giganteum*) native to the Sierra Nevada mountains of the western U.S. Giant sequoias are among Earth's largest organisms, weighing as much as 6,000 tons and reaching an amazing 100 m in height. The large size of sequoias and other trees is based on the presence of **wood**, a tissue composed of numerous pipelike arrays of empty, water-conducting cells whose walls are strengthened by an exceptionally tough polymer known as lignin. These properties enable woody tissues to transport water upward for great distances and also to provide the structural support needed for trees to grow tall and produce many branches and leaves.

In modern seed plants, a special tissue known as the **vascular cambium** produces both thick layers of wood and thinner layers of inner bark. The **inner bark** transports watery solutions of organic compounds. (The structure and function of the vascular cambium, wood, and inner bark are described in more detail in Chapters 36 and 39.) Vascular cambium, wood, and inner bark help gymnosperms and woody angiosperms to compete effectively for light and other resources needed for photosynthesis.

Wood first appeared in a group of ancient seedless plants known as the **progymnosperms** (from the Greek, meaning before gymnosperms). Woody progymnosperms, such as the fossil plant *Archaeopteris*, which lived 370 mya, were the first trees that had leafy twigs (**Figure 32.2**). The vascular tissue of progymnosperms differed from that of earlier vascular plants in being arranged in a ring around a central pith of nonvascular tissue. Seed plants inherited the capacity to make this new arrangement of vascular tissue, which is called a **eustele**. A eustele contains cells that can develop into a vascular cambium as seedlings grow into saplings. The vascular cambium generates wood, allowing saplings to grow into tall trees.

The greatest diversity of gymnosperms occurred during the Mesozoic era, when gymnosperms were the major vegetation present. This period was also known as the Age of Dinosaurs, and gymnosperms are thought to have been the major food for plant-eating dinosaurs during most of their history. Some groups of gymnosperms became extinct before or as a result of the Cretaceous-Paleogene event (K/T event) about 66 mya. Surviving gymnosperm phyla are the cycads (formally, Cycadophyta); *Ginkgo biloba*, the only surviving member of a once large phylum termed Ginkgophyta; and conifers plus Gnetales, which comprise about 800 species. These phyla display distinctive reproductive features and play important roles in ecology and human affairs.

## Cycads Are Endangered in the Wild but Are Widely Used as Ornamentals

Cycads are regarded as the earliest diverging modern gymnosperm phylum, originating more than 300 mya. Nearly 300 cycad species occur today, primarily in tropical and subtropical regions. However, many species of cycads are rare, and their tropical forest homes are increasingly threatened by human activities. Many cycads are listed as endangered, and commercial trade in cycads is regulated by CITES (Convention on International Trade in Endangered Species of Wild Fauna and Flora), a voluntary international agreement between governments to protect such species.

The structure of cycads is so interesting and attractive that many species are cultivated for use in outdoor plantings or as houseplants. The nonwoody stems of some cycads emerge from the ground much like tree trunks, some reaching 15 m in height, whereas the stems of other cycads are not conspicuous because they are subterranean (**Figure 32.3**). Cycads display spreading, palmlike leaves (cycad comes from a Greek word meaning palm). Mature leaves of the African cycad *Encephalartos laurentianus* can reach an astounding 8.8 m in length!

**Figure 32.2 An early forest in which the only trees were the progymnosperm *Archaeopteris*.** This illustration was reconstructed from fossil data.

 **Core Skill: Connections** Look ahead to Figure 54.26a–e. In what way did ancient *Archaeopteris* forests differ from most forests of the present time?

**(a) Emergent cycad stem**

**(b) Submergent cycad stem**

**Figure 32.3 Cycads.** Palmlike foliage and conspicuous seed-producing cones are features of most cycads. **(a)** The stems of some cycads emerge from the ground. **(b)** The stems of other cycads are submerged in the ground, so the leaves emerge at ground level. This image also shows a conspicuous orange conelike structure that bears seeds. a: ©Philippe Psaila/Science Source; b: ©Ed Reschke/Getty Images

Root surface

Cyanobacteria

**(a) Coralloid roots**

**(b) Coralloid root cross section**

**Figure 32.4** **Coralloid roots of cycads.** **(a)** Many cycads produce aboveground branching roots that resemble branched corals. **(b)** This magnified cross section of a coralloid root shows a ring of symbiotic blue-green cyanobacteria, which provide the plant with a form of nitrogen that can be used to make essential cellular compounds.

©Lee W. Wilcox

**Concept Check:** *Why do the coralloid roots grow aboveground?*

In addition to underground roots, which provide anchorage and take up water and minerals, many cycads produce coralloid roots. Such roots extend aboveground and have branching shapes resembling corals (**Figure 32.4a**). Coralloid roots harbor light-dependent, photosynthetic cyanobacteria within their tissues. The cyanobacteria, which form a bright blue-green ring beneath the root surfaces (**Figure 32.4b**), convert atmospheric nitrogen ($N_2$) into ammonia

($NH_3$), providing their plant hosts with the nitrogen that is crucial to growth (see Chapter 38).

Cycad reproduction is distinctive in several ways. Individual cycad plants produce conspicuous conelike structures that bear either ovules and seeds or pollen (see Figure 32.3b). When mature, both types of reproductive structures emit odors that attract beetles. These insects carry pollen to ovules, where the pollen produces tubes that deliver sperm to eggs.

## *Ginkgo biloba* Is the Last Survivor of a Once Diverse Group

The beautiful tree *Ginkgo biloba* (**Figure 32.5a**) is the single remaining species of a phylum that was much more diverse during the Age of Dinosaurs. *G. biloba* takes its species name from the two-lobed shape of its leaves, which have unusual forked veins (**Figure 32.5b**). Widely cultivated modern *Ginkgo* trees are descended from seeds produced by a tree found in a remote Japanese temple garden and brought to Europe by 17th-century explorers.

*G. biloba* trees are widely planted along city streets because they are ornamental and also tolerate cold, heat, and pollution better than many other trees. In addition, these trees are long-lived—individuals can live for more than a thousand years and grow to 30 m in height. Individual trees produce either ovules and seeds or pollen, based on a sex chromosome system much like that of humans. Ovule-producing trees have two X chromosomes; pollen-producing trees have one X and one Y chromosome. Wind disperses pollen to ovules, where pollen grains germinate to produce pollen tubes. These tubes grow through ovule tissues for several months, absorbing nutrients that are used for sperm development. Eventually the pollen tubes burst, delivering flagellate sperm to egg cells. After fertilization, zygotes develop into embryos, and the ovule integument develops into a fleshy, foul-smelling outer seed coat and a hard, inner seed coat (**Figure 32.5c**). For streetside or garden plantings, people usually select the pollen-producing trees to avoid the stinky seeds.

**(a)** *Ginkgo biloba* tree

**b)** *Ginkgo biloba* leaf

**(c)** *Ginkgo biloba* seeds

**Figure 32.5** *Ginkgo biloba.* **(a)** A *Ginkgo biloba* tree; **(b)** fan-shaped leaves with forked veins; and **(c)** seeds with fleshy, foul-smelling seed coats (because of their fleshy, colorful appearance, mature *Ginkgo* seeds are often mistaken for fruits). a: ©Karlene V. Schwartz; b: ©Fancy Photography/Veer/Getty Images; c: ©Topic Photo Agency IN/age fotostock

## Conifers Are the Most Diverse Modern Gymnosperm Lineage

The conifers (**Figure 32.6**) are a lineage of trees named for their seed cones, of which pinecones are familiar examples. Modern conifer families include more than 50 genera. Conifers are particularly common in mountain and high-latitude forests and are important sources of wood and paper pulp.

**Conifer Reproduction**   Conifers produce simple pollen cones and more complex ovule-bearing cones (see step 1 of **Figure 32.7**). The ovule cones, also called seed cones, are composed of many short branch systems that bear ovules. Ovules contain female gametophytes, within which eggs develop (see step 3a of Figure 32.7). The pollen cones of conifers have many leaflike structures, each bearing a microsporangium in which meiosis occurs and pollen grains develop (see step 3b of Figure 32.7).

When conifer pollen is mature, it is released into the wind, which transports it to ovules. When released from pollen tubes, sperm fuse with eggs, generating zygotes that grow into the embryos within seeds (see steps 4–7 of Figure 32.7). Altogether, it takes nearly 2 years for pines (the genus *Pinus*) to complete the processes of male and female gamete development, fertilization, and seed development.

Conifer seeds may also display features that aid in dispersal. For example, the seeds of pines and some other conifers develop wings that aid in wind dispersal (**Figure 32.8a**). Other conifers, such as yew and juniper, produce seeds or cones with bright-colored, fleshy coatings that are attractive to birds, which help to disperse the seeds (**Figure 32.8b** and **c**).

**Conifer Tracheids**   The wood of conifers contains many specialized vascular cells known as tracheids that are adapted for efficient water and mineral conduction even in dry conditions. Like the tracheids of other vascular plants, those of conifers are devoid of cytoplasm and occur in long columns that function like plumbing pipelines (**Figure 32.9a**). Tracheid side and end walls possess many thin-walled, circular **pits** through which water moves both vertically and laterally from one tracheid to another. Conifer pits are unusual in having a porous outer region that lets water flow through and a nonporous, flexible central region called the **torus** (plural, tori) that functions like a valve (**Figure 32.9b**).

If conifer tracheids become dry, a common event in arid or cold habitats, they fill with air and are no longer able to conduct water. In this case, the torus presses against the pit opening, sealing it (**Figure 32.9c**). The torus valve thereby prevents air bubbles from spreading to the next tracheid. This conifer adaptation localizes air bubbles, preventing them from stopping water conduction in other tracheids. The presence of tori in their tracheids helps to explain why conifers have been so successful for hundreds of millions of years.

Conifer wood (and leaves) may also display conspicuous resin ducts, passageways for the flow of syrup-like resin that helps to prevent attack by pathogens and herbivores. Resin that exudes from tree surfaces may trap insects and other organisms, then harden in the air and fossilize, preserving the inclusions in amber.

**Adaptations to Cold Climates**   Many conifers occur in cold climates and thus display numerous adaptations to such environments. Their conical shapes and flexible branches help conifer trees shed snow, preventing heavy snow accumulations from breaking branches. People who use conifers in landscape plantings also value these traits.

Conifer leaf shape and structure are adapted to resist damage from drought that occurs in both summer and winter, when liquid water is scarce. Conifer leaves are often scalelike (**Figure 32.10a**) or needle-shaped (**Figure 32.10b**); these shapes reduce the area of leaf surface from which water can evaporate. In addition, a thick, waxy cuticle coats conifer leaf surfaces (**Figure 32.10c**), retarding water loss and attack by disease organisms.

Many conifers are evergreen; that is, their leaves live for more than 1 year before being shed and are not all shed during the same season. Retaining leaves through winter helps conifers start up

**Figure 32.6 Representative conifers. (a)** Many conifers, such as pines, are not deciduous, meaning that they do not lose all their leaves at the same time in the autumn. **(b)** Some conifers, such as the dawn redwood, are deciduous, meaning that they drop their leaves in the autumn.

a: ©Lee W. Wilcox; b: ©Bryan Pickering/Eye Ubiquitous/Corbis/Getty Images

**(a) Pine (*Pinus ponderosa*)**

**(b) Dawn redwood (*Metasequoia glyptostroboides*)**

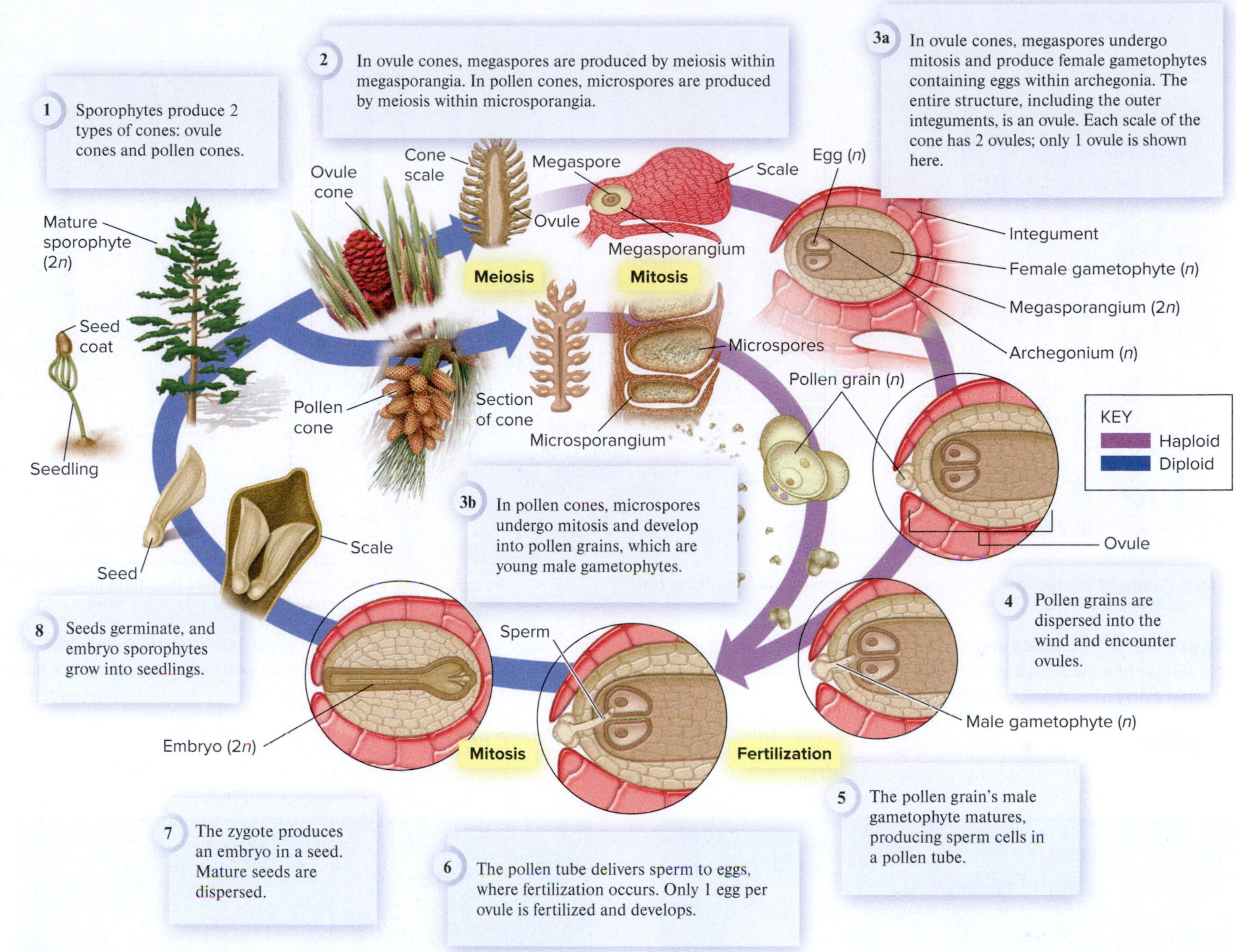

**1** Sporophytes produce 2 types of cones: ovule cones and pollen cones.

**2** In ovule cones, megaspores are produced by meiosis within megasporangia. In pollen cones, microspores are produced by meiosis within microsporangia.

**3a** In ovule cones, megaspores undergo mitosis and produce female gametophytes containing eggs within archegonia. The entire structure, including the outer integuments, is an ovule. Each scale of the cone has 2 ovules; only 1 ovule is shown here.

Mature sporophyte (2*n*)

Seed coat

Seedling

Ovule cone

Pollen cone

Cone scale

Megaspore

Ovule

Scale

Megasporangium

**Meiosis**

**Mitosis**

Section of cone

Microsporangium

Microspores

Egg (*n*)

Integument

Female gametophyte (*n*)

Megasporangium (2*n*)

Archegonium (*n*)

Pollen grain (*n*)

**KEY**
| | |
|---|---|
| | Haploid |
| | Diploid |

Ovule

Scale

Seed

**3b** In pollen cones, microspores undergo mitosis and develop into pollen grains, which are young male gametophytes.

**8** Seeds germinate, and embryo sporophytes grow into seedlings.

Sperm

**4** Pollen grains are dispersed into the wind and encounter ovules.

Male gametophyte (*n*)

Embryo (2*n*)

**Mitosis**

**Fertilization**

**7** The zygote produces an embryo in a seed. Mature seeds are dispersed.

**6** The pollen tube delivers sperm to eggs, where fertilization occurs. Only 1 egg per ovule is fertilized and develops.

**5** The pollen grain's male gametophyte matures, producing sperm cells in a pollen tube.

**Figure 32.7** **The life cycle of the genus *Pinus*.**

👁 **Core Concept: Structure and Function** This diagram illustrates the entire seed-to-seed growth and development cycle of conifers, illustrating structure-function relationships.

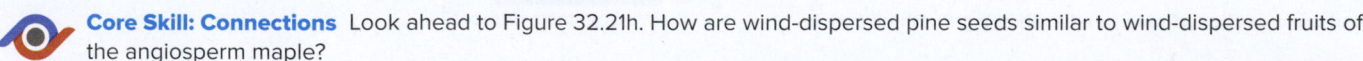

**(a) Pine seed**  **(b) Yew seeds**  **(c) Juniper cones with seeds**

**Figure 32.8** **Conifer seeds.** **(a)** Winged, wind-dispersed seed of the genus *Pinus*. **(b)** Fleshy-coated, bird-dispersed seeds of yew (*Taxus baccata*). **(c)** Fleshy cones of juniper (*Juniperus scopularum*) contain one or more seeds and are dispersed by birds. Juniper seeds are used to flavor gin. a: ©Zach Holmes Photography; b: ©Carmen Hauser/Shutterstock; c: ©Ed Reschke/Getty Images

👁 **Core Skill: Connections** Look ahead to Figure 32.21h. How are wind-dispersed pine seeds similar to wind-dispersed fruits of the angiosperm maple?

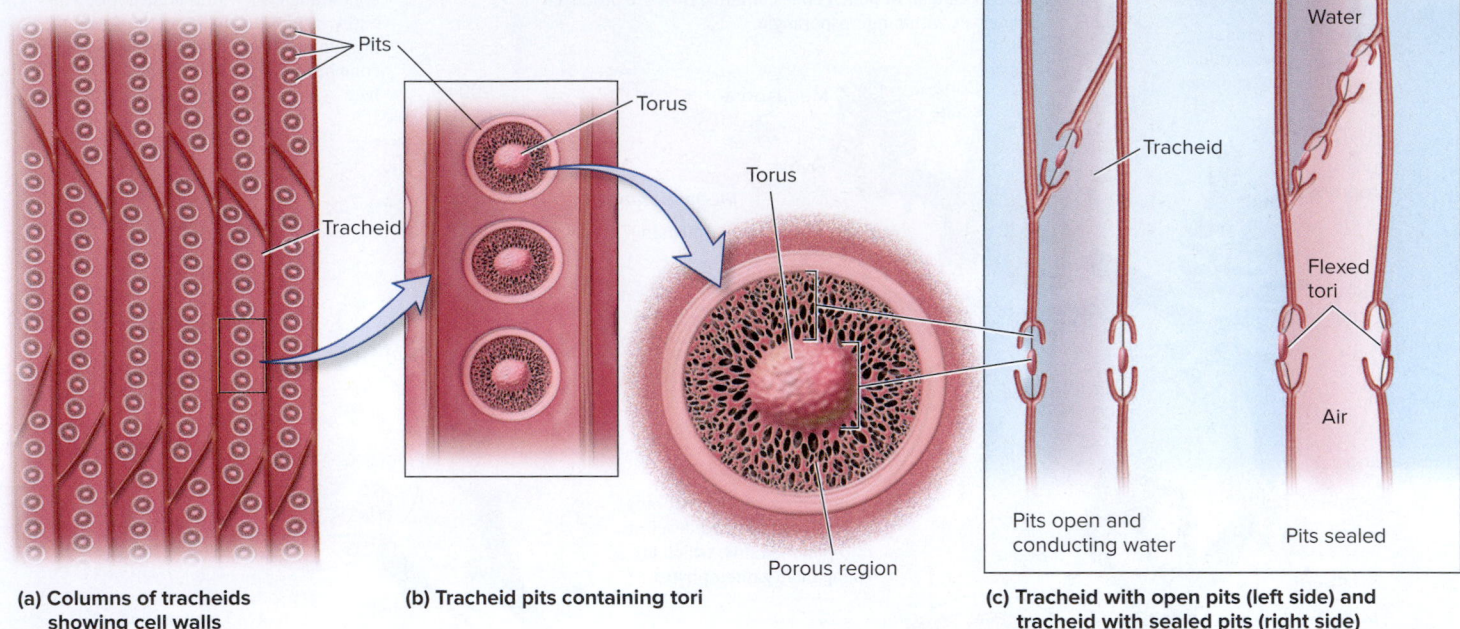

(a) Columns of tracheids showing cell walls

Pits

Tracheid

(b) Tracheid pits containing tori

Torus

Torus

Porous region

(c) Tracheid with open pits (left side) and tracheid with sealed pits (right side)

Water

Tracheid

Flexed tori

Air

Pits open and conducting water

Pits sealed

**Figure 32.9** **Tracheids and tori in conifer wood.** **(a)** The lignin-rich cell walls of the water-conducting cells called tracheids. **(b)** Detailed view of a portion of a tracheid that shows the thin-walled areas known as pits, each with a torus. **(c)** A water-filled tracheid with open pits and an air-filled tracheid with pits sealed by the flexed tori.

 **Core Concept: Structure and Function**  This illustration shows how tori in water-conducting cells of conifers aid survival in arid or cold habitats.

(a) Scale-shaped leaves of Eastern red cedar

(b) Needle-shaped leaves of pine

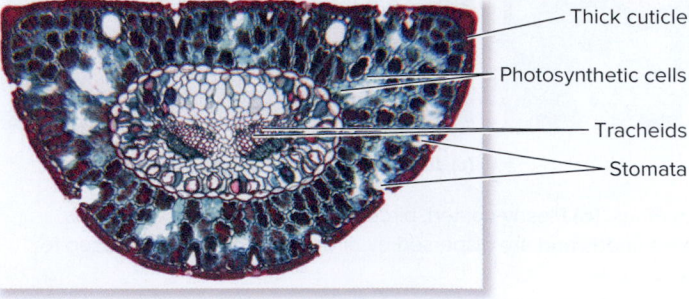

Thick cuticle

Photosynthetic cells

Tracheids

Stomata

(c) Stained cross section of pine needle, showing the thick cuticle

**Figure 32.10** **Conifer leaves.**  The leaves of conifers are typically shaped as small scales or long needles, with similar internal structure. a: ©Steven P. Lynch; b: ©Ken Wagner/Phototake; c: ©Lee W. Wilcox

**Concept Check:**  *In what ways are conifer leaves adapted to resist water loss from their surfaces?*

**(a) Genus Gnetum**

Reproductive structures

Broad leaf

Photosynthetic stem

Tiny scale-like leaves

Reproductive structures

**(b) Ephedra californica**

Reproductive structures

Leaves

**(c) Welwitschia mirabilis**

**Figure 32.11   Gnetales.  (a)** A tropical plant of the genus *Gnetum*, displaying broad leaves and reproductive structures. **(b)** *Ephedra californica* growing in deserts of North America, showing minuscule brown leaves on green, photosynthetic stems and reproductive structures. **(c)** *Welwitschia mirabilis* growing in the Namib Desert of southwestern Africa, showing long, wind-shredded leaves and reproductive structures. a: ©Robert & Linda Mitchell; b: ©2004 James M. Andre; c: ©Wildlife GmbH/Alamy Stock Photo

photosynthesis earlier than deciduous trees, which in spring must replace leaves lost during the previous autumn. Evergreen leaves thus provide an advantage in the short growth season of alpine or high-latitude environments. However, some conifers do lose all their leaves in the autumn. The bald cypress (*Taxodium distichum*) of southern U.S. floodplains, tamarack (*Larix laricina*) of northern bogs, and dawn redwood (*Metasequoia glyptostroboides*; see Figure 32.6b) are examples of deciduous conifers.

**Gnetales**   The conifer clade also includes the Gnetales, an order of three genera, *Gnetum*, *Ephedra*, and *Welwitschia*, that feature distinctive adaptations. *Gnetum* is unusual among modern gymnosperms in having broad leaves similar to those of many tropical plants (**Figure 32.11a**). Such leaves maximize light capture in dim forest habitats. More than 30 species of the genus *Gnetum* occur as vines, shrubs, or trees in tropical Africa or Asia. *Ephedra*, native to arid regions of the southwestern U.S., has tiny brown scalelike leaves and green, photosynthetic stems (**Figure 32.11b**). These adaptations help the plant conserve water by preventing water loss that would otherwise occur from the surfaces of larger leaves. *Ephedra* produces secondary metabolites that aid in plant protection but also affect human physiology. Early settlers of the western U.S. used *Ephedra* to treat colds and other medical conditions. The modern decongestant drug pseudoephedrine is based on the chemical structure of ephedrine, a substance that was named for and originally obtained from *Ephedra*.

*Welwitschia* has only one living representative species. *Welwitschia mirabilis* is a strange-looking plant that grows in the coastal

Namib Desert of southwestern Africa, one of the driest places on Earth (**Figure 32.11c**). A long taproot anchors a stubby stem that barely emerges from the ground. Two very long leaves grow from the stem but are rapidly shredded by the wind into many strips. The plant is thought to obtain most of its water from coastal fog that accumulates on the leaves, explaining how it can grow and reproduce in such a dry place.

## 32.3   The Evolution and Diversity of Modern Angiosperms

### Learning Outcomes:

1. List four flower organs and their functions, and explain how each flower part may have first evolved.
2. Describe how diversification of flowers and fruits enhances seed production and dispersal.
3. Name three major types of angiosperm secondary metabolites, and explain how these affect animals.

More than 124 mya, one extinct gymnosperm group, although it's unclear which one, gave rise to the angiosperms—the flowering plants. Charles Darwin famously referred to the origin of the flowering plants as "an abominable mystery," one that has not been fully solved even today. Recent geological studies indicate that the rise of angiosperms may be related to a global climate change that brought more humid conditions, arising from the breakup of the supercontinent Pangaea (see Chapter 26). Angiosperms have

**Figure 32.12 Angiosperm flowers and fruits.** Citrus plants display the critical innovations of flowering plants: the flowers and fruits shown here and seed endosperm (not shown). ©Bill Ross/Corbis/Getty Images

**Concept Check:** *What other trait occurs widely among angiosperms but rarely among other plants?*

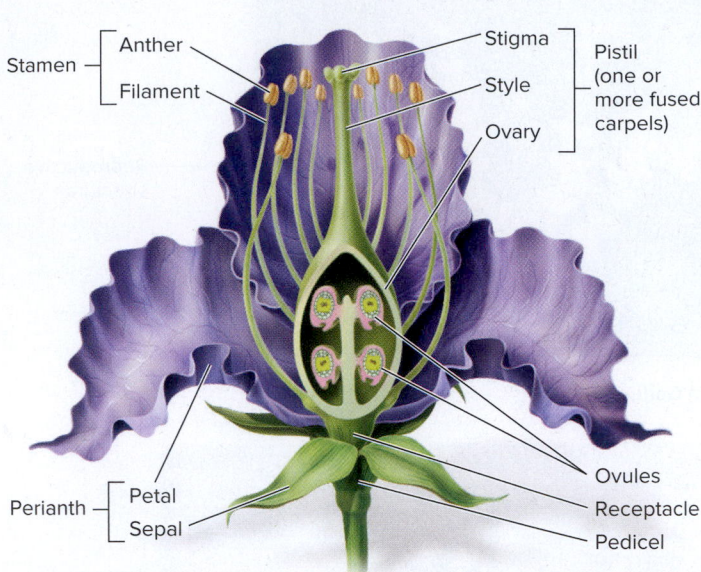

**Figure 32.13 Generalized flower structure.** Although flowers are diverse in size, shape, and color, many have the parts illustrated here.

retained many structural and reproductive features from ancestral seed plants, but have also evolved several distinctive traits.

Flowers and fruits are two of the defining features of angiosperms (**Figure 32.12**). These two features do not occur in other modern plants. The term **angiosperm** is from Greek words meaning enclosed seed, which reflects the presence of seeds within fruits. The seed nutritive material known as endosperm is another defining feature of the flowering plants (see Chapters 31 and 40). Flowers, fruits, and seed endosperm are critical innovations that aid reproduction. Flowers foster seed production, fruits favor seed dispersal, and endosperm food helps embryos within seeds grow into seedlings. In addition, most angiosperms possess distinctive water-conducting cells, known as **vessels**, which are wider than tracheids and therefore increase the efficiency of water flow through plants. Although similar conducting cells occur in some seedless plants and certain gymnosperms, the vessels of angiosperms are thought to have evolved independently.

Although humans obtain wood, medicines, and other valuable products from gymnosperms, we depend even more on the angiosperms. Our food, beverages, and spices—flavored by an amazing variety of secondary metabolites—primarily come from flowering plants. People surround themselves with ornamental flowering plants and decorative items displaying flowers or fruit. We also commonly use flowers and fruit in ceremonies. In this section, we will focus on how flowers, fruits, and secondary metabolites played key roles in angiosperm diversification. We will also examine how features of flowers, fruits, and secondary metabolites are used to classify and identify angiosperm species.

## Flower Organs Evolved from Leaflike Structures

**Flowers** are complex reproductive structures that are specialized for the efficient production of pollen and seeds. The sexual reproduction process of angiosperms depends on flowers. As the flowering plants diversified, flowers of varied types evolved as reproductive adaptations to differing environmental conditions. To understand this

process, we can start by considering the basic flower parts and their roles in reproduction.

***Flower Parts and Their Reproductive Roles*** Flowers are produced at stem tips, and may contain four types of organs: sepals, petals, pollen-producing stamens, and ovule-producing carpels (**Figure 32.13**). These flower organs are supported by tissue known as a **receptacle**, located at the tip of a flower stalk—a **pedicel**. The functioning of several genes that control flower organ development explains why carpels are the most central flower organs, why stamens surround carpels, and why petals and sepals are the outermost flower organs (refer back to Figure 20.24).

Many flowers produce attractive **petals** that play a role in **pollination**, the transfer of pollen among flowers. **Sepals** of many flowers are green and form the outer layer of flower buds. By contrast, the sepals of other flowers look similar to petals, in which case both sepals and petals are known as **tepals**. Sepals and petals of a flower are collectively known as the **perianth**. Most flowers have one or more **stamens**, the structures that produce and disperse pollen. Most flowers also contain a single or multiple **carpels**, structures that produce ovules.

Some flowers lack perianths, stamens, or carpels. Flowers that possess all four types of flower organs are known as **complete flowers**, and flowers lacking one or more organ types are known as **incomplete flowers**. Flowers that contain both stamens and carpels are described as **perfect flowers**, and flowers lacking either stamens or carpels are **imperfect flowers**.

Flowers also differ in the numbers of organs they produce. Some flowers produce only a single carpel, others display several separate carpels, and many possess several carpels that are fused together into a compound structure. Both single and compound carpels are referred to as a **pistil** (from the Latin *pistillum*, meaning pestle) because of a resemblance to the device people use to grind materials to powder in a mortar (see Figure 32.13). Only one pistil is present in flowers that have only one carpel and in flowers with fused carpels. By contrast, flowers possessing several separate carpels display multiple pistils.

Pistil structure can be divided into three regions having distinct functions. The topmost portion of the pistil, known as the **stigma**, receives and recognizes pollen of the appropriate species or genotype. The elongate middle portion of the pistil is called the **style**. The lowermost portion of the pistil is the **ovary**, which encloses and protects ovules.

During the flowering plant life cycle, the stigma allows pollen of the appropriate genetic type to germinate, producing a long pollen tube that grows through the style (see steps 1–4 of **Figure 32.14**). The pollen tube thereby delivers two sperm cells to ovules. In the distinctive angiosperm process known as **double fertilization**, one sperm fuses with the egg to form a zygote, and the other sperm fuses with other haploid cells of the

female gametophyte (see step 5 of Figure 32.14). The latter is the first step in the development of a characteristic angiosperm nutritive tissue known as endosperm. Fed by the endosperm, the zygote develops into an embryo, and the ovule develops into a seed (see steps 6 and 7 of Figure 32.14). Ovaries (and sometimes additional flower parts) develop into fruits.

**Early Flowers**  Fossils of whole plants with recognizable flowers and fruits have been identified from geological deposits that are about 124 million years old, though molecular data and fossil pollen grains suggest that angiosperms may have originated earlier. Flowers were a critical innovation that led to extensive angiosperm diversification. Comparative studies of the structures of modern and fossil flowers suggest how modern stamens and carpels might have arisen.

Structural comparisons and molecular data indicate that stamens are homologous to gymnosperm microsporophylls, leaflike structures

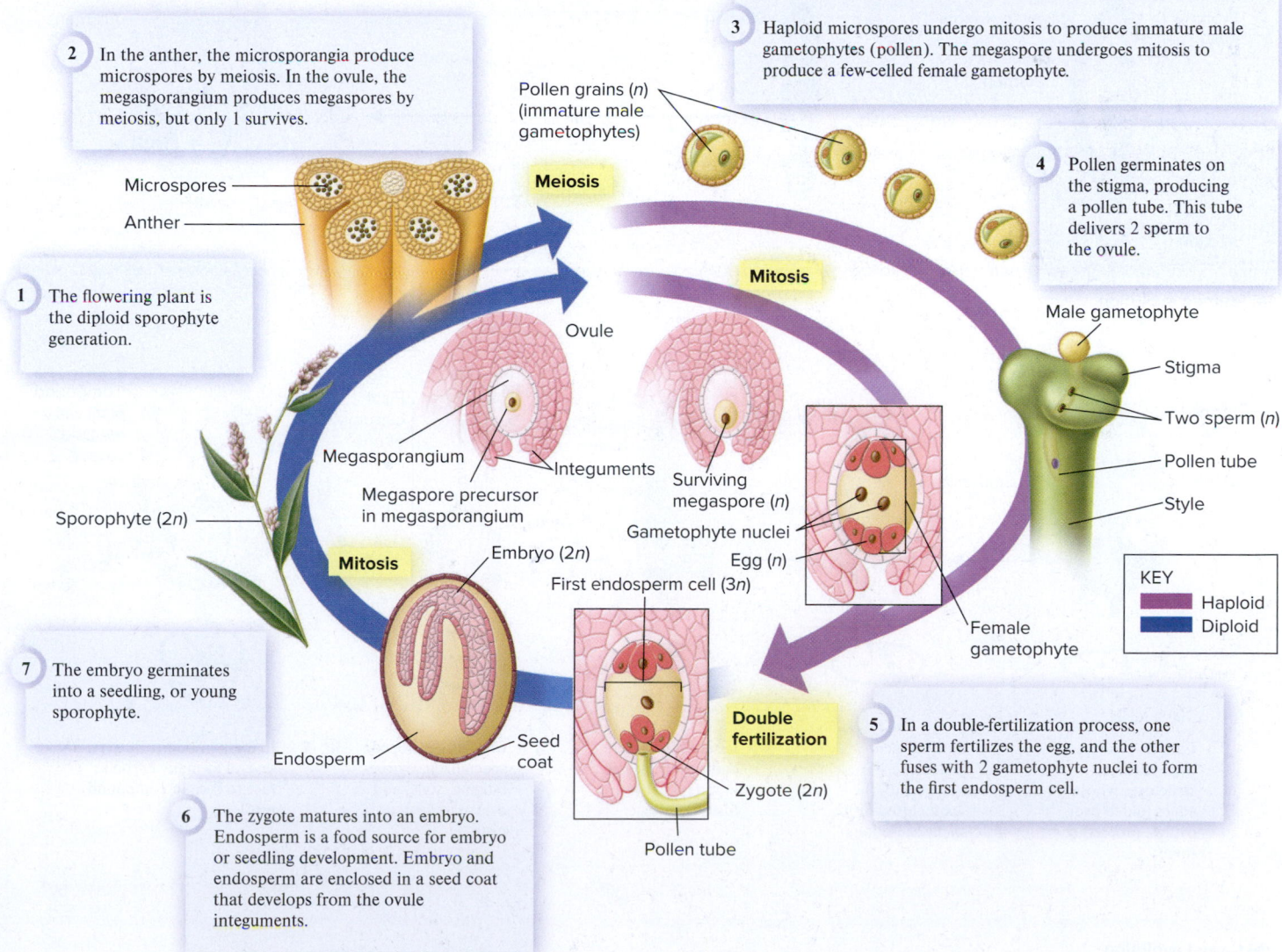

**Figure 32.14**  **The life cycle of a flowering plant, illustrated by the genus *Polygonum*.**  Flowering plant life cycles differ in length and in the number of cells and nuclei occurring in the female gametophyte, with the seven cells and eight nuclei of *Polygonum* being common.

that produce microspores (young pollen). Early fossil flowers and some modern flowers have broad stamens that are leaf-shaped, with elongated, pollen-producing microsporangia on the stamen surface (**Figure 32.15a**). During angiosperm evolution, the stamens of most modern plants have narrowed to form **filaments**, or stalks, that elevate **anthers**, clusters of microsporangia that produce pollen and then open to release it (see steps 2 and 3 of Figure 32.15a). Filaments and anthers are adaptations that foster pollen dispersal.

Plant biologists likewise hypothesize that carpels are homologous to gymnosperm megasporophylls, leaflike structures that bear ovules on their surfaces. In early angiosperms, such leaves folded over ovules, protecting them (see step 1 of **Figure 32.15b**). In support

of this hypothesis is the observation that the carpels of some early-diverging modern plants are leaflike structures that fold over ovules, with the carpel edges stuck together by secretions (see step 2 of Figure 32.15b). During evolution, this folding resulted in carpels that developed specialized regions and completely enclosed ovules (see steps 3 and 4 of Figure 32.15b). Most modern flowers produce carpels whose edges have fused together into a tube whose lower portion (ovary) encloses ovules. Plant biologists hypothesize that such evolutionary change increased ovule protection, which would improve plant fitness.

In contrast, flower sepals and petals have no recognizable homologs in modern gymnosperms. These perianth structures are unique

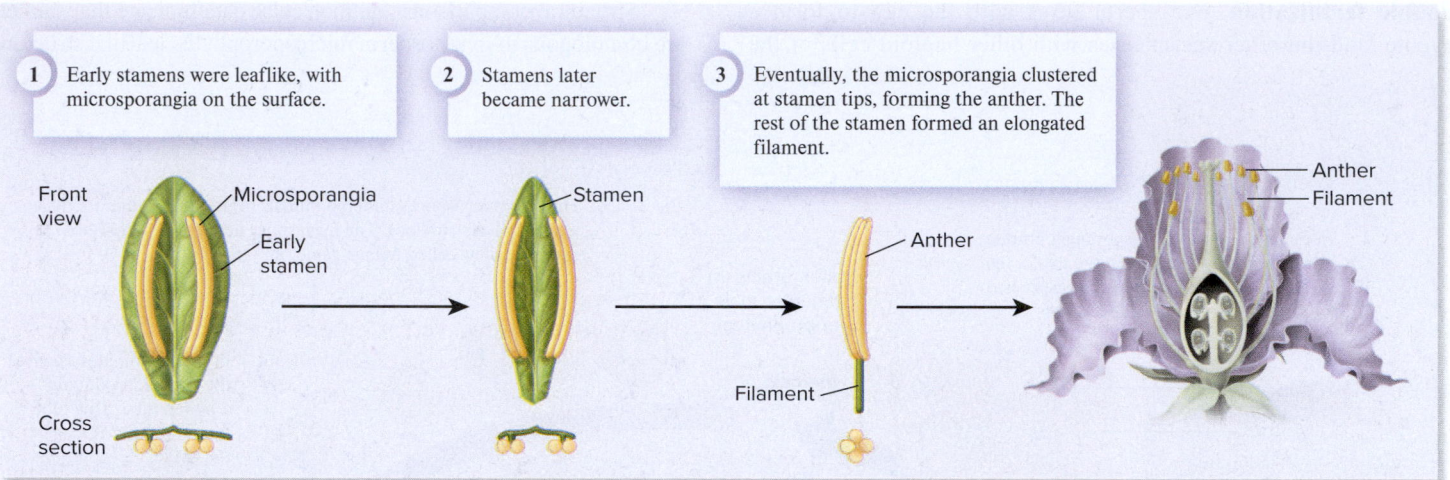

**(a) Stamen evolution**

1. Early stamens were leaflike, with microsporangia on the surface.
2. Stamens later became narrower.
3. Eventually, the microsporangia clustered at stamen tips, forming the anther. The rest of the stamen formed an elongated filament.

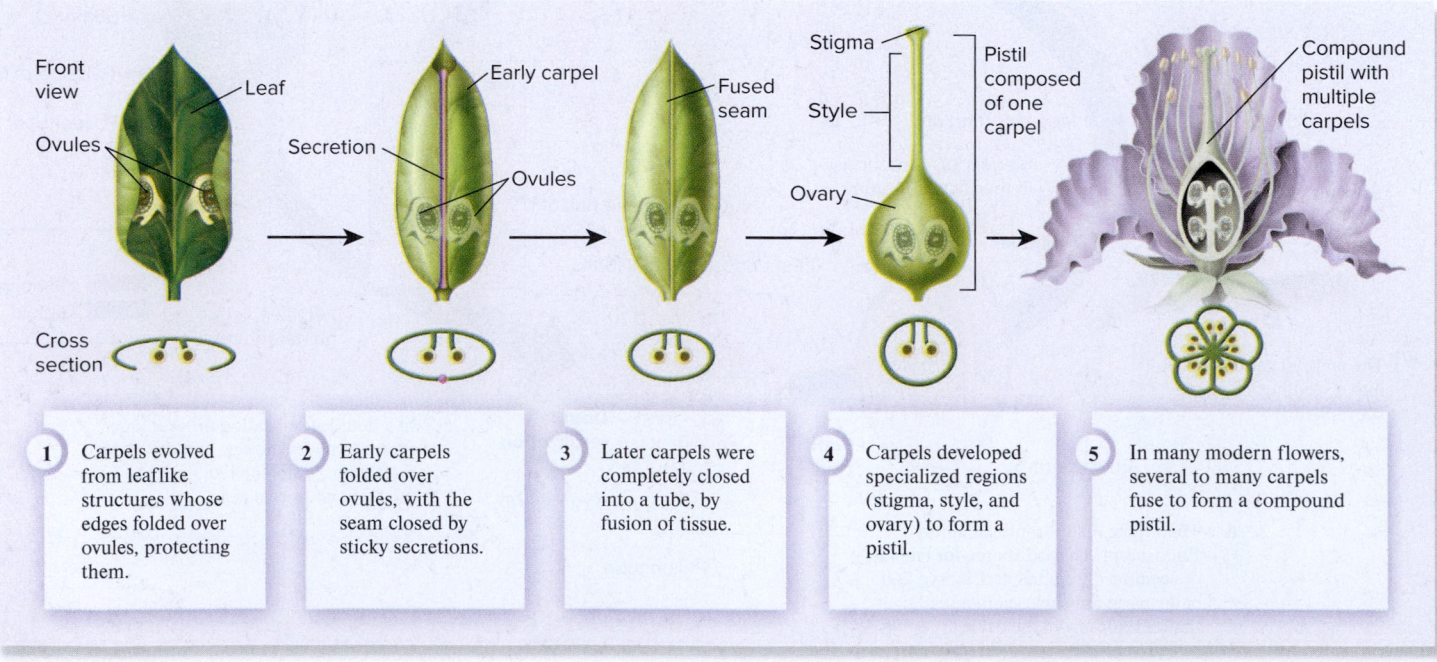

**(b) Carpel evolution**

1. Carpels evolved from leaflike structures whose edges folded over ovules, protecting them.
2. Early carpels folded over ovules, with the seam closed by sticky secretions.
3. Later carpels were completely closed into a tube, by fusion of tissue.
4. Carpels developed specialized regions (stigma, style, and ovary) to form a pistil.
5. In many modern flowers, several to many carpels fuse to form a compound pistil.

**Figure 32.15 Hypothetical evolutionary origin of stamens, carpels, and pistils.** Plant biologists test these models by searching for new fossils or generating additional molecular data.

to angiosperms, so plant biologists have long wondered how sepals and petals arose. Recent analyses indicate that the gene expression patterns of the pollen cones of gymnosperms (see Figure 32.7) share features with flower stamens, as expected, but also with the flower perianth. These data suggest that perianth parts originated from stamen-like structures, by loss of sporangia. The first flowers arose when early stamens, carpels, and perianth parts became aggregated into a single structure.

## Flowering Plants Diversified into Several Lineages, Including Monocots and Eudicots

**Figure 32.16** presents our current understanding of the relationships among modern angiosperm groups. According to gene-sequencing studies, the earliest-diverging modern angiosperms are represented by a single species called *Amborella trichopoda*, a shrub that lives in cloud forests on the South Pacific island of New Caledonia. The flowers of *A. trichopoda* display hypothesized ancient features. For example, the fairly small flowers have stamens with broad filaments and several separate carpels (**Figure 32.17**). *A. trichopoda* also lacks vessels in the water-conducting tissues, but typical angiosperm vessels are present in later-diverging groups of

**Figure 32.17** *Amborella trichopoda* **flower, similar to a hypothesized early flower.** This small flower is only about 3–4 mm in diameter. It displays several central, greenish carpels; nonfunctional stamens; and a pink perianth of tepals. This plant species also produces flowers that lack carpels but have many functional stamens. ©Sangtae Kim, Ph.D.

angiosperms, including water lilies, the star anise plant, and other close relatives (see Figure 32.16). Magnoliids, represented by the genus *Magnolia*, are the next-diverging group. Magnoliids are closely related to two very large and diverse angiosperm lineages: the **monocots** and the **eudicots**.

Monocots and eudicots are named for differences in the number of embryonic leaves called cotyledons. Monocot embryos possess one cotyledon, whereas eudicots possess two cotyledons. Monocots differ from eudicots in several additional ways (look ahead to Table 36.1). For example, monocots typically have flowers with parts numbering three or some multiple of three (**Figure 32.18a**). In contrast, eudicot flower parts often occur in fours, fives, or a multiple of four or five (**Figure 32.18b**).

## Core Concept: Evolution

## Whole-Genome Duplications Influenced the Evolution of Flowering Plants

During evolution, whole-genome duplication has occurred in a wide variety of eukaryotes and has happened on multiple occasions during the evolutionary history of plants. For example, a whole-genome duplication event occurred early in the evolution of seed plants (see the red bar in Figure 32.1). Additional examples include duplication of the entire plant genome before the divergence of *Amborella* (see the blue bar in Figure 32.16) and during eudicot diversification (green bar in Figure 32.16). After such whole-genome duplications, a plant's genome operates as a diploid system. Although genome sizes vary, the number of genes estimated for plants whose genomes have been studied is about 25,000. Whole-genome duplication has the potential to affect species' evolutionary pathways because it offers the opportunity for many genes to diverge, forming gene families.

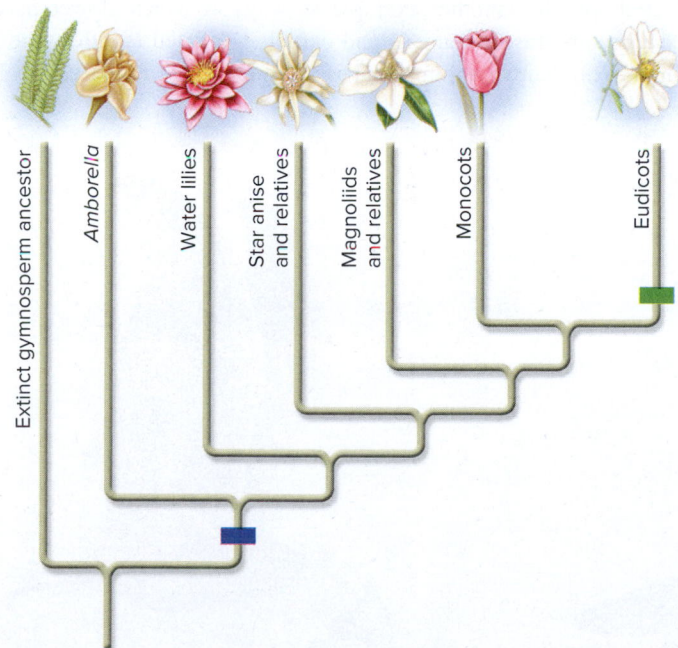

**Figure 32.16** **A phylogenetic tree showing the major modern angiosperm lineages and examples of whole-genome duplication events.** In a whole-genome duplication event, the genome size doubles. Molecular data indicate that the size of plant genomes underwent major increases at different time points and in particular lineages (only some of which are indicated here). For example, a whole genome duplication occurred before the divergence of *Amborella* (blue bar), and the genome size tripled (a whole-genome triplication) during the evolution of eudicots (green bar). Plant evolutionary biologists speculate that these duplication and triplication events strongly influenced the diversification of modern angiosperms.

**Figure 32.18** **One characteristic difference between monocots and eudicots: flower part number.** **(a)** Flowers and buds of a lily (genus *Lilium*), displaying six tepals. **(b)** A flower and buds of apple (genus *Malus*), showing five flower petals. Green sepals are visible around the pink buds. a: ©Dudakova Elena/Shutterstock; b: ©Ed Reschke/Getty Images

**(a) A monocot with six tepals**

**(b) A eudicot with five petals**

## Flower Diversification Has Fostered Efficient Seed Production

During the diversification of flowering plants, flower evolution has involved several types of changes that foster the transfer of pollen from one plant to another. Effective pollination is essential to efficient seed production because it minimizes the amount of energy plants must expend to accomplish sexual reproduction. Fusion of flower organs, clustering of flowers into groups, and reduction in size of the perianth are some examples of changes leading to effective pollination.

Many flowers have fused petals that form floral tubes. Such tubes tend to accumulate sugar-rich nectar that provides a reward for **pollinators**, animals that transfer pollen among plants. The diameters of floral tubes vary among flowers and are evolutionarily tuned to the feeding structures of diverse animals, which range from the narrow tongues of butterflies to the wider bills of nectar-feeding birds (**Figure 32.19**). Nectar-feeding bats stick their heads into even larger tubular flowers to lap up nectar with their tongues. Orchids provide another example of ways in which flower parts have become fused; stamens and carpels are fused together into a single reproductive column that is surrounded by attractive tepals

**(a) Zinnia flower and butterfly**

**(b) Hibiscus flower and hummingbird**

**(c) Saguaro cactus flower and bat**

**Figure 32.19** **Flowers whose perianths form nectar-containing floral tubes of different widths that accommodate different pollinators.** **(a)** This zinnia is composed of an outer rim of showy flowers and a central disc of narrow tubular flowers that produce nectar. Butterflies, but not other pollinators, are able to reach the nectar by means of their narrow tongues. **(b)** The hibiscus flower produces nectar in a floral tube whose diameter corresponds to the dimensions of a hummingbird bill. **(c)** The flower of a saguaro cactus (*Carnegiea gigantea*) forms a floral tube that is wide enough for nectar-feeding bats to get their heads inside. The cactus flower has been drawn here as if it were transparent, to illustrate bat pollination.

(**Figure 32.20a**). This arrangement of flower organs fosters orchid pollination by particular insects and is a distinctive feature of the orchid family.

Many plants produce **inflorescences**, groups of flowers tightly clustered together, which occur in several types. The sunflower family features a type of inflorescence in which many small flowers are

(a) **An orchid flower with fused pistil and stamens**

Tepals

Fused pistil and stamens

Tepals

(b) **A sunflower plant showing inflorescence**

(c) **Grass flowers lacking showy perianth**

**Figure 32.20** **Evolutionary changes in flower structure.** **(a)** An orchid of the genus *Cattleya* has fused stamens and pistil, and six tepals, one of which is specialized to form a lower lip. **(b)** An inflorescence (head) of sunflower (genus *Helianthus*). This inflorescence includes a rim of flowers with conspicuous petals that attract pollinators and an inner disc of flowers that lack attractive perianths. **(c)** Grass flowers of the grass genus *Triticum* lack a showy perianth. a: ©Neil Joy/Science Source; b: ©Pixtal/age fotostock; c: ©blickwinkel/Alamy Stock Photo

**Concept Check:** *What advantage does the nonshowy perianth of grass flowers provide?*

clustered into a head (**Figure 32.20b**). The flowers at the center of a sunflower head function in reproduction and lack showy petals, but flowers at the rim have showy petals that attract pollinators. Flower heads allow pollinators to transfer pollen among a large number of flowers at the same time.

The grass family features flowers with few or no perianths, which explains why grass flowers are not showy (**Figure 32.20c**). This adaptation fosters pollination by wind, since petals would only get in the way of such pollen transfer.

## Diverse Types of Fruits Function in Seed Dispersal

**Fruits** are structures that develop from ovary walls and function in the dispersal of enclosed seeds. Seed dispersal helps to prevent seedlings from competing with their larger parents for scarce resources such as water and light. Dispersal of seeds also allows plants to colonize new habitats. Diverse fruit types illustrate the many ways in which plants have become adapted for effective seed dispersal. Like flower types, fruit types are useful in classifying and identifying angiosperms.

These are just a few examples of the diverse mechanisms that flowering plants use to disperse their seeds.

- Many mature angiosperm fruits, such as cherries, grapes, and lemons, are attractively colored, soft, juicy, and tasty (**Figure 32.21a–c**). Such fruits are adapted to attract animals that consume the fruits, digest the outer portion as food, and eliminate the seeds, thereby dispersing them. Hard seed coats prevent such seeds from being destroyed by the animal's digestive system.

- Strawberries are aggregate fruits, structures consisting of many fruits that all develop from a single flower having multiple pistils (Figure 32.21d). The ovaries of these pistils develop into tiny, single-seeded yellow fruits on a strawberry surface; the fleshy, red, sweet portion of a strawberry develops from a flower receptacle. Aggregate fruits allow a single animal consumer, such as a bird, to disperse many seeds at the same time.

- Pineapples (Figure 32.21e) are juicy multiple fruits that develop when many ovaries of an inflorescence fuse together. Such multiple fruits attract relatively large animals that have the ability to disperse seeds for long distances.

- The plant family informally known as **legumes** is named for its distinctive fruits, dry pods that open down both sides when seeds are mature, thereby releasing them (Figure 32.21f).

- Nuts and grains are additional examples of dry fruits. **Grains** are the characteristic single-seeded fruits of cereal grasses such as rice, corn (maize), barley, and wheat.

- Coconut fruits are adapted for dispersal in ocean currents and can float for months before being cast ashore (Figure 32.21g).

- Maple trees produce dry and thus lightweight fruits having wings, features that foster effective wind dispersal (Figure 32.21h).

- Other plants produce dry fruits with surface burrs that attach to animal fur.

**(a) A fleshy fruit (cherry)**

**(b) A fleshy berry fruit (grape)**

**(c) A fleshy fruit (lemon)**

**(d) An aggregate fruit (strawberry)**

**(e) A multiple fruit (pineapple)**

**(f) Legumes with dry pods (peas)**

**(g) Fruit with husk (coconut)**

**(h) A dry, winged fruit (maple)**

## Angiosperms Produce Diverse Secondary Metabolites That Play Important Roles in Plant Structure, Reproduction, and Protection

Secondary metabolism involves the synthesis of organic compounds that are not essential for cell structure and growth in organisms but aid their survival and reproduction. These molecules, called **secondary metabolites**, are produced by various prokaryotes, protists, fungi, some animals, and all plants, but are most diverse in the angiosperms. About 100,000 different types of secondary metabolites are known, most of which are produced by flowering plants. Because secondary metabolites play essential roles in plant structure, reproduction, and protection, diversification of these compounds has influenced flowering plant evolution. Three major classes of plant secondary metabolites occur: (1) terpenes and terpenoids; (2) phenolics, which include flavonoids and related compounds; and (3) alkaloids (**Figure 32.22**).

About 25,000 types of plant terpenes and terpenoids are constructed from different arrangements of the simple hydrocarbon gas isoprene. Taxol, whose use in the treatment of cancer was noted earlier, is a terpene, as are citronella and a variety of other compounds that repel insects. Rubber, turpentine, rosin, and amber are complex terpenoids that likewise serve important roles in plant biology as well as having useful human applications.

Phenolic compounds are responsible for some flower and fruit colors as well as the distinctive flavors of cinnamon, nutmeg, ginger, cloves, chilies, and vanilla. Phenolics absorb UV radiation, thereby preventing damage to a plant's DNA. They also help to defend plants against insects and disease microbes. Some phenolic compounds found in tea, red wine, grape juice, and blueberries are antioxidants that detoxify free radicals, thereby preventing cellular damage.

Alkaloids are nitrogen-containing secondary metabolites that often ave potent effects on the animal nervous system. Plants produce at least 12,000 types of alkaloids, and certain species produce many alkaloids. Caffeine, nicotine, morphine, ephedrine, cocaine, and codeine are examples of alkaloids that influence the physiology and behavior of humans and are thus of societal concern. Like flower and fruit structures, secondary metabolites are useful in distinguishing among Earth's hundreds of thousands of flowering plant species.

**Figure 32.21** **Representative fruit types.** **(a–c)** The cherry, grape, and lemon are fleshy fruits adapted to attract animals that consume the fruits and excrete the seeds. **(d)** Strawberry is an aggregate fruit, consisting of many tiny, single-seeded fruits produced by a single flower. The fruits are embedded in the surface of a fleshy receptacle that is adapted to attract animal seed-dispersal agents. **(e)** Pineapple is a large multiple fruit formed by the aggregation of smaller fruits, each produced by one of the flowers in an inflorescence. **(f)** Peas produce legumes, fruits that open on two sides to release seeds. **(g)** Coconut fruits possess a fibrous husk that aids dispersal in water. **(h)** Maple trees produce dry fruits with wings adapted for wind dispersal. a–e: ©Lee W. Wilcox; f: ©Gloomerique/Getty Images; g: ©foodanddrinkphotos co/age fotostock; h: ©blickwinkel/Alamy Stock Photo

**Terpene**

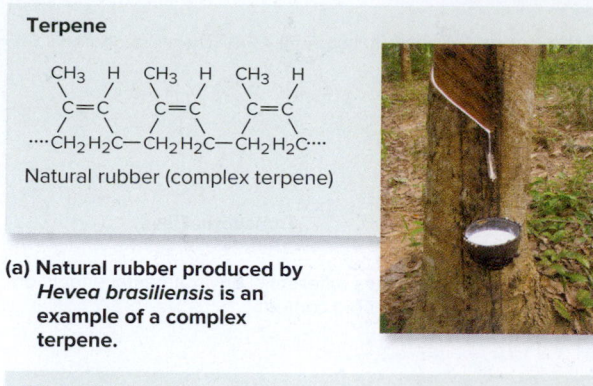

Natural rubber (complex terpene)

**(a) Natural rubber produced by** *Hevea brasiliensis* **is an example of a complex terpene.**

**Phenolic**

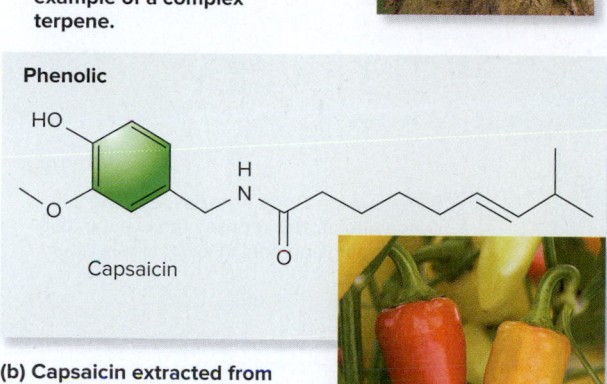

Capsaicin

**(b) Capsaicin extracted from capsicum peppers is an example of a phenolic compound.**

**Alkaloid**

Caffeine

**(c) Caffeine produced by** *Coffea arabica* **is an example of an alkaloid.**

**Figure 32.22  Major types of plant secondary metabolites.** Note that the structures of these plant secondary metabolites differs from that of the primary compounds produced by all cells; primary compounds include sugars and amino acids, described in Chapter 3. The production by plants of terpenes, phenolics, and alkaloids helps to explain how plants survive and reproduce and why plants are useful to humans in so many ways. a: ©Suphatthra China/Shutterstock; b: ©Jonathan Buckley/ GAP Photo/Getty Images; c: ©Science Photo Library/Alamy Stock Photo

 **Core Skill: Process of Science**

# Feature Investigation | Hillig and Mahlberg Analyzed Secondary Metabolites to Explore Species Diversification in the Genus *Cannabis*

The genus *Cannabis* has long been a source of hemp fiber used for ropes and fabric. People have also used *Cannabis* (also known as marijuana) in traditional medicine and as a hallucinogenic drug. *Cannabis* produces THC (tetrahydrocannabinol), a type of alkaloid called a cannabinoid. THC and other cannabinoids are produced in glandular hairs that cover most of the *Cannabis* plant's surface but are particularly rich in leaves located near the flowers. THC mimics compounds known as endocannabinoids, which are naturally produced and act in the animal brain and elsewhere in the body. THC affects humans by binding to receptor proteins in plasma membranes in the same way as natural endocannabinoids do. People sometimes use cannabis to ease pain and other medical conditions.

Because humans have subjected cultivated *Cannabis* plants to artificial selection for so long, plant biologists have been uncertain how cultivated *Cannabis* species are related to those in the wild. In the past, plants cultivated for drug production were often identified as *Cannabis indica*, whereas those grown for hemp were typically known as *Cannabis sativa*. However, these species are difficult to distinguish on the basis of structural features, and the relevance of these names to wild cannabis was unknown. At the same time, species identification has become important for biodiversity studies, agriculture, and law enforcement. For these reasons, plant biologists Karl Hillig and Paul Mahlberg hypothesized that ratios of THC to another cannabinoid known as CBD

(cannabidiol) might aid in defining *Cannabis* species and identifying plant samples at the species level, as shown in **Figure 32.23**.

To test their hypothesis, the investigators began by collecting *Cannabis* fruits (containing seeds) from nearly a hundred diverse locations around the world. As shown in step 1 of Figure 32.23, they used the seeds to grow new plants under uniform conditions in a greenhouse. The investigators next extracted cannabinoids, analyzed them by means of gas chromatography (a laboratory technique used to identify components of a mixture), and determined the ratios of THC to CBD. The results, published in 2004, suggested that the wild and cultivated *Cannabis* samples evaluated in this study could be classified into distinct species: *C. sativa*, displaying relatively low THC levels, and *C. indica*, having relatively high THC levels. More recent genetic studies suggest that the genus *Cannabis* is even more diverse than previously thought.

### Experimental Questions

1. **CoreSKILL »** Designing an experiment requires a plan to achieve an adequate number of samples, in order to allow statistical analysis. Hillig and Mahlburg obtained nearly a hundred *Cannabis* fruit samples from around the world. Why were so many samples needed?

2. Why did Hillig and Mahlberg collect samples from the leaves growing nearest the flowers?

**Figure 32.23** **Hillig and Mahlberg's analysis of secondary metabolites in the genus *Cannabis*.** (top inset): ©Phil Schermeister/National Geographic/ Getty Images; (middle inset): ©Matthew Kellett/Alamy Stock Photo

**GOAL** To determine if cannabinoids aid in distinguishing *Cannabis* species.

**KEY MATERIALS** *Cannabis* fruits obtained from nearly 100 different worldwide sources.

| | Experimental level | Conceptual level |
|---|---|---|
| **1** Grow multiple *Cannabis* plants from seeds under standard conditions in a greenhouse. |  | Eliminates differential environmental effects on cannabinoid content. |
| **2** Extract cannabinoids from leaves surrounding flowers. |  | Extracts were made from tissues richest in cannabinoids; this reduces the chance that cannabinoids present in lower levels would be missed. |
| **3** Analyze cannabinoids by gas chromatography. Determine ratios of THC (tetrahydrocannabinol) to CBD (cannabidiol) in about 200 *Cannabis* plants. | 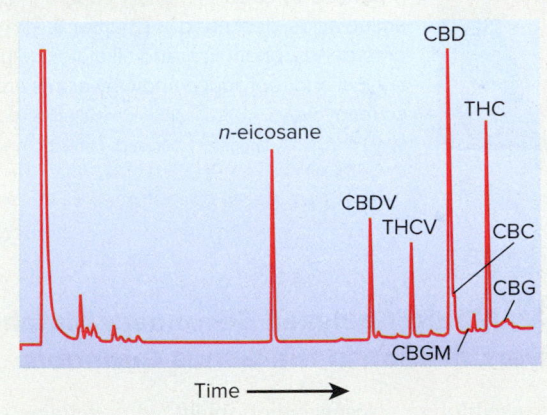 | Previous data suggested that ratios of THC to CBD might be different in separate species. |

**4 THE DATA**

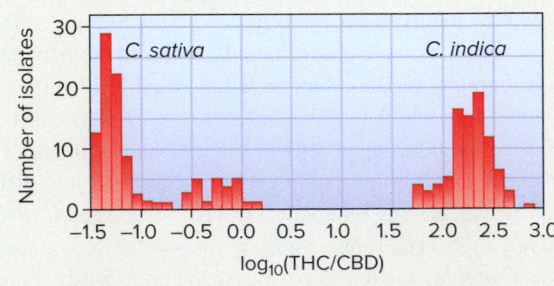

*Cannabis* plants isolated from diverse sources worldwide formed 2 groups—those having relatively high THC to CBD ratios and those having lower THC to CBD ratios.

Plants having low THC to CBD ratios, often used as hemp fiber sources, corresponded to the species *C. sativa*.

Plants having high THC to CBD ratios, often used as drug sources, corresponded to the species *C. indica*.

**5 CONCLUSION** Differing cannabinoid ratios support a concept of 2 *Cannabis* species.

**6 SOURCE** Hillig, K.W., and Mahlberg, P.G. 2004. A chemotaxonomic analysis of cannabinoid variation in Cannabis (Cannabaceae). *American Journal of Botany* 91: 966–975. ©2004 Botanical Society of America. All rights reserved. Used with permission.

**BIO·TIPS**　**THE QUESTION**　*The study by Hillig and Mahlburg revealed that samples of Cannabis plants from diverse sources varied in their ratios of alkaloids known as cannabinoids. Why did the investigators grow plants from seeds in a greenhouse before conducting their analyses of cannabinoid content?*

**T**OPIC　*What topic in biology does this question address?* The topic is the use of secondary metabolites to differentiate species of plants. More specifically, the question addresses the relative effects of genes and environment on plants' production of secondary metabolites.

**I**NFORMATION　*What information do you know based on the question and your understanding of the topic?* You know from earlier chapters of this textbook that the traits organisms express depend on both genes and environment and that genetically determined traits are used for classifying organisms. In this chapter, you have learned that plants, and flowering plants in particular, produce diverse types of secondary metabolites that play important roles in reproduction and protection. Consequently, you might expect that the ratios of cannabinoids in *Cannabis* plants reflect environmental conditions, genetic composition, or both.

**P**ROBLEM-SOLVING **S**TRATEGY　*Design an experiment.* Consider how you might design an experiment to determine how cannabinoid ratios differ among individual plants grown from seeds collected from different sources. Because the seeds came from different sources, they may be from different species of *Cannabis*, defined by genetic characteristics. In your experiment, you would have to control for possible environmental effects, so that such effects would not mask any differences due to genetics.

**ANSWER**　*Growing experimental plants from seeds in a greenhouse under the same (standard) conditions is a way to minimize variation in cannabinoid ratios due to environmental effects. During their development, all of the experimental plants will experience the same conditions of light, moisture, soil minerals, day length, and other factors that affect plant growth and the production of secondary metabolites. Under standardized growth conditions, observed differences in cannabinoid ratios will reflect genetic variation that can be used to classify Cannabis plants into species, as Hillig and Mahlburg were able to do.*

## 32.4 The Role of Coevolution in Angiosperm Diversification

**Learning Outcomes:**

1. Explain the concept of coevolution.
2. List examples of coevolution between plants and animal pollinators.
3. List examples of coevolution between plants and animal seed dispersal agents.

The preceding section described how flowering plants are commonly associated with animals in ways that strongly influence plant evolution. Likewise, plants have influenced animal evolution in a diversity-generating process known as **coevolution**, which is the process by which two or more species of organisms influence each other's evolutionary pathway. During the diversification of flowering plants, coevolution with animals has been a major evolutionary factor. For example, the diversification of bees about 123 million years ago correlates with the diversification of the eudicots, which today make up three-quarters of all angiosperm species. Coevolution is reflected in the diverse forms of most flowers and many fruits and the many ways that plants accomplish effective pollen and seed dispersal. Human attraction to flowers and fruit also is an example of coevolution. This is because human sensory systems are similar to those of various animals that have coevolved with angiosperms.

### Pollination Coevolution Influences the Diversification of Flowers and Animals

Animal pollinators transfer pollen from the anthers of one flower to the stigmas of other flowers of the same species. Pollinators thereby foster genetic variation and enhance the potential for evolutionary change among plants. Insects, birds, bats, and other pollinators learn the characteristics of particular flowers, visiting them preferentially. This animal behavior, known as constancy or fidelity, increases the odds that a flower stigma will receive pollen of the appropriate species. Animal pollinators increase the precision of pollen transfer, which reduces the amount of pollen that plants must produce to achieve pollination. By contrast, wind-pollinated plants must produce much larger amounts of pollen because windblown pollen reaches appropriate flowers by chance.

Flowers attract the most appropriate pollinators by means of attractive colors, odors, shapes, and sizes. Secondary metabolites influence the colors and odors of many flowers. Flavonoids, for example, color many blue, purple, or pink flowers. More than 700 types of chemical compounds contribute to floral odors.

Most flowers reward pollinators with food: sugar-rich nectar, lipid- and protein-rich pollen, or both. In this way, flowering plants provide an important biological service, providing food for many types of animals. However, some flowers "trick" pollinators into visiting or trap pollinators temporarily, thereby achieving pollination without actually rewarding the pollinator. Examples include flowers that look and smell like dead meat, thereby attracting flies, which are fooled but accomplish pollination anyway.

Although many flowers are pollinated by a variety of animals, others have flowers that have become specialized for particular pollinators, and vice versa. These specializations, which have resulted from coevolution, are known as **pollination syndromes** (**Table 32.2**). For example, odorless red flowers, such as those of hibiscus (see Figure 32.19b), are attractive to birds, which can see the color red but have a poor sense of smell. By contrast, bees are not typically attracted to red flowers because bee vision does not extend to the red end of the visible light spectrum. Rather, bees are attracted to blue, purple, yellow, and white flowers having sweet odors. If you are allergic to bee stings or just want to reduce the possibility of being stung, avoid dressing in bee-attracting flower colors and wearing flowery fragrances when in locales frequented by bees.

| Table 32.2 | Pollination Syndromes | |
|---|---|---|
| **Animal features** | **Coevolved flower features** | |
| **Bees** | | |
| Color vision includes ultraviolet (UV), not red | Often blue, purple, yellow, white (not red) colors | |
| Good sense of smell | Fragrant | |
| Require nectar and pollen | Nectar and abundant pollen | |
| **Butterflies** | | |
| Good color vision | Blue, purple, deep pink, orange, red colors | |
| Sense odors with feet | Light floral scent | |
| Need landing place | Landing place | |
| Feed with long, tubular tongue | Nectar in deep, narrow floral tubes | |
| **Moths** | | |
| Active at night | Open at night; white or bright colors | |
| Good sense of smell | Heavy, musky odors | |
| Feed with long, thin tongue | Nectar in deep, narrow floral tubes | |
| **Birds** | | |
| Color vision, includes red | Often colored red | |
| Often require perch | Strong, damage-resistant structure | |
| Poor sense of smell | No fragrance | |
| Feed in daytime | Open in daytime | |
| High nectar requirement | Copious nectar in floral tubes | |
| Hover (hummingbirds) | Pendulous (dangling) flowers | |
| **Bats** | | |
| Color blind | Light, reflective colors | |
| Good sense of smell | Strong odors | |
| Active at night | Open at night | |
| High food requirements | Copious nectar and pollen provided | |
| Navigate by echolocation | Pendulous or borne on tree trunks | |

 **Core Skill: Modeling** The goal of this modeling challenge is to propose a model that shows a series of steps in the process by which a pollinator accomplishes pollination.

**Modeling Challenge:** Table 32.2 lists pollination syndromes, features of animal pollinators and flowers that have co-evolved. Pick one of these pollination syndromes and propose a model that shows a series of steps in the pollination process. The model should begin with the pollinator close to, but not touching the flower, and should end with pollination, pollen attachment to the flower stigma. Your model should answer two key questions: Why does the pollinator visit the flower? How does the pollinator deliver pollen to the stigma?

Pollination syndromes are of practical importance in agriculture and in conservation biology. Fruit growers often import colonies of bees to pollinate flowers of fruit crops and thereby increase

**Figure 32.24** *Brighamia insignis*, **a plant endangered by the loss of its pollinator.** The pollinator that coevolved with *B. insignis* has become extinct, with the result that the plant is unable to produce seeds unless artificially pollinated by humans. ©Garden World Images Ltd/ Alamy Stock Photo

**Concept Check:** *What kind of animal likely pollinated B. insignis?*

crop yields. In recent years, widespread die-offs of bee colonies have become an environmental and agricultural concern. When bee pollinators are not available, growers cannot produce some fruit crops. Some plants have become so specialized to particular pollinators that if the pollinator becomes extinct, the plant becomes endangered. An example is the Hawaiian cliff-dwelling *Brighamia insignis* (**Figure 32.24**), whose presumed moth pollinator has become extinct. Humans that hand-pollinate *B. insignis* are all that stand between this plant and extinction.

## Seed-Dispersal Coevolution Influences the Characteristics of Fruits and Animals

As in the case of pollination, coevolution between plants and their animal seed-dispersal agents has influenced characteristics of both fruits and the seed-dispersing animals. In addition, flowering plant fruits provide food for animals, an important biological service. For example, many of the plants of temperate forests produce fruits that are attractive to resident birds. Such juicy, sweet fruits have small seeds that readily pass through bird guts. Many plants signal fruit ripeness by undergoing color changes from unripe green fruits to red, orange, yellow, blue, or black (**Figure 32.25**). Because birds have good color vision, they are able to detect the presence of ripe fruits and consume them before the fruits drop from plants and rot.

Apples, strawberries, cherries, blueberries, and blackberries are examples of fruits whose seed dispersal adaptations have made them attractive food for humans. By contrast, the lipid-rich fruits of Virginia creeper (*Parthenocissus quinquefolia*) and some other autumn-fruiting plants energize migratory birds but are not tasty to

**Figure 32.25  Fruits attractive to animal seed-dispersal agents.**  Color and odor signals alert coevolved animal species that fruits are ripe, thus favoring the dispersal of mature seeds. ©Beng & Lundberg/naturepl.com

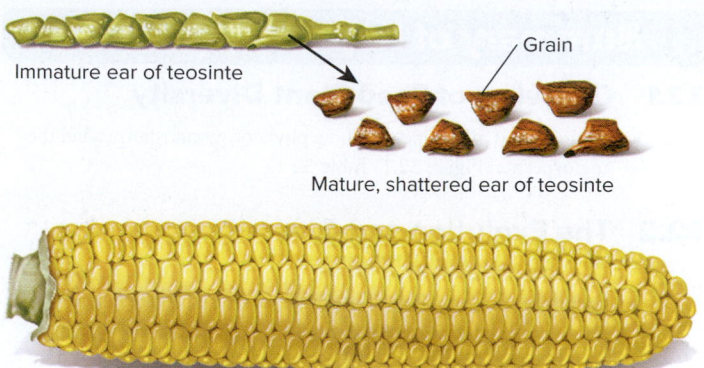

Nonshattering ear of *Z. mays*, subspecies mays.

**Figure 32.26  Ears and grains of modern corn and its ancestor, teosinte.**  This illustration shows that domesticated corn ears are much larger than those of the ancestral grass teosinte. In addition, corn grains are softer and more edible than the grains of teosinte, which are enclosed in a hard casing.

👁 **Core Skill: Science and Society**  The domestication of corn from a wild grass into one of the world's largest production crops is an amazing feat of artificial selection.

humans. The Virginia creeper's leaves often turn fall colors earlier than surrounding plants, thereby signaling the availability of nutritious, ripe fruit to high-flying birds. Such lipid-rich fruits must be consumed promptly because they rot easily, in which case seed dispersal cannot occur.

## 32.5  Human Influences on Angiosperm Diversification

### Learning Outcome:

1. **CoreSKILL »** Describe how molecular information about modern crop plants is used to infer their evolutionary origin by domestication.

By means of the process known as **domestication**, which involves artificial selection for traits desirable to humans, ancient humans transformed wild plant species into new crop species. Cultivated bread wheat (*Triticum aestivum*) was probably among the earliest food crops, having originated more than 8,000 years ago, in what is now southeastern Turkey and northern Syria. Bread wheat originated by a series of steps that included hybridization and whole-genome duplication from wild ancestors (*Triticum boeoticum* and *Triticum dicoccoides*).

Among the earliest changes that occurred during wheat domestication was the loss of **shattering**, the process by which ears of wild grain crops break apart and disperse their grains. A mutation probably caused the ears of some wheat plants to remain intact, a trait that is disadvantageous in nature but beneficial to humans. Nonshattering ears would have been easier for humans to harvest than normal ears. Early farmers probably selected seed stock from plants having nonshattering ears and other favorable traits such as larger grains. These ancient artificial selection processes, together with modern breeding efforts, explain why cultivated wheat differs from its wild relatives in shattering and other properties. The accumulation of these trait differences is why cultivated and wild wheat plants are classified as different species.

About 9,000 years ago, people living in what is now Mexico domesticated one type of the native grass known as teosinte

(formally, *Zea mays* subspecies *parviglumis*). The domestication process produced a new species, *Zea mays* subspecies *mays*, commonly known as corn or maize. The evidence for this pivotal event includes ancient ears that were larger than wild ones and distinctive fossil pollen. Modern ears of corn are much larger than those of teosinte, with many more rows and larger and softer corn grains, and modern corn ears do not shatter, as do those of ancestral teosinte (**Figure 32.26**). These and other trait changes reflect artificial selection accomplished by humans. An analysis of the corn genome, reported in 2005 by Canadian biologist Stephen Wright, American evolutionary biologist Brandon Gaut, and coworkers, suggests that 1,200 corn genes have been affected by artificial selection.

Molecular analyses indicate that domesticated rice (*Oryza sativa*) originated from ancestral wild species of grasses (*Oryza nivara* and/or *Oryza rifipogon*). As in the cases of wheat and corn, domestication of rice involved loss of ear shattering. Researchers have identified a mutation in domesticated strains of rice that alters the amino acid sequence of a protein that regulates shattering. Ancient humans might have unconsciously selected for this mutation while gathering rice from wild populations, because the mutants would not so easily have shed grains during the harvesting process. Eventually, the nonshattering mutant became a widely planted crop throughout Asia, and today it is the food staple for millions of people.

Although humans generated these and other plant species, modern humans have also caused the extinction of plants as the result of habitat destruction and other threats to species. Protecting biodiversity will continue to challenge humans as populations and demands on the Earth's resources increase. Plant biologists are working to identify one or more molecular sequencing tools for use in barcoding plants, a process that is widely used to identify and catalog animals. The ability to bar code plants, which would enable researchers to quickly analyze the DNA of a species and identify it based on existing barcodes, is important to organizations like CITES and others that monitor international trade in endangered plant species.

# Summary of Key Concepts

## 32.1 Overview of Seed Plant Diversity

- Modern seed plants include three phyla of gymnosperms and the angiosperms (Figure 32.1, Table 32.1).

## 32.2 The Evolution and Diversity of Modern Gymnosperms

- Gymnosperms are plants that produce exposed seeds rather than seeds enclosed in fruits. Gymnosperms originated from seedless woody plants known as progymnosperms (Figure 32.2).

- The modern gymnosperms include three phyla: cycads, *Ginkgo biloba*, and the conifers (which include the Gnetales). Nearly 300 species of cycads primarily live in tropical and subtropical regions. Features of cycads include palmlike leaves, nonwoody stems, coralloid roots with cyanobacterial endosymbionts, toxins, and large conelike seed-producing structures (Figures 32.3, 32.4).

- The tree *Ginkgo biloba* is the last surviving species of a phylum that was diverse during the Mesozoic, also known as the Age of Dinosaurs. Individual trees produce ovules and seeds or pollen, with a sex chromosome system much like that of humans (Figure 32.5).

- Conifers have been widespread and diverse members of plant communities for the past 300 million years and are important sources of wood and paper pulp for humans. Reproduction involves simple pollen cones and complex ovule-producing cones. Many conifers display adaptations that aid survival in cold climates. Three genera of Gnetales display distinctive adaptations (Figures 32.6, 32.7, 32.8, 32.9, 32.10, 32.11).

## 32.3 The Evolution and Diversity of Modern Angiosperms

- Angiosperms inherited seeds, the capacity to produce wood, and other features from gymnosperm ancestors, but display distinctive features not found in other land plants, including flowers and fruits (Figure 32.12).

- Flowers foster seed production and are adapted in various ways that aid pollination. The major flower organs are sepals and petals (or tepals), stamens, and carpels, which may occur singly or in fused groups. Both single carpels and compound carpels take a distinctive shape known as a pistil, which displays regions of specialized function. The stigma is a receptive surface for pollen, pollen tubes grow through the style, and ovules develop within the ovary. Pollination is the transfer of pollen from a stamen to a pistil, a process distinct from fertilization. Double fertilization, the production of both a zygote and a nutritive tissue known as endosperm, is a critical innovation of angiosperms. This process allows ovules to develop into seeds containing embryos and endosperm, and ovaries to develop into fruits. Stamens and carpels may have evolved from leaflike structures bearing sporangia (Figures 32.13, 32.14, 32.15).

- Whole-genome duplications have influenced plant evolution, particularly the diversification of the angiosperms. The two largest and most diverse lineages of flowering plants are the monocots and eudicots (Figures 32.16, 32.17, 32.18).

- Flower diversification involved evolutionary changes such as fusion of petals, clustering of flowers into inflorescences, and reduction in size of the perianth. These changes improve the effectiveness of pollination, which enhances seed production (Figures 32.19, 32.20).

- Fruits are structures that enclose seeds and aid in their dispersal. Fruits occur in many types that foster seed dispersal (Figure 32.21).

- Angiosperms produce three main groups of secondary metabolites: (1) terpenes and terpenoids; (2) phenolics, which include flavonoids and related compounds; and (3) alkaloids, which play essential roles in plant structure, reproduction, and defense, respectively (Figure 32.22).

- Hillig and Mahlberg demonstrated the use of particular secondary metabolites in distinguishing species of the societally important genus *Cannabis* (Figure 32.23).

## 32.4 The Role of Coevolution in Angiosperm Diversification

- Coevolutionary interactions between flowering plants and animals that serve as pollen- and seed-dispersal agents played a powerful role in the diversification of both angiosperms and animals (Table 32.2, Figures 32.24, 32.25).

- Human appreciation of flowers and fruits is based on sensory systems similar to those present in the animals with which angiosperms coevolved.

## 32.5 Human Influences on Angiosperm Diversification

- Humans have produced new crop species by domesticating wild plants. The process of domestication involved artificial selection for traits such as nonshattering ears of wheat, corn, and rice (Figure 32.26).

# Assess & Discuss

## Test Yourself

1. What feature(s) must be present for a plant to produce wood?
   a. a type of conducting system in which vascular bundles occur in a ring around pith
   b. a eustele
   c. a vascular cambium
   d. all of the above
   e. none of the above

2. Which sequence of critical innovations reflects the order of their appearance in time?
   a. embryos, vascular tissue, wood, seeds, flowers
   b. vascular tissue, embryos, wood, flowers, seeds
   c. vascular tissue, wood, seeds, embryos, flowers
   d. wood, seeds, embryos, flowers, vascular tissue
   e. seeds, vascular tissue, wood, embryos, flowers

3. How long have ancient and modern groups of gymnosperms been important members of plant communities?
   a. 10,000 years, since the dawn of agriculture
   b. 100,000 years
   c. 300,000 years
   d. 65 million years, since the Cretaceous-Paleogene (K/T) event
   e. 300 million years, since the Coal Age

4. What similar features do gymnosperms and angiosperms possess that differ from other modern vascular plants?
   a. Gymnosperms and angiosperms both produce flagellate sperm.
   b. Gymnosperms and angiosperms both produce flowers.
   c. Gymnosperms and angiosperms both have tracheids, but not vessels, in their vascular tissues.
   d. Gymnosperms and angiosperms both produce fruits.
   e. None of the above statements is true.

5. Which part of a flower receives pollen transported by the wind or a pollinating animal?
   a. perianth          d. pedicel
   b. stigma            e. ovary
   c. filament

6. The primary function of a fruit is to
   a. provide food for the developing seed.
   b. provide food for the developing seedling.
   c. foster pollen dispersal.
   d. foster seed dispersal.
   e. None of the above identifies the primary function of a fruit.

7. Flowers have diversified with regard to
   a. color.
   b. number of flower parts.
   c. fusion of organs.
   d. aggregation into inflorescences.
   e. all of the above.

8. Plants of the genus *Fuchsia* produce deep pink to red flowers that dangle from plants, produce nectar in floral tubes, and have no scent. Based on these features, which animal is most likely to be a coevolved pollinator of these plants?
   a. bee
   b. bat
   c. hummingbird
   d. butterfly
   e. moth

9. Which type of plant secondary metabolite is best known for the antioxidant properties of human foods such as blueberries, tea, and grape juice?
   a. alkaloids
   b. cannabinoids
   c. carotenoids
   d. phenolics
   e. terpenoids

10. What feature(s) of domesticated grain crops might differ from those of wild ancestors?
    a. the degree to which ears shatter, allowing for seed dispersal
    b. grain size
    c. number of grains per ear
    d. softness and edibility of grains
    e. all of the above

## Conceptual Questions

1. Make a diagram that shows how plant biologists think flowers arose.

2. Explain why fruits such as apples, strawberries, and cherries are attractive, nutritious, and harmless foods for humans.

3. **Core Concept: Structure and Function** Compare the structures of an apple flower and a sunflower, explaining how they relate to differences in pollination and seed dispersal.

## Collaborative Questions

1. Where in the world would you have to travel to find wild plants representing all of the gymnosperm phyla, including the three types of Gnetales?

2. How would you go about trying to solve what Darwin called "an abominable mystery," that is, the identity of the seed plant group that was ancestral to the flowering plants?

# An Introduction to Animal Diversity

# 33

**The variety of life forms on Earth is staggering.** Naked mole rats, *Heterocephalus glaber,* are long-lived rodents that remain cancer free. ©John Visser/Photoshot

**N**aked mole rats (*Heterocephalus glaber*) are a species of rodent that live in arid areas of the Horn of Africa, including Ethiopia, Somalia, and Kenya. Their large protruding teeth are used for digging, and their lips are sealed behind the teeth to keep out soil. Naked mole rats have a unique reproductive process in which only one dominant female, the queen, reproduces. But these rats have an even more intriguing claim to fame. Whereas most species of mice and rats live only 4 years on average, naked mole rats can live for at least 30 years. And they are cancer free! Scientists discovered that the rats produce a polysaccharide called high-molecular-mass hyaluronan, or HMM-HA. Secretion of HMM-HA from mole rat cells causes contact inhibition, preventing the cells from overcrowding and forming tumors. This ability is lost in cancer cells. When researchers inhibited the synthesis of HMM-HA by naked mole rat cells, the cells lost their contact inhibition and tumors formed. From these results, biologists are interested in pursuing this line of research to prevent cancer and extend the life in humans. This is just one example in which the study of animal diversity could lead to dramatic improvements in human health.

Animals constitute the most species-rich kingdom. About 1.3 million species have been found and described, and an estimated 2–5 million more species await discovery and classification. Beyond being members of this kingdom ourselves, humans depend on animals. Many different kinds of animals and their products are part of our diet. Humans also enjoy animal species as companions and depend on other species for tests of lifesaving drugs. We share parts of our genome with other organisms such as fruit flies, nematodes, and zebrafish—all of which are used as model organisms for understanding aspects of human molecular and developmental biology.

However, we are also in conflict with animals such as insects that threaten our food supply and transmit deadly diseases. Malaria is transmitted by mosquitoes; sleeping sickness, by tsetse flies; and rabies, by a number of animals, including dogs, raccoons, and bats. With such a huge number and diversity of existing animals and with animals featuring so prominently in our lives, understanding animal diversity is of great importance. Therefore, researchers have spent a great deal of effort in determining the unique characteristics of different taxonomic groups and identifying their evolutionary relationships.

Since the time of Carolus Linnaeus in the 1700s, scientists have classified animals based on their morphology, that is, on their physical structure. In the 1990s, animal classifications based on similarities in DNA and rRNA sequences became more common. Quite often, classifications based on morphology and those based on molecular data were similar, but some important differences arose. In this chapter, we will begin by defining the key characteristics of animals and then take a look at the major features of animal body plans that form the basis of classification. We will explore how new molecular data have enabled scientists to revise and refine the animal phylogenetic tree. As more molecular-based evidence becomes available, systematists will likely continue to redraw the tree of animal life. Therefore, as you read this chapter, keep in mind that the classification of animals is now, and will continue to be, a work in progress.

## 33.1 Characteristics of Animals

**Learning Outcomes:**

1. List the key characteristics of animals that distinguish them from other organisms.
2. Provide a brief overview of the history of animal life on Earth.

The Earth contains a dazzling diversity of animal species, living in environments from the deep sea to the desert and exhibiting an amazing array of characteristics. Most animals move and eat multicellular prey, and therefore, they are loosely differentiated from species in other kingdoms. However, a single definition of an animal is difficult because they are so diverse that biologists can find exceptions to nearly any given characteristic. Even so, a number of key features can help us broadly characterize the group we call animals (**Table 33.1**).

### Animals Are Multicellular Heterotrophs

Animals have several characteristics relating to cell structure, mode of nutrition, movement, and reproduction that collectively distinguish them from other organisms. If we focus on these characteristics, **animals** can be defined as multicellular heterotrophs with cells that lack cell walls,

| Table 33.1 | Common Characteristics of Animals |
|---|---|
| Characteristic | Example |
| Multicellularity | Even relatively simple types of animals such as sponges are multicellular, in contrast to the mostly single-celled eukaryotic microorganisms called protists (see Chapter 28). |
| Heterotrophs | Animals obtain their food by eating other organisms or their products. This contrasts with plants and algae, most of which are autotrophs and essentially make their own food. |
| No cell walls | The cells of plants, fungi, bacteria, archaea, and most protists have a rigid cell wall, but animal cells lack a cell wall and are quite flexible. |
| Nervous tissue | The presence of a nervous system in most animals enables them to respond rapidly to environmental stimuli. |
| Movement | Most animals have a muscle system, which, combined with a nervous system, allows them to move in their environment. |
| Sexual reproduction | Most animals reproduce sexually, with small, mobile sperm uniting with a much larger egg to form a fertilized egg, or zygote. |
| Extracellular matrix | Proteins such as collagen bind animal cells together to give them added support and strength (see Figure 10.1). |
| Characteristic cell junctions | Animals have characteristic cell junctions, called anchoring, tight, and gap junctions (see Figures 10.7, 10.9, 10.11). |
| Special clusters of *Hox* genes | Most animals possess *Hox* genes, which function in patterning the body axis (see Figures 20.16, 24.15). |
| Similar rRNA | Animals all have very similar genes that encode for RNA of the small ribosomal subunit (SSU rRNA; see Figure 12.17). |

the capacity to move at some point in their life cycle, and the ability to reproduce sexually, with sperm fusing directly with eggs.

***Cell Structure***   Like some protists, plants, and most fungi, animals are multicellular. However, animal cells lack cell walls and are flexible. This flexibility facilitates movement. Animal cells gain structural support from an extensive extracellular matrix (ECM) that forms strong fibers outside the cell (refer back to Figure 10.1). Additionally, a group of unique cell junctions—anchoring, tight, and gap junctions—play an important role in holding animal cells in place and allowing communication between cells (refer back to Table 10.3).

***Mode of Nutrition***   All animals are **heterotrophs**; that is, they cannot synthesize their own organic molecules using energy from inorganic substances. Instead, animals must ingest other organisms or their products to sustain life. Many different modes of feeding exist among animals, including suspension feeding (filtering food out of the surrounding water); bulk feeding (eating large food pieces, as done by carnivores and herbivores); and fluid feeding (sucking plant sap or animal body fluids) (**Figure 33.1**). Although fungi and animals both rely on absorptive nutrition—that is, they secrete enzymes that break down complex materials and absorb the resulting small organic molecules—fungi use external digestion to obtain their nutrients. Animals ingest their food into an internal gut and then break it down using enzymes.

***Movement***   Most animals have muscle cells and nerve cells organized into tissues. Muscle tissue is unique to animals, and most animals are capable of some type of locomotion, the ability to move from place to place, in order to acquire food or escape predators. This ability has led to the development of muscular-skeletal systems, systems of sensory structures, and a nervous system that coordinates movement and prey capture. Sessile species such as barnacles, which stay in one place, use bristled appendages to obtain nearby food. However, in many sessile species, although adults are immobile, the larvae can swim.

***Reproduction***   Nearly all members of the animal kingdom reproduce sexually, although certain insects, fish, and lizard species can reproduce asexually. During sexual reproduction, a small, mobile sperm generally unites with a much larger egg to form a fertilized egg, or zygote. Fertilization may occur internally, which is common in terrestrial species, or externally, which is more common in aquatic species. Similarly, embryos develop inside the mother or outside in the mother's environment.

### Animal Life Began More Than a Half Billion Years Ago

The history of animal life spans over 630 million years, starting at the end of the Proterozoic eon, when multicellular animals emerged (refer back to Figure 26.4). The first animals were invertebrates, animals without a vertebral column, or backbone. A profusion of animal phyla appeared during the Cambrian explosion, 533–525 million years ago (mya), including sponges, jellyfish, corals, flatworms, mollusks, annelid worms, the first arthropods, and echinoderms, plus many phyla that no longer exist today (**Figure 33.2**).

(a)

(b)

(c)

**Figure 33.1** **Modes of animal nutrition.** **(a)** Suspension feeders, such as these tube worms, filter food particles from the water column. **(b)** Grizzly bears and other bulk feeders tear off large pieces of their food and chew it or swallow it whole. **(c)** Fluid feeders, such as these aphids, suck fluid from their food source. a: ©waldhaeusl.com/age fotostock; b: ©Enrique R Aguirre Aves/Getty Images; c: ©Bartomeu Borrell/age fotostock

The causes of the sudden increase in animal life at that time are not fully understood, but three explanations have been proposed.

- Species proliferation may have been related to a warm favorable environment. At the same time, atmospheric and aquatic oxygen levels were increasing, permitting increased metabolic rates, and an ozone layer had developed, blocking out harmful UV radiation and allowing complex life to thrive in shallow water and eventually on land.

- The evolution of the *Hox* gene complex may have permitted much variation in morphology.

- As new types of predators evolved, prey developed adaptations that enabled them to avoid their predators, leading to counteradaptations by predators, and so on. This evolutionary "arms race" may have resulted in a proliferation of predator and prey types.

**Figure 33.2** **The profusion of animal life in the Cambrian period, about 520 mya.** This artist's reconstruction of marine life shows many different phyla, some of which are now extinct.
©Publiphoto/Science Source

These hypotheses are not mutually exclusive and may well have operated at the same time.

Around 520 mya, the first vertebrates, fishes, appeared at roughly the same time as the first plants invaded land. The appearance of land plants introduced a viable food source for any organisms that could utilize them. However, the realm of land and air presented organisms with many challenges. For colonization of land to occur, certain species evolved adaptations that prevented them from drying out and enabled them to breathe, move, and reproduce in the new environment, in much the same way as the plant embryo, leaves, seeds, and other adaptations permitted plants to colonize terrestrial habitats (see Chapter 31). For animal species, such features included lungs and internal fertilization. The development of the amniotic egg, which features a tough, protective shell to prevent drying out, enabled animals to be terrestrial for their entire life cycle. The amniotic egg, which is described in Chapter 35, appeared during the Carboniferous period, about 300 mya, and was responsible for the success of the reptiles, which appeared during this period. Reptiles were to dominate the Earth for many millions of years during the rise and fall of the dinosaurs. Mammals appeared at the same time as dinosaurs, although they were not prevalent. The number and diversity of mammals exploded only after the dinosaurs abruptly died out at the end of the Cretaceous era, about 65 mya.

## 33.2 Animal Classification

**Learning Outcomes:**

1. Discuss why choanoflagellates are believed to be the closest living relatives of animals.
2. Describe each of the major morphological and developmental features of animal body plans that form the basis of the classification of animals.

All animals are classified in the domain Eukarya, the supergroup Opisthokonka, and the kingdom Animalia (informally called the animal kingdom) (refer back to Figure 25.1). Although extremely diverse, most biologists agree that the animal kingdom is monophyletic, meaning that all taxa have evolved from a single common ancestor. Today, scientists recognize about 35 animal phyla.

At first glance, many of the animal phyla seem so distantly related to one another (for example, chordates and jellyfish) that making sense of this diversity with a classification scheme seems very challenging. Fortunately, by carefully examining body features and, more recently, by analyzing molecular data such as DNA sequences, evolutionary biologists have been able to propose models that describe the evolutionary relationships among animals. The model shown in **Figure 33.3** describes those relationships for 13 common phyla. In this section, we will explore the major features of animal body plans that form the basis of this animal phylogeny.

## Animals Evolved from a Choanoflagellate-like Ancestor

With the monophyletic nature of the animal kingdom in mind, scientists have attempted to identify the species from which animals most likely evolved. Molecular data indicate that the closest living relative of animals is the flagellated protist known as a choanoflagellate. These tiny, single-celled organisms have a single flagellum surrounded by a collar of cytoplasmic tentacles (refer back to Figure 28.21b).

Some species of choanoflagelles form colonies consisting of many individual organisms on a single stalk. Scientists hypothesize that the first simple animals may have arisen when some of these cells gradually acquired specialized functions—for example, movement or nutrition—while still maintaining coordination with other cells and cell types. As discussed later, evolutionary changes to this simple body plan resulted in critical innovations that led to the more complex body plans found in modern animals.

## Animal Phyla Have Broad Differences Related to Body Plan, Germ Layers, and Features of Embryonic Development

Prior to the use of molecular data in phylogeny, biologists traditionally classified animal diversity in terms of three main morphological and developmental features of animal body plans:

1. Type of body symmetry
2. Number of germ layers
3. Specific features of embryonic development

We will discuss each of these major features of animal body plans next.

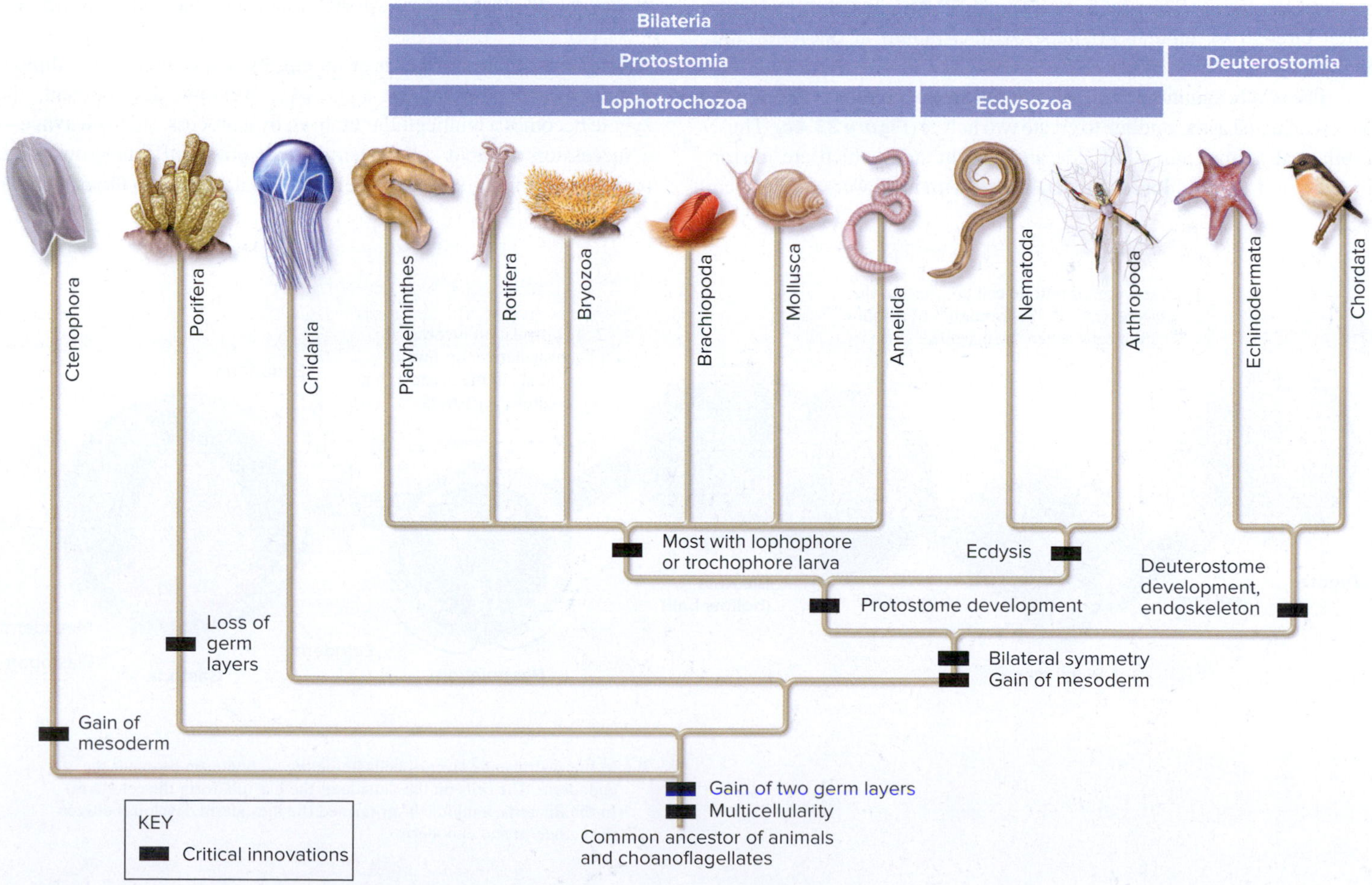

**Figure 33.3  An animal phylogenetic tree based on body plans and molecular data.** Biologists have identified about 35 different animal phyla. We will focus our discussions here and in the next two chapters on the 13 most abundant and recognizable phyla.

 **Core Concept: Evolution** As shown at the bottom of this tree, the first animals evolved from a choanoflagellate-like ancestor.

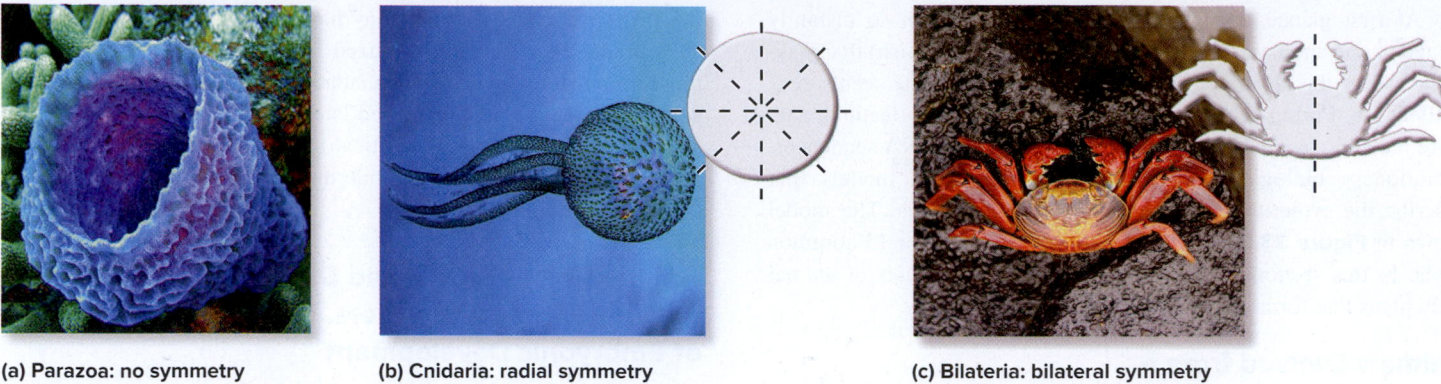

**(a) Parazoa: no symmetry**          **(b) Cnidaria: radial symmetry**          **(c) Bilateria: bilateral symmetry**

**Figure 33.4** **Early divisions in the animal phylogeny.** Animals can be categorized based on body symmetry **(a)** the absence of symmetry (Parazoa, the sponges); **(b)** radial symmetry (the cnidarians); or **(c)** bilateral symmetry (Bilateria, all other animals). a: ©E Teister/age fotostock; b: ©Gavin Parsons/Getty Images; c: ©Jens Kuhfs/Getty Images

*Symmetry*   Animals may be categorized according to the type of symmetry their body displays. Symmetry refers to the existence of balanced proportions of the body on either side of a median plane. Some of the earliest-diverging animals, such as sponges, were asymmetric, meaning they had no plane of symmetry (**Figure 33.4a**). Radially symmetric animals can be divided equally by any longitudinal plane passing through the central axis (**Figure 33.4b**). Such animals are often circular or tubular in shape, with a mouth at one end, and include cnidarians (jellyfish).

Bilaterally symmetric animals, the **Bilateria**, can be divided along a vertical plane at the midline to create two halves (**Figure 33.4c**). Thus, a bilateral animal has a left side and a right side, which are mirror images, as well as a **dorsal** (upper) and a **ventral** (lower) side, which

are not identical, and an **anterior** (head) and a **posterior** (tail) end. Bilateral symmetry is strongly correlated with both the ability to move through the environment and **cephalization**—the localization of sensory structures at the anterior end of the body. Such abilities allow animals to encounter their environment initially with their head, which is best equipped to detect and consume prey and to detect and respond to predators and other dangers. Most animals are bilaterally symmetric.

*Germ Layers*   Fertilization of an egg by a sperm creates a diploid zygote. During the earliest stage of embryonic development, the zygote becomes a multicellular embryo by a process called **cleavage**— a succession of rapid cell divisions with no significant growth that produces a hollow sphere of cells called a **blastula** (**Figure 33.5**).

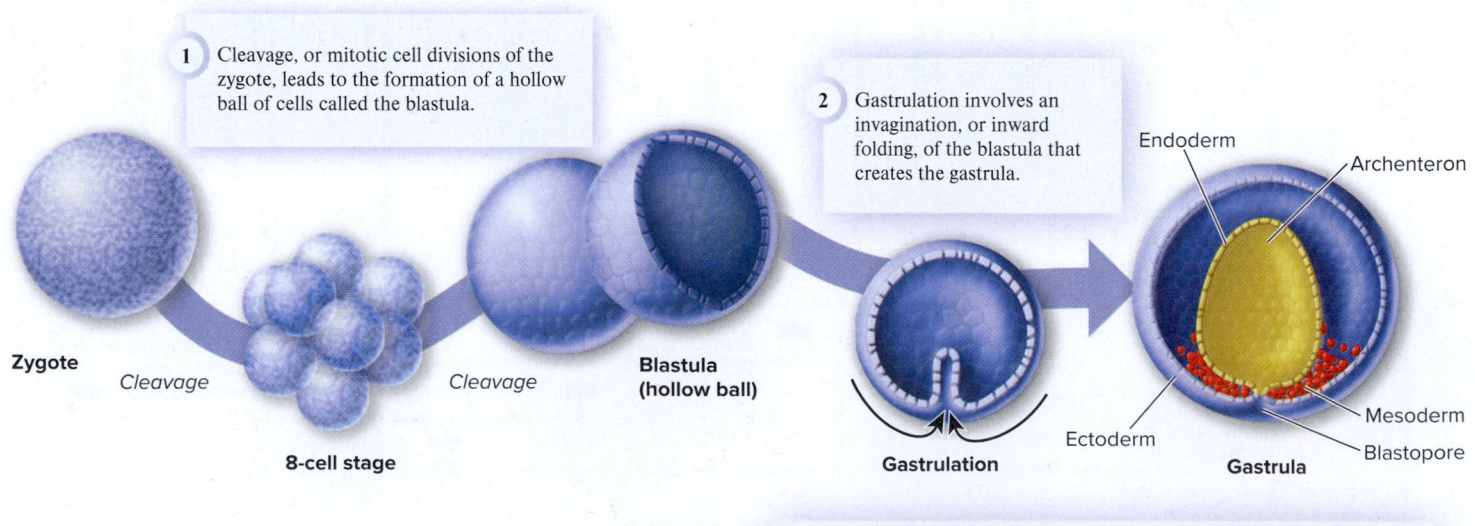

**1** Cleavage, or mitotic cell divisions of the zygote, leads to the formation of a hollow ball of cells called the blastula.

**2** Gastrulation involves an invagination, or inward folding, of the blastula that creates the gastrula.

Endoderm

Archenteron

**Zygote**   *Cleavage*   *Cleavage*   **Blastula (hollow ball)**

**8-cell stage**

Mesoderm

Ectoderm

Blastopore

**Gastrulation**   **Gastrula**

**3** In the gastrula, the layer of cells lining the archenteron becomes the endoderm. The cells on the outside of the blastula form the ectoderm. In the Bilateria, a middle layer termed the mesoderm develops between the ectoderm and endoderm.

**Figure 33.5** **Formation of germ layers.**   Note: Radially symmetric animals (cnidarians) do not form mesoderm.

   **Core Skill: Connections**   Look back to Figure 25.8. Is the existence of three germ layers in triploblastic animals a shared primitive character or a shared derived character?

In all animals except the sponges, the growing embryo then develops different layers of cells, called **germ layers**. During **gastrulation**, an area in the blastula folds inward, or invaginates, creating in the process a structure called a **gastrula**. The inner layer of cells becomes the **endoderm**, which lines the primitive digestive tract. The outer layer, or **ectoderm**, covers the surface of the embryo and differentiates into the epidermis and nervous system.

A key difference between Bilateria and most other animals is that the Bilateria develop a third layer of cells, termed the **mesoderm**, between the ectoderm and endoderm. Mesoderm forms the muscles and most other organs between the digestive tract and the ectoderm. Because the Bilateria have these three distinct germ layers, they are referred to as **triploblastic**, whereas the cnidarians, which have only ectoderm and endoderm, are termed **diploblastic**. Interestingly, the mesoderm of the earliest-diverging animals, the ctenophores, probably originated independently of the mesoderm found in bilaterians.

*Specific Features of Embryonic Development in the Bilateria*    In the Bilateria, a key feature of embryonic development concerns the development of a mouth and anus (**Figure 33.6a**). In gastrulation, the endoderm forms an indentation, the **blastopore**, which is the opening of the archenteron to the outside. In **protostomes** (from the Greek *protos*, meaning first, and *stoma*, meaning mouth) (see Figure 33.3), the blastopore becomes the mouth. If an anus is formed in a protostome species, it develops from a secondary opening. In contrast, in **deuterostomes** (from the Greek *deuteros*, meaning second), the blastopore becomes the anus, and the mouth is formed from a secondary opening.

Protostomes and deuterostomes also differ at the cleavage stage of embryonic development. As mentioned, the earliest stage of embryonic development involves a process known as cleavage (see Figure 33.5). Protostome development is generally characterized by so-called **determinate cleavage**, in which the fate of each embryonic cell is determined very early (**Figure 33.6b**). If one of the cells

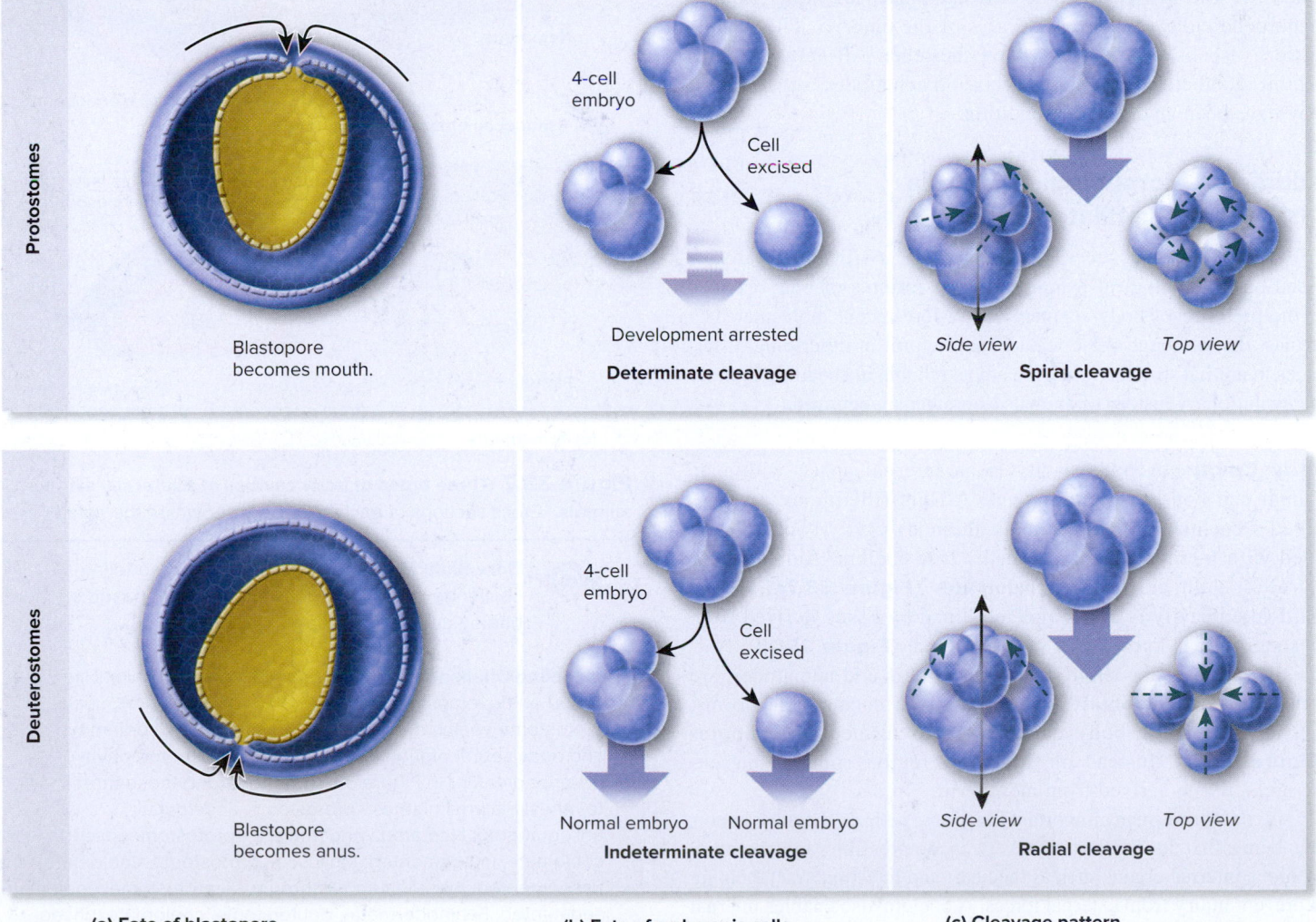

(a) Fate of blastopore          (b) Fate of embryonic cells          (c) Cleavage pattern

**Figure 33.6    Differences in embryonic development between protostomes and deuterostomes.  (a)** In protostomes, the blastopore becomes the mouth. In deuterostomes, the blastopore becomes the anus. **(b)** Protostomes have determinate cleavage, whereas deuterostomes have indeterminate cleavage. **(c)** Many protostomes have spiral cleavage, whereas all deuterostomes have radial cleavage. The dashed arrows indicate the direction of cleavage.

is removed from a four-cell protostome embryo, neither the single cell nor the remaining three-cell mass can form viable embryos, and development is halted. In contrast, deuterostome development in most species is characterized by **indeterminate cleavage**, in which each cell produced by early cleavage retains the ability to develop into a complete embryo. For example, when one cell is excised from a four-cell sea urchin embryo, both the single cell and the remaining three can go on to form viable embryos. Other embryonic cells compensate for the missing cells. In human embryos, if individual embryonic cells separate from one another early in development, identical twins can result.

Another distinguishing feature of the early bilaterian embryo is the cleavage pattern (**Figure 33.6c**). In **spiral cleavage**, the planes of cell cleavage are oblique to the vertical axis of the embryo, resulting in an arrangement in which newly formed upper cells lie centered between the underlying cells. Many protostomes, including mollusks and annelid worms, exhibit spiral cleavage. The coiled shells of some mollusks result from spiral cleavage. Organisms with spiral cleavage are also known as spiralians. In **radial cleavage**, the cleavage planes are either parallel or perpendicular to the vertical axis of the embryo. This results in tiers of cells, one directly above the other. All deuterostomes exhibit radial cleavage, as do insects and nematodes, suggesting it may have been an ancestral condition.

## Additional Morphological Criteria Distinguish the Bilateria

In older phylogenetic trees of animal life, classification was also based on morphological features, such as features of body cavities or the presence of body segmentation. More recent molecular data suggest that although these features are helpful in describing differences in animal structure, they are not as reliable in shedding light on the evolutionary history of animals as previously believed.

***Body Cavity*** A body cavity is an internal space within an animal that houses internal organs. A fluid-filled body cavity is called a **coelom**. In many animals, the body cavity is completely lined with mesoderm and is called a true coelom. Animals with a true coelom are termed **coelomates** (**Figure 33.7a**). If the fluid-filled cavity is not completely lined by tissue derived from mesoderm, it is known as a pseudocoelom (**Figure 33.7b**). Animals with a pseudocoelom, including rotifers and nematodes, are termed **pseudocoelomates**. Some animals, such as flatworms, lack a fluid-filled body cavity and are termed **acoelomates** (**Figure 33.7c**). Instead of fluid, this region contains mesenchyme, a tissue derived from mesoderm.

A coelom has many important functions, perhaps the most important being that its fluid is relatively incompressible and therefore cushions internal organs such as the heart and intestinal tract, helping to prevent injury from external forces. A coelom also enables internal organs to move and grow independently of the outer body wall. Furthermore, in some soft-bodied invertebrates, such as earthworms, the coelom functions as a **hydrostatic skeleton**—a fluid-filled body cavity surrounded by muscles that gives support and shape to the body of organisms. Muscle contractions at one part of the body push this

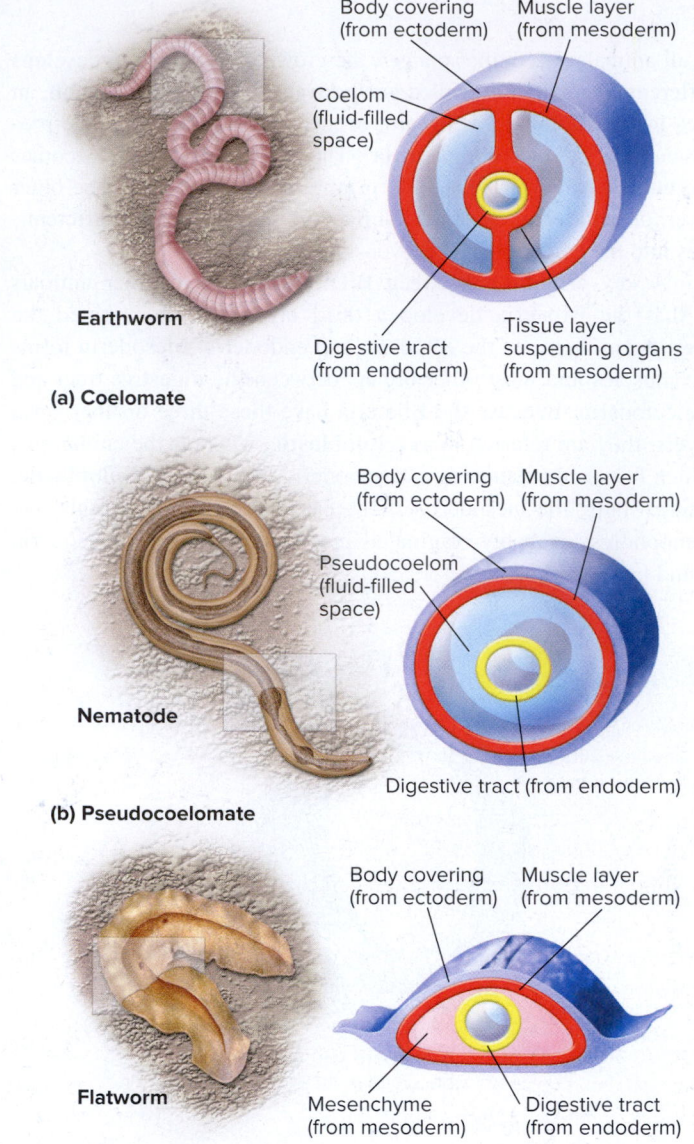

**(a) Coelomate**

**(b) Pseudocoelomate**

**(c) Acoelomate**

**Figure 33.7 Three types of body cavities of bilaterally symmetric animals.** Cross sections of each animal are shown on the right.

**Core Skill: Modeling** The goal of this modeling challenge is to draw a phylogenetic tree based on morphological and developmental information.

**Modeling Challenge:** Older phylogenetic trees of animal life were based on developmental and morphological features, such as protostome versus deuterostome development, coelom type, and body segmentation. Let's consider eight of the phyla described earlier in Figure 33.3 with regard to these three features: Platyhelminthes—protostome, acoelomate, unsegmented; Nematoda and Rotifera—protostome, pseudocoelomate, unsegmented; Mollusca—protostome, coelomate, unsegmented; Annelida and Arthropoda—protostome, coelomate, segmented; Echinodermata—deuterstome, coelomate, unsegmented; and Chordata—deuterostome, coelomate, segmented. Based on these three features, draw a phylogenetic tree for the eight phyla. On your tree, place horizontal black bars to indicate the occurrence of these three critical innovations.

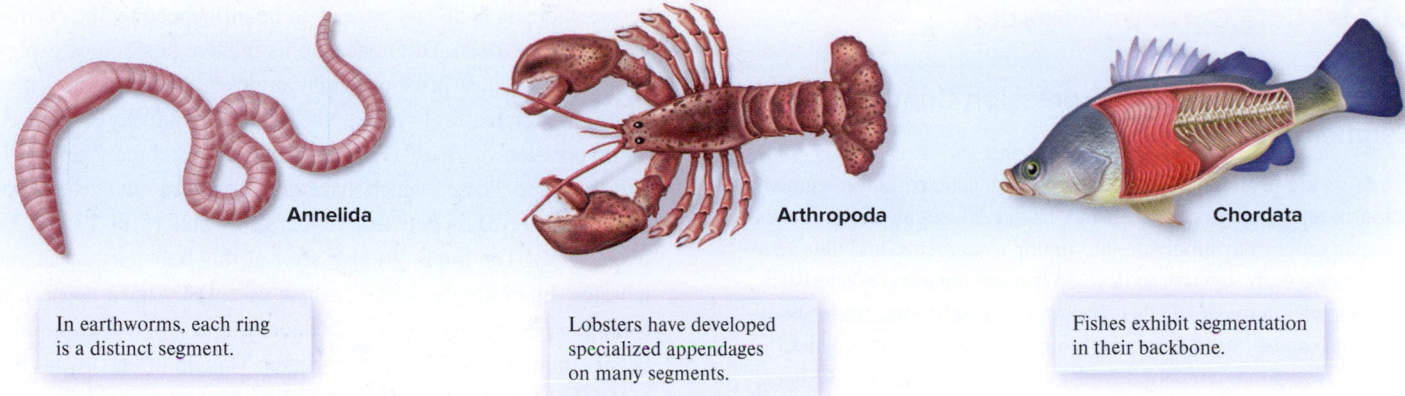

Annelida

Arthropoda

Chordata

In earthworms, each ring is a distinct segment.

Lobsters have developed specialized appendages on many segments.

Fishes exhibit segmentation in their backbone.

**Figure 33.8  Segmentation.** Annelids, arthropods, and chordates all exhibit segmentation.

fluid toward another part of the body. This type of movement can best be observed in an earthworm. Finally, in some organisms, the fluid in the body cavity also acts as a simple circulatory system.

The presence or absence of a coelom or pseudocoelom was previously used in the construction of animal phylogenies. However, scientists now believe this feature may not be useful in classification because animals that once possessed coeloms may have lost them over long periods of evolutionary time, as is true for the ancestors of flatworms. In addition, the coelom may have arisen twice in animal evolution, once in protostomes and once in deuterostomes.

**Segmentation**    Another feature of the animal body plan is the presence or absence of segmentation. In segmentation, the body is divided into regions called segments. Even though segmentation is a common feature of the Bilateria, its presence is more obvious in some phyla compared to others. In annelids (segmented worms), most segments contain the same set of blood vessels, nerves, and muscles (**Figure 33.8**). Some segments may differ, such as those containing the brain or sex organs. Segmentation is also evident in arthropods (such as lobsters and insects), but less so in chordates (such as fish and mammals) (Figure 33.8). Most species of chordata are vertebrates, which possess a series of small bones called vertebrae that form the backbone. The repeating pattern of vertebrae indicates segmentation.

The advantage of segmentation is that it allows specialization of body regions. For example, as we will see in Chapter 34, arthropods exhibit a vast degree of specialization of their segments. Many insects have wings and only three pairs of legs, whereas centipedes have no wings and many legs. Crabs, lobsters, and shrimp have highly specialized thoracic appendages that aid in feeding.

**BIO TIPS**

**THE QUESTION** *Three phyla containing species with obvious segmentation are Annelida, Arthropoda, and Chordata. Are such segmented animals a monophyletic, polyphyletic, or paraphyletic group?*

**T OPIC** *What topic in biology does this question address?* The topic is evolutionary relationships based on segmentation.

**I NFORMATION** *What information do you know based on the question and your understanding of the topic?* In the question, you are reminded that a taxonomic group can be monophyletic, polyphyletic, and paraphyletic. From your understanding of taxonomy and systematics, discussed in Chapter 25, you may remember that a monophyletic group contains a common ancestor and all of its descendants, a paraphyletic group contains a common ancestor but not all of its descendants, and a polyphyletic group contains groups of species with different common ancestors. You should also remember which phyla are segmented.

**P ROBLEM-SOLVING S TRATEGY** *Make a drawing.* One strategy for solving this problem is to draw a simple version of Figure 33.3 with just the names of the phyla and the branches, and then place a star next to each phylum with segmented animals. Next, look at the comparison of monophyletic, polyphyletic, and paraphyletic taxonomic groups shown in Figure 25.5. Decide which pattern best matches your drawing.

**ANSWER** *Because segmented animals have different common ancestors, the group is polyphyletic.*

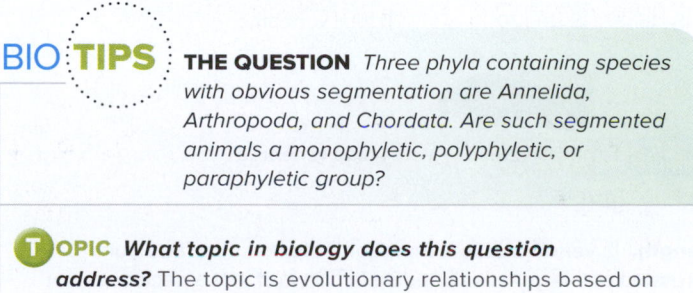

### 🔵 Core Concept: Evolution

## Changes in *Hox* Gene Expression Control Body Segment Specialization

Scientists are beginning to understand the genetic basis for segmentation in animals. As described in Chapter 20, segmentation genes cause an embryo to subdivide into multiple segments, and then *Hox* genes cause each segment to develop its own unique characteristics. Recent studies have shown that changes in specialization among body segments can be traced to relatively simple changes in *Hox* genes.

The *Hox* genes are organized into four clusters of 13 genes, each designated with a number from 1 through 13. Some of these genes are expressed in anterior segments; others are expressed in posterior segments (refer back to Figure 20.17). In the 1990s, Greek molecular biologist Michalis Averof and coworkers showed how relatively simple shifts in the expression patterns of *Hox* genes along the anteroposterior axis can account for the large variation in arthropod appendage types. More recent work by Averof and colleagues (2010) has shown how specific changes in *Hox* expression are linked to changes in crustacean maxillipeds—appendages near the mouth that are used for feeding. Maxillipeds arise in the anterior thoracic segments and display a mixture of locomotory and feeding functions. By knocking out *Hox* genes or expressing *Hox* genes in an abnormal position, the researchers could change maxillipeds into leglike appendages or transform leglike appendages into maxillipeds.

Shifts in the patterns of expression of *Hox* genes in the embryo along the anteroposterior axis are similarly prominent in vertebrate evolution. In vertebrates, the transition from one type of vertebra to another, for example, from cervical (neck) to thoracic (chest) vertebrae, is controlled by particular *Hox* genes. The site of the cervicothoracic boundary appears to be influenced by the *HoxC-6* gene (**Figure 33.9**). Differences in its relative position of expression, which occurs prior to vertebrae development, control neck length in vertebrates. In mice, which have a relatively short neck, the expression of *HoxC-6* begins between vertebrae 7 and 8. In chickens and geese, which have longer necks, the expression begins farther back, between vertebrae 14 and 15 or 17 and 18, respectively. The forelimbs also arise at this boundary in all vertebrates. Interestingly, snakes, which essentially have no neck or forelimbs, do not exhibit this boundary, and *HoxC-6* expression occurs immediately behind their heads. This, in effect, means that snakes got longer by losing their neck and lengthening their chest.

American molecular biologist Sean Carroll has remarked that it is very satisfying to find that the evolution of body forms and novel structures in two of the most successful and diverse animal groups, arthropods and vertebrates, is shaped by the shifting of *Hox* genes. It also reminds us of one of the core concepts of biology—that evolution often involves descent with modification. Much of the diversity in animal phyla can be seen as modifications to a general body plan.

## The Animal Kingdom Encompasses Many Diverse Phyla

**Table 33.2** summarizes the basic characteristics of the major animal phyla. In Chapter 34, we will discuss the Ctenophora, Porifera, Cnidaria, Lophotrochozoa, and Ecdysozoa, and also the invertebrate members of Deuterostomia; these are all the animals without a backbone. In Chapter 35, we will turn our attention to the the phylum Chordata, which is the largest group of deuterostomes. These include fishes, amphibians, reptiles, and mammals, which possess a backbone.

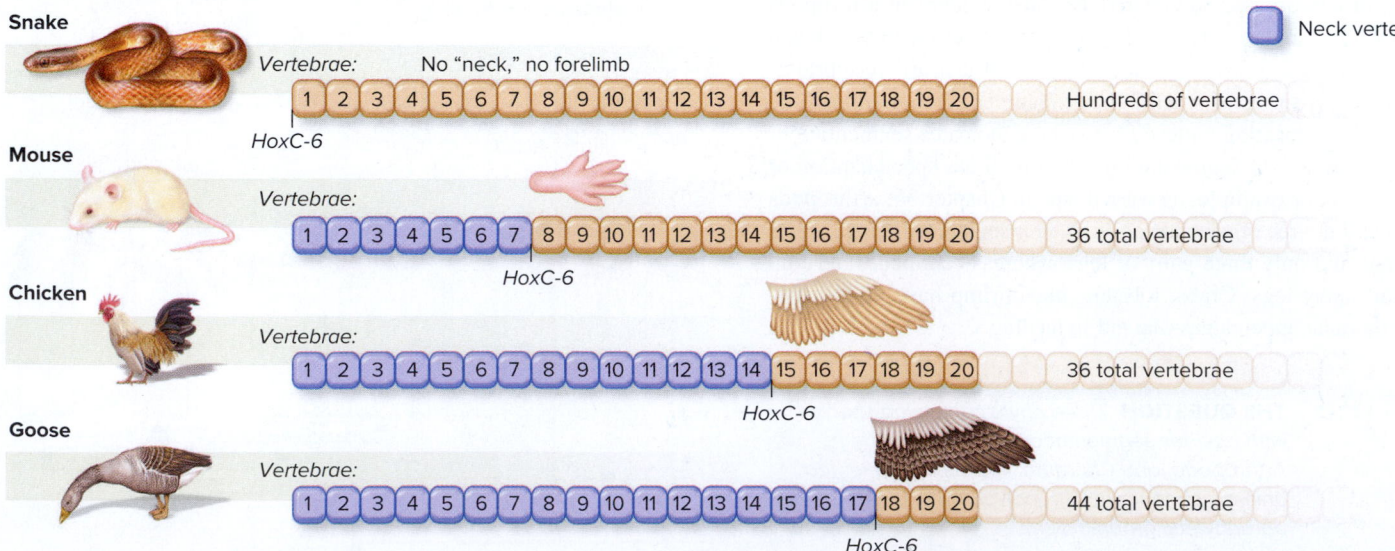

**Figure 33.9** **Relationship between *HoxC-6* gene expression and neck length.** In vertebrates, the transition between neck and trunk vertebrae is controlled by the position of the *HoxC-6* gene. In snakes, the expression of this gene is shifted so far forward that a neck does not develop.

| Table 33.2 | Summary of the Basic Characteristics of the Major Animal Phyla | | | | | | | | | | | |

| Feature | Ctenophora (comb jellies) | Porifera (sponges) | Cnidaria (hydra, anemones, jellyfish) | Platyhel- minthes (flatworms) | Rotifera (rotifers) | Bryozoa and Brachiopoda (bryozoans and brachiopods) | Mollusca (snails, clams, squids) | Annelida (segmented worms) | Nematoda (round- worms) | Arthropoda (insects, arachnids, crustaceans) | Echinoder- mata (sea stars, sea urchins) | Chordata (vertebrates and others) |
|---|---|---|---|---|---|---|---|---|---|---|---|---|
| Estimated number of species | 200 | 8,500 | 9,000 | 20,000 | 2,200 | 4,800 | 110,000 | 18,000 | 25,000 | 1,000,000+ | 7,400 | 69,730 |
| Level of organization | Tissue; lack organs | Cellular; lack tissues and organs | Tissue; lack organs | Organs | Organs | Organs | Organs | Organs | Organs | Organs | Organs | Organs |
| Symmetry | Radial | Absent | Radial | Bilateral | Bilateral | Bilateral | Bilateral | Bilateral | Bilateral | Bilateral | Bilateral larvae, radial adults | Bilateral |
| Cephalization | Absent | Absent | Absent | Present | Present | Reduced | Present | Present | Present | Present | Absent | Present |
| Germ layers | Three | Absent | Two | Three | Three | Three | Three | Three | Three | Three | Three | Three |
| Body cavity, or Coelom | Absent | Absent | Absent | Absent | Pseudo- coelom | Coelom | Reduced Coelom | coelom | Pseudo- Coelom | Reduced coelom | Coelom | Coelom |
| Obvious segmentation in the adult | Absent | Absent | Absent | Absent | Absent | Absent | Absent | Present | Absent | Present | Absent | Present |

## 33.3 The Use of Molecular Data in Constructing Phylogenetic Trees for Animals

**Learning Outcomes:**

1. Discuss how molecular data are used to construct and revise phylogenetic trees.
2. List the morphological features of the Ecdysozoa and the Lophotrochozoa.

In Chapter 25 (see Sections 25.2 and 25.3), we considered how molecular data can be used to construct phylogenetic trees. This approach involves the comparison of genetic data, such as DNA, RNA, and amino acid sequences from different species to estimate their evolutionary relationships based on the degree of similarities between the sequences. More closely related species exhibit fewer sequence differences than distantly related ones.

In Section 33.2, we explored the relationships between major features of animal body plans and animal phylogeny. Early phylogenetic trees for the animal kingdom were based largely on morphological features. In the past few decades, however, major revisions to such trees have occurred when biologists have compared molecular data, such as DNA sequences, with morphological data. The phylogenetic tree shown earlier, in Figure 33.3, is derived from morphological data and also from more recent molecular data.

In this section, we will consider how molecular data are used to refine our understanding of the evolutionary relationships among animals.

### The Sequences of SSU rRNA Genes and *Hox* Genes Are Analyzed to Determine Broad Evolutionary Relationships Among Animals

As discussed in Chapter 25, the sequences of some genes change fairly slowly during evolution. Such slowly changing genes are particularly useful for evaluating broad evolutionary relationships, such as comparing phyla. Scientists have often focused on comparing base sequences in the gene that encodes RNA of the small ribosomal subunit (SSU rRNA) (see Chapter 12). SSU rRNA is universal in all organisms, and its base sequence has changed very slowly over long periods of time. We can appreciate this phenomenon by comparing a very small portion of the sequence of the SSU rRNA gene of a sponge, flatworm, seagull, and paramecium (**Figure 33.10**) in much the same way as we did in Chapter 12 (refer back to Figure 12.17). The three animal sequences are very similar to each other, and all of them differ from that of the paramecium (a protist).

The three animal species shown in Figure 33.10 are members of different phyla. If we compared three different species of sponges with three different species of flatworms, we would find that the sequences from the three species of sponges are more similar to each other than they are to those of the flatworms, and vice versa.

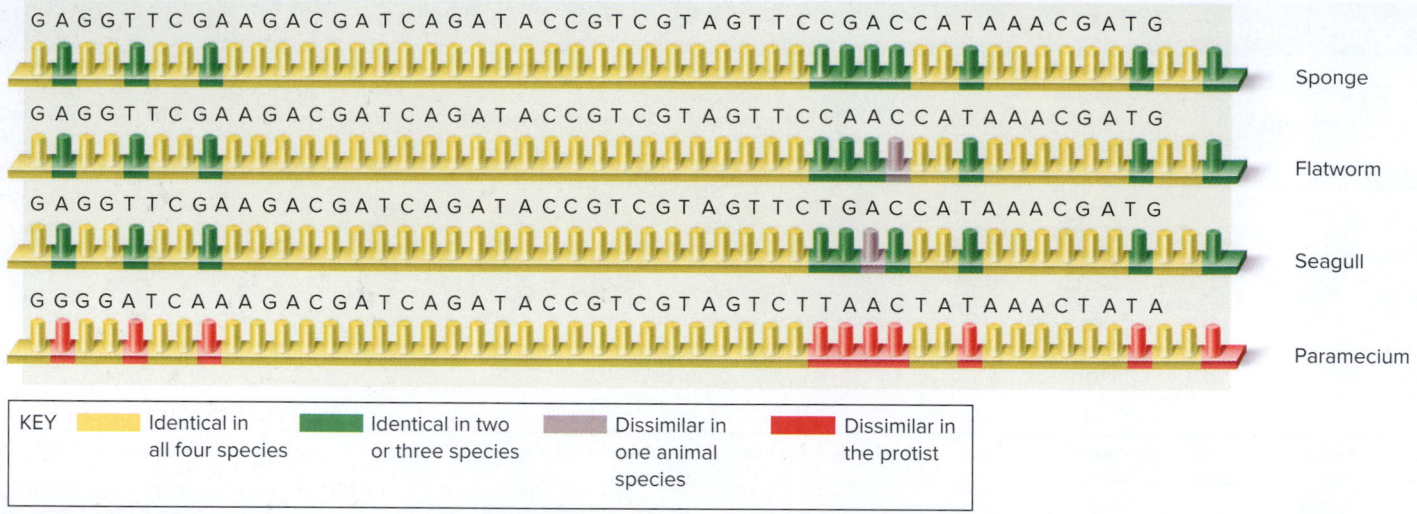

KEY

| Identical in all four species | Identical in two or three species | Dissimilar in one animal species | Dissimilar in the protist |

**Figure 33.10** **Comparison of small subunit (SSU) rRNA gene sequences from three animals and a protist.** Note the similarities between the animals, even though they are very distantly related within the animal kingdom.

 **Core Skill: Connections** Look back at Figure 12.17. Which color represents sequences of bases that are the most evolutionarily conserved?

A second approach for understanding broad evolutionary relationships among animals is to analyze genes that have played a major role in animal diversification. Researchers have studied *Hox* genes, which are found in cnidarians and bilaterians, to study the evolution of body plans (refer back to Figure 24.15). They hypothesize that the duplication of *Hox* genes and gene clusters has led to the evolution of more complex animal body forms. Examination of the genes that regulate early developmental differences has provided insight into the evolution of animal development and the mechanisms by which animal body plans have diversified.

Studies using molecular data have resulted in major revisions to phylogenetic trees that were previously based on morphological data. As an example, **Figure 33.11** compares the phyla in Protostomia. Figure 33.11a is the current view, which was also shown as part of Figure 33.3. Figure 33.11b is a previous model that was based on morphological data. As described next, the revisions that created our current view of the phylogeny of Protostomia are derived from molecular studies.

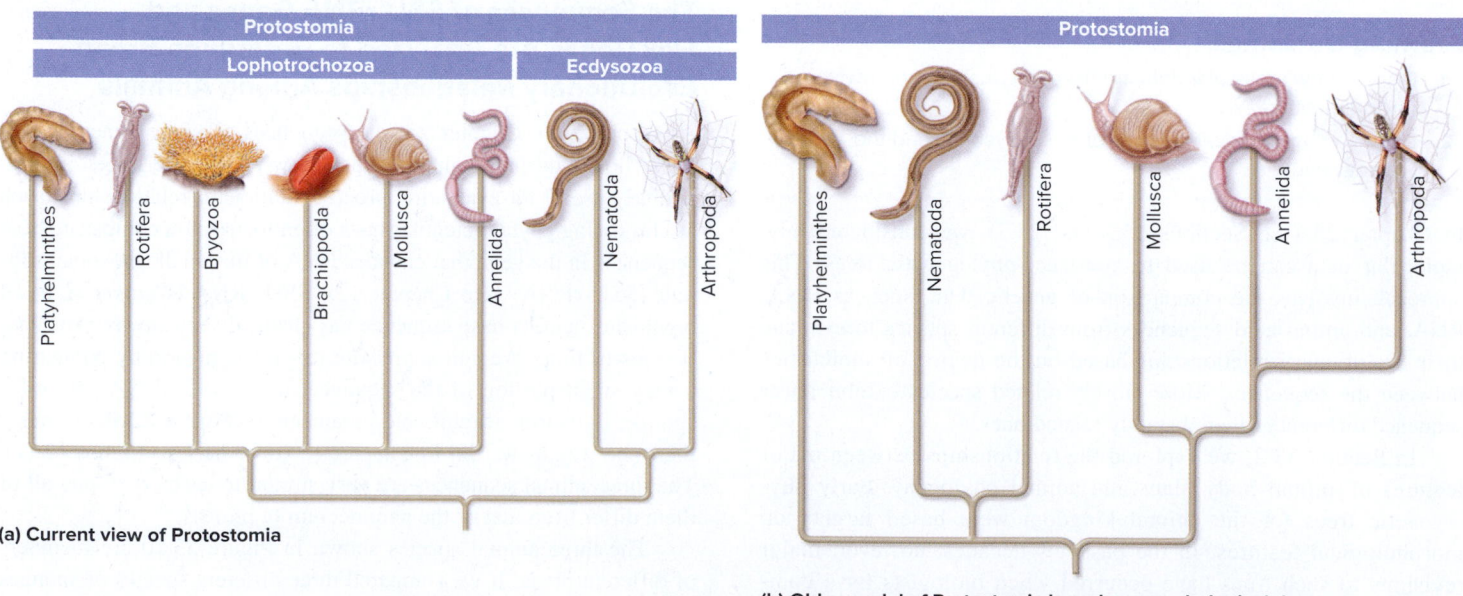

**(a) Current view of Protostomia**

**(b) Older model of Protostomia based on morphological data**

**Figure 33.11** **A comparison of a current phylogenetic tree for Protostomia with an older tree based on morphological data.** The current tree shown in **(a)** is part of the tree shown earlier in Figure 33.3. The model shown in part **(b)** is no longer accepted.

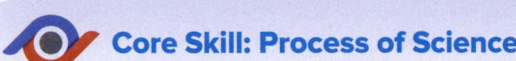

**Core Skill: Process of Science**

**Feature Investigation** | **Aguinaldo and Colleagues Analyzed SSU rRNA Sequences to Determine the Taxonomic Relationships of Arthropods to Other Phyla in Protostomia**

In 1997, American molecular biologists Anna Marie Aguinaldo, James Lake, and colleagues analyzed the relationships of arthropods to other phyla by sequencing the complete gene that encodes SSU rRNA from a variety of representative phyla (**Figure 33.12**). Total genomic DNA was isolated using standard techniques and amplified by polymerase chain reaction (PCR; refer back to Figure 21.6). PCR fragments were then subjected to DNA sequencing, a technique also described in Chapter 21, and the evolutionary relationships among 50 species were examined.

The resulting data indicated the existence of a monophyletic clade—the Ecdysozoa—containing the nematodes and arthropods (see step 3 in Figure 33.12 and Figure 33.11a). The hypothesis that nematodes are more closely related to arthropods than previously thought has important ramifications. First, it implies that

**Figure 33.12**  **A revised animal phylogeny based on a comparison of SSU rRNA genes.**

**GOAL**  To determine the evolutionary relationships among many animal species, especially the relationship of arthropods to other species.

**KEY MATERIALS**  Cellular samples from about 50 animals in different taxa.

| Experimental level | | Conceptual level |
|---|---|---|

**1**  Isolate DNA from animals and subject the DNA to polymerase chain reaction (PCR) to obtain enough material for DNA sequencing. PCR is described in Chapter 21.

For more detail, refer back to Figure 21.6.

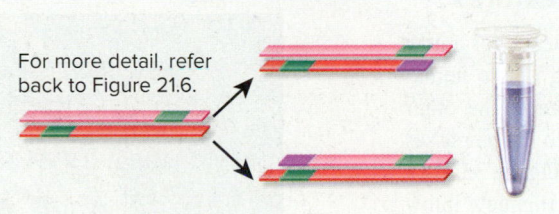

The goal of PCR is to amplify a region in the SSU rRNA gene.

**2**  Sequence the amplified DNA by dideoxy sequencing, also described in Chapter 21.

For more detail, refer back to Figure 21.8.

CACCGTA

Dideoxy sequencing, in which DNA strands are separated according to their lengths by subjecting them to gel electrophoresis, is used to determine the base sequence of DNA.

**3**  Compare the DNA sequences and infer phylogenetic relationships using the cladistic approach described in Chapter 25.

Lophotrochozoa          Ecdysozoa

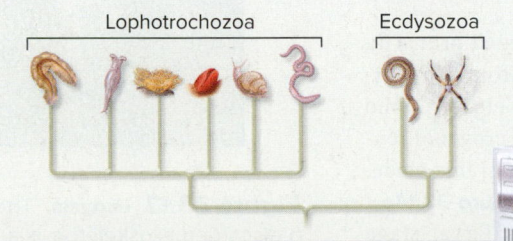

The approach compares traits that are either shared or not shared by different species and creates clades, consisting of a common ancestral species.

**4**  **THE DATA**

This process resulted in a large group of DNA sequences that were then analyzed with the use of computer programs.

two well-researched model organisms, *Caenorhabditis elegans* (a nematode) and the fruit fly *Drosophila melanogaster* (an arthropod), are more closely related than had been believed. Second, morphological classification had assumed that arthropods and annelids were closely related to each other based on the presence of segmentation. Molecular data does not support the previous hypothesis that annelids and arthropods form a clade of segmented animals (see Figure 33.11b).

### Experimental Questions

1. What was the purpose of the study conducted by Aguinaldo and colleagues?
2. **CoreSKILL »** What was the major finding of this particular study?
3. What impact does the new view of nematode and arthropod phylogeny have on other areas of research?

## The Ecdysozoa and Lophotrochozoa Have Some Distinctive Morphological Features

The study by Aguinaldo and colleagues provided evidence for a new clade of molting animals, the **Ecdysozoa**, consisting of the nematodes and arthropods. According to molecular evidence, the other major protostome clade is the **Lophotrochozoa**, which encompasses the mollusks, annelids, and several other phyla (see Figure 33.11a). When some morphologists reviewed their data given this new information, they found morphological support for these new groupings. Let's look at what morphological features make each of these groups unique.

The Ecdysozoa is so named because all of its members secrete a nonliving cuticle, an external skeleton (exoskeleton); think of the hard shell of a beetle or that of a crab. As these animals grow, the exoskeleton becomes too small, and the animal molts, or breaks out of its old exoskeleton, and secretes a newer, larger one (**Figure 33.13**). This molting process is called ecdysis; hence the name Ecdysozoa. Although this group was named for this morphological characteristic, it was first strongly supported as a separate clade by molecular evidence.

Similarly, the Lophotrochozoa clade was organized primarily through analyses of molecular data. Its name stems from two morphological features seen in many organisms of this clade: Lopho is derived from the **lophophore**, a horseshoe-shaped crown of tentacles used for feeding that is present on some phyla in this clade, such as the rotifers, bryozoans, and brachiopods (**Figure 33.14a**); trocho refers to the **trochophore larva**, a distinct larval stage characterized by a band of cilia around its middle that is used for swimming (**Figure 33.14b**). Trochophore larvae are found in several Lophotrochozoa phyla, such as annelid worms and mollusks, indicating their similar ancestry. However, other members of the clade, such as the platyhelminthes, have neither of these morphological features and are classified as lophotrochozoans based strictly on molecular data.

**Figure 33.13 Ecdysis.**  The dragonfly, shown here emerging from a discarded exoskeleton, is a member of the Ecdysozoa—a clade of animals exhibiting ecdysis, the periodic shedding (molting) and re-formation of the exoskeleton. ©Dwight Kuhn

 **Core Concept: Structure and Function**  For animals with exoskeletons, growth and development necessitate molting.

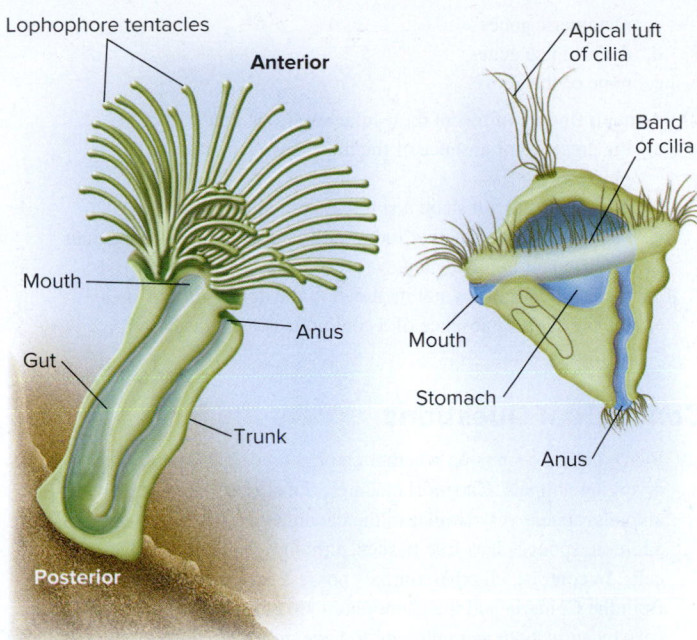

(a) Lophophore of a phoronid worm    (b) Trochophore larva

**Figure 33.14** **Characteristics of the Lophotrochozoa.** **(a)** A lophophore, a crown of ciliated tentacles, generates a current to bring food particles into the mouth. **(b)** The trochophore larval form is found in several animal lineages.

## Summary of Key Concepts

### 33.1  Characteristics of Animals

- Animals constitute a very species-rich kingdom, with a number of characteristics that distinguish them from other organisms, including multicellularity, an extracellular matrix, and unique cell junctions, in addition to heterotrophic feeding and internal digestion and the possession of nervous and muscle tissues (Table 33.1).

- Many different feeding modes are used by animals, including suspension feeding, bulk feeding and fluid feeding (Figure 33.1).

- The history of animal life on Earth spans over 630 million years. A profusion of animal phyla appeared in the Cambrian explosion (533–525 mya). Animals evolved adaptations to deal with the colonization of land, starting about 520 mya, and the number and diversity of mammals exploded after dinosaurs died out at the end of the Cretaceous period, 65 mya (Figure 33.2).

### 33.2  Animal Classification

- The animal kingdom is monophyletic, meaning that all taxa have evolved from a single common ancestor (Figure 33.3).

- Biologists hypothesize that animals evolved from a choanoflagellate-like ancestor.

- Animals can be categorized according to their type of symmetry, whether asymmetric (the sponges), radial (the cnidarians and ctenophores), or bilateral (Bilateria, all other animals) (Figure 33.4).

- The Cnidaria have two embryonic germ layers, the endoderm and the ectoderm, whereas the Bilateria and Ctenophora have a third

germ layer termed the mesoderm, which develops between the endoderm and the ectoderm (Figure 33.5).

- Animals are also classified according to patterns of embryonic development. In protostomes, the blastopore becomes the mouth; in deuterostomes, the blastopore becomes the anus. Most protostomes have spiral cleavage, and all deuterostomes have radial cleavage (Figure 33.6).

- Animals with a coelom, a body cavity that is completely lined with mesoderm, are termed coelomates. Animals that possess a coelom that is not completely lined by tissue derived from mesoderm are called pseudocoelomates. Those animals lacking a fluid-filled body cavity are termed acoelomates (Figure 33.7).

- Segmentation, the division of the body into identical subunits called segments, is an obvious feature of the animal body plan in certain phyla (Figure 33.8).

- Shifts in the pattern of expression of *Hox* genes are prominent in evolution. In vertebrates, the transition from one type of vertebra to another is controlled by certain *Hox* genes (Figure 33.9).

- Each animal phylum shows a distinctive set of general characteristics (Table 33.2).

### 33.3  The Use of Molecular Data in Constructing Phylogenetic Trees for Animals

- Phylogenetic trees are constructed and revised by comparing similarities in DNA, RNA, and amino acid sequences among different species (Figure 33.10).

- Molecular studies resulted in a revision to the animal phylogenetic tree; the protostomes were divided into two major clades: the Ecdysozoa and the Lophotrochozoa (Figures 33.11, 33.12).

- Members of the Ecdysozoa secrete and periodically shed a nonliving cuticle that is typically an exoskeleton, or external skeleton (Figure 33.13).

- The Lophotrochozoa are grouped primarily through analyses of molecular data, but some members are distinguished by two morphological features: the lophophore, a crown of tentacles used for feeding, and the trochophore larva, a distinct larval stage (Figure 33.14).

## Assess & Discuss

### Test Yourself

1. Which of the following is *not* a distinguishing characteristic of animals?
   a. the capacity to move at some point in the life cycle
   b. possession of cell walls
   c. multicellularity
   d. heterotrophy
   e. All of the above are characteristics of animals.

2. Which is the correct hierarchy of divisions in the animal kingdom, from most inclusive to least inclusive?
   a. Protostomia, Ecdysozoa, Bilateria
   b. Ecdysozoa, Protostomia, Bilateria
   c. Bilateria, Protostomia, Ecdysozoa
   d. Protostomia, Bilateria, Ecdysozoa
   e. none of the above

3. Bilateral symmetry is strongly correlated with
   a. the ability to move through the environment.
   b. cephalization.
   c. the ability to detect prey.
   d. a and b.
   e. a, b, and c.

4. In triploblastic animals, the inner lining of the digestive tract is derived from the
   a. ectoderm.
   b. mesoderm.
   c. endoderm.
   d. pseudocoelom.
   e. coelom.

5. Pseudocoelomates
   a. lack a fluid-filled cavity.
   b. have a fluid-filled cavity that is completely lined with mesoderm.
   c. have a fluid-filled cavity that is partially lined with mesoderm.
   d. have a fluid-filled cavity that is not lined with mesoderm.
   e. have an air-filled cavity that is partially lined with mesoderm.

6. Protostomes and deuterostomes can be classified based on
   a. cleavage pattern.
   b. destiny of the blastopore.
   c. whether the fate of the embryonic cells is fixed early during development.
   d. all of the above.

7. Indeterminate cleavage is found in
   a. annelids.
   b. mollusks.
   c. nematodes.
   d. vertebrates.
   e. all of the above.

8. Naturally occurring identical twins are possible only in animals that
   a. have spiral cleavage.
   b. have determinate cleavage.
   c. are protostomes.
   d. have indeterminate cleavage.
   e. have spiral and determinate cleavage and are protosomes.

9. Genes involved in the patterning of the body axis, that is, in determining characteristics such as neck length and appendage formation, are called
   a. small subunit (SSU) rRNA genes.
   b. *Hox* genes.
   c. metameric genes.
   d. determinate genes.
   e. none of the above.

10. A major finding of recent molecular studies is that
    a. the presence or absence of the mesoderm is not important in phylogeny.
    b. all animals do not share a single common ancestor.
    c. body symmetry, whether radial or bilateral, is not an important determinant in phylogeny.
    d. the echinoderms are not included in the deuterostome clade.
    e. the presence or absence of a coelom is not important for classification.

## Conceptual Questions

1. Fierce debate centers on whether ctenophores or sponges are the earliest-diverging animals. Choanoflagellates, the closest living relatives to animals, appear very similar to the choanocyte cells of sponges. In addition, sponges lack true tissues, although they have distinct types of cells. In contrast, all other animals possess one or more types of tissues. Both the Cnidaria and the Ctenophora are radially symmetric and, until very recently, both were thought to have only two germ layers, endoderm and ectoderm. All other animals are bilaterally symmetric and have three germ layers. Draw an animal phylogenetic tree that illustrated this viewpoint, including Cnidaria and Ctenphora as part of the same phyla, and label the critical innovations.

2. Why was the evolution of a coelom important?

3. **Core Concept: Evolution** Brachiopods, bryozoans, mollusks, and annelids all exhibit a lophophore or trochophore larvae. Are these monophyletic, paraphyletic, or polyphyletic groups? Explain your answer. Refer back to Figure 25.5 to remind yourself about these terms.

## Collaborative Questions

1. Discuss the many ways that animals can affect humans, both positively and negatively.

2. Summarize how molecular evidence has enabled scientists to refine their views on animal phylogeny.

# The Invertebrates

# 34

**What is this organism, and how does it feed?**

©Georgie Holland/age fotostock

**T**he organism shown in the chapter opening photograph looks like an underwater plant, complete with long leaflike structures and roots. However, you may be surprised to learn that it's an animal, a type of echinoderm called a feather star, which is related to sea stars. Its long arms catch food particles floating in the ocean current, and tiny tube feet pass these particles into special food gutters that run along the center of each arm and empty into the mouth. The number of arms varies from species to species and may reach 200. Feather stars can creep along the ocean floor by means of rootlike projections called cirri. About 550 species of feather stars are in existence today, but some fossil formations are packed with feather star fragments, showing us how successful the group was in the past.

The history of animal life on Earth has evolved over hundreds of millions of years. Some scientists suggest that changing environmental conditions, such as a buildup of dissolved oxygen and minerals in the ocean or an increase in atmospheric oxygen, eventually permitted higher metabolic rates and increased the activity of a wide range of animals. Others suggest that with the development of sophisticated locomotor skills, a wide range of predators and prey evolved, leading to an evolutionary "arms race" in which predators evolved powerful weapons and prey evolved more powerful defenses against them. Such adaptations and counteradaptations may have led to a proliferation of different lifestyles and taxa. Also, an increase in the number of *Hox* genes may have fostered an increase in animal diversity and complexity.

In this chapter and the next, we will survey the wondrous diversity of animal life on Earth. In this chapter, we will examine the **invertebrates**, animals without a backbone, a category that makes up more than 95% of all animal species. We begin by exploring some of the earliest animal lineages, the ctenophores, sponges, and jellyfish. We will then turn to the Lophotrochozoa and Ecdysozoa, the two groups of protostomes introduced in Chapter 33. Finally, we will examine the deuterostomes, focusing here on the echinoderms and the invertebrate members of the phylum Chordata. The animal classification outlined in Chapter 33 (refer back to Figure 33.3) will serve as the basis for our discussion of animal lineages. A simplified version of Figure 33.3 is shown in **Figure 34.1**. Keep in mind, however, that animal phylogeny is a work in progress, and further revisions,

refinements, and perhaps surprises lay ahead, as the genomes of more and more species are sequenced and compared.

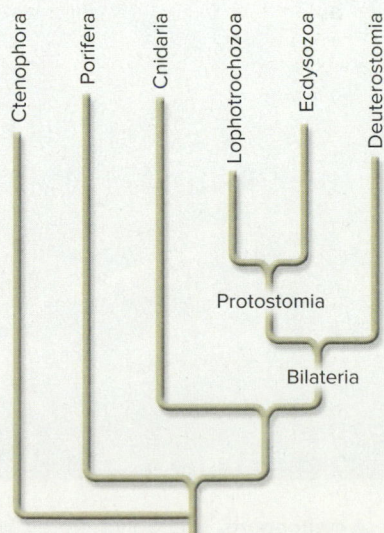

**Figure 34.1** **An animal phylogeny.** This phylogenetic tree summarizes our current understanding about the evolutionary relationships among animal groups.

# 34.1 Ctenophores: The Earliest Animals

## Learning Outcome:

1. Outline the unique features of ctenophores.

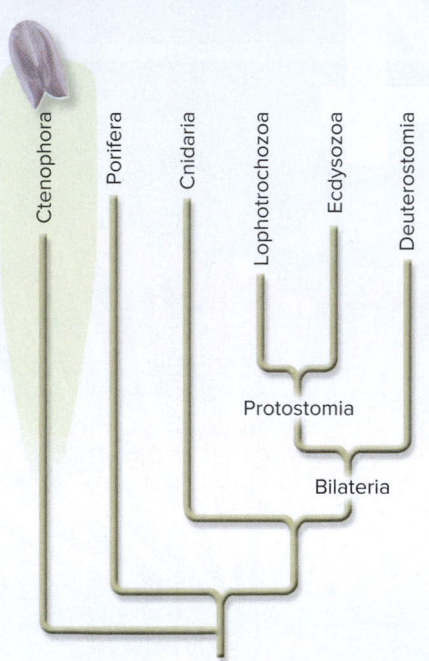

Ctenophores, also known as comb jellies, constitute the earliest-diverging animal lineage. Comb jellies are a small phylum of about 200 species, all of which are marine and look very much like jellyfish (**Figure 34.2**). They have on their surfaces eight rows of cilia that resemble combs. Based on morphological data, the ctenophores were once classified as close relatives to cnidarians, which include jellyfish and corals. However, recent molecular analyses have changed this view and placed them farther apart on the animal evolutionary tree.

The name Ctenophora (pronounced teen-o-for'-ah) comes from the Greek *ktenos,* meaning comb, and *phora,* meaning bearing. The coordinated beating of the cilia propels the ctenophores. Averaging about 1–10 cm in length, comb jellies are probably the largest animals to use cilia for locomotion.

Most comb jellies possess two long tentacles that secrete a sticky substance onto which small prey adhere. The tentacles are then drawn over the mouth. Digestion occurs in a body cavity called a **gastrovascular cavity**, and waste and water are eliminated through two anal pores. Prey are generally small and may include tiny crustaceans called copepods and small fishes.

Comb jellies are often transported around the world in ships' ballast water. *Mnemiopsis leidyi,* a ctenophore species native to the Atlantic coast of North and South America, was accidentally introduced into the Black and Caspian Seas in the 1980s. With a plentiful food supply and a lack of predators, *M. leidyi* underwent a population explosion and ultimately devastated the local fishing industries.

All ctenophores are **hermaphrodites** (from the Greek, for the god Hermes and the goddess Aphorodite), possessing both ovaries and testes, and gametes are shed into the water to unite and eventually form a free-swimming larva that is very similar in form to the adult. Nearly all ctenophores exhibit **bioluminescence**, a phenomenon that results from chemical reactions that give off light. Ctenophores are particularly evident at night. Sometimes, they wash up on shore and make the sand or mud appear luminescent.

Like jellyfish, ctenophores have both muscle and nerve cells organized as a diffuse net centralized at an elementary brain. However, the ctenophore nervous system uses different neurotransmitters than those in bilaterians and jellyfish and has different types of synapses. The presence of muscle cells originating from mesoderm suggests that ctenophores share a three-germ-layer embryonic structure with bilaterians. Even so, recent analyses of the genome of the ctenophore *M. leidyi* suggests that ctenophores lack many of the genes involved in specifying bilaterian mesoderm. In addition, ctenophores lack true *Hox* genes and possess a ctenophore-specific cleavage program. Finally, many bilaterian neuron-specific genes are absent or not expressed in ctenophores. These findings argue against a linear march of evolutionary forms from more simple animals such as ctenophores and sponges to complex bilaterians. Instead, evolutionary studies suggest that ctenophores were the earliest animals to diverge from a choanoflagellate-like ancestor that was multicellular and had a simple nervous system. Later, the ctenophores evolved their own unique way of forming mesoderm, which is different from the bilaterians.

**Figure 34.2  A ctenophore.** Ctenophores are called comb jellies because the eight rows of cilia on their surfaces resemble combs.
©Matthew J. D'Avella/SeaPics.com

# 34.2 Porifera: The Sponges

## Learning Outcomes:

1. Outline the body plan and unique characteristics of sponges.
2. Describe how sponges defend themselves against predators.

Members of the phylum Porifera (from the Latin, meaning pore bearers), are commonly referred to as sponges. Sponges lack true tissues—groups of cells that have a similar structure and function. However, sponges are multicellular and produce different types of specialized

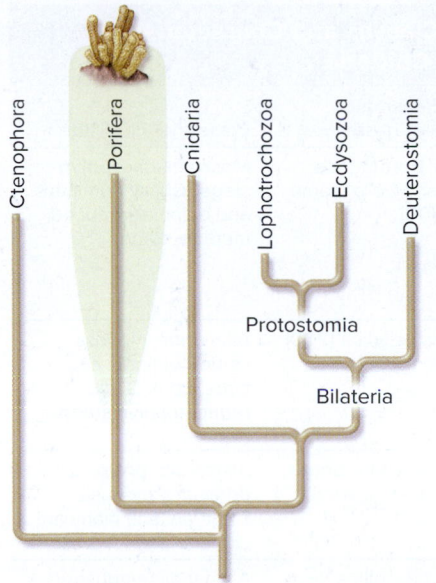

cells. Even though sponges carry most of the genes that are needed for a functioning nervous system, they have lost the ability to produce neurons during evolution. Biologists have identified approximately 8,000 species of sponges, the vast majority of which are marine. Sponges range in size from only a few millimeters across to more than 2 m in diameter. Smaller sponges may be radially symmetric, but most have no apparent symmetry. Some sponges have a low, encrusting growth form, whereas others grow tall and erect (**Figure 34.3a**). Although adult sponges are sessile—that is, anchored in place—the larvae are free-swimming.

## Choanocytes Help Circulate Water

The body of a sponge looks similar to a vase pierced with small holes or pores (**Figure 34.3b**). Water is drawn through these pores into a central cavity, the **spongocoel**, and flows out through the large opening at the top, called the osculum. The water enters the pores by the beating action of the flagella of the **choanocytes**, or collar cells, that line the spongocoel (**Figure 34.3c**). In the process, the choanocytes trap and eat small particulate matter and tiny plankton.

A layer of flattened epithelial cells similar to those making up the outer layer of animals in other phyla protects the sponge body. In between the choanocytes and the epithelial cells lies a gelatinous, protein-rich matrix called the **mesohyl**. Within this matrix are mobile cells called **amoebocytes** that absorb food from choanocytes, digest it, and carry the nutrients to other cells. Thus, considerable cell-to-cell contact and communication exist in sponges. Sponges are unique among the major animal phyla in using intracellular digestion, the uptake of food particles by cells, as a mode of feeding.

## Sponges Have Mechanical and Chemical Defenses Against Predators

Some amoebocytes can form tough skeletal fibers that support the sponge's body. In many sponges, this skeleton consists of sharp **spicules** formed of protein, calcium carbonate, or silica. For example, some deep-ocean species, called glass sponges, are distinguished by having needle-like silica spicules that form elaborate lattice-like skeletons. The presence of such tough spicules

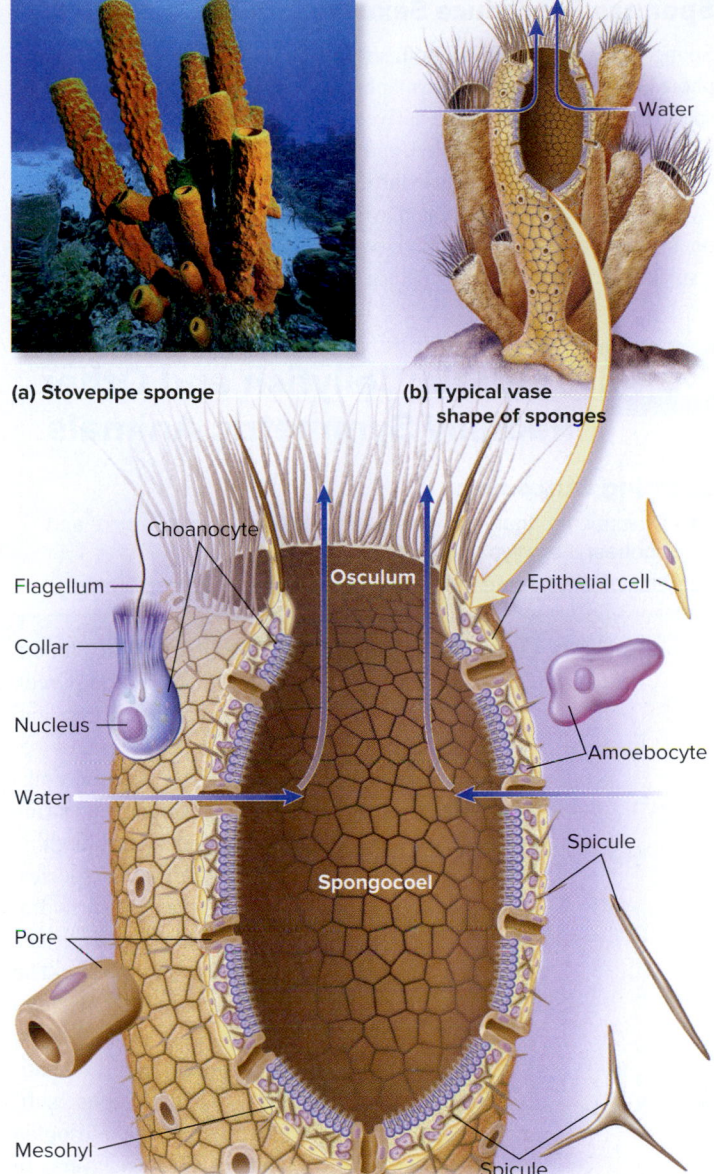

**(a) Stovepipe sponge**

**(b) Typical vase shape of sponges**

**(c) Cross section of sponge morphology**

**Figure 34.3  Sponge body plan. (a)** The stovepipe sponge (*Aplysina archeri*) is a common sponge found on Caribbean reefs. **(b)** Many sponges have a vaselike shape. **(c)** A cross section reveals that sponges are multicellular animals, having various cell types but no distinct tissues. a: ©Norbert Probst/age fotostock

*Concept Check:*  *If sponges are soft and sessile, why aren't they eaten by other organisms?*

may explain why predation of sponges is rare. Other sponges have fibers of a tough protein called **spongin** that lend skeletal support. Spongin skeletons are still commercially harvested and sold as bath sponges. Many species produce toxic defensive chemicals, some of which are being tested as possible anticancer and anti-inflammatory agents in humans.

## Sponges Reproduce Sexually and Asexually

Sponges reproduce through both sexual and asexual means. Like ctenophores, most sponges are hermaphrodites, and thus can produce both sperm and eggs. Gametes are derived from amoebocytes or choanocytes. The eggs remain in the mesohyl, and the sperm are released into the water and carried by water currents to fertilize the eggs of neighboring sponges. Zygotes develop into flagellated swimming larvae that eventually settle on a suitable substrate to become sessile adults. In asexual reproduction, a small fragment or bud may detach and form a new sponge.

## 34.3 Cnidaria: Jellyfish and Other Radially Symmetric Animals

**Learning Outcomes:**

1. Describe the four main classes of cnidarians, and compare and contrast the polyp and medusa body forms.
2. Describe how cnidarians defend themselves and capture prey.

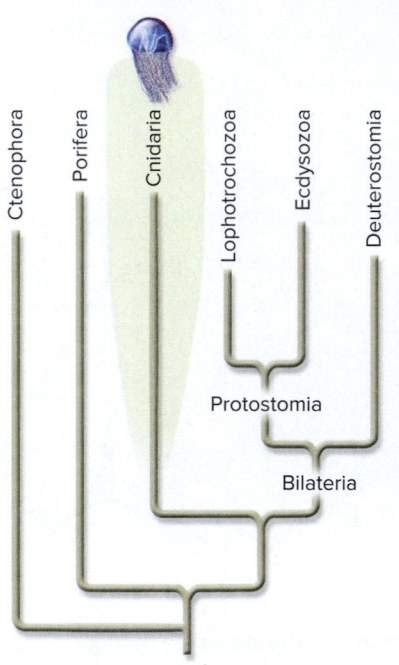

The members of the phylum Cnidaria (from the Greek *knide*, meaning nettle, and *aria*, meaning related to; pronounced nid-air'-e-ah) are mostly found in marine environments, although a few are freshwater species. Cnidaria includes hydra, jellyfish, box jellies, sea anemones, and corals. The cnidarians have only two embryonic germ layers: the ectoderm and the endoderm. A gelatinous substance called the **mesoglea** connects the two layers. In jellyfish, the mesoglea is enlarged and forms a transparent jelly, whereas in hydra and corals, the mesoglea is very thin. Most cnidarians have tentacles around the mouth that aid in prey detection and capture.

The phylum Cnidaria consists of four classes: Hydrozoa (including the Portuguese man-of-war), Scyphozoa (jellyfish), Anthozoa (sea anemones and corals), and Cubozoa (box jellies). The distinguishing characteristics of these classes are shown in **Table 34.1**.

## Cnidarians Exist in Two Different Body Forms

Most cnidarians exist in one of two different body forms with an associated lifestyle: the sessile **polyp** or the motile **medusa** (**Figure 34.4**). For example, corals and sea anemones exhibit only the polyp form, and jellyfish exist predominantly in the medusa form.

The polyp form has a tubular body with an opening at the oral (top) end that is surrounded by tentacles and functions as both mouth and anus

| Table 34.1 | Main Classes and Characteristics of the Cnidaria | |
|---|---|---|
| | **Class and examples (est. number of species)** | **Class characteristics** |
| | Hydrozoa: Portuguese man-of-war, *Hydra*, some corals (2,700) | Mostly marine; polyp stage usually dominant and colonial, reduced medusa stage |
| | Scyphozoa: jellyfish (200) | All marine; medusa stage dominant and large (up to 2 m); reduced polyp stage |
| | Anthozoa: sea anemones, sea fans, most corals (6,000) | All marine; polyp stage dominant; medusa stage absent; many are colonial |
| | Cubozoa: box jellies, sea wasps (20) | All marine; medusa stage dominant; box-shaped |

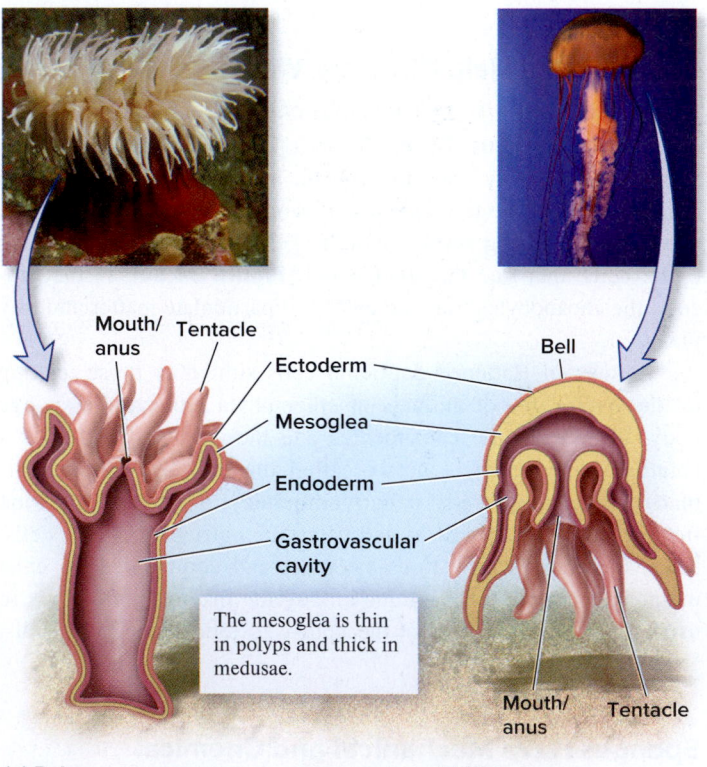

**(a) Polyp**        **(b) Medusa**

**Figure 34.4  Polyp and medusa forms of cnidarians.** Both **(a)** polyp and **(b)** medusa forms have two layers of cells, an outer layer of ectoderm and an inner layer of endoderm. In between is a layer of mesoglea, which is thin in polyps, such as corals, and thick in medusae, such as most jellyfish. a: Source: Linda Snook, NOAA/CBNMS; b: ©Kick Images/Getty Images

**Concept Check:** *What are the body forms of the following types of cnidarians: jellyfish, sea anemone, and Portuguese man-of-war?*

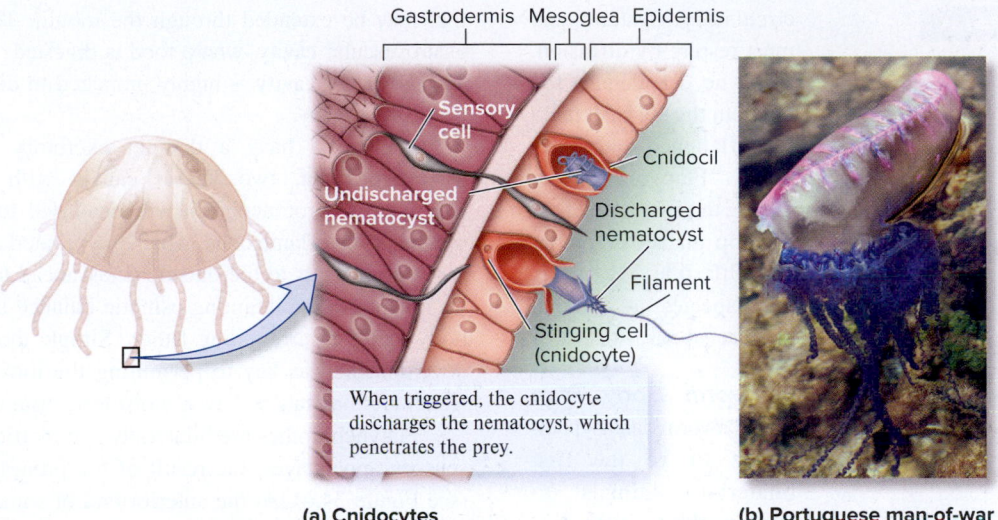

Gastrodermis  Mesoglea  Epidermis

Sensory cell

Undischarged nematocyst

Cnidocil

Discharged nematocyst

Filament

Stinging cell (cnidocyte)

When triggered, the cnidocyte discharges the nematocyst, which penetrates the prey.

**(a) Cnidocytes**

**(b) Portuguese man-of-war**

**Figure 34.5** **Specialized stinging cells of cnidarians, called cnidocytes.** **(a)** Cnidocytes, which contain stinging capsules called nematocysts, are situated in the tentacles. **(b)** The Portuguese man-of-war (*Physalia physalis*) employs cnidocytes that can be lethal to humans. b: ©Nature/UIG/Getty Images

**Concept Check:** *Are cnidocytes recycled for reuse once they have been fired?*

(see Figure 34.4a). The aboral (bottom) end is attached to the substrate. Polyps exist colonially, as in corals, or alone, as in sea anemones. Corals take dissolved calcium and carbonate ions from seawater and precipitate them as limestone underneath their bodies. With some species, this leads to a buildup of limestone deposits. As each successive generation of polyps dies, the limestone remains in place, and new polyps grow on top. Thus, huge underwater limestone deposits called coral reefs are formed (look ahead to Figure 54.26b). The largest of these is Australia's Great Barrier Reef, which stretches over 2,300 km. Many other extensive coral reefs are known, including the reef system along the Florida Keys. All coral reefs occur in warm water, generally between 20°C and 30°C.

The free-swimming medusa form has an umbrella-shaped body with an opening that serves as both mouth and anus located on the concave underside and surrounded by tentacles (see Figure 34.4b). More mobile medusae possess simple sense organs near the bell margin, including organs of equilibrium called statocysts and photosensitive organs known as **ocelli**. When one side of the bell tips upward, the statocysts on that side are stimulated, and muscle contraction is initiated to right the medusa. The ocelli allow medusae to position themselves in particular light levels.

### Cnidarians Have Specialized Stinging Cells

One of the unique and characteristic features of the cnidarians is the existence of stinging cells called **cnidocytes**, which function in defense or the capture of prey (**Figure 34.5a**). Cnidocytes contain **nematocysts**, powerful capsules with an inverted coiled and barbed thread. Each cnidocyte has a hairlike trigger called a **cnidocil** on its surface. When the cnidocil is touched or detects a chemical stimulus, the nematocyst is discharged, and its filament penetrates the prey and injects a small amount of toxin. Small prey are immobilized and passed into the mouth by the tentacles. After discharge, the cnidocyte is absorbed, and a new one grows to replace it. The nematocysts of most cnidarians are not harmful to humans, but those on the tentacles of the larger jellyfish and the Portuguese man-of-war (**Figure 34.5b**) can cause extreme pain or even death.

## 34.4 Lophotrochozoa: The Flatworms, Rotifers, Bryozoans, Brachiopods, Mollusks, and Annelids

**Learning Outcomes:**

1. Describe the unique features of platyhelminthes, rotifers, bryozoans, and brachiopods.
2. Outline the main features and list the major classes of the mollusks.
3. **CoreSKILL »** Analyze the results of Fiorito and Scotto's experiments, and explain how they show that octopuses can learn by watching each another.
4. List the advantages of segmentation in the annelids.

As we explored in Chapter 33 (refer back to Figure 33.3), molecular data suggest three clades of bilateral animals: the Lophotrochozoa and the Ecdysozoa (collectively known as the protostomes) and the Deuterostomia. In this section, we will explore the distinguishing characteristics of the Lophotrochozoa, a diverse group that includes taxa that possess either a lophophore (a crown of ciliated tentacles, seen in Bryozoa and Brachiopoda) or a distinct larval stage called a trochophore (Mollusca and Annelida). Also included in this clade are the Platyhelminthes (some of which have trochophore-like larvae) and the Rotifera (which have a lophophore-like feeding device), both of which share molecular similarities with the other members of the Lophotrochozoa.

### The Phylum Platyhelminthes Consists of Flatworms with No Coelom

Platyhelminthes (from the Greek *platy*, meaning flat, and *helminth*, meaning worm), or flatworms, lack a specialized respiratory or

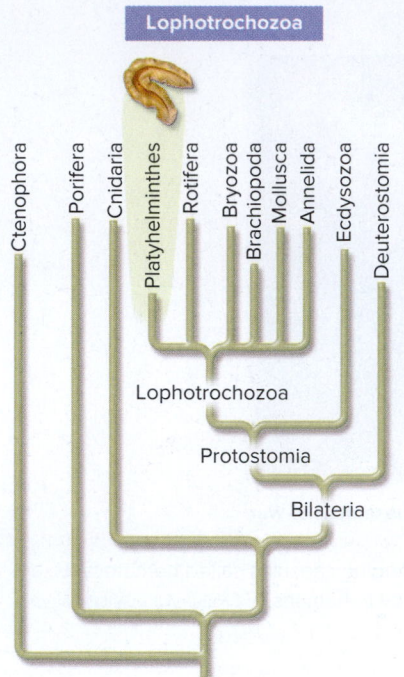

Lophotrochozoa

Ctenophora
Porifera
Cnidaria
Platyhelminthes
Rotifera
Bryozoa
Brachiopoda
Mollusca
Annelida
Ecdysozoa
Deuterostomia

Lophotrochozoa

Protostomia

Bilateria

circulatory system and must respire by diffusion. Thus, no cell can be too far from the surface, making a flattened shape necessary. Flatworms were among the first animals to develop an active predatory lifestyle. However, most species are internal or external parasites.

### Flatworm Body Plan

The flatworms are hypothesized to be the first bilaterian animals to evolve three distinctive embryonic germ layers—ectoderm, endoderm, and mesoderm—with mesoderm replacing the simpler gelatinous mesoglea of cnidarians. For this reason, they are said to be triploblastic. The muscles in flatworms, which are derived from mesoderm, are well developed. The evolution of mesoderm was a critical innovation in animals, leading to the development of more sophisticated organs.

Flatworms lack a coelem—a fluid-filled body cavity in which the gut is suspended. Therefore, they are described as acoelomates. Instead, mesoderm fills the body spaces around the gastrovascular cavity (**Figure 34.6**). The digestive system of flatworms is incomplete, with only one opening, which serves as both a mouth and an anus, as in cnidarians. Most flatworms possess a muscular pharynx

that may be extended through the mouth. The pharynx opens to a gastrovascular cavity, where food is digested. In large flatworms, the gastrovascular cavity is highly branched to distribute nutrients to all parts of the body.

Flatworms have a distinct excretory system consisting of **protonephridia**, two lateral canals with branches capped by **flame cells**. Protonephridia are dead-end tubules lacking internal openings. The flame cells, which are ciliated and waft water through the lateral canals to the outside (look ahead to Figure 49.2), primarily function in maintaining osmotic balance between the flatworm's body and the surrounding fluids. Simple though this system is, its development was key to permitting the movement of animals into freshwater habitats and even moist terrestrial areas.

Platyhelminthes are bilaterally symmetrical with a head bearing sensory appendages, the result of the process called **cephalization** (see Figure 34.6). At the anterior end of some free-living flatworms are light-sensitive eyespots, called ocelli, as well as chemoreceptive and sensory cells that are concentrated in organs called auricles. A pair of **cerebral ganglia**, clusters of nerve cell bodies, receives input from photoreceptors in eyespots and sensory cells. From the ganglia, a pair of lateral nerve cords running the length of the body allow rapid movement of information from anterior to posterior. In addition, transverse nerves form a nerve net on the ventral surface, similar to that of cnidarians. Thus, flatworms show the beginnings of the more centralized type of nervous system seen throughout much of the rest of the animal kingdom.

**The Classes of Flatworms**   The four classes of flatworms are the Turbellaria, Monogenea, Cestoda (tapeworms), and Trematoda (flukes) (**Table 34.2**).

- Turbellarians are the only free-living class of flatworms and are widespread in lakes, ponds, and marine environments (**Figure 34.7a**).

**Figure 34.6   Body plan of a flatworm.** Flatworm morphology is represented by a planarian, a member of the class Turbellaria.

Ocelli (eyespots)
Auricles
Cerebral ganglia
Lateral nerve cords
Protonephridia
Transverse nerve
Gastrovascular cavity
Pharyngeal chamber
Mouth   Pharynx

*Concept Check:*  *How do flatworms breathe?*

| Table 34.2 | Main Classes and Characteristics of Platyhelminthes | |
|---|---|---|
| | **Class and examples (est. number of species)** | **Class characteristics** |
| | Turbellaria: planarian (3,000) | Mostly marine; free-living flatworms; predatory or scavengers |
| | Monogenea: fish flukes (1,000) | Marine and freshwater; usually external parasites of fish; simple life cycle (no intermediate host) |
| | Cestoda: tapeworms (5,000) | Internal parasites of vertebrates; complex life cycle, usually with one intermediate host; no digestive system; nutrients absorbed across epidermis |
| | Trematoda: flukes (11,000) | Internal parasites of vertebrates; complex life cycle with several intermediate hosts |

**(a)**

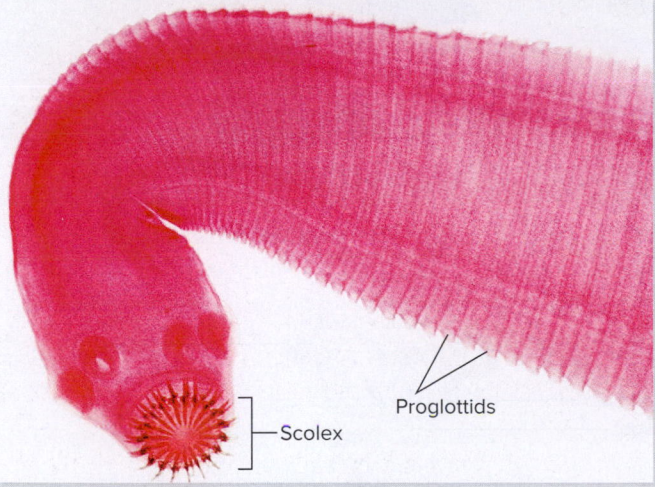

Scolex

Proglottids

**(b)**

**Figure 34.7** **Flatworms.** **(a)** Many free-living marine turbellarians are brightly colored, such as this racing stripe flatworm, *Pseudoceros bifurcus*, from Bali, Indonesia. **(b)** A tapeworm, *Taenia pisiformis*, a member of the class Cestoda. Note the tiny hooks and suckers that make up the scolex. Each segment is a proglottid, which may be filled with eggs.

a: ©Wolfgang Poelzer/age fotostock; b: ©Biophoto Associates/Science Source

 **Core Skill: Science and Society** About 1% of U.S. cattle are infected by beef tapeworms. Consuming beef that is not sufficiently well cooked can lead to infection by these parasites. At least 1,000 hospitalizations a year in the U.S. are due to tapeworm infection, most as a result of eating uncooked pork.

- Monogeneans are relatively simple external parasites with just one host species (a fish).

- Cestodes and trematodes are internally parasitic in humans and other animals and therefore are of great medical and veterinary importance. They possess a variety of organs of attachment, such as hooks and suckers, that enable them to remain embedded within their hosts. For example, cestodes attach to their host by means of an organ at the head end called a scolex

(**Figure 34.7b**). They have no mouth or gastrovascular cavity and absorb nutrients across the body surface.

***Reproduction in Flatworms*** In Platyhelminthes, reproduction is either sexual or asexual. Most species are hermaphroditic but do not fertilize their own eggs. Flatworms can also reproduce asexually by splitting into two parts, with each half regenerating the missing fragment.

Flatworm life cycles can be complex. Cestodes often require two different vertebrate host species, such as pigs or cattle, to begin their life cycle and another host, such as humans, to complete their development. Behind the scolex in cestodes is a long ribbon of identical segments called proglottids, which are segments of sex organs that produce thousands of eggs. The proglottids are continually shed in the host's feces. Human feces passed out onto the ground are eaten with grass by pigs and cattle. Many tapeworms are ingested by humans who consume undercooked, infected meat—hence it is important to cook meat thoroughly.

The life cycle of trematodes is typically more complex than that of cestodes, involving multiple hosts. The first host, called the intermediate host, is usually a mollusk, and the final host, or definitive host, is usually a vertebrate, but often a second or even a third intermediate host is involved. In the case of the Chinese liver fluke (*Clonorchis sinensis*), the adult parasite lives and reproduces in the definitive host, a human (see step 1, **Figure 34.8**). Structures, which are sometimes called eggs, contain encapsulated miracidia; these pass from the host via the feces, and then an intermediate host, such as a snail, eats the miracidia, which transform into sporocysts (see steps 2 and 3). The sporocysts asexually produce more sporocysts called rediae, which develop into a free-swimming life stage called cercariae (see steps 4 and 5). In the last stages of the life cycle, cercariae bore their way out of the snail and infect their second intermediate host, fishes, by entering via the gills. Here, the cercariae develop into juvenile flukes and lodge in fish muscle, which the definitive host will eat (see steps 6 and 7). From the small intestine of the definitive host, the juvenile flukes travel to the liver and grow into adult flukes, and the life cycle begins anew. The probability of each trematode stage reaching a suitable host is low, so trematodes produce large numbers of offspring to ensure that some survive.

Blood flukes, genus *Schistosoma*, are the most common parasitic trematodes infecting humans; they cause the disease known as schistosomiasis. Over 200 million people worldwide, primarily in tropical Asia, Africa, and South America, are infected with schistosomiasis. The inch-long adult flukes may live for years in human hosts, and the release of eggs may cause chronic inflammation and blockage in many organs. Untreated schistosomiasis can result in severe damage to the liver, intestines, and lungs and eventually lead to death. Sewage treatment and access to clean water greatly reduce infection rates.

## Members of the Phylum Rotifera Have a Pseudocoelom and a Ciliated Crown

Members of the phylum Rotifera (from the Latin *rota*, meaning wheel, and *fera*, meaning to bear) get their name from their ciliated crown, or **corona**, which, when beating, looks similar to a rotating wheel

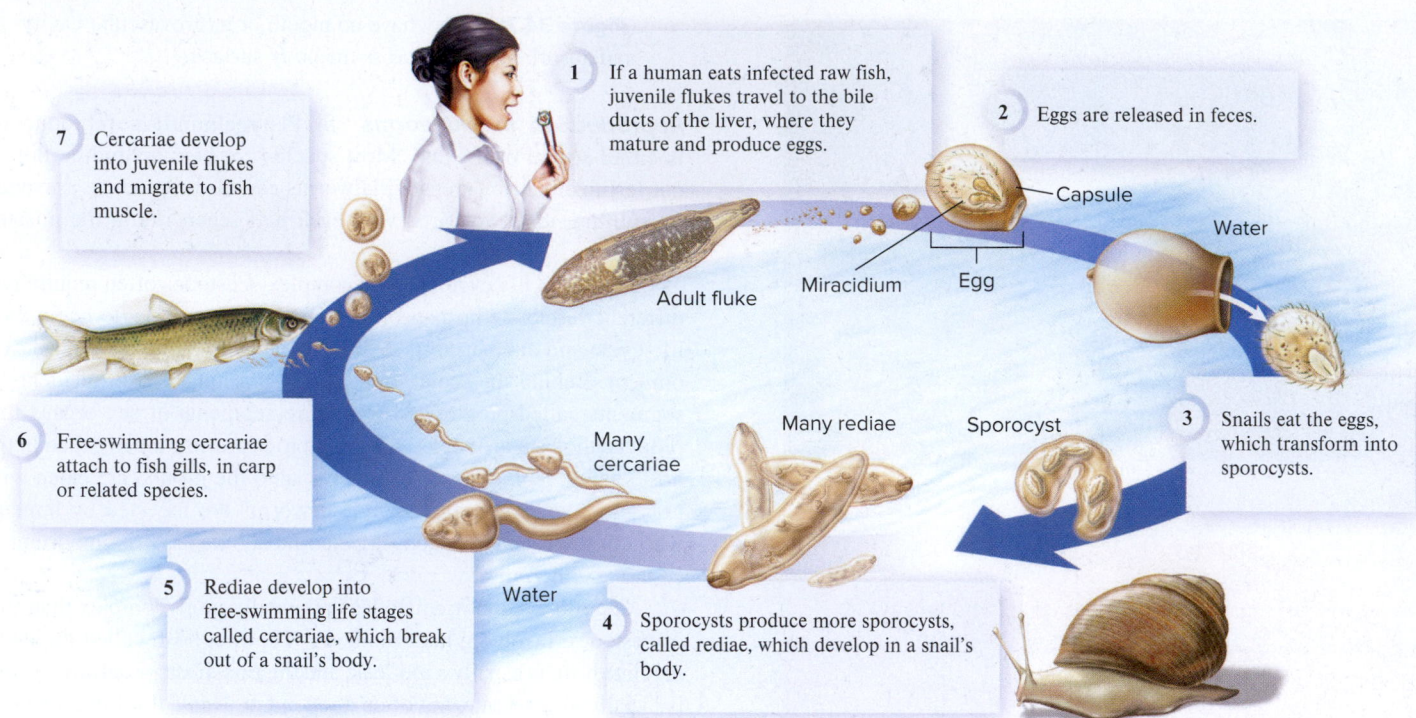

**Figure 34.8   The complete life cycle of a trematode.**   This figure shows the life cycle of the Chinese liver fluke (*Clonorchis sinensis*).

1   If a human eats infected raw fish, juvenile flukes travel to the bile ducts of the liver, where they mature and produce eggs.

2   Eggs are released in feces.

3   Snails eat the eggs, which transform into sporocysts.

4   Sporocysts produce more sporocysts, called rediae, which develop in a snail's body.

5   Rediae develop into free-swimming life stages called cercariae, which break out of a snail's body.

6   Free-swimming cercariae attach to fish gills, in carp or related species.

7   Cercariae develop into juvenile flukes and migrate to fish muscle.

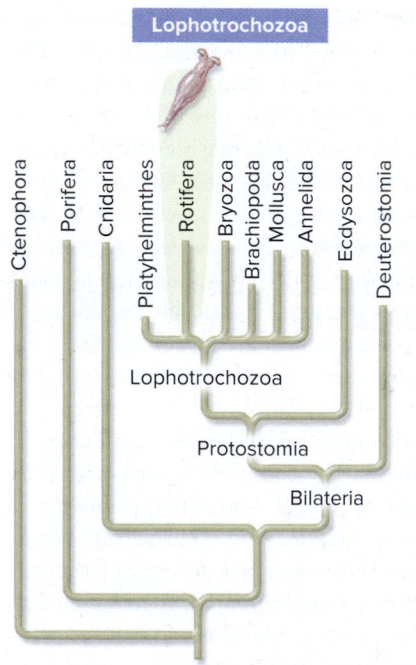

(**Figure 34.9**). Most rotifers are microscopic animals, usually less than 1 mm long, and some have beautiful colors. About 2,200 species of rotifers have been identified. They typically inhabit fresh water, with a few marine and terrestrial species. Most often they are bottom-dwelling organisms, living on a pond floor or along lakeside vegetation. The body of the rotifer bears a jointed foot with one to four toes. **Pedal glands** in the foot secrete a sticky substance that aids in attachment to a substrate.

The internal organs of rotifers lie within a pseudocoelom, a fluid-filled body cavity that is not completely lined with mesoderm. The pseudocoelom serves as a hydrostatic skeleton and as a medium for the internal transport of nutrients and wastes. Rotifers have an alimentary canal, a digestive tract with a separate mouth and anus. For this reason, rotifers can feed more frequently compared to simpler animals that have a single opening to their digestive system, such as cnidarians. The corona of rotifers creates water currents that propel the animal through the water and waft small planktonic organisms or decomposing organic material

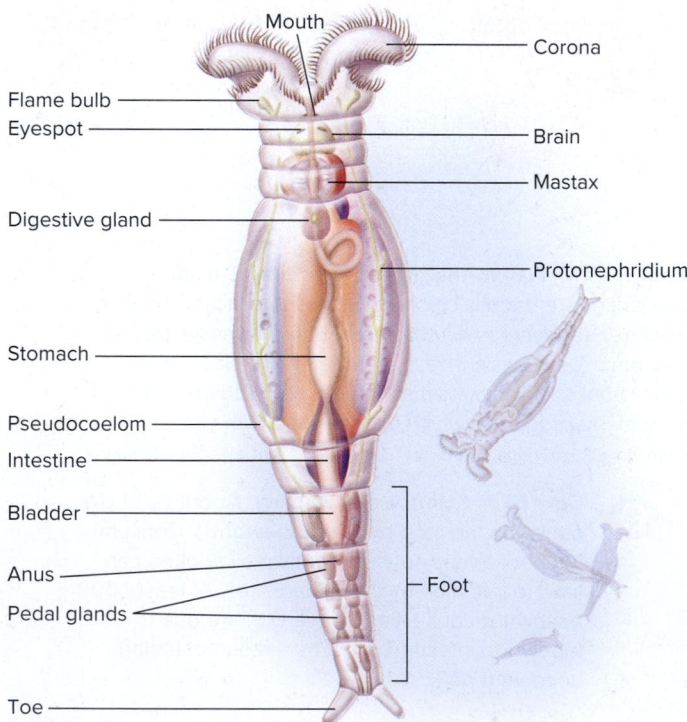

**Figure 34.9   Body plan of a common rotifer, *Philodina genus*.**

toward the mouth. The mouth opens into a circular, muscular pharynx called a **mastax**, which has jaws for grasping and chewing. The mastax, which in some species can protrude through the mouth to seize small prey, is a structure unique to rotifers. They also have a pair of protonephridia with flame bulbs that collect excretory and digestive waste and drain into a bladder, which passes waste to the anus. The nervous system consists of nerves that extend from the sensory organs, especially the eyespots and some bristles on the corona, to the brain.

Reproduction in rotifers is unique. In some species, unfertilized diploid eggs that have not undergone meiotic division develop into females through a process known as **parthenogenesis**. In other species, some unfertilized eggs develop into females, whereas others develop into males that live only long enough to produce and release sperm that fertilize the eggs. The resultant fertilized eggs form zygotes, which have a thick shell and can survive for long periods of harsh conditions, for example, if a water supply dries up, before developing into new females. Because the tiny zygotes are easily transported, rotifers show up in the smallest of aquatic environments, such as birdbaths or roof gutters.

**(a) A bryozoan**

**(b) A brachiopod, the northern lamp shell**

**Figure 34.10**  **Bryozoans and brachiopods.** **(a)** Bryozoans are colonial animals that reside in a nonliving case called a zoecium. **(b)** Brachiopods, such as this northern lamp shell (*Terebratulina septentrionalis*), have dorsal and ventral shells. a: ©G. Guenther/age fotostock; b: ©Gordon MacSkimming/age fotostock

**Concept Check:**  *What are the two main functions of the lophophore?*

## Bryozoa and Brachiopoda Have a Lophophore for Feeding and Gas Exchange

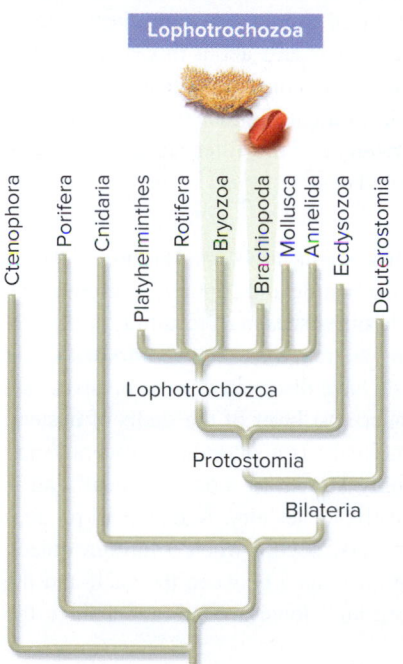

The Bryozoa and the Brachiopoda both possess a lophophore, a ciliary feeding device (refer back to Figure 33.14a), and a true coelom (refer back to Figure 33.7a). The lophophore is a circular fold of the body wall bearing tentacles that draw water toward the mouth. Because a thin extension of the coelom penetrates each tentacle, the tentacles also serve as a respiratory organ. Gases diffuse across the tentacles and into or out of the coelomic fluid and are carried throughout the body. Both phyla have a U-shaped alimentary canal, with the anus located near the mouth but outside of the lophophore.

***Phylum Bryozoa***   The bryozoans (from the Greek *bryon*, meaning moss, and *zoon*, meaning animal) are small colonial animals, most of which are less than 0.5 mm long, that can be found on rocks in shallow aquatic environments. They look very much like plants. Within a colony, each animal secretes and lives inside a nonliving exoskeleton called a zoecium that is composed of chitin or calcium carbonate (**Figure 34.10a**). For this reason, bryozoans have been important reefbuilders. Also, many of them encrust boat hulls and have to be scraped off periodically. About 4,500 species of bryozoans currently exist. They date back to the Paleozoic era, and thousands of fossil forms have been discovered and identified.

***Phylum Brachiopoda***   Brachiopods (from the Greek *brachio*, meaning arm, and *podos*, meaning foot) are marine organisms with two shell halves, much like clams (**Figure 34.10b**). In clams, however, the shell halves are considered to be left and right sides with the plane of symmetry lying parallel to the site at which the shells join. In contrast, brachiopods have a dorsal and ventral shell, with the plane of symmetry perpendicular to the site at which the shells join. The dorsal and ventral shells of brachiopods are of slightly different sizes and shapes. Brachiopods are bottom-dwelling species that attach to the substrate via a muscular pedicle. Although they are now a relatively small group, with about 300 living species, brachiopods flourished in the Paleozoic and Mesozoic eras—about 30,000 fossil species have been identified. Some of these fossil forms represent organisms that reached 30 cm in length, although their modern relatives are only 0.5–8.0 cm long.

## Mollusca Is a Large Phylum Containing Snails, Slugs, Clams, Oysters, Octopuses, and Squids

Mollusks (from the Latin *mollis*, meaning soft) constitute a very large phylum, with over 100,000 living species, including organisms as diverse as snails, clams, octopuses, and chitons. They are an ancient group, as evidenced by the classification of about 35,000 fossil species. Mollusks have considerable economic, aesthetic, and ecological importance to humans. Many serve as sources of food, including scallops, oysters, clams, and squids. A significant industry involves the farming of oysters to produce cultured pearls, and rare and beautiful mollusk shells are extremely valuable to collectors. Snails and slugs can damage vegetables and ornamental plants, and boring mollusks can penetrate wooden ships and wharfs. Mollusks are intermediate hosts to many parasites, and several invasive species have become serious pests. For example, populations of the zebra mussel (*Dreissena polymorpha*) have been introduced into North America from Asia, probably via ballast water from transoceanic ships. Since their introduction, they have spread rapidly throughout the Great Lakes and an increasing number of inland waterways, adversely impacting native organisms and clogging water intake valves of municipal water-treatment plants around the lakes.

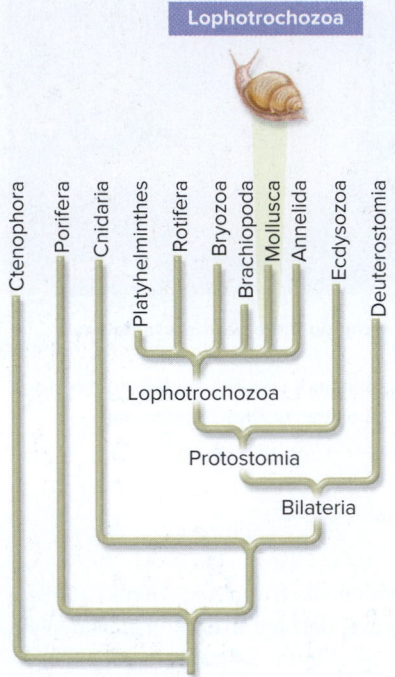

**Mollusk Body Plan**  One common feature of mollusks is their soft body, which in many species is found under a protective external shell. Most mollusks are marine, although some have colonized fresh water. Many snails and slugs have moved onto land, but they survive only in humid areas and where the calcium necessary for shell formation is abundant in the soil. The ability to colonize freshwater and terrestrial habitats has led to a diversification of mollusk body plans. The amazing diversity of mollusks demonstrates how species diversity is related to environmental diversity.

Although great variation in morphology occurs between classes, mollusks have a basic body plan consisting of three parts (**Figure 34.11**).

- A muscular **foot** is usually used for movement, and a **visceral mass** containing the internal organs rests atop the foot.

- The **mantle**, a fold of skin draped over the visceral mass, secretes a shell in those species that form shells.

- The **mantle cavity** houses delicate **gills**, filamentous organs that are specialized for gas exchange. A continuous current of water, often induced by cilia present on the gills or by muscular pumping, flushes out the wastes from the mantle cavity and brings in new oxygen-rich water.

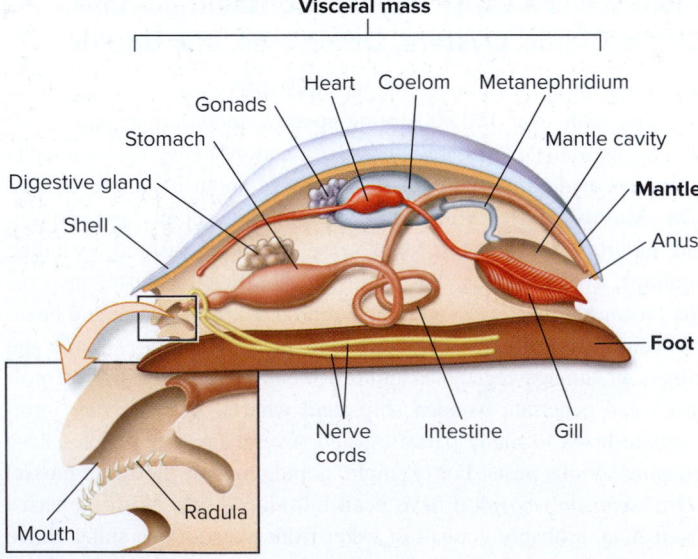

**Figure 34.11  The mollusk body plan.** The generalized body plan of a mollusk includes the characteristic foot, mantle, and visceral mass.

**Concept Check:**  *Do molluscan hearts pump blood?*

Mollusks are coelomate organisms, but the coelom is confined to a small area around the heart. Most species of mollusks have an **open circulatory system** with a heart that pumps a body fluid called hemolymph through vessels and into sinuses. Sinuses are open, fluid-filled cavities between the internal organs. The organs and tissues are therefore continually bathed in hemolymph. The sinuses coalesce to form an open cavity known as the hemocoel (blood cavity). From these sinuses, the hemolymph drains into vessels that take it to the gills and then back to the heart.

Excretory organs called **metanephridia** remove nitrogenous and other wastes. Metanephridia have ciliated funnel-like openings inside the coelom that are connected to ducts that lead to the exterior mantle cavity. The pores from the metanephridia discharge wastes into this cavity. The anus also opens into the mantle cavity. The metanephridial ducts may also serve to discharge sperm or eggs from the gonads. The nervous system varies from simple ganglia and nerve chords in most species to much larger brains and sophisticated organs of touch, smell, taste, and vision in octopuses.

The mollusk's mouth may contain a **radula**, a unique, protrusible, tonguelike organ that has many teeth and is used to eat plants, scrape food particles off rocks, or, if the mollusk is predatory, bore into shells of other species and tear flesh. In the cone shells (genus *Conus*), the radula is reduced to a few poison-injecting teeth on the end of a long proboscis that is cast about in search of prey, such as a worm or even a fish. Some cone shell species produce a neuromuscular toxin that can kill humans. Other mollusks, particularly bivalves, have lost their radula and are filter feeders that strain water brought in by ciliary currents.

**Mollusk Shells**  Most mollusk shells are complex three-layered structures that are secreted by the mantle and continue to grow as the mollusk grows. Shell growth is often seasonal, resulting in distinct growth lines on the shell, much like tree rings (**Figure 34.12a**). Using shell growth patterns, biologists have discovered some bivalves that are over 100 years old. The innermost layer of the shells of oysters, mussels, abalone, and other mollusks is a smooth, iridescent lining called nacre, which is commonly known as mother-of-pearl and is often collected from abalone shells for jewelry. Actual pearl production in mollusks, primarily oysters, occurs when a foreign object, such as a grain of sand, becomes lodged between the shell and the mantle, and layers of nacre are laid down around it to reduce the irritation.

**Reproduction in Mollusks**  Most mollusks species have separate sexes, although some exist as hermaphrodites. Gametes are usually released into the water, where they mix and fertilization occurs. In some snails, however, fertilization is internal, with the male inserting sperm directly into the female. Internal fertilization was a critical innovation enabling some snails to colonize land. In many species, reproduction involves the production of a trochophore larva that develops into a **veliger**, a free-swimming larva that has a rudimentary foot, shell, and mantle (**Figure 34.13**).

**The Major Molluscan Classes**  Of the eight molluscan classes, the four most common are the Bivalvia (clams and mussels),

**(a) A quahog clam, class Bivalvia**

**(c) A snail, class Gastropoda**

**(b) A chiton, class Polyplacophora**

**(d) A nudibranch, class Gastropoda**

**(e) A blue-ringed octopus, class Cephalopoda**

**Figure 34.12  Mollusks.  (a)** A bivalve shell, class Bivalvia, with growth rings. This quahog clam (*Mercenaria mercenaria*) can live over 20 years. **(b)** A chiton (*Tonicella lineata*), a polyplacophoran with a shell made up of eight separate plates. **(c)** A gastropod, the tree snail, *Liguus fasciatus*, from the Florida Everglades showing its characteristic coiled shell. **(d)** A nudibranch (*Phyllidia ocellata*). The nudibranchs are a gastropod subclass whose members have lost their shell altogether. **(e)** The highly poisonous blue-ringed octopus (*Hapalochlaena lunulata*), a cephalopod. a: ©Andrew J. Martinez/Science Source; b: ©Kjell B. Sandved/Science Source; c: ©ImageBROKER/Alamy Stock Photo; d: ©Hal Beral/Corbis /Getty Images; e: ©Richard Merritt FRPS/Getty Images

Polyplacophora (chitons), Gastropoda (snails and slugs), and Cephalopoda (octopuses, squids, and nautiluses) (**Table 34.3**).

- Bivalves are freshwater or marine mollusks whose bodies are enclosed within a hinged shell of two valves, or halves. Prominent members of this class include oysters, clams, mussels, and scallops (**Figure 34.12a**).

- Polyplacophora are marine mollusks with a shell composed of eight separate plates (**Figure 34.12b**). Chitons are common in the intertidal zone, an area above water at low tide and under water at high tide, and they creep along when covered by the tide. Feeding occurs by scraping algae off rock surfaces. When the tide recedes, the muscular foot holds the chiton tight to the rock surface, preventing desiccation.

- The class Gastropoda (from the Greek *gaster*, meaning stomach, and *podos*, meaning foot) is the largest group of mollusks and encompasses about 75,000 living species, including snails, periwinkles, and limpets (**Figure 34.12c**). Most gastropods have a one-piece shell, into which the animal can withdraw to escape predators, However, the class also includes species such as slugs and nudibranchs, whose shells have been greatly reduced or completely lost during their evolution (**Figure 34.12d**). Although gastropods usually occupy marine or freshwater habitats, some species, including snails and slugs, have also colonized land.

The 780 species of Cephalopoda (from the Greek *kephalo*, meaning head, and *podos*, meaning foot) are the most

**Figure 34.13 A snail veliger.** Veligers are free-swimming larval forms of mollusks that look more like adults than the trocophore larvae from which they develop. ©Solvin Zankl/Alamy Stock Photo

| Table 34.3 | Major Classes and Characteristics of Mollusks | |
|---|---|---|
| | **Class and examples (est. number of species)** | **Class characteristics** |
|  | Bivalvia: clams, mussels, oysters, scallops (30,000) | Marine or freshwater; shell with two halves or valves; primarily filter feeders with siphons |
| | Polyplacophora: chitons (860) | Marine; eight-plated shell |
| | Gastropoda: snails, slugs, nudibranchs (75,000) | Marine, freshwater, or terrestrial; most with coiled shell, but shell absent in slugs and nudibranchs; radula present |
| | Cephalopoda: octopuses, squids, nautiluses (780) | Marine; predatory, with tentacles around mouth, often with suckers; shell often absent or reduced; closed circulatory system; jet propulsion via siphon |

morphologically complex of the mollusks and indeed among the most complex of all invertebrates. Most are fast-swimming marine predators that range from organisms just a few centimeters in size to the colossal squid (*Mesonychoteuthis hamiltoni*), which is known to reach over 13 m in length and 495 kg (1,091 lb) in weight. A cephalopod's mouth is surrounded by many long arms commonly armed with suckers. All cephalopods have a beaklike jaw that allows them to bite their prey, and some, such as the blue-ringed octopus (*Hapalochlaena lunulata*), deliver a deadly poison through their saliva (**Figure 34.12e**).

The foot of some cephalopods has become modified into a muscular siphon. Water drawn into the mantle cavity is quickly expelled through the siphon, propelling the organism forward or backward in a kind of jet propulsion. Such vigorous movement requires powerful muscles and a very efficient circulatory system to deliver oxygen and nutrients to the muscles. Cephalopods are the only mollusks with a **closed circulatory system**, in which

blood flows throughout an animal entirely within a series of vessels. One of the advantages of this type of system is that the heart can pump blood through the tissues rapidly, making oxygen more readily available. The blood of cephalopods contains the copper-rich protein hemocyanin for transporting oxygen. Less efficient than the iron-rich hemoglobin of vertebrates, hemocyanin gives the blood a blue color.

Cephalopods have a well-developed nervous system and brain that support their active lifestyle. Their sense organs, especially their eyes, are also very well developed. Many cephalopods (with the exception of nautiluses) have an ink sac that contains the pigment melanin; the sac can be emptied to provide a "smokescreen" to confuse predators. In many species, melanin is also distributed in special pigment cells in the skin, which allows for color changes. Octopuses often change color when disturbed, and they can change color rapidly to blend in with their background and escape detection.

## Core Skill: Process of Science

## Feature Investigation | Fiorito and Scotto's Experiments Showed That Invertebrates Can Exhibit Sophisticated Observational Learning Behavior

The ability to learn by observing the behavior of others has commonly been observed in vertebrates, especially among species that live in social groups. For example, young rhesus macaques (*Macaca mulatta*) that observed their parents fearfully responding to model snakes also developed a fear of snakes and maintained this fear

for 3 months after observing their parent's behavior. In 1992, Italian researchers Graziano Fiorito and Pietro Scotto set out to test the hypothesis that octopuses (invertebrates) can learn by observing the behavior of other octopuses (**Figure 34.14**).

## Figure 34.14 Observational learning in octopuses.

**HYPOTHESIS** Octopuses can learn by observing another's behavior.

**STUDY LOCATION** Laboratory setting with *Octopus vulgaris* collected from the Bay of Naples, Italy.

| | Experimental level | Conceptual level |
|---|---|---|

**1** Train 2 groups of octopuses, one to attack white balls, one to attack red. These are called the demonstrator octopuses.

Reward choice of correct ball (with fish) and punish choice of incorrect ball (with electric shock). Training is complete when octopus makes no "mistakes" in 5 trials.

Conditions a demonstrator octopus to attack a particular color of ball.

**2** In an adjacent tank, allow observer octopus to watch trained demonstrator octopus.

Observer octopus may be learning the correct ball to attack by watching the demonstrator octopus.

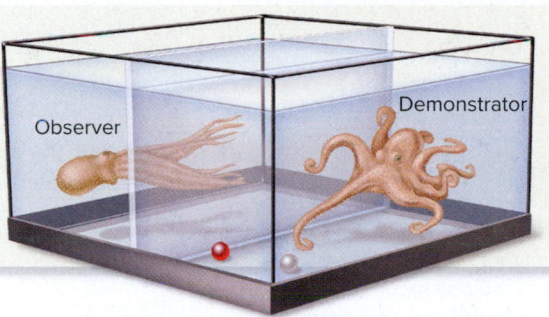

**3** Drop balls into the tank of the observer octopus. Test the observer octopus to see if it makes the same decisions as the demonstrator octopus.

If the observer octopus is learning from the demonstrator octopus, the observer octopus should attack the ball of the same color as the demonstrator octopus was trained to attack.

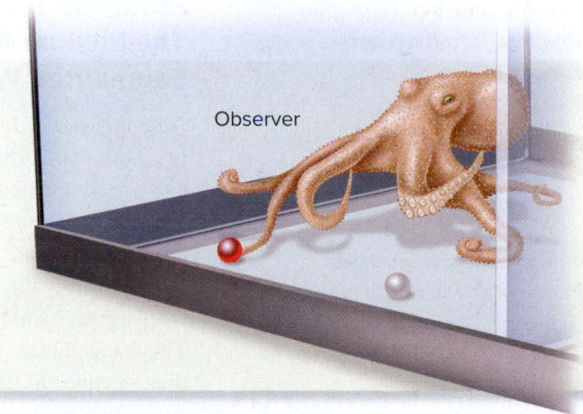

**4** **THE DATA**

| Participant | Color of ball chosen in 5 trials* | |
|---|---|---|
| | Red | White |
| Observers (watched demonstrator attack red) | 4.31 | 0.31 |
| Observers (watched demonstrator attack white) | 0.40 | 4.10 |
| Untrained (did not watch demonstrations) | 2.11 | 1.94 |

*Average of 5 trials; data do not always sum to 5, because some trials resulted in no balls being chosen.

**5** **CONCLUSION** Invertebrate animals are capable of learning from watching other individuals behave, in much the same way as vertebrate species learn from watching others.

**6** **SOURCE** Fiorito, G., and Scotto, P. 1992. Observational learning in Octopus vulgaris. *Science* 256: 545–547.

In their experiments, they used a system of reward (a small piece of fish placed behind a ball so that the octopus could not see it) and punishment (a small electric shock for choosing the wrong ball) to train octopuses to attack either a red or a white ball. This type of learning is called classical conditioning (see Chapter 55). Because octopuses are color blind, they must distinguish between the relative brightness of the balls. Octopuses were considered to be trained when they made no mistakes in five trials.

Observer octopuses in adjacent tanks were then allowed to watch the trained octopuses attacking the balls. In step 3, the observer octopuses were themselves tested, as were untrained octopuses that had never watched the demonstrators. As seen in the data, observers nearly always attacked the same color ball as they had observed the demonstrators attacking. In contrast, the untrained octopuses were equally likely to attack a red or white ball. These results indicate that one octopus can learn by watching the behavior of another octopus. This was a unexpected finding because many researchers thought that such complex learning would not be found in invertebrate species. It is also surprising because *Octopus vulgaris*, the species they studied, lives a solitary existence for most of its life.

### Experimental Questions

1. What was the hypothesis tested by Fiorito and Scotto?
2. **CoreSKILL** » What were the results of the experiment? Did these results support the hypothesis?
3. **CoreSKILL** » Explain the significance of performing the experiment on both observer octopuses and untrained octopuses.

---

## BIO TIPS

**THE QUESTION**  *To determine if the observer octopuses would retain their learning, Fiorito and Scotto conducted follow-up trials. Five days after their initial testing, the observer octopuses were retested for their ability to choose the correct ball. The observers that had watched a demonstrator that was trained to attack a red ball made the following choices of color: red, 3.88; white, 0.50. The observers that had watched a demonstrator trained to attack a white ball made the following choices: red, 0.50; white, 3.70. Were the observer octopuses retaining their learning after 5 days?*

**T**OPIC  ***What topic in biology does this question address?*** The topic is learning; more specifically it is about learning retention in octopuses.

**I**NFORMATION  ***What information do you know based on the question and your understanding of the topic?*** In the data of Figure 34.14, you are given the initial frequency of color choices made by the observers. From the question, you know the frequency of color choices made 5 days later.

**P**ROBLEM-SOLVING **S**TRATEGY  ***Make a calculation.*** To solve this problem, you determine percentages of correct responses, initially (on day 1) and after 5 days, for both groups of observer octopuses.

For the octopuses that watched a demonstrator trained to attack a red ball, the percentages are as follows.

$$\text{Day 1:} \left( \frac{4.31}{4.31 + 0.31} \right) \times 100 = 93.3\%$$

$$\text{Day 5:} \left( \frac{3.88}{3.88 + 0.50} \right) \times 100 = 88.6\%$$

For the octopuses that watched a demonstrator trained to attack a white ball, the percentages are as follows.

$$\text{Day 1:} \left( \frac{4.10}{4.10 + 0.40} \right) \times 100 = 91.1\%$$

$$\text{Day 5:} \left( \frac{3.70}{3.70 + 0.50} \right) \times 100 = 88.1\%$$

**ANSWER**  *The octopuses appeared to retain their learning fairly well, because only a slight drop (less than 5%) was observed after 5 days.*

## The Phylum Annelida Consists of the Segmented Worms

Annelids are a large phylum with about 18,000 species of segmented worms. The members include free-ranging marine worms, tube worms, the familiar earthworm, and leeches. They range in size from less than 1 mm to enormous Australian earthworms that can reach a length of 3 m.

***Annelid Body Plan***  The phylum name Annelida is derived from the Latin *annulus*, meaning little ring. Each ring is a distinct segment of the annelid's body; adjacent segments are separated by septa (**Figure 34.15**). Segmentation in the adult confers three advantages:

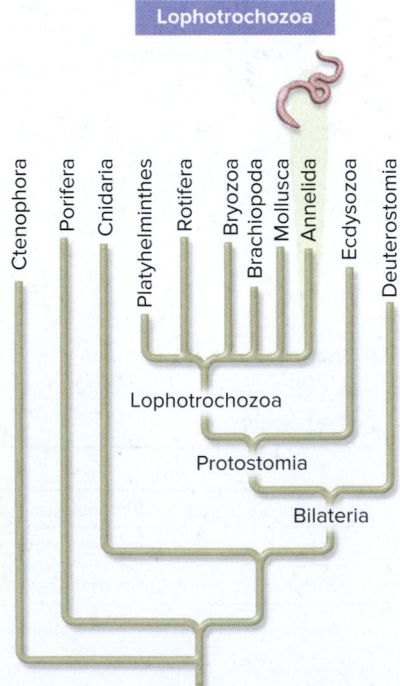

- Many components of the body are repeated in each segment, including blood vessels, nerves, and excretory and reproductive organs. If the components in one segment fail, those of another segment will still function.

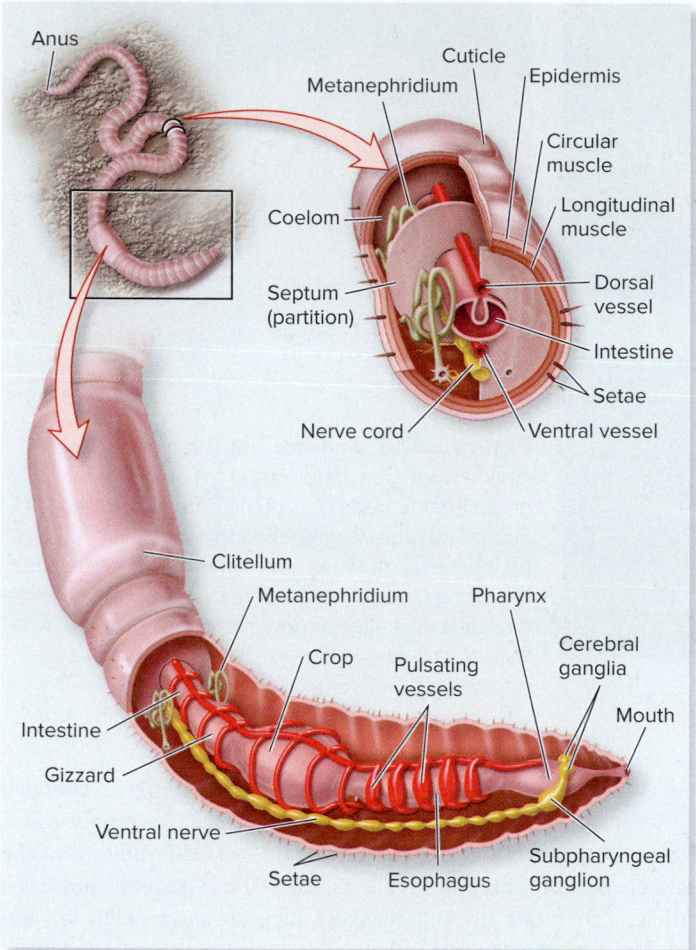

Anus

Cuticle

Metanephridium

Epidermis

Circular muscle

Coelom

Longitudinal muscle

Septum (partition)

Dorsal vessel

Intestine

Nerve cord

Setae

Ventral vessel

Clitellum

Metanephridium

Pharynx

Crop

Pulsating vessels

Cerebral ganglia

Mouth

Intestine

Gizzard

Ventral nerve

Setae

Esophagus

Subpharyngeal ganglion

**Figure 34.15   The segmented body plan of an annelid, as illustrated by an earthworm.**  The segmented nature of the worm is apparent internally as well as externally. Individual segments are separated by septa.

*Concept Check:*  *What are some of the advantages of segmentation?*

- Annelids possess a fluid-filled coelom that acts as a hydrostatic skeleton. In unsegmented coelomate animals, muscle contractions can distort the entire body during movement. However, such distortion is minimized in segmented animals, which allows for more effective locomotion over solid surfaces.

- Segmentation permits specialization of some segments, especially at the annelid's anterior end. Animals with more complex body plans tend to produce a greater variety of specialized segments.

All annelids except the leeches have chitinous bristles, called **setae**, on each segment. In some annelids, these are situated on fleshy, footlike **parapodia** (from the Greek, meaning almost feet) that are pushed into the substrate to provide traction during movement. In others, the setae are held closer to the body. Many annelid species burrow into soil or into muddy marine sediments and extract nutrients from ingested soil or mud. Some annelids also feed on dead or living vegetation, whereas others are predatory or parasitic.

Annelids have a nervous system with a pair of cerebral ganglia that connect to a subpharyngeal ganglion (Figure 34.15). From there,

a large ventral nerve cord runs down the entire length of the body. The ventral nerve cord is unusual because it contains a few very large nerve cells called **giant axons** that facilitate high-speed neuronal conduction and rapid responses to stimuli.

Annelids have an internal transport system in which the circulatory system and the coelomic fluid both carry nutrients, wastes, and respiratory gases. The circulatory system is closed, with dorsal and ventral vessels connected by pairs of pulsating vessels. The blood of most annelid species contains the respiratory pigment hemoglobin. Respiration occurs directly through the permeable skin surface, which restricts annelids to moist environments. The digestive system is complete and unsegmented, with many specialized regions: mouth, pharynx, esophagus, crop, gizzard, intestine, and anus.

**Reproduction in Annelids**  Sexual reproduction in annelids involves two individuals, often of separate sexes, but sometimes hermaphrodites, which exchange sperm via internal fertilization. In some species, asexual reproduction by fission occurs, in which the posterior part of the body breaks off and forms a new individual.

**The Major Annelidan Groups**  In 2011, a study by German evolutionary biologist Torsten Struck and colleagues suggested that the phylum Annelida contains two major groups: the Errantia and the Sedentaria. Members of the Errantia have many long setae bristling out of their body and are supported on footlike parapodia (**Figure 34.16a**). Most of them are free-ranging predators with well-developed eyes and powerful jaws. Many are brightly colored. In turn, most species are important prey for fishes and crustaceans.

In the Sedentaria, setae are in close proximity to the body wall, which facilitates anchorage in tubes and burrows. The more sedentary lifestyle of the Sedentaria is associated with reductions in head appendages. Within this group, three types of lifestyles are apparent: those of tube worms, earthworms, and leeches.

Tube worms are marine sedentarians that exhibit beautiful tentacle crowns for filtering food items, such as plankton, from the water (**Figure 34.16b**). The bulk of the worm remains hidden in a tube deep in the mud or sand.

Earthworms play a unique and beneficial role in conditioning the soil, primarily due to the effects of their burrows and excretion. Earthworms ingest soil and leaf tissue to extract nutrients and in the process create burrows in the earth. As plant material and soil pass through the earthworm's digestive system, it is finely ground in the gizzard into smaller fragments. Once excreted, this material—called castings—enriches the soil (**Figure 34.16c**). Because a worm can eat its own weight in soil every day, worm castings on the soil surface can be extensive. The biologist Charles Darwin was interested in earthworm activity, and his last work, *The Formation of Vegetable Mould, through the Actions of Worms, with Observations on Their Habits*, was the first detailed study of earthworm ecology. In it, he wrote, "All the fertile areas of this planet have at least once passed through the bodies of earthworms."

Leeches are usually found in freshwater environments, but some are marine species and others are terrestrial species that inhabit warm, moist areas such as tropical forests. They have a fixed number of segments, usually 34, though in most species septa are not present. Most leeches are blood-sucking parasites of vertebrates. Unlike

(a)

(b)

(c)

(d)

**Figure 34.16 Annelids. (a)** This free-ranging marine worm from Indonesia is a member of the group Errantia. Members of the group Sedentaria include **(b)** tube worms, **(c)** earthworms, and **(d)** leeches. This leech species, *Hirudo medicinalis*, is sucking blood from a patient to reduce the swelling that can occur after surgery. a: ©WaterFrame/Alamy Stock Photo; b: ©J W Alker/age fotostock; c: ©Colin Varndell/Getty Images; d: ©St. Bartholomew's Hospital/Science Source

cestode and trematode flatworms, which are internally parasitic and host-specific, leeches are generally external parasites that feed on a broad range of hosts, including fishes, amphibians, and mammals. Leeches have powerful suckers at both ends of the body, and the anterior sucker is equipped with razor-sharp jaws that can bore or slice into the host's tissues. The salivary secretion (hirudin) acts as an anticoagulant to stop the prey's blood from clotting and an anesthetic to numb the pain. Leeches can suck up to several times their own weight in blood. They were once used in the medical field in the practice of bloodletting, the withdrawal of often considerable quantities of blood from a patient in the erroneous belief that this would prevent or cure illness and disease. Even today, leeches may be used after surgeries (Figure 34.16d). If the blood vessels are not fully reconnected and excess blood accumulates, a swelling called a hematoma may form. The accumulated blood blocks the delivery of new blood and stops the formation of new vessels. The leeches remove the accumulated blood, and new capillaries are more likely to form.

## 34.5 Ecdysozoa: The Nematodes and Arthropods

### Learning Outcomes:

1. List the distinguishing characteristics of nematodes.
2. Describe the arthropod body plan and its major features.
3. Give examples of the arthropod subphyla Chelicerata, Myriapoda, Hexapoda, and Crustacea.
4. List the features that help to account for the diversity of insect species.
5. **CoreSKILL** » Explain how DNA barcoding can be useful in analyzing and controlling insect populations.

The Ecdysozoa is the sister group to the Lophotrochozoa. Although the separation is supported by molecular evidence, the Ecdysozoa is named for a process called **ecdysis**, or the periodic molting of the exoskeleton (refer back to Figure 33.13). All ecdysozoans possess a **cuticle**, a nonliving covering that both supports and protects the animal. Once formed, however, the cuticle typically cannot increase in size, which restricts the growth of the animal inside. The solution for growth is the formation of a new, softer cuticle under the old one. The old one then splits open and is sloughed off, allowing the new, soft cuticle to expand to a bigger size before it hardens.

The evolution of a cuticle was a critical innovation that led to other changes in ecdysoans. A thick cuticle, as in arthropods, impedes the diffusion of oxygen across the skin. Such species acquire oxygen by lungs, gills, or a set of branching, air-filled tubes called tracheae. The ability to shed the cuticle also opened up developmental options for the ecdysozoans. For example, many species undergo complete metamorphosis, changing from a wormlike larva into a winged adult. Animals with internal skeletons cannot do this because growth occurs only by adding more minerals to the existing skeleton. Another significant adaptation is the development of internal fertilization, which permitted species to live in dry environments. A variety of appendages specialized for locomotion evolved in many species, including legs for walking or swimming and wings for flying.

Because of these innovations, ecdysozoans are an incredibly successful group. Of the eight ecdysozoan phyla, we will consider the two most common: the nematodes and arthropods. The grouping of nematodes and arthropods is a relatively new concept supported by molecular data, and it implies that the process of molting arose only once in animal evolution. In support of this, certain hormones that stimulate molting have been discovered to exist only in both nematodes and arthropods.

## The Phylum Nematoda Consists of Small Pseudocoelomate Worms Covered by a Tough Cuticle

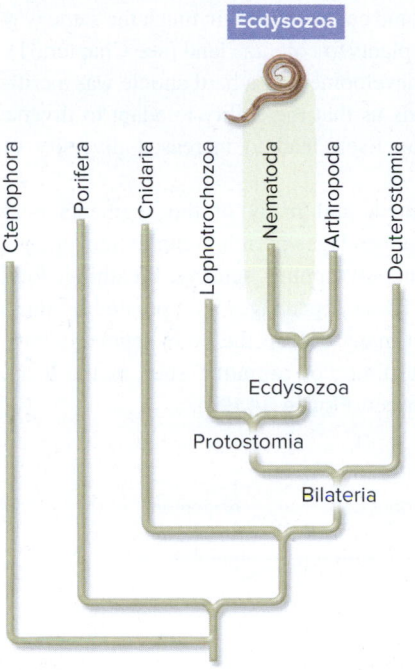

The nematodes (from the Greek *nematos*, meaning thread), also called roundworms, are small, thin worms that range in size from less than 1 mm to about 5 cm (**Figure 34.17**), although some parasitic species measuring 1 m or more have been found in the placenta of sperm whales. Nematodes are ubiquitous organisms that exist in nearly all habitats, from the poles to the tropics. They are found in the soil, in both freshwater and marine environments, and inside plants and animals as parasites. A shovelful of soil may contain a million nematodes. Over 25,000 species are known, but there are probably at least four times as many undiscovered species.

**Key Features of Nematodes** Nematodes have several distinguishing characteristics. A tough cuticle covers the body. The cuticle is secreted by the epidermis and is made primarily of **collagen**, a structural protein also present in vertebrates. The cuticle is shed periodically as the nematode grows. Beneath the epidermis are longitudinal muscles but no circular muscles, which means that muscle contraction results in thrashing of the body rather than smoother wormlike movement. The pseudocoelom functions as both a fluid-filled skeleton and a circulatory system. Diffusion of gases occurs through the cuticle. Nematodes have a complete digestive tract composed of a mouth, pharynx, intestine, and anus. The mouth often contains sharp, piercing organs called **stylets**, and the muscular pharynx functions to suck in food. Excretion of metabolic waste occurs via two simple tubules that have no cilia or flame cells.

**Reproduction in Nematodes** Nematode reproduction is usually sexual, with separate males and females, and fertilization takes place internally. Females are generally larger than males and can produce prodigious numbers of eggs, in some cases, over 100,000 per day. In some species, such as *Caenorhabditis elegans,* both hermaphrodites and males are produced. Hermaphrodites can undergo self fertilization, or they can achieve cross fertilization if they mate with a male. *C. elegans* has become a model organism for researchers to study the process of development (refer back to Figure 20.1b). Development is easily observed because the organism is transparent and composed of relatively few cells, and the generation time is short. An adult *C. elegans* has about 1000 somatic cells.

**Parasitic Nematodes** A large number of nematodes are parasitic in humans and other vertebrates.

- The large roundworm *Ascaris lumbricoides* is a parasite of the small intestine that can reach up to 30 cm in length. Over a billion people worldwide carry this parasite. Although infections are most prevalent in tropical or developing countries, the prevalence of *A. lumbricoides* is relatively high in rural areas of the southeastern U.S. Eggs pass out in feces and can remain viable in the soil for years, although they require ingestion before hatching into an infective stage.

- Hookworms (*Necator americanus*), so named because their anterior end curves dorsally like a hook, are also parasites of the human intestine. The eggs pass out in feces, and recently hatched hookworms can penetrate the skin of a host's foot to establish a new infection. In areas with modern plumbing, these infections are uncommon.

- Pinworms (*Enterobius vermicularis*), although a nuisance, have relatively benign effects on their hosts. The rate of infection in the U.S., however, is staggering: 30% of children and 16% of adults are believed to be hosts. Adult pinworms live in the large intestine and migrate to the anal region at night to lay their eggs, which causes intense itching. The resultant scratching can spread the eggs from the hand to the mouth.

- In the tropics, some 250 million people are infected with *Wuchereria bancrofti*, a fairly large (100 mm) worm that lives in the lymphatic system, blocking the flow of lymph, and, in extreme cases, causing elephantiasis, an extreme swelling of the legs and other body parts (**Figure 34.18**). Females release tiny, live young called microfilariae, which are transmitted to new hosts via mosquitoes.

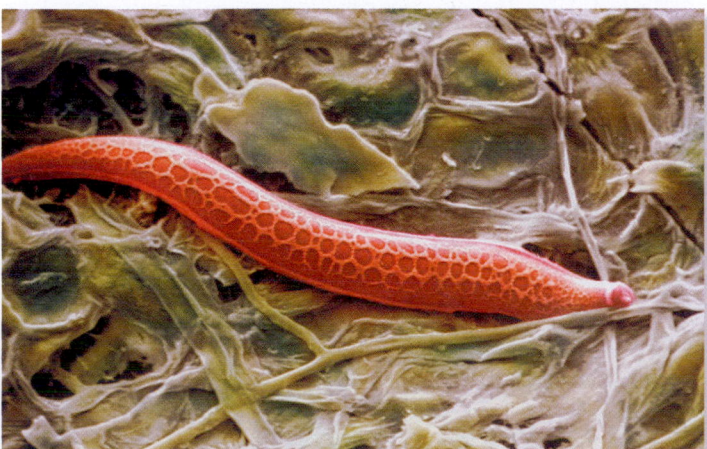

**Figure 34.17** **Scanning electron micrograph of a nematode within a plant leaf.** ©Biophoto Associates/Science Source

**Concept Check:** *Both nematodes and annelids are wormlike in appearance. How are they different?*

## The Phylum Arthropoda Contains Species with Jointed Appendages

The arthropods (from the Greek *arthron*, meaning joint, and *podos*, meaning foot) constitute perhaps the most diverse phylum on Earth, including familiar organisms such as spiders, insects, and crustaceans.

crustaceans, the exoskeleton is reinforced with calcium carbonate to make it extra hard. The exoskeleton provides protection and also a point of attachment for muscles, all of which are internal. It is also relatively impermeable to water, a feature that may have enabled many arthropods to conserve water and colonize land, in much the same way as a tough seed coat allowed plants to colonize land (see Chapter 31). From this point of view, the development of a hard cuticle was a critical innovation. It also reminds us that the ability to adapt to diverse environmental conditions can itself lead to increased diversity of organisms.

Arthropods are segmented, and many of the segments bear jointed appendages. Jointed appendages permit complex movements and functions such as walking, swimming, sensing, breathing, food handling, and reproduction. These appendages are operated by muscles within each segment. In many orders, the body segments have become fused into functional units, or **tagmata**, such as the head, thorax, and abdomen of an insect (**Figure 34.19a**).

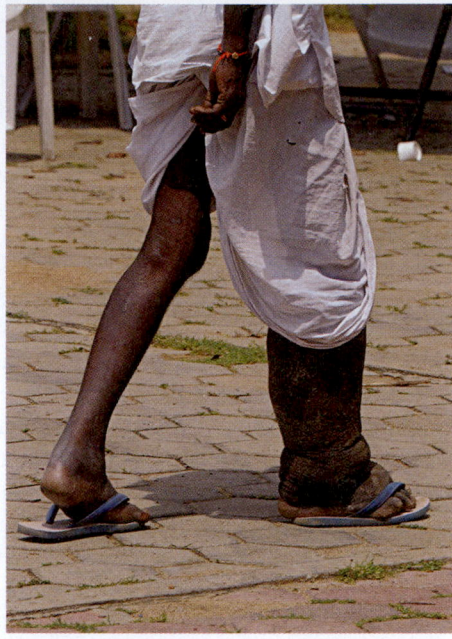

**Figure 34.18** **Elephantiasis in a human leg.** The disease is caused by the nematode parasite *Wuchereria bancrofti*, which lives in the lymphatic system and blocks the flow of lymph. ©Noah Seelam/Stringer/Getty Images

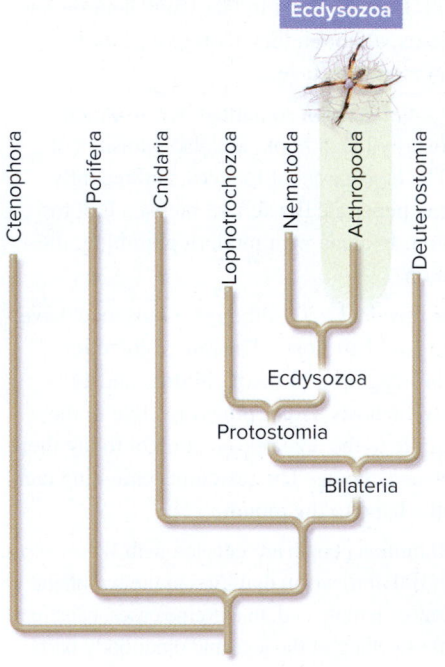

About three-quarters of all described living species present on Earth are arthropods, and scientists have estimated they are also numerically common, with an estimated $10^{18}$ (a billion billion) individual organisms. The huge success of the arthropods, in terms of their sheer numbers and diversity, is related to features that permit these animals to live in all the major areas on Earth, from the poles to the tropics and from marine and freshwater habitats to dry land. These features include an exoskeleton, segmentation, and jointed appendages.

***Arthropod Body Plan*** The body of a typical arthropod is covered by a hard cuticle, an **exoskeleton** (external skeleton), made of layers of chitin and protein. The cuticle can be extremely tough in some parts, as in the shells of crabs, lobsters, and even beetles, yet be soft and flexible in other parts, between body segments and segments of appendages, to allow for movement. In the class of arthropods called

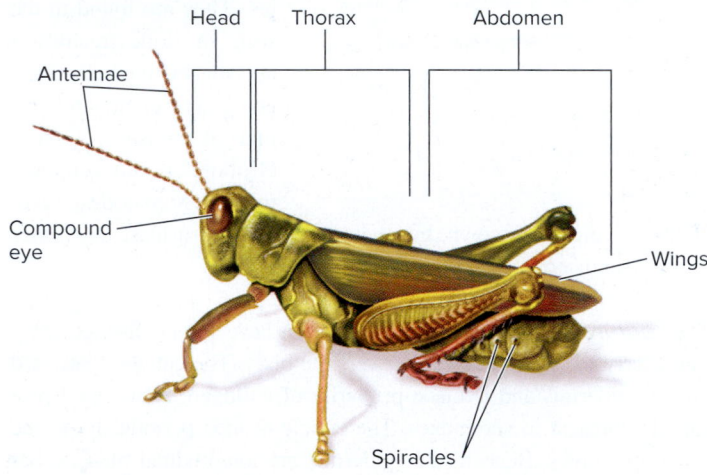

**(a) External anatomy**

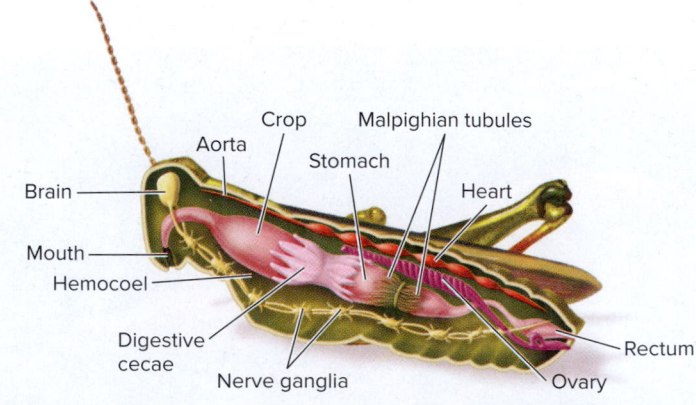

**(b) Internal anatomy**

**Figure 34.19** **Body plan of an arthropod, as represented by a grasshopper.**

 **Core Skill: Connections** Look ahead to Figure 49.5. Why did the Malpighian tubule system play a key role in the colonization of land by insects and other arthropods?

In arthropods, cephalization has resulted in a well-defined head, which includes a brain consisting of two or three cerebral ganglia connected to several smaller ventral nerve ganglia. Arthropods have multiple sensory organs, including organs of sight, touch, smell, hearing, and balance. They have compound eyes composed of many independent visual units called **ommatidia** (singular, ommatidium) (look ahead to Figure 44.13). Together, these lenses render a mosaic-like image of the environment. Some species, particularly some insects, possess additional simple eyes, or ocelli, that are probably only capable of distinguishing light from dark.

Like most mollusks, arthropods have an open circulatory system (look ahead to Figure 48.1a), in which hemolymph is pumped from a tubelike heart into the aorta or short arteries and then into the open sinuses that coalesce to form a cavity called the hemocoel. From the hemocoel, gases and nutrients from the hemolymph diffuse into tissues. The hemolymph flows back into the heart via pores, called ostia, that are equipped with valves.

Because the cuticle impedes the diffusion of gases through the body surface, arthropods require special organs that permit gas exchange. In aquatic arthropods, these consist of feathery gills that have an extensive surface area in contact with the surrounding water. Terrestrial species have a highly developed **tracheal system** (look ahead to Figure 48.18). On the body surface, pores called **spiracles** provide openings to a series of finely branched air tubes within the body called trachea. The tracheal system delivers oxygen directly to tissues and cells, and the circulatory system does not play a role in gas exchange.

The digestive system of arthropods is complex and often includes a mouth, crop, stomach, intestine, and rectum (**Figure 34.19b**). The stomach has glands called digestive cecae that secrete digestive enzymes. Excretion is accomplished by specialized metanephridia or, in insects and some other taxa, by **Malpighian tubules**, extensive tubes that extend from the digestive tract into body cavity, where they are surrounded by hemolymph (look ahead to Figure 49.5). Nitrogenous wastes are absorbed by the tubules and emptied into the gut, where the intestine and rectum reabsorb water and salts and the waste is excreted through the anus. This excretory system, allowing the retention of water, was another critical innovation that permitted the colonization of land by arthropods.

### Major Subphyla of Arthropods

The history of arthropod classification is extensive and active. Although many classifications have been proposed, a 1995 study of the mitochondrial DNA of arthropod species by American geneticist Jeffrey Boore and colleagues suggests a phylogeny with five main subphyla: one now-extinct subphyla, Trilobita (trilobites); and four living subphyla, Chelicerata (spiders and scorpions), Myriapoda (millipedes and centipedes), Hexapoda (insects and relatives), and Crustacea (crabs and relatives) (**Table 34.4**).

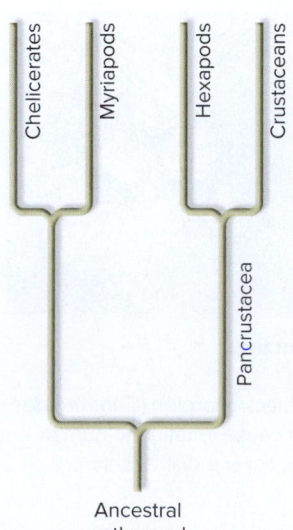

| Table 34.4 | Main Subphyla and Characteristics of Arthropods | |
|---|---|---|
| | **Subphyla and examples (est. number of species)** | **Class characteristics** |
| | Chelicerata: spiders, scorpions, mites, ticks, horseshoe crabs, and sea spiders (74,000) | Body usually with cephalothorax and abdomen only; six pairs of appendages, including four pairs of legs, one pair of fangs, and one pair of pedipalps; terrestrial; predatory or parasitic |
| | Myriapoda: millipedes and centipedes (13,000) | Body with head and highly segmented trunk. In millipedes, each segment with two pairs of walking legs; terrestrial; herbivorous. In centipedes, each segment with one pair of walking legs; terrestrial; predatory, poison jaws |
| | Hexapoda: insects such as beetles, butterflies, flies, fleas, grasshoppers, ants, bees, wasps, termites, and springtails (>1 million) | Body with head, thorax, and abdomen; mouthparts modified for biting, chewing, sucking, or lapping; usually with two pairs of wings and three pairs of legs; mostly terrestrial, some freshwater; herbivorous, parasitic, or predatory |
| | Crustacea: crabs, lobsters, shrimp (45,000) | Body of two to three parts; three or more pairs of legs; chewing mouthparts; usually marine |

Boore's research showed that the Trilobita were among the earliest-diverging arthropods. The lineage then split into two groups. One, often referred to as the Pancrustacea, contains the insects and crustaceans. The other, with no overarching name, contains the myriapods and chelicerates. Molecular evidence suggests that insects are more closely related to crustaceans than they are to spiders or millipedes and centipedes. We will take a closer look at insects and crustaceans due to their relative sizes and importance to humans.

### Subphylum Trilobita: Extinct Early Arthropods

The trilobites were among the earliest arthropods, flourishing in shallow seas of the Paleozoic era, some 500 mya, and dying out about 250 mya. Most trilobites were bottom feeders and were generally 3–10 cm in size, although some reached almost 1 m in length (**Figure 34.20**). They had three main tagmata: the head, thorax, and tail. Trilobites also had two dorsal grooves that divided the body longitudinally into three lobes—an axial lobe and two pleural lobes—a structural characteristic that gave the subphylum its name. Most of the body segments showed little specialization. In contrast, later-diverging arthropods developed specialized appendages on many segments, including appendages for grasping, walking, and swimming.

Head     Thorax     Tail

Pleural lobe

Axial lobe

**Figure 34.20** **A fossil trilobite.** About 4,000 fossil species of these early arthropods, including *Modocia centralis*, shown here, which was about 20 cm long, have been described. ©Sinclair Stammers/Getty Images

### Subphylum Chelicerata: The Spiders, Scorpions, Mites, and Ticks

The Chelicerata consists mainly of the class Arachnida, which contains predatory spiders and scorpions as well as the ticks and mites, some of which are blood-sucking parasites that feed on vertebrates. The two other living classes are the Merostomata, the horseshoe crabs (four species), and the Pycnogonida, the sea spiders (1,000 species), both of which are marine, reflecting the group's marine ancestry. All species have a body consisting of two tagmata: a fused head and thorax, called a cephalothorax, and an abdomen. They also possess six pairs of appendages: the chelicerae, or fangs; a pair of pedipalps, which have various sensory, predatory, or reproductive functions; and four pairs of walking legs.

Spider fangs are supplied with venom from poison glands. Most spider bites are harmless to humans, although they are very effective in immobilizing and/or killing their insect prey. Venom from some species, including the black widow (*Latrodectus mactans*; **Figure 34.21a**) and the brown recluse (*Loxosceles reclusa*), are potentially, although rarely, fatal to humans. The toxin of the black widow is a neurotoxin, which interferes with the functioning of the nervous system, whereas that of the brown recluse is hemolytic,

meaning it destroys red blood cells around the bite. After the spider has subdued its prey, it pumps digestive fluid into the tissues via the fangs and sucks out the partially digested meal.

Spiders have abdominal silk glands, called spinnerets, and many spin webs to catch prey (**Figure 34.22a**). The silk is a protein that stiffens after extrusion from the body because the mechanical shearing causes a change in the organization of the protein's structure. Silk is stronger than steel of the same diameter and is more elastic than Kevlar, the material used in bulletproof vests. Each spider family constructs a characteristic size and style of web and can do it perfectly on its first attempt, indicating that web spinning is an innate (instinctual) behavior (see Chapter 55). Spiders also use silk to wrap up prey and to construct egg sacs. Interestingly, spiders that are fed drugged prey spin their webs differently than do undrugged spiders (**Figure 34.22b** and **c**).

Scorpions (order Scorpionida) are generally tropical or subtropical animals that feed primarily on insects, though they may eat spiders and other arthropods as well as smaller reptiles and mice. Their pedipalps are modified into large claws, and the abdomen tapers into a stinger, which is used to inject venom. Although the venom of most North American species is generally not fatal to humans, that of the *Centruroides* genus from deserts in the U.S. Southwest and Mexico can be deadly. Fatal species are also found in India, Africa, and other countries. Unlike spiders, which lay eggs, scorpions bear live young that the mother then carries around on her back until they have their first molt (**Figure 34.21b**).

In mites and ticks (order Acari), the two main body segments (cephalothorax and abdomen) are fused and appear as one large segment. Many mite species are free-living scavengers that feed on dead plant or animal material. Other mites are serious pests on crops, and some, like chiggers (*Trombicula alfreddugesi*), are parasites of humans that spread diseases such as typhus (**Figure 34.21c**). Chiggers are parasites only on their larval stage. Chiggers do not bore into the skin; their bite and salivary secretions cause skin irritation. *Demodex brevis* is a hair-follicle mite that is common in animals and humans. The mite is estimated to be present on over 90% of adult humans. Although the mite causes no irritation in most humans,

**(a) Black widow spider**

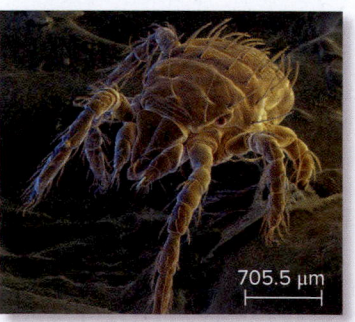

**(b) Scorpion with young**

**(c) Chigger mite**

705.5 μm

**(d) Bont ticks**

**Figure 34.21** **Common arachnids.** **(a)** Female black widow spider (*Latrodectus mactans*). **(b)** The Central American black scorpion (*Centruroides gracilis*) is highly venomous and carries its young on its back. **(c)** SEM of a chigger mite (*Trombicula alfreddugesi*) that can cause irritation to human skin and spread disease. **(d)** These South African bont ticks (*Amblyomma hebraeum*) are feeding on a white rhinoceros. a: ©George Grall/Getty Images; b: ©Mark Smith/Science Source; c: ©David Scharf/Science Source; d: ©Roger De LaHarpe/Gallo Images/Corbis/Getty Images

**Concept Check:** *What is one of the main characteristics distinguishing arachnids from insects?*

(a) Normal web

(b) Web spun by spider fed with prey containing caffeine

(c) Web spun by spider fed with prey containing marijuana

**Figure 34.22** **Spider-web construction by normal and drugged spiders.** a–c: ©NASA/SPL/Science Source

 **Core Concept: Science and Society** Some scientists have suggested using web-spinning spiders to test substances for the presence of drugs or even to indicate environmental contamination.

*Demodex canis* causes the skin disease known as mange in domestic animals, particularly dogs.

Ticks are larger than mites, and all are ectoparasitic, feeding on the body surface of vertebrates. Their life cycle includes attachment to a host, sucking blood until they are replete, and dropping off the host to molt (**Figure 34.21d**). Ticks can carry a variety of viral and bacterial diseases, including Lyme disease, a bacterial disease so named because it was first observed in the town of Lyme, Connecticut, in the 1970s.

**Subphylum Myriapoda: The Millipedes and Centipedes** Myriapods have one pair of antennae on the head and three pairs of appendages that are modified as mouthparts, including mandibles that act like jaws. The millipedes and centipedes, both wormlike arthropods with legs, are among the earliest terrestrial animal phyla known. Millipedes (class Diplopoda) have two pairs of legs per segment, as their class name denotes (from the Latin *diplo*, meaning two, and *podos*, meaning feet), not 1,000 legs, as their common name suggests (**Figure 34.23a**). They are slow-moving herbivores that eat decaying leaves and other plant material. When threatened, the millipede's response is to roll up into a protective coil. Many millipede species

(a) Two millipedes

(b) A centipede

**Figure 34.23** **Millipedes and centipedes.** **(a)** Millipedes have two pairs of legs per segment. **(b)** The venom of the giant centipede (Scolopendra heros) is known to produce significant swelling and pain in humans. a: ©David Aubrey/Corbis/Getty Images; b: ©Larry Miller/Science Source

also have glands on their underside that can eject a variety of toxic, repellent secretions. Some millipedes are brightly colored, warning potential predators that they can protect themselves.

Class Chilopoda (from the Latin *chilo*, meaning lip, and *podos*, meaning feet), or centipedes, are fast-moving carnivores that have one pair of walking legs per segment (**Figure 34.23b**). The head has many sensory appendages, including a pair of antennae and three pairs of appendages modified as mouthparts, including powerful claws connected to poison glands. The venom of some larger species, such as *Scolopendra heros*, is powerful enough to cause pain in humans. Most species do not have a waxy waterproofing layer on their cuticle and so are restricted to moist environments under leaf litter or in decaying logs, usually coming out at night to actively hunt their prey.

**Subphylum Hexapoda: Insects and Relatives** Hexapods are six-legged arthropods. Most are insects, but a few earlier-diverging non-insect hexapods have been identified, including soil-dwelling groups such as collembolans, and molecular studies have shown that these represent a separate but related lineage. Insects are in a class by themselves (Insecta), literally and figuratively. Biologists have classified more species of insects than all other species of animal life combined. Approximately 1 million species of insects have been described thus far, and according to a 2015 estimate by British Entomologist Nigel Stork, 4 million more species await description. At least 90,000 species of insects have been identified in the U.S. and Canada alone. DNA barcoding, which is discussed later in this chapter, can help resolve many taxonomic dilemmas between closely related species.

Insects are the subject of an entire field of scientific study, **entomology**. They are studied in large part because of their significance as pests of the world's agricultural crops and carriers of some of the world's most deadly diseases. Insects live in all terrestrial habitats, and virtually all species of plants are fed upon by at least one, usually tens, and sometimes, in the case of large trees, hundreds of insect species. Because approximately one-quarter of the world's crops are lost annually to insects, researchers are constantly trying to find ways to reduce pest densities. Insect pest reduction often involves chemical control (the use of pesticides) or biological control (the use of living organisms). Many species of insects are also important pests or parasites of humans and livestock, both by their own actions and as vectors of diseases such as malaria and sleeping sickness.

In contrast, insects also provide us with many types of essential biological services. We depend on insects such as honeybees, butterflies, and moths to pollinate our crops. Bees also produce honey, and silkworms are the source of silk fiber. Despite the revulsion they provoke in us, fly larvae (maggots) are important in the decomposition process of both dead plants and animals. In addition, we use insects in the biological control of other insects.

***Key Features of Insects***   Of paramount importance to the success of insects was the evolution of wings, a feature possessed by no other arthropod and indeed no other living animal except birds and bats. Unlike vertebrate wings, however, insect wings are outgrowths of the body wall cuticle and are not true segmental appendages. This means that insects still have all their walking legs. Insects are thus like the mythological horse Pegasus, which sprouted wings out of its back while retaining all four legs. In contrast, birds and bats have one pair of appendages (arms) modified for flight, which leaves them considerably less agile on the ground.

Insects in different orders have also evolved a variety of mouthparts in which the constituent parts, the mandibles and maxillae, are modified for different functions (**Figure 34.24**). Many of these mouthparts are modified walking appendages and are bilaterally paired. As a result, the jaws of many insects, such as grasshoppers, move in a side-to-side motion, rather than up and down as human jaws do.

- Grasshoppers, beetles, dragonflies, and many others have mouthparts adapted for chewing.
- Mosquitoes and many plant pests have mouthparts adapted for piercing and sucking.
- Butterflies and moths have a coiled tongue (**proboscis**) that can be uncoiled, enabling them to drink nectar from flowers.
- Some flies have lapping, spongelike mouthparts that sop up liquid food.

Their varied mouthparts are adaptations that allow insects to specialize their feeding on virtually anything: plant matter, decaying organic matter, and other living animals. The biological diversity of insects is therefore related to environmental diversity, in this case, the variety of foods that insects eat. Parasitic insects attach themselves to other species, and some insect parasites (called hyperparasites) even feed on other parasites, as noted in a verse sometimes attributed to the 18th-century English poet and satirist Jonathan Swift:

Big fleas have little fleas

upon their backs to bite 'em;

and little fleas have lesser fleas

and so, ad infinitum.

***Major Orders of Insects***   The diversity of insects is astounding: Hexapoda is composed of 35 orders, some of which have over 100,000 species. The most common orders are described in **Table 34.5**.

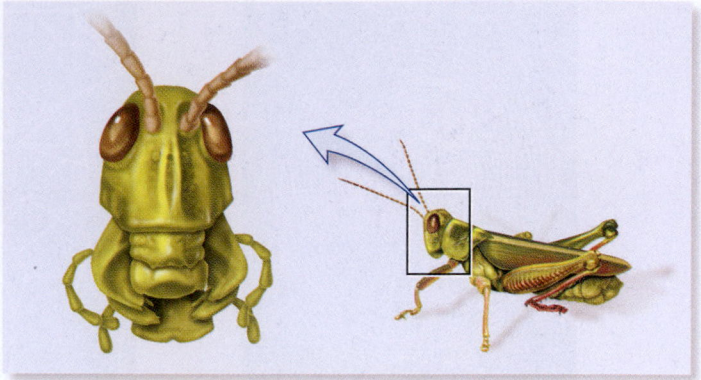

**(a) Chewing (grasshopper)**

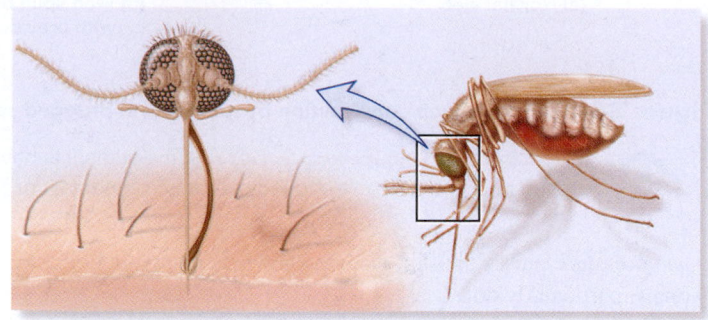

**(b) Piercing and blood sucking (mosquito)**

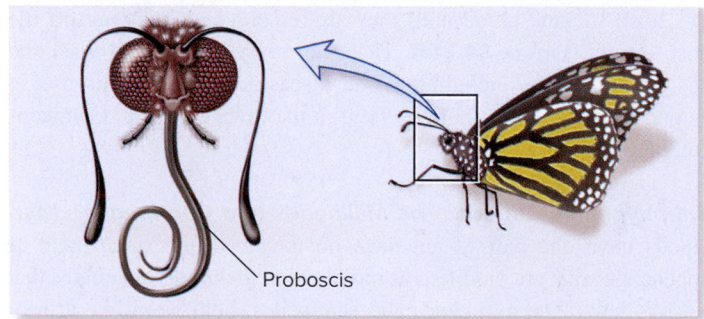

Proboscis

**(c) Nectar sucking (butterfly)**

**(d) Sponging liquid (housefly)**

**Figure 34.24  A variety of insect mouthparts.** Insect mouthparts have become modified in ways that allow insects to feed by a variety of methods, including **(a)** chewing (Orthoptera, Coleoptera, and others), **(b)** piercing and blood sucking (Diptera), **(c)** nectar sucking (Lepidoptera), and **(d)** sponging up liquid (Diptera).

*Concept Check:*  *Insects have a variety of mouthparts. Name two other key insect adaptations.*

| Table 34.5 | **Major Orders and Characteristics of Insects** |
| --- | --- |

| Order and examples (approx. number of described species) | | Order characteristics |
| --- | --- | --- |
| Coleoptera: beetles, weevils (400,000) | | Two pairs of wings (front pair thick and leathery, acting as wing cases, back pair membranous); armored exoskeleton; biting and chewing mouthparts; complete metamorphosis; largest order of insects |
| Hymenoptera: ants, bees, wasps (130,000) | | Two pairs of membranous wings; chewing or sucking mouthparts; many have posterior stinging organ on females; complete metamorphosis; many species social; important pollinators |
| Diptera: flies, mosquitoes (120,000) | | One pair of wings with hind wings modified into halteres (balancing organs); sucking, piercing, or lapping mouthparts; complete metamorphosis; larvae are grublike maggots in various food sources; some adults are disease vectors |
| Lepidoptera: butterflies, moths (150,000) | | Two pairs of colorful wings covered with tiny scales; long tubelike tongue for sucking; complete metamorphosis; larvae are plant-feeding caterpillars; adults are important pollinators |
| Hemiptera: true bugs; assassin bug, bedbug, chinch bug, cicada (82,000) | | Two pairs of membranous wings; piercing or sucking mouthparts; incomplete metamorphosis; many are plant feeders; some are predatory or blood feeders; vectors of plant diseases |
| Orthoptera: crickets, grasshoppers (20,000) | | Two pairs of wings (front pair leathery, back pair membranous); chewing mouthparts; mostly herbivorous; incomplete metamorphosis; powerful hind legs for jumping |
| Odonata: damselflies, dragonflies (5,500) | | Two pairs of long, membranous wings; chewing mouthparts; large eyes; predatory on other insects; incomplete metamorphosis; nymphs aquatic; considered early-diverging insects |
| Siphonaptera: fleas (2,400) | | Wingless, laterally flattened; piercing and sucking mouthparts; adults are bloodsuckers on birds and mammals; jumping legs; complete metamorphosis; vectors of plague |
| Phthiraptera: sucking lice (3,000) | | Wingless ectoparasites; sucking mouthparts; flattened body; reduced eyes; legs with clawlike tarsi for clinging to skin; incomplete metamorphosis; very host-specific; vectors of typhus |
| Isoptera: termites (2,300) | | Two pairs of membranous wings when present; some stages wingless; chewing mouthparts; social species; incomplete metamorphosis |

 **Core Skill: Modeling**  The goal of this modeling challenge is to develop a mathematical model that allows you to estimate the number of insect species on Earth.

**Modeling Challenge:** Some biologists have suggested that we don't know within an order of magnitude how many species exist on Earth. Because insects represent by far the largest taxa on Earth, the answer to this question is dependent on knowing the number of insect species. Of the insects, the best known are the showy butterflies, with 15,000–20,000 species known worldwide. In Britain, insect diversity is almost completely known, with 67 species of butterflies and a total of 24,043 insect species. Create a mathematical model that allows you to estimate the number of insect species on Earth. (Hint: Look ahead to Section 56.1 and read about the mark-recapture technique if you get stuck.) Can you think of some assumptions your model makes?

Although all insects have six legs, different orders have slightly different wing structures, and many of the orders are based on wing type (their names often include the root *pter-*, from the Greek *pteron*, meaning wing).

- In beetles (Coleoptera), only the back pair of wings is functional; the front wings have become protective shell-like coverings under which the back pair folds when not in use.

- Wasps and bees (Hymenoptera) have two pairs of wings that are hooked together and move as one wing.

- Flies (Diptera) possess only one pair of wings (the front pair); the back pair has been modified into a small pair of balancing organs, called halteres, that act like miniature gyroscopes.

- Butterflies (Lepidoptera) have wings that are covered in scales (from the Greek *lepido*, meaning scale); other insects generally have clear, membranous wings.

- In ant and termite colonies, the queen and the drones (males) retain their wings, whereas female individuals called workers have lost theirs. Other orders, such as fleas and lice, are completely wingless.

***Reproduction and Development of Insects***   All insects have separate sexes, and fertilization is internal. During development, the majority (approximately 85%) of insects undergo a change in body form known as **complete metamorphosis** (from the Greek *meta*, meaning change, and *morph*, meaning form) (**Figure 34.25a**). Animals that undergo complete metamorphosis advance through four stages: egg, larva, pupa, and adult. The dramatic body transformation from larva to adult occurs in the pupa stage. The larval stage is often spent in an entirely different habitat from that of the adult, and larval and adult forms use different food sources. Consequently, they do not compete directly for the same resources. The larval stage, such as a caterpillar, is focused on eating and growth, whereas the adult stage involves sexual reproduction. Most adult insects have wings, allowing them to disperse their fertilized eggs over a larger area.

A smaller percentage of insect species undergo **incomplete metamorphosis**, in which morphological changes are more gradual (**Figure 34.25b**). Incomplete metamorphosis has only three stages: egg, nymph, and adult. Young insects, called nymphs, look like miniature adults when they hatch from their eggs, but usually don't have wings. As they grow and feed, they shed their exoskeleton and replace it with a larger one several times, each time entering a new instar, or stage of growth. When the insects reach their adult size, they have also grown wings.

Some insects, such as bees, wasps, ants, and termites, have developed complex social behavior and live cooperatively in underground or aboveground nests. Such colonies exhibit a division of labor, in that some individuals forage for food and care for the brood (workers), others protect the nest (soldiers), and some only reproduce (the queen and drones) (**Figure 34.26**).

***Subphylum Crustacea: Crabs, Lobsters, Barnacles, and Shrimp***
The crustaceans are common inhabitants of marine environments, although some species live in fresh water and a few are terrestrial. Many species, including crabs, lobsters, crayfish, and shrimp, are

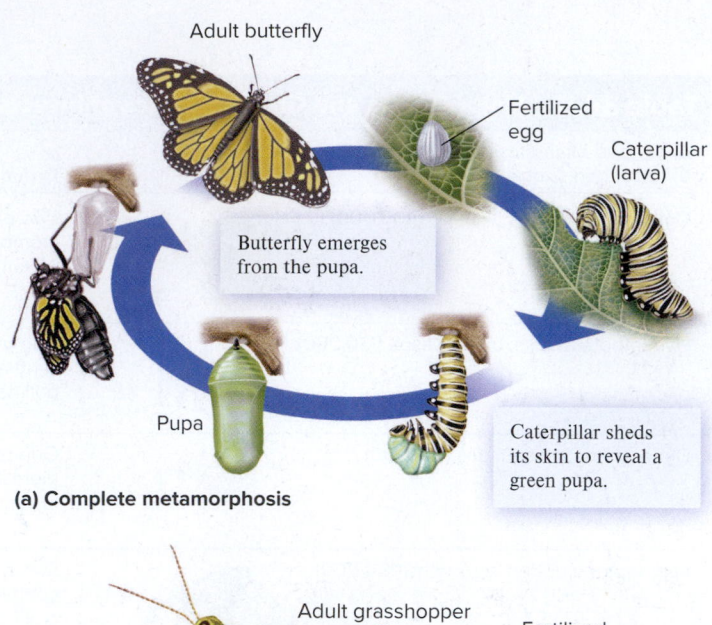

**(a) Complete metamorphosis**

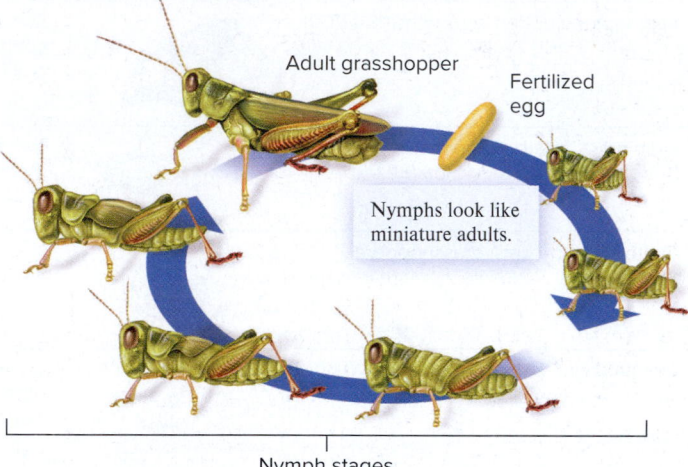

**(b) Incomplete metamorphosis**

**Figure 34.25   Metamorphosis. (a)** Complete metamorphosis, as illustrated by the life cycle of a monarch butterfly. The adult butterfly has a completely different appearance than the larval caterpillar. **(b)** Incomplete metamorphosis, as illustrated by the life cycle of a grasshopper. The eggs hatch into nymphs, essentially miniature versions of the adult.

**(a) Worker and soldier ants**

**(b) Queen ant**

**Figure 34.26   The division of labor in insect societies.** Individuals from the same insect colony may appear very different. Among these army ants (*Eciton burchelli*) from Paraguay, **(a)** workers forage for the colony, soldiers (with large mandibles) protect the colony from predators, and **(b)** the queen produces eggs. a: Source: Alex Wild/myrmecos .net; b: ©Oxford Scientific/Getty Images

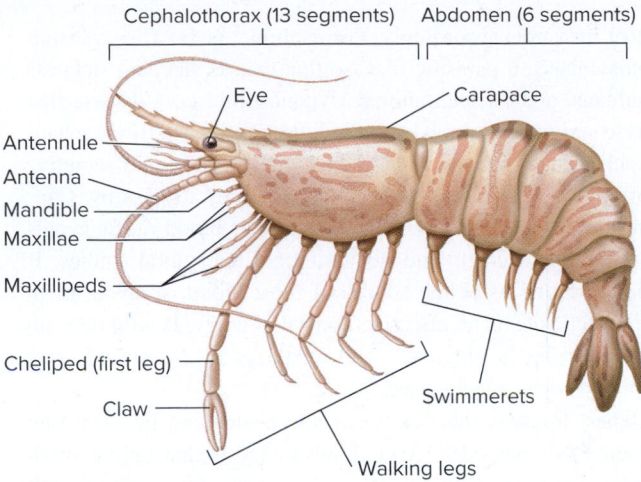

Cephalothorax (13 segments) · Abdomen (6 segments)

Eye · Carapace

Antennule
Antenna
Mandible
Maxillae

Maxillipeds

Cheliped (first leg)

Claw

Swimmerets

Walking legs

**Figure 34.27** **Body plan of a crustacean, as represented by a shrimp.**

 **Core Skill: Connections** Look ahead to Figure 44.8. Where are a crustacean's organs of balance located?

454.5 μm

**Figure 34.28** **Crustacean larva.** The nauplius, a distinct larval stage exhibited by most crustaceans, molts several times before reaching maturity. Many of these larvae are less than 0.01 mm long.
©FLPA/D P Wilson/age fotostock

economically important food items for humans; smaller species are important food sources for other predators.

***Crustacean Body Plan*** The crustaceans are unique among the arthropods in that they possess two pairs of antennae at the anterior end of the body—the antennule (first pair) and antenna (second pair) (**Figure 34.27**). In addition, they have three or more sensory and feeding appendages that are modified mouthparts: the mandibles, maxillae, and maxillipeds. These are followed by walking legs and, often, additional abdominal appendages, called swimmerets, and a powerful tail. In some orders, the first pair of walking legs, or chelipeds, is modified to form powerful claws. The head and thorax are often fused together, forming the cephalothorax. In many species, the cuticle covering the head extends over most of the cephalothorax, forming a hard protective covering called the **carapace**. For growth to occur, a crustacean must shed the entire exoskeleton.

Many crustaceans are predators, but others are scavengers, and some, such as barnacles, are filter feeders. Gas exchange typically

occurs via gills, and crustaceans, like other arthropods, have an open circulatory system. Crustaceans possess two excretory organs: antennal glands and maxillary glands, both modified metanephridia, which open at the bases of the antennae and the maxillae, respectively. Reproduction usually involves separate sexes, and fertilization is internal. Most species carry their eggs in brood pouches under the female's body. Eggs of most species produce larvae that must go through many different molts prior to assuming adult form. The first of these larval stages, called a **nauplius**, is very different in appearance from the adult crustacean (**Figure 34.28**).

***Crustacean Diversity*** Crustacean clades are numerous, but most are small and obscure, although many orders contain important prey items for other marine organisms. For example, copepods are tiny and abundant planktonic crustaceans, which are a food source for filter-feeding organisms and small fish. The clade Cirripedia is composed of the barnacles, crustaceans whose carapace forms calcified plates that cover most of the body (**Figure 34.29a**). Their legs are modified into feathery filter-feeding structures.

(a) Goose barnacles—order Cirripedia

(b) Pill bug—order Isopoda

(c) Coral crab—order Decapoda

**Figure 34.29** **Common crustaceans.** **(a)** Goose barnacles (*Lepas anatifera*). **(b)** Pill bug, or wood louse (*Armadillium vulgare*). **(c)** Coral crab (*Carpilius maculates*). a: ©NHPA/Photoshot; b: ©Miyuki Satake/iStock/Getty Images; c: ©Masa Ushioda/Waterframe/age fotostock

Malacostraca is the largest class of the crustaceans and is divided into many orders. For example, Euphausiacea are shrimplike krill that grow to about 3 cm and provide a large part of the diet of many whales, seals, penguins, fish, and squid. The order Isopoda contains many small species that are parasitic on marine fishes. Terrestrial isopods, better known as pill bugs, or wood lice, retain a strong connection to water and need to live in moist environments such as leaf litter or decaying logs (**Figure 34.29b**). When threatened, they curl up into a tight ball, making it difficult for predators to get a grip on them.

The most familiar Malacostracan order is Decapoda, which includes the crabs and lobsters, the largest crustacean species (**Figure 34.29c**). As their name suggests, these decapods have 10 walking legs (five pairs), although the first pair is invariably modified to support large claws. Most decapods are marine, but many are freshwater species, such as crayfish. In hot, moist tropical areas, some species, called land crabs, are terrestrial.

## Core Concept: Information

### DNA Barcoding: A New Tool for Species Identification

The International Barcoding of Life (IBOL) project, begun in 2003 by Canadian biologist Paul Hebert, is a broad initiative that seeks to create a digital identification system for all life-forms. Hebert made the analogy that the large diversity of products in a grocery store can each be distinguished with a relatively small barcode. Though the diversity of the world's animal species is considerably larger, he reasoned that all species could be distinguished using their DNA. The complete genome would be too large to analyze rapidly, so Hebert suggested analyzing a small piece of DNA of all species. The DNA sequence he proposed for animals is the first 648 base pairs of a gene called *CO1*, for cytochrome oxidase, an enzyme in the electron transport chain of mitochondria (refer back to Figure 7.8). All animals have this gene in their mitochondrial DNA. A key observation is that although this part of the *CO1* gene varies widely between species, it hardly varies at all between individuals of the same species—only 2%.

From a practical perspective, DNA barcoding may be used to analyze and control insect populations For example, about 3,500 species of mosquitoes have been identified, but many of them are hard to tell apart, especially in the field. Some mosquitoes transmit deadly diseases such as malaria and yellow fever and are subject to stringent control measures in many countries. The Mosquito Barcoding Initiative aims to catalog each mosquito species by analyzing the *CO1* gene and thereby build up a DNA barcode database. Field researchers will be able to quickly analyze the DNA of some individuals in a given area and identify them based on existing barcodes. Appropriate control measures can then be instigated against a mosquito population if it contains members of a species that is known to be a disease carrier.

For blood-feeding insects, scientists can also bar code their blood meals, target their feeding preferences, and optimize control measures accordingly. For example, tsetse flies transmit tryptosomiasis, a parasitic disease that causes sleeping sickness in humans, and African animal trypanosomiasis, a disease that leads to serious economic losses of livestock. Tsetse flies are hard to track in nature because of their solitary habits and secretive nature, hiding in bushes and waiting for prey to pass by. Capturing tsetse flies and DNA barcoding their blood meals avoids the necessity of costly and difficult field behavioral studies. If cattle are found to be the source of most blood, spraying them with insecticides is an effective control strategy. If wildlife such as buffalo, giraffe, elephants, and warthogs are the source, then trapping devices may be used.

Hebert foresees the day when all species can be identified by their DNA barcodes. A huge advantage is that only a small sample of cells is necessary. The sample can come from an adult or immature individual, which is a great help since much insect taxonomy is based solely on adults. Many scientists anticipate the day when handheld field barcoding identification devices will be commonly used. At the moment, barcoding involves a laboratory analysis taking about an hour and costing $2.00 per sample.

## 34.6 Deuterostomia: The Echinoderms and Chordates

### Learning Outcomes:

1. Identify the distinguishing characteristics of echinoderms.
2. Describe the four critical innovations in the body plan of chordates.
3. List the two invertebrate subphyla of Chordata, and explain their relationship to the vertebrates.

As discussed in Chapter 33, the deuterostomes are grouped together because they share similarities in patterns of development (refer back to Figure 33.6). Molecular evidence also supports a deuterostome clade. All animals in the phylum Chordata (from the Greek *chorde*, meaning string, referring to the spinal cord), which includes the vertebrates, are deuterostomes. Interestingly, so is one invertebrate group, the phylum Echinodermata, which includes the sea stars, sea urchins, and sea cucumbers. Although there are far fewer phyla and species of deuterostomes than

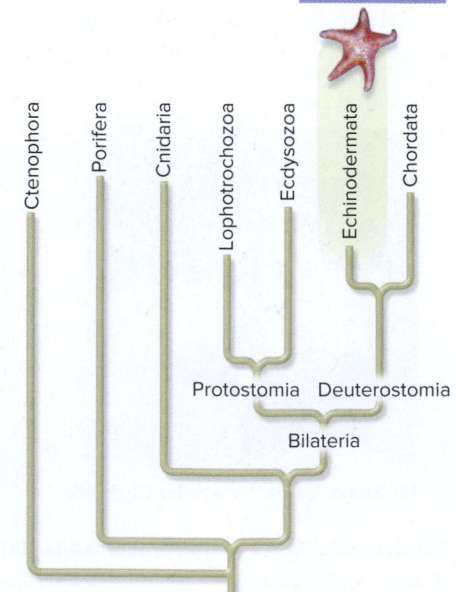

protostomes, the deuterostomes are generally much more familiar to us. After all, we humans are deuterostomes.

We will conclude our discussion of invertebrate biology by turning our attention to the invertebrate deuterostomes. In this section, we will explore the phylum Echinodermata and then introduce the phylum Chordata, looking in particular at its distinguishing characteristics and at its two invertebrate subphyla: the cephalochordates, commonly referred to as the lancelets, and the urochordates, also known as the tunicates. We will discuss the subphylum Vertebrata in Chapter 35.

## The Phylum Echinodermata Includes Sea Stars and Sea Urchins—Species with a Water Vascular System

The phylum Echinodermata (from the Greek *echinos*, meaning spiny, and *derma*, meaning skin) consists of a unique grouping of deuterostomes. A striking feature of all echinoderms is their modified radial symmetry. The body of most species can be divided into five parts pointing out from the center. As a consequence, cephalization is absent in most classes. There is no brain and only a simple nervous system consisting of a central nerve ring from which arise radial branches to each limb. The radial symmetry of echinoderms is secondary, present only in adults. The free-swimming larvae have bilateral symmetry and metamorphose into the radially symmetrical adult form.

**Echinoderm Body Plan** Most echinoderms have an **endoskeleton**, an internal hard skeleton composed of calcareous plates overlaid by a thin skin (**Figure 34.30**). The skeleton is covered with spines and jawlike pincers called pedicellariae, the primary purpose of which is to deter settling of animals such as barnacles. These structures can also have poison glands.

Echinoderms possess a true coelom, and a portion of the coelom has been adapted to serve as a unique **water vascular system**, a network of canals that branch into tiny **tube feet** that function in movement, gas exchange, feeding, and excretion (see inset to Figure 34.30). The water vascular system uses hydraulic power (water pressure generated by the contraction of muscles), which enables the tube feet to extend and contract, allowing echinoderms to move, but only very slowly.

Water enters the water vascular system through the **madreporite**, a sievelike plate on the animal's surface. From there it flows into the **ring canal** in the central disc, into five radial canals, and into the tube feet. At the base of each tube foot is a muscular sac called an **ampulla**, which stores water. Contractions of the ampullae force water into the tube feet, causing them to straighten and extend. When the foot contacts a solid surface, muscles in the foot contract, forcing water back into the ampulla. Sea stars also use their tube feet in feeding, by exerting a constant, strong pressure on bivalves, whose adductor muscles open and close the shell. The adductor muscles eventually tire, allowing the shell to open slightly. At this stage, the sea star everts its stomach and inserts it into the opening. It then digests its prey, using juices secreted from extensive digestive glands. Sea stars also feed on sea urchins, brittle stars, and sand dollars, prey that cannot easily escape them.

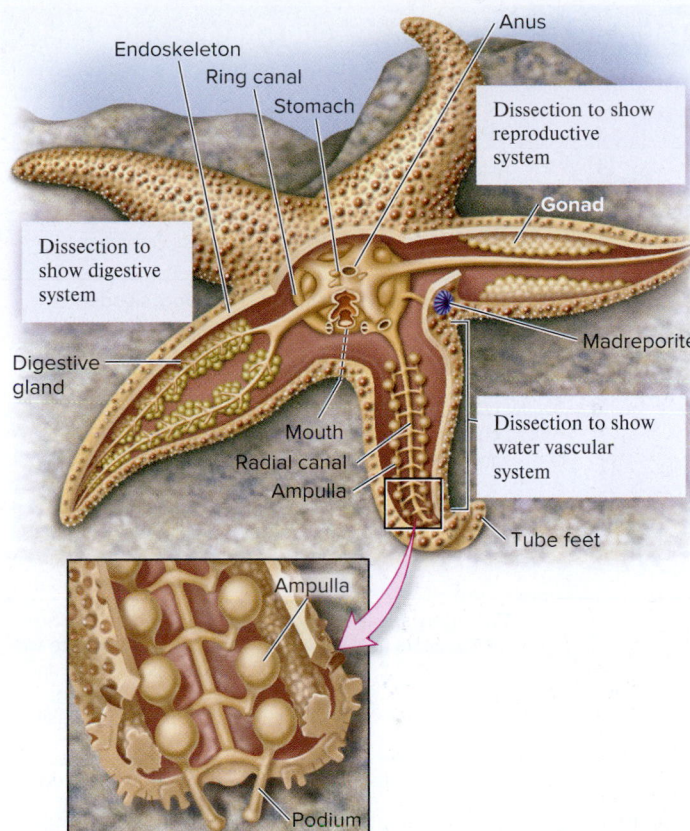

**Figure 34.30** **Body plan of an echinoderm, as represented by a sea star.** The arms of this sea star have been dissected to different degrees to show the echinoderm's various organs. The inset shows a close-up view of the tube feet, part of the water vascular system characteristic of echinoderms.

*Concept Check: Echinoderms and chordates are both deuterostomes. What are three defining features of deuterostomes?*

Echinoderms cannot osmoregulate, so no species have entered freshwater environments. No excretory organs are present. For some species, both respiration and excretion of nitrogenous waste take place by diffusion across their tube feet. Coelomic fluid circulates around the body.

Most echinoderms exhibit **autotomy**, the ability to intentionally detach a body part, such as a limb, that will later regenerate. In some species, a broken limb can even regenerate into a whole animal. Some sea stars regularly reproduce by breaking in two. Most echinoderms reproduce sexually and have separate sexes. Fertilization is usually external, with gametes shed into the water. Fertilized eggs develop into free-swimming larvae, which become sedentary adults.

**The Major Echinoderm Classes** Although over 20 classes of echinoderms have been described from the fossil record, only 5 main classes of echinoderms exist today: the Asteroidea (sea stars), Ophiuroidea (brittle stars), Echinoidea (sea urchins and sand dollars), Crinoidea (sea lilies and feather stars), and Holothuroidea (sea cucumbers). The key features of the echinoderms and their classes are listed in **Table 34.6**, and several members are shown in **Figure 34.31**.

| Table 34.6 | Main Classes and Characteristics of Echinoderms | |
| --- | --- | --- |
| | **Class and examples (est. number of species)** | **Class characteristics** |
| | Asteroidea: sea stars (1,600) | Five arms; tube feet; predatory on bivalves and other echinoderms; eversible stomach |
| | Ophiuroidea: brittle stars (2,000) | Five long, slender arms; tube feet not used for locomotion; no pedicellariae; browse on sea bottom or filter feed |
| | Echinoidea: sea urchins, sand dollars (1,900) | Spherical (sea urchins) or disc-shaped (sand dollars); no arms; tube feet and moveable spines; pedicellariae present; many feed on seaweeds |
| | Crinoidea: sea lilies and feather stars (700) | Cup-shaped; often attached to substrate via stalk; arms feathery and used in filter feeding; very abundant in fossil record |
| | Holothuroidea: sea cucumbers (1,200) | Cucumber-shaped; no arms; spines absent; endoskeleton reduced; tube feet; browse on sea bottom |

## The Phylum Chordata Includes All Vertebrates and Some Invertebrates

The deuterostomes consist of two major phyla: the echinoderms and the chordates. As deuterostomes, both phyla share similar developmental traits. In addition, both have an endoskeleton, consisting in the echinoderms of calcareous plates and in chordates, for the most part, of bone. However, the echinoderm endoskeleton functions in much the same way as the arthropod exoskeleton, in that an important function is providing protection. The chordate endoskeleton serves a very different purpose. In early-diverging chordates, the endoskeleton is composed of a single flexible rod situated dorsally, deep inside the body. Muscles move this rod, and their contractions cause the back

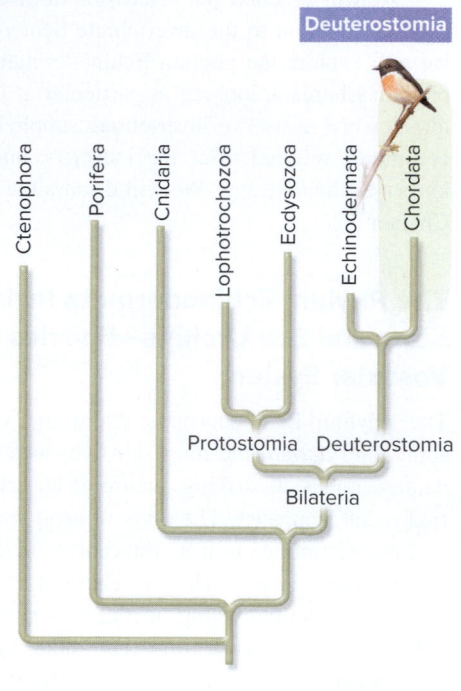

(a) Necklace Sea star, *Fromia monilis*, Baa Atoll, Maldives

(b) Brittle star, *Ophiarachna* spp., Gulf of Mexico

(c) Sea urchin, *Heterocentrotus trigonarius*, Hawaii

(d) Sea Lily, *Proisocrinus ruberrimus*, Indonesia

(e) Bronze-spot sea cucumber, *Holothuria argus*.

**Figure 34.31 Echinoderms. (a)** Sea star. **(b)** Brittle star. **(c)** Sea urchin. **(d)** Sea lily. **(e)** Sea cucumber. a: ©ullstein bild/Getty Images; b: Source: NOAA Okeanos Explorer Program, Gulf of Mexico 2012 Expedition; c: Source: David Burdick/NOAA; d: Source: NOAA Okeanos Explorer Program, INDEX-SATAL 2010; e: ©Poelzer Wolfgang/Alamy Stock Photo

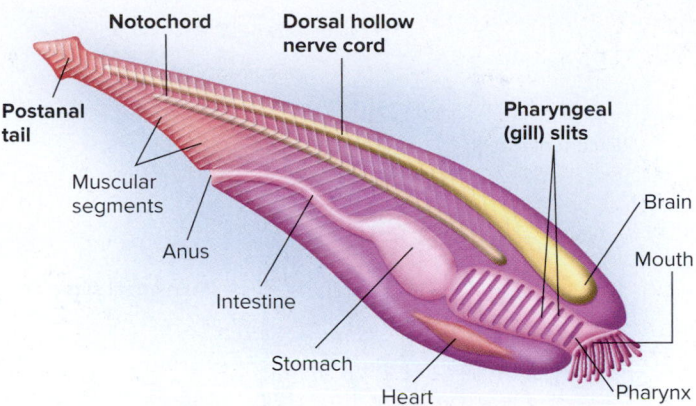

**Figure 34.32 Chordate characteristics.** The generalized chordate body plan has four main features: notochord, dorsal hollow nerve cord, pharyngeal slits, and postanal tail.

and tail end to move from side to side, permitting a swimming motion in water. The endoskeleton becomes more complex in different lineages that develop limbs, as we will see in Chapter 35, but it is always internal, with muscles attached.

Let's take a look at the four critical innovations in the body plan of chordates that distinguish them from all other animal life (**Figure 34.32**):

1. *Notochord*. Chordates are named for the **notochord**, a single flexible rod that lies between the digestive tract and the nerve cord. Composed of fibrous tissue encasing fluid-filled cells, the notochord is stiff yet flexible and provides skeletal support for all early-diverging chordates. In most chordates, such as vertebrates, a more complex jointed backbone usually replaces the notochord; its remnants exist only as the soft material within the discs between each vertebrae.

2. *Dorsal hollow nerve cord*. Many animals have a long nerve cord, but in nonchordate invertebrates, it is a solid tube that lies ventral to the alimentary canal. In contrast, the nerve cord in chordates is a hollow tube that develops dorsal to the alimentary canal. In vertebrates, the dorsal hollow nerve cord develops into the brain and spinal cord.

3. *Pharyngeal slits*. Chordates, like many animals, have a complete gut, from mouth to anus. However, in chordates, slits develop in the pharyngeal region, close to the mouth, that open to the outside. This permits water to enter through the mouth and exit via the slits, without having to go through the digestive tract. In early-diverging chordates, **pharyngeal slits** function as a filter-feeding device, whereas in later-diverging chordates, they develop into gills for gas exchange. In terrestrial chordates, the slits do not fully form, and they become modified for other purposes.

4. *Postanal tail*. Chordates possess a **postanal tail** of variable length that extends posterior to the anal opening. In aquatic chordates such as fishes, the tail is used in locomotion. In terrestrial chordates, the tail may be used for a variety of functions. In virtually all other nonchordate phyla, the anus is at the end of the body.

Although few chordates apart from fishes possess all of these characteristics in their adult life, they all exhibit them at some time during development. For example, in adult humans, the notochord becomes the spinal column, and the dorsal hollow nerve cord becomes the central nervous system. However, humans exhibit pharyngeal slits and a postanal tail only during early embryonic development. All the pharyngeal slits, except one, which forms the auditory (Eustachian) tubes in the ear, are eventually lost, and the postanal tail regresses to form the tailbone (the coccyx).

The phylum Chordata consists of the invertebrate chordates—the subphylum Cephalochordata (lancelets) and the subphylum Urochordata (tunicates)—along with the subphylum Vertebrata. Although the Vertebrata is by far the largest of these subphyla, biologists have focused on the Cephalochordata and Urochordata for clues as to how the chordate phylum may have evolved. Comparisons of gene sequences for the small subunit rRNA (SSU rRNA) show that these two subphyla are our closest invertebrate relatives (**Figure 34.33**).

**Subphylum Cephalochordata: The Lancelets** The cephalochordates (from the Greek *cephalo*, meaning head) look a lot more chordate-like than do tunicates. They are commonly referred to as lancelets, in reference to their bladelike shape and size, about 5–7 cm in length (**Figure 34.34a**). Lancelets are a small subphylum of 26 species, all marine filter feeders, with 4 species occurring in North American waters. Most of them belong to the genus *Branchiostoma*.

The lancelets live mostly buried in sand, with only the anterior end protruding into the water. Lancelets have the four distinguishing chordate characteristics: a clearly discernible notochord (extending well into the head), dorsal hollow nerve cord, pharyngeal slits, and postanal tail (**Figure 34.34b**). They are filter feeders, drawing

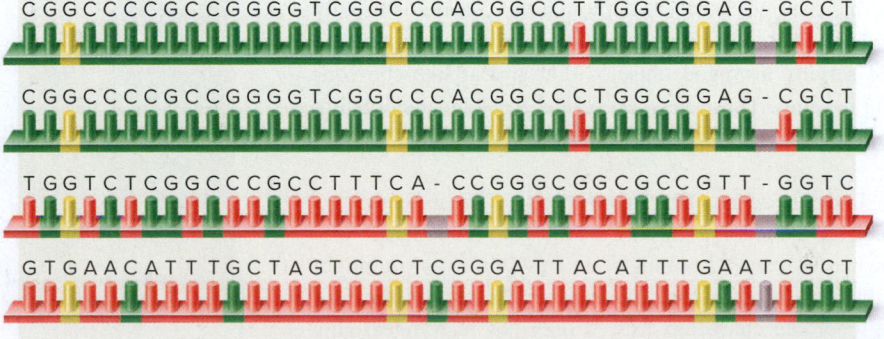

**Figure 34.33 Comparison of SSU rRNA gene sequences of chordate and nonchordate species.** Note the many similarities (yellow) and differences (green and red) among the sequences.

 **Core Concept: Evolution** The DNA sequence similarities between the invertebrate chordates (represented by the lancelet) and the vertebrates (represented by a human) suggest that the former are indeed our closest invertebrate relatives.

**(a) Lancelet in the sand**

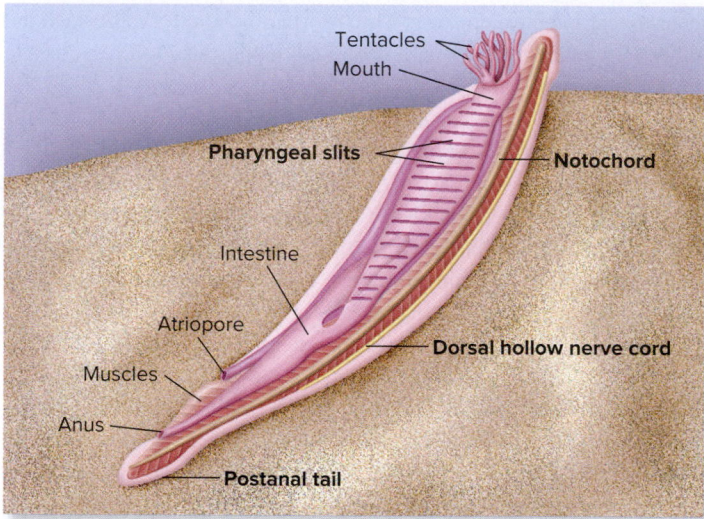

**(b) Body plan of the lancelet**

**Figure 34.34  Lancelets. (a)** A bladelike lancelet. **(b)** The body plan of the lancelet clearly displays the four characteristic chordate features. a: ©Natural Visions/Alamy Stock Photo

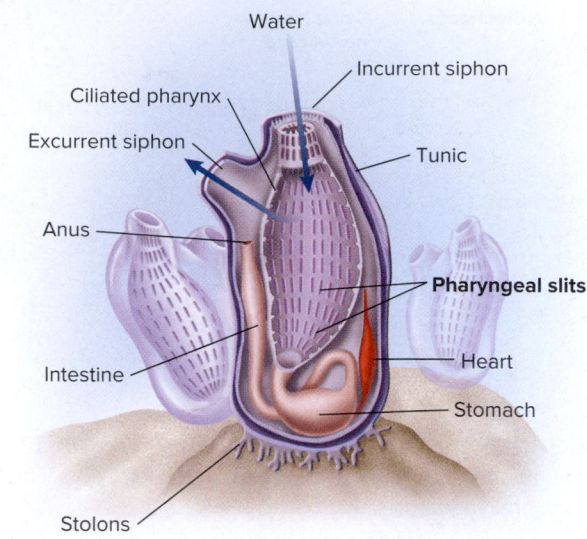

**(a) Adult tunicate**

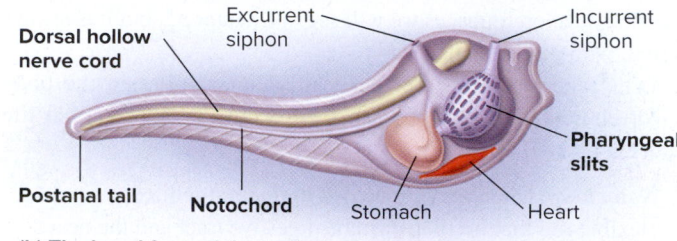

**(b) The larval form of the tunicate**

**(c) Typical tunicate**

**Figure 34.35  Tunicates. (a)** Body plan of the sessile, filter-feeding adult tunicate. **(b)** The larval form, which shows the four characteristic chordate features, has been proposed as a possible ancestor of modern vertebrates. **(c)** The blue tunicate, *Rhopalaea crassa*. c: ©Franco Banfi/Getty Images

water through the mouth and into the pharynx, where it is filtered through the pharyngeal slits. A mucous net across the pharyngeal slits traps food particles, and ciliary action takes the food into the intestine, while water exits via the atriopore. Gas exchange generally takes place across the body surface. Although the lancelet is usually sessile, it can leave its sandy burrow and swim to a new spot, using a sequence of serially arranged muscles that appear like chevrons (<<<<) along its sides. These muscles reflect the segmented nature of the lancelet body and permit a fishlike swimming motion.

**Subphylum Urochordata: The Tunicates**  The urochordates (from the Greek *oura*, meaning tail) are a group of 3,000 marine species also known as tunicates. Looking at an adult tunicate, you might never guess that it is a relative of modern vertebrates. Of the four distinguishing chordate characteristics, it only has pharyngeal slits (**Figure 34.35a**). The larval tunicate, in contrast, looks like a tadpole and exhibits all four chordate hallmarks (**Figure 34.35b**). The larval tadpole swims for only a few days, usually without feeding. Larvae settle on and attach to a rock surface via rootlike extensions called stolons. Here the larvae metamorphose into adult tunicates and in the process lose most of their chordate characteristics.

In 1928, the English marine biologist Walter Garstang suggested modern vertebrates arose from a larval tunicate form that had somehow acquired the ability to reproduce. Analysis of molecular data in 2006 led French evolutionary biologist Frédéric Delsuc and colleagues to propose that tunicates are the closest living relatives of vertebrates. These researchers group the cephalochordates more closely with the echinoderms. This means the common ancestor of living deuterostomes was a free-living, bilaterally symmetrical animal with pharyngeal slits, a segmented body, and a dorsal hollow nerve cord. This ancestral line split into two groups, the echinoderm–cephalochordate group and the tunicate–vertebrate group. Echinoderms lost most of their ancestral features, but cephalochordates did not. According to this view, tunicates lost their segmentation and most became sedentary, whereas vertebrates did not.

Adult tunicates are marine animals, some colonial and others solitary, that superficially resemble sponges or cnidarians. Tunicates are filter feeders that draw in water through an **incurrent siphon**, using a ciliated pharynx, and filter it through extensive pharyngeal slits. The food is trapped on a mucous sheet and then passes via ciliary action to the stomach, intestine, and anus; waste products exit through the excurrent siphon. The whole animal is enclosed in a nonliving **tunic** made of protein and a cellulose-like material called tunicin. Tunicates are also known as sea squirts for their ability to squirt out water from the excurrent siphon when disturbed. They have a rudimentary circulatory system with a heart and a simple nervous system of relatively few nerves connected to sensory tentacles around the incurrent siphon. The animals are mostly hermaphroditic.

## 34.7 A Comparison of Animal Phyla

### Learning Outcome:

1. Compare and contrast the key characteristics of animal phyla.

**Table 34.7** describes the common characteristics of the various animal phyla. In Sections 34.1–34.6, we explored all of the invertebrate phyla and two invertebrate subphyla of the chordates—Cephalochordata (lancelets) and Urochordata (tunicates). In Chapter 35, we will examine the vertebrates.

| Table 34.7 | Summary of the Physical Characteristics of the Major Invertebrate Phyla | | | | | | | | | | | |
|---|---|---|---|---|---|---|---|---|---|---|---|---|
| **Feature** | **Ctenophora (comb jellies)** | **Porifera (sponges)** | **Cnidaria (hydra, anemones, jellyfish)** | **Platyhelminthes (flatworms)** | **Rotifera (rotifers)** | **Bryozoa and Brachiopoda** | **Mollusca (snails, clams, squid)** | **Annelida (segmented worms)** | **Nematoda (roundworms)** | **Arthropoda (insects, arachnids, crustaceans)** | **Echinodermata (sea stars, sea urchins)** | **Chordata (vertebrates and others)** |
| Digestive system | Complete gut | Absent | Gastrovascular cavity | Gastrovascular cavity | Complete gut (usually) | Complete gut | Complete gut | Complete gut | Complete gut | Complete gut | Usually complete gut | Complete gut |
| Circulatory system | Absent | Absent | Absent | Absent | Absent | Absent; open or closed | Open; closed in cephalopods | Closed | Absent | Open | Absent | Closed |
| Respiratory system | Absent | Absent | Absent | Absent | Absent | Absent | Gills | Absent | Absent | Trachae; gills or book lungs (a structure in spiders) | Tube feet; respiratory tree | Gills; lungs |
| Excretory system | Absent | Absent | Absent | Protonephridia with flame cells | Protonephridia | Metanephridia | Metanephridia | Metanephridia | Excretory tubules | Excretory glands resembling metanephridia | Absent | Kidneys |
| Nervous system | Nerve Net | Absent | Nerve net | Brain; cerebral ganglia; lateral nerve chords; nerve net | Brain; nerve cords | No brain; nerve ring | Ganglia; nerve cords | Brain; ventral nerve cord | Brain; nerve cords | Brain; ventral nerve cord | No brain; nerve ring and radial nerves | Well-developed brain; dorsal hollow nerve cord |
| Reproduction | Sexual (hermaphrodite) | Sexual; asexual (budding) | Sexual; asexual (budding) | Sexual (most hermaphroditic); asexual (body splits) | Mostly parthenogenetic; males appear only rarely | Sexual (some hermaphroditic); asexual (budding) | Sexual (some hermaphroditic) | Sexual (some hermaphroditic) | Sexual (some hermaphroditic) | Usually sexual (some hermaphroditic) | Sexual (some hermaphroditic); parthenogenetic; asexual by regeneration (rare) | Sexual; rarely parthenogenetic |
| Support | Mesoglea | Endoskeleton of spicules and collagen | Mesoglea | Parenchyma | Tissue | Exoskeleton | Hydrostatic skeleton and shell | Hydrostatic skeleton | Fluid skeleton | Exoskeleton | Endoskeleton of plates beneath outer skin | Endoskeleton of cartilage or bone |

# Summary of Key Concepts

## 34.1  Ctenophores: The Earliest Animals

- Invertebrates, or animals without a backbone, make up more than 95% of all animal species. Ctenophores are the earliest-diverging animals. They possess a unique nervous system and a mesoderm germ layer. They are predatory, possess a complete gut and use cilia to propel themselves (Figures 34.1, 34.2).

## 34.2  Porifera: The Sponges

- The phylum Porifera, or sponges, lack true tissues, but are multicellular animals possessing several types of cells. They are asymmetric marine filter feeders (Figure 34.3).

## 34.3  Cnidaria: Jellyfish and Other Radially Symmetric Animals

- The phylum Cnidaria includes hydra, jellyfish, box jellies, sea anemones, corals, and the Portuguese man-of-war (Table 34.1). Cnidarians have only two embryonic germ layers: the ectoderm and the endoderm, with a gelatinous substance (mesoglea) connecting the two layers.

- Cnidarians exist in one of two forms: polyp or medusa. A characteristic feature of cnidarians is their stinging cells, or cnidocytes, which function in defense and prey capture (Figures 34.4, 34.5).

## 34.4  Lophotrochozoa: The Flatworms, Rotifers, Bryozoans, Brachiopods, Mollusks, and Annelids

- Most Lophotrochozoa possess either a lophophore or a larval stage called a trochophore. Platyhelminthes, or flatworms, are hypothesized to be the first bilaterian animals to evolve three distinctive embryonic germ layers—ectoderm, endoderm, and mesoderm (Figure 34.6).

- The four classes of flatworms are the Turbellaria, Monogenea, Cestoda (tapeworms), and Trematoda (flukes). Flukes and tapeworms are internally parasitic, with complex life cycles (Figures 34.7, 34.8, Table 34.2).

- Rotifers are microscopic animals that have a complete digestive tract with separate mouth and anus; the mastax, a muscular pharynx, is a structure unique to the rotifers (Figure 34.9).

- The bryozoa and brachiopods both possess a lophophore, a ciliary feeding structure (Figure 34.10).

- The mollusks, which constitute a large phylum with over 100,000 diverse living species, have a basic body plan with three parts—a foot, a visceral mass, and a mantle—and an open circulatory system (Figures 34.11, 34.12, 34.13).

- The four most common mollusk classes are the polyplacophora (chitons), gastropoda (snails and slugs), bivalvia (clams and mussels), and cephalopoda (octopuses, squids, and nautiluses) (Table 34.3).

- Cephalopods are among the most complex of all invertebrates. They are the only mollusks with a closed circulatory system; they have a well-developed nervous system and brain and are believed to exhibit learning by observation (Figure 34.14).

- A striking feature of the annelids is segmentation, in which the body is divided into compartments; specialization of segments is only minimally present at the anterior end (Figure 34.15).

- Annelids are a large phylum containing two main groups: Errantia, which includes free-ranging marine worms, and Sedentaria, which includes tube worms, earthworms, and leeches (Figure 34.16).

## 34.5  Ecdysozoa: The Nematodes and Arthropods

- The ecdysozoans are named for their ability to shed their cuticle, a nonliving covering that provides support and protection. The two most common ecdysozoan phyla are the nematodes and the arthropods.

- Nematodes, which exist in nearly all habitats, have a cuticle made primarily of collagen, a structural protein. The small, free-living nematode *Caenorhabditis elegans* is a model organism. Many nematodes are parasitic in humans (Figures 34.17, 34.18).

- Arthropods are perhaps the most species-rich phylum on Earth. The arthropod body is covered by a cuticle (exoskeleton) made of layers of chitin and protein, and it is segmented, with segments fused into functional units called tagmata (Figure 34.19).

- The five main subphyla of arthropods are Trilobita (trilobites; now extinct), Chelicerata (spiders, scorpions, and relatives), Myriapoda (millipedes and centipedes), Hexapoda (insects and relatives), and Crustacea (crabs and relatives) (Table 34.4, Figures 34.20, 34.21, 34.22, 34.23).

- More insect species are known than all other animal species combined. The development of a variety of wing structures and mouthparts were keys to the success of insects (Figure 34.24, Table 34.5).

- Insects undergo a change in body form during development, either complete metamorphosis or incomplete metamorphosis, and have developed complex social behaviors (Figures 34.25, 34.26).

- Most crustacean orders contain small species and feature prominently in marine food chains. The most well-known order of crustaceans is the Decapoda, which includes the crabs, lobsters, and shrimp (Figures 34.27, 34.28, 34.29).

## 34.6  Deuterostomia: The Echinoderms and Chordates

- The Deuterostomia include the phyla Echinodermata and Chordata. A striking feature of the echinoderms is their radial symmetry, which is present only in adults; the free-swimming larvae are bilaterally symmetrical. Echinoderms possess a unique water vascular system (Figure 34.30).

- Five main classes of echinoderms exist today: the Asteroidea (sea stars), Ophiuroidea (brittle stars), Echinoidea (sea urchins and sand dollars), Crinoidea (sea lilies and feather stars), and Holothuroidea (sea cucumbers) (Table 34.6, Figure 34.31).

- The phylum Chordata is distinguished by four critical innovations: the notochord, dorsal hollow nerve chord, pharyngeal slits, and postanal tail (Figure 34.32).

- The subphylum Cephalochordata (lancelets) and subphylum Urochordata (tunicates) are invertebrate chordates. Genetic studies have shown that tunicates are the closest invertebrate relatives of the vertebrate chordates (subphylum Vertebrata) (Figures 34.33, 34.34, 34.35).

## 34.7    A Comparison of Animal Phyla

- Each of the major animal phyla is distinguished by a unique set of characteristics (Table 34.7).

## Assess & Discuss

### Test Yourself

1. Choanocytes are
   a. a group of protists that are believed to have given rise to animals.
   b. specialized cells of sponges that function to trap and eat small particles.
   c. cells that make up the gelatinous layer in sponges.
   d. cells of sponges that function to transfer nutrients to other cells.
   e. cells that form spicules in sponges.

2. Why aren't sponges eaten more often by predators?
   a. They are protected by silica spicules.
   b. They are protected by toxic defensive chemicals.
   c. They are often eaten; it's just that the leftover cells reaggregate into new, smaller sponges.
   d. Both a and b are correct.
   e. All three explanations, a, b, and c, are correct.

3. Which of the following organisms can produce female offspring through parthenogenesis?
   a. cnidarians          d. rotifers
   b. flukes              e. annelids
   c. choanocytes

4. What organisms survive without a mouth, digestive system, or anus?
   a. cnidarians          d. cestodes
   b. rotifers            e. nematodes
   c. echinoderms

5. Which phylum does not have at least some members with a closed circulatory system?
   a. Lophophorata
   b. Arthopoda
   c. Annelida
   d. Mollusca
   e. All of the above phyla have some members with a closed circulatory system.

6. A defining feature of the Ecdysozoa is a
   a. segmented body.
   b. closed circulatory system.
   c. cuticle.
   d. complete gut.
   e. lophophore.

7. In arthropods, the tracheal system is
   a. a unique set of structures that function in ingestion and digestion of food.
   b. a series of branching tubes extending into the body that allow for gas exchange.
   c. a series of tubules that allow waste products in the blood to be released into the digestive tract.
   d. the series of ommatidia that form the compound eye.
   e. none of the above.

8. Characteristics of the class Arachnida include
   a. two tagmata.
   b. six walking legs.
   c. an aquatic lifestyle.
   d. a lobed body.
   e. both b and d.

9. Incomplete metamorphosis
   a. is characterized by distinct larval and adult stages that do not compete for resources.
   b. is typically seen in arachnids.
   c. involves gradual changes in life stages in which the young resemble the adult stage.
   d. is characteristic of the majority of insects.
   e. always includes a pupal stage.

10. Which clade includes echinoderms?
   a. Protostomia
   b. Bilateria
   c. Ecdysozoa
   d. Lophotrochozoa
   e. Echinoderms are a member of all the above clades.

### Conceptual Questions

1. Compare and contrast the five main feeding methods discussed in the chapter.

2. Why is external fertilization common in aquatic invertebrates but rare in terrestrial species?

3.  **Core Concept: Structure and Function** Explain the difference between complete metamorphosis and incomplete metamorphosis.

### Collaborative Questions

1. Revisit the animal phylogeny outlined in Figure 34.1 and discuss the critical innovations that led to the separation of each of the clades shown.

2. Discuss reasons why insects are the most species-rich taxon.

# The Vertebrates

# 35

**The star-nosed mole (*Condylura cristata*).** This species is a vertebrate, a fascinating group of animals that includes human beings. ©Gary Meszaros/Science Source

**T**he star-nosed mole, *Condylura cristata,* lives in tunnels in wet areas of eastern Canada and the northeastern United States. It is one of the most distinctive mammals anywhere on Earth. The mole lives for the most part in complete darkness and is virtually blind. It feels its way around and finds prey by means of 22 fleshy appendages on its nose, which contain more than 25,000 minute and highly sensitive sensory receptors called Eimer's organs. The mole has a voracious appetite and needs to eat frequently. In fact, the star-nosed mole has been identified as the world's fastest-eating mammal, averaging less than a quarter of a second to identify and consume a food item. Its astoundingly acute sensory abilities more than make up for its poor eyesight. The moles can swim underwater in search of food and smell their prey by exhaling air bubbles then inhaling them to detect scents.

The star-nosed mole is a **vertebrate** (from the Latin *vertebratus,* meaning joint of the spine), an animal with a backbone. Vertebrates range in size from tiny fishes weighing 0.1 g to huge whales

with weights over 100,000 kg. They occupy nearly all of Earth's habitats, from the deepest depths of the oceans to mountaintops and the sky beyond. Throughout history, humans have depended on many vertebrate species for their welfare: by domesticating species such as horses, cattle, pigs, sheep, and chickens; using skin and fur for clothes; and keeping countless species, including cats and dogs, as pets. Many vertebrate species are the subjects of conservation efforts, as we will see in Chapter 60.

In Chapter 34, we discussed two chordate subphyla: the cephalochordates (lancelets) and urochordates (tunicates). The third subphylum of chordates, the Vertebrata, with about 66,000 species, is by far the largest and most dominant group of the phylum. The vertebrates include fishes, amphibians, reptiles, and mammals. In this chapter, we will explore the characteristics of vertebrates and the evolutionary development of the major vertebrate clades.

## 35.1 Vertebrates: Chordates with a Backbone

**Learning Outcome:**

**1.** List the main distinguishing characteristics of vertebrates.

Our current understanding of the relationships between the vertebrate groups is shown in **Figure 35.1**. Nested within the vertebrates are various clades based on morphological characteristics. For example, most vertebrates have jaws and are collectively known as gnathostomes. Many gnathostomes have four limbs for movement and are known as tetrapods.

The vertebrates retain all chordate characteristics outlined in Chapter 34, as well as possessing several additional traits, including the following:

1. *Vertebral column.* During development in vertebrates, the notochord is replaced by a bony or cartilaginous column of interlocking **vertebrae** that provides support and also protects the nerve cord, which lies within its tubelike structure.

2. *Cranium.* The anterior end of the nerve cord elaborates to form a well-developed brain that is encased in a protective bony or

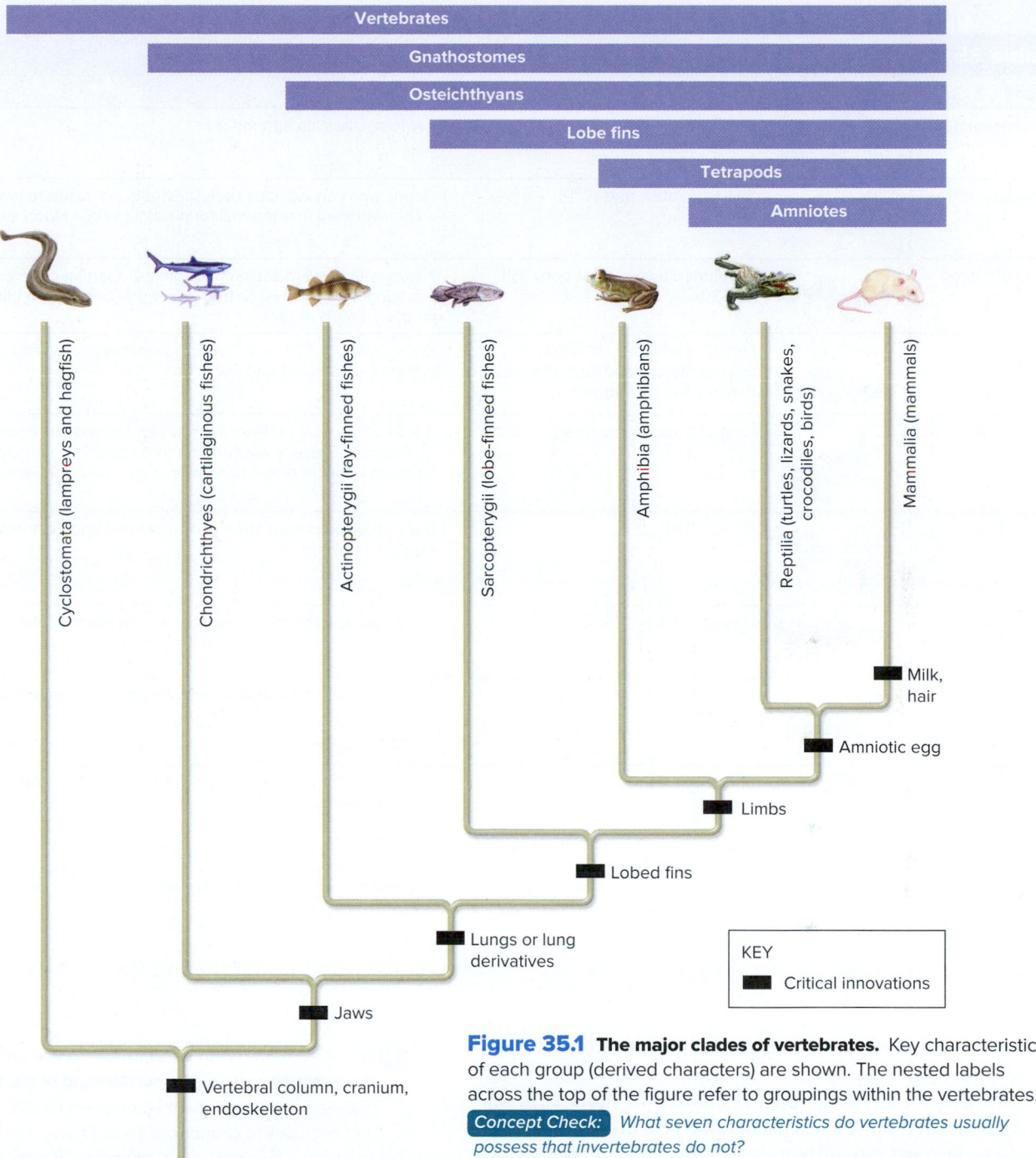

**Figure 35.1** **The major clades of vertebrates.** Key characteristics of each group (derived characters) are shown. The nested labels across the top of the figure refer to groupings within the vertebrates.

*Concept Check:* *What seven characteristics do vertebrates usually possess that invertebrates do not?*

cartilaginous housing called the **cranium**. This continues the trend of cephalization—the development of the head end in animals.

3. ***Endoskeleton of cartilage or bone.*** The cranium and vertebral column are parts of the endoskeleton, the living skeleton of vertebrates that forms within the animal's body. Most vertebrates also have two pairs of appendages, such as fins, legs, or arms. The endoskeleton is composed of either bone or cartilage, materials that are very strong yet more flexible than the chitin found in insects and other arthropods.

Although these are the main distinguishing characteristics of vertebrates, there are others. For example, vertebrates have multiple

clusters of *Hox* genes, compared with the single cluster of *Hox* genes in tunicates and lancelets. These additional gene clusters are believed to have permitted increasingly complex morphologies beyond those possessed by invertebrate chordates. Vertebrates also possess a great diversity of internal organs, including a liver, kidneys, endocrine glands, and a heart with at least two chambers. The liver is unique to vertebrates, and the vertebrate kidneys, endocrine system, and heart are more complex than are analogous structures in invertebrate taxa.

Although these features are exhibited in all vertebrate classes, some classes evolved critical innovations that helped them succeed in specific environments such as on land or in the air. For

## Table 35.1    The Main Clades and Characteristics of Living Vertebrates

| Clade | | Examples (approx. number of species) | Main characteristics |
|---|---|---|---|
| Cyclostomata | | Lampreys and hagfish (100) | Jawless fishes, no appendages |
| Chondrichthyes | | Sharks, skates, rays (970) | Fishes with cartilaginous skeleton; teeth not fused to jaw; no swim bladder; well-developed fins; internal fertilization; single blood circulation |
| Actinopterygii | | Ray-finned fishes, most bony fish (31,830) | Fishes with ossified skeleton; single gill opening covered by operculum; fins supported by rays, fin muscles within body; swim bladder often present; mucous glands in skin |
| Sarcopterygii | | Lobe-finned fishes, of which coelacanths (2) and lungfishes (6) are the only living members | Fishes with ossified skeleton; bony extensions, together with muscles, project into pectoral and pelvic fins |
| Amphibia | | Frogs, toads, salamanders (7,600) | Adults able to live on land; fresh water needed for reproduction; development usually involving metamorphosis from tadpoles; adults with lungs and double blood circulation; moist skin; shell-less eggs |
| Testudines | | Turtles (346) | Body encased in hard shell; no teeth; head and neck retractable into shell; eggs laid on land |
| Squamata | | Lizards, snakes (9,900) | Lower jaw not attached to skull; skin covered in scales |
| Crocodilia | | Crocodiles, alligators (25) | Four-chambered heart; large aquatic predators; parental care of young |
| Aves | | Birds (10,425) | Feathers; hollow bones; air sacs; reduced internal organs; endothermic; four-chambered heart |
| Mammalia | | Mammals (5,500) | Mammary glands; hair; specialized teeth; enlarged skull; external ears; endothermic; four-chambered heart; highly developed brains; diversity of body forms |

example, birds developed feathers and wings, structures that enable most species to fly. Each of the vertebrate clades is distinctly different from one another, as outlined in **Table 35.1**. One of the earliest innovations was the development of jaws. All vertebrates except some early-diverging fishes possessed jaws. Today, the only jawless vertebrates are hagfish and lampreys, which are described in Section 35.2.

## BIO TIPS

**THE QUESTION** *What derived characters (called critical innovations in Figure 35.1) are common to reptiles, amphibians, and lobe-finned fish, but not sharks?*

**T**OPIC **What topic in biology does this question address?** The topic is systematics. More specifically, the question concerns the use of a phylogenetic tree to determine derived characters that are shared among certain taxa.

**I**NFORMATION **What information do you know based on the question and your understanding of the topic?** From the question, you know you are comparing osteichthyans, tetrapods, and amniotes to chondrichthyans. From your understanding of systematics, which is described in Chapter 25, you may remember the definition of shared derived characters. You were given the derived characters (also called critical innovations) in Figure 35.1.

**P**ROBLEM-SOLVING **S**TRATEGY **Compare and contrast.** Look back at Figure 25.9. Construct a table with the derived characters listed on the left side and the four groups being compared across the top. Fill in the table with "Yes" or "No" as done in Figure 25.9a, according to whether a group does or does not possess a derived character.

**ANSWER** *Reptiles, amphibians, and lobe-finned fish all possess lungs or lung derivatives and lobe fins, but sharks do not.*

## 35.2 Cyclostomata: Jawless Fishes

### Learning Outcome:

**1.** Describe the two classes of existing jawless vertebrates.

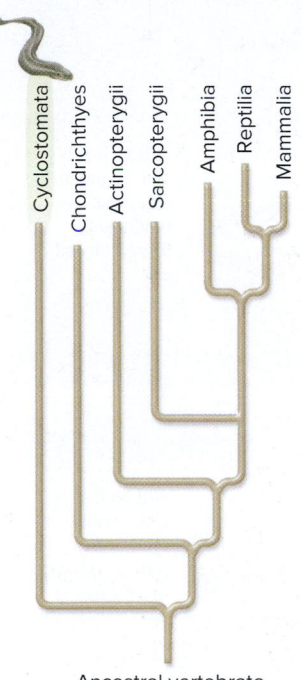

Ancestral vertebrate

Although many species of jawless fishes are known from the fossil record, most became extinct by the end of the Devonian period. Today, the only jawless vertebrates are hagfish and lampreys, together known as the Cyclostomata. Cyclostomes, or "circle mouths," are eel-like animals that do not possess jaws. Sequencing of RNA libraries in 2010, together with genomic surveys, yielded strong support that Cyclostomata is monophyletic. In addition, hagfish and lampreys share a very distinct type of immune system. In this section, we will examine the characteristics of hagfish and lampreys.

### The Hagfish Are the Simplest Living Cyclostomes

The hagfish are entirely marine cyclostomes that lack eyes, jaws, fins and even vertebrae (**Figure 35.2**). The hagfish skeleton consists largely of a notochord and a cartilaginous skull that encloses the brain. The lack of a vertebral column leads to extensive flexibility. So how can hagfish be vertebrates without a vertebral column? The strong molecular support for a cyclostome clade suggests that hagfish anatomy has degenerated to a remarkable degree and only the cranium and diversity of organs provide evidence of vertebrate ancestry.

Hagfish live in the cold waters of northern oceans, close to the muddy bottom, feeding on marine worms and other invertebrates. Essentially blind, hagfish have a very keen sense of smell and are attracted to dead and dying fish. They attach themselves to such fish via toothed plates on the mouth. The powerful tongue then rasps off pieces of tissue. Though the hagfish cannot see approaching predators, they have special glands that produce copious amounts of slime. When provoked, the hagfish increases its slime production dramatically, enough to potentially distract predators or coat their gills and interfere with breathing. Hagfish can sneeze to free their nostrils of their own slime.

### The Lampreys Are Eel-like Animals That Lack Jaws

Lampreys are similar to hagfish because they lack both a hinged jaw and true appendages. However, lampreys do possess a notochord surrounded by a cartilaginous rod that represents a rudimentary vertebral column.

**Figure 35.2   The hagfish.** ©Pat Morris/ardea.com

Lampreys are found in both marine and freshwater environments. Marine lampreys are parasitic as adults. They grasp other fish with their circular mouth (**Figure 35.3a**) and rasp a hole in the fish's side, sucking blood, tissue, and fluids until they are replete (**Figure 35.3b**).

**(a) Jawless mouth of a sea lamprey**

**(b) A sea lamprey feeding**

**Figure 35.3   The lamprey, a modern jawless fish.** **(a)** The sea lamprey (*Petromyzon marinus*) has a circular, jawless mouth. **(b)** A sea lamprey feeding on a fish. a: ©R. Duran/Getty Images; b: ©Jacana/Science Source

Reproduction of all species, whether they live in marine or freshwater environments, is similar. Males and females spawn in freshwater streams, and the resultant larval lampreys bury into the sand or mud, much like lancelets (refer back to Figure 34.34a), emerging to feed on small invertebrates or detritus at night. This stage can last for 3–7 years, at the end of which the larvae metamorphose into adults. In most freshwater species, the adults do not feed at all but quickly mate and die. Young marine lampreys migrate from fresh water to the ocean, and later return to fresh water to spawn and then die.

## 35.3 Gnathostomes: Jawed Vertebrates

### Learning Outcomes:

1. Describe how jaws evolved.
2. Discuss the distinguishing features of sharks.
3. List the three features that distinguish bony fishes from cartilaginous fishes.
4. Outline the differences between the ray-finned fish and the lobe-finned fish.

All vertebrate species that possess jaws are called **gnathostomes** (from the Greek, meaning jaw mouth) (see Figure 35.1). Gnathostomes are a diverse clade of vertebrates that include fishes, amphibians, reptiles, and mammals. The earliest-diverging gnathostomes were fishes. Jawed fishes, which appeared in the mid-Ordovician period (about 470 mya), radiated in both fresh and salt water.

Biologists have identified about 32,800 species of living fishes with jaws, more speices than in any other clade of vertebrates. Most jawed fishes are aquatic, gill-breathing species that usually possess fins and a scaly skin. The three clades of jawed fishes are the Chondrichthyes (cartilaginous fishes), Actinopterygii (ray-finned fishes), and Sarcopterygii (coelacanths and lungfishes) (see Table 35.1). In this section, we will examine the evolution of the vertebrate jaw and then consider these three classes of jawed fishes. The remaining sections of this chapter will explore the characteristics of the other jawed vertebrates.

### A Hinged Jaw Was a Critical Innovation That Aided in Feeding

A hinged jaw was an important evolutionary development that led to a great diversification of vertebrates. It enabled an animal to grip its prey more firmly, thereby increasing its likelihood of capturing the prey and allowing it to attack larger prey. Accompanying the jawed mouth was the development of more sophisticated head and body structures, including two pairs of appendages called fins. Gnathostomes also possess at least two more *Hox* gene clusters than do the cyclostomes (bringing their total to four or more). Developmental biologists speculate that additional *Hox* gene clusters led to increased morphological complexity.

The hinged jaw developed from the gill arches, cartilaginous or bony rods that help to support gills. Similarities between cells that make up jaws and gill arches support this view. Primitive jawless fishes had nine gill arches surrounding the eight gill slits (**Figure 35.4a**). During the late Silurian period (about 417 mya),

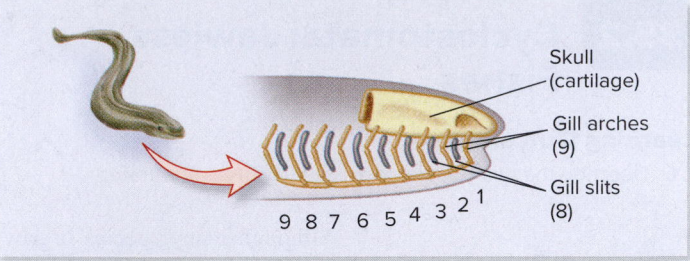

**(a) Primitive jawless fishes**

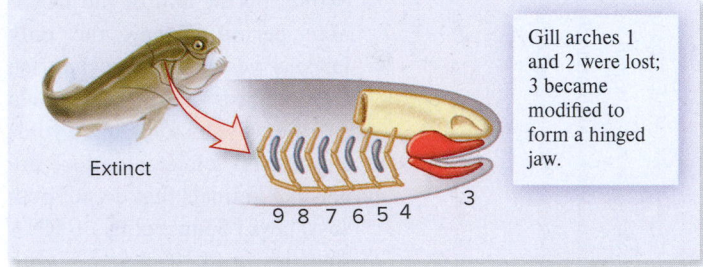

**(b) Early jawed fishes (placoderms)**

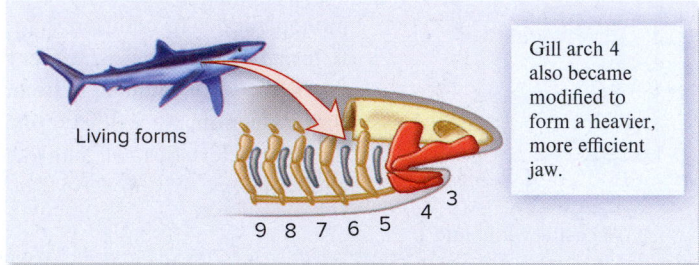

**(c) Modern jawed fishes (cartilaginous and bony fishes)**

**Figure 35.4 The evolution of the vertebrate jaw.** **(a)** Primitive fishes and extant jawless fishes such as lampreys have nine cartilaginous gill arches that support eight gill slits. **(b)** In early jawed fishes such as the placoderms, the first two pairs of gill arches were lost, and the third pair became modified to form a hinged jaw. This left six gill arches (4–9) to support the remaining five gill slits, which were still used in breathing. **(c)** In modern jawed fishes, the fourth gill arch also contributes to jaw support, allowing more powerful bites to be delivered.

 **Core Concept: Structure and Function**  The development of a jaw increased the predatory capabilities of gnathostomes.

some of these gill arches became modified. The first and second gill arches were lost, and the third and fourth pairs evolved to form the jaws (**Figure 35.4b** and **c**). This is how evolution typically works; body features do not appear de novo, but instead, existing features become modified to serve other functions.

By the mid-Devonian period, several classes of jawed fishes were common. Two of them, the Acanthodii (spiny fishes) and Placodermi (armored fishes) died out during a mass extinction late in the Devonian. The reasons for this extinction are not well understood, but other types of jawed fishes present at the same time—the cartilaginous and bony fishes—did not go extinct and flourished in the aftermath of the mass extinction.

## Chondrichthyans Are Fishes with Cartilaginous Skeletons

Members of the clade Chondrichthyes (the **chondrichthyans**)—sharks, skates, and rays—are also called cartilaginous fishes because their skeleton is composed of flexible cartilage rather than bone. The cartilaginous skeleton is not considered an ancestral character but rather a derived character. This means that the ancestors of the chondrichthyans had bony skeletons, but members of this class subsequently lost this feature. This hypothesis is reinforced by the observation that during development, the skeleton of most vertebrates is cartilaginous, and then it becomes bony (ossified) as a hard calcium-phosphate matrix replaces the softer cartilage. Genetic changes in the cartilaginous fishes are believed to prevent the ossification process.

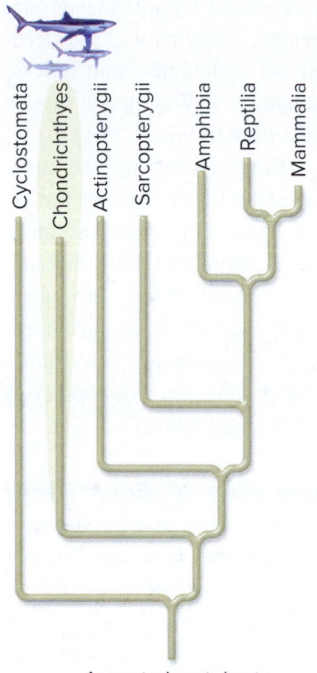

Ancestral vertebrate

***Key Features of Chondrichthyans*** All chondrichthyans are denser than water, which means that they will sink if they stopped swimming. Many sharks never stop swimming and maintain buoyancy via the use of their fins and a large oil-filled liver. Perhaps the most important fin for propulsion is the large and powerful caudal fin, or tail fin, which, when swept from side to side, thrusts the fish forward at great speed (**Figure 35.5a**). For example, great white sharks (*Carcharodon carcharias*) can swim at over 40 km per hour, and Mako sharks (*Isurus oxyrinchus*) have been clocked at nearly 50 km per hour. The paired pelvic fins (at the back) and pectoral fins (at the front) act like flaps on airplane wings, allowing the shark to dive deeper or rise to the surface. They also aid in steering. In addition, the dorsal fin (on the shark's back) acts as a stabilizer to prevent the shark from rolling in the water as the tail fin pushes it forward. During swimming, water continually enters the mouth and is forced over the gills, allowing sharks to extract oxygen and breathe.

Skates and rays are essentially flattened sharks that cruise along the ocean floor, using hugely expanded pectoral fins. In addition, their thin, whiplike tails are often equipped with a venomous barb used in defense (**Figure 35.5b**). Most of the 475 or so species of skates and rays feed on bottom-dwelling crustaceans and mollusks. At times, they may rest on the ocean floor. How do skates and rays breathe when they are not swimming? These species, and a few sharks such as the nurse shark, use a muscular pharynx and jaw muscles to pump water over the gills.

Sharks were among the earliest fishes to develop teeth. Shark teeth evolved from rough scales on the skin that contained dentin and enamel. Although shark's teeth are very sharp and hard, they are not set into the jaw, as are human teeth, so they break off easily. Teeth are

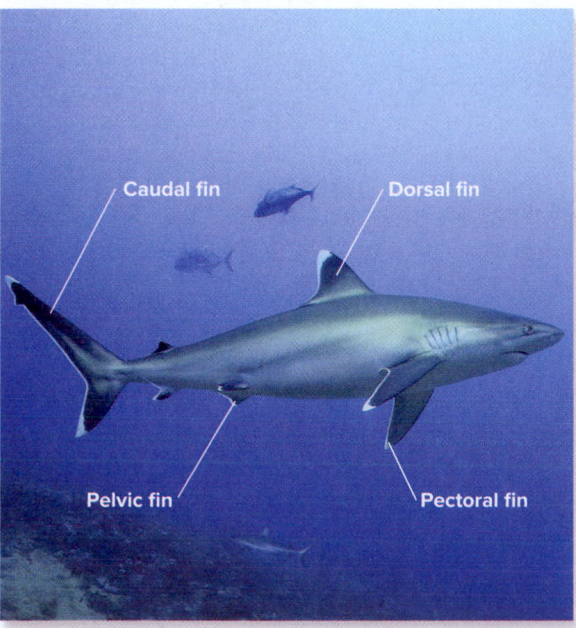

(a) Silvertip shark

(b) Stingray

(c) Rows of shark teeth

(d) Shark egg pouch

**Figure 35.5 Cartilaginous fishes. (a)** The silvertip shark (*Carcharhinus albimarginatus*) is one of the ocean's most powerful predators. **(b)** Stingrays are essentially flattened sharks with very large pectoral fins. **(c)** Close-up of the mouth of a sand tiger shark (*Carcharias taurus*), showing rows of teeth. **(d)** This mermaid's purse (egg pouch) of a dogfish shark (*Scyliorhinus canicula*) is entwined in vegetation to keep it stationary. a: ©Valerie & Ron Taylor/ardea.com; b: ©Bill Curtsinger/National Geographic/Getty Images; c: ©Jeff Rotman/naturepl.com; d: ©Oxford Scientific/Getty Images

 **Core Skill: Connections** Are "fish" a monophyletic group?

continually replaced, row by row (**Figure 35.5c**). Sharks may have 20 rows of teeth, with the front pair in active use and the ones behind ready to grow in as replacements when needed.

In the chondrichthyans, and indeed in all species of fishes, the heart consists of two chambers, an atrium and a ventricle, that contract in sequence. All fishes employ what is known as single circulation, in which blood is pumped from the heart to capillaries in the gills to collect oxygen, and then it flows through arteries to the tissues of the body, before returning to the heart (look ahead to Figure 48.2a).

***Chondrichthyan Senses*** Sharks have a powerful sense of smell, facilitated by sense organs in the nostrils (sharks and other fishes do not use nostrils for breathing). They can see well but cannot distinguish colors. Although sharks have no eardrum, they can detect pressure waves generated by moving objects. All jawed fishes have a row of microscopic organs in the skin, arranged in a line that runs laterally down each side of the body, that can sense movements in the surrounding water. This system of sense organs, known as the **lateral line**, senses pressure waves and sends nervous signals to the inner ear and then on to the brain. Sharks have an extra sense that helps them find and track prey. The ampullae of Lorenzini, vesicles and pores found around the shark's head, are sensory organs that detect electromagnetic fields produced by other organisms.

***Reproduction in Chondrichthyans*** Fertilization is internal in chondrichthyans, with the male transferring sperm to the female via a pair of **claspers**, extensions of the pelvic fins. Some shark species are **oviparous**; that is, they lay eggs, often inside a protective pouch called a mermaid's purse (Figure 35.5d). In **ovoviparous** species, the eggs are retained within the female's body, but there is no placenta to nourish the young. A few species are **viviparous**; the eggs develop within the uterus, receiving nourishment from the mother via a placenta. Both ovoviparous and viviparous sharks give birth to live young.

Sharks never stop growing, so larger individuals are the oldest. Radiocarbon dating of the eye lenses of large Greenland sharks, *Somniosus microcephalus*, revealed the age at sexual maturity to be at least 156 years and showed an average life span of 272 years; the largest animal, at 502 cm long, was determined to be 392±120 years old, making this species the longest-lived vertebrate known.

## The Earliest-Diverging Osteichthyans Are Fishes with Bony Skeletons

Unlike the cartilaginous fishes, all other gnathostomes have a bony skeleton and belong to the clade known as osteichthyes. This term means "bony fish" and was originally proposed for just that group. With the advent of modern phylogenetic systematics, the term **osteichthyans** was expanded to include all vertebrates with a bony skeleton, including tetrapods (refer back to Figure 35.1).

Bony fishes are the most numerous of all types of fishes, with more individuals and more species (about 31,830) than any other group. Most biologists now recognize two living clades: the Actinopterygii (ray-finned fishes) and the Sarcopterygii (lobe-finned fishes).

***Body Plan of Bony Fishes*** Fishes in both clades possess a bony skeleton and scale-covered skin. The skin of bony fishes, unlike the rough skin of sharks, is covered by a thin epidermal layer containing glands that produce mucus, an adaptation that reduces drag during swimming. Just as in the cartilaginous fishes, water is drawn over the gills for breathing, but in bony fishes, a protective flap called an **operculum** covers the gills (**Figure 35.6**). Muscle contractions around the gills and operculum draw water across the gills so that bony fishes do not need to swim continuously to breathe.

Some early bony fishes lived in shallow, oxygen-poor waters and developed lungs as an embryological offshoot of the pharynx. These fish could rise to the water surface and gulp air. As discussed later, modern lungfishes can breathe in this manner. However, in most bony

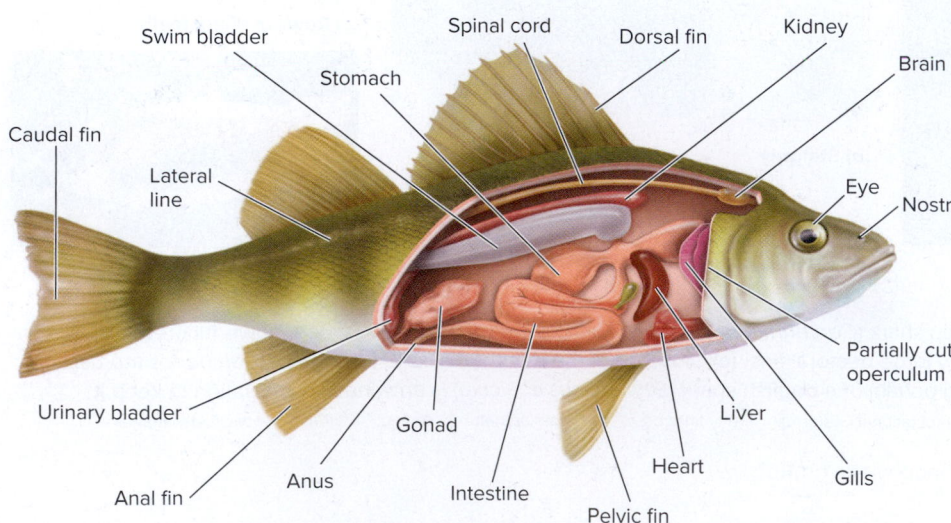

**Figure 35.6 Generalized body plan of a bony fish.** Bony fish possess a swim bladder and an operculum that covers the gills.

**(a) Lionfish (*Pterois volitans*)**

**(b) Whitemouth moray eel (*Gymnothorax meleagris*)**

**(c) Leafy sea dragon (*Phycodurus eques*)**

**Figure 35.7**  **The diversity of ray-finned fishes.** Ray-finned fishes exhibit many different sizes and body shapes. a: Source: Julie Bedford/NOAA; b: ©Andrew Dawson/age fotostock; c: ©Luc Novovitch/Getty Images

*Concept Check:*  *What features distinguish ray-finned fishes from sharks?*

fishes, the lungs evolved into a **swim bladder**, a gas-filled, balloon-like structure that helps the fish remain buoyant in the water even when it is completely stationary. In early-diverging bony fishes, the gut and swim bladder are connected via a duct, and the fishes can fill their swim bladder by gulping air. In later-diverging species, the swim bladder is connected to the circulatory system, and gases are transported in and out of the blood, allowing the fishes to change the volume of the swim bladder and so to rise and sink. Therefore, unlike the sharks, many bony fishes can remain motionless and use a "sit-and-wait" ambush style. These three features—bony skeleton, operculum, and swim bladder—distinguish bony fishes from cartilaginous fishes.

Some fishes, known as bimodal breathers, can breathe through their gills and by gulping air, absorbing oxygen through their digestive tracts or accessory organs. For example, Siamese fighting fish (*Betta splendens*, known as betta), a popular freshwater aquarium fish, is a bimodal breather that is relatively easy to care for, since it can survive without an air pump in its aquarium.

Bony fishes have colonized nearly all aquatic habitats. Following the cooling of the newly formed Earth, water condensed into rain and over a vast period of time filled what are now the oceans. Later, as water evaporated from the oceans and sodium, potassium, and calcium were added via runoff from the land, the oceans became salty. Therefore, most fishes probably evolved in freshwater habitats and secondarily became adapted to marine environments. This, of course, required the development of physiological adaptations to the different osmotic problems seawater presents compared with fresh water (look ahead to Figure 41.17).

**Reproduction in Bony Fishes**  Reproductive strategies of bony fishes vary tremendously, but most species reproduce via external fertilization, with the female shedding her eggs and the male depositing sperm on top of them. Although adult bony fishes can maintain their buoyancy, their eggs tend to sink. This is why many species spawn in shallow, more oxygen- and food-rich waters and why coastal areas are important fish nurseries.

**Actinopterygii, the Ray-Finned Fish**  The most species-rich clade of bony fishes is the Actinopterygii, or **ray-finned fishes**, which includes all bony fishes except the coelacanths and lungfishes. In Actinopterygii, the fins are supported by thin, bony, flexible rays and are moved by muscles on the interior of the body. The clade has a diversity of forms, from lionfish and large predatory moray eels to delicate sea dragons (**Figure 35.7**). Whole fisheries are built around the harvest of species such as cod, anchovies, and salmon.

**Sarcopterygii, the Lobe-Finned Fish**  The Actinistia (coelacanths) and Dipnoi (lungfishes) are both considered Sarcopterygii, or **lobe-finned fishes**. The term Sarcopterygii used to refer solely to the lobe-finned fishes. More recently, evolutionary studies have shown that terrestrial vertebrates (tetrapods) evolved from such fishes. Therefore, the term Sarcopterygii has been expanded to include both lobe-finned fishes and tetrapods (see Figure 35.1). In the lobe-finned fishes, the fins are supported by skeletal extensions of the pectoral and pelvic areas that are moved by muscles within the fins.

The fossil record revealed that the Actinistia, or coelacanths, were a very successful group in the Devonian period, but all fishes of the class were believed to have died off at the end of the Mesozoic era (some 65 mya). You can therefore imagine the scientific excitement when in 1938, a modern coelacanth was discovered as part of the catch of a boat fishing near the Chalumna River in South Africa (**Figure 35.8a**). Intensive searches in the area revealed that coelacanths were living in deep waters off the southern African coast and especially off a group of islands near the coast of Madagascar called the Comoros Islands. Another species was found more recently in Indonesian waters.

Early-diverging lobe-finned fishes probably evolved in fresh water and had lungs, but the coelacanth lost lungs and returned to the sea. One distinctive feature of this group is a special joint in the skull that allows the jaws to open extremely wide and gives the coelacanth a powerful bite. As further evidence of the coelacanth's unusual body plan, its swim bladder is filled with oil rather than gas, although it serves a similar purpose—to increase buoyancy.

The Dipnoi, or **lungfishes**, like the coelacanths, are also not currently a very species-rich group, having just three genera and six species (**Figure 35.8b**). Lungfishes live in oxygen-poor freshwater swamps and ponds. They have both gills and lungs, the latter of which enable them to come to the surface and gulp air. Surprisingly, lungfish will drown if they are unable to breathe air. When ponds dry out, some species of lungfish can dig a burrow and survive in it until the next rain. Because they also have muscular lobe fins, lungfish are often able to successfully traverse quite long distances over shallow-bottomed lakes that may be drying out.

**Figure 35.8** **The Sarcopterygii (lobe-finned fish). (a)** An actinisian, the coelacanth (*Latimeria chalumnae*). **(b)** A dipnoi, the Australian lungfish (*Neoceratodus forsteri*). a: ©Peter Scoones/SPL/Science Source; b: ©D. R. Schrichte/SeaPics.com

**Concept Check:** *How are lungfishes similar to coelacanths?*

The morphological features of coelacanths, lungfishes, and primitive terrestrial vertebrates, together with the similarity of their nuclear genes, suggest to many scientists that lobe-fin ancestors gave rise to three lineages: the coelacanths, the lungfishes, and the tetrapods. In the next section, we will examine the characteristics of tetrapods in more detail.

## 35.4 Tetrapods: Gnathostomes with Four Limbs

**Learning Outcomes:**

1. Describe adaptations that were beneficial for a terrestrial lifestyle.
2. List the amphibian orders, and describe the features that differentiate them.
3. **CoreSKILL »** Explain how relatively simple experiments with mice showed that mutations in just two genes can cause large changes in limb development.

During the Devonian period (from about 417 to 354 mya), a diversity of plants and animals colonized the land. The plants served as both a source of oxygen and a potential food source for animals that ventured out of the aquatic environment. Terrestrial arthropods and vertebrates appeared during the Devonian. The transition to life on land involved

a large number of adaptations. Paramount among these were adaptations preventing desiccation and making locomotion and reproduction on land possible. In this section, we begin by outlining the development of the **tetrapods**, vertebrate animals having four legs or leglike appendages. We will discuss the first terrestrial vertebrates and their immediate descendants, the amphibians. We will then explore the characteristic features and diversity of modern amphibians.

### The Origin of Tetrapods Involved the Development of Four Limbs

The fossil record of the Devonian period demonstrates the evolution of sturdy lobe-finned fishes to fishes with four limbs. The abundance of light and nutrients in shallow waters encouraged a profusion of plant life and the invertebrates that fed on them. The development of lungs enabled lungfishes to colonize these productive yet often oxygen-poor waters. Here, the ability to move in shallow water clogged with plants and debris was more vital than the ability to swim swiftly through open water and may have favored the progressive development of sturdy limbs. As an animal's weight began to be borne more by the limbs, the vertebral column strengthened, and hip bones and shoulder bones were braced against the backbone for added strength.

***Early Transitional Forms*** One of the transitional forms between fish and tetrapods was *Tiktaalik roseae*, nicknamed fishapod (refer back to Figure 22.5). Fishapods had broad skulls with eyes mounted on the top, lungs, and pectoral fins with five finger-like bones. *T. roseae* is an important species, for it represents a transitional form, displaying an intermediate state between an ancestral form and the form of its descendants. Eventually, species more like modern amphibians evolved, species that were still tied to water for reproduction but increasingly lived on land. In these species, the vertebral column, hip bones, and shoulder bones grew sturdier. Such changes were needed as the animal's weight was no longer supported by water but was borne entirely on the limbs.

***Hox Genes and Limb Formation*** The modifications leading to the emergence of tetrapods are the result of changes in the expression of genes, especially *Hox* genes. In tetrapods, *Hox* genes 9–13 work together to specify limb formation from proximal to distal, meaning from close to the point of attachment to the body to the terminal end of the limb (**Figure 35.9**). For example, *Hox9* plays a role in the formation of the scapula, which is attached to the shoulder, whereas *Hox13* plays a role in the formation of metacarpals found in the claws or fingers. As described next in the Feature Investigation, our understanding of *Hox* gene function has come from genetic studies involving *Hox* mutations.

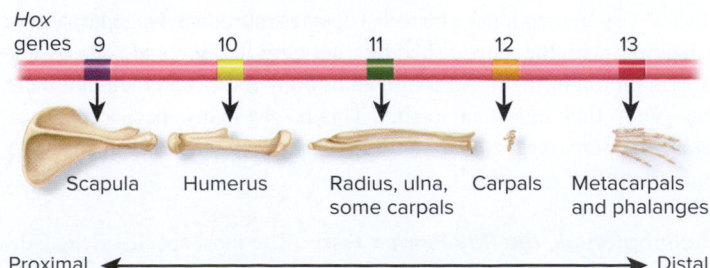

**Figure 35.9** **The roles of *Hox* genes 9–13 in specifying limb formation from proximal to distal.** The axis of limb development in mice is shown, together with the associated genes.

 **Core Skill: Process of Science**

## Feature Investigation | Davis and Colleagues Provided a Genetic-Developmental Explanation for Limb Length in Tetrapods

The development of limbs in tetrapods was a vital step that allowed animals to colonize land. The diversity of vertebrate limb types is amazing, from fins in fish and marine mammals to different wing types in bats and birds to legs and arms in primates. Early in vertebrate evolution, an ancestral gene complex was duplicated twice to give rise to four groups of genes, called *HoxA*, *HoxB*, *HoxC*, and *HoxD*, which control limb development. Among the four groups, 13 different types of *Hox* genes can be found, but any given group does not contain all 13 types (refer back to Figure 20.16).

In 1995, Allen Davis, Mario Capecchi, and colleagues analyzed the effects of mutations in specific *Hox* genes that are responsible for determining limb formation in mice. The vertebrate forelimb is divided into three zones: humerus (upper arm); radius and ulna (forearm); and carpals, metacarpals, and phalanges (digits). The researchers had no specific hypothesis in mind; their goal was to understand the role of *Hox* genes in limb formation. As described in **Figure 35.10**, they began with strains of mice carrying loss-of-function mutations

**Figure 35.10** Relatively simple changes in *Hox* genes control limb formation in tetrapods.

**GOAL** To determine the role of *Hox* genes in limb development in mice.

**KEY MATERIALS** Mice with individual mutations in *HoxA-11* and *HoxD-11* genes.

| Experimental level | Conceptual level |
|---|---|

1 Breed mice with individual mutations in *HoxA-11* and *HoxD-11* genes. (The *A* and *D* refer to wild-type alleles; *a* and *d* are mutant alleles.)

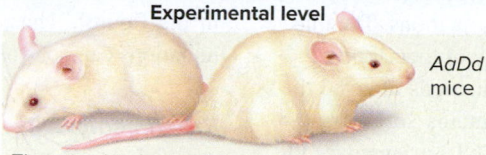

*AaDd* mice

The mice bred were heterozygous for both genes (*AaDd*).

Based on previous studies, researchers expect mutant mice to produce viable offspring, perhaps with altered limb morphologies.

2 Using molecular techniques described in Chapter 20, obtain DNA from the tail and determine the genotypes of offspring.

The resulting genotypes occur in Mendelian ratios, generating mice with different combinations of wild-type and mutant alleles.

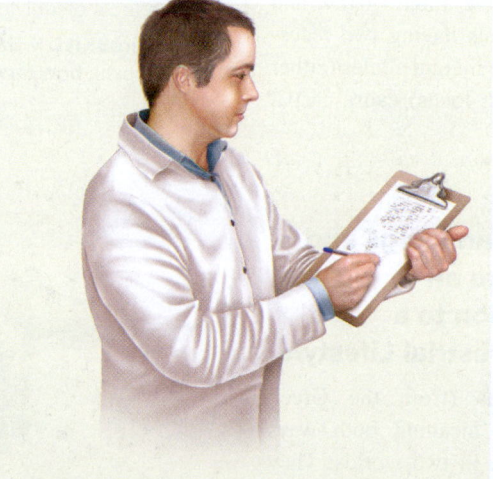

|  | AD | Ad | aD | ad |
|---|---|---|---|---|
| **AD** | AADD | AADd | AaDD | AaDd |
| **Ad** | AADd | AAdd | AaDd | Aadd |
| **aD** | AaDD | AaDd | aaDD | aaDd |
| **ad** | AaDd | Aadd | aaDd | aadd ← Double mutant |

9:3:3:1 phenotypic ratio expected a two-factor cross

3 Stain the skeletons and compare the limb characteristics of the wild-type mice (*AADD*) to those of strains carrying mutant alleles in one or both genes.

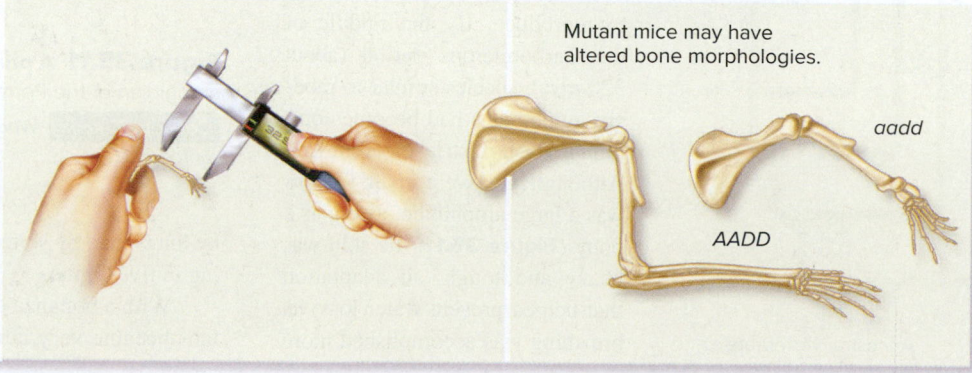

Mutant mice may have altered bone morphologies.

*aadd*

*AADD*

**4    THE DATA**

| Genotype | Carpal bone fusions (% of mice showing the fusion) | | | |
|---|---|---|---|---|
| | Normal (none fused) | NL fused to T | T fused to P | NL fused to T and P |
| AADD | 100 | 0 | 0 | 0 |
| AaDD | 100 | 0 | 0 | 0 |
| aaDD | 33 | 17 | 50 | 0 |
| AADd | 100 | 0 | 0 | 0 |
| AAdd | 0 | 17 | 17 | 67 |
| AaDd | 17 | 17 | 33 | 33 |

**5    CONCLUSION** Relatively simple mutations involving two genes can cause large changes in limb development.

**6    SOURCE** Davis, A.P. et al. 1995. Absence of radius and ulna in mice lacking Hoxa-11 and Hoxd-11. *Nature* 375: 791–795.

in *HoxA-11* or *HoxD-11* that, on their own, did not cause dramatic changes in limb formation. They bred the mice and obtained offspring carrying one, two, three, or four loss-of-function mutations. The mice were then analyzed with regard to the morphology of their limbs.

Taken together, the data indicate that the mutations affected the formation of limbs. For example, the wrist contains seven bones: three proximal carpals—called navicular lunate (NL), triangular (T), and pisiform (P)—and four distal carpals (d1–d4). In mice with the genotypes *aaDD* and *AAdd*, the proximal carpal bones are usually fused together. Individuals having one recessive allele (*AADd* and *AaDD*) do not show this defect, but individuals having two recessive alleles (*AaDd*) often do. Therefore, any two mutant alleles (either in both *HoxA-11* and *HoxD-11* or one in each locus) cause carpal fusions. Deformities became even more severe with three mutant alleles (*Aadd* or *aaDd*) or four mutant alleles (*aadd*) (data not shown in the figure). Thus, the researchers showed that relatively simple mutations can control relatively large changes in limb development.

***Experimental Questions***

1. What was the purpose of the study conducted by Davis and colleagues?

2. **CoreSKILL »** Explain how the researchers were able to study the effects of individual genes.

3. **CoreSKILL »** Summarize the results of the experiment, and explain how they relate to limb development in vertebrates.

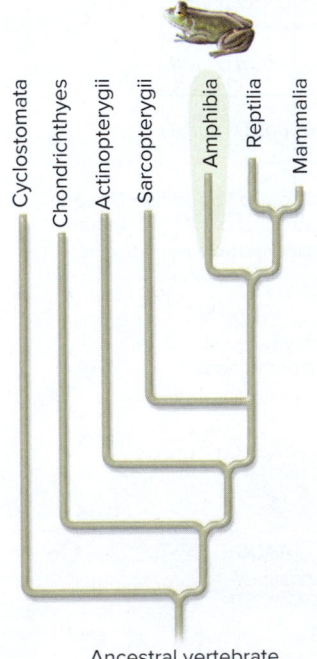

Cyclostomata
Chondrichthyes
Actinopterygii
Sarcopterygii
Amphibia
Reptilia
Mammalia

Ancestral vertebrate

## Amphibian Lungs and Limbs Are an Adaptation to a Semiterrestrial Lifestyle

**Amphibians** (from the Greek *amphibios*, meaning both ways of life) live in two worlds: They have successfully invaded the land, but most must return to the water to reproduce. By the middle of the Carboniferous period (about 320 mya), species similar to modern amphibians had become common in the terrestrial environment. Although most were small, *Cacops* was a large amphibian, as big as a pony (**Figure 35.11**). Its skin was heavy and tough, an adaptation that helped prevent water loss; its breathing was accomplished more

**Figure 35.11 A primitive tetrapod.** *Cacops* was a large, early amphibian of the Permian period.

*Concept Check:* *What were the advantages to animals of moving onto land?*

by lungs than by skin; and it possessed **pentadactyl limbs** (limbs ending in five digits).

With a bonanza of terrestrial arthropods to feast on, the amphibians became very numerous and species rich, and the mid-Permian period (some 260 mya) is sometimes known as the Age of Amphibians. However, most of the large amphibians became extinct at the end of the Permian period. This was the largest known mass extinction in Earth's history, with the extinction of 90–95% of marine species and a large proportion of terrestrial species. Most surviving amphibians were relatively small organisms resembling modern species.

**Key Features of Amphibians**   One of the first challenges terrestrial animals had to overcome was breathing air when on land. Like lungfishes, most amphibians open their mouths to let in air. Alternatively, air may enter through the nostrils. Amphibians then close and raise the floor of the mouth, creating a positive pressure that pumps air into the lungs. This method of breathing is called **buccal pumping**. In addition, the skin of amphibians is much thinner than that of fishes, and amphibians absorb oxygen from the air directly through their outer moist skin or through the skin lining of the inside of the mouth or pharynx. Because the skin of amphibians is usually thin, these animals face the problem of desiccation, or drying out. As a consequence, even amphibian adults are more abundant in damp habitats, such as swamps or rain forests, than in dry areas.

Amphibians have a three-chambered heart, with two atria and one ventricle. One atrium receives blood from the body, and the other receives blood from the lungs. Both atria pump blood into the single ventricle, which pumps some blood to the lungs and some to the rest of the body (look ahead to Figure 48.2b). This form of circulation allows the tissues to receive well-oxygenated blood at a higher pressure than is possible via single circulation, because some of the blood that returns to the heart is directly pumped to the tissues without being slowed down by passage through the lung capillaries. Oxygenated and deoxygenated bloods are kept somewhat separate, which enhances the delivery of nutrients and oxygen to the tissues.

**Reproduction**   In frogs and toads, fertilization is generally external, with males shedding sperm over the gelatinous egg masses laid by the females in water (**Figure 35.12a**). The fertilized eggs lack a shell and would quickly dry out if exposed to the air. The eggs soon hatch into tadpoles (**Figure 35.12b**), small fishlike animals that lack limbs and

breathe through gills. As the tadpole nears the adult stage, the tail and gills are resorbed, and limbs and lungs appear (**Figure 35.12c**). Such a dramatic change in body form, from juvenile to adult, is known as **metamorphosis**. A few species of amphibians do not require water to reproduce. These species are ovoviparous or viviparous—retaining the eggs in the reproductive tract and giving birth to live young.

**Orders of Amphibians**   Approximately 7,600 living amphibian species are known, and the vast majority of these, some 6,700 species, are frogs and toads of the order Anura (from the Greek, meaning tail-less ones) (**Figure 35.13a**). The other two orders are the Apoda (from the Greek, meaning legless ones), the wormlike caecilians; and the Urodela (from the Latin, meaning tailed ones), the salamanders. Global warming is currently threatening many anurans with extinction.

Adult anurans are carnivores, eating a variety of invertebrates by catching them on a long, sticky tongue. In contrast, the aquatic larvae (tadpoles) are primarily herbivores. Frogs generally have smooth, moist skin and long hind legs, making them excellent jumpers and swimmers. In addition to secreting mucus, which keeps their skin moist, some frogs can also secrete poisonous chemicals that deter would-be predators. Some amphibians advertise the poisonous nature of their skin with warning coloration (look ahead to Figure 57.9b). Others use camouflage as a way of avoiding detection by predators. Toads have a drier, bumpier skin and shorter legs than frogs. They are less impressive leapers than frogs, but they can better tolerate drier conditions.

Caecilians (order Apoda) comprise a small order of about 200 species of legless, nearly blind amphibians (**Figure 35.13b**). Most are tropical and burrow in forest soils, but a few live in ponds and streams. They are secondarily legless, which means they evolved from legged ancestors. Caecilians have tiny jaws equipped with teeth and eat worms and other soil invertebrates. In this order, fertilization is internal, and females usually bear live young. The young are nourished inside the mother's body by a thick, creamy secretion known as uterine milk. In most caecilian species, the young grow into adults about 30 cm long, though species up to 1.3 m in length are known.

The salamanders (order Urodela, about 700 species) possess a tail and have a more elongate body than that of anurans (**Figure 35.13c**). During locomotion, they seem to sway from side to side, perhaps reminiscent of how the earliest tetrapods may have walked. Like frogs,

**(a) Gelatinous mass of amphibian eggs**

**(b) Tadpole**

**(c) Tadpole undergoing metamorphosis**

**Figure 35.12   Amphibian development in the wood frog (*Rana sylvatica*). (a)** Amphibian eggs are laid in gelatinous masses in water. **(b)** The eggs develop into tadpoles, aquatic herbivores with a fishlike tail that breathe through gills. **(c)** During metamorphosis, the tadpole loses its gills and tail and develops limbs and lungs. a: ©Don Vail/Alamy Stock Photo; b–c: ©Dwight Kuhn

**(a) Tree frog**

**(b) A caecilian**

**(c) Mud salamander**

**Figure 35.13 Amphibians. (a)** Most amphibians are frogs and toads of the order Anura, including this red-eyed tree frog (*Agalychnis callidryas*). **(b)** The order Apoda includes wormlike caecilians such as this species from Ecuador, *Siphonops annulatus*. **(c)** The order Urodela includes species such as this mud salamander (*Pseudotriton montanus*). a: ©Gregory G. Dimijian/Science Source; b: ©Danita Delimont/Alamy Stock Photo; c: ©Gary Meszaros/Science Source

**Concept Check:** *Do all amphibians produce tadpoles?*

salamanders often have colorful skin patterns that advertise their distastefulness to predators. Salamanders retain their moist skin by living in damp areas under leaves or logs or beneath lush vegetation. They generally range in size from 10 to 30 cm. Fertilization is usually internal, with females using their cloaca, a common opening for the digestive and urogenital tracts, to pick up sperm packets deposited by males. A very few salamander species do not undergo metamorphosis, and the newly hatched young resemble tiny adults. However, some species, such as Cope's giant salamander (*Dicamptodon copei*), retain the gills and tail fins characteristic of the larval stage into adulthood, and mature sexually in the larval stage, a phenomenon known as paedomorphosis.

## 35.5 Amniotes: Tetrapods with a Desiccation-Resistant Egg

**Learning Outcomes:**

1. Diagram the structure of the amniotic egg.
2. Identify the critical innovations of the amniotes.
3. Describe the distinguishing features of the major amniote classes.
4. List the features that allowed birds to fly.

Although amphibians live successfully in a terrestrial environment, they must lay their eggs in water or in a very moist place, so that the shell-less eggs do not dry out on exposure to air. Thus, a critical innovation in animal evolution was the development of a shelled egg that sheltered the embryo from desiccating conditions on land. A shelled egg containing fluids was like a personal enclosed pond for each developing individual. Such an egg evolved in the common ancestor of turtles, lizards, snakes, crocodiles, birds, and mammals—a group of tetrapods collectively known as the **amniotes**. The amniotic egg permitted animals to lay their eggs in a dry place so that reproduction was no longer tied to water. It was truly a critical innovation, untethering animals from water in much the same way as the development of seeds liberated plants from water (see Chapter 31).

In time, the amniotes became very diverse in species and morphology. Mammals are considered amniotes, too, because even though most of them do not lay eggs, they retain other features of amniotic reproduction. In this section, we begin by discussing the structure of the amniotic egg and other adaptations that permitted animal species to become fully terrestrial. We will then discuss the biology of the reptiles, the first group of vertebrates to fully exploit land.

### The Amniotic Egg and Other Innovations Permitted Life on Land

The **amniotic egg** (**Figure 35.14**) contains the developing embryo and the four separate extraembryonic membranes that it produces:

1. The innermost membrane is the **amnion**, which protects the developing embryo in a fluid-filled sac called the amniotic cavity.
2. The **yolk sac** encloses a stockpile of nutrients, in the form of yolk, for the developing embryo.
3. The **allantois** functions as a disposal sac for metabolic wastes.
4. The **chorion**, along with the allantois, provides gas exchange between the embryo and the surrounding air.

Surrounding the chorion is the albumin, or egg white, which also stores nutrients. The **shell** provides a tough, protective covering that is not very permeable to water and prevents the embryo from drying out. However, the shell remains permeable to oxygen and carbon dioxide, so the embryo can breathe. In birds, this shell is hard and calcareous, whereas in reptiles and early-diverging mammals such as the platypus and echidna, it is soft and leathery. In most mammals, however, the embryos embed into the wall of the uterus and receive their nutrients directly from the mother.

Along with the amniotic egg, other critical innovations that enabled animals' conquest of land include the following:

- *Desiccation-resistant skin.* Whereas the skin of amphibians is usually moist and aids in respiration, the skin of amniotes is thicker and water resistant and contains keratin, a tough protein. As a result, most gas exchange takes place through the lungs.
- *Thoracic breathing.* Amphibians use buccal pumping to breathe, contracting the mouth to force air into the lungs. In contrast, amniotes use **thoracic breathing**, in which coordinated contractions of muscles expand the rib cage, creating a negative pressure to suck air in and then forcing it out later. This results in a greater volume of air being displaced with each breath than with buccal pumping.
- *Water-conserving kidneys.* The ability to concentrate wastes prior to elimination and thus conserve water is an important role of the amniotes' kidneys.

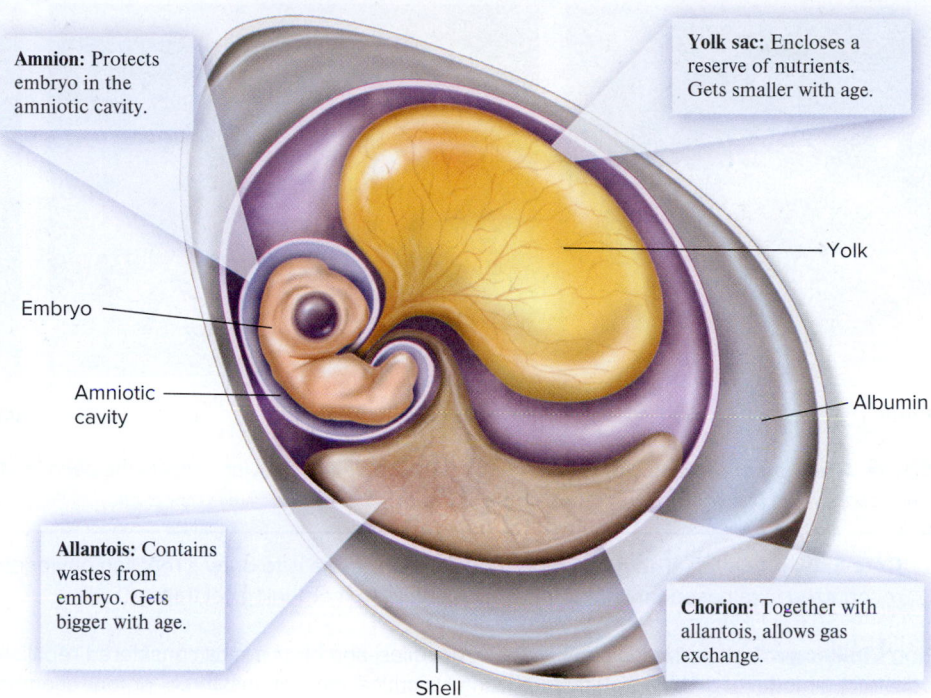

**Amnion:** Protects embryo in the amniotic cavity.

**Yolk sac:** Encloses a reserve of nutrients. Gets smaller with age.

Yolk

Embryo

Albumin

Amniotic cavity

**Allantois:** Contains wastes from embryo. Gets bigger with age.

**Chorion:** Together with allantois, allows gas exchange.

Shell

**Figure 35.14**  **The amniotic egg.**

**Concept Check:** *What are the other critical innovations of amniotes?*

- *Internal fertilization.* Because sperm cannot penetrate a shelled egg, fertilization occurs internally, within the female's body before the shell is secreted. In **internal fertilization,** the male of the species often uses a copulatory organ (penis) to transfer sperm into the female reproductive tract. However, birds usually transfer sperm from cloaca to cloaca.

## Reptiles Include Turtles, Lizards, Snakes, Crocodilians, Dinosaurs, and Birds

Early amniote ancestors gave rise to all modern amniotes that exist today, from lizards and snakes to birds and mammals. The traditional view of amniotes identified three living classes: the reptiles (turtles, lizards, snakes, and crocodilians), birds, and mammals. As we will see later in the chapter, modern systematists have argued that enough similarities exist between birds and the classic reptiles that birds should be considered part of the reptilian lineage. This classification scheme, which is also based on molecular evidence, will be followed in this chapter. The fossil record reveals other reptilian clades, all of which are extinct, including two clades of dinosaurs (ornithischian and saurischian dinosaurs), flying reptiles (pterosaurs), and two clades of ancient aquatic reptiles (icthyosaurs and plesiosaurs).

Cyclostomata
Chondrichthyes
Actinopterygii
Sarcopterygii
Amphibia
Reptilia
Mammalia

Ancestral vertebrate

***Testudines: The Turtles***   Turtles is an umbrella term for terrestrial species, also called tortoises, and aquatic species, sometimes known as terrapins. The turtle lineage is ancient and has remained virtually unchanged for 200 million years. The major distinguishing characteristic of the turtles is a hard protective shell into which the animal can withdraw its head and limbs. In most species, the vertebrae and ribs are fused to form this shell. All turtles lack teeth but have sharp beaks for biting.

Most turtles are aquatic and have webbed feet. The forelimbs of marine species have evolved to become large flippers. All turtles, even the aquatic species, lay their eggs on land, usually in soft sand. The gender of hatchlings is dependent on temperature, with high temperatures producing more females. Marine species often make long migrations to sandy beaches to lay their eggs (**Figure 35.15a**). Most land turtles are quite slow movers, possibly due to a low metabolic rate and a heavy shell. However, they are very long-lived species, often surviving for 120 years or more. Furthermore, turtles do not appear to show reproductive senescence or aging, reproducing continually throughout their lifetime. Most organs such as the liver, lungs, and kidneys of a centenarian turtle function as effectively as do organs in young individuals, prompting genetic researchers to examine the turtle genome for longevity genes. Many turtle species are in danger of extinction, due to egg hunting, harvesting for shells or meat, destruction of habitat and nesting sites, and death in fishing nets.

***Squamata: Lizards and Snakes***   The clade Squamata is a large clade with about 6,270 species of lizards and 3,630 species of snakes. Many species have an elongated body form. One of the defining characteristics of lizards and snakes is a **kinetic skull,** in which the joints between various parts of the skull are extremely mobile. The lower jaw does not join directly to the skull but rather is connected by a multijointed hinge, and the upper jaw is hinged and movable from the rest of the head. This allows the jaws to open relatively wider than other vertebrate jaws, with the

(a) Green turtle

(b) Common collared lizard

(c) Juvenile tree python

**Figure 35.15  A variety of reptiles. (a)** A green turtle (*Chelonia mydas*) laying eggs in the sand in Malaysia. **(b)** The common collared lizard (*Crotaphytus collarus*). **(c)** A juvenile tree python (*Morelia veridis*). a: ©Pat Morris/ardea.com; b: ©Royalty-Free/Corbis; c: ©Mark Kostich/Getty Images

 **Core Skill: Modeling**  The goal of this modeling challenge is to draw a reptilian phylogenetic tree that distinguishes the four recognized taxa based on critical innovations that are morphological traits.

**Modeling Challenge:** Turtles, lizards and snakes, crocodiles, and birds are all considered reptilians, yet we recognize them as distinct taxa. Construct a model of a phylogenetic tree using six critical innovations: three- or four-chambered heart, scales, endothermy, a hard protective shell, a  kinetic skull, and feathers. Be sure to show these critical innovations on your phylogenetic tree with black bars and labels. The four taxa have the following combinations of traits: Crocodilia (four-chambered heart, scales, not endothermic, no protective shell, no kinetic skull, no feathers); Squamata (three-chambered heart, scales, not endothermic, no protective shell, kinetic skull, no feathers); Testudines (three-chambered heart, scales, not endothermic, hard protective shell, no kinetic skull, no feathers); Aves (four-chambered heart, scales, endothermic, no protective shell, no kinetic skull, feathers).

result that lizards and especially snakes can swallow large prey (**Figure 35.16**). Nearly all species are carnivores. Snakes may be venomous, whereas lizards usually are not.

A main difference between lizards and snakes is that lizards generally have limbs, whereas snakes do not (**Figure 35.15b,c**).

**Figure 35.16  The kinetic skull.** In snakes and lizards, both the upper and lower jaw are movable, thereby permitting large prey to be swallowed. This horned bush viper (*Atheris ceratophora*) is swallowing a leaf-folding frog. ©Michele Menegon/ardea.com

**Concept Check:** *Snakes are limbless, so how can they be considered tetrapods?*

Leglessness is a derived character, meaning snake ancestors possessed legs but later lost them.  In tetrapods, the expression of many different genes is needed for limb formation. In addition to *Hox* genes, the *SHH* gene is needed for limb formation and other developmental events. A genetic switch next to this gene is needed for *SHH* gene expression. In 2016, researchers found that pythons, which have tiny little leg bones inside their bodies, have three deletions in this genetic switch, and even more deletions are found in snake species with no leg bones at all. In the same year, a second research team used the CRISPR-Cas system (described in Chapter 21) to modify the expression of the *SHH* gene in mice. When they replaced the genetic switch of mice with the genetic switch from snakes, the resulting mice developed little nubs instead of legs. Taken together, these results indicate that the expression of the *SHH* gene plays an important role in limb formation.

***Crocodilia: The Crocodiles and Alligators***  The Crocodilia is a small clade of large, carnivorous, aquatic animals that have remained essentially unchanged for nearly 200 million years (**Figure 35.17**). Indeed, these animals existed at the same time as the dinosaurs. Most of the 25 recognized species live in tropical or subtropical regions. There are only two extant species of alligators: one living in the southeastern U.S. and one found in China.

Crocodiles have a four-chambered heart, a feature they share with birds and mammals (look ahead to Figure 48.2b). In this regard, crocodiles are more closely related to birds than to any other living reptiles. Their teeth are set in sockets, a feature typical of the dinosaurs and the earliest birds. Similarly, crocodiles care for their young, another trait they have in common with birds. These and other features suggest that crocodiles and

**(a) American alligator**

**(b) American crocodile**

**Figure 35.17  Crocodilians.** The Crocodilia is an ancient clade that has existed unchanged for millions of years. **(a)** Alligators, such as this American alligator (*Alligator mississippiensis*), have a broad snout, and the lower jaw teeth close on the inside of the upper jaw (and thus are almost completely hidden when the mouth is closed). **(b)** Crocodiles, including this American crocodile (*Crocodylus acutus*), have a longer, thinner snout, and the lower jaw teeth close on the outside of the upper jaw (and thus are visible when the mouth is closed). a: ©Warren Jacobi/Corbis; b: ©SteveByland/Getty Images

 **Core Skill: Connections**  Look ahead to Figure 48.2b. In what ways are crocodilians similar to birds and mammals?

birds are more closely related than crocodiles and lizards. As with turtles, the sex of crocodiles' offspring is dependent on nest temperature.

***Ornithischia and Saurischia: The Dinosaurs***  In 1841, English paleontologist Richard Owen coined the term **dinosaur** (from the Greek, meaning terrible lizard) to describe some of the wondrous fossil animals discovered in the 19th century. About 215 mya, dinosaurs were the dominant tetrapods on Earth and remained so for 150 million years, far longer than any other vertebrate. The two main clades were the ornithischians, or bird-hipped dinosaurs, which were herbivores such as *Stegosaurus*; and the saurischians, or lizard-hipped dinosaurs, which were fast, bipedal carnivores such as *Tyrannosaurus rex* (**Figure 35.18**). In contrast to the limbs of lizards, amphibians, and crocodiles, which splay out to the side, the legs of dinosaurs were positioned directly under the body, like pillars, a position that may have helped support their heavy

**(a) Ornithischian (*Stegosaurus*)**        **(b) Saurischian (*Tyrannosaurus*)**

**Figure 35.18  Dinosaurs. (a)** Herbivorous ornithischians included *Stegosaurus*, and **(b)** carnivorous saurischians included bipedal species such as *Tyrannosaurus rex*.

bodies. Because less energy was devoted to lifting the body from the ground, some dinosaurs are believed to have been fast runners. Members of different but closely related clades—the pterosaurs (the first vertebrates to fly) and ichthyosaurs and plesiosaurs (marine reptiles)—were also common at the same time as the ornithishichians and saurischians.

Dinosaurs were the biggest animals ever to walk on the planet, with some weighing up to 50 tonnes (metric tons), or over 100,000 pounds. The variety of the thousands of dinosaur species found in fossil form around the world is staggering. However, perhaps not surprisingly for such long-extinct species, scientists are still hotly debating many details of their lives. For example, an issue still unresolved is whether some dinosaur species were **endothermic**, capable of generating body heat through their own metabolism, as birds and mammals are, or whether they were **ectothermic**, dependent on external heat as the main source of their body heat, as most reptiles are. Another issue is whether dinosaurs exhibited parental care of their young.

All nonavian dinosaurs, and many other animals, died out abruptly during a mass extinction at the end of the Cretaceous period (about 65 mya). Although widely attributed to climatic change brought about by the impact of a meteorite, scientists continue to debate the cause or causes of this mass extinction. We do not yet know why dinosaurs died out, while many other animals, including birds and small mammals, survived.

***Aves: The Birds***  The defining characteristics of birds (Aves, plural of the Latin *avis*, meaning bird) are feathers and nearly all species can fly. As we will see, the ability to fly has shaped nearly every feature of the bird body. The other vertebrates that have evolved the ability to fly, the bats and the now-extinct pterosaurs, used skin stretched tight over elongated limbs to fly. Such a surface can be irreparably damaged, though some holes may heal remarkably quickly. In contrast, birds use feathers, epidermal outgrowths that can be replaced if damaged. Recent research shows that feathers evolved in dinosaurs before the appearance of birds.

## Modern Birds Evolved from Small, Feather-Covered Dinosaurs

To trace the evolution of birds, paleontologists look at transitional forms, the earliest animals that had feathers. One of the first known fossils exhibiting the faint impression of feathers was *Archaeopteryx lithographica* (from the Greek, meaning ancient wings and stone picture), found in a limestone quarry in Germany in 1861. The fossil was dated at 150 million years old, which places it during the Jurassic period. Except for the presence of feathers, *Archaeopteryx* had features similar to those of dinosaurs (**Figure 35.19a**; see also Figure 26.14). First, the fossil included an impression of a long tail with many vertebrae, a dinosaur feature. Some modern birds have long tails, but they are made of feathers, with the actual tailbone being much reduced. Second, the wings had claws halfway down the leading edge, another dinosaur-like character. Among modern birds, only the hoatzin, a South American swamp-inhabiting bird, has claws on its wings, which enable the chicks to climb back into the nest if they fall out. A third dinosaur-like feature is *Archaeopteryx*'s toothed beak. Fourth, the fossil shows that *Archaeopteryx* lacked an enlarged breastbone, a feature that modern birds possess to anchor their large flight muscles, so it likely could not fly.

Structural similarities of the skull, feet, and hind leg bones have led scientists to conclude that *Archaeopteryx* is closely related to **theropods**, a group of bipedal saurischian dinosaurs. The wings and feathers of *Archaeopteryx* may have enabled it to glide from tree to tree, helped to keep it warm, or cut out the glare when folded over its head when hunting, in much the same way as some herons fold their wings over their heads when they are fishing. Later, the wings and feathers may have taken on functions of flight.

In China in the mid-1990s, paleontologists unearthed fossils of about the same age as *Archaeopteryx* that similarly suggest a close kinship between dinosaurs and modern birds. *Caudipteryx zoui* was a dinosaur-like animal with feathers on its wings and tail and a toothed beak (**Figure 35.19b**). *Confuciusornis sanctus* was a small, flightless but completely feathered dinosaur lacking the long, bony tail and toothed jaw found in other theropod dinosaurs. Its large tail feathers may have functioned in courtship displays (**Figure 35.19c**).

These three species—*Archaeopteryx*, *Caudipteryx*, and *Confuciusornis*—help trace a lineage from dinosaurs to birds. By the early Cretaceous period, and only a relatively short period after *Archaeopteryx* evolved, the fossil record shows the existence of a huge array of bird types resembling modern species. These were to share the skies with pterosaurs for 70 million years, before eventually having the airways to themselves.

## Birds Have Feathers, a Lightweight Skeleton, Air Sacs, and Reduced Organs

Modern birds possess many characteristics, including scales on their feet and legs and shelled eggs, that reveal their reptilian ancestry. In addition, however, birds have four features that are unique among living animals and associated with flight.

1. **Feathers.** Feathers are modified scales that keep birds warm and enable flight (**Figure 35.20a**). Soft, downy feathers, which are close to the body, maintain heat, whereas stiffer contour feathers, supported on a modified forelimb, give the wing the airfoil shape it needs to generate lift. Each contour feather develops from a follicle, a tiny pit in the skin. If a feather is lost, a new one can be regrown. The contour feathers consist of many paired barbs, each of which supports barbules that contain hooks that interlock with barbules from neighboring barbs to give the feather its shape (**Figure 35.20b**).

**(a) *Archaeopteryx lithographica***

Avian features:
1. Feathered wings and tail

Reptilian features:
1. Long, bony tail
2. Claws on wings
3. Toothed beak
4. No large breastbone

**(b) *Caudipteryx zoui***

Avian features:
1. Feathered wings and tail

Reptilian features:
1. Short, bony tail
2. Claws on wings
3. Toothed beak
4. No large breastbone

**(c) *Confuciusornis sanctus***

Avian features:
1. Completely feathered
2. No bony tail
3. No teeth in beak

Reptilian features:
1. Flightless
2. Claws on wings
3. No large breastbone

**Figure 35.19 Transitional forms between dinosaurs and birds. (a)** *Archaeopteryx lithographica* was a Jurassic animal with dinosaur-like features as well as wings and feathers. **(b)** *Caudipteryx zoui* was a dinosaur with feathers on its tail and wings. **(c)** *Confuciusornis sanctus* was a birdlike animal with a horny, toothless beak.

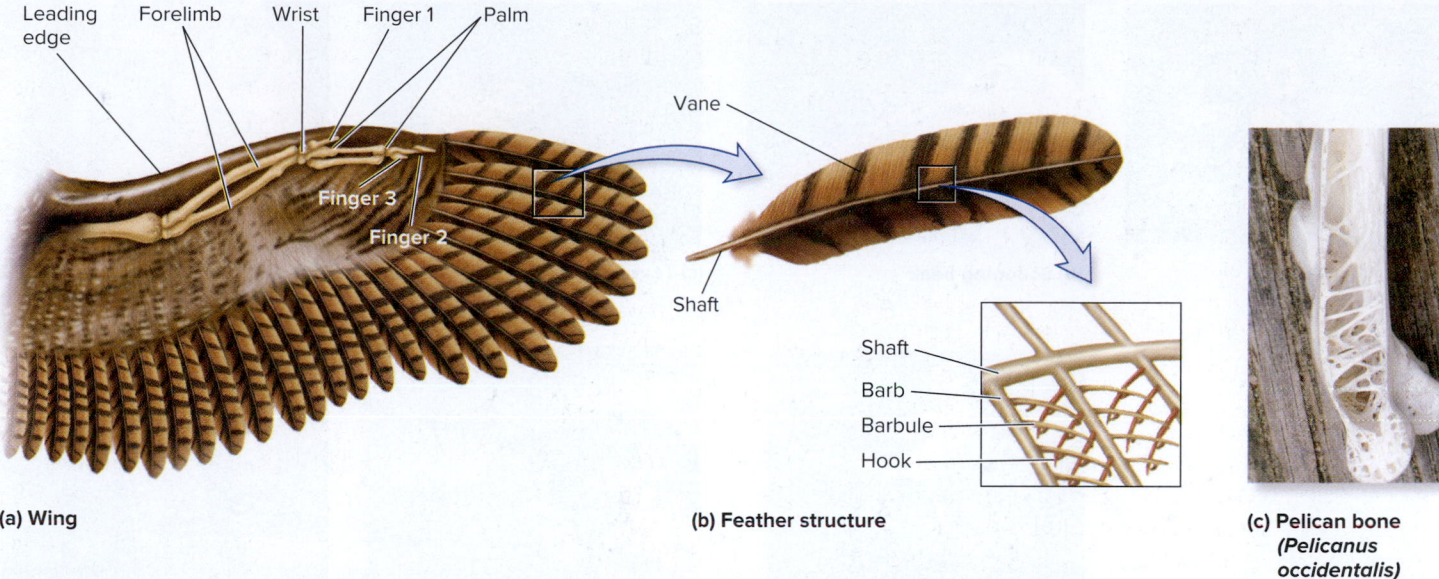

(a) Wing

(b) Feather structure

(c) Pelican bone (*Pelicanus occidentalis*)

**Figure 35.20** **Features of the bird wing and feather.** **(a)** The wing is supported by an elongated and modified forelimb with three extended fingers. **(b)** Each feather has a hollow shaft that supports many barbs, which, in turn, support barbules that interlock with hooks to give the feather its form. **(c)** The bones of a pelican (*Pelicanus occidentalis*) are hollow but crisscrossed with a honeycomb structure that provides added strength.
c: ©Gilbert S. Grant/Science Source

**Concept Check:**  *What adaptations in birds help reduce their body weight to enable flight?*

2. **Air sacs.** Flight requires a great deal of energy generated by an active metabolism that requires abundant oxygen. Birds have nine air sacs—large, hollow sacs that may extend into the bones—that expand and contract when a bird inhales and exhales, while the lungs remain stationary. Air is therefore being constantly moved across the lungs during inhalation and exhalation. Although making bird breathing very efficient, this process also makes birds especially susceptible to airborne toxins (hence, the utility of the canary in the coal mine; the bird's death signaled the presence of harmful carbon dioxide or methane gas that was otherwise unnoticed by miners).

3. **Reduction of organs.** Some organs are reduced in size or are lacking altogether in birds, which reduces the total mass of a bird's body. For example, birds have only one ovary and can carry relatively few eggs. As a result, they lay fewer eggs than most other reptile species. In fact, the gonads of both males and females are reduced, except during the breeding season, when they increase in size. Most birds also lack a urinary bladder. In addition, the lack of teeth reduces weight at the head end.

4. **Lightweight bones.** Most bird bones are thin and hollow and are crisscrossed internally by thin strips of bone, giving them a honeycomb structure (**Figure 35.20c**). An enlarged breastbone, or **sternum**, provides an anchor on which a bird's powerful flight muscles attach. These muscles may contribute up to 30% of the bird's body weight.

Other bird characteristics are related to flight. Rapid flight requires good vision, and bird vision is the best in the vertebrate world. Birds are endotherms. Their body temperatures are generally 40–42°C, considerably warmer than the human body's average of 37°C. This warm temperature ensures rapid metabolism and the quick production of adenosine triphosphate (ATP), which is needed to fuel flight and other activities. Like mammals, birds have a double circulation and a four-chambered heart. This type of circulatory system is more efficient at rapidly providing oxygen to the body, especially to the wings during flight.

Most birds are carnivores, eating insects or other invertebrates. However, some birds, such as parrots, eat just nutrient-rich fruits and seeds. The keratin of bird beaks is tough and malleable, and a wide assortment of beaks has evolved, with the form dependent on the function (**Figure 35.21**).

Bird reproduction involves parental care. Eggs need be kept warm for successful development, which entails brooding by an adult bird. Often, the males and females take turns brooding so that each parent can feed and maintain its strength. Picking successful partners is therefore an important task, and birds often engage in complex courtship rituals (look ahead to Figure 55.21).

## Birds Are Placed into Many Different Orders

Birds are the most species-rich clade of terrestrial vertebrates, with 28 orders, 166 families, and about 10,425 species (**Table 35.2**). Despite this diversity, birds lack the variety of body shapes that exist in the mammals, some of which can swim, others fly, others walk on four legs, and yet others walk only on two legs. Most birds fly, and therefore, most have the same general body shape. The biggest departures from this body shape are the flightless birds, including the cassowaries, emus, and ostriches. These birds have smaller wing bones, and the keel on the breastbone is greatly reduced or absent. Penguins are also flightless birds whose upper limbs are modified as flippers used in swimming.

**(a) Cracking beak**

**(b) Scooping beak**

**(c) Tearing beak**

**(d) Probing beak**

**(e) Nectar-feeding beak**

**(f) Sieving beak**

**Figure 35.21** **A variety of bird beaks.** Birds have evolved a variety of beak shapes used in different types of food gathering. **(a)** Hyacinth macaw (*Anodorhynchus hyacinthinus*)—cracking. **(b)** White pelican (*Pelecanus onocrotalus*)—scooping. **(c)** Verreaux's eagle (*Aquila verreauxii*)—tearing. **(d)** American avocet (*Recurvirostra americana*)—probing. **(e)** Anna's hummingbird (*Calypte anna*)—nectar feeding. **(f)** Roseate spoonbill (*Platalea ajaja*)—sieving. a: ©B. G. Thomson/Science Source; b: ©Jean-Claude Canton/Bruce Coleman Inc./Photoshot; c: ©Morales/age fotostock; d: ©Max Allen/Shutterstock; e: Source: Robert McMorran/USFWS; f: ©Mervyn Rees/Alamy Stock Photo

 **Core Concept: Structure and Function** Each of these beak shapes permits a different method of feeding.

## 35.6 Mammals: Milk-Producing Amniotes

### Learning Outcomes:

1. Identify four features that distinguish mammals from other vertebrate clades.
2. Describe the main orders of mammals.
3. Outline the phylogenetic relationships among the major clades of mammals.

About 225 mya, the first mammals appeared in the mid-Triassic period (refer back to Figure 26.13). They evolved from small mammal-like reptiles that went extinct about 170 mya. Until recently, the earliest mammals were believed to have been small, insect-eating species that lived in the shadows of dinosaurs. However, in 2005, the discovery of two fossils of a 130-million-year-old mammalian genus called *Repenomamus* challenged the notion that all early mammals were small insect eaters. One fossil was of an animal estimated to weigh about 13 kg (30 pounds), about the size of a small dog, which is larger than some dinosaurs living in the same region at the time. The other fossil had the remains of a baby dinosaur in its stomach area.

The extinction of the dinosaurs in the Cretaceous period, some 65 mya, paved the way for mammals to increase in numbers. Today, biologists have identified about 5,500 species of mammals with a diverse array of lifestyles, from fishlike dolphins to birdlike bats, and from small insectivores such as shrews to large herbivores such as giraffes and elephants. The range of sizes and body forms of mammals is unmatched by any other vertebrate group, and mammals are prime illustrations of the concept that organismal diversity is related to environmental diversity. In this section, we will outline the features that distinguish mammals from other taxa and examine the diversity of mammals that exist on Earth.

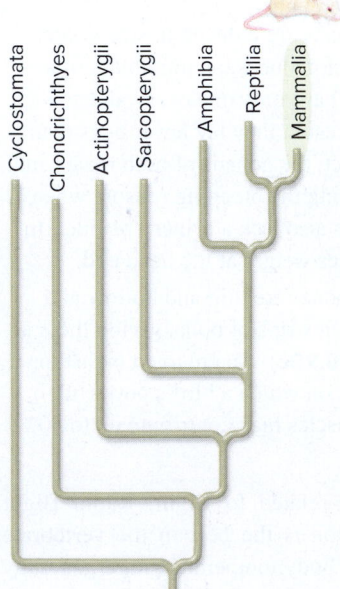

Ancestral vertebrate

### Mammals Have Mammary Glands, Hair, Specialized Teeth, and an Enlarged Skull

Four characteristics distinguish mammals: the possession of mammary glands, hair, specialized teeth, and an enlarged skull.

- *Mammary glands.*
  Mammals, or the clade Mammalia (from the Latin *mamma*, meaning breast), are

| Table 35.2 | | The Main Orders of Birds, in Order of Species Richness | |
|---|---|---|---|
| **Order** | | **Examples (approx. number of species)** | **Main characteristics** |
| Passeriformes | | Robins, starlings, sparrows, warblers (5,900) | Perching birds with perching feet; songbirds |
| Apodiformes | | Hummingbirds, swifts (430) | Fast fliers with rapidly beating wings; small bodies |
| Piciformes | | Woodpeckers, toucans (403) | Large with specialized beaks; two toes pointing forward and two backward |
| Psittaciformes | | Parrots, cockatoos (360) | Large, powerful beaks |
| Charadriiformes | | Gulls, sandpipers (340) | Shorebirds |
| Columbiformes | | Doves, pigeons (310) | Round bodies; short legs |
| Galliformes | | Chickens, pheasants, quail (285) | Often large birds; weak flyers; ground nesters |
| Accipitriformes | | Eagles, hawks, vultures (300) | Large diurnal carnivores; birds of prey; powerful talons; strong beaks |
| Coraciiformes | | Hornbills, kingfishers (206) | Large beaks; cavity nesters |
| Strigiformes | | Owls (205) | Nocturnal carnivores; powerful talons; strong beaks |
| Anseriformes | | Ducks, swans, waterfowl (165) | Able to swim; webbed feet; broad bills |
| Pelecaniformes | | Pelicans, frigate birds (65) | Large, water inhabiting |
| Sphenisciformes | | Penguins (17) | Flightless; wings modified into flippers for swimming; marine; Southern Hemisphere |

named after the female's distinctive mammary glands, which secrete milk. Milk is a fluid rich in fat, sugar, protein, and vital minerals, especially calcium. Newborn mammals suckle this fluid, which helps promote rapid growth.

- *Hair.* All mammals have hair, although some have more than others. Whales have hair in utero, but adults are hairless or retain only a few hairs on their snout. Compared with many mammals, humans are relatively hairless. In some animals, the hair is dense and is referred to as fur. In some aquatic species such as beavers, the fur is so dense it cannot be thoroughly wetted, so the hair underneath remains dry. Mammals are endothermic, and their fur is an efficient insulator. Hair can also take on functions other than insulation. Many mammals, including cats, dogs, walruses, and whales, have sensory hairs called vibrissae (**Figure 35.22a**). Hair can be of many colors, to allow the mammals to blend

into their background (**Figure 35.22b**). In some cases, as in porcupines and hedgehogs, the hairs become long, stiffened, and sharp (quills) and serve as a defense mechanism (**Figure 35.22c**).

- *Specialized teeth.* Mammals are the only vertebrates with highly differentiated teeth—incisors, canines, premolars, and molars—that are adapted for different types of diets (**Figure 35.23**). Although teeth are generally present in all mammalian species, some teeth are larger, smaller, lost, or reduced, depending on diet. Of particular importance to carnivores such as wolves are the piercing canine teeth, whereas herbivorous species such as deer depend on their chisel-like incisors to snip off vegetation and their many molars to grind plant material. Only mammals chew their food in this fashion. Rodent incisors grow continuously throughout life, and species such as beavers wear them down by gnawing tough plant material such as

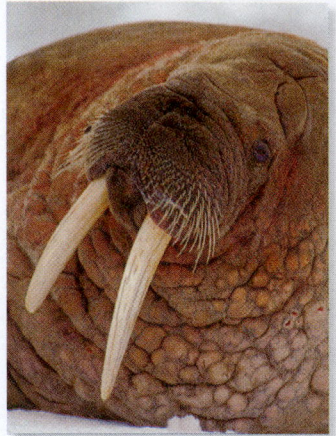
**(a) Sensory hairs**

**(b) Camouflaged coat**

**(c) Defensive quills**

**Figure 35.22  Mammalian hair. (a)** The sensory hairs (vibrissae) of the walrus (*Odobenus rosmarus*). **(b)** The camouflaged coat of a bobcat (*Lynx rufus*). **(c)** The defensive quills of the crested porcupine (*Hystrix africaeaustralis*). a: ©National Geographic Creative/Alamy Stock Photo; b: ©Charles Krebs/Corbis/Getty Images; c: ©Anthony Bannister/Science Source

wood. Mammals that have different types of teeth are called heterodonts; others, such as dolphins, whose teeth are of uniform size and shape, are called homodonts.

- *Enlarged skull.* The mammalian skull differs from other amniote skulls in several ways. First, the brain is enlarged and contained within a relatively large skull. Second, mammals have a single lower jawbone, unlike reptiles, whose lower jaw is composed of multiple bones. Third, mammals have three bones in the middle ear, as opposed to reptiles, which have only one. Fourth, most mammals, except some seals, have external ears.

In addition to those uniquely mammalian characteristics, some, but not all, mammals have the following additional features:

- *The ability to digest plants.* Apart from geese, tortoises and marine iguanas, certain species of mammals are the only large

vertebrates alive today that can exist on a steady diet of grasses or tree leaves; indeed, most large mammals are herbivores. Though mammals cannot digest cellulose, the principal constituent of the cell wall of many plants, some species have a large four-chambered stomach containing cellulose-digesting bacteria. These bacteria can break down the cellulose and make the plant cell contents available to the animal. Other species have an extensive cecum or large intestine where digestion occurs.

- *Horns or antlers.* Mammals are the only living class of vertebrates to possess horns or antlers. Many mammals, especially antelope, cattle, and sheep, have horns, typically consisting of a bony core that is a permanent outgrowth of the skull surrounded by a hairlike keratin sheath, as shown in the large antelope called a kudu (**Figure 35.24a**). Hooves, also made of keratin, protect an animal's toes from the force of impact of its feet against the ground.

**(a) Biting teeth**

**(b) Grinding teeth**

**(c) Gnawing teeth**

**(d) Tusks**

**(e) Grasping teeth**

**Figure 35.23  Mammalian teeth.** Mammals have different types of teeth, according to their diet. **(a)** The wolf has long canine teeth that bite its prey. **(b)** The deer has a long row of molars that grind plant material. **(c)** The beaver, a rodent, has long, continually growing incisors used to gnaw wood. **(d)** The elephant's incisors are modified into tusks. **(e)** Dolphins and other fish or plankton feeders have numerous small teeth used to grasp prey. a: ©Image Source/Corbis; b: ©Sam Camp/Getty Images; c: ©mauritius images GmbH/Alamy Stock Photo; d: ©Gallo Images ROOTS Collection/Getty Images; e: ©Jim Watt/Getty Images

**Figure 35.24 Horns and antlers in mammals.** Mammals have a variety of outgrowths that are used for defense or by males as weapons in contests over females. **(a)** The horns of this male kudu (*Tragelaphus strepsiceros*) are bony outgrowths of the skull covered in a keratin sheath. **(b)** The horns of the black rhinoceros (*Diceros bicornis*) are outgrowths of the epidermis, made of tightly matted hair. **(c)** The antlers of the caribou (*Rangifer tarandus*), also known as reindeer, are made entirely of bone and are grown and shed each year. a: ©Peter Chadwick/ Getty Images; b: Source: Karl Stromayer/USFWS; c: ©Paul A. Souders/Corbis/Getty Images

**(a) Skull outgrowths**   **(b) Epidermal outgrowths**   **(c) Bony antlers**

Rhinoceros horns are outgrowths of the epidermis, consisting of very tightly matted hair (**Figure 35.24b**). In contrast, deer antlers are made entirely of bone (**Figure 35.24c**). Deer grow a new set of antlers each year and shed them after the mating season.

## Mammals Are Morphologically Diverse and Occupy a Wide Range of Environments

Modern mammals are incredibly diverse in morphology and lifestyle (**Table 35.3**). They vary in size from tiny insect-eating bats,

| Table 35.3 | The Main Orders of Mammals, in Order of Species Richness | |
|---|---|---|
| **Order** | **Examples (approx. number of species)** | **Main characteristics** |
| Rodentia | Mice, rats, squirrels, beavers, porcupines (2,277) | Plant eating; gnawing habit, with two pairs of continually growing incisor teeth |
| Chiroptera | Bats (1,157) | Insect or fruit eating; small; have ability to fly; navigate by sonar; nocturnal |
| Eulipotyphla | Shrews, moles, hedgehogs (462) | Insect eaters; primitive placental mammals |
| Primates | Monkeys, apes, humans (437) | Opposable thumb; binocular vision; large brains |
| Carnivora | Cats, dogs, weasels, bears, seals, sea lions (295) | Flesh-eating mammals; canine teeth |
| Artiodactyla | Deer, antelopes, cattle, sheep, goats, camels, pigs (240) | Herbivorous hoofed mammals, usually with two toes, hippopotamus and others with four toes; many with horns or antlers |
| Diprotodontia | Kangaroos, koalas, opossums, wombats (147) | Pouched mammals mainly found in Australia |
| Lagomorpha | Rabbits, hares (92) | Powerful hind legs; rodent-like teeth |
| Cetacea | Whales, dolphins (84) | Marine fishes or plankton feeders; front limbs modified into flippers; no hind limbs; little hair except on snout |
| Perissodactyla | Horses, zebras, tapirs, rhinoceroses (18) | Hoofed herbivorous mammals with odd number of toes, one (horses) or three (rhinoceroses) |
| Monotremata | Duck-billed platypuses, echidna (5) | Egg-laying mammals found only in Australia and New Guinea |
| Proboscidea | Elephants (3) | Long trunk; large, upper incisors modified as tusks |

weighing in at only 2 g, to leviathans such as the blue whale, the largest animal ever known, which tips the scales at 100 tonnes (over 200,000 pounds). Mammalian orders are divided into two distinct subclasses (**Figure 35.25**). The subclass Prototheria contains only the order Monotremata, or **monotremes**, which are found in Australia and New Guinea. There are only five species: the duck-billed platypus (**Figure 35.26a**) and four species of echidna, a spiny animal resembling a hedgehog. Monotremes are early-diverging mammals that lay eggs rather than bear live young, lack a placenta, and have mammary glands with poorly developed nipples. The mothers incubate the eggs, and after hatching, the young simply lap up the milk as it oozes onto the fur.

The subclass Theria contains all remaining live-bearing mammals. The Theria are divided into two clades, the Metatheria and the Eutheria. The clade Metatheria, or the **marsupials**, is a group of seven orders, with about 280 species, including the swamp wallaby pictured in **Figure 35.26b**. Once widespread, members of this order are now largely confined to Australia, although some marsupials exist in South America, and one species—the opossum—is found in North America. Fertilization is internal, and reproduction is viviparous in marsupials. Marsupials have a placenta that nourishes the embryo. Unlike other mammals, however, marsupials are extremely small when they are born (often only 1–2 cm) and make their way to a ventral pouch called a marsupium for further development.

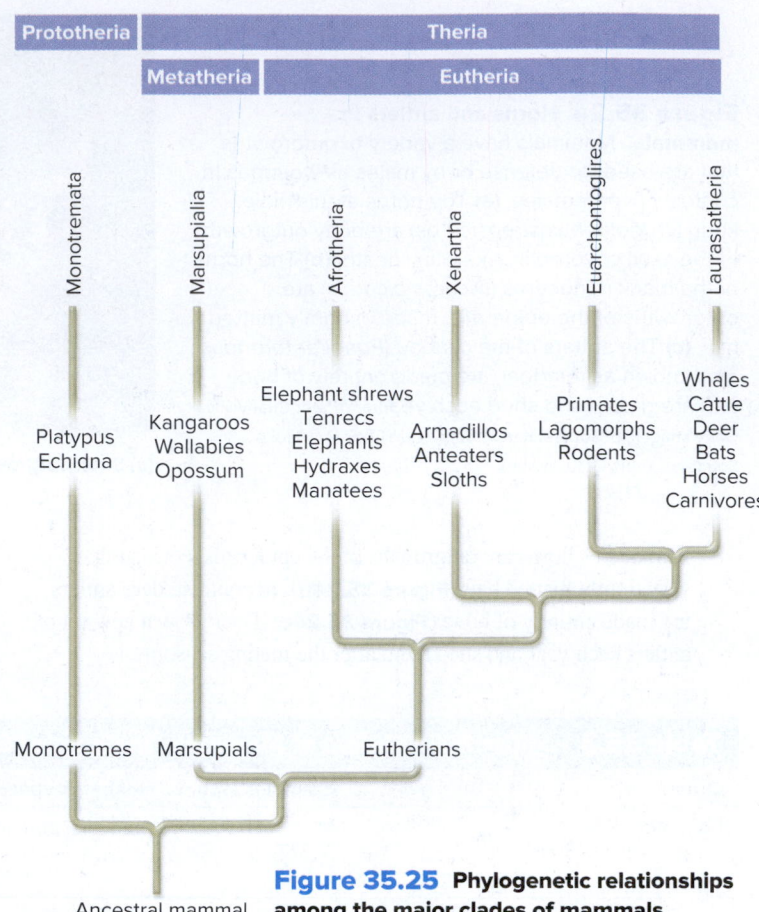

**Figure 35.25** Phylogenetic relationships among the major clades of mammals.

**(a) Prototherian (duck-billed platypus)**

**Figure 35.26** **Diversity among mammals.**
**(a)** Prototherians, such as this duck-billed platypus (*Ornithorhynchus anatinus*), lay eggs, lack a placenta, and possess mammary glands with poorly developed nipples. **(b)** Metatherians, or marsupials, such as this swamp wallaby (*Wallabia bicolor*), feed and carry their developing young, or "joeys," in a ventral pouch. **(c)** Gestation lasts longer in eutherians, and their young are more developed at birth, as illustrated by this young langur monkey (*Presbytis entellus*). a: ©Dave Watts/ naturepl.com; b: ©Nigel Pavitt/Getty Images; c: ©Education Images/Contributor/Getty Images

**(b) Metatherian (swamp wallaby)**

 **Core Skill: Connections** Look ahead to Figure 51.10. The placenta serves as the provisional lungs, intestine, and kidneys of the developing fetus. How much mixing is there of maternal and fetal blood?

**(c) Eutherian (gray langur)**

All other mammalian orders are members of the clade Eutheria and are considered **eutherians**, or placental mammals, such as the langur monkeys shown in **Figure 35.26c**. Eutherians have a long-lived and complex placenta, compared with that of marsupials. In eutherians, fertilization is internal, and reproduction is viviparous, but the developmental period, or gestation, of the young is prolonged. Molecular studies suggest four clades of placental mammals diverged in the Cretaceous. The earliest diverging clade was the Afrotheria, which evolved in the African landmass starting about 110–100 mya. This clade includes the elephant shrews, tenrecs, golden moles, manatees and dugongs, hyraxes, aardvark, and elephants. Shortly thereafter, about 100–95 mya, the Xenartha evolved in South America, where the armadillos, anteaters, and sloths appeared. The other two clades, the Euarchontoglires, containing the primates (refer back to Section 26.3), lagomorphs, and rodents, and the Laurasiatheria, containing the whales, artiodactyla, bats, horses, and carnivores, both evolved in the northern continent of Laurasia and became separate about 95–85 mya. Later, following continental drift, Africa and Arabia collided with Laurasia, and the Isthmus of Panama joined North and South America. These new land bridges facilitated animal movement between once separated continents.

The diversity of mammals is often threatened by human activities such as habitat destruction. In addition, many species are hunted for food. Others, such as wild cats and whales, are hunted for their products (fur and oil, respectively), and still others, such as the oryx, have simply been shot for sport. We will revisit the conservation of such endangered species in Chapter 60.

## Summary of Key Concepts

### 35.1 Vertebrates: Chordates with a Backbone

- Vertebrates have several characteristic features, including a vertebral column, a cranium, and an endoskeleton of cartilage or bone (Figure 35.1, Table 35.1).

### 35.2 Cyclostomata: Jawless Fishes

- Early-diverging vertebrates lacked jaws. Today, the only jawless vertebrates are the hagfish and lampreys (Figures 35.2, 35.3).

### 35.3 Gnathostomes: Jawed Vertebrates

- A critical innovation in vertebrate evolution was the hinged jaw, which first developed in fishes. Gnathostomes are vertebrate species that possess a hinged jaw (Figure 35.4).

- The chondrichthyans (sharks, skates, and rays) have a skeleton composed of flexible cartilage and powerful appendages called fins. They are active predators with acute senses and were among the earliest fishes to develop teeth (Figure 35.5).

- Bony fishes consist of the Actinopterygii (ray-finned fishes, the most species-rich clade), and the Sarcopterygii (lobe-finned fishes). In Actinopterygii, the fins are supported by thin, flexible rays and moved by muscles inside the body (Figures 35.6, 35.7).

- The Sarcopterygii comprise the lobe-finned fishes (Actinistia and Dipnoi) and the tetrapods. In the lobe-finned fishes, the fins are supported by extensions of the pectoral and pelvic areas and are moved by their own muscles (Figure 35.8).

### 35.4 Tetrapods: Gnathostomes with Four Limbs

- The fossil record reveals the evolution of lobe-finned fishes into fishes with four limbs. Recent research has shown that relatively simple mutations control large changes in limb development (Figures 35.9, 35.10, 35.11).

- Amphibians live on land but return to the water to reproduce. The larval stage undergoes metamorphosis, losing gills and tail and gaining lungs and limbs (Figure 35.12).

- The majority of amphibians belong to the order Anura (frogs and toads). Other orders are the Apoda (caecilians) and Urodela (salamanders) (Figure 35.13).

### 35.5 Amniotes: Tetrapods with a Desiccation-Resistant Egg

- The amniotic egg permitted animals to become fully terrestrial. Other critical innovations that enabled the conquest of land included desiccation-resistant skin, thoracic breathing, water-conserving kidneys, and internal fertilization (Figure 35.14).

- Living reptilian clades include the Testudines (turtles), Squamata (lizards and snakes), Crocodilia (crocodiles), and Aves (birds). The Ornithischia and Saurischia are two extinct clades of dinosaurs (Figures 35.15, 35.16, 35.17, 35.18).

- Three species—*Archaeopteryx*, *Caudipteryx*, and *Confuciusornis*—help trace a lineage from dinosaurs to birds (Figure 35.19).

- The four key characteristics of birds are feathers, air sacs, reduced size of organs, and a lightweight skeleton. Birds are the most species-rich clade of terrestrial vertebrates. Many of their unique features are related to flying. Their beak structures reflect varied methods for feeding (Figures 35.20, 35.21, Table 35.2).

### 35.6 Mammals: Milk-Producing Amniotes

- The distinguishing characteristics of mammals are mammary glands, hair, specialized teeth, and an enlarged skull. Mammalian tooth shape varies according to diet. Other distinguishing characteristics of some mammals are the ability to digest plants and horns or antlers (Figures 35.22, 35.23, 35.24).

- Two subclasses of mammals exist: the Prototheria (monotremes) and the Theria (the live-bearing mammals). The live-bearing mammals are, in turn, divided into the Metatheria (marsupials) and Eutheria (placental mammals). The Eutheria have been divided into four different clades (Table 35.3, Figures 35.25, 35.26).

## Assess & Discuss

### Test Yourself

1. Which of the following is *not* a defining characteristic of vertebrates?
   a. cranium      c. vertebral column
   b. hinged jaw      d. endoskeleton

2. Which of the following organisms are considered cyclostomes?
   a. tunicates      d. both a and b
   b. hagfish      e. both b and c
   c. lampreys

3. The presence of a bony skeleton, an operculum, and a swim bladder are all defining characteristics of
   a. Myxini.
   b. lampreys.
   c. Chondrichthyes.
   d. bony fishes.
   e. amphibians.

4. Which type of fish is a lobe-fin?
   a. moray eel
   b. sea dragon
   c. lamprey
   d. stingray
   e. lungfish

5. Organisms that lay eggs are said to be
   a. oviparous.
   b. ovoviparous.
   c. viviparous.
   d. placental.
   e. none of the above.

6. Which clade does not include frogs?
   a. vertebrates
   b. gnathostomes
   c. tetrapods
   d. amniotes
   e. lobe fins

7. In some amphibians, the adult retains certain larval characteristics, a phenomenon known as
   a. metamorphosis.
   b. parthenogenesis.
   c. cephalization.
   d. paedomorphosis.
   e. hermaphrodism.

8. The membrane of the amniotic egg that serves as a site for waste storage is the
   a. amnion.
   b. yolk sac.
   c. allantois.
   d. chorion.
   e. albumin.

9. Which characteristic qualifies lizards as gnathostomes?
   a. a cranium
   b. a skeleton of bone or cartilage
   c. a hinged jaw
   d. the possession of limbs
   e. amniotic eggs

10. Which of the following is *not* a distinguishing characteristic of birds?
   a. amniotic egg
   b. feathers
   c. air sacs
   d. lack of certain organs
   e. lightweight skeletons

## Conceptual Questions

1. How is vertebrate movement similar to arthropod movement, and how is it different?

2. Why aren't all reptiles endothermic given that both birds and mammals are?

3. **Core Concept: Evolution** Are birds considered to be living dinosaurs?

## Collaborative Questions

1. By what means can vertebrates move?

2. Why are amphibians considered good indicator species, which are species whose status provides information on the overall health of an ecosystem?

# UNIT VI

# FLOWERING PLANTS

**Flowering plants**, also known as the angiosperms, are essential to the lives of humans and most other organisms on Earth. Flowering plants provide most of our food, either directly in the form of vegetables and grains and other fruits or indirectly as animal fodder. Cotton, linen, and other fibers that we use for clothing and wood that we use for construction and fuel, as well as powerful cancer drugs and many other medicines, come from flowering plants. This unit reveals molecular, biochemical, structural, evolutionary, and ecological features of the hundreds of thousands of flowering plants that support Earth's life.

The unit begins with Chapter 36, which provides an overview of flowering plant structure and function, focusing on the seed-to-seed life cycle. By comparing plant bodies with those of animals, we will discover how plants are constructed and how they grow. Building on this background, Chapter 37 explains the genetic and physiological bases of plant behavior—plant responses to external stimuli such as day length that are mediated by internally produced hormones. In this chapter, we will see that, like animals, plants have evolved sophisticated sensory systems that monitor environmental conditions and allow plants to respond in predictable ways. Chapter 38 explains the nutritional requirements of plants, a deep understanding of which is critical to human agriculture and our ability to feed our increasing populations. In Chapter 39, we see how plant water transport influences global climate and how plants import organic food into nonphotosynthetic

(36) ©Linda Graham;
(37) ©William D. Bowman;
(38) ©Dwight Kuhn;
(39) ©Barry Mason/Alamy Stock Photo; (40) ©Michael Roach/Alamy Stock Photo

organs and tissues. Chapter 40 focuses on the molecular and cellular bases of plant reproduction, the process that generates seeds and fruits. This chapter ties together key concepts presented throughout the unit: the seed-to-seed life cycle of plants and their distinctively structured bodies, plant development and growth in response to environmental and hormonal influences, and plant acquisition and transport of materials that support growth and reproduction.

 **The following Core Concepts and Core Skills will be emphasized in this unit:**

- *Energy and Matter: Though sunlight is the major source of energy for photosynthetic plants, hundreds of flowering plant species as well as diverse tissues occurring in all flowering plants are heterotrophic and thus require a supply of organic food as a source of energy.*
- *Systems: Some of the ways in which plants detect and respond to light and other environmental factors are similar to those operating in microbes or animals, but others are distinctive.*
- *Energy and Matter: As is the case for animals, it is essential for flowering plants to maintain body water content and energy balance within tolerance limits, or they will die.*
- *Information: Some of the cellular and molecular bases of plant growth and development are also found in microbes and animals, but plants display some unique growth and development modes.*
- *Structure and Function: Have you ever wondered why and how trees become so tall? The unique features of plant bodies as well as variations occurring among plants explain how flowering plants function in nature and in human agriculture.*
- *Process of Science: A modern understanding of flowering plant structure, function, and behavior—such as flower blooming—has been derived from experimental studies, some of which are described in Feature Investigations in this unit.*
- *Modeling: Every chapter has a Modeling Challenge to help you refine this important skill.*

# An Introduction to Flowering Plant Form and Function

# 36

**Cotton is an economically important crop harvested from the seeds of a flowering plant.** ©Linda Graham

This chapter provides an introduction to the flowering plants, focusing on fundamental principles of body form and function—anatomy and physiology. These principles are basic to human efforts to use genetic methods to improve crops in ways that enhance human society, such as developing cotton plants that resist pest attack. The principles of plant form and function are equally important to evolutionary biologists who seek to understand how and why variations in plant body structure arise, and to ecologists who want to know how plants respond to environmental changes.

We will begin by considering the seed-to-seed life of a flowering plant, from seed germination to the production of a new generation. Next, an overview of how plants grow and develop reveals some fascinating similarities to animals, but also intriguing differences. Finally, centering our attention on the adult plant body reveals how adaptation to different environments has generated diverse species whose form and function vary dramatically. This background supports subsequent chapters in this unit that focus on flowering plant behavior, nutrition, transport, and reproduction.

**A**nyone who seeks to improve human life by reducing the effects of disease, producing more food, or improving our environment needs to know something about plant form and function. That's because humans depend on flowering plants not only for nutritious food, fibers such as cotton and linen, wood and paper, medicines, and biofuels, but also for plentiful fresh air and clean water. Knowledge of basic plant form and function, including how plants develop from and produce seeds, is essential to human society.

The flowering plant *Gossypium*, shown in the chapter opening photo, provides an example. Flowering plants are distinguished by flowers that produce fruits containing seeds. As *Gossipium* seeds mature, their surfaces develop a thick blanket of 10,000–20,000 cellulose-rich hairs that may aid dispersal to environments favorable for seedling growth. Such seed hairs form the valuable commodity called cotton, which is harvested from mature *Gossypium* seeds as the fruit opens. Understanding seed hair form and function fosters agricultural production of this renewable organic material, which has many uses.

## 36.1 From Seed to Seed—The Life of a Flowering Plant

**Learning Outcomes:**

1. List ways in which flowering plant reproduction and growth resemble or differ from those of animals.
2. Describe how the two major types of flowering plants—monocots and eudicots—differ in form.
3. **CoreSKILL »** Distinguish among annual, biennial, and perennial plants, and explain how their differences influence human uses of plants as food.

Several major events punctuate the lives of flowering plants, also known as the **angiosperms**. When seeds germinate, dormant embryos begin metabolic activity and start the process of seedling development. Seedlings grow and develop into mature plants capable of reproduction. Finally, flowers produce fruits that disperse the next generation of seeds. In this section, we will briefly survey the life cycle of flowering plants, focusing on the basic structural features of each life stage.

## Seedlings Develop from Embryos in Seeds

**Seeds** are reproductive structures produced by flowering plants and other seed plants, usually as the result of sexual reproduction. Seeds contain embryos that develop into young plants—seedlings—when seeds germinate. As in animals, the embryo is an essential stage in the sexual cycle of plants. The plant sexual cycle explains how embryos typically arise (**Figure 36.1**; see also Figure 32.14).

Unlike animals, sexual reproduction in plants requires two multicellular stages: a gamete-producing **gametophyte** and a spore-producing **sporophyte**. In the life cycle of plants, these two life stages alternate with one another in a process called **alternation of generations**. Flowering plants produce relatively large sporophytes and microscopic gametophytes that grow and develop within flowers (see Figure 36.1). Diploid sporophytes produce haploid spores by the process of meiosis. These spores grow into gametophytes that produce plant gametes—eggs and sperm. Fusion of egg and sperm in the process of fertilization generates a diploid zygote, which undergoes repeated mitotic divisions to form the plant **embryo**.

The plant embryo is a very young sporophyte that lies dormant within a seed, accompanied by a supply of stored food and enclosed by a tough, protective seed coat (**Figure 36.2a**), much like the animal amniotic egg. Like an eggshell, the seed coat protects the delicate plant embryo during the dispersal of seeds from parent plants into the environment. Dispersed seeds may remain dormant in the soil—sometimes for long periods—and then germinate when temperature, moisture, and light conditions are favorable. Such conditions activate embryo metabolism, in which stored food is respired for energy needed for cell division and growth.

As is the case for animals, **growth** of plants is an increase in weight or size, and **development** is a series of changes in the state of a cell, tissue, organ, or organism. Enlarging plant embryos break the seed coat and grow into seedlings (**Figure 36.2b**). If sufficient resources such as water and minerals are available, seedlings may develop into mature sporophyte plants (**Figure 36.2c**).

The angiosperm plant body is simpler in form than most animal bodies and is composed of only three types of organs: stems, leaves, and roots. **Stems** produce leaves and branches and bear the reproductive structures of mature plants. **Leaves** are flattened structures that emerge from stems and are often specialized in ways that enable photosynthesis. Stems and leaves together make up the plant **shoot** (see Figure 36.2b). Mature plants often possess multiple stems bearing many leaves, which together form the **shoot system**. **Roots** provide anchorage in the soil and also foster efficient uptake of water and minerals. The aggregate of a plant's roots make up the **root system** (see Figure 36.2c).

As in animals, the process of body and organ development in plants involves the differentiation of specialized cells having distinctive structures and functions. But unlike animals, plant seedlings and mature plants produce new tissues in specific areas called meristems. A **meristem** (from the Greek *merizein*, meaning to divide) is a region of undifferentiated cells that produces new tissue by cell division. A dormant meristem occurs at the shoot and root tips of seed embryos, and these meristems become active in seedlings (see Figure 36.2a and b). Active meristems also occur at stem and root tips of mature plants. Such meristems are known as shoot and root **apical meristems** because they occur at shoot and root tips, also known as apices.

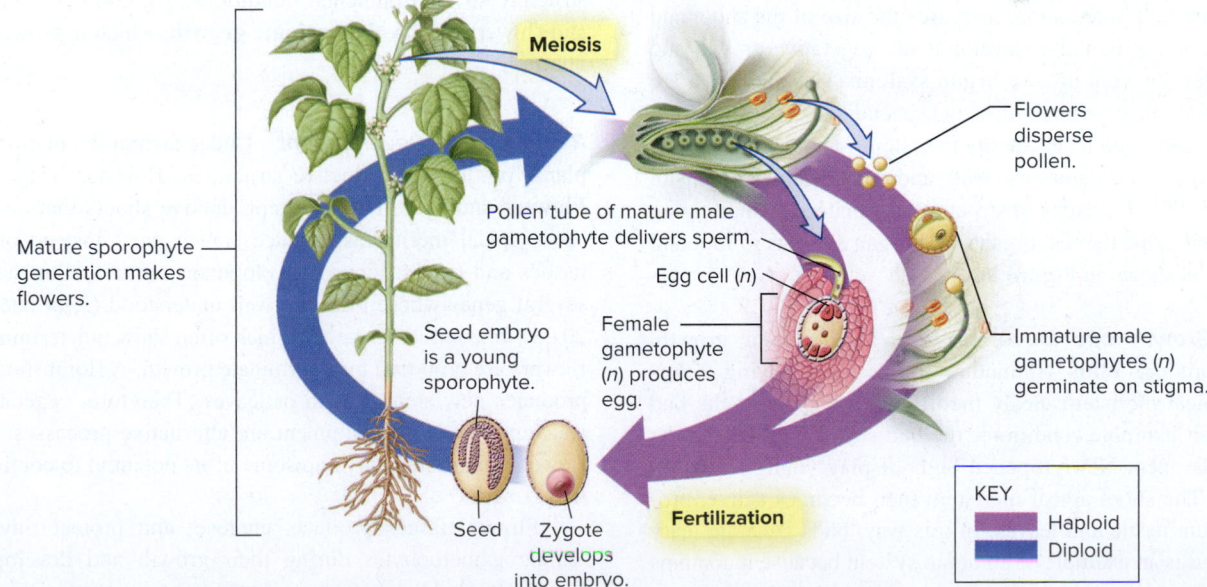

Meiosis

Flowers disperse pollen.

Mature sporophyte generation makes flowers.

Pollen tube of mature male gametophyte delivers sperm.

Egg cell (*n*)

Seed embryo is a young sporophyte.

Female gametophyte (*n*) produces egg.

Immature male gametophytes (*n*) germinate on stigma.

Seed    Zygote develops into embryo.

Fertilization

KEY
Haploid
Diploid

**Figure 36.1  The plant sexual cycle.** The sexual cycle of flowering plants involves alternation of sporophyte and gametophyte generations. In flowering plants, the spore-producing sporophyte is the dominant, conspicuous generation, whereas the tiny gametophytes are mostly hidden within flowers. Flowers produce haploid spores via meiosis. These spores undergo mitosis to grow into male or female gametophytes. Male gametophytes (pollen) contain sperm, and female gametophytes contain egg cells that are fertilized by sperm delivered by male gametophytes. Zygotes develop into embryonic plants that are dispersed in seeds.

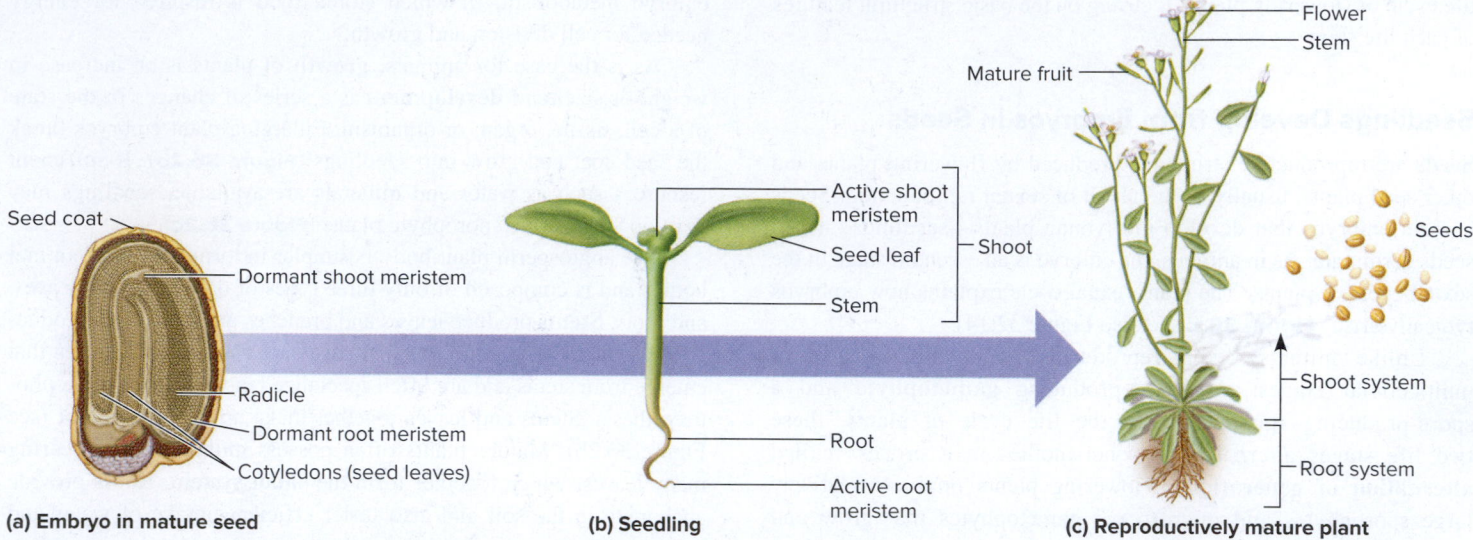

(a) Embryo in mature seed

Seed coat
Dormant shoot meristem
Radicle
Dormant root meristem
Cotyledons (seed leaves)

(b) Seedling

Active shoot meristem
Seed leaf
Stem
Shoot
Root
Active root meristem

(c) Reproductively mature plant

Flower
Stem
Mature fruit
Seeds
Shoot system
Root system

**Figure 36.2**  **The seed-to-seed life cycle of flowering plants, illustrated by *Arabidopsis thaliana*. (a)** Seed embryos possess embryonic leaves, known as cotyledons; a dormant shoot meristem; an embryonic root, known as a radicle; and a dormant root meristem. **(b)** When seeds germinate, the shoot and root meristems become active. Meristem activity allows the radicle to produce the seedling root and the young shoot of the seedling to grow and produce leaves. **(c)** Reproductively mature plants have branched shoot and root systems and bear flowers and fruits that disperse seeds.

 **Core Skill: Process of Science**  The plant *Arabidopsis thaliana* is widely used as a model organism for understanding the genetics of plant form and development.

## Mature Sporophytes Develop from Seedlings

As seedlings develop into mature sporophytes, the shoot typically becomes green and photosynthetic and thus able to produce organic food. Photosynthesis powers the transformation of seedlings into mature plants. The development of mature plants encompasses both **vegetative growth**, a process that increases the size of the shoot and root systems, and reproductive development. Vegetative growth and reproductive development involve **organ systems**, structures that are composed of multiple organs, tissues, and specialized cells. Branches, buds, flowers, seeds, and fruits are organ systems, analogous to organ systems such as the circulatory, skeletal, and reproductive systems in the animal body. The hierarchy of structure in a mature plant, ranging from specialized cells, tissues, organs, and organ systems to root and shoot systems, is shown in **Figure 36.3**.

*Vegetative Growth and Development*  During their growth, plant shoots produce **buds**—miniature shoots, each having a dormant shoot apical meristem. Scaly modified leaves protect the bud contents. Under favorable conditions, the bud scales fall off, and the vegetative buds open. Newly opened buds display young leaves on a short shoot. The shoot apical meristem then becomes active, producing new stem tissue and leaves. In this way, buds generate leafy branches. A bud is an example of an organ system because it contains more than one organ.

Vegetative shoots often display **indeterminate growth**, meaning that apical meristems continuously produce new stem tissues and leaves as long as conditions remain favorable. This process explains how very large plants, such as trees, can develop from seedlings (**Figure 36.4a**). However, plant size is also under genetic control, so

some plants remain small even when they are mature. The tiny floating plants of *Lemna* species, commonly known as duckweeds, which sometimes cover the surfaces of ponds in summer, are examples of plants whose small size is genetically determined (**Figure 36.4b**). Indeterminate growth allows plants to adapt their vegetative body structure to environmental conditions. By contrast, animal bodies and flowers display **determinate growth**, which is growth of limited duration.

*Reproductive Development*  Under favorable conditions, mature plants produce reproductive structures: flowers, seeds, and fruits. **Flowers** and floral buds are reproductive shoots that develop when shoot apical meristems produce flower parts instead of new stem tissues and leaves. Flower development occurs under the control of several genes whose roles are well understood (refer back to Figure 20.24). In contrast to shoots, which often show indeterminate growth, flowers are produced by determinate growth. A floral shoot no longer produces new stem growth or leaves. Therefore, vegetative growth and reproductive development are alternative processes. In order to flower, a plant must give up some of its potential to continue vegetative growth.

Flower tissues produce, enclose, and protect tiny male and female gametophytes during their growth and development (see Figure 36.1). Female gametophytes contain eggs within structures known as ovules, produced in the ovary of a flower pistil. Male gametophytes begin their development within pollen grains produced in the anthers of a flower stamen. Pollen is dispersed to the flower pistil, where the pollen grains may germinate, producing a tube that delivers sperm to eggs. Fertilization generates zygotes, which develop into

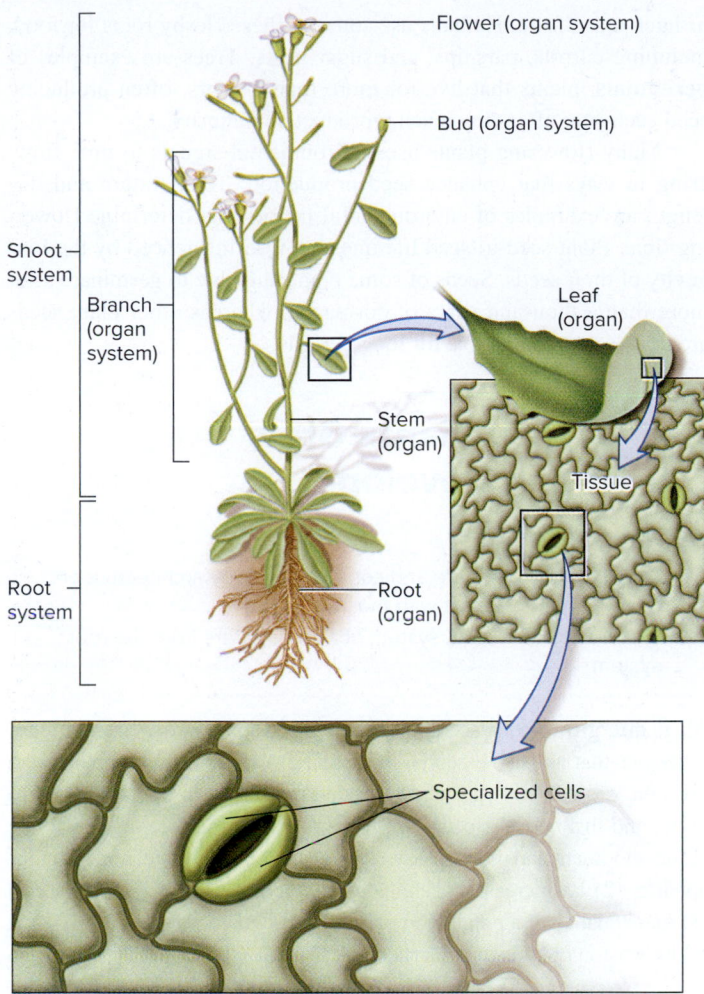

**Figure 36.3** **Levels of biological organization in a plant.** Flowering plant sporophyte bodies consist of a root system and a shoot system. Shoot systems produce organ systems such as buds, flowers, fruits, and seeds, which are composed of organs, tissues, and specialized cells. Root systems are likewise composed of organs, tissues, and specialized cells.

**(a)** *Brachychiton*, a tree native to Australia

**(b)** *Lemna* (common duckweed)

**Figure 36.4** **Plants display indeterminate growth, but vary in size.** A large woody angiosperm **(a)** and tiny duckweed plants **(b)** display indeterminate growth, but genetic differences confer different plant dimensions. a: ©Linda Graham; b: ©Howard Rice/Getty Images

embryos, and also triggers the process by which ovules develop into seeds and flower parts develop into fruits. **Fruits** thus enclose seeds and function in seed dispersal. As noted earlier, flower buds, flowers, fruits, and seeds are organ systems because they consist of more than one organ. For example, flowers typically contain several leafy organs, including sepals and petals, as well as stamens and pistils, which evolved from leaves (refer back to Figure 32.15).

## Flowering Plants Vary in the Structure of Organs and Organ Systems

With some exceptions, flowering plants are classified into two major groups, informally known as the **eudicots** and the **monocots** (refer back to Figure 32.18). These groups take their names from the number of seed leaves (cotyledons) that are present on seed embryos. For example, bean plants and relatives, which possess two (*di*) seed leaves, are examples of eudicots. Most woody trees, shrubs, and vines

are also eudicots. Corn, which has only one (*mono*) seed leaf, is an example of a monocot, as is tiny duckweed (see Figure 36.4b). Eudicots and monocots also vary in the structure of other organs and organ systems. For example, eudicot flowers typically have petals and other parts numbering four, five, or a multiple of those numbers, whereas monocot flower parts usually occur in threes or a multiple of three. Stems, roots, leaves, and pollen of eudicots and monocots also vary in distinctive ways, as shown in **Table 36.1**.

## Flowering Plants Vary in Seed-to-Seed Lifetime

The lifetime of a flowering plant can vary from a few weeks to many years. Plants that die after producing seed during their first year of life are known as **annuals**. Corn and the common bean are examples of annual crops whose nutrient-rich seeds are harvested within a few months after planting and must be replanted at the beginning of each new growing season. Plants that do not reproduce during the first year of life but may reproduce within the following year are known as **biennials**. Such plants often store food in fleshy roots during the first year of growth, and this food fuels reproduction during the second

## Table 36.1 Distinguishing Features of Eudicots and Monocots, Two Major Groups of Flowering Plants

| Feature | Eudicots | Monocots |
|---|---|---|
| Number of seed leaves (cotyledons) | Two | One |
| Number of flower parts | Usually four, five, or multiples of these | Usually three or a multiple of three |
| Stem vascular bundles | Arranged in a ring | Scattered |
| Root system | Branched taproot | Fibrous; adventitious |
| Leaf venation | Netted or branched | Often parallel |
| Pollen | Three pores or slits | One pore or slit |

**Core Skill: Modeling** The goal of this modeling challenge is to sketch a diagram that shows the two major lineages of angiosperms, monocots and eudicots, and helps to explain why nearly three-quarters of flowering plant species are eudicots.

**Modeling Challenge:** Use Figure 32.16, reproduced at the right, to make a simplified sketch that shows only the two major lineages (monocots and eudicots). Use an arrow to indicate an event that may have led to the increased evolutionary diversification of eudicots.

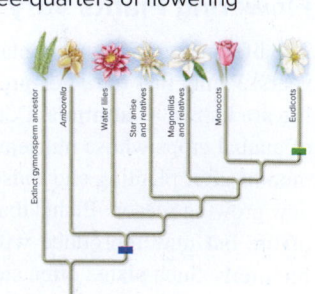

or later year of life. Humans use some of these fleshy roots for food, including carrots, parsnips, and sugar beets. Trees are examples of **perennials**, plants that live for more than 2 years, often producing seed each year after they reach reproductive maturity.

Many flowering plants use environmental signals to time flowering in ways that enhance seed production. Temperature and day length are examples of environmental factors that determine flowering time. Plant seed-to-seed lifetimes are also influenced by the longevity of their seeds. Seeds of some plants are able to germinate after more than a thousand years of dormancy, whereas other plant seeds are unable to remain alive for long periods.

## 36.2 How Plants Grow and Develop

### Learning Outcomes:

1. **CoreSKILL »** Compare and contrast the body architecture and development of plants with those of animals.
2. Explain how the shoot system of a plant differs from the root system.

As plants grow from seedlings, their development depends on four processes that are also essential to animal growth and development: cell division, growth, cell specialization, and programmed cell death. Additional and distinctive aspects of plant growth and development include (1) development and maintenance of a plant-specific architecture throughout life, (2) an increase in length by the activity of apical meristems, (3) maintenance of a population of youthful stem cells in meristems, and (4) expansion of cells in controlled directions, by water uptake.

### Plants Display a Distinctive Architecture

In plant biology, the term apical has two distinct meanings. As we have seen, apical refers to the tips (apices) of shoots and roots, as in shoot apical meristems or root apical meristems. A second meaning for apical is the part of a plant that typically projects upward, which is the top of the shoot. By contrast, the bottom of a root is termed the basal region. So the shoot apical meristem occurs at the apical pole, and the root apical meristem occurs at the basal pole. This property, known as **apical-basal polarity**, explains why plants produce shoots at their tops and roots at their lower regions.

Apical-basal polarity originates during embryo development. As seedlings and maturing plants grow in length by the activity of shoot and root meristems, apical-basal polarity is maintained (**Figure 36.5a**). Animals likewise have anterior and posterior ends, whose development is influenced by *Hox* genes (refer back to Chapter 20). In contrast, plant apical-basal polarity is under the control of genes such as *GNOM*; mutations in these genes result in plant embryos that are cone-shaped or spherical and thus lack normal apical-basal architecture (**Figure 36.5b**).

A second anatomical feature of plants is **radial symmetry**. Plant embryos normally display a cylindrical shape, also known as an axis, which is retained in the stems and roots of seedlings and mature plants. A thin slice or cross section of an embryo, stem, or root is typically circular in shape. Most plants produce new leaves or flower parts in circular whorls, or spirals, around shoot tips (**Figure 36.6**).

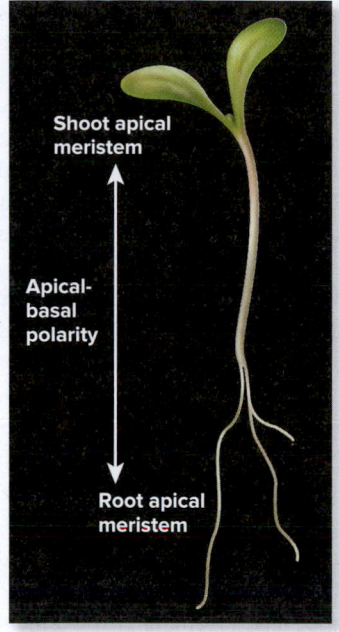

(a) Normal seedling

0.3 mm

(b) Abnormal *GNOM* mutants

**Figure 36.5  Plant apical-basal polarity.  (a)** Normal plants exhibit apical-basal polarity, as shown by this seedling. Growth occurs at two meristems, one at the shoot and one at the root. **(b)** *GNOM* mutants of *Arabidopsis thaliana* lack apical-basal polarity and thus produce abnormal embryos and seedlings. b: ©Prof. Dr. Gerd Jürgens/Universität Tübingen. Image Courtesy Hanno Wolters

 **Core Skill: Connections**  Look back to Figure 20.17 to see the relationship between expression of *Hox* genes and body organization in the mouse, as a model animal. How is animal body polarity similar to that of plants?

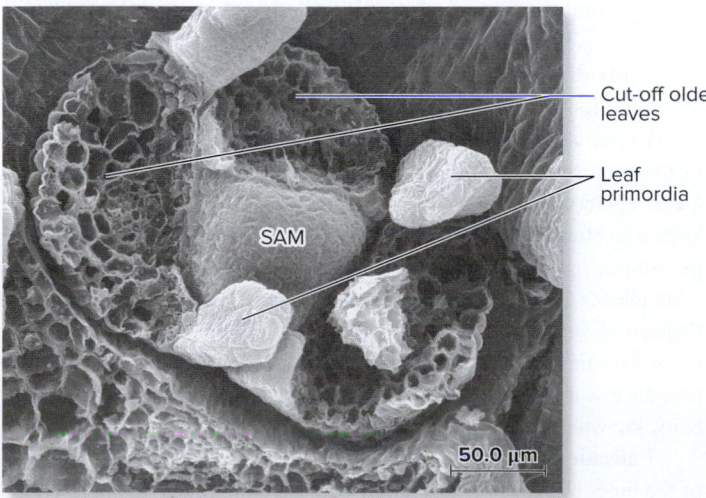

Cut-off older leaves

Leaf primordia

SAM

50.0 μm

**Figure 36.6  Plant radial symmetry.**  This top-down SEM view of a shoot apical meristem (SAM) reveals the radial symmetry of the shoot, indicated by its cylindrical shape. Leaf primordia are produced in circles or spirals around the shoot axis. ©Beth Krizek

Buds and branches likewise emerge from stems in radial patterns, as do lateral roots from a central root axis. Together, apical-basal polarity and radial symmetry explain why diverse plant species have a fundamentally similar architecture. Although radial symmetry also characterizes early-diverging animals, most animals exhibit bilateral symmetry (look back at Figure 33.4).

## Primary Meristems Increase Plant Length and Produce Plant Organs

We previously discussed how plant embryos grow into seedlings by adding new cells from distinctive growth points, the **shoot apical meristem (SAM)** and the **root apical meristem (RAM)**. During plant development, the SAM and RAM of the embryo give rise to many apical meristems located in the buds of shoots and at the tips of roots.

A SAM produces tissues that increase plant length and generate new organs. Such meristems are known as the **primary meristems**, and in a process known as **primary growth**, they ultimately produce primary tissues and organs of diverse types (**Table 36.2**). Tissues differ in their cellular complexity. Simple primary tissues are those composed of only one or two cell types; complex primary tissues are

| Table 36.2 | Examples of Tissues and Specialized Cells Found in Flowering Plants* |
|---|---|
| **Primary Growth** | |
| **Simple primary tissues (composed of one or two cell types)** | **Plant cell types found in those tissues** |
| Parenchyma | Parenchyma cells |
| Collenchyma | Collenchyma cells |
| Sclerenchyma | Fibers and sclereids |
| Root endodermis | Endodermal cells |
| Root pericycle | Pericycle cells |
| **Complex primary tissues (composed of at least two cell types)** | **Plant cell types found in those tissues** |
| Leaf or stem epidermis | Flattened epidermal cells, trichomes, stomatal guard cells |
| Root epidermis | Flattened epidermal cells, root hairs |
| Leaf mesophyll | Spongy parenchyma cells, palisade parenchyma cells |
| Leaf, stem, or root xylem | Tracheids, vessel elements, fibers, parenchyma cells |
| Leaf, stem, or root phloem | Sieve-tube elements, companion cells, fibers, parenchyma cells |
| **Secondary Growth** | |
| **Simple and complex secondary tissues** | **Plant cell types found in those tissues** |
| Secondary xylem (wood) | Tracheids, vessel elements, fibers, parenchyma |
| Secondary phloem (inner bark) | Sieve-tube elements, companion cells, fibers, parenchyma |
| Outer bark | Cork cells |

*This list does not include all of the tissues and cell types found in flowering plants. Some of these examples will be described later in this chapter.

made of more cell types. As described in Section 36.3, the primary meristems of woody plants also give rise to secondary or lateral meristems. In a process known as **secondary growth**, the secondary meristems increase the girth of woody plant stems and roots by producing secondary tissues (Table 36.2).

Plant biologists have discovered that plant cell specialization and tissue development do not depend on the lineage (the parentage) of a cell or tissue. Chemical influences such as hormones, proteins known as transcription factors, and microRNAs (miRNAs) that move through plants are much more important in determining the type of specialized tissue produced by unspecialized plant cells.

*Primary Stem Structure and Development*  New primary stem tissues arise via cell divisions in shoot apical meristems. A layer of outermost tissue known as the **epidermis** develops at the stem surface. The epidermis produces a waxy surface coating known as the **cuticle**, which helps to reduce water loss from the plant surface and to protect plants from damage by ultraviolet (UV) light, animals, and disease-causing microorganisms.

Beneath the epidermis lies the stem **cortex**, which is largely composed of a type of tissue called **parenchyma** (**Figure 36.7a**). This tissue is composed of only one cell type, thin-walled cells known as **parenchyma cells**. These cells often store starch in plastids and

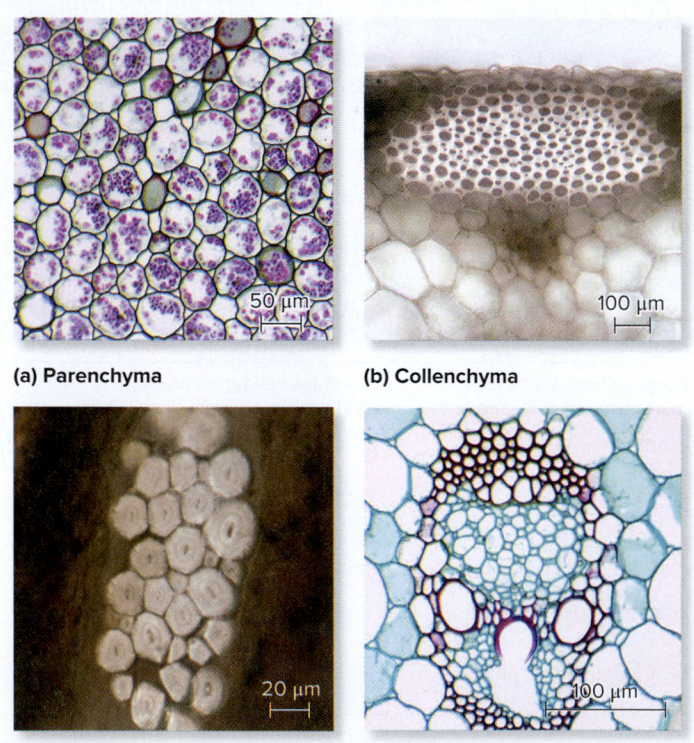

**(a) Parenchyma**

**(b) Collenchyma**

**(c) Sclerenchyma**

**(d) Vascular bundle**

**Figure 36.7**  **Examples of tissues produced by primary shoot meristems.**  **(a)** Parenchyma, **(b)** collenchyma, **(c)** sclerenchyma, and **(d)** a vascular bundle composed of complex xylem and phloem tissues. The purple-stained particles in parenchyma cells are stored starch. The flexible walls of collenchyma cells appear white here, as do the more rigid and thicker walls of sclerenchyma cells. In the vascular bundle, darker-stained and thicker walls of xylem cells contrast with lighter-stained and thinner-walled cells of phloem. a–d: ©Lee W. Wilcox

therefore serve as an organic food reserve. Stem parenchyma also has the ability to undergo cell division (meristematic capacity), which aids wound healing when stems are damaged. The cell division capability of stem parenchyma also explains how people are able to grow new plants from stem cuttings. Stems also contain **collenchyma** (**Figure 36.7b**), tissue composed of flexible **collenchyma cells**, and rigid **sclerenchyma** (**Figure 36.7c**), tissue composed of two types of tough-walled sclerenchyma cells termed **fibers** and **sclereids**. These tough cells provide strength and protection to the plant stem.

New water- and food-conducting tissues develop at the core of a young shoot. These conducting tissues are known as **primary vascular tissues** because they develop from new cells produced by the SAM. Vascular bundles, also known as veins, contain two types of complex tissues—xylem and phloem (see Section 36.3 and Table 36.2). Newly formed stem xylem and phloem connect with older conducting tissues that extend throughout the stem system. Stem xylem and phloem link to vascular tissues of the root system, forming a continuous route for conduction of water, minerals, and organic compounds through the plant.

In contrast to the circulatory system of most animals, the plant conduction system is not closed, but instead is open to the environment. The open plant conduction system is key to the transport of materials within plants (described more completely in Chapter 39). Primary vascular tissues are typically arranged in elongate clusters known as **vascular bundles** that appear round or oval when cross-cut (**Figure 36.7d**). In the primary stems of beans and other eudicots, the vascular bundles are arranged in a ring, which is easily seen in thin slices made across a stem (see Table 36.1). By contrast, in the stems of corn and other monocots, the vascular bundles are scattered.

*Leaf Structure and Development*  Young leaves are produced at the sides of a SAM as small bumps known as **leaf primordia** (see Figure 36.6). As young leaves develop, they acquire vascular tissue that connects to the stem xylem and phloem (**Figure 36.8a**) and leaves also become flattened. Leaf flattening expands the area of leaf surface available for light collection during photosynthesis. For some leaves, thinness is an adaptation that helps them to release excess heat. Leaves also become bilaterally symmetrical, meaning that they can be divided into two equal halves in only one direction, from the leaf tip to its base (**Figure 36.8b**).

Upper and lower leaf tissues develop differently in several ways that foster photosynthesis (**Figure 36.8c**). For example, the more shaded lower epidermis of a leaf usually displays larger numbers of pores, known as **stomata** (from the Greek word *stoma*, meaning mouth), than the sunnier upper surface (refer back to Figure 31.9b). When open, stomata allow $CO_2$ to enter and water vapor and $O_2$ to escape leaf tissues. Closure of stomata helps to prevent excess water loss from plant surfaces. Locating most stomata on the cooler, shadier leaf underside helps to reduce water loss by evaporation. Many plants produce surface leaf hairs, known as trichomes, that also help to reduce water loss.

**Palisade parenchyma** consists of closely packed, elongated cells of the inner leaf that are adapted to absorb sunlight efficiently (see Figure 36.8c). **Spongy parenchyma**, located closer to the lower leaf surface, contains rounder cells separated by abundant air spaces. These air spaces foster $CO_2$ absorption and $O_2$ release by leaves. Together, the palisade and spongy parenchyma are known as the leaf **mesophyll**.

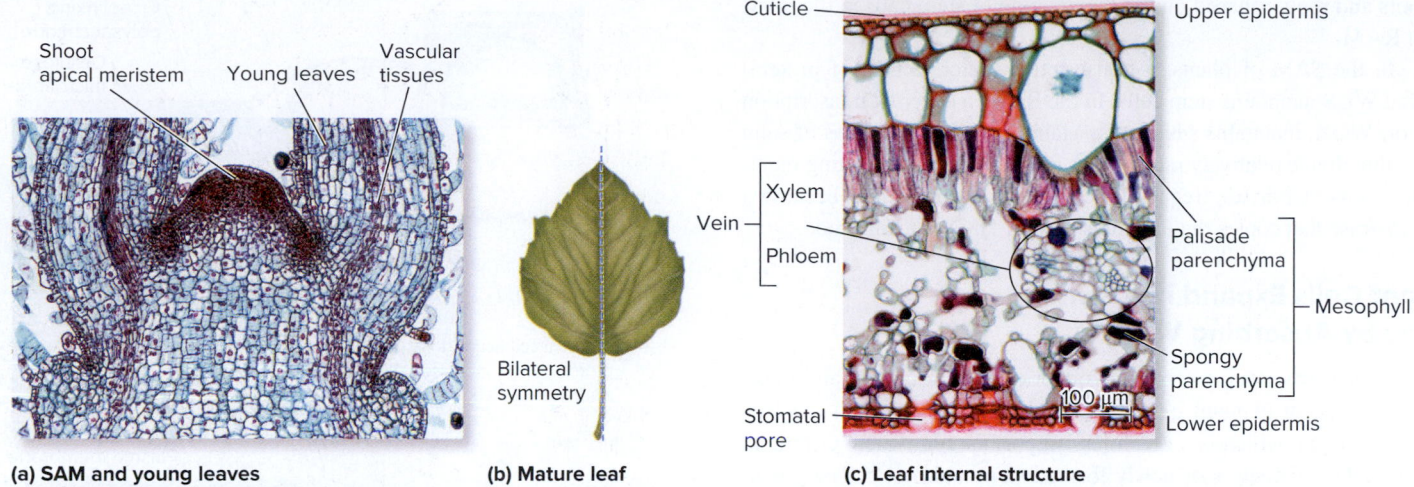

**Figure 36.8  Leaf development and structure. (a)** Young leaves develop at the sides of SAMs, as shown in this thinly sliced, stained shoot tip. The SAM is darkly stained because the cells at this location are small and densely packed. Note that the vascular tissues of young leaves, which also stain darkly, are connected to stem vascular tissues. **(b)** Mature leaves are typically thin and flat and show bilateral symmetry. **(c)** An internal view of a thinly sliced and stained leaf reveals tissue differentiation from the upper to the lower surface. A layer of palisade parenchyma lies just beneath the upper epidermis, which is capped with a waxy cuticle. Veins of conducting tissue (xylem and phloem) are embedded in the photosynthetic mesophyll. Spongy parenchyma lies above the lower epidermis, which displays stomata. These structural features of mature foliage leaves facilitate photosynthesis. a,c: ©Lee W. Wilcox

**Concept Check:**  *What advantage do plant leaves obtain by having stomata located on the lower epidermis?*

**Leaf veins** composed of vascular tissue commonly occur at the junction of palisade and spongy parenchyma or within the spongy parenchyma (see Figure 36.8c). Palisade parenchyma and spongy parenchyma are typically green and active in photosynthesis, a process that requires water, carbon dioxide, and dissolved minerals. Parenchyma cells can only take up carbon dioxide that has first dissolved into water, a feature inherited from ancient aquatic algal ancestors. Therefore, to perform photosynthesis, leaf parenchyma cells must be bathed in water. The xylem tissues of veins conduct water and minerals throughout leaf tissues, fostering photosynthesis. Phloem tissues of leaf veins carry the sugar products of photosynthesis from leaf cells to stem vascular tissues. In this way, sugar produced in leaves can be exported to other parts of the plant.

The leaves of flowering plants have one or more larger main veins, with smaller veins branching from them. The density of veins in angiosperm leaves is about four times that of other vascular plants. This allows plants to conduct materials more efficiently, which helps to explain why flowering plants are the dominant type of vegetation in most habitats today.

***Root System Structure and Development***  In beans and most other eudicots, a main root develops from the embryonic root and then produces branch roots, also known as lateral roots. Such a root system of eudicots is known as a **taproot system**, which has one main root with many branch roots (see Table 36.1). In contrast, the embryonic root of most monocots dies soon after seed germination, and it is replaced by a **fibrous root system** consisting of multiple roots that grow from the stem base (see Table 36.1). Fibrous roots are examples of **adventitious roots**, structures that are produced on the surfaces of stems (and sometimes leaves). They constitute the fibrous root system of most monocots and also can form on eudicots. Roots that develop at the bases of stem cuttings are also adventitious.

As noted earlier, the tips of roots and their branches each possess an apical meristem that adds new cells. Expansion of these new cells allows roots to grow into the soil. As they lengthen, roots produce branches but not from buds, as is the case for stems. Instead, branch roots develop from meristematic tissues located within the root (see Section 36.4). The root system both anchors the plant in the soil and plays an essential role in harvesting water and mineral nutrients. Root tissues are usually not green and photosynthetic, so must rely on organic compounds transported from the shoot. A plant's root system and shoot system depend on each other.

## Plant Meristems Contain Youthful Stem Cells

Plant meristems include undifferentiated cells referred to as **stem cells**. In the late 19th century, the Russian-born American biologist Alexander Maximow coined the term Stammzelle, which is derived from the German words *stamm*, meaning stem, such as a plant stem, and *zelle*, meaning cell. Maximow used the term stem cell to describe animal cells that remain undifferentiated but are able to generate specialized tissues. Animal stem cells are often in the news because of their potential for use in the treatment of human diseases that cause cell or tissue damage. The term stem cell is also widely used for cells located within the plant meristem that likewise remain undifferentiated but can divide and produce cells that generate new tissues. In the context of plant development, the term does not mean any cell located in a plant stem, only the undifferentiated cells located within the meristems of the shoot and root.

When plant stem cells divide, they produce two cells: one that remains young and unspecialized and divides relatively rarely, plus another cell. This second cell may differentiate into various types of specialized cells, but it often continues to divide, thereby adding new cells to shoot and root tips. The indeterminate growth typical of plant shoots and roots is based on the localization of stem cells in the SAM and RAM.

In the SAM of plants, a transcription factor (a type of protein) called WUS maintains stem cells. In the RAM, a different transcription factor, WOX, maintains stem cells. Maintaining apical caches of stem cells that divide relatively rarely helps plants avoid accumulating mutations. Cells that divide frequently tend to acquire mutations because a polymerase that copies the lagging strand of DNA is error-prone.

## Plant Cells Expand in a Controlled Way by Absorbing Water

As we have observed, meristem production of new cells is an important component of plant growth. In addition, plant growth involves cell expansion, which is a much less important component of animal growth. The diameters of newly formed stem and root cells are usually equal in all dimensions, but many soon begin to extend lengthwise, thereby helping shoots and roots to grow longer. Recall that plant cells typically possess a relatively large central vacuole (refer back to Figure 4.24a). Cell extension occurs when water enters the central vacuole by osmosis (**Figure 36.9**). As the central vacuole expands, the cell wall also expands and increases the cell's volume. By taking up water, plant cells can enlarge quickly, allowing rapid plant growth. Bamboo, for instance, can grow taller by 2 m within a week and can grow up to 30 m in less than 3 months! The importance of water uptake in cell expansion helps to explain why plant growth is so dependent on water supply.

Plant cell walls contain cellulose microfibrils that are held together by crosslinking polysaccharides. When plant cells and their vacuoles absorb water, pressure builds on cell walls. In response to this pressure and under acidic conditions, proteins unique to plants—known as expansins—are produced. **Expansins** unzip the crosslinking polysaccharides from cellulose microfibrils so that the cell wall can stretch (**Figure 36.10**). As a result, cells enlarge, often by elongating in a particular direction, which is important to plant form. Some plant cells are able to elongate up to 20 times their original length.

The direction in which a plant cell expands depends on the arrangement of cellulose microfibrils in its cell wall, which is in turn determined by the orientation of cytoplasmic microtubules. These

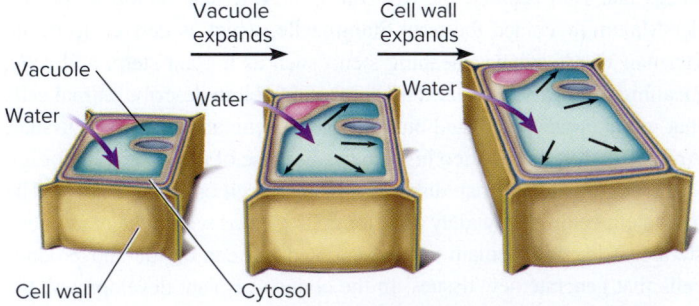

**Figure 36.9** **Plant cells expand by taking up water into their vacuoles.**

 **Core Skill: Connections** Look back at Figure 4.24a to see an electron micrograph of a plant vacuole and at Figures 4.9 and 4.11, which compares generalized animal and plant cells. How do plant cells differ from animal cells in terms of vacuoles?

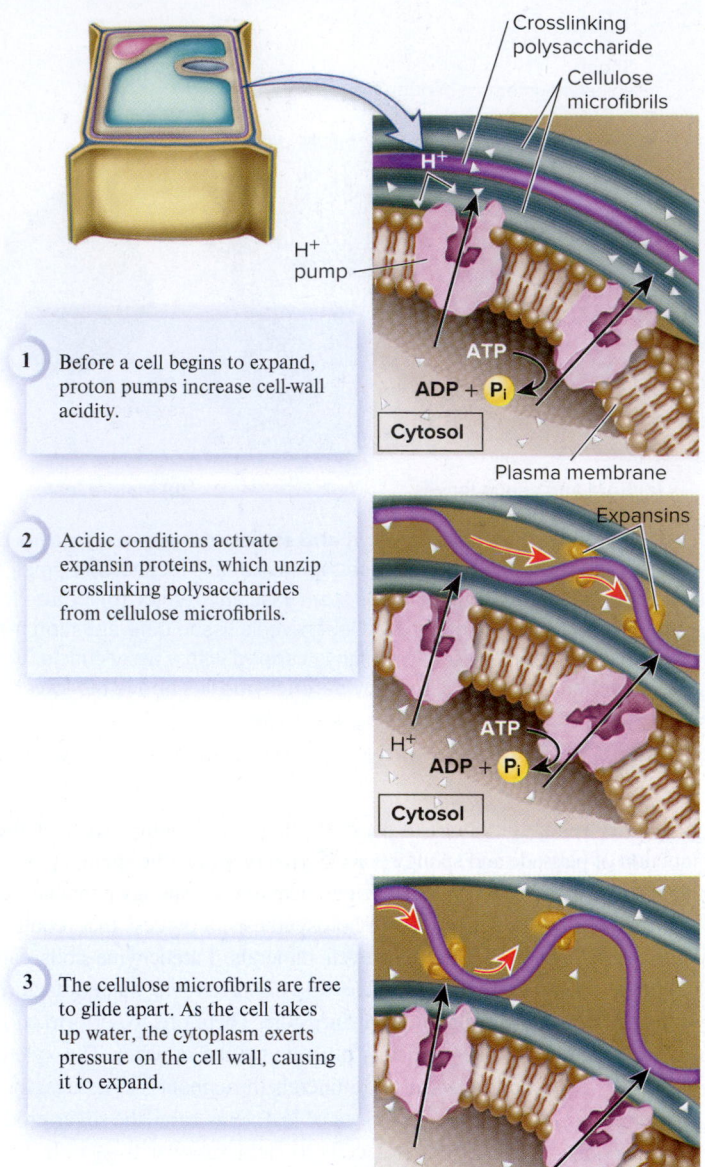

1. Before a cell begins to expand, proton pumps increase cell-wall acidity.

2. Acidic conditions activate expansin proteins, which unzip crosslinking polysaccharides from cellulose microfibrils.

3. The cellulose microfibrils are free to glide apart. As the cell takes up water, the cytoplasm exerts pressure on the cell wall, causing it to expand.

**Figure 36.10** **A hypothetical model of the process of cell-wall expansion.**

microtubules are thought to influence the positions of cellulose-synthesizing protein complexes located in the plant plasma membrane. The protein complexes connect sugars to form cellulose polymers, spinning cellulose microfibrils onto the cell surface to form the cell wall. As a result, cell-wall cellulose microfibrils encircle cells in the same orientation as underlying cytoplasmic microtubules (**Figure 36.11**). Because cellulose microfibrils do not extend lengthwise, plant cell walls expand more easily in a direction perpendicular to the microfibrils.

Microtubules control not only the direction of cell expansion but also the plane of cell division, which is also critical to plant form. Mutation of the *FASS* gene in the model plant *A. thaliana* illustrates the importance of microtubule orientation to plant structure. In cells of the mutant plants, microtubules are randomly arranged, causing cells to divide and grow abnormally, and producing plants with stubby organs.

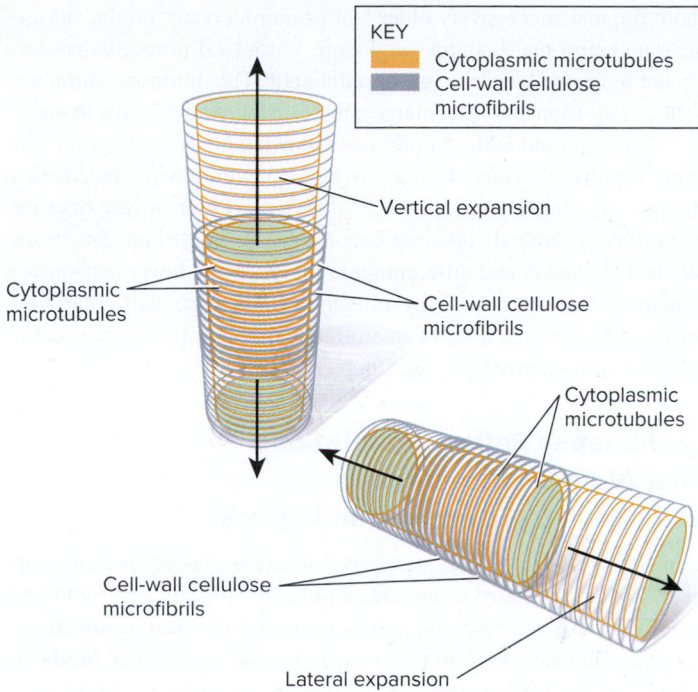

KEY
- Cytoplasmic microtubules
- Cell-wall cellulose microfibrils

Vertical expansion

Cytoplasmic microtubules

Cell-wall cellulose microfibrils

Cytoplasmic microtubules

Cell-wall cellulose microfibrils

Lateral expansion

**Figure 36.11  Control of the direction of plant cell expansion by microfibrils and microtubules.**  Plant cells enlarge in the direction perpendicular to encircling cell-wall cellulose microfibrils, which run parallel to the orientation of underlying cytoplasmic microtubules.

 **Core Skill: Connections**  Look back to Table 4.1 to see an image of microtubules. Are the structures of plant and animal microtubules different?

## 36.3  The Shoot System: Stem and Leaf Adaptations

### Learning Outcomes:

1. Explain why plant shoots are said to have a modular structure.
2. Explain why leaves having different shapes and vein patterns exist in nature.
3. Compare the structure and function of the conducting tissues xylem and phloem.
4. **CoreSKILL »** Create a drawing showing how bark and wood originate in woody plants.

As we have seen, the shoot system includes all of a plant's stems, branches, leaves, and buds. The shoot system also produces flowers and fruits when the plant has reached reproductive maturity. Thus, the shoot system is essential to plant growth, photosynthesis, and reproduction. Features of stems and leaves vary among plants in ways that explain their ecological functions and are useful in distinguishing plant species. In this section, we will examine the general features of shoot systems with an emphasis on stems and leaves.

### Shoot Systems Have a Modular Structure

More than 200 years ago, the German author, politician, and scientist Johann Wolfgang von Goethe realized that plants are modular organisms, composed of repeating units. Shoots are notably modular (**Figure 36.12**).

Each shoot module consists of four parts: a stem node, an internode, a leaf, and an axillary meristem or bud. A **node** is the stem

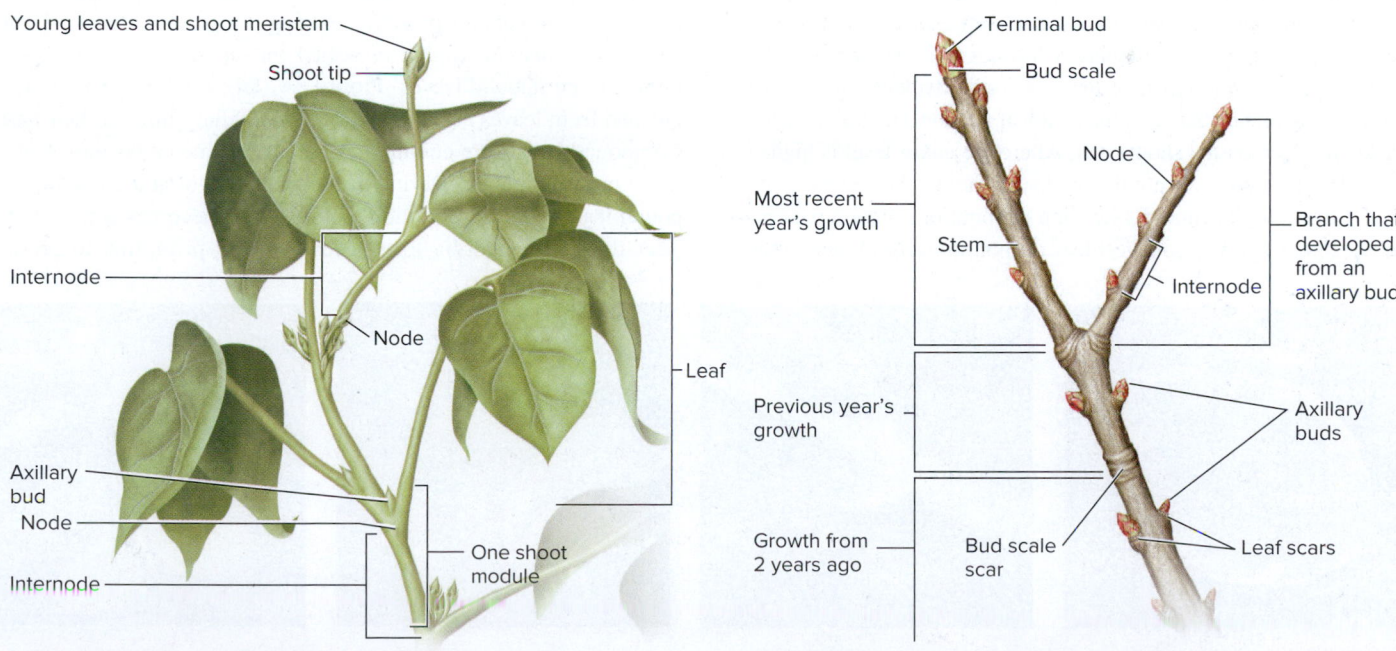

Young leaves and shoot meristem

Shoot tip

Internode

Node

Axillary bud

Node

Internode

Leaf

One shoot module

**(a) Modular structure of herbaceous shoot**

Terminal bud

Bud scale

Node

Most recent year's growth

Stem

Internode

Branch that developed from an axillary bud

Previous year's growth

Axillary buds

Growth from 2 years ago

Bud scale scar

Leaf scars

**(b) Modular structure of woody shoot in winter**

**Figure 36.12  The modular organization of plant shoots.**  **(a)** The top end of an herbaceous stem showing the shoot modules. Each module consists of a node with its associated leaf and axillary meristem or bud and an internode. **(b)** The modular organization shown by a woody stem as it appears during winter. Axillary buds lie above the scars left by leaf fall. Regions between successive sets of bud scale scars mark each year's growth

**Concept Check:**  *If a twig has five sets of bud scale scars, how old is the twig likely to be?*

region from which one or more leaves emerge. An **internode** is the region of stem between adjacent nodes. Differences in numbers and lengths of internodes help to explain why plants differ in height (look back to Figure 36.4).

Each time a young leaf is produced at a SAM, a new meristem develops in the upper angle formed where the leaf emerges from the stem. This angle is known as an axil (from the Greek *axilla*, meaning armpit), and the meristem formed there is called an **axillary meristem**. Such axillary meristems generate **axillary buds**, which can produce flowers or branches known as lateral shoots. Such new branches bear a SAM at their tips. SAMs located at the apices of both main and lateral shoots produce new leaves under the direction of chemical messengers known as hormones.

## Hormones, Mechanical Forces, and miRNAs Influence Leaf Development

As we have noted, leaf primordia are surface bumps of tissue that develop at the sides of a SAM (see Figure 36.6). Production of leaf primordia is under the control of a plant hormone known as an **auxin**. In general, **hormones** are signaling molecules that exert their effects at a site distant from the place where they are produced. Plant hormones are important in coordinating both plant development and plant responses to environmental conditions and are described more completely in Chapter 37.

The outermost epidermal layer of cells at shoot tips produces an auxin, which moves from cell to cell by means of specific membrane transport proteins. The auxin accumulates in particular locations because cells of the shoot apex differ in their ability to import and export the hormone. When auxin accumulates in a particular apical region, it causes increased expression of the gene that encodes expansin. When expansin allows their cell walls to stretch (see Figure 36.10), cells expand by taking up water, thereby forming a tissue bulge—a leaf primordium. The development of leaf primordia depletes auxin from nearby tissue, with the result that the next leaf primordium will develop in a different place on the shoot apex, where the auxin level is higher. Such changes in auxin concentration on the surface of the shoot explain why leaf (and flower) primordia develop in spiral or whorled patterns around the shoot tip. The youngest leaf primordia occur closest to the shoot tip, and successively older leaf primordia occur on the sides of the stem below the shoot tip (see Figure 36.6). Leaf primordia produce a plant hormone known as **gibberellic acid**. This hormone stimulates both cell division and cell enlargement, causing young leaves to grow.

Computer modeling studies have shown that, as the largest major veins rapidly elongate during early leaf development, mechanical stresses play important roles in leaf shaping. Later in leaf development, local chemical signaling becomes more important. For example, leaf flattening and differentiation of upper and lower leaf surface structures are determined by interactions between particular transcription factors and a set of microRNAs (miRNAs) (for more information about microRNAs, see Chapter 13).

## Leaf Shapes Reflect Adaptations That Aid in Photosynthesis and Alleviate Environmental Stress

Leaf flatness facilitates solar energy collection as described in Chapter 8. Leaf shapes also reflect adaptations to stressful environmental conditions. For example, thinness helps leaves to avoid overheating.

The flattened portion of a leaf is known as the leaf **blade**. In beans and most other eudicots, blades are attached to the stem by means of a stalk known as a **petiole**. An axillary bud occurs at the junction of stem and petiole (**Figure 36.13a**). In contrast, corn and other monocots have leaf blades that grow directly from the stem, encircling it to form a leaf sheath (**Figure 36.13b**).

Leaf shape can be simple or compound, each having particular advantages. Simple leaves have only one blade, though the edges may be smooth, toothed, or lobed (**Figure 36.13c**). Simple leaves are advantageous in shady environments because they provide maximal light absorption surface, but they can overheat in sunny environments. As an evolutionary response to heating stress, the blades of some leaves have become highly dissected into leaflets. Such leaves are known as compound leaves (**Figure 36.13d**). Leaflets can be distinguished from leaves because leaflets lack axillary buds at their bases. Compound leaves are common in hot environments because leaflets foster heat dissipation. During the development of at least some compound leaves, the transcription factor KNOX becomes active shortly after the leaf primordia form, causing these primordia to produce

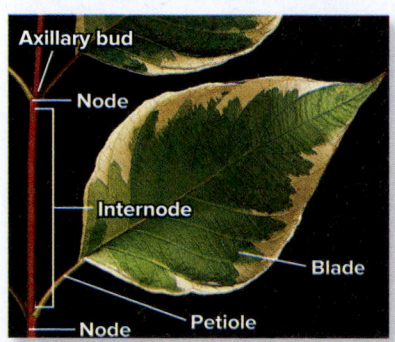

**(a) Eudicot stem with simple leaf having pinnate venation**

**(b) Monocot stem and leaf**

**(c) A simple leaf with palmate venation**

**(d) A compound leaf**

**Figure 36.13 Examples of variation in leaf form.** **(a)** A simple eudicot leaf, showing blade, petiole, and axillary bud. This leaf has a pinnate venation pattern. **(b)** The leaf of a monocot, showing parallel veins. The base of the leaf encircles the stem. **(c)** A simple leaf having palmate venation. **(d)** A compound leaf divided into leaflets. a–d: ©Lee W. Wilcox

multiple growth points that generate the leaflets. By contrast, during simple leaf development, KNOX is not active, because the presence of other proteins suppresses expression of the *KNOX* gene.

## Leaf Vein Patterns Can Be Pinnate or Palmate

Leaf vein patterns are known as **venation**. Eudicot leaves occur in two major venation forms. In **pinnate** venation, one main vein extends from the base to the tip of the leaf and smaller veins branch off the main vein (Figure 36.13a). Alternatively, several main veins may radiate from the base of the leaf like the fingers of your hand, a pattern known as **palmate** venation (Figure 36.13c). In eudicot leaves, small veins connect in a netted pattern, but most monocot leaves have a distinctive parallel venation (Figure 36.13b and Table 36.1).

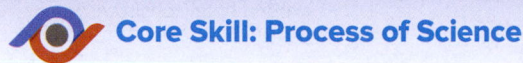

### Core Skill: Process of Science

## Feature Investigation | Sack and Colleagues Showed That Palmate Venation Confers Tolerance of Leaf Vein Breakage

In 2008, Lawren Sack and colleagues studied the adaptive value of leaf venation patterns by comparing water conduction after injury in pinnately and palmately veined leaves (**Figure 36.14**). Leaves with palmate venation have several main veins, whereas leaves with pinnate venation have just one main vein. The investigators hypothesized that the multiple main veins of palmate leaves could confer greater tolerance of vein breakage of the type that would occur during mechanical injury or insect damage; if one main vein were damaged, water flow could continue through the other main veins.

**Figure 36.14** **Sack and colleagues investigated the function of palmate venation.**

**HYPOTHESIS** Palmate venation provides vascular redundancy, which allows leaves to tolerate vein breakage.

**KEY MATERIALS** Seven species of trees or shrubs at Harvard Forest, Petersham, MA.

| Experimental level | Conceptual level |
|---|---|
| **1** Identify 7 species, 4 with pinnately veined leaves and 3 with palmately veined leaves. For each species, the researchers analyzed 10 leaves on 3 different plants. | 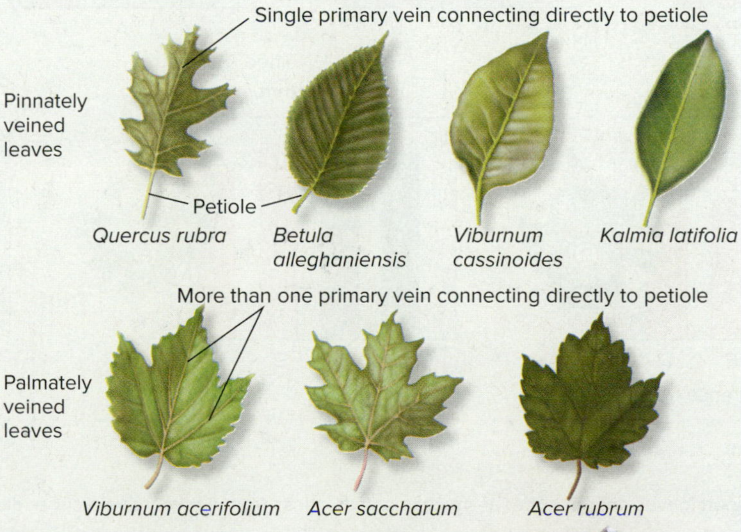 Single primary vein connecting directly to petiole<br><br>Pinnately veined leaves<br><br>Petiole<br>*Quercus rubra*  *Betula alleghaniensis*  *Viburnum cassinoides*  *Kalmia latifolia*<br><br>More than one primary vein connecting directly to petiole<br><br>Palmately veined leaves<br><br>*Viburnum acerifolium*  *Acer saccharum*  *Acer rubrum* | Locate pinnate and palmate leaves for comparison. |
| **2** With the leaf still attached to the plant, use a scalpel to cut across 1 primary vein in 5 experimental leaves but not 5 control leaves. |  | This procedure initially cuts off water supply via 1 primary vein. |

**3** Cover the cuts with medical tape on both top and bottom of leaves. Tape same area of uncut control leaves.

Tape

The tape prevents infection. Controls: Tape is applied to controls for experimental consistency.

**4** Fold cardboard over base of cut leaves, forming a splint. Splint the same area of uncut control leaves.

Splint

The cardboard prevents leaf collapse. Controls: Cardboard is applied to controls for experimental consistency.

**5** Measure water conduction 2–9 weeks after treatment, in 2 regions (A, B) of each pinnate leaf and in 3 regions (A, B, C) of each palmate leaf.

Pinnate leaf
Palmate leaf

After 2–9 weeks, the cuts had healed. Measurement of water conduction will determine the effect of the vein cut.

**6** THE DATA

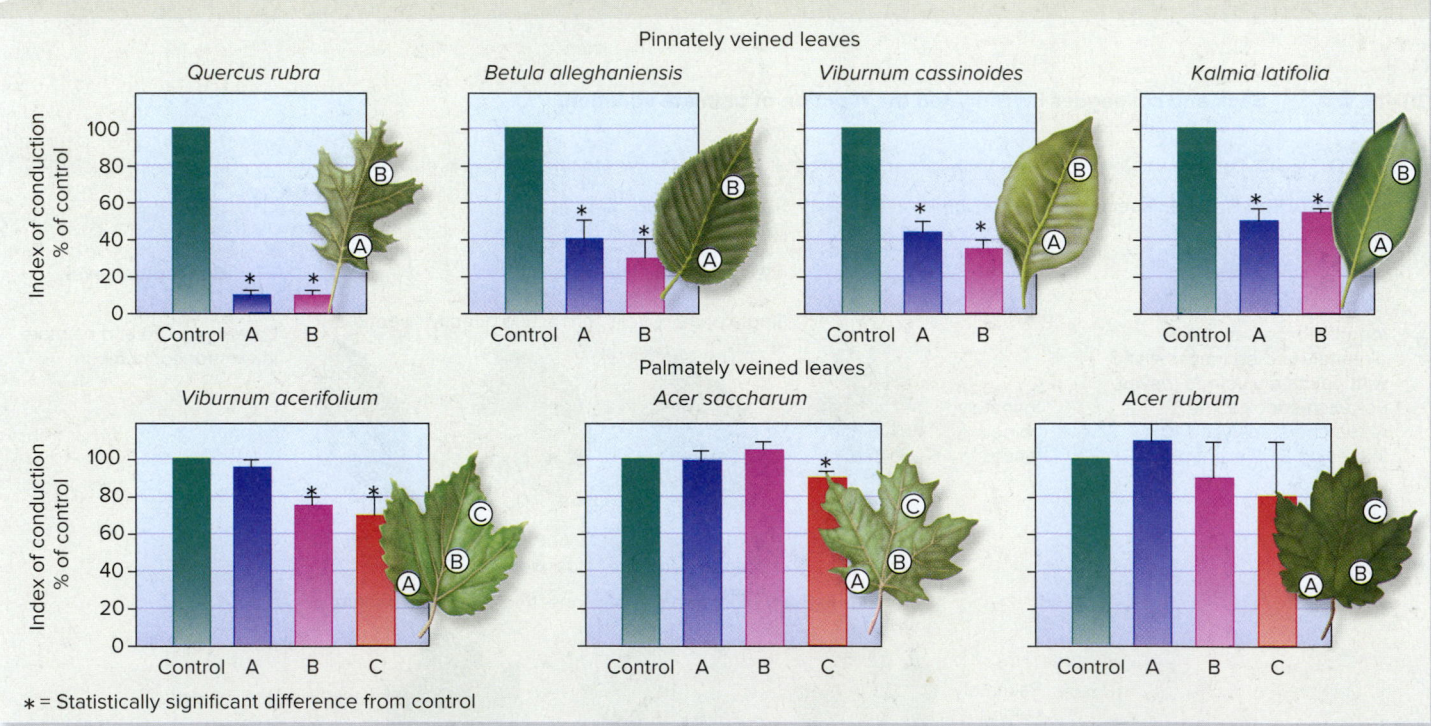

Pinnately veined leaves

*Quercus rubra*    *Betula alleghaniensis*    *Viburnum cassinoides*    *Kalmia latifolia*

Index of conduction % of control

Palmately veined leaves

*Viburnum acerifolium*    *Acer saccharum*    *Acer rubrum*

Index of conduction % of control

* = Statistically significant difference from control

**7** **CONCLUSION** Palmately veined leaves did not suffer as much conduction loss from a primary vein cut as did pinnately veined leaves.

**8** **SOURCE** Sack, L., et al. 2008. Leaf palmate venation and vascular redundancy confer tolerance of hydraulic disruption. *Proceedings of the National Academy of Sciences of the U.S.* 105:1567–1572. Copyright ©2008 National Academy of Sciences, U.S.A. Used with permission.

To test the hypothesis, the investigators experimentally cut a main vein in the leaves of several plants belonging to seven different plant species: four having pinnately veined leaves and three having palmately veined leaves. They conducted the experiments in vivo, that is, in living trees. After the wounds had healed, the investigators measured the water flow within the leaves at two or three places on each leaf (see steps 3–5 of (Figure 36.14). The rate of water flow, called the index of conduction, is expressed as the percentage of water flow of the control leaves, which are uncut leaves. As seen in the data, they found that across all species examined, palmately veined leaves tolerated the disruption in water flow better than pinnately veined leaves.

What are the advantages and disadvantages of these two venation patterns? Although palmate venation provides redundancy in case a main vein becomes damaged, it is more costly in terms of materials needed to construct the additional main veins. Leaves with pinnate venation are less costly to produce and work well when the potential for vein damage is low.

### Experimental Questions

1. Why did Sack and colleagues conduct their studies of palmate venation on plants growing in a forest rather than in a greenhouse?

2. **CoreSKILL »** Why did Sack and colleagues splint leaves having cut veins as well as controls?

3. **CoreSKILL »** Why did Sack and colleagues measure leaf water conduction at two or more places on each leaf?

---

 **BIO TIPS**

**THE QUESTION** *Which species of tree or shrub tested by Sack and colleagues showed the greatest tolerance to damaged primary veins? In other words, in which leaves was water flow least inhibited by leaf damage?*

**T OPIC** *What topic in biology does this question address?* The topic is variation in leaf venation, which many people first encounter as young children making leaf collections for school projects. More specifically, the question is about the relationship between leaf venation and the effects of damage on water flow.

**I NFORMATION** *What information do you know based on the question and your understanding of the topic?* In the question, you are reminded that Sack and colleagues did experiments to study the relationship between leaf venation pattern and the effects of vein damage on water flow. You also have the data in Figure 36.14, which compares the effects of leaf damage among seven different species.

**P ROBLEM-SOLVING S TRATEGY** *Interpret data.* The data in step 6 of Figure 36.14 are presented in the form of histogram bars, showing differences in water conduction among the species studied when leaf veins were cut. Look for the species that displays the least difference between control leaves and experimentally cut leaves. Pay particular attention to the indicator of statistical significance (the asterisk). In displays of numerical data, asterisks are commonly used to indicate significant differences in outcomes between experimentally manipulated organisms and unmanipulated controls.

**ANSWER** *Bars indicating results for leaves of pinnately veined species that were experimentally cut are all noticeably shorter than bars for control leaves, and all are marked with an asterisk,* indicating statistical significance. Among the palmately veined species, the histogram bars for experimentally cut leaves of Viburnum acerifolium are taller than those for pinnately veined species, but shorter than those for the other palmately veined species studied. For the Acer species, the heights of the histogram bars for the Acer saccharum leaves that were experimentally cut are approximately the same as the height of the control bar, except for the bar for leaves cut at point C, which shows a significant difference from the control. By contrast, the heights of the histogram bars for all manipulations of the palmately veined Acer rubrum are similar, and no bars display an asterisk indicating a significant difference from the control. These results indicate that, of the species examined, Acer rubrum showed the greatest tolerance to the experimental leaf cuts.

## Leaf Surface Features Prevent Desiccation, Provide Protection, and Aid in Photosynthesis

Leaf surfaces also show adaptive features. As we have previously discussed, a layer of epidermal tissue occurs at upper and lower leaf surfaces (see Figure 36.8c). These epidermal cells secrete a cuticle composed of protective wax and polyester compounds. The cuticle helps plants to avoid drying in the same way that enclosure in waxed paper keeps food moist. Plants that grow in very arid climates often have thick cuticles, whereas plants native to moist habitats typically have thinner cuticles.

As previously mentioned, some leaf epidermal cells may differentiate into protective spiky or hairlike projections known as **trichomes** (**Figure 36.15**). Broken trichomes of the stinging nettle, for example, release a caustic substance that irritates animals' skin, causing them to avoid consuming these plants. Production of leaf trichomes is under the control of transcription factors; similar proteins control the formation of hairs on seeds, as in cotton plants (see the chapter opening photo).

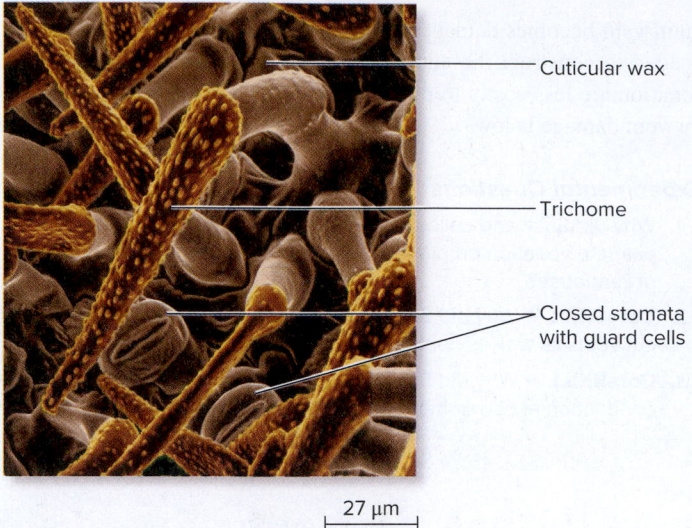

Cuticular wax

Trichome

Closed stomata with guard cells

⊢ 27 μm ⊣

**Figure 36.15 Leaf surface features.** The features shown in this artificially colored SEM are cuticular wax, trichomes, and stomatal pores with guard cells. ©Eye of Science/Science Source

Leaf epidermal cells include pairs of specialized guard cells located on either side of stomata (see Figure 36.15). These **guard cells** allow stomata to be open during moist conditions, which allows a leaf to take up $CO_2$ and release $O_2$ during photosynthesis. Alternatively, the stomata will close when conditions are dry, thereby preventing plants from losing too much water. The genetic basis of guard-cell development is becoming increasingly well understood, as described next.

👁 **Core Concept: Information**

## Genetic Control of Stomatal Guard-Cell Development

The flowering plant *Arabidopsis thaliana* is a model organism that is widely used to explore the genetic basis for plant structure and development. Several features of *A. thaliana* make it suitable for such studies. It is small in size, it has a fast seed-to-seed life cycle, it produces a relatively large number of seeds, and the

genome has been sequenced. Mutants of this plant have been used to identify the genes controlling many aspects of plant structure and development, including the development of specialized stomatal guard cells.

Guard-cell development begins when an unspecialized surface cell divides unequally (Figure 36.16). Unequal cell division is controlled by the protein RBR, which is related to the animal tumor suppressor Rb (Retinoblastoma). Plant cells having mutations in the gene that encodes RBR fail to differentiate normally. The larger of the two leaf cells arising from unequal division becomes a flat, puzzle piece–shaped epidermal cell, and the smaller is called a meristemoid because it functions as a stem cell. Meristemoids undergo one or more unequal cell divisions, producing more puzzle piece–shaped epidermal cells, before finally dividing equally to produce a pair of guard cells. Genetic studies of *A. thaliana* have revealed that the meristemoid secretes a protein that inhibits division by adjacent cells but does not affect cells farther away. This process distributes stomata evenly and prevents too many of them from forming, which could increase the loss of water from plant surfaces.

In 2007, two teams, led by Lynn Pillitteri and Cora MacAlister, respectively, independently reported experiments with *A. thaliana* showing how three closely related genes control guard-cell development at three consecutive steps. A gene called *SPEECHLESS* starts the process by establishing the first unequal cell divisions of meristemoids. A protein encoded by the *MUTE* gene then causes meristemoids to stop dividing unequally and produces the cell that will eventually divide to produce guard cells (guard cells precursor) (see **Figure 36.16**). Disabling mutations of these two genes cause the plant epidermis to completely lack stomatal pores (and so lack epidermal "mouths" and be speechless, or mute). Finally, the gene *FAMA* directs the production of guard cells and their specialization. The proteins encoded by these plant genes are members of a type known as basic helix-loop-helix (bHLH) proteins. Similar bHLH proteins control the development of muscle and nerve cells in animals, such as the protein encoded by *MyoD* (refer back to Figure 20.21).

## Modified Leaves Perform Diverse Functions

Though most leaves function primarily as photosynthetic organs, some plants produce leaves that are modified in ways that allow them to play other roles. For example, threadlike tendrils that help some plants attach to a supporting structure are modified leaves or leaflets (**Figure 36.17a**). The tough scales that protect buds on plants such as the sycamore from winter damage are modified

Leaf primordium

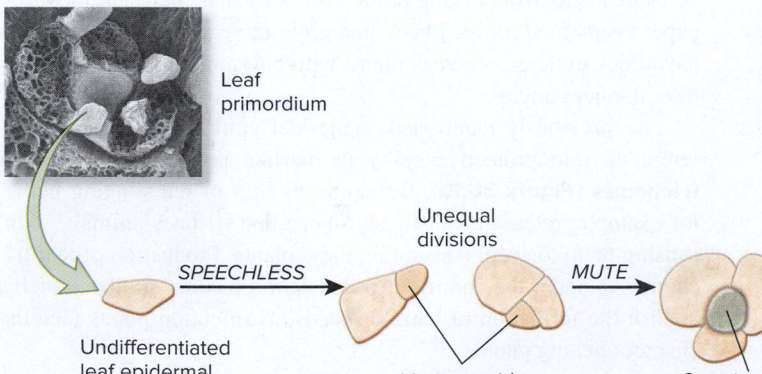

Undifferentiated leaf epidermal cell

*SPEECHLESS*

Unequal divisions

Meristemoids

*MUTE*

Guard cells precursor

Equal division

*FAMA*

Specialized guard cells

**Figure 36.16** The development of stomatal guard cells, controlled by three genes. ©Beth Krizek

**(a) Tendrils**        **(b) Bud scales**        **(c) Bracts**        **(d) Spines**

**Figure 36.17** **Examples of modified leaves.** **(a)** The tendrils of an American vetch plant are modified leaves that help the plant attach to a trellis. **(b)** Bud scales, such as those on this sycamore bud, are modified leaves that protect buds from winter damage. **(c)** The attractive red bracts of poinsettia are modified leaves that function like flower petals to attract pollinator insects to the small flowers. **(d)** Cactus spines, such as these on this giant saguaro, are modified leaves that function in defense. a: ©Ed Reschke/Getty Images; b: ©John Farmar; c: ©Steve Terrill/Corbis/Getty Images; d: ©Don Paulson Photography/Purestock/SuperStock

*Concept Check:* *Cactus leaves are so highly modified for defense that they cannot effectively accomplish photosynthesis, so how do cacti obtain organic compounds?*

leaves (**Figure 36.17b**). Poinsettia "petals" are actually modified leaves known as bracts, which are larger and more brightly colored than the flowers they surround, which helps to attract pollinators (**Figure 36.17c**). Cactus spines are actually modified leaves that have taken on a defensive role, leaving photosynthesis to the cactus stem (**Figure 36.17d**).

## Stems May Contain Primary and Secondary Vascular Systems

Stems, leaves, roots, buds, flowers, and fruits all contain vascular systems composed of xylem and phloem tissues that conduct water, minerals, and organic compounds. **Herbaceous plants** such as corn and bean produce mostly primary vascular tissues. In contrast, **woody plants** produce both primary and secondary vascular tissues. A comparison of primary and secondary vascular tissues will aid in understanding their roles.

*Primary Vascular Tissues*   Primary vascular tissues are composed of primary xylem and phloem. Primary xylem is a complex tissue containing several cell types (see Table 36.2). These include unspecialized parenchyma cells; stiff fibers that provide structural support; and two types of cells that facilitate water transport: narrower tracheids and wider vessel elements (look ahead to Figures 39.11 and 39.12). Arranged in pipeline-like arrays, **tracheids** and **vessel elements** conduct water, along with dissolved minerals, hormones, and some other organic substances. These materials pass from one tracheid or vessel element to another through thin areas in the cell walls known as pits (**Figure 36.18**).

Mature tracheids and vessel elements are no longer living cells, and the absence of cytoplasm facilitates water flow. During development, these cells lose their cytoplasm by programmed cell death, a

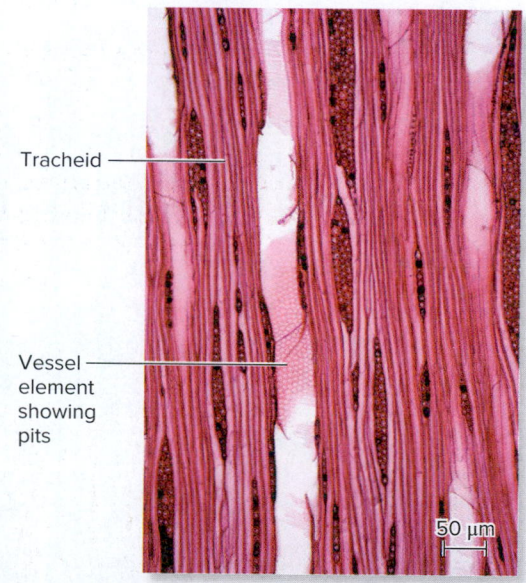

Tracheid

Vessel element showing pits

50 µm

**Figure 36.18** **Water-conducting cells of the xylem.** In this thinly sliced portion of a stem, the lignin-impregnated walls of narrow tracheids and wider vessel elements can be distinguished. In this image, the walls of both tracheids and vessels have been stained with a red dye. ©Lee W. Wilcox

 **Core Concept: Structure and Function** Because lignin strengthens tracheid and vessel walls, they do not collapse as large volumes of water move through them.

process that resembles apoptosis in animals. However, the cell walls of tracheids and vessel elements don't easily break down or collapse because they are impregnated with a tough polymer known as lignin, except at pits. The rigid cell walls of tracheids and vessel elements not only foster water conduction but also help support the plant body. Like the tough plastic used in plumbing pipes, lignin provides the hydrophobic surface needed for water movement, as well as the strength to support trees weighing more than 2,000 metric tons.

In contrast to the conducting cells of xylem, mature conducting cells of phloem are alive. Phloem tissue transports organic compounds such as sugars, amino acids and proteins, hormones, RNA, and certain minerals in a watery solution. Phloem tissue includes **sieve-tube elements**, thin-walled living cells that are arranged end to end to form pipelines (**Figure 36.19**). Pores in the end walls of sieve-tube elements allow the watery solution to move from one cell to another.

Phloem tissue also includes companion cells that aid sieve-tube element metabolism, supportive fibers, and parenchyma cells (see Table 36.2). Phloem fibers are tough-walled sclerenchyma cells that are surprisingly long, 20–50 mm, and valued for their high strength. The phloem fibers of hemp (*Cannabis sativa*), flax (*Linum usitatissimum*), jute (*Corchorus capsularis*), kenaf (*Hibiscus cannabinus*), and ramie (*Boehmeria nivea*) are commercially important in the production of rope, textiles, and paper.

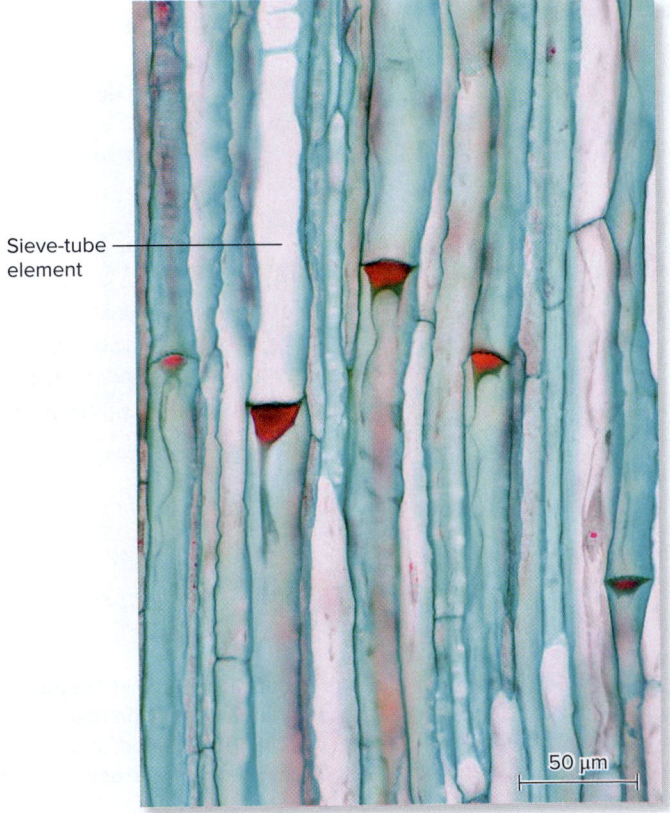

Sieve-tube element

50 μm

**Figure 36.19** **Food-conducting cells of the phloem.** This thinly sliced portion of a stem shows stained, thin-walled sieve-tube elements that conduct a watery solution known as phloem sap. This solution contains sugars and other organic compounds, including amino acids and proteins, hormones, and RNA. When sieve-tube elements are damaged, which occurred when this thin slice was cut, protein plugs form. These plugs, stained red in this image, reduce loss of phloem sap, in much the same way that clotting reduces blood loss in animals. ©Lee W. Wilcox

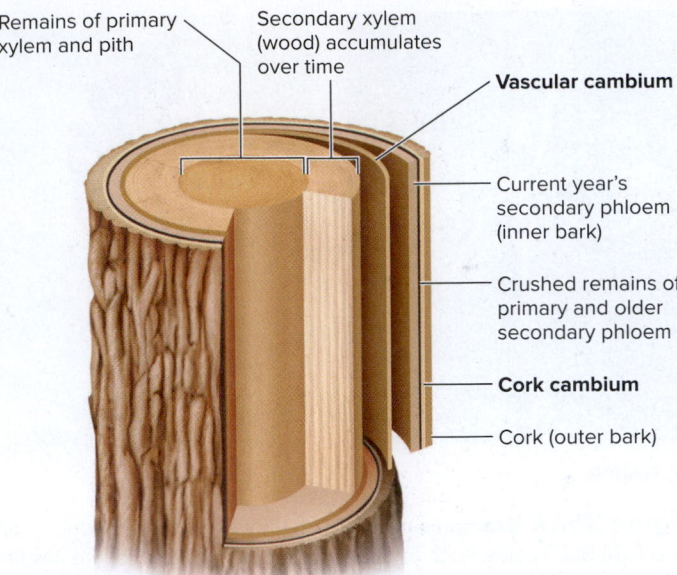

Remains of primary xylem and pith

Secondary xylem (wood) accumulates over time

**Vascular cambium**

Current year's secondary phloem (inner bark)

Crushed remains of primary and older secondary phloem

**Cork cambium**

Cork (outer bark)

**Figure 36.20** **Formation of wood and bark by secondary (lateral) meristems.** The vascular cambium is a thin cylinder of tissue that produces a thick cylinder of wood (secondary xylem) toward the inside of the stem and a thinner cylinder of inner bark (secondary phloem) toward the outside of the stem. The cork cambium forms an outer coating of protective cork (outer bark).

***Secondary Vascular Tissues***   Woody plants begin life as herbaceous seedlings that possess only primary vascular systems. But as these plants mature, they produce secondary vascular tissues and bark. Secondary vascular tissues are composed of secondary xylem and secondary phloem. **Secondary xylem** is also known as **wood**, a component of plants that plays many important roles in human life (**Figure 36.20**). Wood is composed of about 25% lignin, 45% cellulose, and 25% other polysaccharides that are together known as hemicelluloses.

**Secondary phloem** is the **inner bark**. **Outer bark** is composed of protective layers of mostly dead cork cells that cover the outside of woody stems and roots (see Figure 36.20). Therefore, bark includes both inner bark (secondary phloem) and outer bark (cork).

Woody plants produce secondary vascular tissues by means of **secondary meristems**, also known as **lateral meristems**, which form rings of actively dividing cells that encircle the stem. The two types of secondary meristems are vascular cambium and cork cambium, which are derived from primary meristems.

The secondary meristem known as **vascular cambium** is a ring of dividing cells that produces secondary xylem to its interior and secondary phloem to its exterior (see Figure 36.20). Secondary xylem conducts most of a woody plant's water and minerals. Cell divisions that occur in secondary meristems increase the girth of woody stems. During each new growing season, the vascular cambium produces new cylinders of secondary xylem and secondary phloem.

In trees growing in temperate climates, each year's addition of new secondary xylem forms growth rings that can be observed on the cut stem surfaces (**Figure 36.21**). The growth rings of secondary xylem surround the remains of the primary xylem and a central cylinder of parenchyma cells known as the pith. If environmental conditions

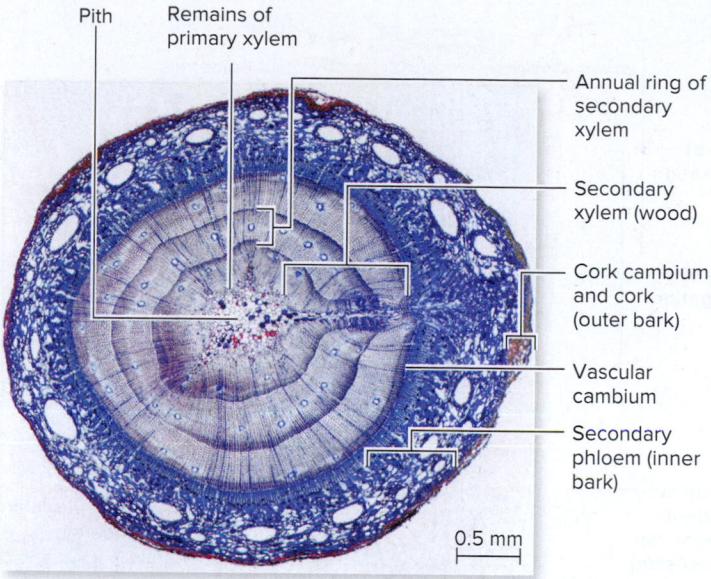

Pith | Remains of primary xylem

Annual ring of secondary xylem

Secondary xylem (wood)

Cork cambium and cork (outer bark)

Vascular cambium

Secondary phloem (inner bark)

0.5 mm

**Figure 36.21** **The anatomy of a woody stem.** Each year, a new cylinder of wood is produced; this yearly wood production appears as annual rings on the cut surface of a woody stem. ©Lee W. Wilcox

*Concept Check:* *Why do tree trunks have a layer of wood (secondary xylem) that is thicker than the layer of inner bark (secondary phloem)?*

favor plant growth, the growth rings formed at that time will be wider than those formed during times of stressful conditions. Climatologists use growth ring widths in samples of old wood to deduce past climatic conditions, and archaeologists use growth ring data to determine the age of wood constructions and artifacts left by ancient cultures.

Secondary xylem may transport water for several years, but usually only the current year's production of secondary phloem is active in food transport. This is because thin-walled sieve elements typically live for only a year. Thus, only a thin layer of phloem, the inner bark, is responsible for most of the sugar transport in a large tree. Deep abrasion of tree bark may damage this thin phloem layer, disrupting a tree's food transport. If a groove is cut all the way around a tree trunk—a practice known as girdling—the tree will die because all of its functional phloem transport routes will have been interrupted.

As a young woody stem begins to increase in diameter, its thin epidermis eventually ruptures and is replaced by outer bark, which is composed of protective cork tissues. Cork is produced by a secondary meristem called the **cork cambium**, another ring of actively dividing cells. The cork cambium surrounds the secondary phloem (see Figures 36.20 and 36.21). Together, the cork cambium, layers of cork tissue produced by the cambium, and associated parenchyma cells are known as a **periderm**. The outer bark becomes thicker as woody stems accumulate multiple periderm layers. The outer bark surface is often interrupted by passages known as **lenticels** that allow inner stem tissues to accomplish gas exchange.

Cork cells are dead when mature, and their walls are layered with suberin, a material that helps to prevent both attack by microbial pathogens and water loss from the stem surface. Cork tissues also produce tannins, compounds that protect against pathogens by inactivating their proteins. The cracked surfaces of tree trunks are dead cork tissues of the outer bark. Commercial cork is sustainably harvested from the cork oak tree (*Quercus suber*) for production of

flooring material, bottle stoppers, and other items. Additional information about the structure and function of primary and secondary xylem and phloem can be found in Chapter 39.

## Modified Stems Display Diverse Forms and Functions

Stems mostly grow upright because light is required for photosynthesis. But some stems, known as rhizomes, occur underground and grow horizontally. For example, potato tubers are the swollen, food-storing tips of rhizomes. Grass stems also grow horizontally, as either rhizomes just beneath the soil surface or stolons, which grow along the soil surface. The leaves and reproductive shoots of grasses grow upward from the point where they are attached to these horizontal stems. Grass blades continue to elongate from their bases even if you cut their tips off, explaining why lawns must be mowed repeatedly during the growing season. The horizontal stems of grasses are adaptations that help to protect vulnerable shoot apical meristems against natural hazards such as fire and grazing animals.

## 36.4  Root System Adaptations

### Learning Outcomes:

1. List ways that root structure has been modified in different plants in response to different habitats.
2. **CoreSKILL »** Make a drawing that shows how branch roots develop.
3. Explain how mobile proteins influence root development.

Roots play the essential roles of absorbing water and minerals, anchoring plants in soil, and storing nutrients. The external form of roots varies among flowering plants, reflecting adaptation to particular life spans or habitats. In contrast, root internal structure is more uniform. In this section, we will first examine the internal structure and development of roots, and then consider a few examples of modified roots that have particular adaptive advantages.

### Root Internal Growth and Tissue Specialization Occur in Distinct Zones

As discussed in Section 36.1, the common bean and other eudicots display an underground taproot system, whereas corn and other monocots have a fibrous root system (see Table 36.1). Studies of gene expression in the eudicot *Arabidopsis thaliana* reveal that roots are amazingly complex in their internal structure, having at least 15 distinct cell types. For our purposes, a simpler microscopic examination of root internal structure reveals three major zones: (1) a root apical meristem (RAM) protected by a root cap, (2) a zone of root elongation, and (3) a zone of maturation in which specialized cells can be observed (**Figure 36.22**).

***Root Apical Meristem and Root Cap***   As discussed earlier, an apical meristem occurs at the tip of each root and its branches. Like the SAM, the RAM contains stem cells, but these are organized differently in root apices. Root stem cells surround a tiny region of cells

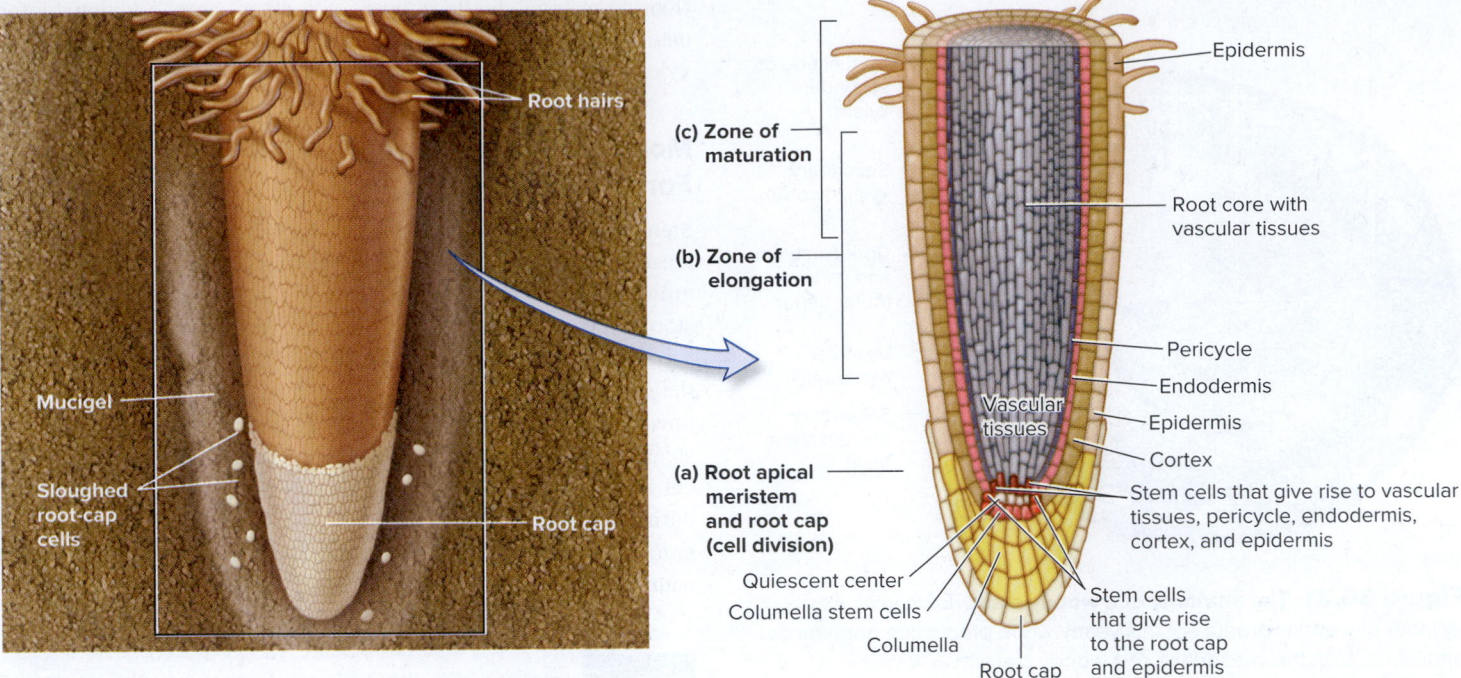

**Figure 36.22  Three zones of root growth.** A longitudinal view of a typical root reveals three major zones: **(a)** a root apical meristem region that includes stem cells, a quiescent zone, columella, and a root cap; **(b)** a zone of elongation; and **(c)** a zone of maturation, characterized by specialized cells and tissues including epidermal root hairs, cylinders of endodermis and pericycle tissue, and a core of vascular tissue.

that rarely divide, known as the quiescent center. Signals emanating from the quiescent center keep nearby stem cells in an undifferentiated state. Root stem cells farther away from the quiescent center produce new cells in multiple directions.

- Toward the root tip, stem cells produce columella cells that sense gravity and touch, which helps roots extend downward into the soil and around obstacles such as rocks.

- At the sides of the quiescent center, stem cells produce a protective root cap and epidermal cells. Root tip epidermal cells secrete a sticky substance called mucigel that lubricates root growth through the soil.

- Toward the shoot, stem cells generate cells that become internal root tissues.

***Zones of Elongation and Maturation*** Above the RAM lies the **zone of elongation**, in which cells extend by water uptake, thereby dramatically increasing root length (see Figure 36.22). The root elongation zone illustrates the general principle that cell expansion in plants is not necessarily linked directly to cell division. (By contrast, animal cell expansion is more closely linked to cell division.)

Above and overlapping with the root zone of elongation is the **zone of maturation**, where most root cell differentiation and tissue specialization occur. Specialized root tissues include mature vascular tissues at the root core, an enclosing cylinder of cells known as the pericycle, another cell cylinder called the endodermis (meaning inside skin), and epidermal cells at the root surface. Relatively unspecialized parenchyma cells form a cortex that lies between the endodermis and the epidermis. Starting with the epidermis and moving

inward, we will take a closer look at these root tissues and factors that control their development.

The zone of maturation can be identified by the presence of numerous microscopic hairs that emerge from the root epidermis. **Root hairs** are specialized epidermal cells that can be as long as 1.3 cm, about the width of your little finger, but are only 10 μm in diameter, less than the width of a finger cell. Their small diameter allows root hairs to obtain water and minerals from soil pores that are too narrow for even the smallest roots to enter. Root hair plasma membranes are rich in transport proteins that use ATP to selectively absorb materials from the soil (look ahead to Figure 39.3).

The production of hairs from root epidermal cells is controlled by the activity of transcription factors that move between cells and also by cell position (**Figure 36.23**). A root hair will develop from epidermal cells lying over the junction of two cortical cells, whereas no hair will develop if an epidermal cell lies over only one cortical cell. Root hairs are so delicate that they are easily damaged by abrasion as roots grow through the soil, and they live for only 4 or 5 days. As a result, root hairs are absent from older regions above the zone of maturation. To compensate, roots must continually produce new root hairs. The average rate of root-hair production has been estimated at more than 100 million per day for some plants. One reason that gardeners use care when transplanting seedlings is to prevent extensive damage to the root hairs.

The epidermis of mature roots encloses a cylinder of parenchyma known as the root cortex (**Figure 36.24**). Much like the cells of the stem cortex, root cortex cells are often rich in starch and therefore serve as a food storage site for plants. The root cortex of some plants contains intercellular air spaces that arise as a result of programmed cell death and provide routes for oxygen diffusion within the root.

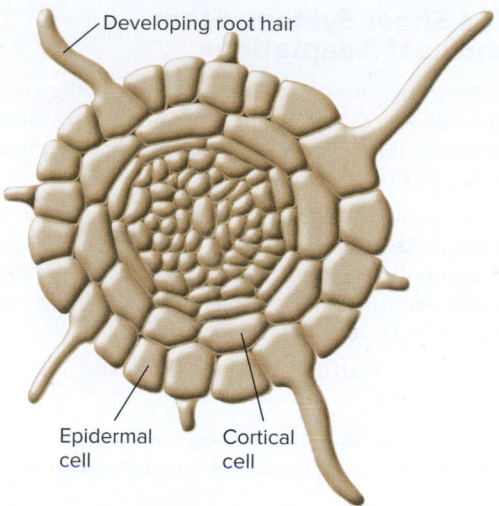

**Figure 36.23** A cross section of a root showing positional influence on root-hair development.

Water and dissolved minerals also diffuse from the environment into roots through spaces between cortex cells, stopping only when they reach a one-cell-thick cylinder of specialized tissue known as **endodermis** (inside skin). This endodermal cell layer is an important component of the mechanism by which roots absorb selected minerals (described further in Chapter 39). Endodermal cells specialize in response to a transcription factor (called SHORT ROOT) that is synthesized in cells of the root core, then transported outward.

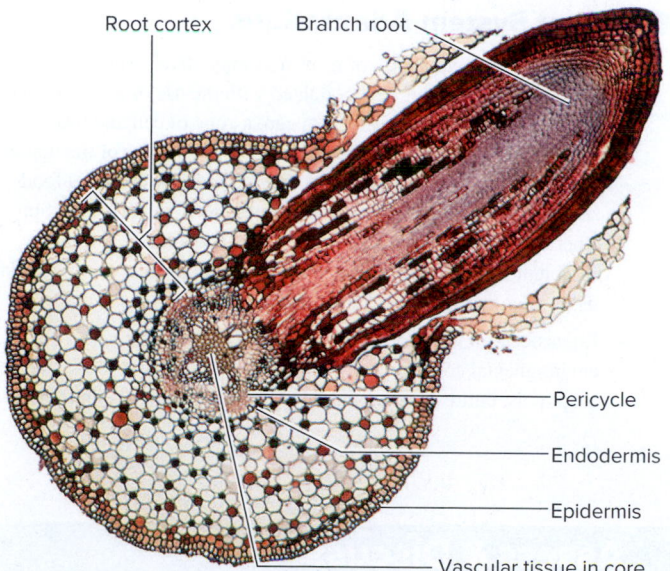

**Figure 36.24** **Cross section of a mature root.** This stained light micrograph shows the epidermis and cortex of a root surrounding a central core of vascular tissue. An inner cortex layer is the endodermis, which surrounds a cylinder of meristematic pericycle tissue. The pericycle has produced a young branch (lateral) root that has grown through the cortex and the epidermis. ©Lee W. Wilcox

*Concept Check:* *Why must lateral roots be produced in the way shown in this figure?*

A cylinder of tissue having cell division (meristematic) capacity, known as the **pericycle**, encloses the root vascular tissue (see Figure 36.24). In response to the hormone auxin, the pericycle produces lateral branch roots that force their way through the cortex to the surface. This process differs from the way that stems produce branches by means of buds. In some roots, the pericycle generates a vascular cambium that produces wood—secondary xylem. Such woody roots also possess a cork cambium that makes a protective covering of suberin-coated cork tissue. Like woody stems, woody roots produce primary vascular tissues in their youth and secondary vascular tissues at maturity. The woody roots of trees are sometimes visible above ground.

## Modified Roots Display Diverse External Forms and Functions

Plants produce several other types of roots that provide adaptive advantages in response to different habitats. For example, corn and many other plants produce supportive prop roots from the lower portions of their stems. Many tropical trees grow in such thin soils that they are vulnerable to being blown down in windstorms. Such trees often produce dramatic aboveground buttress roots that help keep them upright (**Figure 36.25a**). Many mangrove trees that grow along

**(a) Buttress roots**

**(b) Pneumatophores**

**Figure 36.25** **Modified aboveground roots.** **(a)** Buttress roots help to keep tropical trees such as this *Pterocarpus hayesii* from toppling in windstorms. **(b)** Pneumatophores produced by mangroves are roots that extend upward into the air. These roots take up air and then transmit it to underwater roots that grow in oxygen-poor sediments.

a: ©Martin Engelmann/Getty Images; b: ©Peter E. Smith, Natural Sciences Image Library

tropical coasts produce pneumatophores (Greek meaning breath bearers), roots that grow upward into the air (**Figure 36.25b**). Functioning like snorkels, pneumatophores absorb oxygen-rich air, which diffuses to submerged roots growing in oxygen-poor sediments. This mechanism is necessary because roots require a supply of oxygen in order to produce ATP. Roots use the ATP to power their growth and the uptake of mineral nutrients (see Chapter 39).

# Summary of Key Concepts

## 36.1  From Seed to Seed—The Life of a Flowering Plant

- Seed embryos, seedlings, and mature plants are components of the sporophyte generation in the plant sexual cycle; tiny gametophytes develop and grow within flowers (Figure 36.1).

- Plant organs are composed of tissues that contain specialized cells. The basic plant organs are roots, stems, and leaves. Shoot systems include stems and stem branches, and stems produce leaves, buds, flowers, and fruits. Root systems include one or more main roots with branches. Buds, flowers, fruits, and seeds are organ systems, composed of more than one organ (Figures 36.2, 36.3).

- The two major groups of flowering plants—eudicots and monocots—differ in the structure of their seed embryos, flowers, stems, roots, leaves, and pollen. Flowering plants may live for one year (annuals), two years (biennials), or more than two years (perennials) (Figure 36.4, Table 36.1).

## 36.2  How Plants Grow and Develop

- Plant growth and development features the presence of a fundamental architecture based on apical-basal polarity and radial symmetry throughout the life of a plant (Figures 36.5, 36.6).

- Plants grow by producing new cells at meristems and by controlled cell enlargement involving water uptake.

- Shoot apical meristems produce primary meristems that increase plant length and produce organs (Table 36.2).

- The simple plant tissues, containing one or two cell types, include parenchyma, collenchyma, and sclerenchyma. Complex plant tissues include the vascular tissues known as xylem and phloem, and the primary vascular tissues occur in vascular bundles (Figure 36.7).

- Leaves develop from primordia at shoot apices. Foliage leaves have internal and external structure that is adapted for photosynthetic functions (Figure 36.8).

- Meristems include stem cells that divide relatively rarely, thereby reducing the accumulation of harmful mutations. Cells that arise from stem cells may divide more frequently, producing new tissue.

- Plant cells are able to expand under conditions that result in loosening of cell-wall components, and by water uptake into vacuoles. The direction in which plant cells expand is determined by the arrangement of cellulose microfibrils in the cell wall, which is influenced by the orientation of microtubules in the nearby cytosol (Figures 36.9, 36.10, 36.11).

## 36.3  The Shoot System: Stem and Leaf Adaptations

- Shoots are modular systems; each module includes a node, internode, leaf, and axillary meristem or bud. An axillary bud develops in leaf axils; such buds may grow into new branches (Figure 36.12).

- Variations in leaf structure reflect adaptations that aid photosynthesis or protect against stress. For example, Sack and colleagues demonstrated that palmate leaves provide conducting system redundancy that is useful in coping with vein damage (Figures 36.13, 36.14, 36.15).

- Stomatal guard-cell differentiation is controlled by several genes (Figure 36.16).

- Leaves not only function in photosynthesis but also play other roles, including attachment, attraction, and protection (Figure 36.17).

- Herbaceous plants are those whose stems produce little or no wood and are mostly composed of primary vascular tissues. The primary vascular tissues are primary xylem and primary phloem (Figures 36.18, 36.19).

- In addition to primary tissues, woody plants such as trees and shrubs possess secondary meristems that produce wood and bark. The vascular cambium produces secondary xylem (wood) and secondary phloem (inner bark). The cork cambium produces cork tissues that form outer bark (Figures 36.20, 36.21).

- Stems occur in diverse forms that reflect adaptation to environmental conditions. Examples include grass rhizomes, which grow horizontally underground and are therefore better protected from fire and grazing animals (as well as lawnmowers).

## 36.4  Root System Adaptations

- The internal organization of roots is comparatively uniform, and three major zones can be recognized with the use of a microscope: the root apical meristem and root cap, a zone of cell and root elongation, and a zone of tissue maturation. Features of the mature root include epidermal root hairs that aid nutrient uptake, a food-storing cortex, an endodermis that functions in mineral selection, a pericycle that produces lateral (branch) roots (and vascular cambium in the case of woody roots), and an inner core of vascular tissue (Figures 36.22, 36.23, 36.24).

- Roots occur in multiple forms that reflect adaptation to environmental conditions. Examples of aboveground roots include prop roots, buttress roots, and pneumatophores (Figure 36.25).

# Assess & Discuss

## Test Yourself

1. Where would you look to find the gametophyte generation of a flowering plant?
   a. at the shoot apical meristem
   b. at the root apical meristem
   c. in seeds
   d. in flower parts
   e. Flowering plants lack a gametophyte generation.

2.  What is a radicle?
    a.  an embryonic leaf
    b.  an embryonic stem
    c.  an embryonic root
    d.  a mature root system of a monocot
    e.  an organism that has extreme political views

3.  Which type of plant is most likely to have food-rich roots that are useful as human food?
    a.  an annual
    b.  a biennial
    c.  a perennial
    d.  a centennial
    e.  plants that grow along coastal shorelines

4.  Which of the following terms best describes the distinctive architecture of plants?
    a.  radial symmetry and apical-basal polarity
    b.  bilateral symmetry and apical-basal polarity
    c.  radial symmetry and absence of apical-basal polarity
    d.  bilateral symmetry and absence of apical-basal polarity
    e.  absence of symmetry and absence of apical-basal polarity

5.  Which is the most accurate description of how plants grow?
    a.  by the addition of new cells at meristems that include stem cells
    b.  by cell enlargement as the result of water uptake
    c.  by both the addition of new cells and cell expansion
    d.  by addition of fat cells
    e.  by all of the above

6.  Where would you look for leaf primordia?
    a.  at a vegetative shoot tip
    b.  at the root apical meristem
    c.  at the vascular cambium
    d.  at the cork cambium
    e.  in a floral bud

7.  Which leaf tissues display the greatest amount of air space?
    a.  the upper epidermis
    b.  the lower epidermis
    c.  the palisade parenchyma
    d.  the spongy parenchyma
    e.  the vascular tissues

8.  What are adventitious roots?
    a.  roots that develop on plant cuttings that have been placed in water
    b.  buttress roots that grow from tree trunks
    c.  the only kinds of roots produced by monocots, because their embryonic root dies soon after seed germination
    d.  any root that is produced by stem (or sometimes leaf) tissue, rather than developing directly from the embryonic root
    e.  All of the above describe adventitious roots.

9.  During its development, a tracheid elongates in a direction parallel to the shoot or root axis. Based on this information, what can you say about the orientation of cell-wall cellulose microfibrils and cytoplasmic microtubules in the developing tracheid?
    a.  The microfibrils will be oriented perpendicularly (at right angles) to the long axis of the developing tracheid, encircling it, but the cytoplasmic microtubules will be oriented parallel to the direction in which the tracheid is elongating.
    b.  Microfibrils and microtubules will both be oriented perpendicularly (at right angles) to the elongating axis of the tracheid.
    c.  Microfibrils and microtubules will both be oriented parallel to the direction of tracheid elongation.
    d.  Microfibrils will be oriented parallel to the direction of tracheid elongation, but microtubules will be perpendicular (at right angles) to both the microfibrils and the elongating tracheid.
    e.  None of the above is correct.

10. What are examples of woody plants?
    a.  trees
    b.  shrubs
    c.  herbaceous plants
    d.  all of the above
    e.  only a and b

## Conceptual Questions

1.  What would be the consequences if overall plant architecture were bilaterally symmetric?

2.  What would be the consequences if leaves were radially symmetric (shaped like spheres or cylinders)?

3.  **Core Concept: Structure and Function** Why are most tall plants woody, rather than herbaceous?

## Collaborative Questions

1.  Find a tree stump or a large limb that has recently been cut from a tree (or imagine doing so). Which of the following features could you locate with the unaided eye: the outer bark, the inner bark, the secondary xylem, the vascular cambium, annual rings?

2.  Which physical factors would you expect to influence shoot growth most strongly? Which physical factors would you expect to influence underground root growth most strongly?

# Flowering Plants: Behavior

# 37

**Behavior of the snow buttercup.** The snow buttercup (*Ranunculus adoneus*) holds its flowers above the surface of the snow. The flowers move so that they always face the Sun during the day, a behavior known as sun tracking. ©William D. Bowman

T he snow buttercup (*Ranunculus adoneus*) grows in deep snowbanks in the high Rocky Mountains, with flower stems protruding above the snow's surface toward the Sun, as shown in the chapter opening photo. Amazingly, snow buttercup flowers change their position so that they face the Sun throughout the day, a process known as sun tracking. Experiments have demonstrated that sun tracking warms snow buttercup flowers, thereby favoring pollen development and germination. These processes foster fertilization, which leads to more effective seed production. In this way, sun tracking can be understood as an adaptation that increases snow buttercup reproductive fitness. Like sun-tracking solar panels, the leaves of alfalfa, lupine, soybean, common bean, cotton, and other wild and agricultural plants also track the Sun, a process that aids energy storage via photosynthesis.

Sun tracking is but one example of the many ways in which plants display behavior—that is, responses to stimuli. Plants respond not only to changes in the Sun's position, but also to day length, which is the period of daily illumination. In addition to light,

plants respond to gravity, wind, attack by animals and disease-causing microorganisms, and other environmental stimuli. This chapter provides examples of ways in which understanding plant behavior has been key to increasing agricultural productivity and protecting natural ecosystems for the benefit of humans.

We begin with a survey of the diverse types of stimuli that induce plant behavior and review how cells perceive and respond to stimuli by means of signal transduction pathways. Next, we will focus on plant hormones and other major types of mobile internal molecules that influence plant behavior. Finally, we will consider how responses to environmental stimuli foster plant survival and reproduction.

## 37.1 Overview of Plant Behavioral Responses

**Learning Outcomes:**

1. List examples of plant responses to internal and environmental stimuli.
2. Describe the three stages of cell signaling.

**Behavior** is defined as a response of organisms to an internal or external stimulus. Examples of plant behavior include plant movements, some types of which were described in 1880 by Charles Darwin and his son Francis in their book *The Power of Movement in Plants*. Modern time-lapse photography, which makes plant behavior more obvious to humans, reveals that most plants are constantly in motion, bending, twisting, or rotating in dancelike movements known as nutation (**Figure 37.1a**). Some plants display relatively rapid movements, illustrated by the sensitive plant (*Mimosa pudica*), whose leaves quickly fold when touched, then open more slowly (**Figure 37.1b**).

Among the many other important examples of plant behavior that occur are the following:

- Seeds germinate when they detect the presence of sufficient light and moisture for successful seedling growth.
- Plant shoots typically grow toward light and against the pull of gravity.

**(a) Nutation movements**    **(b) Leaf folding**

**Figure 37.1** **Examples of plant movements.** **(a)** Sixteen superimposed photographs of a shoot of the honeysuckle vine *Lonicera japonica*, taken over a period of 2 hours, reveal the circular movement known as nutation. **(b)** Photographs of the sensitive plant (*Mimosa pudica*) made before and shortly after a touch reveal the rapid process of leaf folding. Even if only one leaflet is touched, electrical signals travel throughout the complex leaf, causing the entire organ to fold. The leaves will eventually unfold. a: ©Digital Photography by Ash Kaushesh, University of Central Arkansas, Conway, Arkansas 72035 USA/Image courtesy Botanical Society of America, www.botany.org St. Louis, MO 63110; b (top, bottom): ©Lee W. Wilcox

- Most roots grow toward water and in the same direction as the gravitational force. Roots can also grow around obstacles in the soil.

- Flowers, fruit, and seeds are produced only during the season(s) most favorable for reproductive success.

- Plants take protective actions when they sense attack by disease-causing microbes or hungry animals, thereby preventing excessive damage to their own bodies or those of neighboring plants.

## Plant Behavior Involves Responses to Internal and External Stimuli

To gain a more complete understanding of plant behavior, we will begin by surveying the types of stimuli that cause plant responses. Most people are aware that both internal chemical signals and environmental factors influence animal behavior. Bird nesting behavior in spring, for example, involves hormonal changes triggered by seasonal conditions. Plants likewise respond to internal and environmental stimuli (**Figure 37.2**).

*Internal Stimuli*    Plants respond to two types of internal stimuli: internal biological clocks and mobile chemical signals. Internal biological clocks, known as **circadian rhythms**, occur not only in plants, but also in animals and other organisms. The word circadian comes from the Latin words meaning around and day. A circadian rhythm is any biological process that undergoes a consistent pattern

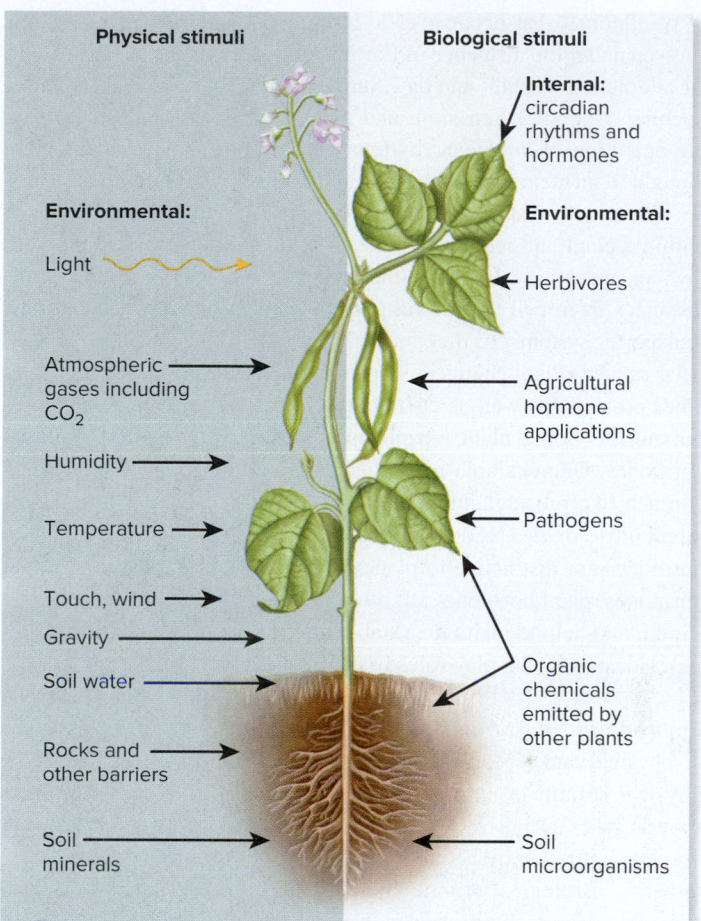

**Figure 37.2** **Types of plant stimuli.** Plants respond to both physical and biological stimuli. Stimuli may be internal to the plant or come from the environment.

 **Core Skill: Modeling** The goal of this modeling challenge is to draw a model that is similar to Figure 37.2 and shows the stimuli experienced by plants in a spacecraft.

**Modeling Challenge:**
Experiments aimed at understanding how plants respond to condition in space have been completed and continue to be performed on the International Space Station. Results of these experiments are helpful in determining how plants can be used to feed space travelers and purify their

Plant growing in the International Space Station. ©NASA Photo/Alamy Stock Photo

air and water. To design effective experiments in space requires understanding the types of environmental stimuli that plants respond to on Earth and how these environmental stimuli and plant responses might differ in space. Draw a model in the same format as Figure 37.2 that shows the stimuli that plants grown on board the International Space Station or other spacecraft would likely experience.

of oscillations that occur over a 24-hour period. Circadian rhythms evolved under the influence of Earth's rotation, which causes the regular alternation of night and day. Sun tracking, leaf movements, flower opening, fragrance emission, and many other behaviors result from the operation of circadian rhythms in plants. Circadian rhythms are thought to influence the timing of about 30% of gene activity in plants.

Plants also respond to internal chemical signals that are produced within a plant and move from one location to another, acting at very low concentrations. In plants, these chemical signals may move short distances from cell to cell via plasmodesmata or by cell membrane transporter systems, or they may move long distances within the vascular system. Plant chemical signals include transcription factors and other proteins, as well as chemically different compounds known as **hormones**. Some plant hormones are similar to particular animal hormones. One example is the plant hormone jasmonic acid, which is much like prostaglandins produced by animals. The gaseous hormone nitric oxide (NO) functions in both animals and plants. Other hormones are distinctive to plants. Even so, as in the case of animal hormones, plant hormones are often produced in response to external stimuli and help to maintain a stable internal environment (homeostasis). Hormones also play roles in plant development.

***Environmental Stimuli*** Plants sense and respond to many types of physical and biological environmental stimuli (see Figure 37.2). Physical stimuli in natural plant environments include light, atmospheric gases such as $CO_2$ and water vapor, temperature, touch, wind, gravity, soil water, rocks and other barriers to root growth, and soil minerals. Biological stimuli include herbivores (animals that consume plant parts), airborne pathogens (disease-causing microbes), organic chemicals emitted from neighboring plants, and beneficial or harmful soil microorganisms. Crop plants also respond to applications of agricultural chemicals, which may include hormones. That plants have evolved such a broad array of sensory capacity is not surprising, because all of the listed environmental influences affect plant survival and reproduction.

***Plant Responses to Environmental Stimuli*** Though plants lack the specialized sense organs typical of animals, receptor molecules located in plant cells sense stimuli and cause responses. When many cells of a tissue receive and respond to the same biological or physical stimuli, entire organs or plant bodies display behavior. For example, houseplants tend to grow toward a light source such as a window. This process, known as positive **phototropism**, involves both a cellular perception of light and a growth response of stem tissue to an internal chemical signal. In general, a **tropism** is a growth response that depends on a stimulus that occurs in a particular direction. In the case of phototropism, the plant senses the direction of light and responds by changing the location of a plant hormone known as an auxin. As a result, the stem bends toward light. We will next review how plant cells more generally receive signals and transmit them intracellularly, a process known as cell signaling that occurs in all cells (refer back to Chapter 9).

## Plant Cell Signaling Involves Receptors, Second Messengers, and Effectors

**Cell signaling** is the process in which a cell receives a physical or chemical signal, thereby switching on an intracellular pathway that leads to a cellular response (**Figure 37.3**). The process of cell signaling often involves three types of molecules: receptors that may

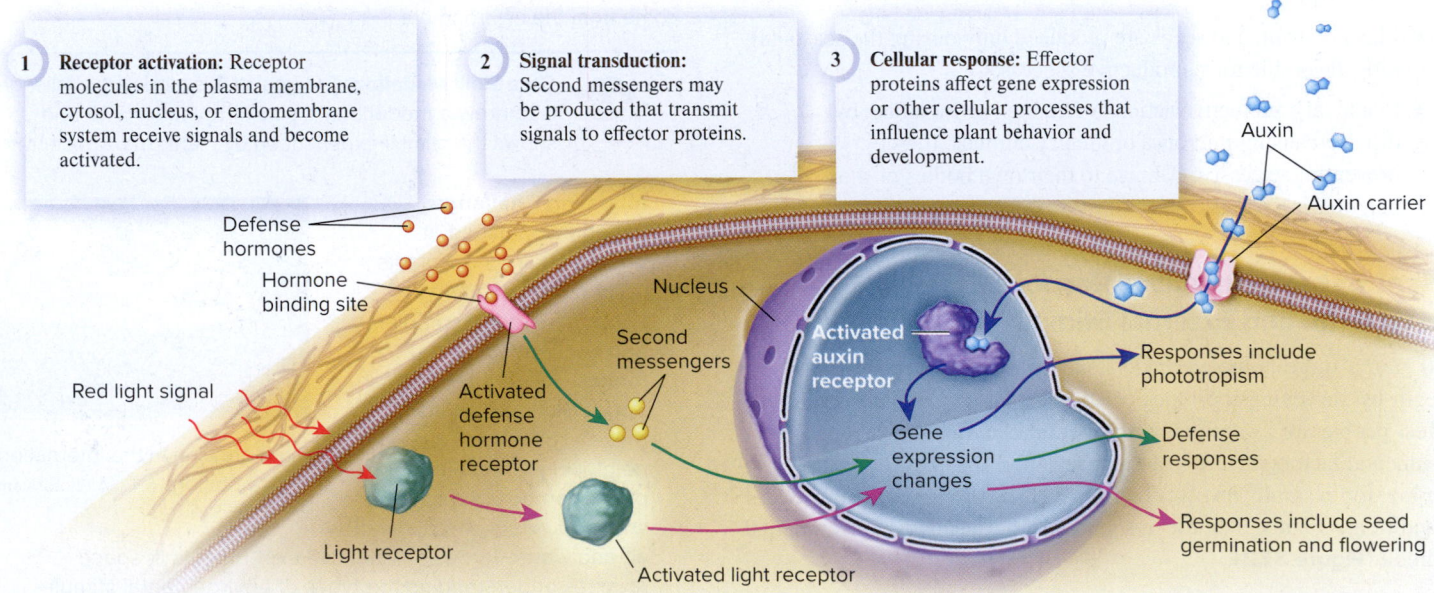

1. **Receptor activation:** Receptor molecules in the plasma membrane, cytosol, nucleus, or endomembrane system receive signals and become activated.

2. **Signal transduction:** Second messengers may be produced that transmit signals to effector proteins.

3. **Cellular response:** Effector proteins affect gene expression or other cellular processes that influence plant behavior and development.

Defense hormones

Hormone binding site

Red light signal

Activated defense hormone receptor

Light receptor

Activated light receptor

Nucleus

Second messengers

Activated auxin receptor

Gene expression changes

Auxin

Auxin carrier

Responses include phototropism

Defense responses

Responses include seed germination and flowering

**Figure 37.3 An overview of plant cell signaling.** Plant cells respond to hormonal signals produced within the plant body, as well as to environmental stimuli. Three different signal transduction processes are shown here: one started by light, one started by a defense hormone having a plasma membrane receptor, and another started by an auxin having a receptor located in the nucleus.

 **Core Concept: Systems** Internal and environmental stimuli are received and elicit responses at the cellular level.

become activated, second messengers that transmit signals, and effectors that cause a cellular response (also refer back to Figure 9.4).

**Receptors** (also known as sensors) are proteins that become activated when they receive a specific type of signal (see Figure 37.3). Receptors occur in diverse cellular locations. Whereas some plant defense receptors are located within the plasma membrane, light receptors may occur in the cytosol and auxin receptors in the nucleus. Some activated receptors directly generate a response, such as an increased flow of ions across a membrane. Other activated receptors bind to signaling molecules that initiate an intracellular signaling pathway. For example, the binding of certain defense hormones to plasma membrane receptors results in the intracellular production of second messenger molecules that transmit the signal.

**Second messengers** transmit messages from many types of activated receptors. Cyclic AMP (cAMP), inositol triphosphate ($IP_3$), and calcium ions ($Ca^{2+}$) are major second messengers in animal and plant cells. $Ca^{2+}$ is a particularly common second messenger in plant cells. Touch and various other stimuli cause $Ca^{2+}$ to flow from storage sites in the lumen of the endoplasmic reticulum (ER) into the cytosol. $Ca^{2+}$ then binds to calmodulin or other calcium-binding proteins. $Ca^{2+}$ binding alters the structure of such proteins, causing them to interact with other cellular proteins or to alter their enzymatic function.

**Effectors** are molecules that directly influence cellular responses. In plants, calcium-dependent protein kinases (CDPKs) are particularly important effector molecules. The last phase of cell signaling occurs when an effector causes a cellular response, such as opening or closing an ion channel or switching the transcription of particular genes on or off. A single activated receptor can dispatch many second-messenger molecules, which, in turn, can activate scores of effectors, leading to many molecular responses within a single cell.

## 37.2  Plant Hormones

**Learning Outcomes:**

1. List some examples of plant hormones.
2. Explain how auxin functions in phototropism.
3. Explain how cytokinins, gibberellins, and ethylene are relevant to agriculture.
4. Explain how abscisic acid and brassinosteroids help plants cope with environmental stress.

Along with proteins and RNA molecules, plant hormones are chemical signals transported within the plant body. When taken up by target cells, the signals elicit responses. Here, we focus on plant hormones, about a dozen types of small molecules that are synthesized in metabolic pathways that also make amino acids, nucleotides, sterols, or secondary metabolites. Auxins, cytokinins, gibberellins, ethylene, abscisic acid, and brassinosteroids are examples of plant hormones (**Table 37.1**).

Individual plant hormones often have multiple effects, and different concentrations or combinations of hormones can produce distinct growth or developmental responses. Several plant hormones are known to act by causing the removal of gene repressors, thereby allowing gene expression to occur. This general mechanism allows hormones to cause relatively rapid responses. A closer look at the major types of plant hormones reveals their multifaceted roles.

## Auxins Are the Master Plant Hormones

Plants produce several types of **auxins**, which are considered to be the master plant hormones because they influence plant structure, development, and behavior in many ways, often working with other hormones. Indoleacetic acid (IAA) is one plant auxin (see Table 37.1), but other natural and artificial compounds have similar structures and effects. In this section, we will refer to this family of related compounds simply as auxin.

Auxin exerts so many effects because it promotes the expression of thousands of genes known as **auxin-responsive genes**. Under low auxin concentrations, proteins called Aux/IAA repressors prevent plant cells from expressing these genes. The repressor proteins prevent gene expression by binding to activator proteins at gene promoters and inhibiting their function. When the auxin concentration is high enough, auxin molecules glue repressors onto a protein complex called TIR1, which causes the breakdown of the repressors. Free of the repressors, the activator proteins are no longer inhibited and enhance the expression of auxin-response genes.

**Auxin Transport**  The way in which auxin is transported into and out of cells is integral to its effects in plants. Auxin is produced in apical shoot tips and young leaves, and it is directionally transported from one living parenchyma cell to another. Auxin in an uncharged form (IAAH) may enter cells from intercellular spaces by means of simple diffusion. However, the negatively charged form (IAA⁻) requires the aid of a plasma membrane protein known as the **auxin influx carrier (AUX1/LAX)**. Several types of proteins, called PIN proteins, transport auxin out of cells. They are named for the pin-shaped shoot apices of plants having mutations in *PIN* genes. Because they transport auxin out of cells, PIN proteins are called **auxin efflux carriers**. They are necessary because auxin occurs as a charged ion in the cytosol that does not readily diffuse out of cells.

In shoots, AUX1/LAX is located at the apical ends of cells, whereas PIN proteins often occur at the basal ends (**Figure 37.4a**). This polar distribution of auxin carriers explains why auxin primarily flows downward in shoots and into roots, a process called **polar transport** (**Figure 37.4b**). However, the locations of auxin carriers can also change within cell plasma membranes, allowing lateral or upward transport of auxin. Differences in the presence and positions of auxin carrier proteins explain variations in auxin concentration within plants. The local auxin concentration allows plant cells to determine their position within the plant body and to respond by dividing, expanding, or specializing.

**Auxin Effects**  In nature, auxin influences plants throughout their lifetimes. Auxin establishes the apical-basal polarity of seed embryos, induces vascular tissue to differentiate, mediates phototropism, promotes formation of adventitious roots, and stimulates fruit development. Many of auxin's effects are also of practical importance to humans. Auxin is used to produce some types of seedless fruit, retard premature fruit drop in orchards, and stimulate root development on stem cuttings. Although we still have much to learn about auxin's function, its role in phototropism has been elucidated by a series of experiments, described next.

## Table 37.1   Examples of Plant Hormones

| Type of plant hormone | Chemical structure of an example | Functions* |
|---|---|---|
| Auxins | Indoleacetic acid (IAA) | Establish apical-basal polarity, induce vascular tissue development, mediate phototropism, promote formation of adventitious roots, inhibit leaf and fruit drop, and stimulate fruit development |
| Cytokinins | Zeatin | Promote cell division, influence cell specialization and plant aging, activate secondary meristem development, promote adventitious root growth, and promote shoot development on callus |
| Gibberellins | Gibberellic acid | Stimulate cell division and cell elongation, stimulate stem elongation and flowering, and promote seed germination |
| Ethylene | Ethylene $H_2C = CH_2$ | Promotes seedling growth, induces fruit ripening, plays a role in leaf and petal aging and drop, coordinates defenses against osmotic stress and pathogen attack |
| Abscisic acid | Abscisic acid | Slows or stops metabolism during environmental stress, induces bud and seed dormancy, prevents seed germination in unfavorable conditions, and promotes stomatal closing |
| Brassinosteroids | Brassinolide | Promote cell expansion, stimulate shoot elongation, retard leaf drop, stimulate xylem development, and promote stress responses |

*The lists of functions are only partial lists.

**The Role of Auxin in Phototropism**   In the 1880s, Charles Darwin and his son Francis were the first to publish results of experiments on plant phototropism. The Darwins performed their experiments on cereal seedlings, whose tips are protected by a sheath of tissue called a coleoptile. In a simple but elegant experiment, the Darwins covered either the tips or lower portions of coleoptiles with shading materials such as blackened glass tubes, left other seedlings uncovered, and removed the tips of some seedlings. They then compared how those seedlings responded to illumination from the side. The seedlings whose tips were left uncovered grew toward the light, whereas seedlings whose tips were covered or removed did not. The Darwins concluded that seedling tips transmit some "influence" to lower portions,

causing them to bend toward the light. You can probably guess what this influence was, but technology available at the time did not allow the Darwins to determine this.

Three decades later, in the 1910s, Danish botanist Peter Boysen-Jensen confirmed the Darwins' results and demonstrated that the influence was a chemical substance that diffused from the tips of the seedlings to other parts. To do this, Boysen-Jensen cut off the tips of oat seedlings and placed either a porous layer of gelatin or a nonporous material such as a sheet of the mineral mica on the cut surface. Then he replaced the tips. Oat seedlings layered with porous gelatin displayed a normal phototropic response, bending toward the light, but those layered with nonporous mica did not.

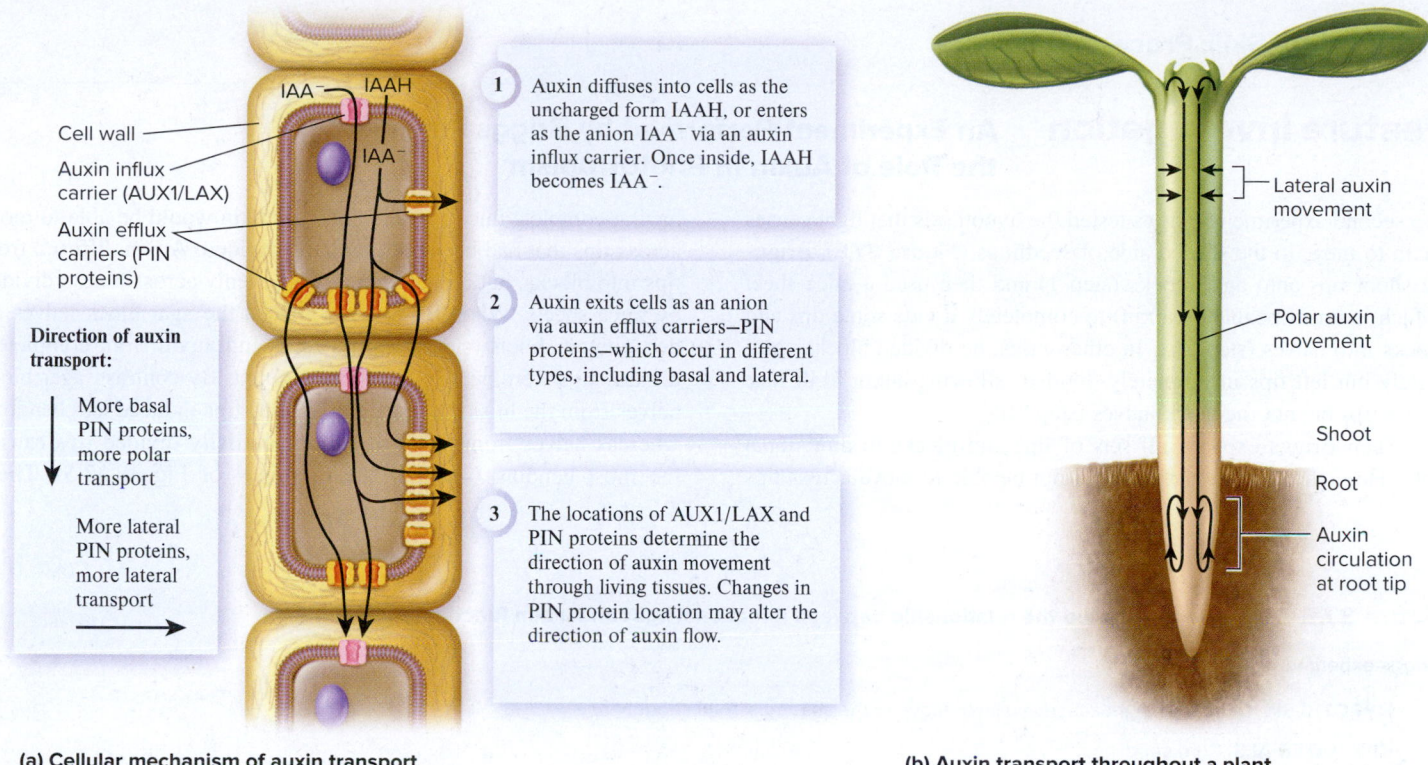

**(a) Cellular mechanism of auxin transport**

Cell wall

Auxin influx carrier (AUX1/LAX)

Auxin efflux carriers (PIN proteins)

IAA⁻ IAAH

IAA⁻

**1** Auxin diffuses into cells as the uncharged form IAAH, or enters as the anion IAA⁻ via an auxin influx carrier. Once inside, IAAH becomes IAA⁻.

**Direction of auxin transport:**

More basal PIN proteins, more polar transport

More lateral PIN proteins, more lateral transport

**2** Auxin exits cells as an anion via auxin efflux carriers—PIN proteins—which occur in different types, including basal and lateral.

**3** The locations of AUX1/LAX and PIN proteins determine the direction of auxin movement through living tissues. Changes in PIN protein location may alter the direction of auxin flow.

**(b) Auxin transport throughout a plant**

Lateral auxin movement

Polar auxin movement

Shoot

Root

Auxin circulation at root tip

**Figure 37.4** **Auxin transport.** **(a)** Polar and lateral auxin transport is controlled by the distribution of auxin efflux carriers located in the plasma membrane. When efflux carriers primarily occur at the basal ends of cells, auxin will flow downward. Auxin may flow laterally when auxin efflux carriers occur at the sides of cells. **(b)** In a whole plant, auxin primarily flows downward from shoot tips to root tips, where it then flows upward for a short distance.

**Concept Check:** *How could auxin carriers be organized to allow auxin to move upward in roots?*

Boysen-Jensen's experiment demonstrated that the phototropic substance was a diffusible chemical, but exactly which one, and how it worked, remained unknown. A series of additional experiments provided some answers.

In the 1920s, Dutch plant physiologist Frits Went named the substance discovered by Boysen-Jensen auxin (from the Greek word *auxein*, meaning to increase). Although the chemical structure of auxin was not determined until 1934, Went performed experiments that helped explain how auxin works. In a first step, Went cut the tips off oat seedlings and placed these tips onto agar blocks. Agar, a complex polysaccharide derived from red algae, forms a mesh capable of holding considerable water and dissolved compounds. Agar's permeability to auxin is similar to that of the protein gelatin used by Boysen-Jensen, but agar is much more stable at room temperature and more resistant to microbial breakdown and therefore is easier to use in laboratory experiments. In Went's experiment, the auxin diffused from cut seedling tips into these agar blocks. In the next steps, he treated decapitated seedlings in one of four ways: (1) placed auxin-laden agar blocks off-center on some, (2) placed auxin-laden blocks evenly on others, (3) placed plain agar blocks off-center on some, and (4) left some uncapped. All seedlings were then kept in the dark throughout the experiment. Only seedlings that were capped off-center with an auxin-laden block grew in the direction away from the agar block.

This experiment demonstrated that auxin application could substitute for the directional light stimulus and suggested that asymmetric auxin distribution is the mechanism by which light causes plants to bend.

Subsequently, Went and N. O. Cholodny independently proposed that light causes auxin to move to the unlit side of seedling tips, causing cells on that side to elongate more, which results in bending. But other scientists argued that bending could result if light destroys auxin on the illuminated side of a seedling. In the 1950s, American plant biologist Winslow Briggs designed two experiments to test these alternate hypotheses.

In his first experiment, which tested the hypothesis that auxin might be destroyed by light, Briggs first grew corn seedlings in the dark. Then he cut off their tips, put the tips on agar blocks, and exposed some to darkness and others to directional light. During this process, auxin from tips diffused evenly into the agar blocks. If auxin were destroyed by light, agar blocks under lighted tips should receive less auxin than blocks under tips kept in the dark. The auxin-destruction hypothesis also predicts that when agar blocks from lighted tips are placed on one side of decapitated seedlings, they should cause less bending than will blocks from tips kept in the dark. However, Briggs discovered that both types of agar blocks caused the same amount of shoot bending. This result is not consistent with the hypothesis that light destroys auxin.

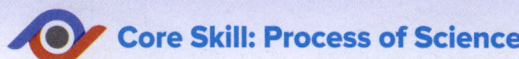

**Core Skill: Process of Science**

# Feature Investigation | An Experiment Performed by Briggs Revealed the Role of Auxin in Phototropism

In a second experiment, Briggs tested the hypothesis that light causes auxin to move to the shaded side of seedlings (**Figure 37.5**). Briggs set shoot tips onto agar blocks (step 1) and then used a mica sheet (which is impermeable to auxin) to completely divide some tips and blocks into halves (step 2A). In other cases, he divided blocks completely but left tips incompletely divided, allowing auxin to diffuse across tips but not the block halves (step 2B).

Then Briggs exposed all sets of tips and blocks to directional light. He predicted that auxin would not be able to move across tips having complete mica barriers but that auxin would be able to move across tips that had been only partially divided. Auxin diffused from tips into blocks, but it could not diffuse evenly across blocks divided by mica sheets. When Briggs later placed the agar block halves on decapitated shoots (step 3), those receiving auxin from completely divided tips were bent by the same amount. By contrast, agar block halves from the lit side of partially divided tips induced less bending, whereas halves from the unlit side of partially divided tips caused the most bending (see the data in step 4 of Figure 37.5). These

**Figure 37.5** Briggs demonstrated the relationship between directional light and auxin function.

Briggs experiment

**HYPOTHESIS** Directional light causes auxin to move to the shaded side of shoot tips.

**KEY MATERIALS** Corn seedlings.

| | Experimental level | Conceptual level |
|---|---|---|
| **1** Place shoot tips on agar blocks. | 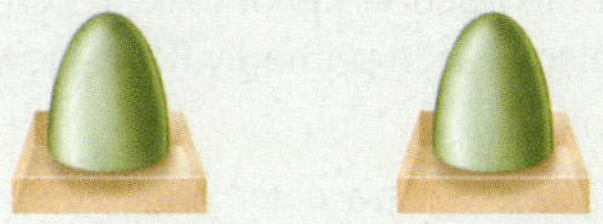 | |
| **2** Divide some tip/block combinations completely with a mica sheet, which prevents diffusion between the 2 halves of the tip and agar block. Divide some tip/block combinations only partially with a mica sheet. This allows auxin diffusion across the tip, but not across the agar block. Expose both to directional light. | 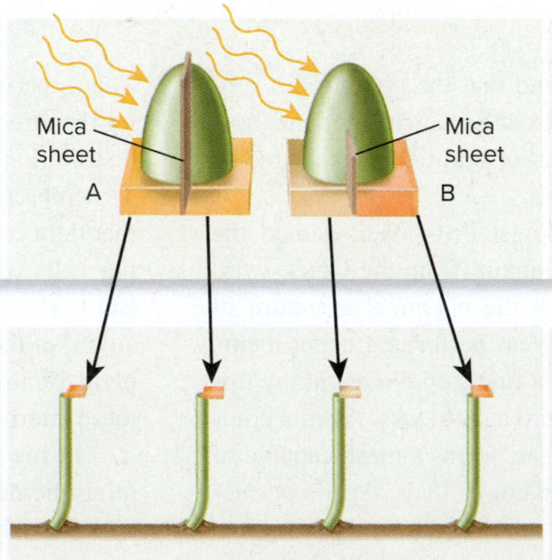 Mica sheet — A    Mica sheet — B | If directional light causes auxins to move to shaded side of shoot tips, agar block in B will contain more auxin on right side.  |
| **3** Remove agar block halves from tips. Place agar halves onto right sides of shoots, which have their tips removed. | | If directional light causes auxins to move laterally, the block half beneath the left side of the partially divided tip shown in B should cause the least shoot bending, whereas the block half beneath the right side of B should cause the greatest amount of bending. |

**4  THE DATA**

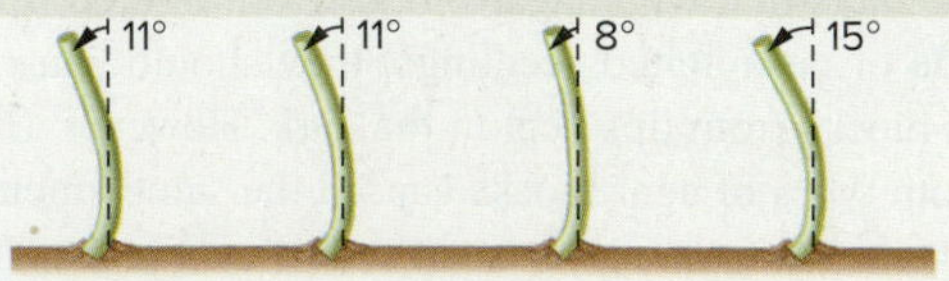

11°   11°   8°   15°

**5  CONCLUSION** In A, the mica sheets prevented auxin from moving to the shaded side of tips. In B, auxin was able to move in response to directional light. Agar block pieces from the shaded side of tips contained more auxin and therefore caused greater shoot bending. Hypothesis is correct.

**6  SOURCE** Briggs, W. R. 1963. Mediation of phototropic responses of corn coleoptiles by lateral transport of auxin. *Plant Physiology* 38(3): 237-247.

experimental results support the hypothesis that unidirectional light causes auxin to accumulate on the shaded side. Modern plant scientists would explain such auxin movement as the result of lateral transport involving PIN proteins.

How might auxin accumulation cause phototropic bending? One widely held hypothesis is that auxin accumulation on the shaded side of a plant shoot causes plasma membrane proton pumps located there to work at a faster rate. In response, the cell wall becomes more acidic, which activates expansins—proteins that break crosslinks between cellulose microfibrils and allow cells to elongate (refer back to Figure 36.10). This process might explain how auxin accumulation in cells located on the shaded side of shoot tips causes them to

elongate more than do cells on the sunny side, causing the tip to bend toward the light.

### Experimental Questions

1. What is the current hypothesized mechanism by which auxin accumulation causes shoot bending in response to directional light?

2. **CoreSKILL »** Figure 37.5 illustrates experiments performed with four seedling tips. Would this number really be enough to allow conclusions to be made about how such seedling tips would generally respond?

---

## BIO·TIPS

**THE QUESTION** *Show how Briggs tested the auxin-destruction hypothesis: that coleoptile bending in response to directional light occurs because light destroys auxin.*

**T**OPIC *What topic in biology does this question address?* The topic is how plants sense and respond to directional light. More specifically, the question asks you to show the procedure Briggs used to investigate the possibility that bending toward the light occurs because light destroys auxin.

**I**NFORMATION *What information do you know based on the question and your understanding of the topic?* You learned earlier in this chapter that phototropism in plants is initiated by directional light and followed by a growth response of stem tissue to an internal chemical signal. The signal was discovered to be the asymmetric distribution of the plant hormone auxin. The first experiment of Briggs is described in the paragraph that precedes the Feature Investigation. However, Briggs' first experiment is not illustrated.

**P**ROBLEM-SOLVING **S**TRATEGY *Make a drawing.* Creating a diagram is an excellent way to achieve understanding and communicate that understanding to others. Use the text description of Briggs's experiment to make diagrams that show how he conducted his first experiment.

**ANSWER** *Start by drawing two cut coleoptile tips placed onto agar blocks. Add arrows on one of the drawings that indicate how auxin is expected to flow into the agar block if the coleoptile tip is evenly illuminated. Add arrows to the other drawing to indicate how auxin is expected to flow into the agar block if directional light destroys auxin. These contrasting drawings illustrate the beginning hypothesis. Now draw diagrams that show how coleoptiles whose tips have been removed respond when agar blocks are placed on one side of them. This is the experimental test. Finally, draw these coleoptiles after they have responded by bending, which is the experimental result. Lack of a difference in bending of the experimental coleoptiles is evidence that directional light does not cause phototropism by destroying auxin.*

## Cytokinins Stimulate Cell Division

Like auxins, the plant hormones known as **cytokinins** play varied and important roles throughout the lives of plants. The name of these hormones reflects their major effect—an increase in the rate of plant cytokinesis, or cell division. Root tips are major sites of cytokinin production, but shoots and seeds also make this plant hormone. Transported in the xylem to meristems and other plant parts, cytokinins bind to receptors thought to be located in the plasma membrane. At shoot and root tips, cytokinins influence meristem size, stem cell activity, and vascular tissue development. Cytokinins are also involved in root and shoot growth and branching, the production of flowers and seeds, and leaf aging.

*Plant Tissue Culture*    In the laboratory, cytokinin and auxin are essential to cloning plants. Cloning involves a process, known as **plant tissue culture**, which is used commercially to produce thousands of identical plants having the same desirable characteristics.

- Plant tissue culture begins with pieces of stem, leaf, or root that have been removed from a plant, and have had their surfaces sterilized to prevent growth of microbes (**Figure 37.6**, step 1).

- The cleaned plant pieces are then placed into dishes containing nutrients (minerals, vitamins, and sugar) and various proportions of auxin and cytokinin. If the proportions of auxin and cytokinin are about the same (1:1), plant cells undergo division, forming a mass of white tissue known as a callus (step 2).

- If the callus is then transferred to a new dish containing the same nutrients, with auxin-to-cytokinin proportions greater than 10:1, the callus will form roots (step 3).

- After root formation has occurred, the proportion of auxin-to-cytokinin can be changed to less than 10:1. This causes the callus to develop green shoots (step 4).

- By altering the ratios of auxin and cytokinin, entire plants can be regenerated from a callus. A single callus can be divided into many pieces and each piece treated with these hormones, thereby producing many hundreds of identical new plants.

## Gibberellins Stimulate Cell Division and Elongation

The **gibberellins** (such as gibberellic acid, or GA; see Table 37.1) are another group of plant hormones. Gibberellins are produced

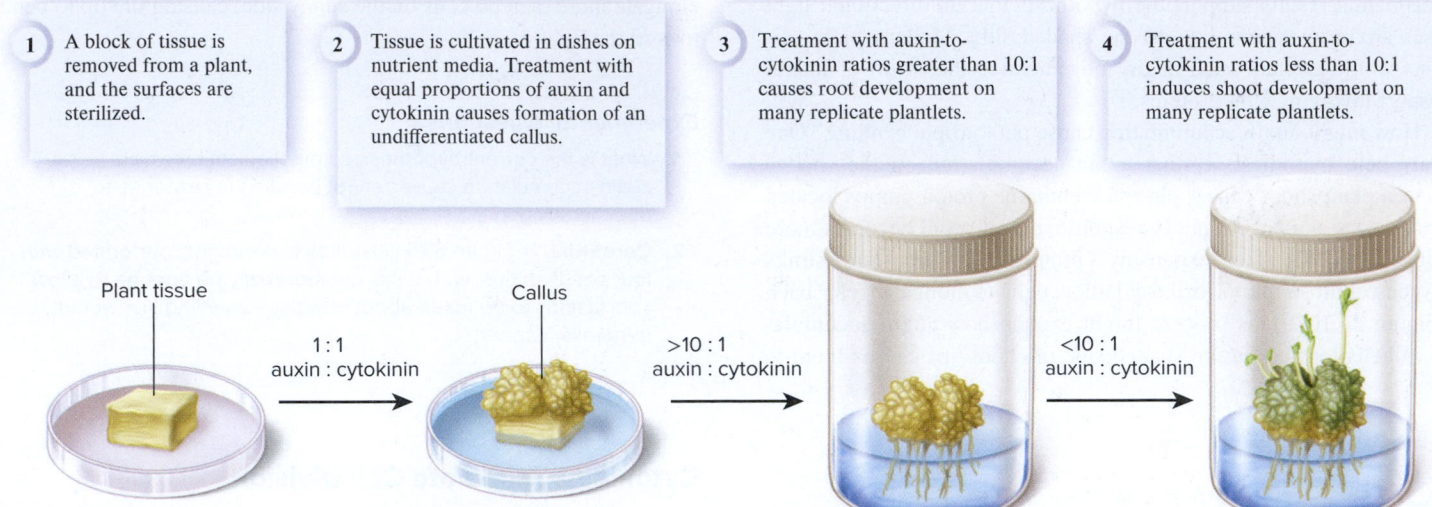

1   A block of tissue is removed from a plant, and the surfaces are sterilized.

2   Tissue is cultivated in dishes on nutrient media. Treatment with equal proportions of auxin and cytokinin causes formation of an undifferentiated callus.

3   Treatment with auxin-to-cytokinin ratios greater than 10:1 causes root development on many replicate plantlets.

4   Treatment with auxin-to-cytokinin ratios less than 10:1 induces shoot development on many replicate plantlets.

Plant tissue

Callus

1 : 1
auxin : cytokinin

>10 : 1
auxin : cytokinin

<10 : 1
auxin : cytokinin

**Figure 37.6**   **The process of plant tissue culture.** Plant tissue culture illustrates the effect of different proportions of auxin and cytokinin on plant organ development.

*Concept Check:*   *How do commercial growers use plant tissue culture to produce many identical plants?*

in apical buds, roots, young leaves, and seed embryos. In addition to promoting shoot development on laboratory calluses, gibberellins interact with light and other hormones to foster seed germination and enhance stem elongation and flowering in nature. Gibberellins also retard leaf and fruit aging. These multiple effects largely arise from the hormones' stimulatory effects on cell division and elongation.

More than a hundred different forms of gibberellin have been found. Many kinds of dwarf plants are short because they produce less gibberellin than taller varieties of the same species. The dwarf strain of pea plants Mendel used in some of his breeding experiments is an example. When dwarf varieties of plants are experimentally sprayed with gibberellin, their stems grow to normal heights. However, dwarf wheat and rice crops are valued in agriculture because they can be more productive and less vulnerable to storm damage than taller varieties. Since the discovery of gibberellin, plant scientists have discovered how this plant hormone works at the molecular level and how gibberellin regulation of plant growth evolved, discussed next.

### Core Concept: Evolution

## Gibberellin Function Arose in a Series of Stages During Plant Evolution

In flowering plants, gibberellin works by helping to liberate repressed transcription factors. In the absence of gibberellin, certain proteins (known as DELLA proteins) bind to particular transcription factor proteins needed for the expression of gibberellin-responsive genes (**Figure 37.7a**). In this way,

**(a) Gibberellin absent**

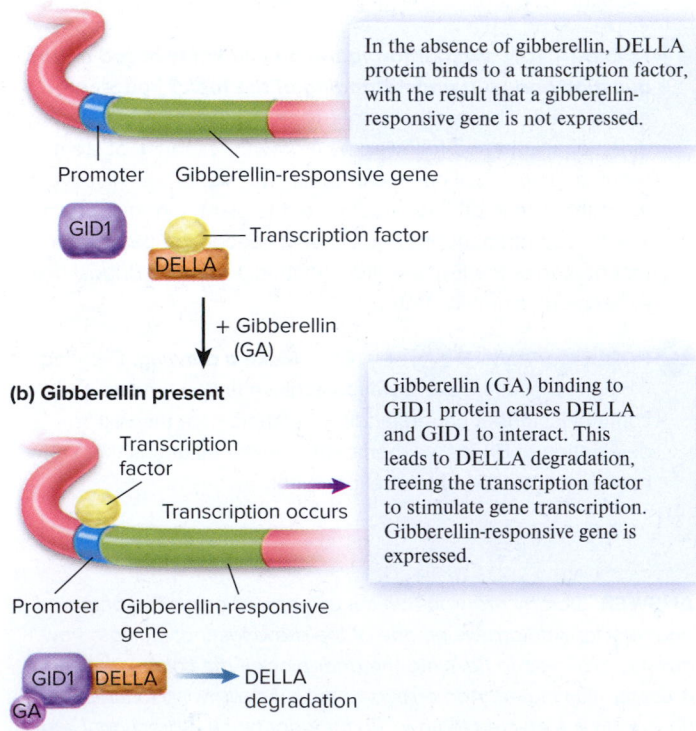

In the absence of gibberellin, DELLA protein binds to a transcription factor, with the result that a gibberellin-responsive gene is not expressed.

Promoter    Gibberellin-responsive gene

GID1

Transcription factor

DELLA

+ Gibberellin
(GA)

**(b) Gibberellin present**

Transcription factor

Transcription occurs

Gibberellin (GA) binding to GID1 protein causes DELLA and GID1 to interact. This leads to DELLA degradation, freeing the transcription factor to stimulate gene transcription. Gibberellin-responsive gene is expressed.

Promoter    Gibberellin-responsive gene

GID1   DELLA  → DELLA degradation

GA

**Figure 37.7**   **Gibberellin works by releasing trapped transcription factors.** **(a)** In the absence of gibberellin, DELLA binds transcription factors, and so gibberellin-responsive genes are not expressed. **(b)** When gibberellin binds to the protein GID1, GID1 can bind DELLA proteins, which causes DELLA proteins to be degraded. As a result, transcription factors that had been bound to DELLA proteins are released and can bind to gene promoters, thereby inducing gene expression.

DELLAs function as brakes that restrain cell division and expansion. When sufficient gibberellin is present, it releases the brakes by binding to a protein known as GID1 (**Figure 37.7b**). Gibberellin binding to GID1 causes DELLAs to release transcription factor proteins and fosters the destruction of DELLAs. In the absence of DELLA proteins, transcription factor proteins are able to bind to the promoter regions of gibberellin-responsive genes, allowing their expression. As a result, cell division and expansion occur, leading to growth.

In 2007, Yuki Yasumura, Nicholas Harberd, and their colleagues reported that the components of the gibberellin-DELLA mechanisms in flowering plants arose from features present in more ancient lycophyte and bryophyte lineages. Modern bryophytes and lycophytes possess GID1 as well as DELLA and the transcription factor proteins it binds, but these substances do not affect plant growth. This observation suggests that DELLA-mediated repression of plant growth evolved after the divergence of lycophytes but prior to the appearance of the first gymnosperms and flowering plants (**Figure 37.8**). Though the necessary components (DELLAs and GID1 proteins) were present earlier, only later did they assemble into a growth regulation system.

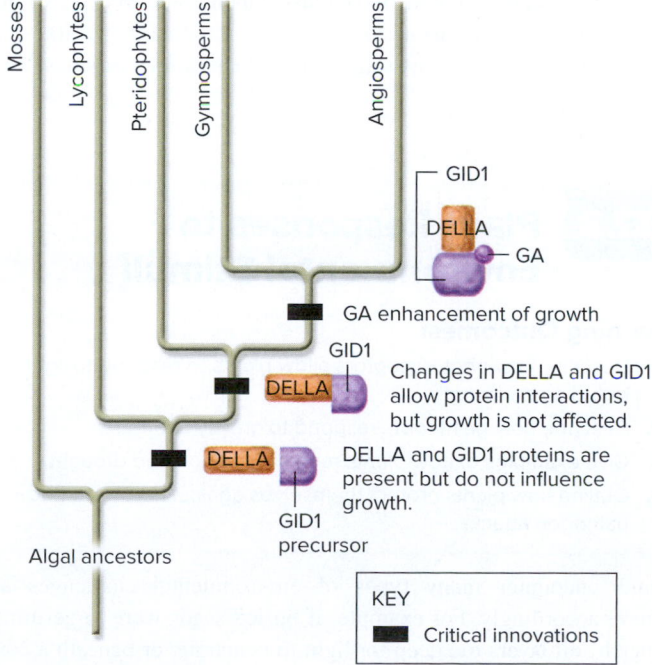

**Figure 37.8** **Evolution of the gibberellin-DELLA system.** Although DELLA and GID1 proteins are present in a moss, they do not interact, nor does gibberellin bind to GID1. In lycophytes, GID1 interacts with DELLA, but the system does not influence growth. The growth-enhancing responses of gibberellin apparently evolved after the divergence of lycophytes.

## Ethylene's Effects Include Cell Expansion

The plant hormone **ethylene** is particularly important in coordinating plant developmental and stress responses. Ethylene is a simple hydrocarbon gas produced during seedling growth, flower development, and fruit ripening (see Table 37.1). In the root tip, ethylene determines how many stem cells remain inactive in the quiescent center and how many cells undergo divisions. This hormone also plays important roles in defense against osmotic stress and pathogen attack, and leaf and petal aging and drop. As a gas, ethylene is able to diffuse through the plasma membrane and cytosol to bind to ethylene receptors localized in the endoplasmic reticulum. When activated by ethylene binding, these receptors inactivate a protein kinase known as CTR1. This action ultimately enables transcription factors to induce the transcription of various genes.

People first noticed the effects of ethylene gas on plants in the 1800s when they observed that street-side trees exposed to leaking street lanterns unexpectedly lost their leaves. A 17-year-old student in St. Petersburg, Russia, Dimitry Neljubov, performed the first experiments to explore the effects of illumination gas on plants. He exposed pea seedlings grown in the laboratory to illumination gas and noticed that the pea seedlings grew sideways rather than upward. Then he tested the individual chemical components of illumination gas for the same effect. After conducting many experiments, in 1901, Neljubov reported that ethylene was the only component of illumination gas that caused the seedlings to grow horizontally and that ethylene was effective in very low concentrations (as low as 0.06 parts per million in air). Later, scientists established that ethylene influences cell expansion, often in association with auxin. Ethylene does this by increasing the disorder of microtubules within cells, thereby causing random orientation of cell-wall microfibrils. As a result, cells exposed to ethylene tend to expand in all directions rather than elongating.

To understand how ethylene affects cell growth, researchers have exposed growing seedlings to varying concentrations of ethylene in the dark. (**Figure 37.9**). At low concentrations, ethylene prevents the seedling stem and root from elongating. At moderate concentrations, the hormone induces the stem and root to swell radially, thereby increasing in thickness. Together, these responses strengthen the seedling stem and root. At even higher concentrations, the seedling stem bends so that embryonic leaves and the delicate meristem grow horizontally rather than vertically; this is the sideways growth response that Neljubov first observed. Taken together, the three effects of ethylene on seedling growth is called the triple response.

What is the biological significance of the triple response? These experimental results provide insight regarding the growth of seedlings while they are still within the soil. The tender apical meristems of seedlings could be easily damaged during their growth through crusty soil. Ethylene helps seedlings avoid damage by thickening the stem and root. The bent portion of the stem, known as a hook, then pushes up through the soil. The hook forms as the result of an imbalance of auxin across the stem axis, which causes cells on one side of the stem to elongate faster than cells on the other side. Ethylene drives this auxin imbalance.

Knowledge of the effects of ethylene on fruit has been very useful commercially. Ripe fruit can be easily damaged during transit,

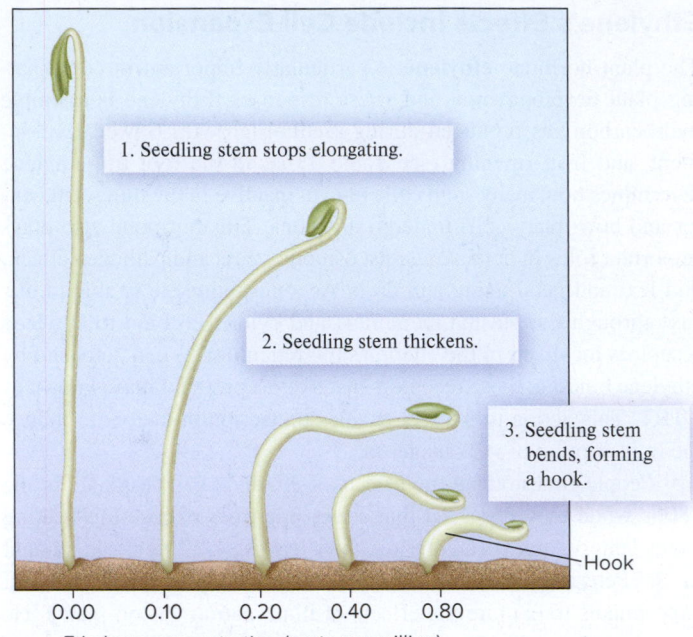

1. Seedling stem stops elongating.

2. Seedling stem thickens.

3. Seedling stem bends, forming a hook.

Hook

0.00    0.10    0.20    0.40    0.80

Ethylene concentration (parts per million)

**Figure 37.9  Seedling growth showing the triple response to ethylene.** Ethylene applications at increasing concentrations cause seedlings to cease elongation, swell radially, and bend to form a hook that can push upward through the soil. Ethylene produced naturally within seedlings causes the same response.

**Concept Check:** *What adaptive advantage does this seedling behavior provide?*

but tomatoes and apples can be picked before they ripen for transport with minimal damage. At their destination, such fruit can be ripened by treatment with ethylene. However, fruit that becomes overripe may exude ethylene, which hastens ripening in nearby, unripe fruit. For this reason, fruit that must be stored for extended periods is kept in ethylene-free environments.

## Several Hormones Help Plants Cope with Environmental Stresses

Several plant hormones share the property of helping plants respond to environmental stresses such as flooding, drought, high salinity, cold, heat, and attack by disease-causing microorganisms and animal herbivores. These protective hormones include the major plant hormones known as **abscisic acid** and **brassinosteroids** (see Table 37.1).

*Abscisic Acid*   Abscisic acid (ABA) was named at a time when plant biologists thought that it played a role in leaf or fruit drop, also known as abscission. Later, they discovered that ethylene actually causes leaf and fruit abscission, whereas abscisic acid slows or stops plant metabolism when growing conditions are poor.

ABA has a variety of effects on plant growth and function. For example, ABA may induce bud and seed dormancy. Dormant buds and seeds resume growth only when specific environmental signals reveal the onset of conditions suitable for survival. In preparation

for winter, ABA stimulates the formation of tough, protective scales around the buds of perennial plants. Seed coats of apple, cherry, and other plants also accumulate ABA, which prevents seeds from germinating unless temperature and moisture conditions are favorable for seedling growth. Water-stressed roots also produce ABA, which is then transported to shoots, where (together with ABA produced by water-stressed leaf mesophyll) it helps to prevent water loss from leaf surfaces by inducing leaf pores (stomata) to close.

To exert its effects, ABA binds to ABA receptors, which are soluble proteins in the cytoplasm. The binding of ABA to its receptors promotes the transcription of ABA-responsive genes, thereby leading to a cellular response.

*Brassinosteroids*   Brassinosteroids are named after the cruciferous plant genus *Brassica* (which includes cabbage and broccoli), in which they were first identified. However, seeds, fruit, shoots, leaves, and flower buds of all types of flowering plants contain brassinosteroids. These plant hormones induce water uptake by vacuoles and influence enzymes that alter cell-wall carbohydrates, thereby fostering cell expansion. Mutations that affect brassinosteroid synthesis cause plants to exhibit dwarfism. Such plants have small, dark green cells because their tissues are unable to expand. Brassinosteroids also impede leaf drop, help grass leaves to unroll, and stimulate xylem development. They can be applied to crops to help protect plants from heat, cold, high salinity, and herbicide injury.

Brassinosteroids are chemically related to animal steroid hormones, such as human sex hormones. However, unlike animal steroid hormones, which bind to receptors in the nucleus or cytosol, brassinosteroids bind to receptors in the plasma membrane. When they bind brassinosteroids, the membrane receptors initiate a signal transduction pathway that activates transcription factors for brassinosteroid-responsive genes.

## 37.3  Plant Responses to Environmental Stimuli

### Learning Outcomes:

1. Describe how photoreceptors allow plants to respond to light, including day length.
2. Describe how plant roots respond to gravity and touch.
3. Give examples of how plants respond to flood and drought.
4. Outline how plants protect themselves against herbivore and pathogen attacks.

Plants encounter many types of environmental challenges and behave accordingly. For example, if buried seeds were to germinate beneath soil layers too deep for light to penetrate, or beneath a cover of established plants, seedlings would not be able to obtain sufficient light for photosynthesis and would die. A related reproductive challenge for plants is to flower at times of the year that are most beneficial for achieving pollination or seed dispersal. In this section, we will examine how plants determine if there is enough light for seeds to germinate and for seedlings to grow and how they determine when to flower.

## Plants Detect Light via Photoreceptors

Many activities in plants, such as photosynthesis and flowering, require the proper amount of light and must occur during the correct time of year. To sense the amount and direction of light, plants produce light sensors known as **photoreceptors**, of which there are several types. Some types of plant photoreceptors also occur in animals, whereas others are particular to plants. Each type of photoreceptor has a light-absorbing component as well as other regions that respond to light absorption by switching on signal transduction pathways. Responses by many cells in a tissue or organ cumulatively result in behaviors such as sun tracking, phototropism, flowering, and seed germination.

***Blue-Light Receptors***   Cryptochrome and phototropin are two types of blue-light receptors—molecules that absorb and respond to blue light. Experiments suggest that **cryptochrome** helps young seedlings determine if their environment has enough light for adequate photosynthesis. If not, seedlings continue to elongate through the soil, toward the light. Cryptochromes also function as light sensors in animals.

   **Phototropin** is a blue-light sensor so named because one of its roles is to promote positive phototropism, the tendency of a plant to grow toward a light source. The light-activated form of this sensor has two components: a protein that has a kinase domain and a flavin pigment that can absorb blue light.

   As you may recall from Chapter 9, a protein kinase domain catalyzes the attachment of phosphate to proteins. In the dark, flavin is not covalently bound to the phototropin protein and its kinase domain is inactive. However, when flavin absorbs blue light, it changes conformation and covalently binds to the protein. Flavin binding, in turn, changes the conformation of the phototropin protein, allowing it to phosphorylate itself by means of the protein kinase domain. Therefore, when a plant is exposed to directional blue light, phototropin becomes phosphorylated. Though the steps are not entirely understood, phosphorylation is thought to initiate a series of events that alters the distribution of auxin and thereby causes the plant to grow toward the light. In this way, a light signal is converted into a chemical signal that influences plant growth.

***Phytochrome, the Red-Light and Far-Red-Light Receptor***
Many plant growth and developmental processes are influenced by **phytochrome**, a red-light and far-red-light receptor. Phytochrome operates much like a light switch, flipping back and forth between two conformations: $P_r$ absorbs red light, and $P_{fr}$ absorbs only far-red light (light having a wavelength longer than that of red light) (**Figure 37.10**). When red light is abundant, as in full sunlight, $P_r$ absorbs red light and changes to $P_{fr}$, which activates cellular responses such as seed germination. When left in the dark for a long period, $P_{fr}$ slowly transforms into $P_r$.

   The role of phytochrome as a plant "light switch" has been shown experimentally in studies of lettuce-seed germination. Researchers have found that water-soaked lettuce seeds will not germinate in darkness, but they will germinate if exposed to as little as 1 minute of red light (**Figure 37.11**). This amount of light exposure is sufficient to transform a critical amount of $P_r$ to the active $P_{fr}$ conformation, which stimulates germination. However, if this brief red-light treatment is followed by a few minutes of treatment with far-red light, the lettuce seeds will not germinate. This short period of far-red illumination is enough to convert seed $P_{fr}$ back to the inactive $P_r$ conformation.

   The most recent light exposure determines whether the phytochrome occurs in the active or inactive conformation. In nature, if seeds are close enough to the surface that their phytochrome is switched on by red light, the seeds will germinate. But if seeds are buried too deeply for red light to penetrate, they will not germinate. In this way, seeds can sense if they are close enough to the surface to begin the germination process.

   In the dark, phytochrome molecules in the $P_r$ state reside in the cytosol. After exposure to red light, activated phytochrome ($P_{fr}$)

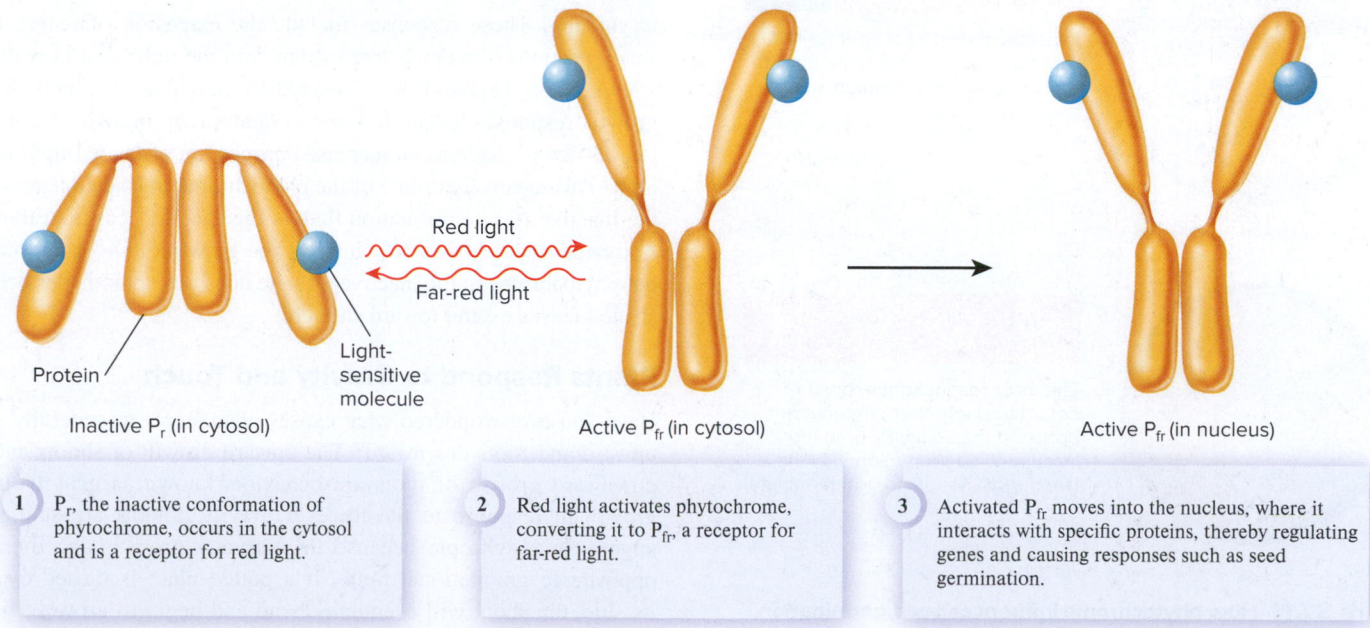

Red light

Far-red light

Protein

Light-sensitive molecule

Inactive $P_r$ (in cytosol)

Active $P_{fr}$ (in cytosol)

Active $P_{fr}$ (in nucleus)

1  $P_r$, the inactive conformation of phytochrome, occurs in the cytosol and is a receptor for red light.

2  Red light activates phytochrome, converting it to $P_{fr}$, a receptor for far-red light.

3  Activated $P_{fr}$ moves into the nucleus, where it interacts with specific proteins, thereby regulating genes and causing responses such as seed germination.

**Figure 37.10   How phytochrome acts as a molecular light switch.**

**Figure 37.11** **How phytochrome influences seed germination.**

*Concept Check:* *Describe the change in phytochrome that would occur if a deeply buried seed were uncovered enough to receive sunlight.*

molecules move from the cytosol to the nucleus. Within the nucleus, $P_{fr}$ interacts with a transcription factor protein known as PIF3 (phytochrome-interacting factor 3). PIF3 binds to the regulatory elements of several phytochrome-responsive genes, functioning as a positive regulator of some of the genes and as a negative regulator of others.

***Photoperiodism*** A plant's ability to measure and respond to the amount of light and day length is a process called **photoperiodism**, which influences the timing of dormancy and flowering. Phytochromes also play a critical role in photoperiodism. Flowering plants can be classified as long-day, short-day, or day-neutral plants. When scientists named these groups, they thought that plants measured the amount of daylight. Researchers later discovered that plants actually measure night length.

Lettuce, spinach, radish, beet, clover, gladiolus, and iris are examples of **long-day plants** because they flower in spring or early summer, when the night period is shorter (and thus the day length is longer) than a defined length (**Figure 37.12**). In contrast, asters, strawberries, dahlias, poinsettias, potatoes, soybeans, and goldenrods are examples of **short-day plants** because they flower only when the night length is longer than a defined period. Such night lengths occur in late summer, fall, or winter, when days are short. As shown in Figure 37.12, when plants are given an experimental light flash in the middle of a long dark period, the long-day plants flower but the short-day plants do not. These results indicate that both types of plants measure night length. Roses, snapdragons, cotton, carnations, dandelions, sunflowers, tomatoes, and cucumbers flower regardless of the night length, as long as day length meets the minimal requirements for plant growth, and are thus known as **day-neutral plants**.

Ornamental plant growers manipulate night length to produce flowers for market during seasons when they are not naturally available. For example, chrysanthemums are short-day plants that usually flower in the fall, but growers use light-blocking shades to increase night length in order to produce flowering plants at any season.

***Shading Responses*** Phytochrome also mediates plant responses to shading. These responses include the extension of leaves from shady portions of a dense tree canopy into the light, and growth that allows plants to avoid being shaded by neighboring plants. These growth responses occur by the elongation of branch internodes. Leaves detect shade as an increased proportion of far-red light to red light. This means that more of the phytochrome in shaded leaves is in the inactive ($P_r$) conformation than is the case for leaves in the sun. Activated phytochrome ($P_{fr}$) inhibits the growth of shoot internodes, but phytochrome in the inactivated state does not, so branches bearing shaded leaves extend toward sunlight.

## Plants Respond to Gravity and Touch

Have you ever wondered what causes plant stems to generally grow upward and roots downward? The upward growth of shoots and the downward growth of roots are behaviors known as **gravitropism**, growth in response to the force of gravity. Shoots are said to be negatively gravitropic because they usually grow in the direction opposite to gravitational force. If a potted plant is turned over on its side, the shoot will eventually bend and begin to grow vertically again (**Figure 37.13**). Most roots are said to be positively gravitropic because they grow in the same direction as the gravitational force.

**Figure 37.12  Flowering and photoperiodism.**  Iris is a long-day plant that flowers in response to the short nights of late spring and early summer, whereas goldenrod is a short-day plant that flowers in response to the longer nights of autumn. The length of night is the critical factor, as shown by the effects of light flashes. (top left, bottom left): ©Ray Bulson/Newscom; (top right, bottom right): ©Lee W. Wilcox; (middle left): ©Garden Picture Library/Getty Images; (middle right): ©Comstock Images/Getty Images

**Concept Check:**  *What would happen if you gave flowering plants a brief exposure to darkness in the middle of the daytime?*

**Figure 37.13  Negative gravitropism in a shoot.**  This tomato shoot system has resumed upward growth after being placed on its side. Upward growth started about 4 hours after the plant was turned sideways; this photo was taken 20 hours later. Shoots sense gravity by means of starch-heavy statoliths present in stem tissue near the central vascular tissue. ©Lee W. Wilcox

Both roots and shoots detect gravity by means of starch-heavy plastids known as **statoliths**, which are located in specialized gravity-sensing cells called statocytes. In shoots, statocytes are located in a tissue known as the endodermis, which forms a sheath around

vascular tissues. In roots, gravity-sensing cells primarily occur in the center of the root cap (**Figure 37.14**).

Gravity causes the relatively heavy statoliths to sink, which affects the levels of calcium ion second messengers that influence the direction of auxin transport. When roots are vertical, auxin moves upward equally on both sides and the root continues to elongate in the downward direction. However, a change in the direction of root growth will alter the relative position of the statoliths. For example, in a root that becomes oriented horizontally, the statoliths are pulled by gravity to the lower sides of statocytes (see right side of Figure 37.14). The change in statolith position causes more auxin to move to cells on the lower sides of roots. In roots, auxin inhibits cell elongation (in contrast to its action in shoots). Therefore, root growth slows on the lower side, while cell elongation continues normally on the upper side. This process causes the root to bend, so that it eventually grows downward again.

Recent studies suggest that gravity responses are related to touch responses, known as **thigmotropism** (from the Greek *thigma*, meaning touch). For example, when roots encounter rocks or other barriers to their downward growth in the soil, they display a touch response that temporarily supersedes their response to gravity. Such roots grow horizontally until they get around the barrier, whereupon downward growth in response to gravity resumes. Plant shoots also respond to touch; examples include vines with tendrils that wind around or clasp supporting structures. Wind also induces touch responses. In very

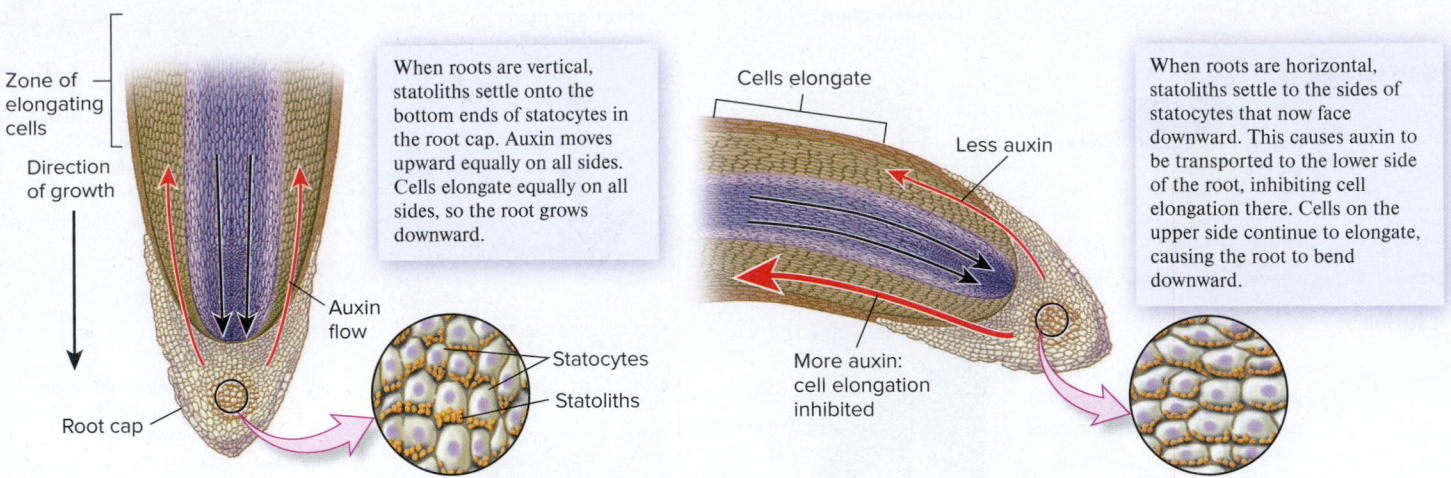

**Figure 37.14** **Positive gravitropism in a root.** Root-tip cells sense gravity by means of starch-heavy statoliths present in cells at the center of the root cap.

**Concept Check:** *Is there any other environmental signal that roots could use to achieve downward growth?*

windy places, trees tend to be shorter than normal, giving them the advantage of being less likely to be blown over than taller trees. In the laboratory, plant scientists have simulated natural touch responses by rubbing plant stems and found that this treatment can result in shorter plants. Touch causes the release of calcium ion second messengers that influence gene expression.

More rapid responses to touch, such as leaf folding by the sensitive plant (see Figure 37.1b), are based on changes in the water content of cells within a structure known as a pulvinus (plural, pulvini), a swelling located at the base of attachment of each pair of leaflets in

complex leaves. A pulvinus consists of a thick layer of parenchyma cells that surrounds a core of vascular tissue (**Figure 37.15**). When the leaflet of a sensitive plant is touched, an electrical signal called an action potential opens ion channels in parenchyma cells near the lower surfaces of the pulvini. These cells expel potassium and chloride ions, causing water to flow out and the cells to become flattened. This bends the leaflets together, starting the leaflet-folding process. The action potential generated at the touch site also flows through the leaflet, causing many or all of the leaflets to also bend, with the result that the entire leaf folds. Reversal of this process allows the leaf to unfold.

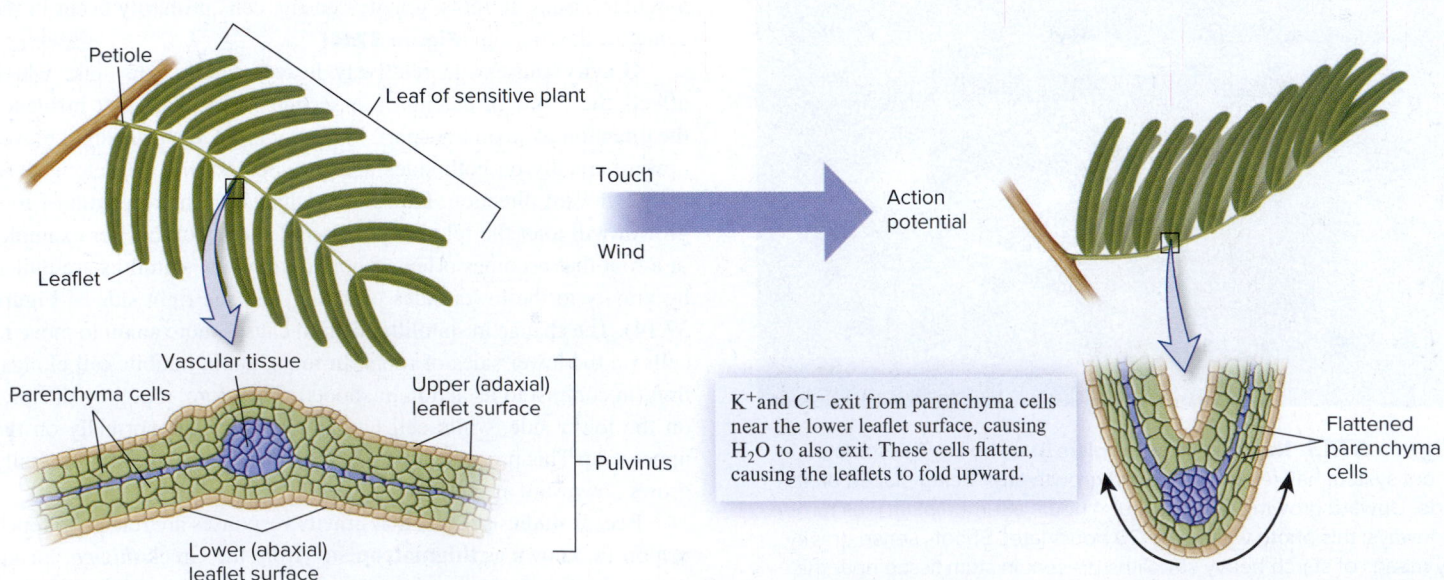

**Figure 37.15** **Leaf folding in the sensitive plant: How pulvini change the positions of leaflets.** The electrical signals known as action potentials result from the rapid flow of ions through membrane ion channels. Electrical signals spread from one cell to another through plasmodesmata. Cells near the lower surface of the pulvinus respond to ion flow by losing water, causing them to flatten.

 **Core Skill: Connections** Action potentials are also key to the function of the nervous systems of animals (see Chapter 42).

The action of pulvini also explains some plant movements that are unrelated to touch, including sleep movements and sun tracking. Sleep movements are changes in leaf position that occur in response to day-night cycles. Sun tracking, as described at the beginning of this chapter, is the movement of leaves or flowers in response to the Sun's position.

## Plants Respond to Physical Stresses Such as Flooding and Drought

Plants display many types of adaptations that help them cope with unfavorable growth conditions, such as flooding and drought. These responses are often mediated by hormones.

In most plants, roots get oxygen from air that is located in the spaces between soil particles. The major harmful effect of flooding is that too much water makes roots unable to obtain sufficient oxygen to fuel cellular respiration. Without oxygen from the air, roots cannot produce the ATP needed to absorb minerals from soil.

Many plants reduce the effects of flooding by producing **aerenchyma**, a tissue containing large, snorkel-like air channels that allow more oxygen to flow from shoots to the submerged roots (**Figure 37.16**). In some plants, aerenchyma formation is developmentally programmed. For example, aerenchyma develops in the roots of many plants native to wetland habitats even when the soil is not wet. Alternatively, aerenchyma formation may be a response to a change in environmental conditions. The roots of cultivated plants, such as corn, develop aerenchyma in response to flooding. Aerenchyma formation is regulated by the action of ethylene, which leads to programmed cell death, followed by cell collapse.

During a drought, plants are likely to receive too little water. The amount of water content in a plant is also affected by other environmental stresses—high salinity, heat, and cold—that reduce the amount of liquid water present in plant cells. Most plants that lose half or more of their water are unable to recover. To prevent this from happening, plants have acquired diverse adaptations that reduce water loss, such as the closure of stomata and, in extreme cases, wilting. The hormone abscisic acid triggers these responses.

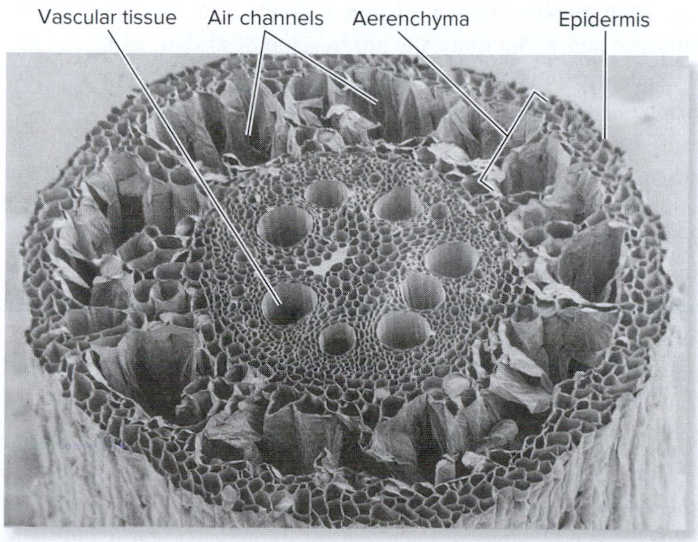

## Plants Respond to Biological Stresses Such as Herbivore and Pathogen Attacks

Plants are vulnerable to attack by animal herbivores and pathogens—disease-causing microorganisms. Structural features such as cuticles, epidermal trichomes, and outer bark, described in Chapter 36, help to reduce infections and herbivore attacks. These structures, together with chemical defense compounds, explain why remarkably little natural vegetation is lost to herbivore or pathogen attack. However, agricultural crops can be more vulnerable to attack than their wild counterparts. This increased vulnerability arose because some protective adaptations were lost during crop domestication as the result of genetic changes that increase edibility. For this reason, crop scientists are particularly interested in understanding plant defense, with the goal of being able to breed or genetically engineer crop plants that are better protected from pests.

***Plant Responses to Herbivores***    Plants produce diverse herbivore-defense compounds including secondary metabolites—terpenes and terpenoids, phenolics, and alkaloids (refer back to Figure 32.22)—and also poisonous hydrogen cyanide and other molecules. Some of these substances act directly on herbivores, making plants taste bad so that herbivores learn to avoid them. Other chemical compounds function indirectly. For example, when attacked by insect caterpillars, cruciferous plants release terpenoids that attract the bodyguard wasp (*Cotesia rubecula*), which attacks the caterpillars.

Some wounded plants can communicate with nearby plants, helping them to resist attack by herbivores (**Figure 37.17**). For example, when cutworms wound tomato plants, damaged cells release a volatile alcohol, hexenol. Nearby plants convert the hexenol into a compound that reduces the cutworms' ability to attack them.

Plants can also receive signals warning of herbivore attack from other plants by means of connecting parasitic plants. Non-photosynthetic dodders (*Cuscuta* species) (see Figure 38.20) form natural grafts to the stems of photosynthetic host plants, sometimes connecting many host plants of the same or different species. These parasite grafts allow the transmission of metabolites, proteins, mRNAs, and warning compounds from plants that have suffered herbivore attack to plants that have not yet been attacked, allowing the latter to take preemptive defensive action.

***Plant Responses to Pathogen Attack***    Every year, about 15% of crop production is lost to diseases caused by certain bacteria, fungi, protists, or viruses. During their evolution, plants have evolved mechanisms for detecting and responding to such microbes, but microbes can rapidly evolve new disease processes. Agricultural scientists aim to understand how pathogenic microbes attack plants and how plants are able to prevent disease, in order to develop disease-resistant crop varieties.

Vascular tissue    Air channels    Aerenchyma    Epidermis

**Figure 37.16  A plant response to flooding.** This image of a slice of a root shows air channels within aerenchyma. Courtesy Dr. Malcolm Drew, Texas A&M University

**Core Concept: Structure and Function**  Air channels within aerenchyma allow air to flow from shoots to roots even when the plant is partially submerged.

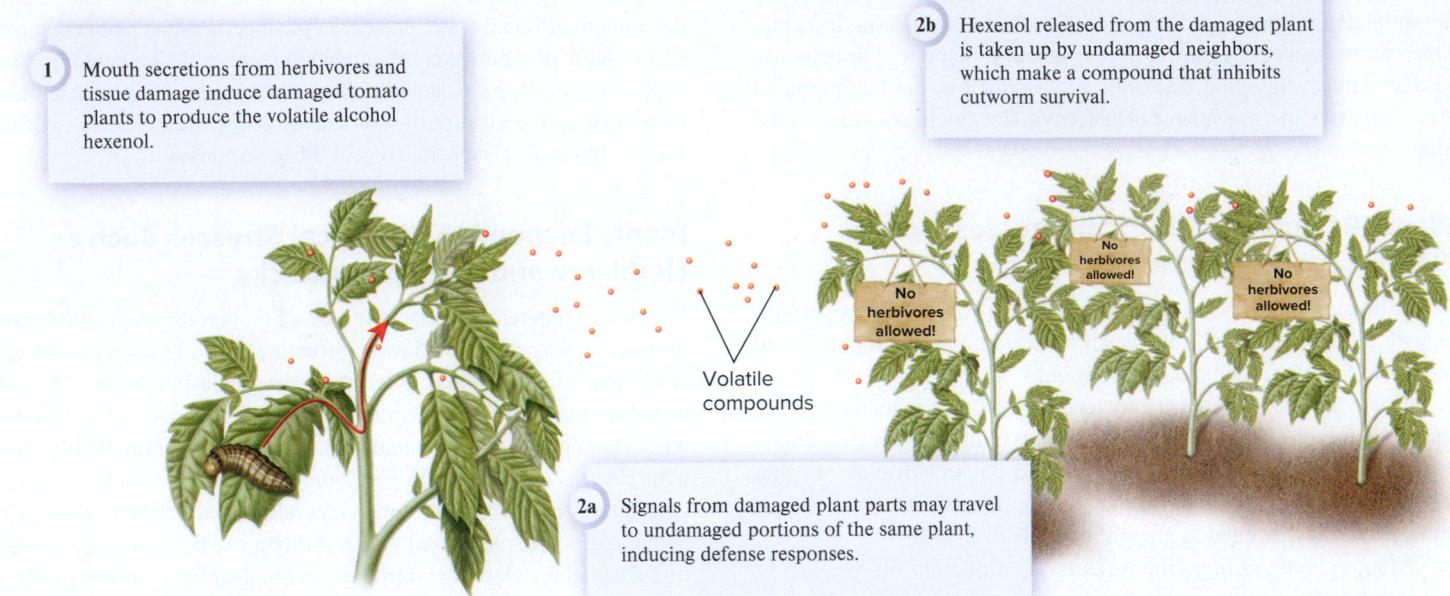

**1** Mouth secretions from herbivores and tissue damage induce damaged tomato plants to produce the volatile alcohol hexenol.

**2b** Hexenol released from the damaged plant is taken up by undamaged neighbors, which make a compound that inhibits cutworm survival.

Volatile compounds

No herbivores allowed!

**2a** Signals from damaged plant parts may travel to undamaged portions of the same plant, inducing defense responses.

**Figure 37.17** **Plant responses to herbivore attack.** **(1)** Leaf damage induces defensive responses such as the release of volatile compounds. **(2a)** These compounds trigger defenses in undamaged parts of the same plant. **(2b)** Such volatile compounds may also induce defenses in neighboring plants, which then become less vulnerable to herbivores.

Plant pathogens produce compounds known as **elicitors**. Such elicitors promote **virulence**, which is the ability of a pathogen to infect its host and cause disease. **Avirulence genes (*Avr* genes)** are microbial genes that encode elicitors or encode enzymes that are needed to synthesize elicitors (**Figure 37.18**, step 1). Some elicitors bind to the surface of plant cells whereas others are injected into the cytosol. For example, certain bacteria inject elicitors into plant cells by means of syringe-like systems (see Figure 27.18). Fungal pathogens often deliver elicitors into plant cells and absorb nutrients from them by means of penetration structures known as haustoria (from a Latin word, meaning to drink; see Figure 29.23).

To detect elicitors, plant species have evolved first and second lines of defense (see step 2 of Figure 37.18). The first line of defense consists of plasma membrane receptors, typically protein kinases, which specifically bind pathogen-associated elicitors. Bacterial elicitors include lipopolysaccharides, the flagellar protein flagellin, and the cell wall material peptidoglycan. Fungal elicitors include the cell wall compound chitin. A second line of defense involves receptors that bind elicitors in the cytosol. Together, these two signal detection systems allow plants to detect elicitors produced by pathogens and arm themselves against attack.

In plants, many types of **resistance genes (*R* genes)** encode receptor proteins that recognize elicitors. If a plant is genetically unable to produce a receptor that can recognize a pathogen's presence or elicitor, disease may result. By contrast, plants successfully resist disease when the product of a dominant *R* gene (a receptor) recognizes a pathogen's presence by characteristic signal compounds or its dominant *Avr* gene product (an elicitor). Diverse alleles of *R* genes occur in plant populations, providing plants with a large capacity to cope with different types of pathogens. As described next, these receptors, together with plant responses—the hypersensitive response and systemic acquired resistance—represent the plant immune system.

***The Hypersensitive Response to Pathogen Attack*** The binding of elicitors to receptors results in the production of chemical defense signals: hydrogen peroxide ($H_2O_2$), the gas nitrous oxide (NO), and mobile hormones (see steps 3 and 4 of Figure 37.18). The plant **hypersensitive response (HR)** is a local reaction to pathogen attack that limits the progression of disease. $H_2O_2$ can kill pathogens and also helps strengthen the cell wall by promoting the formation of cross-links in cell-wall polymers. The gaseous hormone NO works with $H_2O_2$ to stimulate the synthesis of hydrolytic enzymes, defensive secondary metabolites, defense hormones, and tough lignin in the cell walls of nearby tissues. NO also induces programmed cell death as a way of limiting the spread of infection. Necrotic spots are brown patches on plant organs that reveal where pathogens have attacked plants and where plant tissues have battled back. In some plants, bacterial infection causes the cells to avoid secreting sugars, which reduces food available to the pathogen. Cellular miRNAs also may help to destroy the nucleic acids of pathogenic viruses.

***Systemic Acquired Resistance*** Defense hormones, which are made at or near a site of pathogen infection, send alarm signals to other parts of the plant, which respond by preparing defenses. Whereas the HR is a local reaction, **systemic acquired resistance (SAR)** is a defensive response of the whole plant induced by a pathogen attack (**Figure 37.19**). SAR immunizes the plant not only from the inducing pathogen, but also many others for weeks to months. Plant SAR is somewhat similar to the immune systems present in animals.

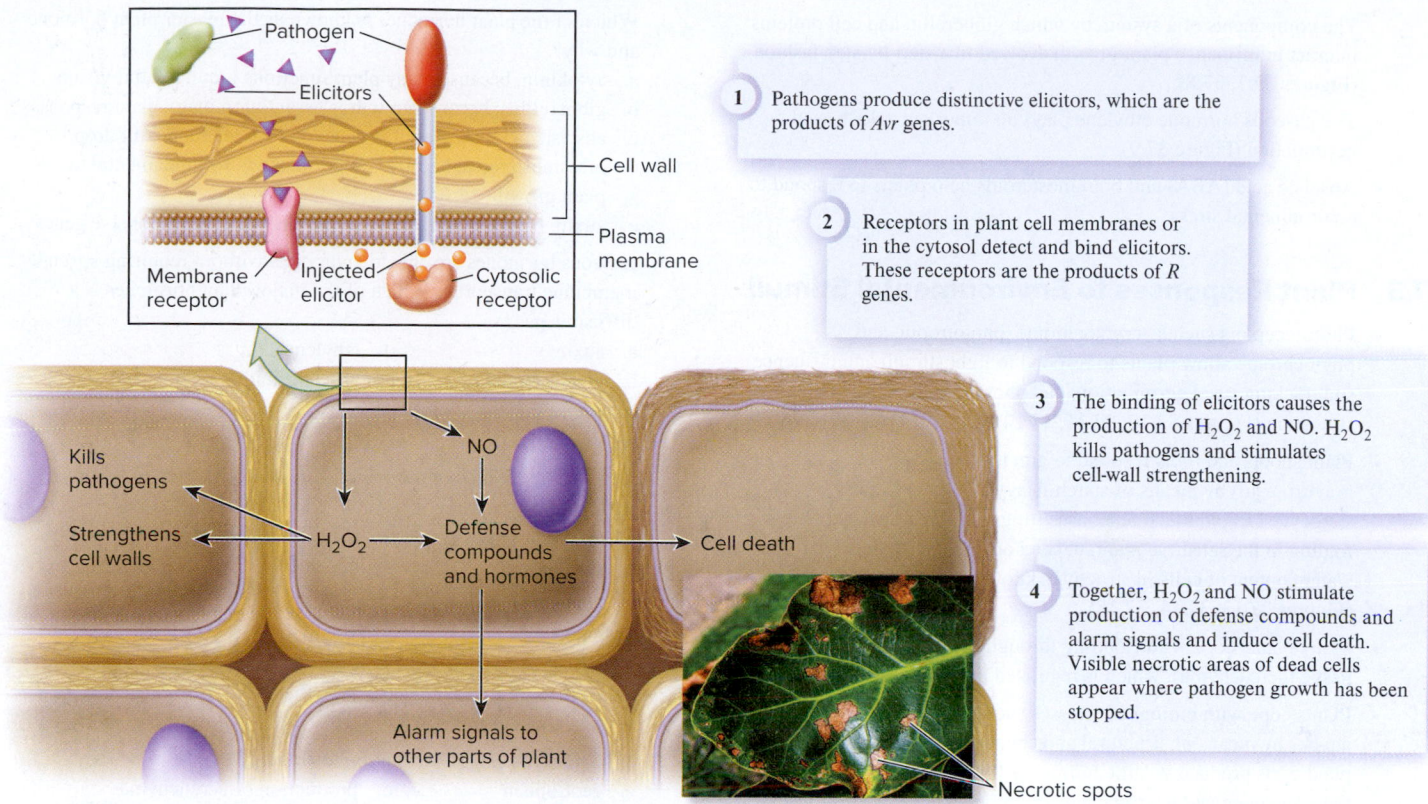

1  Pathogens produce distinctive elicitors, which are the products of *Avr* genes.

2  Receptors in plant cell membranes or in the cytosol detect and bind elicitors. These receptors are the products of *R* genes.

3  The binding of elicitors causes the production of $H_2O_2$ and NO. $H_2O_2$ kills pathogens and stimulates cell-wall strengthening.

4  Together, $H_2O_2$ and NO stimulate production of defense compounds and alarm signals and induce cell death. Visible necrotic areas of dead cells appear where pathogen growth has been stopped.

**Figure 37.18** **Pathogen/plant interactions and the hypersensitive response to pathogen attack.** (inset): ©G. R. "Dick" Roberts/Natural Sciences Image Library

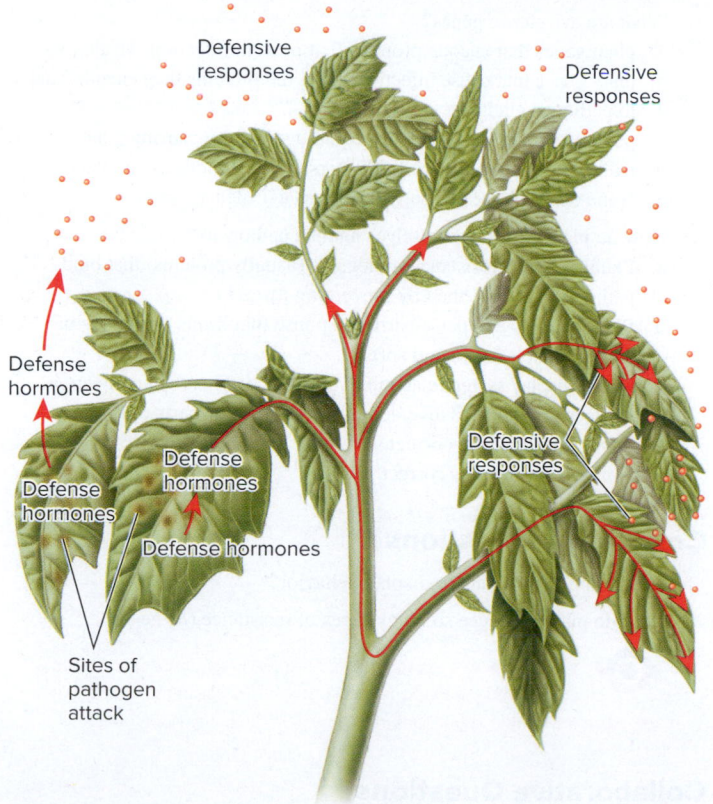

**Figure 37.19** **Systemic acquired resistance to pathogen attack.**

**Concept Check:** *In what way is systemic acquired resistance in plants similar to the immune systems of animals?*

## Summary of Key Concepts

### 37.1 Overview of Plant Behavioral Responses

- Plants sense and respond to diverse internal and external stimuli and thus display behavior (Figures 37.1, 37.2).

- During the process of plant cell signaling, cellular receptors respond to environmental stimuli as well as to internal hormonal signals. The process involves receptor activation, signal transduction by second messengers, and a cellular response (Figure 37.3).

- Cellular responses include changes in gene expression and ion channels that influence plant growth, development, reproduction, chemistry, and movements such as sun tracking and leaf folding.

### 37.2 Plant Hormones

- Plant hormones interact with environmental stimuli to control plant development, growth, and behavior (Table 37.1).

- Auxin plays an important role in many aspects of plant behavior, including phototropism, as demonstrated by the classic experiments of the Darwins, Went, Briggs, and others. Auxin can be transported downward or upward (polar transport) and sideways (lateral transport) in the plant. The positioning of auxin influx and efflux carriers determines the direction of auxin transport (Figures 37.4, 37.5).

- The major effect of cytokinins is to increase the rate of cell division. Auxin and cytokinin are used in laboratory and commercial plant tissue culture (Figure 37.6).

- Gibberellin function illustrates the general principle that plant hormones often act to release cellular brakes on gene expression.

The components of a system by which gibberellin and cell proteins interact to influence plant growth evolved in a step-by-step fashion (Figures 37.7, 37.8).

- The gaseous hormone ethylene plays an important role in seed germination (Figure 37.9).

- Abscisic acid (ABA) and brassinosteroids help plants to respond to environmental stress.

## 37.3   Plant Responses to Environmental Stimuli

- Photoreceptors such as cryptochrome, phototropin, and phytochrome allow plants to respond to light stimuli and influence sun tracking, seed germination, and photoperiodic control of flowering (Figures 37.10, 37.11, 37.12).

- Plant shoots and roots respond to gravity (behaviors known as gravitropism) by means of starch-heavy statoliths located within statocytes. Rapid touch responses (thigmotropism), such as leaf folding in the sensitive plant, depend on changes in the water content of cells in structures known as pulvini (Figures 37.13, 37.14, 37.15).

- Many plants cope with flooding through the production of special tissue (aerenchyma), which is regulated by ethylene (Figure 37.16).

- Plants cope with biological stresses such as herbivore and pathogen attacks by means of structural and chemical adaptations. Injured plant parts produce volatile hormones that signal other parts of the same plant and nearby plants to produce defensive responses. The hypersensitive response (HR) is a local defensive response to pathogen attack, whereas systemic acquired resistance (SAR) is a whole-plant immune response (Figures 37.17, 37.18, 37.19).

## Assess & Discuss

### Test Yourself

1. Examples of plant hormones include
   a. cyclic AMP, $IP_3$, and calcium ions.
   b. calcium, CDPKs, and DELLA proteins.
   c. auxin, cytokinin, and gibberellin.
   d. cryptochrome, phototropin, and phytochrome.
   e. statoliths, pulvini, and aerenchyma.

2. Phototropism is the
   a. production of flowers in response to a particular day length.
   b. production of flowers in response to a particular night length.
   c. growth response of a plant, organ system, or organ to directional light.
   d. growth response of a plant, organ system, or organ to gravity.
   e. growth response of a plant, organ system, or organ to touch.

3. What is the most accurate order of events during signal transduction?
   a. first, receptor activation; then, messenger signaling; and last, an effector response
   b. first, an effector response; then, messenger signaling; and last, receptor activation
   c. first, messenger signaling; then, receptor activation; and last, an effector response
   d. first, an effector response; then, receptor activation; and last, messenger signaling
   e. none of the above

4. Which of the plant hormones is known as the master plant hormone and why?
   a. cytokinin, because many plant functions require cell division
   b. gibberellins, because growth is essential to many plant responses
   c. abscisic acid, because it is necessary for leaf and fruit drop
   d. brassinosteroids, because water uptake is so fundamental to plant growth
   e. auxin, because there are many different auxin-responsive genes

5. Gaseous hormones are able to enter cells without requiring special membrane transporter. Which of the major plant hormones is a diffusible gas?
   a. auxin                  d. ethylene
   b. gibberellin            e. abscisic acid
   c. cytokinin

6. Photoreceptor molecules allow plant cells to detect light of particular wavelengths. Which of these molecules is considered to be a plant photoreceptor?
   a. cryptochrome           d. a, b, and c
   b. phototropin            e. none of the above
   c. phytochrome

7. Thigmotropism is a plant response to
   a. light.                 d. gravity.
   b. cold.                  e. drought.
   c. touch.

8. Which response is an adaptation to flooding?
   a. geotropism             d. production of aerenchyma
   b. stomatal closure       e. opening aquaporins
   c. photoperiodism

9. What are avirulence genes?
   a. plant genes that encode proteins that prevent infection (virulence)
   b. plant genes that cause infection when the proteins they encode bind to pathogen elicitors
   c. pathogen genes that prevent the pathogens from causing plant disease
   d. pathogen genes that encode elicitors that foster disease in plants
   e. None of the above correctly describes avirulence genes.

10. How do plants defend themselves against pathogens?
    a. Plants produce resistance molecules (usually proteins) that bind pathogen elicitors, thereby preventing disease.
    b. Plants display a hypersensitive response that limits the ability of pathogens to survive and spread.
    c. Plants display systemic acquired resistance, whereby an infection induces immunity to diverse pathogens in other parts of a plant.
    d. All of the above are correct.
    e. None of the above is correct.

### Conceptual Questions

1. Why can plants be said to display behavior?

2. Why do plants produce so many types of resistance (R) genes?

3. **Core Concept: Systems**  Because diverse plants exude volatile compounds in response to herbivore or pathogen attack, some experts have written about "talking plants." Is there any such thing?

### Collaborative Questions

1. Why are most wild plants distasteful, and some even poisonous, to people?

2. How could you increase the resistance of a crop plant species to particular types of herbivores?

# Flowering Plants: Nutrition

# 38

The tentacled leaves of the sundews (*Drosera rotundifolia and Drosera intermedia*), shown with a trapped fly, are a plant adaptation for the acquisition of nutrients. ©Dwight Kuhn

**M**any types of fascinating carnivorous (meat-eating) plants grow abundantly in wetlands around the world, even though the soils in these places are infertile. How is this possible? Like most plants, carnivorous plants are photosynthetic and thus produce their own organic food from carbon dioxide and water, using sunlight as an energy source. These resources are abundant in wetlands, but wetland soils are low in other nutrients, such as nitrogen, that are needed for plant growth. Carnivorous plants, such as the sundews shown in the chapter opening photo, have adapted by obtaining nutrients from the bodies of trapped insects and other small animals. Carnivorous plants lure animals with enticing fragrances, brightly colored leaves, or glistening sugar-rich drops of nectar. The unsuspecting prey fall into deep, water-filled pitchers; become ensnared by gluelike mucilage; or are trapped within the walls of leafy jails whose doors suddenly snap shut. Decomposition of the animal bodies releases nutrients that plant leaves quickly absorb. Other wild and cultivated plants face similar nutritional challenges and likewise display adaptations that help them acquire sufficient resources for growth and reproduction.

This chapter focuses on plant nutrition, the processes by which plants obtain essential resources. We will begin by describing the resources needed by plants for completion of their seed-to-seed life cycle in good health. Next, we will explore the role of soil as an essential resource for plants. Last, we will examine the biological sources of plant nutrients, focusing on nutritional associations between plants and microorganisms, sources of nutrients for carnivorous plants, and the ways some plants obtain nutrients from other plants. An understanding of these topics is crucial for those who seek ways to grow more plant-derived food or biofuels for humans without causing environmental harm. Plant nutritional information is also useful to people who tend gardens or houseplants or who want to restore degraded habitats.

## 38.1 Plant Nutritional Requirements

### Learning Outcomes:

1. List the major nutritional resources that most plants need for healthy growth.
2. Describe how some plants have adapted to light limitation in shady habitats.
3. **CoreSKILL »** Distinguish between plant macronutrients and micronutrients, using quantitative comparisons.

Carnivorous plants illustrate the concept that all plants—like all animals—have nutritional requirements. A **nutrient** is a substance that is metabolized by or incorporated into an organism. Photosynthetic plants require carbon dioxide ($CO_2$), water ($H_2O$), and more than a

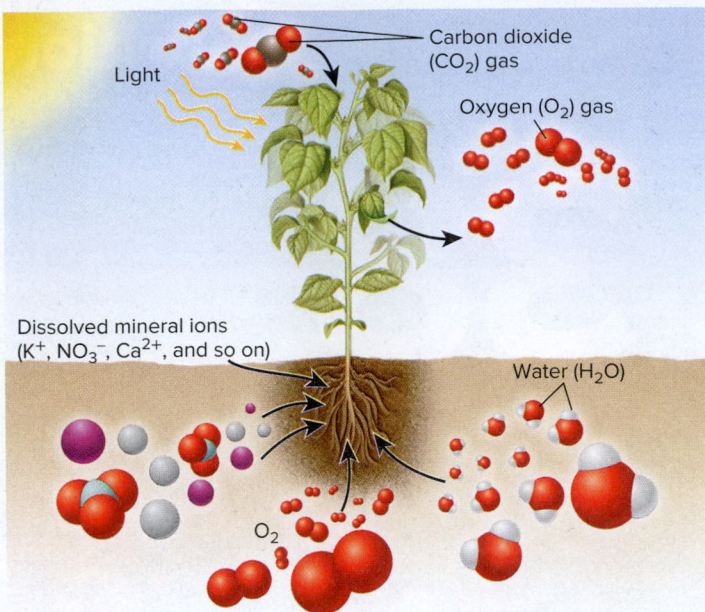

**Figure 38.1** **The major types of plant nutrients and their sources.** Elements that are required in relatively large amounts are known as macronutrients, whereas elements required in smaller amounts are known as micronutrients. A mineral is a naturally occurring inorganic compound. Many minerals dissolve in water to form ions.

dozen elements, such as potassium (as $K^+$), nitrogen (in the form of $NH_4^+$ or $NO_3^-$), and calcium (as $Ca^{2+}$) to produce organic food by means of photosynthesis. $CO_2$ is primarily absorbed from air, whereas $H_2O$ and elements are primarily taken up from soil in the form of dissolved ions (**Figure 38.1**). As is the case for animals, deficiency symptoms develop in plants that receive too little of these substances. The environmental scarcity of nutrients selects for adaptations that help plants to acquire them, illustrated by the existence of carnivorous plants.

**Essential elements** are chemical elements that are required by plants for survival and play many roles in plant metabolism, often functioning as enzyme cofactors (**Table 38.1**). Elements that are generally required in amounts of at least 1 g/kg of plant dry mass are known as **macronutrients**. In contrast, elements that are needed in amounts equal to or less than 0.1 g/kg of plant dry mass are known as **micronutrients**, or trace elements. Because insufficient amounts of light, carbon dioxide, water, and other mineral nutrients can limit the extent of green plant growth, these resources are known as **limiting factors**. If you were in charge of a garden, greenhouse, farm, or forest used to generate products such as wood or paper, or if you were overseeing an environmental restoration project, you would want to understand the conditions that foster or limit plant growth. In this section, we will focus on plant resource requirements, starting with light.

## Light Is an Essential Resource for the Growth of Green Plants

All photosynthetic plants require light energy for the formation of the covalent bonds of organic compounds that make up the plant body. Green plants' use of light energy as an essential resource parallels animals' nutritional requirements for organic food as a source of chemical energy. Several hundred species of plants, such as the Indian

| Table 38.1 | Plant Essential Nutrients | | | |
|---|---|---|---|---|
| Element (chemical symbol) | Percentage of plant dry mass | Major source | Form taken up by plants | Function(s) |
| **Macronutrients** | | | | |
| Carbon (C) | 45 | Air | $CO_2$ | Component of all organic molecules |
| Oxygen (O) | 45 | Air, soil, water | $CO_2$, $O_2$, $H_2O$ | Component of organic molecules |
| Hydrogen (H) | 6 | Water | $H_2O$ | Component of all organic molecules; protons used in chemiosmosis and cotransport |
| Nitrogen (N) | 1.5 | Soil | $NO_3^-$, $NH_4^+$ | Component of proteins, nucleic acids, chlorophyll, coenzymes, and alkaloids |
| Potassium (K) | 1.0 | Soil | $K^+$ | Has essential role in cell ionic balance |
| Calcium (Ca) | 0.5 | Soil | $Ca^{2+}$ | Component of cell walls; messenger in signal transduction |
| Magnesium (Mg) | 0.2 | Soil | $Mg^{2+}$ | Component of chlorophyll; activates some enzymes |
| Phosphorus (P) | 0.2 | Soil | $HPO_4^{2-}$ | Component of nucleic acids, ATP, phospholipids, and some coenzymes |
| Sulfur (S) | 0.1 | Soil | $SO_4^{2-}$ | Component of proteins, some coenzymes, and defense compounds |
| **Micronutrients** | | | | |
| Chlorine (Cl) | 0.01 | Soil | $Cl^-$ | Required for water splitting in photosystem II; cell ion balance |
| Iron (Fe) | 0.01 | Soil | $Fe^{3+}$, $Fe^{2+}$ | Enzyme cofactor; component of cytochromes |
| Manganese (Mn) | 0.005 | Soil | $Mn^{2+}$ | Enzyme cofactor; required for water splitting in photosystem II |
| Boron (B) | 0.002 | Soil | $B(OH)_3$ | Enzyme cofactor; component of cell walls |
| Zinc (Zn) | 0.002 | Soil | $Zn^{2+}$ | Enzyme cofactor |
| Sodium (Na) | 0.001 | Soil | $Na^+$ | Required to generate PEP in $C_4$ and CAM plants* |
| Copper (Cu) | 0.0006 | Soil | $Cu^+$, $Cu^{2+}$ | Enzyme cofactor |
| Molybdenum (Mo) | 0.00001 | Soil | $MoO_4^{2-}$ | Enzyme cofactor |
| Nickel (Ni) | 0.000005 | Soil | $Ni^{2+}$ | Enzyme cofactor |

*PEP stands for phosphoenolpyruvate; CAM stands for crassulacean acid metabolism.

**Figure 38.2** The heterotrophic flowering plant, Indian pipe (*Monotropa uniflora*). ©Robert Ziemba/age fotostock

 **Core Concept: Energy and Matter** The nongreen heterotrophic plant *M. uniflora* lacks photosynthetic capacity and therefore must absorb organic compounds for use as an energy source. By contrast, nearby autotrophic green plants use sunlight as a source of energy.

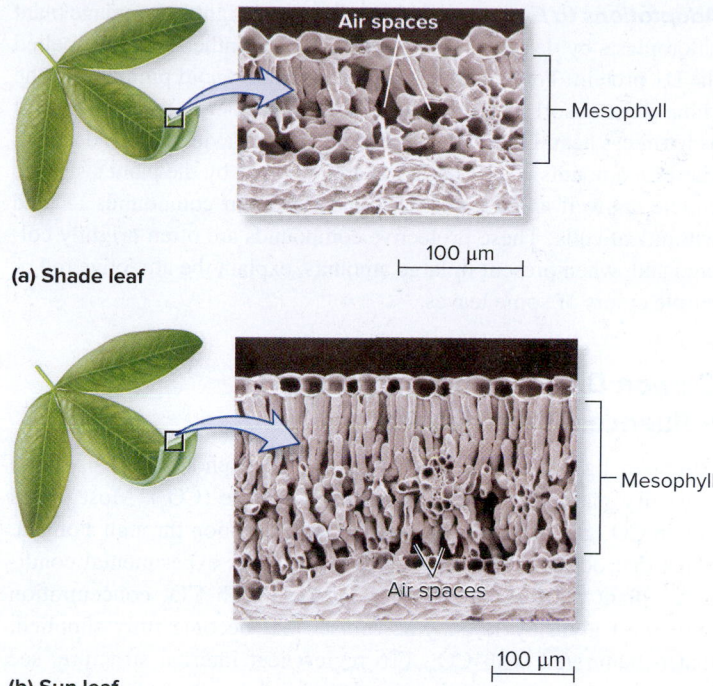

(a) Shade leaf

(b) Sun leaf

**Figure 38.3** Shade and sun leaves. a–b: ©Raymon Donahue

 **Core Concept: Structure and Function** These scanning electron micrographs of cut leaves reveal the thinner mesophyll and greater number of air spaces in (a) shade leaves as compared with (b) sun leaves.

pipe (*Monotropa uniflora*), have become heterotrophic and lost their photosynthetic capacity. In heterotrophic plants, essential nutrients include absorbed organic compounds that replace light as a source of energy (**Figure 38.2**).

In nature, plants must adapt to environments with varying amounts of light and shade. For example, in forests, light availability limits the growth of tree seedlings and other small plants that are shaded by the leafy tree canopy overhead. Conversely, plants growing in deserts or on mountains often experience light so intense that it can damage the photosynthetic components. Plants have evolved adaptations that help them cope with environments that have too little or too much light.

**Adaptations to Shade** Plants can adapt to shading by producing thin leaves that allow some light to pass through to other leaves, producing more chlorophyll, or producing distinctive sun and shade leaves (**Figure 38.3**). Sun leaves have a thicker layer of chlorophyll-containing mesophyll and are able to harvest more of the bright sunlight that penetrates deeply into the leaf. In contrast, shade leaves have a thinner mesophyll layer, with more air spaces than do sun leaves. The arrangement of leaves on plants and stem-branching patterns also reflect adaptations that reduce shading. For example, many tropical forest trees that must compete for light with closely crowded neighbors are extremely tall, and they produce branches and leaves only at their very tops. Shorter plants native to the shady interiors of tropical rain forests are so well adapted to these moist but dim conditions that they make excellent houseplants (**Figure 38.4**).

**Figure 38.4** The tropical houseplant *Monstera deliciosa*. *M. deliciosa* makes an attractive houseplant because its large, deep green leaves are adapted to the moist and shady conditions present in the interiors of tropical forests. ©Royal Botanical Gardens Kew

***Adaptations to Excessive Light***    Too much light can damage plant chloroplasts by destroying an essential photosynthetic protein, called the D1 protein. To avoid damage, specific carotenoid pigments in the chloroplast absorb some of the excess light energy and dissipate it as harmless heat. Other adaptations prevent ultraviolet (UV) damage. Harmful amounts of UV radiation are absorbed by the plant's surface cuticle, as well as by carotenoid and flavonoid compounds located within leaf cells. These protective compounds are often brightly colored and, when present in large amounts, explain the attractive red or purple colors of some leaves.

## Carbon Dioxide Concentration Influences Plant Growth

Although light provides the energy for plant photosynthesis, most plant dry mass originates from carbon dioxide ($CO_2$). Most plants obtain $CO_2$ gas from the atmosphere by absorption through stomata, pores that occur in the plant epidermis. Under experimental conditions, plant photosynthetic rates increase with $CO_2$ concentration until the Calvin cycle enzyme rubisco has become fully supplied, that is, saturated with $CO_2$. (To review leaf internal structure, see Figure 36.8c; to review the role of rubisco in the Calvin cycle of photosynthesis, see Figure 8.13.)

***$CO_2$ Limitation***    In nature, plants are often unable to obtain enough $CO_2$ for maximal photosynthesis. As a result, $CO_2$ limits agricultural crop productivity. This is partly because the modern atmospheric concentration of $CO_2$ is only 390 µL/L (a small fraction of atmospheric content), whereas considerably more $CO_2$ would be required to maximize the rate of photosynthesis in most plants. Evidence that plant photosynthesis can be limited by $CO_2$ availability is provided by studies of crops grown in greenhouses, where atmospheric gas content can be controlled. When supplied with air enriched with $CO_2$, tomatoes, cucumbers, leafy vegetables, and some other crops can double their growth rate. In nature, plants that experience hot, dry conditions are particularly vulnerable to low $CO_2$ levels when their stomata close to conserve water. When stomata are closed, plants cannot absorb $CO_2$, which limits photosynthesis. Many plants possess structural and biochemical adaptations that help them cope with $CO_2$ limitation by improving $CO_2$ absorption.

***$CO_2$-Absorption Adaptations***    Many plants that live in arid or hot environments display an adaptation known as $C_4$ photosynthesis, which improves productivity by aiding $CO_2$ absorption (refer back to Figure 8.17). About 30,000 plant species utilize $C_4$ photosynthesis, which is thought to have evolved from ancestral $C_3$ photosynthesis on more than 40 separate occasions. Today, $C_4$ plants dominate extensive warm grassland habitats of the world, contributing about 25% of the total terrestrial photosynthesis.

$C_4$ photosynthesis relies on specializations in the internal structure of leaves. In $C_4$ plants, the leaf mesophyll harvests light, but does not contain Calvin-cycle enzymes that bind $CO_2$ to form sugar. Instead, $C_4$ leaf mesophyll produces the enzyme phosphoenolpyruvate (PEP) carboxylase, which binds $CO_2$ more readily than does rubisco. PEP carboxylase adds $CO_2$ to PEP, a three-carbon molecule, to produce the four-carbon compound oxaloacetate. The leaves of

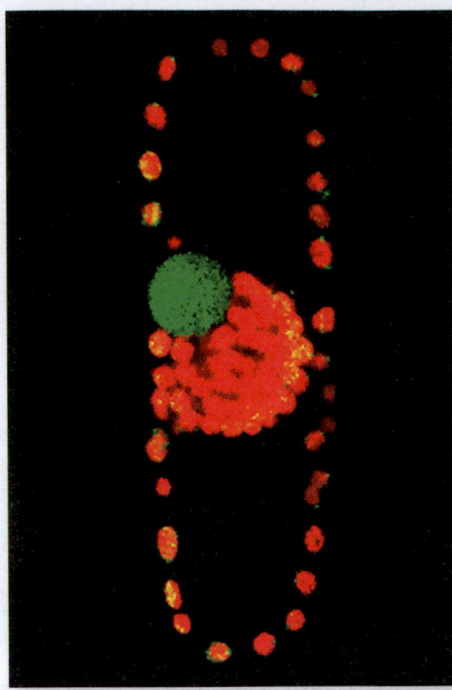

**Figure 38.5   A $CO_2$ absorption adaptation.** In contrast to most $C_4$ plants, *Bienertia cycloptera* is able to conduct $C_4$ photosynthesis within the confines of a single cell. In this fluorescence photograph of a *B. cycloptera* cell, the red plastids at the cell periphery function like the mesophyll cells of most $C_4$ plants. In contrast, the red plastids clustered near the green nucleus function much like the bundle-sheath cells of most $C_4$ plants. The plastids appear red because chlorophyll emits red light when it fluoresces. Confocal fluorescence micrograph courtesy Simon D.X. Chuong

**Concept Check:** *Which plastids should contain more of the enzyme rubisco, plastids at the cell periphery or those clustered near the nucleus?*

$C_4$ plants also have increased vein density, and specialized bundle-sheath cells surround each vein. In the leaves of most $C_4$ plants, such as maize (corn), the four-carbon compounds produced in mesophyll move through intracellular connections into the bundle-sheath cells. Within the bundle-sheath cells, enzymes release $CO_2$ from the $C_4$ compounds, and the $CO_2$ is incorporated into organic carbon by Calvin cycle enzymes.

Some plants native to arid regions have evolved a type of $C_4$ photosynthesis that occurs within a single cell. Photosynthetic cells of these plants have two types of plastids, one type containing PEP carboxylase and the other containing rubisco (**Figure 38.5**). $C_4$ photosynthesis is particularly valuable to plants that occur in hot, dry environments because it allows the plants to absorb sufficient $CO_2$ without losing too much water.

## Water Is an Essential Plant Resource

Water is essential to plants for several reasons. As a nutrient, water is the source of most of the hydrogen atoms and some of the oxygen atoms in organic compounds (see Table 38.1). For example, oxygen is incorporated into organic molecules during hydrolysis reactions (refer back to Figure 3.4b). Water is also the solvent for mineral nutrients and is the main transport medium in plants, allowing movement of minerals and other solutes throughout the plant body via the vascular tissues. Cytoplasmic and vacuolar water also helps to support plants by maintaining hydrostatic pressure on the cell wall.

Though plants vary in water content, water typically makes up about 90% of the weight of living plants. Most plants die when their water content falls below half of the amount normal for that particular species. Although many types of desiccation-resistant plants display adaptations that allow them to survive for extended periods in fairly dry conditions, all plants require an adequate supply of water for active metabolism and growth.

## Deficiency Syndromes Occur When Plants Receive an Inadequate Supply of Certain Nutrients

In addition to light, $CO_2$, and water, plants require additional elements for growth and survival (see Table 38.1). These elements occur naturally in water and soil or they can be added in the form of fertilizers. Plant biologists have quantitatively analyzed the elemental requirements of plants by growing them hydroponically, that is, by bathing plant roots in a water solution to which elements in the form of dissolved minerals are added in various combinations and amounts. Hydroponic studies reveal that when plants lack an adequate supply of an essential elemental nutrient, they display characteristic deficiency symptoms. Such symptoms include failure to reproduce, tissue death, and changes in leaf color. Yellowing of leaves, known as **chlorosis**, is a common mineral deficiency symptom, because many different elements are needed for chlorophyll production (**Figure 38.6**).

**Figure 38.6  Chlorosis as a symptom of mineral deficiency.** This camellia plant is suffering from an iron deficiency, as revealed by the yellow leaves, a symptom known as chlorosis. ©Geoff Kidd/SPL/Science

**Concept Check:** *Does chlorosis always indicate iron deficiency in plants?*

## 38.2  The Role of Soil in Plant Nutrition

### Learning Outcomes:

1. List the benefits of soil organic matter for plant growth.
2. Explain why plants require a source of fixed nitrogen and how they obtain it.
3. Describe various plant adaptations that increase the ability to obtain phosphorus.

Soil is an essential resource for most wild and cultivated plants, providing water and other essential nutrients. For this reason, extensive loss of soil by wind and water erosion is of wide concern. Soils vary greatly in fertility, that is, their ability to support plant growth. Thus, plants of many types have had to adapt to the challenges of obtaining nutrients from poor soils. In this section, we will explore soil structure and chemistry from the perspective of plant growth and examine how plants take up nutrients from the soil.

### The Physical Structure of Soils Affects Their Aeration, Water-Holding Capacity, and Fertility

Natural soils display layers, known as soil horizons (**Figure 38.7a**). The remains of plants that have recently died and other organisms form a layer of litter above the **topsoil**; the topsoil is also known as the A horizon. Many of the inorganic minerals and organic materials that enrich high-quality topsoil arise from the activities of microorganisms that decompose the litter. In this way, the minerals contained in living plants are eventually recycled to subsequent generations. Beneath the topsoil lie layers called subsoil and soil base, which are largely composed of mineral materials. Bedrock is the bottom layer that supports the soil horizons. Plant roots play an important role in conveying deep-lying minerals to the surface, thereby helping to enrich the topsoil.

Soil horizons vary in composition and thickness, depending on various factors—including climate, vegetation, bedrock type, and human influences. For example, natural grasslands produce deep, rich topsoil that is used for cropland in many regions of the world (**Figure 38.7b**). In dramatic contrast, tropical rain forests often have only thin layers of topsoil; their low-fertility soils are composed mostly of inorganic materials that are not useful to plants (**Figure 38.7c**). Farmers cope with reduced soil fertility by adding organic or inorganic fertilizers. The proportions of organic to inorganic materials and the sizes of inorganic particles are used to classify soils into different types. Soils also display variation in their amount of aeration, water-holding capacity, pH, and mineral content. All of these soil properties affect plant growth. For these reasons, we will take a closer look at soil structure and the role of fertilizers.

***Soil Organic Matter***  Soil organic matter, also known as humus, is largely derived from plant detritus, the dead and decaying remains of plants, although animal wastes and decayed animal bodies also contribute to the organic content of soils. Soil organic matter provides many benefits. Organic-rich soils, containing 8% or so organic matter, are less likely to erode—wash or blow away with water or wind. Soil organic matter also binds mineral nutrients, thereby fostering soil fertility, and gives soil a soft consistency that fosters plant root growth and farmers' ability to cultivate. Gardeners often produce their

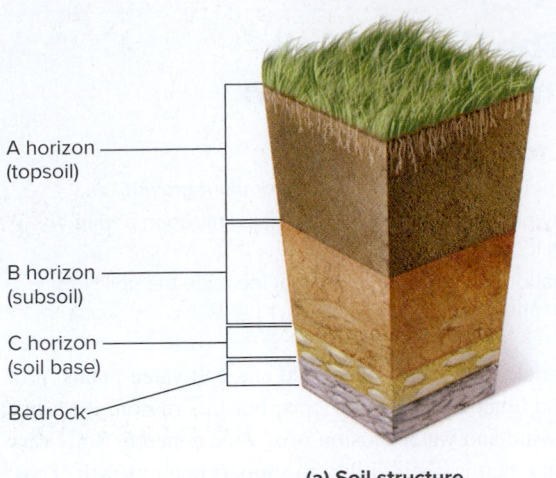

A horizon
(topsoil)

B horizon
(subsoil)

C horizon
(soil base)

Bedrock

**(a) Soil structure**

**(b) Thick layer of
topsoil in cropland**

**(c) Thin layer of topsoil in rain forest**

**Figure 38.7** **Soil horizons, the structural layers of soil.** **(a)** Diagram showing general soil structure. **(b)** A vertical view of an agricultural soil, showing a relatively deep layer of dark topsoil. **(c)** A vertical view of a tropical rain forest soil, showing a thin layer of dark topsoil. b: Courtesy of C.A. Stiles, University of Wisconsin; c: ©Ruddy Gold/age fotostock

own organic-matter-rich compost by layering vegetable waste from the kitchen and yard waste with soil, and turning the pile occasionally to introduce the oxygen necessary for decomposition (**Figure 38.8**). Mature compost is then mixed with garden soil to improve its fertility.

***Inorganic Soil Constituents*** Inorganic materials in soil are derived from the physical and chemical breakdown of rock, a process known as **weathering**. Rock, which is an aggregate of two or more minerals, is physically weathered by changes such as cycles of freezing and thawing. Lichens and plant roots may produce organic acids that contribute to chemical weathering of rocks. During chemical weathering, soluble salts are washed out, and minerals are hydrolyzed or oxidized. **Leaching** is the dissolution and removal of inorganic ions as water percolates through materials. Heavy rainfall can reduce the fertility of soils by leaching large amounts of nutrients from them.

The leached elements often end up in natural bodies of water, where they can foster the growth of cyanobacteria, algae, and aquatic plants.

Inorganic soil materials occur as particles that can be categorized according to their size as sand, silt, or clay (**Figure 38.9**). Sand grains range from 2 mm to 62.5 μm in diameter. Particles of silt range from 20 to 2 μm in diameter, and clay particles are even smaller. Soils can be classified according to their relative content of coarse and fine materials. For example, soils that contain 45% or more sand and 35% or less clay are classified as sandy soils. The other main types of soil are silt, clay, and loam. **Loam** contains a mixture of sand, silt, and clay and is ideal for the cultivation of most plants.

Because of their size differences, sand, silt, and clay particles confer different properties on soils. The relatively large size and irregular shapes of sand particles allow air and water to move rapidly through sandy soils, which are said to be porous (**Figure 38.10**).

**Figure 38.8** **Composting.** Gardeners produce compost by layering small amounts of soil with vegetable waste from the kitchen and yard waste and periodically turning the pile to introduce the oxygen needed for decomposition. Compost can be used to increase the organic and mineral content of garden soils. Source: USDA

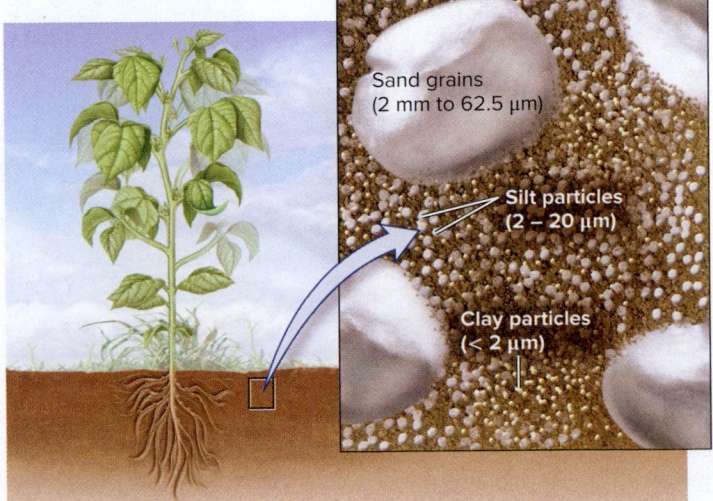

Sand grains
(2 mm to 62.5 μm)

Silt particles
(2 – 20 μm)

Clay particles
(< 2 μm)

**Figure 38.9** **The relative sizes of inorganic soil components.**

Silt and clay particles fit closely together, so soils containing larger amounts of these materials are less porous than sandy soils. Water percolates less easily through silty and clay soils, which therefore retain more ionic mineral nutrients than do sandy soils. Clay particles have negative charges on their surfaces that electrostatically bind positively charged ions (cations) such as $NH_4^+$, $Ca^{2+}$, and $Fe^{2+}$ (**Figure 38.11a**). Cations having higher valence numbers (such as $Fe^{2+}$) are bound more tightly than ions having lower valence numbers.

In order to be available to plants, cations must be detached from clay particles. Hydrogen ions ($H^+$, protons) are able to replace mineral cations on the surfaces of clay particles in a process known as **cation exchange** (**Figure 38.11b**). Cation exchange releases cations to soil water, making them available for uptake by plant roots. However, free ions are also more easily washed out of soil. If the $H^+$ concentration becomes too high, large numbers of mineral ions are released and can be leached from the soil by heavy rainfall. Such leached minerals may include heavy metals such as aluminum ($Al^{3+}$) that would otherwise be bound in soil. Cation exchange is the mechanism by which acid rain, which adds $H^+$ to soil, causes loss of soil fertility and the pollution of streams with toxic substances, such as aluminum, that can harm human health.

Despite their water- and mineral-retention features, silt- and clay-rich soils may be poorly aerated and therefore unfavorable to root growth. Gardeners often mix organic materials and sand into silt or clay-rich soils to improve their aeration properties. Loam is the preferred soil for agriculture because it combines the aeration provided by sand with the mineral and water retention capacity of silt and clay. Even with a good soil mix, fertilizers may be needed to achieve maximum crop production.

***The Role of Fertilizers*** **Fertilizers** are soil additions that enhance plant growth by providing essential elements. The addition of fertilizer to soils can compensate for deficiencies in soil organic matter or mineral content and thus improve soil fertility. Fertilizers occur in organic and inorganic forms. Organic fertilizers are those in which

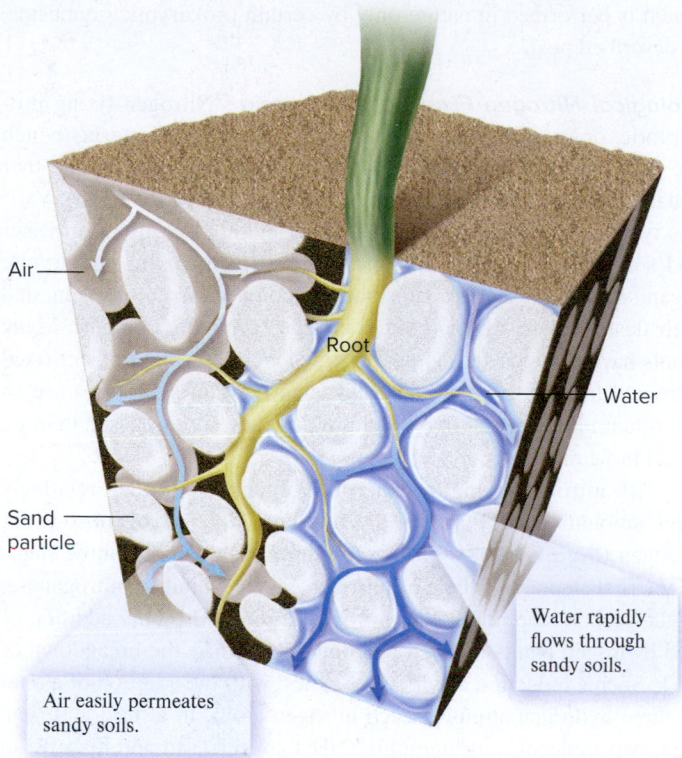

**Figure 38.10** **The movement of air and water in sandy soil.**

**Concept Check:** *In which type of soil is nutrient leaching more of a problem, sandy or clay soil?*

Sandy soils are well aerated, which is favorable to the growth of plant roots, which require oxygen for cellular respiration. However, sandy soils hold less water than the same volume of clay, and rapid percolation of water through sandy soils both reduces the amount of water available to the roots and leaches minerals from the soil.

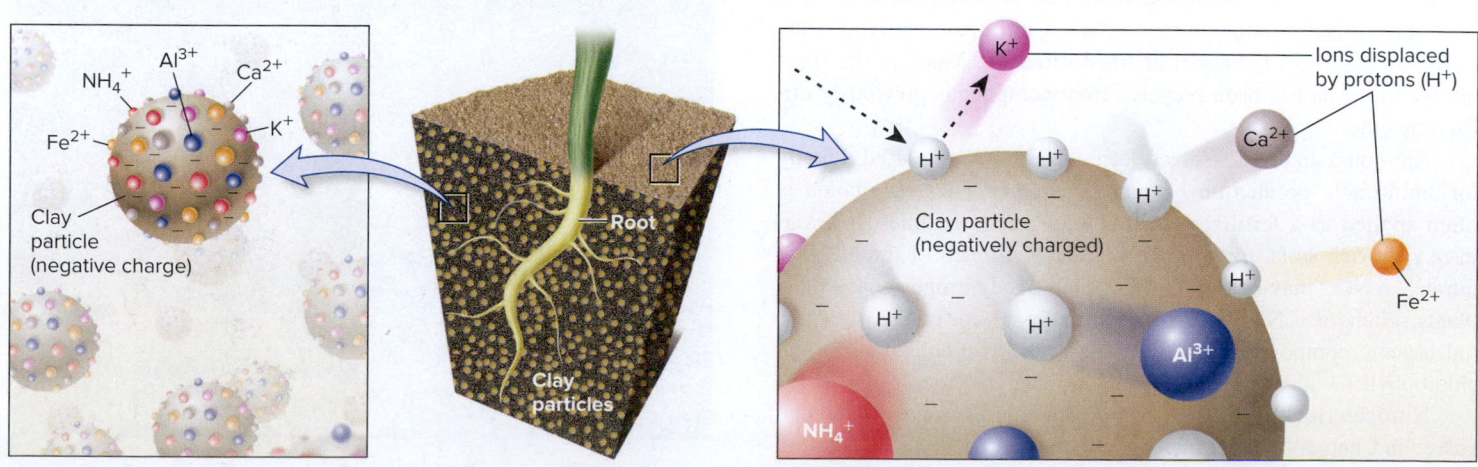

(a) **Electrostatic attraction between clay particles and mineral ions**

(b) **Cation exchange**

**Figure 38.11** **Cation binding and exchange.** (a) Clay and organic particles in the soil display negative electrostatic surface charges that bind cations. Bound cations include not only plant mineral nutrients such as ammonium ($NH_4^+$) but also cations that are not plant nutrients, such as $Al^{3+}$. (b) Cation exchange occurs when protons ($H^+$) displace other cations, releasing them to soil water. This process makes cations more available for uptake by plant roots, but it also increases the potential for cations to leach away during rains or floods.

most of the minerals are bound to organic molecules and are thus released relatively slowly. Organic fertilizers play an important role in **organic farming**, the production of crops without the use of inorganic fertilizers, growth substances, and pesticides. Manure and compost are examples of organic fertilizers.

In contrast, inorganic fertilizers consist largely of inorganic minerals, which are immediately useful to plants but can more easily leach from soils during heavy rainfall. Nitrogen (N), phosphorus (P), and potassium (K) are the mineral nutrients that most frequently limit crop growth. For this reason, these minerals are the main components of the most common type of commercial inorganic fertilizers. Such fertilizers are available with different ratios of minerals, which are optimal for different types of plants.

Excessive application of fertilizers to fields and lawns is undesirable because minerals not taken up by plant roots are easily washed by rain into waterways, where they can fuel large growths of algae and aquatic plants that can harm other aquatic life-forms. For example, large areas of the Gulf of Mexico and other coastal regions are now known as "dead zones" because microbial decomposition of large algal populations has depleted oxygen from the water, suffocating the animal life. These large populations of algae are fostered by fertilizers that wash from farm soils into rivers, such as the Mississippi River, that drain into coastal oceans. More careful application of fertilizers and the planting of vegetation that absorbs mineral nutrients along stream and river edges can help to reduce or prevent dead zones.

## Plants Require Fixed Nitrogen

Nitrogen is frequently limiting to plant growth in nature and in crop fields, because large amounts of it are required by plants to synthesize amino acids, nucleotides, and alkaloids, among many other cellular constituents. Nitrogen is the largest component of plants by mass after carbon, oxygen, and hydrogen. Although the Earth's atmosphere is 78% nitrogen gas ($N_2$), plants cannot utilize nitrogen in this form. To be of use to plants, soil nitrogen must occur in another form, such as ammonia ($NH_3$), ammonium ion ($NH_4^+$), or nitrate ion ($NO_3^-$); any of these substances is known as **fixed nitrogen**. Much of the fixed nitrogen in soils has been recycled from compounds previously utilized by other organisms.

Ammonia and its dissolved form—$NH_4^+$—can be used directly for amino acid production by plants, explaining why ammonia is often applied as a fertilizer to farm fields in springtime. However, in oxygen-rich soils, microorganisms oxidize much of the $NH_4^+$ to nitrate, so $NO_3^-$ may be the form in which fixed nitrogen enters most plants. Plants use $NH_4^+$ and $NO_3^-$ to make a wide range of essential organic compounds, including amino acids, nucleic acids, and chlorophyll.

Nitrogen flows through the environment in a nitrogen cycle, discussed in Chapter 59 (look ahead to Figure 59.15). **Nitrogen fixation** is the process by which atmospheric $N_2$ is combined with hydrogen to produce $NH_3$. New fixed nitrogen can be added to soils by the action of lightning, fire, or air pollution, as well as by biological and industrial nitrogen fixation. The nitrogen in inorganic fertilizers is produced by **industrial nitrogen fixation**, a human activity. Most of the fixed nitrogen in soils is produced by **biological nitrogen fixation**, which is performed in nature only by certain prokaryotic organisms, as described next.

***Biological Nitrogen Fixation by Bacteria***  Nitrogen-fixing prokaryotic organisms include many types of cyanobacteria, which are photosynthetic organisms that occur in oceans, lakes, and other aquatic systems, as well as in surface soil crusts (**Figure 38.12**). Various types of nonphotosynthetic bacteria and archaea living in water and soil are also able to fix nitrogen. Nitrogen-fixing prokaryotic organisms often excrete a substantial amount of fixed nitrogen, and their death makes still more fixed nitrogen available to plants. Many plants have nitrogen-fixing, prokaryotic symbionts that transfer fixed nitrogen directly to plant cells. Nitrogen-fixation symbioses are so important in nature and in agriculture that they are discussed in more detail in Section 38.3.

All nitrogen-fixing prokaryotic organisms utilize relatively large amounts of ATP and an enzyme known as **nitrogenase** to fix nitrogen (**Figure 38.13**). The fixation process occurs in three steps. In the first step, a molecule of nitrogen gas ($N_2$) binds to nitrogenase. In the second step, the bound nitrogen is reduced by the addition of two hydrogen atoms (2 H), a reaction powered by the breakdown of ATP. Such a reduction occurs three times, with the addition of a total of three hydrogen atoms to each nitrogen atom. In a third and final step, two molecules of ammonia ($NH_3$) are released and dissolve in cell water to form $NH_4^+$. The nitrogenase enzyme is then free to bind more $N_2$.

Because the $O_2$ molecule resembles $N_2$, oxygen can bind to the active site of nitrogenase. Oxygen binding disables nitrogenase, thereby stopping nitrogen fixation. For this reason, many nitrogen-fixing microorganisms exist in anaerobic environments or they turn off the expression of the nitrogenase gene when oxygen is present. To increase nitrogen availability, crop scientists are working to genetically engineer nitrogen-fixation capacity into crop plants such as rice

**Figure 38.12  A soil surface crust that includes nitrogen-fixing, soil-enriching cyanobacteria and lichens.** Such crusts are widespread in grasslands and other arid regions. Source: Jayne Belnap, U.S. Geological Survey

*Concept Check:*  *How might soil crusts influence the ecology and economy of regions in which grazing is important?*

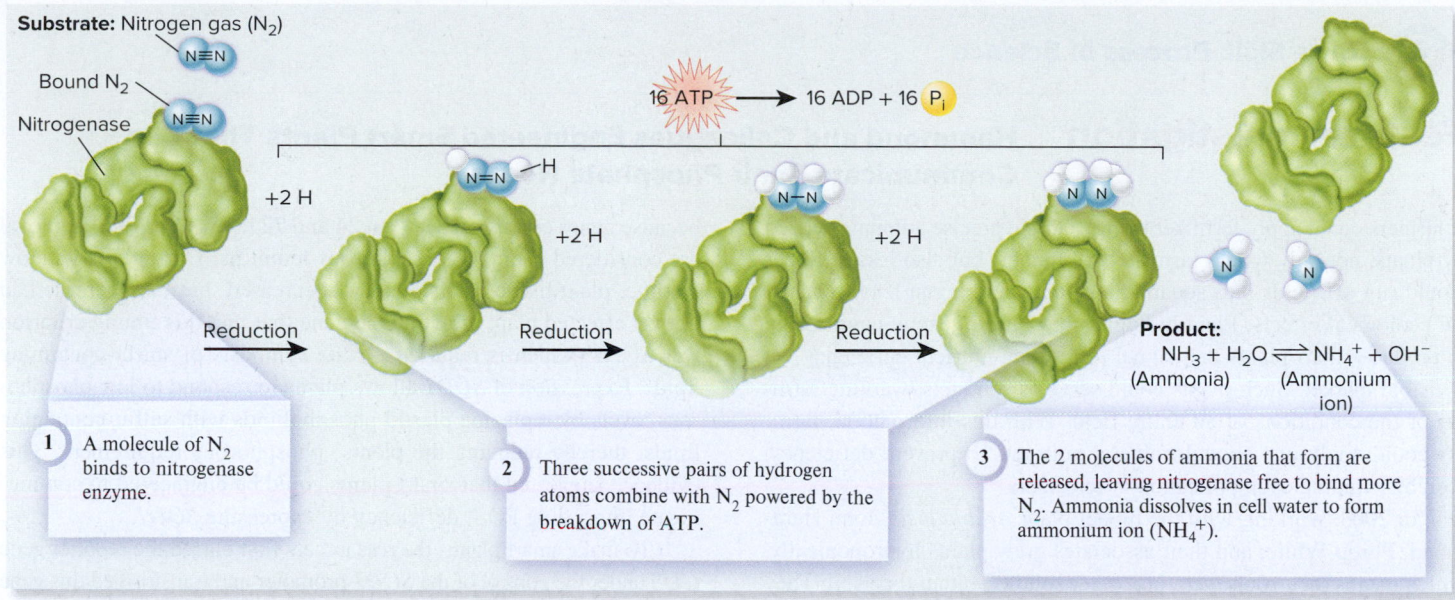

**Figure 38.13** **The biological process of nitrogen fixation.**

*Concept Check:* *What common substance inactivates the enzyme nitrogenase by binding to its active site?*

and maize (corn). However, the vulnerability of nitrogenase to oxygen means that this enzyme may need to be altered so that it binds oxygen less readily, or plants must be engineered with some mechanism that protects nitrogenase from oxygen.

*Industrial Nitrogen Fixation* Worldwide, farmers apply more than 80 million metric tons of nitrogen fertilizer per year. The fixed nitrogen found in fertilizer is produced industrially from $N_2$ by means of a procedure invented by German chemists Fritz Haber and Carl Bosch in 1909. The reduction of $N_2$ gas to $NH_3$ is energetically favorable at room temperature, but the activation energy is very high, so the reaction occurs extremely slowly. Using an iron catalyst, temperatures of 400–650°C (752–1,202°F), and high pressures (150–400 atmospheres), the Haber-Bosch process generates $NH_3$ rapidly. However, because of its high energy requirements, industrial nitrogen fixation can be costly from the perspective of many of the world's farmers. The high cost of fertilizers helps to explain why agricultural scientists are so interested in the possibility of genetically engineering biological nitrogen fixation into crop plants.

## Plants Display Adaptations for Acquiring Phosphorus

Phosphorus (P) is another soil mineral that often limits plant growth. Plants obtain phosphorus from the ion known as phosphate ($PO_4^{3-}$), which occurs in the soil in three dissolved forms: $H_3PO_4$, $H_2PO_4^-$, and $HPO_4^{2-}$. $HPO_4^{2-}$ (called hydrogen phosphate ion) is the form most commonly absorbed by plants. Phosphate uptake involves ATP-requiring proton cotransport, as is the case for the uptake of other anions such as nitrate ($NO_3^-$) and sulfate ($SO_4^{2-}$). The uptake of these nutrients at the root hair surface is discussed further in Chapter 39.

Although phosphorus can be abundant in soil, it is often unavailable to plants. One reason is that $PO_4^{3-}$ forms tightly bound complexes with clay, iron and aluminum oxides, and calcium carbonate in soils. In addition, soil microbes incorporate $PO_4^{3-}$ into organic compounds that are not taken up by plants.

Because plants need large supplies of phosphorus for a variety of cell processes, they have evolved various adaptations that increase their ability to obtain $PO_4^{3-}$ from soil. A common adaptation for acquiring $PO_4^{3-}$ is the symbiotic association of plant roots with various types of fungi (see Section 38.3). In addition, plants that grow in soils having low phosphorous content may produce more highly branched roots and more and longer root hairs. Plant roots also secrete protons and organic acids such as citrate and malate into the soil, which help release phosphorus from inorganic complexes. For example, the plant *Lupinus alba* releases as much as 25% of its total photosynthetic carbon into the soil as organic acids—a high price to pay but apparently one that is essential for the plant to obtain sufficient phosphorus. Plants may also secrete phosphatase enzymes from roots. These enzymes release phosphorus from organic compounds in the soil. The general process by which P, N, $CO_2$, and other minerals are released from organic compounds is called **mineralization**.

Farmers and gardeners apply phosphate-rich fertilizers to crop fields and gardens as a way of preventing phosphorous deficiencies, which reduce yields. Phosphorous fertilizers are obtained from phosphate-rich mineral deposits, but experts have warned that inexpensive sources of $PO_4^{3-}$ will be exhausted within the next 90 years. Consequently, there is much interest in devising ways to maximize the efficiency by which plants are able to take up and use phosphorus. Genetic engineering to produce "smart plants" that can sense the levels of nutrients in the soil may offer some options.

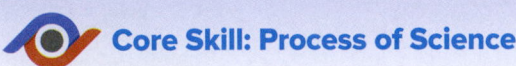

## Core Skill: Process of Science

## Feature Investigation | Hammond and Colleagues Engineered Smart Plants That Can Communicate Their Phosphate Needs

If farmers could apply fertilizer to crops in the precise amounts needed by plants, not only would farmers save money, but also less fertilizer would run off fields into aquatic habitats, where it can lead to harmful ecological effects. Plant biologists have used genetic engineering to produce smart plants that signal impending nutrient deficiency via a visible marker. Such plants could serve as sentinels, warning farmers of the conditions of an entire field. With this information, farmers could apply just enough mineral nutrients to prevent deficiency, thereby avoiding overapplication of fertilizers.

In 2003, working with the model plant *Arabidopsis*, John Hammond, Philip White, and their associates grew plants hydroponically, which means their roots were in a water solution rather than soil. They identified some of the genes whose expression changes when plants are transferred from nutrient solutions containing sufficient phosphorus to solutions lacking phosphorus. They found that some genes were turned on quickly after phosphorus removal, but other genes took much longer, up to 100 hours or more. This timing is important

because genes expressed between 24 and 72 hours after $PO_4^{3-}$ removal are considered useful as phosphorous monitors. During this window of time, plant tissue levels of $PO_4^{3-}$ decreased, but the decrease had not yet affected plant growth. One gene that met this timing criterion was *SQD1*, which is required for the synthesis of sulfur-containing lipids. Expression of *SQD1* allows plants to respond to low phosphorous levels by replacing plastid phospholipids with sulfur-containing lipids, thereby reducing the plants' phosphorous requirement. This evidence suggested that smart plants could be engineered to communicate impending $PO_4^{3-}$ deficiency by expressing *SQD1*.

To make smart plants, the researchers first placed the reporter gene *GUS* under the control of the *SQD1* promoter and transformed this gene into *Arabidopsis* plants (**Figure 38.14**). The researchers then grew these genetically engineered plants for various time periods in hydroponic solutions of differing $PO_4^{3-}$ levels. After different time periods, they removed leaves and chemically treated them with a compound that produces a light blue color when the *GUS* gene is expressed. Some leaves were

**Figure 38.14** The experiment of Hammond and colleagues showed that plants can be engineered to communicate changes in the level of nutrients.

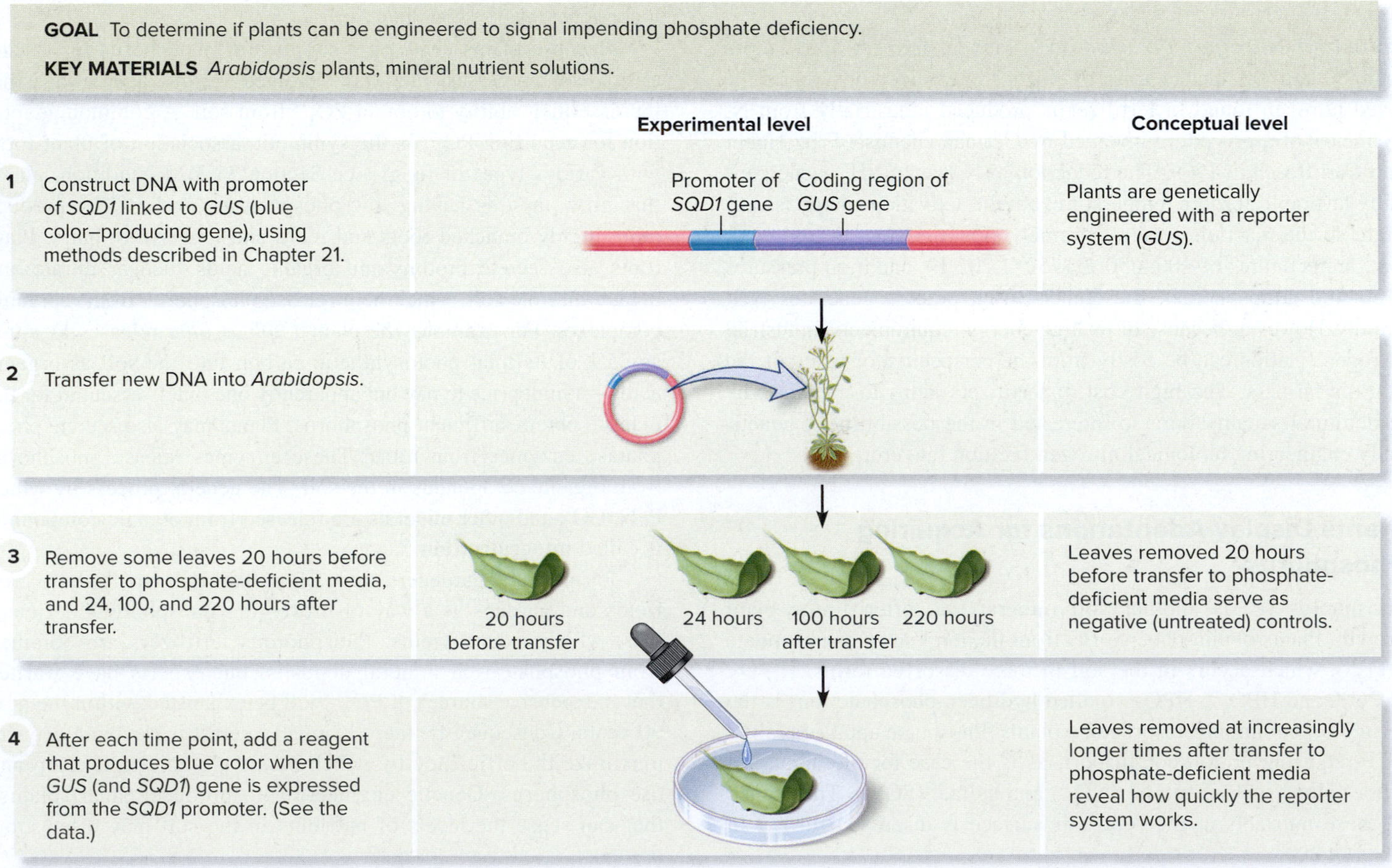

| | | |
|---|---|---|
| **GOAL** To determine if plants can be engineered to signal impending phosphate deficiency. | | |
| **KEY MATERIALS** *Arabidopsis* plants, mineral nutrient solutions. | | |

| | Experimental level | Conceptual level |
|---|---|---|
| **1** Construct DNA with promoter of *SQD1* linked to *GUS* (blue color–producing gene), using methods described in Chapter 21. | Promoter of    Coding region of<br>*SQD1* gene    *GUS* gene | Plants are genetically engineered with a reporter system (*GUS*). |
| **2** Transfer new DNA into *Arabidopsis*. | | |
| **3** Remove some leaves 20 hours before transfer to phosphate-deficient media, and 24, 100, and 220 hours after transfer. | 20 hours before transfer    24 hours   100 hours   220 hours after transfer | Leaves removed 20 hours before transfer to phosphate-deficient media serve as negative (untreated) controls. |
| **4** After each time point, add a reagent that produces blue color when the *GUS* (and *SQD1*) gene is expressed from the *SQD1* promoter. (See the data.) | | Leaves removed at increasingly longer times after transfer to phosphate-deficient media reveal how quickly the reporter system works. |

**5   THE DATA**

Control    24 hours    100 hours    220 hours

**6   CONCLUSION** Plants can be genetically engineered to express color signals in time for farmers to apply fertilizer sufficient to prevent nutrient deficiency.

**7   SOURCE** Hammond, J. P., et al. 2003. Changes in gene expression in Arabidopsis shoots during phosphate starvation and the potential for developing smart plants. *Plant Physiology* 132: 578–596.

removed before transfer to the phosphate-deficient solution and served as controls. Because the *GUS* gene was under the control of the *SQD1* promoter, leaves from plants that were developing PO$_4^{3-}$ deficiency turned blue! These smart plants were able to communicate impending phosphorous deficiency in time for a farmer to apply fertilizer.

In a later and more extensive study of gene expression in *Arabidopsis*, other investigators discovered that PO$_4^{3-}$ induces the activity of about 612 genes and represses the activity of about 254 genes. Some of these genes may encode proteins useful in monitoring plant phosphorous status. If smart plant technology can be developed for crop plants, farmers may be able to monitor and fertilize fields with much greater precision.

### Experimental Questions

1. What advantage do plants obtain when the *SQD1* gene is expressed?

2. **CoreSKILL »** How were the investigators able to identify potential sentinel plants that were starting to experience phosphorous deficiency?

**BIO TIPS**

**THE QUESTION** *Explain why Hammond and colleagues sought to identify plant genes whose expression changed during the period from 24 to 72 hours after the imposition of a phosphorus limitation.*

**T OPIC** *What topic in biology does this question address?* The topic is plant nutrition, or more specifically, plant responses to nutrient limitation and their relevance to developing technological monitoring systems that allow more efficient applications of fertilizer to crops.

**I NFORMATION** *What information do you know based on the question and your understanding of the topic?* Phosphorus is an element that is essential to living things, including plants. Table 38.1 briefly reviews its role in plant nutrition. This chapter has explained how nutrients affect plant growth, and you have also learned that soil phosphorus is often bound to inorganic and organic materials that reduce its availability to plant roots. Thus, phosphorus is often applied to fields to enhance crop yields. You also know that plants (like other living organisms) respond to environmental change and that such responses include changes in gene expression.

**P ROBLEM-SOLVING S TRATEGY** Think about Hammond and colleagues' desire to find genes that would react to phosphorous limitation in a rather short period and whose response could be observed. Realize that a system for monitoring plant nutrient limitation needs to reveal the problem

before it causes irreversible effects. Understand that farmers do not want to wait so long to apply fertilizer that crop plants will suffer in the meantime.

**ANSWER** *Monitoring changes in plant gene expression at different times after the onset of phosphorous limitation allowed the investigators to identify genes that could be used to reveal the point at which the plants started to respond to nutrient limitation but before they incurred serious damage.*

## 38.3 Biological Sources of Plant Nutrients

### Learning Outcomes:

1. Explain the importance of mycorrhizal fungi to plants.

2. List the major types of prokaryotic organisms that occur in symbioses with plants, thereby helping them to acquire nitrogen.

3. Distinguish carnivorous plants from parasitic plants, giving examples.

This section focuses on several fascinating ways in which plants use other organisms as a means to obtain nutrients. Biological mechanisms by which plants obtain nutrients include maintaining symbiotic relationships with fungi or bacteria, capturing animal prey (by carnivorous plants), and serving as hosts for nonphotosynthetic plant parasites.

## Mycorrhizal Associations Help Most Plants Obtain Mineral Nutrients

At least 80% of seed plants have symbiotic associations with fungi that live within the tissues of plant roots or that envelop root surfaces (Figures 30.13–30.15). These associations are termed **mycorrhizae**; the prefix *myco* refers to fungi, and *rhiza* means root, so the term literally means fungus root.

In mycorrhizal associations, soil fungi obtain organic food from the roots of photosynthetic plant hosts, and the fungi supply the plants with water and mineral nutrients. Due to the extensive mycelia that fungi produce within the soil, these fungus root associations provide an exceptionally efficient way for plants to harvest water and minerals, especially phosphate, from a much larger volume of soil than is available to roots by themselves. The presence of lush vegetation on thin, infertile tropical rain forest soils is largely due to the ability of mycorrhizae to rapidly absorb mineral nutrients released by decaying organisms and transmit the nutrients directly to plant roots (**Figure 38.15**). In many tropical rain forests, mineral nutrients occur within the bodies of living organisms, rather than accumulating in the soil where they could easily be leached away by heavy, frequent rains.

Various species of ghostly pale plants have lost their photosynthetic pigments (see Figure 38.2). Such heterotrophic plants have become dependent on organic compounds supplied by fungi that form

**Figure 38.15** **Nutrient acquisition via mycorrhizae.**

 **Core Concept: Systems** In all forests, but particularly those of tropical regions, mycorrhizal fungi rapidly collect soil minerals released from decaying organisms and transport them directly to plant roots. Such efficient nutrient cycling bypasses the soil, from which mineral ions can be easily leached by heavy rainfall. This system of material movement among diverse organisms explains how lush forests can grow on thin, infertile soils.

mycorrhizal associations with a photosynthetic host, such as a nearby tree. In this process, known as mycoheterotrophy, the fungus serves as an underground conduit for the flow of organic nutrients from a green, photosynthetic plant to a heterotrophic plant. Many plant seedlings that grow in the shade of taller plants also use mycoheterotrophy to survive until they are able to obtain enough light for photosynthesis.

## Plant-Bacteria Symbioses Provide Some Plants with Fixed Nitrogen

Soils contain large populations of thousands of bacterial species, many of which aid plant growth. Some soil bacteria produce plant hormones that affect root structure, others help plants to tolerate drought and other stresses, and some provide plants with nutrients, notably fixed nitrogen. In nitrogen-fixation symbioses, the plants provide organic nutrients to the bacteria, and the bacteria supply the plants with a much higher supply of fixed nitrogen than the plants could obtain from most soils. Representatives of three types of nitrogen-fixing bacteria—cyanobacteria, actinobacteria, and proteobacteria—are symbiotically associated with specific types of plants. (For more information about the characteristics of these bacterial groups, see Chapter 27.)

**Plant-Cyanobacteria Symbioses** Although most cyanobacteria are themselves photosynthetic, organic compounds supplied by plant partners subsidize the high energy costs of nitrogen fixation. This subsidizing allows the cyanobacteria to fix more nitrogen than they require, secreting the excess to plant partners. Nitrogen-fixing cyanobacteria form symbioses with some bryophytes, ferns, and gymnosperms, as well as the flowering plant *Gunnera*. This plant, commonly known as the giant rhubarb or prickly rhubarb, can produce leaves almost 3 m across (**Figure 38.16**). Nitrogen-fixing symbionts are advantageous to *Gunnera* because this large plant grows in nitrogen-poor habitats, such as volcanic slopes in Hawaii. *Gunnera* harbors cyanobacteria within stems and leaf petioles. In these locations, the cyanobacteria can use cyclic electron flow to transform light energy into ATP, needed to produce fixed nitrogen. The presence of nitrogen-fixing cyanobacteria helps to explain why *Gunnera* can grow to dramatic size on poor soils.

**Woody Plant–Actinobacteria Symbioses** In contrast to photosynthetic cyanobacteria, actinobacteria are heterotrophic nitrogen-fixing bacteria. The actinobacterial genus *Frankia* occurs in nodules formed on the underground roots of certain shrubs or trees, such as alder (*Alnus*) and myrtle (*Myrica*). These plants receive fixed nitrogen from their bacterial partners, which, in turn, obtain organic nutrients. Woody plants, such as shrubs of the genus *Ceanothus*, that have *Frankia* symbionts are able to grow abundantly even in places where soil nitrogen is low. This symbiosis helps to explain why *Ceanothus* covers extensive areas in mountainous regions of the western U.S.

**Legume-Rhizobia Symbioses** The nitrogen-fixation symbioses most important in nature and to agriculture involve certain species of proteobacteria that are collectively known as **rhizobia** (from the Greek *rhiza*, meaning root). Rhizobia live within root cells of wild and cultivated legumes, forming legume-rhizobia symbioses. In nature, legume plants are important sources of fixed nitrogen for other plants. When legumes die, they generate soil organic matter that is enriched

**Figure 38.16** *Gunnera* **growing on nitrogen-poor soil.** Nitrogen-fixing cyanobacteria that live within cavities in this plant's leaf petioles provide the plant with fixed nitrogen, which explains how such a large plant can grow on infertile soils. ©J. Hyvönen; (inset): ©Photo by Birgitta Bergman, Department of Botany, Stockholm University (Sweden)

with fixed nitrogen. Consequently, wild legumes are regarded as particularly valuable members of natural plant communities.

Important legume crops include soybeans, peas, beans, peanuts, clover, and alfalfa. Foods produced from soybeans, peas, beans, and peanuts are valued for their high protein content. Clover and alfalfa are used for animal food and to enrich fields with the fixed nitrogen needed by subsequent food crops. The value of these crops arises from their fixed-nitrogen content. The amount of ammonia produced by legume-rhizobia symbioses nearly equals the world's entire industrial production.

### 👁 Core Concepts: Systems, Information

### Development of Legume-Rhizobia Symbioses

Rhizobia can live independently in the soil, but are also found within lumpy **nodules** that form on legume roots (**Figure 38.17**). Different species of rhizobia preferentially form symbioses with particular plant species. Because of their agricultural importance, these legume-rhizobia symbioses have been extensively studied, and a great deal is now known about the molecular basis of their development. This information is potentially useful in efforts to genetically engineer nitrogen-fixation capacity into nonlegume crops.

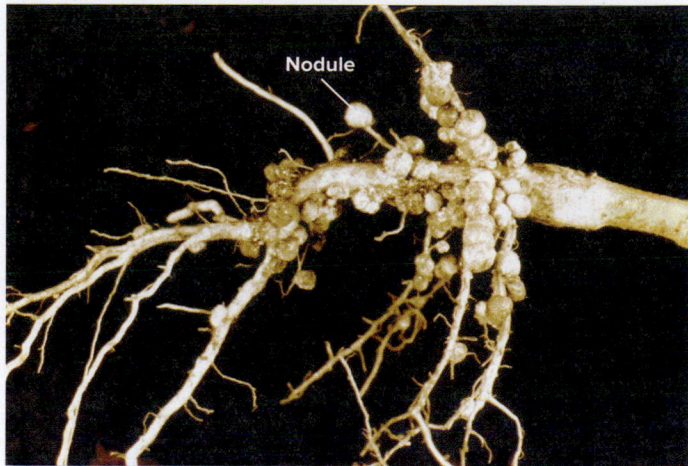

**Figure 38.17** **Legume root nodules.** The cells of nodules on the roots of this soybean plant (*Glycine max*) and other legumes contain nitrogen-fixing bacteria known as rhizobia. ©Scimat/Science Source

Nodule development involves a series of chemical signals sent back and forth between rhizobia and their host plants (**Figure 38.18**). Legumes start this exchange by secreting particular flavonoid compounds from their roots. Recall that flavonoids are phenolic secondary metabolites that play essential roles in plant structure, reproduction, and protection (see Figure 32.22b). These flavonoids bind to receptors in the plasma membranes of compatible soil rhizobia (Figure 38.18, step 1). In response, the rhizobia typically secrete **Nod factors** (nodulation factors). Each rhizobial species produces Nod factors with distinctive structural variations that can be recognized by the preferred host species. These Nod factors function something like keys that unlock doors, allowing bacteria to enter roots via root hairs. The factors bind to receptors in the membranes of root hair cells in the host plant (step 2).

Within minutes after its membrane receptors bind Nod factors, the root hair plasma membrane allows an influx of calcium ions, and a few minutes later, root hair calcium ion concentrations start oscillating rapidly. Root hairs respond to these calcium changes by swelling at their tips and curling around the rhizobia (step 3). The rhizobia then inject infection proteins into root hairs. In response, the cell wall at the root hair tip changes so that bacterial enzymes can erode a small hole in the wall, allowing the bacterial cells to enter. The plasma membrane forms a tubular infection thread through which rhizobia move into the root cortex. The tip of the infection thread fuses with the plasma membrane of a cortex cell, then the rhizobia are released into the cell's cytoplasm, each bacterial cell enclosed by the host membrane (step 4).

Meanwhile, plants produce proteins known as **nodulins** that foster root nodule development. Within 18–30 hours after the initial infection, root cortex cells start to divide to form root nodules. Environmental conditions in developing nodules cause rhizobia to undergo changes in their structure and gene expression patterns. These modified rhizobia are known as **bacteroids** (step 5). Bacteroid respiration provides the large amounts of ATP that are necessary for nitrogen fixation.

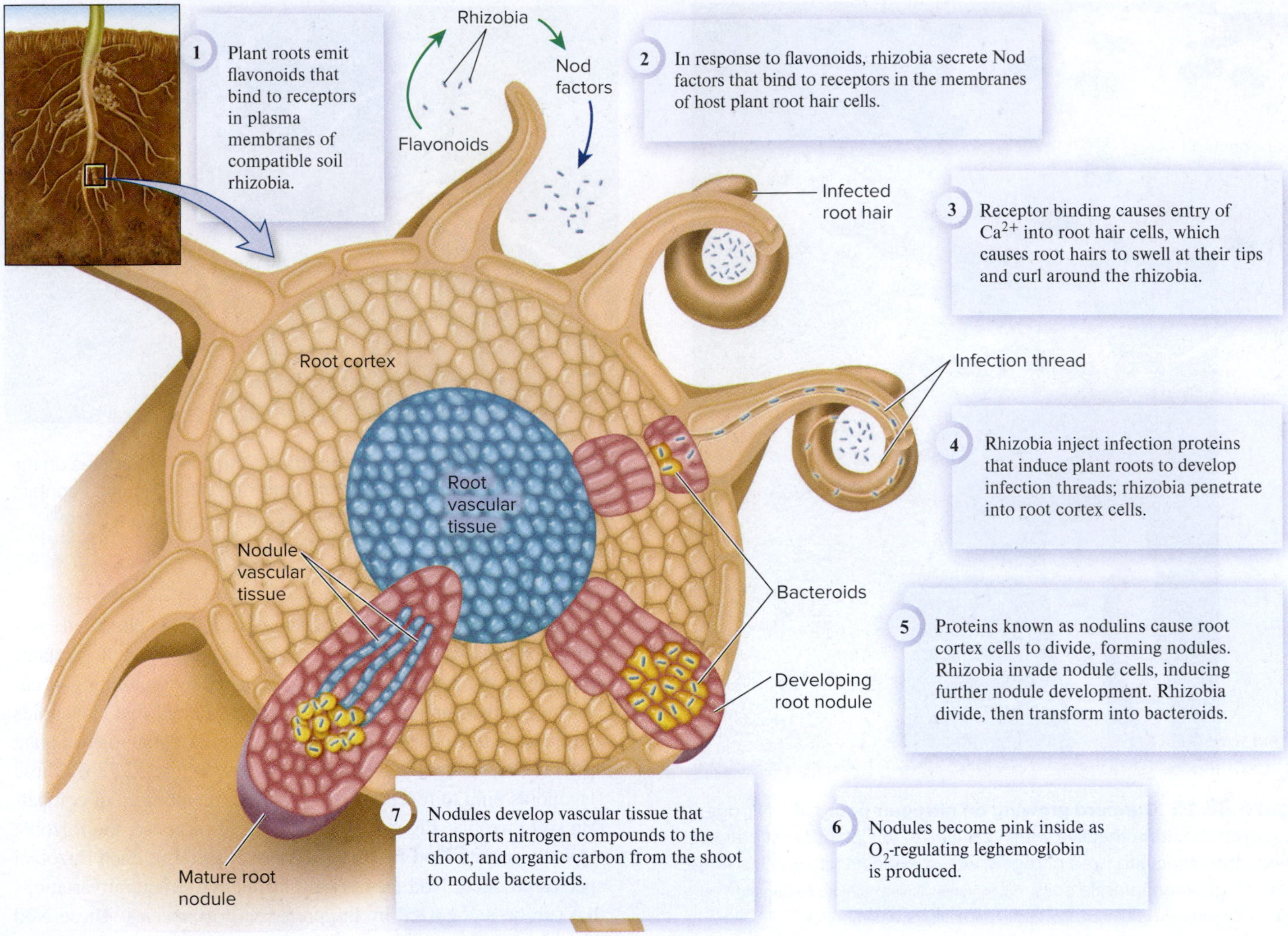

**1** Plant roots emit flavonoids that bind to receptors in plasma membranes of compatible soil rhizobia.

Rhizobia

Flavonoids

Nod factors

**2** In response to flavonoids, rhizobia secrete Nod factors that bind to receptors in the membranes of host plant root hair cells.

Infected root hair

**3** Receptor binding causes entry of $Ca^{2+}$ into root hair cells, which causes root hairs to swell at their tips and curl around the rhizobia.

Root cortex

Infection thread

**4** Rhizobia inject infection proteins that induce plant roots to develop infection threads; rhizobia penetrate into root cortex cells.

Root vascular tissue

Nodule vascular tissue

Bacteroids

**5** Proteins known as nodulins cause root cortex cells to divide, forming nodules. Rhizobia invade nodule cells, inducing further nodule development. Rhizobia divide, then transform into bacteroids.

Developing root nodule

Mature root nodule

**7** Nodules develop vascular tissue that transports nitrogen compounds to the shoot, and organic carbon from the shoot to nodule bacteroids.

**6** Nodules become pink inside as $O_2$-regulating leghemoglobin is produced.

**Figure 38.18** **Root nodule development.** The process of root nodule development involves a chemical "conversation" between the legume plant and nitrogen-fixing bacteria.

**Core Skill: Modeling** The goal of this modeling challenge is to use information provided in Figure 38.18 to describe how root nodule development might differ in a mutant plant lacking nodulin proteins.

**Modeling Challenge:** Figure 38.18 is a mechanistic model illustrating a series of steps involved in the development of legume root nodules that house beneficial nitrogen-fixing bacteria known as rhizobia. Assume that each step in the model depends on successful completion of the previous step. Let's suppose a mutant plant was unable to produce nodulins. Propose a revised model that describes what will happen when rhizobia infect such a plant.

Legume nodules typically produce **leghemoglobin** (legume hemoglobin), a pink protein that helps to regulate local oxygen concentrations, transporting enough oxygen to bacteroids to support respiration but preventing oxygen from disabling nitrogenase (step 6). Mature nodules also produce vascular tissue that moves nitrogen fixed by bacteroids to the root vascular system for transport throughout the plant. These nodule vascular tissues also supply organic food produced by the legume to their bacteroid partners (step 7). Nearly 5,000 gene expression changes are associated with the legume-rhizobia symbiosis.

## Carnivorous Plants Are Autotrophs That Obtain Mineral Nutrients from Animals

About 600 species of flowering plants have adapted to low-nitrogen environments by evolving mechanisms for trapping and digesting animals and are therefore known as carnivorous plants. Their leaves are modified in ways that allow them to capture animal prey, primarily insects, though larger animals are sometimes snared as well. Carnivorous plants are photosynthetic autotrophs that supply their own organic compounds; prey animals are primarily sources of nitrogen. The experimental use of radiolabeled prey insects has revealed

**(a) Pitcher plant (genus *Nepenthes*)**

**(b) Venus flytrap (*Dionaea muscipula*)**

**Figure 38.19** **Carnivorous plants.** **(a)** A pitcher plant passively captures animals that accidentally fall into its water-filled pitcher. **(b)** The Venus flytrap has an active trap that is stimulated by the touch of its prey, in this case, a fly. a: ©DeAgostini/Getty Images; b: ©Rafael Ben-Ari/Chameleons Eye/Photoshot/Newscom

that carnivorous plants obtain as much as 87% of their nitrogen from animals.

The trapping mechanisms used by carnivorous plants are classified as passive or active. Plants with passive trapping mechanisms depend on the prey to fall or wander into a trap. For example, tropical pitcher plants (genus *Nepenthes*) have leaves that are folded and partially fused to form tubes that collect rainwater (**Figure 38.19a**). The interior walls of these pitchers are slippery and have downward-pointing hairs. Insects and other small animals such as lizards and frogs that fall into the pitchers are unable to climb out. Eventually, the trapped animals drown and are digested by microbes living within the pitchers.

Plants with active mechanisms, such as Venus flytraps and sundews (see the chapter opening photo), have traps that are stimulated by touch. Charles Darwin, who was fascinated by carnivorous plants, was one of the first to study the trapping mechanisms of Venus flytraps and sundews. The Venus flytrap (*Dionaea muscipula*) has an active trap formed by a two-lobed leaf that is edged with lance-shaped teeth (**Figure 38.19b**). The leaf surface produces modified hairs, usually three per leaf lobe. If a single hair is touched—perhaps by wind or rain, or debris—and another touch does not occur soon thereafter, nothing happens. But when a fly or similar insect prey lands on the leaf and brushes against the same hair twice, or touches a second hair within 20–40 seconds, the leaf lobes snap shut around it.

Experimental studies indicate that a traveling electrical wave, known as an action potential, develops in the stimulated hairs of Venus flytrap. The electrical signal travels from cell to cell along plasma membranes, via plasmodesmata, at about 10 cm/second. This signal causes leaf cells to take up ions and water so that the leaf enlarges and changes shape, springing the trap. Action potentials also cause leaf gland cells to secrete hydrogen ions and digestive enzymes into the trap, forming what has been called a "green stomach." Trap cells use transporter proteins to take up materials released by digestion of the prey. Digestion is typically complete within 10 days, whereupon the trap may reopen. Trap leaves can go through three or four digestive cycles during their lifetime.

Sundews (such as *Drosera rotundifolia*) have leaves bearing glandular hairs whose sticky tips glisten in the sunlight. Insects that land on sundew leaves get mired in the sticky mucilage exuded by these hairs, as shown in the chapter opening photo. As the insects struggle to get away, they become covered with more mucilage and

eventually smother as their breathing pores become clogged. Darwin discovered that sundew leaves bend after being touched and that glandular hairs not originally in contact with the insects also bend, folding over the prey as you would fold your fingers over an object in your palm. Later, investigators discovered that this bending involves the plant hormone auxin. In response to touch, auxin accumulates in sundew leaf tips and then flows downward, stimulating the cell expansion that causes the leaf bending. The glandular hairs also produce enzymes that digest the prey.

## Parasitic Plants Obtain Nutrients from Photosynthetic Plants

More than 4,500 species of plants live as complete or partial **parasites**, organisms that obtain all or much of their water, minerals, and organic compounds from another organism. Dodder and witchweed are prominent examples of plants that are completely parasitic.

Dodder (*Cuscuta pentagona*) lacks roots and does not grow from the soil. Instead, all of the 150 species of this parasite live aboveground (**Figure 38.20**). These parasites twine their yellow or orange stems around green plant hosts, into which they sink peg-shaped, absorptive structures known as haustoria. These haustoria tap into the host plant's vascular system, gaining water, minerals, and sugar, which the parasite uses for growth and reproduction. The long, flexible stems of dodder often loop from one plant to another, allowing an individual dodder plant to tap into many different host plants at the same time. Dodder reproduces very rapidly by means of broken-off stem fragments and seeds. A single dodder plant can produce more than 16,000 seeds. In consequence, dodder is a widespread agricultural pest that attacks citrus, tomatoes, and many other fruit, vegetable, forage, and flower crops. Although dodder can harm host plants, recent research has revealed that dodder can help plant hosts defend against animal herbivores (see Chapter 37).

Another group of parasitic plants, the witchweeds (genus *Striga*), are serious problems for agriculture worldwide, because these parasites attack major cereal crops: corn, sorghum, rice, and millet. Witchweed seeds lie dormant in soil until secretions from host plant roots stimulate their germination. Genetic engineers are working to find ways to protect crops from the debilitating effects of these crop parasites.

**Figure 38.20** **A parasitic plant.** Dodder (*Cuscuta pentagona*) is an example of a parasitic plant that obtains all of its water, minerals, and organic compounds from one or more green plant hosts. Source: Charles T. Bryson, USDA Agricultural Research Service, Bugwood.org

 **Core Skill: Connections** Look back to Figure 34.7b to see an example of a parasitic animal—a tapeworm—and Table 34.2, which describes characteristics of tapeworms. What is similar about food acquisition in dodder and tapeworms?

## Summary of Key Concepts

### 38.1 Plant Nutritional Requirements

- The nutrients required by green plants are $CO_2$, $H_2O$, and several types of elements absorbed from soil or water (Figure 38.1, Table 38.1).

- Green plants require light energy as an essential resource for growth. Like animals and fungi, heterotrophic plants obtain chemical energy by metabolizing organic compounds absorbed from their environment (Figure 38.2).

- Plants display many adaptations that allow them to cope with insufficient or excess amounts of light and inadequate $CO_2$ (Figures 38.3, 38.4, 38.5).

- Inadequate supplies of certain nutrients can limit plant growth and cause mineral deficiency symptoms (Figure 38.6).

### 38.2 The Role of Soil in Plant Nutrition

- Natural soils display layers known as soil horizons. Soils are composed of organic material and inorganic minerals. Soil organic matter is largely derived from plant detritus, animal wastes, and decayed animal bodies (Figures 38.7, 38.8).

- Inorganic soil components occur as particles that can be categorized according to their size as sand, silt, or clay. Cation exchange releases cations to soil water, making cations available for uptake by plant roots (Figures 38.9, 38.10, 38.11).

- Biological or industrial processes convert atmospheric nitrogen gas into fixed nitrogen that plants can utilize. Biological nitrogen fixation can be performed only by certain prokaryotes (Figures 38.12, 38.13).

- Plants display several types of adaptations to cope with phosphate deficiency. Genetically modified smart plants can signal impending phosphate deficiency (Figure 38.14).

### 38.3 Biological Sources of Plant Nutrients

- Mycorrhizal fungi, which are associated with the roots of most plants, provide plants with water, phosphorus, and other minerals (Figure 38.15).

- Nitrogen-fixing prokaryotes living within the tissues of some plants provide them with fixed nitrogen. Legume-rhizobia associations are particularly important in nature and in agriculture (Figures 38.16, 38.17, 38.18).

- Carnivorous plants obtain mineral nutrients from the digested bodies of trapped animals. Parasitic plants obtain water, minerals, and organic compounds from green plant hosts (Figures 38.19, 38.20).

## Assess & Discuss

### Test Yourself

1. Which of the following substances can limit plant growth in nature?
   a. sunlight
   b. water
   c. carbon dioxide
   d. fixed nitrogen
   e. all of the above

2. In what form do plants take up most soil minerals?
   a. as ions dissolved in water
   b. as neutral salts
   c. as mineral-clay complexes
   d. linked to particles of organic carbon
   e. None of the above describes how plants take up minerals.

3. Why do plants need sulfur?
   a. for the construction of cell walls
   b. as an essential component of chlorophyll
   c. to produce proteins and some coenzymes
   d. for all of the above
   e. for none of the above

4. Soil organic matter provides the benefit of
   a. allowing water to percolate rapidly through soil.
   b. making soil softer in consistency.
   c. increasing the aluminum content of soil.
   d. causing minerals to be leached more rapidly from soil.
   e. none of the above.

5. Which environments are conducive to heavy leaching of minerals from soils?
   a. those having soils that are composed primarily of sand particles
   b. those having acidic soils
   c. those impacted by acid rain
   d. regions characterized by heavy rainfall
   e. all of the above

6. Which property is *not* characteristic of clay-rich soils?
   a. high mineral nutrient retention
   b. high water retention
   c. high aeration
   d. lower amounts of sand than clay
   e. All of the above characterize clay-rich soils.

7. Which of the plants listed below is heterotrophic (has to obtain organic food from the environment)?
   a. a green houseplant
   b. a legume plant such as bean
   c. a carnivorous plant such as the sundew
   d. ghostly white *Monotropa*
   e. none of the above

8. What kinds of organisms occur in nitrogen-fixing symbioses with plants?
   a. cyanobacteria
   b. actinobacteria
   c. rhizobia
   d. all of the above
   e. none of the above

9. How do legume roots attract rhizobia?
   a. They secrete flavonoids.
   b. They secrete carotenoids.
   c. They secrete alkaloids.
   d. They secrete Nod factors.
   e. None of the above is a means of attracting rhizobia.

10. Which plant uses a passive trap to obtain animal prey as a source of mineral nutrients?
    a. the Indian pipe (*Monotropa uniflora*)
    b. the tropical pitcher plant (*Nepenthes* spp.)
    c. the Venus flytrap (*Dionaea muscipula*)
    d. dodder (*Cuscuta* spp.)
    e. all of the above

## Conceptual Questions

1. Why are agricultural experts and ecologists concerned about overfertilization of crop fields?

2. Draw a diagram showing how rhizobia and legume roots communicate chemically during nodule formation.

3.  **Core Concept: Systems** How can lush tropical forests grow on relatively thin soils from which nutrients have been leached by frequent, heavy rains?

## Collaborative Questions

1. Imagine that you have bought a farm and want to start growing a crop to sell at a local market. How could you go about determining if the soil needs to be fertilized and with what mineral nutrients?

2. **Core Skill: Science and Society** Imagine that you own a large farm with a trout stream running through it. How would you protect the water quality of the stream?

# Flowering Plants: Transport

# 39

**A shade tree.** The evaporation of water from plant leaves cools them and us, and even affects local and global climate.

©Barry Mason/Alamy Stock Photo

O n hot days, people naturally gravitate to the cool shade beneath trees, as shown in the chapter opening photo. But most people do not realize that trees are not only sun umbrellas. Plants actually cool the air around them as water evaporates from their surfaces. That's why grass feels cool when you walk barefoot on it, even on a hot day.

Plants benefit from this evaporation process—known as transpiration—because it cools their surfaces and enables the movement of a continuous stream of water from the soil, through roots and stems, to leaves. This evaporation process not only helps to distribute water throughout the plant body, but also aids the movement of dissolved minerals, organic compounds such as sugars and hormones, and other organic materials over long distances within plants. Transport is therefore crucial for the functions of plant growth, behavior, and nutrition, which we discussed in the preceding three chapters.

In addition, plant transport plays a critical role in Earth's global climate. On a worldwide basis, plant transpiration annually moves $3.2 \times 10^4$ billion tons of water from the soil into the atmosphere as water vapor. Plant transpiration provides important ecological services. For example, plant-produced atmospheric water vapor is the source of 30% of rain, with the rest originating from evaporation at the surfaces of oceans and freshwater bodies. Along with other atmospheric gases (including carbon dioxide and methane), water vapor also works as a greenhouse gas that helps to warm Earth's climate by absorbing the Sun's heat. Plant transport processes are also relevant to agriculture, as humans seek to improve crop productivity and the efficiency of water and nutrient use. Conservation biologists appreciate that plant transport processes are important to the preservation and restoration of natural environments.

To understand plant transport more fully, in this chapter, we first survey the materials that move through plants and the general directions of such movements. Next, we will focus more closely on water and solute uptake by plant cells. We will then examine how these materials are moved within plants over short and long distances and explore some of the plant adaptations that allow such transport to be as efficient as possible in a variety of environments. In the process, we will learn why plants are so cool!

## 39.1 Overview of Plant Transport

**Learning Outcomes:**

1. Describe how plants transport water and minerals from roots to leaves.
2. Describe how plants transport organic molecules from leaves to nonphotosynthetic parts.

In Chapters 36–38, we surveyed the structure, behavior, and nutrition of flowering plants (the angiosperms) and discussed the interdependence of plant root and shoot systems. We have seen that in most plants, the root system absorbs water and dissolved minerals from the soil and that the shoot system takes up carbon dioxide ($CO_2$) from the atmosphere via stomata, pores that occur in the surfaces of leaves and other aboveground structures

(**Figure 39.1**). Photosynthetic cells use these materials to produce sugar and other organic compounds needed for overall plant growth and reproduction. Nonphotosynthetic plant cells, such as those of roots and flowers, depend on organic food produced by green tissues.

The growth and survival of plants depends on a two-way transport system. Organic food is transported from photosynthetic to nonphotosynthetic parts. In most plants, this transport is generally downward from the leaves to the roots. In addition, water and minerals are transported upward from roots to shoots. Since plants can grow to sizable heights, the tallest trees being over 100 m tall, transport of materials often occurs over long distances.

The long-distance transport of water, dissolved minerals, and sugar throughout the plant body occurs within a continuous system of conducting tissues. Recall that the complex tissues of vascular plants that primarily conduct water and dissolved minerals are known as the xylem, and those that conduct mostly organic substances in a watery sap are termed the phloem. These conducting tissues are key to the ability of vascular plants to thrive in terrestrial habitats, which can sometimes be quite arid. To fully understand how plants accomplish long-distance transport, we will begin by reviewing the processes by which minerals, organic compounds, and water are taken up and move at the cellular level.

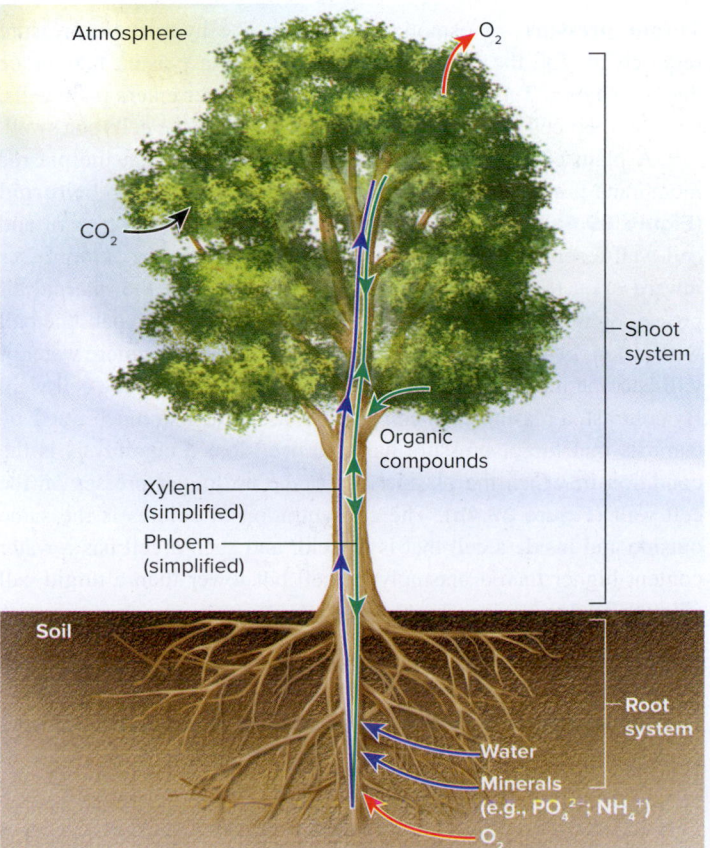

**Figure 39.1 Overview of material uptake and long-distance transport processes in plants.**

## 39.2 Uptake and Movement of Materials at the Cellular Level

### Learning Outcomes:

1. Describe the differences among turgid, flaccid, and plasmolyzed plant cells.
2. **CoreSKILL »** Explain how water potential and relative water content are calculated.
3. List ways in which plants cope with cellular osmotic stress.

Chapter 5 described how all cells use both passive and active processes to import or export materials. Here, we briefly review these processes, illustrating how they work in flowering plants.

### Passive Transport Does Not Require the Input of Energy

Recall that water, gases, and certain small, uncharged compounds can diffuse across plasma membranes in the direction of their concentration gradients. **Passive transport** is the movement of materials into or out of cells down a concentration gradient without the expenditure of energy in the form of ATP. Passive transport across plasma membranes occurs in two ways: by simple or facilitated diffusion. **Simple diffusion** into or out of cells is the movement of molecules through a phospholipid bilayer down a concentration gradient. **Facilitated diffusion** is the transport of molecules across a plasma membrane down a concentration gradient with the aid of membrane transport proteins (**Figure 39.2a**).

The two main types of membrane transport proteins that function in facilitated diffusion are channels and transporters. **Channels** are membrane pores formed by proteins that allow movement of ions and molecules across membranes (see Figure 39.2). **Transporters** are proteins that transport molecules by binding them on one side of the membrane and then changing conformation so that the molecule is released to the other side of the membrane (refer back to Figure 5.17). Transporters increase the rate at which specific mineral ions and organic molecules are able to enter or leave plant cells and vacuoles.

Recall that osmosis is the diffusion of water across a selectively permeable membrane in response to differences in solute concentrations. In the case of plants, water moves from a solution that has a lower solute concentration (soil) to one of higher solute concentration (root cells). Osmotic water uptake into living plant cells is essential to photosynthesis, as well as to cell expansion and structural support. However, simple diffusion of water does not occur rapidly enough to supply the water needs of rapidly expanding plant cells. In this case, facilitated diffusion of water occurs through protein channels known as **aquaporins**, which occur widely in living cells. Thirty-five distinct aquaporin genes have been identified in the genome of the model plant *Arabidopsis*. Aquaporins increase the rate at which water flows into expanding plant cells and their vacuoles. Similarly, many other types of plasma membrane protein channels and transporters facilitate the diffusion of specific mineral ions and organic molecules into and out of plant cells and vacuoles.

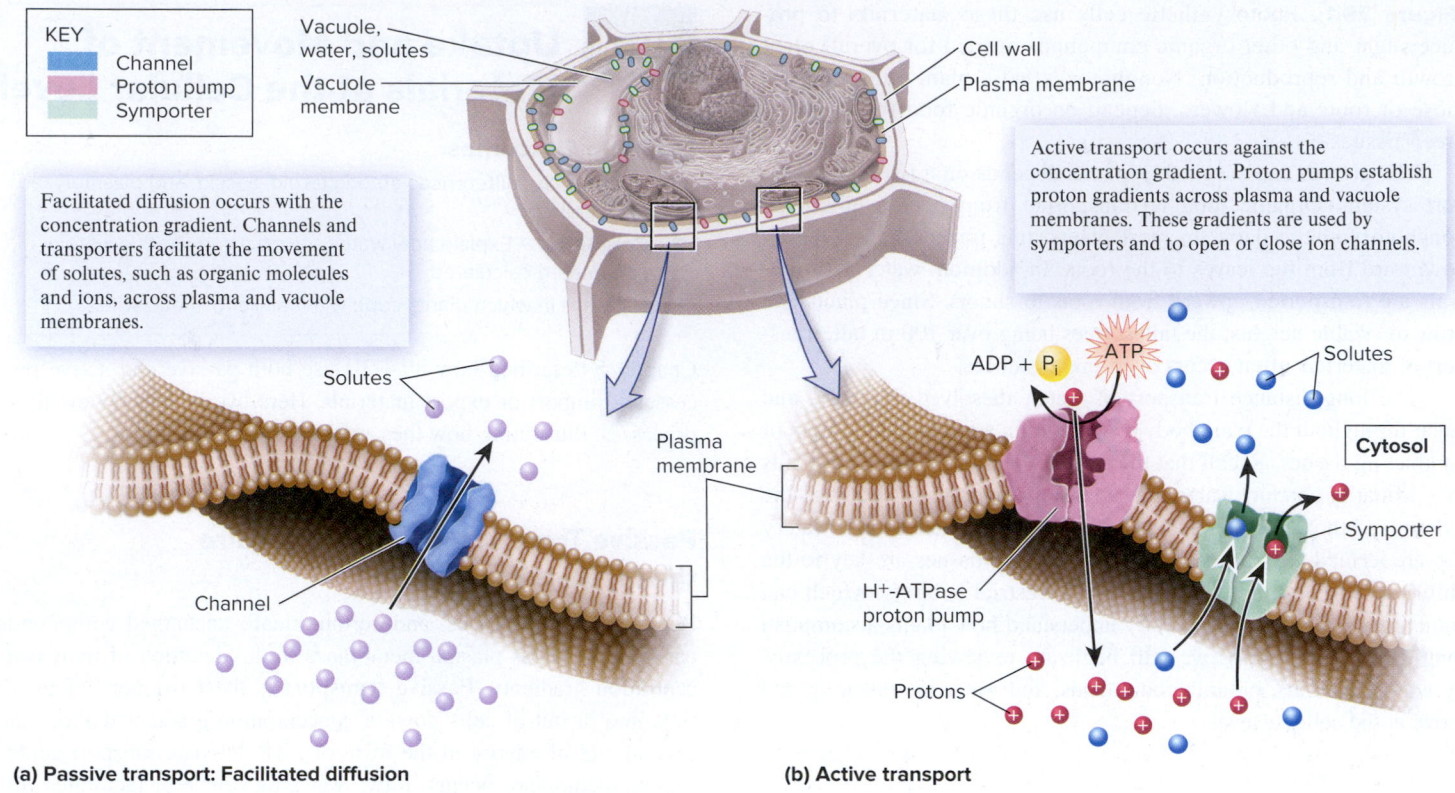

KEY
- Channel
- Proton pump
- Symporter

Vacuole, water, solutes

Vacuole membrane

Cell wall

Plasma membrane

Facilitated diffusion occurs with the concentration gradient. Channels and transporters facilitate the movement of solutes, such as organic molecules and ions, across plasma and vacuole membranes.

Active transport occurs against the concentration gradient. Proton pumps establish proton gradients across plasma and vacuole membranes. These gradients are used by symporters and to open or close ion channels.

Solutes

Plasma membrane

Channel

ADP + $P_i$    ATP

Solutes

Cytosol

$H^+$-ATPase proton pump

Symporter

Protons

(a) Passive transport: Facilitated diffusion

(b) Active transport

**Figure 39.2** Passive and active transport.

## ATP Hydrolysis Powers Active Transport

If a substance must be transported across a plasma membrane against its concentration gradient, work must be performed, in the process known as active transport. During **active transport**, membrane transport proteins use energy to move substances against their concentration gradients. An example is the $H^+$-ATPase proton pump, found in the plasma membranes of plant cells, which uses ATP to pump $H^+$, which are protons, against a gradient (**Figure 39.2b**). This concentration gradient generates an electrical charge difference across the membrane, which is known as a **membrane potential**. Energy is released when protons pass back across the plasma membrane, in the direction of their proton gradient. This energy can then be used to power other active transport of ions or organic materials. For example, it might be used to open or close ion channels or in the functioning of a proton-solute **symporter**, a protein that transports two substances in the same direction across a membrane. Symporters are needed for the uptake of organic solutes such as sugars, amino acids, and nucleotide bases.

Active transport proteins are particularly abundant in root cell membranes (**Figure 39.3**), allowing root cells to concentrate dissolved mineral nutrients to more than 75 times their abundance in soil. As a result, soil water flows into root cells by osmosis. We next take a closer look at osmotic water movement into and out of plant cells.

## Cellular Water Content Is Influenced by Solute Content and Turgor Pressure

The water content of plant cells depends on osmosis, and osmosis depends on two factors: solute content and turgor pressure.

**Turgor pressure**, or osmotic pressure, is the hydrostatic pressure required to stop the net flow of water across a plasma membrane due to osmosis. Turgor pressure increases as water enters plant cells, because their cell walls restrict the extent to which the cells can swell.

A plant cell whose cytosol is so full of water that the plasma membrane presses right up against the cell wall is said to be **turgid** (**Figure 39.4a**). The pressure relationship between the cytosol and cell wall resembles the way that a soccer ball's leather skin presses inward upon the air within, while at the same time the internal air presses on the ball's cover. If you add more air to a limp ball, the ball will stiffen. In the same way, if a nonturgid cell absorbs more water, it will become more rigid as the water exerts pressure on the cell wall. By contrast, a plasmolyzed cell is one that has lost so much water by osmosis that turgor pressure has also been lost. **Plasmolysis** is the condition in which the plasma membrane no longer presses on the cell wall (**Figure 39.4b**). The concentration of solutes is the same outside and inside a cell that is **flaccid**, and such a cell has a water content higher than a plasmolyzed cell but lower than a turgid cell (**Figure 39.4c**).

## Water Potential Affects the Movement of Water Into and Out of Cells and a Plant's Relative Water Content

Together, solute concentration and the presence of a cell wall influence an important plant cell property known as **water potential**, the potential energy of water. Water moves from a region of higher water potential to a region of lower water potential. A good analogy is a waterfall, in which the water has high gravitational potential energy

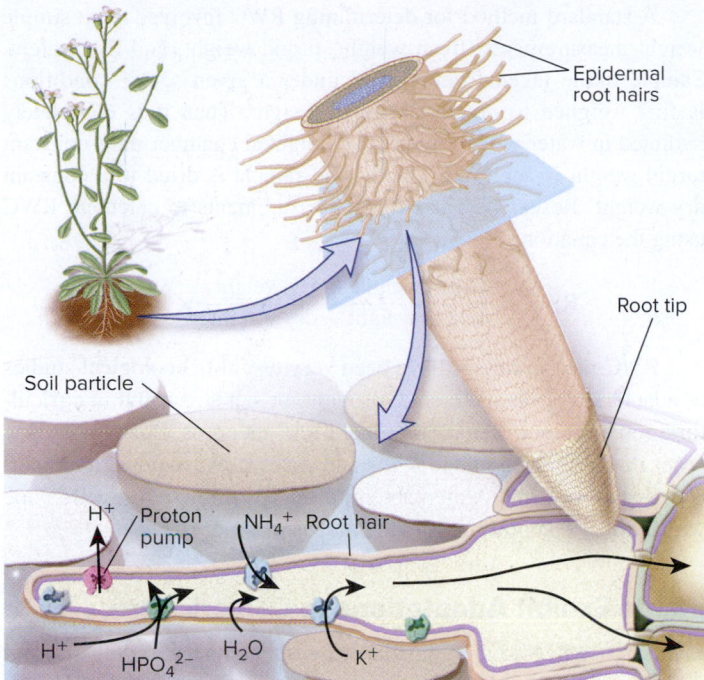

**Figure 39.3** **Ion uptake at root-hair membranes.** The H$^+$-ATPase proton pump establishes an electrochemical gradient that drives the active uptake of solutes. The resulting increase in intracellular solute concentration also drives the osmotic diffusion of water into the cell.

**Core Concept: Energy and Matter** When mineral ion concentrations in the soil are lower than those within cells, root-hair plasma membranes take up nutrient ions by active transport, which requires energy in the form of ATP.

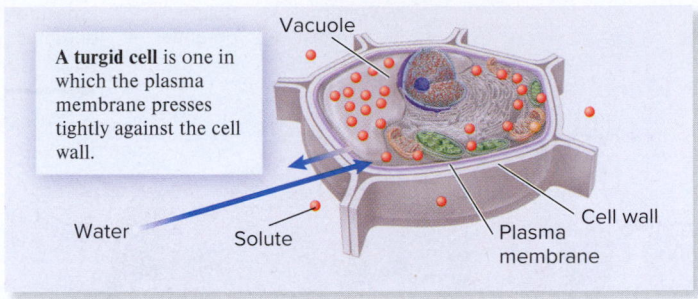

**(a) Turgid cell in a hypotonic solution**

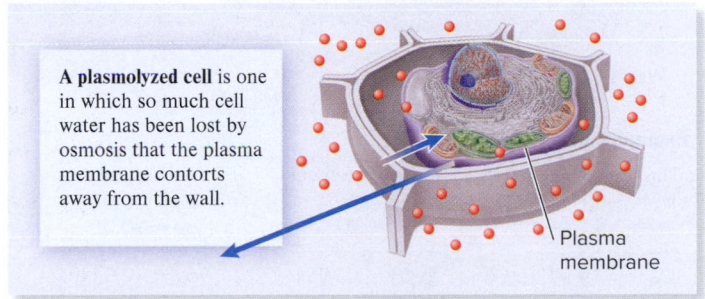

**(b) Plasmolyzed cell in a hypertonic solution**

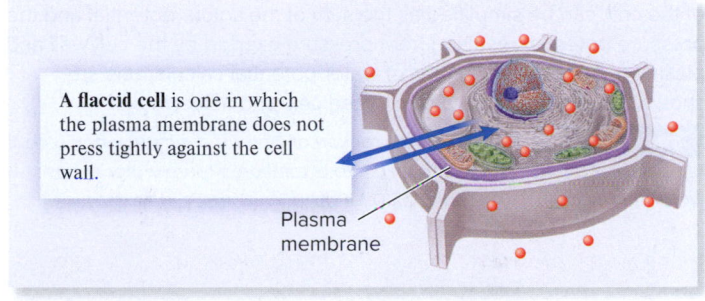

**(c) Flaccid cell in an isotonic solution**

**Figure 39.4** **Turgid, plasmolyzed, and flaccid plant cells.** **(a)** When the concentration of solutes inside a cell is greater than that outside (the cell is surrounded by a hypotonic solution), more water may enter the cell than will leave it. As a result, the plant cell may become swollen, or turgid. **(b)** When the concentration of solutes outside a cell is greater than within it (the cell is surrounded by a hypertonic solution), more water will leave the cell than will enter it. As a result, a plant cell will become plasmolyzed. **(c)** When a cell is bathed in an isotonic solution, it will be flaccid.

at the top. Pressure also influences water potential; a waterfall would flow upward if pressure greater than the force of gravity were applied. Solutes and some other factors also affect water potential.

Water potential is measured in pressure units known as megapascals (MPa) (a pascal is equal to 1 newton per square meter). One MPa is equal to 10 times the average air pressure at sea level, about the same pressure that occurs within an inflated bicycle tire. As another reference point, 1 MPa is several times the pressure in typical home plumbing pipes, which you experience when turning on a water faucet.

In the study of plants, the concept of water potential is used to understand the movement of water into and out of cells (cellular water potential) and between entire plants and their environment. The concept of relative water content is used to gauge the water status of whole plants or organs.

***Cellular Water Potential*** A water potential equation can be used to predict the direction of water movement, given information about the solute concentrations inside and outside of plant cells and a measure of pressure at the cell-wall–membrane interface. In this equation, cellular water potential is symbolized by the Greek letter psi ($\psi$) with the subscript W for water: $\psi_W$. In its simplest form, total $\psi_W$ is calculated as:

$$\psi_W = \psi_S + \psi_P$$

where $\psi_S$ is solute potential and $\psi_P$ is pressure potential.

**Solute potential** is the component of water potential due to the presence of solute molecules. As you might expect, solute potential is proportional to the concentration of solutes in a solution. The solute potential of pure water open to the air, at sea level and room temperature, is defined as zero. When solutes are added, they interact with water molecules, thereby diluting the water and affecting its disorder. As a result, fewer free water molecules are present, which reduces the potential energy of water. Thus, in the absence of a pressure potential, water that contains solutes always has a negative solute potential. The higher the concentration of dissolved solutes, the lower (more negative) the solute potential.

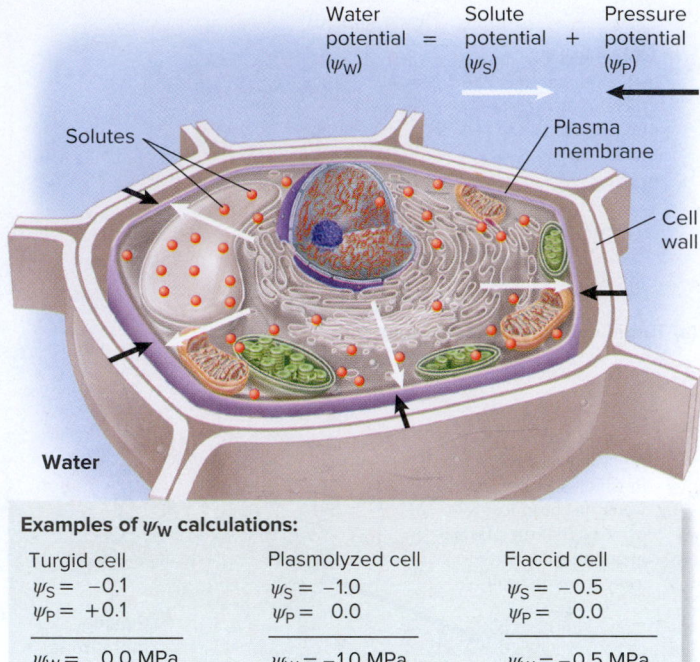

Water potential = Solute potential + Pressure potential
$(\psi_W)$         $(\psi_S)$          $(\psi_P)$

Solutes

Plasma membrane

Cell wall

**Water**

**Examples of $\psi_W$ calculations:**

| Turgid cell | Plasmolyzed cell | Flaccid cell |
|---|---|---|
| $\psi_S = -0.1$ | $\psi_S = -1.0$ | $\psi_S = -0.5$ |
| $\psi_P = +0.1$ | $\psi_P = 0.0$ | $\psi_P = 0.0$ |
| $\psi_W = 0.0$ MPa | $\psi_W = -1.0$ MPa | $\psi_W = -0.5$ MPa |

**Figure 39.5** **Plant-cell water potential.** The water potential of a plant cell, which predicts the direction of water movement into or out of the cell, can be simplified as the sum of the solute potential and the pressure potential resulting from pressure exerted by the cell wall and plasma membrane. Examples of water potential calculations are shown for a turgid cell, a plasmolyzed cell, and a flaccid cell.

**Concept Check:** *Describe the direction of water movement when a cell of each of the three types is placed into a solution of pure water (whose water potential, $\psi_w$, is defined as 0 MPa).*

**Pressure potential** $(\psi_P)$ is the component of water potential due to hydrostatic pressure. In plant cells, the hydrostatic pressure is determined in part by the resistance provided by the cell wall. Because of this resistance, the value for pressure potential can be either positive or negative. For example, a turgid cell has a positive pressure potential, which typically measures about 1 MPa. This high pressure inside turgid plant cells is a testimony to the strength of their cellulose-rich cell walls. In contrast to turgid cells, both flaccid and plasmolyzed cells have a pressure potential of zero. Plants or plant organs having many cells with low turgor pressure appear wilted. If your houseplants become wilted, watering will enable the cells to increase their pressure potential, restoring cell turgor and normal plant appearance. Therefore, the water content of an entire organ or plant is influenced by the water potential of its component cells (**Figure 39.5**).

*Relative Water Content*   The property known as **relative water content (RWC)** is often used to gauge the water content of a plant organ or entire plant and is easy to measure. RWC integrates the water potential of all cells within an organ or plant and is thus a measure of relative turgidity. Measurements of RWC can be used to predict a plant's ability to recover from the wilted condition. An RWC of less than 50% spells death for most plants, but some plants can tolerate lower water content for substantial time periods.

A standard method for determining RWC involves three simple weight measurements: fresh weight, turgid weight, and dry weight. Sample tissue taken from a plant under a given set of conditions is first weighed to obtain the fresh weight. Then it is completely hydrated in water within an enclosed, lighted chamber until constant turgid weight is achieved. Finally, the sample is dried to a constant dry weight. Researchers use these measurements to calculate RWC using the equation

$$RWC = \frac{(\text{fresh weight} - \text{dry weight})}{(\text{turgid weight} - \text{dry weight})} \times 100$$

RWC measurements have been very useful in ecological studies of adaptation of plants to cold, drought, or salt stress and in agricultural research for developing drought-tolerant crops. Developing new crops that are better able to withstand water stress requires an understanding of not only water potential but also how plant cells cope with cellular osmotic stress, which leads to water stress.

## Plants Exhibit Adaptations to Osmotic Stress

Plants native to cold, dry, or saline environments have evolved many different adaptations that allow them to cope with low water content. For example, plants often increase the solute concentrations of their cells' cytosol, a process known as **osmotic adjustment**. Increased amounts of the amino acid proline, sugars, or sugar alcohols such as mannitol decrease the cells' water potential, thus drawing water into cells. By increasing the concentration of solutes inside cells, cold-resistant plants prevent water from moving out of their cells when ice crystal formation in intercellular spaces lowers the water potential outside the cells. The additional solutes also lower the freezing point of the cytosol, in the same way that adding antifreeze to a car's radiator in winter keeps the radiator fluid from freezing.

Plants of arid lands often possess adaptations that help them survive water stress. Many can survive in a nearly dry state for as much as 10 months of the year, growing and reproducing only after the rains come. The cytosol of such desiccation-tolerant plants is typically rich in sugars that bind to phospholipids to form a glasslike structure. This helps to stabilize the cellular membranes, preventing them from becoming damaged during plasmolysis (see Figure 39.4b). Plant cells under water stress may also increase the number of plasma membrane aquaporins. These additional protein channels increase the rate of water uptake, allowing cells to recover turgor more quickly when water becomes available.

## 39.3 Tissue-Level Transport

**Learning Outcomes:**
1. Define transmembrane transport, symplastic transport, and apoplastic transport.
2. Explain how the root endodermis functions as a diffusion barrier.

Thus far, we have examined how water, dissolved minerals, and organic compounds enter or leave plant cells. In this section, we will consider short-distance transport within and among nearby tissues.

## Solutes Can Move Within Tissues via Transmembrane, Symplastic, and Apoplastic Transport

Tissue-level transport occurs in three forms: transmembrane transport, symplastic transport, and apoplastic transport (**Figure 39.6**).

**Transmembrane transport** involves the export of a material from one cell via membrane proteins, followed by import of the same substance by an adjacent cell (Figure 39.6a). One prominent example of transmembrane transport is the movement of the plant hormone auxin downward in shoots. Auxin travels from one phloem parenchyma cell to another in a linear series with the aid of carrier proteins (refer back to Figure 37.4). This process explains how auxin produced in one part of the plant body can influence more distant tissues.

**Symplastic transport** is the movement of a substance from the cytosol of one cell to the cytosol of an adjacent cell via membrane-lined channels called plasmodesmata (Figure 39.6b). Plasmodesmata are large enough in diameter to allow transport of proteins and nucleic acids, as well as smaller molecules. Together, all of a plant's protoplasts (the cell contents without the cell walls) and plasmodesmata form the **symplast**. Because plant cells are interconnected via plasmodesmata, symplastic transport has the potential to move a wide variety of molecules between cells and within the tissues of the plant body.

The **apoplast** is the continuum of water-filled cell walls and intercellular spaces (Figure 39.6c). **Apoplastic transport** is the movement of solutes along cell walls and in the spaces between cells. Water and dissolved minerals often move through plant tissues for short distances by apoplastic transport.

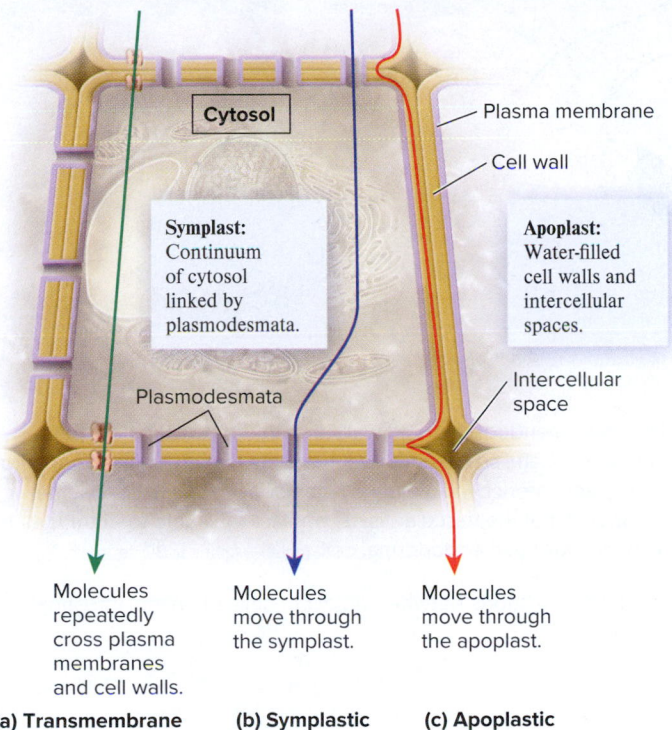

(a) **Transmembrane**   (b) **Symplastic**   (c) **Apoplastic**

**Figure 39.6** **Three routes of tissue-level transport in plants: (a) transmembrane, (b) symplastic, and (c) apoplastic.**

## Roots Take Up Minerals via Symplastic Transport or by Apoplastic and Transmembrane Transport

Both symplastic and apoplastic transport play important roles in mineral nutrient transport through the outer tissues of roots (**Figure 39.7**). The plasma membranes of epidermal root hair cells are rich in channels and transporters that selectively absorb essential mineral ions from soil water. As shown by the blue arrow in Figure 39.7, absorbed ions can move symplastically from the cytosol of root hairs, cortex, and endodermis directly to xylem parenchyma cells. Plasmodesmata make such cell-to-cell, tissue-level transport possible.

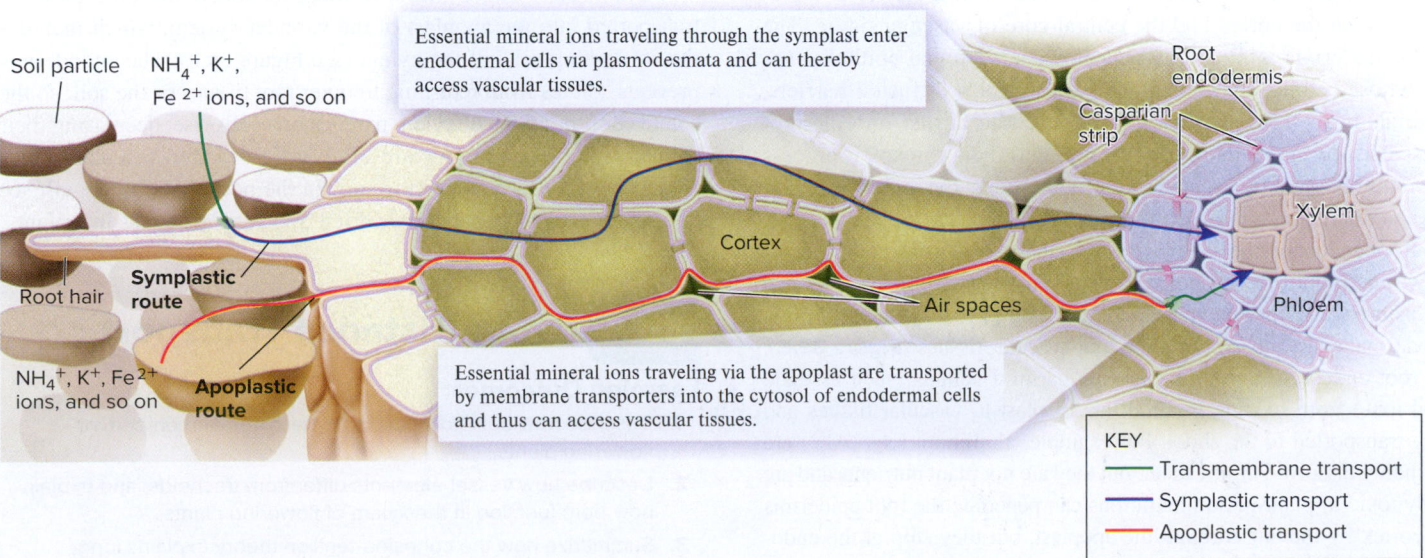

**Figure 39.7** **Symplastic and apoplastic transport of mineral ions in roots.**

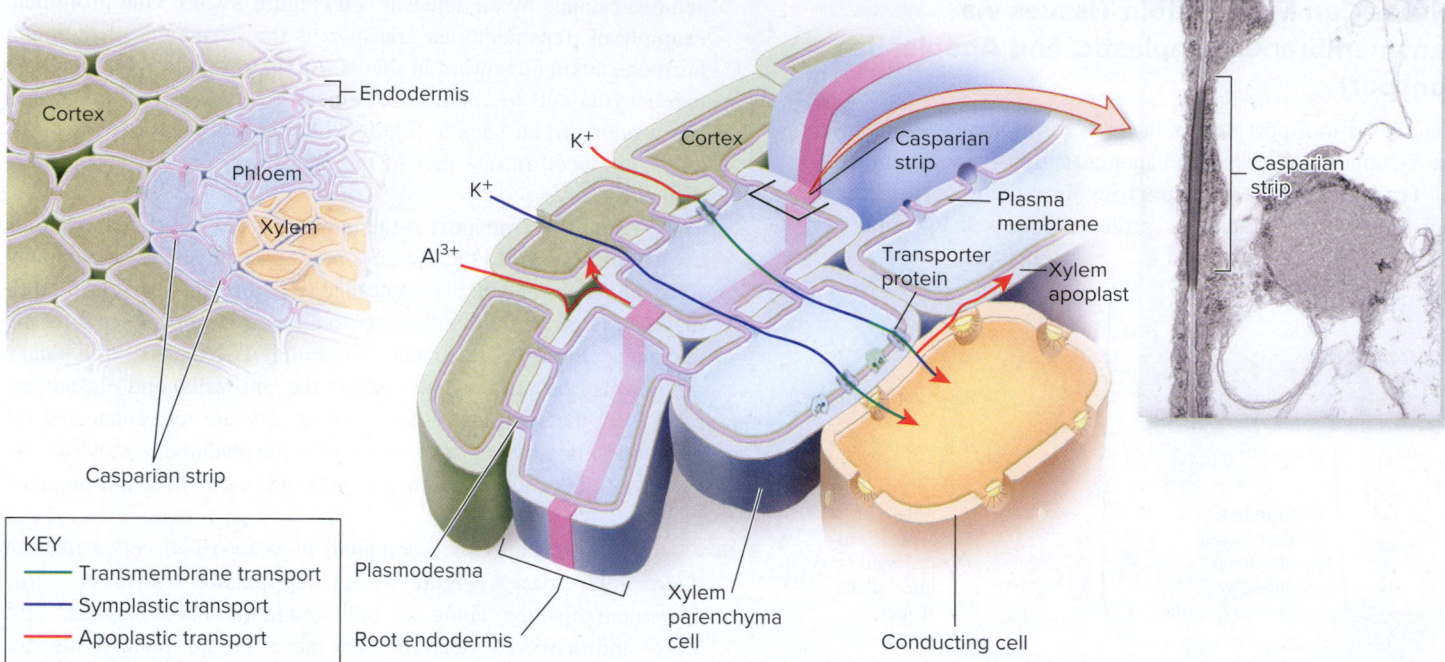

**KEY**
- Transmembrane transport
- Symplastic transport
- Apoplastic transport

**Figure 39.8 Ion transport pathways across the root endodermis.** Casparian strips in endodermal cell walls prevent apoplastic transport across the root endodermis, limiting entry of harmful soil minerals such as $Al^{3+}$ and exit of useful solutes. Mineral nutrients that are transported into the cytosol of endodermal cells are able to pass through the endodermal barrier to xylem parenchyma cells via plasmodesmata. Once past the endodermis, nutrient ions such as $K^+$ are moved across plasma membranes to the apoplast of the vascular tissue and are thus able to enter the xylem. Inset shows a transmission electron micrograph (TEM) of a Casparian strip in the wall of an endodermal cell. (right): ©James S. Busse

**Core Skill: Connections** Look back to Figure 10.9 to see a diagram of tight junctions between adjacent cells of animal intestinal epithelium. How is the root endodermis similar in structure and function?

Apoplastic transport can also move water and dissolved minerals into root epidermal and cortex tissues. However, apoplastic movement of water and minerals stops at the **endodermis**, a term meaning inside skin. In roots, the endodermis is a thin cylinder of tissue whose close-fitting cells and specialized cell walls form a barrier to diffusion between the cortex and the central core of vascular tissue. The plant endodermis is functionally analogous to animal epithelial tissues whose cellular tight junctions likewise form diffusion barriers. Materials in the root apoplast cannot penetrate farther into the root unless endodermal cells transport them into their cytosol, a process that requires specific transporter proteins (see the green junction between the red and blue arrows in Figure 39.7).

Root endodermal cell walls possess ribbon-like strips composed of waxy, waterproof suberin and phenolic polymers. These ribbons, known as **Casparian strips**, prevent apoplastic transport of solutes through endodermal cell walls and into the root vascular tissues (**Figure 39.8**). The root endodermis thereby prevents harmful solutes (such as toxic metal ions) from moving through the apoplast to vascular tissues and being transported to the shoot. For example, aluminum ions ($Al^{3+}$) are commonly dissolved in soil water, but they are not plant nutrients and are highly toxic to plants. Aluminum ions can penetrate the root epidermis and cortex by moving through the apoplast, but they stop at the endodermis because they are unable to enter the cytosol of endodermal cells.

Endodermal plasma membranes possess specific channels and transporters for essential mineral nutrients (such as $K^+$), which allows

them to enter the cytosol of root endodermal cells. By moving through endodermal cytosol, symplastically transported essential minerals are able to bypass the Casparian strips. Therefore, the root endodermis functions as a molecular filter that allows the passage of beneficial solutes.

Once solutes have moved through endodermal cells, they are transported into the apoplast of the vascular system, which includes the conducting cells of the xylem (see Figure 39.8). The endodermis prevents solutes from returning to outer root tissues or the soil, so the solute concentrations of xylem parenchyma cells rise, decreasing their cellular water potential. As a result, water flows into vascular tissues from outer root tissues and the soil. In the next section, we will see how water and solutes are transported over long distances in a plant.

## 39.4 Long-Distance Transport

**Learning Outcomes:**

1. Describe how bulk flow occurs in the xylem and phloem of flowering plants.
2. Describe how vessel elements differ from tracheids, and explain how both function in the xylem of flowering plants.
3. Summarize how the cohesion-tension theory explains long-distance water movement in plants.
4. Explain how stomata and leaf abscission help reduce transpirational water loss.

5. Describe how sieve-tube elements and companion cells work together to form a transport system.

6. **CoreSKILL »** Create a diagram showing how and why phloem sap moves from source to sink.

Plants rely on long-distance transport to move water and dissolved materials between roots and shoots and among organs. Tall trees are able to transport water and minerals to astounding heights, more than 110 m in some cases. This is possible because plants possess an extensive, branched, long-distance vascular system composed of xylem and phloem tissues. Watery solutions move through these tissues by **bulk flow**, the mass movement of liquid caused by pressure, gravity, or both. Plant conducting tissues are specialized in ways that foster bulk flow and aid plants in adapting to water stress. In this section, we will take a closer look at bulk flow and the major factors involved in long-distance transport by this process.

## Bulk Flow Is Water Movement Under the Influence of Pressure and Gravity

Bulk flow (also known as mass flow) occurs when molecules of liquid all move together from one place to another as the result of differences in pressure and/or gravity. One example of bulk flow is leaching, the movement of water and dissolved minerals downward through soil layers as the result of gravity (Chapter 38). Bulk flow is one way in which mineral ions can move through soil toward plant roots. Likewise, once inside the plant, minerals and other dissolved solutes can move through xylem and phloem conducting tissues via bulk flow, which is much faster than diffusion. For example, phloem sap, which contains sugars and other dissolved solutes, moves by bulk flow up to 1 m per hour. The bulk flow of water and solutes within xylem and phloem results from differences in water pressure. However, the pressure differences originate from different processes in these two types of conducting tissues.

In xylem, the bulk flow of water and dissolved minerals is driven by **transpiration** (from the French *transpirer*, meaning to perspire), the process in which water evaporates from the aerial parts of plants, usually via the stomata of leaves (**Figure 39.9**). The extent to which

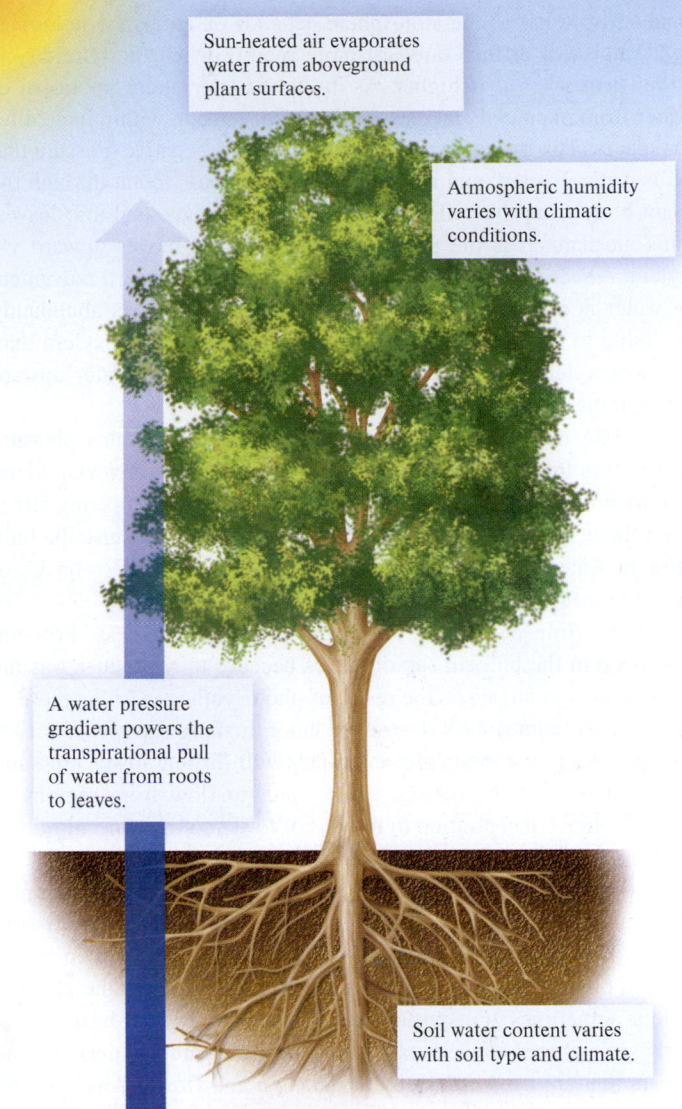

Sun-heated air evaporates water from aboveground plant surfaces.

Atmospheric humidity varies with climatic conditions.

A water pressure gradient powers the transpirational pull of water from roots to leaves.

Soil water content varies with soil type and climate.

**Figure 39.9  Upward transport of water in xylem.**  Water pressure differences between moist soil and drier air drive the upward movement of water in plants.

---

 **Core Skill: Modeling**  The goal of this modeling challenge is to show how gravity might impact plant size by influencing transport.

**Modeling Challenge:** Figure 39.9 is a model showing the relationship between evaporation at a tree's leaf surfaces and the water pressure gradient that occurs in the tree's water transport system (xylem), given adequate soil water availability. This model assumes constant gravity, which on Earth averages 9.8 m/s² (32.2 ft/s²) for a standard latitude of 45° 32'33".

In recent years, diverse planets have been detected outside of our solar system, some occurring in orbits that might allow the existence of liquid water and life. These include three of the seven planets (labeled e, f, and g) in the TRAPPIST-1 system. Let's suppose a particular tree species on Earth attains a maximal height of 100 m. Imagine that the seeds from such trees could be transported to other planets having an orbit favorable to the presence of liquid water and to tree growth.

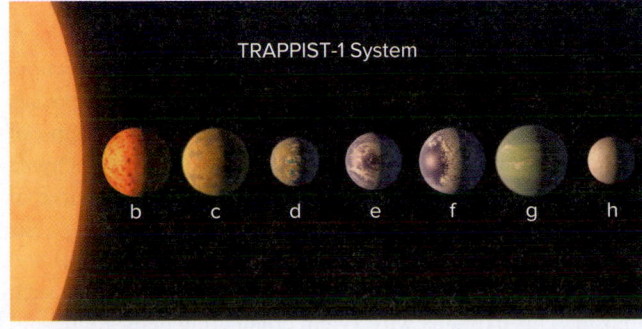

TRAPPIST-1 System

b    c    d    e    f    g    h

Source: NASA/JPL-Caltech

Sketch three trees showing differences in the maximal height that you expect to occur on three planets, one having much lower gravity than occurs on Earth, one with gravity similar to Earth's, and one with gravity much higher than Earth's. On your sketch, indicate the height of each tree and explain the differences in height.

plants lose water to the atmosphere depends on atmospheric humidity; water will diffuse outward more readily when the humidity is lower than when it is higher. As discussed later, the evaporation of water from plant cells increases the surface tension within intercellular spaces. This rise in surface tension creates a negative pressure that pulls water upward, and it moves as a continuous stream through the plant body from the soil, into roots, through stems, and into leaves. This mechanism is the primary way that water moves upward via xylem. A second mechanism that can promote the upward movement of water is related to root pressure. When soil water is abundantly available to roots, water pressure will be higher in root xylem than in shoot xylem. This positive pressure tends to push water upward through the xylem.

While the evaporation of water from leaves creates pressure differences that facilitate the upward movement of water via xylem, gravity promotes a downward movement. These two competing forces limit the heights that terrestrial plants can achieve. Because the bulk flow of water in plant xylem is affected by gravity, trees on Earth rarely exceed 100 m in height.

Bulk flow in phloem arises by a different process. Pressure builds up in the phloem sap of leaves because they produce organic solutes such as sugars as the result of photosynthesis. An increase in sap organic solutes causes water to enter, making sap pressure rise. Phloem sap that contains fewer solutes will display lower pressure. Such differences in pressure cause water to flow from regions of higher solute concentration to regions of lower solute concentrations, such as developing fruit and roots.

Although xylem serves as the primary transport system for water and minerals, and phloem for organic compounds dissolved in water, the transport functions of xylem and phloem overlap somewhat. Phloem can aid in the distribution of certain minerals, and xylem sometimes transports organic compounds. For example, in early spring, trees convert starch stored in stem parenchyma cells into sugars that are used during bud expansion and flower development. These sugars are transported in xylem sap. Maple trees produce a copious flow of sugar-rich xylem sap that people have long tapped to make maple syrup (**Figure 39.10**).

## Xylem Is Adapted for Long-Distance Transport of Water and Minerals

Xylem's structure plays an essential role in its transport function. The xylem of flowering plants contains several types of specialized cells, some of which remain alive at maturity, and some of which are dead when they are fully functional. Xylem parenchyma cells are alive, but thick-walled supportive fibers may be alive or dead at maturity. Two types of specialized water-conducting cells are always dead and empty of cytosol when mature: tracheids and vessel elements. Bulk flow would be impeded by obstructions such as cytoplasm. This explains why xylem conducting cells are devoid of cytoplasm.

Together, tracheids and vessel elements are known as **tracheary elements**. During the development of tracheary elements, a secondary wall is deposited in patterns, such as spirals or rings, on the inside of the primary cell wall. This secondary wall is rich in a plastic-like polymer known as lignin. Because lignin is

**Figure 39.10 Sugar-maple tapping.** In early spring, xylem transports sugar from storage sites to the shoot buds of woody plants such as this sugar maple (*Acer saccharum*). ©Andre Jenny/Alamy Stock Photo

 **Core Skill: Science and Society** People tap sugar maples by boring holes into the tree trunks and collecting the xylem sap (shown here in buckets). The sap is boiled to evaporate some of the water, leaving the concentrated product—maple syrup.

resistant to compression, microbial decay, and water infiltration, it confers strength, durability, and waterproofing. Like the plumbing pipes of a building, lignin-reinforced tracheary elements do not readily collapse as water moves through them under tension. In contrast, the xylem of lignin-deficient mutant plants is more collapsible. These characteristics explain how tracheary elements contribute to structural support of the plant body as well as functioning in transport.

**Tracheids** Long and narrow in shape, **tracheids** typically have slanted end walls that fit together to form long tubes (**Figure 39.11a**). The end walls of tracheids are not lignified, nor are large areas of the side walls of tracheids that occur in plant tissues that are still growing. Such tracheids are extensible because they have spirals or rings of lignin that allow them to continue elongating (**Figure 39.11b**). In contrast, tracheids that develop in tissues that have already expanded have more lignin, which makes them rigid and unable to elongate any more. Tracheid walls that are extensively lignified display numerous small, lignin-free cell-wall regions known as **pits**. At such pits, the thin primary wall of the tracheid remains readily permeable to water. Water moves from one tracheid to another both vertically and laterally through pits.

**Vessels and Vessel Elements** Mature **vessel elements** are a second type of water-conducting cell present in xylem tissue. Vessel elements are aligned in pipeline-like structures known as **vessels** (**Figure 39.12a**). Flowering plants are distinguished from other plant groups by the abundance of vessels; nonflowering plants primarily rely on tracheids for water conduction. Vessel elements are larger in diameter and longer than tracheids, conferring greater capacity for

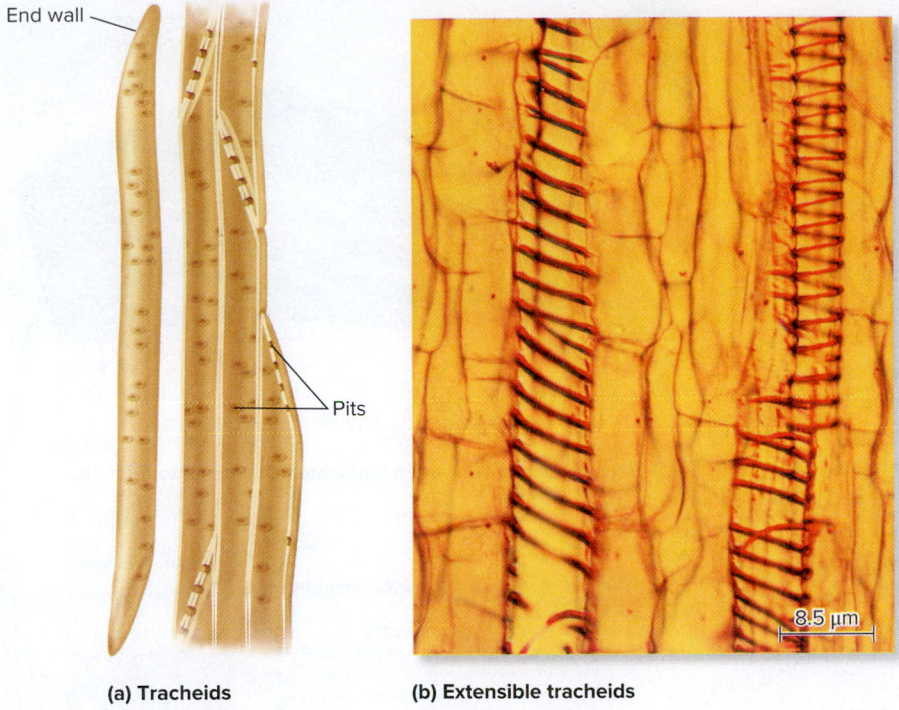

**(a) Tracheids**    **(b) Extensible tracheids**

**Figure 39.11  Tracheid cells in xylem tissue. (a)** Tracheids are long, tubular cells with slanted end walls. Water and ions move from cell to cell through the pits. **(b)** Light micrograph of extensible tracheids from the xylem of pumpkin. b: ©Astrid & Hanns-Frieder Michler/Science Source

*Concept Check:* *If you applied a stain specific for lignin to tracheids present in a longitudinal slice of a plant stem that is still growing in length, then observed the cells with a light microscope, what portions of the tracheids would be stained, and what parts would not be stained?*

bulk flow; they therefore represent one of the many ways in which flowering plants are particularly well adapted to life on land.

Development of vessel elements resembles that of tracheids in some ways. For example, lignified secondary walls are deposited in spirals or rings on the inside of the primary cell wall. Vessel elements also have numerous pits in their side walls. In contrast to tracheids, the end walls and some side walls of vessel elements are extensively perforated, meaning that all cell-wall material is removed from some areas (**Figure 39.12b** and **c**). This allows water to flow faster from one vessel element to another than it can flow from one tracheid to another.

Because the perforated end walls and large diameter of vessels allow them to transport more water at a faster rate than tracheids can, you might wonder why flowering plants possess two types of water-conducting cells. The answer is that vessels can be more vulnerable than tracheids to embolism, meaning blockage by air bubbles. Once an embolism forms in a vessel element, an air bubble can move through the large end-wall perforations into another element, thereby blocking the flow through an entire vessel. Just as air bubbles can cause disruption of blood circulation in people, sometimes leading to death, such bubbles also disrupt water transport in plants, sometimes severely.

An embolism can form within a xylem vessel as the result of physical damage, drought, or repeated cycles of freezing and thawing. Air bubbles form frequently during winter, because air does not dissolve

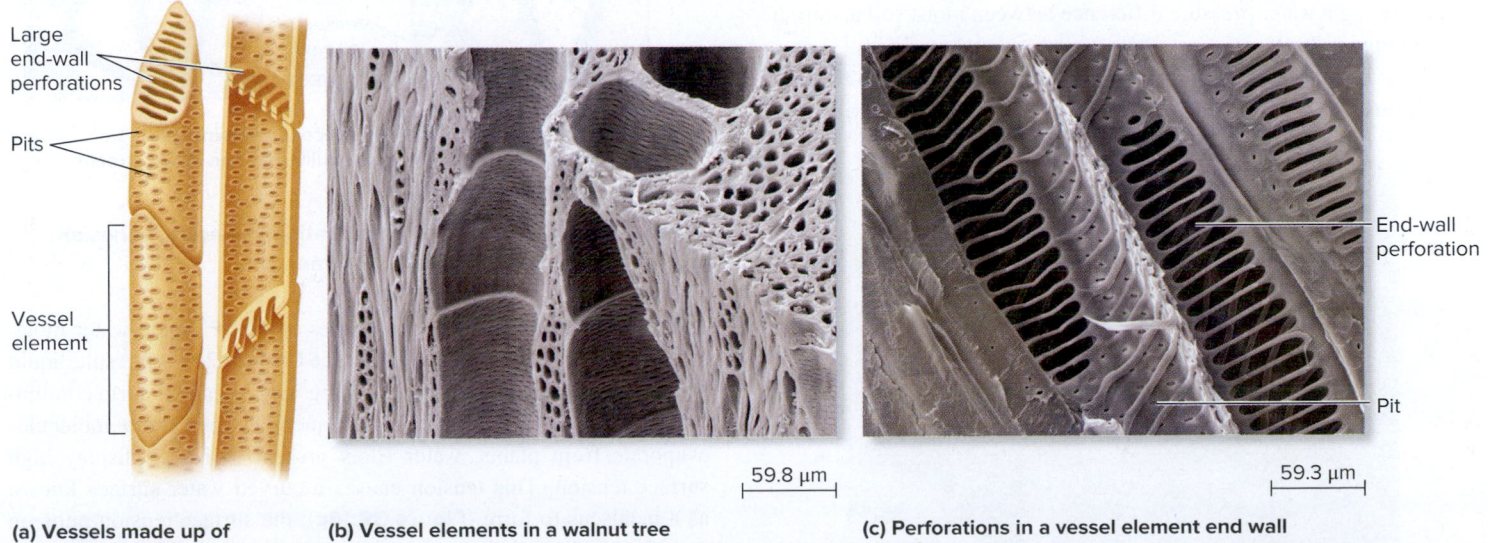

**(a) Vessels made up of vessel elements**    **(b) Vessel elements in a walnut tree**    **(c) Perforations in a vessel element end wall**

**Figure 39.12  Vessels composed of vessel elements in xylem tissue. (a)** This illustration shows the wide diameter of vessel elements with many pits and end-wall perforations. **(b)** SEM of vessels in the wood of the walnut tree (genus *Juglans*). **(c)** SEM of a perforated vessel element end wall from the tulip tree (*Liriodendron tulipifera*). b–c: ©Power and Syred/Science Source

*Concept Check:* *Which structural features of vessel elements explain the vulnerability of vessels to embolism, that is, blockage by air bubbles?*

in ice. By the end of a cold winter, the functional vessels of many woody plants have become almost completely blocked by air. Blocked vessels in trees often cease to function in water transport and must be replaced by new growth in the spring. Fortunately, even if vessels become blocked, water conduction can still occur via tracheids. This is because tracheid pits are so small that they do not allow air bubbles to move to other tracheids. Thus, an air bubble tends to be confined to the single tracheid in which it first formed, and water continues to flow through nearby tracheids. Tracheids thereby provide a fail-safe conduction route when vessels have become blocked by embolism.

Some plants are able to refill embolized vessels by means of a process known as **root pressure**. At night, the xylem of roots may accumulate high concentrations of ions that are not immediately transported upward to shoots. In this case, the root acts much like a cell rich in solutes, with the result that water gushes in so rapidly that it pushes upward to leaves. Evidence of this process can be observed in the early morning as droplets of water at the edges of leaves, a phenomenon known as **guttation** (**Figure 39.13**). As the water rushes upward, it can dislodge air bubbles or dissolve them, thereby reversing an embolism. Root pressure refilling has been observed to occur in nonwoody plants such as corn (*Zea mays*) and in some woody plants, including the sugar maple (*Acer saccharum*).

## Cohesion-Tension Theory Explains the Role of Transpiration in Long-Distance Water Transport

As mentioned, transpiration is the process by which water is lost from the aerial parts of plants, usually via stomata on the surfaces of leaves (**Figure 39.14a**). Evaporation occurs more quickly under conditions of low humidity and/or high temperature. Transpiration is capable of pulling water via bulk flow up to the tops of the tallest trees and is the primary way in which water is transported for long distances in plants. Plants expend no energy to transport water and minerals by transpiration. Rather, the Sun's energy indirectly powers this process by generating a water pressure difference between moist soil and drier air (see Figure 39.9).

How does evaporation at plant surfaces influence long-distance water transport in the xylem? To answer this question, we must

**Figure 39.13** **Guttation, the result of root pressure.** ©Custom Life Science Images/Alamy Stock Photo

**Concept Check:** *What function does root pressure serve in plants?*

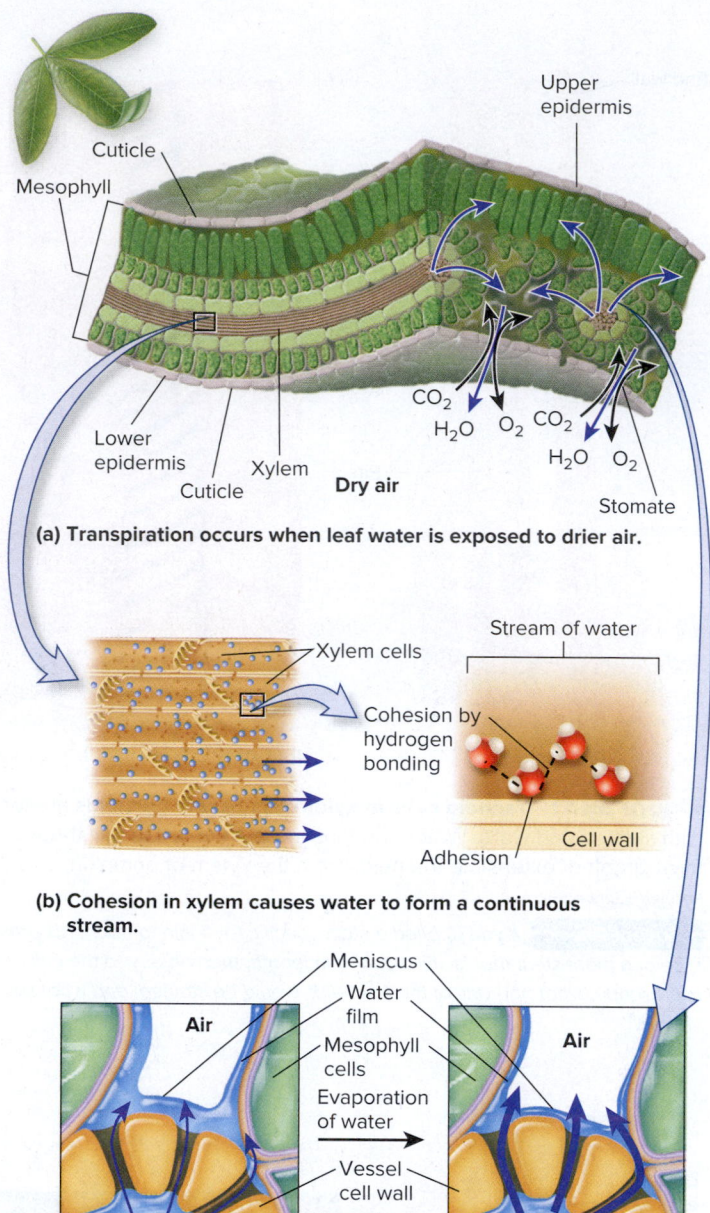

**(a)** Transpiration occurs when leaf water is exposed to drier air.

**(b)** Cohesion in xylem causes water to form a continuous stream.

**(c)** When water evaporates, the surface tension increases in the intercellular spaces of cells, pulling on the water stream in xylem.

**Figure 39.14** **The roles of transpiration, cohesion, adhesion, and tension in long-distance water transport.**

consider the unique physical properties of water. Liquid water molecules are linked by hydrogen bonds (see Chapter 2). As a result, liquid water is amazingly cohesive, explaining why it tends to form continuous streams (**Figure 39.14b**). Consequently, when water molecules evaporate from plants, water films present in leaves display high surface tension. This tension causes a curved water surface known as a meniscus to form (**Figure 39.14c**); the surface tension pulls on neighboring liquid water molecules and eventually on water in the nearest vein, which is connected to the plant's entire water supply. As the result of water's cohesion and the tension exerted on water at the plant's surface, a continuous stream of water can be pulled up through the plant body from the soil, into roots, through stems, and into leaves. This explanation for long-distance water movement in plants is known as the **cohesion-tension theory**. (Recall from Chapter 1 that a scientific theory is a well-established concept, not just a hypothesis.)

Plant transpiration moves huge amounts of water from the soil to the atmosphere. About 99% of the water that enters plants via roots is generally lost as water vapor during transpiration. Each crop season, a single corn plant (*Zea mays*) loses more than 200 L of water, which is more than 100 times the corn plant's mass. A typical tree loses 400 L of water per day! On a regional and global basis, plant transpiration has enormous climate effects. For example, an estimated one-half to three-quarters of rainfall received by the Amazon tropical rain forest actually originates from plant-transpired water vapor, often visible as mist (**Figure 39.15**). Furthermore, about half of the solar heat received by Amazonian plants is dispersed to the atmosphere during transpiration. This heat dispersal has a cooling effect on regional ground temperature, which would be much higher in the absence of plant transpiration. Such cooling effects result from water's unusually high heat of vaporization, the amount of heat needed to isolate water molecules from the liquid phase and move them to the vapor phase. Most of this energy is needed to break the large numbers of hydrogen bonds that occur in liquid water. The evaporation of large amounts of water from plant surfaces effectively dissipates heat, explaining how plants cool themselves and their environments.

Although evaporation of water from plant surfaces plays an essential role in bulk flow through xylem, plants will die if they lose too much water. Plant surfaces, including those of leaves, are typically covered by a cuticle, a wax-containing layer that retards water loss. Only about 5% of water evaporated from plant surfaces emerges through the cuticle. More than 90% of the water that evaporates from plants is lost through stomata, surface pores that can be closed to retain water or opened to allow the entry of $CO_2$ needed for photosynthesis.

Plants face a constant dilemma: whether to open their stomata for $CO_2$ intake and suffer the effect of reduced water content or to close the stomata to retain water, thereby preventing $CO_2$ uptake. When the stomata are open, $O_2$ also exits the plant, as does water vapor when the atmospheric humidity is relatively low. Stomata are often abundantly located on the lower surfaces of leaves. Tobacco leaves, for example, possess an estimated 12,000 stomata per square centimeter of leaf surface! As described next, the regulation of stomatal opening and closing is a key way for plants to avoid excessive water loss.

## Closing of Stomata Helps to Reduce Transpirational Water Loss

Depending on their environment, almost all plants experience water stress, or water deficit, which is an inadequate amount of water. Water stress is common for plants of the world's arid regions, and their growth is often limited by water availability. Even plants of moist, forested regions of the world experience water stress during drier or colder seasons, under windy conditions, or during a drought. The leaves at the tops of tall trees are generally under considerable water stress because gravity has a substantial effect on their water potential.

Earlier, we considered examples of plant cellular adaptations to deal with osmotic stress. Plants have evolved additional ways to prevent excessive loss of water by transpiration. One example is the regulation of stomatal opening and closing. Plant stomata close to conserve water under conditions of water stress and open when the stress has been relieved, allowing air exchange with the leaf's spongy mesophyll. Stomata are bordered by a pair of **guard cells**, which are sausage-shaped chloroplast-containing cells attached to one another at their ends (**Figure 39.16a**). The distinctive structural features of guard cells explain how they are able to open and close a pore.

As guard cells become fully turgid, their volume expands by 40–100%. This expansion does not occur evenly, however, because the innermost cell walls of guard cells are thicker and less extensible than are other parts of the cell walls. In addition, the guard cells expand primarily in the lengthwise direction because bands of radially oriented cellulose microfibrils prevent lateral expansion (**Figure 39.16b**). When guard cells are turgid, a stomatal pore opens between them, allowing air exchange with the leaf's spongy mesophyll. Conversely, when the guard cells lose their turgor, their volume decreases, and the stomatal pore closes.

What causes the change in turgor that opens or closes stomatal pores? In flowering plants, stomata often open early in the morning, in response to sunlight. This response makes sense, given that light, water, and carbon dioxide are all required for photosynthesis. Blue light stimulates $H^+$-ATPase proton pumps in guard cells, leading to the uptake of ions, especially potassium ($K^+$), and other solutes. As a result of increases in solute concentrations inside guard cells, osmotic water uptake occurs via plasma membrane aquaporins, resulting in cell expansion and stomatal opening (**Figure 39.17a**).

At night, the reverse process closes the stomata of flowering plants. Potassium and other solutes are pumped out of guard cells,

**Figure 39.15  Plant-transpired water vapor rising as mist from a tropical rain forest.** This mist visually illustrates the enormous amount of water that is transpired from the surfaces of plants into the atmosphere. Water vapor derived from plant transpiration is an important source of rainfall, and the process of evaporation cools plant surfaces and affects the local and global climate. ©Adalberto Rios Szalay/Sexto Sol/Stockbyte/Getty Images

**Concept Check:** *Why does evaporation of water have such a powerful cooling effect?*

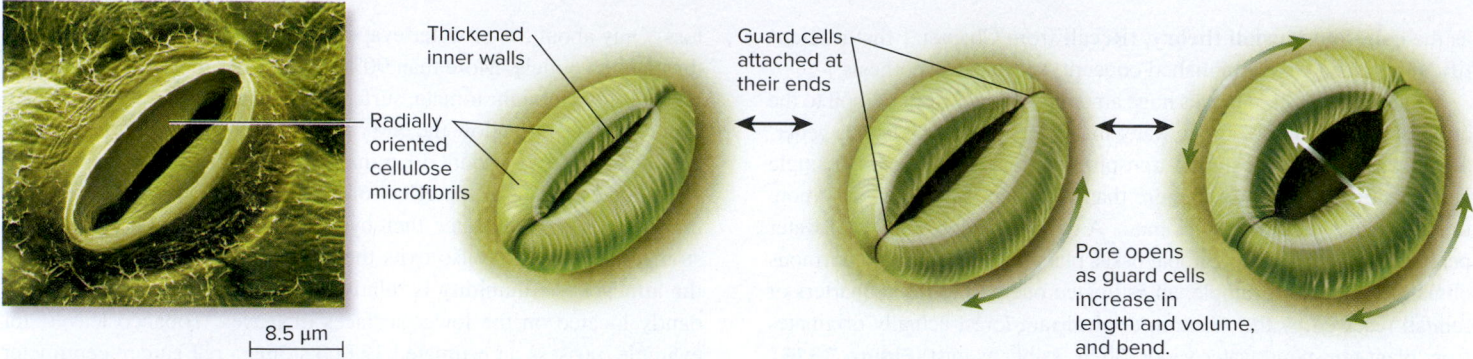

(a) Stomatal guard cells

8.5 μm

(b) The roles of radial orientation of cellulose microfibrils and thickened inner walls in opening or closing guard cells

**Figure 39.16    The structure of guard cells.**  When flaccid, guard cells close stomata. Turgid guard cells produce a stomatal opening. **(a)** An SEM of a stomate in a rose leaf, showing the two guard cells bordering a partly open pore. **(b)** Thickened inner cell walls and radial orientation of cellulose microfibrils in the guard-cell walls explain why they separate when turgid, forming a pore. a: ©Andrew Syred/Science Source

*Concept Check:*    *How could you make a physical model that would illustrate how guard-cell structure affects function?*

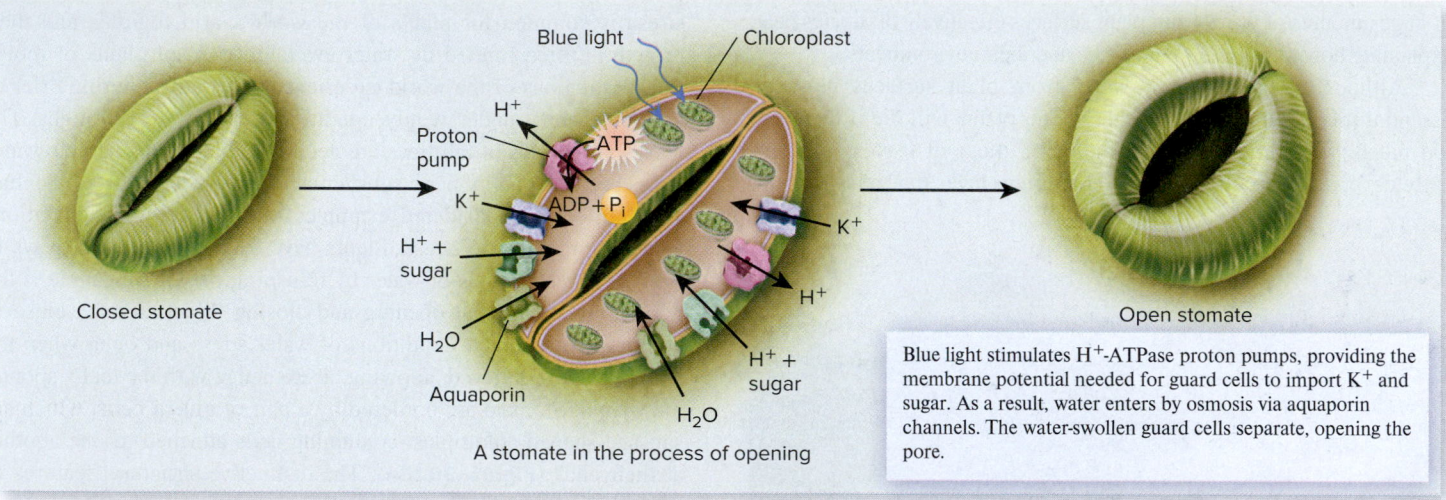

Blue light stimulates H⁺-ATPase proton pumps, providing the membrane potential needed for guard cells to import K⁺ and sugar. As a result, water enters by osmosis via aquaporin channels. The water-swollen guard cells separate, opening the pore.

(a) The process of stomate opening

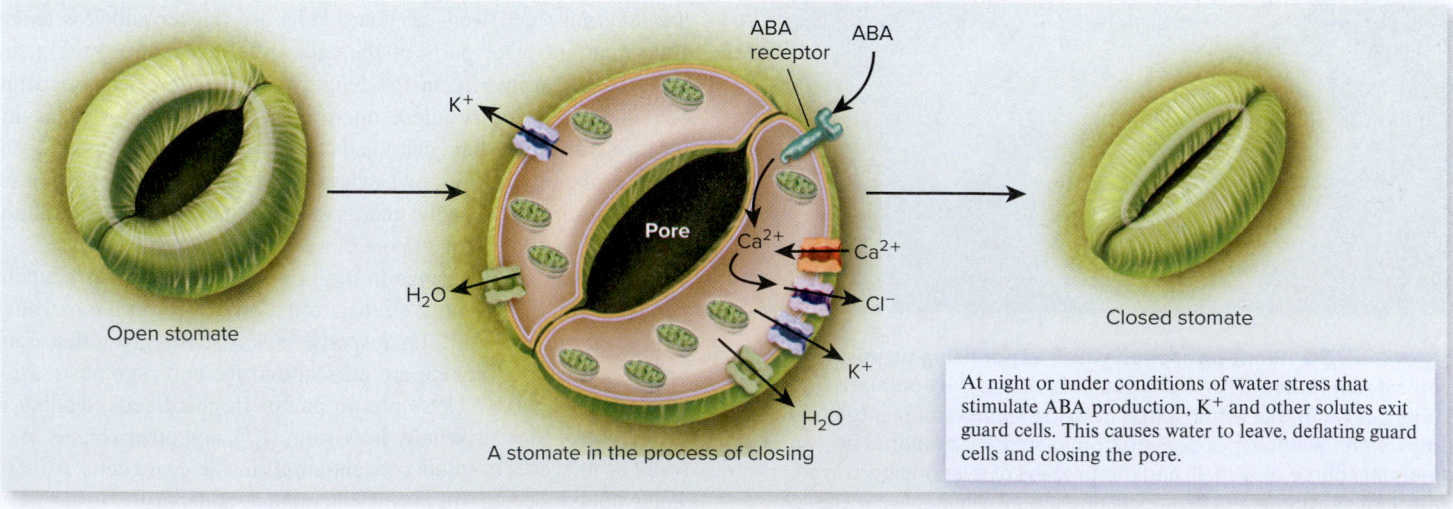

At night or under conditions of water stress that stimulate ABA production, K⁺ and other solutes exit guard cells. This causes water to leave, deflating guard cells and closing the pore.

(b) The process of stomate closing

**Figure 39.17    How stomata of flowering plants open and close.  (a)** Angiosperm stomata usually open in response to sunlight. **(b)** Angiosperm stomata usually close in response to lack of sunlight. They can also close during the day under conditions of water stress, which induces plants to produce more of the hormone abscisic acid (ABA). Guard-cell plasma membranes possess ABA receptors, which receive the drought signal.

causing water to exit, deflating the cells and resulting in pore closure. Flowering plants also close their stomata during the daytime under conditions of water stress, a process mediated by the stress hormone abscisic acid (ABA). Water stress causes a 50-fold increase in ABA, which is transported in the xylem sap to guard cells. ABA then binds to a receptor, which elicits a $Ca^{2+}$ second messenger, causing the guard cells to lose solutes and deflate (**Figure 39.17b**). An understanding of the process of stomatal closure suggests strategies by which crop plants might be genetically engineered to more effectively withstand drought, as described next.

 **Core Skill: Process of Science**

# Feature Investigation | Park, Cutler, and Colleagues Genetically Engineered an ABA Receptor Protein to Foster Crop Survival During Droughts

Global climate change has increased the frequency and length of droughts that endanger agricultural production of human food. Increasing crop resistance to water stress is one way in which plant biologists are trying to improve agricultural productivity. The plant hormone abscisic acid (ABA) is key to drought response in seed plants, so agricultural biologists have been working to identify methods that farmers could use to stimulate ABA activity in crop plants.

A team of biochemists led by Sang-Youl Park and Sean Cutler hypothesized that some chemicals already approved for use in protecting crops (known as agrochemicals) might be able to bind to mutant versions of the ABA receptor and thereby activate it. To test this hypothesis, they created a large collection of *Arabidopsis* plants containing different mutations in the ABA receptor gene (known as *PYR1*). The collection included all of the 475 single substitutions that could possibly occur in the 25 amino acids at the ABA binding site.

The team then examined the responses of these mutant plants to a set of widely used agrochemicals. This process, which involved more than 7,000 individual tests of the binding of different agrochemicals to the ABA receptor, revealed that some of the mutants responded to mandipropamid, an agrochemical widely used to control pathogens called oomycetes, which are protists that cause diseases in diverse species of natural plants and crops. The researchers discovered that, in the responsive mutants, the ability of the agrochemical to fit within the ABA binding site was improved, and hydrophobic interactions and hydrogen bonding occurred between the agrochemical and the receptor.

With this fundamental information in hand, the research team next tried to determine if such mutations had applicability to crop species. To do so, the investigators created tomato plants that carried the same kind of mutations that caused the ABA receptor to recognize mandipropamid in *Arabidopsis* (**Figure 39.18**). They then

**Figure 38.18**  **The Park team's experiment with a genetically engineered ABA receptor protein.** (thermal images): ©Sean Cutler and Sang-Youl Park

> **HYPOTHESIS**  Tomato plants bearing a genetically engineered ABA receptor will close their stomata in response to treatment with an agrochemical.
>
> **KEY MATERIALS**  Genetically engineered tomato plants, wild-type (nonengineered) tomato plants, agrochemical, controlled environment culture chamber, thermal imaging equipment.

| | **Experimental level** | **Conceptual level** |
|---|---|---|
| **1** Grow replicate wild-type (WT) and genetically engineered (E) plants of the same age together for 3 weeks. |  | Replicates provide reproducibility, and controlled growth conditions reduce the effects of environment. |
| **2** Obtain thermal images before experimental treatment. | 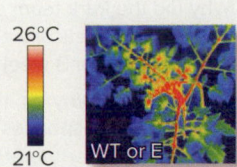  26°C  21°C  WT or E | Plant temperature indicates degree of evaporation through open stomata. Blue indicates open stomata; red indicates closed stomata. |

**3** Treat all plants with the agrochemical.

WT plants with the agrochemical should not respond to treatment by closing their stomata, but engineered plants should do so.

**4** Obtain thermal images after experimental treatment.

Comparison of thermal images reveals differences in surface temperatures resulting from stomata closure.

**5   THE DATA**

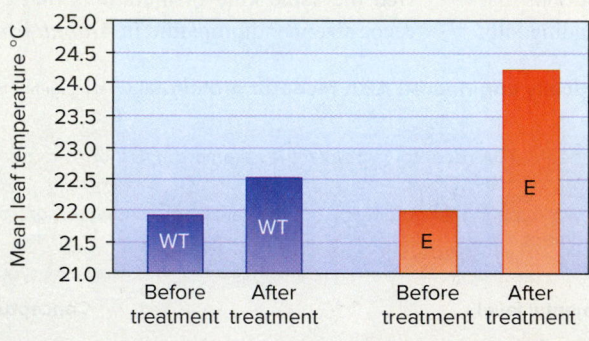

**6   CONCLUSION**  Farmers might be able to use common agrochemicals to foster survival of genetically engineered crops during drought periods.

**7   SOURCE**  Park, S.-Y., Peterson, F. C., Mosquna, A., Yao, J., Volkman, B.F., and Cutler, S.R. 2015. Agrochemical control of plant water use using engineered abscisic acid receptors. *Nature*. 520: 545–548.

used measurements of leaf temperature to assess the responsiveness of the ABA receptors in these plants to the agrochemical mandipropamid. A higher leaf temperature indicates stomatal closing. As seen in the data, the genetically engineered plants (E) appeared to close their stomata in response to mandipropamid. This result suggests that genetic engineering can be used to create crop varieties that farmers can treat with agrochemicals to increase drought resistance.

*Experimental Questions*

**1.** Why did the Park team grow replicate wild-type (WT) tomato plants and genetically engineered (E) tomato plants of the same age together for 3 weeks?

**2. CoreSKILL »** Why did the Park team use thermal images to compare the responses of the WT and E plants before and after experimental treatment with the agrochemical?

## BIO TIPS

**THE QUESTION** *The data in step 5 of Figure 39.18 reveal that genetically engineered (E) tomato plants showed a higher level of response to application of an agrochemical than did wild-type (WT) plants. Did the WT plants also show any response to the experimental treatment, and if so, why might this response have occurred?*

**T OPIC** *What topic in biology does this question address?* The topic is leaf stomatal responses of angiosperms. More specifically, the question concerns how genetic engineering can improve a crop's ability to close its stomata when exposed to an agrochemical.

**I NFORMATION** *What information do you know based on the question and your understanding of the topic?* In the question, you are reminded of the data in Figure 39.18. From this chapter and Chapter 36, you know that stomata are pores that occur on plant surfaces, particularly the undersides of leaves. Figure 39.17 shows how the binding of ABA to ABA receptors causes stomatal closure, thereby reducing water loss by transpiration. From Section 9.2, you may recall that hormones bind to receptors in a very specific way to form a ligand-receptor complex.

**P ROBLEM-SOLVING S TRATEGY** *Interpret data. Compare and contrast.* Compare the bar graphs in step 5 of Figure 39.18 and the thermal images collected before and after the experimental treatment (in steps 2 and 4). Think about the structure of the ABA receptor in WT plants (those not genetically engineered) versus in mutant plants, and consider how the agrochemical mandipropamid might interact with the ABA receptor in these different strains.

**ANSWER** *Agrochemical treatment did result in a small increase in leaf temperature in WT tomato plants. This result may indicate that WT plants close at least some of their stomata in response to agrochemical treatment, though not as many as in the E plants. One explanation is that the agrochemical binds to the ABA receptor in wild-type plants, but not as well as to the ABA receptor in the engineered plants. Genetic engineering improved the ability of a particular agrochemical to fit within the receptor-binding site and activate it. For this reason, the ABA receptors of E plants bind the agrochemical more effectively than do those of WT plants. The results suggest that WT plants respond by closing some stomata, but E plants respond more strongly.*

## Leaf Abscission Also Prevents Water Loss

A second way that plants can prevent water loss is by dropping their leaves, a process known as **leaf abscission**. Angiosperm trees and shrubs of seasonally cold habitats experience water stress every winter, when evaporation from plant surfaces occurs, yet soil water is frozen and therefore unavailable for uptake by roots. Desert plants also experience water stress conditions, but at less predictable times and for much of the year. Both types of plants are adapted to cope with water stress by dropping their leaves. This process lets these plants avoid very low leaf water potentials and the consequent danger

(a) Ocotillo with leaves          (b) Ocotillo without leaves

**Figure 39.19   Leaf abscission as a drought adaptation.** The ocotillo (*Fouquieria splendens*), a plant native to North American deserts, is known for its ability to respond to intermittent rain and drought by producing and dropping leaves multiple times within a year. a: ©Dan Suzio/Science Source; b: ©Matthew Heinrichs/Alamy Stock Photo

*Concept Check:* *Why does the ocotillo not drop its leaves at a single predictable time each year, as do temperate angiosperm trees and shrubs?*

of xylem embolism. Leaf abscission also reduces the amount of root mass that plants must produce to obtain water under arid conditions. The ocotillo (*Fouquieria splendens*) of North American deserts can produce leaves after sporadic rains and then drop all of its leaves as a direct response to drought as many as six times a year (**Figure 39.19**).

The sugar maple (*Acer saccharum*) is an example of the many types of temperate forest trees or shrubs that drop their leaves each autumn and are thus known as deciduous plants. Deciduous plants contrast with evergreen conifers, whose needle- or scale-shaped leaves are adaptations that reduce the area of leaf surface from which water can evaporate, which helps these gymnosperms cope with water stress during the cold season (refer back to Figure 32.10). The broader, thinner leaves produced by many angiosperms are well adapted for efficient light-capture, but more vulnerable to the stresses caused by cold. During their evolution, temperate zone angiosperm trees and shrubs have acquired the genetic capacity to predict the onset of cold, dry winter conditions and respond with pre-emptive leaf abscission. In contrast to the case of the ocotillo, autumn leaf drop in temperate angiosperms is not directly induced by drought.

Leaf abscission is a highly coordinated developmental process. The hormone ethylene stimulates an abscission zone to develop at the bases of leaf petioles, where they join stems (**Figure 39.20a**). The abscission zone contains two types of tissues: a separation layer of short, thin-walled cells and an underlying protective layer of cork cells (**Figure 39.20b**). The walls of the cork cells contain both water-proofing wax and phenolic polymers that retard microbial attack. As the abscission layer develops across the vein linking the petiole with the stem, this layer eventually cuts off the water supply to the leaf. Chlorophyll in the leaves degrades, revealing colorful orange and yellow carotenoid and xanthophyll pigments that were previously

**(a) LM of leaf abscission zone at the junction of a petiole and stem, stained with dyes**

Petiole

Conducting tissue

Abscission zone

Stem

25 µm

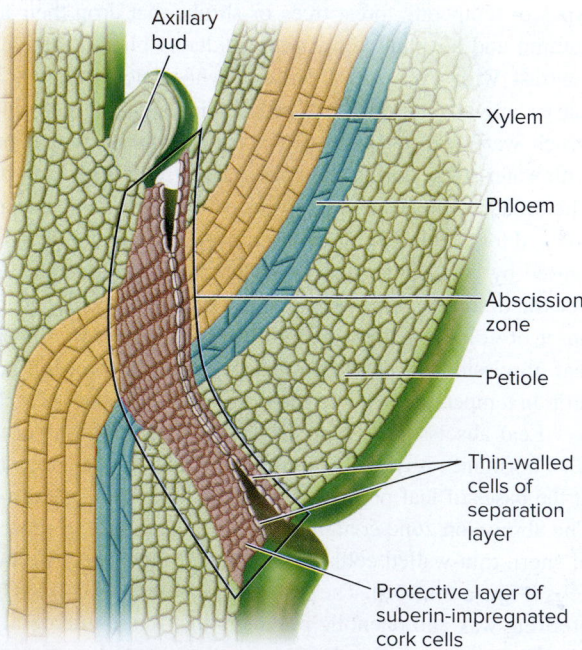

Axillary bud

Xylem

Phloem

Abscission zone

Petiole

Thin-walled cells of separation layer

Protective layer of suberin-impregnated cork cells

**(b) Tissues involved in leaf abscission**

**Figure 39.20 Leaf abscission.** Leaf abscission helps maintain plant hydration. a: ©Lee W. Wilcox

masked by the chlorophyll. In addition, some plants synthesize red and reddish-blue pigments in response to changing environmental conditions. The presence of these pigments explains colorful autumn vegetation in temperate zones. Enzymes eventually break down the cell-wall components of the separation layer, causing the petiole to break off the stem. The underlying protective layer forms a leaf scar that seals the wound, helping to protect the plant stem from water loss and pathogen attack. When environmental conditions change for the better, new branches and leaves arise from axillary buds that occur just above leaf scars (see Figure 39.20b).

## Long-Distance Transport of Organic Molecules Occurs in the Phloem

Phloem plays an essential role in long-distance transport of organic molecules and some minerals in the plant body. Phloem often transports sugars from where they are produced to other sites where they are also used. Recall that primary phloem occurs in the vascular bundles of herbaceous plants and secondary phloem occurs as the inner bark of woody plants (refer back to Figure 36.20). In contrast to xylem, whose transport tissues are dead and empty of cytoplasm at maturity, mature phloem tissues remain alive and retain at least some cytoplasmic components. A closer look at phloem structure and function will help to illuminate these differences.

***Phloem Structure*** Phloem tissues of flowering plants include supporting fibers, parenchyma cells, sieve-tube elements, and adjacent companion cells. **Sieve-tube elements** are arranged end to end to form transport pipes (**Figure 39.21**), analogous to the way that the xylem's vessel elements are aligned to form longitudinal vessels.

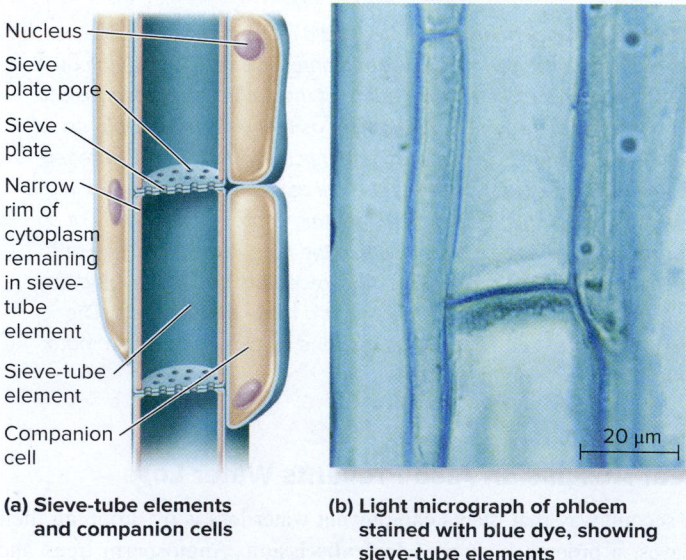

Nucleus

Sieve plate pore

Sieve plate

Narrow rim of cytoplasm remaining in sieve-tube element

Sieve-tube element

Companion cell

20 µm

**(a) Sieve-tube elements and companion cells**

**(b) Light micrograph of phloem stained with blue dye, showing sieve-tube elements**

**Figure 39.21 Sieve-tube elements and companion cells of phloem.** b: ©Lee W. Wilcox

**Core Skill: Connections** Look back to Figure 36.16, which illustrates the development of guard cells. In what way are guard-cell development and companion-cell development similar?

Together, the sieve-tube elements and companion cells form a system for the transport of soluble organic substances made during photosynthesis. In addition, phloem sap may contain hormones such as auxin and abscisic acid, as well as many types of proteins and RNA.

Each pair, consisting of a sieve-tube element and its companion cell, has a common origin. The two cells are produced by an unequal division of a single precursor cell and are therefore linked by plasmodesmata formed at cytokinesis. The smaller of the two cells develops into a companion cell, whose name reflects its life-support function, and the larger of the two cells develops into a sieve-tube element. The sieve-tube element loses its nucleus and most of its cytoplasm in an adaptation that reduces obstruction to bulk flow. Mature sieve-tube elements retain only a thin film of cytoplasm near the cell wall that includes some endoplasmic reticulum, plastids, and mitochondria. The end walls of developing sieve-tube elements become perforated by the action of wall-digesting enzymes that enlarge existing plasmodesmata. The perforated end walls of mature sieve-tube elements are known as **sieve plates**, and the numerous perforations are known as **sieve plate pores**. Phloem sap passes through these plates from one sieve-tube element to another.

Mature sieve-tube elements are not dead. However, because they lack a nucleus, they depend on their neighboring companion cell for messenger RNAs (mRNAs) and proteins, which are supplied via plasmodesmata. For example, when a plant's conducting system is damaged, a short-term wound response occurs that involves a protein known as **P protein** (for phloem protein). Large amounts of this protein accumulate along sieve plates, preventing loss of phloem sap (**Figure 39.22**). This protein accumulation functions much like a clot that helps reduce blood loss from wounded animals. P protein also binds to the cell walls of pathogens, thereby helping to prevent infection at wounds. However, sieve-tube elements cannot produce P protein by themselves. The companion cells provide either P protein mRNA or the protein itself to sieve-tube elements. In a longer term response, plants deposit the carbohydrate callose as a sealant at the wound site.

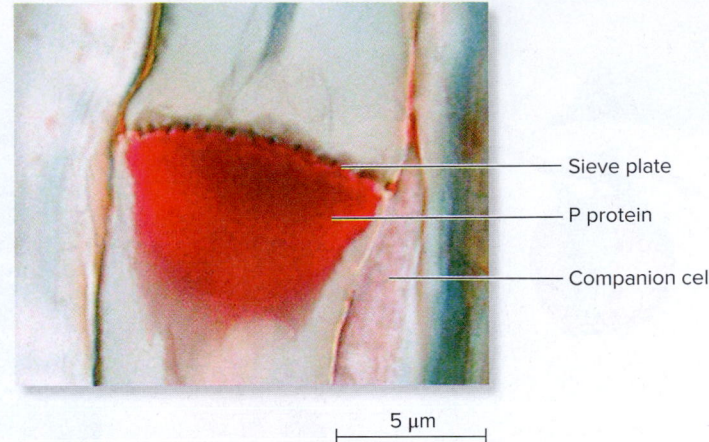

Sieve plate

P protein

Companion cell

5 μm

**Figure 39.22  Phloem wound response.** When phloem is damaged, the cytoplasm of a sieve-tube element surges toward the sieve plate, depositing P protein, stained red in this light micrograph. In this location, P protein helps to prevent infection and leakage of solutes. ©Lee W. Wilcox

*Phloem Loading*  Companion cells also play an essential role in moving sugars into sieve-tube elements for long-distance transport, a process known as **phloem loading**. Although glucose and some other monosaccharides can occur in phloem, the disaccharide sucrose is the main form in which most plants transport sugar over long distances. Plant biologists think that sucrose is less vulnerable to metabolic breakdown en route than are monosaccharides.

Two types of phloem loading occur: symplastic and partly apoplastic. Many woody plants transport sucrose from sugar-producing cells of the leaf mesophyll to companion cells and then to sieve-tube elements via plasmodesmata, a process known as symplastic phloem loading (**Figure 39.23a**). The advantage of symplastic loading is that

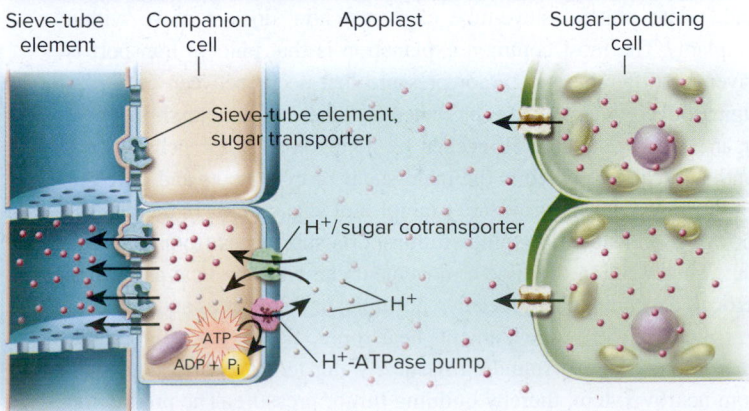

In some plants, sugar moves from sugar-producing cells into companion cells and sieve-tube elements via plasmodesmata.

Sieve-tube element | Companion cell | Sugar-producing cell

Sucrose

Plasmodesmata

**(a) Symplastic phloem loading**

In some plants, sugar moves from sugar-producing cells into the apoplast. ATP is required for active transport into companion cells. Sugar moves into sieve-tube elements via transporters and via plasmodesmata.

Sieve-tube element | Companion cell | Apoplast | Sugar-producing cell

Sieve-tube element, sugar transporter

$H^+$/sugar cotransporter

$H^+$

ATP

ADP + $P_i$

$H^+$-ATPase pump

**(b) Partly apoplastic phloem loading**

**Figure 39.23  Symplastic and partly apoplastic phloem loading.**

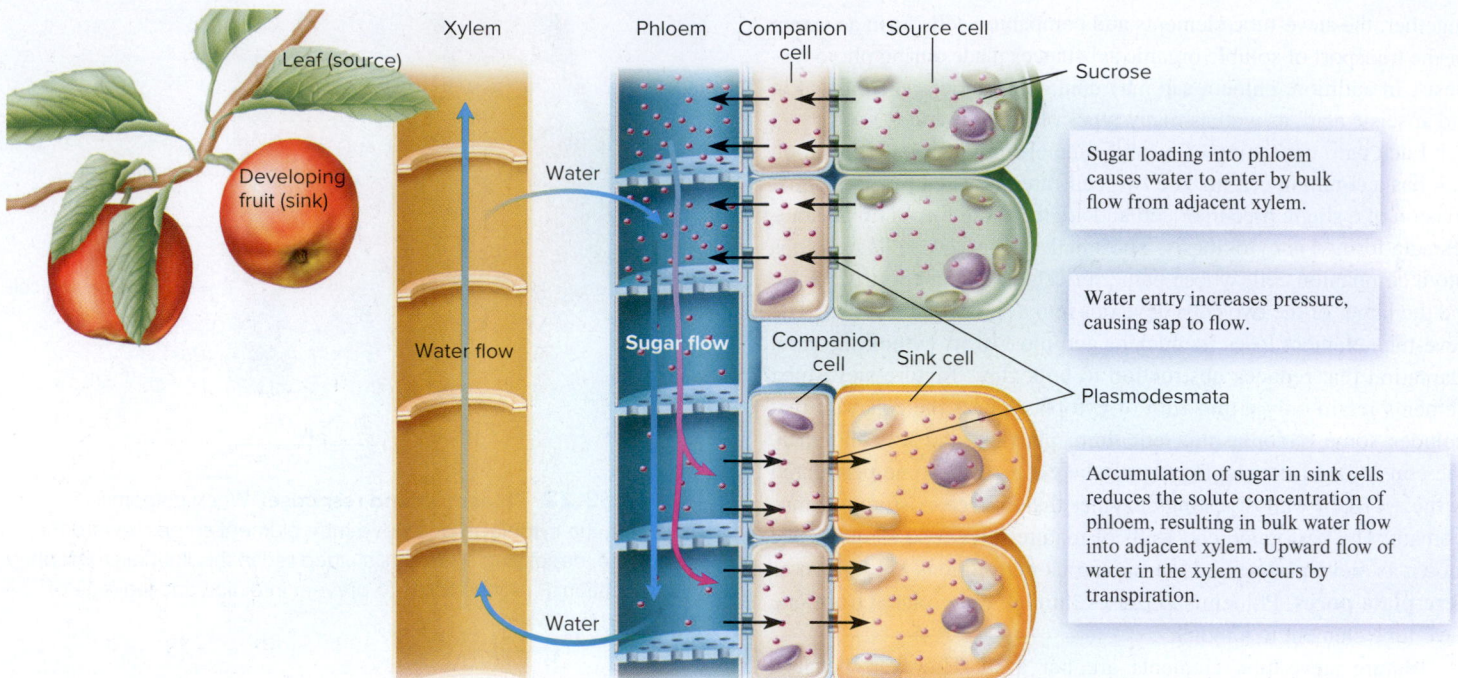

**Figure 39.24** Pressure-flow hypothesis for phloem transport.

it does not require ATP; by moving through plasmodesmata, the sugar does not have to cross plasma membranes.

In contrast, most herbaceous plants, including important crop plants and the model plant *Arabidopsis,* load sugar into sieve-tube elements or companion cells from intercellular spaces, often against a concentration gradient. ATP must be used to move the sugar across a plasma membrane into a companion cell or sieve-tube element (**Figure 39.23b**). Therefore, this second type of phloem loading is partly apoplastic and partly a transmembrane process.

### The Pressure-Flow Hypothesis Helps Explain Transport in Phloem Tissues

We have learned that transpiration, driven by the Sun's energy, moves water in plant xylem by cohesion and tension. Once sugar has been loaded into phloem sieve-tube elements, how does it move within the plant? The most common explanation is that phloem transport is driven by differences in turgor pressure that occur between cells of a **sugar source**, where sugar is produced, and those of a **sugar sink**, where sugar is not made, but still needed by nonphotosynthetic cells. Photosynthetic leaf mesophyll is the main sugar source. Roots and developing leaves, seeds, and fruits are examples of sugar sinks. In the process known as **translocation**, phloem transports substances from source to sink. The direction of phloem movement can be horizontal as well as vertical, depending on the relative positions of the sources and sinks.

Because sieve-tube elements near source tissues have higher solute contents than surrounding tissues, water tends to rush into them from nearby xylem, thereby building turgor pressure. The production of turgor pressure requires an intact plasma membrane. This explains why mature phloem sieve-tube elements must be alive in order to function.

In contrast to sieve-tube elements near source tissues, sieve-tube elements near sink tissues have lower solute concentrations. The resulting water pressure difference drives the bulk flow of phloem sap from source to sink tissues. This explanation for translocation is known as the **pressure-flow hypothesis** (**Figure 39.24**). At sink tissues, sugar is typically unloaded through plasmodesmata (see Figure 39.24). Because plasmodesmata are very narrow in comparison with sieve-tube elements, they slow the flow of phloem sap from sieve-tube elements into sink tissues. This reduction in flow rate helps to equalize the distribution of phloem sap, preventing delivery of too much sap to any single sink. When the solute concentration of phloem sap has been sufficiently reduced, water flows from the phloem back into the xylem, where upward transport occurs. The reliance of phloem bulk flow on water supplied by the xylem explains the close proximity of phloem and xylem tissues in vascular bundles and woody stems.

## Summary of Key Concepts

### 39.1 Overview of Plant Transport

- The plant root system takes up water and minerals, and the shoot system absorbs carbon dioxide from the air. Photosynthetic cells use these materials to produce organic compounds. Xylem transports water and minerals from roots to shoots, and phloem transports organic compounds from photosynthetic to nonphotosynthetic tissues (Figure 39.1).

### 39.2 Uptake and Movement of Materials at the Cellular Level

- Facilitated diffusion is the transport of molecules across plasma membranes down a concentration gradient with the aid of membrane transport proteins, typically channels and transporters. Active transport—the transport of substances across plasma membranes against concentration gradients—usually requires ATP. Plasma membrane proton pumps use the energy released by ATP hydrolysis to move protons from the cytosol into the intercellular space, generating a membrane potential. Potential energy released by the flow of protons back into the cell can be coupled to the transport of ions and solutes (Figures 39.2, 39.3).

- Turgor pressure increases as water enters plant cells because cell walls restrict the extent to which cells can swell. A cell that is so full of water that the plasma membrane presses closely against the cell wall is turgid. A cell that contains so little water that the plasma membrane pulls away from the cell wall is plasmolyzed. A flaccid cell has a water content between these extremes (Figure 39.4).

- Solute potential and pressure potential arising from the presence of a cell wall are major factors that determine cellular water potential. Water moves from a region of higher water potential to a region of lower water potential (Figure 39.5).

- Plants display a variety of adaptations that help them cope with osmotic stress.

## 39.3 Tissue-Level Transport

- Tissue-level transport occurs in three forms. Transmembrane transport involves the movement of materials from one cell to another from intercellular spaces, across plasma membranes, and into the cytosol. Symplastic transport allows materials to move from one cell to another through the symplast, the cells' cytosol and plasmodesmata, without crossing plasma membranes. In apoplastic transport, water and solutes move through the apoplast, the water-filled cell walls and intercellular spaces of tissues (Figures 39.6, 39.7).

- In roots, waxy Casparian strips in cell walls of endodermal tissue function as diffusion barriers that reduce both the entry of harmful soil minerals into and the exit of useful solutes from the vascular tissues (Figure 39.8).

## 39.4 Long-Distance Transport

- Water and solutes move long distances by bulk flow within the xylem and phloem. Plant vascular tissues are adapted in ways that reduce resistance to bulk flow. Bulk flow of water upward in xylem is powered by the water pressure difference between moist soil and drier air, the latter resulting from solar heating. Xylem is the main conduit for water and dissolved mineral nutrients, but it may also transport certain organic compounds (Figures 39.9, 39.10).

- The water-conducting cells of xylem, tracheids and vessel elements (together known as tracheary elements), are dead and empty of cytoplasm at maturity. Pits in tracheary element walls allow water entry and exit, and narrow or constrict in response to xylem sap solute content. Vessel elements are wider than tracheids but are more vulnerable to blockage by air bubbles, or embolism. Root pressure, the effects of which include guttation, the appearance in the morning of water drops on leaf tips, helps some plants to refill embolized vessels (Figures 39.11, 39.12, 39.13).

- Transpiration is the evaporative loss of water from plant surfaces. The cohesion-tension theory proposes that as the result of water's cohesion and the tension exerted on water at the plant's surface by evaporation, a continuous stream of water is pulled up through the plant body from the soil, into roots, through stems, and into leaves, from which water is evaporated into the atmosphere (Figures 39.14, 39.15).

- Regulation of stomatal opening and closing helps plants prevent excessive water loss by transpiration. Expansion of guard cells causes stomata to open, allowing $CO_2$ intake. Guard-cell deflation causes pores to close, limiting water loss. Abscisic acid (ABA) stimulates stomatal closure in response to drought, a process that suggests strategies for genetically engineering crops that have greater drought tolerance. Plants under existing or predicted water stress often drop their leaves in a process known as abscission, an adaptive response that lets plants avoid very low water potentials and the threat of embolism (Figures 39.16, 39.17, 39.18, 39.19, 39.20).

- Organic solutes and minerals are transported in phloem sap as the result of osmosis. Phloem sap moves within sieve-tube elements, which are living when mature, but lack a nucleus and are thus dependent on companion cells. Phloem loading, the movement of sugars into sieve-tube elements for long-distance transport, occurs by symplastic or partly apoplastic transport (Figures 39.21, 39.22, 39.23).

- In the process known as translocation, phloem transports substances from source to sink. The pressure-flow hypothesis helps to explain translocation as a process driven by differences in turgor pressure that occur between a sugar source (for example, leaves) and a sugar sink (for example, developing fruit) (Figure 39.24).

## Assess & Discuss

### Test Yourself

1. An aquaporin is
   a. a channel protein that allows the influx of $K^+$ into cells, causing water to also flow in between the phospholipids of the plasma membrane.
   b. a type of blue-colored pore in the epidermal surfaces of plants.
   c. a protein channel in plasma membranes that facilitates the diffusion of water.
   d. a transport protein, or transporter, in plasma membranes that uses protons to cotransport water.
   e. none of the above.

2. Why is turgor pressure a property of plant cells?
   a. Plant cells possess the necessary chloroplasts.
   b. Plant cells possess a cell wall, necessary for formation of turgor.
   c. Plant cells possess mitochondria, which provide the ATP needed for turgor.
   d. All of the above are true.
   e. none of the above is true.

3. How do plant cells avoid losing too much water in very cold, dry, or saline habitats?
   a. They balance the osmotic condition of their cytosol with that of the environment.
   b. Their epidermal cells are coated with waxy cuticle.
   c. They stabilize their membranes with sugars.
   d. They produce more aquaporin water channels to take maximum advantage of available moisture.
   e. All of the above help plants avoid water loss.

4. What are ways in which plants accomplish tissue-level transport?
   a. transmembrane transport of solutes from one cell to another
   b. symplastic transport of materials from one cell to another via plasmodesmata
   c. apoplastic transport of water and dissolved solutes through cell walls and intercellular spaces
   d. All of the above are used for tissue-level transport.
   e. None of the above is used for tissue-level transport.

5. A root endodermis is
   a. an innermost layer of cortex cells that display characteristic Casparian strips.
   b. a layer of cells just inside the epidermis of a root.
   c. a layer of cells just outside the epidermis of a root.
   d. a group of specialized cells that occur within the root epidermis.
   e. none of the above.

6. Which of the following statements best explains how water enters root cells from the soil?
   a. Roots accumulate sugars from shoots, thereby increasing root cell ability to absorb water by osmosis.
   b. Roots actively pump water from the soil using the chemical energy of ATP.
   c. Water enters the spongy spaces between cells and within cell walls.
   d. Membrane-level transport of ions from soil into root hair cells increases the osmotic flow of water into them.
   e. Both c and d are correct.

7. What features of water explain how it can be drawn up a tall tree from roots to leaves?
   a. cohesion, the result of extensive hydrogen bonding
   b. adhesion, water's tendency to stick to surfaces such as the inner walls of tracheid and vessels
   c. high surface tension that develops when water evaporates from intercellular leaf spaces
   d. all of the above
   e. none of the above

8. What feature of vascular plants contributes to their ability to maintain relatively stable internal water content?
   a. a waxy surface cuticle
   b. an extensive root system that mines water from soil
   c. specialized water-conducting tracheary elements composed of dead cells
   d. epidermal pores that open and close
   e. All of the above are contributing features.

9. What structural features of guard cells underlie their ability to form an open pore in plant epidermal surfaces?
   a. thickened inner cell walls and radially oriented microfibrils
   b. thickened outer cell walls and radially oriented microfibrils
   c. thickened inner cell walls and longitudinal microfibrils
   d. thickened outer cell walls and longitudinal microfibrils
   e. uniform thickness of cell walls and randomly arranged microfibrils

10. What substances plug wounded sieve-tube elements, thereby preventing the leakage of phloem sap?
    a. X protein and callose
    b. C protein and callose
    c. P protein and callose
    d. P protein and sucrose
    e. none of the above

## Conceptual Questions

1. Why is it a bad idea to overfertilize your houseplants? If the amount recommended on the package is good, wouldn't more be better?

2. Why is it a bad idea for subsistence farmers (those barely able to grow enough crops to feed themselves) to allow livestock to graze natural vegetation to the point that it disappears?

3.  **Core Concept: Energy and Matter** Imagine two plants that have similar mineral nutrient concentrations in their root tissues. If one plant is planted in soil of lower mineral concentration and the other is planted in soil of higher mineral concentration, which plant will likely need to use more energy in the form of ATP to take up additional soil minerals at the root hair plasma membranes?

## Collaborative Questions

1. **Core Skill: Science and Society** Take a look outside or imagine a forest or grassland. What can you deduce about the availability of soil water from the types of plants that occur?

2. Imagine that you are part of a team assigned to determine what environmental conditions best suit a new crop so that the crop can be recommended to farmers in appropriate climate regions. What features of the crop plants might you investigate?

# Flowering Plants: Reproduction

## 40

**The reproductive success of dandelions.** ©Michael Roach/Alamy Stock Photo

D andelions sometimes seem to be taking over the world, growing abundantly in open, sunny areas. Bright yellow flower heads, as shown in the chapter opening photo, are one of the secrets of dandelions' success. If you pull a dandelion flower head apart, you can see that it is actually a bouquet of 200 or so small flowers. Each flower produces a tiny, one-seeded fruit equipped with a "parachute" for effective long-distance dispersal by wind. Each dandelion plant can produce up to 5,000 fruits during its lifetime, which explains how dandelions can spread so rapidly across a landscape.

Though most flowering plants produce seeds by means of sexual reproduction, dandelions and some other plants are able to produce seeds by asexual reproduction, a process that does not involve the fusion of gametes. As a result, the traits of asexually reproducing dandelion parents and their progeny are uniform. Asexual reproduction could be very usefully applied in agriculture, because most crops reproduce only sexually. Each year many U.S. farmers buy and plant hybrid seeds that develop into mature plants having uniform and desirable trait combinations. Such farmers typically do not use seed from one year's crop to plant the next, because sexual reproduction mixes genes into diverse combinations present among the resulting seeds. For this reason, plants that grow from sexually produced seeds do not uniformly express the desirable trait combinations present in their hybrid parents. But if hybrid crop plants could be engineered to produce seed asexually, as dandelions do, farmers might be able to use seed from one crop to plant the next and continue to harvest uniformly desirable crops. Because reproduction is so important to agriculture, plant biologists aim to better understand both sexual and asexual reproduction in flowering plants.

Flowers, fruits, and seeds are essential reproductive features of the diverse types of flowering plants found in nature. This chapter begins with an overview of the reproductive cycle of flowering plants that describes how flowers, fruits, and seeds function. This overview provides essential background for a closer look at flower structure and development and some of the genes that control flower production and appearance. We will also examine the sexual reproductive processes by which plants produce gametes and accomplish fertilization, thereby producing zygotes, embryos, seeds, fruits, and seedlings. Finally, we will take a closer look at the ways in which dandelions and some other flowering plants reproduce without using the sexual process.

## 40.1 An Overview of Flowering Plant Reproduction

**Learning Outcomes:**

1. Explain how the sexual life cycle of flowering plants differs from that of earlier evolved mosses.
2. List the four organs found in many flowers and the functions of each organ.
3. Describe the phenomenon of double fertilization.
4. Describe the parts of an angiosperm seed, and explain how each part is produced and functions.

Most flowering plants display **sexual reproduction**, the process by which two gametes fuse to produce offspring that have unique combinations of genes. Flowering plants, also known as angiosperms, inherited their sexual life cycle, known as alternation of generations, from ancestors extending back to the earliest land plants (see Chapters 31 and 32). Though all plants share the same basic life cycle, flowering plants display unique reproductive features. In this section, we will first review the general features of alternation of generations and then consider more specific features of the angiosperm life cycle.

## Flowering Plants Display Alternation of Generations

All groups of land plants produce two multicellular life cycle stages, in essence, two distinct plants. These two life cycle stages are the diploid, spore-producing **sporophyte** and the haploid, gamete-producing **gametophyte**. In all groups of plants, haploid spores are typically produced by diploid sporophytes via meiosis. These spores undergo mitotic cell divisions to produce multicellular gametophytes. Certain cells within the gametophytes differentiate into gametes. In contrast to meiosis in animals, plant meiosis does not directly generate the gametes.

The processes of meiosis and fertilization form the transitions between the plant sporophyte and gametophyte life stages and link them in a cycle (**Figure 40.1**). The land plant life cycle is known as **alternation of generations** because it involves the cycling between distinct sporophyte and gametophyte generations.

During the evolutionary diversification of land plants, the sporophyte generation has become larger and more complex, while the gametophyte generation has become smaller and less complex. To see this change, let's compare the life cycle stages of mosses to those of angiosperms. Mosses arose early in the history of land plants, whereas angiosperms appeared much later. During the intervening

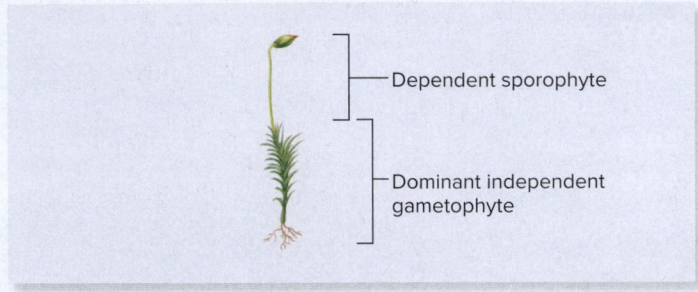

**(a) Gametophyte-dominant bryophyte (moss)**

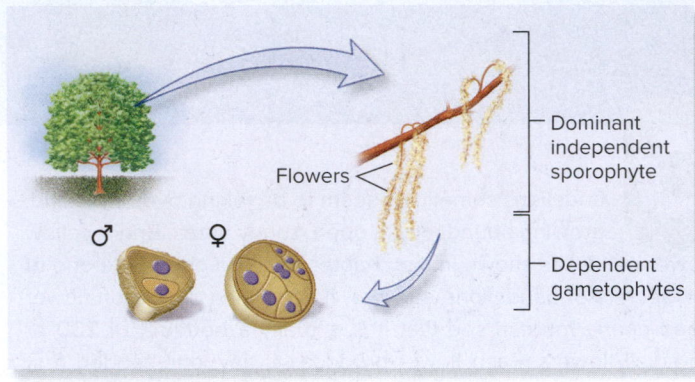

**(b) Sporophyte-dominant flowering plant (oak)**

**Figure 40.2  Evolutionary shift in life cycle stage dominance in plants. (a)** In mosses, the gametophyte is the dominant life cycle stage, and the sporophyte depends on the gametophyte for resources. **(b)** In flowering plants such as oak trees, the sporophyte life cycle stage is dominant. Microscopic flowering plant gametophytes develop and grow within sporophyte flowers and depend completely on sporophytes.

time, the relative sizes and dependence of the sporophyte and gametophyte generations changed dramatically. Moss sporophytes are small structures that always grow attached to larger, photosynthetic gametophytes, because moss sporophytes are incapable of independent life (**Figure 40.2a**). Moss sporophytes depend on gametophytes to supply them with essential nutrients.

In contrast, flowering plant sporophytes are notably larger and more complex than gametophytes. A tall oak tree, for example, is a single sporophyte. Oak gametophytes are few-celled, microscopic structures that develop and grow within flowers (**Figure 40.2b**). In addition, photosynthetic oak seedlings and trees grow independently, but nonphotosynthetic oak gametophytes depend completely on the sporophyte generation for their nutrition. A closer look at flower structure will provide a more complete view of angiosperm gametophyte structure and function.

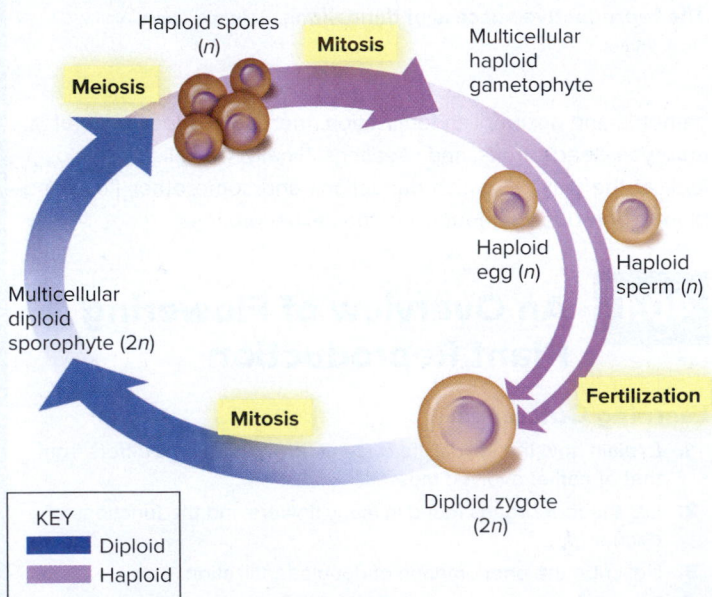

**Figure 40.1  Alternation of generations, the plant life cycle.**

 **Core Skill: Connections** Look back to Figure 32.14 to see a more detailed illustration of the flowering plant life cycle. How can you recognize and where can you find the gametophyte generation of a flowering plant?

## Flowers Produce and Nurture Male and Female Gametophytes

The literary wit Gertrude Stein famously paraphrased Shakespeare ("a rose by any other name would smell as sweet") when she wrote, "A rose is a rose is a rose." Everyone knows that a rose is a flower, but what, exactly, is a flower? A **flower** is defined as a reproductive shoot, a stem branch that produces reproductive organs instead of leaves. Flowers are organ systems because several different organs typically occur within a flower (see Chapter 36). Flower organs are produced by shoot apical meristems much like those that generate leaves and are thought to have evolved from leaflike structures by descent with modification (refer back to Figure 32.15). Most flowering plants are classified into one of two major groups, eudicots or monocots (Chapter 32).

Most eudicot flowers contain four types of organs: sepals, petals, stamens, and carpels (**Figure 40.3**). **Sepals** often function to protect the unopened flower bud. **Petals** usually serve to attract insects or other animals for pollen transport (refer back to Figures 32.19 and 32.20). Monocot flowers typically have tepals, which are outer flower parts that are not differentiated into petals and sepals. Monocots and eudicots alike produce stamens and/or carpels at the flower's center. **Stamens** and **carpels** each produce distinctive types of spores by the process of meiosis. From these spores, tiny multicellular

gametophytes develop, and certain gametophytic cells become specialized gametes.

***Stamens***    Stamens produce **male gametophytes** and foster their early development. Most stamens display an elongate **filament**, which is topped by an anther (see Figure 40.3). Filaments contain vascular tissue that delivers nutrients from the parental sporophyte to the anthers. Each **anther** is a group of four sporangia, structures in which spores are produced. Within the anther's sporangia, many diploid cells undergo meiosis, each producing four tiny, haploid spores. Because they are so small, generally 25–50 μm in diameter, the spores produced within anthers are known as **microspores**. Immature male gametophytes, known as **pollen grains**, develop from microspores. The term pollen comes from a Latin word meaning fine flour, reflecting the small size of pollen grains. Pollen grains are eventually dispersed through pores or slits in the anthers. At the time of dispersal, the pollen grain is a two- or three-celled immature male gametophyte produced by mitotic division. During a later phase of development, a mature male gametophyte produces **sperm cells**.

***Carpels***    Carpels are vase-shaped structures that produce, enclose, and nurture **female gametophytes**. Carpels contain veins of vascular tissue that deliver nutrients from the parent sporophyte to the developing gametophytes. The term **pistil** (named for its resemblance to the pestle used to grind materials to a powder) refers to a single carpel or several fused carpels (see Figure 40.3). The topmost portion of a pistil, known as a **stigma** (Greek, meaning mark), receives pollen grains. The **style** is the middle portion of the pistil, and an ovary is at the bottom of the pistil. The **ovary** produces and nourishes one or more ovules.

An **ovule** of flowering plants consists of a spore-producing structure (a sporangium) and enclosing tissues consisting of modified leaves, known as **integuments**. Within an ovule, a diploid cell produces four haploid **megaspores** by meiosis, three of which die. The surviving megaspore generates a female gametophyte by mitosis. The female gametophytes of flowering plants typically consist of seven cells, one of which is the female gamete, the **egg**. This basic information about male and female gametophytes helps in understanding how they function to produce a young sporophyte within a seed.

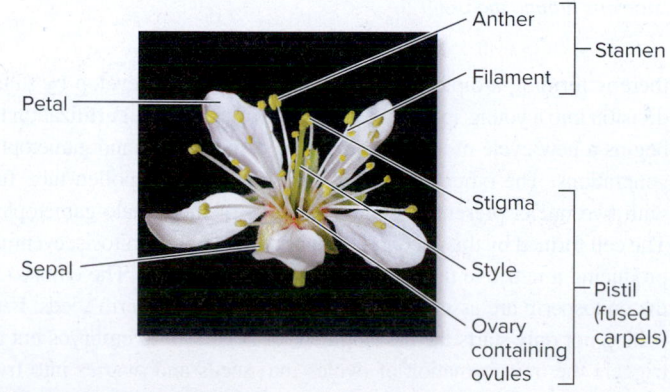

**(a) Flower parts**

**(b) *Prunus americana* (plum)**

**Figure 40.3**  **The structure of a typical eudicot flower.** b: ©Lee W. Wilcox

**Concept Check:** *Do all flowers have all of the structures illustrated here?*

## Fertilization Triggers the Development of Embryonic Sporophytes, Seeds, and Fruits

In flowering plants, fertilization leads to the production of a young sporophyte that lies within a seed, completing the life cycle (**Figure 40.4**). Prior to fertilization, pollen grains released from anthers first find their way to the stigma of a compatible flower, a process known as **pollination**. Some plants display **self-pollination**, in which pollen from the anthers of a flower is transferred to the stigma of the same flower or between flowers of the same plant. **Cross-pollination**, which occurs when a stigma receives pollen from a different plant of the same species, is also common. Many flowers are attractive to insects or other animals that transport pollen, whereas oak flowers and those of some other angiosperms are adapted for pollen transport by wind. A few plants move pollen by means of water currents. Flowers are adapted for effective pollination by diverse pollination mechanisms (refer back to Table 32.2).

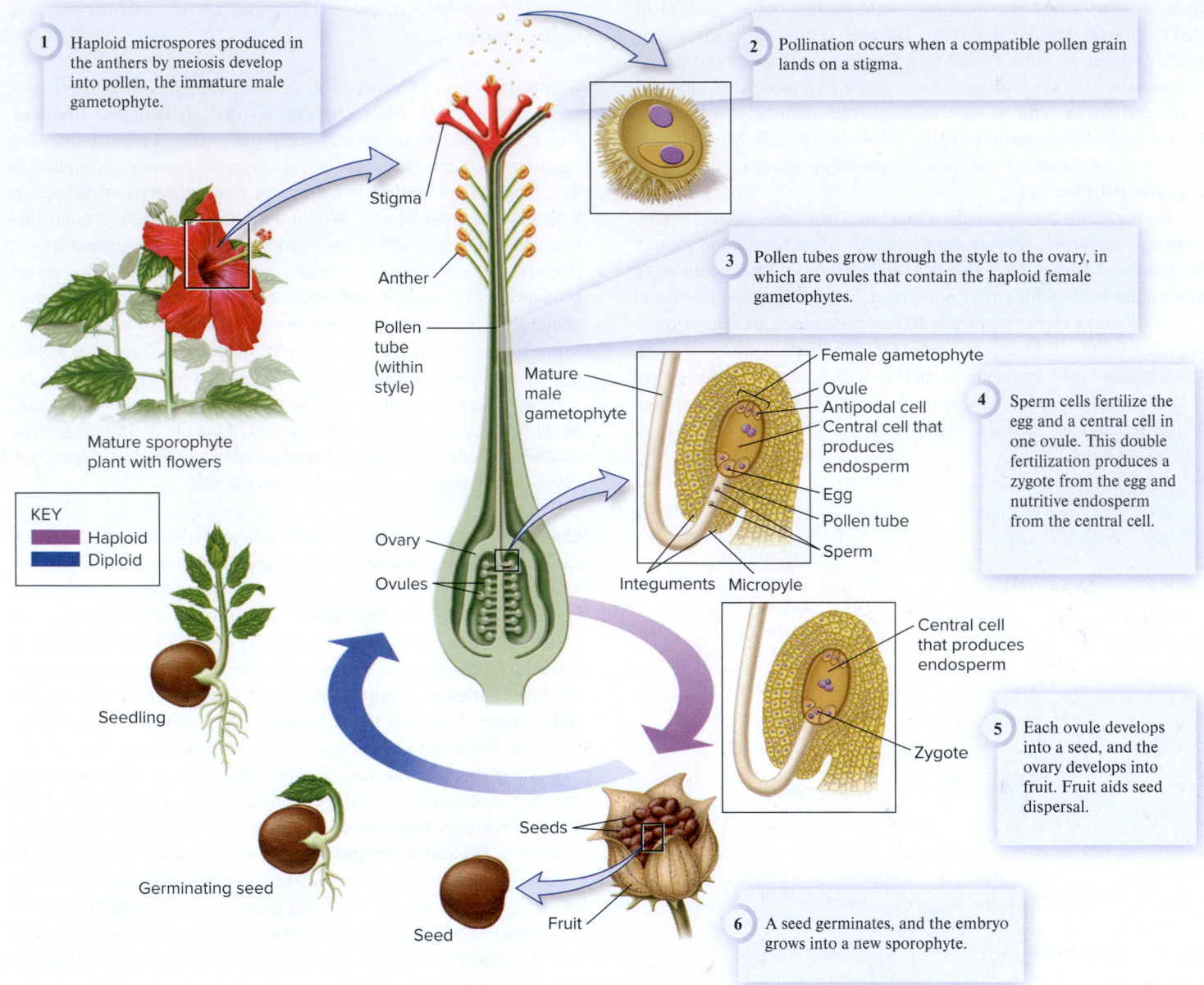

**1** Haploid microspores produced in the anthers by meiosis develop into pollen, the immature male gametophyte.

**2** Pollination occurs when a compatible pollen grain lands on a stigma.

**3** Pollen tubes grow through the style to the ovary, in which are ovules that contain the haploid female gametophytes.

**4** Sperm cells fertilize the egg and a central cell in one ovule. This double fertilization produces a zygote from the egg and nutritive endosperm from the central cell.

**5** Each ovule develops into a seed, and the ovary develops into fruit. Fruit aids seed dispersal.

**6** A seed germinates, and the embryo grows into a new sporophyte.

Stigma
Anther
Pollen tube (within style)
Mature male gametophyte
Ovary
Ovules
Mature sporophyte plant with flowers
Seedling
Germinating seed
Seed
Fruit
Seeds

Female gametophyte
Ovule
Antipodal cell
Central cell that produces endosperm
Egg
Pollen tube
Sperm
Integuments    Micropyle

Central cell that produces endosperm
Zygote

KEY
Haploid
Diploid

**Figure 40.4  The life cycle of a flowering plant.** The plant reproductive cycle is illustrated here by the eudicot hibiscus.

*Concept Check:* *What advantage does the hibiscus flower gain by clustering its stamens around the pistil?*

***Pollen Germination***   When pollen grains land on the stigma, the stigma functions as a gatekeeper, allowing only pollen of appropriate genotype to germinate. During germination, a pollen grain produces a long, thin **pollen tube** that contains two sperm cells. The pollen tube grows through the style toward the ovary. Upon reaching the ovules, the pollen tube grows through the **micropyle**, an opening in the ovule, and delivers sperm to the female gametophyte (see Figure 40.4). These sperm unite with haploid cells of the female gametophyte in the process of **fertilization**. Note that pollination and fertilization are distinct processes in flowering plants.

***Double Fertilization***   Angiosperms display a phenomenon known as **double fertilization**; that is, two different fertilization events occur. One of the two sperm cells delivered by a pollen tube fertilizes the egg cell,

thereby forming a diploid **zygote**. This zygote may develop by mitotic division into a young sporophyte, known as an **embryo**. Fertilization thus begins a new cycle of alternation between sporophyte and gametophyte generations. The other sperm delivered by the same pollen tube fuses with two nuclei present in one of the cells of the female gametophyte. The cell formed by this second fertilization undergoes mitosis, eventually producing a nutritive tissue known as the **endosperm**. The embryo and the endosperm are essential parts of maturing angiosperm seeds. Fertilization not only starts the development of zygotes into embryos but also triggers the transformation of ovules into seeds and ovaries into fruits. Embryo, seed, and fruit development occur at the same time.

***Embryos and Seeds***   An embryo is a young, multicellular, diploid sporophyte that develops from a single-celled zygote by mitosis.

Because they are not yet capable of photosynthesis, embryos depend on sporophytes for organic food and other materials. Therefore, embryo development occurs within developing seeds located in a flower ovary. Seeds develop from fertilized ovules. Each developing seed contains an embryo and nutritive endosperm tissue, enclosed and protected by a **seed coat** that develops from the ovule's integuments. When embryos and the seed coat have fully matured, they undergo drying, and the seed enters a phase of metabolic slowdown known as **dormancy**. Fully mature, dormant seeds are ready to be dispersed.

*Fruit and Seed Dispersal*   A **fruit** is a structure that encloses and helps to disperse seeds (see Figure 40.4). Seed dispersal benefits plants by reducing competition for resources among seedlings and parental plants, and it allows plants to colonize new sites.

Fruits develop from the flower's ovary and sometimes include other flower parts. Young fruits bearing immature seeds are typically small and green. During the time that embryos and seeds are developing, the fruit also matures. The ovary wall changes into a fruit wall known as a **pericarp** (from the Greek, meaning surrounding the fruit).

Mature fruits vary greatly among plant species in size, shape, color, and water content. These variations represent adaptations for seed dispersal in different ways. For example, single-seeded dandelion fruits are dry and lightweight and bear a fluffy "parachute" derived from the flower's sepals (see the chapter opening photo). These features foster dispersal by wind. In contrast, coconut fruits feature an airy husk (the pericarp) that keeps them afloat in ocean currents so that they can be carried from one tropical shore to another. Inside the coconut fruit is a single, large seed loaded with liquid and solid endosperm, which people consume as coconut milk and coconut meat, respectively (**Figure 40.5a**). These large amounts of endosperm provide nutrients that sustain coconut seedling growth on infertile, sandy shores. Fruit variation is also extremely important to wild animals and in human agriculture. For example, most fruit crops are juicy and sweet, with relatively small seeds (**Figure 40.5b**). In nature, these features foster dispersal by birds and other animals that feed on such fruits.

*Seed Germination and Seedlings*   If a dispersed seed encounters favorable conditions, including sufficient sunlight and water, it will undergo **germination**. During seed germination, the embryo absorbs water, becomes metabolically active, and grows out of the seed coat, producing a seedling. If the seedling obtains sufficient nutrients from the environment, it grows into a mature sporophyte capable of producing flowers. In the next section, we will focus on flower production, structure, and development.

## 40.2  Flower Production, Structure, and Development

**Learning Outcomes:**

1. List some examples of particular genes that control flower production or shape.
2. Give an example of how gene expression affects flowering time or flower appearance.
3. **CoreSKILL »** Explain recent research findings on how flowers bloom, that is, open up from a closed bud.

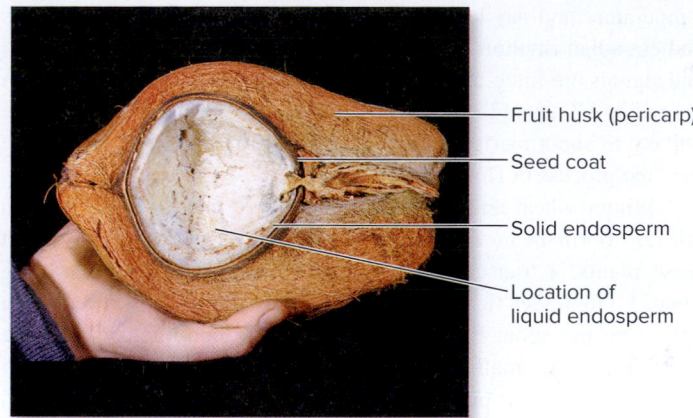

**(a) Coconut fruit and seed**

Labels: Fruit husk (pericarp); Seed coat; Solid endosperm; Location of liquid endosperm

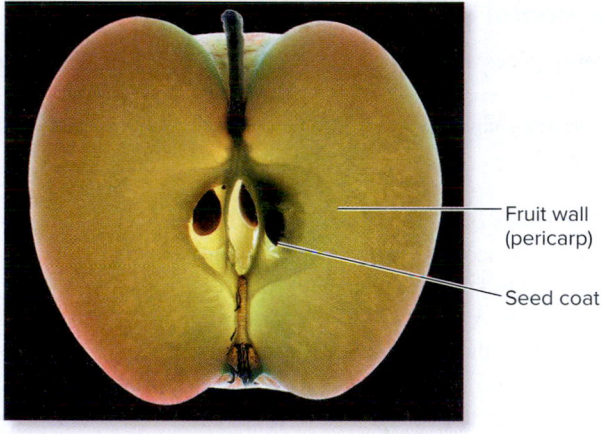

**(b) Apple fruit and seed**

Labels: Fruit wall (pericarp); Seed coat

**Figure 40.5**  **Fruit adaptations for seed dispersal.** **(a)** The coconut fruit's outer wall, called a husk, allows the fruit to float and thus disperse its seed among tropical shores. **(b)** The apple is a juicy, sweet fruit that attracts animals to consume it, thereby helping to disperse its seeds. a: ©Natural Sciences Image Library; b: ©Lee W. Wilcox

Flowers are essential sources of food for many animal pollinators. As the result of coevolutionary relationships with such animals, flowers occur in a spectacular array of colors and forms that attract particular pollinators (refer back to Table 32.2). Flowers also attract humans because we possess sensory systems much like those of animal pollinators. We give bouquets to show love and appreciation; decorate homes, workplaces, and objects with flowers; display flower arrangements on ceremonial occasions; and make perfume from flowers. Consequently, many types of flowers are grown for the florist and perfume industries. Flowers are also necessary for the production of grain and other fruit crops. For these and other reasons, biologists investigate how flower development is controlled by environmental signals and changes in gene expression.

### Environmental Signals Interact with Genes to Control Flower Production

You've probably noticed that different plants flower at particular times of the year. How do plants know when to flower? Flowering time is controlled by the integration of environmental information such as temperature and day length (photoperiod) with hormonal influences and circadian rhythms (see Chapter 37). These stimuli are perceived and signals are integrated by leaves, which produce a mobile protein known as FT (for Flowering locus Time). FT protein travels in the phloem to shoot meristems and interacts with other proteins there to start the process of flower production.

Winter wheat and some other plants are planted and sprout in fall, are dormant in winter, and flower in the following spring. In these plants, a transcription factor known as FLC (for Flowering Locus C) represses flowering genes, thereby preventing flowers from appearing too soon. Exposure to cold winter conditions causes the production of a small RNA molecule that leads to the silencing of FLC, allowing these plants to flower when the appropriate day length occurs in spring. The process by which cold exposure allows plants to flower in spring is known as vernalization.

## Developmental Genes Control Flower Structure

Organ identity genes specify the four basic flower organs that occur in eudicot flowers—sepals, petals, stamens, and carpels. Other genes determine flower shape, color, odor, or grouping into bunches known as inflorescences.

***The Genetic Basis of Flower Organ Identity***    Sepals, petals, stamens, and carpels occur in four concentric rings known as **whorls**. Sepals (collectively known as the calyx) form the outermost whorl, and petals (together known as the corolla) form an adjacent whorl. Stamens (together, the androecium) create a third whorl, and carpels (the gynoecium) form the innermost whorl (**Figure 40.6**). The **perianth** consists of the calyx plus the corolla. In monocot flowers, such as those of lilies, the perianth consists of tepals rather than petals and sepals. You may recall that *A*, *B*, *C*, and *E* genes encode transcription factors that control the production and arrangement of these whorls (refer back to Figure 20.24).

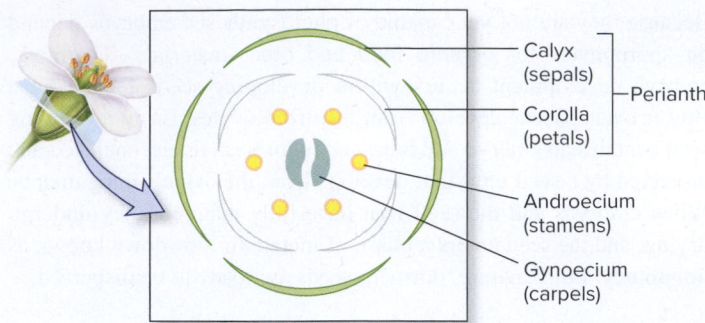

**Figure 40.6**  **The occurrence of eudicot flower parts in concentric whorls.**

***Variation in Number of Whorls***    Eudicot flowers that possess all four types of flower whorls—calyx, corolla, androecium (stamens), and gynoecium (one or more carpels)—are known as **complete flowers**. Complete monocot flowers possess tepals, androecium, and gynoecium. In contrast, flowers that lack one or more flower whorls are described as **incomplete flowers**. Flowers having both stamens and carpels are said to be **perfect flowers**, whereas flowers lacking stamens or carpels are described as **imperfect flowers**. An imperfect flower that produces only carpels is known as a carpellate flower (or pistillate flower). Imperfect flowers that produce only stamens are described as staminate flowers.

Corn produces both imperfect staminate and carpellate flowers on an individual plant (**Figure 40.7**). The flowers of corn start to develop as perfect flowers, but in carpellate flowers, the stamens stop developing. After pollination and fertilization, each carpellate flower produces one of the kernels on a cob of corn. In contrast, staminate flowers of corn, which are found in corn tassels, produce the pollen. Corn is termed **monoecious** (meaning one house) because it produces staminate and carpellate flowers on the same plant. Holly and willow also produce staminate and carpellate flowers, though on separate plants, and are examples of plants described as **dioecious** (meaning two houses).

**Figure 40.7**  **Imperfect flowers of corn, a monoecious plant. (a)** Staminate flowers lack carpels, and **(b)** carpellate flowers lack stamens, but both types of flowers occur on a single corn plant. In contrast, dioecious plants produce staminate and carpellate flowers on separate plants. These features foster cross-pollination. a: ©Scott Sinklier/age fotostock; b: ©Robert and Jean Pollock

*Concept Check:*  *What inference can you draw from the observation that flowers of corn lack showy petals?*

**(a) Staminate flowers of *Zea mays* (corn)**

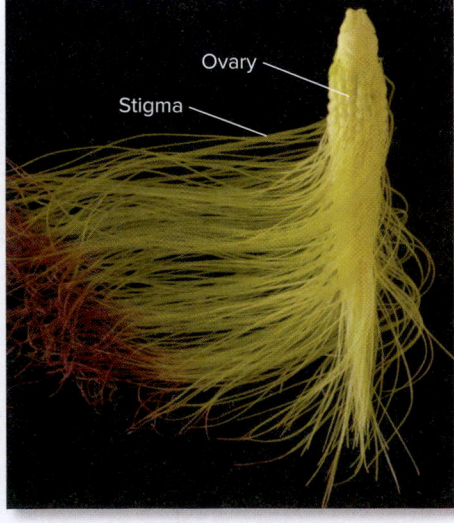

**(b) Carpellate flowers of *Zea mays* (corn)**

***Variation in Flower Organ Number*** In addition to variation in whorls, flowers vary in number of organs. Eudicots and monocots commonly differ in number of flower organs (refer back to Table 36.1). Eudicot flower organs usually occur in fours or fives or a multiple of either of these numbers. By contrast, monocot flower organs typically occur in threes or a multiple of three. Some flowers possess relatively few organs. For example, a minuscule flower of the tiny floating aquatic plant *Lemna gibba* has no perianth and only two stamens.

Many plants sold for use in gardens have been bred so that the flowers produce multiple organs. Garden roses, for example, typically have many more whorls of petals than do wild roses. This change results from a mutation that causes organs that would have become stamens to instead develop into additional petals.

***Variation in Flower Color*** Flowers and their parts may vary in color. The calyx and corolla of monocot flowers such as tulip and lily are often similar in appearance and attractive function. In contrast, eudicot flowers tend to have green, leaflike sepals (see Figure 40.3) that are quite distinct from petals, which are often colorful and fragrant. Color variations arise from differences in gene action that influence the pathways by which plants synthesize pigments.

Differences in flower color were key to recognizing one of the first cases of gene silencing in plants. Gene silencing occurs when miRNA and siRNA block gene expression (see Chapter 13). In an attempt to produce flowers of deeper color, researchers introduced an extra copy of a pigment-producing gene into petunia plants. Surprisingly, the extra gene copy sometimes produced flowers whose petals had white patches or were completely white (**Figure 40.8**). Adding the extra gene had caused the plants to produce a small-interfering RNA that not only silenced the expression of the extra gene, but also silenced the natural pigment-producing genes, causing white patches on the petals.

***Variation in Flower Fragrance*** The fragrances of flowers result from secondary metabolites that diffuse into the air from petals and other flower organs. Recently, RNA interference (RNAi) (see Chapter 13) was used to show that petunia flowers use a transporter protein to actively transport fragrance compounds into the air. This process allows flowers to communicate with pollinator animals more effectively than if fragrances were emitted by passive diffusion alone. Plants produce chemically diverse fragrances to attract particular types of animal pollinators. Because humans possess sensory systems similar to those of many pollinators, we are also attracted to many of these fragrances.

***Flower Shape Variation Resulting from Organ Fusion*** During their development, many flowers undergo genetically controlled fusion of whorls or fusion of the organs within a whorl. For example, pistils are often composed of two or more fused carpels. In addition, stamen filaments often partially fuse with the carpel or form a tube surrounding the pistil, a feature displayed by hibiscus flowers (see Figure 40.4). Each small dandelion flower has five petals that are fused at their sides to form a single strap-shaped structure. Some flowers have petals that are fused together to form a tube that holds nectar consumed by animal pollinators (see Figure 32.19).

***Variations in Flower Symmetry and Aggregation*** Flower shape variation can also result from changes in symmetry. Flowers that possess radial symmetry are described as regular, actinomorphic, or polysymmetric flowers. Flowers having radial symmetry can be divided into two equal parts by more than one plane inserted through the center of the flower. In contrast, flowers that display bilateral symmetry are known as irregular, zygomorphic, or monosymmetric flowers. Flowers having bilateral symmetry can be divided into two equal parts by only a single plane inserted through the center.

Symmetry, like other flower features, is under genetic control. The production of flowers having bilateral symmetry is controlled by transcription factors such as those encoded by the *CYCLOIDEA*

**(a) Normal purple petunia flowers**    **(b) Petunia flower affected by siRNA**

**Figure 40.8 Gene silencing and flower color in petunias. (a)** Normal purple petunia (*Petunia hybrida*) flowers produced by the expression of all genes involved in flavonoid synthesis. **(b)** Flower of a genetically engineered plant displays petal tissues in which expression of one of the genes needed for production of purple pigment production has been silenced (suppressed) by siRNA. a: ©Michael Davis/Getty Images; b: ©Richard Jorgensen

**(a) Normal snapdragon flower**        **(b) Snapdragon flower with *CYCLOIDEA* mutation**

**Figure 40.9** **Genetic control of flower symmetry.** **(a)** Normal snapdragon flowers, with functioning CYCLOIDEA genes, are bilaterally symmetric. **(b)** Snapdragon plants carrying mutations in the CYCLOIDEA gene produce flowers that have radial symmetry.

a–b: Courtesy of John Innes Centre

 **Core Concept: Information** A comparison of the structure of a normal snapdragon flower and a flower from a plant having a mutation in the *CYCLOIDEA* gene illustrates the effect of genetic information on phenotype.

gene. For example, snapdragon (*Antirrhinum majus*) flowers are normally bilaterally symmetric, but a loss-of-function mutation in the *CYCLOIDEA* gene causes these flowers to display radial symmetry (**Figure 40.9**).

The cut-flower crop plant *Gerbera hybrida* provides an additional example of genetic influences on symmetry. *Gerbera* flowers are actually **inflorescences**, groups of flowers clustered together. Within the same inflorescence, *Gerbera* flowers occurring at the rim differ from flowers at the center (**Figure 40.10**), as in the case of sunflowers (see Figure 32.20b). During development of this inflorescence, a transcription factor similar to that encoded by the *CYCLOIDEA* gene is expressed more strongly at the rim than at the center. As a result, rim flowers develop bilateral symmetry, whereas the center flowers have radial symmetry. The center flowers produce stamens that supply pollen. By contrast, stamens do not develop in the rim flowers, and their petals fuse to form larger colorful structures that are specialized to attract pollinators.

**Figure 40.10** **An inflorescence.** This *Gerbera* daisy illustrates the occurrence of flowers of differing symmetry within the same inflorescence. ©Burke/Triolo Productions/Getty Images

**Concept Check:** *List several ways in which the flowers at the rim of the Gerbera inflorescence differ from those at the center.*

---

 **Core Skill: Process of Science**

## Feature Investigation | Liang and Mahadevan Used Time-Lapse Video and Mathematical Modeling to Explain How Flowers Bloom

Like the curled limbs of a human embryo within the womb, concave and overlapping petals occupy minimal space within a flower bud. But when flowers bloom, petals rapidly become convex, spreading outward from each other. Various hypotheses had been proposed to explain petal movements during the opening of flowers. One was that growth started at the midrib running along the center of each petal; another proposed that the blooming process was driven by differential growth rates of top and bottom petal surfaces. In 2011, Haiyi Liang and Lakshminarayanan Mahadevan reported a new explanation for blooming based on their studies of the Asiatic lily, *Lilium casablanca* (**Figure 40.11**).

The investigators placed cut stems bearing young green buds into water, in an environment of constant humidity, temperature, and light.

To evaluate growth during blooming, they first painted small black dots 1 cm apart along the edges and centers of many tepals. (Recall that the flowers of monocot plants such as lily have similarly structured perianth parts known as tepals rather than separate petals and sepals.) By marking many tepals, the researchers repeated, or replicated, the experiment. Replication is key to the experimental process because it helps researchers avoid making erroneous conclusions that might be reached on the basis of just a few observations. The investigators then used time-lapse video to track changes in positions of the black dots during the opening of young green buds, until the flowers had fully bloomed $4\frac{1}{2}$ days later. This process allowed the investigators to measure and graph growth strain at the petal edge versus at the center. Strain is the

**Figure 40.11** Liang and Mahadevan used time-lapse video and mathematical modeling to explain how flowers bloom.

(6): Haiyi Liang and L. Mahadevana 2011. Growth, geometry, and mechanics of a blooming lily, *PNAS* 108: 5516–21, Fig. 2c

**GOAL** To better understand flower blooming.

**KEY MATERIALS** Asiatic lily, *Lilium casablanca*, cut stems with flower buds.

| | | **Experimental level** | **Conceptual level** |
|---|---|---|---|

**1** Paint small black dots at 1 cm intervals along edges and centers of perianth parts on replicate flower buds.

Painted dots

Compare rate of growth at center and edges of tepals.

**2** Maintain cut stems at constant temperature, humidity, and light conditions for 4.5 days, until flowers open.

Avoid confounding the results due to environmental changes.

**3** Set up an automatic time-lapse video system that records blooming at intervals of 1 minute.

Generates a record of changes in the positions of dots on tepals.

**4** Use equations for growth of thin films to model tepal shape changes.

Compare model to video-based observations.

**5** **THE DATA**

(1) The strain of longitudinal growth at tepal edges increases more rapidly than at centers.

(2) Simulation of the blooming process based on mathematical modeling matches observations made with time-lapse video.

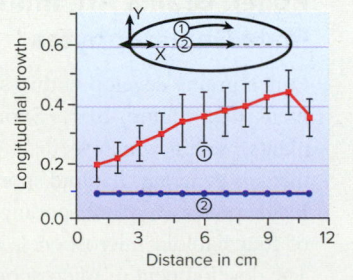

**6** **CONCLUSION** Lily blooming is caused by faster growth at the edges than at centers of tepals.

Wrinkled tepals

The blooming process suggests new ways to engineer the shapes of plastic sheets.

**7** **SOURCE** Liang, H., and Mahadevan, L. 2011. Growth, geometry, and mechanics of a blooming lily. *Proceedings of the National Academy of Sciences (USA)* 108: 5516–5521.
Copyright ©2011 by PNAS. All rights reserved. Used with permission.

ratio of the change in length of a material to the initial unstressed length. These measurements also enabled them to mathematically model changes in tepal shape, starting with equations that describe changes in the shapes of thin sheets (see Figure 40.11).

The biologists observed that by the end of the fourth day, each bud had absorbed one-fifth of a liter of water, increased in length by 10% and width by 20%, and turned white. The inner three tepals had become wrinkled, especially at the edges, a key clue to the mechanism underlying the blooming process. The wrinkling reflected faster growth at the edges than in the center. This difference in growth-generated strain values is large enough to cause the flower to bloom rapidly as individual tepals simultaneously reversed curvature and bent outward, a process predicted by mathematical modeling. The study revealed that flower blooming is based on a different mechanism than previously thought and illustrates the value of applying physics and mathematical models to biological phenomena.

### Experimental Questions:

1. Why do you think Liang and Mahadevan used *Lilium casablanca* for their analysis of flower blooming?
2. How did time-lapse video improve data gathering in this study?

## BIO TIPS

**THE QUESTION** *By employing time-lapse video, Liang and Mahadevan determined that the blooming of lily flowers resulted from faster growth of the tepals at their edges than at their centers. Consider the data from their study in step 5 of Figure 40.11. How much faster was the longitudinal growth at the edges of the tepals than the longitudinal growth at the centers?*

**T OPIC** *What topic in biology does this question address?* The topic is flowering plant reproduction, or, more specifically, how flowers bloom, that is, open fully from the bud stage. Most people have observed flowers in stages of opening, but few have watched the process from start to finish. For this reason, the physical mechanism underlying flower blooming had not been clear until this experiment was performed.

**I NFORMATION** *What information do you know based on the question and your understanding of the topic?* In the question, you are reminded of the data in Figure 40.11. From your understanding of the topic, you may recall that monocots such as lilies produce flowers whose perianth parts are known as tepals because they are not specialized into outer sepals and inner petals, as they are in eudicots. The outermost parts of the lily bud transform into petal-like tepals as the flower blooms. The investigators focused on the longitudinal (lengthwise) growth of the tepals.

**P ROBLEM-SOLVING S TRATEGY** *Interpret data.* The data are presented in the form of line graphs based on video recordings of changes in the positions of black dots marked on the surfaces of lily buds. Compare the red line showing longitudinal growth at the tepal edges to the blue line showing longitudinal growth at the tepal centers.

**ANSWER** *Longitudinal growth of the tepals always occurred more rapidly at the edges than at the centers, and the maximal difference was more than 4 times (a ratio of 0.45/0.1).*

## 40.3 Male and Female Gametophytes and Double Fertilization

### Learning Outcomes:

1. Explain how gamete production in plants differs from that in animals.
2. Explain how the pistil controls pollen germination.
3. Describe the structure of the female gametophyte.
4. Outline the different fates of the two sperm cells transmitted by each pollen tube.

In animals, the cells that will undergo meiosis to produce gametes are known as the germ line, and they are set aside from other body cells during early development. By contrast, plants do not establish gamete-producing cells during early development and meiosis does not generate plant gametes. Even so, similar proteins of the Rb (retinoblastoma) family are involved in starting the process of gamete development in both animals and plants. As we have seen, plant gametes are produced by the life cycle stage known as gametophytes. In this section, we will focus on plant male and female gametophytes, gametes, and fertilization.

### Pollen Grains Are Immature Male Gametophytes

Pollen grains develop within sporangia located in the anthers of stamens (see Figure 40.4). Sporangia are structures produced by all plants, within which meiosis generates spores. (Note that in plants, meiosis generates haploid spores, not haploid gametes as in animals.) Inside protective plant sporangia, diploid body cells produce a cluster of four haploid microspores, each having a thin cellulose cell wall. The development of microspores into pollen grains involves two processes that occur at the same time: (1) microspore division to produce a young male gametophyte, and (2) development of a tough pollen wall that protects the gametophyte during pollen transport. Both of these processes are completed before the anthers release pollen.

Each microspore nucleus undergoes one or two mitotic divisions to form a young male gametophyte. The first division gives rise to two specialized cells: a tube cell and a generative cell suspended within the tube cell (**Figure 40.12a**). The **generative cell** divides to produce two sperm cells, either before or (more commonly) after pollination. The **tube cell** produces the pollen tube, which delivers sperm to the female gametophyte.

A mature pollen grain has a tough wall, and each plant species produces pollen whose wall has a distinctive sculptural shape (**Figure 40.12b**). The wall, which surrounds the plasma membrane of the tube cell, is composed largely of a tough polymer known as sporopollenin. Named for its presence on the surfaces of mature spores and pollen, sporopollenin protects spores and pollen from damage.

Development of the pollen grain wall starts with deposition of a blanket of the carbohydrate callose around each cluster of four microspores after they form by meiosis. The callose blanket isolates the microspores from the influences of adjacent sporophyte tissues, thereby aiding pollen differentiation. Callose also provides a surface pattern for sporopollenin deposition and holds microspores together until an anther enzyme degrades the callose, freeing pollen grains from each other.

As pollen grains mature, anther cells secrete a pollen coat, a layer of material that covers the sporopollenin-rich pollen wall. Coat materials include additional sporopollenin, pigments that give pollen its typically yellow, orange, or brown coloration, and lipids and proteins that aid in pollen attachment to carpels. Certain of these pollen coat compounds are responsible for allergic reactions in people exposed to particular types of airborne pollen. About 10% of flowering plants are wind-pollinated, and such plants produce copious amounts of pollen.

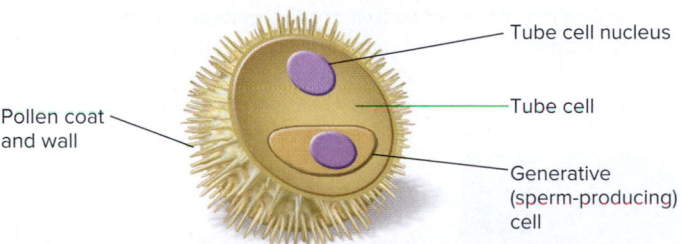

**(a) A cut pollen grain showing immature male gametophyte**

Tube cell nucleus
Tube cell
Pollen coat and wall
Generative (sperm-producing) cell

16.7 µm

**(b) SEM of whole pollen grains showing distinctive sculptural shapes**

**Figure 40.12  Pollen grains.  (a)** Diagram of cut pollen grain. **(b)** SEM of whole pollen grains of different species. b: ©Phanie/Alamy Stock Photo

**Concept Check:** *What is the maximum number of cells in a mature male gametophyte of a flowering plant?*

For example, ragweed plants (genus *Ambrosia*), which are commonly associated with allergies, each produce an estimated 1 billion pollen grains during a year!

## After Pollination, the Pistil Controls Pollen Germination

Recall that pollination is the process by which pollen is delivered to surfaces of the stigma, the uppermost part of the flower pistil. However, even if a pollen grain reaches the stigma of the right flower species, it may not be able to germinate and produce a sperm-delivery tube. The stigma and the style determine whether or not pollen grains germinate and pollen tubes grow toward ovules. How is this accomplished?

About half of plant species can serve as both mother and father to their progeny, because pollen produced by those plants is able to germinate on pistils of the same plants. Such self-pollinating plants are said to display **self-compatibility (SC)**. By contrast, plants that prevent the germination of pollen that is too genetically similar to the pistil are said to display **self-incompatibility (SI)**. Similar to human cultural practices that prevent mating between close relatives, SI helps to decrease the likelihood of recessive disorders in offspring. In many plants, SI involves the *S* gene locus, which encodes S proteins. Each locus contains genetic sequences that determine pollen compatibility traits and pistil compatibility traits. Multiple *S* alleles for both genes occur in plant populations.

Two major types of SI are known: gametophytic SI and sporophytic SI. Gametophytic SI occurs when one or more *S* genes within pollen determine compatibility by encoding S proteins located in the pollen cytosol. When the gametophyte controls compatibility, tubes may start to grow from incompatible pollen, but S proteins encoded by pistil cells enter the tubes and destroy the pollen tube RNA. This halts tube growth. In contrast, S proteins within genetically compatible pollen bind the pistil-produced proteins, thereby preventing destruction of the pollen tube RNA and allowing tube growth to continue (**Figure 40.13a**).

Sporophytic SI occurs when pollen compatibility is determined by the sporophyte that produces the pollen. This control is exerted when anthers deposit proteins into the pollen coat. When the sporophyte controls compatibility, pollen cannot germinate when S proteins in the plasma membranes of stigma cells recognize (bind) the S proteins of incompatible pollen. In this case, protein binding leads to signal transduction processes that prevent pollen germination. However, pollen can germinate if stigma proteins are unable to bind genetically distinct S proteins in its coat (Figure 40.13b).

## A Female Gametophyte Develops Within Each Ovule

In flowering plants, each ovule produces a single female gametophyte that often consists of seven cells and eight nuclei (**Figure 40.14**). One of these cells is an egg cell, which lies wedged between two cells known as **synergids**. These synergids help move nutrients from the larger sporophyte to the nonphotosynthetic female gametophyte. Synergids also secrete small proteins called LURES that act as attractants for pollen tube growth. The LURES cause pollen tubes to deposit specific receptor proteins in the membranes of tube tip cells lying closest to the LURES. Then, the binding of LURES to the receptors steers pollen tube growth

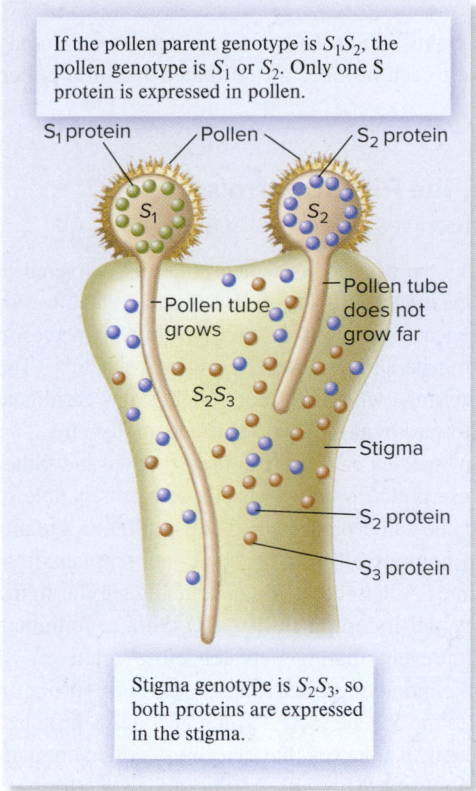

If the pollen parent genotype is $S_1S_2$, the pollen genotype is $S_1$ or $S_2$. Only one S protein is expressed in pollen.

S₁ protein — Pollen — S₂ protein

Pollen tube grows

Pollen tube does not grow far

$S_2S_3$

Stigma

S₂ protein

S₃ protein

Stigma genotype is $S_2S_3$, so both proteins are expressed in the stigma.

**(a) Gametophytic SI: If pollen S allele does not match either stigma allele, pollen will germinate.**

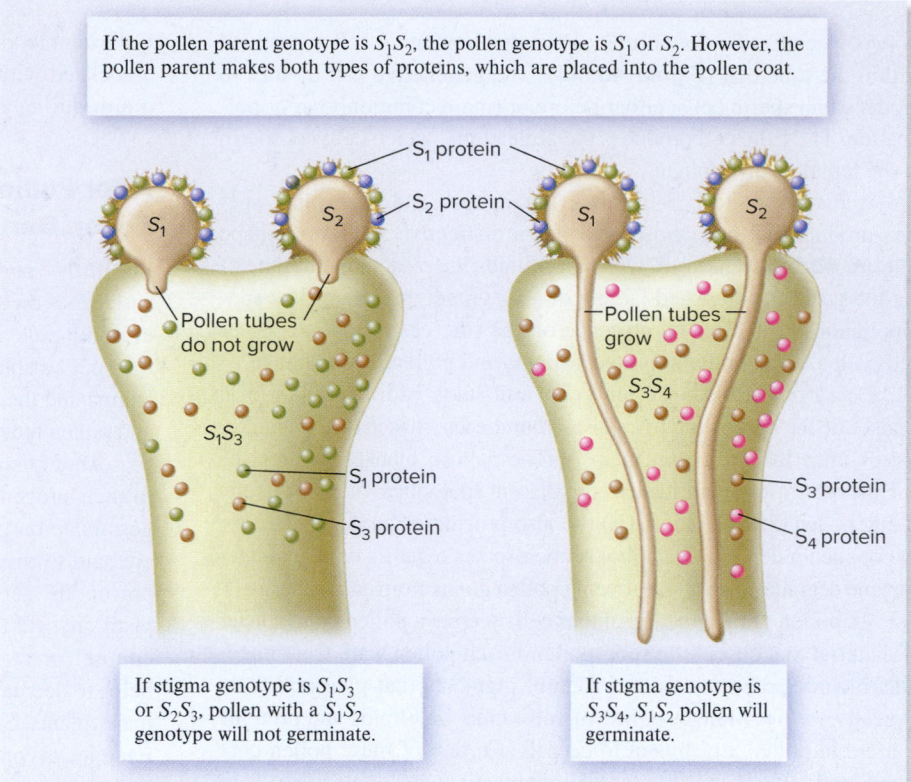

If the pollen parent genotype is $S_1S_2$, the pollen genotype is $S_1$ or $S_2$. However, the pollen parent makes both types of proteins, which are placed into the pollen coat.

S₁ protein

S₂ protein

Pollen tubes do not grow

$S_1S_3$

S₁ protein

S₃ protein

Pollen tubes grow

$S_3S_4$

S₃ protein

S₄ protein

If stigma genotype is $S_1S_3$ or $S_2S_3$, pollen with a $S_1S_2$ genotype will not germinate.

If stigma genotype is $S_3S_4$, $S_1S_2$ pollen will germinate.

**(b) Sporophytic SI: If pollen coat S proteins do not match either stigma S protein, pollen tubes will grow.**

**Figure 40.13 Self-incompatibility.** Self-incompatibility helps plants to avoid combinations of gametes that are too genetically similar. It is controlled by interactions between chemical constituents of pollen and the pistil. **(a)** In gametophytic self-incompatibility, compatibility between pollen and pistil is determined by the haploid genotype of the pollen. In this case, pollen S protein is located in the cytosol. **(b)** In sporophytic self-incompatibility, compatibility between pollen and pistil is determined by the sporophyte that produced the pollen and contributed S proteins to its coat.

through the micropyle opening to reach the female gametophyte (see step 4 in Figure 40.4).

The other four cells of the angiosperm female gametophyte consist of three antipodal cells, whose functions are not well understood, and a large **central cell** that contains two nuclei. This central cell and the egg cell are involved in the process of double fertilization.

## Pollen Tubes Deliver Sperm Cells That Accomplish Double Fertilization

As we have seen, when pollen grains germinate successfully, the tube cell produces a long pollen tube. Within the tube, the generative cell divides by mitosis to produce two sperm cells. To deliver sperm to female gametophytes, the tube must grow at its tip from the stigma, through the style, to reach the ovule (**Figure 40.15**). Tube growth, controlled by the tube cell nucleus, is accomplished by adding new cytoplasm and cell-wall material to the elongating tip. Pollen tubes have been observed to grow toward ovules at about 0.5 mm per hour, commonly taking from 1 hour to 2 days to reach their destination.

When a pollen tube encounters an ovule, it enters through the micropyle and penetrates a female synergid. The tube then stops growing, and its thin tip wall bursts, releasing the sperm. The bursting process propels the two sperm toward the egg and the central cell. One sperm nucleus fuses with the egg cell to produce a zygote, the

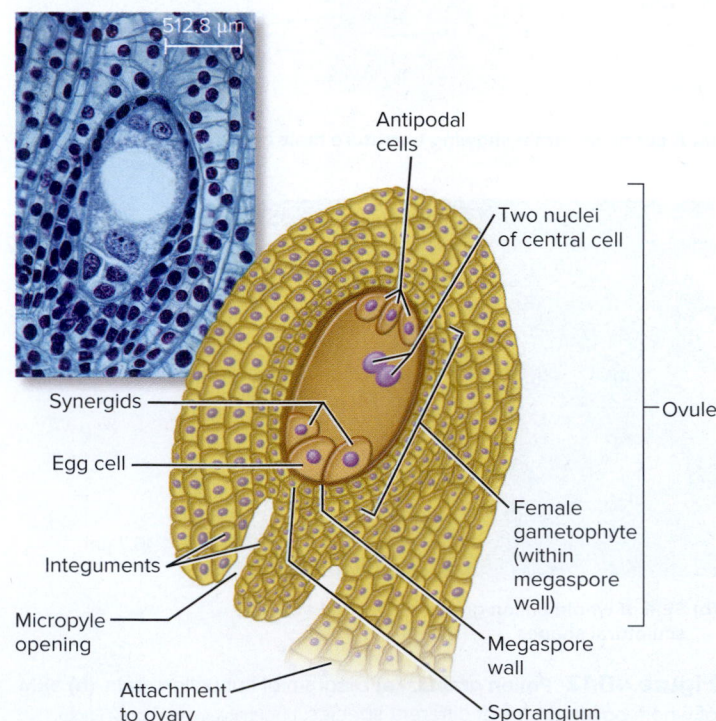

512.8 μm

Antipodal cells

Two nuclei of central cell

Synergids

Egg cell

Ovule

Integuments

Female gametophyte (within megaspore wall)

Micropyle opening

Megaspore wall

Attachment to ovary

Sporangium

**Figure 40.14 Angiosperm female gametophyte within an ovule.**
©Ed Reschke/Getty Images

*Concept Check:* *How do female gametophytes obtain nutrients?*

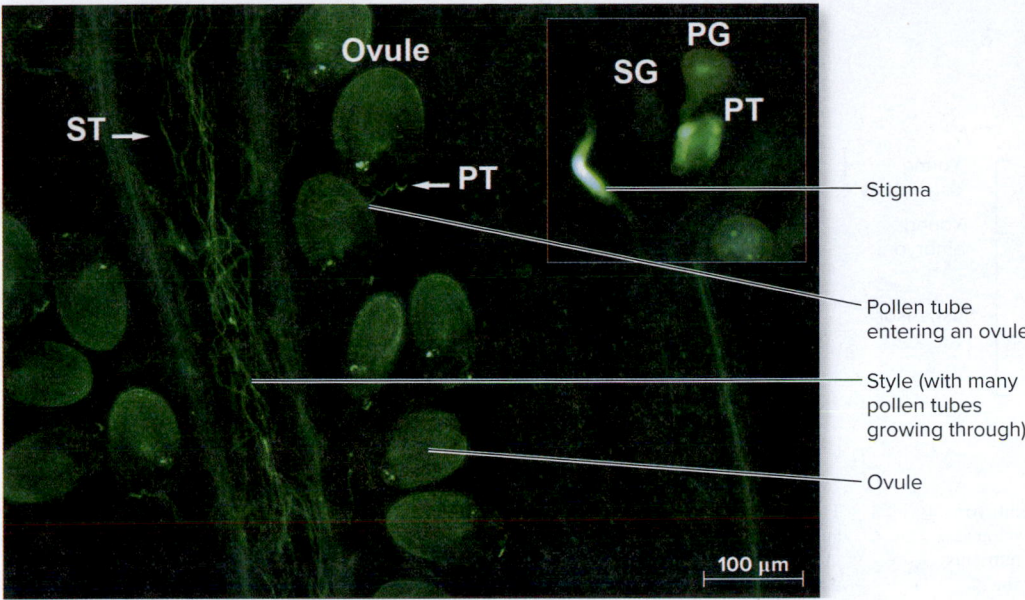

**Figure 40.15 Pollen tubes delivering sperm to ovules.** This fluorescence microscopic view shows pollen tubes (PT) growing through the style (ST) toward ovules. Courtesy J. M. Escobar Restrepo

 **Core Skill: Connections** Look ahead to Figure 51.5, which illustrates the male sexual organ of humans, an adaptation that allows internal fertilization. How is the pollen tube of plants like a human penis?

first cell of a new sporophyte generation (see step 4 in Figure 40.4). The other sperm fuses with the two nuclei of the central cell to form the first endosperm cell (see step 5 in Figure 40.4). The term double fertilization arises from these two processes.

The zygote and fertilized central cell have dramatically different fates. The endosperm absorbs protein, lipid, carbohydrates, vitamins, and minerals from the mother plant and stores these nutrients. The stored nutrients provide material and energy needed by the zygote to develop into an embryo and by the embryo to develop into a seedling. The nutritive role of endosperm explains why a large percentage of human and animal food comes from seed endosperm of grain crops; corn, wheat, rice, and other grain crops generate more than 380 billion pounds of endosperm per year in the U.S. alone.

You might wonder why the embryo and endosperm resulting from double fertilization have such different fates despite similar genetic composition. Part of the answer is that during its development, plant endosperm undergoes genomic imprinting by DNA methylation, a process also known to occur in mammals (see Figure 18.2). The imprinting process causes gene expression to occur differently in plant endosperm than it does in the embryo.

## 40.4 Embryo, Seed, Fruit, and Seedling Development

### Learning Outcomes:

1. Outline the major stages in angiosperm embryo development.
2. List examples of environmental and internal factors that influence seed germination.
3. Explain how seed germination varies among plants.

Seeds and fruits are major features of plant reproduction and are essential to the nutrition of animals, including humans. Seeds contain dormant plant embryos that may develop into seedlings under favorable conditions. As discussed earlier, fruits aid seed dispersal, which allows plants to colonize new sites. Embryos, seeds, and fruits mature simultaneously, and their development is coordinated by hormonal signals that were introduced in Chapters 36 and 37. Seedling development is also hormonally regulated.

### Embryos Develop from Zygotes

Fueled by endosperm nutrients, angiosperm embryos undergo development in a series of stages known as **embryogenesis** (**Figure 40.16**). Sometime within a period of days to several weeks following fertilization, a zygote begins to divide. At this point, a zygote is blanketed with a layer of callose, which helps to seal it off from the environment, thereby fostering embryo-specific gene expression. The zygote's first cell division is unequal, producing a smaller cell and a larger cell (Figure 40.16, step 1), similar to the asymmetric division of animal zygotes in certain species. In plants, this unequal division helps to establish the apical-basal (top-bottom) polarity of the embryo, which persists through the life of the plant.

After unequal cell division of the plant zygote, the smaller cell develops into the embryo, whose radial symmetry is established at this point and continues in adult plants. The larger cell develops into a **suspensor**, a short chain of cells anchored near the micropyle at the ovule entrance (Figure 40.16, step 2). The suspensor channels nutrients and hormones into the young embryo, which absorbs them at its surfaces.

Young eudicot embryos are spherical, but they soon become heart-shaped as the seedling leaves, called **cotyledons**, start to develop. At this point, an auxin and a mobile transcription factor are involved in establishing the young shoot and root at the apical and basal poles, respectively. This process is also influenced by the TOPLESS protein, which helps to repress auxin-responsive genes that would otherwise promote root development at the apical pole. (Mutants that have lost normal *TOPLESS* function have roots at both apical and basal poles, but no shoots, and are thus topless.) Eudicot embryos such as *Arabidopsis* then become torpedo-shaped, and as the cotyledons grow, they often curl to fit within the developing seed (Figure 40.16, step 4). In contrast, mature monocot embryos are cylindrical, with a single cotyledon and a side notch where the apical meristem forms.

### Mature Seeds Contain Dormant Embryos

As seeds mature, they undergo changes leading to dormancy, an adaptation that prevents them from germinating when environmental

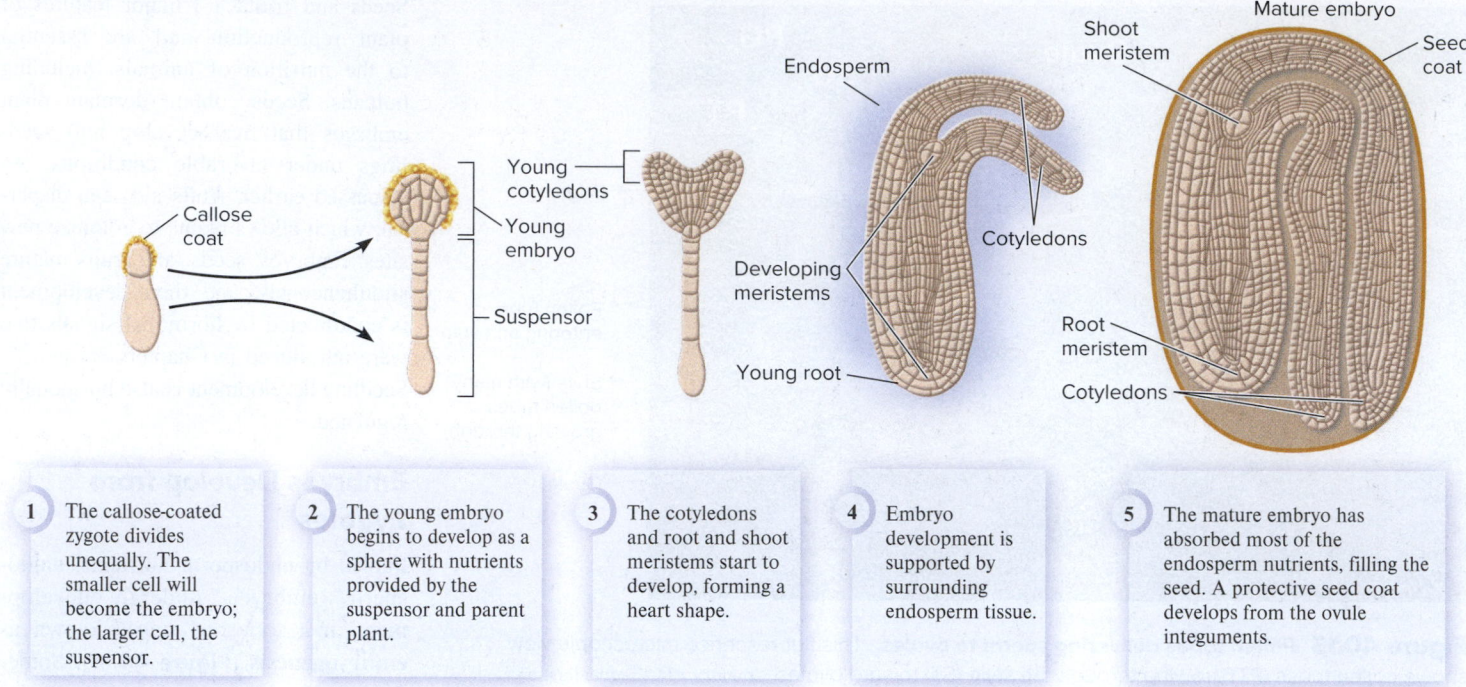

1. The callose-coated zygote divides unequally. The smaller cell will become the embryo; the larger cell, the suspensor.

2. The young embryo begins to develop as a sphere with nutrients provided by the suspensor and parent plant.

3. The cotyledons and root and shoot meristems start to develop, forming a heart shape.

4. Embryo development is supported by surrounding endosperm tissue.

5. The mature embryo has absorbed most of the endosperm nutrients, filling the seed. A protective seed coat develops from the ovule integuments.

**Figure 40.16** **Embryogenesis in the eudicot *Arabidopsis*.**

**Concept Check:** *How would the eudicot embryo differ if its TOPLESS gene were nonfunctional?*

conditions are not suitable for seedling growth. Seed maturation includes transformation of the ovule's integuments into a tough seed coat (see Figure 40.16, step 5). The seed coat restrains seedlings from growing and contains suberin, a lipid-rich material that prevents the entry of water and oxygen, thereby maintaining low seed metabolism. In addition, the coats of some seeds are darkly colored with pigments that may help to prevent damage by UV radiation or microbial attack.

A key change leading to seed dormancy is the gradual, controlled loss of water from the embryo and other seed tissues. During this process, the embryo becomes relatively dry, but is able to survive. The water content of dispersed seeds is only 5–15%. Abscisic acid (ABA) is a hormone that induces the activity of genes that help embryo tissues to survive the drying process. Some of these desiccation-tolerance genes encode proteins that form loose coils enclosing cell contents, thereby preventing damage as the cytoplasm becomes almost completely dry. When the seeds of flowering plants are dry and ready for dispersal, they are released from the plant while enclosed in a fruit or released when the fruit breaks open.

The structure of mature monocot and eudicot seeds differs. Within eudicot seeds, mature embryos often display an **epicotyl**, the portion of an embryonic stem with two tiny leaves in a first bud that is located above the point of attachment of the cotyledons (**Figure 40.17a**). The **hypocotyl** is the portion of an embryonic stem located below the point of attachment of the cotyledons. An embryonic root, the **radicle**, extends from the hypocotyl. Much of the endosperm has been absorbed into the large cotyledons. In contrast, mature monocot embryos, such as those of corn, feature an epicotyl with a first bud enclosed in a protective sheath known as the **coleoptile**. The young monocot root is enclosed within a protective envelope known as the **coleorhiza** (**Figure 40.17b**).

## Fruits Develop from Ovaries and Other Flower Parts

All fruits develop from ovaries and sometimes other flower parts. They occur in diverse forms that aid seed dispersal. Some fruits are dry, whereas others are moist and juicy; some open to release seeds, and others do not. Fruits also display a wide variety of sizes, colors, and fragrances. These variations result from differences in the process of fruit development. Plant hormones, including auxins, gibberellic acid, and cytokinins, control this transformation. ABA stimulates cell expansion, and ethylene influences fruit ripening. For instance, ethylene helps to ripen nuts, a type of dry fruit, by inducing plasma membranes to rupture, causing water loss. Under the influence of plant hormones, the pericarp (ripened ovary wall) of peaches, plums, and related fruits swells and softens, and orange or red chromoplasts replace green chloroplasts. As fruits mature, the outer protective cuticle often becomes very thick, contributing to peel toughness, which helps to prevent microbe attack. In addition, many maturing fruits increase their sugar and acid content, which produces the distinctive tastes of ripe fruits. Many fruits also produce fragrant volatile compounds.

Differences in the shape, color, fragrance, and moisture content of wild fruits reflect evolutionary adaptation for effective seed dispersal. Though many fruits and seeds are dispersed by wind or water or by attaching to animal fur, others are consumed by fruit-eating animals that are attracted by the colors and fragrances. Blackberries provide a good example of fruits adapted for animal dispersal. Blackberry flowers produce many separate pistils, each containing a single ovule (**Figure 40.18a**). Following pollination and fertilization, the ovary of each pistil develops into a sweet, juicy fruitlet containing a single seed. As the individual fruitlets develop, they fuse together at the sides.

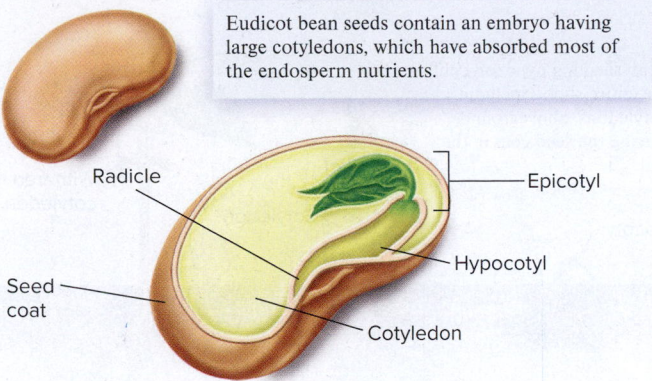

Eudicot bean seeds contain an embryo having large cotyledons, which have absorbed most of the endosperm nutrients.

Radicle

Epicotyl

Seed coat

Hypocotyl

Cotyledon

**(a) Eudicot bean seed, showing embryo with epicotyl, hypocotyl, and radicle**

Monocot corn kernels are single-seeded fruits. The seed embryo has one cotyledon that presses against the nutritive endosperm. The embryonic shoot tip is protected by a tissue sheath—the coleoptile. The embryonic root is protected by a tissue sheath—the coleorhiza.

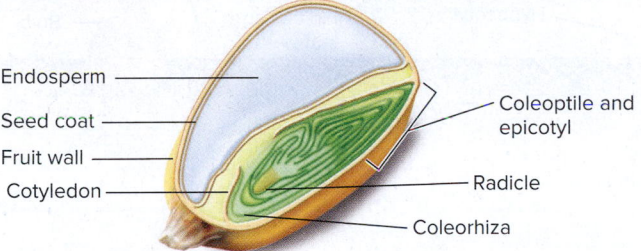

Endosperm

Seed coat

Coleoptile and epicotyl

Fruit wall

Cotyledon

Radicle

Coleorhiza

**(b) Monocot corn seed, showing an embryo protected by coleoptile and coleorhiza**

**Figure 40.17**   **Structure of mature seeds and embryos.**  Note: You may want to look ahead to Figure 40.20a to see the relative locations of some of these structures in young seedlings.

**Concept Check:**  *Why do mature seeds of eudicots lack extensive amounts of endosperm?*

Consequently, the many fruitlets produced by a single blackberry flower are dispersed together (**Figure 40.18b**). Attracted by the color, birds consume the whole aggregate and excrete the seeds, thereby dispersing many at a time. Many other types of fruits occur and these likewise represent adaptations that foster seed dispersal (refer back to Figure 32.21). Although a fruit is usually defined as a mature ovary containing seeds, seedless fruits such as watermelon are produced commercially by genetic modification or treatment with artificial auxin.

## Environmental and Internal Factors Influence Seed Germination

Seeds vary greatly in their ability to germinate after dispersal. Small seeds such as those of dandelions and lettuces germinate quickly if light is available. Other seeds require a period of dormancy before germination occurs. Some seeds can remain dormant for amazingly long time periods. For example, a lotus (*Nelumbo nucifera*) seed collected from a lake bed in China germinated at the age of 1,300 years, as determined by radiocarbon dating. In 2005, plant scientists germinated a 2,000-year-old date seed found in Israel.

Water is generally required to rehydrate seeds so that embryos can resume their metabolic activity. Water absorption also swells seeds, helping to break the seed coat and allowing embryonic organs to emerge. In some cases, rainfall of sufficient duration to leach germination-inhibiting compounds out of seeds is required. The optimal temperature for germination of most seeds lies between 25°C and 30.25°C (77°F and 86.25°F). This explains why gardeners wait until the soil is warm before planting seeds outdoors in spring. However, some seeds need a period of cold treatment or seed coat abrasion before they will germinate. Such physical stimuli induce the activity of more than 2,000 genes associated with seed germination.

When grass seeds rehydrate, the young shoot secretes the hormone gibberellic acid from the seed cotyledon into the outermost endosperm layer, known as the aleurone. In response, the aleurone secretes digestive enzymes into the central endosperm, releasing sugars from stored starch (**Figure 40.19**). The seedling uses these sugars for growth. This

Pistils
Ovary with ovule

**(a)** *Rubus allegheniensis* **(common blackberry) flower**

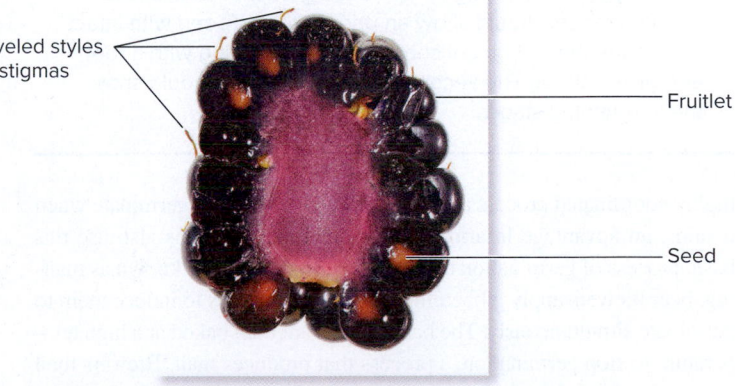

Shriveled styles and stigmas

Fruitlet

Seed

**(b) Blackberry fruit**

**Figure 40.18**  **Blackberry flower and fruit.**  **(a)** Each of the many separate pistils in a blackberry flower is able to produce a single one-seed fruit (called a fruitlet) if fertilization occurs. **(b)** Together, the individual fruitlets of the blackberry compose an aggregate fruit that allows many seeds to be efficiently dispersed at the same time by the same animal agent. a–b: ©Lee W. Wilcox

  **Core Concept: Systems**  The dispersal by birds of many blackberry seeds at the same time is an example of the interconnection of living systems.

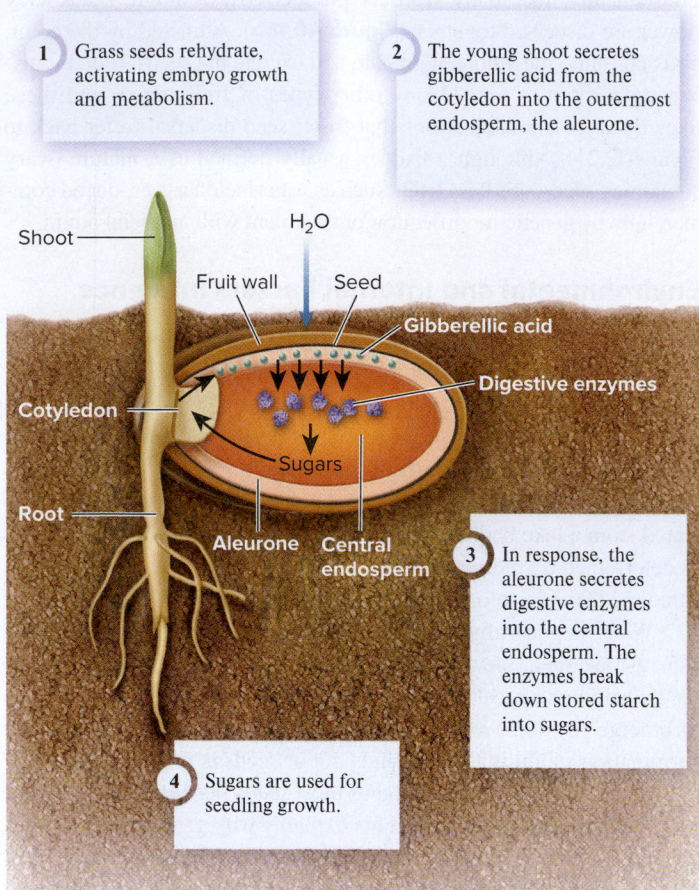

1. Grass seeds rehydrate, activating embryo growth and metabolism.

2. The young shoot secretes gibberellic acid from the cotyledon into the outermost endosperm, the aleurone.

3. In response, the aleurone secretes digestive enzymes into the central endosperm. The enzymes break down stored starch into sugars.

4. Sugars are used for seedling growth.

**Figure 40.19** Germination of grass seeds.

 **Core Skill: Modeling** The goal of this modeling challenge is to elaborate on the seed germination model shown in Figure 40.19 by making a flow diagram.

**Modeling Challenge:** Make a flow diagram that expands the model in Figure 40.19 into four successive models, each showing a separate stage in the seed germination process. The first model should show an ungerminated seed with intact seed coat; the last model should show a seedling with a root and green shoot. The second and third models should show the intervening stages.

highly coordinated process allows grass seeds to quickly germinate when it rains, an advantage in arid grassland habitats. Humans also use this basic process of germination to make beer. In the process known as malting, beer brewers apply gibberellic acid to barley seeds to induce them to germinate simultaneously. The barley seeds are then baked at a high temperature to stop germination, a process that produces malt. Brewers then treat malt with water and heat, add the dried flowers of the hop plant (the genus *Humulus*), and add yeasts to ferment the plant sugars to alcohol.

Once seeds have germinated, plants vary in the process by which the embryonic shoot emerges. When bean and onion seeds germinate, the hypocotyl forms a hook that first breaches the soil surface and then straightens, thereby pulling the rest of the seedling and cotyledons aboveground (**Figure 40.20a** and **b**). In contrast, when

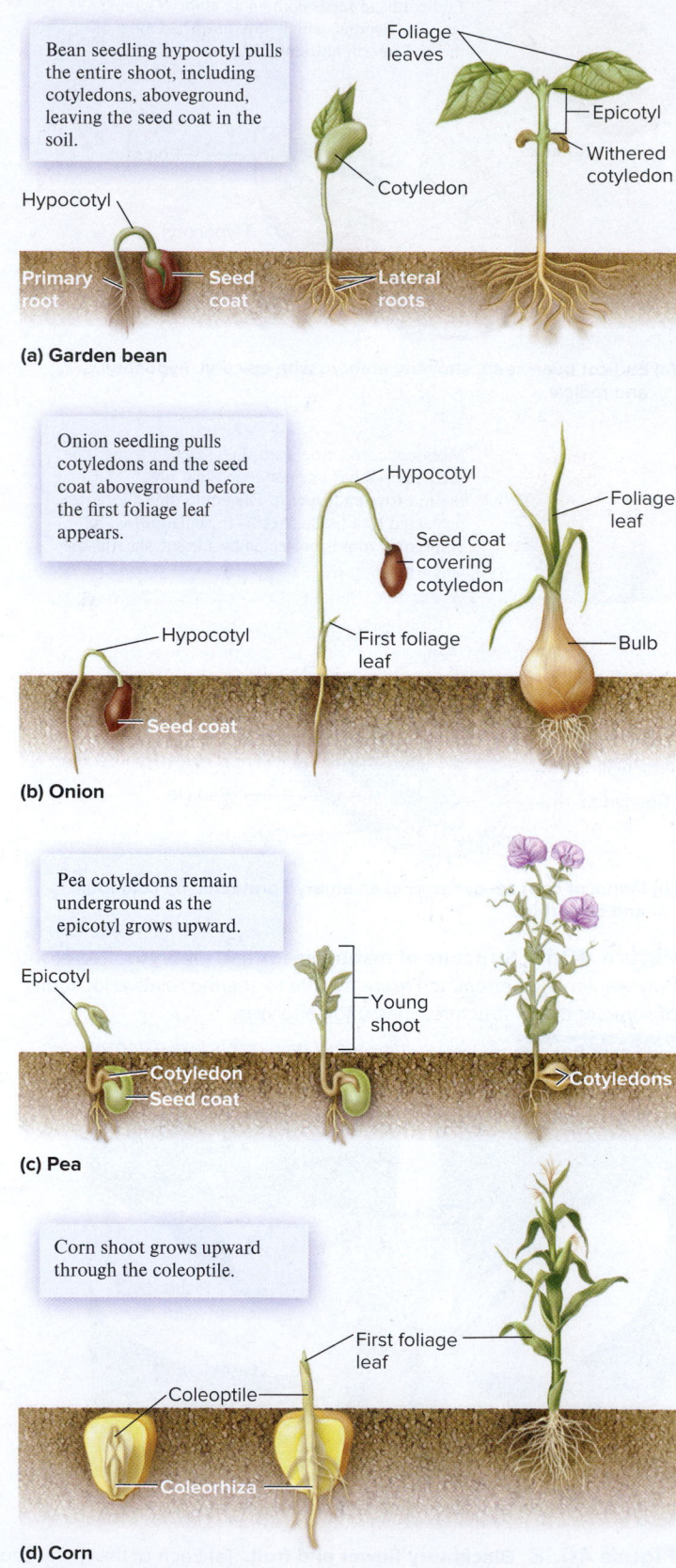

**(a) Garden bean**

**(b) Onion**

**(c) Pea**

**(d) Corn**

**Figure 40.20** Variations in seed germination and seedling growth patterns.

pea seeds germinate, the epicotyl forms a hook that pulls the shoot tip out of the ground, leaving the cotyledons beneath the soil surface (**Figure 40.20c**). In both cases, the tough hook cells bear the brunt of passage through hard surface soil crusts, thereby protecting the delicate shoot tips. The plant hormone ethylene controls seedling hook formation, as described in Chapter 37 (refer back to Figure 37.9). However, not all seedlings form hooks. For example, as they grow through the soil, the shoot tips of corn seedlings and those of other grasses are protected by the coleoptile, a protective sheath that encloses the first foliage leaves (**Figure 40.20d**).

**Figure 40.21** **Asexual reproduction via somatic embryogenesis.** The leaves of this *Kalanchoë* plant bear small plantlets around the edges. When mature, these plantlets drop off and, under the right conditions, grow into new plants. ©Lee W. Wilcox

## 40.5 Asexual Reproduction in Flowering Plants

**Learning Outcomes:**

1. Explain the benefits of asexual reproduction in flowering plants.
2. List several examples of ways in which plant asexual reproduction is important to agriculture.

Many plants rely on sexual reproduction. However, a wide variety of angiosperms reproduce primarily by asexual means, and other plants commonly utilize both sexual and asexual reproduction. **Asexual reproduction** is the production of new individuals from a single parent without the occurrence of fertilization.

Although sexual reproduction provides beneficial genetic variation, asexual reproduction can be advantageous in other ways. For example, asexual reproduction maintains favorable gene combinations that allow faster population growth in stable environments. Asexual reproduction is also advantageous in stressful habitats where pollinators or mates can be rare, because it allows a single individual to start a new population. Finally, asexual reproduction allows some plants to persist for very long periods of time. Among the oldest known plants are creosote bushes that are asexual clones of a parent that grew from a seed about 12,000 years ago! The propagation of plants from cuttings is a form of asexual reproduction that is widely used commercially and by home gardeners. In this section, we will explore the three main mechanisms of plant asexual reproduction: specialized reproductive structures, somatic embryogenesis, and apomixis.

### Vegetative Reproduction Generates Offspring from Nonreproductive Organs

Certain forms of asexual reproduction are referred to as **vegetative reproduction** because they involve the nonreproductive parts of plants. Roots, stems, and leaves are vegetative plant organs that may function as asexual reproductive structures.

***Specialized Reproductive Structures*** Biologists have identified many cases in which vegetative structures function in asexual reproduction. For example, root sprouts, such as those produced by aspens, can generate entire groves of genetically identical trees. Sucker shoots, such as those appearing at the bases of banana plants and date palms, and the pieces of tuber with "eyes" (which are buds) that develop into potato plants are examples of vegetative plant organs that have agricultural importance. Attractive horticultural varieties of African violets and other plants can be propagated from leaf cuttings. Such asexual offspring

grow into adult plants that have the same valued properties as their parent plant. In contrast, seeds from these plants would produce diverse progeny, not all of which would have the economically prized properties.

***Somatic Embryogeneis*** In our previous examples of vegetative reproduction, offspring were produced from vegetative structures, but the offspring themselves did not originate from embryos. **Somatic embryogenesis** is the production of plant embryos from body (somatic) cells. Embryos can develop from many types of plant cells, such as leaf cells. Somatic embryos develop normally to the torpedo stage but do not dehydrate and become dormant, as is normal for zygotic embryos. Rather, somatic embryos produce root and shoot systems and develop into mature plants.

Somatic embryogenesis occurs naturally in many species including citrus, mango, onion, and tobacco plants. The common houseplant *Kalanchoë daigremontiana* has leaves bearing many tiny plantlets at their edges. When these detach, they are able to take root and grow into new individuals (**Figure 40.21**). As discussed shortly, recent molecular studies have revealed that embryogenesis genes are involved in this process and its evolution.

Agricultural scientists have also made use of somatic embryogenesis. In the 1950s, British plant biologist F. C. Steward and associates were the first researchers to successfully clone a complex organism, the carrot plant, by means of somatic embryogenesis. They used differentiated cells from carrot roots, grew the cells in conditions that caused some cells to lose their specialized properties and develop into embryos, and then cultivated each embryo in conditions that favored development into a mature carrot plant. Many types of plants are now cloned by somatic embryogenesis, which allows commercial growers to produce large numbers of genetically identical individuals.

**Core Concept: Evolution**

### The Evolution of Plantlet Production in *Kalanchoë*

*Kalanchoë daigremontiana* is informally known as mother of thousands because it produces many plantlets at the edges of its leaves. Helena Garces, Neelima Sinha, and their colleagues investigated the evolution of this type of asexual reproduction by studying genes that are involved in the development of organs and embryos in four species of *Kalanchoë*. In addition to *K. daigremontiana*, they investigated *K. marmorata*, which does not produce plantlets; *K. pinnata*, which produces plantlets only under certain stressful conditions; and *K. gastonis-bonnieri*, which produces plantlets both normally and when stressed. In leaves of these species, the biologists looked for expression of the gene *STM*, which encodes a key regulator of leaf production at the shoot meristem. They found that *STM* is expressed in cells at the leaf margins of all *Kalanchoë* species that produce leaf plantlets, but not those of the species that do not produce plantlets. Then the investigators checked for the expression in leaves of two genes that are involved in embryo development (*LEC1* and *FUS3*). They discovered that these embryo-linked genes were expressed only in the leaf margins of species that normally form plantlets, not the species in which plantlet formation is induced by stress.

In a survey of a larger number of species, these investigators also discovered that *Kalanchoë* species that normally produce plantlets have an altered LEC1 protein. The normal form of LEC1 protein is essential to the process by which embryos become dry and thus tolerant of arid conditions. LEC1 is therefore necessary for the production of viable seeds, those able to germinate. *Kalanchoë* species that produce plantlets only under stressful conditions and those that do not produce plantlets at all were able to produce viable seeds. By contrast, the seeds of plantlet-producing species were not viable.

Together, these data allowed the investigators to infer that the evolution of plantlet formation began when certain leaf cells of some species gained the ability to function like a shoot meristem, thereby producing structures resembling small shoots. In some of the descendants of these species, normal LEC1 function in seeds was lost, as was the ability to produce viable seeds. Some species adapted by expressing the embryo-development process in leaf margin cells, a process that allowed them to produce plantlets. Such plantlets are not affected by loss of LEC1 function because, unlike seeds, they do not undergo a drying process during development.

## Apomixis Is Seed Production Without Fertilization

**Apomixis** (from the Greek, meaning away from mixing, that is, genetic mixing) is a natural asexual reproductive process in which fruits and seeds are produced within flowers in the absence of fertilization. More than 300 species of flowering plants, including hawkweeds, dandelions, and some types of citrus, are able to reproduce asexually by apomixis. Dandelions and some other apomictic plants require pollination to stimulate seed development, but others do not. Agricultural scientists are interested in apomixis as a potential method for producing genetically uniform seeds, propagating hybrids, and removing the need for fertilization in crop plants.

Most studies of apomixis have been carried out with dandelions, because they are widespread and populations are composed mainly of individuals that reproduce by apomixis. In these plants, meiosis produces microspores, but most pollen grains have abnormal chromosomes. Such grains produce pollen tubes but not sperm cells, so that fertilization does not follow pollination. However, female gametophytes and the eggs they produce are diploid. This condition arises because during the preceding meiotic divisions that generate megaspores,

homologous chromosomes do not pair and meiosis II does not occur. The diploid eggs of apomictic dandelions develop into normal embryos without fertilization and an endosperm develops from the unfertilized central cell. This explains why apomictic dandelions can produce their single-seeded fruits despite the absence of gamete fusion.

## Summary of Key Concepts

### 40.1 An Overview of Flowering Plant Reproduction

- Flowering plants display a sexual life cycle known as alternation of generations. The gamete-producing male and female gametophytes of flowering plants are very small and depend entirely on nurturing sporophytic tissues. Plant gametes arise by the process of mitosis (Figures 40.1, 40.2).

- Flowers are reproductive shoots that develop from a shoot apical meristem. The role of flowers is to promote seed production. A flower shoot generally produces four types of organs: sepals and petals (or tepals in the case of monocots), stamens, and carpels. Stamens produce pollen grains, which are immature male gametophytes. Carpels produce, enclose, and nurture female gametophytes (Figure 40.3).

- Mature male gametophytes, or pollen tubes, each produce two sperm and deliver them to ovules within the ovary. Flowering plants display double fertilization: One of the two sperm released from a pollen tube combines with an egg cell to form a zygote, and the other fuses with two nuclei located in a central cell of the female gametophyte, producing the first cell of endosperm tissue. Endosperm is a nutritive tissue that supports development of an embryonic sporophyte (Figure 40.4).

- Seeds are reproductive structures that contain a dormant embryo enclosed by a protective seed coat that develops from ovule integuments. Fruits are structures that contain seeds and foster seed dispersal. Like flowers and endosperm, fruits are unique features of flowering plants (Figure 40.5).

### 40.2 Flower Production, Structure, and Development

- Plants flower in response to environmental stimuli, such as temperature and day length, by the conversion of a leaf-producing shoot into a flowering shoot.

- Flowers vary in the types of whorls present, the number of flower organs, color, fragrance, organ fusion, symmetry, and arrangement (whether single or in inflorescences). These variations are related to pollination mechanisms and are genetically controlled (Figures 40.6, 40.7, 40.8, 40.9, 40.10).

- Flower blooming involves dramatic changes in the shape of petals or tepals caused by faster growth at their edges than at their centers (Figure 40.11).

### 40.3 Male and Female Gametophytes and Double Fertilization

- Pollen grains are immature male gametophytes protected by a tough wall composed largely of sporopollenin. After pollination, interactions between proteins of pistil cells and those of pollen determine pollen germination (Figures 40.12, 40.13).

- Female gametophyte development occurs within an ovule. Mature female gametophytes include an egg and two synergids, a central cell with two nuclei, and three antipodal cells (Figure 40.14).

- Germinated pollen delivers two sperm to female gametophytes by means of a long pollen tube. The style plays a role in the guidance, nutrition, and fate of the pollen tube. One sperm nucleus fuses with the egg to produce a zygote, the first cell of a new sporophyte generation. The other sperm nucleus fuses with the two nuclei of the central cell, generating the first cell of the nutritive endosperm tissue (Figure 40.15).

## 40.4 Embryo, Seed, Fruit, and Seedling Development

- Unequal division of a zygote leads to development of a nutritive suspensor and an embryo. Young eudicot embryos are heart-shaped as two embryonic leaves (cotyledons) develop. The embryo assumes a torpedo shape as the embryonic root forms. Monocot embryos have only a single cotyledon; the embryonic shoot tip is protected by the coleoptile, and the embryonic root is protected by the coleorhiza (Figures 40.16).

- Mature seeds contain embryos that become dry and are protected by desiccation-resistance proteins and a tough seed coat. These adaptations enable seeds to withstand long periods of dormancy, germinating only when conditions are favorable for seedling survival. Mature fruits develop from ovaries and aid in seed dispersal (Figures 40.17, 40.18).

- Seed germination is influenced by environmental and internal factors. The embryonic root (radicle) is the first organ to emerge, an adaptation that allows rapid water uptake, essential for seedling development (Figures 40.19, 40.20).

## 40.5 Asexual Reproduction in Flowering Plants

- Asexual reproduction is the production of new individuals from a single parent without the occurrence of fertilization. Vegetative reproduction is the development of whole plants from nonreproductive organs. Somatic embryogenesis is the production of embryos from individual body cells. Apomixis is a mechanism by which some plants produce seeds from flowers without fertilization (Figure 40.21).

# Assess & Discuss

## Test Yourself

1. Where do the pollen grains of flowering plants develop?
   a. in the anthers of a flower
   b. in the carpels of a flower
   c. while being dispersed by wind, water, or animals
   d. within ovules
   e. within pistils

2. Where do mature male gametophytes of flowering plants primarily develop?
   a. in the anthers of a flower
   b. in the carpels of a flower
   c. while being dispersed by wind, water, or animals
   d. within ovules
   e. on the surfaces of leaves

3. Where would you find female gametophytes of a flowering plant?
   a. in the anthers of a flower
   b. at the stigma of a pistil
   c. in the style
   d. within ovules in a flower's ovary
   e. in structures that are dispersed by wind, water, or animals

4. How does double fertilization occur in flowering plants?
   a. The two sperm in a pollen tube fertilize the two egg cells present in each female gametophyte.
   b. One of the two sperm in a pollen tube fertilizes the single egg in a female gametophyte, and the other fuses with the two nuclei present in the central cell.

   c. Two sperm, one contributed by each of two different pollen tubes, fertilize the two egg cells in a single female gametophyte.
   d. Two sperm contributed by separate pollen tubes enter a single female gametophyte; one of the sperm fertilizes the egg cell, and the other fertilizes the central cell.
   e. None of the above is correct.

5. A seed is
   a. an embryo produced by the fertilization of an egg, which is protected by a seed coat.
   b. a structure that germinates to form a seedling under the right conditions.
   c. an embryo produced by parthenogenesis that is enclosed by a seed coat.
   d. all of the above.
   e. none of the above.

6. What is the chemical stimulus of flowering that is produced by leaves and transported to the shoot apical meristem?
   a. the hormone auxin
   b. the protein STM
   c. the carbohydrate callose
   d. the mineral ion $K^+$
   e. None of the above is correct.

7. How many whorls of organs occur in complete flowers?
   a. two          d. eight
   b. four         e. ten
   c. six

8. If an ovary contains eight ovules, how many seeds could potentially result if pollen tubes reach all eight ovules?
   a. one          d. more than 20
   b. four         e. None of the above is correct.
   c. eight

9. What function(s) does the carbohydrate callose have in the reproduction of flowering plants?
   a. Callose forms a coat that isolates young embryos during their early development.
   b. Callose forms a coat that isolates groups of four microspores during their early development into pollen grains.
   c. Callose helps to form the patterns on the sporopollenin walls of pollen grains.
   d. All of the above are functions of callose.
   e. None of the above is a function of callose.

10. From what structure does a fruit pericarp primarily develop?
    a. the style          d. a group of fused sepals
    b. a stamen filament  e. the stigma
    c. the ovary wall

## Conceptual Questions

1. Why are pollen grain walls composed largely of sporopollenin?

2. Why are seed coats often tough?

3.  **Core Concept: Evolution** Modern flowering plants likely possessed a single common ancestor. Why, then, do flowers occur in such a diversity of shapes and colors?

## Collaborative Questions

1. Observe orchid flowers or view images of them. Are these flowers bilaterally symmetric or radially symmetric? Are these flowers more likely to be wind-pollinated or pollinated by animals? What gene might be involved in the production of orchid flower shape?

2. How do plants prevent the production of many offspring expressing deleterious recessive traits?

# UNIT VII
# ANIMALS

Despite the amazing diversity of animal life, fundamental similarities link the millions of animal species. We will explore many of these similarities in this unit. The basic features of animal bodies and the maintenance of homeostasis will be introduced in Chapter 41. Chapters 42–44 will discuss major principles of nervous systems. The ability of animals to move through their environments will be covered in Chapter 45. The digestion and absorption of nutrients and the control of metabolism are the subjects of Chapters 46 and 47. Circulatory, respiratory, excretory, and endocrine systems are then covered in Chapters 48–50. The unit continues with Chapters 51–52, which focus on animal reproduction and development and immune systems, and concludes with Chapter 53, which describes the integrated responses of animals' organ systems to a life-threatening challenge to homeostasis.

 **The following Core Concepts and Core Skills will be emphasized in this unit:**

- *Evolution:* *The fundamental principles of evolution are key to understanding virtually all aspects of animal biology. This core concept is covered in several places in all the chapters. For example, we will look at the evolution of nervous systems in Chapter 43 and at an ancient family of proteins that provide immunity in animals in Chapter 52.*

- *Energy and Matter:* *We will see how many of the processes that achieve and maintain homeostasis require energy. For example, it takes energy to maintain blood flow (Chapter 48) and a stable concentration of sodium ions in an animal's body fluids (Chapter 49).*

- *Information:* *How genes play a role in determining key traits in animals will be discussed throughout this unit.*

- *Structure and Function:* *The relationship between structure and function in animals at the levels of cells, tissues, organs, and whole bodies will be evident throughout the unit. This core concept of animal biology is particularly relevant to our discussion of nervous systems (Chapters 42–44) and digestive systems (Chapter 46).*

- *Systems.* *The organ systems of animal bodies interact in many ways to maintain normal functioning of processes such as blood pressure and digestion. In Chapter 53, we will explore an integrated response to a major challenge to homeostasis.*

- *Process of Science:* *Most chapters have a Feature Investigation that describes a classic or recent experiment notable not only for its innovation and creativity, but also for moving the field of animal biology forward in a significant way.*

- *Modeling:* *Each chapter has a Modeling Challenge to help you refine this important skill.*

- *Science and Society:* *Most chapters include a section titled "Impact on Public Health," in which basic science from the chapter is applied to the human condition, with emphasis on how society at large is impacted by disease.*

# Animal Bodies and Homeostasis

# 41

**Perspiring and drinking water are both mechanisms that help achieve homeostasis—a stable internal body environment.**
©John Rowley/Getty Images

L ook at the woman in the chapter opening photo. What is happening in her body at that moment? Obviously, it is a hot day, and the woman is perspiring and thirsty; it appears that she has just been exercising outdoors. During such exertion, her body must prevent an excessive buildup of heat, because high body temperature can alter the activities of enzymes and damage cellular membranes. One way some animals, including humans, can eliminate heat from the body is to perspire. Recall from Chapter 2 that water has a high heat of vaporization, and therefore the evaporation of water from the body surface helps eliminate body heat. This benefit comes with a cost, however, because perspiration depletes some of the body's water, potentially leading to changes in ion concentrations in body fluids, blood pressure, and other critical physiological variables. Consequently, structures in the brain and other organs that are sensitive to the body's fluid level and ion concentrations trigger the sensation of thirst, and the woman drinks water to replenish what was lost through perspiration. The process through which the different aspects of an animal's internal environment (in this case, temperature and body fluids) are maintained within normal limits, even in the face of changing circumstances or external challenges, is known as **homeostasis**.

Maintenance of homeostasis is one of the fundamental principles of virtually all aspects of animal biology. Before we can fully appreciate what homeostasis means to animals and how it is achieved, however, we first need to understand some basic features of animal bodies. We begin this chapter with a discussion of the organization of cells into tissues and tissues into organs. Chapter 10 considered the organization of cells into tissues from the perspective of cell biology. Here, we look at tissues and organs from the perspective of the whole animal, examining how the properties of life arise from the complex interactions of its components. Next, we will discuss a core concept of biology that also helps us understand homeostasis, the concept that structure (form) determines function. We then link these principles together with a detailed look at what homeostasis means for different animals, how it may be challenged, and how it is restored or maintained. The final section of the chapter considers the important topic of homeostatic control of internal fluids.

## 41.1 Organization of Animal Bodies

**Learning Outcomes:**

1. List the different categories of animal tissues, providing general functions and specific examples of each.
2. Name the various organ systems found in many animals, list the components of each, and describe their general functions.

All animals share similarities in the ways in which they exchange materials with their surroundings, obtain energy from organic nutrients, synthesize complex molecules, reproduce themselves, and detect and respond to signals in their environment. Animals typically begin life as a single cell—most commonly a fertilized egg—which divides to create two cells, each of which divides in turn, resulting in four cells, and so on. If cell division were the only event

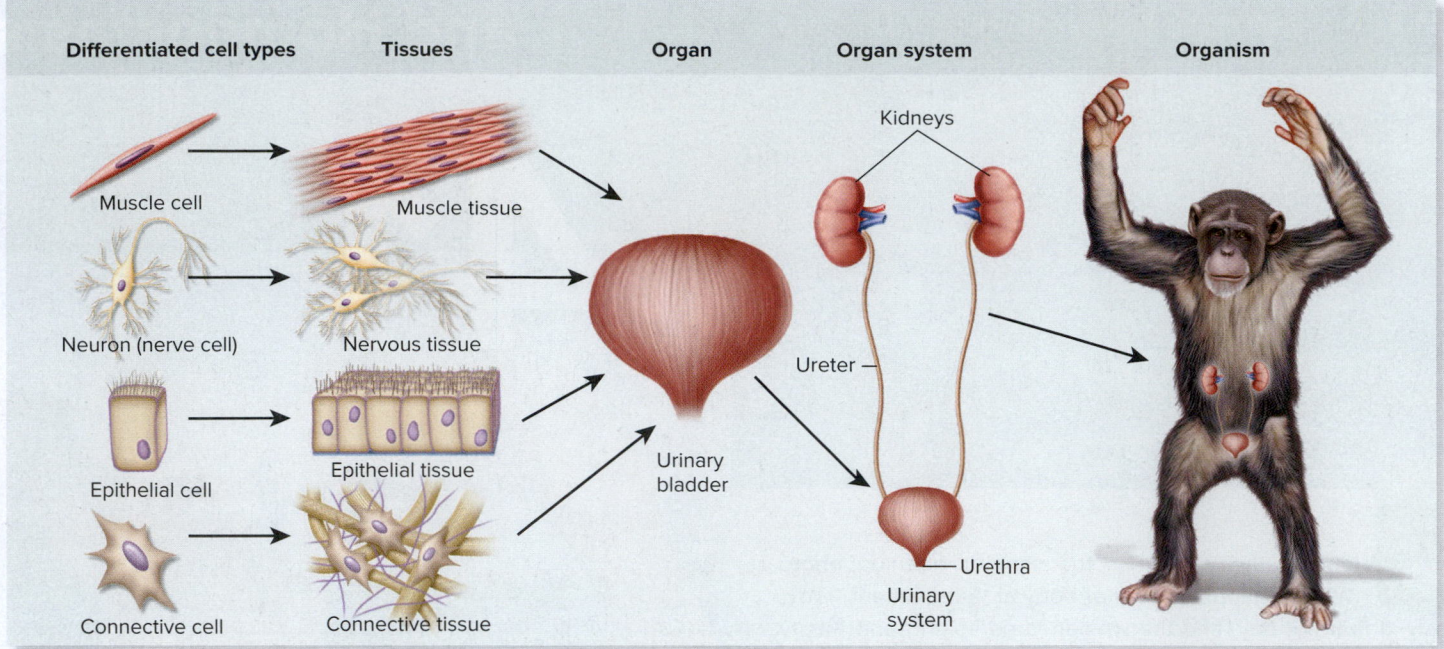

| Differentiated cell types | Tissues | Organ | Organ system | Organism |

Muscle cell

Muscle tissue

Neuron (nerve cell)

Nervous tissue

Epithelial cell

Epithelial tissue

Connective cell

Connective tissue

Urinary bladder

Kidneys

Ureter

Urethra

Urinary system

**Figure 41.1  The internal organization of cells, tissues, organs, and an organ system in a mammal.** The urinary system of mammals is part of the larger excretory system that includes all the structures that function to remove soluble wastes from animals' bodies.

 **Core Concept: Systems** New properties of life emerge from complex interactions. By themselves, none of the four tissues that constitute a bladder or a kidney could perform the functions of those organs, but when combined in precise ways, the result is a functional organ system capable of removing soluble wastes from the fluids of an animal's body.

occurring, the end result would be a spherical mass of identical cells. As we will see in Chapter 51, however, cells become specialized during development to perform a particular function (that is, they differentiate). Examples of differentiated cells are muscle and blood cells. As **Figure 41.1** shows for the urinary system of mammals, the cells of an animal's body are organized into progressively more complex structures, including tissues, organs, and organ systems. In this section, we will explore this organization in greater detail.

## Specialized Cells Are Organized into Tissues

A **tissue** is an association of many cells that have a similar structure and function. The tissues in a typical animal's body can be classified into four types, according to their locations and the functions they perform: muscle, nervous, epithelial, and connective tissues. Within each of these functional categories, subtypes of tissues perform variations of the given function, as illustrated by the three types of muscle tissue.

***Muscle Tissues*** **Muscle tissues** consist of cells specialized to shorten, or contract, generating the mechanical forces that may produce body movement, decrease the diameter of a tube, or exert pressure on a fluid-filled cavity. Three types of muscle tissue are found in animals: skeletal, smooth, and cardiac (**Figure 41.2**).

**Skeletal muscles** are generally attached to the skeleton of an animal. When skeletal muscles are stimulated by signals from the nervous system, they generate force that leads to the contraction of the muscle (see Chapter 45). Contraction of these muscles may be under voluntary control and produce the movements required for locomotion, such as extending limbs or flapping wings.

**Smooth muscles** surround hollow tubes and cavities inside the body's organs, so their contraction can move the contents of those organs. For example, the contraction of smooth muscle in the stomach wall propels partially digested food into the intestines, where it can be digested. Smooth muscle also surrounds and forms part of small blood vessels and airway tubes (bronchioles). Contraction in those regions reduces blood or air flow, respectively. In the circulatory system, this contraction helps direct blood to regions of the body that most require it at any given time. In the airways of the respiratory system, it helps direct air to the healthiest parts of a lung. Contraction of all smooth muscle is involuntary—that is, it occurs without conscious control.

In the third type of muscle tissue, **cardiac muscle**, physical and electrical connections between individual cells enable many cells to contract almost simultaneously. Like smooth muscle, cardiac muscle cannot be contracted voluntarily. It is found only in the heart, where it provides the force that generates sufficient pressure to pump blood through an animal's body.

***Nervous Tissues*** **Nervous tissues** are composed of a complex network of cells called **neurons** that are specialized to receive, generate, and conduct electrical signals from one part of an animal's body to another part (**Figure 41.3**). Depending on where it is generated in an animal's body, an electrical signal produced in one neuron may stimulate or inhibit other neurons to initiate new electrical signals,

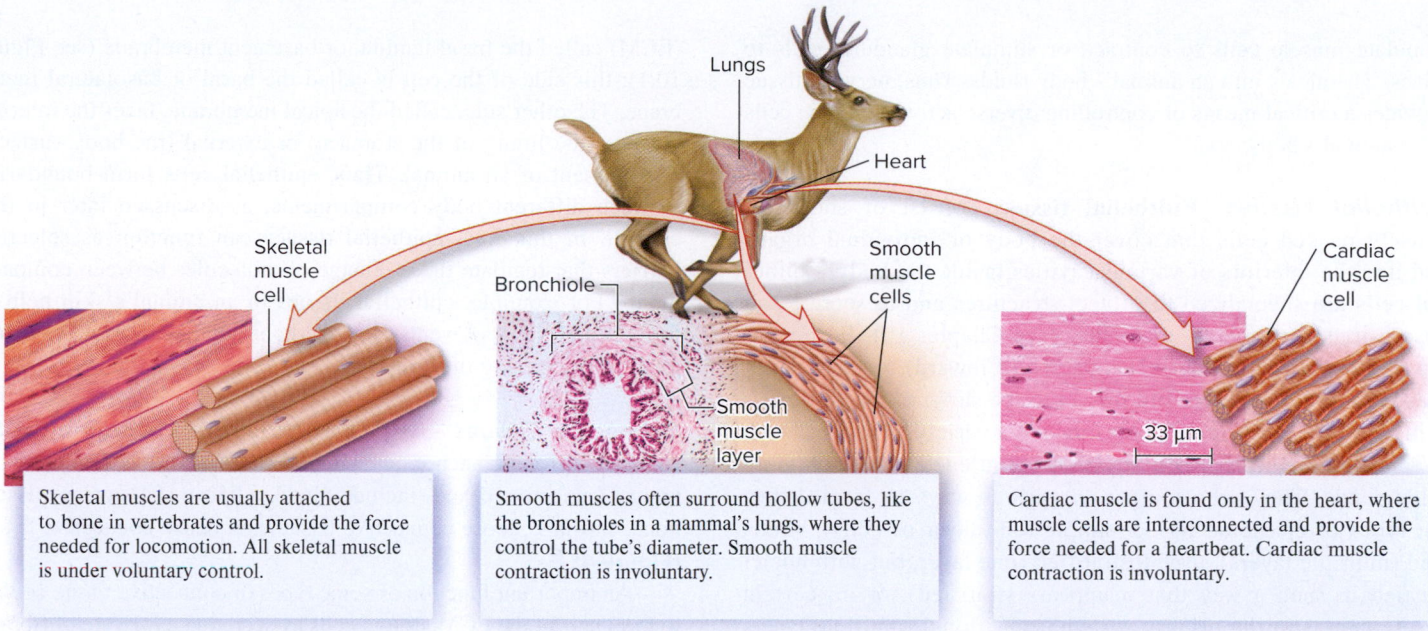

Skeletal muscles are usually attached to bone in vertebrates and provide the force needed for locomotion. All skeletal muscle is under voluntary control.

Smooth muscles often surround hollow tubes, like the bronchioles in a mammal's lungs, where they control the tube's diameter. Smooth muscle contraction is involuntary.

Cardiac muscle is found only in the heart, where muscle cells are interconnected and provide the force needed for a heartbeat. Cardiac muscle contraction is involuntary.

**Figure 41.2** **Three types of muscle tissue: skeletal, smooth, and cardiac.** All three types produce force, but they differ in their appearance and in their locations within animals' bodies. (left): ©Michael Abbey/Science Source; (middle): ©Sinclair Stammers/Science Source; (right): ©Innerspace Imaging/Science Source

 **Core Skill: Modeling** The goal of this modeling challenge is to draw a model showing how smooth muscle contraction affects the width of a tube such as a bronchiole.

**Modeling Challenge:** Smooth muscle often surrounds tubelike structures in animal bodies, where it controls the degree to which the tubes are opened. In the disease asthma, smooth muscle tissue in the small air tubes called bronchioles is sometimes in a highly contracted state. Shown below is a model of a fully opened bronchiole in a human. Draw another model of the bronchiole as you think it will appear when the person is experiencing asthmatic symptoms (contracted smooth muscle tissue). Referring to your model, explain why individuals with asthma have difficulty breathing when their symptoms flare up.

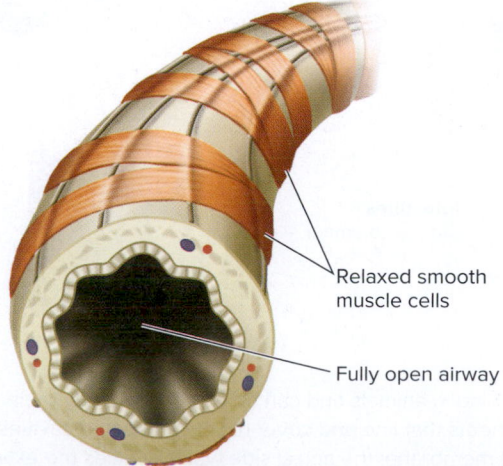

**Open bronchiole**

A healthy bronchiole in a human lung. Note that the smooth muscle tissue is in a relaxed state, and the bronchiole is fully opened.

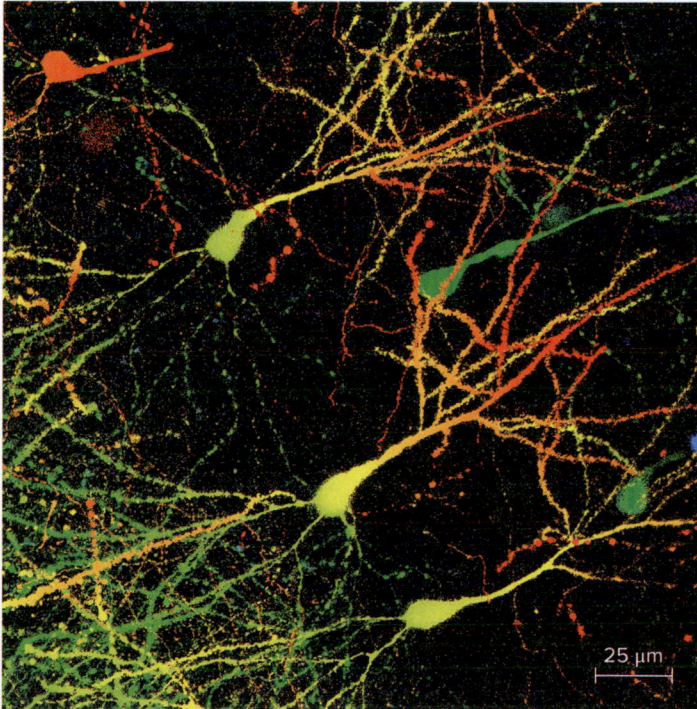

**Figure 41.3** **Nervous tissue in the brain of a vertebrate.** Nervous tissue consists of neurons with extensive cell-cell contacts, as shown in this confocal micrograph of a section from a human brain. The cells are labeled with fluorescent markers; different colors signify different depths in the section. ©Dr. Gopal Murti/SPL/Science Source

 **Core Concept: Systems** Animals interact with their environments in many ways. These interactions largely depend on the functioning of nervous tissue, which allows animals to sense and respond to changes in the environment.

stimulate muscle cells to contract, or stimulate glandular cells to release chemicals into an animal's body fluids. Thus, nervous tissue provides a critical means of controlling diverse activities of the cells in an animal's body.

**Epithelial Tissues** **Epithelial tissues** consist of sheets of densely packed cells that cover the body or individual organs and line the interiors of various cavities inside the body. Epithelial cells are specialized to protect structures and to secrete and absorb ions and organic molecules (see Chapter 10). For example, epithelial tissue can invaginate (fold inward) to form sweat glands that secrete water and ions onto the surface of an animal's skin. Epithelial cells in animals come in a variety of shapes, such as cuboidal (cube-shaped), squamous (flattened), and columnar (elongated). They are arranged in various ways to form different types of epithelial tissue: simple (one layer of cells), stratified (multiple layers), pseudostratified (one layer, but with nuclei located in such a way that it appears stratified), or, in certain cases such as in the urinary system, transitional (multiple layers with the ability to expand and contract) (**Figure 41.4**).

Regardless of their shape, organization into tissues, or location, all epithelial cells are asymmetric, or polarized. This means that one side of such a cell is anchored to or faces an extracellular matrix (ECM) called the basal lamina, or basement membrane (see Figure 10.1); this side of the cell is called the basal or basolateral membrane. The other side, called the apical membrane, faces the internal (such as the lining of the stomach) or external (the body surface) environment of an animal. Thus, epithelial cells form boundaries between different body compartments, as discussed later in this chapter. In this way, epithelial tissues can function as selective barriers that regulate the exchange of molecules between compartments. For example, epithelial tissues in an animal's skin help to form a barrier that prevents most substances in the external environment from entering the body.

**Connective Tissues** As their name implies, **connective tissues** connect, surround, anchor, and support the structures of an animal's body. Connective tissues include blood, adipose (fat-storing) tissue, bone, cartilage, loose connective tissue, and dense connective tissue (**Figure 41.5**).

An important function of some types of connective tissue cells is to form part of the ECM around cells by secreting a mixture of fibrous proteins and carbohydrates, such as glycosaminoglycans. These carbohydrates may covalently attach to proteins to form proteoglycans (refer back to Figure 10.4). In some cases, the ECM is rich in minerals. The final characteristics of any type of connective tissue are determined

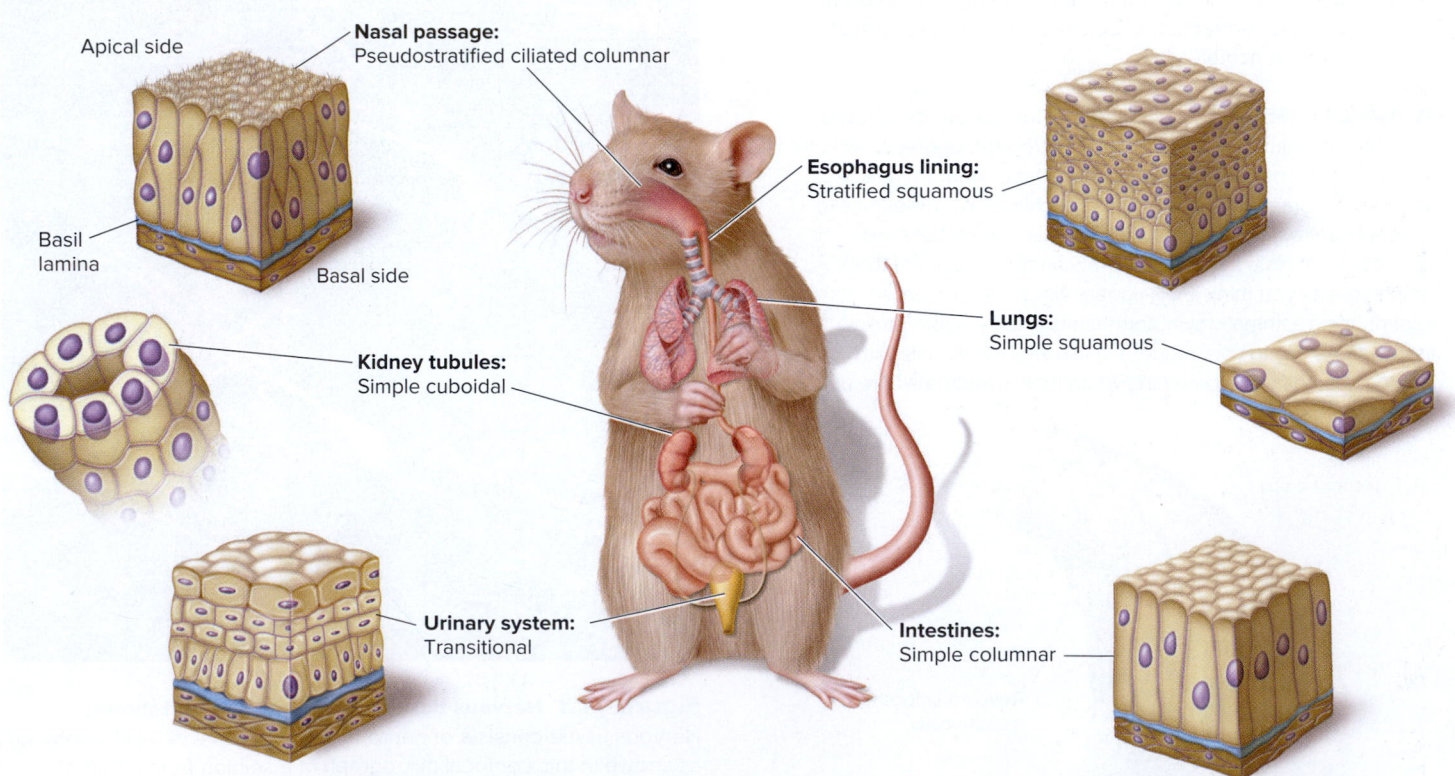

**Figure 41.4 Examples of epithelial tissue.** Several types of epithelial tissue are found in animals and can be distinguished by their appearance. Epithelial tissue is used to construct body coverings and the protective sheets that line and cover hollow tubes and cavities. The epithelial cells that make up epithelial tissues have an apical and basal (or basolateral) membrane; the apical side typically faces the exterior of the body or the lumen of a structure such as the intestine.

 **Core Concept: Structure and Function** Note how different types of epithelial cells arranged in different ways form tissues with different functions. For example, simple cuboidal cells arranged as tubules in the kidney permit the passage of filtered body fluids, and pseudostratified ciliated columnar cells lining the nasal passage act as filters of airborne particles and debris.

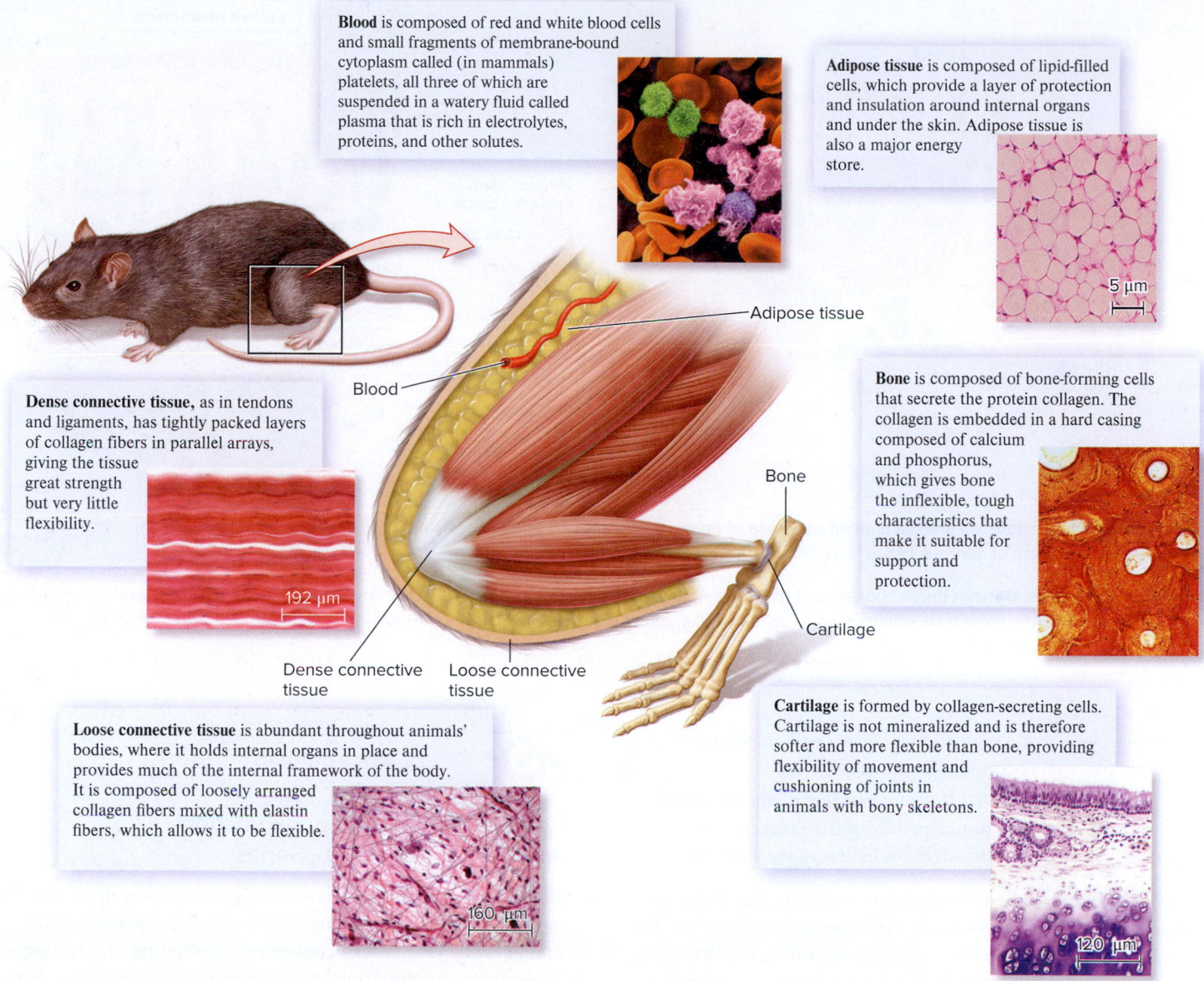

**Blood** is composed of red and white blood cells and small fragments of membrane-bound cytoplasm called (in mammals) platelets, all three of which are suspended in a watery fluid called plasma that is rich in electrolytes, proteins, and other solutes.

**Adipose tissue** is composed of lipid-filled cells, which provide a layer of protection and insulation around internal organs and under the skin. Adipose tissue is also a major energy store.

5 μm

Adipose tissue

Blood

**Dense connective tissue,** as in tendons and ligaments, has tightly packed layers of collagen fibers in parallel arrays, giving the tissue great strength but very little flexibility.

192 μm

Bone

**Bone** is composed of bone-forming cells that secrete the protein collagen. The collagen is embedded in a hard casing composed of calcium and phosphorus, which gives bone the inflexible, tough characteristics that make it suitable for support and protection.

Cartilage

Dense connective tissue

Loose connective tissue

**Cartilage** is formed by collagen-secreting cells. Cartilage is not mineralized and is therefore softer and more flexible than bone, providing flexibility of movement and cushioning of joints in animals with bony skeletons.

**Loose connective tissue** is abundant throughout animals' bodies, where it holds internal organs in place and provides much of the internal framework of the body. It is composed of loosely arranged collagen fibers mixed with elastin fibers, which allows it to be flexible.

160 μm

120 μm

**Figure 41.5 Examples of connective tissue in mammals.** Connective tissues connect, surround, anchor, and support other tissues and may exist as a suspension of cells (blood), clumps of cells (fat), or tough, rigid material (bone and cartilage). The samples have been stained or the micrographs have been colorized to reveal connective tissue. (blood): ©Dennis Kunkel Microscopy, Inc./Phototake; (adipose tissue): ©Ed Reschke/Getty Images; (bone): ©Innerspace Imaging/Science Source; (cartilage): ©Victor P. Eroschenko; (loose connective tissue, dense connective tissue): ©McGraw-Hill Education/Al Telser, photographer

in part by the relative proportions and types of proteins, proteoglycans, and minerals secreted into the ECM. The ECM serves several general functions, which include (1) providing a scaffold to which cells attach and organize themselves into more complex structures, (2) protecting and cushioning parts of the body, (3) providing mechanical strength, and (4) cell signaling—transmitting information to the cells that helps regulate their activity, migration, growth, and differentiation.

The proteins of the ECM consist mainly of two types. The first type is insoluble fiber-like proteins such as collagen and the rubber-band-like protein elastin; these proteins are often referred to as fibers. A second category is adhesive proteins (fibronectin and laminin) that serve to organize the protein and carbohydrate components of the ECM (refer back to Table 10.1).

## Different Tissue Types Combine to Form Organs and Organ Systems

An **organ** is composed of two or more kinds of tissues arranged in various proportions and patterns, such as sheets, tubes, layers, bundles, or strips. For example, the vertebrate stomach (**Figure 41.6**) consists of the following layers:

- an outer covering of simple squamous epithelial tissue;
- connective tissue layers covering and cementing the organ together;
- layers of smooth muscle tissue, the contractions of which mechanically break up food and propel it through the stomach and into the small intestine;

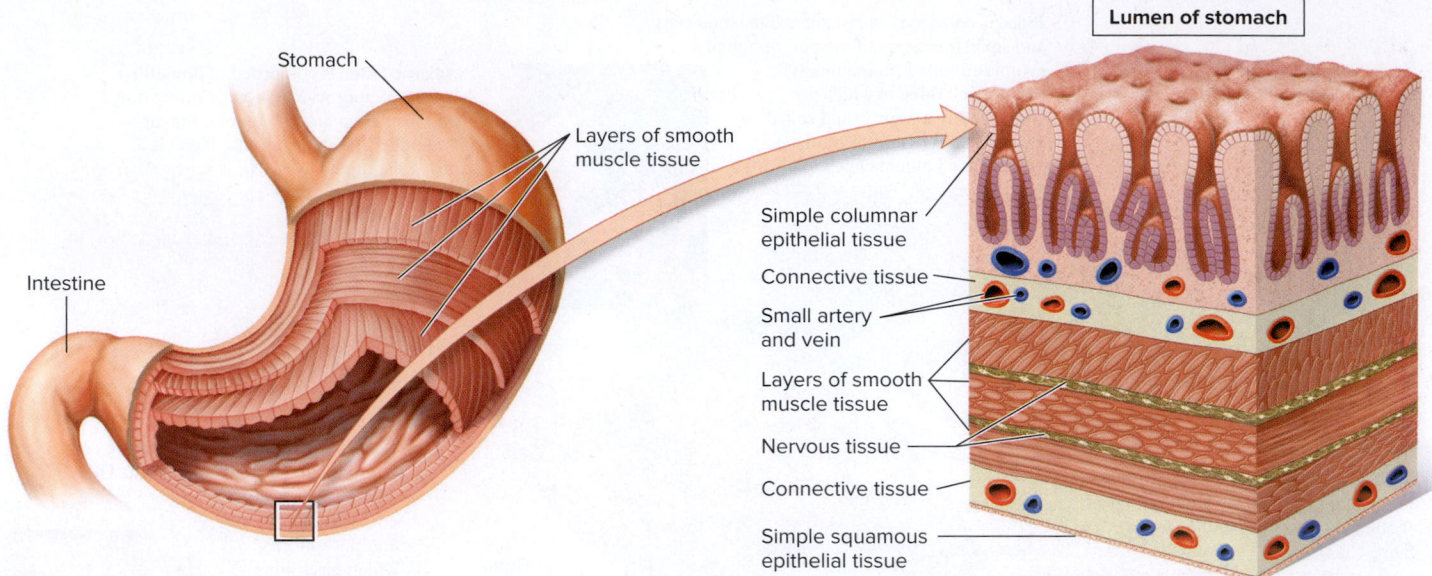

**Figure 41.6** **The vertebrate stomach as an example of an organ composed of all four tissue types.** In this illustration, the thickness and appearance of the layers of nervous tissue have been considerably exaggerated for visual clarity.

 **Core Skill: Connections** Look ahead to Figure 46.6 and compare the structure of the small intestine shown there with the structure of the stomach shown in Figure 41.6. What can you conclude about common functions of the stomach and intestine from this comparison?

- nervous tissue that comes in close contact with the smooth muscle tissue and helps regulate its activity; and

- an inner lining of simple columnar epithelial tissue that secretes enzymes and acid (important in the digestive process) and protective mucus into the cavity, or lumen, of the stomach.

In an **organ system**, different organs work together to perform an overall function or functions. In the example just described, the stomach is part of the digestive system, along with other structures, such as the mouth, esophagus, small and large intestines, and anus. In another familiar example, the vertebrate circulatory system includes the heart, blood vessels and blood. The organ systems found in animals are described in **Table 41.1**.

Organ systems should not be considered as functioning in isolation. Instead, they frequently influence each other and depend on each other in many ways. For example, signals from the nervous, circulatory, and endocrine systems strongly influence how much water the mammalian kidney retains as it forms urine, an adaptation that can be lifesaving under certain circumstances.

The spatial arrangement of organs into organ systems is part of the overall body plan of animals. Organ systems develop at specific times and locations within the body and, in bilateral animals, along the anteroposterior body axis, as do other structures, such as limbs, tentacles, antennae, and other animal appendages. Scientists have long questioned how the layout of animal bodies is determined during the period when an embryo is developing. Remarkably, organ development in most animals appears to be under the control of a highly conserved family of body-plan genes with homologs in most animals, as described next.

 **Core Concept: Information**

## Organ Development and Function Are Controlled by *Hox* Genes

In previous chapters, you have learned about a family of genes called *Hox* genes that are found in most animal phyla. In bilateral animals, *Hox* genes determine the formation of structures along the anteroposterior body axis during development. For example, we saw in Chapter 20 how these genes determine the number and position of legs and wings in *Drosophila*. *Hox* genes play a similar role in determining the spatial patterning of the vertebrate body and appendages. Recently, scientists have begun exploring the role of *Hox* genes in the development and spatial patterning of the organs that make up animals' organ systems.

By generating mutant mice that fail to express one or more *Hox* genes, researchers have discovered that these genes have the important function of determining where within the vertebrate body particular organs form. Recall from Chapter 20 (refer back to Figures 20.16 and 20.17) that mouse *Hox* genes are arranged in four clusters, designated A–D, with multiple genes per cluster. Homologous genes (for example, *HoxA-3, HoxB-3,* and *HoxD-3*) that are found within a single species are called paralogs. Such paralogs typically act in concert to regulate similar developmental processes. For example, *HoxA-3* is important for development of anterior parts of the body, including

| Table 41.1 | Organ Systems Found in Animals | |
|---|---|---|
| Organ system | Major components* | Major functions |
| Circulatory | Contractile element (heart or vessel); distribution network (blood vessels); blood or hemolymph | Distributes solutes (nutrients, gases, wastes, and so on) to all parts of an animal's body |
| Digestive | Ingestion structures (mouth, mouthparts); storage structures (crop, stomach); digestive and absorptive structures (stomach, intestines); elimination structures (rectum, anus); accessory structures (pancreas, gallbladder) | Breaks complex foods into absorbable units; absorbs organic nutrients, ions, and water; eliminates solid wastes |
| Endocrine | All cells, tissues, organs, or glands that secrete hormones | Regulates and coordinates growth, development, metabolism, mineral balance, water balance, blood pressure, behavior, and reproduction |
| Excretory | All organs including respiratory structures (e.g., gills and lungs) that are involved in removing soluble wastes from the body; the vertebrate urinary system is a part of the excretory system and includes the kidneys, ureters, bladder and urethra | Eliminates soluble metabolic wastes; regulates body fluid volume and solute concentrations |
| Immune and lymphatic | Circulating white blood cells (leukocytes); lymph organs, lymph vessels and nodes | Defends against pathogens |
| Integumentary | Body surfaces (skin) | Protects from dehydration and injury; defends against pathogens; in some animals, plays a role in regulation of body temperature |
| Muscular-skeletal | Force-producing structures (muscles); support structures (bones, cartilage, exoskeleton); connective structures (tendons, ligaments) | Produces locomotion; generates force; propels materials through body organs; supports body |
| Nervous | Processing (brain); sensory structures; signal delivery (spinal cord, peripheral nerves and ganglia, sense organs) | Regulates and coordinates movement, sensation, organ functions, and learning |
| Reproductive | Gonads and associated structures | Produces gametes (sperm and egg); in some animals, provides nutritive environment for embryo and fetus |
| Respiratory | Gas-exchange sites (gills, skin, trachea, lungs) | Exchanges oxygen and carbon dioxide with the environment; regulates blood pH |

*Selected examples only; these do not necessarily pertain to all animals.

the neck. When this gene is knocked out, mouse embryos show defects in neck structure. Notably, the organs within the neck—including the thymus, thyroid, and parathyroid glands—do not develop normally. When two or more paralogs of this group are knocked out, certain neck organs fail to form at all. Experimental deletion of *Hox* genes associated with other body segments does not affect the development and function of neck glands and organs. Likewise, investigators have uncovered vital roles of different *Hox* genes in lung development within the thorax and in the proper positioning and development of the vertebrate kidneys in the abdomen.

Of particular interest is the discovery that *Hox* genes are important not only for spatial patterning of organs but also for their growth, development, and function. Paralogs of *Hox* 1 and 3, for instance, help determine the final branching patterns of the airways of the lungs, the final size of the lungs, and the ability of the lungs to produce secretions that are important for breathing air after birth. Other *Hox* genes have been shown to control cell proliferation, shape changes, apoptosis, cell migration, and cell-cell adhesion within various organs. Similar results have been found in invertebrates, such as the leech, where *Hox* genes are first expressed during organ formation, and in *Drosophila*, where the final shape and size of the heart are partly controlled by *Hox* genes.

## 41.2 The Relationship Between Structure and Function

### Learning Outcomes:

1. Provide an example of how the structure of an animal's tissues or organs can help predict the function.

2. **CoreSKILL »** Describe the quantitative relationship between the surface area of an object and its volume, and explain the importance of this relationship to animal form and function.

A key principle of biology emphasized throughout this unit is that form (structure) determines function. The appearance or structure of an animal's tissues and organs can often help us predict their functions. For example, let's compare the respiratory systems of an insect and a mammal (**Figure 41.7**). The respiratory systems of animals exchange oxygen from the environment with carbon dioxide generated by the body. Although many important differences exist between the respiratory systems of insects and mammals, notably the presence of lungs in mammals, certain structural similarities suggest that both systems serve similar functions. In both cases, for example, a series of internal branching tubes composed of epithelial and connective tissues arises from one or more openings that connect with the outside environment (the mouth and nose in the mammal, and the body surface pores called spiracles in the insect). These tubes become smaller and smaller as they continue to branch, eventually terminating in narrow structures that are only one cell thick.

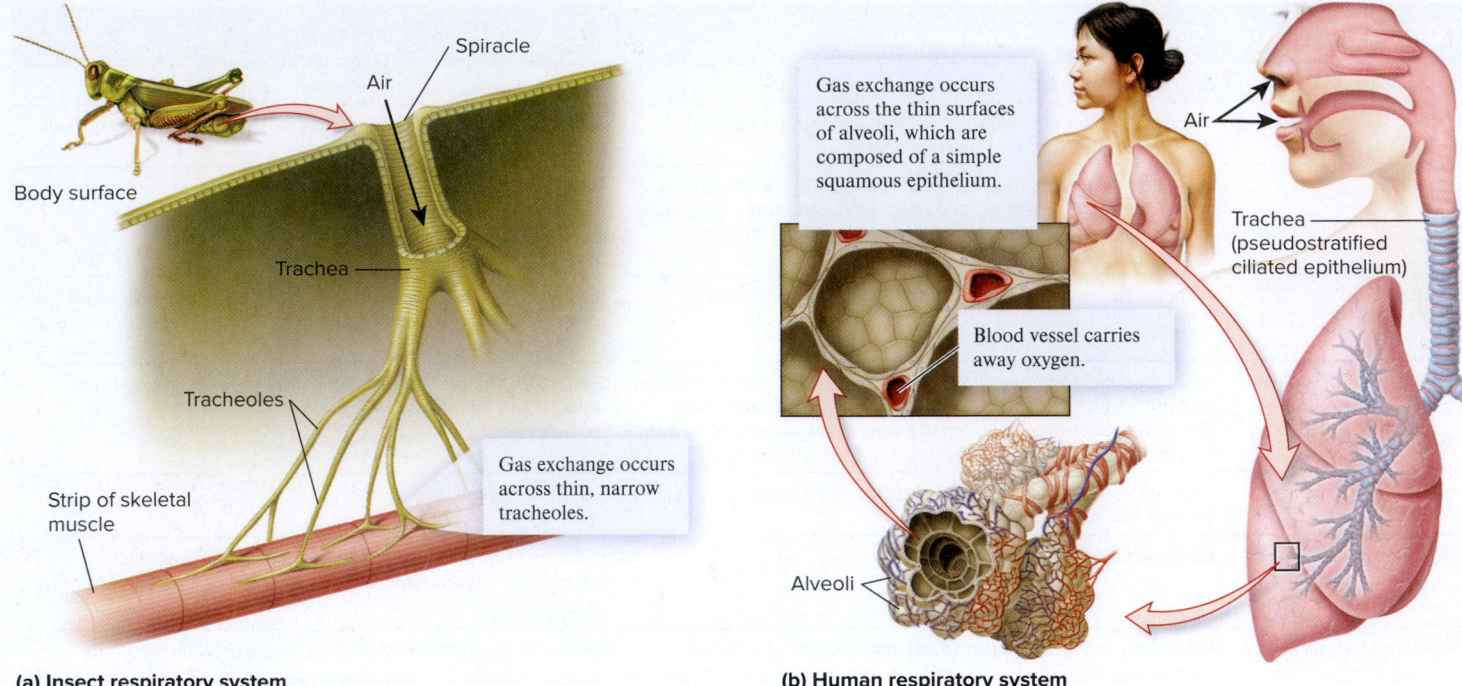

**(a) Insect respiratory system**

Spiracle

Air

Body surface

Trachea

Tracheoles

Strip of skeletal muscle

Gas exchange occurs across thin, narrow tracheoles.

**(b) Human respiratory system**

Gas exchange occurs across the thin surfaces of alveoli, which are composed of a simple squamous epithelium.

Air

Trachea (pseudostratified ciliated epithelium)

Blood vessel carries away oxygen.

Alveoli

**Figure 41.7** **Comparison of the branching air tubes in (a) an insect and (b) a mammal.** Note the similar features of highly branching, internalized hollow tubules that connect to the outside air, suggesting that these systems perform similar functions.

In both cases, these branching tubes serve as conduits for air to flow back and forth between the environment and the internal spaces of the animal. In the insect, the ends of the branching tubes (called tracheoles) are where oxygen diffuses from the air to the fluid around individual cells (and from there to intracellular fluid) (Figure 41.7a). In the mammal, the ends of the tubes form saclike structures called alveoli across which oxygen diffuses into the blood (Figure 41.7b).

If we examine the mammalian lung in greater detail, we see that the alveoli are composed of extremely thin, squamous epithelial cells. The shape of the cells provides a clue to their function. Their flat, thin structure permits rapid diffusion of gases across them. Imagine the resistance to oxygen diffusion if the cells were thick or scaly, like the cells of the body surface of many animals, for example. Therefore, both the gross and microscopic anatomy of the gas-exchange surfaces of respiratory systems facilitates their functions.

An additional structural similarity found in essentially all respiratory surfaces, including gills, is an extensive surface area. This structural similarity applies to all cells, tissues, and organs that mediate diffusion or absorption of a solute from one compartment to another or that require extensive cell-to-cell contacts. Consider, for instance, the finger-like projections of the small intestine of a human, the skin folds of some high-altitude frogs, the cellular extensions of neurons of a mouse, and the feathery antennae of a moth (**Figure 41.8**). What do these structures have in common? They all have a large surface area, which maximizes their ability to absorb nutrients (intestine), obtain oxygen by diffusion from the environment (frog skin), communicate with other cells (neurons), or detect airborne molecules (moth antennae).

The relationship between a structure's surface area and its volume is called the **surface area/volume (SA/V) ratio** (see Figure 4.12). A high SA/V ratio is ideal for exchange of heat, solutes, gases, and water across a surface without contributing greatly to the mass or volume of a body part. This concept will apply throughout this unit as we explore the ways in which animals obtain energy, regulate their metabolism and body temperature, obtain oxygen, and eliminate wastes.

A large increase in surface area of a structure, however, comes at the expense of greatly increasing volume if the shape of the structure is not changed (refer back to Figure 4.12). As a spherical object enlarges, its volume grows relatively more than its surface area, because its surface area increases by a power of 2, whereas its volume increases by a power of 3. For example, if the radius of a sphere is increased by a factor of 10, its surface area increases 100 times, but its volume increases 1,000 times.

The relationship between surface area and volume also applies to an animal's organs and could create certain disadvantages. For example, the ability to obtain sufficient oxygen from water requires a great amount of surface area on the structures that make up a fish's gills. If such structures were spherical, the gills would need to be extremely large to accomplish the needed gas exchange. The challenge of packaging an extensive surface area into a confined space is overcome by variations in shape. The gills of fish are comprised of many flattened disc- or platelike structures. Water flows over both sides of each disc (look ahead to Figure 48.17a). The flattened discs greatly increase surface area and thereby facilitate the function of gas exchange. At the same time, the flattened discs minimize the volume of the gill. Similarly, the inner surface of the human intestine in Figure 41.8a folds inward to form finger-like extensions, thereby increasing the surface area without greatly increasing the volume.

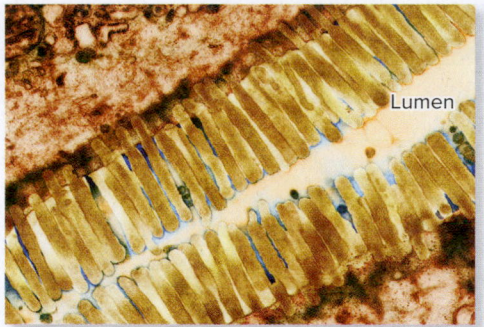

**(a) Human intestine**

**(b) Frog skin**

**Figure 41.8** **Examples of structures in which extensive surface area is important for function.** A large surface area allows **(a)** high rates of transport of nutrients across the intestine of a human, **(b)** increased diffusion of oxygen across the folds of skin of a frog living at high altitude, where O$_2$ is less available, **(c)** extensive communication between neurons in a mouse's brain, and **(d)** detection of airborne chemicals by moth antennae. a: ©Biophoto Associates/Science Source; b: ©Dante Fenolio/Science Source; c: ©Thomas Deerinck, NCMIR/Science Source; d: ©Anthony Bannister/Science Source

*Concept Check:* *Is an extensive surface area important only for animals, or could it also provide advantages to other living organisms?*

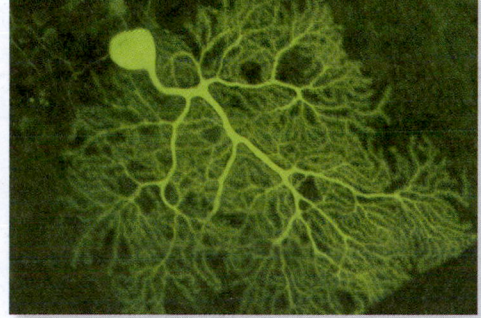

**(c) Mouse brain neuron stained with a fluorescent marker**

**(d) Moth antennae**

# 41.3 General Principles of Homeostasis

## Learning Outcomes:

1. Explain how the concept of homeostasis applies to the internal environment of animals.
2. List several variables that are regulated within a homeostatic range in vertebrate animals.
3. Name the four components of a homeostatic control system, and describe the importance of each to the regulation of an animal's internal environment.
4. **CoreSKILL »** Contrast negative feedback, positive feedback, and feedforward regulation, and explain how they do or do not contribute to the maintenance of homeostasis in animals.
5. Explain the importance of paracrine and hormonal signaling to homeostasis, and provide examples of each.

The environmental conditions in which animals live are rarely, if ever, constant. Animals are exposed to fluctuations in air and water temperatures, nutrient and water supplies, pH, and, in some cases, oxygen availability. Any one of these environmental changes could be harmful or even fatal if an animal is unable to respond appropriately. However, as you might expect from the incredible diversity of environments in which they exist, animals can adjust in many ways to their surroundings and thrive.

The process of maintaining a relatively stable internal environment despite changes in the external surroundings is known as homeostasis (from the Greek *homoios*, meaning similar, and *stasis*, meaning to stand still). The term was coined in the 20th century by

American physiologist Walter Cannon, but the concept itself originated in the 19th century with French physiologist Claude Bernard, who postulated that a constant *milieu interieur* (internal environment) was a prerequisite for good health.

## Vertebrates Maintain Most Physiological Variables Within a Narrow Range

In vertebrates, the common physiological variables—concentrations of blood-borne solutes such as minerals, glucose, and oxygen, for example—are usually maintained within a certain range despite fluctuating external environmental conditions (**Table 41.2**). At first glance, homeostasis may appear to be a state of stable balance of physiological variables. However, this simple description cannot capture the scope of homeostasis. For example, no physiological function is constant for very long, which is why we call them variables. Some variables may fluctuate around an average value during the course of a single day yet still be considered in balance. Homeostasis is a dynamic process, not a static one.

Consider an example of a physiological variable in your own body. Normally, blood sugar (glucose) remains at fairly steady and predictable concentrations in any healthy individual. After a meal, however, the concentration of glucose in your blood can increase quickly, especially if you have just eaten something sweet. Conversely, if you skip a few meals, your blood glucose concentration may decrease slightly (**Figure 41.9**). Such fluctuations above and below the normal value might suggest that blood glucose concentration is not homeostatically controlled, but this is incorrect. Once blood glucose increases or decreases, homeostatic mechanisms restore the concentration back toward normal. In the case of glucose, the nervous and endocrine systems are primarily responsible for this quick

| Table 41.2 | Selected Examples of Homeostatic Variables in Animals | |
|---|---|---|
| **Variable** | **Factors that influence homeostasis** | **Examples of functions** |
| ***Minerals*** | Eating food; excreting wastes | |
| $Na^+$ and $K^+$ | | Establish resting membrane potentials across plasma membranes in all cells and transmit electrical signals in excitable tissues (muscles and nervous tissue) |
| $Ca^{2+}$ | | Important for muscle contraction; neuron function; skeleton and shell formation |
| $Fe^{2+}$ | | Binds and transports oxygen in blood or body fluids (some invertebrates use copper instead of iron) |
| ***Energy sources*** | Eating food; expending energy | |
| Glucose | | Broken down to provide energy for use by all cells, especially brain cells |
| Fat | | Provides an alternate source of energy, particularly for cells not in the nervous system; major component of plasma membranes |
| ATP | | Provides energy to drive most chemical reactions and body functions; modifies function of many proteins by transferring a phosphate group to proteins |
| ***Body temperature*** | Rate of energy expenditure; environmental temperature; behavioral mechanisms (look ahead to Chapter 47) | Determines the rate of chemical reactions in an animal's body |
| ***pH of body fluids*** | Hydrogen ion transporters in cells; buffers in body fluids; rates of energy expenditure; breathing rate | Affects enzymatic activity in all cells |
| ***Other variables*** | | |
| Oxygen and carbon dioxide | Movement of air or water across respiratory surfaces (for example, lungs and gills); metabolic rate | Oxygen circulates in body fluids and enters cells, where it is used during the production of ATP; carbon dioxide is a waste product that is eliminated to the environment, but it is also a key factor that regulates the rate of breathing. |
| Water | Drinking, eating, excretion of wastes, perspiration, osmosis across body surface (skin or gills) | Numerous biological functions including participating in chemical reactions; helping to regulate body temperature; acting as a solvent for biologically important molecules (refer back to Chapter 2) |

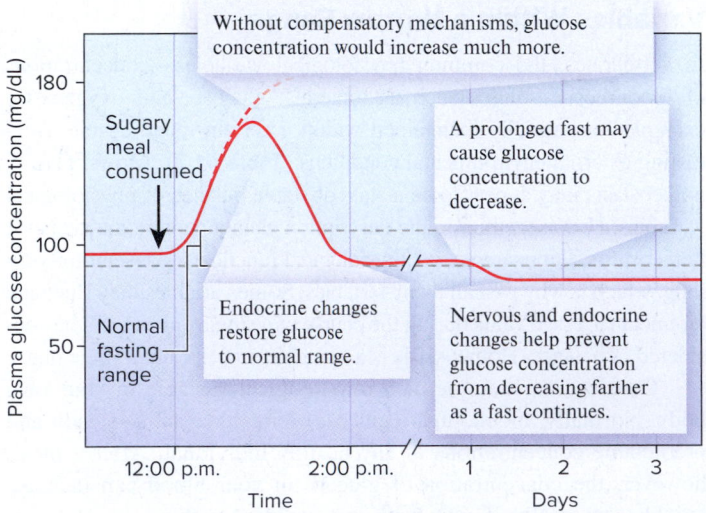

**Figure 41.9** **An example of a homeostatically controlled physiological variable, glucose concentration in human blood.** Note that glucose concentration in the plasma may increase or decrease, depending on whether an animal has recently eaten. However, even after a sugary meal or a prolonged fast, homeostatic mechanisms either return glucose concentration to normal or enable it to remain within the range required for survival. Traditional units for glucose concentration used in the U.S. appear on the vertical axis; as a reference, a value of 100 mg/dL is equal to 5.5 mM.

adjustment, but in other examples, a wide variety of control systems may be initiated. In later chapters, we will see how every organ and tissue of an animal's body contributes to homeostasis, sometimes in multiple ways, and usually in concert with each other.

Homeostasis, then, does not imply that a given physiological function or variable is rigidly constant. Instead, homeostasis means that a variable fluctuates within a certain normal range and that once it deviates from that range, compensatory mechanisms restore the variable toward normal.

## Homeostatic Control Systems Maintain the Internal Environment

The activities of cells, tissues, and organs must be regulated and coordinated with each other so that any change in the extracellular fluid—the internal environment—initiates a response to correct the change. These compensating regulatory responses are performed by homeostatic control systems. A **homeostatic control system** must have several components:

- a **set point**, which is the normal value for a controlled variable;
- a **sensor**, which monitors the level or activity of a particular variable;
- an **integrator**, which compares signals from the sensor with the set point; and
- an **effector**, which compensates for any deviation between the actual value and the set point.

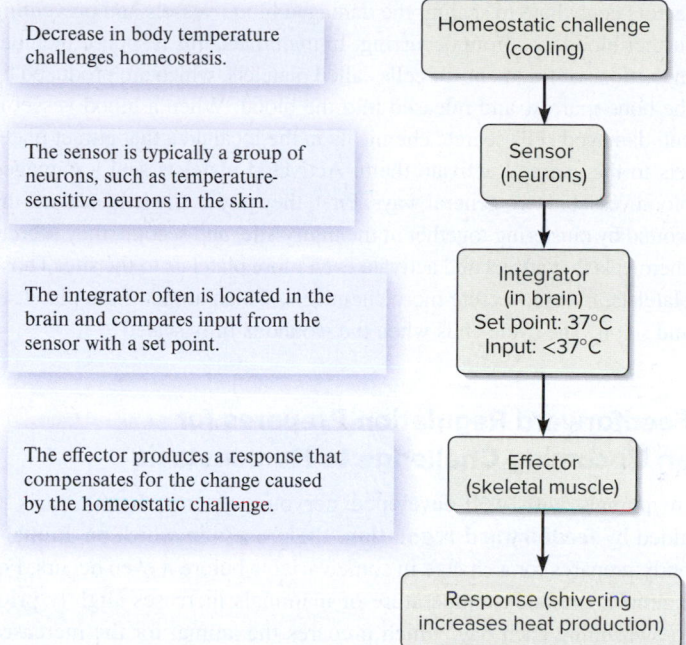

**Decrease in body temperature challenges homeostasis.**

**The sensor is typically a group of neurons, such as temperature-sensitive neurons in the skin.**

**The integrator often is located in the brain and compares input from the sensor with a set point.**

**The effector produces a response that compensates for the change caused by the homeostatic challenge.**

Homeostatic challenge (cooling)

Sensor (neurons)

Integrator (in brain)
Set point: 37°C
Input: <37°C

Effector (skeletal muscle)

Response (shivering increases heat production)

**Figure 41.10    An example of a homeostatic control system.**  The mechanisms for responding to a decrease in body temperature are shown. Different homeostatic control systems have different sensors, integrators, and effectors.

**Figure 41.10** shows an example of a homeostatic control system that regulates body temperature in mammals. This system is somewhat analogous to the heating system of a home. In that case, a sensor and integrator within the thermostat compare the actual room temperature with the set point temperature that was determined by setting the thermostat to a given temperature. If the room temperature becomes cooler than the thermostat setting, the effector (furnace) is activated and adds heat to the room. In a mammal, the sensors are temperature-sensitive neurons in the skin and brain, whereas the integrator is a collection of neurons within the brain. Signals from this part of the brain are sent along nerves to the effectors, which include skeletal muscles. If body temperature decreases, the muscles contract vigorously in response to these signals, resulting in shivering—a key way in which mammals' bodies generate heat. We will discuss other heat-conserving and heat-generating mechanisms that contribute to this important homeostatic control system in Chapter 47.

 **Core Skill: Modeling**

BIO **TIPS**    **THE QUESTION**  *A mouse emerges from its burrow on a cold day to forage for food. Its body temperature quickly decreases, but then fails to increase back to normal. Looking at the model of a homeostatic control system in Figure 41.10, how might you account for this failure to respond?*

**T**OPIC  *What topic in biology does this question address?*
The topic is homeostatic control systems. Specifically, the question asks you to explain how such a system might fail.

**I**NFORMATION  *What information do you know based on the question and your understanding of the topic?*
From the question, you know that the mouse had a homeostatic challenge and failed to respond to it. From your understanding

of the topic, you know that there are several components to any homeostatic control system.

**P**ROBLEM-SOLVING **S**TRATEGY  *Sort out the steps in a complicated process.* Figure 41.10 indicates that in a homeostatic control system that regulates body temperature in a mammal, several responses must occur in sequence for the body temperature to be maintained when the mammal is in a cold environment. These include: (1) sensing the change in the variable (body temperature), (2) comparing it to a set point in the brain, and (3) inducing a compensatory change in temperature by activating effectors (skeletal muscles). To answer the question, consider which, if any, of these responses could have failed to occur in the mouse.

**ANSWER**  *In the mouse, any of these three responses could have failed to function properly. Perhaps the temperature-sensitive neurons were not functioning, or the brain cells responsible for comparing the input to a set point were diseased or not functioning. Finally, the output to the effectors or the effectors themselves may not have been functioning normally. Without further data, none of these possibilities can be ruled out.*

## Negative Feedback Is a Key Feature of Homeostasis

As you have seen in Figure 41.10, homeostatic mechanisms can move a variable back toward its set point. Such mechanisms must be controlled so that a homeostatic response does not overcompensate. This form of regulation is termed a **negative feedback loop**, or simply negative feedback. As an example, **Figure 41.11** considers a negative feedback loop involving homeostatic changes to blood pressure. When the blood pressure of an animal decreases due to blood loss, pressure sensors in the heart and certain blood vessels detect the change in pressure and send the information to the integrator—the brain (Figure 41.11). In the brain, the signal is compared with the normal set point for blood pressure. The brain responds to the deviation from the set point in two ways. First, signals are sent along nerves to the effectors—in this case, the kidneys, heart, and blood vessels. Second, the brain stimulates the release of certain hormones into the blood; these hormones provide an additional signal to the effectors. The result is that the heart beats more rapidly and forcefully, the kidneys produce less urine and thereby retain more water in the body, and the blood vessels direct blood to the most vital organs such as the brain. These responses raise the animal's blood pressure back toward the set point.

To prevent overcompensation, that is, to prevent the blood pressure from becoming higher than normal, the return of blood pressure to its set point removes the stimulus from the sensor (see the dashed arrow in Figure 41.11). This negative feedback, in turn, shuts off further production of the hormonal and neural responses. If negative feedback did not occur, the blood pressure would not only rebound back to the set point but might continue to increase to abnormally high and possibly dangerous levels.

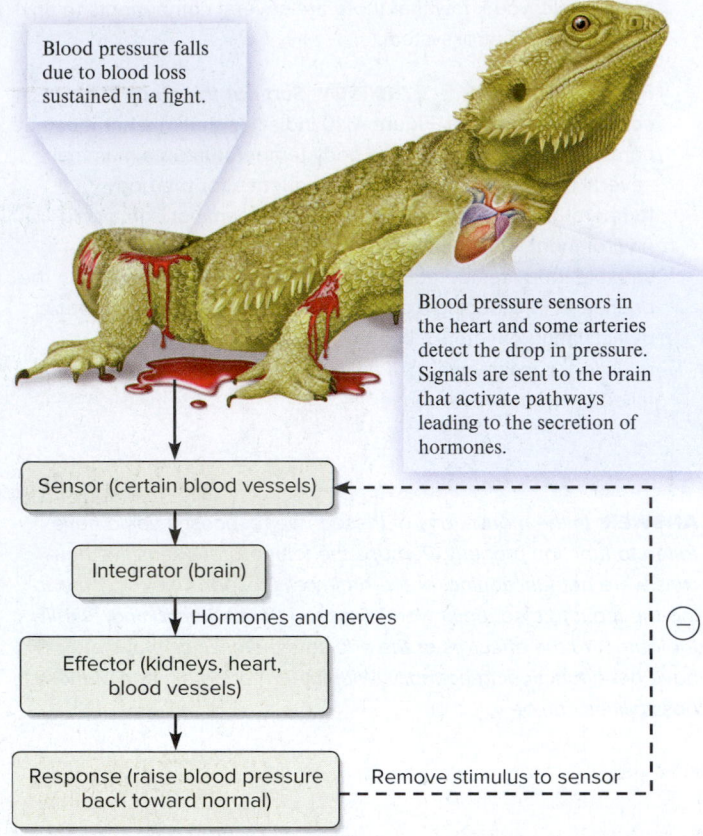

Blood pressure falls due to blood loss sustained in a fight.

Blood pressure sensors in the heart and some arteries detect the drop in pressure. Signals are sent to the brain that activate pathways leading to the secretion of hormones.

Sensor (certain blood vessels)

Integrator (brain)

Hormones and nerves

Effector (kidneys, heart, blood vessels)

Response (raise blood pressure back toward normal)

Remove stimulus to sensor

**Figure 41.11  A negative feedback loop as a mechanism by which homeostatic control systems operate.** In this example, loss of blood results in a drop in blood pressure, which could be life-threatening if not corrected. Effectors such as the kidneys, heart, and blood vessels help restore blood pressure toward normal. They do not increase blood pressure above normal, however, because of negative feedback (as denoted by the minus sign next to the dashed arrow).

 **Core Skill: Connections** Can negative feedback occur at levels other than the organ system level shown in this figure? Refer back to Figure 6.13 for help.

## Positive Feedback Does Not Achieve Homeostasis

Thus far, we have considered homeostatic mechanisms that occur via negative feedback. A **positive feedback loop**, or simply positive feedback, accelerates or amplifies a process (think of an avalanche that begins with a small snowball rolling down a steep hill). A positive feedback loop moves a system away from homeostasis, because a change in a variable or process leads to events that amplify that change. Look again at Figure 41.10. Imagine what might happen if shivering not only generated heat, but also in some way stimulated the sensor neurons to send additional stimulatory signals to the effectors. In such a circumstance, heat production would continue without any means of stopping, leading to a dangerously high body temperature. This is contrary to the principle of homeostasis, in which large fluctuations in a variable are minimized and reversed.

One example of positive feedback occurs in the process of blood clotting in mammals (**Figure 41.12**). If an animal receives a wound that results in bleeding, as shown in Figure 41.11, various blood-borne factors contribute to sealing the damaged blood vessels and preventing further blood loss from occurring. In mammals, this response includes the actions of fragments of cells called platelets, which are produced by the bone marrow and released into the blood. When a blood vessel is cut, damaged cells secrete chemicals in the local area that attract platelets to the site and activate them. Activated platelets seal a damaged blood vessel in two general ways. First, they physically help seal off the wound by clustering together at the injury site, and second, they secrete chemicals that attract and activate even more platelets to the site. Those platelets, in turn, secrete more chemicals, which attract more platelets, and so on. The cycle ends when the wound is fully sealed.

## Feedforward Regulation Prepares for an Upcoming Challenge to Homeostasis

In animals with well-developed nervous systems, homeostasis is aided by **feedforward regulation**, the process in which an animal's body prepares for a change in some variable before it even occurs. For example, the body temperature of mammals increases slightly prior to awakening each day, which prepares the animal for the increased metabolic demands of being awake and active. A famous example of feedforward regulation, first characterized by Russian physiologist Ivan Pavlov in the 1890s, involves the changes that occur when a hungry dog smells or sees food. First, the dog starts to salivate, and its stomach begins to churn and produce acid. Salivation and activity of the stomach are important components of the digestive process, yet at this stage, the animal has not actually eaten any food. Instead, its digestive system is already preparing for the arrival of food in order to maximize digestive efficiency, speed the flow of nutrients into the blood, and minimize the time required for active cells to replenish energy stores.

In the preceding example, feedforward regulation uses sensory detectors that recognize odors and sights. Many examples of such regulation, including the example described by Pavlov, result from, or are modified by, the phenomenon called learning. The result of this process is that the nervous system learns to anticipate a homeostatic challenge. Familiar examples are the increased heart rate and breathing rate that occur just before an athletic competition—demonstrated, for example, in trained racehorses before the start of a race (**Figure 41.13**). The process of training, in which a horse's body learns to prepare for the exertion of the ensuing race, prevents any delay between the start of exercise and the adequate flow of blood and nutrients to skeletal muscle.

## Local and Long-Distance Chemical Signals Coordinate Homeostatic Responses

A common thread that links all homeostatic processes together is communication between cells, whether the cells are close to each other or in different parts of an animal's body. Some homeostatic responses are highly localized, occurring only in the area of a disturbance. For example, damage to an area of skin causes cells in the injured area to release molecules that help contain the injury, prevent infections, and promote tissue repair in the immediate vicinity (see Chapter 52). Local responses provide areas of an animal's body with mechanisms for local self-regulation. It is

**Figure 41.12** **An example of positive feedback.** When a blood vessel is cut, platelets help seal the damaged site by forming a clot. As platelets are attracted to the site and activated by secretions from damaged cells, they secrete their own chemicals that attract and activate more platelets. Those platelets continue the cycle until the wound is finally sealed.

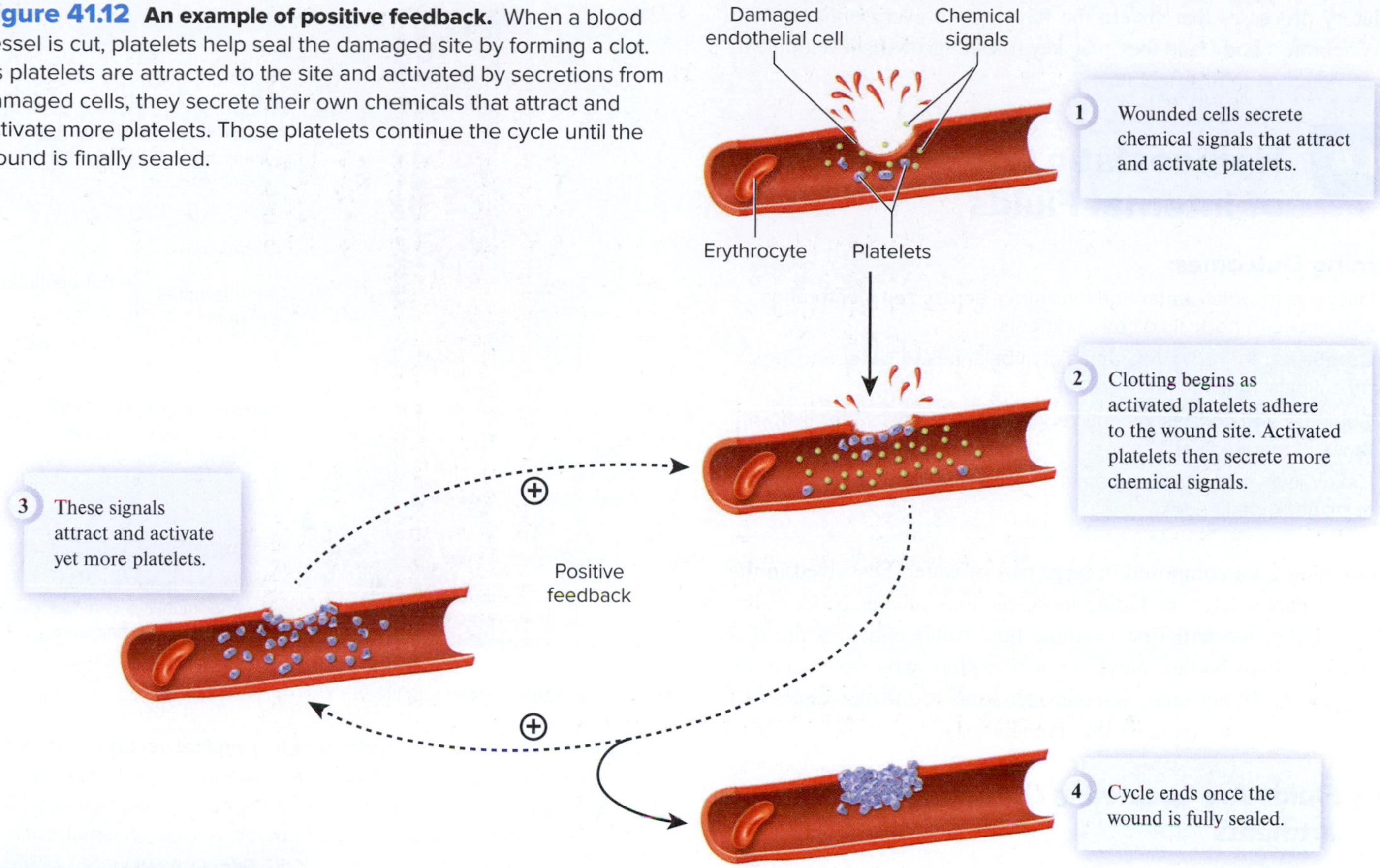

Damaged endothelial cell

Chemical signals

Erythrocyte   Platelets

**1** Wounded cells secrete chemical signals that attract and activate platelets.

**2** Clotting begins as activated platelets adhere to the wound site. Activated platelets then secrete more chemical signals.

**3** These signals attract and activate yet more platelets.

Positive feedback

⊕

⊕

**4** Cycle ends once the wound is fully sealed.

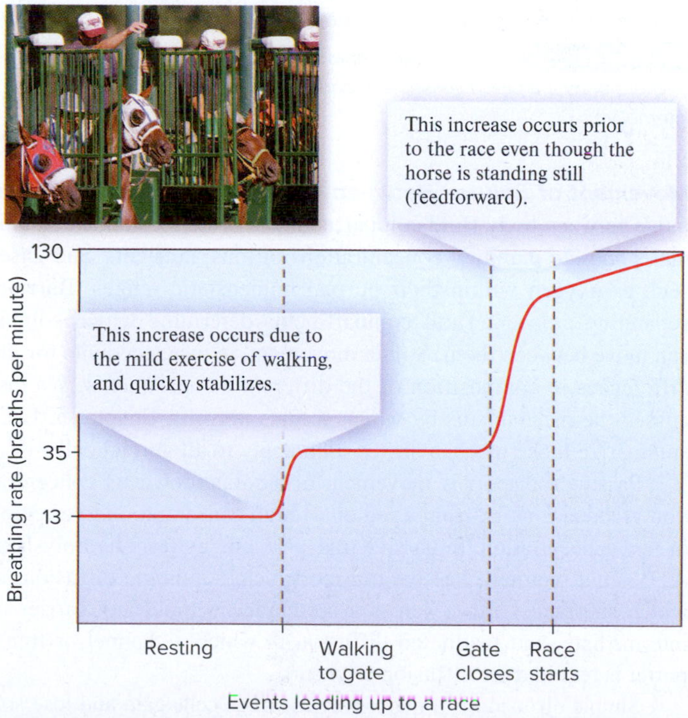

This increase occurs prior to the race even though the horse is standing still (feedforward).

This increase occurs due to the mild exercise of walking, and quickly stabilizes.

Resting   Walking to gate   Gate closes   Race starts

Events leading up to a race

Breathing rate (breaths per minute)

130

35

13

**Figure 41.13** **Feedforward regulation of breathing rate in an animal trained for athletic exercise.** Feedforward regulation prepares an animal's body for an ensuing challenge or event, such as a race. ©Mitch Wojnarowicz/The Image Works

*Concept Check:* *What similar feedforward process might occur in an animal in nature?.*

no benefit to an animal to promote tissue repair in regions of the body that are not injured. This type of cellular communication—in which molecules are released into the extracellular fluid and act on nearby cells—is called **paracrine signaling** (refer back to Figure 9.3d).

Another example of extremely localized signaling occurs between neurons. A common way in which neurons communicate is through the release of neurotransmitters, small signaling molecules that are synthesized and stored in neurons. When a neuron releases neurotransmitters, they diffuse and then bind to receptor proteins on an adjacent neuron (or in some cases a muscle or gland cell), altering the activity of that cell. This type of cell-to-cell communication is typically very rapid, finishing within milliseconds. Consequently, neurotransmitter responses can make immediate homeostatic adjustments, like those associated with reflexes. These are just two of the many types of localized signaling that occur in animals' bodies and that will be described in subsequent chapters in this unit.

In addition to using paracrine signaling and neurotransmitter release, cells can communicate over long distances by releasing chemical messenger molecules into the blood. This type of signaling is mediated by **hormones**—chemical messengers produced by the endocrine system of animals. A hormone released in response to a homeostatic disturbance, such as the decrease in blood pressure described earlier, can influence the activities of many different cells, tissues, and organs simultaneously because the hormone is carried throughout the entire blood circulation. Some hormones act quickly—within seconds—whereas others take minutes or even hours for their effects to occur. In subsequent chapters, we will see that hormones are a key part of the regulatory processes that govern the functions of every organ system in a vertebrate's body, and they play key roles in growth, development, and reproduction in invertebrates.

## 41.4  Homeostatic Control of Internal Fluids

**Learning Outcomes:**

1. List ways in which water and ions move across cell membranes and between body fluid compartments.
2. **CoreSKILL »** Predict outcomes of imbalances of water and ions on animals' function and survival.
3. Compare and contrast osmotic adaptations of freshwater fish with those of marine fish.
4. List two ways of classifying animals according to how they adapt to osmotic challenges.

Animal bodies are composed in large part of water. Dissolved in the water are many solutes, including inorganic ions such as $Na^+$ and $K^+$. In this section, we will first examine how water and ions are distributed in animal bodies. Next, we will explore why water and ion homeostasis is so important and consider some of the challenges that must be overcome to maintain that homeostasis.

### Body Fluids Are Located in Different Fluid Compartments

Most of the water in an animal's body is contained inside its cells; this fluid is called **intracellular fluid** (from the Latin *intra*, meaning inside of). The rest of the water in the body exists outside of the cells; this fluid is called **extracellular fluid** (from the Latin *extra*, meaning outside of). Plasma membranes separate the intracellular fluid from the extracellular fluid.

In vertebrates and some invertebrates, extracellular fluid is composed of the watery (noncellular) part of blood, called **plasma**, and the fluid that fills the spaces that surround cells, called **interstitial fluid** (from the Latin *inter*, meaning between) (**Figure 41.14**). In such animals, plasma and interstitial fluid are kept separate, with plasma contained within blood vessels in a closed circulatory system. The interstitial fluid and the plasma are separated by the walls of vessels (arteries, capillaries, and veins). In many invertebrates with open circulatory systems (see Chapter 48), however, plasma and interstitial fluid are intermingled in a single fluid called hemolymph.

The locations of fluids depicted in Figure 41.14 are called body fluid compartments. In a typical vertebrate, the total water volume in the three compartments (intracellular fluid, plasma, and interstitial fluid) accounts for about two-thirds of body weight, with solids comprising the rest. Of the total body water, up to two-thirds is intracellular and one-third extracellular, with the majority of the latter located in the interstitial compartment.

The solute composition of the extracellular fluid is very different from that of the intracellular fluid. Maintaining differences in solute composition across the plasma membrane is an important way in which animal cells regulate their own activity. For example, many different proteins that are important in regulating cellular events such as mitosis, cytokinesis, and metabolism are confined to the intracellular fluid.

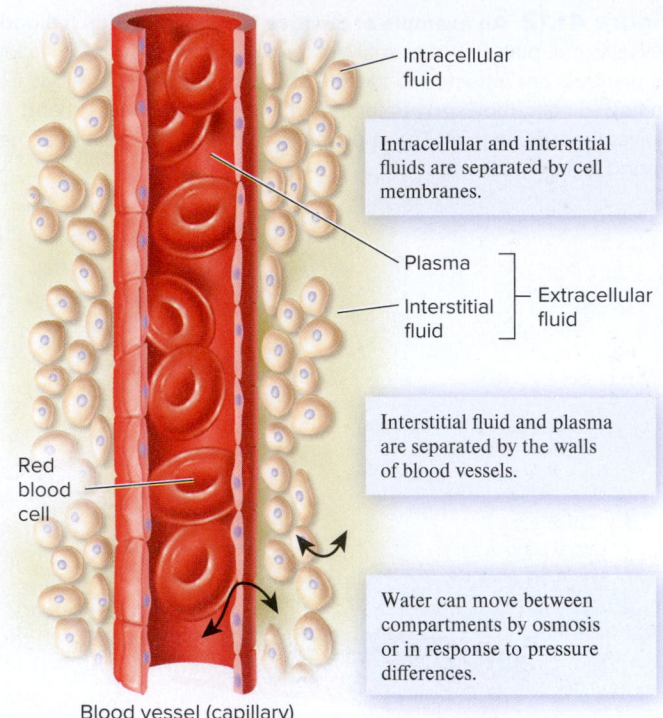

Intracellular fluid

Intracellular and interstitial fluids are separated by cell membranes.

Plasma

Interstitial fluid — Extracellular fluid

Interstitial fluid and plasma are separated by the walls of blood vessels.

Red blood cell

Water can move between compartments by osmosis or in response to pressure differences.

Blood vessel (capillary)

**Figure 41.14  Fluid compartments in a typical vertebrate.** Most of the fluid within an animal's body exists within cells (intracellular fluid). Extracellular fluid is that portion of the body's fluid that lies outside cells (interstitial fluid) or within blood vessels (plasma), such as the capillary shown here. Arrows indicate directions of water movement between adjacent compartments.

**Concept Check:** *What would happen to the distribution of water in an animal's body if a blood vessel was damaged and leaked its contents internally?*

***Movement of Solutes Between Compartments***   Solutes must move between body fluid compartments in order for cells in an animal's body to maintain concentrations of ions, nutrients, and gases such as oxygen within their normal homeostatic ranges. Barriers separating adjacent fluid compartments determine which solutes can move between them. Solute movement, in turn, accounts for the differences in composition of the different compartments. We discussed the mechanisms by which solutes move in Chapter 5. Let's summarize those mechanisms, which apply to all animal cells.

Passive transport is movement of a solute down its concentration gradient, that is, from a region of high concentration to a region of low concentration. In passive transport, energy from hydrolysis of ATP is not required. Passive transport includes simple diffusion, in which substances move across a membrane without any carrier or intermediate, and facilitated diffusion in which a channel or transporter is required for diffusion to occur.

Simple diffusion is a major way in which cells gain and lose solutes. Molecules that can cross phospholipid bilayers are able to passively diffuse into or out of a cell. Examples include nonpolar molecules such as many lipids, and gases such as oxygen and carbon dioxide.

The rate of simple diffusion depends on several factors, notably the concentration gradient of the solute and the area across which it is diffusing. The rate of simple diffusion of a solute across a membrane of given thickness can be calculated using a modified form of Fick's first law of diffusion adapted for movement across a membrane:

$$J = KA(C_1 - C_2)$$

where $J$ is the rate of simple diffusion, $K$ is a constant that includes temperature, $A$ is the cross-sectional area of the barrier across which diffusion is occurring, and $C_1$ and $C_2$ are the concentrations of the solute at two locations (for example, inside and outside a cell). This equation is used to determine how changes in solute concentrations, temperature, or area can influence the rate at which a substance moves across a plasma membrane. For example, breathing a gas mixture from a tank that is enriched in oxygen will increase the amount of oxygen entering the blood of a mountain climber at high altitude, where oxygen is limited. According to Fick's first law, the difference between $C_1$ (oxygen in the inhaled gas mixture) and $C_2$ (oxygen in the blood) will be increased by breathing from the tank. Therefore, we can predict that $J$, the rate of diffusion of oxygen into the blood, will also be increased, an important survival mechanism at very high altitudes.

Most polar molecules and ions, however, can move through a plasma membrane only with the help of a transport protein, as in facilitated diffusion. In one case, the membrane has channels that permit the solute to diffuse down its concentration gradient through the bilayer. Examples of substances that diffuse through channels are ions such as Na⁺. In a second case, proteins in the membrane bind a solute and shuttle it down its concentration gradient across the lipid membrane. An example of a common solute that moves across membranes in this way is glucose. In both of these forms of diffusion, solutes move down their concentration gradients and hydrolysis of ATP is not required for their diffusion.

By contrast, in active transport, energy is required to move a solute against a concentration gradient (refer back to Figure 5.19). Typically, we will encounter this type of transport in this unit when discussing how animal cells maintain different concentrations of various ions

across their plasma membranes and how such concentration differences relate to a cell's ability to function. One example of a function dependent on differences in ion concentration is the generation of electrical gradients across the membranes of muscle cells and neurons.

***Movement of Water Between Compartments***   Water can readily move between adjacent compartments in an animal's body, because barriers such as plasma membranes tend to be highly permeable to water, due to the presence of water channels called aquaporins (see Figure 5.16). This movement depends on pressure differences in the fluids of each compartment or on differences in solute concentrations that lead to osmosis (see Chapter 5), in which water moves from a region of lower solute concentration to one of higher solute concentration.

To function properly, cells require a relatively stable internal composition, including ion and protein concentrations, cellular volume, and pH. A decrease in solute concentration outside a cell, for example, will cause water to move by osmosis from outside the cell to inside. In this case, osmosis redistributes water from the interstitial to the intracellular compartment. This movement will cause a cell to become deformed as it swells due to the influx of water. In contrast, an increase in extracellular solute concentration will lead to osmosis of water from inside the cell to outside, causing the cell to shrink. In either case, a swollen or shrunken animal cell generally is more fragile than a normal cell and will be destroyed if its membrane ruptures. **Figure 41.15** shows examples of mammalian red blood cells (known as erythrocytes) in which intracellular fluid levels have been altered. This can occur, for example, if the cells are exposed to extracellular fluids with either a higher or lower solute concentration than the fluid inside the cell. When erythrocytes swell, they may burst, a phenomenon called hemolysis. Shrinkage of erythrocytes is called crenation and is also potentially destructive (Figure 41.15, middle panel).

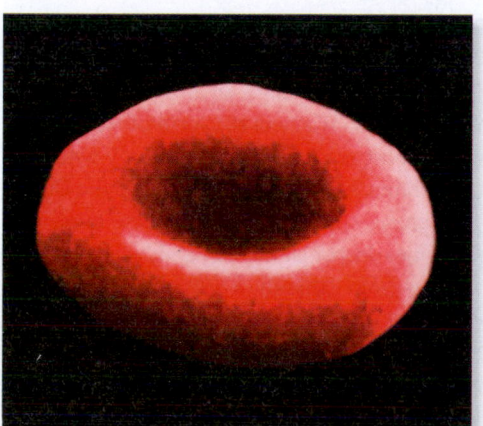

Erythrocyte in a solution of normal solute concentration

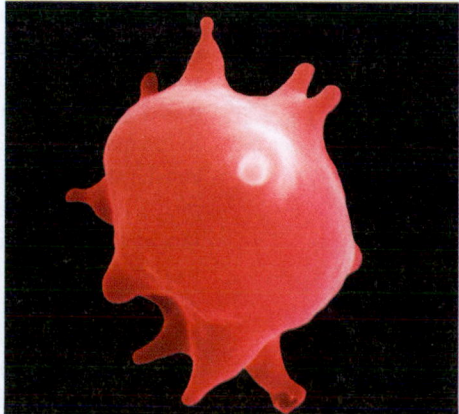

Erythrocyte that has lost intracellular fluid when placed in a solution of higher than normal solute concentration

Erythrocyte that has gained intracellular fluid when placed in a solution of lower than normal solute concentration

**Figure 41.15   Changes in cell shape due to alterations in intracellular fluid volume.**  Alterations in intracellular fluid volume can have drastic effects on cell shape, as shown by these SEM images of erythrocytes. Large changes in shape like those in the middle and right are usually lethal for cells. Each cell is approximately 5–8 μm in diameter. (left): ©Mary Martin/Science Source; (middle, right): ©David M. Phillips/Science Source

*Concept Check:*   *What effect will changes in intracellular fluid volume have on intracellular solute concentration?*

## A Balance of Water and Ions Is Critical for Survival

Maintenance of normal levels of body water is of great importance for all animals. Not only is water the major portion of an animal's body mass, it is also the solvent that permits solutes to participate in chemical reactions. As described in Chapters 2 and 3, water itself participates in important chemical reactions, notably hydrolysis reactions. In addition, water is the transport vehicle that brings $O_2$ and nutrients to cells and removes wastes generated by metabolism.

When an animal's water volume is reduced below the normal range, we say the animal is dehydrated. In terrestrial animals, dehydration may occur if sufficient drinking water is not available or when water is lost by evaporation (through perspiring or panting). Dehydration can be a serious, potentially life-threatening condition. For example, because blood is roughly 50% water (plasma), blood volume tends to decrease in dehydrated animals. Decreased blood volume compromises the ability of the circulatory system to move nutrients and wastes throughout the body and to assist in the regulation of body temperature on hot days.

Ion balance is also very important for animals. A change of only a few percentage points in the extracellular fluid concentration of potassium ions ($K^+$), for example, can trigger changes in nerve, heart, and skeletal muscle function by altering their electrical activities. Other ions, such as calcium ($Ca^{2+}$), magnesium ($Mg^{2+}$), phosphate ($PO_4^{3-}$), and sulfate ($SO_4^{2-}$), also participate in various biological activities. Their functions include serving as cofactors for enzyme activation, participating in bone formation, forming part of the extracellular matrices around cells, and activating cellular events such as exocytosis and muscle contraction. An imbalance in any of these ions can seriously disrupt cellular activities.

The solute concentration of an aqueous solution is known as the solution's **osmolarity**, expressed in milliosmoles/liter (mOsm/L). The number of dissolved solute particles determines a solution's osmolarity. For example, a 150 mM NaCl solution has an osmolarity of 300 mOsm/L, because each NaCl molecule dissociates into two ions, one $Na^+$ and one $Cl^-$ ($2 \times 150 = 300$). The value of 300 mOsm/L is well within the range of typical osmolarities of animal body fluids. Solutions with an osmolarity greater than normal are called hyperosmotic solutions; those with an osmolarity less than normal are hypo-osmotic solutions. An iso-osmotic solution is one that has the same osmolarity as a typical animal cell.

## Many Exchanges of Ions and Water with the Environment Are Obligatory

Many vital processes—eliminating nitrogenous wastes, obtaining $O_2$ and eliminating carbon dioxide ($CO_2$), consuming and metabolizing food, and regulating body temperature—have the potential to disturb ion and water homeostasis. Therefore, these processes require additional energy expenditure to minimize or reverse the disturbance. Exchanges of ions and water with the environment that occur as a consequence of such vital processes are called obligatory exchanges (because the animal is obligated to make them) (**Figure 41.16**).

### Exchanges Due to Elimination of Nitrogenous Wastes
When carbohydrates and fats are metabolized by animal cells, the major waste product is $CO_2$, which is exhaled or, in some animals, diffuses across

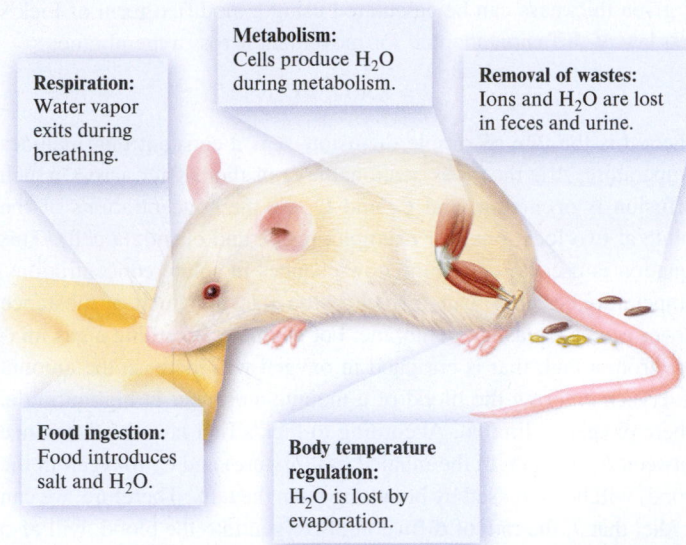

**Respiration:** Water vapor exits during breathing.

**Metabolism:** Cells produce $H_2O$ during metabolism.

**Removal of wastes:** Ions and $H_2O$ are lost in feces and urine.

**Food ingestion:** Food introduces salt and $H_2O$.

**Body temperature regulation:** $H_2O$ is lost by evaporation.

**Figure 41.16 Types of obligatory ion and water exchanges in a terrestrial animal.** Obligatory exchanges with the environment occur as the result of necessary life processes.

**Concept Check:** *Can animals completely avoid all the losses resulting from obligatory exchanges?*

the body surface. By contrast, proteins and nucleic acids contain nitrogen; when these molecules are broken down and metabolized, nitrogenous wastes are generated. Nitrogenous wastes are molecules that include nitrogen from amino groups (—$NH_2$). These wastes are toxic at high concentrations and must be eliminated from the body but, unlike $CO_2$, cannot be eliminated by exhaling or diffusion. As we will see in Chapter 49, the excretion of nitrogenous wastes is carried out by excretory organs such as kidneys and often requires body water.

### Exchanges Due to Respiration
The requirements for respiration and for water and ion balance present different challenges to air- and water-breathing animals. To ventilate its lungs, an air-breathing animal moves air in and out of its airways. Water in the form of water vapor in the mouth, nasal cavity, and upper airways exits the body with each exhalation. As an animal becomes more active, it requires more $O_2$ and produces more $CO_2$. These changes are met by an increase in respiratory activity. Breathing becomes deeper and more rapid, which, in turn, increases the rate of water loss from the body. Therefore, respiration in animals with lungs is associated with significant water loss, as you can observe in cold weather when you can "see your breath."

Also, as described in Chapter 48, small, active animals with high metabolic rates usually have faster breathing rates than do larger, less active animals. Consequently, the potential for water loss due to respiration is relatively greater in small animals, particularly in endotherms (birds and mammals). A hummingbird for example, may have 15–20 times the water loss per gram of body mass than does a large goose.

In water-breathing animals, the challenge of water and ion homeostasis is more complex, because such animals move water, not air, over their respiratory organs (gills). Gills, like all respiratory organs, are thin structures with large amounts of surface area and an

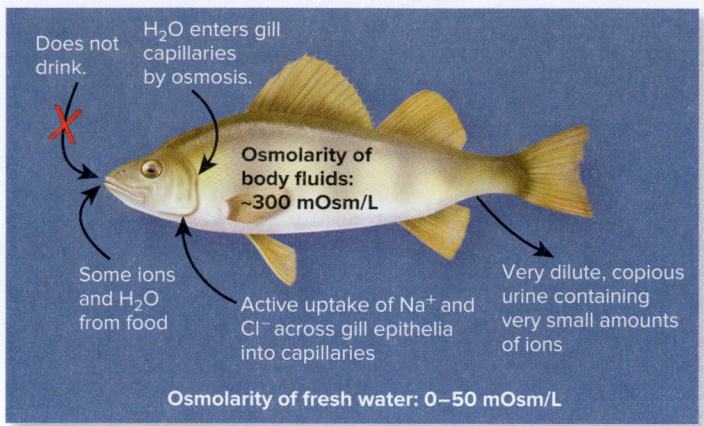

**(a) Freshwater fish**

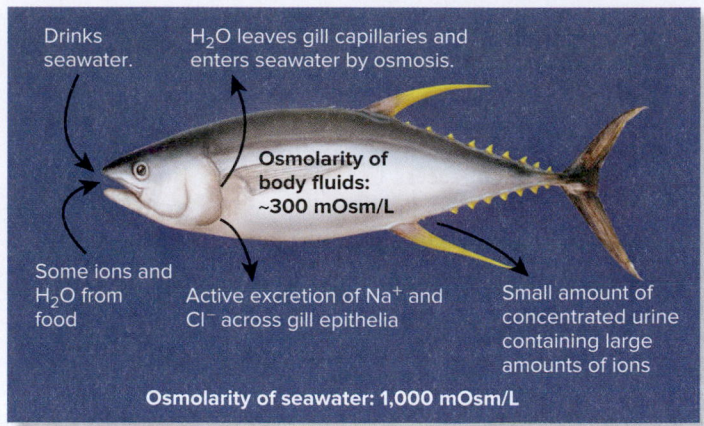

**(b) Saltwater fish**

**Figure 41.17** **Ion and water balance in water-breathers.** Water breathing creates osmoregulatory challenges due to diffusion of ions and osmosis of water across gills. These challenges differ between **(a)** freshwater and **(b)** saltwater fishes and are addressed by behaviors (drinking or not drinking water), by active transport of ions across the gills, and by alterations in urine output.

 **Core Concept: Energy and Matter** Energy from the hydrolysis of ATP is required for all of the active transport processes illustrated in Figure 41.17. This energy is required to maintain homeostatic body fluid osmolarities despite the osmotic challenges imposed by very different environments.

extensive network of blood vessels. Although these features make gills ideal for gas exchange by diffusion between the blood and the surrounding water, they also make them ideal for ion and water movement by diffusion and osmosis, respectively.

When differences occur in ion concentration between a water-breathing animal's body fluids and the surrounding water, breathing via the gills has the potential to disrupt ion and water balance. Fishes or other water-breathing animals that live in fresh water and those that live in salt water face opposite challenges in maintaining this balance (**Figure 41.17**). The internal fluid osmolarity of most fishes is usually within the range of 225–400 mOsm/L, similar to that of most other vertebrates. Because freshwater lakes and rivers have very little ion content (usually <25 mOsm/L), this high concentration gradient for ions could promote the loss of ions from a fish's body into the fresh water. Likewise, a high osmotic gradient favors the movement of water from the lake or river into the body fluids of a freshwater fish. Freshwater fishes, therefore, gain water and lose ions when ventilating their gills (Figure 41.17a). If left uncorrected, these changes would cause a dangerous decrease in blood ion concentrations.

Freshwater fishes maintain water and ion balance via two different mechanisms. First, their kidneys are adapted to producing copious amounts of dilute urine—up to 30% of their body mass per day (an amount that would be equivalent to about 25 L per day in an average-sized human!). Second, specialized gill epithelial cells actively transport Na+ and Cl− from the surrounding water into the fish's blood. Thus, these two important ions are recaptured from the water. Freshwater fishes rarely, if ever, drink water, except for any that might be swallowed with food.

Saltwater fishes have the opposite problem. They tend to gain ions and lose water across their gills, because seawater has a much higher osmolarity (about 1,000 mOsm/L) than that of their body fluids (Figure 41.17b). The gain of ions and the loss of water from the

body are only partly offset by the kidneys, which in marine fishes produce very little urine so that as much water as possible can be retained in the body. The urine that is produced has a higher ion concentration than that of freshwater fishes. To prevent dehydration from occurring, marine fishes must drink. However, the only water available to them is the hyperosmotic seawater, which has a very high ion content. Paradoxically, therefore, marine fish drink seawater to replenish the water lost by osmosis through their gills. What does the fish do with all of the ions it ingested? The ingested ions must be eliminated, and this process is accomplished by gill epithelial cells. In contrast to the gills of freshwater fishes, which pump ions from the water into the fluids of the fish, the gills of marine fishes pump ions out of the fish and into the ocean. Thus, marine fishes drink seawater to replace the water lost through their gills by osmosis and then expend energy to transport the excess ions out of the body.

***Exchanges Due to Feeding*** Because foods contain salts and water, eating also involves obligatory exchanges of these substances. Some plant products are over 95% water by weight, and other foods may contain high amounts of Na+ or other minerals. Therefore, the type of diet an animal consumes determines how much salt and water it ingests.

Once food has been digested and absorbed, the unusable parts of food are excreted as solid wastes. Some ions and water are lost by this route in most animals, but exceptions exist. Desert-dwelling kangaroo rats such as *Dipodomys panamintensis* produce fecal pellets that are almost completely dry, which helps these animals conserve water.

When food molecules are metabolized to provide energy that will be stored in the chemical bonds of ATP, oxygen captures electrons and combines with hydrogen ions, thereby making water (refer back to Figure 7.8). This water is sometimes called metabolic water to indicate its origin.

As noted earlier, marine fishes drink seawater. Other animals besides marine fishes may also drink seawater, either because fresh water is unavailable or because they ingest some with the food they eat. Many marine reptiles and birds also ingest seawater when consuming prey or, in some cases, when they spend prolonged periods at sea and have no access to fresh water for drinking. These animals have specialized epithelial cells that line structures called salt glands, located in groups around the nostrils, mouth, and eyes (**Figure 41.18**). Ions (notably $Na^+$ and $Cl^-$) move from the blood into the interstitial fluid, and from there, they are actively transported by the epithelial cells of the salt glands into the tubules of the gland. The ions and a small amount of fluid then collect into a central duct and are excreted as highly concentrated solutions. In general, vertebrates without salt glands cannot survive by drinking seawater, because they have no means of creating and excreting such a highly concentrated salt solution. Some marine mammals have been observed to occasionally drink small amounts of seawater, but most appear to never drink at all. These animals get their water from the food they eat.

***Exchanges Due to Evaporation of Water*** Endotherms (animals that generate their body heat) use body water to cool off when they are active or in a hot environment. For example, sweating and panting are used to cool the body. These activities use the evaporation of water to draw heat out of the body. In the process, however, the animal loses water and, in sweat, some ions. You know from tasting sweat that it is salty, but the saltiness of sweat and that of blood are not the same. Sweat is a hypo-osmotic solution compared with blood; that is, it has a lower concentration of solutes. Thus, the fluid left behind in the body after perspiration has both a lower volume and a higher solute concentration than normal.

Other than perspiration and panting, very little water is gained or lost directly across the body surface of most terrestrial vertebrates, because their skin is impermeable to water; exceptions include amphibians. In invertebrates, the rate of water loss across the body surface depends on whether the animal is soft-bodied, like worms, or covered in a waxy, water-impermeable cuticle, like most insects.

The significance of obligatory exchanges and their effects on homeostasis was dramatically illustrated by a long-term investigation by a research team at the University of Florida, described next. Their discovery led to a revolution in our understanding of exercise physiology in humans.

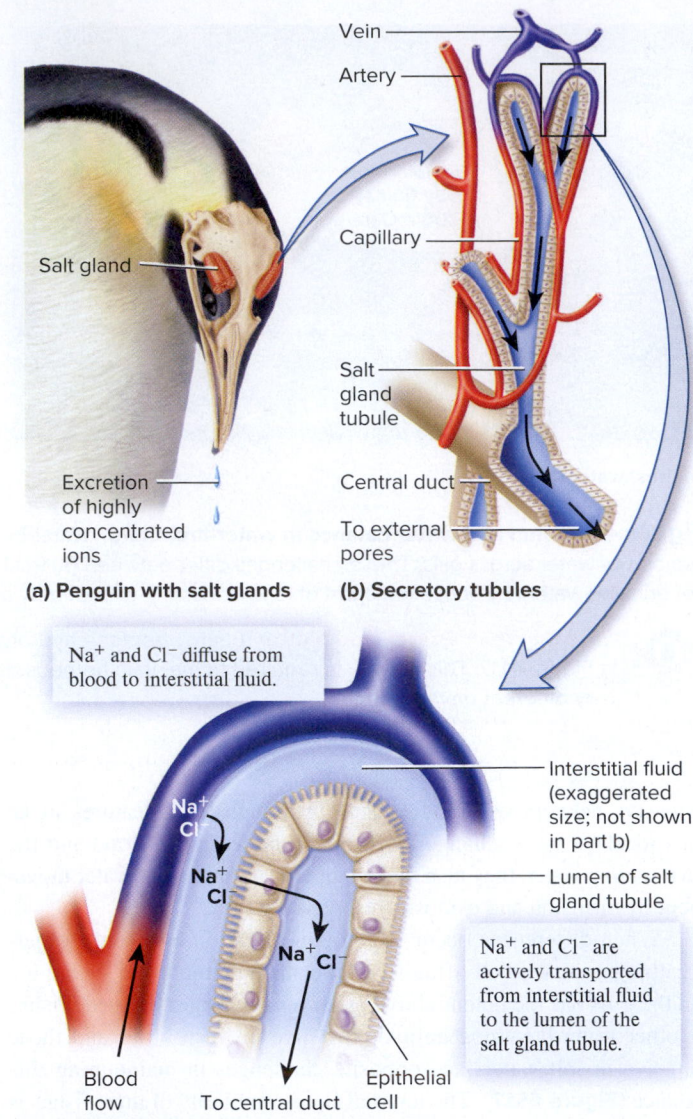

**(a) Penguin with salt glands** **(b) Secretory tubules**

$Na^+$ and $Cl^-$ diffuse from blood to interstitial fluid.

**(c) Collection of salt solution in the tubule**

**Figure 41.18 Salt glands as an adaptation for marine life.** Many marine birds and reptiles have salt glands, which contain a network of secretory tubules that actively transport $Na^+$ and $Cl^-$ from the interstitial fluid into the tubule lumen. The viscous solution then moves through a central duct and to the outside environment through pores in the nose, around the eyes, and in other locations. The black arrows indicate direction of flow of blood or salt gland excretions.

*Concept Check:* *Why can't humans survive by drinking seawater?*

## Core Skill: Process of Science

## Feature Investigation | Cade and Colleagues Discovered Why Athletes' Performances Wane on Hot Days

On a typically hot summer day in the mid-1960s in Gainesville, Florida, the University of Florida football team was practicing in full equipment. The players were rapidly becoming dehydrated and, unbeknownst to them, the osmolarity of their body fluids was increasing as their bodies produced copious amounts of dilute sweat in an effort to maintain body temperature. The athletes became aware of two things. First, they discovered that they did not need to urinate for long periods after a strenuous practice session, and, second, their performance on the field suffered as they became increasingly fatigued and more susceptible to severe muscle cramps. Occasionally, players would require medical treatment for their symptoms. In extreme cases, athletes exercising in these conditions have been known to occasionally develop seizures—uncontrolled activity of neurons in the brain. This situation did not escape the notice of the team physicians and, notably, university faculty member and kidney specialist Robert Cade.

Many of the symptoms experienced by the players could be readily explained. The fatigue was directly related to loss of water from the body, which put a strain on the circulatory system and reduced the amount of blood flow to muscles and other organs. It was worsened by a slight decrease in blood glucose concentration during the long periods of strenuous activity without food. The muscle cramps and even the occasional seizures arose from an imbalance in extracellular ions—notably $Na^+$ and $K^+$—which are secreted out of the body by sweat glands in the process of perspiration. The resulting imbalance in extracellular fluid ion concentrations caused a change in the electrical properties of muscle cells and neurons, which triggered the spasms. Lastly, decreased urine production is one of the body's mechanisms for retaining fluid when body water is decreasing.

The key question was: How could these effects of strenuous exercise best be reversed or prevented? The answer was simple and clever. Cade and his colleagues rejected the prevailing view that drinking any fluids during heavy exercise somehow contributed to cramps and other problems. Instead, they hypothesized that the best way to maintain ion and water homeostasis in a profusely sweating person is to restore to the body exactly what was lost; that is, the person should drink a solution that resembles sweat!

The first thing Cade needed to do was analyze precisely how much $Na^+$, $K^+$, and other ions are actually present in sweat. Fortunately, he had an abundance of human sweat at his disposal to analyze. Once the players left the field, their jerseys were wrung out into a container, and the composition of the collected sweat was determined with an ion analyzer, the flame spectrophotometer shown in **Figure 41.19**. The concentrations were then compared with known values of ion concentrations in human blood. Today, we know that the composition of human sweat can change under certain conditions and can vary among people, but Cade's results were typical. The athletes' sweat contained mostly $Na^+$, $K^+$, and $Cl^-$ at concentrations that indicated the solution was dilute compared with blood. Once Cade completed this analysis,

**Figure 41.19** **Cade and colleagues discovered a way to improve athletic performance and prevent ion and water imbalance during strenuous exercise.**

**HYPOTHESIS** Athletic performance can be enhanced by maintaining the body's ion and $H_2O$ balance during exercise.

**KEY MATERIALS** Supply of human sweat for analysis, ion analyzer, salt solution.

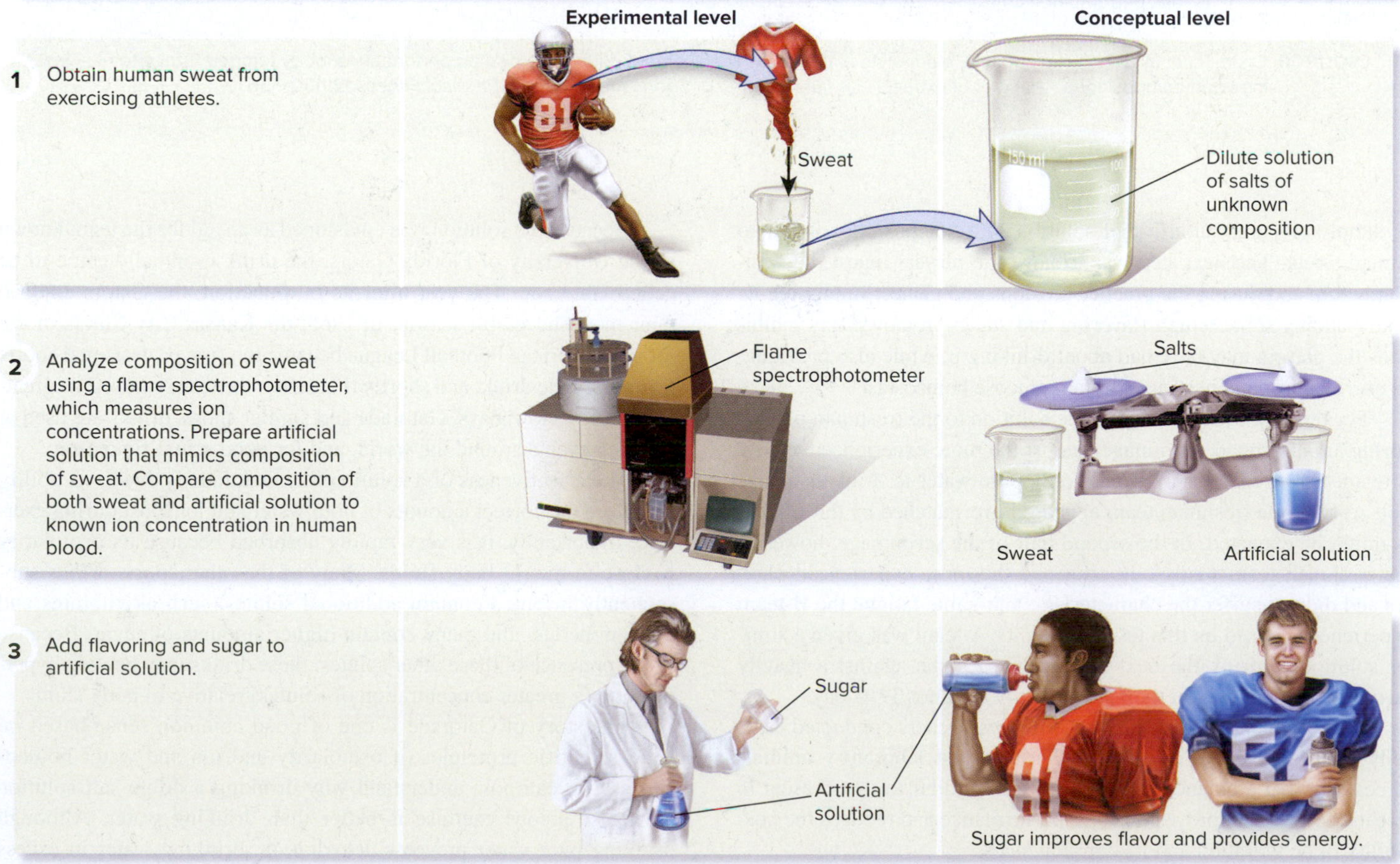

Experimental level | Conceptual level

1 Obtain human sweat from exercising athletes.

Sweat

Dilute solution of salts of unknown composition

2 Analyze composition of sweat using a flame spectrophotometer, which measures ion concentrations. Prepare artificial solution that mimics composition of sweat. Compare composition of both sweat and artificial solution to known ion concentration in human blood.

Flame spectrophotometer

Salts

Sweat          Artificial solution

3 Add flavoring and sugar to artificial solution.

Sugar

Artificial solution

Sugar improves flavor and provides energy.

**4** Provide freshman team with the artificial solution and varsity B-team with water. Hold scrimmage.

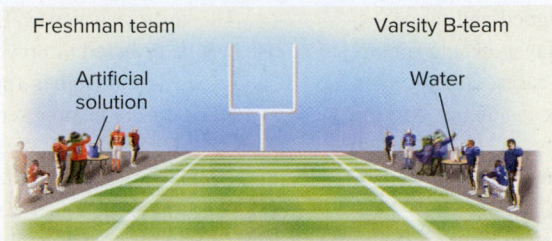

Freshman team    Varsity B-team

Artificial solution    Water

**5** THE DATA

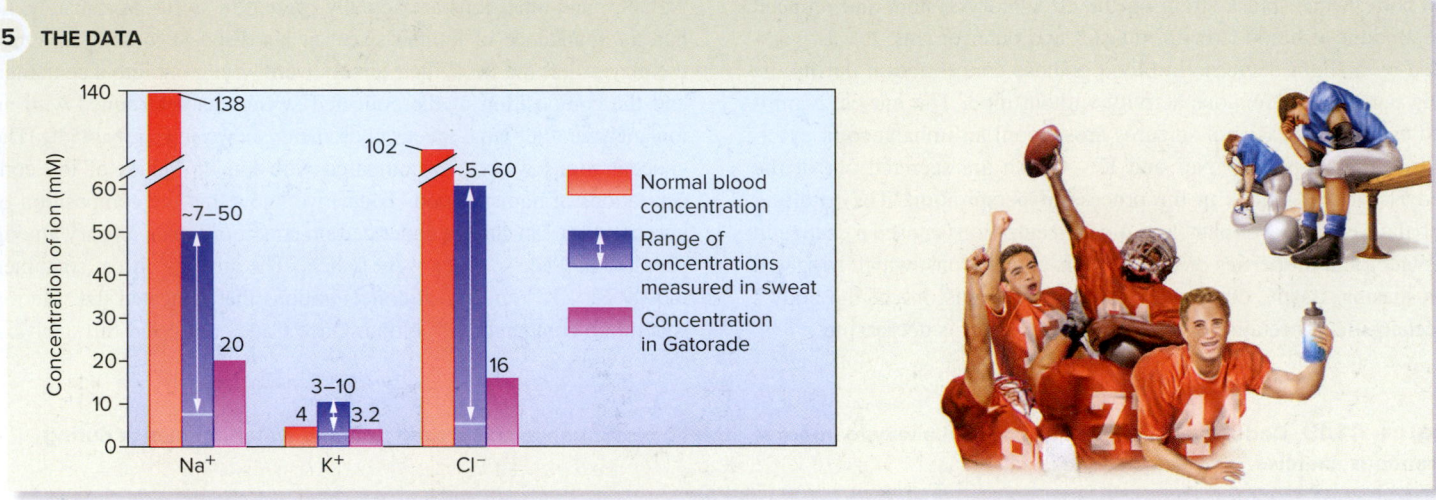

**6** **CONCLUSION** Replacement of ions and lost water using a fluid with solute concentrations similar to those found in human sweat improves athletic performance compared to water replacement alone.

**7** **SOURCE** Cade, R., et al. 1972. Effect of fluid, electrolyte and glucose replacement during exercise on performance, body temperature, rate of sweat loss, and compositional changes of extracellular fluid. *Journal of Sports Medicine and Physical Fitness* 12: 150–156.

he simply prepared an artificial solution of a composition similar to human sweat. The next step was to have the players ingest the solution before and during the practice sessions and games. Improving its taste—adding some lemon flavoring and sugar—removed any inhibitions the players may have had about drinking it, while also providing an energy boost and helping maintain glucose homeostasis.

For the first trial, Cade gave the solution to the freshman players during an intrasquad scrimmage against the more experienced varsity B-team, whose members received only pure water to drink as a control. At first, the freshman team appeared overmatched by the varsity, as might be expected. In the second half of the scrimmage, however, the freshman team vastly outperformed the more experienced players and did not suffer the characteristic late-game fatigue the B-team experienced. Based on this test, the varsity A-team was given a similar solution to drink the next day during a game against a heavily favored opponent, whom they beat handily on a hot 39°C day.

In subsequent years, Cade and other researchers conducted carefully controlled experiments with humans and laboratory animals to confirm that a balanced solution of ions similar to that present in sweat effectively improves exercise performance and reduces the possibility of dehydration and its consequences.

Because the solution was envisioned as an aid for the team known as the University of Florida Gators, the drink eventually came to be called Gatorade. The year after its introduction, the Gators enjoyed their most successful season. In 1965, the Kansas City Chiefs of the former American Football League became the first professional sports team to try the drink, and shortly thereafter, the team enjoyed its greatest success. Nowadays, Gatorade and similar sports drinks are used at sporting events around the world, and for good scientific reason.

The effectiveness of a solution like Gatorade is due to its ability to restore the correct amounts of both water and ions lost during exercise. Importantly, it is very rapidly absorbed because its osmolarity is close to that of body fluids. Many of the other sports drinks subsequently invented contain additional solutes, such as vitamins and other minerals, and many contain higher amounts of sugar. Because of the presence of these other solutes, these drinks may be very hyperosmotic (a greater concentration of solutes) relative to body fluids.

The story of Gatorade is one of good common sense based on solid scientific principles of osmolarity and ion and water homeostasis. You can now understand why drinking a dilute salt solution during strenuous exercise is better than drinking water. Although drinking pure water prevents dehydration, drinking water in excess will actually decrease plasma ion concentrations to below normal. In other words, it will replace one type of ion imbalance with another.

## Animals Adapt to Osmotic Challenges by Regulating or Conforming

Animals adapt to osmotic challenges posed by the environment in one of two major ways. Some animals regulate their internal osmolarity at a very stable level, whereas others conform to the osmolarity of their environment (for example, the sea). Animals that maintain very stable internal ion concentrations and osmolarities, even when living in water that has an osmolarity that is very different from that of their body fluids or living on land, are called **osmoregulators**. Such animals drink or excrete water and ions as necessary to maintain an internal osmolarity that is generally about 300 mOsm/L, or about one-third that of seawater and at least 10 times that of fresh water. All terrestrial animals are osmoregulators, as are all freshwater animals and many marine animals, including bony fishes and some crustaceans. Osmoregulators maintain stable cellular levels of ions and water, but this requires considerable expenditure of energy, primarily to pump ions into and out of epithelial cells.

Most marine invertebrates and some vertebrates—notably sharks—use a different means to control body fluid composition. In this case, the osmolarity of extracellular and intracellular fluids is matched with seawater. These animals are called **osmoconformers**, because their osmolarity conforms to that of their environment. The osmolarity of blood and other fluids of marine osmoconformers is like that of seawater, around 1,000 mOsm/L. An advantage of having body fluids conform to the osmolarity of the surrounding seawater is that there is much less tendency to gain or lose water by osmosis across the skin or gills. Thus, sharks and other osmoconformers expend less energy to compensate for water gain or loss than do other aquatic animals. However, osmoconformers are generally limited to the marine environment.

Vertebrate osmoconformers have a high concentration of uncharged molecules dissolved in their extracellular fluids. This allows the extracellular fluids and seawater to have similar osmolarities, while preventing an excessive accumulation of ions in the body. The body fluids of sharks and other osmoconformers contain sugars, amino acids, and metabolic waste products that produce an osmolarity very similar to that of seawater, even though the concentrations of $Na^+$, $K^+$, and other critical ions in these fluids are similar to those of osmoregulators.

Vertebrate osmoconformers cannot tolerate high ion concentrations in their body fluids any better than can osmoregulators. One reason is because a proper ion balance is required for normal electrical signaling in neurons and muscle cells. In addition, very high ion concentrations tend to disrupt the three-dimensional structure of many proteins, rendering them inactive. Consequently, the body fluids of vertebrate osmoconformers are less salty—that is, they have fewer ions—than seawater, as is also the case for all osmoregulators. Therefore, vertebrate osmoconformers such as sharks tend to gain ions by diffusion across their gills. The excess ions are eliminated by the kidneys and a type of salt gland called the rectal gland.

## Summary of Key Concepts

### 41.1 Organization of Animal Bodies

- In an animal's body, differentiated cells with similar properties associate with each other to form tissues, which combine with other types of tissues to form organs. Organs are functionally and in some cases anatomically linked to form organ systems (Figure 41.1).

- Muscle tissues consist of cells specialized to contract. The three categories of muscle tissue are skeletal, smooth, and cardiac. Nervous tissues initiate and conduct electrical signals from one part of an animal's body to another part. Epithelial tissues are specialized to protect structures and to secrete and absorb ions and organic molecules. Connective tissues connect, surround, anchor, and support the structures of an animal's body (Figures 41.2, 41.3, 41.4, 41.5).

- An organ is composed of two or more kinds of tissues. In an organ system, different organs work together to perform an overall function (Figure 41.6, Table 41.1).

- The development, spatial positioning, and functions of many body organs are under the control of *Hox* genes in vertebrates.

### 41.2 The Relationship Between Structure and Function

- The structure of an animal's tissues and organs are related to the function of those structures (Figure 41.7).

- Extensive surface area maximizes the ability of a tissue or organ to absorb solutes, exchange oxygen and carbon dioxide with the environment, communicate with other cells, and receive sensory information from the environment. The ratio between a structure's surface area and its volume is called the surface area/volume (SA/V) ratio (Figure 41.8).

### 41.3 General Principles of Homeostasis

- Homeostasis is the process of maintaining a relatively stable internal environment despite changes in the external environment. Some animals conform to their environment, and others regulate internal processes in response to their environment.

- Vertebrates maintain most physiological variables within a certain range despite variations in external environmental conditions (Table 41.2, Figure 41.9).

- Homeostatic control systems regulate the activities of cells, tissues, and organs. Negative feedback loops minimize changes in a variable and prevent homeostatic responses from overcompensating (Figures 41.10, 41.11).

- Positive feedback loops accelerate a process and do not achieve homeostasis (Figure 41.12).

- Feedforward regulation prepares an animal's body for an upcoming challenge to homeostasis (Figure 41.13).

- Chemical communication between cells is essential to homeostasis. Local and long-distance chemical signals coordinate homeostatic processes.

## 41.4 Homeostatic Control of Internal Fluids

- The body fluids of many animals are located in fluid compartments and are of three types: intracellular fluid, plasma, and interstitial fluid. Alterations in intracellular fluid volume can have drastic effects on cell shape (Figures 41.14, 41.15).

- Exchanges of ions and water with the environment resulting from vital processes, such as respiration, feeding, and the elimination of wastes, are called obligatory exchanges (Figure 41.16).

- The solute concentration of a solution is known as the solution's osmolarity. Fishes and other water-breathing animals that live in fresh water and those that live in salt water face opposite osmoregulatory challenges (Figures 41.17, 41.18).

- Robert Cade and his colleagues discovered that fluid replacement during strenuous exercise is particularly beneficial if the fluid contains solutes at concentrations resembling those in human sweat (Figure 41.19).

- Animals that maintain constant internal ion concentrations and osmolarities are called osmoregulators. Animals in which internal osmolarity conforms to the osmolarity of the environment are called osmoconformers.

## Assess & Discuss

### Test Yourself

1. Tissue that is specialized to conduct electrical signals from one structure in the body to another structure is_____tissue.
   - a. epithelial
   - b. connective
   - c. nervous
   - d. muscle

2. Structures composed of two or more tissue types arranged in various proportions and patterns are
   - a. cells.
   - b. tissues.
   - c. organs.
   - d. organ systems.
   - e. organisms.

3. The extracellular matrix (ECM) is partly formed by some types of connective tissue cells and
   - a. contains fibrous proteins that provide structural support to cells.
   - b. provides a scaffolding for the cells.
   - c. plays a role in cellular communication.
   - d. does all of the above.
   - e. does a and b only.

4. From an examination of the structure of many animal organs,
   - a. it is apparent that an organ's surface area increases more than its volume as an organ enlarges.
   - b. a general function can sometimes be predicted based on the structural adaptations.
   - c. it can be seen that different tissues do not come into contact with each other within an organ.
   - d. all four tissues are equally represented in all organs.
   - e. None of the above is correct.

5. Most of the water in an animal's body
   - a. lacks any type of dissolved ions or other solutes.
   - b. is found in the spaces between cells.
   - c. is contained inside the cells.
   - d. is located in the extracellular fluid.
   - e. is unable to move between body compartments.

6. The folds, convolutions, or extensions found in many animal structures result in
   - a. decreased level of activity in that particular structure.
   - b. interruption in the normal functioning of the structure.
   - c. increased surface area for absorption, communication, or exchange.
   - d. increased volume without a change in surface area.
   - e. none of the above.

7. Adapting to changes in the external environment and maintaining internal variables within physiological ranges is
   - a. equilibrium.
   - b. a conditioned response.
   - c. positive feedback.
   - d. homeostasis.
   - e. both c and d.

8. Which of the following statements regarding negative feedback is *false*?
   - a. It helps regulate variables such as body temperature and blood pressure.
   - b. It is the mechanism by which platelets seal a wound in a blood vessel in mammals.
   - c. It is a major feature of homeostatic control systems.
   - d. It prevents homeostatic responses from overcompensating.
   - e. It may occur at the organ, cellular, or molecular level.

9. A change in ion concentrations in an animal's body may result in
   - a. altered membrane potentials that disrupt normal cell function.
   - b. disruption of certain biochemical processes that occur in the cell.
   - c. movement of water between fluid compartments.
   - d. a and b only.
   - e. a, b, and c.

10. Which statement is *false*?
    - a. Sweat has a lower solute concentration than internal body fluids such as blood.
    - b. Some animals can survive without drinking water.
    - c. A solution with an osmolarity of 1,000 mOsm/L is hyperosmotic relative to the body fluids of most fishes.
    - d. Marine fishes, but not freshwater fishes, drink the water they swim in.
    - e. A 300 mOsm/L solution of NaCl contains 300 mOsm/L $Na^+$ and 300 mOsm/L $Cl^-$.

### Conceptual Questions

1. Describe the relationship between structure and function in animals, and explain how surface area and volume are related.

2. Define homeostasis, and give examples of homeostatically regulated physiological variables in animals.

3. **Core Concept: Energy and Matter** How are energy use and the maintenance of homeostasis related?

### Collaborative Questions

1. Define the terms negative feedback loop and positive feedback loop, and explain how they differ.

2. Discuss the organization of animal bodies from the cellular to the organ system level. Given that organ systems interact extensively with each other, should each system really be considered distinct, or is it possible to think of an animal's body as being one large, integrated "system"?

# Neuroscience I: Cells of the Nervous System

# 42

**Fluorescent stain of a section through a mammalian brain.** The nervous systems of animals contain signaling cells called neurons with different structures and functions. In this two-photon fluorescent micrograph of a section of a mammalian brain, the red-colored cells are known to function in controlling an animal's ability to move. ©NIGMS/Yinghua Ma/Timothy Vartanian/Cornell University /Science Source

**A**s you begin reading, stop and think. Can you describe everything you are doing right now? Your eyes are sensing light reflected off this page, and your brain is interpreting the meanings of the words you are reading. You may be hearing sounds and, in some cases, choosing to ignore them as you concentrate on reading. Your digestive system may be sending signals about hunger, or perhaps if you've just eaten, you feel full and are digesting what you ate. You are breathing, perspiring, feeling the chair on which you're sitting, and your heart is beating. All of these processes and many others are under the control of the **nervous system**, coordinated circuits of cells that sense internal and environmental changes and transmit signals that enable you to respond in an appropriate way. Our nervous system helps us exert control over our bodies. It also allows us, like other animals, to sense what is going on in the outside world, initiate actions that influence events and respond to demands, and regulate internal processes. You are conscious of some of these functions, such as reading this text or feeling hungry. Many others, however, such as maintaining your body temperature and controlling your heart rate, occur without your awareness.

**Neuroscience** is the scientific study of nervous systems. Neuroscientists are interested in topics such as the structure and function of the brain and the biological basis of consciousness, memory, learning, and behavior. Neuroscience is an interdisciplinary field that interfaces with other disciplines such as cell and molecular biology, chemistry, physics, psychology, and linguistics. It is experiencing an unprecedented level of new discoveries and rapid growth.

In this chapter, our focus on neuroscience will be at the cellular level. Nervous systems are composed of circuits of **neurons**, highly specialized cells that communicate with each other and with other types of cells by electrical or chemical signals (see the chapter opening photo). In many animals, such as ourselves, neurons become organized into a central processing area of the nervous system called a **brain**. The brain sends commands to and receives signals from various parts of the body through **nerves**—bundles of neuronal cell extensions encased in connective tissue and projecting to and from various tissues and organs.

We begin by investigating the special features of neurons that make them suited for rapid communication between cells. We will look at electrical and chemical gradients across the neuronal plasma membrane and how these gradients produce a way for neurons to convey signals. Next, we will explore how neurons send and receive signals. We will consider general features of the main classes of signaling molecules called neurotransmitters and end by examining the various effects of therapeutic, recreational, and illicit drugs on neurotransmitter action. Chapters 43 and 44 will explore how nervous systems evolved and are organized, how the brain functions, and how animals use their nervous systems to sense the world around them.

**42.1** # Cellular Components of Nervous Systems

## Learning Outcomes:

1. Explain the difference between the central nervous system (CNS) and the peripheral nervous system (PNS).

2. List the cellular components of a nervous system, and describe the function of each cell type.

3. **CoreSKILL »** Identify the different parts of a typical neuron, and explain how the structure of each relates to its function.

4. Describe the functional relationships among sensory neurons, motor neurons, and interneurons.

5. **CoreSKILL »** Explain what a reflex is, and propose reasons why reflexes are important in animals.

The organization of the nervous system permits extremely rapid responses to changes in an animal's external or internal environment. In many animals, the **central nervous system (CNS)** consists of a brain and a nerve cord, which in vertebrates extends from the brain through the vertebral column and is called the **spinal cord**. The **peripheral nervous system (PNS)** consists of all neurons and projections of their plasma membranes that are outside of but connect with the CNS, such as projections that end on muscle and gland cells. In certain invertebrates with simple nervous systems, the distinction between a CNS and PNS is less clear or not present.

The evolution of nervous systems has given animals the ability to receive information about the environment via their PNS, transmit that information along nerves to a CNS where the information is interpreted, and, if necessary, initiate a behavioral response via their PNS (**Figure 42.1**). For example, if a hungry hyena receives stimuli such as the smell and taste of food, odor-sensing cells in the nose and taste-sensing cells in the tongue act as receptors for the stimuli and then send signals along nerves to the brain. There, the signals are interpreted and recognized. The brain then sends a signal via nerves that stimulate gland cells in the mouth, which respond by producing saliva in preparation for the arrival and swallowing of food. In this section, we will survey the general properties of the cells of the central and peripheral nervous system.

## Cells of the Nervous System Are Specialized to Receive and Send Signals

Nervous systems transfer signals from one part of the body to another and direct the activities of cells, tissues, organs, and glands. Although these are complex tasks, nervous systems have only two unique classes of cells: neurons, which function in signaling, and glia, which have numerous support roles.

*Neurons*   All animals except sponges have neurons, cells that send and receive electrical and chemical signals to and from each other and other types of cells throughout the body. The number of neurons in the nervous systems of different species varies widely, partly as a function of the size of an animal's head and brain, but also as a function of the complexity of its behavior. As a comparison, the tiny, short-lived nematode *Caenorhabditis elegans* has 302 neurons in its nervous system, compared with several thousand in a wasp, several

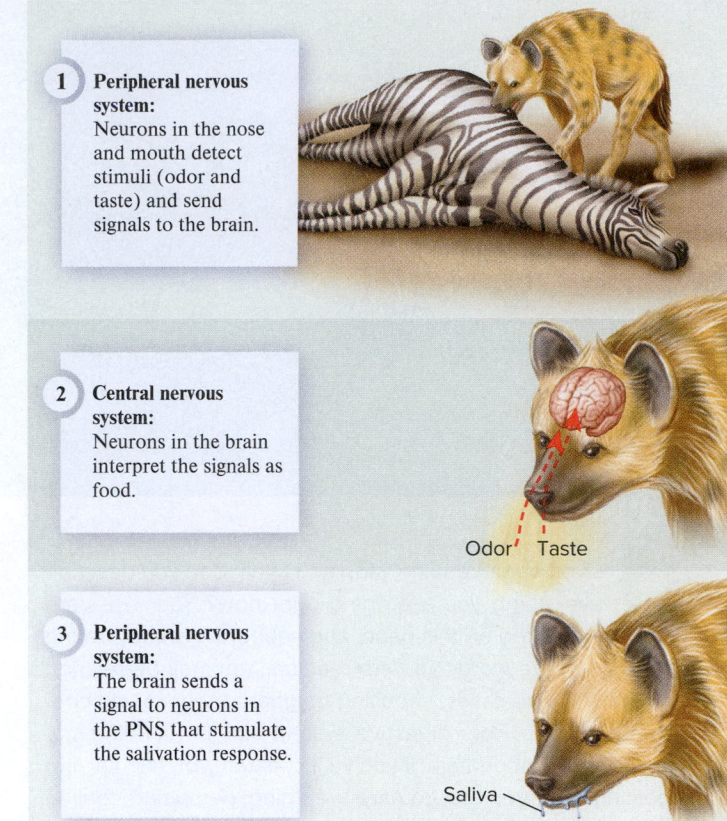

**1**  **Peripheral nervous system:** Neurons in the nose and mouth detect stimuli (odor and taste) and send signals to the brain.

**2**  **Central nervous system:** Neurons in the brain interpret the signals as food.

Odor  Taste

**3**  **Peripheral nervous system:** The brain sends a signal to neurons in the PNS that stimulate the salivation response.

Saliva

**Figure 42.1 Functional relation between the central and peripheral nervous systems.** In this example, a hungry hyena senses odor and taste, and the brain interprets these signals as a potential food source. This initiates a biological response (salivation) that prepares the hyena for eating.

hundred thousand in a salamander, 300 million in an octopus, and over 100 billion in a human!

Regardless of their total number, neurons in one animal species look and function much like neurons from any other species. A neuron is composed of a **cell body** (sometimes called the soma), which contains the cell nucleus and other organelles (**Figure 42.2a** and **b**). Two types of extensions or projections arise from the cell body: dendrites and the axon. **Dendrites** (from the Greek word *dendron*, meaning tree) may be single projections of the cell body but more commonly are elaborate treelike structures with numerous branching extensions that provide a large surface area for contacts with other neurons. Electrical and chemical messages from other neurons are received by the dendrites, and electrical signals generated in the dendrites move toward the cell body (**Figure 42.2c**). The cell body processes these signals, as well as others that it receives directly, and may generate an outgoing signal to a structure called an axon.

An **axon** is an extension of the cell body that transmits signals along its length and eventually to other cells. An axon may be only a few micrometers long or as long as 2 m, such as those in very large or long-limbed animals. A typical neuron has a single axon, which may have branches and which may be wrapped in an insulating layer of glia cell tissue called myelin (described shortly). The part of the axon

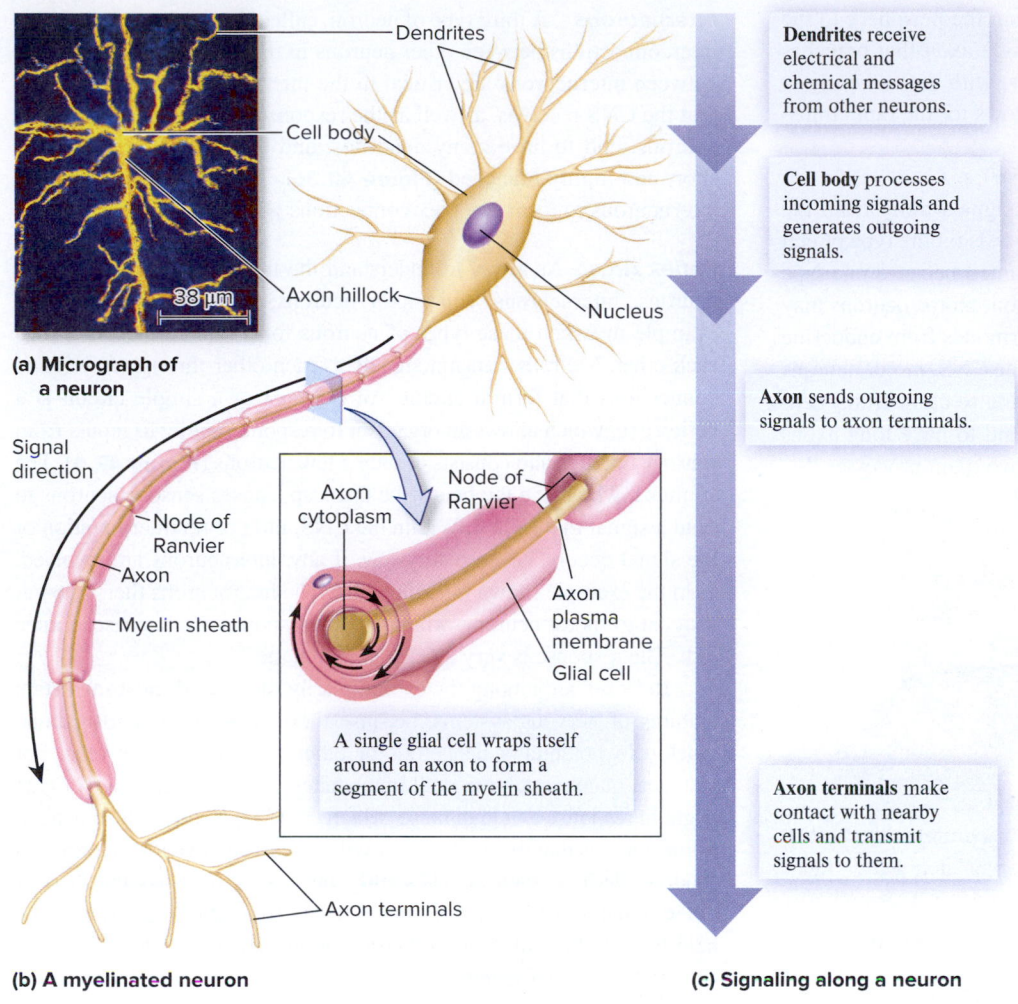

(a) Micrograph of a neuron

38 μm

Dendrites

Cell body

Axon hillock

Nucleus

Signal direction

Node of Ranvier

Axon

Myelin sheath

Axon cytoplasm

Node of Ranvier

Axon plasma membrane

Glial cell

A single glial cell wraps itself around an axon to form a segment of the myelin sheath.

Axon terminals

(b) A myelinated neuron

**Dendrites** receive electrical and chemical messages from other neurons.

**Cell body** processes incoming signals and generates outgoing signals.

**Axon** sends outgoing signals to axon terminals.

**Axon terminals** make contact with nearby cells and transmit signals to them.

(c) Signaling along a neuron

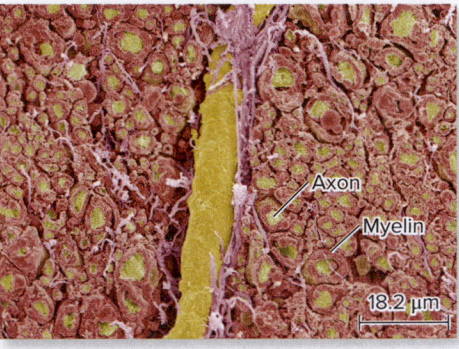

Axon

Myelin

18.2 μm

(d) Cross section of a nerve

**Figure 42.2 Structure and basic function of a typical vertebrate neuron and associated glial cells. (a)** A stained neuron seen at high magnification (confocal fluorescence microscopy). **(b)** A diagrammatic representation of a peripheral neuron with associated glial cells—in this instance, a type called Schwann cells. The Schwann cells wrap their membranes around the axon at regular intervals, creating a structure called a myelin sheath that is interrupted by bare patches called nodes of Ranvier. **(c)** The structures involved in signaling by a neuron are identified; signals travel in the direction shown by the arrow. **(d)** Cross section of a nerve as seen in a false-color SEM. Axons are light green and surrounded by myelin sheaths (red). Connective tissue fibers (purple) and a portion of a blood vessel (yellow) can also be seen. a: ©James Cavallini/BSIP/Phototake; d: ©Steve Gschmeissner/Science Source

closest to the cell body is named the **axon hillock**. As we will see later, the axon hillock is important in the generation of the electrical signals that travel along an axon. At the other end of the axon are one or more **axon terminals**, which convey electrical or chemical messages to other cells, such as other neurons or muscle or gland cells.

Within an animal's body, many axons tend to run in parallel bundles to form nerves, within and around which are protective layers of connective tissue (**Figure 42.2d**). Nerves enter and leave the CNS and transmit signals in both directions between the PNS and the CNS. Along the way, the axon terminals communicate with particular cells of the body.

*Glia* Glia (from the Greek, meaning glue) are cells that surround neurons and perform numerous functions. One type of glia, called astrocytes, provide metabolic support for neurons and are also involved in forming the blood-brain barrier, which is a physical barrier between blood vessels and most parts of the CNS. This barrier prevents the passage of toxins and other damaging chemicals from the blood into the extracellular fluid around neurons in the CNS, but allows passage of nutrients and gases. Astrocytes also help to maintain a stable concentration of ions in the extracellular fluid. Other glia, called microglia, participate in immune functions and remove cellular debris produced by damaged or dying cells.

In vertebrates, specialized glial cells wrap around certain axons at regular intervals to form an insulating layer called a **myelin sheath** (see Figure 42.2b and d). The sheath is periodically interrupted by noninsulated gaps called **nodes of Ranvier**. In the vertebrate brain and spinal cord, the myelin-producing glial cells are called **oligodendrocytes**. **Schwann cells** are the glial cells that form myelin on axons that travel outside the brain and spinal cord. As we will see later, myelin and the nodes of Ranvier increase the speed with which electrical signals pass along the axon.

## Sensory Neurons, Motor Neurons, and Interneurons Form Pathways in a Nervous System

Neurons can be categorized into three main types: sensory neurons, motor neurons, and interneurons. The structures of each type reflect their specialized functions.

*Sensory Neurons* As their name suggests, **sensory neurons** detect or sense information from the outside world, such as light, odors, touch, or heat. In addition, sensory neurons detect internal body conditions such as blood pressure or body temperature. Sensory neurons are also called afferent (from the Latin, meaning to bring toward) neurons because they transmit information from the periphery to the CNS. Many sensory neurons have a long, single axon that branches into a peripheral process and a central process, with the cell body in between (**Figure 42.3a**). This arrangement allows for the rapid transmission of a sensory signal to the CNS.

***Motor Neurons***    **Motor neurons** transmit signals away from the CNS and elicit some type of response that depends on the type of cell receiving the signal. Motor neurons are so named because one type of response they cause is movement. In addition, motor neurons may cause other effects such as the secretion of hormones from endocrine glands. Because they send signals away from the CNS, motor neurons are also called efferent (from the Latin, meaning to carry from) neurons. Like sensory neurons, motor neurons tend to have long axons (**Figure 42.3b**), but these do not branch into two main processes.

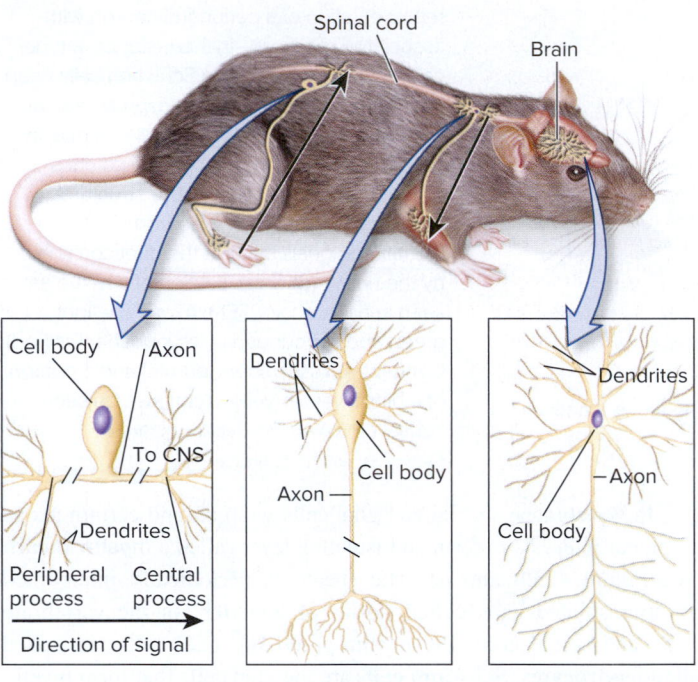

**(a) Sensory neuron**   **(b) Motor neuron**   **(c) Interneuron**

**Figure 42.3   Types of neurons. (a)** Vertebrate sensory neurons are afferent neurons with an axon that bypasses the cell body and projects to the CNS. **(b)** Motor neurons are efferent neurons that transmit signals away from the CNS and usually have long axons that enable them to act on distant cells. In (a) and (b), the hatch marks indicate that the axons are not to scale, but are in fact much longer. **(c)** Interneurons are usually short neurons that connect two or more other neurons within the CNS. Although short, the axons and dendrites may have extensive branches, allowing them to receive many inputs and transmit signals to many neurons.

 **Core Concept: Structure and Function**  Cells are the simplest units of any organ system, including the nervous system. However, the structural variety and complexity of cells within an organ system can be great, as seen here. These structural differences are responsible for the different functions of neurons.

***Interneurons***    A third type of neuron, called the **interneuron**, forms interconnections between other neurons in the CNS. The signals sent between interneurons are critical in the interpretation of information that the CNS receives, as well as the response that it may elicit. Interneurons tend to have many dendrites, and their axons are typically short and highly branched (**Figure 42.3c**). This arrangement allows interneurons to form complex connections with many other cells.

***Reflex Arcs***    As a way to understand the interplay between sensory neurons, interneurons, and motor neurons, let's consider a simple example in which these types of neurons form interconnections with each other. Neurons transmit signals to each other through a series of connections that form a circuit. An example of a simple circuit is a **reflex arc**, which allows an organism to respond rapidly to inputs from sensory neurons and consists of only a few neurons (**Figure 42.4**). The stimulus, which is a tap below the kneecap, causes sensory neurons to send a signal to the CNS. Within the CNS, little or no interpretation of the signal occurs, because very few, if any, interneurons are involved, as in the example shown in Figure 42.4. The interneurons then transmit a signal to motor neurons, which elicit a response, in this case, a knee jerk. The response is very quick and automatic.

Reflexes are among the evolutionarily oldest and most important features of nervous systems, because they allow animals to respond quickly to potentially dangerous or otherwise important events. For instance, many vertebrates will immediately cringe, jump, leap, or take flight in response to a loud noise, which could represent sudden danger. Some animals that live in the water will reflexively dive in response to a shadow overhead, which could signify the presence of a passing shark or other predator. Many infant primates have strong grasping reflexes that help them hold onto their mother as she moves about. Other reflexes, such as the patellar tendon reflex just described, are important for postural changes and locomotion. Countless examples of useful reflexes are found in animals, and their importance is evident from the observation that they arose early in evolution and exist in nearly all animals.

## 42.2 Electrical Properties of Neurons and the Resting Membrane Potential

**Learning Outcomes:**

1. Explain the meaning of membrane potential.
2. Describe how the resting membrane potential is established and maintained.
3. Describe how an electrochemical gradient determines the direction in which an ion will move.
4. **CoreSKILL »** Write the Nernst equation, and use it to predict the direction in which different ions move across a plasma membrane.

In the late 18th century, Italian scientists Luigi Galvani and Alessandro Volta examined the ways in which frog leg muscles could be stimulated to contract. They dissected the muscles along with their associated nerves and placed them in a saline (NaCl) solution. The saline solution approximated the ion concentrations normally found in frog plasma, and helped keep the muscles and nerves alive for a short time after removal from the animal. The two scientists discovered that when they stimulated the nerve

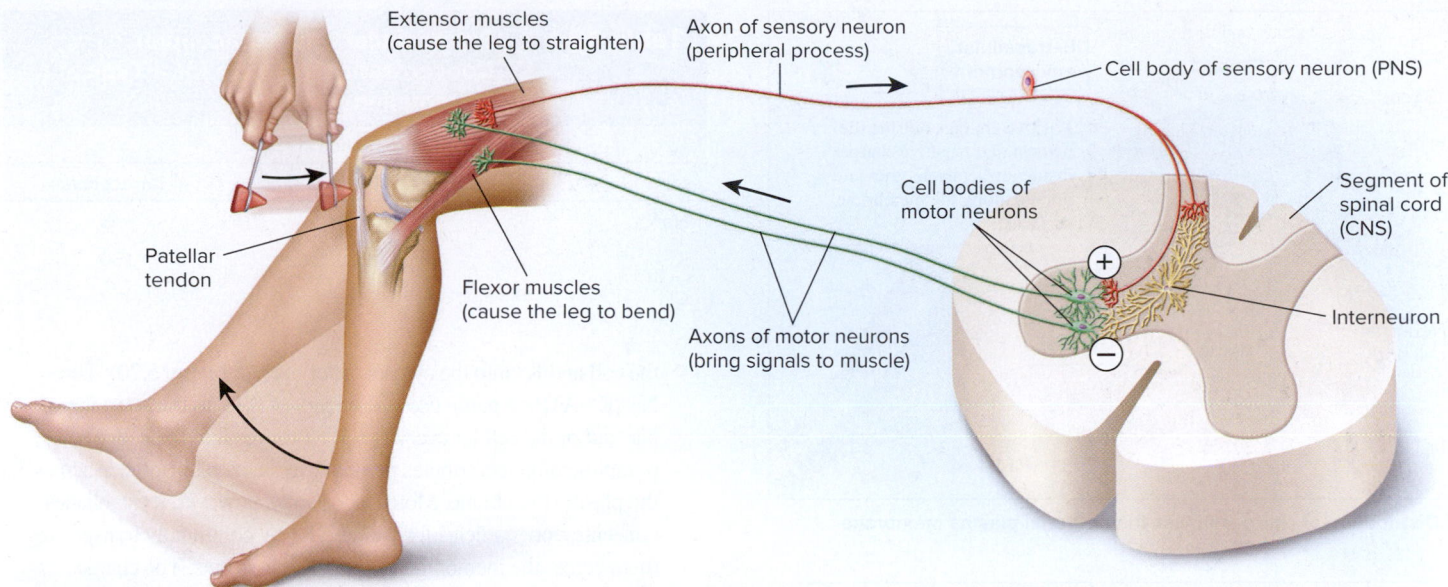

**Figure 42.4   A reflex arc.** The knee-jerk response is an example of a reflex arc. A tap below the kneecap (also known as the patella) stretches the patellar tendon, which acts as a stimulus for a sensory neuron. This stimulus initiates a reflex arc that activates (+) a motor neuron that causes the extensor muscle on top of the thigh to contract. At the same time, an interneuron inhibits (–) the motor neuron of the flexor muscle, causing it to relax.

**Concept Check:** *Animals have many types of reflexes. Once initiated, must all reflexes occur to completion, or do you think that in some cases they may be overridden or partially suppressed?*

that was attached to a muscle or directly stimulated the muscle itself with a source of electric current, the muscle contracted. They concluded that electricity was required for muscle contraction to occur, whether it originated at the nerve or the muscle itself. Eventually, Galvani postulated that electric current could somehow be generated by the nerve and muscle themselves, something he called "animal electricity."

Today, we know that Galvani's animal electricity comes from neurons, which use electrical signals to communicate with other neurons, muscle cells, or gland cells. These signals, often called nerve impulses but properly called action potentials, involve changes in the amount of electric charge across a neuron's plasma membrane. In this section, we will first examine the electrical and chemical gradients across the plasma membrane of neurons. Later, we will explore how such gradients provide a way for neurons to conduct action potentials. (In Chapter 45, we will discuss how muscle cells, too, can generate action potentials.)

## Neurons Establish Differences in Ion Concentration and Electric Charge Across Their Membranes

Like all cell membranes, the plasma membrane of a neuron acts as a barrier that separates charges. Ion concentrations differ between the interior and exterior of a cell, and this sets the stage for the establishment of differences in the net charge across a membrane. Such differences in charge act as an electrical force measured in **millivolts** (mV), named after Alessandro Volta. Analogous to a battery, neurons have negative and positive poles, but these are the inside and outside surfaces, respectively, of the plasma membrane. For this reason, a neuron is said to be electrically polarized. The difference between

the electric charges along the inside and outside surfaces of a cell membrane is called a potential difference, or **membrane potential**. The **resting membrane potential** refers to the membrane potential of an unstimulated cell that is not generating action potentials.

Let's begin our discussion of electrical signaling by examining how the resting membrane potential is established and maintained. When investigators first measured the resting membrane potential of neurons, they registered a voltage of about −70 mV inside the cell with respect to the outside. This means that the interior of the cell had a more negative charge than the exterior, which turns out to be typical of animal cells in their resting state. For comparison, a resting potential of −70 mV is tiny compared with the voltages used to provide electric current in a home (approximately 120 V), or even that of a small 1.5 V battery. Nonetheless, this tiny difference in charge across the membrane of a neuron is sufficient to provide the means for generating an action potential that can travel from one end of a neuron to the other, as we will see later in this chapter.

The resting membrane potential is determined by the ions located along the inner and outer surfaces of the plasma membrane (**Figure 42.5a**). Ions of opposite charges align on either side of the membrane because they are attracted to each other due to electrical forces. Negative ions within the cell are drawn to positive ions arrayed on the outer surface of the plasma membrane. Although there are more positive charges along the outside surface of a neuron and more negative charges inside, the actual number of ions that contribute to the resting membrane potential is extremely small compared with the total number of ions inside and outside the cell. **Table 42.1** lists the ions that are important in establishing the resting potential and their typical intracellular and extracellular concentrations in mammals and many other vertebrates. The ions that are critical for establishing the resting membrane potential are $Na^+$ and $K^+$ and, to a lesser extent, $Cl^-$.

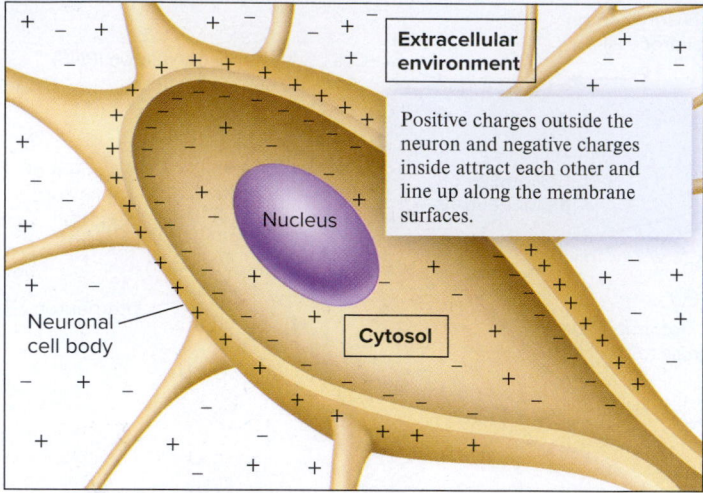

**(a) Distribution of charges across the neuronal plasma membrane**

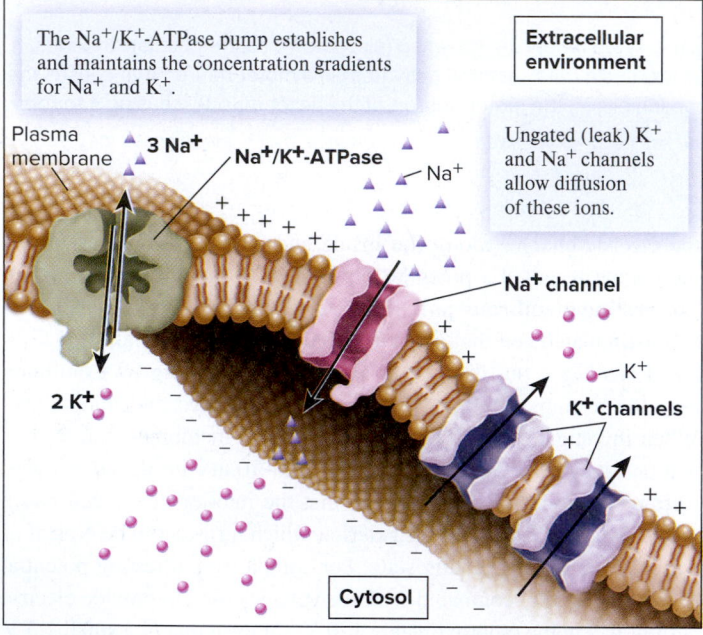

**(b) Two major factors that influence the resting membrane potential**

**Figure 42.5  The resting membrane potential.** The slight excess of negative charges inside and positive charges outside the cell membrane is shown in part **(a)**. In part **(b)**, the two major factors that contribute to this charge distribution are shown: the Na⁺/K⁺-ATPase pump that establishes ion concentration gradients, and ungated ion channels that permit diffusion of Na⁺ and especially K⁺ across the membrane.

**Core Concept: Energy and Matter** Animals use energy for many processes. The Na⁺/K⁺-ATPase pump uses the energy stored in the bonds of ATP to transport ions across the neuronal plasma membrane. Establishing and maintaining resting membrane potentials accounts for a significant fraction of the energy consumed per day in most animals.

Two factors are primarily responsible for determining the resting membrane potential (**Figure 42.5b**).

1. *Establishment of ion concentration gradients.* The Na⁺/K⁺-ATPase pump within the plasma membrane continually moves Na⁺ out of

**Table 42.1   Extracellular and Intracellular Concentrations of Ions for a Typical Mammalian Neuron**

| Ion | Concentration (mM) | |
|-----|-----|-----|
| | Extracellular | Intracellular |
| Na⁺ | 145 | 15 |
| K⁺ | 5 | 150 |
| Cl⁻ | 110 | 7 |

the cell and K⁺ into the cytosol (refer back to Figure 5.20). The Na⁺/K⁺-ATPase pump uses the energy of ATP to transport three Na⁺ out of the cell for every two K⁺ it moves into the cell. The pump therefore contributes modestly to a charge difference across the plasma membrane. More importantly, however, it establishes concentration gradients for Na⁺ and K⁺ by continually transporting them across the membrane in opposite directions. The charge difference and concentration gradients will determine the directions in which the ions will move by diffusion either into or out of the cell.

2. *Unequal membrane permeabilities to different ions.* The plasma membrane contains ion-specific channels that affect the permeability of the membrane to Na⁺ and K⁺. Ungated channels that are specific for Na⁺ or K⁺ influence the resting potential by allowing the passive movement of these ions. An ungated channel, sometimes referred to as a leak channel, is one that is open at rest and that does not require a stimulus such as ligand binding to open. Most neurons have about 10 to 100 times more ungated K⁺ channels than ungated Na⁺ channels. Therefore, at rest, the membrane is more permeable to K⁺ than to Na⁺. As you might predict from their concentration gradients shown in Table 42.1, Na⁺ tends to diffuse into cells and K⁺ diffuses out of cells through their respective channels when the cell is at rest. The greater number of leak channels for K⁺, therefore, means that diffusion of K⁺ makes a more significant contribution to resting membrane potential than does diffusion of Na⁺, and excess positive charges exit the cell.

Now let's see how this movement of ions occurs, and how it produces the resting membrane potential.

## An Electrochemical Gradient Governs the Movement of Ions Across a Membrane

The direction in which an ion diffuses depends on the **electrochemical gradient** for that ion, which is the combined effect of both an electrical and a chemical (concentration) gradient. **Figure 42.6** considers the concept of an electrochemical gradient for K⁺, using two compartments separated by a semipermeable membrane that permits the diffusion of only K⁺. Figure 42.6a illustrates an electrical gradient. In this case, the concentration of K⁺ is equal on both sides of the membrane, but the concentrations of other ions (Na⁺ and Cl⁻) are unequal on opposite sides of the membrane and thereby produce an electrical gradient. Because K⁺ is positively charged, it is attracted to the side of the membrane with more negative charge. Figure 42.6b shows a concentration gradient in which K⁺ concentration is higher on one side than the other. In this scenario, K⁺ diffuses from a region of high to low concentration. Finally, Figure 42.6c shows a balance of forces creating

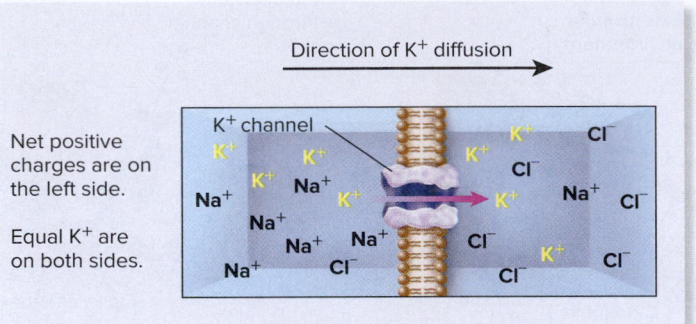

**(a) Electrical gradient, no concentration gradient for K⁺**

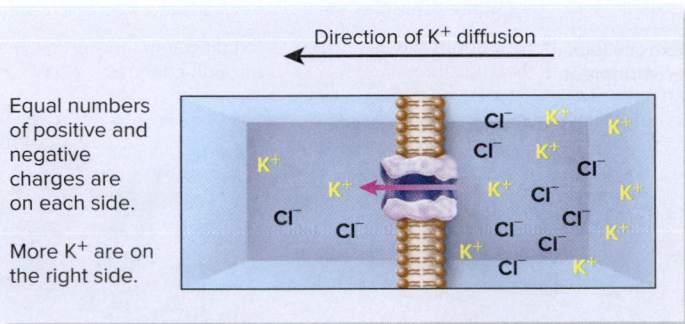

**(b) Concentration gradient for K⁺, no electrical gradient**

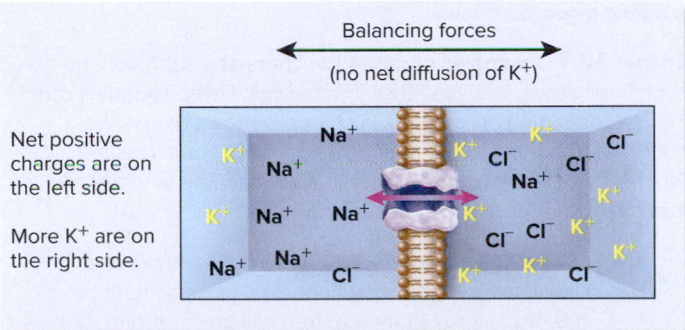

**(c) Electrochemical equilibrium**

**Figure 42.6  Electrical and chemical (concentration) gradients.** This hypothetical example depicts two compartments separated by a membrane that is permeable only to K⁺. **(a)** In this example, the compartments initially contain equal concentrations of K⁺, but an electrical gradient exists due to an unequal distribution of Na⁺ and Cl⁻. Potassium ions are attracted to the higher amount of negative charge on the right side of the membrane. **(b)** In this case, there is initially no electrical gradient across the membrane, and the left compartment contains a lower concentration of K⁺ than in the right compartment. Under these conditions, K⁺ diffuses down its chemical concentration gradient from right to left. **(c)** This example illustrates opposing electrical and concentration gradients. The right compartment contains a higher concentration of K⁺, and the left side has a higher net amount of positive charge. These gradients balance each other, so no net movement of K⁺ occurs and a state of electrochemical equilibrium is reached.

 **Core Skill: Connections** What are some biological functions of ion electrochemical gradients in living organisms? Look back at Table 5.3 for help.

an electrochemical equilibrium. The electrical gradient favors the movement of K⁺ from left to right, but the concentration gradient favors movement from right to left. These opposing forces create an electrochemical equilibrium in which there is no net diffusion of K⁺ in either direction. In living cells, the membrane potential at which this occurs for a particular ion at a given concentration gradient is referred to as that ion's **equilibrium potential**.

With two different forces—electrical and concentration gradients—acting on a given ion, is it possible to predict the direction that ion will move across a membrane at any concentration gradient? In other words, can we compare the relative strengths of the electrical and chemical gradients and predict their net effect? By measuring the membrane potential of isolated neurons in the presence of changing concentrations of extracellular ions, scientists have deduced a mathematical formula that relates electrical and concentration gradients to each other. This formula, named the **Nernst equation** after German chemist and Nobel laureate Walther Nernst who derived it, gives the calculated equilibrium potential for an ion at any given concentration gradient. For monovalent cations such as Na⁺ and K⁺ at 37°C, the Nernst equation can be expressed as

$$E = 60 \text{ mV} \log_{10} ([X_{extracellular}]/[X_{intracellular}])$$

where $E$ is the equilibrium potential; $[X]$ is the concentration of an ion, outside or inside the cell; and 60 mV is a value that depends on temperature, valence, and other factors. (For anions, the value is −60 mV.)

The Nernst equation allows researchers to predict when an ion is or is not in electrochemical equilibrium. To understand the usefulness of this equation, consider two examples. First, let's suppose the resting membrane potential of a neuron is found to be −88.6 mV, and the K⁺ concentrations are 5 mM outside and 150 mM inside the neuron. If we enter these concentrations into the Nernst equation, we obtain the equilibrium potential for K⁺:

$$E = 60 \text{ mV} \log_{10} (5/150) = 60 \text{ mV} (-1.48)$$

$$= -88.6 \text{ mV}$$

Under these conditions, where the calculated K⁺ equilibrium potential equals the actual resting membrane potential, K⁺ is in electrochemical equilibrium, and no net diffusion of K⁺ occurs, even when many K⁺ channels are open.

As a second example, let's suppose that the membrane potential is at a typical resting value of −70 mV and the Na⁺ concentration is 100 mM outside and 10 mM inside. If we enter these concentrations into the Nernst equation, we get

$$E = 60 \text{ mV} \log_{10} (100/10) = 60 \text{ mV} (1)$$

$$= 60 \text{ mV}$$

At a resting membrane potential of −70 mV, the value of +60 mV tells us that Na⁺ is not in electrochemical equilibrium. When the equilibrium potential for a given ion—calculated by the Nernst equation—and the actual resting membrane potential do not match, there will be a driving force for that ion to diffuse across the membrane. In this example, Na⁺ will diffuse into the cell, because both the electrical and concentration gradients favor an inward flow. However, if the membrane potential was +60 mV instead of −70 mV, Na⁺ would be in electrochemical equilibrium at those concentrations, and therefore, no net diffusion of Na⁺ would occur in either direction.

By establishing electrochemical gradients and maintaining them with the Na⁺/K⁺-ATPase pump, neurons have the ability to quickly allow ions to move across the plasma membrane by opening additional (gated) channels that were previously closed. The movement of a charged ion down its electrochemical gradient results in an electric current (a current is any net unidirectional flow of electric charge). This small current provides the necessary electrical signal that neurons use to communicate with one another, as we will see next.

## 42.3 Generation and Transmission of Electrical Signals Along Neurons

### Learning Outcomes:

1. Distinguish between ligand-gated and voltage-gated ion channels.
2. Compare and contrast graded potentials and action potentials.
3. **CoreSKILL »** Describe the steps involved in the generation of an action potential, beginning with the spread of excitation to the axon hillock.
4. Describe the events in the propagation of an action potential.
5. Explain how myelination influences the propagation of an action potential.

Communication between neurons begins when one cell receives a stimulus and sends an electrical signal along its plasma membrane via currents generated by ion movements. This signal then influences the next neuron in a circuit. Each signal is brief (only a few milliseconds), but a neuron may receive and transmit millions of signals in its lifetime. In this section, we will survey the amazing ability of neurons to send and receive rapid, brief, and repeated signals.

### Signaling by a Neuron Occurs Through Changes in the Membrane Potential

Recall that a cell is polarized because of the separation of charge across its membrane. At rest, the inside of the membrane is more negative than the outside. Changes in the membrane potential, therefore, are changes in the degree of polarization. **Depolarization** occurs when the cell membrane becomes less polarized, that is, less negative inside the cell relative to the surrounding fluid. As described later, when a neuron is stimulated, one or more types of membrane ion channels open. Often, these are Na⁺ channels; when they open, Na⁺ ions diffuse into the cell, bringing with them their positive charge. This makes the new membrane potential somewhat less negative than the resting membrane potential. Consequently, the membrane is said to be depolarized. In contrast, **hyperpolarization** occurs when the cell membrane becomes more polarized, that is, more negative on the inside relative to the extracellular fluid. For example, opening more K⁺ channels would result in an increase in the rate of diffusion of K⁺ out of a cell. This would make the charge along the inside of the cell membrane more negative than it normally is while at the resting membrane potential.

Whereas all cells in an animal's body have a membrane potential, neurons (and muscle cells) are called excitable cells because they can generate electrical signals by changing their membrane potentials. This is accomplished by gated ion channels, so called because they open and close in a manner analogous to a gate in a fence (**Figure 42.7**). **Voltage-gated ion channels** open and close in response to changes in

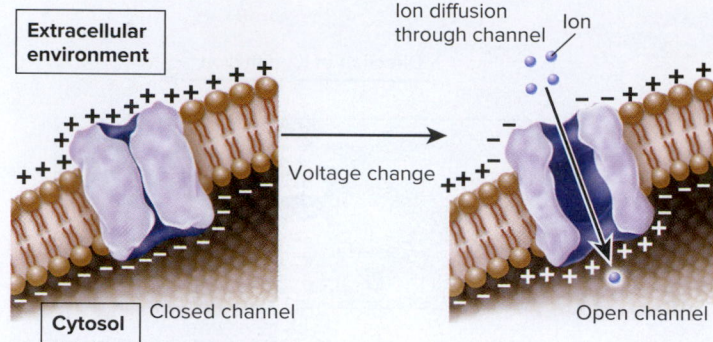

**(a) Voltage-gated ion channel**

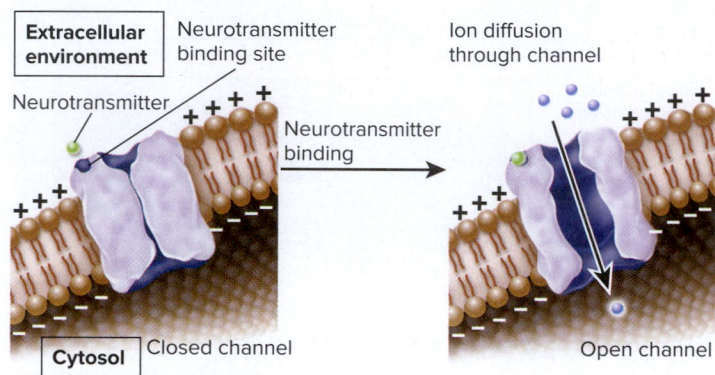

**(b) Ligand-gated ion channel**

**Figure 42.7 Examples of gated ion channels. (a)** A voltage-gated ion channel allows ions to diffuse into the cell. These channels open or close depending on changes in charge (voltage) across the membrane. **(b)** A ligand-gated ion channel opens or closes in response to ligand binding. In the example here, the binding of a neurotransmitter opens the channel.

 **Core Skill: Connections** Are ions the only substances that can move through channels in plasma membranes? Refer back to Figure 5.16 to recall an important molecule that diffuses through membranes via channels.

voltage across the membrane. **Ligand-gated ion channels** open or close when ligands, molecules such as neurotransmitters, bind to them. As we will discuss in detail later, neurotransmitters are small molecules that are secreted from an axon terminal of one neuron and diffuse to the membrane of another neuron (or other type of cell). There, they bind to and activate membrane receptors that are themselves ion channels, or are linked in some way to ion channels. The opening and closing of ligand-gated and voltage-gated ion channels are responsible for two types of changes in a neuron's membrane potential: graded potentials and action potentials.

### Graded Potentials Vary in Size Depending on the Strength of a Stimulus

A **graded potential** is any depolarization or hyperpolarization that varies depending on the strength of the stimulus. A large change in membrane potential occurs when a strong stimulus opens many channels, whereas a weak stimulus causes a small change because only a small number of channels open (**Figure 42.8**).

Graded potentials occur locally on a particular area of the plasma membrane, such as on dendrites or the cell body, where an electrical or chemical stimulus opens ion channels. From this area, a graded

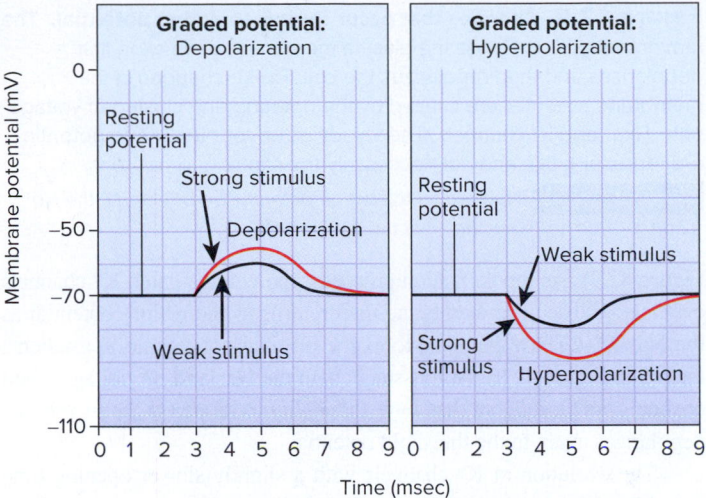

**Figure 42.8  Graded membrane potentials.** Within a limited range, the change in the membrane potential occurs in proportion to the intensity of the stimulus. If the membrane potential becomes less negative following a stimulus, the resulting change is called a depolarization. If the membrane potential becomes more negative, it is called hyperpolarization.

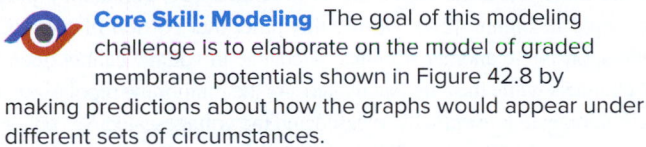 **Core Skill: Modeling** The goal of this modeling challenge is to elaborate on the model of graded membrane potentials shown in Figure 42.8 by making predictions about how the graphs would appear under different sets of circumstances.

**Modeling Challenge:** Later in this chapter, you will learn how a neuron may receive both depolarizing and hyperpolarizing stimuli, sometimes at the same time. Draw a graph like the ones in Figure 42.8 using the following information: At 2 milliseconds, a neuron simultaneously receives a strong depolarizing (SD) stimulus and also an equally strong hyperpolarizing (SH) stimulus. At 10 milliseconds, the neuron simultaneously receives an SD stimulus and a weak hyperpolarizing (WH) stimulus. You should assume that any graded potential lasts about 3.5 milliseconds, as indicated in **Figure 42.8**.

potential spreads a small distance across a region of the plasma membrane. In a short time, the membrane potential returns to the resting potential because ion pumps restore the ion concentration gradients, and the ion channels close again.

Graded potentials occur on all neurons and are particularly important for the function of sensory neurons, which must distinguish between strong and weak environmental stimuli. As discussed next, graded potentials can act as triggers for the long distance type of electrical signal, the action potential.

## Action Potentials Occur in an All-or-None Manner

**Action potentials** are the electrical events that carry a signal along an axon. In contrast to a graded potential, an action potential is always a large depolarization. All action potentials in a given neuron have a very similar amplitude, which is the degree to which an action potential changes the membrane potential away from its resting state. Once an action potential has been triggered, it occurs in an all-or-none

manner. In other words, it cannot be graded. Unlike a graded potential, an action potential is actively propagated along the axon, regenerating itself as it travels. Action potentials travel rapidly down the axon from the axon hillock to the axon terminals, where they initiate a response at a junction with another cell.

***Depolarization Phase of an Action Potential*** **Figure 42.9** shows the electrical changes that happen in a localized region of an axon when an action potential is occurring. Voltage-gated $Na^+$ and $K^+$ channels are both present in very high numbers from the axon hillock to the axon terminals. An action potential begins when a graded potential is large enough to spread to the axon hillock without dying out, and depolarizes the membrane there to a value called the threshold potential (step 2 in Figure 42.9). The **threshold potential** is the membrane potential, typically around −55 to −50 mV, that is sufficient to open large numbers of voltage-gated $Na^+$ channels and trigger an action potential.

Voltage-gated $Na^+$ channels open rapidly when the membrane potential changes from the resting membrane potential (for example, around −70 mV) to the threshold potential. The opening of voltage-gated $Na^+$ channels involves a charge-mediated change in the conformation of the membrane-spanning region of the channel (see step 2 in Figure 42.9). When the protein changes shape, the central pore opens, and $Na^+$ rapidly diffuses across the membrane into the cell down its electrochemical gradient (recall that a typical equilibrium potential for $Na^+$ is around +60 mV). This influx of charges further depolarizes the cell, causing even more voltage-gated $Na^+$ channels to open, resulting in the spike in membrane potential that characterizes the action potential. This positive feedback process is so rapid that the membrane potential reaches its peak positive value in less than 1 millisecond (msec)!

***Repolarization Phase of an Action Potential*** The action potential approaches but does not reach the equilibrium potential for $Na^+$. When the membrane potential becomes sufficiently positive, a second, delayed conformational change in the $Na^+$ channel blocks the continued flow of $Na^+$ into the cell. This conformational change involves the **inactivation gate**, a string of amino acids that juts out from the channel protein into the cytosol (step 3 in Figure 42.9). The inactivation gate swings into the channel pore, thereby preventing any further movement of $Na^+$ through the channel into the cell. Thus, the inactivation gate terminates the depolarization phase of an action potential. Under these conditions, the $Na^+$ channel is said to be inactivated. The inactivation gate does not swing out of the channel until the membrane potential returns to a value that is close to the resting potential.

Once $Na^+$ channels are inactivated, the continued action of the $Na^+/K^+$-ATPase pump begins the process of repolarization by transporting 3 $Na^+$ out of the cell for every 2 $K^+$ pumped in. Of greater importance, however, is a change that occurs in membrane permeability to $K^+$. Voltage-gated $K^+$ channels are also opened by the change in voltage to the threshold potential, but they open about 1 msec later than $Na^+$ channels. When $K^+$ channels open, some $K^+$ ions leave the cell, moving down the electrochemical gradient, and the membrane potential becomes more negative again as a result. The membrane even exhibits a brief period of hyperpolarization because the $K^+$ channels do not close as quickly as do the $Na^+$ channels. Instead, they remain open for a short time; this allows the membrane potential to approach the equilibrium potential for $K^+$, which is usually more negative than the resting potential (step 4 in

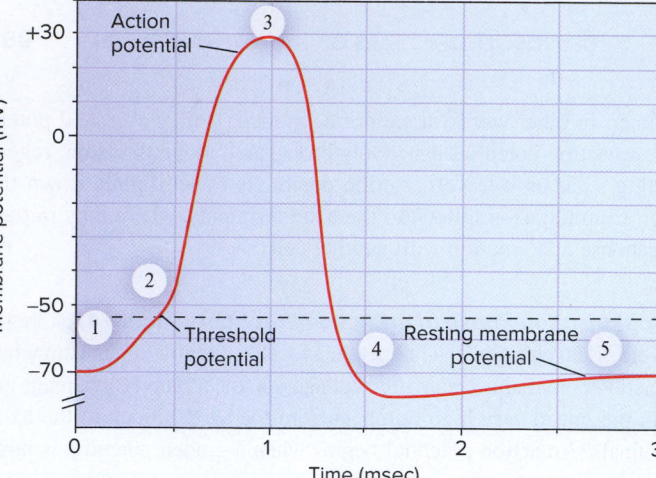

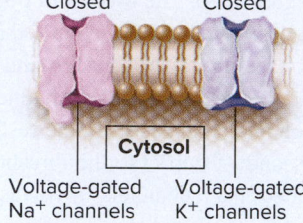

1. **Resting membrane potential:** The membrane is at the resting membrane potential. Leak channels are open but voltage-gated channels are closed.

2. **Depolarization to threshold:** Voltage-gated Na⁺ channels open. Na⁺ diffuses into the cell and depolarizes the membrane. An action potential is triggered if the cell is sufficiently stimulated so that the threshold potential of about –55 to –50 mV is reached.

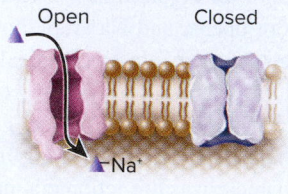

3. **Peak of action potential:** At about +30 mV, voltage-gated Na⁺ channels are inactivated, and voltage-gated K⁺ channels open. K⁺ exits the cell and repolarizes the membrane. At this time, the membrane is in its absolute refractory period.

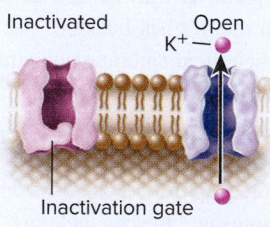

4. **Repolarization:** Voltage-gated Na⁺ channels change from inactivated to closed. Voltage-gated K⁺ channels remain open, causing a hyperpolarization of the membrane. The membrane is now in its relative refractory period.

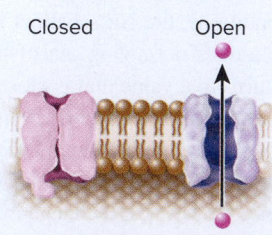

5. **Restoration of resting membrane potential:** Voltage-gated K⁺ channels close. The resting potential of the membrane is restored by the Na⁺/K⁺-ATPase pump and the leak channels.

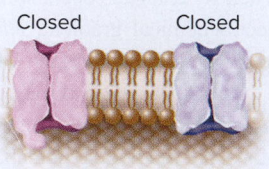

**Figure 42.9 Changes that occur during an action potential.** The movement of ions across the plasma membrane of an axon first depolarizes and then repolarizes the cell. These changes in the membrane potential are caused by the opening and closing of voltage-gated Na⁺ and K⁺ channels. The values given for membrane potential are representative and not necessarily the same in all neurons.

**Concept Check:** *How would the action potential be affected if the Na⁺ channels did not have their inactivation gates?*

Figure 42.9). As repolarization proceeds, the voltage-gated K⁺ channels eventually close. The membrane then returns to the resting potential as the Na⁺/K⁺-ATPase pump restores the original concentration gradients (step 5 in Figure 42.9). At this stage, both the Na⁺ and K⁺ voltage-gated channels are closed, but they have the ability to reopen if the membrane depolarizes again to the threshold potential.

The evolution of K⁺ channels with a slightly slower opening time than Na⁺ channels was a key event that led to the formation of nervous systems. Imagine what would happen if the voltage-gated Na⁺ and K⁺ channels opened simultaneously: As Na⁺ entered the cell down its electrochemical gradient, K⁺ would leave the cell down its electrochemical gradient, and they would negate each other's effects on membrane potential.

***Refractory Periods*** During the times when the voltage-gated Na⁺ channels are inactivated, the membrane is in its **absolute refractory period** (see step 3 in Figure 42.9). During this time, that portion of membrane is unresponsive to another stimulus. A change in voltage cannot open the Na⁺ channels while they are inactivated. As the membrane repolarizes, the inactivation gate is eventually released and the voltage-gated Na⁺ channels revert to the closed state. While the voltage-gated K⁺ channels are still open, however, the membrane enters a brief **relative refractory period** (see step 4 in Figure 42.9). During this time the membrane is hyperpolarized; a new action potential may be generated but only in response to an unusually large stimulus that can counteract the continued efflux of K⁺. Therefore, the refractory periods place limits on the frequency with which a neuron can generate and transmit action potentials. As we will see, the refractory periods also ensure that the action potential does not "retrace its steps" by moving backward toward the cell body.

**Core Skill: Modeling**

BIO **TIPS** **THE QUESTION** *Imagine that just before the repolarization phase of an action potential begins, the voltage-gated K⁺ channels become blocked and are unable to open. How might such an action potential look on a graph like that in Figure 42.9?*

**T** **OPIC** *What topic in biology does this question address?* The topic is an action potential. Specifically, the question addresses how voltage-gated K⁺ channels can affect the shape of a graph of an action potential.

**I** **NFORMATION** *What information do you know based on the question and your understanding of the topic?* From the question, you know that an action potential has begun, but the voltage-gated K⁺ channels cannot open. From your understanding of the topic and from Figure 42.9, you know that voltage-gated K⁺ channels normally open at the peak of an action potential; the movement of K⁺ through these channels out of the cell is what causes repolarization of the cell. You also know that the Na⁺/K⁺–ATPase pump contributes to repolarization.

**P**ROBLEM-SOLVING **S**TRATEGY *Make a drawing.* Referring to Figure 42.9 as a guide, draw a similar graph of an action potential and imagine how the second half of it—that is, the portion beginning at step 3 in the figure—would appear if K⁺ ions were not able to leave the cell through their voltage-gated channels.

**ANSWER** *The action potential would be prolonged because repolarization would be slower. Repolarization would still occur, because the Na⁺/K⁺–ATPase pump would eventually restore the ion concentration gradients. Without the rapid and large efflux of K⁺ out of the cell through the voltage-gated channels, however, repolarization would take much longer. Note that there would be no hyperpolarization, because this normally occurs as K⁺ leaves the cell through voltage-gated channels that temporarily remain open while Na⁺ channels are closed.*

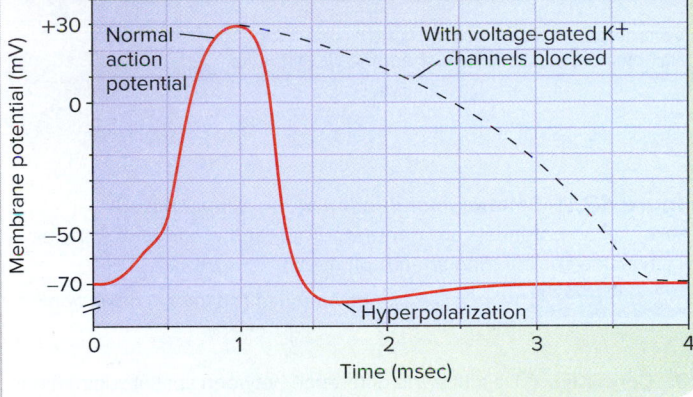

## Action Potentials Are Propagated Along the Axon to the Axon Terminals

Thus far, we have considered the electrical changes that happen when an action potential is initiated. We will now consider how such potentials are conducted down an axon from the axon hillock to the axon terminals (**Figure 42.10**). Let's begin at the axon hillock. When a neuron receives stimuli from other cells, this causes a graded potential in the cell body of the neuron that reaches the axon hillock. If the change in membrane potential is sufficient to reach threshold, an action potential will be triggered. An action potential first occurs with the abrupt opening of several voltage-gated Na⁺ channels just beyond the axon hillock, where voltage-gated Na⁺ channels are abundant. This action potential, in turn, triggers the opening of nearby Na⁺ channels farther along the axon, which allows even more Na⁺ to flow into the neuron and depolarize a region closer to the axon terminals, leading to another action potential. In this way, the sequential opening of Na⁺ channels along the axon membrane conducts a wave of depolarization from the axon hillock to the axon terminals.

*Direction of Propagation* Why doesn't the action potential move backward from the terminals to the axon hillock? Experimentally, if an axon is stimulated in its middle, action potentials can travel in both directions, toward the cell body and the terminals. The reason why this doesn't ordinarily happen is the inactivation state of the Na⁺ channel, which contributes to the absolute refractory period. Again, let's begin at the hillock.

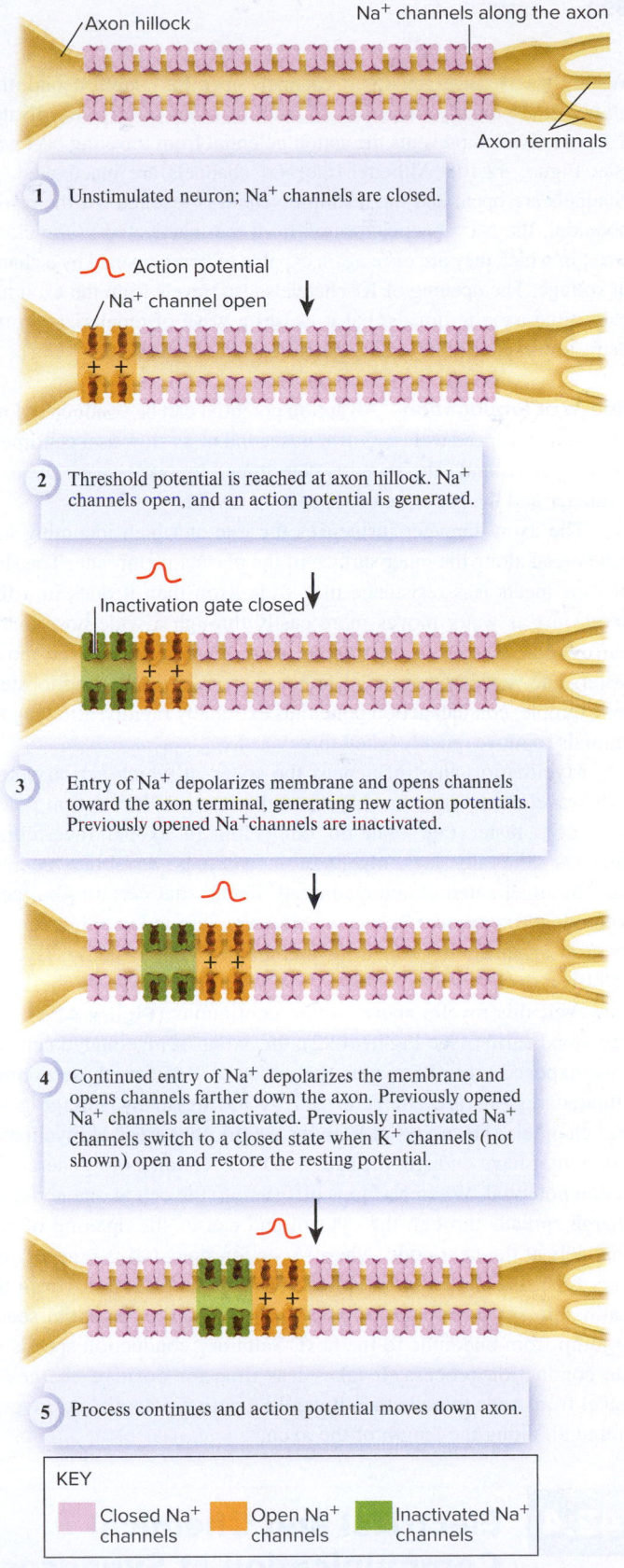

1. Unstimulated neuron; Na⁺ channels are closed.

2. Threshold potential is reached at axon hillock. Na⁺ channels open, and an action potential is generated.

3. Entry of Na⁺ depolarizes membrane and opens channels toward the axon terminal, generating new action potentials. Previously opened Na⁺ channels are inactivated.

4. Continued entry of Na⁺ depolarizes the membrane and opens channels farther down the axon. Previously opened Na⁺ channels are inactivated. Previously inactivated Na⁺ channels switch to a closed state when K⁺ channels (not shown) open and restore the resting potential.

5. Process continues and action potential moves down axon.

**KEY**

| Closed Na⁺ channels | Open Na⁺ channels | Inactivated Na⁺ channels |

**Figure 42.10 Conduction of the action potential along an axon.** All Na⁺ channels shown are voltage-gated channels.

 **Core Concept: Energy and Matter** Transmission of electrical signals generated by the movements of ions along an axon is a very rapid mechanism of cell-to-cell communication.

When these voltage-gated Na$^+$ channels open for 1 millisecond, they allow Na$^+$ to rapidly enter the cell, and then they become inactivated. This inactivation prevents the action potential from traveling backward (see Figure 42.10). Although the Na$^+$ channels are inactivated, K$^+$ channels are open, and the resting potential is restored. At the resting potential, the Na$^+$ channels switch from the inactivated to the closed state, in which they are once again capable of being opened by a change in voltage. The opening of K$^+$ channels also travels from the axon hillock to the axon terminals, but it causes a wave of repolarization that helps to reestablish the resting potential.

***Speed of Propagation***   An action potential can be conducted down the axon at a speed as fast as 100 m/second or as slow as a centimeter or two per second. The speed is determined by two factors: the axon diameter and the presence or absence of myelin.

The axon diameter influences the rate at which incoming ions can spread along the inner surface of the plasma membrane. The flow of ions meets less resistance in a wide axon than it does in a thin axon, just as water moves more easily through a wide hose than a narrow one. Therefore, in a wider axon, the action potential moves faster. The very large axons of motor neurons of squids and lobsters, for example, conduct action potentials extremely rapidly, allowing the animals to move quickly when threatened.

Myelination also influences the speed at which action potentials travel along an axon. Myelinated axons conduct action potentials at a faster rate than do unmyelinated axons. Invertebrate neurons generally lack myelination, whereas vertebrate neurons may be myelinated or unmyelinated. Recall that certain glial cells (oligodendrocytes and Schwann cells) wrap around vertebrate axons to form an insulating sheath of membrane. The insulating layer of myelin reduces charge leakage across the membrane of the axon. However, this myelin sheath is not continuous (**Figure 42.11**). As described earlier (see Figure 42.2), the axons of myelinated neurons have exposed areas known as the nodes of Ranvier; these nonmyelinated regions are characterized by having many voltage-gated Na$^+$ channels. The nodes of Ranvier are the only areas of myelinated axons that have enough voltage-gated Na$^+$ channels to generate an action potential. When Na$^+$ ions diffuse into the cell at one node, the charge spreads through the cytosol and causes the opening of Na$^+$ channels at the next node, where an action potential is regenerated. This type of conduction is called **saltatory conduction** (from the Latin *saltare*, meaning to leap) because the action potential seems to jump from one node to the next. Saltatory conduction speeds up the conduction process. It takes less time for positive charges to travel from node to node than it would if action potentials were generated all along the length of the axon.

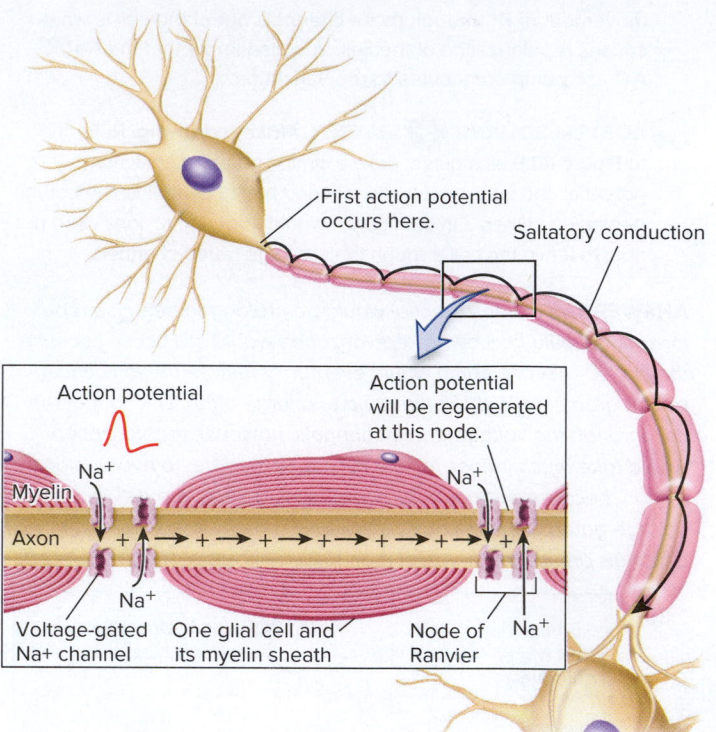

**Figure 42.11** **Saltatory conduction along a myelinated axon.** Action potentials are generated only at the nodes of Ranvier, which lack a surrounding sheath of myelin.

**Concept Check:**   *What is a major advantage of saltatory conduction?*

3. **CoreSKILL »** Describe the difference between spatial summation and temporal summation, and make predictions about postsynaptic potentials in a cell receiving multiple inputs from other cells.
4. List the classes of neurotransmitters, and provide brief descriptions of their generalized functions.
5. Describe the two general types of postsynaptic membrane receptors.

Neurons communicate with other cells at a **synapse**, which is a junction where an axon terminal meets another neuron, muscle cell, or gland cell. At a synapse, an electrical or chemical signal passes from an axon terminal to the next cell. A synapse includes an axon terminal of the neuron that is sending the signal, the nearby plasma membrane of the receiving cell, and in certain cases the **synaptic cleft**, or extracellular space between the two cells. The **presynaptic cell** sends the signal, and the **postsynaptic cell** receives it. A given cell may be presynaptic to one cell, and postsynaptic to another (**Figure 42.12**).

## Synapses May Be Electrical or Chemical

By studying neurons from both invertebrates and vertebrates, researchers have identified two types of synapses: electrical and chemical. The first type, the **electrical synapse**, directly passes electric current from the presynaptic to the postsynaptic cell. The electrical signal passes through this type of synapse extremely rapidly, because the plasma membranes of adjacent cells are connected by gap junctions (refer back to Figure 10.11) that allow electric charge to move directly from one cell to the other. An electrical synapse does not have a synaptic

# 42.4   Electrical and Chemical Communication at Synapses

**Learning Outcomes:**

1. Describe the structural features of a synapse.
2. Distinguish between two types of potentials produced in postsynaptic cells: excitatory postsynaptic potentials (EPSPs) and inhibitory postsynaptic potentials (IPSPs).

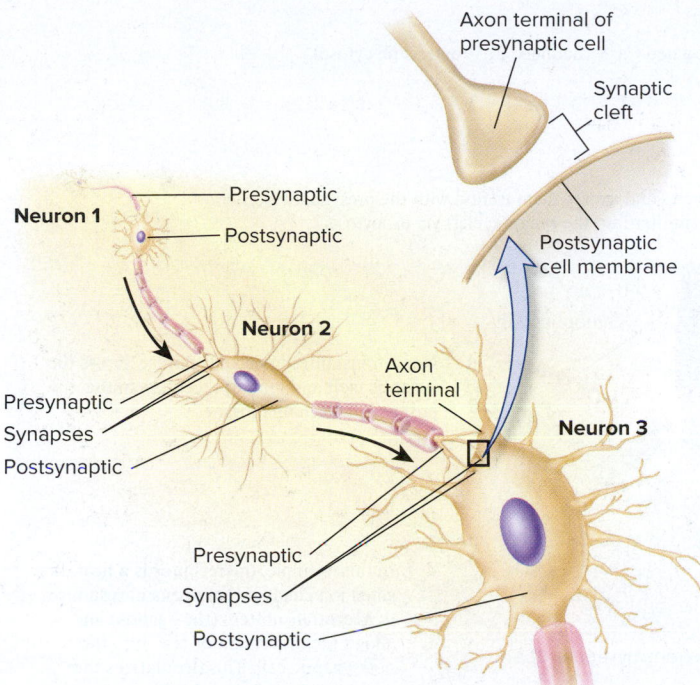

**Figure 42.12 Presynaptic and postsynaptic cells.** The arrows show the direction of signal transmission from one neuron to the next. Note that neuron 2 is postsynaptic with respect to neuron 1 and presynaptic with respect to neuron 3. The example shown here is a chemical synapse.

cleft, but a small intercellular gap is found between the presynaptic and postsynaptic membranes. Recent research indicates that electrical synapses are more widespread among taxa, including vertebrates, than originally thought. The most well-studied examples, however, occur in some aquatic invertebrates such as leeches. In those animals, electrical synapses occur in parts of the body where a group of neurons must fire rapidly and synchronously, such as when an animal must coordinate a number of muscles to swim or escape danger.

The second type of synapse is a **chemical synapse**, in which a neurotransmitter is released from an axon terminal and acts as a signal from the presynaptic to the postsynaptic cell. Chemical synapses appear to be more common than electrical synapses, particularly in vertebrates. Some neurons release only one type of neurotransmitter, and some neurons can release two or more different ones. Chemical synapses are slower than electrical synapses. However, unlike electrical synapses, a major advantage of these synapses is that they allow for complex modulation of the responses of postsynaptic cells, as we will see next.

## At Chemical Synapses, Neurotransmitters Generate Excitatory or Inhibitory Signals in Postsynaptic Cells

**Figure 42.13** shows the steps that occur when two cells communicate via a chemical synapse. An axon terminal of a presynaptic cell contains vesicles—small, membrane-enclosed packets, each containing thousands of molecules of neurotransmitter. The membranes of axon terminals contain voltage-gated $Ca^{2+}$ channels. When an action potential arrives at an axon terminal, the voltage change from the action potential opens these channels, allowing $Ca^{2+}$ to diffuse down

its electrochemical gradient into the cell. Calcium binds to a protein associated with the vesicle membrane. This triggers exocytosis, in which the vesicle fuses with the presynaptic membrane, thereby releasing its neurotransmitter molecules into the synaptic cleft. The neurotransmitter molecules diffuse across the 10- to 20-nm-wide synaptic cleft and bind to ligand-gated ion channels or other receptor proteins in the postsynaptic cell membrane.

The binding of neurotransmitter molecules opens or closes ligand-gated ion channels, thereby changing the membrane potential of the postsynaptic cell. In some cases, neurotransmitters directly bind to ion channels and cause the channels to open or close. In other cases, neurotransmitters bind to receptor proteins on the postsynaptic cell membrane, which leads to the accumulation of second messengers in the cytosol of the cell (see Chapter 9). These messengers, in turn, open or close ion channels by any of several mechanisms.

Some neurotransmitters are called excitatory, because they depolarize the postsynaptic membrane; the response is an **excitatory postsynaptic potential (EPSP)**. It is called excitatory because the depolarization of the postsynaptic cell membrane brings the membrane potential closer to the threshold potential that would trigger an action potential. An EPSP is a graded potential that can be caused by the opening of $Na^+$ channels in a neuronal membrane or by the closing of $K^+$ channels. In both cases, positive charges accumulate inside the cell.

By contrast, an inhibitory neurotransmitter usually hyperpolarizes the postsynaptic membrane, producing an **inhibitory postsynaptic potential (IPSP)**. This reduces the likelihood of an action potential by moving the membrane further away from threshold. An IPSP is a graded potential that can be caused by, for example, the opening of $Cl^-$ channels. The equilibrium potential for $Cl^-$ is typically close to or more negative than the resting membrane potential. Therefore, when $Cl^-$ channels are opened, $Cl^-$ moves into cells, making them more negatively charged and bringing them closer to the equilibrium potential for $Cl^-$. Hyperpolarization is the most common way in which neurons are inhibited.

To end the stimulation of the postsynaptic cell, neurotransmitter molecules in the synaptic cleft must be broken down by enzymes or transported back into the terminal of the presynaptic cell and repackaged into vesicles for reuse. The latter event, called reuptake, is an efficient mechanism for recapturing and reusing excess neurotransmitters that were released into the synaptic cleft. It also prevents these neurotransmitters from diffusing away from the synapse and possibly interacting with other distant cells. As we will see in the next section, drugs that block the reuptake process are used to treat people with disorders of neurotransmission, including depression.

## Neurons Respond to Multiple Synaptic Inputs

A neuron may receive inputs from many synapses on its dendrites or cell body (**Figure 42.14a**), and some of these synapses may release neurotransmitters onto the neuron at the same or nearly the same time. Certain neurotransmitters are excitatory, and others are inhibitory. When do these different inputs lead to an action potential? The effect of a single synapse is usually far too weak to elicit an action potential in a postsynaptic neuron. However, when multiple EPSPs are generated at one time at many synapses along different regions of the dendrites and cell body, their depolarizations sum together. The

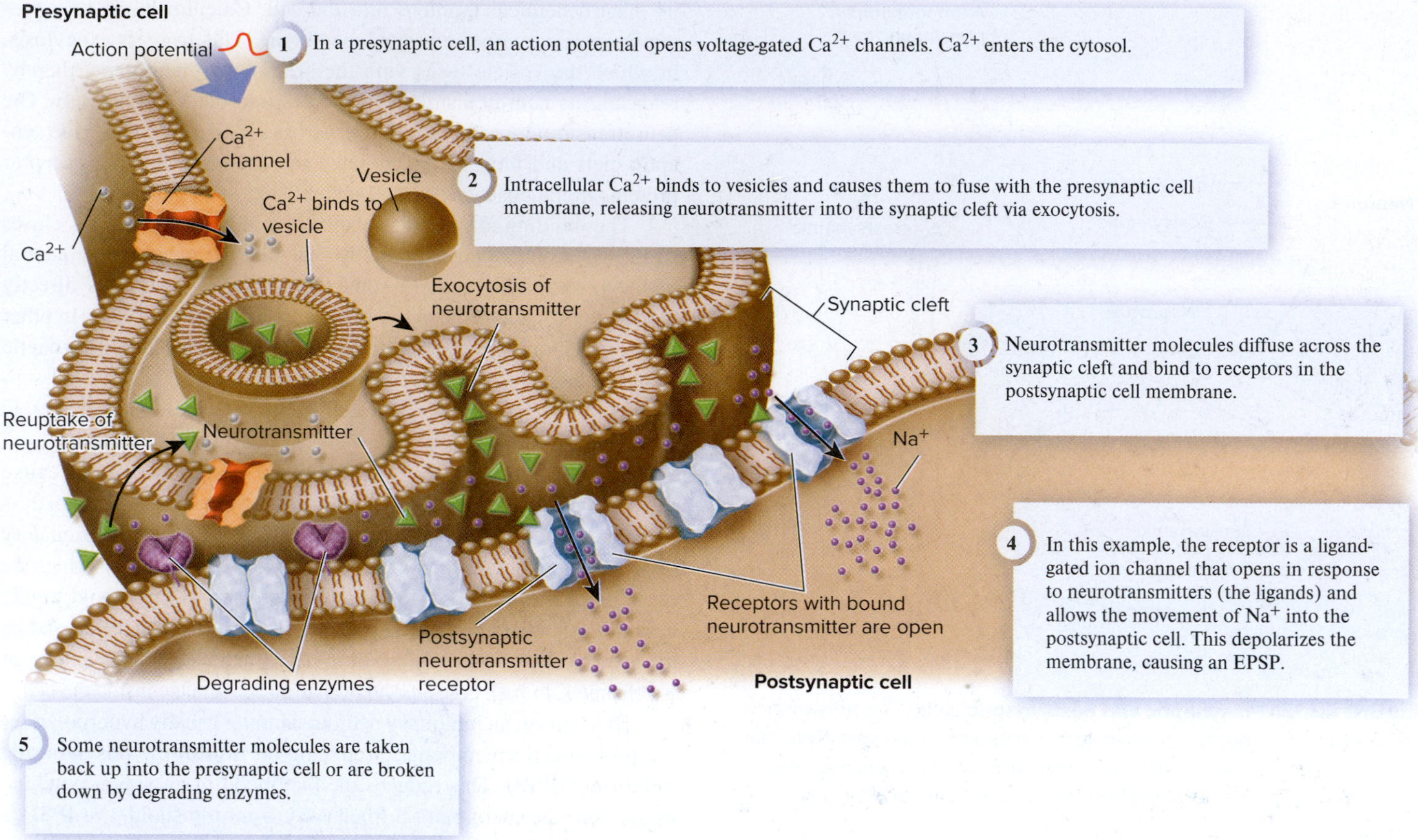

**Presynaptic cell**

Action potential

**1** In a presynaptic cell, an action potential opens voltage-gated $Ca^{2+}$ channels. $Ca^{2+}$ enters the cytosol.

$Ca^{2+}$ channel

Vesicle

$Ca^{2+}$ binds to vesicle

$Ca^{2+}$

**2** Intracellular $Ca^{2+}$ binds to vesicles and causes them to fuse with the presynaptic cell membrane, releasing neurotransmitter into the synaptic cleft via exocytosis.

Exocytosis of neurotransmitter

Synaptic cleft

**3** Neurotransmitter molecules diffuse across the synaptic cleft and bind to receptors in the postsynaptic cell membrane.

Reuptake of neurotransmitter

Neurotransmitter

$Na^+$

**4** In this example, the receptor is a ligand-gated ion channel that opens in response to neurotransmitters (the ligands) and allows the movement of $Na^+$ into the postsynaptic cell. This depolarizes the membrane, causing an EPSP.

Degrading enzymes

Postsynaptic neurotransmitter receptor

Receptors with bound neurotransmitter are open

**Postsynaptic cell**

**5** Some neurotransmitter molecules are taken back up into the presynaptic cell or are broken down by degrading enzymes.

**(a) Events occurring at a chemical synapse**

Axon terminal   Synapse

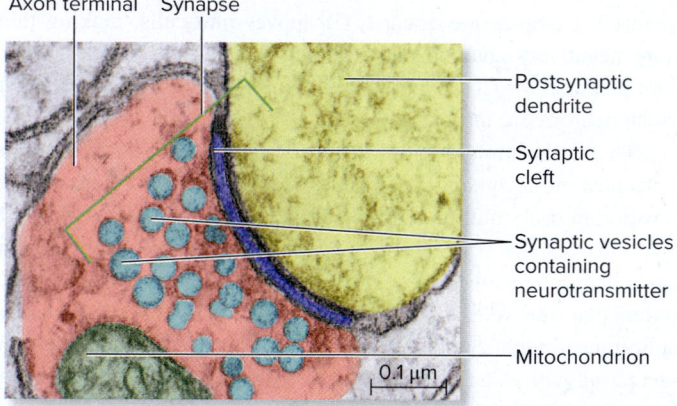

Postsynaptic dendrite

Synaptic cleft

Synaptic vesicles containing neurotransmitter

Mitochondrion

0.1 μm

**(b) False-color electron micrograph of a chemical synapse**

**Figure 42.13 Structure and function of a chemical synapse.**
**(a)** In response to an action potential, $Ca^{2+}$ enters the presynaptic neuron axon terminal. This results in fusion of the vesicle with the plasma membrane, which releases neurotransmitter molecules into the synaptic cleft. The neurotransmitter molecules then bind to receptors in the plasma membrane of the postsynaptic cell. This causes ion channels to open ($Na^+$ in this example) or close, which, in turn, changes the membrane potential of the postsynaptic cell. **(b)** False-color electron micrograph of a chemical synapse from a rat brain. b: ©McGraw-Hill Education/Al Telser, photographer

resulting larger depolarization may bring the membrane potential at the axon hillock to the threshold potential, initiating an action potential. When an action potential is initiated in this manner, the process is called **spatial summation** (**Figure 42.14b**).

Alternatively, two or more EPSPs may arrive at the same location in quick succession, such that the first EPSP has not yet decayed away when the next EPSP arrives. In that case, the depolarizations sum and may reach threshold upon arrival at the axon hillock. This process is called **temporal summation** (**Figure 42.14c**).

A third possibility is that EPSPs and IPSPs may arrive together at a postsynaptic cell. In this case, the two types of signals will cancel

each other out, and no action potential is elicited (**Figure 42.14d**). From this discussion, you might deduce that when two or more IPSPs arrive together, their hyperpolarizations might sum, too. That is exactly what happens. The membrane potential of the cell receiving multiple IPSPs moves farther away from threshold (**Figure 42.14e**).

In addition to the number of synapses that stimulate the postsynaptic membrane, the location of the synapses is also important. Synapses that occur far from the axon hillock are less effective than synapses on the cell body nearer the axon hillock.

Collectively, the various possible synaptic inputs shown in Figure 42.14a represent a major advantage of chemical synapses, allowing an animal to discern between competing signals. As one simple example, imagine a hungry fish confronted with a worm (food) and a predator (danger). IPSPs sent to hunger-sensitive neurons in the brain suppress those neurons, while other neurons that respond to danger cues receive EPSPs and thus generate action potentials that control, among other things, motor function. As a result, the fish ignores the worm and swims to safety.

## Neurotransmitters Can Be Categorized According to Size or Structure

Neuroscientists have identified more than 100 different neurotransmitters in animals. Generally, neurotransmitters are categorized by size or structure (**Table 42.2**). The changing balance between excitatory and inhibitory neurotransmitters controls the state of nervous system circuits at any one time. To understand how neurotransmitters work, imagine driving a car with one foot on the gas pedal and one foot on the brake. To speed up, you could press down on the gas pedal, ease up on the brake, or both, whereas to slow down, you could do the opposite. All nervous systems operate in this way, with combined excitatory and inhibitory actions of neurotransmitters.

Next, we consider major features of the different chemical classes of neurotransmitters found in animals, which include acetylcholine,

biogenic amines, amino acids, neuropeptides, and gaseous neurotransmitters. With the exception of acetylcholine, all of these classes contain several different neurotransmitters that are similar in chemical structure but may have different functions in nervous systems.

***Acetylcholine*** Acetylcholine (ACh), one of the most widespread neurotransmitters in animals, is released at the synapses of **neuromuscular junctions**, where a neuron contacts a skeletal or cardiac muscle cell. It is also released at synapses within the brain and elsewhere. Acetylcholine acts as an excitatory neurotransmitter in the brain and on skeletal muscle cells and certain gland cells, but it is inhibitory when released from neurons that control cardiac muscle contraction. As we will see later, certain neurotransmitters such as ACh can exert both excitatory and inhibitory effects because they typically have more than one type of receptor to which they bind on different cells. The receptors, in turn, can be linked with different signaling mechanisms.

***Biogenic Amines*** The biogenic amines are compounds containing amine groups that are formed from amino acids. Common biogenic amines include the catecholamines—dopamine, norepinephrine, and epinephrine—and serotonin and histamine. The catecholamines are formed from the amino acid tyrosine, serotonin is formed from tryptophan, and histamine from histidine.

In addition to widespread physiological effects such as control of heart and lung function, catecholamines and serotonin are psychoactive; that is, they affect mood, attention, behavior, and learning. In humans, for example, abnormally high or low levels of catecholamines and serotonin have been associated with a variety of mental illnesses, including schizophrenia, attention-deficit disorder, and depression. Histamine is well known as a component of allergic reactions in people, but this is not related to its neurotransmitter functions. In the brain, neurons that produce histamine are important in modulating sleep. They are most active during waking and are nearly inactive during sleep. This explains why certain antihistamines (drugs that block the ability of histamine to bind to its receptor, thereby inhibiting its action) used to treat colds and allergies also induce drowsiness.

Dendrites of postsynaptic neuron

Presynaptic neurons

Cell body of postsynaptic neuron

Axon hillock of postsynaptic neuron

**KEY**

Inputs that are excitatory (E)

Inputs that are inhibitory (I)

**(a) A single neuron receiving many inputs**

**Figure 42.14 Integration of synaptic inputs.** Membrane potential changes in a postsynaptic membrane are shown under several different circumstances. When multiple excitatory (E) or inhibitory (I) inputs arrive simultaneously or nearly so, the subsequent EPSPs and IPSPs may add (summate) or cancel each other out, depending on the situation.

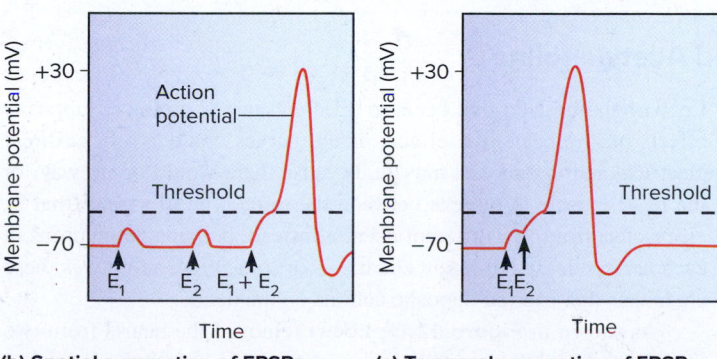

**(b) Spatial summation of EPSPs**

**(c) Temporal summation of EPSPs**

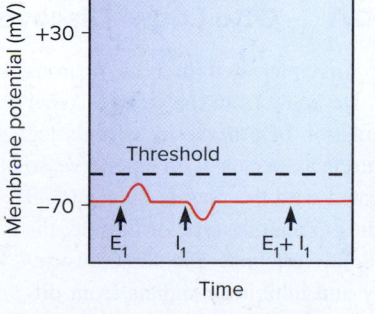

**(d) Cancellation of EPSP and IPSP**

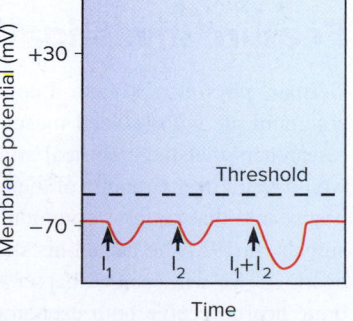

**(e) Spatial summation of IPSPs**

| Table 42.2 | Classes of Neurotransmitters and Some Representative Examples and Functions |
|---|---|
| **Transmitter** | **Some major functions*** |
| Acetylcholine | CNS: Stimulates the brain; important in memory, motor control, and many other functions |
| | PNS: Stimulates skeletal muscle at neuromuscular junctions; inhibits cardiac muscle; promotes digestion |
| ***Biogenic amines*** | |
| Catecholamines: dopamine, norepinephrine, epinephrine | CNS: Regulate mood, attention, learning, behavior |
| Serotonin | PNS: Stimulates cardiac muscle; improves lung function; helps animals respond to stressful situations |
| Histamine | CNS: Helps to maintain awake state |
| ***Amino acids*** | |
| Excitatory amino acids: glutamate, aspartate | CNS: Widespread mediators of activity in all areas of CNS; the major "on" signal of the CNS |
| Inhibitory amino acids: γ-aminobutyric acid (GABA), glycine | CNS: Hyperpolarize neurons; act as a "brake" on the nervous system |
| ***Neuropeptides*** | |
| Opiate peptides: endorphin | CNS: Modulate postsynaptic cell response to neurotransmitters; play a role in mood, behavior, appetite, pain perception, and many other functions |
| ***Gases*** | |
| Nitric oxide, carbon monoxide | CNS: Possible role in memory and odor sensation |
| | PNS: Relax smooth muscle, especially in blood vessels |

*Note: CNS stands for central nervous system; PNS stands for peripheral nervous system.

**Amino Acids**   The amino acids glutamate, aspartate, glycine, and γ-aminobutyric acid (GABA) function as neurotransmitters. Glutamate is the most widespread excitatory neurotransmitter found in animal nervous systems, whereas GABA is the most common inhibitory neurotransmitter. GABA hyperpolarizes the postsynaptic membrane by opening Cl⁻ channels, allowing negatively charged Cl⁻ to diffuse into the cell (see Table 42.1). In this way, GABA brings neurons further away from the threshold potential required to generate an action potential, and thus it acts as the major "brake" on the CNS.

***Neuropeptides***   Neuropeptides are short polypeptides containing from 2 to about 15 amino acids. Like the other neurotransmitters discussed, neuropeptides can be excitatory or inhibitory. Neuropeptides are often called **neuromodulators**, because they can alter or modulate the response of the postsynaptic neuron to other neurotransmitters. For example, a neuropeptide may stimulate synthesis of receptors for another neurotransmitter, which makes a cell more responsive to that neurotransmitter. One group of neuropeptides is called the opiate peptides because opium-like drugs, such as morphine, bind to their receptors. Opiate peptides include the endorphins, a group of peptides that decrease pain and cause natural feelings of euphoria (extreme happiness and sense of invulnerability).

***Gaseous Neurotransmitters***   Certain gases such as nitric oxide (NO) and carbon monoxide (CO) act locally in many tissues and sometimes function as neurotransmitters. Unlike other neurotransmitters, these molecules are not sequestered into vesicles and are produced locally as required. Gaseous neurotransmitters are short-acting and influence other cells by diffusion from a presynaptic cell. In humans, NO is responsible for relaxing the smooth muscle surrounding blood vessels, including those in the penis. When a male becomes sexually aroused, NO levels increase in this tissue, dilating the vessels and increasing blood flow into the penis, producing an erection. Several drugs used to treat male sexual dysfunction enhance erections by increasing or mimicking the action of NO on smooth muscle. The functions of CO are still uncertain, but scientists think it may act as a neurotransmitter in pathways that mediate the sense of smell in some animals, such as the terrestrial mollusk *Limax maximus*.

The discovery that the actions of neurons are mediated in large part by neurotransmitters was one of the most significant achievements in the history of neuroscience. This discovery laid the foundation for our understanding of how chemical synapses function, and also provided insight regarding the basis of some neurological and muscular diseases. Amazingly, the existence of neurotransmitters was demonstrated in a remarkably simple experiment that arose from a dream, as we see next.

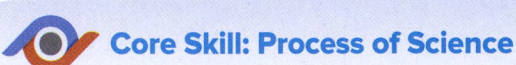

**Core Skill: Process of Science**

# Feature Investigation | Otto Loewi Discovered Acetylcholine

German physiologist Otto Loewi was interested in how neurons communicate with skeletal muscle. He knew from the work of other researchers that the electrical stimulation of a nerve in a frog's leg would result in contraction of the muscle associated with that nerve, so it appeared that neurons communicated with the muscle by electrical signals. In 1921, he turned his studies to another type of muscle, the heart. As we will see in Chapter 48, all vertebrate and some invertebrate hearts receive both excitatory and inhibitory signals from different nerves that regulate the rhythm and intensity of the heartbeat.

Loewi hypothesized that because different nerves produced opposite effects on the heart, the effects of the nerves could not be a direct electrical action on heart muscle, because there would be no way for the heart muscle to discern between the same type of signal (that is, electricity) from two different nerves. Instead, perhaps the neurons in each nerve released different chemicals of some type, and it was these chemicals that exerted opposite actions on the heart.

As shown in **Figure 42.15**, Loewi removed the hearts from two frogs and placed the hearts in baths containing a solution of ions and

**Figure 42.15**  **Loewi's experimental discovery of chemical neurotransmission.**

**HYPOTHESIS**  Neurons release chemical substances that influence the activity of the heart.

**KEY MATERIALS**  Two frog hearts, saline solution, and stimulating and recording electrodes.

| Experimental level | Conceptual level |
|---|---|

1  Dissect hearts from 2 frogs and place in saline solution. Heart 1 still has its vagus nerve attached.

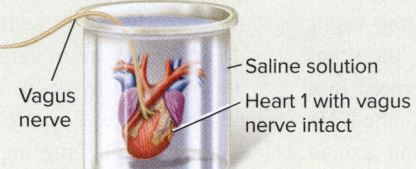

Vagus nerve — Saline solution — Heart 1 with vagus nerve intact

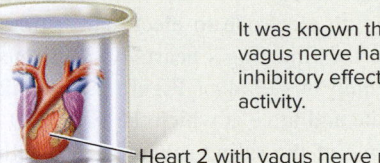

It was known that the vagus nerve has an inhibitory effect on heart activity.

Heart 2 with vagus nerve removed

2  Electrically stimulate vagus nerve of heart 1.

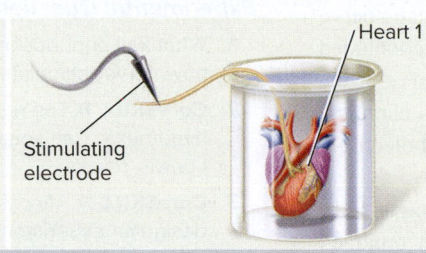

Heart 1 — Stimulating electrode

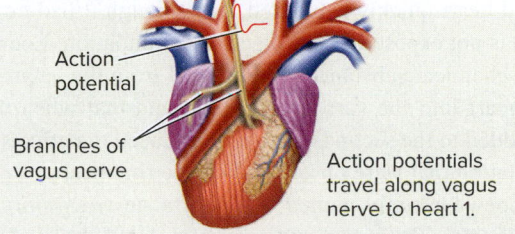

Action potential

Branches of vagus nerve

Action potentials travel along vagus nerve to heart 1.

3  Record strength and number of beats in heart 1 before and after electrical stimulation of vagus nerve. Next, remove a sample of the saline solution in and around heart 1, and transfer to heart 2. Record activity of heart 2. This was done using mercury manometers that were connected to each heart. The manometers measure pressure, which is due to the contractile force of the heart beating.

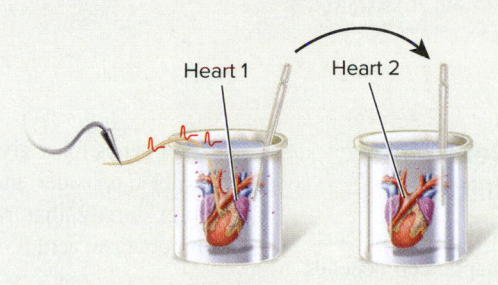

Heart 1    Heart 2

If stimulation of vagus nerve resulted in the release of chemicals onto heart 1, then these same chemicals (some of which may diffuse into the saline solution) should have an identical effect on heart 2.

4  **THE DATA**

Stimulation of vagus nerve of heart 1

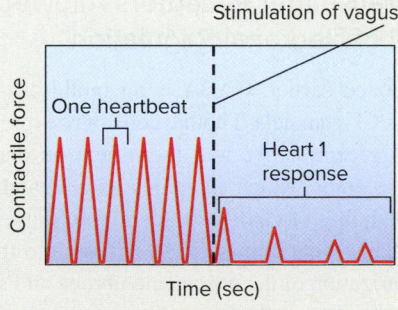

One heartbeat — Heart 1 response

Contractile force

Time (sec)

Addition of saline solution from within and around heart 1

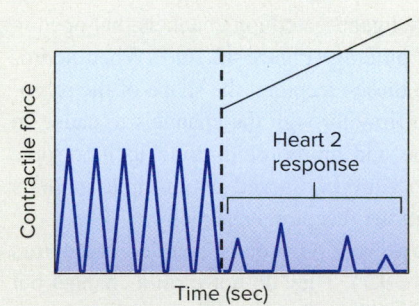

Heart 2 response

Contractile force

Time (sec)

5  **CONCLUSION**  Electrical stimulation causes the vagus nerve to secrete chemicals that decrease heart contractions.

6  **SOURCE**  Loewi, O. 1921. On humoral transmission of the action of heart nerves. Pflugers archives. *European Journal of Physiology* 189: 239–242.

other substances that helped keep the organ alive. When a frog's heart is maintained in this solution, it will continue to beat for several hours before it eventually stops.

Initially, Loewi began by examining the major inhibitory nerve of the vertebrate heart, called the vagus nerve. When Loewi dissected the hearts from the two frogs, he left the associated vagus nerve intact in one heart, but removed it from the second heart. Next, Loewi used an electrode to electrically stimulate the vagus nerve attached to the first frog's heart. As illustrated by the schematic data presented in step 4 of Figure 42.15, this resulted in a decrease in the rate and force at which the first heart contracted. He then removed some of the saline solution from within and around the first heart and transferred it to the solution that was bathing the second (unstimulated) heart. The rate and force of beating of the second heart quickly decreased, even though it had no vagus nerve and was not exposed to any electrical stimulation. Loewi concluded that a chemical substance was released from the vagus nerve of the first heart into the surrounding fluid and that when this chemical was added to the second heart, it reproduced the effects of electrical stimulation that were observed with the first heart.

Loewi initially named this substance *vagusstoff* (vagus substance), after the vagus nerve he stimulated. It was later renamed acetylcholine when its chemical nature (acetic acid bonded to choline) was determined. Acetylcholine was the first neurotransmitter

discovered. Loewi's research opened the door to what we now know about chemical transmission at synapses, and benefited the now-enormous pharmaceutical industry, which builds on this knowledge to treat neurological disorders (diseases of the nervous system).

Interestingly, as Loewi described later, the idea for his experiment, which would be largely responsible for his earning a share of the 1936 Nobel Prize in Physiology or Medicine, came to him in a dream. He woke in the middle of the night, scribbled down his idea, and returned to sleep. The next morning, he despaired to find that he couldn't read his sleepy scribbling! Incredibly, he had the dream again the following night and, not taking any chances, got up and went directly to his laboratory.

### Experimental Questions

1. What key prior observation led Loewi to develop his hypothesis of how nerves stimulate or inhibit heart muscle contractions?

2. **CoreSKILL »** The results of Loewi's experiment supported his hypothesis. Can you think of an alternative hypothesis based on Loewi's results?

3. **CoreSKILL »** After you finish reading the rest of this chapter, design an experiment to test the hypothesis that acetylcholine is Loewi's proposed *vagusstoff*, using a heart (without its vagus nerve), some acetylcholine, and an acetylcholine receptor blocker.

## Postsynaptic Membrane Receptors Determine the Type of Response to Neurotransmitters

As mentioned earlier, some neurotransmitters can have both excitatory and inhibitory effects. The response of the postsynaptic cell depends on the type of receptor present in the postsynaptic membrane. The two major types of postsynaptic membrane receptors are ionotropic and metabotropic, and many neurotransmitters, such as acetylcholine, act on both (**Figure 42.16**). Neurotransmitter molecules bind to the extracellular portions of these receptors.

**Ionotropic receptors** are ligand-gated ion channels that open in response to neurotransmitter binding (Figure 42.16a). When neurotransmitter molecules bind to these receptors, the shape of the receptor changes, allowing ions to flow through the channels to cause an EPSP or IPSP. Acetylcholine and amino acids bind to ionotropic receptors. Ionotropic receptors are composed of multiple subunits that associate in a ring to form the receptor's channel.

**Metabotropic receptors** are G-protein-coupled receptors (GPCRs) (refer back to Figure 9.7). They do not form a channel but instead are coupled to an intracellular signaling pathway that initiates changes in the postsynaptic cell (Figure 42.16b). A common type of response is the phosphorylation of plasma membrane ion channels for sodium, potassium, or calcium ions. One example of a function of metabotropic receptors is the activation of sensory cells that respond to visual and other stimuli.

Not only do many neurotransmitters bind to more than one type of receptor, but receptors that are composed of subunits may exist in multiple forms made up of different combinations of the subunits. How could an organism benefit from expressing such a variety of

receptor types for a given neurotransmitter? To understand the benefits, let's consider the amazing complexity of one well-studied receptor family that recognizes the amino acid neurotransmitter γ-aminobutyric acid (GABA).

 **Core Concepts: Evolution, Information**

### The Evolution of Varied Subunit Compositions of Neurotransmitter Receptors Allowed for Precise Control of Neuronal Regulation

As mentioned earlier, GABA is an inhibitory neurotransmitter that opens $Cl^-$ channels. Though cells may express different types of GABA receptors, we will focus here on one type called the $GABA_A$ receptor, which functions as a ligand-gated ion channel. The binding of GABA to this ionotropic receptor opens the channel. This event allows $Cl^-$ to diffuse into the cell, causing a hyperpolarization of the plasma membrane and shifting the membrane potential toward the equilibrium potential for $Cl^-$ (usually between about −70 to −90 mV) (**Figure 42.17**). In this way, the binding of GABA to this receptor decreases the likelihood that a neuron will generate an action potential.

The $GABA_A$ receptor is a good example that illustrates how different combinations of receptor subunits can influence a postsynaptic response to a neurotransmitter. $GABA_A$ receptors are usually composed of five polypeptide subunits (designated α, β, γ, and so on). All animals with a nervous system express one or more forms

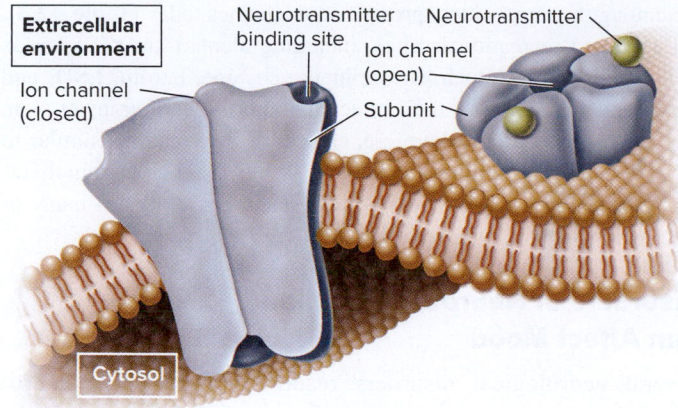

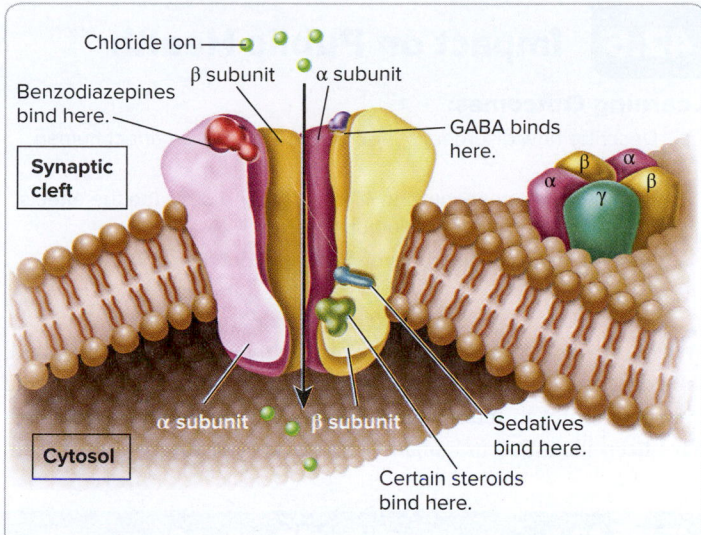

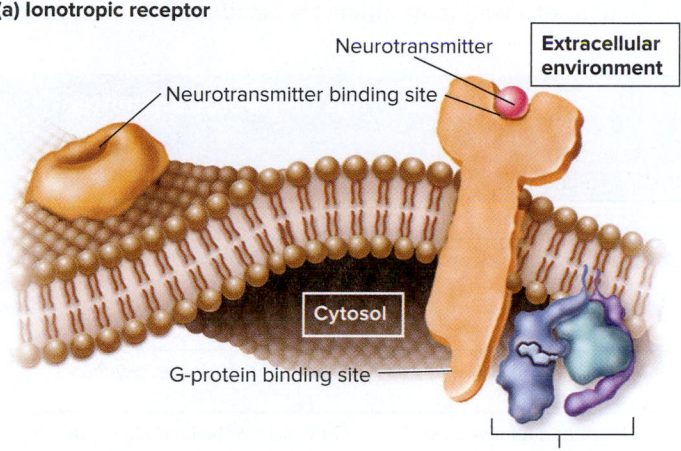

(a) **Ionotropic receptor**

(b) **Metabotropic receptor**

**Figure 42.16  The two major types of postsynaptic receptors.** **(a)** Ionotropic receptors are ligand-gated ion channels that have several subunits. Neurotransmitters bind to ionotropic receptors and directly open ion channels in the membrane. **(b)** Metabotropic receptors are G-protein-coupled receptors, which are discussed in Chapter 9. Neurotransmitters bind to metabotropic receptors and initiate a signaling pathway that opens or closes ion channels.

**Figure 42.17  The structure and function of the GABA$_A$ receptor.** Each type of GABA$_A$ receptor has five subunits that form an ion channel allowing the passage of chloride ions. When GABA binds to the receptor, chloride ions move through the open channel and hyperpolarize the cell. Various other molecules bind to different sites on the receptor. These include molecules that naturally occur in animals' bodies, such as certain steroid hormones, and drugs, such as sedatives and benzodiazepines, used to treat anxiety and other disorders.

**Core Concept: Structure and Function** The three-dimensional structure of the GABA$_A$ receptor determines its ability to bind a particular neurotransmitter and, as shown in Figure 42.17, various other ligands that can modulate the receptor's activity.

of the GABA$_A$ receptor. For example, the genomes of humans and other mammals have a group of homologous genes that encode at least 19 different GABA$_A$ receptor subunits. In addition, subunit variation can be further increased by alternative splicing (refer back to Figure 14.21). This amazing variety in subunits allows cells to potentially express dozens of different kinds of GABA$_A$ receptors.

What selective advantages led to the evolution of such a variety of different GABA$_A$ receptor subunits? Though the answer is not entirely understood, each type of subunit has its own unique properties that can fine-tune the function of the GABA$_A$ receptor so that it works optimally in the neuron in which it is expressed. For example, the various subunits may differ in their affinity for GABA and the rate of Cl$^-$ movement through the channel.

More recently, neuroscientists have determined that the various subunits may differ in their ability to recognize molecules other than GABA (see Figure 42.17). This work has shown that the subunits of the GABA$_A$ receptor bind a variety of other molecules, including naturally occurring ones such as certain steroid hormones. The binding of these molecules enhances or reduces the effectiveness of GABA in activating the receptor. This knowledge has proven beneficial in understanding how certain drugs exert their actions. For example, ethanol—found in alcoholic drinks—binds to one of the GABA$_A$ receptor subunits expressed in brain and motor neurons and enhances the actions of GABA. This may explain in part why alcohol depresses the activity of the brain and impairs motor coordination, among other effects. Other subunits of the GABA$_A$ receptor bind drugs, including benzodiazepines such as alprazolam (Xanax), which are used to treat chronic or severe anxiety. The inhibitory effects of GABA may be part of the mechanism for achieving a balance between anxiety and calmness. The ability of the receptor to bind numerous ligands and the many different combinations of subunits in the receptor provide an enormous degree of control over precisely how this neurotransmitter system regulates the activity of the brain.

## 42.5 | Impact on Public Health

### Learning Outcomes:

1. Describe how disorders of neurotransmission can affect human health.
2. List several recreational and illicit drugs, and describe the effects these drugs have on the nervous system.
3. Give examples of disorders of neural conduction, and describe how they arise.

When neurons fail to develop properly or their function is impaired, the consequences can be devastating, affecting mood, behavior, and even the ability to think or move. Over 100 neurological disorders have been identified in humans, and therapeutic drugs to treat them are among the most widely prescribed medicines today (**Table 42.3**). All so-called recreational drugs, including alcohol and tobacco, as well as illicit drugs such as marijuana, cocaine, heroin, LSD, and methamphetamine, exert their effects by altering neurotransmission. The use of these drugs, therefore, can result in symptoms similar to those of neurological disorders. Note in Table 42.3 that certain therapeutic drugs used to treat neurological disorders also exert many of their actions on neurotransmission.

### Disorders of Neurotransmission Can Affect Mood

Several neurological disorders result from disrupted neurotransmission between cells. Genetic processes involved in the production of neurotransmitters or malfunction of synaptic

| Table 42.3 | Representative Effects of Common Therapeutic, Illicit, or Recreational Drugs on Neurotransmitter Action and Mood | | |
|---|---|---|---|
| Name of drug | Actions on neurotransmission | Effects on mood | Effects of abuse or overdose |
| **Illicit or Recreational*** | | | |
| Alcohol (ethanol) | Enhances inhibitory GABA transmission; increases dopamine transmission; inhibits glutamate transmission | Relaxation; euphoria; sleepiness | Liver damage; brain damage |
| Amphetamines ("uppers," "crystal meth," "speed") | Stimulate the release of dopamine and norepinephrine | Euphoria; increased activity | High blood pressure; psychosis |
| Cocaine | Blocks norepinephrine and dopamine reuptake | Intense euphoria followed by depression | Convulsions; hallucinations; death from overdose |
| LSD (lysergic acid diethylamide) | Binds to serotonin receptors | Hallucinations; sensory distortions | Unpredictable and irrational behavior |
| Marijuana (tetrahydrocannabinol) | Binds to receptors for natural cannabinoids | Increased sense of well-being; decreased short-term memory; decreased goal-directed behavior; increased appetite | Delusions; paranoia; confusion |
| Narcotics: heroin, morphine, fentanyl, oxycodone | Bind to opiate receptors | Pain relief; euphoria; sedation | Slowed breathing; death from overdose |
| Nicotine | Initially stimulates but then depresses activity in adrenal medulla and neurons in the peripheral nervous system; increases dopamine in brain | Increased attention; decreased irritability | Heart disease and lung disease |
| PCP (phencyclidine, or "angel dust") | Blocks channel for excitatory amino acid neurotransmitters; increases dopamine activity | Violent behavior; feelings of power; numbness; disorganized thoughts | Psychosis; convulsions; coma; death |
| **Therapeutic** | | | |
| Tricyclic antidepressants (for example, Elavil, Anafranil) | Block the reuptake of norepinephrine from synapses | Relieve depression and obsessive-compulsive disorder | Drowsiness; confusion |
| Selective serotonin reuptake inhibitors (SSRIs) (for example, Prozac, Zoloft, Paxil, Lexapro) | Block the reuptake of serotonin from synapses | Relieve depression and obsessive-compulsive disorder | Insomnia; anxiety; headache |
| Monoamine oxidase inhibitors (for example, Parnate, Nardil) | Block the breakdown of biogenic amine neurotransmitters | Relieve depression | Liver damage; hyperexcitability |
| Antianxiety drugs: benzodiazepines [for example, Xanax, Valium, Librium, Rohypnol ("date rape drug," "roofies")] | Bind to GABA receptors and increase inhibitory neurotransmission | Relieve anxiety; cause sleepiness and in some cases amnesia | Drowsiness; memory loss in some cases |
| Antipsychotic drugs: phenothiazines (for example, Thorazine, Mellaril, Stelazine) and atypical antipsychotics (Abilify, Risperdal) | Block dopamine receptors | Ease schizophrenic symptoms | Decreased control of movement |

*Some of these illicit or recreational drugs have some therapeutic value under certain conditions.

events can increase or decrease activity at synapses, which, in turn, affects emotions and behavior. The most common mood disorder is **major depressive disorder** or, simply, depression. This illness results in prolonged periods of sadness, despair, and lack of interest in daily activities without alternating episodes of euphoria. Depression affects 5–12% of men and 10–25% of women at some time during their lives. This condition is thought to result from decreased activity of synapses that release biogenic amines, such as serotonin, which changes neuronal activity within specific areas in the brain involved in processing emotion. Drugs used to treat major depression include the **selective serotonin reuptake inhibitors (SSRIs)**, such as Prozac, Zoloft, and Paxil, which reduce the reuptake of serotonin into the presynaptic terminal after it is released. This allows serotonin to accumulate in the synaptic cleft, counteracting the deficit that causes the alteration in mood.

In many respects, the use of such drugs to treat depression reflects a profound change in the public's attitude toward mental illness. Historical attitudes toward mood disorders held that individuals who were depressed lacked the ability to cope with stressful events in their lives. Only relatively recently has it become accepted that mood disorders are typically caused by changes in the balance of neurotransmitters in the brain. Drugs can be very effective in treating these disorders. Patients taking SSRIs often report decreased sadness, increased energy, and a greater interest in daily activities.

## Many Illicit Drugs Disrupt Normal Neurotransmission

Many illicit drugs work at the synapse to either enhance or interfere with the normal mechanisms of neurotransmission (see Table 42.3). At the presynaptic terminal, such drugs can decrease neurotransmitter release by reducing $Ca^{2+}$ entry into the cell or by preventing the exocytosis of vesicles containing stored neurotransmitters. In the synaptic cleft, drugs can slow the rate at which the neurotransmitter is broken down into an inactive form or taken back up into the presynaptic neuron, thereby prolonging the action of the neurotransmitter in the synaptic cleft. Some substances act on the postsynaptic membrane by either preventing the neurotransmitter from binding to its receptor or by acting as a substitute for the neurotransmitter by stimulating the receptor.

In effect, these drugs produce changes or imbalances in neurotransmission similar to those observed in some neurological disorders. These substances can induce euphoria, increase activity, alter mood, and produce hallucinations. They can also have potentially life-threatening effects and may be highly addictive.

Some drugs, such as cocaine, block the removal of dopamine and norepinephrine from the synaptic cleft by preventing their reuptake into the presynaptic terminal. Morphine and marijuana mimic the actions of biological substances already in the brain, binding to receptors on the postsynaptic membrane. With these drugs, the resulting effects are much stronger than are the effects of natural neurotransmitters. It is no surprise that many of these drugs are mind-altering. They do, after all, change the ways in which neurons communicate with each other.

## Disorders of Conduction May Result in Motor Problems and Abnormal Neuronal Development

Some human diseases are caused by the inability of certain axons to properly conduct an action potential. This occurs most commonly because an axon fails to become myelinated or because a myelinated axon becomes demyelinated.

In **congenital hypothyroidism**, axons fail to become wrapped with myelin during fetal development, which leads to slow conduction speeds and abnormal connections between brain neurons. This results in profound mental defects that cannot be reversed unless treatment begins immediately after birth. Congenital hypothyroidism is caused by a deficiency of thyroid hormone in the fetus. Among its many actions, thyroid hormone stimulates the formation of myelin during fetal development. However, thyroid hormone cannot be synthesized without the element iodine, which is part of its structure. The iodine in the fetus comes from the mother's diet. If a mother's dietary intake of iodine is too low, the fetus will not have enough iodine to make its own thyroid hormone, and therefore will not be able to make normal amounts of myelin. Congenital hypothyroidism is rare in the U.S. and many other countries since the advent of iodized table salt. However, it is a serious public health concern in many parts of the world.

Unlike congenital hypothyroidism, **multiple sclerosis (MS)** is a myelin-related disease that usually begins between the ages of 20 and 50 in individuals with apparently healthy nervous systems. With MS, a person's own immune system, for reasons unknown, attacks and destroys myelin as if it were a foreign substance. Eventually, these repeated attacks leave multiple scarred (sclerotic) areas of tissue in the nervous system (**Figure 42.18**) and impair the function of myelinated neurons that control movement, speech, memory, and emotion. Multiple sclerosis is a serious and unpredictable disease, characterized by flare-ups

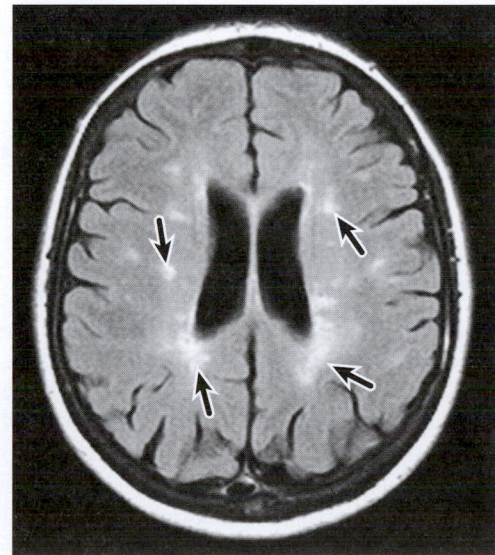

**Figure 42.18  MRI scan of a brain from a person with multiple sclerosis.** The arrows point to a few of numerous regions of damaged nervous tissue (white patches). ©Du Cane Medical Imaging Ltd./Science Source

 **Core Skill: Science and Society** Research into the treatment and prevention of human (and animal) disease has been greatly facilitated by technological advances such as MRI and other imaging techniques.

followed by periods of remission in which symptoms are reduced or absent. No cure is currently available, but certain drugs may slow its progression and reduce the severity of symptoms. This disease affects roughly 2.5 million people worldwide, about 75% of them women.

## Summary of Key Concepts

### 42.1  Cellular Components of Nervous Systems

- The central nervous system (CNS) is composed of a brain and a nerve cord. The peripheral nervous system (PNS) consists of all neurons and their projections that are outside of and connect with the CNS. Nerves transmit signals between the PNS and CNS (Figure 42.1).

- The two major classes of cells in nervous systems are neurons and glia. In neurons, signals flow from dendrites to the cell body and then to the axon and axon terminal. Types of neurons include sensory neurons, motor neurons, and interneurons. A neuron's function is a reflection of its structure (Figures 42.2, 42.3).

- The most basic neural circuit is a reflex arc, which acts rapidly in response to inputs from sensory neurons and consists of only a few neurons (Figure 42.4).

### 42.2  Electrical Properties of Neurons and the Resting Membrane Potential

- Neuronal membranes are electrically polarized. The membrane potential is determined by the differential distribution and differential permeability of ions across the plasma membrane. The resting membrane potential is the membrane potential of a cell that is not sending electrical signals. Neurons use electrical signals to communicate with other neurons, muscle cells, or gland cells. These signals involve changes in the amount of electric charge on either side of a cell's plasma membrane (Figure 42.5, Table 42.1).

- Diffusion of ions through membrane channels occurs as a result of the concentration gradient of an ion across the membrane and the electric charge difference across the membrane. Ions move in response to an electrochemical gradient (Figure 42.6).

- The Nernst equation gives the equilibrium potential for an ion at any given concentration gradient.

### 42.3  Generation and Transmission of Electrical Signals Along Neurons

- Gated ion channels enable a cell to communicate by changing its membrane potential. The opening and closing of voltage-gated and ligand-gated ion channels cause two types of changes in the neuron's membrane potential—graded potentials and action potentials (Figures 42.7, 42.8).

- Graded potentials can trigger an action potential, an event that carries an electrical signal along an axon, from the axon hillock to the axon terminal. Axon diameter and myelination influence the rate of propagation of an action potential (Figures 42.9, 42.10, 42.11).

### 42.4  Electrical and Chemical Communication at Synapses

- In an electrical synapse, electric current is conducted from one cell to another via gap junctions. In a chemical synapse, a neurotransmitter carries the signal from the presynaptic to the postsynaptic cell. Many excitatory postsynaptic potentials (EPSPs) generated at one time can sum together and bring the membrane potential to the threshold potential, initiating an action potential (Figures 42.12, 42.13, 42.14).

- Chemical classes of neurotransmitters found in animals include acetylcholine, biogenic amines, amino acids, neuropeptides, and gaseous neurotransmitters. The discovery that the actions of neurons are mediated in large part by neurotransmitters was one of the most significant discoveries in the history of neuroscience (Table 42.2, Figure 42.15).

- The receptors of the postsynaptic neuron determine the types of signals that pass from one neuron to the other. The two major types of postsynaptic receptors are ionotropic and metabotropic (Figures 42.16, 42.17).

### 42.5  Impact on Public Health

- Most neurological conditions can be classified as disorders of either neurotransmission or conduction. Mood disorders caused by disrupted neurotransmission include major depressive disorder. Drugs used in the treatment of neurological disorders and many recreational and illicit drugs usually alter neurotransmission (Table 42.3).

- Some neurological conditions are caused by the inability of the axon to conduct an action potential. This occurs most commonly because axons fail to become myelinated (as in congenital hypothyroidism) or because myelinated axons become demyelinated (as in multiple sclerosis) (Figure 42.18).

## Assess & Discuss

### Test Yourself

1. In vertebrates, the brain and the spinal cord constitute
   a. the peripheral nervous system.
   b. the efferent division of the nervous system.
   c. the central nervous system.
   d. the autonomic nervous system.
   e. the central and peripheral nervous systems.

2. The structures of a neuron that function mainly in receiving signals from other neurons are the
   a. myelin sheaths.
   b. axons.
   c. axon terminals.
   d. dendrites.
   e. $K^+$ channels.

3. The myelin sheath
   a. is produced by neurons in the peripheral nervous system.
   b. is formed only around neurons in the brain.
   c. is present on all neurons.
   d. is generally present around long axons in either the CNS or the PNS.
   e. significantly slows transmission along neurons.

4. Neurons that function mainly in connecting other neurons in the central nervous system are
   a. sensory neurons.
   b. efferent neurons.
   c. motor neurons.
   d. afferent neurons.
   e. interneurons.

5. The difference in charges across the plasma membrane of an unstimulated neuron is called
   a. an EPSP.
   b. the resting membrane potential.
   c. an IPSP.
   d. the graded potential.
   e. the action potential.

6. Which of the following contribute(s) to the resting membrane potential?
   a. the relative leakiness of the membrane to $Na^+$ and $K^+$
   b. active transport of ions across the membrane
   c. concentration of $Na^+$ and $K^+$ inside and outside of the cell
   d. all of the above
   e. b and c only

7. A neuron has reached a threshold potential when it has depolarized to the point where
   a. the voltage-gated $Na^+$ channels have become inactivated.
   b. sufficient numbers of voltage-gated $Na^+$ channels open to initiate a positive feedback cycle, contributing to further depolarization.
   c. voltage-gated $K^+$ channels close.
   d. voltage-gated $Na^+$ channels close.
   e. both b and c occur.

8. The speed of transmission of an action potential along an axon is influenced by
   a. the presence of myelin.
   b. an increased concentration of $Ca^{2+}$.
   c. the diameter of the axon.
   d. all of the above.
   e. a and c only.

9. Gap junctions are characteristic of
   a. electrical synapses.
   b. chemical synapses.
   c. acetylcholine synapses.
   d. GABA synapses.
   e. synapses between motor neurons and muscle cells.

10. The response of the postsynaptic cell at a chemical synapse is determined by
    a. the type of neurotransmitter released at the synapse.
    b. the type of receptors the postsynaptic cell has.
    c. whether or not an axon is myelinated.
    d. whether the synapse is on a dendrite or directly on the cell body of the postsynaptic cell.
    e. a and b, both.

## Conceptual Questions

1. Describe the difference between graded and action potentials.

2. In certain diseases, such as kidney failure, the $Na^+$ concentration in the body's extracellular fluid can be altered. What effect might a high extracellular $Na^+$ concentration have on neurons?

3. **Core Concept: Structure and Function** How can the core concept that structure determines function be applied to neurons?

## Collaborative Questions

1. Describe the difference between an electrical synapse and a chemical synapse. What advantage is provided by chemical synapses?

2. Name the parts of a neuron, and give a brief description of their major characteristics.

# Neuroscience II: Evolution, Structure, and Function of the Nervous System

# 43

**Three-dimensional reconstruction of the brain of a fruit fly.** The brains of animals are organized into anatomical structures with specialized functions. For example, the green structures in this image are important for learning and memory, the purple structures for olfaction, and the large orange structures on either side for processing visual information. Courtesy Ann-Shyn Chiang, Tsing Hua Chair Professor/Brain Research Center & Institute of Biotechnology/National Tsing Hua University

I t will take you approximately 2–3 seconds to read this sentence. During that time, many of the 100 billion or so neurons of your brain will have collectively fired off millions of action potentials. Some of those signals will help process the visual information reaching your eyes as you scan the lines. Others will activate centers of learning and memory to allow you to understand the meanings of the words you've read. Still other signals will help filter out extraneous inputs—such as background noise—that might distract you from your task. The complexity of the seemingly simple task of reading a single sentence indicates the enormous level of activity that goes on continually in the brain, even at rest.

The wonder of the brain lies in its incredible complexity. The human brain, for example, has several thousand miles of interconnected neurons and hundreds of trillions of synapses, resulting in a total surface area that if spread out, would cover more than four soccer fields. The brain allows us to move, think, and experience sensation and emotion. Groups of neurons also coordinate homeostatic functions such as breathing, blood circulation, and body temperature. When we examine the way that groups of neurons communicate, we begin to understand the complex mental functions of nervous systems, including learning, memory, and motivation.

Neuroscience—the study of nervous systems—is an area of intense research activity worldwide. In 2013, the administration of President Barack Obama announced the formation of the BRAIN Initiative (Brain Research through Advancing Innovative Neurotechnologies), a 10-year, federally funded effort to map the location, structure, and function of all the 100 billion or so neurons in the human brain. This initiative has the potential to revolutionize our understanding of how the human brain functions in both health and disease, and it launches a thrilling new era of neuroscience research. As one recent example, the Human Connectome Project—an international consortium of researchers at several institutions funded largely by the National Institutes of Health—has provided fascinating new insight into the organization and interconnectedness of pathways throughout the human brain.

In this chapter, we will first survey a variety of nervous systems, which allow animals to sense and respond to environmental changes. We will then examine the nervous system of humans. However, keep in mind that we still have much to learn about the organization, connectivity, and functions of nervous system structures. Our own nervous system is fascinating and mysterious, and the study of how it functions will ultimately tell us much about what makes us human.

## 43.1 The Evolution and Development of Nervous Systems

### Learning Outcomes:

1. List the different types of nervous systems found in animals.
2. Describe the general anatomical organization of the brain in vertebrates.
3. Relate the structural changes in brain complexity that accompanied the evolution of mammals to brain function.

Animal nervous systems are the products of hundreds of millions of years of evolution. They provide advantages to animals that promote survival and reproductive success. For example, nervous systems allow animals to sense their environment and respond to changes in an appropriate way. In addition, nervous systems form connections with muscles and facilitate movement, which allows some animals to travel distances to obtain food. Likewise, nervous systems help animals avoid predation and other environmental dangers; form social bonds that enhance the chances of survival for both the individual and the group; and even perform the complex tasks of thinking, learning, remembering, and planning.

Studying the evolution and development of nervous systems helps us understand how particular nervous systems are adapted to different functions. At the structural level, the organization of nervous systems ranges from a relatively simple network of a small number of cells to the complexity of the human brain. The characteristics of an animal's nervous system determine the behaviors that it displays. In this section, we will survey some comparative features of the nervous systems of invertebrates and vertebrates.

## The Evolution of Nervous Systems Gave Animals the Ability to Sense and Respond to Changes in the Environment

Precisely when nervous systems first arose and whether or not the nervous systems of most or all animals can be traced back to a common ancestor are questions of active investigation by neuroscientists. For example, recent genetic studies have uncovered remarkable similarities in the expression and activation of genes coding for proteins that regulate neuronal development across taxa in bilaterally symmetric animals. Those studies suggest that the patterning of nervous system development in these bilaterians may be traced to a common ancestor that lived more than 500 mya!

Today, all animals except sponges have a nervous system. Interestingly, though, researchers have discovered that sponges express dozens of genes that are similar to genes expressed in human neurons, particularly those that encode proteins that regulate synaptic function. The functions of the sponge genes are uncertain, but the proteins encoded by the genes interact with each other in ways that are reminiscent of human synaptic proteins. Thus, the origin of nervous systems almost certainly can be traced to genes of evolutionarily ancient organisms. As animal species evolved, these genes were modified and formed the basis of all future nervous systems.

The simplest nervous system is the **nerve net** of the radially symmetric cnidarians (jellyfish, hydras, and anemones; **Figure 43.1a**). The neurons are arranged in a network of connections between the inner and outer body layers of the animals. A characteristic feature of nerve nets is that activation of neurons in any one region leads to activation of most or all other neurons, with the excitation spreading in all directions at once. Many of these neurons stimulate contractile cells to contract. This allows the organism to move large areas of its body simultaneously, thereby coordinating simple movements such as swimming. Recent research has identified regions of specialized function in the nerve nets of some cnidarians, such as local sensory neurons in the outer body wall. These findings push the origin of specialized nervous system structures and function further back in evolutionary

time than previously thought. Some cnidarians, such as the jellyfish, have two nerve nets: one for moving tentacles and one for swimming.

Sea stars and other echinoderms also have a simple nervous system, but it is slightly more complex than that of cnidarians. A nerve ring surrounds the mouth and is connected to larger radial nerves extending into the arms (**Figure 43.1b**). This arrangement allows the mouth and arms to operate independently.

During the evolution of animals, more complex body types have been associated with **cephalization** (from the Greek *cephalo*, meaning head). This term refers to the concentration of sense organs at the anterior end of the body, forming an increasingly complex **brain** that controls sensory and motor functions of the entire body. Within the brain, neuronal pathways provide the integrative functions necessary for an animal to make more sophisticated responses to its environment. Brains are found in all vertebrates and most invertebrates, and they are usually composed of more than one anatomical and functional region with considerable complexity.

Platyhelminthes (flatworms) was the first animal phylum to evolve a brain with defined regions exhibiting many synaptic connections. In these animals, different regions of the brain appear to integrate inputs from sense organs, such as the eyes, and control motor output, such as movements involved in swimming (**Figure 43.1c**). Two nerve cords extend along the ventral surface of the animal from the anterior end to the posterior end and are connected to each other by transverse nerves.

In annelids (segmented worms), ganglia and nerves are present at each body segment, where they coordinate local sensory and motor activities (**Figure 43.1d**). Ganglia are collections of neuron cell bodies with limited processing ability, limited synapses, and few to no subdivisions like those found in a brain. Ganglia are present in most animals and often serve to coordinate local signaling in a body part.

In the simpler types of mollusks, such as the snail, the nervous system is very similar to that of the annelids. The head contains a pair of anterior ganglia; paired nerve cords extend from these ganglia and send branches to the eyes, muscular foot, and digestive system.

In arthropods, such as *Drosophila* (**Figure 43.1e** and the chapter opening photo), the brain has several subdivisions with distinct, well-defined anatomical borders and functions, such as a region devoted to learning and memory. Some mollusks, such as the squid and octopus, have complex brains with subdivisions that allow these animals to coordinate the sophisticated visual sensing and motor behaviors necessary for their predatory lifestyle (**Figure 43.1f**).

In the embryos of chordates, a dorsally located nerve cord is present that, in vertebrates, develops into the brain and **spinal cord**. The brain and spinal cord constitute the **central nervous system (CNS)** (**Figure 43.1g**). Nerves from the **peripheral nervous system (PNS)** relay signals into and out of the CNS at separate regions along the spinal cord.

### Brains of Vertebrates Have Three Basic Divisions

Development of the vertebrate brain begins with the formation of a central fold in the embryo called the neural tube, or dorsal nerve cord. This hollow tube is the structure from which the entire nervous system develops (look ahead to Figure 51.19). Increased cell proliferation leads to bending and folding of the neural tube during embryonic development, resulting in bulges that become separate divisions of the nervous system. The anterior end develops into the brain, while the posterior portion becomes the spinal cord.

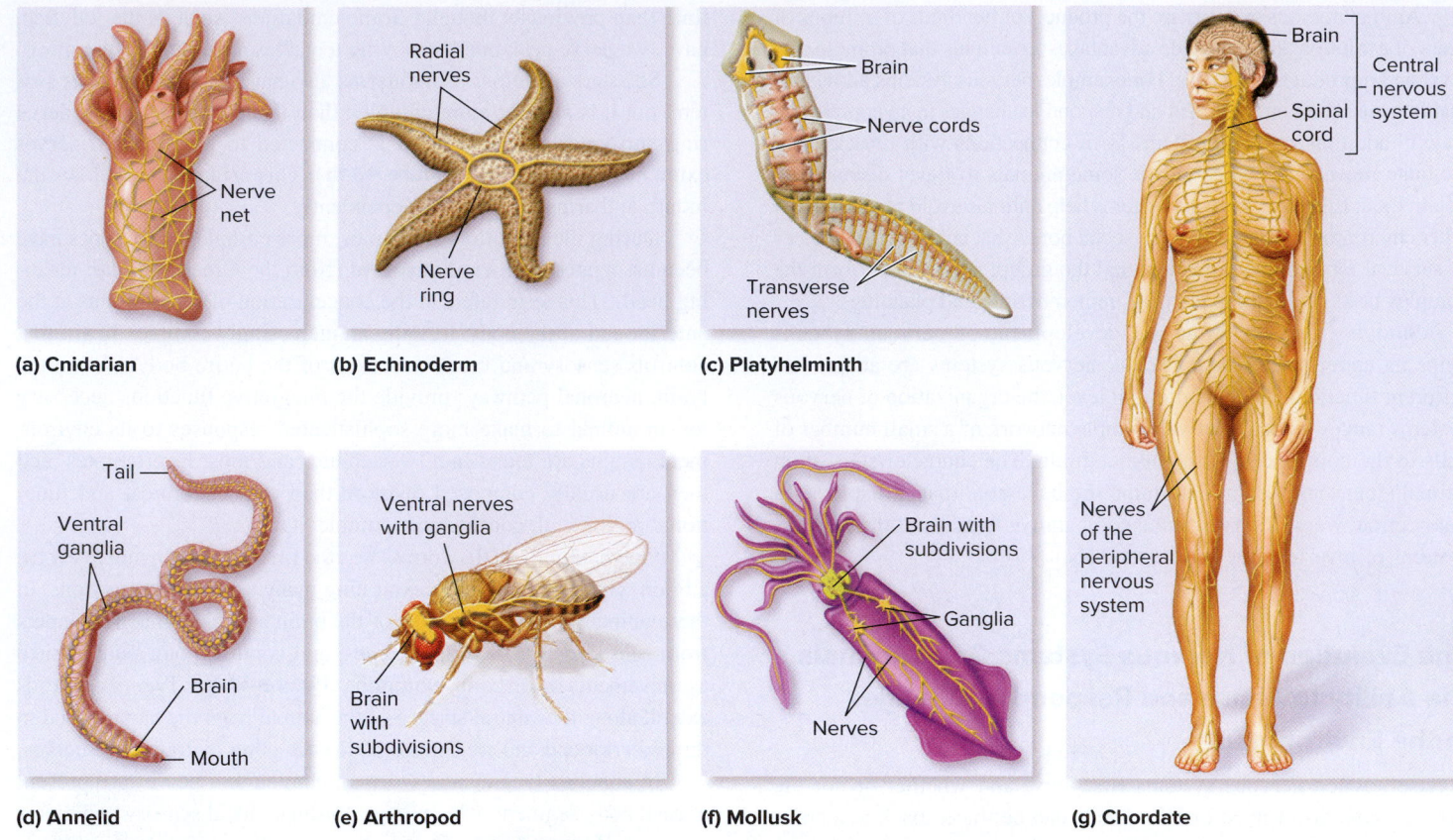

**(a) Cnidarian** — Nerve net

**(b) Echinoderm** — Radial nerves, Nerve ring

**(c) Platyhelminth** — Brain, Nerve cords, Transverse nerves

**(d) Annelid** — Tail, Ventral ganglia, Brain, Mouth

**(e) Arthropod** — Ventral nerves with ganglia, Brain with subdivisions

**(f) Mollusk** — Brain with subdivisions, Ganglia, Nerves

**(g) Chordate** — Brain, Central nervous system, Spinal cord, Nerves of the peripheral nervous system

**Figure 43.1** **Representative nervous systems throughout the animal kingdom.**

 **Core Skill: Connections** Nervous systems are one of the defining features of animals. In Chapter 33, you learned that several other features define animals and distinguish them from other living organisms. What are some of these features?

In vertebrates, the brain has three major anatomic divisions: **hindbrain**, **midbrain**, and **forebrain** (**Figure 43.2**). Fossils of jawless fishes that lived 400 mya show that their brains were already organized into the three basic divisions that have been retained in all modern vertebrates.

Let's look at the development of the human brain. At 4 weeks, the human embryo already exhibits the hindbrain, midbrain, and forebrain (Figure 43.2a). Just a week later, the hindbrain and forebrain have each formed two separate subdivisions (Figure 43.2b). The hindbrain subdivides into the metencephalon and the myelencephalon. The forebrain subdivides into the telencephalon and the diencephalon. The midbrain, by contrast, does not subdivide and is termed the mesencephalon. By the time the human brain is fully developed, some of these structures have further divided and specialized (Figure 43.2c; their functions will be described in Section 43.2). The development of brain subdivisions increases the capacity of the brain to perform complex, distinct functions.

## Evolution of Increased Brain Complexity Involved a Larger, Highly Folded Forebrain

As evolution resulted in animals with more complex nervous systems, the size of the forebrain and its major subdivision, the **cerebrum**, also increased, making up a greater proportion of the brain. Many of the important functions of the forebrain are carried out by neurons in the outer surface of the cerebrum called the **cerebral cortex**. Therefore, increased complexity of the brain is also correlated with an increased surface area of the cerebral cortex (see Figure 43.2). During the evolution of mammals, this increase in surface area occurred more rapidly than an expansion in the size of the skull. How could this occur? The answer is that the external surface of the forebrain in animals with increasingly complex brains is highly convoluted, forming many folds and grooves. Compare the relatively smooth-looking surface of the brain of a rat with the highly folded one of a dolphin or a primate in **Figure 43.3**.

As body size increases across the animal kingdom, you might expect that brain mass would increase proportionately—that the brain of an elephant would be proportionately larger than that of a bat, for instance. That is generally the case, with a few important exceptions (**Figure 43.4**). In particular, the masses of the human and dolphin brains are considerably greater than would be expected on the basis of body mass (note the logarithmic scale on the vertical axis).

Brain mass and the amount of folding of forebrain structures are correlated with more complex behaviors. Why is this so? As we will see in the next section, the outer surface of the brain, the cerebral cortex, plays a key role in conscious thought, reasoning, and learning. Greater size and increased folding provide more surface area to support a larger number of neurons and synapses, which, in turn, facilitates processing and interpretation of information. Even so, evidence does not suggest that people with small differences in brain size differ

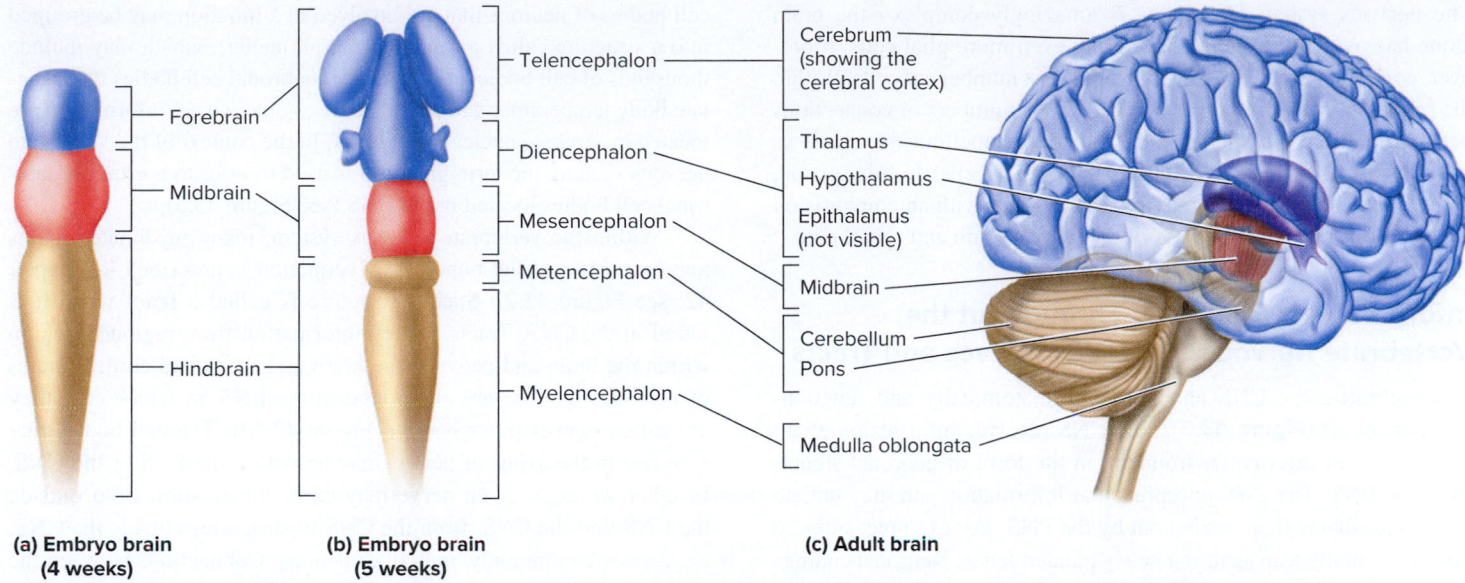

**(a) Embryo brain**
**(4 weeks)**

**(b) Embryo brain**
**(5 weeks)**

**(c) Adult brain**

**Figure 43.2 Development of the human brain.** Vertebrate brains begin as three major divisions, which then develop into additional subdivisions. The structures shown here that occur during embryonic development at **(a)** 4 weeks and **(b)** 5 weeks are compared with **(c)** their appearance in the adult brain. Some forebrain structures beneath the cerebrum are not shown. The functions of all of these structures will be described in Section 43.2.

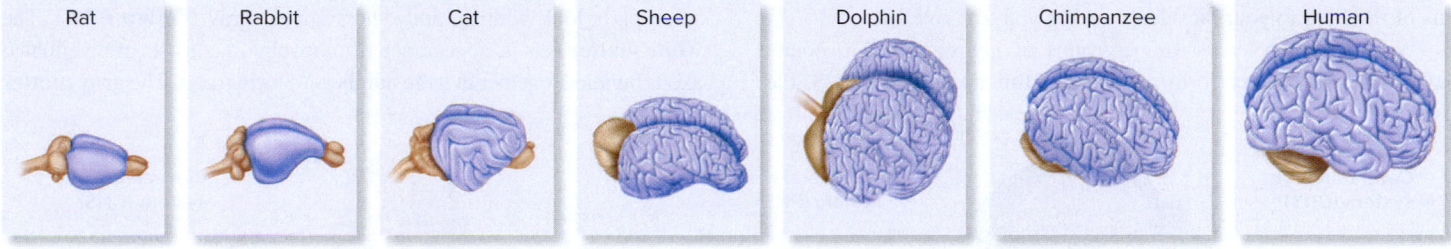

**Figure 43.3 The degree of cerebral cortex folding in different mammalian species.** The brains are not shown to scale.

**Core Concept: Structure and Function** The greater amount of folding of the cerebral cortex of certain mammals increases the surface area of this part of the brain, allowing for more neuronal connections and thus more complex behaviors.

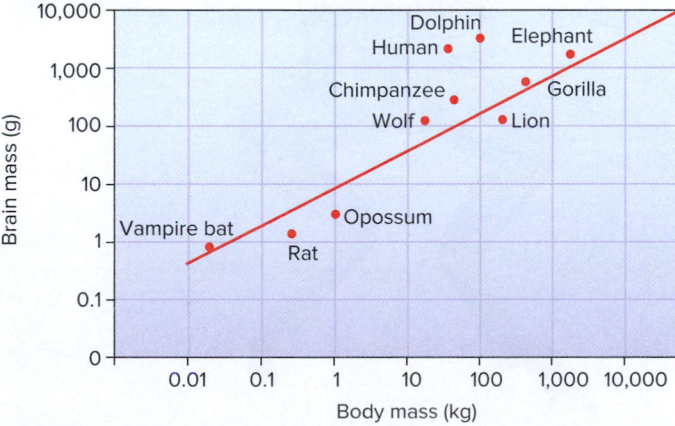

**Figure 43.4 Brain mass as a function of body mass in mammals.** For most mammals, brain mass is in proportion to body mass. However, humans and dolphins have a much greater brain mass relative to their body mass than do other mammals.

**Concept Check:** *Scientists have determined that the brain mass of Homo neanderthalensis (Neanderthals) was greater than that of our own. Does this mean that Neanderthals were more intelligent and capable of more complex behaviors than we are?*

in intelligence. Also, it would be wrong to assume that an animal with a small brain is profoundly limited in its behavioral repertoire. A bat with a 0.9-g brain and an elephant with a 2,500-g brain can both perform a great variety of interesting and complex behaviors, such as navigating across great distances and interacting with fellow members of their species.

**43.2  Structure and Function of the Nervous Systems of Humans and Other Vertebrates**

**Learning Outcomes:**

1. Outline the anatomical organization of the human nervous system.
2. Describe the organization of the peripheral nervous system.
3. Identify the differences between the somatic and autonomic nervous systems.
4. Briefly describe the major structures and functions of the human hindbrain, midbrain, and forebrain.

The nervous system of humans is amazingly complex—the brain alone has over 100 billion neurons and even more glial cells. Moreover, complexity is defined by more than just numbers of cells. Within the human brain, for example, are enormous numbers of connections between neurons—a single neuron in the cerebellum may have as many as 100,000 or more synapses with other cells! In this section, we will examine the vertebrate nervous system, with an emphasis on the functions of the major parts of the human brain and spinal cord.

## Information Is Conveyed Throughout the Vertebrate Nervous System by Nerves and Tracts

In vertebrates, the CNS and PNS are anatomically and functionally connected (**Figure 43.5**). The CNS receives information about the internal or external environment in the form of neuronal signals from the PNS. The CNS interprets that information and may initiate a response that is then carried out by the PNS. For example, suppose you accidentally lean against a newly painted fence. Neuronal endings in your skin, which are part of the PNS, would transmit tactile (touch) information through axons that bring information directly into the spinal cord. From there, the information travels to your brain, where the sensation is analyzed and identified as something sticky. Signals are sent from your brain, down your spinal cord, and through the neurons of the PNS to your muscles, causing you to move away.

Within the nervous system, groups of neurons may associate with each other and perform a particular function. In the CNS, the cell bodies of neurons that are involved in a function may be grouped into a structure called a **nucleus** (plural, nuclei), which may include thousands of cell bodies. For instance, neuronal cell bodies that regulate body temperature and those that recognize visual information are located in separate nuclei in the brain. In the context of the vertebrate nervous system, the term ganglion is used to refer to a group of neuronal cell bodies located in the PNS (see Figure 43.5).

Within the vertebrate nervous system, many myelinated axons may occur in parallel bundles. (Myelination is described in Chapter 42; see Figure 42.2.) Such a structure is called a **tract** when it is found in the CNS. Tracts convey information from region to region within the brain and between the brain and the spinal cord. Bundles of myelinated axons are also found in the PNS, in which case they are called **nerves** (refer back to Figure 42.2d). The cell bodies that give rise to the axons of nerves may be within the PNS or the CNS. In other words, a given nerve may carry information from outside the CNS into the CNS, from the CNS to structures outside the CNS, or, as occurs commonly, in both directions. Connections between the PNS and the CNS occur at the brain or spinal cord. **Cranial nerves** are directly connected to the brain, primarily to sites within the hindbrain and midbrain. By comparison, **spinal nerves** are connections between the PNS and spinal cord (see Figure 43.5).

One of the most obvious characteristics of the CNS is that some parts look whitish, and others appear gray (**Figure 43.6**). The **white matter** gets its appearance from myelin; it consists of myelinated axons bundled together in large numbers to form tracts. The **gray matter**

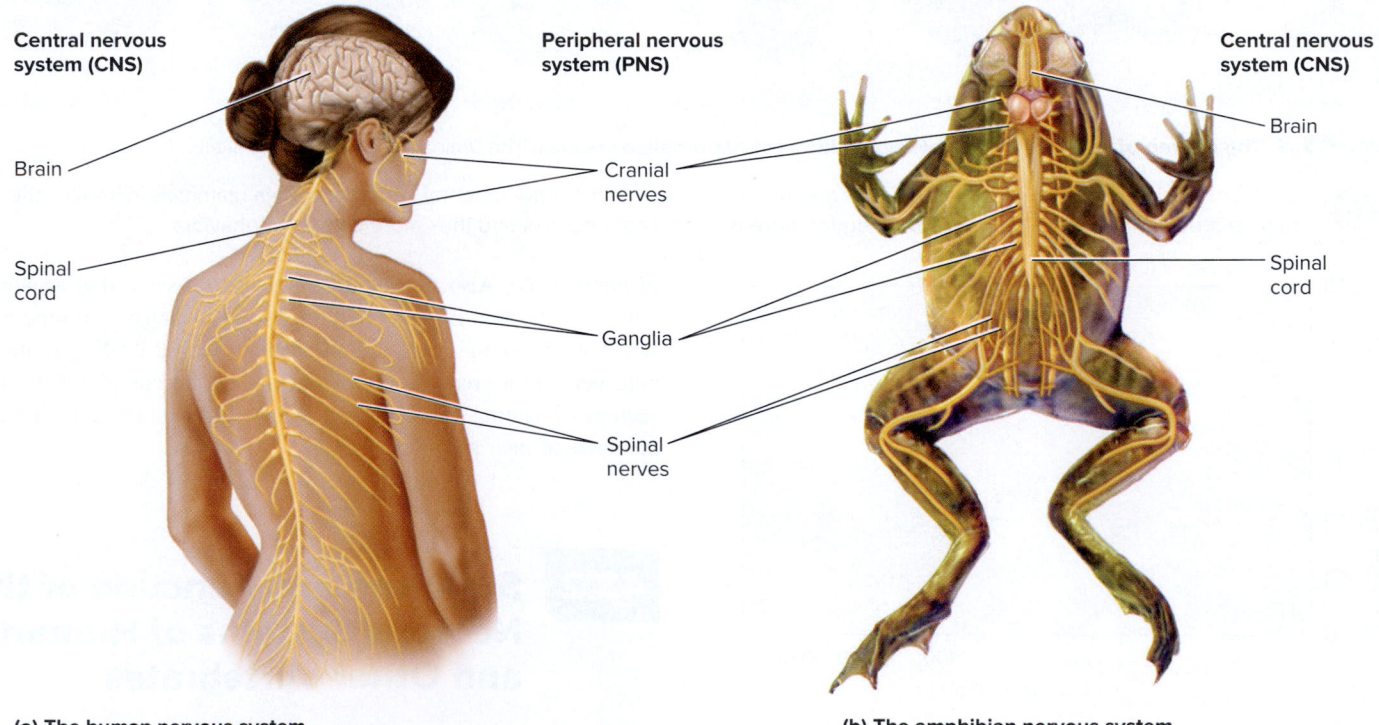

**Central nervous system (CNS)**
- Brain
- Spinal cord

**Peripheral nervous system (PNS)**
- Cranial nerves
- Ganglia
- Spinal nerves

**Central nervous system (CNS)**
- Brain
- Spinal cord

(a) The human nervous system

(b) The amphibian nervous system

**Figure 43.5** **Organization of the vertebrate nervous system.** The CNS consists of the brain and spinal cord, both of which are encased in bone (not shown). The PNS includes cranial nerves, ganglia, and spinal nerves, which carry information to and from the CNS, and many other neurons throughout the body. Note the similarities between two widely divergent vertebrates, **(a)** humans and **(b)** frogs.

 **Core Concept: Evolution** The similarities between the nervous systems of these two very different-looking animals point to a common ancestor.

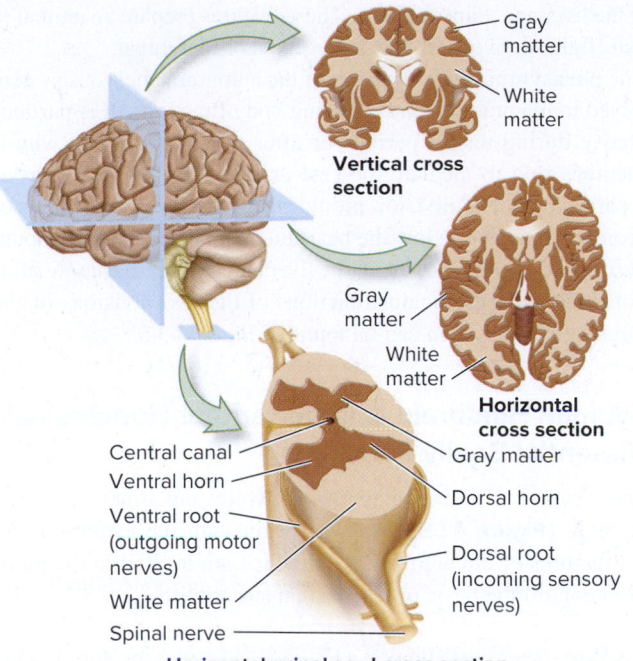

**Vertical cross section**

Gray matter
White matter

Gray matter
White matter

**Horizontal cross section**

Central canal
Ventral horn
Ventral root (outgoing motor nerves)
White matter
Spinal nerve

Gray matter
Dorsal horn
Dorsal root (incoming sensory nerves)

**Horizontal spinal cord cross section**

**(a) Gray and white matter in the brain and spinal cord**

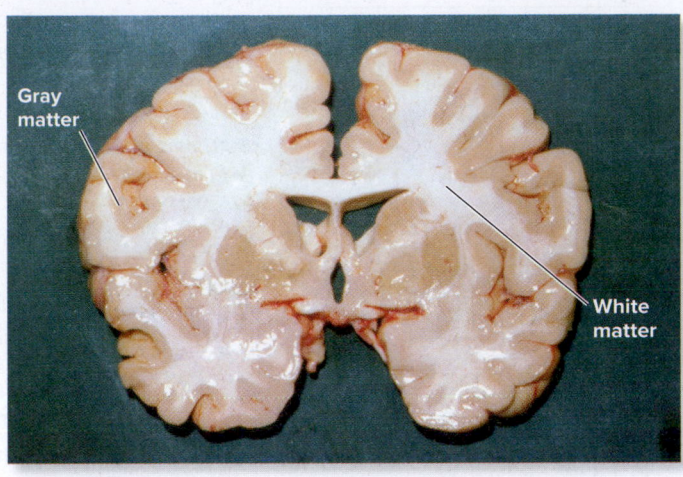

Gray matter
White matter

**(b) Photograph of a vertical cross section of the human brain**

**Figure 43.6** **Gray matter and white matter in the CNS.** **(a)** The gray matter is composed of cell bodies, dendrites, and unmyelinated axons. The white matter consists of tracts of myelinated axons. **(b)** Photograph of a vertical cross section through an adult human brain. In these images, gray matter is darkened for better visibility. b: ©Biophoto Associates/Science Source

**Concept Check:** *Is a spinal nerve composed of axons of afferent or efferent neurons, or both?*

is darker in appearance and consists of neuronal cell bodies, dendrites, and some unmyelinated axons. The cerebral cortex is composed of gray matter that sits on top of a large collection of white matter pathways. In the spinal cord, the gray matter is located in the center and forms two dorsal extensions, or horns, and two ventral horns (Figure 43.6a). Each dorsal horn connects to a dorsal root, which is part of a spinal nerve. Dorsal roots receive incoming information from sensory (afferent) nerves of the PNS. The ventral horn connects to the ventral root, which is also part of a spinal nerve that transmits outgoing information to motor (efferent) nerves. A central canal runs through the spinal cord, carrying a nutritive and protective fluid, as described shortly.

## The CNS Is Encased in Protective Structures

Unlike the PNS, the CNS is encased in protective structures including bone (the skull and vertebrae) and three layers of sheathlike membranes called **meninges** (**Figure 43.7**). The outermost membrane, the dura mater (from the Latin, meaning hard mother), is a thick protective layer that lies just inside the skull and vertebrae. The middle membrane is called the arachnoid mater (from the Latin, meaning spidery mother) because it has numerous weblike tissue connections to the innermost membrane, the pia mater (from the Latin, meaning thin mother). The pia mater is a very thin membrane that lies on the surface of the brain and spinal cord, folding with the brain's surface.

Between the arachnoid mater and pia mater is the subarachnoid space. This space is filled with **cerebrospinal fluid**, which surrounds the exterior of the brain and spinal cord and absorbs physical shocks to the brain that result from sudden movements or blows to the head. The cerebrospinal fluid contains nutrients, hormones, and other substances that are taken up by cells of the brain. The fluid is also a reservoir for metabolic waste products that are then carried away by the circulatory system. In addition to the subarachnoid space, the cerebrospinal fluid

also fills a series of connected cavities called the ventricles that lie deep within the brain and connect to the central canal that extends the length of the spinal cord (see Figure 43.6). These fluid-filled structures provide a cushion of support and protection for the CNS.

## The PNS Consists of the Somatic and Autonomic Nervous Systems

The PNS of vertebrates is subdivided into two major functional and anatomical components: the somatic nervous system and the autonomic nervous system. Both divisions have sensory (afferent) nerves and motor (efferent) nerves.

*Somatic Nervous System*   The major functions of the **somatic nervous system** are to sense the external environment and control skeletal muscles. The afferent sensory neurons of the somatic nervous system receive stimuli, such as heat, light, odors, chemicals (in food), sounds, and touch, and transmit signals to the CNS. The efferent motor neurons of the somatic nervous system control skeletal muscles. The cell bodies of these motor neurons are located within the CNS. The axons from these cells leave the spinal cord and project directly onto skeletal muscle without any intermediary synapses along the way.

Many of the responses of the somatic nervous system can be controlled consciously. For example, we use our somatic nervous system to walk and hold a pencil. However, not all responses are voluntary. An example is a reflex arc, such as the knee-jerk response, which is automatic (refer back to Figure 42.4).

*Autonomic Nervous System*   The **autonomic nervous system** regulates homeostasis and organ function. For example, it is involved in regulating heart rate, blood pressure, glucose homeostasis, and the

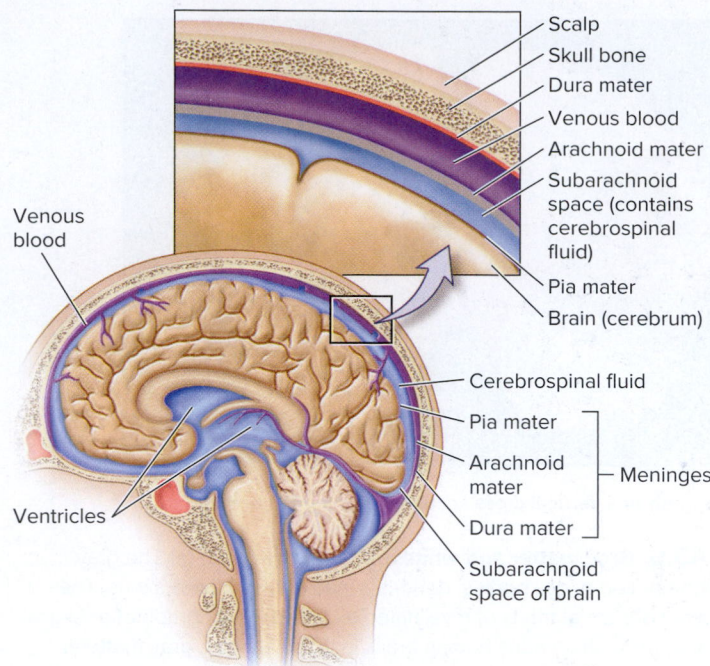

Scalp
Skull bone
Dura mater
Venous blood
Arachnoid mater
Subarachnoid space (contains cerebrospinal fluid)
Pia mater
Brain (cerebrum)

Venous blood

Cerebrospinal fluid
Pia mater
Arachnoid mater — Meninges
Dura mater
Ventricles
Subarachnoid space of brain

**Figure 43.7** **The meninges and ventricles of the CNS.** The thicknesses of the meninges are exaggerated for illustrative purposes. Note that the cerebrospinal fluid encases the entire CNS and also fills the ventricles.

**Concept Check:** *In a procedure known as a lumbar puncture (commonly referred to as a spinal tap), physicians use a needle to withdraw a small amount of cerebrospinal fluid from the bottom of the spine to help diagnose specific illnesses. Considering that this fluid encases the brain, what effects might this procedure have on a patient?*

amount of stomach acid secreted in response to a meal. The autonomic nervous system is predominantly composed of efferent motor neurons. For the most part, the autonomic nervous system is not subject to voluntary control. For example, we usually cannot consciously change our heart rate or blood pressure.

The efferent pathways of the autonomic nervous system involve sets of two motor neurons. The cell body of the first neuron is within the CNS and synapses on a second neuron in ganglia outside the spinal cord; these ganglia, therefore, are part of the PNS. This second neuron sends its axon to an effector cell, where it alters that cell's function. These neurons control smooth muscles, cardiac muscle, and glands.

The efferent nerves of the autonomic nervous system are subdivided into the sympathetic and parasympathetic divisions (**Figure 43.8**). Both divisions of the autonomic system act on the same organs and usually have opposing actions. The **sympathetic division** is responsible for rapidly activating systems that prepare the body for danger or stress. Imagine, for example, the physiological responses that would occur if a person was hiking and came upon a grizzly bear. They are collectively called the **fight-or-flight response**, which is characterized by increased heart rate, stronger pumping action of the heart, relaxed (dilated) airways and faster breathing, inhibition of digestive activity, increased blood flow to skeletal muscles, and increased secretion into the blood of energy-supplying substances such as glucose and

fats by the liver and adipose tissue. These features prepare an animal to confront (fight) or avoid (flight) a perceived or real threat.

The **parasympathetic division** of the autonomic nervous system is involved in maintaining and restoring body functions. It is particularly active during restful periods or after a meal, which is why it is sometimes said to mediate the **rest-or-digest response**. Neurons of the parasympathetic division promote digestion and absorption of food from the intestines, slow the heart rate, and decrease the amount of fuel supplied to the blood from the liver and adipose tissue. A summary of these and other major functions of the two divisions of the autonomic nervous system can be found in Figure 43.8.

## The Human Hindbrain Is Important for Homeostasis and Essential Bodily Functions

Let's now turn our attention to the structure and function of the human brain (**Figure 43.9**). We will begin with the evolutionarily oldest structures of the brain, some of which are located in the hindbrain and control the basic processes that sustain life.

*Cerebellum* The **cerebellum** is a large structure that sits dorsal to the rest of the hindbrain and receives sensory inputs from the cerebral cortex and the auditory and visual areas of the brain. It also receives inputs from the spinal cord and inner ears that convey information about the position of the limbs and head, respectively, and thereby helps maintain balance and coordinate hand-eye movements. In addition, the cerebellum helps control the use of multiple muscles at one time and synchronizes fine motor activities such as texting, making a jump shot in basketball, or touching the fingers to the tip of the nose with your eyes closed. When the cerebellum is damaged or injured, such as in an accident, a person finds it difficult to maintain balance and fine-tune motor functions. Although historically scientists have thought that the cerebellum does not function in learning, memory, and conscious thought, recent evidence has strongly suggested that the cerebellum may provide significant cognitive functions, the full extent of which remains to be discovered.

*Pons* The **pons** sits anterior to the medulla oblongata and ventral to the cerebellum. Major tracts involved in motor function pass through the pons into and out of the cerebellum, so the pons serves as a relay between the cerebellum and other areas of the brain. In addition to this integrative motor function, the pons contains nuclei that have a very important role in regulating the rate and depth of breathing. The pons is also the origin of some cranial nerves.

*Medulla Oblongata* The **medulla oblongata** is located between the pons and the anterior part of the spinal cord. It coordinates many processes that maintain homeostasis. It is involved in the control of heart rate, breathing, blood pressure, digestion, swallowing, and vomiting, and it is the origin of several cranial nerves.

## The Human Midbrain Processes Sensory Inputs and Maintains Alertness

The midbrain lies anterior to the pons. It processes several types of sensory inputs, including vision, olfaction, and audition. It has tracts that pass this information to other parts of the brain for further processing

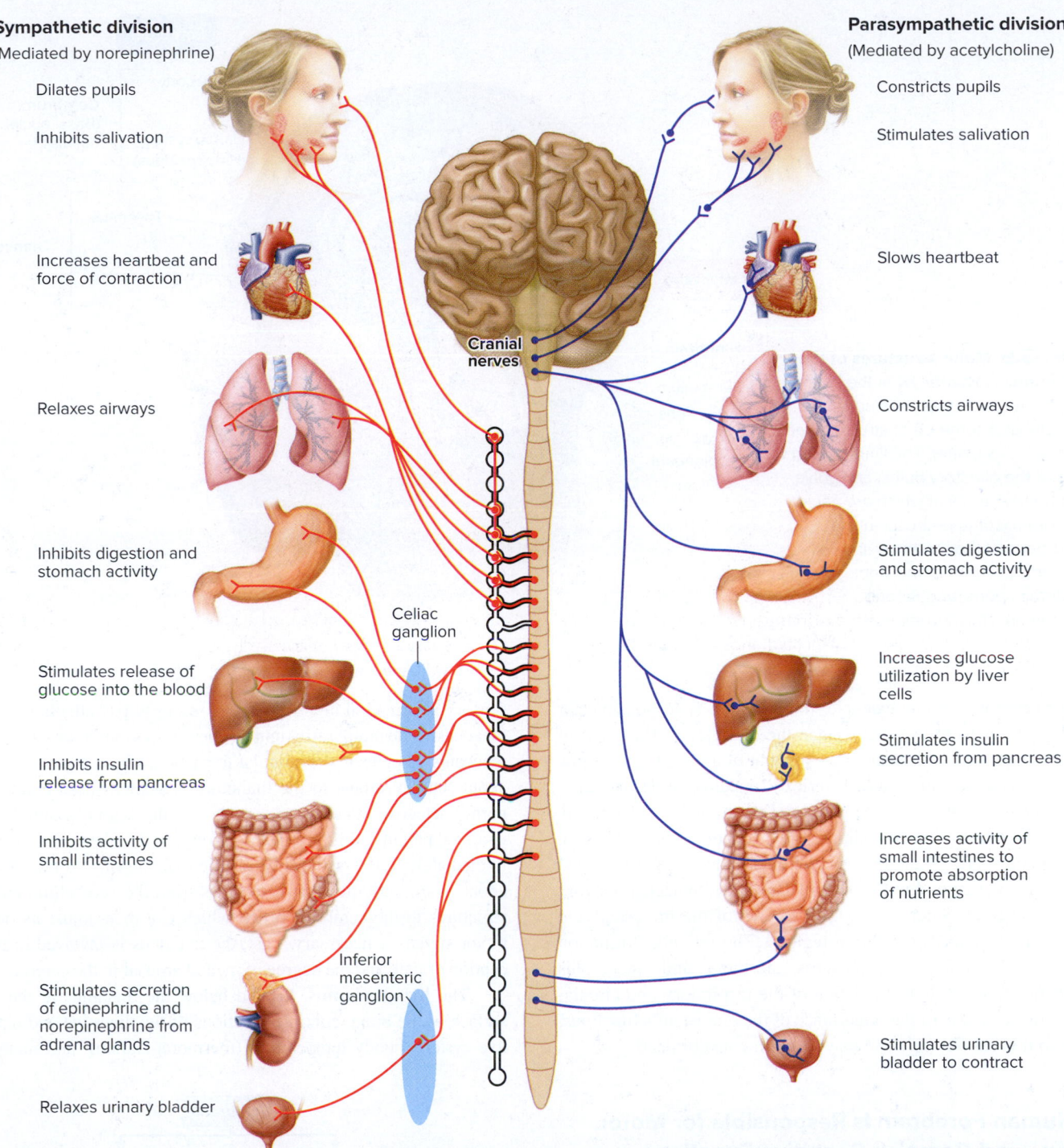

**Sympathetic division**
(Mediated by norepinephrine)

Dilates pupils

Inhibits salivation

Increases heartbeat and force of contraction

Relaxes airways

Inhibits digestion and stomach activity

Stimulates release of glucose into the blood

Inhibits insulin release from pancreas

Inhibits activity of small intestines

Stimulates secretion of epinephrine and norepinephrine from adrenal glands

Relaxes urinary bladder

Celiac ganglion

Inferior mesenteric ganglion

Cranial nerves

**Parasympathetic division**
(Mediated by acetylcholine)

Constricts pupils

Stimulates salivation

Slows heartbeat

Constricts airways

Stimulates digestion and stomach activity

Increases glucose utilization by liver cells

Stimulates insulin secretion from pancreas

Increases activity of small intestines to promote absorption of nutrients

Stimulates urinary bladder to contract

**Figure 43.8 Sympathetic and parasympathetic divisions of the autonomic nervous system.** For simplicity, only some of the major functions of each division are shown in this figure. The sympathetic and parasympathetic systems tend to have opposite effects, and most parts of the body receive inputs from both divisions. Nerves from the sympathetic division make connections with a chain of ganglia, most, but not all, of which are alongside the spinal cord. Nerves from the parasympathetic division make connections in ganglia near or in their targets.

**Core Concept: Systems** The autonomic nervous system is important for maintaining homeostasis. Many organ systems in the body are controlled in opposite ways by the two divisions of this branch of the nervous system. Therefore, the functions of these structures can be modulated in two directions. For example, heart rate can be accelerated or slowed to match an animal's immediate metabolic requirements.

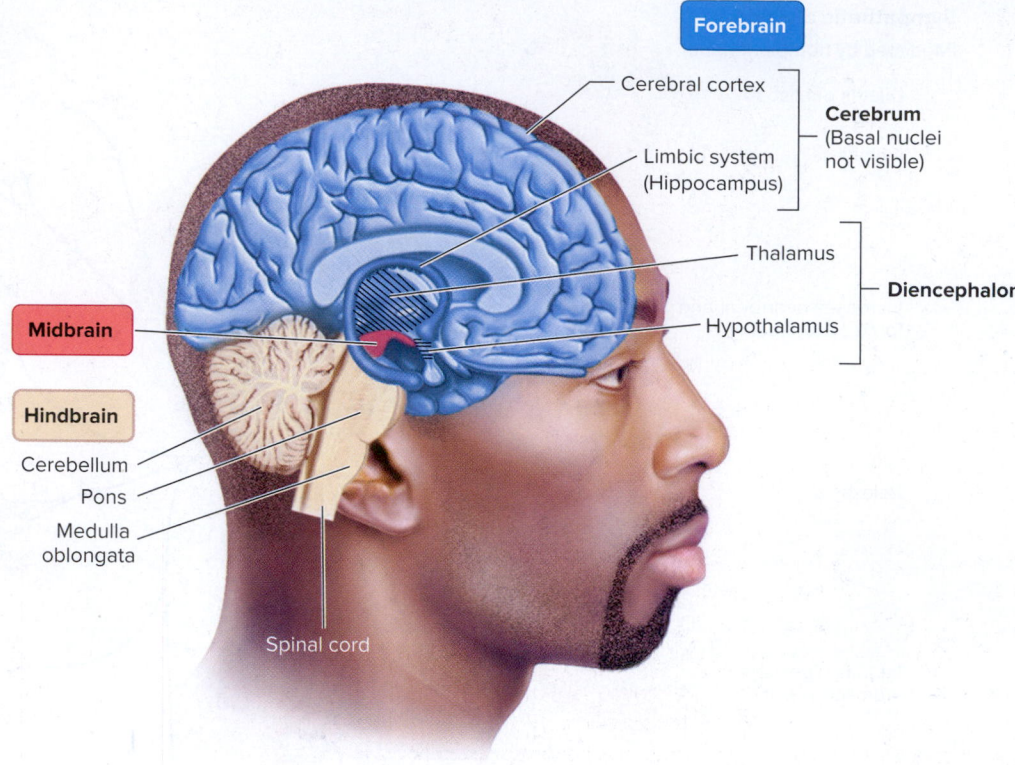

**Figure 43.9 Major structures of the human brain.** An overview of the brain, showing several internal structures (the basal nuclei and some other structures are not visible in this plane). The limbic system consists of the olfactory bulbs, amygdala, and hippocampus, all of which are part of the cerebrum (some neuroscientists consider parts of the thalamus and hypothalamus as part of the limbic system). The midbrain, pons, and medulla oblongata collectively comprise the brainstem.

and interpretation. As one example, the midbrain is responsible for activating neural pathways that change the diameter of the pupil of the eye in response to a change in the amount of ambient light. If the midbrain is damaged, this pupillary reflex is impaired or destroyed.

The medulla oblongata, the pons, and the midbrain collectively constitute the **brainstem**. In addition to the functions just described, all three major parts of the brainstem contain additional nuclei that together form the **reticular formation**. This interconnected network of nuclei and tracts extends throughout much of the brainstem and sends signals to many other brain regions. The reticular formation maintains and controls consciousness, alertness, and sleep, plus essential functions such as regulation of the respiratory and circulatory systems. Because of the importance of the brainstem's functions, damage to it is catastrophic and may result in coma or death.

## The Human Forebrain Is Responsible for Motor, Sensory, and Complex Cognitive Functions

The human forebrain comprises the diencephalon and cerebrum (also known as the telencephalon) (**Figure 43.10**). The diencephalon is made up of the thalamus, hypothalamus, and epithalamus. The cerebrum consists of the cerebral cortex, basal nuclei, and limbic system.

***Diencephalon (Thalamus, Hypothalamus, and Epithalamus)*** In all vertebrates including humans, the **thalamus** functions in relaying sensory information to appropriate parts of the cerebrum and, in turn, directing outputs from the cerebrum to other parts of the brain. It receives input from all sensory systems except olfaction. One type of processing performed by neurons in the thalamus is filtering out

sensory information in a way that allows us to pay attention to important cues while temporarily ignoring less important ones. This filtering mechanism begins in the reticular formation, which relays information about sensory inputs to the thalamus. Together, these brain regions permit selective attention to certain stimuli. A good example of this is a new parent's ability to sleep through a thunderstorm but awaken immediately to the cry of a baby. The thalamus also directs feedback about motor activities that it receives from the cerebellum and other structures to the cerebral cortex, which can then adjust its outgoing motor signals if necessary. Last, the thalamus is involved in the perception of pain and the degree of mental arousal in the cerebral cortex.

The **hypothalamus**, located below the thalamus at the ventral surface of the brain, controls functions of the digestive and reproductive systems, body temperature (thermoregulation), and many basic

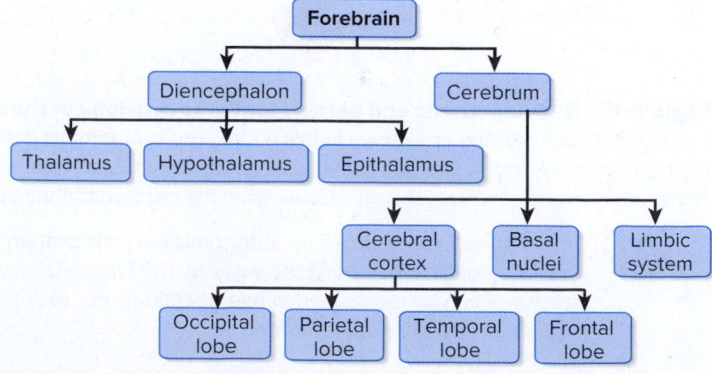

**Figure 43.10 Relationships among structures that make up the human forebrain.**

behaviors such as eating and drinking. This area has great importance for homeostasis and the control of behavior. Though small in size, it is composed of many nuclei, each with its own vital functions. Certain of these nuclei produce and release hormones, which travel through blood vessels to the pituitary gland located just beneath the brain. The pituitary gland, in turn, regulates hormone secretion from other glands in the body, including the thyroid, gonads, and adrenal glands. In this way, the hypothalamus acts as a link between the nervous and endocrine systems. In addition to producing hormones, the hypothalamus is sensitive to the actions of other hormones. For example, certain hormones produced by cells in the stomach, intestine, adipose tissue, gonads, and elsewhere act within the hypothalamus to facilitate feeding, drinking, sexual, and aggressive behaviors. Finally, a small pair of hypothalamic nuclei called the suprachiasmatic nuclei act as the master clock of the CNS, establishing circadian rhythms, which control behavioral, physiological, and hormonal rhythms over the 24-hour day.

The **epithalamus** is a collection of structures that have varied functions in the control of food and water intake, the integration of olfactory and visceral inputs with emotion and memory centers of the brain, and in some vertebrates, rhythmic and seasonal behaviors. One of these structures, the pineal gland, is located in the center of the brain and secretes a hormone called melatonin into the blood. Production of melatonin is regulated by the length of the light period in each day. Although still debated, melatonin has been suggested to function in daily cycles such as our sleep/wake cycle.

***Cerebrum: Hemispheres***   As mentioned earlier in this section, the cerebrum consists of the cerebral cortex, basal nuclei, and limbic system. One of the most recognizable features of the cerebrum, however, is its division into two halves, or **hemispheres**. Each hemisphere is connected to the other by a major tract called the **corpus callosum** (**Figure 43.11a**). In the 1950s, American neuroscientists Roger Sperry and Ronald Meyers examined the separate functions of the hemispheres in laboratory animals by performing split-brain surgeries in which they severed the corpus callosum. The animals that underwent such surgery maintained their overall health and functioning. Therefore, the surgery was thought to be safe for humans. In 1961, split-brain surgery was used for the first time to treat patients with severe epilepsy, a disorder characterized by uncontrolled electrical activity (seizures) that begins in one hemisphere and can spread via the corpus callosum to the other side. Cutting the connection between the hemispheres decreased the severity of epileptic seizures by reducing their spread.

Split-brain surgery also provided an opportunity for the researchers to make critical observations about the importance of communication between the two hemispheres. Patients who had undergone such surgery generally show normal behavior and intellectual function, because both hemispheres can function fairly independently. However, psychological tests revealed that the two sides of the brain also process different types of information. One study demonstrated that the left hemisphere produces a descriptive word for an object but does not identify certain characteristics of that object, such as its shape and texture (**Figure 43.11b**). The right hemisphere, in contrast, cannot use words to name the object but can identify other qualities. Sperry, Meyers, and other neuroscientists have concluded that the left hemisphere is involved in understanding language and producing speech in most people. Therefore, the left hemisphere is said to be

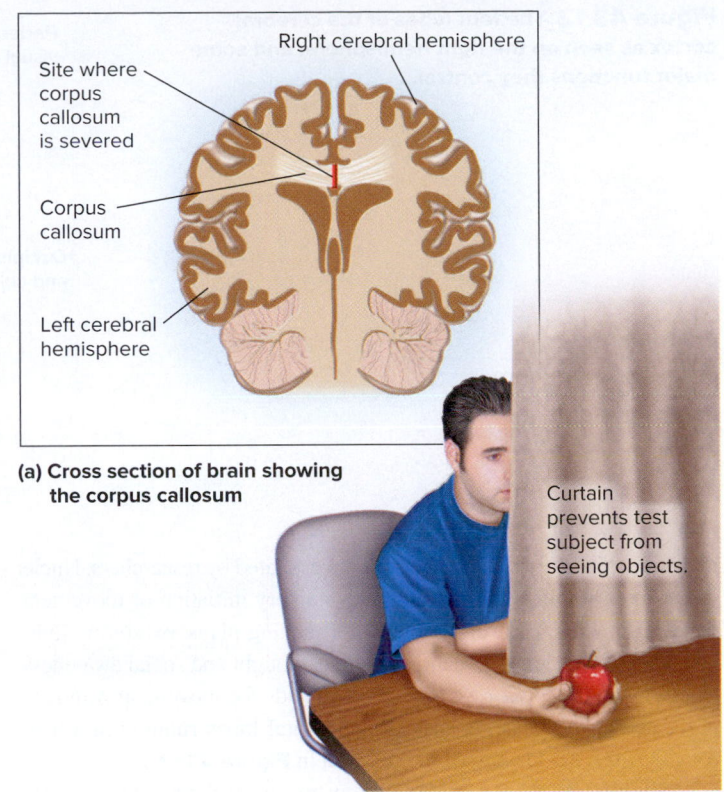

(a) **Cross section of brain showing the corpus callosum**

(b) **Testing of split-brain patient**

**Figure 43.11   The hemispheres of the human brain.** **(a)** The cerebral hemispheres and their connection by the corpus callosum. (Note: The left hemisphere controls the right side of the body, and the right hemisphere controls the left side.) **(b)** Split-brain patient being tested for hemispheric dominance. By using this apparatus, Roger Sperry and his collaborators showed that the left and right cerebral hemispheres have different capabilities. When a split-brain patient held an object in his right hand but could not see it or touch it with his left hand, he could give it a name (for example, "an apple"). When he held another object in his left hand, he could describe it (for example, "smooth"), but could not name it.

**Concept Check:**  *With her eyes closed, a split-brain patient was given a rock to hold, and she described it as a rock. Which hand was it in?*

dominant for those functions. The right hemisphere is dominant for nonverbal memories, recognizing faces, and interpreting emotions. In 1981, Sperry received the Nobel Prize in Physiology or Medicine for his insight regarding specialization in each hemisphere of the brain.

***Cerebrum: Cerebral Cortex***   As mentioned earlier, the cerebral cortex is the surface layer of gray matter that covers the cerebrum (see the darkly shaded outer rim of the brain hemispheres in Figure 43.11a). Within the cortex are identifiable regions with neurons that function in sensory, motor, or other functions. Although the cerebral cortex is only a few millimeters thick, it contains about 10% of all the neurons in the human brain.

The cerebral cortex is broadly divided into four lobes in each hemisphere of the brain: the frontal, parietal, occipital, and temporal lobes named for the bones that overlie those regions during embryonic development (**Figure 43.12**). Each lobe has a number of functions,

**Figure 43.12** The four lobes of the cerebral cortex as seen on the right hemisphere, and some major functions they control.

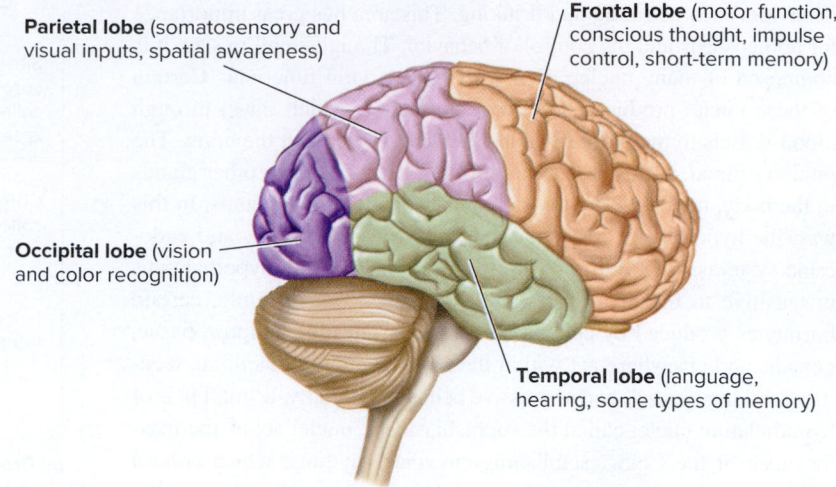

**Parietal lobe** (somatosensory and visual inputs, spatial awareness)

**Frontal lobe** (motor function, conscious thought, impulse control, short-term memory)

**Occipital lobe** (vision and color recognition)

**Temporal lobe** (language, hearing, some types of memory)

many of which are still being actively investigated by researchers. Nuclei in the **frontal lobe** are important for voluntary initiation of movement, decision making, controlling impulses, making plans, exhibiting judgment, short-term memory, and conscious thought and social awareness. The primary motor cortex, where commands for movement originate, is located at the posterior part of the frontal lobes running in a band roughly from ear to ear (see the red area in **Figure 43.13**).

Slightly posterior to the motor cortex and at the beginning of the **parietal lobe** is the region known as the somatosensory cortex (blue region in Figure 43.13). The somatosensory cortex and parietal lobe receive and interpret sensory input from somatic pathways, including touch from the surface of the body. In addition, the parietal lobe has an important role in spatial awareness, that is, our ability to use visual cues to orient ourselves in space.

The **occipital lobe** controls many aspects of visual perception and color recognition. The **temporal lobe** is necessary for language, hearing, and some types of memory.

An amazing finding is that sensory inputs enter and motor outputs exit the cerebral cortex in a pattern that forms a map of the body (see Figure 43.13). The regions of the body are represented in proportion to the amount of cortical area devoted to them. For instance, a larger part of the cerebral cortex is devoted to sensory inputs from the lips than from other areas of the face. The lips have more neuron endings and are more sensitive to touch than these other areas. Other cortical functions are also mapped in this way. For example, a map that reflects different sound frequencies (the pitch of sound) exists in the temporal lobes. The organization of the cerebral cortex may not be permanent, however, because the map may change depending on the amount of use or disuse of a given part of the body, as discussed in the Feature Investigation later in this chapter.

**Cerebrum: The Basal Nuclei**    The **basal nuclei** (or, as they have been historically but inaccurately referred to, basal ganglia) are a group of nuclei that surround the thalamus and lie beneath the cerebral cortex. Like the cerebellum, the basal nuclei are involved in planning, learning, and fine-tuning movements. They also function via a complex circuitry to initiate or inhibit movements.

Parkinson's disease is a relatively common neurological disorder that affects the basal nuclei. People with Parkinson's disease have

trouble initiating movement, such as beginning to move their legs when they wish to walk. They are capable of walking once movement has begun, but they move slowly with muscle tremors and a shuffling, jerky gait. These symptoms result from the gradual deterioration of dopamine-releasing neurons in an area of the midbrain called the substantia nigra, the neurons of which send axons to the basal nuclei. People in the early stages of Parkinson's disease can be treated with L-dopa, a substance that enters the blood and travels to the basal nuclei. There, axon terminals from remaining healthy cells originating in the substantia nigra take up the L-dopa and convert it into dopamine, which is then released onto cells of the basal nuclei. L-Dopa, therefore, increases the amount of dopamine in the basal nuclei and reduces the symptoms of Parkinson's disease.

**Cerebrum: The Limbic System**    The **limbic system** refers to a collection of evolutionarily older structures that form an inner layer at the base of the forebrain. These include structures such as the **olfactory bulbs** (which process information about smells), **amygdala**, and **hippocampus**. Some neuroscientists also consider parts of the diencephalon as part of the limbic system, because of the extensive connections between these regions. The limbic system is primarily involved in the formation and expression of emotions, and it plays an important role in learning, long-term memory, and the perception and recognition of smells. In humans, the expression of emotions occurs early in childhood before the more advanced functions of the cerebral cortex are evident. Thus, even very young babies can express fear, distress, and anger as well as bond emotionally with their parents.

Deep within the brain, the amygdala is one of the limbic system structures critical for understanding and remembering emotional situations. This structure also is involved in the ability to recognize emotional expression in others. Emotions are not unique to humans, however, and some are clearly present in other mammals. Being able to express and detect emotions imparts a selective advantage by enabling animals to establish and maintain relationships. Emotions such as fear help an animal defend itself against danger by avoiding conflict. Likewise, anger is associated with aggression, a key behavior by which many animals defend themselves or their territories.

Adjacent to the amygdala and forming a loop within the medial regions within the brain, the hippocampus is composed of several layers

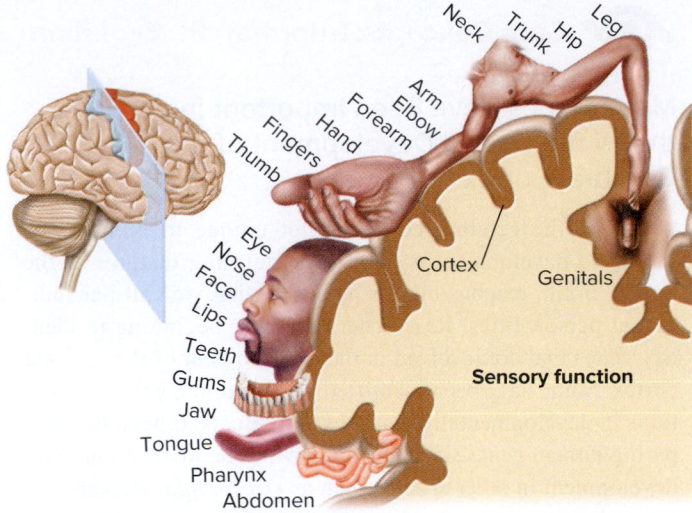

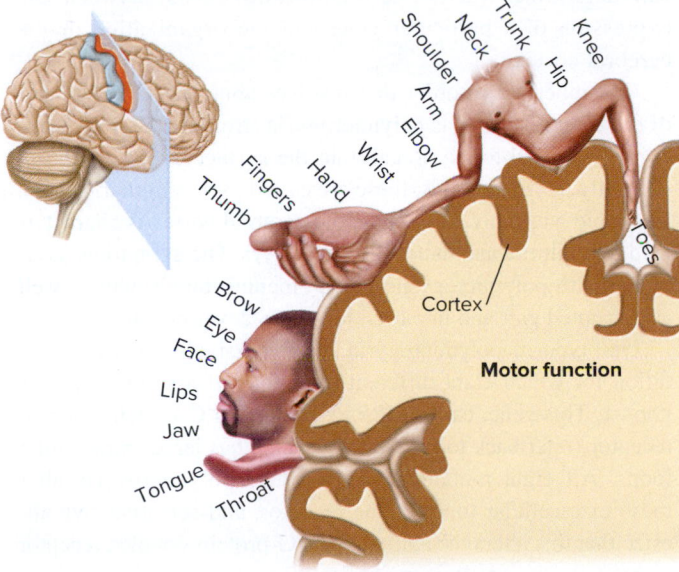

**Figure 43.13** **Maps of human body parts along the cerebral cortex.** These maps represent how the cortex interprets sensory information from these body parts and controls body movements of these parts (motor function). The relative sizes of body parts reflect the relative amount of cortex devoted to them. The blue region is the somatosensory cortex; the red region is the primary motor cortex.

of cells that are connected in a circuit. Its main function appears to be establishing memories for spatial locations, facts, and sequences of events. Damage to certain parts of the hippocampus in humans results in an inability to form new memories, a devastating condition that prevents recognition of other people or even an awareness of daily events.

Experiments with laboratory animals have demonstrated the importance of the hippocampus for memory and learning in other mammals. In a particularly well-studied example, rats are placed into a pool of milky water containing a hidden platform. The animals swim until they find the platform, on which they can safely stand. The time it takes to find the platform in subsequent trials is shorter as they learn and remember its whereabouts. This type of spatial

learning depends on activity in the hippocampus. Rats with parts of their hippocampus destroyed fail to improve their times with repeated trials. The hippocampus also receives extensive inputs from the olfactory bulbs, which may explain why smells are such potent triggers of memory in humans and why many animals use their sense of smell as a major way to learn and remember aspects of their environments.

Some of the major functions of the hindbrain, midbrain, and forebrain are summarized in **Table 43.1**.

| Table 43.1 | Major Functions of Brain Regions in Humans | |
|---|---|---|
| **Region** | | **Major Functions** |
| **Hindbrain** | | |
| | *Medulla oblongata and pons* | Coordinate homeostatic functions such as breathing, heart rate, digestion; form part of reticular formation that controls sleep and alertness; origins of cranial nerves |
| | *Cerebellum* | Fine-tuning of complex body movements; maintenance of balance |
| **Midbrain** | | Processes visual, auditory, and olfactory sensory inputs; forms part of reticular formation |
| **Forebrain** | | |
| *Diencephalon* | | |
| | *Thalamus* | Routes sensory information (except olfaction) to discrete parts of cerebrum; filters irrelevant sensory information; directs outgoing motor information from cerebral cortex to spinal cord; involved in pain perception and mental arousal |
| | *Hypothalamus* | Regulates activities of gastrointestinal and reproductive systems; controls function of pituitary gland; regulates body temperature, appetite, thirst, aggressive behavior, sexual behavior, and body rhythms |
| | *Epithalamus* | Produces cerebrospinal fluid; plays a role in food and water intake; contains the pineal gland, which may regulate sleep/wake cycle and body rhythms |
| *Cerebrum* | | |
| | *Cerebral cortex* | Voluntary motor control; perception of sensory inputs; attention; integration of sensory and motor information; generation of speech; decision making; impulse control; judgment; planning; conscious thought; learning; memory; and emotion |
| | *Basal nuclei* | Planning, fine-tuning, initiating, inhibiting and learning movements |
| | *Limbic system* | Formation and expression of emotions; perception of odors; learning and memory |

## Core Skill: Modeling

**BIO TIPS** **THE QUESTION** *Animal brains have numerous anatomical and functional structures. Many of these structures are comprised of smaller structures, which in turn contain even smaller structures. Draw a hierarchical model that describes the relationships among the larger and smaller structures of the human brain that contain the region responsible for visual perception and color recognition.*

**T**OPIC *What topic in biology does this question address?*

The topic is the structure of animal brains. More specifically, the question asks you to identify a hierarchy of structure in a region of the human brain.

**I**NFORMATION *What information do you know based on the question and your understanding of the topic?*

From the question, you know that the human brain contains numerous structures that contain smaller structures within them. The question also indicates that the region of the brain responsible for visual perception and color recognition is part of progressively larger structures. From your understanding of the topic, you may remember that the three major divisions of the brain are the forebrain, midbrain, and hindbrain, and that sensory processing (such as vision) occurs within the forebrain. You also need to recall which part of the forebrain is responsible for visual processing.

**P**ROBLEM-SOLVING **S**TRATEGY *Make a drawing.* Draw the major divisions of the forebrain in a hierarchy, with the forebrain at the top, and continuing down to the region responsible for visual processing. Refer to Figures 43.9 and 43.10 for help.

**ANSWER** *Visual processing occurs in the occipital lobe of the cerebral cortex, which is part of the cerebrum, which in turn is one of the two largest divisions of the forebrain, one of the three major divisions of the brain*

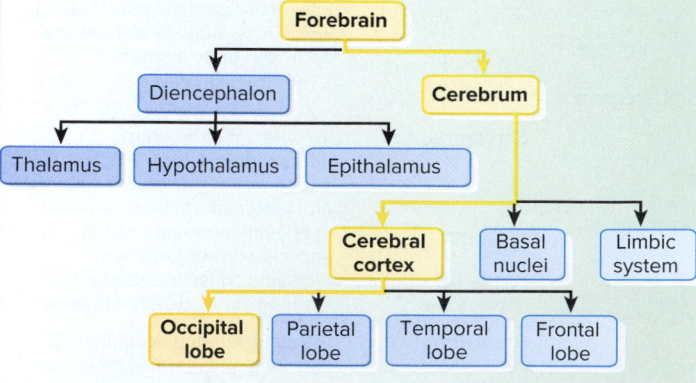

## Core Concepts: Information, Evolution

### Many Genes Have Been Important in the Evolution and Development of the Cerebral Cortex

Although the cerebral cortex is not unique to humans, its extensive development is one of the defining features of the human brain, responsible for much of what we call our individual personalities. Researchers are now beginning to identify genes that are involved in the development of the cerebral cortex. Some have been identified by examining genetic mutations in developmentally disabled individuals; others, by comparing human genes with genes known to be involved in brain development in other species such as *Drosophila*. Researchers have also compared these genes in many species that show notable differences in cerebral structure. This last approach can determine whether a relationship exists between the expression of a particular gene and the organization of the cerebral cortex.

One inherited disorder that involves abnormal development of the cerebral cortex is polymicrogyria (from the Greek, meaning many small folds). Recall that the surface of the cerebrum normally has many folds; these are called gyri (singular, gyrus). In people with polymicrogyria, the cerebral cortex is characterized by multiple and unusually small gyri. The symptoms associated with polymicrogyria include mental impairment as well as disrupted gait and impaired language development.

One type of polymicrogyria is a recessively inherited condition for which eight different mutations of a single gene are known. This gene, called *GPR56*, encodes a G-protein-coupled receptor (refer back to Figure 9.7), which has large extracellular loops. All eight mutations that produce polymicrogyria alter these extracellular loops of the receptor, and scientists hypothesize that this alters the ability of the G-protein-coupled receptor to bind its ligand. Recent studies using transgenic mice have demonstrated that preventing expression of the *GPR56* gene results in decreased neurogenesis during embryonic life, whereas overexpression of the gene has the opposite effect, resulting in increased proliferation of neurons.

Several other genes, including microcephalin (*MCPH1*) (from the Greek, meaning small head) and *ASPM* (*abnormal spindle-like microcephaly-associated*), have been shown to be determinants of brain size. For example, mutations of these genes in the human population produce individuals with much smaller frontal lobes. Interestingly, the sequences of these genes in several primates, including humans, as well as in other mammals such as dogs and sheep, have shown that the proteins produced by the normal *MCPH1* and *ASPM* genes have undergone greater changes in humans and great apes than in other species. Therefore, these genes may have been under greater selection pressure in animals with larger cerebral cortexes, suggesting that the genes have played a key role in the evolution of the cerebral cortex.

**43.3** ## 43.3  Cellular Basis of Learning and Memory

### Learning Outcomes:

1. Define the terms learning and memory, and describe their relationship to one another.

2. Describe how memory is related to changes in the strength of connections between neurons.

3. **CoreSKILL »** Describe the evidence that shows that the brain is capable of neurogenesis, and predict how neurogenesis might be important for human learning and memory.

4. Summarize the similarities and differences in the technologies of CT, MRI, and fMRI.

In the past few decades, an exciting advance in neuroscience has occurred—researchers have begun to understand complex behaviors, such as learning and memory, at the cellular level. Though it is difficult to separate the two concepts, **learning** can be defined as the process by which new information is acquired. Learning is an evolutionary adaptation that allows past experiences to affect ongoing and future behavior. **Memory** is the ability to retain, retrieve, and use information that was previously learned. Memory connects an animal's experiences throughout life. Our own behavior is largely controlled by what we have learned and remember from past experiences. Neuroscientists want to understand how the brain learns and how it captures memories. In this section, we will examine some current ideas about how this may be achieved at the cellular level and consider experimental approaches that researchers employ when investigating such complicated phenomena.

### Learning and Memory Occur via Changes Within Neurons and in Their Connections with Each Other

Beginning in the 1960s, research along two fronts led to key insights regarding the cellular basis of memory. Norwegian neuroscientist Terje Lømo and British researcher Timothy Bliss focused their efforts on the hippocampus. As described earlier, this is a key region of the brain involved with learning and memory. Lømo and Bliss conducted experiments on anesthetized rabbits to monitor signal transmission across particular regions of the hippocampus. Their key discovery involved the effects of multiple stimuli. Experimentally, a series of short, electrical stimulations to a neuron was shown to strengthen, or potentiate, its communication at a synapse with an adjacent cell for minutes or hours. Such multiple stimuli caused neurons to communicate more readily; responses were stronger and more prolonged. This phenomenon was termed **long-term potentiation (LTP)**. LTP is the long-lasting strengthening of the connection between neurons. Later work showed that LTP occurs naturally in the hippocampus and can last from hours to days, and even years.

Austrian-born American neuroscientist Eric Kandel also was interested in learning and the formation of memory. In the 1960s, however, he took a different approach by studying a simpler organism called the California sea slug, or sea hare (*Aplysia californica*). He chose this organism for several key reasons. First, it has only about 20,000 neurons, making it easier to identify pathways that are involved in specific types of behavior. Second, some of the neurons in this organism are extremely large, which facilitated the study of action potentials via microelectrodes. In addition, the large size of the neurons made it technically simpler to inject substances into them and study their effects. Finally, another advantage is that Kandel and colleagues could isolate proteins and mRNA from these large neurons and identify the biochemical and genetic changes that occur when the animal responds to a stimulus.

Much of Kandel's work focused on one type of learning affecting the gill-withdrawal reflex, a simple protective reflex that is thought to involve less than 100 neurons in the CNS. The gill and siphon are organs involved in respiration, located in the animal's mantle cavity and protected by muscular appendages called parapodia (**Figure 43.14a**). When the siphon is gently touched with a fine probe, the sea slug closes the siphon and retracts its gills into the mantle cavity for protection (**Figure 43.14b**). Though a reflex, this behavior is subject to learning. For example, if the touching of the siphon is accompanied with a brief electrical shock to the tail, the sea slug can learn to withdraw its gill in response to a subsequent shock without the siphon being touched. Kandel's study of the sea slug is similar in some ways to the famous conditioning experiments of Ivan Pavlov. Interestingly, a single tail shock paired with a touch of the siphon will result in conditioning that lasts for a few minutes. Amazingly, though, multiple

**(a) Sea slug**

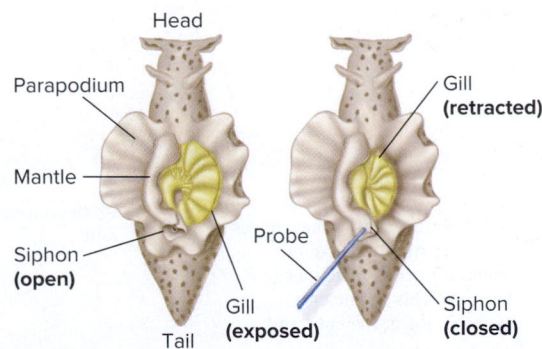

**(b) Gill-withdrawal reflex**

**Figure 43.14   The gill-withdrawal reflex in the sea slug.** (a) Photo of *Aplysia punctata* (closely related to *A. californica*) in its natural habitat; note the parapodium and tail. (b) When touched with a probe, the siphon closes and the gill retracts. In this drawing, the parapodia are moved apart for a better view of the gill. a: ©Premaphotos/Alamy Stock Photo

 **Core Skill: Process of Science**  The use of animals such as *Aplysia* as model organisms has revealed numerous fundamentally important properties that evolved in the nervous systems of all animals. Despite the obvious differences in complexity between a sea slug and a human, much of what is learned in these relatively simple animal models is fully applicable to the function of human neurons and the nervous system.

trials over several days result in a lasting memory—a shock given 3 weeks later (without siphon touch) still results in the gill-withdrawal reflex!

Over the course of many years, the work of Kandel and colleagues revealed many clues regarding the cellular basis of learning and memory. As in vertebrates, memory in the sea slug occurs in two forms: short-term memory and long-term memory (**Figure 43.15**). Short-term

memory lasts for minutes or hours. This type of memory is typically caused by a single stimulus. Kandel found that short-term memory does not require the synthesis of new proteins. Rather, a single stimulus activates intracellular second-messenger pathways that make it easier for neurons involved in a particular behavior to communicate with each other. For example, as shown in Figure 43.15a, a single stimulus may lead to the activation of protein kinases such as protein kinase A (PKA) in the presynaptic (sensory) cell. PKA, in turn, can phosphorylate proteins such as ion channels, which leads to release of an increased amount of neurotransmitter. These changes enhance the transmission of a signal between the presynaptic and postsynaptic cells.

Kandel and colleagues also discovered that repeated stimuli result in long-term memory, which lasts days or weeks. Such repeated stimuli require the synthesis of new proteins (Figure 43.15b). Long-term memory involves the activation of genes in the presynaptic cell, which leads to the synthesis of mRNA and the translation of the encoded proteins. Once made, such proteins cause the formation of additional synaptic connections. These connections also allow the presynaptic and postsynaptic cells to communicate with each other more readily. Such a change in strength of the connection between two neurons, which occurs as a result of learning, is termed **synaptic plasticity**.

Kandel's work provides a foundation for our ability to understand how learning and memory may occur at the cellular level. Short-term memory may involve changes in pre-existing cellular proteins that make it easier for neurons to communicate. Long-term memory results in protein synthesis that causes physical changes in the synapse itself, also affecting communication. Later studies by Kandel and others showed that such changes also occur in vertebrates such as the mouse. For his work on learning and memory, Kandel was awarded a share of the Nobel Prize in Physiology or Medicine in 2000.

Recently, researchers have made astonishing progress toward understanding the cellular networks that encode memories in the mammalian brain. For example, using the technique called optogenetics, in which light is used to activate ion channels in genetically altered neurons of the hippocampus of mice, investigators have for the first time mapped a cellular network that stores a specific memory, in this case, one that is associated with a stimulus that induces fear. They have even been able to induce behaviors associated with that memory by reactivating the network at a later time in response to a nonfearful stimulus! In other words, the investigators have created a false memory in mice.

## Neurogenesis May Also Contribute to Learning and Memory

Until fairly recently, neuroscientists had thought that the brain of vertebrates did not produce new neurons in adulthood, that is, that the brain was incapable of **neurogenesis**, the production of new neurons by cell division. However, in 1983, a study demonstrating neurogenesis in an adult vertebrate was carried out by Argentinean biologist Fernando Nottebohm and colleagues. Their research revealed that an increase in the number of neurons in certain brain areas of the canary occurred during the mating season. In the late 1990s, evidence also revealed that the primate and human CNS, like other parts of the body, contain stem cells, cells with the potential to differentiate into a variety of cells. For example, American researchers Elizabeth Gould and Bruce McEwen demonstrated the appearance of new neurons in the hippocampus and olfactory bulbs of adult marmosets and rhesus monkeys.

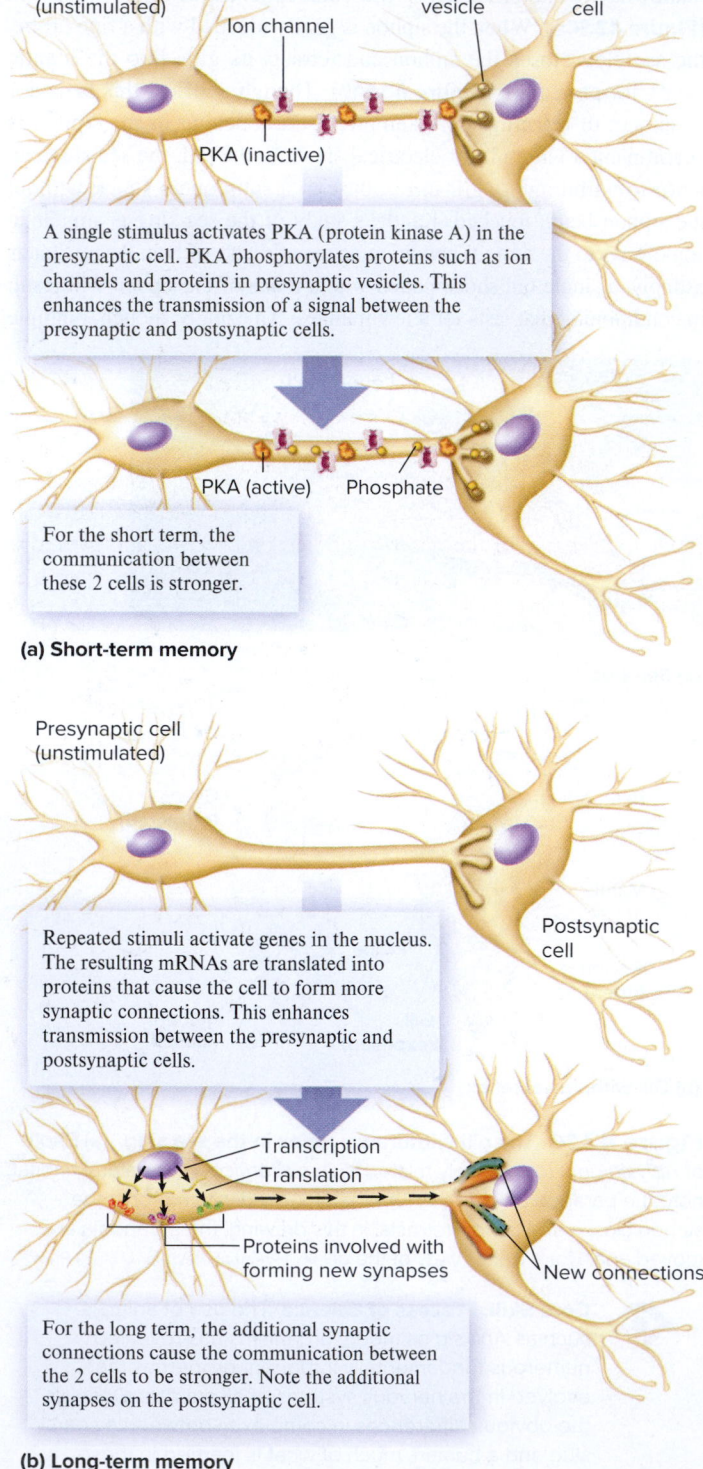

Presynaptic cell (unstimulated)　Ion channel　Presynaptic vesicle　Postsynaptic cell

PKA (inactive)

A single stimulus activates PKA (protein kinase A) in the presynaptic cell. PKA phosphorylates proteins such as ion channels and proteins in presynaptic vesicles. This enhances the transmission of a signal between the presynaptic and postsynaptic cells.

PKA (active)　Phosphate

For the short term, the communication between these 2 cells is stronger.

**(a) Short-term memory**

Presynaptic cell (unstimulated)

Postsynaptic cell

Repeated stimuli activate genes in the nucleus. The resulting mRNAs are translated into proteins that cause the cell to form more synaptic connections. This enhances transmission between the presynaptic and postsynaptic cells.

Transcription
Translation

Proteins involved with forming new synapses　New connections

For the long term, the additional synaptic connections cause the communication between the 2 cells to be stronger. Note the additional synapses on the postsynaptic cell.

**(b) Long-term memory**

**Figure 43.15** Cellular changes associated with short-term and long-term memory in the sea slug.

In 1998, American researcher Fred Gage and Swedish physician Peter Eriksson made a key discovery: They found evidence of recent mitotic activity in neurons in the hippocampus of deceased adult cancer patients. Those patients had previously been treated with a drug called bromodeoxyuridine (BrdU) to combat their cancer. BrdU is taken up by cancerous cells, but also by any other cells that are actively dividing. Its presence in cells can be detected with special stains on sections of brain or other tissue. Gage conducted such staining procedures on the brains of the deceased patients and observed the presence of BrdU in the hippocampus, suggesting recent neurogenesis.

A key question is whether the neurogenesis observed in adult brains is involved in learning and memory. This question is hotly debated and not resolved. However, some evidence suggests that it could play a role. For example, studies have shown that the hippocampus of adult monkeys grows new neurons when the animals are placed in socially enriching environments, and the formation of these neurons slows when animals are chronically stressed. Also, other studies of rats suggested that new neurons are retained in the hippocampus in response to training in particular tasks that require hippocampus function.

## Imaging Studies Help Scientists Understand the Functions of Brain Regions

Because of the enormous numbers of neurons and connections between neurons in the vertebrate brain, a key challenge in neuroscience is to understand how such complexity results in sophisticated forms of learning, memory, and responses to environmental conditions. Several imaging techniques allow doctors and researchers to examine the structure and activity level of the human brain without anesthesia or surgery. The earliest technique to be developed was computerized tomography (CT). A **CT scan** involves the use of an X-ray beam and a series of detectors that rotate around the head, producing slices of images that are reconstructed into three-dimensional images based on differences in the density of brain tissue. CT scans can easily visualize the ventricles and differences between white and gray matter, but they cannot examine the brain in great detail.

A more sensitive method, called **magnetic resonance imaging (MRI)**, was developed in the 1980s. The patient is placed in a device that contains a magnet powerful enough to generate a magnetic field many thousands of times greater than that of the Earth. This stabilizes the spinning, or resonance, of atomic nuclei (usually those of hydrogen atoms in water molecules) so that most of the nuclei align with the magnetic field. When body tissue is stimulated with a beam of radio waves, its atoms absorb the energy of the waves and the resonance of their nuclei changes, thereby altering their alignment with the magnetic field. When the radio wave pulse stops, the atoms release their energy, which is recorded by a detector. This information is analyzed by a computer, and an image is produced. MRI images allow detection of structures as small as 0.10 mm. For example, they can provide information about abnormal tissue, such as brain tumors, which respond to magnetic and radio frequency pulses differently than normal tissue. MRIs are widely used in medicine to check for injured tissue, cancers, and other abnormalities throughout the body. In 2003, American chemist Paul Lauterbur and British physicist Peter Mansfield received the Nobel Prize in Physiology or Medicine for their work in developing this technique.

With certain modifications, MRI can be used to assess the functional activity of areas within the brain. This technique, which is widely used by neuroscientists, is called **functional MRI (fMRI)**. It takes advantage of the observation that blood flow, and therefore oxygen delivery, increases to areas where neurons are more active. This increase in oxygenation is detected via fMRI. In this way, fMRI determines which neurons in particular areas of the brain are active when an individual performs certain intellectual or motor tasks (**Figure 43.16**). The principle is similar to that applied in standard MRI, except that the increased oxygen use of active tissue alters the resonance of local hydrogen atoms.

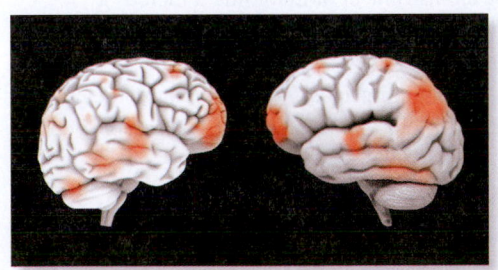 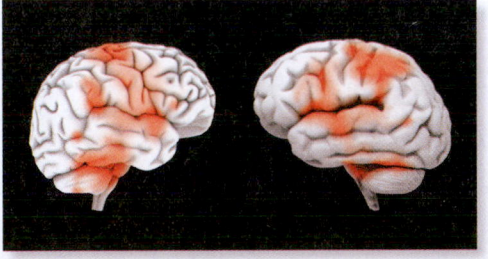

(a) Brain activity of a person thinking about a task that requires finger movements

(b) Brain activity when the same person is performing this task

**Figure 43.16** Exploring the functional activity of brain regions using fMRI scans. Red shading indicates greater O$_2$ use; both hemispheres are shown. (brain on left): ©Science Photo Library/Alamy Stock Photo; (brain on right): ©Simone Brandt/Getty Images

 **Core Skill: Modeling** The goal of this modeling challenge is to create a model that shows the regions of the human brain that become activated by environmental stimuli, and propose experiments to identify which stimuli are activating particular regions.

**Modeling Challenge:** Imagine a person undergoing an fMRI procedure while viewing a video of a nature scene such as a forest, including all of the usual environmental sounds. The subject is told to think about the images and try to remember them for a brief time. Trace an image of the human brain from Figure 43.16. Next, refer back to Figure 43.12. Based on the information in Figure 43.12 make a model of the areas in the brain that you predict would be activated in an fMRI image during the test period. In your model, the areas of the brain that use more oxygen should be shaded in a color, such as orange. Then, if your model shows more than one shaded region, explain how you might redo the experiment to distinguish which regions responded to the visual images, which to auditory cues, and which to other functions such as conscious thought and short-term memory.

The use of fMRI has revealed many fascinating aspects of the activities of different brain regions, notably in people who have suffered brain damage or loss of sensory inputs. For example, individuals who are blind from birth might be predicted to have occipital lobes that are less functional or active than are those in sighted persons (recall that the occipital lobes have the major role in visual processing). However, the work of American researcher Harold Burton and others has revealed with fMRI that the occipital lobes of blind persons are active but have become adapted to other sensory functions such as tactile signals from the fingers, including those arising from Braille reading. Amazingly, this reassignment of occipital function occurs to some extent even in individuals who have lost their vision later in life. Most likely, this does not represent a new function of the occipital lobes, but rather an expansion of an existing function that remains relatively minor in sighted persons.

The plasticity of the brain revealed by the work of Burton and coworkers is not restricted to clinical situations, as just described. MRI and fMRI are also revealing differences in brain structure and function in individuals due to the types of activities in which they regularly engage, as described next.

## Core Skill: Process of Science

# Feature Investigation | Gaser and Schlaug Discovered That the Sizes of Certain Brain Structures Differ Between Musicians and Nonmusicians

MRI and fMRI have been extremely useful in revealing which brain areas are involved in a particular function. They have also shown that the human brain is surprisingly adaptable. A number of studies have been carried out on musicians, because they practice a particular skill extensively throughout their lives, enabling researchers to study the effects of repeated use on brain function.

American neuroscientist Christian Gaser and German neuroscientist Gottfried Schlaug used MRI to examine the sizes of brain structures in three groups of people—professional musicians, amateur musicians, and nonmusicians. The researchers hypothesized that repeated exposure to musical training would increase the size of brain areas associated with visual, motor, and auditory skills, because each of these activities is used to read, make, and interpret music.

As shown in **Figure 43.17**, individuals were assigned to one of three groups based on their reported history of musical training: professional musicians with over 2 hours of musical practice time each day, amateur musicians who played a musical instrument regularly but not professionally (practicing about 1 hour/day), and those who never played a musical instrument regularly. After controlling for factors such as age and other characteristics, the researchers conducted MRIs on the three groups (see steps 2 and 3 of Figure 43.17). As seen in the data, brain areas involved in hearing, moving the fingers, and coordinating movements with vision and hearing were larger in professionals than in amateur musicians, and larger in amateurs than in nonmusicians. The region of the brain that controls finger movements was particularly well developed in the professional musicians, an interesting finding because all of the musicians in this study played keyboard instruments such as the piano.

In subsequent studies, Schlaug and colleagues examined the brains of people of different ages who either did or did not have musical training beginning in childhood. The researchers were able to correlate the degree of training with the sizes of different brain regions. They identified one particular region within the temporal lobes that was highly correlated with the extent of musical training in the course of one's life. This region, called the planum temporale, is believed to be especially important for recognizing and interpreting sound, identifying its source, and translating auditory signals (such as music) into motor processes (such as playing an instrument or humming a melody).

In a related study, American researchers Vincent Schmithorst and Scott Holland used fMRI to determine if musicians' brains were activated differently than nonmusicians' brains when they heard music. These researchers found that one area of the cerebral cortex

**Figure 43.17** Gaser and Schlaug's study of the size of visual, motor, and auditory nuclei in the brains of musicians and nonmusicians.

**HYPOTHESIS** Musical training is associated with structural differences in the brain.

**KEY MATERIALS** Volunteer subjects with different degrees of musical training.

| | Experimental level | Conceptual level |
|---|---|---|
| **1** Establish 3 groups of subjects with different musical backgrounds. | Interview subjects for musical history and assign to 1 of 3 groups.  | **Controls (nonmusicians)** No musical training. <br><br> **Amateur musicians** Play an instrument about 1 hr/day and are not employed as musicians. <br><br> **Professional musicians** Employed as musicians and practice their instrument > 2 hr/day. |

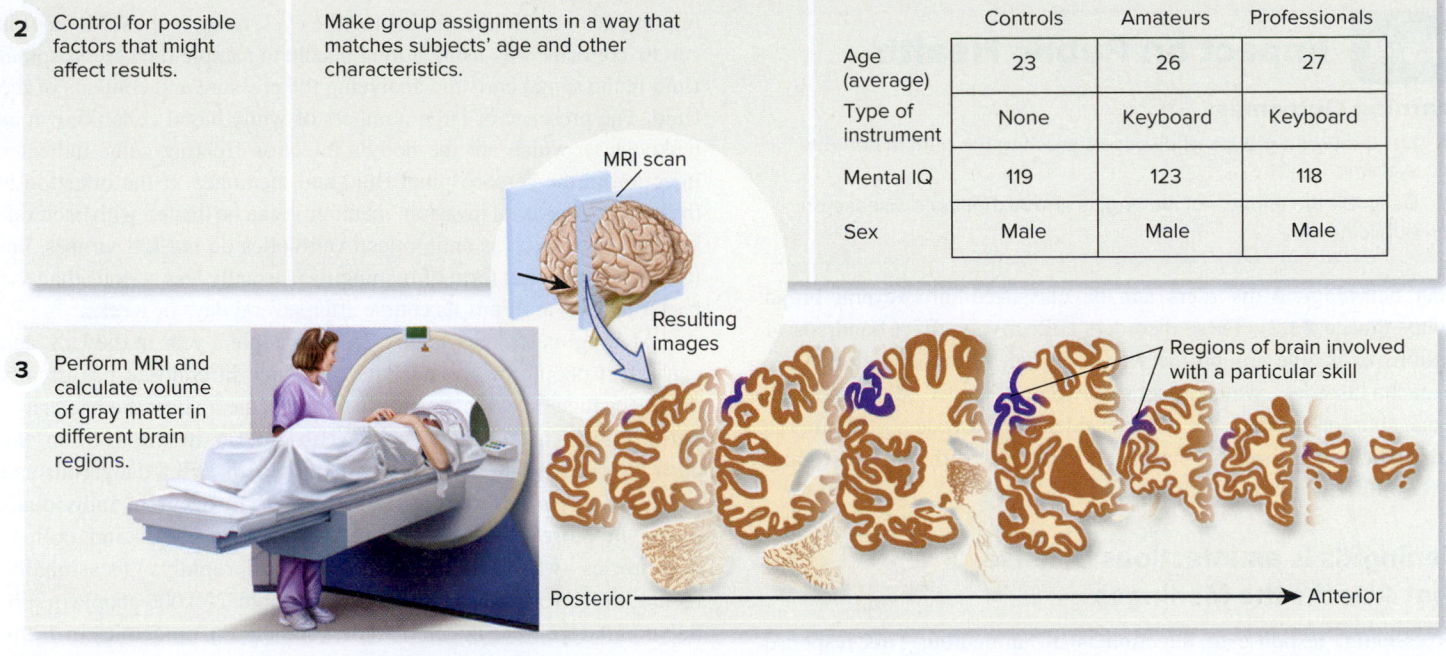

**2** Control for possible factors that might affect results.

Make group assignments in a way that matches subjects' age and other characteristics.

| | Controls | Amateurs | Professionals |
|---|---|---|---|
| Age (average) | 23 | 26 | 27 |
| Type of instrument | None | Keyboard | Keyboard |
| Mental IQ | 119 | 123 | 118 |
| Sex | Male | Male | Male |

MRI scan

Resulting images

**3** Perform MRI and calculate volume of gray matter in different brain regions.

Regions of brain involved with a particular skill

Posterior ———————————————→ Anterior

**4 THE DATA**

Results from step 3*:

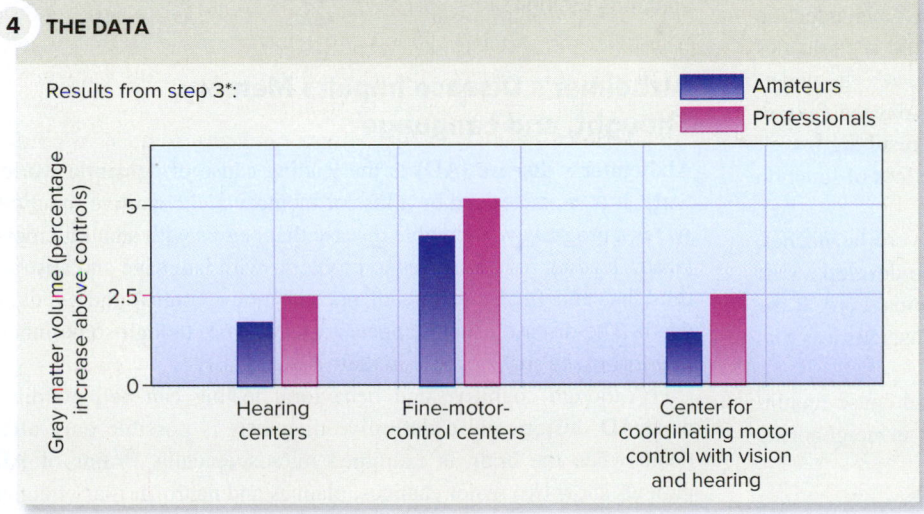

Gray matter volume (percentage increase above controls)

- Amateurs
- Professionals

Hearing centers | Fine-motor-control centers | Center for coordinating motor control with vision and hearing

*Controls are not shown separately because the data are expressed relative to controls.

**5 CONCLUSION** Musical training is associated with increased volumes of brain regions involved in hearing, fine-motor control, and the coordination of motor and sensory information.

**6 SOURCE** Gaser, C., and Schlaug, G. 2003. Brain structures differ between musicians and non-musicians. *Journal of Neuroscience* 23: 9240–9245.

was selectively activated by melodies only in musicians. This study differed from that of Gaser and Schlaug, because Schmithorst and Holland examined the activity of brain areas as well as their sizes. Their results showed that listening to music activates certain neurons and pathways in the brains of musicians but not in nonmusicians.

The human studies of Gaser, Schlaug, Schmithorst, and Holland have not determined the underlying reason(s) for increased size of certain brain structures. One possibility is that people with increased size of these regions are more likely to become musicians. Alternatively, musical training may actually cause certain regions of the brain to grow larger and alter their neuronal pathways. In other research studies involving experimental animals, groups of animals have been randomly separated into those learning a task versus controls which do not learn the task. Such experiments have shown increases in the size of brain regions that are associated with learning and memory. The increased size may result from formation of new synapses,

growth of blood vessels to the region, and/or production of more glial cells.

### Experimental Questions

**1.** What was the hypothesis proposed by Gaser and Schlaug? How did Gaser and Schlaug test this hypothesis? What were the results of their experiment?

**2. CoreSKILL »** How did the research of Schmithorst and Holland differ from that of Gaser and Schlaug? Were their results generally supportive of Gaser and Schlaug's hypothesis?

**3. CoreSKILL »** Based on the results described here and what you have learned in this chapter, predict general differences that might be found between the brain of a professional athlete, such as a tennis player, and that of a nonathlete. What about the brain of a person who has been deaf from birth compared to that of a person with normal hearing?

# 43.4 | Impact on Public Health

**Learning Outcomes:**

1. List the broad groups of diseases affecting the human nervous system.
2. Describe the impacts of meningitis and Alzheimer's disease on public health.

Most neurological disorders can be classified into several broad groups (Table 43.2). These disorders collectively affect hundreds of millions of people around the world. By far, the most common are headache disorders, which affect nearly half of the world's population at one time or another. We will consider two disorders—meningitis and Alzheimer's disease—that result from very different causes and affect millions of individuals worldwide.

## Meningitis Is an Infectious Disease That Attacks the Meninges

An essential response to infection is inflammation. This response increases the permeability of blood vessels in infected areas, allowing immune cells to be delivered to the site of an infection. When infection occurs within the meninges (causing **meningitis**), fluid accumulates in the subarachnoid space. This accumulation compresses the underlying brain tissue and its blood vessels, interrupting oxygen flow to the neurons of the cerebral cortex. If not treated, the resulting loss of oxygen (and nutrients) causes neuronal death and the loss of function of brain regions associated with those neurons.

The initial symptoms of meningitis include severe headaches, fever, or seizures. Many patients with meningitis also develop a stiff neck because the inflammation proceeds down the spinal cord. If the infection progresses untreated, it may lead to unconsciousness and even death within hours.

Several different viruses or bacterial species can cause meningitis. It usually results from an untreated infection in neighboring regions, such as the sinuses behind the eyes, nose, or ears. Meningitis can be confirmed by using a long needle to sample the cerebrospinal fluid in the spinal cord and analyzing the pressure and contents of the fluid. The presence of large numbers of white blood cells (known as leukocytes), which are the body's infection-fighting cells, indicates infection in the cerebrospinal fluid and meninges. If the infection is the result of bacterial invasion, meningitis can be treated with bacteria-killing agents such as antibiotics. Antibiotics do not kill viruses, but fortunately the viral form of meningitis is usually less serious than the bacterial form and runs its course after several days or weeks.

Meningitis strikes roughly 25,000 people a year in the U.S. and can affect people of any age. Its incidence in children has greatly declined since the widespread use of a vaccine against the bacterium *Haemophilus influenzae* type b (Hib) began in the U.S. in the early 1990s. Despite the vaccine, meningitis is still a dangerous and prevalent disease worldwide, and it tends to occur in individuals living in close quarters, such as military barracks and college dormitories, where infections may spread rapidly. Occasionally, meningitis can become epidemic. For example, 250,000 people in sub-Saharan Africa were infected and 25,000 died in epidemics in 1996, and nearly 75,000 people in Southeast Asia died of a meningitis epidemic in 2004.

## Alzheimer's Disease Impairs Memory, Thought, and Language

**Alzheimer's disease (AD)** is the leading cause of dementia worldwide. It is characterized by a loss of memory and cognitive function. AD is a progressive incurable disease that begins with small memory lapses, leading in later stages to problems with language and abstract thinking, and finally to loss of normal motor control and eventual death. The disease usually appears after age 65, though some inherited forms can strike people in their 30s and 40s.

Although cognitive and behavioral testing can help to diagnose AD, historically a definitive diagnosis is possible only after death when the brain is examined microscopically. Brains of AD patients show two major changes: plaques and neurofibrillary tangles (Figure 43.18). Plaques are extracellular deposits of a misfolded protein, β-amyloid, that forms large, sticky aggregates. These plaques were first noted in 1906 by German physician Alois Alzheimer, after whom the disease was named. Neurofibrillary tangles are intracellular, twisted accumulations of cytoskeletal fibers. Scientists are unsure how these changes influence intellectual function and memory. AD is also associated with the degeneration and death of neurons, particularly in the hippocampus and parietal lobes, which is why it is considered a neurodegenerative disease.

Researchers have identified variation in a few genes whose products are associated with the likelihood of developing AD later in life, but the underlying changes that result in the expression of these and other possible AD-related genes are still the subject of considerable research. Although genetics undoubtedly plays a role in AD, it is not the only possible cause. For example, when one identical twin develops AD, the other appears to be at increased risk but does not always develop the disease, even if he or she survives to very old age. Moreover, evidence suggests that severe head injuries, metabolic diseases such as diabetes, and heart and blood vessel disease may predispose a person to AD in later life.

| Table 43.2 | Categories of Diseases and Disorders Affecting the Human Central Nervous System |
|---|---|
| **Category** | **Examples** |
| Infectious | Meningitis, encephalitis |
| Neurodegenerative | Alzheimer's disease |
| Movement | Parkinson's disease |
| Seizure | Epilepsy |
| Sleep | Sleep apnea (brain fails to regulate breathing during sleep) |
| Tumors | Glioma (a tumor arising from glial cells) |
| Headache | Migraine (severe recurring headache) |
| Mood | Major depressive disorder; bipolar disorder (see Chapter 42) |
| Demyelinating | Multiple sclerosis (see Chapter 42) |
| Injury-related | Brain and spinal cord injuries due to accidents |

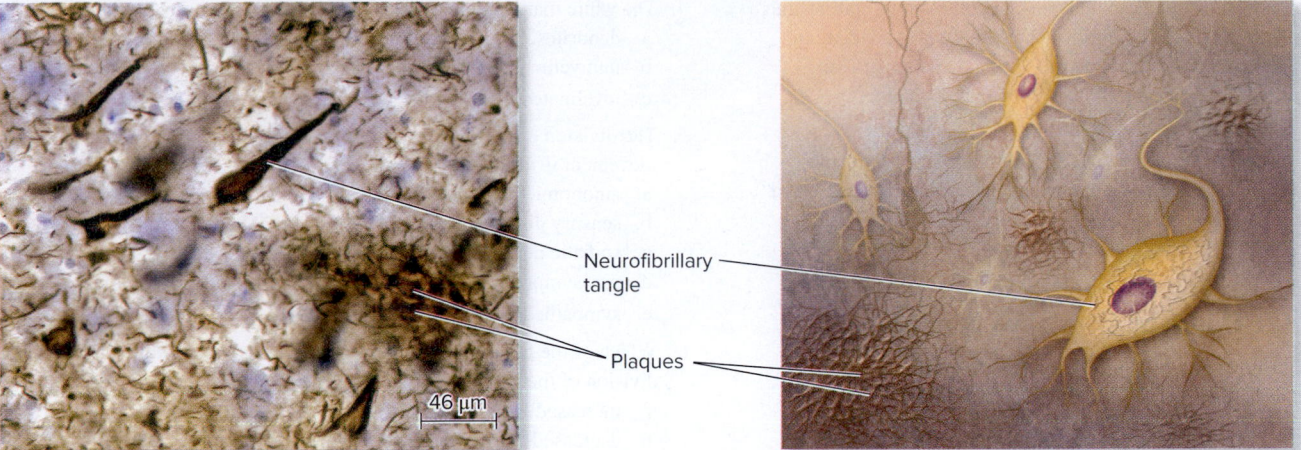

**Figure 43.18** **Cellular section from the brain of a person who died from Alzheimer's disease.** The section has been stained for visualization of proteins found in plaques and neurofibrillary tangles. An illustration of plaques and tangles is shown for comparison.
(left): ©McGraw-Hill Education/Al Telser, photographer.

Currently, AD cannot be prevented or cured. However, four major clinical approaches to prevent or slow down its progression are currently being tested or employed. These approaches are designed to (1) induce a person's immune system to destroy β-amyloid as soon as it is formed, (2) prevent the formation of β-amyloid with drugs that block its synthesis, (3) prevent the accumulation of β-amyloid into large aggregates using antiaggregation drugs, or (4) restore concentrations of certain neurotransmitters that decline in concentration in the brains of people with AD. Each of these approaches holds great promise but is still unproven. Exciting recent research has demonstrated that neuronal stem cell transplants into the hippocampus of old mice can reverse the learning and memory deficits associated with a mouse model of Alzheimer's disease, providing hope for potential future therapies for humans.

Until a cure for AD is found, its impact on public health remains enormous. Currently, about 4–5 million Americans have AD, and this number is expected to grow to nearly 16 million by 2050. The prevalence of the disease is about 3% for people between the ages of 65 and 74, and 25–50% for people older than 85. Estimated costs associated with providing health care and housing for AD patients (30% of whom live in nursing homes), as well as lost productivity in the workplace, total a staggering $100 billion per year in the U.S. This number will rise substantially now that the oldest members of the population spike known as the baby boom generation are older than 70.

## Summary of Key Concepts

### 43.1 The Evolution and Development of Nervous Systems

- All multicellular animals except sponges have a nervous system. Simple nervous systems include the nerve net of cnidarians. As animal bodies become more complex, cephalization occurs, with the formation of regionally subdivided and specialized brain that is capable of more functions (Figure 43.1).

- In all vertebrates, the three major divisions of the brain are the hindbrain, midbrain, and forebrain. Human embryos develop these divisions by 4 weeks (Figure 43.2).

- Additional folding of the brain and increased brain mass allow for expansion of regions associated with conscious thought, reasoning, and learning (Figures 43.3, 43.4).

### 43.2 Structure and Function of the Nervous Systems of Humans and Other Vertebrates

- In humans and other vertebrates, the brain and spinal cord are the central nervous system (CNS). The neurons and all axons outside the CNS, including the cranial and spinal nerves, constitute the peripheral nervous system (PNS). The CNS receives sensory input from the PNS, and the PNS acts on commands from the CNS (Figure 43.5).

- The gray matter of the CNS is composed of dendrites, cell bodies, and unmyelinated axons. The white matter consists of tracts of myelinated axons. The meninges are protective coverings of the CNS. Cerebrospinal fluid fills the subarachnoid space and ventricles (Figures 43.6, 43.7).

- The PNS can be subdivided into the somatic and autonomic nervous systems. The somatic nervous system senses external environmental conditions and controls skeletal muscles and skin. The autonomic nervous system senses internal body conditions and controls homeostasis. The efferent part of the autonomic nervous system is divided into two components: sympathetic (fight-or-flight response) and parasympathetic (rest-or-digest response) (Figure 43.8).

- The evolutionarily oldest structures of the brain, some of which are located in the hindbrain, control the basic processes that sustain life. These structures include the medulla oblongata, cerebellum, and pons (Figure 43.9).

- The midbrain processes several types of sensory inputs, including vision, olfaction, and audition. The medulla oblongata, the pons, and the midbrain collectively constitute the brainstem. Some nuclei of these three structures form the reticular formation, a network of nuclei and tracts that sends signals to many other brain regions (Figure 43.9).

- The forebrain is made of the thalamus, hypothalamus, epithalamus (diencephalon), and the cerebrum. The cerebrum is divided into two hemispheres. Each hemisphere is specialized to perform certain aspects of behavior and can operate independently (Figures 43.10, 43.11).

- The cerebrum consists of the basal nuclei, limbic system, and cerebral cortex. Each side of the human cerebral cortex is divided into four lobes, each of which has a number of functions (Figures 43.12, 43.13, Table 43.1).

## 43.3 Cellular Basis of Learning and Memory

- Learning is the process by which new information is acquired. Memory is the ability to retain, retrieve, and use information that was previously learned.

- Repeated stimuli result in long-term potentiation, in which the connections between adjacent neurons become stronger. Studies of the sea slug indicate that short-term memory is caused by a single stimulus that activates second-messenger pathways. Long-term memory is caused by repeated stimuli that activate genes, which results in stronger synaptic connections, a phenomenon called synaptic plasticity (Figures 43.14, 43.15).

- Imaging techniques such as CT scans, MRI, and fMRI allow neuroscientists and physicians to examine the structure and activity of the brain (Figures 43.16, 43.17).

## 43.4 Impact on Public Health

- Disorders of the human central nervous system can be placed into several broad categories (Table 43.2).

- Meningitis is a potentially life-threatening infectious disease in which the meninges become inflamed. Alzheimer's disease is a progressive disorder characterized by the formation of plaques and neurofibrillary tangles in brain tissue. Both are examples of neurological disorders with a large impact on public health (Figure 43.18).

## Assess & Discuss

### Test Yourself

1. A nerve net consists of
   a. bilateral neurons that extend from the head of the animal to the tail.
   b. a group of neurons that are interconnected and are activated all at once.
   c. a single nerve cord with ganglia in each body segment.
   d. a central nervous system with peripheral nerves associated with different body structures.
   e. none of the above.

2. The division of the vertebrate brain that includes the cerebellum is the
   a. hindbrain.        d. forebrain.
   b. telencephalon.    e. diencephalon.
   c. midbrain.

3. In general, the brains of more complex vertebrates
   a. are smaller.
   b. have fewer neurons but with more connections.
   c. have more folds in the cerebral cortex.
   d. use less oxygen.
   e. Both a and c are true.

4. The white matter of the CNS is composed of
   a. dendrites.             d. cell bodies.
   b. unmyelinated axons.    e. a and b only.
   c. myelinated axons.

5. The division of the nervous system that controls voluntary muscle movement is the
   a. autonomic nervous system.
   b. sensory division.
   c. somatic nervous system.
   d. parasympathetic division.
   e. sympathetic division.

6. Which of the following is *not* a response to activation of the sympathetic division of the autonomic nervous system?
   a. increased breathing rate
   b. decreased heart rate
   c. increased blood flow to the skeletal muscles
   d. increased blood glucose levels
   e. inhibition of digestion

7. The_____acts as a relay for the cerebrum.
   a. medulla          d. midbrain
   b. pons             e. thalamus
   c. hypothalamus

8. The_____is a portion of the limbic system that is important for memory formation.
   a. amygdala         d. epithalamus
   b. hippocampus      e. mesencephalon
   c. pons

9. In humans, the_____hemisphere of the cerebrum is dominant in nonverbal processing.
   a. right
   b. left
   c. The hemispheres contribute equally to nonverbal processing.

10. _____ is a progressive disease that causes a loss of memory and intellectual and emotional function.
    a. Meningitis
    b. Parkinson's disease
    c. Amnesia
    d. Alzheimer's disease
    e. Stroke

### Conceptual Questions

1. One of the most important and fundamental functions of all nervous systems is the reflex, several of which were described in this chapter and in Chapter 42. Describe why reflexes are adaptive.

2. Explain the differences between white matter and gray matter.

3. **Core Concept: Systems** New properties of life emerge from complex interactions, such as those between cells. How is this principle evident in the structure and function of animal nervous systems?

### Collaborative Questions

1. Discuss how the two parts of the nervous system of many animals—the central and peripheral nervous systems—interact with each other.

2. List the three major divisions of the brain of vertebrates, and briefly describe the function of each in humans.

# Neuroscience III: Sensory Systems

# 44

**Compound eyes of a robber fly (family Asilidae).**
©Gustavo Mazzarollo/Moment/Getty Images

**W**hat does the world look, sound, smell, and feel like to other animals? We can never know exactly how an animal perceives its environment. Biologists can perform experiments, however, to determine the capabilities of an animal's sensory systems. For instance, researchers have examined the structure of the goldfish eye, conducted behavioral studies to determine if a goldfish can discriminate between different colors, and measured the electrical responses of neurons in the animal's visual system to different visual stimuli. From such studies, we know that goldfish and probably most animals can discriminate between light of different wavelengths.

Despite the prevalence of such abilities as light, odor, sound, taste, and touch detection across animal taxa, the sensory experience of different animals may differ radically from our own. We do not see the color patterns of flowers produced by the reflection of UV light the way honeybees do, or hear the very low frequency sounds (such as those produced by earthquakes) that elephants, whales, and alligators hear. We cannot detect the presence of chemicals using our entire body surface, the way an earthworm can as it seeks food. We cannot detect the electric field generated by excitable tissues of marine animals, although many elasmobranchs and catfish can.

Senses allow living organisms to perceive their environments. In neuroscience, a **sense** is broadly defined as a system that consists of specialized cells that detect a specific type of chemical or physical stimulus (also known as a modality) and send signals to the central nervous system (CNS) for interpretation. The senses allow animals to perceive subtle and complex aspects of their environments. They are the windows through which animals experience the world around them. The nervous systems of most animals also have the ability to sense signals arising from within the body, such as hunger or pain.

In this chapter, we will examine how nervous systems collect incoming sensory information and how membrane potentials of specialized neurons change in response to sensory inputs. We will then learn that other structures of the nervous system may modify or enhance this neural activity before sending it to the brain, where it is interpreted. Finally, we will discuss how problems with sensory systems—in particular vision and hearing deficits—can affect human health.

## 44.1 An Introduction to Sensation

**Learning Outcomes:**

1. Describe the relationship between sensory transduction and perception.
2. Explain how sensory receptors transmit the intensity of a stimulus.
3. List the classes of sensory receptors and the stimuli to which they respond.

Sensory systems convert chemical or physical stimuli from an animal's body or the external environment into a signal that causes a change in the membrane potential of sensory neurons. **Sensory transduction** is the process by which incoming stimuli are converted into neural (electrical) signals. Sensory transduction involves cellular changes, such as opening of ion channels, which cause either graded potentials or action potentials in neurons.

**Perception** is an awareness of sensations that are experienced. For instance, touching a hot object generates a thermal sensation, which initiates a neuronal response in the brain, giving us the perception that this stimulus is hot. Not all sensations are consciously perceived by an organism. Most of the time, for example, we are not aware of the touch of our clothing. The brain also processes sensory information in areas that do not generate conscious thought. For instance, certain neurons constantly monitor blood pressure and the concentrations of oxygen, glucose, and other substances in the blood, but we are not aware that this is occurring.

We begin our study of sensory systems by examining the specialized cells that receive sensory inputs. A **sensory receptor** is in some cases a neuron. In other cases, it may be a specialized epithelial cell that synapses with a neuron referred to as a sensory neuron. In both cases, the sensory receptor recognizes an internal or external (environmental) stimulus and initiates sensory transduction by creating graded potentials (described in Chapter 42) in itself or an adjacent cell (**Figure 44.1**). If a response is strong enough, sensory receptors initiate electrical responses to stimuli, such as chemicals, light, heat, and sound, which lead to action potentials that are sent to the CNS.

## A Strong Stimulus Generates More Frequent Action Potentials

How do sensory receptors transmit the intensity of a stimulus? Let's consider an example involving weak and strong stimuli to the sense of touch (**Figure 44.2**). Sensory transduction begins when the specialized endings of a sensory receptor respond to a stimulus. Such a stimulus—in this case, the touch of a glass rod—opens ion channels that allow sodium ions ($Na^+$) to diffuse down their electrochemical gradient into the cell, depolarizing the sensory receptor. The amount of depolarization is directly related to the intensity of the stimulus, because a stronger stimulus opens more ion channels.

The first response of a sensory receptor is usually a graded change in the membrane potential of the cell body that is proportional to the intensity of the stimulus (see Figure 44.2). The membrane potential, known as a **receptor potential** in these cells, becomes less and less negative as the strength of the stimulus increases. When a stimulus is strong enough, it depolarizes the membrane to the threshold potential at the axon hillock and produces an action potential in a sensory neuron (refer back to Figure 42.9 for a description of the action potential).

Recall from Chapter 42 that action potentials proceed in an all-or-none manner, regardless of the nature or strength of the stimulus that elicits them. How, then, can action potentials provide information about the intensity of a stimulus? The answer is that the strength of the stimulus determines the frequency of action potentials generated. A strong stimulus generates many action potentials in a short period of time, because it can overcome the membrane's relative refractory period (see Figure 42.9). As a result, the frequency of action potentials is higher when the stimulus is strong than when it is weak. The action potentials are transmitted into the CNS and carried to the brain for interpretation. The brain interprets a higher frequency of action potentials as a more intense stimulus.

## The CNS Processes Each Sense Within Its Own Pathway

Different stimuli produce different sensations because they activate specific neural pathways that are dedicated to processing only that

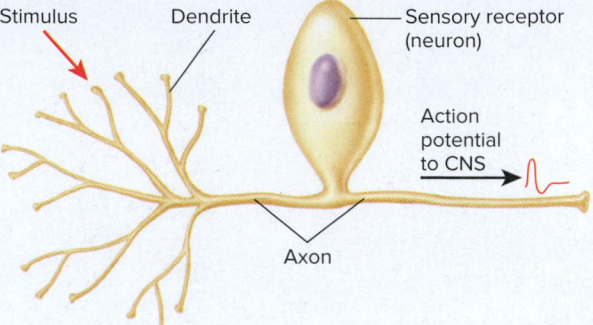

**(a) A neuron as a sensory receptor**

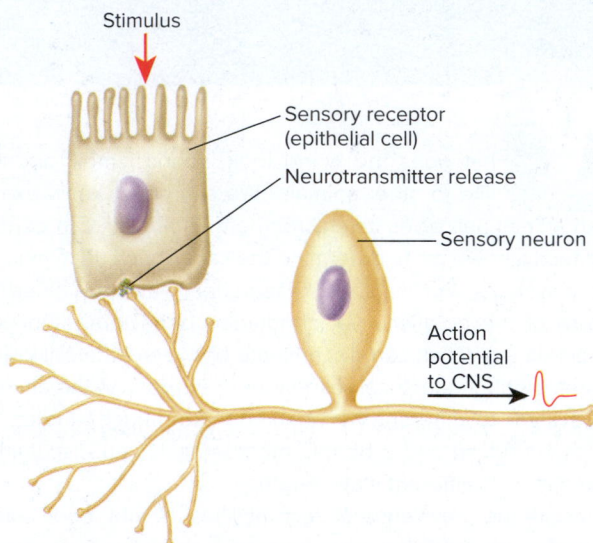

**(b) A specialized epithelial cell as a sensory receptor**

**Figure 44.1 Sensory receptors. (a)** Many sensory receptors are neurons that directly sense stimuli. **(b)** Other sensory receptors are specialized epithelial cells that sense stimuli and release neurotransmitter that stimulates nearby neurons called sensory neurons. In both cases, when stimulated, the neurons send action potentials to the CNS, where the signals are interpreted.

 **Core Skill: Connections** The term receptor is used in more than one way in biology. What is the difference between the sensory receptors described in this chapter and the membrane receptors described in Chapter 9?

type of stimulus. We know that we are seeing light because the signals generated by visual sensory receptors in the eye are transmitted along a neural pathway that sends action potentials into areas of the brain that are devoted to processing vision. For this reason, the brain interprets such signals as visual stimuli. The brain can separate and identify each sense because each one uses its own dedicated pathway.

Sensory receptors can be divided into general classes based on the type of stimulus, or modality, to which they respond. Each class of receptor uses a different mechanism to detect stimuli and to transmit the information to different regions of the CNS.

- **Mechanoreceptors** transduce mechanical energy such as touch, pressure, stretch, movement, and sound.

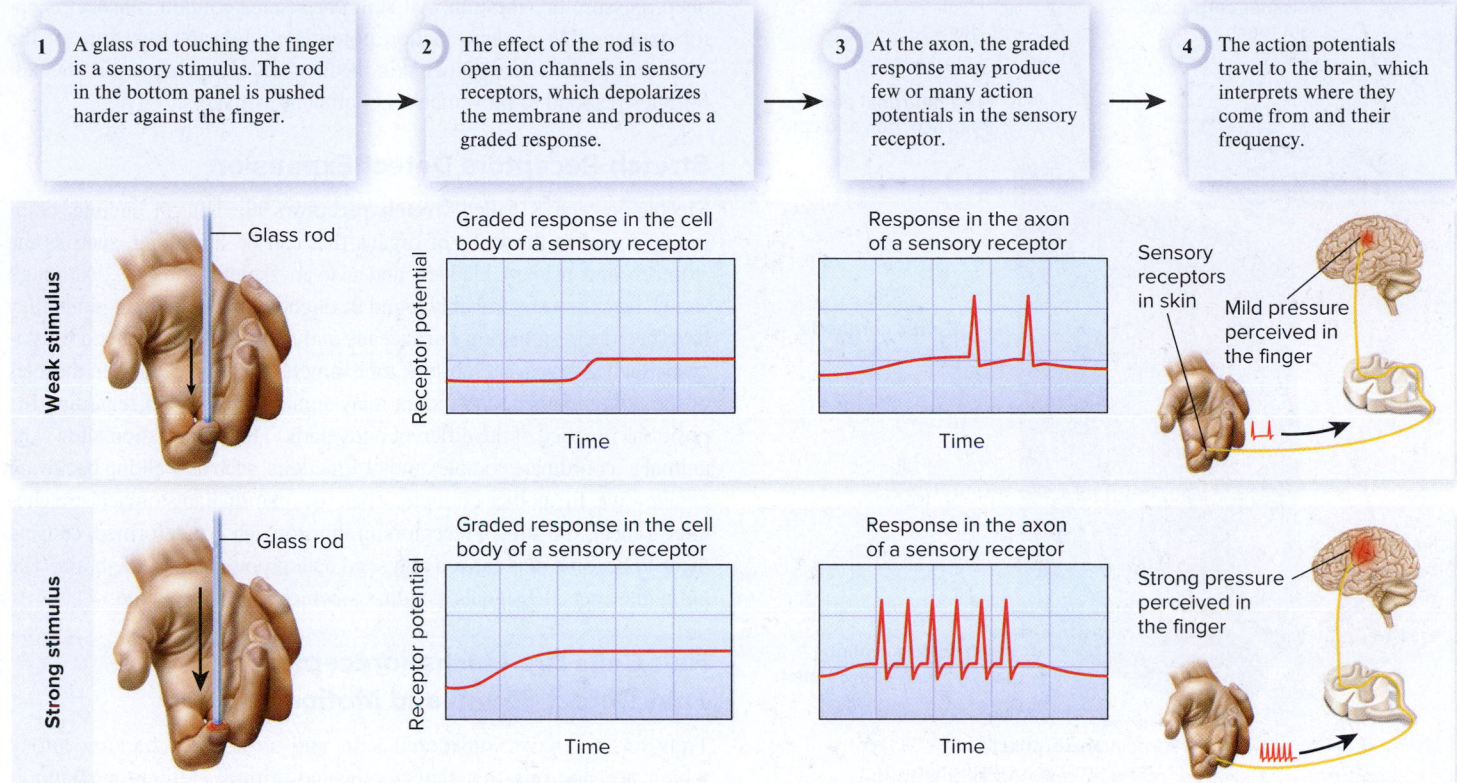

**Figure 44.2** **Transduction of a sensory stimulus of two different intensities.** In this example, the sensory receptor is a neuron that is embedded within deformable structures composed of connective tissue. Note the faster and larger graded response following the stronger stimulus.

- **Thermoreceptors** detect cold and heat.

- **Nociceptors**, or pain receptors, detect extreme heat, cold, and pressure, as well as certain potentially damaging molecules such as acids.

- **Electromagnetic receptors** sense radiation within a portion of the electromagnetic spectrum, including visible, UV, and infrared light, as well as electrical and magnetic fields in some animals.

- **Photoreceptors** are a type of electromagnetic receptor that detect visible light.

- **Chemoreceptors** recognize specific chemical compounds in the air, water, body fluids, or food.

In most of the remaining sections of this chapter, we will examine the structures and functions of these types of sensory receptors and the organs in which they are found.

## 44.2 Mechanoreception

**Learning Outcomes:**

1. List the types of mechanoreceptors that detect touch, stretch, or movement, and describe how their structures relate to the functions of hearing and balance.

2. Describe the structure of the mammalian ear, and explain how mechanical forces move through it.

3. Give examples of adaptations for hearing in animals that inhabit different environments.

4. Describe how body position and movement are detected by sense organs.

Mechanoreceptors are cells that detect physical stimuli such as touch. Physically touching or deforming a mechanoreceptor cell opens ion channels in its plasma membrane (see Figure 44.2). As discussed in this section, some mechanoreceptors are neurons that send action potentials to the CNS in response to physical stimuli. Other mechanoreceptors are specialized epithelial cells that contain hairlike structures that bend in response to mechanical forces.

### Skin Receptors Detect Touch and Pressure

Several types of receptors in the skin of many animals detect touch, deep pressure, or the bending of hairs on the skin. Some of these specialized receptors consist of neuronal dendrites covered in dense connective tissue. In mammals, these receptors are located at different depths below the surface of the skin, which makes them suitable for responding to different types of stimuli (**Figure 44.3**). For example, **Meissner corpuscles** lie just beneath the skin surface and sense touch and light pressure. They are found throughout the skin but are concentrated in areas sensitive to light touch, such as the fingertips, lips, eyelids, and genitals. In contrast, **Pacinian corpuscles** and **Ruffini corpuscles** are located much deeper beneath the surface, particularly in the soles of the feet and the palms of the hands. These corpuscles respond best to

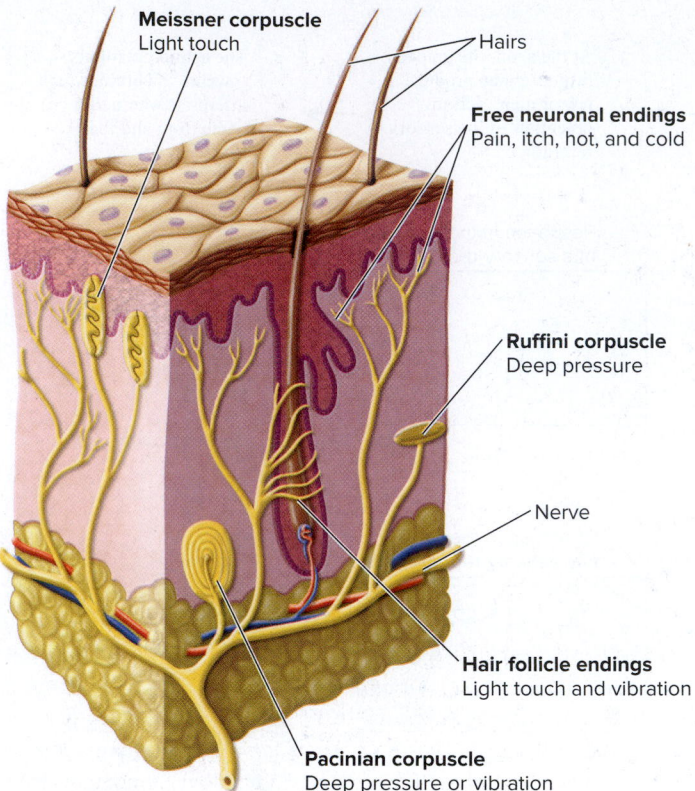

**Meissner corpuscle**
Light touch

Hairs

**Free neuronal endings**
Pain, itch, hot, and cold

**Ruffini corpuscle**
Deep pressure

Nerve

**Hair follicle endings**
Light touch and vibration

**Pacinian corpuscle**
Deep pressure or vibration

**Figure 44.3** **Examples of sensory receptors in the skin of mammals.**

**Concept Check:** *The several types of touch receptors in the skin respond to different stimuli. What different touch sensations are you aware of?*

deep pressure or vibration. All skin corpuscles contain sensory receptor neurons that generate action potentials when the structure of the corpuscle is deformed. Other skin mechanoreceptors located in the hair follicles respond to movements of hairs and whiskers.

## Stretch Receptors Detect Expansion

Mechanoreceptors called **stretch receptors** are neuron endings commonly found in the walls of organs that can be distended, such as the stomach and urinary bladder, and also in skeletal muscles. Although stretch receptors are probably found throughout the animal kingdom, they have been best studied in crustaceans and mammals. In decapod crustaceans such as crabs and lobsters, for example, stretch receptors in muscles of the tail, abdomen, and thorax relay signals to the brain regarding the positions in space of the different body parts. This information allows the animal to coordinate complex motor functions, such as walking backward or sideways. In another example, when the mammalian stomach stretches after a meal, the stretch receptors in the stomach are deformed, causing them to become depolarized and send action potentials to the brain. The brain interprets the signals as fullness, which inhibits appetite.

## Hair Cells Are Mechanoreceptors That Detect Sound and Motion

Thus far, we have considered skin and stretch mechanoreceptors, which are neurons that detect physical stimuli. Other mechanoreceptors are specialized epithelial cells called **hair cells**, which have deformable projections called **stereocilia**. The stereocilia are different from true cilia (see Figure 4.17) because they do not contain motor proteins in their structure. Instead, they are displaced by movements of fluid or other physical stimuli (**Figure 44.4**). The stereocilia on hair cells allow animals to detect movements and sound waves.

| At rest (unstimulated) | Excited | Inhibited |
|---|---|---|

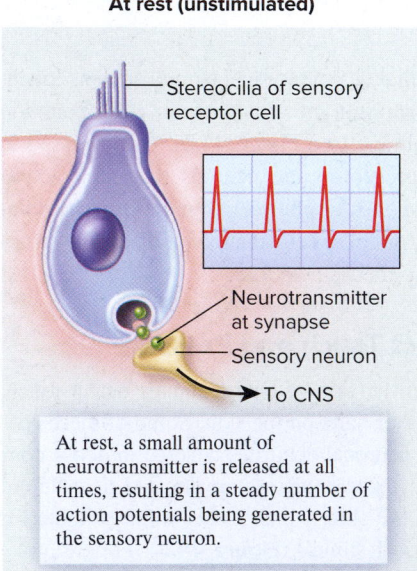

Stereocilia of sensory receptor cell

Neurotransmitter at synapse
Sensory neuron
To CNS

At rest, a small amount of neurotransmitter is released at all times, resulting in a steady number of action potentials being generated in the sensory neuron.

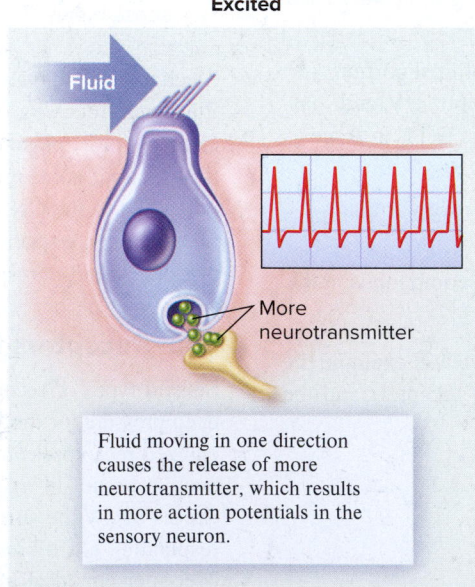

Fluid

More neurotransmitter

Fluid moving in one direction causes the release of more neurotransmitter, which results in more action potentials in the sensory neuron.

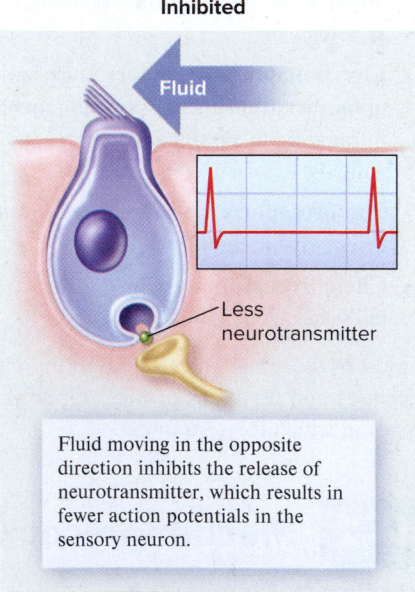

Fluid

Less neurotransmitter

Fluid moving in the opposite direction inhibits the release of neurotransmitter, which results in fewer action potentials in the sensory neuron.

**Figure 44.4** **The response of hair cells to mechanical stimulation.** The stereocilia inside these hair cells are hairlike projections of the plasma membrane that contain actin filaments.

**Core Skill: Connections** How are stereocilia different from cilia, which are described in Chapter 4?

Hair cells contain ion channels that open or close when the stereocilia bend. This leads to a change in the cell's membrane potential. When the plasma membrane depolarizes, voltage-gated $Ca^{2+}$ channels open, resulting in the release of neurotransmitter molecules from the hair cells. The neurotransmitters then bind to protein receptors in adjacent sensory neurons and may initiate action potentials that are sent to the CNS. When unstimulated, the stereocilia are not bent, and the hair cells release only a small amount of neurotransmitter onto nearby sensory neurons, resulting in a resting rate of action potentials from the sensory neurons. In the example shown in Figure 44.4, bending of the stereocilia in one direction in response to fluid movement increases the release of neurotransmitter from the hair cell, exciting the sensory receptors, whereas bending in the other direction decreases the release of the same neurotransmitter, inhibiting the sensory receptors. The result is an increase or decrease, respectively, in the rate of action potentials produced in the sensory neurons.

Hair cells provide a rich array of sensory capabilities in many animal species. For example, these cells are found in the hearing and equilibrium (balance) organs of many invertebrates and vertebrates, where they detect sound or changes in head position. They are also found along the body surface of fishes and some amphibians, where they detect external water currents, as described later.

## Audition (Hearing) Involves the Detection of Sound Waves

The sense of hearing, called **audition**, is the ability to detect and interpret sound waves. This sense is critical for the survival and reproduction of many types of animals. For example, a mother seal locates her pup by hearing its calls, and a male bird sings an elaborate song to attract a mate. Hearing is also important for detecting the approach of danger—a predator, a thunderstorm, an automobile—and locating its source.

Sound travels through air or water in waves. The distance from one peak of the wave to the next peak is a **wavelength**. The number of complete wavelengths that occur in 1 second is called the **frequency** of the sound, which is measured in the number of waves per second, or hertz (Hz), after German physicist and pioneer of radio wave research, Heinrich Hertz. The length and frequency of sound waves impart certain characteristics to the stimulus. Short wavelengths have high frequencies that are perceived as a high pitch, or tone, and long wavelengths have lower frequencies and a lower pitch. The human hearing range is about 20–20,000 Hz.

The sense of hearing is present in vertebrates and arthropods, but not in other phyla. Arthropods do not appear to have more than a general sensitivity to sound, although some exceptions exist. For example, some species of moths have sound-sensitive membranes that detect the high frequencies emitted by their chief predators, bats. The sense of hearing, however, is especially well developed in many vertebrates (notably birds and mammals). We turn now to a detailed discussion of the mammalian ear and the mechanism by which it detects sound, including the importance of hair cells for hearing.

***Structure of the Mammalian Ear*** The mammalian ear has three main compartments: the outer, middle, and inner ears (**Figure 44.5**). The outer ear consists of the external ear, or pinna (plural, pinnae), and the auditory canal. The outer ear is separated from the middle ear by the tympanic membrane (eardrum). The middle ear contains three small bones called ossicles (named the malleus, incus, and stapes) that link movements of the eardrum with the oval window. The oval window is another membrane similar to the eardrum that separates the middle ear from the inner ear. The inner ear is composed of the **cochlea** (from the Latin, meaning snail)—a coiled chamber of bone containing hair cells and the membrane-like round window—and the vestibular system, which functions in balance, as described later.

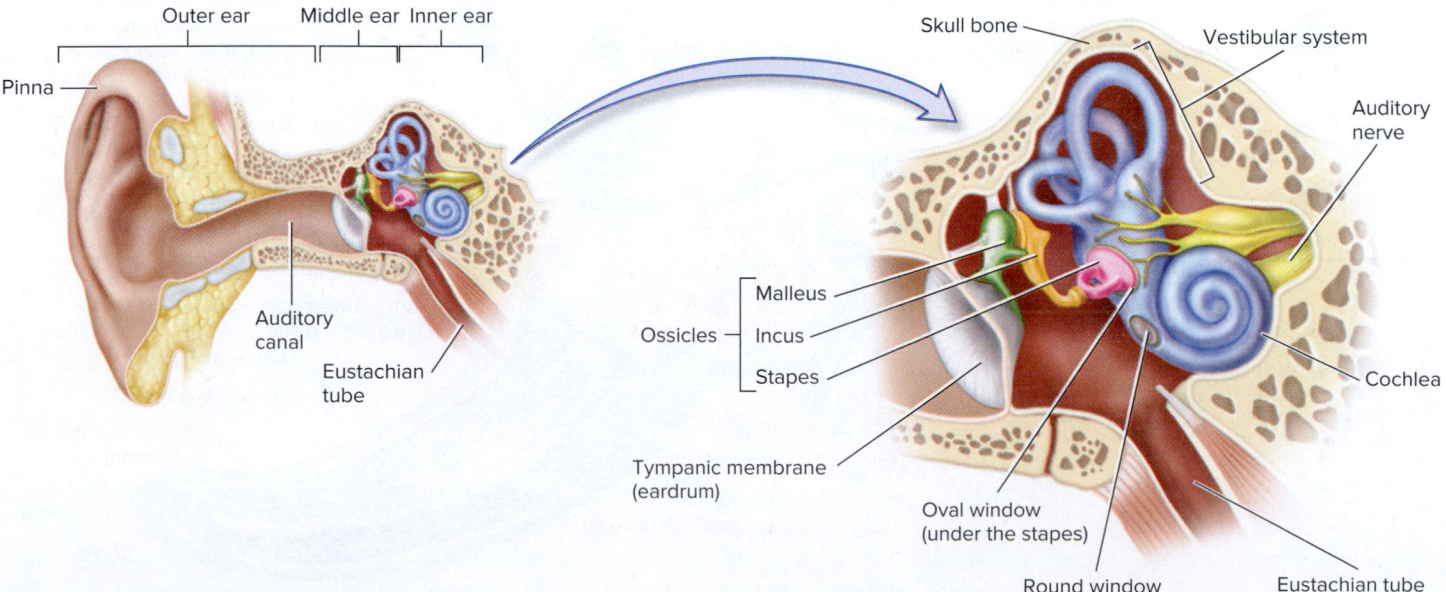

**Figure 44.5 The structure of the human ear.** The three main compartments are the outer, middle, and inner ear; the latter two are shown in more detail in the inset. (The eustachian tube shown here is a structure that connects the middle ear to the pharynx and functions to equalize air pressures in the ear.)

These structures in the inner ear generate the signals that travel via the auditory nerve to the auditory cortex of the brain.

***Generation of Electrical Signals in the Mammalian Ear***  To understand how mammals hear, let's first consider how mechanical forces move through the ear. Sound waves entering the outer ear cause the tympanic membrane to vibrate back and forth (**Figure 44.6**). The malleus, incus, and stapes transfer the vibration of the tympanic membrane to the oval window, causing it to vibrate against the cochlea. This vibration sends pressure waves through a fluid called perilymph. Perilymph is found within two narrow passages in the cochlea called the vestibular and tympanic canals, which are separated by a tube called the cochlear duct. The waves travel from the vestibular canal to the tympanic canal and eventually strike the round window, where they dissipate. Along the way, the waves cause the vibration of a sheath-like structure called the **basilar membrane**, which is formed from elastic fibers tensed across the cochlear duct. Sounds of very low frequency (longer wavelength) create pressure waves that take the complete route through the vestibular and tympanic canals (see the green arrows in Figure 44.6). Sounds of higher frequency (shorter wavelength) produce pressure waves that follow a different route, passing from the vestibular canal through the cochlear duct (as shown by the blue arrows in Figure 44.6). They then pass through the basilar membrane, before reaching the tympanic canal.

Within the cochlea, mechanical vibrations are transduced into electrical signals. This happens in a structure called the **organ of Corti** which rests on top of the basilar membrane. To understand how this works, we need to look at a cross section through the cochlea (**Figure 44.7**). The organ of Corti contains supporting cells and rows of hair cells. The stereocilia of the hair cells are embedded in a gelatinous tectorial membrane. The vibration of the basilar membrane bends the stereocilia in one direction and then the other. When bent in one direction, the hair cells depolarize and release neurotransmitter, which activates adjacent sensory neurons that then send action potentials to the CNS via the auditory nerve. When bent in the other direction, the hair cells hyperpolarize and stop releasing neurotransmitter. In this way, the frequency of action potentials generated by the sensory neurons is determined by the up-and-down vibration of the basilar membrane.

The basilar membrane is lined with protein fibers that span its width. These fibers function much like the strings of a guitar. The fibers near the oval and round windows at the base of the cochlea are short and rigid, and they vibrate in response to high-frequency waves. Longer and more resilient fibers are near the other end of the cochlea and vibrate to lower-frequency waves. For this reason, hair cells closer to the oval and round windows respond to high-pitched sounds, whereas those at the opposite end are triggered by lower-pitched sounds. When we hear a great number of sound frequencies at once, such as at a musical concert, the waves traveling through the cochlea activate hair cells all along the basilar membrane in a physical representation of the music! These cells stimulate sensory neurons, which send multiple action potentials to the auditory areas of the brain for processing. The most incredible feature of this process, however, is that the mammalian ear and brain can "tune in" to all of these frequencies simultaneously.

## Adaptations for Hearing Provide Survival Advantages for Animals

The range of audible pitches varies among different species of animals. As noted, humans can hear between about 20 and 20,000 Hz (conversation averages 90–300 Hz). Insectivorous bats, toothed whales, and

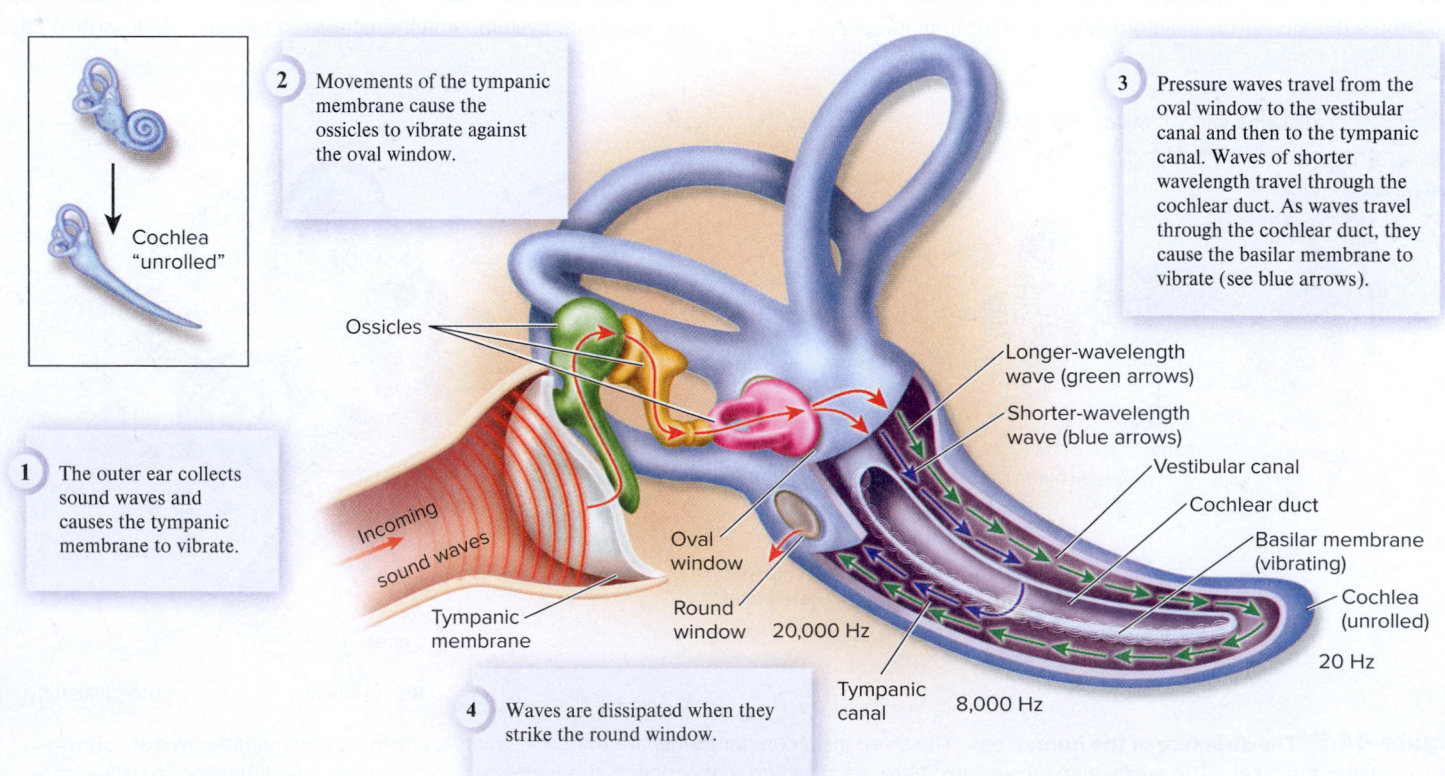

**Figure 44.6 Movement of pressure waves through the human ear.**

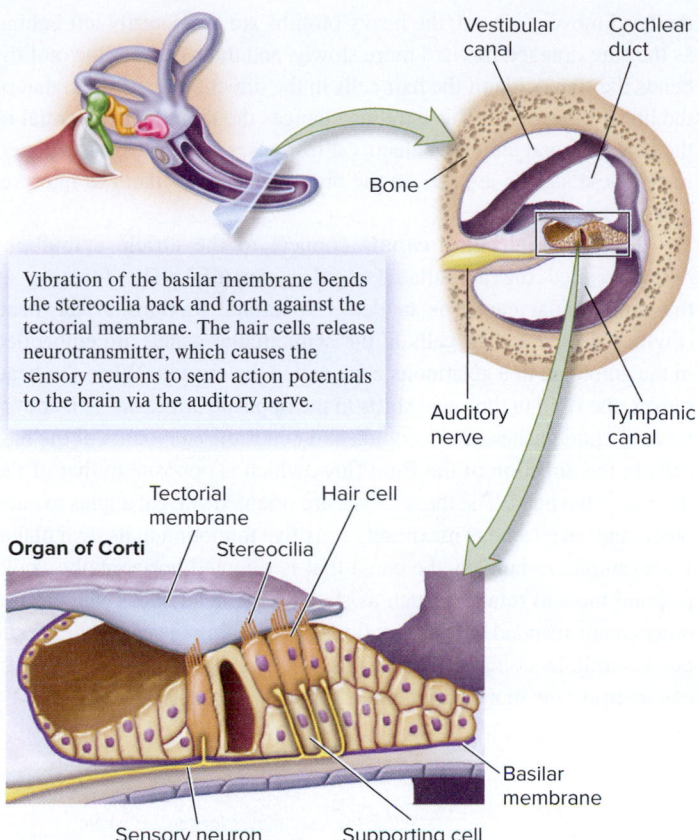

Vibration of the basilar membrane bends the stereocilia back and forth against the tectorial membrane. The hair cells release neurotransmitter, which causes the sensory neurons to send action potentials to the brain via the auditory nerve.

**Figure 44.7** **Transduction of mechanical vibrations to action potentials in the organ of Corti.**

**Concept Check:** *What causes vibration of the basilar membrane?*

some species of moths may have the highest-frequency sensitivity (to 100,000–240,000 Hz), and baleen whales and elephants may have the lowest-frequency sensitivity (to nearly 1 Hz). These adaptations increase the animals' ability to communicate and survive. For instance, low-frequency sounds carry great distances through water or air and hearing them is especially useful for animals with large territories.

A vital feature of hearing is the ability to determine the direction from which a sound originates. For example, this ability may make the difference between a successful and an unsuccessful predator. How does an animal locate a sound? Under most circumstances, sound does not arrive at both ears simultaneously. Sound waves coming from the right, for example, excite the sensory receptors in an animal's right ear first and the left ear some milliseconds later, and therefore the brain receives action potentials from the auditory nerves of each ear at slightly different times. The brain interprets the time difference to determine the direction from which a sound came.

Animals such as owls that rely on hearing to pinpoint prey, particularly in the dark, tend to be extremely good at identifying the direction of a sound. An interesting experiment demonstrated this by outfitting owls in a dark room with small headphones. Just as in a human hearing test, sounds could be sent to either headphone or to both. If the investigator sent a high-pitched noise (that mimicked the sounds of a mouse) first to the left headphone and then a single millisecond later to the right headphone, the owl turned its head to the left, because the owl's brain perceived the sound to be coming from that direction. If

the noises reached both headphones simultaneously, the owl behaved as if the signal was coming from directly in front of its head.

Bats in the air, whales and dolphins in the sea, and shrews in underground tunnels generate high-frequency sound waves to determine the location of an object. In this phenomenon, called **echolocation**, the sound waves bounce off a distant object and return to the animal, like an echo. The time it takes for the sound to return to each ear indicates the distance and direction of the object. Echolocation is especially useful in situations where vision is limited, such as in the dark.

## The Sense of Balance Is Mediated by Statocysts in Invertebrates and by the Vestibular System in Vertebrates

Let's now turn our attention to another form of mechanoreception, the sense of balance, also called equilibrium. Balance is part of a broader sense called proprioception, which is an animal's ability to sense the position, orientation, and movement of its body. Being able to sense body position is vital for the survival of animals. This is how a lobster, for example, rights itself when flipped over by a predator or how a bird maintains its balance while flying.

***Statocysts*** Many aquatic invertebrates have sensory organs called **statocysts** that send information to the brain about the position of the animal in space (**Figure 44.8**). Statocysts are small round structures consisting of an outer sphere of hair cells and one or more **statoliths**, which are tiny granules of sand or other dense objects. When the

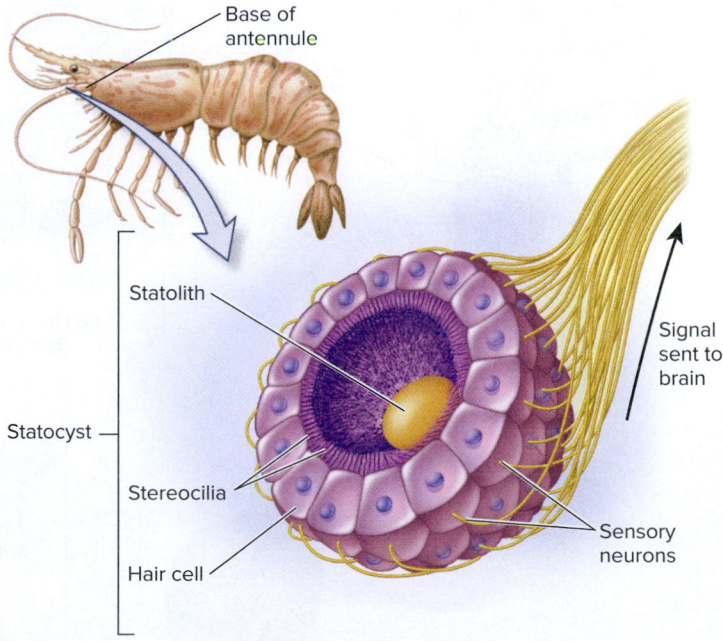

**Figure 44.8** **Sensing of balance in aquatic invertebrates.** Statocysts located near the antennae consist of a sphere of sensory hair cells surrounding one or more stony statoliths. When the animal moves, gravity shifts the statolith and stimulates the hair cells beneath it.

 **Core Skill: Connections** Is the use of statoliths unique to animals? Look back to Figure 37.14 for help.

animal moves, gravity alters the statoliths' position. If the animal turns on its side, for example, the movement of statoliths stimulates a new set of hair cells to release neurotransmitter, generating action potentials in sensory neurons that inform the brain of the change in body position.

Several experiments have demonstrated the importance of statoliths. In one particularly dramatic example, researchers replaced the statoliths of crayfish with iron filings. Moving a magnet to different positions around the animal displaced the filings, causing the animal to change its position, and even to swim upside-down when the magnet was placed directly above its head.

**The Vestibular System** The organ of balance in vertebrates, known as the **vestibular system**, is located in the inner ear next to the cochlea (**Figure 44.9**). The vestibular system is composed of a series of fluid-filled sacs and tubules, which provide information about either linear or rotational movements. The utricle and saccule, the two sacs nearest the cochlea, detect linear movements of the head (Figure 44.9a), such as those that occur when an animal runs, jumps, or changes its posture. The hair cells within these structures are embedded in a gelatinous substance that contains granules of calcium carbonate called **otoliths** (from the Latin, meaning ear stones), which are analogous to statoliths. When

the head moves forward, the heavy otoliths are temporarily left behind as they are dragged forward more slowly, and the weight of the otoliths bends the stereocilia of the hair cells in the direction opposite to that of the linear movement. This bending changes the membrane potential of the hair cells and alters the electrical responses of nearby sensory neurons. These signals are sent to the brain, which uses them to interpret how the head has moved.

Three **semicircular canals** connect to the utricle at bulbous regions called the ampullae (singular, ampulla). The function of the semicircular canals is to detect rotational motions of the head (Figure 44.9). The hair cells in the semicircular canals are embedded in the ampullae in a gelatinous cone called the **cupula**. When the head moves, the fluid in the canal shifts in the opposite direction. This movement of fluid pushes on the cupula and bends the stereocilia of the hair cells in the direction of the fluid flow, which is opposite to that of the motion of the head. The three canals are oriented at right angles to each other, and each canal is maximally sensitive to motion in its own plane. For example, in humans the canal that is oriented horizontally would respond most to rotations such as shaking the head "no," whereas the other canals respond to "yes" motions or to tipping the ear to the shoulder. Overall, by comparing the signals from the three canals, the brain can interpret the motion of the head in three dimensions.

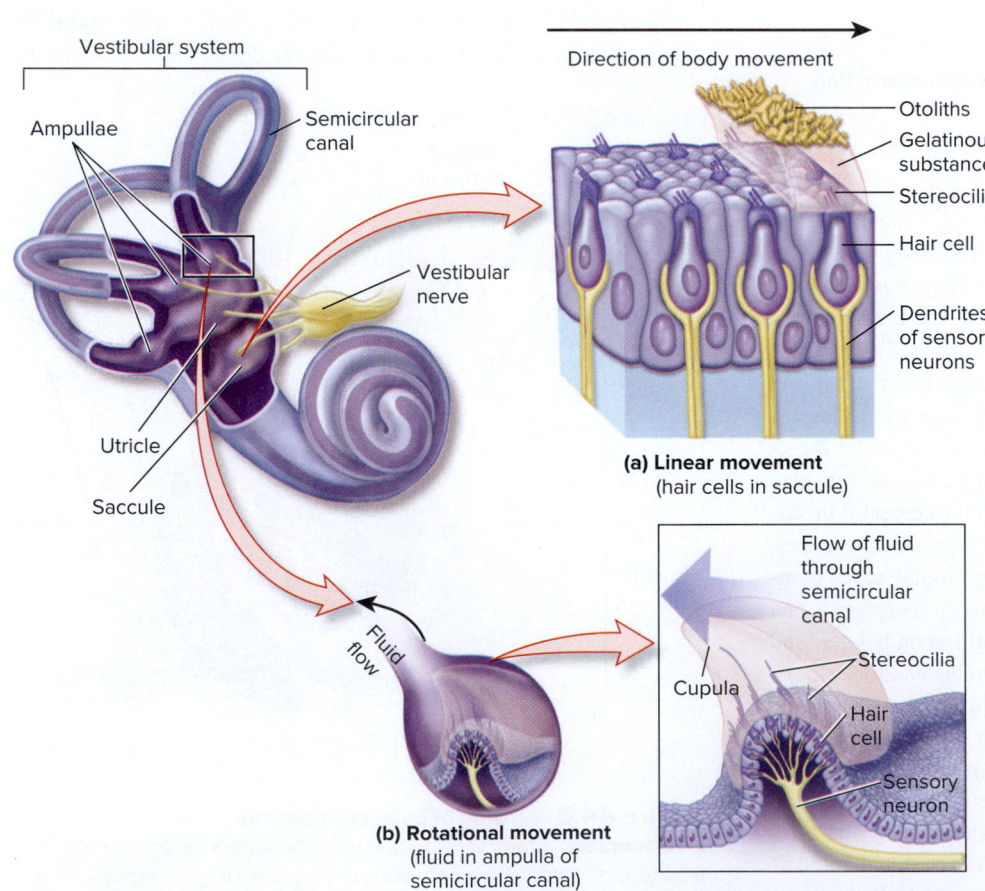

(a) **Linear movement**
(hair cells in saccule)

(b) **Rotational movement**
(fluid in ampulla of semicircular canal)

When the head moves forward, as when we walk or run, the otoliths move forward more slowly, and inertia bends the stereocilia on the hair cells. Electrical signals are sent to the brain via the vestibular nerve, which interprets the stimulus as linear motion.

The semicircular canals are at right angles to each other. The fluid in a canal will move opposite the direction of motion, bending the stereocilia within the cupula. This sends a signal via the vestibular nerve. The brain interprets the signals from all three canals as various rotational movements of the head.

**Figure 44.9** **The vertebrate vestibular system.**

*Concept Check:* *Note the orientation of the three semicircular canals with respect to each other. Why are the canals oriented in three different planes?*

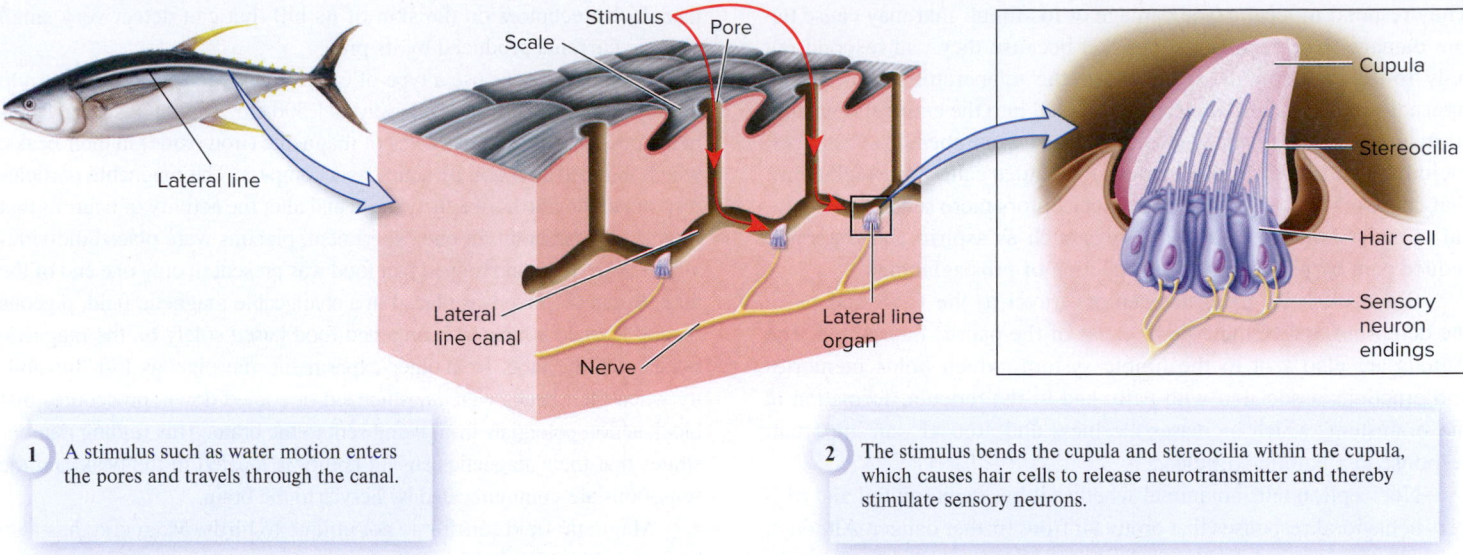

Lateral line

Stimulus    Pore

Scale

Cupula

Stereocilia

Hair cell

Lateral
line canal

Lateral line
organ

Sensory
neuron
endings

Nerve

**1** A stimulus such as water motion enters the pores and travels through the canal.

**2** The stimulus bends the cupula and stereocilia within the cupula, which causes hair cells to release neurotransmitter and thereby stimulate sensory neurons.

**Figure 44.10** Mechanoreceptors in the lateral line system of fishes that detect changes in water movement.

The vestibular system of vertebrates provides conscious information about body position and movement. It also supplies unconscious information for reflexes that maintain normal posture, control head and eye movements, and assist in locomotion. Researchers have discovered correlations between the types of locomotion in which an animal engages, and the size of its semicircular canals relative to the animal's body mass. Among primates, for example, agile animals with jerky forms of locomotion, such as leaping tarsiers, have much larger canals than do lorises, which are quadripedal and move slowly.

## Mechanoreceptors in the Lateral Line System Detect Movements in Water

Fishes and some toads detect changes in their environment through a **lateral line system** (**Figure 44.10**). This sensory system has hair cells that detect changes in water currents brought about by waves, nearby moving objects, and low-frequency sounds traveling through the water. The lateral line system runs along both sides of the body and the head of the animal. Small pores let water enter into a lateral line canal. The stereocilia of hair cells of each lateral line organ protrude into a gelatinous structure called a cupula (similar to the cupula of the vertebrate vestibular system). When the cupula is moved by the water, the stereocilia bend, causing the release of neurotransmitter from the hair cell. This stimulates a response in sensory neurons at the base of the hair cells. The response provides information to the brain about changes in water movement, such as the approach of a predator.

## 44.3 Thermoreception and Nociception

### Learning Outcome:

1. **CoreSKILL** » Distinguish between thermoreception and nociception, and propose reasons why they are vital to the safety and survival of an animal.

The perception of temperature and pain enables animals to respond effectively to potentially dangerous changes in their environments. As described in this section, these sensory stimuli are related in that their receptors are located in some of the same areas (for example, the skin; see Figure 44.3), share similar physical features, and under certain conditions result in similar perceptions.

## Thermoreceptors Detect Temperature

Sensing the outside temperature is important for animals because their body temperature is affected by the external temperature. This is particularly true for ectotherms, animals whose body temperature changes with the environmental temperature. Animals can survive only if their body temperature remains within certain limits, because cell membranes and the proteins in cells function optimally only within a particular temperature range. There are two types of thermoreceptors: those that respond to hot and those that respond to cold. Both types are sensory neurons without any structural specializations or associated structures, unlike most mechanoreceptors, for example. The peripheral endings of these sensory neurons respond to cold or hot temperatures by activating or inhibiting enzymes within their plasma membranes, which alters membrane ion channels.

In addition to skin thermoreceptors that sense the outside temperature, thermoreceptors in various organs and the brains of some animals also detect changes in core body temperature. Activation of thermoreceptors triggers physiological and behavioral adjustments that help maintain homeostasis. These adjustments, described in Chapter 47, include shivering, changes in blood flow to or away from the skin, and behaviors such as seeking shade or sunlight. In addition, thermoreceptors are often linked with reflexive behaviors, such as when an animal steps on a hot surface and pulls its foot away.

## Nociceptors Warn of Pain

Like thermoreceptors, nociceptors are sensory neurons with free peripheral endings in the skin and internal organs (see Figure 44.3).

They respond to local tissue damage or to stimuli that may cause tissue damage. Nociceptors are unusual because they can respond not only to external stimuli, such as extreme temperatures, but also to internal stimuli, such as molecules released into the extracellular fluid from injured cells. Damaged cells release a number of substances, including acids and small signaling molecules called prostaglandins, that cause inflammation and make nociceptors more sensitive to painful stimuli. Anti-inflammatory drugs such as aspirin and ibuprofen reduce pain by preventing the production of prostaglandins.

Signals arising from nociceptors travel to the CNS and reach the cerebrum, where the type or cause of the pain is interpreted. The signals are also sent to the limbic system, which holds memories and emotions associated with pain, and to the reticular formation in the brainstem, which increases alertness and arousal—an important response to a painful stimulus.

Nociception tells an animal whether it has been injured and triggers behavioral responses that protect it from further danger. Although in many cases we cannot know whether or how animals perceive pain, nociceptors have been identified in all classes of vertebrates and in many invertebrates.

## 44.4 Electromagnetic Reception

### Learning Outcome:

1. Identify ways that animals use electromagnetic receptors to sense their environments.

Electromagnetic receptors detect radiation within a wide range of the electromagnetic spectrum, including those wavelengths that correspond to visible light, UV light, and infrared light, as well as electrical and magnetic stimuli. Photoreceptors are specialized electromagnetic receptors that respond to light and are described in Section 44.5. Here, we will examine the ability of some animals to sense electric and magnetic fields and also heat in the form of infrared radiation.

The ability to detect the presence of nearby prey or predators can be especially challenging in animals that inhabit low-light environments. The more ways an animal has to detect other animals, the better it can avoid danger or obtain food. One mechanism found widely in fishes is electroreception, in which specialized sensory structures detect electric fields in the environment. There are two general types of electroreception. First, many fishes living in dark waters can detect the weak electric field generated by the activity of excitable tissues such as the muscles and nerves of other animals. To do this, they use exquisitely sensitive electroreceptors located in pores in the head region. These sensory receptors are as heavily innervated as the eyes of these fishes, suggesting the importance of this sense for their survival. Sharks and rays in particular can detect even the tiny electrical signals generated by the beating hearts of prey hiding beneath a layer of sand on the ocean floor. In a second type of electroreception, some fishes generate their own electric fields with a special organ derived from excitable tissue. As a fish swims through its environment, nearby objects will disturb this field and this disturbance can be sensed by the fish. This might happen, for example, if a potential prey or predator moved close to the animal. Although electroreception is primarily found in fishes, it is not unique to them. The platypus, a mammal that lives in the murky waters of streams and ponds,

has electroreceptors on the skin of its bill that can detect very small electric currents produced by its prey.

Homing pigeons use a type of electromagnetic sensing to return to their starting points from as far away as 1,500 km. This navigational feat is made possible by small particles of magnetite (iron oxide) in their beaks, which indicate direction by acting as a compass. The magnetite particles respond to the Earth's magnetic field and alter the activity of neurons that project into the brain. In one experiment, pigeons were placed individually in large tubes and trained that food was present in only one end of the tube. When the tube was placed in a changeable magnetic field, pigeons readily learned which end contained food based solely on the magnetic polarity of the tube. In another experiment, the pigeons lost this ability when their beaks were anesthetized or cooled down, procedures that block action potentials from being sent to the brain. This finding demonstrates that their magnetic sensing ability is located in the beak and the sensations are communicated by nerves to the brain.

Magnetic field sensing is not unique to birds. Magnetite has also been found in the heads of migratory fishes such as rainbow trout. However, this probably does not entirely explain the extraordinary ability of migratory animals to navigate great distances, because other cues, such as smell and visual recognition of landmarks, also appear to play roles in this process.

Venomous snakes known as pit vipers (a group that includes copperheads and rattlesnakes) can localize prey in the dark with detectors that sense the heat emitted from animals as infrared radiation. These detectors are located in pits on each side of the head between the eyes and nostrils (**Figure 44.11**). Within the pit, a thin, nerve-rich, temperature-sensitive membrane becomes activated in response to infrared waves emitted by live animals. When the snake detects the heat of the animal, it localizes its prey by moving its head back and forth until both pits detect the same intensity of radiation. This indicates that the prey is centered in front of the snake.

Electrical, magnetic, and infrared sensing are adaptations for long-distance migration or low-light environments. When light is plentiful, however, photoreception, discussed in the next section, becomes a dominant sensory ability in many animals.

Pit—contains heat sensors

**Figure 44.11 Infrared sensing.** Sensory pits enable a white-lipped pit viper (*Cryptelytrops albolabris*) to detect the heat given off by its prey. ©Daniel Heuclin/Science Source

*Concept Check:* *What advantage does having sensory pits on both sides of the head provide to a pit viper?*

## 44.5 Photoreception

**Learning Outcomes:**

1. Describe the structure of invertebrate visual organs.
2. Describe the structure of the vertebrate (single-lens) eye, and explain how its structure is related to its ability to form images.
3. Compare and contrast the structure and function of rods and cones.
4. Describe the steps involved in the mechanisms by which photoreceptors respond to light in a single-lens eye.
5. Outline the neural pathway by which visual signals travel to reach the brain.

Although it is a form of electromagnetic reception, photoreception is such an important and widespread sense that we will cover it separately here. Visual systems employ specialized neurons called photoreceptors, which detect photons of light arriving from the Sun or other light sources or reflecting off an object. A photon is the fundamental unit of electromagnetic radiation and has the properties of both a particle and a wave. The properties of light are described in Chapter 8. In this section, we will examine the organs found in animals, usually called **eyes**, that detect light and send signals to the brain. The amazing features of these organs reflect the importance of vision in the animal world.

### Eyecups and Compound Eyes Are Found in Many Invertebrates

The ability to sense light is an ancient adaptation found even in many unicellular organisms, where it may provide a selection advantage for photosynthesis or protection. Typically, such organisms may have a single eyespot that contains a small number of light-sensitive molecules, but they lack the ability to discern the direction of a light source or to interpret a visual image.

**Eyecups**    A slightly more sophisticated visual organ is found in free-living flatworms such as *Planaria*. These animals have two concave structures called **eyecups** (**Figure 44.12**). Each eyecup contains the endings of photoreceptor cells and a layer of pigment cells that shields the photoreceptors from one side. The left and right eyecups receive light from different directions. This allows the eyecups to detect not only the presence or absence of light, but also its direction. The nervous system compares the amount of light detected by each eyecup, and the flatworm moves toward darkness, a behavior that protects it from predators. However, this type of photoreceptor does not form visual images of the environment.

**Compound Eyes**    In contrast, arthropods and some annelids have image-forming **compound eyes** (**Figure 44.13a**), which consist of several hundred to more than 10,000 light detectors called **ommatidia** (singular, ommatidium) (**Figure 44.13b**). Each ommatidium makes up one facet of the eye. Within the ommatidium, a two-part **lens**—composed of an outer region called the cornea and an inner crystalline cone—focuses light onto a long central structure called a rhabdom (**Figure 44.13c**). The rhabdom is a column of light-sensitive microvilli that project from the cell membranes of the photoreceptor cells of the ommatidium (**Figure 44.13d**). The light-sensitive molecules required for vision are located in the microvilli. The extensive surface area imparted by the microvilli provides the eye with increased sensitivity to light. Pigmented cells surrounding the photoreceptor cells absorb

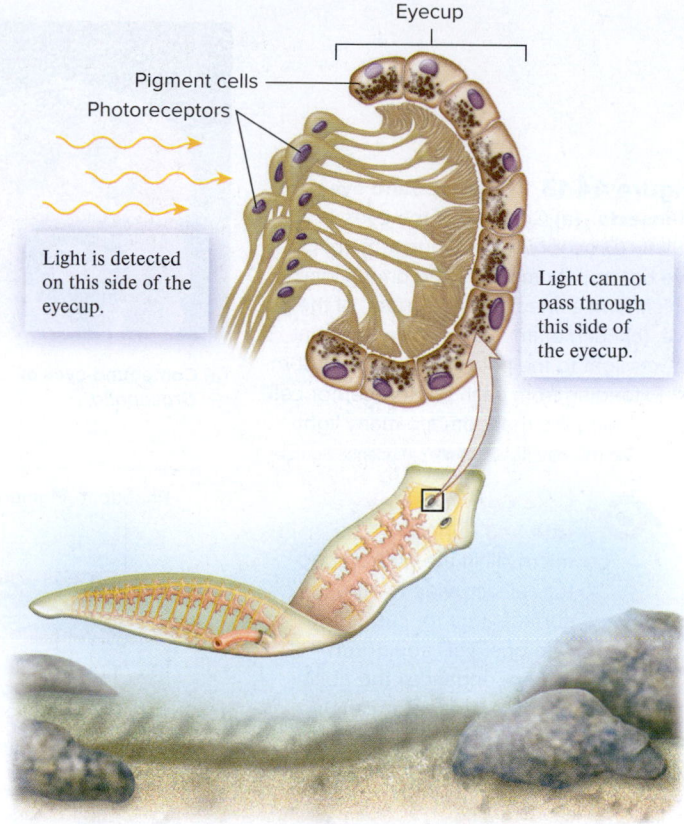

**Figure 44.12   The eyecup of a flatworm.** The orientation of the eyecup allows light to stimulate photoreceptors from primarily one direction. This type of visual organ senses the presence or absence of light and its direction, but does not form visual images.

excess light and thereby isolate each ommatidium from its neighbors. Thus, each ommatidium is pointed at one narrow area in space.

Each ommatidium senses the intensity and color of light. Although each ommatidium receives light from only a very narrow field, collectively they provide animals with a wide viewing area. Combining the different inputs from neighboring ommatidia, the compound eye is believed to form a mosaic-type image that the brain interprets. Animals such as bees and fruit flies, with large numbers of ommatidia, presumably have sharper vision and a wider field of vision than do those with fewer sensory cells, such as grasshoppers.

As anyone who has tried to swat a fly knows, the compound eye is extremely sensitive to movement as an object moves across successive ommatidia. This helps flying insects evade birds and other predators. Behavioral studies have shown, however, that the resolving power of even the best compound eye is considerably less than that of the single-lens eye, which we consider next.

### Vertebrates and Some Invertebrates Have a Single-Lens Eye

**Structure of the Single-Lens Eye**    **Single-lens eyes** are found in vertebrates and also in some mollusks, such as squid, octopuses, and some snails, and in some annelids. In such eyes, different patterns of light emitted from or reflected off objects in the animal's field of view are transmitted through a small opening, or **pupil**, through a single focusing lens, to a sheetlike layer of photoreceptors in the **retina** at the back of the eye (**Figure 44.14a**). The light inputs form a

**Figure 44.13** **The compound eye of insects.** **(a)** Close-up of the eyes of a fruit fly (*Drosophila melanogaster*). **(b)** Each eye has approximately 1,000 ommatidia, which form a sheet on the surface of the eye. **(c)** Each ommatidium has a lens that directs light to the photosensitive rhabdom. **(d)** Extending from each photoreceptor cell and forming the rhabdom are many light-sensitive microvilli. a: ©Omikron/Science Source

 **Core Concept: Structure and Function** The microvilli in the photoreceptor cells provide a large surface area for capturing photons. This structural feature enhances the ability of a photoreceptor cell to function as a light detector.

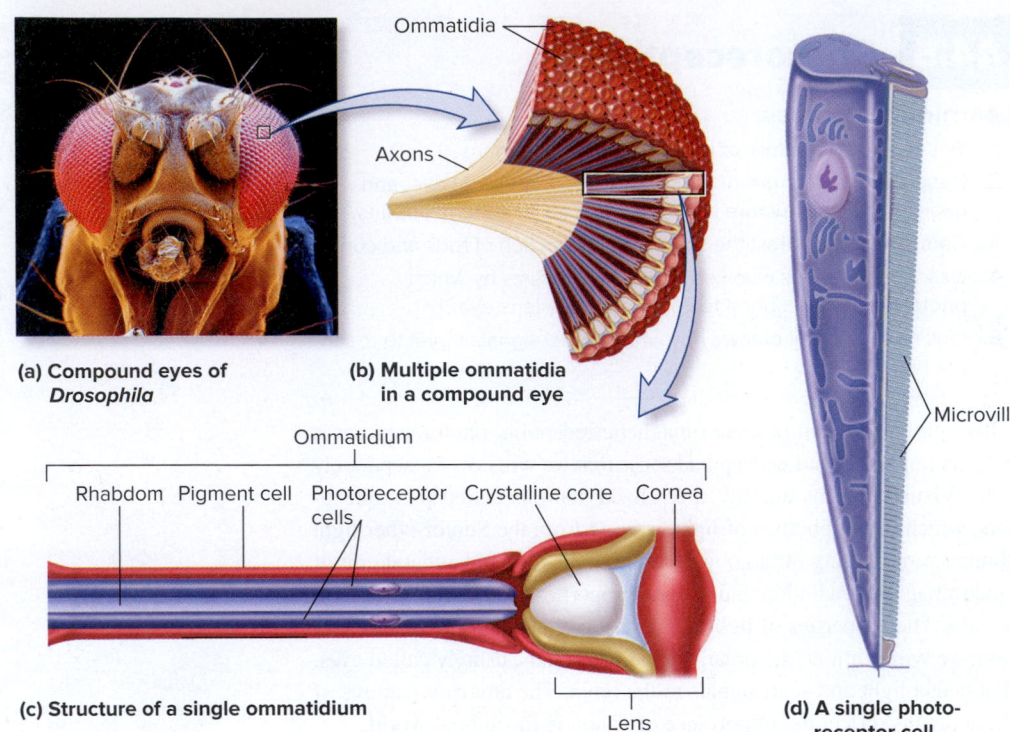

**(a) Compound eyes of *Drosophila***

**(b) Multiple ommatidia in a compound eye**

**(c) Structure of a single ommatidium**

Ommatidium

Rhabdom   Pigment cell   Photoreceptor cells   Crystalline cone   Cornea

Lens

**(d) A single photoreceptor cell**

Microvilli

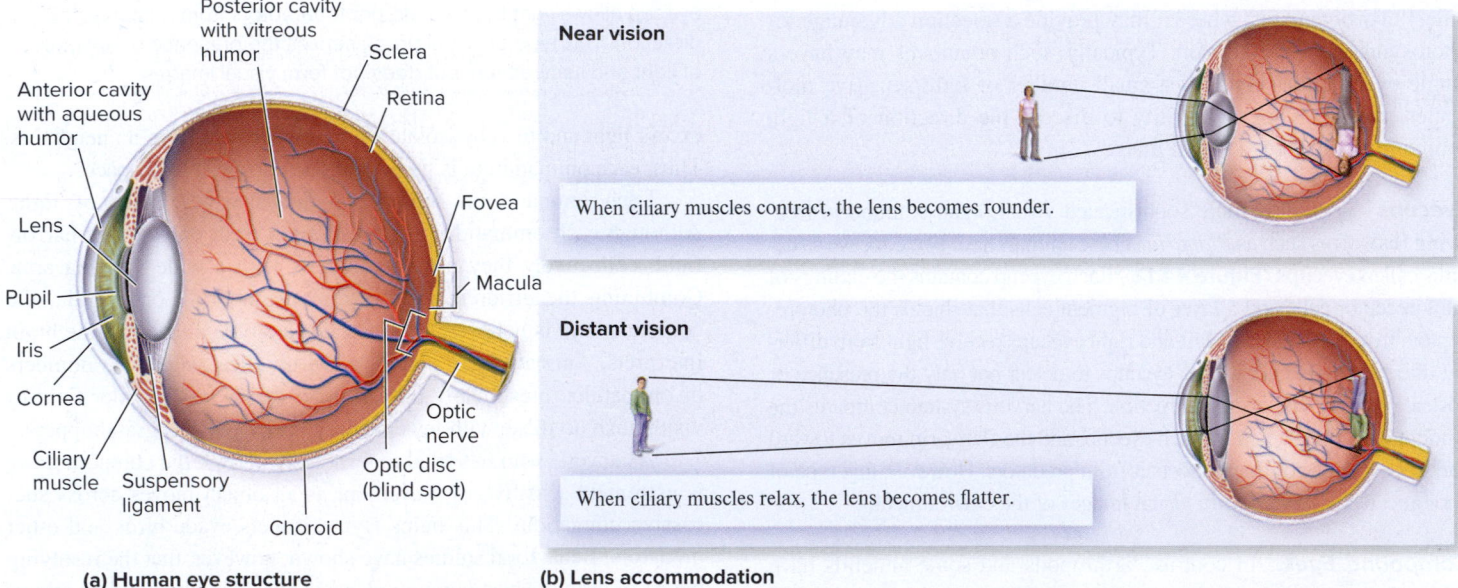

**(a) Human eye structure**

Posterior cavity with vitreous humor
Sclera
Retina
Anterior cavity with aqueous humor
Lens
Fovea
Pupil
Macula
Iris
Cornea
Optic nerve
Ciliary muscle
Suspensory ligament
Optic disc (blind spot)
Choroid

**(b) Lens accommodation**

**Near vision**

When ciliary muscles contract, the lens becomes rounder.

**Distant vision**

When ciliary muscles relax, the lens becomes flatter.

**Figure 44.14** **The vertebrate single-lens eye.** **(a)** The structure of the human eye. **(b)** Changes in lens shape during accommodation. When an object is near, the ciliary muscles contract and the lens becomes rounder, causing light to bend more. When the object is far away, the ciliary muscles relax and the lens flattens. **(c)** Demonstration of the blind spot. First, position this picture in front of your face. Next, close your left eye and stare at the black spot with your right eye while you move the picture toward and away from your face. At some point, light reflecting off the plus (+) sign will fall directly on your optic disc, and it will seem to disappear. It will then reappear as you continue moving the picture.

**(c) Demonstration of blind spot**

visual image of the environment on the retina. The activation of these photoreceptors triggers electrical changes in neurons that pass out of the eye through the **optic nerves**, which carry the signals to the brain. The brain then interprets the visual image that was transmitted.

As illustrated in Figure 44.14a, the vertebrate eye has a tough outer sheath called the sclera (the white of the eye). Between the sclera and the retina is a layer of blood vessels called the choroid. At the front of the eye, the sclera is continuous with a thin, clear layer known as the **cornea**. As in compound eyes, the cornea functions partly to focus light in single-lens eyes, and it also plays a protective role. In single-lens eyes, however, the lens plays the major role in focusing light onto photoreceptors.

Within the vertebrate eye are two cavities, the anterior and posterior cavities. The anterior cavity is the part of the eye between the lens and the cornea. Within this cavity is the **iris**—the circle of pigmented smooth muscle responsible for eye color. The anterior cavity is largely filled with a thin liquid called the aqueous humor that helps maintain eye pressure and shape, and may serve a nutritive function. The larger posterior cavity between the lens and the retina contains the thicker vitreous humor, which further helps maintain the shape of the eye.

The hole in the center of the iris is the pupil. The size of the pupil changes when the smooth muscles of the iris reflexively relax or contract to allow more or less light to enter the eye.

*Light Focusing*   Because light radiates in all directions from a light source, light must be bent (refracted) inward toward the photoreceptors at the back of the eye. This is accomplished by the cornea and the lens. Whenever light passes from one medium to another medium of a different density, light waves will bend (try looking at a pencil in a glass partly filled with water). The cornea, which is at the interface between the air and the aqueous humor, initially refracts the light. The light then passes through the lens, where it is refracted again and focused onto the layer of photoreceptors, the retina, at the back of the posterior chamber. The bending of the incoming light results in an upside-down and laterally inverted image on the retina, but the brain adjusts for this, and the image is perceived correctly.

The lens is adjusted to focus light that originates from different distances. In fishes and amphibians, the lens is moved forward or backward. In the avian and mammalian eye, the lens remains stationary but changes shape to become more or less round. When the lens is stretched, it flattens, and light passing through it bends less than when it is round. Contraction and relaxation of the ciliary muscles adjust the lens according to the angle at which light enters the eye, a process called **accommodation** (**Figure 44.14b**).

How does the retina form sharp images? The region on the retina directly in line with the pupil and lens is called the macula. Near the center of the macula is the **fovea**, which contains the highest density of photoreceptors for color. The fovea is responsible for the sharpness with which we and many other animals see in daylight. However, the retina also has limitations in forming images. In the eye, as mentioned, the image initiates signals that travel from the retina to the brain through neurons in the optic nerves that exit the eye. In vertebrates, the point on the retina where the optic nerve leaves the eye is called the **optic disc**. The optic disc does not have any photoreceptors, forming a blind spot where light does not activate a response (**Figure 44.14c**). Invertebrates with single-lens eyes do not have a blind spot, because

the photoreceptors in their eyes are at the front of the retina. Therefore, the optic nerve does not pass through the layer of photoreceptors before leaving the eye.

## Rods and Cones Are Photoreceptor Cells

Vertebrates have two types of photoreceptors with names that are derived from their shapes: rods and cones (**Figure 44.15**). **Rods** are very sensitive to low-intensity light and can respond to as little as one photon, but they do not discriminate different colors. Rods are useful mostly at night, and they send signals to the brain that generate a black-and-white visual image. **Cones** are less sensitive to low levels of light but, unlike rods, are sensitive to wavelengths of light that allow animals to perceive color.

Rods and cones are cells with three functional parts: the outer segment, the inner segment, and the synaptic terminal. The **outer segment** of the cell contains folds of membranes that form stacks like discs (**Figure 44.16**). These discs contain the pigment molecules that absorb light. The **inner segment** of the cell contains the cell nucleus and other cytoplasmic organelles. Rods and cones do not have axons but have synaptic terminals with neurotransmitter-containing vesicles, which synapse with neurons within the retina.

Nocturnal animals (those active predominantly at night) rely primarily on rod vision, though some do have limited color vision. In diurnal animals (active predominantly by day) with both rods and cones, such as humans, the rods are located around the periphery of the retina away from the fovea. Therefore, it is easiest to see low-intensity light if it comes into the eye at an angle. You can easily verify this. In early evening, before many stars are visible, look at the sky until you notice a star out of the corner of your eye. Now shift your gaze to where you thought you saw the star. You will probably not be able to locate it anymore. When you look away again so that light from the dim star enters your eye at an angle, it will reappear. This demonstrates that under low-light conditions, your vision is better when the light is directed to the part of the retina that contains only rods.

Cones are used in daylight by most diurnal vertebrate species and by some insects such as the honeybee, which can detect the

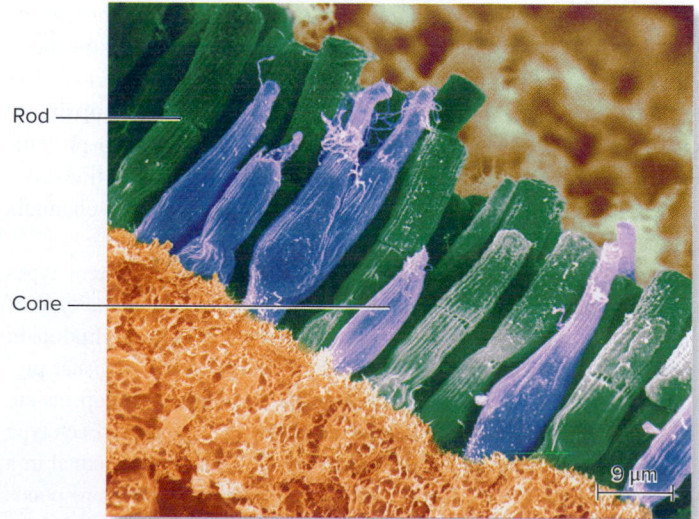

Rod

Cone

9 μm

**Figure 44.15  Rod and cone photoreceptors.** Rods are shown as green and cones as blue in this false-color scanning electron micrograph (SEM). ©Eye of Science/Science Source

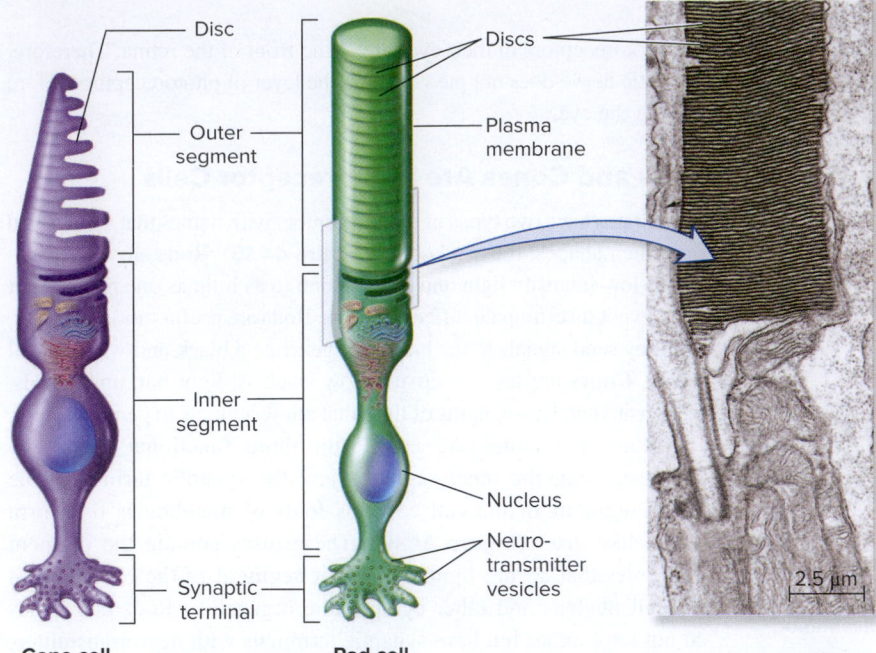

**Figure 44.16  Structure of a cone and a rod.**  The illustration shows the structure of a cone and rod photoreceptor and the appearance of a rod in a transmission electron micrograph (TEM). Note the stacks of multiple membranous discs in the outer segments of the cells. (right) ©Don W. Fawcett/Science Source

**Core Skill: Connections**  Refer back to Figure 44.13d. What structural similarity do the insect and vertebrate photoreceptors share?

yellow color of pollen. Compared to rods, the human retina has fewer cones, which are clustered in and around the fovea. Cones provide sharp images because of their density at the fovea. Although they are less sensitive to light than rods, this is less critical in daylight because the amount of light reaching the eyes at this time far exceeds what is needed to stimulate any photoreceptor cell.

## Rods and Cones Contain Visual Pigments That Absorb Light

Visual pigments are molecules that absorb light; they are found embedded in the disc membranes of the outer segments of rods and cones. In the mid-20th century, American biologist and Nobel Laureate George Wald discovered that these pigments consist of two components bonded together. The first component is **retinal**, a derivative of vitamin A that is capable of absorbing light energy. The discovery of retinal in the visual pigment explains the need for vitamin A in the human diet and its importance in vision. The second component of visual pigments is a protein called **opsin**, of which there are several types. Opsins are examples of G-protein-coupled receptors (see Figure 9.7), which trigger a signal transduction pathway that changes the permeability of membrane channels to ions.

Rods and cones have visual pigments containing different types of opsin proteins. These pigments are named according to the type of opsin they contain. In rods, the visual pigment is named **rhodopsin** (**Figure 44.17**). Cones contain any one of several types of visual pigments called cone pigments, or **photopsins**. In humans, photopsins are composed of retinal plus one of three possible opsin proteins. Each type of opsin protein determines the wavelength of light that the retinal in a cone can absorb. For example, each cone pigment in humans responds best to red, green, or blue light. Any given cone cell makes only one type of cone pigment. Many different shades of these colors can be perceived, however, because the brain uses information about the proportion of each type of cone that was stimulated to generate perceptions of all other colors.

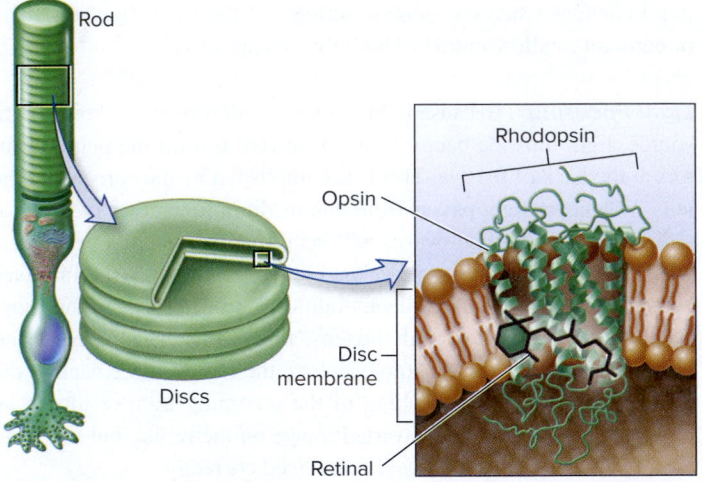

**Figure 44.17  A visual pigment.**  The visual pigment rhodopsin is found within the membrane of the rod photoreceptor discs. It is composed of a transmembrane protein, opsin, that is bonded to a molecule of retinal, a derivative of vitamin A that is capable of absorbing light energy.

**Core Concept: Structure and Function**  The stack of discs in the outer segment of a photoreceptor is an excellent example of a structural adaptation that increases surface area without significantly increasing volume, thereby allowing for greater light-capturing ability and improving the function of the cell as a sensory cell.

**Core Concept: Evolution**

## Color Vision Is an Ancient Adaptation in Animals

Color vision requires the presence of at least two types of opsins with optimal sensitivities to light of different wavelengths. Genome analyses and behavioral and neurophysiological testing of many animals have confirmed that color vision is widespread across invertebrate and vertebrate taxa, and arose early in animal evolution. It is debatable why color vision provided a selective advantage to animals. Some investigators hypothesize that it produced more acute vision due to an overall improved contrast and was not originally related to commonly observed phenomena today, including colorful plumage displays or brightly colored flowers. Coloration in plants may have evolved in response to the color sensitivity in animals, not vice versa.

Evolutionary studies indicate that the primordial color sensitivity was to short wavelengths in the ultraviolet/blue region of visible light (refer back to Figure 8.4). From there, additional opsins evolved with sensitivities to medium and longer wavelengths. Most vertebrates have four types of color-sensitive photoreceptors. Some insects, including the butterfly *Papilio xuthus*, have up to six types of light receptors, and certain crustaceans (for example, mantis shrimp) have 12! Such animals presumably see shades of colors we cannot. A well-studied model is the honeybee, which has four color-sensitive photoreceptors that respond to wavelengths ranging from 300 nm to 600 nm (refer back to Figure 8.4). Honeybees can perceive ultraviolet light that is not seen by humans. A rainbow would appear very differently to a honeybee than it does to humans. A rainbow seen by a bee would extend past the blue edge and stop short of the red edge.

As mentioned, most vertebrates have four types of color-sensitive photoreceptors. However, most modern species of mammals are dichromatic; that is, they only have two types of opsins and therefore see a more limited color palette than do other vertebrates. One possible explanation for dichromatic vision may be related to the evolution of mammals. The earliest mammals were probably nocturnal animals. Recall that cones are less sensitive to low levels of light and therefore color vision is limited at night. For this reason, natural selection may have favored a loss of some color-sensitive photoreceptors, thereby allowing for more rods and improving night vision.

Compared with other mammals, primates are an exception to dichromatic color vision. A relatively recent gene duplication event led to the reappearance of a third opsin during the course of primate evolution, one that is sensitive to middle (green) wavelengths. The evolution of this third opsin may have imparted an advantage that allowed fruit and leaf-eating primates to better discern orange, red, and yellow fruits against a background of green leaves, but this is uncertain.

About 92% of human males and over 99% of females have trichromatic color vision. However, deviations from trichromatic color vision may result from defects in the cone pigments arising from mutations in the opsin genes. The most common is red-green color blindness, which occurs predominantly in men (1 in 12 males compared with 1 in 200 females). Individuals with red-green color blindness either lack the red or green cone pigments entirely or, more commonly, have one or both of them in an abnormal form. In one form of red-green color blindness, for example, an abnormal green pigment responds to red light as well as green, making it difficult to discriminate between the two colors.

Researchers have determined that color blindness results from a recessive mutation in one or more genes encoding the opsins. Genes encoding the red and green opsins are located on the X chromosome, but the gene encoding the blue opsin is located on a different chromosome. In males, the presence of only one X chromosome means that a single recessive allele from the mother results in red-

green color blindness, even though the mother herself may not be color blind (**Figure 44.18**). Although there is no cure or treatment for color blindness in humans, in 2009 American researcher Katherine Mancuso and colleagues were able to restore trichromatic color vision in adult squirrel monkeys who had been dichromatic (red-green color blind) since birth. This was accomplished using gene therapy in which the missing opsin gene was introduced into each eye, where it was then expressed in existing photoreceptors. Behavioral testing revealed that the monkeys could, for the first time, distinguish red and green colors. Incredibly, the visual centers of the adult brain were able to perceive this new input immediately, despite not having been "wired" for red-green signals during early life.

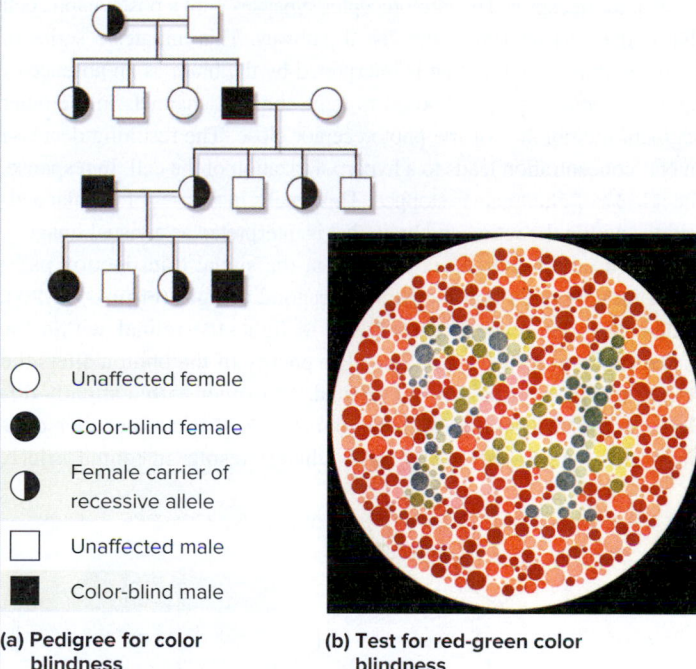

○ Unaffected female

● Color-blind female

◐ Female carrier of recessive allele

□ Unaffected male

■ Color-blind male

**(a) Pedigree for color blindness**

**(b) Test for red-green color blindness**

**Figure 44.18  Color blindness.** **(a)** A pedigree for red-green color blindness showing all possible offspring. **(b)** A standard eye test to screen for red-green color blindness. People with red-green color blindness will not see the number 74 hidden in this pattern.

b: ©Steve Allen/Getty Images

**Concept Check:** *Why is red-green color blindness rare in females?*

 **Core Skill: Modeling** The goal of this modeling challenge is to make predictions about the transmission of color blindness to offspring of parents with genotypes that are different from those in the pedigree shown in Figure 44.18.

**Modeling Challenge:** Based on your understanding of red-green color blindness (see Figure 44.18), create a Punnett square to predict the ratios of genotypes and phenotypes of offspring from a color-blind mother and a father with normal color vision. Refer back to Section 17.1 for help in setting up a Punnett square, which is a type of model used to predict the outcomes of crosses. Let $X^C$ represent the allele for normal color vision, and $X^c$ the recessive allele that causes color blindness. Remember that only the X chromosome carries this gene.

## Photons Change Photoreceptor Activity by Altering the Conformation of Visual Pigments

Photoreceptors differ from other sensory receptor cells because at rest in the dark their membrane is slightly depolarized, whereas in response to a light stimulus, it becomes hyperpolarized (**Figure 44.19**). In the dark, the cell membranes of the outer segments of resting cells are highly permeable to sodium ions. $Na^+$ diffuses into the cytosol of the cell through open $Na^+$ channels in the outer segment membrane. The $Na^+$ channels are gated by intracellular cyclic guanosine monophosphate (cGMP). In the dark, cytosolic concentrations of cGMP are high, keeping $Na^+$ channels open and depolarizing the cell. This depolarization results in a continuous release of the neurotransmitter glutamate from the synaptic terminal of the photoreceptor. The photoreceptor synapses with a postsynaptic cell that is the next neuron in the visual pathway. This initiates a series of events within the retina that is interpreted by the brain as an absence of light. In contrast, when exposed to light, the $Na^+$ channels in the outer segment membranes of the photoreceptor close. The resulting decrease in $Na^+$ concentration leads to a hyperpolarization of the cell. In response, the release of glutamate is stopped. This results in a series of cellular activations within the retina and brain that is interpreted as a visual image.

Let's take a more detailed look at the signal transduction pathway that allows a photoreceptor to respond to light (**Figure 44.20**). When the photoreceptor is exposed to light, the retinal within the visual pigment absorbs a photon. The energy of the photon alters the retinal from *cis*-retinal to *trans*-retinal, an isomer with a slightly different conformation due to a rotation at one of the molecule's double bonds (see Figure 44.20). This change results in retinal briefly

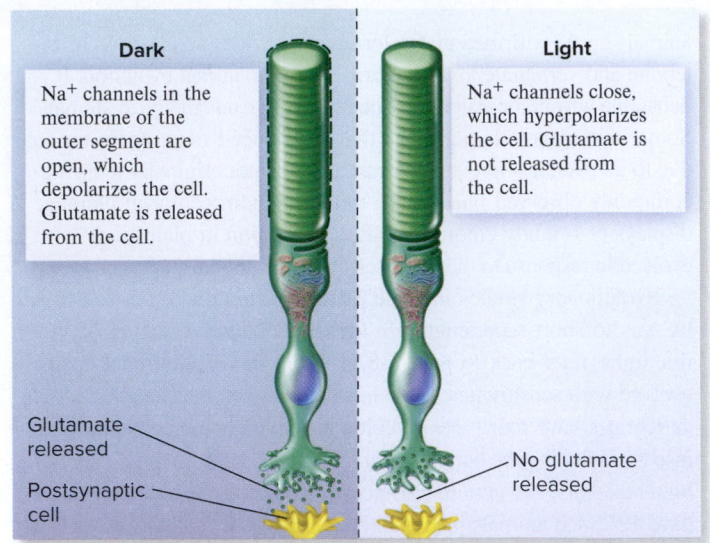

**Figure 44.19** Membrane potential response of a photoreceptor to dark and light.

dissociating from the opsin protein, causing the opsin to change its three-dimensional shape and activate a G protein called transducin, located in the disc membrane. The activated transducin, in turn, activates another disc protein, the enzyme phosphodiesterase.

The action of phosphodiesterase results in the closure of $Na^+$ channels in the outer segment membrane. Remember that in the dark, these channels are kept open by intracellular cGMP. However, when

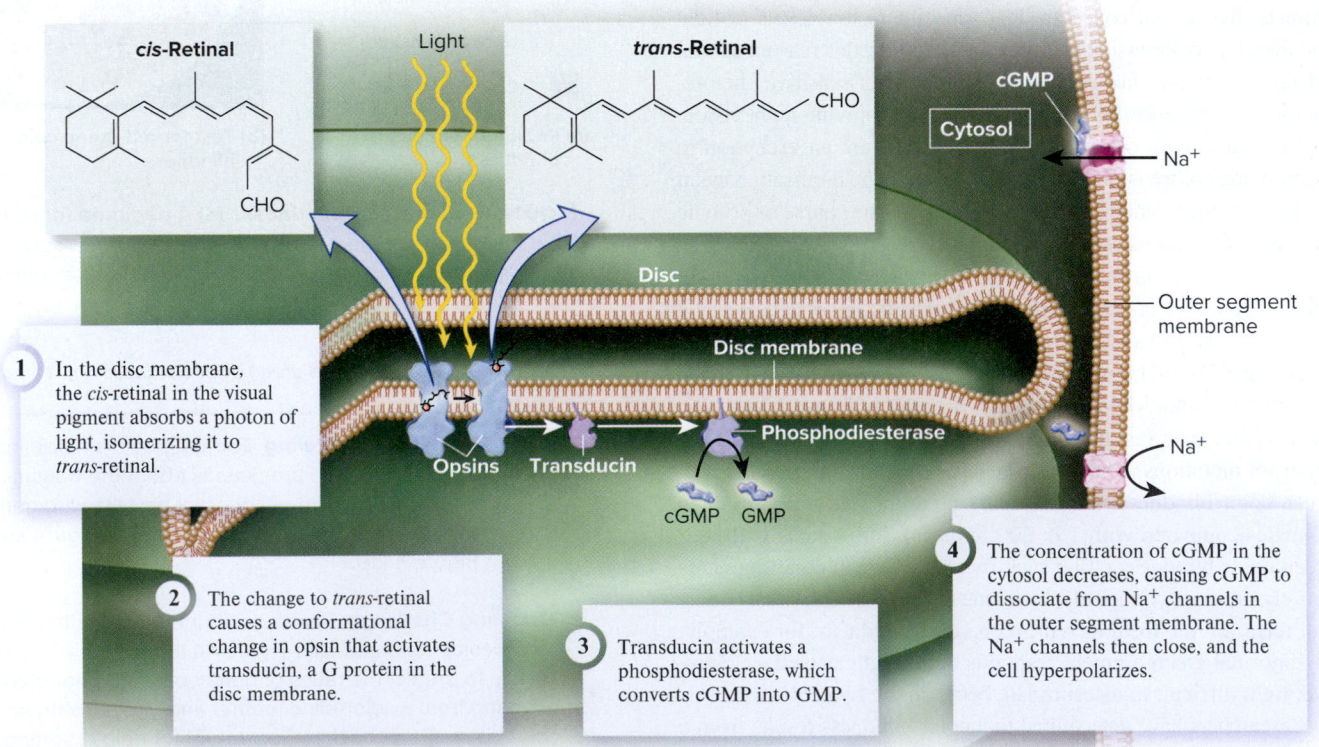

**Figure 44.20** Signal transduction pathway in a photoreceptor (rod) cell in response to light.

 **Core Concept: Energy and Matter** The energy of light alters the configuration of the retinal molecule, causing it to isomerize. This results in a conformational change in the opsin associated with that retinal.

phosphodiesterase is activated, it decreases the concentration of cytosolic cGMP by converting that molecule to GMP. This results in the dissociation of cGMP from the channels, so the channels close, and there is no longer any net diffusion of Na$^+$ into the cell. The membrane potential of the cell becomes less positive than it was in the dark. Therefore, the response of the cell is a hyperpolarization that is proportional to the intensity of the light. The final result is a decrease in glutamate release from the photoreceptor (see Figure 44.19), ultimately leading to a visual image. The sequential activation of enzymes following activation of a single photoreceptor results in an amplification of the original signal (refer back to Figure 9.14). Because of this property, extremely low levels of light are detectable.

## The Visual Image Is Refined in the Retina

Thus far, we have considered the structure of the vertebrate eye and how photoreceptors transduce light. We will now turn our attention to the neural pathway through which the visual signal travels to reach the brain. To do so, we must consider the cellular organization of the retina. The vertebrate retina has several layers of cells (**Figure 44.21**). The photoreceptors (rods and cones) form the deepest layer, closest to the

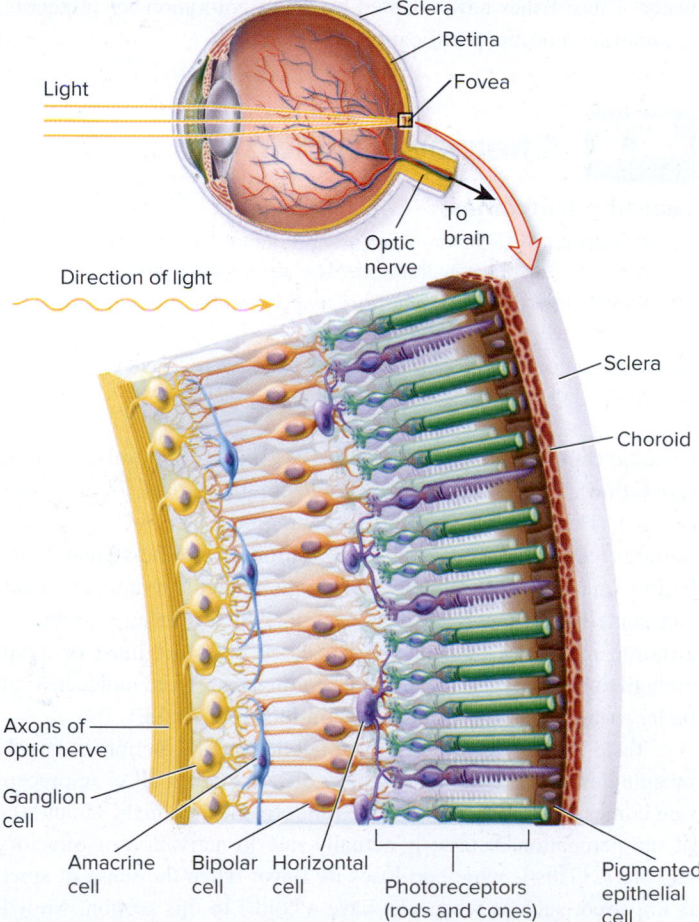

**Figure 44.21 The arrangement of cells in the retina.** Light passes through layers of cells before it reaches the photoreceptors. Amacrine and horizontal cells integrate the responses of the bipolar and ganglion cells. The ganglion cells generate action potentials that carry the information to the brain.

*Concept Check:* *With what cells do photoreceptors form synapses?*

sclera. Immediately behind the photoreceptors is a pigmented epithelium that absorbs light that missed the photoreceptors; this absorption prevents scattering of light within the retina, which could degrade the sharpness of vision. Because the photoreceptors are positioned at the back of the retina, light must pass through two transparent layers of cells before it reaches them. The middle layer contains **bipolar cells**, so named because one end (or "pole") of the cell synapses with the photoreceptors, and the other end relays responses to a top layer of cells, the **ganglion cells**. The axons of the ganglion cells extend out of the eye into the optic nerve. In addition, two other types of cells, horizontal and amacrine cells, are interspersed across the retina.

The pathway for light reception begins at the rods and cones. These photoreceptor cells release neurotransmitter molecules that affect the membrane potential of bipolar cells. The membrane potential of the bipolar cells determines the amount of neurotransmitter that they release, which, in turn, controls the membrane potential of ganglion cells. When a threshold potential is reached in ganglion cells, action potentials are sent out of the eye via the optic nerve to the brain. These signals travel along pathways that include the thalamus, brainstem, cerebellum, and the cerebral cortex. Visual information is further refined and interpreted within the vision centers of the cortex. The cortex responds to such characteristics of the visual scene as whether something is moving, how far away it is, how one color compares with another, and the nature of the image (for example, a face). The cortex does not form a picture in the brain, but forms a spatial and temporal pattern of electrical activity that is perceived as an image.

Horizontal and amacrine cells modify electrical signals as they pass from the photoreceptors to the ganglion cells. These cells adjust the signals significantly, enhancing an animal's ability to visualize a scene by emphasizing the differences between images. Horizontal cells make connections between photoreceptors and help to define the boundaries of an image. Amacrine cells are important in adjusting the eye to different light intensities and increasing the sensitivity of the eye to moving images. The ability of the retina to refine the image is especially well developed in birds and reptiles. These animals have complex retinas that process the image extensively before it is interpreted in the brain.

## Vertebrate Eyes Are Adapted to Environmental Conditions and Life Histories

Many vertebrates have modifications of their visual systems that are the result of evolutionary adaptations to environmental conditions. Other adaptations have occurred as a result of behavioral requirements for obtaining food or attracting a mate. For instance, raptors can resolve images while flying at speeds close to 150 mph. In another example, cats and certain other animals have reflective surfaces at the backs of their retinas that help direct light onto photoreceptors even in low light conditions such as at night.

***Differences in Eye Placement*** Except for some of the ray-finned fishes, blunt-headed cetaceans, and most amphibians, vertebrate animals have some degree of **binocular (or stereoscopic) vision**. Animals with both eyes located at or near the front of the head, such as primates and raptors, have greater binocular vision, because the overlapping images coming into both eyes are processed together in the brain to form one perception (**Figure 44.22**). Binocular vision provides excellent depth perception because the images come into each eye from

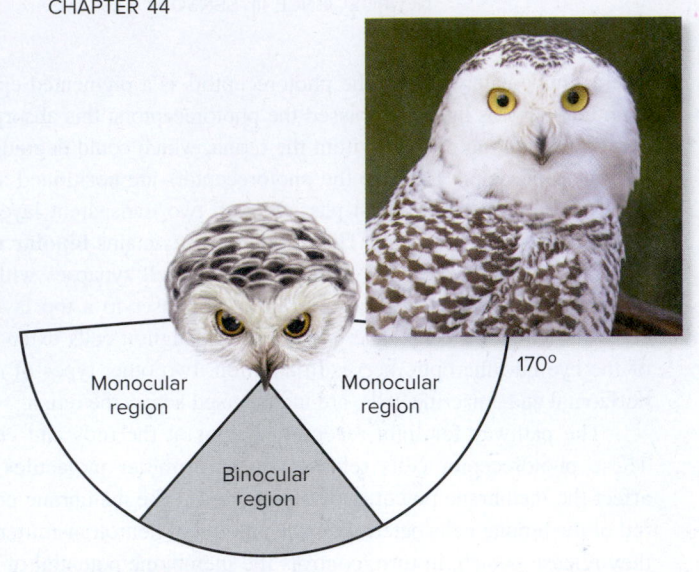

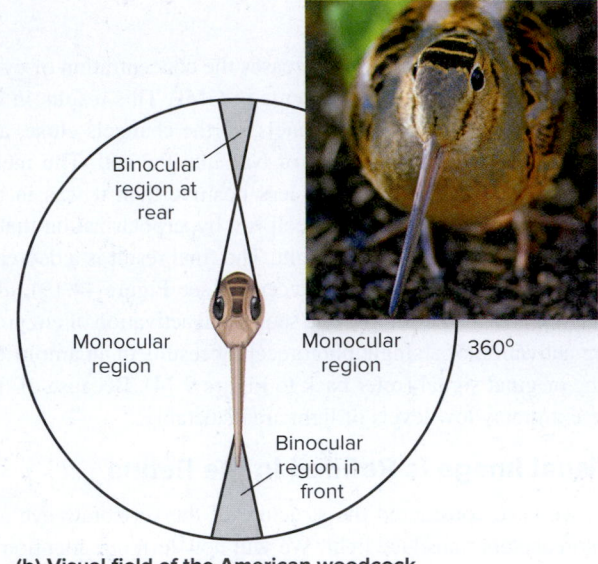

**(a) Visual field of the snowy owl**

**(b) Visual field of the American woodcock**

**Figure 44.22 Examples of binocular and monocular fields of vision.** Visual fields are shown for **(a)** the snowy owl (*Bubo scandiacus*) and **(b)** the American woodcock (*Scolopax minor*). Monocular regions are white; binocular regions are shaded. a: ©Enjoylife2/Getty Images; b: ©Cal Vornberger/Getty Images

slightly different angles. The brain processes those tiny differences to determine where an object is relative to other objects in its environment. Predators benefit from binocular vision because it helps them judge distance and determine the location of their prey. Binocular vision is present in predatory birds such as the snowy owl (Figure 44.22a) and mammals, particularly in arboreal animals that must judge distances between tree limbs. It is not unique to vertebrates, however. Predatory insects such as mantids also have binocular vision.

In contrast, animals with eyes on the sides of the head, such as most fishes, blunt-headed cetaceans, amphibians, herbivorous mammals, and insects, have strictly or primarily **monocular vision**. Monocular vision allows an animal to see a wide area at one time, at the cost of reduced depth perception. Many prey species have monocular vision, which may have evolved because it helps them scan for predators across a wide field of vision. The placement of the eyes in the American woodcock (*Scolopax minor*) actually permits a field of vision of 360°, most of which is monocular (Figure 44.22b). In other words, these and similar birds can see directly behind themselves, even when digging in the dirt for worms!

***Vision in the Deep Sea*** Fishes and other deep-sea vertebrates have color vision that is limited primarily to the color blue. Light with longer wavelengths that would be seen as red or orange does not usually penetrate more than ~6 m into the water, whereas the higher-energy, shorter-wavelength light, which we perceive as blue, can penetrate to greater depths. Aquatic animals that live in the deep sea are usually capable of seeing only blue, because they generally have only one opsin, which is responsive to blue light. Deep-dwelling fishes tend to be drab-colored, because most wavelengths of light do not penetrate that far and therefore could not reflect off the surface of the fishes. Thus, there was no selection pressure for evolving bright coloration. In those deep-dwelling fishes with more than one type of opsin, the additional visual pigments detect the bioluminescence (self-generated light, like that of a firefly) produced by their own or other species.

By contrast, fishes that live near the water's surface sometimes have four or five different opsins, giving them excellent color vision. Not surprisingly, shallow water and surface-dwelling fishes are often very colorful, because light of all wavelengths penetrates shallow water. These fishes have adapted by using coloration for protection (camouflage) or for identification.

## 44.6 Chemoreception

**Learning Outcomes:**

1. **CoreSKILL** » Describe olfaction and gustation in animals, and propose reasons why these senses are advantageous.
2. Explain how olfactory receptors respond to the binding of odor molecules.
3. Outline how receptor cells within taste buds respond to the binding of food molecules.

Chemoreception includes the senses of smell (**olfaction**) and taste (**gustation**), both of which involve detecting chemicals in air, water, or food. These chemicals bind to chemoreceptors, which, in turn, initiate electrical responses in other neurons that send signals to the brain. Amazingly, the binding of a single molecule to a receptor cell can sometimes be perceived as an odor! In terrestrial vertebrates, airborne molecules that bind to olfactory receptors must be small enough to be carried in the air and into the nose. Taste molecules can be larger because they are conveyed in food and liquid.

Taste and smell are closely related. The distinction is largely meaningless for aquatic animals, because for them all chemoreception comes through the water. Even in terrestrial animals, about 80% of the perception of taste is actually due to activation of olfactory receptors. (This is why food loses its flavor when the sense of smell is impaired, such as when you have a cold.) In this section, we will explore chemoreception in invertebrates and vertebrates, focusing on insects and mammals since they provide well-studied examples.

### Olfaction and Taste in Insects Involve Chemoreceptors in Sensory Hairs

Insects rely on odor and taste for finding food and mates. In insects, chemoreceptors are neurons that are located on sensory hairs on the

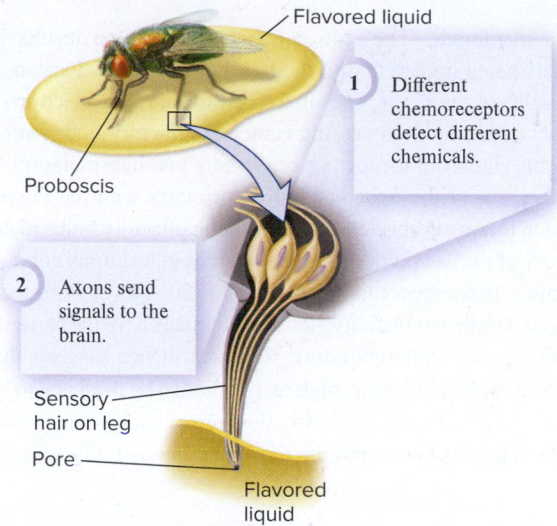

1. Different chemoreceptors detect different chemicals.

2. Axons send signals to the brain.

Flavored liquid

Proboscis

Sensory hair on leg

Pore

Flavored liquid

**(a) Chemoreception in the blowfly**

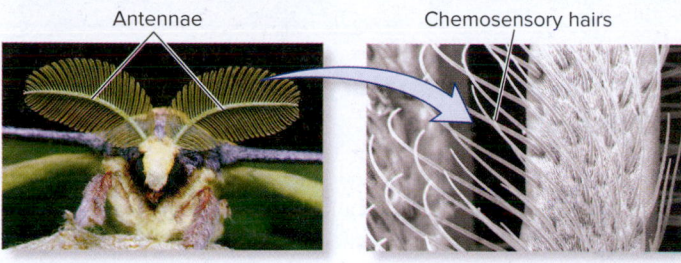

Antennae

Chemosensory hairs

**(b) Chemosensory hairs on antennae of the male moth detect the odor molecules.**

**Figure 44.23  Chemoreception in insects. (a)** Chemosensory cells of the blowfly located on the proboscis, legs, and feet sense different chemicals. **(b)** Chemosensory hairs on a male moth's antennae bind odor molecules secreted by females. b (left): ©Anthony Bannister/Science Source; b (right): Courtesy of Louisa Howard, Dartmouth College

proboscis (coiled tongue), legs, feet, and antennae. Each sensory hair on the proboscis and feet has a pore at the tip through which the substance passes. As one example, **Figure 44.23a** shows a blowfly with four separate chemoreceptors within each hair. Each of these neurons responds to different molecules. Receptors on dendrites of the chemoreceptor cells inside the pore bind to the molecules and initiate a sensory transduction pathway that opens ion channels in the membrane. This depolarizes the plasma membrane of the chemoreceptor cell and generates action potentials, which are sent to the brain for interpretation.

In certain moths, males have elaborate antennae that can sense pheromones, extremely potent signaling molecules given off by a female. The female secretes a sex-attractant pheromone into the air from an abdominal gland. The chemosensory hairs on the male's antennae (**Figure 44.23b**) can detect extremely low concentrations of the pheromone from several kilometers away. This highly sensitive detection system enables the male to locate the female in the dark or at a distance.

## Mammalian Olfactory Receptors Respond to the Binding of Odor Molecules

The olfactory sensitivity of mammals varies widely depending on their supply of olfactory receptor cells, which ranges from 5 or 6 million in

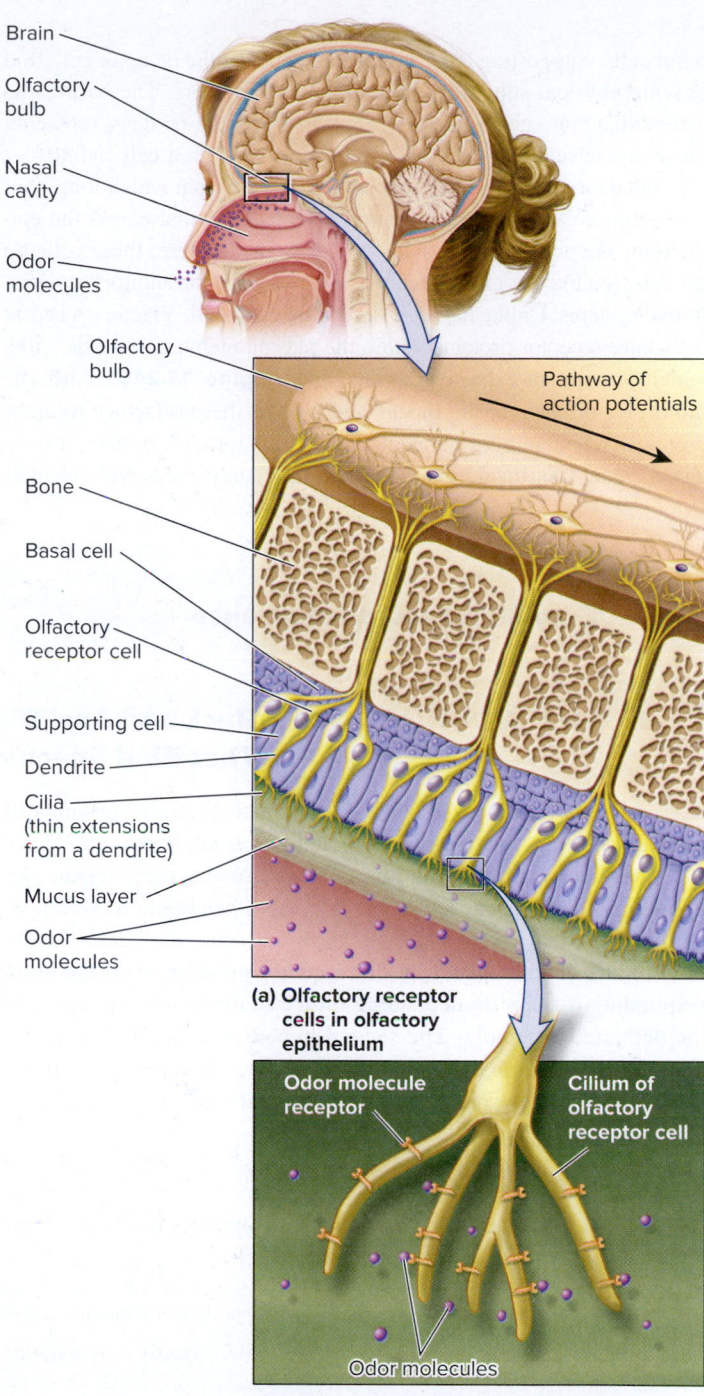

Brain

Olfactory bulb

Nasal cavity

Odor molecules

Olfactory bulb

Pathway of action potentials

Bone

Basal cell

Olfactory receptor cell

Supporting cell

Dendrite

Cilia (thin extensions from a dendrite)

Mucus layer

Odor molecules

**(a) Olfactory receptor cells in olfactory epithelium**

Odor molecule receptor

Cilium of olfactory receptor cell

Odor molecules

**(b) Cilia from dendrite with odor molecule receptors**

**Figure 44.24  Olfactory structures in the human nose.** Odor molecules dissolve in a layer of mucus that coats the olfactory receptor cells. The molecules bind to protein receptors in the membranes of cilia that extend from the olfactory receptor cells. Action potentials in the olfactory receptor cells are conducted to cells in the olfactory bulb, and from there to the brain for interpretation. Basal cells periodically differentiate into new olfactory receptor cells, replacing dead or damaged cells.

humans to 100 million in rabbits and 220 million in dogs. Mammalian olfactory sensory receptors are neurons that are located in the epithelial tissue at the upper part of the nasal cavity (**Figure 44.24a**). These cells are surrounded by two additional cell types: supporting cells and basal cells. Supporting cells are located between the receptor cells and provide physical support for the olfactory receptors. The basal cells differentiate into new olfactory receptors every 30–60 days, replacing those that have died after prolonged exposure of their cell endings.

Olfactory sensory receptors have dendrites from which long, thin extensions called cilia extend into a mucous layer that covers the epithelium. Despite the superficial similarity in structure, these cells do not function like the mechanoreceptor hair cells of the auditory and vestibular systems. Unlike hair-cell stereocilia that bend, olfactory receptor cells have receptor proteins within the plasma membranes of the cilia, which provide an extensive surface area (**Figure 44.24b**). Airborne molecules dissolve in the mucus and bind to these olfactory receptor proteins. When an odor molecule binds to its receptor protein, it initiates a signal transduction pathway that ultimately opens Na$^+$ channels

in the plasma membrane. The subsequent depolarization results in action potentials being transmitted to the next series of cells located in the olfactory bulbs of the brain. The olfactory bulbs are a collection of neurons that act as an initial processing center of olfactory information and relay it to the cerebrum for further processing and interpretation.

The relative size of the olfactory bulbs correlates with the importance of olfaction to a given species. In humans, the olfactory bulbs make up only about 5% of the weight of the brain, whereas in nocturnal animals like rats and mice, they can comprise as much as 20%. Even with their relatively limited olfactory sensitivity, however, humans have the capacity to detect 10,000 or more different odors. Recent evidence suggests that this number could in principle be as high as 1 trillion! The mechanism by which mammals detect so many different odors remained a mystery until 1991, when two scientists uncovered the molecular basis of olfaction.

## 👁 Core Skill: Process of Science

## Feature Investigation | Buck and Axel Discovered a Family of Olfactory Receptor Proteins That Bind Specific Odor Molecules

How does the olfactory system discriminate among thousands of different odors? American neuroscientists Linda Buck and Richard Axel set out to answer this question. When they began, two hypotheses were proposed to explain this phenomenon. One possibility was that many different types of odor molecules might bind to one or just a few types of receptor proteins, with the brain responding differently depending on the number or distribution of the activated receptors. The second hypothesis was that olfactory receptor cells can make many different types of receptor proteins, each type binding a particular odor molecule or group of structurally related odor molecules.

To begin their study, Buck and Axel made the logical assumption that olfactory receptor proteins would be highly expressed in olfactory sensory receptor cells, but not in other parts of the body. Based on previous work, they also postulated that the receptor proteins would be members of the large family of G-protein-coupled receptors (GPCRs). As shown in **Figure 44.25**, they isolated olfactory sensory receptor cells from rats and then used a homogenizer to break open the cells to release their mRNA. The mRNA was purified and then used to make complementary DNA (cDNA) using the enzyme reverse transcriptase. This generated a large collection of cDNAs, representing all of the genes that were expressed in the receptor cells at the time of mRNA collection.

**Figure 44.25** Buck and Axel identified olfactory receptor proteins in olfactory receptor cells.

**HYPOTHESES** 1. Many different types of odor molecules bind to just a few types of receptor proteins. 2. Odor molecules are detected by many specific olfactory receptor proteins belonging to the family of G-protein-coupled receptors (GPCRs).

**KEY MATERIALS** Laboratory rats (*Rattus norvegicus*), PCR reagents, DNA-sequencing gels.

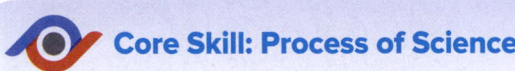

**Experimental level**          **Conceptual level**

1   Dissect and homogenize olfactory epithelium from laboratory rats.

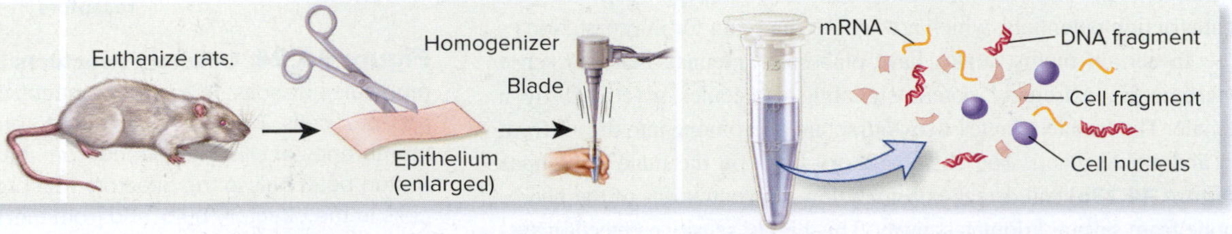

Euthanize rats.

Homogenizer

Blade

Epithelium (enlarged)

mRNA — DNA fragment

Cell fragment

Cell nucleus

2   Purify mRNA. Make cDNA (described in Chapter 21) from the mRNA, using reverse transcriptase.

Add mRNA and reverse transcriptase.

Many double-stranded cDNAs

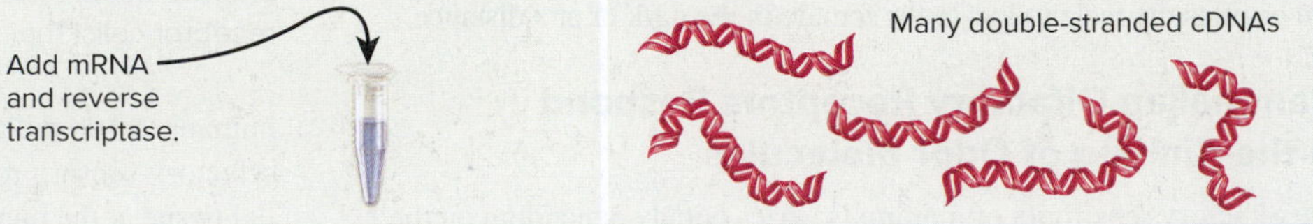

**3** Add primers that bind specifically to genes that encode GPCRs. Subject to PCR as described in Chapter 21.

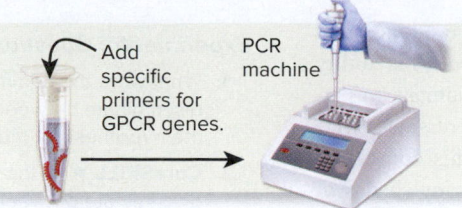

Add specific primers for GPCR genes.

PCR machine

Primers will hybridize only with cDNA that codes for proteins in the GPCR family and amplify those genes. Many different PCR products are obtained, each corresponding to a different gene.

**4** Subject each PCR product to DNA sequencing, also described in Chapter 21.

Output from automated sequencing (example)

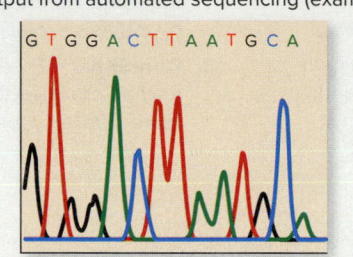

GTGGACTTAATGCA

Different GPCRs will have slightly different DNA sequences.

**5** **THE DATA**

At least 100 different GPCRs were uniquely expressed in olfactory sensory receptor cells. Analysis of their predicted amino acid sequences revealed significant variability in the putative ligand-binding regions (transmembrane domains 3–5).

○ Amino acids that were the same in all the olfactory GPCRs.

● Amino acids that were different among the olfactory GPCRs.

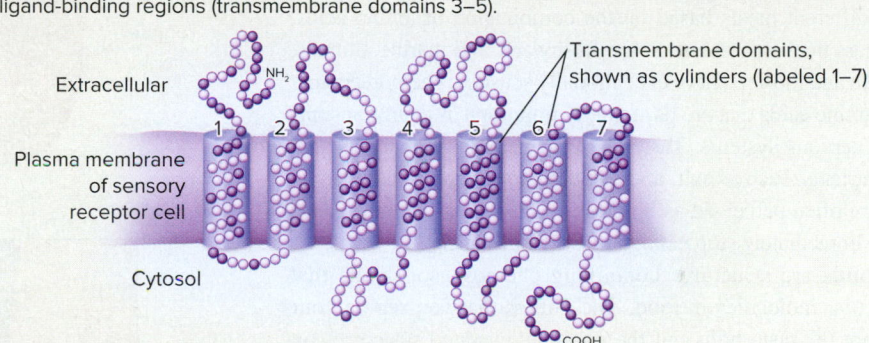

Extracellular

Plasma membrane of sensory receptor cell

Cytosol

Transmembrane domains, shown as cylinders (labeled 1–7)

**6** **CONCLUSION** Olfactory receptor cells express many different receptor proteins that account for an animal's ability to detect a wide variety of odors.

**7** **SOURCE** Buck, L., and Axel, R. 1991. A novel multigene family may encode odorant receptors: A molecular basis for odor recognition. *Cell* 65: 175–187.

To determine if any of these cDNAs encoded GPCRs, the researchers used primers that recognized conserved regions within previously identified genes that encoded GPCRs. A conserved region is a DNA sequence that rarely changes among different members of a gene family. The primers were used in PCR to amplify cDNAs that encoded GPCRs. This produced many PCR products that were then subjected to DNA sequencing.

As shown in the data in step 5 of Figure 44.25, Buck and Axel initially identified at least 100 different genes, each encoding a GPCR with a slightly different amino acid sequence, as predicted from the DNA sequences. Significant variability in the predicted amino acid sequences for the GPCRs was found in the putative ligand-binding region of the molecules (believed to be transmembrane domains 3–5). Further research showed that these genes were expressed only in olfactory cells, and not in other parts of the body. These results were consistent with the second hypothesis, namely, that organisms produce a large number of distinct olfactory receptor proteins, each type binding a particular odor molecule or a group of related odor molecules.

Since these studies, researchers have determined that this family of olfactory genes in mammals is surprisingly large. In humans, roughly 400 genes encode olfactory receptor proteins. The diversity of olfactory receptor proteins is further increased by alternative splicing, which is described in Chapter 14 (refer back to Figure 14.21). Each olfactory receptor cell is thought to express only one type of GPCR that recognizes its own specific odor molecule or group of closely related molecules. Most odors that an animal encounters, however, are due to multiple chemicals that activate many different types of odor receptors at the same time. We perceive odors based on the combination of receptors that become activated. This ability to combine different inputs allows animals such as humans to perceive at least thousands of different odors despite expressing only a few hundred different olfactory receptor proteins. This combining of inputs is similar to the way vertebrates perceive many shades of color despite having only a few types of photoreceptors, as described earlier.

The research of Buck and Axel explained, in part, how animals detect a myriad of odors. In 2004, the two neuroscientists received the Nobel Prize in Physiology or Medicine for this pioneering work.

**Experimental Questions**

1. What were the two major hypotheses to explain how animals discriminate between different odors? How did Buck and Axel test the hypothesis of multiple olfactory receptor proteins?

2. **CoreSKILL** » Of the two hypotheses explaining how animals discriminate between different odors (see question 1), which one was supported by the results of this experiment? With the evidence presented by Buck and Axel, what is the current hypothesis explaining the discrimination of odors in animals?

3. **CoreSKILL** » Predict the survival or reproductive advantages that may be provided to animals by the ability to sense thousands or more different types of odors.

## Taste Buds Detect Food Molecules

Chemical senses are present in all animal phyla, although not all animals have taste-sensing organs. Many animals use their taste (gustation) sense to select appropriate foods. For example, butterflies select nectar based on the sugars found in a particular flower, and carnivorous animals detect the taste of different meats based on the combination of amino acids, fats, and sugars that are present. Some freshwater and marine animals, such as catfish and lobsters, have exceptionally sensitive chemoreceptors for specific amino acids that are particularly important as neurotransmitters in their nervous systems. Taste can also help an animal seek out necessary nutrients, such as salt, and avoid poisonous chemicals. Toxic substances are often perceived as bitter or distasteful, which may cause an animal to immediately stop eating any such substance.

**Taste buds** are structures containing chemosensory cells that detect particular molecules in food. The bumps that you see on your tongue are not the taste buds but the papillae, elevated structures on the tongue that collect food molecules in depressions called taste pores. These pores contain many taste buds with sensory receptor cells (**Figure 44.26**). The sensory receptor cells, along with several supporting cells, form a complex structure that is organized like the wedges of an orange. The tips of the sensory receptor cells have microvilli that extend into the taste pore (another example of an anatomical sensory adaptation that increases surface area). Here, molecules in food that have dissolved in saliva bind to receptor proteins. This binding triggers intracellular signals that alter ion permeability and membrane potentials. The sensory receptor cells then release neurotransmitters onto underlying sensory neurons. Action potentials travel from these neurons to the thalamus and other regions of the forebrain, where the taste is perceived.

There are a number of different types of taste cells, and they are distributed on the tongue in different areas that overlap considerably. Each type of cell has a specific transduction mechanism that allows it to detect specific chemicals present in ingested foods and fluids. Their activation results in the perception of sweet, sour, salty, and bitter tastes. In addition, a recently recognized fifth taste perception called umami (after a Japanese word for delicious) is associated with the presence in ingested food of glutamate and other similar amino acids, and is usually described as making food flavorful. It may account for the widely recognized effects of monosodium glutamate (MSG) in enhancing the flavor of food.

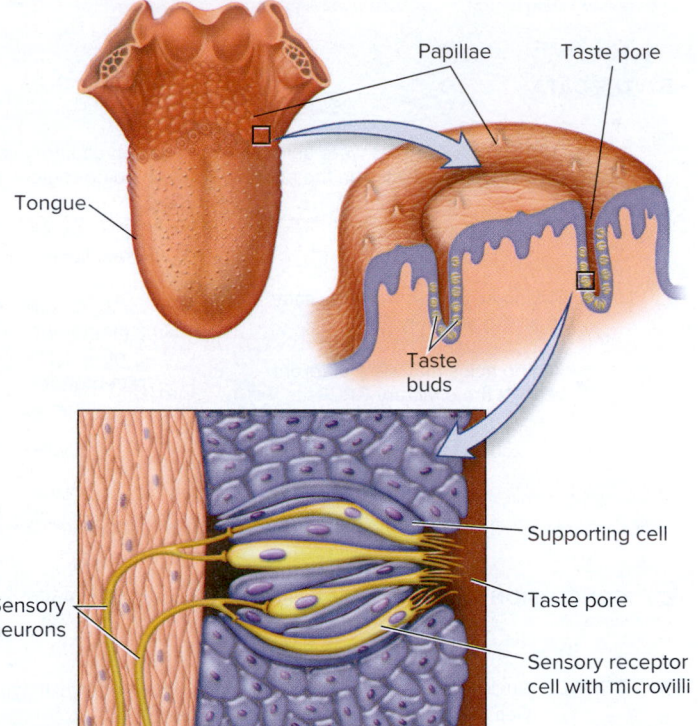

**Figure 44.26 Structures involved in the sense of taste.** This sense occurs in taste buds, which contain the sensory receptor cells that respond to dissolved food molecules.

*Concept Check:* *What are some advantages of having the ability to perceive tastes that are salty, sweet, sour, bitter, and umami?*

The senses of taste and smell are enhanced when we are hungry, a phenomenon that most likely occurs in other animals as well. Once we have eaten, we are less aware of the smell and taste of food. The importance of this for survival is clear: A hungry animal needs to eat. An enhanced sense of smell aids in locating food, and a heightened sense of taste encourages an animal to eat. Afterward, these senses become temporarily dampened so as not to distract an animal from its other needs. The mechanism by which these changes occur is uncertain, but it may involve a temporary alteration in the number of smell and taste receptors, or in their ability to bind ligands.

## Core Skill: Connections

### BIO TIPS

**THE QUESTION** *Look at Figures 44.13d, 44.16, 44.23, 44.24, and 44.26. Despite their many differences, the sensory receptor cells depicted in these figures share at least one important similarity in their general structures. Can you identify that structural similarity?*

**T OPIC** *What topic in biology does this question address?*
The topic is sensory receptor cells in animals. More specifically, the question asks you to find a structural similarity shared by many of these receptors.

**I NFORMATION** *What information do you know based on the question and your understanding of the topic?*
From the question, you know that there is at least one important structural feature that is common to these sensory receptor cells: the insect photoreceptor, the vertebrate photoreceptors, the invertebrate chemosensory cell, the vertebrate olfactory receptor cell, and the vertebrate gustatory sensory receptor cell. From your understanding of the topic, you know that sensory receptor cells are highly specialized structures that have evolved to detect and discriminate among particular types of stimuli.

**P ROBLEM-SOLVING S TRATEGY** *Compare and contrast. Relate structure and function.* To answer this question, look carefully at the structural features of these various sensory receptors, and compare those features to identify a similarity they share.

**ANSWER** *In each of these sensory receptors, structural adaptations are present that greatly increase the total surface area of the cell. These adaptations include microvilli, cilia, and hairs. Despite their differences in overall appearance and in the stimuli to which they respond, many animal sensory receptors have evolved structures with extensive surface areas to optimize their ability to detect environmental signals such as light and chemicals.*

## 44.7 Impact on Public Health

### Learning Outcome:

1. Identify common types of visual and hearing deficits in humans, the causes of each, and their impacts on public health.

Sensory disorders are among the most common neurological problems found in humans and range from mild (needing eyeglasses or a hearing aid) to severe (blindness or deafness). This section presents an overview of a few representative sensory disorders that have a major impact on public health.

### Visual Disorders Include Glaucoma, Macular Degeneration, and Cataracts

Visual disorders affect an enormous number of people. In the U.S. alone, more than 10 million people have severe eye problems that cannot be corrected by eyeglasses, and more than 1 million people are blind (42 million worldwide). The costs to the U.S. economy associated with blindness and severe visual loss amount to nearly $4 billion yearly. Although vision loss has many causes, three disorders account for over half of all cases: glaucoma, macular degeneration, and cataracts.

*Glaucoma* Normally, the fluid that makes up the aqueous humor in the eye is produced and reabsorbed (drained) in a circulation that keeps the fluid level nearly constant. In **glaucoma**, drainage of aqueous humor becomes blocked, and the pressure inside the eye increases as the fluid level rises. If untreated, this eventually damages cells in the retina and leads to irreversible loss of vision (**Figure 44.27a**). The cause of glaucoma is not always known, but in some cases it is due to trauma or severe infection of the eye, chronic use of certain medicines, blood vessel disease of the eyes, or diseases such as diabetes. Some forms of glaucoma may have a hereditary component. Glaucoma can be treated with eye drops that contain a drug that decreases fluid production in the eye, or with laser surgery to reshape and improve the drainage structures in the eye. The American Foundation for the Blind estimates that 400,000 new cases of glaucoma are diagnosed each year in the U.S., and up to 10 million individuals have elevated eye pressure. Glaucoma accounts for roughly 10% of all cases of blindness in the U.S.

*Macular Degeneration* In **macular degeneration**, photoreceptor cells in and around the macula (the region that contains the fovea) are lost. Because cones are most densely packed in this region, this condition is associated with loss of sharpness and color vision (**Figure 44.27b**). Although macular degeneration does not usually occur in people under 60 years old, it can occur at any age. Risk factors include heredity, smoking, hypertension, elevated blood cholesterol, and obesity. Macular degeneration is the leading cause of blindness in the U.S., accounting for roughly 25% of all cases, but its causes remain obscure. There is currently no cure, but some treatment options may slow the progression of the disease.

*Cataracts* **Cataracts**, regions of accumulation of protein in the lens, cloud the lens and cause blurring, poor night vision, and difficulty focusing on nearby objects (**Figure 44.27c**). By age 65, as many as 50% of individuals have one or more cataracts in either eye, and this jumps to 70% by age 75. Many cataracts are small enough not to affect vision. Many people do not even realize they have cataracts until they undergo an eye exam. The causes of cataracts are not all known but include trauma, chronic use of certain medicinal drugs such as adrenal steroids, hypertension, diabetes, and heredity. Cataracts are also associated with alcohol abuse, excessive exposure to UV radiation from the sun, and smoking. The treatment, when needed, is usually to have the affected lens surgically removed and replaced with an artificial lens. Nearly 1.4 million such surgeries are performed each year in the U.S. Without the natural lens, however, which normally helps protect the retina by absorbing some of the high-energy UV light from the Sun, the retina must be protected during the day by wearing dark sunglasses.

(a) How the world is seen by a person with glaucoma

(b) How the world is seen by a person with macular degeneration

(c) How the world is seen by a person with cataracts

**Figure 44.27** **Approximation of the appearance of the visual field in a person with advanced (a) glaucoma, (b) macular degeneration, and (c) cataracts.** a–c: Source: National Eye Institute, National Institutes of Health

 **Core Skill: Science and Society** As researchers learn more and more about the risk factors and causes of different eye diseases, treatments for these diseases are becoming more effective. For example, many people with glaucoma can now be treated with a simple laser procedure that requires only a few minutes and results in little or no discomfort. Such individuals may be effectively cured of the disease for long periods.

## Damaged Hair Cells Within the Cochlea Can Cause Deafness

**Deafness** (partial or complete hearing loss) is usually caused by damage to the hair cells within the cochlea, although some cases result from functional problems in brain areas that process sound or in the nerves that carry information from the hair cells to the brain. When the hair cells are damaged, noises have to be louder to be detected, and an affected person may require the use of hearing aids to amplify incoming sounds.

Hearing loss may be mild or severe and may result from many causes, including injury to the ear or head, hereditary defects of the inner ear, and exposure to certain diseases (for example, rubella) or toxins during fetal life. By far, however, the most significant cause of hearing loss is repeated, long-term exposure to loud noise.

The amplitude of a sound wave (that is, the distance between the peak and the trough of the wave) determines its loudness. As the amplitude of a sound wave gets larger, the perceived loudness of the sound increases. The loudness of sound is measured on a logarithmic scale with units of decibels (dB; technically, decibels measure the intensity of the pressure of the sound wave, which we perceive as loudness). Normal conversation is about 60 dB, a chainsaw is 108 dB, and a jet plane taking off can reach 150 dB. Estimates of noise-induced hearing loss, which usually results from job-related activities, range from 7% to 21% of all cases of hearing loss worldwide; this makes it the leading occupational disorder in the U.S. It is, therefore, one of the most significant disabilities in the U.S. in terms of numbers of people affected and costs to society. Scientists estimate that nearly 40 million people are exposed daily to dangerous noise levels in the U.S. as a result of their occupation.

Recent research has led to a better understanding of the mechanism by which noise impairs hearing. Chronic exposure to loud sounds appears to produce a state of metabolic exhaustion in the hair cells of the cochlea. As a result, the cells become fatigued and are unable to maintain normal biochemical processes. One consequence of this is a buildup of free radicals. These chemical species oxidize lipids in cellular membranes, damaging the membranes in the process. Mitochondrial membranes appear to be particularly susceptible to free radicals. Once mitochondria are destroyed, a cell's ability to produce the ATP needed to fulfill its energy demands is compromised, and the cell dies. As hair cells die, the ear becomes less sensitive to sound.

Researchers are investigating drugs that might prevent the formation of free radicals in the cells of the ear, but such drugs are not yet available. If a cochlea is severely damaged, it can be surgically replaced with an artificial cochlear implant. In response to sound waves, these devices generate electrical signals that can stimulate the auditory nerve, which communicates with the brain. Cochlear implants cannot restore hearing to normal, but they can make it possible to hear conversations.

## Summary of Key Concepts

### 44.1 An Introduction to Sensation

- Sensory transduction is the process by which incoming stimuli are converted to neural (electrical) signals. Perception is an awareness of sensations that are experienced. Sensory receptors are either neurons or specialized epithelial cells that respond to stimuli and begin the process of sensory transduction. A sensory receptor often responds to a stimulus by eliciting a graded response proportional to the intensity of the stimulus (Figures 44.1, 44.2).

- Sensory receptors can be divided into the general classes of mechanoreceptors, thermoreceptors, nociceptors, electromagnetic receptors (including photoreceptors), and chemoreceptors.

## 44.2 Mechanoreception

- Mechanoreceptors respond to physical stimuli such as touch, pressure, stretch, movement, and sound. In mammals, several types of receptors in the skin detect stimuli such as touch and pressure, including Meissner corpuscles, Ruffini corpuscles, and Pacinian corpuscles (Figure 44.3).

- Hair cells have projections called stereocilia that respond to movements of fluid or other stimuli and release neurotransmitter that may result in action potentials in an adjacent sensory neuron (Figure 44.4).

- Audition (hearing) is the ability to sense sound waves and is well developed in vertebrates. The mammalian ear has three main compartments: the outer, middle, and inner ears (Figure 44.5).

- Sound waves move through outer ear, tympanic membrane, malleus, incus, and stapes to the cochlea of the ear, where they cause the basilar membrane to vibrate. This vibration causes the stereocilia of overlying hair cells to bend, which elicits action potentials that are sent to the brain, leading to the perception of sound (Figures 44.6, 44.7).

- Many adaptations for hearing improve animals' ability to sense sounds in diverse environments.

- Statocysts in certain invertebrates allow these animals to sense body position. The vestibular system in vertebrates allows animals to sense linear and rotational movement (Figures 44.8, 44.9).

- The lateral line system of fishes consists of hair cells that detect water movements (Figure 44.10).

## 44.3 Thermoreception and Nociception

- Thermoreceptors in the skin and brain allow an animal to sense external and internal temperatures, respectively. Nociceptors in the skin and internal organs sense pain, an important survival adaptation in animals.

## 44.4 Electromagnetic Reception

- Some animals can detect electrical and magnetic stimuli; in some cases, this ability is used to locate prey (Figure 44.11).

## 44.5 Photoreception

- The eyecup in flatworms is a simple eye that detects light and its direction but does not form an image. The compound eye found in many invertebrates consists of many ommatidia that focus light (Figures 44.12, 44.13).

- The single-lens eye is found in vertebrates and certain invertebrates. In such an eye, objects in the animal's field of view emit or reflect patterns of light that are transmitted through the pupil and the lens to a layer of photoreceptors called the retina. Activation of photoreceptors triggers electrical changes in neurons that pass out of the eye through the optic nerves, carrying the signals to the brain, which then interprets the incoming signals to form a visual image. Stretching and flattening of the lens aids in focusing on objects at varying distances (Figure 44.14).

- Rods and cones are photoreceptors found in the vertebrate eye. Rods do not discriminate colors; cones can detect color. The visual pigment in rods and cones consists of retinal and one of several proteins called opsin (Figures 44.15, 44.16, 44.17).

- Red-green color blindness is due to a defect in a type of opsin found in cone pigments (Figure 44.18).

- Absorption of photons by retinal triggers a signal transduction pathway that causes hyperpolarization of the cell leading to a visual image (Figures 44.19, 44.20).

- The retina is composed of layers of cells, including photoreceptors, that receive light input and convey electrical signals to the visual centers of the brain via the optic nerve (Figure 44.21).

- The vertebrate eye has several adaptations that aid in vision for particular animals in different environments (Figure 44.22).

## 44.6 Chemoreception

- Invertebrates have taste receptors on their proboscis, legs, feet, and antennae. Some insects can detect pheromones (Figure 44.23).

- In mammals, olfactory receptors are neurons located in epithelial tissue at the upper part of the nasal cavity. They contain cilia with protein receptors that bind specific odor molecules. The binding initiates a signal transduction pathway that ultimately depolarizes the plasma membrane and transmits action potentials to the olfactory bulb of the brain. Buck and Axel discovered that olfactory sensory receptors have many different types of protein receptors for odor molecules (Figures 44.24, 44.25).

- Taste buds contain chemosensory cells that detect molecules in food. There are different types of taste cells, the activation of which results in the perception of five tastes: sweet, sour, salty, bitter, and umami (Figure 44.26).

## 44.7 Impact on Public Health

- Common visual disorders include glaucoma, macular degeneration, and cataracts. Prolonged exposure to intense sound can damage hair cells within the cochlea and lead to deafness (Figure 44.27).

## Assess & Discuss

### Test Yourself

1. The process in which incoming sensory stimulation is converted to electrical signals in neurons is called
   a. an action potential.
   b. sensory reception.
   c. perception.
   d. sensory transduction.
   e. perceptual transduction.

2. Vibrations of the _____, located within the _____, bend stereocilia in the hair cells of the mammalian ear and activate sensory neurons.
   a. tympanic membrane, cochlear duct
   b. round window, middle ear
   c. basilar membrane, organ of Corti
   d. ossicle, outer ear
   e. ampulla, semicircular canals

3. _____ sense pain; _____ sense heat or cold; and _____ sense touch.
   a. Mechanoreceptors; thermoreceptors; nociceptors
   b. Nociceptors; thermoreceptors; mechanoreceptors
   c. Nociceptors; thermoreceptors; stretch receptors
   d. Mechanoreceptors; nociceptors; stretch receptors
   e. Nociceptors; photoreceptors; mechanoreceptors

4. Statocysts are sensory organs for
   a. hearing found in many invertebrates.
   b. equilibrium found in mammals.
   c. equilibrium found in many invertebrates.
   d. water current changes found in fish.
   e. hearing found in vertebrates.

5. In which process(es) are hair cells involved?
   a. balance in vertebrates and invertebrates
   b. hearing in mammals
   c. vision in animals with compound eyes
   d. heat sensing in pit vipers
   e. both a and b

6. Which statement about compound eyes is *true*?
   a. They do not contain a lens or lenslike structure.
   b. They cannot sense color.
   c. They have one ommatidium per eye.
   d. They are found in insects and also many vertebrates.
   e. They probably have less resolving power than single-lens eyes.

7. In the mammalian eye, light from near or far objects is focused on the retina when
   a. the lens moves forward or backward.
   b. the lens rounds up or flattens.
   c. the eyeball changes shape.
   d. the cornea changes shape.
   e. the retina changes shape.

8. The amount of glutamate released from photoreceptors of the vertebrate eye is highest when
   a. an animal is in full sunlight.
   b. an animal is in a completely dark place.
   c. an animal is in a dimly lit place.
   d. $Na^+$ channels of the photoreceptors are closed.
   e. Both a and d are true.

9. Cone pigments discriminate between different wavelengths of light due to
   a. their location in the retina.
   b. the amount of light they absorb.
   c. the type of retinal they have.
   d. the type of opsin protein they have.
   e. interactions with bipolar cells.

10. The stimulation for olfaction involves odorant molecules
    a. bending the cilia of olfactory sensory receptor cells.
    b. binding to protein receptors of olfactory sensory receptor cells.
    c. entering the cytoplasm of olfactory sensory receptor cells.
    d. opening $K^+$ channels of olfactory sensory receptor cells.
    e. binding to cells located in the olfactory bulbs.

## Conceptual Questions

1. Distinguish between sensory transduction and perception.

2. Despite the differences in their appearances, can you identify structural and functional similarities in the compound and single-lens eyes found in animals?

3. **Core Concept: Systems** Which sense do you think is the least important to human survival?

## Collaborative Questions

1. Discuss the several types of sensory stimuli that animals can detect, and describe some general features of the sensory receptors that are adapted for each of these stimuli.

2. Explain how the structures of the mammalian ear are adapted to detect and distinguish different frequencies of sound.

# Muscular-Skeletal Systems and Locomotion

# 45

Red-eyed tree frog (*Agalychnis callidryas*) using its muscular-skeletal system to leap from one plant to another. ©Scott Linstead/Science Source

In 1991, Olympic athlete Mike Powell established a long-jump record of 8.95 m, roughly 5 times the height of a person. A jump of nearly 9 m sounds impressive, and it is for us, but how do humans compare with other animals? Red kangaroos can hop up to 12 m. Some tree frogs, like the one shown in the chapter opening photo, can leap distances up to 1.4 m without a running start, yet the frog is only 4.5 cm long and weighs only 8 g! This is a sensational leap for an animal that size, roughly 30 times its body length. The jumping abilities of frogs are related to their relatively large leg muscle mass, the elongation of the bones in their legs and feet, and elastic elements in the connective tissue associated with their muscle and bone. Indeed, the jumping ability of frogs makes them excellent model organisms for the study of muscle function in vertebrates.

Muscles are composed of highly specialized cells that have the ability to contract in response to stimuli. The three types of muscle tissue—skeletal, smooth, and cardiac—were introduced in Chapter 41. This chapter will explore the structure and function of skeletal muscle and the mechanism by which it controls movements, such as those required for **locomotion**, the movement of an animal from place to place.

For skeletal muscles to produce locomotion, they must exert a force on an animal's skeleton. We begin the chapter, therefore, with an overview of animal skeletons. Then we will examine the structure and function of skeletal muscle and see how the interaction of two muscle proteins, actin and myosin, produces muscle contraction. Next, we will consider various modes of locomotion in animals. We will conclude with a consideration of important bone and muscle diseases in humans.

## 45.1  Types of Animal Skeletons

**Learning Outcomes:**

1. Distinguish between exoskeletons and endoskeletons.
2. **CoreSKILL »** Relate the structure of an exoskeleton to its function, and propose reasons why such a skeleton provides advantages to an animal.

3. List the major functions of the vertebrate skeleton.
4. Describe the composition of vertebrate bone.

When we think of the word skeleton, an image of the vertebrate system of bones usually comes to mind. However, invertebrates possess a skeleton as well, although it is not made of bone. Therefore, a broader definition of a **skeleton** is a structure that serves one or more functions related to support, protection, and locomotion. Using this definition, the two major types of skeletons found in animals are exoskeletons and endoskeletons. A third type of skeleton that broadly fits the definition is called a hydroskeleton (which will not be discussed in this chapter). This type is found in some soft-bodied invertebrates that use water pressure to propel their bodies.

### Exoskeletons Are on the Outside of Animal Bodies

Arthropods have an **exoskeleton**, an external skeleton that surrounds and protects most of the body surface (**Figure 45.1a**). Exoskeletons provide support for the body, protection from the environment and predators, and protection for internal organs. The arthropod skeleton is made of a polysaccharide called chitin, and in crustaceans such as lobsters and shrimp, it is sometimes strengthened with calcium and other minerals. Exoskeletons are often tough, durable, and

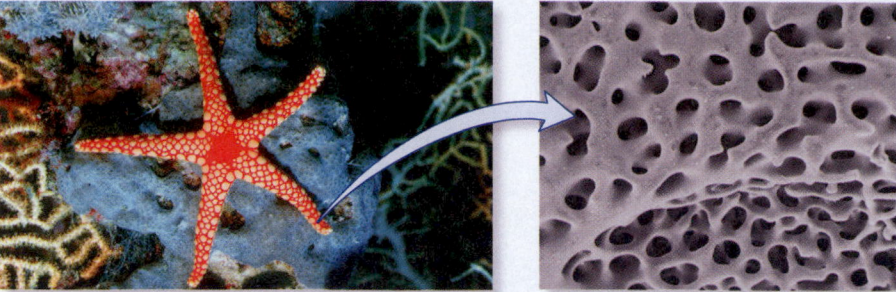

(a) An arthropod next to its recently shed exoskeleton

(b) Echinoderm (sea star) and (right) SEM of its endoskeleton

**Figure 45.1  Types of skeletons. (a)** Exoskeleton. An arthropod's skeleton covers and protects its body, but it must be periodically shed and replaced to allow growth of the animal, as shown in this photo. **(b)** Endoskeleton. Echinoderms such as the sea star (starfish) have endoskeletons of bony plates made of calcium carbonate ($CaCO_3$), shown in detail in the SEM. Vertebrate endoskeletons will be described later. a: ©Tom McHugh/Science Source; b (left): ©Georgette Douwma/Science Source; b (right): ©The Natural History Museum, London

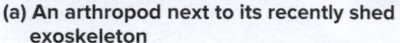
**Concept Check:** *Unlike exoskeletons, endoskeletons do not provide protection for the body surface of animals. How can the lack of an external skeleton be an advantage for animals?*

segmented to allow for flexibility and movement. However, to allow growth, they must be periodically shed, regrown, and strengthened again, a process called ecdysis, or molting (see Figure 45.1a). A disadvantage of exoskeletons is that when an animal is molting, its new exoskeleton is temporarily soft, making the animal more vulnerable to predators and the environment.

Exoskeletons vary enormously in their complexity, thickness, and durability. The differences in exoskeletons are usually adaptations that enhance an animal's survival. Think, for example, of the difference between the body surfaces of a butterfly and a lobster. A butterfly's exoskeleton must be light enough for the animal to fly, whereas the thick, tough exoskeleton of the lobster provides a very effective defense against predators. Exoskeletons may seem primitive compared to the endoskeletons of vertebrates (discussed next), particularly because of the requirement for molting. Even so, arthropods are among the most successful of all animal phyla living today, having survived for hundreds of millions of years and inhabiting nearly every possible ecological niche on the planet. Clearly, exoskeletons have been advantageous for one of the planet's greatest success stories.

## Endoskeletons Are Internal Support Structures

Like exoskeletons, **endoskeletons** provide support and protection. Unlike exoskeletons, however, endoskeletons are internal structures and do not protect the body surface. Some endoskeletons do, however, protect internal organs such as those in the thorax of vertebrates.

Various types of endoskeletons are found in some species of sponges and all echinoderms and vertebrates. Minerals including $Ca^{2+}$, $Mg^{2+}$, $PO_4^{2-}$, and $CO_3^{2-}$ supply the hardening material that gives an endoskeleton its firm structure. The endoskeletons of sponges consist of spiky networks of proteins and minerals, while those of echinoderms are made up of mineralized platelike structures. Beneath the body surface of echinoderms, for example, arrays of mineralized plates made largely of $CaCO_3$ extend into the spines and arms that radiate from the main body (**Figure 45.1b**).

Vertebrate endoskeletons, by contrast, are composed of either cartilage (refer back to Figure 41.5), as in cartilaginous fishes (sharks,

rays, and skates), or both cartilage and bone, as in bony fishes, amphibians, reptiles, birds, and mammals.

## Bone Consists of a Mixture of Organic and Mineral Components

**Bone** is a living, dynamic tissue with both organic and mineral components. Organic materials include cells that form bone, called osteoblasts and osteocytes, and cells that break it down, called osteoclasts. The organic part of bone is secreted by osteoblasts and osteocytes and consists largely of the protein collagen, which has a unique triple helical structure that gives bone both strength and flexibility (refer back to Figure 10.2). The mineral part of bone is composed of a crystalline mixture of primarily $Ca^{2+}$ and $PO_4^{2-}$ and other ions that provide rigidity. These ions must be obtained in an animal's diet, absorbed into the blood, and deposited in bone.

A proper proportion of organic and mineral components is required for normal bone function. Bone lacking sufficient mineral, for example, is easily fractured. Bone is formed at high rates during an animal's growth periods, but even in adulthood bone is continuously formed, broken down, and re-formed. The skeleton is continually changing—the one in your body right now is completely remodeled from the one that was in your body a few years ago. Similarly, the skeleton you will have a few years from now will be different from the one you have today.

## The Vertebrate Skeleton Performs Several Important Functions

In the vertebrate skeleton, bones are connected in ways that allow for support, protection of internal structures, and movement. The vertebrate skeleton is often considered in two parts: the axial and appendicular skeletons (**Figure 45.2**). The axial skeleton is composed of the bones that form the main longitudinal axis of an animal's body, including the skull, vertebrae, sternum, and ribs. The appendicular skeleton consists of the limb bones and the bones that connect them to the axial skeleton. A **joint** is formed where two or more bones come together. Some joints permit free movement (for example, the joints in human shoulders).

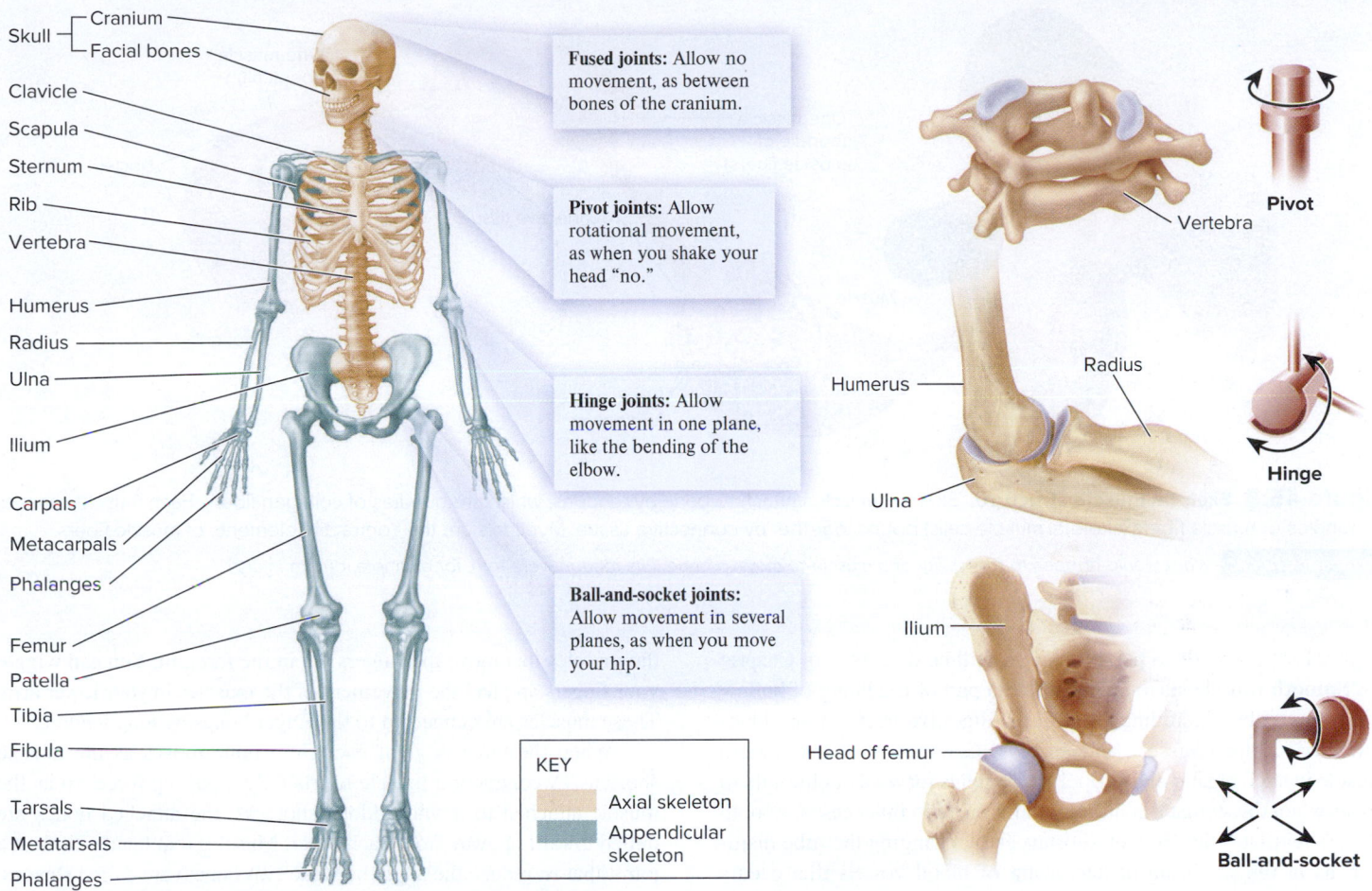

**Figure 45.2  The adult human skeleton.** This diagram shows the axial (beige) and appendicular (green) parts of the skeleton in an adult human, an animal with an endoskeleton. The adult human skeleton consists of 206 separate bones. Three examples of movable joints—pivot, hinge, and ball-and-socket—are shown.

Figure 45.2 illustrates three types of joints that allow different types of movements: pivot, hinge, and ball-and-socket joints. Other joints do not allow movement (fused joints like those interlocking the skull bones) or allow only limited movement (such as those of the vertebral column).

The skeleton of vertebrates serves several other functions in addition to support, protection, and movement. For example, blood cells and platelets, the latter of which help blood to clot (see Chapter 48), are formed within the soft, fatty interior (called the marrow) of certain bones including the ilia, the vertebrae, and the ends of the femurs. In addition, homeostasis of $Ca^{2+}$ and $PO_4^{2-}$ levels in the blood is achieved in large part through exchanges of these ions between bone and blood. For example, if dietary intake of $Ca^{2+}$ is low, $Ca^{2+}$ is removed from bone and added to the blood, so that all of the vital cellular activities that depend on $Ca^{2+}$, such as neuron signaling and muscle contraction, can continue to function normally. If dietary $Ca^{2+}$ is restored to normal, any available excess $Ca^{2+}$ is redeposited in bone. This $Ca^{2+}$ cycling is under the control of hormones such as parathyroid hormone produced by the parathyroid glands (see Chapter 50). About 99% of all the $Ca^{2+}$ in a typical vertebrate's body exists in bone. This represents a huge reservoir of $Ca^{2+}$ for the blood.

Bones cannot move by themselves but instead provide the scaffold on which skeletal muscles cause body movement. We turn now to skeletal muscle and the mechanism by which it generates force.

## 45.2  Skeletal Muscle Structure and the Mechanism of Force Generation

**Learning Outcomes:**

1. List the three types of muscle tissue found in vertebrates, and describe where they are found in the body.
2. Describe how antagonistic muscles function at a joint.
3. Identify the structural components of a muscle down to the level of the sarcomere.
4. Explain the sliding filament mechanism of muscle contraction.
5. Explain how tropomyosin and troponin help regulate muscle contraction.
6. **CoreSKILL »** Predict how skeletal muscle contraction will be affected by changes in electrical activity in motor neurons.
7. Describe the structural features of a neuromuscular junction and the role of acetylcholine in mediating skeletal muscle excitation.

Vertebrates have three types of muscle tissue that are classified according to their structure, function, and control mechanisms. **Cardiac muscle** is found only in the heart and provides the force

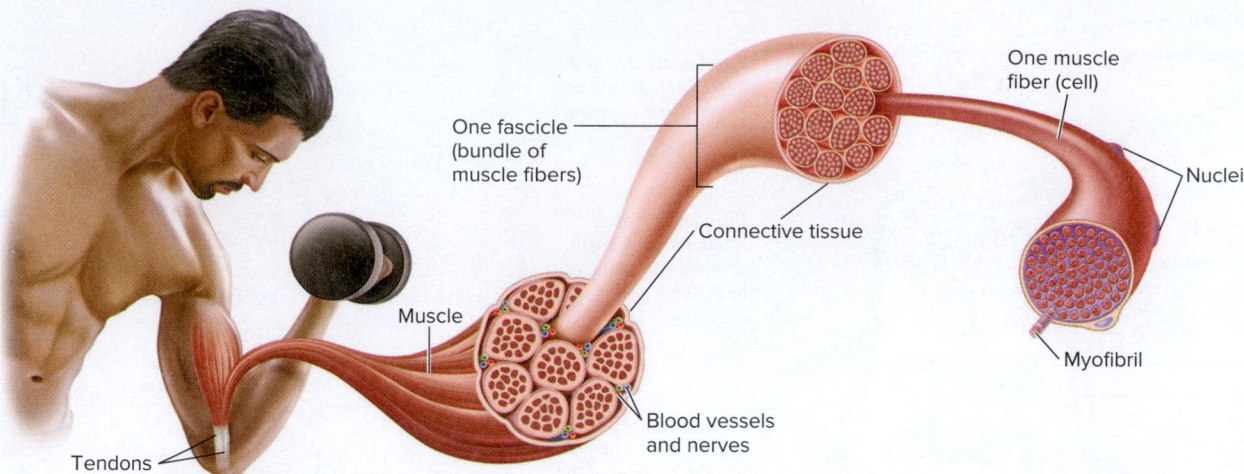

**Figure 45.3** **Skeletal muscle structure.** Skeletal muscles attach to bone by tendons, which are bundles of collagen fibers. Each muscle consists of bundles of muscle fibers (skeletal muscle cells) bound together by connective tissue. Myofibrils are the contractile elements of muscle fibers.

**Concept Check:** *What would happen to the ability of a muscle to move a bone if its tendon were torn, for example, due to injury?*

required for the heart to pump blood; it will be discussed in Chapter 48. **Smooth muscle** surrounds and forms part of the lining of hollow organs and tubes, including those of the digestive tract, urinary bladder, uterus, blood vessels, and airways. Contraction of the smooth muscle in such organs may propel the contents forward or churn them up, as when the stomach contracts after a meal. In other cases, smooth muscle regulates the flow of substances by changing the tube diameter, as in the widening or narrowing of blood vessels that occurs when different parts of the body require more or less nutrients and oxygen. Smooth muscle contraction is not under voluntary control. Instead, it is controlled by the autonomic nervous system, hormones, and local chemical signals.

**Skeletal muscle** is found throughout the body and is directly involved in locomotion. In vertebrates, but not invertebrates, skeletal muscle is electrically excitable—it can generate action potentials in response to a stimulus (invertebrate skeletal muscle cells have graded membrane potentials but do not have action potentials). The action potentials of vertebrate skeletal muscle cells result in an increased concentration of $Ca^{2+}$ in the cytosol, which triggers force generation. Before seeing how this is possible, however, let's begin with an overview of skeletal muscle structure and function.

## A Skeletal Muscle Is a Contractile Organ That Supports and Moves Bones

A skeletal muscle such as the bicep is an organ comprised of cells—called **muscle fibers**—bound together in bundles (called fascicles) by a succession of connective tissue layers (**Figure 45.3**). Skeletal muscles are usually linked to bones by bundles of collagen fibers known as tendons. The transmission of force from contracting muscle to bone can be likened to a number of people pulling on a rope attached to a heavy object. Each person corresponds to a single muscle fiber, the rope corresponds to the tendons, and the bone is the heavy object.

Some tendons are very long, with the site of their attachment to bone far removed from the end of the muscle. For example, some of

the muscles that move the fingers are in the forearm. You can wiggle your fingers and feel the movement of the muscles in your lower arm. These muscles are connected to the finger bones by long tendons.

When the force is great enough, a bone moves as the muscle shortens. A contracting muscle exerts only a pulling force, so as the muscle attached to it via tendons shortens, the attached bones are pulled toward or away from each other. Muscles that bend a limb at a joint (that is, reduce the angle between two bones) are called **flexors**, whereas muscles that straighten a limb (increase the angle between two bones) are called **extensors**. Two or more muscles that produce oppositely directed movements at a joint are known as **antagonists**. For example, in **Figure 45.4**, we can see that contraction of the hamstrings flexes the leg at the knee joint, relaxing the quadriceps, whereas contraction of the quadriceps causes the leg to extend and the hamstrings to relax. Both antagonistic muscles exert only a pulling force when they contract; where they connect to the shin bones determines whether the leg is flexed or extended.

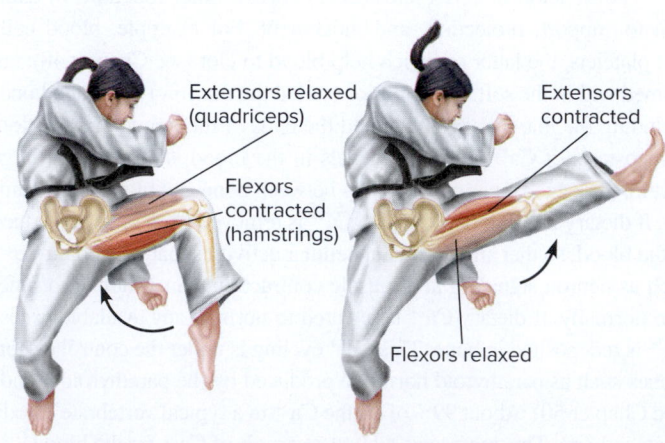

**Figure 45.4** **Actions of flexors and extensors.** The figure shows how skeletal muscles cause flexion or extension of a limb. When the flexor muscle contracts, the extensor relaxes, and vice versa.

## Muscle Fibers Contain Myofibrils Composed of Arrays of Filaments

Each skeletal muscle fiber arises from several cells that fuse to form a single mature cell with multiple nuclei (see Figure 45.3). Each fiber contains numerous cylindrical bundles known as **myofibrils** (**Figure 45.5**). Myofibrils extend from one end of the fiber to the other and are linked to the tendons at the ends of the fiber. Each myofibril contains numerous functional structures called thick and thin filaments. These filaments are arranged in a repeating pattern running the length of the myofibril. One complete unit of this repeating pattern is known as a **sarcomere** (from the Greek *sarco*, meaning muscle, and *mer*, meaning part). The **thick filaments** are composed almost entirely of the protein **myosin**. Myosin is a motor protein that hydrolyzes ATP as a source of energy. The **thin filaments**, which are about half the diameter of the thick filaments, contain the cytoskeletal protein **actin** and associated proteins. Because the arrangement of thick and thin filaments looks like a series of light and dark bands when viewed by a microscope, skeletal muscle is also known as striated muscle.

The components of a sarcomere, shown in Figure 45.5, have the following features:

- The A band is formed by the thick filaments located in the middle of each sarcomere, where their orderly parallel arrangement produces a wide, dark band. A portion of the thin filaments overlaps the thick filaments in this band.

- The Z line is a network of proteins to which thin filaments are attached. Two successive Z lines define the boundaries of one sarcomere.

- The I band lies between the A bands of two adjacent sarcomeres. Each I band contains those portions of the thin filaments that do not overlap the thick filaments, and each I band is bisected by a Z line.

- The H zone is a narrow region in the center of the A band. It corresponds to the space between the two sets of thin filaments in each sarcomere.

- The M line is in the center of the H zone and is composed of proteins that link the central regions of adjacent thick filaments.

- The spaces between overlapping thick and thin filaments are bridged by projections known as **cross-bridges**, which are regions of myosin molecules that extend from the surface of the thick filaments toward the thin filaments (see Figure 45.5).

## Skeletal Muscle Shortens When Thin Filaments Slide Past Thick Filaments

Movement requires shortening of muscles to pull against attached tendons and bones. However, a muscle can generate force, or contract, without producing movement. Holding a heavy weight at a constant position, for example, requires muscle contraction, but not muscle shortening. Therefore, as used in muscle physiology, the term contraction refers to activation of the cross-bridges within muscle fibers, which initiates the generation of force. When the activating

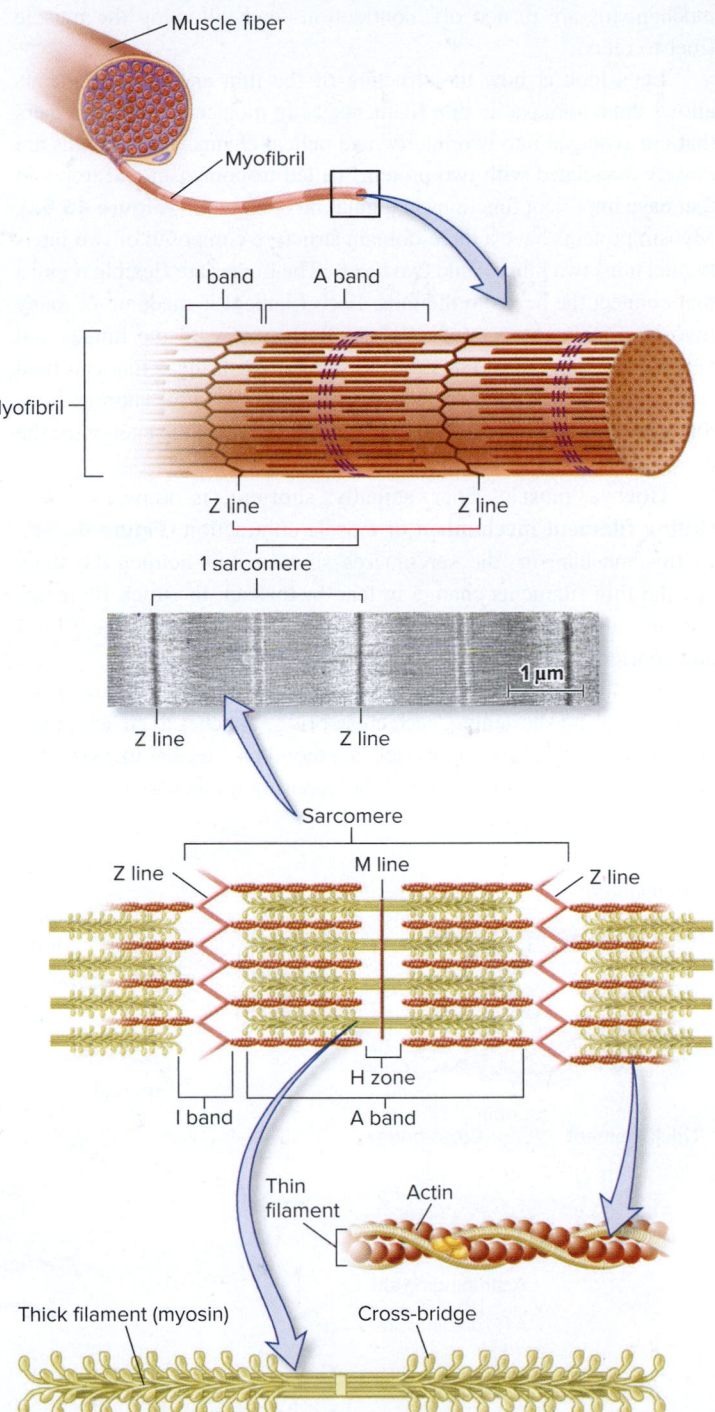

**Figure 45.5 Myofibril structure.** Each muscle fiber consists of numerous myofibrils containing thick and thin filaments. Their arrangement produces a striated banding pattern that can be seen with microscopy (in the inset). A myofibril is composed of repeating units called sarcomeres. ©Dr. H. E. Huxley

 **Core Concept: Structure and Function** A muscle (an organ) consists of many cells called fibers arranged in parallel. As the cells change their shape by shortening, the shape of the entire organ also changes; that is, the muscle contracts.

mechanisms are turned off, contractions end, allowing the muscle fiber to relax.

Let's look at how the structure of the thin and thick filaments allows them to move. In thin filaments, actin molecules form polymers that are arranged into two intertwined helical chains. These chains are closely associated with two proteins called tropomyosin and troponin that have important functions in regulating contraction (**Figure 45.6a**). Myosin proteins have a three-domain structure composed of two intertwined tails, two hinges, and two heads. The hinges are flexible regions that connect the heads to the tails. Each filament is made up of many myosin proteins associated in a parallel array, with the hinges and heads extending out to the sides, forming cross-bridges that can bind to actin. Each head contains two binding sites—one for actin and one for ATP. The hydrolysis of ATP by myosin provides the energy for the cross-bridge to move via a bending motion at the hinge.

How a muscle fiber actually shortens is known as the **sliding filament mechanism** of muscle contraction (**Figure 45.6b**). In this mechanism, the sarcomeres shorten, but neither the thick nor the thin filaments change in length. Instead, the thick filaments remain stationary while the thin filaments slide, pulling on the Z lines and shortening the sarcomere.

The sliding filament movement is propelled by the myosin cross-bridges. During shortening, each cross-bridge attaches to an actin molecule in a thin filament and moves in a motion somewhat like your fist bending at your wrist. Because of the opposing orientation of the thick filaments, the movement of a cross-bridge forces the thin filaments toward the center of the sarcomere (the M line in the H zone), thereby narrowing the H zone and shortening the sarcomere (see Figure 45.6b). One stroke of a cross-bridge produces only a very small movement of a thin filament relative to a thick filament. As long as a muscle fiber continues to be stimulated to contract, however, each cross-bridge repeats its motion many times, resulting in continued sliding of the thin filaments. Thus, the ability of a muscle fiber to generate force and movement depends on the amount of interaction between actin and myosin.

The protein myosin was first discovered in extracts of frog leg muscle in the 1860s by German physiologist Wilhelm Kühne. We now know that eukaryotic genomes encode a family of related myosin proteins, and that at least one of them may have made a significant contribution to human evolution, as described next.

 **Core Concept: Evolution**

### Myosins Are an Ancient Family of Proteins

Myosins are among the most ancient eukaryotic proteins and are found throughout the animal kingdom. Small differences in the sequences of genes that arose from a single primordial myosin gene have led to the tissue-specific expression of numerous

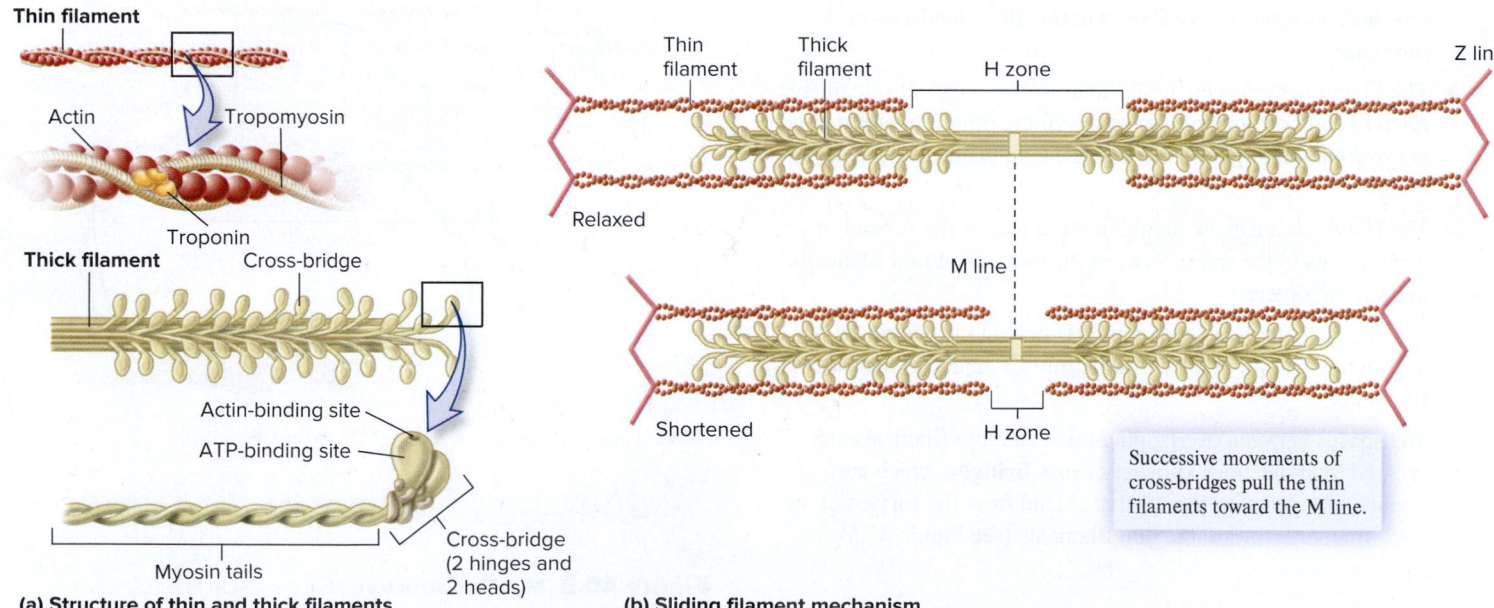

**(a) Structure of thin and thick filaments**

**(b) Sliding filament mechanism**

**Figure 45.6 Structure and function of the thin and thick filaments.** **(a)** Thin filaments are composed of two intertwined actin molecules and their associated proteins, tropomyosin and troponin. Thick filaments are made of the protein myosin, which has two intertwined tails, two hinges, and two heads that contain an actin-binding site and an ATP-binding site. The end of a myosin molecule that contains the actin-binding site is bent at an angle to form the cross-bridge. **(b)** When the cross-bridges on myosin molecules bind to actin, the thin filaments are pulled toward the M line, shortening the sarcomere. The sliding of thin filaments past the overlapping thick filaments shortens the sarcomere, but does not change the lengths of the filaments themselves.

 **Core Concept: Structure and Function** The precise structural arrangement of the proteins that make up thick and thin filaments gives them their function. Without this structure, actin and myosin would not produce a force that results in the shortening of the muscle cell.

different myosin proteins, each with a characteristic ability to bind actin and hydrolyze ATP. In skeletal muscle, myosin is important for generating force, but in other types of cells, various myosins function in organelle trafficking, cell locomotion, mitosis and other activities. One member of the large family of myosin genes, called *MYH16*, encodes a myosin polypeptide that is expressed mainly in the skeletal muscles of the jaw in primates. In 2004, American researcher Hansell Stedman and colleagues discovered that this gene, although present in humans, did not encode a functional polypeptide in humans because of a mutation that deleted two base pairs. This mutation was present in 100% of people tested worldwide but was not found in eight nonhuman primate species tested, including chimpanzees.

Based on genome comparisons and estimates of genetic divergence among species, the researchers estimated that the mutation occurred approximately 2.4 mya. Significantly, the genus *Homo*, with its smaller jaw and larger braincase, is also thought to have first appeared about 2.4 mya, leading Stedman to suggest that the loss of the MYH16 protein led to the smaller, less muscular jaws characteristic of modern humans (**Figure 45.7**). Because jaw muscles are attached to skull bones, massive jaw muscles may have placed a mechanical constraint on skull growth in early hominins (as well as in modern nonhuman primates). Smaller jaw-closing muscles would place less constraint on bone formation of the skull (in particular, at the areas of muscle attachment obscuring the sutures or growth plates). This in turn would prevent the early cessation of brain growth seen in the nonhuman primates. This may have eliminated a major constraint on the evolution of the larger brains seen in modern humans.

Stedman's hypothesis is compelling, given that the myosin mutation affects the jaw-closing muscles and is present only in humans. Another research group, using statistical analyses of additional gene sequences made available through the Chimpanzee Sequencing and Analysis Consortium, concluded that the mutation may have arisen earlier—perhaps 4.0–5.4 mya, before the appearance of small-jawed species of *Homo* in the fossil record. Stedman's hypothesis will require confirmation by future research. Scientists are certain that a human-specific mutation in *MYH16* arose at some point in the evolution of hominins. Whether the mutation facilitated expansion of the cranium, which in turn may have permitted enlargement of the brain, remains unknown. The hypothesis, however, does provide an intriguing piece of the puzzle of the evolution of modern humans.

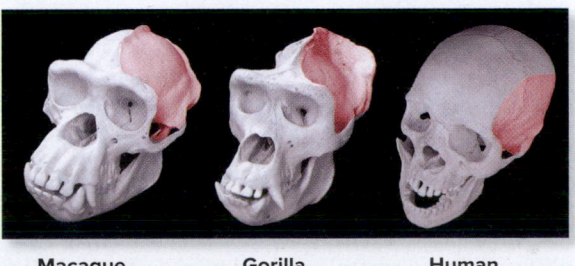

**Macaque          Gorilla          Human**

**Figure 45.7** **Comparison of jaw size and area of muscle attachment (shown in pink) in modern primates.** Note: Skulls are not to scale. ©Hansell H. Stedman, M.D.

## The Cross-Bridge Cycle Requires ATP and Ca²⁺

Let's now turn our attention to how actin and myosin interact to promote muscle contraction and shortening. The sequence of events that occurs between the time when a cross-bridge binds to a thin filament and when it is set to repeat the process is known as a **cross-bridge cycle**.

Cross-bridge cycling occurs when the $Ca^{2+}$ concentration exceeds a critical threshold in the cytosol. This usually occurs when neural input results in the release of $Ca^{2+}$ from intracellular storage sites (described in detail shortly). In other words, the contraction of skeletal muscle fibers is under nervous control.

**Figure 45.8** illustrates the events that occur during the four steps of the cross-bridge cycle: (1) cross-bridge binding, (2) power stroke (moving), (3) detaching, and (4) resetting. As the cycle begins, the myosin cross-bridges are in an energized state, which is produced by the hydrolysis of their bound ATP to ADP and inorganic phosphate ($P_i$). The ADP and $P_i$ remain bound to the cross-bridge until step 2. The sequence of storage and release of energy by myosin is analogous to the operation of a mousetrap: Energy is first stored in the trap by cocking the spring (ATP hydrolysis) and is then released by the springing of the trap (head binds to actin and moves in power stroke). Let's look at each step of the process more closely.

*Step 1: Ca²⁺ concentration increases, triggering the cross-bridge to bind to actin.* When the $Ca^{2+}$ concentration in the cytosol increases, an energized myosin cross-bridge, along with its associated ADP and $P_i$, binds to an actin molecule on a thin filament.

*Step 2: Release of $P_i$ fuels the power stroke. The cross-bridge and thin filaments move.* The binding of an energized myosin cross-bridge to actin in step 1 triggers the release of $P_i$. This release causes a conformational change in the hinge of the myosin molecule. The change in conformation causes the cross-bridge to rotate toward the M line in the H zone at the center of the sarcomere, as myosin returns to its lower energy conformation. This step is known as the **power stroke**, which moves the actin filament. At this time, ADP is released from the cross-bridge.

*Step 3: ATP binds to myosin, causing the cross-bridge to detach.* During the power stroke, myosin is bound very firmly to actin. After the power stroke is completed, this linkage must be broken to allow the cross-bridge to be reenergized and repeat the cycle. The binding of a new molecule of ATP to the myosin cross-bridge alters myosin's conformation and breaks the link between actin and myosin. ATP is not hydrolyzed in this step. Instead, ATP functions at this stage as an allosteric modulator of the myosin head, weakening the binding of myosin to actin and leading to their dissociation.

*Step 4: ATP hydrolysis re-energizes and resets the cross-bridge.* After actin and myosin dissociate, the ATP bound to myosin is hydrolyzed by the ATPase activity of myosin. This hydrolysis re-energizes myosin, causing it to reset to the position that allows actin binding. If $Ca^{2+}$ is still available at this time, the cross-bridge can reattach to a new actin molecule in the thin filament, and the cross-bridge cycle will repeat, causing the muscle fiber to shorten further.

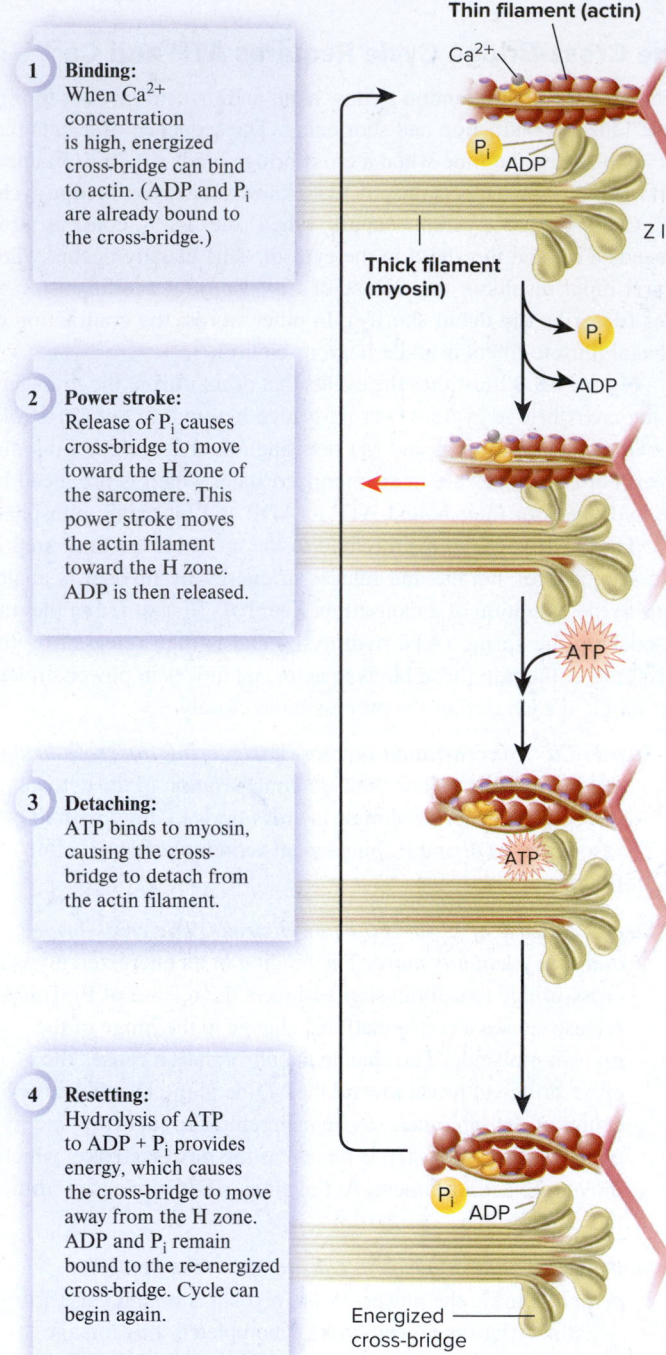

**1  Binding:**
When Ca²⁺ concentration is high, energized cross-bridge can bind to actin. (ADP and P$_i$ are already bound to the cross-bridge.)

**2  Power stroke:**
Release of P$_i$ causes cross-bridge to move toward the H zone of the sarcomere. This power stroke moves the actin filament toward the H zone. ADP is then released.

**3  Detaching:**
ATP binds to myosin, causing the cross-bridge to detach from the actin filament.

**4  Resetting:**
Hydrolysis of ATP to ADP + P$_i$ provides energy, which causes the cross-bridge to move away from the H zone. ADP and P$_i$ remain bound to the re-energized cross-bridge. Cycle can begin again.

Thin filament (actin)
Ca²⁺
P$_i$  ADP
Z line
Thick filament (myosin)
P$_i$
ADP
ATP
ATP
P$_i$  ADP
Energized cross-bridge

**Figure 45.8  The four steps of cross-bridge cycling in skeletal muscle.**

As we have seen, ATP performs two different roles in the cross-bridge cycle. First, the energy released from ATP hydrolysis provides the energy for cross-bridge movement. Second, binding of ATP to myosin breaks the link formed between actin and myosin, allowing a new cycle to start. Although the precise mechanisms may differ slightly between vertebrates and invertebrates, biologists think that all skeletal muscle functions via steps similar to those just described.

## The Regulation of Muscle Contraction by Ca²⁺ Is Mediated by Tropomyosin and Troponin

How does the presence of Ca²⁺ in the cytosol of muscle cells regulate the cycling of cross-bridges? The answer requires a closer look at the two additional thin filament proteins mentioned earlier, tropomyosin and troponin.

**Tropomyosin** is a rod-shaped molecule composed of two intertwined protein subunits (**Figure 45.9a**). Tropomyosin proteins are arranged end to end along the thin filament. In the absence of Ca²⁺, they partially cover the myosin-binding site on each actin molecule, thereby preventing cross-bridges from making contact with actin. Each tropomyosin molecule is held in this blocking position by **troponin**, a smaller, globular-shaped protein with three subunits that is bound to both tropomyosin and actin. In this way, troponin and tropomyosin block access to myosin-binding sites on actin molecules in the relaxed muscle fiber.

One of the three subunits of troponin is capable of binding Ca²⁺. The binding of Ca²⁺ produces a change in the shape of troponin, which—through troponin's linkage to tropomyosin—causes tropomyosin to move away from the myosin-binding site on each actin molecule (**Figure 45.9b**). This movement allows cross-bridge cycling to occur. Conversely, release of Ca²⁺ from troponin reverses the process, blocking the myosin-binding site and turning off contractile activity.

Thus, the concentration of cytosolic Ca²⁺ determines whether or not Ca²⁺ is bound to troponin molecules, which, in turn, determines the number of actin sites available for cross-bridge binding. The cytosolic concentration of Ca²⁺, however, is very low in resting muscle. Let's see how the Ca²⁺ concentration is increased so that contraction can occur.

## Ca²⁺ Concentration and Contraction of Skeletal Muscle Fibers Are Coupled with Electrical Excitation

Like neurons, vertebrate skeletal muscle cells generate and propagate action potentials in response to an appropriate stimulus. The propagation of action potentials causes an increase in the concentration of cytosolic Ca²⁺, which triggers contraction of a muscle fiber. This sequence of events by which an action potential in the plasma membrane of a muscle fiber leads to cross-bridge activity is called **excitation-contraction coupling**. The electrical activity in the plasma membrane does not act directly on the contractile proteins but instead acts as a stimulus to increase cytosolic Ca²⁺ concentration. The increased Ca²⁺ concentration continues to activate the contractile apparatus long after electrical activity in the membrane has ceased.

The source of the increased cytosolic Ca²⁺ that occurs following a muscle action potential is the muscle fiber's **sarcoplasmic reticulum**, which acts as a Ca²⁺ reservoir. The sarcoplasmic reticulum, which is a specialized form of the endoplasmic reticulum, is composed of interconnected sleevelike compartments and sacs around each myofibril (**Figure 45.10**). Separate tubular structures, the **transverse tubules (T-tubules)**, are invaginations of the plasma membrane that lie close to the sarcoplasmic reticulum. The T-tubules conduct action potentials from the outer surface of the muscle fiber to the myofibrils. An action potential causes the opening of Ca²⁺ channels in the lateral sacs of the sarcoplasmic reticulum, which allows Ca²⁺ to diffuse into the cytosol and bind to troponin, initiating cross-bridge cycling.

A contraction continues until Ca²⁺ is removed from troponin and the cytosol. This is achieved by ATP-driven Ca²⁺ transporters in the sarcoplasmic reticulum that decrease the Ca²⁺ concentration in the cytosol back to its resting concentration.

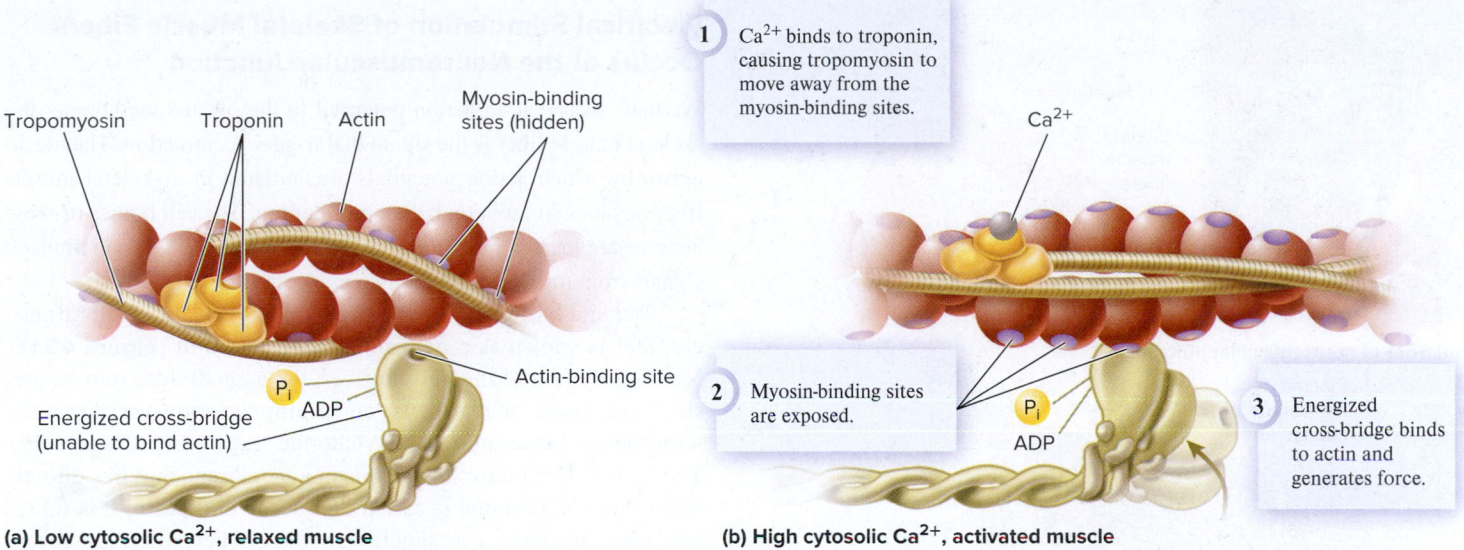

**1** $Ca^{2+}$ binds to troponin, causing tropomyosin to move away from the myosin-binding sites.

Tropomyosin  Troponin  Actin  Myosin-binding sites (hidden)

$Ca^{2+}$

$P_i$  ADP  Actin-binding site

Energized cross-bridge (unable to bind actin)

**(a) Low cytosolic $Ca^{2+}$, relaxed muscle**

**2** Myosin-binding sites are exposed.

$P_i$  ADP

**3** Energized cross-bridge binds to actin and generates force.

**(b) High cytosolic $Ca^{2+}$, activated muscle**

**Figure 45.9** **Function of $Ca^{2+}$, tropomyosin, and troponin in cross-bridge cycling.** **(a)** When cytosolic $Ca^{2+}$ is low, the myosin-binding sites on actin are blocked by tropomyosin. **(b)** When cytosolic $Ca^{2+}$ increases, $Ca^{2+}$ binds to troponin, which, in turn, causes tropomyosin to move away from the myosin-binding sites on actin.

Opening of transverse tubule to extracellular fluid

Muscle fiber plasma membrane

**1** Action potentials propagate along the plasma membrane and down the transverse tubules (T-tubules).

T-tubule  Muscle fiber plasma membrane  Sarcoplasmic reticulum

T-tubule

**2** The depolarization produced by the action potentials opens voltage-gated $Ca^{2+}$ channels in the membranes of the sarcoplasmic reticulum, out of which $Ca^{2+}$ diffuses into the cell cytosol.

$Ca^{2+}$

$Ca^{2+}$ (binds to troponin)

$Ca^{2+}$

ATP  ADP + $P_i$

$Ca^{2+}$ channel

Cytosol

Myofibrils

Cytosol

Mitochondrion

**3** $Ca^{2+}$ then binds to troponin in the myofibril, initiating muscle fiber contraction.

**4** $Ca^{2+}$ is then pumped back into the sarcoplasmic reticulum by ATP-driven $Ca^{2+}$ transporters. This results in muscle fiber relaxation.

Transverse tubules

Sarcoplasmic reticulum

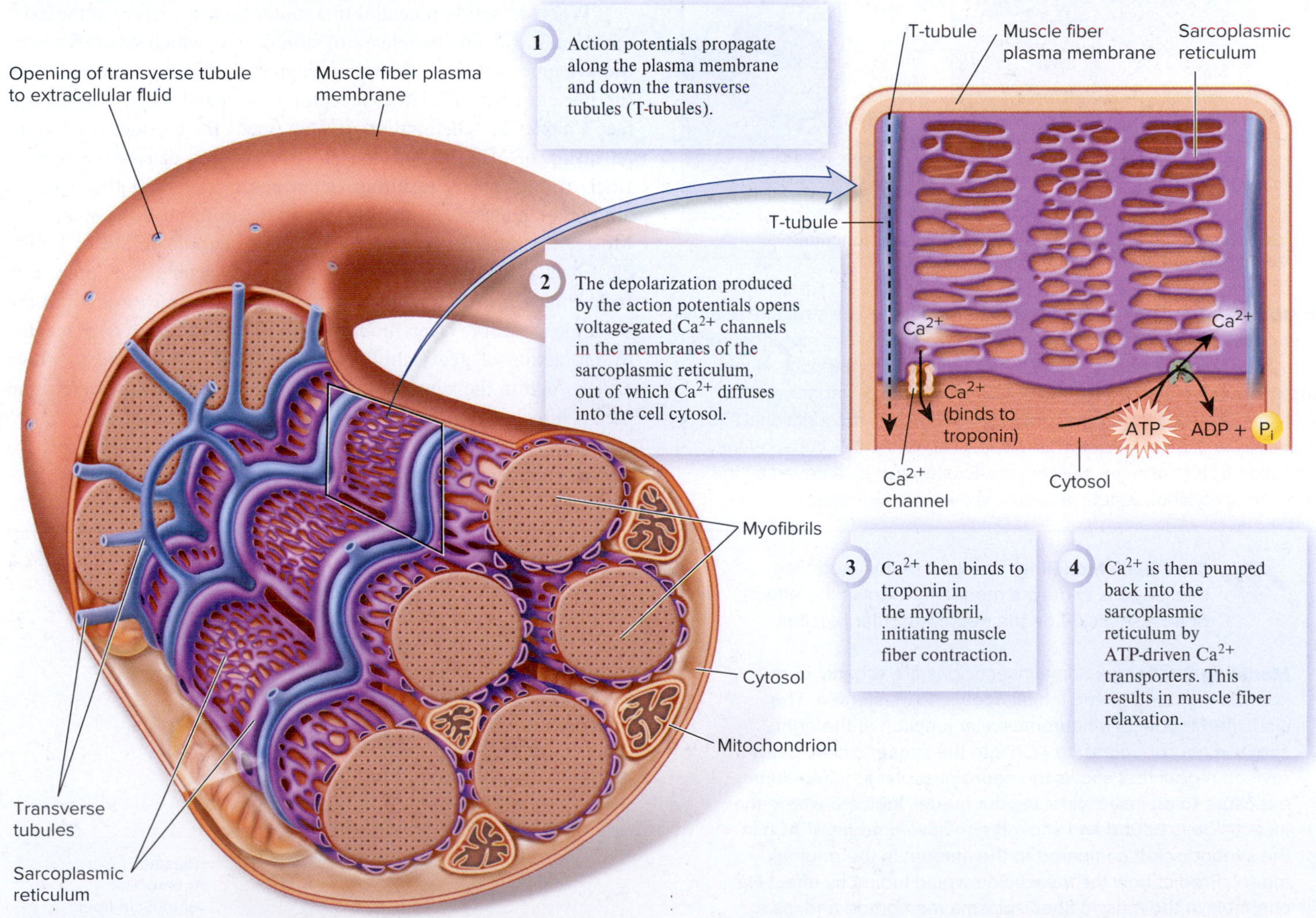

**Figure 45.10** **Structure and function of the sarcoplasmic reticulum, transverse tubules (T-tubules), and myofibrils in a skeletal muscle fiber.**

 **Core Skill: Connections** What have you learned earlier about the role of voltage-gated $Ca^{2+}$ channels in cell-to-cell communication? Look back at Figure 42.13 for help.

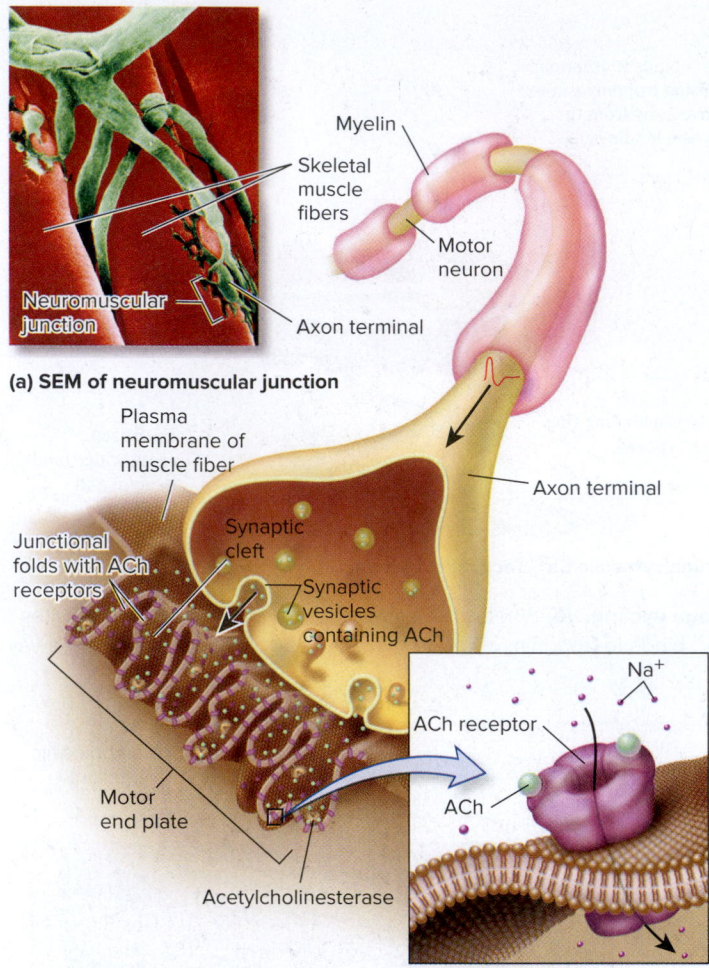

**(a) SEM of neuromuscular junction**

**(b) Structures of, and events at, the neuromuscular junction (only part of the motor neuron is shown)**

**Figure 45.11 The neuromuscular junction. (a)** The structure of a neuromuscular junction as seen in a colorized SEM. The motor neurons are green, and the muscle fibers are red. **(b)** Action potentials in the motor neuron cause exocytosis of ACh-containing synaptic vesicles. ACh binds to receptors in the plasma membrane in the junctional folds of the skeletal muscle fiber. This initiates Na⁺ entry and, consequently, an action potential in the muscle fiber. Excess ACh is removed by the enzyme acetylcholinesterase. a: ©Don W. Fawcett/Science Source

## Electrical Stimulation of Skeletal Muscle Fibers Occurs at the Neuromuscular Junction

We have seen that an action potential in the plasma membrane of a skeletal muscle fiber is the signal that triggers contraction. The mechanism by which action potentials are initiated in a skeletal muscle fiber involves stimulation by a motor neuron. The cell bodies of these neurons are located in the central nervous system (CNS) and transmit signals from the CNS to directly control muscles.

The site where a motor neuron's axon synapses with a muscle fiber is known as a **neuromuscular junction** (**Figure 45.11**). Near the surface of the muscle fiber, the axon divides into several short processes, or terminals, containing synaptic vesicles filled with the neurotransmitter acetylcholine (ACh) (see Chapter 42). The region of the muscle fiber plasma membrane that lies directly under an axon terminal is called the **motor end plate**; it is folded into what are known as junctional folds, where the ACh receptors are located. These folds increase the total surface area available for the membrane to respond to ACh. The extracellular space between the axon terminal and the motor end plate is called the synaptic cleft.

When an action potential in a motor neuron arrives at the axon terminal, it triggers the release of stored ACh, which diffuses across the synaptic cleft and binds to receptors in the junctional folds of the muscle fiber. The ACh receptor is a ligand-gated ion channel (see Chapter 5). The binding of ACh opens the channel and causes an influx of Na⁺ into the muscle fiber, which causes the muscle fiber to depolarize, resulting in an action potential that spreads along the membrane of the muscle fiber and through the T-tubules. Most neuromuscular junctions are located near the middle of a muscle fiber, and newly generated muscle action potentials propagate from this region toward the ends of the fiber and throughout the T-tubule network. Overstimulation of a muscle fiber is prevented by the action of **acetylcholinesterase**. This enzyme breaks down excess ACh in the synaptic cleft to inactive forms that cannot bind the ACh receptor.

Now let's explore how skeletal muscle is adapted to meet the varied functional demands of vertebrates.

**Core Skill: Modeling** The goal of this modeling challenge is to draw a model that shows the effects of an insecticide on the neuromuscular junction.

**Modeling Challenge:** Some insecticides are organic compounds that bind to and inhibit acetylcholinesterase. The simplified model of a neuromuscular junction at the right shows a neuron releasing ACh into the synaptic cleft. Draw a revised model that shows the neuromuscular junction during exposure to an insecticide. In your model, indicate where the insecticide is bound and show the relative amount of ACh in the synaptic cleft compared to the amount in the original model. Predict how the insecticide would indirectly affect Na⁺ channels in the muscle fiber's plasma membrane and, as a result, muscle function.

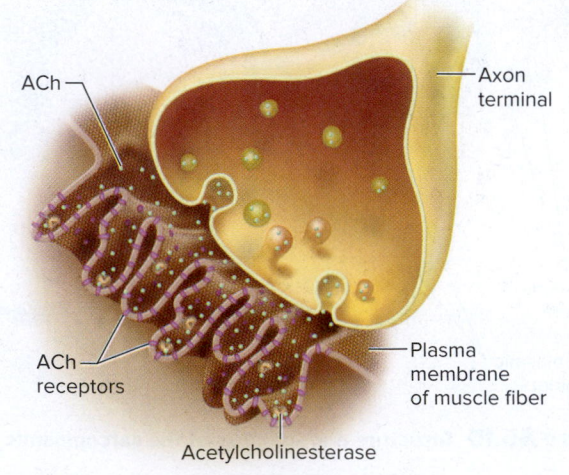

Neuromuscular junction without insecticide

## 45.3 Types of Skeletal Muscle Fibers and Their Functions

**Learning Outcomes:**

1. Outline the general characteristics of the three types of skeletal muscle fibers.
2. **CoreSKILL »** Predict how skeletal muscles may adapt to exercise.

Animals use skeletal muscle for a wide variety of different activities, such as locomotion, stretching, chewing, breathing, and maintaining posture and balance, to name a few. Therefore, it is not surprising that not all skeletal muscle fibers share identical mechanical and metabolic characteristics. In this section, we will consider how different types of fibers can be broadly classified on the basis of their rates of shortening (as either fast or slow) and the way in which they produce the ATP required for contraction (as oxidative or glycolytic).

### Skeletal Muscle Fibers Are Adapted for Different Types of Movement

Recall that there are many types of myosin, each with slightly different characteristics but all sharing an ability to hydrolyze ATP and bind actin. Different skeletal muscle fibers contain forms of myosin that differ in the maximal rates at which they can hydrolyze ATP. This, in turn, determines the maximal rates of cross-bridge cycling and muscle shortening. Fibers containing myosin with low ATPase activity are called **slow fibers**. Those containing myosin with higher ATPase activity are classified as **fast fibers**. Although the rate of cross-bridge cycling is about four times faster in fast fibers than in slow fibers, the maximal force produced by both types of cross-bridges is approximately the same.

The second means of classifying skeletal muscle fibers is based on the type of metabolic pathways available for synthesizing ATP. Fibers that contain numerous mitochondria and have a high capacity for oxidative phosphorylation are classified as **oxidative fibers**. Most of the ATP production by such fibers depends on blood flow delivering oxygen and nutrients to the muscle. Not surprisingly, therefore, these fibers are surrounded by many small blood vessels. They also contain large amounts of **myoglobin**, an oxygen-binding protein that increases the availability of oxygen in the fiber by providing an intracellular reservoir of the gas. The large amounts of myoglobin present in oxidative fibers give these fibers a dark-red color. For this reason, oxidative fibers are often referred to as red muscle fibers. The benefit of red muscle fibers is they can maintain sustained action over a long period of time without fatigue.

By contrast, **glycolytic fibers** have fewer mitochondria but possess a high concentration of the proteins involved in glycolysis (glycolytic enzymes; refer back to Figure 7.3) and large stores of glycogen, the storage form of glucose. Corresponding to their limited use of oxygen, these fibers are surrounded by relatively few blood vessels and contain little myoglobin. The lack of myoglobin is responsible for the pale color of glycolytic fibers and their designation as white muscle fibers.

On the basis of these two characteristics, three major types of skeletal muscle fibers have been distinguished:

1. **Slow-oxidative fibers** have low rates of myosin ATPase activity but have the ability to make large amounts of ATP. These fibers are useful for prolonged, regular types of movement, such as steady flight over a period of time, long-distance swimming, or the maintenance of posture. These muscles, for example, are what give the red color to the dark meat of ducks, which use the muscles for flight. Long-distance runners have a high proportion of these fibers in their leg muscles. These types of activities require muscles that do not fatigue easily.

2. **Fast-oxidative fibers** have high myosin ATPase activity and can make large amounts of ATP. Like slow-oxidative fibers, these fibers do not fatigue quickly and can be used for long-term actions. They are also particularly suited for rapid actions, such as the rapid trilling sounds made by the throat muscles in songbirds or the shaking of a rattlesnake's tail that produces a clicking sound.

3. **Fast-glycolytic fibers** have high myosin ATPase activity but cannot make as much ATP as oxidative fibers, because their source of ATP is glycolysis. These fibers are best suited for rapid, intense actions, such as a cheetah's short sprint at maximum speed. Sloths, by contrast, have few or no fast-glycolytic fibers in their leg muscles, which is not surprising given a sloth's very sedentary lifestyle. Fast-glycolytic fibers fatigue more rapidly than fast-oxidative fibers. The breast meat of chickens, for example, appears white because, unlike ducks, chickens do not fly except for very short distances and therefore do not require oxidative pectoral muscles. The fast-glycolytic muscles of chickens, however, are ideal for short flights in the air that help them quickly escape predators. When they land, chickens use slow-oxidative fibers in their leg muscles to run long distances as they continue to elude a predator.

Different muscle groups within an animal's body have different proportions of each type of fiber interspersed with one another; many activities require the action of all three types of fibers at once. This is important when you consider the wide range of animal activities related to locomotion alone, including walking, climbing, running, swimming, flying, crawling, crouching, jumping, and maintaining balance and posture. Depending on the requirements of an animal at any given moment, the motor nerve inputs can be adjusted to stimulate different ratios of fiber types. When you lift a heavy weight for a brief time, the fast-glycolytic fibers in your arm muscles are activated in large numbers. When a crab uses its pincers to grab prey, fast-glycolytic muscles snap the claws closed quickly, but then slow-oxidative fibers maintain a tight grip for as long as required. The characteristics of the three types of skeletal muscle fibers are summarized in **Table 45.1.**

### Muscles Adapt to Exercise

The regularity with which a muscle is used, as well as the duration and intensity of the activity and whether it includes resistance, affects the properties of the muscle. For example, increased amounts of resistance exercise—such as weight lifting—results in hypertrophy (increase in size) of muscle fibers. Because the number of fibers in a muscle does not normally change significantly throughout adult

| Table 45.1 | Characteristics of the Three Types of Skeletal Muscle Fibers | | |
|---|---|---|---|
| | Slow-oxidative | Fast-oxidative | Fast-glycolytic |
| Primary source of ATP production | Oxidative phosphorylation | Oxidative phosphorylation | Glycolysis |
| Mitochondria | Many | Many | Few |
| Blood supply | High | High | Moderate |
| Myoglobin content | High (red) | High (red) | Low (white) |
| Rate of fatigue | Slow | Intermediate | Fast |
| Myosin ATPase activity | Low | High | High |
| Rate of contraction | Slow | Fast | Fast |

life, the increases in muscle size that occur with resistance exercise result primarily from increases in the size of each fiber. These fibers undergo an increase in diameter due to the increased synthesis of actin and myosin filaments, which form more myofibrils.

Exercise of relatively low intensity but long duration—popularly called aerobic exercise, including running and swimming—increases the number of mitochondria in the fibers that are required in this type of activity. In addition, the number of blood vessels around these fibers increases to supply the greater energy demands of active muscle. All of these changes increase endurance.

By contrast, short-duration, high-intensity exercise, such as weight lifting, primarily affects fast-glycolytic fibers, which are used during strong contractions. In addition, glycolytic activity is enhanced by increased synthesis of glycolytic enzymes. The results of such high-intensity exercise are the increased strength and bulging muscles of a conditioned weight lifter. Such muscles, although very powerful, have little capacity for endurance and therefore fatigue rapidly.

A decline or cessation of muscular activity results in the condition called **atrophy**, a reduction in the size of the muscle. Likewise, if the neurons to a skeletal muscle are destroyed or the neuromuscular junctions become nonfunctional, the denervated muscle fibers will become progressively smaller in diameter. This condition is known as denervation atrophy. Even with an intact nerve supply, a muscle can atrophy if it is not used for a long period of time, as when a broken limb is immobilized in a cast.

The mechanism by which changes occur in skeletal muscle during exercise is an active area of research, but recent discoveries have provided intriguing clues, an example of which is described next.

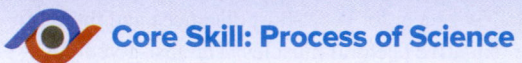 **Core Skill: Process of Science**

## Feature Investigation | Evans and Colleagues Activated a Gene to Produce "Marathon Mice"

In the course of investigating possible ways to reverse or prevent obesity in humans, American biologist Ron Evans and his colleagues discovered one way in which the proportions of oxidative and glycolytic fibers change in skeletal muscle. Evans was interested in a gene that encodes a transcription factor called PPAR-δ. Activation of this protein results in the expression of genes that enable skeletal muscle or other cells to more efficiently burn fat instead of glucose for energy. Evans hypothesized that mice in which PPAR-δ was chronically activated at high concentrations would lose weight due to increased fat burning, as shown in **Figure 45.12**.

To test this hypothesis, Evans created mice that carried the *PPAR-δ* gene in a modified form (see Chapter 21). The modified gene had a promoter that caused the gene to be expressed only in skeletal muscle cells. The region of the gene that encoded PPAR-δ was linked to a region of another gene that encoded a viral protein domain called VP16. This domain facilitates gene activation. The researchers expected the combination of PPAR-δ and VP16 to strongly activate genes that enable cells to preferentially metabolize fat instead of glucose for energy.

In the first part of the experiment, Evans monitored the body weights of the transgenic mice after they reached adulthood. The transgenic mice gained significantly less weight than the wild-type mice when fed high-fat diets (which normally cause mice—like humans—to gain weight), confirming Evans's hypothesis. As is sometimes the case in scientific discovery, an unexpected finding arose from this study. When Evans examined several tissues in these mice under the microscope, he observed that the skeletal muscle of the transgenic mice showed a noticeable shift from glycolytic fibers to slow-oxidative fibers. The skeletal muscle in transgenic mice appeared redder than it did in wild-type mice. It contained more myoglobin and mitochondria, and had higher concentrations of oxidative enzymes capable of providing the cells with sustained levels of ATP. These changes occurred even though the mice had not been subjected to exercise training.

Based on these observations, Evans tested a second hypothesis that the transgenic mice would have a greater capacity for prolonged exercise than wild-type mice. When the transgenic mice were challenged with an endurance exercise test, Evans discovered that they outperformed age- and weight-matched wild-type mice by a factor of nearly twofold! They could sustain a high level of activity on a miniature treadmill for nearly twice as long as wild-type mice (hence the nickname "marathon mice"). This effect occurred in transgenic mice even without prior exercise training. In other words, simply increasing the ratio of oxidative to glycolytic fibers gave the mice greater ability to sustain aerobic activity.

The results of these experiments indicate that increasing the amount of activated PPAR-δ facilitates an oxidative state in skeletal muscle fibers that somehow signals them to convert to types that are best suited for oxidative metabolism. Therefore, the switch in fiber type that occurs in exercise training may not require exercise per se, and it may be mediated in part by proteins that activate or induce *PPAR-δ* expression. These results may have important implications for enhancing physical endurance in humans, as well as for possible treatments for various muscle diseases and for prevention of obesity.

**Figure 45.12** **Evans and colleagues' activation of a gene to produce "marathon mice."** (mice on treadmills): ©Mark Richards/PhotoEdit

**HYPOTHESES** 1. Increased expression of genes that lead to increased fat oxidation in skeletal muscle cells will prevent obesity in mice.
2. Transgenic mice have a greater capacity for prolonged exercise than do wild-type mice.

**KEY MATERIALS** Mice, light and electron microscopes, motorized treadmills.

| | Experimental level | Conceptual level |
|---|---|---|

**1** Prepare a modified gene containing a skeletal muscle–specific promoter and a coding sequence that links *VP16* and *PPAR-δ*. See Chapter 21 for gene cloning methods.

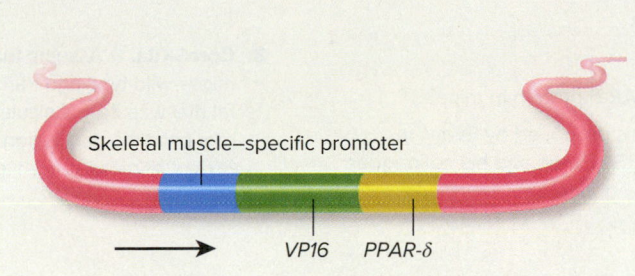

Skeletal muscle–specific promoter

VP16   PPAR-δ

Skeletal muscle–specific promoter ensures gene is turned on only in skeletal muscle.

*VP16* encodes a domain that always activates transcription.

*PPAR-δ* encodes a transcription factor that specifically activates genes that allow cells to efficiently burn fat.

**2** Make transgenic mice expressing the *VP16–PPAR-δ* gene.

See Chapter 21 for a discussion of gene cloning.

All of the cells will carry this gene, but only skeletal muscle cells will express the gene.

**3** Perform the following tests:

(a) Feed wild-type control mice and transgenic mice a normal-fat diet (4%) and then switch to a high-fat diet (35%). Weigh mice weekly.
(b) Examine the muscle fibers in the mice.
(c) Test their endurance on a treadmill.

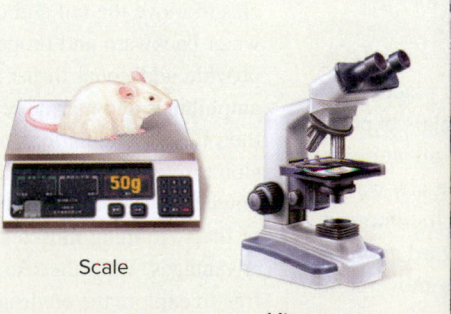

Scale

Microscope

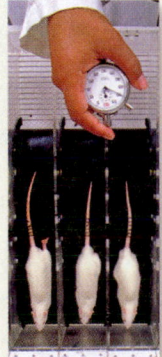

Treadmills

(a) Eating a high-fat diet is known to cause obesity in mice and other mammals.
(b) The appearance of skeletal muscle can be examined by light and electron microscopy.
(c) The treadmills are motorized to keep mice moving until they become exhausted, at which time the treadmills are stopped.

**4** **THE DATA**

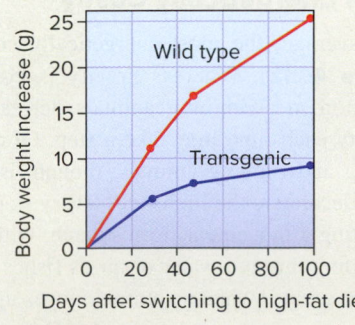

**(a) Weight gain**

Characteristics of skeletal muscle in transgenic mice:

– redder than wild type
– more myoglobin
– more mitochondria
– more slow-oxidative fibers

**(b) Difference in skeletal muscle**

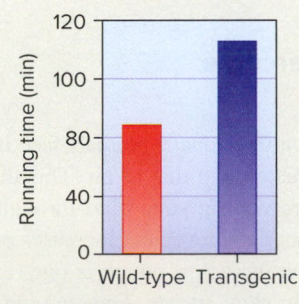

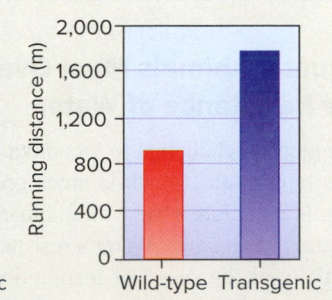

**(c) Muscle endurance**

**5**    **CONCLUSION** PPAR-δ contributes to both weight loss and endurance in mice. The fiber-type switching associated with exercise does not require exercise, because increasing fat oxidation in skeletal muscle cells resulted in more oxidative fibers even without exercise training.

**6**    **SOURCE** Wang, Y.X., et al. 2004. Gene targeting turns mice into long-distance runners. *Public Library of Science Biology* 2: 322.

### Experimental Questions

1. What is the normal function of the PPAR-δ protein in mice?

2. **CoreSKILL** ≫ What was the hypothesis proposed by Evans in relation to PPAR-δ and obesity? How did Evans and his colleagues test this hypothesis, and what did they observe?

3. **CoreSKILL** ≫ Assume that the mean weight of both groups of mice—wild-type and transgenic—prior to being switched to the high-fat diet was 24 g. Calculate the approximate mean weights of the two groups at the conclusion of the study. How does a diet containing 35% of its calories from fat compare to a typical human diet?

## 45.4    Animal Locomotion

### Learning Outcomes:

1. Describe the mechanisms of animal locomotion in water, on land, and in air.

2. **CoreSKILL** ≫ Compare the quantitative differences in energy costs of swimming, running, and flying.

3. Explain how evolution has shaped the structures used for locomotion.

Locomotion, the movement of an animal from place to place, may take many forms, as described earlier. In all cases, animals experience certain constraints to locomotion. For example, all animals must overcome frictional forces (drag) generated by the air, water, or surface of the ground. In addition, all forms of locomotion require energy to provide thrust, defined as the forward motion of an animal in any environment, and/or lift, which is movement against gravity.

Although the precise mechanism may differ among animals, locomotion with few exceptions (such as the rhythmic beating of cilia in ctenophores) results from muscular contractions that exert force on one of the types of skeletons discussed at the beginning of this chapter. In this section, we will examine the similarities and differences between locomotion in water, on land, and in air.

### Aquatic Animals Must Overcome the Resistance of Water

The greatest challenge to locomotion that aquatic animals face is the density of water, which is much greater than that of air. This difference is apparent when you compare waving your hand through the air and under water. Water's resistance to movement increases exponentially as the speed of locomotion increases, which is one reason why many fishes swim at relatively slow speeds. Overcoming this resistance requires considerable muscular effort. Most swimming animals, including fishes, amphibians, reptiles, diving birds, and marine mammals, have evolved streamlined bodies that reduce drag and so make swimming more efficient.

Although the density of water creates challenges to locomotion, it also provides certain benefits. An energetic advantage to swimming is that fishes and other swimmers do not need to provide as much lift to overcome gravity. Because the density of water is similar to that of an animal's body, water provides buoyancy, which helps support the animal's weight.

The mechanism of swimming is similar among many different vertebrates. Most fishes, for example, contract posterior skeletal muscles to move the tail end of the animal from side to side. This pushes water backward and propels the fish forward. Other muscles and fins provide additional thrust and enable changes in direction. Likewise, amphibians and marine reptiles rely predominantly on their hind legs, their tail, or undulations of the posterior parts of the body for propulsion through the water. Cetaceans (whales and dolphins) use up-and-down thrusts of their tail flukes to provide propulsion. Confining most of the swimming muscles to the rear of an animal's body has certain advantages. With the rear end devoted to movement, the front end is free to explore the environment, fight off aggressors, or find food.

Energetically, swimming is the "cheapest" form of locomotion in animals that are adapted to it, due to streamlining, the relatively slow speed of most swimmers, and the buoyancy of water. By contrast, terrestrial animals face considerable energetic costs associated with locomotion, as we examine next.

### Locomotion on Land Is Energetically Costly

Locomotion on land is, on average, the most energetically costly means of locomotion (**Figure 45.13**). Whereas gravity is not an important factor for locomotion in swimming animals, terrestrial animals must overcome gravity each time they take a step. Of even greater importance to walking and running animals, though, is the requirement to accelerate and decelerate the limbs with every step. In essence, each step is like starting a movement from scratch, without the luxury of occasionally gliding through water or air as fishes and birds do. This challenge is even greater when an animal moves uphill or over rough terrain.

Apart from terrestrial gastropods, which move along the surface of the ground on a layer of secreted mucus, and snakes, which undulate along the ground on a portion of their ventral body surface, most

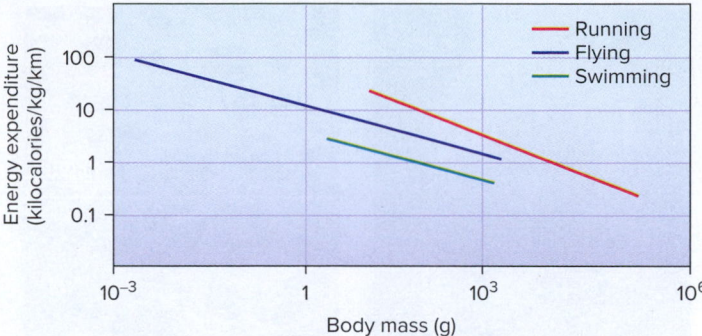

**Figure 45.13 Energy costs of locomotion.** The energy costs of three different modes of locomotion for animals of different sizes are shown. The vertical axis measures the energy cost as kilocalories expended per kilogram (kg) per kilometer (km). Energy costs are highest for runners compared with similarly sized fliers and swimmers. Note: Only a portion of the full range of body sizes of swimmers is shown.

terrestrial animals limit the amount of contact with the ground while moving, thereby minimizing the amount of friction they encounter. Tetrapods usually have only two feet on the ground at any time when walking, and for brief moments, an animal such as a horse galloping at full speed has all four feet off the ground.

Having fewer legs touching the ground at any one time helps increase speed but can compromise stability. Arthropods, for example, have at least six legs. This apparently provides excellent stability but reduces maximal speed (although cockroaches can attain rapid speeds by running on only two legs). At the other extreme are animals that move by jumping—fleas, certain spiders, click beetles, grasshoppers, frogs, and kangaroos.

## Flying Has Evolved in Four Different Lineages

Flying is a highly successful means of locomotion and is hypothesized to have evolved in four different lineages: pterosaurs (extinct reptiles that were the first vertebrates to fly), insects, birds, and mammals (bats). Flying provides numerous advantages: Animals can escape land-based predators, scan their surroundings over great distances, and inhabit environments such as high cliffs that may be inaccessible to nonflying animals. The mechanics of flying, however, require animals to overcome gravity and air resistance, which makes flying more energetically costly than swimming but still less costly than running on land (see Figure 45.13). Many migratory birds can travel hundreds of miles daily for a week or longer. As with swimming, resistance to movement in flight is decreased by streamlined bodies. However, earthbound animals have one advantage over flying animals—they can grow to much larger sizes than animals that fly. The vast majority of flying animals have a mass between about 1 mg and 1 kg. Only a few large birds have masses exceeding 10 kg. Although this represents a wide range, it falls far short of the sizes achieved by earthbound or aquatic animals.

In flying vertebrates, lift and thrust are provided by pectoral and other muscles that move the wings. The pectoral muscles are so powerful and massive that they constitute as much as 15–20% of a bird's total body mass and up to 30% in hummingbirds, which

use their wings not only to fly but also to hover. The requirement for large, strong pectoral muscles is one reason why the body mass of flying vertebrates is limited. The extinct pterosaurs would seem to be an exception because some species were known to have had wingspans of nearly 10 m. However, scientists think that these large animals were unable to generate the force required to lift their massive bodies off the ground and instead glided from trees or cliffs to fly.

In birds and bats, the wings are modifications of the forelimbs. In general, bat wings are far more maneuverable than bird wings, because unlike birds, bats have digits at the end of their forelimbs/wings. These allow bats to precisely alter the shape of their wings and perform rapid, fine-tuned changes in direction, even at high speeds. The largest birds, such as hawks and eagles, are able to glide because of the great surface area of their large wings. By using a bird's momentum to propel it forward, gliding provides a considerable energy savings. Bats and small birds, however, can glide for only very brief moments.

 **Core Skill: Quantitative Reasoning**

**BIO TIPS** **THE QUESTION** *Look again at Figure 45.13. Which flying animal expends more total energy over a distance of 1 km: one that weighs 1 g or another that weighs 1 kg?*

**T**OPIC *What topic in biology does this question address?* The topic is animal locomotion. More specifically, the question concerns the relationship of body mass to energy expenditure in flying animals.

**I**NFORMATION *What information do you know based on the question and your understanding of the topic?* From the question, you know you need to compare two animals that differ in body mass by a factor of 1,000 (1 g versus 1 kg) with regard to their total energy expenditure while flying an equivalent distance (1 km). From your understanding of the topic, based on Figure 45.13, you know that smaller animals expend more energy than larger ones when energy expenditure is normalized to a standard mass (such as 1 kg).

**P**ROBLEM-SOLVING **S**TRATEGY *Interpret data. Compare and contrast.* To answer this question, you must interpret the relationship between body mass and energy expenditure for flying animals (Figure 45.13). First, find the approximate energy values for flying animals with body masses of 1 g and 1 kg (1,000 g), by moving upward from those masses on the horizontal axis to the line for flying animals and then moving left from there to the vertical axis to get the energy values. Second, consider that the question asks for *total* energy expenditure. The data shown are for energy expenditure per kilogram of body mass. Thus, you must account for the fact that the animals differ in mass by a factor of 1,000.

## 45.5 Impact on Public Health

**Learning Outcomes:**

1. Describe the impacts of the bone diseases rickets and osteoporosis on public health.
2. Recognize the main symptoms and causes of muscular dystrophy.

In this chapter, we have seen how skeletal muscles and, in vertebrates, bones function together to provide animals with protection and enable them to move around in their environments. A number of diseases affect bone structure and function in humans. Bone disease may involve defects in either the mineral or organic components of bone. Poor bone formation and structure may result from inadequate nutrition, hormonal imbalances, aging, or skeletal muscle atrophy, to name just a few of the common causes.

In addition, many diseases or disorders directly affect the contraction of skeletal muscle. Some of them are temporary and not serious, such as muscle cramps, whereas others are chronic and severe, such as the disease muscular dystrophy. Also, some muscle diseases result from defects that originate in parts of the nervous system that control contraction of the muscle fibers rather than from defects that originate in the fibers themselves. One example is amyotrophic lateral sclerosis, a degenerative disease in which the destruction of motor neurons leads to skeletal muscle atrophy that may result in death from respiratory failure.

Other diseases that affect skeletal muscle function result when normal processes go awry. For example, the immune system that normally protects the body from pathogens may turn on itself, or faulty genes may produce an abnormal protein. In this section, we will look at a few of these conditions in more detail.

### Rickets and Osteoporosis Affect the Bones of Millions of People

Bone diseases are fairly common, particularly among individuals older than 50. Two major abnormalities can occur in bone. The first is improper mineral deposition in bone, usually due to inadequate dietary calcium intake or inadequate absorption of $Ca^{2+}$ from the small intestine. Without adequate minerals, bone becomes soft and easily deformed, as occurs in the weight-bearing bones of the legs of children with **rickets** (or **osteomalacia**, as it is called in adults) (**Figure 45.14a**). These disorders are best prevented or treated with vitamin D, because this vitamin is the most important factor in promoting absorption of ingested $Ca^{2+}$ from the small intestine.

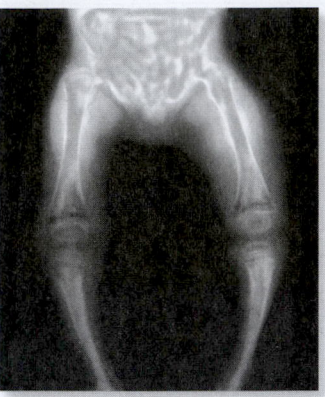
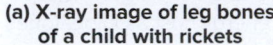
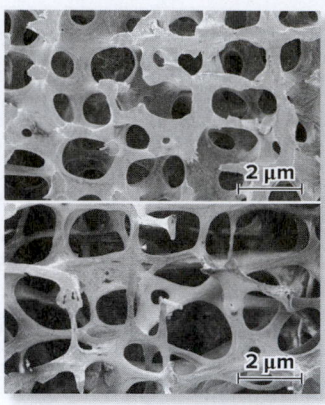

(a) X-ray image of leg bones of a child with rickets

(b) Histologic appearance of normal bone (top) and bone from a person with osteoporosis (bottom)

**Figure 45.14  Human bone diseases.** a: ©Dr. LR/Science Source; b: ©Tim Arnett, University College London

A second major abnormality is a more common disease called **osteoporosis**, in which both the mineral and organic portions of bone are decreased (**Figure 45.14b**). This disease, which affects four times as many women as men, occurs when the normal balance between bone formation and bone breakdown is disrupted.

One cause of osteoporosis is prolonged disuse of muscles. In ways that are not completely clear, the force produced by active skeletal muscle contractions helps maintain bone mass. When muscles are not or cannot be used—due to paralysis or long-term immobilizing illnesses—bone mass declines.

More commonly, osteoporosis may result from hormonal imbalances. Some hormones—for example, estrogen—stimulate bone formation. When estrogen concentrations decline after menopause (the time when a woman's reproductive cycles cease), bone density may decline, increasing the risk of bone fractures. Men can get osteoporosis, too, but since they typically have more bone mass to start, and do not have a pronounced drop in sex hormone concentrations like women, the onset of the disease is usually later and less severe than in women.

In contrast to estrogen, some hormones—such as parathyroid hormone (see Chapter 50)—act to demineralize bone, releasing $Ca^{2+}$ into the extracellular fluid as part of the way in which the body normally maintains $Ca^{2+}$ homeostasis in the blood. If such hormones are present in excess, however, they can cause enough demineralization of bone to result in osteoporosis. This may happen in rare cases when the glands that make these hormones malfunction and overproduce the hormones.

Osteoporosis can be minimized with adequate calcium and vitamin D intake and weight-bearing exercise programs. In some cases, postmenopausal women may be given estrogen to replace what their bodies are no longer producing, but this therapy is controversial due to the potential adverse effects of estrogen on cardiovascular health and possible increased risk of developing breast cancer. Osteoporosis is the most prevalent bone disease in the U.S., affecting up to 15–30 million individuals. It results in annual national expenditures of approximately $15–20 billion in hospital and other medical costs.

## Muscular Dystrophy Is a Rare Genetic Disease That Causes Muscle Degeneration

The group of diseases collectively called muscular dystrophy affect 1 of every 3,500 American males; it is much less common in females. **Muscular dystrophy** is associated with the progressive degeneration of skeletal and cardiac muscle fibers, weakening the muscles and leading ultimately to death from heart failure and other causes. The signs and symptoms become evident at about 2–6 years of age, and most affected individuals do not survive beyond the age of 30.

The most common form of muscular dystrophy, called **Duchenne muscular dystrophy**, is an X-linked recessive disorder resulting from a defective gene on the X chromosome. Because females have two X chromosomes, and males only one (plus one Y), a heterozygote female with one abnormal and one normal allele will not generally develop Duchenne muscular dystrophy. If she passes the abnormal allele to a son, however, he will have the disease. The affected gene encodes a protein known as dystrophin, which is absent in patients with the disease. Dystrophin is a large protein that links cytoskeletal proteins to the plasma membrane. Scientists think it is involved in maintaining the structural integrity of the plasma membrane in muscle fibers. In the absence of dystrophin, the plasma membrane of muscle fibers is disrupted, causing extracellular fluid to enter the cell. Eventually, the cell ruptures and dies.

One focus of interest in attempts to develop a treatment for muscular dystrophy is a protein called myostatin, which is one of a large number of proteins that control muscle growth in mammals. Myostatin functions to prevent overgrowth of skeletal muscle by inhibiting maturation of stem cells into new muscle fibers and by preventing excessive growth of mature fibers. Researchers have identified animals with mutations in the gene that encodes myostatin. The mutations result in the production of an inactive protein, and such animals show astonishing muscle development (**Figure 45.15**). Whether

**Figure 45.15  An animal with a mutation in the gene that encodes myostatin, which resulted in inactive myostatin protein and overdevelopment of skeletal muscle.** McPherron, A. C., and Lee, S. 1997. Double muscling in cattle due to mutations in the myostatin gene. *PNAS* 94: 12,457-12,461.

or not targeting the myostatin protein will lead to a treatment that can reverse muscle loss in patients with muscular dystrophy and other diseases remains to be determined.

## Summary of Key Concepts

### 45.1  Types of Animal Skeletons

- Skeletons are structures that provide support and protection and also function in locomotion.

- Two types of skeletons are commonly found in animals. In arthropods, exoskeletons are protective external structures that must be shed to accommodate an animal's growth. Endoskeletons—found in some species of sponges and in all echinoderms and vertebrates—are internal structures that grow with an animal but do not protect its body surface (Figure 45.1).

- Vertebrate endoskeletons are considered to have two parts: axial and appendicular skeletons. A joint is formed where two or more bones of a vertebrate endoskeleton come together (Figure 45.2).

- In addition to the functions of support, protection, and locomotion, the vertebrate skeleton produces blood cells and constitutes a reservoir for ions crucial to homeostasis. A proper proportion of organic to mineral components is required for normal bone function.

### 45.2  Skeletal Muscle Structure and the Mechanism of Force Generation

- A skeletal muscle is a grouping of cells, called muscle fibers, bound together by connective tissue layers. Muscles that bend a limb at a joint are called flexors, whereas muscles that straighten a limb are called extensors. Two or more muscles that produce oppositely directed movements at a joint are known as antagonists (Figures 45.3, 45.4).

- Within each muscle cell are cylindrical bundles known as myofibrils, each of which contains thick filaments of myosin and thin filaments composed of actin and two other proteins, arranged in repeating units called sarcomeres. Regions of the thick filaments that extend toward the thin filaments are called cross-bridges (Figure 45.5).

- During muscle contraction, the sarcomeres shorten by a process known as the sliding filament mechanism. In muscle contraction, the thick filaments remain stationary while the thin filaments slide past them propelled by the action of cross-bridges (Figure 45.6).

- Mutations in a myosin gene expressed mainly in the jaw muscle may have allowed the human brain to become larger (Figure 45.7).

- In the cross-bridge cycle, the binding of a myosin cross-bridge to actin causes a change in the shape of the myosin molecule. As a result, the two filaments slide past each other, shortening the sarcomere and contracting the muscle fiber. Release of the cross-bridge from actin and return of myosin to its original conformation require ATP binding and hydrolysis (Figure 45.8).

- Tropomyosin and troponin, the two proteins associated with actin, play a critical role in the regulation of muscle contraction. The binding of $Ca^{2+}$ to troponin allows tropomyosin to move away from the myosin-binding sites, initiating cross-bridge binding (Figure 45.9).

- The concentration of cytosolic $Ca^{2+}$ determines the number of actin sites available for cross-bridge binding. The source of the cytosolic $Ca^{2+}$ involved in a muscle fiber's action potential is the fiber's sarcoplasmic reticulum. Transverse tubules (T-tubules) conduct action potentials from the plasma membrane at the outer surface of the muscle fiber to the interior of the cell, allowing $Ca^{2+}$ to be released from the sarcoplasmic reticulum into the cytosol (Figure 45.10).

- Electrical stimulation of skeletal muscle occurs at a neuromuscular junction, the location where a motor neuron's axon terminal synapses with a muscle fiber (Figure 45.11).

## 45.3   Types of Skeletal Muscle Fibers and Their Functions

- Three major types of skeletal muscle fibers have been distinguished. Slow-oxidative fibers have low rates of myosin ATPase activity; they do not fatigue easily and are used for prolonged, regular activities. Fast-oxidative fibers have high myosin ATPase activity, do not fatigue quickly, and are particularly suited for rapid, long-term actions. Fast-glycolytic fibers have high myosin ATPase activity but cannot make as much ATP as oxidative fibers; they are best suited for rapid, short-term actions (Table 45.1).

- Increased expression of the *PPAR-δ* gene in mice results in increased slow-oxidative muscle, greater exercise endurance, and weight loss (Figure 45.12).

## 45.4   Animal Locomotion

- Locomotion, the movement of an animal from place to place, takes many forms, including swimming, walking, running, crawling, hopping, and flying.

- Due to streamlining, the relatively slow speed of most swimmers, and the buoyancy of water, swimming is energetically the most efficient form of locomotion. Locomotion on land is, on average, the most energetically costly means of locomotion. The energy expenditure required for flight is intermediate between those for swimming and land-based locomotion (Figure 45.13).

## 45.5   Impact on Public Health

- Several health conditions affect bone or muscle structure and function in humans. Rickets (osteomalacia in adults) is characterized by soft, deformed bones, usually resulting from insufficient dietary intake of calcium or inadequate absorption of $Ca^{2+}$. In osteoporosis, bone density is reduced when bone formation fails to keep pace with normal bone breakdown. Muscular dystrophy is an ultimately fatal genetic disease associated with the progressive degeneration of skeletal and cardiac muscle fibers (Figures 45.14, 45.15).

## Assess & Discuss

### Test Yourself

1. A disadvantage of an exoskeleton is that it
   a. cannot protect an animal's internal organs.
   b. must be periodically shed, leaving the animal in a vulnerable state.
   c. does not provide any flexibility for the ease of movement of an animal.
   d. is a soft, easily damaged structure.
   e. cannot protect the outside of the body surface.

2. Which, if any, of the following is *not* a function of the vertebrate skeleton?
   a. structural support
   b. protection of internal organs
   c. $Ca^{2+}$ reserve
   d. blood cell production
   e. All of the above are functions of the vertebrate skeleton.

3. The protein that provides strength and flexibility to bone is
   a. actin.
   b. myosin.
   c. myoglobin.
   d. collagen.
   e. elastin.

4. Which of the following statements is *true*?
   a. A muscle fiber is a collection of cells embedded in connective tissue.
   b. A sarcomere contains both actin and myosin molecules arranged in a parallel fashion.
   c. The function of $Ca^{2+}$ in contraction is to bind to tropomysin.
   d. Myofibrils are individual muscle cells.
   e. The I band of a sarcomere is the region where thin and thick filaments overlap.

5. The function of ATP during muscle contraction is to
   a. cause an allosteric change in myosin so it detaches from actin.
   b. provide the energy necessary for the movement of the cross-bridge.
   c. expose the myosin-binding sites on the thin filaments.
   d. do all of the above.
   e. do a and b only.

6. The function of $Ca^{2+}$ in skeletal muscle contraction is to
   a. cause an allosteric change in myosin so that it detaches from actin.
   b. provide the energy necessary for the movement of the cross-bridge.
   c. expose the myosin-binding sites on the thin filaments.
   d. stimulate an action potential in the muscle fiber.
   e. do a and c only.

7. Stimulation of a muscle fiber by a motor neuron occurs at
   a. the neuromuscular junction.
   b. the transverse tubules.
   c. the myofibril.
   d. the sarcoplasmic reticulum.
   e. none of the above.

8. Muscle fibers that have a large number of mitochondria, contain large amounts of myoglobin, and exhibit low rates of ATPase activity are called _____ fibers.
   a. slow-glycolytic
   b. fast-glycolytic
   c. intermediate
   d. fast-oxidative
   e. slow-oxidative

9. Which of the following statements about movement and locomotion is *false*?
   a. Terrestrial animals and flying animals expend energy to provide lift.
   b. Swimming animals typically expend energy to provide thrust but not lift.
   c. Flexors and extensors are examples of muscles called agonists.

d. Flexors cause bending at a joint.

e. Extensors cause straightening of a limb.

10. For animals adapted to it, swimming is energetically the cheapest form of locomotion because of

a. the streamlined body forms of aquatic animals.

b. the slow speed of locomotion of many swimmers.

c. the buoyancy of water.

d. a and c only.

e. a, b, and c.

## Conceptual Questions

1. Compare the structural and functional features of exoskeletons and endoskeletons and their advantages and disadvantages.

2. Describe as many types of animal locomotion as you can, and discuss the benefits and challenges presented by each in terms of energy usage.

3. **Core Concept: Energy and Matter** Explain the role of energy with regard to the mechanism by which skeletal muscle cells contract.

## Collaborative Questions

1. List and briefly describe the steps in the cross-bridge cycle.

2. Discuss the three types of muscle tissues found in vertebrates, and identify distinguishing functional features of each.

# Nutrition and Animal Digestive Systems

# 46

**A balanced meal containing many nutrients, including carbohydrates, lipids, proteins, vitamins, minerals, and water.**

©Richard Hutchings/PhotoEdit

**W**hat would 50,000,000 calories worth of food look like if it were assembled in one place? (Try to imagine about 170,000 cheeseburgers or slices of pizza.) That is roughly the number of calories you will consume in your lifetime (in food labeling, 1 food calorie is actually equivalent to 1 kilocalorie, sometimes written as 1 Calorie with a capital C). All that energy is required for the trillions of body cells to perform varied activities such as synthesizing proteins, making cellular organelles, and maintaining concentration gradients of ions across cellular membranes. These and other activities are necessary for homeostasis and require a lifelong supply of energy and chemical building blocks. These materials must be consumed by animals in the form of nutrients. A **nutrient** is any organic or inorganic substance that is taken in by an organism and is required for survival, growth, development, tissue repair, or reproduction. The process of consuming and using nutrients is called **nutrition**. All organisms require nutrients to survive. Animals receive their nutrients by consuming food, such as the nutritionally balanced meal in the chapter opening photo.

Food processing in animals occurs in four phases: ingestion, digestion, absorption, and egestion (**Figure 46.1**). **Ingestion** is the act of taking food into the body via a structure such as a mouth. From there, the food moves into a digestive cavity, known in most animals as an **alimentary canal**. If the nutrients in food are in a form that cannot be directly used by cells, they must be broken down into smaller molecules, a process known as **digestion**. This is followed by the process of **absorption**, in which ions, water, and small molecules diffuse or are transported from the alimentary canal into an animal's body fluids. **Egestion** (or defecation) is the process by which animals pass undigested material and other wastes out of the body.

In this chapter, we will begin by looking at the types of nutrients that animals require. Next, we will discuss the mechanisms by which animals digest and absorb those nutrients and consider how these processes are regulated. We will end by considering some of the common ways in which the function of the digestive system may go awry in humans.

## 46.1 Animal Nutrition

**Learning Outcomes:**

1. List the major categories of nutrients consumed by animals, and describe some of their general functions.
2. Identify four groups of essential nutrients, listing several examples of each.
3. **CoreSKILL »** Predict some of the consequences of a deficiency of a given nutrient.

Animals require both organic (carbon-containing) and inorganic nutrients. Organic nutrients fall into five categories: carbohydrates, proteins, lipids, nucleic acids, and vitamins. These provide energy and act as the building blocks of new molecules or serve as coenzymes in enzymatic reactions. Inorganic nutrients include water and minerals such as calcium, copper, and iron. The importance of water to life was

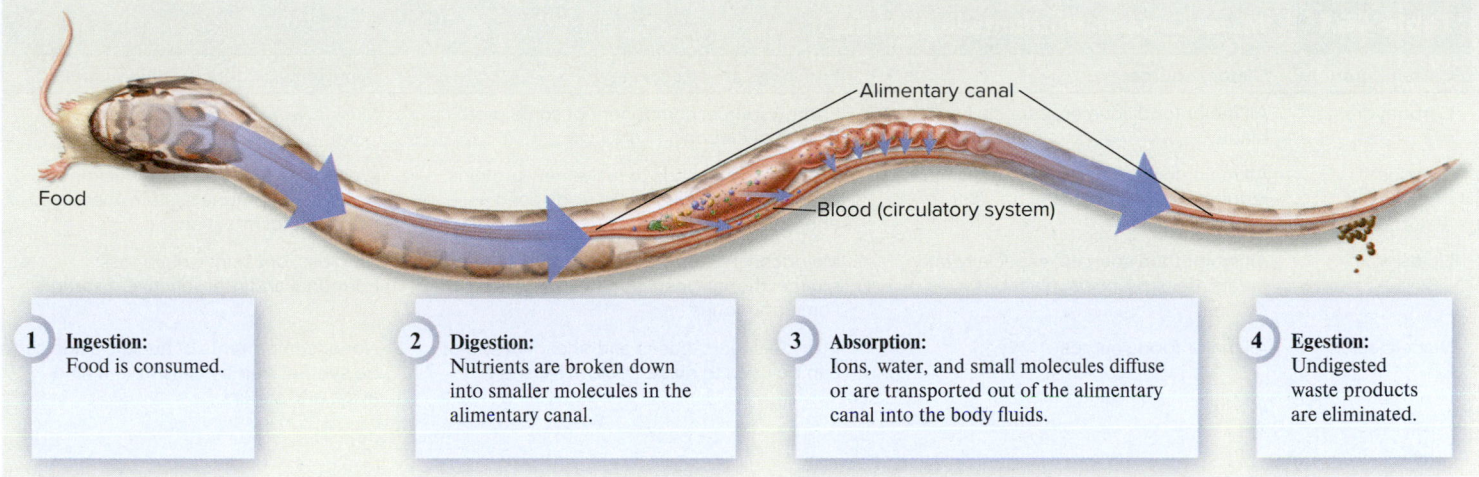

1 **Ingestion:** Food is consumed.

2 **Digestion:** Nutrients are broken down into smaller molecules in the alimentary canal.

3 **Absorption:** Ions, water, and small molecules diffuse or are transported out of the alimentary canal into the body fluids.

4 **Egestion:** Undigested waste products are eliminated.

**Figure 46.1** An overview of the four phases of food use in animals, as shown in a snake that has consumed a rat.

covered in Chapter 2. Minerals serve many functions, including acting as cofactors in enzymatic reactions.

Because of similarities in the ways their organ systems function, animals share many nutritional requirements. However, the types of foods eaten by animals differ depending on an animal's physiology. The digestive systems of some **herbivores**, animals that eat only vegetation, contain microorganisms that assist in the digestion of cellulose, so these animals are well adapted to subsist on plants. By contrast, **carnivores** are primarily adapted to consume animal flesh or fluids, and **omnivores**, such as humans, eat both plant and animal products.

Although broadly useful, the three dietary categories are limited when describing the diversity of animal feeding habits. For example, some animals eat protists, algae, and/or fungi as a major source of food. Also, other animals are almost strictly carnivores at one time of year but herbivores at other times. Many nonmigratory birds, for example, feed on insects and worms during the summer but switch to eating whatever vegetation, buds, or seeds they can find during the winter.

The amount of each type of nutrient required by an animal may also differ, depending on the animal's activity level, or metabolic rate, which we will discuss in Chapter 47. Generally, highly active and energetic animals, such as most birds and mammals, require a proportionally greater amount of nutrients each day than do relatively inactive or sedentary animals, such as nonmotile invertebrates. In this section, we will examine the various organic and inorganic nutrients consumed by animals and some of their major functions.

## Animals Require Nutrients for Energy and the Synthesis of New Molecules

Although different species consume a wide variety of foods, all animals require the same fundamental organic molecules (**Table 46.1**; see also Chapter 3 for a discussion of the chemical nature of organic molecules). Ingested organic molecules are used for two general purposes: to provide energy and to make new molecules. Carbohydrates supply energy-yielding glucose and the carbon required for building organic molecules. Proteins supply amino acids that, in addition to being building blocks for new proteins, can also be used as an energy source. Lipids supply components for membrane-building and thermal

insulation and also provide energy. Nucleic acids supply some of the components required for DNA, RNA, and ATP synthesis. The other category of organic nutrients—vitamins—are required in small quantities and function as coenzymes in various chemical reactions.

As described in Chapter 6, the bonds of organic molecules may be broken to release energy. This energy can be used in the synthesis of ATP, a primary energy source of all cells. Indirectly, therefore, organic molecules provide the energy required for most of the chemical reactions that occur in animals' bodies. Alternatively, organic molecules serve as building blocks to synthesize new cellular molecules. For example, animals use amino acids obtained from food to make the specialized proteins that their cells require.

## Essential Nutrients Must Be Obtained from the Diet

Animal cells can synthesize many organic molecules, but certain compounds cannot be synthesized from any ingested or stored precursor molecule. These **essential nutrients** must be obtained in the diet in their complete form. The word essential refers to the fact that these nutrients must come from food. It does not mean that other nutrients are less important, because many synthesized nutrients are required for an animal's survival. The essential nutrients can be classified into four groups: essential amino acids, essential fatty acids, vitamins, and minerals.

*Essential Amino Acids*   Nine **essential amino acids** are required in the diet of humans and many, but not all, other animals—isoleucine, leucine, lysine, methionine, phenylalanine, histidine, threonine, tryptophan, and valine. These amino acids are required for building proteins but cannot be synthesized by an animal's cells, unlike the other 11 amino acids that also make up proteins. Also, animal cells do not store amino acids. Therefore, without a recurring supply of these nine amino acids, protein synthesis in each cell in an animal's body would slow down or stop completely. Carnivores and omnivores readily obtain all of the essential amino acids, because meat (animal muscle) contains all 20 amino acids. Unlike animal meat, most plants do not contain every essential amino acid in sufficient quantities to supply a human's nutritive requirements. Therefore, people who

| Table 46.1 | Major Organic Nutrients in Animals | | |
|---|---|---|---|
| Class of nutrient | Dietary sources | Functions in vertebrates | Symptoms of deficiency in humans |
| Carbohydrates | All major food sources, especially starchy plants | Energy source; component of some proteins; source of carbon | Muscle weakness; weight loss |
| Proteins | All major food sources, especially meat, legumes, cereals, roots | Provide amino acids to make new proteins; build muscle; some amino acids used as energy source | Weight loss; muscle loss; weakness; weakened immune system; increased likelihood of infections |
| Lipids | All major food sources, especially fatty meats, dairy products, plant oils | Major component of cell membranes; energy source; thermal insulator; building blocks of some hormones | Hair loss; dry skin; weight loss; hormonal and reproductive disorders |
| Nucleic acids | All major food sources | Provide sugars, bases and phosphates that can be used to make DNA, RNA, and ATP | None; components of nucleic acids can be synthesized by cells from amino acids and sugars |

follow a strict vegetarian diet must find ways to balance the protein content of the plant matter they eat. By contrast, some herbivores, such as cows, have evolved the capacity to synthesize the essential amino acids, which allows them to subsist entirely on a diet of plants.

***Essential Fatty Acids***   The **essential fatty acids** are certain unsaturated fatty acids, such as linoleic acid (see Figure 3.9), that cannot be synthesized by animal cells. Linoleic acid is vital to an animal's health because it is converted in cells to another fatty acid, called arachidonic acid. This fatty acid is the precursor required for production of several compounds that are important in many aspects of animal physiology. Such compounds include the prostaglandins, which function in pain, blood clotting, and smooth muscle contraction. Some animals—such as felines—cannot synthesize arachidonic acid from linoleic acid, so arachidonic acid is an essential fatty acid in those species. Unsaturated fatty acids are found primarily in plants, which provide a dietary source of essential fatty acids for both herbivores and omnivores. Strict carnivores, however, obtain their essential fatty acids from fishes or from the adipose (fat) tissue of birds and mammals.

***Vitamins***   **Vitamins** are important organic nutrients that serve as coenzymes for many metabolic and biosynthetic reactions. The two categories of vitamins are water-soluble and fat-soluble. Water-soluble vitamins, such as vitamin C, are not stored in the body and must be regularly ingested. Fat-soluble vitamins, such as vitamin A, are stored to some degree in adipose tissue. Not all animals require the same vitamins in their diet, however. Among vertebrates, for example, only primates and guinea pigs cannot synthesize their own vitamin C and must therefore consume it in the diet. **Table 46.2** summarizes the vitamins required by animals, their dietary sources, some of their important functions, and some health consequences associated with their deficiencies in humans.

***Minerals***   **Minerals** are inorganic ions required by animals for normal functioning of cells. Minerals such as iron and zinc are required as cofactors for or constituents of some enzymes and other proteins. Other minerals such as calcium are required for bone, muscle, and nervous system function. Still others—notably sodium and potassium—contribute to changes in electrical differences across plasma membranes and therefore are especially critical for heart, skeletal muscle, and neuronal activity. **Table 46.3** summarizes some of the most important minerals and their functions. Many minerals are

required in only trace amounts, far less than 1 mg/day in a relatively large mammal such as a human. Nonetheless, without regular consumption of these small amounts, serious health problems arise.

Regardless of the type of food consumed by an animal, the useful parts of the material that is eaten must in many cases be digested into molecules that the animal's cells can absorb or transport. Next, we will consider how animals digest their food and absorb the nutrients.

## 46.2 General Principles of Digestion and Absorption of Nutrients

**Learning Outcomes:**

1. Compare the processes of intracellular and extracellular digestion, and explain why one is far more common than the other.
2. **CoreSKILL »** Make predictions about the functions of parts of the alimentary canal based on their structural adaptations.
3. Distinguish between passive and active absorption of food.

Once food has been ingested, some of the macromolecules from the food must be broken down (digested) so that the nutrients can be absorbed by the cells of the digestive tract. In this section, we will examine some of the major principles of digestion and absorption in animals, beginning with where digestion takes place.

### Digestion Usually Occurs Extracellularly

Food is digested either inside cells (intracellularly) or, more commonly, outside cells (extracellularly). Intracellular digestion occurs only in some very simple invertebrates such as sponges and single-celled organisms and to a limited extent in cnidarians. It involves using phagocytosis to bring food particles directly into a cell, where the food is segregated from the rest of the cytoplasm in food vacuoles. Once inside these vacuoles, the macromolecules in the food are digested by hydrolytic enzymes into monomers (the building blocks of polymers), which then are moved out of the vacuole to be used directly by that cell. Intracellular digestion cannot support the metabolic demands of an active animal, because only extremely tiny bits of food can be phagocytosed at one time. This form of digestion also does not provide a mechanism for storing large quantities

| Table 46.2 | Vitamins Required by Animals* | | |
|---|---|---|---|
| **Class of nutrient** | **Dietary sources** | **Functions in vertebrates** | **Symptoms of deficiency in humans** |
| *Water-soluble vitamins* | | | |
| Biotin | Liver; legumes; soybeans; eggs; nuts; mushrooms; some green vegetables | Coenzyme for gluconeogenesis and fatty acid and amino acid metabolism | Skin rash; nausea; loss of appetite; mental disorders (depression or hallucinations) |
| Folic acid | Green vegetables; nuts; legumes; whole grains; organ meats (especially liver, kidney, heart) | Coenzyme required for synthesis of nucleic acids | Anemia (a lower than normal number of erythrocytes in the blood); depression; birth defects |
| Niacin | Legumes; nuts; milk; eggs; meat | Involved in many oxidation-reduction reactions | Skin rashes; diarrhea; mental confusion; memory loss |
| Pantothenic acid | Nearly all foods | Part of coenzyme A, which is involved in numerous synthetic reactions, including formation of cholesterol | Burning sensation in hands and feet; gastrointestinal symptoms; depression |
| Vitamin $B_1$ (thiamine) | Meats; legumes; whole grains | Coenzyme involved in metabolism of sugars and some amino acids | Beriberi (muscular weakness, anemia, heart problems, loss of weight) |
| Vitamin $B_2$ (riboflavin) | Dairy foods; meats; organ meats; cereals; some vegetables | Respiratory coenzyme; required for metabolism of fats, carbohydrates, and proteins | Seborrhea (excessive oil secretion from skin glands resulting in skin lesions) |
| Vitamin $B_6$ (pyridoxine) | Meats; liver; fish; nuts; whole grains; legumes | Coenzyme for over 100 enzymes that participate in amino acid metabolism, lipid metabolism, and heme synthesis | Seborrhea; nerve disorders; depression; confusion; muscle spasms |
| Vitamin $B_{12}$ | Meats; liver; eggs; some shellfish; dairy foods | Required for erythrocyte formation | Anemia; nervous system disorders leading to sensory problems; balance and gait problems; loss of bladder and bowel control |
| Vitamin C (ascorbic acid) | Citrus fruits; green vegetables; tomatoes; potatoes | Antioxidant and free-radical scavenger; aids in iron absorption; helps maintain healthy connective tissue and gums | Scurvy (connective tissue disease associated with skin lesions, weakness, poor wound healing, tooth decay); bleeding gums |
| *Fat-soluble vitamins* | | | |
| Vitamin A (retinol) | Liver; green and yellow vegetables; some fruits in small amounts | Component of visual pigments; regulatory molecule affecting transcription; important for reproduction and immunity | Night blindness due to loss of visual ability; skin lesions; impaired immunity |
| Vitamin D | Fish oils; fish; egg yolk; liver; synthesized in skin via sunlight | Required for calcium and phosphorus absorption from intestine; bone growth | Rickets (weakened, deformed bones) in children; osteomalacia (weak bones) in adults |
| Vitamin E | Meats; vegetable oils; grains; nuts; seeds; small amounts in some fruits and vegetables | Antioxidant; inhibits prostaglandin synthesis | Visual disturbances; possibly skeletal muscle atrophy; peripheral nerve disorders |
| Vitamin K | Legumes; green vegetables; some fruits; some vegetable oils (olive oil, soybean oil); liver; synthesized by bacteria in the large intestine | Component of blood clotting mechanism | Reduced blood clotting ability |

*Not all animals require each of these vitamins. Many mammals, for example, can synthesize vitamin C.

of food so that an animal can digest it slowly while engaging in its other activities.

Most animals digest food via extracellular digestion in a cavity of some sort. Extracellular digestion protects the interior of the cells from the actions of hydrolytic enzymes and allows animals to consume large prey or large pieces of vegetation. Food enters the digestive cavity, where it is stored, slowly digested, and absorbed gradually over long periods of time, ranging from hours (for example, after a human eats a pizza) to weeks (after a python eats a gazelle).

In the simplest form of extracellular digestion—seen in invertebrates such as flatworms and cnidarians—the digestive cavity has one opening that serves as both an entry and an exit port (**Figure 46.2**). The digestive cavity of these animals is called a **gastrovascular cavity**, because not only does digestion occur within it, but fluid movements in the cavity also serve as a circulatory—or vascular—system to distribute digested nutrients throughout the animal's body. Food within a gastrovascular cavity is partially digested by enzymes that are secreted into the cavity by the cells facing the cavity. As the food particles become small enough, they are phagocytosed by the lining cells

## Table 46.3    Minerals Required by Animals

| Mineral | Dietary sources | Functions in vertebrates | Symptoms of deficiency in humans |
|---|---|---|---|
| Calcium (Ca) | Dairy products; cereals; legumes; whole grains; green leafy vegetables; bones (eaten by some animals) | Bone and tooth formation; exocytosis of stored secretions in nerves and other cells; muscle contraction; blood clotting | Muscular disorders; loss of bone; reduced growth in children |
| Chlorine (Cl) | Meats; dairy foods; blood; natural deposits of salt | Participates in electrical, acid-base, and osmotic balance across cell membranes, notably those of neurons and heart cells | Muscular and nerve disorders |
| Chromium (Cr) | Liver; seafood; some nuts; meats; mushrooms; some vegetables | Required for proper glucose metabolism, possibly by aiding the action of the hormone insulin | Disorders of lipid and glucose balance in blood |
| Copper (Cu) | Fish; shellfish; nuts; legumes; liver and other organ meats | Required for hemoglobin production and melanin synthesis; required for connective tissue formation; serves as oxygen-binding component in some invertebrates | Anemia; bone changes |
| Iodine (I) | Seaweed; seafood; milk; iodized salt | Required for formation of thyroid hormones | Inability to make thyroid hormones, resulting in enlarged thyroid gland |
| Iron (Fe) | Liver and other organs; some meats; eggs; legumes; leafy green vegetables | Oxygen-binding component of hemoglobin; cofactor for some enzymes | Anemia |
| Magnesium (Mg) | Hay; grasses; whole grains; green leafy vegetables | Cofactor for many enzymes that use ATP as a substrate | Changes in nervous system function |
| Manganese (Mn) and molybdenum (Mo) | Nuts; whole grains; legumes; vegetables; liver | Cofactors for many enzymes | Poor growth; abnormal skeletal formation; nervous system disorders (convulsions) |
| Phosphorus (P) | Dairy foods; grains; legumes; nuts; meats | Bone and tooth formation; component of DNA, RNA, and ATP | Bone loss; muscle weakness |
| Potassium (K) | Meats; fruits; vegetables; dairy foods; grains | Participates in electrical, acid-base, and osmotic balance across cell membranes, notably those of neurons and heart cells | Muscle weakness; serious heart irregularities; GI symptoms |
| Selenium (Se) | Seafood; eggs; chicken; soybeans; grains | Antioxidant; cofactor for some enzymes | Keshan disease (damage to and loss of heart muscle) |
| Sulfur (S) | Proteins from any source | Component of two amino acids (methionine and cysteine) | Inability to synthesize many proteins |
| Sodium (Na) | Many fruits; vegetables; meats; table salt | Participates in electrical, acid-base, and osmotic balance across cell membranes, notably those of neurons and heart cells | Muscle cramps; changes in nerve activities |
| Zinc (Zn) | Widely found in meats; fish; shellfish (oysters); grains | Many functions related to tissue repair; sperm development; cofactor for many metabolic enzymes; required for certain transcription factors to bind to DNA | Stunted growth; loss of certain sensations like taste; impaired immune function; skin lesions |

and further digested intracellularly. Undigested material that remains in the gastrovascular cavity is expelled.

## Most Digestive Cavities Are Alimentary Canals with Specialized Regions and Openings at Opposite Ends

In contrast to the gastrovascular cavities of certain invertebrates, the digestive systems of all other animals consist of a single elongated tube, or alimentary canal, with an opening at both ends through which food passes from one end to the other (see Figure 46.1). Along its length, the tube usually contains smooth muscle. The contractions of this muscle help churn up the ingested food so that it is mechanically broken into smaller fragments.

The canal is lined on its interior surface by a layer of epithelial cells. These cells synthesize and secrete digestive enzymes and other factors into the hollow cavity (or lumen, as such cavities are called) of the alimentary canal, and they secrete certain hormones into the blood that help

regulate digestive processes. The cells are also involved with transporting digested material out of the canal and into an animal's body fluids. In addition, a diverse collection of microorganisms, called a microbiome, inhabit the alimentary canal. The microbiome plays a key role in digestion and even affects the growth of young animals (see Chapter 30).

Along its length, an alimentary canal has several specialized regions that vary according to species. Because of these specializations, digestive processes requiring acidic conditions, for example, can be segregated from those requiring neutral pH, and undigested food can be stored in one region while digestion continues in another area. The ability of some animals to store food in a stomach, for example, allows them to eat less frequently, leaving time for other activities.

## Absorption of Food May Be Passive or Active

Once food is digested, the nutrients must be absorbed by the epithelial cells that line specialized portions of the alimentary canal. This occurs in different ways, either by simple or facilitated diffusion or

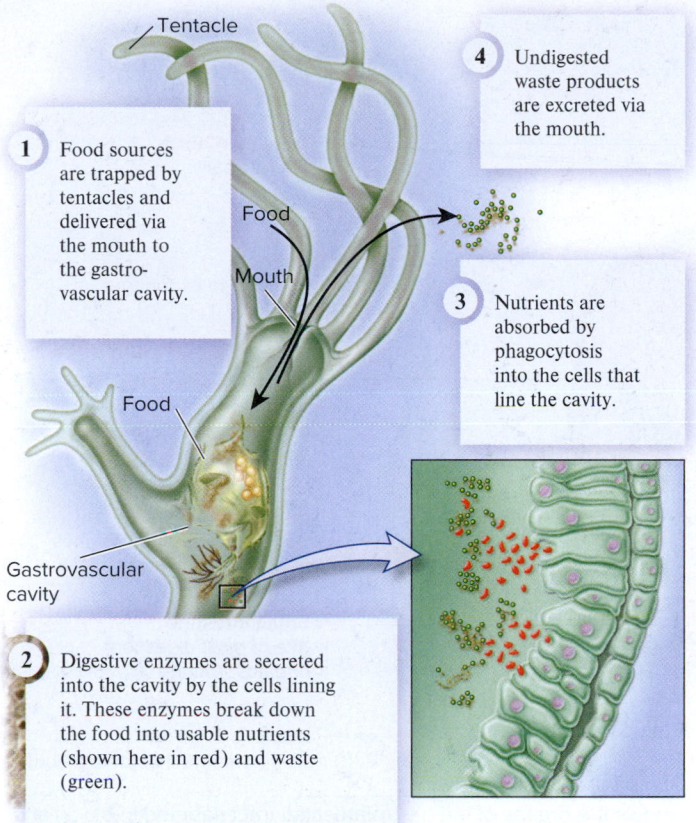

① **Food sources are trapped by tentacles and delivered via the mouth to the gastro-vascular cavity.**

④ **Undigested waste products are excreted via the mouth.**

③ **Nutrients are absorbed by phagocytosis into the cells that line the cavity.**

② **Digestive enzymes are secreted into the cavity by the cells lining it. These enzymes break down the food into usable nutrients (shown here in red) and waste (green).**

**Figure 46.2** **Extracellular digestion in a gastrovascular cavity.** In animals with a gastrovascular cavity, such as the cnidarian *Hydra* illustrated here, most digestion occurs extracellularly within the cavity.

by active transport. Small, hydrophobic molecules such as fatty acids diffuse down concentration gradients across the epithelium. Ions and other molecules are transported by facilitated diffusion or active transport. Minerals are ions and therefore, like all polar substances, do not readily cross plasma membranes. Instead, they are usually actively transported across the membranes of epithelial cells by ATP-dependent ion pumps. In other cases, small, hydrophilic organic nutrients are transported by secondary active transport, usually with $Na^+$.

After nutrients enter the epithelial cells of the alimentary canal, the cells use some of the nutrients for their own requirements. In animals with a closed circulatory system, most of the nutrients are transported across the basolateral surface of the epithelial cells (the side opposite the lumen), where they can enter into nearby blood or lymph vessels and circulate to the other cells of the body. Thus, nutrients enter the alimentary canal in food, are digested within the canal into monomers or other small fragments that can be transported into epithelial cells, and from there are released into the blood, where they can reach all of the body's cells. In the special case of water, osmotic gradients established by the transport of ions and other nutrients out of the epithelial cells draw water by osmosis from the canal, across the epithelial cells, and from there into the blood.

The mechanisms that activate and control the digestive and absorptive functions of the alimentary canal have been extensively studied in vertebrates and have great importance for human health. In the next sections, we will explore the structure and function of the digestive systems of vertebrates.

## 46.3 Overview of Vertebrate Digestive Systems

### Learning Outcomes:

1. Describe the general structure of the vertebrate digestive system.
2. Describe how food moves through regions of the alimentary canal, and explain how each region contributes to the processes of digestion and absorption.
3. **CoreSKILL »** Predict problems that might arise if saliva or acid could not be secreted into the mouth or stomach, respectively.
4. Outline how microorganisms can help digest cellulose in ruminants and other herbivores.

The vertebrate **digestive system** consists of the alimentary canal—also known as the gastrointestinal (or GI) tract—plus several accessory organs (**Figure 46.3**). As illustrated in Figure 46.3, the human alimentary canal consists of the oral cavity (mouth), pharynx, esophagus, stomach, small and large intestines, and anus. The accessory organs, not all of which are found in all vertebrates, are the salivary glands, liver, gallbladder, and pancreas. The differences between the digestive systems of various vertebrates reveal much about their respective diets. In this section, we will look at a few of the most important differences as we discuss the structure and function of each part of the vertebrate digestive system.

### The Vertebrate Alimentary Canal is Divided into Functional Regions

The alimentary canal is one continuous tube that varies in structure and function along its length, with three general sections. The first section, at the anterior end, functions primarily in the ingestion of food. It contains the mouth (with its accessory organs, the salivary glands), pharynx (throat), and esophagus. The middle section, which functions in the storage and initial digestion of food, contains one or more food storage or digestive organs, including the crop, gizzard, and stomach, depending on the species. This section also contains the upper part of the small intestine—where most of the digestion and absorption of food takes place—and accessory organs that connect with the intestine, including the pancreas, liver, and gallbladder. The third section, the posterior part of the canal, functions in final digestion and absorption and the elimination of nondigestible wastes. It consists of the remainder of the small intestine and, in most vertebrates other than fishes, a large intestine. Undigested material is defecated through an opening called an anus or, in many amphibians, reptiles, and birds, a cloaca (a common opening for the digestive and urogenital tracts).

From the midpoint of the esophagus to the anus or cloaca, the GI tract has the same general structure, with a lumen that is lined by a layer of epithelial and glandular cells. Included in the epithelial layer are secretory cells that release a protective coating of mucus into the lumen of the tract, and for this reason this layer of cells is also referred to as the mucosa. Other cells in the mucosal layer release hormones into the blood in response to the presence of food. Passing through the mucosal layer are ducts from secretory glands that release acid, enzymes, water, and ions into the lumen. From the stomach onward, the epithelial cells are linked along the edges of their apical surfaces by tight junctions that prevent digestive enzymes and undigested food from moving between the cells and out of the alimentary

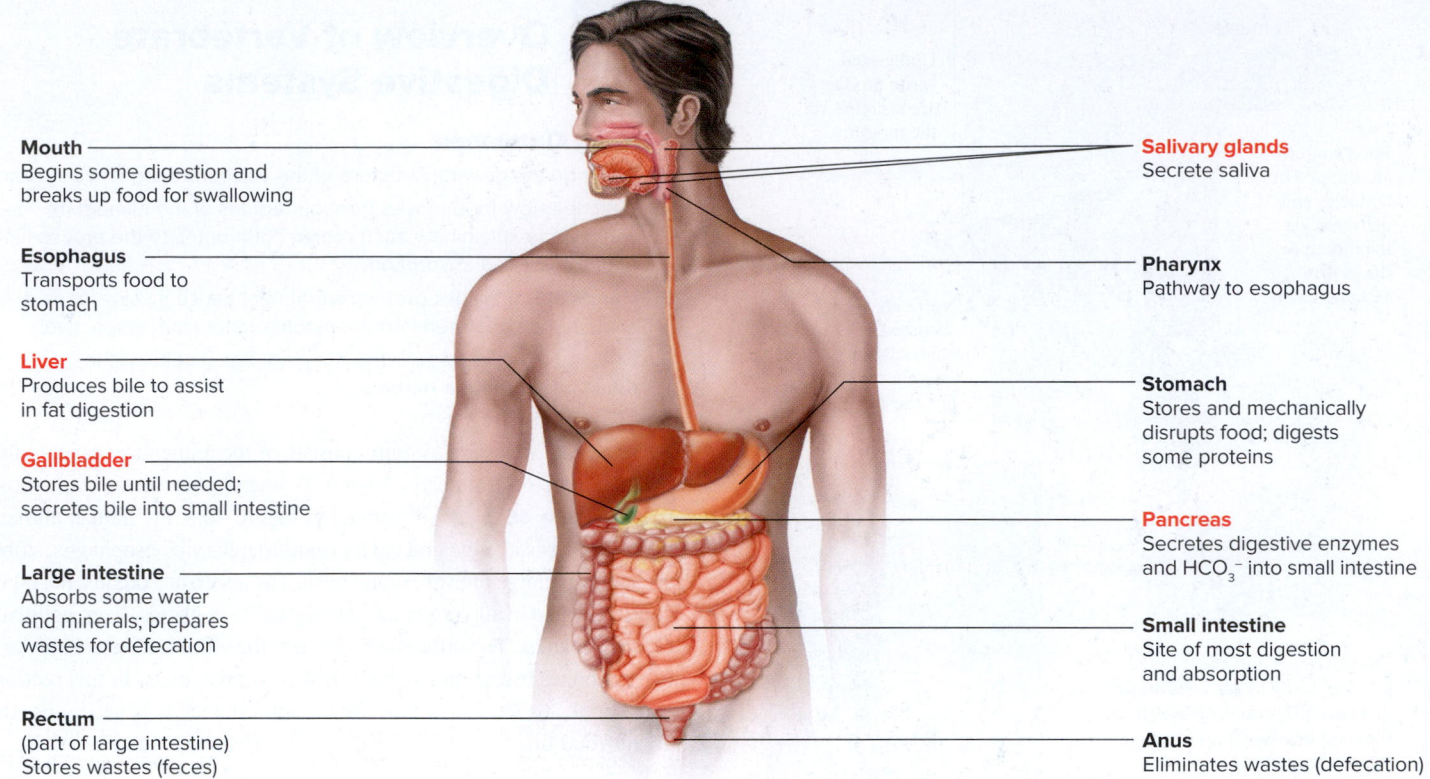

**Mouth**
Begins some digestion and
breaks up food for swallowing

**Esophagus**
Transports food to
stomach

**Liver**
Produces bile to assist
in fat digestion

**Gallbladder**
Stores bile until needed;
secretes bile into small intestine

**Large intestine**
Absorbs some water
and minerals; prepares
wastes for defecation

**Rectum**
(part of large intestine)
Stores wastes (feces)

**Salivary glands**
Secrete saliva

**Pharynx**
Pathway to esophagus

**Stomach**
Stores and mechanically
disrupts food; digests
some proteins

**Pancreas**
Secretes digestive enzymes
and $HCO_3^-$ into small intestine

**Small intestine**
Site of most digestion
and absorption

**Anus**
Eliminates wastes (defecation)

**Figure 46.3  A vertebrate digestive system, that of a human.** This figure shows the organs of the gastrointestinal tract (labeled in black) and the major accessory organs (labeled in red). Not all vertebrates share identical features of the digestive system. For example, some fishes lack a stomach, and many birds and some mammals lack a gallbladder.

 **Core Concept: Systems**  Organ systems such as the digestive system are typically composed of numerous different organs that may or may not be physically connected but function together.

canal. In some places, such as the small intestine, the apical surface of the mucosa is highly convoluted, a feature that increases the surface area available for digestion and absorption.

The epithelial cell layer is surrounded by layers of tissue made of smooth muscles, neurons, connective tissue, and blood vessels. The neurons are activated by signals originating from regions of the central nervous system that respond to the sight and smell of food. The neurons of the alimentary canal are also activated directly by the presence of food in the tract. Contraction of the smooth muscles is controlled by these neurons and results in mechanical mixing of the contents within the stomach and intestine. This helps speed up digestion and also brings digested foods into contact with the epithelium to facilitate absorption.

## Food Processing and Carbohydrate Digestion Begin in the Mouth

Ingestion and the start of digestion begin once food enters the mouth. In terrestrial vertebrates, the presence of food stimulates salivary glands in and around the mouth, cheeks, tongue, and throat to produce a flow of saliva—a watery fluid containing proteins, mucus, and antibacterial substances that keeps the mouth moist and clean. Unlike terrestrial vertebrates, fishes, which lack true salivary

glands, secrete mucus from specialized cells in their mouth and pharynx.

Saliva has several functions, not all of which pertain to all vertebrates:

- to moisten and lubricate food to facilitate swallowing;
- to dissolve food particles to facilitate the ability of specialized chemical-sensing structures called taste buds to taste food (refer back to Figure 44.26);
- to kill ingested bacteria with a variety of antibacterial compounds, including antibodies; and
- to initiate digestion of carbohydrates through the action of a secreted enzyme called **amylase**.

Digestion is the least important of these functions in most vertebrates, few of which produce salivary amylase. In humans and other primates, salivary amylase is present and accounts for only a very small percent of total carbohydrate digestion.

The other functions of saliva, however, are very important. For example, imagine trying to swallow food with a perfectly dry mouth. Also, the antibiotic properties of saliva help keep the mouth free of infection. In people who have had cancerous salivary glands removed, the teeth and gums often become so diseased that tooth loss occurs.

## Peristalsis Moves Swallowed Food Through the Esophagus to a Storage Organ

As food is swallowed, it moves into the next segments of the alimentary canal, the **pharynx** and **esophagus**. These structures, although not contributing to digestion or absorption, serve as a pathway to a storage organ.

*Pharynx and Esophagus*    The smooth muscles of the pharynx and esophagus contribute to swallowing. In the pharynx, swallowing begins as a voluntary action but continues in the esophagus by the process of **peristalsis**—rhythmic, spontaneous waves of smooth muscle contraction that begin near the mouth and end at the stomach. When most vertebrates eat, the mouth and stomach are roughly horizontal with respect to each other. Indeed, the head of a terrestrial grazing animal is even lower than its stomach when swallowing. The wavelike action of peristalsis ensures that swallowed food is moved toward the stomach and does not remain in the esophagus or even move backward into the pharynx if the head is lowered.

*The Crop*    In some animals, food moves directly from the upper esophagus to a storage organ called the **crop**, which is a dilation of the lower esophagus (**Figure 46.4**). Crops are found in most birds (and are also found in many invertebrates, including insects and some worms). Food is stored and softened by watery secretions in the crop, but little or no digestion occurs there. Because they process large amounts of tough food, birds that eat primarily grains and seeds have larger crops than do birds that eat only insects and worms. The material that birds regurgitate to their young comes from the crop. In some

species, such as pigeons and doves, the epithelial cells that line the crop lumen secrete a lipid-rich watery solution called crop milk or pigeon milk into the material to be regurgitated.

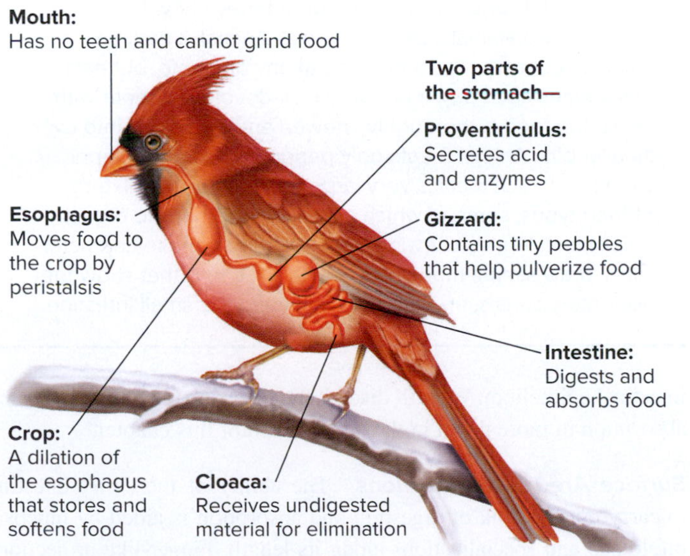

## Food Processing Continues and Protein Digestion Begins in the Stomach

Once food passes through the esophagus (and crop, if present), it reaches a storage organ, the stomach.

*General Features of Stomachs*    A **stomach** is a muscular, saclike organ that most likely evolved as a means of storing food. Stomach-like organs are found in all vertebrate classes, but a true stomach (defined as one that produces hydrochloric acid, HCl) is absent in many species of teleost fishes that eat very small amounts of food at once. In addition to its storage function, the muscular nature of the stomach helps mechanically break up large chunks of food into smaller, more easily digestible fragments. Lastly, the stomach partially digests some of the macromolecules in food and regulates the rate at which its contents empty into the small intestine.

Glands within the stomach mucosa secrete HCl and an inactive protein called pepsinogen into the stomach lumen. One function of the acid is to convert pepsinogen into the active enzyme **pepsin**, which is a protease and begins the digestion of protein. Why do stomach gland cells secrete pepsinogen instead of pepsin? The answer is that if the cells produced active pepsin, they would digest their own cellular proteins. The epithelium of the stomach is coated with a layer of alkaline mucus that protects the stomach lining from the effects of HCl.

In addition to its function in activating pepsin, HCl kills many of the microorganisms that may have been ingested with food, and also helps dissolve the particulate matter in food. In addition, the acid environment in the stomach (or gastric) lumen alters the ionization of polar molecules, especially proteins. This disrupts the structural framework of the tissues in food and makes the proteins more accessible to pepsin, which partially digests some proteins into smaller polypeptide fragments. By contrast, no significant digestion of carbohydrates or lipids occurs in the stomach.

*Specialization Among Vertebrates*    In birds, the stomach is divided into two parts: the proventriculus and the gizzard (see Figure 46.4). The **proventriculus** is the glandular portion of the stomach that secretes acid and pepsinogen. Partially digested and acidified food then moves to the **gizzard**, a muscular structure with a rough inner lining that grinds food into smaller fragments. This increases the surface area available to digestive enzymes that act in the small intestine.

The gizzard contains sand or tiny stones swallowed by the bird. The gritty sand and stones take the place of teeth and help mash and grind ingested food. Eventually, the pebbles in the gizzard become smaller as they are worn away, and they are eliminated. Thus, birds must occasionally restock the gizzard with new grinding stones. Grain-eating birds, particularly chickens and other fowl, many passerines (perching birds), and pigeons and doves, generally have more muscular gizzards than do insectivorous birds, because of the difficulty in breaking down plant matter. Gizzards, incidentally, are not unique to birds. Certain reptiles that are closely related to birds, such as crocodiles, also contain muscular gizzards. In addition, some species of herbivorous fishes (for example, members of the family

**Mouth:**
Has no teeth and cannot grind food

**Two parts of the stomach—**

**Proventriculus:**
Secretes acid and enzymes

**Esophagus:**
Moves food to the crop by peristalsis

**Gizzard:**
Contains tiny pebbles that help pulverize food

**Intestine:**
Digests and absorbs food

**Crop:**
A dilation of the esophagus that stores and softens food

**Cloaca:**
Receives undigested material for elimination

**Figure 46.4    The alimentary canal of birds.** The avian alimentary canal contains specialized regions for storing and softening food (the crop) and pulverizing food (the gizzard). The gizzard and proventriculus constitute the stomach. Undigested material is eliminated through the cloaca.

**Concept Check:**    *Smooth, polished stones have been found in the stomach region of fossilized skeletons of ancient sauropod dinosaurs. What does this discovery suggest about the alimentary canal of such animals?*

Acanthuridae) ingest quantities of inorganic grit with their meals, which helps to grind up food in a portion of the stomach that is modified into a strong, muscular grinding organ like a gizzard.

Digestive actions of the stomach reduce food particles to **chyme**, a solution that contains water, ions, molecular fragments of proteins, nucleic acids, and carbohydrates, droplets of fat, and various other small molecules. Very few of these molecules, except water, can cross the epithelium of the stomach. Therefore, little or no absorption of nutrients occurs in the stomach.

***Specializations in Ruminants***   Cellulose, the main macromolecule of plant cell walls (refer back to Figures 3.7 and 10.5), is a very important part of the diet of herbivores and many omnivores. In the human diet, cellulose is also called fiber or roughage, and is useful in helping eliminate solid wastes. The human digestive system is not equipped with the enzymes to digest cellulose, and it therefore passes through the system intact. Herbivores called **ruminants** (sheep, goats, llamas, and cows) also lack the enzymes, but they are able to digest cellulose with the help of their microbiome. Microorganisms living within their digestive tracts break down the cellulose into monosaccharides that can be absorbed along with other by-products of microbial digestion, such as fatty acids and some vitamins. In this way, bacteria and protists predigest the food, and the animal absorbs the broken-down cellulose and uses its sugar as a food source.

Ruminants have a complex esophagus and stomach consisting of several chambers, beginning with three outpouchings of the lower esophagus called the rumen, reticulum, and omasum, in that order (**Figure 46.5**). The rumen and reticulum contain the microorganisms that digest cellulose, and the omasum absorbs some of the water and ions released from the chewed and partially digested food (called the cud). The cud is occasionally regurgitated, rechewed, and swallowed again. Eventually, the partially digested food, the microorganisms, and the by-products of microbial digestion (including useful products such as fatty acids) reach the true stomach, the abomasum, which contains the acid and proteases typical of the vertebrate stomach. From the abomasum, the material passes to the intestines, where digestion and absorption are completed. The microorganisms are killed by the secreted stomach acid, and digested or eliminated. Some microorganisms survive in the rumen, however, and quickly multiply to replenish their populations, ensuring that a well-balanced mutualistic relationship is maintained.

## Most Digestion and Absorption Occur in the Small Intestine

Nearly all digestion of food and absorption of nutrients and water occur in the **small intestine**, the portion of the alimentary canal that leads from the stomach to the large intestine or, in animals without a large intestine, directly to the anus or cloaca. Hydrolytic enzymes break down macromolecules of organic nutrients into smaller monomers. Some of these hydrolytic enzymes are on the apical membrane of the intestinal epithelial cells; others are secreted by the pancreas and enter the intestinal lumen through a connecting duct. The products of digestion are absorbed across the epithelial cells and enter the blood. Vitamins and minerals, which do not require enzymatic digestion, are also absorbed in the small intestine. Water is absorbed by osmosis from the small intestine in response to the movement of nutrients across the

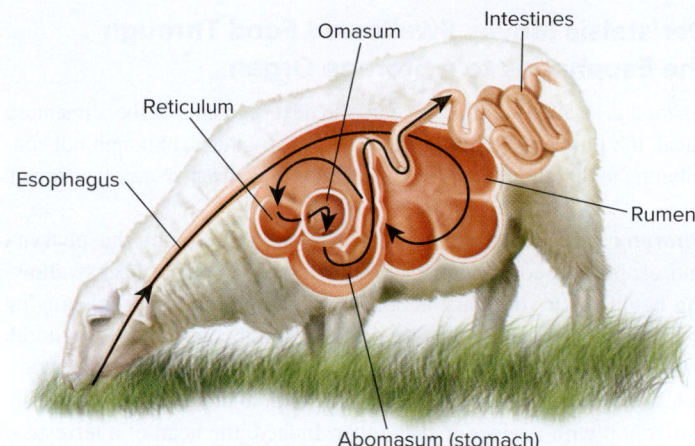

**Figure 46.5  Digestive tract of a ruminant.** Ruminants have a complex arrangement of three modified pouches arising from the esophagus: the rumen, reticulum, and omasum. The rumen and reticulum act as storage and processing sites (in large ruminants, the rumen may store up to 95 L of undigested food); the omasum absorbs some water and ions. Digestion by acid and pepsin takes place in the abomasum, which connects with the intestines.

 **Core Skill: Modeling**   The goal of this modeling challenge is to predict how the structure of the alimentary canal may differ in two animals with different eating patterns, diets, and teeth.

**Modeling Challenge:** Refer to the models of the alimentary canals of a human, a bird, and a ruminant in Figures 46.3, 46.4, and 46.5. Note the structural differences in the esophagus and stomach in these three vertebrates, reflecting their different diets and also whether or not they chew food. Let's consider two animals, called A and B. Animal A eats only a small bit of easily digestible food at any one time, but eats many times each day. It has very well-developed teeth with which the food is thoroughly chewed and torn apart into even smaller bits. Animal B eats only periodically (say, once per week), but each meal is very large and contains a mixture of food types, some of which are tough and hard to digest. This animal has no chewing teeth. Based on these characteristics, draw simple models for animals A and B that show their alimentary canals, up to but not including the small intestine.

intestinal epithelium. We will discuss the mechanisms of digestion and absorption in more detail in the next section of this chapter.

***Surface Area Specializations***   The ability of the small intestine to carry out the bulk of digestion and absorption is aided by mucosal infoldings and specializations along its length. Finger-like projections known as **villi** (singular, villus) extend into the lumen of the vertebrate small intestine (**Figure 46.6**). The surface of each villus is covered with a layer of epithelial cells, the plasma membranes of which form small projections called **microvilli**, known collectively as a **brush border**. The combination of folded mucosa, villi, and microvilli increases the small intestine's surface area about 600-fold above that of a flat-surfaced tube having the same length and diameter. The small intestine is small in diameter compared with the large intestine, but it is very long—3 m

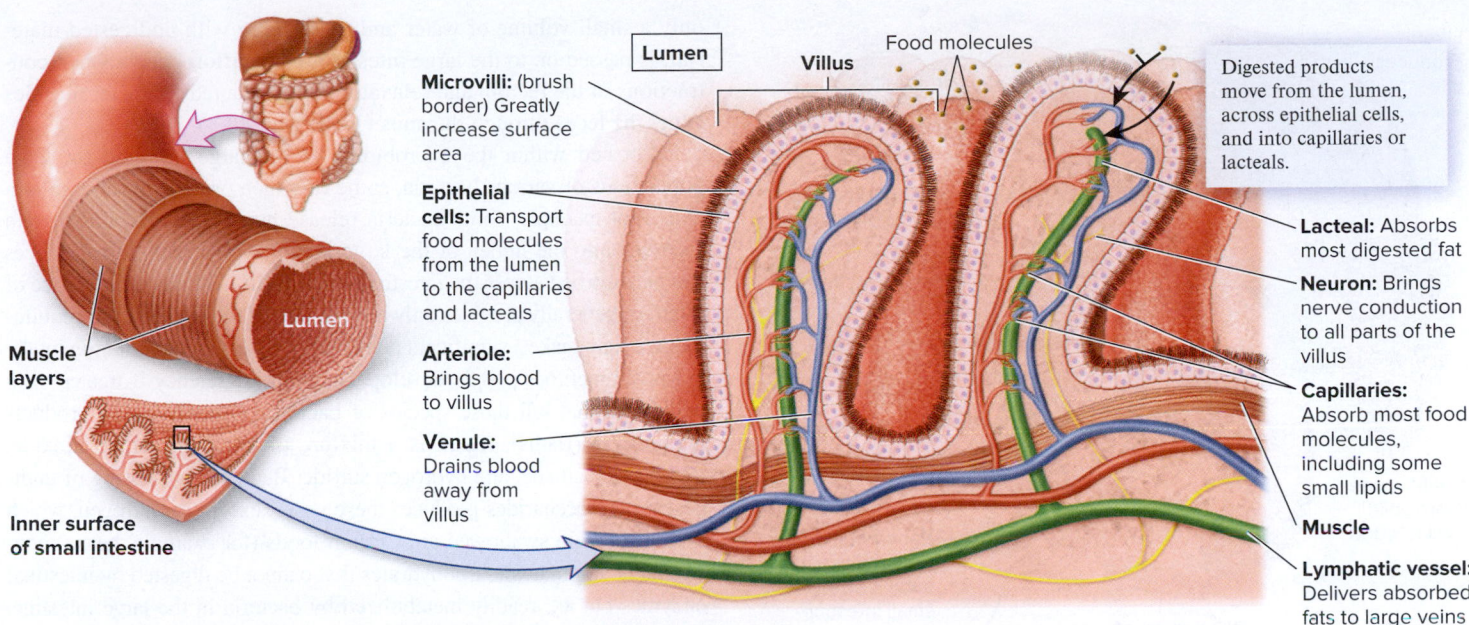

**Lumen**

**Villus**

Food molecules

**Microvilli:** (brush border) Greatly increase surface area

**Epithelial cells:** Transport food molecules from the lumen to the capillaries and lacteals

Digested products move from the lumen, across epithelial cells, and into capillaries or lacteals.

**Lacteal:** Absorbs most digested fat

**Neuron:** Brings nerve conduction to all parts of the villus

**Arteriole:** Brings blood to villus

**Venule:** Drains blood away from villus

**Capillaries:** Absorb most food molecules, including some small lipids

**Muscle**

**Lymphatic vessel:** Delivers absorbed fats to large veins

**Muscle layers**

**Lumen**

**Inner surface of small intestine**

**Figure 46.6  The specialized arrangement of tissues in the small intestine.** The inner surface of the small intestine is folded into numerous villi, which increase the surface area for digestion and absorption. Within each villus are capillaries and a lymphatic vessel called a lacteal, into which absorbed nutrients are transported. The epithelial cells of the villi have extensions from their surface called microvilli. The microvilli constitute the brush border of the intestine and greatly add to the total surface area.

 **Core Concept: Structure and Function**  The specializations along the inner surface of the small intestine create an enormous surface area that greatly enhances the rate at which the intestine is able to digest large molecules and absorb nutrients.

in an adult human (if removed from the abdomen, the small intestine can expand to almost twice as long as its normal length, because the muscles relax). This brings the total surface area of the human small intestine to about 300 m²—roughly the size of a tennis court! This enormous surface area means that the likelihood of an ingested food particle encountering a digestive enzyme and being absorbed across the epithelium is very high, so digestion and absorption proceed rapidly.

***Absorption into Blood and Lymph***  The center of each intestinal villus is occupied by capillaries, the smallest blood vessels in the body, and by a special type of larger vessel called a **lacteal**, which is part of the lymphatic system (see Figure 46.6). Most of the fat absorbed in the small intestine exists as bulky protein-coated particles that are too large to enter capillaries. Consequently, absorbed fat enters the larger, wider lacteals. Material absorbed by the lacteals eventually empties into the circulatory system. Other nutrients are absorbed directly into the capillaries and from there into veins.

## Accessory Organs Secrete into the Small Intestine Substances That Aid Digestion

As the chyme moves through the small intestine, three organs—the pancreas, liver, and gallbladder—secrete substances that flow via ducts into the first portion of the intestine, which is called the **duodenum** (**Figure 46.7**).

***Pancreas***  The **pancreas**, a complex organ located behind and below the stomach in humans (see Figure 46.3), has several functions, but in this chapter, we will focus on those that are directly involved in digestion. The gland secretes digestive enzymes and a fluid rich

in bicarbonate ions ($HCO_3^-$). The $HCO_3^-$ neutralizes the acidity of chyme, which would otherwise inactivate the pancreatic enzymes in the small intestine and could also damage the intestinal epithelium.

***Liver***  The **liver** is the site of bile production. **Bile** contains $HCO_3^-$, cholesterol, phospholipids, a number of organic wastes, and a group of amphipathic substances that are derived from cholesterol and are known as **bile salts**. The $HCO_3^-$, like that from the pancreas, helps neutralize acid ($H^+$) from the stomach, and the phospholipids and bile salts break up dietary fat droplets, thereby increasing their accessibility to digestive enzymes.

***Gallbladder***  The liver secretes bile into small ducts that join to form the common hepatic duct. Between meals, secreted bile is stored in the **gallbladder**, a small sac underneath the liver. During a meal, the smooth muscles in the gallbladder contract, secreting the bile solution into a connecting duct called the common bile duct (see Figure 46.7). The opening of a sphincter allows the bile to flow into the lumen of the small intestine. The gallbladder, therefore, is a storage organ that allows the release of large amounts of bile to be precisely timed to the consumption of fats. However, many animals such as horses and doves that secrete bile do not have a gallbladder. In humans, the gallbladder can be surgically removed without impairing bile secretion by the liver or its flow into the intestinal tract. People without a gallbladder can still digest fat but may need to limit the amount of fat they eat at one time because bile secretion can no longer be well timed to a meal.

The digested nutrients, along with water, are absorbed across the plasma membranes of the brush border cells. Peristalsis slowly propels the remaining contents through the posterior two portions of

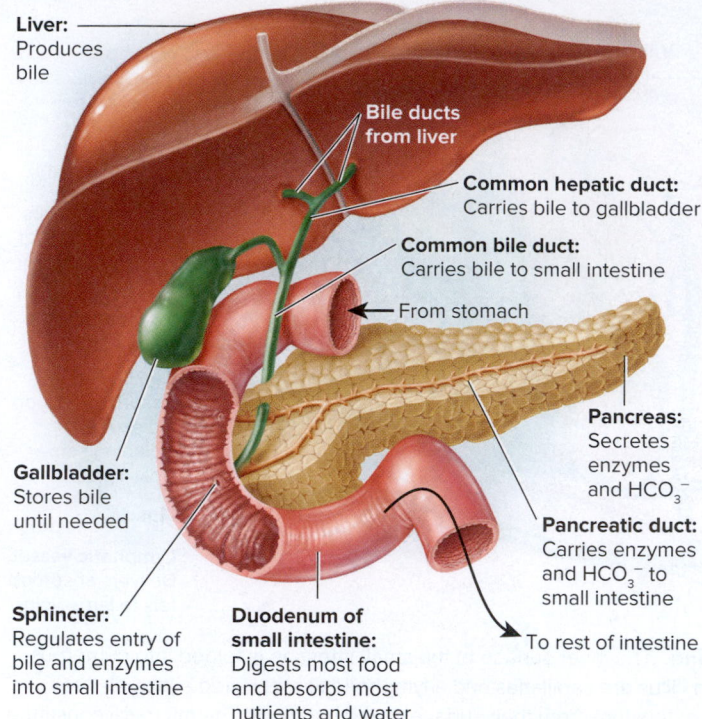

**Liver:**
Produces
bile

**Bile ducts
from liver**

**Common hepatic duct:**
Carries bile to gallbladder

**Common bile duct:**
Carries bile to small intestine

From stomach

**Gallbladder:**
Stores bile
until needed

**Pancreas:**
Secretes
enzymes
and $HCO_3^-$

**Pancreatic duct:**
Carries enzymes
and $HCO_3^-$ to
small intestine

**Sphincter:**
Regulates entry of
bile and enzymes
into small intestine

**Duodenum of
small intestine:**
Digests most food
and absorbs most
nutrients and water

To rest of intestine

**Figure 46.7  The arrangement and functions of the vertebrate liver, gallbladder, pancreas, and small intestine.** Bile drains from the liver into the gallbladder through the common hepatic duct. During a meal, bile is secreted from the gallbladder and enters the duodenum of the small intestine through the common bile duct. Simultaneously, secretions from the pancreas travel through the pancreatic duct, which joins with the common bile duct from the gallbladder and empties into the small intestine. The secretions from the two ducts into the small intestine are regulated by a muscular sphincter, a ring of smooth muscle that relaxes to allow secretions through or contracts to prevent their entry.

**Concept Check:**  *What advantage does an animal gain by having a gallbladder?*

the small intestine, called the jejunum and the ileum, where further absorption occurs. Finally, depending on the animal, the remaining material enters the large intestine or the anus or cloaca.

## The Large Intestine Concentrates Undigested Material, Which Is Then Egested via the Anus

The size of the large intestine varies greatly among different vertebrates, and the organ is vestigial or even absent in many animals, notably fishes. In humans, the large intestine is a tube about 6 cm in diameter and 1–1.5 m long. Its first portion, the **cecum**, forms a small pouch from which extends the appendix, a finger-like projection having no certain essential function but may contribute to the body's immune defense mechanisms. The next part of the large intestine in humans and other mammals is called the **colon**. The terminal portion of the colon empties into the **rectum**, a short segment of the large intestine that ends at the **anus**, the opening at the posterior end of the alimentary canal.

The primary functions of the large intestine are to store and concentrate fecal material before defecation and to absorb some of the remaining ions and water that were not absorbed in the small intestine. Because most substances are absorbed in the small intestine,

only a small volume of water and ions, along with undigested material, is passed on to the large intestine. **Defecation** occurs when contractions of the rectum and relaxation of associated sphincter muscles expel the feces through the anus (see Figure 46.3).

Located within the microbiome of the large intestine are large populations of various bacteria, some of which provide benefits to animals. For example, some bacteria release by-products such as certain vitamins into the lumen of the large intestine, where these substances can be absorbed across the intestinal epithelium. Although this source of vitamins generally provides only a small part of the normal daily requirement, it may make a significant contribution when dietary vitamin intake is low. Sometimes people develop a vitamin deficiency if treated with antibiotics that kill these species of bacteria. Other bacterial products include gas (flatus), which is a mixture of nitrogen, carbon dioxide, hydrogen, methane, and hydrogen sulfide. Bacterial processing of undigested polysaccharides produces these gases, except for nitrogen, which is derived from swallowed air. Certain foods (for example, beans) contain large amounts of carbohydrates that cannot be digested by intestinal enzymes but are readily metabolized by bacteria in the large intestine, producing gas. The bacteria are continually eliminated in the feces along with undigested material, so their populations normally stay in check.

## 46.4  Mechanisms of Digestion and Absorption in Vertebrates

### Learning Outcomes:

1. Describe the mechanisms of digestion and absorption of carbohydrates, proteins, and fats in vertebrates, and explain the importance of enzymes in some of these processes.
2. Explain why some nutrients do not require digestion prior to being absorbed.
3. **CoreSKILL »** Propose a hypothesis explaining why most human adults are lactose-intolerant to some degree.

The preceding sections provided an overview of nutrition and the basic features of digestive systems. We turn now to a more detailed description of the mechanisms by which carbohydrates, proteins, and lipids are processed in the vertebrate digestive system and how the end products of digestion are absorbed across intestinal cells. Nucleic acids are handled via mechanisms similar to those for these other nutrients and will not be discussed further.

### Carbohydrates Are Digested by Amylase and Brush Border Enzymes and Absorbed in the Small Intestine

In omnivores such as humans, most of the ingested carbohydrates are the polysaccharides (polymers of glucose) starch and cellulose from plants and glycogen from animals. The remainder consists of simple carbohydrates, such as the monosaccharides fructose and glucose in fruit, and disaccharides (combinations of two monosaccharides), such as lactose in milk. Humans also add the disaccharide sucrose (table sugar) to their food. Certain other animals consume sucrose from sources such as maple sap and sugarcane.

Although a very small amount of polysaccharide is digested in the mouth by salivary amylase, almost all polysaccharide digestion takes place in the small intestine due to the action of amylase secreted into the

**Figure 46.8 Digestion and absorption of carbohydrates in the small intestine.** Digestion and absorption occur in the same cells but are shown separately here for clarity. For simplicity, the microvilli making up the brush border are not shown.

 **Core Skill: Connections** Are the transport processes shown in Figure 46.8, facilitated diffusion and secondary active transport, unique to animal cells? Look back at Figure 37.3 for help.

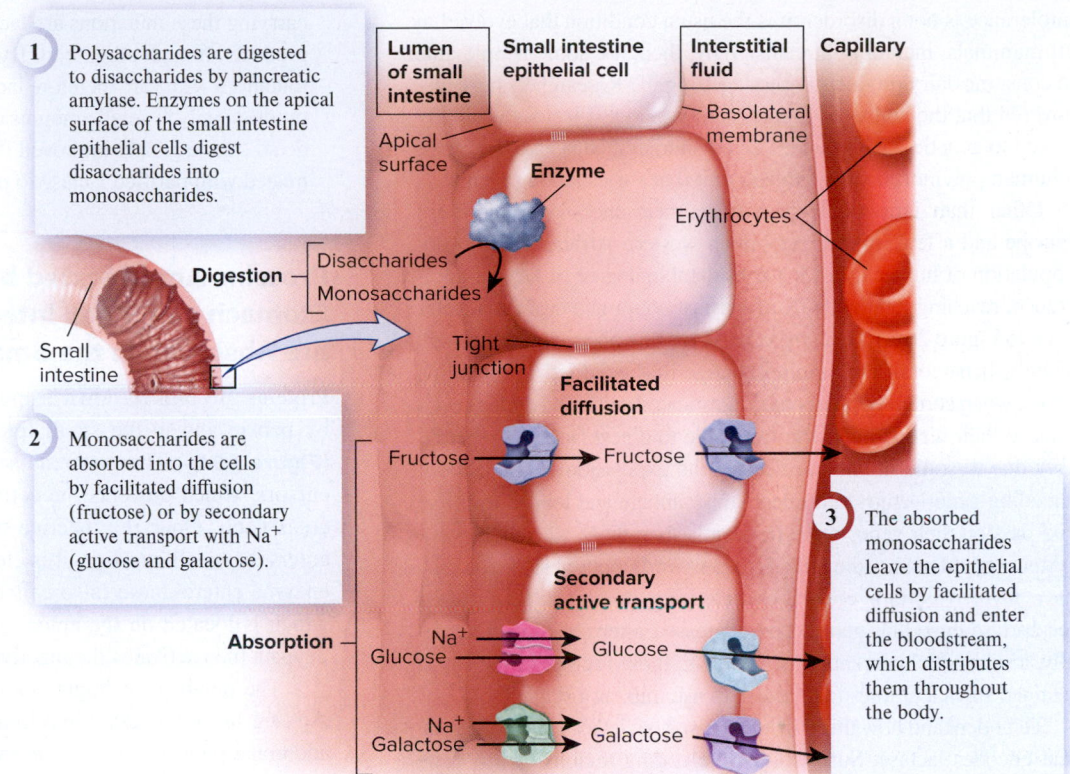

① Polysaccharides are digested to disaccharides by pancreatic amylase. Enzymes on the apical surface of the small intestine epithelial cells digest disaccharides into monosaccharides.

② Monosaccharides are absorbed into the cells by facilitated diffusion (fructose) or by secondary active transport with Na+ (glucose and galactose).

③ The absorbed monosaccharides leave the epithelial cells by facilitated diffusion and enter the bloodstream, which distributes them throughout the body.

intestine by the pancreas. The products of starch digestion via amylase are molecules of the disaccharide maltose. Maltose and any ingested sucrose and lactose are broken down into monosaccharides—fructose, glucose, and galactose—by enzymes located on the brush border of the small intestine epithelial cells (**Figure 46.8**). The monosaccharides are then absorbed into the epithelial cells. Fructose crosses the apical membrane (the surface facing the lumen) of the epithelial cells by facilitated diffusion, whereas glucose and galactose undergo secondary active transport (see Figure 5.19b) coupled to Na+. Monosaccharides then leave the epithelial cells by way of facilitated diffusion through transporters located in the basolateral membrane of the epithelial cells and enter the blood. The circulatory system distributes the monosaccharides and other absorbed nutrients to the cells of the body.

Do you feel ill after drinking milk or eating dairy products? If so, you are among the majority of people who cannot adequately digest lactose, the chief disaccharide in milk. A small percentage of humans, however, retain the ability to fully digest lactose throughout life. Next, we'll examine this phenomenon and consider its genetic and evolutionary basis.

 **Core Concept: Evolution**

## Evolution and Genetics Explain Lactose Intolerance

One of the defining features of mammals is that they produce and consume milk, a nutrient-rich solution containing all the macro- and micronutrients required by mammalian offspring. With rare exceptions, the milk of all mammals contains the disaccharide lactose. Lactose is digested by the intestinal brush border enzyme lactase, which cleaves lactose into glucose and galactose. After

infancy, humans with lactose intolerance cannot adequately digest lactose because their lactase is either inactive, absent, or present only in small amounts.

Milk is the sole food of most mammals shortly after birth, and the primary food for various lengths of time thereafter until weaning, the transition from consuming mother's milk to eating a diet of solid foods. Once weaned, mammals never again drink milk, except, of course, for humans. The popular notion of adult cats lapping up milk from a bowl is a misconception. Most adult cats are lactose-intolerant. Some visits to the veterinarian are related to GI symptoms caused by well-meaning owners who regularly feed milk to their adult pets.

Because the only dietary source for lactose is milk, it is not surprising that older mammals lose the ability to digest this disaccharide. This occurs because the gene that encodes the enzyme lactase is shut off at the age of weaning or shortly thereafter. The developmental mechanisms that turn off lactase production and activity are not firmly established, but they are known to involve decreased transcription of the lactase-encoding gene.

If an adult mammal were to drink milk, the undigested lactose would remain in the intestine. As a result, water that normally would be absorbed by osmosis with the digested monosaccharides formed from lactose would also remain in the intestine. Farther along the alimentary canal, microorganisms in the large intestine digest some of the lactose for their own use and, in the process, release by-products such as hydrogen and other gases. The combination of water retention and bacterial action results in GI symptoms such as diarrhea, gas, and cramps.

An estimated 90% of the world's human population cannot fully digest lactose after early childhood. In other words, lactose intolerance is not a disorder, it is the usual condition that evolved in all mammals, including humans. Why, then, are some people able to consume dairy products without getting ill? Researchers have discovered that the ability of human adults to digest lactose is clearly linked to genetic background. Lactose intolerance is an example of a human polymorphism, a genetic trait that varies among people.

Other than individuals who trace their ancestry to northern Europe and a few isolated regions in western Africa, nearly every population of humans shows a considerable degree of lactose intolerance, reaching nearly 100% in most of south and east Asia (refer back to Figure 26.23). One hypothesis for this phenomenon appears to be a behavioral and cultural change that occurred in Neolithic times, when certain populations domesticated cattle and added cow's milk to their diet. Adults who were able to digest lactose—presumably due to some mutation affecting the expression of the lactase-encoding gene—enjoyed a selective advantage and tended to thrive and pass on their genes more frequently than those whose digestive systems could not handle milk. In other words, natural selection may have favored certain populations carrying mutations that caused the lactase-encoding gene to be expressed after weaning. Eventually, a substantial percentage of people in those regions of the world retained sufficient intestinal lactase to use milk as a staple food.

To understand how this trait was passed on, let's examine the gene that encodes lactase. Surprisingly, the coding regions of this gene and its core promoter are identical whether individuals are lactose-tolerant or lactose-intolerant. Recently, however, Finnish investigators uncovered two single-nucleotide changes located in presumed regulatory sites that control the expression of the lactase-encoding gene. These changes are associated with prolonged expression of the gene, allowing it to continue to be active after weaning. People carrying these mutations are lactose-tolerant and can consume milk products through adulthood. By comparison, adults who lack these mutations—and are therefore lactose-intolerant—can consume dairy products only in small amounts or not at all. However, their ability to do so can be greatly improved if the product has been commercially treated with purified lactase to predigest the lactose.

## Proteins Are Digested by Proteases in the Stomach and Small Intestine and Amino Acids are Absorbed in the Small Intestine

Proteins are broken down to polypeptide fragments in the stomach by pepsin and in the small intestine by proteases such as **trypsin** (**Figure 46.9**). The pancreas secretes its enzymes as inactive precursors, which prevents the active enzymes from digesting the pancreas itself. Once the inactive form of trypsin (called trypsinogen) enters the small intestine, it is cleaved into the active molecule by the enzyme enterokinase (also called enteropeptidase), the active site of which is located on the apical surfaces of intestinal epithelial cells. Trypsin then activates the inactive forms of other pancreatic enzymes.

The polypeptide fragments produced by trypsin and other proteases are further digested into individual amino acids by the actions of additional proteases located on the brush border in the small intestine. Together, these proteases cleave off one amino acid at a time from the N-terminus and C-terminus of polypeptide fragments. Individual amino acids, coupled to $Na^+$, then enter the epithelial cells by secondary active transport. Amino acids leave these cells and enter the blood by facilitated diffusion across the basolateral membrane. Like carbohydrates, proteins are almost completely digested and absorbed in the duodenum of the small intestine.

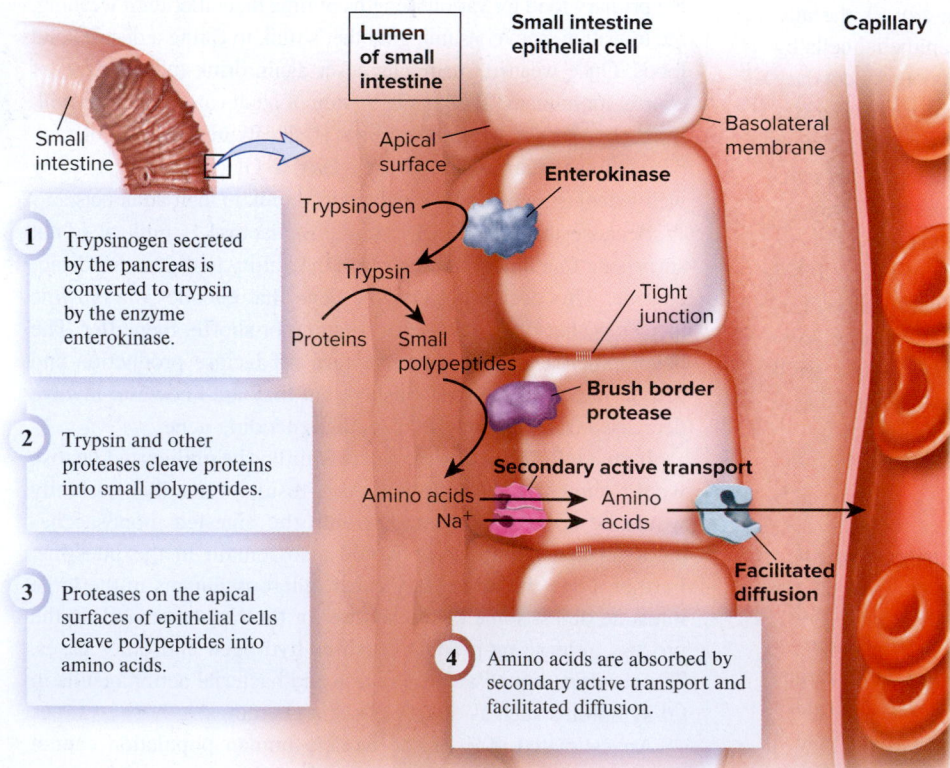

**1** Trypsinogen secreted by the pancreas is converted to trypsin by the enzyme enterokinase.

**2** Trypsin and other proteases cleave proteins into small polypeptides.

**3** Proteases on the apical surfaces of epithelial cells cleave polypeptides into amino acids.

**4** Amino acids are absorbed by secondary active transport and facilitated diffusion.

**Figure 46.9** Digestion of proteins and absorption of amino acids in the small intestine.

## Lipids Are Digested by Lipases and Absorbed in the Small Intestine

Most ingested lipids are in the form of triglycerides (fats). Fat digestion occurs almost entirely in the small intestine. The major digestive enzyme in this process is **lipase**, secreted by cells of the pancreas into the small intestine. The lipase reaction catalyzes the splitting of bonds in a triglyceride, producing two free fatty acids and a monoglyceride:

$$\text{Triglyceride} \rightarrow 2 \text{ Free fatty acids} + 1 \text{ Monoglyceride}$$

***Emulsification***    Fats are poorly soluble in water and aggregate into large droplets, as you can see if you shake a salad dressing made of oil and vinegar. Because lipase is a water-soluble enzyme, its digestive action in the small intestine can take place only at the surface layer of a lipid droplet.

The rate of digestion is substantially increased by the process of **emulsification**, which disrupts the large lipid droplets into many tiny droplets, increasing their total surface area and therefore providing greater exposure to lipase (**Figure 46.10a**). Emulsification occurs in two steps. First, the muscular contractions of the stomach and small intestine mechanically break up large lipid droplets into many smaller ones. Second, the nonpolar portions of amphipathic molecules from bile (bile salts and phospholipids) dissolve in a lipid droplet; the polar (charged)

regions dissolve in the surrounding watery chyme. The charged hydrophilic groups of bile salts and phospholipids repel each other; this prevents the small lipid droplets from coalescing into larger ones. The resulting suspension of small lipid droplets is called an emulsion.

***Formation of Micelles***    Although emulsification speeds up digestion of fats, absorption of the poorly soluble products of the lipase reaction is facilitated by a second action of bile salts and phospholipids, the formation of **micelles** (see Figure 46.10a). Micelles consist of bile salts, phospholipids, fatty acids, and monoglycerides clustered together. Temporarily storing breakdown products of lipid digestion in micelles allows the products to remain in solution, so they can slowly diffuse out of the micelle and across the apical membrane of the epithelial cell. Thus, the micelles keep most of the insoluble fat digestion products in small soluble aggregates while gradually releasing very small quantities of lipids to diffuse into the intestinal epithelium. Note in Figure 46.10a that it is not the micelle that is absorbed but rather the individual lipid molecules that are released from the micelle.

***Formation of Chylomicrons***    During their passage through the epithelial cells, fatty acids and monoglycerides are resynthesized into triglycerides in the smooth endoplasmic reticulum (SER). This process decreases the concentration of cytosolic free fatty acids and monoglycerides in the epithelial cells, and so maintains a diffusion gradient for these molecules from the lumen into the cells. The resynthesized triglycerides aggregate into **chylomicrons**, large droplets coated with amphipathic proteins that perform an emulsifying function similar to

**Figure 46.10    Digestion and absorption of emulsified fat in the small intestine.** (Note: Fatty acids and other molecules are not drawn to scale but are enlarged for clarity.)

Small intestine

Emulsified small lipid droplet

Bile salt

Triglyceride

Phospholipid

Small intestine epithelial cell

Apical membrane

Tight junction

**Lipase (digestion)**

Triglycerides are broken down by lipase into fatty acids and monoglycerides. These small lipids, along with cholesterol and vitamins, form micelles.

Diffusion (absorption) of small lipids

Fatty acid

Monoglyceride

Bile salt

The small lipids gradually leave the micelles and diffuse into the epithelial cells of the intestine.

Micelle (greatly enlarged)

**(a) Digestion of emulsified fats into micelles, and absorption into intestinal cells**

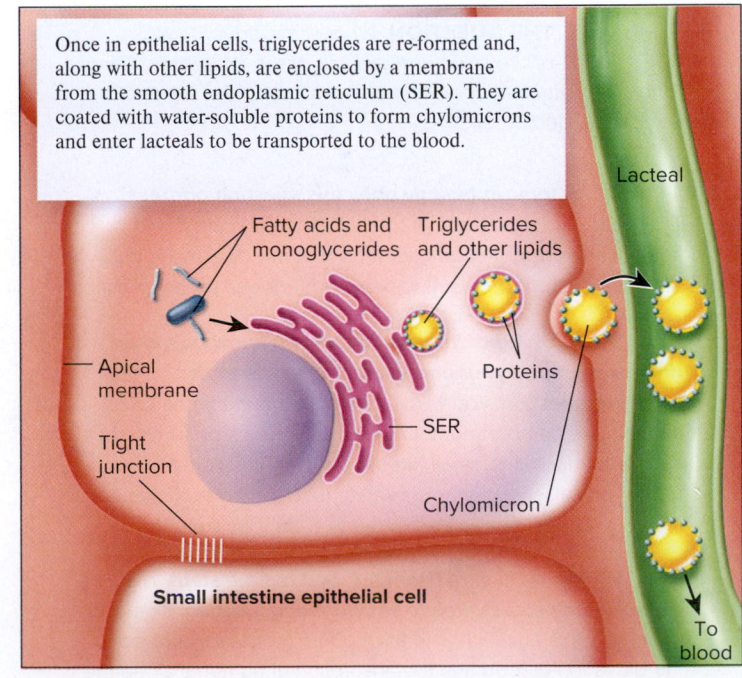

Once in epithelial cells, triglycerides are re-formed and, along with other lipids, are enclosed by a membrane from the smooth endoplasmic reticulum (SER). They are coated with water-soluble proteins to form chylomicrons and enter lacteals to be transported to the blood.

Lacteal

Fatty acids and monoglycerides

Triglycerides and other lipids

Apical membrane

Proteins

Tight junction

SER

Chylomicron

Small intestine epithelial cell

To blood

**(b) Absorption of triglycerides via the formation and secretion of chylomicrons**

that of bile salts (**Figure 46.10b**). In addition to triglycerides, chylomicrons contain phospholipids, cholesterol, and fat-soluble vitamins that have been absorbed by the same process that led to fatty acid and monoglyceride movement into the epithelial cells of the small intestine.

Chylomicrons are released by exocytosis from the epithelial cells and enter lacteals (see Figure 46.6). The fluid from the lacteals eventually empties into the blood.

## Vitamins, Minerals, and Water Are Not Digested and Are Absorbed Mostly in the Small Intestine

As stated earlier, vitamins, minerals, and water do not require digestion, and they are absorbed in their complete form. Most water-soluble vitamins are absorbed by diffusion or active transport in the small intestine. The fat-soluble vitamins—A, D, E, and K—follow the pathway for lipid absorption just described (see Figure 46.10). Any interference with the secretion of bile or the action of bile salts in the intestine decreases the absorption of fat-soluble vitamins.

Water is the most abundant substance in chyme. Small amounts of ingested water are absorbed in the stomach, but the stomach has a small surface area available for diffusion and lacks the solute-absorbing mechanisms that create the osmotic gradients necessary for water absorption. The great majority of water absorption occurs in the small intestine. The epithelial membranes of the small intestine are very permeable to water, which diffuses across the epithelium whenever an osmotic gradient is established by the active absorption of solutes, particularly $Na^+$, $Cl^-$, and $HCO_3^-$. Other minerals, such as $K^+$, $Mg^{2+}$, and $Ca^{2+}$, are present in smaller concentrations and are also absorbed by this mechanism, as are trace elements such as zinc and iodine. The mechanisms of absorption of these molecules generally involve transport proteins.

**BIO TIPS** **THE QUESTION** *What is the advantage of the structural division of the alimentary canal of animals into multiple compartments, including the stomach, small intestine, and so on?*

**T** **OPIC** *What topic in biology does this question address?*
The topic is the anatomy of the alimentary canal. More specifically, the question concerns how its subdivision into different compartments may be advantageous for an animal.

**I** **NFORMATION** *What information do you know based on the question and your understanding of the topic?*
From the question, you know that the alimentary canals of animals are compartmentalized into discrete structures, such as the stomach and small intestine, which has some advantage to the animals. From your understanding of the topic, you may recall what these structures are and what their functions are.

**P** **ROBLEM-SOLVING** **S** **TRATEGY** *Sort out the steps in a complicated process. Relate structure and function.* One way to answer this question is to think about the steps involved in the digestion of food and the absorption of nutrients along the alimentary canal. Which functions occur in which parts of the canal, and why? Next, think about the structural differences along the canal, and how those differences relate to the various functions.

**ANSWER** *Having discrete structures along the alimentary canal allows for the separation of functions. For example, food is acidified in the stomach, which is well suited to withstand the effects of low pH. The presence of a stomach allows for the slow processing of chyme as it enters the small intestine, allowing animals to eat large amounts of food and store much of it in the stomach without overwhelming the ability of the intestines to digest and absorb the food. Digestive enzymes can be released directly into the small intestine, where they are required, and not secreted throughout the entire canal. It would be wasteful to secrete digestive enzymes into the large intestine, where little or no digestion occurs. Absorptive structures, such as the brush border of the small intestine, are adapted for this function and possess digestive enzymes, transport proteins, and a large surface area. Such adaptations would have no value in the esophagus, for example, which acts primarily to move food to the stomach.*

## 46.5 Neural and Endocrine Control of Digestion

**Learning Outcomes:**
1. Explain how the nervous system controls different features of the digestive process.
2. Name three major hormones important for the regulation of digestion in vertebrates, and describe the role of each.

The digestive systems of animals are regulated by the nervous and endocrine systems. Neurotransmitters from neurons and hormones from endocrine glands control the volume of saliva produced, the amount of acid produced in the stomach, the timing and amount of secretions from the gallbladder and pancreas, and the rate and strength of muscle contractions along the alimentary canal. In this section, we will examine the major mechanisms by which the nervous and endocrine systems control the activity of the vertebrate digestive system.

### The Nervous System Controls Muscle and Secretory Activity

The nervous system can affect the activities of the digestive system in two major ways: (1) local control of muscle and glandular activity by the neurons within the alimentary canal and (2) long-distance regulation by the brain.

Within the walls along the length of the alimentary canal is a highly branched, interconnected collection of neurons that function in local control of the digestive system. These neurons interact with nearby smooth muscles, glands, and epithelial cells. Stimulation of neuronal activity at one point along the alimentary canal can lead to signals that are transmitted up and down the length of the canal. When food enters the small intestine, for example, the intestine is expanded. This directly activates stretch-sensitive neurons in the intestinal wall. Signals are sent from these neurons to the smooth muscles of the stomach, where they decrease the strength of the contractions of those muscles. This slows the rate at which chyme moves from the stomach into the small intestine, giving the intestine sufficient time for digestion and absorption of the amounts of chyme it receives. In this way, the alimentary canal can regulate its own function independent of the brain.

In long-distance regulation, the brain communicates with neurons in the walls of the stomach and intestines and thereby influences the movement and secretory activity of the alimentary canal. For example, emotional stress, a brain-related event, can affect digestive processes. Likewise, the sight, smell, and taste of food activate digestive functions even before food reaches the stomach. These stimuli when processed by the nervous system act via nerves from the brain as a feedforward mechanism (see Figure 41.13), so saliva production, stomach activity, and digestion are ready to begin immediately once food is ingested.

## Hormones Regulate the Rate of Digestion

Hormones are chemical messengers secreted by specialized cells into the blood, where they travel to all parts of an animal's body and act on various target cells (see Figure 9.3e). The hormones that control the digestive system are secreted mainly from cells scattered throughout the epithelium of the stomach and small intestine. One surface of each hormone-producing cell is exposed to the lumen of the alimentary canal. At this surface, chemical substances in chyme stimulate cells in the stomach epithelium to release a hormone called **gastrin**, which reaches all the parts of the stomach through the bloodstream (**Figure 46.11**). The presence of gastrin stimulates smooth muscle contraction in the stomach, which helps move chyme into the small intestine. Gastrin also stimulates acid production by stomach epithelial cells. In the small intestine, the arrival of chyme stimulates release of the hormones **cholecystokinin (CCK)** and **secretin** from intestinal epithelial cells. Cells of the pancreas respond to CCK by secreting digestive enzymes and to secretin by secreting acid-neutralizing bicarbonate ions ($HCO_3^-$) into the small intestine. CCK also stimulates contraction of the gallbladder and therefore bile release.

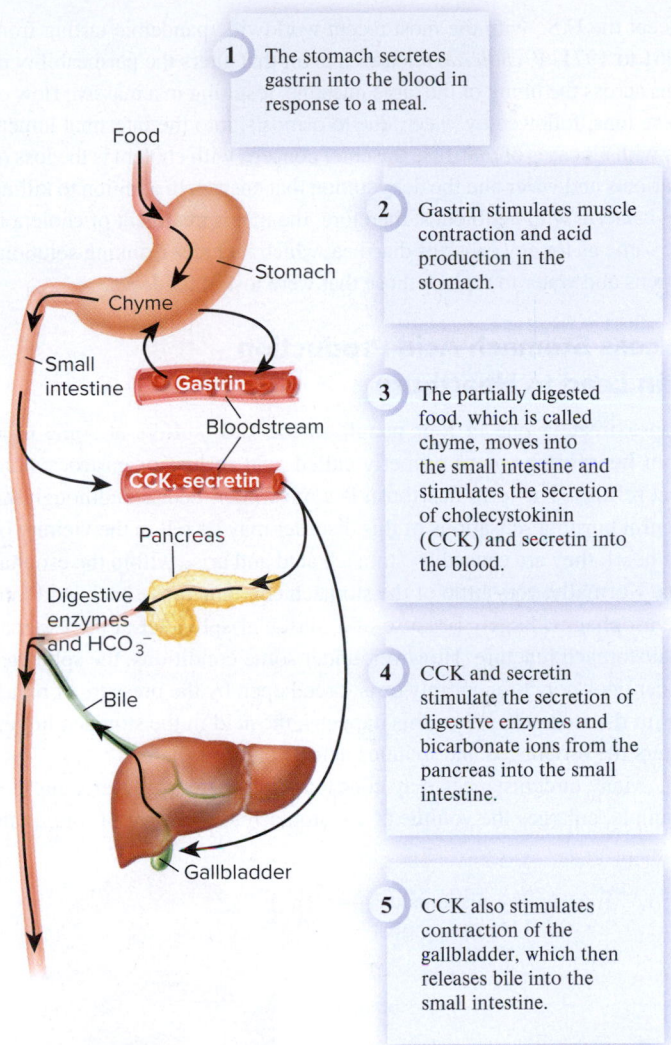

1  The stomach secretes gastrin into the blood in response to a meal.

2  Gastrin stimulates muscle contraction and acid production in the stomach.

3  The partially digested food, which is called chyme, moves into the small intestine and stimulates the secretion of cholecystokinin (CCK) and secretin into the blood.

4  CCK and secretin stimulate the secretion of digestive enzymes and bicarbonate ions from the pancreas into the small intestine.

5  CCK also stimulates contraction of the gallbladder, which then releases bile into the small intestine.

**Figure 46.11  Hormonal regulation of digestion in the stomach and small intestine.**

 **Core Skill: Connections**  In Chapter 41, you learned about the process of negative feedback, which helps maintain homeostasis in animals. Can you make a prediction about the possible feedback effects, if any, of CCK on the stomach? Do you think CCK might stimulate or inhibit smooth muscle activity and acid production in the stomach?

---

## 46.6  Impact on Public Health

### Learning Outcomes:

1.  Describe the impact of diarrhea, heartburn, and ulcers on public health.
2.  Identify the primary cause of ulcers.

---

As we have seen, the functioning of the vertebrate digestive system is extraordinarily fine-tuned by the nervous and endocrine systems. When people are well nourished, therefore, you might assume that digestive problems would be rare. However, each year in the U.S. alone, GI complaints account for approximately 40 million visits to doctors and 10 million visits to hospitals and emergency rooms. Among the most common GI problems registered by hospitals and physicians are diarrhea, heartburn, and ulcers.

### Diarrhea Is the Most Common Gastrointestinal Disorder Worldwide

According to the World Health Organization, worldwide there are over 2 billion cases of **diarrhea**—loose, watery stools occurring at least three times per day—every year. Most episodes of diarrhea resolve within a few days. Typically they result from infection with a pathogen, such as a virus or bacterium. Other times, however, diarrhea may be caused by food sensitivities (such as lactose intolerance,

discussed earlier), reactions to medications, stress-related disorders, or parasites that inhabit the rectum and colon.

Most cases of pathogen-related diarrhea in the U.S. result from exposure to one of several related types of bacteria, including those of the genera *Salmonella*, *Shigella*, and *Escherichia*. Often, these infections are eliminated by the body's immune system within a day or two, but sometimes medical attention is required. One cause of diarrhea stands apart as particularly dangerous, however. Cholera is a disease caused by the bacterium *Vibrio cholerae*, usually ingested by consuming contaminated food or water. Each year at least 2,000 people die of cholera worldwide, with another 100,000 people contracting the disease but surviving. Nearly all cases of cholera occur in Africa, parts of Asia (notably China), and India, although scattered outbreaks have occurred nearly everywhere except the U.S., with the most recent worldwide pandemic lasting from 1961 to 1971. *V. cholerae* releases a toxin that alters the permeability of ions across the lining of the large intestine, resulting in a massive flow of these ions, followed by water, due to osmosis, into the intestinal lumen. As with all cases of diarrhea, the chief concern with cholera is the loss of nutrients and water and the dehydration that ensues. In addition to killing the bacteria with antibiotics, therefore, the major treatment of cholera is the same as for any cause of diarrhea, which includes drinking solutions of ions and water to replace those that were lost in the feces.

## Excess Stomach Acid Production Can Lead to Heartburn

Approximately one in four people in the U.S. suffers at some time from **heartburn** (more properly called acid reflux, or gastroesophageal reflux). The term heartburn is a misnomer, because although the painful burning sensations of this disorder may be felt in the vicinity of the heart, they are caused by stomach acid and arise within the esophagus. Normally, very little of the stomach contents move backward into the esophagus, largely because of a muscular sphincter at the esophageal/stomach juncture. However, under some conditions, the sphincter either does not close entirely or is forced open by the pressure of material in the stomach. When this happens, the acid in the stomach lumen enters the esophagus and irritates neuron endings there.

Many circumstances may contribute to heartburn. Overeating, for example, enlarges the volume of the stomach to the point of forcing its contents through the esophageal sphincter. Lying down after a big meal removes the effect of gravity on food in the stomach and may allow some acid to leak backward. Heartburn is also associated with smoking and consumption of alcohol, citrus fruits (which are acidic), chocolate (which contains theobromine, an alkaloid related to caffeine that relaxes the esophageal sphincter), and fatty foods (which take longer to digest than other foods and therefore slow the rate at which the stomach empties). One of the most common causes is pregnancy. Toward the latter third of pregnancy, the growing fetus pushes up on the abdominal contents, which tends to force material from the stomach into the esophagus.

Common antacids contain calcium carbonate, which buffers the acid in the stomach and esophagus. In severe cases, heartburn can damage the walls of the esophagus enough to cause a chronic cough and pain, or even perforate the esophagus. Antacids may not be sufficient to treat these individuals, who are instead given drugs that inhibit the stomach's ability to produce acid. These drugs are among the most widely prescribed medications in the U.S.

## Erosion of the Walls of the Alimentary Canal Causes an Ulcer

Erosion of any portion of the wall of the alimentary canal due to any cause is called an **ulcer**. Most ulcers occur in the stomach (gastric ulcers) and the duodenum, and occasionally also in the lower esophagus, because these sites have the greatest exposure to acid (**Figure 46.12**). Ulcers are typically less than an inch wide; if left

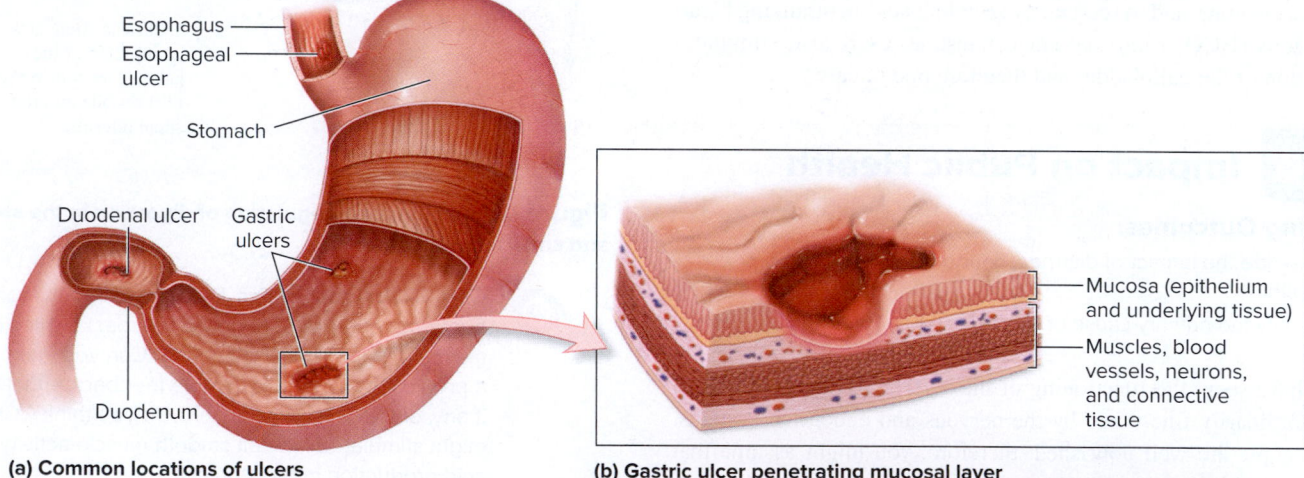

(a) **Common locations of ulcers**

(b) **Gastric ulcer penetrating mucosal layer**

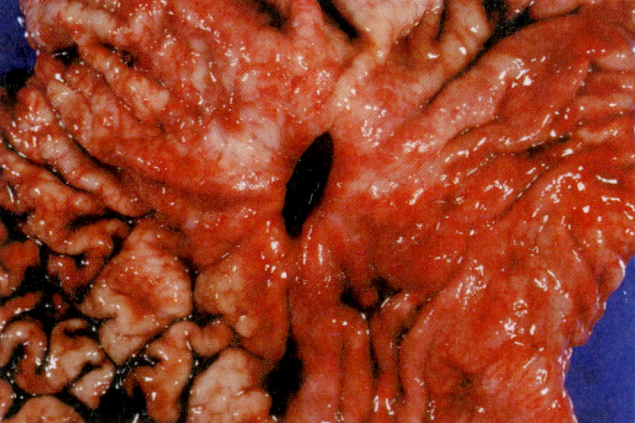

(c) **Human gastric ulcer**

**Figure 46.12 Gastric and duodenal ulcers. (a)** Common locations of ulcers in humans. **(b)** Illustration of a gastric ulcer eroding through the mucosal epithelium. **(c)** Photograph of an actual gastric ulcer that has penetrated through the stomach wall. c: ©Javier Domingo/Phototake

untreated, contents of the lumen may leak into the surrounding body cavity, where enzymes and acids from the stomach can do considerable damage. As many as 20 million Americans have an ulcer. Each year, 40,000 patients require surgery to repair tissue damaged by ulcers, and around 6,000 die due to complications from the disease.

Acid is essential for ulcer formation, but we now know that its overproduction is not usually the primary cause. Many patients with ulcers have perfectly normal rates of acid production. Contrary to popular belief, stress or spicy food is not a major cause of ulcers either. So what is the main cause of ulcers? In the 1980s, two Australian scientists, Barry Marshall and J. Robin Warren, proposed that most stomach ulcers arise from a bacterial infection. This idea did not gain quick acceptance because many scientists believed that bacteria could not survive the acidic conditions in the stomach. As we see next, however, Marshall and Warren were able to provide support for their hypothesis with compelling experimental data.

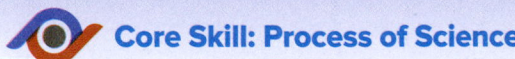

## Core Skill: Process of Science

## Feature Investigation | Marshall and Warren and Coworkers Demonstrated a Link Between Bacterial Infection and Ulcers

For many years, the conventional wisdom regarding ulcers was that they occurred as a consequence of overproduction of stomach acid, which in turn was due to factors such as stress, eating spicy foods, and smoking, among others. In the early 1980s, Barry Marshall and J. Robin Warren obtained evidence that some individuals with gastritis (inflammation of the stomach lining) or ulcers had active colonies of a bacterium in the stomach (initially thought to be *Campylobacter pylori* but later identified as *Heliobacter pylori*). This was a startling observation, because it had been assumed that no organism could survive the harsh pH and enzymes present in the stomach. In preliminary studies, Marshall and Warren further observed that treatment of such individuals with compounds that kill bacteria could reduce colonies of *H. pylori* and provide relief from gastritis. Based on these observations, they hypothesized that symptoms of gastritis and ulcers in certain individuals were caused by *H. pylori* infection and that eradication of the infection would cure the disease (**Figure 46.13**).

To test their hypothesis, the investigators recruited individuals who were being treated for ulcers at a local clinic. First, the presence of *H. pylori* and an ulcer was confirmed using visual inspection, biopsies, enzymatic tests, and bacterial cultures in vitro. On the basis of these exams and tests, 100 patients were chosen for the study. They were assigned to one of four groups based on the treatment they were to be given: antacid plus placebo, antacid plus antibiotic, bismuth plus placebo, or bismuth plus antibiotic. Bismuth is known to have antibacterial properties, but its mechanism of action is unclear. The antacid used in the study temporarily relieves the symptoms of

**Figure 46.13** Marshall and Warren and coworkers demonstrate that *H. pylori* infection is a cause of ulcers in humans.

(inset): ©Juergen Berger/SPL/Science Source

| HYPOTHESIS | *H. pylori* infection is a cause of ulcers in humans. |
| --- | --- |
| KEY MATERIALS | Endoscope, bacterial culture plates and medium, histologic stains, biopsy instruments. |

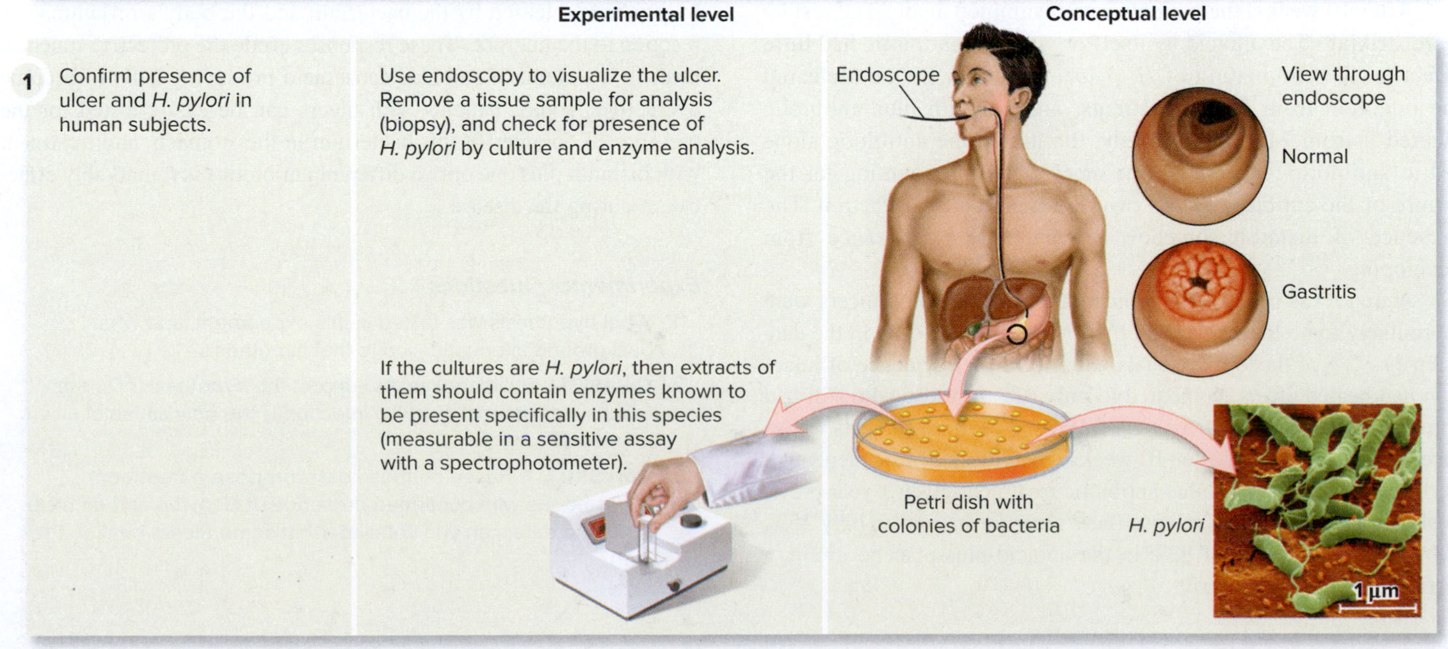

|  | Experimental level | Conceptual level |
| --- | --- | --- |
| **1** Confirm presence of ulcer and *H. pylori* in human subjects. | Use endoscopy to visualize the ulcer. Remove a tissue sample for analysis (biopsy), and check for presence of *H. pylori* by culture and enzyme analysis. | Endoscope — View through endoscope / Normal / Gastritis |
|  | If the cultures are *H. pylori*, then extracts of them should contain enzymes known to be present specifically in this species (measurable in a sensitive assay with a spectrophotometer). | Petri dish with colonies of bacteria / *H. pylori* / 1 μm |

**2**  Randomly assign subjects to 1 of 4 treatment groups.

Treatment Group 1: Antacid + Placebo
Treatment Group 2: Antacid + Antibiotic
Treatment Group 3: Bismuth + Placebo
Treatment Group 4: Bismuth + Antibiotic
↓ 10 weeks
Re-evaluate for ulcers and *H. pylori*
↓ 1 year
Follow-up exams

Group 1 is a control, because antacids help the symptoms of ulcers, but do not cure them. A placebo is an ineffective treatment with no function. An antibiotic is a compound that kills bacteria. Bismuth is an element that also has bacteria-killing properties.

**3**  THE DATA

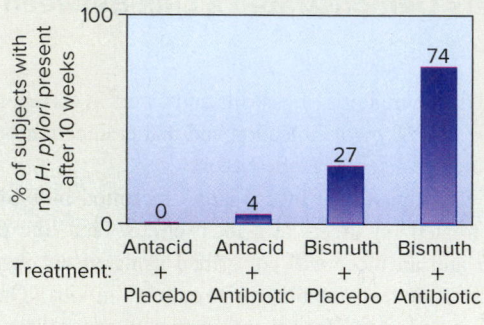
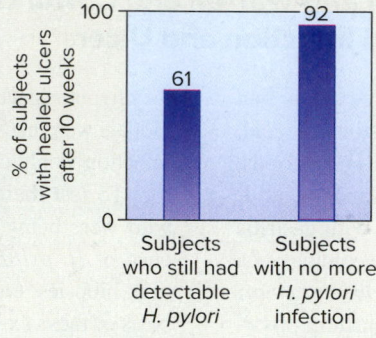
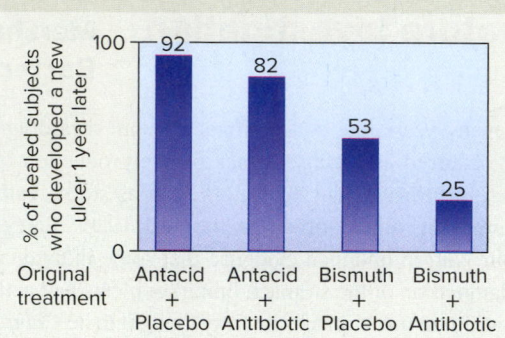

**4**  **CONCLUSION**  Infection with *H. pylori* is a significant cause of ulcers in humans.

**5**  **SOURCE**  Marshall, B. J., et al. 1988. Prospective double-blind trial of duodenal ulcer relapse after eradication of *Campylobacter pylori*. *The Lancet* 332: 1437–1442.

an ulcer by neutralizing acid. It does not cure ulcers, however, and therefore served as a control.

After 10 weeks, the patients were examined again. The results were striking: The antacid by itself or with the antibiotic had little effect on the elimination of *H. pylori* infection. Bismuth cleared the infection from 27% of patients, and bismuth plus antibiotic cleared it from 74%. Interestingly, the use of the antibiotic alone led to antibiotic-resistant strains of *H. pylori*, accounting for the failure of the antibiotic on its own to eliminate the infection. The presence of bismuth somehow prevented this resistance from developing.

A trend was observed in which patients with cured ulcers were more likely to no longer have *H. pylori* infection. As seen in the data (step 3), 92% of those individuals with no further evidence of infection had healed ulcers, compared to only 61% of those who still had signs of infection. Finally, a 1-year follow-up of those subjects whose ulcers had fully healed after 10 weeks (regardless of treatment) was performed. The bismuth-plus-antibiotic treatment given 1 year previously reduced the rate of relapse (new ulcer formation) to only 25%, compared with a high of 92% in the antacid-plus-placebo group, a statistically significant result.

The investigators concluded that *H. pylori* infection is a common cause of ulcers in humans. The mechanism seems to involve toxic by-products released by the bacterium and the body's inflammatory reaction to the microbe. These responses erode the protective mucosal layer in the stomach and also cause acid-producing cells to become overactive. Today, patients with ulcers can be easily tested for the presence of colonies of this bacterium in the stomach, and treatment with bismuth plus one or two different antibiotics is remarkably effective at curing the disease.

***Experimental Questions***

**1.** What hypothesis was tested in this experiment, and what background information led to that hypothesis?

**2.** **CoreSKILL »** Did the results support the hypothesis? Do the results indicate that *H. pylori* infection is the only cause of ulcers in people?

**3.** **CoreSKILL »** Based on these data, propose a treatment for someone with confirmed presence of *H. pylori* and an ulcer. From the data, can you conclude that some ulcers heal on their own?

## Summary of Key Concepts

- The four phases of food use in animals are ingestion, digestion, absorption, and egestion. Animals require organic nutrients—carbohydrates, proteins, lipids, nucleic acids, and vitamins—and inorganic nutrients including water and minerals. Differences in nutritional demands reflect an animal's physiology and environment (Figure 46.1).

### 46.1 Animal Nutrition

- Herbivores preferentially eat plants, carnivores consume animal flesh or fluids, and omnivores eat both plant and animal products.

- Essential nutrients include nine amino acids that cannot be synthesized by humans and many other animals; vitamins, which are organic nutrients that serve as coenzymes for metabolic and biosynthetic reactions; certain fatty acids; and minerals, which are inorganic ions required for many cellular functions (Tables 46.1, 46.2, 46.3).

### 46.2 General Principles of Digestion and Absorption of Food

- Intracellular digestion occurs in single-celled organisms and simple invertebrates. Most animals digest food via extracellular digestion. Flatworms and cnidarians have a gastrovascular cavity with one opening that serves as both entry and exit port (Figure 46.2).

- Nearly all other animals have an alimentary canal, open at the mouth and at the anus or cloaca and segregated into specialized regions, through which food passes from one end to the other.

- Once food has been digested, the nutrients can be absorbed via simple diffusion, facilitated diffusion, or primary or secondary active transport.

### 46.3 Overview of Vertebrate Digestive Systems

- The vertebrate digestive system consists of the alimentary canal plus several associated organs. The anterior portion of the canal contains the oral cavity, pharynx (throat), and esophagus. The middle portion contains food storage or digestive organs (crop, gizzard, and/or stomach), and the upper part of the small intestine (duodenum). The posterior part of the alimentary canal contains the remainder of the small intestine, the large intestine, the rectum, and the anus or cloaca (Figure 46.3).

- Saliva facilitates swallowing, dissolves food particles, and initiates digestion. Peristalsis moves food through the pharynx and esophagus to the stomach, where digestion begins (Figure 46.4).

- The stomach stores food, partially digests proteins, and regulates the rate at which chyme empties into the small intestine. Ruminants have a complex esophagus and stomach with several chambers that facilitates cellulose digestion (Figure 46.5).

- Nearly all digestion and absorption occur in the small intestine. The combination of villi and microvilli increases the small intestine's surface area and maximizes the efficiency of digestion and absorption (Figure 46.6).

- The pancreas secretes digestive enzymes and a bicarbonate-rich fluid that neutralizes the acidic chyme. The liver secretes bile, which aids in the digestion of fats (Figure 46.7).

- The large intestine concentrates undigested material, which is then eliminated from the anus.

### 46.4 Mechanisms of Digestion and Absorption in Vertebrates

- Carbohydrate digestion occurs in the small intestine, where maltose and other disaccharides are digested into monomers by enzymes and transported across epithelial cells into the blood (Figure 46.8).

- In humans, the ability to digest lactose after weaning is linked to a genetic mutation.

- Proteins are broken down to peptide fragments in the stomach by pepsin and further broken down into amino acids in the small intestine by trypsin and other proteases (Figure 46.9).

- Lipid digestion and absorption occurs entirely in the small intestine by the action of pancreatic lipase (Figure 46.10).

- Most water-soluble vitamins are absorbed by diffusion or active transport. Fat-soluble vitamins follow the pathway for lipid absorption.

### 46.5 Neural and Endocrine Control of Digestion

- The digestive systems of animals are regulated primarily by the nervous and endocrine systems. The nervous system affects the digestive system in two major ways: (1) local control by neurons in the alimentary canal and (2) long-distance regulation by the brain.

- Several hormones function together to regulate the rate of digestion (Figure 46.11).

### 46.6 Impact on Public Health

- Diarrhea, heartburn, and ulcers are common disorders involving the alimentary canal. Marshall and Warren and coworkers demonstrated that most ulcers are due to bacterial infection (Figures 46.12, 46.13).

## Assess & Discuss

### Test Yourself

1. The process of enzymatically breaking down large molecules into smaller molecules that can be used by cells is
   a. absorption.          d. digestion.
   b. secretion.           e. egestion.
   c. ingestion.

2. The term essential nutrients refers to
   a. all the carbohydrates, proteins, and lipids ingested by an organism.
   b. nutrients that an animal cannot manufacture from other molecules.
   c. nutrients that must be obtained from the diet in their complete form.
   d. all organic nutrients.
   e. Both b and c are true of essential nutrients.

3. Which of the following statements is *false*?
   a. Trypsin digests proteins in the small intestine.
   b. Enzymes on the brush border of the small intestine complete the digestion of carbohydrates.
   c. Free fatty acids are absorbed across the apical membranes of small intestinal epithelial cells.

d. Approximately 10% of the world's adult population is lactose-intolerant.

e. Bile is a mixture of wastes, $HCO_3^-$, cholesterol, and amphipathic compounds.

4. Which of the following statements is *true*?
   a. Intracellular digestion commonly occurs in vertebrates.
   b. Absorption of nutrients always requires active transport.
   c. Alimentary canals have two openings, whereas gastrovascular cavities have only one.
   d. Extracellular, but not intracellular, digestion requires hydrolytic enzymes.
   e. Most minerals are absorbed by simple diffusion.

5. To which part of the alimentary canal is the pancreas connected?
   a. esophagus         d. cecum
   b. stomach           e. large intestine
   c. small intestine

6. Which of the following statements regarding the vertebrate stomach is *false*?
   a. Its cells secrete the protease pepsin in its active form.
   b. It is a saclike organ that may have evolved to store food.
   c. Its cells secrete hydrochloric acid.
   d. It is the initial site of protein digestion.
   e. Little or no absorption of nutrients occurs there.

7. Absorption in the small intestine is increased by
   a. the many villi present on the inner surface of the small intestine.
   b. the brush border formed by microvilli on the cells of the villi.
   c. the presence of numerous transporter molecules on the epithelial cells.
   d. all of the above.
   e. a and b only.

8. In birds, the secretion of acid and pepsinogen occurs in the
   a. crop.              d. cloaca.
   b. gizzard.           e. gallbladder.
   c. proventriculus.

9. In many vertebrates, bile is produced by the _____ and stored in the _____.
   a. liver, small intestine
   b. gallbladder, liver
   c. pancreas, small intestine
   d. small intestine, gallbladder
   e. liver, gallbladder

10. Which of the following is *true* of the large intestine?
    a. It contains microorganisms that may produce useful by-products that can be absorbed.
    b. It stores and concentrates fecal material.
    c. Its cells absorb ions and water that remain in chyme after it leaves the small intestine.
    d. It varies considerably in size and may even be absent in some vertebrates.
    e. All of the above are true of the large intestine.

## Conceptual Questions

1. Distinguish between digestion and absorption.

2. Explain the functions of the crop and gizzard in birds. Can you propose a reason why a crop and gizzard did not evolve in humans?

3. **Core Concept: Structure and Function** Describe the various structural adaptations of the vertebrate small intestine from the organ to the cellular level, and explain how each is related to the functions of digestion and absorption.

## Collaborative Questions

1. Discuss the role of enzymes in digestion. What is their major function, and where are they produced and secreted?

2. Define nutrient. On the basis of that definition, do you consider water a nutrient? Briefly discuss the essential nutrients, vitamins, and minerals that animals must obtain through their diet.

# Control of Energy Balance, Metabolic Rate, and Body Temperature

## 47

**A genetically obese, leptin-deficient mouse and a wild-type mouse.** ©The Rockefeller University/AP Images

In 1997, two young cousins being treated for extreme obesity were brought to the attention of researchers studying a newly discovered hormone called **leptin**. Leptin had been demonstrated to inhibit appetite in laboratory rodents. The researchers hypothesized that the children—who weighed approximately 30 kg by age 2, and in one case, 86 kg by age 9—were not producing leptin. This hypothesis was confirmed by molecular analyses and traced to a mutation in the leptin-encoding gene. Subsequently, several other individuals with mutations affecting leptin production were identified, leading in each case to extreme childhood obesity. Treatment of these individuals with leptin has proven beneficial in restoring normal body weight. Although leptin deficiency is rare and is only one of many possible causes of obesity in humans, animal models of leptin deficiency, such as the mouse shown in the chapter opening photo, represent a remarkable breakthrough in our understanding of the genetic bases of the control of appetite and metabolism. How nutrients that supply energy are processed, stored, and used by animals' bodies in times of food abundance or fasting is a focus of this chapter.

In Chapter 46, we saw that all animals require nutrients to assemble the macromolecules that make up body tissues. Some of the ingested nutrients, such as carbohydrates, lipids, and proteins, also represent a form of energy that can be used to synthesize ATP within cells. Most animals, however, do not have a constant supply of nutrients. For example, insects, cephalopods, and all vertebrates sleep or have periods of greatly reduced activity for part of the day, during which time they do not eat. In addition, environmental changes may reduce the food supply, leading to long periods of fasting. As a consequence of the irregular and sometimes unpredictable flow of nutrients into the body, animals have evolved an array of mechanisms to adequately maintain concentrations of important fuel molecules in the blood or other body fluids, even when fasting. In the first two sections of this chapter, we will explore how ingested nutrients are stored in the body for such times of need and the mechanisms by which these stores are tapped.

**Metabolism** refers to all the bodily activities and chemical reactions in an organism that maintain life. **Metabolic rate** is the rate at which an organism uses energy to power these reactions.

Animals may have widely different metabolic rates, which determine the amount of nutrients they require. The greater an animal's metabolic rate, the more heat it generates as a by-product of breaking down nutrients and using the energy of their chemical bonds to synthesize ATP. Some of this heat escapes to the environment, and some is used to warm an animal's body. Heat, in turn, can speed up metabolism. Metabolism and body temperature are therefore closely related, and we will examine this relationship in the next two sections of the chapter. We will also discuss energy balance, the balance between energy consumption and expenditure. The chapter will conclude with a discussion of the public health impact of human disorders associated with metabolism, including obesity.

## 47.1 Use and Storage of Energy

**Learning Outcomes:**

1. Describe the processes by which glucose, triglycerides, and amino acids are absorbed during the absorptive state.

2. Outline the two major ways that vertebrates can increase their blood glucose concentration during the postabsorptive state.

3. Define glucose sparing, and explain why it is important.

Once nutrients have been ingested, they are either used or, in many cases, stored. The handling of nutrients can be divided into two alternating phases. The **absorptive state** occurs when ingested nutrients enter the blood from the alimentary canal. The **postabsorptive state** occurs when the alimentary canal is empty of nutrients and the body's own stores must supply energy. During the absorptive state, some ingested nutrients supply the immediate energy requirements of the body. The rest are added to the body's energy stores to be called upon during the next postabsorptive state. An average meal in a human requires about 4 hours for complete absorption. Therefore, our usual three-meal-a-day pattern places us in the postabsorptive state during the late morning and afternoon and part of the night. Total-body energy stores are adequate for the average human to withstand a fast of several weeks. By contrast, some animals can barely survive a single missed meal—particularly if they have low energy reserves—because their relative metabolic requirements are much greater than our own. In this section, we will focus on nutrient absorption, use, and storage in vertebrates, with a closer look at mammals.

## In the Absorptive State, Nutrients Are Absorbed and Used or Stored

The categories of nutrients that are absorbed either intact or after digestion during the absorptive state include carbohydrates, lipids, proteins, nucleic acids, vitamins, minerals, and water. We will look in depth at the first three of these; **Figure 47.1** gives an overview of what happens to these nutrients. Digested carbohydrates are absorbed as monosaccharides, including glucose (refer back to Figure 46.8).

Triglycerides are absorbed after first being digested into fatty acids and monoglycerides. Proteins are broken down into amino acids, which are then absorbed (refer back to Figure 46.9).

***Absorbed Carbohydrates*** The chief carbohydrate monomer absorbed from the alimentary canal of vertebrates is glucose, which is one of the two major energy sources during the absorptive state (triglycerides being the other). Much of the absorbed glucose enters all cells and is enzymatically broken down, resulting in the formation of hydrogen ions, carbon dioxide, and water and, in the process, providing the energy required to synthesize ATP from ADP and inorganic phosphate (Figure 47.1a). Because skeletal muscle makes up a large fraction of body mass in most vertebrates, it is a major consumer of glucose, particularly when an animal is active. In all vertebrates, skeletal muscle also incorporates some of the glucose into the polymer glycogen, which is stored in the muscle cells for future use. If more glucose is absorbed into the blood than is required for immediate energy demands, a portion of the excess is incorporated into glycogen in the liver, and the remainder is broken down to provide substrates for synthesizing triglycerides in adipose cells. The structures of glycogen and triglycerides are described in Figures 3.7 and 3.8, respectively.

***Absorbed Triglycerides*** Triglycerides are too large to diffuse across the plasma membranes of the intestinal epithelial cells. As described in Chapter 46 (refer back to Figure 46.10), triglycerides are digested into fatty acids and monoglycerides in the lumen of the small intestine, and then these breakdown products are resynthesized into triglycerides once they have diffused into the intestinal epithelial

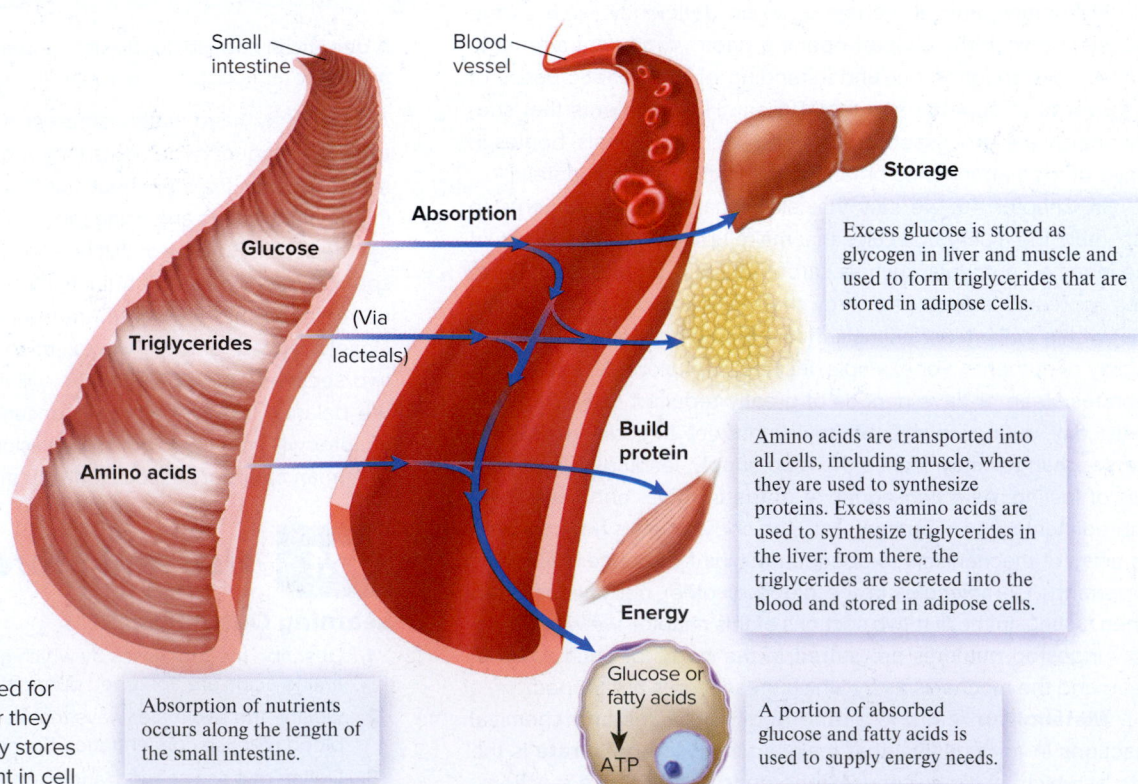

**Figure 47.1 Events of the absorptive state.** The products of digestion are absorbed into the blood along the length of the small intestine. (Note that triglycerides are absorbed into lacteals and from there into the blood; refer back to Figures 46.6 and 46.10 for details.) These nutrients are used for immediate energy demands, or they are deposited in cells as energy stores or as macromolecules important in cell function, such as for building proteins.

Small intestine

Blood vessel

Absorption

Glucose

Triglycerides

(Via lacteals)

Amino acids

Storage

Excess glucose is stored as glycogen in liver and muscle and used to form triglycerides that are stored in adipose cells.

Build protein

Amino acids are transported into all cells, including muscle, where they are used to synthesize proteins. Excess amino acids are used to synthesize triglycerides in the liver; from there, the triglycerides are secreted into the blood and stored in adipose cells.

Energy

Glucose or fatty acids

→ ATP

Absorption of nutrients occurs along the length of the small intestine.

A portion of absorbed glucose and fatty acids is used to supply energy needs.

cells. The triglycerides and other ingested lipids (for example, cholesterol) are packaged into chylomicrons, which enter lymph and from there the blood. As blood moves through adipose tissue, a blood vessel enzyme called lipoprotein lipase releases the fatty acids from the triglycerides in the chylomicrons. The released fatty acids then diffuse into adipose cells to re-form triglycerides (Figure 47.1b). These triglycerides are stored in adipose cells until an animal requires additional energy.

As with glucose, some of the ingested fatty acids are not stored but are used by most organs other than the brain during the absorptive state to provide energy. The relative amounts of carbohydrate and fat used for energy during the absorptive state depend largely on the composition of a meal.

***Absorbed Amino Acids***    Amino acids are taken up by all body cells, where they are used to synthesize proteins (Figure 47.1c). All cells require a regular supply of amino acids, because proteins are constantly being synthesized and degraded. However, unlike excess glucose and fatty acids, which are stored as glycogen and triglycerides, respectively, excess amino acids that are ingested are not stored as protein. Instead, excess amino acids are enzymatically broken down and their products are used in the synthesis of fatty acids, which then get incorporated into triglycerides. The triglycerides are then packaged and released into the blood to be taken up and stored in adipose cells. Therefore, eating large amounts of protein does not increase stores of body protein.

## In the Postabsorptive State, Stored Nutrients Are Released and Used

As the postabsorptive state begins, synthesis of glycogen and triglycerides slows, and the breakdown of these substances begins. During this state, macromolecules formed during the absorptive state are broken down to supply monomers that can be used for energy. No glucose is available to be absorbed from the intestines during this time, yet the blood glucose concentration must be maintained because the cells of the central nervous system (CNS) normally rely almost entirely on glucose for energy. A large decrease in blood glucose concentration can disrupt CNS functions, ranging from subtle impairment of mental function to seizures, coma, or even death.

The events that maintain the blood glucose concentration fall into two categories: (1) reactions that provide glucose to the blood and (2) cellular use of fatty acids for energy, thus sparing glucose for the CNS. Let's look at each of these.

***Production of Glucose from Glycogen and Other Sources***    Vertebrates can produce glucose during the postabsorptive state in two major ways: by breaking down glycogen and by synthesizing new glucose. First, the glycogen that was formed during the absorptive state can be broken back down into molecules of glucose by hydrolysis, a process known as **glycogenolysis** (**Figure 47.2a**). This process occurs primarily in the liver, from which the glucose is released into the blood, where it can travel to all cells. Skeletal muscle glycogen is

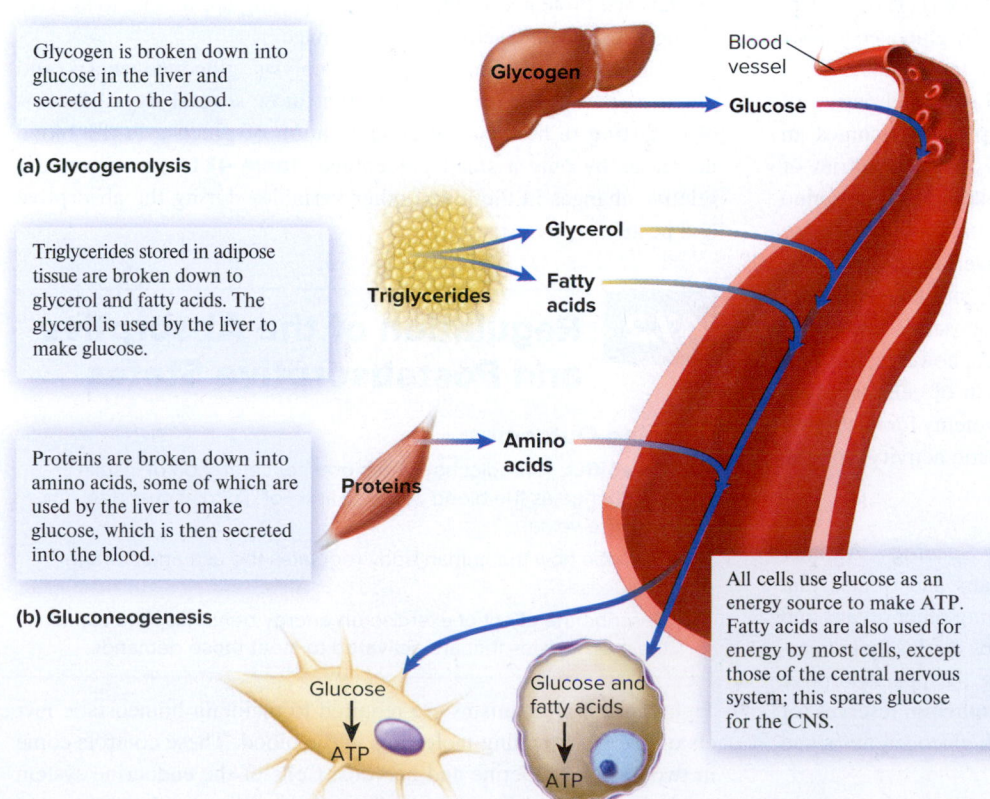

Glycogen is broken down into glucose in the liver and secreted into the blood.

**(a) Glycogenolysis**

Triglycerides stored in adipose tissue are broken down to glycerol and fatty acids. The glycerol is used by the liver to make glucose.

Proteins are broken down into amino acids, some of which are used by the liver to make glucose, which is then secreted into the blood.

**(b) Gluconeogenesis**

Glycogen

Blood vessel

Glucose

Glycerol

Fatty acids

Triglycerides

Amino acids

Proteins

All cells use glucose as an energy source to make ATP. Fatty acids are also used for energy by most cells, except those of the central nervous system; this spares glucose for the CNS.

Glucose

ATP

CNS

Glucose and fatty acids

ATP

All other cells

**Figure 47.2  Events of the postabsorptive state.** In the postabsorptive state, macromolecules formed and stored during the absorptive state are broken down into smaller molecules that can be released into the blood and used for energy. **(a)** This begins with glycogenolysis—the breakdown of glycogen into glucose. **(b)** In gluconeogenesis, the breakdown products of triglycerides and proteins, namely, glycerol, fatty acids, and certain amino acids, are used to synthesize glucose in the liver.

| Table 47.1 | Relative Changes in the Use and Generation of Energy Sources During the Absorptive and Postabsorptive States | | | | | |
|---|---|---|---|---|---|---|
| | Glucose absorption from alimentary canal | Glucose use by cells | Synthesis of triglycerides | Use of fatty acids for energy by cells | Breakdown of glycogen in muscle and liver | Blood concentration of glucose |
| Absorptive state | High | High | High | Low/moderate | Low | Normal |
| Postabsorptive state | Absent | Moderate (some glucose is spared for CNS use) | Low | High | High | Normal |

also broken down into glucose by glycogenolysis, but this glucose is used exclusively by the muscle cells and not secreted into the blood.

The amount of liver glycogen available to provide glucose during the postabsorptive state varies among animals, but it is generally sufficient to maintain blood glucose concentration for only a brief time, such as an overnight fast. Therefore, a second mechanism for maintaining blood glucose concentration is required if the postabsorptive state continues longer. In the process of **gluconeogenesis** (literally, creation of new glucose), enzymes in the liver synthesize glucose from noncarbohydrate precursors. The glucose is then secreted into the blood (Figure 47.2b). This process occurs in all vertebrates but appears to be especially important in mammals.

A major precursor for gluconeogenesis is glycerol, which is released from triglycerides in adipose tissue by the breakdown process called **lipolysis**. In lipolysis, lipase enzymes within adipose cells hydrolyze triglycerides into fatty acids and glycerol, both of which enter the bloodstream. The fatty acids diffuse into cells, where they are used as an alternative to glucose as an energy source (except in the CNS, which continues to require primarily glucose). The glycerol is taken up by the liver, where enzymes process it by gluconeogenesis to synthesize glucose, which is then released back into the blood.

If the postabsorptive state continues for an extended period of time—as when an animal fails to find food—protein becomes an increasingly important source of blood glucose. Large quantities of protein in muscle and other tissues can be broken down to amino acids without serious tissue damage or loss of function. The amino acids enter the blood and are taken up by the liver. In the liver, the amino group is removed from each amino acid, and the remainder of the molecule is used for the synthesis of glucose by a stepwise series of enzyme-catalyzed reactions. This process, however, has limits. Continued protein loss can result in the death of cells throughout an animal's body because they depend on proteins for such vital processes as plasma membrane function, enzymatic activity, and the formation of organelles.

***Lipid Metabolism by Other Tissues: Glucose Sparing***   Another way that glucose is made available to the organs and tissues that require it the most—such as the brain—is by having other organs and tissues decrease their dependence on glucose. They do this by increasing their use of fat as an energy supply during the postabsorptive state. This metabolic adjustment, called **glucose sparing**, reserves (or spares) the glucose produced by the liver through glycogenolysis and gluconeogenesis for use by the CNS.

The essential step in glucose sparing is lipolysis, the breakdown of adipose tissue triglycerides, which, as stated earlier, liberates fatty

acids and glycerol into the blood. In vertebrates, the circulating fatty acids are taken up and used to provide energy by almost all tissues, excluding the central nervous system, whose cells do not express the enzymes required to break down fatty acids for energy.

Of the vertebrate body's tissues and organs, the liver is unique in that most of the fatty acids entering it during the postabsorptive state are not used by that organ for energy. Instead they are processed into three small compounds collectively called **ketones**. Ketones are released into the blood during prolonged fasting. They provide an important energy source for the many cells, including those of the brain, that are able to oxidize these compounds via the citric acid cycle.

The use of fatty acids and ketones during fasting provides energy for the body, sparing the available glucose for the brain. Moreover, as just mentioned, the brain can use ketones for energy, and it does so increasingly as ketones build up in the blood during the first few days of a fast. The survival value of this phenomenon is significant. When the brain decreases its glucose requirement by using ketones, much less protein breakdown is required to supply amino acids for gluconeogenesis. Protein stores last longer, enabling the animal to survive a longer fast without serious tissue damage.

The combined effects of glycogenolysis, gluconeogenesis, and glucose sparing are so effective, that after several days of complete fasting in humans the concentration of glucose in the blood decreases by only a small percentage. **Table 47.1** summarizes the relative changes in these and other variables during the absorptive and postabsorptive states.

## 47.2 Regulation of the Absorptive and Postabsorptive States

**Learning Outcomes:**

1. **CoreSKILL »** Predict how the blood concentration of insulin will change as the blood concentration of glucose changes, and vice versa.
2. Describe how the human body regulates the use and storage of glucose.
3. Describe the effect of exercise on energy demands, and explain the mechanisms that are activated to meet those demands.

Tight control mechanisms are required to maintain homeostatic levels of energy-providing molecules in the blood. These controls come in two forms: endocrine and nervous. Cells of the endocrine system produce the blood-borne long-distance signaling molecules called hormones. Several hormones function together with signals arising from cells of the nervous system in these control mechanisms. In this section, we will see that one common function of the endocrine and nervous systems is to regulate the processes of glycogenolysis and gluconeogenesis so that glucose is made available to cells at all times.

## Insulin Is a Key Regulator of Metabolism

The blood concentration of **insulin**, a polypeptide hormone made by the pancreas, increases during the absorptive state and decreases during the postabsorptive state. Insulin regulates metabolism primarily by regulating the blood glucose concentration. It does this by promoting the transport of glucose from extracellular fluid into cells, where it can be used for metabolism. Glucose is a polar molecule that cannot cross plasma membranes without the aid of a transport protein.

Insulin stimulates glucose uptake by binding to a cell-surface receptor and stimulating an intracellular signaling pathway. This pathway increases the availability of transport proteins called glucose transporters (GLUTs) in the plasma membrane (**Figure 47.3**). These GLUTs are located within preformed vesicles stored in the cytosol of cells. When these vesicles are stimulated by the insulin signaling pathway, they fuse with the plasma membrane, making more GLUTs available to transport glucose into the cell. Consequently, insulin functions to decrease the blood glucose concentration because it increases uptake of glucose into cells.

Insulin exerts its effects mainly on skeletal and cardiac muscle cells and adipose cells, because these cells have insulin receptors in their plasma membranes. However, animal cells have many types of GLUTs, but only one requires insulin for its activity, as described next.

**Core Concept: Evolution**

## A Family of GLUT Proteins Transports Glucose in All Animal Cells

All animal cells use glucose for energy and thus require transporters to move glucose across their plasma membranes. GLUTs and the genes that encode them are evolutionarily ancient, and their structure is very similar across phyla. For example, almost 70% of the sequence of a gene that encodes a GLUT in *Drosophila* is identical to that of a gene encoding one of the GLUTs in humans. In more closely related phyla, such as birds and mammals, GLUTs are even more similar, with up to 95% of their amino acid sequences being the same. These and other considerations indicate that the GLUTs arose by accumulated mutations of a common ancestral gene. Over the course of evolution, some GLUTs acquired differences in substrate specificity (for example, some transport fructose rather than glucose) and in regulatory capacity (such as whether or not their expression is regulated by insulin).

In mammals, GLUTs make up a family of at least 14 related proteins that share similar structures but are expressed in different tissues. The different GLUTs vary in their ability to bind glucose. For example, some GLUTs have high affinity for glucose, and others have low affinity. High affinity means the protein can bind glucose even at very low concentrations of glucose. Let's look at the properties of three GLUTs, named GLUT1, GLUT3, and GLUT4. In mammals, skeletal muscle and adipose cells express the protein GLUT4, which has a low affinity for glucose, but one that is sufficient for the concentration of glucose normally

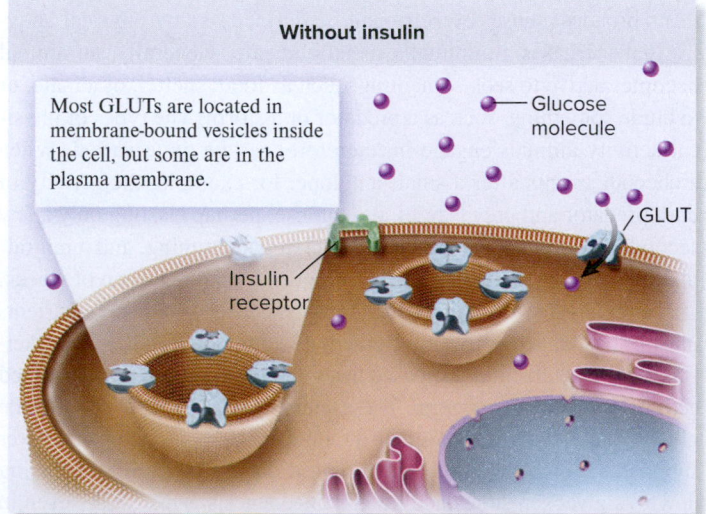
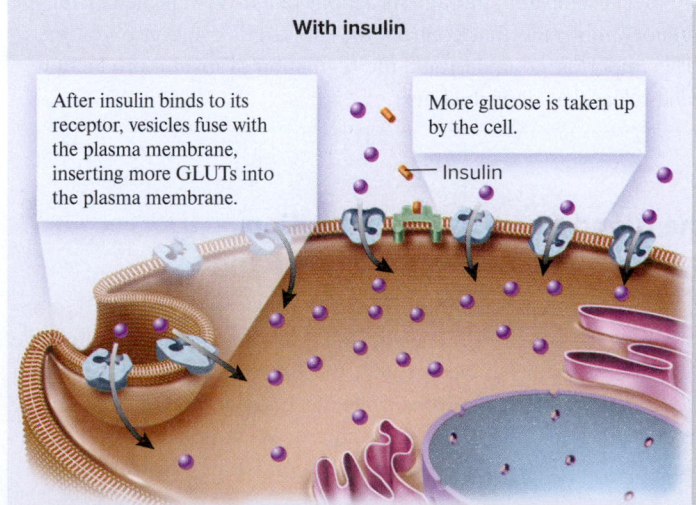

**Figure 47.3** **The effect of insulin on glucose transporters (GLUTs).** In adipose cells and skeletal and cardiac muscle cells, the presence of insulin results in the fusion of intracellular vesicles containing GLUTs with the plasma membrane, where they facilitate glucose uptake into the cell.

 **Core Skill: Connections** In what other contexts does the fusion of intracellular vesicles with a cell's plasma membrane occur? (For one example, look back at Figure 5.21.)

found in mammalian blood. This is also the only GLUT whose movement to the plasma membrane requires that the cell be stimulated by insulin. Consequently, as the glucose concentration increases in the blood after a meal, the blood concentration of insulin increases. The increased insulin recruits more GLUT4 molecules from vesicles in the cytosol to the plasma membrane of skeletal muscle and adipose cells.

By comparison, GLUT1 and GLUT3 are found predominantly in the brain, where they act in concert to mediate the transport of glucose from blood vessels to the interstitial fluid of the brain, and from there into brain cells. GLUT1 and GLUT3 have much higher affinity for glucose than do other GLUTs. This means that neurons of the brain can transport glucose into their cytosol even when the concentration of glucose in the extracellular fluid is very low, a clear survival advantage for the brain. In addition, insulin is not required for GLUT1 and GLUT3 to be present in the plasma membrane, unlike the situation for GLUT4. Instead, GLUT1 and GLUT3 are always present in the plasma membranes of cells expressing them. As a result of these properties of GLUT1 and GLUT3, neurons of the brain still receive adequate energy for survival even if an animal's blood glucose concentration decreases significantly, perhaps due to disease or starvation. Insulin-dependent cells, in contrast, would not be able to take up glucose under these circumstances, thus sparing glucose for the brain.

Expressing multiple types of the same functional class of protein means that different parts of an animal's body can meet their own particular metabolic demands. Moreover, these demands may change during development. For example, high-affinity GLUTs such as GLUT1 and GLUT3 are present in high numbers in the plasma membranes of many embryonic cells, which have a greater requirement for glucose, but are present in smaller numbers during other stages of life.

Because insulin, through its actions on GLUT4, is the key regulatory molecule that controls the blood glucose concentration, it is important to understand the regulation of insulin production and release, as described next.

## The Blood Glucose Concentration Is Maintained Within a Normal Range

To maintain homeostasis, the blood glucose concentration is controlled by a system of checks and balances. In the absorptive state, after a meal has been eaten, the blood glucose concentration increases; in the postabsorptive state, depending on how long it lasts, the concentration may begin to decrease. What homeostatic mechanisms keep the blood glucose concentration within a normal range?

### Increased Glucose in the Blood in the Absorptive State
The primary factor controlling the secretion of insulin from the pancreas is the blood glucose concentration (**Figure 47.4a**). An increase in an animal's blood glucose concentration directly stimulates cells of the pancreas to secrete insulin roughly in proportion to the amount of glucose in the blood. Later, after insulin promotes glucose uptake into its target cells, the resulting decrease in the blood glucose concentration removes the signal for insulin secretion. This is an example of a negative feedback loop, as described in Figure 41.11.

In addition to the blood glucose concentration, inputs from the nervous system to the pancreas also play a role in the regulation of insulin secretion. During a meal, signals from the parasympathetic division of the autonomic nervous system (recall the rest-or-digest response discussed in Chapter 43) stimulate the secretion of insulin into the blood.

### Decreased Glucose in the Postabsorptive State
Several factors act in concert to prevent blood glucose from decreasing below the normal homeostatic range, even during a short fast. Otherwise, glucose could decrease so much—a condition called hypoglycemia—that despite the high-affinity GLUT1 and GLUT3 in neuronal plasma membranes, there would not be enough glucose to keep these neurons alive.

If for any reason—such as a prolonged fast—the blood glucose concentration decreases below the normal homeostatic range for an animal, neurons within the hypothalamus in the brain that respond to changes in the extracellular concentration of glucose are activated (Figure 47.4b). Signals from the hypothalamus then stimulate the production of glucose-elevating factors. These include numerous hormones, notably **glucagon**, a protein hormone that is also secreted from the pancreas and that stimulates the processes of glycogenolysis, gluconeogenesis, and ketone synthesis in the liver. In addition, certain other hormones from various endocrine glands, as well as the neurotransmitter norepinephrine released from neurons of the sympathetic division of the autonomic nervous system, stimulate adipose tissue to release fatty acids into the blood. The fatty acids diffuse across plasma membranes and provide another source of energy for the synthesis of ATP. The overall effect is to maintain or increase the blood concentrations of glucose, fatty acids, and ketones during the postabsorptive state or during a prolonged fast.

## More Energy Is Required During Physical Activity

We think of exercise as something humans do for fun or fitness, but in its broadest sense, **exercise** can be defined as any physical activity that increases an animal's metabolic rate. Generally, an animal becomes active to seek something, such as food, shelter, or a mate, or to elude something, such as a predator or a storm. The types of physical activity animals engage in, therefore, can be quite varied. When a cheetah sprints after a small antelope, for example, the activity of both predator and prey is brief and intense, perhaps lasting only a few seconds. By contrast, a tuna may never stop swimming, and a migrating bird may fly a hundred miles a day or more over a span of weeks.

For all types of activity, including exercise in humans, nutrients must be available to provide the energy required for such things as skeletal muscle contraction, increased heart and lung activity, and increased activity of the nervous system. These energy-providing nutrients include glucose and fatty acids as well as the muscle's own glycogen. The liver supplies the blood with the glucose used during exercise by breaking down its glycogen stores and by gluconeogenesis. This occurs even in the absorptive state, and thus the blood glucose concentration increases above normal when an animal is active at such times. In addition, an increase in adipose tissue lipolysis releases fatty acids into the blood, which provides an additional source of energy for the exercising muscle.

These events are mediated by the same hormones and nerves responsible for the regulation of the postabsorptive state. For example,

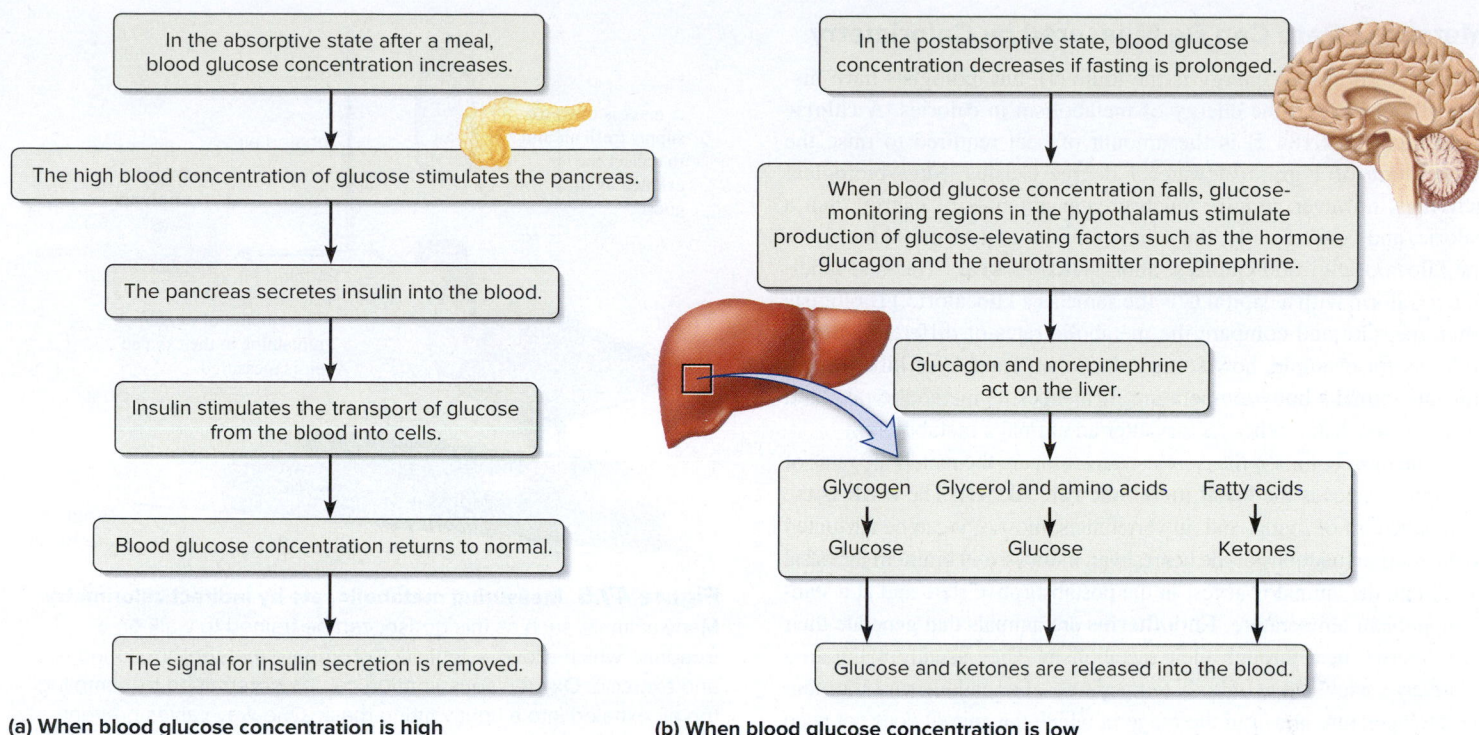

In the absorptive state after a meal, blood glucose concentration increases.

↓

The high blood concentration of glucose stimulates the pancreas.

↓

The pancreas secretes insulin into the blood.

↓

Insulin stimulates the transport of glucose from the blood into cells.

↓

Blood glucose concentration returns to normal.

↓

The signal for insulin secretion is removed.

**(a) When blood glucose concentration is high**

In the postabsorptive state, blood glucose concentration decreases if fasting is prolonged.

↓

When blood glucose concentration falls, glucose-monitoring regions in the hypothalamus stimulate production of glucose-elevating factors such as the hormone glucagon and the neurotransmitter norepinephrine.

↓

Glucagon and norepinephrine act on the liver.

↓

| Glycogen | Glycerol and amino acids | Fatty acids |
| ↓ | ↓ | ↓ |
| Glucose | Glucose | Ketones |

↓

Glucose and ketones are released into the blood.

**(b) When blood glucose concentration is low**

**Figure 47.4** **The homeostatic control of blood glucose concentration within a normal range by hormones and the nervous system.**

 **Core Skill: Modeling**  The goal of this modeling challenge is to make predictions about glucose homeostasis in a mammal whose cellular sensitivity to insulin is abnormally decreased.

**Modeling Challenge:** Mammals sometimes develop a disease called type 2 diabetes mellitus, in which cells that are normally sensitive to insulin no longer are able to respond properly to the hormone. Revise the model of glucose homeostasis shown in Figure 47.4a to show how glucose and insulin concentrations in the blood will be altered in a mammal with this form of diabetes. You will need to change the text in the last three boxes.

inputs from the sympathetic division to the pancreas inhibit insulin secretion during exercise. Consequently, glucose transport into muscle and adipose cells is decreased, which tends to increase the blood glucose concentration (this is part of the fight-or-flight response described in Chapter 43). Because the brain does not depend on insulin for glucose transport across neuronal membranes, as just described, more of the body's supply of glucose is available to the brain at such times, while other cells can use fatty acids for energy. Therefore, the body uses all available forms of energy in response to fasting or exercise.

<table><tr><td>**47.3**</td><td></td></tr></table>

## 47.3 Energy Balance and Metabolic Rate

**Learning Outcomes:**

1. Define basal metabolic rate (BMR), and explain how it is measured.
2. **CoreSKILL** » Predict the effects of various factors on metabolic rate.
3. Describe how hormones regulate metabolic rate and appetite.

Animals have a wide range of metabolic demands that depend on numerous factors. Active animals, such as migrating birds, use energy at a greater relative rate than do inactive animals, such as hibernating mammals. Likewise, juveniles typically use energy at a greater rate relative to mature animals. Recall that the amount of energy an organism uses in a given period of time to power its metabolic requirements is called its metabolic rate.

A fundamental characteristic of energy is that it can be neither created nor destroyed, but it can be converted from one form to another (the first law of thermodynamics; see Chapter 6). The breakdown of organic molecules liberates energy in their chemical bonds, which is transferred to the bonds in ATP. This is the energy that cells harness to perform various biological activities such as muscle contraction, active transport, and molecular synthesis. We refer to these functions as work. Not all of the energy liberated from the breakdown of organic molecules is used to do work, however. Some of it appears as heat, which contributes to an animal's body temperature or is dissipated to the environment. In this section, we examine how metabolic rate is measured in animals; how metabolic rate is influenced by activity, digestion, and body mass; and how a balance is achieved between energy consumption and expenditure.

## Metabolic Rate Can Be Measured by Calorimetry

The standard unit of energy is the joule (J), but biologists have historically quantified the energy of metabolism in calories. A **calorie** (equivalent to 4.184 J) is the amount of heat required to raise the temperature of 1 gram of water 1 degree Celsius. Most biological activities, however, require much greater amounts of energy than a calorie, and consequently, the more common unit of measurement is the kilocalorie (1,000 calories, abbreviated as **kcal**). (In food labeling, a Calorie with a capital C is the same as a kilocalorie.) Biologists often measure and compare the metabolic rates of different animals to learn, for example, how some animals are capable of hibernating, how an animal's body temperature influences its metabolic rate, and how hormones and other factors alter an animal's metabolism.

The most common measure used to compare the metabolic rates of different species is the **basal metabolic rate (BMR)**. The BMR is the metabolic cost of living, and, in vertebrates, most of it can be attributed to the routine functions of the heart, liver, kidneys, and brain. In the basal condition, the animal is at rest in the postabsorptive state and at a standard ambient temperature. **Endotherms** are animals that generate their own internal heat through their metabolism. They usually maintain a relatively narrow range of body temperatures. For endotherms, the standard temperature is within the range in which the animal does not need to generate additional heat (for example, by shivering) or lose excess heat (for example, by perspiration). This temperature range is called an animal's thermoneutral zone. The BMR of **ectotherms**—animals that acquire their heat from the environment—must be measured at a standard temperature for each species that approximates the average temperature that the species normally encounters. In this case, the term standard metabolic rate (SMR) is used instead of BMR, because the basal condition in ectotherms is harder to define than for endotherms.

The usual method for measuring metabolic rate, **indirect calorimetry**, is based on the principle that animals require oxygen to metabolize foodstuffs. The more fuel being metabolized—that is, the greater the metabolic rate—the more oxygen must be consumed by the animal. Measuring the rate at which a resting animal uses oxygen, therefore, provides a good estimate of BMR. Indirect calorimetry can also be used to compare the metabolic rates of an animal during rest and activity, when oxygen consumption increases (**Figure 47.5**). One limitation to this method is that a small percentage of fuel is metabolized anaerobically—that is, without oxygen—and thus indirect calorimetry underestimates the actual metabolic rate.

## Activity, Digestion, and Body Mass Influence Metabolic Rate

Not all tissues in the body use oxygen and produce heat at the same rate. Some structures, such as skin, consume relatively little oxygen under resting conditions, whereas others, such as the brain, heart, and liver, have high rates of metabolism even when an animal is sleeping. Also, the metabolic rates of different tissues can vary depending on their activity. For example, the metabolism of the alimentary canal increases when food is being digested, and that of skeletal muscle increases during exercise.

***Physical Activity***    The primary factor that increases metabolic rate is altered skeletal muscle activity. Even small increases in muscle

A mask is fitted to supply fresh air and to collect the air exhaled by the goose.

Air

Inhaled air

**Oxygen analyzer**

Exhaled air

The amount of oxygen remaining in the exhaled air is measured.

Angled treadmill

**Figure 47.5  Measuring metabolic rate by indirect calorimetry.** Many animals, such as this goose, can be trained to walk on a treadmill, which allows scientists to compare metabolism during rest and exercise. Oxygen consumption can be determined by sampling the air exhaled into a tightly fitting mask. One-way valves prevent inhaled and exhaled air from mixing.

**Core Skill: Process of Science**  This figure illustrates how biologists can learn about general physiological principles using a variety of animal models. For example, many geese and other birds often fly for long periods at very high altitudes where oxygen availability is limited. Using experimental procedures like the one shown here allows scientists to understand how such animals can function under conditions that would be very challenging for humans. For example, the percentage of oxygen in the inhaled air can be adjusted to match that found at high altitude, or the degree of activity can be adjusted by altering the treadmill.

contraction significantly increase metabolic rate; strenuous activity increases it even more. For example, the total daily expenditure of kilocalories may vary for a healthy adult human from approximately 1,350 kcal for a small person at rest to more than 7,000 kcal for a cyclist competing in the Tour de France. Metabolic rate is also slowed during sleep, due partly to decreased muscle activity, and increased during exposure to cold temperatures, due to increased muscle activity from shivering.

***Digesting Food***    Eating and digesting food also increase the metabolic rate. Particularly in mammals that eat meat, this may increase metabolic rate (and associated heat production) by 10–50% for a few hours after eating. You may have noticed this **food-induced thermogenesis** after consuming a large meal. Ingested protein produces the greatest effect, whereas carbohydrate and fat produce less. The increased heat is believed to result partly from the processing of the absorbed nutrients by the liver and from the energy expended by the alimentary canal in digestion and absorption. Food-induced thermogenesis is observed in nearly all vertebrates, but it is most notable in certain reptiles such as snakes that eat infrequent but very large meals. Body temperature can increase by several degrees Celsius in such animals, and persist for days and even weeks depending on the size of the meal.

*Body Mass*   Another factor affecting metabolic rate is body mass. In general, a large animal uses greater amounts of energy than does a small animal because the large animal has more mass and more cells, all of which consume fuel and generate heat. The metabolic rate and heat generation of an elephant are clearly greater than those of a mouse, for instance. However, when the metabolic rate of an elephant and a mouse are scaled to their respective body masses, we find that the energy expenditure per gram of body mass in a mouse is much higher than the comparable calculated value for an elephant. **Mass-specific BMR** is the amount of energy expended per gram of body mass in the resting condition.

Mass-specific BMR is a relative term that allows scientists to compare basal metabolic rates among animals of different sizes. Research has shown that the relationship between mass-specific BMR and body mass is exponential (**Figure 47.6**). One possible explanation is that the ratio of an animal's surface area to its volume or body mass is greater in smaller animals than in larger animals (refer back to Figure 4.12). Therefore, smaller animals lose heat more rapidly than larger ones. According to this hypothesis, smaller animals must generate more heat per gram of body mass than larger animals to compensate for their heat loss. However,

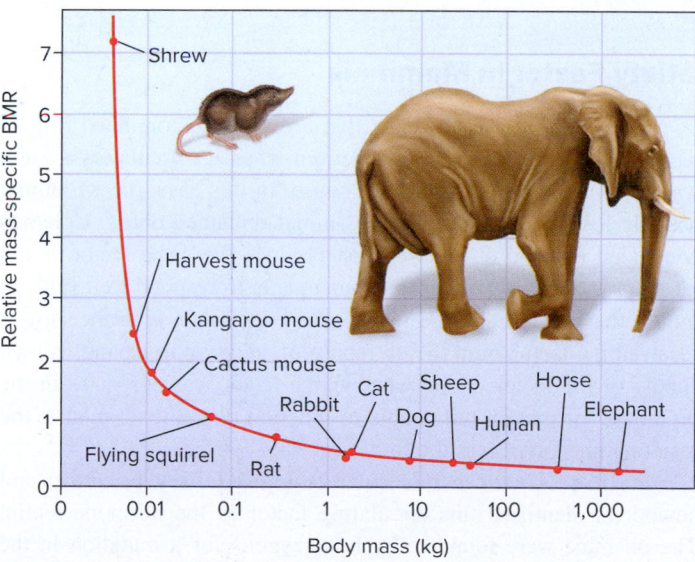

**Figure 47.6  Metabolic rates of animals that differ in size.** Metabolism can be scaled to body mass by measuring oxygen consumption and normalizing it to the animal's body mass (mass-specific BMR). Note that when expressed in this way, the mass-specific BMR of a shrew is greater than that of an elephant, even though the total oxygen consumption and heat output of the elephant are much greater. The values on the vertical axis are relative units of metabolism.

*Concept Check:*  *Can the relationship between body size and metabolic rate be used to propose hypotheses about metabolic rates of extinct animals?*

although this hypothesis appears to provide an explanation for the relationship between metabolism and body size in endotherms, it does not explain the observation that the same relationship exists in almost all animals, including ectotherms.

## Hormones and the Nervous System Control Food Intake

When the daily amount of energy within the food that an animal consumes is equal to the amount of energy it expends, the animal's body weight remains stable. Tipping the balance in either direction causes weight gain or loss; that is, the total body mass increases or decreases. Normally, energy is stored in the form of fat in adipose tissue.

Body weight in an adult animal is usually regulated around a predetermined set point that differs among species and between individuals. Body weight is maintained by adjusting caloric intake and energy expenditure in response to changes in body weight. This mechanism usually works very precisely in those animals in which it has been studied. For example, if given the opportunity, a mammal that eats less one day will eat more the next day to compensate for the previous day's deficit. Similarly, if an animal is overfed one day, it may eat less the next day.

Short-term control of feeding generally involves a feeling of **satiety**, that is, fullness. As an animal's stomach and small intestine stretch to accommodate food, nerves send inhibitory signals from these structures to the hunger center in the hypothalamus. At the same time, the stomach and small intestine release into the blood hormones that reach the hypothalamus and suppress hunger. These satiety signals remove the sensation of hunger and set the time period before it returns again.

Long-term control of food intake is mediated by many different molecules in the brain, by hormones, and by emotional state, particularly in humans. One hormone that has received considerable attention in recent years for its ability to control appetite and metabolic rate is leptin (from the Greek *leptos*, meaning thin). Leptin has been identified in all classes of vertebrates but has been most extensively studied in mammals. In these animals, leptin is produced by adipose cells in proportion to fat mass: As more fat is stored in adipose cells, more leptin is secreted into the blood. Leptin acts on the hypothalamus to decrease appetite and increase metabolic rate (**Figure 47.7a**). In this way, the brain is made aware of how much fat is stored in the body at all times, and it can adjust appetite and metabolic rate appropriately if fat stores decline or increase. If an animal fasts for a period of time, its adipose cells shrink as they release their stored fat into the blood. The decrease in leptin secretion resulting from the decreased adipose mass results in a decrease in BMR and an increase in appetite. This may be the true evolutionary significance of leptin, namely that its disappearance from the blood lowers the BMR, consequently prolonging life during periods of starvation (**Figure 47.7b**).

Leptin was discovered in 1994, but its existence was postulated decades before that by the pioneering work of Douglas Coleman, who investigated the nature of mutations in mice that result in obesity. His observations continue to have important implications for human health today.

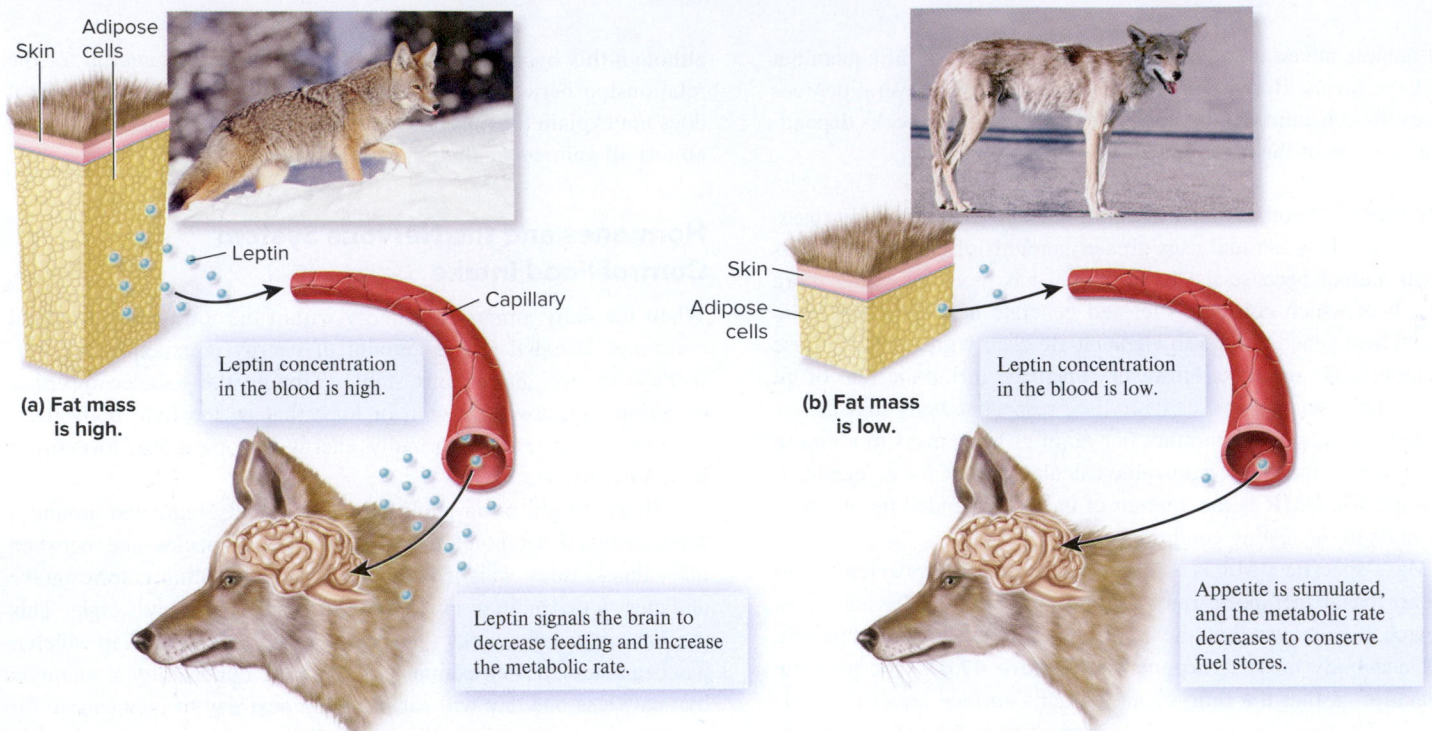

**Figure 47.7   The role of leptin in regulating appetite and metabolic rate.**  In animals such as the coyote, changes in the blood leptin concentration result directly from changes in fat mass. Animals with more fat make more leptin. a: ©Paul McCormick/Getty Images; b: ©William S. Clark/ Frank Lane Picture Agency/Corbis/Getty Images

---

## Core Skill: Process of Science

# Feature Investigation | Coleman Revealed a Satiety Factor in Mammals

For many years, scientists wondered how most animals regulate their body mass around a predetermined set point, despite fluctuations in their food supply. They postulated that other parts of the body somehow communicated with the brain to signal when energy stores were above or below normal. In the 1970s, Canadian-American researcher Douglas Coleman tested this hypothesis in an experiment involving parabiosis, the surgical connection of the abdominal walls of two animals, such that the blood supply from one animal intermixes with that of the other.

Coleman used two strains of mice called ob and db mice, which carry different mutations that result in inherited forms of obesity, characterized by the excessive accumulation of body fat (an example of an ob mouse is shown in the chapter opening photo). Coleman first connected a wild-type (wt) mouse, one that lacked these mutations, with either an ob mouse or a db mouse, as shown in **Figure 47.8**. He discovered that when the circulatory system of the ob mouse was in contact with that of the wt mouse, the ob mouse ate less and gained less weight than usual. This suggested that the blood of the wt mouse contained a circulating factor that signals the brain when an animal has sufficient fat stored in its body and adjusts appetite accordingly. The ob mouse was deficient in this factor, but when exposed to it through the wt mouse's circulation, it responded in the appropriate way. The wt mouse of the parabiosis pair apparently retained a sufficient amount of the factor in its blood, because it maintained its body weight at a normal level.

Coleman noticed, however, that a db mouse continued to gain weight at an abnormally high rate even when its circulatory system was in contact with that of a wt mouse. In this case, the wt animal actually lost weight while the db animal remained obese. Coleman concluded that the db mouse must produce the same factor as the wt mouse, but for some reason, was unable to respond to it. The wt mouse that was parabiosed to the db mouse lost weight because it received the factor from the db mouse, in addition to having its own supply of the factor. Thus, whether the factor was absent as in the ob mouse, or present but unable to function as in the db mouse, the resulting phenotype was the same—obesity.

In 1994, American molecular biologist Jeffrey Friedman and coworkers identified this circulating factor as the hormone leptin. The ob mice were found to be homozygous for a mutation in the leptin gene, which produced an inactive leptin molecule, whereas db mice produced leptin but did not respond to it. The db mice were found to produce even greater amounts of leptin than wt mice, which explained why the wt mouse in Coleman's experiments lost weight when parabiosed with a db mouse. Friedman and others later showed that adipose cells produce leptin in direct proportion to the total fat mass of an animal, as stated earlier.

At first, the work of Coleman and Friedman generated considerable excitement that leptin might be useful to treat obesity in humans, but this has thus far proven difficult. Why? Recent research

**Figure 47.8**  Coleman's parabiosis experiments revealed a satiety factor in wild-type mice that was absent in genetically obese mice.

**HYPOTHESIS**  Body weight is controlled by a factor that circulates in the blood. This factor is absent in strains of mice that have an inherited form of obesity.

**KEY MATERIALS**  Two different strains of genetically obese mice, normal (wild-type) mice.

| Experimental level | Conceptual level |
|---|---|

**1** Surgically connect the abdominal walls of a genetically obese and normal (wt; wild-type) mouse. After a few days, blood vessels from each mouse cross to the other mouse. Monitor changes in body weight. Note: 2 different strains of obese mice were tested, called ob and db mice.

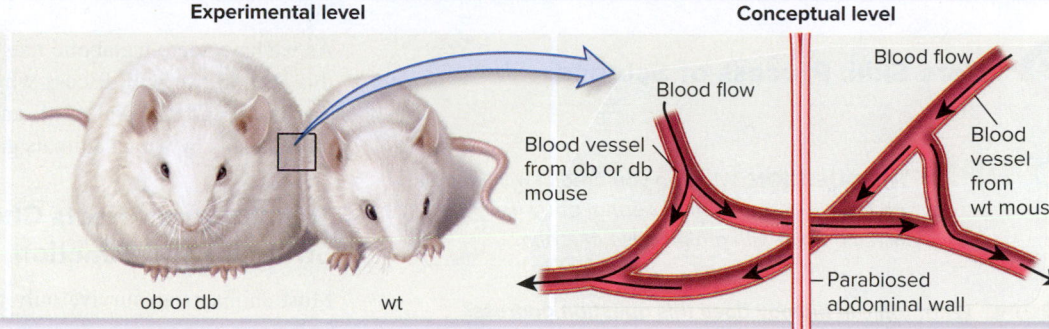

**2** Feed mice a normal diet for several weeks, then visually inspect and weigh each pair.

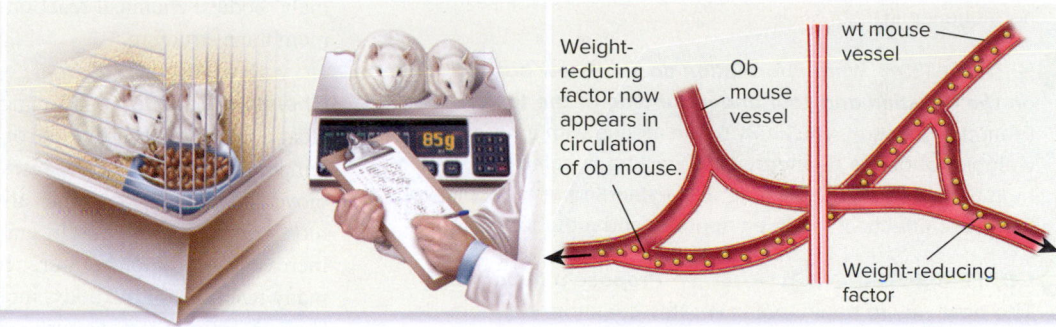

**3  THE DATA**

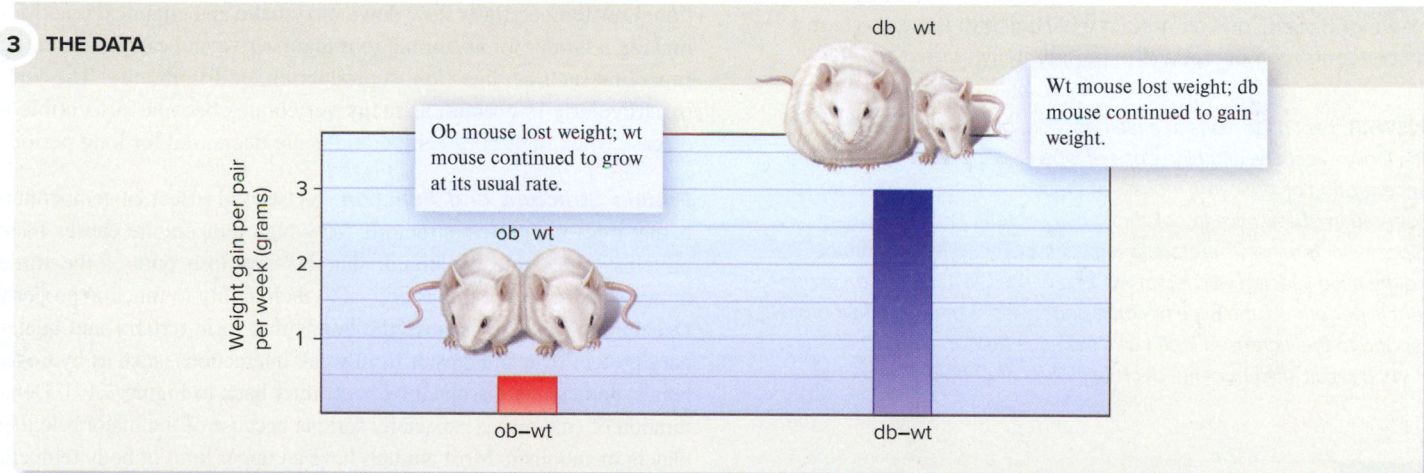

**4  CONCLUSION**  Wild-type mice secrete a blood-borne factor that decreases body weight. The factor is absent from ob mice but present in db mice. Ob mice retain the ability to respond to the factor, unlike db mice, which cannot respond to it.

**5  SOURCE**  Coleman, D. L. 1973. Effects of parabiosis of obese with diabetes and normal mice. *Diabetologia* 9: 294–298.

has revealed that most obese humans are more like the db mice than the ob mice. That is, they produce leptin but fail to respond adequately to it, and therefore, simply increasing the concentration of leptin in the blood may not have a significant effect on body weight. However, other studies have shown that leptin normally acts in nonobese humans in a manner much like it does in wt rodents.

As noted in the chapter introduction, researchers have identified rare individuals in whom leptin is not produced due to a mutation in the leptin gene. These individuals are extremely obese and respond well to injections of leptin, losing considerable weight. The body weight disorders of such individuals, therefore, are reminiscent of the condition in ob mice.

*Experimental Questions*

1. What observation led to the experiments conducted by Coleman?

2. **CoreSKILL »** How did the experimental linking of the bloodstreams of the wild-type mice and the mutant mice affect the body weight of both strains of mutants? Why did the db mice fail to lose weight when parabiosed with the wt mice?

3. **CoreSKILL »** Predict what will happen to the respective body weights of ob and db mice if they are parabiosed to one another.

 **Core Skill: Process of Science**

 **THE QUESTION** *What do you predict will happen if a healthy mouse with a normal body weight is injected daily with a high dose of leptin?*

**T** **OPIC** *What topic in biology does this question address?*
The topic is the control of body weight in a mouse by the hormone leptin.

**I** **NFORMATION** *What information do you know based on the question and your understanding of the topic?*
From the question, you know that a mouse with a normal body weight (not obese or underweight) will be injected daily with a high dose of leptin. From your understanding of the topic, you know the effects of leptin on appetite and metabolic rate.

**P** **ROBLEM-SOLVING** **S** **TRATEGY** *Propose a hypothesis.*
The best way to begin to answer this question is to consider how appetite and metabolic rate are related to body weight; then you can predict what will happen if the blood concentration of leptin is experimentally increased.

**ANSWER** *Leptin functions to inhibit appetite and increase metabolic rate. Daily injections of a high dose of leptin will increase the blood concentration of leptin in a mouse, perhaps to a level similar to those observed in obese mice (recall that leptin secretion increases as adiposity increases). Therefore, a logical hypothesis is that normal, healthy mice injected with leptin will lose weight. When such an experiment is performed, the mice become underweight because their brains respond to the increased leptin as if the mice had an excess of body fat. As a result, their appetite decreases and metabolism increases.*

## 47.4 Regulation of Body Temperature

**Learning Outcomes:**

1. Provide examples of how changes in temperature affect chemical reactions, protein functions, and membrane structure.
2. List and define the four terms used to categorize organisms based on their source of heat and ability to maintain body temperature.
3. Identify the four main mechanisms animals use to exchange heat with the environment.
4. **CoreSKILL »** Identify several mechanisms by which animals can alter the rate of heat gain or loss, and describe some ways in which body structures facilitate these mechanisms.

As we have seen, metabolic rate and body temperature are linked. In this section, we will discuss why body temperature is important for the health and survival of all animals and consider the homeostatic mechanisms by which animals gain or lose heat.

### Temperature Affects Chemical Reactions, Protein Structure and Function, and Membrane Structure

Most animals can survive only in a relatively narrow range of temperatures. Temperature has an effect on three vital features of animals' bodies: chemical reactions, protein structure and function, and membrane structure.

*Chemical Reactions*    Chemical reactions depend on temperature. Heat accelerates the motion of molecules, so as an animal's body temperature increases, the rates at which the molecules in its body move and contact each other also increase. Consequently, the rate of most chemical reactions in animals increases significantly with an increase in body temperature. In addition, enzymes, which catalyze many reactions in the body, including those involved in metabolism, have an optimal temperature range for their maximal catalytic function. Low temperatures slow down enzymatic and chemical reactions, making it harder for an animal to remain active and carry out internal functions such as digestion, reproduction, and immunity. The latter is particularly important, as many vertebrates become susceptible to disease when their body temperatures are decreased for long periods.

*Protein Structure and Function*    A second effect of temperature is that it affects protein structure. Very high temperature causes many proteins to become denatured; that is, they lose part of the three-dimensional structure that is crucial to their ability to function properly. Denaturation occurs because the bonds that form tertiary and quaternary protein structures result from weak interactions, such as hydrogen bonds, and can be disrupted by heat (refer back to Figure 3.17). Denaturation of enzymes is especially serious because of the major role they play in metabolism. Most animals have an upper limit of body temperature at which they can survive. Mammals generally have a resting body temperature of 35–38°C (95–100°F). In humans, a body temperature of 41°C (106°F) inhibits protein function and proper signaling within the nervous system, and a body temperature of 42–43°C (107–109°F) is usually fatal.

*Membrane Structure*    A third effect of temperature is that it affects the structures of the plasma membrane and intracellular membranes. At low temperatures, membranes become less fluid and more rigid, primarily due to changes in their phospholipids. Rigid membranes are less able to perform biological functions, such as transporting ions and binding extracellular molecules to receptors on the membrane surface. Alternatively, if the temperature becomes too high, membranes can become leaky.

## Ectotherms and Endotherms May Have Fluctuating or Stable Body Temperatures

Biologists classify animals according to both the source of heat used to warm their bodies and their ability to maintain body temperature. Recall that ectotherms depend on external heat sources to warm their bodies, whereas endotherms use their own metabolically generated heat to warm themselves. **Homeotherms** have body temperatures that are maintained within a narrow range, whereas **heterotherms** have body temperatures that vary widely in response to the environmental temperature (**Figure 47.9**). Most animals can be categorized as either endotherms or ectotherms and as either homeotherms or heterotherms. Generally, birds and mammals are endothermic and homeothermic. Other vertebrates and most invertebrates are ectothermic and heterothermic.

Ectotherms are usually heterotherms because most environments on Earth have fluctuating temperatures over short periods of time. However, this is not always the case. For example, a fish living in deep ocean waters is an ectotherm but also a homeotherm because the temperature of the water—and therefore of its body—remains relatively constant for an extended period of time. Fishes that live in waters with fluctuating temperatures, by contrast, are ectothermic and heterothermic.

Even endothermic homeotherms do not have truly constant body temperatures. They have a narrow range of body temperatures that increases or decreases slightly in extreme climates, during physical activity, or during sleep. The important feature is that birds and mammals can quickly adjust the body's mechanisms for retaining or releasing heat so that body temperature remains within the optimal narrow range. This regulation provides the advantage that the body's chemical reactions are at optimal levels even when the environment imposes extreme challenges. The metabolic rate of a resting mammal, for example, is roughly six times greater than that of a comparably sized reptile. A suddenly awakened mammal is instantly capable of maximal activity even on a winter day, but an icy-cold reptile could be at the mercy of a predator because of the time required to warm itself in order to flee.

Endothermy does have three major disadvantages, however:

1. **Requirement for larger amounts of energy.** To produce sufficient heat by metabolic processes, endotherms must consume larger amounts of food to provide the nutrients used by cells in the formation of ATP, during which heat is generated. Small endotherms with high mass-specific BMRs, such as shrews (see Figure 47.6), must eat almost continuously and may die if deprived of food for as little as a day. By contrast, many ectotherms, such as snakes, can regularly live for weeks without eating.

2. **Risk of overheating.** Endotherms have a greater risk of hyperthermia, or overheating, during periods of intense activity, even in cold weather.

3. **Loss of body fluids.** As described shortly, the prevention of overheating often requires the evaporation of bodily fluids (and thus a need for replenishment of fluid). Therefore, many endotherms are restricted to environments where fresh water is plentiful.

## Animals Exchange Heat with the Environment in Four Ways

The surface of an animal's body can lose or gain heat from the external environment via four mechanisms: radiation, evaporation, convection, and conduction (**Figure 47.10**).

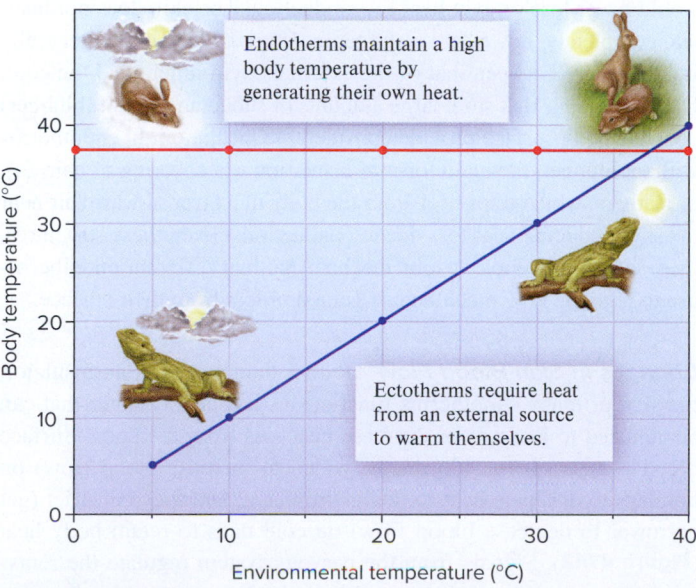

**Figure 47.9  Body temperature and environmental temperature in endotherms and ectotherms.** A rabbit is an endotherm but also a homeotherm because its body temperature doesn't change much. A lizard is an ectotherm but also a heterotherm because its body temperature changes considerably.

Endotherms maintain a high body temperature by generating their own heat.

Ectotherms require heat from an external source to warm themselves.

*Body temperature (°C)* / *Environmental temperature (°C)*

**Concept Check:** *Into what thermoregulatory categories do humans fit?*

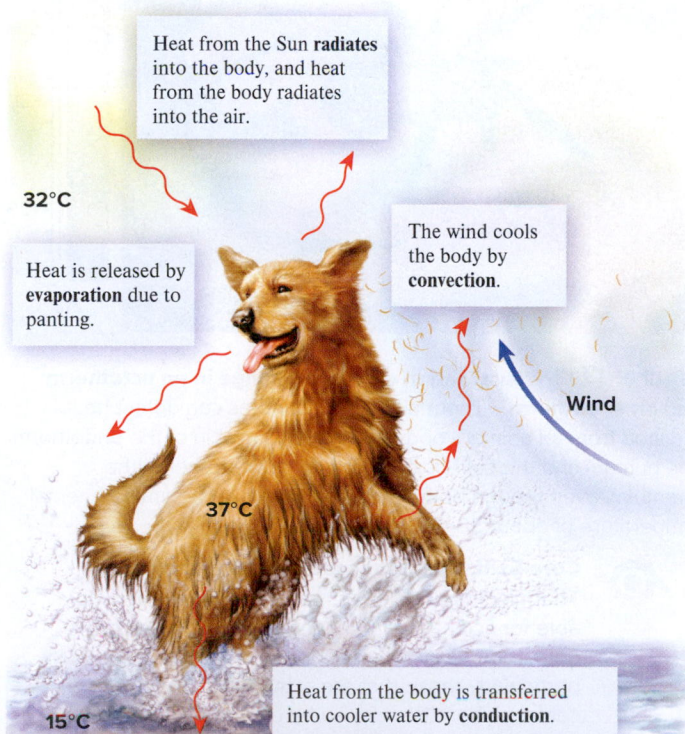

Heat from the Sun **radiates** into the body, and heat from the body radiates into the air.

32°C

Heat is released by **evaporation** due to panting.

The wind cools the body by **convection**.

Wind

37°C

15°C

Heat from the body is transferred into cooler water by **conduction**.

**Figure 47.10  Types of heat exchange.** The four ways in which animals exchange heat with the environment are radiation, evaporation, convection, and conduction.

 **Core Concept: Energy and Matter** Heat is a form of energy and can be exchanged between an animal's body and its environment.

- **Radiation** is the emission of electromagnetic waves from the surface of an object or organism. The rate of emission is determined by the temperature of the radiating surface. Thus, if the surface of an animal's body is warmer than the environment, the body loses heat at a rate that depends on the temperature difference. If the outside temperature is warmer than body temperature, the body gains heat by radiation, for instance from sunlight. We can observe radiated heat from an animal's body with imaging devices that detect infrared light, the wavelength at which thermal energy is radiated from animal bodies (**Figure 47.11**).

- **Evaporation** is the conversion of water from the liquid to the gaseous state. Animals can lose body heat through evaporation of water from the skin and membranes lining the respiratory tract, including the surface of the tongue. A large amount of energy in the form of heat is required to transform water from liquid to gas. Whenever water vaporizes from the body's surface, the heat required to drive the process is conducted from the surface, thereby cooling the animal.

- **Convection** is the transfer of heat by the movement of air or fluid next to the body. For example, the air close to an endotherm's body is heated by conduction. Because warm air is less dense than cold air, the warm air near the body rises and carries away heat by convection. Convection is aided by creating currents of air around an animal's body. Humans may do this by sitting near fans, but other animals can create cooling air currents by other means, such as when an elephant waves its ears.

- **Conduction** is the process by which the body surface loses or gains heat through direct contact with cooler or warmer substances. The greater the temperature difference, the greater is the rate of heat transfer. Different materials have different abilities to absorb heat, however. As we saw in Chapter 2, water has a higher specific heat than air, meaning that at any temperature, water will retain greater amounts of heat than will air. Consequently, aquatic animals in water that is 10°C lose considerably more heat in a short time than terrestrial animals lose in air that is 10°C. Even on a hot day, terrestrial animals can lose heat by immersing themselves in water.

The four mechanisms of heat transfer just described can be regulated in animals in such a way that heat is retained within the body at some times and lost from the body at other times, as we see next.

## Several Mechanisms Can Alter Rates of Heat Gain or Loss in Endotherms

For purposes of temperature control, think of an endotherm's body as a central core surrounded by a shell consisting of skin and subcutaneous (just below the skin) tissue. Depending on the species, the temperature of the central core of endotherms is regulated at approximately 35–42°C (95–108°F), but the temperature of the outer surface of the skin varies considerably. If the skin were a perfect insulator, the body would never lose or gain heat by conduction. The skin does not insulate completely, however, so the temperature of its surface generally is somewhere between that of the external environment and the core. Only in animals that store large amounts of subcutaneous fat (blubber) does the body surface provide considerable insulation. In endotherms without blubber, the main form of insulation is a covering of hair, fur, or feathers, which traps heat from the body in a layer of warm air near the skin, reducing heat loss due to conduction. Given these structures, then, let's take a look at four mechanisms that different endotherms use to regulate how much heat is gained or lost from their surface.

***Changes in Skin Blood Flow*** Rather than acting as an insulator, the skin of many endotherms functions as a heat exchanger that can be adjusted to increase or decrease heat loss from the body. Surface blood vessels of the skin dilate (widen to increase blood flow) on hot days to dissipate heat to the environment, and they constrict (get narrower to decrease blood flow) on cold days to retain body heat (**Figure 47.12**). Signals from the nervous system regulate the relaxation or contraction of the smooth muscles that control the diameter of these blood vessels. Diving birds and diving mammals are good examples of animals that use this mechanism. Ducks, seals, and walruses greatly decrease the amount of blood flowing to the skin when they dive in cold waters. This allows them to retain body heat that would otherwise be conducted into the water.

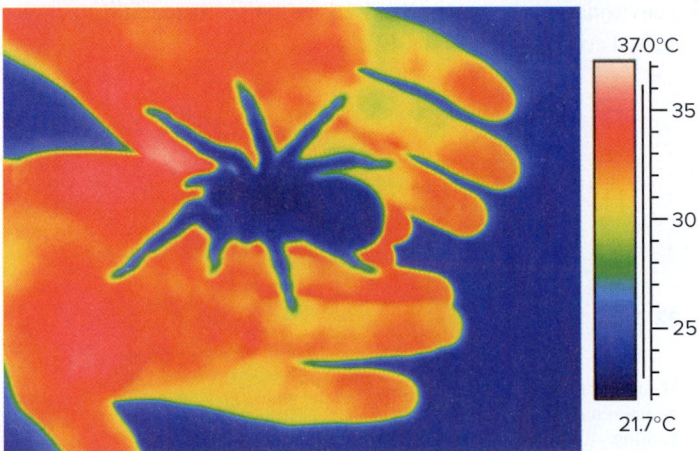

37.0°C
— 35
— 30
— 25
21.7°C

**Figure 47.11** **Visualization of heat exchange in an ectotherm and an endotherm.** Thermal-imaging cameras can detect heat radiated from an animal's body. Note the warm skin of the endotherm (the human) and the cold body surface of the ectotherm (the tarantula), even though both animals are at the same environmental temperature of about 20–25°C. ©Nutscode/T Service/Science Source

 **Core Skill: Communication and Collaboration**

Thermal-imaging cameras have proven to be valuable for scientists attempting to gather quantitative data about large populations of endothermic animals, particularly those that are active only at night. The cameras were developed by engineers; biologists use them for census-taking and work closely with statisticians who help interpret the data. One example involves obtaining an accurate census of large colonies (up to several million individuals) of insectivorous bats in North America, many of whom have suffered enormous losses due to a fungal disease called white-nose syndrome.

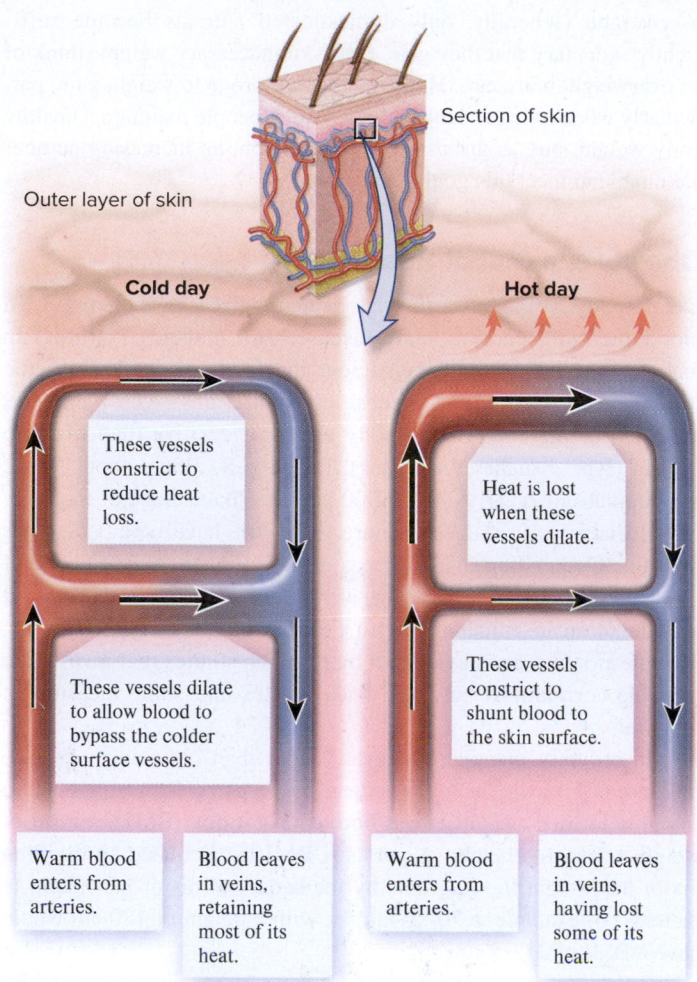

**Figure 47.12  Regulation of heat exchange in the skin.** As shown in this schematic illustration, the skin functions as a variable heat exchanger. The arrows in the blood vessels indicate the direction and relative amount of blood flow.

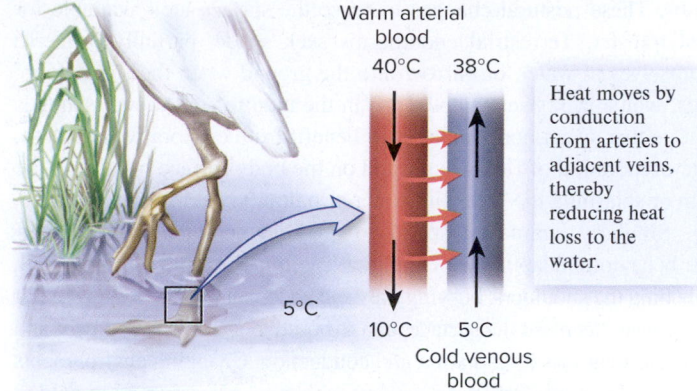

**(a) Countercurrent heat exchange in the leg of an endotherm**

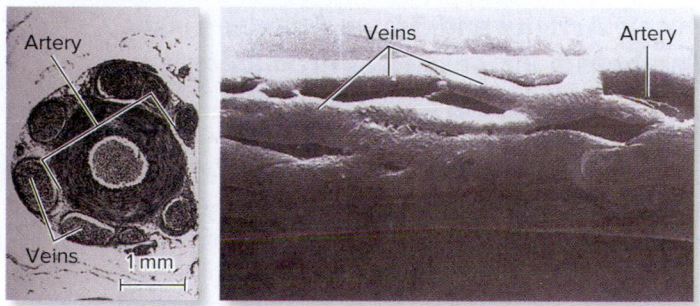

**(b) Cross section and surface view of veins covering an artery**

**Figure 47.13  Countercurrent exchange. (a)** Countercurrent exchange retains heat in the leg of an endotherm such as this bird. Black arrows in vessels indicate direction of blood flow. **(b)** A micrograph of the arrangement of veins surrounding an artery in a wading bird's leg. The artery is almost completely covered by overlying veins, allowing efficient heat exchange between the vessels.
b (left, right): Courtesy of Uffe Midtgård, University of Copenhagen

**Countercurrent Exchange**  Many endotherms and ectotherms regulate heat loss to the environment through **countercurrent exchange**, in which heat is transferred between fluids flowing in opposite directions. Countercurrent exchange regulates heat loss to the environment by returning heat to the body's core and keeping the core much warmer than the extremities. In endotherms, countercurrent exchange occurs primarily in the extremities—the flippers of dolphins, for example, or the legs of birds and certain other terrestrial animals (**Figure 47.13a**). As warm blood travels from the core through arteries down a wading bird's leg, for example, heat moves by conduction from the artery to adjacent veins that carry cooler blood from the feet in the opposite direction (**Figure 47.13b**). By the time the arterial blood reaches the tip of the leg, its temperature has dropped considerably, reducing the amount of heat lost to the environment and returning the heat via the veins to the body's core.

**Evaporative Heat Loss**  Recall that some animals lose body heat through evaporation of water from the skin and membranes lining the respiratory tract. Heat exchange in some mammals is regulated by changing the rate of water evaporation through perspiration. Nerves to the sweat glands stimulate the production of sweat, a dilute solution

containing $Na^+$ and $Cl^-$. The most important factor determining evaporation rate—and therefore heat loss—is the water vapor concentration, or humidity, of the air. The discomfort you feel on a humid day is due to the slow rate of evaporation. Your sweat glands continue to secrete, but most of the sweat simply remains on your skin and so your body temperature remains elevated, especially during exercise.

In endotherms that lack sweat glands, such as birds, or those that have very few such glands, such as dogs and cats, panting (short, rapid breaths with the mouth open) promotes evaporation of water from the tongue surface. Panting has advantages over sweating, because no ions are lost, and panting provides the air current that promotes heat exchange by convection. However, the surface area of the mouth and tongue is relatively small, which limits the rate at which heat can be eliminated. Interestingly, many reptiles also pant on hot days, suggesting that panting evolved prior to endothermy.

**Behavioral Adaptations**  Behavioral mechanisms can also alter heat loss by radiation, conduction, and evaporation. Two such behaviors involve changing exposed surface area and changing surroundings. On hot days, birds may ruffle their feathers and raise their wings, whereas many mammals will reduce their activity and spread their limbs. These postural changes increase the surface area available for heat transfer. Terrestrial endotherms seek shade, partially immerse themselves in water, or burrow into the ground when the sun is high. Pigs, which lack sweat glands, roll in the mud to cool down. Animals that neither sweat nor pant can still benefit from evaporative heat loss. The evaporation of fluids deposited on the body surface by licking the skin or splashing the skin with water also draws heat from the body.

Similarly, animals respond to cold temperatures with numerous behavioral adaptations. Huddling in groups, curling up into a ball, hunching the shoulders, burying the head and feet in feathers, and similar maneuvers decrease the surface area exposed to a cold environment and decrease heat loss by radiation and conduction. Changing environments is also a common strategy for coping with cold. Migration from cold to warmer regions occurs in numerous species of birds and mammals.

## Muscle Activity and Brown Adipose Tissue Metabolism Increase Heat Production

We have discussed how heat is gained or lost to the environment and how heat can be retained by reducing blood flow to the skin on a cold day. Body temperature, however, is a balance between these factors and heat production. Changes in muscle activity constitute a major control of heat production for temperature regulation in endotherms.

When an endotherm is in its thermoneutral zone, no significant adjustments are necessary to maintain core body temperature. When exposed to temperatures below the thermoneutral zone, however, core body temperature begins to decrease. The primary response to decreasing temperatures is to decrease the flow of blood to regions that permit conduction of heat. If this does not adequately decrease heat loss, skeletal muscle contraction is increased. This leads to shivering, which consists of rapid muscle contractions without any locomotion. Virtually all of the energy liberated by the contracting muscles appears as internal heat, a process known as **shivering thermogenesis**. Many birds that remain in cold climates during the winter shiver almost continuously.

In many mammals, chronic cold exposure also induces **nonshivering thermogenesis**, an increase in the metabolic rate and therefore heat production that is not due to increased muscle activity. Nonshivering thermogenesis occurs primarily in **brown adipose tissue** (also called brown fat), a specialized tissue in small mammals such as hibernating bats, small rodents living in cold environments, and many newborn mammals, including humans. Brown adipose tissue is responsive to hormones and signals from the nervous system, which are activated when body temperature decreases. Unlike the adipose tissue discussed previously, which stores energy in the form of fat, brown adipose tissue metabolizes fat and generates heat as a by-product.

## 47.5 | Impact on Public Health

### Learning Outcome:

1. Define body mass index (BMI), and explain how it is used to assess health risks associated with being overweight and obesity.

As we have seen, most animals, when provided adequate nutrients, maintain their body mass around a set point that is normal for their species. We rarely observe healthy animals in nature that are overweight. Generally, only domesticated animals become sufficiently sedentary that they gain excess, unnecessary weight (think of an overweight housecat). Humans, too, are prone to weight gain, particularly when living sedentary lives. Many people maintain a healthy body weight, but, as discussed in this section, an increasing number are unable to meet this goal.

### Obesity Is a Global Health Issue

Excess body fat increases the risk of many diseases, including high blood pressure, cancer, heart disease, and diabetes mellitus. In diabetes mellitus, either insufficient insulin is available from the pancreas to control blood glucose concentrations (type 1 diabetes mellitus) or the cells of the body are less sensitive than usual to insulin (type 2 diabetes mellitus). In the U.S. alone, about 8% of the population—nearly 24,000,000 people—have diabetes mellitus. Of all diabetics in the U.S., more than 90% have type 2 diabetes mellitus. Compelling evidence has directly linked the incidence of this type of diabetes with being overweight. At what point does fat accumulation in humans start to pose a health risk? Historically, this question has been evaluated by research studies that investigate possible correlations between disease rates and some measure of body fat.

In ordinary practice, however, rather than obtaining a precise measure of body fat, a simple indicator of a person's potential health risk due to their weight is the **body mass index (BMI)**, a ratio of weight relative to height. A person's BMI is calculated by dividing his or her weight in kilograms by the square of his or her height in meters. For example, a 70-kg human with a height of 180 cm would have a BMI of 21.6 kg/m²:

$$BMI = 70 \text{ kg}/(1.8)^2 \text{ m}$$
$$= 21.6 \text{ kg/m}^2$$

The BMI is not a measure of body fat. It gives an estimate of how overweight a person may be, but does not account for such things as a person's muscle development, which could also increase weight and BMI. Nonetheless, current National Institutes of Health guidelines categorize BMIs of 25 or more as overweight, that is, as having increased health risk. BMIs of 30 or greater are considered obese, with a greatly increased health risk. Data compiled by the Centers for Disease Control and Prevention in Atlanta, Georgia, and other U.S. federal agencies indicate that two-thirds or more of U.S. adults age 20–74 are now overweight or obese (**Figure 47.14**). One of the more troubling statistics is that the percentage of adults who are overweight but not obese has remained relatively unchanged since 1960, at about one-third. However, the percentage of obese adults has risen during that time from about 13% to the current level of nearly 40%. Since as recently as the early 1990s, the CDC estimates that the average body weight of Americans has risen by 10 pounds. Even more troubling, the rate of childhood obesity has also risen. In the U.S., the incidence of obesity in children age 6–11 has increased from 2–3% in 1960 to the current estimate of approximately 18.5%.

The increase in obesity is not confined to the U.S., but has become a worldwide trend. According to the World Health Organization, more than 1 billion adults globally are overweight and 300 million are obese.

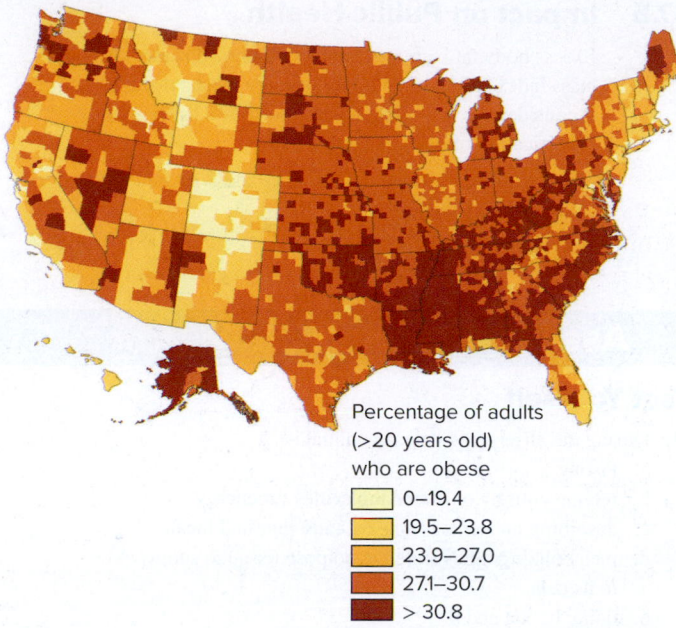

Percentage of adults
(>20 years old)
who are obese

| | |
|---|---|
| | 0–19.4 |
| | 19.5–23.8 |
| | 23.9–27.0 |
| | 27.1–30.7 |
| | > 30.8 |

**Figure 47.14** **Obesity in U.S. adults.** The Centers for Disease Control and Prevention estimate that many areas of the country experience obesity rates over 30% in adults (the darkest colored areas). Source: CDC.

 **Core Skill: Science and Society** Research into the causes of obesity and its consequences for human health have led to significant changes in the way we view the importance of maintaining a healthy body mass, but unfortunately obesity continues to place a significant burden on the health care systems of much of the western world.

Some studies indicate that genetic factors play an important role in obesity. Identical twins separated soon after birth and raised in different households have strikingly similar body weights as adults. Researchers hypothesize that natural selection favored the evolution of so-called thrifty genes, which boosted our ancestors' ability to store fat from each feast in order to sustain them through the next famine. Given today's abundance of high-fat and high-carbohydrate foods in many countries, what was once a survival mechanism may now be a liability.

The methods and goals of treating obesity are undergoing extensive rethinking. An increase in body fat is generally due to an excess of energy intake over energy expenditure, and overweight people have traditionally been advised to follow a low-calorie diet. However, such diets alone have limited effectiveness, because over 90% of obese people regain most or all of their lost weight within 5 years. This disturbing phenomenon may be related to the observation that metabolic rate decreases as the concentration of leptin in the blood decreases. Metabolic rate may decrease sufficiently to prevent further weight loss on a diet of as little as 1,000 kcal per day.

Research indicates that crash diets are not an effective long-term method for controlling weight. Instead, caloric intake should be set at a realistic level that can be maintained for the rest of one's life. This reduction in caloric intake should lead to a slow, steady weight loss

of no more than 1 pound per week until body weight stabilizes at a new, lower level. Most important, any program of weight loss should include increased physical activity. The exercise itself burns calories, but more importantly, it partially offsets the tendency for the metabolic rate to decrease. As a bonus, the combination of exercise and caloric restriction causes a person to lose more fat and less protein than with caloric restriction alone.

The impact of obesity on public health is enormous, accounting for many illnesses requiring hospitalization and chronic drug therapy, and well over 100,000 premature deaths per year. Its impact on the economy is far-reaching as well. The economic toll of obesity-related illnesses is felt in the loss of worker-hours in the workplace and in the costs of hospital stays, physician office visits, nursing home care, and medications. Current estimates by the Centers for Disease Control and Prevention are that as much as 20% of all U.S. health-care expenditures are directly or indirectly related to obesity! Obesity can affect society in unexpected ways. As an example, the increasing weight load of passengers forces airplanes to burn 350 million additional gallons of fuel each year, compared with just 25 years ago! This translates into nearly 4 million tons of additional pollution released into the atmosphere every year. In another recent example, for safety reasons, the U.S. Coast Guard has begun downgrading the allowable number of passengers on large state ferries due to the increasing average weight of U.S. adults.

## Summary of Key Concepts

### 47.1  Use and Storage of Energy

- An animal's utilization of nutrients has two states: the absorptive state, during which ingested nutrients are entering the blood from the alimentary canal, and the postabsorptive state, during which the GI tract is empty of nutrients and the body's own stores must supply energy.

- Glucose and fatty acids are the two major energy sources during the absorptive state. Much of the absorbed glucose immediately enters cells and is enzymatically broken down, providing energy required to synthesize ATP. Most absorbed triglycerides are stored in adipose cells until an animal requires additional energy. Amino acids are taken up by all body cells and used to synthesize proteins (Figure 47.1).

- The events that maintain blood glucose concentration in the postabsorptive state fall into two categories: (1) the reactions of glycogenolysis and gluconeogenesis, which provide glucose to the blood, and (2) cellular use of fatty acids for energy, which spares glucose for use by the nervous system (Figure 47.2, Table 47.1).

### 47.2  Regulation of the Absorptive and Postabsorptive States

- Tight control mechanisms, in the form of several hormones and the nervous system, maintain homeostatic concentrations of fuel molecules in an animal's blood. The hormone insulin acts on certain cells to facilitate the diffusion of glucose from blood into the cell cytosol via glucose transporters (GLUTs). All animal cells use GLUTs to transport glucose across their plasma membranes (Figure 47.3).

- In vertebrates, an increase in blood glucose concentration in the absorptive state stimulates the cells of the pancreas to secrete insulin; a decrease in that concentration removes the signal for secretion. In the postabsorptive state, when the blood glucose concentration decreases, glucose-monitoring regions in the hypothalamus stimulate production of glucose-elevating factors such as glucagon and norepinephrine (Figure 47.4).

- Exercise or any type of physical activity increases an animal's metabolic rate. Exercise increases an animal's requirement for nutrients, including glucose and fatty acids, to provide energy.

## 47.3   Energy Balance and Metabolic Rate

- An animal's metabolic rate refers to the amount of energy it uses in a given period of time to power all of its metabolic requirements.

- The most common measure for comparing metabolic rates of different species is the basal metabolic rate (BMR). Most of the basal metabolism is due to the routine functions of the heart, liver, kidneys, and brain (Figure 47.5).

- Many factors affect metabolism, including skeletal muscle activity, whether an animal has recently eaten, and body mass (Figure 47.6).

- When the daily amount of energy in consumed foods equals the amount of energy expended, body weight remains stable. Tipping the balance in either direction causes weight gain or loss by increasing or decreasing total body energy content.

- Short-term control of feeding generally involves satiety signals that remove the sensation of hunger and set the time period before hunger returns again. Experiments by Coleman and Friedman investigated the hormone leptin as a satiety factor in mammals. Leptin has since been found in all classes of vertebrates (Figures 47.7, 47.8).

## 47.4   Regulation of Body Temperature

- Most animals can survive only in a relatively narrow temperature range that allows chemical reactions to proceed, maintains the structures of membranes, and avoids denaturing proteins.

- Animals can be classified according to their source of heat and their ability to maintain body temperature. Ectotherms depend on external heat sources to warm their bodies, whereas endotherms use their own metabolically generated heat to warm themselves. Homeotherms maintain their body temperature within a narrow range, but heterotherms have body temperatures that vary with the environmental temperature (Figure 47.9).

- The surface of an animal's body can lose or gain heat from the external environment via four mechanisms: radiation, evaporation, convection, and conduction (Figures 47.10, 47.11).

- The skin can function as a variable heat exchanger; blood vessels near the skin surface dilate to dissipate heat or constrict to retain it. Both endotherms and ectotherms regulate heat loss through countercurrent exchange, which retains heat by returning it to the body's core and keeping the core warmer than the extremities. Heat exchange can also be regulated by changing the rate of water evaporation via perspiration. Behavioral mechanisms can alter heat loss by radiation, conduction, and convection (Figures 47.12, 47.13).

- Muscle activity (shivering thermogenesis) and brown adipose tissue metabolism (nonshivering thermogenesis) increase the production of heat.

## 47.5   Impact on Public Health

- Excess body fat increases the risk of many diseases. A body mass index (BMI) of 25 kg/m² or more means that a person is considered overweight, and a value of 30 kg/m² gives a classification of obese (Figure 47.14).

- Obesity can have serious health risks and is treated with caloric restriction and exercise.

## Assess & Discuss

### Test Yourself

1. During the absorptive state, an animal is
   a. fasting.
   b. relying entirely on stored molecules for energy.
   c. absorbing nutrients from a recently ingested meal.
   d. metabolizing lipids stored in adipose tissue to supply ATP to its cells.
   e. doing both a and b.

2. Gluconeogenesis
   a. occurs when the liver synthesizes glucose from noncarbohydrate precursors.
   b. is the process by which glycogen is broken down to glucose.
   c. occurs primarily when an animal is in the absorptive state.
   d. occurs when triglycerides are being formed and stored in adipose cells.
   e. occurs primarily in skeletal muscle.

3. In the process of _____, most tissues of the vertebrate body metabolize fat instead of glucose to ensure that _____ tissue has an adequate supply of glucose.
   a. gluconeogenesis, muscle
   b. glucose sparing, epithelial
   c. glycogenolysis, nervous
   d. glucose sparing, nervous
   e. gluconeogenesis, epithelial

4. Ketones are compounds that are derived from _____ and are synthesized primarily in the _____ state.
   a. glucose, absorptive
   b. glycogen, absorptive
   c. fatty acids, postabsorptive
   d. amino acids, postabsorptive
   e. triglycerides, absorptive

5. Insulin primarily regulates the blood glucose concentration by
   a. stimulating the recruitment of GLUTs from the cytosol to the plasma membrane for transport of glucose from extracellular to intracellular fluid.
   b. stimulating gluconeogenesis.
   c. suppressing glucose uptake by muscle tissue.
   d. stimulating the release of glucose from glycogen reserves in the liver.
   e. inhibiting the synthesis of new GLUTs.

6. The rate at which an animal uses energy is called
   a. the body mass index.
   b. the animal's energy consumption.
   c. the metabolic rate.
   d. nonshivering thermogenesis.
   e. shivering thermogenesis.

7. Which factor(s) may increase metabolic rate?
   a. shivering
   b. decreased muscle activity
   c. sleeping
   d. consumption of a meal
   e. both a and d

8. Which molecule acts on brain centers to decrease appetite in mammals and other vertebrates?
   a. GLUT4          d. glucagon
   b. glycogen       e. a ketone
   c. leptin

9. Animals that have body temperatures that are maintained within a narrow range are
   a. endotherms.    d. heterotherms.
   b. ectotherms.    e. both b and d.
   c. homeotherms.

10. The rate of heat loss in a mammal is regulated by
    a. the degree of blood flow at the surface of the skin.
    b. the amount of perspiration.
    c. behavioral adaptations.
    d. air currents near the animal's body.
    e. all of the above.

## Conceptual Questions

1. Explain the functions of insulin. Why do you think a hormone such as insulin is required to carry out these functions?

2. Explain how appetite is controlled by the brain. What is the benefit of having a hormone released from adipose cells in proportion to total fat mass?

3. **Core Concept: Structure and Function** How does this core concept apply to countercurrent exchange?

## Collaborative Questions

1. Discuss the differences between being ectothermic and endothermic and between being heterothermic and homeothermic.

2. Discuss four ways that animals exchange heat with their environment.

# Circulatory and Respiratory Systems

# 48

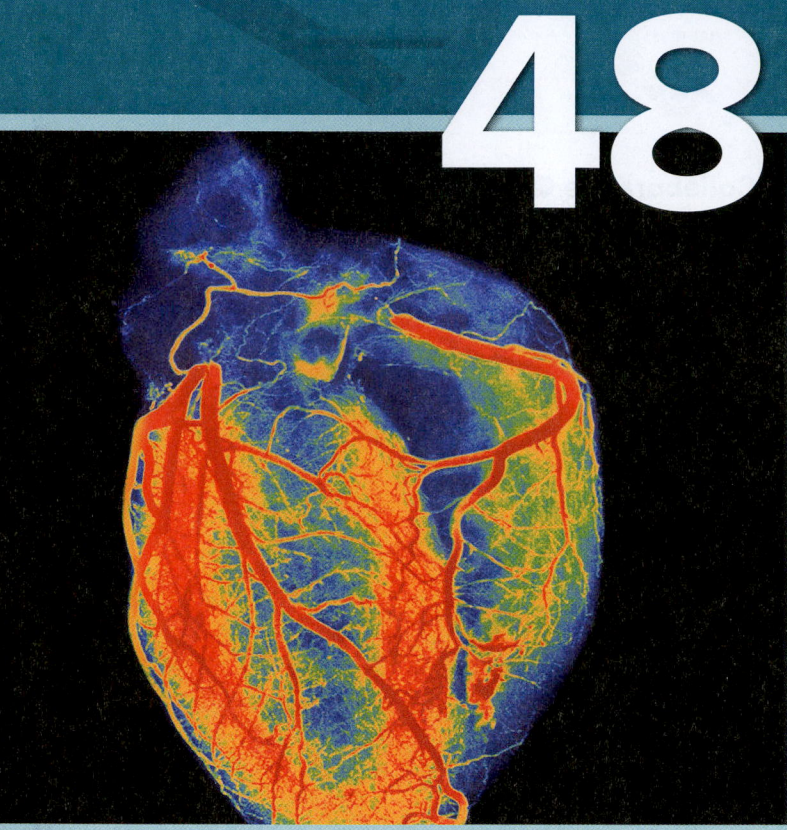

**Image from cardiac angiography of a human heart (contrast dye injection made the coronary blood vessels [in red] easier to visualize).** ©SPL/Science Source

**A**n overweight 62-year-old woman decided to get herself "back in shape" after years of a sedentary lifestyle. Rather than gradually increasing her activity level, however, she embarked on a rigorous, demanding exercise program at a local gym. While running at 7 mph on a 10% inclined treadmill, she felt an acute, crushing sensation in her chest. She became extremely anxious and stopped exercising, but the pain did not go away. She felt light-headed, short of breath and nauseated, and called for help. An ambulance arrived, and the paramedic determined from her vital signs that the woman was experiencing a heart attack. She was transported to the emergency department of a local hospital and remained there for further treatment. Eventually, she was released and placed on several medications to control her blood pressure, prevent blood clots from forming, and improve the function of the healthy parts of her heart. She also received counseling from a nutritionist and a physical therapist, who provided her with a sensible plan for improving her overall health.

Cardiovascular disease—disease of the heart or blood vessels—is the leading cause of death in the U.S. and in much of the rest of the developed world. In the U.S. alone, an estimated 80 million or more people have one or more diseases that affect the heart and blood vessels. About 1–1.5 million of those individuals will have a heart attack in the next year, and between 400,000 and 600,000 of them will die as a result.

The heart is a muscular pump that requires considerable nutrients and oxygen to sustain its unceasing muscular effort. The chapter opening photo shows the extensive network of blood vessels coursing through a typical mammalian heart. Should any of those vessels become diseased, the regions of the heart supplied by them can die. That is what happens during a heart attack. When this occurs, the damaged heart may not be able to pump blood forcefully enough to generate the pressure required for sufficient blood to reach all the cells of the body.

The heart and blood vessels that function to transport blood or, in certain animals, a fluid called hemolymph to all regions of an animal's body comprise the **circulatory system**. Circulatory systems transport necessary materials to cells and transport waste products away from cells to be released into the environment. Two of the most important solutes carried in the blood of many animals are oxygen ($O_2$) and carbon dioxide ($CO_2$). Oxygen enters an animal's body via respiratory organs such as gills or lungs. It then dissolves into body fluids and is distributed to cells. As described in Chapter 7, $O_2$ is used by mitochondria during the formation of ATP. One of the waste products of those reactions is $CO_2$, which diffuses out of cells. In vertebrates, $CO_2$ is returned by the circulatory system to the respiratory organ and is then released into the environment. The process of moving $O_2$ and $CO_2$ in opposite directions between the environment, body fluids, and cells is called **gas exchange**, or **respiration**. (The latter term differs from cellular respiration, in which cells generate energy from metabolism and produce ATP, as described in Chapter 7.) A **respiratory system** includes all of an animal's structures that contribute to gas exchange.

This chapter explores the mechanisms that control the circulation of body fluids and the means by which animals obtain and transport $O_2$, rid themselves of $CO_2$, and cope with the challenges imposed by changing metabolic demands.

## 48.1    Types of Circulatory Systems

**Learning Outcomes:**

1. Compare and contrast open and closed circulatory systems.
2. Describe the differences between single and double circulations in vertebrates.

Except for the simple type of circulation that exists in certain invertebrates that have a gastrovascular cavity (look back to Figure 46.2), two basic types of animal circulatory systems exist: open systems and closed systems. Here, we compare and contrast some of the key features of the two major types of circulatory systems found in animals.

### In Open Circulatory Systems, Hemolymph Enters the Body Cavity

As illustrated in **Figure 48.1a**, an **open circulatory system** is characterized by a fluid that is pumped by one or more contractile hearts into the body cavity (hemocoel) of an animal. Therefore, the fluid in the blood vessels and the interstitial fluid that surrounds cells mingle in one large body compartment, rather than being located in separate body compartments. The mixed fluid is called **hemolymph**. Nutrients and wastes are exchanged by diffusion between the hemolymph and body cells. The hemolymph is eventually returned to the heart through vessels or through small openings called ostia. Oxygen and $CO_2$ are not transported in hemolymph. In

animals with open circulatory systems, these gases are exchanged with the environment by a different mechanism, which we will examine later.

Open circulatory systems are found in most invertebrates and have certain advantages. They are metabolically inexpensive because little energy is needed to pump hemolymph into the hemocoel. Hemolymph simply enters the hemocoel through open-ended vessels. In addition, open circulatory systems can adapt to changes in an animal's metabolic demands. As an insect takes flight, for example, its flight muscles contract more forcefully and rapidly, which acts to expand and compress the animal's thorax. This movement helps propel hemolymph throughout the hemocoel and into and out of the hearts. In other words, as the animal's physical activity increases, its circulation becomes more effective, recharging the metabolically active cells with nutrients. The ability to adjust the circulation to meet an animal's requirements is one of the most important features of any circulatory system.

Despite the phenomenal success of species with open circulatory systems, this type of circulation has certain limitations. For instance, because the hemolymph empties in bulk into the hemocoel, it cannot be selectively delivered to individual regions of the body that may have increased metabolic requirements relative to other regions. By contrast, during periods of increased physical activity in vertebrates, larger amounts of blood are delivered to skeletal muscles and away from less metabolically active tissues and organs, an extremely useful adaptation that results in greater endurance. This selective delivery is possible because these animals have a closed circulatory system, as we explore next.

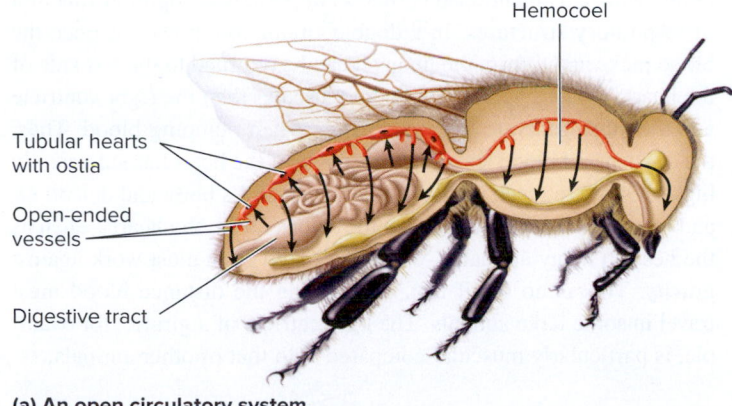

Hemocoel

Tubular hearts with ostia

Open-ended vessels

Digestive tract

**(a) An open circulatory system**

**Figure 48.1  Types of circulatory systems.  (a)** In open circulatory systems, one or more muscular, tubular hearts pump hemolymph through open-ended vessels, where it percolates throughout the body. In arthropods such as this honeybee, hemolymph re-enters the heart through ostia. The arrows show the movement of the hemolymph. **(b)** In closed circulatory systems, such as that of this earthworm, blood remains within vessels and hearts, and it recirculates without emptying into the body cavity. Arrows indicate the direction of blood flow.

**Concept Check:** *It is incorrect to think of open circulatory systems as primitive compared with closed systems. Why?*

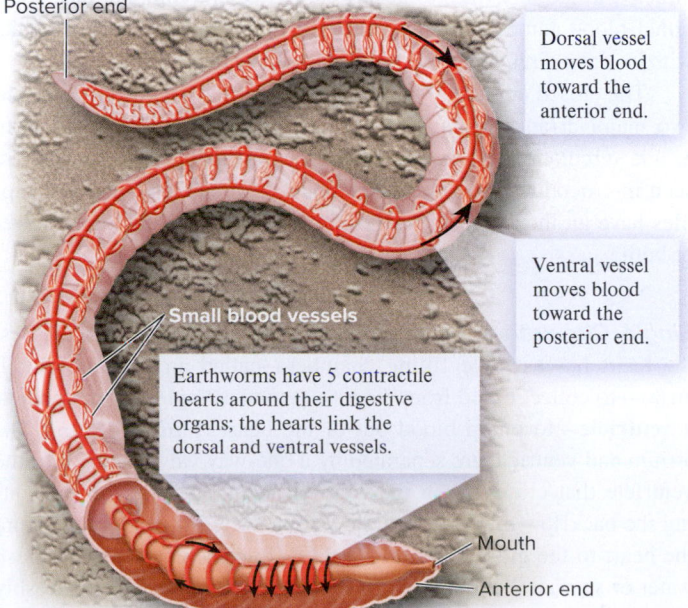

Posterior end

Dorsal vessel moves blood toward the anterior end.

Ventral vessel moves blood toward the posterior end.

Small blood vessels

Earthworms have 5 contractile hearts around their digestive organs; the hearts link the dorsal and ventral vessels.

Mouth

Anterior end

**(b) A closed circulatory system.**

## In Closed Circulatory Systems, Blood Remains in Blood Vessels

In a **closed circulatory system**, blood and interstitial fluid are physically separated and differ in their components and solute composition. Closed circulatory systems evolved independently several times and are found in earthworms, cephalopods (squids and octopuses), and all vertebrates (**Figure 48.1b**). Despite some differences in structure among species, closed circulatory systems share certain key features, which will be described throughout the rest of this chapter:

- They transport **blood**—a fluid connective tissue containing a mixture of cells and solutes under pressure that is generated by one or more contractile, muscular hearts. Both $O_2$ and $CO_2$ are transported in the blood of closed circulatory systems.

- Except when injury occurs, blood remains at all times within the blood vessels that distribute it throughout the body. Vessels called **arteries** carry blood away from the heart, and **veins** return blood back to the heart. **Capillaries** link the arterial and venous vessels and are the sites where water and solutes are exchanged between blood and interstitial fluid.

- The activity of closed circulatory systems can be adjusted to match an animal's metabolic demands.

- Within limits, closed circulatory systems generally can repair themselves when injured.

- Closed circulatory systems grow in size as an animal grows.

A closed circulatory system offers several advantages. First, animals can grow to a larger size, because blood can be directed to every cell of an animal's body, no matter how large the body. Nearly all body cells are within one or two cell widths of a blood vessel. Second, blood flow can be selectively increased or decreased to supply different parts of the body with the precise amount of blood required at any given moment. After a meal, for example, more blood can be directed to the intestines to absorb nutrients, and on a hot day, additional blood can be routed to the skin to dissipate heat. An animal with an open circulatory system cannot make these adjustments.

The closed circulatory system of vertebrates can be divided into two major arrangements: single circulation and double circulation. Single circulations are seen in fishes, and double circulations are seen in crocodiles, birds, and mammals. Amphibians and most reptiles have an intermediate type of circulation that combines features of both.

***Single Circulation: Fishes***   In the single circulation of fishes, the heart has a single filling chamber—called an **atrium** (plural, atria)—to collect blood from the tissues, and a single exit chamber—a **ventricle**—to pump blood out of the heart (**Figure 48.2a**). The atrium and ventricle are separated by a one-way valve facing into the ventricle that closes when the ventricle contracts, thereby preventing the backflow of blood into the atrium. Arteries carry blood from the heart to the gills, which pick up $O_2$ from the environment (fresh water or seawater) and unload $CO_2$ into the environment. The freshly oxygenated blood then circulates through other arteries to the rest of the body. There, $O_2$ and nutrients are delivered to cells, and $CO_2$ and other wastes diffuse from cells into the blood. Finally, the partially

deoxygenated blood is returned to the heart via veins, where it is pumped back to the gills for another load of oxygen.

An important feature of all respiratory surfaces is that they function best when the blood flowing through them is maintained at a low pressure. Thus, the fish heart does not generate high pressure when it pumps blood to the gills, thereby protecting the delicate gill tissue and facilitating the exchange of gases. As a consequence, however, the pressure of blood leaving the gills will also be low, and this limits the rate at which oxygenated blood can be delivered to the body's cells.

***Double Circulation: Crocodiles, Birds, and Mammals***   In a double circulation, oxygenated and deoxygenated blood are completely separated into two distinct circuits, the **systemic circulation** and the **pulmonary circulation** (**Figure 48.2b**). All animals with double circulations have four-chambered hearts: a left atrium and ventricle and a right atrium and ventricle. The two atria are separated from each other by a connective tissue septum, as are the two ventricles. Another septum separates the each atrium from its corresponding ventricle. Thus, each chamber is filled separately and the two circulations do not mix.

The pulmonary circulation receives partially deoxygenated blood from the right ventricle and delivers fully oxygenated blood from the lungs to the left atrium, which then passes it on to the left ventricle. The left ventricle pumps its blood to all the body's tissues via the systemic circulation. The systemic circulation then returns the partially deoxygenated blood from the tissues to the right atrium, which then sends it to the right ventricle, and the cycle starts again. Each ventricle fills with blood and then contracts simultaneously with the other.

A major advantage of a double circulation is that the two ventricles can function as if they were, in effect, two hearts, each with its own ability to pump blood under different pressures. The blood from the right ventricle can be pumped under low pressure to the lungs, which, as mentioned earlier, is important for optimal function of respiratory structures. In a double circulation, however, once the blood picks up $O_2$ from the lungs, it can be returned to the left side of the heart. The left ventricle is more muscular than the right ventricle and therefore generates higher pressure when pumping blood. Thus, the oxygenated blood leaving the left side of the heart has sufficiently high pressure to reach all the cells of an animal's body and deliver $O_2$ and nutrients at a high rate, even to regions above the heart—such as the head in many animals—where the blood flow must work against gravity. This is no small feat, considering the distance blood must travel in some large animals. The left ventricle of a giraffe, for example, is particularly muscular compared with that of other animals.

 **Core Concept: Evolution**

### A Four-Chambered Heart Evolved from Simple Contractile Tubes

The heart is the first vertebrate organ to become functional during embryonic development, and the factors responsible for its development have been extensively studied. Interestingly, at least one of these developmental factors has also recently shed light on how a heart develops four chambers in crocodiles, birds, and mammals.

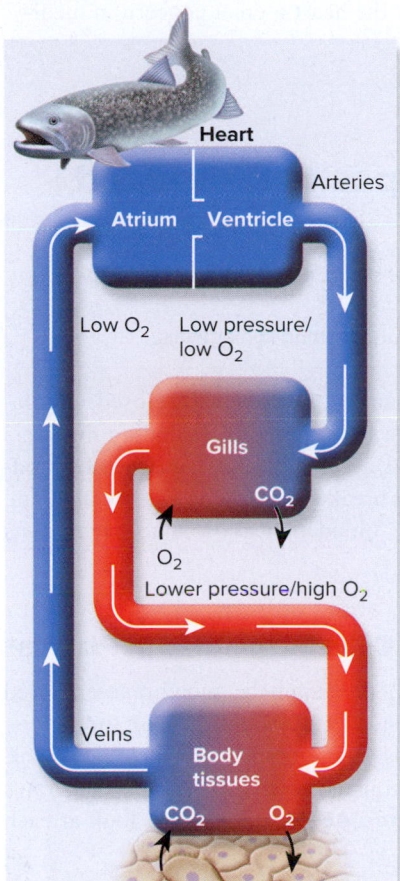

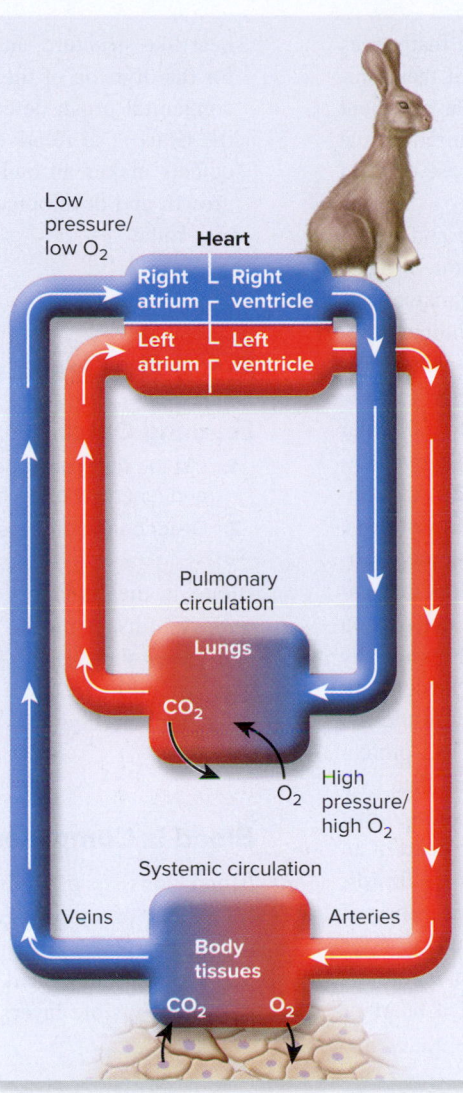

**Figure 48.2** **Representative vertebrate circulatory systems.** **(a)** Fishes have a single circulation in which blood is pumped from the heart to the gills, from which it circulates to the rest of the body tissues, and finally returns to the heart. **(b)** Crocodiles, birds, and mammals have a double circulation, in which oxygenated blood is pumped under high pressure to the body's tissues, and deoxygenated blood is pumped under low pressure to the lungs. A connective tissue septum (shown as while lines) separates the two atria and the two ventricles, and also separates the atria from the ventricles. Valves are represented as breaks in the septums between atria and ventricles.

KEY
- $O_2$-rich blood
- $O_2$-poor blood
- Direction of blood flow

**(a) Single circulation in fishes**

**(b) Double circulation in crocodiles, birds, and mammals**

---

 **Core Skill: Modeling**  The goal of this modeling challenge is to revise the model shown in Figure 48.2b to reflect the existence of a hole between the atria.

**Modeling Challenge:** As shown in the model in Figure 48.2b, animals with double circulations have septa that divide the two atria from each other and the two ventricles from each other. On rare occasions, humans are born with a hole in the septum between the two atria. Draw a model that incorporates such a hole and then color the blood in the two circulations, starting with the color conventions shown in the key. You will have to think carefully how the colors may change from those shown in Figure 48.2b.

---

The first heartlike organ appears to have evolved at least 500 mya. Ancestral hearts may have resembled those observed today in cephalochordates and urochordates—a simple, linear tube with wavelike contractile properties but no valves or chambers (refer back to Figure 34.35). A valvelike structure first appears in the tubular hearts of some arthropods, such as *Drosophila*, but such hearts do not have clearly defined chambers. Analyses of animal genomes have identified numerous genes that are critical for the development of the chambered

vertebrate heart, including a core set of five highly conserved genes. Interestingly, however, one or more of these five genes are also expressed in linear tube hearts such as those just described and even in contractile structures in animals without hearts. For example, cnidarians do not have a heart or blood vessels, but they have muscle cells in their contractile bell that not only produce locomotion but in the process circulate water through the animal's gastrovascular cavity. These muscle cells express at least three of these heart developmental genes. Nematodes also do not have a heart but, instead, have a contractile pharynx that serves part of the same function. The pharyngeal muscles of these animals express four of the five key genes for heart development that are found in all vertebrates. Thus, the genes required for the formation of a contractile, multichambered heart arose early in the evolution of animals.

Why are some vertebrate hearts four-chambered and others are not? In a 2009 study, American researcher Benoit Bruneau and colleagues compared the expression of one of the major heart development genes, called *Tbx5*, in the hearts of amphibians, reptiles, birds, and mammals. (The *Tbx5* gene encodes Tbx5 protein, a transcription factor.) The researchers discovered that the expression of *Tbx5* became increasingly restricted to particular locations within the hearts of animals displaying greater separation of the ventricles into two chambers (**Figure 48.3**). Amphibians have two atria but only a single ventricle. In these animals, *Tbx5* was uniformly expressed throughout the developing ventricle. In lizards, in which a septum only partially divides the ventricle into two chambers, *Tbx5* expression was absent in a portion of the right half of the ventricle. In birds and mammals, with fully partitioned ventricles, Tbx5 protein was not made in the developing right ventricle, but was present in large amounts in the left ventricle. This finding suggests that a gradient of Tbx5 protein is required for the ventricle to form two chambers and thereby a complete double circulation. The investigators also showed that if *Tbx5* expression was experimentally induced in the heart of an embryonic mouse in a pattern that mimicked that seen in lizards, the mouse heart developed only three clearly defined chambers—there was no distinction between left and right ventricles!

From these and other studies, the expression of a core set of genes appears to be vital for the formation of a heart or heartlike structure, and a subset of genes, like *Tbx5*, is critical for the division of the heart into chambers. The most common congenital organ defects in humans are those associated with the heart. The relatively high incidence of heart-related birth defects makes an understanding of the genetic control of the growth and development of the heart a great concern in medicine today.

## 48.2 The Composition of Blood

**Learning Outcomes:**

1. List the four components of blood, and describe the composition and functions of each.
2. Describe the process of blood clotting in mammals.

Blood is the transport medium of animals with closed circulatory systems. It moves necessary materials—including nutrients and gases such as $O_2$—to all cells and takes away waste products, including $CO_2$ and other breakdown products of metabolism. What are the components of blood that allow it to perform its functions?

### Blood Is Composed of Cells Suspended in Plasma

Blood consists of cells and, in mammals, membrane-enclosed fragments of cytosol, suspended in a solution containing dissolved nutrients, proteins, gases, and other molecules. If we collect a blood sample and spin it in a centrifuge, the blood separates into three visible layers (**Figure 48.4**). Let's take a look at each of these.

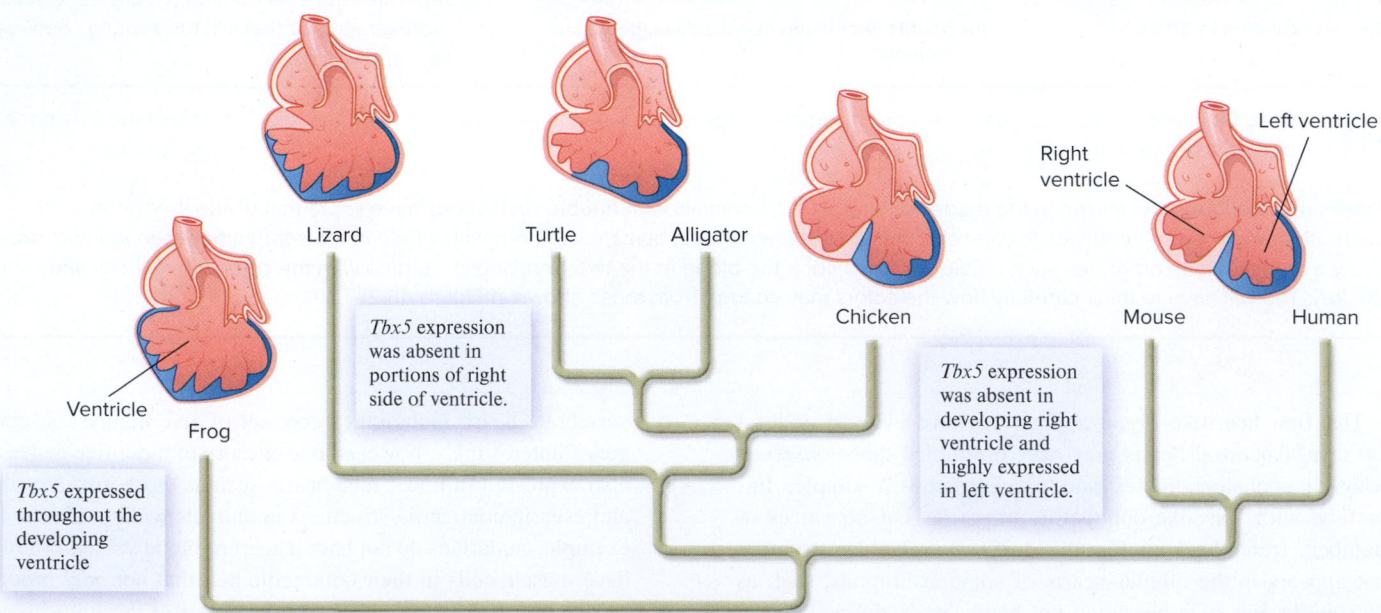

**Figure 48.3 Expression of the *Tbx5* gene in embryonic vertebrate hearts.** The blue color indicates where in the developing heart muscle the *Tbx5* gene was expressed. Note: Structural details have been omitted for simplicity. Source: Kazuko Koshiba-Takeuchi et al., 2009. Reptilian heart development and the molecular basis of cardiac chamber evolution. *Nature* 461: 95–98.

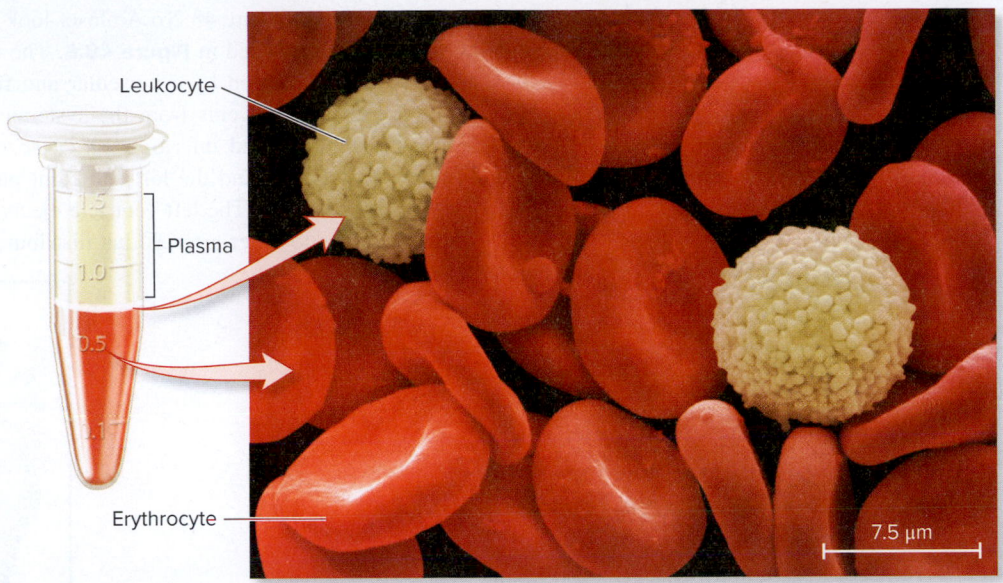

**Figure 48.4  Components of blood.** When a blood sample is centrifuged, it forms three visible layers: plasma, leukocytes, and erythrocytes. Leukocytes, shown in the scanning electron micrograph, are the white blood cells that make up part of the immune system. Erythrocytes are red blood cells, which carry oxygen. Note: The leukocyte layer is enlarged and not to scale, for illustrative purposes. (right) ©Power and Syred/Science Source

 **Core Skill: Connections** Leukocytes are part of the immune system of animals. Are immune defenses unique to animals? (Refer back to Figures 37.17 through 37.19 for help.)

***Plasma***  The top layer of the centrifuged blood sample is a yellowish solution called **plasma** (see Figure 48.4). Plasma typically makes up about half of the total volume of blood in most vertebrates. It is made up of water and dissolved nutrients, oxygen, waste products of metabolism, and many other molecules released by cells, such as hormones and proteins. These proteins serve numerous functions, such as helping in the formation of blood clots, which seal off wounds to blood vessels.

***Leukocytes***  Beneath the plasma in the centrifuged blood sample is a narrow white layer of **leukocytes**, also known as white blood cells (see Figure 48.4). Leukocytes develop from a specialized connective tissue (the marrow) of certain bones in vertebrates. Although there are several types—which we describe further in Chapter 52—all leukocytes perform vital functions that defend the body against infection and disease and thus are key components of an animal's immune system.

***Erythrocytes***  The bottom visible layer of the centrifuged blood sample consists of **erythrocytes**, also called red blood cells because of their color (see Figure 48.4). The term **hematocrit** refers to the volume of blood (expressed as percentage) that is composed of erythrocytes, usually between 35% and 65% among vertebrates. Erythrocytes serve the critical function of transporting oxygen throughout the body. There are approximately a thousand times more erythrocytes than leukocytes in the circulation. Like leukocytes, erythrocytes are derived from cells in the bone marrow. In most vertebrates, mature erythrocytes retain their nuclei and other cellular organelles, but in all mammals (and a few species of fishes

and amphibians), the nuclei are lost upon maturation. The lack of a nucleus and many other organelles in the mammalian erythrocyte increases the cell's oxygen-carrying capacity and contributes to its characteristic biconcave shape (see Figure 48.4). The biconcave shape of the mammalian erythrocyte increases its surface area relative to the flattened disc or oval shape seen in most other vertebrates. This is believed to increase the efficiency of gas exchange between the erythrocyte and the body fluids.

Oxygen is poorly soluble in plasma. Consequently, the amount of oxygen that dissolves in plasma usually cannot support a vertebrate's basal metabolic rate, let alone more strenuous activity. Within the cytosol of erythrocytes, however, are large amounts of the protein **hemoglobin**, which can reversibly bind to oxygen. Hemoglobin greatly increases the reservoir of oxygen in the blood and enables animals to be more active. In a later section, we will consider the mechanisms by which hemoglobin binds and releases oxygen.

***Platelets***  Vertebrate blood has a fourth component not visible in a centrifuged sample of blood, called **platelets** in mammals and thrombocytes in other vertebrates. Platelets, which are derived from bone marrow cells, are cell fragments that lack a nucleus. They serve a crucial role in the formation of blood clots that limit blood loss after injury. The formation of a blood clot requires several steps, two of which include platelets (**Figure 48.5**). First, when a blood vessel is injured, platelets secrete substances that cause them to clump together and bind to collagen fibers in the surrounding connective tissue at the wound site. This forms a plug that prevents continued blood loss. Second, other platelet secretions interact with plasma proteins to cause the precipitation from solution of a fibrous protein called

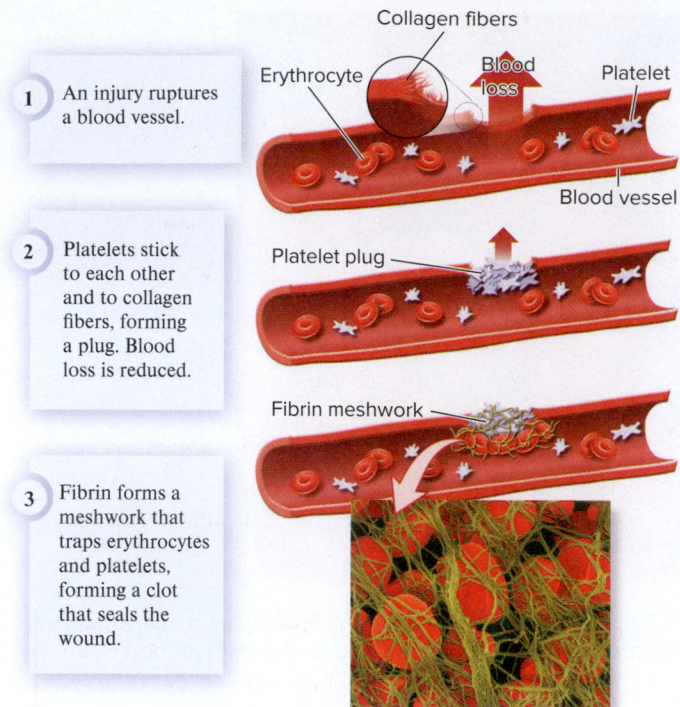

**Figure 48.5  Platelets and the process of blood clot formation.**  A blood clot forms in two major steps: A platelet plug reduces initial blood loss, and a fibrin clot then seals the wound. An example of a fibrin clot is shown in the scanning electron micrograph.

(bottom) ©Dennis Kunkel Microscopy, Inc./Phototake

fibrin. Fibrin forms a meshwork of threadlike fibers that wrap around and between platelets and erythrocytes, enlarging and thickening the plug to form a clot. Blood clotting begins within seconds and helps prevent injured animals from losing too much blood. Eventually, the body absorbs the clot as the injured vessel heals.

## 48.3 The Vertebrate Heart and Its Function

### Learning Outcomes:

1. Describe the structure of the vertebrate heart.
2. **CoreSKILL »** Outline the sequence of events of the cardiac cycle, and draw conclusions regarding the functional relationship between electrical and contractile events.
3. **CoreSKILL »** Interpret the meaning of the tracings on an electrocardiogram.

In Section 48.1, we considered some of the general features of circulatory systems in animals. In this section, we will examine the structure and function of the vertebrate heart, focusing on the human heart, which carries out a double circulation.

### Vertebrate Hearts That Carry Out a Double Circulation Have Two Atria and Two Ventricles

As described earlier, animals with a double circulation, such as mammals, have a heart with a left atrium and ventricle and a right atrium

and ventricle (see Figure 48.2b) A closer look at the structure of the human heart is provided in **Figure 48.6**. The two sides of the heart are physically separated by a muscular and fibrous septum. Blood enters the atria from veins from the systemic or pulmonary circulation; the superior and inferior vena cavae return blood from the systemic circulation, and the left and right pulmonary veins return blood from the lungs. The left ventricle ejects blood into the **aorta**, which branches to other arteries that distribute the blood. The right

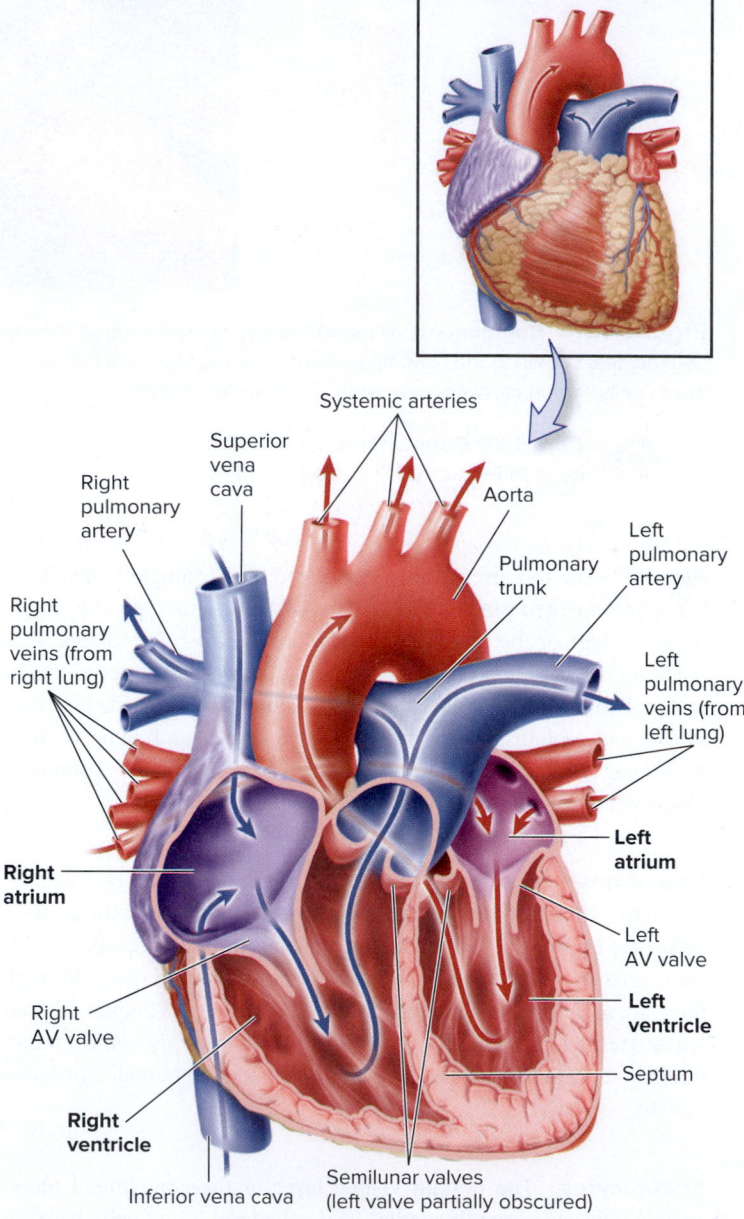

**Figure 48.6  The mammalian heart and circulation.**  This cutaway illustration of a human heart shows the major blood vessels entering and leaving the heart, and the relationships of the four chambers (see the inset for the intact heart). Regions that contain oxygenated blood are shown in red, deoxygenated in blue. Note that the pulmonary veins carry oxygenated blood because they return blood from the lungs, whereas veins from the systemic circulation carry deoxygenated blood.

**Concept Check:**  *Which vessel or vessels leading from the mammalian heart carry fully oxygenated blood?*

ventricle ejects blood into the pulmonary trunk, which divides into **pulmonary arteries** that lead to the right and left lungs.

Valves control the direction that blood flows through the heart. Blood normally can flow in only one direction through a valve. One-way valves called **atrioventricular (AV) valves** control the movement of blood between the atria and ventricles. Blood flows down a pressure gradient through the AV valves from the atria into the ventricles. One-way **semilunar valves** are found between each ventricle and the artery into which a ventricle sends blood (either the aorta or the pulmonary trunk). The right ventricle pumps deoxygenated blood to the lungs through the right semilunar valve, and the left ventricle pumps oxygenated blood to the rest of the body through the left semilunar valve.

## Vertebrates Have a Myogenic Heart

What causes the heart to beat so steadily? Animals cannot consciously initiate heart contractions. The beating of the heart is initiated either by nerves or by intrinsic activity of the heart muscle cells themselves. Many arthropods, for example, have a neurogenic heart that will not beat unless it receives regular electrical impulses from the nervous system. All vertebrates, however, have a myogenic heart, in which the signaling mechanism that initiates contraction resides within the cardiac muscle itself.

In myogenic hearts, cardiac muscle is distinguished by the interconnectedness between individual cardiac muscle cells, or cardiac myocytes. Each myocyte has membrane extensions that form interlocking networks with other myocytes. Within these networks, called intercalated discs, are many gap junctions (refer back to Figure 10.11)

that electrically couple the myocytes. The large number of gap junctions permits the rapid spread of electric current from cell to cell, so that all parts of the heart are rapidly stimulated at nearly the same time. This electric current is the signal that increases the intracellular concentration of $Ca^{2+}$ within the myocytes. Although the details are different, the increase in $Ca^{2+}$ triggers contraction in a manner that is generally similar to skeletal muscle contraction (see Chapter 45).

Myogenic hearts are electrically excitable and generate their own action potentials. The rate and forcefulness of the beating of myogenic hearts can, however, be regulated by the nervous system. For example, the autonomic nervous system increases heart rate in vertebrates during the fight-or-flight response (see Chapter 43). Nonetheless, myogenic hearts continue to beat on their own if dissected out of an animal and placed in a nutrient bath, even with no nerves present.

## Excitation of the Heart Begins in the Atria and Spreads to the Ventricles

As with skeletal muscle (see Chapter 45), contraction of the vertebrate heart cannot occur until the muscle has been electrically excited. The electrical excitation of the vertebrate heart has two phases: atrial and ventricular. In atrial excitation, electrical signals are generated within the wall of the right atrium at the **sinoatrial (SA) node**, or **pacemaker** (**Figure 48.7**). The SA node is a collection of modified myocytes that have an inherently unstable resting membrane potential. Ion channels in the membranes of these cells are opened spontaneously and allow the influx of positively charged ions into the cytosol, thereby depolarizing the cell. These depolarizations produce action potentials in the

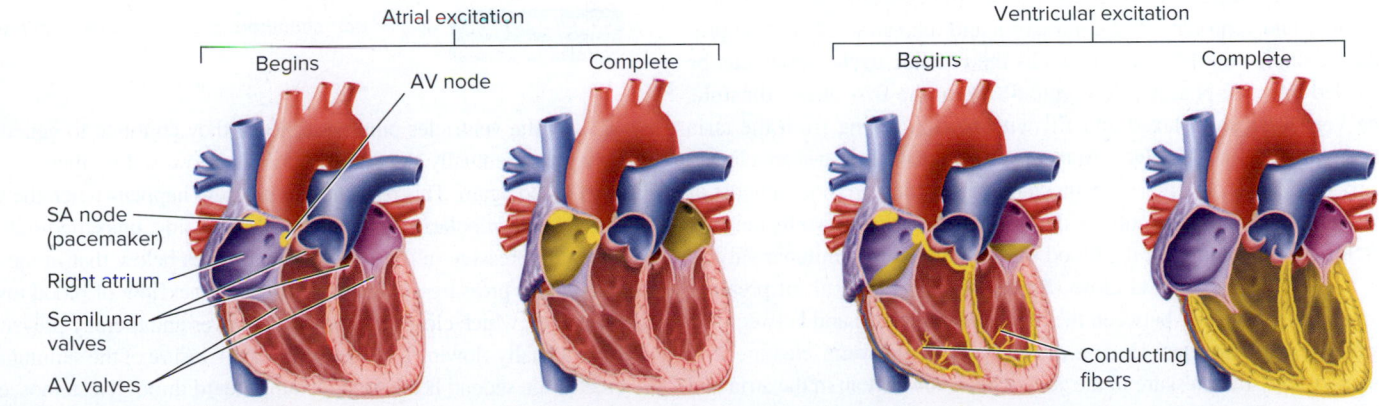

| | Atrial excitation | | Ventricular excitation | |
| --- | --- | --- | --- | --- |
| | Begins | Complete | Begins | Complete |
| **AV valves** | OPEN | OPEN | Beginning to close | CLOSED |
| **Semilunar valves** | CLOSED | CLOSED | CLOSED | OPEN |
| **Phase of cardiac cycle** | Diastole | Diastole | Systole beginning | Systole |
| **Atria** | Relaxed | Contracted | Beginning to relax | Relaxed |
| **Ventricles** | Relaxed | Relaxed | Beginning to contract | Contracted |
| **Chambers with highest pressure** | All are low pressure | Atria (slightly) | Ventricles | Ventricles |

**Figure 48.7  Electrical activity in the mammalian heart and the events of the cardiac cycle.**  The spread of electrical activity, which is shown in yellow, begins in the sinoatrial (SA) node and quickly spreads through the atria to the atrioventricular (AV) node. Branches from the AV node transmit electrical activity throughout the ventricles via conducting fibers. The atria and ventricles do not contract until they have become electrically excited. Changes in pressure gradients between the atria and ventricles, and between the ventricles and the aorta or pulmonary trunk, are the forces that open or close the two sets of heart valves. Note: The left semilunar valve is only partly visible in this orientation.

SA node cells with a range of frequencies that are characteristic for a given species. The frequency determines an animal's heart rate.

Once the SA node cells generate action potentials, the potentials quickly spread across both atria through the gap junctions described earlier. The action potentials trigger an influx of $Ca^{2+}$ through voltage-gated channels into the muscle cell cytosol, which activates contraction. Because the impulses spread very rapidly across the atria, both atria contract together almost as if they were one large muscle cell. Atrial contraction assists in providing a final push of blood through the open AV valves into the ventricles.

To begin ventricular excitation, action potentials initiated in the SA node must somehow move to the ventricles, but the septum separating the atria and ventricles does not contain gap junctions. However, another node of specialized myocytes, the **atrioventricular (AV) node,** is located at the junction between the atria and ventricles and conducts the electrical signals from the atria to the ventricles. Like the SA node, the AV node is electrically excitable, but its cells require a longer time to become excited than do the cells of the SA node. This electrical delay allows time for the atria to contract before the ventricles do. Fibers branching from the AV node spread electrical impulses along a conducting system of cardiac muscle cells that extends throughout the muscular walls of the ventricles. This conducting system ensures that all of the ventricular muscle gets depolarized quickly and nearly simultaneously (see Figure 48.7). In this way, the entire mass of both ventricles contracts in a coordinated way in response to depolarization.

## The Cardiac Cycle Has Two Phases: Diastole and Systole

Each beat of the vertebrate heart requires the coordinated activities of the atria and ventricles. The contraction and relaxation events that produce a single heartbeat are known as the **cardiac cycle**, which can be divided into two phases (see Figure 48.7). In the first phase, **diastole**, the ventricles are relaxed and fill with blood coming from the atria through the open AV valves. At the end of diastole, the atria are electrically stimulated, and therefore they contract, which provides a boost of additional blood to the ventricles. In the second phase, **systole**, the ventricles contract and eject the blood through the open semilunar valves.

The valves open and close (**Figure 48.8**) as a result of pressure gradients established between the atria and ventricles and between the ventricles and arteries. During diastole, when the ventricles are in a relaxed state, the pressure in the ventricles is lower than in the atria and the arteries. Therefore, the AV valves are open, but the semilunar valves are closed. Once the ventricles are electrically excited, they begin contracting, and the pressure within the ventricles rapidly increases. This marks the beginning of systole. Because the ventricles are thicker and stronger than the atria, ventricular contractions generate much greater pressure. When the pressure in the ventricles exceeds that in the atria, the AV valves are forced closed and no more blood enters the ventricles. Valve closure also prevents blood from flowing backward into the atria. Closure of the AV valves makes the "lub" sound of the familiar "lub-dub" heard through a stethoscope. Ventricular pressure continues to increase as the ventricles continue contracting. Eventually the pressure in the ventricles exceeds the pressure in the arteries, which causes the semilunar valves to open (see Figure 48.8b). Blood is then ejected through the open semilunar valves into the arteries.

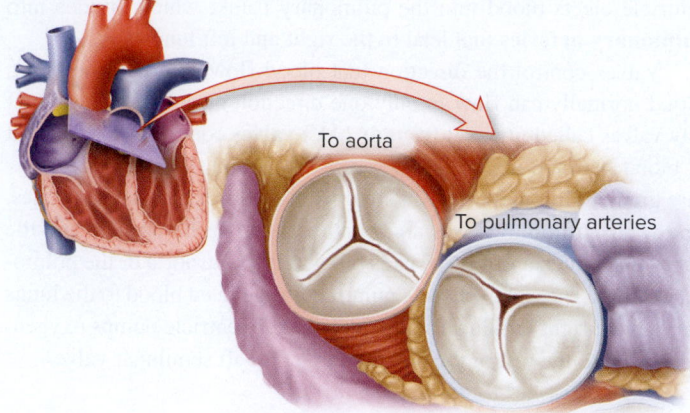

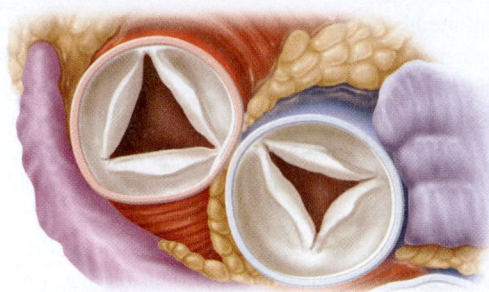

**(a) Semilunar valves in nearly closed position**

**(b) Semilunar valves in opened position**

**Figure 48.8   Appearance and function of heart valves.** The heart valves consist of two or three flaps of tissue that open in one direction. Shown in this illustration are the semilunar valves of a human heart.

*Concept Check:*  *Which heart chambers pump blood through the semilunar valves?*

As the ventricles pump out blood, they continue to generate pressure, but eventually the pressure must be lowered so that they can fill with blood again. This reduction in pressure happens when the ventricle muscle cells repolarize and return to their resting, unexcited state. At this time, the pressure in the ventricles decreases below that in the arteries. The higher pressure in the arteries causes backflow of blood toward the ventricles, which closes the semilunar valves and thereby prevents blood from actually flowing back into the heart. Closure of the semilunar valves creates the second heart sound, "dub," heard through a stethoscope.

Throughout systole, meanwhile, the atria continue to fill with blood from the veins, which slowly increases the pressure in the atria as they expand. As diastole begins, the pressure in the atria exceeds that in the relaxing ventricles, and the AV valves open again, bringing a new volume of blood into the ventricles and starting a new cycle.

**Blood pressure**, which is defined as the force exerted by blood on the walls of blood vessels, is highest in the arteries during systole, while the ventricles are generating pressure, and lowest during diastole, when the ventricles are relaxing. For this reason, blood pressure is measured with two numbers, the systolic and diastolic pressures. For historical reasons, the units are usually given in millimeters of mercury (mmHg). Among vertebrates, blood pressures are highest in mammals and birds and lowest in fishes. For example, a typical healthy blood pressure in humans is around 120/80 mmHg (systolic/diastolic).

## An ECG Tracks Electrical Events During the Cardiac Cycle

An **electrocardiogram** (**ECG or EKG**, with the latter abbreviation from the original German spelling) is a medical test used to evaluate the function of the heart. An ECG is a record of the summed electrical potentials between various positions on the body. These potentials arise from the electrical signals generated by myocytes during the cardiac cycle. Sensitive electrodes are placed on the surface of the body to monitor the wave of electric current initiated by the SA node. This wave spreads in sequence through the atria, AV node, and ventricles. This procedure works because the body fluids that surround the heart conduct electricity, even the relatively weak impulses generated by a beating heart.

The tracing of an ECG reveals several waves of electrical excitation known as the P, QRS, and T waves (**Figure 48.9**):

- The P wave begins when the SA node generates action potentials and ends when the two atria are completely depolarized. In the cardiac cycle, the P wave is followed by atrial contraction, while the ventricles are in diastole and filling with blood through the open AV valves.

- The QRS wave (or complex) begins when the branches from the AV node excite the ventricles and ends when both ventricles depolarize completely. In the cardiac cycle, the QRS complex is followed by ventricular contraction during systole. The complex shape of this wave reflects the various routes by which electrical activity courses through the two ventricles.

- The T wave results from the repolarization of the ventricles back to their resting state. In the cardiac cycle, this causes ventricular relaxation as diastole begins again. (No wave is visible for atrial repolarization because it occurs simultaneously with the large QRS complex.)

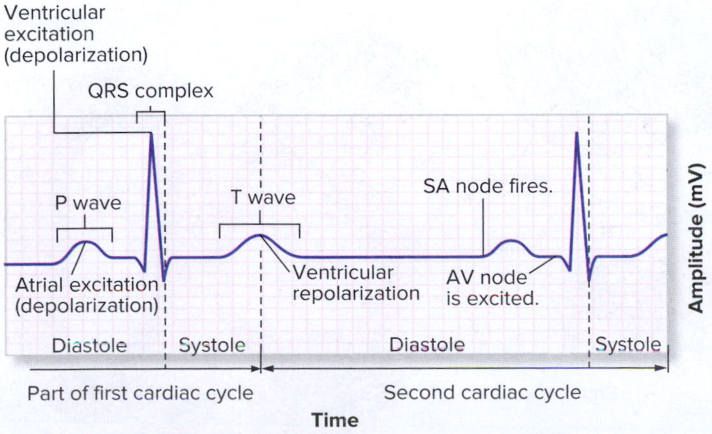

**Figure 48.9 An electrocardiogram (ECG).** Electrodes placed on the skin detect electrical impulses occurring in the heart, and an ECG is a recording of them. The resultant waveform is a useful indicator of cardiac health. Note that the ventricular wave (QRS complex) is taller than the atrial wave (P wave). This is because the ventricles are larger than the atria and generate more electrical activity.

The ECG record displays both the amplitude (strength) of the electrical signal and the direction in which it is moving within the chest. From this information, physicians can determine whether a person's heart signals have a normal frequency, strength, duration, and pattern.

### BIO TIPS

**THE QUESTION** *Place the following events of the vertebrate cardiac cycle in their proper sequence: contraction of the ventricles, QRS complex on the ECG, the first heart sound, and the ejection of blood into the arteries.*

**T OPIC** *What topic in biology does this question address?*
The topic is the cardiac cycle of the vertebrate heart. More specifically, the question asks you to put four events from that cycle into their correct order.

**I NFORMATION** *What information do you know based on the question and your understanding of the topic?*
The question lists electrical and mechanical events that must occur in a sequential way during the cardiac cycle. From your understanding of the topic, you know that electrical events in the heart's chambers must precede mechanical ones (that is, contraction). You also know what event causes the first heart sound (closure of the atrioventricular valves).

**P ROBLEM-SOLVING S TRATEGY** *Sort out the steps in a complicated process.* First, look again at Figure 48.9, and then try to determine where the different waves on an ECG best fit into Figure 48.7. In other words, try to match the electrical events of Figure 48.9 with the mechanical ones in Figure 48.7.

**ANSWER** *The correct sequence is as follows: QRS complex on the ECG, ventricular contraction, the first heart sound, ejection of blood into the arteries. The QRS complex is the electrical event that stimulates ventricular contraction. Once that contraction begins, the pressure in the ventricles increases above that in the atria, shutting the AV valves (making the first heart sound). Very soon after that, the increasing pressure opens the semilunar valves and blood is ejected into the arteries.*

## 48.4 Blood Vessels

### Learning Outcome:

1. Explain the structural and functional distinctions among arteries, arterioles, capillaries, venules, and veins.

In the previous section, we discussed the vertebrate heart, which pumps blood through a closed circulatory system. **Figure 48.10** provides an overview of the organization of blood vessels as blood flows in a closed circulatory system. Blood is pumped by the heart to large arteries and then flows to branching small arteries and eventually to the smallest arteries, which are called arterioles. Arterioles bring blood to the smallest vessels, the capillaries.

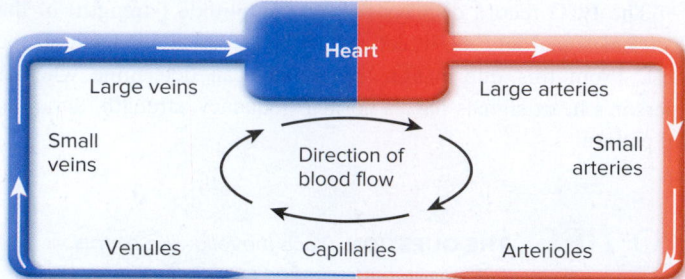

**Figure 48.10** **The sequence of blood flow in a closed circulatory system.** Overview of blood flow through vessels in a closed circulatory system. Regions of gas exchange with the environment (such as the lungs), have been omitted for simplicity.

Gas and nutrient exchange occurs between the blood in the capillaries and the cells surrounding the capillaries. From capillaries, the blood then flows back to the heart through the smallest veins, called venules, to small veins and finally to large veins. In this section, we will examine the structures and functions of these vessels.

## Arteries Distribute Blood to Organs and Tissues

Arteries are thick-walled vessels that consist of layers of smooth muscle and connective tissue wrapped around a single-celled inner layer of specialized epithelial cells called an endothelium. The endothelium forms a smooth lining in contact with the blood (**Figure 48.11a**). Because thick layers of tissue surround the endothelium, most dissolved substances cannot diffuse across arteries. Instead, arteries act as conducting tubes that distribute blood leaving the heart to all the organs and tissues of an animal's body.

In vertebrates, the walls of the largest arteries, such as the aorta, also contain one or more layers of elastin, a protein with elastic properties (refer back to Figure 10.3). As the aorta stretches to accommodate blood arriving from the heart, the elastin layers also stretch. The thick layers of tissue in the aorta and other large arteries prevent them from stretching more than a small amount, however. When the heart relaxes as it readies for another beat, the elastin layers in the aorta and large arteries recoil to their original state, something like the release of a stretched rubber band. The recoiling vessels generate a force on the blood within them. This force helps prevent blood pressure from decreasing too much while the heart is refilling during diastole.

## Arterioles Distribute Blood to Capillaries

As arteries carry blood away from the heart, they branch repeatedly and become smaller in diameter. Eventually, the vessels are little more than a single-celled layer of endothelium with one or two layers of smooth muscle and connective tissue (see Figure 48.11a). These are the arterioles, which deliver blood to regions of the body in proportion to metabolic demands. The adjustment of blood flow is accomplished by changing the diameter of arterioles. They widen, or dilate, in areas of high metabolic activity and narrow, or constrict, in inactive regions. Arterioles dilate when their smooth muscle cells relax and constrict when these cells contract.

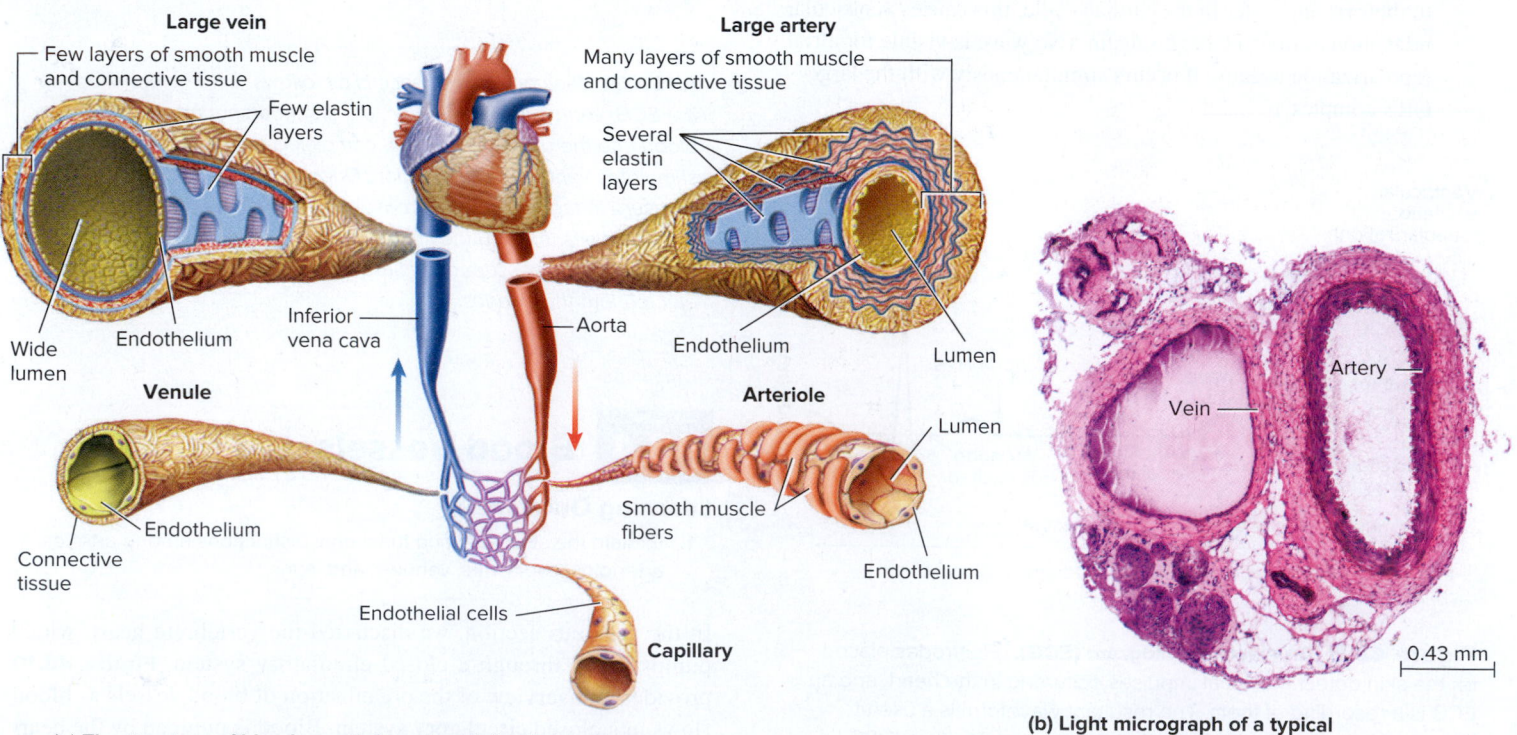

**(a) The structure of blood vessels in a closed circulatory system**

**(b) Light micrograph of a typical artery and vein in a mammal**

**Figure 48.11** **Comparative features of blood vessels.** **(a)** Structures of blood vessels (not drawn to scale). **(b)** Light micrograph of a small artery near a vein. Note the differences between the two vessels in the thickness and composition of the walls. b: ©National Geographic Creative/Alamy Stock Photo

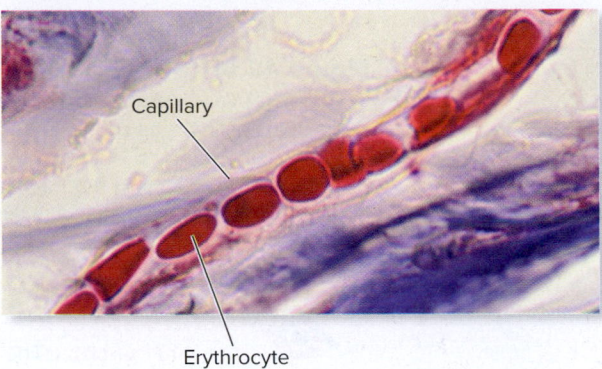

**Figure 48.12** **Erythrocytes moving through a capillary.** Note that the cells move along in single file, seen here through a light microscope. The diameter of the capillary is approximately 7 μm.

©Ed Reschke/Getty Images

**Concept Check:** *Do erythrocytes enter and exit through pores in capillaries?*

## Capillaries Are the Site of Gas and Nutrient Exchange

Arterioles branch into the tiny, thin-walled capillaries. Capillaries are tubes composed of a single-celled layer of endothelium resting on a layer of extracellular matrix called a basal lamina (refer back to Figure 41.4). Capillaries are the narrowest blood vessels in animals; essentially every cell in an animal's body is near one. The diameter of a capillary is about the same as the width of erythrocytes, which move through the capillaries in single file (**Figure 48.12**).

With the exception of large proteins, most solutes readily diffuse between the plasma in a capillary and the surrounding interstitial fluid and intracellular fluid. Nutrients, hormones, and other solutes diffuse from the plasma and into cells; waste products generated by cells diffuse in the opposite direction. Gases are exchanged across capillaries between the blood and the environment via a respiratory organ, such as the lungs, and between the blood and tissues (**Figure 48.13**).

Blood enters capillaries under pressure that is created by the beating of the heart. This pressure forces some of the fluid in blood out through tiny openings in capillaries into the interstitial fluid. These openings are wide enough to permit water but not cells to leave the capillary. The movement of fluid between capillaries and interstitial fluid is important for maintaining a normal distribution of water between the body fluid compartments.

If the fluid that leaves a capillary were to remain in the interstitial fluid, the volume of plasma in the blood would decrease, and the interstitial fluid would swell. Most of the fluid that leaves at the beginning of a capillary, however, is recaptured at the capillary's end. Any excess fluid is picked up by lymphatic vessels (which will be described in Chapter 52) and eventually returned to veins.

## Venules and Veins Return Blood to the Heart

Once blood travels through capillaries, picking up any substances secreted or diffusing from the cells of the body, it enters the venules, which are small, thin-walled extensions of capillaries. The venules empty into veins that return blood to the heart. The walls of veins are much thinner, less muscular, and more easily distended, or stretched, than those of arteries (see Figure 48.11a,b).

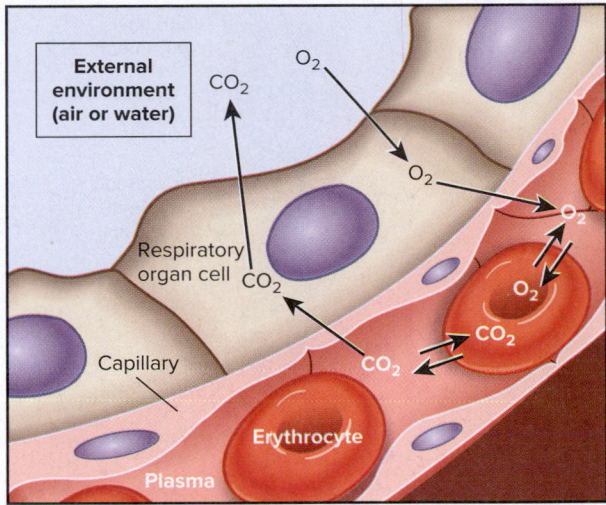

**(a) Gas exchange between the environment and a respiratory organ**

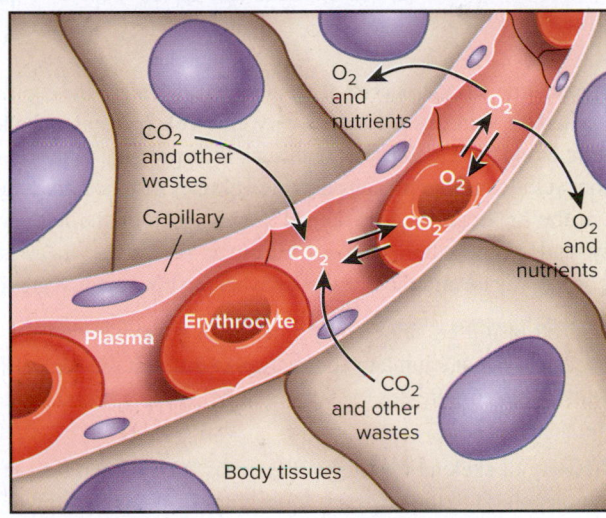

**(b) Gas, nutrient, and waste exchange between the blood and tissues**

**Figure 48.13** **Overview of gas and other solute exchange between capillaries and cells.** **(a)** In vertebrates, oxygen ($O_2$) diffuses from the environment across the cells of a respiratory organ into the blood (plasma and erythrocytes); carbon dioxide ($CO_2$) diffuses in the opposite direction. **(b)** From the blood, $O_2$ and nutrients diffuse into tissue cells, where they are used during the synthesis of ATP and other activities. Cells generate $CO_2$ and other waste products, which diffuse out of cells and into blood. Note: Interstitial fluid is omitted for simplicity.

 **Core Concepts: Systems and Energy and Matter**
A fundamental principle of biology is that living organisms maintain homeostasis. Keep this principle in mind as you read the rest of this chapter. The exchange of $O_2$ and $CO_2$ illustrated here is absolutely essential for all homeostatic processes in animals, in part because oxygen is required for the production of much of an animal's ATP. The ATP, in turn, provides the energy required to sustain processes that contribute to homeostasis. Two organ systems, the respiratory and circulatory systems, must function together to make this happen.

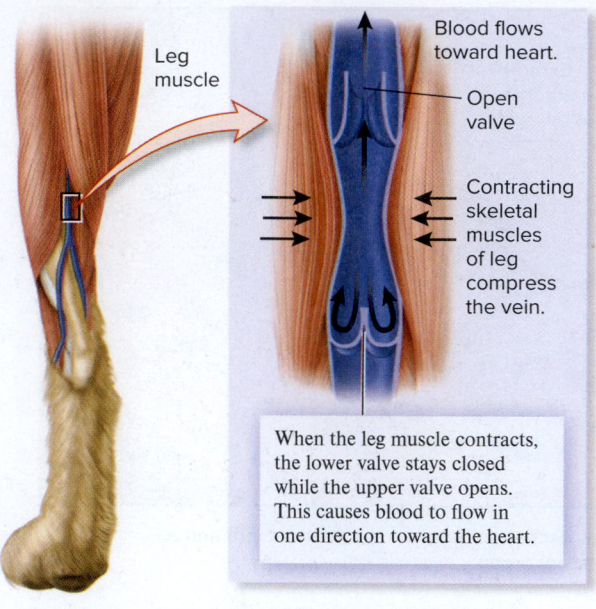

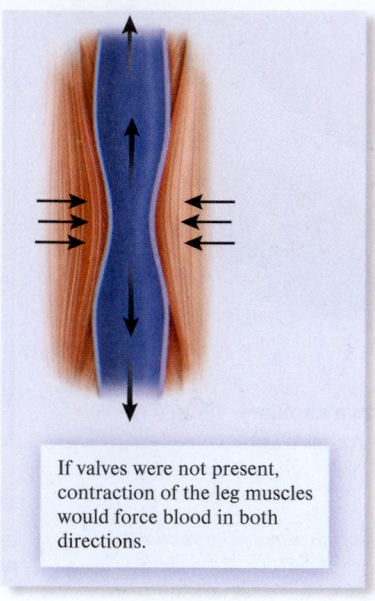

Leg muscle

Blood flows toward heart.

Open valve

Contracting skeletal muscles of leg compress the vein.

When the leg muscle contracts, the lower valve stays closed while the upper valve opens. This causes blood to flow in one direction toward the heart.

If valves were not present, contraction of the leg muscles would force blood in both directions.

**(a) Vein with one-way valves**

**(b) Vein without valves**

**Figure 48.14** **One-way valves in veins. (a)** Valves are typically present in the veins of limbs, as shown in this dog's leg, where they assist the return of blood to the heart against the force of gravity. **(b)** Blood moving in a vein without valves would flow in both directions if the vein were compressed.

 **Core Concept: Structure and Function** The structure of valves in the circulatory system permits them to open in one direction only. This structure, in turn, determines the function of the valves. In this example, they facilitate the movement of blood toward the heart.

By the time blood has been distributed through all of the capillaries and reached the veins, its pressure is very low. This presents a challenge in getting blood back to the heart with such little pressure to propel it. In addition, veins can fill with considerable volumes of blood, particularly veins in the lower parts of an animal's body, such as the legs, where gravity tends to cause blood to pool. However, several factors assist the blood on its way toward the heart, even when it flows against gravity:

- Neurotransmitters released by neurons of the sympathetic nervous system stimulate contraction of smooth muscles of leg veins. This compresses the veins and helps force blood back to the heart.

- Activity of skeletal muscles in the limbs assists the return of venous blood to the heart, by squeezing the veins that travel through them (**Figure 48.14**).

- One-way valves inside veins ensure that blood returning from below the heart moves in only one direction, toward the heart (see Figure 48.14). The valves can open toward the heart, but not in the other direction. When a leg vein, for example, is compressed by the mechanisms just mentioned, the blood forces open the valves and blood moves in one direction. When the vein is relaxed, the valves close again, preventing blood from flowing backward, away from the heart. (By contrast, veins located above the heart, like those in the necks of bipeds and some quadrupeds, lack valves because gravity is sufficient to return blood toward the heart.)

Many people can easily observe the effects of gravity on venous blood flow. When the arms are held down by the sides, the veins are visible on the backs of the hands. When the arms are raised above the head, the bulging veins quickly lose blood and become less visible. When blood from the veins returns to the heart, it must travel against gravity when the arms are at a person's sides and with gravity when the arms are elevated. Blood drains from veins much more effectively when gravity works in its favor.

## 48.5 Relationship Among Blood Pressure, Blood Flow, and Resistance

### Learning Outcomes:

1. **CoreSKILL** » Be able to relate blood pressure, blood flow, and resistance in a mathematical way.
2. Distinguish between the effects of vasodilation and vasoconstriction on blood flow.
3. **CoreSKILL** » Predict the effects of changes in resistance and cardiac output on blood pressure.

Blood pressure—the force exerted by blood on vessel walls—is responsible for blood flow, the movement of blood through the vessels. **Resistance ($R$)** refers to the tendency of blood vessels to slow the flow of blood through their lumens. Blood pressure varies through an animal's body in part because of differences in resistance. Such differences can be considered on a local level, as in an exercising muscle, or on a systemic level, as in the resistance the heart must overcome to pump blood throughout the entire body. In this section, we will explore the relationship among blood pressure, blood flow, and resistance.

### Blood Pressure, Blood Flow, and Resistance Are Mathematically Related

The relationship among blood pressure, blood flow, and resistance is stated by Poiseuille's law, which was derived in the 1840s by Jean Marie Louis Poiseuille, a French physician and physiologist. His law is simplified here:

$$\text{Flow}(F) = \Delta\text{Pressure}(P)/\text{Resistance}(R)$$

Stated mathematically, blood flow through a blood vessel is directly proportional to the difference ($\Delta$) in pressure of the blood between the beginning and end of the vessel, and inversely proportional to the resistance created by that vessel. The equation can be rearranged as $\Delta P = F \times R$, which demonstrates that blood pressure depends on both blood flow and resistance. Poiseuille's law applies to blood flow through a single vessel, an organ, or the entire body.

## Resistance to Flow Depends on the Radius of Arterioles

A change in arteriolar resistance is the major mechanism for increasing or decreasing blood flow to a region. The relationship between arteriolar radius and resistance is not linear. Resistance is inversely proportional to the radius of the vessel raised to the fourth power: $R \propto 1/r^4$, where $\propto$ means "proportional to" and $r$ is the radius of the arteriolar lumen. Let's consider an arteriole with a radius that increases by a factor of 2. This would occur if the smooth muscles of the arteriole relaxed sufficiently to allow the vessel to dilate and double its original radius. Because resistance is inversely proportional to the fourth power of vessel radius, an increase in radius by a factor of 2 will result in a decrease in resistance of $2^4$, or 16-fold.

**Vasodilation** refers to an increase in blood vessel radius, and **vasoconstriction** refers to a decrease in blood vessel radius. The signals that control arteriolar radius come from three sources:

- *Local factors.* Locally, metabolic by-products such as carbon dioxide, lactic acid, and other substances secreted by metabolically active tissues cause nearby arterioles to vasodilate, thereby decreasing their resistance to blood flow. This lowering of resistance permits more blood flow to the active region, facilitating oxygen and nutrient delivery and waste removal.

- *Hormones.* Hormones secreted by glands throughout the body can also regulate arteriolar radius. For example, during the fight-or-flight response, some hormones cause the arterioles that deliver blood to the small intestine to vasoconstrict, routing blood away from the intestine and to more immediately vital areas such as the heart and skeletal muscles.

- *Nervous system.* Signals from the autonomic nervous system cause contraction or relaxation of arteriolar smooth muscle; this effect is similar to that described for hormones.

## Cardiac Output and Resistance Determine Blood Pressure

Because blood vessels provide resistance to blood flow, the heart must beat forcefully enough to overcome that resistance throughout the whole body. **Cardiac output (CO)** is the amount of blood the heart pumps per unit of time, usually expressed in units of liters per minute (L/min). Poiseuille's law can be adapted to the whole body. In this case, pressure refers to arterial blood pressure (BP), flow refers to CO, and resistance refers to the total resistance provided by all the arterial vessels of the systemic circulation (**total peripheral resistance**, or **TPR**). Thus, we have BP = CO × TPR. The values of CO and TPR determine the pressure the blood exerts in the arterial vessels of a closed circulatory system.

Cardiac output depends on how often an animal's heart beats each minute, and how much blood it ejects with each beat. Each beat, or stroke, of the heart ejects an amount of blood known as the

| Table 48.1 | Comparative Features of Representative Mammalian Hearts* | | | |
|---|---|---|---|---|
| Animal | Body mass (kg) | Heart mass (kg) | Stroke volume (L) | Heart rate (bpm)[†] |
| Shrew[††] | 0.0024 (2.4 g) | 0.000035 (35 mg) | 0.000008 (8 μl) | 835 |
| Rat | 0.20 | 0.001 | 0.0002 | 360 |
| Rabbit | 2 | 0.012 | 0.0013 | 189 |
| Small dog | 5 | 0.030 | 0.007 | 120 |
| Large dog | 30 | 0.180 | 0.040 | 88 |
| Human | 75 | 0.380 | 0.075 | 70 |
| Horse | 450 | 3.50 | 0.90 | 38 |
| Elephant | 4,000 | 25 | 4.0 | 25 |
| Blue whale | 100,000 | 600 | 100 | 10 |

*Values are based on average body masses and resting conditions. In some cases, stroke volumes are estimates based on heart size.
[†]bpm = beats per minute.
[††]The shrew reported here is the Etruscan shrew, one of the smallest known mammals. Its heart is somewhat larger than would be predicted for its body mass. Note its heart rate; at 835 bpm, the heart beats 14 times per second!

**stroke volume (SV)**, which is roughly proportional to the size of the heart (**Table 48.1**). Thus, if we know the stroke volume of a heart and can measure the heart rate (HR, the number of beats per minute, or bpm), we can determine the CO. Simply put, CO = SV × HR. As one example from Table 48.1, the CO of a blue whale is

$$100 \, \text{L/beat} \times 10 \, \text{beats/min} = 1{,}000 \, \text{L/min}$$

To put that number in perspective, 1,000 L/min is roughly equivalent to 250 gal/min, or 25 ordinary fish aquaria!

Of course, the CO of a blue whale is far greater than that of a human, a dog, or a shrew. The heart of a typical shrew, for example, is the size of a small pea, whereas the heart of a blue whale is as large as a cow. Typically, heart size varies in proportion to body mass within a given class of vertebrates, with birds and mammals having larger hearts than similarly sized fishes, amphibians, or reptiles. Note in Table 48.1 that heart rate decreases as mammals get larger, but heart mass and stroke volume increase roughly in proportion with body mass. Smaller animals have smaller hearts and, therefore, smaller stroke volumes. However, small animals have faster heart rates than do large animals. A shrew's resting heart rate may be more than 800 bpm, whereas a blue whale's heart may beat only 10 times per minute (although the volume of blood ejected with each of those beats is enormous!). The faster heart rates of small animals give them a greater cardiac output than predicted from the size of their hearts, which helps them meet the extraordinary oxygen and nutrient demands of their relatively high metabolisms.

The greater the cardiac output and resistance to blood flow, the higher the blood pressure will be. Imagine that the circulatory system is like a faucet (the heart) connected to a garden hose (the arteries and arterioles) (**Figure 48.15**). If the faucet is fully open (analogous to maximal cardiac output) and the hose is not blocked (analogous to low resistance), the amount of water rushing through the hose will be high, and so will the water pressure, representing blood pressure (Figure 48.15a). If the faucet is only partially open, the water pressure will be lower

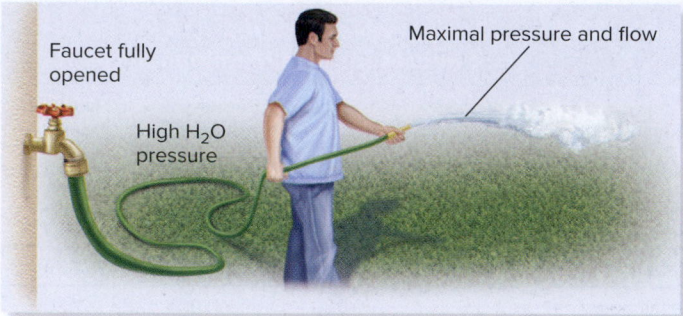

**(a) Maximal cardiac output with low resistance**

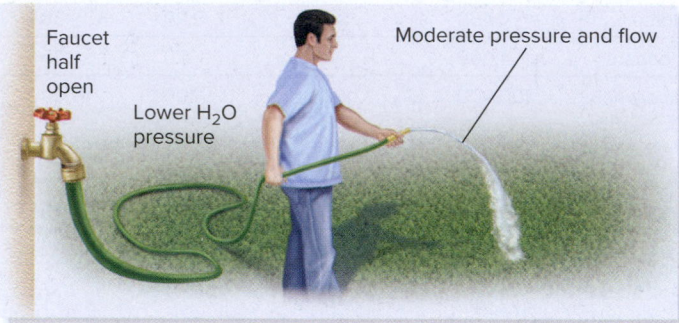

**(b) Moderate cardiac output with low resistance**

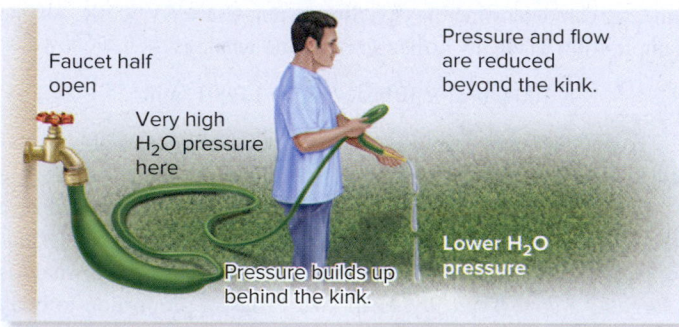

**(c) Moderate cardiac output with high resistance**

**Figure 48.15** **The relationship among cardiac output, resistance, and blood pressure.** A hose analogy shows the way in which cardiac output (the faucet) and resistance (constriction of the hose) affect blood pressure (the water leaving the hose).

 **Core Skill: Quantitative Reasoning** Cardiac output, resistance and blood pressure are mathematically related. Based on that relationship, it is possible to make accurate predictions about the degree to which blood pressure will change, for example, if cardiac output increases by some amount. In humans, this is important when considering certain medical drugs used to treat heart disease.

(Figure 48.15b). However, now imagine that the faucet is partially open but a region of the hose is kinked, or constricted, representing a region of high resistance. In that case, the pressure of the water in the hose will increase between the faucet and the point where the hose is kinked, and will decrease beyond the constriction, as will the flow of water (Figure 48.15c).

Arterial blood pressure, therefore, is a function of how hard the heart is working and how constricted or dilated the various arterioles are. Blood pressure must be high enough for blood to reach all body tissues even at the farthest extremities, but not high enough to damage blood vessels or force excess plasma out of capillaries.

## 48.6 Physical Properties of Gases

**Learning Outcomes:**

1. List the relative amounts of the major gases that make up the air we breathe.
2. Define partial pressure, and explain how it is calculated.
3. Describe the three factors that determine how much gas dissolves in a solution.

We will now turn our attention to the process of gas exchange between the environment and the blood and other fluids of an animal's body. In many animal species, gas exchange involves an interplay between the circulatory system and the respiratory system. The exchange of $O_2$ and $CO_2$ depends on the solubility of these gases in water and their rate of diffusion. Air is composed of about 21% $O_2$, 78% nitrogen ($N_2$), and roughly 1% $CO_2$ and other gases. From a respiratory standpoint, $N_2$ can usually be ignored because it plays no role in energy production nor is it created as a waste product of metabolism. In this section, we begin by examining some of the basic properties of the two major gases that are important in respiration.

### Gases Exert Pressure, Which Depends on Altitude

The gases in the air exert pressure on the body surfaces of terrestrial animals, although the pressure is not perceptible (unless it changes suddenly, as when your ears "pop" in a descending airplane). This pressure is called **atmospheric pressure**. It can be measured by noting how high a column of mercury is forced upward by the air pressure in a device called a mercury manometer. The traditional unit of gas pressure, therefore, is millimeters of mercury (mmHg)—the same as for blood pressure.

At sea level, atmospheric pressure is 760 mmHg. It decreases at higher elevations because the gravitational pull of the Earth decreases; consequently, gases expand and fewer gas molecules are in a given volume of air at high altitude. To visualize why gas pressure decreases at higher altitudes, think about how hydrostatic pressure decreases from the ocean floor to the ocean surface. Now, imagine that you are standing at the bottom of an "ocean" of air. The closer you get to the "ocean's" surface (the top of the atmosphere), the lower the pressure exerted on your body.

Atmospheric pressure is the sum of the pressures exerted by each gas in air, in exact proportion to their amounts. The individual pressure of each gas is its **partial pressure**, symbolized by a capital P and a subscript depending on the gas. At sea level, the partial pressure of oxygen ($P_{O_2}$) in the air we breathe is 21% of the atmospheric pressure at sea level, or 160 mmHg ($0.21 \times 760$ mmHg). The percentage of oxygen and other gases in air remains the same regardless of altitude, but the lower the atmospheric pressure, the lower the partial pressure of oxygen in air. In Denver, for example, where the atmospheric pressure is only about 640 mmHg, the $P_{O_2}$ is 21% of 640 mmHg, or 134.4 mmHg.

The partial pressure of oxygen in the environment provides the driving force for its diffusion from air (or, as we will see, water) across an animal's respiratory surface and into its blood. All gases diffuse from regions of higher pressure to regions of lower pressure. Consequently, the rate of oxygen diffusion into the blood of a terrestrial animal decreases when the animal moves from sea level to a higher altitude, where the $P_{O_2}$ is lower. Thus, humans who climb to high altitudes often need to breathe from a tank containing gas that is enriched in oxygen to compensate for the lower $P_{O_2}$.

## Pressure, Temperature, and Other Solutes Influence the Solubility of Gases

Gases dissolve in water, including fresh water, seawater, and all body fluids. Gases such as $O_2$ exert their biological effects while in solution. However, most gases dissolve rather poorly in water. Less $O_2$ is in a given volume of water than in air, for example, which places constraints on how active many aquatic organisms can be. Among the factors that influence the solubility of a gas in water, the following are particularly important.

- *Pressure of the gas.* As the pressure of a gas that comes into contact with water increases, more of that gas will dissolve, up to a limit that is specific for each gas at a given temperature. The partial pressure of gas in water is given in the same units as atmospheric pressure (mmHg). For example, if water is in contact with air that has a $P_{O_2}$ of 160 mmHg, the $P_{O_2}$ of the water is also 160 mmHg.

- *Temperature of water.* More gas dissolves in a given volume of cold water than in warm water. At higher temperatures, gases in solution have more thermal energy and are therefore more likely to escape from the liquid. This means that animals inhabiting warm waters generally have less $O_2$ available to them than do animals in cold waters.

- *Presence of other solutes.* Ions and other solutes decrease the amount of gas that dissolves in water. Thus, plasma—which contains many solutes—dissolves less $O_2$ than does pure water. Likewise, less $O_2$ dissolves in seawater than in fresh water at any given temperature and pressure. Animals living in cold freshwater lakes, therefore, generally have more $O_2$ available to them than animals living in warm, salty seas.

## 48.7 Types of Respiratory Systems

### Learning Outcomes:

1. Describe the different ways in which animals obtain oxygen from the environment.
2. **CoreSKILL »** Diagram the countercurrent exchange that occurs in gill ventilation.
3. State Boyle's law, and explain how it relates to ventilation in air-breathing vertebrates.

A **respiratory system** is all of the components of an animal's body that contribute to the exchange of $O_2$ and $CO_2$ between the external environment and cells of the body. Four major types of gas-exchange organs are found in animals: the body surface, gills, tracheae, or lungs. **Ventilation** is the process of bringing oxygenated water or air

into contact with a gas-exchange (respiratory) organ. In this section, we will examine the mechanisms of ventilation used by animals with different respiratory systems.

## Some Animals Exchange Gases Across the Body Surface

In those invertebrates that are only a few cell layers thick, such as cnidarians and platyhelminthes, $O_2$ and $CO_2$ can diffuse directly across the body surface (refer back to Figures 34.4 and 34.6). In this way, $O_2$ reaches all the interior cells, in some cases without any specialized circulatory system.

Even in some vertebrates, the body surface may be permeable to gases. Amphibians have moist, permeable skin. On land, amphibians rely primarily on their lungs, but when they are under water and thus cannot breathe air, $O_2$ and $CO_2$ can diffuse from the water across the skin. In some species, skin folds increase surface area and thereby increase the rates of diffusion of gases (refer back to Figure 41.8b). The ability to exchange gases across the skin is an adaptation that permits amphibians to spend prolonged times under water.

## Water-Breathing Animals Use Gills for Gas Exchange

Water-breathing animals use specialized respiratory structures called **gills**, which can be either external or internal. External gills are uncovered extensions from the body surface, found in many invertebrates and the larval forms of some amphibians. Internal gills, which occur in fishes, are enclosed in a protective cavity.

*External Gills* External gills vary widely in appearance, but all have a large surface area, often in the form of elaborate projections (**Figure 48.16**). In many cases, external gills are ventilated by being

**Figure 48.16** **Example of an animal (a nudibranch) with external gills.** ©Hal Beral/V&W/The Image Works

 **Core Concept: Structure and Function** Note how the gills of this nudibranch have extensive surface modifications to increase the total area available for gas exchange. This structural feature is not unique to external gills; all gas-exchange surfaces have structural modifications that maximize their function.

moved back and forth through the water. The ability to move external gills is particularly important for sessile invertebrates, which must otherwise rely on sporadic local water currents or muscular efforts of their bodies to create local currents for ventilation.

Despite the success of marine invertebrates, external gills have several limitations. First, they are unprotected and therefore are susceptible to damage from the environment. Second, because water is much denser than air, considerable energy is required to continually wave the gills back and forth through the water (think of the difference between waving your hand through air and waving it through water). Finally, their appearance and motion may draw the attention of predators.

**Internal Gills**    By contrast, fishes have internal gills, which are covered by a bony plate called an operculum (**Figure 48.17a**, also refer back to Figure 35.6). Fish gills are confined within the opercular

cavity—the space beneath the operculum—which protects the gills and helps streamline the body.

The main support structures of gills are the gill arches, from which project gill filaments composed of numerous platelike structures called **lamellae** (**Figure 48.17b**). Blood vessels run the length of the filaments. Oxygen-poor blood travels through a vessel called the afferent vessel along one side of the filament, and oxygen-rich blood travels through another vessel called the efferent vessel along the other side. Within the lamellae are numerous capillaries, all oriented with blood flowing from the oxygen-poor vessel to the oxygen-rich one.

Water enters a fish's mouth and flows between the lamellae in the opposite direction from blood flowing through the lamellar capillaries. This arrangement of water and blood flow is an example of **countercurrent exchange** (refer back to Figure 47.13). As oxygenated water encounters the lamellae, it comes into close

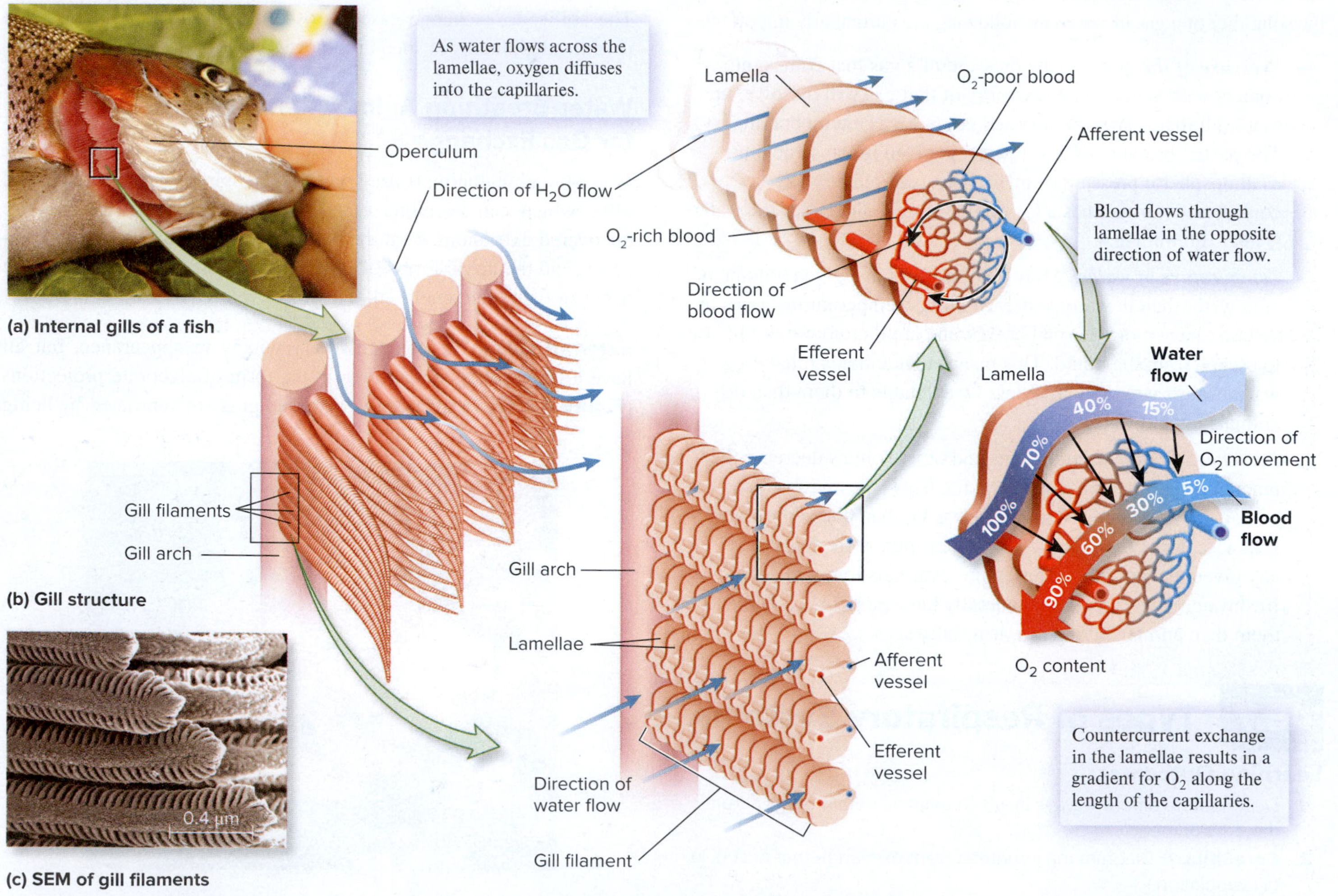

**(a) Internal gills of a fish**

Operculum

Direction of H₂O flow

**(b) Gill structure**

Gill filaments

Gill arch

**(c) SEM of gill filaments**

0.4 μm

As water flows across the lamellae, oxygen diffuses into the capillaries.

Lamella

O₂-poor blood

Afferent vessel

O₂-rich blood

Direction of blood flow

Efferent vessel

Blood flows through lamellae in the opposite direction of water flow.

Gill arch

Lamellae

Afferent vessel

Efferent vessel

Direction of water flow

Gill filament

Lamella

Water flow

40%    15%

70%

100%    60%    30%    Direction of O₂ movement

90%    5%    Blood flow

O₂ content

Countercurrent exchange in the lamellae results in a gradient for O₂ along the length of the capillaries.

**Figure 48.17    Structure of fish gills. (a)** The operculum, which has been lifted up in this photo, protects the gills underneath. **(b)** The gills are composed of gill arches, from which numerous pairs of filaments arise. Thin, platelike lamellae are arrayed along the filaments. Blood flows through the capillaries of the gill filaments in the opposite direction from water flowing between lamellae, a mechanism called countercurrent exchange. **(c)** Several filaments with their lamellae, as revealed in a scanning electron micrograph (SEM). a: ©Sarah Ahrens/Alamy Stock Photo; c: ©Electron Microscopy Unit, Royal Holloway, University of London

 **Core Skill: Connections** You have learned about countercurrent exchange in other contexts (for example, refer back to Figure 47.13). How does it relate to heat exchange in vertebrates?

proximity to blood in the gill capillaries. Gases such as $O_2$ always diffuse along a pressure gradient from a region of higher pressure to one of lower pressure. Thus, $O_2$ diffuses from the water into the capillaries of the lamellae. As water continues to flow across the lamellar surface, it encounters regions of capillaries that have not yet picked up $O_2$—in other words, even as $O_2$ begins to diffuse from the water into the gill capillaries, a sufficient pressure gradient remains along the lamellae to permit diffusion of more of the remaining $O_2$ from the water. This is an extremely efficient way to remove as much $O_2$ from the water as possible before the water passes out of the operculum.

Different fishes can ventilate their gills in three possible ways:

- Actively drawing water in through the mouth and out the operculum

- Swimming with the mouth open so that water continually moves across the gills

- Facing into a current of water while resting, but keeping the mouth open

Each of these ventilation mechanisms is a flow-through system—water moves only in one direction so that the gills are constantly in contact with fresh, oxygenated water. This improves gas exchange and is an important adaptation for water-breathing, considering the generally lower oxygen content of water compared to air.

## Insects Use Tracheal Systems to Exchange Gases with the Air

Air-breathing probably evolved as an adaptation in aquatic animals inhabiting regions that were subject to periodic drought. One of the major mechanisms that animals evolved to breathe air is the **tracheal system** found in insects. Along the surface of an insect's body are tiny openings to the outside called **spiracles**. Arising from the spiracles are sturdy tubes called **tracheae** (singular, trachea) (**Figure 48.18**). Tracheae branch extensively into ever-smaller tubes called tracheoles, which eventually become small enough that their tips contact virtually every cell in the body. At their tips, tracheoles are filled with a small amount of fluid. Air flowing into the tracheoles comes into contact with this fluid. Oxygen from the air dissolves in the fluid, and from there it diffuses across the tracheoles and into nearby cells. Carbon dioxide diffuses in the opposite direction, from cells into the tracheoles, and from there to the environment.

When an insect's oxygen demands increase due to increased activity, muscular movements of its abdomen and thorax draw air into and out of the tracheae a little like a bellows. An insect's muscles and tracheal system match ventilation with the animal's exercise intensity and $O_2$ requirements. This is particularly important in flying insects, which have very great metabolic demands.

As discussed earlier, the open circulatory system of insects does not participate in gas exchange. Oxygen diffuses directly from air to trachea to tracheoles and finally to body cells. This mechanism of ventilation and $O_2$ delivery is very effective. The relative metabolic rate of insect flight muscles is among the highest known of any tissue

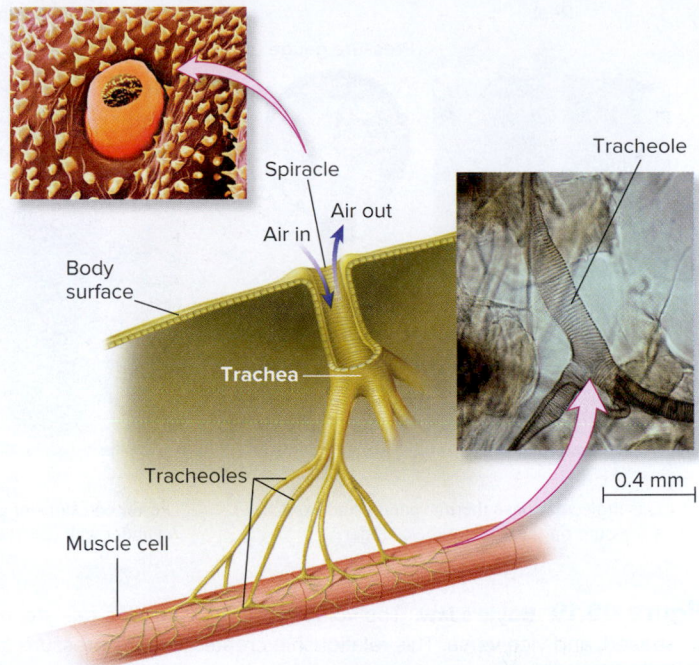

**Figure 48.18** **The tracheal system of insects.** Air enters holes on the body surface called spiracles. Oxygen diffuses directly from the fluid-filled tracheole tips to cells that come into contact with the tips. The circulatory system plays no role in gas exchange. The micrographs show a single spiracle and a branching tracheole. (left) ©Microfield Scientific Ltd/Science Source; (right) ©Ed Reschke/Getty Images

*Concept Check:* *How might the structure of the respiratory system of insects be related to their relatively small size?*

in any animal, and the tracheal system supplies enough $O_2$ to meet those enormous demands.

## Air-Breathing Vertebrates Use Lungs to Exchange Gases

Except for some amphibians such as lungless salamanders, all air-breathing terrestrial vertebrates use lungs to bring $O_2$ into the circulatory system and remove $CO_2$. **Lungs** are internal, paired structures that arise from the pharynx during embryonic life. All lungs receive deoxygenated blood from the heart and return oxygenated blood to the heart.

Most vertebrates ventilate their lungs by a process called **negative pressure filling**, in which the pressure of air in the lungs is decreased below that of the environment in order to create a pressure gradient that draws air into the lungs. Boyle's law states that the pressure and volume of a gas are inversely related (**Figure 48.19**). For example, when the volume in which a gas is contained increases, the pressure of the gas in that container decreases. By expanding its lungs (inhaling), an animal creates a pressure gradient for air to move from the atmosphere (higher pressure) into its lungs (lower pressure). When an animal exhales, the lungs become compressed, increasing the pressure of the air inside them. This increased pressure causes air to leave the lungs.

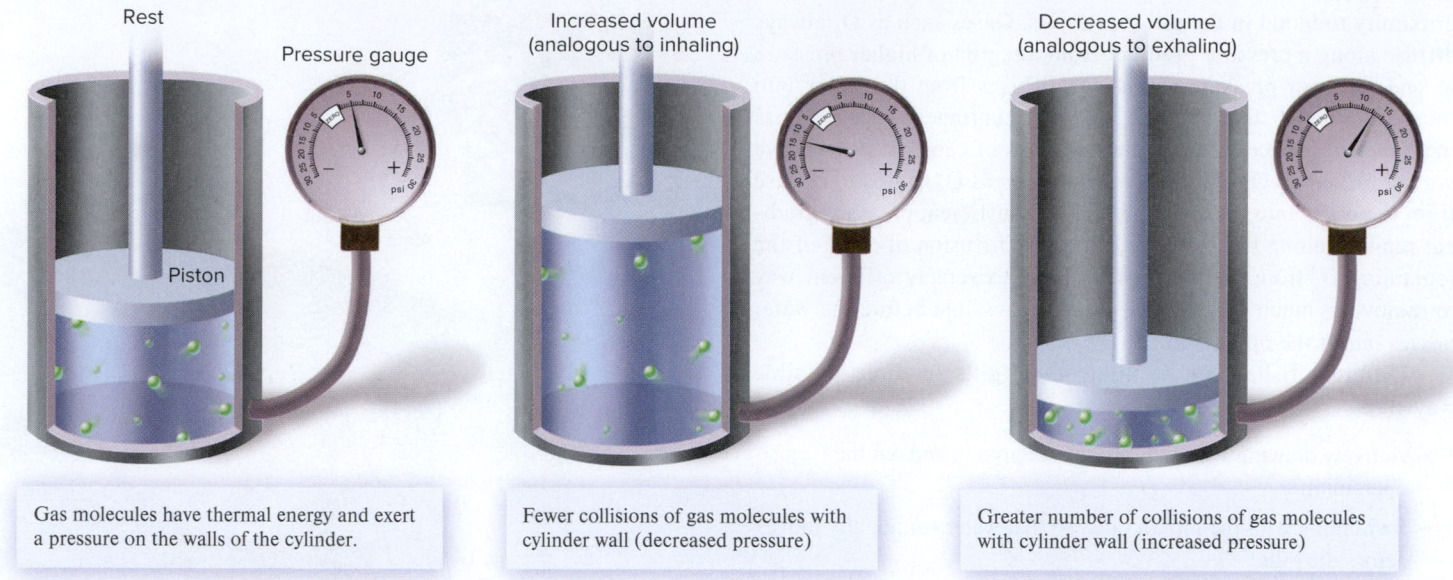

| Rest | Increased volume (analogous to inhaling) | Decreased volume (analogous to exhaling) |
| --- | --- | --- |
| Pressure gauge | | |
| Piston | | |
| Gas molecules have thermal energy and exert a pressure on the walls of the cylinder. | Fewer collisions of gas molecules with cylinder wall (decreased pressure) | Greater number of collisions of gas molecules with cylinder wall (increased pressure) |

**Figure 48.19 Boyle's law.** The volume and pressure of a gas are inversely related; that is, when the volume is increased, the pressure is decreased, and vice versa. This relationship creates the gas pressure gradients that ventilate the vertebrate lungs.

**Concept Check:** *As lungs expand during inhalation, what happens to the pressure of the air inside them?*

## 48.8 Structure and Function of the Mammalian Respiratory System

**Learning Outcomes:**

1. Describe the components of the mammalian respiratory system and the structure of the mammalian lung.
2. Outline the process of ventilation in mammalian lungs.

In this section, we will examine in detail the structures of the mammalian respiratory system and the mechanisms by which mammals ventilate their lungs.

### During Ventilation, Air Flows Through a Series of Branching Tubes

In mammals, the respiratory system includes the nose, mouth, pharynx, larynx, trachea, branching tubes, lungs, and muscles and connective tissues that encase these structures within the thoracic (chest) cavity (**Figure 48.20a**). When humans and other mammals breathe, air first enters the nose and mouth, where it is warmed and humidified. These effects protect the lungs from drying out. While in the nose, the air is partially purified as it flows over a coating of sticky mucus in the nasal cavity. The mucus and hairs in the nasal cavity trap some of the larger dust and other particles that are inhaled with air. These are then removed by the body's immune cells or swallowed.

The inhaled air from the mouth and nose converges at the back of the throat, or **pharynx**, a common passageway for air and food. From there, air passes through the **larynx**, which contains the vocal cords. Air flows from the larynx into the **trachea**, a tube that leads to the lungs.

The trachea is partially ringed by cartilage that provides rigidity and ensures that the trachea always remains open. Inhaled air flows

down the trachea as it branches into two smaller tubes, called **bronchi** (singular, bronchus), which lead to each lung. The bronchi branch repeatedly into smaller and smaller tubes, eventually becoming thin-walled **bronchioles** surrounded by circular rings of smooth muscle (**Figure 48.20b**). Bronchioles can dilate or constrict in a manner analogous to that of arterioles, the small blood vessels that deliver blood to capillaries (look back at Figure 48.11).

The **alveoli** (singular, alveolus) are the saclike regions of the lungs where gas exchange occurs (**Figure 48.20c**). The alveoli are highly adapted for gas exchange and consist of two major types of cells. Gases diffuse across type I cells, whereas type II cells are secretory cells (described later). The alveoli are only one cell thick and resemble extremely thin sacs, appearing like bunches of grapes on a stem. Deoxygenated blood pumped from the right ventricle of the heart flows to the many capillaries surrounding the alveoli. Oxygen diffuses from the lumen of each alveolus across the very thin type I alveolar cells, through the interstitial space outside the cells, and into the capillaries (see Figure 48.20b). Carbon dioxide diffuses in the opposite direction. The newly oxygenated blood from the lungs then flows to the left atrium of the heart and from there enters the left ventricle, where it is pumped out through the aorta to the rest of the body.

### The Pleural Sacs Protect the Lungs

The lungs are soft, delicate tissues that could easily be damaged by the surrounding bone, muscle, and connective tissue of the thorax if not protected. Each lung is encased in a **pleural sac**, a double layer of thin, moist connective tissue. Between the two layers is a microscopically thin layer of water that acts as a lubricant and makes the two tissue layers adhere to each other.

In addition to protecting the lungs, the inner pleural sac adheres to its lung, and the outer pleural sac adheres to the chest wall. In this way, movements of the chest wall result in similar movements of the lungs. This is important because the lungs are not muscular and,

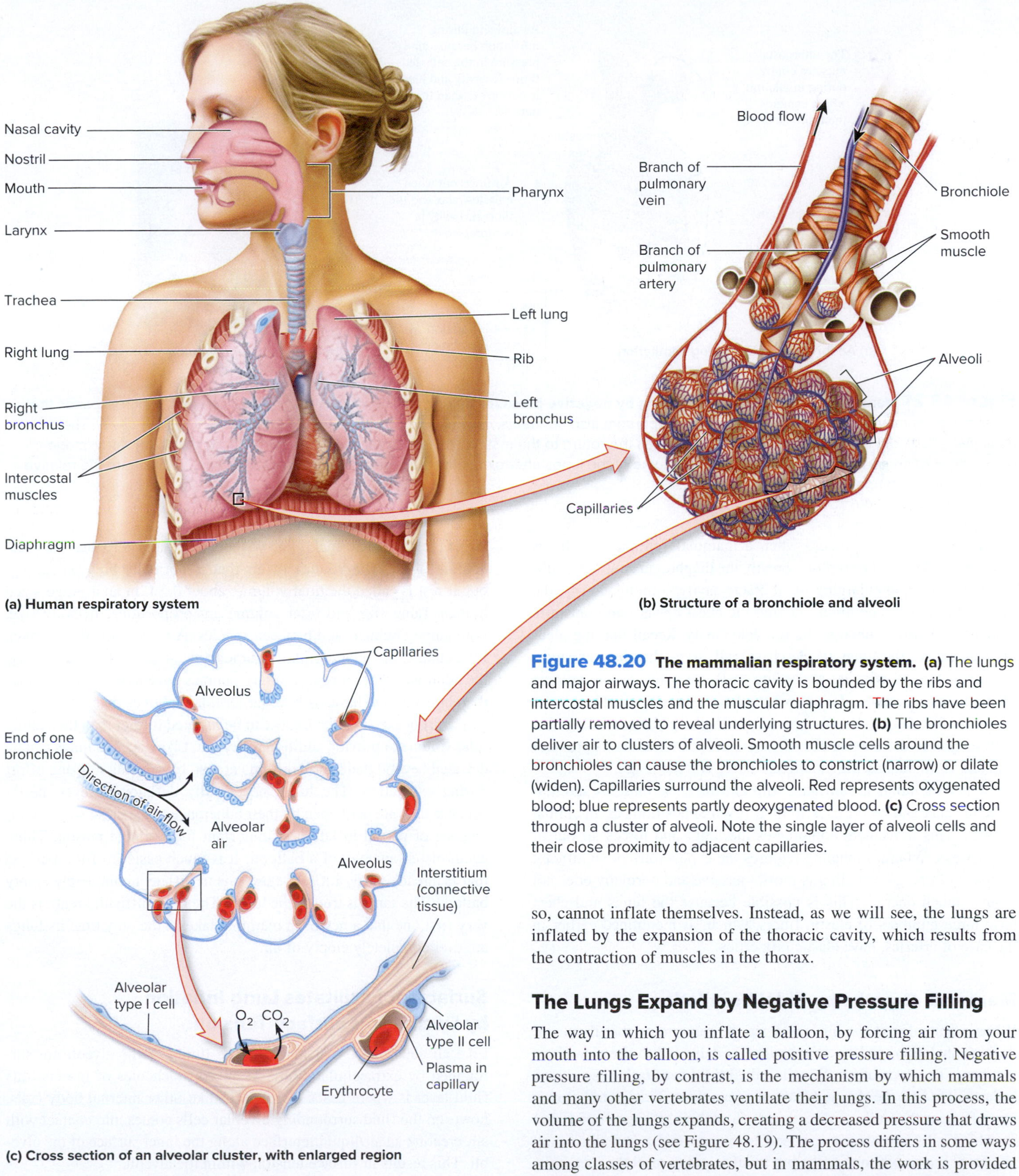

## (a) Human respiratory system

Nasal cavity

Nostril

Mouth

Larynx

Trachea

Right lung

Right bronchus

Intercostal muscles

Diaphragm

Pharynx

Left lung

Rib

Left bronchus

## (b) Structure of a bronchiole and alveoli

Blood flow

Branch of pulmonary vein

Branch of pulmonary artery

Capillaries

Bronchiole

Smooth muscle

Alveoli

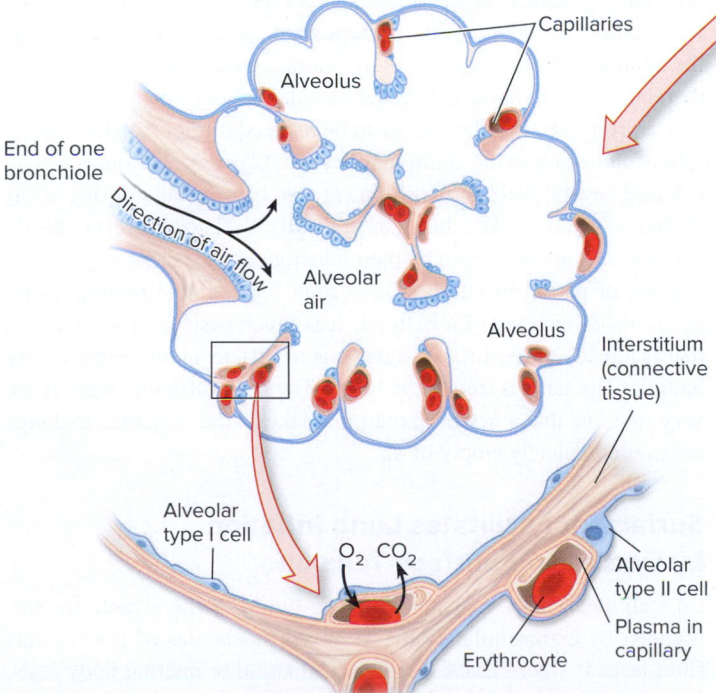

**(c) Cross section of an alveolar cluster, with enlarged region**

Capillaries

Alveolus

End of one bronchiole

Direction of air flow

Alveolar air

Alveolus

Interstitium (connective tissue)

Alveolar type I cell

$O_2$  $CO_2$

Alveolar type II cell

Plasma in capillary

Erythrocyte

**Figure 48.20   The mammalian respiratory system.   (a)** The lungs and major airways. The thoracic cavity is bounded by the ribs and intercostal muscles and the muscular diaphragm. The ribs have been partially removed to reveal underlying structures. **(b)** The bronchioles deliver air to clusters of alveoli. Smooth muscle cells around the bronchioles can cause the bronchioles to constrict (narrow) or dilate (widen). Capillaries surround the alveoli. Red represents oxygenated blood; blue represents partly deoxygenated blood. **(c)** Cross section through a cluster of alveoli. Note the single layer of alveoli cells and their close proximity to adjacent capillaries.

so, cannot inflate themselves. Instead, as we will see, the lungs are inflated by the expansion of the thoracic cavity, which results from the contraction of muscles in the thorax.

## The Lungs Expand by Negative Pressure Filling

The way in which you inflate a balloon, by forcing air from your mouth into the balloon, is called positive pressure filling. Negative pressure filling, by contrast, is the mechanism by which mammals and many other vertebrates ventilate their lungs. In this process, the volume of the lungs expands, creating a decreased pressure that draws air into the lungs (see Figure 48.19). The process differs in some ways among classes of vertebrates, but in mammals, the work is provided by the intercostal muscles, which surround and connect the ribs in the chest, and a large muscle called the **diaphragm** (see Figure 48.20a), which divides the thoracic cavity from the abdomen.

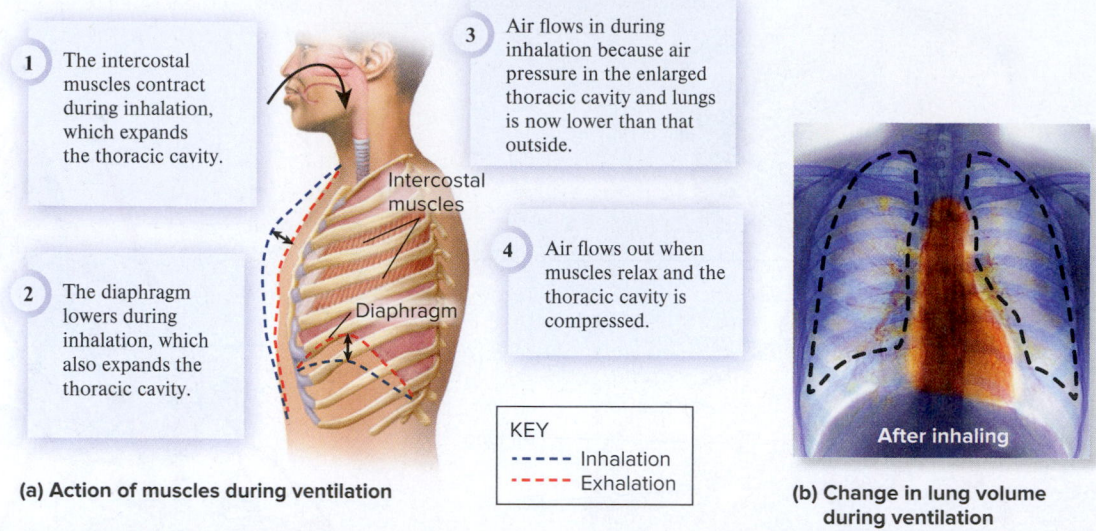

1. The intercostal muscles contract during inhalation, which expands the thoracic cavity.

2. The diaphragm lowers during inhalation, which also expands the thoracic cavity.

3. Air flows in during inhalation because air pressure in the enlarged thoracic cavity and lungs is now lower than that outside.

4. Air flows out when muscles relax and the thoracic cavity is compressed.

Intercostal muscles

Diaphragm

KEY
- - - - Inhalation
- - - - Exhalation

After inhaling

**(a) Action of muscles during ventilation**

**(b) Change in lung volume during ventilation**

**Figure 48.21  Ventilation of the mammalian lung by negative pressure filling. (a)** The intercostal muscles contract, which expands the chest cavity by moving the ribs up and out. The diaphragm also contracts, causing it to pull downward, further expanding the cavity. The muscular efforts of inhalation require energy, whereas the return to the resting state by exhaling occurs primarily by recoil. **(b)** X-ray image of the chest of an adult man after inhaling. The volume of the lungs after exhaling is superimposed using dashed lines to illustrate the relative change in lung volume. b: ©Pr. M. Brauner/Science Source

Let's follow the process when a mammal ventilates its lungs (**Figure 48.21**). At the start of a breath, the diaphragm contracts, pulling downward and enlarging the thoracic cavity. Simultaneously, the intercostal muscles contract, moving the chest upward and outward, which also helps to enlarge the thoracic cavity. Recall that the pleural sacs adhere the lungs to the chest wall, so as the chest expands, the lungs expand with it. According to Boyle's law, as the volume of the lungs increases, the pressures of the gases within them must decrease. In other words, the pressure in the lungs becomes negative with respect to the outside air. Air, therefore, flows down its pressure gradient from outside the mouth and nose, into the lungs.

Once the lungs are inflated with air, the chest muscles and diaphragm relax and recoil back to their original positions as an animal exhales. This movement compresses the lungs and forces air out of the airways. Whereas inhaling requires the expenditure of significant amounts of energy, exhaling is mostly passive and normally does not require much energy. This is possible because the lungs and chest have large numbers of elastin fibers, which, as mentioned earlier in the case of arteries, have elastic properties.

## Mammals Breathe by Tidal Ventilation

When mammals exhale, air leaves via the same route that it entered during inhalation, and no new oxygen is delivered to the airways at that time. This type of breathing is called **tidal ventilation**, so named because it is like the ebb and flow of ocean tides. Tidal ventilation is less efficient than the unidirectional, flow-through system of fishes in which gills are always exposed to oxygenated water during all phases of the respiratory cycle, but is sufficient to meet the metabolic demands of terrestrial vertebrates.

As you likely know from experience, the lungs are neither fully inflated nor deflated at rest. For example, you could easily take a larger breath than normal if you wished, or exhale more than the usual amount of air. The volume of air that is normally breathed in and out at rest is called the tidal volume, about 0.5 L in an average-sized human. Lung size and tidal volume are proportional to body size, both among humans and between species. A 6-foot-tall adult human, for example, has a larger tidal volume than a 4-foot-tall child because the adult has larger lungs. Similarly, horses have larger tidal volumes than humans, and humans have larger tidal volumes than dogs.

During exertion, the lungs can be inflated further than the resting tidal volume to provide additional oxygen. Likewise, the lungs can be deflated beyond their normal limits at rest, by exerting a strong effort during exhalation. The lungs never fully deflate, however, partly because they are held open by their adherence to the chest wall. Maintenance of partial inflation is important for a simple reason. Think again of our analogy of a balloon. It is much easier to fill a balloon that is already partly inflated than it is to inflate a completely empty balloon. The same is true of the lungs. The most difficult breath is the very first one that a newborn mammal takes—the only time its lungs are ever completely empty of air.

## Surfactant Facilitates Lung Inflation by Decreasing Surface Tension

Like all cells, those that make up the lining of the alveoli are surrounded by extracellular fluid. As in the tracheoles of insects, this fluid layer is where gases dissolve. Unlike other internal body cells, however, the fluid surrounding alveolar cells comes into contact with air, creating an air/liquid interface along the inner surface of the alveoli. This results in surface tension within the alveoli.

Surface tension results from the attractive forces between water molecules at an air/liquid interface and partly explains why droplets of water form beads. This tension produces a force that tends to collapse alveoli as water molecules lining the alveolar inner surfaces are attracted to each other. If many or all of the alveoli collapsed, however, the amount of surface area available for gas exchange in the lungs would be greatly reduced. What prevents them from collapsing? The type II cells of the alveoli produce **surfactant**, a mixture of proteins and amphipathic lipids (that is, lipids with both polar and nonpolar regions), and secrete it into the alveolar lumen. Surfactant molecules dissolve in the fluid layer inside the alveoli and remain at the fluid-air interface. They are believed to increase the distance between water molecules at the fluid surface. This effect reduces surface tension in the alveolar walls, allowing them to remain open.

Surface tension is particularly important in the transition from fetal to postnatal life in mammals. Most mammalian fetuses are encased in fluid within the uterus. Consequently, their lungs do not have an air/liquid interface, and they do not start producing surfactant until the final stages of fetal development. In humans, surfactant production begins around week 26 of gestation but does not begin to approach final levels until after week 33 (normal pregnancy length is about 40 weeks). If a human baby is born prematurely (defined as prior to week 37 of gestation), sufficient surfactant may not be available, and consequently, many alveoli may collapse after birth. This condition, known as respiratory distress syndrome (RDS) of the newborn, can be partially alleviated by inserting a tube in the trachea and injecting a natural or synthetic form of surfactant. Each year in the U.S., approximately 500,000 babies are born prematurely. Of those, roughly 25,000–40,000 will be born sufficiently premature as to be diagnosed with RDS.

The discovery of surfactant and how it could be used to treat infants with RDS was among the greatest scientific achievements of the 20th century. The treatment continues to save many thousands of lives each year, as we see next.

### Core Skill: Process of Science

## Feature Investigation | Fujiwara and Colleagues Demonstrated the Effectiveness of Administering Surfactant to Newborns with RDS

In the early part of the 20th century, researchers noted that isolated lungs from experimental animals were easier to inflate if they were first filled with fluid, which eliminated the air/liquid interface that exists in vivo. Later, investigators discovered the presence of substances in mammalian lungs (including those of humans) that had surface-active (surfactant) properties. As described earlier, surfactant is any substance that decreases surface tension in a liquid. Surfactants typically contain amphipathic molecules; that is, a portion of such a molecule is hydrophilic (soluble in water) and another portion is hydrophobic.

The hydrophilic region of a surfactant molecule dissolves in water in the alveoli fluid, while the hydrophobic region is excluded from the water and projects into the air. As mentioned, one hypothesis for how surfactants work is that by inserting themselves between water molecules at the interface between a liquid and the air, the surfactant molecules increase the distance between water molecules. Recall from Chapter 2 that each water molecule has a region of partial negative charge and another of partial positive charge. These charges produce an attractive force between water molecules. A principle of physics is that electrical forces decrease with the square of the distance between charges; thus, as surfactant increases the distance between water molecules at the surface of a liquid, the attractive forces between the water molecules decreases exponentially. This is what is meant when it is said that surface tension is decreased in the presence of surfactant.

As noted earlier, respiratory distress syndrome (RDS) is an extremely serious condition that occurs in many newborns born prematurely. In such infants, the lungs have not matured to the point that they produce sufficient amounts of surfactant to maintain a normal surface tension in the alveoli. The reason surfactant is not produced early in fetal life is there is no air/liquid interface in the fetal lungs. Instead, the lungs of fetuses are filled with amniotic fluid (the fluid in which a fetus is suspended within the uterus). Indeed, the lungs are not used for gas exchange at this time, because a fetus receives oxygen via the umbilical circulation. If a baby is born before sufficient surfactant is available, surface tension in the lungs will be too high to allow the alveoli to remain open, and they will all collapse within a few breaths of air. Recall that it is easier to inflate a balloon that is partially inflated than one that is fully deflated; in a premature infant with RDS, all of the millions of alveoli are analogous to deflated balloons. Under such conditions, it becomes impossible for a newborn to fully inflate the lungs with each breath, and he or she quickly becomes very ill. If surfactant can be administered to such babies, their chances of survival should be greatly increased because they will be better able to ventilate their lungs.

How can surfactant be administered to newborns? Prior to 1980, experiments in animals suggested that administering animal lung extracts that contained surfactant to prematurely born experimental animals improved the lung function and chances of survival of those animals. In 1980, Tetsuro Fujiwara and coworkers tested this in human newborns for the first time (**Figure 48.22**). Their hypothesis was that administering a modified form of surfactant (obtained from cows) to very premature infants with RDS would improve lung function and increase the chances that such infants would survive. Prior to the administration of surfactant, RDS was confirmed by determining the blood $O_2$ and $CO_2$ levels of the newborns. The babies were placed in ventilators with a mixture of air that was greatly enriched in $O_2$ during this time as part of the standard treatment for premature infants. As shown in the data in step 4 of Figure 48.22, the newborns' blood $O_2$ levels were lower than normal despite the oxygen-rich air they were breathing, and their $CO_2$ levels were also higher than normal; both measurements indicated poor lung function. The infants were then administered surfactant through a tube placed within the trachea (called an endotracheal tube). Next, their blood gases were re-examined at intervals over the next few hours. As seen in the data, the arterial $P_{O_2}$ of each infant increased dramatically within an hour or so of surfactant administration! Other signs of improved overall health were also noted, such as heart, kidney, and gastrointestinal function (all of these organs are adversely affected when a baby has RDS and is not receiving sufficient oxygen). Within 3 hours after surfactant administration, the physicians were able to decrease the amount of oxygen supplied via the ventilator from approximately 80% to 38%, and eventually nearly to normal (which, you will recall, is 21%). In other words, by that time, the lungs of the infants were functioning almost normally, and the requirement for oxygen supplementation was greatly reduced.

Of the 12 newborns treated, 10 survived and 2 died (but from causes not directly attributable to RDS). While it is, of course, tragic that any infant failed to survive, consider that in 1979, just prior to

**Figure 48.22**  **The pioneering study of Fujiwara and colleagues established that treatment with surfactant greatly improved the survival rate of premature infants with RDS.**

**HYPOTHESIS** Administration of surfactant to infants born with RDS will improve their lung function and chances for survival.

**KEY MATERIALS**   Ventilator to supply air mixture containing up to 80% oxygen, blood gas analyzer, surfactant extracted from cow lungs.

**Experimental level**                                        **Conceptual level**

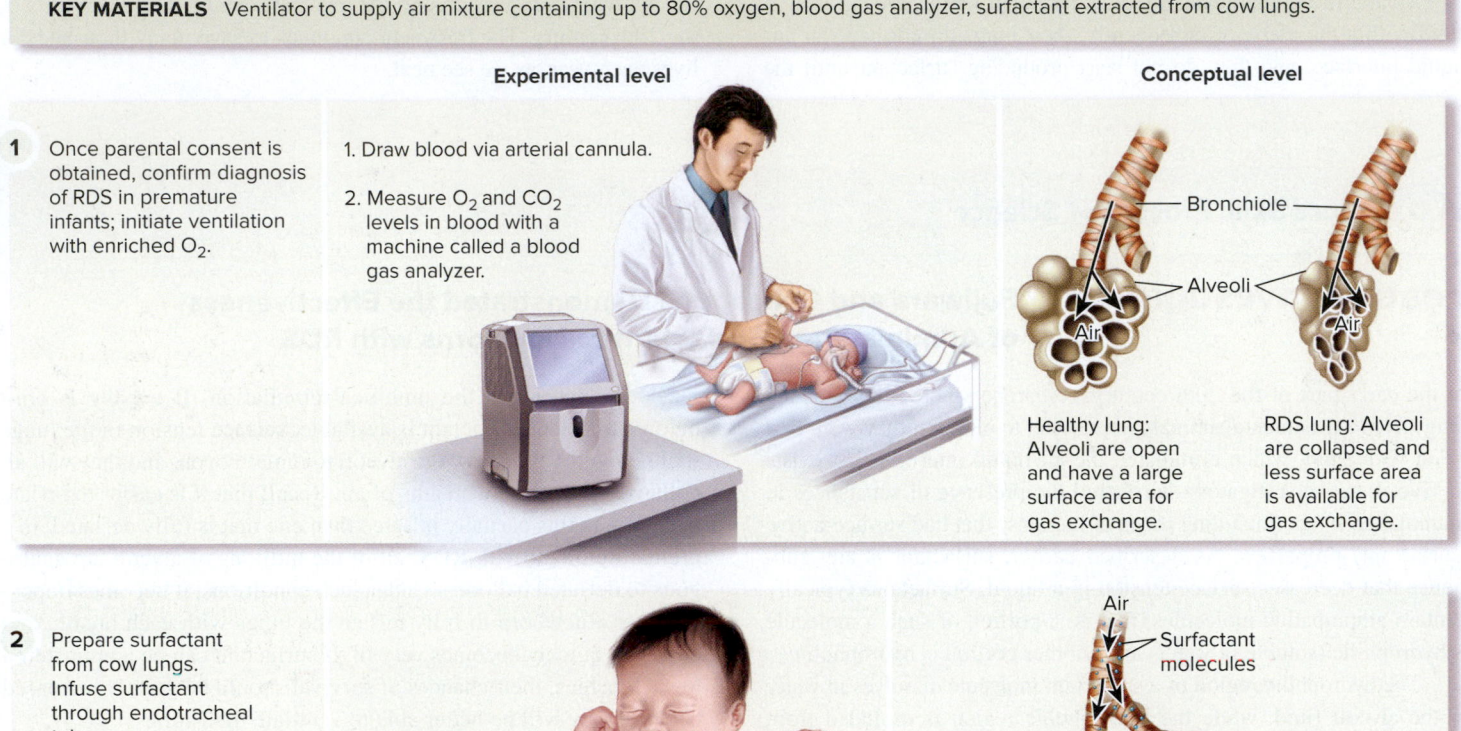

1   Once parental consent is obtained, confirm diagnosis of RDS in premature infants; initiate ventilation with enriched $O_2$.

1. Draw blood via arterial cannula.

2. Measure $O_2$ and $CO_2$ levels in blood with a machine called a blood gas analyzer.

Bronchiole

Alveoli

Air

Air

Healthy lung: Alveoli are open and have a large surface area for gas exchange.

RDS lung: Alveoli are collapsed and less surface area is available for gas exchange.

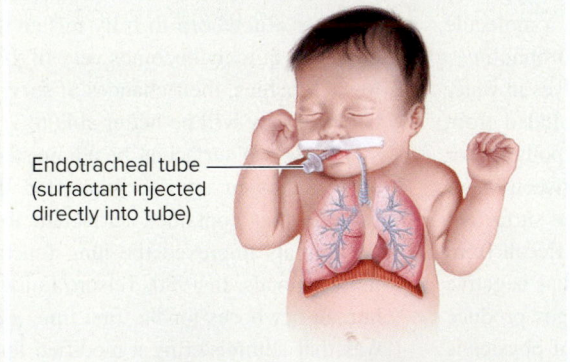

2   Prepare surfactant from cow lungs. Infuse surfactant through endotracheal tube.

Endotracheal tube (surfactant injected directly into tube)

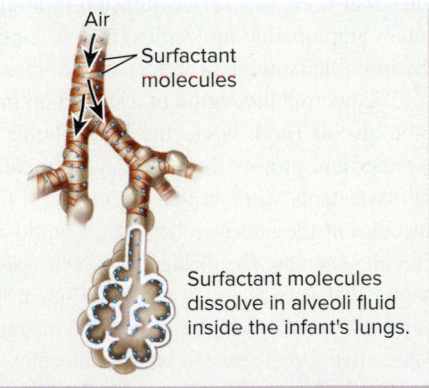

Air

Surfactant molecules

Surfactant molecules dissolve in alveoli fluid inside the infant's lungs.

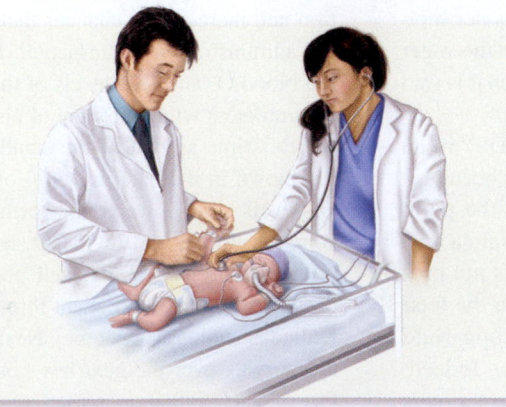

3   Monitor blood gases for next several hours; check overall health of infants.

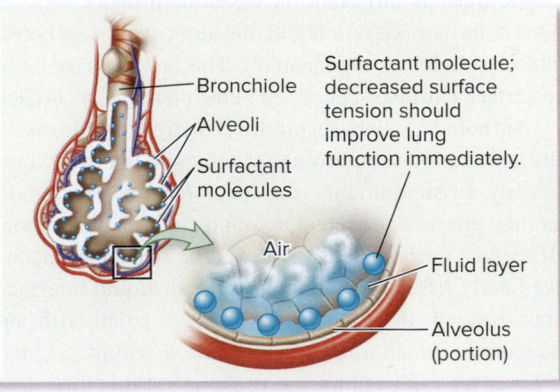

Bronchiole

Alveoli

Surfactant molecules

Surfactant molecule; decreased surface tension should improve lung function immediately.

Air

Fluid layer

Alveolus (portion)

**4** **THE DATA**

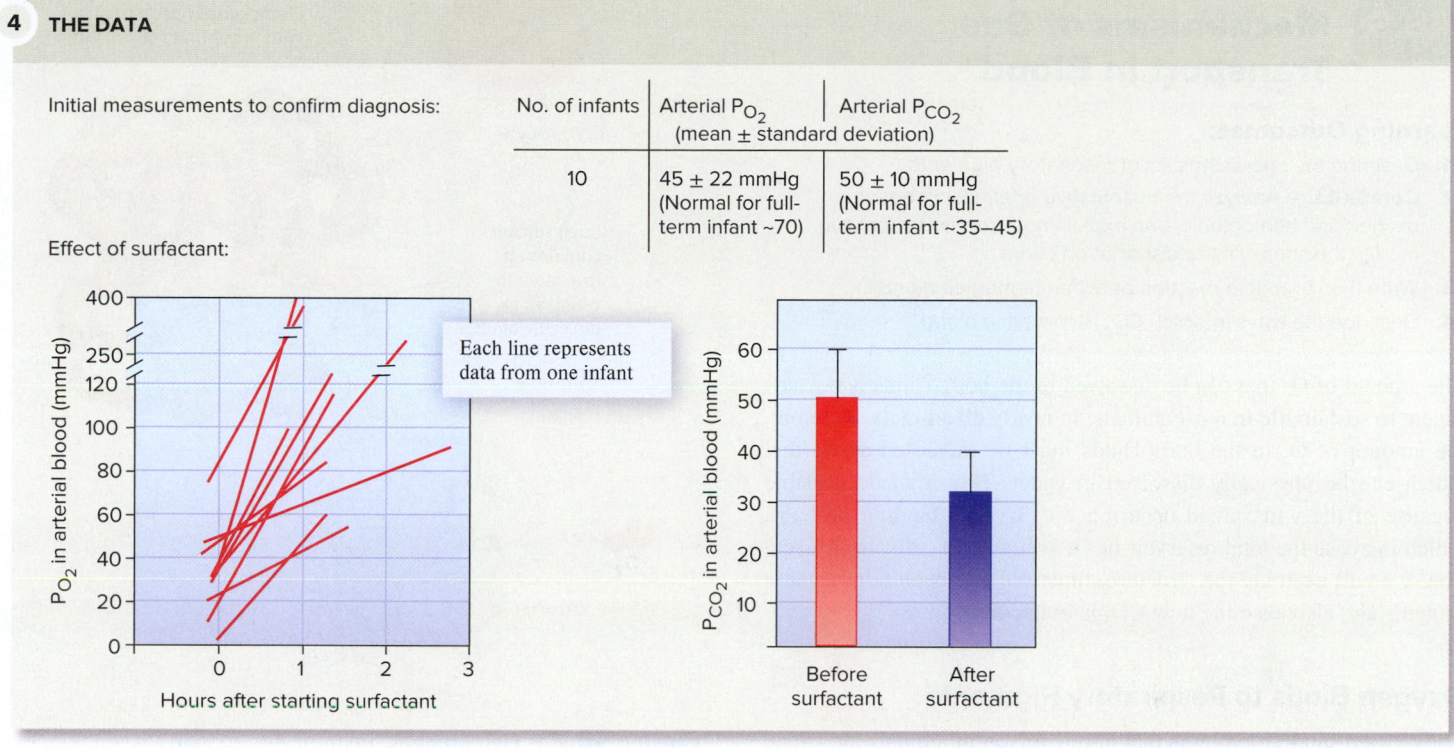

| Initial measurements to confirm diagnosis: | No. of infants | Arterial $P_{O_2}$ (mean ± standard deviation) | Arterial $P_{CO_2}$ |
|---|---|---|---|
| | 10 | 45 ± 22 mmHg (Normal for full-term infant ~70) | 50 ± 10 mmHg (Normal for full-term infant ~35–45) |

Effect of surfactant:

Each line represents data from one infant

**5** **CONCLUSION** Treatment with surfactant quickly improved lung function as evidenced by greatly increased blood $O_2$ levels and decreased $CO_2$ levels, to within the normal range. Surfactant therapy improves overall health and chances for survival in premature infants.

**6** **SOURCE** Fujiwara, T. et al. 1980. Artificial surfactant therapy in hyaline-membrane disease. *The Lancet* 315: 55–59.

the advent of surfactant therapy, RDS was the second leading cause of death in infants in the U.S. Following Fujiwara's pioneering study in 1980 and subsequent improvements in the ways in which surfactant was produced and administered, RDS fell to being the fourth leading cause of death in infants by 1988, and to ninth in 2010. Prior to the 1980s, approximately 25,000 infants died from RDS in the U.S. each year. Today, despite no decrease in the number of premature births annually, fewer than 900 RDS infants do not survive. Whereas that number is still tragically high, it represents an astounding improvement, and ranks the development of the surfactant treatment among the greatest scientific achievements of modern times.

Perhaps you noticed something unusual about the experimental design in Fujiwara's study. Did you notice that there were no control groups (infants who received a vehicle solution without surfactant)? Today, the gold standard for clinical studies is a randomized control trial (RCT). However, at the time of Fujiwara's study, not all clinical experiments were controlled as they are today. In some cases, this was due to practical considerations of cost and the limited availability of subjects. In others, ethical considerations may have taken precedence, such as the immediate need to save the lives of the infants in Fujiwara's study. Nonetheless, since Fujiwara's study, the results have been replicated and expanded upon many times in

animal models and in humans using RCT protocols, and surfactant therapy is now routinely used in hospitals around the world to save countless lives.

### Experimental Questions

1. **CoreSKILL** » Glucocorticoids are hormones made by the adrenal glands. They are known to stimulate surfactant production in the lungs. If a woman was told that due to complications of her pregnancy, she needed to have her baby delivered very prematurely, explain how our understanding of glucocorticoids could be used to help prevent RDS in the newborn of this woman.

2. **CoreSKILL** » Refer to the arterial blood $P_{O_2}$ measurements for the newborns in Fujiwara's study. A normal, healthy oxygen level in a full-term (not premature) newborn is about 70 mmHg (for various reasons, this is lower than in adults). Did some of the infants in Fujiwara's study have oxygen levels much greater than 70 mmHg after surfactant therapy? Why might this be? (Hint: How were the babies treated in addition to receiving surfactant?)

3. **CoreSKILL** » Prior to treatment, arterial blood $P_{O_2}$ levels in the newborns in this study were very low. Propose a hypothesis to explain why the newborns' blood $P_{CO_2}$ levels were higher than normal.

## 48.9 Mechanisms of Gas Transport in Blood

**Learning Outcomes:**

1. Describe the characteristics of respiratory pigments.
2. **CoreSKILL »** Analyze the quantitative relationship between oxygen and hemoglobin, and explain how certain factors can modify the shape of the dissociation curve.
3. Write the reversible reaction between hemoglobin and $O_2$.
4. Describe the ways in which $CO_2$ is carried in blood.

The amount of $O_2$ that can be dissolved in the body fluids is not sufficient to sustain life in most animals. In nearly all animals, therefore, the amount of $O_2$ in the body fluids must be increased above that which can be physically dissolved in water. This is made possible because of the widespread occurrence of oxygen-binding proteins, which increase the total reservoir of $O_2$ available to cells. In this section, we will examine the structure, function, and evolution of these proteins and also examine how $CO_2$ is transported.

### Oxygen Binds to Respiratory Pigments

The oxygen-binding proteins that have evolved in animals are called **respiratory pigments** because they have a color (blue or red). In vertebrates, the pigments are contained within erythrocytes, whereas many invertebrates have these pigments in their hemolymph. Respiratory pigments are proteins containing one or more metal atoms that bind to oxygen. In vertebrates and many marine invertebrates, the metal is typically iron ($Fe^{2+}$). As mentioned earlier, hemoglobin is the major iron-containing pigment and gives blood its red color. In decapod crustaceans, arachnids, and many mollusks, the metal is copper ($Cu^{2+}$). The copper-containing pigment **hemocyanin** gives the blood or hemolymph a bluish tint.

Hemoglobin gets its name because it is a globular protein—which refers to the shape and water solubility of a protein—and because it contains a chemical group called a heme in its core. An atom of iron is bound within the heme group. In vertebrates, hemoglobin consists of four polypeptide subunits, each with its own heme and iron atom to which a molecule of $O_2$ can bind. Thus, a single hemoglobin protein can bind up to four molecules of oxygen (**Figure 48.23**).

Respiratory pigments such as hemoglobin share certain characteristics that make them ideal for transporting $O_2$. First, they all have a high affinity for $O_2$. Second, the binding between the pigment and $O_2$ is noncovalent and reversible. The reversibility of $O_2$ binding allows respiratory pigments to unload $O_2$ to cells that require it. The reaction that describes the reversible binding of $O_2$ to hemoglobin (Hb) is

$$Hb + O_2 \rightleftharpoons HbO_2$$

where $HbO_2$ is oxyhemoglobin (hemoglobin with bound $O_2$), and the double arrows ($\rightleftharpoons$) indicate that the reaction is reversible.

The amount of pigment present in the blood is sufficient to provide enough $O_2$ to meet most oxygen demands, except for those that occur during the most strenuous exertion. In humans, for example, the presence of hemoglobin gives blood about 45 times more $O_2$-carrying capacity than plasma alone would have.

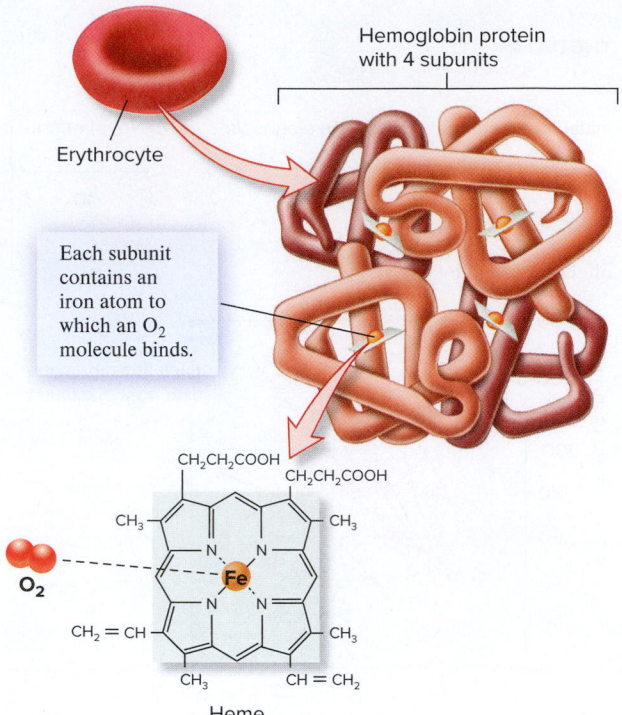

**Figure 48.23 Hemoglobin.** Erythrocytes contain large amounts of the protein hemoglobin. Oxygen binds reversibly to iron atoms in the heme portion of each subunit of hemoglobin.

**Core Concept: Structure and Function** The precise quaternary structure of hemoglobin permits its association with heme groups, which contain the iron atoms that bind oxygen molecules.

### The Amount of Oxygen Bound to Hemoglobin Depends on the $P_{O_2}$ of Blood

The partial pressure of $O_2$ ($P_{O_2}$) in blood is a measure of its dissolved concentration. When $P_{O_2}$ is high, more $O_2$ binds to hemoglobin, whereas fewer $O_2$ molecules will be bound when $P_{O_2}$ is low. **Figure 48.24** shows the relationship between $O_2$ binding and $P_{O_2}$, known as an **oxygen-hemoglobin dissociation curve**, for humans. At a $P_{O_2}$ of 100 mmHg, which is typical of the oxygenated blood leaving the lungs, nearly every hemoglobin molecule is bound to four $O_2$ molecules (at the far right in the graph). Under these conditions, hemoglobin is nearly 100% saturated with $O_2$. The $P_{O_2}$ of blood leaving the tissue capillaries of other parts of the body is lower and depends on metabolic activity. At rest, the average $P_{O_2}$ of blood capillaries in these other parts of the body is typically around 40 mmHg. At this $P_{O_2}$, hemoglobin releases some $O_2$ molecules, decreasing to about 75% saturation with $O_2$. During strenuous exercise, $P_{O_2}$ in the capillaries drops even further (as low as 20 mmHg). Consequently, during exercise hemoglobin releases even more $O_2$ and becomes less saturated. In this way, hemoglobin performs its role of $O_2$ delivery. In the lungs, it binds $O_2$, and elsewhere it releases $O_2$ as required.

The curve in Figure 48.24 is not linear but S-shaped (sigmoidal). This shape results because the subunits of hemoglobin cooperate with each other in binding $O_2$. Once a molecule of $O_2$ binds to one subunit's iron atom, the shape of the entire hemoglobin protein changes,

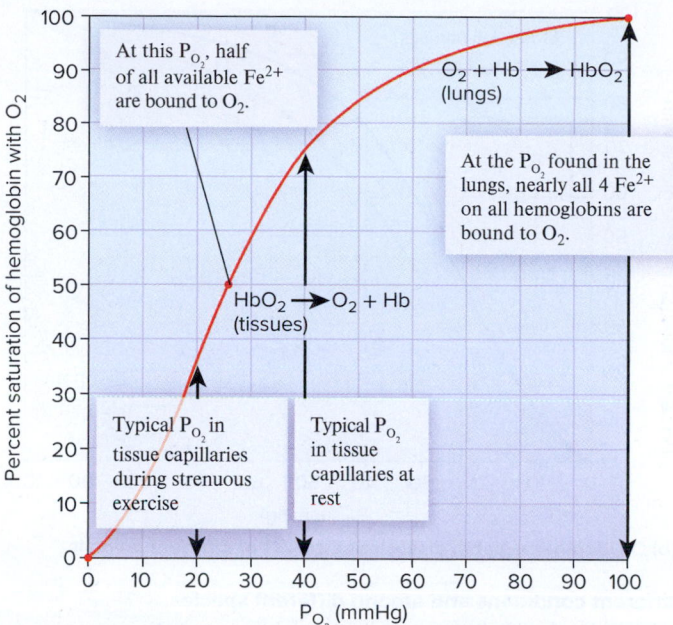

**Figure 48.24 The human oxygen-hemoglobin dissociation curve.** Depending on the partial pressure of oxygen ($P_{O_2}$), oxygen is either loaded onto hemoglobin, as in the lungs, or unloaded from hemoglobin, as in the rest of the body tissues. When $P_{O_2}$ is high, more $Fe^{2+}$ ions are bound to $O_2$, and therefore the hemoglobin is more saturated with $O_2$.

making it easier for a second $O_2$ to bind to the next subunit, and so on. Thus, the relationship between $P_{O_2}$ and the amount of $O_2$ bound to hemoglobin becomes very steep in the portion of the curve that represents the pressures that occur in the tissue capillaries throughout the body. This steepness allows $O_2$ release from hemoglobin to be very sensitive to even small decreases in $P_{O_2}$ generated by the diffusion of $O_2$ from capillaries to cells. To visualize this, follow the curve from right to left and see how hemoglobin saturation decreases as oxygen pressure decreases. In the other direction, the curve levels off at high oxygen pressures as 100% saturation is approached.

## The Affinity of Hemoglobin for Oxygen Is Decreased by Factors Such as Temperature, $CO_2$, and pH

One of the remarkable features of the oxygen-hemoglobin binding relationship is that it is influenced by metabolic waste products such as $CO_2$ and $H^+$ and by heat (temperature). **Figure 48.25a** shows three curves, one obtained under normal resting conditions and the others in the presence of low or high levels of $CO_2$. Carbon dioxide binds to amino acids in the hemoglobin protein (not to the iron, as $O_2$ does), and when it does, it decreases the affinity of the hemoglobin for $O_2$. This is an example of allosteric regulation, as described in Chapter 6 (look back at Figure 6.7c). Note how an increase in $CO_2$ shifts the curve to the right, such that at any $P_{O_2}$, less $O_2$ is bound to hemoglobin. Another way of saying this is that at any $P_{O_2}$, more $O_2$ has been released from hemoglobin, thus becoming available to cells.

A similar shift in the curve occurs with an increase in acidity (increased $H^+$ concentration), because $H^+$ can also bind to hemoglobin

and alter its oxygen-carrying capacity; the effect of $CO_2$ and $H^+$ on the oxygen-hemoglobin dissociation curve is known as the **Bohr effect**. Elevated temperature also reduces the affinity of hemoglobin for $O_2$, resulting in a right-shifted curve.

Cells generate each of these products—$CO_2$, $H^+$, and heat— when they are actively metabolizing nutrients such as glucose. The metabolic products enter the surrounding capillaries and diffuse into erythrocytes, where they alter the shape of hemoglobin, causing it to release more of its $O_2$ than would normally occur at that $P_{O_2}$. This phenomenon is a way in which individual body tissues obtain more $O_2$ from the blood to match their higher metabolic demands. Thus, when an animal increases its physical activity, the skeletal muscles generate more $CO_2$, $H^+$, and heat than do some other tissues. Therefore, the hemoglobin in muscle capillaries releases more $O_2$ to muscle cells compared to the hemoglobin in the capillaries of these other tissues.

The shift in the oxygen-hemoglobin dissociation curve occurs in all classes of vertebrates (although not in all species), but it has different magnitudes in different species. Not surprisingly, perhaps, animals with relatively high metabolic rates, such as mice, show a greater Bohr effect than do animals with lower relative metabolic rates. Recall that these same waste products of metabolism also cause local vasodilation of arterioles. The more metabolically active a tissue is, therefore, the more blood flow it receives, which means more oxygen-bound hemoglobin. Moreover, the oxygen is unloaded from hemoglobin more readily due to the shift in the curve. This is an excellent example of how adaptive changes in circulatory and respiratory functions often are complementary.

Regardless of their sensitivity to $CO_2$ and other factors, the hemoglobins of metabolically active animals also have a lower affinity for oxygen (**Figure 48.25b**) even in the absence of a greater than normal amount of $CO_2$, $H^+$, and heat. In small, active animals, the curves are displaced to the right relative to the human curve (in other words, the hemoglobin $P_{50}$ value is higher in small animals than it is in larger animals with lower relative metabolic rates). In contrast, larger animals with slower relative metabolic rates, such as the elephant, have curves shifted to the left compared with that of humans.

At high oxygen pressures, such as those that occur in the lungs, these shifts have little relevance because nearly all the hemoglobin is bound to oxygen at those pressures. At lower $P_{O_2}$, however, such as those that would occur in the capillaries of metabolically active tissues, the difference in the curves becomes significant. For example, look at the three curves at a $P_{O_2}$ of 40 mmHg, a typical value found in tissues that are using oxygen at a resting rate. The mouse hemoglobin has less oxygen bound (its hemoglobin is less saturated) at that pressure than does the human hemoglobin, which in turn has less $O_2$ bound than does the elephant hemoglobin. In other words, the mouse hemoglobin has released more of its $O_2$ to the active tissues than have the other animal hemoglobins. The different properties of hemoglobin among species result from changes in the sequence of hemoglobin, the result of numerous evolutionary changes in the genes encoding hemoglobin subunits in vertebrates.

## Carbon Dioxide Is Transported in the Blood in Three Forms

As with oxygen, only a limited amount of $CO_2$ physically dissolves in blood. An additional amount of $CO_2$ is carried bound to hemoglobin,

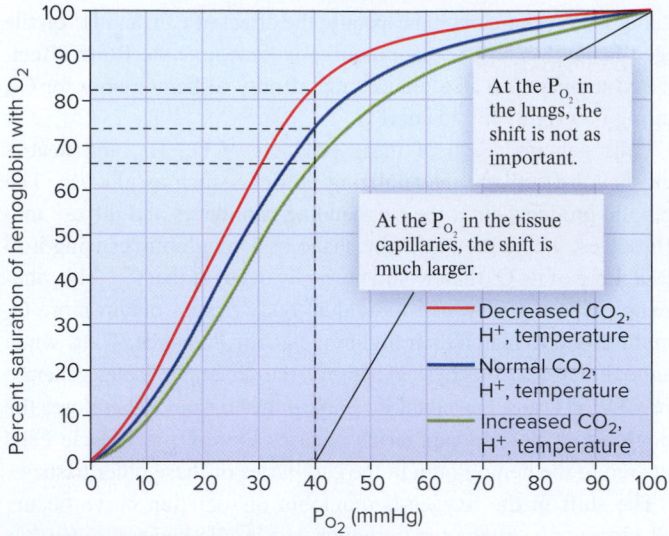

**(a) Shifts in the oxygen-hemoglobin dissociation curve**

**(b) Oxygen-hemoglobin dissociation curves of different animals**

**Figure 48.25** **Changes in oxygen-hemoglobin dissociation curves under different conditions and among different species.**
**(a)** Increasing or decreasing the amounts of $CO_2$ or $H^+$ (pH) and the temperature of the blood shifts the oxygen-hemoglobin dissociation curve. Metabolically active tissues generate more of these products, including heat. The change in affinity of hemoglobin for oxygen ($O_2$) allows different tissues to obtain $O_2$ in proportion to their metabolic requirements. **(b)** Oxygen-hemoglobin dissociation curves for three mammals with low (elephant), moderate (human), or high (mouse) relative metabolic rates. For any $P_{O_2}$, such as the one selected in the graph (40 mmHg), which is typical of the $P_{O_2}$ of tissue capillaries, less $O_2$ is bound to mouse hemoglobin than to human or elephant hemoglobin, and less $O_2$ is bound to human hemoglobin than to elephant hemoglobin. Therefore, $O_2$ is unloaded from hemoglobin more readily in smaller animals.

**Concept Check:** *What would happen to the position of the middle curve in Figure 48.25a following infusion of an alkaline compound such as bicarbonate ions ($HCO_3^-$) into the blood of a resting, healthy individual?*

as noted earlier. The majority of $CO_2$, however, is converted into highly soluble bicarbonate ions ($HCO_3^-$). This conversion is achieved by the following reactions, where $H_2CO_3$ is a short-lived compound called carbonic acid that immediately dissociates to an $H^+$ and an $HCO_3^-$:

$$CO_2 + H_2O \rightleftharpoons H_2CO_3 \rightleftharpoons H^+ + HCO_3^-$$

Note that one $H^+$ is formed for every $CO_2$ that enters this reaction. The resulting pH change is part of what makes $CO_2$ a dangerous waste product, because the activities of most enzymes in an animal's body are very sensitive to changes in pH.

These reactions are readily reversible. The first step is catalyzed in both directions by the enzyme carbonic anhydrase, which is present in high amounts in erythrocytes. As you learned in Chapter 2, the concentrations of reactants and products affect the rate of a chemical reaction. For example, as tissues release $CO_2$ into capillaries, the forward reaction (from left to right, as written here) will be favored, and the pH of the blood will decrease because the concentration of $H^+$ will increase. Conversely, when $CO_2$ diffuses out of lung capillaries and is exhaled, the reactions will proceed from right to left, and the pH of the blood will increase as $H^+$ combines with $HCO_3^-$.

# 48.10 Control of Ventilation

## Learning Outcome:

1. Describe the role of respiratory centers and chemoreceptors in the regulation of ventilation.

In the previous sections, we examined different ways that animals ventilate their respiratory organs. Now let's look at how the mechanisms

of breathing are controlled in mammals. Unlike the heart, lungs are neither muscles nor electrically excitable tissue. Therefore, they cannot initiate or regulate their own expansion. Nonetheless, lungs require a mechanism to rhythmically expand and recoil because animals cannot consciously control breathing at all times; for example, such control cannot be maintained during sleep. In this section, we examine the ways in which the nervous system and chemoreceptors control ventilation in mammals.

## The Nervous System Contains the Control Center for Ventilation

The control center that initiates rhythmic expansion of the lungs is a collection of nuclei in the central nervous system. In mammals, these **respiratory centers** are located in the pons and medulla oblongata of the brainstem (**Figure 48.26**). Neurons within these regions rhythmically generate action potentials, somewhat analogously to the way the SA node of the heart generates a rhythm. These electrical signals travel from the brainstem through two sets of nerves. The first set stimulates the intercostal muscles, and the second set stimulates the diaphragm. When the lungs expand in response to the contraction of these muscles, stretch-sensitive neurons in the lungs and chest send signals to the respiratory centers, informing them that the lungs are inflated. The respiratory centers then temporarily turn off the stimulating signals until the animal exhales, whereupon new signals are sent to the breathing muscles.

Although the brainstem automatically generates a steady rhythm of breathing, it can be modified or overridden. For example, animals that dive underwater—including humans when we swim—can

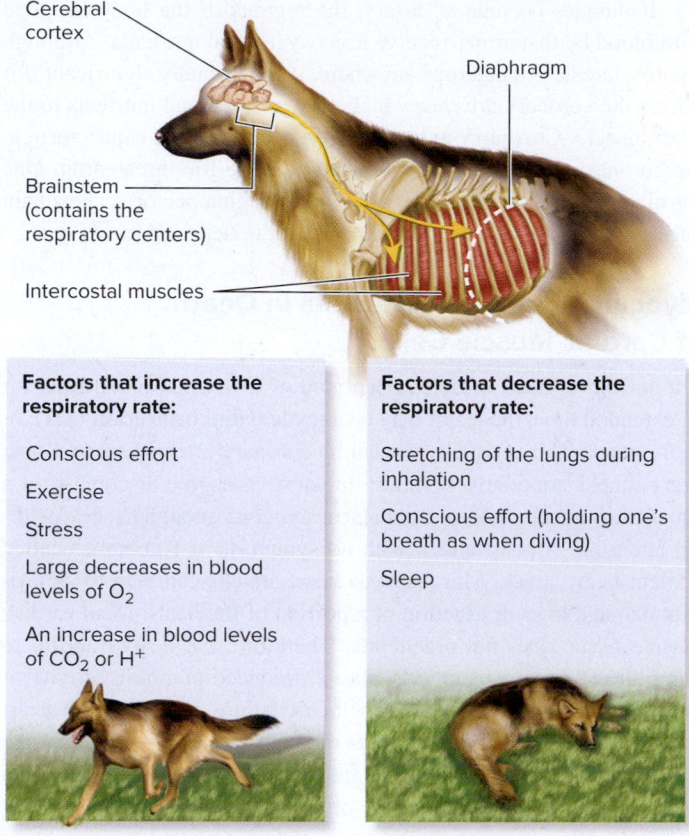

Cerebral cortex

Diaphragm

Brainstem (contains the respiratory centers)

Intercostal muscles

**Factors that increase the respiratory rate:**

Conscious effort

Exercise

Stress

Large decreases in blood levels of $O_2$

An increase in blood levels of $CO_2$ or $H^+$

**Factors that decrease the respiratory rate:**

Stretching of the lungs during inhalation

Conscious effort (holding one's breath as when diving)

Sleep

**Figure 48.26** **The control of breathing via respiratory centers in the mammalian brain.** Neurons in the brainstem send action potentials along neurons in nerves that stimulate the intercostal muscles and diaphragm. The factors listed here can modulate the rate at which action potentials are generated and therefore the respiratory rate.

 **Core Skill: Connections** The brainstem of vertebrates was described in Chapter 43. What parts of the mammalian brain comprise the brainstem?

temporarily hold their breath. The respiratory centers decrease their activity during sleep and increase it during stress. In addition, increased breathing occurs in response to physical activity. At such times, a variety of neural, endocrine and metabolic factors converge on the respiratory centers to increase the rate of signals to the breathing muscles, resulting in faster and deeper breaths.

As with cardiovascular systems, respiratory activity varies with body mass, as seen for mammals in **Table 48.2**. Comparing Tables 48.1 and 48.2, we can see that in mammals, circulatory and respiratory systems evolved similarly with respect to the metabolic demands of animals of different body mass. Smaller animals have proportionally smaller lungs (as evidenced by tidal volumes) and hearts than do larger animals, but they have faster breathing rates and heart rates. These adaptations allow smaller animals to deliver oxygen to tissues at a rate sufficient for their relatively high metabolic demands.

## Chemoreceptors Modulate the Activity of the Respiratory Centers

The respiratory centers are influenced by the partial pressures of oxygen and carbon dioxide in the arteries, as well as the concentration

| Table 48.2 | Respiratory Characteristics of Representative Mammals* | | | |
|---|---|---|---|---|
| Animal | Body mass (kg) | Tidal volume (L) | Breaths/ min | $P_{50}$ value |
| Shrew | 0.0024 | 0.00003 | 700 | 37 |
| Rat | 0.20 | 0.0016 | 85 | 35 |
| Dog | 25 | 0.27 | 20 | 29 |
| Human | 75 | 0.50 | 12 | 26 |
| Horse | 450 | 6.50 | 9 | 24 |

*Note: All values are averages from resting animals. Tidal volume is the volume of air breathed in with each breath. Note that tidal volume increases as an animal's mass (and therefore lung size) increases. By contrast, breathing rate is higher in smaller animals. Similar relationships occur in other vertebrates, notably birds. The $P_{50}$ value is the oxygen pressure at which an animal's hemoglobin is 50% saturated with $O_2$. Higher $P_{50}$ values correspond to lower affinities of hemoglobin for $O_2$ (in other words, animals with high $P_{50}$ values unload $O_2$ from hemoglobin more readily than do animals with low $P_{50}$ values; see Figure 48.25).

of hydrogen ions (in other words, the pH of the blood). Recall from Chapter 44 that chemoreceptors recognize specific chemicals in the air, water, body fluids, or food. Chemoreceptors located in the aorta, carotid arteries, and the brainstem detect the circulating levels of $O_2$, $CO_2$, and $H^+$ and relay that information through nerves or interneurons to the respiratory centers.

If the arterial $P_{O_2}$ decreases well below normal, as might occur at high altitude or in certain respiratory diseases, the chemoreceptors signal the respiratory centers to increase the rate and depth of breathing to increase ventilation of the lungs. The increase in ventilation brings more oxygen into the blood. Similarly, a buildup of $CO_2$ in the blood, which would occur if an animal's ventilation were lower than normal (again, often the result of respiratory disease), signals the respiratory centers to stimulate breathing. The increased ventilation not only brings in more $O_2$, but also helps eliminate more $CO_2$. Finally, an increased concentration of $H^+$ in the blood (such as during physical activity) activates chemoreceptors that signal the brain that the blood is too acidic. These signals lead to an increase in the rate of breathing, which eliminates additional $CO_2$, thereby allowing more $H^+$ to bind to $HCO_3^-$.

## 48.11 Impact on Public Health

### Learning Outcomes:

1. Define hypertension and atherosclerosis, and explain their impacts on human health.
2. List the causes, symptoms, and current medical treatments for myocardial infarction (MI), or heart attack.
3. Outline the causes, symptoms, and current treatments of asthma.
4. Discuss the impact of smoking tobacco on respiratory health.

Diseases of the heart and blood vessels account for more deaths each year in the U.S. than any other cause. Why is cardiovascular disease so devastating? One reason is that damage to structures of the circulatory system often occurs slowly, over many years, and without warning symptoms until the disease has reached late stages. Cardiovascular disease not only has a dramatic impact on the health of many Americans but also has a staggering impact on the economy, including expenditures for health-care services, medications, and lost productivity.

Respiratory diseases of all kinds (including lung cancer) afflict as many as 10% of the U.S. population and result in an estimated 300,000–400,000 deaths per year, making lung disease among the top three causes of death in the U.S. The economic impact of respiratory diseases on the U.S. economy is immense, with recent estimates of up to $150 billion per year in health-related costs and lost productivity. Many of these diseases are chronic—once they appear, they last for the rest of a person's life. Lifestyle factors, such as smoking tobacco and exposure to air pollution, cause some respiratory disorders or make existing conditions worse. In this section, we will examine the nature and some common causes of cardiovascular and respiratory diseases in humans.

## Hypertension and Atherosclerosis Contribute to Heart and Blood Vessel Disease

**Hypertension**, or high blood pressure, refers to an arterial blood pressure that is chronically above normal. The normal range of blood pressure in healthy humans varies from about 90/60 to 120/80 mmHg. Values above 120/80 mmHg are considered elevated, and a resting blood pressure of 140/90 mmHg or higher indicates hypertension. Many researchers and physicians today believe that the threshold for defining hypertension should even be lower. Hypertension has many causes, including obesity, smoking, aging, kidney disease, excess male hormones, and genetic factors, although in many cases, the cause is unknown. It can often be treated with diet and exercise and with drugs that cause vasodilation, thereby reducing total peripheral resistance.

Hypertension rarely has any noticeable symptoms, which is why it is often referred to as a "silent killer." For this reason, it is important to have blood pressure checked regularly. Without treatment, hypertension can damage arteries, contributing to the formation of **plaques**—accumulations of lipids, fibrous tissue, and smooth muscle cells—along the inner surfaces of arterial walls. Plaques may lead to **atherosclerosis**, in which plaques cause the arteries to narrow and harden (**Figure 48.27**). Large plaques may occlude (block) the lumen of an artery entirely. Arterial plaques may arise from a variety of factors in addition to hypertension, including calcium and fat deposits, and are also correlated with obesity, high blood cholesterol concentrations, and smoking.

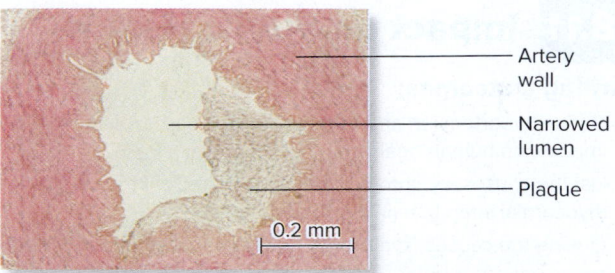

**Figure 48.27** **An atherosclerotic plaque in a small artery.** Compare this with the image of a healthy artery shown in Figure 48.11b. ©Biophoto Associates/SPL/Science Source

 **Core Skill: Science and Society** Heart and blood vessel disease are the leading killers among adults in the U.S. Intensive research into the causes of atherosclerosis has helped educate people about behaviors that increase the risk of this disease, such as smoking.

If plaques occlude an artery, the regions of the body supplied with blood by that artery receive less oxygen and nutrients. Although atherosclerosis is dangerous anywhere, it is especially significant if it affects the coronary arteries, which carry oxygen and nutrients to the heart muscle. **Coronary artery disease** occurs when plaques form in the coronary vessels—a condition that can be life-threatening. One warning sign of coronary artery disease is angina pectoris, chest pain during exertion due to the heart muscle being deprived of oxygen.

## Myocardial Infarction Results in Death of Cardiac Muscle Cells

If a portion of heart muscle is deprived of its normal blood flow for an extended time, the result may be a **myocardial infarction** (MI), or heart attack. This is usually caused by coronary artery disease. Some heart attacks are relatively minor. In some cases, the discomfort of a small heart attack may not even alarm someone enough to seek medical attention. A heart attack with no symptoms is sometimes called a silent heart attack. More serious heart attacks can lead to significant damage to or destruction of a portion of the heart. Dead cardiac muscle tissue does not regenerate. Therefore, the heart's ability to pump blood is permanently decreased. Reduced pumping activity of the heart can result in congestive heart failure, in which the heart cannot pump enough blood to meet the body's needs. As noted in the chapter introduction, each year in the U.S. between 1 and 1.5 million people suffer a heart attack, many of which are fatal.

Preventing a heart attack from occurring at all is obviously the best strategy for long-term health. Procedures are available that allow physicians to monitor the status of the coronary vessels in people thought to have heart disease. In the procedure called **cardiac angiography**, the coronary arteries can be visualized by injecting a dye into a person's veins and then taking an X-ray image of the chest (see the chapter opening image). The resulting image helps a physician determine if the vessels are narrowed by disease.

If a blockage is found, several common treatments can restore blood flow through a coronary artery. One is **balloon angioplasty**, in which a thin tube with a tiny, inflatable balloon at its tip is threaded through the artery to the diseased area. Inflating the balloon compresses the plaque against the arterial wall, widening the lumen. In most cases, a wire-mesh device called a stent is inserted into the artery after angioplasty has expanded it, providing a sort of lattice to hold the artery open (**Figure 48.28**). A treatment for more serious coronary artery disease is a **coronary artery bypass**, in which a small piece of healthy blood vessel is removed from one part of the body and surgically grafted onto the coronary circulation in such a way that blood bypasses the diseased region of the unhealthy artery.

## Asthma Is a Disease of Hyperreactive Bronchioles

We saw previously that bronchioles resemble arterioles in that both are thin tubes with circular rings of smooth muscles that can relax or contract. In the disease **asthma**, however, the smooth muscles of the bronchioles are hyper-reactive to many stimuli and contract more than usual (**Figure 48.29**). Contraction of these muscles narrows the bronchioles, causing bronchoconstriction. This constriction makes it difficult to move air into and out of the lungs, because resistance to airflow increases when the diameter of the airways decreases

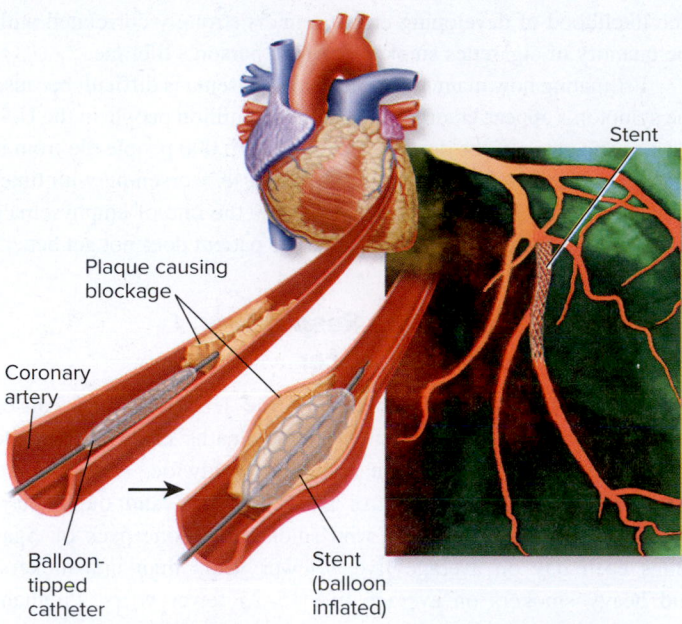

**Figure 48.28  A treatment for blocked blood vessels.** Balloon angioplasty can widen diseased arteries, followed by insertion of a stent. The inset shows a stent placed in a coronary artery of a human patient. *(inset)* ©Sovereign/ISM/Phototake

 **Core Skill: Science and Society**  Biologists investigating the mechanisms of blood flow, heart function, and blood pressure in animals can use that knowledge to develop treatments for heart disease in humans.

(analogous to what you learned earlier about blood vessels). Often, the resistance to airflow can be so great that the movement of air creates a characteristic wheezing sound.

Asthma tends to run in families and therefore likely has a genetic basis in many individuals with the disease. Several known triggers can elicit bronchoconstriction, including exercise, cold air, and allergic reactions. The last of these—allergic reactions—is of interest because asthma is believed to be partly the result of an imbalance in the immune system, which controls inflammation and other allergic responses. During flare-ups of asthma, inflammation results in the secretion of a viscous, mucus-like fluid into the lumen of bronchioles and other airway tubes. This fluid inhibits airflow and worsens symptoms.

The symptoms of asthma can be alleviated by inhaling an aerosol mist containing **bronchodilators**, compounds that bind to receptors located on the plasma membranes of smooth muscle cells of bronchioles. These compounds, which are related to the neurotransmitter norepinephrine, cause bronchiolar smooth muscle cells to relax. This, in turn, allows the bronchioles to dilate (widen). To help reduce the inflammation of the lungs, the medication may also contain an adrenocortical steroid hormone with anti-inflammatory actions. Currently, there is no cure for asthma, but with regular treatment and the avoidance of known triggers, most people with this disease lead normal lives with only few restrictions.

## Emphysema Causes Permanent Lung Damage

Unlike asthma, in which the major problems are periodically inflamed airways and hyper-reactive bronchioles, **emphysema** is a progressive disease that involves extensive lung damage (**Figure 48.30**).

**Figure 48.29**  Comparison of healthy and asthmatic bronchioles in a human.

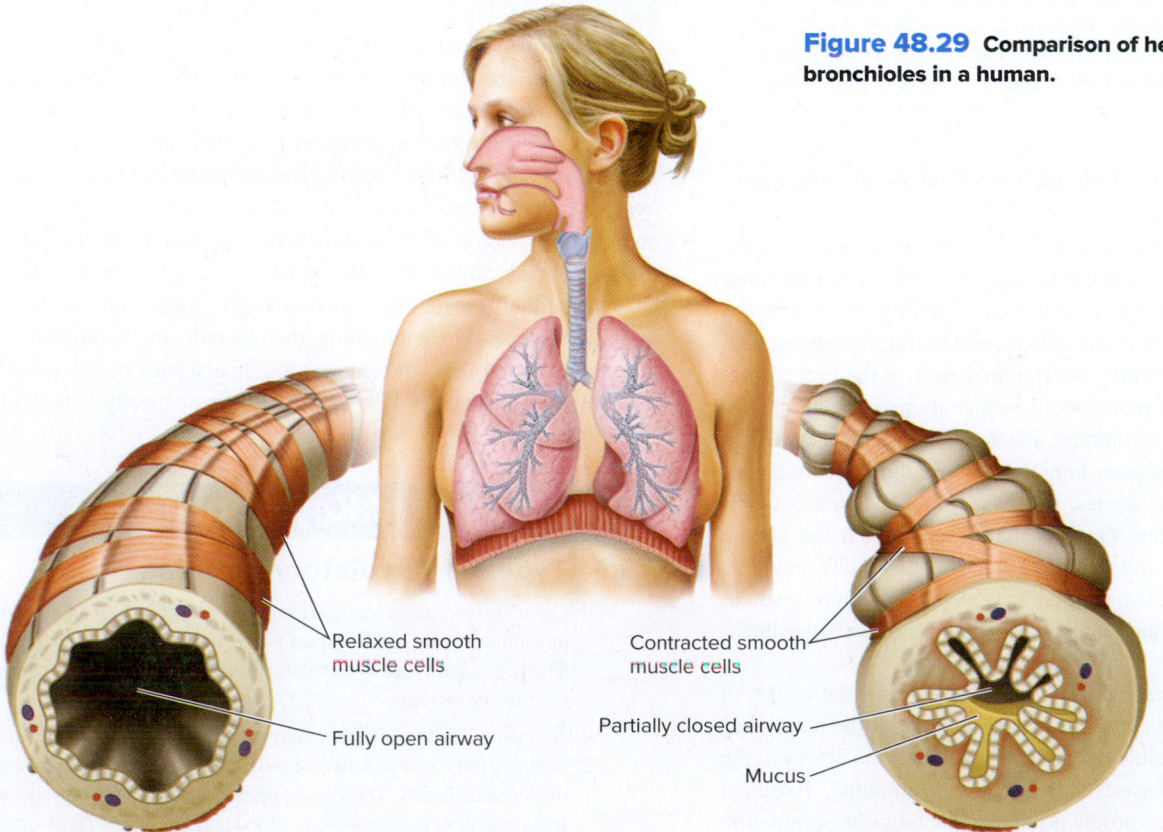

Relaxed smooth muscle cells

Fully open airway

Contracted smooth muscle cells

Partially closed airway

Mucus

**Healthy bronchiole**

**Asthmatic bronchiole (constricted)**

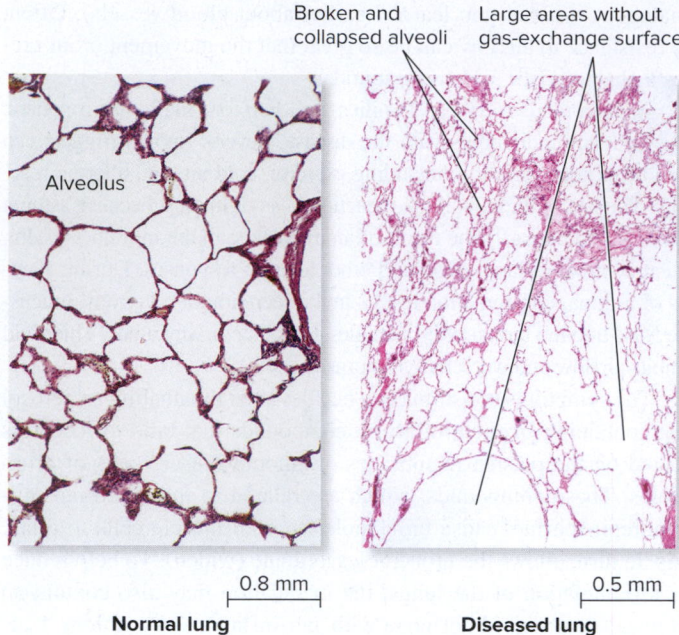

Broken and collapsed alveoli

Large areas without gas-exchange surfaces

Alveolus

0.8 mm
**Normal lung**

0.5 mm
**Diseased lung**

**Figure 48.30  The effects of emphysema.** These light micrographs compare a section of a normal, healthy lung (left) with that of a lung from a person who died of emphysema (right). The collapse and destruction of alveoli caused by this disease reduces the surface area available for gas exchange in the lungs. (left) ©Astrid & Hanns-Frieder Michler/SPL/Science Source; (right) ©McGraw-Hill Education/Al Telser, photographer

 **Core Skill: Science and Society** Lung disease affects millions of people. Biologists study the causes, mechanisms, and possible treatments of human diseases such as emphysema and asthma. Such research has improved and extended the lives of many millions of individuals.

The disease decreases the total surface area of the alveoli, which consequently decreases the rate of oxygen diffusion from the lungs into the circulation. Consequently, one sign of emphysema is a $P_{O_2}$ in the arteries that is lower than normal. It is also physically harder for those with emphysema to exhale because the recoil ability of the lungs is decreased due to the loss of elastin fibers, and therefore, arterial $CO_2$ concentrations increase. Finally, the terminal ends of the bronchioles are often damaged, which increases resistance to airflow and creates one of the disease's main symptoms, shortness of breath.

Reduced blood $O_2$ and poor lung function limit a person's ability to function, and in its late stages, emphysema results in a patient's being essentially bedridden. Oxygen therapy, in which the person breathes a mixture of air and pure $O_2$ from a portable gas tank, can provide some relief. The extra oxygen increases the pressure gradient for $O_2$ from the alveoli to the lung capillaries, promoting better diffusion of $O_2$ into the blood.

In some cases, emphysema results from an enzyme deficiency in the lungs that destroys the elastic protein that provides the recoil during exhalation or from chronic exposure to air pollution. However, the overwhelming majority of cases, 85%, are due to smoking. Toxins in cigarettes and other tobacco products damage the lungs by stimulating leukocytes to release proteolytic enzymes that degrade lung tissue.

The likelihood of developing emphysema is strongly correlated with the quantity of cigarettes smoked during a person's lifetime.

Estimating how many people have emphysema is difficult because the symptoms appear gradually. More than 3 million people in the U.S. have severe cases of the disease, and at least 15,000 people die from it each year. Emphysema is a progressive disease, worsening with time. Although medical care can sometimes slow the rate of emphysema's progress, the disease is not curable and the patient does not get better.

## Tobacco Smoke Causes Respiratory Health Problems and Cancer

Smoking tobacco products is one of the leading global causes of death, contributing to about 430,000 deaths each year in the U.S. alone and over 5 million annually worldwide. According to the Centers for Disease Control and Prevention and the American Lung Association, people who smoke up to one pack of cigarettes each day on average live 7 fewer years than nonsmokers, and heavy smokers on average live 15–25 fewer years. Pregnant women who smoke run a high risk of their baby being born underweight, a potentially serious condition that may affect the newborn's long-term health.

Up to 85% of all new cases of lung cancer diagnosed each year are attributable to smoking, making lung cancer the leading cause of preventable death. Equally important, however, is that smoking is estimated to be responsible for nearly 30% of all cancers, including cancer of the mouth and throat, esophagus, bladder, pancreas, and ovaries. Smoking is also a leading cause of cardiovascular disease, high blood pressure, and stroke. Smoking only a few cigarettes per day significantly increases the risk of heart disease.

Because tobacco smoke is inhaled directly into the lungs, the chemicals in the smoke can do considerable damage to lung tissue. Even adolescents who have only recently started smoking have increased mucus (phlegm) production in their airways, shortness of breath, and reduced lung growth. Thousands of chemicals, including over 40 known cancer-causing compounds, have been identified in cigarette smoke. Some of these chemicals—such as formaldehyde—are toxic to all cells. Others, like the odorless gas carbon monoxide (CO), have harmful effects on lung function in particular. CO competes with $O_2$ for binding sites in hemoglobin, thereby reducing hemoglobin saturation. Heavy smokers who smoke more than a pack of cigarettes each day may have as much as 15% less $O_2$-carrying capacity in their blood.

## Summary of Key Concepts

### 48.1  Types of Circulatory Systems

- Circulatory systems transport necessary materials to all cells of an animal's body and transport waste products away from cells. The two major types of circulatory systems are open and closed circulatory systems.

- An open circulatory system does not transport $O_2$ and $CO_2$. It contains three components: hemolymph, open-ended vessels, and one or more hearts. The vessels open into the animal's body cavity. In a closed circulatory system, blood and interstitial fluid are physically separated at all times by blood vessels (Figure 48.1).

- The closed circulatory systems of vertebrates transport $O_2$ and $CO_2$ and may be single or double circulations. In a double circulation, blood circulates at two different pressures through the systemic circulation and pulmonary circulation (Figures 48.2, 48.3).

## 48.2 The Composition of Blood

- Blood is a fluid connective tissue consisting of cells suspended in a solution containing dissolved nutrients, proteins, gases, and other molecules. Blood has four components: plasma, leukocytes, erythrocytes, and (in vertebrates) platelets or thrombocytes (Figure 48.4).

- Platelets contribute to clot formation in two ways: by binding together to form part of the clot itself and by secreting substances that lead to precipitation of the protein fibrin (Figure 48.5).

## 48.3 The Vertebrate Heart and Its Function

- Blood enters the atria from veins of the systemic or pulmonary circulation, flows down a pressure gradient through the AV valves into the ventricles, and is pumped out through the semilunar valves into the arteries of the systemic and pulmonary circulations (Figure 48.6).

- Excitation of the atria begins at the sinoatrial (SA) node; excitation of the ventricles begins at the atrioventricular (AV) node. The electrical impulses spread through the atria and ventricles and stimulate the heart muscle to contract. The cardiac cycle has two phases: diastole and systole (Figures 48.7, 48.8).

- The electrical activity of the heart can be monitored using an electrocardiogram (Figure 48.9).

## 48.4 Blood Vessels

- Arteries carry blood away from the heart. Arterioles distribute blood to capillaries, which are the site of gas and nutrient exchange. Blood from the capillaries flows into venules and then into veins, which return it to the heart (Figures 48.10, 48.11, 48.12, 48.13).

- Muscle activity and one-way valves help move venous blood toward the heart against gravity (Figure 48.14).

## 48.5 Relationship Among Blood Pressure, Blood Flow, and Resistance

- Blood pressure, which is the force exerted by blood on the walls of blood vessels, is responsible for moving blood through the vessels. Resistance refers to the tendency of blood vessels to slow the flow of blood through their lumens.

- Local blood flow through a vessel is directly proportional to the pressure of the blood entering the vessel and inversely proportional to the resistance created by that vessel. Arteriole radius is the major factor that regulates resistance.

- Cardiac output overcomes resistance to generate systemic blood pressure and depends on the size of an animal's heart, how often it beats each minute, and how strongly it contracts with each beat (Table 48.1).

- Cardiac output (CO) and total peripheral resistance (TPR) determine blood pressure (BP): BP = CO × TPR (Figure 48.15).

## 48.6 Physical Properties of Gases

- The partial pressure of oxygen ($P_{O_2}$) in the environment provides the driving force for its diffusion. Atmospheric pressure decreases at higher elevations.

- Three factors—the pressure of the gas, temperature of the water, and presence of any other solutes—affect the solubility of a gas in water.

## 48.7 Types of Respiratory Systems

- A respiratory system is all of the components of an animal's body that contribute to the exchange of gases between the external environment and cells of the body. Ventilation is the process of bringing oxygenated water or air into contact with a respiratory organ.

- The body surface is permeable to gases in some invertebrates and in amphibians. Water-breathing animals use external or internal gills for gas exchange (Figures 48.16, 48.17).

- In insects, air in the tracheoles comes into contact with fluid at the tracheole tips. Oxygen from the air dissolves in this fluid and diffuses across the tracheole wall and into nearby cells (Figure 48.18).

- Air-breathing vertebrates generally use lungs to obtain $O_2$ and eliminate $CO_2$. All lungs receive deoxygenated blood from the heart and return oxygenated blood to the heart. Most vertebrates fill their lungs by negative pressure filling, which follows the principles of Boyle's law (Figure 48.19).

## 48.8 Structure and Function of the Mammalian Respiratory System

- The mammalian respiratory system includes the nose, mouth, airways, lungs, muscles, and connective tissues that encase these structures within the thoracic cavity (Figure 48.20).

- Mammals ventilate their lungs by negative pressure filling. The work is provided by the intercostal muscles and diaphragm. Surfactant decreases the surface tension of water helping to keep the alveoli open (Figures 48.21, 48.22).

## 48.9 Mechanisms of Gas Transport in Blood

- A limited amount of any gas can dissolve in water. Such limits are overcome in part by respiratory pigments that provide additional oxygen-carrying capacity. Hemoglobin is the major iron-containing pigment in vertebrates and many invertebrates (Figure 48.23).

- The amount of oxygen bound to hemoglobin depends on the partial pressure of $O_2$ ($P_{O_2}$) in the blood. Metabolic products such as $CO_2$ and $H^+$ and heat can influence oxygen-hemoglobin binding (Figures 48.24, 48.25).

- $CO_2$ is transported dissolved in solution, bound to hemoglobin, and in the form of $HCO_3^-$.

## 48.10 Control of Ventilation

- In mammals, respiratory centers in the brainstem initiate the rhythmic expansion of the lungs (Figure 48.26).

- Chemoreceptors detect arterial blood levels of $H^+$, $CO_2$, and $O_2$. The chemoreceptors relay this information to the respiratory centers, which in turn affect the breathing rate.

## 48.11 Impact on Public Health

- Cardiovascular disease accounts for more deaths each year in the U.S. than any other cause. Cardiovascular disorders include hypertension, atherosclerosis, coronary artery disease, angina pectoris, and myocardial infarction (heart attack) (Figure 48.27).

- Cardiovascular diagnostic techniques and treatments include cardiac angiography, balloon angioplasty, and coronary artery bypass (Figure 48.28).

- In asthma, the muscles of the bronchioles contract more than usual, increasing resistance to airflow (Figure 48.29).

- Smoking tobacco products is one of the leading global causes of death. Smoking is strongly linked to cancer, cardiovascular disease, stroke, and emphysema, in which lung tissues are severely damaged (Figure 48.30).

# Assess & Discuss

## Test Yourself

1. Hemolymph differs from blood in that it
   a. does not contain blood cells.
   b. is a mixed fluid found in the hemocoel.
   c. circulates through closed circulatory systems only.
   d. functions only in defense of the body and not transport.
   e. does not pass through a heart.

2. A typical hematocrit value for a human is around 42%. This means that
   a. the typical fluid portion of blood is about 42% of the total volume.
   b. the leukocytes make up 42% of the blood volume.
   c. the erythrocytes make up 42% of the blood volume.
   d. the leukocytes and erythrocytes together make up 58% of the blood volume.
   e. the erythrocytes alone make up 58% of the blood volume.

3. A major advantage of a double circulation is that
   a. blood can be pumped to the upper portions of the body by one circuit and to the lower portions of the body by the other circuit.
   b. each circuit can pump blood with differing pressures to optimize the function of each.
   c. the oxygenated blood can mix with the deoxygenated blood before being pumped to the tissues of the body.
   d. less energy is required to provide nutrients and oxygen to the tissues of the body.
   e. All of the above are advantages of a double circulation.

4. The function of erythrocytes is to
   a. transport oxygen throughout the body.
   b. defend the body against infection and disease.
   c. transport chemical signals throughout the body.
   d. secrete the proteins that form blood clots.
   e. do both a and d.

5. For blood flow through a closed circulation, which is the correct sequence of vessels beginning at the heart?
   a. arteriole, artery, capillary, vein, venule
   b. artery, capillary, arteriole, venule, vein
   c. vein, venule, capillary, arteriole, artery
   d. artery, arteriole, capillary, venule, vein
   e. artery, arteriole, capillary, vein, venule

6. Carbon dioxide is considered a harmful waste product of cellular respiration because it
   a. lowers the pH of the blood.
   b. lowers the $H^+$ concentration in the blood.
   c. competes with oxygen for transport in the blood.
   d. does all of the above.
   e. does a and b only.

7. The countercurrent exchange mechanism in fish gills
   a. maximizes oxygen diffusion into the bloodstream.
   b. is a less efficient mechanism for gas exchange than that used in mammalian lungs.
   c. occurs because the flow of blood is in the same direction as water flowing across the gills.
   d. facilitates diffusion of carbon dioxide into the blood of the fish.
   e. facilitates diffusion of oxygen to the environment.

8. The tracheal system of insects
   a. consists of several tracheae that connect to multiple lungs within the different segments of the body.
   b. consists of extensively branching tubes that are in close contact with all the cells of the body.
   c. allows oxygen to diffuse directly across the thin exoskeleton of the insect to the bloodstream.
   d. cannot function without constant movement of the wings to move air into and out of the body.
   e. provides oxygen that is carried through the animal's body in hemolymph.

9. Which of the following factors does *not* increase the rate of breathing by influencing the chemoreceptors?
   a. an increase in $P_{CO_2}$ in the arterial blood
   b. an increase in $P_{O_2}$ in the arterial blood
   c. a decrease in the pH of the arterial blood
   d. a decrease in $P_{O_2}$ in the arterial blood
   e. an increase in the $H^+$ concentration in the arterial blood

10. The majority of oxygen is transported in the blood of vertebrates
    a. by binding to plasma proteins.
    b. by binding to hemoglobin in erythrocytes.
    c. as dissolved gas in the plasma.
    d. as dissolved gas in the cytoplasm in the erythrocytes.
    e. by binding to hemoglobin in the plasma.

## Conceptual Questions

1. Explain the differences between closed and open circulatory systems. What advantages does a closed circulatory system provide?

2. Explain why it is an advantage for $CO_2$, $H^+$, and heat to be major factors in decreasing the affinity of hemoglobin for oxygen.

3. **Core Concepts: Structure and Function** A core concept of biology is that structure determines function. How is this concept related to what you have learned in this chapter about the hemoglobin molecule?

## Collaborative Questions

1. Describe the cardiac cycle, and explain why heart valves must open in only one direction.

2. List the components of the mammalian respiratory system, and describe the major functions of each.

# Excretory Systems

# 49

I f you have ever noticed how quickly the water in an aquarium or a swimming pool becomes dirty when the filter is not functioning, you will have a good idea of the importance of filtering the soluble wastes from an animal's body fluids. The human kidneys, for example, are remarkable filtration devices. Although each one is only about the size of a computer mouse, the kidneys are able to filter blood at a rate of 150–200 L/day. Considering that a typical adult has only 5 L or so of blood, that is an astonishingly effective filtration mechanism. By the time a person reaches 50 years of age, his or her kidneys have filtered roughly 3,000,000 L of blood! As they perform this function, the kidneys not only remove soluble waste products of metabolism from the filtered blood, but also recapture from it useful substances such as sodium ions and water.

The ability of organisms to maintain homeostasis is one of the fundamental principles of biology and a common theme of the previous several chapters. Excretory systems are critical not only for removing soluble wastes from body fluids, but also for homeostasis, particularly of ion and water balance. An **excretory system** includes all of an animal's organs (such as gills, lungs, kidneys, and, in some animals, the body surface) that function to remove soluble wastes generated from metabolism. These wastes include such substances as $CO_2$ and the nitrogenous wastes from protein and nucleic acid metabolism.

As described in Chapter 48, the elimination of $CO_2$ from the body is carried out by respiratory organs such as gills and lungs, which is why these structures can also be considered part of an animal's excretory system. In this chapter, we examine how different excretory organs participate in eliminating nitrogenous and other soluble wastes from animal bodies. We then highlight some of the major features of the mammalian kidney, examine how the kidneys eliminate wastes and, in the process, regulate ion and water balance. We conclude by considering how kidney disease affects human health.

**Colorized scanning electron micrograph of the filtration apparatus of a vertebrate kidney.** As blood flows through bundles of capillaries known as a glomerulus (shown in blue) in the kidneys, some of the fluid in the blood leaves and enters underlying tubules called nephrons. Soluble wastes can be eliminated from the body via this mechanism. ©Steve Gschmeissner/SPL/Getty Images

## 49.1 Excretory Systems in Different Animal Groups

**Learning Outcomes:**

1. Describe the forms of nitrogenous wastes generated by animals.
2. Describe the general processes of filtration, reabsorption, secretion, and excretion.
3. Identify several invertebrate osmoregulatory organs, and compare and contrast the process of elimination in each.
4. Recognize the general structural and functional features of kidneys, which are common to all vertebrates.

Animals use one or more organs to rid themselves of metabolic wastes, excess water and ions, and toxins from their environment. At high concentrations, metabolic wastes and ions can be very harmful to cell function. Most excretory organs contain tubular structures lined with epithelial cells that have the capacity to transport ions and other solutes across their membranes. Wastes are excreted out of the body by means of these tubes. In this section, we will consider the structure and function of invertebrate and vertebrate excretory organs. First, however, let's examine the major forms of nitrogenous wastes generated in animals by the metabolism of proteins (amino acids) and nucleic acids.

## Nitrogenous Wastes Exist in Three Major Forms

Nitrogenous wastes are usually produced in one of three forms—ammonia (and ammonium ions), urea, or uric acid (**Figure 49.1**). Different animal groups produce a particular form of such wastes, depending on the species and the environment in which it lives. All nitrogenous wastes originate from the metabolism of amino acids or nucleic acids.

- Ammonia ($NH_3$) and ammonium ions ($NH_4^+$) are the most toxic of the nitrogenous wastes because they disrupt the pH of body fluids, ion electrochemical gradients, and many chemical reactions that involve oxidations and reductions. Animals that excrete wastes in this form typically live exclusively or most of the time in water. In marine invertebrates, $NH_3$ and $NH_4^+$ are continually excreted across the skin and gills, whereas in freshwater and most saltwater fishes, these wastes are excreted via the gills and kidneys. Because $NH_3$ is so toxic, aquatic animals excrete it as quickly as it is formed. The chief advantage of excreting nitrogenous wastes as $NH_3$ and $NH_4^+$ is that energy is not required for their conversion to a less toxic product, as is the case for urea and uric acid. A disadvantage is that considerable body water is required to excrete them at a rate that maintains dilute, safe concentrations within the animal.

- **Urea** is produced from $NH_3$ by all mammals, most amphibians, some marine fishes, some reptiles, and some terrestrial invertebrates. One advantage of producing urea is that it is less toxic than $NH_3$, and thus animals can tolerate some accumulation of it in their body fluids. Another advantage is that urea does not require large volumes of water for its excretion. Thus, producing urea conserves water, removes the necessity for constant excretion, and reduces the likelihood of toxicity. One drawback of producing urea is that the metabolic synthesis of urea from $NH_3$ requires a moderate expenditure of ATP and thus consumes part of an animal's total daily energy budget.

- **Uric acid** or other nitrogenous compounds called purines are produced by birds, insects, and most reptiles. Like urea, these compounds are less toxic than ammonia, but they are even more energetically costly to synthesize from $NH_3$. However, because they are poorly soluble in water, they are not excreted in a watery urine but instead are packaged with other waste products and excess salts into a semisolid, partly dried precipitate that is excreted. The energy investment required to produce uric acid, therefore, is balanced against the water conserved by excreting nitrogenous wastes in this form.

## Excretory Systems Use Four Processes to Remove Nitrogenous and Other Soluble Wastes from the Body

Most excretory systems function by using one or more of the following processes: filtration, reabsorption, secretion, and excretion (**Figure 49.2**).

1. **Filtration.** In filtration, an organ functions like a sieve or filter, removing some of the water and its small solutes from the blood, interstitial fluid, or hemolymph, while excluding blood

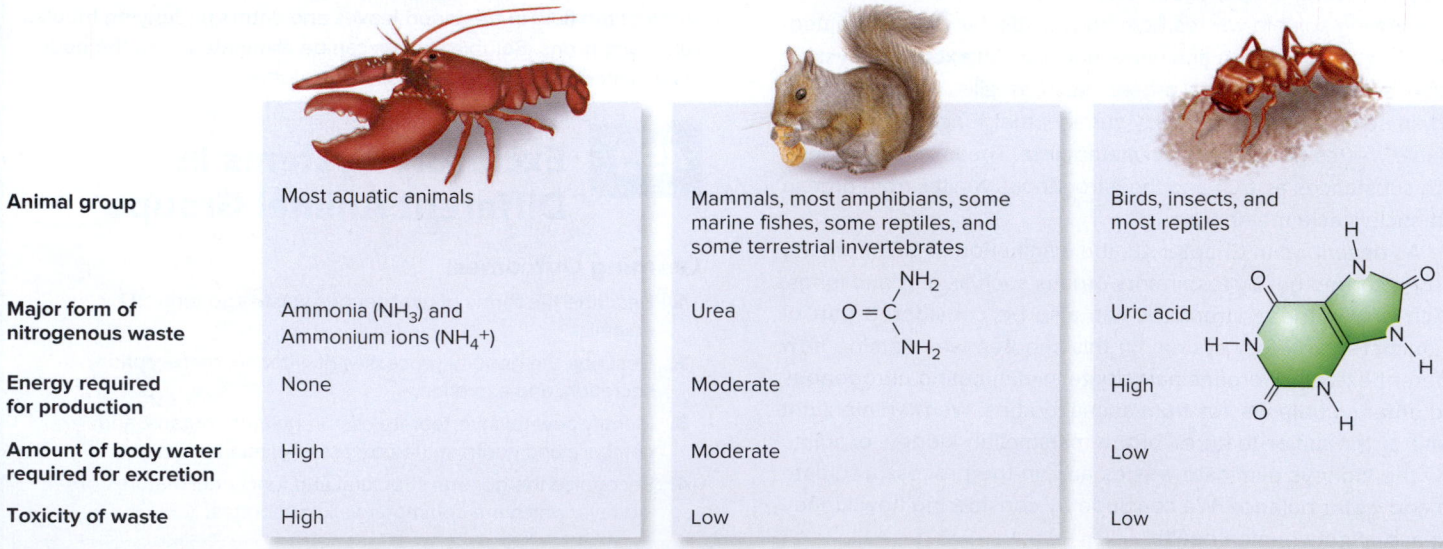

| Animal group | Most aquatic animals | Mammals, most amphibians, some marine fishes, some reptiles, and some terrestrial invertebrates | Birds, insects, and most reptiles |
|---|---|---|---|
| Major form of nitrogenous waste | Ammonia ($NH_3$) and Ammonium ions ($NH_4^+$) | Urea | Uric acid |
| Energy required for production | None | Moderate | High |
| Amount of body water required for excretion | High | Moderate | Low |
| Toxicity of waste | High | Low | Low |

**Figure 49.1 Nitrogenous wastes produced by different animal groups.** The three major forms of nitrogenous wastes, which are derived from the breakdown of proteins or nucleic acids, have different properties.

**Core Concept: Energy and Matter** Energy that is stored in the bonds of ATP is required to metabolize nitrogenous wastes into urea and uric acid. Many animals, therefore, must expend energy each day to rid themselves of nitrogenous wastes.

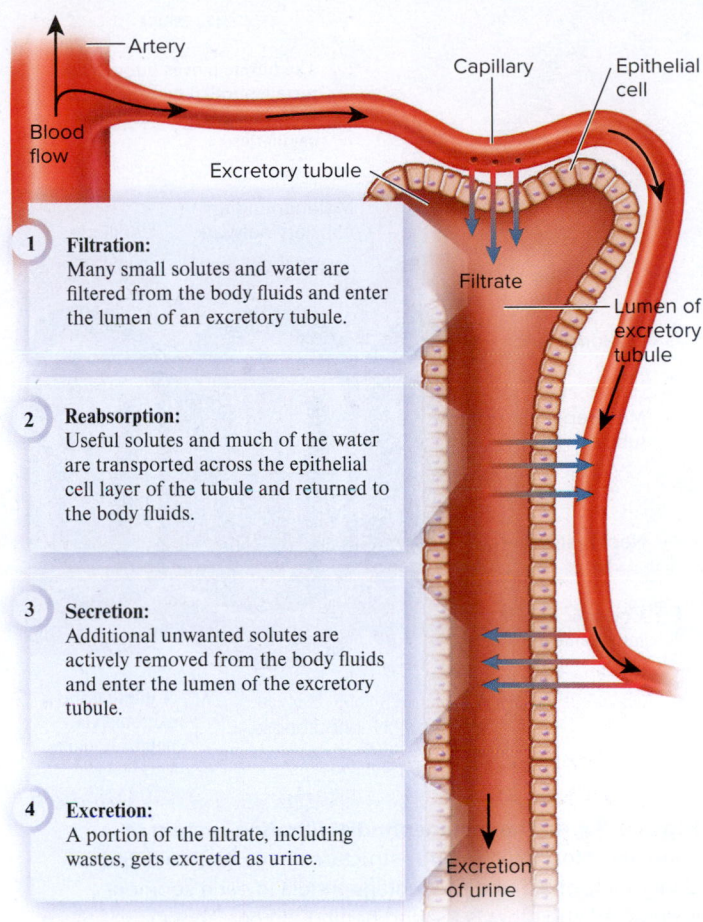

**Figure 49.2** **Basic features of the function of many excretory systems.**

*Concept Check:* *What is the benefit of secreting substances into the tubules?*

cells and large solutes such as proteins. A typical filtration system is found in the mammalian kidney, in which a portion of the plasma component of the blood is forced under pressure through leaky capillaries and into the kidney tubules. The material that passes through the filter and enters the excretory organ for further processing or excretion is called a **filtrate**.

2. **Reabsorption**. In reabsorption, some of the material in the filtrate is recaptured and returned to the blood. This is an important feature of many excretory organs, because the formation of a filtrate is not selective, apart from the exclusion of proteins and blood cells. In other words, in order to filter the blood and remove soluble wastes, important solutes such as ions, sugars, and amino acids also get filtered in the process. Reabsorbing these useful solutes requires active transport pumps or other transport systems. Much of the filtered water also gets reabsorbed.

3. **Secretion**. In some cases, solutes may be excreted from the body in quantities greater than those found in the initial filtrate. How is this possible? Some solutes are actively transported from the interstitial fluid surrounding the epithelial cells of the tubules into their lumens. This process, called secretion, supplements the

amount of a solute that can normally be removed by filtration alone. Excretory organs often eliminate particularly toxic compounds from an animal's body in this way. Secretion is a very effective means of removing wastes from body fluids. Some marine fishes, for example, use secretion as the sole means of removing soluble wastes from the blood. These animals do not form a filtrate at all.

4. **Excretion**. The process of expelling soluble waste or harmful materials from the body is called excretion. In animals that form a filtrate, the part of the filtrate that remains after reabsorption has been completed and that gets excreted is called **urine**.

## Many Invertebrates Remove Soluble Wastes from Their Body Fluids by Filtration

Invertebrates use a variety of mechanisms to filter their body fluids. The simplest filtration mechanism in invertebrates is the protonephridia system of flatworms (**Figure 49.3**). **Protonephridia** (singular, protonephridium) are a series of branching tubules that filter fluids from the body cavity by means of ciliated cells that cap the ends of the tubule branches. The beating of the cilia bears some resemblance to a flickering flame, which is why these cells are known as flame cells. As fluid is drawn through slitlike openings of the flame cells and into the lumen, it percolates through the tubule, where most solutes are reabsorbed back into the extracellular fluid. Excess water that

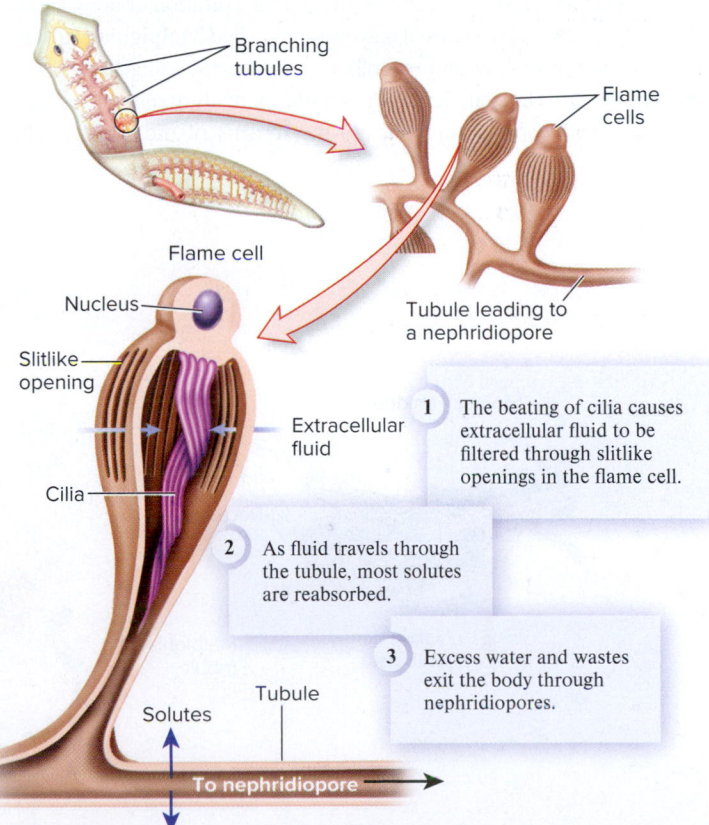

**Figure 49.3** **The protonephridial filtration system of flatworms.** As the filtrate moves along the tubules, most solutes are reabsorbed. The final excreted fluid is usually hypo-osmotic relative to body fluids, a typical adaptation found in freshwater animals that helps eliminate excess water.

entered the flatworm's body from the environment and some wastes travel through the tubules and exit the body through tiny openings in the body wall called nephridiopores. Much of the nitrogenous waste in flatworms actually diffuses across the body surface into the surrounding water. Therefore, the protonephridia are primarily osmoregulatory organs. The urine of flatworms is hypo-osmotic (more dilute) relative to the body fluids, an adaptation for life in fresh water.

Annelids, such as earthworms, use a different filtration mechanism, called a metanephridial system (**Figure 49.4**). Pairs of **metanephridia** (singular, metanephridium) are located in each body segment and consist of a tubular network that begins with a funnel-like structure called a nephrostome. The nephrostomes collect coelomic fluid, which contains nitrogenous wastes, through tiny pores that exclude large solutes. Na+, Cl−, and other useful solutes are reabsorbed by active transport along the length of the tubules that extend from the nephrostomes, and then diffuse into nearby capillaries. The nitrogenous wastes remain behind in the tubules and are excreted through nephridiopores in the body wall. Many annelids live in watery environments and like flatworms, excrete hypo-osmotic urine.

## Insects Remove Wastes from Their Body Fluids by Secretion

The insect excretory system is quite different from that of other invertebrates, because it involves secretion rather than filtration of body fluids. In insects, an extensive series of narrow tubes called **Malpighian tubules** arises from the midgut and extends into the surrounding hemolymph (**Figure 49.5**). The cells lining the tubules actively transport ions and uric acid from the hemolymph into the lumens of the tubules. This

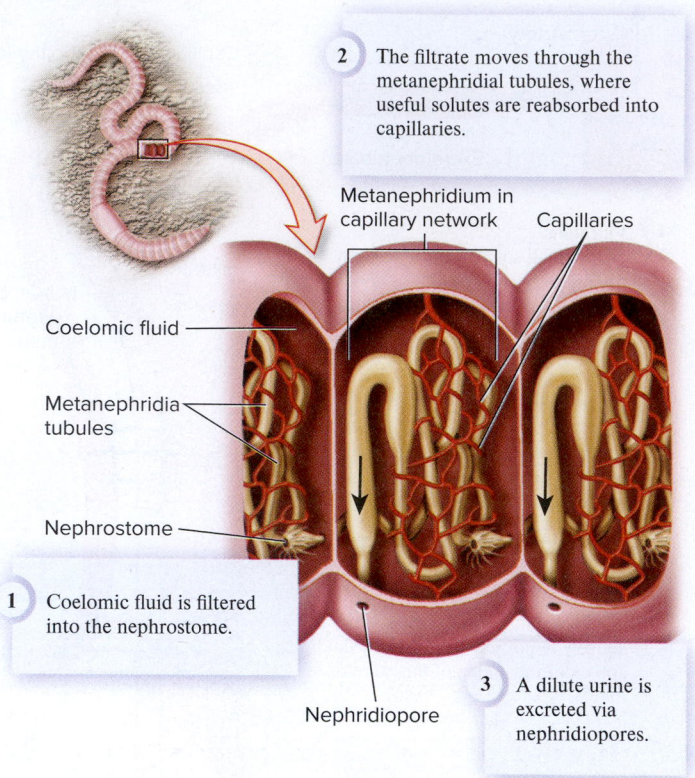

**2** The filtrate moves through the metanephridial tubules, where useful solutes are reabsorbed into capillaries.

Metanephridium in capillary network — Capillaries

Coelomic fluid

Metanephridia tubules

Nephrostome

**1** Coelomic fluid is filtered into the nephrostome.

Nephridiopore

**3** A dilute urine is excreted via nephridiopores.

**Figure 49.4  The metanephridial filtration system of annelids.** Most internal body structures have been omitted for clarity. Only one of the two metanephridia in each segment is depicted; the other is located behind the one shown.

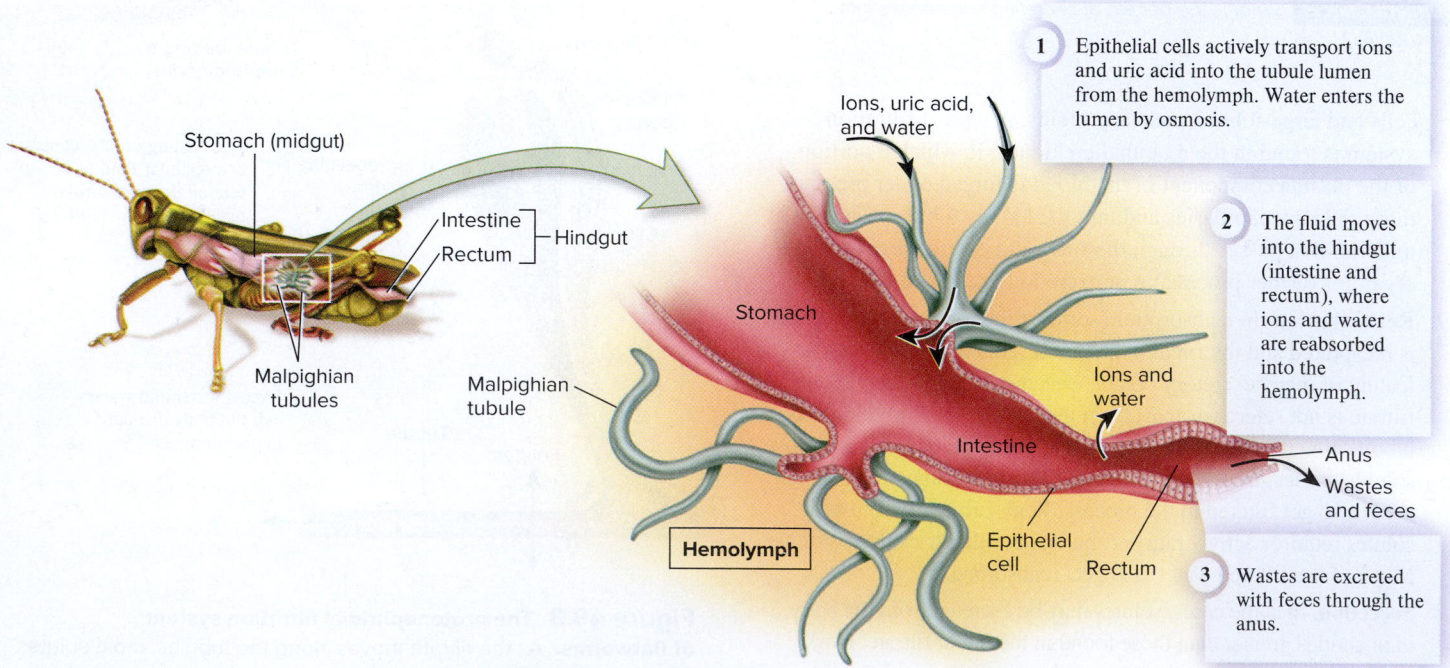

Stomach (midgut)

Intestine
Rectum — Hindgut

Malpighian tubules

Malpighian tubule

Ions, uric acid, and water

**1** Epithelial cells actively transport ions and uric acid into the tubule lumen from the hemolymph. Water enters the lumen by osmosis.

**2** The fluid moves into the hindgut (intestine and rectum), where ions and water are reabsorbed into the hemolymph.

Stomach

Ions and water

Intestine

Anus

Wastes and feces

Hemolymph

Epithelial cell

Rectum

**3** Wastes are excreted with feces through the anus.

**Figure 49.5  Malpighian tubules form the secretory system of an insect.** The tubules, which are longer and more convoluted than shown in this simplified illustration, extend into the body cavity, where they are surrounded by hemolymph.

secretion process creates an osmotic gradient that draws water into the tubules. The fluid moves from the tubules into the hindgut—the intestine and rectum—where much of the useful solutes and water is reabsorbed. The nitrogenous wastes, any excess ions, and other waste compounds are excreted together with the feces through the anus.

Unlike other invertebrates, most terrestrial insects, apart from blood-sucking ones, excrete urine that is either iso-osmotic (has the same total solute concentration) or hyperosmotic (is more concentrated) relative to body fluids. This difference in urine concentration reflects the general principle that life in dry environments is associated with a risk of dehydration and a need to conserve water.

## Vertebrates Remove Wastes from Their Body Fluids by Filtration and Secretion

The major excretory organ found in all vertebrates is the **kidney**. The kidneys of all vertebrates have many features in common. They typically contain specialized tubules composed of epithelial cells that participate in the excretion of nitrogenous wastes and other solutes. In addition, kidneys have a critical role in both ion and water homeostasis by promoting active transport of $Na^+$, $K^+$, and other ions across their membranes. In response to an animal's changing requirements, these processes can be controlled; that is, they can be stimulated or inhibited, usually by the actions of nerves and hormones. Most vertebrate kidneys are filtration kidneys, with the exception of purely secretory kidneys found in some marine fishes that need to minimize water loss (refer back to Figure 41.17). Filtration in the kidneys is controlled by mechanical forces, such as the hydrostatic pressure exerted by blood entering the capillaries of the kidneys. The mammalian kidney has been especially well studied and is examined in detail in the next section.

## 49.2 Structure and Function of the Mammalian Kidney

### Learning Outcomes:

1. Name the primary components of the urinary system in mammals, and describe the major anatomical features of the human kidney.
2. Describe how the structural features of different parts of a nephron relate to specific functions in the formation of urine.
3. Explain how the actions of two hormones, aldosterone and antidiuretic hormone (ADH), mediate the final composition of urine.
4. **CoreSKILL »** Predict the effects of changes in the blood concentrations of aldosterone and antidiuretic hormone on the osmolarity of blood.

In mammals, the two kidneys lie in the abdominal cavity (**Figure 49.6a**). The urine formed in each kidney collects in a central area called the renal pelvis. (The adjective renal means pertaining to the kidneys.) From there, urine flows through tubes called ureters into the urinary bladder, where it is temporarily stored. Urine is excreted via the urethra. Collectively, the kidneys, ureters, urinary bladder, and urethra constitute a part of the excretory system called the urinary system in mammals.

Each kidney has an outer portion called the renal cortex and an inner portion called the renal medulla (**Figure 49.6b**). The cortex is the primary site of blood filtration. In the medulla, the filtrate becomes concentrated by the reabsorption of water back into the blood. In this section, we will examine the structural features of kidneys that allow them to function as a filtration system and to regulate water and ion homeostasis.

## The Functional Units of the Kidney Are Called Nephrons

Depending on its size, a mammalian kidney contains as many as several million **nephrons**, which are the structural and functional units of a kidney and only a single cell thick. Each nephron is composed of an initial portion that filters the blood and a tubular region that modifies the filtrate.

As shown in **Figure 49.6c**, the filtering process begins at a region of the nephron called the **renal corpuscle**. The renal corpuscle forms a filtrate from blood that is free of cells and large proteins. This filtrate then leaves the renal corpuscle and passes through the remainder of the nephron called the **renal tubule**. Each renal tubule has three major segments with specialized functions, as shown in Figure 49.6c. As the filtrate flows through the renal tubule, it is modified by substances being reabsorbed from it or secreted into it. Ultimately the filtrate remaining at the end of each renal tubule enters a collecting duct, which brings it to the pelvis. When it reaches the renal pelvis, the filtrate is called urine. From the renal pelvis, urine flows into the urinary bladder. In many vertebrates, the urinary bladder plays a role in water balance by regulating how much water is reabsorbed from the urine before it is excreted. In mammals, however, the bladder functions only as a storage site. Let's look more closely at each part of the nephron and its associated structures.

***The Renal Corpuscle*** Each renal corpuscle contains a cluster of capillaries called the **glomerulus** (plural, glomeruli) (**Figure 49.7a**; also refer back to the chapter opening photo). Each glomerulus is supplied with blood under pressure by an **afferent arteriole**, and blood exits the glomerulus via an **efferent arteriole** (afferent means toward, and efferent means away from). The glomerulus protrudes into a fluid-filled space called **Bowman's capsule**, after English physiologist William Bowman who first described it in 1841. The combination of a glomerulus and a Bowman's capsule constitutes a renal corpuscle. The glomerular capillaries contain fenestrations, tiny holes in their walls that permit flow of plasma out of the capillaries. Cells called podocytes encase the capillaries and form filtration slits that allow the passage of small solutes but are believed to play a role in helping prevent large proteins from entering the filtrate (**Figure 49.7b**).

***The Renal Tubule*** The renal tubule, which is continuous with Bowman's capsule, is made of a single layer of epithelial cells joined by tight junctions and resting on a basement membrane. The basolateral surfaces of the epithelial cells have numerous $Na^+/K^+$-ATPase pumps in their membranes that, as we will see, function in reabsorption. The epithelial cells differ in structure and function along the tubule's length. Three major segments of the renal tubule are recognized (see Figure 49.6c):

- The **proximal tubule** is the segment of the renal tubule that drains Bowman's capsule.
- The **loop of Henle** arises from the proximal tubule; it is a long, hairpin-shaped loop consisting of a descending limb that comes

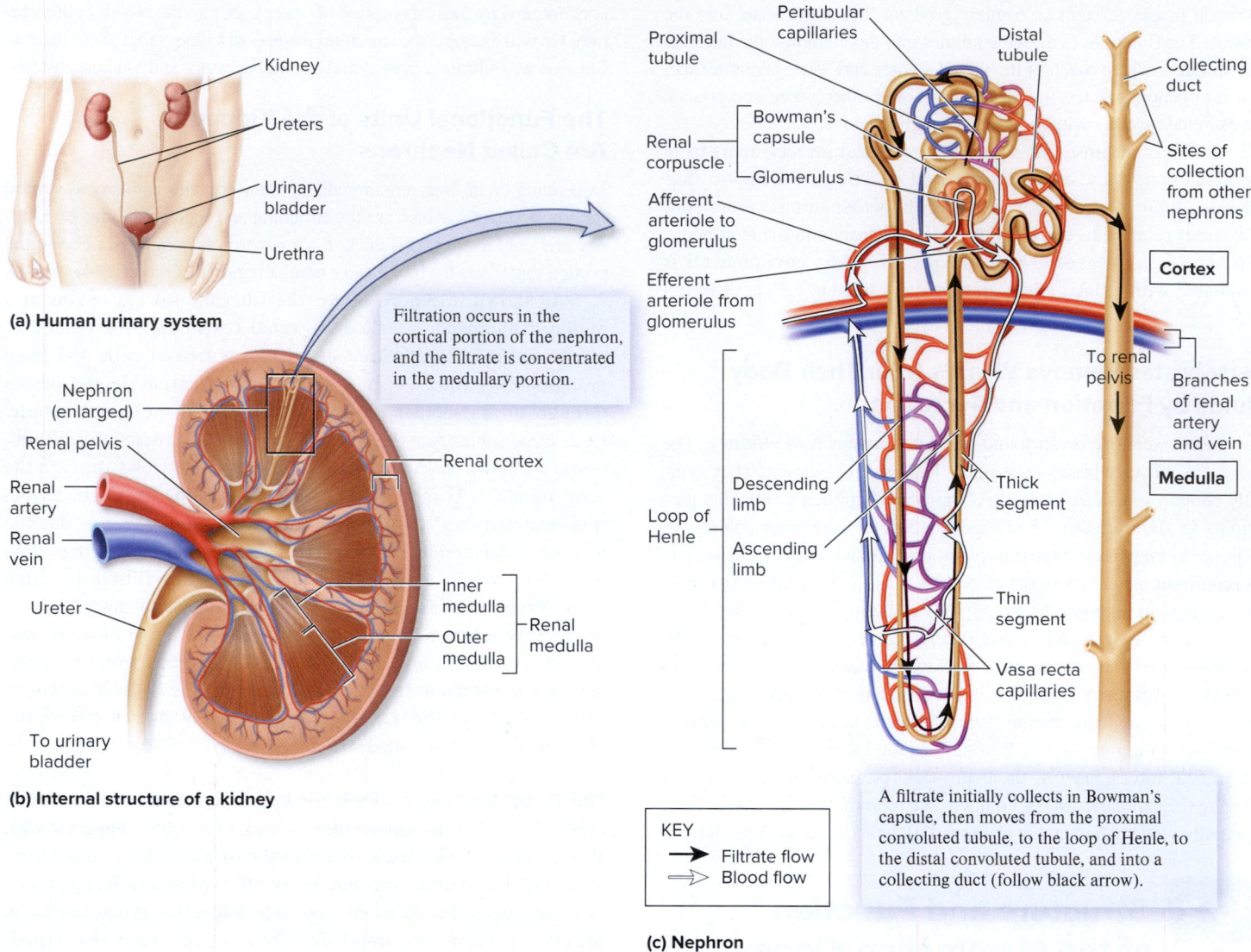

**(a) Human urinary system**

Filtration occurs in the cortical portion of the nephron, and the filtrate is concentrated in the medullary portion.

**(b) Internal structure of a kidney**

**KEY**
→ Filtrate flow
⇒ Blood flow

A filtrate initially collects in Bowman's capsule, then moves from the proximal convoluted tubule, to the loop of Henle, to the distal convoluted tubule, and into a collecting duct (follow black arrow).

**(c) Nephron**

**Figure 49.6** **The mammalian urinary system, including the basic functional unit of the nephron.** (a) The human urinary system in a female. In the male, the urethra passes through the penis. (b) View of a section through a kidney, showing the locations of the major internal structures and a single nephron (enlarged; nephrons are microscopic). (c) Structure of a nephron. The nephron begins at Bowman's capsule, continues along the renal tubule, and empties into a collecting duct. The major segments of the renal tubule are shown (proximal tubule, loop of Henle, and distal tubule). One segment, the loop of Henle, has additional segments identified here. Many nephrons empty into a given collecting duct. Surrounding the nephron are capillaries, called peritubular capillaries in the cortex and vasa recta capillaries in the medulla.

from the proximal tubule and an ascending limb. The ascending limb has two segments—a thick segment and a thin segment—with different properties.

- The **distal tubule** arises from the thick segment of the loop of Henle. The filtrate flows from the distal tubule into a collecting duct.

***Capillaries of the Nephron*** All along its length, each renal tubule is surrounded by capillaries. These include the peritubular capillaries in the cortex and the vasa recta capillaries in the medulla (see Figure 49.6c). Both sets of capillaries carry away reabsorbed solutes and water from the filtrate in the nephron or collecting ducts, and return them to the bloodstream.

## A Filtrate Is Produced by Hydrostatic Pressure in Bowman's Capsule

Filtration begins as blood flows through the glomerulus and a portion of the plasma leaves the glomerular capillaries and filters into Bowman's capsule. Only a small portion of the plasma is filtered from the blood as it circulates through the glomerulus. Most of the plasma, therefore, exits the glomerulus by an efferent arteriole (see Figure 49.7a). Proteins and blood cells are prevented from leaving the glomerular capillaries because of the small diameter of the fenestrations and the presumed action of the filtration slits, mentioned earlier.

The rate at which the filtrate is formed in the glomeruli is called the **glomerular filtration rate (GFR)**. The GFR can be increased by dilation (widening) of the afferent arterioles. When an afferent

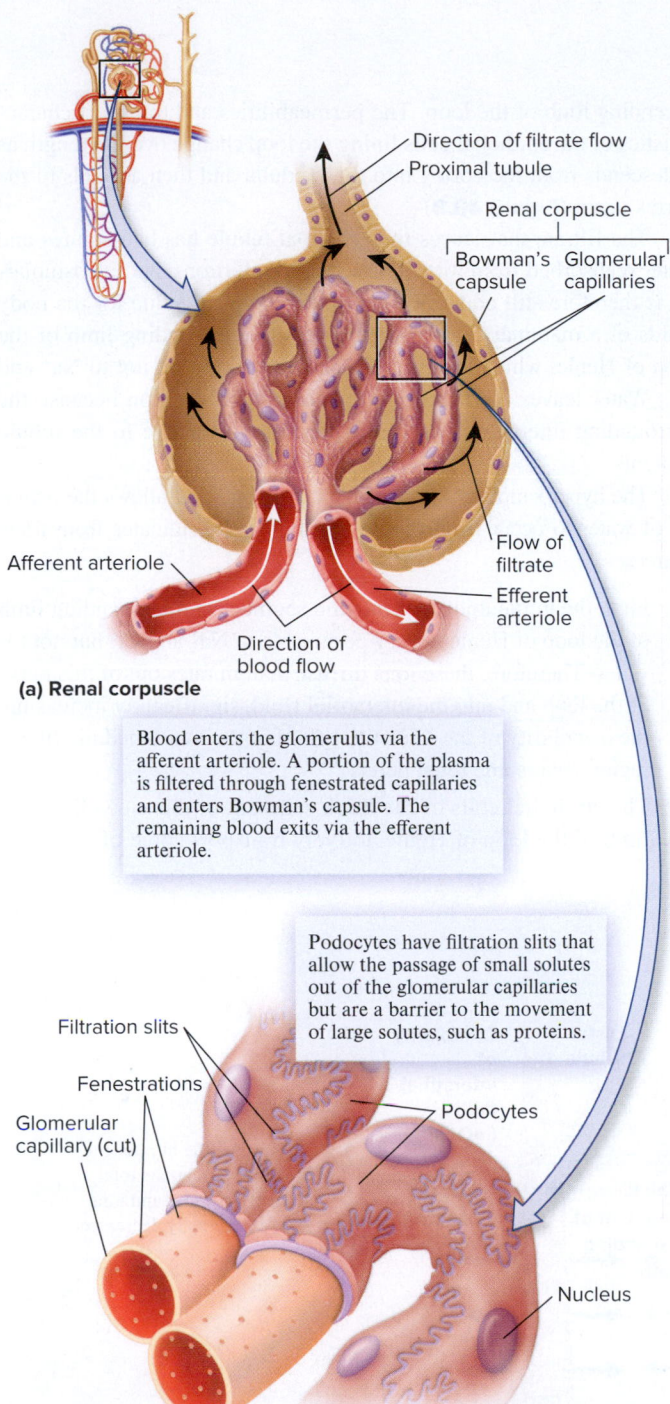

Direction of filtrate flow
Proximal tubule
Renal corpuscle
Bowman's capsule
Glomerular capillaries
Flow of filtrate
Efferent arteriole
Direction of blood flow
Afferent arteriole

**(a) Renal corpuscle**

Blood enters the glomerulus via the afferent arteriole. A portion of the plasma is filtered through fenestrated capillaries and enters Bowman's capsule. The remaining blood exits via the efferent arteriole.

Podocytes have filtration slits that allow the passage of small solutes out of the glomerular capillaries but are a barrier to the movement of large solutes, such as proteins.

Filtration slits
Fenestrations
Glomerular capillary (cut)
Podocytes
Nucleus

**(b) Glomerular capillaries with podocytes**

**Figure 49.7** **The structure and function of the renal corpuscle.** **(a)** A renal corpuscle comprises Bowman's capsule and the glomerular capillaries that make up the glomerulus. It is here that the filtrate is first formed. **(b)** The glomerular capillaries are completely encased in podocytes, specialized epithelial cells that support the glomerulus and are believed to act in part as a filtration barrier.

 **Core Concept: Structure and Function** The structure of the capillaries in the renal glomerulus is suited to their function. Fenestrations permit the passage of plasma, but not blood cells, out of the capillaries. The structure of the podocytes is also suited to their function, in that filtration slits prevent the passage of plasma proteins into the filtrate. Damage to the structure of the podocytes might cause proteins to leak into the filtrate and be lost in the urine.

arteriole dilates, more blood enters the glomerulus, increasing the hydrostatic pressure in its capillaries and forcing more plasma through the fenestrations in the capillaries and into Bowman's capsule. This might happen, for example, when nerves or hormones signal that there is excess water in the body and that more water must be excreted in the urine.

By contrast, constriction (narrowing) of the afferent arterioles will decrease the amount of blood entering the glomerular capillaries and therefore decrease the GFR. This might occur in response to dehydration or a loss of blood due to a severe injury. In such situations, decreasing the GFR results in less urine production, which, in turn, minimizes how much water is lost from the body. The tradeoff, however, is that the blood is less effectively cleared of wastes.

## Useful Solutes Are Reabsorbed from the Filtrate in the Proximal Tubule

The filtrate flows from the renal corpuscle to the proximal tubule. Unwanted substances remain in the filtrate. By contrast, anywhere from two-thirds to all of a particular useful solute is reabsorbed from the filtrate in the proximal tubule. These solutes include $Ca^{2+}$, $Na^+$, $K^+$, $Cl^-$, $HCO_3^-$ (bicarbonate ion), and organic molecules such as glucose, vitamins, and amino acids. Some ions diffuse through channels in the membranes of the epithelial cells that form the proximal tubule. Others are actively transported across the tubule epithelium. Organic molecules generally are reabsorbed by being coupled to the transport of ions such as $Na^+$. The reabsorption of solutes and water is enhanced by microvilli that extend into the lumen from the apical membrane of the epithelial cells of the proximal tubule. This anatomical adaptation, called a **brush border**, creates an enormous surface area for the positioning of transporters and channels (**Figure 49.8**).

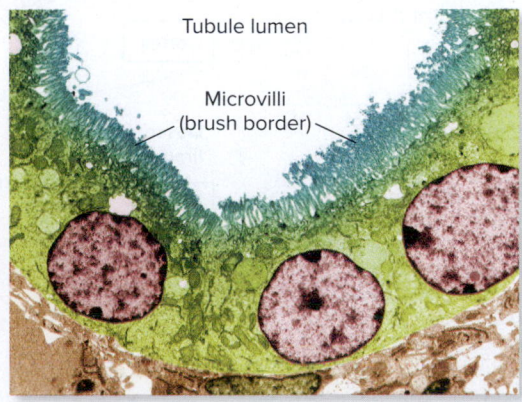

Tubule lumen
Microvilli (brush border)

**Figure 49.8** **Electron micrograph of cuboidal epithelial cells of the proximal tubule of a rat nephron.** Note the extensive brush border of microvilli on the apical membranes of the cells. ©Steve Gschmeissner/SPL/Science Source

 **Core Skill: Connections** In what other organ is a brush border found (refer back to Figure 46.6)? What general conclusions can you draw about the function of such specialized epithelia?

Most of the water in the filtrate is reabsorbed by osmosis as the ions and organic molecules are transported from the lumen of the proximal tubule to the interstitial fluid. From the interstitial fluid, the solutes and water diffuse into peritubular capillaries to return to the blood.

Some solutes that are not required by an animal or that are potentially toxic at high concentrations diffuse out of the peritubular capillaries and are actively transported across the renal tubule epithelium into the lumen of the proximal tubule. Examples of such substances include drugs (for example, penicillin), naturally occurring toxins, nucleoside metabolites, and ions such as $K^+$ and $H^+$ (if they are in excess). These solutes are excreted in the final urine.

By the time the filtrate leaves the proximal tubule, its volume and composition have changed considerably. The amount of solutes and water it contains is much reduced, and the useful organic molecules normally have all been removed and returned to the blood.

## Additional Water and Ions Are Reabsorbed from the Filtrate Along the Loop of Henle

In the loop of Henle, the filtrate flows down the descending limb of the loop, makes a U-turn at the bottom, and then moves back up the ascending limb of the loop. The permeabilities and transport characteristics of the epithelial cells lining the loop change over its length as it descends from the cortex into the medulla and then ascends to the cortex again (**Figure 49.9**).

The filtrate that leaves the proximal tubule has had solutes and water reabsorbed from it in about equal proportions, and its osmolarity is therefore still about 300 mOsm/L, a typical value for the body fluids of a mammal. This filtrate enters the descending limb of the loop of Henle, which is very permeable to water but not to $Na^+$ and $Cl^-$. Water leaves the filtrate by osmosis in this region because the surrounding interstitial fluid is hyperosmotic relative to the tubule contents.

The hyperosmolarity of the interstitial fluid that allows the osmosis of water to occur from the descending limb originates from three sources:

1. First, the initial upturn of the thin segment of the ascending limb of the loop of Henle is very permeable to $Na^+$ and $Cl^-$ but not to water. Therefore, these ions diffuse at high rates out of this part of the loop and into the interstitial fluid, significantly increasing the osmolarity of the interstitial fluid of the inner medulla (the region nearest the renal pelvis).

2. The epithelial cells of the thick segment of the ascending limb of the loop of Henle actively transport some of the

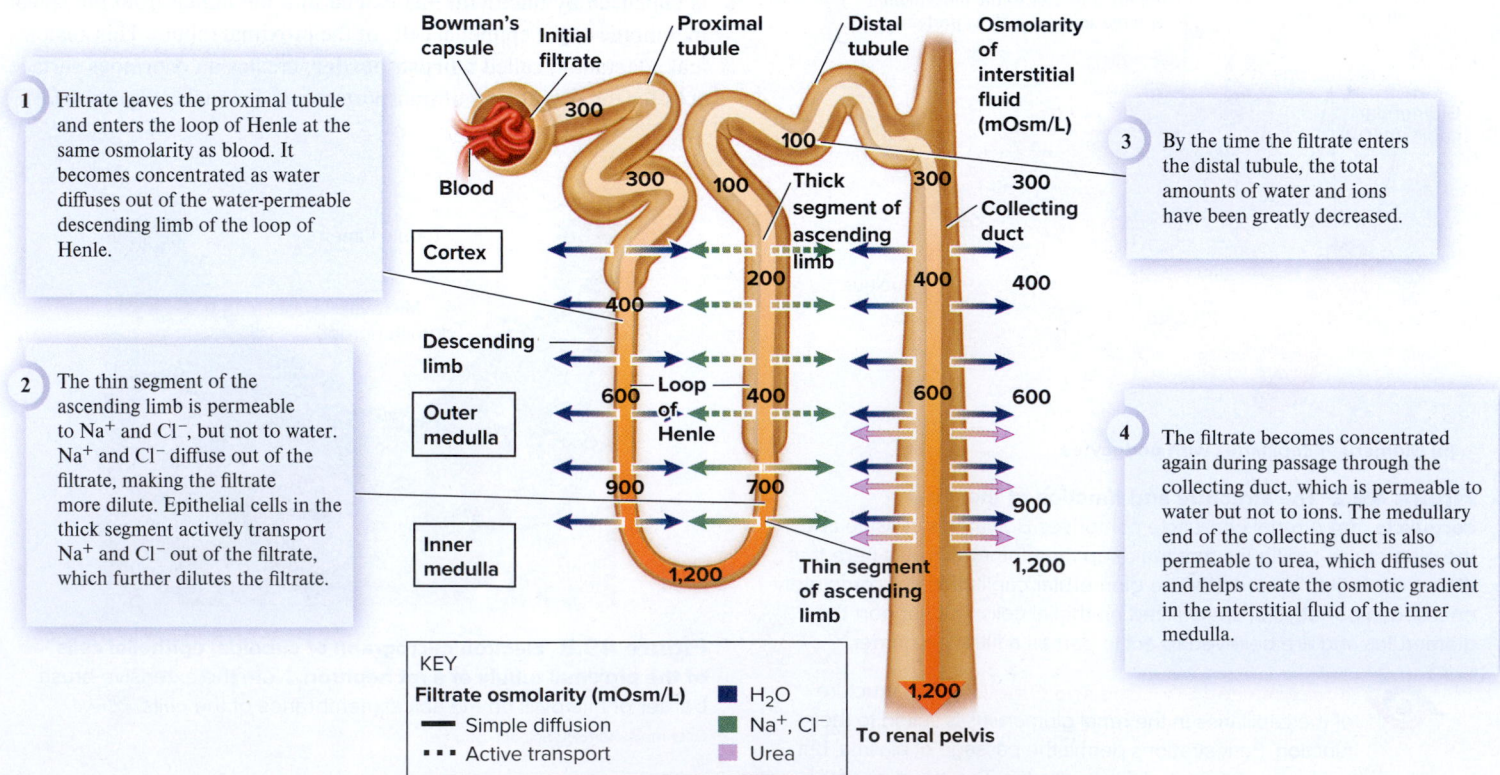

**1** Filtrate leaves the proximal tubule and enters the loop of Henle at the same osmolarity as blood. It becomes concentrated as water diffuses out of the water-permeable descending limb of the loop of Henle.

**2** The thin segment of the ascending limb is permeable to $Na^+$ and $Cl^-$, but not to water. $Na^+$ and $Cl^-$ diffuse out of the filtrate, making the filtrate more dilute. Epithelial cells in the thick segment actively transport $Na^+$ and $Cl^-$ out of the filtrate, which further dilutes the filtrate.

**3** By the time the filtrate enters the distal tubule, the total amounts of water and ions have been greatly decreased.

**4** The filtrate becomes concentrated again during passage through the collecting duct, which is permeable to water but not to ions. The medullary end of the collecting duct is also permeable to urea, which diffuses out and helps create the osmotic gradient in the interstitial fluid of the inner medulla.

**KEY**
**Filtrate osmolarity (mOsm/L)**
— Simple diffusion
••• Active transport
■ $H_2O$
■ $Na^+$, $Cl^-$
■ Urea

**Figure 49.9 Tubule permeabilities and osmolarities of the filtrate in the nephron and collecting duct.**

remaining Na⁺ and Cl⁻ out of the filtrate and into the interstitial fluid of the outer medulla (the region near the cortex).

3. Some of the urea that is present in the filtrate does not get excreted in the urine, but instead diffuses out of the lower ends of the collecting ducts and into the interstitial fluid of the inner medulla, contributing to the total solute concentration and, therefore, osmolarity.

Collectively, these solutes create the osmotic force that draws water out of the filtrate in the descending limb. The water then enters local capillaries and rejoins the blood circulation.

As water diffuses out of the filtrate in the descending limb of the loop of Henle, the filtrate becomes more and more concentrated, and its osmolarity increases from 300 mOsm/L to about 1,200 mOsm/L (or even higher in some species). During its passage up the ascending limb, however, the osmolarity of the filtrate decreases to about 200 mOsm/L as ions diffuse or are transported out of the tubule into the interstitial fluid, as just mentioned. As noted earlier, the thin segment of the ascending limb is permeable to Na⁺ and Cl⁻, but not water, so Na⁺ and Cl⁻ diffuse out of the filtrate in this segment; the exit of these ions begins to dilute the filtrate. The thick segment is also impermeable to water, but in this segment, epithelial cells actively transport Na⁺ and Cl⁻ out of the filtrate, and the filtrate becomes increasingly more dilute.

As a consequence of ion movement out of the ascending limb, the osmolarity of the kidney interstitial fluid increases in a gradient from the renal cortex to the inner medulla. This extracellular osmolarity gradient is what allows water to diffuse by osmosis from the descending limb of the loop of Henle all along its length. In other words, this is an example of a countercurrent exchange system, like those that operate in heat and gas exchange in some animals (refer back to Figures 47.13 and 48.17). A major difference between those countercurrent exchange systems and the one in the loop of Henle, however, is that the latter requires energy-dependent ion pumps to maintain the necessary concentration gradient. Because energy is used to increase—or multiply— the gradient, the loop of Henle system is also referred to as a **countercurrent multiplication system**.

The chief advantage provided to an animal by the loop of Henle is that the final volume of urine produced has been reduced by the reabsorption of water along the osmotic gradient. This reabsorption of water is especially important in animals in which total body water stores are regularly in danger of being depleted. For example, desert mammals such as the kangaroo rat tend to have longer loops of Henle than other mammals. The extra length of the loop provides for a very large osmotic gradient in the medulla and, therefore, more water-reabsorbing capacity. At the other extreme, animals that generally must eliminate excess water have shorter or absent loops of Henle. Freshwater fishes, for example, do not have a loop of Henle in their nephrons. In their case, it is advantageous to excrete as much water as possible to compensate for the large amounts of water that constantly enter the blood by osmosis across their gills when they breathe.

## The Concentrations of Ions Are Fine-Tuned in the Distal Tubule and Cortical Collecting Duct

By the time the filtrate reaches the distal tubule and cortical collecting duct (the part of the collecting duct that extends up into the kidney cortex), most of the reabsorbed ions and other solutes, along with much of the water, have already been restored to the blood, and the filtrate has been diluted to an osmolarity of about 100 mOsm/L (see Figure 49.9). However, the remaining water and the concentrations of certain ions in the filtrate, notably Na⁺ and K⁺, can still be fine-tuned to precisely match an animal's requirements for water and ion homeostasis. This process is mediated by the actions of two hormones called aldosterone and antidiuretic hormone (ADH).

***Aldosterone: Na⁺ Reabsorption and K⁺ Secretion*** Aldosterone is a steroid hormone produced by the adrenal glands, two small glands that sit atop each kidney. It acts on epithelial cells of the distal tubule and cortical collecting duct, stimulating the active transport of three Na⁺ out of the filtrate into the extracellular fluid (reabsorption) for every two K⁺ it pumps into cells from the interstitial fluid and ultimately into the filtrate (secretion) (**Figure 49.10**). This 3:2 proportion of ions exiting versus entering creates an osmotic gradient. Water from the filtrate follows the Na⁺ by osmosis into the extracellular fluid and blood. This loss of water makes the remaining filtrate that moves on to the lower, medullary part of the collecting duct a bit more concentrated than it was before, with an osmolarity of about 400 mOsm/L. Aldosterone concentrations increase in the blood whenever the Na⁺ concentration of the blood is lower than normal or the K⁺ concentration is higher than normal. Such imbalances might occur, for example, due to dietary changes. Through its actions on the nephron, aldosterone corrects such imbalances, while also increasing the amount of water reabsorbed from the filtrate.

***ADH: Water Reabsorption*** The lower part of the collecting duct, in the inner medulla, is the final place where urine composition can be altered in mammals. The osmolarity of the filtrate increases during passage along the collecting duct, which is permeable to water but not to ions, allowing water to diffuse out of the duct by osmosis. As previously described, the cells of the collecting duct in the inner medulla are permeable to urea, which enters the interstitial fluid and contributes to the osmotic gradient there.

In addition, however, the permeability of the epithelial cells of the collecting ducts to water can be regulated independently of ion permeability, depending on an animal's requirement at any given moment for retaining or excreting water. This happens under the influence of the polypeptide hormone **antidiuretic hormone (ADH)**, also known as vasopressin. When ADH is present in the blood, it acts to increase the number of water channels called aquaporins (from the Latin, meaning water pores; refer back to Chapter 5) in the apical

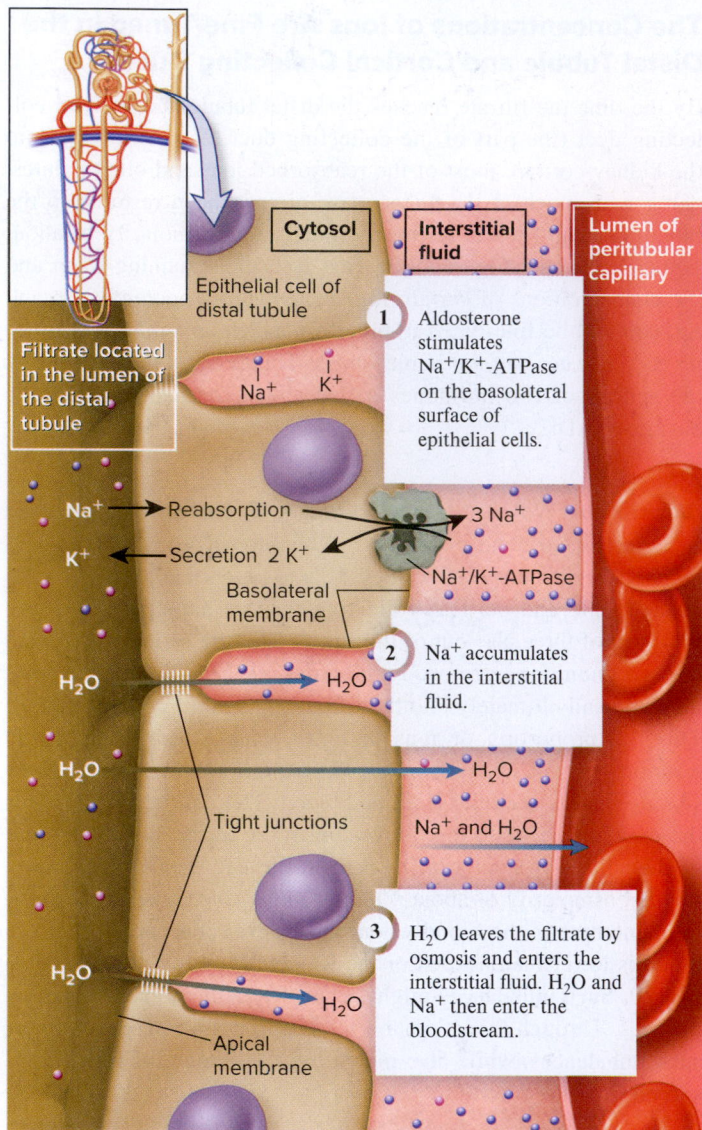

**Figure 49.10** **Effect of aldosterone on ion and water movement.** Aldosterone stimulates Na$^+$/K$^+$-ATPase pumps on the surfaces of epithelial cells of the distal tubule and cortical collecting duct. These pumps are located only on the side of the cells facing the interstitial fluid (the basolateral surface). This activity creates an osmotic gradient, as three Na$^+$ are reabsorbed from the filtrate for every two K$^+$ secreted into it. Water then leaves the tubule by osmosis. The pump also creates a diffusion gradient for Na$^+$ entry and K$^+$ exit on the apical side of the cell. The net effect is reabsorption of Na$^+$ and water, and secretion of K$^+$, which gets excreted in the urine. Na$^+$ and water enter the peritubular capillaries and are carried away in the bloodstream.

 **Core Skill: Connections** What special property of epithelial cells is demonstrated by their multiple responses to aldosterone? Look back at Figure 41.4 for help.

membranes of the collecting duct cells (**Figure 49.11**). It does this by stimulating intracellular signaling mechanisms that promote the fusion of cytosolic vesicles containing aquaporins with the apical membranes of the duct epithelial cells, resulting in the insertion of aquaporins into those membranes.

Because the collecting ducts are located within the hyperosmotic medulla of the kidney, an osmotic gradient draws water from the filtrate in the duct into the duct epithelial cells. The water exits the cells on the other side (the basolateral surface), where another set of aquaporins is present. These latter aquaporins are always present in the basolateral membrane; unlike those on the apical side of the cells, they do not require the presence of ADH to be inserted into the membrane. Once the water moves from the cells into the interstitial fluid, it enters the vasa recta capillaries and the blood. In this way, as the filtrate travels through the collecting ducts, it becomes greatly concentrated (hyperosmotic) compared to blood—as much as four to five times more concentrated in humans, for example.

ADH concentrations increase in the blood when a mammal needs to conserve water, such as during periods of dehydration. The ability of the kidneys to produce hyperosmotic urine is a major determinant of an animal's ability to survive in conditions where water availability is limited. By contrast, when there is excess water in the body, the concentration of ADH in the blood decreases. This decrease results in endocytosis of portions of the apical membranes, along with their aquaporins. This, in turn, decreases the water permeability of the collecting ducts. In such a case, urine increases in volume and becomes more dilute because less water is reabsorbed from the collecting ducts.

BIO **TIPS**

**THE QUESTION** *The proximal tubule is the only region of the vertebrate nephron that contains a brush border. Can you propose an explanation for why this is the case?*

**T OPIC** *What topic in biology does this question address?*
The topic is the purpose of a brush border, specifically one found in the proximal tubules.

**I NFORMATION** *What information do you know based on the question and your understanding of the topic?*
From the question, you know that the proximal tubule contains a brush border but the rest of the nephron does not. From your understanding of the topic, you may remember that a brush border consists of epithelial cells with microvilli extending from their apical membranes into a lumen.

**P ROBLEM-SOLVING S TRATEGY** *Relate structure to function.* Recall the functions that are associated with the structural adaptation provided by a brush border, and then consider the different functions that are carried out along a nephron.

**ANSWER** *Brush borders increase the surface area available for many transport processes, such as the reabsorption of solutes from the lumen of a proximal tubule into the blood. More than two-thirds of most useful solutes are reabsorbed in the proximal tubule of the nephron. The structure of the epithelial cells of the proximal tubule, therefore, reflects their major contribution to reabsorption. Because comparatively little reabsorption of solutes occurs along the distal tubule, there would have been little selective advantage to evolve a brush border there.*

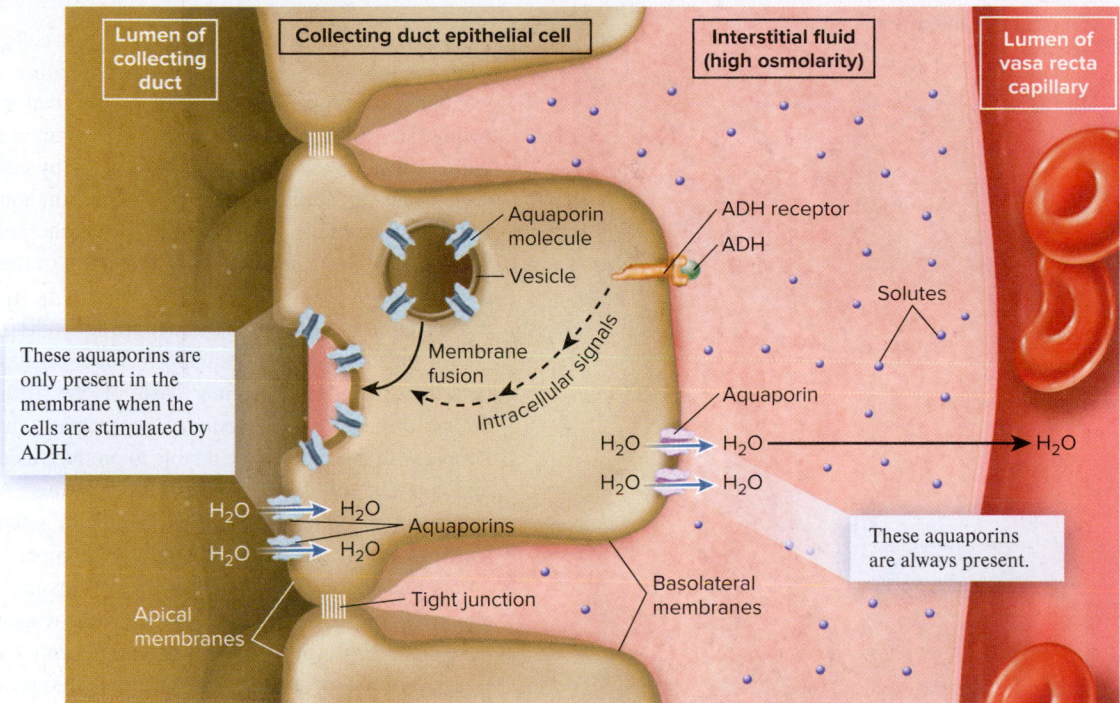

**Figure 49.11** **The effect of antidiuretic hormone (ADH) on water reabsorption in the collecting ducts of the kidney.** Water molecules diffuse more rapidly across a membrane when the membrane contains aquaporins. The epithelial cells of collecting ducts have two sets of aquaporins. One set is always present on the basolateral side of the cells; the other set inserts into the apical membrane only in the presence of ADH. ADH activates cell-signaling mechanisms that stimulate intracellular storage vesicles containing aquaporins to fuse with the apical membrane. Water moves by osmosis across the cell and into the interstitial fluid, and from there enters the vasa recta capillaries to be transported into the circulation. When ADH concentrations in the blood decrease (for example, when an animal is fully hydrated), the aquaporins on the apical membrane are returned to the intracellular storage vesicles by the process of endocytosis.

 **Core Skill: Modeling**  The goal of this modeling challenge is to draw a model of a cell that depicts the location of aquaporins in the absence of ADH.

**Modeling Challenge:** Figure 49.11 shows a model of an epithelial cell that is being stimulated by ADH. Draw a new model of this cell that depicts the location of aquaporins in the absence of ADH. How would this change in the location of the aquaporins affect the movement of water?

 **Core Concept: Evolution**

## Aquaporins in Animals Are Part of an Ancient Superfamily of Channel Proteins

Biological membranes such as the plasma membranes of animal cells are composed of a lipid bilayer that inhibits the movement of water (refer back to Figure 5.1). The questions of how water moves rapidly across membranes like those of the collecting ducts and how ADH controls the amount of water reabsorption were not solved until the early 1990s. At that time, American physiologist Peter Agre and colleagues discovered the first of what would eventually be recognized as a new family of proteins, called aquaporins, a discovery that earned Agre the Nobel Prize in Chemistry in 2003 (see Figure 5.16). Scientists now know that aquaporins are a subfamily of an even larger, ancient family of

proteins that function as membrane channels and transporters and are found in all domains of living species.

Based on DNA sequencing, aquaporins have been classified into numerous evolutionary groups. For example, there are five major groups of aquaporins in plants, and four in mammals. Each major group typically contains more than one closely related aquaporin. For example, mammals have at least 13 individual aquaporins, several of which are expressed in the kidney. The differences between aquaporin-encoding genes led to proteins with slightly different functions. As one important example, some aquaporins can transport other small molecules beside water, such as glycerol and urea. In addition, the various aquaporins in an organism can be distinguished on the basis of their cell-specific expression and whether and how they are regulated.

Aquaporins are proteins with six transmembrane domains and two short loops in the membrane (**Figure 49.12**). In animals,

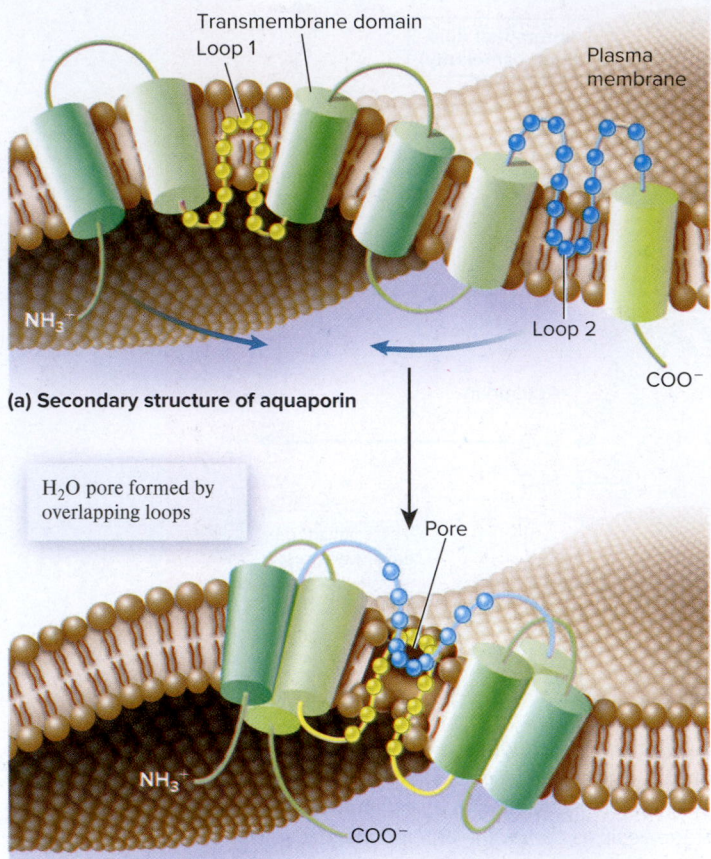

**(a) Secondary structure of aquaporin**

H₂O pore formed by overlapping loops

Pore

**(b) Three-dimensional (tertiary) structure of aquaporin**

**Figure 49.12 Schematic aquaporin structure.** All proteins of the aquaporin family share a similar structure, with six membrane-spanning domains (represented as cylinders) and two loops that come together to form a channel. **(a)** Secondary structure of aquaporin. This highly schematic representation highlights the two separate regions of the molecule that come together to form the water pore. **(b)** Model of the tertiary structure of aquaporin showing pore formation.

the two short loops appear to come together to form the three-dimensional core of the water channel. The importance of the loops is consistent with the observation that the amino acid sequences of these portions are very similar among widely divergent species. Water must pass through a zone of constriction, created by the loops, that reduces the channel opening to a pore that is about 30 picometers ($30 \times 10^{-12}$ m) wide, or just about the width of a water molecule. Scattered along the inner part of the channel are arginines, amino acids that are positively charged. The charged arginines participate in hydrogen bonding with water molecules, facilitating their single-file movement through each channel at rates that have been estimated to be up to billions of water molecules per second!

Within the various extracellular and intracellular domains of aquaporin proteins are sites that can be modified by enzymes, such as protein kinases. These sites suggest that the opening and closing of these channels may be gated by stimuli, like the

gating of ion channels in neurons and other cells. In addition, the promoter region of certain aquaporin-encoding genes contains a site that is recognized by transcriptional activator proteins that are responsive to the presence of cAMP, a common intracellular signaling molecule and one that is generated by cells stimulated by ADH. Thus, in addition to its rapid effect on aquaporin insertion into cell membranes (see Figure 49.11), another mechanism by which ADH promotes osmosis of water out of the renal collecting ducts, and thereby reduces urine volume, is by stimulating the transcription of one or more aquaporin-encoding genes.

An understanding of aquaporin function has allowed scientists to explain the molecular basis of one form of an inherited human disease called hereditary nephrogenic diabetes insipidus. People with this disease are unable to produce a concentrated urine and consequently lose large amounts of water. A mutation in an aquaporin-encoding gene (*AQP2*) results in a form of the protein showing any of several abnormalities: improper folding, impaired ability to enter the plasma membrane, or impaired ability to form a channel core. AQP2 is the aquaporin that is responsive to the presence of ADH. Because of this, water reabsorption from the kidneys is greatly impaired in individuals with mutations in the *AQP2* gene.

Agre's discovery may have widespread implications for other areas of biology and human health. For example, certain types of plant diseases are associated with abnormal aquaporin expression in roots. In addition, some scientists are investigating the possibility that drugs that inhibit bacterial aquaporins may someday be useful antibiotics. The use of drugs to inhibit one class of aquaporins has even been suggested as a possible antiperspirant!

## 49.3 Impact on Public Health

**Learning Outcomes:**

1. Describe several common health issues related to diseases of the kidneys.
2. **CoreSKILL »** Diagram the process of hemodialysis, and predict its benefits and limitations.

Diseases and disorders of the kidney are a major cause of illness in humans. According to statistics released by the Centers for Disease Control and Prevention, up to 20 million people suffer from kidney disease in the U.S., with approximately 15,000–20,000 individuals receiving kidney transplants each year. This section gives an overview of kidney diseases and disorders in humans and also discusses some of the available treatments for these conditions.

### Kidney Disease Disrupts Homeostasis

Many diseases affect the kidneys. Diabetes, bacterial infections, allergies, congenital defects, kidney stones (accumulation of mineral deposits in nephron tubules), tumors, and toxic chemicals are some possible sources of kidney damage or disease. A buildup of fluid pressure due to obstruction of the urethra or a ureter may damage one or both kidneys and increase the likelihood of a bacterial infection and eventual renal failure. The symptoms of renal failure include fatigue, weakness, swelling in the abdomen and other regions, changes in the amount and composition of urine, anemia, ion imbalances, nausea, and muscle disturbances, among others. These symptoms are similar regardless of the cause of the disease, and all stem from the condition known as **uremia**, the retention of urea and other waste products in the blood.

Assuming that a person with poorly functioning kidneys continues to ingest a normal diet containing the usual quantities of nutrients and ions, what problems arise? Potentially toxic waste products that would normally enter the nephron tubules in large amounts by filtration instead build up in the blood, because kidney damage significantly decreases the number of functioning nephrons. In addition, the excretion of $K^+$ is impaired because too few nephrons remain capable of normal tubular secretion of this ion. Increased $K^+$ in the blood is an extremely serious condition, because of the importance of stable extracellular concentrations of $K^+$ to the control of heart and neuron function.

The kidneys are still able to perform their homeostatic functions reasonably well as long as at least 20% or so of the nephrons are functioning normally. The remaining nephrons undergo alterations in function—filtration, reabsorption, and secretion—to partially compensate for the missing nephrons. For example, each remaining nephron increases its rate of $K^+$ secretion so that the total amount of $K^+$ excreted by the kidneys can be maintained at or near homeostatic levels. The kidneys' compensatory abilities are limited, however. If, for example, someone with severe renal disease were to eat a diet high in $K^+$, the remaining nephrons might not be able to secrete enough $K^+$ to prevent its concentration from increasing in the extracellular fluid.

## Kidney Disease May Be Treated with Hemodialysis or Transplantation

Diseased kidneys may eventually reach a point where they can no longer excrete and reabsorb water and ions at rates that maintain homeostasis, or excrete waste products as fast as they are produced. Adjusting a person's diet can help reduce the severity of these problems. For example, decreasing how much $K^+$ one consumes decreases the amount of $K^+$ that must be excreted. However, such alterations may not eliminate the problems. In that case, doctors must use various procedures to artificially perform the kidneys' excretory functions.

*Hemodialysis*  The most common of these procedures is **hemodialysis**. The general term dialysis means to separate substances in solution using a porous membrane. In hemodialysis, blood from one of the patient's arteries is redirected through a dialysis machine, which is called a dialyzer (**Figure 49.13**). Within the dialyzer, blood flows through cellophane tubing that is surrounded by a special dialysis fluid. The tubing is highly permeable to most solutes but relatively impermeable to protein and completely impermeable to blood cells. These characteristics are designed to be similar to those of the kidneys' glomerular capillaries. The dialysis fluid has ion concentrations similar to those in plasma but contains no urea or other substances that are to be completely removed from the plasma.

As blood flows through the tubing in the dialyzer, small solutes diffuse out into the dialysis fluid until an equilibrium is reached. If, for example, the patient's plasma $K^+$ concentration is above normal, $K^+$ diffuses out of the blood across the cellophane tubing and into

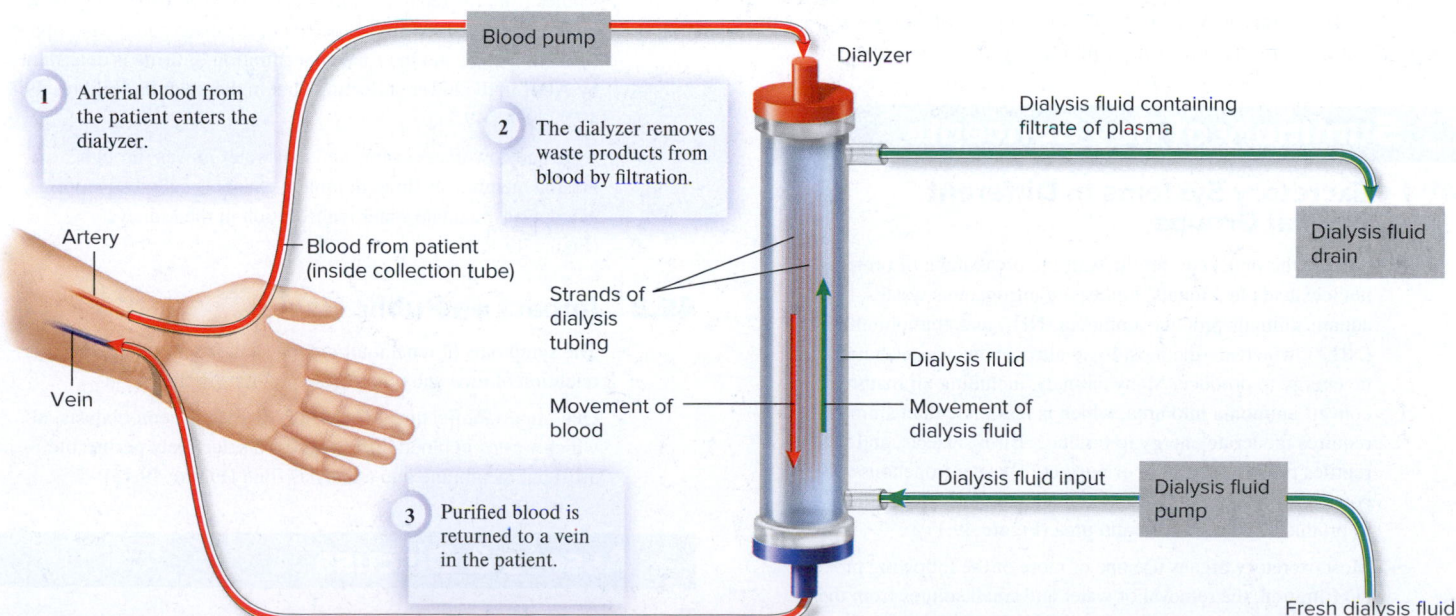

**Figure 49.13  Simplified diagram of hemodialysis.**  The dialyzer is composed of many strands of very thin, sievelike tubing. In the dialyzer, blood within the dialysis tubing and the dialysis fluid bathing the tubing move in opposite directions (in a countercurrent), which maximizes diffusion of substances out of the blood. The dialyzer provides a large surface area for diffusion of waste products out of the blood and into the dialysis fluid.

**Core Skills: Connections, Science and Society**  Note that the medical procedure of dialysis uses a feature found in many animals—countercurrent exchange. What other examples of countercurrent exchange have you learned about? (Refer back to Figures 47.13, 48.17, and 49.9 for help.)

the dialysis fluid. Similarly, waste products and excess amounts of other substances also diffuse into the dialysis fluid and thus are eliminated from the body. Note in Figure 49.13 that blood and dialysis fluid move in opposite directions through the dialyzer. This pattern of movement establishes an artificial countercurrent exchange system that increases the efficiency with which waste products are removed from the blood. The dialyzed blood is then returned to one of the patient's veins through another type of tubing that leaves the dialyzer.

Some patients with reversible, temporary forms of kidney disease may require hemodialysis for only days or weeks. However, patients with chronic, irreversible kidney disease require treatment for the rest of their lives, unless they receive a kidney transplant. Such patients undergo hemodialysis several times a week. Each year nearly 400,000 Americans undergo some type of dialysis.

*Kidney Transplantation*    The treatment of choice for most patients with permanent kidney disease is kidney transplantation. Rejection of the transplanted kidney by the recipient's body is a potential problem, but great strides have been made in decreasing the frequency of rejection. Many people who might benefit from a transplant, however, do not receive one, because the number of people needing a transplant far exceeds the number of donors. Currently, the major source of kidneys for transplanting is from recently deceased persons. Improved public understanding may lead many more individuals to give permission to have their kidneys and other organs used following their death. Recently, donation from a living, related donor has become more common, particularly with improved methods for preventing rejection. As noted earlier, the mammalian kidney can perform its functions with only a fraction of its nephrons intact. Due to this large safety margin, a person who donates one of his or her kidneys can function quite normally with only one kidney.

## Summary of Key Concepts

### 49.1    Excretory Systems in Different Animal Groups

- Among the important products of the breakdown of proteins and nucleic acids in animals' bodies are nitrogenous wastes. Most aquatic animals produce ammonia ($NH_3$) and ammonium ions ($NH_4^+$), which are the most toxic nitrogenous wastes but require no energy to produce. Many animals, including all mammals, convert ammonia into urea, which is less toxic than ammonia but requires moderate energy to produce. Birds, insects, and most reptiles produce uric acid or purines. These nitrogenous wastes conserve water and are less toxic but energetically costlier to produce than ammonia and urea (Figure 49.1).

- Most excretory organs use one or more of the following processes: (1) filtration, the removal of water and small solutes from the body fluids; (2) reabsorption, in which useful filtered solutes are returned to the body fluids via transport systems; (3) secretion, in which unnecessary or harmful solutes are actively transported from the extracellular fluid and into the excretory tubule; and (4) excretion, in which waste is passed out of the body (Figure 49.2).

- In the protonephridial system of flatworms, a series of branching tubules filters fluids from the body cavity into the tubule lumens via the actions of ciliated flame cells (Figure 49.3).

- In the metanephridial system of annelids, pairs of metanephridia located in each body segment filter interstitial fluid and dilute urine is excreted via nephridiopores in the body wall (Figure 49.4).

- In insects, cells lining the Malpighian tubules actively transport ions and uric acid from the hemolymph into the lumens of the tubules; water follows by osmosis. After useful ions and water are reabsorbed into the hemolymph, wastes are excreted from the body (Figure 49.5).

### 49.2    Structure and Function of the Mammalian Kidney

- The urinary system in humans consists of the kidneys, ureters, urinary bladder, and urethra. Urine is excreted through the urethra. Each kidney is composed of an outer renal cortex and an inner renal medulla (Figure 49.6).

- Nephrons, the functional units of the kidney, are composed of a filtering component, called the renal corpuscle, and a renal tubule that empties into a collecting duct. Each renal corpuscle contains a cluster of capillaries called the glomerulus within a structure called Bowman's capsule. Each glomerulus is supplied with blood under pressure by an afferent arteriole. Each tubule of a nephron is composed of a proximal tubule, a loop of Henle, and a distal tubule. Renal tubules are surrounded by peritubular capillaries in the cortex and by the vasa recta capillaries in the medulla (Figure 49.7).

- Different portions of the renal tubule have different permeabilities to solutes and water. Most reabsorption of useful solutes occurs in the proximal tubule. Water and ion reabsorption continues along the loop of Henle, using a countercurrent exchange system. The reabsorption of solutes by the vasa recta minimizes the loss of solutes from the renal medulla. Fine-tuning of urine composition by the hormone aldosterone occurs in the distal tubule and upper collecting duct, and the final concentration of urine is determined by ADH in the lower collecting duct in the medulla (Figures 49.8, 49.9, 49.10, 49.11).

- Agre and coworkers discovered that water moves through plasma membranes through protein channels called aquaporins. Aquaporins regulate water reabsorption in the kidneys (Figure 49.12).

### 49.3    Impact on Public Health

- The symptoms of renal malfunction stem from uremia, the retention of urea and other waste products in the blood.

- One important treatment for kidney disease is hemodialysis, in which wastes in blood diffuse across a selectively permeable artificial membrane into a dialysis fluid (Figure 49.13).

## Assess & Discuss

### Test Yourself

1. If a person were to drink an excess of water, what would happen as a result?
   a. The GFR would decrease.
   b. The concentration of ADH in the blood would decrease.
   c. The concentration of ADH in the blood would increase.
   d. GFR would increase.
   e. Both b and d would occur.

2. Nitrogenous wastes are the by-products of the metabolism of
   a. carbohydrates.          d. proteins.
   b. lipids.                 e. both c and d.
   c. nucleic acids.

3. Which of the following is *not* an example of a filtration mechanism for removing soluble wastes from the body fluids of animals?
   a. Malpighian tubules
   b. the vertebrate kidney
   c. the excretory system of annelids
   d. metanephridia

4. The active transport of solutes into the lumen of an excretory organ is called
   a. filtration.
   b. reabsorption.
   c. secretion.
   d. excretion.

5. The excretory organs found in insects are composed of
   a. protonephridia.          d. a filtration kidney.
   b. metanephridia.           e. a secretory kidney.
   c. Malpighian tubules.

6. In the mammalian kidney, filtration is mainly driven by the
   a. solute concentration in the tubular filtrate.
   b. solute concentration in the blood.
   c. osmolarity of the blood.
   d. osmolarity of the filtrate in the distal tubule.
   e. hydrostatic pressure in the blood vessels of the glomerulus.

7. In the mammalian urinary system, the urine formed in the kidneys is carried to the urinary bladder by the _____ and from the bladder to the outside of the body by the _____.
   a. collecting duct, renal pelvis
   b. nephron, collecting duct
   c. renal pelvis, urethra
   d. ureters, urethra
   e. urethra, ureters

8. Which of the following causes an increase in $Na^+$ reabsorption in the distal tubule and cortical collecting duct?
   a. an increase in aldosterone concentration
   b. an increase in antidiuretic hormone concentration
   c. a decrease in aldosterone concentration
   d. a decrease in antidiuretic hormone concentration
   e. Both b and c cause this increase in reabsorption.

9. Which would be expected to decrease the osmolarity of a mammal's body fluids?
   a. an increase in the plasma aldosterone concentration
   b. a decrease in the plasma ADH concentration
   c. dehydration
   d. loss of blood due to injury
   e. an increase in the plasma ADH concentration

10. Aquaporins are
    a. ion channels.
    b. water channels.
    c. receptors for ADH.
    d. small pores in the fenestrated capillaries of the glomerulus.
    e. part of a family of proteins encoded by genes found only in animals.

## Conceptual Questions

1. Define nitrogenous waste, and list the main types produced in animals. What are some advantages and disadvantages of excreting different types of nitrogenous wastes?

2. List and describe the three processes involved in urine production. Do each of these processes occur for every substance that enters an excretory organ such as the kidney?

3.  **Core Concept: Systems** Organ systems often interact and function together. Which organ systems participate in the elimination of wastes from animal bodies?

## Collaborative Questions

1. Describe as many mechanisms as you can by which soluble wastes are removed from the body fluids of animals.

2. Briefly discuss the parts and functions of the nephron in the mammalian kidney. Which functions occur in multiple regions of the nephron, and which occur in only one portion?

# Endocrine Systems

# 50

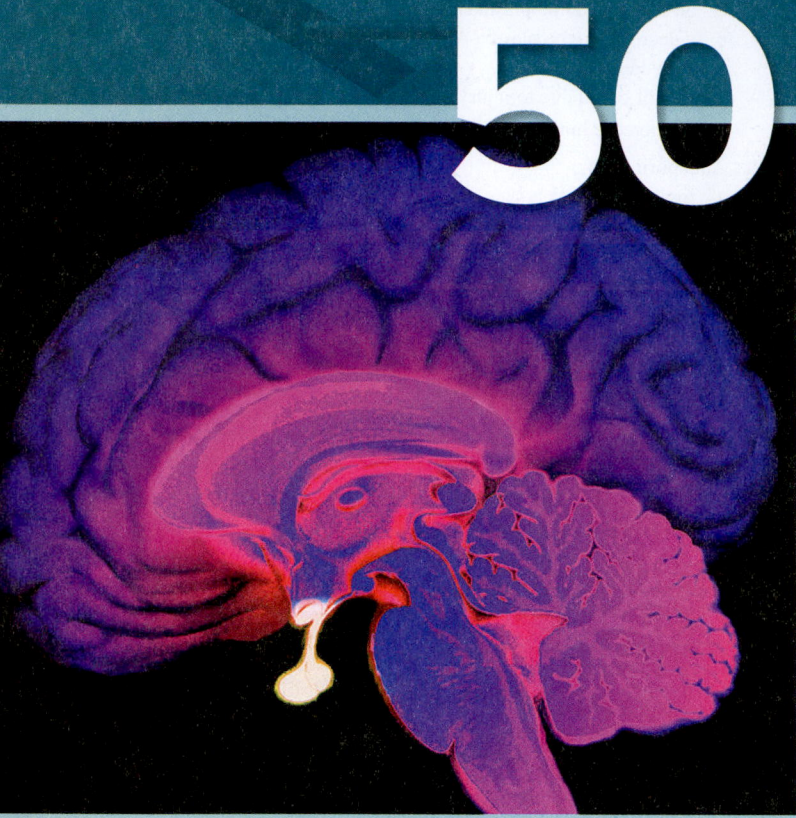

**A section through a human brain, highlighting the pituitary gland and its connection to the hypothalamus (white).** Both of these structures secrete numerous hormones. (Image is a three-dimensional MRI.) ©Sovereign/ISM/Phototake

A 22-year-old man was seen by his physician because of a complaint of hair loss and the appearance of severe acne over much of his face, neck, back, and shoulders. He also reported feelings of irritability and aggression. Upon examination, the man was found to have additional symptoms, including hypertension, an elevated plasma cholesterol concentration, an increased hematocrit (red blood cell count; see Chapter 48), and, alarmingly, shrunken testes. When questioned, the patient admitted that for 6 months he had been taking an oral form of the illegal drug stanozolol in an effort to improve his physique by building more muscle mass. Stanozolol is a synthetic version of a steroid hormone known as an androgen. A **hormone** is a chemical signal produced by cells that is secreted into the extracellular fluid. Unlike paracrine substances, which act on nearby cells, hormones travel through the blood or hemolymph to one or more distant target tissues to alter their functions.

In males, androgens are produced by the testes, which are reproductive organs that contain hormone-producing cells. The most well-known androgen is testosterone. As we will discuss in this chapter, in normal amounts, androgens have important functions in the physiological events associated with puberty and reproduction, including the increase in skeletal muscle mass that accompanies male puberty. This young man hoped to increase his muscle mass by consuming large amounts of a synthetic androgen that mimics testosterone. At high concentrations, however, androgens promote increased erythrocyte production, which, in turn, forces the heart to work harder and can lead to hypertension. Likewise, high concentrations of androgens increase the amount of cholesterol in the blood, promote fluid retention, damage the liver, cause hair loss, and increase the activity of the sebaceous glands. At such concentrations, androgens also strongly inhibit the activity of the testes. The body senses that sufficient androgens are already available in the blood. In response, the testes do not need to make more androgens and thus shrink in size as they become less and less active.

Fortunately, this individual was educated by his doctor about the consequences of misuse of these powerful hormones, and he discontinued the practice. Had he continued to take the synthetic androgen, all of his symptoms would have worsened, and he would have run the risk of serious and irreversible damage to his heart, liver, and other organs, while increasing his risk of cancer and other diseases.

Androgens and other hormones are found in all vertebrates and many invertebrates. Hormones are often produced in response to developmental changes, as in puberty, and to homeostatic challenges, such as a change in an animal's blood pressure or body temperature. A chief function of hormones is to counter these challenges and maintain homeostasis. Hormones affect a wide range of body functions, including gastrointestinal activity, blood pressure regulation, cholesterol balance, fluid and mineral balance, and reproduction.

Hormones are made by cells in numerous organs of an animal's body. In addition, hormone-producing cells are often found in specialized glands, called **endocrine glands**, the primary function of which is hormone synthesis and secretion. An example is the

## 50.1 Types of Hormones and Their Mechanisms of Action

### Learning Outcomes:

1. List some of the major endocrine glands found in animals and the hormones they secrete.
2. List and give examples of the three different chemical classes of hormones, and describe some of their structural differences.
3. Describe the cellular locations of receptors for lipid-soluble hormones and receptors for water-soluble hormones.

**Figure 50.1** is an overview of the major vertebrate endocrine glands and their hormones; you should take a moment and review this figure as preparation for the rest of the chapter. Throughout this chapter, we will use the human as a representative species when describing the location and appearance of vertebrate endocrine glands. All vertebrates share similar glands, although their locations and histological structures may show certain differences. We will occasionally point out important functional differences in various hormones that are present in all vertebrates. In this section, we will examine some of the general characteristics of hormones that are shared by all or most animals, including how they act and how they are controlled. Keep in mind that endocrine signaling is just one type of cell-to-cell communication. Refer back to Figure 9.3 for a review of the different types of cell-signaling mechanisms found in animals.

### Hormones Are Classified According to Their Structure

Hormones fall into three broad classes: amines, polypeptides, and steroid hormones (**Table 50.1**). The amines and the polypeptides share similar chemical properties and mechanisms of action, whereas the steroid hormones act very differently from the other two classes.

The amine hormones are derived from an amino acid, either tyrosine or tryptophan. Tyrosine is the precursor for the hormones epinephrine and norepinephrine, which are produced in the adrenal medulla and are important in a vertebrate's response to stress (the fight-or-flight response; see Chapter 43). It is also the precursor for dopamine, a hormone that is made by the brain and functions in the control of the anterior pituitary gland. Tyrosine is also the molecular backbone of thyroid hormone, which is made by the thyroid gland and is an important regulator of metabolic rate, growth, and development. The major hormone derived from tryptophan is melatonin, which is produced within an endocrine gland called the pineal gland, located in the vertebrate brain. Melatonin is important in some species for controlling circadian rhythms, daily cycles of physiological or biochemical processes. In some mammals such as sheep and hamsters that breed only at certain times of year, melatonin controls seasonal cycles of reproduction.

Polypeptides are the most abundant hormones. These hormones participate in numerous body functions such as metabolism, mineral balance, growth, and reproduction. Examples include insulin and glucagon made by the pancreas, and leptin made by adipose tissue.

Amine and polypeptide hormones are generally synthesized and secreted at a steady rate or in a circadian rhythm, until an additional amount is required. In that case, transcription factors within the cell direct the increased transcription of the gene encoding the hormone in question. Conversely, when less hormone is required, gene transcription is slowed or stopped. These hormones are too large and hydrophilic to diffuse from the cytosol across the plasma membrane and then into the extracellular fluid. Instead, they are packaged into secretory vesicles in much the same way as neurotransmitters are packaged in neuron axon terminals. This packaging provides a ready means of secreting the hormones by exocytosis and establishes a reservoir of stored hormone available for immediate release when required. If a cell is stimulated to secrete more than the usual amount of a stored amine or polypeptide hormone, it also is typically stimulated to synthesize new hormone molecules to replace them.

Steroid hormones are synthesized from cholesterol, as shown in **Figure 50.2**, and thus all steroid hormones are lipids, unlike the other classes of hormones. Steroid hormones, unlike water-soluble hormones, are not packaged into secretory vesicles because a lipophilic steroid could diffuse across the lipid membrane of the vesicle. Instead, steroid hormones are made on demand, and no significant amount of them is stored.

Steroids are less soluble in water than are amines or polypeptide hormones. Due to this limited solubility, steroids are usually bound to large, soluble proteins in the blood that serve as carriers. By combining with these proteins, steroids can reach high concentrations in the blood. They can then be released from their binding proteins into the plasma in small amounts at a time. From there, they diffuse into cells.

### Hormones Act Through Plasma Membrane or Intracellular Receptors

The amine and polypeptide hormones are generally water-soluble, and the steroid hormones are lipid-soluble. An exception is thyroid hormone, which is a lipophilic amine hormone. All of the other amine and polypeptide hormones are not able to cross plasma membranes

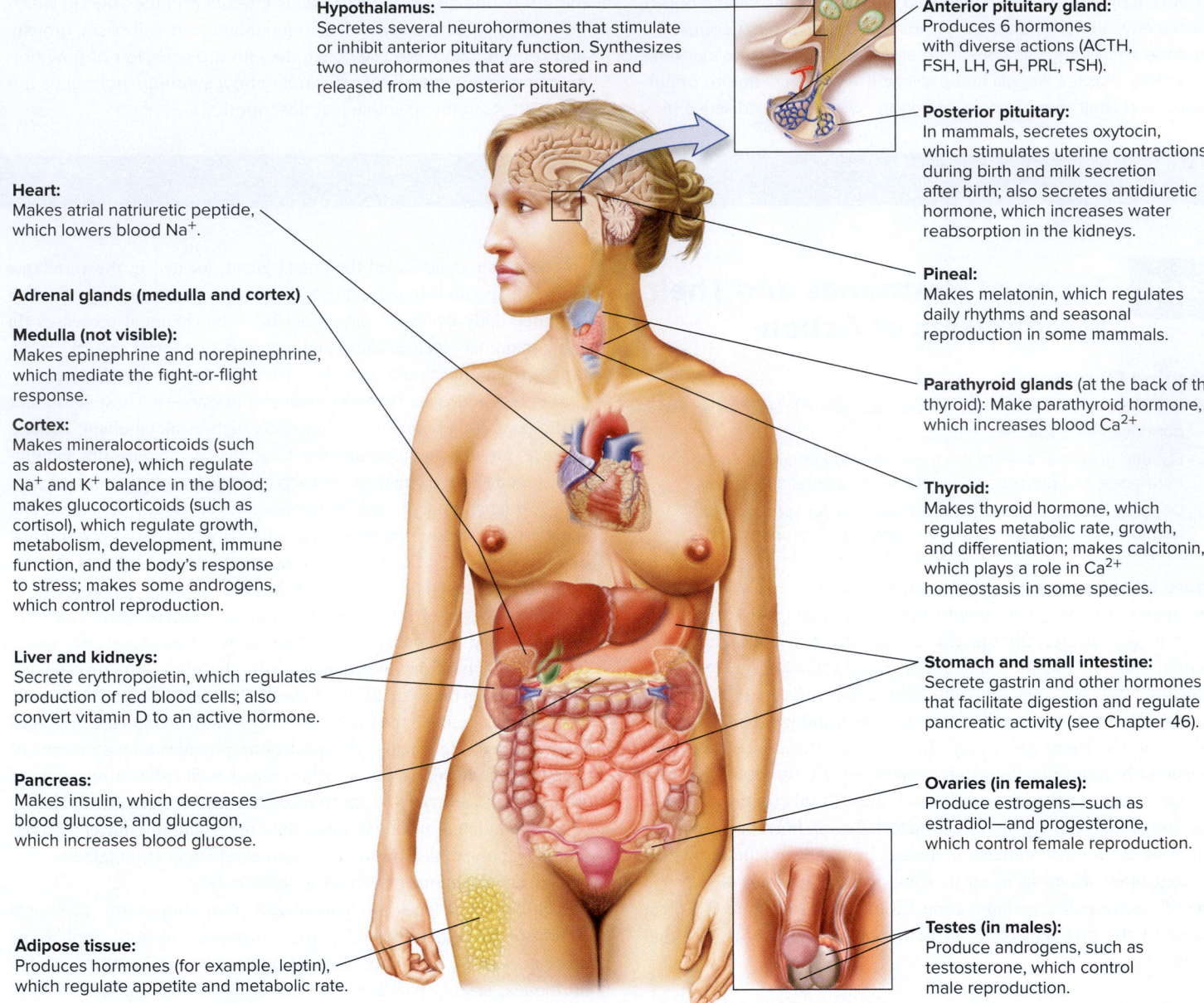

**Hypothalamus:** Secretes several neurohormones that stimulate or inhibit anterior pituitary function. Synthesizes two neurohormones that are stored in and released from the posterior pituitary.

**Anterior pituitary gland:** Produces 6 hormones with diverse actions (ACTH, FSH, LH, GH, PRL, TSH).

**Posterior pituitary:** In mammals, secretes oxytocin, which stimulates uterine contractions during birth and milk secretion after birth; also secretes antidiuretic hormone, which increases water reabsorption in the kidneys.

**Heart:** Makes atrial natriuretic peptide, which lowers blood $Na^+$.

**Pineal:** Makes melatonin, which regulates daily rhythms and seasonal reproduction in some mammals.

**Adrenal glands (medulla and cortex)**

**Medulla (not visible):** Makes epinephrine and norepinephrine, which mediate the fight-or-flight response.

**Parathyroid glands** (at the back of the thyroid): Make parathyroid hormone, which increases blood $Ca^{2+}$.

**Cortex:** Makes mineralocorticoids (such as aldosterone), which regulate $Na^+$ and $K^+$ balance in the blood; makes glucocorticoids (such as cortisol), which regulate growth, metabolism, development, immune function, and the body's response to stress; makes some androgens, which control reproduction.

**Thyroid:** Makes thyroid hormone, which regulates metabolic rate, growth, and differentiation; makes calcitonin, which plays a role in $Ca^{2+}$ homeostasis in some species.

**Liver and kidneys:** Secrete erythropoietin, which regulates production of red blood cells; also convert vitamin D to an active hormone.

**Stomach and small intestine:** Secrete gastrin and other hormones that facilitate digestion and regulate pancreatic activity (see Chapter 46).

**Pancreas:** Makes insulin, which decreases blood glucose, and glucagon, which increases blood glucose.

**Ovaries (in females):** Produce estrogens—such as estradiol—and progesterone, which control female reproduction.

**Adipose tissue:** Produces hormones (for example, leptin), which regulate appetite and metabolic rate.

**Testes (in males):** Produce androgens, such as testosterone, which control male reproduction.

**Figure 50.1** **Overview of the vertebrate endocrine system.** This figure shows many of the major endocrine glands and other structures that constitute the vertebrate endocrine system (as seen in a human) and gives the major functions of the hormones produced by those glands.

| Table 50.1 | Chemical Classes of Hormones | | | |
|---|---|---|---|---|
| **Class** | **Chemical properties** | **Location of target cell receptors** | **Mechanism of action** | **Examples** |
| Amines | Derived from tyrosine or tryptophan; small, water-soluble (except thyroid hormones, which are lipophilic) | Plasma membrane (except thyroid hormones, which act via intracellular receptors) | Stimulate second-messenger pathways (except thyroid hormone, which acts via changes in gene transcription) | Epinephrine, norepinephrine, dopamine, thyroid hormone, melatonin |
| Polypeptides | Water-soluble | Plasma membrane | Stimulate second-messenger pathways | Insulin, glucagon, leptin |
| Steroids | Synthesized from cholesterol; lipid-soluble | Cytosol or nucleus | Usually stimulate gene transcription directly | Aldosterone, cortisol, androgens, estrogens |

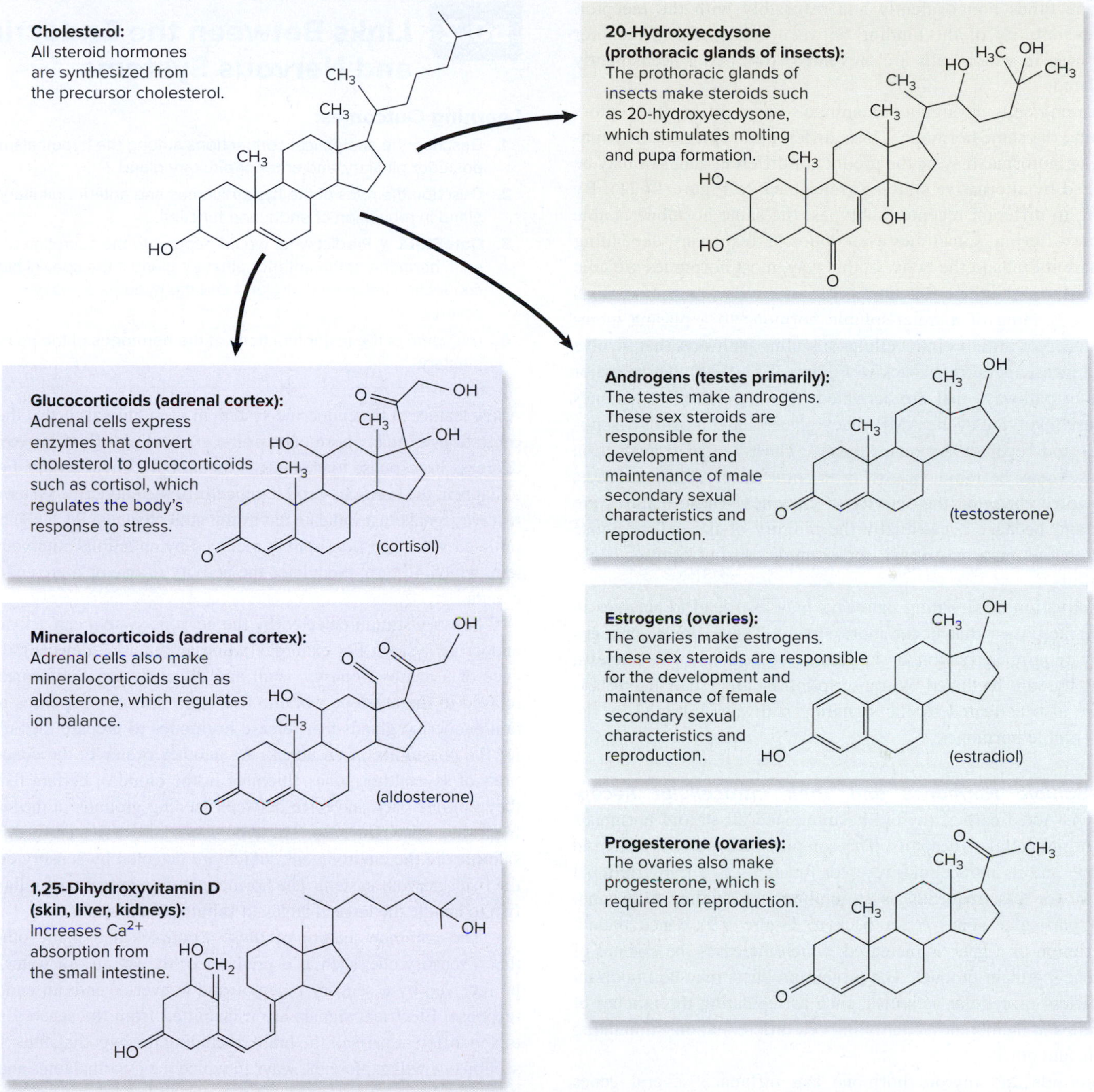

**Cholesterol:**
All steroid hormones are synthesized from the precursor cholesterol.

**20-Hydroxyecdysone (prothoracic glands of insects):**
The prothoracic glands of insects make steroids such as 20-hydroxyecdysone, which stimulates molting and pupa formation.

**Glucocorticoids (adrenal cortex):**
Adrenal cells express enzymes that convert cholesterol to glucocorticoids such as cortisol, which regulates the body's response to stress.

(cortisol)

**Androgens (testes primarily):**
The testes make androgens. These sex steroids are responsible for the development and maintenance of male secondary sexual characteristics and reproduction.

(testosterone)

**Mineralocorticoids (adrenal cortex):**
Adrenal cells also make mineralocorticoids such as aldosterone, which regulates ion balance.

(aldosterone)

**Estrogens (ovaries):**
The ovaries make estrogens. These sex steroids are responsible for the development and maintenance of female secondary sexual characteristics and reproduction.

(estradiol)

**1,25-Dihydroxyvitamin D (skin, liver, kidneys):**
Increases $Ca^{2+}$ absorption from the small intestine.

**Progesterone (ovaries):**
The ovaries also make progesterone, which is required for reproduction.

**Figure 50.2  Synthesis and functions of the major steroid hormones in animals.**

 **Core Concept: Structure and Function**  Notice how slight changes in the arrangements of side groups attached to the four-ring skeleton derived from cholesterol create products with completely different functions. Keep in mind that the three-dimensional shape of a molecule may be changed much more by such chemical modifications than is visible in a two-dimensional depiction.

and must bind to a receptor protein on the surface of a target cell. Steroid hormones and thyroid hormone, however, being lipophilic, diffuse across the target cell's plasma membranes. They bind to a receptor protein located in the cytosol or nucleus. Hormones that act through plasma membrane receptors tend to elicit fast responses (in seconds to minutes), whereas those that act via intracellular receptors generally act much more slowly (within hours), as we see next.

*Water-Soluble Hormones and Their Plasma Membrane Receptors*  All amine and polypeptide hormones other than thyroid hormone act by binding to a receptor protein located in the plasma membrane (refer back to Figure 9.5). Only cells having the proper receptors on their surfaces can respond to these hormones. Thus, although a given hormone travels throughout the entire circulatory system, it activates only specific cells. The hormone binds noncovalently and reversibly with the receptor. The reversibility of the binding between hormone and receptor is one way in which cells are prevented from being permanently stimulated.

Among cells throughout an animal's body, different receptors may bind the same hormone. These different receptors, called subtypes, or isoforms, may be the product of different genes or may be produced by alternative splicing (refer back to Figure 14.21). By binding to different receptor subtypes, the same hormone is able to elicit differing, sometimes even opposite, responses, depending on where it binds in the body. In this way, most hormones are able to serve more than one function.

The binding of a water-soluble hormone to a plasma membrane receptor initiates intracellular signaling pathways that involve second messengers (refer back to Figures 9.12–9.14). Three major signaling pathways that are activated by water-soluble hormones are those involving cyclic AMP, diacylglycerol and inositol trisphosphate, and receptor tyrosine kinases. These signal transduction processes may be rapid, occurring in some cases within seconds, and involve changing the activity of enzymes. These features are important, because occasionally the rapidity of the cell response to a hormone may be critical, for example, during fight-or-flight conditions.

Activation of signaling pathways may also lead to changes in cellular activities that occur more slowly. These slower changes usually require activation or inhibition of genes in the nucleus, actions that are mediated by transcription factors (see Chapter 14) that are also activated by the signaling pathways initiated by the water-soluble hormones.

### Lipid-Soluble Hormones and Their Intracellular Receptors

As noted earlier, thyroid hormone and all steroid hormones bind to intracellular receptors. The complex of a steroid or thyroid hormone and its intracellular receptor functions as a transcriptional activator (or less frequently as an inhibitor) by binding to enhancers of particular genes (refer back to Figure 9.9). Once bound, transcription of a gene is increased, which increases the amount of that gene's protein product. The protein products may be important in a variety of cellular activities, such as regulating the number of ion pumps in membranes, controlling cell differentiation, stimulating growth, and others.

Steroids and thyroid hormone can influence several genes within a single cell or in different cells. In this way, one hormone can exert a variety of actions throughout the body. The numerous and varied physical changes that accompany puberty in mammals result from the actions of two steroid hormones—androgens in males and estrogens in females—and are among the most striking and commonly recognized examples of this kind of widespread action.

Generally, hormone concentrations in the blood remain within a relatively narrow range, but they can be increased or decreased beyond that range if required. One of the ways in which changes in hormone concentrations are initiated is through sensory input to an animal's brain. As we see next, the nervous system and endocrine system are functionally linked in many animals, including all vertebrates.

## 50.2 Links Between the Endocrine and Nervous Systems

### Learning Outcomes:

1. Describe the anatomical connections among the hypothalamus, posterior pituitary, and anterior pituitary gland.

2. Describe the roles of the hypothalamus and anterior pituitary gland in regulation of endocrine function.

3. **CoreSKILL »** Predict what would happen to the secretion of each hormone of the anterior pituitary gland if the special blood connection between that gland and the hypothalamus was severed.

4. List some of the major functions of the hormones of the posterior pituitary.

A key feature of the endocrine system in most animals is that the concentrations of many hormones in the extracellular fluid increase and decrease in response to changes in an animal's environment. For this to happen, the hormone-producing cells of the endocrine system must receive signals that indicate environmental changes. Such signals are initiated when sensory input is received by an animal's nervous system, which, in turn, modulates the activity of one or more endocrine glands.

Sensory stimuli detected by the nervous system can activate the endocrine system. For example, when an antelope detects the presence of a nearby lioness, visual and olfactory sense information is relayed to the antelope's brain. The brain initiates responses in certain endocrine glands that release hormones to prepare the antelope for the possibility of an attack. As another example, the concentrations of several hormones fluctuate in the blood of certain fishes as they migrate back and forth between feeding grounds in the sea and freshwater spawning sites. These hormones are activated by different salinities in the environment, which are detected by sensory cells of the fish's nervous system. The hormones act to prepare the gills of the fish to handle the large changes in salinity of the water.

The common feature of these examples and many others is that a sensory cue, such as a predator or the salinity of water, must be received by a sensory receptor and converted into an endocrine response. Electrical signals are transmitted from the sensory receptors to different parts of the brain, including the hypothalamus. In this section, we will explore the ways in which the hypothalamus and pituitary gland in vertebrates link the nervous and endocrine systems. The brain/endocrine link in invertebrates will be covered in Section 50.5.

### The Hypothalamus and Pituitary Gland Are Physically Connected

As described in Chapter 43, the **hypothalamus** is a collection of several nuclei that are located at the ventral surface of the vertebrate brain (**Figure 50.3a**). Neurons in these nuclei receive signals from most parts of the brain. In turn, they are connected to an endocrine gland sitting directly below the hypothalamus, called the **pituitary gland** (see the bottom part of the highlighted structure in the chapter opening photo). The pituitary gland in humans is made up of two lobes, the anterior and posterior lobes. Some vertebrate species have an

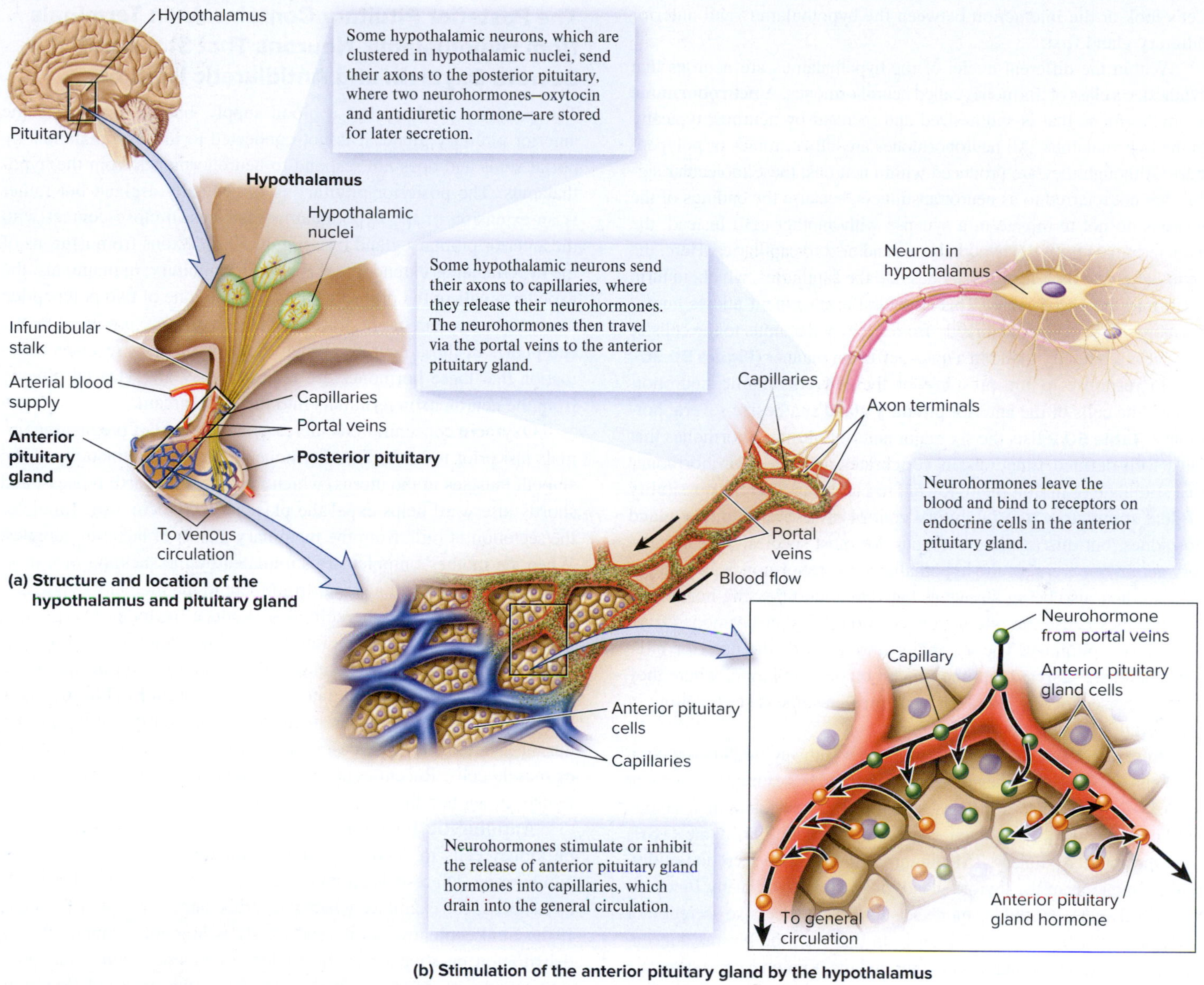

**Hypothalamus**

Pituitary

Some hypothalamic neurons, which are clustered in hypothalamic nuclei, send their axons to the posterior pituitary, where two neurohormones—oxytocin and antidiuretic hormone—are stored for later secretion.

**Hypothalamus**

Hypothalamic nuclei

Some hypothalamic neurons send their axons to capillaries, where they release their neurohormones. The neurohormones then travel via the portal veins to the anterior pituitary gland.

Neuron in hypothalamus

Infundibular stalk

Arterial blood supply

Capillaries

Axon terminals

**Anterior pituitary gland**

Capillaries

Portal veins

**Posterior pituitary**

Neurohormones leave the blood and bind to receptors on endocrine cells in the anterior pituitary gland.

To venous circulation

**(a) Structure and location of the hypothalamus and pituitary gland**

Portal veins

Blood flow

Anterior pituitary cells

Capillaries

Neurohormone from portal veins

Capillary

Anterior pituitary gland cells

Neurohormones stimulate or inhibit the release of anterior pituitary gland hormones into capillaries, which drain into the general circulation.

Anterior pituitary gland hormone

To general circulation

**(b) Stimulation of the anterior pituitary gland by the hypothalamus**

**Figure 50.3** **Relationship between the hypothalamus and the pituitary gland in a human.**

 **Core Skill: Connections** Recall the definition of neurotransmitter from Chapter 42 and look back at Figure 42.13. What determines whether a molecule such as dopamine is considered a neurotransmitter or a hormone?

intermediate lobe that secretes hormones that function in such things as seasonal changes in coat color in certain mammals. This lobe is believed to be vestigial in many other species, including primates.

The hypothalamus and pituitary are connected by a thin piece of tissue called the **infundibular stalk** and also by a system of blood vessels called portal veins. **Portal veins** differ from ordinary veins in that they not only collect blood from capillaries—like all veins do—but they also form another set of capillaries, instead of returning the blood directly to the heart like other veins. The portal veins extend through the length of the infundibular stalk. Within the anterior lobe of the pituitary gland—often simply called the anterior pituitary gland—the portal veins empty into a second set of capillaries. This arrangement of blood vessels bypasses the general circulation and

allows the hypothalamus to communicate directly with the anterior pituitary gland. Let's now explore the nature of this communication.

## The Hypothalamus and Anterior Pituitary Gland Have Integrated Functions

As described in Chapter 43, the nuclei of the vertebrate hypothalamus are vital for such diverse functions as reproduction, circadian rhythms, appetite, metabolism, and responses to stress. The hypothalamus has such wide-ranging effects in part because it acts as a master control, signaling the pituitary gland when to produce and secrete its various hormones. However, the hypothalamus communicates differently with the anterior and posterior lobes of the pituitary gland.

Let's look at the interaction between the hypothalamus and anterior pituitary gland first.

Within the different nuclei of the hypothalamus are neurons that synthesize a class of hormones called neurohormones. A **neurohormone** is any hormone that is synthesized and secreted by neurons, typically in the hypothalamus. All neurohormones are either amines or polypeptides. Although they are produced within neurons, these molecular signals are not referred to as neurotransmitters, because the endings of the neurons do not terminate in a synapse with another cell. Instead, the axon terminals from the hypothalamus end next to capillaries. Here, the neurons secrete their neurohormones into the capillaries, which, in turn, drain into the portal veins. This structural arrangement allows neurohormones to be delivered directly from the hypothalamus to the cells of the anterior pituitary gland in a quick, efficient manner (**Figure 50.3b**).

In response to the presence of these hypothalamic neurohormones, the cells of the anterior pituitary gland synthesize several hormones. **Table 50.2** lists the six major anterior pituitary hormones that have well-defined functions in vertebrates and the neurohormones that stimulate or inhibit them. Recent research suggests the possibility of dual stimulatory and inhibitory control of several pituitary gland hormones, but this remains uncertain for most species. The stimulatory action of several of the hypothalamic neurohormones led historically to their also being known as hypothalamic-releasing hormones, because they cause the release, or secretion, of other hormones from the anterior pituitary. The six major hormones of the anterior pituitary gland are secreted into the general blood circulation, where they act on other endocrine glands or structures, in some cases resulting in the secretion of yet other hormones.

In several cases, the final hormone in a pathway inhibits secretion of a neurohormone via a negative feedback loop. For example, looking at Table 50.2, we see that the neurohormone thyrotropin-releasing hormone (TRH) stimulates the secretion of thyroid-stimulating hormone (TSH) from the anterior pituitary. TSH, in turn, stimulates the secretion of thyroid hormone from the thyroid gland. In addition to its many functions (described later), thyroid hormone acts to prevent excessive secretion of TRH and TSH by negative feedback. In this way, the plasma concentrations of anterior pituitary hormones and the hormones they stimulate other glands to secrete are usually kept within an optimal range.

## The Posterior Pituitary Contains Axon Terminals from Hypothalamic Neurons That Store and Secrete Oxytocin and Antidiuretic Hormone

The posterior pituitary has a blood supply, but, in contrast to the anterior pituitary gland, it is not connected to the hypothalamus by portal veins and does not respond to neurohormones from the hypothalamus. The posterior pituitary is not actually a gland but rather is an extension of the hypothalamus that lies in close contact with the anterior pituitary gland (see Figure 50.3). Axons from neurons of the hypothalamus extend into the posterior pituitary. In mammals, the axon terminals in the posterior pituitary store one of two polypeptide hormones, oxytocin or antidiuretic hormone, that are produced by the cell bodies of those neurons. When the hypothalamus receives information that these hormones are required, they are released directly from the neuron axon terminals into the bloodstream.

**Oxytocin** concentrations increase in the blood of pregnant mammals just prior to birth. This hormone stimulates contractions of the smooth muscles in the uterus, which facilitates the birth process, and shortly afterward helps expel the placenta. Oxytocin also stimulates the secretion of milk from the mammary glands of lactating females. When the mother's nipples are stimulated by the suckling of a newborn, neurons transmit a signal from there to the mother's hypothalamus, which stimulates the cells that produce oxytocin. Oxytocin is released from the posterior pituitary into the blood where it travels to smooth muscle cells surrounding the secretory components of the mammary glands. This stimulates the release of milk. Thus, oxytocin has two important and different functions, one during birth and one during lactation. In both cases, the hormone stimulates contraction of muscle cells. Recent evidence also supports a role of oxytocin in mother-infant bonding and prosocial behavior in some species.

**Antidiuretic hormone (ADH)** gets its name because it acts on kidney cells to decrease urine production—a process known as antidiuresis. (Diuresis is an increased excretion of water in the urine, as happens, for example, when you drink large amounts of fluids.) If the fluid content of an animal's body is low, for example, during dehydration or after a significant loss of blood, volume and pressure sensors in structures of the circulatory system send signals to

| Table 50.2 | Hormones of the Vertebrate Anterior Pituitary Gland and Hypothalamus | | |
|---|---|---|---|
| **Anterior pituitary gland hormone** | **Stimulatory neurohormone from hypothalamus** | **Inhibitory neurohormone from hypothalamus** | **Major functions** |
| Adrenocorticotropic hormone (ACTH) | Corticotropin-releasing hormone (CRH) | None known | Stimulates adrenal cortex to make glucocorticoids |
| Follicle-stimulating hormone (FSH) | Gonadotropin-releasing hormone (GnRH) | None known | Stimulates germ cell development and sex steroid production in gonads |
| Luteinizing hormone (LH) | GnRH | None known | Stimulates release of eggs in females; stimulates sex steroid production from gonads |
| Growth hormone (GH) | Growth hormone-releasing hormone (GHRH) | Somatostatin | Promotes linear growth; regulates glucose and fatty acid balance in blood |
| Thyroid-stimulating hormone (TSH) | Thyrotropin-releasing hormone (TRH) | None known | Stimulates thyroid gland to make thyroid hormone |
| Prolactin (PRL) | TRH and other factors have been suggested as stimulators of PRL in some species, but this is not certain. | Dopamine | Stimulates milk formation in mammals; participates in mineral balance in other vertebrates |

the hypothalamus. This results in an increase in the amount of ADH that is secreted into the blood from the posterior pituitary. ADH acts to increase the number of water-channel proteins called aquaporins that are present in the apical membranes of cells in the collecting ducts of the kidneys (refer back to Figure 49.11). Water is reabsorbed from the filtrate through these aquaporins. Minimizing the volume of water lost in the urine is an adaptation that conserves body water when necessary.

At high concentrations, ADH also increases blood pressure by stimulating vasoconstriction of blood vessels, a function that accounts for the other common name of ADH, vasopressin. Like oxytocin, therefore, ADH has more than one function. The two major functions of ADH are related in that they contribute to maintaining blood pressure and fluid levels in the body.

Oxytocin and ADH are well-studied examples of the evolution of hormones. Although these two hormones are found only in mammals, many invertebrates and all nonmammalian vertebrates secrete one or more polypeptides that are chemically similar to oxytocin and ADH but are not identical to the mammalian hormones. One of these is **vasotocin**, which combines some of the chemical structure of oxytocin and ADH. Research indicates that an ancestral vasotocin-encoding gene duplicated at some point, and then those two genes evolved into the genes for oxytocin and ADH found in mammals. Because only mammals lactate, the functions of vasotocin in birds, fishes, and other vertebrates must be different from that of oxytocin in mammals. Research has shown that vasotocin is responsible for regulating ion and water balance in the blood of nonmammalian vertebrates. The observation that members of the vasotocin/oxytocin/ADH gene family arose early in animal evolution also suggests that oxytocin and ADH may have additional, unrecognized actions that have been retained in mammals. For example, human males have oxytocin in their blood, but its role cannot be the same as in females because, of course, only females give birth and lactate. The major physiological functions of oxytocin in male humans and other animals have not been unequivocally identified.

## 50.3 Hormonal Control of Metabolism and Energy Balance

### Learning Outcomes:

1. Explain how the structure of the thyroid gland is linked to its function (synthesis of thyroid hormone).
2. Describe the important anatomical features of the thyroid gland and the major function of thyroid hormone in adult animals.
3. List the hormones produced by the pancreas and adrenal glands, and provide a brief description of the major metabolic functions of each.
4. **Core Skill »** Identify one or more ways in which type 1 and type 2 diabetes mellitus can be distinguished if an individual shows symptoms of diabetes.

An important function of the endocrine system is to regulate an animal's metabolic rate and energy balance. Hormones are partly responsible for regulating energy balance by modulating appetite,

digestion, absorption of nutrients, and the concentration of glucose in the blood and its transport into cells. Although many hormones are involved in these processes, those from the thyroid gland, adrenal glands, pancreas, and adipose tissue have particularly important functions in vertebrates, as described in this section.

## Thyroid Hormone Stimulates Metabolism

Thyroid tissue does not always form a single, distinct gland in vertebrates. In fishes, for example, thyroid tissue is scattered throughout the body. In tetrapods, it is more consolidated into paired glands in the neck or, in mammals, in a single bilobed gland straddling the trachea just below the larynx (**Figure 50.4a**). Regardless of its appearance or location, however, the thyroid consists of many small, spherical structures called follicles, each consisting of a shell of epithelial cells called follicular cells and a core of a gel-like substance called the colloid. The colloid consists primarily of large amounts of the protein thyroglobulin (TG), which plays a major role in the synthesis of thyroid hormone.

***Synthesis of Thyroid Hormone*** Thyroid hormone is produced when thyrotropin-releasing hormone (TRH; see Table 50.2), which is made in the hypothalamus, stimulates the anterior pituitary gland to secrete thyroid-stimulating hormone (TSH) into the blood. TSH stimulates the follicular cells of the thyroid gland to begin the process of making thyroid hormone. Thyroid hormone has a negative feedback effect on the hypothalamus and anterior pituitary, preventing them from producing too much TRH and TSH, thereby keeping the circulating concentration of thyroid hormone in check.

**Figure 50.4b** shows the pathway leading to thyroid hormone synthesis. First, iodide—converted in the intestine from dietary iodine—diffuses from the bloodstream into the interstitial fluid, from where it is transported across the basolateral membrane of the thyroid follicular cells. The iodide then diffuses through the apical membrane and enters the colloid, where it is oxidized back to iodine and bonds to tyrosine side chains in thyroglobulin. When the thyroid follicular cells are stimulated by TSH, the apical membranes undergo endocytosis, bringing colloid with its iodinated thyroglobulin into the cell. The endocytotic vesicles fuse with lysosomes. There, lysosomal enzymes cleave the iodinated tyrosines from thyroglobulin to form **thyroxine**, also called $T_4$, because it contains four iodines and **triiodothyronine**, or $T_3$, which has three iodines. Both $T_4$ and $T_3$ then diffuse out of the follicular cells across the basolateral membrane into the blood. These molecules are carried throughout the body and diffuse into cells. Inside cells, most of the $T_4$ is converted by enzymes that remove one iodine, forming $T_3$. It is $T_3$ that actually binds to cellular receptor proteins. $T_4$ can be considered, therefore, as a circulating reservoir of the active hormone $T_3$. From here on, we will for simplicity refer only to $T_3$ as thyroid hormone.

***Stimulation of Metabolism*** A major action of thyroid hormone in adult animals is to stimulate energy expenditure by many different cell types. This occurs in large part by increasing the number and activity of the $Na^+/K^+$-ATPase pumps in plasma membranes. As these pumps hydrolyze ATP, the cellular concentration of ATP decreases. This decrease releases feedback inhibition on the cell's metabolism

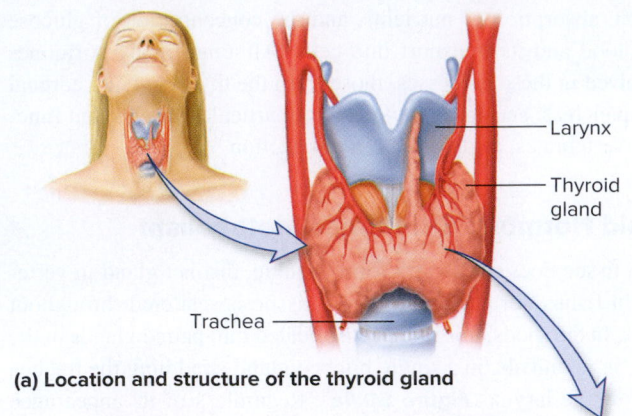

**Larynx**

**Thyroid gland**

**Trachea**

**(a) Location and structure of the thyroid gland**

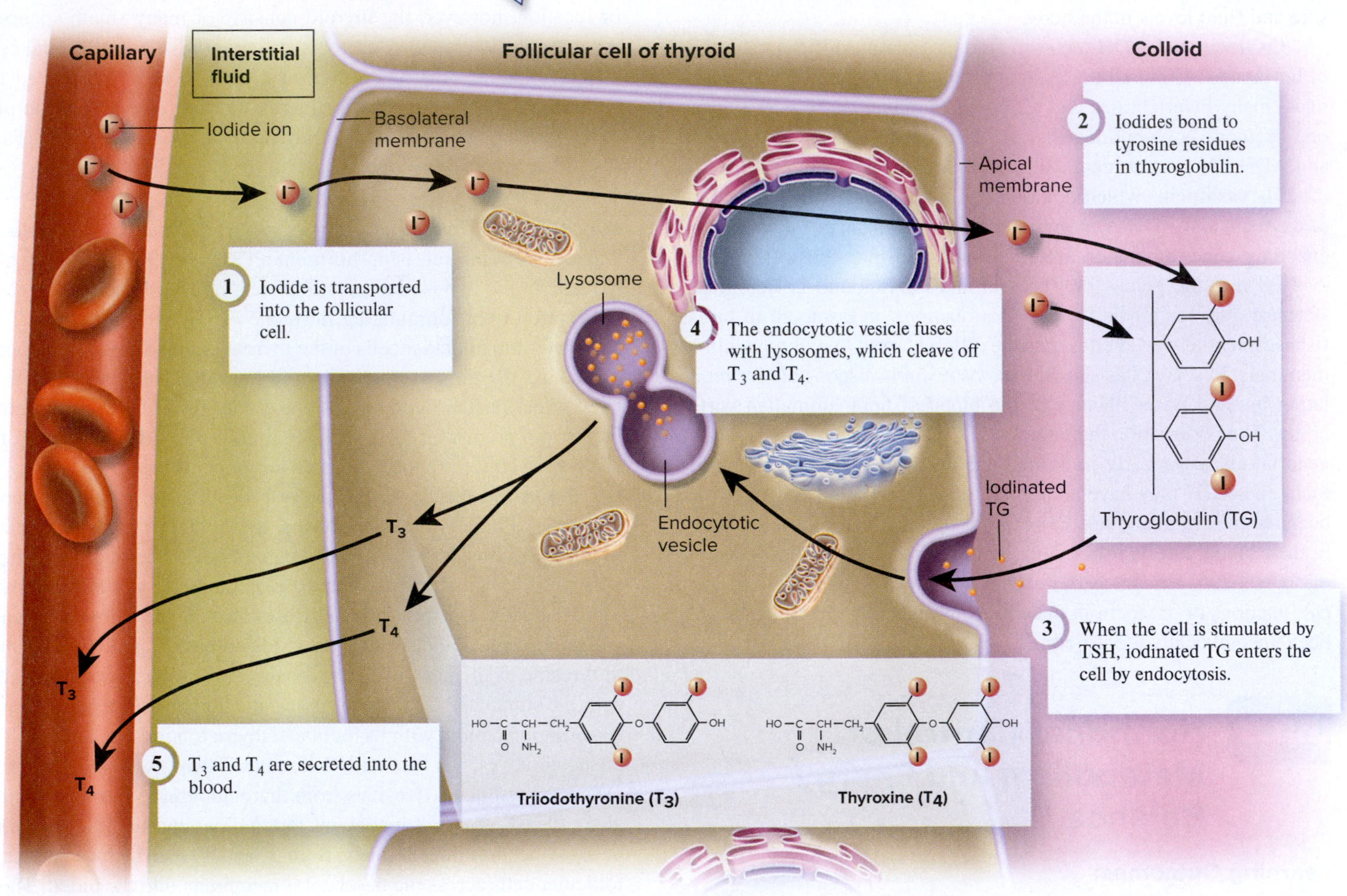

**Figure 50.4** **The thyroid gland and synthesis of thyroid hormone.** **(a)** Location and structure of the gland in a human. Vertebrates usually have thyroid tissue in the neck region, but it is not always consolidated into a single structure as shown here. **(b)** The steps involved in production of thyroid hormone. Shown is a cross section through a small part of a single follicle in a thyroid gland, with a nearby capillary (not to scale). The blood delivers iodide to the gland and picks up thyroid hormone secreted by the gland. Although $T_4$ and $T_3$ are both synthesized in the thyroid gland, $T_3$ is the active form of thyroid hormone. $T_4$ circulates in the blood and gets converted by cells to $T_3$.

**Capillary** | **Interstitial fluid** | **Follicular cell of thyroid** | **Colloid**

Iodide ion

**Basolateral membrane**

**Apical membrane**

**2** Iodides bond to tyrosine residues in thyroglobulin.

**1** Iodide is transported into the follicular cell.

**Lysosome**

**4** The endocytotic vesicle fuses with lysosomes, which cleave off $T_3$ and $T_4$.

**Thyroglobulin (TG)**

Iodinated TG

$T_3$

**Endocytotic vesicle**

$T_4$

**3** When the cell is stimulated by TSH, iodinated TG enters the cell by endocytosis.

$T_3$

$T_4$

**5** $T_3$ and $T_4$ are secreted into the blood.

**Triiodothyronine (T$_3$)**

**Thyroxine (T$_4$)**

**(b) Synthesis of thyroid hormone**

of glucose. As a result, glucose metabolism is increased, providing the energy required to produce more ATP. Whenever metabolism is increased, heat production is increased. Consequently, a person with a hyperactive thyroid gland (hyperthyroidism; from the Greek *hyper*, meaning over or above) generally feels warm, whereas the opposite condition (hypothyroidism; from the Greek *hypo*, meaning under or below) results in a sensation of coldness. Biologists have estimated that up to 70% of the heat produced by some endotherms is attributable solely to the actions of thyroid hormone on metabolic rate. Therefore, it is not surprising that humans with hyperthyroidism often lose weight, while weight gain is typically a problem in hypothyroidism.

***Effect of Iodine Deficiency*** The observation that thyroid hormone cannot be made without iodine leads to some interesting and unique consequences. The availability of iodine in the diet of most animals varies widely. As a consequence, the ability to store large amounts of thyroglobulin in the colloid of the thyroid was an important evolutionary adaptation. In this way, during times when iodine ingestion is sufficient, many thyroglobulin molecules have their tyrosine residues bound to iodines, one of the first steps in forming $T_3$. During times of low iodine availability, this reservoir of iodinated tyrosines in thyroglobulin molecules can be tapped. Humans, for example, have at least a 2-month supply of thyroid hormones even if dietary iodine

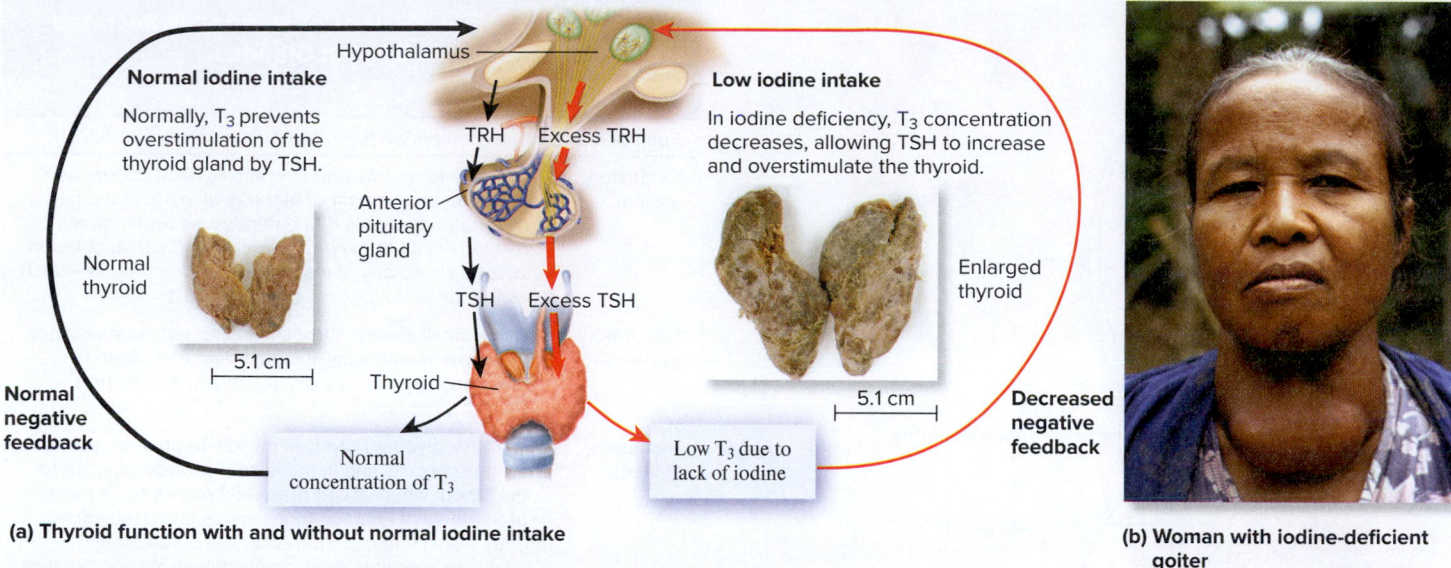

(a) Thyroid function with and without normal iodine intake

(b) Woman with iodine-deficient goiter

**Figure 50.5** **Consequences of inadequate iodine in the diet.** **(a)** With normal iodine intake, as shown on the left, $T_3$ controls TSH secretion by negative feedback. Without adequate iodine, as shown on the right, less $T_3$ is synthesized, which decreases the negative feedback effect of $T_3$. Consequently, the TSH concentration increases and the thyroid gland enlarges. **(b)** An extreme example of an enlarged thyroid gland, or goiter, due to iodine deficiency. a (left inset, right inset): ©S. Goodwin & Dr. Max Hincke, Division of Clinical and Functional Anatomy, University of Ottawa; b: ©Bruce Coleman Inc./ AlamyStock Photo

 **Core Concept: Systems** Recall from Chapter 41 that a key way in which homeostasis is maintained is via negative feedback. Negative feedback participates in the control mechanisms that regulate most organ systems, including the endocrine system. Notice in Figure 50.5 how a change in dietary intake of iodine results in decreased negative feedback and subsequent overgrowth of the thyroid gland.

becomes unavailable. In most industrialized countries, iodine deficiency is rarely a problem due to the introduction of iodized salt in the mid-20th century. However, in some regions of the world, it is still a major health problem.

The left side of **Figure 50.5a** shows the pattern of $T_3$ production from a healthy thyroid gland when iodine ingestion is adequate. The right side of the figure shows the consequences when $T_3$ is not produced in normal amounts, for example, due to a lack of iodine in the diet. Recall that hormones that are secreted in response to the actions of anterior pituitary gland hormones often exert a check on their own blood concentrations by negative feedback inhibition. If the plasma concentration of $T_3$ becomes lower than normal due to inadequate iodine ingestion, there will be less negative feedback on the hypothalamus and anterior pituitary gland, resulting in increased TRH and, consequently, TSH concentrations. The thyroid gland responds to the increased TSH by increasing the cellular machinery needed to produce more and more thyroglobulin, even though in the absence of iodine, no additional $T_3$ can be synthesized. What results is an overgrown gland that still lacks the resources to make $T_3$. This condition is known as an **iodine-deficient goiter** (**Figure 50.5b**). In humans, the problem can be alleviated either by adding iodine to the diet or by taking pills that contain thyroid hormone. Goiters are not unique to humans. Iodine deficiency is relatively common among vertebrates, and goiters are found frequently in reptiles and birds, particularly those that subsist on all-seed diets, which are generally low in iodine.

## Hormones of the Adrenal Glands and Pancreas Regulate the Concentration of Energy-Yielding Molecules in the Blood

Thyroid hormone regulates an animal's metabolism. For metabolism to proceed normally, however, cells must have adequate sources of energy available, usually in the form of glucose and fatty acids. The brain, in particular, must have a constant supply of glucose because brain cells have relatively limited storage capacity for fuel. Regulation of energy availability to cells is in large part accomplished by the hormones of the adrenal glands and the pancreas.

**Adrenal Glands** The adrenal glands, so named because they sit on top of the kidneys (from the Latin *ad*, meaning toward, and *renis*, meaning kidney), are multifunctional glands that contain an inner region called the adrenal medulla and an outer region called the adrenal cortex (**Figure 50.6**). These two regions produce different hormones that have widespread effects in animals, including effects on metabolism (see **Table 50.3** for a summary of many of the metabolic and nonmetabolic actions of these hormones).

When an animal is fasting, the sympathetic nervous system stimulates the cells of the adrenal medulla to secrete the amine hormones epinephrine and norepinephrine. Together, these two hormones are responsible for most of the physiological reactions of the fight-or-flight response that were described in Chapter 43 and Table 9.1. These reactions include increased production of glucose by the liver

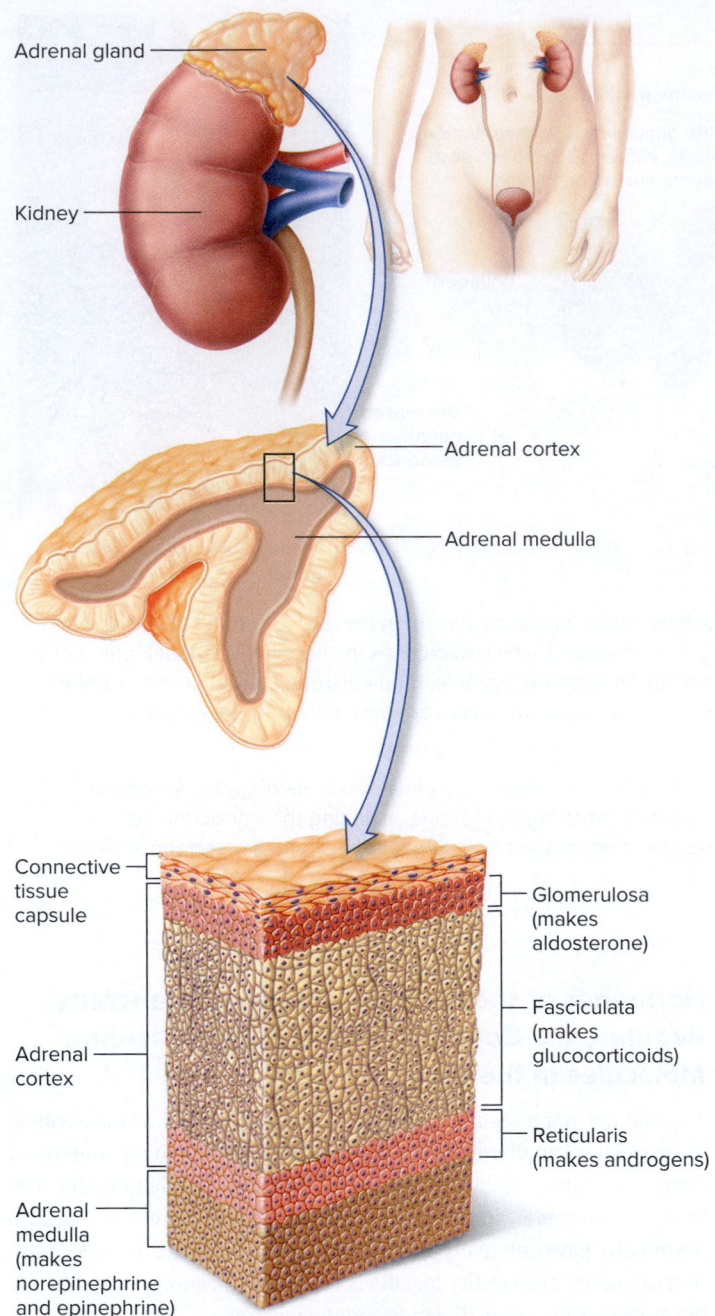

**Figure 50.6** Location, structure, and function of the adrenal glands.

| Table 50.3 | Major Actions of Hormones Secreted by the Adrenal Glands |
|---|---|
| **Site of action** | **Action(s)** |
| *Epinephrine and norepinephrine (secreted by the adrenal medulla)* | |
| Circulatory system | Increase heart rate and strength of heart contractions to maximize pumping of blood to all parts of the body; dilate blood vessels that enter tissues requiring more oxygen—such as skeletal muscle—and constrict blood vessels to regions of less immediate importance—such as the gut and kidneys |
| Respiratory system | Dilate small airways (bronchioles) to reduce resistance to airflow in mammals; increase rate and depth of breathing to maximize oxygen intake and carbon dioxide elimination |
| Metabolic system | Increase glycogenolysis in muscle to provide glucose for muscle cells and in the liver to provide glucose to the blood, where it can reach all body cells; increase breakdown of adipose triglycerides into usable fuel (fatty acids) that can then enter the bloodstream; stimulate secretion of glucagon, which acts on the liver to promote gluconeogenesis |
| Nervous system | Increase arousal and alertness; inhibit nonessential functions such as appetite |
| *Glucocorticoids (secreted by the adrenal cortex)* | |
| Liver | Stimulate gluconeogenesis, thus providing glucose to the blood |
| Adipose tissue | Stimulate breakdown of triglycerides into fatty acids and glycerol for fuel |
| Muscle and adipose tissue | Inhibit sensitivity to insulin, making more glucose available to brain cells, which do not require insulin to transport glucose across their plasma membranes |
| Bone | Inhibit bone growth and formation, because such processes require large amounts of nutrients that could be used to combat stress instead |
| Lungs | Stimulate lung maturation in the fetus |
| Immune | Suppress immune system function and reduce inflammation |
| Other | Regulate $Na^+$ and $Cl^-$ balance in migratory fishes; stimulate nervous system development in most vertebrates; stimulate protein breakdown to provide amino acids to the liver for gluconeogenesis; inhibit reproduction |

via stimulation of glycogenolysis (release of glucose from glycogen) and gluconeogenesis (synthesis of glucose from other sources) (refer back to Figure 47.2).

The outer part of the adrenal gland, the cortex, is subdivided into three zones: the glomerulosa, the fasciculata, and the reticularis (see Figure 50.6). The outer zone, the glomerulosa, is the region that makes the hormone aldosterone, which acts to maintain $Na^+$, $K^+$, and water balance (described later). The innermost cortical zone, known as the reticularis, functions in humans to make certain androgens, but its function in other animals is not as clear. The bulk of the cortex is the middle zone, the fasciculata, which produces glucocorticoid hormones such as **cortisol**.

Glucocorticoids are a group of related catabolic hormones that promote the breakdown of molecules and macromolecules. For example, they act on bone, immune, muscle, and adipose tissue to break down proteins and lipids to provide energy for the body's organs—in particular, vital organs such as the heart and brain. This source of energy is not only important for an animal that is fasting, but also for one that is facing an acute stress (traditionally defined in biology as a real or perceived threat to homeostasis), because feeding and digestion both stop during the fight-or-flight response. Consequently, internal stores become the only source of energy. These adrenal steroids are called glucocorticoids because one of their major actions is to promote gluconeogenesis in the liver during times of stress, thus providing glucose to the blood. Glucocorticoid release is stimulated by the anterior pituitary hormone adrenocorticotropic hormone (ACTH), the release of which in turn is stimulated by the neurohormone corticotropin-releasing hormone (CRH) (refer back to Table 50.2).

The actions of these two hormones are opposite with respect to each other—insulin decreases and glucagon increases blood glucose concentrations.

Maintaining homeostatic concentrations of glucose and other nutrients in the blood is a vital process that keeps cells functioning optimally. When an animal has not eaten for some time, its energy stores become depleted, and the blood glucose concentration begins to decrease. Under these conditions, glucagon is secreted into the blood, where it acts on the liver to stimulate glycogenolysis within seconds (**Figure 50.8**, right side). A second action of glucagon on the liver is particularly important for responses to prolonged fasting. In that case, glucagon stimulates gluconeogenesis. Due to the combined actions of cortisol, epinephrine, and glucagon, the concentration of glucose in the blood of a fasting animal is prevented from decreasing significantly below the normal homeostatic range.

***Hormonal Changes After a Meal*** In contrast to fasting, after an animal eats a meal, the concentrations of glucose and other nutrients in the blood become elevated. Restoring the normal blood concentrations of glucose, fats, and amino acids is almost exclusively under the control of insulin, one of the few hormones that is absolutely essential for survival in animals. The secretion of insulin is directly stimulated by an increased concentration of glucose in the blood (Figure 50.8, left side). Once in the blood, insulin acts on plasma membrane receptors located primarily in cells of adipose and skeletal and cardiac muscle tissues to facilitate the transport of glucose across the plasma membrane into the cytosol. Glucose is then metabolized for cellular functions or used to synthesize stored energy forms such as fat.

As discussed in Chapter 47, the process of glucose transport involves the actions of proteins called glucose transporters (GLUTs; refer back to Figure 47.3). These proteins are located in cytosolic membrane-bound vesicles. The major function of insulin is to stimulate fusion of the vesicles with the plasma membrane. Once a vesicle with its GLUTs fuses with the plasma membrane, the GLUTs begin transporting glucose from the extracellular fluid into the cell. When the blood glucose concentration returns to normal, the stimulus for insulin secretion disappears, and the insulin concentration in the blood decreases. This, in turn, results in a decrease in the number of GLUTs in plasma membranes, because the GLUTs are subjected to endocytosis and thereby return to membrane vesicles in the cytosol. The action of insulin is not limited to glucose transport. It also stimulates the transport of amino acids into cells and promotes fat deposition in adipose tissue.

In the absence of sufficient insulin, as occurs in the disease **type 1 diabetes mellitus (T1DM)**, less extracellular glucose can cross plasma membranes, and consequently, glucose accumulates to a very high concentration in the blood. T1DM occurs in many mammals and other vertebrates. It is caused when the body's immune system mistakenly attacks and destroys the insulin-producing cells of the islets of Langerhans. One consequence of the disease is that muscle and adipose cells cannot receive the normal amount of glucose to produce the ATP they require. In addition, the unusually large amount of glucose in the blood overwhelms the kidneys' ability to reabsorb it from the kidney filtrate, and glucose appears in the urine. Fortunately, this form of the disease

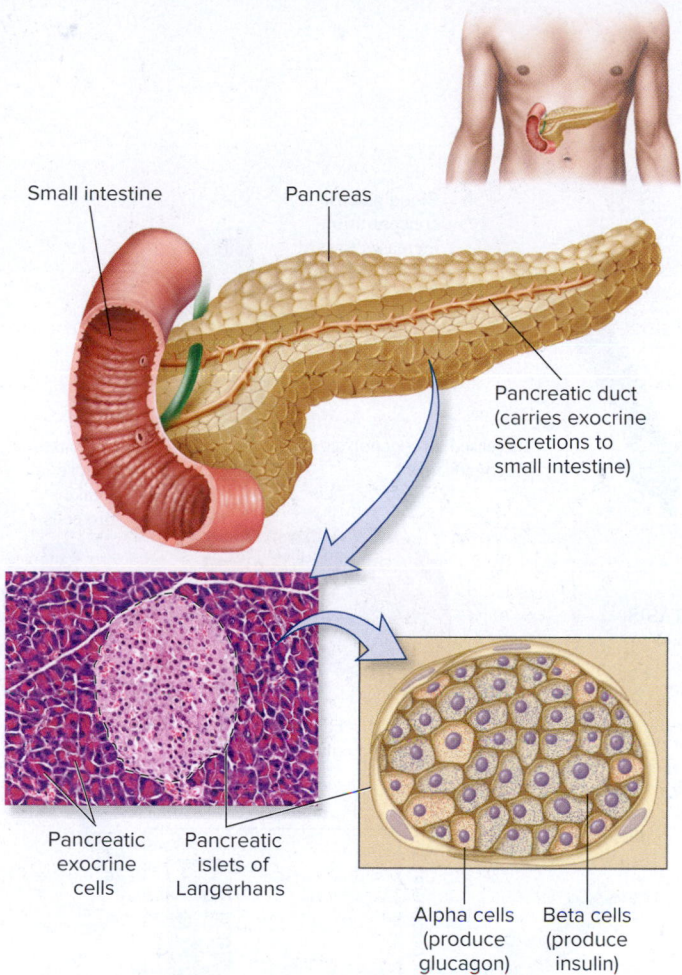

Small intestine    Pancreas

Pancreatic duct
(carries exocrine
secretions to
small intestine)

Pancreatic
exocrine
cells

Pancreatic
islets of
Langerhans

Alpha cells    Beta cells
(produce      (produce
glucagon)     insulin)

**Figure 50.7** **Location, appearance, and internal structure of the mammalian pancreas.** Within the exocrine pancreas are scattered islets of Langerhans, which are endocrine tissue. Only the exocrine products are secreted into the intestine. The hormones from the islets of Langerhans are secreted into the blood (not shown). (bottom left):
©Cultura RM/Alamy Stock Photo

 **Core Skill: Connections** The pancreas contains both exocrine and endocrine tissue. Is this property ever observed in other organs? (Hint: Think about the processes associated with digestion and absorption of food described in Chapter 46.)

***Pancreas*** The pancreas is a complex organ that is both an exocrine and endocrine gland (**Figure 50.7**). An **exocrine gland** is one in which epithelial cells secrete chemicals into a duct, which carries those molecules directly to another structure or to the outside surface of the body. Familiar examples of exocrine glands include the sweat glands and salivary glands. Most of the mass of the pancreas consists of exocrine cells. The secretions of the exocrine pancreas empty into the small intestine, where they aid digestion.

The nonexocrine portion of the pancreas consists of endocrine cells that produce polypeptide hormones. Spherical clusters of endocrine cells called **islets of Langerhans** are scattered in large numbers throughout the pancreas. Within the islets are alpha cells, which make **glucagon**, and beta cells, which make **insulin**.

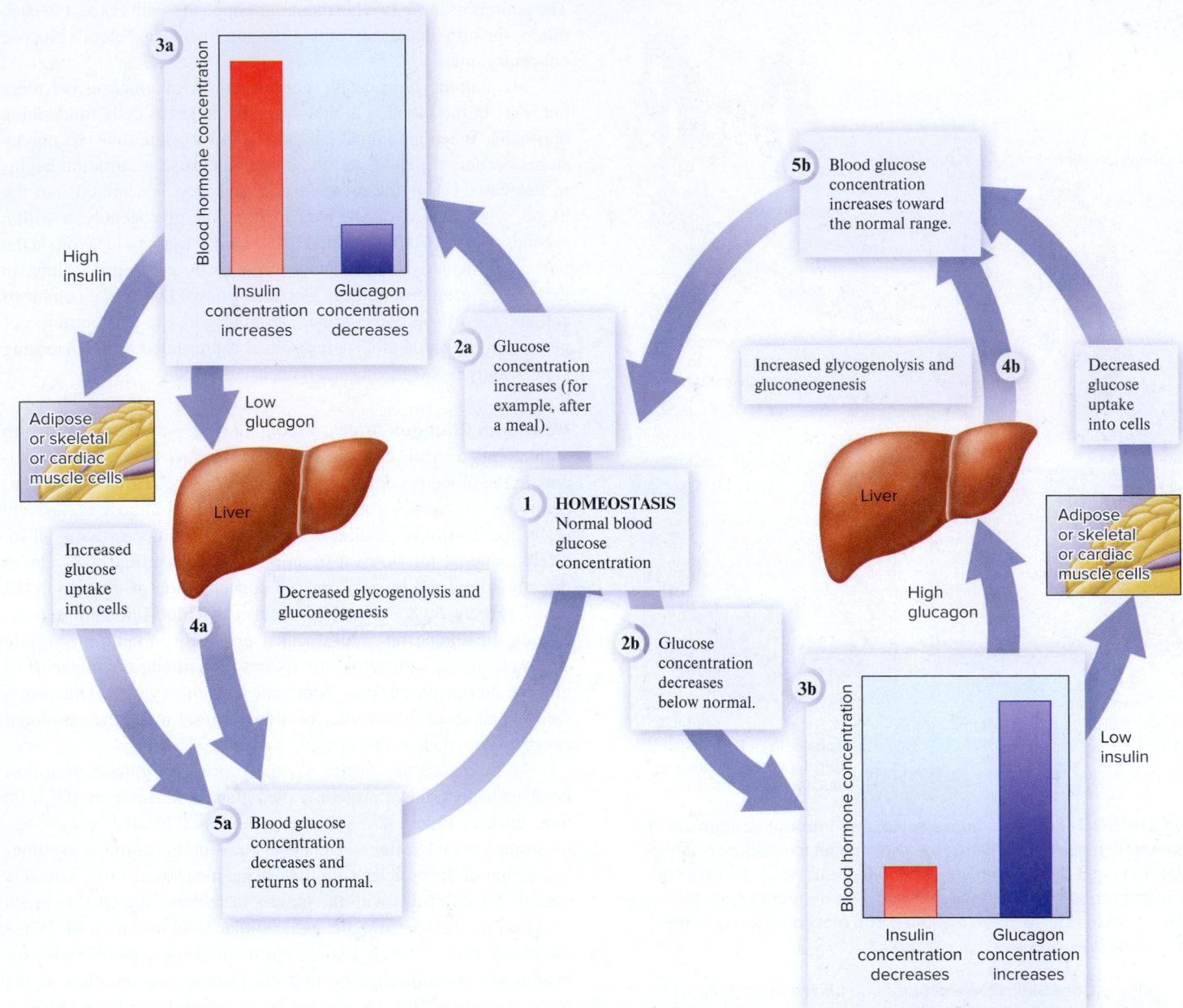

**Figure 50.8** **The functions of insulin and glucagon in glucose homeostasis in the blood.** Glucagon and insulin respond in opposite ways to changes in the plasma glucose concentration. If the plasma glucose concentration decreases below normal (right), the insulin concentration decreases, and glucagon increases. If the plasma glucose concentration increases above normal (left), the insulin concentration increases, and glucagon decreases. Both of these mechanisms return the plasma glucose concentration to normal. In addition, through their action on the liver, hormones from the adrenal medulla and adrenal cortex play significant roles in elevating glucose concentration (not shown).

is treatable in humans and animals (for example, common pets) with regular monitoring of blood glucose concentration and daily administration of insulin.

In humans, the more common form of diabetes, which accounts for roughly 90–95% of all cases (estimates range between 15 and 20 million people in the U.S. alone), is **type 2 diabetes mellitus (T2DM)**. In T2DM, the pancreas may function relatively normally and is not attacked by the immune system. However, the cells of the body lose much of their ability to respond to insulin for reasons that are still unclear. T2DM

is linked to obesity and can often be prevented or reversed with weight control. Drugs are also available that improve the ability of cells to respond to insulin. Formerly, T2DM was known as adult-onset diabetes, because it usually appeared in middle age. This name is no longer used, because the rising tide of childhood obesity in developed countries like the U.S. has unfortunately sparked a sharp increase in the number of young people with the disease.

The discovery of insulin and its application to the human condition was one of the greatest and most influential achievements in the history of medical research, as described next.

## Core Skill: Process of Science

# Feature Investigation | Banting, Best, MacLeod, and Collip Were the First to Isolate Active Insulin

In the 19th century, scientists discovered that in addition to the exocrine part of the pancreas, the organ contains spherical clusters of cells that are not associated with its exocrine digestive functions. These clusters were called islets by their discoverer, German physiologist Paul Langerhans. By the early 20th century, German scientists had discovered that in dogs, complete removal of the pancreas immediately resulted in diabetes. Researchers assumed, therefore, that cells of the islets of Langerhans produced a factor of some kind that prevented diabetes—in other words, it helped maintain glucose homeostasis by preventing the blood glucose concentration from increasing uncontrollably. The factor, however, proved impossible to isolate using the chemical purification methods available at the time. Researchers assumed that during the process of grinding up a pancreas to produce an extract, the factor was destroyed by the digestive enzymes of the exocrine pancreas.

This problem was eventually solved in 1921 by a team of Canadian scientists: surgeon Frederick Banting, medical student Charles Best, and biochemist James Collip. The work was performed in the laboratory of John MacLeod, a renowned expert in carbohydrate metabolism at the time. Banting had read a paper in a medical journal that described a deceased patient in whom the pancreatic duct, which carries digestive juices to the small intestine (see Figure 46.7), had become clogged due to calcium deposits. The closed duct caused pressure to build up behind the blockage, which eventually caused the exocrine part of the pancreas to atrophy and die. The islets of

Langerhans, however, survived intact. Banting hypothesized that if he were to tie off, or ligate, the pancreatic duct of an experimental animal, and then wait a sufficient time for the exocrine pancreas to die, he could more easily obtain an active glucose-lowering factor from the remaining islets without the problem of contamination by digestive enzymes.

Banting and Best proceeded to ligate the pancreatic ducts of several dogs, as shown in **Figure 50.9**. After waiting 7 weeks, an amount of time they previously had determined was sufficient for the exocrine part of the pancreas to atrophy, they prepared extracts of the remaining parts of the pancreas, including its islets of Langerhans. This was done by removing the atrophied pancreas from each dog and grinding it up with a mortar and pestle in an acid solution. The extract was then injected into a second group of dogs in which diabetes had previously been induced by surgically removing the pancreas. The researchers discovered that the extract was capable of keeping the diabetic dogs healthy for a brief time. However, they were unable to isolate sufficient quantities of the active factor in the extract to keep the experiments going longer than a day or so, nor were they able to obtain sufficiently pure factor to prevent side effects such as infection and fever.

This is where Collip's expertise came in. Collip developed a method to precipitate contaminating proteins from the extract by adding alcohol to the acid (step 3, Figure 50.9). Different proteins will precipitate from a solution in response to different concentrations

---

**Figure 50.9**  **The isolation of insulin by Banting, Best, MacLeod, and Collip.**

  **Core Concept: Science and Society**  The discovery of insulin and develpment of the methods to manufacture it in large amounts have improved and saved the lives of many millions of people. Today, dozens of human (and animal) diseases, including diabetes, are treatable with the use of hormones.

**HYPOTHESIS**  Ligation of pancreatic ducts will cause atrophy of the exocrine pancreas, allowing extraction of an active glucose-lowering factor from the remaining portion of the pancreas.

**KEY MATERIALS**  One group of dogs for ligation experiments; second group of dogs made diabetic by having pancreas removed.

Experimental level | Conceptual level

1 Ligate pancreatic ducts in one group of dogs by tying threads around the base of the ducts and pulling them tight.

Surgeon operating on midgut region of a dog

Ligated ducts block flow of digestive juices, which damages exocrine pancreas.

Pancreas

Islets of Langerhans

Small intestine

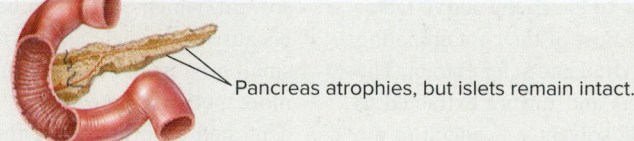

2 Allow 7 weeks for atrophy of pancreas.

Pancreas atrophies, but islets remain intact.

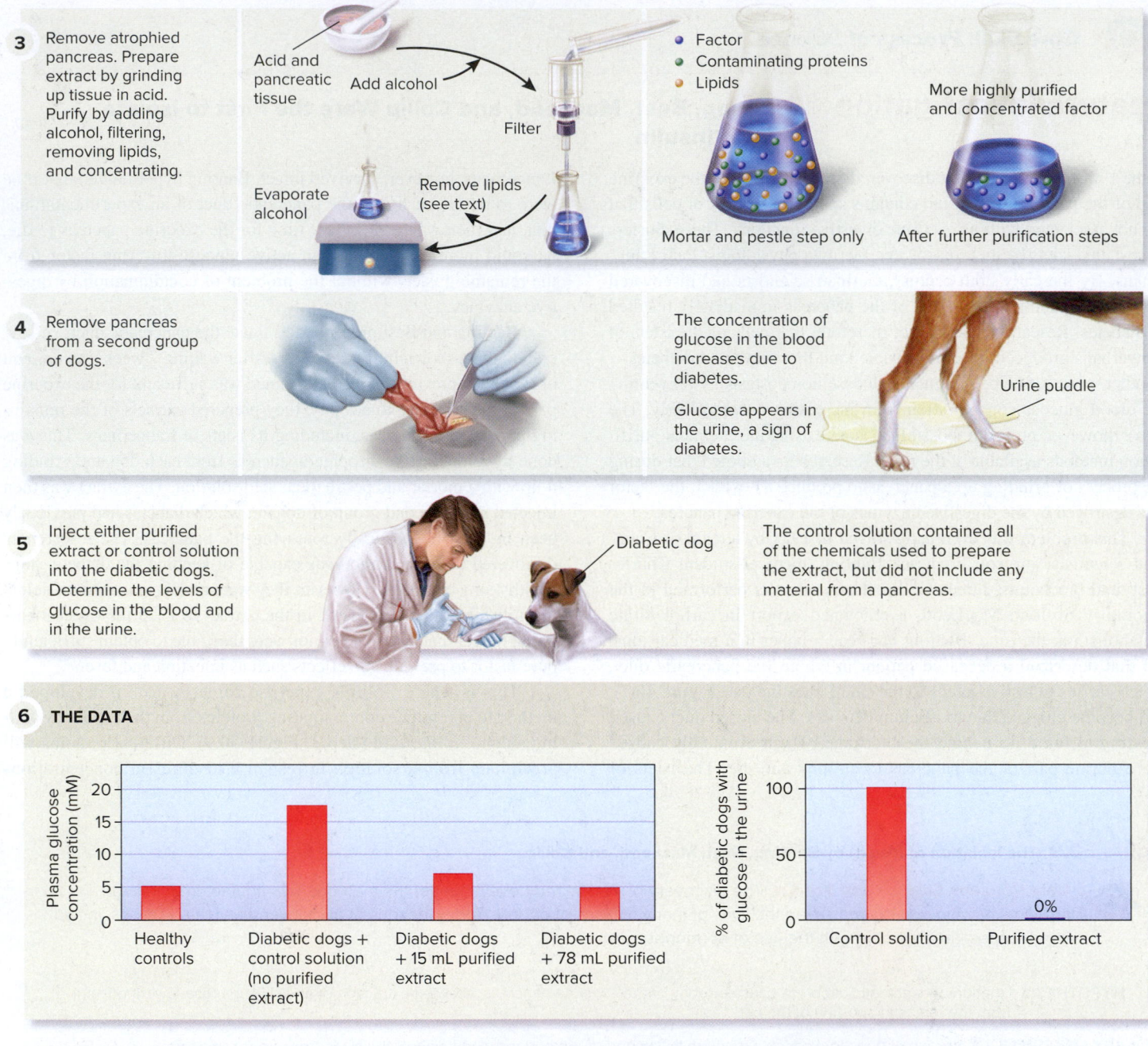

**3** Remove atrophied pancreas. Prepare extract by grinding up tissue in acid. Purify by adding alcohol, filtering, removing lipids, and concentrating.

Acid and pancreatic tissue

Add alcohol

Filter

Evaporate alcohol

Remove lipids (see text)

- Factor
- Contaminating proteins
- Lipids

More highly purified and concentrated factor

Mortar and pestle step only

After further purification steps

**4** Remove pancreas from a second group of dogs.

The concentration of glucose in the blood increases due to diabetes.

Glucose appears in the urine, a sign of diabetes.

Urine puddle

**5** Inject either purified extract or control solution into the diabetic dogs. Determine the levels of glucose in the blood and in the urine.

Diabetic dog

The control solution contained all of the chemicals used to prepare the extract, but did not include any material from a pancreas.

**6** THE DATA

Plasma glucose concentration (mM)

- Healthy controls
- Diabetic dogs + control solution (no purified extract)
- Diabetic dogs + 15 mL purified extract
- Diabetic dogs + 78 mL purified extract

% of diabetic dogs with glucose in the urine

- Control solution
- Purified extract — 0%

**7** **CONCLUSION** Atrophy of the exocrine portion of the pancreas eliminated digestive enzymes that would have degraded the glucose-lowering factor (insulin) during its purification from the pancreas. Insulin was effective in reversing the major symptoms of diabetes in dogs.

**8** **SOURCE** Banting, F. G., and Best, C. H. 1922. The internal secretion of the pancreas. *Journal of Laboratory and Clinical Medicine* 7:256–271.

of alcohol. Using a concentration too low to cause the active factor to precipitate, Collip was able to remove most of the contaminating proteins, leaving behind a much more purified and safer extract. The extract was then filtered to remove debris and further extracted to remove lipids. In this step, a hydrophobic solvent was added to the partially purified extract. The lipids in the extract dissolved in the solvent, which could then be decanted and removed. This was done because the researchers correctly hypothesized that the factor was a small protein and that by removing lipids, they would obtain an even more purified preparation. Finally, the purified extract was concentrated by evaporating the alcohol, which increased its potency. When increasing amounts of this purified factor were injected into a second group of diabetic dogs, their blood glucose concentrations decreased and even returned to normal, and there was no longer any glucose in their urine. Diabetic dogs that received only a control solution that did not contain the factor continued to show a high glucose concentration in the blood and urine, compared with healthy controls.

In later experiments, two subsequent innovations enabled the researchers to obtain larger amounts of the factor and more accurately assess its potency without using the ligation procedure. First, the researchers chose the very large pancreases of cows, obtained at a local slaughterhouse, for the starting material from which to prepare the extracts. Second, Collip developed a highly sensitive assay to more precisely measure the concentration of glucose in the blood of an animal before and after injection of the purified factor. The combination of the improved chemical purification steps, larger amounts of starting material, and improved assays for testing the extract proved to be the keys that enabled the team to test the factor, which they

eventually named insulin, on human patients. The first successful test came in 1922 on a 14-year-old boy in Toronto, who was seriously ill from T1DM. The success of the team in rapidly isolating insulin and proving its effectiveness was so significant that Banting and MacLeod were awarded the 1923 Nobel Prize in Medicine or Physiology. The Nobel committee felt that Banting and MacLeod were the leaders of the project and that it was they who deserved the prize, but the two scientists disagreed. Banting shared the monetary portion of his award with Best, and MacLeod shared his with Collip.

### Experimental Questions

1. How did Banting and Best propose to obtain the glucose-lowering factor produced by the pancreas?

2. What was Collip's contribution to the isolation of insulin?

3. **CoreSKILL »** Propose an explanation as to why glucose appears in the urine of untreated diabetic animals or humans.

## Adipose Tissue Secretes the Hormone Leptin, Which Regulates Appetite

Hormones also contribute to energy balance by exerting effects on appetite and, as a consequence, on food consumption. As you might imagine from its importance as the major storage site for energy-yielding fat, a chief source of appetite-regulating hormones is adipose tissue. One such hormone, introduced in Chapter 47, is the protein leptin, which was first characterized in 1994 in rodents but has since been observed in all vertebrate classes. Leptin is released by adipose cells into the blood in direct proportion to the amount of adipose tissue in the body, and it acts on the hypothalamus to inhibit appetite and increase metabolic rate (refer back to Figure 47.7). When adipose stores are low, however, leptin secretion from adipose cells decreases, resulting in a decreased blood concentration of leptin. This removes the inhibitory effect of leptin on appetite, resulting in an increase in appetite and increased food consumption by the animal, along with a decrease in metabolic rate. In these ways, the amount of energy stored in an animal's body in the form of fat is communicated to the brain via leptin to regulate how hungry an animal feels.

We have seen that hormones influence energy homeostasis by regulating nutrient concentrations in the blood, transport of nutrients into cells, metabolic rate, and appetite. In the next section, we will consider how hormones control another feature of homeostasis, regulating concentrations of key minerals in the blood.

## 50.4 Hormonal Control of Mineral Balance

### Learning Outcomes:

1. Describe the hormonal control of $Ca^{2+}$ homeostasis.

2. Explain the regulation in the kidneys of water and $Na^+$ and $K^+$ balance by antidiuretic hormone, aldosterone, and atrial natriuretic factor.

All animals must maintain a proper balance of minerals such as $Ca^{2+}$, $Na^+$, and $K^+$ in their cells and extracellular fluids. These ions

participate in numerous functions common throughout much of the animal kingdom. For example, the partitioning of ions across plasma membranes determines in part the electrical properties of neurons and muscle cells. In this section, we will see how maintaining a homeostatic balance of these ions in vertebrates is coordinated in large part by hormones.

## Vitamin D and Parathyroid Hormone Regulate $Ca^{2+}$ Concentration in the Blood of Vertebrates

Calcium ions have critical roles in neuronal transmission, heart function, muscle contraction, bone formation, and numerous other events. Therefore, the concentration of $Ca^{2+}$ in the blood is among the most tightly regulated variables in an animal's body.

***1,25-Dihydroxyvitamin D*** Calcium is obtained from the diet and absorbed by the small intestine. Animals obtain calcium from the bones of prey, many types of vegetation, fruits, and even, in some cases, the water they drink. Like all charged particles, $Ca^{2+}$ cannot readily cross plasma membranes, including those of the epithelial cells of the small intestine, and thus its transport must be regulated. This process is controlled by the action of a derivative of **vitamin D**. In most mammals that receive regular exposure to sunlight, skin cells produce vitamin D from a precursor called 7-dehydrocholesterol, a reaction that requires the energy of UV light (**Figure 50.10**). In addition, many animals obtain vitamin D from food, and people may obtain the vitamin from milk and other foods to which it has been added as a supplement. This supplementation is especially important for people who live at latitudes that receive little sunlight for much or part of the year.

Before it can act, vitamin D must first be modified by two enzymes that each add one hydroxyl group to specific carbon atoms; the first enzyme acts in the liver and the second in the kidney. The final active product is called 1,25-dihydroxyvitamin D. This molecule is a hormone that is secreted into the blood by an organ—the kidneys—and then acts on a distant target tissue—the small intestine. The major function of 1,25-dihydroxyvitamin D in the intestine is

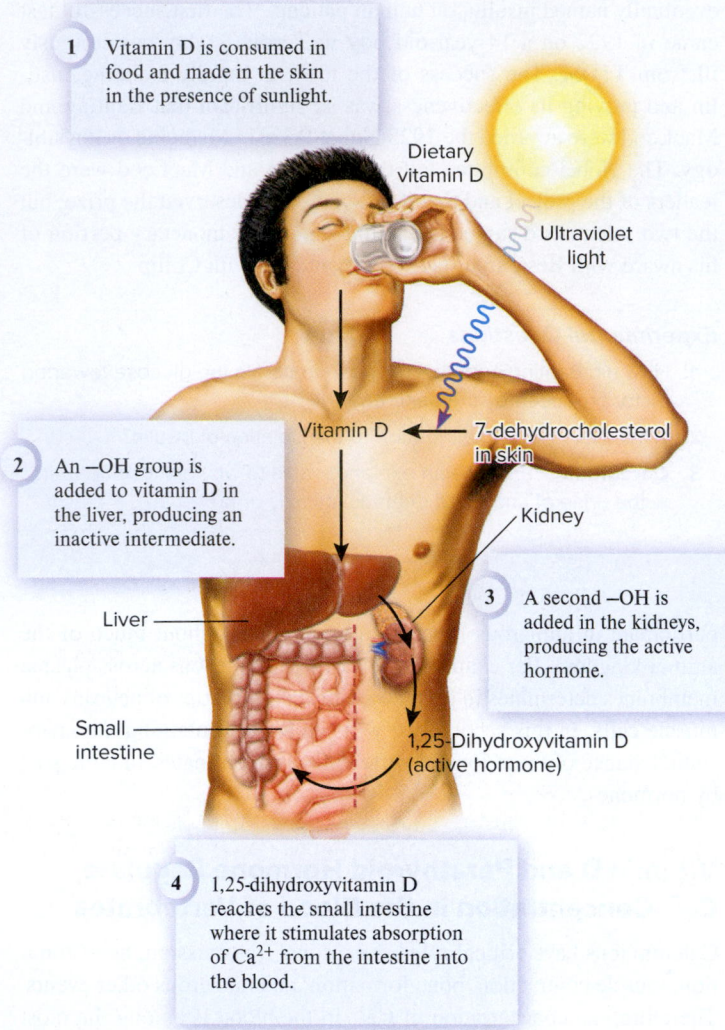

1. Vitamin D is consumed in food and made in the skin in the presence of sunlight.

Dietary vitamin D

Ultraviolet light

Vitamin D ← 7-dehydrocholesterol in skin

2. An —OH group is added to vitamin D in the liver, producing an inactive intermediate.

Kidney

Liver

3. A second —OH is added in the kidneys, producing the active hormone.

Small intestine

1,25-Dihydroxyvitamin D (active hormone)

4. 1,25-dihydroxyvitamin D reaches the small intestine where it stimulates absorption of Ca²⁺ from the intestine into the blood.

**Figure 50.10** **Synthesis of the active hormone formed from vitamin D or its precursor in the skin.**

*Concept Check:* *Would you predict that all mammals synthesize the active form of vitamin D using the energy of sunlight?*

to activate expression of a $Ca^{2+}$ pump in the intestinal epithelium, thereby stimulating the absorption of $Ca^{2+}$. The $Ca^{2+}$ then enters the blood, which delivers the ions to tissues for such activities as building bone and maintaining nerve, muscle, and heart functions. If the active form of vitamin D is not present in the blood at a sufficient level, the bones lose $Ca^{2+}$ and become weakened.

**Parathyroid Hormone** Even when $Ca^{2+}$ is not present in the diet, or when 1,25-dihydroxyvitamin D is not formed in normal amounts because of insufficient exposure to sunlight, the blood concentration of $Ca^{2+}$ does not normally decrease significantly, because of a hormone called **parathyroid hormone (PTH)**. In all tetrapods, this hormone is secreted from several small glands in the neck called parathyroid glands, which in humans are located behind the thyroid gland (**Figure 50.11a**). Fishes produce parathyroid hormone from their gills and do not have parathyroid glands; recent evidence suggests that the gills of fishes and parathyroid glands of tetrapods are evolutionarily related. Cells of the parathyroid glands (or gills)

express receptors in their plasma membranes that bind extracellular $Ca^{2+}$. The binding of $Ca^{2+}$ to these receptors inhibits the secretion of PTH. Thus, a decrease in the blood concentration of $Ca^{2+}$ ends this inhibition and stimulates the cells of the parathyroid glands to secrete more PTH.

PTH acts on bone to stimulate the activity of cells that dissolve the mineral part of bone. This activity releases $Ca^{2+}$, which then enters the blood (**Figure 50.11b**). Typically, only a very small fraction of the total $Ca^{2+}$ in bone is removed in this way. Therefore, besides providing a skeletal framework for the vertebrate body, bone also serves as an important reservoir of $Ca^{2+}$. PTH also acts to increase reabsorption of $Ca^{2+}$ from the filtrate in the kidneys, so less $Ca^{2+}$ is excreted in the urine. If the blood $Ca^{2+}$ concentration increases, PTH secretion is inhibited. Without PTH, $Ca^{2+}$ homeostasis is not possible.

**Calcitonin** In addition to 1,25-dihydroxyvitamin D and PTH, another hormone called **calcitonin** functions in $Ca^{2+}$ homeostasis in some vertebrates, notably fishes and possibly in some mammals. Calcitonin is a polypeptide produced in and secreted from cells in the thyroid gland. Its function is the opposite in many respects of that of PTH. Calcitonin promotes excretion of $Ca^{2+}$ via the kidneys and deposition of $Ca^{2+}$ into bone, thereby decreasing the blood $Ca^{2+}$ concentration. This function is especially important in marine fishes, because of the high $Ca^{2+}$ content of seawater and the entry of this ion into the body fluids of the animals when they drink (refer back to Figure 41.17).

**BIO TIPS** : **THE QUESTION** *What might happen to the blood concentrations of Ca²⁺ and parathyroid hormone (PTH) if the kidney enzyme responsible for the formation of 1,25-dihydroxyvitamin D was absent, nonfunctional, or inhibited for some reason?*

**T OPIC** *What topic in biology does this question address?*
The topic is the relationship between the blood concentrations of $Ca^{2+}$ and those of two hormones that are known to help regulate $Ca^{2+}$ homeostasis. Specifically, how does a change in the amount of one hormone affect the blood concentrations of the other and of $Ca^{2+}$?

**I NFORMATION** *What information do you know based on the question and your understanding of the topic?*
From the question, you know that 1,25-dihydroxyvitamin D is not being produced in normal amounts. From your understanding of the topic, you may remember that this hormone acts in the small intestine to stimulate absorption of $Ca^{2+}$ into the blood. You also know the function of PTH and that its secretion is controlled by the blood concentration of $Ca^{2+}$.

**P ROBLEM-SOLVING S TRATEGY** *Sort out the steps in a complicated process. Predict the outcome.* Based on your knowledge of the relationships among the blood concentrations of $Ca^{2+}$, 1,25-dihydroxyvitamin D, and PTH, you can predict the sequence of events that will follow a decrease in the concentration of 1,25-dihydroxyvitamin D.

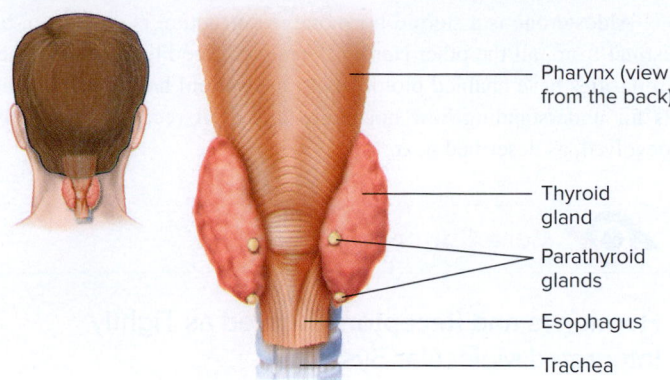

**(a) Location of the parathyroid glands**

Pharynx (view from the back)

Thyroid gland

Parathyroid glands

Esophagus

Trachea

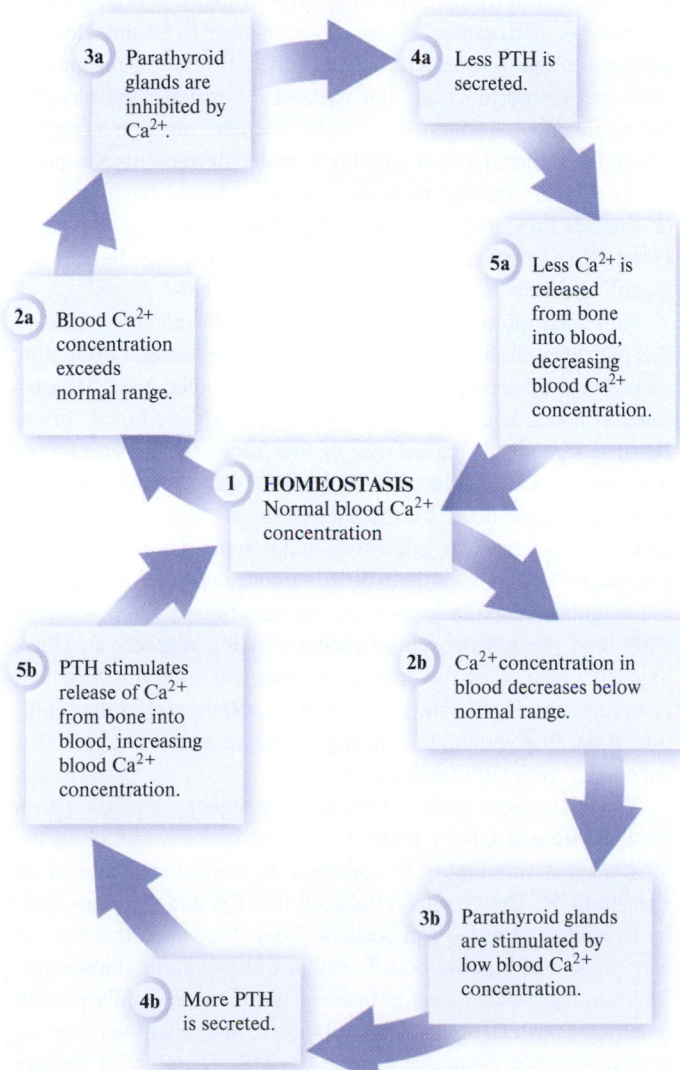

**3a** Parathyroid glands are inhibited by $Ca^{2+}$.

**4a** Less PTH is secreted.

**2a** Blood $Ca^{2+}$ concentration exceeds normal range.

**5a** Less $Ca^{2+}$ is released from bone into blood, decreasing blood $Ca^{2+}$ concentration.

**1  HOMEOSTASIS** Normal blood $Ca^{2+}$ concentration

**5b** PTH stimulates release of $Ca^{2+}$ from bone into blood, increasing blood $Ca^{2+}$ concentration.

**2b** $Ca^{2+}$ concentration in blood decreases below normal range.

**3b** Parathyroid glands are stimulated by low blood $Ca^{2+}$ concentration.

**4b** More PTH is secreted.

**(b) Homeostatic control of blood $Ca^{2+}$ concentration**

**Figure 50.11  The role of parathyroid hormone in calcium homeostasis. (a)** In humans, four small parathyroid glands are located behind the thyroid. **(b)** The action of PTH: Steps 2a–5a occur when $Ca^{2+}$ is in excess in the blood; steps 2b–5b occur when the concentration of $Ca^{2+}$ in the blood is below normal.

**ANSWER**  *The sequence of events is as follows:*

1. *If the blood concentration of 1,25-dihydroxyvitamin D is decreased, the ability to absorb $Ca^{2+}$ from the intestine will be decreased.*
2. *This will lead to a decreased blood concentration of $Ca^{2+}$ as most of the dietary calcium will not get absorbed.*
3. *A decrease in blood concentration of $Ca^{2+}$ will trigger an increase in the secretion of PTH from the parathyroid glands.*
4. *PTH, in turn, will act on bone to resorb some $Ca^{2+}$ that will then enter the blood and help to prevent the $Ca^{2+}$ concentration from decreasing further.*

## Several Hormones Regulate the Extracellular Fluid Concentrations of Na⁺ and K⁺ in Vertebrates

Like $Ca^{2+}$, concentrations of $Na^+$ and $K^+$ in the body fluids of most animals are tightly regulated, because these ions are the basis of membrane potential formation and action potential generation (see Chapter 42), among other functions. Like $Ca^{2+}$, $Na^+$ and $K^+$ are ingested in the diet and excreted in the urine at rates that maintain homeostatic concentrations in the body fluids. Unlike $Ca^{2+}$ in vertebrates, however, no large reservoirs exist in the body for $Na^+$ and $K^+$. One of the key mechanisms that regulates the concentrations of these ions in the blood is altering the rate of $Na^+$, $K^+$, and water reabsorption from the urine as it is being formed in the kidneys. This is accomplished in large part by the actions of three hormones: ADH, aldosterone, and atrial natriuretic peptide (**Figure 50.12**). All three of these hormones exert their effects on the kidneys.

The vertebrate kidney normally reabsorbs most $Na^+$ and $K^+$ from the fluid filtered through the glomeruli (refer back to Figure 49.9). However, dietary and other changes can alter the concentrations of these ions in the blood. When this happens, the kidney works to restore the ions to their normal concentrations. For example, if the blood $Na^+$ concentration increases above normal, the osmolarity of the blood will increase. Osmoreceptors in the brain detect this and stimulate ADH secretion from the posterior pituitary.

As mentioned earlier, ADH acts on the kidneys to reabsorb water from the forming urine. Increasing the amount of ADH available to act on the kidneys results in less water being excreted in the urine (and more being reabsorbed into the blood). In addition, the

 **Core Skill: Modeling**  The goal of this modeling challenge is to make a model that shows the effects on $Ca^{2+}$ homeostasis if cellular sensitivity to PTH is decreased.

**Modeling Challenge:** A type of endocrine disorder in many vertebrates, including mammals, occurs when target cells lose their ability to respond to a particular hormone. For example, target cells may lose their ability to express a functional receptor for a hormone. Consider the bottom part of the model in Figure 50.11b, with the steps labeled 1 through 5b. Revise that model to show how $Ca^{2+}$ homeostasis would be affected if bone cells were unable to respond to PTH.

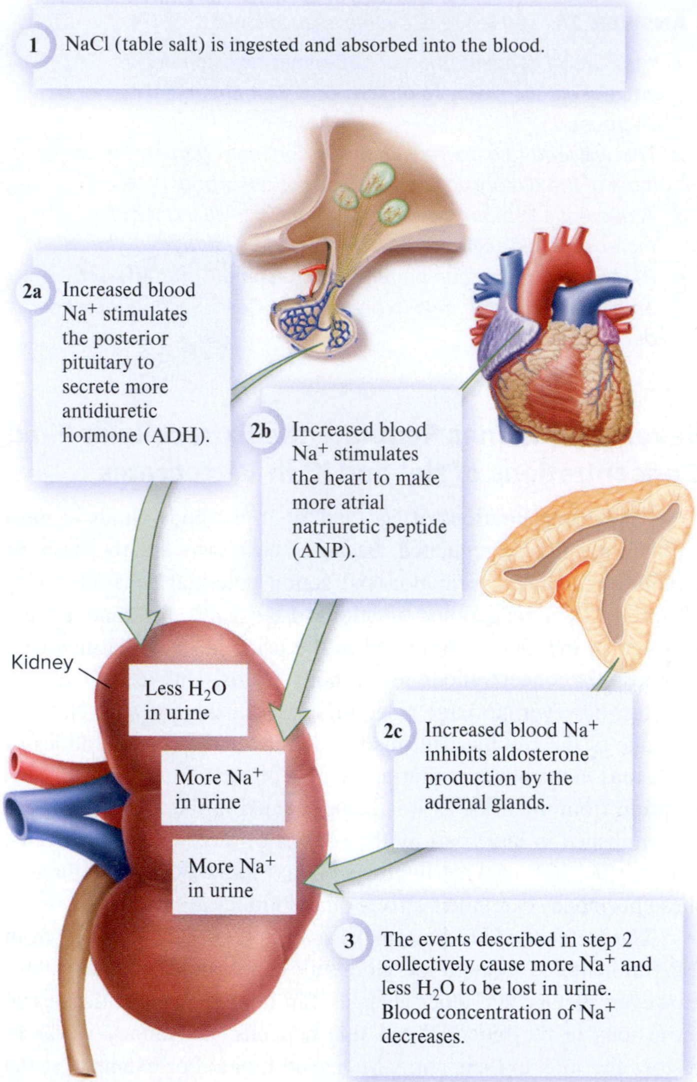

1. NaCl (table salt) is ingested and absorbed into the blood.

2a. Increased blood Na⁺ stimulates the posterior pituitary to secrete more antidiuretic hormone (ADH).

2b. Increased blood Na⁺ stimulates the heart to make more atrial natriuretic peptide (ANP).

Kidney

Less $H_2O$ in urine

More Na⁺ in urine

2c. Increased blood Na⁺ inhibits aldosterone production by the adrenal glands.

More Na⁺ in urine

3. The events described in step 2 collectively cause more Na⁺ and less $H_2O$ to be lost in urine. Blood concentration of Na⁺ decreases.

**Figure 50.12** **An example of Na⁺ balance achieved by the coordinated actions of three hormones.**

*Concept Check:* *Given the importance of these ions in numerous aspects of animal biology, is it surprising that there is more than one hormone that regulates Na⁺ and K⁺ balance?*

synthesis and secretion of the adrenal steroid hormone **aldosterone** are directly inhibited in response to an increase in Na⁺ concentration. Normally, aldosterone increases Na⁺ reabsorption in the kidney, and therefore, its absence results in more Na⁺ excretion in the urine than normal. (Aldosterone also regulates the extracellular concentration of K⁺ by promoting its secretion into the urine; see Figure 49.10.)

In addition, **atrial natriuretic peptide (ANP)** is secreted from the atria of the heart whenever the blood concentration of Na⁺ increases. ANP causes natriuresis (a loss of Na⁺ in the urine; from the neo-Latin *natrium*, meaning sodium) by decreasing Na⁺ reabsorption. Thus, ANP and aldosterone have opposite effects on Na⁺ balance in the body, which is why their concentrations in the blood tend to change in opposite directions. The combined effect of decreasing water loss in the urine and reabsorbing less Na⁺ is to decrease the concentration of Na⁺ in blood and other body fluids, returning its concentration to normal.

Aldosterone is a steroid hormone. Its structure is similar to, but distinct from, all the other steroid hormones (see Figure 50.2). Their similarities have enabled biologists to use steroid hormones as models for understanding how hormones and their receptors may have coevolved, as described next.

## Core Concept: Evolution

### Hormones and Receptors Evolved as Tightly Integrated Molecular Systems

We have explored how hormones act by binding to receptor proteins located on the cell surface or in the cytosol or nucleus. This raises an intriguing question: Which evolved first, a given hormone or its receptor? There would appear to be no selection pressure to evolve one without the other. Without a receptor, a hormone has no function, and without a hormone, the receptor has no function. Hormones and their receptors function as tightly integrated molecular systems. Each molecule requires the other for biological activity. How such systems could have evolved, or whether they evolved together, has long been a major puzzle. However, recent research has generated intriguing new hypotheses regarding the evolution of these signaling systems.

American biologist Joseph Thornton and colleagues tackled this puzzle by examining how two structurally related steroid hormones evolved separate activities and distinct receptors. Aldosterone, as noted, is a steroid hormone made by the adrenal cortex. Because it regulates the balance of two minerals (Na⁺ and K⁺) in the body, it is known as a mineralocorticoid. Cortisol is another steroid hormone made by the adrenal cortex and is known as a glucocorticoid because one of its major functions is to regulate glucose balance. The actions of these hormones are mediated by intracellular receptors known as the mineralocorticoid receptor (MR) and the glucocorticoid receptor (GR), respectively. Phylogenetic analysis of the gene sequences that code for these two receptors suggests that they arose at least 450 mya, after the evolution of the first vertebrates, by duplication of an ancient gene that encoded a corticoid receptor (CR).

The researchers analyzed the known sequences of the genes for the MR and GR of many vertebrate species, including the most ancient vertebrates, to deduce a theoretical sequence of an ancestral CR. They then synthesized this CR and tested its ability to bind aldosterone and cortisol. They discovered that the CR was capable of binding both hormones, particularly aldosterone. This finding was surprising, because aldosterone is only present in tetrapods, which arose around 300 mya, long after the proposed gene duplication event that created the MR. Therefore, it appears that a receptor with high affinity for aldosterone was present long before animals acquired the capacity to synthesize aldosterone. The receptor seems to have evolved to bind other steroids. Around 450 mya, the gene for the CR duplicated, and then the two resulting genes seem to have further mutated such that one receptor gained high affinity primarily for mineralocorticoids and the other for glucocorticoids.

These studies suggest that the ability of an animal to respond to aldosterone evolved long before aldosterone did, because a receptor with high affinity was already in place in animals. When aldosterone evolved in tetrapods, it was able to use the MR derivative of the ancestral CR. In this example, the answer to "which came first?" appears to be that the receptor evolved first and the hormone later.

# 50.5 Hormonal Control of Growth and Development

## Learning Outcomes:

1. List and describe the functions of the polypeptide hormones involved in the regulation of growth and development in vertebrates.

2. Identify the functions of the three major hormones that control growth and development in insects.

Hormones play a nearly universal role in controlling growth and development in animals. Growth can occur slowly over long periods or in brief spurts, with some animals exhibiting both types of growth. Many mammals, for example, grow slowly but steadily until puberty, then experience a period of rapid growth, followed by slower rates of growth, and finally cessation of growth. Many insects, however, grow and develop in spurts during molting periods. Although growth is determined by many factors, notably adequate nutrition, it is regulated in large part by the endocrine system.

Growth is distinguished from development, which is the process by which cells form tissues with specific functions, and tissues develop into larger and more complex structures. In this section, we will examine how both processes depend in part on the endocrine system.

## Vertebrates Require a Balance of Several Hormones for Normal Growth

Several hormones stimulate growth in vertebrates. The anterior pituitary gland produces **growth hormone (GH)**, which is under the control of the hypothalamus (see Table 50.2). GH acts on the liver to produce another hormone, called **insulin-like growth factor-1 (IGF-1)**. In mammals, IGF-1 stimulates the elongation of bones, especially during puberty, when mammals become reproductively mature. This growth is further accelerated by the steroid hormones of the gonads, resulting in the rapid growth spurt in puberty. Eventually, however, the gonadal hormones cause the growth regions of bone to seal, preventing any further bone elongation.

GH, IGF-1, and gonadal steroid hormones continue to be produced in adult vertebrates—including humans—even though growth has ceased. In adulthood, GH serves metabolic functions, such as helping to regulate the concentrations of glucose and fatty acids in the blood. The gonadal steroids are important for reproduction in adults (see Chapter 51).

In rare cases, a tumor of the GH-secreting cells of the anterior pituitary gland produces excess GH. If this occurs during childhood

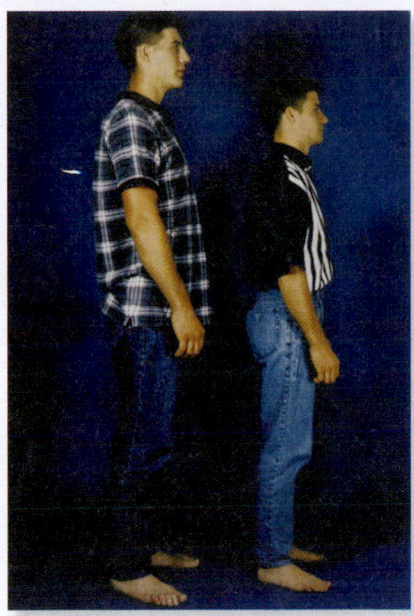

**Figure 50.13** **Pituitary gigantism in one individual from a pair of identical twins.** This disorder is caused by a high concentration of growth hormone and insulin-like growth factor-1, which leads to increased height. If the excess of growth hormone continues after growth stops, it results in enlarged bones, including those of the face, hands, and feet. Courtesy Dr. Robert F. Gagel

**Concept Check:** *Did the individual on the left in Figure 50.13 develop a growth hormone disorder before or after puberty?*

in humans, a person with this disorder can grow exceedingly tall and is known as a **pituitary giant** (**Figure 50.13**). Soon after puberty, growth stops. If a tumor causes a high GH concentration only after puberty when growth has ceased, the excess GH causes many bones, such as those of the face, hands and feet, to thicken and enlarge, a condition known as **acromegaly**. People with acromegaly are generally treated with a synthetic form of somatostatin, which is the inhibitor of GH made by the hypothalamus (see Table 50.2). In many cases, however, the tumor must be surgically removed. A person may be both a pituitary giant and an acromegalic, if the excess GH production began before puberty but then continued for a time after growth had stopped and medical treatment was not begun.

By contrast, if the pituitary fails to make adequate amounts of GH during childhood in a human, the concentrations of GH and IGF-1 in the blood will be lower than normal. In such cases, growth is stunted, resulting in one of the many possible causes of **short stature**. Individuals with this condition can be treated with injections of recombinant human GH and will grow to relatively normal height, as long as treatment begins before puberty is completed, so that the bones can still elongate.

Hormones are also required during fetal life, not for growth but for development of the brain, lungs, and other organs into fully mature, functional structures. For example, cortisol from the fetal adrenal gland of mammals is vital for proper lung formation. In humans, premature babies born before the adrenal glands have matured have lungs that are not capable of inflating properly. Such babies require hospitalization to survive (refer back to Figure 48.22 for details about the treatment of respiratory distress syndrome in newborns).

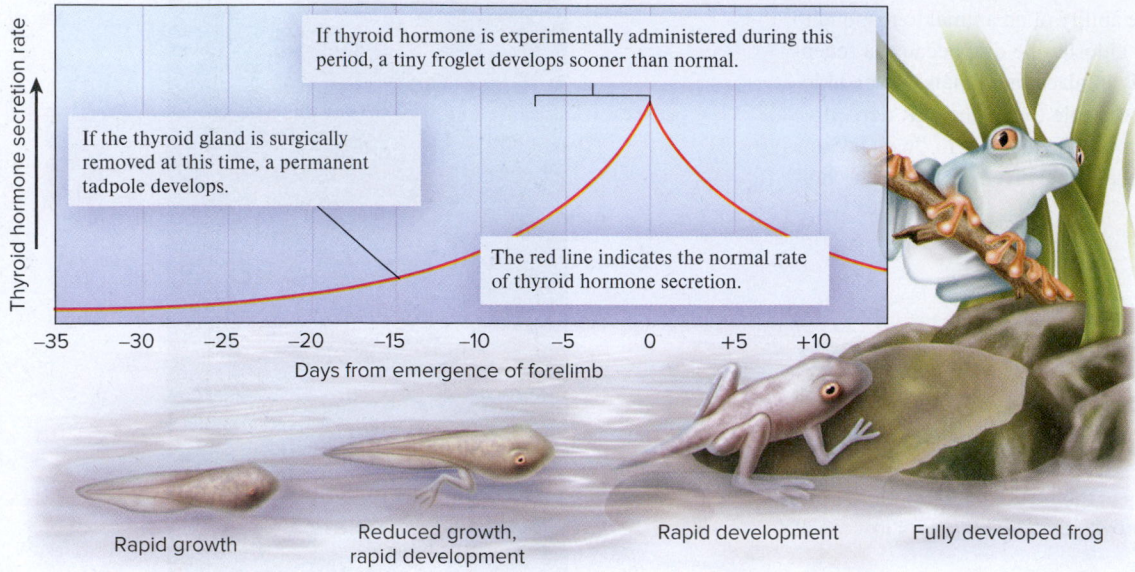

If thyroid hormone is experimentally administered during this period, a tiny froglet develops sooner than normal.

If the thyroid gland is surgically removed at this time, a permanent tadpole develops.

The red line indicates the normal rate of thyroid hormone secretion.

Thyroid hormone secretion rate

Days from emergence of forelimb

−35  −30  −25  −20  −15  −10  −5  0  +5  +10

Rapid growth

Reduced growth, rapid development

Rapid development

Fully developed frog

**Stages of normal development**

**(a) The effect of thyroid hormones on tadpole development**

**(b) Japanese flounder**

**Figure 50.14 Effects of thyroid hormone on animal development. (a)** Experimental manipulation of thyroid hormone concentration can slow down or accelerate development of tadpoles. **(b)** The eyes of a bottom-dwelling flounder (*Platichthys flesus*) migrate to one side of the head during development, partly in response to thyroid hormone.
b: ©blickwinkel/Alamy Stock Photo

 **Core Concept: Evolution** The regulation of animal development is an ancestral function of thyroid hormone. In the course of evolution, this hormone has gained many other functions, but has retained a developmental role in all vertebrates, including humans.

As another example, thyroid hormone influences development among all vertebrates. In amphibians, thyroid hormone has a critical function in metamorphosis, notably in tadpoles, where it promotes the resorption of the tail and development of the legs (**Figure 50.14a**). This effect can be dramatically demonstrated by experimentally altering a tadpole's thyroid hormone levels. Decreasing the amount of thyroid hormone results in a tadpole that does not transform into a frog, whereas increasing the hormone causes a tadpole to undergo metamorphosis sooner than normal, resulting in a tiny froglet.

In fishes, thyroid hormone has an equally critical function in development. For example, among species of flatfish such as flounder, thyroid hormone is responsible for the characteristic change in appearance that occurs in these species as they settle into a sedentary existence on the ocean bottom. The fins and gill covers migrate to the dorsal surface facing the water; the dorsal body surface becomes pigmented; and, most remarkably, the eyes migrate to the same side of the head so that one eye is not unused on the side facing the ocean floor (**Figure 50.14b**). Each of these metamorphic events is under the direct control of thyroid hormone.

## Invertebrates Grow and Develop in Spurts Under the Control of Three Major Hormones

Like vertebrate endocrine pathways, hormonal control systems in insects and other invertebrates often involve multiple glands and neural structures acting in concert. In insects, for example, the endocrine system is critical for the growth and development of larvae and their eventual development into pupae (**Figure 50.15**). In the case of larval growth and metamorphosis, specialized neurosecretory cells in the brain periodically secrete a hormone called **prothoracicotropic hormone** (**PTTH**), which then stimulates a pair of endocrine glands called the prothoracic glands. These glands,

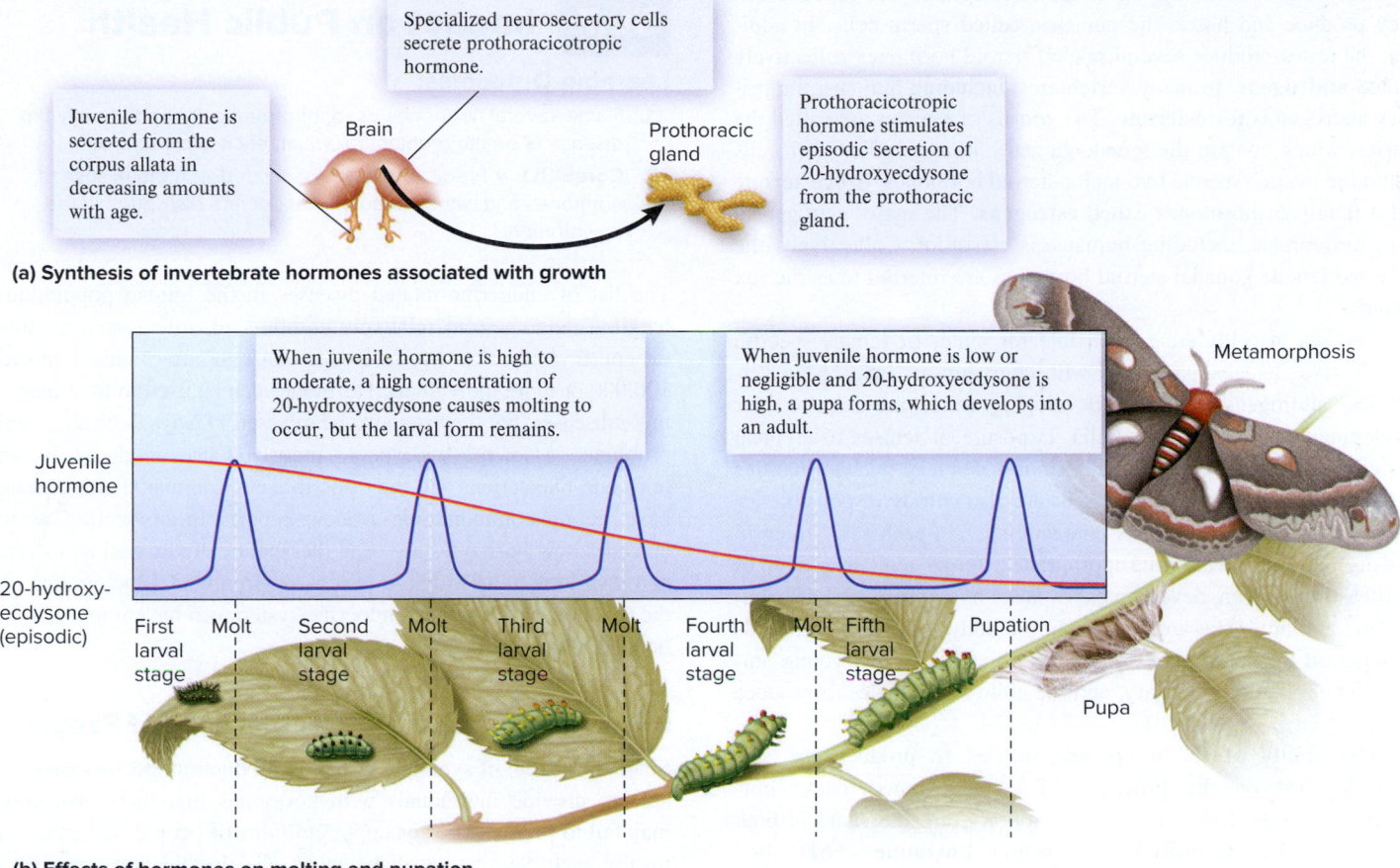

Specialized neurosecretory cells secrete prothoracicotropic hormone.

Juvenile hormone is secreted from the corpus allata in decreasing amounts with age.

Brain

Prothoracic gland

Prothoracicotropic hormone stimulates episodic secretion of 20-hydroxyecdysone from the prothoracic gland.

**(a) Synthesis of invertebrate hormones associated with growth**

When juvenile hormone is high to moderate, a high concentration of 20-hydroxyecdysone causes molting to occur, but the larval form remains.

When juvenile hormone is low or negligible and 20-hydroxyecdysone is high, a pupa forms, which develops into an adult.

Metamorphosis

Juvenile hormone

20-hydroxy-ecdysone (episodic)

First larval stage | Molt | Second larval stage | Molt | Third larval stage | Molt | Fourth larval stage | Molt | Fifth larval stage | Pupation

Pupa

**(b) Effects of hormones on molting and pupation**

**Figure 50.15  Hormonal control of insect development.**  Development of insects requires the coordinated actions of three hormones. Note: The relative concentrations and secretion patterns of 20-hydroxyecdysone and juvenile hormone in this figure are schematic and not representative of all insects.

*Concept Check:*  *In what part of a cell do you predict the receptor for the steroid hormone 20-hydroxyecdysone is located?*

located in the thorax, synthesize and secrete into the hemolymph a steroid hormone called **20-hydroxyecdysone**. This hormone is secreted only periodically. In response to each burst of secretion, the larva undergoes a rapid development and molts (sheds its cuticle). It then begins a new developmental period until it molts again in response to another episode of secretion. The molting process is known as ecdysis, from which the name of the hormone is derived (refer back to Figure 33.13).

Throughout larval development, paired neurosecretory structures behind the brain, called the corpus allata, secrete another hormone, a protein called **juvenile hormone (JH)**. Although 20-hydroxyecdysone induces molting, JH prevents metamorphosis into an adult (hence the name juvenile, which reflects the fact that JH fosters the larval stage). As a larva progresses through different stages, however, the amount of JH it produces gradually declines until the concentration is nearly zero (see Figure 50.15). During this time, 20-hydroxyecdysone continues to be periodically secreted. The decline in JH below a certain concentration results in the transition from larva to pupa in response to a burst of 20-hydroxyecdysone. The near absence of JH is a prerequisite for the final step of metamorphosis into an adult.

## 50.6  Hormonal Control of Reproduction

**Learning Outcomes:**

1. List the vertebrate male and female sex steroids, their sites of production, and their functions in reproductive development.
2. **CoreSKILL »** Propose a mechanism by which hormones can link nutrition with reproduction.

The topic of reproduction is covered in detail in Chapter 51, but it is worth noting here that in all vertebrates and probably most invertebrates, reproduction is closely linked with endocrine function. In this section, we will briefly discuss the most common reproductive hormones and some of their major actions in male and female vertebrates.

### The Gonads Secrete Sex Steroids That Influence Most Aspects of Reproduction

Hormones produced by the gonads of vertebrates have vital functions in nearly all aspects of reproduction, from reproductive behaviors to the ability to produce offspring. In males, the gonads are called testes. They produce and house the gametes, called sperm cells. In addition, the testes produce several related steroid hormones collectively termed **androgens**. In many vertebrates, including humans, the primary androgen is **testosterone**. The gonads of females are called the ovaries, which contain the female gametes, or egg cells. Other cells within the ovaries secrete two major steroid hormones, **progesterone** and a family of hormones called **estrogens**. The major estrogen in many vertebrates, including humans, is **estradiol**. Collectively, the male and female gonadal steroid hormones are referred to as the sex steroids.

The sex steroids are responsible for male- or female-specific reproductive changes associated with courtship and mating. In vertebrates, androgens from the developing testes are required for development of the male genitalia. Exposure of fetuses to atypical concentrations of male or female hormones can lead to abnormalities in genitalia morphology. Sex steroids are also chiefly responsible for development of secondary sex characteristics in each sex, which in humans include growth of the appropriate external genitals, growth of the breasts in women, development of facial hair in men, and distribution and amount of fat and muscle in the body. Finally, sex steroids are required for maturation of gametes, the transition of young animals to reproductive maturity, and the ability of females to produce young.

The ability of the testes and ovaries to produce sex steroids depends on the presence of gonadotropins. These hormones are secreted by the anterior pituitary, are the same in both sexes, and include **follicle-stimulating hormone (FSH)** and **luteinizing hormone (LH)** (see Table 50.2). Therefore, identical anterior pituitary gland hormones control the gonads in both male and female animals.

## Nutrition and Reproduction Are Linked Through Hormones

An interesting feature of the endocrine control of reproduction is the observation that puberty is delayed in mammals, including humans, that are very undernourished. Similarly, fertility—the ability to produce offspring—is reduced in women and other adult female mammals under such conditions. This makes sense, because supporting the nutritional demands of a growing fetus would be difficult without adequate nutrition and energy stores. In such a case, it is more advantageous for an animal to delay reproduction until sufficient food is available.

How does the brain of a female mammal determine when sufficient energy is stored in her body to support pregnancy? The answer appears to be partly the result of hormonal signals. For example, if the amount of fat in a female mammal's body decreases, so does the concentration of leptin in her blood, as described earlier. Leptin has been demonstrated to stimulate synthesis and secretion of reproductive hormones such as FSH and LH. Consequently, undernourishment that causes a decrease in leptin also results in decreased production of reproductive hormones, thereby contributing to a loss of fertility. Therefore, leptin acts as a link between energy stores and the reproductive system.

## 50.7 Impact on Public Health

**Learning Outcomes:**

1. Name several hormones used therapeutically, and identify the disease or health problem for which each is prescribed.
2. **CoreSKILL »** Predict the risks associated with the misuse of hormones and with exposure to endocrine disruptors in the environment.

The list of endocrine-related diseases in the human population is lengthy, ranging from relatively common disorders such as diabetes and thyroid disease to rare ones that may affect only 1 in every 100,000 or more individuals. Hormones can be used to treat many of these diseases, but they can also be misused in ways that cause health problems. Also, the widespread industrial use of chemicals with hormone-like actions and their possible environmental consequences have become common topics of news reports. In this section, we will consider how hormones are used therapeutically as well as misused, and then look at an example of how environmental factors that alter the function of the human endocrine system can have important public health implications.

### Hormones Are Used to Treat Millions of People

Since the advent of synthetic hormone production, doctors have been able to provide individuals with hormones that their own bodies may fail to produce. For example, millions of people self-administer insulin each day to treat diabetes or take thyroid hormone supplements to treat hypothyroidism. Many women take hormones to help induce pregnancy or control the symptoms of menopause. A growing number of men, too, are administered gonadal hormones if the levels produced by their bodies decline significantly in later life. Other examples of therapeutic hormone use include recombinant human growth hormone to treat abnormally short stature in prepubertal children, epinephrine inhalers to treat asthma, and glucocorticoids to treat inflammation, lung disease, and skin disorders, to name just a few.

### Hormone Misuse Can Have Disastrous Consequences

As we saw at the beginning of this chapter, some individuals, typically those in competitive sports, self-administer hormones such as androgens. This practice may increase muscle mass and improve athletic performance. However, the price is steep. First, androgens exert negative feedback effects on secretion of FSH and LH by the anterior pituitary gland. As a consequence, the anterior pituitary gland stops producing and secreting FSH and LH while the user is taking supplements containing androgens. In males, this causes the testes to shrink, as they no longer are making sperm, and the man becomes infertile; this is what happened to the young man described at the beginning of the chapter. Androgen administration has also been linked to extreme aggressive behavior, cardiovascular disease and heart attacks, skin problems, and certain cancers. Women using androgens face health risks similar to those of men but, in addition, develop masculinizing traits, including increased growth of body hair and thinning scalp hair.

Another example of misuse occurs when a hormone is used to boost the number of erythrocytes in the circulation in order to increase the oxygen-carrying capacity of the blood (a practice often called blood-doping). The frequency of this practice has increased in recent years among athletes participating in long-distance aerobic activities, such as competitive cycling and cross-country skiing. The hormone used is **erythropoietin (EPO)**, which acts by stimulating the maturation of erythrocytes in the bone marrow and their release into the blood. EPO is primarily made by the kidneys as part of the homeostatic control of erythrocyte production. Its secretion is increased above normal in response to any situation where additional blood cells are required, for example, following blood loss or when a person lives at high altitudes, where the oxygen pressure is low. When EPO is used abusively, however, the number of erythrocytes can reach such a high level that the blood becomes much more viscous than normal. This increased viscosity puts a serious strain on the heart, which must work harder to pump the thickened blood. Since the 1990s, the international cycling community has been rocked by an alarming number of world-class European cyclists who died of heart attacks in the prime of their lives. These individuals had been using EPO to gain an unfair and, as it turns out, unwise advantage over their peers. Testing continues to detect injected EPO in the blood of some cyclists and other endurance athletes.

## Synthetic Compounds May Act as Endocrine Disruptors

A recent and disturbing phenomenon is the growing prevalence of so-called **endocrine disruptors** in lakes, streams, ocean water, and soil exposed to pollution runoff. These chemicals are often derived from industrial waste and have molecular structures that in some cases sufficiently resemble estrogen to bind to estrogen receptors. They are also widespread in common household goods including nearly all plastic products. If these compounds make their way into drinking water or food, they can exert estrogen-like actions or inhibit the actions of the body's own estrogen. Such effects can lead to significant consequences for fertility and the development of embryos and fetuses.

The extent of the risk from endocrine disruptors is hotly debated, but the number of mature, functional germ cells produced in animals as diverse as mollusks and human males has declined substantially during the past 50 years in the U.S. In addition, researchers throughout the world have noted feminization of freshwater fishes downstream of wastewater facilities. For example, male fishes that were exposed to such conditions during development show increased production of proteins normally made by females bearing eggs. They also show changes in gonadal structures such that there is a resemblance to the female structures. Further research is being conducted to gain more information on the consequences of these contaminants for animal and human endocrine systems.

The U.S. Environmental Protection Agency currently lists about 10,000 chemicals that are recommended for screening or have already been documented to exert endocrine disruptive actions. Other than reproductive hormones, the hormones that are most commonly affected by different types of endocrine disruptors are thyroid hormone and, to a lesser extent, glucocorticoids.

## Summary of Key Concepts

### 50.1 Types of Hormones and Their Mechanisms of Action

- The endocrine glands and other organs with hormone-secreting cells constitute the endocrine system. Endocrine glands contain epithelial cells that secrete hormones into the bloodstream, where they circulate throughout the body. Although slower than the electrical signaling in nervous systems, chemical signaling complements nervous system regulation through its varying actions in multiple locations across widely ranging time frames (Figure 50.1).

- Hormones fall into three broad classes: amines, polypeptides and steroids. Water-soluble hormones (amines and polypeptides) act on receptor molecules located in the plasma membrane, whereas lipid-soluble hormones (steroids) act on intracellular receptors (Table 50.1).

- Synthesis of amine hormones is mediated by enzymatic conversion of either tyrosine or tryptophan into their respective hormones. The steroid hormones are derived from cholesterol (Figure 50.2).

### 50.2 Links Between the Endocrine and Nervous Systems

- Sensory input from an animal's nervous system modulates the activity of certain endocrine glands and influences blood concentrations of many hormones.

- The hypothalamus in vertebrates is physically connected to the pituitary gland, which consists of anterior and posterior lobes (Figure 50.3).

- Within the hypothalamus are numerous neurons that synthesize neurohormones and stimulate the anterior pituitary. In response, the anterior pituitary gland synthesizes six different hormones that respond to the presence of hypothalamic neurohormones. They are adrenocorticotropic hormone (ACTH), follicle-stimulating hormone (FSH), luteinizing hormone (LH), growth hormone (GH), thyroid-stimulating hormone (TSH), and prolactin (PRL) (Table 50.2).

- In mammals, the neuron terminals in the posterior pituitary store and secrete one of two hormones: oxytocin or antidiuretic hormone (ADH).

### 50.3 Hormonal Control of Metabolism and Energy Balance

- Hormones are partly responsible for regulating energy use by cells, such as modulating appetite, digestion, absorption of nutrients, and blood concentration of glucose. Although many hormones are involved in these processes, one from the thyroid gland ($T_3$), three from the adrenal glands (cortisol, epinephrine, and norepinephrine), two from the pancreas (insulin and glucagon), and one from adipose tissue (leptin) have especially important functions in vertebrates.

- The thyroid gland makes $T_3$, which contains iodine. A major action of $T_3$ in adult animals is to stimulate energy consumption by many different cell types (Figures 50.4, 50.5).

- The adrenal glands produce epinephrine and norepinephrine in the medulla and glucocorticoids in the cortex; these hormones increase the blood glucose concentration. The endocrine pancreas produces insulin and glucagon, which have opposite effects on blood glucose concentration (Figures 50.6, 50.7, Table 50.3).

- Maintaining normal glucose and other nutrient concentrations in the blood is a vital process that keeps cells functioning optimally. The combined short-term and long-term actions of insulin, glucagon, epinephrine, and cortisol help maintain normal blood glucose concentrations during fasting (Figure 50.8).

- Ground-breaking research by Banting, Best, MacLeod, and Collip isolated insulin for therapeutic use in treating diabetes mellitus (Figure 50.9).

- Adipose tissue is an important source of appetite-regulating hormones, including leptin, which acts on the hypothalamus to inhibit appetite.

## 50.4 Hormonal Control of Mineral Balance

- Because of the important functions that $Ca^{2+}$ has in neuronal transmission, heart function, muscle contraction, and numerous other physiological processes, the concentration of $Ca^{2+}$ in the blood is among the most tightly regulated variables in an animal's body. Vitamin D and parathyroid hormone, the latter produced by the parathyroid glands, regulate the blood concentration of $Ca^{2+}$ (Figures 50.10, 50.11).

- $Na^+$ and $K^+$ play crucial roles in membrane potential formation, action potential generation, and other functions. A key mechanism that regulates blood concentrations of these ions is altering the rate of $Na^+$, $K^+$, and water reabsorption from the urine as it is being formed in the kidneys. This regulation is accomplished in large part by the actions of ADH, aldosterone, and atrial natriuretic peptide (Figure 50.12).

## 50.5 Hormonal Control of Growth and Development

- Hormones play a crucial role in regulating growth and development. In vertebrates, normal growth depends on a balance between growth hormone, insulin-like growth factor 1, and gonadal hormones. Thyroid hormones affect development among all vertebrates (Figures 50.13, 50.14).

- In insects, prothoracicotropic hormone, 20-hydroxyecdysone, and juvenile hormone control the growth of larvae and their development into pupae (Figure 50.15).

## 50.6 Hormonal Control of Reproduction

- Hormones produced by the gonads play vital roles in nearly all aspects of reproduction. The ability of the testes and ovaries to produce the sex steroids depends on the gonadotropins, which are the same in both sexes and include follicle-stimulating hormone and luteinizing hormone.

## 50.7 Impact on Public Health

- Hormones are used therapeutically to treat a variety of human disorders, including diabetes, infertility, growth disorders, asthma, and inflammation.

- Androgen misuse can disrupt normal hormone concentrations and cause health risks such as cardiovascular disease, skin problems,

cancer, infertility in men, and masculinizing traits in women. Blood-doping with erythropoietin can make blood dangerously thick.

- Endocrine disruptors, such as chemicals derived from industrial waste, may bind to estrogen receptors in animals' bodies. They may exert estrogen-like actions or inhibit the actions of the body's own estrogen.

## Assess & Discuss

### Test Yourself

1. Which is the defining feature of hormones?
   a. They are only produced in endocrine glands.
   b. They are secreted into the blood, where they may reach one or more types of distant target cells, thereby altering cell function throughout the body.
   c. They are released only by neurons.
   d. They are never released by neurons.
   e. They are secreted into ducts, where they diffuse to another nearby gland or other structure.

2. Steroid hormones are synthesized from _____ and bind to _____.
   a. proteins; membrane receptors
   b. fatty acids; membrane receptors
   c. tyrosine; intracellular receptors
   d. proteins; intracellular receptors
   e. cholesterol; intracellular receptors

3. Which of the following statements about polypeptide hormones is *false*?
   a. They bind to receptors located on the cell membrane.
   b. Most of them are lipophilic.
   c. They are the most abundant class of hormones.
   d. They normally activate second messengers.
   e. They bind reversibly to receptors.

4. Chronic deficiency of iodine in a vertebrate's diet will lead to
   a. increased secretion of TRH, decreased secretion of TSH, decreased $T_3$ concentration in the blood, and a goiter.
   b. decreased secretion of TRH and TSH, decreased $T_3$ concentration in the blood, and a goiter.
   c. increased $T_3$ concentration in the blood, decreased secretion of TRH and TSH, but no goiter.
   d. decreased $T_3$ concentration in the blood, increased secretion of TRH and TSH, and a goiter.
   e. decreased secretion of TRH, increased secretion of TSH, decreased $T_3$ concentration in the blood, and a goiter.

5. The hypothalamus and the pituitary gland are physically connected by
   a. arteries.
   b. the infundibular stalk and portal veins.
   c. the adrenal medulla.
   d. the spinal cord.
   e. the intermediate lobe.

6. Antidiuretic hormone (ADH)
   a. increases water reabsorption in the kidneys.
   b. regulates blood pressure by constricting arterioles.
   c. decreases the volume of urine produced by the kidneys.
   d. probably arose from the same ancestral gene as that of oxytocin.
   e. does all of the above.

7. Which of the following pairs of hormones are involved in the regulation of blood $Ca^{2+}$ concentration in vertebrates?
   a. aldosterone and ANP
   b. insulin and glucagon

c.  parathyroid hormone and 1,25-dihydroxyvitamin D
d.  prolactin and oxytocin
e.  thyroxine and TSH

8.  In invertebrates, molting of larvae is stimulated by
    a.  growth hormone.
    b.  cortisol.
    c.  juvenile hormone.
    d.  20-hydroxyecdysone.
    e.  aldosterone.

9.  Which of the following is *true* of the adrenal glands?
    a.  They produce insulin.
    b.  They produce hormones that control ion balance and maintain glucose homeostasis.
    c.  They produce only steroid hormones.
    d.  They are inhibited by the pituitary hormone ACTH.
    e.  They chiefly regulate $Ca^{2+}$ balance.

10. Endocrine disruptors are
    a.  chemicals released by the nervous system to override the endocrine system.
    b.  chemicals released by the male of a species to decrease the fertility of other males.
    c.  drugs used to treat overactive endocrine structures.
    d.  chemicals derived from industrial waste that may alter endocrine function.
    e.  all of the above.

## Conceptual Questions

1.  What is the function of leptin, and what is the benefit to an animal of an adipose-derived signaling molecule? What function does an appetite serve in animals?

2.  Distinguish between type 1 and type 2 diabetes mellitus.

3.  **Core Concept: Systems**  Organ systems often exert dual control over variables such as blood pressure, respiration, and growth. How do the opposing actions of insulin and glucagon maintain glucose homeostasis? When are these hormones released into the blood? What might happen to a nonfasting mammal that was injected with a high dose of glucagon?

## Collaborative Questions

1.  Discuss the roles of hormones in growth and development.

2.  Discuss the functions of the different steroid hormones, and indicate where they are produced in the mammalian body.

# Animal Reproduction and Development

# 51

**Most animals reproduce by sexual reproduction.** These oysters, shown here releasing their sperm and eggs into the water where they will combine and form new organisms, reproduce by sexual reproduction, which favors genetic variation in species. ©Robert F. Sisson/National Geographic/Getty Images

**A**n 18-year-old woman joined her college's cross-country track team after competing for 2 years in high school. Gradually, her training increased from about 25 km/week to a very demanding 90 km/week. Recognizing the increased ability and competitiveness of college athletes relative to those at her high school, and at the urging of her track coach, she restricted her daily food intake to become as lean as possible. As her training and diet restriction progressed, she became increasingly preoccupied with monitoring what she was eating. Eventually, her weight dropped from 59 kg to 46 kg. This was followed by the cessation of her menstrual periods for several months. Alarmed that something was wrong, she visited her physician. After a battery of tests ruled out a variety of possible causes for missing her periods, she was diagnosed with secondary amenorrhea. This is a general term for the absence of menstrual cycles in a person who had previously been cycling normally. In her case, it was clear that the cause of her condition was undernourishment combined with intense exercise. She was encouraged to receive mental health

counseling in conjunction with her parents and coach, in order to address her eating disorder, and to work with a dietitian to tailor an appropriate meal plan to support her desired level of activity.

Secondary amenorrhea is common among elite female athletes in sports and activities like cross-country track, gymnastics, and ballet, where physical activity is demanding and a lean body mass is encouraged. It is a warning sign that a person is overdoing things and needs to correct a nutritional imbalance. When a woman's body is stressed in this way, her reproductive function ceases. From a physiological perspective, this makes sense, as the likelihood of supporting a healthy pregnancy in such circumstances is greatly reduced and the body requires whatever energy is available just to survive.

This woman's case illustrates the delicate control of **reproduction**—the processes by which organisms produce offspring. The biological mechanisms that favor successful reproduction in the animal kingdom are extraordinarily diverse. Many of the observable differences in animal behavior and anatomy are the result of adaptations that increase an animal's chances of reproducing.

In this chapter, we will examine the diverse means of reproduction throughout the animal kingdom, including asexual and sexual reproduction. We will consider the control mechanisms of reproduction in mammals, including humans. We then will cover some of the general events in embryonic development and will conclude with a discussion of some key issues related to fertility in the human population today, including those that affected the young woman described in this introduction.

## 51.1 Overview of Sexual and Asexual Reproduction

**Learning Outcomes:**

1. Describe several differences between sexual and asexual reproduction.
2. **CoreSKILL »** Propose a hypothesis explaining why most animals reproduce by sexual reproduction.

**Sexual reproduction** is the production of a new individual by the joining of two haploid reproductive cells called

**gametes**—typically one from each parent. This usually produces offspring that are genetically different from both parents. **Asexual reproduction** in animals occurs when offspring are produced from a single parent without the fusion of gametes. The offspring are clones of the parent organism. In this section, we will consider the processes of sexual and asexual reproduction, as well as their advantages and disadvantages.

## Sexual Reproduction Occurs in Most Animals

Sexual reproduction occurs in the vast majority of animal species and involves the joining of two gametes. The gametes are spermatozoa (usually shortened to **sperm**) from the male and **egg cells**, or ova (singular, ovum), from the female. (In some species, sperm and egg cells are both produced by a single organism called a **hermaphrodite**.) When a sperm unites with an egg—the process called **fertilization**—each haploid gamete contributes its set of chromosomes to produce a diploid cell called a fertilized egg, or **zygote**. As the zygote undergoes cell divisions and begins to develop, it is called an **embryo**.

## Asexual Reproduction Takes Several Forms

Asexual reproduction occurs in some invertebrates and in a small number of vertebrate species. These animals reproduce in one of three ways: budding, regeneration, or parthenogenesis. **Budding**, which is seen in cnidarians, occurs when a portion of the parent organism pinches off to form a complete, new individual (**Figure 51.1a**). In this process, cells from the parent undergo mitosis and differentiate into specific types of structures before the new individual breaks away from the parent. At any one time, a parent organism may have one, two, or multiple buds forming simultaneously. Budding continues throughout such an animal's lifetime.

   Certain species of sponges, echinoderms, and worms reproduce by **regeneration**, in which a complete organism can be formed from a fragment of the animal's body. In some sea stars and starfish, for example, an arm removed by injury or predation can grow into an entirely new individual (**Figure 51.1b**). Similarly, a flatworm bisected into two pieces can regenerate into two distinct individuals.

   **Parthenogenesis** is the development of offspring from an unfertilized egg. Animals produced by parthenogenesis are usually diploid but in some cases may be haploid. This form of asexual reproduction occurs in several invertebrate classes and in a few species of fishes and reptiles.

## Sexual and Asexual Reproduction Have Advantages and Disadvantages

Asexual reproduction provides a relatively simple way for an organism to produce many copies of itself, whereas sexual reproduction usually requires two individuals to produce offspring. What are the advantages and disadvantages of each method?

   Asexual reproduction has certain advantages over sexual reproduction. First, an animal can reproduce asexually even if it is isolated from others of its own species—for example, because the animal is nonmotile for much of its life or because it rarely encounters another member of its species. Second, individuals can reproduce rapidly at any time because they need not seek out, attract, and mate with an

**(a) Budding**

**(b) Regeneration**

**Figure 51.1  Examples of asexual reproduction. (a)** *Hydra* with a single bud. **(b)** A starfish regenerating a new body from a single arm.
a: ©Clouds Hill Imaging Ltd./Corbis/Getty Images; b: ©WaterFrame/Alamy Stock Photo

individual of the opposite sex. Asexual reproduction, therefore, is an effective way of generating large numbers of offspring. Although many kinds of animals reproduce asexually, it is more prevalent in species that live in very stable environments, where there is little selection pressure for genetic diversity in the population.

   Compared with asexual reproduction, sexual reproduction is associated with unique costs. Two types of gametes (sperm and eggs) must be made (often in large numbers), males and females require specialized body parts to mate with each other, and the two sexes must be able to find each other. Yet given that the vast majority of eukaryotic species reproduce sexually, the following question has intrigued biologists since the time of Darwin: What is the advantage of sexual reproduction for the perpetuation of a species?

   In the context of species survival, the major difference between asexual and sexual reproduction is that sexual reproduction allows for greater genetic variation due to genetic recombination. Only certain alleles from each parent are passed on, and when a set of genes from one parent mixes with a different set from the other parent, the offspring are never exactly like either of their parents. Thus, a key consequence of sexual reproduction is greater genetic variation within a population.

   One prevalent hypothesis about the advantage of sexual reproduction is that it allows a more rapid adaptation to environmental changes than does asexual reproduction. In particular, sexual reproduction allows alleles within a species to be redistributed via crossing over and independent assortment across many generations. As a result, some offspring carry combinations of alleles that promote survival and reproduction, whereas other offspring may carry less favorable combinations. As described in Chapter 23, natural selection is the process in which certain alleles or allele combinations that promote greater reproductive success become more prevalent, whereas those that result in lower fitness become less prevalent. By comparison, the alleles of asexual organisms are not reassorted from generation to generation. As a result, it is more difficult to accumulate potentially beneficial alleles within individuals of these species. Also, as described next, sexual reproduction may facilitate elimination of harmful alleles from a population.

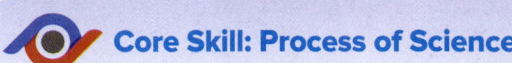

## Core Skill: Process of Science

## Feature Investigation | Paland and Lynch Provided Evidence That Sexual Reproduction May Promote the Elimination of Harmful Mutations in Populations

Evolutionary biologists have suggested that the inability of asexual species to reassort alleles may be a key disadvantage for them, compared with sexually reproducing species. To investigate this question, American researchers Susanne Paland and Michael Lynch studied the persistence of mutations in populations of *Daphnia pulex*, a freshwater organism commonly known as the water flea. The researchers chose this organism because some natural populations reproduce asexually, and others reproduce sexually.

In their experiment, shown in **Figure 51.2**, Paland and Lynch studied the sequences of several mitochondrial genes in 14 sexually reproducing and 14 asexually reproducing populations of *D. pulex*. The researchers hypothesized that asexual populations would be less able to eliminate harmful mutations. As discussed in Chapter 15, random gene mutations that change the amino acid sequence of an encoded protein are much more likely to be harmful than beneficial. Because the alleles of sexually reproducing populations can be

**Figure 51.2** **Paland and Lynch demonstrated the importance of sexual reproduction in reducing the frequency of harmful genetic mutations.**

**HYPOTHESIS** Sexual reproduction allows for greater mixing of alleles of different genes and thereby may prevent the accumulation of detrimental alleles in a population.

**KEY MATERIALS** The researchers collected samples of *Daphnia pulex* from many natural populations. A total of 14 sexual populations and 14 asexual populations were studied.

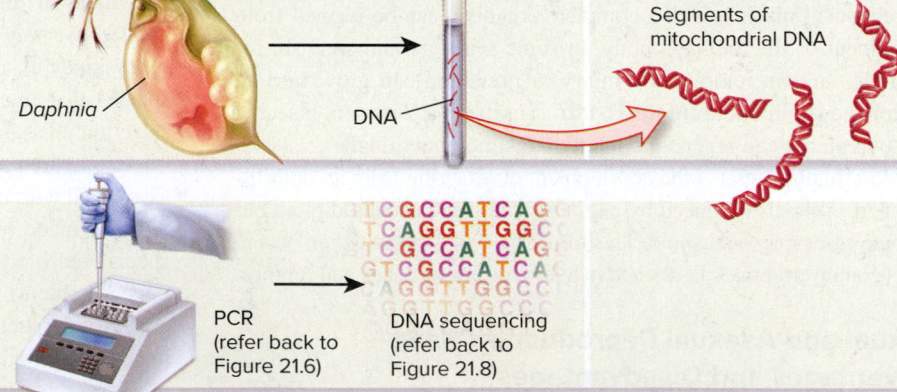

|  | Experimental level | Conceptual level |
|---|---|---|
| 1 | Isolate mitochondrial DNA from members of 28 populations of *D. pulex*. This involves breaking open cells and extracting the DNA (refer back to Chapter 21). | *Daphnia*    DNA | Segments of mitochondrial DNA |
| 2 | Amplify regions of mitochondrial genes, using PCR. Subject the regions to DNA sequencing. The techniques of PCR and DNA sequencing are described in Chapter 21 (refer back to Figures 21.6 and 21.8). | PCR (refer back to Figure 21.6)    DNA sequencing (refer back to Figure 21.8) |  |

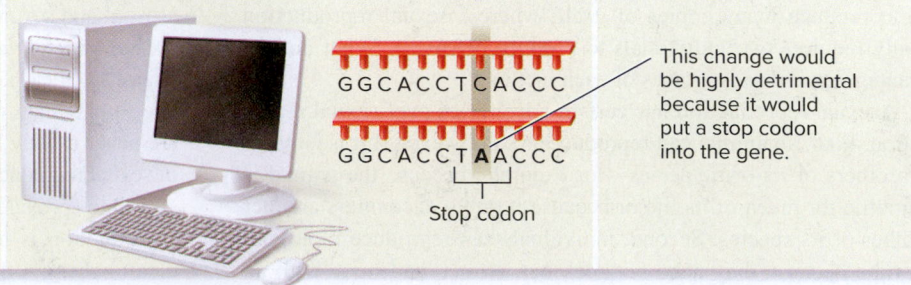

3  Using computer technology, align the sequences and determine the number of DNA changes that would cause amino acid substitutions. These amino acid changes were categorized as those that would be highly deleterious, moderately deleterious, mildly deleterious, or neutral for protein function. Compare the sexual and asexual populations for the persistence of these types of changes.

GGCACCTCACCC

GGCACCTAACCC

Stop codon

This change would be highly detrimental because it would put a stop codon into the gene.

**4    THE DATA**

Results from step 3:

| Types of amino acid substitutions (The amino acid substitutions were due to rare mutations that occurred in the natural populations of *D. pulex*.) | % of total amino acid substitutions | Allowed to persist in: | |
|---|---|---|---|
| | | Sexual populations | Asexual populations |
| Highly deleterious | 73.2 | No | No |
| Moderately deleterious | 13.3 | No | Yes |
| Mildly deleterious | 4.4 | Yes | Yes |
| Neutral | 9.1 | Yes | Yes |

**5    CONCLUSION**  Moderately deleterious mutations are less likely to persist in populations of animals that reproduce sexually.

**6    SOURCE**  Paland, S., and Lynch, M. 2006. Transition to asexuality results in excess amino acid substitutions. *Science* 311: 990–992.

reassorted from generation to generation, some offspring will not inherit such detrimental alleles.

As you can see in step 4 of Figure 51.2, the researchers discovered that both the sexual and asexual populations could eliminate highly deleterious mutations. Organisms harboring such mutations probably died and did not reproduce. In addition, the sexual and asexual populations both retained mildly deleterious and neutral mutations. However, moderately deleterious mutations were eliminated from the sexual populations but not from the asexual ones. One interpretation of these data is that sexual reproduction allowed for the reassortment of beneficial and detrimental alleles, making it easier for sexually reproducing populations to eliminate those mutations that are moderately detrimental.

### Experimental Questions

1. **CoreSKILL »** How did Paland and Lynch test the hypothesis that sexual reproduction reduced the frequency of deleterious mutations in a population?

2. **CoreSKILL »** Approximately how many times more often did deleterious mutations persist in the asexually reproducing populations compared to the sexually reproducing ones?

3. What is the proposed evolutionary benefit of sexual reproduction?

## 51.2  Gametogenesis and Fertilization

### Learning Outcomes:

1. Outline the processes of spermatogenesis and oogenesis.
2. **CoreSKILL »** Propose reasons why internal fertilization is advantageous.

The formation of gametes and the event of fertilization are common features of sexual reproduction. In this section, we will explore how most animals produce gametes and how two gametes join to form a new organism.

### Sperm and Eggs Are Produced During the Process of Gametogenesis

In most species of animals, male and female gametes are formed within **gonads**—the **testes** (singular, testis) and the **ovaries** (singular, ovary). In species with separate sexes, males have testes and females have ovaries. However, some animal species are hermaphroditic, meaning that a single individual produces sperm and egg cells. Familiar examples of hermaphroditic species include earthworms and most snails.

The formation of gametes—**gametogenesis**—begins with cells called germ cells. Germ cells multiply by mitosis, resulting in diploid cells (carrying two sets of chromosomes, denoted as 2*n*) called **spermatogonia** (singular, spermatogonium) and **oogonia** (singular, oogonium) (**Figure 51.3**). Some of these cells become **primary spermatocytes** or **primary oocytes** that may begin the process of meiosis. (See Chapter 16 for a review of the processes of mitosis and meiosis.) Until this point, the development of sperm and eggs is similar. From then on, gametogenesis differs between the two types of gamete.

***Spermatogenesis***  The formation of haploid sperm from a diploid germ cell is called **spermatogenesis**. As shown in Figure 51.3a, primary spermatocytes begin this process by undergoing the first meiotic division (meiosis I) to produce two haploid (*n*) cells called **secondary spermatocytes**. These cells also undergo meiosis II, producing four haploid **spermatids** that eventually differentiate into mature haploid sperm cells. Gametogenesis in males, therefore, results in four gametes from each spermatogonium.

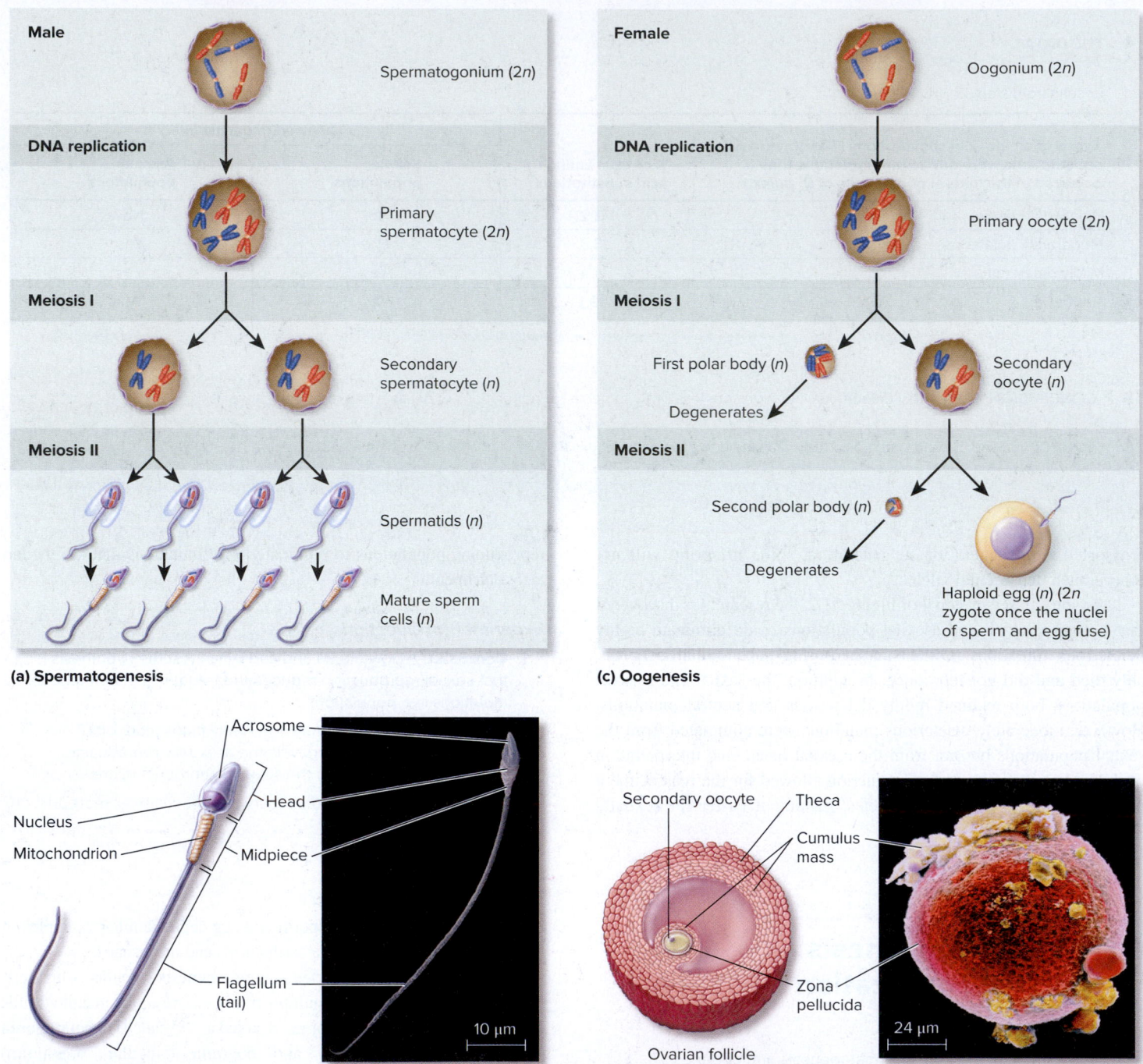

**(a) Spermatogenesis**

**(c) Oogenesis**

**(b) Mature human sperm**

**(d) Mature human follicle and oocyte**

**Figure 51.3  Gametogenesis and gametes in human males and females.  (a)** In the process of spermatogenesis, male diploid ($2n$) germ cells undergo two meiotic divisions to produce mature haploid ($n$) sperm. **(b)** The characteristic head, midpiece, and flagellum (tail) of a mature human sperm, as seen in a drawing and accompanying scanning electron micrograph (SEM). **(c)** The process of oogenesis in females, which produces a haploid secondary oocyte that enters but does not complete meiosis II until it is fertilized. **(d)** Mature follicle and oocyte. The drawing depicts a secondary oocyte within its follicle; the SEM shows an isolated human oocyte covered by its zona pellucida and remnants of the cumulus mass. b: ©Eye of Science/Science Source; d: ©P. M. Motta & G. Familiari/Univ. La Sapienza/Science Source

**Core Concept: Structure and Function**  The structure of a sperm cell is a good example at the cellular level of this core concept of biology. Each of the three major structural elements of the sperm cell is suited for a particular function. When all three elements function together, the sperm is capable of fertilizing an egg.

The most striking change in each spermatocyte as it differentiates into a sperm is the formation of a flagellum, also called the tail (Figure 51.3b). The movements of the tail require energy and make the sperm motile. The sperm also has a head region, which contains the nucleus that carries the chromosomes. At the tip of the head is a special structure called the **acrosome**, which contains proteolytic enzymes that help break down the protective outer layers surrounding an egg cell. The head and tail are separated by a midpiece that typically contains as many as several dozen mitochondria, depending on the species, that produce the ATP required for tail movements.

*Oogenesis* Whereas spermatogenesis produces four gametes from each primary spermatocyte, **oogenesis** results in the production of a single haploid egg from each primary oocyte (Figure 51.3c). In mammals, oogenesis begins in the ovary of a female embryo. Oogonia divide by mitosis to form primary oocytes that enter meiosis I but stop the process, remaining in an arrested state after birth. Meiosis I does not resume until puberty, the time after which a mammal first becomes capable of reproducing. Beginning at puberty, meiosis I is completed in some primary oocytes, producing from each one a haploid **secondary oocyte** plus a smaller cell, called a polar body, that eventually degenerates. Meiosis II begins in the secondary oocyte but stops at metaphase. Once the secondary oocyte is released from the ovary in the process of **ovulation**, it can become fertilized if sperm are available. Meiosis II is not complete until the oocyte is fertilized by a sperm. If the oocyte does not encounter a sperm, it never undergoes the second meiotic division. Once a haploid egg nucleus fuses with a haploid sperm nucleus, a diploid zygote is produced.

Each oocyte undergoes growth and development within an ovarian structure called a follicle. In mammals, a layer of glycoproteins called the **zona pellucida** surrounds the surface of the secondary oocyte in the follicle. In addition, a layer of cells called the cumulus mass, which provides protection and nutritive support, is outside the zona pellucida. An outer cellular layer called the theca produces hormones that control oocyte growth (Figure 51.3d). These layers have important roles in fertilization, as discussed later.

## Fertilization Involves the Union of Sperm with Egg

Fertilization is the process by which the haploid male and female gametes unite and become a diploid zygote. Several important cellular and molecular processes must occur before the nuclei of the gametes can fuse. The mechanism by which the egg and sperm make contact has been studied extensively in sea urchins, and evidence suggests that a similar process occurs in humans and other mammals. Sea urchin eggs emit chemical attractants that bind to nearby sperm. This increases cellular respiration (that is, the breakdown of nutrients to synthesize ATP) within the sperm, which helps increase sperm motility. The sperm then swim toward the egg by following the concentration gradient of the attractants.

For a sperm to physically contact the egg, it must first penetrate the layers surrounding the egg's plasma membrane (**Figure 51.4**). In sea urchins, as in many animals, the sperm must penetrate a jelly-like layer consisting of glycoproteins and polysaccharides before contacting the plasma membrane of the egg. The sperm is able to do this because of the **acrosomal reaction**, in which proteases and other hydrolytic enzymes are released from the acrosome onto the jelly coat of the egg (see Figure 51.4a). These enzymes dissolve a localized region of the jelly coat, allowing the sperm head to bind to proteins in the egg's plasma membrane. Binding is followed by fusion of the sperm head membrane with the egg membrane, and shortly thereafter by penetration of the sperm head and release of its nucleus into the egg.

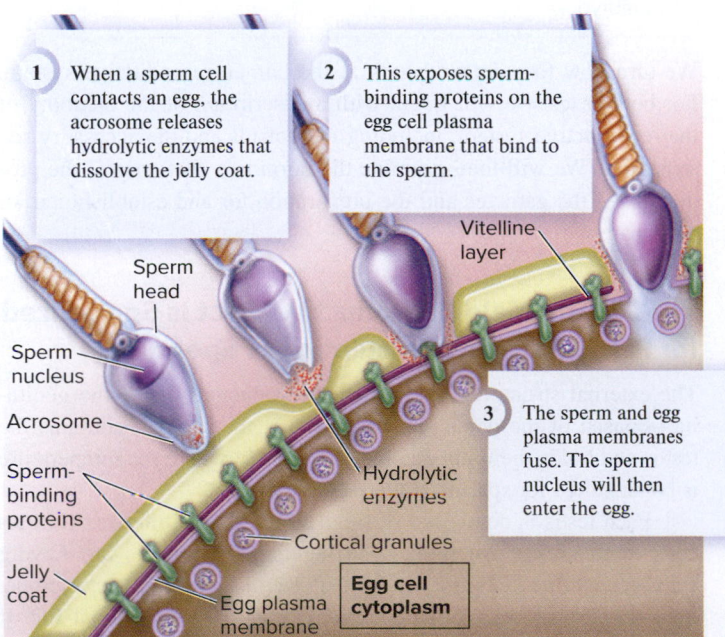

**1** When a sperm cell contacts an egg, the acrosome releases hydrolytic enzymes that dissolve the jelly coat.

**2** This exposes sperm-binding proteins on the egg cell plasma membrane that bind to the sperm.

Vitelline layer

Sperm head

Sperm nucleus

Acrosome

Sperm-binding proteins

Jelly coat

Hydrolytic enzymes

Egg plasma membrane

Cortical granules

**3** The sperm and egg plasma membranes fuse. The sperm nucleus will then enter the egg.

Egg cell cytoplasm

**(a) Acrosomal reaction**

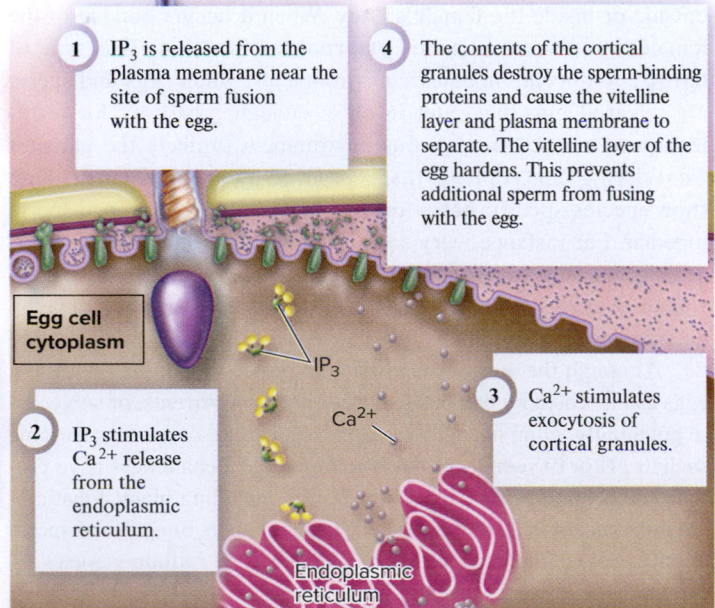

**1** IP$_3$ is released from the plasma membrane near the site of sperm fusion with the egg.

**4** The contents of the cortical granules destroy the sperm-binding proteins and cause the vitelline layer and plasma membrane to separate. The vitelline layer of the egg hardens. This prevents additional sperm from fusing with the egg.

Egg cell cytoplasm

IP$_3$

Ca$^{2+}$

**2** IP$_3$ stimulates Ca$^{2+}$ release from the endoplasmic reticulum.

**3** Ca$^{2+}$ stimulates exocytosis of cortical granules.

Endoplasmic reticulum

**(b) Cortical reaction**

**Figure 51.4** **The events during fertilization of an egg by a sperm.** **(a)** Acrosomal reaction. The contact of a sperm with an egg initiates a series of events that permits the head of the sperm to bind to the plasma membrane of the egg. **(b)** Cortical reaction. Sperm fusion leads to an increased concentration of cytosolic Ca$^{2+}$ that ultimately causes the vitelline layer of the egg to harden, preventing additional sperm from fusing.

The acrosomal reaction is followed by the **cortical reaction** (Figure 51.4b). The cytosolic $Ca^{2+}$ concentration in eggs, as in most cells, is kept low by several mechanisms, including the transport of $Ca^{2+}$ by an ATP-dependent pump out of the cytosol and into the endoplasmic reticulum. When the sperm binds to the egg, a small signaling molecule called inositol trisphosphate ($IP_3$) is released from the region of the plasma membrane nearest to the sperm entry point. $IP_3$ then binds to nearby sites on the endoplasmic reticulum (ER) and opens $Ca^{2+}$ channels. Within seconds after a sperm cell binds to an egg, $Ca^{2+}$ is released from the lumen of the ER and into the cytosol.

The release of $Ca^{2+}$ in the cortical reaction has several important effects. First, it causes exocytosis of membrane-bound vesicles in the egg's cytosol called cortical granules. The cortical granules release enzymes and other substances that inactivate the sperm-binding proteins on the plasma membrane. At that time, also, the outer coating of the egg cell, known as the vitelline layer in sea urchins or the zona pellucida in vertebrates (see Figure 51.3), becomes hardened and begins to separate from the plasma membrane. These events create a barrier to additional sperm fusing with the egg. Additionally, the burst of cytosolic $Ca^{2+}$ leads to the activation of molecular signaling pathways that initiate the first cell cycle and triggers an increase in protein synthesis and metabolism within the egg cell.

Shortly afterward, the nucleus of the sperm fuses with the nucleus of the egg, creating a diploid zygote. The first cell division of the zygote occurs approximately 90 minutes after fertilization in sea urchins and amphibians, but it can take up to 24 hours in mammals.

## In a Given Species, Fertilization Occurs Either Outside or Inside the Female

For sperm to fertilize eggs, the two gametes must physically come into contact. In species with separate sexes, this can occur either outside or inside the female's body. When it occurs outside of the female, the process is called **external fertilization**. This type of fertilization occurs in aquatic environments, when eggs and sperm are released into the water in close enough proximity for fertilization to occur. The aqueous environment protects the gametes from drying out. Animals that reproduce by external fertilization show species-specific behaviors that help bring the eggs and sperm together. For instance, very soon after a female fish lays her eggs, a male deposits his sperm cells in the water nearby, where they spread over the clump of eggs. The fertilized eggs then develop outside the mother's body.

Although the aqueous environment protects against desiccation, eggs can be eaten by predators, washed away by currents, or subjected to potentially lethal changes in water temperature or another variable such as pH or oxygen level. Such environmental challenges have provided selection pressure for some species, including many aquatic or amphibious animals, to release very large numbers of eggs and sperm at once. For example, oysters may release several million gametes, as shown in the chapter opening photo.

In contrast to external fertilization, most terrestrial animals and some aquatic animals reproduce by **internal fertilization**, in which sperm are deposited within the reproductive tract of the female during the act called copulation. Internal fertilization protects the zygotes from dessication, environmental hazards, and predation, and

guarantees that sperm are placed and remain in very close proximity to eggs. Once fertilization occurs within the female, the zygotes then develop into offspring.

The behaviors and anatomical structures involved in achieving internal fertilization are extremely varied among species. Typically, mating involves accessory sex organs, which are reproductive structures other than the gonads. The external accessory sex organs involved in copulation include certain types of genitalia (for example, the **penis** and the **vagina** in mammals), which are used to physically join the male and female so that sperm can be deposited directly into the female's reproductive tract. A penis or analogous structure is present in most insects, reptiles, some species of birds (ratites and waterfowl), and all mammals. However, males of other vertebrate species—including most birds—that reproduce by internal fertilization lack a structure that can be inserted into the female, instead depositing sperm in the female by cloacal contact. The cloaca is a common opening for the reproductive, digestive, and excretory systems in these animals.

## 51.3 Human Reproductive Structure and Function

### Learning Outcomes:

1. Describe the structures of the human male reproductive tract.
2. Outline the process of hormonal control of the male reproductive system.
3. Describe the structures of the human female reproductive tract.
4. Diagram the events of the ovarian cycle.
5. Outline the process of hormonal control of the reproductive cycle of the human female.
6. Explain how maternal hormones prepare the uterus to accept the embryo.

We turn now to a detailed look at the human reproductive system. For both sexes, we will begin with a description of the anatomy of the reproductive system, including the gonads and the accessory sex structures. We will then examine the hormones that control the production of the gametes and the preparation for and establishment of pregnancy.

### The Human Male Reproductive Tract Is Specialized for Production and Ejaculation of Sperm

The external structures of the male reproductive tract—the genitalia—consist of the penis and the scrotum, the sac that contains the testes and holds them outside the body cavity where the temperature is better suited for spermatogenesis (**Figure 51.5**).

Each testis is composed of tightly packed **seminiferous tubules** encased in connective tissue. Surrounding the tubules are Leydig cells—endocrine cells that secrete the steroid hormone testosterone. Spermatogenesis begins at puberty and continues throughout life. It occurs all along the walls of the seminiferous tubules. Cells at the earliest stages of spermatogenesis—the spermatogonia—are located nearest the outer surface of the wall. Cells of more advanced stages are located progressively inward, so that the mature sperm are released into the tubule lumen. Cells within the seminiferous tubules are continuously

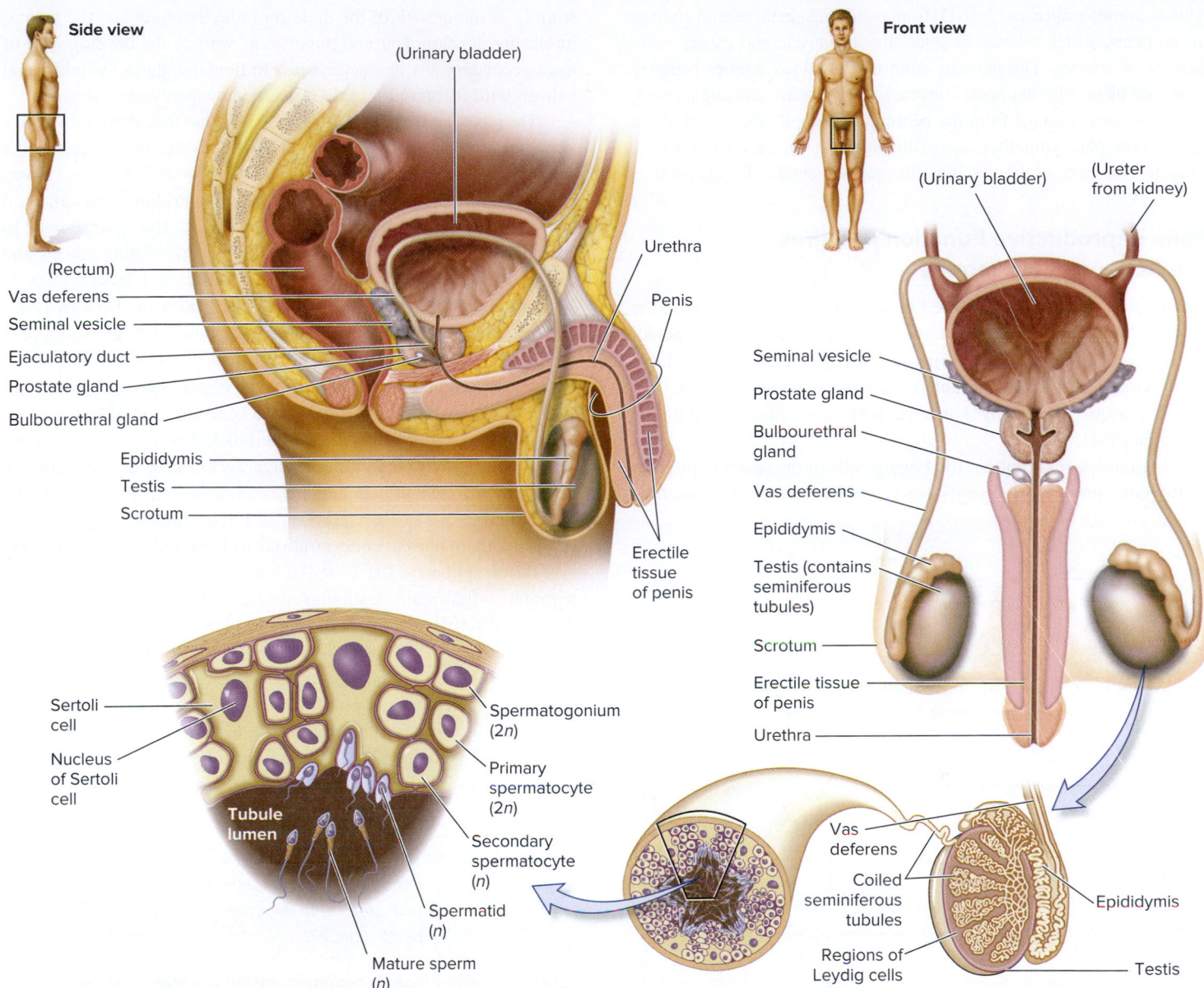

**Side view**

(Urinary bladder)

(Rectum)
Vas deferens
Seminal vesicle
Ejaculatory duct
Prostate gland
Bulbourethral gland

Urethra
Penis

Epididymis
Testis
Scrotum

Erectile
tissue
of penis

**Front view**

(Urinary bladder)
(Ureter
from kidney)

Seminal vesicle
Prostate gland
Bulbourethral
gland
Vas deferens
Epididymis
Testis (contains
seminiferous
tubules)
Scrotum
Erectile tissue
of penis
Urethra

Sertoli
cell
Nucleus
of Sertoli
cell

**Tubule
lumen**

Spermatogonium
(2n)
Primary
spermatocyte
(2n)
Secondary
spermatocyte
(n)
Spermatid
(n)
Mature sperm
(n)

Vas
deferens
Coiled
seminiferous
tubules
Regions of
Leydig cells

Epididymis

Testis

**Figure 51.5  Male reproductive structures in humans.**  Side and front views of the male reproductive system (nonreproductive structures are identified in parentheses for orientation purposes). The enlargement (at bottom) shows the internal structure of a testis and associated structures and indicates the stages of spermatogenesis.

developing from spermatogonia into spermatocytes and eventually to sperm, so at any one time, all types of cells are present along the seminiferous tubule. Support cells, called Sertoli cells, surround the developing spermatogonia and spermatocytes, providing them with nutrients and protection and playing a role in their maturation into sperm.

Sperm moving out of the seminiferous tubules are emptied into the epididymis, a coiled, tubular structure located on the surface of the testis (see Figure 51.5). Here the sperm complete their differentiation by becoming motile and gaining the capacity to fertilize an egg.

Sperm leave the epididymis through the vas deferens, a muscular tube leading to the ejaculatory duct, which then connects to the urethra (see Figure 51.5). The urethra originates at the bladder and extends to the end of the penis. In males, the urethra not only conducts urine but also carries **semen**, a mixture containing fluid and sperm, that is released during **ejaculation**—the movement of

semen through the urethra by contraction of muscles at the base of the penis.

The liquid components of semen are important for the survival and movement of sperm through the female reproductive tract. This liquid is formed by three accessory glands that secrete substances into the urethra to mix with the sperm: the seminal vesicles, the prostate gland, and the bulbourethral glands. The seminal vesicles secrete the monosaccharide fructose, the main nutrient for sperm, as well as other factors that enhance sperm motility and survival. The prostate gland and bulbourethral glands secrete a thin, alkaline fluid that protects sperm from acidic fluids in the urethra and within the female reproductive tract.

Introduction of sperm into the female reproductive system during copulation is made possible by erection of the penis. Erection occurs when blood fills spongy erectile tissue located along the length of the penis (see Figure 51.5). Sexual arousal stimulates release of the gaseous neurotransmitter nitric oxide (NO) from parasympathetic neuron endings in the penis, which relaxes vascular smooth muscle and causes vaso-dilation of arteries. The pressure from the distended arteries bringing increased blood into the penis constricts nearby veins, causing a reduction in venous drainage from the penis, engorging it with blood. After ejaculation, parasympathetic signaling and NO release are decreased, causing a reversal of the vascular changes responsible for erection.

## Male Reproductive Function Requires the Actions of Hormones

Recall from Chapter 50 that the hypothalamus is a structure on the ventral surface of the brain that synthesizes neurohormones, including gonadotropin-releasing hormone (GnRH) (refer back to Table 50.1). GnRH stimulates the anterior pituitary gland to release two gonadotropins: luteinizing hormone (LH) and follicle-stimulating hormone (FSH).

In males, LH stimulates the Leydig cells of the testes to produce androgens, particularly testosterone (**Figure 51.6**). Testosterone

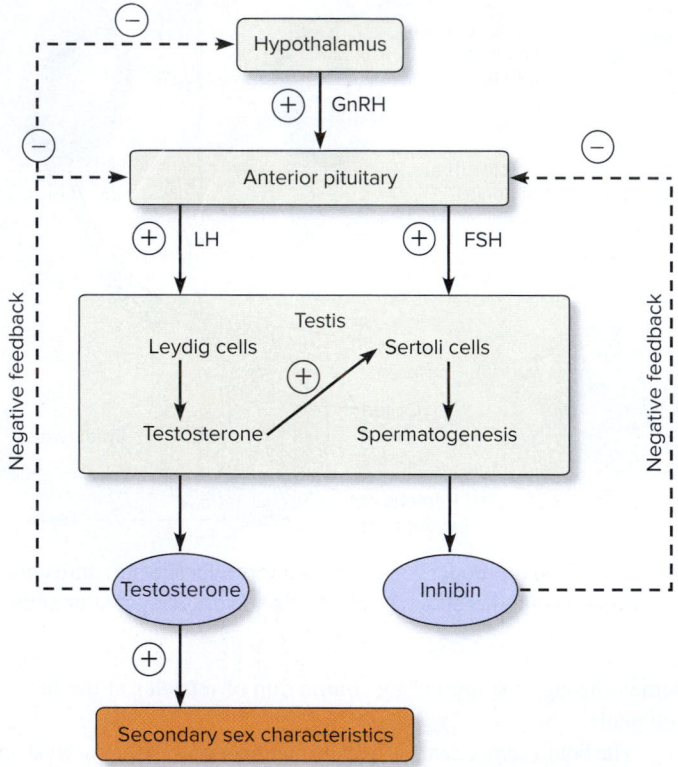

stimulates the growth of the male reproductive tract and the genitalia during development and puberty, as well as the development of male secondary sex characteristics. In humans, these include facial hair growth, increased muscle mass, and deepening of the voice.

The other pituitary gonadotropin, FSH, functions along with testosterone to stimulate spermatogenesis. FSH does this by stimulating the activity of the Sertoli cells within the seminiferous tubules (see Figures 51.5 and 51.6). The Sertoli cells provide the nutritional and structural support necessary for development of the sperm. They also respond to testosterone produced by the Leydig cells, by stimulating mitosis and meiosis of the germ cells associated with them in the tubules.

Production of sperm and testosterone is kept in check by negative feedback mechanisms that control the amount of gonadotropins produced (see Figure 51.6). Testosterone at high concentrations inhibits the secretion of GnRH from the hypothalamus, so both LH and FSH are inhibited when the blood concentration of testosterone is high. Testosterone also directly inhibits LH secretion by the anterior pituitary gland. In addition, when Sertoli cells are activated by FSH, they secrete a protein hormone called inhibin, which enters the blood and inhibits further secretion of FSH. These feedback mechanisms maintain normal concentrations of FSH and LH in the blood.

Before puberty, LH is not released in sufficient amounts to stimulate significant testicular production of testosterone, and the reproductive system is quiescent—spermatogenesis does not occur. Although the mechanisms that initiate puberty in mammals are still not completely understood, research has shown that increased GnRH production at that time initiates increased LH and FSH secretion from the pituitary. The testosterone induced by LH stimulates development of adult male characteristics. Testosterone is also responsible for an increased sex drive (libido) at this time.

**Figure 51.6  The hormonal control of male reproduction.** In response to LH, Leydig cells in the testes secrete testosterone, which, along with FSH, acts on Sertoli cells to facilitate spermatogenesis. Negative signs indicate inhibitory effects via negative feedback, and plus signs indicate stimulatory effects.

 **Core Concept: Systems** Organ systems, including the reproductive and endocrine systems, have regulatory mechanisms that monitor and adjust system activity. In this example, negative feedback is an important way in which the secretion of reproductive hormones is controlled.

 **THE QUESTION** *What will happen to spermatogenesis in a man taking supplements of testosterone to increase muscle mass and athletic performance?*

**T** **OPIC** *What topic in biology does this question address?* The topic concerns the effects of abnormally increased amounts of testosterone on spermatogenesis in men.

**I** **NFORMATION** *What information do you know based on the question and your understanding of the topic?* From the question, you can assume that the testosterone supplements increase the man's plasma concentration of testosterone above the normal range. From your understanding of the topic, you may recall that the blood concentrations of testosterone and the hypothalamic hormone GnRH are linked with spermatogenesis, as shown in Figure 51.6.

**P** **ROBLEM-SOLVING** **S** **TRATEGY** *Sort out the steps in a complicated process. Predict the outcome.* Consider the sequence of events beginning with an increase in the blood concentration of testosterone. What happens next to the concentrations of GnRH and the gonadotropins? In turn, what follows from the changes in those hormones? Based on this sequence of steps, predict what happens to sperm production.

**ANSWER** *Using Figure 51.6 as a guide, let's follow the sequence of events. (1) Taking testosterone supplements will increase the blood concentration of testosterone above normal. (2) This, in turn, will inhibit (by negative feedback) the secretion of GnRH from the hypothalamus and that of the gonadotropins LH and FSH from the anterior pituitary. (3) The blood concentrations of LH and FSH decrease as a consequence. (4) With decreased LH and FSH, there will be less stimulation of the Leydig and Sertoli cells of the testes. (5) With less stimulation of these two cell types, spermatogenesis will decrease or even stop. This is, in fact, what happens with chronic use of androgens such as testosterone in men.*

## The Female Reproductive Tract Is Specialized for Production and Fertilization of the Egg and Development of the Embryo

The female genitalia are composed of the larger, hair-covered outer folds called the labia majora and smaller, inner folds, called the labia minora, which surround the external opening of the reproductive tract (**Figure 51.7**). At the anterior part of the labia minora is the clitoris, which is erectile tissue of the same embryonic origin as the penis. Like the penis, the clitoris becomes engorged with blood during sexual arousal and is very sensitive to sexual stimulation. Unlike males, however, the openings of the reproductive tract and the urethra are separate in females. The opening of the urethra is located between the clitoris and the opening of the reproductive tract.

The external opening of the reproductive tract leads to the vagina, a tubular, smooth muscle structure into which sperm are deposited during copulation. At the end of the vagina is the cervix, the opening to the **uterus**, a pear-shaped organ that holds and nourishes the growing embryo. It consists of an inner lining of glandular and secretory cells, called the endometrium, and a thick, muscular layer. We will discuss the functions of the uterus later in the chapter.

Oocytes develop within one of the two ovaries (see Figure 51.7), which are suspended within the abdominal cavity by connective tissue. In humans, each ovary is typically a little larger than an almond. During ovulation, a secondary oocyte is released by an ovary and is quickly drawn into a thin tube, the **oviduct** (also called the Fallopian tube), by the actions of undulating fimbriae (singular, fimbria), finger-like projections of the oviduct that extend out to the ovary.

The secondary oocyte is moved down the length of the oviduct by cilia on the oviduct's inner surface. For fertilization to take place, sperm must travel through the cervix and uterus and then into the oviduct, where fertilization typically occurs. Upon contact with a sperm, the secondary oocyte completes meiosis II, and the union of sperm and the secondary oocyte creates a fertilized egg, or zygote. The zygote undergoes several cell divisions to become a **blastocyst**, a ball of approximately 32–150 cells that enters the uterus, where embryonic development will proceed, as we will examine in detail later.

## Gametogenesis in Females Is a Cyclical Process Within the Ovaries

In contrast to males, in which spermatogenesis continues throughout postpubertal life in the testes, most female mammals, including humans, appear to be born with all the primary oocytes they will ever have. At birth, each ovary in a human female has about 1 million

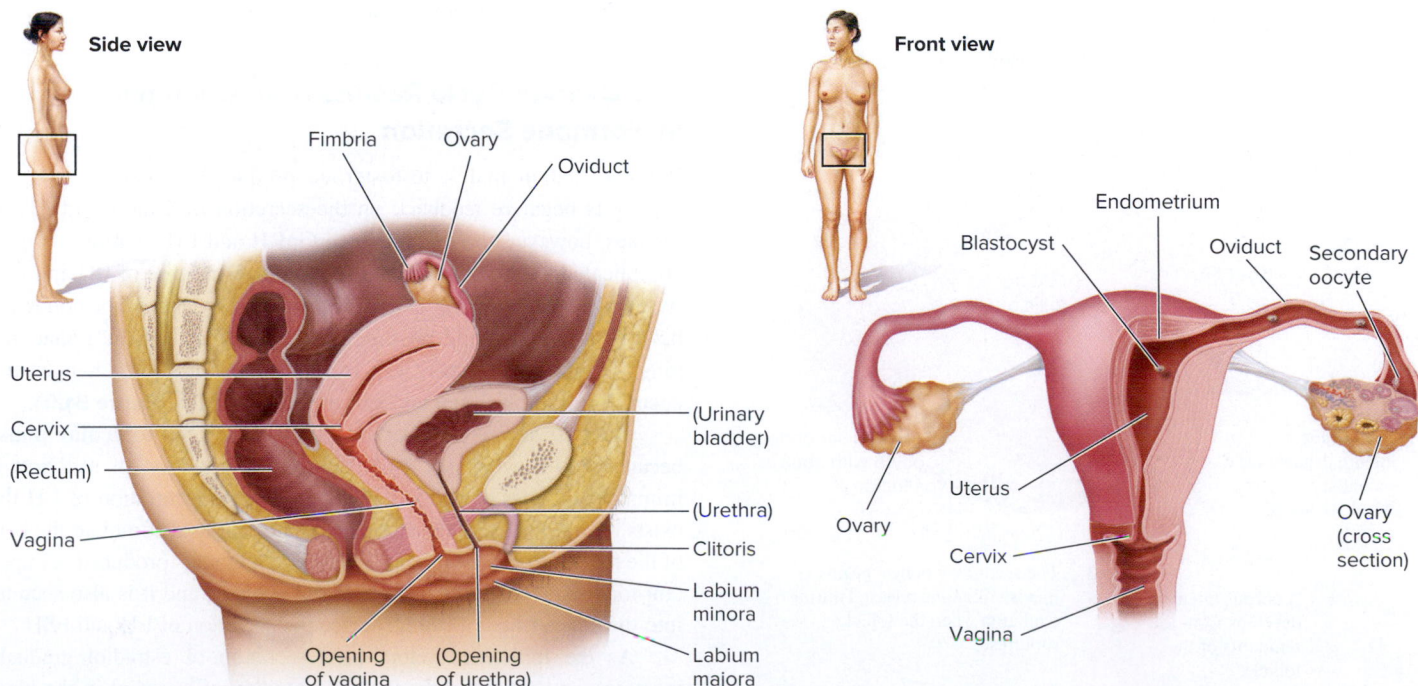

**Figure 51.7** **Female reproductive structure and function in humans.** Side and front views of the female reproductive system (nonreproductive structures are identified in parentheses for orientation purposes). An oocyte moves from the ovary into the oviduct (or Fallopian tube), where it may be fertilized and develop into a blastocyst. Subsequently, the blastocyst enters the uterus, where it may implant in the endometrium, the inner lining of the uterus.

primary oocytes, which are arrested in prophase of meiosis I. Most of these degenerate before the onset of puberty, when each ovary contains about 200,000 primary oocytes. Other than this degeneration, the ovaries are quiescent until puberty, when they begin to show cyclical activity.

As described in Chapter 50, cells within the ovaries secrete a family of steroid hormones called estrogens. The major estrogen is **estradiol**, which has a critical role in ovulation and influences the secondary sex characteristics of females. The secondary sex characteristics, which begin to develop at puberty, include development of breasts, widening of the pelvis (an adaptation for giving birth), and a characteristic pattern of fat deposition.

The **ovarian cycle** involves the development of an ovarian follicle, the release of a secondary oocyte, and the formation and subsequent regression of the empty follicle (**Figure 51.8**). During the first week of the ovarian cycle in humans, several primary oocytes that have been maturing for several months, each within its own follicle, now begin their final maturation steps (see step 1, Figure 51.8). By the beginning of the second week, all but one of these growing follicles and their primary oocytes degenerate. The mechanisms by which one follicle becomes dominant and survives are still uncertain. The single remaining follicle continues to develop and enlarge (while a new crop of immature follicles begins their slow growth period for future ovarian cycles). During that time, the primary oocyte of the enlarging follicle completes meiosis I, becomes a secondary oocyte, and begins meiosis II (see steps 2 and 3, Figure 51.8).

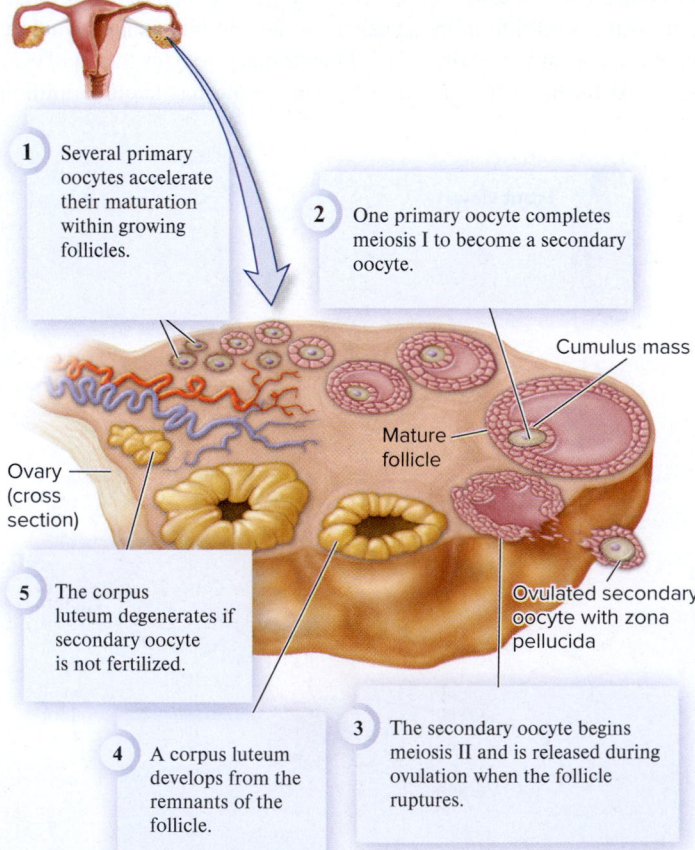

**1** Several primary oocytes accelerate their maturation within growing follicles.

**2** One primary oocyte completes meiosis I to become a secondary oocyte.

Cumulus mass

Mature follicle

Ovary (cross section)

**5** The corpus luteum degenerates if secondary oocyte is not fertilized.

Ovulated secondary oocyte with zona pellucida

**4** A corpus luteum develops from the remnants of the follicle.

**3** The secondary oocyte begins meiosis II and is released during ovulation when the follicle ruptures.

**Figure 51.8 Follicle and oocyte development in the ovarian cycle.** Development of an oocyte and corpus luteum within the ovary are events that occur during a single ovarian cycle.

The developing secondary oocyte is surrounded by cells of the cumulus mass and theca, which protect and nurture it and secrete estradiol. The estradiol is secreted into the blood, where it functions to control the secretion of LH and FSH from the anterior pituitary gland, as we'll discuss shortly. Some estradiol is also secreted into the follicle, where it stimulates fluid secretion into the inner core of the follicle. As the follicle continues to grow, the fluid pressure inside the follicle increases, until it begins to form a bulge. Eventually, ovulation occurs as the follicle ruptures, and the secondary oocyte, the zona pellucida, and some surrounding supportive cells of the cumulus mass are released from the ovary (see step 3 of Figure 51.8).

Cells in the empty follicle subsequently undergo pronounced anatomical and physiological changes, differentiating into a structure called the **corpus luteum** (see step 4, Figure 51.8). In humans, the corpus luteum is active for approximately the second half of the ovarian cycle. It is responsible for secreting hormones that stimulate the development of the uterus required for sustaining an embryo in the event of a pregnancy. If pregnancy does not occur, the corpus luteum degenerates, and a new group of immature follicles with their primary oocytes develops (see step 5, Figure 51.8).

In humans, a typical ovarian cycle lasts approximately 28 days. As a result, of the 200,000 or so primary oocytes that were in each ovary at the onset of puberty, only about 300 to 500 secondary oocytes are ovulated over a woman's 30- to 40-year reproductive lifetime. In addition, degeneration of other primary oocytes continues at a rate of about 1000 per month throughout much of adulthood. The mechanisms that cause this cell death are still being investigated. Eventually, the oocytes become depleted and a woman stops having ovarian cycles, an event called **menopause**. The average onset of menopause in the U.S. is approximately 51 years of age. After menopause, a woman is no longer capable of ovulation.

## The Ovarian Cycle Results from Changes in Hormone Secretion

We saw that in males, testosterone produced by cells in the testes exerts negative feedback on the secretion of GnRH and LH. In females, however, the regulation of GnRH and LH secretion is more complicated. Although GnRH also stimulates release of LH and FSH in females, the resulting estradiol produced by these hormones can have both negative and positive feedback effects on the gonadotropins. To understand this, let's examine the hormonal changes that occur during the ovarian cycle in a human female (**Figure 51.9**).

The first half of the ovarian cycle is called the **follicular phase**, because this is when the growth and differentiation of a cohort of immature follicles occur. The relatively low concentration of LH that exists during follicular development is sufficient to stimulate the cells of the follicle to make estradiol. The estradiol that is produced is important for enlargement and growth of the oocytes, and it is also secreted into the blood where it can influence the secretion of LH and FSH.

As the follicles develop, their secretion of estradiol gradually increases, and consequently, the concentration of estradiol in the blood begins to increase (see Figure 51.9). Another ovarian steroid, progesterone, is also secreted into the blood in very low amounts at this time. Initially, estradiol and progesterone exert a negative feedback action on the secretion of LH and FSH from the pituitary gland. As a result, LH and

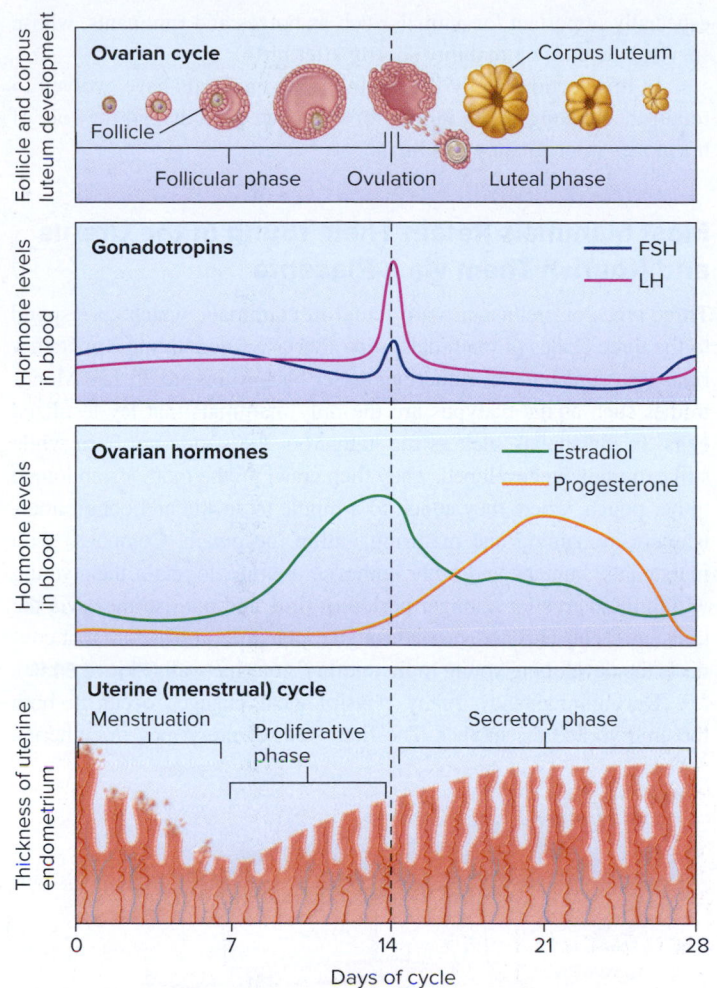

**Figure 51.9** **The ovarian and uterine cycles in a human female.** The ovarian cycle is divided into the follicular and luteal phases. The uterine cycle is divided into menstruation and the proliferative and secretory phases.

 **Core Skill: Modeling** The goal of this modeling challenge is to predict the pattern of blood concentrations of LH and FSH in a woman taking a daily oral contraceptive that contains estradiol and progesterone.

**Modeling Challenge:** As described later in Section 51.6, a common method of preventing ovulation in women who do not wish to become pregnant is taking a daily pill that contains an estrogen, such as estradiol, and progesterone. This pill is called a contraceptive, which is any medication or device that acts to decrease the likelihood of a pregnancy. The ovarian steroids are given at a sufficient dose that they exert strong negative feedback on LH and FSH secretion, but do not exert positive feedback. In a model based on Figure 51.9, first draw the normal pattern for the blood concentrations of LH and FSH during a cycle. Then, on the same model, draw a predicted pattern for the blood levels of LH and FSH over the course of 28 days in a woman taking a contraceptive pill containing estradiol and progesterone.

FSH concentrations in the blood are kept at a relatively low level, just high enough to continue promoting follicle development. As the cycle proceeds, all but the largest of the developing follicles eventually die, as discussed earlier. As the remaining follicle nears full development and gets ready for ovulation, its production of estradiol increases, causing the blood concentration of estradiol to increase sharply. At that time, the feedback action of estradiol on LH and FSH switches from negative to positive, by mechanisms that involve increased GnRH secretion from the hypothalamus. This event results in a sudden, sharp surge in LH and FSH in the blood. The LH released from the pituitary as a result of positive feedback by estradiol induces rupture of the follicle; that is, it induces ovulation.

Ovulation marks the end of the follicular phase and the beginning of the **luteal phase** of the ovarian cycle, named after the corpus luteum. During this phase, the high concentration of LH in the blood initiates development of the corpus luteum. The corpus luteum secretes large amounts of progesterone, the dominant ovarian hormone of the luteal phase, plus some estradiol. The increasing concentration of progesterone in the blood inhibits LH and FSH secretion by negative feedback and is essential to prepare the uterus to receive and nourish the embryo. If fertilization of the secondary oocyte does not occur, the corpus luteum degenerates after 2 weeks and progesterone release decreases, allowing LH and FSH to initiate development of a new set of oocytes and their follicles. However, if fertilization does occur, the blastocyst develops a surrounding layer of cells that secrete an LH-like hormone, called **chorionic gonadotropin**, which maintains the corpus luteum and its ability to secrete progesterone. Chorionic gonadotropin is only present in pregnant women. Therefore, the detection of this hormone is the basis for some pregnancy tests.

## Maternal Hormones Prepare the Uterus to Accept the Embryo

In humans, the ovarian cycle occurs in parallel with changes in the lining of the uterus during the **uterine cycle**, or **menstrual cycle**. The hormones produced by the ovarian follicle influence the development of the endometrium, the glandular inner layer of the uterus. As depicted at the bottom left of Figure 51.9, a period of bleeding called **menstruation** marks the beginning of the uterine cycle and the follicular phase of the ovarian cycle. During menstruation, the endometrium is sloughed off and released from the body.

By about the end of the first week of the menstrual cycle, the endometrium is ready to grow again in response to the newly increasing concentration of estradiol secreted by a developing follicle. This phase of the menstrual cycle, which corresponds to the latter part of the ovarian follicular phase, is called the proliferative phase (see Figure 51.9) because uterine cells begin to grow and divide. During this time, the endometrium becomes thicker and more vascularized. During the subsequent luteal phase of the ovarian cycle, progesterone from the corpus luteum initiates further endometrial growth, including the development of glands that secrete nutritive substances that sustain the embryo during its first 2 weeks in the uterus. This part of the menstrual cycle is called the secretory phase because of these glandular secretions. If fertilization does not occur, degeneration of the corpus luteum and the associated decrease in progesterone and estradiol concentrations initiate menstruation and the beginning of the next uterine cycle. If fertilization does occur, however, the blastocyst becomes embedded in the endometrium and pregnancy begins, as described in the next section.

## 51.4    Pregnancy and Birth in Mammals

**Learning Outcomes:**

1. Explain the relationship between the fetal and maternal structures of the placenta.
2. Outline the hormonal control of the birth process.

**Pregnancy**, or gestation, is the time during which a developing embryo or fetus grows within the uterus of the mother. Physiologically, pregnancy is considered to begin not at fertilization but when the embryo is established in the uterine lining. This occurs within days of fertilization in animals with short gestation lengths but may take weeks in large animals with long gestations.

In mammals, gestation length varies widely and is roughly related to the size of adults in a particular species. Small animals such as hamsters and mice have gestation periods of 16 to 21 days, canids have longer pregnancies of about 50 to 75 days, humans average about 268 days, and the Asian elephant carries its fetus up to 660 days. The advantages of prolonging prenatal development are twofold: The embryo is protected while it is developing in the uterus, and the offspring can be more fully developed at birth. Being more fully developed at birth is

especially important for animals such as horses and ruminants, whose survival depends on mobility shortly after birth.

In this section, we will examine how mammals have evolved to retain their young in the uterus for extended periods and the role of hormones in pregnancy and birth.

### Most Mammals Retain Their Young in the Uterus and Nourish Them via a Placenta

Three types of pregnancies are found in mammals, which correspond to the three clades of mammals: prototherians (monotremes), metatherians (marsupials), and eutherians (refer back to Figure 35.25). Monotremes such as the platypus are the only mammals that lay fertilized eggs. In marsupials such as the kangaroo, the young are born while still extremely undeveloped. They then crawl up the mother's abdomen to her pouch, where they attach to a nipple to suckle and obtain nourishment, remaining and maturing within the pouch. Compared with marsupials, humans and other eutherian mammals retain their young within the uterus for a longer period of time and nourish them via the transfer of nutrients and gases through a **placenta**, a structure that connects the developing young to the mother's uterine wall (**Figure 51.10**).

During pregnancy, many physiological changes occur in both the embryo and the mother. The first event of pregnancy in eutherian

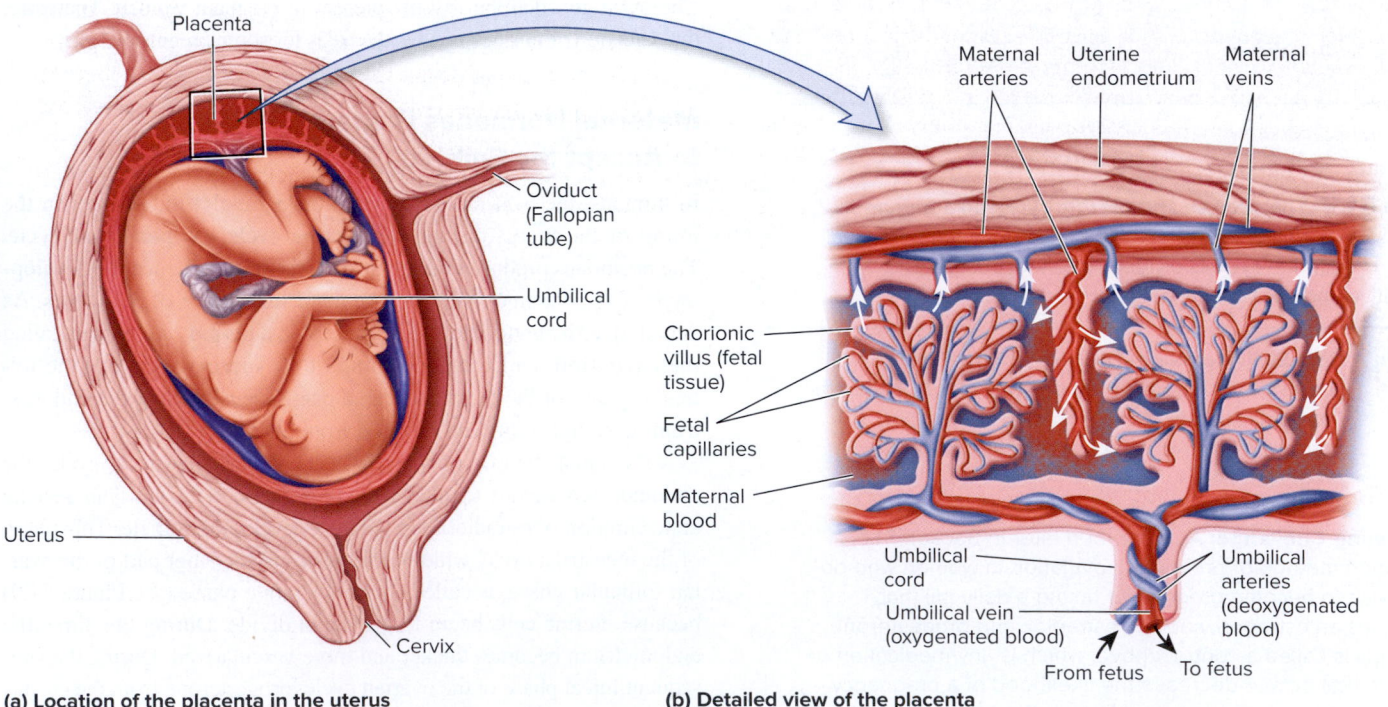

**(a) Location of the placenta in the uterus**

**(b) Detailed view of the placenta**

**Figure 51.10    The structure of the placenta.** In mammals, the placenta is composed of both maternal and embryonic or fetal tissues. **(a)** Overview of placental location and structure in the human. **(b)** Enlarged view of the placenta showing the relationship between fetal and maternal structures. Note that in humans, blood in the fetal and maternal circulations does not mix. The white arrows indicate diffusion of solutes such as nutrients and wastes; black arrows indicate direction of blood flow.

 **Core Skill: Connections** Note that the umbilical arteries are shown in blue to signify that they carry deoxygenated blood, and the umbilical vein is shown in red to signify it carries oxygenated blood. Normally, arteries carry oxygenated blood, and veins carry deoxygenated blood, except in the pulmonary circulation (refer back to Figure 48.2b). Why is the circulation through the placenta like that of the pulmonary circulation?

(placental) mammals is **implantation**, when the blastocyst embeds within the uterine endometrium, which typically occurs in humans around 8 to 10 days after fertilization. As mentioned earlier, the implanted blastocyst initially receives nutrients directly from endometrial glands. However, shortly after implantation, newly developing embryonic tissues merge with the endometrium to form the placenta, which remains in place and grows larger as the embryo matures into a **fetus**. (In humans, an embryo is called a fetus after the eighth week of gestation.) The placenta has a maternal portion and an embryonic or fetal portion.

The placenta is rich in blood vessels from both the mother and the fetus. The maternal and fetal sets of vessels lie in close proximity. The fetal portion of the placenta, called the chorion, contains convoluted structures called chorionic villi that provide a large surface area containing capillaries where nutrients, gases, and other solutes are exchanged. Nutrients and oxygen from the mother are carried through maternal arteries, where her blood pools in large areas of the fetal placenta surrounding the fetal capillaries. Solutes diffuse from the maternal blood into fetal capillaries, and from there flow into the umbilical vein, part of the fetal circulation. In turn, carbon dioxide and other waste products from the fetus are carried through the umbilical artery to the placenta, where they diffuse into the mother's circulation and are eventually excreted. Because of this placental organization, the maternal and fetal blood do not mix.

Prenatal development in humans is generally described as having three trimesters, each of which lasts about 3 months. At the end of the first trimester, the rudiments of the organs are present, and the developing fetus is about an inch long. The second trimester is an extremely rapid phase of growth. During the third trimester, the lungs of the fetus mature so that they are ready to function as gas-exchange organs when the fetus makes the transition to breathing air.

## Core Concept: Evolution

### The Evolution of the Globin Gene Family Has Been Important for Internal Gestation in Mammals

As discussed in Chapter 21, genes can be duplicated, which sometimes creates gene families. Gene families have been important in the evolution of complex traits because the various members of a gene family enable the expression of complex, specialized forms and functions. An interesting example is the globin gene family in animals. Globin genes encode polypeptides that are subunits of proteins that function in oxygen binding. Hemoglobin, which is present in erythrocytes, carries oxygen throughout the body in all vertebrates and many invertebrates, delivering oxygen to all of the body's cells. In humans, the globin gene family is composed of several homologous genes that were originally derived from a single ancestral globin gene (refer back to Figure 21.14).

All of the globin polypeptides are subunits of proteins that function in oxygen binding, but the various family members tend to have specialized functions. For example, certain globin

| Table 51.1 | Globin Gene Expression During Mammalian Development | | |
|---|---|---|---|
| Stage of development | Globin genes expressed | Hemoglobin composition | Oxygen affinity (P50)* |
| Embryo | ε-globin and ζ-globin | Two ε-globin and two ζ-globin subunits | 5–13.5 mmHg |
| Fetus | γ-globin and α-globin | Two γ-globin and two α-globin subunits | 19.5 mmHg |
| Birth to adult | β-globin and α-globin | Two β-globin and two α-globin subunits | 26.5 mmHg |

*$P_{50}$ values represent the partial pressure of oxygen required to half-saturate hemoglobin (refer back to Figure 48.24): A lower $P_{50}$ indicates a higher affinity of hemoglobin for oxygen. The value for embryos is an estimate based on in vitro experiments. All values are for human hemoglobins.

genes are expressed only during particular stages of embryonic development (refer back to Figure 14.3). This phenomenon has special importance in placental mammals, because the oxygen demands of a growing embryo and fetus are quite different from the demands of its mother. These different demands are met by the differential expression of hemoglobin genes during prenatal development.

Altogether, five globin genes—designated α, β, γ, ε, and ζ—encode the major subunits that are found in hemoglobin proteins at different developmental stages. During embryonic development, the ε-globin and ζ-globin genes are turned on, resulting in embryonic hemoglobin with a very high affinity for oxygen (**Table 51.1**). At the fetal stage, these genes are turned off, and the α-globin and γ-globin genes are turned on, producing fetal hemoglobin with slightly lower (but still high) affinity for oxygen. Finally, just before birth, expression of the γ-globin gene decreases, and the β-globin gene is turned on, resulting in adult hemoglobin, which has a lower affinity for oxygen than either the embryonic or fetal forms. The higher affinities of embryonic and fetal hemoglobins enable the embryo and fetus to remove oxygen from the mother's bloodstream and use that oxygen to meet their own metabolic demands. Therefore, the evolution of different globin genes each of which is expressed at particular stages of development enables placental mammals to develop in the uterus without breathing on their own or being exposed to atmospheric oxygen.

## Birth Is Dependent on Hormones That Elicit a Positive Feedback Loop

Birth—also called parturition—is initiated by the actions of several hormones and other factors secreted by the placenta and glands of the mother (**Figure 51.11**). Toward the end of pregnancy, hormones from the fetus stimulate the placenta to start secreting large amounts of estrogens such as estradiol into the maternal circulation.

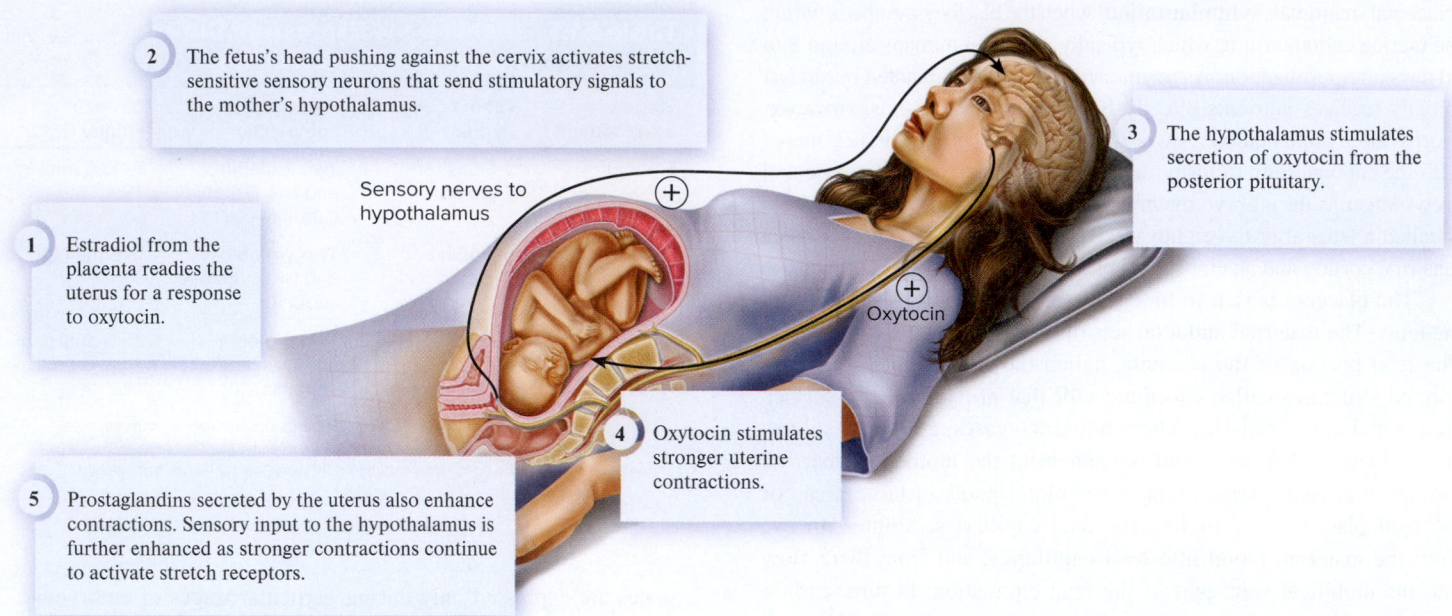

**2** The fetus's head pushing against the cervix activates stretch-sensitive sensory neurons that send stimulatory signals to the mother's hypothalamus.

**3** The hypothalamus stimulates secretion of oxytocin from the posterior pituitary.

Sensory nerves to hypothalamus

**1** Estradiol from the placenta readies the uterus for a response to oxytocin.

Oxytocin

**4** Oxytocin stimulates stronger uterine contractions.

**5** Prostaglandins secreted by the uterus also enhance contractions. Sensory input to the hypothalamus is further enhanced as stronger contractions continue to activate stretch receptors.

**Figure 51.11 Hormonal control of parturition.** Birth relies on neural signals from the uterus and maternal hormones that act on the uterus. In response to sensory neural input arising from the push of the fetus on the cervix, the maternal posterior pituitary releases oxytocin into the blood, which stimulates uterine smooth muscle contractions. The secretion of prostaglandins by the uterus also increases the strength of the contractions. Sensory receptors in the uterus detect the more forceful contractions and signal the mother's posterior pituitary to secrete more oxytocin, thus completing a positive feedback loop that further strengthens the contractions.

 **Core Skill: Connections** In what other context have you learned about positive feedback in female reproduction? See Figure 51.9 for help.

Estrogens have at least two major effects on uterine tissue at this time. First, they promote the formation of gap junctions between uterine smooth muscle cells, which enables coordinated uterine contractions. Second, estrogens enhance uterine sensitivity to the hormone oxytocin.

Recall from Chapter 50 that oxytocin is a posterior pituitary hormone that stimulates contraction of uterine muscle. The high concentration of estradiol in the mother's blood near the end of pregnancy stimulates the production of oxytocin receptors in smooth muscle cells of the uterus, thereby making the uterus more sensitive to oxytocin. At the same time, the fetus usually positions itself with its head above the cervix in preparation for birth. The pressure of the fetus's head pressing on the cervix stretches the smooth muscle of the lower uterus including the cervix. This stretch is detected by neurons in these structures. Signals from the stretch-sensitive neurons are sent to the mother's hypothalamus, triggering the release of oxytocin into the blood.

Binding of oxytocin to its receptors initiates the strong uterine muscle contractions that are the hallmark of **labor**. In addition to its direct action on uterine muscle, oxytocin stimulates uterine secretion of small signaling molecules called prostaglandins that act with oxytocin to further increase the strength of the muscle contractions. The stronger contractions elicit even greater stimulation of stretch receptors, resulting in more oxytocin release from the mother's pituitary, which causes yet stronger contractions. This positive feedback loop continues until the baby is born.

Labor occurs in three stages (**Figure 51.12**). In stage one, dilation and thinning of the cervix occur, which makes it easier for the fetus to pass out of the uterus. In stage two, uterine contractions get stronger

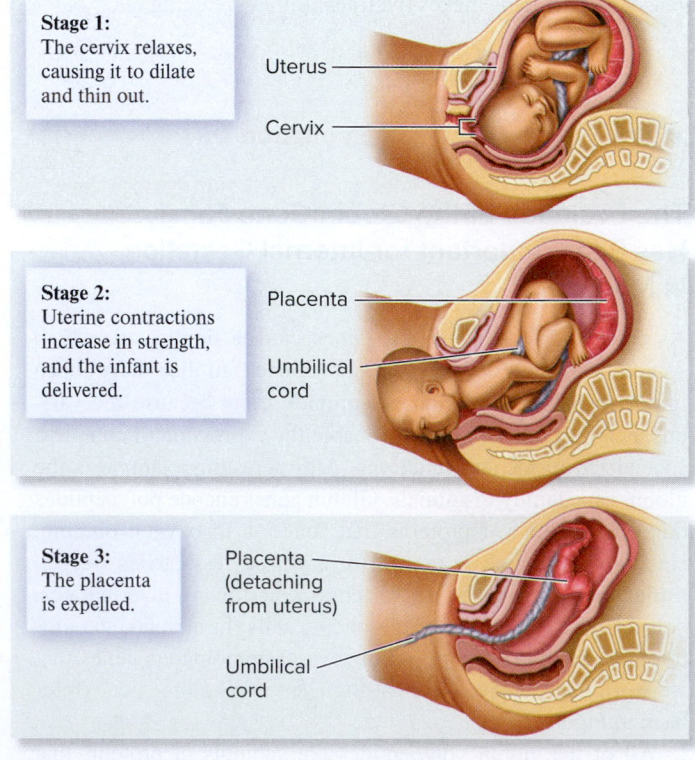

**Stage 1:** The cervix relaxes, causing it to dilate and thin out.

Uterus

Cervix

**Stage 2:** Uterine contractions increase in strength, and the infant is delivered.

Placenta

Umbilical cord

**Stage 3:** The placenta is expelled.

Placenta (detaching from uterus)

Umbilical cord

**Figure 51.12 The three stages of labor.**

*Concept Check:* *Many female mammals consume the placenta after giving birth. What is the benefit of this behavior?*

and more frequent. The fetus is pushed, usually headfirst, through the cervix and the vagina and out into the world. In stage three, the contractions continue for a short while. Blood vessels within the placenta and umbilical cord contract and block further blood flow, making the newborn independent from the mother. As the oxytocin-induced contractions continue for a short period, the placenta detaches from the uterine wall and is delivered a few minutes after the birth of the baby.

## 51.5 General Events of Embryonic Development

### Learning Outcomes:

1. Describe the general events of embryonic development, and identify the end result of each.
2. Describe the process of cleavage, beginning at fertilization and leading into gastrulation.
3. **CoreSKILL »** Predict the fates of cells in a gastrula as they differentiate into the three major germ layers.
4. Outline the early development of the nervous system during neurulation.
5. Describe the migration and fates of neural crest cells.

The process by which a fertilized egg (that is, a zygote) is transformed into an animal with distinct physiological systems and body parts is called **embryonic development**. As an animal develops, cells arrange themselves in coordinated ways that lead to the establishment of a body plan. The final, adult body plan of most animals is organized along three axes: the dorsoventral axis, the anteroposterior axis, and the left-right axis. Along these axes are often separate sections, or body segments, each containing specific body parts such as a wing or leg.

To establish the correct body plan, each cell in a developing animal must receive information regarding its relative position within the body. Such information determines where a cell should move to, whether or not it should divide (or die), and what types of functions it will ultimately perform. Each cell receives the required positional information from its neighboring cells. This information is provided in a variety of ways, including intracellular and extracellular signaling molecules and by cell-to-cell contacts. A cell responds to positional information by dividing (cell division), dying (cell death, or apoptosis), or migrating from one region of the embryo to another (cell migration). Last, in the process of **cell differentiation**, different cells within a developing organism acquire specialized forms and functions, due to the expression of cell-specific genes.

Embryonic development follows a similar pattern in most animals. Most modern animals are triploblastic; that is, they develop from embryos with three distinct cell layers called the ectoderm, mesoderm, and endoderm (refer back to Figure 33.5). Triploblasts include vertebrates and most invertebrates other than sponges and cnidarians. Development in these animals can be categorized into five general events: (1) fertilization (discussed earlier; see Figure 51.4), (2) cleavage, (3) gastrulation, (4) neurulation, and (5) organogenesis (**Figure 51.13**). In this section, we will examine the key aspects of the general events of animal development beginning with cleavage. However, it should be noted that many species also have an additional event called metamorphosis, which is a transition from a feeding larval form to an adult (refer back to Figure 34.25). Metamorphosis occurs after organogenesis and facilitates the rapid growth of young organisms into mature ones. Examples of metamorphosis include the transformation of a caterpillar into a butterfly and that of a tadpole into a frog.

### Cell Divisions Without Cell Growth Create a Cleavage-Stage Embryo

The initial cell cycles of embryos are unique because they involve repeated cell divisions without cell growth. The process by which

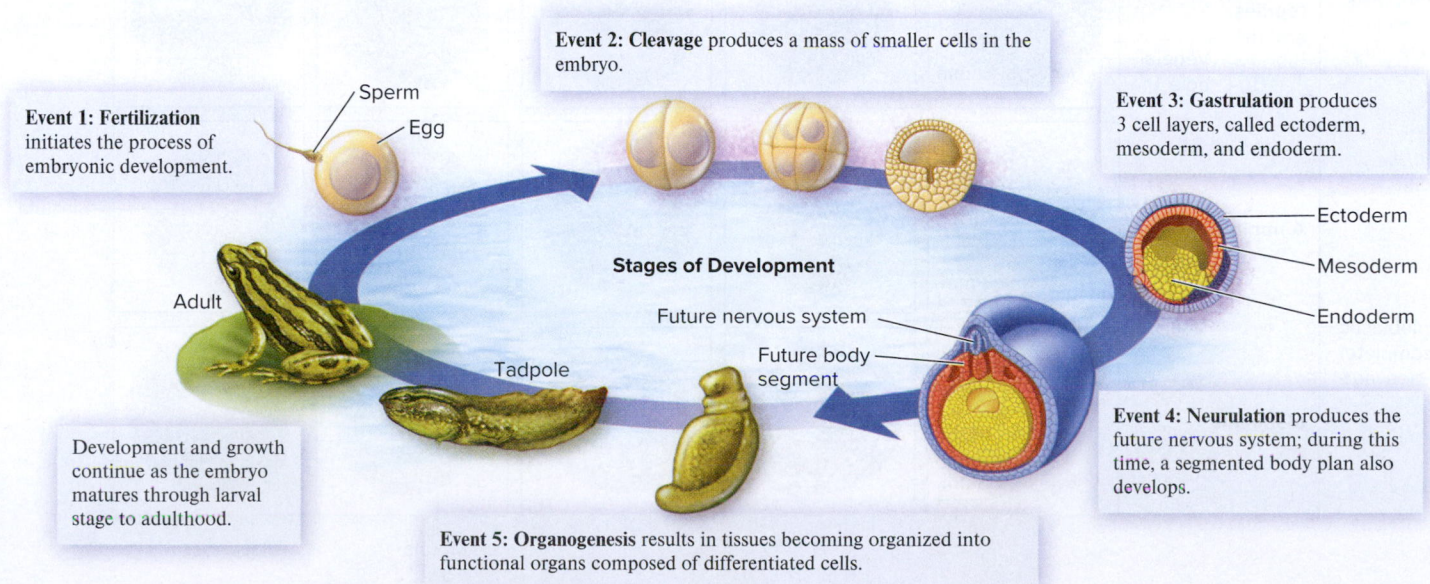

**Event 1: Fertilization** initiates the process of embryonic development.

Sperm

Egg

**Event 2: Cleavage** produces a mass of smaller cells in the embryo.

**Event 3: Gastrulation** produces 3 cell layers, called ectoderm, mesoderm, and endoderm.

Ectoderm

Mesoderm

Endoderm

**Stages of Development**

Future nervous system

Future body segment

**Event 4: Neurulation** produces the future nervous system; during this time, a segmented body plan also develops.

Adult

Tadpole

Development and growth continue as the embryo matures through larval stage to adulthood.

**Event 5: Organogenesis** results in tissues becoming organized into functional organs composed of differentiated cells.

**Figure 51.13 Overview of events of embryonic development.** This figure shows the general developmental events of all vertebrate embryos, using a frog as an example.

these cell cycles occur is called **cleavage**. The embryonic cells repeatedly split in two, resulting in several generations of daughter cells that are roughly half the size of the cells that gave rise to them.

In most species in which development occurs outside the mother, cell division during cleavage represents one of the fastest cell cycles found in nature. This minimizes the time in which eggs or early embryos could be eaten by predators. The cell cycle during cleavage in amphibians, for example, requires only 20 minutes. During each 20-minute cell cycle, complete genome replication, mitosis, and duplication of the nuclear envelope are followed by cytokinesis. In placental mammals, in which development occurs within the protective environment of the mother's body, cell divisions during cleavage are slower, requiring about 12 hours to complete.

The daughter cells produced during the cleavage stage of development are known as **blastomeres**. Individual blastomeres are bound together, and an outer, single-cell layer of blastomeres forms a sheet of epithelial cells that separates the embryo from its environment. After formation of the outer epithelial layer, the embryos of many animals take up water and form a cavity called a **blastocoel**. The embryo at this stage is called a **blastula**. The blastocoel provides a space into which cells will migrate to form the digestive tract and other structures of the embryo.

***Animal and Vegetal Poles of Cleavage-Stage Embryos*** Among triploblastic organisms, cleavage-stage embryos can vary considerably in size and appearance. This variation is in part related to whether or not the egg contains yolk and, if so, the location and amount of that yolk. Yolk is a nutrient-rich food store that is used by the developing embryo. The eggs of birds, some fishes, and some other vertebrates have large amounts of yolk. In the eggs and early embryos of these species, yolk is most concentrated toward one end—or pole—called the **vegetal pole**. Much less yolk, and much more cytoplasm, is concentrated near the

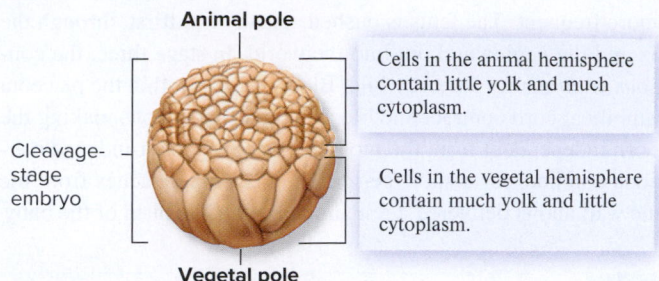

**Animal pole**

Cells in the animal hemisphere contain little yolk and much cytoplasm.

Cleavage-stage embryo

Cells in the vegetal hemisphere contain much yolk and little cytoplasm.

**Vegetal pole**

**Figure 51.14** **Polarity in an amphibian cleavage-stage embryo.**

opposite pole, called the **animal pole** (**Figure 51.14**). These poles form the apices of the vegetal and animal hemispheres, which determine in part the future anteroposterior (head-tail) and dorsoventral (back-front or top-bottom, depending on the species) axes of the embryo.

***Meroblastic Cleavage: Birds, Reptiles, and Fishes*** In some species that exhibit animal and vegetal poles, the cleavage process is called **meroblastic cleavage**, or incomplete cleavage, because only the region of the embryo located within the animal hemisphere undergoes cell division (**Figure 51.15**). Instead of forming a ball of cells (a blastula), in this type of cleavage, a flattened disc of blastomeres known as a **blastoderm** develops on top of the yolk mass.

***Holoblastic Cleavage: Amphibians and Mammals*** In animals whose eggs have smaller amounts of yolk, cleavage during the first cell division is complete and bisects the entire zygote into two equal-sized blastomeres. This complete cleavage, or **holoblastic cleavage**, occurs in amphibians and mammals (see Figure 51.15). In amphibians, cleavage-stage embryos form a blastula, as previously noted. In mammals, however, cleavage-stage embryos undergo a process called compaction, in

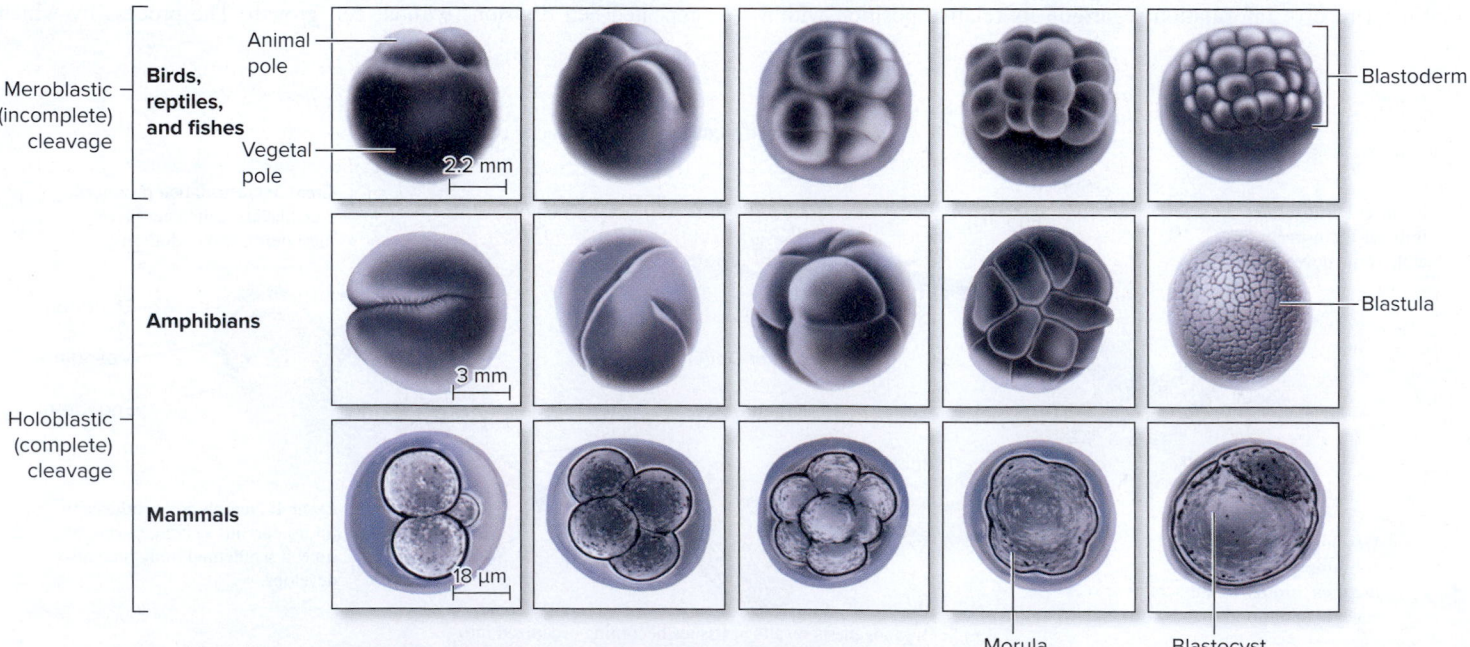

**Figure 51.15** **Meroblastic and holoblastic cleavage.** Early embryos of birds, reptiles, and many fishes undergo incomplete (meroblastic) cleavage, whereas most amphibian and mammalian embryos undergo complete (holoblastic) cleavage. The amount of yolk in the egg (not visible in these illustrations) contributes to many of these morphological differences observed in various species.

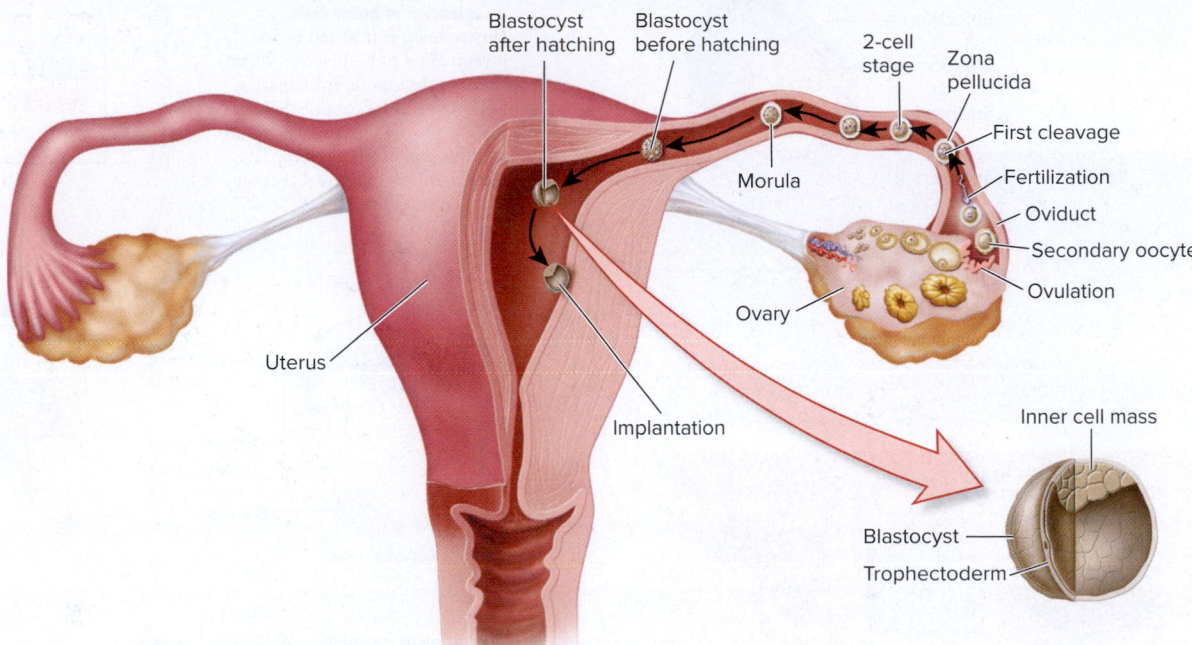

**Figure 51.16** **The sites of early embryonic development in mammals.** After an ovulated secondary oocyte is fertilized, initial cleavage and development of the resulting embryo occur in the oviduct of the mother. The blastocyst hatches from the zona pellucida before implanting into the inner lining of the uterus. A blastocyst is composed of an outer epithelial layer, called the trophectoderm, which gives rise to extraembryonic tissues such as the placenta, and an inner cell mass, which develops into the embryo.

which the amount of physical contact between cells is maximized. At this stage, the embryo in these species is called a **morula**. The morula then proceeds to form a blastocyst, the mammalian counterpart of a blastula.

***Cleavage and Implantation in Mammals***   In mammals, the events of fertilization and cleavage occur in the oviduct (**Figure 51.16**). The blastocyst has a morphological appearance that differs from that of the blastula or blastoderm in nonmammalian species, and shows no animal-vegetal polarity that is analogous to that of other chordates. The blastocyst consists of an outer epithelial layer called the **trophectoderm**, which gives rise to the placenta, and an inner layer called the inner cell mass, which develops into the embryo. Upon reaching the uterus, the blastocyst hatches from the zona pellucida, the layer of glycoproteins that surrounds the secondary oocyte and is retained up to this time to prevent premature adhesion of the blastocyst to the oviduct. The blastocyst then implants in the endometrium of the mother's uterus (see Figure 51.16). In humans, this entire process takes about 8 to 10 days.

Toward the end of cleavage, cell cycles become less synchronous, and the embryo begins to express its own genes. The embryo's shift from existing exclusively on maternal factors to developing in response to products derived from its own genome begins 6–24 hours after fertilization in vertebrates. This shift is followed by the next general event of development, called gastrulation.

## Gastrulation Establishes the Three Germ Layers in the Embryo

**Gastrulation** is one of the most dramatic events of embryonic development in animals because of the major cell movements that occur.

During gastrulation, the hollow ball of cells that makes up a blastula or blastocyst develops into a highly organized structure called a **gastrula** (refer back to Figure 33.5). A key event is the formation of germ layers, which are primary layers of cells that form during gastrulation. In the gastrula-stage embryo, three germ layers—**ectoderm**, **mesoderm**, and **endoderm**—become clearly established. These distinct germ layers are partially differentiated tissues that occupy discrete regions of the embryo, with an outer ectoderm, a middle mesoderm, and an inner endoderm layer. Each type of germ layer eventually gives rise to different structures. The organization that emerges during gastrulation is most evident by the clear establishment of the digestive tube and body axes. Gastrulation is the first time when both the anteroposterior and dorsoventral body axes are clearly evident in the embryo.

Each germ layer gives rise to different types of cells and body structures (**Figure 51.17**). The ectoderm in the gastrula forms the epidermis and nervous system in the later embryo. The mesoderm gives rise to muscles, kidneys, blood, heart, limbs, connective tissues, and notochord; the last is a key feature of all chordates, described later in this section. The endoderm becomes the epithelial lining of the pancreas, thyroid, lungs, digestive tract, liver, and urinary bladder.

Some of the most detailed descriptions of the events in gastrulation come from the study of frog and other amphibian embryos. The major events of gastrulation are depicted in **Figure 51.18** and described next.

***Invagination and Involution: Formation of Germ Layers and Archenteron***   Prior to gastrulation, the blastula of amphibians is enclosed in a simple, spherical epithelial cell layer. Gastrulation

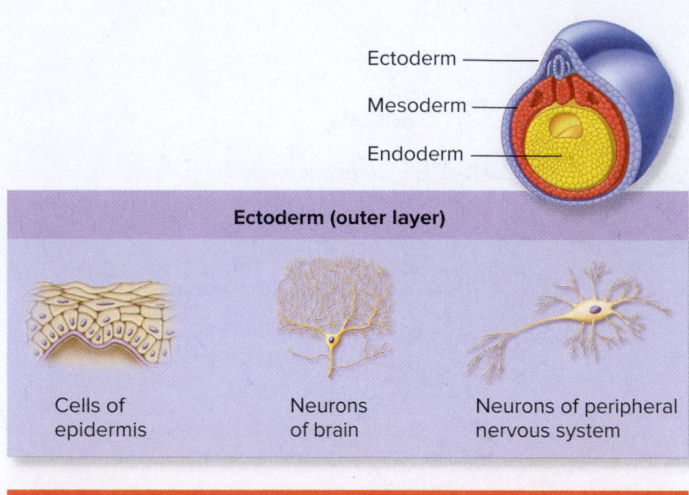

Ectoderm
Mesoderm
Endoderm

### Ectoderm (outer layer)

| Cells of epidermis | Neurons of brain | Neurons of peripheral nervous system |

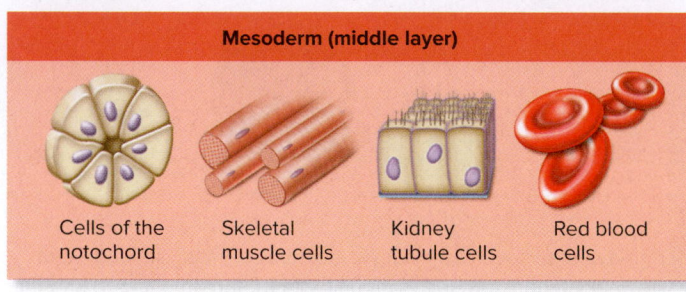

### Mesoderm (middle layer)

| Cells of the notochord | Skeletal muscle cells | Kidney tubule cells | Red blood cells |

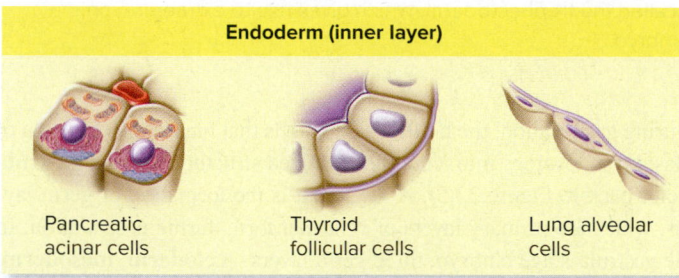

### Endoderm (inner layer)

| Pancreatic acinar cells | Thyroid follicular cells | Lung alveolar cells |

**Figure 51.17** **Examples of cell types derived from ectoderm, mesoderm, and endoderm.**

 **Core Concept: Structure and Function** The formation of specific cell types with different structures and functions is an example of development. Growth entails an increase in the numbers and/or size of each of these cell types.

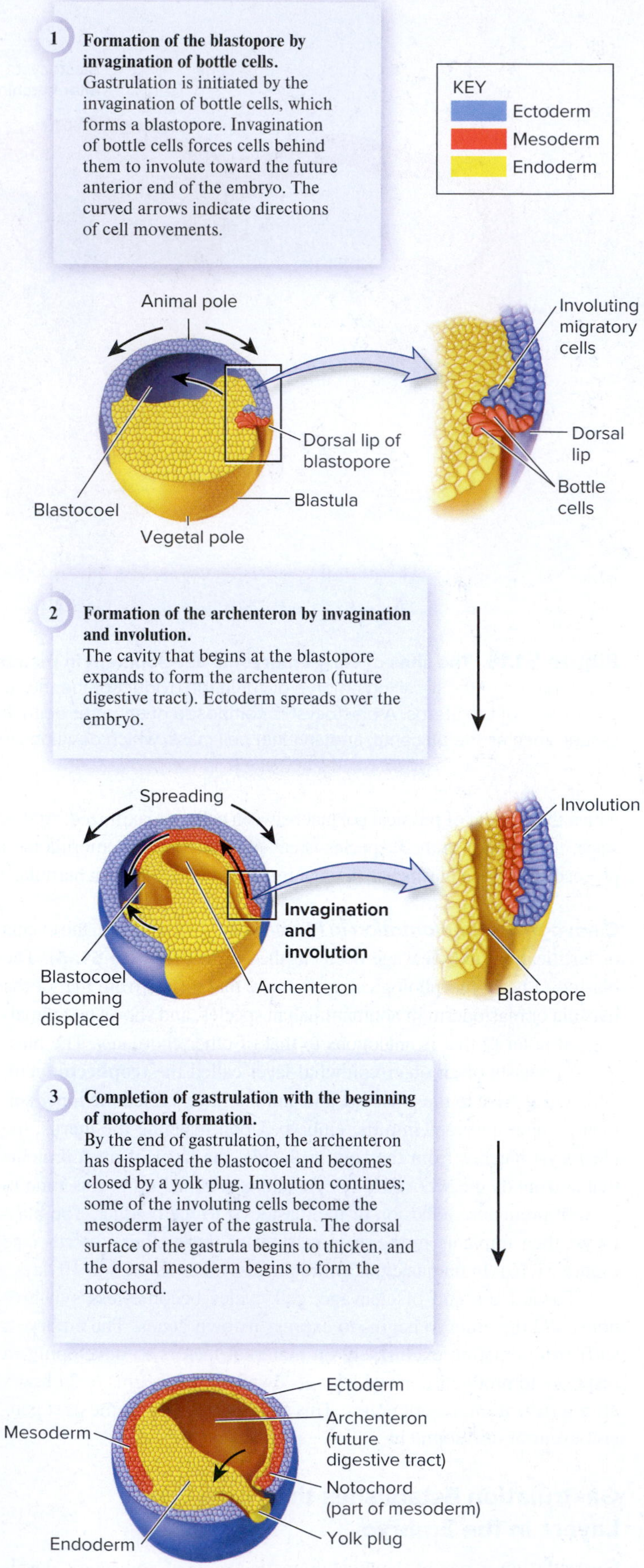

begins when a band of epithelial cells located at the vegetal hemisphere of the blastula—called bottle cells—invaginates (pinches in), pushing cells from the outside of the embryo to the inside (see Figure 51.18, step 1). This process creates a small opening called the **blastopore**, which defines the anteroposterior axis of the animal. The initiating site of invagination becomes what is called the dorsal lip of the blastopore. This change in morphology of only a few key cells in the embryo initiates the gastrulation process in amphibians.

Once the bottle cells change their shape and push into the interior of the embryo, other cell movements occur, and together these orchestrated movements establish the mesoderm and endoderm of the organism, including its future digestive tract. Just before invagination begins, cells of the animal hemisphere spread out and move downward. When they arrive at the blastopore, they enter the opening and

**1** **Formation of the blastopore by invagination of bottle cells.** Gastrulation is initiated by the invagination of bottle cells, which forms a blastopore. Invagination of bottle cells forces cells behind them to involute toward the future anterior end of the embryo. The curved arrows indicate directions of cell movements.

**KEY**
Ectoderm
Mesoderm
Endoderm

Animal pole
Involuting migratory cells
Dorsal lip of blastopore
Dorsal lip
Blastocoel
Blastula
Bottle cells
Vegetal pole

**2** **Formation of the archenteron by invagination and involution.** The cavity that begins at the blastopore expands to form the archenteron (future digestive tract). Ectoderm spreads over the embryo.

Spreading
Involution
Invagination and involution
Blastocoel becoming displaced
Archenteron
Blastopore

**3** **Completion of gastrulation with the beginning of notochord formation.** By the end of gastrulation, the archenteron has displaced the blastocoel and becomes closed by a yolk plug. Involution continues; some of the involuting cells become the mesoderm layer of the gastrula. The dorsal surface of the gastrula begins to thicken, and the dorsal mesoderm begins to form the notochord.

Mesoderm
Ectoderm
Archenteron (future digestive tract)
Notochord (part of mesoderm)
Endoderm
Yolk plug

**Figure 51.18** **The events of gastrulation in amphibians.**

subsequently migrate upward along the roof of the blastocoel, toward the animal pole of the embryo. This folding back of sheets of surface cells into the interior of the embryo is called involution (see Figure 51.18, step 2). After involution, dorsal mesodermal cells migrate toward the animal pole by crawling along the roof of the blastocoel, with endoderm following closely behind.

As the opening from the blastopore extends into the embryo, a new cavity called the **archenteron** displaces the existing blastocoel (see Figure 51.18, steps 2 and 3). The archenteron becomes the organism's future digestive tract. The blastopore opening remains sealed with a yolk-rich piece of tissue called the yolk plug until later in development. In chordates and echinoderms, the opening formed by the blastopore ultimately becomes the anus of the organism (refer back to Figure 33.6). Meanwhile, during involution, surface cells spread from the animal hemisphere to surround the entire vegetal hemisphere to become the future ectoderm. The result of these cellular rearrangements is an embryo with three distinct germ layers.

***Notochord Formation*** A distinguishing anatomical feature that begins forming at the end of gastrulation in all chordates is the **notochord**—a structure derived from mesoderm that provides rigidity along the anteroposterior axis in the dorsal side of the gastrula (see Figure 51.18, step 3). The presence of a notochord defines the phylum Chordata. The notochord persists in the trunk and tail of fishes and amphibians. In birds and mammals, the notochord disappears by the time vertebrae have formed. By the time the notochord has formed, the dorsal ectoderm overlying the notochord begins to thicken, which initiates the next general event in development, called neurulation.

## Neurulation Involves Formation of the Central Nervous System and Segmentation of the Body

By studying development in several different vertebrate species, researchers are beginning to understand some of the fundamental steps in the formation of the central nervous system (CNS)—the brain and spinal cord—in vertebrates. The multistep embryological process responsible for initiating CNS formation is called **neurulation** (**Figure 51.19**). Neurulation occurs just after gastrulation and involves the formation of the **neural tube** from ectoderm located dorsal to the notochord. All neurons and their supporting cells in the CNS originate from neural precursor cells derived from the neural tube.

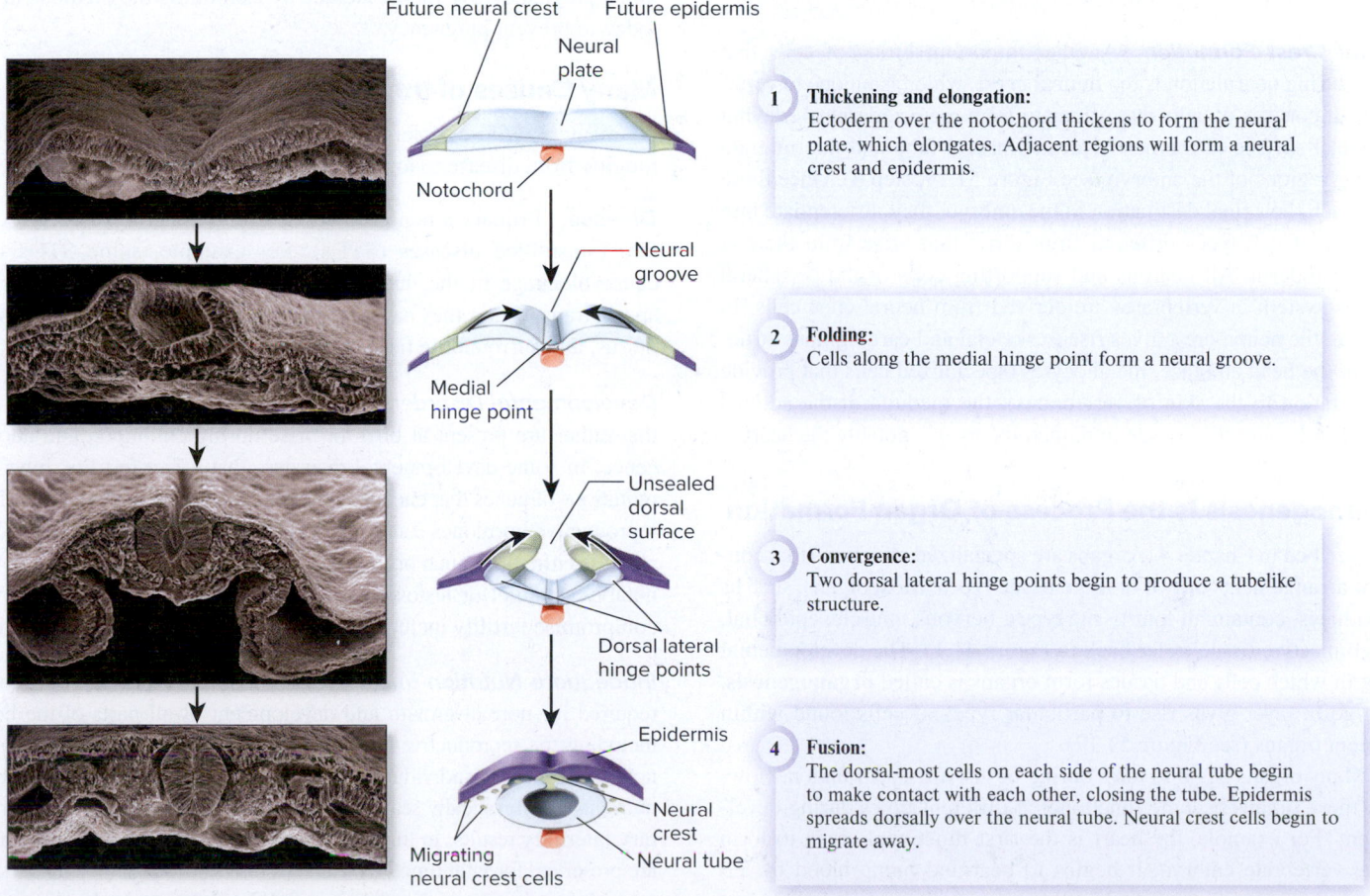

Future neural crest    Future epidermis

Neural plate

Notochord

**1** Thickening and elongation:
Ectoderm over the notochord thickens to form the neural plate, which elongates. Adjacent regions will form a neural crest and epidermis.

Neural groove

Medial hinge point

**2** Folding:
Cells along the medial hinge point form a neural groove.

Unsealed dorsal surface

Dorsal lateral hinge points

**3** Convergence:
Two dorsal lateral hinge points begin to produce a tubelike structure.

Epidermis

Neural crest

Migrating neural crest cells    Neural tube

**4** Fusion:
The dorsal-most cells on each side of the neural tube begin to make contact with each other, closing the tube. Epidermis spreads dorsally over the neural tube. Neural crest cells begin to migrate away.

**Figure 51.19** **Neurulation and the beginning of neural crest formation in vertebrates.** The major steps of neurulation are (1) thickening and elongation; (2) folding, which creates the neural groove; (3) convergence, in which the neural tube begins to take shape; and (4) fusion, in which the neural tube is completed. In a later event, cells migrate away from the neural crest to form several other structures, including the neurons of the peripheral nervous system. (1–4) *Courtesy of Kathryn Tosney*

### Neural Tube Formation

Neurulation in vertebrates occurs in several major steps, as shown in Figure 51.19.

1. In the first step, ectoderm overlaying the notochord thickens to form the neural plate. Adjacent regions will eventually form the epidermis and a structure called the neural crest (discussed shortly). The neural plate then elongates, resulting in the formation of a single, dorsal, elongated epithelial cell layer that is aligned with the animal's anteroposterior axis.

2. Next, a column of cells along the midline of the neural plate—the medial hinge point—initiates folding of the neural plate, leading to the formation of the neural groove.

3. After folding, bilateral columns of cells in the dorsal lateral hinge points then undergo morphological changes that lead to convergence of the two sides of the neural groove and generation of a tubelike structure that is not yet sealed on the dorsal surface.

4. In the next step of neurulation, called fusion (see Figure 51.19, step 4), the dorsal-most cells on either side of the neural tube are released from adjacent ectoderm and make contact with each other, thereby closing the neural tube. At the same time, ectoderm on either side of the neural tube moves toward the centerline, then up and over the neural tube, where it forms the dorsal epidermis of the embryo.

**Neural Crest Formation**    Another important group of cells that arises during neurulation is the **neural crest**, which is unique to vertebrates. It consists of cells that originate from the ectoderm overlaying the dorsal surface of the newly formed neural tube and that migrate to other regions of the embryo (see Figure 51.19, step 4). Once these cells reach their final destination in the embryo, they differentiate into a variety of cell types different from those that arise from the rest of the ectoderm. All neurons and supporting cells of the peripheral nervous system in vertebrates are derived from neural crest cells. In addition, the neural crest gives rise to skeletal and cartilaginous structures in the head and face, melanocytes (specialized cells that provide pigmentation to the skin of vertebrates), the medulla of the adrenal glands, and connective tissue in numerous organs, notably the heart.

## Organogenesis Is the Process of Organ Formation

As described in Chapter 41, organs are specialized structures that consist of arrangements of two or more tissue types. Most organs, such as the kidneys, contain all four tissue types: nervous, muscle, epithelial, and connective tissue (refer back to Figure 41.1). The developmental event in which cells and tissues form organs is called **organogenesis**. Each germ layer gives rise to particular types of cells found within different organs (see Figure 51.17).

Many organs begin to form during or just after neurulation. However, these organs become functional at different times during development. For example, the heart is the first functional organ to form in the vertebrate embryo. It begins to beat and pump blood by 2.5 days after fertilization in chicks, 9 days in mice, and about 22 days in humans. By contrast, the lungs of mammals do not acquire the ability to function until shortly before birth. As we saw in Chapter 41, the development of different organs in animals is controlled by genes in the embryo—notably the *Hox* genes. *Hox* genes are important

for establishing structures along the anteroposterior axis. Many of the genes controlling the processes of gastrulation, neurulation, and organogenesis encode secreted proteins or growth factors that induce cells in their local vicinity to differentiate along a specific developmental pathway. For example, the notochord produces many signaling proteins that help establish tissue patterns in the embryo. Proteins produced within it induce segment-specific expression of the *Hox* genes in subsequent stages of development.

## 51.6    Impact on Public Health

**Learning Outcomes:**

1. Explain the most common causes of human infertility.
2. Compare the different types of birth control, and explain their mechanisms of pregnancy prevention.

Approximately 5–10% of individuals of reproductive age in the U.S. are infertile; that is, they cannot reproduce. In men and women alike, this problem can be caused by a variety of factors. In this section, we discuss some of the common causes of **infertility**—the inability of a man to produce sufficient numbers or quality of sperm to impregnate a woman, or the inability of a woman to become pregnant or maintain a pregnancy. We then conclude by examining the methods in use today to prevent pregnancy.

## Many Causes of Infertility Have Been Identified

As many as 75% of infertility cases have some identifiable cause, ranging from disease to toxin exposure.

**Disease**    Primary among the diseases that affect fertility are sexually transmitted diseases (STDs). For example, some STDs may cause blockage in the ducts of the testes, thus preventing normal sperm transport, or they can cause permanent damage to the oviducts, uterus, and surrounding tissues.

**Developmental Disorders**    Developmental disorders are conditions that either are present at birth or arise during childhood and adolescence. In some developmental disorders that affect fertility, inherited mutations of genes that encode enzymes involved in the biosynthesis of reproductive hormones cause abnormal expression of those genes. The result is either too much or too little of one or more of these hormones, notably estradiol or testosterone. Other developmental disorders that compromise fertility include malformations of the cervix or oviducts.

**Inadequate Nutrition and Other Stressors**    Adequate nutrition is required for normal growth and development of all parts of the body, including the reproductive system. Because the reproductive system is not essential for an individual's survival, it often becomes inactive when nutrients are chronically scarce—such as during starvation—and temporary infertility results. In this way, precious stores of energy in the body are preserved for vital functions, such as those of the brain and heart.

Malnutrition can also delay sexual development and puberty during late childhood and early adolescence. Undernourished children may begin puberty several years later than normal. The brains of all mammals, including humans, contain a center that monitors the body's fat stores. One of the triggers that initiate puberty may be a signal—such as the hormone leptin—from adipose tissue to the brain (see Chapter 50). Very low fat stores in undernourished girls, for example, result in decreased leptin in the blood. This low level of leptin signals the brain that the body does not contain sufficient energy to support the energetic demands of pregnancy, and thus puberty is delayed.

Starvation or poor nutrition is considered a type of stress, defined as any real or perceived threat to an animal's homeostasis. Physical and psychological forms of stress affect fertility in humans. Many nonessential functions, including the maintenance of menstrual cycles in women, can be suppressed by chronic stress. The reproductive consequences of stress are much greater in females than in males, likely because only females bear the energetic cost of pregnancy. Interestingly, from a reproduction viewpoint, the human body responds to long-term strenuous physical exercise in the same way that it responds to long-term stress. This is why many young ballerinas and gymnasts experience delayed puberty and why female long-distance runners (like the one in the chapter introduction) may have menstrual cycles that are abnormal or absent (secondary amenorrhea).

***Other Causes***    When the causes of infertility cannot be determined, a variety of factors come under suspicion. Among these possible causes are ingestion of toxins (for example, certain heavy metals such as cadmium), tobacco smoking, marijuana use, and injuries to the gonads.

***Treatments***    Among several currently available treatments for increasing the likelihood of pregnancy in infertile couples are hormone therapy for the woman to increase egg production and a collection of procedures known as **assisted reproductive technologies (ART)**. In the most common ART procedure, in vitro fertilization, sperm and eggs collected from a man and a woman are placed together in culture dishes. Once the sperm have fertilized the eggs and the resulting zygotes have undergone several cell divisions, one or more of the embryos are inserted into a woman's uterus with the goal that one will implant. When this procedure was first used in 1978, the children born as a result came to be known as "test-tube babies." Since then, several million children have been born using this technology.

## Contraception Usually Prevents Pregnancy

The use of methods to prevent fertilization or the implantation of a fertilized egg is termed **contraception**. Methods of contraception can be either permanent or temporary.

***Permanent Methods***    The permanent forms of contraception surgically prevent the transport of gametes through the reproductive tract (**Figure 51.20a**). **Vasectomy** is a surgical procedure in men that severs the vas deferens, thereby preventing the release of sperm at ejaculation (however, semen is still released). In women, **tubal ligation** involves the cutting and sealing of the oviducts. This procedure prevents the movement of the egg from the oviduct into the uterus. These procedures are considered permanent, because it is difficult—sometimes impossible—to reverse the surgery.

***Temporary Methods***    Temporary methods of preventing fertilization include barrier methods, which prevent sperm from reaching an egg (**Figure 51.20b**). Barrier methods include **vaginal diaphragms,**

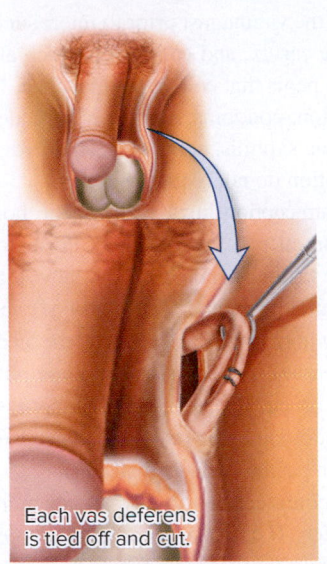
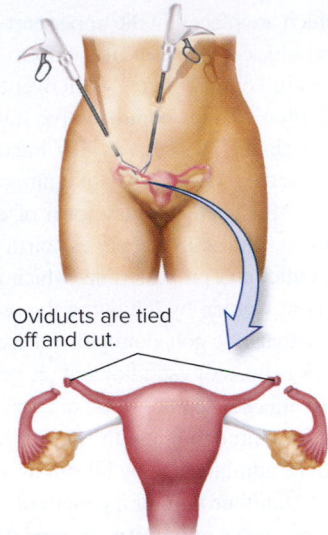

Oviducts are tied off and cut.

Each vas deferens is tied off and cut.

**Vasectomy (<1.0%)**
**(a) Permanent methods**

**Tubal ligation (<1.0%)**

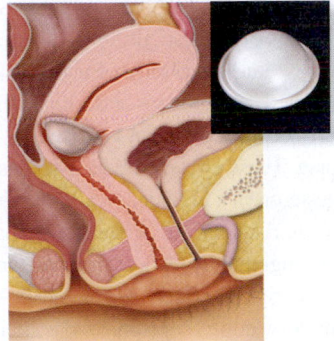
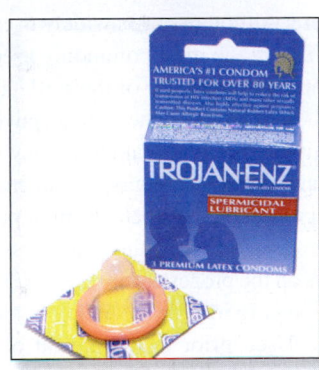

**Diaphragm (5–20%)**

**Condoms (male) (2–15%)**

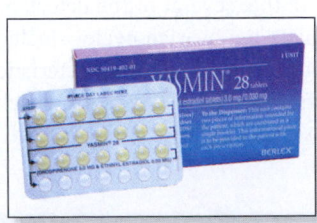
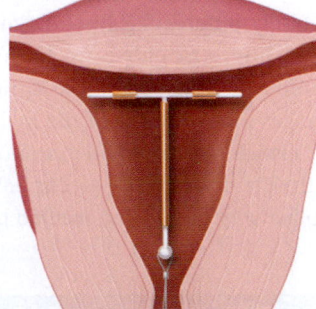

**Oral contraceptive (1–2%)**
**(b) Temporary methods**

**Intrauterine device (IUD) (1–2%)**

**Figure 51.20  Examples of contraceptive methods.** These methods may be used by men or women to **(a)** permanently or **(b)** temporarily prevent pregnancy. The estimated first-year failure rates for each method are given in parentheses (collected from data published by the U.S. Food and Drug Administration and other organizations). A failure rate of 10% means that 10 of every 100 women using that method of contraception will become pregnant in the first year of use. The large failure rate for use of condoms and diaphragms is due to improper use of these devices by many people. Female condoms are also available and have a failure rate of approximately 20%. (b diaphragm, condoms, oral contraceptive) ©McGraw-Hill Education/Jill Braaten, photographer

which are placed in the upper part of the vagina just prior to intercourse and block movement of sperm to the cervix, and **condoms**, which are sheathlike membranes worn over the penis that collect the ejaculate. In addition to their contraceptive function, condoms significantly reduce the risk of STDs such as HIV infection, syphilis, gonorrhea, chlamydia, and herpes. Other types of contraception do not reduce this risk.

Another temporary form of contraception involves synthetic hormones. Oral contraceptives (birth control pills) are synthetic forms of estradiol and progesterone, which are taken by mouth to prevent ovulation in women by inhibiting the secretion of pituitary LH and FSH and hypothalamic gonadotrophin releasing hormone (GnRH) (recall the feedback actions of estradiol and progesterone). The hormones in these pills also affect the composition of cervical mucus such that sperm cannot easily pass through it into the uterus. In addition to the oral route, hormones can be administered by injections, skin patches, and vaginal rings.

Another temporary method of contraception involves placement in the uterus of an **intrauterine device (IUD)**, a small object that prevents fertilization in part by inhibiting sperm movement and survival in the uterus. Some IUDs may also induce local inflammation in the endometrium and make it less likely that a blastocyst could implant there. Although not as widely used as other contraceptives in the U.S., IUDs are the most commonly used means of contraception by women worldwide because of their effectiveness and simplicity of use.

In addition to the contraceptive methods used before or during intercourse, any of several emergency contraception pills can be taken by a woman within 72 hours after intercourse. These pills include a high dose of progestin (a synthetic form of progesterone), a high dose of combined estrogen and progestin, and a dose of ulipristal acetate, a substance that acts on the progesterone receptor. These ingredients work to prevent ovulation or to impede the ability of sperm to reach and fertilize an egg.

Used prior to the advent of modern contraception, and still in use today by individuals who prefer not to use contraceptives, is the rhythm method, which involves abstaining from sexual intercourse near the time of ovulation. Its main drawback is the difficulty in precisely pinpointing the time of ovulation, which can occur at any time within a roughly 2-week window of the 28-day cycle. Predicting the time of ovulation is also difficult in that several of the detectable changes characteristic of the midpoint of the ovarian cycle—including a small rise in body temperature and changes in the characteristics of the cervical mucus—occur only after ovulation. These drawbacks explain why the rhythm method has a relatively high failure rate.

## Summary of Key Concepts

### 51.1 Overview of Sexual and Asexual Reproduction

- Asexual reproduction occurs when offspring are produced from a single parent, without the fusion of genetic material from two parents (Figure 51.1).

- Sexual reproduction is the production of a new individual by the joining of two haploid gametes: a sperm from the male and an egg from the female.

- Paland and Lynch showed that sexual reproduction was more effective than asexual reproduction in eliminating deleterious mutations from populations of *Daphnia pulex* (Figure 51.2).

### 51.2 Gametogenesis and Fertilization

- Male and female gametes are formed within the gonads—the testes in males and the ovaries in females. Within the ovaries, each oocyte undergoes growth and development within a follicle before it leaves the ovary in a process called ovulation (Figure 51.3).

- Major events in fertilization include the acrosomal and cortical reactions (Figure 51.4).

- In external fertilization, sperm and eggs are released into an aquatic environment, where they unite. In internal fertilization, sperm are deposited within the reproductive tract of the female during copulation or by cloacal contact.

### 51.3 Human Reproductive Structure and Function

- Sperm production requires testosterone. Sperm move out of the seminiferous tubules and into the epididymis, which leads into the vas deferens and the ejaculatory duct. The urethra conducts semen, a mixture containing fluid and sperm, during ejaculation. The fluid components of semen are produced in the seminal vesicles, the prostate gland, and the bulbourethral glands (Figures 51.5, 51.6).

- In the ovary, primary oocytes develop into secondary oocytes. If the secondary oocyte is fertilized by a sperm, the fertilized egg undergoes several cell divisions to become a blastocyst, a ball of cells that enters the uterus (Figures 51.7, 51.8).

- In females, changes in hormone secretion produce the ovarian cycle and control the uterine (or menstrual) cycle. The cessation of ovarian cycles is called menopause (Figure 51.9).

### 51.4 Pregnancy and Birth in Mammals

- Pregnancy is the time during which a developing embryo or fetus grows within the uterus of the mother. Implantation is the embedding of the blastocyst within the uterine endometrium. Most mammals retain and nourish their young within the uterus via transfer of nutrients and gases through the placenta (Figure 51.10).

- The evolution of the globin gene family contributed to the ability of placental mammals to develop inside the mother's uterus (Table 51.1).

- Birth is initiated by hormones produced by the placenta and by the mother's endocrine system. Oxytocin stimulates the strong uterine muscle contractions that are the hallmark of the three-stage process called labor (Figures 51.11, 51.12).

### 51.5 General Events of Embryonic Development

- The process by which a fertilized egg is transformed into an organism with distinct physiological systems and body parts is called embryonic development. The process by which different cells within a developing organism acquire specialized forms and functions, due to the expression of cell-specific genes, is called cellular differentiation.

- Development in many animals, including vertebrates, involves five general events: fertilization, cleavage, gastrulation, neurulation, and organogenesis (Figure 51.13).

- Cleavage involves cell divisions without cell growth and results in daughter cells called blastomeres. Cleavage-stage embryos in triploblasts have animal and vegetal hemispheres. The hemispheres determine, in part, the future anteroposterior and dorsoventral axes of the embryo (Figure 51.14).

- Incomplete, or meroblastic, cleavage occurs in vertebrates whose eggs contain large amounts of yolk. Complete, or holoblastic, cleavage occurs in animals whose eggs have smaller amounts of yolk. In mammals, cleavage occurs in the oviduct and the blastocyst implants in the uterine wall (Figures 51.15, 51.16).

- During gastrulation, the hollow ball of cells that makes up a blastula develops into a gastrula, containing the three germ layers: endoderm, mesoderm, and ectoderm. Each germ layer gives rise to specific structures (Figures 51.17, 51.18).

- Neurulation is the multistep embryological process responsible for initiating formation of the nervous system. Neurulation involves the formation of the neural tube and neural crest (Figure 51.19).

- Organogenesis, the developmental event during which cells and tissues form organs, begins during or just after neurulation. However, organs become functional at different times during development.

## 51.6   Impact on Public Health

- Infertility is the inability of a man to produce sufficient numbers or quality of sperm to impregnate a woman, or the inability of a woman to become pregnant or to maintain a pregnancy. Many causes of infertility are known.

- The use of methods to prevent fertilization or implantation of a fertilized egg is termed contraception. Methods of contraception include vasectomy and tubal ligation, vaginal diaphragms, condoms, oral contraceptives, and IUDs (Figure 51.20).

## Assess & Discuss

### Test Yourself

1. The development of offspring from unfertilized eggs is
   a. budding.
   b. cloning.
   c. fragmentation.
   d. parthenogenesis.
   e. implantation.

2. Which is considered an advantage of sexual reproduction?
   a. necessity to locate a mate
   b. increased energy expenditure in producing gametes that may not be used in reproduction
   c. increased genetic variation
   d. decreased genetic variation
   e. both a and b

3. Spermatogonia
   a. are haploid germ cells.
   b. are mature haploid cells.
   c. are mature male gametes.
   d. have flagella.
   e. are diploid germ cells.

4. Compared with external fertilization, in internal fertilization,
   a. male gametes have a higher chance of coming into close proximity to female gametes.
   b. gametes are less protected against predation or other harmful environmental factors.
   c. gametes typically become dessicated.
   d. gametes come into contact only outside the mother's reproductive tract.
   e. Both b and c occur.

5. Which sequence correctly describes the pathway followed by sperm in mammals?
   a. epididymis→ vas deferens→ seminiferous tubules→ ejaculatory duct→ urethra
   b. seminiferous tubules→ vas deferens→ epididymis→ ejaculatory duct→ urethra
   c. vas deferens→ seminiferous tubules→ epididymis→ ejaculatory duct→ urethra
   d. seminiferous tubules→ epididymis→ vas deferens→ ejaculatory duct→ urethra
   e. epididymis→ seminiferous tubules→ ejaculatory duct→ vas deferens→ urethra

6. The fructose in semen is secreted by
   a. the epididymis.
   b. the seminiferous tubules.
   c. the seminal vesicles.
   d. the prostate gland.
   e. the bulbourethral glands.

7. A major function of FSH is to
   a. stimulate the development of the gonads during early development.
   b. stimulate spermatogenesis in males and oocyte maturation in females.
   c. increase the secretion of testosterone by the testes.
   d. regulate the secretion of the bulbourethral glands.
   e. inhibit the activity of Sertoli cells in the testes.

8. During the human ovarian cycle, ovulation is stimulated by
   a. a decrease in FSH secretion.
   b. an increase in progesterone secretion.
   c. an increase in LH secretion.
   d. the presence of semen in the vagina.
   e. a decrease in estradiol concentration in the bloodstream.

9. In vertebrates, the digestive tract forms from
   a. the blastopore.
   b. the dorsal lip.
   c. the archenteron.
   d. the mesoderm.
   e. both a and d.

10. Cells of the neural crest
   a. give rise to the central nervous system.
   b. originate from ectoderm.
   c. migrate to different areas of the body and differentiate into a variety of cells, including neurons of the peripheral nervous system.
   d. do all of the above.
   e. do b and c only.

## Conceptual Questions

1. What disadvantages are associated with external fertilization, and what is one major way in which animals overcome those disadvantages?

2. How does the hypothalamus influence vertebrate reproduction?

3. **Core Concept: Matter and Energy** What are some of the energy costs associated with sexual reproduction? What outweighs those costs and accounts for the observation that most animals reproduce sexually?

## Collaborative Questions

1. Describe the major events in animal development beginning with cleavage.

2. Describe the events of the ovarian and uterine cycles, and explain their relative timing.

# Immune Systems

# 52

In an immune system response, a macrophage engulfs numerous rod-shaped bacteria (image is a false-color SEM). ©SPL/Science Source

In 2001, as many as 10 million domestic sheep and cattle had to be destroyed in England due to an epidemic of hoof-and-mouth disease, a viral disease that infected animals across the country. A similarly severe outbreak resulted in millions of diseased pigs and cattle being killed in South Korea and Japan in 2010–2011. Beginning in 2006, as many as 6 million bats in North America have died as a result of an epidemic of a fungal disease called white-nose syndrome, reducing the populations of some species by as much as 99%. Since the 1990s, thousands of gorillas and chimpanzees have died from outbreaks of the ebola virus in parts of Africa, decimating their populations. Countless numbers of crows and other birds have died from West Nile virus in recent years, a virus that, like ebola, can be transmitted to humans. These and other epidemic illnesses highlight the importance of understanding how animals protect themselves from harmful microorganisms and their secretions, as well as against other internal threats, such as cancer.

The ability of an animal to ward off these threats—an animal's **immunity**, or immune defenses—is the subject of this chapter. The cells and organs within an animal's body that contribute to immune defenses collectively constitute that animal's **immune system**. The study of immunity is called immunology. Immunologists examine the processes by which the immune system protects an animal from foreign matter, whether living or nonliving. In these processes, immune defenses recognize the body's own molecules as "self" and attack anything that is foreign, or "nonself."

The two major types of immune defenses are innate and adaptive. **Innate immunity** refers to the body's defenses that are present at birth and that act against any foreign material in much the same way, regardless of the specific identity of the invading material. Innate immunity includes a set of cellular and chemical defenses that oppose substances that breach the external barriers (skin and mucous membranes) of an animal's body. An example of an innate immune response is seen in the chapter opening image, in which a cell known as a macrophage is engulfing numerous bacteria. All animals have innate immune defenses.

In contrast, **adaptive immunity** (also called acquired immunity) develops only after an animal's body is exposed to foreign substances. This type of immunity is characterized by the ability of certain cells of the immune system to recognize a particular foreign substance and initiate a response that targets that substance specifically. Another feature distinguishing adaptive immunity is that repeated exposure to a foreign substance elicits greater and greater defense responses. In contrast, in innate immunity, each exposure to the foreign material elicits the same magnitude of defense responses. Adaptive immunity has been identified in all vertebrates except for the jawless fishes, but has not been unequivocally identified in invertebrates.

We begin the chapter with a brief overview of the different pathogens that cause disease in animals. We will then consider the mechanisms that provide animals with innate and adaptive immunity against harmful pathogens. We will conclude with a discussion of the public health implications of some immunity-related conditions in humans.

## 52.1 Types of Pathogens

**Learning Outcome:**

1. List the three main types of pathogens that elicit immune responses, and briefly describe how each damages its host organism.

An animal's immune defenses must protect against a variety of foreign materials, but most important among them are disease-causing viruses and microorganisms, or **pathogens**. Pathogens exist in nearly every ecological niche on Earth. Both terrestrial and aquatic animals, including invertebrates and vertebrates, encounter each of the three major types of pathogens: certain bacteria, viruses, and eukaryotic parasites. A fourth type of infectious agent—proteins called prions—were considered in Chapter 19 (refer back to Figure 19.8) and will not be discussed here.

### Bacteria are Prokaryotes Responsible for Many Animal Diseases

As described in Chapter 27, bacteria are prokaryotic organisms that lack a true nucleus. Bacteria can either damage tissues at an open wound site on an animal or release toxins that enter the bloodstream and disrupt functions in other parts of an animal's body. Bacteria are responsible for many diseases and infections in animals and humans, including typhoid fever, strep throat, skin infections, ear infections, anthrax, plague, cholera, and food poisoning. The major ways in which bacteria gain entry into an animal's body are through direct bodily contact, open wounds, inhalation through the respiratory tract, and ingestion with food or water that has been contaminated with fecal matter. This last situation may arise because many infectious bacteria enter the intestines of animals and are excreted in feces, which may be deposited near food or water sources used by other animals.

### Viruses May Cause Illness or Cancer

All viruses contain nucleic acid (DNA or RNA) within a protein coat (see Chapter 19). Unlike bacteria, viruses lack the metabolic machinery to synthesize the proteins they require to replicate themselves. Instead, they must infect a host cell and use its biochemical and genetic machinery, including nucleotides and energy sources, to make more viruses. The viral nucleic acid directs the host cell to synthesize the proteins required for the virus's replication. After entering a cell, some viruses, such as the common cold virus, multiply rapidly, kill the cell, and then infect other cells. Other viruses can lie dormant within host cells before suddenly undergoing rapid replication, which causes cell damage or death. Finally, certain viruses can transform their host cells into cancerous cells. Viruses are responsible for a great variety of illnesses, including ebola, swine fever, West Nile encephalitis, hoof-and-mouth disease, influenza, and some sexually transmitted diseases, such as that caused by the human immunodeficiency virus (HIV). Like bacterial infections, viral infections can spread rapidly among animals and can be lethal. Viruses typically enter an animal's body through the respiratory tract or through an open wound.

### Eukaryotic Parasites Can Enter Animal Bodies in Several Ways and Cause Widespread Illness

Eukaryotic parasites—whether protists, fungi, or worms—damage a host by using the host's nutrients for their own growth and reproduction or by secreting toxic chemicals. In humans, parasites account for an enormous number of cases of disease annually. For example, several hundred million people are infected each year with one of the mosquito-borne protists of the genus *Plasmodium* that cause malaria. Parasitic infections may enter a host through the bite of an infected insect, as in malaria; by ingestion of food or water containing parasitic organisms, such as roundworms; or in some cases, by penetrating the skin, as with blood flukes.

## 52.2 Innate Immunity

**Learning Outcomes:**

1. Identify the general characteristics of innate defense mechanisms in animals.
2. Explain how an animal's body surface provides protection from pathogens.
3. Describe the process of phagocytosis, and explain its importance in innate immune responses.
4. List the cell types involved in innate immunity, and identify the functions of each.
5. Identify the correct sequence of the events in the inflammatory response.
6. Describe the role of Toll proteins in innate immunity.

Innate immune defenses protect against foreign cells or substances without having to recognize the invaders' specific identities. This type of defense mechanism is called innate because animals inherit the ability to perform these protective functions and because this type of immunity does not require prior exposure to invaders. Instead of distinguishing among foreign materials, innate defenses recognize some general property marking the invader as foreign, such as a particular class of carbohydrate or lipid present in the cell walls of many different kinds of microbes. For this reason, innate immunity is sometimes referred to as nonspecific immunity.

In this section, we will consider the innate immune defenses. These include the actions of phagocytic cells, the response to injury or infection known as inflammation, and various proteins secreted by cells of the immune system that facilitate the destruction of pathogens. We begin, however, with a brief look at barrier defenses that can help prevent pathogens from entering an animal's body fluids and cells.

### An Animal's Body Surface Is an Initial Line of Defense

Although not part of an animal's innate immunity, an animal's initial defense against pathogens is the cellular, chemical, and anatomical barrier provided by a surface exposed to the external environment. Very few microorganisms can penetrate the intact skin or body surface of most animals, particularly the tough, thick, or scaly skin characteristic of many vertebrates or the exoskeleton of many arthropods. In addition, glands in the body surfaces of many invertebrates and vertebrates secrete a variety of antimicrobial molecules onto the body surface, including mild acids and enzymes such as lysozyme that destroy bacterial cell walls.

The mucus secreted from cells in the mucous membranes lining the respiratory and upper digestive tracts of vertebrates also contains antimicrobial molecules. More importantly, mucus is sticky—microbes that become stuck in it are prevented from penetrating the mucous membrane barrier. They are either swept up by cilia into the pharynx and then swallowed, or they are engulfed by cells that are present in both tracts. Pathogens ingested with food are often destroyed by the acidic environment of an animal's stomach.

If, however, a pathogen is able to penetrate a barrier and gain entry into an animal's internal tissues and fluids, innate defense mechanisms are activated. These mechanisms are mediated by several types of cells that reside in the body fluids and tissues, as described next.

## In Innate Immunity, Phagocytic Cells Defend Against Pathogens That Enter the Body

Several different types of cells have important functions in innate immunity in vertebrates. As noted in **Figure 52.1**, some of these cell types are **phagocytes**—cells capable of phagocytosis. **Phagocytosis** is a type of endocytosis in which a cell engulfs particulate matter, which usually is then destroyed by proteases or oxidizing compounds such as hydrogen peroxide. Phagocytes are found in the body fluids, such as hemolymph and blood, and also within various tissues and organs. They are present in all classes of animals and are among the most fundamental and ancestral forms of immune defenses.

In vertebrates, most phagocytes are a type of blood cell called **leukocytes** (see Figure 52.1). All leukocytes are derived from a common type of stem cell called adult hematopoietic stem cells (refer back to Figure 20.20), which in mammals and birds are found in the bone marrow. These stem cells give rise to several types of leukocytes and other immune cells that have specialized functions. The leukocytes involved in innate immunity include the following:

- **Neutrophils** are the most abundant phagocytes. They are found in blood and some may enter tissues during inflammation. After neutrophils engulf bacteria by phagocytosis, the bacteria are destroyed within endocytotic vacuoles by proteases, oxidizing compounds, and antibacterial proteins called defensins. The production and release of neutrophils from bone marrow are greatly stimulated during the course of an infection.

- **Eosinophils** are found in the blood and in the mucosal surfaces lining the gastrointestinal, respiratory, and urinary tracts, where they fight off parasitic invasions. Eosinophils act in some cases by releasing toxic chemicals that kill parasites and in other cases by phagocytosis.

- **Monocytes** are phagocytes that circulate in the blood for a short time, after which they migrate into tissues and organs and develop into macrophages.

- **Macrophages** are strategically located where they will encounter foreign matter, for example, in epithelia in contact with the external environment, such as skin and the linings

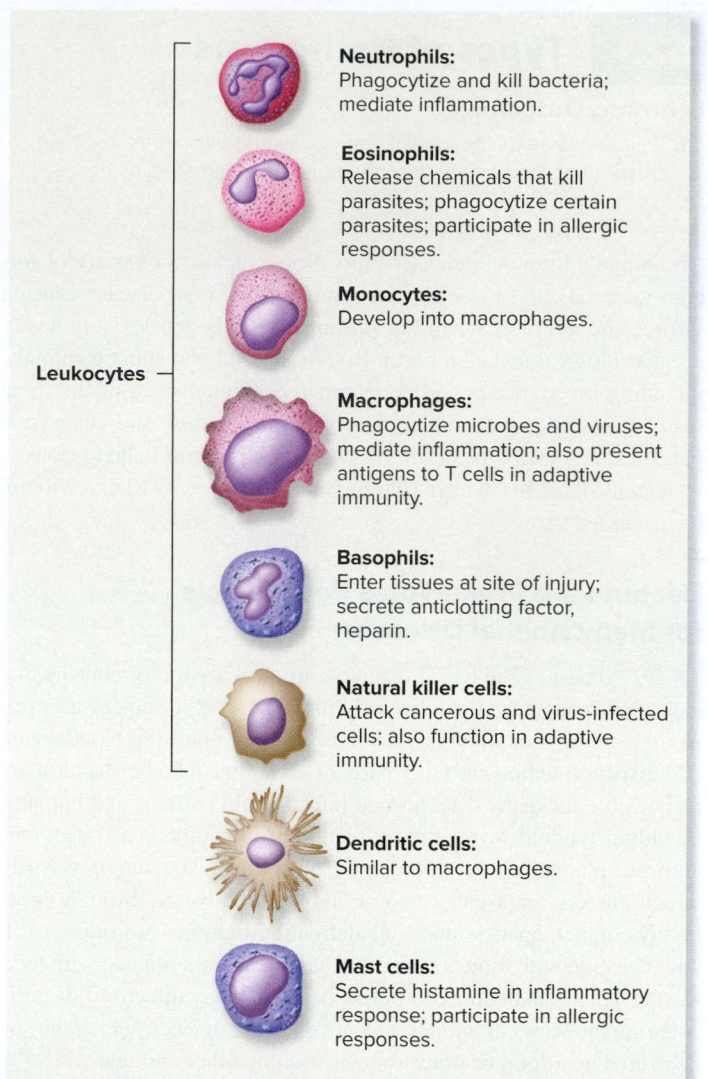

**Figure 52.1** **Cells involved in innate immunity in vertebrates.** Six of these types of cells are leukocytes, and several are phagocytes. Some of these cells also participate in adaptive immunity; a few of those actions are included here for reference and will be described in a later section.

of respiratory and digestive tracts. Macrophages are large phagocytes capable of engulfing viruses and bacteria, as shown in the chapter opening image. As will be described later, macrophages also function in adaptive immunity.

- **Basophils** are secretory cells, not phagocytes. They secrete an anticlotting factor called heparin at the site of an infection, which helps the circulation flush out the infected site. Basophils also secrete histamine, which attracts infection-fighting cells and proteins to the site.

- **Natural killer (NK) cells** are another kind of leukocyte called lymphocytes, of which there are several types. We will discuss the other types of lymphocytes later, because they play the major role in adaptive immunity. Like macrophages, NK cells participate in both innate and adaptive immunity. These cells are part of an animal's innate defenses because they recognize general features on the surface of cancerous cells or virus-infected cells. They act by releasing chemicals into the vicinity of cancerous or virus-infected cells, thereby killing those cells.

In addition to leukocytes, two other types of cells derived from bone marrow stem cells have important functions in innate immunity. **Dendritic cells** are scattered throughout most tissues, where they perform various macrophage-like functions. **Mast cells** are found throughout connective tissues, particularly beneath the epithelial surfaces of the body. Mast cells secrete many locally acting molecules, including histamine. Histamine and other substances are involved in inflammation, a fundamental component of the innate defense mechanism, which we now examine.

## Inflammation Is an Innate Response to Infection or Injury

**Inflammation** is an innate local response to infection or injury. The functions of inflammation are to destroy or inactivate pathogens, to clear the infected region of dead cells and other debris, and to set the stage for tissue repair. The key cellular components of this process are neutrophils, macrophages, dendritic cells, and mast cells.

The events of inflammation are induced and regulated by chemical mediators. These include a family of proteins called **cytokines** that function in both innate and adaptive immune defenses. Cytokines provide a chemical communication network that synchronizes the components of the immune response. Most cytokines are secreted by more than one type of immune cell and also by certain nonimmune cells such as fibroblasts and the endothelial cells of blood vessels.

The sequence of events in a typical inflammatory response to a bacterial infection is summarized in **Figure 52.2**. A tissue injury such as that caused by a splinter begins the inflammatory process, which results in the familiar signs and symptoms of local redness, swelling, heat, and pain.

Substances secreted into the extracellular fluid from injured tissue cells, mast cells, and neutrophils contribute to the inflammatory response. For example, histamine from mast cells and nitric oxide from endothelial cells (Figure 52.2, step 1) cause vasodilation (widening) of the small blood vessels in the infected and damaged area, which induces the vessels to leak (step 2).

These vascular changes provide two benefits. First, the increased blood flow to the inflamed area, which accounts for the redness and heat, speeds the delivery of beneficial proteins and leukocytes and increases local metabolism to facilitate healing. Second, the increased vascular permeability ensures that the plasma proteins that participate in inflammation can gain entry to the interstitial fluid. The swelling in an inflamed area also results from this increased leakiness of blood vessels.

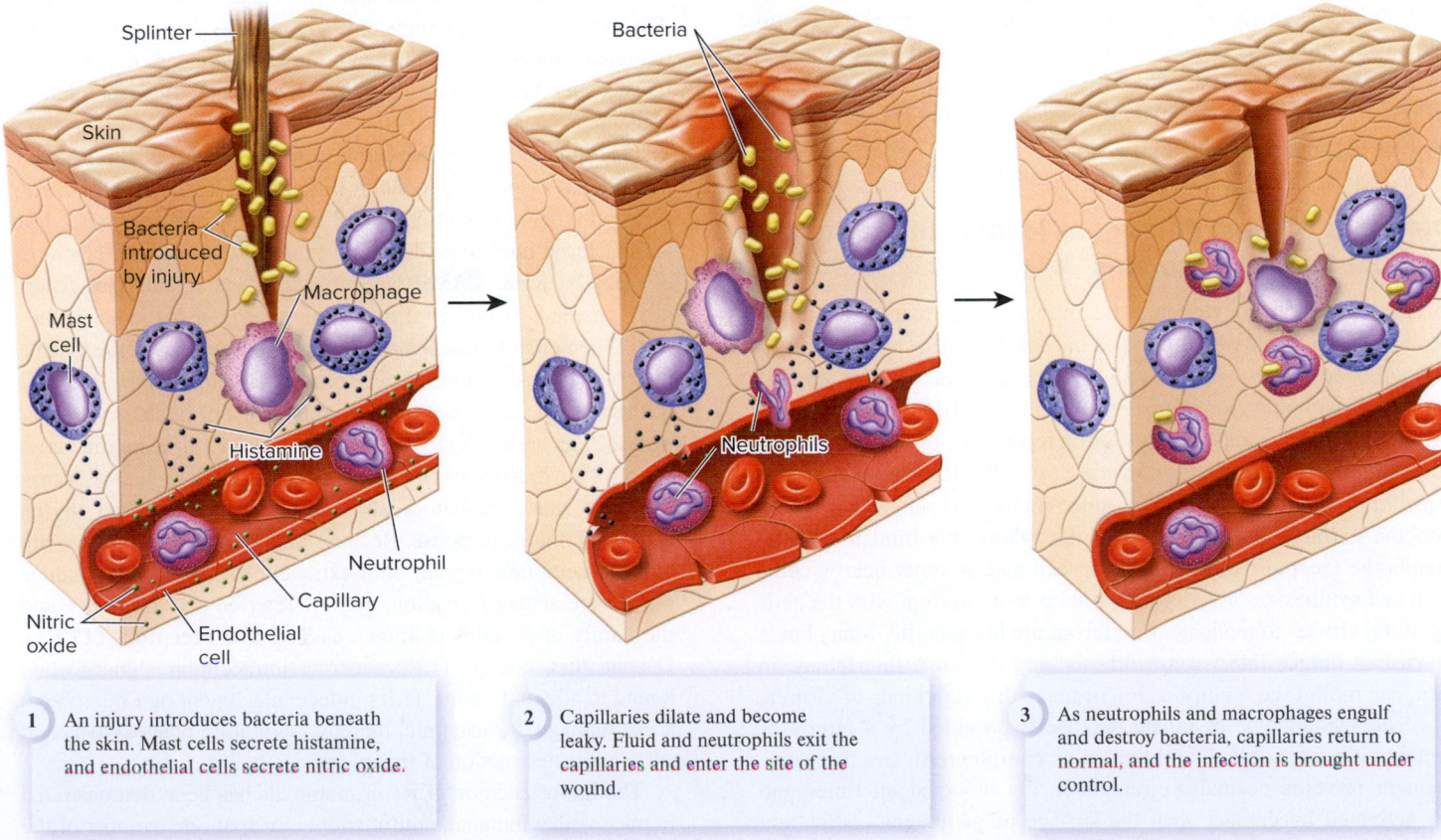

1. An injury introduces bacteria beneath the skin. Mast cells secrete histamine, and endothelial cells secrete nitric oxide.

2. Capillaries dilate and become leaky. Fluid and neutrophils exit the capillaries and enter the site of the wound.

3. As neutrophils and macrophages engulf and destroy bacteria, capillaries return to normal, and the infection is brought under control.

**Figure 52.2** **The events in inflammation.** Shown are the initial stages of inflammation in response to a penetrating wound that introduces bacteria beneath the skin.

**Concept Check:** *Inflammation is often associated with swelling of the inflamed area. Could this swelling have an adaptive value?*

Once neutrophils and macrophages arrive at the site of the injury, they begin the process of phagocytizing and destroying the bacteria (Figure 52.2, step 3). The initial step in phagocytosis involves the interaction of phagocyte surface receptors with certain carbohydrates or lipids in the bacterial cell walls. Subsequently, the neutrophils and macrophages also release antimicrobial substances into the extracellular fluid that can destroy bacteria even before phagocytosis occurs. Other secreted substances such as nitric oxide function as inflammatory mediators. The result is positive feedback: Once phagocytes enter the area and encounter bacteria, they release inflammatory mediators that bring even more phagocytes into the area.

Inflammation may sometimes be accompanied by **fever**, an increase in an animal's body temperature that results from an infection. This is distinguished from hyperthermia, an increase in body temperature resulting from any number of causes, such as overexertion. In humans, a body temperature greater than 38°C (100.4°F) is considered to be a fever (normal is about 37°C [98.6°F]).

The precise mechanisms by which a fever arises are not completely understood. In mammals, cytokines released into the circulation by activated macrophages act within the hypothalamus to raise the body's set point for temperature, just as when you turn up a thermostat in a room. The result is an increase in heat generation. Whereas a sustained high fever requires medical attention because of the damaging effects of high temperature on membrane function, enzyme activity, and other processes (see Chapter 47), it appears from animal models that a small or moderate increase in temperature stimulates leukocyte activity and proliferation and provides a less hospitable environment for at least some types of pathogens. For example, fever promotes the uptake of iron from the blood into the liver, thus preventing microbes from obtaining this element that is important for their survival and reproduction.

## Defenses Against Pathogens Include Interferons and Complement Proteins

In addition to phagocytes and the inflammatory response, animal bodies have at least two other types of innate defenses against invading pathogens. These defenses allow for extracellular destruction of pathogens without prior exposure to them. The first of these innate defenses is used against viruses. **Interferons** are proteins that generally inhibit viral replication inside host cells. In response to viral infection, many types of cells produce interferons and secrete them into the extracellular fluid. When the interferons bind to plasma membrane receptors on the secreting cell and on other nearby cells, each cell synthesizes a variety of proteins that interfere with the ability of the viruses to replicate. Interferons are not specific. Many kinds of viruses induce interferon synthesis, and the same interferons, in turn, can inhibit the multiplication of many different kinds of viruses.

The second type of innate defense is provided by a family of proteins present in blood and known as **complement**. Inactive complement proteins normally circulate in the blood at all times and are activated by contact with the surface of pathogens. Activation of complement proteins results in a cascade of events. Among their many actions, complement proteins stimulate the release of histamine from mast cells, thereby increasing permeability of local blood vessels, as described earlier. Five of the active proteins generated

in the complement cascade form a multi-unit protein called the **membrane attack complex (MAC)**, which, by embedding itself in the pathogen's cell membrane or envelope (in the case of some viruses), creates pores in the membrane. Water and ions enter the pathogen through the channels, and the pathogen is killed or destroyed.

 **Core Concept: Evolution**

### Innate Immune Responses Require Proteins That Recognize Features of Many Pathogens

As we have seen, innate immunity is not specific. It often depends on an immune cell recognizing some general molecular feature common to many types of pathogens. These features are called **pathogen-associated molecular patterns (PAMPs)**. Before 1985, however, it was not known how that recognition was accomplished. At that time, German biologist Christiane Nüsslein-Volhard and American biologist Eric Wieschaus were interested in how animal embryos differentiate into adults. In the course of their studies, they discovered a protein they named Toll (now called Toll-1) that was required for the proper dorsoventral orientation of the body of the fruit fly, *Drosophila melanogaster*. In a landmark study in 1996, it was discovered that Toll-1 also conferred on adult flies the ability to fight off fungal infections (see the following Feature Investigation). It has since become clear that a family of Toll proteins exists in animals from nematodes to mammals, including humans, and these proteins are found in the plasma membranes and endosomal membranes of macrophages and dendritic cells, among others.

One function of the Toll proteins in vertebrates is to recognize and bind to ligands that contain PAMPs, such as lipopolysaccharide and other common microbial lipids and carbohydrates; viral and bacterial nucleic acids; and a protein found in the flagellum of many bacteria. PAMPs are molecular features that are generally considered to be vital to the survival of a particular pathogen. When one of these ligands binds a Toll protein on the plasma membrane of an immune cell, second messengers are generated in the immune cell, triggering secretion of inflammatory mediators such as various cytokines. These in turn stimulate the activity of other leukocytes involved in the innate immune response. Some of these mediators also activate cells involved in the adaptive immune response. Because many of the Toll proteins are plasma-membrane-bound, bind extracellular ligands, and induce second-messenger formation, they are referred to as receptors, and the family of proteins is known as **Toll-like receptors (TLRs)**. Despite this, not all TLRs generate intracellular signals when bound to a ligand. Some TLRs induce attachment of a microbe to a macrophage, for example, thereby facilitating phagocytosis and subsequent destruction of the microbe.

The importance of TLRs in mammals has been demonstrated in mice with a mutated, nonfunctional form of one member of the family called Toll-4. These mice are hypersensitive to the effects of injections of lipopolysaccharide (to mimic a bacterial infection) and are less able to ward off bacterial infections. In humans, recent studies suggest that certain naturally occurring variants in a specific TLR are associated with increased risk of certain diseases.

TLRs are currently an active area of investigation among biologists because of their significance in the immune response. Certain domains of these receptors have also been identified in plants, where they may also be involved in disease resistance. In 2018, Israeli researchers Shany Doron and colleagues reported that the intracellular domain of a vertebrate TLR shares similarity to that found in newly discovered defense proteins in prokaryotes. The genes that encode these proteins appear to confer nonspecific protection for bacteria against a wide variety of phages. Therefore, TLRs may be among the first immune defense mechanisms to have evolved in living organisms. For their investigations of embryonic development, which included the discovery of Toll proteins, Nüsslein-Volhard and Wieschaus were awarded the 1995 Nobel Prize in Physiology or Medicine.

## Core Skill: Process of Science

## Feature Investigation | Lemaitre and Colleagues Identify an Immune Function for Toll Protein in *Drosophila*

As you've just learned, Toll protein was initially identified as a critical molecule directing the proper dorsoventral development of *Drosophila* embryos. Interestingly, however, certain aspects of the sequence and function of Toll share similarities with at least one important immune protein found in vertebrates. For example, both Toll and the receptor for a cytokine known as interleukin-1 are transmembrane proteins with similar amino acid sequences in their cytosolic domains. Moreover, both proteins activate a similar intracellular signaling pathway once they have bound an extracellular ligand. These findings led to the hypothesis proposed by French researcher Bruno Lemaitre and coworkers, working in the laboratory of Jules Hoffmann, that Toll protein may also function in immunity in *Drosophila*.

To test this hypothesis, the researchers compared wild-type flies to flies in which the *Toll* gene was underexpressed due to a mutation (**Figure 52.3**). They then exposed the flies to a variety of pathogens, by pricking them with a needle that had been dipped into water (as a control) or a concentrated solution of fungal spores. Survival of the flies was monitored over the next several days, and compared with the survival of wild-type flies that had normal levels of Toll protein. What they found was remarkable: Within days of being infected with the fungus *Aspergillus fumigatus*, 100% of the mutant flies died, but only about 30% of the wild-type flies died (see the data in step 4 of Figure 52.3). The mutant flies were observed to be covered in germinating hyphae of the fungus, clearly demonstrating that the flies were unable to fight the infection. Moreover, analysis of mRNA from the infected and control (uninfected) flies revealed that, in contrast to the wild-type flies, the infected mutant flies were unable to increase expression of a key antifungal gene that encoded a polypeptide called drosomycin.

Since then, these and other researchers have identified much of the mechanism by which Toll promotes immunity to certain microbes in *Drosophila* and, presumably, other insects. Activation

**Figure 52.3** **Lemaitre and colleagues identified the importance of Toll protein in immunity in *Drosophila*.** ©Daniel L. Geiger/SNAP/Alamy Stock Photo

**HYPOTHESIS** The Toll signaling pathway is required for the immune response in *Drosophila*.

**KEY MATERIALS** Wild-type *Drosophila*; mutant strain of *Drosophila* that underexpresses the receptor for Toll protein; spores of the fungus *A. fumigatus*; cDNA probe to detect mRNA of antifungal polypeptide drosomycin.

| Experimental level | Conceptual level |
|---|---|

**1** Prick flies with a clean needle that was previously dipped in water (control) or a solution of *A. fumigatus* spores.

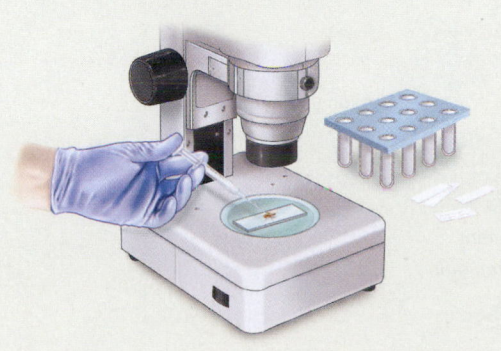

*A. fumigatus* is a common pathogen that *Drosophila* is normally able to defend against.

2.7 μm

**2** Follow survival rates of flies for 6 days.

Stopper (allows for air flow)

Flies grow in plastic vials for 6 days.

Flies

Nutrient mix

Dead flies

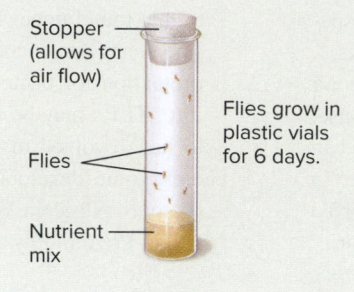

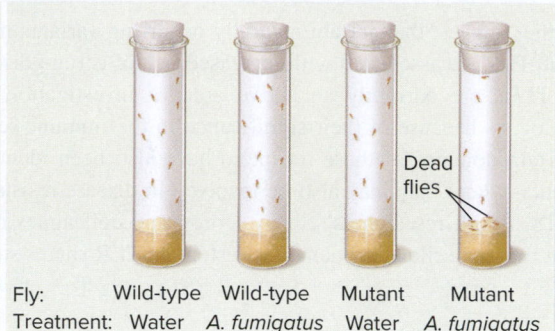

| Fly: | Wild-type | Wild-type | Mutant | Mutant |
|---|---|---|---|---|
| Treatment: | Water | A. fumigatus | Water | A. fumigatus |

**3** Examine expression of antipathogen genes in wild-type and mutant flies, using Northern blot analysis:

1. Extract mRNA from flies.
2. Separate individual mRNAs by gel electrophoresis.
3. Transfer mRNAs from the gel to a nylon membrane.
4. Probe membrane with a radioactively labeled fragment of cDNA complementary to a region of drosomycin mRNA.
5. Expose membrane to X-ray film.

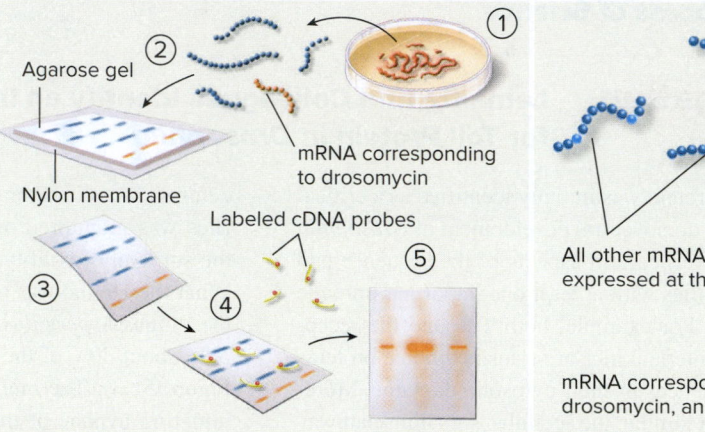

Agarose gel

Nylon membrane

mRNA corresponding to drosomycin

Labeled cDNA probes

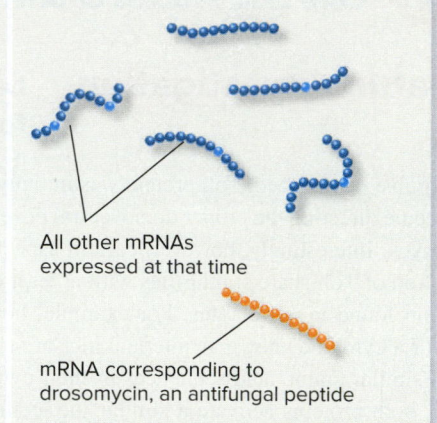

All other mRNAs expressed at that time

mRNA corresponding to drosomycin, an antifungal peptide

**4** THE DATA

Survival test results:

Percent survival

Days after challenge

Wild-type/water
Mutant/water
Wild-type/A. fumigatus

Mutant/A. fumigatus

Northern blot results:

Drosomycin mRNA

| Strain: | Wild-type | Wild-type | Mutant |
|---|---|---|---|
| Treatment: | Water | A. fumigatus | A. fumigatus |

**5** CONCLUSION Toll protein receptor is essential for survival of *Drosophila* exposed to a common fungal pathogen and for induction of antifungal genes.

**6** SOURCE Lemaitre, B., Nicolas, E., Michaut, L., Reichhart, T., and Hoffmann, J. A. 1996. The dorsoventral regulatory gene cassette spätzel/toll/cactus controls the potent antifungal response in *Drosophila* adults. *Cell* 86: 973–983.

of Toll protein induces expression of several genes that encode anti-microbial molecules. Unlike TLRs, however, in *Drosophila* Toll does not directly recognize PAMPs associated with microbial membranes. Rather, microbial infection induces activation of extracellular signaling molecules in *Drosophila* that, in turn, bind to Toll that is located on cell membranes. This binding induces intracellular signals that activate gene expression. Despite this difference, the identification of an immune function of an ancestral Toll protein paved the way for a clearer understanding of TLRs and innate immunity in all animals. In 2011, Jules Hoffmann received a share

of the Nobel Prize in Physiology or Medicine for this pioneering research.

---

**Experimental Questions**

1. What led the investigators to hypothesize that Toll protein may serve an immune function in *Drosophila*?

2. Is *Drosophila* Toll protein a receptor that recognizes PAMPs?

3. **CoreSKILL »** Based on the results of their survival studies of infected flies, was the researchers' hypothesis supported?

---

## 52.3   Adaptive Immunity

### Learning Outcomes:

1. Describe the components of the immune system that provide adaptive immunity, including the major classes of lymphocytes.

2. Outline the three stages of the adaptive immune response.

3. Compare and contrast the humoral immune response with the cell-mediated immune response.

4. **CoreSKILL »** Explain how the structures of different immunoglobulins allow predictions to be made about their functions.

5. Explain how genetic rearrangement creates a diverse set of antibodies in mammals.

6. Describe how B and T cells that recognize self molecules are destroyed.

7. **CoreSKILL »** Predict the benefit of secondary immune responses, and explain the function of memory cells in these responses.

In adaptive immunity, cells of the immune system first encounter and later recognize a specific foreign cell or protein to be destroyed or eliminated, as opposed to recognizing some general molecular feature common to many different pathogens. Any molecule that can trigger an adaptive immune response is called an **antigen**. An antigen, therefore, is any molecule that the host does not recognize as self. Most antigens are either proteins or very large polysaccharides. Antigens include the protein coats of viruses, bacterial surface proteins, specific macromolecules on pollens and other allergens, membrane proteins of cancerous or transplanted cells, toxins, and vaccines.

Adaptive immunity is found in all jawed vertebrates. Jawless fishes possess some but not all of the features of adaptive immunity found in other vertebrates. Adaptive immunity is considered to be absent in invertebrates. Recently, however, scientists who study the evolution of immune systems have uncovered several interesting features of invertebrate immune function that suggest some invertebrates have a limited ability to adapt immune activity to a specific invader. Despite this intriguing finding, however, we will consider adaptive immune responses in the context in which they are best understood, the jawed vertebrates.

### Adaptive Immune Defense Mechanisms Include Lymphoid Organs, Tissues, and Cells

The cells of the immune system that are responsible for adaptive immunity are a type of leukocyte called **lymphocytes**, mentioned in the preceding section. At any given time, some lymphocytes circulate in the

blood, but most of them reside in a group of organs and tissues that constitute the **lymphatic system** (**Figure 52.4a**). The system is composed primarily of a network of lymphatic vessels. As described in Chapter 48, not all of the fluid that exits a capillary is returned to that capillary. The residual fluid mixes with the interstitial fluid and is called lymph. The lymphatic vessels collect the lymph and return it to the blood and circulatory system. Various lymphoid organs and tissues are located throughout the lymphatic system. They are grouped into primary lymphoid organs, in which lymphocytes differentiate into mature cells, and secondary lymphoid organs in which lymphocytes multiply and function.

*Primary Lymphoid Organs*   The primary lymphoid organs are the structures in which lymphocytes differentiate into mature immune cells. These are the bone marrow in birds and mammals and the thymus gland in all vertebrates. In animals without extensive bone marrow, specialized regions of other organs such as the kidney and liver serve as primary lymphoid organs. Destruction of or damage to a primary lymphoid organ usually results in a severe inability to fight off infections.

*Secondary Lymphoid Organs*   The primary lymphoid organs supply mature lymphocytes to secondary lymphoid tissues and organs, where the lymphocytes multiply and function. These include the lymph nodes of mammals (**Figure 52.4b**), the spleen (found in all jawed vertebrates and the largest secondary lymphoid structure), the tonsils (small, rounded lymphoid organs in the pharyngeal region of mammals), and scattered lymphocyte accumulations in the linings of the intestinal, respiratory, genital, and urinary tracts. The loss of any of the secondary lymphoid organs, although not as serious, nevertheless increases the risk of local or systemic infections throughout an animal's life. For example, in humans, the spleen must occasionally be surgically removed due to injury or disease. Such individuals must be carefully monitored for the rest of their lives because of their increased vulnerability to infection.

*Recirculation of Lymphocytes*   After leaving the bone marrow or thymus gland, lymphocytes circulate between the secondary lymphoid organs, blood, lymph, and all the tissues of the body. Lymphocytes from all the secondary lymphoid organs continually leave those structures and are carried to the bloodstream. Simultaneously, some circulating lymphocytes leave venules all over the body to enter interstitial fluid. From there, they re-enter lymphatic vessels and are carried back to secondary lymphoid organs. This constant recirculation of lymphocytes increases the likelihood that any given lymphocyte will encounter an antigen it is specifically programmed to recognize.

**Primary lymphoid organs**          **Secondary lymphoid organs**

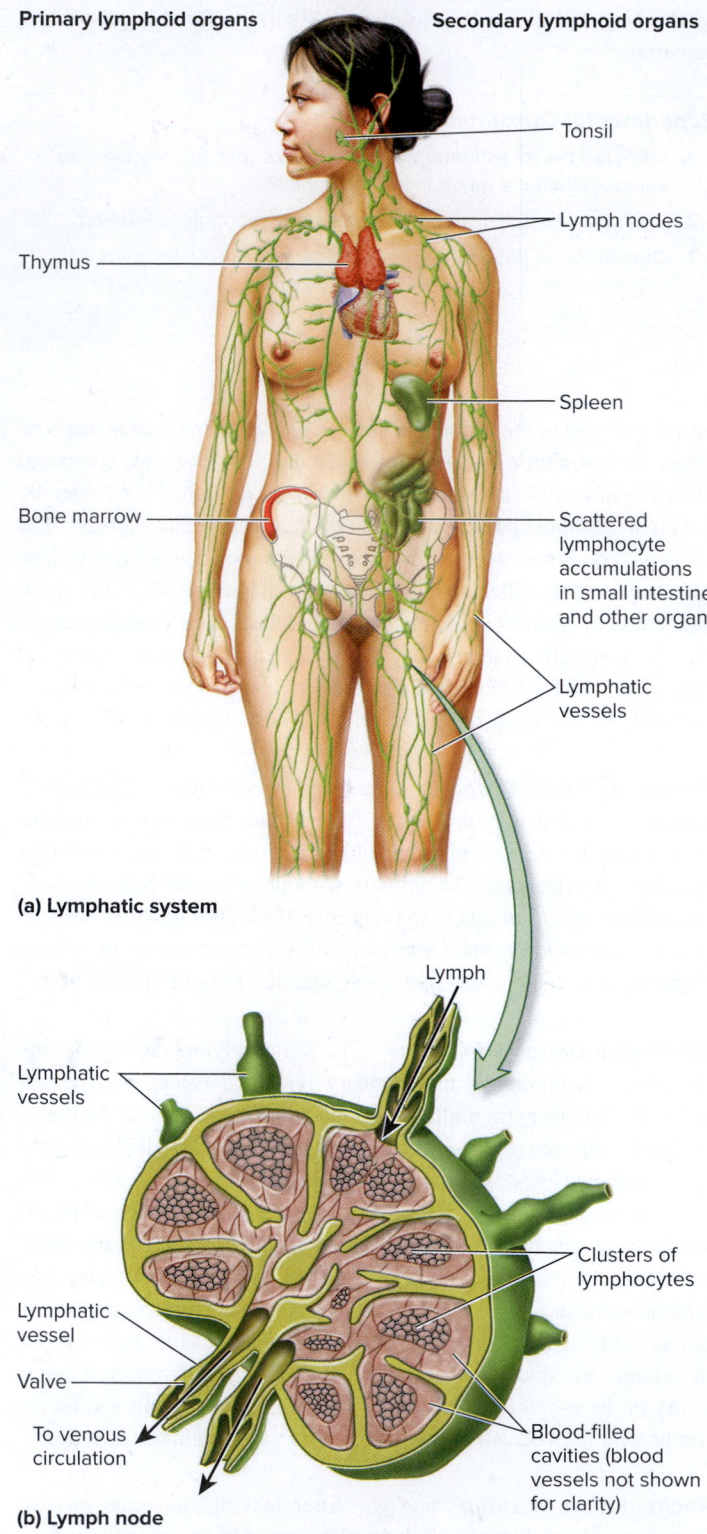

(a) Lymphatic system

(b) Lymph node

**Figure 52.4** **The lymphatic system in humans. (a)** The major components of the human lymphatic system. Primary lymphoid organs are shown in red, and secondary lymphoid organs are shown in green. In adult humans, the primary lymphoid organs in bone are found in the sternum, ribs, parts of the skull, small regions of the femur and humerus, and, as shown here, the hip bones.
**(b)** The structure of a lymph node. Lymph nodes occur along the course of lymphatic vessels, which drain lymph from tissues and return it to the venous circulation. Within a lymph node, lymph percolates through open cavities containing clusters of lymphocytes.

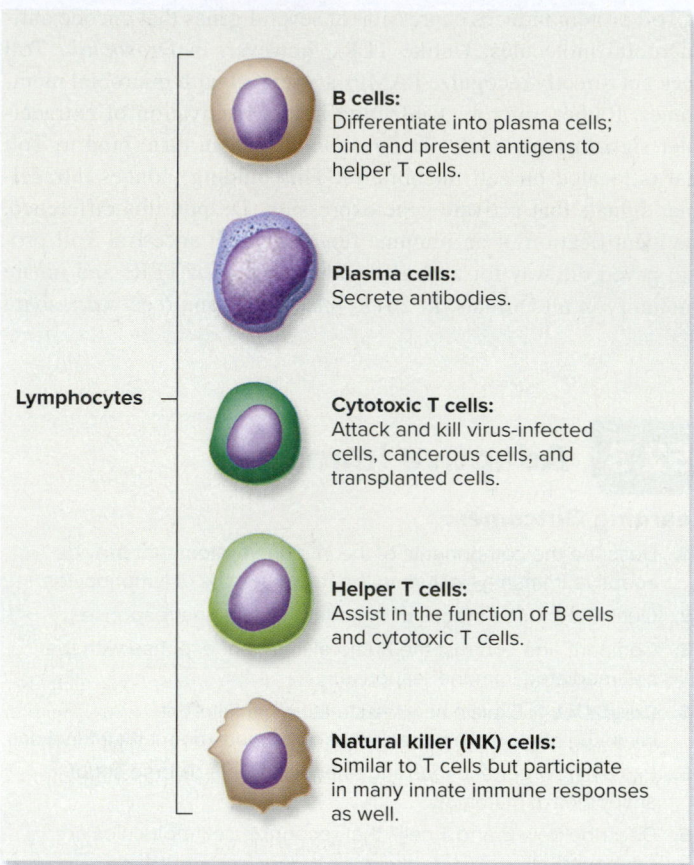

**Figure 52.5** **Types of lymphocytes involved in adaptive immunity in vertebrates.** Some important functions of each type of lymphocyte are noted. The shape and color conventions shown in this figure will be used throughout this chapter.

## Lymphocytes Include B Cells and T Cells

Different kinds of lymphocytes participate in specific coordinated immune system responses (**Figure 52.5**). In addition to NK cells, described earlier, the two major types of lymphocytes are **B cells** and **T cells**. B cells were first observed to mature in an avian organ called the bursa of Fabricius, and therefore were named B cells. In mammals, B cells mature within bone marrow. If stimulated by antigen, some B cells differentiate further into **plasma cells**, which synthesize and secrete **antibodies**, proteins that bind to and help destroy foreign molecules, as described later.

T cells may directly kill infected, mutated, or transplanted cells. They are called T cells because they mature within the thymus gland.

 **Core Concept: Systems** The lymphatic system functions to protect an animal against pathogens. Like most organ systems, its function is dependent on other systems. For example, lymph arises from fluid that exits the capillaries of the circulatory system and mixes with interstitial fluid. After this fluid passes through the lymphatic vessels, it is returned to the circulatory system.

T cells that mediate defense mechanisms include two major types of lymphocytes: cytotoxic T cells and helper T cells. **Cytotoxic T cells** travel to the location of their targets, bind to the targets by recognizing an antigen, and kill the targets via secreted chemicals. In addition, responses mediated by cytotoxic T cells are directed against body cells that have become cancerous or infected by pathogens. NK cells also destroy such cells by secreting toxic chemicals.

As their name implies, **helper T cells** do not themselves function as effector cells. Instead, they assist in the activation and function of B cells and cytotoxic T cells. With only a few exceptions, B cells and cytotoxic T cells cannot function adequately unless they are stimulated by cytokines secreted from helper T cells.

An additional type of T cell that won't be considered further here is the regulatory T cell. These cells function to suppress other T cell activities and to help prevent destruction of self proteins. They serve as a check on immune activity.

## Adaptive Immune Responses Include Humoral and Cell-Mediated Immunity and Occur in Three Stages

Lymphocytes function by recognizing antigens such as those found on viruses, bacteria, and the surface of cancerous cells. The ability of lymphocytes to distinguish one antigen from another has a central role in adaptive immunity. There are two types of adaptive immunity. In **humoral immunity**, plasma cells secrete antibodies that bind to antigens. The adjective humoral denotes communication by way of soluble chemical messengers (the word humors was once used to refer to bodily fluids). In **cell-mediated immunity**, cytotoxic T cells directly encounter and destroy infected body cells, cancerous cells, or transplanted cells.

An adaptive immune response can usually be divided into three stages (**Figure 52.6**): recognition of antigen, activation and proliferation of lymphocytes, and attack against the recognized antigen.

**Stage 1: Recognition of Antigen** During its development, each lymphocyte synthesizes a type of membrane receptor that can bind to a specific antigen. If the lymphocyte subsequently encounters that antigen, the antigen becomes bound to the receptor. This specific binding is the immunological meaning of the word recognize: Antigens that bind to a lymphocyte receptor are said to be recognized by the lymphocyte. The ability of lymphocytes to distinguish one antigen from another, therefore, is determined by the nature of their plasma membrane receptors. Each lymphocyte is specific for just one type of antigen.

**Stage 2: Activation and Proliferation of Lymphocytes** In the second stage, the binding of an antigen to a receptor on a lymphocyte activates that lymphocyte. Upon activation, the lymphocyte undergoes multiple cycles of cell division. The result is the formation of

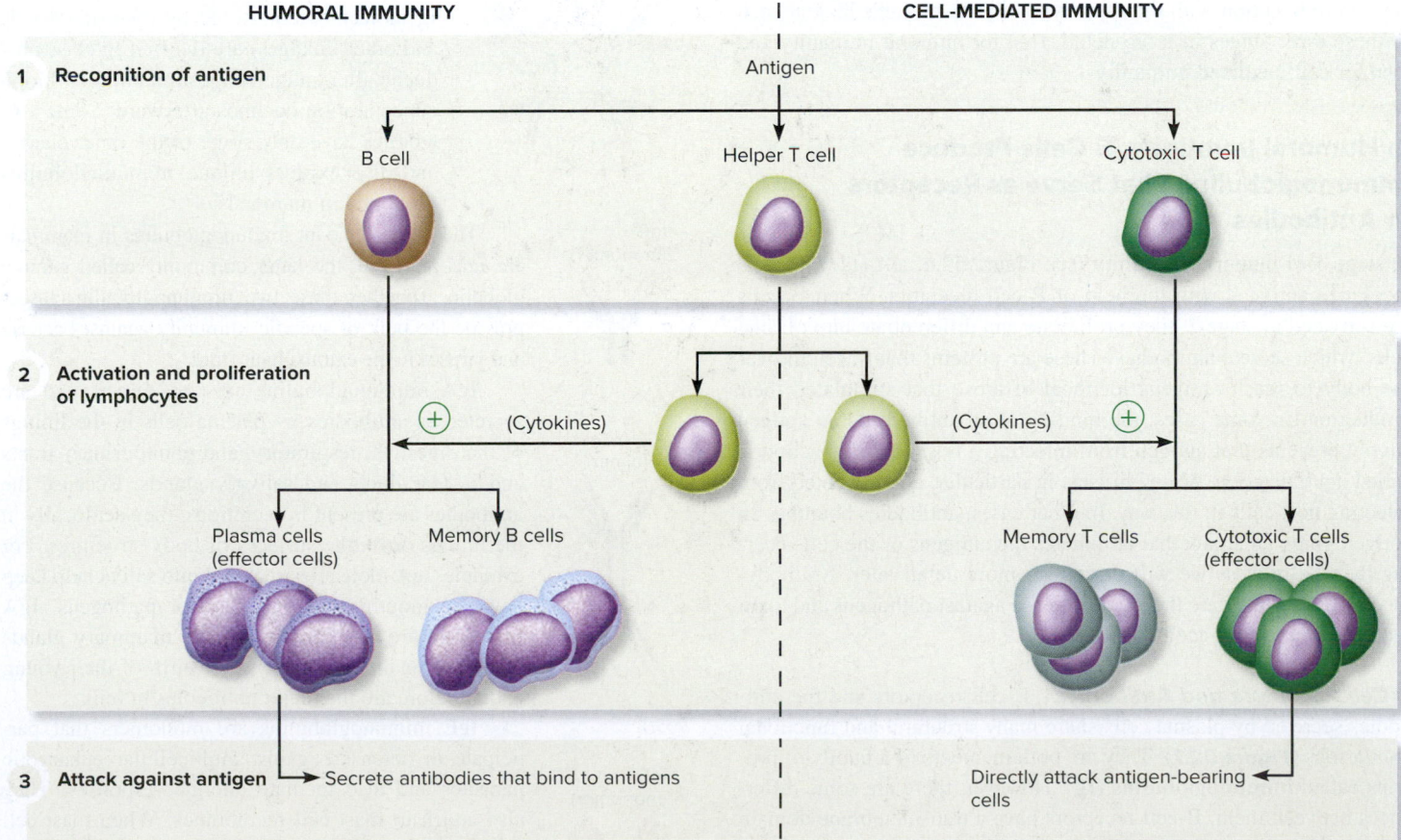

**Figure 52.6 The three stages of an adaptive immune response.** All three cell types in step 1 recognize the same antigen. Helper T cells secrete cytokines that activate B cells and cytotoxic T cells, as indicated by the ⊕ symbols. Both B cells and T cells undergo cell division to form clones when activated, and in both cases, some of the cloned cells are set aside as memory cells to fight off a future infection of the same type.

many identical cells called clones that express the same receptor as the receptor that first recognized the antigen. The structure of the antigen determines which individual lymphocytes will be activated to form clones. This process requires the function of helper T cells, which divide when activated and then secrete the cytokines that promote further cell division. Some of the cloned lymphocytes become **effector cells**, the plasma cells and cytotoxic T cells that carry out the attack response. Others are stored as **memory cells**, which remain poised to recognize the antigen if it returns in the future.

The size of the lymphocyte population is staggering. For example, 100 million different lymphocytes, each with the ability to recognize a unique antigen, are estimated to exist in a human's immune system. This vast population explains why our bodies are able to recognize so many different antigens as foreign and eventually destroy them.

### Stage 3: Attack Against Antigen

In the third stage, the effector cells attack all antigens of the kind that initiated the immune response. Plasma cells carry out a humoral response by secreting antibodies into the blood. These antibodies then recruit and guide other molecules and cells that perform the actual attack. Activated cytotoxic T cells, by contrast, carry out cell-mediated immunity. They directly attack and kill the cells bearing the antigens.

Once the attack is successfully completed, the great majority of plasma cells and cytotoxic T cells that participated in it die by apoptosis. However, memory cells persist even after the immune response has been successfully completed, so they can recognize and fight off any future infection with the same type of antigen. Let's look at each of these three stages in more detail, first for humoral immunity and then for cell-mediated immunity.

## In Humoral Immunity, B Cells Produce Immunoglobulins That Serve as Receptors or Antibodies

In stage 1 of humoral immunity (see Figure 52.6, left side), B cells recognize antigens with the help of B-cell receptors. When B cells are activated in stage 2, they proliferate and differentiate into plasma cells, which secrete antibodies. These are proteins that travel all over the body to reach antigens identical to those that stimulated their production. In some cases, the binding of an antibody to an antigen simply prevents that antigen from infecting a host cell; this action is called neutralization. Many viruses, in particular, are prevented from infecting host cells in this way. In other cases, antibodies bound to an antigen guide an attack that eliminates the antigens or the cells bearing them, a process we will discuss in more detail later. Antibody-mediated responses are the major defense against pathogens and toxin molecules in the extracellular fluid.

### B-Cell Receptors and Antibodies

B-cell receptors and the antibodies secreted by plasma cells share many structural and functional similarities (**Figure 52.7**). They are both members of a family of proteins called **immunoglobulins (Ig)**. However, there are some differences between them. B-cell receptors have a transmembrane domain that anchors them in the plasma membrane of the B cell (Figure 52.7a). Antibodies are soluble proteins that are secreted from plasma cells (Figure 52.7b). Interestingly, B-cell receptors and the antibodies

made by plasma cells are encoded by the same genes. In plasma cells, the pre-mRNA is alternatively spliced, a phenomenon described in Chapter 14 (see Figure 14.21), so the transmembrane domain is not present in the protein. For this reason, the B-cell receptor in a particular B cell and secreted antibodies from the resulting plasma cells recognize the exact same antigen.

### Immunoglobulin Classes and Structure

Each immunoglobulin molecule is composed of four interlinked polypeptides: two long heavy chains and two short light chains (see Figure 52.7). A hinge region that provides the molecule with flexibility separates the light chains and upper parts of the heavy chains from the lower parts of the heavy chains. One portion of an immunoglobulin chain is called the **constant region**. The amino acid sequence of the constant region is identical for all immunoglobulins of a given class and is what distinguishes the classes from each other. The constant regions are important for the binding of antibodies to immune cells and to complement proteins, and thus contribute to the eventual destruction of antigen that is bound to an antibody. A defining feature of immunoglobulins, however, is their **variable region**, which gets its name because its amino acid sequence varies among different B cells. The variable region is the site that specifically recognizes a particular antigen.

IgM
(pentamer)

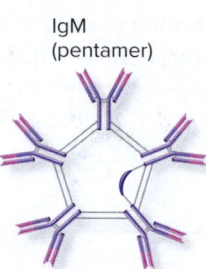

Mammals have five classes of immunoglobulins, designated IgM, IgG, IgA, IgE, and IgD. All vertebrates have IgM molecules. These pentamers (made of five Ig molecules connected by disulfide bonds and other linkages) are the first Ig class produced after antigen exposure, but their blood concentration declines afterward. Some vertebrates have only some of the other classes and also express unique immunoglobulins not found in mammals.

The most abundant immunoglobulins in mammals are IgM and IgG, the latter commonly called gamma globulin. Together these two immunoglobulin classes provide the bulk of specific immunity against bacteria and viruses in the extracellular fluid.

IgG
(monomer)

IgA immunoglobulins exist as dimers and are secreted as antibodies by plasma cells in the linings of the digestive, respiratory, and genitourinary tracts and in tear ducts and salivary glands. Because the antibodies are present in secretions, they act locally in the linings or on the surfaces of body structures. For example, IgA molecules secreted into saliva help keep animals' mouths relatively free of pathogens. IgA molecules are also secreted by the mammary glands of female mammals shortly after birth of their young and therefore are the major antibodies in milk.

IgA
(dimer)

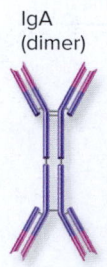

IgE immunoglobulins are monomers that participate in defenses against multicellular eukaryotic parasites and also mediate allergic responses. They also attach to mast cell membranes. When mast cell IgE molecules bind antigen, the mast cell secretes its histamine into the extracellular fluid, causing vasodilation and contributing to the allergic response. In people

IgE
(monomer)

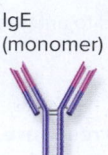

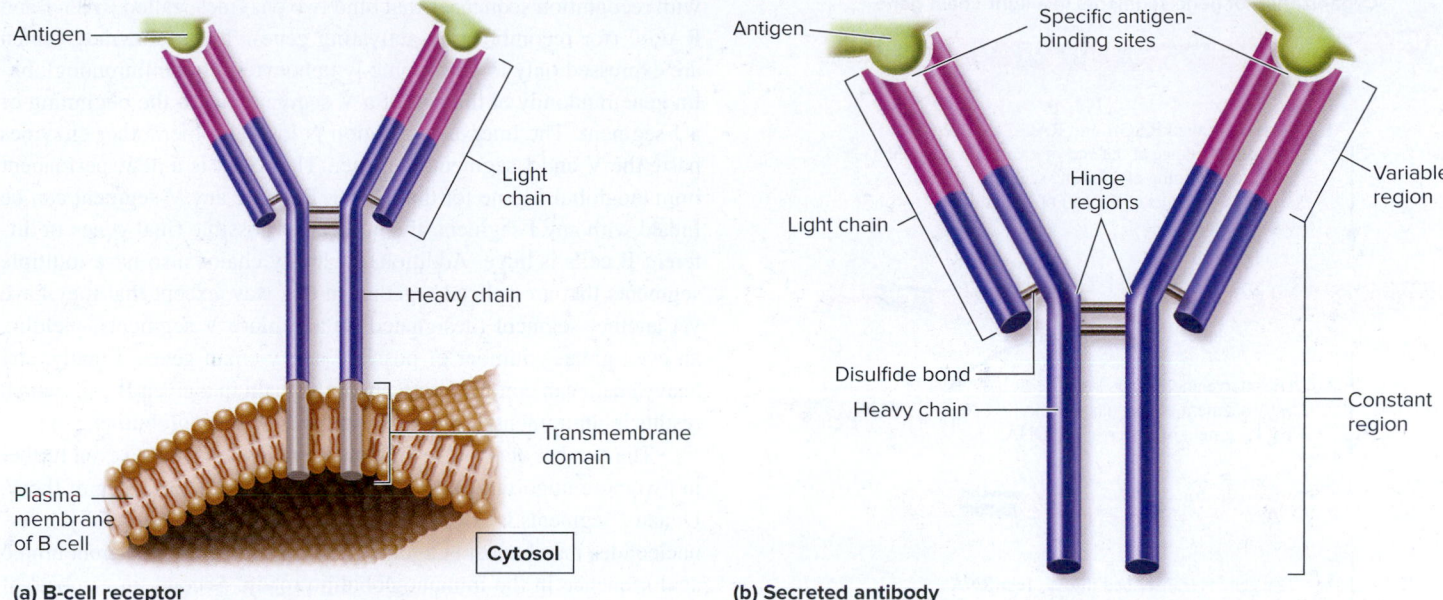

**(a) B-cell receptor**

Antigen — Light chain — Heavy chain — Transmembrane domain — Plasma membrane of B cell — Cytosol

**(b) Secreted antibody**

Antigen — Specific antigen-binding sites — Hinge regions — Variable region — Light chain — Disulfide bond — Heavy chain — Constant region

**Figure 52.7  Immunoglobulins. (a)** B-cell receptor and **(b)** secreted antibody. Immunoglobulins are composed of two heavy chains and two light chains. Disulfide bonds hold the chains together. A B-cell receptor is anchored to the plasma membrane by a transmembrane domain, whereas an antibody is secreted into the extracellular fluid. Within both types of immunoglobulins, the constant regions of the heavy and light chains (shown in blue) have identical amino acid sequences. In contrast, the antigen-binding sites formed by the light- and heavy-chain variable regions (shown in purple) have unique amino acid sequences and give each receptor or antibody its specificity for a particular antigen.

**Core Skill: Modeling**  The goal of this modeling challenge is to make a model that depicts the binding of an immunoglobulin to an antigen that has a different structure than the one shown in Figure 52.7.

**Modeling Challenge:** The structures of proteins determine their functions. This is true of all immunoglobulins. Using the model of a secreted antibody shown in Figure 52.7b as a starting point, draw a new model of an antibody that binds to the antigen depicted to the right.

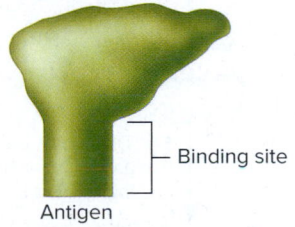

Binding site

Antigen

**An antigen with its binding site indicated**

who are particularly sensitive to allergens, this response is easily demonstrated by a pinprick injection of an antigen—such as one of the proteins associated with hay fever—into a small region under the skin, or by applying the antigen topically for an extended period of time (called patch testing). The resultant local inflammation and reddening of the skin are mediated in large part by IgE molecules.

IgD (monomer)

The functions of IgD are still unclear. However, IgD molecules are present both in blood and on the surface of B cells, and they are known to bind antigens on B cells, thus possibly contributing to B-cell activation.

The amino acid sequences of the variable regions vary widely from immunoglobulin to immunoglobulin in a given Ig class. The enormous number of variable sequences results in countless unique structures within each class of immunoglobulins. Thus, each of the five classes of immunoglobulins contains up to millions of unique molecules each capable of combining with only one specific antigen or, in some cases, with several antigens with very similar structures. How did animals evolve the ability to make all of these

different immunoglobulins? The explanation for this remarkable array was first proposed in the 1970s, as we see next.

***Recombination and Immunoglobulin Structure***  The human genome contains about 200 genes that encode immunoglobulins. This raises an intriguing question: How can an animal such as a human produce millions of different immunoglobulin proteins if there are only 200 immunoglobulin genes? The answer is that the 200 genes undergo a unique process involving gene rearrangements. This phenomenon was discovered by Japanese scientist Susumu Tonegawa and others in the 1970s. Tonegawa was recognized for this achievement in 1987 with the Nobel Prize in Physiology or Medicine.

Along the length of a typical human immunoglobulin gene are numerous gene segments that encode a portion of the final immunoglobulin protein (**Figure 52.8**). In light chains, these gene segments are of three types, called variable, joining, and constant segments. A total of 40 variable (V) segments encode the antigen-binding site. These are next to four joining (J) segments and a single constant (C) segment. Each segment along the length of the gene is associated

**Organization of gene segments in a light-chain gene**

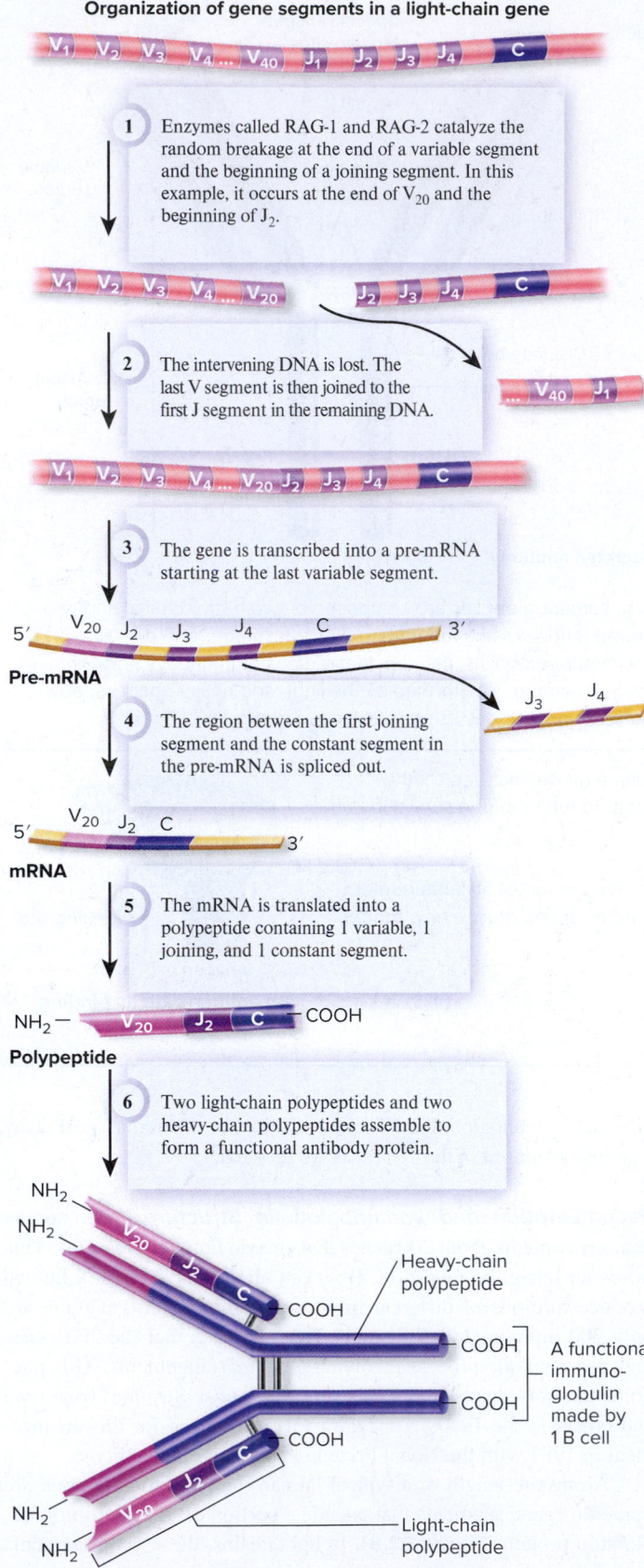

1  Enzymes called RAG-1 and RAG-2 catalyze the random breakage at the end of a variable segment and the beginning of a joining segment. In this example, it occurs at the end of $V_{20}$ and the beginning of $J_2$.

2  The intervening DNA is lost. The last V segment is then joined to the first J segment in the remaining DNA.

3  The gene is transcribed into a pre-mRNA starting at the last variable segment.

**Pre-mRNA**

4  The region between the first joining segment and the constant segment in the pre-mRNA is spliced out.

**mRNA**

5  The mRNA is translated into a polypeptide containing 1 variable, 1 joining, and 1 constant segment.

**Polypeptide**

6  Two light-chain polypeptides and two heavy-chain polypeptides assemble to form a functional antibody protein.

Heavy-chain polypeptide

A functional immuno-globulin made by 1 B cell

Light-chain polypeptide

**Figure 52.8  The mechanism of immunoglobulin diversity.** Although this figure shows events for a light-chain, events similar to those depicted here also occur in the production of the heavy chains, creating even more structural diversity.

with recognition sequences that bind two enzymes, called RAG-1 and RAG-2 (for recombination-activating gene). These enzymes, which are expressed only in developing lymphocytes, cut an immunoglobulin gene randomly at the end of a V segment and at the beginning of a J segment. The intervening region is lost, and then other enzymes paste the V and J segments together. The result is a new, permanent immunoglobulin gene for that B cell. Because any V segment can be linked with any J segment, the number of possible final genes in different B cells is huge. Additionally, heavy chains also have multiple segments that are spliced together in this way, except that they have yet another segment (designated D) and more V segments, yielding an even greater number of possible heavy-chain genes. Finally, any heavy chain can combine with any light chain in a given B cell, which results in an immense number of possible immunoglobulins.

The number of possible immunoglobulins is increased even further in two more important ways within B cells. First, the joining of the V, D, and J segments is not always a precise process. Occasionally, a few nucleotides may be lost at a joining end, resulting in a different amino acid sequence in the immunoglobulin protein. Second, in a subset of activated B cells, the DNA encoding the variable antigen-binding sites of immunoglobulins undergoes a unique process known as hypermutation, which primarily produces point mutations. The result is a hypervariable region of the light and heavy chains of all immunoglobulins.

The three processes of gene recombination, imprecise joining of gene segments, and hypermutation cause each lymphocyte within an individual's body to produce a unique type of immunoglobulin. The evolution of adaptive immunity resulted in an incredibly diverse array of immunoglobulins capable of recognizing an incredibly diverse array of antigens. Nearly any foreign antigen that is taken into the body is recognized by some lymphocytes in this large population.

## Activated B Cells Produce Plasma Cells That Secrete Antibodies

In stages 2 and 3 of the humoral immune response, B cells are activated and differentiate into either plasma or memory cells. The plasma cells secrete antibodies that bind the antigen that was detected (see Figure 52.6, left side). Let's take a closer look at these processes.

***Clonal Selection***    B cells are activated by a specific antigen, with the aid of a helper T cell (a process we will discuss later). When an antigen-stimulated lymphocyte divides and replicates itself, the progeny of this lymphocyte—all of which express the same receptor—are clones. The process by which these clones are formed is called **clonal selection** (**Figure 52.9**). This term emphasizes that lymphocyte proliferation is selected by exposure to an antigen.

***Antibody Destruction of Pathogens via Opsonization***    The antibodies secreted from the plasma cells circulate through the lymphatic system and the blood. Eventually, the antibodies combine with the antigen that initiated the immune response. These antibodies then direct the attack against the pathogen to which they are now bound. Thus, immunoglobulins have two distinct roles in humoral immune responses. First, during antigen recognition, immunoglobulins (B-cell receptors) on the surface of B cells bind to antigen brought to them. Second, immunoglobulins (antibodies) secreted by the resulting plasma cells bind to pathogens bearing the same antigens, marking them as the targets to be attacked.

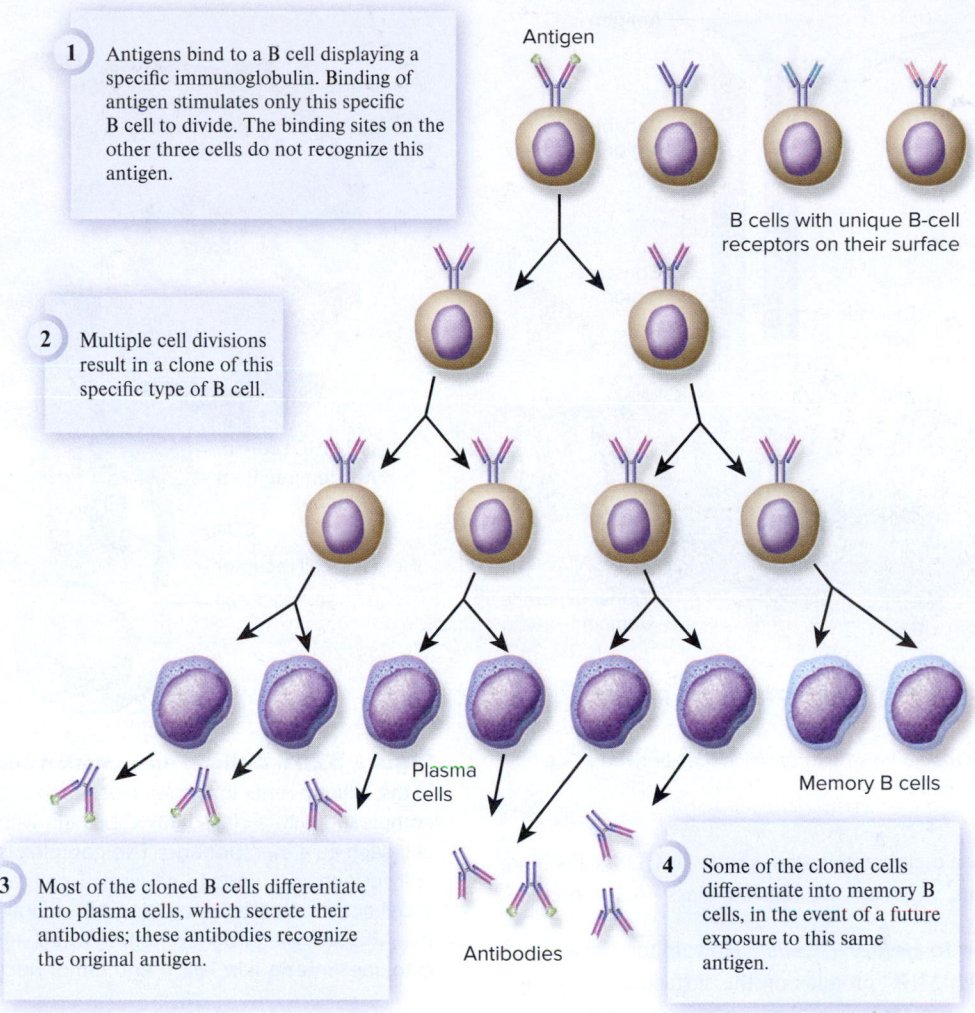

**1** Antigens bind to a B cell displaying a specific immunoglobulin. Binding of antigen stimulates only this specific B cell to divide. The binding sites on the other three cells do not recognize this antigen.

Antigen

B cells with unique B-cell receptors on their surface

**2** Multiple cell divisions result in a clone of this specific type of B cell.

Plasma cells

Memory B cells

**3** Most of the cloned B cells differentiate into plasma cells, which secrete their antibodies; these antibodies recognize the original antigen.

Antibodies

**4** Some of the cloned cells differentiate into memory B cells, in the event of a future exposure to this same antigen.

**Figure 52.9  Clonal selection.** In this example, a B cell with a specific immunoglobulin on its surface recognizes an antigen and is stimulated to divide into a clone of identical cells.

Instead of directly destroying the pathogens, antibodies bound to antigen on the pathogen surface inactivate the pathogens in various ways. Antibodies may physically link the pathogens to phagocytes (neutrophils and macrophages), complement proteins, or NK cells. This linkage—called **opsonization**—triggers the attack mechanism and ensures that only the pathogens, and not nearby body cells, are destroyed.

In a second mechanism, antibodies that recognize toxins produced by bacterial pathogens in the extracellular fluid bind to the toxins, thereby preventing them from harming susceptible body cells. The antibody-antigen complexes that are formed are then destroyed by phagocytes.

In a similar way, antibodies that recognize certain viral surface proteins bind to the viruses in the extracellular fluid, preventing them from attaching to the plasma membranes of potential host cells. As with bacterial toxins, the antibody-virus complexes that are formed are subsequently phagocytized.

## In Cell-Mediated Immunity, T Cells Recognize Antigens Complexed with Self Proteins

In cell-mediated immunity, T cells recognize and are activated by antigens (see Figure 52.6, right side). T-cell receptors for antigens have specific regions that differ from one T cell to another. As shown in **Figure 52.10**, they are composed of two polypeptides, each with a variable and a constant region, along with a transmembrane domain. The variable regions recognize an antigen. As in B-cell development, multiple DNA rearrangements occur during T-cell maturation, leading to millions of distinct types of T cells, each with a receptor of unique specificity. For T cells, this maturation occurs as they develop in the thymus gland.

A T-cell receptor cannot bind to an antigen unless the antigen is already complexed with a protein that is found on the surface of another cell, such as a macrophage. Such proteins are encoded by a gene family known as the **major histocompatibility complex (MHC)**, and thus the proteins are called MHC proteins. In humans, two major classes of MHC proteins are known. Class I MHC proteins are found on the surface of all human body cells except erythrocytes (that is, all nucleated cells). Class II MHC proteins are found primarily on the surface of macrophages, B cells, and dendritic cells.

The two different types of T cells have different MHC requirements. Cytotoxic T cells require that antigen be associated with class I MHC proteins, whereas helper T cells require an association with class II MHC proteins. One reason for this difference stems from the presence of different proteins on the surfaces of T cells; helper T cells can be identified by a unique membrane protein called CD4, and cytotoxic

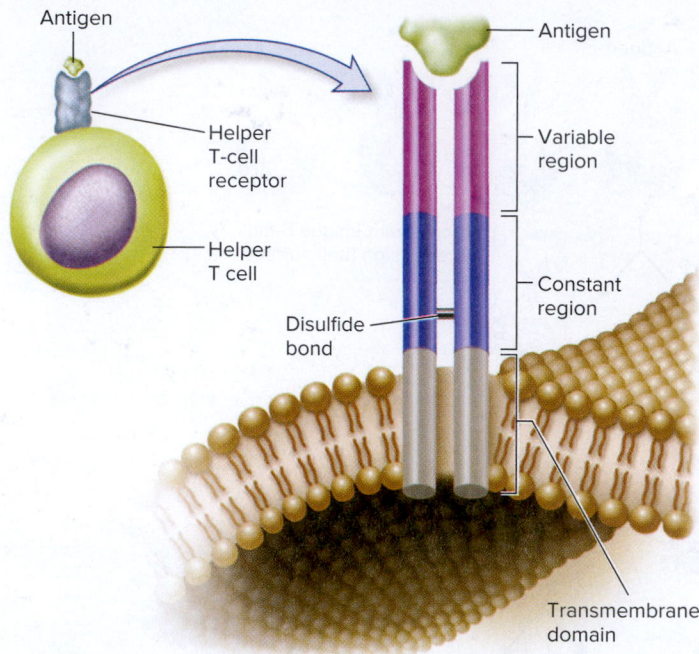

**Figure 52.10**  **Structure of a T-cell receptor in a plasma membrane.**

> *Concept Check:* *What are some structural similarities between T-cell receptors and B-cell receptors?*

T cells are identified by a membrane protein known as CD8. CD4 binds to class II MHC proteins, whereas CD8 binds to class I MHC proteins.

***Antigen Presentation to Helper T Cells***    How do foreign antigens become complexed with MHC proteins on the surface of the body's own cells? The answer involves the mechanism known as antigen presentation. As previously noted, helper T cells can bind antigen only when the antigen appears on the plasma membrane of a host cell complexed with the cell's class II MHC proteins. Cells bearing these complexes, therefore, function as **antigen-presenting cells (APCs)**. Because only macrophages, B cells, and dendritic cells express class II MHC proteins, only these cells can function as APCs for helper T cells.

Let's consider the function of macrophages as APCs for helper T cells (**Figure 52.11**). After a microbe or noncellular antigen has been phagocytized by a macrophage in an innate immune response, antigens, such as proteins, are partially broken down into smaller polypeptide fragments by the macrophage's proteolytic enzymes within endosomes. The resulting digested fragments then bind in the endosomes to class II MHC proteins synthesized by the macrophage. Each fragment-MHC complex is then transported to the cell surface, where it is displayed on the plasma membrane. A specific helper T-cell receptor then binds this entire complex on the cell surface of the macrophage. The CD4 protein helps link the two cells. The molecule that is complexed to MHC proteins and presented to the helper T cells is not the intact antigen but instead a polypeptide fragment of the antigen— called an antigenic determinant, or **epitope**. Even so, it is customary to call this antigen presentation rather than epitope presentation.

B cells process antigen and present it to helper T cells in essentially the same way as macrophages do. The ability of B cells to present antigen to helper T cells is a second function of B cells in

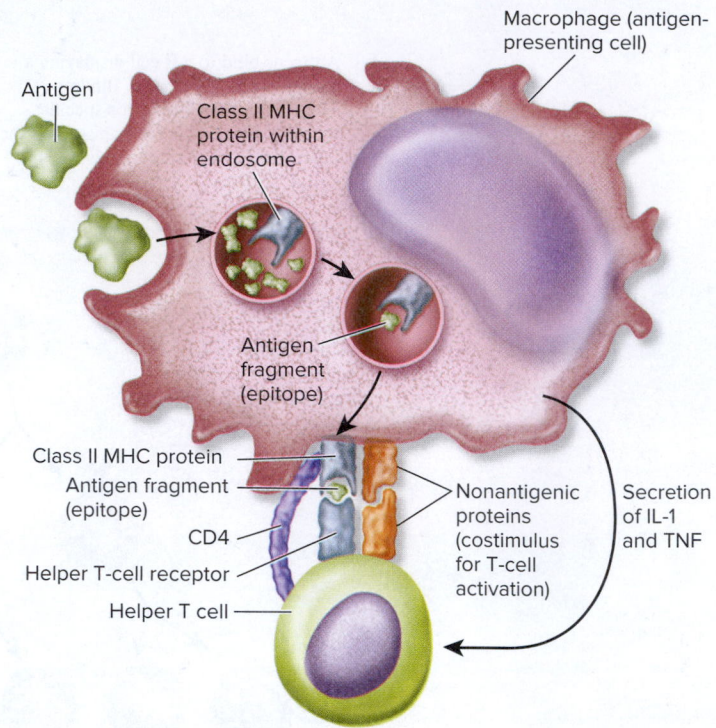

**Figure 52.11**  **Antigen presentation and helper T-cell activation.** In the initial events in helper T-cell activation, an antigen fragment is complexed with a class II MHC protein within an antigen-presenting cell such as a macrophage. The complex is then displayed on the cell surface and binds to a helper T-cell receptor. Also required for T-cell activation are the binding of nonantigenic proteins between the APC and the attached helper T cell, and the actions of the cytokines interleukin 1 (IL-1) and tumor necrosis factor (TNF).

response to antigenic stimulation, in addition to their differentiation into antibody-secreting plasma cells.

The binding between the helper T-cell receptor and the complex of antigen fragment and class II MHC protein on an APC is the essential antigen-specific event in helper T-cell activation. However, by itself, this specific binding does not result in T-cell activation. In addition, interactions occur between nonantigenic proteins on the surfaces of the attached helper T cell and the APC. These interactions comprise a necessary costimulus for T-cell activation (see Figure 52.11).

Finally, the antigen-dependent binding of the APC to the helper T cell plus the costimulus induces the APC to secrete large amounts of two cytokines—interleukin 1 (IL-1) and tumor necrosis factor (TNF). These molecules also stimulate the attached helper T cell.

Thus, the APC participates in the activation of a helper T cell in three ways: (1) presenting antigen, (2) providing a costimulus, and (3) secreting cytokines. Activated helper T cells then secrete additional cytokines of their own that stimulate B cells and cytotoxic T cells.

***Helper T Cells and B-Cell Activation***    Now we can reconsider how B cells are activated by the actions of helper T cells. This process begins when a helper T cell specific for a particular antigen binds to a complex of that antigen and a class II MHC protein on an APC, activating the helper T cell. Along with other signals, this binding induces the activated helper T cell to undergo many cycles of cell division. Some of the resulting activated helper T cells then bind to B cells that display the same antigen on their surfaces. This binding, along with additional cytokines, stimulates the B cell to go through the process of clonal selection. Thus, helper T cells are so named because they help activate B cells that have bound antigen, in addition to their participation in activation of cytotoxic T cells and antigen presentation.

**Antigen Presentation to Cytotoxic T Cells**  Unlike helper T cells, cytotoxic T cells require class I MHC proteins for activation. This distinction helps explain the major function of cytotoxic T cells—destruction of any of the body's own altered cells that have become cancerous or infected with viruses.

How do antigens that activate cytotoxic T cells arise? In viral infections, once a virus has entered a host cell, the expression of viral genes results in the synthesis of viral proteins, which are foreign to the cell. Cancer cells accumulate mutations, some of which alter the amino acid sequence of proteins. Such abnormal proteins act as antigens.

In both virus-infected and cancerous cells, cytosolic enzymes hydrolyze some of the antigenic proteins into polypeptide fragments, which are transported into the endoplasmic reticulum. There the fragments are complexed with the host cell's class I MHC proteins and then shuttled by the secretory pathway to the plasma membrane, where a cytotoxic T cell specific for the antigen/MHC protein complex can bind to it. Once binding occurs, cytotoxic T cells release chemicals that kill the infected or cancerous cell, as discussed next.

## Activated Cytotoxic T Cells Kill Infected or Cancerous Cells

So far, this section has described how immune responses provide long-term defenses against bacteria, viruses, and foreign molecules that enter the body's extracellular fluid. We now examine how an animal's own cells that have become infected by viruses or transformed into cancerous cells are destroyed by cell-mediated immune responses (see Figure 52.6, right side).

What is the value of destroying virus-infected host cells? First, and most importantly, such destruction prevents cells from making more viruses. Second, for cells that already are making mature viruses, the lysis of infected cells releases the viruses into the extracellular fluid, where they can be neutralized by circulating antibody.

**Functions of Cytotoxic T Cells**  A typical cytotoxic T-cell response triggered by viral infection of a vertebrate's body cells is summarized in **Figure 52.12**. The response triggered by a cancerous cell would be similar. A virus-infected cell produces foreign proteins, viral antigens that are processed and presented on the plasma membrane of the cell complexed with class I MHC proteins. Cytotoxic T cells specific for the particular antigen bind to the complex (Figure 52.12, step 1). As with B cells, binding to antigen alone does not cause activation of the cytotoxic T cell. Cytokines from nearby activated helper T cells are also required.

Macrophages phagocytize extracellular viruses (or, in the case of cancer, antigens released from the surface of cancerous cells) and then process and present antigen, in association with class II MHC proteins, to the helper T cells (step 2 of Figure 52.12). In addition, the macrophages provide a costimulus and also secrete IL-1 and TNF. The activated helper T cell releases IL-2 and other cytokines, which stimulate proliferation of the helper T cell.

IL-2 and other cytokines also act on the cytotoxic T cell bound to the surface of the virus-infected or cancerous cell, stimulating this attack cell to proliferate. Why is proliferation important if a cytotoxic T cell has already located and bound to its target? The answer is that there is rarely just one virus-infected or cancerous cell. By expanding the population of cytotoxic T cells capable of scanning the entire body and recognizing the particular antigen, the likelihood is greater that the other virus-infected or cancerous cells will be encountered by an appropriate cytotoxic T cell.

The cytotoxic T cells specific for that virus then find and bind to other virus-infected cells (Figure 52.12, step 3). There are two major ways in which cytotoxic T cells kill their target cells. The first way involves the release by exocytosis of secretions from the T cell. The second mechanism involves binding of a membrane protein on the T cell to surface receptors on the target cell. Both mechanisms induce apoptosis in the target cell. Let's examine the first mechanism in detail. Each cytotoxic T cell releases the contents of its secretory vesicles directly into the extracellular space between itself and the target cell to which it is bound (thereby ensuring that other nearby host cells will not be killed). These vesicles contain proteases and a protein called perforin, which is similar in structure to the proteins of the complement system's membrane attack complex. Perforin inserts into the target cell's membrane and forms pores (perforations) in it (step 4). These pores allow proteases secreted by a cytotoxic T cell to enter the attacked cell and induce apoptosis. Perforin can also cause a virus-infected cell to take up so much water that it bursts, a second way to kill it. The cytotoxic T cell is not harmed by this process and can then continue to kill other virus-infected cells (step 5).

Viruses can be eliminated from an animal's body in two ways: through the humoral actions of antibodies in body fluids and through the cell-mediated killing of virus-infected cells by cytotoxic T cells. Although cytotoxic T cells have an important role in the attack against such cells, they are not the only mechanisms. NK cells also destroy virus-infected and cancerous cells by secreting toxic chemicals. As mentioned earlier in this chapter, NK cells can recognize general features on the surface of such cells and participate in innate immunity. In addition, in a cell-mediated immune response, NK cells can be linked to such target cells by antibodies and then can destroy them by release of toxic molecules.

## Summary: Example of an Adaptive Immune Response

Let's bring together our discussion of the adaptive immune system by looking in detail at one example in which a humoral immune response results in the destruction of bacteria. The sequence of events, which is quite similar to the humoral response to a virus in the extracellular fluid, is summarized in **Figure 52.13**. For this example, we consider the response in mammals, in which lymph nodes are present. Many features of the response, however, are similar in other vertebrates.

This process starts the same way as for innate responses, with the bacteria penetrating one of the body's linings through an injury and entering the interstitial fluid (Figure 52.13, step 1). The bacteria then move with lymph into the lymphatic system and are carried to lymph nodes (step 2). Within the lymph node, a macrophage and a B cell recognize the bacteria as foreign and bind to them.

As we have discussed, the process of B-cell activation usually requires the activation of helper T cells. The helper T cell binds to a

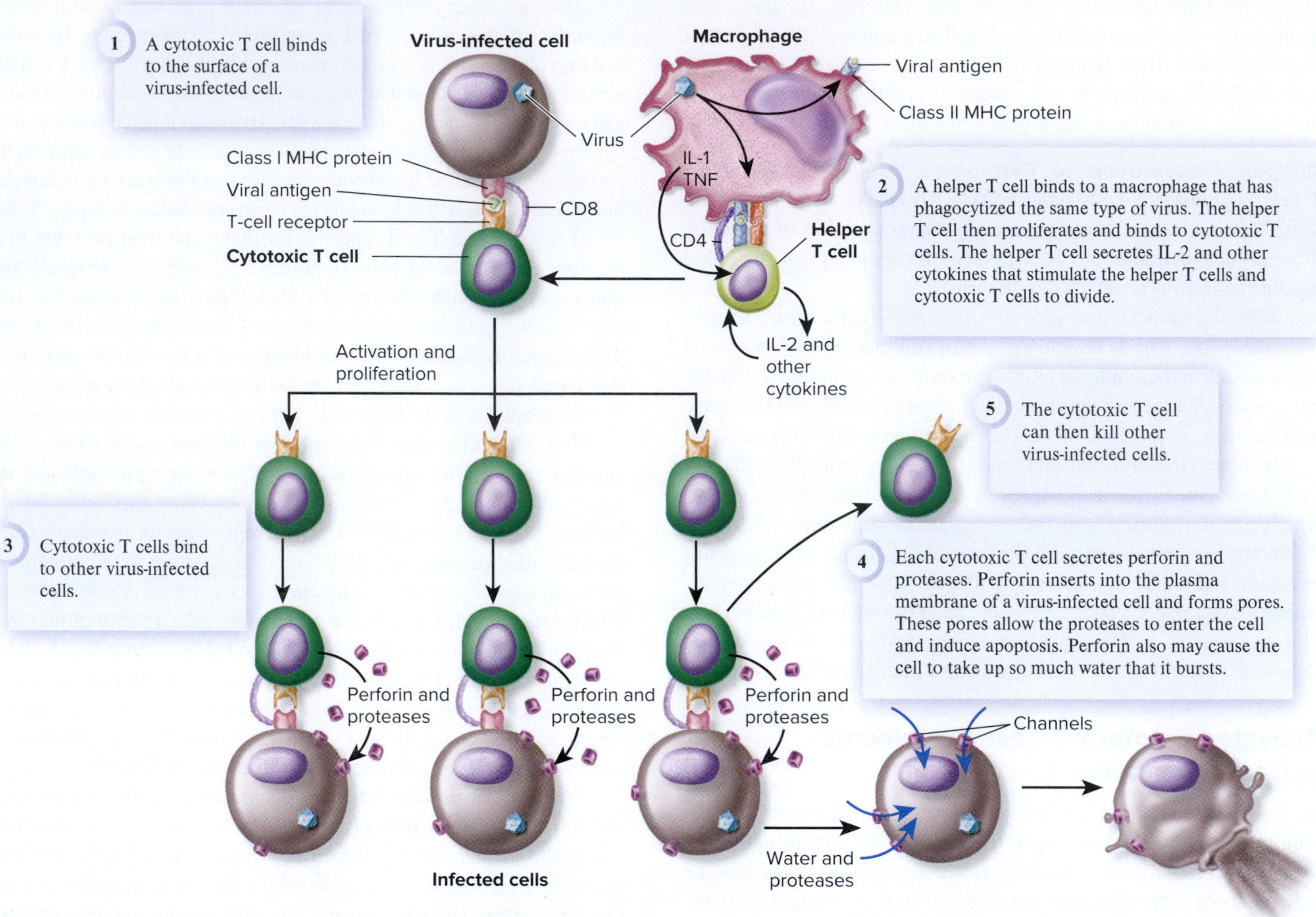

1. A cytotoxic T cell binds to the surface of a virus-infected cell.

**Virus-infected cell**

**Macrophage**

Viral antigen

Class II MHC protein

Virus

Class I MHC protein

Viral antigen

T-cell receptor

**Cytotoxic T cell**

CD8

IL-1 TNF

CD4

**Helper T cell**

2. A helper T cell binds to a macrophage that has phagocytized the same type of virus. The helper T cell then proliferates and binds to cytotoxic T cells. The helper T cell secretes IL-2 and other cytokines that stimulate the helper T cells and cytotoxic T cells to divide.

IL-2 and other cytokines

Activation and proliferation

5. The cytotoxic T cell can then kill other virus-infected cells.

3. Cytotoxic T cells bind to other virus-infected cells.

4. Each cytotoxic T cell secretes perforin and proteases. Perforin inserts into the plasma membrane of a virus-infected cell and forms pores. These pores allow the proteases to enter the cell and induce apoptosis. Perforin also may cause the cell to take up so much water that it bursts.

Perforin and proteases

Perforin and proteases

Perforin and proteases

Channels

**Infected cells**

Water and proteases

**Figure 52.12  Summary of events in the killing of virus-infected cells by cytotoxic T cells.** The sequence is similar for cancerous cells attacked by cytotoxic T cells.

 **Core Skill: Science and Society**  Viruses cause illness in animals and humans alike. In many cases, a viral illness can be transmitted from animals to humans. Many viral illnesses cause only temporary symptoms, such as those of the common cold. Other viruses, however, can cause very serious or life-threatening illnesses, such as the dengue virus, which infects as many as 100 million people per year and causes tens of thousands of deaths annually.

complex of processed antigen and class II MHC protein on an APC (Figure 52.13, step 3). In this case, the APC is a macrophage that has phagocytized the bacterium, hydrolyzed its proteins into polypeptide fragments, complexed the fragments with class II MHC proteins, and displayed the complexes on its surface. Once a helper T cell specific for the complex binds to it, the helper T cell becomes activated. The macrophage helps this process in two other ways: It provides a costimulus, and it secretes the cytokines IL-1 and TNF.

IL-1 and TNF stimulate the helper T cell to secrete another cytokine, IL-2. IL-2 stimulates the activated helper T cell to divide, which leads eventually to the formation of a clone of activated helper T cells (Figure 52.13, step 4). The activated helper T cells bind to B cells that display the antigen, and also secrete IL-2 and other cytokines (step 5). Some of these cytokines provide additional signals that are usually required to activate nearby antigen-bound B cells to proliferate (step 6). These cells differentiate into memory cells, which help ward off possible future attacks by the same antigen, and plasma cells, which

secrete specific antibodies (step 7). The antibodies enter the bloodstream and bind to bacterial cells, which are then destroyed (step 8).

## B Cells and T Cells That Recognize Self Molecules Must Be Killed or Inhibited

As we have seen, the lymphocytes responsible for the adaptive immune response in vertebrates are very capable killers of pathogens—so capable, in fact, that it raises a question: Why don't these cells attack and kill normal self cells? In other words, how does the body distinguish between self and nonself components and develop what is called **immune tolerance**, or tolerance of its own molecules?

Recall that the huge diversity of lymphocyte receptors is ultimately the result of multiple random DNA cutting and recombination processes. It is virtually certain, therefore, that every animal possessing adaptive immune defenses will have lymphocytes with receptors that can bind to that individual's own proteins. The continued existence and functioning

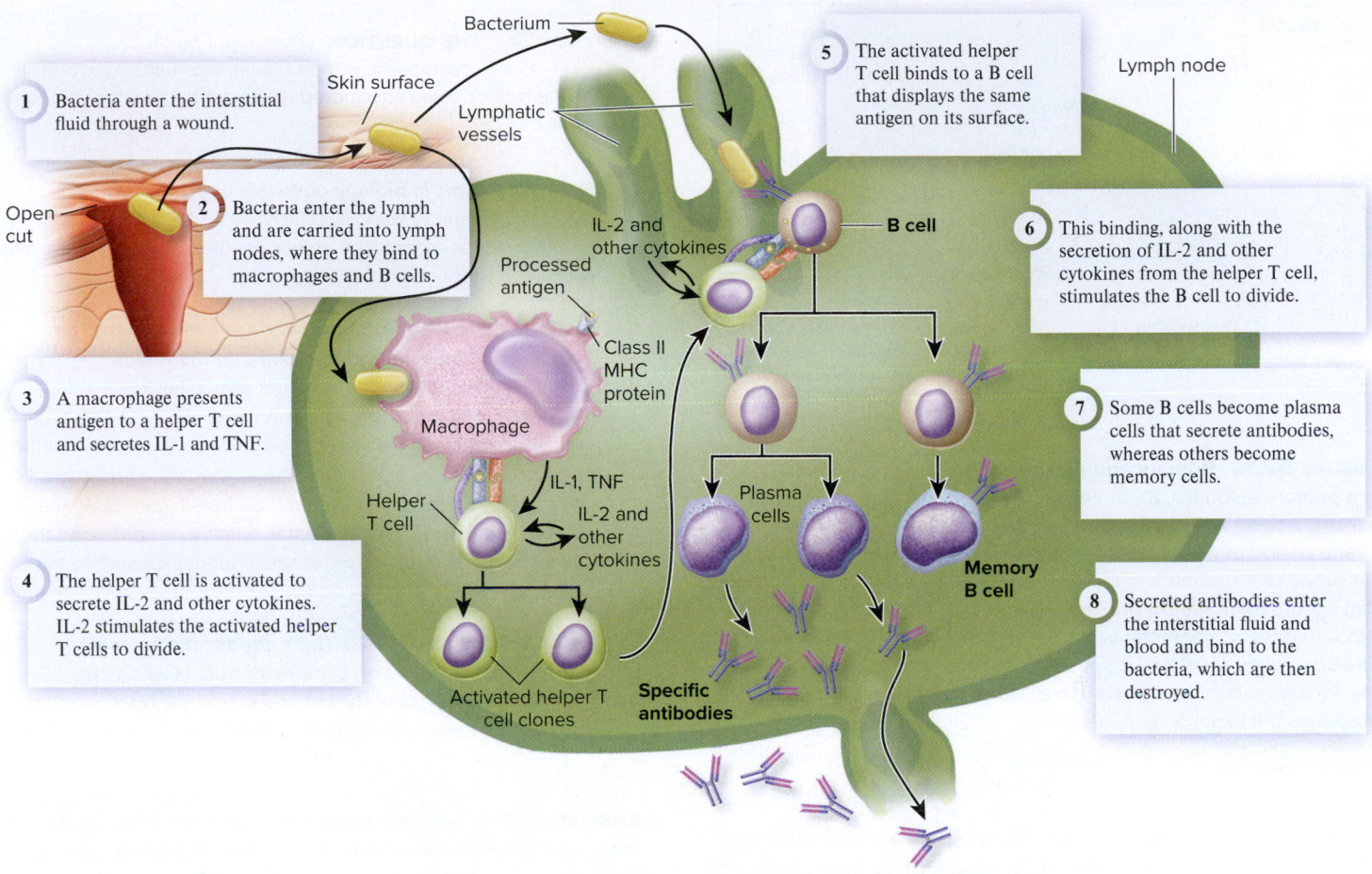

**Figure 52.13** **Summary of events in a typical humoral immune response.** Most of the events depicted occur within a lymph node.

of such lymphocytes would be disastrous, because such binding would launch an immune attack against all body cells expressing these proteins.

At least two mechanisms explain why individuals normally lack active lymphocytes that respond to self components. First, during early development in vertebrates, T cells are exposed to a wide mix of self proteins in the thymus gland. Those T cells with receptors capable of binding self proteins are destroyed by apoptosis in a process termed **clonal deletion**. The second mechanism, termed **clonal inactivation**, occurs outside the thymus gland and causes potentially self-reacting T cells to become nonresponsive. B cells undergo similar processes. The mechanisms by which these two events occur are still under investigation.

Occasionally, however, these mechanisms fail, and an animal's immune cells attack the body's own cells. When this happens, it produces an autoimmune disease. **Autoimmune diseases** are conditions in which the body's normal state of immune tolerance somehow breaks down, with the result that both humoral and cell-mediated attacks are directed against the body's own cells and tissues. A growing number of animal and human diseases are being recognized as autoimmune in origin. Examples in humans include multiple sclerosis, in which myelin around neurons is attacked; myasthenia gravis, in which the receptors for acetylcholine on skeletal muscle cells are the targets; rheumatoid arthritis, in which joints are damaged; systemic lupus erythematosus, in which numerous organs are damaged; and type 1 diabetes mellitus, in which the insulin-producing cells of

the pancreas are destroyed. Treatments for autoimmune disease range from treating the symptoms (for example, administering insulin to individuals with diabetes) to suppressing the immune system with drugs.

## Immunological Memory Is an Important Feature of Adaptive Immunity

As we have learned, the magnitude of the adaptive immune response to a given antigen depends on whether or not the body has previously been exposed to that antigen. Consider, for example, the humoral immune response. In mammals, antibody production in response to the first contact with an antigen occurs slowly, over a few weeks. This response to an initial antigen exposure is termed a **primary immune response** (**Figure 52.14**). Any subsequent infection by the same pathogen elicits a rapid and heightened production of additional specific antibodies against that particular antigen, a reaction termed a **secondary immune response**.

In the case of humoral immunity, this secondary response occurs more quickly, is stronger, and lasts longer because memory B cells that were produced in the primary response are quickly stimulated to multiply and differentiate into thousands of plasma cells. These cells then produce large amounts of specific antibodies. The immune system's ability to produce this secondary response is called **immunological memory**.

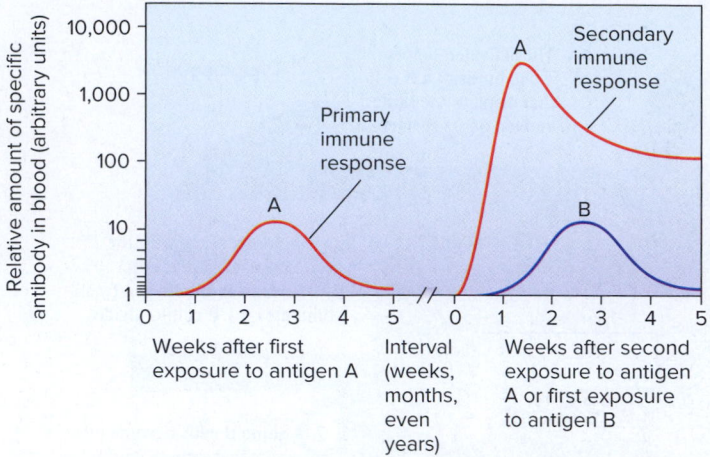

**Figure 52.14 Primary and secondary immune responses.** In a primary response, as shown on the left of this graph, an initial exposure to an antigen produces modest levels of specific antibody over a period of weeks. In a secondary response, subsequent exposure to the same antigen results in greater antibody production that occurs more rapidly and lasts longer than a primary response. (Note that the scale on the vertical axis of the graph is logarithmic.) The secondary response is specific for that antigen. During a secondary response, exposure to a different antigen for the first time produces the usual primary response.

**Concept Check:** *What is the advantage of a secondary immune response?*

Immunological memory explains why humans and other animals are able to fight off many illnesses, such as many common childhood diseases, after having been previously exposed to them. The adaptive response to exposure to any type of antigen is known as **active immunity**. Active immunity not only results from natural exposure to antigens, but also is the basis for exposures to antigens that occur in vaccinations. In **vaccinations**, small quantities of living, dead, or altered microbes, small quantities of toxins, or harmless antigenic molecules derived from a pathogen or its toxins are injected into the body, resulting in a primary immune response, including the production of memory cells. Subsequent natural exposure to the immunizing antigen results in a rapid, effective response that can prevent or reduce the severity of disease.

In contrast to active immunity, another type of adaptive immunity, called **passive immunity**, confers protection against disease through the direct transfer of antibodies from one individual to another. Passive immunity can occur naturally, as when IgG molecules cross the mammalian placenta to protect a fetus from various pathogens, or when a newborn mammal receives antibodies from breast milk. It can also occur artificially, as when a person is given an injection of IgG molecules shortly after being exposed to hepatitis viruses. Recent advances in the creation of highly specific and pure antibodies, called **monoclonal antibodies** because they are derived from a single clone of cells prepared in a laboratory, have paved the way for the use of passive immunity to combat certain types of cancer. Because antibodies are proteins and are eventually broken down and removed from the body, the protection afforded by the transfer of antibodies in passive immunity is relatively short-lived, usually lasting only a few weeks or months.

**BIO TIPS**

**THE QUESTION** *What might be the consequences if an animal was unable to form memory cells after being vaccinated against a particular pathogen or antigen?*

**T** **OPIC** *What topic in biology does this question address?* The topic is memory cells. Specifically, you are asked to predict the outcome for an animal that was unable to form memory cells following a vaccination.

**I** **NFORMATION** *What information do you know based on the question and your understanding of the topic?* From the question, you know that an animal has received a vaccination to help protect it from some disease caused by a particular pathogen or antigen. From your understanding of the topic, you may recall that vaccinations rely on the processes of active immunity. You also have learned from Figure 52.14 that adaptive immunity produces a small primary response and then a larger secondary response on subsequent exposure to the same antigen.

**P** **ROBLEM-SOLVING** **S** **TRATEGY** *Predict the outcome.* First, consider the function of memory cells: How do they contribute to the secondary immune response? Next, predict what might happen without that contribution.

**ANSWER** *Artificial exposure to small quantities of a pathogen or antigen (vaccination) results in a small primary immune response, but also causes the differentiation of activated T and B cells into memory cells. These cells are activated by subsequent (natural) exposure to the same pathogen or antigen, resulting in a large secondary immune response. Without the production of memory cells, the secondary immune response would be greatly diminished, resembling the blue curve labeled B on the right side in Figure 52.14.*

## 52.4 Impact on Public Health

**Learning Outcomes:**

1. **CoreSKILL »** Predict some influences of lifestyle on immunity.
2. Explain the role of the immune system in organ transplant rejection and allergic reactions.
3. Describe the effects of HIV on a human immune system, and describe a current method of treating HIV infection.

In this section, we will consider a few ways in which the functioning of the immune system can be affected by lifestyle, medical interventions, allergies, and destruction of immune cells. Collectively, the effects of disorders of the immune system have an almost immeasurable impact on public health in terms of worker productivity, healthcare resources, and the economy.

### Lifestyle Has an Important Influence on Immunity

Adequate nutrition is essential for good health. Protein-calorie malnutrition in particular is the single greatest contributor to decreased resistance to infection worldwide. When adequate amino acids for synthesizing essential proteins are not available, immune function is impaired. Deficits of certain nonprotein nutrients can also lower resistance to infection.

Both stress and state of mind can affect resistance to infection and to cancer. The immune system can alter neural and endocrine function, and, in turn, neural and endocrine activity modifies immune function. For example, lymphoid tissue receives input from nerves, and immune cells have receptors for certain hormones. Conversely, immune cells release cytokines that have important effects on the brain and endocrine system. In addition, lymphocytes secrete several hormones that are also produced by endocrine glands. The multiple brain-body interactions that affect disease resistance are the subject of a field of study called psychoneuroimmunology.

Of the hormones associated with stress, the adrenal hormone cortisol has received the most attention due to its powerful suppressive effect on inflammation and adaptive immunity. Among other things, cortisol inhibits production of inflammatory mediators, reduces capillary permeability in injured areas, and suppresses the growth and activity of certain types of leukocytes. In this way, it acts as a sort of brake on the immune system, suppressing its activity. This should not be surprising, since all organ systems are generally under dual control, with stimulatory signals and inhibitory signals creating a homeostatic balance. During chronic stress or when cortisol is used to treat certain illnesses for long periods of time, however, this inhibitory action may be severe enough to cause immunosuppression. This inhibition is a key link between stress and health. Chronic stress may lead to a chronically increased concentration of cortisol in the blood that, by suppressing the body's immune responses, lowers resistance to infection.

Another feature of a person's lifestyle that appears to affect immune function is exercise. The influence of physical exercise on the body's resistance to infection and cancer has been debated for decades. Evidence now suggests that the intensity, duration, regularity, and psychological stress of the exercise all have important influences—both positive and negative—on a variety of immune functions, such as the numbers of circulating NK cells. Although evidence suggests that too much intense exercise can impair immunity, most experts currently believe that moderate exercise and physical conditioning have net beneficial effects on the immune system and on disease resistance. Recent studies suggest that exercise may be particularly beneficial in helping to ward off the onset of breast cancer, one of the most common types of cancer in women.

## Organ Transplants Are Medical Procedures That Can Cause Serious Immune Reactions

Organ transplants have saved numerous lives. However, they carry the possibility of provoking immune reactions that can threaten the life of the recipient. Since the mid-20th century, organ transplants from a healthy or recently deceased donor to a recipient have become widespread. The United Network for Organ Sharing reports that approximately 28,500 organs are transplanted in the U.S. each year, with kidney (17,000), liver (6,000), heart (2,000), and lung (1,800) transplants accounting for most of these procedures. The major obstacle to successful transplantation of tissues and organs is a reaction called graft rejection, in which a person's immune system recognizes the transplant

(also called a graft) as foreign and attacks it as it would any foreign cells. Although B cells and macrophages play some role in graft rejection, cytotoxic T cells and helper T cells are mainly responsible. To minimize this possibility, transplant patients are given drugs that suppress immune function, a procedure that is associated with some risk.

Except for grafts from identical twins, the class I MHC proteins on graft cells differ from those on the recipient's cells, as do the class II MHC proteins present on macrophages in the graft. Consequently, the recipient's T cells recognize the MHC proteins in the graft as foreign, and cytotoxic T cells (with the aid of helper T cells) destroy the graft cells.

## Allergies Affect the Quality of Life of Millions of People

An **allergy** (one type of a group of related immune disorders known as hypersensitivities) is a condition in which immune responses to environmental antigens cause inflammation and damage to body cells. Antigens that induce allergic reactions are called allergens. Common examples of allergens include ragweed pollen and animal dander. Most allergens themselves are relatively or completely harmless. It is the immune responses to them that cause the damage. In essence, then, allergy is immunity gone awry, for the response is of inappropriate strength and duration for the stimulus. In the U.S. alone, as many as 40 million people (about 13% of the population) suffer from allergies.

For any allergy to develop, a genetically predisposed person must first be exposed to the allergen—a process called sensitization. Subsequent exposures elicit the damaging immune responses we recognize as an allergy. Hypersensitivities can be broadly classified according to the speed of the response. Those that take up to several days to develop are considered delayed hypersensitivities. The skin rash that appears after contact with poison ivy is an example. More common are reactions considered immediate hypersensitivities, which can develop in minutes. Allergies fall into this category.

In immediate hypersensitivity, sensitization to the allergen leads to the production of specific antibodies and a clone of memory B cells. In individuals who are genetically susceptible to allergies, antigens that elicit immediate hypersensitivity reactions stimulate the production of IgE antibodies. Upon their release from plasma cells, these IgE molecules circulate throughout the body and become attached to mast cells in connective tissue (**Figure 52.15**). When the same antigen enters the body at some future time and binds with IgE that is bound to mast cells, the mast cell is stimulated to secrete many inflammatory mediators, including histamine, which then initiate an inflammatory response.

The signs and symptoms of allergies reflect both the effects of inflammatory mediators and the body site in which the binding of antigen to the IgE bound to a mast cell occurs. When, for example, a previously sensitized person inhales ragweed pollen, the antigen combines with the variable region of IgE, and the constant region of IgE binds to mast cells in the airways. The mast cells release their contents, which induce increased mucus secretion, increased blood flow, swelling of the epithelial lining, and contraction of the smooth muscle surrounding airways. These effects produce the congestion, runny nose, sneezing, and, in some persons, difficulty in breathing characteristic of hay fever. Antihistamines are drugs taken by people to block the action of histamine that is released during these allergic responses. These drugs prevent histamine from binding to its receptor

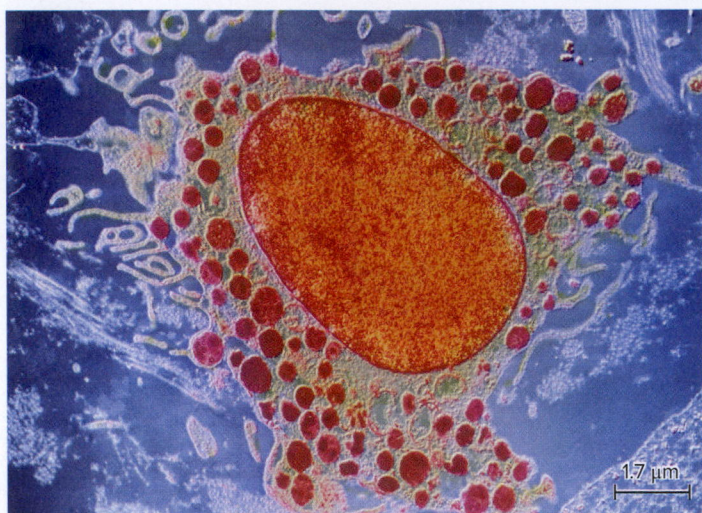

**Figure 52.15** Colorized electron micrograph of a mast cell, showing numerous secretory vesicles filled with histamine, a major mediator of allergic responses. ©CNRI/Science Source

protein on its target cells, thereby preventing or relieving some of the symptoms of allergy.

## Acquired Immunodeficiency Syndrome (AIDS) Results from HIV Infection

Acquired immunodeficiency syndrome (AIDS) is caused by the human immunodeficiency virus (HIV), which weakens the immune system by preferentially killing helper T cells. HIV is a retrovirus, a virus that contains RNA as its genetic material. Once inside a helper T cell, HIV uses the enzyme reverse transcriptase to transcribe its RNA into DNA, which is then integrated

into the chromosomal DNA of the host's T cells (refer back to Figure 19.4b). Later, viral replication within the T cell results in the death of the cell.

HIV infects helper T cells because the CD4 protein in their plasma membranes acts as a receptor for an HIV capsid protein. However, binding to CD4 is not sufficient to enable HIV to enter the helper T cell. Another T-cell surface protein, which normally acts as a receptor for certain cytokines, must serve as a coreceptor. Interestingly, individuals possessing a mutation in this cytokine receptor are highly resistant to HIV infection, so much research is now focused on the possible therapeutic use of chemicals that can bind to and block this coreceptor.

HIV not only directly kills helper T cells, but it also indirectly causes additional helper T-cell death by inducing cytotoxic T cells to kill HIV-infected helper T cells. In addition, by still poorly understood mechanisms, HIV causes the death of many uninfected helper T cells by apoptosis. Without adequate numbers of helper T cells, neither B cells nor cytotoxic T cells can function normally. Both humoral and cell-mediated immunity are compromised. Many individuals with AIDS die from infections and cancers that ordinarily would be readily prevented by a fully functional immune system.

AIDS, first described in 1981, has since reached pandemic proportions although its prevalence has been leveling off for several years. About 36 million people worldwide are currently living with HIV infection, and an estimated 5,000–10,000 new infections occur each day (**Figure 52.16**). The major routes of HIV transmission are (1) unprotected sexual intercourse with an infected partner; (2) transfer of contaminated blood or blood products between individuals, such as the sharing of needles among intravenous drug users, or, less commonly, as a result of a blood transfusion; (3) transfer from an infected mother to her child across the placenta or during delivery; or (4) transfer via breast milk during nursing.

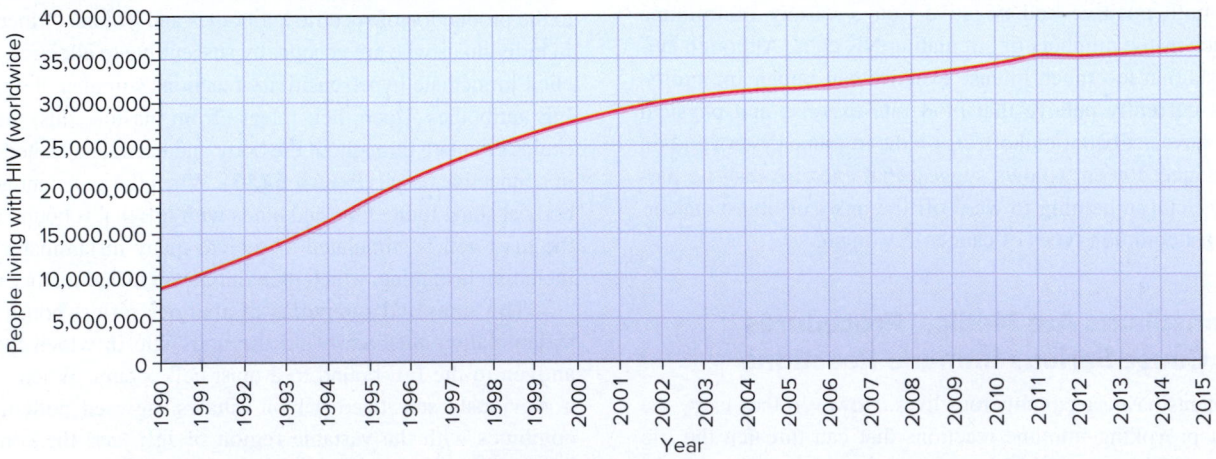

**Figure 52.16** Worldwide incidence of people living with HIV/AIDS. Data are from the *2016 World Aids Day Report* published by UNAIDS (The Joint United Nations Programme on HIV/AIDS). Source: UNAIDS, *2016 World Aids Day Report,* Geneva: UNAIDS, 2016.

 **Core Skills: Science and Society, and Communication and Collaboration** Although the number of people living with HIV has not decreased, research into its causes, prevention, and treatment has greatly slowed the rate at which it is increasing in the human population (compare the slope of the curve in the early 1990s to that in recent years). These advances have arisen from a worldwide effort by scientists and policy makers. Note that the data illustrated in this figure were collected and published not in a scientific journal but by UNAIDS, which is associated with the United Nations.

The great majority of individuals now infected with HIV show no signs of AIDS. Their infections are diagnosed by the presence of anti-HIV antibodies or HIV RNA in the blood. However, if left untreated, HIV infection commonly develops into AIDS in about 10 years. During the first 5 years, killed helper T cells are typically replaced by new cells, so T-cell concentrations remain normal, and the individual remains asymptomatic. Over the next 5 years, T-cell concentrations begin to decline, until at some point, AIDS reveals itself in the form of opportunistic viral, bacterial, and fungal infections. Certain unusual cancers, such as Kaposi sarcoma, also occur with high frequency. In untreated individuals, death usually occurs within 2 years after the onset of AIDS symptoms.

Treatment for HIV-infected individuals has two components: one directed against the virus itself to delay progression of the disease, and one to prevent or treat the opportunistic infections and cancers that ultimately cause death. One current antiviral approach involves administering a combination of four drugs, known as HAART (highly active antiretroviral therapy). Two of the drugs inhibit the action of reverse transcriptase in converting viral RNA into DNA within the host cell, a third drug inhibits an HIV enzyme required for assembling new viruses, and a fourth class of drugs called fusion inhibitors prevent the virus from entering T cells. These treatments have been demonstrated to be effective in slowing the rate at which infection with HIV leads to AIDS. Unfortunately, however, the HAART regimen is associated with numerous side effects, including nausea, vomiting, diarrhea, metabolic disturbances, and liver damage. Much research is under way to find better treatments and ultimately to cure this disease. For example, drugs have been developed to block the integration of viral DNA (formed after viral RNA has been reverse transcribed into DNA) into host DNA or to increase production of new helper T cells. The possibility of using gene therapy to alter the structure of the coreceptor described earlier, thereby preventing the entry of HIV into helper T cells, is currently being investigated. Finally, the use of specific activators of a subtype of cannabinoid receptor has also shown promise as a means of inhibiting HIV replication in infected cells.

## Summary of Key Concepts

- An animal's cells and organs that collectively contribute to its immunity constitute the animal's immune system.
- In innate immunity, the body's defenses are present at birth and act against foreign materials in much the same way regardless of the specific identity of the invading material. Adaptive immunity develops only after the body is exposed to foreign substances and targets those foreign substances specifically.

### 52.1 Types of Pathogens

- Three major types of pathogens elicit an immune response: bacteria, viruses, and eukaryotic parasites.

### 52.2 Innate Immunity

- An important innate defense is carried out by phagocytes—cells capable of phagocytosis. In vertebrates, most phagocytes are a type of blood cell called leukocytes. The leukocytes involved in

innate immunity include neutrophils, eosinophils, monocytes, macrophages, basophils, and natural killer (NK) cells (Figure 52.1).

- Inflammation is an innate local response to infection or injury characterized by local redness, swelling, heat, and pain. The events of inflammation are induced and regulated by chemical mediators called cytokines (Figure 52.2).

- Antimicrobial proteins include interferons, which inhibit viral replication, and complement proteins, which kill microbes without prior phagocytosis. Activation of the complement proteins results in the formation of a membrane attack complex (MAC), which creates pores in the microbial plasma membrane and kills the microbe.

- Toll-like receptors (TLRs) are evolutionarily ancient proteins that recognize common molecular features of many pathogens (Figure 52.3).

### 52.3 Adaptive Immunity

- A foreign molecule that the host does not recognize as self and that triggers an adaptive immune response is an antigen.

- Leukocytes called lymphocytes are responsible for adaptive immune responses. Most lymphocytes reside in a group of organs and tissues that constitute the lymphatic system (Figure 52.4).

- Lymphocytes responsible for adaptive immunity are B cells and T cells. B cells differentiate into antibody-producing cells called plasma cells. T cells include cytotoxic T cells, which directly kill target cells, and helper T cells, which assist in the activation and function of B cells and cytotoxic T cells (Figure 52.5).

- Immunologists recognize two types of adaptive immunity. In humoral immunity, plasma cells secrete antibodies that bind to antigens. In cell-mediated immunity, cytotoxic T cells directly attack and destroy abnormal body cells.

- Adaptive immune responses occur in three stages: The first stage is recognition of an antigen; the second is activation and proliferation of lymphocytes; and the third is attack against the antigen (Figure 52.6).

- In humoral immunity, B cells recognize antigens with B-cell receptors. When B cells are activated, they proliferate and differentiate into plasma cells, which secrete antibodies.

- Both B-cell receptors and antibodies belong to a family of proteins called immunoglobulins. Immunoglobulins contain a constant region, which is identical for all immunoglobulins of a given Ig class, and a variable region that serves as the antigen-binding site (Figure 52.7).

- The random joining of V, D, and J domains via recombination is crucial in enabling plasma cells to produce a diverse array of antibodies capable of recognizing many different antigens. In addition, imprecise end joining and hypermutation contribute to antibody diversity (Figure 52.8).

- B cells that are activated by an antigen differentiate into plasma cells by a process called clonal selection. Antibodies combine with the antigen that activated the B cell and guide an attack that eliminates the antigen or the cells bearing it (Figure 52.9).

- Major histocompatibility complex (MHC) proteins are cellular "identity tags" that serve as genetic markers of self. Class I MHC proteins are found on the surface of all human body cells except erythrocytes. Class II MHC proteins are found only on the surface of macrophages, B cells, and dendritic cells.

- The binding between a helper T-cell receptor and an antigen bound to class II MHC proteins on an APC is essential to helper T-cell activation. Once activated, helper T cells can help to activate both B cells and cytotoxic T cells (Figure 52.10).

- Antigen-presenting cells (APCs) are cells bearing fragments of antigen, called antigenic determinants, or epitopes, complexed with the cell's MHC proteins (Figure 52.11).

- Cell-mediated immune responses are mediated by cytotoxic T cells, which directly kill virus-infected and cancerous cells via secreted chemicals. Humoral immune responses are mediated by B cells and plasma cells. In both types of responses, helper T cells are required (Figures 52.12, 52.13).

- The process by which the body distinguishes between self and nonself components is called immune tolerance. Individuals normally lack active lymphocytes that respond to self components because of two mechanisms. T cells with receptors capable of binding self proteins are destroyed by apoptosis in a process termed clonal deletion. Clonal inactivation causes potentially self-reacting lymphocytes to become nonresponsive. When the body's immune cells attack the body's own cells, the result is an autoimmune disease.

- Upon initial exposure to an antigen, the body produces a primary immune response. Any subsequent exposure to the same antigen elicits an immediate and heightened response termed a secondary immune response. The immune system's ability to produce this secondary response is called immunological memory (Figure 52.14).

- The acquired response to exposure to any type of antigen is known as active immunity. The artificial exposures to antigen that occur in vaccinations and immunizations also induce active immunity. In contrast, passive immunity confers protection against disease through the direct transfer of antibodies from one individual to another.

## 52.4 Impact on Public Health

- Factors that cause malfunction of the immune system include lifestyle; organ transplants; allergies; and acquired immune deficiency syndrome (AIDS), caused by the human immunodeficiency virus (HIV). AIDS reduces the body's immunity by killing helper T cells (Figures 52.15, 52.16).

## Assess & Discuss

### Test Yourself

1. Which of the following is *not* an example of a barrier defense in animals?
   a. skin
   b. secretions from skin glands
   c. exoskeleton
   d. mucus
   e. antibodies

2. The leukocytes that are found in mucosal surfaces and that play a role in defending an animal's body against parasitic infections are
   a. neutrophils.
   b. eosinophils.
   c. basophils.
   d. monocytes.
   e. NK cells.

3. The vascular changes of inflammation
   a. lead to an increase in bacterial cells at the injury site.
   b. decrease the number of leukocytes at the injury site.
   c. allow plasma proteins to move easily from the bloodstream to the injury site.
   d. decrease the number of antibodies at the injury site.
   e. activate lymphocytes.

4. Which statement about adaptive immunity is *true*?
   a. Adaptive immunity only requires the presence of helper T cells to function properly.
   b. Adaptive immunity allows recognition of nonspecific molecular markers on many types of pathogens.
   c. Adaptive immunity is triggered by exposure to a specific antigen.
   d. Adaptive immunity includes inflammation.
   e. All of the above are true.

5. Memory B cells are
   a. cloned lymphocytes that are active in fighting subsequent infections.
   b. cloned lymphocytes that are active during a primary infection.
   c. NK cells that recognize cancer cells and destroy them.
   d. cells that produce antibodies.
   e. macrophages that have recognized self antigens.

6. Which of the following are phagocytes?
   a. mast cells
   b. all lymphocytes
   c. all leukocytes
   d. plasma cells
   e. dendritic cells

7. The antigen-binding site of an antibody is the
   a. constant region.
   b. variable region.
   c. complete heavy chain.
   d. complete light chain.
   e. hinge region.

8. A major difference between the activation of B cells and the activation of cytotoxic T cells is that
   a. cytotoxic T cells must interact with antigens bound to plasma membranes.
   b. B cells do not interact with any other type of lymphocyte.
   c. B cells are suppressed by helper T cells.
   d. cytotoxic T cells produce antibodies.
   e. only cytotoxic T cells express immunoglobulins on their membranes.

9. Cells that process foreign proteins and complex them with their MHC proteins are called
   a. cytotoxic T cells.
   b. plasma cells.
   c. NK cells.
   d. antigen-presenting cells.
   e. helper T cells.

10. HIV causes immune deficiency because the virus
    a. destroys all the cytotoxic T cells.
    b. preferentially destroys helper T cells that regulate the immune system.
    c. directly inactivates plasma cells.
    d. causes mutations that lead to autoimmune diseases.
    e. causes helper T cells to multiply uncontrollably.

## Conceptual Questions

1. Distinguish between innate and adaptive immunity.

2. Living organisms interact with their environments. Such interactions can involve potentially threatening environmental factors such as pathogens. Distinguish three types of pathogens that the immune system of animals protects against.

3. **Core Concept: Structure and Function** Describe the basic structure of an immunoglobulin, and explain how the structure relates to its ability to recognize a specific antigen.

## Collaborative Questions

1. Explain the function of cytotoxic T cells.

2. List the different types of lymphocytes, and briefly describe a key structure or function of each type.

# Integrated Responses of Animal Organ Systems to a Challenge to Homeostasis

# 53

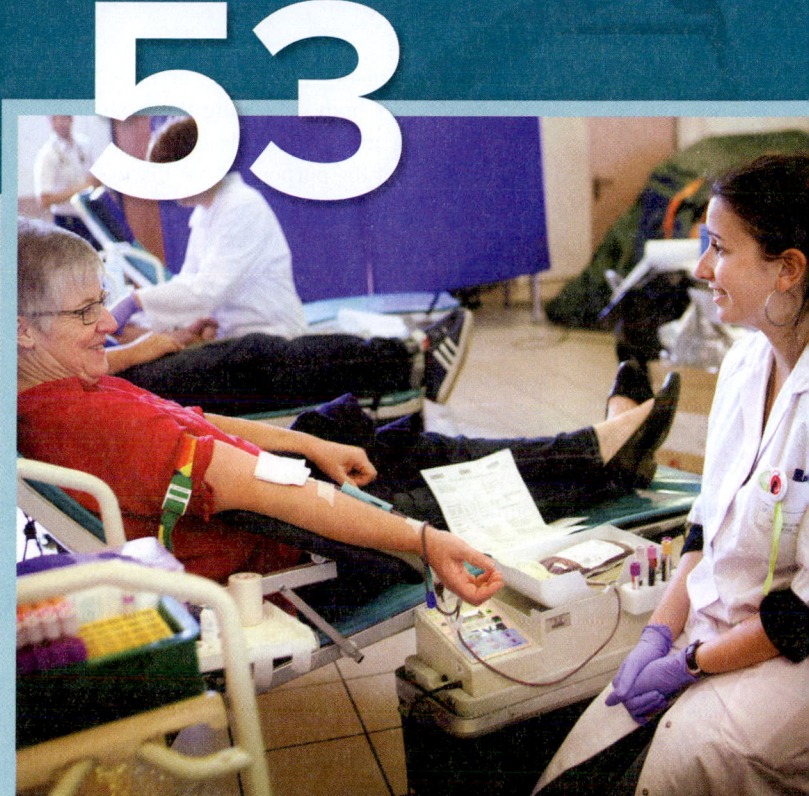

**A person donating blood.** Donations such as these improve the chances for survival of individuals who have lost significant amounts of blood due to injury. Whether blood is donated or lost due to injury, however, homeostatic mechanisms help prevent blood pressure from falling to a dangerous level. ©BSIP SA/Alamy Stock Photo

**S**hortly into a high school soccer match, a 16-year-old girl falls to the ground after tripping over an opponent's foot. A teammate running behind the girl cannot stop herself in time and accidentally kicks the girl with a hard, cleated shoe in the lower back, near the region of the girl's left kidney. The injury is not terribly painful, and the girl is able to get up on her own and walk off the field, where she sits down and rests. By the end of the match, however, she is complaining of feeling "a little woozy." She becomes agitated yet remains alert and responsive. Nonetheless, her concerned coach has her transported by ambulance to a local hospital. A paramedic records her blood pressure as 108/66 mmHg—which, according to the girl's mother, is a little lower than usual. Her heart rate is 101 beats per minute (bpm), which is abnormally rapid. She is breathing at a rate of 20 breaths/minute, which is above normal.

The kick she received has ruptured one or more large blood vessels, causing her to bleed internally. The loss of blood through a ruptured blood vessel is called a **hemorrhage**. Without prompt treatment, a significant hemorrhage runs the risk of leading to a condition known as **shock**, in which the ability of the circulatory system to provide nutrients and oxygen to vital organs decreases to the point that those organs can no longer function properly and their cells begin to die.

A fundamental principle of biology is that living organisms maintain homeostasis. Throughout this unit, you have seen numerous examples of how this principle applies to animal biology. Recall from Chapter 41 that organ systems in animals do not function in isolation. That is, changes in the activity of one organ system often result in changes in other organ systems. Many times, two or more organ systems function together to control an important homeostatic variable such as blood pressure. In this chapter, we will take a detailed look at how a significant challenge to homeostasis results in changes in the activities of all organ systems, which function together to meet the challenge of restoring homeostasis. Our example of a homeostatic challenge will be hemorrhage, which may be internal, as in the girl just described, or external, as when a cut causes an animal to bleed to the outside. Keep in mind, however, that there are many homeostatic challenges that result in the coordinated activation of more than one organ system; hemorrhage is just one well-studied example. The nature or magnitude of the responses may differ for different situations, but the general concept of an integrated whole-body response remains the same.

Hemorrhage can have life-threatening consequences if not compensated for by homeostatic mechanisms. A significant loss of blood decreases an animal's blood pressure, which slows the delivery of nutrients and oxygen to vital organs such as the brain and heart. The result can be catastrophic and even fatal if pressure is not corrected. The compensatory mechanisms for hemorrhage have been studied in all classes of vertebrates but most thoroughly in mammals, and thus we will use mammals as our example in this chapter. We will begin by describing these compensatory mechanisms and then, at the end of the chapter, return to the case of the young soccer player and discover how an understanding of homeostatic control mechanisms helped her recover.

# 53.1 Effects of Hemorrhage on Blood Pressure and Organ Function

**Learning Outcomes:**

1. Explain why a decrease in blood pressure can be dangerous.
2. **CoreSKILL »** Predict changes that may occur to an animal's blood pressure when its blood volume is increased or decreased, and explain why.

In any type of hemorrhage, no matter how small or large, one of the first responses of an animal's body is sealing the wound. This is accomplished by the clotting mechanisms described in Chapter 48 (refer back to Figure 48.5). For the purpose of our discussion, we will assume that an animal suffering a hemorrhage has normal clotting mechanisms and that they are initiated immediately following the hemorrhage and continue for some time until the wound is closed. In this section, we will focus on understanding why hemorrhage is so dangerous and how blood volume and pressure are linked.

## Decreased Blood Pressure May Lead to Cell Death

Blood circulates in an animal's body under pressure, as described in Chapter 48. The source of that pressure is the beating of the heart. Blood flows through arteries when it leaves the heart and pushes against the inner surfaces of the vessels. This force is what we refer to when we measure a person's blood pressure using an arm cuff, for example. In most mammals, blood pressure is roughly similar to that of humans, but different animals face different challenges in getting blood to all parts of the body. Think of the distance blood must travel from the heart to the brain in a giraffe. Giraffes have evolved an exceptionally strong heart that moves blood up their long neck. Even in humans, the distance from the heart to the brain results in a decrease in blood pressure as the blood flows upward against gravity. You may have experienced a sense of light-headedness on occasion when standing suddenly; it takes a second or two to generate the pressure required to overcome the effect of gravity on the blood circulating to your head. This temporary light-headedness illustrates the importance of blood pressure. All cells require uninterrupted delivery of nutrients and oxygen to perform the metabolic and other activities required for optimal functioning and survival. Very active cells, such as those of the brain, are the first to show signs of malfunction if the rate of nutrient and oxygen delivery is decreased.

Because of its vital importance for cell function, let's focus for the moment on oxygen delivery. The rate of oxygen delivery to any tissue in an animal's body depends on at least four factors, which were described in detail in Chapter 48. These include

- the rate of blood flow to a region, which in turn is influenced by the degree to which vessels are dilated (widened) or constricted (narrowed);
- the number of erythrocytes in a given volume of blood;
- the percent saturation of hemoglobin in the erythrocytes; and
- the effectiveness of hemoglobin releasing its oxygen to cells.

Following a hemorrhage, the first of these factors—blood flow—is immediately affected, as blood pressure begins to decrease. Very quickly, unless compensatory events occur, this alone can be sufficient to cause widespread cell death. Adjustments to all four factors, however, can help prevent such cell death from happening.

## Blood Volume and Blood Pressure Are Linked

Refer again to the chapter opening photo of a person donating blood. This procedure removes about 500 mL of blood, which is roughly 10% of the total blood volume of a typical adult human. This controlled procedure has been adapted by investigators in order to understand how an animal's body responds to a hemorrhage. For instance, using a thin, flexible, hollow tube (called a cannula) inserted into a vein of an experimental animal to withdraw carefully controlled amounts of blood, investigators can determine the relationship between the degree of hemorrhage and its effects on different organ systems. Then, investigators can monitor how these organ systems respond in an integrated, coordinated way—first, to prevent blood pressure from continuing to decrease and, second, to help restore blood volume and pressure to normal.

The results of such experiments indicate that mammals can usually cope well with a 10% hemorrhage, as humans can when donating blood. Symptoms of a 10% hemorrhage are relatively mild and, in some people, are barely noticeable. Blood pressure may remain normal or may decrease only slightly. Part of the reason that symptoms are minor can be attributed to the large volume of blood that is present in the large veins, such as those in the legs. Veins are expandable and can accommodate more blood than can arteries. This blood can serve as a reservoir for times when an animal becomes more active and requires greater cardiac output (refer back to Figures 48.10, 48.14, and 48.15). After a hemorrhage, this reservoir can also be tapped to help maintain heart function and blood pressure. However, larger hemorrhages—for example, 20% and greater of an animal's blood volume—do result in a significant decrease in blood pressure. Remarkably, though, animals may survive even such a large hemorrhage (**Figure 53.1**).

**Figure 53.1**  **An adult zebra that survived an attack by a lioness.** These deep wounds have begun healing. Note that there is no bleeding at this time. Nonetheless, these injuries would initially have caused considerable hemorrhage. The animal's homeostatic responses to the loss of blood may have helped save its life. ©Images of Africa Photobank/Alamy Stock Photo

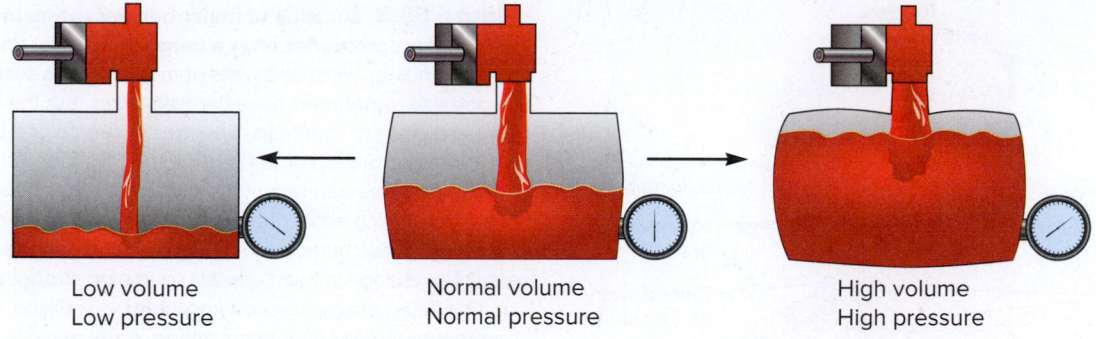

Low volume
Low pressure

Normal volume
Normal pressure

High volume
High pressure

**Figure 53.2** **The relationship between blood volume and blood pressure.** In a system such as the artificial one shown here, an increase in fluid volume will increase the pressure that the fluid exerts on the walls of the chamber. In a similar way, increasing blood volume will increase the pressure of the blood in vessels such as arteries. A hemorrhage will decrease volume and thus decrease pressure.

What is the link between blood volume and blood pressure? Mammals have a closed circulatory system (refer back to Figures 48.1b and 48.2b). In such a system, volume and pressure are closely related. Look at the artificial closed system shown in **Figure 53.2**. When volume is normal, so is pressure. However, a decrease in volume decreases pressure, whereas an increase in volume causes the pressure to rise.

Recall that the blood pressure of a mammal that has lost about 10% of its blood volume does not change much. What is helping to maintain blood pressure? How is the venous reservoir of blood returned to the heart, and what other changes occur to improve circulatory function? As we will see next, numerous organ systems function together to help prevent a dangerous drop in blood pressure under these circumstances. These responses occur in two phases, one within seconds to minutes and one requiring hours to days or even weeks.

## 53.2 The Rapid Phase of the Homeostatic Response to Hemorrhage

### Learning Outcomes:

1. Define baroreceptors, and describe where they are located in mammals.
2. **CoreSKILL »** Predict what will happen to the firing rate of baroreceptors if blood volume is increased or decreased.
3. Describe the baroreceptor reflex in mammals, its dependence on the nervous system, and the changes it induces in the function of the circulatory system.
4. Compare the distribution of blood to different organs before and after hemorrhage.
5. Explain the mechanisms that cause redistribution of blood between organs.
6. Describe how the structures of the respiratory system contribute to restoring pressure following hemorrhage.

Compensatory responses to hemorrhage occur across a timeline, and consist of both very rapid responses and more delayed responses. Although many of the responses overlap in time, those that begin within seconds can be distinguished from those that occur after hours, days or even weeks. The initial, nearly instantaneous homeostatic

response to a decrease in blood volume and pressure is mediated by the nervous system. When a hemorrhage occurs, the nervous system must receive information that blood pressure and volume are decreasing, then respond appropriately. Refer back to Figure 41.9, in which a typical homeostatic control system is described. Two key elements of such a system are a sensor, which detects a variable, and an integrator, which receives input from the sensor and compares that input to a set point (which in our example is the normal value for blood pressure). For monitoring blood volume and pressure, the sensors are located in blood vessels and the heart. The integrator is located in the brainstem. In this section, we will examine how these control elements help minimize a fall in blood pressure after a hemorrhage has occurred.

### Baroreceptors Immediately Sense Changes in Blood Pressure and Initiate a Compensatory Reflex

Special pressure-sensitive regions exist within the heart and certain blood vessels in all vertebrates. For simplicity, we will focus on those that exist in the walls of certain large arteries in mammals. These include the carotid arteries in the neck, which supply blood to the brain, and the arch of the aorta, the first artery that emerges from the left ventricle. These regions contain the endings of neurons, known as **baroreceptors** (from the Greek *baros,* meaning weight or pressure), which are in constant communication with the medulla oblongata of the brainstem (**Figure 53.3a**). They mediate an important mechanism by which blood pressure is regulated in vertebrates called the **baroreceptor reflex** (**Figure 53.3b**). This response to a change in blood pressure is termed a reflex, because, like all reflexes, it is involuntary and rapid.

Let's first examine what happens if blood pressure decreases below normal, as in hemorrhage. Normally, each time the heart contracts and ejects blood, the carotid arteries and aorta are stretched, and consequently, so are the baroreceptors in those vessels. This stretching opens ion channels in the baroreceptors, depolarizing them and initiating action potentials that are sent to the brain. If the blood pressure decreases below the normal range, the arteries are less stretched, and ion channels are opened less frequently. Consequently, the baroreceptors send action potentials at a decreased frequency compared to normal. In the terminology used in Chapter 41, baroreceptors are the sensors for blood pressure, and the brainstem is the integrator. The set point for blood pressure corresponds to a baseline frequency of action

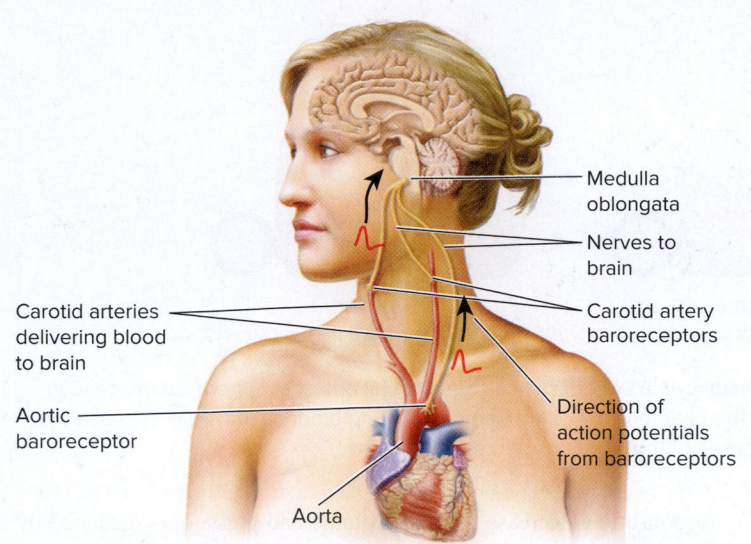

**(a) Location of baroreceptors in a human**

Carotid arteries delivering blood to brain

Aortic baroreceptor

Medulla oblongata

Nerves to brain

Carotid artery baroreceptors

Direction of action potentials from baroreceptors

Aorta

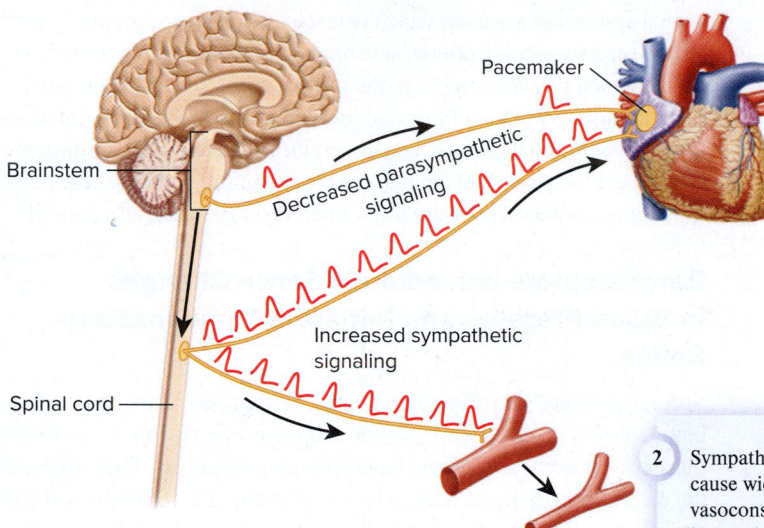

**(b) The baroreceptor reflex after hemorrhage**

Pacemaker

Brainstem

Decreased parasympathetic signaling

Increased sympathetic signaling

Spinal cord

**1** Sympathetic signals increase heart rate after hemorrhage. A decrease in parasympathetic signals also increases heart rate.

**2** Sympathetic signals cause widespread vasoconstriction after hemorrhage.

**Figure 53.3  Location of major baroreceptors in a human, and the baroreceptor reflex after a hemorrhage.** **(a)** Baroreceptors are neuron endings within the walls of major arteries such as the aorta and carotids. All vertebrates have baroreceptors, but the locations shown here are those of mammals. Changes in stretch of a blood vessel that contains a baroreceptor will change the frequency of action potentials generated in the baroreceptor. These signals are relayed to the brainstem, which elicits outgoing signals via the sympathetic and parasympathetic branches of the autonomic nervous system. The result is a change in heart rate and pumping strength as well as a change in peripheral vasoconstriction or vasodilation. **(b)** The baroreceptor reflex after hemorrhage. A decrease in pressure leads to decreased signals from the baroreceptors to the brainstem. In turn, this leads to increased activity of nerves of the sympathetic nervous system exiting the spinal cord to the heart and blood vessels. There is also a decrease in signals to the heart from nerves of the parasympathetic nervous system that exit from the brainstem. Note: Details of autonomic nervous system anatomy are omitted for simplicity.

**Core Skill: Connections** In what other organs are stretch-sensitive receptors found? Refer back to Figure 42.4 and Section 44.2 for help, and think about structures of the digestive and urinary systems (Chapters 46 and 49).

potentials arriving in the brainstem when blood pressure is normal. A lower frequency of action potentials than the set point is interpreted by the medulla oblongata as a lower blood pressure than normal.

The effectors in this control system are nerves exiting the brain and spinal cord via the autonomic nervous system (refer back to Figure 43.8). The response to decreased activation of baroreceptors is an increase in the amount of norepinephrine released from neurons of the sympathetic nervous system onto cardiac sinoatrial node (pacemaker) cells, cardiac ventricle muscle cells, and vascular smooth muscle cells (see Figure 53.3b). In addition, there is a decrease in the release of acetylcholine from neurons of the parasympathetic branch of the autonomic nervous system onto the pacemaker cells. Pacemaker cells control the frequency of heart contractions; norepinephrine stimulates their activity and acetylcholine slows them down. Together, these changes result in an increased heart rate (HR). Norepinephrine also increases the force of contraction of the ventricles.

The latter effect increases the stroke volume (SV) of the heart, or the amount of blood ejected with each beat. Recall from Chapter 48 that the cardiac output (CO), or the volume of blood pumped from the heart per minute, is the product of HR and SV.

$$\underset{\text{(mL/min)}}{\text{CO}} = \underset{\text{(beats/min)}}{\text{HR}} \times \underset{\text{(mL/beat)}}{\text{SV}}$$

Therefore, the baroreceptor reflex in this example results in an increased CO. As described in Chapter 48, the relationship between CO and blood pressure (BP) is:

$$\text{BP} = \text{CO} \times \text{TPR}$$

where TPR is the total peripheral resistance determined by the degree of vasoconstriction of blood vessels, notably the arterioles. Because CO is directly related to blood pressure, an increase in CO increases pressure. In addition, though, stimulation of the baroreceptors also results in vasoconstriction in many parts of the body, again due to increased norepinephrine release from neurons of the sympathetic nervous system. The smaller the radius of a blood vessel, the more resistance it imparts to the movement of blood (refer back to Figure 48.15). Consequently, the reflexive increase in CO and vasoconstriction (increased TPR) together help to compensate for a decrease in blood pressure.

By contrast, if blood pressure increases above normal, the walls of the aorta and carotid arteries are stretched more than normal. In such a situation, the baroreceptors send a greater than normal rate of action potentials to the brainstem, which interprets this as higher than normal blood pressure. The result is a sequence of events that are opposite to those just described. In this case, the sympathetic neurons are inhibited and the parasympathetic neurons are stimulated. Thus, less norepinephrine and more acetylcholine are secreted, thereby decreasing cardiac output and causing vasodilation (decreased TPR). Together, these events decrease blood pressure toward normal.

When a mild hemorrhage occurs, the baroreceptor response is usually sufficient to restore a normal blood pressure. However, when a more serious hemorrhage happens, the baroreceptor reflex can prevent pressure from continuing to fall but cannot return it all the way back to normal. The baroreceptor reflex occurs within seconds and can be considered the first and most critical line of defense against hemorrhage-induced low blood pressure.

 **Core Concept: Evolution**

## Baroreceptors May Have Evolved to Minimize Increases in Blood Pressure in Vertebrates

A baroreceptor reflex appears to be present in some form in all vertebrates. However, the location and function of the baroreceptors have changed during evolution. Although this chapter primarily describes the baroreceptor reflex that follows a decrease in blood pressure in mammals, the selection pressure for the evolution of this reflex was probably related to the dangers of acute or chronic elevations in blood pressure. Factors that can increase blood pressure above normal in different vertebrates include such things as acute stress, immune cell secretions, mineral imbalances, and body temperature changes.

To understand why baroreceptors may have evolved to protect against elevations in blood pressure, let's consider the different circulatory challenges imposed on aquatic and terrestrial vertebrates. In water, fishes are essentially weightless due to water's buoyancy, which equalizes pressure around all parts of their bodies. There is no significant gravitational effect on the circulation of blood in a typical fish's body. In addition, its brain and heart are roughly horizontal with respect to each other. When animals evolved the ability to live on land, however, the effect of gravity became much more important. This is particularly apparent when considering animals such as horses, giraffes, deer, humans or any mammal with its head and brain located above its heart.

Without the buoyant effect of water, and particularly in a vertically oriented body, blood tends to pool in the lower part of the body. This makes it harder to return blood to the heart, and to

pump it from there to the brain. From this discussion, you might conclude that a terrestrial animal such as a human would require greater blood pressure than that of a fish in order to counter the effects of gravity, and this is true.

Another key factor is the complexity of the circulatory system. Recall from Figure 48.2 that fishes have a single circulation. Crocodiles, birds, and mammals have double circulations, and amphibians and most reptiles have a circulation with features of both types. In the single circulation of fishes, blood leaves the heart under pressure and travels through an artery directly to the gills. After exchanging gases with the environment, the blood leaves the gills and travels to the rest of the body, dissipating some pressure along the journey. Consequently, even though a fish is not significantly affected by gravitational effects, its single circulation means that the location with highest blood pressure is its respiratory organ. As was described in Chapter 48, all respiratory organs such as gills and lungs are delicate structures that are particularly vulnerable to damage when exposed to high blood pressure. Unlike the major baroreceptors in mammals that are found in the arteries that lead to the brain, the baroreceptors in fish are located in the arteries that supply the gills! Similarly, amphibian baroreceptors are located in an artery that delivers blood to its lungs and skin (both of which serve as respiratory organs). Most reptiles, too, have their major baroreceptor in vessels leading to the lungs. All of these animals, because they do not have a complete double circulation, must generate higher pressure in the vessels that lead to the respiratory organs. The evolution of baroreceptors that could respond to an increase in blood pressure by triggering a reflex that lowers the pressure to normal provided a clear selection advantage by protecting the respiratory organs. The primary response in such animals is a rapid decrease in heart rate mediated by the neurotransmitter acetylcholine.

Why, then, are some of the baroreceptors of mammals, as shown in Figure 53.3, located in the carotid arteries? Animals with double circulations protect their lungs by having two different pressures in the circulation. The weaker, right side of the heart pumps blood to the lungs under very low pressure, and the stronger, left side pumps blood to the rest of the body (including the brain) under much higher pressure that is sufficient to counter the effects of gravity. A double circulation, therefore, would seem to eliminate the problem of dangerously high blood pressure in respiratory organs. By contrast, however, even with a double circulation, blood pressure may still decrease suddenly when an animal changes posture, for example, or as described in this chapter has suffered a hemorrhage. Unfortunately, the brain is the organ farthest from the heart that must contend with gravity! Consequently, mammals in particular evolved baroreceptors in their carotid arteries so that any decrease in pressure of blood traveling to the brain could be quickly corrected for by an increase in heart rate and other variables. The baroreceptor response shifted over the course of evolution from primarily an acetylcholine-mediated decrease in heart rate in fishes and other vertebrates, which protects the respiratory organs from high pressure, to a bidirectional effect on the mammalian heart mediated by acetylcholine and norepinephrine, which protects the brain from low pressure.

## Core Skill: Process of Science

# Feature Investigation | Cowley and Colleagues Determined the Function of Baroreceptors in the Control of Blood Pressure in Mammals

From the discussion thus far about hemorrhage, you might imagine that baroreceptors evolved in vertebrates to compensate for large, sudden swings in blood pressure that could otherwise be life-threatening. Researchers considered, however, that such circumstances are not typically encountered in the daily lives of animals. Therefore, an alternative hypothesis is that the evolution of baroreceptors provides a selective advantage to animals in controlling their minute-to-minute blood pressure. American researchers Allen Cowley, Jr., Jean Liard, and Arthur Guyton tested this hypothesis by recording blood pressure in dogs who either had their arterial baroreceptors denervated (cut) or left intact (**Figure 53.4**).

One group of anesthetized animals had their carotid and aortic baroreceptors surgically denervated. After the denervation procedure, the nerves do not grow back. Therefore, the denervated baroreceptors could not transmit information about blood pressure to the brainstem, where the control centers for the baroreceptor reflex are located. Control animals experienced the same surgical procedures but the nerves were not cut.

After the surgery, the dogs were trained to rest peacefully in a comfortable sling in a quiet testing room for up to a few hours. The dogs were also trained to lie down or stand up on command. Blood pressure measurements were continuously recorded on a machine from

**Figure 53.4** Cowley and coworkers determined the function of arterial baroreceptors in the control of blood pressure in dogs.

**HYPOTHESIS** Arterial baroreceptors function to minimize minute-to-minute fluctuations in blood pressure in a mammal.

**KEY MATERIALS** The researchers studied adult dogs, some of which had their baroreceptors surgically denervated (cut). Key materials included blood pressure recorders, surgical implements, and a quiet testing room for the dogs.

| Experimental level | Conceptual level |
|---|---|

1. Surgically denervate arterial baroreceptors in anesthetized dogs. Control dogs subjected to the same procedures but baroreceptors are left intact.

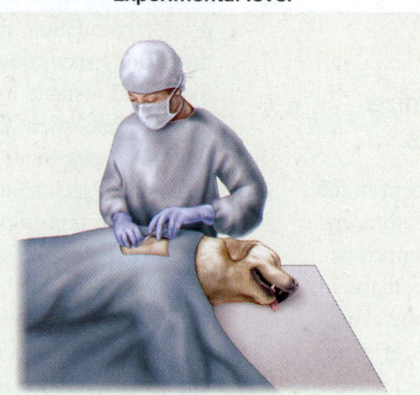

Signals from baroreceptors (shown here for aorta) cannot reach brainstem.

Nerve
Brainstem
Aorta
Aortic baroreceptors
Carotid arteries

2. Allow dogs to acclimate to the quiet testing room for several days. Record baseline blood pressure using a catheter (a hollow plastic tube) inserted in a leg artery. Continue recordings for several hours. In some experiments, record blood pressure in dogs that are either standing or lying down.

Dogs rest comfortably in a sling that permits slight movements, lying down, or standing up, but prevents turning around or jumping. This protects the dogs from injury and minimizes the likelihood of damaging the catheter or pressure recording instruments.

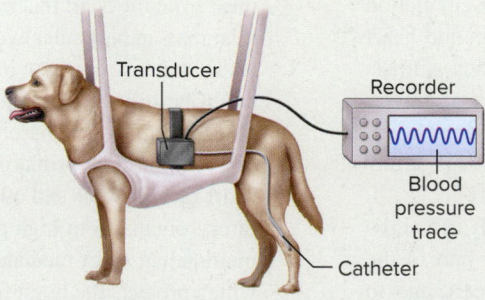

Transducer
Recorder
Blood pressure trace
Catheter

Quiet room minimizes stress to the dog which enables a more accurate recording of resting, baseline blood pressure. A blood pressure transducer is a device that contains a flexible membrane that moves with each pulse of pressure coming from the arterial catheter. This movement activates an electric current that leads to a recording device.

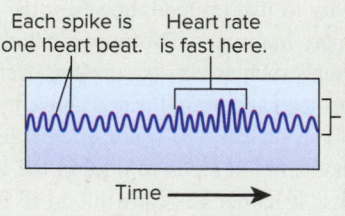

Each spike is one heart beat. Heart rate is fast here.

The height of each spike is proportional to blood pressure. Higher spikes indicate higher blood pressure.

Time →

**3** **THE DATA**

Portions of representative records from 2 dogs:

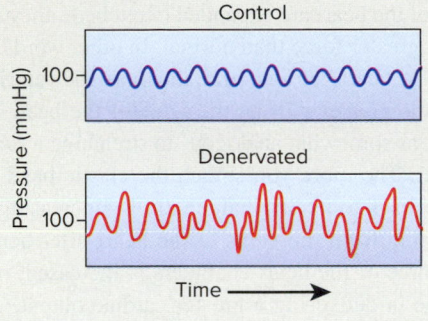

Summary of records from 12–15 dogs over 4 hours:

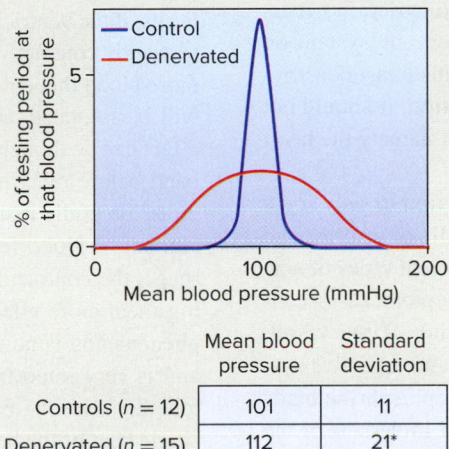

| | Mean blood pressure | Standard deviation |
|---|---|---|
| Controls (n = 12) | 101 | 11 |
| Denervated (n = 15) | 112 | 21* |

*Significantly greater than control standard deviation (in other words, denervated dogs have significantly more variable baseline blood pressure)

Effects of a postural change from lying prone to standing up:

| | Control | Denervated |
|---|---|---|
| Decrease in blood pressure upon standing (mmHg) | 15 ± 1 | 30* ± 1 |
| Mean recovery time to baseline blood pressure (seconds) | 21 | 81* |
| Increase in heart rate upon standing (beats per min) | 34 ± 4 | 24* ± 1 |

*Significantly different from control value

**4** **CONCLUSION** Baroreceptors function in the minute-to-minute control of blood pressure in dogs. Without functioning baroreceptors, blood pressure fluctuates significantly even at rest. Moreover, a simple postural change that normally causes a slight and transient decrease in blood pressure, is compensated for less well without baroreceptors. Upon standing, blood pressure decreases to a greater exent in the absence of baroreceptor input, takes nearly four times longer to recover, and elicits a smaller increase in heart rate.

**5** **SOURCE** Cowley, A. W., Liard, J. F., and Guyton, H. C. 1973. Role of the baroreceptor reflex in daily control of arterial blood pressure and other variables in dogs. *Circulation Research* 32: 564–575.

a catheter in a leg artery that was connected to a device called a pressure transducer.

As shown in step 3 in Figure 53.4, the blood pressure of the control dogs was stable over a given time period, but that of the denervated dogs varied considerably. The mean pressure of the two groups was slightly different, but the most obvious difference was in the variability during several hours of recording. In addition, the resting heart rates of the two groups showed a similar pattern. A significantly greater variability was observed in the resting heart rates of denervated dogs compared to control animals (not shown in step 3).

When a mammal such as a dog or human changes posture from the supine to a standing position, a transient decrease in blood pressure typically occurs due to gravitational effects. In the table shown in step 3, note that the posture-induced decrease in blood pressure of denervated dogs was twice that of control dogs. Moreover, the denervated animals required significantly longer to recover to normal blood pressure, and they had a smaller compensatory increase in heart rate than did control animals. In other experiments not shown here, even the mild excitement of seeing the handler enter the room caused pronounced fluctuations in blood pressure in the denervated dogs but not in the control dogs.

From these results, the researchers concluded that the major minute-to-minute function of the arterial baroreceptors is to chronically maintain an animal's blood pressure within a narrow range, in part by quickly compensating for the effects of posture, excitement, and many other physical or psychological factors. These results agree well with current thinking about the evolution of baroreceptors in vertebrates, as discussed earlier.

---

*Experimental Questions*

1. **CoreSKILL** ≫ Note from the data in step 3 that, although the carotid and aortic baroreceptors were denervated, the blood pressure of dogs did eventually return to normal after standing up. Propose a hypothesis to explain this result. Look back at Figure 48.13 for help and recall how skeletal muscle contraction contributes to blood circulation.

2. What was the purpose of testing the animals in a quiet, isolated environment?

3. **CoreSKILL** ≫ How could the researchers have used a similar experimental procedure to test the hypothesis that the most important contribution to the baroreceptor reflex arose specifically from the carotid artery baroreceptors?

## The Nervous System Stimulates Redistribution of Blood to Vital Organs When Pressure Is Low

As you have just learned, part of the baroreceptor reflex following a decrease in blood pressure involves vasoconstriction in blood vessels in different parts of an animal's body. This vasoconstriction occurs in part through the actions of the sympathetic nervous system on smooth muscle cells of arterioles. However, because vasoconstriction limits blood flow past the site of the constriction, it should not occur in parts of the body that are vital for survival, namely the heart and brain.

How does an animal's body limit vasoconstriction to only some regions? The answer is that the smooth muscle cells of arteries and arterioles in different parts of the body express different types of norepinephrine receptors. Binding to one type of receptor causes contraction of the muscles, resulting in vasoconstriction of the vessel. This binding occurs in the skin, skeletal muscles, and digestive system, among other locations. By contrast, when norepinephrine binds to the other type of norepinephrine receptor, found in vessels of the heart and brain, it helps keep those vessels dilated to ensure sufficient blood flow to those regions (**Figure 53.5**). This redistribution of blood away from organs with less immediately vital functions to those with more important functions is an effective and extremely rapid means of helping to compensate for a decreased total blood volume.

In addition to their actions on the arterioles, the neurons of the sympathetic nervous system also have endings on the smooth muscles of many of the large veins, such as those in the legs. When stimulated by norepinephrine, these smooth muscles contract, and this helps squeeze blood up through the veins toward the heart. Recall from Chapter 48 that venous blood pressure is very low compared to arterial pressure. Compressing the veins in this way helps provide an important flow of additional blood from the venous reservoir.

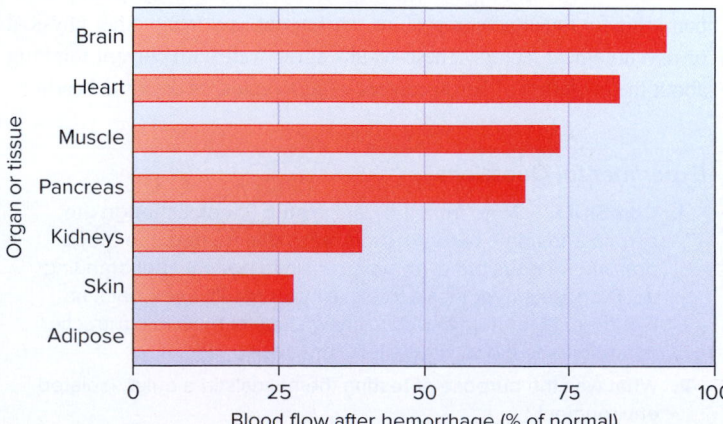

**Figure 53.5 Changes in the distribution of total blood volume after hemorrhage.** Redistributing blood away from areas that are not immediately important ensures that the heart and brain still receive sufficient blood despite hemorrhage. The idealized values shown here are based on research derived from numerous mammalian species.

There is an additional benefit from the increased venous blood returning to the heart. In the early 20th century, German physiologist Otto Frank and British physiologist Ernest Starling, working independently to investigate the mechanisms by which heart muscle cells contract, recognized an intriguing property of that type of muscle. Within limits, if the ventricles of the heart are expanded (stretched), they subsequently contract with greater force than normal. In other words, the more blood that enters the heart before a beat, the more the ventricles will be stretched, and the greater will be the force of the beat. This effect can be thought of as somewhat analogous to stretching a rubber band before releasing it. The more you stretch the elastic band, the more forcefully it snaps back to its original length. By increasing the amount of blood returning from the veins to the heart after hemorrhage, the contractile force of the heart chambers is increased, making them more effective in delivering a greater cardiac output. This phenomenon is now known as the **Frank-Starling law of the heart** and is very important in the early response to blood loss. It occurs together with the baroreceptor reflex and can be observed within a single beat of the heart.

## The Respiratory System Aids in Circulating Blood and Delivering Oxygen

In addition to the nervous system, the respiratory system also participates in the rapid response to hemorrhage. When oxygen delivery to any region of an animal's body is decreased, the cells in that region depend more on fermentation to produce the ATP required for survival. This increase in fermentation occurs within seconds to minutes of insufficient blood flow and oxygen delivery. A by-product of fermentation is lactic acid, which is released into the blood and decreases its pH.

Recall from Chapter 48 that chemoreceptors are present in certain blood vessels of the circulatory system of animals. In mammals, the chemoreceptors are associated with the same vessels that contain the baroreceptors. They are sensitive to changes in the amount of oxygen, carbon dioxide, and acidity of the blood (**Figure 53.6**). After a hemorrhage, the buildup of lactic acid (decreased pH) stimulates the chemoreceptors, which then activate associated neurons that send signals to the brainstem centers that control breathing (refer back to Figure 48.26). Breaths become deeper and more rapid, thereby maximizing the saturation of hemoglobin with oxygen. In addition, the mechanical action of the chest and diaphragm creates a siphon-like effect that draws blood up from veins below the diaphragm, allowing greater return of venous blood to the heart. This **thoraco-abdominal pump**—or respiratory pump, as it is sometimes called—is another important way in which the Frank-Starling effect is generated (**Figure 53.7**). In this way, the muscular-skeletal system makes a contribution to the restoration of blood pressure.

The acidic pH of the blood has another benefit, one that is not related to control of blood pressure or volume. As you learned in Chapter 48 (refer back to Figure 48.25), a decreased pH results in a decreased affinity of hemoglobin for oxygen (recall that this is called the Bohr effect). The lactic acid generated by poorly oxygenated tissues causes hemoglobin to unload a greater amount of oxygen, which is an important benefit that helps offset the decreased oxygen delivery to the tissues resulting from decreased blood flow.

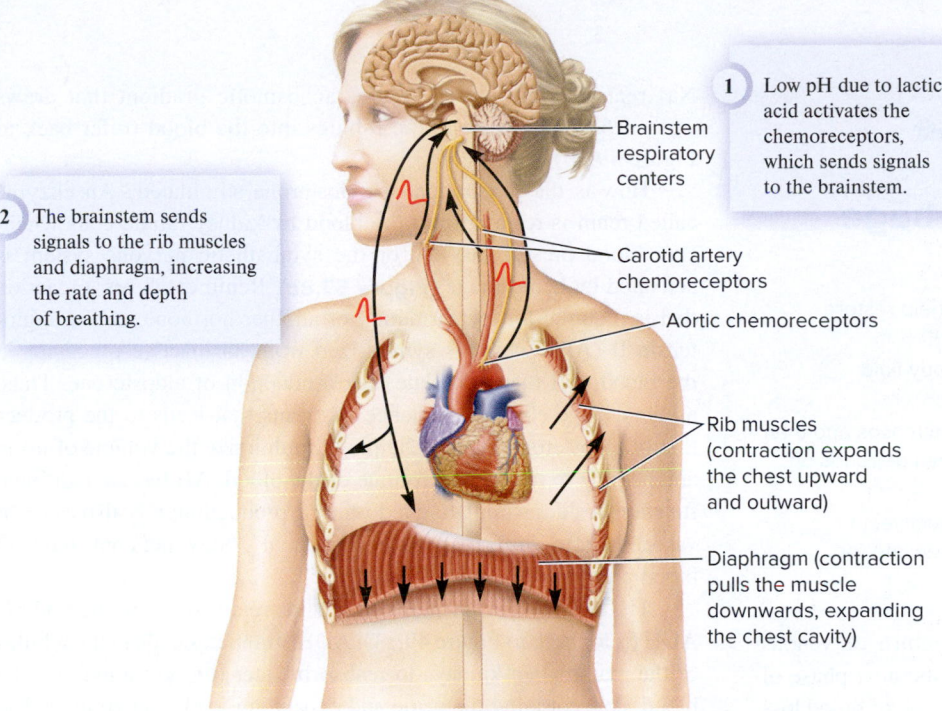

② The brainstem sends signals to the rib muscles and diaphragm, increasing the rate and depth of breathing.

① Low pH due to lactic acid activates the chemoreceptors, which sends signals to the brainstem.

Brainstem respiratory centers

Carotid artery chemoreceptors

Aortic chemoreceptors

Rib muscles (contraction expands the chest upward and outward)

Diaphragm (contraction pulls the muscle downwards, expanding the chest cavity)

**Figure 53.6** **Location of peripheral chemoreceptors and reflex activation of breathing in a human.** Nearby the baroreceptors are specialized cells sensitive to the amount of oxygen, carbon dioxide, and acid (pH) in the blood (not shown are additional chemoreceptors located within the brain). Activation of neurons associated with these cells results in signals that travel to the brainstem respiratory centers. Effector nerves leave the central nervous system and travel to the muscles of respiration, including the diaphragm and rib (intercostal) muscles, increasing the rate and depth of breathing.

*Concept Check:* *What is the significance of the location of chemoreceptors and baroreceptors (see Figure 53.3)?*

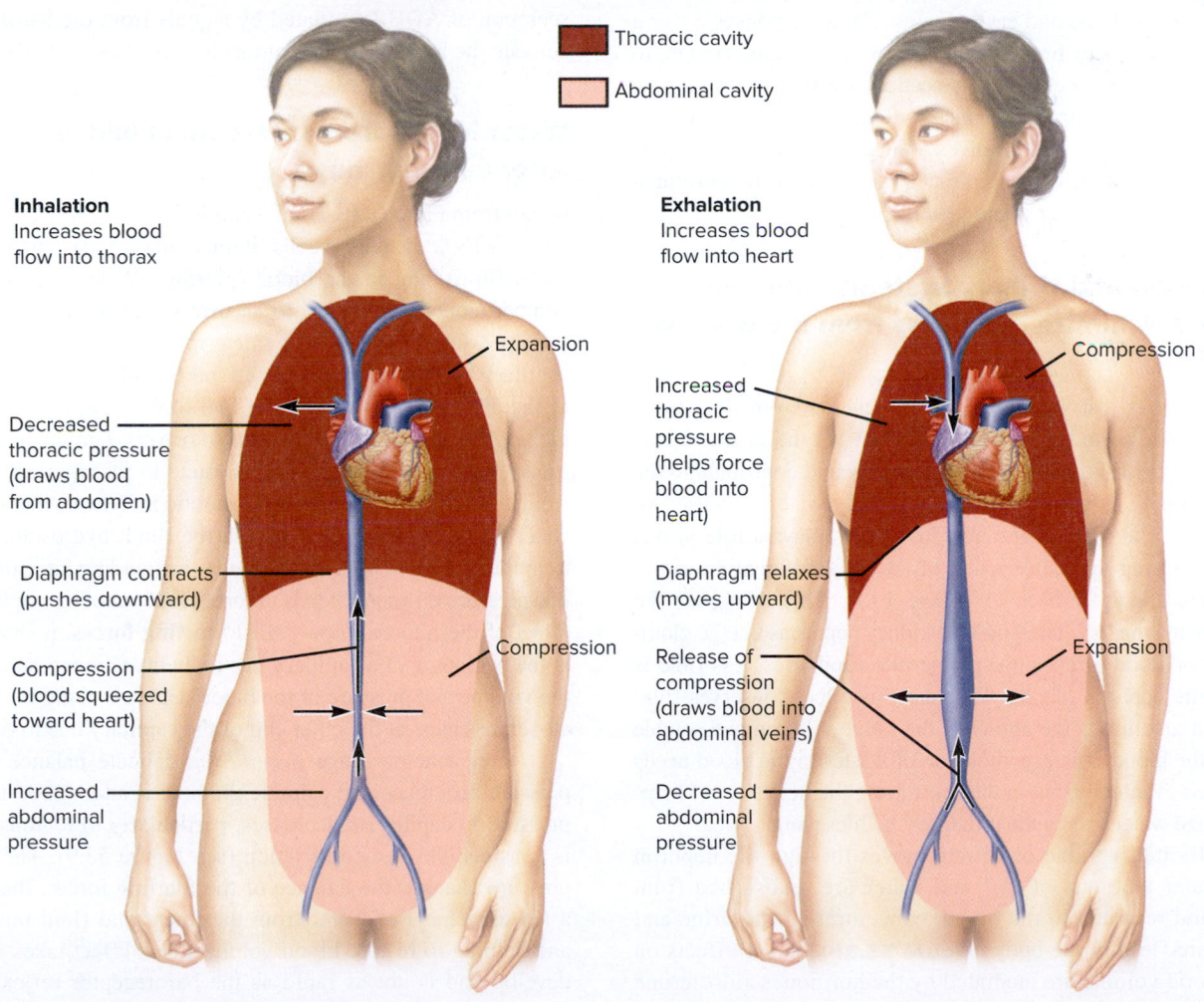

■ Thoracic cavity

□ Abdominal cavity

**Inhalation**
Increases blood flow into thorax

Expansion

Decreased thoracic pressure (draws blood from abdomen)

Diaphragm contracts (pushes downward)

Compression (blood squeezed toward heart)

Compression

Increased abdominal pressure

**Exhalation**
Increases blood flow into heart

Compression

Increased thoracic pressure (helps force blood into heart)

Diaphragm relaxes (moves upward)

Release of compression (draws blood into abdominal veins)

Expansion

Decreased abdominal pressure

**Figure 53.7** **The thoraco-abdominal pump mechanism.** Blood is drawn toward the heart from the abdomen, through the diaphragm, and into the thoracic cavity by the actions of skeletal muscles. When an individual is inhaling, the thorax expands, decreasing the pressure within it. In addition, as the diaphragm pushes down on the abdominal contents, the pressure in the abdomen slightly increases. Because veins are readily collapsed or expanded, these pressure differences tend to draw blood from the abdomen into the thorax. Fast, deep breathing, as occurs when chemoreceptors are activated after a hemorrhage, facilitates venous return to the heart by this mechanism. The increased venous return contributes to the Frank-Starling effect, which enables the heart to pump more forcefully.

# 53.3 The Secondary Phase of the Homeostatic Response to Hemorrhage

## Learning Outcomes:

1. Explain how the endocrine and urinary systems help restore blood volume.

2. Describe how and why water moves between body fluid compartments after a hemorrhage.

3. Describe the mechanism by which hematocrit decreases and then returns to normal, and explain how this mechanism depends on the endocrine and muscular-skeletal systems.

4. List some changes in the functions of the reproductive, integumentary, and digestive systems that compensate for hemorrhage.

After the baroreceptor reflex and the increased return of venous blood contribute to stabilizing the blood pressure, the next phase of the compensatory response to the homeostatic challenge of blood loss is to restore fluid volume and erythrocytes. These responses cover a time period of anywhere from an hour or so after a hemorrhage, to several days and even weeks. In this section, we will examine how this process largely requires the concerted actions of the endocrine and urinary systems but also involves a special feature, the way in which fluid compartments in an animal's body can be redistributed without any input from other organ systems.

## The Endocrine and Urinary Systems Function Together to Minimize Fluid Loss from the Body via the Urine

An animal that has lost significant blood benefits from decreasing the amount of body fluid that is normally lost via the urine. Rather than being lost, this fluid can remain in the body and contribute to the total blood volume. In such circumstances, body water is conserved in several ways. The sympathetic nervous system plays a role in this response, by causing vasoconstriction of the renal afferent arterioles (**Figure 53.8a** and refer back to Figure 49.7). This decreases the amount of plasma that is filtered into the kidney nephrons via the glomeruli. With less blood entering the glomerular capillaries, less fluid is filtered into the nephron tubules. Whereas this effect conserves fluid in the body, it also limits the ability of the kidneys to remove soluble wastes from the blood. The accumulation of wastes in the blood needs to be corrected eventually, but in the short term, the net effect of helping to minimize water loss in the urine may be lifesaving.

As the limited volume of filtrate moves through the nephron tubules, greater amounts of $Na^+$ and water are reabsorbed from the filtrate and returned to the blood, concentrating the urine and retaining more fluid in the body (**Figure 53.8b**). These effects on urine and blood volume are mediated by the hormones aldosterone and antidiuretic hormone (ADH) produced by the endocrine system. The effects are first seen within an hour of the hemorrhage and may continue for many hours. Aldosterone, a steroid hormone produced by the adrenal glands, is secreted whenever blood volume is decreased. It acts on distal regions of nephrons to increase

$Na^+$ reabsorption, which creates an osmotic gradient that draws water from the filtrate in the tubules into the blood (refer back to Figure 49.10).

How is the production of aldosterone stimulated? An enzyme called renin is released into the blood by kidney tubule cells whenever blood pressure is low or the sympathetic nervous system is activated by hemorrhage (**Figure 53.8c**). Renin catalyzes a reaction that is required for the formation of another hormone, called angiotensin II (AII), which is synthesized from an inactive precursor in the blood. AII then stimulates the production of aldosterone. Thus, a hemorrhage elicits a sequence of events that leads to the production of aldosterone, thereby helping to minimize the volume of urine and helping to restore the volume of the blood. AII has an additional function besides stimulating aldosterone production; it is also a potent vasoconstrictor in many parts of an animal's body and contributes to blood pressure in that way, too.

The other major hormone that helps retain water in the body is ADH (refer back to Figure 49.11). ADH stimulates cells in the tubules of the mammalian kidneys to reabsorb water (but not ions) into the blood, concentrating the urine and expanding the blood volume. The secretion of ADH is initiated by signals from the baroreceptors that activate the hypothalamic neurons that synthesize ADH.

## Water Moves from Interstitial Fluid to the Plasma After a Hemorrhage

Recall from Figure 41.14 that water is found in three compartments of an animal's body: within cells (intracellular fluid), outside cells (interstitial fluid), and in the blood (plasma). Water will move between compartments in response to an osmotic gradient or a difference in fluid pressure. Ernest Starling determined that water moves across a capillary between the plasma and interstitial fluid compartments due to the difference between these two forces (**Figure 53.9**). In mammals, plasma and interstitial fluid are very similar in composition, except that much more protein is found in plasma than in interstitial fluid. This difference creates an osmotic imbalance that tends to draw water into a capillary. By contrast, the fluid (hydrostatic) pressure in a capillary is much higher than in the surrounding interstitial fluid, and this pressure difference tends to force fluid out of the capillary. The net effect of these forces, now called **Starling forces**, is that fluid leaves at the beginning of a capillary due to hydrostatic pressure and then, as the volume and pressure of the blood decrease along the length of the vessel, re-enters at the other end of the capillary due to osmosis.

After a hemorrhage occurs, the delicate balance between the pressure difference and osmotic gradient is altered because the blood pressure in capillaries decreases, particularly in regions where there is considerable vasoconstriction (see Figure 53.9). The lower blood pressure changes the balance of the Starling forces, thereby causing a net movement of water from the interstitial fluid into the plasma and helping to restore blood volume. This effect takes some time to develop and is not as rapid as the baroreceptor reflex or even the endocrine responses just mentioned, but it is a very important part of the longer-term response to hemorrhage.

What effect might this redistribution of body fluid have on the hematocrit—the percentage of an animal's blood volume that is occupied by erythrocytes (refer back to Figure 48.4)? For a few hours after

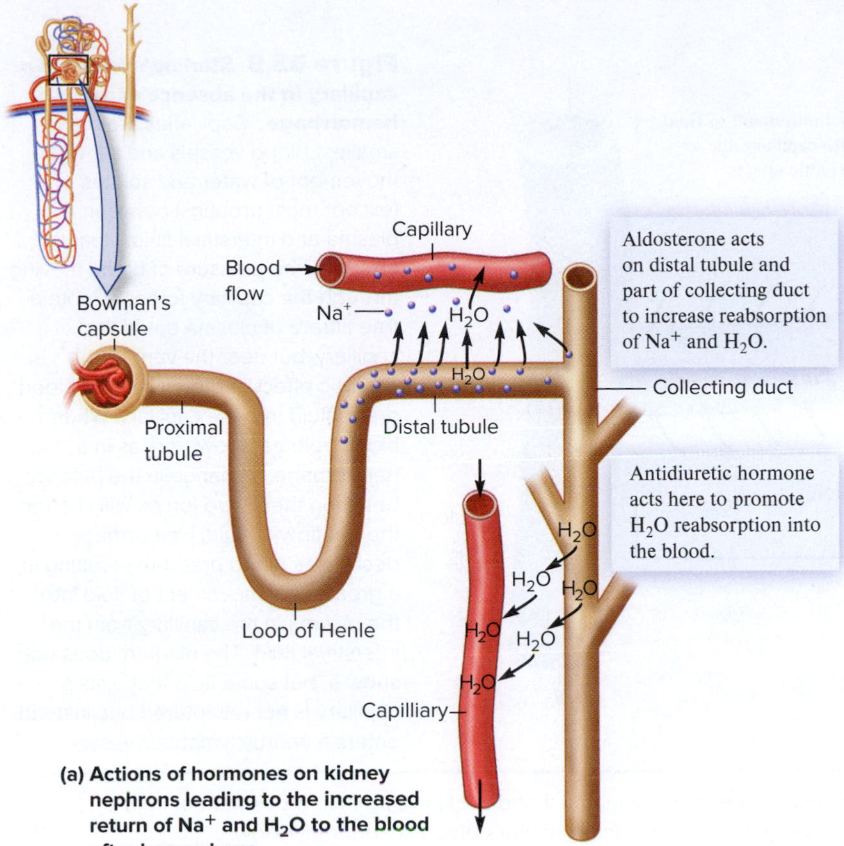

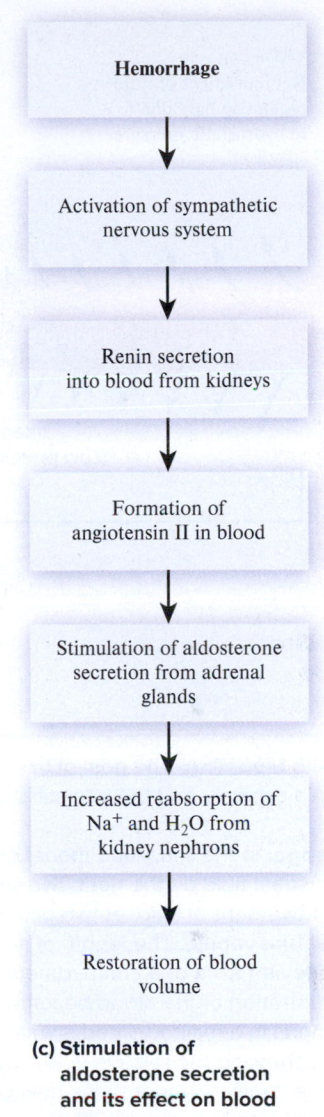

Capillary

Blood flow

Bowman's capsule

$Na^+$ — $H_2O$

Aldosterone acts on distal tubule and part of collecting duct to increase reabsorption of $Na^+$ and $H_2O$.

Proximal tubule

$H_2O$

Distal tubule

Collecting duct

Antidiuretic hormone acts here to promote $H_2O$ reabsorption into the blood.

$H_2O$ $H_2O$ $H_2O$ $H_2O$ $H_2O$

Loop of Henle

Capilliary

**(a) Actions of hormones on kidney nephrons leading to the increased return of $Na^+$ and $H_2O$ to the blood after hemorrhage**

Efferent arteriole

Glomerulus

Proximal tubule

**Increased volume of filtrate**

Afferent arteriole

Dilated

Blood flow

Normal

**Normal volume of filtrate**

Constricted (hemorrhage)

**Decreased volume of filtrate**

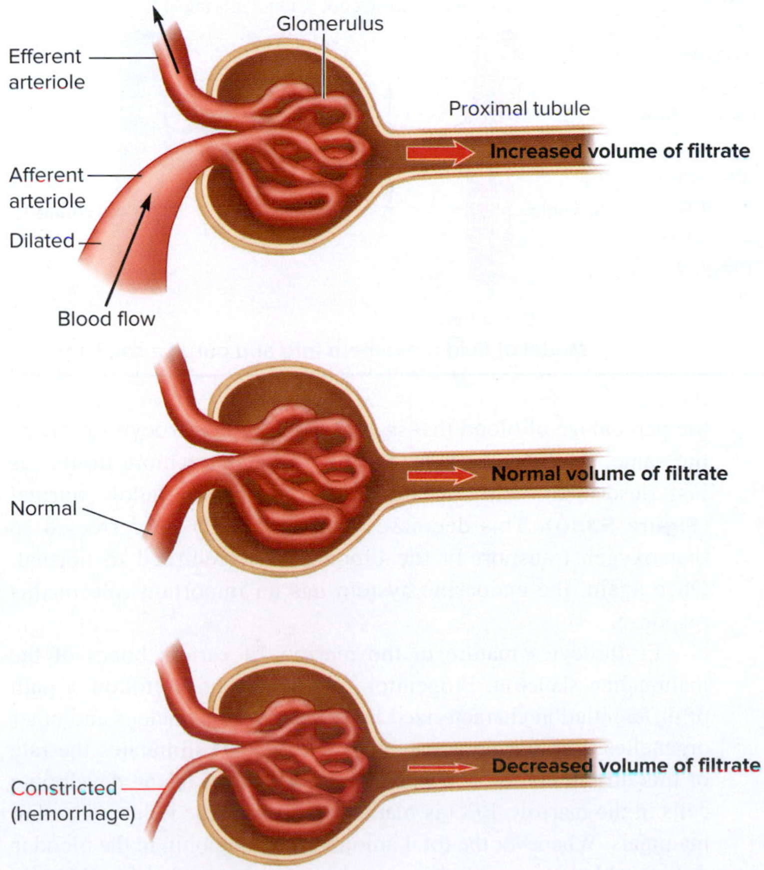

**(b) Effect of vasodilation and vasoconstriction of afferent arteriole on the amount of filtrate formed in a nephron**

**Hemorrhage**

↓

Activation of sympathetic nervous system

↓

Renin secretion into blood from kidneys

↓

Formation of angiotensin II in blood

↓

Stimulation of aldosterone secretion from adrenal glands

↓

Increased reabsorption of $Na^+$ and $H_2O$ from kidney nephrons

↓

Restoration of blood volume

**(c) Stimulation of aldosterone secretion and its effect on blood volume**

**Figure 53.8** **The renal response to hemorrhage. (a)** Schematic illustration of the effect of changing the diameter of afferent arterioles on the filtration of fluid out of a glomerulus capillary into a nephron tubule. The afferent arteriole brings blood into a glomerulus associated with a kidney nephron. The relative amount of filtrate (red arrows) in the proximal tubule of the nephron is indicated by the thickness of the arrow. Following a hemorrhage, vasoconstriction of an afferent arteriole will decrease the amount of fluid that gets filtered into the tubule, thereby preserving fluid in the blood. **(b)** Additional changes along the distal tubule of the nephron and collecting ducts promote greater reabsorption of $Na^+$ and water into the blood. The result is a decreased volume of urine that is more concentrated than normal, as well as a return of additional fluid to the circulation. **(c)** The production of the hormones angiotensin II (AII) and aldosterone, along with some major actions. Intermediate steps in AII synthesis are not shown for simplicity.

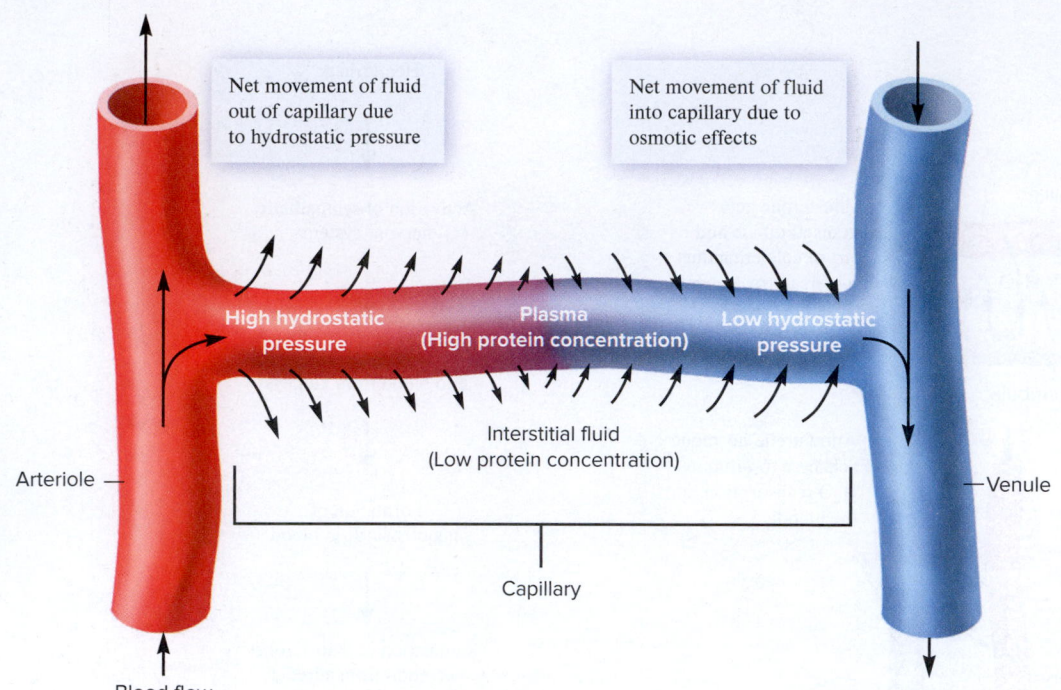

**Figure 53.9 Starling's forces in a capillary in the absence of a hemorrhage.** Capillaries are the smallest blood vessels and allow the movement of water and solutes (except most proteins) between the plasma and interstitial fluid. Near the arteriole, the pressure of blood moving through the capillary forces a protein-free filtrate of plasma out of the capillary, but near the venule, the osmotic effect of proteins in the blood draws fluid into the capillary. When blood volume is lowered, as in a hemorrhage, a change in the balance between these two forces will change the net flow of fluid. Hemorrhage decreases blood pressure resulting in a greater net movement of fluid into the plasma in the capillary from the interstitial fluid. The diagram does not show it, but some fluid that exits a capillary is not recaptured but, instead, enters a nearby lymphatic vessel.

 **Core Skill: Modeling** The goal of this modeling challenge is to make a model that depicts the relative movement of fluid between a capillary and the interstitial fluid when the amount of protein in an animal's blood is abnormally low.

**Modeling Challenge:** In the simplified model of fluid movement between a capillary and interstitial fluid on the right, two sets of arrows are used to indicate the net movement of fluid at two places along a capillary: near the arteriole and near the venule. The length of an arrow represents the relative amount of fluid moving from one compartment to the other. Let's assume that the protein concentration of the blood becomes lower than normal (this could happen, for example, in a malnourished animal). Draw a model like the one given with arrows showing the relative fluid movements at the arterial and venule ends of the capillary under these new conditions. Remember to use the length of an arrow to indicate the relative amount of fluid movement; a longer arrow represents a greater amount of fluid moving into or out of the capillary.

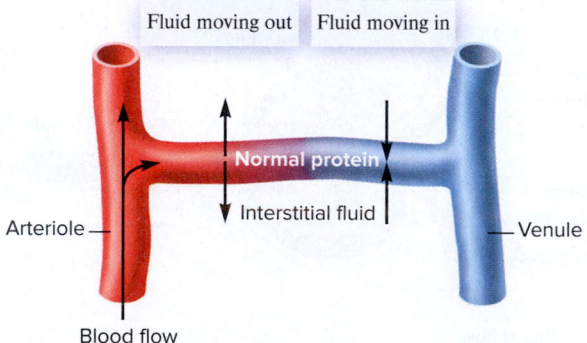

Model of fluid movement into and out of a capillary

a hemorrhage, the hematocrit does not change, because whole blood is lost from the injured vessel. However, the hematocrit will eventually decrease below normal as the blood becomes diluted by the movement of interstitial fluid into capillaries due to the Starling forces. As you have learned throughout this unit, homeostatic responses to one challenge often produce new challenges that must, in turn, be addressed at a later time. In this case, restoring blood volume so that pressure can be maintained is of more immediate importance than is bringing the hemocrit back to normal. As we will see next, however, hematocrit is eventually restored during later stages of the secondary phase.

## Hematocrit Changes in a Biphasic Manner After a Hemorrhage

Recall that initially hematocrit does not change when blood is lost following a hemorrhage. The total blood volume decreases, but

the percentage of blood that is comprised of erythrocytes remains the same. As time goes on, the blood becomes more dilute (as just described), and the hematocrit decreases below normal (**Figure 53.10**). This decrease must eventually be corrected so that oxygen transport in the blood can be returned to normal. Once again, the endocrine system has an important role in this response.

Erythrocytes mature in the marrow of certain bones of the mammalian skeleton. Progenitor cells in the bones follow a path of differentiation characterized by the loss of the nucleus and other organelles. The hormone erythropoietin (EPO) stimulates the rate of this differentiation and increases the survival of the developing cells in the marrow. EPO is made primarily by the kidneys in adult mammals. Whenever the total amount of hemoglobin in the blood is decreased below normal for any reason (such as a decreased hematocrit), the EPO-producing cells of the kidneys are activated to produce and secrete EPO into the blood. The effect of the hormone

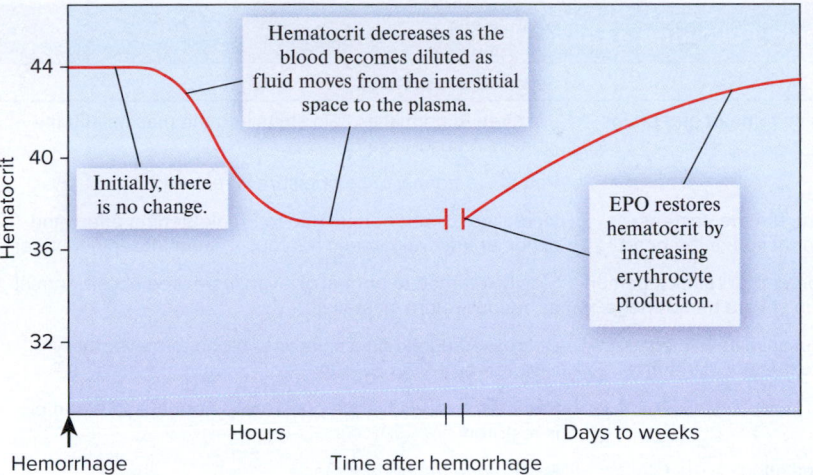

Initially, there is no change.

Hematocrit decreases as the blood becomes diluted as fluid moves from the interstitial space to the plasma.

EPO restores hematocrit by increasing erythrocyte production.

**Figure 53.10 The effect of fluid shifts and erythrocyte production on hematocrit following a hemorrhage.** Note that hematocrit decreases as fluid moves from the interstitial fluid into the blood vessel. This is later compensated for by the production of more erythrocytes in the bone marrow and their release into the blood under the control of the hormone erythropoietin (EPO).

 **Core Concept: Systems** The endocrine system secretes erythropoietin, which stimulates erythrocyte maturation within certain bones of the muscular-skeletal system. The circulatory system also plays a role in fluid shifts between body fluid compartments. Taken together, these systems all play roles in the changes observed in hematocrit after a hemorrhage.

on the number of circulating erythrocytes is not apparent for at least several hours, and it does not fully restore hematocrit for up to several days or even weeks, depending on the magnitude of the hemorrhage.

Interestingly, some mammals such as horses, seals, and pigs contain a large reservoir of mature erythrocytes in their spleen, an organ of the immune system involved in the production of leukocytes (refer back to Figure 52.4). In response to activation by the sympathetic nervous system, smooth muscle cells in the spleen contract, causing the spleen to eject its reservoir of erythrocytes into the circulation. This happens much more quickly than the slow EPO-induced restoration of hematocrit and provides a clear survival advantage after hemorrhage. Under normal conditions, the spleen response gives such animals a boost in the oxygen-carrying capacity of the blood during periods of increased activity or, in the case of seals, when diving under water.

## Other Organ Systems Indirectly Help Compensate for Hemorrhage

As we have seen, the circulatory, nervous, respiratory, endocrine, and urinary systems have a variety of important functions that contribute to an animal's ability to survive a serious hemorrhage. In addition, all other organ systems are affected to some degree by hemorrhage, and they all are involved at least indirectly in the overall compensatory response that restores blood volume and blood pressure.

You have already learned how the respiratory system contributes to the return of blood from the veins to the heart. The muscular-skeletal system may also assist the flow of blood from the limbs to the heart. Recall that one-way valves within limb veins help move blood against gravity toward the heart (refer back to Figure 48.14). Each time the skeletal muscles of the limbs contract, the veins coursing through those muscles are squeezed and blood is pushed through the valves. The anxiety that may be triggered by the injury and hemorrhage may encourage such muscular activity, and this has the benefit of contributing to the Frank-Starling effect.

The integumentary system also contributes indirectly as the arterioles that normally deliver blood to the skin constrict, redirecting

blood to vital organs. This also happens throughout much of the alimentary canal of the digestive system. In addition, the smooth muscles of the canal relax, due to inhibitory actions of the sympathetic nervous system. This effect occurs even if an animal has just eaten a meal prior to the hemorrhage, and it results in a considerable energy savings. In such emergency situations, energy supplies tend to be spared for the heart, the brain, and a few other organs, such as the placenta in a pregnant female.

In major homeostatic challenges such as hemorrhage, energy is also redirected away from the reproductive system. At such times, it is more important that the individual survives so that it may reproduce another day.

Finally, even the immune and lymphatic systems respond to hemorrhage. We have already learned that a major immune system organ, the spleen, may contribute to the maintenance of hematocrit in certain mammals. In addition, if the injury has resulted from an external wound, part of the response to hemorrhage is to prevent infection. The cells and secretions of the innate immune response trigger a local inflammatory reaction (refer back to Figure 52.2) at the site of the wound, and the adaptive immune response responds to any specific pathogens that have entered the circulation or local tissues.

**Table 53.1** summarizes some of the major events of the homeostatic response to hemorrhage in a mammal. Not all homeostatic challenges engage such widespread responses, but typically the activities of more than one organ system are affected directly or indirectly. There are many common examples of this fundamental feature of homeostasis. For example, when you are ill, you may have a fever caused by secretions of the cells of the immune system. Fever induces sweating, which in turn decreases the amount of fluid and alters the concentrations of $Na^+$ and other ions in the blood. The urinary and endocrine systems help to restore fluid and ion balance. If you happen to miss a meal or two, the endocrine and nervous systems together will help maintain glucose homeostasis in the blood. If you move to a high altitude or go mountain-climbing, the respiratory system will help maintain the amount of oxygen in your blood by stimulating your breathing rate. That will, however, decrease the amount of carbon dioxide in the blood, which in turn will increase the pH of

| Table 53.1 | Summary of Key Homeostatic Responses Following a Hemorrhage | |
|---|---|---|
| **Variable** | **Rapid responses** | **Secondary responses** |
| Rate of blood flow to organs | Decreased to most organs, but less so to heart and brain | Restored to normal as fluid shifts into the plasma and the kidneys reabsorb water |
| Heart rate (HR) | Increased | Returns to normal once pressure is restored |
| Stroke volume (SV) | Decreased due to less venous return, but the decrease is minimized by increased force of contraction of the heart | Increased toward normal due to Frank-Starling effect and action of the sympathetic nervous system (norepinephrine) |
| Cardiac output (CO = HR × SV) | Maintained near normal by the baroreceptor reflex if hemorrhage is not severe, but decreased in severe hemorrhage | Slowly returned to normal or even increased above normal as venous return increases |
| Hematocrit | Not usually a rapid response except in animals with contractile spleens that release erythrocytes into the bloodstream | Decreased due to fluid shifts into blood, but eventually restored by erythropoietin |
| Breathing rate and depth | Increased (improves venous return) | Returned to normal once pressure is normal and blood pH is restored |
| Hemoglobin (Hb) saturation | Increased toward 100% by hyperventilation | Restored to the normal range |
| Unloading of oxygen from Hb | Increased by Bohr effect due to lactic acid | Restored to normal once blood flow to organs returns to normal and pH is normalized |
| Urine production | Decreased due to vasoconstriction of afferent arterioles | Further decreased due to effects of ADH and aldosterone |

the blood. The urinary system will then help restore pH to normal by excreting bicarbonate ions in the urine and reabsorbing more hydrogen ions into the blood. These and many other examples demonstrate once again that organ systems do not function in isolation but are integrated with each other to ensure homeostasis of all of an animal's physiological functions.

**BIO·TIPS**   **THE QUESTION** *Many animals encounter periods of dehydration due, for example, to lack of available water to drink or to disease. One such disease in humans is cholera, which is induced by the bacterium Vibrio cholerae and involves a massive loss of body water via the gastrointestinal tract. What do you predict would happen to the baroreceptor reflex following severe dehydration in a person? How might other organ systems respond to dehydration?*

**T**OPIC *What topic in biology does this question address?* The topic is dehydration. Specifically, you are asked to predict the organ system responses in a person who is severely dehydrated.

**I**NFORMATION *What information do you know based on the question and your understanding of the topic?* From the question, you know that a person has lost considerable body water. From your understanding of the topic, you may remember that blood is more than 50% water by volume, and that blood volume and blood pressure are directly related. You may also recall how the human body responds to a lower blood volume, such as from a hemorrhage.

**P**ROBLEM-SOLVING **S**TRATEGY *Predict the outcome.* First, remember the events that occur in the body following a loss of blood (hemorrhage). Now think how a loss of water and a hemorrhage are related.

**ANSWER** *Dehydration results in a loss of volume in all three body compartments, including the blood. Thus, if dehydration is severe enough, you can predict that the body's responses to it would in many ways be similar to those described in this chapter for hemorrhage. In fact, this is true, and dehydration is associated with increased baroreceptor activity followed by many of the events listed in Table 53.1. Heart rate increases, peripheral resistance increases, particularly in less vital regions of the body, and hormones are activated to help retain water in the body and decrease urine production.*

## 53.4 Impact on Public Health

### Learning Outcomes:
1. Explain the relation between the degree of hemorrhage and its consequences in a human.
2. Describe procedures used to help restore normal circulatory function in humans after a large hemorrhage.

Let's now return to the case of the girl described at the beginning of the chapter. Her initial symptoms were light-headedness, agitation without any sign of mental confusion, rapid breathing rate, a slightly lower blood pressure than normal for her, and a very rapid heart rate. All of these suggested significant internal bleeding due to the rupture of one or more blood vessels, most likely near her kidney on the side where she was kicked. Once in the ambulance, she was laid on her back with her feet slightly elevated. Gravity helped to return blood from her leg veins to the right atrium of her heart, which was important because the amount of blood ejected with each heartbeat is roughly proportional to the amount of blood that fills the heart (recall the Frank-Starling law of the heart). She was anxious but still conscious and alert, suggesting that only a moderate hemorrhage had occurred.

A decision was made to not infuse whole blood but, instead, to simply supply fluids. A cannula was placed into a vein in her hand and attached to a solution of ions, which was allowed to rapidly enter her circulation at an initial rate of about 1 liter/10 minutes. The fluid was approximately isotonic with her blood. Recall from Chapter 41 (and refer back to Figure 5.13) that cells can be deformed and even ruptured when exposed to extracellular fluid with a tonicity that is very different from normal.

After sufficient fluids were infused, the girl's blood pressure rose to 114/71 mmHg, which was within her normal range. Her heart rate was still somewhat elevated at 84 bpm, compared to her usual resting rate of about 72 bpm. This higher rate, however, could mainly be attributed to her continued anxiety, not to the baroreceptor reflex. Her breathing rate by this time was a steady 15 breaths per minute, normal for a girl her age, and her hematocrit was 34 (normal is about 38–46 for females). A blood sample was drawn to test for indicators of kidney damage. Later analyses of the sample, along with a CT scan of her thorax and abdomen, revealed that the kidneys and other major organs were functioning properly and undamaged, and thus it appeared that fluid replacement was sufficient in this case. Her hematocrit was low for some time, but it eventually recovered due to the actions of erythropoietin.

Had the hemorrhage been more severe—for example, resulting in a loss of greater than 20% of her total blood volume—the outcome might have been very different. Under such conditions, shock ensues, with a variety of symptoms, including very low blood pressure, rapid heart rate, and in extreme cases a loss of consciousness and multiple organ failure. During shock, the baroreceptor reflex and endocrine events described earlier may be insufficient to prevent pressure from continuing to fall. In such situations, whole blood is sometimes infused, because the total amount of hemoglobin in the person's blood may be dangerously low. In addition, hormones such as epinephrine may be infused to facilitate heart action and vasoconstriction.

Even with these treatments, however, a person who has progressed too far along in shock may not recover. One reason is that as organs begin to fail due to the profound lack of oxygen and nutrient delivery, their ability to contribute to the mechanisms compensating for the blood loss decreases. Another reason is a positive feedback cycle that can develop. As cells die in large numbers, they often release inflammatory substances that result in vasodilation (a normal response to inflammation; refer back to Figure 52.2). This can offset the vasoconstriction mediated by the sympathetic nervous system and thereby decrease blood pressure even more, which further decreases the perfusion of organs with blood, causing more cell death, and so on (**Figure 53.11**).

Roughly every 3 minutes, a person dies in the U.S. from a traumatic injury, such as those incurred in car accidents or from operating heavy machinery. Hemorrhage is the leading cause of death due to traumatic injuries in the U.S. Each year, several hundred thousand individuals in the U.S. also suffer hemorrhaging due to gastrointestinal diseases or abnormally weakened blood vessels (aneurysms). Hemorrhage is also a very serious concern in battlefield injuries. For these reasons, investigators continue to research improved means of preventing or reversing shock. The girl in this case was fortunate that she received prompt treatment, and she was able to resume her athletic pursuits after a recovery period of nearly 4 weeks.

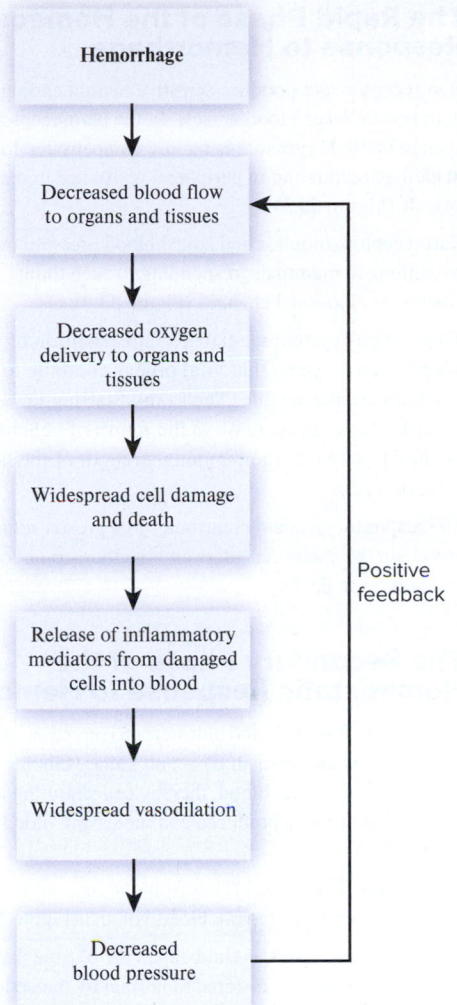

**Figure 53.11 A positive feedback cycle that worsens shock.** As blood flow to tissues decreases, cells begin to die and release their contents, including inflammatory mediators (as described in detail in Chapter 52). Inflammation, in turn, tends to cause small blood vessels to dilate, which decreases resistance and therefore lowers blood pressure even further.

 **Core Skill: Connections** In what other context have you learned about positive feedback? Refer back to Figures 51.9 and 51.11 for help.

## Summary of Key Concepts

### 53.1 Effects of Hemorrhage on Blood Pressure and Organ Function

- Hemorrhage is a loss of blood from a damaged blood vessel. It may decrease delivery of oxygen and nutrients to cells (Figure 53.1).

- Blood volume and blood pressure are closely related. A decrease in volume due to hemorrhage can lead to a fall in pressure. Smaller hemorrhages may produce no obvious symptoms, but large hemorrhages can have serious consequences (Figure 53.2).

## 53.2 The Rapid Phase of the Homeostatic Response to Hemorrhage

- Baroreceptors are pressure-sensitive neuron endings associated with certain large blood vessels. In the baroreceptor reflex, a change in blood pressure is rapidly compensated for by a change in cardiac output and in peripheral resistance in many blood vessels (Figure 53.3).

- Baroreceptors monitor and adjust blood pressure under baseline conditions in mammals, responding to such things as changes in posture or emotional changes (Figure 53.4).

- The nervous system mediates a redistribution of blood after hemorrhage so that vital organs continue to receive as much blood as possible. The Frank-Starling law of the heart describes how an increase in the return of venous blood to the heart increases the pumping strength of the heart muscle (Figure 53.5).

- The respiratory system contributes to a greater return of venous blood via the chemoreceptors and the thoraco-abdominal pump (Figures 53.6, 53.7).

## 53.3 The Secondary Phase of the Homeostatic Response to Hemorrhage

- Antidiuretic hormone and aldosterone (secretion of which is stimulated by angiotensin II) act on kidney tubules to help retain water and ions in the blood, thereby concentrating and decreasing the amount of urine produced and increasing blood volume (Figure 53.8).

- After hemorrhage, fluid flows from the interstitial space to the plasma, which helps restore blood volume (Figure 53.9).

- Hematocrit decreases as fluid enters the plasma due to Starling forces, then is slowly restored to normal by the action of erythropoietin on immature erythrocytes in bone marrow (Figure 53.10).

- All organ systems are affected by hemorrhage to some degree and directly or indirectly contribute to the restoration of homeostasis.

## 53.4 Impact on Public Health

- The treatment for hemorrhage depends on the magnitude of blood loss. For relatively moderate hemorrhages, iso-osmotic fluid replacement is usually sufficient. For severe hemorrhage, whole blood and vasoactive drugs may be administered (Figure 53.11).

- Hemorrhage is the leading cause of death due to traumatic injuries in the United States.

## Assess & Discuss

### Test Yourself

1. How does the respiratory system contribute to the restoration of blood pressure following a hemorrhage?
   a. by increasing the saturation of hemoglobin with oxygen
   b. by helping to increase the return of blood from veins to the heart
   c. by increasing the heart rate
   d. by activating the baroreceptor reflex
   e. through the Bohr effect

2. Following hemorrhage, blood flow decreases to the _____ but is maintained as close to normal as possible in the _____.
   a. brain and heart, digestive system
   b. skin, digestive system
   c. kidneys and heart, brain and skin
   d. skin and digestive system, brain and heart
   e. digestive system and liver, skin and kidneys

3. Which of the following will increase the hematocrit after hemorrhage in a human?
   a. an intravenous infusion of erythropoietin
   b. Starling forces
   c. intravenous infusion of iso-osmotic NaCl solution
   d. intravenous infusion of a hyperosmotic NaCl solution
   e. both a and b

4. The frequency of action potentials sent from baroreceptors to the brain _____ when blood volume is decreased, resulting in _____ heart rate.
   a. increases, an increased
   b. increases, a decreased
   c. decreases, an increased
   d. decreases, a decreased

5. An animal's normal stroke volume is 9 mL/beat and its normal heart rate is 125 beats/min. Immediately after a hemorrhage, its heart rate increases to 161 beats/min and its stroke volume does not change. What is its new cardiac output?
   a. 1.45 L/min          d. 17.9 L/min
   b. 0.145 L/min         e. 0.055 L/min
   c. 17.9 mL/min

6. For the animal in question 5, what could explain why the stroke volume remained unchanged?
   a. Stroke volume is not usually affected by hemorrhage.
   b. The heart muscle cannot contract more forcefully than it normally does at rest.
   c. The hemorrhage may have been so severe that venous return to the heart cannot be significantly increased above normal.
   d. Constant stroke volume is a compensatory response to prevent blood pressure from increasing too much.
   e. Stroke volume increases only when the heart rate does not change from normal.

7. Antidiuretic hormone
   a. causes vasodilation of blood vessels.
   b. stimulates aldosterone production by the adrenal glands.
   c. stimulates reabsorption of both Na+ and water from the kidneys.
   d. is stimulated by AII.
   e. increases water but not ion reabsorption in the kidneys.

8. Which of the following act(s) as a vasoconstrictor?
   a. norepinephrine          d. aldosterone
   b. acetylcholine           e. both a and c
   c. AII

9. Imagine that the baroreceptor reflex was *not* functional in an animal that had experienced a severe hemorrhage. What would happen?
   a. Heart rate would increase significantly, but the strength of each heartbeat would not.
   b. The initial phase of recovery would be normal, but the delayed responses would be absent.
   c. The animal would probably develop shock very quickly as its pressure rapidly fell.
   d. Heart rate would decrease immediately after the hemorrhage but then slowly return to normal.
   e. There would be little significant effect on the overall compensatory response to hemorrhage.

10. A day after an animal suffered a moderate hemorrhage, what would be *true*?
    a. The animal's blood pressure would be normal, but its stamina (ability to be very active for long periods) would be decreased.
    b. The concentration of erythropoietin in the animal's blood would be increasing.
    c. The animal's respiration rate would be increased above normal.
    d. The flow of blood to the animal's vital organs would be decreased.
    e. Both a and b would be true.

## Conceptual Questions

1. Try to imagine the integrated responses of an animal's body to other major homeostatic challenges. Don't limit yourself to mammals. For example, what are some of the challenges a bird might encounter while flying at a high altitude over a mountain range?

2. Some people get momentarily light-headed after donating blood and then getting up from the chair or cot on which they were reclining. Why? Despite the light-headedness, few people actually faint at this time. What prevents them from fainting?

3. **Core Concept: Energy and Matter** A core concept of biology is that living organisms use energy. In what ways does the integrated response of the body to restore homeostasis after a hemorrhage use energy?

## Collaborative Questions

1. Identify some secondary problems that arise as a result of the integrated response to hemorrhage in a mammal.

2. Describe the nature of the baroreceptor reflex that occurs if blood pressure is *increased* above normal.

# UNIT VIII
# ECOLOGY

Ecology is the study of interactions among organisms and between organisms and their environment. These interactions govern the number of species in an area and their population densities. Ecologists work at the largest scales of any biologists.

Chapter 54 introduces the field of ecology and discusses the effects of physical variables such as temperature and moisture. At the largest scales, variation in temperature and moisture create distinct large-scale habitats, called biomes. Chapter 55 discusses behavioral ecology and explores how behavior contributes to the fitness of organisms. The chapter begins by investigating how different behaviors arise and ends by examining group behavior and mating systems. The next two chapters examine population growth and the constraints to growth provided by competitors and natural enemies. Chapter 56 introduces the demographic tools needed to study population growth and provides simple mathematical models of growth. Chapter 57 discusses the effects of competition, mutualism, predation, herbivory, and parasitism on populations. Chapter 58 focuses on communities and ecosystems and considers the factors that influence the number of species in a community, and the flow of energy and nutrients through the living and nonliving components of the environment. Chapter 59 covers the recent rapid growth of the human population and the impacts humans have on natural systems. These impacts include global warming and climate change, pollution, habitat destruction, overexploitation and invasive species. Finally, Chapter 60 focuses on the conservation of life on Earth and the various strategies used to protect genetic, species, and ecosystem diversity.

(54): ©Dante Fenolio/Science Source; (55): ©FLPA/Alamy Stock Photo; (56): ©Mike Lockhart; (57): ©bastianbodyl/iStock/Getty Images; (58): ©Gary Braasch/Corbis Historical/Getty Images; (59): Source: Photo courtesy of National Park Service, Everglades National Park; (60): ©Patrick Pleul/dpa/picture-alliance/Newscom

## The following Core Concepts and Core Skills will be emphasized in this unit:

- *Energy and Matter:* We discuss the flow of energy in Chapter 58 and nutrient cycles in Chapter 59.
- *Information:* In Chapter 55, we will examine the influence of learning on behavior and see that the acquisition of information is critical in local and long-range movement and in foraging and communication.
- *Science and Society:* Humans have profound impacts on ecological systems, from global warming, climate change, and pollution, through habitat destruction, overexploitation and invasive species, all of which are detailed in Chapter 59.
- *Systems:* Living organisms interact with their environment (Chapter 54). Populations of different species interact in many different ways, such as competition, mutualism, predation, herbivory, and parasitism (Chapter 57) to form complex ecological communities and ecosystems (Chapter 58).
- *Process of Science:* All chapters have a Feature Investigation, which illustrates how science is performed by describing an important and pivotal experiment that expanded the boundaries of ecological knowledge.
- *Modeling:* Each chapter has a Modeling Challenge to help you refine this important skill.

# An Introduction to Ecology and Biomes

# 54

**Diminishing and disappearing populations.** Population sizes of the Panamanian golden frog (*Atelopus zeteki*) have decreased greatly over the past 20 years, and populations of many other species of harlequin frogs have disappeared entirely. Ecologists are investigating the reasons for this decline. ©Dante Fenolio/Science Source

 **Core Skill: Connections** Look ahead to Chapter 59. What are the main threats to species?

I n 2006, a study led by J. Alan Pounds of the Monteverde Cloud Forest Preserve in Costa Rica reported that two-thirds of the 110 species of harlequin frogs in mountainous areas of Central and South America had become extinct over the previous 20 years. The researchers noted that populations of other species, such as the Panamanian golden frog (*Atelopus zeteki*), had been greatly reduced (see the chapter opening photo). The question was why. The culprit was identified as a disease-causing fungus, *Batrachochytrium dendrobatidis*, but Pounds's study implicated global warming—a gradual increase in the average temperature of the Earth's atmosphere—as the agent causing outbreaks of the fungus. One effect of climate change due to global warming is to increase the cloud cover, which reduces daytime temperatures and raises nighttime temperatures. Researchers believe that this combination has created favorable conditions for the spread of *B. dendrobatidis* and other pathogens, which thrive in cooler daytime temperatures. Pounds, the team's lead researcher and an ecologist, was quoted as saying, "Disease is the bullet killing frogs, but climate change is pulling the trigger."

**Ecology** is the study of interactions among organisms and between organisms and their environments. Interactions among organisms are called **biotic** interactions, and those between organisms and their nonliving environment are termed **abiotic** interactions. These interactions, in turn, govern the numbers of species in an area and their population densities. This first chapter of this unit on ecology will introduce the four broad areas of ecology: organismal, population, community, and ecosystems ecology. Next, we will explore how ecologists approach and conduct their work. We will then turn our focus to abiotic interactions and examine the effects of factors such as temperature, water, light, pH, and salt concentrations on the distributions of organisms. We will conclude with a consideration of climate and its large influence on **biomes,** the major types of habitats where organisms are found.

Before 1960, the field of ecology was dominated by taxonomy, natural history, and speculation about observed patterns. An ecologist's tools of the trade included sweep nets, quadrats (small, measured plots of land used to sample living things), and specimen jars. Since that time, the number of ecological studies has exploded, and ecologists have become active in investigating environmental change on local, regional, and global scales. Ecologists have embraced experimentation and adapted concepts and methods derived from biochemistry, genetics, mathematics, and physiology. Their tools have kept pace with technological innovations. Now an ecologists' equipment is just as likely to include laptops, satellite-generated images, and chemical autoanalyzers.

Ecological studies have important implications in the real world, as will be amply illustrated by examples discussed throughout this unit. However, there is a distinction between ecology and **environmental science,** the application of ecology to real-world problems. To use an analogy, ecology is to environmental science as physics is to engineering. Both physics and ecology provide the theoretical framework on which more applied studies are based. Engineers rely on the principles of physics to build bridges. Environmental scientists rely on the principles of ecology to solve environmental problems.

(a) A single organism

(b) A population of zebras

(c) An African grassland community

(d) Nutrient flow in an African grassland community

**Figure 54.1** **The scale of ecology.** **(a)** Organismal ecology. What is the temperature tolerance of this zebra? **(b)** Population ecology. What factors influence the growth of zebra populations in Africa? **(c)** Community ecology. What factors influence the number of species in African grassland communities? **(d)** Ecosystem ecology. How do water, energy, and nutrients flow among plants, zebras, and other herbivores and carnivores in African grassland communities? a: ©PHOTOCREO Michal Bednarek/Shutterstock; b: ©Alain Pons/Biosphoto; c: ©Paul Springett/Alamy Stock Photo; d: ©Art Wolfe/Science Source

## 54.1 The Scale of Ecology

**Learning Outcome:**

1. Describe and differentiate between the different scales on which ecologists work.

Ecology ranges in scale from the study of an individual organism, to studying populations, communities, and ecosystems (**Figure 54.1**). This section introduces the broad areas of organismal, population, community, and ecosystem ecology and presents an investigation that helps illuminate the field of population ecology.

### Organismal Ecology Investigates How Adaptations and Choices by Individuals Affect Their Reproduction and Survival

**Organismal ecology** is the study of the ways in which individual organisms meet the challenges of their abiotic and biotic interactions within their environments. It can be divided into two subdisciplines. The first, **physiological ecology**, investigates how organisms are physiologically adapted to their environment and how the environment impacts the distribution of species. Much of this chapter discusses physiological ecology. The second subdiscipline, **behavioral ecology**, focuses on how the behavior of individual organisms contributes to their survival and reproductive success, which, in turn, eventually affects the population density of the species. This is the topic of Chapter 55.

### Population Ecology Describes How Populations Grow and Interact with Other Species

**Population ecology** focuses on groups of interbreeding individuals, called populations. A primary goal of population ecology is to understand the factors that affect a population's growth and determine its size and density. Although the attention of a population ecologist may be aimed at studying the population of a particular species, the relative abundance of that species is often influenced by its interactions with other species. Thus, population ecology includes the study of **species interactions**, such as predation, competition, and parasitism. Knowing what factors affect populations can help us lessen species endangerment, stop extinctions, and control invasive species.

 **Core Skill: Process of Science**

## Feature Investigation | Callaway and Aschehoug's Experiments Showed That the Secretion of Chemicals Gives Invasive Plants a Competitive Edge over Native Species

One important topic in the area of population ecology concerns **introduced species**, species that are moved from a native location to another location, usually by humans. Such species sometimes spread so aggressively that they crowd out native organisms, in which case they are considered **invasive species**. We will examine the effect of invasive species on native species more extensively in Chapter 59. Plants have often become invasive because they have escaped their natural enemies, primarily insects that occur in the country of origin and not in the new locale. One way of controlling these species, therefore, has been to import the plant's natural enemies. This method is known as **biological control**. However, an investigation of the population ecol-

ogy of diffuse knapweed (*Centaurea diffusa*), a Eurasian plant that has established itself in many areas of North America, suggests a different reason for the success of invasive species.

American researchers Ragan Callaway and Erik Aschehoug hypothesized that the roots of this particular species secrete powerful toxins, called **allelochemicals**, that kill the roots of other species, allowing *C. diffusa* to proliferate. To test their hypothesis, Callaway and Aschehoug collected seeds of three native Montana grasses, *Koeleria cristata*, *Festuca idahoensis*, and *Agropyron spicata*, and grew each of them with or without the introduced *Centaurea* species (**Figure 54.2**). As hypothesized, *C. diffusa* depressed the biomass of

**Figure 54.2** **Experimental evidence of the effect of allelochemicals on plant production.**

**HYPOTHESIS** Introduced plants from Eurasia outcompete native Montana grasses by secreting allelochemicals from their roots.

**KEY MATERIALS** Seeds of *Centaurea diffusa and other native grasses* from Eurasia plus seeds of native Montana grasses.

|  | Experimental level | Conceptual level |
|---|---|---|
| **1** Collect seeds of native Montana grasses and plant with and without seeds of invasive *C. diffusa* from Eurasia. Three months after sowing seeds, the plants are harvested, dried, and weighed. | | *C. diffusa* significantly reduces biomass of native Montana grasses. |
| **2** Collect seeds of grasses from Eurasia of the same three genera as the Montana grasses and plant with and without *C. diffusa*. Three months after sowing seeds, the plants are harvested, dried, and weighed. | | *C. diffusa* doesn't depress the biomass of grasses native to Eurasia as much. |

**3   THE DATA***

*The biomass is that of the genus noted at the top of each graph.

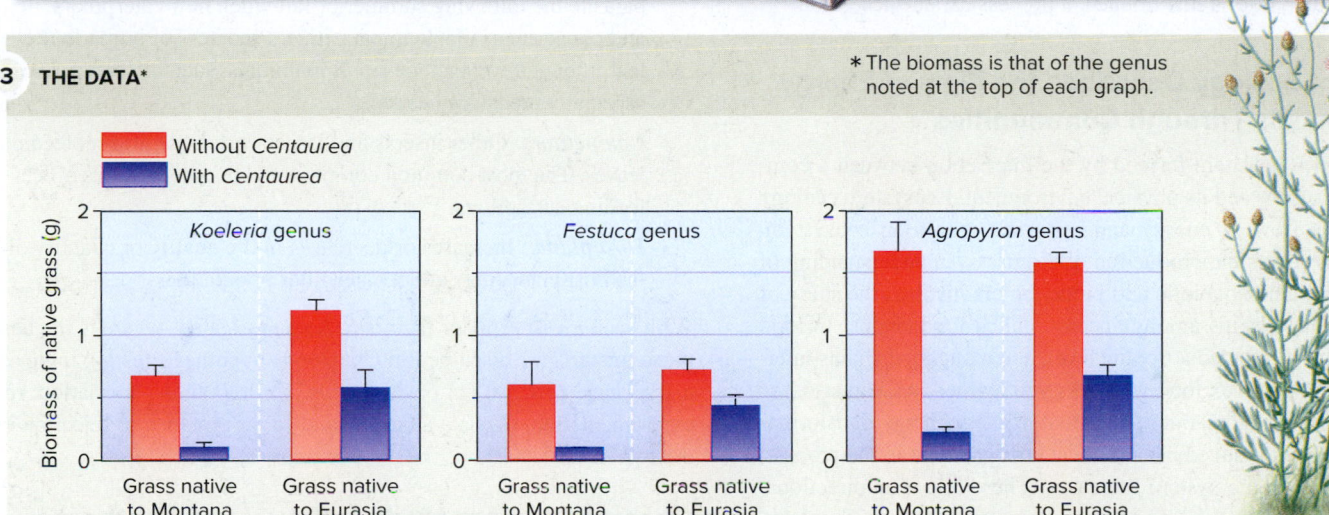

**4   CONCLUSION** *Centaurea diffusa*, a Eurasian grass, is invasive in the U.S. because it secretes allelochemicals, which inhibit the growth of native plants.

**5   SOURCE** Callaway, R. M., and Aschehoug, E. T. 2000. Invasive plants versus their old and new neighbors. *Science* 290: 521–523.

the native grasses. When the experiments were repeated with grasses native to Eurasia, *Koeleria laerssenii, Festuca ovina,* and *Agropyron cristatum,* the growth of each species was inhibited, but to a significantly lesser degree than the growth of the species from Montana.

In other experiments not described in Figure 54.2, Callaway and Aschehoug added activated carbon to the soil, which absorbs the allelochemical(s) excreted by the *C. diffusa* roots. With activated carbon added, the Montana grass species increased in biomass compared with the previous experiments. The researchers concluded that *C. diffusa* outcompetes Montana grasses by secreting one or more allelochemicals and that Eurasian grasses are not as susceptible to the chemical's effect because they coevolved with *C. diffusa* and may have developed some resistance to its allelochemical(s). If the reason for the success of invasive plants can be attributed to the chemicals they secrete, this calls into question the effectiveness of biological control of invasive plants

by importation of their natural enemies. This study on the population biology of an invasive plant changed our perspective on why such species succeed and could affect attempts to control them in the future.

*Experimental Questions*

1. Prior to Callaway and Aschehoug's study, what was the prevailing hypothesis on why invasive species succeed in new environments?

2. Briefly describe the evidence collected to support the allelochemical hypothesis.

3. What was the function of the activated carbon used in a subsequent test of the hypothesis?

4. **CoreSKILL »** Explain what else could have been measured in these experiments to further investigate the interactions between *C. diffusa* and native grasses.

## Community Ecology Focuses on the Factors That Influence the Number of Species in a Given Area

**Community ecology** studies how populations of species interact and form functional communities. For example, a forest is a community of trees, herbs, shrubs, grasses, the herbivores that eat them, and the carnivores that prey on the herbivores.

Community ecology focuses on why certain areas have high numbers of species (that is, are species-rich), but other areas have low numbers of species (that is, are species-poor). Although ecologists are interested in species richness for its own sake, a link also exists between species richness and community function. Ecologists generally believe that species-rich communities perform better than species-poor communities. As discussed in Chapter 58, more species may make a community more stable, that is, more resistant to disturbances such as invasive species. Community ecology also considers how species composition and community structure change over time and, in particular, after a disturbance, a process called succession.

## Ecosystem Ecology Describes the Flow of Energy and Chemicals Through Communities

An **ecosystem** is a system formed by the interaction between a community of organisms and its physical environment. **Ecosystem ecology** deals with the flow of energy and materials within an ecosystem, which in turn, affects the production of biomass. An understanding of the interactions among abiotic and biotic factors involves the study of the feeding relationships among species, called food chains. In food chains, each level is called a trophic level, and many food chains interconnect to form complex food webs. As you learned in Chapter 6, the second law of thermodynamics states that in every energy transformation, free energy is reduced because heat energy is lost in the process, and the entropy of the system increases. Therefore, a unidirectional flow of energy occurs through an ecosystem, with energy dissipated at every step. An ecosystem needs a recurring input of energy from an external source—in most cases, the Sun—to sustain itself.

## 54.2 Ecological Methods

### Learning Outcome:

1. **CoreSKILL »** Explain how the five steps of hypothesis testing discussed in Chapter 1 can be applied to an ecological research project.

How do ecologists go about studying their subject? As an example, let's consider an insect pest of apple trees. The oak winter moth, *Operophtera brumata*, is a small moth whose larvae feed on a variety of trees and shrubs in Europe and Asia. Egg hatch often coincides with the opening of buds, so young caterpillars can feed on soft new foliage. Sometimes, however, eggs hatch before leaves appear, and, rather than starve, caterpillars "balloon" away on silken threads in an attempt to reach trees whose leaves have already appeared. Although this is a risky business, being able to feed on a variety of food plants maximizes the caterpillars' chances of survival. In the early 20th century, oak winter moths were accidentally introduced into North America, where larvae became a problem in apple orchards in British Columbia and the northwestern United States, including Oregon. How could these moths be controlled? A thorough understanding of their biology was needed.

First, ecologists drew up a web of interactions between the oak winter moth and the factors that could affect its population size (**Figure 54.3**). These numerous and varied factors include the following:

- *Abiotic factors such as temperature and rainfall:* Warm temperatures can accelerate an egg hatch, meaning that larvae will appear before buds open and leaves are available to feed on.

- *Natural enemies, including bird predators of adult moths and caterpillars, insect parasites, bacterial parasites, and pupal predators:* Bird predators consume relatively few adult moths and caterpillars. Bacterial parasites tend to kill larvae before they pupate. Insect parasites usually lay their eggs inside the caterpillars or pupae, and the developing parasite gradually eats the caterpillar or pupa from the inside out, emerging as an adult parasite the following summer or fall when new caterpillars and pupae are available to parasitize. Predators of pupae in the leaf litter and soil include small mammals, such as shrews, and especially predatory beetles.

- *Competitors:* Other insects and larger vertebrate grazers feed on leaves. The most common competitors are other species of leaf-feeding caterpillars.

- *Host plants:* Increases or decreases in the quality or quantity of the host plants may affect caterpillar populations.

With such a vast array of factors to be investigated, where is the best place to start? As discussed in Chapter 1, hypothesis testing involves a five-stage process: (1) observations, (2) hypothesis formation, (3) experimentation, (4) data analysis, and (5) acceptance or rejection of the hypothesis.

## Observations Are Made to Develop Hypotheses

In Canada, size fluctuations in populations of oak winter moths, which are pests on apple trees, were observed for many years to determine if the population sizes were affected by fluctuations in abiotic or biotic interactions, such as levels of parasitism, predator abundance, or the quality of available leaves. These factors caused relatively little mortality among the moths. However, in the 1950s, *Cyzenis albicans*, a parasite of oak winter moth larvae, was imported from Europe and released. Within 5 years of this release, oak winter moth populations were dramatically reduced.

Ecologists noticed that in areas where the moths were more abundant, parasitism of their larvae by *C. albicans* increased (**Figure 54.4a**). In areas with few moths, such parasitism was low. This finding meant that the parasites were killing more caterpillars in areas where the caterpillars were more common, which tended to control caterpillar numbers. Using statistical methods, ecologists may find a line of best fit, a straight line that represents a summary of the relationship between two variables, as shown in Figure 54.4a. Alternatively, if the data points had not been tightly clustered, as in **Figure 54.4b**, ecologists would have had little confidence that parasitism affects the moth population size.

Many statistical tests are used to determine whether or not two variables are significantly related to one another. The type of relationship shown in Figure 54.4a is called a significant correlation.

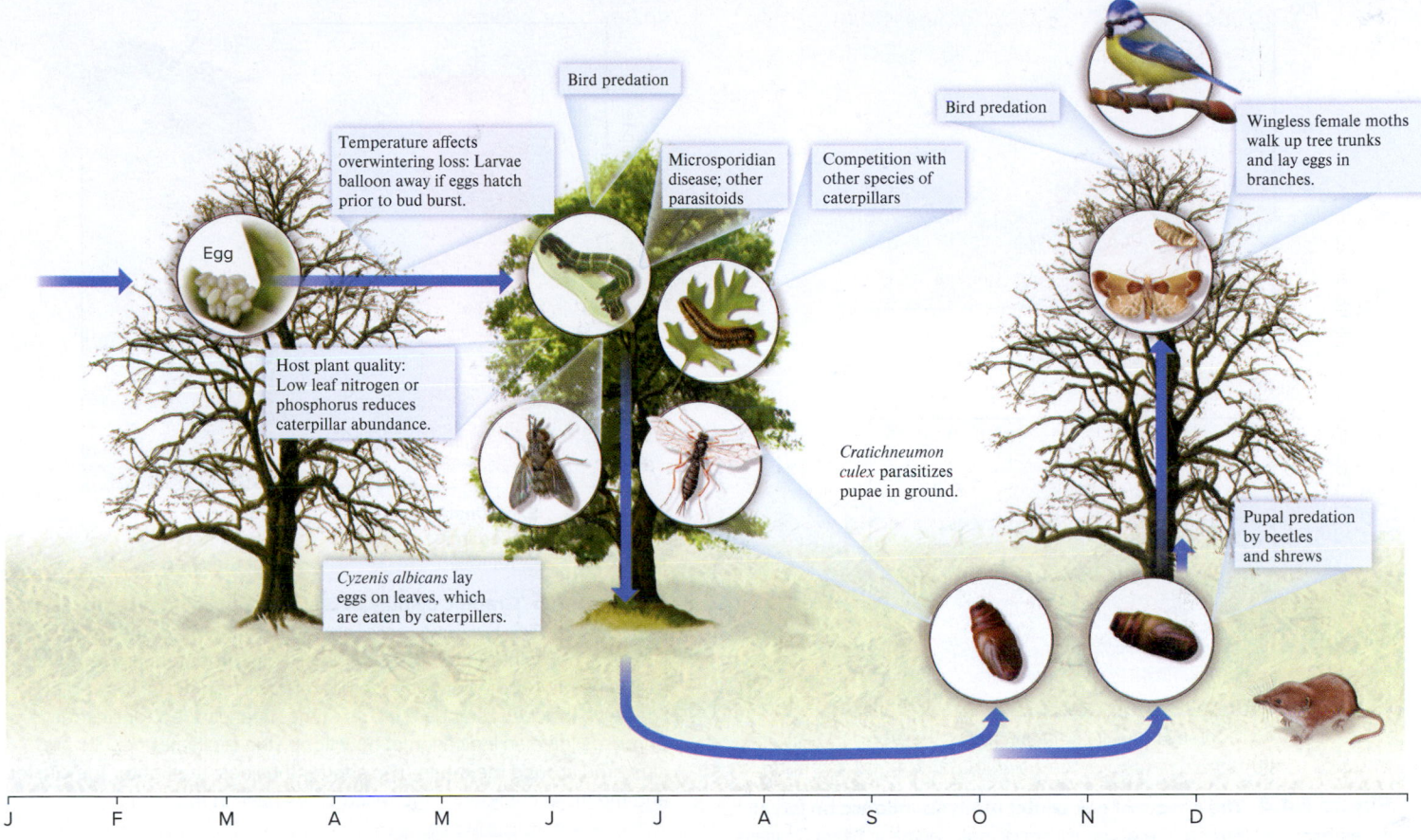

J   F   M   A   M   J   J   A   S   O   N   D

**Figure 54.3** **Web of interactions showing factors that might influence populations of oak winter moths.** Eggs hatch in early April, coincident with the opening of buds (bud burst). If leaves are not available, caterpillars balloon away on silken threads to try to reach trees with leaves. Caterpillars feed on leaves and may compete with other leaf feeders. Eggs of the parasite *Cyzenis albicans*, which are laid on the foliage, are unwittingly devoured by the caterpillars and the parasites hatch inside them. Caterpillars drop to the ground to pupate in the soil and leaf litter. Adult *Cratichneumon culex* directly parasitize pupae. Pupal predation is performed by soil-dwelling beetles and shrews. Adult, wingless, female oak winter moths climb up trees to lay their eggs in November and December.

Ecologists have to be cautious when forming conclusions based on correlations. For this reason, after conducting observations, ecologists usually turn to experiments to test their hypotheses.

## Experiments Are Conducted to Test Hypotheses

In Canada, after oak winter moth populations were reduced by parasitism due to *C. albicans*, additional control of the population sizes was thought to arise from pupal predation. High numbers of generalist beetle and shrew predators were observed in the field.

An experiment to test the hypothesis that pupal predation impacts oak winter moth abundance might involve removing pupal predators and examining subsequent survival rates of the moth pupae. Removing pupal predators might be achieved by putting out traps to catch predatory beetles and small mammals or by applying a pesticide to the soil in July to kill the predators prior to the moths' pupation in the fall. Ecologists could then measure survival rates of the pupae during the fall. If predators are having a significant effect, then removing them should cause moth pupal populations to substantially increase. The experiment would be conducted using two groups with equal numbers of trees: a group of trees in which the predators had been removed in the ground beneath (the experimental group), and a group of trees with predators still present (the control group). Any differences in oak winter moth pupal survival over the fall would likely be due to differences in predation.

## Data Analysis Permits Rejection or Acceptance of a Hypothesis

Performing an experiment several times is called **replication**. Ecologists might replicate the experiment involving the removal of predators of the oak winter moths five or ten times, or even more. At the end of the replications, the researchers would sum the total number of emerging moths, divide the sum by the number of replications, and calculate the mean. As a hypothetical example, let's suppose that the surviving numbers of moth pupae per tree are 5, 4, 7, 8, 12, 15, 13, 6, 8, and 10; the mean number of surviving pupae is then 8.8. For the control group, where the predators were not removed, the surviving numbers of pupae per tree are 2, 4, 7, 5, 3, 6, 11, 4, 1, and 3, with a mean of 4.6. Without predators, the mean number of pupae surviving is almost double the mean number surviving with predators. When subjected to a statistical analysis, the differences in the means are not likely to have occurred due to random chance. This analysis would allow ecologists to accept the hypothesis that predators were the cause of the changes in oak winter moth pupal abundance.

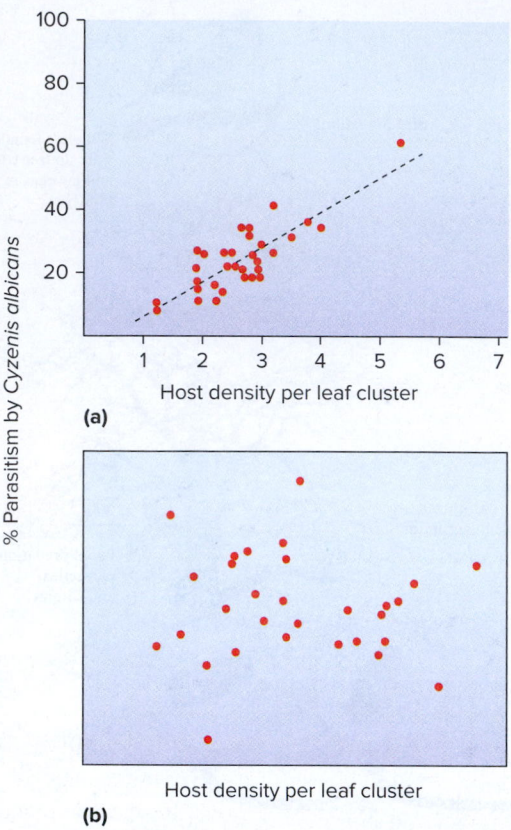

**(a)**

**(b)**

**Figure 54.4** **The effects of oak winter moth abundance on larval parasitism. (a)** Percentage of parasitism of moth larvae in Nova Scotia is dependent on the number of caterpillars per leaf cluster of 4–5 leaves. Data from an apple orchard in Nova Scotia in 1983. **(b)** In this hypothetical graph, there is no apparent relationship between the two variables.

 **Core Skill: Modeling** The goal of this modeling challenge is to make a graph that describes a different type of parasite/host relationship. Based on your graph, explain how this relationship could come about, and how it will affect the host population size.

**Modeling Challenge:** Although the percentage of parasitism of moth larvae by *C. albicans* increases with larval density on leaf clusters, other parasites and hosts may show a different outcome. Make a graph in which the percentage of parasitism decreases with host density. How could such a relationship come about? How would this relationship affect the size of the host population?

The results of such a predator removal experiment are displayed in the bar graph in **Figure 54.5**. Researcher Jens Roland applied the pesticide Diazinon to the soil prior to oak winter moth pupation. This treatment killed larvae and adults of predator beetles but not the moth pupae, because the pesticide had been degraded by the time the moth larvae dropped to the ground. Results showed that pupal predation was indeed important in influencing the numbers of surviving moth pupa.

Ecologists use a variety of tests to determine whether the differences between experimental and control groups are statistically significant. We will not cover the mechanics of these tests, but when experimental and control groups are described in this unit as differing, the results are statistically significant differences unless stated otherwise.

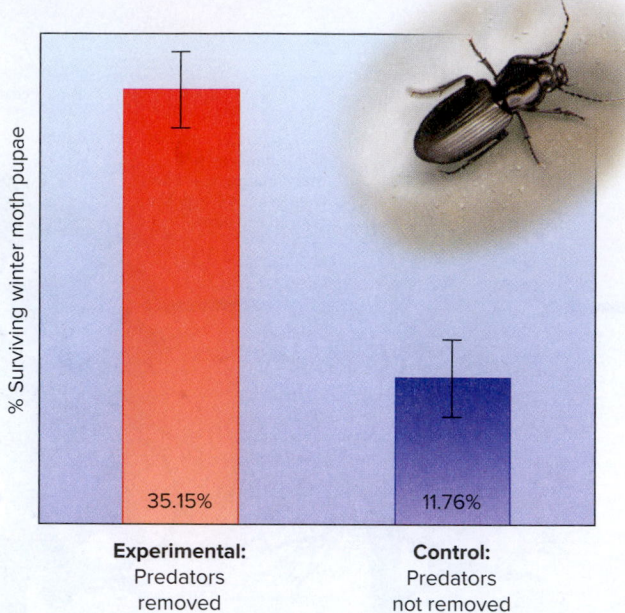

**Figure 54.5** **Graphic display of results of a predator removal experiment.** The two bars represent the average percentages of surviving oak winter moth pupae when beetle predators are removed (experimental) and when they are not removed (control). The vertical lines with brackets (standard errors of the mean; added for illustrative purposes) give an indication of how tightly the replicated results are clustered around the mean. The shorter the lines, the tighter the cluster and the more confidence the researchers have in the result.

BIO **TIPS** **THE QUESTION** *Let's suppose that over a period of several years, continual removal of the beetles that prey on the oak winter moths yielded the opposite results from those shown in Figure 54.5—that is, numbers of surviving moths increased where predators were not removed. How would you interpret these results?*

**T** **OPIC** *What topic in biology does this question address?* The topic is interpreting the results of field experiments, more specifically, the results of a predator removal experiment.

**I** **NFORMATION** *What information do you know based on the question and your understanding of the topic?* In the question, you are reminded that beetle predators eat oak winter moth pupae. From your understanding of the topic, you may remember that most predators feed on a variety of types of prey. Thus, beetle predators likely eat a variety of insect larvae and pupae, probably including pupae of other moths whose caterpillars eat the same kind of vegetation as oak winter moth pupae.

**P** **ROBLEM-SOLVING** **S** **TRATEGY** *Predict the outcome.* Think about the web of interactions that are possible among the oak winter moth, its beetle predators, and other competing caterpillar larvae.

**ANSWER** *One possible explanation is that predator removal increases the number of competing moths. In turn, increased competition might then reduce the number of oak winter moth caterpillars. In this scenario, the enemy (beetle) of the moth's enemy (competing caterpillar species) is the moth's friend.*

## 54.3 The Environment's Effect on the Distribution of Organisms

**Learning Outcomes:**

1. Give examples of how extremes of temperature, both low and high, drastically affect the distribution and abundance of organisms.

2. Explain how other environmental factors, such as wind, water availability, light availability, salt concentration, and pH of soil and water, can affect the distributions of organisms.

As mentioned in Section 54.2, in addition to natural enemies, physical (abiotic) factors have powerful effects on population sizes in most ecological systems. Both the distribution patterns of organisms and their abundance are limited by physical features of the environment such as temperature, wind, availability of water and light, salinity, and pH (**Table 54.1**). In this section, we will examine how these environmental factors affect natural populations.

### Temperature Has an Important Effect on the Distribution of Plants and Animals

Temperature is perhaps the most important factor in the distribution of organisms because of its effect on biological processes and because of the inability of most organisms to regulate their body temperature precisely. For example, the organisms that form coral reefs secrete a calcium carbonate shell. Shell formation and coral deposition are accelerated at high temperatures but are suppressed in cold water. Coral reefs are therefore abundant only in warm water, and a close correspondence is observed between the 20°C isotherm for the average daily temperature during the coldest month of the year and the limits of the distribution of coral reefs (**Figure 54.6**). (An isotherm is a line on a map connecting points of equal temperature.) Coral reefs are located between the two 20°C isotherm lines that are formed above and below the equator.

***Low Temperatures*** Frost is probably the single most important factor limiting the geographic distribution of tropical and subtropical

**Table 54.1  Selected Abiotic Factors and Their Effects on Organisms**

| Factor | Effect |
|---|---|
| Temperature | Low temperatures freeze many plants; high temperatures denature proteins. Some plants require fire for germination. |
| Wind | Wind amplifies effects of cool temperatures (wind chill) and water loss; creates pounding waves. |
| Water | Insufficient water limits plant growth and animal abundance; excess water drowns plants and other organisms. |
| Light | Insufficient light limits plant growth, particularly in aquatic environments. |
| Salinity | High salinity generally reduces plant growth in terrestrial habitats; affects osmosis in marine and freshwater environments. |
| pH | Variations in pH affect decomposition and nutrient availability in terrestrial systems; directly influence mortality in both aquatic and terrestrial habitats. |

plants. Cold temperatures can be lethal because cells may rupture if the water they contain freezes. Disruption of cells via freezing is especially lethal to plants that produce poisonous chemical defenses against herbivores. One form of white clover, *Trifolium repens*, produces cyanide as a defense against herbivores. However, frost injury is lethal to these plants because when cell membranes are disrupted, the toxin is released into other tissues. As a result, the cyanide-producing form is less common in colder regions of Europe (**Figure 54.7**).

The geographic range of endothermic animals are also affected by temperature. For example, the eastern phoebe (*Sayornis phoebe*), a small bird, has a northern winter range that coincides with an average minimum January temperature above 4°C. Such limits are probably related to the energy demands associated with cold temperatures. Cold temperatures mean higher metabolic costs, which, in turn, require high feeding rates. Below 4°C, the eastern phoebe cannot feed fast enough or, more likely, find enough food to keep warm.

***High Temperatures*** High temperatures are also limiting for many plants and animals because relatively few species can survive internal

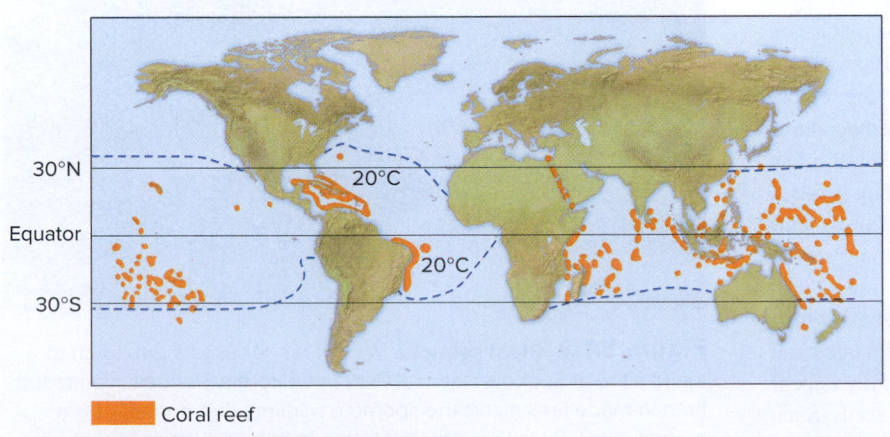

**Coral reef**

**(a) Worldwide distribution of coral reefs**

**(b) A coral reef**

**Figure 54.6  Worldwide locations of coral reefs. (a)** Coral reef formation is limited to waters bounded by the 20°C isotherm (dashed line), a line where the average daily temperature is 20°C during the coldest month of the year. **(b)** Coral reef in the Pacific Ocean. b: ©Michael McCoy/Science Source

**Concept Check:** *Why are coral reefs limited to warm water?*

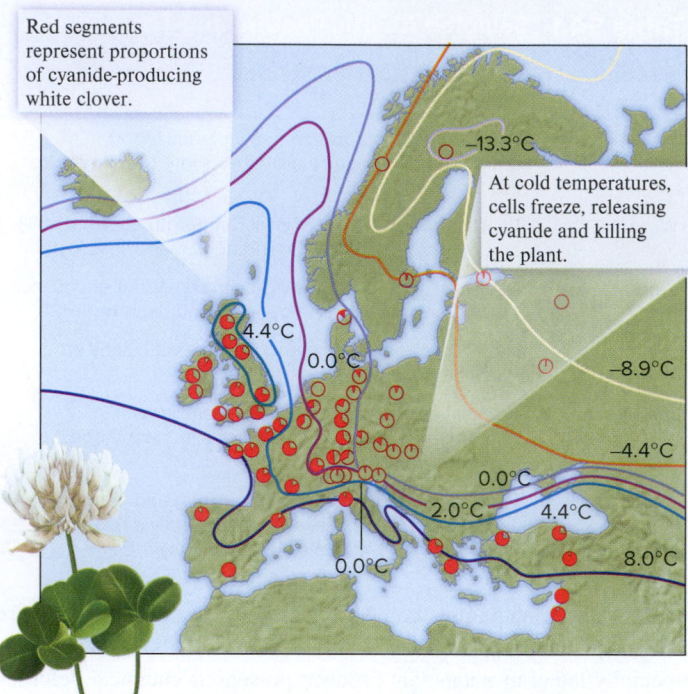

Red segments represent proportions of cyanide-producing white clover.

−13.3°C

At cold temperatures, cells freeze, releasing cyanide and killing the plant.

4.4°C

0.0°C

−8.9°C

−4.4°C

0.0°C

2.0°C    4.4°C

0.0°C    8.0°C

**Figure 54.7** **Frequency of the cyanide-producing form in populations of white clover, *Trifolium repens*, is affected by temperature.** The proportion of the cyanide-producing form is represented by the red section of each circle. This form is more common in warmer regions of Europe. Lines are January isotherms.

 **Core Concept: Systems** In many cases, the geographical ranges of organisms are determined by the physical environment.

temperatures more than a few degrees above their metabolic optimum. We have noted that corals are sensitive to low temperatures; however, they are sensitive to very high temperatures as well. When temperatures are too high, the symbiotic algae that live within coral die and are expelled, causing a phenomenon known as coral bleaching. Once bleaching occurs, the coral tissue loses its color and turns a pale white (**Figure 54.8**). El Niño is a weather phenomenon characterized by a major increase in the water temperature of the equatorial Pacific Ocean. In 2016, a massive wave of coral bleaching struck the Great Barrier Reef after an El Niño event brought abnormally warm waters to the region. Over 60 percent of the coral reefs were bleached. Elevated temperatures of only 2°C above the expected summertime maximum, for just a few weeks, were enough to kill over 50 percent of the reef-building corals.

The ultimate high temperatures that many terrestrial organisms face are the result of fire. However, some species depend on frequent low-intensity fires for their reproductive success. The jack pine (*Pinus banksiana*) of the northern U.S. produces serotinous cones, which remain sealed by pine resin until the heat of a fire melts them open and releases the seeds. In the west, giant sequoia trees are similarly dependent on periodic low-intensity fires for germination of their seeds. Such fires both enhance the release of seeds and clear out competing vegetation at the base of the tree so that seeds can germinate and grow. Fire-suppression practices that attempt to protect forests from fires can actually have undesirable results by preventing the regeneration of

**Figure 54.8** **Coral bleaching. Mantanani Island, Malaysia.**
©Jonathan Bird/Getty Images

fire-dependent species. Furthermore, fire prevention can result in an accumulation of vegetation beneath the canopy (the understory) that may later fuel hotter and more damaging fires. The U.S. Forest Service uses controlled human-made fires to mimic the natural disturbance of periodic fires and maintain fire-dependent forest species (**Figure 54.9**).

**Figure 54.9** **Giant sequoia.** A park ranger uses a drip torch to ignite a fire at Sequoia National Park in California. Periodic, controlled human-made fires mimic the sporadic wildfires that normally burn natural areas. Such fires are vital to the health of giant sequoia populations, because they serve to open the pine cones and release the seeds. ©Raymond Gehman/Corbis Documentary/Getty Images

*Concept Check:* *Why are some fires very destructive to natural systems?*

## Core Concept: Information

### Temperature Tolerance May Be Manipulated by Genetic Engineering

In this section, we have considered how low and high temperatures affect natural populations. The effects of temperature are also important in agriculture. For example, below-freezing temperatures can be very damaging to plant tissue, either killing the plant or greatly reducing its productivity. Frost injury causes losses to agriculture of more than $1 billion annually in the U.S. Frost has been considered an unavoidable result of subfreezing temperatures, but genetic engineering is beginning to change this view.

Between 0°C and –40°C, pure water will be liquid unless provided with an ice nucleus or template on which an ice crystal can be built. Any number of ions or molecules within an organism's cells can act as a template. Researchers discovered that some bacteria commonly found on leaf surfaces act as ice nuclei, triggering the formation of ice crystals and eventually causing frost damage. The genes that confer ice nucleation resistance have been identified, isolated, and deactivated in a genetically engineered strain of the bacteria *Pseudomonas syringae*. When this strain is allowed to colonize strawberries, frost damage is greatly reduced, and plants can withstand an additional 5°C drop in temperature before ice forms.

As another way to avoid frost damage and other types of damage due to cold temperatures, researchers are studying plants that grow in cold regions and identifying genes that confer cold tolerance. For example, certain genes encode enzymes that catalyze the synthesis of osmoprotectants, which are usually small, electrically neutral molecules that help to stabilize proteins and membranes against abiotic stresses such as cold temperatures. With regard to agriculture, a goal is to transfer such cold-tolerance genes to agriculturally important species, and thereby confer better resistance to cold and other abiotic stresses. Although the number of transgenic crops is relatively small, this approach has been applied to some varieties of tomato, tobacco, rice, maize, alfalfa, cotton, canola and flax. One drawback is possible side effects of the introduced gene. For example, transgenic tomatoes that are less sensitive to frost may exhibit dwarfism and reduced fruit set.

At the other end of the temperature spectrum, heat shock proteins (HSPs) help organisms cope with the stress of high temperatures. At high temperatures, proteins may denature, that is, either unfold or bind to other proteins to form misfolded aggregations. HSPs act as molecular chaperones, proteins that help in the proper folding of other proteins, to prevent these types of events from taking place (refer back to Figure 4.34). HSPs normally constitute only about 2% of a cell's soluble protein content, but this can increase to 20% when a cell is stressed, whether by heat, cold, drought, or other conditions. The genes that encode HSPs are extremely common and are found in the genomes of all organisms, from bacteria to plants and animals. In 2006, several genes responsible for inducing the synthesis of HSPs were identified in tomato and maize. Researchers have produced transgenic tobacco plants that grow better than normal plants under higher temperatures.

In the tropics, high temperatures can substantially decrease the growth rates and productivity of many crop species. There is now substantial interest in identifying crop strains with naturally high HSP levels for use in crop-breeding programs. Given the projected continuation of global warming, such research seems particularly timely.

## Wind Can Amplify the Effects of Temperature

Wind is created by temperature gradients. As air heats up, it becomes less dense and rises. As hot air rises, cooler air rushes in to take its place. For example, hot air rising in the tropics is replaced by cooler air flowing in from more temperate regions, thereby creating northerly or southerly winds.

Wind affects living organisms in a variety of ways. It increases the rate of heat loss by convection, the transfer of heat by the movement of air next to the body (the wind chill factor). Wind also contributes to water loss in organisms by increasing the rate of evaporation in animals and transpiration in plants.

Winds can also intensify oceanic wave action, with resulting effects for aquatic organisms. On the ocean's rocky shore, seaweeds survive heavy surf by relying on a combination of holdfasts and flexible structures. The animals of this zone have powerful organic glues and muscular feet to hold them in place (**Figure 54.10**).

## The Availability of Water Has Important Effects on the Abundance of Organisms

As discussed in Chapter 2, water performs many vital functions in all living organisms. It acts as a solvent for chemical reactions, takes part in hydrolysis and dehydration reactions, is the means by which animals eliminate wastes, and is used for support in plants and in some invertebrates as part of a hydrostatic skeleton.

The distribution patterns of many plants are limited by available water. For example, the density of creosote bushes in the Mojave Desert increases in wetter areas. In cold climates, water can be present but locked up as permafrost and, therefore, unavailable. Alpine trees stop growing at a point on the mountainside where they cannot take up enough moisture to offset transpiration losses. This point, known as the timberline, is readily apparent on many mountainsides.

Animals face problems of water balance, too, and their distribution and population density are strongly affected by water availability. Because most animals depend ultimately on plants for food, their distribution is intrinsically linked to those of their food sources. Such a phenomenon regulates the number of buffalo (*Syncerus caffer*) in the Serengeti area of Africa. In this area, grass productivity is related to the amount of rainfall in the previous month. Buffalo density is governed by grass availability, so a significant correlation is found between buffalo density and rainfall (**Figure 54.11**). The only exception occurs in the vicinity of Lake Manyara, where groundwater promotes plant growth.

**(a) Brown alga with a holdfast**

**(b) A mussel with byssal threads**

**Figure 54.10** **Animals and plants of the intertidal zone adhering to their rocky surface.** **(a)** The brown alga (*Laminaria digitata*) has a holdfast that enables it to cling to the rock surface. **(b)** The mussel (*Mytilus edulis*) attaches to the surface of a rock by proteinaceous threads (byssal threads) that extend from the animal's muscular foot. a: ©FLPA/D P Wilson/age fotostock; b: ©Biophoto Associates/Science Source

## Light Can Be a Limiting Resource for Plants and Algae

Because light is necessary for photosynthesis, it can be a limiting resource for plants. However, what may be sufficient light to support the growth of one plant species may be insufficient for another. Many plant species, such as eastern hemlock (*Tsuga canadensis*), grow best in shady conditions. Hemlock saplings grow in the understory below the forest canopy, reaching maximal photosynthesis at one-quarter of full sunlight. Other plants, such as sugarcane (*Saccharum officinarum*), continue to increase their photosynthetic rate as light intensity increases.

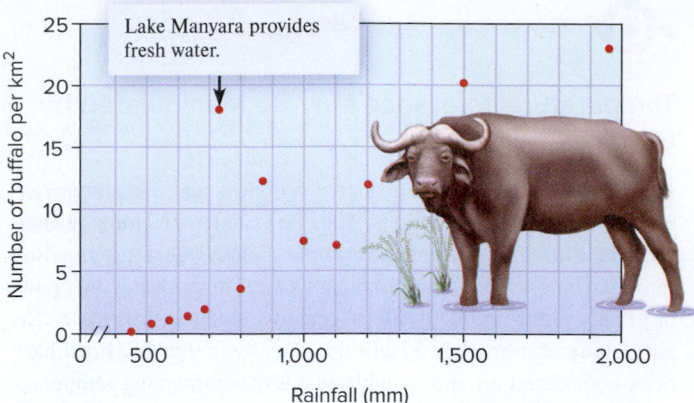

**Figure 54.11** **The relationship between the amount of rainfall and the density of buffalo.** In the Serengeti area of Africa, buffalo density is dependent on grass availability, which itself depends on annual rainfall. An exception is near bodies of water, such as Lake Manyara. Greater water availability leads to greater grass growth and buffalo densities.

In aquatic environments, light may be an even more limiting factor because water absorbs light, preventing photosynthesis at depths greater than 100 m. Most aquatic plants and algae are therefore limited to a fairly narrow zone close to the surface, where light is sufficient to allow photosynthesis to occur. This zone is known as the **photic zone**. In marine environments, seaweeds at greater depths have wider thalli (leaf-like light-gathering structures) than those nearer the surface, because wide thalli can collect more light. In addition, in aquatic environments, plant color changes with depth. At the surface, plants and algae appear green, as they are in terrestrial conditions, because they absorb red and blue light, but not green (**Figure 54.12a**). At greater depths, red light is mostly absorbed by water, leaving predominantly blue-green light. Red algae found in deeper water possess pigments that enable them to utilize blue-green light efficiently and that reflect red light (**Figure 54.12b**).

## The Concentration of Salts in Soil or Water Can Be Critical

Salt concentrations vary widely in aquatic environments and have a great effect on osmotic balance in animals. Oceans contain considerably more dissolved minerals than rivers because oceans continually receive the nutrient-rich waters of rivers, and the Sun evaporates pure water from ocean surfaces, making concentrations of minerals such as salts even higher.

The phenomenon of osmosis influences how living organisms cope with different environments. Freshwater fishes cannot live in salt water, and saltwater fishes cannot live in fresh water. Each employs different mechanisms to maintain an osmotic balance with their environment (refer back to Figure 41.17). Freshwater fishes are hyperosmotic (having a greater concentration of solutes) relative to their environment and tend to gain water by osmosis as it diffuses through the thin tissue of the gills and mouth. To counter this, the fish continually eliminate water in the urine. However, to avoid losing all dissolved ions, many ions are reabsorbed into the bloodstream at the kidneys. Many marine fishes are hypo-osmotic (having a lower concentration of solutes) relative to their environment and tend to lose water as seawater passes over the mouth and gills. They drink water

**(a) Green algae at the ocean surface**

**(b) Red algae at a greater depth**

**Figure 54.12**  **Algae growing at different ocean depths.** **(a)** In the eastern Pacific Ocean, off the coast of California, these giant kelp floating at the ocean surface are green, just like terrestrial plants. **(b)** In contrast, at 75-m depth, in the McGrail Bank off the coast of Texas in the Gulf of Mexico, most seaweeds are pink and red because their pigments absorb the blue-green light that reaches such depths. a: ©Gregory Ochocki/Science Source; b: Source: FGBNMS/UNCW-NURC/NOAA

to compensate for this loss, but the water contains a higher concentration of salt, which must then be excreted at the gills and kidneys.

Salt in the soil also affects the growth of plants. In arid terrestrial regions, salt accumulates in soil where water settles and then evaporates. Salt accumulation can be of great significance to agriculture in arid environments where continued watering, together with the addition of salt-based fertilizers, greatly increases salt concentration in the soil and reduces crop yields. A few terrestrial plants are adapted to live in saline soil along seacoasts. Here the vegetation consists largely of **halophytes**, species that can tolerate higher salt concentrations in their cell sap than regular plants. Species such as mangroves and *Spartina* grasses have salt glands that exude salt onto the surface of the leaves, where it forms tiny white salt crystals (**Figure 54.13**).

## The pH of Soil or Water Can Limit the Distribution of Organisms

As discussed in Chapter 2, the pH of water can be acidic, alkaline, or neutral. Variation in pH can have a major effect on the distribution of organisms. Normal rainwater has a pH of about 5.6, which is slightly

**Figure 54.13**  **Plant adaptations for salty conditions.** Special salt glands in the leaves of *Spartina* exude salt, enabling this grass to exist in saline intertidal conditions. ©Virginia P. Weinland/Science Source

 **Core Skill: Connections**  Look back to Section 38.2. Why can't most plants grow in salty habitats?

acidic because the absorption of atmospheric carbon dioxide ($CO_2$) and sulfur dioxide ($SO_2$) into rain droplets forms carbonic and sulfuric acids, respectively. However, most plants grow best at a soil water pH of about 6.5, a value at which soil nutrients are most readily available to plants. Only a few genera, such as rhododendrons and azaleas (genus *Rhododendron*), can live in soils with a pH of 4.0 or less. Furthermore, at a pH of 5.2 or less, nitrifying bacteria do not function properly, which prevents organic matter from decomposing. In general, alkaline soils containing chalk and limestone have a higher pH and sustain a much richer flora (and associated fauna) than do acidic soils (**Figure 54.14**).

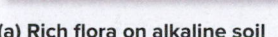

**(a) Rich flora on alkaline soil**

**(b) Sparse flora on acidic soil**

**Figure 54.14**  **Species-rich flora of chalk grassland compared with species-poor flora of acid soils.** **(a)** At Mount Caburn in England, the lime-rich chalk hills foster a much greater variety of plant and animal species than are observed at **(b)** a heathland site elsewhere in England. Heathlands are a product of thousands of years of human clearance of natural forest areas and are characterized by acidic, nutrient-poor soils. a: ©Peter Wakely/English Nature; b: ©G. A. Matthews/ SPL/Science Source

**Concept Check:**  *Why do acidic soils support fewer species of plants and animals than lime-rich soils?*

Likewise, the number of fishes and other species also decreases in acidic waters. The optimal pH for most freshwater fishes and bottom-dwelling invertebrates is between 6.0 and 9.0. Acidity in lakes increases the amount of toxic metals, such as mercury, aluminum, and lead, which can leach into the water from surrounding soil and rock. Too much mercury or too much aluminum can interfere with gill function, causing fishes to suffocate.

The susceptibility of both aquatic and terrestrial organisms to changes in pH explains why ecologists are so concerned about **acid rain**, precipitation with a pH of less than 5.6. Acid rain results from the burning of fossil fuels such as coal, oil, and natural gas, which releases $SO_2$ and nitrogen oxide ($NO_2$) into the atmosphere. These react with oxygen in the air to form sulfuric acid and nitric acid, which falls to the Earth's surface in rain or snow. Such precipitation is most harmful to aquatic environments. Acid rain can make rivers and especially lakes more acidic, diminishing their ability to sustain fishes and other aquatic life.

## 54.4 Climate and Its Relationship to Biological Communities

### Learning Outcomes:

1. Explain how global temperature differentials drive atmospheric circulation.
2. Explain how both mountains and large bodies of water can change local temperature and precipitation patterns.

Temperature, wind, precipitation, and light are components of **climate**, the prevailing weather pattern in a given region. As we have seen, the distribution and abundance of organisms are influenced by these factors. Therefore, to understand the patterns of abundance of life on Earth, ecologists need to study the global climate. In this section, we will examine global climate patterns, focusing on how temperature variation drives atmospheric circulation and how features such as elevation and landmass can alter these patterns.

### Atmospheric Circulation Is Driven by Global Temperature Differentials

Substantial differences in temperature occur over the Earth, mainly due to latitudinal variations in the incoming solar radiation. At higher latitudes, such as in northern Canada and Russia, the Sun's rays hit the Earth obliquely and are spread out over more of the planet's surface than they are in equatorial areas (**Figure 54.15**). More heat is also lost in the atmosphere of higher latitudes because the Sun's rays travel a greater distance through the atmosphere, allowing more heat to be dissipated by cloud cover. The result is that 40% less solar energy strikes polar latitudes than equatorial areas. Generally, temperatures increase as the amount of solar radiation increases (**Figure 54.16**). However, in the tropics, both cloudiness and rain reduce average temperature, so temperatures do not continue to increase toward the equator.

Global patterns of atmospheric circulation and precipitation are influenced by solar energy. In 1735, English meteorologist George Hadley made the initial contribution to a model of general

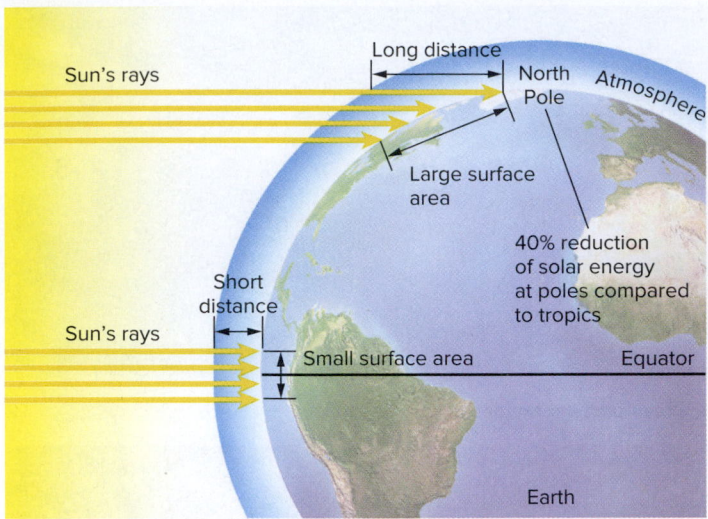

**Figure 54.15  The intensity of solar radiation at different latitudes.** In polar areas, the Sun's rays strike the Earth at an oblique angle and deliver less energy than at tropical locations. In tropical areas, the energy is concentrated over a smaller surface and travels a shorter distance through the atmosphere.

atmospheric circulation. In his model, high temperatures at the equator cause the surface equatorial air to heat up and rise vertically into the atmosphere. The vertical rising of the hot air cools the land by convection (look back to Figure 47.10). As the warm air rises away from its source of heat, it cools and becomes less buoyant, but this cool air does not sink back to the surface because of the warm air underneath it. The warm air rising near the equator forms towers of cumulus clouds that provide rainfall, which, in turn, maintains the lush vegetation of the equatorial rain forests. As the upper flow

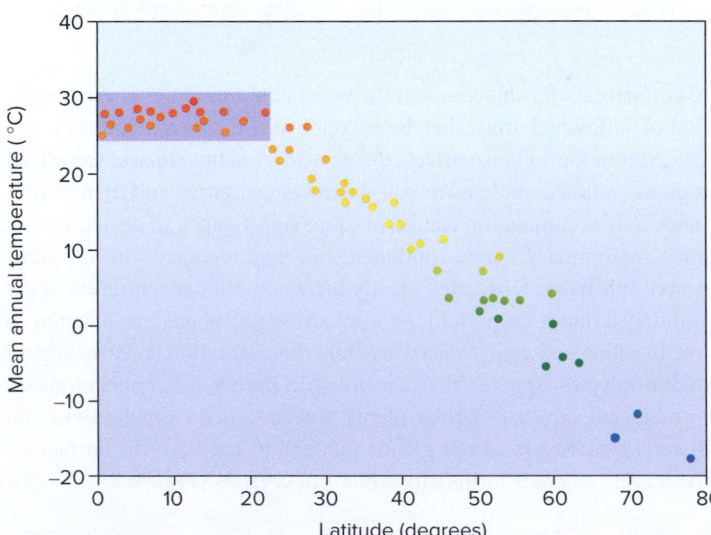

**Figure 54.16  Variation of the Earth's temperature.** The temperatures shown in this figure were measured at moderately moist continental locations of low elevation.

*Concept Check:* *Why is there a wide band of similar temperatures at the tropics? (See the orange and red dots highlighted by the purple rectangle.)*

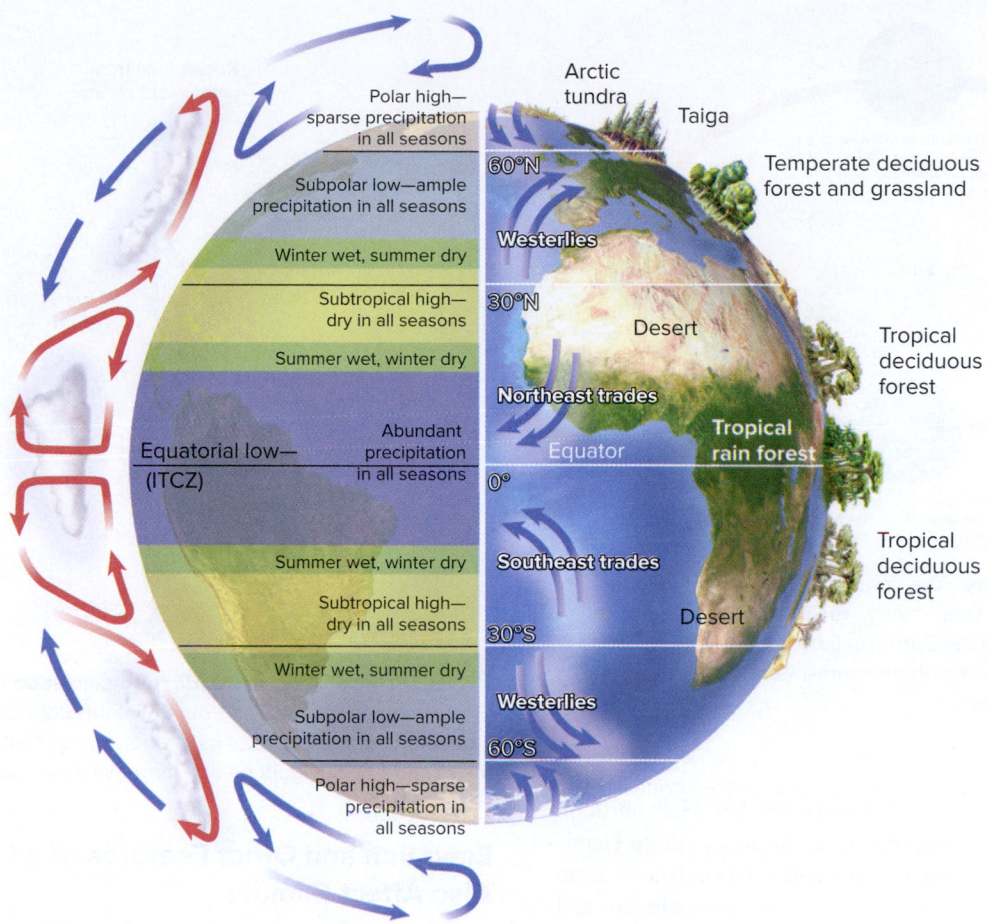

**Figure 54.17 Global circulation based on a modified three-cell model.** Tropical forests exist mainly in a band around the equator, where it is hot and rainy. At latitudes around 20° to 30° north and south, the air is hot and dry, and deserts exist. A secondary zone of precipitation exists at latitudes around 45° to 55° north and south, where temperate forests are located. The polar regions are generally cold and dry. The term high refers to areas of high pressure resulting from falling air. Low refers to areas of low pressure resulting from rising air. Northerly or southerly air movements are deflected west or east and are shown as westerlies, or so-called northeast/southeast trade winds, which allowed sailing ships to explore the world. The ITCZ is the intertropical convergence zone.

of air moves toward the poles, it begins to subside, or fall back to Earth, at latitudes of about 20–30° north and south of the equator. These **subsidence zones** are areas of high pressure and are the sites of the world's tropical deserts, because the subsiding air is relatively dry, having released all of its moisture over the equator (**Figure 54.17**).

From the centers of the subsidence zones, the surface flow splits into two directions, one of which flows toward the pole and the other toward the equator. The equatorial flow from both hemispheres meets near the equator in a region called the intertropical convergence zone (ITCZ). In the three-cell model, the circulation between latitudes 30° and 60° is opposite that of the cell nearest the equator because the net surface flow is poleward. Additional zones of high precipitation occur in this cell, usually between latitudes 45° and 55°. In the final circulation cell, at the poles, the air has cooled and descends, but it has little moisture left, explaining why many high-latitude regions are desert-like. The distributions of the major biomes discussed in Section 54.5 are largely determined by temperature differences and

the wind and rainfall patterns they generate. Hot, tropical forest blankets the tropics, where rainfall is high. At latitudes of about 20° to 30°, the air cools and descends, but it is without moisture, so hot deserts occur in that area. The middle cell of the circulation model shows us that at latitudes of about 45° to 55°, the air has warmed and gained moisture, so it ascends, dropping rainfall over the wet, temperate forests of the Pacific Northwest and Western Europe in the Northern Hemisphere and New Zealand and Chile in the Southern Hemisphere.

## The Tilt and Rotation of the Earth Also Affect Climate

The three-cell model provides a good understanding of global circulation, but it is oversimplified. Overlaid on these patterns is a seasonality caused by the tilt of the Earth. The Earth's axis of rotation is tilted at 23.5° from the vertical for the full 365 days it takes to orbit the Sun (**Figure 54.18**). The solar equator, the area receiving

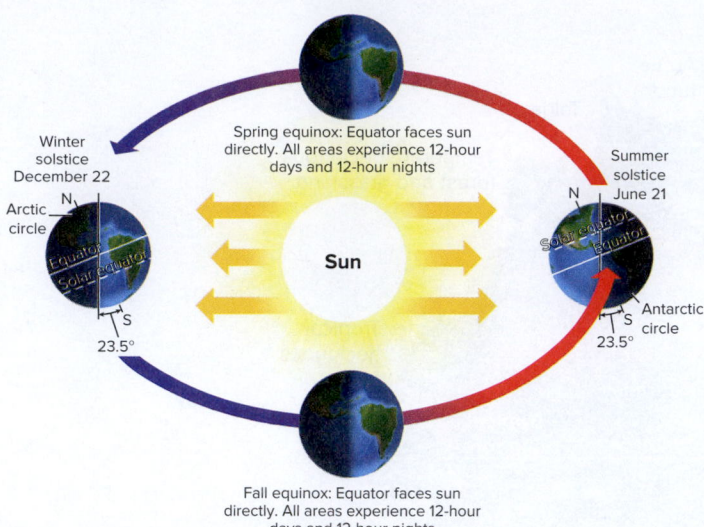

**Figure 54.18** **Seasonality.** The constant 23.5° tilt of the Earth's axis causes changes in the solar energy striking the Earth during its 365-day procession around the Sun. The times of the solstices and equinoxes are shown for the Northern Hemisphere and are reversed for the Southern Hemisphere.

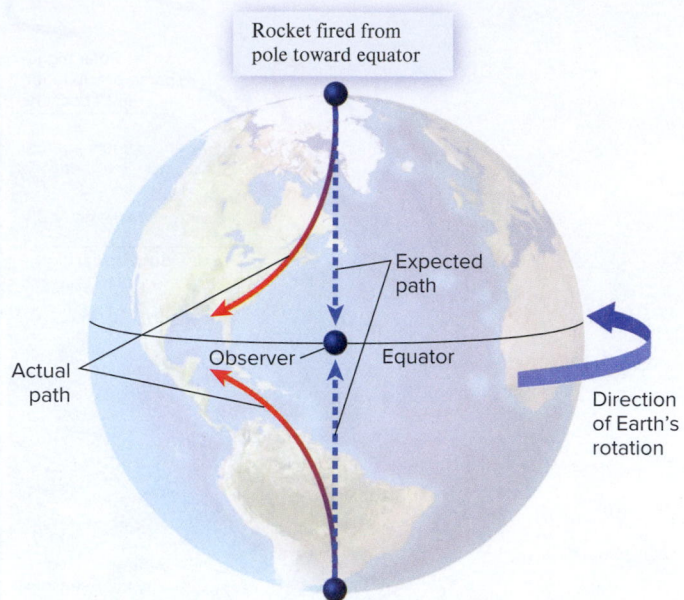

**Figure 54.19** **Diagrammatic representation of the Coriolis force.** Even though a rocket fired from the North Pole flies due south, by the time it reaches the equator its intended target would have moved and the landing point would be at a more westward spot.

the most solar energy, varies seasonally and reaches 23.5° north on June 21 and 23.5° south on December 21. In the Northern Hemisphere, these dates are called the summer and winter solstices (from the Latin *sol,* sun, and *sistit,* stands). For several days before and after each solstice, the noontime elevation of the Sun appears in the same place. On March 21 and September 22, the so-called spring and autumn equinoxes, all locations in the Northern and Southern Hemispheres have approximately equal amounts of solar radiation. This means that for half of the year the Northern Hemisphere receives more solar energy, and for the other half of the year the Southern Hemisphere receives more solar energy. At 60° north, during the northern winter, temperatures in Siberia may average only –12°C, whereas in the summer they may average 16°C, a difference of 28°C. In contrast, tropical temperatures vary relatively little, perhaps 2–3°C, year round. Southern Hemisphere temperatures also vary seasonally, but the large expanses of open water moderate the temperature extremes.

In addition to seasonal changes in temperatures, the wind direction is deflected by the rotation of the Earth, a phenomenon called the Coriolis force, after French scientist Gaspard Gustave de Coriolis. Imagine a rocket fired from the North or South Pole toward the equator. By the time the rocket reaches the equator, the target would have moved to the east. The rocket always travels due south or north, but to an observer at the equator it would bend some distance to the west before landing (**Figure 54.19**). A similar phenomenon occurs with winds and aircraft, though pilots compensate for Coriolis forces when flying. The opposite phenomenon happens when a rocket is fired from the equator either north or south. The Coriolis force deflects wind directions east or west, gives spin to storm systems, and is a reason hurricanes spin counterclockwise in the Northern Hemisphere but clockwise in the Southern Hemisphere.

## Elevation and Other Features of a Landmass Can Also Affect Climate

Thus far, we have considered how global temperatures and wind patterns affect climate. The geographic features of a landmass can also have an important effect. For example, the elevation of a region greatly influences its temperature range. On mountains, temperatures decrease with increasing elevation. This decrease is a result of a process known as **adiabatic cooling**, in which increasing elevation leads to a decrease in air pressure. When air is blown across the Earth's surface and up over mountains, it expands because of the reduced pressure. As it expands, it cools at a rate of about 10°C for every 1,000 m in elevation, as long as no water vapor or cloud formation occurs. (Adiabatic cooling is also the process applied in the function of a refrigerator, in which refrigerant gas cools as it expands coming out of the compressor.) A vertical ascent of 600 m produces a temperature change roughly equivalent to that brought about by an increase in latitude of 1,000 km. This explains why mountaintop vegetation, even in tropical areas, can have the characteristics of a colder biome.

Mountains can also influence patterns of precipitation. When warm, moist air encounters the windward side of a mountain, it flows upward and cools, releasing precipitation in the form of rain or snow. On the side of the mountain sheltered from the wind (the leeward side), drier air descends, producing what is called a **rain shadow**, an area where precipitation is noticeably less (**Figure 54.20a**). For example, the western side of the Cascade Range in Washington State receives more than 500 cm of annual precipitation, whereas the eastern side receives only 50 cm.

The proximity of a landmass to a large body of water can affect climate because land heats and cools more quickly than the sea.

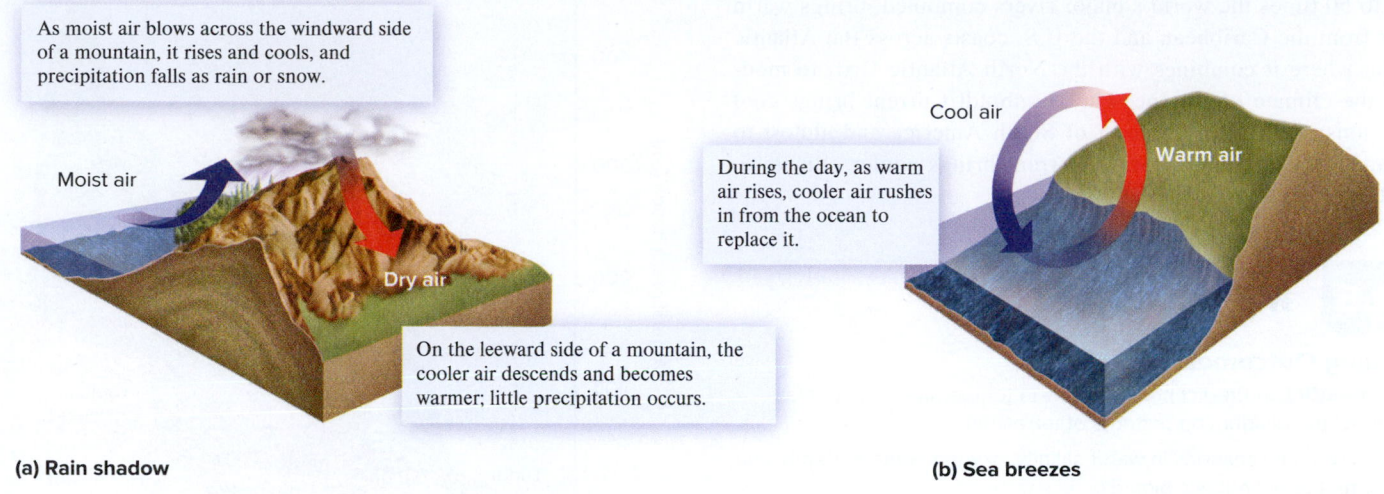

(a) Rain shadow

As moist air blows across the windward side of a mountain, it rises and cools, and precipitation falls as rain or snow.

Moist air

Dry air

On the leeward side of a mountain, the cooler air descends and becomes warmer; little precipitation occurs.

(b) Sea breezes

Cool air

Warm air

During the day, as warm air rises, cooler air rushes in from the ocean to replace it.

**Figure 54.20**  **The influence of elevation and proximity to water on climate.**

Recall from Chapter 2 that water has a very high specific heat—the amount of energy required to raise the temperature of 1 gram of a substance by 1°C. The specific heat of land is much lower than that of water, allowing the land to warm quicker than the water. During the day, the warmed air rises and cooler air flows in to replace it. This pattern creates the familiar onshore sea breezes in coastal areas (**Figure 54.20b**). At night, the land cools quicker than the sea, and so the pattern is reversed, creating offshore breezes. The sea, therefore, has a moderating effect on the temperatures of coastal regions and especially islands. The climates of coastal regions may differ markedly from those of their climatic zones. Many never experience frost, and fog is often evident. Thus, along coastal areas, vegetation patterns may differ from those in areas farther inland. Some areas of the U.S. would be deserts were it not for the warm water of the sea and the moisture-laden clouds that form above them.

Together with the rotation of the Earth, winds also create ocean currents. The major ocean currents act as "pinwheels" between continents, running clockwise in the ocean basins of the Northern Hemisphere and counterclockwise in those of the Southern Hemisphere (**Figure 54.21**). The Gulf Stream, equivalent in

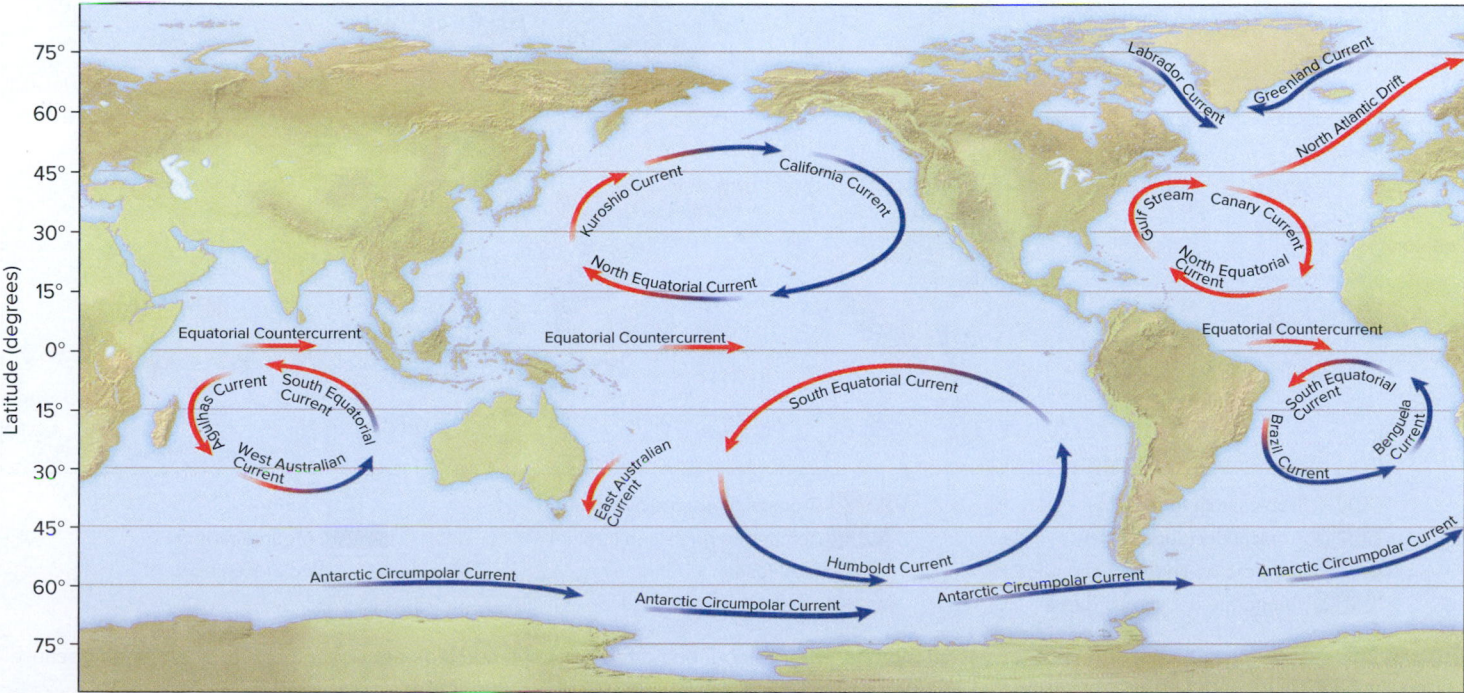

**Figure 54.21**  **Major ocean currents of the world.**  The red arrows represent warm water; the blue arrows, cold water.

flow to 50 times the world's major rivers combined, brings warm water from the Caribbean and the U.S. coasts across the Atlantic Ocean, where it combines with the North Atlantic Drift to moderate the climate of Europe. The Humboldt Current brings cool conditions to the western coast of South America and almost to the equator, and the California Current brings cooler climate to the Hawaiian Islands.

## 54.5   Major Biomes

### Learning Outcomes:

1. **CoreSKILL »** Predict how changes in temperature and rainfall affect the distribution patterns of terrestrial biomes.

2. Discuss how changes in water salinity, oxygen content, depth, and current affect aquatic biomes.

Differences in climate on Earth help to define its different **biomes**, which are major types of habitat characterized by distinctive plant and animal life. Biomes can be subdivided into terrestrial and aquatic. Many types of classification schemes are used for mapping the geographic extent of terrestrial biomes, but one of the most useful was developed by American ecologist Robert Whittaker, who classified terrestrial biomes according to the physical factors of average annual precipitation and temperature (**Figure 54.22**). This classification scheme recognizes 10 terrestrial biomes (**Figure 54.23**). Aquatic biomes are generally

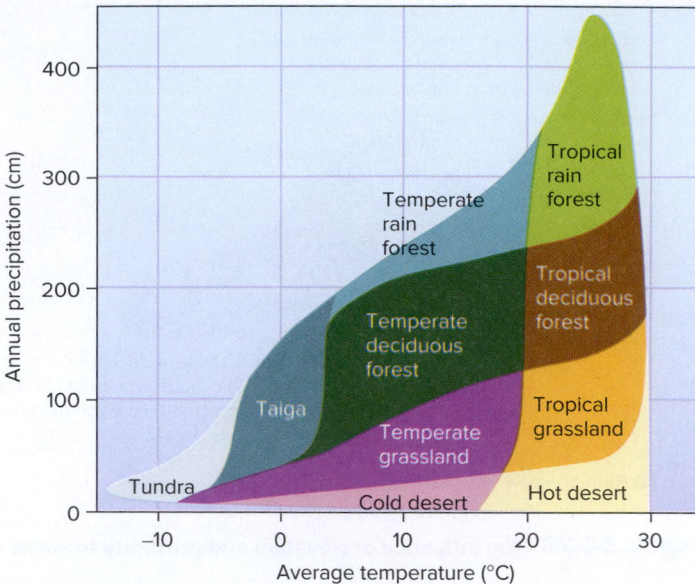

**Figure 54.22** **The relationship between the world's terrestrial biome types and temperature and precipitation patterns.**

*Concept Check:* *What other factors may influence biome types?*

differentiated by water salinity, current strength, water depth, oxygen content, and light availability. In this section, we will explore the main characteristics of Earth's major terrestrial and aquatic biomes.

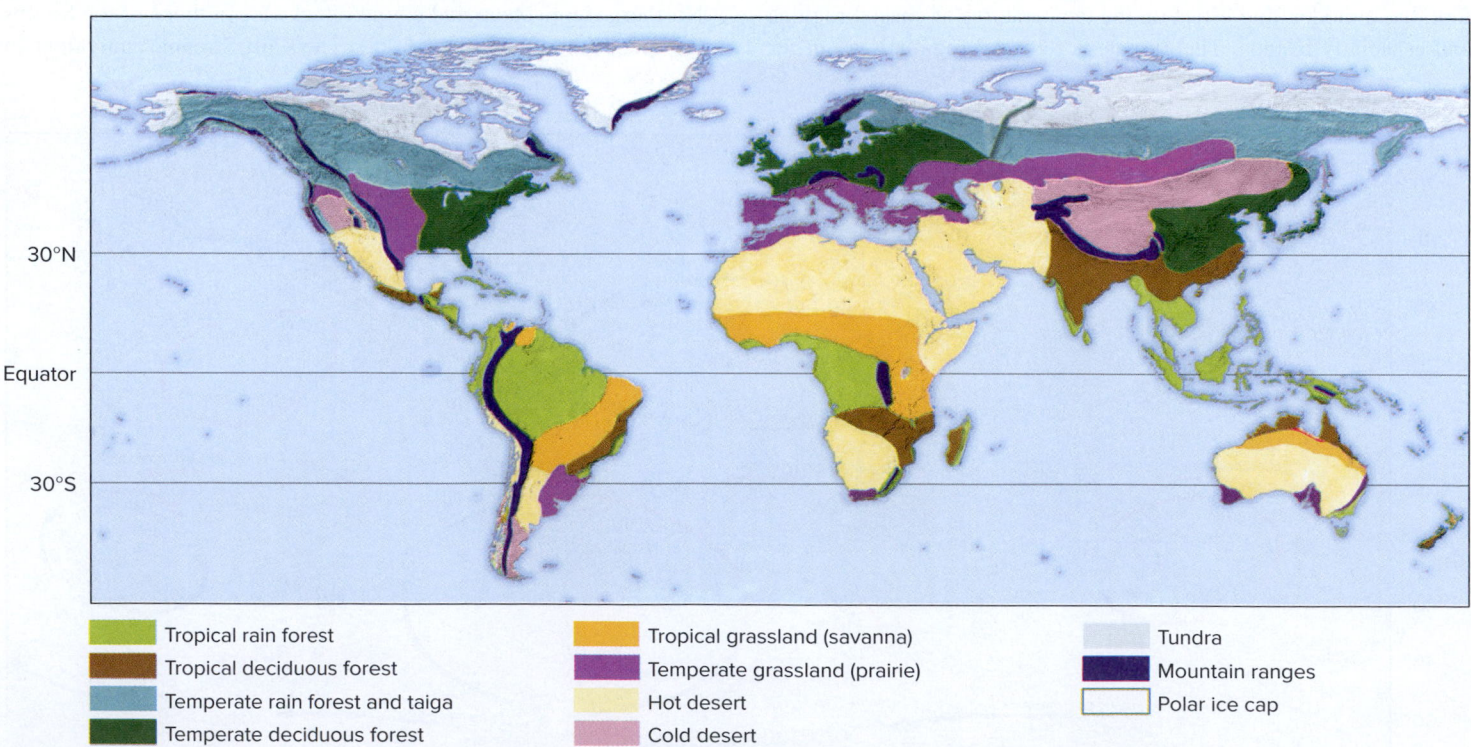

| Tropical rain forest | Tropical grassland (savanna) | Tundra |
| Tropical deciduous forest | Temperate grassland (prairie) | Mountain ranges |
| Temperate rain forest and taiga | Hot desert | Polar ice cap |
| Temperate deciduous forest | Cold desert | |

**Figure 54.23** **Geographic locations of terrestrial biomes.** The distribution patterns of taiga and temperate rain forest are combined because of their similarity in tree species and because temperate rain forest is actually limited to a very small area.

*Concept Check:* *In a globally warmed world, what biome might expand into areas currently occupied by tundra?*

# Terrestrial Biomes

**Figure 54.24a–j** illustrates the 10 major terrestrial biomes and identifies their main characteristics. Although broad terrestrial biomes are a useful way of defining the main types of communities on Earth, ecologists acknowledge that not all communities fit neatly into one of these ten major biome types. Also, one biome type often grades into another, as seen on mountain ranges (**Figure 54.24k**). Soil conditions can also influence biome type. In California, serpentine soils, which are dry and nutrient-poor, support only sparse vegetation. In the eastern U.S., most of New Jersey's coastal plain, called the Pine Barrens, consists of sandy, nutrient-poor soil that cannot support the surrounding deciduous forest and instead contains grasses and low shrubs growing among open stands of pygmy pitch pine and oak trees.

## Tropical Rain Forest

Figure 54.24a

### Tropical rain forest in Java

**Physical Environment:** Rainfall exceeds 230 cm per year, and the temperature is hot year round, averaging 25–29°C. Soils are often shallow and nutrient-poor.

**Location:** This biome is found in equatorial regions. Tropical forests cover much of northern South America, Central America, western and central Africa, Southeast Asia, and various islands in the Indian and Pacific oceans.

**Plant Life:** The numbers of plant species found in tropical forests can be staggering, often reaching as many as 100 tree species per square kilometer. Leaves often narrow to drip-tips at the apex so that rainwater drains quickly. Many trees have large buttresses that help support their shallow root systems. Little light penetrates the **canopy**, the uppermost layer of tree foliage, and the ground cover is often sparse. Vines and epiphytes, plants that live perched on trees and are not rooted in the ground, are common.

**Animal Life:** Animal life in the tropical rain forests is diverse; insects, reptiles, amphibians, and mammals are well represented. Large mammals, however, are not common. Because many of the plant species are widely scattered in tropical forests, plants do not typically rely on wind for pollination or to disperse their seed. Instead, animals are important in pollinating flowers and dispersing fruits and seeds. Mimicry and bright protective coloration, warning of bad taste or the existence of toxins, are common.

**Effects of Humans:** Humans are affecting tropical forests greatly by logging and by clearing the land for agriculture. Many South American tropical forests are cleared to create grasslands for cattle.

©Vladislav T. Jirousek/Shutterstock

# Terrestrial Biomes (continued)

## Tropical Deciduous Forest

Figure 54.24b

©Theo Allofs/theoallofs.com

**Tropical deciduous forest in Bandhavgarh National Park, India**

**Physical Environment:** Rainfall is substantial, at around 130–280 cm a year, and temperatures are hot year round, averaging 25–39°C. This biome experiences a distinct dry season that often lasts 2–3 months or longer. Shortages of water in the soil can occur in the dry season.

**Location:** This biome exists in equatorial regions where rainfall is more seasonal than it is in tropical rain forests. Much of India consists of tropical deciduous forest, containing teak trees. Brazil, Thailand, and Mexico also contain tropical deciduous forest. At the wet edges, this biome may grade into tropical rain forests; at the dry edges, it may grade into tropical grasslands or savannas.

**Plant Life:** Because of the distinct dry season, many of the trees in tropical deciduous forests shed their leaves, just as they do in temperate forests, and an understory of herbs and grasses may grow during this time. Where the dry season is 6–7 months long, tropical deciduous forests may contain shorter, thorny plants such as acacia trees, whose thorns deter moisture-seeking animals, and the forest is then referred to as a tropical thorn forest.

**Animal Life:** The diversity of animal life is high, and species such as monkeys, antelopes, wild pigs, and tigers are present. However, as with plant diversity, animal diversity is less than that of tropical rain forests. Tropical thorn forests may contain more browsing mammals; hence, the evolution of plant thorns as a defense.

**Effects of Humans:** The soil of tropical deciduous forests is more fertile than that of tropical rain forests. Land is increasingly being logged and cleared for agriculture and a growing human population.

## Temperate Rain Forest

Figure 54.24c

### Hoh Rain Forest in Olympic National Park, Washington

**Physical Environment:** Rainfall is abundant, usually exceeding 200 cm a year. The condensation of water from dense coastal fogs augments the normal rainfall. Temperatures seldom drop below freezing in the winter, and summer temperatures rarely exceed 27°C.

**Location:** The area of this biome type is small, consisting of a thin strip along the northwest coast of North America from northern California through Washington State, British Columbia, and into southeastern Alaska (where it is called tongass). It also exists in southwestern South America along the Chilean coast. It is found only in coastal locales because of the moderating influence of the ocean on air temperature.

**Plant Life:** The dominant vegetation type, especially in North America, consists of large evergreen trees such as western hemlock, Douglas fir, and Sitka spruce. The high moisture content allows epiphytes to thrive. Cool temperatures slow the activity of decomposers, so the litter layer is thick and spongy.

**Animal Life:** In North America, the temperate rain forest is rich in species such as mule deer, elk, squirrels, and numerous birds such as jays and nuthatches. Because of the abundant moisture and moderate temperatures, reptiles and amphibians are also common.

**Effects of Humans:** This biome is a prolific producer of wood and supplies much timber; logging threatens the survival of the forest in some areas.

©RGB Ventures/SuperStock/Alamy Stock Photo

## Temperate Deciduous Forest

Figure 54.24d

### Temperate deciduous forest in Maryland

**Physical Environment:** Annual rainfall is generally between 75 and 200 cm. Temperatures fall below freezing each winter but not usually below –12°C.

**Location:** Large tracts of temperate deciduous forest are evident in the eastern U.S., Western Europe, and eastern Asia. In the Southern Hemisphere, eucalyptus forests occur in Australia, and stands of southern beech are found in southern South America, New Zealand, and Australia.

**Plant Life:** Species diversity is much lower in temperate deciduous forests than in tropical forests, with only about three to four tree species per square kilometer. Several tree genera may be dominant in a given locality—for example, oaks, hickories, and maples are usually dominant in the eastern U.S. Commonly, leaves are shed in the fall and reappear in the spring. Many herbaceous plants flower in spring before the trees leaf out and block the light. Even in the summer, though, the forest is not as dense as in tropical forests, so ground cover is abundant.

**Animal Life:** Animals are adapted to the vagaries of the climate; many mammals hibernate during the cold months, birds migrate, and insects enter diapause, a condition of dormancy passed usually as a pupa. Reptiles, which depend on solar radiation for heat, are relatively uncommon. Mammals include squirrels, wolves, bobcats, foxes, bears, and mountain lions.

**Effects of Humans:** Logging has eliminated much of the temperate deciduous forest from populated portions of Europe and North America. Because the annual leaf drop promotes high soil nutrient levels, soils are rich and easily converted to agriculture. Much of the human population lives in the regions where temperate deciduous forest is found, and both agriculture and development are threats to the biome.

©Marie-Ann Daloia/123RF

## Temperate Coniferous Forest (Taiga)

Figure 54.24e

### Temperate coniferous forest in Canada

**Physical Environment:** Precipitation is generally between 30 and 100 cm and often occurs in the form of snow. Temperatures are very cold, often below freezing for long periods of time.

**Location:** The biome of temperate coniferous forest, known commonly by its Russian name, taiga, lies north of the temperate-zone forests and grasslands. Vast tracts of taiga exist in North America and Russia. In the Southern Hemisphere, little land area occurs at latitudes at which extensive taiga could exist.

**Plant Life:** Most of the trees are evergreens or conifers with tough needles, hence the similarity of taiga to temperate rain forest. In this biome, spruces, firs, and pines generally dominate, and the number of tree species is relatively low. Many of the conifers have conical shapes to reduce bough breakage from heavy loads of snow. As in tropical forests, the understory is sparse because the dense year-round canopies prevent sunlight from penetrating. Soils are poor because the fallen needles decay so slowly in the cold temperatures that a layer of needles builds up and acidifies the soil, reducing the numbers of understory species.

**Animal Life:** Reptiles and amphibians are rare because of the low temperatures. Insects are strongly periodic but may often reach outbreak proportions in times of warm temperatures. Mammals that inhabit this biome, such as bears, lynxes, moose, beavers, and squirrels, are heavily furred.

**Effects of Humans:** Humans have not extensively settled this biome, but it has been quite heavily logged. Exploration and development of oil and natural gas reserves are also a threat.

(Top) ©John E Marriott/Getty Images

## Tropical Grassland (Savanna)

**Figure 54.24f**

(Top) ©Jon Arnold Images Ltd/Alamy Stock Photo

### Tropical grassland of the Masai Mara Game Reserve in Kenya

**Physical Environment:** This biome includes hot, tropical areas, with a low or seasonal rainfall between 50 and 130 cm per year. There is often an extensive dry season. Temperatures average 24–29°C.

**Location:** Extensive savannas occur in Africa, South America, and northern Australia.

**Plant Life:** Wide expanses of grasses dominate savannas, but occasional thorny trees, such as acacias, may occur. Fire is prevalent in this biome, so most plants have well-developed root systems that enable them to resprout quickly after a fire.

**Animal Life:** The world's greatest assemblages of large mammals occur in the savanna biome. Herds of antelope, zebra, and wildebeest are found, together with their associated predators: cheetah, lion, leopard, and hyena. Termite mounds dot the landscape in some areas. The extensive herbivory of large grazers, together with frequent fires, may help maintain savannas and prevent their development into forests.

**Effects of Humans:** Savanna soils are often poor because the occasional rain leaches nutrients. Nevertheless, conversion of this biome to agricultural land is rampant, especially in Africa. Overstocking of land for pasturage of domestic animals can greatly reduce grass coverage through overgrazing, turning the area desert-like. This process is known as **desertification**.

## Temperate Grassland (Prairie)

**Figure 54.24g**

### Temperate grassland in Wyoming

**Physical Environment:** Annual rainfall is generally between 25 and 100 cm, too low to support a forest but higher than in deserts. Temperatures in the winter sometimes fall below –10°C, whereas summers may be very hot, approaching 30°C.

**Location:** Temperate grasslands include the prairies of North America, the steppes of Russia, the pampas of Argentina, and the veldt of South Africa. In addition to the limiting amounts of rain, fire and grazing animals may also prevent the establishment of trees in the temperate grasslands. Where temperatures rarely fall below freezing and most of the rain falls in the winter, chaparral, a fire-adapted community featuring shrubs and small trees, occurs. Chaparral is seen at around 30° latitude, where cool ocean waters moderate the climate, as along the coasts of California, South Africa, Chile, and southwest Australia and in countries surrounding the Mediterranean Sea. Some ecologists recognize chaparral as a distinct biome type.

**Plant Life:** From east to west in North America and from north to south in Asia, grasslands show differentiation along moisture gradients. In Illinois, with an annual rainfall of 80 cm, tall prairie grasses such as big bluestem and switchgrass grow to about 2 m high. Along the eastern base of the Rockies, 1,300 km to the west, where rainfall is only 40 cm, prairie grasses such as buffalo grass and blue grama rarely exceed 0.5 m in height. Similar gradients occur in South Africa and Argentina.

**Animal Life:** Where the grasslands remain, large mammals are the most prominent members of the fauna: bison and pronghorn in North America, wild horses in Eurasia, and large kangaroos in Australia. Burrowing animals such as North American gophers and African mole rats are also common.

**Effects of Humans:** Prairie soil is among the richest in the world, having 12 times the humus layer of a typical forest soil. Worldwide, most prairies have been converted to agriculture, and original temperate grassland habitats are among the rarest biomes in the world.

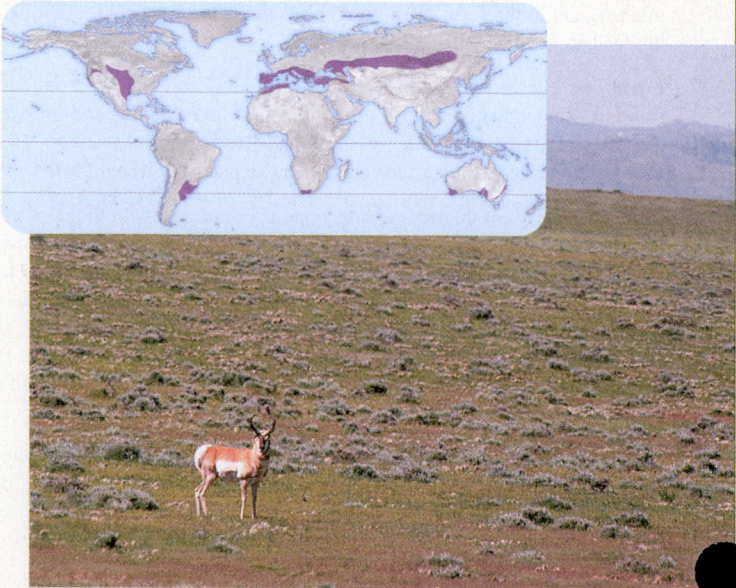

©D. Robert & Lorri Franz/Corbis Documentary/Getty Images

## Hot Desert

Figure 54.24h

### The Namib Desert, Namibia

**Physical Environment:** Rainfall is generally less than 30 cm per year. Temperatures are variable, from below freezing at night to as high as 50°C in the day.

**Location:** Hot deserts are found around latitudes of 20–30° north and south. Prominent deserts include the Sahara of North Africa, the Kalahari and Namib of southern Africa, the Atacama of Chile, the Sonoran of northern Mexico and the southwest U.S., and the Simpson of Australia.

**Plant Life:** Three forms of plant life are adapted to deserts: annuals, succulents, and desert shrubs. Annuals circumvent drought by growing only when there is rain. Succulents, such as the saguaro cactus and other barrel cacti of the southwestern deserts, store water. Desert shrubs, such as the spraylike ocotillo, have short trunks, numerous branches, and small, thick leaves that can be shed in prolonged dry periods. In many plants, spines or volatile chemical compounds serve as a defense against water-seeking herbivores.

**Animal Life:** To conserve water, desert plants produce many small seeds, and animals that eat those seeds, such as ants, birds, and rodents, are common. Reptiles are numerous, because high temperatures permit these ectothermic animals to

©Eye Ubiquitous/Contributor/Getty Images

maintain a warm body temperature. Lizards and snakes are important predators of seed-eating mammals.

**Effects of Humans:** Ambitious irrigation schemes and the prolific use of underground water have allowed humans to develop deserts and grow crops there. Salinization, a buildup in the salt content of the soil that results from irrigation in areas of low rainfall, is prevalent. Off-road vehicles can disturb the fragile desert communities.

## Cold Desert

Figure 54.24i

### The Gobi Desert of Mongolia

**Physical Environment:** Precipitation is less than 25 cm a year and is often in the form of snow. Rainfall usually comes in the spring. In the daytime, temperatures can be high in the summer, 21–26°C, but average around freezing, –2 to 4°C, in the winter.

**Location:** Cold deserts are found in dry regions at middle to high latitudes, especially in the interiors of continents and in the rain shadows of

mountains. Cold deserts are found in North America (the Great Basin Desert), in eastern Argentina (the Patagonian Desert), and in central Asia (the Gobi Desert).

**Plant Life:** Cold deserts are relatively poor in terms of numbers of plant species. Most plants are small in stature, being only between 15 and 120 cm tall. Many species are deciduous and spiny. The Great Basin Desert in Nevada, Utah, and bordering states is a cold desert dominated by sagebrush.

**Animal Life:** As in hot deserts, large numbers of plants produce small seeds on which numerous ants, birds, and rodents feed. Many species live in burrows to escape the cold. In the Great Basin Desert, pocket mice, jackrabbits, kit foxes, and coyotes are common.

**Effects of Humans:** Agriculture is hampered because of low temperatures and low rainfall, and human populations are not extensive. If the top layer of soil is disturbed by human intrusions, such as by off-road vehicles, erosion occurs rapidly and even less vegetation can exist.

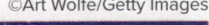

©Art Wolfe/Getty Images

## Tundra

Figure 54.24j

©Bruce Coleman/Photoshot

### Denali National Park in Alaska

**Physical Environment:** Precipitation is generally less than 25 cm per year and is often locked up as snow and unavailable for plants. Deeper water can be locked away for a large part of the year in **permafrost**, a layer of permanently frozen soil. The growing season is short, only 50–60 days. Summer temperatures are only 3–12°C, and even during the long summer days, the ground thaws to less than 1 m in depth. Midwinter temperatures average –32°C.

**Location:** Tundra (from the Finnish *tunturia*, meaning treeless plain) exists mainly in the Northern Hemisphere, north of temperate coniferous forest, because very little land area occurs in the Southern Hemisphere at the latitude where tundra would occur.

**Plant Life:** With so little available water, trees cannot grow. Vegetation occurs in the form of fragile, slow-growing lichens, mosses, grasses, sedges, and occasional shrubs, which grow close to the ground. Plant diversity is very low. In some places, desert conditions prevail because so little moisture falls.

**Animal Life:** Animals of the arctic tundra have adapted to the cold by having good insulation. Many birds, especially shorebirds and waterfowl, migrate. The fauna is much richer in summer than in winter. Many insects spend the winter at immature stages of growth, which are more resistant to cold than the adult forms. Larger animals include such herbivores as musk oxen and caribou in North America (the latter are called reindeer in Europe and Asia). Smaller animals include hares and lemmings. Common predators include arctic foxes, wolves, snowy owls, and polar bears near the coast.

**Effects of Humans:** Though this area is sparsely populated, mineral extraction, especially of oil, has the potential to significantly affect this biome. Ecosystem recovery from such damage would be very slow.

## Mountain Ranges

Figure 54.24k

### Rocky Mountains of Colorado

**Physical Environment:** Mountain ranges must be viewed differently from other biomes. On mountains, temperature decreases with increasing elevation through adiabatic cooling, as discussed earlier in this section. Thus, precipitation and temperature may change dramatically, depending on elevation and whether the mountainside is to windward or leeward.

**Location:** Mountain ranges exist in many areas of the world, but among the largest are the Himalayas in Asia, the Rockies in North America, and the Andes in South America.

**Plant Life:** A variety of biomes can be found on a single mountain range. Biome type may change from temperate forest through taiga and into tundra on an elevation gradient in the Rocky Mountains, and even from tropical forest to tundra on the highest peaks of the Andes in tropical South America. In tropical regions, daylight averages 12 hours per day throughout the year. Instead of a period of intense productivity seen in arctic tundra, vegetation in the tropical alpine tundra exhibits slow but steady rates of photosynthesis and growth all year.

**Animal Life:** The animals of this biome are as varied as the number of habitats it contains. Generally, more species of plants and animals are found at lower elevations than at higher ones. At higher elevations, animals such as bighorn sheep and mountain goats climb the craggy slopes and have skidproof pads on their hooves. Birds of prey, such as eagles, are frequent predators of the furry rodents found at higher elevations, including guinea pigs and marmots.

©McPHOTO/age fotostock

**Effects of Humans:** Logging and agriculture at lower elevations can cause habitat degradation. Because of the steep slopes, mountain soils are often well drained, thin, and especially susceptible to erosion following clearing for agriculture.

## Aquatic Biomes Consist of Marine and Freshwater Regions

The ecology of freshwater habitats is governed largely by the unusual properties of water. Water is at its most dense at 4°C and becomes less dense as it warms or cools. At 0°C water freezes and is in its least dense state, so ice floats on unfrozen water. This explains why lakes and rivers freeze from the top down and why free-flowing water is at the bottom of a frozen lake or river (**Figure 54.25a**). From a fish's point of view, this property is advantageous, because if ice sank, all temperate lakes would freeze solid in winter, and no fish would exist in lakes outside the tropics. Oxygen content is depleted toward the bottoms of lakes by the respiration of bottom-dwelling organisms.

In the spring, ice melts, water warms and spring storms mix the water layers, creating uniform conditions of temperature and oxygen (**Figure 54.25b**). This mixing is termed the **spring overturn**. In deeper temperate lakes, in the summer, three layers are present (**Figure 54.25c**). An upper layer, called the **epilimnion**, is warmed by the sun and mixed well by the wind. Below this lies a transition zone known as the **thermocline**, where the temperature declines rapidly. Lower still is the **hypolimnion**, a cool layer too far below the surface to be much warmed and with low light levels where photosynthesis is absent and oxygen supply is low. In the fall the upper layers cool, and as their density increases they sink. Storms cause the fall overturn (**Figure 54.25d** ), in which the water in the lake is thoroughly mixed and the thermocline disappears.

Coastal areas are influenced mainly by tides and waves. The gravitational pull of the Moon is 2.2 times greater than that of the Sun. As the Earth turns, each area of the globe is closer to the Moon once a day. At the equator, oceans are pulled toward the Moon at this time, creating high tides at the equator and low tides at higher latitudes. Similarly, when an ocean is on the opposite side of the Earth away from the Moon, the tide is high. This is because the Earth is itself pulled more toward the Moon at this point, leaving the water behind, causing the water to rise relative to the Earth. Thus, most areas of the Earth have two high tides per day.

Waves can range in size from small ripples to huge swells. As the wind blows, the friction between the air and the water creates small ripples. Once the ripples have formed, the wind has something to push against and the waves may increase in size. Four factors influence wave size: wind speed, fetch, duration of time of wind, and water depth. Fetch is the distance of open water over which the wind can blow. Shorter fetches in lakes reduce wave size, while long fetches of open ocean may allow the formation of 4- to 5-m waves.

Within aquatic environments, several different biome types are recognized, including distinct marine aquatic biomes (intertidal zone, coral reef, and open ocean) and freshwater biomes (lakes, rivers, and wetlands). These biomes are distinguished primarily by differences in salinity, oxygen content, depth, current strength, and availability of light (**Figure 54.26a–f**). Freshwater habitats are traditionally divided into **lentic**, or standing-water habitats (from the Latin *lenis*, meaning calm), and **lotic**, or running-water habitats (from the Latin *lotus*, meaning washed).

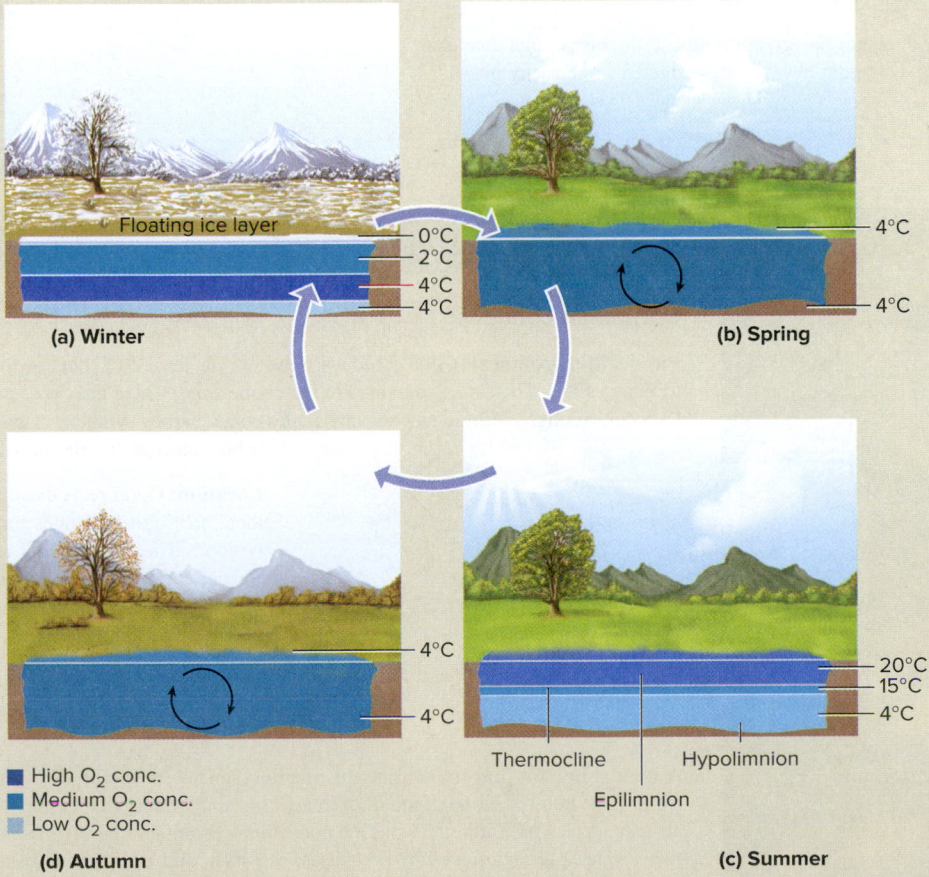

**Figure 54.25  Annual cycle of a temperate lake.**  Cross section of a temperate lake with temperature profiles according to depth for each season. **(a)** The lake surface freezes in winter. **(b)** When the ice melts in the spring, the cold water again sinks and mixes the lake. **(c)** In the summer, the warmest water occurs at the surface, and water temperature decreases with depth. **(d)** Cold air temperatures in the fall cool the upper layers and this dense cold water sinks, thoroughly mixing the lake.

# Aquatic Biomes (continued)

## Intertidal Zone

Figure 54.26a

### Olympic Coast National Marine Sanctuary in Washington State

**Physical Environment:** The **intertidal zone**, the area where the land meets the sea, is alternately submerged and exposed by the daily cycle of tides. The resident organisms are subject to huge daily variations in temperature, light intensity, and availability of seawater.

**Location:** Throughout the world, the area where the land meets the sea consists of sandy shore, mudflats, or rocky shore.

**Plant Life:** Plant life may be quite limited because the sand or mud is constantly shifted by the tide. Mangroves may colonize mudflats in tropical areas, and salt marsh grasses may colonize mudflats in temperate locations. On the rocky shore, green algae and seaweeds predominate.

**Animal Life:** Animal life may be quite diverse. On the rocky shore, sea anemones, snails, hermit crabs, and small fishes live in tide pools. On the rock face, there may be a variety of limpets, mussels, sea stars, sea urchins, snails, sponges, tube worms, whelks, isopods, and chitons. At low tides, organisms may be dry and vulnerable to predation by a variety of animals, including birds and mammals. High tides bring predatory fishes. Sandy or muddy shores may contain burrowing marine worms, crabs, and small isopods.

**Effects of Humans:** Human development has greatly reduced the beach area available to shorebirds and breeding turtles. Oil spills have greatly affected some rocky intertidal areas.

©Nature Picture Library/Alamy Stock Photo

## Coral Reef

Figure 54.26b

### Caribbean coral reef

**Physical Environment:** Corals need warm water of at least 20°C but less than 30°C to survive (refer back to Figure 54.6). They are also limited to the photic zone, where light penetrates and allows photosynthesis to occur. Sunlight is important because many corals harbor symbiotic algae, or dinoflagellates, that contribute nutrients to the animals and that require light to live.

**Location:** Coral reefs exist in warm tropical waters where there are solid substrates for attachment and water clarity is good. The largest coral reef in the world is the Great Barrier Reef off the Australian coastline, but other coral reefs are found in the Atlantic Ocean, the Red Sea, and the Pacific and Indian Oceans.

**Plant Life:** Dinoflagellate algae live within the coral tissue, and a variety of red and green algae live on the coral reef surface.

**Animal Life:** An immense variety of microorganisms, invertebrates, and fishes live among the coral, making the coral reef one of the most interesting and species-rich biomes on Earth. Probably 30–40% of all fish species on Earth are found on coral reefs. Prominent herbivores include snails, sea urchins, and fishes. These are consumed by octopuses, sea stars, and carnivorous fishes. Many species are brightly colored, warning predators of their toxic nature.

**Effects of Humans:** Collectors have removed many corals and fishes for the aquarium trade, and marine pollution threatens water clarity in some areas. Perhaps the greatest threat to coral reefs is global warming. Water temperatures that are too high (over 30°C) and high pH caused by elevated $CO_2$ levels both contribute to coral bleaching.

©Stephen Frink/Corbis Documentary/Getty Images

## The Open Ocean

Figure 54.26c

### Manta ray in the open ocean

**Physical Environment:** In the open ocean, sometimes called the **pelagic zone**, water depth averages 4,000 m. Nutrient concentrations are typically low, though the waters may be periodically enriched by ocean **upwelling**, the circulation of cold, mineral-rich nutrients from deeper water to the surface. Pelagic waters are mostly cold, only warming near the surface.

**Location:** Across the globe, covering 70% of the Earth's surface.

**Plant Life:** In the photic zone, many microscopic photosynthetic organisms (**phytoplankton**) grow and reproduce while drifting in ocean currents. Phytoplankton account for nearly half the photosynthetic activity on Earth and produce much of the world's oxygen.

**Animal Life:** Open-ocean organisms include **zooplankton**, drifting animals such as small worms, copepods (tiny shrimplike creatures), small jellyfish, and small invertebrate and fish larvae that graze on the phytoplankton. Heterotrophic protists are also included in zooplankton. The open ocean also includes free-swimming animals collectively called **nekton**, which can swim against the currents to locate food. Nekton includes large squids, fishes, sea turtles, and marine mammals. Only a few of these organisms live at any great depth. In some areas, a unique assemblage of animals is associated with deep-sea hydrothermal vents that spew hot (350°C) water rich in hydrogen sulfide. Large worms and other chemoautotrophic organisms exist together in this dark, oxygen-poor environment (refer back to Figure 22.3).

**Effects of Humans:** Oil spills and a long history of garbage disposal have polluted the ocean floors of many areas. Overfishing has caused many fish populations to crash, and the whaling industry has greatly reduced the numbers of most species of whales.

©Jeffrey L. Rotman/Corbis Documentary/Getty Images

## Lentic Habitats

Figure 54.26d

Source: NPS Photo, R.Illiescu

### Everglades National Park, Florida

**Physical Environment:** The lentic habitat consists of still, often deep water. Its physical characteristics depend greatly on the surrounding land, which dictates what nutrients collect in the lake. Young lakes often start off clear and with little plant life. Such lakes are called **oligotrophic**. With age, the lake becomes richer in dissolved nutrients from erosion and runoff from surrounding land, with the result that cyanobacteria and algae spread, reducing the water clarity. Such lakes are termed **eutrophic**. The process of eutrophication occurs naturally but can be sped up by human activities (see Chapter 59).

**Location:** Throughout all the continents of the world.

**Plant Life:** In addition to phytoplankton, lentic habitats may have rooted vegetation, which often extends above the water surface (emergent vegetation), such as cattails, plus deeper-dwelling aquatic plants and algae.

**Animal Life:** Animals include fishes, frogs, turtles, crayfish, insect larvae, and many species of insects. In tropical and subtropical lakes, alligators and crocodiles are common.

**Effects of Humans:** Agricultural runoff, including fertilizers and sewage, can greatly increase lake nutrient levels and speed up the process of eutrophication, resulting in phytoplankton blooms and fish kills. In some areas, invasive species of invertebrates and fishes are outcompeting native species.

## Lotic Habitats

**Figure 54.26e**

### Fast-flowing river in the Pacific Northwest

**Physical Environment:** In lotic habitats, flowing water prevents nutrient accumulations and phytoplankton blooms. The current also mixes water thoroughly, providing a well-aerated habitat of relatively uniform temperature. The current, oxygen level, and clarity are greater at the source of a stream (its headwaters) than in the lower reaches. Nutrient levels are generally less in the headwaters.

**Location:** On all continents except Antarctica.

**Plant Life:** In slow-moving streams and rivers, algae and rooted plants may be present; in faster-moving rivers, leaves from surrounding forests are the primary food source for animals.

**Animal Life:** Lotic habitats have a fauna completely different from that of lentic waters. Animals are adapted to stay in place despite an often-strong current. Many of the smaller organisms are flat and attach themselves to rocks to avoid being swept away. Others live on the underside of large boulders, where the current is much reduced. Fish such as trout may be present in rivers with cool temperatures, high oxygen, and clear water. In warmer, murkier waters, catfish and carp may be abundant.

**Effects of Humans:** Animals of lotic systems are not well adapted for low-oxygen environments and thus are particularly susceptible to oxygen-reducing pollutants such as sewage. Dams across rivers have prevented the passage of migratory species such as salmon.

©Ron Crabtree/Getty Images

## Wetlands

**Figure 54.26f**

### Yellow Waters River, Kakadu National Park, Northern Territory, Australia

**Physical Environment:** At the margins of both lentic and lotic habitats, wetlands may develop. Wetlands are areas regularly saturated by surface water or groundwater. They range from marshes (treeless areas where herbaceous species predominate), to swamps (wet areas dominated by trees), and bogs (depressions dominated by marshes). Many wetlands are seasonally flooded when rivers overflow their banks or lake levels rise. Some wetlands also develop along estuaries, areas where river water merges with ocean water, and high tides can flood the land. Because of generally high nutrient levels, oxygen levels are fairly low. Temperatures vary substantially with location.

**Location:** Worldwide, except in Antarctica.

**Plant Life:** Wetlands are among the most productive, species-rich areas in the world. In North America, floating plants such as lilies and rooted species such as sedges, cattails, cypress, and gum trees predominate.

**Animal Life:** Most wetlands are rich in animal species. Wetlands are a prime habitat for wading and diving birds. In addition, they are home to a profusion of insects, from mosquitoes to dragonflies. Vertebrate predators include many amphibians, reptiles, otters, and alligators.

**Effects of Humans:** Long mistakenly regarded as wasteland by humans, many wetlands have been drained and developed for housing and industry. Wetlands play a valuable role in protecting coastal communities from hurricanes, and the loss of wetlands in Louisiana contributed to the severity of effects of Hurricane Katrina in 2005.

©Larry Mulvehill/Science Source

**Learning Outcomes:**

1. Describe how the distribution of species on Earth may result from continental drift.
2. Explain the concept of biogeographical regions.

**Biogeography** is the study of the geographic distribution of extinct and living species. An understanding of evolution and geological change over long time periods helps explain some of the patterns we see in modern biomes. For example, South America, Africa, and Australia all have similar biomes, ranging from tropical to temperate, yet each continent has distinctive animal and plant life. South America is inhabited by sloths, anteaters, armadillos, and monkeys with prehensile tails. Africa possesses a wide variety of antelopes, zebras, giraffes, lions, baboons, the okapi, and the aardvark. Australia, which has no native placental mammals except bats, is home to a variety of marsupials such as kangaroos, koala bears, Tasmanian devils, and wombats, as well as the egg-laying monotremes, namely, the duck-billed platypus and four species of echidnas.

Most continents also have distinct species of plants. For example, eucalyptus trees are native only in Australia. In South American deserts, succulent plants belong to the family Cactaceae, the cacti. In Africa, they belong to the genus *Euphorbia*, the spurges. In North America, the pines, *Pinus* spp., and firs, *Abies* spp., are common, but they do not occur south of the mountains of central Mexico. In contrast, palms are common in South America and do not generally occur north of the mountains of central Mexico, except for several genera in southern California and Florida.

A plausible explanation for differences in species distributions is that abiotic factors are of paramount importance and that each continent supports the species best adapted to it. However, the spread of introduced species has proved this explanation incorrect: European rabbits introduced into Australia proliferated rapidly, and eucalyptus from Australia grows well in California. Based on biogeography, a more plausible explanation is that the independent evolution of separate, unconnected populations has generated different species in different places. In this section, we will examine how biogeography impacts our understanding of species distributions.

### Continental Drift Has Affected the Distribution of Extinct and Modern Species

An important topic in biogeography is the phenomenon of **continental drift**, which has involved major changes in the relative location of continents due to the slow movement of the Earth's surface plates (refer back to Figure 26.5). Continental drift can explain the discovery of similar fossils on different continents, and it can sometimes explain the similarities and differences of modern species that now inhabit these continents.

With regard to fossils, paleontologists have noted the occurrence of similar fossils in South America, Africa, India, Antarctica, and Australia. Many of these fossils were of large land animals, such as the Triassic reptiles *Lystrosaurus* and *Cynognathus*, that could not have easily dispersed among continents, or of plants whose seeds were not likely to be dispersed far by wind, such as the fossil fern *Glossopteris*. These discoveries can be explained by two observations. First, such species existed prior to the continental drift that created the modern continents. And second, their ranges encompassed regions within two or more different continents (**Figure 54.27**).

Continental drift can also explain similarities and differences among modern species on different continents. For example, many species of mammals are found on different continents. This observation is consistent with evolutionary studies indicating that mammals arose about 200 to 225 million years ago, prior to the separation of the major continents. However, after the continents drifted apart, many new mammalian species have arisen and their ranges are often limited to a single continent. Such range limitation is more common for those species that are unable to travel great distances.

### Long-Distance Migrations and Extinctions Affect the Distributions of Species

For species that can travel great differences, biogeography reveals a how species can evolve into two or more closely related species that are widely separated geographically, called disjunct distributions. The distributions of many present-day species are relics of once much broader distributions. For example, currently four living species of tapir are known: three in Central and South America and one in Malaysia (**Figure 54.28**). Fossil records reveal a much more widespread distribution over much of Europe, Asia, and North America. The oldest fossils of the ancestral *Paleotapirus* were identified in Europe, making it likely that this was the center of origin of tapirs. Migrations of later-evolving *Protapirus* resulted in a more widespread distribution. Cooling resulted in the demise of tapirs in all areas except the tropical locations.

Another well-known example of a disjunct distribution is the restricted distribution of monotremes and marsupials. These animals were once plentiful over North America and Europe. They spread into the rest of the world, including South America and Australia, at the end of the Cretaceous period when, although the continents were separated, land bridges existed between them. Later, placental mammals evolved in North America and displaced the marsupials there, apart from a few species such as the opossum. However, placental mammals could not invade Australia because by then the land bridge was broken.

Elephants and camels also have disjunct distributions. Elephants originated in Africa and subsequently dispersed through Eurasia and across the Bering land bridge from Siberia to North America, where many are found as fossils. They subsequently became extinct everywhere except Africa and India. Camels

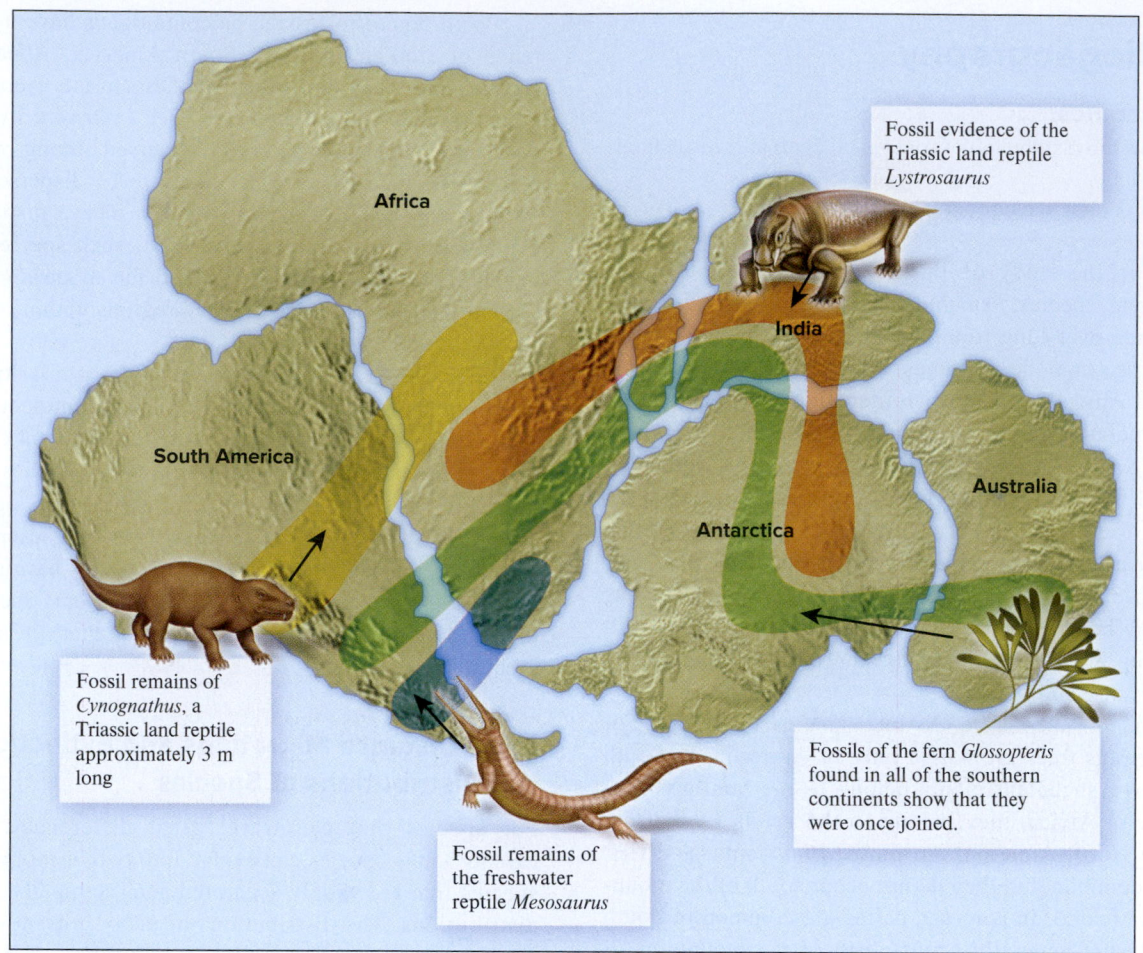

Fossil evidence of the Triassic land reptile *Lystrosaurus*

Africa

India

South America

Australia

Antarctica

Fossil remains of *Cynognathus*, a Triassic land reptile approximately 3 m long

Fossils of the fern *Glossopteris* found in all of the southern continents show that they were once joined.

Fossil remains of the freshwater reptile *Mesosaurus*

**Figure 54.27 Examples of fossil plants and animals found on different continents.** South America, Africa, India, Antarctica, and Australia were once united as Gondwana (refer back to Figure 26.5). The locations of the fossil remains of the four species shown here can be explained by their ranges prior to the drifting of the continents away from each other.

evolved in North America and made the reverse trek across the Bering land bridge into Eurasia; they also crossed into South America via the Central American isthmus. They have since become extinct everywhere except Asia, North Africa, and South America.

## Ecologists Recognize Biogeographic Regions

In the late 19th century, British naturalist Alfred Russel Wallace was one of the earliest scientists to realize that certain plant and animal taxa were restricted to certain geographic areas of the Earth. For example, the distribution patterns of guinea pigs, anteaters, and many other groups are confined to Central and South America, from central Mexico southward. The whole area was distinct enough for Wallace to proclaim it the Neotropical region. Wallace went on to divide the world's biota into six major **biogeographic regions**: Nearctic, Palearctic, Neotropical,

Ethiopian, Oriental, and Australian (**Figure 54.29**). These regions are still widely accepted today, though debate continues about the exact locations of the boundary lines.

Biogeographical regions correspond largely to continents but more exactly to areas bounded by major barriers to dispersal, like the Himalayas and the Sahara Desert. Within these regions, areas of similar climates are often inhabited by species with similar appearance and habits but from different taxonomic groups. For example, the kangaroo rats of North American deserts, the jerboas of central Asian deserts, and the hopping mice of Australian deserts look similar and occupy similar hot, arid environments, but they arose from different lineages, belonging to the families Heteromyidae, Dipodidae, and Muridae, respectively. As noted in Chapter 22, this phenomenon, called convergent evolution, has led to the emergence of similar species that have evolved from different ancestors.

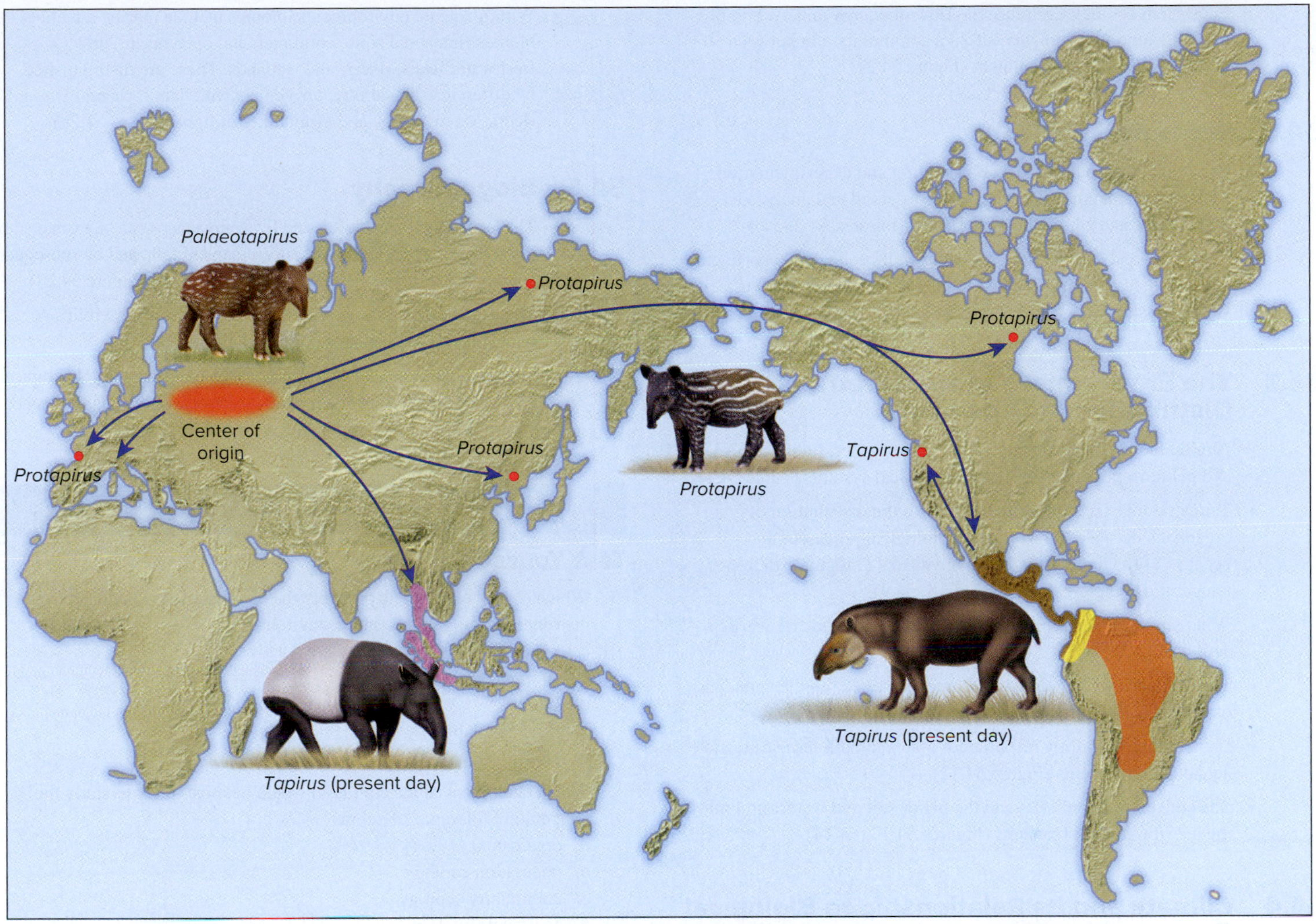

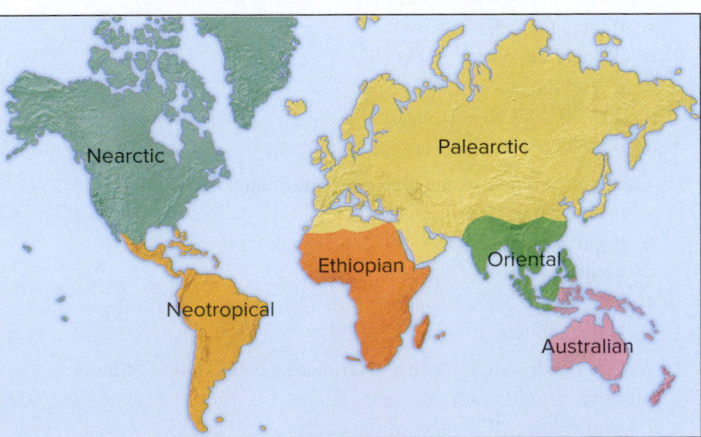

| Tapirus indicus | Tapirus pinchaque | Tapirus terrestris | Tapirus bairdi |

**Figure 54.28  Tapir distribution.** Of the four living tapir species, three are in Central and South America and one is in Malaysia. Fossil evidence suggests a European origin of the ancestral *Paleotapirus* and a dispersal of later-evolving *Protapirus*. A more widespread distribution followed, with tapirs dying out in other regions (marked with a red dot) possibly due to climate change.

 **Core Concept: Evolution**  Knowing that all tapir species share a common European ancestor makes it easier to explain their current distribution pattern.

**Figure 54.29  The biogeographic regions proposed by Wallace.** Note that the borders do not always demarcate continents.

# Summary of Key Concepts

## 54.1  The Scale of Ecology

- Ecologists study the interactions among organisms and between organisms and their environments. The field of ecology is subdivided into broad areas of organismal, population, community, and ecosystem ecology (Figure 54.1).

- Organismal ecology considers how individuals are adapted to their environment and how the behavior of an individual organism contributes to its survival and reproductive success and the population density of the species. Population ecology explores those factors that influence a population's growth, size, and density, including species interactions such as competition, predation, and parasitism. Community ecology studies how populations of species interact and form functional communities.

Ecosystem ecology examines the flow of energy and cycling of nutrients among organisms within a community and between organisms and the environment (Figure 54.2).

## 54.2   Ecological Methods

- Ecological methods focus on observation and experimentation. Interactions among species are often observed and analyzed graphically, and a hypothesis is formed (Figures 54.3, 54.4).

- Ecologists test their hypotheses using well-replicated experiments. The results are often presented graphically and analyzed via a variety of statistical tests (Figure 54.5).

## 54.3   The Environment's Effect on the Distribution of Organisms

- Abiotic factors such as temperature, wind, water, light, salinity, and pH have powerful effects on ecological systems (Table 54.1).

- Temperature exerts important effects on the distribution of organisms because of its effect on biological processes and because of the inability of many organisms to regulate their body temperature (Figures 54.6, 54.7, 54.8, 54.9).

- Wind amplifies the effects of temperature and modifies wave action (Figure 54.10).

- The availability of water has an important effect on the abundance of organisms (Figure 54.11).

- Light can be a limiting resource for plants in both terrestrial and aquatic environments (Figure 54.12).

- The concentration of salts and the pH of soil and water can limit the distribution of organisms (Figures 54.13, 54.14).

## 54.4   Climate and Its Relationship to Biological Communities

- Global temperature differentials are caused by variations in incoming solar radiation and patterns of atmospheric circulation (Figures 54.15, 54.16, 54.17).

- The tilt and rotation of the Earth also affect climate, causing seasonality (Figure 54.18).

- The Coriolis force deflects the direction of northerly and southerly winds, affecting climate (Figure 54.19).

- Elevation and the proximity of a landmass to a large body of water can similarly affect climate (Figures 54.20, 54.21).

## 54.5   Major Biomes

- Climate has a large effect on biomes, which are major types of habitats characterized by distinctive plant and animal life (Figures 54.22, 54.23).

- Terrestrial biomes are generally named for their climate and vegetation type and include tropical rain forest, tropical deciduous forest, temperate rain forest, temperate deciduous forest, temperate coniferous forest (taiga), tropical grassland (savanna), temperate grassland (prairie), hot and cold deserts, and tundra. In mountain ranges, biome type may vary with elevation (Figure 54.24).

- Aquatic biomes are affected by tides and waves and by the fact that ice floats on water; deep lakes go through seasonal changes (Figure 54.25).

- Within aquatic environments, biomes include marine aquatic biomes (intertidal zone, coral reef, and open ocean) and freshwater lakes, rivers, and wetlands. These are distinguished by differences in salinity, oxygen content, depth, current strength (lentic versus lotic), and availability of light (Figure 54.26).

## 54.6   Biogeography

- The location of many fossils and living organisms can be explained by the origin of a species on one supercontinent followed by subsequent continental drift and evolution of related species (Figure 54.27).

- The current distribution patterns of some species are relics of once much broader distributions (Figure 54.28).

- Six major biogeographical regions—Nearctic, Palearctic, Neotropical, Ethiopian, Oriental, and Australian—each of which contains distinct groups of species, are widely recognized today (Figure 54.29).

## Assess & Discuss

### Test Yourself

1. Which of the following is probably the most important factor in the distribution of organisms in the environment?
   a. light
   b. temperature
   c. salinity
   d. water availability
   e. pH

2. Which ecological subdiscipline(s) would be most likely to study the temperature tolerance of zebras?
   a. organismal ecology
   b. population ecology
   c. community ecology
   d. ecosystem ecology
   e. both a and b

3. Physics is to engineering as ecology is to
   a. biology.
   b. environmental science.
   c. chemistry.
   d. mathematics.
   e. statistics.

4. The world's major subsidence zones occur at the latitudes
   a. 0° and 45°–55°.
   b. 30° and the poles.
   c. 0° and the poles.
   d. 20°–30° and 45°–55°.
   e. 0° and 20°–30°.

5. The most common biome type, in terms of area occupied, is
   a. open ocean.
   b. tropical rainforest.
   c. tundra.
   d. hot desert.
   e. lentic habitats.

6. What is the driving force that determines the circulation of the atmospheric air?
   a. temperature differences of the Earth
   b. winds
   c. ocean currents

d.  mountain ridges

e.  all of the above

7.  In this biome, rainfall is between 25 cm and 100 cm and temperatures vary between −10°C in winter and 30°C in summer. Which biome is it?

a.  tropical rainforest

b.  tropical deciduous forest

c.  savanna

d.  prairie

e.  temperate deciduous forest

8.  What characteristic(s) (is) are commonly used to identify the terrestrial biomes of the Earth?

a.  temperature

b.  precipitation

c.  vegetation

d.  all of the above

e.  a and b only

9.  Young lakes are often clear and with little plant life. Such lakes are called

a.  oligotrophic.

b.  eutrophic.

c.  lotic.

d.  lentic.

e.  pelagic.

10.  The unique group of marsupial species found in Australia is largely a result of

a.  adiabatic cooling.

b.  climate.

c.  continental drift and evolution.

d.  rain shadows.

e.  biogeographic regions.

## Conceptual Questions

1.  If mountains are closer to the Sun than valleys, why aren't they hotter?

2.  Why are fires generally more frequent in prairies than in hotter, drier deserts?

3.  **Core Concept: Systems** In most locations on Earth at about 20–30° latitude, air cools and descends, and hot deserts occur. Florida is situated between 31°N and 24°N. Why does it not support a desert biome?

## Collaborative Questions

1.  The so-called Telegraph Fire, near Yosemite National Park, in 2008, was one of the worst in California that year, burning more than 46 square miles covered by timber that had not burned in over 100 years. What could be done to prevent such a catastrophic fire in the park itself?

2.  Based on your knowledge of biomes, identify the biome in which you live. List and describe the organisms that you have observed in your biome. Why might your observations not fit the biome predicted to occur in your area from temperature/precipitation profiles?

# Behavioral Ecology

# 55

**Killdeer (*Charadrius vociferus*) removing an eggshell from its nest.** What is the selective advantage of this behavior? ©FLPA/Alamy Stock Photo

**A**fter their young hatch, nesting birds often pick up the empty eggshells and carry them away from the nest. One might think that they are being neat and tidy or are minimizing the risk of bacterial infection to the chicks, but there is more to the behavior than this. The chicks and unhatched eggs are well camouflaged in the nest, but the white color of the empty eggshell quickly attracts the attention of predators such as crows that would kill and eat the chicks or remaining eggs. By removing the old eggshells, the parents are increasing the chances that their offspring—and thus their genes—will survive. This behavior, which is likely to be an instinctive activity, has been fostered by natural selection: Birds that perform this activity have a higher rate of chick survival than those that don't, which increases the likelihood that the expression of genes that promote this behavior are passed on to future generations.

**Behavior** is the observable response of organisms to an external or internal stimulus. In this chapter, we focus our attention on the field of **behavioral ecology**, the study of how behavior contributes to the differential survival and reproduction of organisms. Contemporary behavioral ecology builds on earlier work that focused primarily on how organisms behave. In the early 20th century, scientific studies of animal behavior, termed **ethology** (from the Greek *ethos*, meaning habit or manner), focused on the specific genetic and physiological mechanisms of behavior. These factors are called **proximate causes**. For example, we could hypothesize that male deer rut or fight with other males in the fall because a change in day length stimulates the eyes, brain, and pituitary gland and triggers hormonal changes in their bodies. The founders of ethology, ethologists Karl von Frisch, Konrad Lorenz, and Niko Tinbergen, shared the 1973 Nobel Prize in Physiology or Medicine for their pioneering discoveries concerning the proximate causes of behavior.

However, we could also hypothesize that male deer fight to determine which deer get to mate with the most female deer and pass on their genes. This hypothesis leads to a different answer than the one that is concerned with changes in day length. This answer focuses on the adaptive significance of fighting to the deer, that is, on the effect of a particular behavior on reproductive success. These factors are called **ultimate causes** of behavior. Since the 1970s, behavioral ecologists have placed a greater emphasis on understanding the ultimate causes of behavior.

In this chapter, we will explore the role of both proximate and ultimate causes of behavior. We begin by investigating how behavior is achieved, examining the roles of both genetics and the environment. In doing so, we will examine the important contributions of von Frisch, Lorenz, and Tinbergen. We will consider how different behaviors are involved in movement, gathering food, and communication. Later, we will investigate how organisms interact in groups, whether an organism can truly behave in a way that benefits others at a cost to itself, and how behavior shapes different mating systems. The chapter focuses on animal behavior, because the behavior of other organisms is more limited and less well understood.

## 55.1 The Influence of Genetics and Learning on Behavior

**Learning Outcomes:**

1. Describe the differences among innate behavior, conditioning, and learning.
2. Distinguish between classical conditioning and operant conditioning.
3. Give examples of how genetics and learning influence most behaviors.

Behavior is controlled by genetics and the environment, and in this chapter, we will discuss the influence of both. Determining to what degree a behavior is influenced by genes versus the environment depends on the particular genes and environment in question. Genes that control behavior in complex animals, such as vertebrates, typically act on the development of the nervous system and musculature—physical traits that evolve through natural selection. The expression of many genes is required for behaviors in animals. Even so, ethologists have identified examples in which the alteration in a single gene may dramatically change a particular behavior. To use the analogy of baking a cake, a change in one ingredient of the recipe may change the taste of the cake, but that does not mean that the one ingredient is responsible for the entire cake.

In this section, we begin by examining how genes can affect behavior and consider several examples of genetically programmed behaviors. Later, we will explore several types of learned behaviors, including behaviors established by classical and operant conditioning and cognitive learning, and conclude by exploring an example of the interaction of genetics and learning on behavior.

### Fixed Action Patterns Are Genetically Programmed

Behaviors that seem to be genetically programmed are referred to as **innate** (also called instinctual). Most individuals will exhibit the same behavior without prior training. For example, a spider will spin a specific web without ever seeing a member of its own species build one. The courtship behaviors of many bird species are so stereotyped as to be virtually identical.

A classic example of innate behavior is the egg-rolling response in geese (**Figure 55.1**). If an incubating goose notices an egg out of the nest, she will extend her neck toward the egg, get up, and then roll the egg back to the nest using her beak. Such behavior functions to improve fitness because it increases the survival of offspring. Eggs that roll out of the nest get cold and fail to hatch. Geese that fail to exhibit the egg-rolling response would pass on fewer of their genes to future generations.

Egg-rolling behavior is an example of what ethologists term a **fixed action pattern (FAP)**, a behavior that, once initiated, continues until completed. For example, if the egg is removed while the goose is in the process of rolling it back toward the nest, the goose still completes the FAP, as though she were rolling back the now-absent egg to the nest. The stimulus to initiate this behavior is obviously a strong one, which ethologists term a **sign stimulus**. The sign stimulus for the goose is that an egg has rolled out of the nest. According to ethologists, this stimulus acts on the goose's central nervous system, which provides a neural stimulus to initiate the motor program, or FAP. Interestingly, any round object, from a wooden egg to a volleyball, can elicit the egg-rolling response. Although sign stimuli usually have certain key components, they are not necessarily very specific.

Niko Tinbergen's study of male stickleback fish provides another classic example of a FAP. Male sticklebacks, which have a characteristic red belly, will attack other male sticklebacks that invade their territory. Tinbergen found that sticklebacks attacked small, unrealistic model fish having a red ventral surface (the sign stimulus), while ignoring a realistic male stickleback model that lacked a red underside (**Figure 55.2**).

 **Core Concept: Evolution**

### Some Behavior Results from Simple Genetic Influences

In 1964, biologist W. C. Rothenbuhler studied the effect of genes on the behavior of honeybees. Some strains of bees were observed to be hygienic; that is, they detect and remove diseased larvae from the nest. This behavior involves two distinct maneuvers: uncapping

| 1 | The female goose extends her neck toward the egg. | 2 | The goose gets up from the nest and approaches the egg. | 3 | The goose places her neck above the egg. | 4 | The goose rolls the egg back to the nest with her beak and neck. |

**Figure 55.1 A fixed action pattern as an example of innate behavior.** A female goose retrieves an egg that has rolled outside the nest through a set sequence of movements. The goose completes this entire sequence even if a researcher takes the egg away before the goose has rolled it back to the nest.

the wax cells and then discarding the dead larvae. Other strains are not hygienic and do not exhibit such behavior. Using genetic crosses, Rothenbuhler demonstrated that a recessive allele of one gene (*u*) controls cell uncapping and a recessive allele of a different gene (*r*) controls larval removal. Strains with the genotype, *uurr,* are hygienic strains, whereas *UURR* strains are nonhygienic. When the two strains were crossed, all the $F_1$ hybrids were nonhygienic (*UuRr*). When the $F_1$ hybrids were crossed with the hygienic strain (*uurr*), four different genotypes were produced: one-quarter of the offspring were hygienic (*uurr*), one-quarter were nonhygienic and showed neither behavior (*UuRr*), one-quarter uncapped the cells but failed to remove the larvae (*uuRr*), and one-quarter removed the larvae but only if the cells were uncapped for them (*Uurr*).

More recently, in 2004, American neuroscientist Barry Richmond and colleagues showed how the work ethic of monkeys is affected by a gene expressed in a region of the brain called the rhinal cortex. Most primates, humans and monkeys included, tend to work harder when a deadline looms. Richmond's team trained four monkeys to release a lever at the exact moment a spot on a computer screen changed color from red to green. The monkeys had to complete this task three times, but only on the third trial did they receive a food reward, regardless of how they performed on the first two trials. As an indication of how many trials were left, the monkeys could see a gray bar on the screen. As the bar became brighter, the monkeys knew they were reaching the last trial, and they worked more diligently for the reward. In the first two trials, the monkeys made more errors than in the last trial. Next, the team switched off the gene known to be involved in processing reward signals. To do this, the researchers injected a short strand of DNA into the monkeys' brains. The effects were only temporary, 10–12 weeks, but during that time the monkeys were unable to determine how many trials were left before the reward was given, and they worked vigilantly to receive the reward on every trial, making few errors even on trials one and two.

## Conditioning Occurs When a Relationship Between a Stimulus and a Response Is Learned

Although many of the behavioral patterns exhibited by animals are largely innate, sometimes animals can make modifications to their behavior based on previous experience, a process that involves **learning**. Perhaps the simplest form of learning is **habituation**, in which an organism learns to ignore a repeated stimulus. For example, animals in African safari parks become habituated to the presence of vehicles containing tourists; these vehicles are neither a threat nor a benefit to them. Birds can become habituated to the presence of a scarecrow, resulting in damage to crops. Habituation can be a problem at airports, where birds eventually ignore the alarm calls designed to scare them away from the runways.

Habituation is a form of nonassociative learning, a change in response to a repeated stimulus without association with a positive or negative reinforcement. Alternatively, an association may gradually develop between a stimulus and a response. Such a change in behavior is termed **associative learning**. In associative learning, a behavior is changed or

**Figure 55.2   A fixed action pattern elicited by a sign stimulus.** The sign stimulus for male sticklebacks to attack other males entering their territory is a red ventral surface. In experiments, male sticklebacks attacked all models that had a red underside, while ignoring a realistic model of a stickleback that lacked the red belly.

conditioned through the association. The two main types of associative learning are termed classical conditioning and operant conditioning.

In **classical conditioning**, an involuntary response comes to be associated positively or negatively with a stimulus that did not originally elicit the response. This type of learning was investigated by Russian psychologist Ivan Pavlov. In his original experiments in the 1920s, Pavlov restrained a hungry dog in a harness and presented small portions of food at regular intervals (refer back to Section 41.3). The dog would salivate whenever it smelled the food. Pavlov then began to sound a metronome when presenting the food. Eventually the dog would salivate at the sound of the metronome, whether or not the food was present. Classical conditioning is widely observed in animals. For example, many insects quickly learn to associate certain flower odors with nectar rewards and other flower odors with no rewards. In humans, the sound of a dentist's drill is enough to produce a feeling of uneasiness, tension, and sweaty palms.

In **operant conditioning**, an animal's behavior is reinforced by a consequence, either a reward or a punishment. The classic example of operant conditioning comes from the work of the American psychologist B. F. Skinner, who placed laboratory animals, usually rats, in a specially devised cage with a lever that came to be known as a Skinner box. If the rat pressed on the lever, a small amount of food would be dispensed. At the beginning of the experiment, the rat would often bump into the lever by accident, eat the food, and continue exploring its cage. Later, it would learn to associate the lever with obtaining food. Eventually, if it was hungry, the rat would almost continually press the lever. Operant conditioning, also called trial-and-error learning, is common in animals. Often it is associated with negative rather

**(a) Blue jay eating monarch**  **(b) Vomiting reaction**

**Figure 55.3** **Operant conditioning, also known as trial-and-error learning.** **(a)** A young blue jay will eat a monarch butterfly, not knowing that it is noxious. **(b)** After the first experience of vomiting after eating a monarch, a blue jay will avoid such insects in the future.
a–b: ©L.P. Brower, Sweet Briar College

*Concept Check:* *What's the difference between operant conditioning and classical conditioning?*

 **Core Skill: Connections** Look ahead to Figure 57.9f. How might operant conditioning be related to the similar appearance of king snakes and coral snakes?

than positive reinforcement. For example, toads eventually refuse to strike at insects that sting, such as wasps and bees, and birds will learn to avoid bad-tasting butterflies (**Figure 55.3**). In humans, giving children a reward for completing homework is a positive reinforcer.

## Cognitive Learning Involves Conscious Thought

**Cognitive learning** refers to the ability to solve problems with conscious thought and includes activities such as perception, analysis, judgment, recollection, and imagining. In the 1920s, German psychologist Wolfgang Köhler conducted a series of classic experiments with chimpanzees, and the results suggested that animals other than humans can exhibit cognitive learning. In the experiments, a chimpanzee was left in a room with bananas hanging from the ceiling and out of reach (**Figure 55.4**). Also present in the room were several wooden boxes. At first, the chimp tried in vain to jump up and grab the bananas. After a while, however, it began to arrange the boxes one on top of another underneath the fruit. Eventually, the chimp climbed the boxes and retrieved the fruit.

Many other examples of such behavior have been observed. Chimps strip leaves off twigs and use the twigs to poke into ant nests, withdrawing the twig and licking the ants off. Captive ravens have been shown to retrieve meat suspended from a branch by a string, even though they have never encountered the problem before. They pull up on the string, step on it, and then pull up on the string again, repeating the process until the meat is within reach.

## Both Genetics and Learning Influence Most Behaviors

Much of the behavior we have discussed so far has been presented as either innate or learned, but the behavior we observe in nature is usually a mixture of both. Bird songs provide a good example. Many birds learn their songs as juveniles, when they hear their parents sing. If juvenile white-crowned sparrows are raised in isolation, their adult songs do not resemble the typical species-specific song (**Figure 55.5**). If they hear only the song of a different species, such as the song sparrow, they again sing a poorly developed adult song. However, if they hear the song of the white-crowned sparrow, they will learn to sing a fully developed white-crowned sparrow song. The birds are genetically programmed to learn, but they will sing the song correctly only if the appropriate instructive program is in place to guide learning.

**Figure 55.4** **Cognitive learning involving problem-solving ability.** This chimp has devised a solution to the problem of retrieving bananas that were initially out of its reach. (all): ©Lilo Hess/The LIFE Images Collection/Getty Images

 **Core Skill: Connections** Look back at Figure 43.17. Does learning affect brain structure?

| Song heard by juvenile | | Song sung by juvenile | |
|---|---|---|---|
| No song heard | | Abnormal song | |
| Song of song sparrow | | Abnormal song | |
| Song of white-crowned sparrow | | Normal song | |

**Figure 55.5  The interaction between genetics and learning.** The lines represent the different sound frequencies produced by the birds over a short time interval. The juvenile white-crowned sparrow will sing an abnormal song if it is kept in isolation or hears only the song of a different species. However, the juvenile will sing the normal white-crowned sparrow song if exposed to it. (left): ©Robert Shantz/Alamy Stock Photo

*Concept Check:* *Cuckoos lay their eggs in other birds' nests, so their young are reared by parent birds of a different species. However, unlike the white-crowned sparrow, adult cuckoos always sing their own distinctive song, not that of the host species they hear as juveniles. How is this possible?*

For some types of behavior, learning can be coupled with innate behavior only during a limited time period of development, which is called a **critical period**. An example is **imprinting**. During this process, learning occurs during a brief critical period and establishes a long-lasting behavioral response to a specific object or individual, such as recognition and bonding to a parent. Imprinting was studied by Austrian ethologist Konrad Lorenz in the 1930s. Lorenz noted that young birds of some species imprint on their mother during a critical period that is usually within a few hours after hatching. This behavior serves them well, because in many species of ducks and geese, it would be hard for the mother to keep track of all her offspring as they walk or swim. After imprinting takes place, the offspring keep track of the mother.

The survival of young ducks or geese requires that they quickly learn to follow their mother's movements. Lorenz raised greylag geese from eggs, and soon after they hatched, he used himself as the model for imprinting. As a result, the young goslings imprinted on Lorenz and followed him around (**Figure 55.6**). For the rest of their life, they preferred the company of Lorenz and other humans to geese. Studies have shown that even an object as foreign as a black box, watering can, or flashing light can be imprinted on if it is the first moving object the chick sees during the critical period. In nature, if young geese are not provided with any stimulus during the critical period, they will fail to imprint on anything, and without parental care, they will almost certainly die.

Other animals imprint in different ways. Newborn shrews imprint on the scent of their mother. Mothers also can imprint on their own young within a few hours. For example, a relatively common trick used in sheep farming is to disguise a lamb whose mother has died or abandoned it by wrapping it in the fleece of another ewe's stillborn lamb. That second ewe will then care for the abandoned lamb because it smells like her own. In these situations, the innate behavior is the ability to imprint soon after birth, and the factors in the environment are the stimulus to which the imprinting is directed.

**Figure 55.6  Lorenz being followed by his imprinted geese.** Newborn geese follow the first object they see after hatching and later will follow that particular object only. They normally follow their mother but can be induced to imprint on humans. The first thing these young geese saw after hatching was ethologist Konrad Lorenz. ©Nina Leen/The LIFE Picture Collection/Getty Images

## 55.2  Local Movement and Long-Range Migration

**Learning Outcomes:**

1. Distinguish between kinesis and taxis, two different types of local movement.
2. Describe the three mechanisms animals use during migration.
3. **CoreSKILL »** Analyze experiments that determine whether animals learn using visual or olfactory clues.

Organisms need to find their way, both locally and over what can be extremely long distances. Locally, organisms continually need to locate sources of food, water, mates, and perhaps nesting sites. Migration involves the longer-distance seasonal movement of animals, usually between overwintering areas and summer breeding sites; these are often hundreds or even thousands of kilometers apart. Several different types of behavior may be involved in these movements.

In this section, we begin by exploring local movement and animals' use of landmarks to guide their movements. We will then consider migration and examine the possible mechanisms used by migrating animals to find their way.

## Local Movement Can Involve Kinesis, Taxis, and Memory

The simplest forms of movement are mere responses to stimuli. A **kinesis** is a movement in response to a stimulus, but one that is not directed toward or away from the source of the stimulus. A simple experiment often done in classrooms is to observe the activity levels of woodlice, sometimes called sow bugs or pill bugs, in dry areas and moist areas. The woodlice move faster in drier areas, and they slow down when they reach moist environments. This behavior tends to keep them in damper areas, which they prefer in order to avoid desiccation.

A **taxis** is a more directed type of movement response, either toward (positive taxis) or away from (negative taxis) an external stimulus. Cockroaches exhibit negative phototaxis, meaning they tend to move away from light. Under low-light conditions, the photosynthetic unicellular flagellate *Euglena gracilis* shows positive phototaxis and moves toward a light source. Sea turtle hatchings are also strongly attracted to light. On emerging from their nests, they crawl toward the brightest location, traditionally the reflected moonlight on the ocean's surface. Lighted houses on the shore can disorient the hatchlings, however, and lead them to wander away from the ocean and succumb to dehydration, exhaustion, and predation. This is why beachfront property owners are requested to turn their lights down in turtle-hatching season. Male silk moths orient themselves in relation to wind direction (anemotaxis). If the air current carries the scent of a female moth, they will move upwind to locate it. Some freshwater fishes orient themselves to the currents of streams. Many fishes exhibit positive rheotaxis (from the Greek *rheos*, meaning current), in that they swim toward the oncoming water current. This taxis aids in respiration because more water passes through their gills and also helps them from being washed downstream.

Sometimes memory and landmarks may be used to aid in local movements. Dutch-born ethologist Niko Tinbergen showed how the female digger wasp uses landmarks to relocate her nests, as described next.

 **Core Skill: Process of Science**

## Feature Investigation | Tinbergen's Experiments Showed That Digger Wasps Use Landmarks to Find Their Nests

In the sandy, dry soils of Europe, the solitary female digger wasp (*Philanthus triangulum*) digs four to five nests in which to lay her eggs. Each nest stretches obliquely down into the ground for 40–80 cm. The wasp follows the digging by performing a sequence of apparently genetically programmed events. She catches and stings a honeybee, which paralyzes it; returns to the nest; drags the bee into the nest; and lays an egg on it. The egg hatches into a larva, which feeds on the paralyzed bee. However, the larva needs to ingest five to six bees before it is fully developed. This means the wasp must catch and sting four to five more bees for each larva. She can carry only one bee at a time. After each visit, the wasp must seal the nest with soil, find a new bee, relocate the nest, open it, and add the bee.

How does the wasp relocate the nest after spending considerable time away? Niko Tinbergen observed the wasps hover and fly around the nest each time they took off. He hypothesized that they were learning the nest position by creating a mental map of the landmarks in the area.

To test his hypothesis, Tinbergen experimentally adjusted the landmarks around the burrow that the wasps might be using as cues (**Figure 55.7**). First, he put a ring of pinecones around the

**Figure 55.7** How Niko Tinbergen discovered the digger wasp's nest-locating behavior.

**Concept Check:** *How would you test what type of spatial landmarks are used by female digger wasps?*

**HYPOTHESIS** Digger wasps (*Philanthus triangulum*) use visual landmarks to locate their nests.

**STARTING LOCATION** The female digger wasp excavates an underground nest, to which she returns daily, bringing food to the larvae located inside.

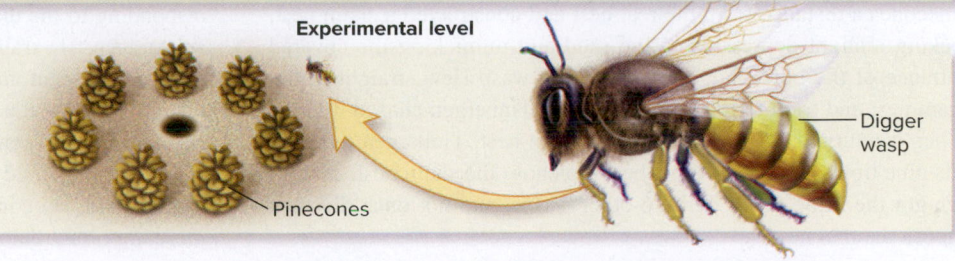

Experimental level

1 Place a ring of pinecones around the nest to train the wasp to associate pinecones with the nest.

Pinecones

Digger wasp

**2** After the wasp leaves the nest to hunt, move the pinecones 30 cm from the real nest. The wasp returns and flies to the center of the pinecone circle instead of the real nest. Repeated experiments yield similar results (see data), indicating that the wasp uses landmarks as visual cues.

Move pinecones 30 cm from the nest.

**3** To test whether it is the shape or the smell of the pinecones that elicits the response, perform the same experiment as above, except use pinecones with no scent and add 2 small pieces of cardboard coated with pine oil.

Pine oil

Cardboard

**4** After the wasp leaves the nest, move the pinecones 30 cm from the nest, but leave the scented cardboard at the nest. The wasps again fly to the pinecone nest (see data), indicating that it is the arrangement of cones, not their smell, that elicits the learning.

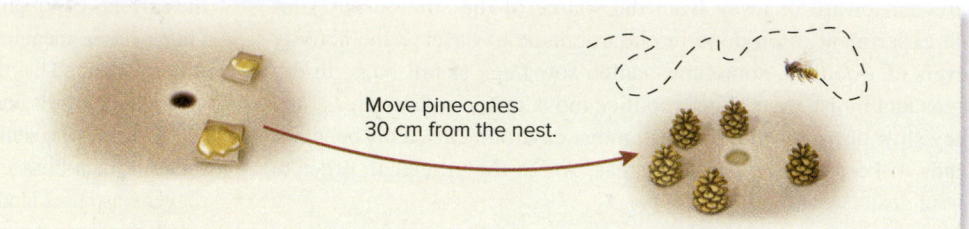

Move pinecones 30 cm from the nest.

**5** **THE DATA***

**Results from steps 1 and 2:**

| Wasp # | Number of return visits per wasp to real nest without pinecones | Number of return visits per wasp to sham nest with pinecones |
|---|---|---|
| 1–17 | 0 | ~9 |

**Results from steps 3 and 4:**

| Wasp # | Number of return visits per wasp to real nest with scented cardboard | Number of return visits per wasp to sham nest with pinecones |
|---|---|---|
| 18–22 | 0 | ~6 |

*Seventeen wasps, numbered 1–17, were studied as described in steps 1 and 2. Five wasps, numbered 18–22, were studied as described in steps 3 and 4.

**6** **CONCLUSION** Digger wasps remember the positions of visual landmarks and use them as aids in local movements.

**7** **SOURCE** Tinbergen, N. 1951. *The study of instinct*. Clarendon Press, Oxford.

nest entrance to train the wasp to associate the pinecones with the nest. Then, when the wasp was out hunting, he moved the circle of pinecones a distance from the real nest and constructed a sham nest, making a slight depression in the sand and mimicking the covered entrance of the burrow. On returning, the wasp flew straight to the sham nest and tried to locate the entrance. Tinbergen chased it away. When it returned, it again flew to the sham nest. Tinbergen repeated this nine times, and every time the wasp chose the sham nest. Tinbergen got the same result with 16 other wasps, and not once did they choose the real nest.

Next, Tinbergen investigated the type of stimulus that might be eliciting the learning. He hypothesized that the wasps could be responding to the distinctive scent of the pinecones rather than their appearance. He trained the wasps by placing a circle of pinecones that had no scent and two small pieces of cardboard coated in pine oil around the real nest. He then moved the cones to surround a sham nest and left the scented cardboard around the real nest. The returning wasps again ignored the real nest with the scented cardboard and flew to the sham. He concluded that for the wasps, sight was apparently more important than smell in determining landmarks.

*Experimental Questions*

1. What observations were important for the development of Tinbergen's hypothesis explaining how digger wasps located their nests?

2. **CoreSKILL** » How did Tinbergen test the hypothesis that the wasps were using landmarks to relocate the nest? What were the results?

3. **CoreSKILL** » Did the Tinbergen experiment rule out any other cue the wasps may have been using besides the sight of pinecones?

---

 ## Core Skill: Quantitative Reasoning

BIO **TIPS** **THE QUESTION** *Tinbergen also investigated whether digger wasps use visual and/or olfactory cues to locate their honeybee prey. He attached a number of dead bees to thread tethers so that they blew in the wind. Some of them were descented with alcohol. He also tethered bee-sized pieces of twigs to threads, and scented some of these by shaking them with dead bees. Finally, he tied dead bees to twigs, which did not move in the wind. He presented these five types of "dummies" to digger wasps and observed their behavior. The wasps either ignored the prey, hovered in place before it, pounced on or grasped it, or attempted to sting or capture it. The resultant numbers of events in 100 hours of observing are shown in the table below. Do digger wasps rely on visual or olfactory cues, or both, to locate their prey?*

| Experimental treatment | Hover | Pounce or grasp | Sting or capture |
|---|---|---|---|
| Tethered bees | 26 | 14 | 26 |
| Alcohol-treated tethered bees | 20 | 9 | 0 |
| Bee-sized tethered twigs | 14 | 3 | 0 |
| Bee-sized scented, tethered twigs | 34 | 20 | 9 |
| Bees tied to twigs | 0 | 0 | 0 |

**T**OPIC *What topic in biology does this question address?* The topic is animal behavior. More specifically, the question concerns how digger wasps locate suitable prey.

**I**NFORMATION *What information do you know based on the question and your understanding of the topic?* From the question, you know that digger wasps may use visual or olfactory cues, or both, to locate their prey. From your understanding of the topic that you gained from Figure 55.7, you know that digger wasps use visual landmarks to locate their nests.

**P**ROBLEM-SOLVING **S**TRATEGY *Interpret data.* Add two columns to the table of results, with one column labeled "Visual," which includes both accurate bee shape and movement, and the other "Olfactory," which corresponds to bee scent. Place a check mark in each of these columns if the results of the experiment indicate that digger wasps may be responding to bee shape, bee movement, or bee scent. Place a minus if they are not.

**ANSWER**

| Experimental | Hover | Pounce or grasp | Sting or capture | Visual Bee shape/ Movement | Olfactory |
|---|---|---|---|---|---|
| Tethered bees | 26 | 14 | 26 | ✓ / ✓ | ✓ |
| Alcohol-treated tethered bees | 20 | 9 | 0 | ✓ / ✓ | – |
| Bee-sized tethered twigs | 14 | 3 | 0 | – / ✓ | – |
| Bee-sized scented, tethered twigs | 34 | 20 | 9 | – / ✓ | ✓ |
| Bees tied to twigs | 0 | 0 | 0 | – / – | ✓ |

*With regard to visual cues, an accurate bee shape was not necessary; the wasps responded to bee-sized tethered twigs. The visual cue of prey movement was very important, because bees tied to twigs did not elicit a response. The olfactory cue also appeared important, because scented bees or scented bee-sized twigs elicited a greater response compared to those that were unscented. However, the olfactory cue alone was not sufficient to elicit a response because bees tied to twigs were not recognized as prey. Taken together, a visual cue (movement) and an olfactory cue (scent) are used by digger wasps to locate their prey.*

## Migration Involves Long-Range Movement and More Complex Spatial Navigation

The activities of some animal species involve **migration**, which is a long-range seasonal movement. Migrations are usually linked to temperature changes, availability of food, and suitable breeding areas. For example, nearly half the bird species of North America migrate to South America to escape the cold winters and feed, returning to North America in the spring to breed. Arctic terns that breed in Arctic Canada and Asia in summer migrate to the Antarctic to feed in the winter and then return to breed. This staggering journey involves up to a 40,000-km (25,000-mile) round-trip, most of it over the open ocean, during which the birds must stay airborne for days at a time!

The monarch butterfly of North America migrates to overwinter in California, Mexico, and possibly south Florida and Cuba (**Figure 55.8**). An interesting point about the northward journey of the monarch is that it involves several generations of butterflies to complete. On their way back to the northern U.S. and Canada, the butterflies lay eggs and die. The caterpillars develop on milkweed plants, and the resultant adults continue to journey farther north.

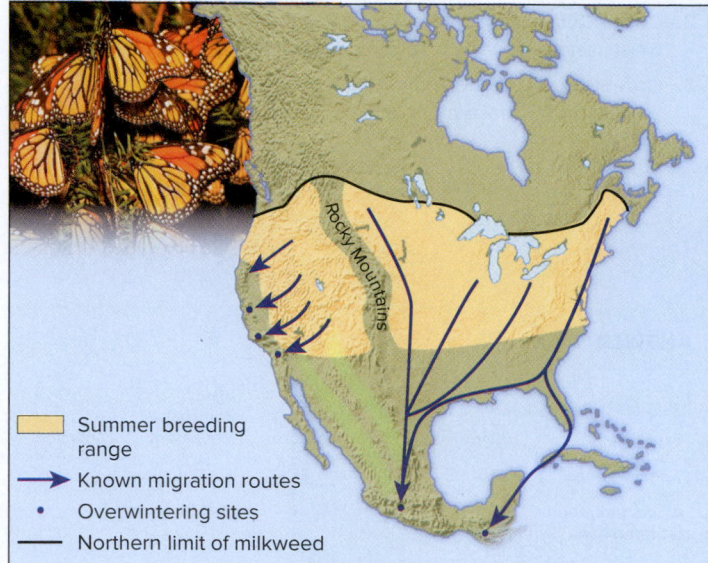

**Figure 55.8** **Monarch butterfly migration.** Many monarch butterflies east of the Rocky Mountains migrate to a small area in Mexico to avoid the cold northern weather. Here they roost together in large numbers in fir trees (inset). Some butterflies may stay in Florida and Cuba. Butterflies west of the Rockies overwinter in mild coastal California locations. ©Jodi Jacobson/Getty Images

**Concept Check:** *Why is the seasonal movement of monarchs an unusual example of migration?*

This cycle happens several times in the course of the return journey. The northward and southward migrations are unique in that none of the individuals has ever been to the destinations before. Therefore, the ability to migrate must be an innate behavior.

How do migrating animals find their way? Three mechanisms may be involved: piloting, orientation, and navigation. In **piloting**, an animal moves from one familiar landmark to the next. For example, many whale species migrate between summer feeding areas and winter calving grounds. Gray whales migrate between the Bering Sea near Alaska to coastal areas of Mexico. Features of the coastline, including mountain ranges, and rivers, may aid in navigation. In **orientation**, animals have the ability to follow a compass bearing and travel in a straight line. **Navigation** involves the ability not only to follow a compass bearing but also to set or adjust it.

An experiment with starlings helps illuminate the difference between orientation and navigation (**Figure 55.9**). European starlings breed in Scandinavia and northeastern Europe and migrate in a southwest direction toward coastal France and southern England to spend the winter. Migrating starlings were captured and tagged in the Netherlands and then transported south to Switzerland and released. Juvenile birds, which had never made the trip before, flew southwest in their migration and were later recaptured in Spain. Adult birds, with more experience, returned to their normal wintering range by adjusting their course by approximately 90°. This result indicates that the adult birds can actually navigate, whereas the juveniles rely on orientation.

Many species use a combination of navigational reference points, including the position of the Sun, the stars (for nighttime travel), and

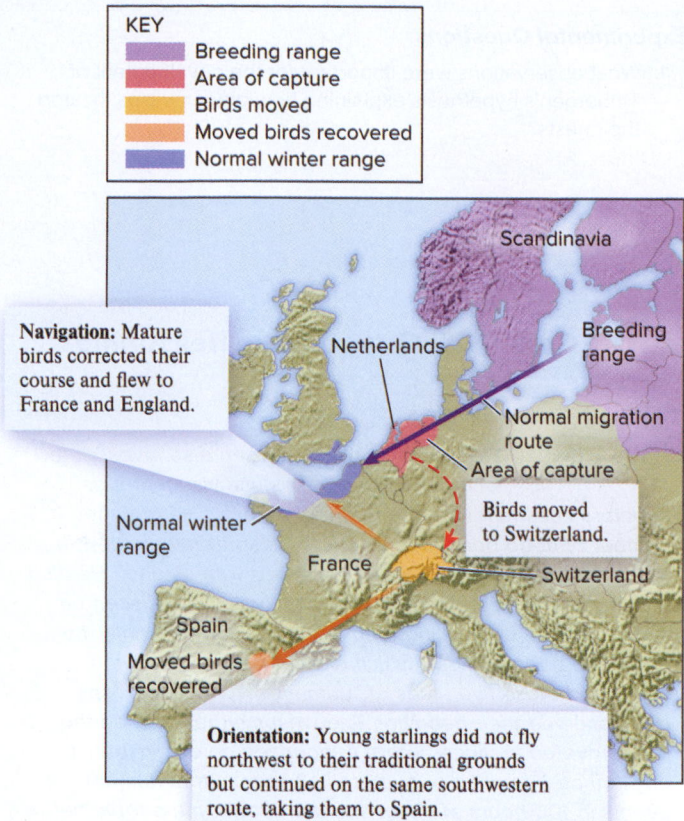

**Figure 55.9** **Orientation versus navigation.** Starlings normally migrate from breeding grounds in Scandinavia and northeastern Europe through the Netherlands and northern Germany to overwintering sites in France and England. This involves a southwest flight. When juveniles were captured in the Netherlands and moved to Switzerland, they continued on in a southwestern direction and ended up in Spain. When adult birds were captured and moved, they changed course and flew to their normal overwintering areas.

Earth's magnetic field. Homing pigeons have magnetite in their beaks that acts as a compass to indicate direction (refer back to Section 44.4 on electromagnetic reception). Navigation by the Sun or the stars also requires the use of a timing device to compensate for the ever-changing position of these reference points. Many migrants, therefore, possess the equivalent of an internal clock. Pigeons integrate their internal clock with the position of the Sun. Researchers have altered the internal clock of pigeons by keeping them under artificial lights for certain periods of time. When the pigeons are released, they display predictable deviations in their flight. For every hour that their internal clock is shifted, the orientation of the birds shifts about 15°.

Not all examples of animal migration are well understood. Green sea turtles feed off the coast of Brazil yet swim east for 2,300 km (1,429 miles) to lay their eggs on Ascension Island, an 8-km-wide island in the center of the Atlantic Ocean between Brazil and Africa. It is not known why the turtles lay their eggs on this speck of an island or how they succeed in finding it. Perhaps fewer predators exist on Ascension than on other beaches. A combination of magnetic orientation and chemical cues may help them find it. Thus, although scientists have made many discoveries about animal navigation, much remains to be learned about how animals acquire a map sense.

## 55.3 Foraging Behavior and Defense of Territory

**Learning Outcomes:**

1. Describe and give examples of optimal foraging.
2. Explain how the risk of predation prevents optimal foraging in some species.
3. Outline the costs and benefits of defending a territory.

Food gathering, or foraging, often involves decisions about whether to remain at a resource patch and look for more food or look for a completely new patch. The analysis of these decisions is often viewed in terms of **optimality theory**, which predicts that an animal should behave in a way that maximizes the benefits of a behavior minus its costs. In this case, the benefits are the nutritional or caloric value of the food items, and the costs are the energetic or caloric costs of movement. Optimality theory can also be applied to other behavioral issues such as how large a territory to defend. Too small a territory would contain insufficient resources, such as food and mates, and too large a territory would be too energetically costly to defend. Theoretically, then, there is an optimal territory size for a given individual to defend. In this section, we will explore examples that involve optimal foraging for food and optimal defense of territory.

### Optimal Foraging Entails Maximizing the Benefits and Minimizing the Costs of Food Gathering

**Optimal foraging** is the concept that in a given circumstance, an animal seeks to obtain the most energy possible with the least expenditure of energy. The underlying assumption of optimal foraging is that natural selection favors animals that are maximally efficient at propagating their genes and at performing all other functions that serve this purpose. In this model, the more net energy an individual gains in a limited time, the greater the reproductive success.

Shore crabs (*Carcinus maenas*) eat many different-sized mussels but tend to feed preferentially on intermediate-sized mussels, which give them the highest rate of energy return (**Figure 55.10**). Very large mussels yield more energy, but they take so long for the crab to open that they are actually less profitable, in terms of energy yield per unit time spent, than smaller sizes. Very small mussels are easy to crack open but contain so little food that they are not worth the effort. This leaves intermediate-sized mussels as the preferred size. Of course, the intermediate-sized mussels may take a longer time to locate, because more crabs are looking for them, so crabs eat some less profitable but more frequently encountered sizes of mussels. The result is that the diet consists of mussels in a range of sizes around the preferred optimal size.

In some cases, animals do not forage optimally. Many animals seek not only to maximize food intake but also to minimize the risk of predation. Some species may only dart out to take food from time to time. For example, many species of ants display this behavior because they can be attacked by parasitic flies that hover around ant colonies. Although ants are relatively quick, some of them cannot avoid these fast-flying parasites, which lay an egg on the ant's head. Once the parasite's eggs hatch, the larvae bore their way into the ant's head, eventually killing it.

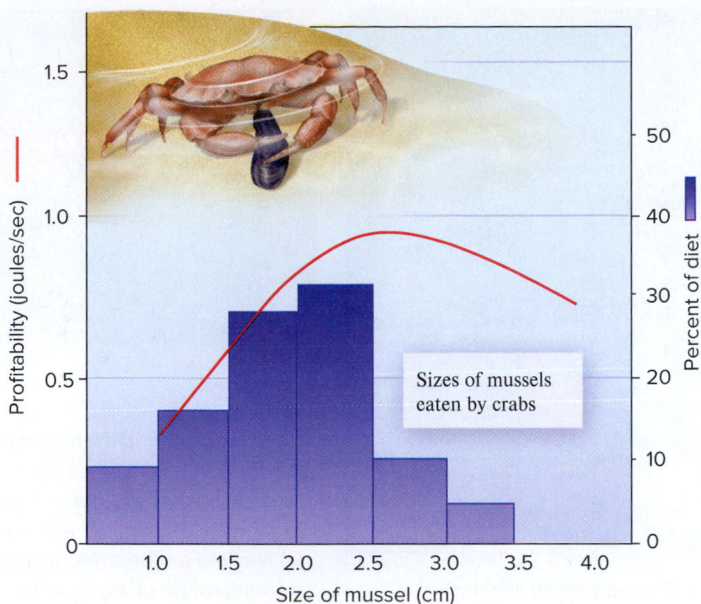

**Figure 55.10** **Optimal foraging behavior in shore crabs.** When offered a choice of equal numbers of each size mussel, shore crabs (*Carcinus maenas*) prefer intermediate-sized mussels that provide the highest rate of energy return. Profitability is the energy yield (joules) per second of time used in breaking open the shell.

In the tropics, leafcutter ants, *Atta cephalotes*, are subject to attack by the fly *Neodohrniphora curvinervis* (**Figure 55.11a**). The larger worker ants are more likely to be attacked, because smaller workers' heads are too small to allow proper development of the parasite flies' larvae. When Matthew Orr compared the size

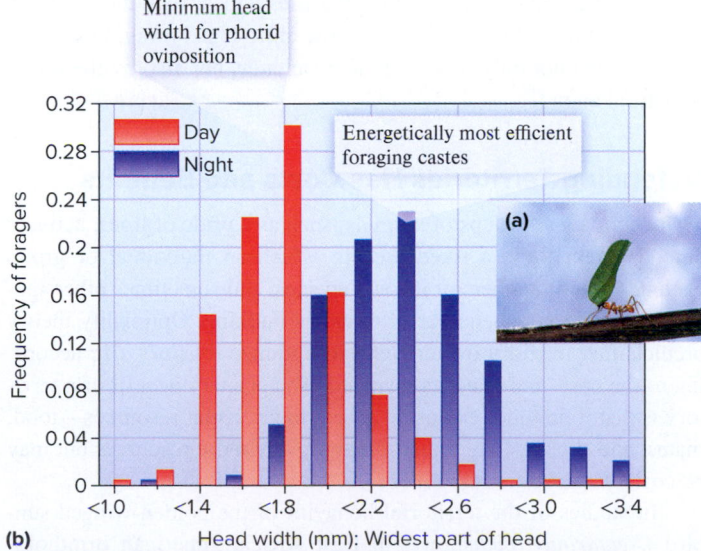

**Figure 55.11** **Changes in foraging strategies in the presence of parasites.** **(a)** The tropical leafcutter ant, *Atta cephalotes*. **(b)** The most efficient leafcutter ants are larger individuals. However, larger individuals are restricted to foraging at night because of the activity of parasitic flies during the day. During the day, only small ants forage, as these are not subject to the same levels of parasitism. a: Source: Photo by Scott Bauer/USDA

**(a) Golden-winged sunbird**

**(b) Cheetah**

**(c) Nesting gannets**

**Figure 55.12** **Differing territory sizes among animals.** **(a)** The golden-winged sunbird of East Africa (*Nectarinia reichenowi* ) has a medium territory size that depends on the number of flowers it can obtain resources from and defend. **(b)** Cheetahs (*Acinonyx jubatus*) hunt over large areas and can have extensive territories. This male is urine-marking part of his territory in the southern Serengeti, near Ndutu, Tanzania. **(c)** Nesting gannets (*Morus bassanus*) have much smaller territories, in which each bird is just beyond the pecking range of its neighbor. a: ©Tony Camacho/Science Source; b: ©Gregory G. Dimijian/Science Source; c: ©Getty Images

distributions of the leafcutter ants that foraged during the day versus those that foraged at night, he discovered a large difference in the size distributions (**Figure 55.11b**). Most foraging by larger workers is done at night, when it is too dark for the flies to see properly and their activity ceases. The lesser amount of foraging in the day is performed by smaller ants that are less susceptible to fly attack. Although the data are not shown in Figure 55.11, Orr extended the activity of *N. curvinervis* by positioning lights outside the leafcutter nests. In these cases, the foraging activities of the leafcutters were thrown into disarray. Ants were backed up around the nest, many rising on their hind legs to snap at the flies which, with the extended light, continued to try to parasitize the ants. Thus, foraging activity is influenced not only by energetic efficiency, but also by the threat of natural enemies.

## Defending Territories Has Costs and Benefits

Many animals or groups of animals, such as a pride of lions, actively defend a **territory**, a fixed area in which an individual or group excludes other members of its own species, and sometimes other species, by aggressive behavior or territory marking. Optimality theory predicts that territory owners tend to optimize territory size according to the costs and benefits involved. The primary benefit of a territory is that it provides exclusive access to particular resources—food, mates, and shelter. Larger territories provide more resources but may be costly to defend.

In studies of the territorial behavior of the golden-winged sunbird (*Nectarinia reichenowi*) in East Africa, American ornithologists Frank Gill and Larry Wolf measured the energy content of nectar as the benefit of maintaining a territory and compared it to the energy costs of activities such as perching, flying, and fighting (**Figure 55.12a**). Defending the territory ensured that other sunbirds did not take nectar from available flowers, thus increasing the amount of nectar in each flower. In defending a territory, the sunbird gained 780 Calories (kilocalories) a day in extra nectar content. However,

the sunbird also spent 728 Calories in defense of the territory, yielding a net gain of 52 Calories a day and making territorial defense advantageous.

Territory size differs considerably among species. Male cheetahs defend relatively large territories, often 40 km² (about 15 square miles). These territories are usually located where densities of prey are relatively high. Males warn intruders to stay away from their territories by scent marking (**Figure 55.12b**). In contrast, territories set up solely to defend areas for mating or nesting are often relatively small. For example, male sea lions defend small areas of beach. The preferred areas contain the largest number of females and are controlled by the largest breeding bulls. The size of the territory of some nesting birds, such as gannets, is determined by how far the bird can reach to peck its neighbor without leaving its nest (**Figure 55.12c**).

## 55.4 Communication

**Learning Outcome:**

1. Give examples of how animals use chemical, auditory, visual, and tactile communication.

**Communication** is the use of specially designed visual, chemical, auditory, or tactile signals to modify the behavior of others. It may be used for many purposes, including defending territories, maintaining contact with offspring, courtship, and contests between males. The use of different forms of communication between organisms depends on the environment in which they live. For example, visual communication plays little role in the signals of nocturnal animals. Similarly, for animals in dense forests, sounds are of prime importance. Sound, however, is a temporary signal. Scent can last longer and is often used to mark the large territories of some mammals. In this section, we will examine the various types of communication—chemical, auditory, visual, and tactile—that occur among animals.

## Chemical Communication Is Often Used to Mark Territories or Attract Mates

The chemical marking of territories is common among animals, especially among members of the canine and feline families (see Figure 55.12b). Scent trails are often used by social insects to recruit workers to help bring prey to the nest. Fire ants (genus *Solenopsis*) attack large, living prey, and many ants are needed to drag the prey back to the nest. The scout that finds the prey lays down a scent trail from the prey back to the nest. The scent excites other workers, which follow the trail to the prey. The scent marker is very volatile, and the trail effectively disappears in a few minutes to avoid mass confusion over old trails.

Animals frequently use chemicals to attract mates. Female moths attract males by powerful chemical attractants called **pheromones**. Male moths have receptors that can detect as little as a single molecule. Among social organisms, some individuals use pheromones to manipulate the behavior of others. For example, a queen bee releases pheromones that suppress the development of the reproductive system of workers, which ensures that she is the only reproductive female in the hive.

## Auditory Communication Is Often Used to Attract Mates and to Deter Competitors

Many species communicate by making sounds. Because the ground can absorb sound waves, sound travels farther in the air, which is why many birds and insects perch on branches or leaves when singing. Air is on average 14 times less turbulent at dawn and dusk than during the rest of the day, so sound carries farther then, which helps explain the preference of most animals for calling at these times. Birds living near airports advance their dawn chorus to reduce overlap with aircraft noise. Some insects utilize the very plants on which they feed as a medium of song transmission. Many male leafhopper and planthopper insects vibrate their abdomens on leaves and create species-specific courtship songs that are transmitted by adjacent vegetation and are picked up by nearby females of the same species.

Although many males use auditory communication to attract females, some females use calls to attract the attention of males.

Female elephant seals scream loudly when approached by a nondominant male. This sound attracts the attention of the dominant male, which drives the nondominant male away. In this way, the female is guaranteed a mating with the strongest male. Sound production can attract predators as well as mates. Some bats listen for the mating calls of male frogs to find their prey. Parasitic flies detect and locate chirping male crickets and then deposit larvae on or near them. The larvae latch onto and penetrate the cricket and eventually kill it. Sound may also be used by males during competition over females. In many animals, lower-pitched sounds come from larger males, so by calling to one another, males can gauge the size of their opponents and decide whether it is worth fighting.

## Visual Communication Is Often Used in Courtship and Aggressive Displays

In courtship, animals use a vast number of visual signals to identify and select potential mates. Competition among males involving displays to attract females has led to elaborate coloration and extensive ornamentation in some species. For example, peacocks and males of many bird species have developed elaborate plumage to attract females.

Male fireflies display light flashes that are species specific with regard to number and duration of flashes (**Figure 55.13a**). Females respond with a flash of their own. Such bright flashes are also bound to attract predators. Some female fireflies use mimicry to their advantage. Female fireflies of the species *Photuris versicolor* mimic the flashing responses normally given by females of other species, such as *Photinus tanytoxus*, in order to lure the males of those species close enough to eat them!

Visual signals are also used to resolve disputes over territories or mates. Deer and antelope have antlers or horns that they use to display and spar over territory and females. Most of these matches never develop into outright fights, because the males gauge their opponent's strength by the size of these ornaments (look ahead to Figure 55.22b). Among insects, the "horns" of rhinoceros beetles and the eye stems of stalk-eyed flies send similar signals (**Figure 55.13b**).

**(a) Firefly flashing**

**(b) Male hercules beetles fighting**

**Figure 55.13**   **Visual communication. (a)** Communication between fireflies is conducted by species-specific light flashes emitted by organs located on the underside of the abdomen. **(b)** The horns of these hercules beetles provide a signal about the strength of their owners. a: ©Darwin Dale/Science Source; b: ©Taylor Weidman/LightRocket/Getty Images

  **Core Concept: Evolution**   In both of the cases illustrated in Figure 55.13, morphological features influence animal behavior, which, in turn, affects reproductive success.

**(a)** Bees clustering around a recently returned scout, seen in the center vibrating her abdomen.

**(b)** Round dance

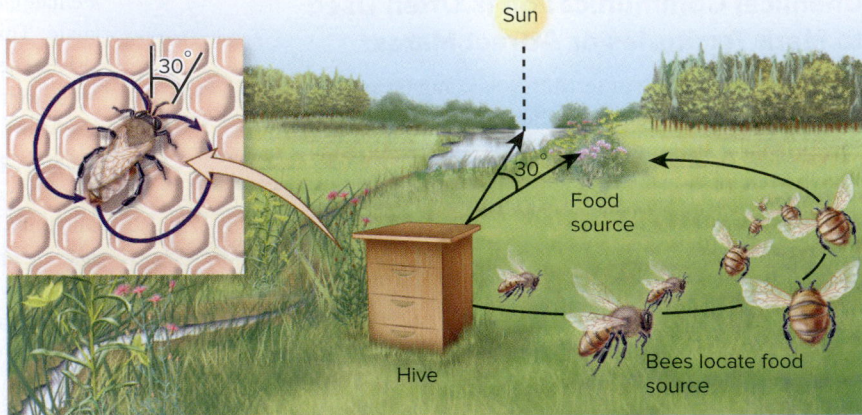

**(c) Waggle dance:** The angle of the waggle to the vertical orientation of the honeycomb corresponds to the angle of the food source from the Sun.

**Figure 55.14** **Tactile communication among honeybees regarding food sources.** **(a)** Bees gather around a newly returned scout to receive information about nearby food sources. **(b)** If the food is less than 50 m away, the scout performs a round dance. **(c)** If the food is more than 50 m away, the scout performs a waggle dance, which conveys information about its location. If the dance is performed at a 30° angle to the right of the hive's vertical plane, then the food source is located at a 30° angle to the right of the Sun. a: ©Scott Camazine/ Alamy Stock Photo

## Tactile Communication Is Used to Strengthen Social Bonds and to Convey Information About Food

Animals often use tactile communication to establish bonds between group members. Primates frequently groom one another, and canines and felines may nuzzle and lick each other. Many insects use tactile communication to convey information on the whereabouts of food. Members of the ant genus *Leptothorax* feed on prey such as dead insects. When a scouting ant encounters such prey, it usually needs an additional worker to help bring it back to the nest. Rather than laying a scent trail, which is energetically costly, the scout ant recruits a helper and physically leads it to the food source. The helper runs in tandem with the scout, its antennae touching the scout's abdomen.

One of the most fascinating examples of tactile communication among animals is the dance of the honeybee, studied by German ethologist Karl von Frisch in the 1940s. Bees commonly live in large hives; in the case of the European honeybee (*Apis mellifera*), the hive consists of 30,000–40,000 individuals. The flowering plants on which the bees forage can be located long distances from the hive and are distributed in a patchy manner, with any given patch usually containing many flowers that store more nectar and pollen than an individual bee can carry back to the nest.

A scout bee that locates a resource patch returns to the hive and recruits more workers to join it (**Figure 55.14a**). Because the inside of the hive is dark, a visual signal will not be effective. Instead, the scout uses a tactile signal. The scout dances on the vertical side of a honeycomb, and the dance is monitored by other bees, which follow and touch her to interpret the message. If the food is relatively close to the hive, less than 50 m away, the scout performs a round dance, rapidly moving in a circle, first in one direction and then the other. The other bees know the food is relatively close at hand,

and the smell of the scout tells them what flower species to look for (**Figure 55.14b**).

If the food is more than 50 m away, the scout will perform a different type of dance, called a waggle dance. In this dance, the scout traces a figure 8, in the middle of which she waggles her abdomen and produces bursts of sound. Again, the other bees maintain contact with her. Occasionally, the scout regurgitates a small sample of nectar so the bees know the type of food source they are looking for. The truly amazing part of the waggle dance is that the angle at which the central part of the figure 8 deviates from the vertical direction of the comb represents the same angle at which the food source deviates from the point at which the Sun hits the horizon (**Figure 55.14c**). The direction is always up-to-date, because the bee adjusts the dance as the Sun moves across the sky.

## 55.5 Living in Groups

**Learning Outcome:**

**1.** Outline the costs and benefits of living in groups.

As we have seen, much of animal behavior is directed at other animals. Some of the more complex behavior occurs when animals live together in groups such as flocks or herds. A central concern of ecology is to explain the distribution patterns of organisms, and a very important related task is to understand the reason for variation in the degree of group living. One way to approach this question is to assess the costs and benefits involved. Although group living increases competition for food and the spread of disease, it also has benefits that compensate for the costs involved. Many of these benefits relate to locating food sources, assistance in rearing offspring, access to mates, and group defense against predators. Group living can reduce predator success in at least two ways: through increased vigilance and through protection in numbers.

## Living in Large Groups May Reduce the Risk of Predation Because of Increased Vigilance

For many predators, success depends on the element of surprise. If an individual is alerted to an attack, the predator's chance of success is lowered. A wood pigeon (*Columba palumbus*) in a flock takes to the air when it spots a goshawk (*Accipiter gentilis*). Once one pigeon takes flight, the other members of the flock are alerted and follow suit. If each individual in a group occasionally scans the environment for predators, the larger the group, the less time an individual forager needs to devote to vigilance and the more time it can spend feeding. This explanation is referred to as the **many-eyes hypothesis** (**Figure 55.15**). Of course, cheating is a possibility, because some birds might never look up, relying on others to keep watch while they keep feeding. However, the individual that happens to be scanning when a predator approaches is most likely to escape, a fact that tends to discourage cheating.

## Living in Groups Offers Protection by the Selfish Herd

Group living also provides protection in sheer numbers. Typically, predators take one prey animal per attack. In any given attack, an individual antelope in a herd of 100 has a 1 in 100 chance of being selected, whereas a single individual has a 1 in 1 chance. Large herds may be attacked more frequently than a solitary individual, but a herd is unlikely to attract 100 times more attacks than an individual, often because of the territorial nature of predators. Furthermore, large numbers of prey

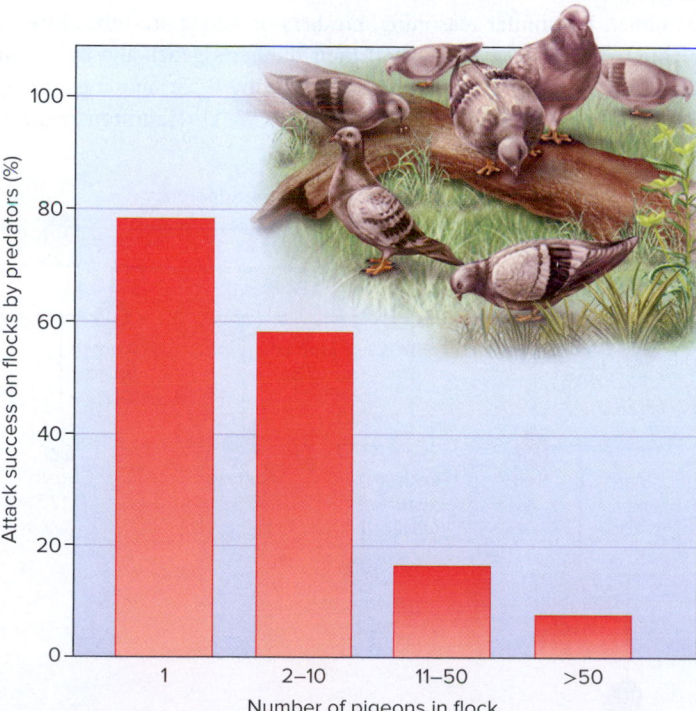

**Figure 55.15 Living in groups and the many-eyes hypothesis.** The larger the number of wood pigeons, the less likely an attack will be successful.

*Concept Check:* *What other advantage do individuals in large groups have when the group is being attacked by a predator?*

are able to defend themselves better than single individuals, which usually choose to flee. For example, groups of nesting black-headed gulls mob a crow, thereby reducing the crow's ability to steal the gulls' eggs.

Research has shown that within a group, each individual can minimize the danger to itself by choosing the location that is as close to the center of the group as possible. This was the subject of a famous paper, "The Geometry of the Selfish Herd," by British evolutionary biologist W. D. Hamilton. The explanation of this type of defense is that predators are likely to attack prey on the periphery because they are easier to isolate visually. Many animals in herds tend to bunch close together when they are under attack, making it physically difficult for the predator to get to the center of the herd.

Overall, group size may be the result of a trade-off between the costs and benefits of group living. Although much group behavior serves to reduce predation, other complex behavior occurs in groups, including grooming behavior and behavior that appears to benefit the group at the expense of the individual. For example, a honeybee stings a potential hive predator to discourage it. The bee's stinger is barbed, and once it has penetrated the predator's skin, the bee cannot withdraw it. The bee's only means of escape is to tear away part of its abdomen, leaving the stinger behind and dying in the process. In the next section, we will explore the reasons for such apparent altruistic behavior, in which an individual incurs costs to itself for the benefit of others.

## 55.6 Altruism

**Learning Outcomes:**

1. **CoreSKILL »** Evaluate the arguments for and against the concept of group selection.
2. Describe how the concept of kin selection can explain altruistic behavior.
3. **CoreSKILL »** Predict which types of relatives are more or less likely to be the recipients of apparent altruistic acts.
4. Explain why eusociality is a form of altruism.

In Chapter 23, we learned that natural selection is a process in which certain individuals are more likely to pass on their genes, yet we see many instances in which some individuals forego reproducing altogether, apparently to benefit the group. How do ecologists explain **altruism**, a behavior that appears to benefit others at a cost to oneself? In this section, we begin by discussing whether such behavior evolved for the good of the group or for the good of the individual. As we will see, most altruistic acts serve to benefit the individual's close relatives. We will explore the concept of kin selection, which argues that acts of self-sacrifice indirectly promote the spread of an organism's genes, and see how this plays out in an extreme form in the genetics of social insect colonies. Last, we will examine reciprocal altruism, instances of altruism among nonkin.

## In Nature, Selfish Behavior Is More Likely Than Altruism

One of the first attempts to explain the existence of altruism was called **group selection**, the premise that natural selection produces outcomes beneficial for the whole group or species. In 1962, British ecologist V. C.

Wynne-Edwards argued that a group containing altruists, each willing to subordinate its interests for the good of the group, would have a survival advantage over a group composed of selfish individuals. In concept, the idea of group selection seemed straightforward and logical: A group that consisted of selfish individuals would overexploit its resources and die out, but the fitness of a group with altruists would be enhanced.

In the late 1960s, the idea of group selection came under severe criticism. Leading the charge was American evolutionary biologist George C. Williams, who argued that evolution acts through the individual; that is, adaptive traits generally are selected for because they benefit the survival and reproduction of the individual rather than the group. Some of Williams's arguments against group selection follow.

*Mutation*    Individuals carrying mutations that allow them to readily use resources for themselves or their offspring have an advantage in a population in which individuals limit their resource use. Consider a species of bird in which a pair lays only two eggs; that is, it has a clutch size of two, and the resources are not overexploited for the good of the group. Laying two eggs ensures a replacement of the parent birds but prevents a population explosion. Suppose a mutant bird that lays three eggs arises. If the population is not overexploiting its resources, sufficient food may be available for all three young to survive. If this happens, the three-egg genotype eventually becomes more common than the two-egg genotype.

*Immigration*    Even in a population in which all pairs laid two eggs and no mutations occurred to increase clutch size, selfish individuals that laid more could still immigrate from other areas. In nature, populations are rarely sufficiently isolated to prevent immigration of selfish mutants from other populations.

*Resource Prediction*    Group selection assumes that individuals are able to assess and predict future food availability and population density within their own habitat. There is little evidence that they can. For example, it is difficult to imagine that songbirds would be able to predict the future supply of the caterpillars that they feed to their young and adjust their clutch size accordingly.

Most ecologists accept individual gain as a more plausible result of natural selection than group selection. Population size is more often controlled by competition in which individuals strive to command as much of a resource as they can. Such selfishness can cause some seemingly surprising behaviors. For example, male Hanuman langurs (*Semnopithecus entellus*) fight mothers and kill infants when they take over groups of females from other males (**Figure 55.16**). The reason for the behavior is that when they are not nursing their young, females become sexually receptive much sooner, hastening the day when the male can father his own offspring. Infanticide ensures that the male can father more offspring, and the genes governing this tendency spread by natural selection.

## Apparent Altruistic Behavior in Nature Is Often Associated with Kin Selection

If individual selfishness is more common than group selection, how do we account for what appear to be examples of altruism in nature? Some propose that the answer lies in a concept known as **kin selection**, selection for behavior that lowers an individual's own fitness but enhances the reproductive success of a relative. Because

**Figure 55.16  Langur monkeys fighting.** Male Hanuman langurs (*Semnopithecus entellus*) may act aggressively toward mothers and even kill their young, hastening the day when the females become sexually receptive and the males can father their own offspring.
©Andrew Parkinson/Stockbyte/Getty Images

relatives share many of the same genes, altruist individuals increase the likelihood that their genes are passed along to future generations by enhancing the reproductive success of their relatives.

The probability that any two individuals will share a copy of a particular gene is a quantity, $r$, called the **coefficient of relatedness**. During meiosis in a diploid species, any given copy of a gene has a 50% chance of segregating into an egg or sperm. A mother and father are on average related to their children by an amount $r = 0.5$, because half of a child's genes come from its mother and half from its father. By similar reasoning, brothers or sisters are related by an amount $r = 0.5$ (they share half their mother's genes and half their father's); grandchildren and grandparents, by 0.25; and cousins, by 0.125 (**Figure 55.17**). In 1964, ecologist W. D. Hamilton realized

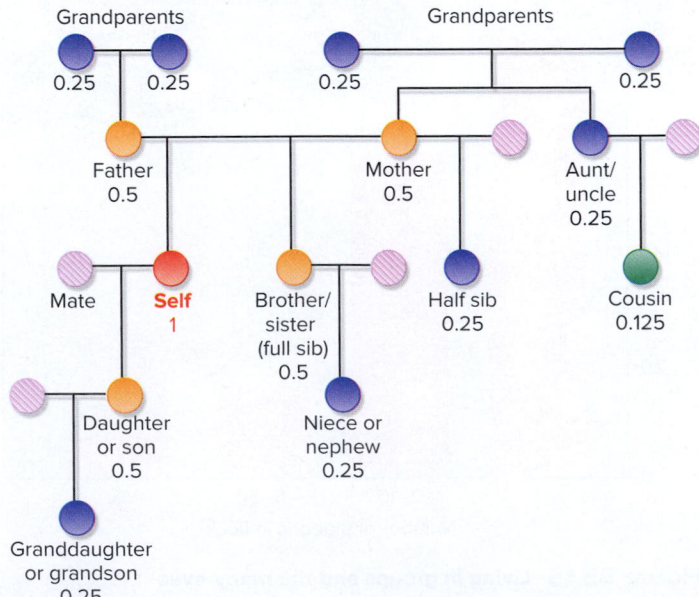

**Figure 55.17  Degree of genetic relatedness to self in a diploid organism.** Pink hatched circles represent completely unrelated individuals.

the implication of the coefficient of relatedness for the evolution of altruism. An organism not only can pass on its genes by having offspring, but also can pass them on via the reproductive success of siblings, nieces, nephews, and cousins. Thus, an organism has a vested interest in protecting its brothers and sisters, and even their offspring.

The term **inclusive fitness** is used to designate the total number of copies of genes passed on through one's relatives, as well as one's own reproductive output. Hamilton proposed that an altruistic gene is favored by natural selection when

$$rB > C$$

where $r$ is the coefficient of relatedness of donor (the altruist) to the recipient, $B$ is the benefit received by the recipient of the altruism, and $C$ is the cost incurred by the donor. This relationship is known as **Hamilton's rule**. For example, let's suppose an altruist act caused an altruist to be killed, but saved the lives of its three sisters. If this act prevented the altruist from producing two offspring, but allowed the three sisters to produce two offspring each for a total of six, then $r = 0.5$, $B = 6$, and $C = 2$.

Let's consider a situation involving altruism within a group of insects. Many insect larvae, especially caterpillars, are soft-bodied creatures. They rely on a bad taste or toxin to deter predators and advertise this condition with bright warning colors. For example, noxious *Datana ministra* caterpillars, which feed on oaks and other trees, have bright red and yellow stripes and adopt a specific posture with head and tail ends upturned when threatened (**Figure 55.18**). After a predator kills and eats one of the caterpillars, the bright warning colors help the predator learn to avoid similar individuals in the future.

How are warning colors related to kin selection? Animals with warning colors often aggregate in kin groups because they hatch from the same egg mass. Therefore, the death of one individual is likely to benefit its siblings, which are less likely to be attacked by the same predator in the future, and thus its genes are more likely to be passed on to the next generation. This explains why the genes for bright color and a warning posture are passed on from generation to generation. Let's suppose that the death of a caterpillar by a predator prevented that caterpillar from eventually producing 100 offspring, but saved the lives of 5 siblings, which were able to produce 100 offspring each. In this case, $r = 0.5$, $B = 500$, and $C = 100$, and the benefit ($0.5 \times 500 = 250$) is greater than the cost (100).

A common example of altruism in social animals occurs when a sentry raises an alarm call in the presence of a predator. This behavior has been observed in Belding's ground squirrels (*Spermophilus beldingi*). The squirrels feed in groups, with certain individuals acting as sentries and watching for predators. As a predator approaches, the sentry typically gives an alarm call, and the group members retreat into their burrows. Similar behavior occurs in prairie dogs (*Cynomys* spp.) (**Figure 55.19**). In drawing attention to itself, the sentry is at a higher risk of being attacked by the predator. However, in many groups, those closest to the sentry are most likely to be offspring or brothers or sisters. Thus, the altruistic act of alarm calling is thought to be favored by kin selection. Supporting this idea is the observation that most alarm calling is done by females, because they are more likely to stay in the colony where they were born and have kin nearby, whereas the males are more apt to disperse far from the colony.

**Figure 55.19** **Alarm calling, a possible example of kin selection.** This prairie dog sentry is emitting an alarm call to warn other individuals, which are often close kin, of the presence of a predator. It is believed that by doing so, the sentry draws attention away from the others but becomes an easier target itself. ©Danita Delimont/Alamy Stock Photo

**Figure 55.18** **Altruistic behavior or kin selection?** *Datana ministra* caterpillars exhibit a bright, striped warning pattern to advertise their bad taste to predators. ©Peter Stiling

*Concept Check:* *Why do D. ministra caterpillars congregate in clusters?*

 **Core Concept: Evolution** The similarities in DNA between kin promote behaviors whereby some animals act to save the lives of their close relatives, making it more likely to pass those genes to future generations.

**THE QUESTION** *From the perspective of inclusive fitness, should you sacrifice your life to save two sisters and/or nine cousins? Assume that your sacrifice prevents you from having two children, but allows all of your sisters or all of your cousins to have two children each.*

**T**OPIC  *What topic in biology does this question address?* The topic is animal behavior. More specifically, the question concerns how the genetic relatedness of individuals influences apparently altruistic behavior.

**I**NFORMATION  *What information do you know based on the question and your understanding of the topic?* From the question, you may realize that you have to compare the relatedness between you and a sister with that between you and a cousin. From your understanding of the topic and from Figure 55.17, you know that the coefficient of relatedness for a full sibling (a sister) is $r = 0.5$. The coefficient of relatedness for a cousin is $r = 0.125$. You also may remember how to use Hamilton's rule ($rB > C$) to calculate the answer (where $r$ is the coefficient of relatedness of donor to recipient, $B$ is the benefit received by the recipient of the altruism, and $C$ is the cost incurred by the donor).

**P**ROBLEM-SOLVING **S**TRATEGY  *Make a calculation. Compare and contrast.* Use Hamilton's rule to determine for which set of relatives it would be more beneficial (from the perspective of natural selection) to sacrifice your life. In the first case, $r = 0.5$, $B = 2 \times 2$, and $C = 2$. Therefore, $rB = 2$ and $C = 2$. In the second case, $r = 0.125$, $B = 9 \times 2$, and $C = 2$. So $rB = 2.25$ and $C = 2$.

**ANSWER** *From the perspective of inclusive fitness, you should sacrifice your life to save nine cousins but not two sisters.*

## Altruism in Eusocial Animals Arises Partly from Genetics and Partly from Lifestyle

Perhaps the most extreme form of altruism is the evolution of sterile castes in social animals, in which the vast majority of females, known as workers, do not reproduce but instead help one reproductive female (the queen) to raise offspring, a phenomenon called **eusociality**. In insects, the explanation of eusociality lies partly in the genetics of most social insect reproduction. Females develop from fertilized eggs and are diploid, the product of fertilization of an egg by a sperm. Males develop from unfertilized eggs and are haploid.

Such a system of sex determination is called the **haplodiploid system** (refer back to Figure 17.14d). If they have the same parents, each daughter receives an identical set of genes from her haploid father. The other half of a female's genes comes from her diploid mother. The coefficient of relatedness ($r$) of sisters is 0.50 (from father) + 0.25 (from mother) = 0.75. The result is that females are more related to their sisters (0.75) than they would be to their

own offspring (0.50). From the perspective of natural selection, these numbers suggest that it is evolutionarily advantageous for females to stay in the nest or hive and care for other female offspring of the queen, which are their full sisters.

## Unrelated Individuals May Engage in Altruistic Acts If the Altruism Is Likely to Be Reciprocated

Even though ecologists have suggested that kin selection can explain altruism, examples of altruism between unrelated individuals have also been observed. What drives this type of behavior appears to be a "You scratch my back, I'll scratch yours" form of reciprocal altruism, in which the cost to the animal of behaving altruistically is offset by the likelihood of a return benefit. This occurs in nature, for example, when unrelated chimps groom each other.

American biologist Gerald Wilkinson has noted that female vampire bats exhibit reciprocal altruism via food sharing. Vampire bats can die after 60 hours without a blood meal, because they can no longer maintain their correct body temperature. Adult females share their food with their young, the young of other females, and other unrelated females that have not fed. The females and their dependent young roost together in groups of 8 to 12. A hungry female will solicit food from another female by approaching and grooming her. The female being groomed then regurgitates part of her blood meal for the other. The roles of blood donor and recipient are often reversed, and Wilkinson showed that unrelated females are more likely to share with those that had recently shared with them. The probability of a female not reciprocating is decreased, because the roost consists of individuals that remain associated with each other for long periods of time.

## 55.7  Mating Systems

**Learning Outcomes:**
1. Compare and contrast promiscuous, monogamous, polygynous, and polyandrous mating systems.
2. **CoreSKILL** » Predict the circumstances that favor polygyny.

In nature, males produce millions of sperm, but females produce far fewer eggs. It would seem that the majority of males are superfluous because one male could easily fertilize all the females in a local area. If one male can mate with many females, why in most species does the sex ratio remain at approximately 1 to 1? The answer lies with natural selection. Let's consider a hypothetical population that contains 10 females to every male; each male mates, on average, with 10 females. A parent whose children were exclusively sons could expect to have 10 times the number of grandchildren of a parent with the same number of daughters. Under such conditions, natural selection would favor the spread of genes for male-producing tendencies, and males would become prevalent in the population. If the population were mainly males, females would be at a premium, and natural selection would favor the spread of genes for female-producing tendencies. Such constraints operate on the numbers of both male and

Many

Number of female partners

One

| Polygyny (elephant seal) | Promiscuity (bonobos) |
| Monogamy (birds) | Polyandry (pipefish) |

One          Many

Number of male partners

**Figure 55.20** **The four different animal mating strategies.**

*Megaptera novaeangliae*, are widely scattered throughout their breeding range, and genetic studies suggest that they have a promiscuous mating system.

## In Monogamous Mating Systems, Males and Females Are Paired for at Least One Reproductive Season

In **monogamy**, each individual mates exclusively with one partner over at least a single breeding cycle and sometimes longer. Several hypotheses explain the existence of monogamy. The first is the **mate-guarding hypothesis**, which suggests that males stay with a female to protect her from being fertilized by other males. Such a strategy may be advantageous when receptive females are widely scattered and difficult to find.

The **male-assistance hypothesis** maintains that males remain with females to help them rear their offspring. Monogamy is common among birds, about 70% of which are socially monogamous; that is, the pairings remain intact during at least one breeding season. According to the male-assistance hypothesis, monogamy is prevalent in birds because eggs and chicks take a considerable amount of parental care. Most eggs need to be incubated continuously if they are to hatch, and chicks require almost continual feeding. It is therefore in the male's best interest to help raise his young, because he would have few surviving offspring if he did not.

The **female-enforced monogamy hypothesis** suggests that females stop their male partners from being polygynous. Male and female burying beetles (*Nicrophorus defodiens*) work together to bury small, dead animals, which provide a food resource for their developing offspring. Males release pheromones to attract other females to the site. However, while an additional female might increase the male's fitness, the additional developing offspring might compete with the offspring of the first female, decreasing her fitness. As a result, on smelling these pheromones, the first female interferes with the male's attempts at signaling, preserving the monogamous relationship.

Recent research by American neuroscientists Larry Young and Elizabeth Hammock has shown that social behavior such as fidelity may have a genetic basis. These researchers found that fidelity of male voles depends on the length of a short tandem repeat sequence (STR) in a gene that codes for a key hormone receptor. Adult male voles with the long version of the STR were more apt to form pair bonds with female partners and nurture their offspring than were voles with the short version.

female offspring, keeping the sex ratio at about 1:1. This idea was developed in 1930 by British geneticist Ronald Fisher and has come to be known as Fisher's principle.

Even though the sex ratio is fairly even in most species, that doesn't mean that one female always mates with one male, or vice versa. Four different types of mating systems occur in nature (**Figure 55.20**). In some species, mating is promiscuous, with each female and each male mating with multiple partners within a breeding season. In monogamy, each individual mates exclusively with one partner over at least a single breeding cycle and sometimes for longer. In contrast, polygamy is a system in which either males or females mate with more than one partner in a breeding season. In polygyny (Greek for many females), one male mates with more than one female, but females mate only with one male. In polyandry (Greek for many males), one female mates with several males, but males mate with only one female. In this section, we will examine these four types of mating systems.

## In Promiscuous Mating Systems, Each Male or Female Mates with Multiple Partners

Chimpanzees and bonobos are somewhat **promiscuous**; each male mates with many females, and vice versa. Here sex alleviates conflict within the social group, but sometimes promiscuity is favored in unpredictable environments. Females that maximize the genetic diversity of their offspring are more likely to have at least some offspring that will survive in a changing world. Intertidal and terrestrial mollusks are also usually promiscuous. Individuals copulate with several partners, and eggs are fertilized with sperm from several different individuals. These mollusks are slow moving and risk desiccation when searching for a mate. The risk of not finding a mate is believed to promote promiscuous mating. Humpback whales,

## In Polygynous Mating Systems, One Male Mates with Many Females

In **polygyny** (Greek, meaning many females), one male mates with more than one female in a single breeding season. Physiological constraints often dictate that female organisms must care for the young. Because of these constraints, at least in many organisms with internal fertilization, such as mammals and some fishes, males are able to mate with and then desert several females. Polygynous systems are therefore associated with uniparental care of young, with males contributing little. Sexual maturity is often delayed in males that fight because of the considerable time it takes to reach a sufficiently large size to compete for females.

Polygyny is influenced by the temporal or spatial distribution of breeding females and by the availability of resources. In cases when all females are sexually receptive within the same narrow period of time, little opportunity exists for a male to garner all the females for himself. When female reproductive receptivity is spread out over weeks or months, however, males have a greater opportunity to mate with more than one female. Where some critical resource is patchily distributed and in short supply, certain males may dominate the resource and breed with more than one visiting female. For example, the major source of nestling death in the lark bunting (*Calamospiza melanocorys*), which lives in North American grasslands, is overheating from too much exposure to the Sun. Prime territories are therefore those with abundant shade, and some males with shaded territories attract two females, even though the second female can expect no help from the male in the process of rearing young. Males in some exposed territories remain bachelors for the season. From the dominant male's point of view, polygyny is advantageous; from the female's point of view, there may be costs. Although by choosing dominant males, a female may be gaining access to good resources, she will have to share these resources with other females.

Sometimes males defend a group of females without commanding a resource-based territory. This pattern is more common when females naturally congregate in groups or herds, perhaps to avoid predation, as with horses, zebras, and some deer, and where space is limited, as with southern elephant seals. Usually the largest and strongest males command most of the matings, but being a dominant male is usually so exhausting that males may only manage to remain the strongest male for a year or two.

Polygynous mating can occur where neither resources nor groups of females are defended. In some instances, particularly in birds and mammals, males display in designated communal courting areas called **leks** (**Figure 55.21**). Females come to these areas specifically to find a mate, and they choose a prospective mate after the males have performed elaborate displays. Most females seek to mate with the best male, so a few of the flashiest males perform the vast majority of the matings. At a lek of the white-bearded manakin (*Manacus manacus*) of South America, one male accounted for 75% of the 438 matings even though there were as many as 10 males. A second male mated 56 times (13% of matings), but six others mated only a total of 10 times.

## In Polyandrous Mating Systems, One Female Mates with Many Males

**Polyandry** (Greek, meaning many males), in which one female mates with several males, is rarer than polygyny. Nevertheless, it occurs in many species of mammals, birds, reptiles, and insects. In the Arctic tundra, the summer season is short but very productive, providing a bonanza of insect food for 2 months. The productivity of the breeding grounds of the spotted sandpiper (*Actitis macularia*) is so high

**Figure 55.21   Male birds at a lek.** Black grouse (*Tetrao tetrix*) congregate at a moorland lek in Scotland in April. Females visit the leks, and males display to them. ©Chris Knights/ardea.com

that the female becomes rather like an egg factory, laying up to five clutches of four eggs each in 40 days. Her reproductive success is limited not by food but by the number of males she can find to incubate the eggs, and females compete for males, defending territories where the males sit.

Polyandry is also seen in some species where egg predation is high, and males are needed to guard the nests. For example, in the pipefish (*Syngnathus typhle*), males have brood pouches that provide eggs with safety and a supply of oxygen- and nutrient-rich water. Females produce enough eggs to fill the brood pouches of two males and may mate with more than one male.

## Mating Systems Tend to Differ in the Degree of Sexual Dimorphism

**Sexual dimorphism** is a pronounced difference in the morphologies of the two sexes within a species. Although males and females differ in all mating systems, the degree to which they differ tends to vary. Though exceptions occur, species that follow a monogamous mating system tend to be similar in body size and structure (**Figure 55.22a**). By comparison, sexual dimorphism is often dramatic in polygamous mating systems. In polygyny, males usually develop a much larger body size to boost success in competition over mates (**Figure 55.22b**). By comparison, in polyandry, the females are typically the larger of the sexes (see **Figure 55.22c**).

(a) **Monogamous species**

(b) **Polygynous species**

(c) **Polyandrous species**

**Figure 55.22** **Sexual dimorphism in body size and mating system.** **(a)** In monogamous species, such as these Manchurian cranes, *Grus japonensis*, males and females do not exhibit pronounced sexual dimorphism and appear very similar. **(b)** In polygynous species, such as elk, *Cervus canadensis*, males are bigger than females, and male elk have large horns with which they engage in combat over females. **(c)** In polyandrous species, females are usually bigger, as is the case for these golden silk spiders, *Nephila clavipes*. a: ©Masahiro Iijima/ardea.com; b: ©Wildlife GmbH/Alamy Stock Photo; c: ©Millard H. Sharp/Science Source

 **Core Skill: Modeling** The goal of this modeling challenge is to draw a set of bar graphs illustrating the relative body sizes of males and females in different mating systems: monogamy, polygyny, and polyandry. You should be able to explain your graphical model.

**Modeling Challenge:** Draw a bar graph with three sets of two bars each to illustrate what you think are the relative body-size ratios of males:females in monogamous, polygynous, and polyandrous species, and explain your graph. Scale the *y*-axis from 0 to 2.0, with 1.0 representing equal body sizes for the two sexes.

## Summary of Key Concepts

### 55.1 The Influence of Genetics and Learning on Behavior

- Behavior is usually due to the interaction of an organism's genes and the environment.

- Genetically programmed behaviors are termed innate and often involve a sign stimulus that initiates a fixed action pattern (FAP) (Figures 55.1, 55.2).

- Organisms can often make modifications to their behavior based on previous experience, a process called learning. Some forms of learning include habituation, classical conditioning, operant conditioning, and cognitive learning (Figures 55.3, 55.4).

- Much behavior is a mixture of innate and learned behaviors. A good example of this occurs in a process called imprinting, in which animals develop strong attachments that influence subsequent behavior (Figures 55.5, 55.6).

### 55.2 Local Movement and Long-Range Migration

- The simplest forms of local movement involve kinesis, taxis, and memory (Figure 55.7).

- Many animals undergo long-range seasonal movement called migration in order to feed or breed. They do this using three mechanisms: piloting (the ability to move from one landmark to the next), orientation (the ability to follow a compass bearing), and navigation (the ability to set, follow, and adjust a compass bearing) (Figures 55.8, 55.9).

### 55.3 Foraging Behavior and Defense of Territory

- Animals use complex behavior in food gathering or foraging. Optimality theory views foraging behavior as a compromise between the costs and benefits involved. The theory of optimal foraging assumes that animals modify their behavior to keep the ratio of their energy uptake to energy expenditure high. The risk of predation has an influence on foraging behavior (Figures 55.10, 55.11).

- The size of a territory, a fixed area in which an individual or group excludes other members of its own species, tends to be optimized according to the costs and benefits involved (Figure 55.12).

### 55.4 Communication

- Communication is a form of behavior. The use of different forms of communication between organisms depends on the environment in which they live.

- Chemical communication often involves marking territories; auditory and visual forms of communication are often used to attract mates. A fascinating form of tactile communication involves the dance of the honeybee (Figures 55.13, 55.14).

## 55.5 Living in Groups

- Many benefits of group living relate to defense against predators, offering protection through sheer numbers and through what is called the many-eyes hypothesis or the geometry of the selfish herd (Figure 55.15).

## 55.6 Altruism

- Altruism is behavior that benefits others at a cost to oneself. One of the first hypotheses to explain altruism, called group selection, suggested that natural selection produced outcomes beneficial for the group. Biologists now believe that most apparently altruistic acts are associated with outcomes beneficial to those most closely related to the individual, a concept termed kin selection (Figures 55.16, 55.17, 55.18, 55.19).

- Altruism among eusocial animals may arise from the unique genetics of the animals.

- Altruism is known to exist among nonrelated individuals that live in close proximity for long periods of time.

## 55.7 Mating Systems

- Four types of mating systems are found among animals: promiscuity, monogamy, polygyny, and polyandry (Figure 55.20).

- Polygynous mating can often occur in situations when males dominate a resource, defend groups of females, or display in common courting areas called leks (Figure 55.21).

- Relative body size of males and females depends on the species' mating system (Figure 55.22).

# Assess & Discuss

## Test Yourself

1. What is the proximate cause of male deer fighting over females?
   a. To determine their supremacy over other males
   b. To injure other males so that those males cannot mate with the females
   c. To maximize the number of genes the male deer pass on
   d. Because changes in day length stimulate this behavior
   e. Because fighting helps rid the herd of weaker individuals

2. Geotaxis is a response to the force of gravity. Fruit flies placed in a vial will move to the top of the vial. This is an example of _____ geotaxis.
   a. positive            d. negative
   b. neutral             e. learned
   c. innate

3. Certain behaviors do not require past training. Such behaviors tend to be the same in all individuals and are referred to as _____ behaviors.
   a. genetically programmed
   b. instinctual
   c. innate
   d. All of the above refer to such behaviors.
   e. Only b and c refer to such behaviors.

4. Patrick has decided to teach his puppy a few new tricks. Each time the puppy responds correctly to Patrick's command, the puppy is given a treat. This is an example of
   a. habituation.          d. imprinting.
   b. classical conditioning.   e. orientation.
   c. operant conditioning.

5. Whales have magnetite in their retinas, which aids in navigation during migration by
   a. piloting.
   b. locating the position of the Sun.
   c. use of the Earth's magnetic fields.
   d. locating the positions of the stars.
   e. doing none of the above.

6. For group living to evolve, the benefits of living in a group must have been greater than the costs. Which of the following is an example of a benefit of living in a group?
   a. reduced spread of disease and/or parasites
   b. increased food availability
   c. reduced competition for mates
   d. decreased risk of predation
   e. all of the above

7. The modification of behavior based on prior experience is called
   a. a fixed action pattern.      d. adjustment behavior.
   b. learning.                    e. innate.
   c. navigation.

8. When an individual behaves in a way that reduces its own fitness but increases the fitness of others of its species, it is exhibiting
   a. kin selection.       d. selfishness.
   b. group selection.     e. ignorance.
   c. altruism.

9. In ants, which employ a haplodiploid mating system, the coefficient of relatedness, $r$, for fathers and sons is equal to
   a. 0.           d. 0.5.
   b. 0.125.       e. 0.75.
   c. 0.25.

10. In a polygynous mating system,
   a. one male mates with one female.
   b. one female mates with many different males.
   c. one male mates with many different females.
   d. many different females mate with many different males.

## Conceptual Questions

1. Some male spiders are eaten by the females after copulation. How can this act be seen to benefit the males?

2. Male parental care occurs in only 7% of fishes and amphibian species with internal fertilization but in 69% of species with external fertilization. Propose an explanation for why this is so.

3. **Core Concept: Evolution**  When no relatives are present, female Belding's ground squirrels produce alarm calls less than 20% of the time when predators approach. When sisters or daughters are present, alarm calls are made in nearly 75% of instances when predators approach. Explain these observations.

## Collaborative Questions

1. Whooping cranes (*Grus americana*) are an endangered species and are bred in captivity to increase their numbers. One problem with this approach is that these cranes are migratory. In the absence of other cranes, can you think of an innovative way researchers might have used crane behavior to ensure safe passage of the birds to overwintering sites?

2. If a bumblebee queen mates with two males, instead of one, how would this influence her colony's eusociality?

# Population Ecology

## 56

A group of black-footed ferrets in Meeteetse, Wyoming. ©Mike Lockhart

I n 1981, the last known population of black-footed ferrets, *Mustela nigripes*, was discovered near Meeteetse, Wyoming. Shortly thereafter, all but 18 of the 100 known ferrets in Meeteetse died of canine distemper. The remainder were captured between 1985 and 1987, inoculated against distemper, and bred in captivity, with the intent of re-establishing the population in the wild later on. Since then, populations have been established in Arizona, Colorado, Montana, South Dakota, Utah, Wyoming, and Chihuahua, Mexico.

In Wyoming, an area called Shirley Basin was one of those targeted for reintroduction of captive-born ferrets. From 1991 to 1994, Shirley Basin received 228 ferrets, but distemper again triggered a decline in the population size. By 1997, only 5 ferrets were found. Extinction of this population seemed imminent. Monitoring efforts, which might disturb the animals, decreased. Surprisingly, by 2003, a total of 52 animals were found, and by 2006, 223 were present. A goal of population ecology is to understand such things as the rapid increase in the number of ferrets in Shirley Basin.

In the case of sexually reproducing species, a **population** can be defined as a group of interbreeding individuals of the same species occupying the same area at the same time. Thus, we can think of a population of water lilies in a particular lake, the lion population in the Ngorongoro crater in Africa, or the human population of New York City. The boundaries of a population can be a little difficult to define, though they may correspond to geographic features such as the edges of a lake or forest or the area of a mountain valley or a certain island. Individuals may enter or leave a population, such as the human population of New York City or the deer population in North Carolina. Thus, populations are often fluid entities, with individuals moving into (immigrating) or out of (emigrating) an area.

This chapter explores **population ecology**, the study of the factors that affect population size and how these factors change over space and time. To study populations, ecologists need to employ some of the tools of **demography**, the study of birth rates, death rates, age distributions, and the sizes of populations. In this chapter, we begin our discussion by examining the ways that ecologists measure and categorize populations. We will explore characteristics of populations and how growth rates are determined by the number of reproductive individuals in the population and their fertility rate. These data are used to construct simple mathematical models that allow ecologists to analyze population growth, such as that of the black-footed ferrets, and predict future growth. We will also look at the factors that limit the growth of populations.

## 56.1 Understanding Populations

**Learning Outcomes:**

1. List the different techniques ecologists use to measure population density.
2. **CoreSKILL »** Calculate population size using mark-recapture data, and explain the limitations of this technique.
3. Identify the three main patterns of dispersion observed in nature.
4. Describe the difference between semelparity and iteroparity—two different reproductive strategies.

Within their areas of distribution, organisms occur in varying numbers. We recognize this pattern by saying a plant or animal is rare in one place and common in another. For added precision, ecologists quantify distribution further and talk in terms of **population density**—the number of organisms of a given species in a given unit area or volume. Population growth affects population density, and knowledge of both can help in making decisions about the management of species. How long will it take for a population of an endangered species to recover to a healthy level if we protect it from its most serious threats? For example, how quickly will the black-footed ferret populations increase in Wyoming? A knowledge of population growth rates and population densities would allow us to predict future ferret population sizes.

In this section, we will examine techniques that are used to measure population sizes and densities and other characteristics of populations within their habitats. We will also discuss the different reproductive strategies organisms use and see how ecologists assign individuals to different groups called age classes.

## Ecologists Use Many Different Methods to Quantify Population Size and Density

The simplest method for measuring population size is to visually count the number of organisms in a given area. We can reasonably do this only if the area is small and the organisms are relatively large. For example, we can readily determine the number of gumbo limbo trees (*Bursera simaruba*) on a small island in the Florida Keys. Normally, however, population ecologists calculate the density of plants or animals in a small area and use this figure to estimate the total abundance over a larger area.

For plants, algae, or other sessile organisms such as intertidal animals, it is fairly easy to count numbers of individuals per square meter or, for larger organisms such as trees, numbers per hectare (an area of land equivalent to 2.471 acres). However, many plant individuals are clonal; that is, they grow in patches of genetically identical individuals. Rather than counting individuals, ecologists can also use the amount of ground covered by plants as an estimate of vegetation density.

***Quadrats Can Be Used to Quantify Population Densities of Plants and Sessile Species*** Plant ecologists use a sampling device called a **quadrat**, a square frame that often, but not always, measures $50 \times 50$ cm and encloses an area of 0.25 m² (**Figure 56.1a**). The ecologists count the numbers of plants of a given species inside the quadrat to obtain a density estimate per square meter. For example, if you counted densities of 20, 35, 30, and 15 plants in four quadrats, you could reliably say that the density of this species was 25 individuals per 0.25 m², or 100/m². For larger plants, such as trees, a quadrat would be ineffective. To count such organisms, many ecologists perform a **line transect**, in which a long piece of string is stretched out and any tree along its length is counted. For example, to count tree species across a large area, we could lay out a 100-m line transect and count all the trees within 1 m on either side of the transect. In effect, this transect is little more than a long, thin quadrat encompassing 200 m². By performing five such transects, we could obtain estimates of tree density per 1,000 m² and then extrapolate that to a number per hectare or per island.

***Traps Are Used to Study More Mobile Species*** Several different sampling methods exist for quantifying the population density of animals, which are more mobile than plants. Suction traps, like giant aerial vacuum cleaners, can suck flying insects from the sky. Pitfall traps set into the ground can catch species such as spiders, lizards, or beetles wandering over the surface (**Figure 56.1b**). Sweep nets can be passed over vegetation to dislodge and capture the insects feeding there. Mist nets—very fine netting spread between trees—can entangle flying birds and bats (**Figure 56.1c**). Baited snap traps, such as mouse traps, or live traps can snare terrestrial animals (**Figure 56.1d**). Population density can thus be estimated as the number of animals caught per trap or per unit area where a given number of traps are set, for example, 10 traps per 100 m² of habitat.

***The Mark-Recapture Technique Can Be Used to Estimate Population Sizes*** Sometimes population biologists capture animals and then tag and release them (**Figure 56.2**). The rationale behind

**(a) Quadrat**  **(b) Pitfall trap**  **(c) Mist net**  **(d) Live trap**

**Figure 56.1 Sampling techniques. (a)** Quadrats are frequently used to count the number of plants per unit area. **(b)** Pitfall traps set into the ground catch wandering species such as beetles and spiders. **(c)** Mist nets consist of very fine mesh to entangle birds or bats. **(d)** Baited live traps catch terrestrial animals, including this lion tamarin, *Leontopithecus rosalia*, in Brazil. a: ©Paul Glendell/Alamy Stock Photo; b: ©Nigel Cattlin/Science Source; c: Source: U.S. Fish & Wildlife Service/Donna Dewhurs; d: ©Jami Tarris/Corbis/Getty Images

**Figure 56.2  The mark-recapture technique for estimating population size.** A leg tag identifies this Red Knot (*Calidiris canutus*). Recapture, or visual relocation, of such marked animals permits estimates of population size. Source: U.S. Fish & Wildlife Service/Greg Breese

the **mark-recapture technique** is that after the tagged animals are released, they mix freely with unmarked individuals and within a short time are randomly mixed within the population. The population is resampled, and the numbers of marked and unmarked individuals are recorded. We assume that the ratio of marked to unmarked individuals in the second sample is the same as the ratio of marked individuals in the first sample to the total population size, that is:

$$\frac{\text{Number of individuals marked in first catch}}{\text{Total population size, } N} = \frac{\text{Number of marked recaptures in second catch}}{\text{Total number of second catch}}$$

Let's say we catch 50 largemouth bass in a lake and mark them with colored fin tags. A week later, we return to the lake and catch 40 fish and 5 of them were previously tagged fish. If we assume that no immigration or emigration has occurred, which is quite likely in a closed system like a lake, and that there have been no births or deaths of fish, then the total population size is given by rearranging the equation:

$$\text{Total population size, } N = \frac{\substack{\text{Number of marked individuals in first catch} \\ \times \text{ Total number of second catch}}}{\substack{\text{Number of marked recaptures} \\ \text{in second catch}}}$$

Using the data for the bass, we have

$$N = \frac{50 \times 40}{5} = \frac{2,000}{5} = 400$$

From this calculation, we estimate that the lake has a total population size of 400 largemouth bass. This information could be useful for game and fish personnel who wish to know the total size of a fish population in order to set catch limits.

However, the mark-recapture technique can have drawbacks. Some animals that have been marked may learn to avoid the traps.

Recapture rates will then be low, resulting in an overestimate of population size. Imagine that instead of 5 tagged fish out of 40 recaptured fish, we get only 2 tagged fish. Now our population size estimate is 2,000/2 = 1,000, a dramatic increase in our population size estimate. On the other hand, some animals can become "trap-happy," particularly if the traps are baited with food. This effect would result in an underestimate of the population size.

Because of the limitations of the mark-recapture technique, ecologists also use other methods to estimate population density. For some larger terrestrial or marine species, captured animals can be fitted with radio collars and followed remotely, using an antennal tracking device. Their home ranges can be determined and population estimates developed based on the area of available habitat. Unmanned aircraft systems (UAS), or drones, have emerged as a safe, low-cost method to document wildlife abundance and have been used to provide animal counts in relatively inaccessible places, including counts of walruses in rough water areas and Steller sea lions in the outer Aleutian Islands, as well as flamingoes, orangutans, and rhinoceroses. For many species with valuable pelts, ecologists can track population densities through time by examining pelt records taken from trading stations. They can also estimate relative population density by examining catch per unit effort, which is especially valuable in commercial fisheries. Ecologists can't easily expect to count the number of fishes in an area of ocean, but can count the number caught, say, per 100 hours of trawling. For frogs or birds, they can count chorusing or singing individuals.

## Core Skill: Quantitative Reasoning

**BIO TIPS**   **THE QUESTION** *Suppose we capture and mark 110 Rocky Mountain goats in a population and later recapture 100 goats, 20 of which have ear tags. What is the estimate of the total population size?*

**T**OPIC   ***What topic in biology does this question address?*** The topic is estimating population size, specifically, using the mark-recapture technique.

**I**NFORMATION   ***What information do you know based on the question and your understanding of the topic?*** From the question, you know how many goats were marked, how many were recaptured, and how many of the recaptured goats were also marked. From your understanding of the mark-recapture technique, you may recall the formula used to estimate population size.

**P**ROBLEM-SOLVING **S**TRATEGY   ***Make a calculation.*** Insert the relevant data into the formula given in the text:

$$\text{Total population size, } N = \frac{\substack{\text{Number of marked individuals in first catch} \\ \times \text{ Total number of second catch}}}{\substack{\text{Number of marked recaptures in second} \\ \text{catch}}}$$

$$N = \frac{110 \times 100}{20} = 550$$

**ANSWER** *The estimated total population size is 550 goats.*

(a) Clumped      (b) Uniform      (c) Random

**Figure 56.3** **Three types of dispersion.** **(a)** A clumped distribution pattern, as for these plants clustered around an oasis, often results from the uneven distribution of a resource, in this case, water. **(b)** A uniform distribution pattern, as for these nesting black-browed albatrosses (*Diomedea melanophris*) on the Falkland Islands, may be a result of competition or social interactions. **(c)** A random distribution pattern, as for these sheep, *Ovis aires*, in a meadow in the Basque Country, Spain., is the least common form of spacing. a: ©Priakhin Mikhail/Alamy Stock Photo; b: ©Fritz Polking/Corbis Documentary/Getty Images; c: ©Pixtal/AGE Fotostock

**Concept Check:** *What is the distribution pattern of students in a half-empty classroom?*

 **Core Skill: Connections** Look back to Figure 55.12b. What is the likely dispersion pattern of cheetahs in the wild?

## Populations Show Different Degrees of Spacing Among Individuals

Individuals within a population show different patterns of **dispersion**; that is, they can be clustered together or spread out to varying degrees. The three basic kinds of dispersion patterns are clumped, uniform, and random.

The type of dispersion observed in nature can tell us a lot about what processes shape group structure. The most common dispersion pattern is **clumped**, because resources in nature tend to be clustered. For example, certain plants may do better in moist conditions, and moisture is greater in low-lying areas (**Figure 56.3a**). Social behavior among animals that aggregate into flocks or herds reflects a clumped pattern.

On the other hand, competition may cause a **uniform** dispersion pattern among individuals, as among trees in a forest. At first, the pattern of trees and seedlings may appear random as seedlings develop from seeds dropped at random, but competition among roots may cause some trees to be outcompeted by others, causing a thinning out and resulting in a relatively uniform distribution. Thus, the dispersion pattern starts out random but ends up uniform. Uniform dispersions may also result from social interactions, as among some nesting birds, which tend to keep an even distance from one other (**Figure 56.3b**).

Perhaps the rarest dispersion pattern is **random**, in which the probability of finding an individual at any point in an area is equal, because resources in nature are rarely randomly spaced. Where resources are common and abundant, as in moist, fertile soil, the dispersion patterns of plants may lack a pattern as plants germinate from randomly dispersed wind-blown seeds. Grazing animals may also exhibit a random distribution across areas of grassland (**Figure 56.3c**).

## Reproductive Strategies May Differ Among Species

To better understand how populations grow in size, let's consider their reproductive strategies. Some organisms produce all of their offspring in a single reproductive event. This pattern, called **semelparity** (from the Latin *semel*, meaning once, and *parere*, meaning to bear), is common in insects and invertebrates and also occurs in organisms such as salmon, bamboo grasses, and agave plants (**Figure 56.4a**). These individuals reproduce once only and die. Semelparous organisms such as agaves may live for many years before reproducing, or they may be annual plants that develop from seed, flower, and drop their own seed within a year. Semelparous organisms often produce groups of same-aged young called **cohorts** that grow at similar rates.

Other organisms reproduce in successive years or breeding seasons. The pattern of repeated reproduction at intervals throughout the life cycle is called **iteroparity** (from the Latin *itero*, meaning to repeat). Iteroparous organisms generally have many young of different ages. This reproductive pattern is found in most vertebrates, nonwoody perennial plants, and trees. Among iteroparous organisms, much variation occurs in the number of reproductive events and in the number of offspring per event. Many species, such as birds or trees in temperate areas, have distinct breeding seasons (seasonal iteroparity) that lead to distinct generations (**Figure 56.4b**). For a few species, individuals reproduce repeatedly and at any time of the year. This pattern is termed continuous iteroparity and is exhibited by some tropical species, many parasites, and many primates (**Figure 56.4c**).

Why do species reproduce in a semelparous or iteroparous mode? The answer may lie in part in environmental uncertainty. If the survival rate of juveniles is very low and unpredictable, then selection

Birth |————— Agave lifetime —————⬤→| Death

Birth |—— Blue tit lifetime ⬤—⬤—⬤—⬤—⬤—⬤—⬤ →| Death

Birth |—— Chimpanzee lifetime ⬤—⬤⬤—⬤⬤—⬤ →| Death

⬤ Reproductive event

**(a) Semelparity**          **(b) Iteroparity (seasonal)**          **(c) Iteroparity (continuous)**

**Figure 56.4** **Differences in reproductive strategies.** Species such as **(a)** agave plants (*Agave shawii*) are semelparous, meaning they breed once in their lifetime and then die. This contrasts with **(b)** blue tits (*Parus caeruleus*) and **(c)** chimpanzees (*Pan troglodytes*), which are iteroparous and breed more than once in their lifetime. a: ©Doug Sherman/Geofile; b: ©John Foxx/Getty Images; c: ©M. Watson/ardea.com

favors repeated reproduction and a long reproductive life to increase the chance that juveniles will survive in at least some years. If the environment is stable, then selection favors a single act of reproduction, because the organism can devote all its energy to making offspring, not maintaining its own body. Under favorable circumstances, annual plants produce more seeds per unit biomass than trees, which have to invest a lot of energy in maintenance. However, when the environment becomes stressful, annuals run the risk of their seeds not germinating. They must rely on some seeds successfully lying dormant and germinating after the environmental stress has ended.

## 56.2 Demography

**Learning Outcomes:**

1. Describe the difference between information summarized in a life table and information summarized in a survivorship curve.
2. Differentiate among type I, II, and III survivorship curves, and give examples of organisms that exhibit those types of curves.
3. **CoreSKILL »** Use survivorship and age-specific fertility data to predict a population's growth.

As mentioned in the chapter opening, demography is the study of birth rates, death rates, age distributions, and the sizes of populations. One way to predict how the size of a population will change is to examine a cohort of individuals from birth to death. For most animals and plants with a relatively short generation time, this involves marking a group of individuals in a population as soon as they are born or germinate and following their fate through their lifetime. After determining the relative numbers of juveniles and mature individuals, researchers can construct a **life table**—a table that provides data on the numbers of living individuals in various age classes in a population and their relative fertilities. In this section, we will explore how ecologists construct life tables, plot survivorship curves, and use age-specific fertility data to predict changes in population sizes.

## A Life Table Summarizes Survival and Fertility Patterns

With regard to demography, let's consider the North American beaver (*Castor canadensis*). Prized for their pelts, by the mid-19th century, these animals had been hunted and trapped to near extinction. Beavers began to be protected by laws in the 20th century, and populations recovered in many areas, often growing to what some considered to be nuisance status. In Newfoundland, Canada, legislation supported trapping as a management technique.

From 1964 to 1971, trappers provided mandibles from which teeth were extracted for age classification. If many mandibles were obtained from, say, 1-year-old beavers, then such animals were probably common in the population. If the number of mandibles from 2-year-old beavers was low, the mortality is high for the 1-year-old age class. From the mandible data, researchers constructed a life table in which age classes represent 1 year (**Table 56.1**). Only females are included in these data, because only females produce offspring. This table also includes age-specific fertility rates, which will be discussed later in this section.

The number of individuals alive at the start of the time period (in this case, a year) is referred to as $n_x$, where $n$ is the number and $x$ refers to the particular age class. By subtracting the value of $n_x$ from the number alive at the start of the previous year, we can calculate the number dying in a given age class or year, $d_x$. Thus, $d_x = n_x - n_{x+1}$. For example, in Table 56.1, 273 beavers were alive at the age of 5 years ($n_5$), and only 205 were alive at the age of six years ($n_6$); thus, 68 died between the age of 5 and 6: $d_5 = n_5 - n_6$, or $d_5 = 273 - 205 = 68$.

## Survivorship Curves Reveal Different Patterns of Survival

A **survivorship curve** is a graphical plot of the numbers of surviving individuals for each age class in a population. Ecologists may construct survivorship curves for two common reasons. First, a survivorship curve may reveal unexpected changes in survivorship

**Table 56.1**    Life Table for the Beaver (*Castor canadensis*) in Newfoundland, Canada

| Age (years), $x$ | *Number alive at start of year, $n_x$ | Number dying during year, $d_x$ | Proportion alive at start of year, $l_x$ | Age-specific fertility, $m_x$ | $l_x m_x$ |
|---|---|---|---|---|---|
| 0–1 | 3,695 | 1,995 | 1.000 | 0.000 | 0 |
| 1–2 | 1,700 | 684 | 0.460 | 0.315 | 0.145 |
| 2–3 | 1,016 | 359 | 0.275 | 0.400 | 0.110 |
| 3–4 | 657 | 286 | 0.178 | 0.895 | 0.159 |
| 4–5 | 371 | 98 | 0.100 | 1.244 | 0.124 |
| 5–6 | 273 | 68 | 0.074 | 1.440 | 0.107 |
| 6–7 | 205 | 40 | 0.055 | 1.282 | 0.071 |
| 7–8 | 165 | 38 | 0.045 | 1.280 | 0.058 |
| 8–9 | 127 | 14 | 0.034 | 1.387 | 0.047 |
| 9–10 | 113 | 26 | 0.031 | 1.080 | 0.033 |
| 10–11 | 87 | 37 | 0.024 | 1.800 | 0.043 |
| 11–12 | 50 | 4 | 0.014 | 1.080 | 0.015 |
| 12–13 | 46 | 17 | 0.012 | 1.440 | 0.017 |
| 13–14 | 29 | 7 | 0.007 | 0.720 | 0.005 |
| 14+ | 22 | 22 | 0.006 | 0.720 | 0.004 |

Net reproductive rate, $\Sigma l_x m_x = 0.938$

due to predation or human activities. Second, such graphs provide information about the patterns of survival among different species.

**Figure 56.5** shows a survivorship curve for the North American beaver. The value of $n_x$, the number of individuals, is typically expressed on a log scale. Ecologists use a log scale to examine rates of change with time, not changes in absolute numbers. Also, the use of a log scale makes it easier to plot a wide range of population sizes.

Another advantage of a log scale is that it allows ecologists to compare populations that differ in their relative sizes. For example, if an age class begins with 1,000 individuals and 500 are lost in year 1, the log of the decrease is

$$\log_{10} 1,000 - \log_{10} 500 = 3.0 - 2.7 = 0.3 \text{ per year}$$

If an age class begins with 100 individuals and 50 are lost, the log of the decrease is similarly

$$\log_{10} 100 - \log_{10} 50 = 2.0 - 1.7 = 0.3 \text{ per year}$$

In both cases, 50% of the individuals have been lost and the rates of change are identical, even though the absolute numbers are different. Plotting the data on a log scale, as in Figure 56.5, ensures that regardless of the relatives sizes of the age classes, the rate of change of one survivorship curve, that is, the slope of the curve, can easily be compared with that of another curve.

Among different species, survivorship curves generally fall into one of three patterns (**Figure 56.6**). With a type I curve, the rate of loss for

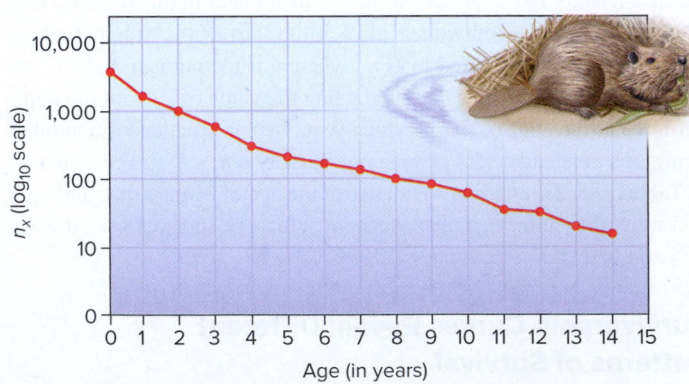

**Figure 56.5  Survivorship curve for the North American beaver.** The survivorship curve is generated by plotting the number of surviving individuals, $n_x$, from any given cohort of young, usually measured on a log scale, against age. This survivorship curve shows a fairly uniform rate of decline through time.

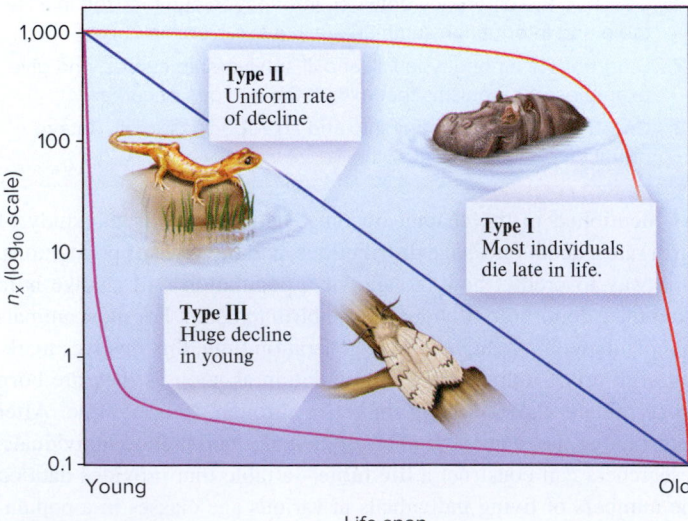

**Figure 56.6  Idealized survivorship curves.**

*Concept Check:* Which type of survivorship curve would you expect for (a) mussels and (b) turtles?

juveniles is relatively low, and most individuals are lost later in life, as they become older and more prone to sickness and predators (see the Feature Investigation). Organisms that exhibit type I survivorship curves have relatively few offspring but invest much time and resources in raising their young. Many large mammals, including humans, exhibit type I curves.

At the other end of the scale is a type III curve, in which the rate of loss for juveniles is relatively high, and the survivorship curve flattens out for those organisms that have avoided early death. Many fishes and marine invertebrates fit this pattern. Most of the juveniles die or are eaten, but a few reach a favorable habitat and thrive. For

example, once they find a suitable rock face on which to attach themselves, barnacles grow and survive very well. Many insects and plants also fit the type III survivorship curve, because they lay many eggs or release hundreds of seeds, respectively.

Type II curves represent a middle ground, with fairly uniform death rates over time. Species with type II survivorship curves include many birds, small mammals, reptiles, and some annual plants. The North American beaver population exhibits this survivorship curve. Keep in mind, however, that these are generalized curves and that few populations fit them exactly.

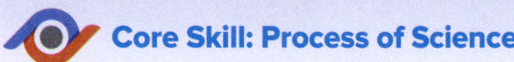

## Core Skill: Process of Science

## Feature Investigation | Murie's Construction of a Survivorship Curve for Dall Mountain Sheep Suggested That the Youngest and Oldest Sheep Were Most Vulnerable to Predation by Wolves

The Dall mountain sheep (*Ovis dalli*) lives in mountainous regions, including the Arctic and sub-Arctic regions of Alaska. In the late 1930s, the U.S. National Park Service was bombarded with public concerns that wolves were responsible for a sharp decline in the population of Dall mountain sheep in Denali National Park (then Mt. McKinley National Park). Shooting the wolves was advocated as a way of increasing the number of sheep. Because meaningful data on sheep mortality were not available, the Park Service enlisted American biologist Adolph Murie to determine whether the wolves killed enough sheep to justify controlling the wolf population. In addition

to spending many hours observing interactions between wolves and sheep, Murie also collected sheep skulls, determining the sheep's age at death by counting annual growth rings on the horns.

In 1947, American ecologist Edward Deevey put Murie's data in the form of a life table that listed each age class and the number of skulls in it (**Figure 56.7**). Although Murie had collected 608 skulls, Deevey expressed the data per 1,000 individuals to allow for comparison with other life tables. From the data, Deevey constructed a survivorship curve. For the Dall mountain sheep in Denali National Park, there was a slight initial decline in survivorship as young lambs

**Figure 56.7** **Examining the survivorship curve of a Dall mountain sheep population reveals information on the cause of death.**

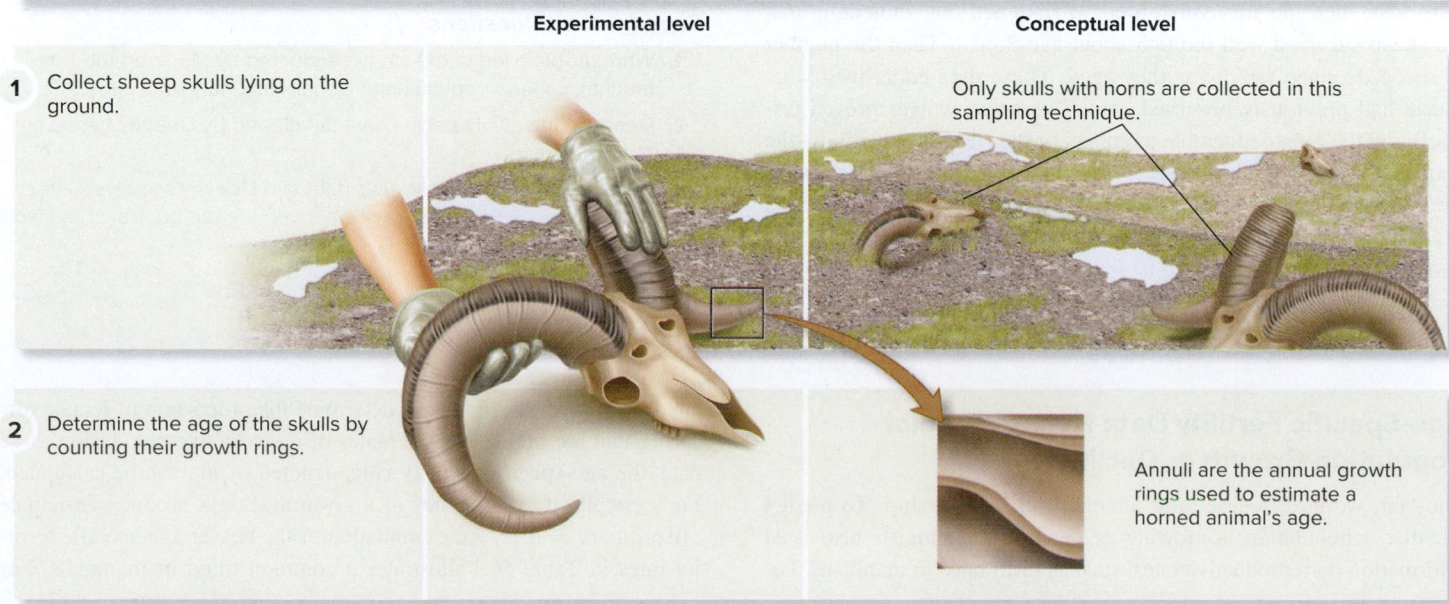

**HYPOTHESIS** Culling the wolf population would protect reproductively active adults in the Dall mountain sheep population.

**STARTING LOCATION** Denali National Park (formerly known as Mt. McKinley National Park) in Alaska, where wolf predation of sheep is common.

| Experimental level | Conceptual level |
| --- | --- |

1  Collect sheep skulls lying on the ground.

Only skulls with horns are collected in this sampling technique.

2  Determine the age of the skulls by counting their growth rings.

Annuli are the annual growth rings used to estimate a horned animal's age.

**3** Organize the data into a life table (see step 4), and construct a survivorship curve using the data.

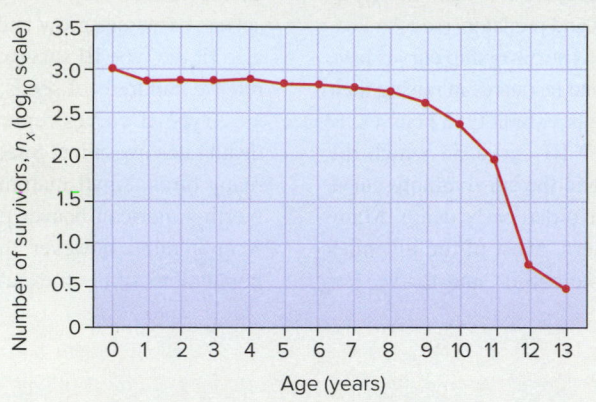

Survivorship curve for the Dall mountain sheep shows the number of sheep alive in each age class on a log scale, plotted against age in years.

**4    THE DATA**

Results used in step 3:

| Age class | Number alive, $n_x$ | $\log_{10} n_x$ | Age class | Number alive, $n_x$ | $\log_{10} n_x$ |
|---|---|---|---|---|---|
| 0–1 | 1,000 | 3.00 | 7–8 | 640 | 2.81 |
| 1–2 | 801 | 2.90 | 8–9 | 571 | 2.76 |
| 2–3 | 789 | 2.90 | 9–10 | 439 | 2.64 |
| 3–4 | 776 | 2.89 | 10–11 | 252 | 2.40 |
| 4–5 | 764 | 2.88 | 11–12 | 96 | 1.98 |
| 5–6 | 734 | 2.86 | 12–13 | 6 | 0.78 |
| 6–7 | 688 | 2.84 | 13–14 | 3 | 0.48 |

**5    CONCLUSION** Most Dall mountain sheep die when very young or very old. Culling the wolf population would not greatly increase sheep survival.

**6    SOURCE** Deevey, E. S., Jr. 1947. Life tables for natural populations of animals. *Quarterly Review of Biology* 22:283–314.

were lost; then the survivorship curve flattened out, indicating that the sheep survived well through about age 7 or 8. Then the number of sheep declined rapidly as they aged. These data underlined what Murie had previously observed, which was that wolves preyed primarily on the most vulnerable members of the sheep population—the youngest and, to a much greater extent, the oldest. Such predation would not be expected to dramatically reduce the sheep population because the reproductively active adults were not being taken. The Park Service ultimately ended its limited wolf-control program.

*Experimental Questions*

1. What problem led to the study conducted by Murie on the Dall mountain sheep population of Denali National Park?

2. Describe the survivorship curve developed by Deevey based on Murie's data.

3. **CoreSKILL »** How did Murie's data and Deevey's analysis affect the decision of the Park Service concerning the control of the wolf population?

## Age-Specific Fertility Data Help to Predict Population Growth or Decline

Thus far, we have focused our attention on survivorship. To predict whether a population is growing or declining, ecologists also need information on reproductive rates, such as birth rates in mammals. For any given age class, they must determine the proportion of female offspring that are produced by females of reproductive age. Using these data, the **age-specific fertility rate**, denoted by $m_x$, can be calculated. For example, if 100 females of a given age class produce 75 female offspring, $m_x = 0.75$. An examination of the beaver age-specific fertility rates in Table 56.1 illustrates a common trend in mammals. For this beaver population in particular, and for many mammalian species in general, no offspring are born to young females, that is, $m_x = 0$. As females mature sexually, age-specific fertility goes up and remains fairly high until later in life until females reach a post-reproductive age. Females in the intermediate age classes have the highest fertility rates.

The number of offspring produced by females of any given age class depends on two factors: the number of females in that age class and their age-specific fertility rate. Let's return to the life table in Table 56.1 and calculate the reproductive rate of this beaver population. First, we use the survivorship data to find the proportion of females alive at the start of any given age class. This age-specific survivorship, termed $l_x$, equals $n_x/n_0$, where $n_0$ is the number alive at time 0, the start of the study, and $n_x$ is the number alive at the beginning of age class $x$. For example, the proportion of the original beaver population still alive at the start of the sixth age class, $l_5$, equals $n_5/n_0 = 273/3,695$, or 0.074. This means that 7.4% of the original beaver population survived to age 5. Next, we multiply the data in the two columns, $l_x$ and $m_x$, for each row, to give us the values in the last column, $l_x m_x$, which represent the average number of offspring per female age class. These values indicate the contribution of each age class to the overall population reproductive rate.

The overall reproductive rate per generation is the number of offspring born to all females of all ages, where a generation is defined as the mean period between the birth of females and the birth of their offspring. Therefore, to calculate the generational growth rate, we sum all the values of $l_x m_x$; that is, we find $\Sigma l_x m_x$, where the $\Sigma$ symbol means "sum of." This summed value, $R_0$, is called the **net reproductive rate**.

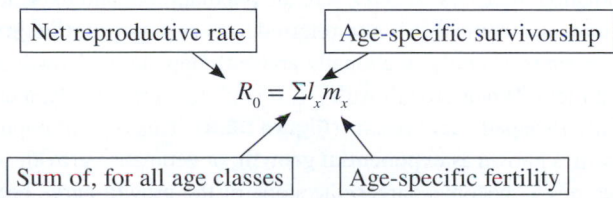

To calculate the population size in the next generation, we simply multiply the number of individuals in the population by the net reproductive rate. Thus, the population size in the next generation, $N_{t+1}$, is determined by the number in the population now, at time $t$, which is given by $N_t$, multiplied by $R_0$.

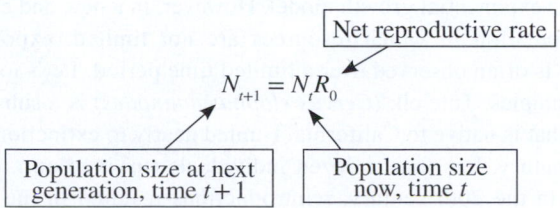

Let's consider an example in which the number of beavers alive now, $N_t$, is 1,000, and $R_0 = 1.1$. This means that the beaver population is reproducing at a rate that is 10% greater than simply replacing itself. The size of the population in the next generation, $N_{t+1}$, is given by

$$N_{t+1} = N_t R_0$$
$$N_{t+1} = 1,000 \times 1.1$$
$$= 1,100$$

Therefore, the number of beavers in the next generation is 1,100, and the population will have grown larger.

Population growth depends on the value of $R_0$. If $R_0 > 1$, then the population will grow. If $R_0 < 1$, the population is in decline. If $R_0 = 1$, the population size stays the same, and we say it is at **equilibrium**. In the case of the beavers, Table 56.1 reveals that $R_0 = 0.938$, which is less than 1, and, therefore, the population is declining. This is valuable information, because it tells us that at that time, the beaver population in Newfoundland needed some form of protection (perhaps bans on trapping and hunting) in order to attain a population level at equilibrium.

Because of the effort involved in calculating $R_0$, the net reproductive rate, ecologists sometimes use a shortcut to predict population growth. Imagine a bird species that breeds annually. To measure population growth, ecologists count the number of birds in the population, $N_0$. Let's say $N_0 = 100$. The next year, ecologists count 110 birds in the same population, so $N_1 = 110$. The **finite rate of increase**, $\lambda$, is the ratio of the population size from one year to the next, calculated as

$$\lambda = N_1/N_0$$

In this case, $\lambda = 1.10$, or 10%. $\lambda$ is often given as percent annual growth, and $t$ is a number of years.

Let's consider a population of birds growing at a rate of 5% per year. To calculate the size of the population after 5 years, we substitute $\lambda$ for $R_0$:

$$N_t = 100, \lambda = 1.05, \text{ and } t = 5$$
$$\text{therefore, } N_{t+5} = 100(1.05)^5 = 127.6$$

What's the difference between $R_0$ and $\lambda$? $R_0$ represents the net reproductive rate per generation. $\lambda$ represents the finite rate of population change over some time interval, often a year. When species are annual breeders that live 1 year, such as annual plants, $R_0 = \lambda$. For species that breed for multiple years, $R_0 \neq \lambda$. The size of a given population can be predicted in two ways:

$$N_t = N_0 R_0{}^t, \text{ where } t = \text{a number of generations}$$
$$\text{and } N_t = N_0 \lambda^t, \text{ where } t = \text{a number of time intervals}$$

Populations grow when $R_0$ or $\lambda > 1$; populations decline when $R_0$ or $\lambda < 1$; and they are at equilibrium when $R_0$ or $\lambda = 1$.

## 56.3 How Populations Grow

**Learning Outcomes:**

1. **CoreSKILL »** Calculate the population growth of continuously breeding organisms using the per capita growth rate (*r*).
2. Distinguish between exponential growth and logistic growth.
3. Explain how density-dependent factors and density-independent factors regulate population size.
4. Compare and contrast life history strategies of *r*-selected and *K*-selected species.

The demographic data provided in life tables can be used to predict how populations grow from generation to generation. However, other population growth models can provide valuable insights into how populations grow over shorter time periods. The simplest of these assumes that for any given time interval, a population will grow if the number of offspring produced is greater than the number of deaths. In this section, we will examine two different types of these simple models. The first assumes that resources are not limiting, and the result is prodigious growth. The second, and perhaps more biologically realistic model, assumes that resources are limiting; population growth is limited and results in stable population sizes. We will then consider how other factors, such as natural enemies, might limit population growth and discuss the overall life history strategies exhibited by different species.

## The Per Capita Growth Rate Is Used to Predict How Populations Will Grow

The change in population size over any time period can be written as the number of new offspring produced per unit time interval minus the number of deaths per unit time interval. In the case of mammals that give birth to live young, this change can be viewed as the number of births minus the number of deaths. As an example, let's consider a population of 1,000 rabbits. Over the course of 1 year, this population experienced 100 births and 50 deaths, resulting in 1,050 rabbits the following year. We can write this formula mathematically as

$$\frac{\text{Change in numbers}}{\text{Change in time}} = \text{Births} - \text{Deaths}$$

or

$$\frac{\Delta N}{\Delta t} = B - D$$

The Greek letter delta, $\Delta$, indicates change, so $\Delta N$ is the change in number and $\Delta t$ is the change in time; $B$ is the number of births per time unit; and $D$ is the number of deaths per time unit.

The numbers of births and deaths can also be expressed per individual in the population. In the case of the rabbit population, the birth of 100 rabbits to a population of 1,000 represents a per capita birth rate, $b$, of 100/1,000, or 0.10. The death of 50 rabbits in a population of 1,000 is a per capita death rate, $d$, of 50/1,000, or 0.05. The rate of change in a population size is given by the following equation:

$$\frac{\Delta N}{\Delta t} = bN - dN$$

For our rabbit population,

$$\frac{\Delta N}{\Delta t} = 0.10 \times 1,000 - 0.05 \times 1,000 = 50$$

If $\Delta t = 1$ year, the rabbit population increases by 50 individuals in a year.

Ecologists often simplify this formula by representing $b - d$ as $r$, the **per capita growth rate**. Thus, $bN - dN$ can be written as $rN$. Because ecologists are also interested in population growth rates over very short time intervals, so-called instantaneous growth rates, instead of writing

$$\frac{\Delta N}{\Delta t}$$

they write

$$\frac{dN}{dt}$$

which is the notation of differential calculus. The equations essentially mean the same thing, except that $dN/dt$ reflects very short time intervals. Thus, for the rabbit population,

$$\frac{dN}{dt} = rN = (0.10 - 0.05)N = 50$$

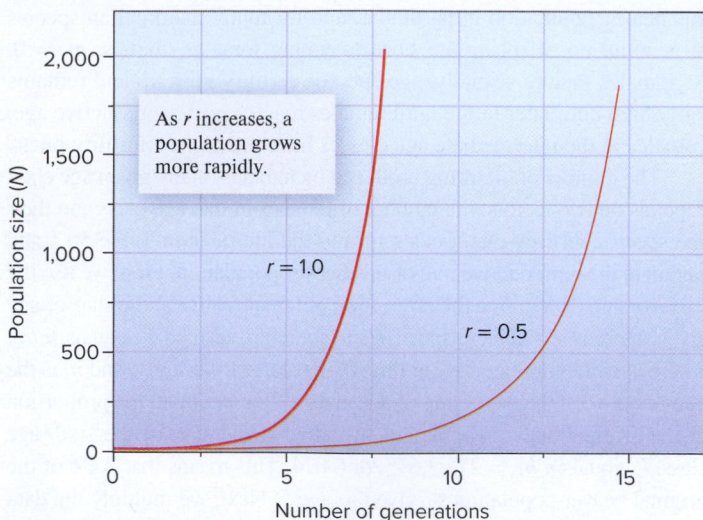

**Figure 56.8** **Exponential population growth.** As the value of $r$ increases, the slope of the curve gets steeper.

## Exponential Growth Occurs When Resources Are Not Limiting

Why do populations grow or decline in numbers? When the value of the per capita growth rate, $r$, is less than zero, the population decreases; when $r = 0$, the population remains constant; and when $r > 0$, the population increases. When $r = 0$, the population is said to be at equilibrium, a situation that is also referred to as **zero population growth**.

Even if $r$ is only fractionally above 0, population growth can be quite rapid. When growth with $r > 0$ is plotted graphically, a characteristic J-shaped curve results (**Figure 56.8**). This type of population growth is known as **exponential growth**, or geometric growth. As the value of $r_{max}$ becomes larger, the slope of the growth curve becomes steeper. If conditions are optimal for population growth, $r$ is at its maximum and is called the **intrinsic rate of increase** (denoted $r_{max}$). Thus, the rate of population growth under optimal conditions is $dN/dt = r_{max}N$.

How do field data fit this simple model for exponential growth? Population growth cannot go on forever, as envisioned with the exponential growth model. However, in a new and expanding population in which resources are not limited, exponential growth is often observed over a limited time period. Let's look at a few examples. Tule elk (*Cervus elaphus nannodes*) is a subspecies of elk that is native to California. Hunted nearly to extinction in the 19th century, less than a dozen individuals survived on a private ranch. In the 20th century, reintroductions resulted in the recovery of tule elk to around 3,500 individuals. One reintroduction was made in 1978 at Point Reyes National Seashore in California, where 10 animals—2 males and 8 females—were released. By 1993, the herd had reached 214 individuals, and it continued to grow in an exponential fashion until 1998, when the herd size stood at 549 (**Figure 56.9a**). This was deemed an excessive number for the size of the available habitat, and animals were removed to begin herds in other locations. Since then, herd size at Point Reyes has been maintained at around 350.

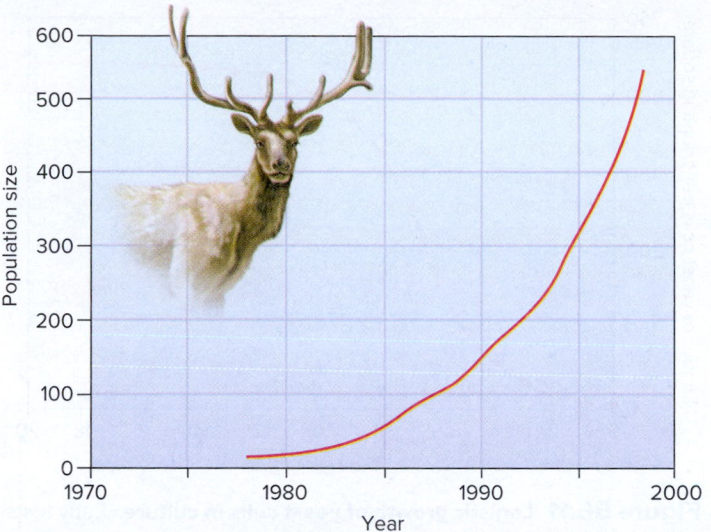

**(a) Tule elk**

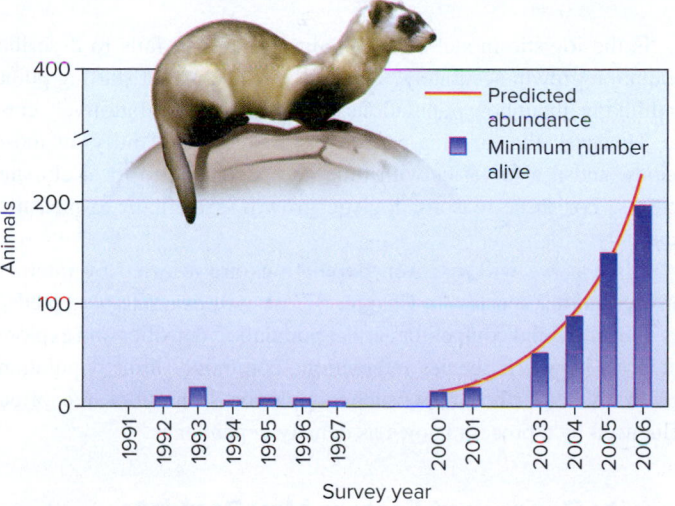

Predicted abundance

Minimum number alive

**(b) Black-footed ferret**

**Figure 56.9** **Exponential growth following reintroduction of a population into a habitat.** **(a)** A population of tule elk (*Cervus elaphus nannodes*) reintroduced to Point Reyes National Seashore in 1978 fits a pattern of exponential growth. **(b)** Black-footed ferrets (*Mustela nigripes*), reintroduced to Shirley Basin, Wyoming, in 1991, showed exponential growth after 2000. No survey was conducted in 2002.

b: Stiling, P. D. (2014). *Ecology*. New York, NY: McGraw-Hill. Fig. No.: 10.2(b) Page No.: 197.

The growth of the recovering black-footed ferret population in Wyoming, mentioned at the beginning of the chapter, also fits the exponential growth pattern (**Figure 56.9b**). The ferrets had been reintroduced in 1991, but the population declined for many years, so that by 1997, only five were living. However, from 2000 to 2006, the population grew in an exponential fashion. A value of $r = 0.47$ was calculated for the increase in population size of the ferrets during those years. Ecologists have noted that, in some cases, populations remain at low levels for an extended period of time before conditions become favorable for exponential growth.

The growth of some introduced species also seems to fit the pattern of exponential growth. The rapid expansion of rabbits after their introduction into southern Australia in the late 19th century is a case in point. In 1859, British immigrant Thomas Austin received two dozen European rabbits from England. Rabbit gestation lasts a mere 31 days, and in southern Australia, each female rabbit could produce up to 10 litters of at least six young each year. The rabbits had essentially no enemies and ate the grass used by sheep and other grazing animals. Even when two-thirds of the population was shot for sport, which was the purpose of the initial introduction, the population grew into the millions within a few short years. By 1875, rabbits were reported on the west coast of Australia, having moved over 1,760 km across the continent despite the deployment of huge, thousand-kilometer-long fences (rabbit-proof fences) meant to contain them.

Finally, one of the most prominent examples of exponential growth is the growth of the global human population. Because of its great importance, we will examine human populations separately in Chapter 59.

## Logistic Growth Occurs When Resources Are Limited

Despite its applicability to certain rapidly growing populations, the exponential growth model is not appropriate in most situations, because it assumes unlimited resources. Although resources can be unlimited when population sizes are small, they eventually become limiting if populations continue to grow. When this occurs, the per capita growth rate decreases. The upper boundary for the population size in a given environment is known as the **carrying capacity (K)**. A more realistic equation to describe population growth, one that takes into account the amount of available resources, is

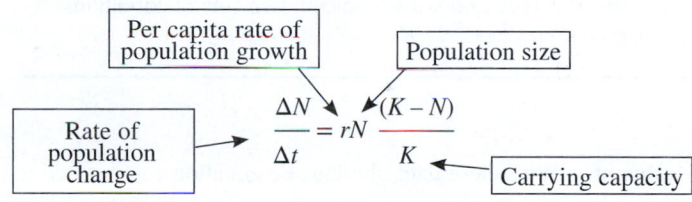

where $(K - N)/K$ represents the proportion of the carrying capacity that is unused by the population. This equation is called the **logistic equation**.

As the population size, $N$, grows, it moves closer to the carrying capacity, $K$, with fewer available resources for population growth. At large values of $N$, the value of $(K - N)/K$ becomes small, and population growth is small. If $K = 1,000$, $N = 900$, and $r = 0.1$, then

$$\frac{dN}{dt} = (0.1)(900) \times \frac{(1,000-900)}{1,000}$$

$$\frac{dN}{dt} = 9$$

In this instance, population growth is 9 individuals per unit of time.

Let's consider how an ecologist could use the logistic equation. First, the value of $K$ would come from intense field and laboratory work in which researchers would determine the amount of resources, such as food, needed by each individual and then determine the amount of available food in a given area in the wild. Field censuses in the wild would be conducted to determine $N$, and field censuses of births and deaths per unit time would be carried out to calculate

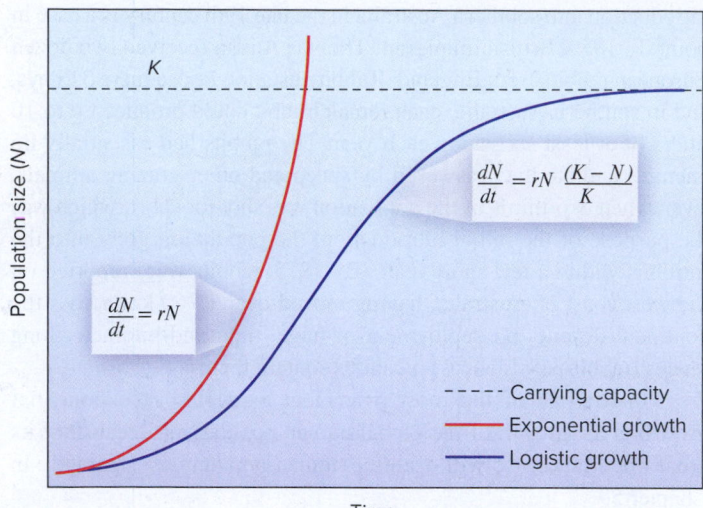

**Figure 56.10 Exponential versus logistic growth.** Exponential (J-shaped) growth occurs in an environment with unlimited resources, whereas logistic (S-shaped) growth occurs in an environment with limited resources.

👁 **Core Skill: Modeling** The goal of this modeling challenge is use different values for $r$, $N$, and $K$ to create graphical models based on the logistic equation.

**Modeling Challenge:** In the growth curve for logistic growth in Figure 56.10 (see the blue line), the value for $r$ is greater than 1 in the starting population (at time zero) and the value for $N$ is much less than $K$. Based on the logistic equation, draw two growth curves under the following two sets of conditions: $r < 0$, $N \ll K$ and $r > 0$, $N = K$.

the value of $r$. With these data, the logistic equation can be used to predict changes in population size as a function of time. As seen in **Figure 56.10**, if $r > 0$ and $N$ is much less than $K$ for the starting population at time zero, an S-shaped growth curve results. This pattern, in which a population initially grows rapidly but then grows more slowly as $N$ approaches $K$, is called **logistic growth**.

Does the logistic growth model provide a better fit to growth patterns of plants and animals in the wild than the exponential model? In some instances, such as laboratory cultures of bacteria and yeasts, the logistic growth model provides a very good fit (**Figure 56.11**). In nature, however, variations in temperature, rainfall, or resources can cause changes in carrying capacity and thus in population size. The uniform conditions of temperature, moisture, and resource levels of the laboratory do not usually exist in the outside world. In addition, time lags may occur between changes in carrying capacity and changes in reproduction. For instance, pregnant females are still likely to give birth even when resources are declining. This kind of time lag can lead to temporary overshoots of population density beyond the carrying capacity. Therefore, relatively few exact fits of the logistic growth model to population growth have been documented in the field. Instead, populations tend to fluctuate around the limits suggested by the logistic model, with frequent overshoots and undershoots.

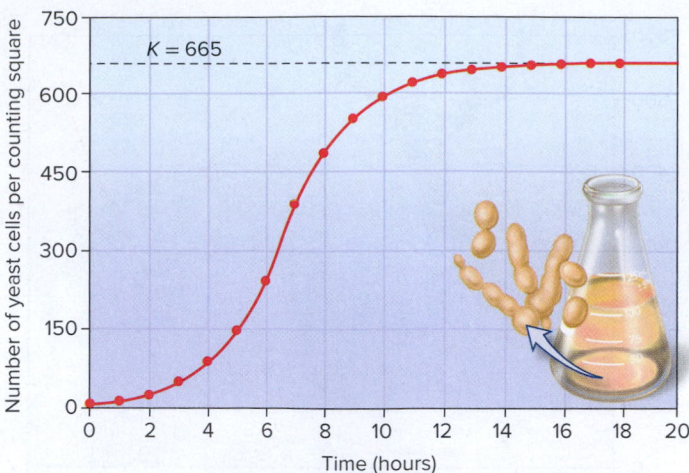

**Figure 56.11 Logistic growth of yeast cells in culture.** Early tests of the logistic growth curve were validated by growth of yeast cells in laboratory cultures. These populations showed the typical S-shaped growth curve.

Is the logistic model of little value because it fails to describe population growth accurately? Not really. It is a useful starting point for thinking about how populations grow, and it seems intuitively correct. However, the carrying capacity is difficult to identify for most species, and it also varies with time and according to local climate patterns. For these reasons, logistic growth is difficult to measure accurately.

Also, as we will discover, populations are affected by interactions with other species. In Chapter 57, we will examine how predators, parasites, and competitors affect population densities and explore situations in which species interactions commonly limit population growth. As described next, such population limitations are often influenced by a process known as density dependence.

## Density-Dependent Factors May Regulate Population Sizes

A **density-dependent factor** is a mortality factor whose influence increases with the density of the population. Parasitism, predation, and competition are some of the many density-dependent factors that may reduce the population densities of living organisms and stabilize them at equilibrium levels. Such factors can be density-dependent when their effect depends on the density of the population; they are responsible for the deaths of relatively more individuals in a population when densities are higher and fewer individuals when densities are lower. For example, many predators develop a visual search image for a particular prey. When a prey is rare, predators tend to ignore it and kill relatively few. When a prey is common, predators key in on it and kill relatively more. In England, for example, predatory shrews kill proportionately more moth pupae in leaf litter when the pupae are common compared with when they are rare. Density-dependent mortality may also occur as population densities increase and competition for scarce resources increases, reducing offspring production or survival. Parasitism may also act in a density-dependent manner. Parasites are able to pass from host to host more easily as the host's population density increases.

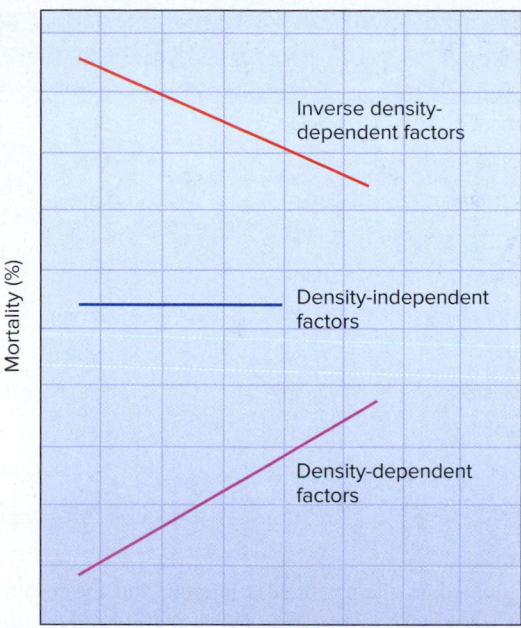

**Figure 56.12** **Three ways that factors affect mortality in response to changes in population density.** For a density-dependent factor, mortality increases with population density; for a density-independent factor, mortality remains unchanged. For an inverse density-dependent factor, mortality decreases as a population increases in size.

*Concept Check:* *Which types of factors tend to stabilize populations at equilibrium levels?*

Density dependence can be evaluated by plotting mortality, expressed as a percentage, as a function of population density (**Figure 56.12**). If mortality increases with density, a positive slope is observed. The factor tends to have a greater effect on dense populations than on sparse ones and is clearly acting in a density-dependent manner.

A **density-independent factor** is a mortality factor whose influence is not affected by changes in population size or density. When mortality due to such a factor is plotted as a function of density, a flat line results. Physical factors are commonly density-independent, including drought, freezes, floods, and disturbances such as fire. For example, in hard freezes, the same proportion of organisms such as birds or plants are usually killed, no matter how large the population size.

Finally, a mortality factor whose influence decreases with increasing population size is considered an **inverse density-dependent factor**. In this case, a negative slope results when mortality is plotted as a function of density. For example, a territorial predator such as a lion may always kill the same number of wildebeest prey, regardless of wildebeest density. The lion is acting in an inverse density-dependent manner, because it is taking a smaller proportion of the population at higher density.

Determining which factors act in a density-dependent or density-independent fashion has large practical implications. Foresters, game managers, and conservation biologists alike are interested in maintaining populations. For example, if a specific disease were to act in a density-dependent manner on white-tailed deer, the removal of deer predators by game managers might not result in an increase in deer population size, because proportionately more deer would be killed by disease.

## Life History Strategies Incorporate Traits Relating to Survival and Competitive Ability

The population parameters we have discussed—including iteroparity versus semelparity, exponential versus logistic growth, and density-dependent versus density-independent factors—have important implications for how populations grow and for the reproductive success of populations and species. These reproductive strategies can be viewed in the context of a much larger picture of life history strategies, sets of physiological and behavioral features that incorporate not only reproductive traits but also survivorship, length of life, habitat type, and competitive ability, among other characteristics.

When comparing many different species, life history strategies follow a continuum. At the one end are species, termed *r*-**selected species**, that have a high rate of per capita population growth (*r*), but poor competitive ability. An example is a dandelion, which produces huge numbers of tiny seeds and therefore has a high value of *r* (**Figure 56.13a**). Weeds exist in disturbed habitats such as gaps in a forest canopy where trees have blown down, and also in areas disturbed by humans such as agricultural fields or backyard gardens. An *r*-selected species such as a weed grows quickly and reaches reproductive age early, devoting much energy to producing a large number of seeds that disperse widely. These weed species generally remain small, and individuals do not live long. In the animal world, insects are mostly *r*-selected species that produce many young and have short life cycles.

At the other end of the continuum are species, termed *K*-**selected species**, whose populations are relatively stable and often exist at or near the carrying capacity (*K*), of the environment. An example is an oak tree that exists in a mature forest (**Figure 56.13b**). Oak trees grow slowly and reach reproductive age late, having to devote much energy to growth and maintenance. A *K*-selected species like a tree grows large and shades out *r*-selected species like weeds, eventually outcompeting them. Such trees live a long time and produce seeds repeatedly every year when mature. These seeds are bigger than those of *r*-selected species, but do not disperse widely. Acorns contain a large food reserve that helps them grow, whereas dandelion seeds must rely on whatever nutrients they can gather from the soil where they land. Mammals, such as elephants, that grow slowly, have few young, and reach large sizes are typical of *K*-selected animal species. **Table 56.2** compares the general characteristics of *r*- and *K*-selected species.

In a human-dominated world, almost every life history feature of a *K*-selected species sets it at risk of extinction. First, *K*-selected species tend to be larger, so they need more habitat in which to live. For example, Florida panthers need huge tracts of land to establish their territories and hunt for deer (look ahead to Figure 60.10c). *K*-selected species tend to have fewer offspring, so their populations cannot recover as fast from disturbances such as fire or overhunting. California condors, for example, produce only a single chick

- Small size
- Rapid growth
- Short life span

- Many small seeds
- Good seed dispersal

**(a) *r*-selected species**

- Large size
- Slow growth
- Long life span

- Few large seeds
- Poor seed dispersal

**(b) *K*-selected species**

**Figure 56.13 Life history strategies.** Differences in traits of a dandelion **(a)** and an oak tree **(b)** illustrate some of the differences between *r*- and *K*-selected species.

 **Core Skill: Connections** Look back at Figure 40.21. What is unusual about the reproductive strategy of the mother of thousands plant?

every other year. *K*-selected species breed at a later age, and the time required for the population to grow from a small to a large size is long. Gestation time in elephants is 22 months, and elephants take at least 7 years to become sexually mature. Large trees, such as the giant sequoia; large terrestrial mammals, such as elephants, rhinoceroses, and grizzly bears; and large marine mammals, such as blue whales and sperm whales, all run the risk of extinction.

What are the advantages to being a *K*-selected species? In a world not disturbed by humans, *K*-selected species fare well. However, in a human-dominated world, many *K*-selected species are selectively

| Table 56.2 | Characteristics of *r*- and *K*-Selected Species | |
|---|---|---|
| Life history feature | *r*-selected species | *K*-selected species |
| Development | Rapid | Slow |
| Reproductive rate | High | Low |
| Reproductive age | Early | Late |
| Body size | Small | Large |
| Length of life | Short | Long |
| Competitive ability | Weak | Strong |
| Survivorship | Type III | Type I |
| Population size | Variable | Fairly constant |
| Dispersal ability | Good | Poor |
| Habitat type | Disturbed | Not disturbed |
| Parental care (animals) | Low | High |

logged or hunted, or their habitat is altered, and the resulting small population sizes make extinction a real possibility. Interestingly, the coast redwood seems to be an exception, a fact perhaps attributable to its unusual genome, as described next.

 **Core Concept: Evolution**

## Hexaploidy Increases the Growth of Coast Redwood Trees

Besides being home to the world's most massive tree, the giant sequoia (*Sequoiadendron giganteum*), California is also the location of the world's tallest tree, the coast redwood (*Sequoia sempervirens*), a towering giant that can grow to over 90 m and can live for up to 2,000 years (**Figure 56.14**). These trees are currently confined to a relatively small 700-km strip along the Pacific coast from California to southern Oregon, an area characterized by year-long moderate temperatures, heavy winter rains, and dense summer fog. Interestingly, because this climate was far more common in an earlier era, these trees were once dispersed throughout the Northern Hemisphere.

How is this huge species different from other tree species? In 1948, researchers made the startling discovery that the tree is hexaploid; that is, each of its cells contains six sets of chromosomes, with 66 chromosomes in total. (Keep in mind that humans have two sets of chromosomes in most cells.) Although hexaploidy is relatively common in grasses and shrubs, it is unusual in trees, particularly gymnosperms. The coast redwood is the only known hexaploid conifer. Hexaploidy means that each tree may have several different alleles for any given gene, which leads to a very genetically diverse population. American molecular biologist Chris Brinegar has found that hardly any two trees have exactly the same genetic constitution. Such genetic diversity allows greater adaptation to environmental conditions and more defensive adaptations against insect or fungal pests. Indeed, living redwoods have no known lethal diseases, and pests do not cause

**Figure 56.14** **The coast redwood (*Sequoia sempervirens*)—a hexaploid conifer.** The coast redwood can grow to over 90 m, and the oldest living trees are over 2,000 years old. Their great genetic variation may help explain their incredible growth and longevity. ©Fuse/Getty Images

 **Core Concept: Information** Despite being large, *K*-selected species, coast redwoods grow faster than any other known conifer, and this rapid growth may be due to their unusual hexaploid genome.

significant damage. What's more, with six sets of genes, trees also have the potential for great variety in their gene products, the proteins, which may help explain their prodigious growth. The coast redwood grows faster than any other conifer on Earth, and this is why it is an exception to most *K*-selected species.

## Summary of Key Concepts

- In the case of sexually reproducing species, a population can be defined as a group of interbreeding individuals of the same species occupying the same area at the same time.

- Population ecology is the study of how populations grow or decline and what factors promote and limit growth.

### 56.1  Understanding Populations

- Ecologists measure population density, the numbers of organisms in a given unit area, in different ways (Figures 56.1, 56.2).

- Populations show different patterns of dispersion, including clumped (the most common), uniform, and random. Species also exhibit different reproductive strategies, and those that reproduce only once often have cohorts of offspring (Figures 56.3, 56.4).

### 56.2  Demography

- Demography is the study of birth rates, death rates, age distributions, and the sizes of populations.

- Life tables summarize data on numbers of living individuals in various age classes in a population and their relative fertilities (Table 56.1).

- Survivorship curves plot the numbers of surviving individuals at different ages. (Figures 56.5, 56.6, 56.7).

- Age-specific fertility and survivorship data help determine the overall growth rate per generation, or the net reproductive rate ($R_0$).

### 56.3  How Populations Grow

- The per capita growth rate ($r$) is used to determine how populations grow over any time period. When $r$ is > 0 and resources are unlimited, exponential growth occurs (Figures 56.8, 56.9).

- Logistic (S-shaped) growth takes into account the upper boundary for a population, called the carrying capacity, and occurs in an environment where resources are limited (Figures 56.10, 56.11).

- Density-dependent factors are mortality factors whose influence varies with population density. Density-independent factors are those whose influence does not vary with population density (Figure 56.12).

- Life history strategies are a set of features including reproductive traits, survivorship, length of life, habitat type, and competitive ability. Life history strategies follow a continuum, with *r*-selected species (those with a high rate of population growth but poor competitive ability) at one end and *K*-selected species (those with a lower rate of population growth but better competitive ability) at the other (Figures 56.13, 56.14, Table 56.2).

## Assess & Discuss

### Test Yourself

1. A student decides to use the mark-recapture technique to estimate the population size of mosquitofish in a small pond near his home. In the first catch, he marked 45 individuals. Two weeks later, he captured 62 individuals, of which 8 were marked. What is the estimated size of the population based on these data?

   a. 134
   b. 349
   c. 558
   d. 1,016
   e. 22,320

Questions 2–4 refer to the following life table:

| Age | $n_x$ | $d_x$ | $l_x$ | $m_x$ | $l_x m_x$ |
|---|---|---|---|---|---|
| 0 | 100 | 35 | 1.00 | 0 | 0 |
| 1 | 65 | ? | 0.65 | 0 | 0 |
| 2 | 45 | 15 | ? | 3 | 1.35 |
| 3 | 30 | 20 | 0.30 | 1 | ? |
| 4 | 10 | 10 | 0.10 | 1 | 0.10 |
| 5 | 0 | 0 | 0.00 | 1 | 0.0 |

2. How many individuals in this population die between their first and second birthday?

   a. 65
   b. 45
   c. 35
   d. 25
   e. 20

3. What proportion of newborns survive to age 2?
   a. 0.55
   b. 0.45
   c. 0.35
   d. 0.20
   e. 0.15

4. What is the net reproductive rate?
   a. 5
   b. 2.5
   c. 1.75
   d. 1.45
   e. 0.80

5. _____ survivorship curves are usually associated with organisms that have high mortality rates in the early stages of life.
   a. Type I
   b. Type II
   c. Type III
   d. Types I and II
   e. Types II and III

6. If $\Sigma l_x m_x$ is equal to 0.5, what prediction can we make about the population?
   a. This population is essentially not changing in numbers.
   b. This population is in decline.
   c. This population is growing.
   d. This population is in equilibrium.
   e. None of the above is correct.

7. Under what condition(s) will a population grow?
   a. $R_0 = 0.5$
   b. $r = 0.4$
   c. $\lambda = 0.8$
   d. conditions a and b
   e. conditions b and c

8. The maximum number of individuals a certain area can sustain is known as the
   a. intrinsic rate of growth.
   b. resource limit.
   c. carrying capacity.
   d. logistic equation.
   e. equilibrium size.

Questions 9 and 10 refer to the following generalized growth patterns, plotted on arithmetic scales. In each case, choose the correct pattern from the four options.

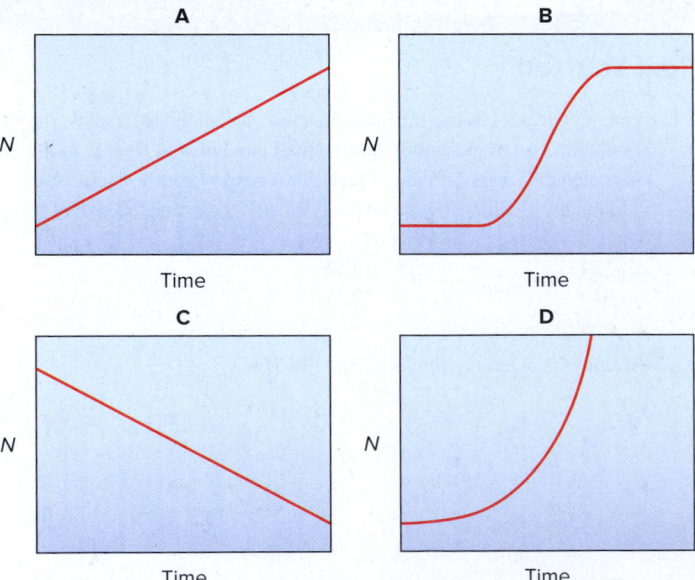

Each pattern may be used once, more than once, or not at all.

9. Which pattern occurs when a population exhibits a constant per capita growth rate?
   a. A
   b. B
   c. C
   d. D
   e. none of the above

10. Which pattern is observed when a population is heading toward extinction?
   a. A
   b. B
   c. C
   d. D
   e. none of the above

## Conceptual Questions

1. As a researcher, you are using the mark-recapture technique (see Figure 56.2) to study a population of largemouth bass. Say you do a poor job tagging the fish, and 20% of the tags fall off. How does this influence your estimate of population size? Is it too high or too low?

2. Using the logistic equation, calculate population growth when $K = 1,000$, $N = 100$, and $r = 0.1$. Compare the result with that shown in Section 56.3, where $K = 1,000$, $N = 900$, and $r = 0.1$, and with the result when $K = 1,000$, $N = 500$, and $r = 0.1$.

3. **Core Concept: Systems** Imagine two types of ponds. One type dries out when there is little rain, and the other type has a water level that fluctuates but some water is always present. Contrast the reproductive strategies of organisms that might live in each pond.

## Collaborative Questions

1. Describe where students on campus might show each type of dispersion pattern, and explain why these patterns might occur.

2. You survey an annually breeding butterfly population and discover 1,000 females. The next year, you count 1,200 females in the same area. What is the finite rate of increase, $\lambda$? What will the population be five years from now if the rate of increase is the same each year?

# Species Interactions

# 57

In this species interaction, a shark is feeding on a ray, which in turn feeds on bay scallops. ©bastianbodyl/iStock/Getty Images

I n 2007, marine biologist Ransom Myers and his colleagues showed that overfishing had severely depleted the numbers of 11 shark species that occur along the eastern seaboard of the U.S. Several shark species had declined by over 99% since the 1950s. Because of strong interactions between the sharks and other marine species, this drastic reduction of the shark population had at least two other effects. First, there was a large increase in the main prey species of the sharks, rays and skates. Second, the increase in rays and skates reduced the densities of their prey—bay scallops (*Argopecten irradians*). Such losses contributed to the closure of the bay scallop industry in North Carolina.

In this chapter, we turn from considering populations on their own to investigating how they interact with populations of other species that live in the same locality. Such **species interactions** can take a variety of forms (**Table 57.1**). **Competition** is defined as an interaction that affects all the interacting species negatively (–/–), as when two species compete over food or other resources. Sometimes an interaction is one-sided, being detrimental to one species and neutral to the other, an interaction called **amensalism** (–/0). **Predation**, **herbivory**, and **parasitism** all have a positive effect on one species and a negative effect on the other (+/–). **Mutualism** is an interaction in which both species benefit (+/+), whereas **commensalism** benefits one species and leaves the other unaffected (+/0).

To illustrate how species interact in nature, let's consider a rabbit population in a grassland community (**Figure 57.1**). To determine what factors influence the size and density of the rabbit population, we need to understand each of the possible species interactions. For example, the rabbit population could be limited by the quality of available food. It is also likely that other species, such as deer, use the same food resource and thus compete with the rabbits. The rabbit population could be limited by predation from foxes or by the virus that causes the disease myxomatosis, which is usually spread by fleas and mosquitoes. It is also possible that other interactions, such as mutualism or commensalism, may occur.

In this chapter, we will example these various types of species interactions, beginning with competition. We will conclude with a discussion of two different models, called bottom-up and top-down, which present opposing viewpoints regarding the species that are most influential in controlling population densities of several different species within ecological systems.

| Table 57.1 | Summary of the Types of Species Interactions | |
|---|---|---|
| **Nature of interaction** | **Species 1*** | **Species 2*** |
| Competition | – | – |
| Amensalism | – | 0 |
| Predation, herbivory, parasitism | + | – |
| Mutualism | + | + |
| Commensalism | + | 0 |

*+ = positive effect; – = negative effect; 0 = no effect.

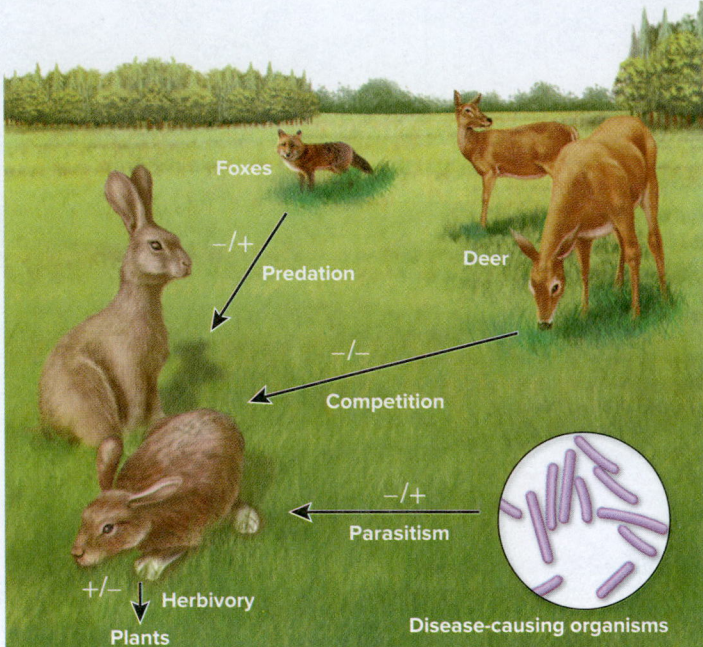

**Figure 57.1** **Species interactions.** These rabbits can interact with a variety of species, experiencing predation by foxes, competition with deer for food, and parasitism from various disease-causing organisms. Herbivory occurs when rabbits feed on various plants. The effects of each species on the other are identified by the labels associated with the arrows, as discussed in the text.

## 57.1    Competition

### Learning Outcomes:

1. Describe the different types of competition that occur in nature.
2. **CoreSKILL »** Predict how the competitive exclusion principle can lead to resource partitioning among species.
3. Give examples of how morphological differences may allow species to coexist.

In this section, we will examine different types of competition in nature and consider how ecologists have analyzed competition among two or more species for environmental resources. Although species may compete for resources, we will also explore how differences in lifestyle or morphology can reduce the overlap in their habitat use, thus allowing them to coexist.

### Different Types of Competition Occur in Nature

Ecologists have identified several different types of competition in nature (**Figure 57.2**). Competition may be **intraspecific**, which occurs between individuals of the same species. In this section, we will focus on competition that is **interspecific**, or between individuals of different species. Competition can also be characterized by the mechanism by which it occurs. In **exploitation competition**, organisms compete indirectly through the consumption of a limited resource, with each obtaining as much as it can. For example, when fly maggots compete in a mouse carcass, not all the individuals can

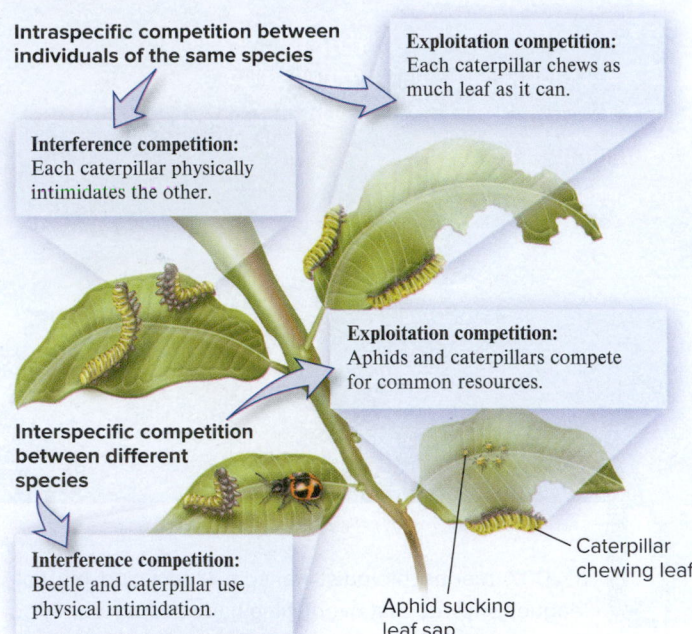

**Figure 57.2** **The different types of competition in nature.**

**Concept Check:** *How would you classify competition between vultures feeding on roadkill?*

command enough of the resource to survive and become adult flies. In **interference competition**, individuals interact directly with one another by physical force or intimidation. Often this force is ritualized into aggressive behavior associated with territoriality. In these cases, strong individuals survive and take the bulk of the resources, and weaker ones perish or, at best, survive under suboptimal conditions.

Competition between species is not always equal. Many plant species compete for natural resources such as water and sunlight. To give themselves an advantage, some species produce and secrete chemicals from their roots that inhibit the growth of other species. For example, diffuse knapweed, an introduced species, secretes root chemicals called allelochemicals into the surrounding environment that kill the roots of native grass species—a phenomenon called **allelopathy** (refer back to Figure 54.2).

### Field Studies Show That Competition Occurs Frequently in Nature

In a 1983 review of field studies by American ecologist Joseph Connell, competition was found in 55% of 215 species surveyed, demonstrating that it is indeed frequent in nature. Generally in studies of single pairs of species utilizing the same resource, competition is almost always reported (90%), whereas in studies involving more species, the frequency of competition drops to 50%. Why should this be the case? Imagine a resource such as a series of different-sized grains that has four species—ants, beetles, mice, and birds—feeding on it (**Figure 57.3a**). The ants feed on the smallest grains, the beetles and mice on the intermediate sizes, and the birds on the largest. If only adjacent species compete with each other, competition is expected to occur only between ant and beetle, beetle and mouse, and mouse and bird. Thus, competition occurs in only three out of the six possible

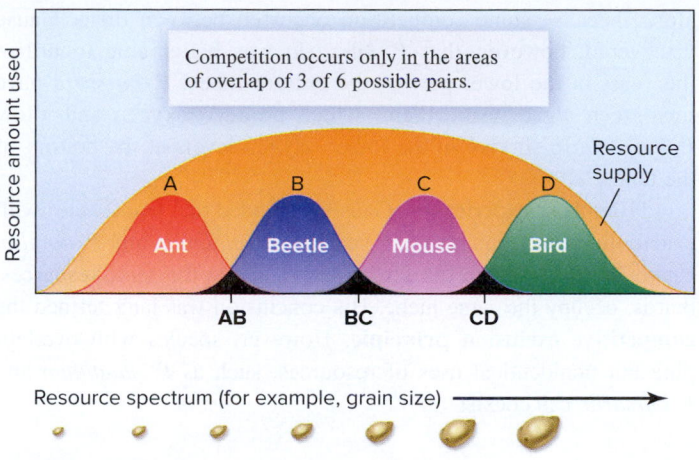

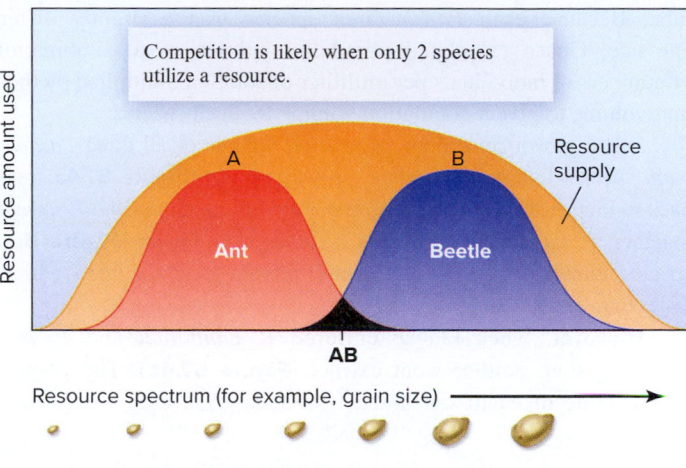

**(a) Competition among 4 species for a resource**

**(b) Competition among 2 species for a resource**

**Figure 57.3  The frequency of competition according to the number of species involved. (a)** Resource supply and utilization curves of four species, A, B, C, and D, along the spectrum of a hypothetical resource such as grain size. If competition occurs only between species with adjacent resource utilization curves, competition is expected between three of the six possible pairings: A and B; B and C; and C and D. **(b)** When only two species utilize a resource set, competition is nearly always expected between them.

species pairs (50%). Naturally, the percentage will vary according to the number of species on the resource spectrum. If only three species occur along the spectrum, we expect competition in two of the three pairs (67%). If just two species utilize the resource spectrum, however, we expect competition in almost 100% of the cases (**Figure 57.3b**).

Some other general patterns were evident from Connell's review. Plants showed a high degree of competition, perhaps because they are rooted in the ground and cannot easily escape or perhaps because they are competing for the same set of limiting nutrients—water, light, and minerals. Marine organisms tended to compete more than terrestrial ones, perhaps because many of the species studied lived in the intertidal zone and were attached to the rock face, and thus remain stationary like plants. Because the area of the rock face is limited, competition for space is quite important.

**THE QUESTION** If five species utilized the resource spectrum in Figure 57.3, what percentage of the interactions would be competitive?

**T OPIC  What topic in biology does this question address?**
The topic is species interactions. More specifically, the question asks you to determine what proportion of species that utilize a resource spectrum are likely to compete.

**I NFORMATION  What information do you know based on the question and your understanding of the topic?**
From the question, you know that there are five species utilizing the same resource spectrum. From Figure 57.3 and the text discussion, you may recall that only adjacent species usually compete.

**P ROBLEM-SOLVING S TRATEGY  Make a calculation. Interpret data.** A good way to start to solve this problem is to construct a table of possible species interactions, with the first potential competing species in columns and the second potential competing species in rows, as shown below.

| First potential competing species | Second potential competing species | | | | |
|---|---|---|---|---|---|
| | A | B | C | D | E |
| A | X | AB | AC | AD | AE |
| B | | X | BC | BD | BE |
| C | | | X | CD | CE |
| D | | | | X | DE |
| E | | | | | X |

Next, determine the total number of possible species interactions (which is 10). Then, divide the number of cases where adjacent species compete (for example, A and B or B and C) by the total number of possible species interactions and multiply that result by 100.

**ANSWER** For this group of species, the percentage of interactions that are likely to be competitive is 40%.

## Species May Coexist If They Do Not Occupy Identical Niches

Although competition is common, researchers have proposed several mechanisms by which two competing species can coexist. One states that similar species can coexist if they occupy different niches. A **niche** is the unique set of habitat resources a species requires as well as its effect on an ecological system.

In 1934, the Russian microbiologist Georgyi Gause began to study competition between three protist species, *Paramecium aurelia*, *Paramecium bursaria*, and *Paramecium caudatum*, all of which feed on bacteria and yeast, which, in turn, feed on an oatmeal medium in a culture tube in the laboratory. The bacteria tended to be located in the oxygen-rich upper part of the culture tube, and the yeast in the oxygen-poor lower part of the tube. Because each *Paramecium* species was a slightly different size, Gause calculated population growth as a combination of numbers of individuals per milliliter of solution multiplied by their unit volume to give a population volume for each species.

When grown separately, population volume of all three *Paramecium* species followed a logistic growth pattern (**Figure 57.4a**; refer back to Figure 56.11). When Gause cultured *P. caudatum* and *P. aurelia* together, *P. caudatum* eventually went extinct (**Figure 57.4b**). Both species utilized bacteria as food, but *P. aurelia* grew at a rate six times faster than *P. caudatum*.

However, when Gause cultured *P. caudatum* and *P. bursaria* together, neither went extinct (**Figure 57.4c**). The population volume of each was much less than when they were grown alone, because some competition occurred between them. Gause discovered, however, that *P. bursaria* was better able to utilize the yeast in the lower part of the culture tubes. *P. bursaria* have tiny green algae inside them, which produce oxygen and allow *P. bursaria* to survive in the lower oxygen levels at the bottom of the tubes.

From these experiments, Gause concluded that two species with exactly the same requirements, such as *P. caudatum* and *P. aurelia*, cannot live together in the same place and use the same resources, that is, occupy the same niche. His conclusion was later termed the **competitive exclusion principle**. However, species with overlapping but nonidentical uses of resources, such as *P. caudatum* and *P. bursaria*, can coexist.

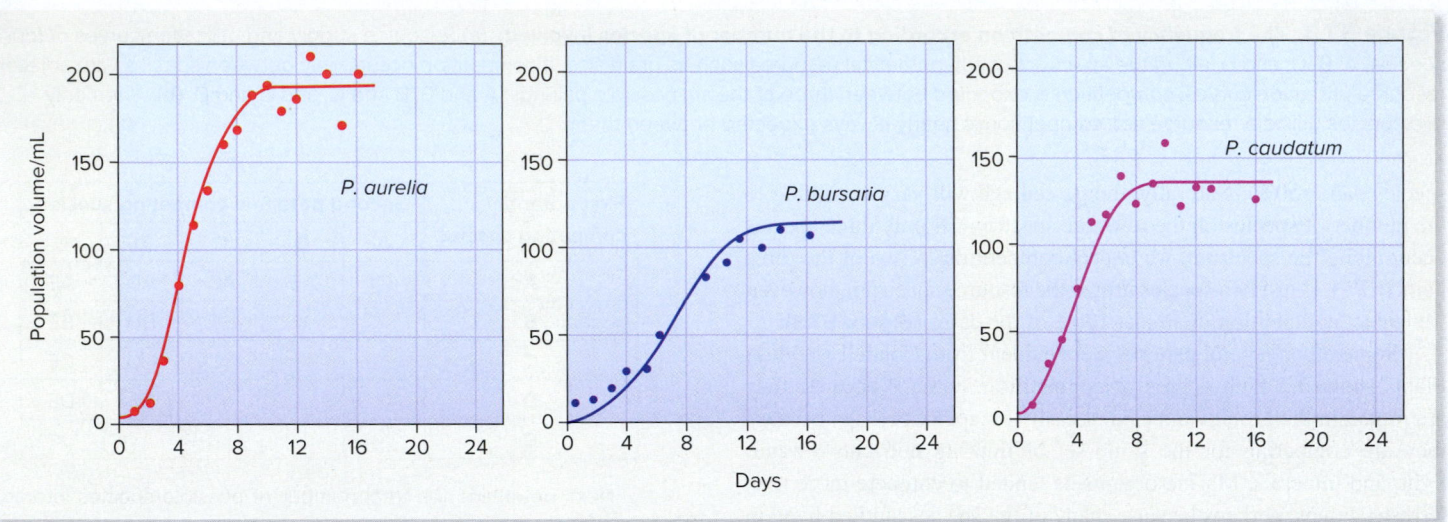

**(a) Each *Paramecium* species grown alone**

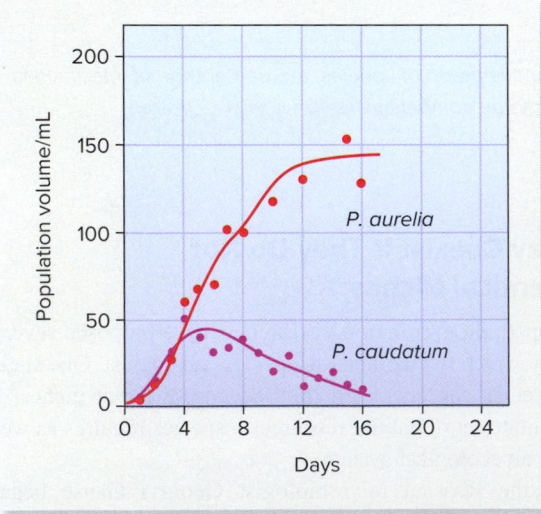

**(b) Competition between *P. aurelia* and *P. caudatum***

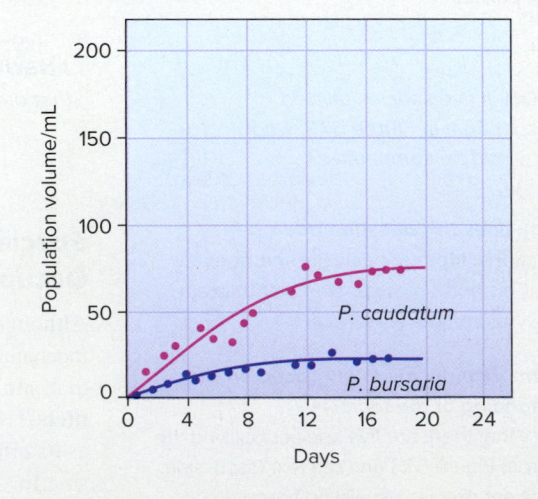

**(c) Competition between *P. caudatum* and *P. bursaria***

**Figure 57.4 Competition among *Paramecium* species. (a)** When grown alone, each of three species, *Paramecium aurelia, Paramecium bursaria*, and *Paramecium caudatum*, grows according to the logistic model. **(b)** When *P. aurelia* is grown with *P. caudatum*, the density of *P. aurelia* is lower than when grown alone, and *P. caudatum* goes extinct. **(c)** When *P. caudatum* is grown with *P. bursaria*, the population densities of both are lowered, but they coexist.

## Resource Partitioning Allows Different Species to Coexist

If competition over the same niche drives one species to local extinction, at least in the laboratory, how different must two species be to coexist, and in what features do they usually differ? To address such questions, in 1958, American ecologist Robert MacArthur examined coexistence between five species of warblers feeding in spruce trees in New England. Since all of these belonged to the genus *Dendroica* and were therefore closely related, they would be expected to compete strongly, possibly sufficiently strongly to cause extinctions. MacArthur found that the species occupied different heights and locations in the tree, and therefore, each probably fed on a different range of insects (**Figure 57.5**). In addition, the Cape May warbler fed on flying insects and tended to remain on the outside of the trees.

The term **resource partitioning** refers to the differentiation of niches, both in space and time, that enables similar species to coexist in a community, just as the five species of warblers feeding in different parts of a spruce tree. We can think of resource partitioning as reflecting the results of past competition, in which competition leads the inferior competitor to eventually occupy a different niche.

British ornithologist David Lack examined competition and coexistence among about 40 species of British passerines, or perching birds (**Figure 57.6**). As a group, these birds had fairly similar lifestyles. Most segregated according to some resource factor, with habitat being the most common one. For example, although all the passerines fed on insects, some would feed exclusively in grasslands, others in forests, some low to the ground, and others high in trees, so the insects present in these locations would be likely to differ somewhat. Birds also segregated by size—so bigger species would take different-sized food from that used by smaller species—and by feeding habit—with some feeding on insects on foliage, others on tree trunks, and so on. Some species also fed in different winter ranges, whereas others occurred in

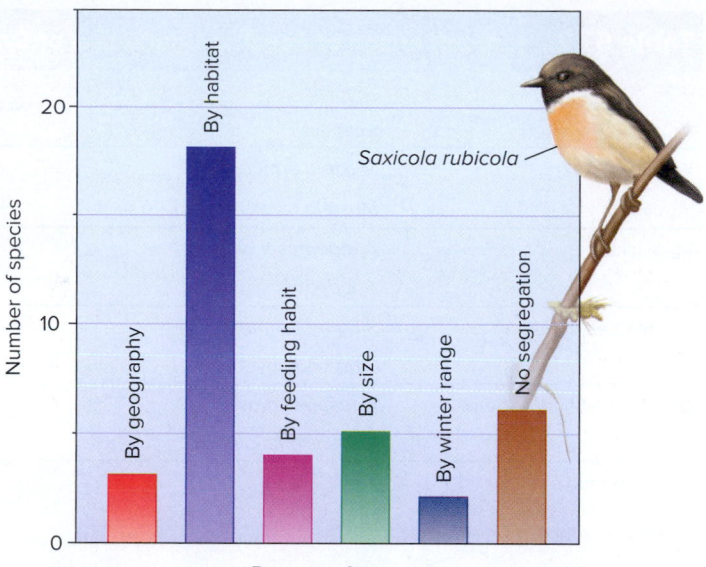

**Figure 57.6  Segregation of 40 bird species according to resource factor.** Among 40 species of passerine birds, most segregation was by habitat, followed by size, feeding habit, geography, and type of winter range. In about 15% of cases, no obvious segregation was observed. More than half of all bird species, including *Saxicola rubicola*, are passerines, also known as perching birds.

**Concept Check:** *Do you think these results for passerine birds are typical for most other species? For example, do most other species segregate by habitat?*

different parts of the country (separation by geography). About 15% of bird species showed no segregation at all.

## Character Displacement Also Allows Species to Coexist

In the case of the five species of *Dendroica* warblers, resources were partitioned in different parts of a tree. Another way for different species to coexist involves variation in morphological features that affects their ability to utilize resources. In 1959, British-born American biologist G. Evelyn Hutchinson examined the sizes of mouthparts or other body parts important in feeding and compared their sizes across species that were **sympatric** (occurring in the same geographic area) and **allopatric** (occurring in different geographic areas). Hutchinson hypothesized that when two species are sympatric, each species will tend to specialize on different types of food. This specialization may be associated with differences in the size of body parts used for feeding, also called feeding characters. The tendency for two species to diverge in morphology and thus resource use because of competition is called **character displacement**. Alternatively, when species are allopatric, there tends to be no selective pressure to specialize on a particular food source, so the size of the feeding character does not evolve to become larger or smaller. Rather, the feeding character is expected to retain a "middle of the road" size that allows the species to exploit the largest range of food types.

One of the classic cases of character displacement involves a study of Galápagos finches, several closely related species of

**Figure 57.5  Resource partitioning.** Among five species of warblers feeding in North American spruce trees, each species prefers to feed at a different height and in a different part of the tree, thus reducing competition.

## Table 57.2    Comparison of Feeding Characters of Sympatric and Allopatric Species

| Animal (character) | Species | Measurement (mm) when | | Ratio* when | |
|---|---|---|---|---|---|
| | | Sympatric | Allopatric | Sympatric | Allopatric |
| Weasels (skull) | *Mustela erminea* | 50.4 | 46.0 | 1.28 | 1.07 |
| | *Mustela nivalis* | 39.3 | 42.9 | | |
| Mice (skull) | *Apodemus flavicollis* | 27.0 | 26.7 | 1.09 | 1.04 |
| | *Apodemus sylvaticus* | 24.8 | 25.6 | | |
| Nuthatches (beak) | *Sitta tephronota* | 29.0 | 25.5 | 1.23 | 1.02 |
| | *Sitta neumayer* | 23.5 | 26.0 | | |
| Galápagos finches (beak) | *Geospiza fortis* | 12.0 | 10.5 | 1.43 | 1.13 |
| | *Geospiza fuliginosa* | 8.4 | 9.3 | | |
| Average ratio | | | | 1.26 | 1.06 |

*Ratio of larger to smaller character

finches Charles Darwin discovered on the Galápagos Islands (refer back to Table 22.1). When two species, *Geospiza fortis* and *Geospiza fuliginosa*, are sympatric, their beak sizes (bill depths) are different: *G. fortis* has a larger bill depth, which enables it to feed on bigger seeds, whereas *G. fuliginosa* has a smaller bill depth, which enables it to crack small seeds more efficiently. However, when these species are allopatric, that is, existing on different islands, their bills are more similar in depth. Researchers studying *Geospiza* concluded that the bill depth differences evolved in ways that minimized competition.

How great must differences between feeding characters be in order to permit coexistence? Hutchinson noted that the ratio between sizes of feeding characters when species were sympatric (and thus competed) averaged about 1.3 (**Table 57.2**). In contrast, the ratio between sizes of feeding characters when species were allopatric (and did not compete) was closer to 1.0. Hutchinson proposed that the

value of 1.3, a roughly 30% difference, could be used as an indication of the amount of difference necessary to permit two species to coexist.

## The Realized Niche Is Smaller Than the Fundamental Niche Due to Competition

Most species perform best over a physiologically optimal range of conditions called the **fundamental niche**. However, if some part of the fundamental niche is occupied by competitors, the range of an organism may be limited to an area known as the **realized niche**, where the competitor is absent. Researchers have established that one of the best methods of determining an organism's fundamental niche is to temporarily remove one of the competing species and examine the effect on the other species. A now-classic example of this method involved a study of the interactions between two species of barnacles conducted on the west coast of Scotland, described next.

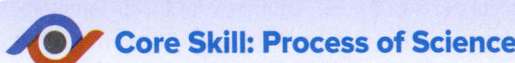

### Core Skill: Process of Science

## Feature Investigation | Connell's Experiments with Barnacle Species Revealed Each Species' Fundamental and Realized Niches

*Chthamalus stellatus* and *Semibalanus balanoides* (formerly known as *Balanus balanoides*) are two species of barnacles that dominate the Scottish coastline. Each organism's realized niche on the intertidal zone is well defined. *Chthamalus* occurs in the upper intertidal zone, and *Semibalanus* is restricted to the lower intertidal zone. Joseph Connell sought to determine if the range of *Chthamalus* adults would be expanded in the absence of competition from *Semibalanus* (**Figure 57.7**).

To do this, Connell obtained rocks from high on the rock face, just below the high-tide level, where only *Chthamalus* grew. These rocks already contained young and mature *Chthamalus*. He then moved the rocks into the *Semibalanus* zone, fastened them down with screws, and allowed *Semibalanus* to also colonize them. Once *Semibalanus* had colonized these rocks, he took the rocks out, removed

all the *Semibalanus* organisms from one side of the rocks with a needle, and then returned the rocks to the lower intertidal zone, screwing them down once again. As seen in the data, the mortality of *Chthamalus* on rock halves with *Semibalanus* was fairly high. On the *Semibalanus*-free halves, however, *Chthamalus* survived well.

In other studies, Connell also monitored survival of natural patches of both barnacle species where both occurred on the intertidal zone at the upper margin of the *Semibalanus* distribution. In a period of unusually low tides and warm weather, when no water reached any barnacles for several days, desiccation became a real threat to both species' survival. During this time, young *Semibalanus* suffered a 92% mortality rate, and older individuals, a 51% mortality rate. At the same time, young *Chthamalus* experienced a 62% mortality rate compared with a rate of only 2% for more-resistant older individuals. Clearly, *Semibalanus* is not

**Figure 57.7** **Connell's experimental manipulation of species indicated the presence of competition.**

**HYPOTHESIS** *Chthamalus stellatus* is being competitively excluded from the lower intertidal zone by the species *Semibalanus balanoides*.

**STARTING LOCATION** The intertidal zone of the rocky shores of the Scottish coast, where the two species of barnacles occur.

Experimental level

**1** Transfer rocks containing young and mature *Chthamalus* from the upper intertidal zone to the lower intertidal zone, and fasten them down in the new location with screws.

**2** Allow *Semibalanus* to colonize the rocks.

**3** After the colonization period is over, remove *Semibalanus* from half of each rock with a needle (leaving the other half undisturbed). Return the rocks to the lower intertidal zone, and fasten them down once again.

**4** Monitor the survival of *Chthamalus* on both sides of the rocks.

*Chthamalus* grows on the side where *Semibalanus* has been removed, indicating that *Semibalanus* may exclude *Chthamalus* from certain habitats.

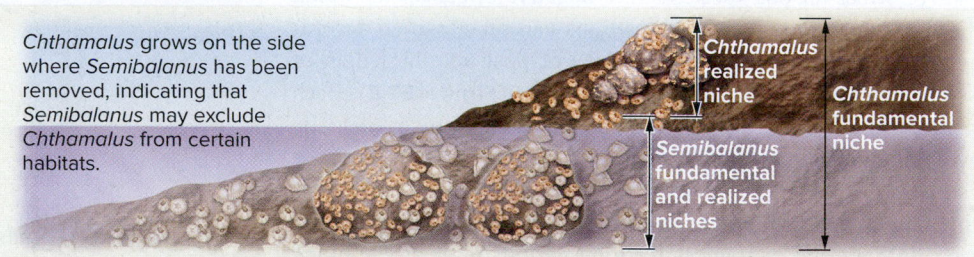

**5** **THE DATA**

| Rock No. | Side of rock | % *Chthamalus* mortality over 1 year | |
| --- | --- | --- | --- |
| | | **Young barnacles** | **Mature barnacles** |
| 13b | *Semibalanus* removed | 35 | 0 |
| | *Semibalanus* not removed | 90 | 31 |
| 12a | *Semibalanus* removed | 44 | 37 |
| | *Semibalanus* not removed | 95 | 71 |
| 14a | *Semibalanus* removed | 40 | 36 |
| | *Semibalanus* not removed | 86 | 75 |

> **6** **CONCLUSION** The data from this study indicate that *Chthamalus* is not found on the lower rock face because of competition with *Semibalanus*. Other studies indicate that *Chthamalus* occupies the upper rock face because it is more resistant to desiccation.
>
> **7** **SOURCE** Connell, J. H. 1971. The influence of interspecific competition and other factors on the distribution of the barnacle *Chthamalus stellatus*. *Ecology* 42: 710–732.

as resistant to desiccation as *Chthamalus* and could not survive in the upper intertidal zone where *Chthamalus* occurs. *Chthamalus* is more resistant to desiccation than *Semibalanus* and can be found higher in the intertidal zone. Thus, whereas the lower limit of *Chthamalus* was set by competition with *Semibalanus*, the upper limit was controlled by desiccation. Although the potential distribution, the fundamental niche, of *Chthamalus* extends over the entire intertidal zone, its actual distribution, the realized niche, is restricted to the upper zone.

**Experimental Questions**

1. Describe the realized niches for the two species of barnacles used in Connell's experiment.

2. Outline the procedure Connell used in the experiments.

3. **CoreSKILL** » How can you explain the presence of *Chthamalus* in the upper intertidal zone if *Semibalanus* was shown to outcompete the species in the first experiment?

## 57.2 Predation, Herbivory, and Parasitism

**Learning Outcomes:**

1. List and describe strategies that animals use to avoid predation.

2. Give examples of the effects of predators on prey populations.

3. Explain why plants and herbivores are said to be in an "evolutionary arms race."

4. Describe variations on the parasitic lifestyle.

Predation, herbivory, and parasitism are interactions that have a positive effect for one species and a negative effect for the other. These categories of species interactions can be classified according to how lethal they are for the prey or host and the length of association between the consumer and prey (**Figure 57.8**). Each has particular characteristics that set it apart. Herbivory usually involves nonlethal predation on plants, whereas predation generally results in the death of the prey. Parasitism, like herbivory, is typically nonlethal and differs from predation in that the adult parasite typically lives and reproduces for long periods in or on the living host (refer back to Figure 34.8). Parasitoids, insects that lay eggs in living hosts, have features in common with both predators and parasites. They always kill their prey, as predators do, but unlike predators, which immediately kill their prey, parasitoids kill the host more slowly. Parasitoids are common in the insect world and include parasitic wasps and flies that feed on many other insects such as caterpillars.

In this section, we will begin by looking at antipredator strategies and see how, despite such strategies, predation remains a key factor affecting the density of prey. We will survey the strategies plants use to deter herbivores and, in turn, the strategies herbivores use to overcome host plant defenses. Finally, we will investigate parasitism, which may be the predominant lifestyle on Earth.

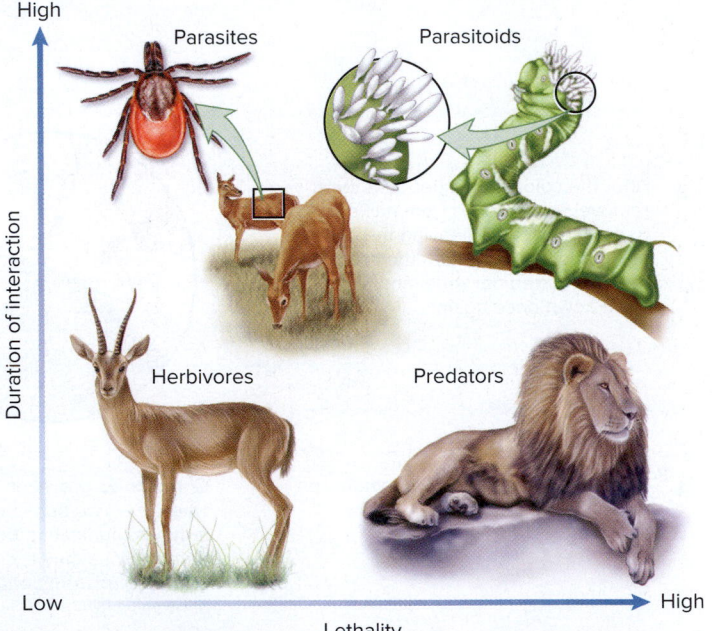

**Figure 57.8 Possible interactions between species.** Lethality represents the probability that an interaction results in the death of the prey. Duration represents the length of the interaction between the consumer and the prey.

*Concept Check:* *Where might omnivores fit in this figure?*

### Animals Have Evolved Many Antipredator Strategies

The variety of strategies that animals have evolved to avoid being eaten suggests that predation is a strong selective force. Common strategies include chemical defense, camouflage, mimicry, displays of intimidation, and armor and weaponry.

***Chemical Defense*** A great many species have evolved chemical defenses against predation. A dramatic example of a chemical defense is that employed by the bombardier beetle (*Stenaptinus insignis*), which has been studied by German-born American entomologist Tom Eisner and coworkers. These beetles possess a reservoir of hydroquinone and hydrogen peroxide in their abdomen. When threatened, they eject the chemicals into an "explosion chamber," where the subsequent release of oxygen causes the whole mixture to be violently ejected as a hot spray (about 88°C, or 190°F) that can be directed at the beetle's attacker (**Figure 57.9a**). Many other arthropods, such as millipedes, also have chemical sprays, and this type of defense is also found in vertebrates, as anyone who has had a close encounter with a skunk can testify.

(a) This bombardier beetle (*Stenaptinus insignis*) directs its hot, stinging spray at a forceps "attacker."

(b) Aposematic coloration advertises the poisonous nature of this blue poison arrow frog (*Dendrobates azureus*) from South America.

(c) Camouflage allows this pygmy sea horse (*Hippocampus bargibanti*) from Bali to blend in with its background.

(d) In a display of intimidation, this porcupine fish (*Diodon hystrix*) puffs itself up to look threatening to its predators.

(e) Müllerian mimicry. Viceroy (left) and monarch (right) butterflies are both noxious and have similar color patterns.

(f) In this example of Batesian mimicry, an innocuous scarlet king snake (*Lampropeltis elapsoides*) (left) mimics the venomous eastern coral snake (*Micrurus fulvius*) (right).

(g) Sable antelope have horns that may be used as predator defense.

**Figure 57.9 Antipredator adaptations.** a: ©CB2/ZOB/Supplied by WENN/Newscom; b: ©Hans D. Dossenbach/ardea.com; c: ©Thomas Aichinger/V&W/The Image Works; d: ©Paul Springett/Alamy Stock Photo; e (left): ©Rick & Nora Bowers/Alamy Stock Photo; e (right): ©Eric Carr/Alamy Stock Photo; f (left): ©Suzanne L. & Joseph T. Collins/Science Source; f (right): ©Jason Ondreicka/Alamy Stock Photo; g: ©cd123/123RF

**Concept Check:** *According to the classification of species interactions in Table 57.1, how would you classify Batesian and Müllerian mimicry?*

 **Core Skill: Connections** Refer back to Figure 34.12. Which types of antipredator adaptations are possessed by mollusks?

Often associated with a chemical defense is **aposematic coloration**, or warning coloration, which advertises an organism's unpalatable taste. For instance, the ladybird beetle's bright red color warns of the toxic defensive chemicals it exudes when threatened, and many tropical frogs have bright warning coloration that calls attention to their skin's lethality (**Figure 57.9b**). Monarch butterfly caterpillars feed exclusively on milkweed, which contains toxic chemicals called cardiac glycosides that pass into the caterpillars. In the 1960s, American entomologist Lincoln Brower and coworkers showed that after inexperienced blue jays ate a monarch butterfly and suffered a violent vomiting reaction, they learned to associate the striking orange-and-black appearance of the butterfly with a noxious reaction (refer back to Figure 55.3).

**Camouflage**   Camouflage is the blending of an organism with the background of its habitat and is a common method of avoiding detection by predators. For example, many grasshoppers are green and blend in with the foliage on which they feed. Stick insects mimic branches and twigs with their long, slender bodies. In most cases, these animals stay perfectly still when threatened, because movement alerts a predator. Camouflage is prevalent in the vertebrate world, too. Many sea horses adopt a body shape and color pattern similar to the environment in which they are found (**Figure 57.9c**).

**Displays of Intimidation**   Some animals put on displays of intimidation in an attempt to discourage predators. For example, a cat arches its back, a frilled lizard extends its collar, and a porcupine fish inflates itself when threatened in order to appear larger (**Figure 57.9d**). All of these animals use displays to deceive potential predators about the ease with which they can be eaten.

**Mimicry**   Mimicry, the resemblance of a species (the mimic) to another species (the model), also secures protection from predators. There are two major types of mimicry. In **Müllerian mimicry**, two or more toxic species converge to look the same, thus reinforcing the basic distasteful design. One example is the black-and-yellow-striped bands of several different types of bees and wasps. The viceroy butterfly (*Limenitis archippus*) and the monarch butterfly (*Danaus plexippus*) are examples of Müllerian mimicry. Both species are unpalatable and look similar, but the viceroy can be distinguished from the monarch by a black line that crosses its hindwings (**Figure 57.9e**).

   **Batesian mimicry** is the mimicry of an unpalatable species (the model) by a palatable one (the mimic). Some of the best examples involve flies, especially hoverflies of the family Syrphidae, which are striped black and yellow and resemble stinging bees and wasps but are themselves harmless. Among vertebrates, the nonvenomous scarlet king snake (*Lampropeltis elapsoides*) mimics the venomous coral snake (*Micrurus fulvius*), thereby gaining protection from would-be predators (**Figure 57.9f**).

**Armor and Weaponry**   Physical defenses that prey use against predators are varied and numerous. The shells of tortoises and freshwater turtles are an effective means of defense against most predators. Though many animals developed horns and antlers for sexual selection, they can also be used in defense against predators (**Figure 57.9g**). Invertebrate species often have powerful claws, pincers, or, in the case of scorpions, venomous stingers that can be used in defense as well as offense.

## Despite the Impressive Array of Defenses, Predators Can Still Affect Prey Densities

Research studies have shown that predators can have a significant effect on prey populations. Considerable data exist on the interaction of the Canada lynx (*Lynx canadensis*) and its prey, snowshoe hares (*Lepus americanus*), because of the value of the pelts of both animals. In 1942, British ecologist Charles Elton analyzed the records of furs traded by trappers to the Hudson's Bay Company in Canada over a 100-year period. Analysis of the records showed that a dramatic 9- to 11-year cycle existed for as long as records had been kept (**Figure 57.10**). As hare density increases, there is an increase in density of the lynx, which then depresses hare numbers. This is followed by a decline in the number of lynx, and the cycle begins again. Using radio collars to track individual hares, researchers were able to determine that 90% of individuals died of predation.

   Ecologists have found that in nearly 1,500 predator-prey studies, over two-thirds (72%) showed a large depression of prey density by predators. Thus, we can conclude that in the majority of cases, predators influence the abundance of their prey in their native environment. The variety of antipredator mechanisms discussed earlier also shows how predation is important enough to select for the evolution of chemical defenses, camouflage, and mimicry in prey. Taken together, these data indicate that predation is a powerful force in nature.

## Plants and Herbivores May Be Engaged in an Evolutionary Arms Race

Herbivory involves the consumption of plant material or the material of similar life-forms such as algae. Herbivory can be lethal to plants, especially for small species. However, it is often nonlethal, because many plant species, particularly larger ones, can regrow. Plants present a massive food resource to any organism versatile enough to use it, so why don't herbivores eat more of the food available to them? After all, unlike most animals, plants cannot move to escape being eaten.

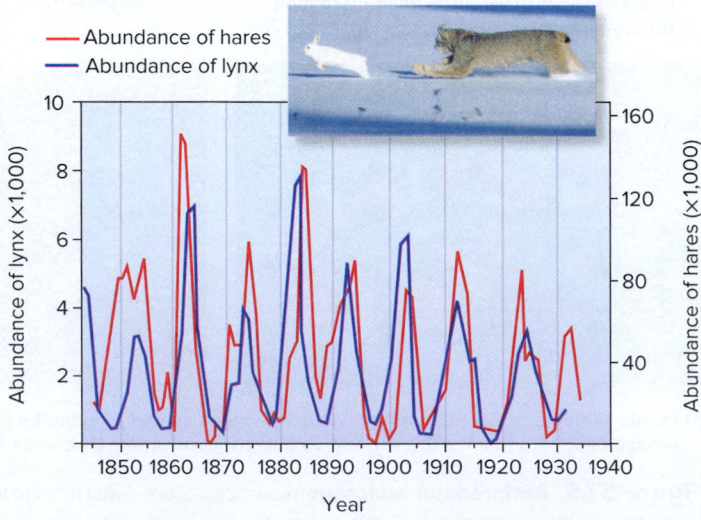

**Figure 57.10   Relationship between a predator and a prey population.**   The 9- to 11-year oscillation in the abundance of the snowshoe hare (*Lepus americanus*) and the Canada lynx (*Lynx canadensis*) was revealed from pelt trading records of the Hudson's Bay Company. ©Alan & Sandy Carey/Science Source

Two hypotheses have been proposed to answer the question of why more plant material is not eaten. First, predators and parasites may keep herbivore numbers low, thereby sparing the plants. The many examples of the strength of predation provide evidence for this view. Second, the plant world is not as helpless as it appears. The sea of green is armed with defensive spines, tough cuticles, noxious chemicals, and more. Let's take a closer look at plant defenses against herbivores and the ways herbivores attempt to overcome them.

***Plant Defenses Against Herbivores*** A plant's first line of defense is mechanical. Many plants have effective mechanical defenses, such as thorns and spines (**Figure 57.11a**). In addition, tough fibers discourage herbivore feeding, and grasses and palms sequester silica, making their foliage difficult to chew. Furthermore, an array of unusual and powerful chemicals is present in plants, including alkaloids (nicotine in tobacco, morphine in poppies, cocaine in coca, and caffeine in coffee), phenolics (lignin in wood and tannin in leaves), and terpenoids (in peppermint) (**Figure 57.11b–d**). Such compounds are not part of the primary metabolic pathway that plants use to obtain energy and are therefore referred to as **secondary metabolites**. Most of these chemicals are bitter tasting or toxic, thereby deterring herbivores from feeding. The staggering variety of secondary metabolites in plants, over 25,000, may be testament to the large number of organisms that feed on plants. In an interesting twist, many of these compounds have medicinal properties that have proved to be beneficial to humans.

An understanding of plant defenses is of great use to agriculturalists. The better a crop plant can defend against pests, the higher the crop yield. The ability of plants to prevent herbivory via either chemical or mechanical defenses is also known as **host plant resistance**. One serious problem associated with commercial development of host plant resistance is that it may take a long time to breed into plants—between 10 and 15 years. This time frame results because of the time it takes to identify the responsible chemicals and develop the resistant genetic lines. Also, resistance to one pest may come at the cost of increasing susceptibility to other pests. Finally, some pest strains can overcome the plant's mechanisms of resistance.

Despite these problems, host plant resistance is a good tactic for farmers. Host plant resistance reduces the need for chemical insecticides and is less environmentally harmful, generally having few side effects on other species in the community. About 75% of cropland in the U.S. utilizes pest-resistant plant varieties, most of these being resistant to plant pathogens. For example, Bt corn is a variety of corn that has been genetically modified by incorporating a gene from the soil bacterium *Bacillus thuringiensis* (Bt) that encodes a protein, Bt toxin, that is toxic to some insects. Genetic engineers have also produced Bt cotton, Bt tomato, and genetically modified varieties of many other crop species.

***Overcoming Host Plant Resistance*** Herbivores can sometimes overcome plant defenses. They detoxify poisons mainly by two chemical pathways: oxidation and conjugation. Oxidation, the most important of these mechanisms, occurs in the liver of mammals and in the midgut of insects. It involves catalysis of the secondary metabolite to a corresponding alcohol by a group of enzymes known as mixed-function oxidases (MFOs). Conjugation, often the next step in detoxification, occurs by uniting the harmful compound or its oxidation product with another molecule to create an inactive and readily excreted product.

**(a) Thorns on rose stems**

**(b) Alkaloids in tobacco**

**(c) Phenolics in tea**

**(d) Terpenoids in peppermint**

**Figure 57.11** **Defenses against herbivory.** **(a)** Mechanical defenses include plant spines and thorns, as on this shrub, *Rosa multiflora*. Plants also possess an array of unusual and powerful chemicals, including **(b)** alkaloids, such as nicotine in tobacco, **(c)** phenolics, such as tannins in tea leaves (near Mount Fuji, Japan), and **(d)** terpenoids in peppermint leaves. a: ©John Dudak/Phototake; b: ©Glen Allison/The Image Bank/Getty Images; c: ©Toyofumi Mori/The Image Bank/Getty Images; d: ©Gilbert S. Grant/Science Source

*Concept Check:* *Of the defenses shown here, which type would be most effective in deterring invertebrate herbivores?*

 **Core Skill: Connections** Which species of plant produces the alkaloid capsaicin? Refer back to Figure 32.22.

## Herbivores May Have Dramatic Effects on Plant Populations

A good method for estimating the effects of herbivory on plant populations is to remove the herbivores and examine subsequent plant growth. A conclusion from such experiments is that invertebrate herbivores, such as insects, usually have a stronger effect on plant populations than vertebrate herbivores such as mammals, at least in terrestrial systems. Thus, although large grazers like bison in North America or antelopes in Africa might seem to be of huge importance in grasslands, it is more likely that grasshoppers are the more significant herbivores because of their sheer weight of numbers. In forests,

**(a) Before biological control**   **(b) After biological control**

**Figure 57.12** **The effects of an insect herbivore on an invasive plant pest.** The prickly pear cactus, *Opuntia stricta*, in Chinchilla, Australia, **(a)** before control by the cactus moth, *Cactoblastis cactorum*, in 1928, and **(b)** after control, in 1929. a: Source: State Library of Queensland, neg. no. API-101-01-0001r; b: Source: State Library of Queensland, neg. no. API-101-01-0002r

 **Core Skill: Science and Society** Control of prickly pear by the cactus moth cleared the cacti from hundreds of thousands of hectares of rangeland, allowing sheep to graze the area and farming to thrive.

invertebrate grazers such as caterpillars have greater access to canopy leaves than vertebrates and are also likely to have a greater effect.

An example of the strong effects of insect herbivores on host plants was provided by changes in the population size of the prickly pear cactus, *Opuntia stricta*. In 1839, this cactus was imported from the Americas to Australia to start a dye industry. Small cochineal insects, which fed on the cactus, could be crushed to collect a red dye. Unfortunately, the cactus spread and became a serious pest. By 1925, it occupied 240,000 km$^2$ of rangeland, which was rendered useless for sheep and cattle grazing. In 1925, control measures were initiated by introducing the cactus moth, *Cactoblastis cactorum*, from South America. By 1932, the stands of prickly pear had been destroyed (**Figure 57.12**).

## Parasitism Usually Involves a Long-Term, Nonlethal Relationship Between a Parasite and Its Host

A **parasite** feeds on another organism, called the host, for a relatively long time, but does not normally kill it outright. Some parasites remain attached to their hosts for most of their life. For example, tapeworms spend their entire adult life inside the host's alimentary canal and even reproduce within their host. Others, such as the Chinese liver fluke, have more complex life cycles that require multiple hosts (refer back to Figure 34.8). To facilitate transmission, many parasites induce changes in the behavior of one host, making that host more susceptible to being eaten by a second host. For example, rodents infected with the brain parasite, *Toxoplasma gondii*, are more active and less fearful of cats and their smells, which increases the chances that infected rodents will be eaten by a cat, the parasite's definitive host.

The malaria parasite, a single-celled species in the genus *Plasmodium*, has a complex life cycle involving two hosts, mosquitoes and

vertebrates (refer back to Figure 28.29). *Plasmodium* interferes with the ability of the mosquito to draw blood from its vertebrate hosts. This increases the number of attacks the infected mosquitoes make in order to try and obtain enough blood. Increased attack rates maximize the transmission rates of the *Plasmodium* itself. British epidemiologist Jacob Koella and colleagues showed that most uninfected mosquitoes generally feed on just one human host at night, and only 10% bite more than one person. Multiple biting of different hosts increased to 22% in malaria-infected mosquitoes. In addition, the saliva of the infected mosquitoes was changed, making the host's blood flow less freely into the mouthparts. Similar behavior is exhibited by leishmaniasis parasites in sand flies and bubonic plague parasites in fleas.

Some flowering plants are parasitic on other plants. **Holoparasites** lack chlorophyll and are totally dependent on host plants for their water and nutrients. One famous holoparasite is *Rafflesia arnoldii*, which lives most of its life within the body of its host, a *Tetrastigma* vine, which grows in tropical rain forests (**Figure 57.13**). Only the *Rafflesia* flower develops externally. It is a massive flower, 1 m in diameter, and the largest known in the world. **Hemiparasites** are usually able to carry out photosynthesis, but depend on their hosts for water and mineral nutrients. Mistletoe (*Viscum album*) is a hemiparasite that grows on the stems of trees. Hemiparasites typically have a broader range of hosts than do holoparasites, which may be confined to a single or a few host species.

Parasites that feed on one species or just a few closely related hosts are termed **monophagous**. By contrast, **polyphagous** species can feed on many different host species, often from more than one family. We can also distinguish parasites as **microparasites** (for example, pathogenic bacteria), which multiply within their hosts, sometimes within the host cells, and **macroparasites** (such as schistosomes), which live in the host but release infective juvenile stages outside the host's body. Usually, the host has a strong immunological response to microparasitic infections. For macroparasitic infections, however,

**Figure 57.13** **A holoparasite.** *Rafflesia arnoldii*, the world's biggest flower, is produced by a holoparasite in Indonesian rain forests. ©A & J Visage/Alamy Stock Photo

 **Core Skill: Connections** More than 4,500 species of plants live as complete or partial parasites. What is an example of another important parasitic plant that was discussed in Chapter 38? Refer back to Figure 38.20.

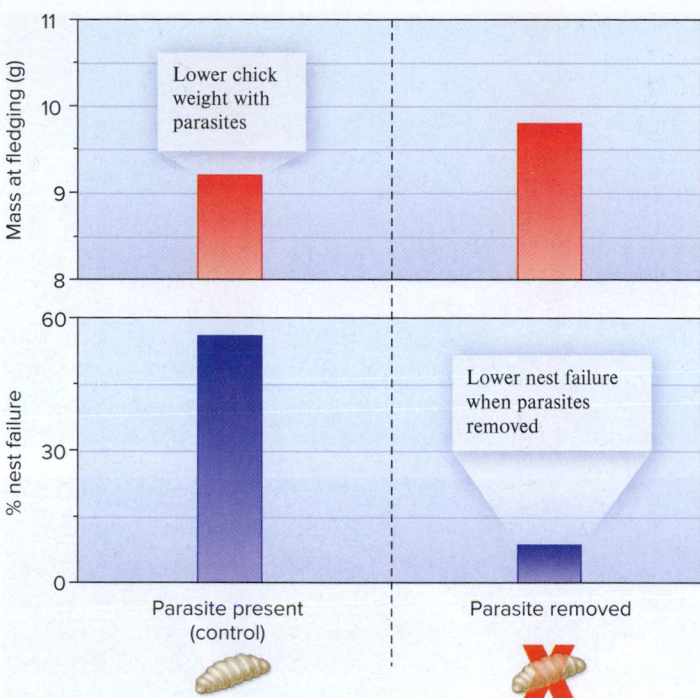

**Figure 57.14  Parasite removal experiments.** The left side shows the results when blowfly larvae were present in the nests of blue tits. The right side shows the results when these parasites were removed.

the immunological response is short-lived. Such infections tend to be persistent, and the hosts are subject to continual reinfection.

Last, we can distinguish **ectoparasites**, such as ticks and fleas, which live outside of the host's body, from **endoparasites**, such as pathogenic bacteria and tapeworms, which live inside the host's body. Problems of definition arise with regard to plant parasites, which seem to straddle both camps. For example, some parasitic plants, such as mistletoe, exist partly outside of the host's body and partly inside. Outgrowths called haustoria penetrate inside the host plant to tap into nutrients. Being endoparasitic on a host seems to require greater specialization than ectoparasitism. Therefore, ectoparasitic animals such as leeches feed on a wider variety of hosts than do endoparasites such as liver flukes.

Parasitism is a common way of life. Vast numbers of species are parasites, including bacteria, protozoa, flatworms (flukes and tapeworms), nematodes, and various arthropods (ticks, mites, and fleas). Most plant and animal species harbor many parasites. For example, on average, each mammalian species hosts two cestode species, two trematodes, four nematodes, and one acanthocephalan. For birds, the numbers are even higher. With the exception of single-celled microorganisms, a free-living organism that does not harbor multiple parasites is a rarity.

## Parasitism Has Strong Effects on Host Populations

As in studies of other species interactions, a direct method for determining the effect of parasites on their host population is to remove the parasites and to re-examine the population. However, this is difficult to do, primarily because of the small size and unusual life histories of many parasites, which make them difficult to remove from a host completely. The few cases of experimental removal confirm that parasites

can reduce host population densities. For example, the nests of birds such as blue tits (*Cyanistes caeruleus*) are often infested with parasitic blowfly larvae that feed on the blood of nestlings. In 1997, French biologist Sylvie Hurtrez-Bousses and colleagues experimentally reduced blowfly larval parasites of young blue tits in nests in Corsica. Parasite removal was cleverly achieved by taking the nests from 145 nest boxes, removing the young, microwaving the nests to kill the parasites, and then returning the nests and chicks to the wild. The success of chicks in microwaved nests was compared with that in nonmicrowaved (control) nests. The parasite-free blue tit chicks had greater body mass at fledging, the time when feathers first grow (**Figure 57.14**). Perhaps more important was the observation that complete nest failure, that is, death of all chicks, was much higher in control nests than in treated nests.

## 57.3  Mutualism and Commensalism

### Learning Outcomes:

1. Distinguish between the different types of mutualism in nature.
2. Give examples of the species interaction called commensalism.

In this section, we will examine interactions that are beneficial to at least one of the species involved. In mutualism, both species gain from the interaction. For example, in mutualistic pollination systems, the plant benefits by the transfer of pollen, and the pollinator typically gains a nectar meal. In commensalism, one species benefits, and the other remains unaffected. For example, in some forms of seed dispersal, barbed seeds are transported to new germination sites in the fur of mammals. The plant species benefit by dispersing their progeny, but the mammals are generally unaffected.

In the field of agriculture, humans have entered into mutualistic relationships with many species. For example, the association of humans with plants has resulted in some of the most far-reaching ecological changes on Earth. Humans have planted huge areas of the Earth with crops, allowing these plant populations to reach densities they never would attain on their own. In return, the crops have led to expanded human populations because of the increased amounts of food they provide.

### Mutualism Is an Association Between Two Species That Benefits Both

Different types of mutualisms occur in nature. In **resource-based mutualisms**, both species receive a benefit in the form of resource transfer of energy and nutrients. In **defensive mutualisms**, one species receives food or shelter in return for defending another species; they often involve an animal defending a plant or an herbivore. **Dispersive mutualisms** are interactions in which a species receives food in return for transporting the pollen or seeds of its partner.

***Resource-Based Mutualism***  Leafcutter ants of the group Attini, of which there are about 210 species, enter into a mutualistic relationship with a fungus. A typical colony of about 9 million ants has the collective biomass of a cow and harvests the equivalent of a cow's daily requirement of fresh vegetation. Instead of consuming the leaves directly, however, the ants chew them into a pulp, which they store underground as a substrate on which the fungus grows

(a)

(b)

**Figure 57.15  Resource-based mutualism. (a)** Leafcutter ants, *Atta cephalotes*, cut leaves and chew them to a pulp underground. **(b)** The ants feed on structures produced by fungi, which develop in the pulp. a: ©Bryan Mullennix/The Image Bank/Getty Images; b: ©Ralph Clevenger/Corbis Documentary/Getty Images

**Core Skill: Systems** The mutualism shown in Figure 57.15 benefits both species.

(**Figure 57.15a**). The ants shelter and tend the fungus, helping it reproduce and grow and weeding out competing fungi. In return, the fungus produces specialized structures known as gongylidia, which serve as food for the ants (**Figure 57.15b**). In this way, the ants circumvent the chemical defenses of the leaves, which are digested by the fungus.

*Defensive Mutualism*   A commonly observed mutualism occurs between ants and aphids. Aphids are fairly defenseless creatures and are easy prey for most predators. The aphids feed on plant sap and have to process a significant amount of it to get their required nutrients. In doing so, they excrete a lot of fluid, and some of the sugars still remain in the excreted fluid, which is called "honeydew." The ants drink the honeydew and, in return, protect the aphids from an array of predators, such as ladybird beetle larvae, by driving the predators away. In some cases, the ants herd the aphids like cattle, moving them from one area to another (**Figure 57.16a**).

In other cases, ants enter into a mutualistic relationship with a plant itself. One of the most famous examples involves acacia trees in Central America, whose large thorns provide food and nesting sites for ants (**Figure 57.16b**). In return, the ants bite and discourage both insect and vertebrate herbivores from feeding on the trees. They also trim away foliage from competing plants and kill neighboring plant shoots, ensuring more light, water, and nutrient supplies for the acacias.

**Ants defending an acacia plant in exchange for food and shelter**

**Figure 57.16  Defensive mutualism. (a)** This red carpenter ant, *Camponotus ferrugineus*, tends aphids feeding on a twig. The ants consume the sugar-rich honeydew produced by the aphids and in return protect the aphids from predators. **(b)** Ants, usually *Pseudomyrmex ferruginea*, make nests inside the large, hornlike thorns of the bull's horn acacia and defend the plant against insects and mammals. In return, the acacia (*Acacia collinsii*) provides two forms of food to the ants: protein-rich granules called Beltian bodies and nectar from extrafloral nectaries (nectar-producing glands that are located away from the flower). a: ©Ed Reschke/Photolibrary/Getty Images; b: ©Gregory G. Dimijian, M.D./Science Source

**Concept Check:** *Is the relationship between ants and bull's horn acacia an example of facultative or obligatory mutualism?*

*Dispersive Mutualism*   Many examples of plant-animal mutualisms involve pollination and seed dispersal. In these cases, neither species can live without the other, a concept called **obligatory mutualism**. This contrasts with **facultative mutualism**, in which the interaction is beneficial but not essential to the survival and reproduction of either species.

From a plant's perspective, an ideal pollinator would be a specialist, moving quickly among individuals but retaining a high fidelity to a plant species. Two ways that plant species in an area promote the pollinator's species fidelity is by synchronized flowering within a species and by sequential flowering of different species through the year. Synchronized flowering within a species means that all flowers of one species are available over a relatively short time period whereas sequential flowering ensures that only one species of flower

**Figure 57.17** **Dispersive mutalism.** This blackbird (*Turdus merula*) is an effective seed disperser. ©Mike Wilkes/naturepl.com

 **Core Skill: Connections** Refer back to Table 32.2. Birds have excellent color vision and are also involved in the pollination of flowers. What is a common color of the flowers of bird-pollinated plants?

is available to pollinators at any one time. The plant should provide just enough nectar to attract a pollinator's visit. From the pollinator's perspective, it would be best to be a generalist and obtain nectar and pollen from as many flowers as possible in a small area, thus minimizing the energy spent on flight between patches. This suggests that although dispersive mutualisms are beneficial to both species, the species' optimal needs are quite different.

Mutualistic interactions are also common in the seed-dispersal systems of plants. Fruits provide organic nutrients, minerals, and water. In return for this juicy meal, animals unwittingly disperse the enclosed seeds, which pass through the digestive tract unharmed. Fruits eaten by birds and mammals often have attractive colors (**Figure 57.17**). Those that attract nocturnal bats are not brightly colored but instead give off a pungent odor.

## In Commensalism, One Partner Receives a Benefit While the Other Is Unaffected

Commensalism is an interaction between species in which one benefits and the other is neither helped nor harmed. Such is the case when orchids or other epiphytes grow in forks of tropical trees. The tree is unaffected, but the orchid gains support and increased exposure to sunlight and rain. Cattle egrets feed in pastures and fields among cattle, whose movements stir up insect prey for the birds. The egrets benefit from the association, but the cattle generally do not. One of the best examples of commensalism involves **phoresy**, in which one organism uses a second organism for transportation. Hummingbird flower mites feed on the pollen of flowers and travel between flowers in the nostrils (nares) of hummingbirds. The flowers that the mites inhabit live only a short while before dying, so the mites relocate by scuttling into the nares of visiting hummingbirds and hitching a ride to the next flower. When the hummingbird visits a new flower, the mites disembark. Presumably, the hummingbirds are unaffected.

Some commensalisms involve one species "cheating" on the other without harming it. In the bogs of Maine, the grass-pink orchid (*Calopogon pulchellus*) produces no nectar, but it mimics

**(a) An orchid without nectar mimicking a female bee**

**(b) Seed dispersal via hooked seeds**

**Figure 57.18** **Commensalisms.** (a) Bee orchids (*Ophrys apifera*) mimic the shape of a female bee. Male bees copulate with the flowers, transferring pollen but getting no nectar reward. (b) Hooked seeds of burdock (*Arctium minus*) have lodged in the fur of a white-footed mouse (*Peromyscus leucopus*). The plant benefits from the relationship by the dispersal of its seeds, and the animal is not affected. a: ©E. A. Janes/age fotostock; b: ©Dwight Kuhn

the nectar-producing rose pogonia (*Pogonia ophioglossoides*) and is therefore still visited by bees. Another example involves bee orchids (*Ophrys apifera*) that mimic the appearance and scent of female bees. Males try to copulate with the flowers and in the process pick up and transfer pollen (**Figure 57.18a**). The stimuli of the bee orchid flowers are so effective that male bees prefer to mate with them even in the presence of actual female bees! Many plants have essentially cheated their potential mutualistic seed-dispersal agents out of a meal by developing seeds with barbs or hooks that lodge in the animals' fur or feathers rather than their stomachs (**Figure 57.18b**). In these cases, the plants receive free seed dispersal, and the animals receive nothing, except perhaps minor annoyance. This type of relationship is fairly common; most hikers and dogs have at some time gathered spiny or sticky seeds as they wandered through woods or fields.

## 57.4 Bottom-Up and Top-Down Control

### Learning Outcome:

1. Explain the differences between bottom-up control and top-down control as conceptual models of how species interactions limit population sizes.

In this chapter, we have seen that interactions between species, such as competition, predation, and parasitism, are important in nature. Let's return to a question posed in Chapter 56: How can ecologists determine which factors, along with abiotic factors such as temperature and moisture, are the most important in affecting population sizes? The question is also asked by many applied biologists, such as foresters, marine biologists, and conservation biologists, who are interested in managing a population's size.

Some ecologists stress the importance of so-called bottom-up factors, such as plant quality and abundance in determining the sizes

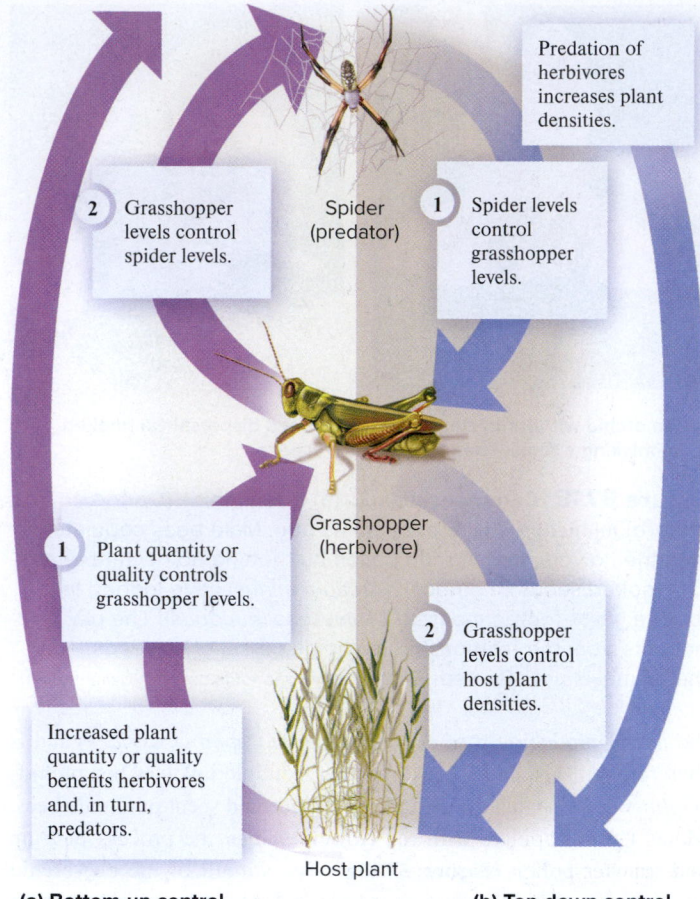

**Figure 57.19** **Bottom-up control versus top-down control.**
**(a)** Bottom-up control proposes that host plant quantity or quality limits the density of herbivores, such as grasshoppers, which, in turn, sets limits on the abundance of predators, such as spiders. **(b)** Top-down control proposes that predators limit the number of herbivores, which, in turn, increases host plant density.

> *Concept Check:* *You apply fertilizer to a bush, and this increases spider density on the bush. What type of control is occurring?*

of herbivore populations and the populations of predators that feed on them. In the example of **Figure 57.19a**, plant quantity or quality limits the density of herbivores, such as grasshoppers, which, in turn, sets limits on the abundance of predators, such as spiders. Other ecologists stress the importance of top-down factors (**Figure 57.19b**). In this case, the spiders play a major role in determining the size of grasshopper populations, which control the size of the plant population.

Current thinking is that both bottom-up and top-down control are important in affecting population sizes. Their relative importance depends on the environment and the types of species interactions that are involved. In this section, we will briefly discuss some of the evidence for bottom-up and top-down control.

## Bottom-Up Control Suggests That Food Limitation Influences Population Densities

At least two lines of evidence suggest that bottom-up control is important in limiting population sizes. First, as discussed in Chapter 58, not all of the energy in food resources can be used by the consumer. For example,

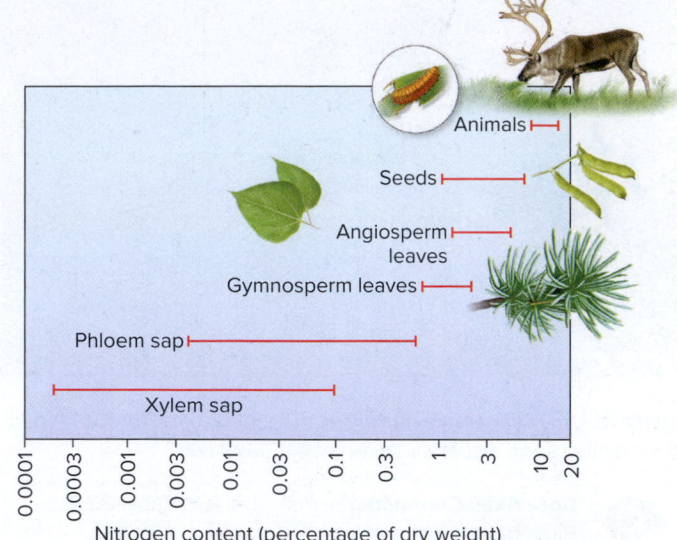

**Figure 57.20** **Nitrogen content of plants and animals.**

when a herbivore eats plant material, some of the energy is lost in the form of heat. This phenomenon, based on the thermodynamic properties of energy transfer, suggests that the quantity and quality of plants may be limiting for the population sizes of all other species that rely on them.

Second, much evidence supports the **nitrogen-limitation hypothesis** that organisms select food in terms of the nitrogen content of the tissue. This is largely due to the different proportions of nitrogen in plants and animals (**Figure 57.20**). Animal tissue generally contains about 10 times as much nitrogen as plant tissue. For this reason, herbivores favor high-nitrogen plants. The addition of nitrogen-containing fertilizer has repeatedly been shown to benefit herbivores. Nearly 60% of 186 studies investigating the effects of fertilization on herbivores reported that increasing a plant's tissue nitrogen concentration through fertilization had strong positive effects on herbivore population sizes, survivorship, growth, and fecundity. Taken together, both lines of evidence indicate that plant densities influence herbivore densities, which in turn influence the densities of carnivores that feed on herbivores, and the carnivores that feed on other animals.

## Top-Down Control Suggests That Natural Enemies Influence Population Densities

Top-down models suggest that predators control populations of their prey (ultimately, herbivores) and that these herbivores control plant populations. Supporting evidence comes from studies of predator removal and addition. Wolves, *Canus lupus*, are being reintroduced into many areas of the United States, and their populations are growing. In a 2005 study, Mark Hebblewhite and colleagues observed colonization of certain areas in Banff National Park in Canada by wolves and compared these areas to those closer to human settlements, which wolves avoid. The wolves reduced the density of elk, *Cervus elaphus*, which in turn promoted the growth of two major elk food plants: aspen, *Populus tremuloides*, and willow, *Salix* spp. (**Figure 57.21a**). Increased plant availability also increased the abundance of songbirds, especially obligate willow specialists, such as the American redstart, *Setophaga ruticilla* (**Figure 57.21b–e**). Beaver abundance also increased due to a greater availability of willow.

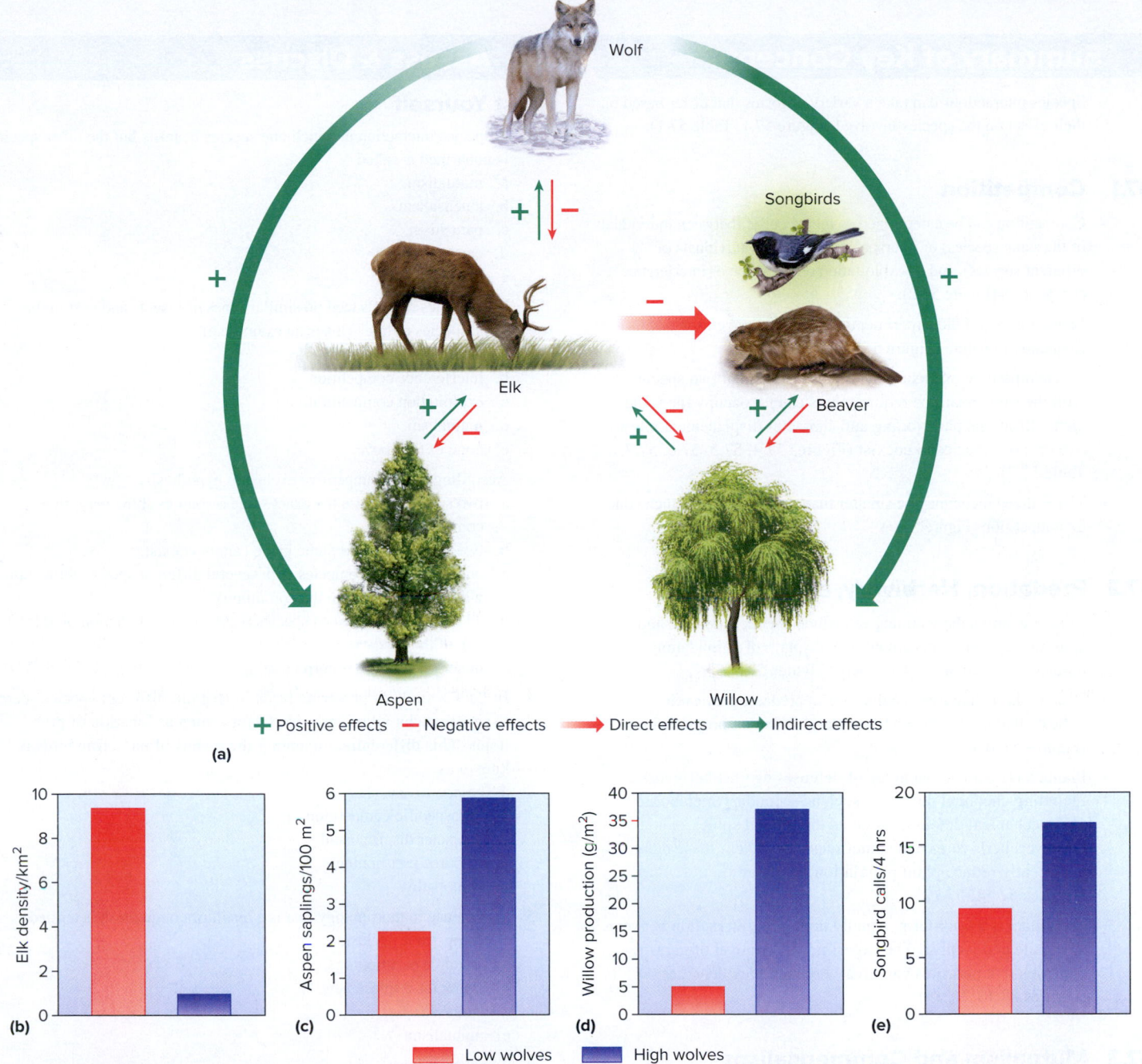

**Positive effects**     **Negative effects**     **Direct effects**     **Indirect effects**

(a)

(b) Elk density/km$^2$

(c) Aspen saplings/100 m$^2$

(d) Willow production (g/m$^2$)

(e) Songbird calls/4 hrs

**Low wolves**     **High wolves**

**Figure 57.21 Studies on wolf addition provide support for the top-down model.** (a) The simplified trophic interactions of wolves, elk, plants, and other organisms in Banff National Park, Canada. Increased wolf abundance caused (b) decreased elk abundance, (c) increased abundance of aspen trees and (d) willow trees, and (e) increased songbird abundance.

**Core Skill: Modeling** The goal of this modeling challenge is to use your understanding of top-down control to construct a simple bar graph that predicts the relative abundances of plants and herbivores under three different conditions: plants alone, plant plus herbivores, and plants plus herbivores and carnivores that prey on the herbivores.

**Modeling Challenge:** Let's assume that a group of plants, herbivores, and carnivores that feed on those herbivores, are subject to top-down control. An ecologist studies the relative numbers of plants in three different areas of the same size. In one area, both the herbivores and carnivores have been removed. In a second area, only the carnivores have been removed. In a third area, the plants, herbivores, and carnivores are all present. Draw a series of bar graphs that predict the relative population sizes of the plant and herbivore populations in these three different areas. The y-axis should be labeled "Relative population size," and three pairs of bars, one pair for each of the three areas, should appear along the x-axis. Within each pair of bars, the left bar should indicate the size of the plant population and the right bar should show the size of the herbivore population. Note: The y-axis does not need to have numbers on it; the goal is to predict the relative population sizes, not the actual numbers.

## Summary of Key Concepts

- Species interactions can take a variety of forms that differ based on their effect on the species involved (Figure 57.1, Table 57.1).

### 57.1 Competition

- Competition can be categorized as intraspecific (between individuals of the same species) or interspecific (between individuals of different species), and as exploitation competition or interference competition (Figure 57.2).

- Laboratory and field experiments show that competition occurs frequently in nature (Figure 57.3).

- The competitive exclusion hypothesis states that two species with the same resource requirements cannot occupy the same niche. Resource partitioning and character displacement allow two or more species to coexist (Figures 57.4, 57.5, 57.6, 57.7, Table 57.2).

- The realized niche may be smaller than the fundamental niche due to competition (Figure 57.7).

### 57.2 Predation, Herbivory, and Parasitism

- Common antipredator strategies include chemical defense and aposematic coloration, camouflage, displays of intimidation, mimicry, and armor and weaponry (Figures 57.8, 57.9).

- Despite these defenses, oscillations in predator-prey cycles indicate that predators can have a large effect on prey densities (Figures 57.10).

- Plants have evolved an array of defenses against herbivores, including chemical defenses, such as secondary metabolites, and mechanical defenses, such as thorns and spines. However, herbivores can sometimes circumvent these defenses and greatly reduce plant population densities (Figures 57.11, 57.12).

- Parasitism is a long-term, usually nonlethal interaction between a parasite and its host. The experimental removal of parasites confirms that parasites can greatly reduce prey densities (Figures 57.13, 57.14).

### 57.3 Mutualism and Commensalism

- Mutualism is an association between two species that benefits both. In a resource-based mutualism, both species receive a benefit in the form of resources; a defensive mutualism typically involves an animal defending either a plant or herbivore; and a dispersive mutualism involves animals dispersing a plant's pollen or seeds (Figures 57.15, 57.16, 57.17).

- In a commensal relationship, one partner receives a benefit while the other is not affected (Figure 57.18).

### 57.4 Bottom-Up and Top-Down Control

- Bottom-up control suggests that plant quality or quantity regulates the abundance of all herbivore and predator species; top-down control suggests that the abundance of predators controls herbivore and plant densities (Figures 57.19, 57.20, 57.21).

## Assess & Discuss

### Test Yourself

1. A species interaction in which one species benefits but the other species is unharmed is called
   a. mutualism.
   b. amensalism.
   c. parasitism.
   d. commensalism.
   e. mimicry.

2. Two species of birds feed on similar types of insects and nest in the same species of tree. This is an example of
   a. intraspecific competition.
   b. interference competition.
   c. exploitation competition.
   d. mutualism.
   e. none of the above.

3. According to the competitive exclusion hypothesis,
   a. two species that use the exact same resources show very little competition.
   b. two species with the same niche cannot coexist.
   c. one species that competes with several different species for resources will be excluded from the community.
   d. all competition between species results in the extinction of at least one of the species.
   e. none of the above is correct.

4. In Lack's study of passerine birds in Britain, different species seem to segregate based on resource factors, such as location of prey items. This differentiation among the niches of passerine birds is known as
   a. competitive exclusion.
   b. intraspecific competition.
   c. character displacement.
   d. resource partitioning.
   e. allelopathy.

5. Divergence in morphology that is a result of competition is termed
   a. competitive exclusion.
   b. resource partitioning.
   c. character displacement.
   d. amensalism.
   e. mutualism.

6. Tapeworms have
   a. low lethality and low duration of interaction.
   b. low lethality and high duration of interaction.
   c. high lethality and low duration of interaction.
   d. high lethality and high duration of interaction.
   e. none of the above.

7. Ticks are regarded as
   a. monophagous endoparasites.
   b. monophagous ectoparasites.
   c. polyphagous endoparasites.
   d. polyphagous ectoparasites.
   d. none of the above.

8. Batesian mimicry differs from Müllerian mimicry in that
   a. in Batesian mimicry, both species possess the chemical defense.
   b. in Batesian mimicry, one species possesses the chemical defense.
   c. in Müllerian mimicry, one species has several different mimics.

d. in Müllerian mimicry, one species has several different chemical defenses.

e. in Batesian mimicry, cryptic coloration is always found.

9. Poppies are protected from herbivores by the
    a. alkaloid nicotine.       d. phenolic lignin.
    b. alkaloid morphine.       e. terpenoid caffeine.
    c. phenolic tannin.

10. Parasitic plants that rely solely on their host for nutrients are called
    a. hemiparasites.           d. monophagous.
    b. fungi.                   e. polyphagous.
    c. holoparasites.

## Conceptual Questions

1. Can the removal of ectoparasites from the coat of one primate by another primate (grooming) be viewed as selfish behavior, as discussed in Chapter 55? Why or why not?

2.  **Core Skill: Science and Society** Crop pests cost millions of dollars to control annually. What factors do you think might limit such losses?

## Collaborative Questions

1. Explain how the reintroduction of wolves in Yellowstone National Park might be beneficial.

2. Detail several antipredator strategies that animals have evolved.

3. Can you think of examples of mimicry used by predators to catch prey rather than used by prey to avoid being eaten? Look back to Figure 55.13a and the accompanying text discussion.

# Communities and Ecosystems: Ecological Organization on Large Scales

# 58

**Mount St. Helens erupting in 1980.** Ecologists have been monitoring gradual change in the species composition of the area since the disturbance. ©Gary Braasch/Corbis Historical/Getty Images

The term **ecosystem** was first used in 1935 by British plant ecologist A. G. Tansley to describe the system formed by the interaction between a community of organisms and its physical environment. **Ecosystem ecology** addresses the flow of energy and the production of **biomass**, which is the total mass of living matter in a given area, usually measured in grams or kilograms per square meter. In Section 58.5, we will explore energy flow, the movement of energy through an ecosystem. In examining energy flow, we will consider the complex networks of feeding relationships among species, which are represented by food webs. Finally, we will focus on the measurement of biomass within ecosystems.

## 58.1 Patterns of Species Richness and Species Diversity

**Learning Outcomes:**

1. Identify the latitudinal gradient of species richness.
2. List and describe three hypotheses for observed patterns of species richness.
3. **CoreSKILL »** Calculate the Shannon diversity index.

**A** t 8:32 a.m. on May 18, 1980, Mount St. Helens, in the Washington Cascades, erupted. The blast felled trees over a 600-km² area, and the landslide that followed destroyed everything in its path, killing nearly 60 people. For nearly 40 years, ecologists have studied the recovery of plant and animal species in that area and noted how the appearance of some species has facilitated the recovery of others.

The assemblage of populations of different species that live in the same place at the same time is known as a **community**. **Community ecology** explores the factors that influence the number and abundance of species in a community. In this chapter, we will begin by considering patterns of species richness and diversity. We will examine why, on a global scale, the number of species is usually greatest in the tropics and declines toward the poles. Next, we will discuss why the recovery of communities following a disturbance such as a fire or a volcanic eruption tends to occur in a predictable sequence—a process termed succession—which may be determined by the balance between the rates of immigration and extinction.

Community ecology is concerned with the factors that influence the number of different species in a community, or **species richness**. Globally, the number of species of most taxa varies along a latitudinal gradient, generally increasing from polar to temperate areas and reaching a maximum in the tropics. For example, the species richness of North American birds increases from Arctic Canada to Panama (**Figure 58.1**). A similar pattern exists for mammals, amphibians, reptiles, and plants. Although the latitudinal gradient of species richness is an important pattern, species richness is also influenced by topographical variation. More mountains mean more hilltops, valleys, and differing habitats. Thus, the number of birds is greater in the U.S. mountainous West. Species richness is also reduced by the peninsular effect, in which the number of species decreases as a function of distance from the main body of land.

Many hypotheses for the latitudinal gradient in species richness have been advanced. In this section, we will consider three hypotheses for patterns of species richness. The key factors are evolutionary

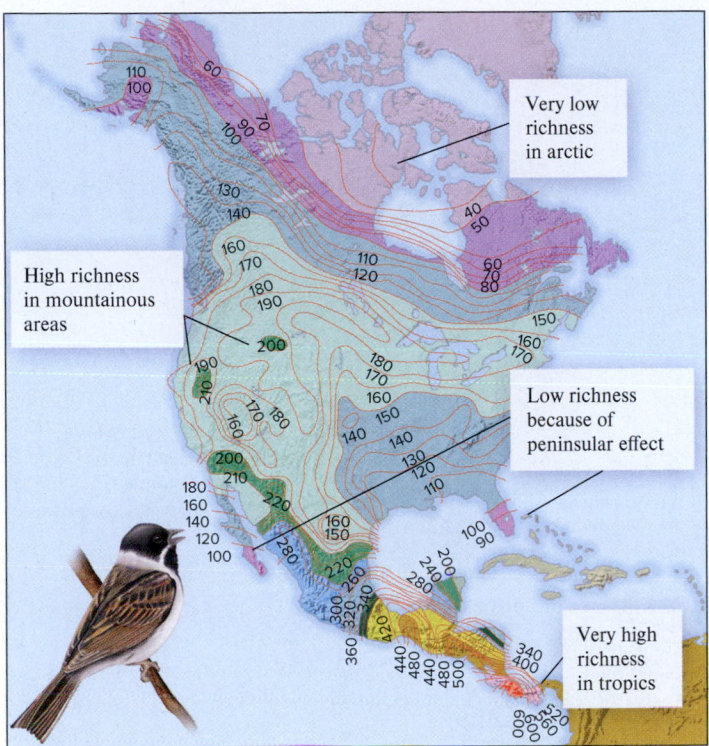

**Figure 58.1  Species richness of birds in North America.** The values indicate the numbers of different species in a given area. Contour lines show equal numbers of bird species, with colors indicating incremental changes. Note the pronounced latitudinal gradient toward the tropics and the high diversity in California, a region of considerable topographical variation and habitat diversity.

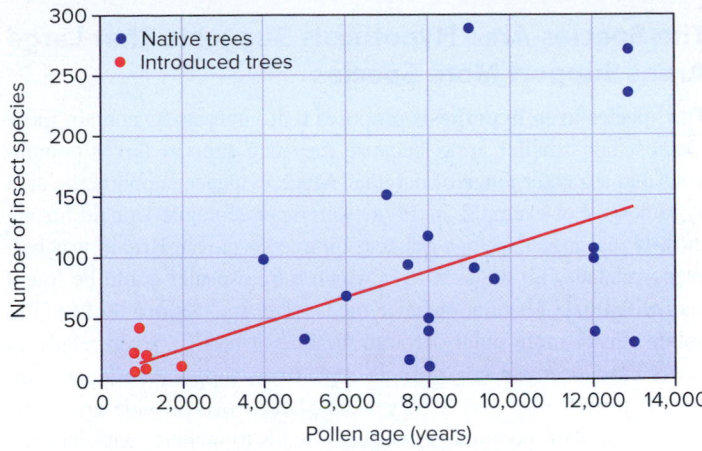

**(a) Insect species richness increases on older tree species.**

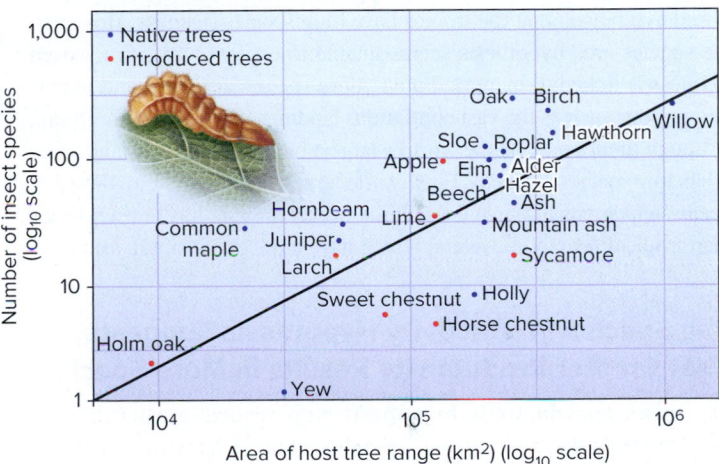

**(b) Insect species richness increases on more widely occurring tree species.**

**Figure 58.2  Relationship between species richness on British host trees and both evolutionary time and area. (a)** Insect species richness is greater on evolutionary older tree species, which supports the species-time hypothesis. **(b)** A positive correlation is also found between insect species richness and the area of the host tree's range, in square kilometers (km²), which supports the species-area hypothesis. Note the log scales in part (b).

time, area, and productivity; each factor can influence species richness to some degree. Although they are discussed separately here, these hypotheses are not mutually exclusive. All three factors can potentially contribute to patterns of species richness.

## The Species-Time Hypothesis Suggests That Communities Diversify with Age

Many ecologists argue that communities diversify, or gain species, with time. Tropical communities are usually older than temperate communities, because the species in temperate regions are periodically wiped out by glaciers. The **species-time hypothesis** proposes that temperate regions have less species-rich communities than tropical ones because they are younger. According to this idea, Ice Ages have driven many species extinct in temperate regions, and it takes time for remaining species to evolve and diversify in those regions after the glaciers have retreated. More than half of the families of flowering plants have no temperate representatives.

In support of the species-time hypothesis, British ecologist H. John Birks found a significant correlation between the numbers of species of insects on various British trees and the evolutionary ages of those tree species (**Figure 58.2a**). Many of the tree species in Britain are relatively recent colonists, having appeared following the departure of the glaciers that covered most of the islands after the last Ice Age. Birks used radiocarbon dating of pollen collected from

deep lake sediments to estimate the length of time a tree species had been present in Britain. No tree species had been present for longer than 13,000 years. He then gathered information on numbers of insect species present on trees from lists provided by other experts who had been examining the insect fauna of trees in Britain for many years. The significant relationship between pollen age and the total number of insects indicated that older tree species support more insect species.

However, ecologists recognize drawbacks to the species-time hypothesis. For example, this hypothesis may help explain variations in the species richness of terrestrial organisms, but it has limited applicability to marine organisms. Although we might not expect terrestrial species, particularly plants, to redistribute themselves quickly following a glaciation—especially if there is a physical barrier, like the English Channel, to overcome—there seems to be no reason that marine organisms couldn't relatively easily shift their distribution patterns during glaciations, yet the latitudinal gradient of species richness still exists in marine habitats.

## The Species-Area Hypothesis Suggests That Large Areas Support More Species

The **species-area hypothesis** proposes that larger areas contain more species than smaller areas because they can support larger populations and a greater range of habitats. Much evidence supports the area hypothesis. For example, in 1974, American ecologist Donald Strong showed that insect species richness on tree species in Britain was better correlated with the area over which a tree species could be found than with time of habitation since the last Ice Age (**Figure 58.2b**). The points cluster more tightly around the line of best fit. Even relatively newly introduced species, such as apple trees, supported a large number of insect herbivores if they were planted over a wide area. The observation that the number of species tends to increase with increasing area is called the **species-area effect**.

The large, climatically similar area of the tropics has been proposed as a reason that the tropics have high species richness. However, the species-area hypothesis seems unable to explain why, if increased richness is linked to increased area, more species are not found in certain regions such as the vast contiguous landmass of Asia. Furthermore, although tundra may be the world's largest biome in terms of landmass, it has low species richness. Finally, the largest marine system, the open ocean, which has the greatest volume of any habitat, has fewer species than tropical nearshore waters, which have a relatively small volume.

## The Species-Productivity Hypothesis Suggests That Greater Productivity Results in More Species

The **species-productivity hypothesis** proposes that greater production by plants results in greater overall species richness. An increase in plant productivity, the total mass of plant material produced over time, leads to an increase in the number of herbivores and hence an increase in the number of predator, parasite, and scavenger species. Productivity itself is influenced by factors such as temperature and rainfall, because many plants grow better where it is warm and wet. For example, in 1987, Canadian biologist David Currie and colleagues showed that the species richness of trees in North America is best predicted by the **evapotranspiration rate**, the rate at which water moves into the atmosphere through the processes of evaporation from the soil and other surfaces and transpiration of plants, which are influenced by the amount of solar energy (**Figure 58.3**).

Once again, however, exceptions are observed. In 1993, American researchers Robert Latham and Robert Ricklefs showed that although patterns of tree species richness in North America support the species-productivity hypothesis, the pattern does not hold for broad comparisons between continents. For example, the temperate forests of eastern Asia support substantially higher numbers of tree species (729) than do climatically similar areas of North America (253) or Europe (124). These three areas have different evolutionary histories and different neighboring areas from which species might have invaded.

## Microbial Diversity Can Be Analyzed by DNA Sequencing

An accurate determination of species richness depends on detailed knowledge of which and how many of each species are present. This

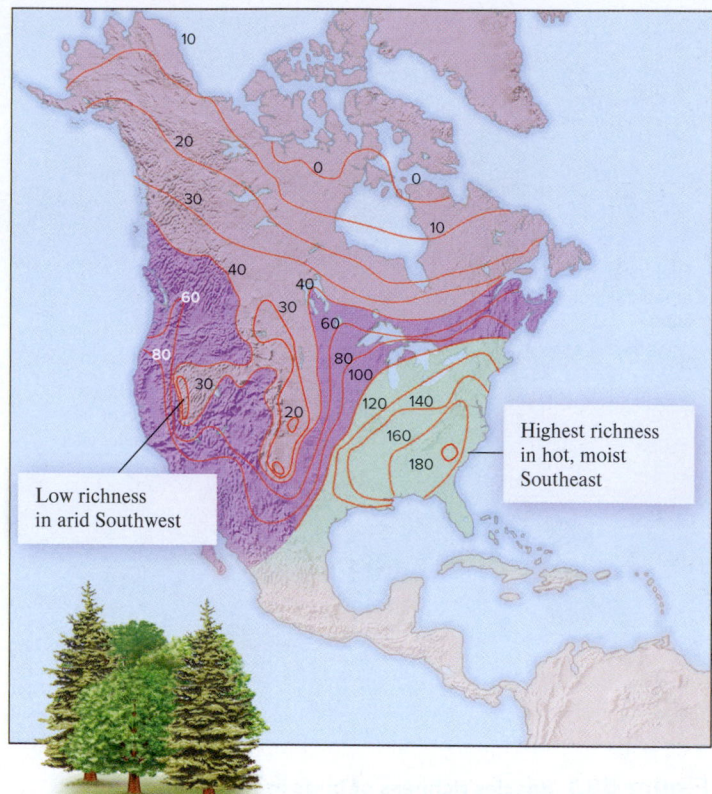

**Figure 58.3** **Tree species richness in North America.** Contour lines show equal numbers of tree species, with colors indicating incremental changes. Tree species richness and evapotranspiration rates are highest in the Southeast.

*Concept Check:* *Why doesn't the species richness of trees increase in mountainous areas of the West, as is true for birds?*

information is relatively easy to determine for communities of vertebrates and some invertebrates, but it is much more difficult for microbial communities, which includes bacteria, archaea, and many species of protists and fungi. The two main obstacles for identifying microbial species are that they are very small and researchers have been unable to devise methods for growing most of them in the laboratory. These obstacles make it difficult to study the characteristics of most microbes and thereby identify new microbial species. Yet knowledge of microbial communities is of great importance, because microbes carry out vital functions such as nitrogen fixation and decomposition. As described in Chapter 30, the species richness of microbial samples can be analyzed using DNA sequencing methods (refer back to Figure 30.3).

## The Shannon Diversity Index Is a Measure of Species Diversity

So far, we have discussed communities in terms of variations in species richness. However, ecologists need to take into account not only the number of species in a community but also their frequency of occurrence, or **relative abundance**. For example, consider two hypothetical communities, A and B, both with two species and 100 total individuals.

|  | Number of individuals of species 1 | Number of individuals of species 2 |
|---|---|---|
| Community A | 99 | 1 |
| Community B | 50 | 50 |

| Species | Abundance | $p_i$ | $\ln p_i$ | $p_i \ln p_i$ |
|---|---|---|---|---|
| 1 | 50 | 0.5 | −0.693 | −0.347 |
| 2 | 30 | 0.3 | −1.204 | −0.361 |
| 3 | 10 | 0.1 | −2.302 | −0.230 |
| 4 | 9 | 0.09 | −2.408 | −0.217 |
| 5 | 1 | 0.01 | −4.605 | −0.046 |
| Total | 5 | 100 | 1.00 | $\sum p_i \ln p_i$ = −1.201 |

The species richness of community B equals that of community A, because they both contain two species. However, community B is considered more diverse than A because the distribution of individuals between species is more even. You would be much more likely to encounter both species in community B than in community A, where one species dominates. **Species diversity** is a measure of the diversity of an ecological community that incorporates both the number of species and their relative abundance.

To measure the species diversity of a community, ecologists calculate what is known as a diversity index. Although many different indices are available, the most widely used is the **Shannon diversity index** ($H_S$), which is calculated using the formula

$$H_S = -\sum p_i \ln p_i$$

where $p_i$ is the proportion of individuals belonging to species $i$ in a community, ln is the natural logarithm, and $\Sigma$ indicates summation. For example, for a species of which there are 50 individuals out of a total of 100 in the community, $p_i$ is 50/100, or 0.5. The natural log of 0.5 is −0.693. For this species, $p_i \ln p_i$ is then 0.5(−0.693) = −0.347. For a hypothetical community with 5 species and 100 total individuals, the Shannon diversity index is calculated as follows:

Remember that, in the formula, the negative sign in front of the summation changes the summed value to positive, so the index calculated in the above table becomes 1.201, not −1.201.

Values of the Shannon diversity index for real communities often fall between 1.5 and 3.5, and the higher the value, the greater the diversity. **Table 58.1** calculates the diversity of two bird communities in Indonesia with similar species richness but differing species abundance. The bird communities were surveyed in a pristine unlogged forest and in a selectively logged lowland forest. To document diversity, British biologist Stuart Marsden established census stations in the two forests and recorded the type and number of all bird species for a number of 10-minute periods. A greater number of individual birds was seen in the logged areas (2,345) than in the unlogged ones (1,824), but a high proportion of the individuals in the logged areas (0.386) belonged to just one species, *Nectarinia jugularis*. Calculation of the Shannon diversity index showed a

## Table 58.1 Shannon Diversity Index for Bird Species on Logged and Unlogged Sites in Indonesia

| Species | Unlogged N | Unlogged $p_i$ | Unlogged $p_i \ln p_i$ | Logged N | Logged $p_i$ | Logged $p_i \ln p_i$ |
|---|---|---|---|---|---|---|
| *Nectarinia jugularis*, olive-backed sunbird | 410 | 0.225 | −0.336 | 910 | 0.386 | −0.367 |
| *Ducula bicolor*, pied imperial pigeon | 230 | 0.126 | −0.261 | 220 | 0.093 | −0.221 |
| *Philemon subcorniculatus*, grey-necked friarbird | 210 | 0.115 | −0.249 | 240 | 0.102 | −0.233 |
| *Nectarinia aspasia*, black sunbird | 190 | 0.104 | −0.235 | 120 | 0.051 | −0.152 |
| *Dicaeum vulneratum*, ashy flowerpecker | 185 | 0.101 | −0.232 | 280 | 0.119 | −0.253 |
| *Ducula perspicillata*, white-eyed imperial pigeon | 170 | 0.093 | −0.221 | 180 | 0.076 | −0.196 |
| *Phylloscopus borealis*, arctic warbler | 160 | 0.088 | −0.214 | 140 | 0.059 | −0.167 |
| *Eos bornea*, red lory | 88 | 0.048 | −0.146 | 73 | 0.031 | −0.108 |
| *Ixos affinis*, golden bulbul | 76 | 0.042 | −0.133 | 31 | 0.013 | −0.056 |
| *Geoffroyus geoffroyi*, red-cheeked parrot | 44 | 0.024 | −0.089 | 54 | 0.023 | −0.087 |
| *Rhyticeros plicatus*, Papuan hornbill | 24 | 0.013 | −0.056 | 27 | 0.011 | −0.050 |
| *Cacatua moluccensis*, Moluccan cockatoo | 12 | 0.007 | −0.035 | 1 | 0.001 | −0.007 |
| *Tanygnathus megalorynchos*, great-billed parrot | 9 | 0.005 | −0.026 | 11 | 0.005 | −0.026 |
| *Eclectus roratus*, electus parrot | 7 | 0.004 | −0.022 | 0 | 0 | 0 |
| *Macropygia amboinensis*, brown cuckoo-dove | 6 | 0.003 | −0.017 | 7 | 0.003 | −0.017 |
| *Cacomantis sepulcralis*, ruby-breasted cuckoo | 3 | 0.002 | −0.012 | 0 | 0 | 0 |
| *Trichoglossus haematodus*, rainbow lorikeet | 0 | 0 | 0 | 64 | 0.027 | −0.097 |
| **Total** | **1,824** | **1.0** |  | **2,345** | **1.0** |  |
| **Shannon diversity index** |  |  | **2.284** |  |  | **2.037** |

higher diversity of birds in the unlogged area, 2.284 versus 2.037, which is a sizable difference, considering the logarithmic nature of the index.

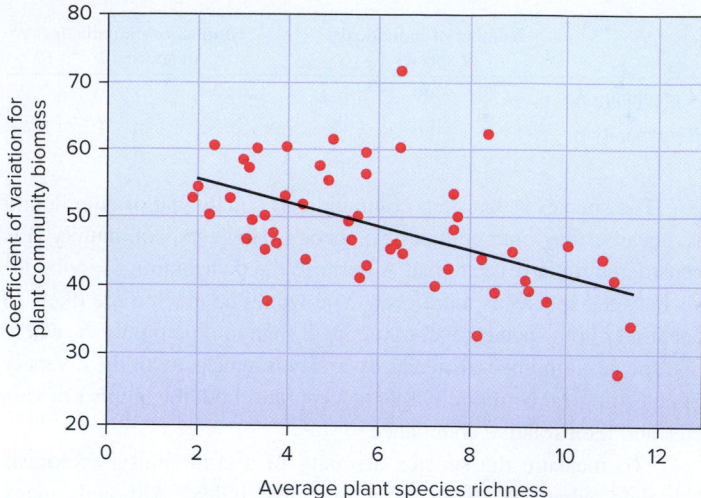

**Figure 58.4   Biomass variation and species richness.** Tilman's 11-year study of grassland plots in Minnesota revealed that year-to-year variability in community biomass was lower in species-rich plots. Each dot represents an individual plot. Only the plots from one field are graphed.

## 58.2   Species Richness and Community Stability

### Learning Outcome:

1. Describe the diversity-stability hypothesis, and explain the evidence supporting it.

A community is often seen as stable when little to no change can be detected in the number of species and their abundances over a given time period. Such a community may also be said to be in equilibrium. Community stability is an important consideration to ecologists. A decrease in the stability of a community over time may alert ecologists to a possible problem. In this section, we will consider the relationship between species richness and community stability, using evidence from a field experiment.

### The Diversity-Stability Hypothesis States That Species-Rich Communities Are More Stable Than Those with Fewer Species

The link between species richness and stability was first explicitly proposed by British ecologist Charles Elton in the 1950s. He suggested that a disturbance in a species-rich community would be cushioned by large numbers of interacting species and would not produce as drastic an effect as would a disturbance in a species-poor community. For example, an introduced predator or parasite would be more likely to cause extinctions in a species-poor community compared to a more species-rich community, where its effects would be buffered by interactions with more species. Elton argued that outbreaks of pests are often found on cultivated land or land disturbed by humans, both of which are species-poor communities with few naturally occurring species. His idea became known as the **diversity-stability hypothesis**.

However, some ecologists challenged Elton's association of diversity with stability. They pointed out many examples of introduced species that have assumed pest proportions in species-rich areas, including rabbits in Australia and pigs in North America. They also suggested that disturbed or cultivated land may suffer from pest outbreaks not because of its low number of species, but because introduced species often have no natural enemies in their new environment. In contrast, native species and their natural enemies have had long associations. For example, in Europe and North America, rabbit species and their predators, such as fox species, have coevolved based on a predator-prey relationship over the course of many generations. This relationship prevents the rabbit populations from increasing to pest proportions.

### Field Studies Have Linked Stability to Richness

To test the diversity-stability hypothesis, research was needed in which the stability of communities was measured and then compared to their species richness. In 1996, American ecologist David Tilman reported the relationship between species richness and community stability from an 11-year study of 207 grassland plots in Minnesota that varied in their species richness. At the end of every year, he measured the biomass of every plant species in each plot. He then analyzed how much this biomass varied from year to year through a statistical calculation called the coefficient of variation, which is a measure of relative variability. (It is the ratio of the standard deviation to the mean.) Less variation in biomass indicates community stability. Year-to-year variation in plant community biomass was significantly lower in plots with greater plant species richness (**Figure 58.4**). These results are consistent with the diversity-stability hypothesis—higher species richness enhances community stability.

Tilman suggested that species-rich communities are more likely to contain disturbance-resistant species that, in the event of a disturbance, could grow and compensate for the loss of disturbance-sensitive species. For example, when a change in climate such as drought decreased the abundance of competitively dominant species that thrived under normal conditions, drought-resistant species increased in mass and replaced them. Thus, declines in the number of sensitive species were compensated for by increases in other species, which acted to stabilize total community biomass.

## 58.3   Succession: Community Change

### Learning Outcomes:

1. Distinguish between primary and secondary succession.
2. Compare and contrast facilitation, inhibition, and tolerance as mechanisms of succession.

The term **succession** describes the gradual and continuous change in species composition of a community following a disturbance. **Primary succession** refers to succession on a newly exposed site that has no biological legacy in terms of plants, animals, or microbes, such as bare ground caused by a volcanic eruption or the sediment created by the retreat of glaciers. In primary succession on land, the plants must often build up the soil, and thus a long time—even hundreds of years—may be required for the process. Only a tiny proportion of the Earth's surface is currently undergoing primary succession, including the area around Mount St. Helens and on new lava flows around the volcanoes in Hawaii and off the coast of Iceland, and behind retreating glaciers in Alaska and Canada.

**Secondary succession** refers to succession on a site that has previously supported life but has undergone a disturbance such as a fire, tornado, hurricane, or flood. In terrestrial areas, soil is already present. Clearing a natural forest and farming the land for several years is an example of a severe forest disturbance that does not kill all native species. Some plants and many soil bacteria, nematodes, and insects are still present. Secondary succession occurs if farming has ended. The secondary succession in abandoned farmlands can lead to a pattern of vegetation quite different from one that develops after primary succession following glacial retreat. For example, the plowing and added fertilizers, herbicides, and pesticides may have caused substantial changes in the soil of an abandoned field, allowing species that require a lot of nitrogen to colonize. These species would not be present for many years in newly created glacial soils.

American plant ecologist Frederic Clements is often viewed as the founder of successional theory. His work in the early 20th century emphasized succession as proceeding through several stages to a distinct end point, or **climax community**. Although disturbance can return a community from a later stage to an earlier stage, generally the community progresses in one direction. Clements's depiction of succession focused on a process termed facilitation, but two other mechanisms of succession—inhibition and tolerance—have since been described. In this section, we will examine the evidence for each of them.

## Facilitation Assumes That Each Invading Species Creates a More Favorable Habitat for Succeeding Species

A key assumption of Clements's view of succession is that each colonizing species makes the local environment a little different, such as a little shadier or a little richer in soil nitrogen, so that it becomes more suitable for other species, which then invade and outcompete the earlier residents. This process, known as **facilitation**, continues until the most competitively dominant species have colonized, when the community is at climax. The composition of the climax community for any given region is determined by climate, soil condition, and frequency of disturbance.

Succession following the gradual retreat of Alaskan glaciers is often used as a specific example of facilitation as a mechanism of succession. Over the past 200 years, the glaciers in Glacier Bay have undergone a dramatic retreat of nearly 100 km (**Figure 58.5**). Succession in Glacier Bay has followed a distinct pattern of vegetation. As glaciers retreat, they leave moraines—deposits of stones, pulverized rock, and debris that serve as soil. In Alaska, the bare soil has a low nitrogen content and little organic matter.

In the pioneer stage, the soil is first colonized by a black crust of cyanobacteria, mosses, lichens, horsetails (*Equisetum variegatum*), with the occasional river beauty (*Epilobium latifolium*) (**Figure 58.6a**). Because the cyanobacteria are nitrogen fixers, the soil nitrogen increases a little, but soil depth and litterfall (fallen leaves, twigs, and

(a) Glacier Bay, Alaska

(b) Glacial retreat

**Figure 58.5** **The degree of glacier retreat at Glacier Bay, Alaska, since 1794.** (a) Primary succession begins on the bare rock and soil evident at the edges of the retreating glacier. (b) The lines reflect the position of the glacier in 1794 and its subsequent retreat northward.
a: ©Charles D. Winters/Science Source

*Concept Check:* *Why might an ecologist think of walking the coastline of Glacier Bay as the equivalent of walking back in time?*

other plant material) are still minimal. At this stage, there are rare instances of seeds and seedlings of dwarf shrubs of the rose family, commonly called mountain avens (*Dryas drummondii*); alders (*Alnus sinuata*); and spruce. After about 40 years (the *Dryas* stage), mountain avens dominate the landscape (**Figure 58.6b**). Soil nitrogen increases, as does soil depth and litterfall, and alder trees begin to invade.

At about 60 years, alders form dense, close thickets (**Figure 58.6c**). Alders have nitrogen-fixing bacteria that live mutualistically in their roots and convert nitrogen from the air into a biologically useful form. Soil nitrogen dramatically increases, as does litterfall. Sitka spruce trees (*Picea sitchensis*) begin to invade at about this time. After about 75 to 100 years, the spruce trees begin to overtop the alders, shading them out. The litterfall is still high, and the large volume of needles turns the soil acidic. The shade causes competitive exclusion of many of the original understory species, including alder, and only mosses carpet the ground. At this stage, seedlings of western hemlock (*Tsuga heterophylla*) and mountain hemlock

| Stage | Pioneer | *Dryas* | Alder | Spruce-hemlock |
|---|---|---|---|---|
| Time (years) since glacial retreat | 5 | 40 | 60 | 200 |
| Soil depth (cm) | 5.2 | 7.0 | 8.8 | 15.1 |
| Soil N (g/m$^2$) | 3.8 | 5.3 | 21.8 | 53.3 |
| Soil pH | 7.2 | 7.3 | 6.8 | 3.6 |
| Litterfall (g/m$^2$/yr) | 1.5 | 2.8 | 277 | 261 |

Cyanobacteria
Moss
Lichens

Mountain avens
(*Dryas drummondii*)

Alder
(*Alnus sinuata*)

Spruce
(*Picea sitchensis*)
Western hemlock
(*Tsuga heterophylla*)

**(a) Pioneer stage**  **(b) *Dryas* stage**  **(c) Alder stage**  **(d) Spruce-hemlock stage**

**Figure 58.6**  **The pattern of primary succession at Glacier Bay, Alaska.** **(a)** The first species to colonize the bare ground following retreat of the glaciers are small species such as cyanobacteria, moss, and lichens. **(b)** Mountain avens (*Dryas drummondii*) is a flower common in the *Dryas* stage. **(c)** Soil nitrogen and litterfall increase rapidly as alder (*Alnus sinuata*) invades. Note also the appearance of a few spruce trees higher up the valley. **(d)** Sitka spruce (*Picea sitchensis*) and hemlock (*Tsuga heterophylla*) trees make up a climax spruce-hemlock forest at Glacier Bay, with moss carpeting the ground. Two hundred years ago, glaciers occupied this spot. a: ©Leon Werdinger/Alamy Stock Photo; b: ©James Hager/age fotostock; c: ©Accent Alaska.com/Alamy Stock Photo; d: ©Craig Lovell/Eagle Visions P/Newscom

(*Tsuga mertensiana*) may also occur. After 200 years, a mixed spruce-hemlock climax forest occupies the location (**Figure 58.6d**).

Other studies also provide evidence of facilitation. Early primary succession on Mount St. Helens shows that decomposition of fungi allows mosses and other fungi to colonize the soil, providing evidence of facilitation. Succession on sand dunes also supports the facilitation model; pioneer plant species stabilize the sand dunes and facilitate the establishment of subsequent plant species. The foredunes, those nearest the shoreline, are the most frequently disturbed and are maintained in a state of early succession, whereas more stable communities develop farther away from the shoreline.

Succession also occurs in aquatic communities. Although soils do not develop in marine environments, facilitation may still occur when one species enhances the quality of settling and establishment sites for another species. When experimental test plates used to measure settling rates of marine organisms were placed in Delaware Bay, researchers discovered that certain cnidarians enhanced the attachment of tunicates, and both facilitated the attachment of mussels, the dominant species in the community. In this experiment, the smooth surface of the test plates prevented many species from colonizing, but once the surface became rougher, because of the presence of the cnidarians, many other species were able to colonize. In a similar fashion, early colonizing bacteria, which create biofilms on rock surfaces, can facilitate succession of other organisms.

## Inhibition Implies That Early Colonists Can Prevent Later Arrivals from Replacing Them

Although data on succession in some communities fit the facilitation model, researchers have proposed alternative hypotheses concerning how succession may operate. In the mechanism known as **inhibition**, early colonists prevent colonization by other species. For example, removing the litter of *Setaria faberi*, an early successional plant species in abandoned New Jersey farm fields, causes an increase in the biomass of a later species, *Erigeron annuus*. The release of toxic compounds from decomposing *S. faberi* litter or physical obstruction by the litter itself blocks the establishment of *E. annuus*. Without the litter present, however, *E. annuus* dominates and reduces the biomass of *S. faberi*.

Inhibition has been seen as the primary method of succession in the marine intertidal zone, where space is limited. In this habitat, early successional species are at a great advantage in maintaining possession of valuable space. In 1974, American ecologist Wayne Sousa created an environment for testing how succession works in the intertidal zone by scraping rock faces clean of all algae or putting out fresh boulders or concrete blocks. The first colonists of these areas were the green algae of the genus *Ulva*. By removing those algae from the substrate, Sousa showed that the large red alga *Chondracanthus canaliculatus* was able to colonize more quickly (**Figure 58.7**).

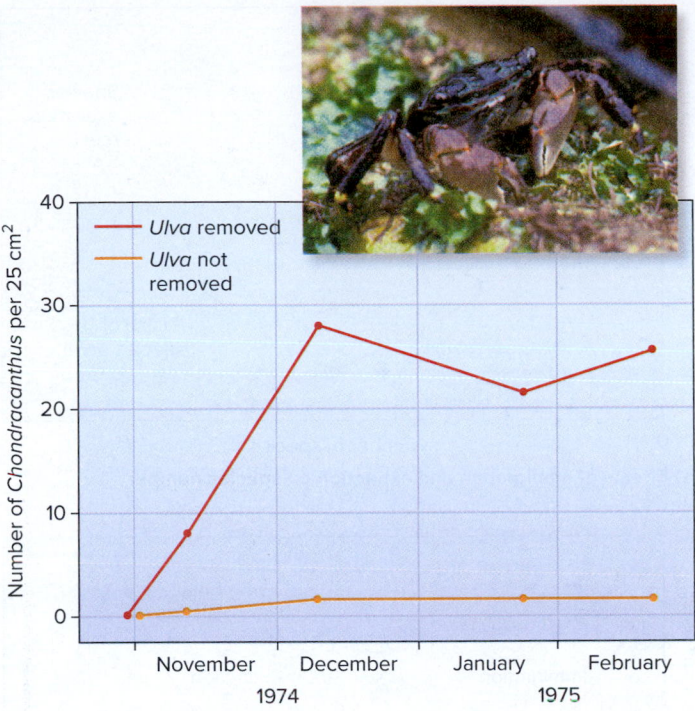

**Figure 58.7** **Inhibition as a method of succession in the marine intertidal zone.** Removing *Ulva* from intertidal rock faces allowed colonization by *Chondracanthus canaliculatus*. The inset shows *Ulva* on a rock face with the striped shore crab *Pachygrapsus crassipes,* a herbivore. ©Wayne Sousa/University of California, Berkeley

The results of Sousa's study indicate that early colonists can inhibit rather than facilitate the invasion of subsequent colonists. Succession may eventually occur because early-colonizing species, such as *Ulva*, are more susceptible than later successional species, such as *C. canaliculatus*, to the rigors of the physical environment and to attacks by herbivores, such as crabs (*Pachygrapsus crassipes*).

### Tolerance Suggests That Early Colonists Neither Facilitate Nor Inhibit Later Colonists

In 1977, researchers Joseph Connell and Ralph Slatyer proposed a third mechanism of succession, which they termed **tolerance**. In this process, any species can start the succession, but the eventual climax community is reached in a somewhat orderly fashion. The species that establish themselves and remain do not change the environment in ways that either facilitate or inhibit subsequent colonists. Species have differing tolerances to the intensity of competition as more species accumulate. Relatively competition-intolerant species are more successful early in succession, when the intensity of competition is low and resources are abundant. Relatively competition-tolerant species appear later in succession and at climax.

Connell and Slatyer found the best evidence for the tolerance model in American plant ecologist Frank Egler's earlier work on floral succession. In the 1950s, Egler showed that succession in plant communities is determined largely by species that already exist in the ground as buried seeds or old roots. Whichever species germinates first or regenerates from roots initiates the succession sequence. Germination or root regeneration, in turn, depends on the timing

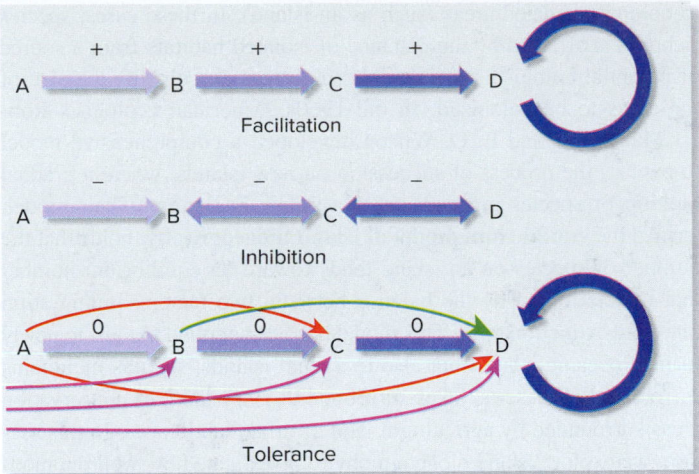

**Figure 58.8** **Three models of succession.** A, B, C, and D represent four stages, with D representing the climax community. An arrow means "is replaced by," + stands for "facilitation," – stands for "inhibition," and 0 indicates no effect. The facilitation model is the classic model of succession. In the inhibition model, early-arriving species outcompete later-arriving species. The tolerance model depends on which species gets there first. The colored arrows show that succession may bypass some stages in the tolerance model.

*Concept Check:* *Inhibition implies that competition exists between species, with early-arriving species tending to outcompete later arrivals, at least for a while. Does competition or mutualism feature more prominently in facilitation?*

of a disturbance. For example, an early-season tree fall would promote early-germinating species to grow in the subsequent light gap, whereas a late-season tree fall would promote the growth of late-germinating species. As succession proceeds, species that are earlier in germination may be outcompeted by different species.

### The Three Models of Succession Differ in How One Species Affects Colonization by a Different Species

The key distinction between the three models of succession is in the manner in which one species affects the colonization of another species. In the facilitation model, species replacement is facilitated by previous colonists; in the inhibition model, it is inhibited by the action of previous colonists; and in the tolerance model, species may be affected by previous colonists, but they do not require them (**Figure 58.8**).

## 58.4 Island Biogeography

### Learning Outcomes:

1. **CoreSKILL »** Interpret a graph that illustrates the equilibrium model of island biogeography.
2. List the three predictions of the model, and discuss how well the evidence supports each one.

In some newly formed habitats such as volcanic islands, succession may be affected not only by facilitation, inhibition, or tolerance but also by the ability of species from neighboring areas (such as a mainland) to colonize isolated areas (such as an island). In these cases, species richness is affected by the distance of isolated habitats from a source of potential colonists from neighboring areas and also by the size of the areas to be colonized. In the 1960s, American ecologists Robert MacArthur and E. O. Wilson developed a comprehensive model to explain the process of succession on new islands, where a gradual buildup of species proceeds from a sterile beginning. Their model, termed the **equilibrium model of island biogeography**, holds that the number of species on an island tends toward an equilibrium number that is determined by the balance between two factors: immigration rates and extinction rates. This model has been applied not just to newly formed oceanic islands but also to virtual islands, such as mountains surrounded by deserts, lakes surrounded by dry land, or conservation areas surrounded by agricultural land or urban landscapes. In this section, we explore island biogeography to investigate how well the model's predictions are supported by data.

## The Island Biogeography Model Suggests That During Succession, Gains in Immigration Are Balanced by Losses from Extinction

MacArthur and Wilson's model of island biogeography suggests that species repeatedly arrive on an island and either thrive or become extinct. The rate of immigration of new species is highest when no species are present on the island. As the number of species accumulates, the immigration rate decreases, since subsequent immigrants are more likely to represent species already present on the island. The rate of extinction is low at the time of first colonization, because few species are present and many have large populations. With the addition of new species, the population sizes of some species diminish, so the probability of extinction increases. Over time, the number of species tends toward an equilibrium value, $\hat{S}$, in which the rates of immigration and extinction are equal. Species may continue to arrive and go extinct, but the number of species on the island remains approximately the same.

MacArthur and Wilson reasoned that when plotted graphically, both the immigration and extinction lines would be curved, for several reasons (**Figure 58.9a**). First, species arrive on islands at different rates. Some organisms, including plants with seed-dispersal mechanisms and winged animals, are more mobile than others and arrive quickly. Other organisms arrive more slowly. This pattern causes the immigration curve to start off steep but get progressively shallower. On the other hand, extinctions rise at accelerating rates, because as later species arrive, competition increases and more species are likely to go extinct.

A strength of the island biogeography model is that it generates several testable predictions:

1. *Species-area relationships.* The number of species should increase with increasing island size (area), a concept known as the species-area effect (see Figure 58.2b). Extinction rates should be lower on larger islands because population sizes are larger and less susceptible to extinction (**Figure 58.9b**).

2. *Species-distance relationships.* The number of species should decrease with increasing distance of the island from the

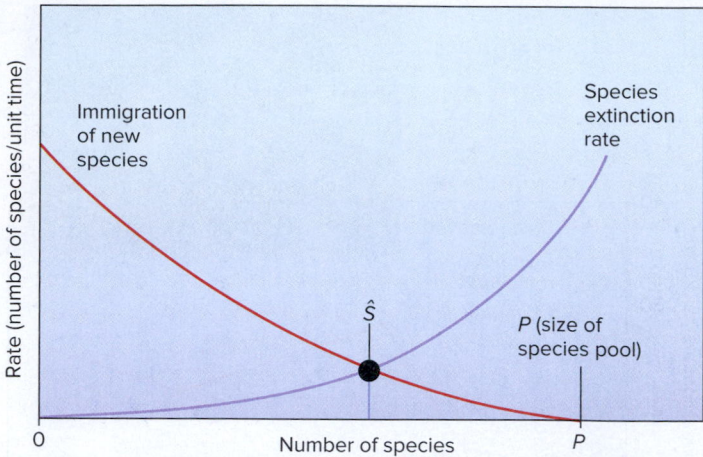

**(a) Effects of immigration and extinction on species number**

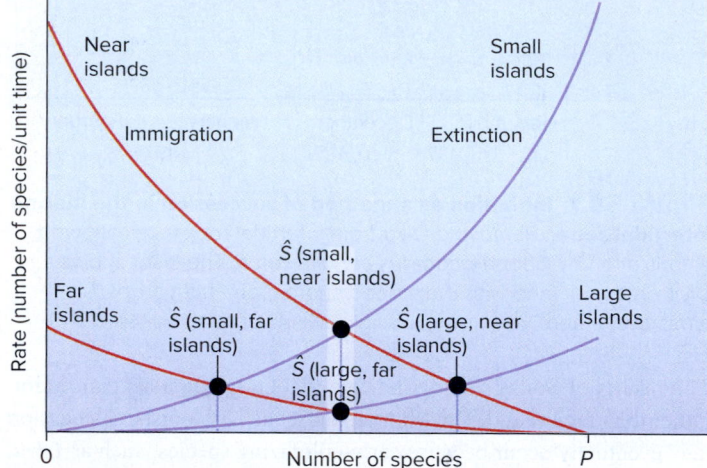

**(b) Added effects of island size and proximity to the mainland on species number**

**Figure 58.9  MacArthur and Wilson's equilibrium model of island biogeography.** **(a)** The opposing effects of immigration and extinction produce an equilibrium number of species on an island, $\hat{S}$. This number can vary from 0 species to $P$ species, the total number of species available to colonize. **(b)** $\hat{S}$ varies according to the island's size and distance from the mainland. An increase in distance (near to far) lowers the immigration rate. An increase in island area (small to large) lowers the extinction rate.

 **Core Skill: Connections**  Look ahead to Figure 60.8. How might the equilibrium model of island biogeography be useful in the design of nature reserves?

mainland, that is, from the **source pool**, the pool of potential species available to colonize the island. Immigration rates should be greater on islands near the source pool because species do not have as far to travel (see Figure 58.9b).

3. *Species turnover.* The turnover of species should be considerable. While the number of species on an island might remain relatively constant, the identity of the species should vary over time as new species colonize the island and others become extinct. Eventually, turnover is reduced when a

climax community is reached and it becomes difficult for new immigrants to colonize the island.

Let's examine the three predictions of the island biogeography model and see how well the data support each one.

**Species-Area Relationships**   The West Indies has traditionally been a key location for ecologists studying island biogeography. The physical geography and the plant and animal life of the islands are well known. Furthermore, all of the Lesser Antilles, from Anguilla in the north to Grenada in the south, enjoy a similar climate and are surrounded by deep water (**Figure 58.10a**). In 1999, Robert Ricklefs and Irby Lovette summarized the available data on the richness of species of four groups of animals—birds, bats, reptiles and amphibians, and butterflies—across 19 islands that varied in area over two orders of magnitude (13–1,510 km²). In each case, a positive correlation occurred between area and species richness (**Figure 58.10b**).

**Species-Distance Relationships**   In studies of the numbers of lowland forest bird species in Polynesia, MacArthur and Wilson found that the number of species decreased with the distance from the source pool of New Guinea (**Figure 58.11**). They expressed the richness of bird species on the islands as a percentage of the number of bird species found on New Guinea. A significant decline in this percentage was observed with increasing distance. More-distant islands were inhabited by lower numbers of species than nearer islands. This research substantiated the prediction of species richness declining with increasing distance from the source pool.

**Species Turnover**   Studies involving species turnover on islands are difficult to perform because detailed and complete species lists are needed over long periods of time, usually many years and often decades. The lists that do exist are often compiled in a casual way and are not usually suitable for comparison with more modern data. In 1980, British researcher Francis Gilbert reviewed 25 investigations carried out to analyze turnover and found a lack of this type of rigor in nearly all of them. Furthermore, most of the observed turnover in these studies, usually less than 1% per year, or less than one species per year, appeared to be due to immigrants that never became established rather than to the extinction of well-established species. More recent studies have revealed similar findings, suggesting that the rates of turnover are low rather than high, giving little conclusive support to the third prediction of the equilibrium model of island biogeography. Even the most rigorous study, by E. O. Wilson and his student Daniel Simberloff, showed negligible turnover, as described next.

(a) **Lesser Antilles Islands**

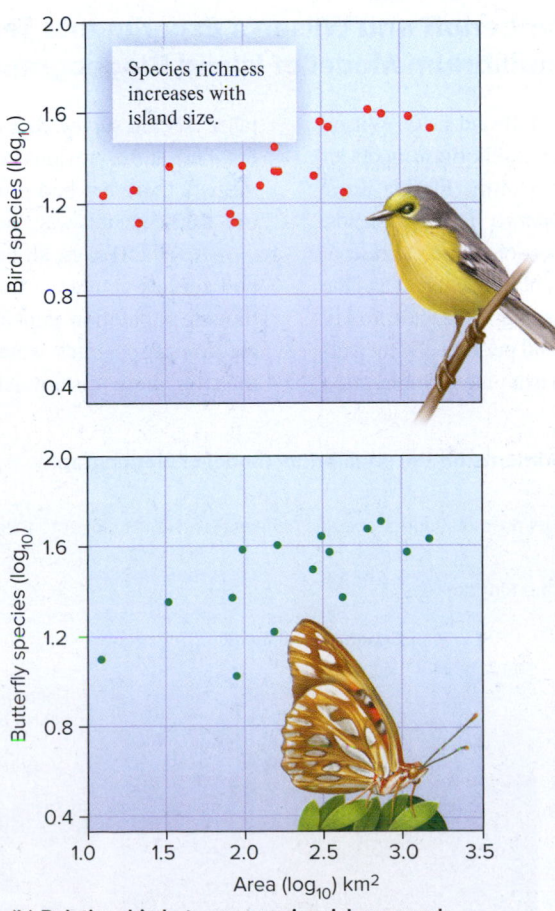

(b) **Relationship between species richness and island size**

**Figure 58.10   Species richness and island size. (a)** The Lesser Antilles extend from Anguilla in the north to Grenada in the south. **(b)** On these islands, the number of bird and butterfly species increases with the area of an island. Note that these relationships are traditionally plotted on a double logarithmic scale, a so-called log-log plot, in which the horizontal axis is the logarithm to the base 10 of the area and the vertical axis is the logarithm to the base 10 of the number of species. A linear plot of the area versus the number of species would be difficult to produce because of the wide range of areas and of variation in species richness. Logarithmic scales condense this variation to manageable limits.

**Concept Check:** *Calculate the approximate change in bird species richness across islands in the Lesser Antilles.*

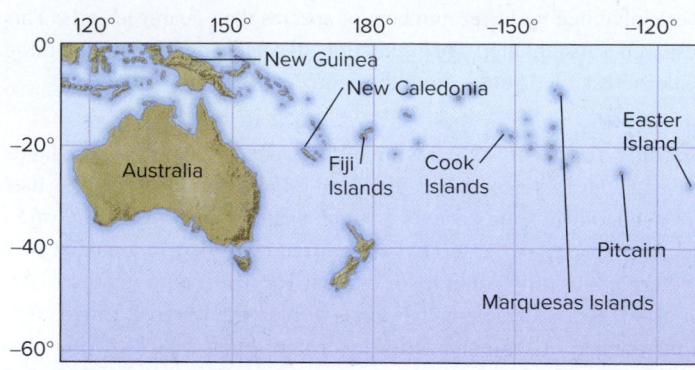

(a) New Guinea and neighboring islands

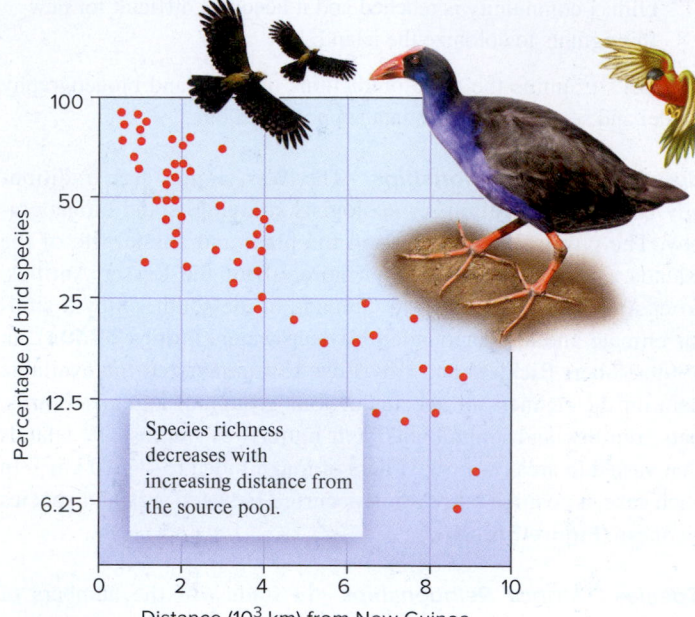

(b) Relationship between species richness and distance from source

**Figure 58.11  Species richness and distance from the source pool. (a)** Map of Australia, New Guinea, and these Polynesian Islands: New Caledonia, Fiji Islands, Cook Islands, Marquesas Islands, Pitcairn, and Easter Island. **(b)** The number of bird species on the islands decreases with increasing distance from the source pool, New Guinea. The species richness is expressed as the percentage of bird species on New Guinea.

---

 **Core Skill: Process of Science**

## Feature Investigation | Simberloff and Wilson's Experiments Tested the Predictions of the Equilibrium Model of Island Biogeography

In the 1960s, American ecologists Daniel Simberloff and E. O. Wilson conducted possibly the most rigorous test of the equilibrium model of island biogeography ever performed, using islands in the Florida Keys. First, they surveyed small red mangrove (*Rhizophora mangle*) islands, 11–25 m in diameter, taking a census of the numbers of all their terrestrial arthropods. Then they enclosed each island with a plastic tent and had the islands fumigated with methyl bromide, a short-acting insecticide, to kill all arthropods on them. The tents were removed, and periodically thereafter Wilson and Simberloff surveyed the islands to examine recolonization rates. At each survey, they counted all species present, noting any species not there at the previous census and the absence of others that were previously there but had presumably gone extinct (see the data of **Figure 58.12**). In this way, they estimated turnover of species on the islands.

After 250 days, all but one of the islands had a number of arthropod species similar to the number observed before fumigation, even though population densities were still low. The data indicated that recolonization rates were higher on islands nearer to the mainland than on more distant islands—as the island biogeography model

**Figure 58.12  Simberloff and Wilson's experiments on the equilibrium model of biogeography.** *(photos)* Courtesy Dr. D. Simberloff, University of Tennessee

| | |
|---|---|
| **HYPOTHESIS** | The island biogeography model predicts higher species richness for islands closer to the mainland and significant turnover of species on islands. |
| **STARTING LOCATION** | Mangrove islands in the Florida Keys. |

|  | Experimental level | Conceptual level |
|---|---|---|
| **1** Take initial census of all terrestrial arthropods on 4 mangrove islands. Erect a framework over each mangrove island. |  | 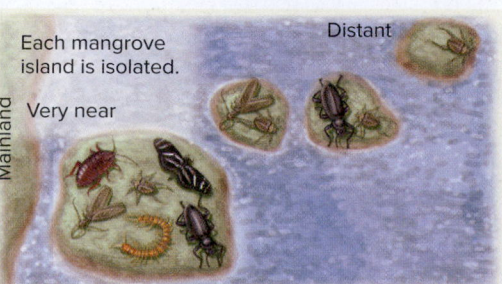 |

**2**  Cover each framework with a tent and fumigate with methyl bromide to kill all arthropod species.

Methyl bromide is a short-acting insecticide that at low levels will not kill plant life.

Mainland

Very near

Distant

**3**  Remove the tents and conduct censuses every month to monitor recolonization of arthropods and to determine extinction rates.

Mangrove islands are recolonized.

Mainland

Very near

Distant

**4   THE DATA**  Island E2 was closest to the mainland and supported the highest number of species both before and after fumigation. E3 and ST2 were at an intermediate distance from the mainland, and E1 was the most distant.

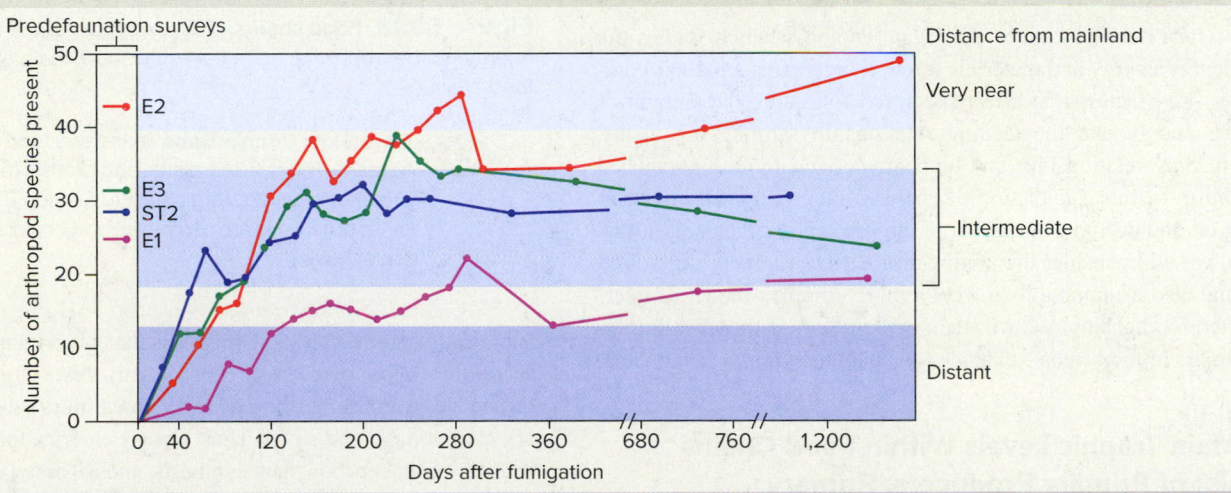

**5   CONCLUSION**  Island distance from the mainland influences species richness on mangrove islands in the Florida Keys. However, species turnover is minimal, and species richness changes little following initial recolonization.

**6   SOURCE**  Simberloff, D. S. 1978. Colonization of Islands by Insects: Immigration, Extinction and Diversity, pp. 139–153 in L. A. Mound and N. Waloff (eds.), *Diversity of Insect Faunas*. Blackwell Scientific Publications, Oxford, U.K.

predicts. However, the data, which consisted of lists of species on islands before and after extinctions, provided little support for the prediction of substantial turnover. Rates of turnover were low, only 1.5 extinctions per year, compared with the 15–40 species found on the islands within a year. Simberloff and Wilson concluded that turnover probably involves only a small subset of transient or less common species, with the more common species remaining permanent after colonization.

**Experimental Questions**

1. What was the purpose of Simberloff and Wilson's study?

2. Why did the researchers conduct a thorough species survey of arthropods before experimental removal of all the arthropod species?

3. **CoreSKILL »** What did the researchers conclude about the relationship between island proximity to the mainland and species richness and turnover?

The equilibrium model of island biogeography has stimulated much research to confirm the strong effects of area and distance on species richness. However, species turnover appears to be low rather than considerable, which suggests that succession on most islands is a fairly orderly process. This conclusion means that colonization is not a random process and that the same species seem to colonize first and other species gradually appear in the same order.

The principles of island biogeography have been applied to wildlife preserves, which are essentially islands in a sea of developed land consisting of agricultural fields or urban sprawl. Conservationists have therefore utilized the model of island biogeography in the design of nature preserves, a topic we will return to in Chapter 60.

## 58.5 Food Webs and Energy Flow

### Learning Outcomes:

1. Compare and contrast food chains and food webs, and distinguish primary producers and primary, secondary, and tertiary consumers.

2. **CoreSKILL » Calculate** the efficiency of consumers as energy transformers in two ways.

3. List and describe the different types of ecological pyramids.

We now turn our attention to ecosystem ecology, which studies the movement of energy and materials through organisms and their communities. Key factors that affect species richness are the amount of available energy and the feeding relationships among organisms. These feeding relationships can be characterized by an unbranched **food chain**—a linear depiction of energy flow, with each organism feeding on and deriving energy from the preceding organism. In this section, we will consider the unidirectional flow of energy in a food chain and also examine a food web, a more complex model of interconnected food chains. We will then explore two of the most important features of food webs—chain length and the pyramid of numbers.

### The Main Trophic Levels Within Food Chains Consist of Primary Producers, Primary Consumers, and Secondary Consumers

Each level in a food chain is called a **trophic level** (from the Greek *trophos*, meaning feeder), with different species feeding at different levels. In a food-chain diagram, an arrow connects each trophic level with the one above it (**Figure 58.13**). Food chains typically consist of organisms that obtain energy in different ways. **Autotrophs** harvest light or chemical energy and store that energy in carbon compounds. Most autotrophs, which include plants, algae, and photosynthetic bacteria, use sunlight for this process. These organisms, called **primary producers**, form the base of the food chain. They produce the energy-rich organic molecules upon which nearly all other organisms depend.

Organisms that consume organic molecules from their environment to sustain life and thus receive their nutrition by eating other organisms or products of organisms are termed **heterotrophs**. Heterotrophs that obtain their food by consuming primary producers are termed **primary consumers** (also called herbivores) and include most protists, most animals, and even some plants such as mistletoe, which

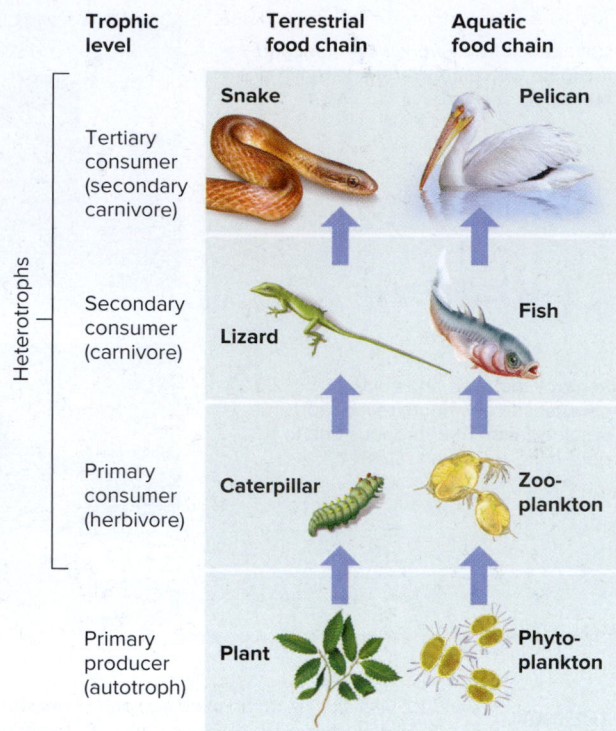

**Figure 58.13 Food chains.** Two examples of the flow of food energy up the trophic levels: a terrestrial food chain and an aquatic food chain.

 **Core Skill: Connections** In the two food chains illustrated in Figure 58.13, plants and protists (phytoplankton) are the producers. Refer back to Section 27.4. What other organisms are producers and could also support food chains?

is parasitic on other plants. Organisms that eat primary consumers are **secondary consumers** (also called carnivores). Organisms that feed on secondary consumers are **tertiary consumers** (also called secondary carnivores), and so on. Thus, energy enters a food chain through primary producers, via photosynthesis, and is passed up the food chain to primary, secondary, and tertiary consumers (see Figure 58.13).

At each trophic level, many organisms die before they are eaten. Most energy from the first trophic level, such as plants, goes unconsumed by herbivores. Instead, unconsumed plants die and decompose in place. This material, along with dead remains of animals and waste products, is called **detritus**. Consumers that get their energy from detritus, called **detritivores**, break down dead organisms from all trophic levels. For example, carrion beetles feed on the dead bodies of other animals. Some detritivores, which do not ingest their food but feed by absorbing it on a molecular scale, are known as decomposers. The most important are fungi and bacteria. In terrestrial systems, detritivores probably consume 80–90% of plant matter, with different groups such as earthworms and fungi working in concert to extract most of the energy. Detritivores may, in turn, support a community of predators that feed on them.

In nature, feeding relationships are usually more complex than simple food chains. For example, many different herbivore species may feed on the same plant species. Also, each species of herbivore may feed on several different plant species. For instance, on the African savanna, cheetahs, lions, and hyenas all eat a variety of prey, including wildebeest, impala, and Thompson's gazelle. These herbivores, in turn, eat a variety of trees and grasses. Such relationships between plants and animals are depicted as a **food web**, a complex model of interconnected food chains in which multiple links occur among different species (**Figure 58.14**).

## In Most Food Webs, Chain Lengths Are Short

Let's examine some of the characteristics of food webs in more detail. The chain length is sum of the number of links between the trophic levels involved. For example, if a lion feeds on a zebra, and a zebra feeds on grass, the chain length is two. In many food webs, chain lengths tend

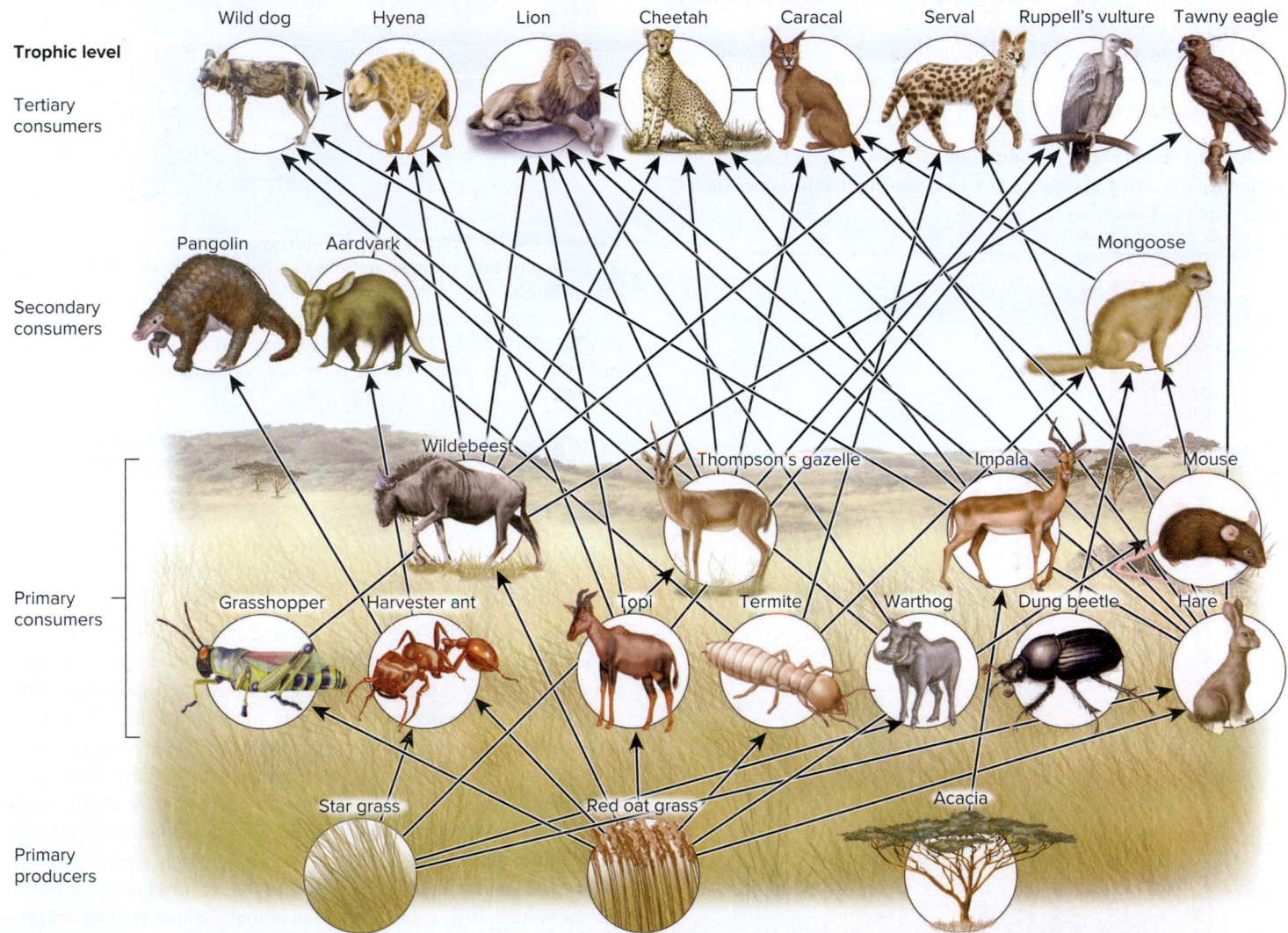

**Figure 58.14** **A food web depicting the feeding relationships of an African savanna ecosystem.** Each trophic level is occupied by different species. Generally, each species feeds on, or is fed upon by, more than one species.

*Concept Check:* *At which trophic level(s) do decomposers feed?*

---

**Core Skill: Modeling** The goal of this modeling challenge is to draw a food web that depicts some of the main species that make up the food web in Banff National Park.

**Modeling Challenge:** Take a look back at Figure 57.21, which shows some species interactions. Beginning with this group of species, create a food web that also includes the following species: cottonwood (a tree), wheatgrass, mouse, mule deer, bison, and coyote. Remember that autotrophs, such as plants, are placed at the bottom of the food web, the herbivores are placed on the next level, and the carnivores are placed above them. The arrows point from lower trophic levels to higher trophic levels. In your food web, aspen, cottonwood, and willow are all eaten by beaver, elk, and mule deer. Wheatgrass is eaten by bison, mouse, and elk. Coyotes eat mouse and beaver, and wolves eat coyote, elk, mule deer, and bison. You may omit the songbirds from your food web.

to be short, usually five or fewer, because of two main factors. First, many organisms cannot digest all their prey. They take only the easily digestible plant leaves or animal tissue such as muscles and internal organs, leaving the hard wood or energy-rich bones behind. Second, much of the energy assimilated by animals is used in maintenance and is lost from the organism as heat. Both of these factors acting together means that, on average, only about 10% of available energy is transferred from one trophic level to another. Because energy is lost at each link, after a few links, most of the available energy has been expended and relatively little energy is available for higher trophic levels (**Figure 58.15**).

As described next, ecologists evaluate the efficiency of consumers as energy transformers in two ways: production efficiency and trophic-level transfer efficiency.

***Production Efficiency*** **Production efficiency** is defined as the percentage of energy assimilated by an organism that becomes incorporated into new biomass.

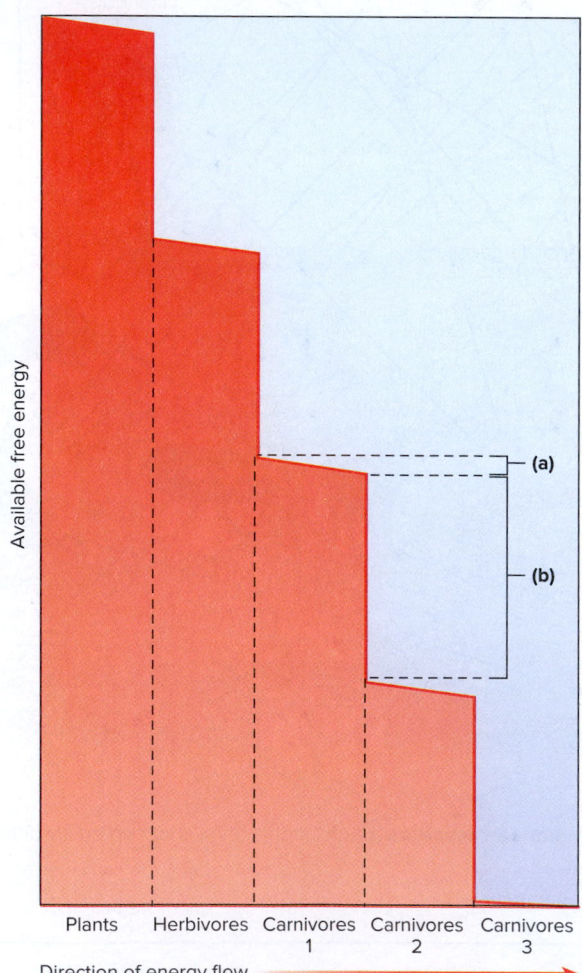

**Figure 58.15 Energy flow through a food web.** This graph of energy flow through a food web shows five trophic levels with four links between the trophic levels. **(a)** Energy lost as heat in a single trophic level. **(b)** Energy lost in the transfer of energy from one trophic level to another.

**Core Concept: Energy and Matter** Within trophic levels, energy is lost to maintenance, and between trophic levels, energy is lost due to imperfect efficiency of transfer.

Energy derived from food 1,000 J

Growth 16 J

Cellular respiration 807 J

Feces 177 J

**Figure 58.16 Production efficiency.** The production efficiency of this squirrel, a mammal, is relatively low. If a mouthful of food contains 1,000 joules (J) of energy, about 807 J is used in cellular respiration to fuel metabolic processes (80.7%), and 177 J (17.7%) is lost in feces. About 16 J of the 823 J assimilated is converted into biomass, a production efficiency of 1.9%.

 **Core Concept: Energy and Matter** For the squirrel, maintaining a constant body temperature reduces its production efficiency.

$$\text{Production efficiency} = \frac{\text{Net productivity}}{\text{Assimilation}} \times 100$$

Here, net productivity is the energy, stored in biomass, that has accumulated over a given time span, and assimilation is the total amount of energy taken in by an organism over the same time span. Invertebrates generally have relatively high production efficiencies that average about 10–40%. Microorganisms also have relatively high production efficiencies. Vertebrates tend to have lower production efficiencies than invertebrates, because they devote more energy to sustaining their metabolism than to new biomass production (**Figure 58.16**). Even within vertebrates, much variation occurs. Fishes, which are ectotherms, typically have production efficiencies of around 10%, and birds and mammals, which are endotherms, have production efficiencies in the range of 1–2%. In large part, this difference reflects the energy cost of maintaining a constant body temperature.

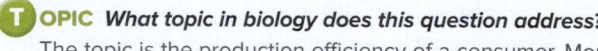

BIO·TIPS    **THE QUESTION** *What is the production efficiency of a caterpillar if a period of feeding delivers 1,000 J of energy of which 500 J is lost in feces, 320 J is used in cellular respiration, and 180 J goes to growth (is converted to biomass)?*

**T OPIC** *What topic in biology does this question address?*
The topic is the production efficiency of a consumer. More specifically, the question asks you to calculate the production efficiency of a caterpillar.

**I** NFORMATION *What information do you know based on the question and your understanding of the topic?* From the question, you know that if a caterpillar eats food that provides 1,000 J of energy, 500 J is lost in feces, 320 J is used in cellular respiration, and 180 J is used for growth. From your understanding of the topic, you may recall that the production efficiency is calculated using this formula:

$$\text{Production efficiency} = \frac{\text{Net productivity}}{\text{Assimilation}} \times 100$$

**P** ROBLEM-SOLVING **S** TRATEGY *Interpret data. Make a calculation.* Insert the data given in the question into the formula. Net productivity is the amount of energy used for growth, or 180 J. Assimilation is the energy used for growth plus the energy used in cellular respiration, or 320 J + 180 J = 500 J.

$$\text{Production efficiency} = \frac{180 \text{ J}}{500 \text{ J}} \times 100 = 36\%$$

**ANSWER** *The caterpillar's production efficiency is 36%.*

***Trophic-Level Transfer Efficiency*** A second way to measure the efficiency of consumers as energy transformers is **trophic-level transfer efficiency**, which is the amount of energy at one trophic level that is acquired by the trophic level above and incorporated into biomass. Calculating this value provides a way of examining energy flow between trophic levels, not just energy flow for an individual species. Trophic-level transfer efficiency is calculated as follows:

$$\begin{array}{l}\text{Trophic-level transfer} \\ \text{efficiency}\end{array} = \frac{\text{Production at trophic level } n}{\text{Production at trophic level } n - 1} \times 100$$

For example, recall from Chapter 54 that zooplankton consists of small drifting animals that graze on microscopic photosynthetic organisms called phytoplankton. If a lake produced 14 g/m² of zooplankton (trophic level *n*) and 100 g/m² of phytoplankton (trophic level *n* − 1), the trophic-level transfer efficiency between these levels would be 14%. Trophic-level transfer efficiency tends to average around 10%, though there is much variation.

Trophic-level transfer efficiency is generally low for two reasons. First, many organisms cannot digest all of their prey. They take only the easily digestible plant leaves or animal tissue such as muscles and guts, leaving the hard wood or energy-rich bones behind. Second, much of the energy assimilated by animals is used in maintenance, so most energy is lost from the system as heat. The 10% average transfer rate of energy from one trophic level to another also necessitates short food webs of no more than four or five levels. Relatively little energy is available for the higher levels.

## Ecological Pyramids Describe the Distribution of Numbers, Biomass, or Energy Between Trophic Levels

The abundance of organisms, biomass, or available energy at each trophic level of a food web can be expressed graphically as an ecological pyramid. One of the best-known ecological pyramids, described by British ecologist Charles Elton in 1927, is the **pyramid of numbers**, in which the number of individuals decreases at each trophic level, with a large

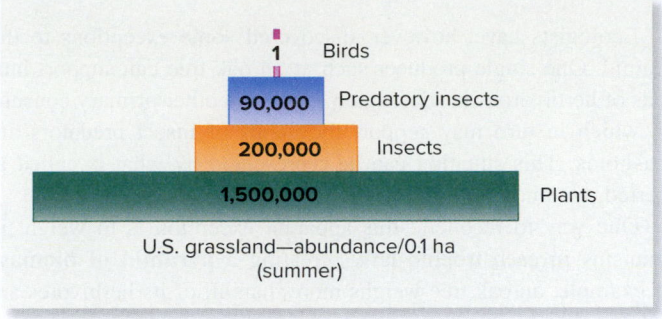

**(a) Pyramid of numbers**

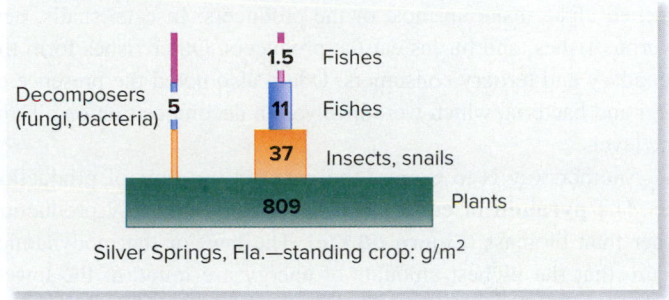

**(b) Pyramid of biomass**

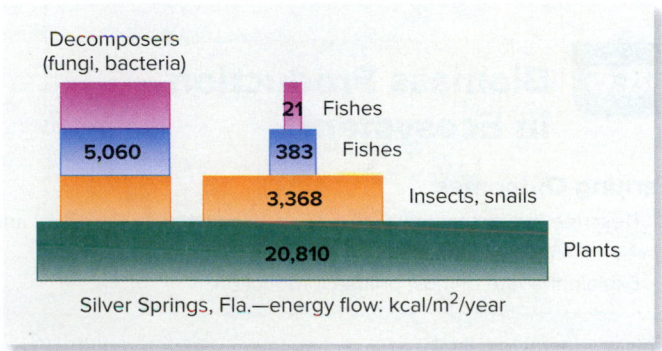

**(c) Pyramid of energy**

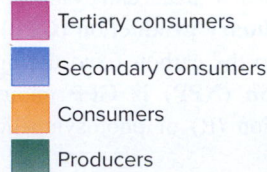

**Figure 58.17 Ecological pyramids in food webs. (a)** In this pyramid of numbers, the abundance of species in an American grassland decreases with increasing trophic level. **(b)** In a pyramid of biomass, the amount of biological material is represented instead of numbers of individuals. Note the presence of decomposers that decompose material at all trophic levels. **(c)** A pyramid of energy for the ecosystem of Silver Springs, Florida. Note the large energy transfers of decomposers, despite their small biomass.

number of individuals at the base and fewer individuals at the top. For example, in a grassland, there may be millions of individual plants per hectare, hundreds of thousands of insects that feed on the plants, tens of thousands of spiders feeding on the insects, and one or two birds that feed on the spiders (**Figure 58.17a**).

Ecologists have, however, discovered some exceptions to this pyramid. One single producer such as an oak tree can support hundreds of herbivorous beetles, caterpillars, and other primary consumers, which in turn may support thousands of insect predators and parasitoids. This situation can be represented by what is called an inverted pyramid of numbers.

One way to reconcile this apparent exception is to weigh the organisms in each trophic level, creating a **pyramid of biomass**. For example, an oak tree weighs more than all of its herbivores and predators combined. American ecologist Howard Odum measured the pyramid of biomass for a freshwater ecosystem, Silver Springs, in Florida (**Figure 58.17b**). Beds of eelgrass (genus *Sagittaria*) and attached algae make up most of the producers. Insects, snails, herbivorous fishes, and turtles eat the producers. Other fishes form the secondary and tertiary consumers. Odum also noted the presence of fungi and bacteria, which were involved in decomposition on all trophic levels.

Another way is to express the pyramid in terms of production rate. The **pyramid of energy** shows the rate of energy production rather than biomass (**Figure 58.17c**). The laws of thermodynamics ensure that the highest amounts of energy are found at the lowest trophic levels. The energy pyramid for Silver Springs also shows that large amounts of energy pass through decomposers, despite their relatively small biomass.

## 58.6 Biomass Production in Ecosystems

**Learning Outcomes:**

1. Describe the factors that limit primary production in terrestrial and aquatic ecosystems.
2. Explain the fate of most primary production.

In this section, we will take a closer look at biomass production in ecosystems. Because the bulk of the Earth's biosphere, 99.9% by mass, consists of primary producers, when we measure ecosystem biomass production, we are primarily interested in plants, algae, and cyanobacteria. Their production is called **gross primary production (GPP)**. Gross primary production is equivalent to the carbon fixed during photosynthesis. **Net primary production (NPP)** is GPP minus the energy used during cellular respiration (R) of photosynthetic organisms.

$$NPP = GPP - R$$

NPP is thus the amount of energy available to primary consumers. Unless otherwise noted, the term **primary production** refers to NPP.

### Primary Production in Terrestrial Ecosystems Is Influenced by Water, Temperature, and Nutrient Availability

In terrestrial ecosystems, water is a major determinant of primary production, and primary production shows an almost linear increase with annual precipitation, at least in arid regions. Likewise, temperature,

which affects production primarily by slowing or accelerating plant metabolic rates, is also important. A lack of **nutrients**, key elements in usable form, particularly nitrogen and phosphorus, can also limit primary production in terrestrial ecosystems, as farmers know only too well. Fertilizers are commonly used to boost the production of annual crops.

In 1984, Susan Cargill and Robert Jefferies showed how a lack of both nitrogen and phosphorus limited production in salt marsh sedges and grasses in Hudson Bay, Canada (**Figure 58.18**). Of these two nutrients, nitrogen was the **limiting factor**—the one in the shortest supply; without it, the addition of phosphorus did not increase production. However, once nitrogen was added and was no longer limiting, phosphorus became the limiting factor. The addition of nitrogen and phosphorus together increased production the most. This result supports a principle known as **Liebig's law of the minimum**, named for Justus von Liebig, a 19th-century German chemist, which states that species' biomass or abundance is limited by the scarcest factor. This factor can change, as the Hudson Bay experiment showed. When sufficient nitrogen is available, phosphorus becomes the limiting factor. If both nitrogen and phosphorus are present in sufficient amounts, then productivity will be limited by another nutrient.

### Primary Production in Aquatic Ecosystems Is Limited Mainly by Light and Nutrient Availability

Of the factors limiting primary production in aquatic ecosystems, the most important are the availability of sufficient light and nutrients. Light is particularly likely to be in short supply because water readily absorbs light. At a depth of 1 m, more than half the solar radiation has been absorbed. By 20 m, only 5–10% of the radiation has not been absorbed. The decrease in light is what limits the depth of algal growth.

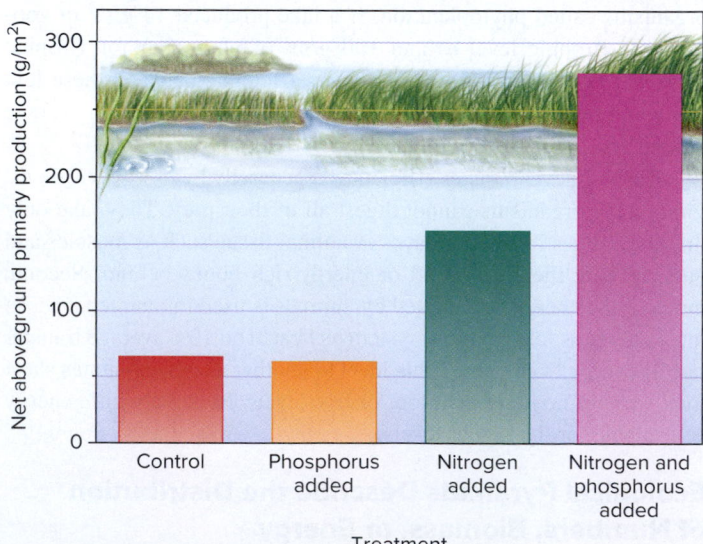

**Figure 58.18** **Limitation of primary production by nitrogen and phosphorus.** Net aboveground primary production of a salt marsh sedge (*Carex subspathacea*) in response to nutrient addition. Nitrogen is the limiting factor. After nitrogen is added, phosphorus becomes the limiting factor.

The most important nutrients affecting primary production in aquatic systems are nitrogen and phosphorus, because they occur in very low concentrations. Whereas soil contains about 0.5% nitrogen, seawater contains only 0.00005% nitrogen. Enrichment of the aquatic environment by the addition of nitrogen and phosphorus occurs naturally in areas of **upwellings**—places where cold, deep, nutrient-rich water containing sediment from the ocean floor is brought to the surface by strong currents, resulting in very productive ecosystems and plentiful fishes. Some of the largest areas of upwelling occur in the Antarctic and along the coasts of Peru and California.

However, too much nutrient supply can be harmful to aquatic systems, resulting in large, unchecked growths of algae called algal blooms. When the algae die, they are consumed by bacteria that, as they respire, deplete the surrounding water of oxygen, causing dead zones with little oxygen to support other aquatic life. Such dead zones are prominent along coastal areas where fertilizer-rich rivers discharge into the oceans. The largest of these is the 22,000-km$^2$ (8,500-mile$^2$) area in the Gulf of Mexico where the Mississippi River dumps high loads of nutrients.

## Primary Production Varies Across the Earth

Knowing which factors limit primary production helps ecologists understand why the mean net primary production varies across the Earth. Modern methods of estimating productivity use orbiting satellites to measure differences in the electromagnetic radiation reflected back from the different vegetation types on Earth (**Figure 58.19**). When we look at the oceans, bright greens, yellows, and reds indicate high chlorophyll concentrations. Some of the highest marine chlorophyll concentrations occur at continental margins, where river nutrients pour into the oceans. Upwellings along coasts also bring nutrient-rich water to the surface. Northern oceans, and to a lesser extent southern oceans, are also very productive, because seasonal storms and temperature changes allow vertical mixing of water, bringing nutrient-rich water to the surface. In the spring, the increased amount of light and nutrients permit rapid phytoplankton growth

until the nutrients are all used up. Many other marine areas, including tropical oceans, are highly unproductive.

Over land, the productivity of forests in all parts of the world, from the tropics to northern and southern temperate areas, is high, but production is often higher in temperate rather than tropical forests. This matches the pattern of productivity observed in the oceans. Although tropical forests enjoy warm temperatures and abundant rainfall, such conditions weather soils rapidly. Tropical soils are low in available forms of most plant nutrients because of loss through leaching. In contrast, temperate soils tend to have much greater concentrations of essential nutrients because of lower rates of nutrient loss and more frequent grinding of fresh minerals by the cycles of continental glaciations over the past 3 million years. Prairies and savannas are also highly productive because their plant biomass usually dies and decomposes each year, returning a portion of the nutrients to the soil, and temperatures and rainfall are not limiting. Deserts and tundra have low productivity because of a lack of water and low temperatures, respectively. Wetlands tend to be extremely productive, primarily because water is not limiting and nutrient levels are high.

## Most Primary Production Is Consumed by Detritivores

A strong relationship exists between primary production and secondary production, usually measured as the biomass of herbivores. Typically, more plant biomass, and thus more primary production, leads to an increased biomass of consumers. However, it has been shown that in ecosystems as diverse as forests and salt marshes most primary production goes to detritivores, not herbivores.

In 1962, American ecologist John Teal examined energy flow in a Georgia salt marsh (**Figure 58.20**). In salt marshes, most of the energy from the Sun goes to two types of organisms: *Spartina* plants and marine algae. The *Spartina* plants are rooted in the ground, whereas the algae float on the water's surface or live on the mud or on *Spartina* leaves at low tide. These photosynthetic organisms absorb about 6% of the sunlight. Most of this energy, 77.6%, is used in plant and

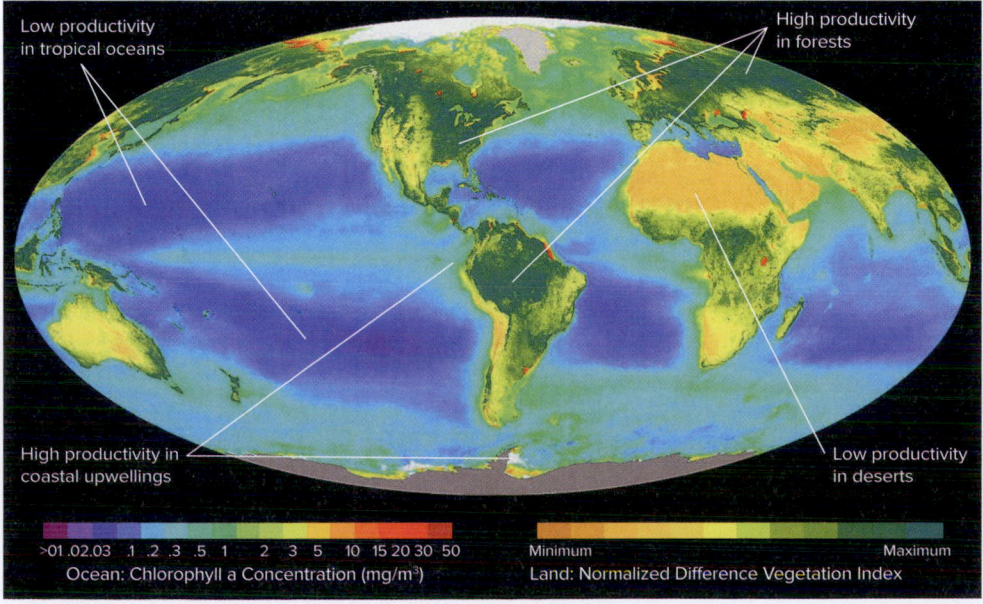

**Figure 58.19  Primary productivity across Earth as measured via satellites.** Ocean productivity is determined by measuring chlorophyll concentrations at the ocean's surface. The normalized difference vegetation index (NDVI) quantifies terrestrial vegetation by measuring the difference between near-infrared (which vegetation strongly reflects) and red light (which vegetation absorbs).

Source: Provided by the SeaWiFS Project, NASA/Goddard Space Flight Center, and ORBIMAGE

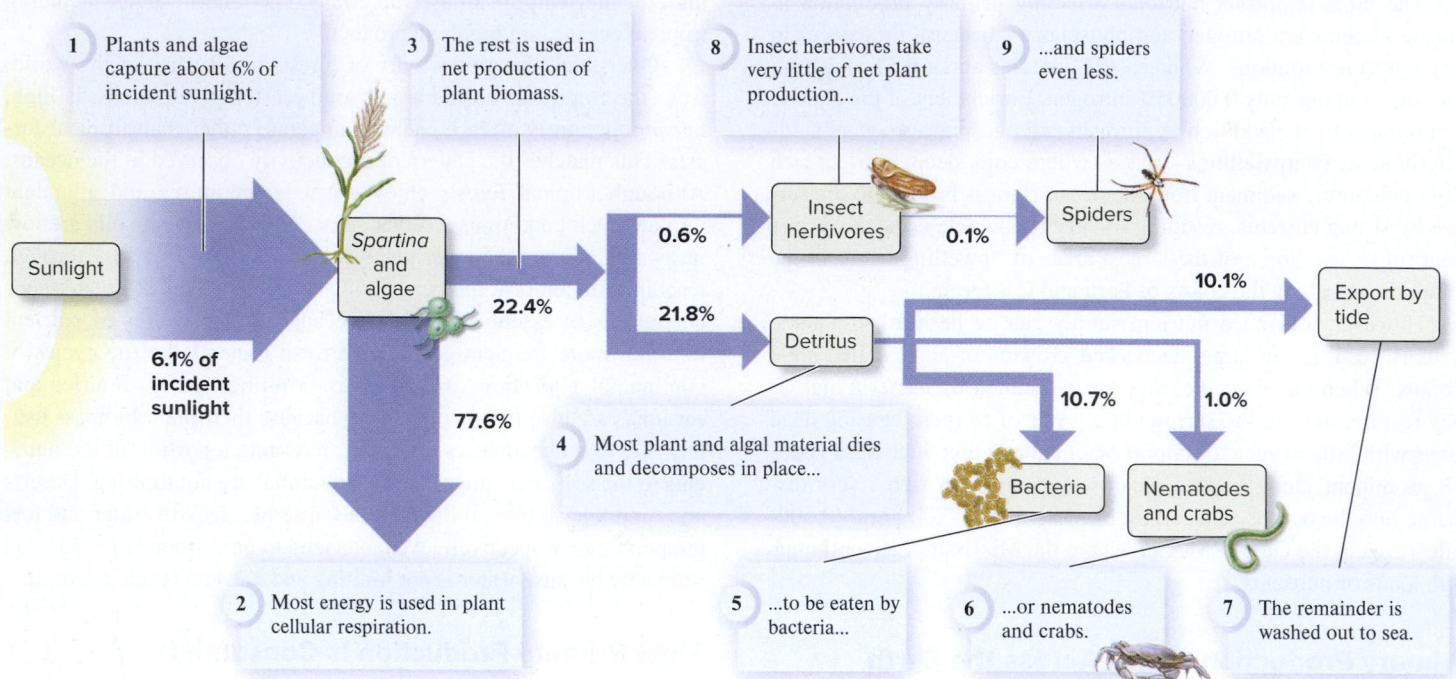

1  Plants and algae capture about 6% of incident sunlight.

3  The rest is used in net production of plant biomass.

8  Insect herbivores take very little of net plant production...

9  ...and spiders even less.

Sunlight

*Spartina* and algae

Insect herbivores

Spiders

0.6%

0.1%

6.1% of incident sunlight

22.4%

21.8%

Detritus

10.1%

Export by tide

77.6%

10.7%

1.0%

Bacteria

Nematodes and crabs

4  Most plant and algal material dies and decomposes in place...

2  Most energy is used in plant cellular respiration.

5  ...to be eaten by bacteria...

6  ...or nematodes and crabs.

7  The remainder is washed out to sea.

**Figure 58.20  Energy-flow diagram for a Georgia salt marsh.** Numbers represent the percentage of gross primary production that flows into different trophic levels or is used in plant respiration.

algal cellular respiration. The energy that is accumulated in biomass is 22.4%, and most of the biomass dies in place and rots on the muddy ground, to be consumed by bacteria. Bacteria are the major detritivores in this system, followed distantly by nematodes and crabs, which feed on tiny food particles as they sift through the mud. Some of the dead material is also removed from the system (exported) by the tide. The herbivores take very little of the plant production, around 0.6%, eating only a small proportion of the *Spartina* and none of the algae. A fraction of herbivore biomass is then consumed by spiders. Overall, if we view the species in ecosystems as transformers of energy, then plants, algae, and cyanobacteria are by far the most important organisms on the planet, other species of bacteria are next, and animals are a distant third.

## Summary of Key Concepts

- The assemblage of many populations of different species in the same location is known as a community. Community ecology explores the factors that influence the number and abundance of these species. An ecosystem is the system formed between communities and their physical environment. Ecosystem ecology addresses the flow of energy and the production of biomass within ecosystems.

### 58.1  Patterns of Species Richness and Species Diversity

- The number of species of most taxa varies according to geographic location, generally increasing from polar areas to tropical areas (Figure 58.1).

- Different hypotheses for the variation in species richness have been advanced, including the species-time hypothesis, the

species-area hypothesis, and the species-productivity hypothesis (Figures 58.2, 58.3).

- Species diversity takes into account both species richness and species abundance. The most widely used measure of species diversity is called the Shannon diversity index (Table 58.1).

### 58.2  Species Richness and Community Stability

- Community stability is an important concept in ecology. The diversity-stability hypothesis maintains that species-rich communities are more stable than communities with fewer species. Tilman's field studies, which showed that year-to-year variation in plant biomass decreased with increasing species richness, established a link between diversity and stability (Figure 58.4).

### 58.3  Succession: Community Change

- Succession describes the gradual and continuous change in community structure over time. Primary succession refers to succession on a newly exposed site with no prior biological legacy. Secondary succession refers to succession on a site that has supported life but has undergone a disturbance (Figures 58.5, 58.6).

- Three mechanisms have been proposed for succession. With facilitation, each species facilitates, or makes the environment more suitable for, subsequent species. With inhibition, initial species inhibit later colonists. With tolerance, any species can start the succession, and species replacement is unaffected by previous colonists (Figures 58.7, 58.8).

## 58.4 Island Biogeography

- In the equilibrium model of island biogeography, the number of species on an island tends toward an equilibrium number determined by the balance between immigration and extinction rates (Figure 58.9).

- The model predicts that the number of species increases with increasing island size, that the number of species decreases with distance from the source pool, and that turnover is high. Support exists for the first two predictions of the model, but experiments on islands in the Florida Keys refuted the third prediction (Figures 58.10, 58.11, 58.12).

## 58.5 Food Webs and Energy Flow

- Ecosystem ecology studies the movement of energy and materials through organisms and their communities. Organisms that obtain energy from light or chemicals are primary producers (or autotrophs). Organisms that feed on primary producers are called primary consumers (or herbivores). Organisms that feed on primary consumers are called secondary consumers (or carnivores). Consumers that get their energy from the remains and waste products of organisms are called detritivores (Figure 58.13).

- Food webs are complex models of interconnected food chains in which multiple links occur between species. Food webs tend to have five or fewer links between top and bottom trophic levels (Figures 58.14, 58.15).

- Production efficiency measures the percentage of energy assimilated by an organism that becomes incorporated into new biomass. Trophic-level transfer efficiency measures the energy available at one trophic level that is acquired by the level above. These efficiencies can be expressed in the form of ecological pyramids, of which the best known is the pyramid of numbers (Figures 59.16, 59.17).

## 58.6 Biomass Production in Ecosystems

- Net primary production (NPP) is gross primary production (GPP) minus the energy released during respiration via photosynthetic organisms. NPP in terrestrial ecosystems is limited primarily by temperature and the availability of water and nutrients. In aquatic ecosystems, it is limited mainly by the availability of light and nutrients (Figures 58.18, 58.19).

- Secondary production is limited by available primary production, but most primary production goes to detritivores (Figure 58.20).

## Assess & Discuss

### Test Yourself

1. A community with many individuals but few different species exhibits
   a. low abundance and high species complexity.
   b. high stability.
   c. low species richness and high abundance.
   d. high species diversity.
   e. high abundance and high species richness.

2. Lake Baikal in Siberia is an ancient unglaciated temperate lake and contains 580 species of bottom-dwelling invertebrates. Great Slave Lake, a comparably sized lake that is at the same latitude in northern Canada and was once glaciated, contains only four bottom-dwelling invertebrate species. This observation supports which of the following hypotheses?
   a. species-time
   b. species-area
   c. species-productivity
   d. both a and b
   e. both b and c

3. Which evidence suggests that more diverse communities are more stable than less diverse communities?
   a. Agricultural land, with fewer species, undergoes less frequent pest outbreaks than natural prairies containing more species.
   b. In Australia, introduced rabbits frequently assume pest proportions. In Europe, coevolved predators such as foxes prevent rabbit populations from reaching pest proportions.
   c. Long-term studies of American grasslands show fields with high numbers of plant species vary less in biomass from year to year.
   d. Both a and b are correct.
   e. Both b and c are correct.

4. The process of primary succession occurs
   a. around a recently erupted volcano.
   b. on a newly plowed field.
   c. on a hillside that has suffered a mudslide.
   d. on a recently flooded riverbank.
   e. on none of the above.

5. Which of the following represent(s) secondary succession?
   a. plants growing in cracks in the pavement of a quiet street
   b. the recovery of vegetation following the 2004 Indonesian tsunami
   c. the colonization of new sand by beach plants
   d. the recovery of forests following a wildfire
   e. both b and d

6. Which is part of MacArthur and Wilson's equilibrium theory of island biogeography?
   a. $\hat{S}$ is increased by distance from the source pool.
   b. $\hat{S}$ is decreased by island size.
   c. $\hat{S}$ is a balance between immigration and extinction.
   d. Island size influences immigration rates.
   e. Distance from the source pool influences extinction rates.

7. As we learned in Chapter 27, some bacteria and archaea are able to use energy from the oxidation of sulphur, iron, or hydrogen. These organisms can be classified as
   a. heterotrophs.        d. both a and b.
   b. autotrophs.          e. both b and c.
   c. producers.

8. Which organisms are the most important consumers of energy in a Georgia salt marsh?
   a. *Spartina* grass and algae
   b. insects
   c. spiders
   d. crabs
   e. bacteria

9. Primary production in aquatic systems is limited mainly by
   a. temperature and moisture.
   b. temperature and light.
   c. temperature and nutrients.
   d. light and nutrients.
   e. light and moisture.

10. The most highly productive terrestrial ecosystems are
    a. deserts.
    d. savannas.
    b. prairies.
    e. tundra.
    c. forests.

## Conceptual Questions

1. Forest A has 5 tree species with 100 individuals and forest B has 5 tree species with 10 individuals. What is the Shannon diversity index for both forests? Which forest has the higher diversity? What do your answers tell you about the limitations of the Shannon diversity index?

2. At what trophic level does a carrion beetle feed?

3. **Core Concept: Systems** In the nutrient-poor heathlands of Europe, scotch heather (*Calluna vulgaris*) and cross-leaved heath (*Erica tetralix*) are gradually replaced by variegated purple moor grass (*Molinia caerulea*) and wavy hair grass (*Deschampsia flexuosa*). Adding *C. vulgaris* litter or nitrogen fertilizer to the soil speeds up this process. Explain this phenomenon, and identify which mechanism of succession is supported.

## Collaborative Questions

1. List some possible ecological disturbances, their likely frequency in natural communities, and the severity of their effects.

2. Calculate the species diversity (the Shannon diversity index) of the following four communities. Which community has the highest diversity? What is the maximum diversity each community could have?

| Community | Relative abundance of species | | | $H_s$ | Maximum possible diversity |
|---|---|---|---|---|---|
| | Species 1 | Species 2 | Species 3 | | |
| 1 | 90 | 10 | — | | |
| 2 | 50 | 50 | — | | |
| 3 | 80 | 10 | 10 | | |
| 4 | 33.3 | 33.3 | 33.3 | | |

# The Age of Humans

# 59

**Burmese python in the Florida Everglades.** This species of snake was introduced into the Florida Everglades by humans and has dramatically decreased the populations of several native species. Source: Photo courtesy of National Park Service, Everglades National Park

In January 2016, British geoscientist Colin Waters and colleagues proposed formal recognition of a new geological epoch, the Anthropocene, the last and latest epoch of the Quaternary period (refer back to Figure 26.4). Some scientists have argued that the Anthropocene is a pop-culture term and a political statement, based on the word **anthropogenic**, meaning resulting from human activity, rather than a new epoch. But support is growing to embrace the Anthropocene, a term coined in the 1980s by American ecologist Eugene Stoermer and later popularized by Dutch atmospheric chemist Paul Crutzen. In the 1970s and 1980s, Crutzen showed how human pollutants can damage the ozone layer, a discovery for which he won the Nobel Prize in Chemistry in 1995, along with Mexican chemist Mario Molina and American chemist Frank Rowland.

While agreement is widespread that humans have a profound effect on the global environment, the formal recognition of a new geological time unit is subject to specific criteria. Global-scale changes must be recorded in geological stratigraphic (layered) material, usually rock, and in ice or marine sediments. Waters and colleagues noted the appearance of manufactured items in these sediments, including plastics, concrete, aluminum, and particulates from fossil-fuel combustion. Perhaps the most distinctive new identifiable items in sediments were radioactive elements from the 500 aboveground nuclear blasts that occurred between 1945 and 1963, which created globe-encircling debris.

Whether or not the Anthropocene gains wide acceptance in the scientific literature, there is no disputing the immense impacts of humans on natural systems. In this chapter, we will examine some of these impacts, which include global warming, effects on biogeochemical cycles, habitat destruction such as deforestation, the overexploitation of various species via overfishing and overhunting, and the introduction of invasive species to all corners of the globe (see the chapter opening photo of a Burmese python in the Everglades). We will begin the chapter by addressing the huge numbers of humans on the planet and our prodigious rate of population growth.

## 59.1 Human Population Growth

**Learning Outcomes:**

1. Describe how differences in age structure and human fertility across different countries affect human population growth.

2. Explain the concept of an ecological footprint.

In 2015, the world's population was estimated to be increasing at the rate of 150 people every minute: about 2 per minute in developed nations and 148 per minute in less-developed nations. In this section, we will examine human population growth trends in more detail and discuss how knowledge of the human population's age structure and fertility levels can help predict its future growth. We will then investigate the carrying capacity of the Earth for humans and explore how the concept of an ecological footprint, which measures human resource use, can help us determine this carrying capacity.

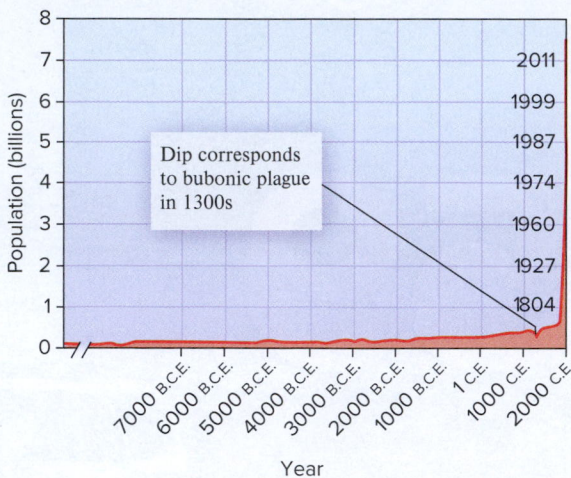

**Figure 59.1** The growth pattern of the human population through history.

## Human Populations Show Extreme Recent Growth

Until the beginning of agriculture and the domestication of animals, around 10,000 B.C.E., the average rate of human population growth was very low. With the establishment of agriculture, the world's population grew to about 300 million by 1 C.E. and to 800 million by the year 1750. Between 1750 and 2011, a relatively short period of human history, the world's human population surged from 800 million to 7 billion (**Figure 59.1**). In 2017, the number of humans was estimated at 7.5 billion. Scientists are very interested in determining when and at what size the human population will level off.

## Knowledge of a Population's Age Structure Can Help Predict Its Future Growth

The age structure of a population can help to predict future population growth. In all populations, **age structure** refers to the relative numbers of individuals of each defined age group. This information is commonly displayed as a population pyramid. In West Africa, for example, children younger than age 15 make up nearly half of the population, creating a pyramid with a wide base and narrow top (**Figure 59.2a**). Even if fertility rates decline, the population is expected to significantly increase as these young people move into childbearing age. The age structure of Western Europe is much more uniform (**Figure 59.2b**). Even if the fertility rate of young women in Western Europe increases to a level higher than that of their mothers, the annual numbers of births will still be relatively low because of the low number of women of childbearing age.

## Human Population Fertility Rates Vary Widely Around the World

Global population growth can be estimated by determining the **total fertility rate (TFR)**, the average number of live births a woman has during her lifetime (**Figure 59.3**). The total fertility rate differs considerably from one geographic area to another. In Africa, the total fertility rate of 4.6 in 2010 has declined substantially since the 1970s, when it was around 6.7 children per woman. In Latin America and Southeast Asia, the rates declined even more from the 1970s and are now at around 2.1. Canada and most countries in Europe have a TFR of less than 2.0. The TFR was about 1.86 in the U.S. in 2016. In Russia,

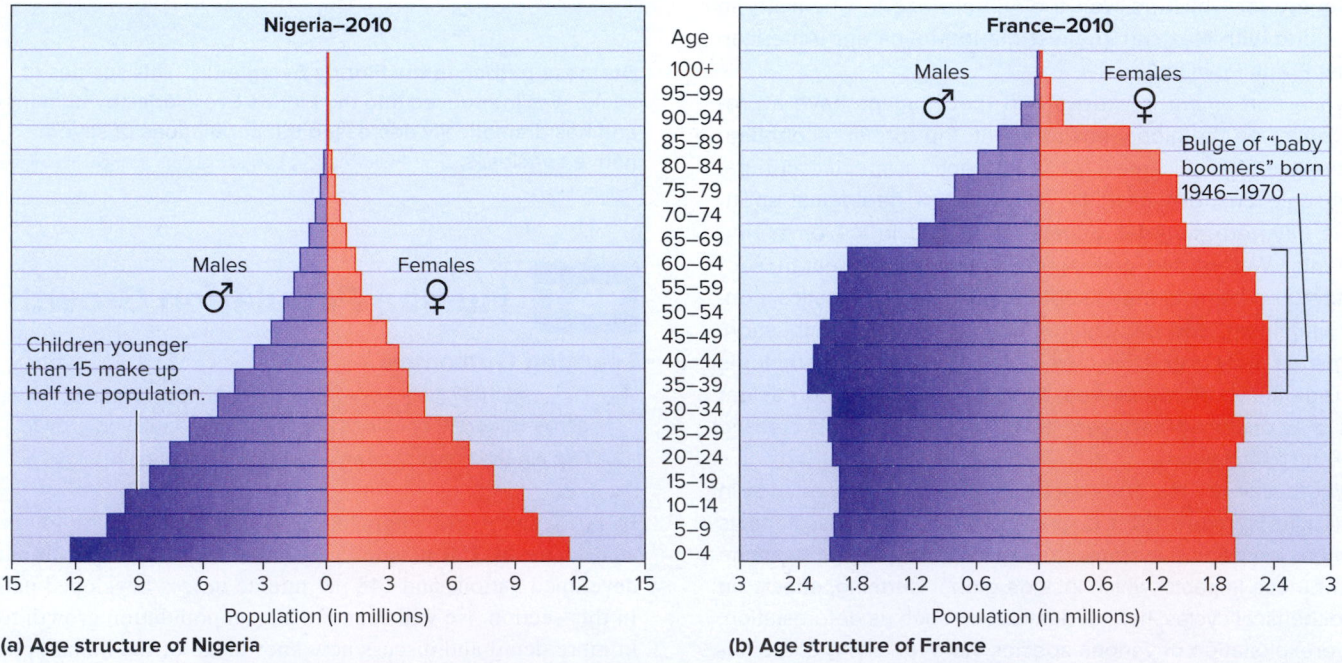

**(a) Age structure of Nigeria**

**(b) Age structure of France**

**Figure 59.2** The age structure of human populations in Nigeria and France, as of 2010. **(a)** In developing areas of the world such as Nigeria, children comprise the most abundant age group. Population growth is rapid. **(b)** In the developed countries of Western Europe, the age structure is more evenly distributed. The bulge represents those born in the post–World War II "baby boom," when birth rates climbed due to stabilization of political and economic conditions. Population growth in developed countries is close to zero.

**Concept Check:** *If the population pyramid in Figure 59.2a was inverted, what would you conclude about the age structure of the population?*

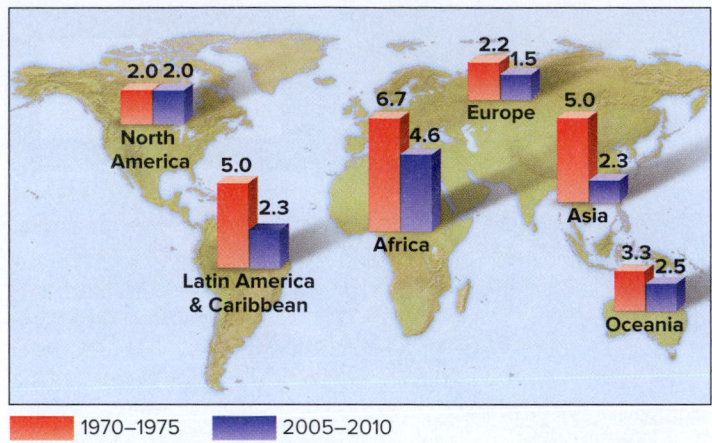

1970–1975    2005–2010

**Figure 59.3** **Total fertility rates (TFRs) among major regions of the world.** Data refer to the average number of children born to a woman during her lifetime.

fertility rates have dropped to 1.61. In China, although the TFR is only 1.6, the population there will still continue to increase until at least 2025 because of the large number of women of reproductive age.

Although the global TFR declined from 4.47 in the 1970s to 2.42 in 2016, this is still greater than the average of 2.3 needed for zero population growth. The replacement rate is slightly higher than 2.0, the number necessary to replace a mother and father, due to natural mortality prior to reproduction. The replacement rate varies globally, from 2.1 in developed countries to between 2.5 and 3.3 in developing countries.

The wide variation in fertility rates makes it difficult to predict future population growth. A 2010 United Nations report presented world population projections to the year 2100 for three different growth scenarios: low, medium, and high (**Figure 59.4**). The three scenarios are based on three different assumptions about fertility rate. Using a low fertility rate estimate of only 1.5 children per woman, the population would reach a maximum of about 8 billion people by

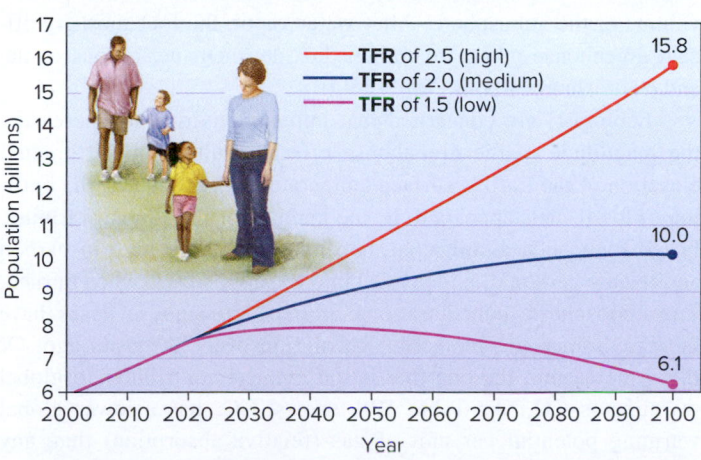

**Figure 59.4** **Population predictions for 2000–2100, using three different total fertility rates (TFRs).**

 **Core Skill: Science and Society** How TFR is calculated has a great influence on assumptions about how human global population size will change over the next 80 years.

2050. A more realistic assumption may be to use the fertility rate estimate of 2.0 or even 2.5; with these rates, the population would continue to rise to 10 billion or almost 16 billion, respectively.

## The Concept of an Ecological Footprint Helps Estimate Carrying Capacity

Recall from Chapter 56 that carrying capacity refers to the maximum population size that can be sustained by an environment. What is the Earth's carrying capacity for the human population, and when will it be reached? Estimates vary widely. Much of the speculation on the upper boundary of the world's population size centers on lifestyle. To use a simplistic example, if everyone on the planet ate meat extensively and drove large cars, then the carrying capacity would be a lot less than if people were vegetarians and used bicycles as their main means of transportation.

In the 1990s, Swiss researcher Mathis Wackernagel and his coworkers calculated how much land is needed to support each person on Earth. Everybody has an effect on the Earth, because we consume the land's resources, including crops, wood, fossil fuels, minerals, and so on. Thus, each person has an **ecological footprint**—the amount of productive land needed to support a person. The average footprint size for everyone on the planet is about 3 hectares (1 ha = 10,000 m²), but wide variation is found around the globe (**Figure 59.5**). The ecological footprint of the average Canadian is 7.5 hectares, and it is about 10 hectares for the average American.

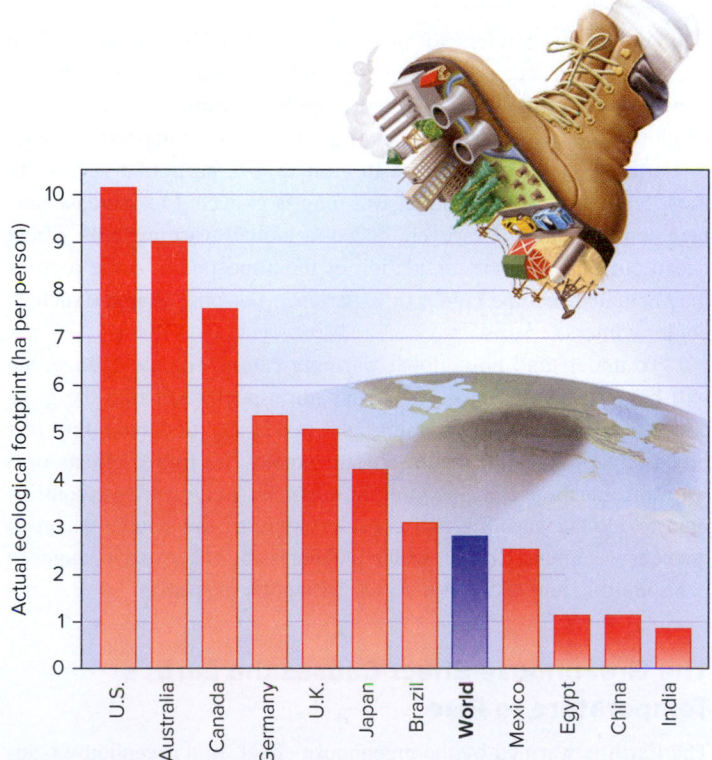

**Figure 59.5** **Ecological footprints of different countries.** The term ecological footprint refers to the amount of productive land needed to support the average individual of that country.

**Concept Check:** *What is your ecological footprint?*

In most developed countries, the largest component of land use is for energy, followed by food and then forestry. Much of the land needed to provide energy serves to absorb the $CO_2$ emitted by the use of fossil fuels. If everyone required 10 hectares, as the average American does, we would need three Earths to provide us with the needed resources. Many people in less-developed countries use far fewer resources.

Globally, humans are already beyond the Earth's carrying capacity if we were to live in a sustainable manner. How has this happened? Many people currently live in an unsustainable manner, using more resources than can be regenerated in any given year. Furthermore, a rapidly growing human population contributes to environmental changes on a broad scale including habitat loss for wildlife, overhunting and overfishing, and pollution. As described next, pollution is thought to contribute to climate change on a global scale.

## 59.2 Global Warming and Climate Change

**Learning Outcomes:**

1. Describe how the greenhouse effect contributes to global warming.
2. Explain how global warming may affect sea levels and precipitation patterns.
3. Predict how climate change will change species distribution patterns.

**Global warming** refers to an increase in Earth's average surface temperature. In recent years, ecologists have been investigating how human activities have contributed to global warming. A key impact of global warming is **climate change**, which is a long-term change in Earth's climate or a change in climate in a particular region. In most regions on Earth, global warming is expected to increase surface temperatures. However, because global warming can affect ocean currents and the circulation of the atmosphere, some regions may actually become colder or experience seasonal changes such as colder winters.

To understand how global warming causes climate change, we will begin by exploring how Earth's atmosphere acts like the glass panes in a greenhouse, trapping heat inside. By understanding this process, ecologists are better able to predict the future effects of a warming Earth. We will explore how human activities have contributed to global warming and consider some of the potential consequences of climate change, which include rising sea levels, changes in precipitation, and changes in the distributions of species.

### The Greenhouse Effect Causes the Earth's Temperature to Rise

The Earth is warmed by the greenhouse effect. In a greenhouse, sunlight penetrates the glass and heats the plants inside. The heat is radiated by the plants but the glass acts to trap the heat inside. Similarly, solar radiation in the form of short-wave energy passes through the atmosphere to heat the surface of the Earth. This energy is then radiated from the Earth's warmed surface into the atmosphere, but in the

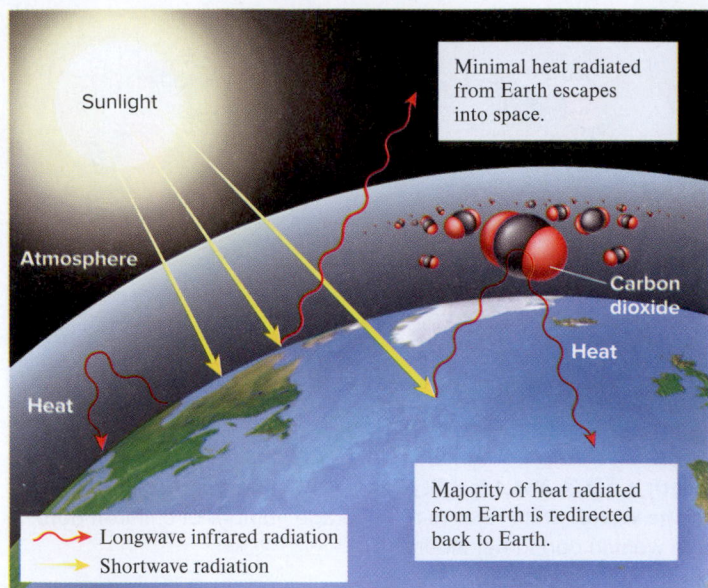

**Figure 59.6** **The greenhouse effect is caused by the insulating effect of atmospheric carbon dioxide.** Solar radiation, in the form of short-wave energy, passes through the atmosphere to heat the Earth's surface. Long-wave infrared energy is radiated back into the atmosphere. Most infrared energy is redirected back to Earth by atmospheric gases, including carbon dioxide molecules, causing global temperatures to rise.

form of long-wave infrared radiation. Atmospheric gases absorb much of this infrared energy and radiate it a second time to the Earth's surface, causing its temperature to rise further (**Figure 59.6**). The greenhouse effect is important to life on Earth. Without the greenhouse effect, global temperatures would be much lower than they are, perhaps averaging only –17°C compared with the existing average of +15°C.

The greenhouse effect is caused by a small group of gases, mainly water vapor, that together make up less than 1% of the total volume of the atmosphere. After water vapor, the four most significant greenhouse gases are carbon dioxide, methane, nitrous oxide, and chlorofluorocarbons (**Table 59.1**).

Ecologists are concerned that human activities are increasing the magnitude of the greenhouse effect, resulting in the gradual elevation of the Earth's surface temperature. In particular, the burning of fossil fuels appears to be the main culprit of global warming. Fossil fuels such as oil, coal, and natural gas are high in carbon and release carbon dioxide ($CO_2$) into the atmosphere when burned. The atmospheric concentrations of most greenhouse gases have increased since the Industrial Revolution over 200 years ago. Of those increasing, the one that is the greatest contributor to global warming is $CO_2$. As Table 59.1 shows, $CO_2$ has a lower global warming potential per unit of gas (relative absorption) than any other major greenhouse gas, but its concentration in the atmosphere is much higher. An analysis of air trapped in glaciers shows that concentrations of atmospheric $CO_2$ increased from about 280 ppm (parts per million) in the pre-industrial 19th century to 400 ppm in 2015 (**Figure 59.7a**).

Since 1957, air samples have been collected directly at Mauna Loa, Hawaii, a relatively unpolluted site. The data show a 25%

| Table 59.1 | The Major Greenhouse Gases and Their Contribution to Global Warming | | | |
|---|---|---|---|---|
| | Carbon dioxide ($CO_2$) | Methane ($CH_4$) | Nitrous oxide ($N_2O$) | Chlorofluorocarbons (CFCs) |
| Relative absorption in ppm of increase* | 1 | 21 | 310 | 10,000 |
| Atmospheric concentration (ppm†) | 400 | 1.75 | 0.315 | 0.0005 |
| Contribution to global warming | 73% | 7% | 19% | 1% |
| Percent from natural sources; type of source | 20–30%; volcanoes | 70–90%; swamps, gas from termites and ruminants | 90–100%; soils | 0% |
| Major human-made sources | Fossil-fuel use, deforestation | Rice paddies, landfills, biomass burning, coal and gas exploitation | Cultivated soil, fossil-fuel use, automobiles, industry | Previously manufactured products (for example, aerosol propellants) but now banned in the U.S. and the E.U. |

*Relative absorption is the warming potential per unit of gas.
†ppm = parts per million

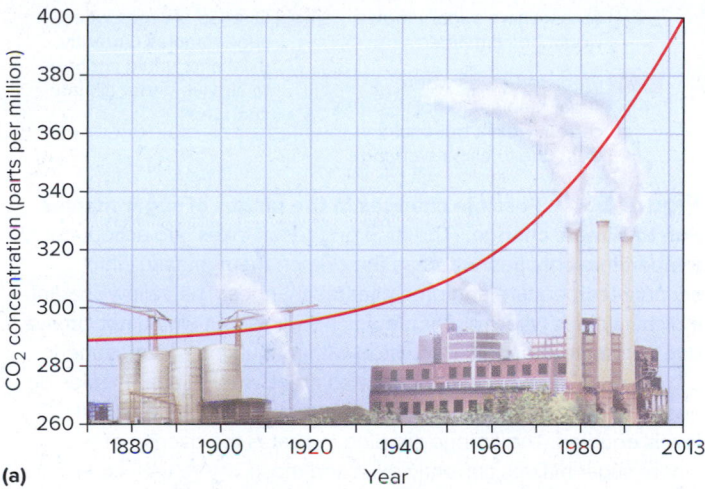

(a)

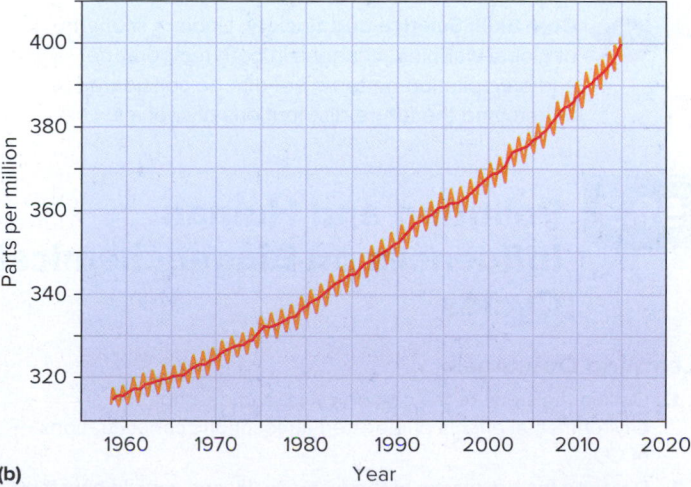

(b)

**Figure 59.7**  **The increase of atmospheric carbon dioxide.**
**(a)** Increases in the burning of fossil fuels have caused an increase in atmospheric $CO_2$ since the late 19th century. **(b)** In air samples taken directly at Mauna Loa, Hawaii, since 1957, $CO_2$ levels have shown consistent increases.

*Concept Check:* *Why does the amount of $CO_2$ fluctuate seasonally in the graph in Figure 59.7b?*

increase in atmospheric $CO_2$ levels in just 58 years, from 313 ppm to 400 ppm during 1957–2015 (**Figure 59.7b**). In addition, seasonal oscillations in $CO_2$ occur, as seen in the orange line of Figure 59.7b. These types of oscillations differ in the two hemispheres of Earth. The Northern Hemisphere, where Hawaii is located, has greater land area and plant biomass than the Southern Hemisphere. In the northern summer, more $CO_2$ is absorbed by plants and atmospheric $CO_2$ level declines slightly. In the northern winter, less $CO_2$ is absorbed and atmospheric $CO_2$ levels increase.

## Global Warming Is Expected to Cause Sea Levels to Rise

To predict the effects of global warming, most scientists focus on a future point, about 2100, when the concentration of atmospheric $CO_2$ will have doubled—that is, increased to about 700 ppm compared with the late-20th-century level of 350 ppm. The 2014 report of the Intergovernmental Panel on Climate Change (IPCC) suggested that if greenhouse gas emissions continue to rise at the same rate, this would lead to a 2–4°C warming by about the end of this century. This increase in temperature might not seem like much, but it is comparable to the warming that ended the last Ice Age.

One consequence of global warming is its effect on sea levels due to the melting of glaciers. The term **sea level** refers to the average level for the surface of one or more of Earth's oceans. With a 2–4°C warming, sea levels are expected to rise by about 50–60 cm. Sea levels have already risen 10–25 cm over the past century, and seawater entry into coastal forests has killed trees in many areas.

An alarming problem is the long time lag between warming temperatures and the time it takes for ice to melt. Even if future temperature increases were prevented, glaciers would continue to melt for many years. For example, the ice sheets of Greenland will take hundreds of years to melt, even though the increase in temperature needed to melt them could be reached in 50–100 years. You might envision this as being similar to an ice cube on a kitchen counter—it takes a while for the ice cube to melt, even though room temperature is considerably above freezing.

Therefore, even if no greenhouse gases were added to the atmosphere after 2020, at least a quarter of the sea-level rise would still be expected by 2100.

## One Effect of Global Warming on Climate Change Is to Alter Precipitation Patterns on Land

An increase in global temperatures is also expected to promote climate change and alter global precipitation. Higher temperatures increase evaporation, and warmer air holds more water vapor than cooler air. Global warming will generally lead to more frequent and heavier precipitation. Over the period 1986–2000, the average increase in precipitation over land was 3.5 mm per year. However, over normally dry areas, global warming is expected to increase evaporation of water from Earth's surface, but the level of moisture in the atmosphere over dry regions may not be high enough to facilitate rain. Therefore, the increased temperatures will likely lead to widespread droughts in desert areas.

We can make the analogy that the atmosphere is like a sponge; warm, moist air allows the sponge to absorb more water, but in dry conditions nothing more can be rung out of the sponge. Northern and southern temperate areas have already shown precipitation increases, as have tropical and subtropical areas. However, desert areas such as the Sahel region south of the Sahara have become drier. In essence, the wet get wetter and the dry get drier.

## Climate Change Is Expected to Alter the Distribution of Species

Assuming that the previous scenario of climate change is accurate, scientists are beginning to consider the consequences on natural and human-made ecosystems. As most regions on Earth become warmer and a few regions become colder, most plant species cannot easily disperse and move north or south into newly created climatic regions that will be suitable for them. Many tree species take hundreds, even thousands, of years for seed dispersal.

Paleobotanist Margaret Davis was among the first to investigate how species' ranges would be changed in a globally warmed world. She predicted that in the event of a $CO_2$ doubling, many tree species in North America, such as sugar maples, would suffer range contractions because their southern ranges would be eliminated due to higher temperatures (see the yellow areas in **Figure 59.8a**). This contraction in the trees' distribution could be offset by the creation of new favorable habitats in Canada (see the red areas in Figure 59.8a). However, most scientists hypothesize that the climatic zones would shift toward the poles faster than the tree species could either migrate via seed dispersal or evolve to become better adapted to their altered climate. Therefore, the areas shown in red in Figure 59.8a are hypothetical and may not be achieved. When Margaret Davis mapped out the future distribution of the sugar maple, *Acer saccharum,* in a globally warmed world and also took into account changed precipitation patterns, the predicted future area of distribution was considerably less than the prediction taking into account only temperature changes (**Figure 59.8b**).

**Current and projected ranges of sugar maple**

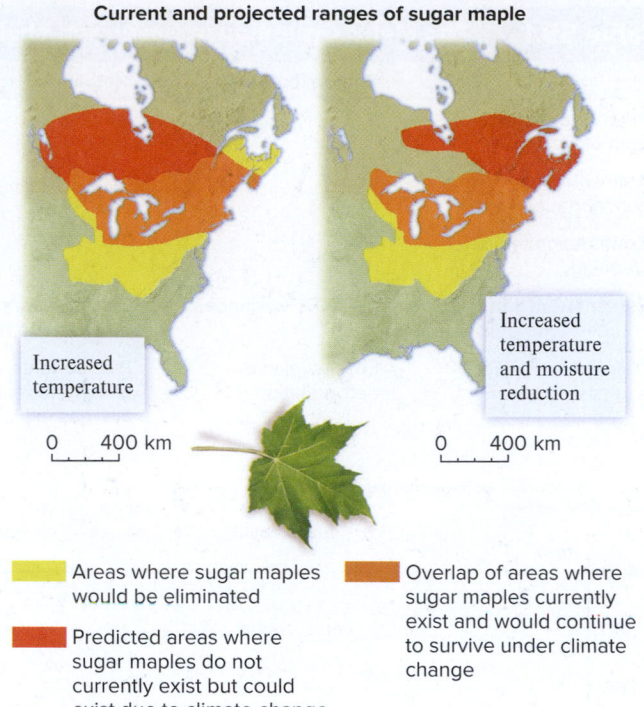

Increased temperature

Increased temperature and moisture reduction

0    400 km

0    400 km

Areas where sugar maples would be eliminated

Predicted areas where sugar maples do not currently exist but could exist due to climate change

Overlap of areas where sugar maples currently exist and would continue to survive under climate change

**Figure 59.8  Possible changes in the ranges of sugar maples due to climate change.** The map on the left takes into account only temperature changes, whereas the one on the right takes into account temperature and precipitation changes. The yellow shading indicates areas where sugar maples now exist but would not survive due to climate changes associated with a doubling of atmospheric $CO_2$ levels. The red shading indicates areas where sugar maples do not currently exist but would have a favorable climate to exist if $CO_2$ levels doubled. The orange shading indicates the overlap of areas where sugar maples currently exist and those where they could also exist due to climate change.

**Core Skill: Science and Society**  Under a scenario of global warming, changes in both temperature and precipitation patterns need to be considered to understand the future distributions of species.

## 59.3  Pollution and Human Influences on Biogeochemical Cycles

### Learning Outcomes:

**1.** Outline the steps of the carbon cycle, and describe the environmental effects of elevated atmospheric concentrations of $CO_2$.

**2.** Describe the processes of the water cycle, and explain how they are affected by humans.

**3.** Describe the role of phosphorus in lake eutrophication.

**4.** List the five main steps of the nitrogen cycle, and identify the human influences on it.

A unit of energy moves through an ecosystem only once, passing through the trophic levels of a food web from producer to consumer and dissipating as heat. In contrast, chemical elements such as carbon or nitrogen follow a cycle, moving from the physical environment to organisms and back to the environment, where the cycle begins again. Because the movements of chemicals through ecosystems involve biological, geological, and chemical transport mechanisms, they are termed **biogeochemical cycles**. Biological mechanisms involve the absorption of chemicals by living organisms and their subsequent release back into the environment. Geological mechanisms include weathering and erosion of rocks and transporting of elements by surface and subsurface drainage. Chemical transport mechanisms include dissolved matter in rain and snow, atmospheric gases, and dust blown by the wind.

In addition to the basic building blocks of carbon, hydrogen, and oxygen, the elements required in the greatest amounts by living organisms are phosphorus and nitrogen. In this section, we take a detailed look at the cycles of these nutrients, which can be divided into two broad types: (1) local cycles, such as the phosphorus cycle, which involve elements with no atmospheric mechanism for long-distance transfer, and (2) global cycles, which involve an interchange between the atmosphere and the ecosystem. In our discussion of biogeochemical cycles, we will take a close look at how these cycles are affected by human activities, such as the burning of fossil fuels and the use of fertilizers.

## The Carbon Cycle Is Affected by Human Activities Such as the Burning of Fossil Fuels and Deforestation

The movement of carbon from the atmosphere into organisms and back again is known as the carbon cycle. The rates of the processes shown in **Figure 59.9** vary greatly. Some carbon becomes locked up in reservoirs with a low turnover rate. For example, a fraction of the material from primary producers is transformed into deposits of coal, natural gas, and oil, which are collectively known as **fossil fuels**. In addition, much carbon is incorporated into the shells of marine organisms, which eventually form huge limestone deposits on the ocean floor or in terrestrial rocks, where turnover is extremely slow. As a result, rocks and fossil fuels contain the world's largest reserves of carbon.

In contrast, much of the carbon in phototrophs, which include plants, algae, and cyanobacteria, turns over much more rapidly. These organisms acquire $CO_2$ from the atmosphere and incorporate it into the organic matter of their own biomass via photosynthesis. Each year, plants, algae, and cyanobacteria remove approximately one-seventh of the $CO_2$ from the atmosphere. At the same time, respiration and the decomposition of phototrophs recycle a similar amount of carbon back into the atmosphere as $CO_2$. Other natural sources also release significant amounts of $CO_2$ into the atmosphere. These include volcanoes, hot springs, and fires. Animals can return some $CO_2$ to the atmosphere by respiration and decomposition, but the amount of carbon flowing through this part of the cycle is minimal.

Human activities, primarily the burning of fossil fuels, are increasingly causing large amounts of $CO_2$ to enter the atmosphere. In addition, deforestation contributes to elevation of atmospheric $CO_2$ because there is less vegetation to absorb $CO_2$ from the atmosphere. Direct measurements over the past five decades show a steady rise in atmospheric $CO_2$ (see Figure 59.7), a pattern that shows no sign of slowing.

What are the consequences of elevated levels of atmospheric $CO_2$? As discussed in Section 59.2, ecologists have identified a link

between a rise in atmospheric $CO_2$ and global warming. Elevated atmospheric $CO_2$ can also increase the acidity of the oceans, mainly near the surface, because carbonic acid is formed when $CO_2$ dissolves in water. Higher acidity inhibits shell growth in marine mollusks, crabs, corals and echinoderms, and can cause reproductive disorders in some fish. Also, as described next, higher $CO_2$ levels may boost plant growth but lower the amount of herbivory.

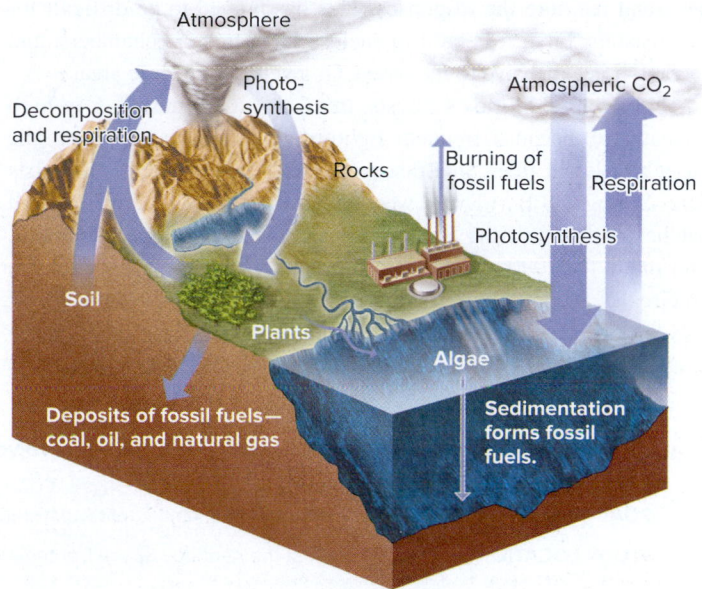

**Figure 59.9** **The carbon cycle.** Each year, plants and algae remove about one-seventh of the $CO_2$ in the atmosphere. Because the carbon flow due to animal respiration and decomposition is so small, it is not represented. The widths of the arrows indicate the relative contribution of each process to the cycle.

*Concept Check:* *Where are the world's greatest stores of carbon?*

 **Core Skill: Connections** Refer back to Table 2.2. Carbon is one of just four elements that account for the vast majority of atoms in living organisms. What are the other three, and, therefore, what biogeochemical cycles might be the most important to us?

 **Core Skill: Modeling** The goal of this modeling challenge is to create a model of the carbon cycle by linking reservoirs and processes.

**Modeling Challenge:** Figure 59.9 is a pictorial model of the carbon cycle. In this modeling challenge, you will create a more schematic model of that cycle. Your model should have two types of components. First, reservoirs for carbon should be drawn as boxes with a label in each. For example, one box in your model will be labeled "Atmosphere." The second component of your model will be arrows that connect the boxes. These arrows represent processes and should also be labeled. For example, one arrow will be labeled "Burning of fossil fuels." Note: Any given box can be linked to another box by multiple arrows.

## Core Skill: Process of Science

## Feature Investigation | Stiling and Drake's Experiments with Elevated CO₂ Showed an Increase in Plant Growth but a Decrease in Herbivore Survival

How will forests of the future respond to elevated $CO_2$? To begin to answer such a question, ecologists ideally would enclose large areas of forests with chambers, increase the $CO_2$ content within the chambers, and measure the responses. This has proved to be difficult for two reasons. First, it is hard to enclose large trees in chambers, and second, it is expensive to increase $CO_2$ levels over a large area.

In much of Florida's forests, trees are small, only 3–5 m high at maturity, because frequent lightning-initiated fires prevent the growth of larger trees. In a discovery-based investigation, ecologists Peter Stiling and Bert Drake were able to increase $CO_2$ levels around patches of forest at the Kennedy Space Center in Cape Canaveral, Florida. In the 1990s, they teamed up with NASA engineers to create 16 circular, open-topped chambers (**Figure 59.10**). In eight of these, they increased atmospheric $CO_2$ to double the ambient level, from around 350 ppm to 700 ppm, the latter of which is the atmospheric

concentration predicted by the end of the 21st century. The experiments commenced in 1996 and lasted until 2007.

The experiment shown in Figure 59.10 focuses on the effects of elevated $CO_2$ on a type of herbivore called a leaf miner, the most common type of herbivore at this site. Because the chambers were open-topped, these insect herbivores could come and go. Censuses were conducted on all species of leaf miners (step 2) and leaves were also examined for damage by leaf miners (step 3). Leaf miner larvae are small enough to burrow between the surfaces of plant leaves, creating leaf mines. These mines result in brown areas, as shown in Figure 59.10, step 3. The analysis of a leaf mine reveals whether a larva survived and emerged, died of nutritional inadequacy, was eaten by predators, or was attacked by parasitoids.

As seen in the data in step 4 of Figure 59.10, the mortality of insect herbivores, namely leaf miners, was higher under conditions

**Figure 59.10** **The effects of elevated atmospheric CO₂ on herbivore survival.** (photos) ©Peter Stiling

GOAL To determine the effects of elevated $CO_2$ on a forest ecosystem; effects on herbivore survival are highlighted here.

STUDY LOCATION Patches of forest at the Kennedy Space Center in Cape Canaveral, Florida.

| | Experimental level | Conceptual level |
|---|---|---|
| **1** Erect 16 open-top chambers around native vegetation. Increase $CO_2$ levels from 350 ppm to 700 ppm in half of them. | 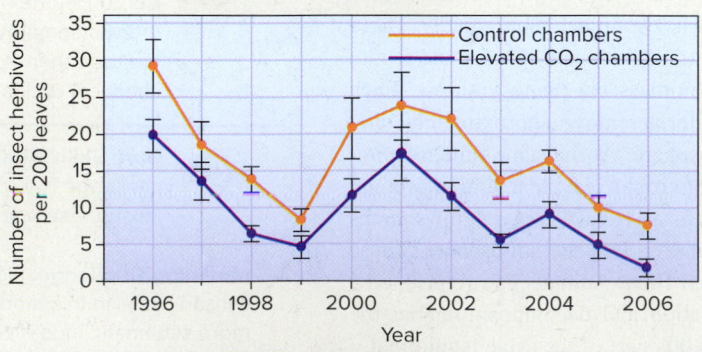 | Expected atmospheric $CO_2$ level is 700 ppm by end of the 21st century. Open-top chambers allow movement of herbivores in and out of chambers. |
| **2** Conduct a yearly count of numbers of insect herbivores per 200 leaves in each chamber. | | |
| **3** Analyze the leaf mines created by leaf miners to determine if the leaf miners survived and emerged, or if they died due to nutritional inadequacy or by the action of predators or parasitoids. |  | Elevated $CO_2$ reduces foliar nitrogen, inhibits normal insect development, and prolongs the feeding time of herbivores, allowing natural enemies greater opportunities to attack them. |

**4 THE DATA**

| Source of mortality* | Elevated $CO_2$ (% mortality) | Control (% mortality) |
|---|---|---|
| Nutritional inadequacy | 10.2 | 5.0 |
| Predators | 2.4 | 2.0 |
| Parasitoids | 10.0 | 3.2 |

*Data refer only to mortality of larvae within leaves and do not sum to 100%. Mortality of eggs on leaves, pupae in the soil, and flying adults is unknown.

**5 CONCLUSION** Elevated $CO_2$ increases herbivore mortality.

**6 SOURCE** Stiling, P., and Cornelissen, T. 2007. How does elevated carbon dioxide ($CO_2$) affect plant-herbivore interactions? A field experiment and meta-analysis of $CO_2$-mediated changes on plant chemistry and herbivore performance. *Global Change Biology* 13: 1823–1842.

of elevated $CO_2$. This decrease in herbivore survival occurred even though the plants produced more biomass in elevated $CO_2$, because $CO_2$ is limiting to plant growth. (Note: The data for plant biomass increases are not shown in the figure.) Lower leaf miner densities could be attributed to higher mortality rates due to nutritional inadequacy, predators, or parasitoids (see step 3).

Stiling and Drake hypothesized that part of the reason for the decline was that even though plants increased in mass, the existing soil nitrogen was diluted over a greater volume of plant material, so the nitrogen level in leaves decreased. This could have resulted in increased insect mortality by two means. First, poorer leaf quality could directly increase insect death because leaf nitrogen levels may have been too low to support the normal development of the leaf miners. Second, lower leaf quality is expected to increase the amount of time insects need to feed to gain sufficient nitrogen. Increased feeding times, in turn, may lead to increased exposure to natural enemies,

such as predatory spiders and ants and parasitoids (see Figure 57.8), so mortality from natural enemies also increased (see the data of Figure 59.10). Thus, in a world of elevated $CO_2$, plant growth may increase, and herbivory may decrease.

*Experimental Questions*

1. What was the hypothesis of Stiling and Drake's experiment?

2. **CoreSKILL** » Explain the purpose of increasing the $CO_2$ levels in only half of the chambers in the experiment and not all of the chambers.

3. **CoreSKILL** » Imagine that the mortality rates from parasitoids in the eight elevated $CO_2$ chambers were 6%, 11%, 9%, 8%, 13%, 12%, 14%, and 7%, and the rates in the ambient chambers were 1%, 4%, 5%, 2%, 3%, 6%, 1.6%, and 3%. Perform a statistical test to see if mortality from parasitoids was different between elevated and ambient $CO_2$ chambers.

## Global Warming and the Interruption of Water Flow by Humans Affect the Water Cycle

The water cycle, also called the hydrological cycle, differs from the cycles of other nutrients in that very little of the water that cycles through ecosystems is chemically changed by any of the cycle's components (**Figure 59.11**). This cycle is a physical process, fueled by the Sun's energy, rather than a chemical one, because it consists of essentially two phenomena: evaporation and precipitation. Even so, the water cycle has important biological components. Over land, 90% of the water that reaches the atmosphere is moisture that has passed through plants and exited from the leaves via evapotranspiration. Only about 2% of the total volume of Earth's water is found in the bodies of organisms or is held frozen or in the soil. The rest cycles from bodies of water, to the atmosphere, and then to the land and back to bodies of water again.

Human activities have altered the water cycle in different ways. As discussed in Section 59.2, climate change may have the greatest effect on the water cycle by altering precipitation patterns around the

world and causing glaciers to melt. On a smaller scale, humans have built structures that affect the flow of moving bodies of water. To increase the amount of available water and to create hydroelectric power, humans have interrupted the water cycle in many ways, most prominently through the use of dams to create reservoirs. Such dams can greatly interfere with the migration of fishes such as salmon and affect their ability to reproduce and survive. Other activities, such as tapping into underground water supplies, or **aquifers**, for drinking water removes more water than is put back by rainfall and can cause shallow ponds and lakes to dry up and sinkholes to develop, exacerbating local shortages.

## Fertilizers and Other Pollutants Affect the Phosphorus Cycle

Phosphorus has no gaseous phase and thus no atmospheric component; that is, it is not moved by wind. As a result, the phosphorus cycle is viewed as a local cycle even though phosphorus can be transported over long distances by moving water (**Figure 59.12**). The Earth's

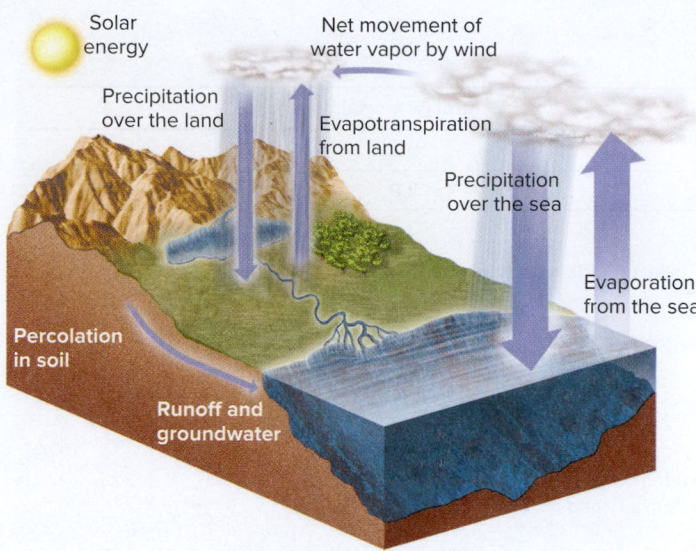

**Figure 59.11 The water cycle.** This cycle is primarily a physical process, not a chemical one. Solar energy drives the water cycle, causing evaporation of water from the ocean and evapotranspiration from the soil and land plants. The next step is condensation of water vapor into clouds followed by precipitation. The widths of the arrows indicate the relative contribution of each process to the cycle.

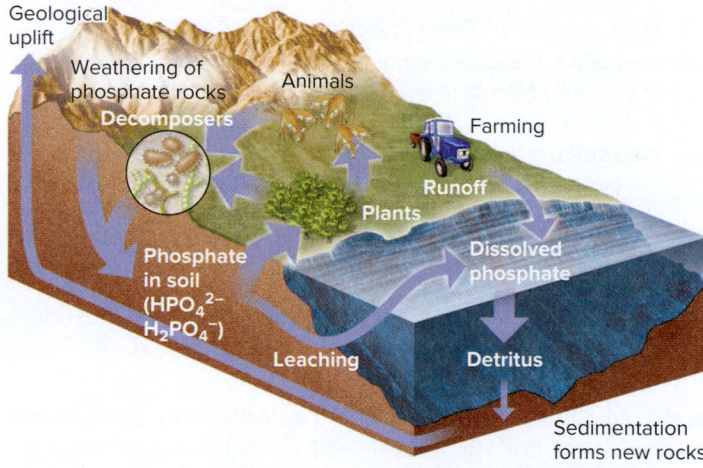

**Figure 59.12 The phosphorus cycle.** Unlike other major biogeochemical cycles, the phosphorus cycle does not have an atmospheric component and thus cycles only locally. The widths of the arrows indicate the relative contribution of each process to the cycle.

crust is the main storehouse for this element. Weathering and erosion of rocks release phosphorus into the soil. Plants have the metabolic means to absorb dissolved ionized forms of phosphorus, the most important of which occurs as phosphate ($HPO_4^{2-}$ or $H_2PO_4^-$). Herbivores obtain their phosphorus only from eating plants, and carnivores obtain it by eating herbivores or other carnivores. When plants and animals excrete wastes or die, the phosphorus becomes available to decomposers, which release it back to the soil. Leaching and runoff eventually wash much phosphate into aquatic systems, where algae utilize it. Phosphate that is not taken up into the food chain settles to the ocean floor or lake bottom, forming sedimentary rock.

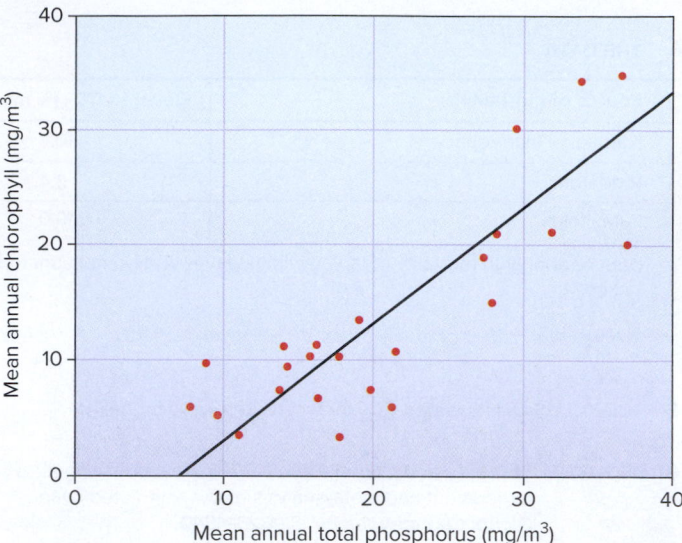

**Figure 59.13 The relationship between primary production and total phosphorus concentration.** As shown in this graph, primary production (measured by chlorophyll concentration) increases linearly with an increase in phosphorus. Each dot represents a different lake.

Phosphorus is commonly a limiting nutrient in aquatic ecosystems. When more phosphorus is added, the growth of algae and aquatic plants can dramatically increase. In lakes, primary production and total phosphorus concentration appear to follow a linear relationship (**Figure 59.13**). What is the consequence of high primary production? When the algae and plants die, they sink to the bottom, where bacteria decompose them and consume the dissolved oxygen in the water. Dissolved oxygen concentrations can then drop too low for fishes to breathe, killing them. The process by which elevated nutrient levels lead to an overgrowth of algae and the subsequent depletion of water oxygen concentrations is known as **eutrophication**.

Eutrophication is frequently due to the enrichment of water with nutrients derived from human activities, such as fertilizer use and sewage dumping. For example, Lake Erie became eutrophic in the 1960s due to the runoff of fertilizer rich in phosphorus from farms and to the industrial and domestic pollutants released from the many cities along its shores (**Figure 59.14a**). Fish species such as white fish and lake trout became severely depleted. The U.S. and Canada teamed together to reduce the levels of discharge by 80%, primarily through eliminating phosphorus in laundry detergents and maintaining strict controls on the phosphorus content of wastewater from sewage treatment plants. Fortunately, lake systems have great potential for recovery after phosphorus inputs are reduced, and Lake Erie has experienced fewer algal blooms, clearer water, and a restoration of fish populations (**Figure 59.14b**).

## Fertilizers and Industrial Pollutants Affect the Nitrogen Cycle

Nitrogen is an essential component of proteins, nucleic acids, and chlorophyll. Because 78% of the Earth's atmosphere consists of nitrogen gas ($N_2$), it may seem that nitrogen should not be in short supply for organisms. However, nitrogen is often a limiting factor in

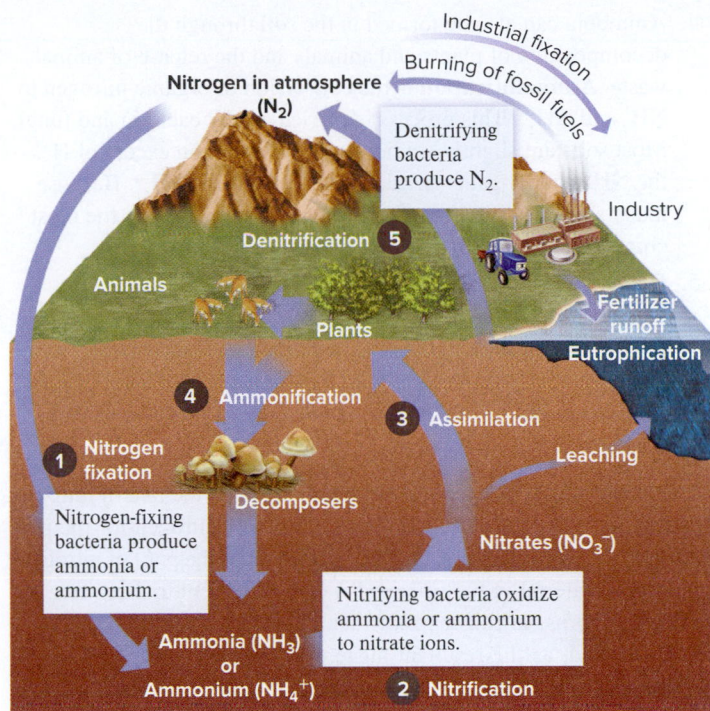

**Figure 59.15** **The nitrogen cycle.** The five main parts of the nitrogen cycle are (1) nitrogen fixation, (2) nitrification, (3) assimilation, (4) ammonification, and (5) denitrification. The recycling of nitrogen from dead plants and animals into the soil and then back into plants is of paramount importance because this is the main pathway for nitrogen to enter the soil. The widths of the arrows indicate the relative contribution of each process to the cycle.

**(a) Polluted (eutrophication)**

**(b) Cleared up**

**Figure 59.14** **Phosphorus pollution in Lake Erie.** The lake **(a)** in the 1960s, when eutrophic and polluted by industrial effluent and fertilizer runoff, and **(b)** in 2007, after eutrophication was reversed by pollution control laws. a: ©JK Enright/Alamy Stock Photo; b: ©Rolf Hicker Photography/Alamy Stock Photo

ecosystems because $N_2$ molecules must be broken apart before the individual nitrogen atoms can combine with other elements. Because there is a triple bond between its two nitrogen atoms, $N_2$ is very stable, and only certain bacteria can break it apart into usable forms such

as ammonia ($NH_3$). This process, called nitrogen fixation, is a critical component of the five-part nitrogen cycle (**Figure 59.15**):

1. A few species of bacteria can accomplish **nitrogen fixation**, that is, convert atmospheric $N_2$ to forms usable by other organisms. The bacteria that fix nitrogen are fulfilling their own metabolic needs, but in the process, they release ammonia ($NH_3$) or ammonium ($NH_4^+$), which can be used by some plants. An important group of nitrogen-fixing bacteria, known as rhizobia, live in nodules on the roots of legumes, including peas, beans, lentils, and peanuts, and of some woody plants. Cyanobacteria are important nitrogen fixers in terrestrial and aquatic systems (refer back to Figure 38.12).

2. In the process of **nitrification**, soil bacteria convert $NH_3$ or $NH_4^+$ to nitrate ($NO_3^-$), a form of nitrogen commonly used by plants. Bacteria of the genera *Nitrosomonas* and *Nitrococcus* first oxidize the forms of ammonia to nitrite ($NO_2^-$), after which bacteria of the genus *Nitrobacter* convert $NO_2^-$ to $NO_3^-$.

3. **Assimilation** is the process by which inorganic substances are incorporated into organic molecules. In the nitrogen cycle, organisms assimilate nitrogen by taking up $NH_3$, $NH_4^+$, and $NO_3^-$ formed through nitrogen fixation and nitrification and incorporating them into organic molecules. Plants take up these forms of nitrogen through their roots, and animals assimilate nitrogen from the plant tissues they ingest.

4. Ammonia can also be formed in the soil through the decomposition of plants and animals and the release of animal waste. **Ammonification** is the conversion of organic nitrogen to $NH_3$ and $NH_4^+$. This process is carried out by bacteria and fungi. Most soils are slightly acidic, and, because of an excess of $H^+$, the $NH_3$ rapidly gains an additional $H^+$ to form $NH_4^+$. Because many soils lack nitrifying bacteria, ammonification is the most common pathway for nitrogen to enter the soil.

5. **Denitrification** is the reduction of $NO_3^-$ to $N_2$. Denitrifying bacteria, which are anaerobic and use $NO_3^-$ in their metabolism instead of $O_2$, perform the reverse of their nitrogen-fixing counterparts by delivering $N_2$ to the atmosphere. This process contributes only a relatively small amount of nitrogen to the atmosphere.

Human activities have approximately doubled the rate of nitrogen input to the nitrogen cycle. Industrial fixation of nitrogen for the production of fertilizer makes a major contribution to the pool of nitrogen-containing material in the soils and waters of agricultural regions. As with phosphorus, fertilizer runoff can cause eutrophication of rivers and lakes, and, as the resultant algae die, decomposition by bacteria depletes the oxygen level of the water, resulting in fish kills.

Excess $NO_3^-$ in surface or groundwater systems used for drinking water is also a health hazard, particularly for infants. In the body, $NO_3^-$ is converted to $NO_2^-$, which then combines with hemoglobin to form methemoglobin, a type of hemoglobin that does not carry oxygen. In infants, the production of large amounts of $NO_2^-$ can cause methemoglobinemia, a dangerous condition in which the level of $O_2$ carried through the body decreases.

Finally, burning fossil fuels releases not only carbon but also nitrogen in the form of nitrous oxide ($N_2O$), which contributes to air pollution. $N_2O$ can react with rainwater to form nitric acid ($HNO_3$), a component of acid rain, which decreases the pH of lakes and streams and increases fish mortality. The release of sulfur from the burning of fossil fuels also contributes to acid rain since the sulfur reacts with rainwater to form highly acidic sulfuric acid.

## 59.4 Pollution and Biomagnification

### Learning Outcome:

1. Define biomagnfication, and explain its relationship to pollution.

In Section 59.3, we considered how pollution generated by humans can affect biogeochemical cycles. Certain types of pollutants can also accumulate within the bodies of animals and plants. The tendency of certain chemicals to concentrate in organisms at higher trophic levels in food chains, which is called **biomagnification**, can cause health and reproductive problems for certain organisms.

The passage of dichlorodiphenyltrichloroethane (DDT), an insecticide used against mosquitoes and agricultural pests, through food chains provides a startling example of biomagnification. DDT was first synthesized by chemists in 1874. In 1939, its insecticidal properties were recognized by Paul Müller, a Swiss scientist who won the 1948 Nobel Prize in Physiology or Medicine for his discovery and subsequent research on

the uses of the chemical. The first important application of DDT was in human health programs during and after World War II, particularly as a means of controlling mosquito-borne malaria; at that time, its use in agriculture also began. The global production of DDT peaked in 1970, when 175 million kilograms of the insecticide was manufactured.

DDT has several chemical and physical properties that profoundly influence its ecological effect. First, DDT is persistent in the environment. It is not rapidly degraded to other, less toxic chemicals by microorganisms or by physical agents such as light and heat. The typical persistence in soil of DDT is about 10 years, which is two to three times longer than the persistence of many other insecticides. Another important characteristic of DDT is its low solubility in water and its high solubility in fats, or lipids. In the environment, most lipids are present in living tissue. Therefore, because of its high lipid solubility, DDT tends to concentrate in biological tissues.

Because biomagnification occurs at each step of a food chain, organisms at higher trophic levels can accumulate especially high concentrations of DDT in their lipids. A typical pattern of biomagnification is illustrated in **Figure 59.16**, which shows the relative amounts of DDT found in a Lake Michigan food chain. The highest concentration of the insecticide was found in gulls, tertiary consumers that feed on fishes, which are the secondary consumers that eat small insects. An unanticipated effect of DDT on bird species was its interference with the metabolic process of eggshell formation. The result

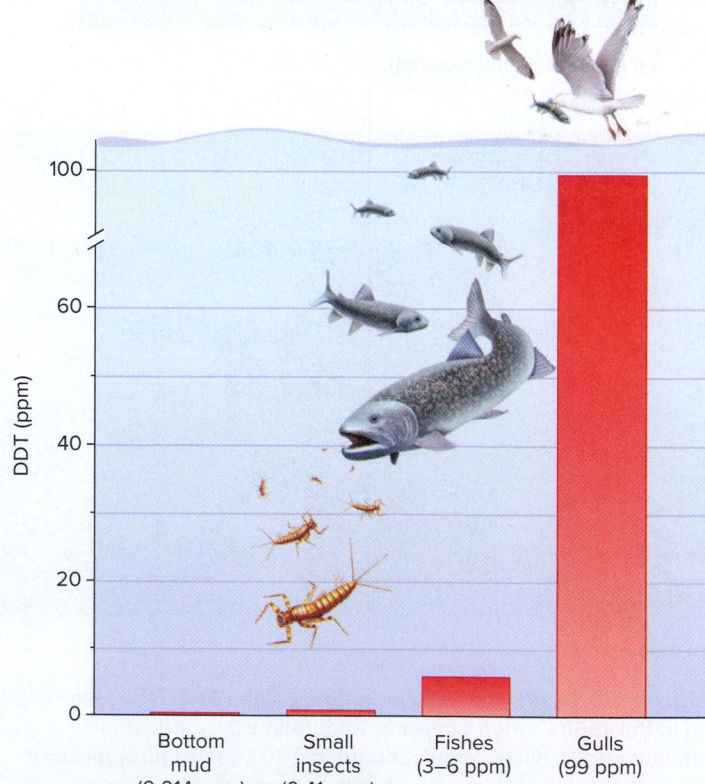

**Figure 59.16 Biomagnification in a Lake Michigan food chain.** The DDT tissue concentration in gulls, a tertiary consumer, was about 240 times that in the small insects sharing the same environment. The biomagnification of DDT in lipids causes its concentration to increase at each successive level in the food chain. The unit ppm is equivalent to 1 mg of DDT per kg of biological material.

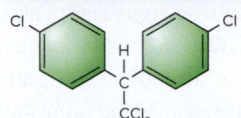

**DDT (dichlorodiphenyltrichloroethane)**
- Persists in environment
- High solubility in lipids
- Found in high concentrations at higher trophic levels

**Figure 59.17  Thinning of eggshells caused by DDT.** Ibis eggs, Texas Gulf Coast. ©George Silk/The LIFE Picture Collection/Getty Images

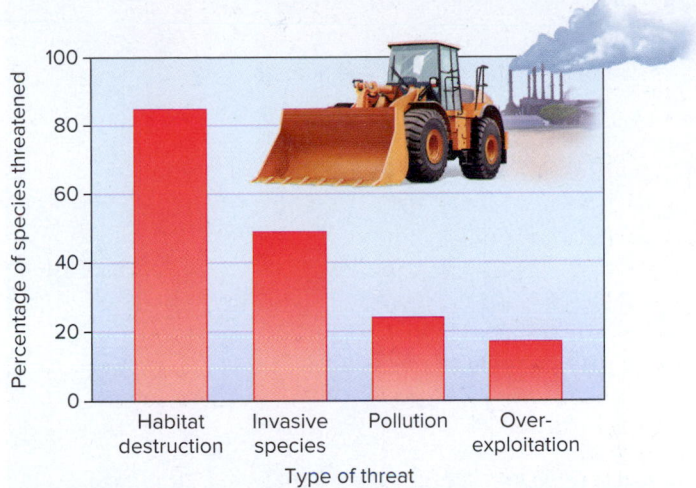

**Figure 59.18  Percentages of plant and animal species threatened by various causes in the U.S.** Species can suffer from multiple threats, so categories do not sum to 100%.

**Core Skill: Science and Society** The effects of humans greatly increase the percentage of species threatened with extinction.

was thin-shelled eggs that often broke under the weight of incubating birds (**Figure 59.17**). DDT was responsible for a dramatic decrease in the populations of many birds due to failed reproduction. Relatively high levels of the chemical were also found to be present in some game fishes, which, as a result, became unfit for human consumption.

Because of growing awareness of the adverse effects of DDT, most industrialized countries, including the U.S., banned the use of the chemical by the early 1970s. The good news is that following the outlawing of DDT, populations of the most severely affected bird species have recovered. However, had scientists initially possessed a more thorough knowledge of how DDT accumulates in food chains, some of the damage to the bird populations might have been prevented.

Biomagnification occurs for other types of molecules that do not break down quickly in the environment and are not excreted efficiently by living organisms. Such substances include mercury, which is emitted from coal-fired power plants, and persistent organic pollutants, or POPs, that are used in herbicides and pesticides. Underwater mining of the ocean floor to extract minerals releases sulfide and selenium that can also biomagnify in food chains.

## 59.5  Habitat Destruction

### Learning Outcomes:

1. Identify the main causes of habitat destruction by humans.
2. Explain why tropical deforestation is a particularly destructive form of habitat loss.
3. Describe how the impact of agriculture causes ecological changes.

**Habitat destruction** is usually a human-driven process in which a natural habitat is altered in a way that prevents it from supporting the species that were originally present. Organisms that previously occupied the habitat are displaced or unable to survive, thereby reducing biodiversity. Habitat destruction is the primary cause of species extinction, and a high percentage of existing species are threatened

by this process (**Figure 59.18**). Other human-driven practices that are threatening the survival of many species include pollution, which has already been discussed in this chapter, and overexploitation and invasive species, which are described in the last two sections.

Habitat destruction includes deforestation, conversion of habitat to agricultural land, urbanization, strip mining, quarrying, and many other forms of land modification. Urbanization, the development of cities on previously natural or agricultural areas, is the most human-dominated and fastest-growing type of land use worldwide, and it devastates the land more severely than nearly any other form of habitat destruction. Freshwater habitats have also suffered via dam construction and river channelization. Wetlands have been drained for agricultural purposes and have been filled in for urban or industrial development. In the U.S., as much as 90% of freshwater marshes have disappeared in states such as Iowa and California, though the national average is approximately 53%. In this section, we will examine the two most widespread and interrelated types of habitat destruction, at least in terms of their influence on species extinctions. These are deforestation and the conversion of land to agricultural purposes.

## Deforestation by Humans Threatens the Existence of Many Species

**Deforestation**, the conversion of forested areas to nonforested land, is a prime cause of the extinction of species (**Figure 59.19**). About one-third of the world's land surface is covered with forests, and much of this area is at risk of deforestation.

Many species live in forests. For example, among North American terrestrial wildlife, about one-quarter of the bird species (272 species) and more than 10% of mammalian species (49 species) have an obligatory relationship with forest cover, meaning that they depend on trees for food and nesting sites. In terms of wildlife use, oaks are among the

**Figure 59.19 Deforestation.** Cascade Mountains near Seattle, Washington, 1906. Source: Library of Congress Prints and Photographs Division [LC-USZ62-67666]

most valuable trees in North America. At least 100 species of birds and mammals include acorns in their diets, and for many species of wildlife, the annual acorn crop is a major determinant of their abundance. Most woodpeckers, as well as many other types of birds, nest in holes that they excavate in trees, and their food usually consists of insects collected on or in trees. The ivory-billed woodpecker (*Campephilus principalis*), the largest known woodpecker in North America and a former inhabitant of wetlands and forests of the southeastern United States, is presumed to have gone extinct in the 1950s due to destruction of its habitat by heavy logging (**Figure 59.20**).

**Figure 59.20 Habitat destruction as an important cause of species extinction.** The ivory-billed woodpecker, the third-largest woodpecker in the world, is thought to have gone extinct in the southeastern U.S. because of habitat destruction, though some unconfirmed sightings have occurred in the past two decades. ©James T. Tanner/Science Source

Tropical forests, primarily rain forests but also deciduous forests, are among the most species-rich terrestrial habitats and exist primarily in three areas of the globe: Africa, Asia, and Latin America. Latin America contains more than the other two areas combined, with Brazil having the greatest amount of tropical forest of any nation. The Amazon rain forest has been termed "the lungs of the planet" because it produces more than 20% of the world's oxygen. Tropical forests, which once covered 14% of the Earth's dry land, now cover only about 7%. The good news is that rates of tropical deforestation decreased from 0.18% per year in the 1990s to 0.08% annually between 2010 and 2015.

Although tropical deforestation occurs for different reasons, clearing land for agriculture has been identified as the prime cause of forest loss. Logging in excess of regrowth is also a significant cause of forest loss, particularly in Asian forests. Collecting wood for fuel can also be important but generally is more of a problem in lightly wooded areas. Finally, the construction of mines, dams, and oil installations is a minor cause of direct deforestation, but the discharge of chemicals and silt into rivers can cause much damage indirectly. The roads built into the regions where logging is being done often open them up to further development.

Conservation of tropical forests would not only save many rare species from extinction, it would also benefit humans in several ways. First, many of the world's crops, including oranges, lemons, bananas, cacao (chocolate), coffee, and vanilla, evolved in tropical rain forests. Rain forests remain a depository of genetic variation that could be used in future breeding of these crops. In addition, rain forests are particularly rich in plants with unusual chemicals that they use in defense (refer back to Chapter 57). Many of these chemicals also have medicinal value. Although many prescription drugs sold worldwide come from plant-derived sources, less than 1% of tropical plants have been tested for their medicinal properties. One estimate has proposed lifetime values for tropical forests at $6,330 per hectare if renewable and sustainable resources such as fruit, rubber, and nuts are harvested. In comparison, forests are worth only $1,000 per hectare for timber and $2,960 per hectare when converted to agricultural grazing land. For these reasons, it seems to make good economic and ecological sense to slow the rate of tropical deforestation.

## The Development of Land for Agriculture Causes Adverse Ecological Changes

No other human endeavor on Earth requires as much land as agriculture. More land has been converted to agriculture since 1945 than in the 18th and 19th centuries combined. The average area of land under crop cultivation worldwide is 11.5%, with an additional 25.8% given over to rangeland for grazing. However, this amount varies substantially among regions (**Table 59.2**).

The planting of crops and the grazing of livestock have produced far-reaching ecological effects on Earth (**Figure 59.21**). The changes to the land may result in soil erosion, a decrease in soil fertility, flooding, silting of rivers, desertification, and the loss of wildlife habitat. Also, as we have seen, runoff water from agricultural land often contains high levels of nitrogen and phosphorus from fertilizers as well as residual pesticides and manure, all of which contaminate streams and lakes. Some

| Table 59.2 | Percentage of Land Area Used for Agricultural Purposes | | |
|---|---|---|---|
| **Continental area** | **% Cropland** | **% Pastures** | **Total** |
| World | 11.5 | 25.8 | 37.3 |
| Asia | 20.0 | 33.0 | 53.0 |
| Central America and the Caribbean | 15.6 | 36.7 | 52.3 |
| Europe | 13.1 | 7.9 | 21.0 |
| Middle East and North Africa | 7.6 | 28.1 | 35.6 |
| North America | 11.4 | 12.5 | 23.9 |
| Oceania | 6.2 | 48.2 | 54.4 |
| South America | 6.8 | 28.8 | 35.6 |
| Sub-Sahara and Africa | 8.1 | 33.9 | 42.0 |

(a)

(b)

**Figure 59.21 Modern agriculture and its impact on the environment.** Modern agriculture has converted vast areas of land into fields for **(a)** the growing of crops and **(b)** the grazing of livestock. This conversion results in the destruction of habitat for native species and has a variety of negative ecological effects on soils and bodies of water. a: Source: USDA; b: ©Martial Colomb/Photographer's Choice/Getty Images

4 billion tons of topsoil are washed into U.S. waterways each year, severely impacting the organisms that live there. Topsoil can also be whipped away by wind.

## 59.6 Overexploitation

**Learning Outcomes:**

1. Outline the main causes of overexploitation of plant and animal populations.
2. Describe examples of the overexploitation of animals and plants.
3. **CoreSKILL »** Calculate the maximum sustainable yield of a population based on its carrying capacity and per capita rate of increase.

In ecology, **overexploitation** is the practice in which humans harvest a particular species at a rate that is unsustainable, based on its natural rate of mortality and capacity for reproduction. Overexploitation, particularly the hunting of animals, has been the cause of many extinctions in the past. In this section, we will consider several examples of overexploitation and then examine how ecologists make calculations to determine if overexploitation is occurring.

### Many Species Have Gone Extinct or Are Currently Threatened Due to Overexploitation

With regard to animals, hunting and fishing have been the main practices that facilitate overexploitation. For other groups, such as plants, the ability to identify valuable species and remove them from their native habitats has led to overexploitation. In 2015, a study by Canadian conservation biologist Chris Darimont and colleagues examined the effects of humans as hunters of terrestrial mammals and fishers of marine fishes. They compared 2,125 estimates of exploited animal populations and showed that humans kill adult prey at rates up to 14 times higher than other predators (**Figure 59.22**). Darimont and colleagues termed humans "super predators" and suggested that, in the long run, this level of hunting and fishing would not be sustainable. As seen in Figure 59.22, humans are predators of species that are considered the top predators in their native environment. For example, humans have greatly decreased the populations of wolves, which are successful predators in many different habitats.

Let's now consider a few examples in which humans have overexploited animals and plants.

***Land Mammals*** The list of mammals threatened with extinction or actually driven extinct by hunting is long. In the prairies of North America, hunting diminished populations of buffalo from around 70 million in the 18th century to 1,150 by 1899. Since then, buffalo numbers have rebounded to about 350,000. In Russia, another inhabitant of the prairies, the Eurasian wild horse, also known as the tarpan, (*Equus ferus*), had been hunted to extinction by the 1860s.

Even today, poachers threaten many land mammals. Rhinoceroses have a horn and elephants have tusks that some people believe will cure everything from cancer to hangovers. The black market value for rhino horn in Southeast Asia is about $65,0000 per kilogram, making it more expensive by weight than gold, diamonds, or

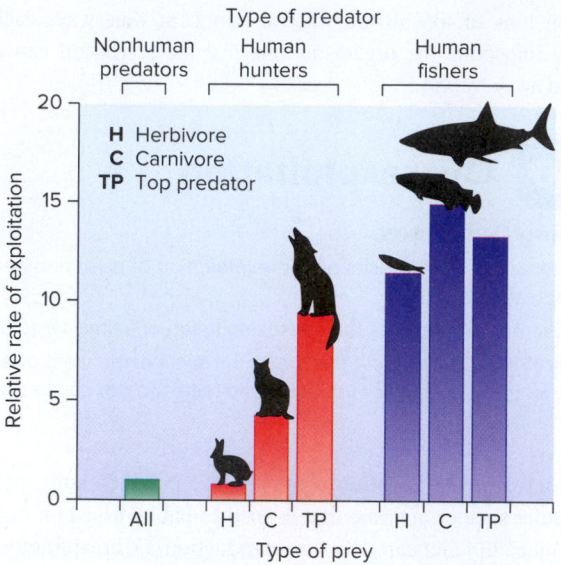

Type of predator

Figure 59.22 **Humans as hunters.** The rates at which humans exploit adult land mammals and marine fish vastly exceed the impacts of other predators. Marine fish of all trophic levels are similarly affected. In contrast, land predators are exploited at much higher rates than herbivores. The term top predator refers to the most successful predator in a given habitat, which also can be a prey to humans.

**Core Skill: Science and Society** Human hunting and fishing massively impacts populations of herbivores, carnivores, and top predators, compared to the effects of non-human predators.

cocaine. Roughly 100,000 African elephants were poached across the continent between 2010 and 2012. In 2011 alone, poachers killed roughly 1 in every 12 African elephants. Many experts believe that the last of the great woolly mammoth populations were hunted to extinction at the end of the last glacial period about 12,000 years ago.

***Whales*** A 2014 estimate of the number of whales killed by industrial harvesting during the last century, 1900–1999, by Roberta Rocha, Director of Science at the New Bedford Whaling Museum in Massachusetts, revealed that nearly three million whales were harvested, in what was the largest cull, in terms of biomass, in human history. The history of whaling in general, and Antarctic whaling in particular, has been characterized by a progression from more valuable or more easily caught species to less attractive ones, as populations of the original targets were depleted (**Figure 59.23**). In the Antarctic, blue whales (*Balaenoptera musculus*) dominated the catches through the 1930s, but by the middle 1950s, few were being taken, although the species was not legally protected until 1965. As the populations of blue whales diminished, attention turned to the fin whale (*B. physalus*) which was originally the most abundant of all whales in the Southern Ocean. By the 1960s, numbers of this species had diminished rapidly. Sei whales (*B. borealis*) were almost ignored by whalers until the bigger species were no longer available. They were hardly taken at all until about 1958, but then catches increased rapidly and reached a peak of about 20,000 in 1964–1965. Catches declined rapidly thereafter, this time due to the introduction of a catch limit of

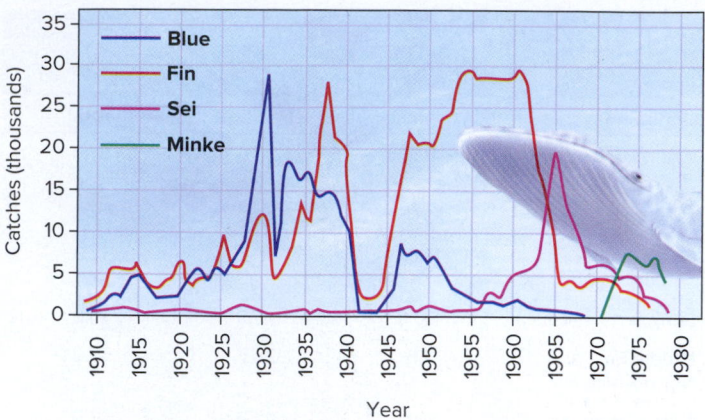

**Figure 59.23** **Sequential decline of whale catches in the Antarctic, showing the strong effect of human predators.**

10. Then the relatively small minke whales (*B. acutorostrata*), which were ignored in the Southern Ocean until 1971–1972, began to be taken. Since that time, minkes have been the largest component of the southern baleen whale catch. The story has been similar in the Northern Hemisphere.

In 1982, the International Whaling Commission (IWC) voted for a moratorium on all commercial whaling. The IWC proposed an end to commercial whaling in 1985–1986, a proposal that did not actually take effect until 1988. The good news is that following the moratorium, the populations of some whales have increased. Blue whales appear to have quadrupled their numbers off the California coast during the 1980s, showing the impact that protection from a predator can have. Steve Palumbi's laboratory estimated past population sizes of North Atlantic whales by examining mitochondrial DNA sequence variation. Population size estimates for fin and humpback whales suggested that pre-whaling populations were 6–20 times higher than present-day population estimates and that full recovery of whale populations will take another 70–100 years.

***Birds*** A poignant example of overexploitation was the dodo (*Raphus cucullatus*), a flightless bird that was native only to the island of Mauritius and had no known predators. A combination of overexploitation and introduced species led to its extinction within 200 years of the arrival of humans. Sailors hunted it for its meat, and the rats and pigs they brought to the island, the latter as a food source, destroyed the dodos' eggs and chicks in their ground nests.

Two abundant species of North American birds, the passenger pigeon and the Carolina parakeet, had been hunted to extinction by the early 20th century. The passenger pigeon (*Ectopistes migratorius*) was once the most common bird in North America, probably accounting for over 40% of the entire number of birds (**Figure 59.24**). Flock sizes were estimated to be over 1 billion birds. It may seem improbable that the most common bird on the continent could be hunted to extinction for its meat, but that is just what happened. The flocking behavior of the birds made them relatively easy targets for hunters, who used special firearms to harvest the birds in quantity. In 1876, in Michigan alone, over 1.6 million birds were killed and sent to markets in the eastern U.S. Similarly, the Carolina

**Figure 59.24 Overexploitation has caused species extinctions.** The passenger pigeon, *Ectopistes migratorius*, which was once among the most abundant bird species on Earth, was hunted to extinction for its meat. ©Topham/The Image Works

parakeet (*Conuropsis carolinensis*), the only species of parrot native to the eastern U.S., had been similarly hunted to extinction by the early 1900s.

***Fishes*** Many species of fish are harvested to the degree that the rate of removal exceeds the rate of reproduction. Eventually, these fish populations crash and the fishery collapses. In the case of the Canadian cod fishery, overfishing and collapse came in the early 1990s after hundreds of years of fishing (**Figure 59.25**). Since that time, the species has not been economically fished. There is hope that the Canadian cod fisheries will eventually recover, but the recovery will be slow because the fish do not spawn until 7 years of age. A 2010 study showed that populations near Newfoundland and Labrador were still only 10% of their original sizes.

In the Florida Keys, historical photographs document the reduction of large trophy fish taken over the period 1956–2007 (**Figure 59.26**). The mean fish size declined from an estimated

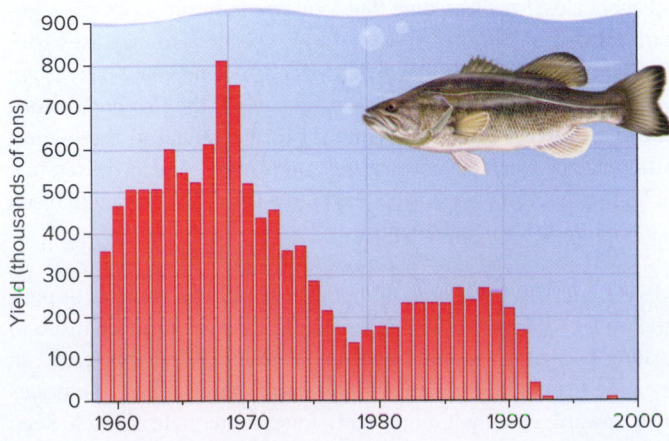

**Figure 59.25 Commercial cod fisheries.** Changes in Canadian cod yields from 1960 to 2000.

(a) 1957

(b) early 1980s

(c) 2007

**Figure 59.26 Change in the sizes of trophy fish caught on Key West charter boats.** **(a)** 1957, **(b)** early 1980s, and **(c)** 2007. a: Source: From the archives of Monroe County Public Library, Key West, Florida. Photo by Wil-Art Studio/Art Stickel; b: Source: From the archives of Monroe County Public Library, Key West, Florida. Photo by Dale McDonald; c: Photo courtesy Loren McClenachan, Ph.D.

**Figure 59.27** **Rothschild's orchid, *Paphiopedilum rothschildianum*.** Overcollection in northern Borneo has pushed this species close to extinction in the wild. ©Matthew Lambley/Alamy Stock Photo

19.9 kg in 1956 to 2.3 kg in 2007! A shift was observed in the types of species caught. In 1956, large groupers (*Epinephellus* spp.) were commonly landed together with large sharks greater than 2 m. In 2007, small snappers (*Lutjanus campechanus* and *Ocyurus chrysurus*) with an average length of 34.4 cm and sharks less than 1 m long were landed.

**Plants** Many species of valuable plants have also been severely overexploited for human use, including West Indian mahogany (*Swietenia mahogani*) in the Bahamas, and Lebanese cedar (*Cedrus libani*), which in Lebanon has been reduced to a few scattered forest remnants. Rare cacti and orchids have also been threatened by collectors, who seek to own a rare organism or to profit from its sale. Rare orchids are very valuable. A single stem of Rothschild's orchid (*Paphiopedilum rothschildianum*), also known as the Gold of Kinabalu, is reported to sell for $5,000 (**Figure 59.27**). No wonder that after its discovery, this orchid was stripped from the wild in northern Borneo by orchid smugglers, pushing it close to extinction.

Plants may also be collected for their medicinal value, real or supposed. In the U.S., American ginseng (*Panax quinquefolicus*), native to eastern North America, has been sought after to treat the common cold. Plants are so enthusiastically collected by "sang hunters," who also sell it to Chinese traders, that ginseng has become threatened or endangered in some states. The Botanic Gardens Conservation International Organization has stated that over 400 medical plant species are at risk of extinction because of overcollection and deforestation, including many species of yew tree (genus *Taxus*), whose bark is used for cancer drugs.

## Quantitative Analysis

### Ecologists Make Calculations to Determine If Overexploitation Is Occurring

For a population that is being harvested by humans, the **yield** is the number of individuals harvested in a given unit of time. **Maximum sustainable yield (MSY)** is the largest number of individuals that can be harvested without causing long-term decreases in the population. If the maximum sustainable yield is exceeded on a consistent basis, overexploitation is occurring.

According to the logistic equation described in Chapter 56, change in the number of individuals in a population over time, *dN/dt*, occurs according to the following equation:

$$\frac{dN}{dt} = rN\frac{(K - N)}{K}$$

where

$N$ is the number of individuals,

$t$ is a unit of time,

$r$ is the per capita rate of population growth, and

$K$ is the carrying capacity.

The greatest population growth, and thus the maximum sustainable yield, occurs at the midpoint of the logistic curve, at $K/2$ (refer back to Figure 56.10).

MSY can thus be estimated as

$$MSY = \frac{rK}{2}$$

As an example, if $K/2 = 10{,}000$ and $r = 0.14$, then $MSY = 1{,}400$.

MSY has been extensively used for fisheries management. However, the use of the model has some drawbacks. For example, biologists cannot easily go under water and count the number of individuals. The model also ignores the age and reproductive status of the individuals being harvested. As an alternative, simple models based on data collection from fisheries may be used, which compare costs and revenues and consider MSY (**Figure 59.28**). The maximum profit (red line) is greater than the profit obtained at MSY (blue line).

Why is MSY important to fisheries (and the harvesting of other species)? Let's assume that total economic costs increase linearly as fishing effort increases. Total revenue also increases with fishing effort, up to a point. After MSY is reached, any increase in effort would result in a decrease in revenue as the fish populations became overfished. The maximum profit is equal to the biggest difference between the cost and revenue curves, and as the red line in Figure 59.28 shows, this occurs below the effort needed to reach the maximum sustainable yield.

Why do fisheries become overfished? Most estimates of MSY are nearly always too high, because of overestimates of population size. Furthermore, incremental improvements are made to fishing gear over the years and the same level of effort results in increased catches. Finally, as fish populations decline, the prices inch upward, making it more likely for commercial fishing to keep harvesting them. In the case of the cod industry, the introduction

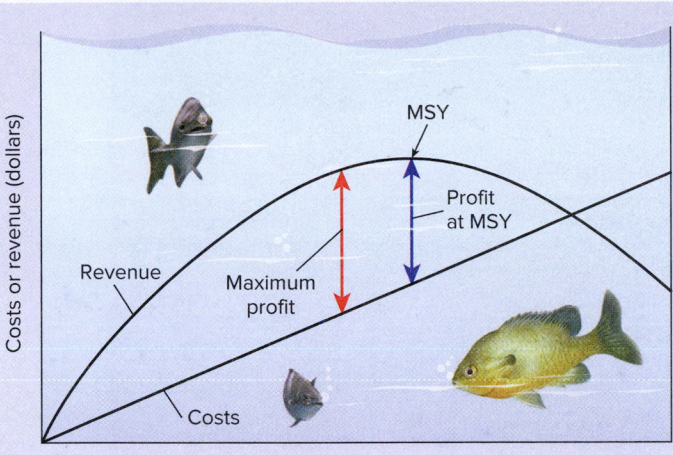

**Figure 59.28 Economic fisheries model based on revenue and costs.** The model assumes that costs increase linearly with effort but revenue reaches a maximum at maximum sustainable yield (MSY). Maximum profit, however, occurs below MSY.

of radar and sonar allowed crews to pursue fish over huge areas. The crews also caught enormous numbers of noncommercial fish, which were discarded. Among these were important prey species for cod, such as capelin. Finally, although undersized cod, which cannot spawn, were returned to the ocean, such discards do not always survive. As the cod populations went into a tailspin, increased effort resulted in more undersized fish being caught and more discarded.

## 59.7 Invasive Species

### Learning Outcomes:

1. Define introduced species and invasive species.
2. Compare and contrast invasive species as competitors, as predators, and as pathogens.

As noted in the Feature Investigation of Chapter 54, **introduced species** are those species moved by humans from their native habitat to another location. Most often, the species are introduced for agricultural or landscaping purposes or as sources of timber, meat, or wool, and they may need humans for their continued survival. In some cases, as with plants, insects, or aquatic species, organisms are unintentionally transported via the movement of cargo by ships or planes. Regardless of the way they have been transported, some introduced species become **invasive species**, spreading naturally and impacting native species. Why do invasive species become successful in their new habitat? In most cases, they have one or more of the following characteristics.

- Invasive species may reproduce rapidly.
- They may not have natural enemies in their new habitat.

- Certain native species may lack defenses against the invasive species.
- Invasive species may compete aggressively for resources.
- They may tolerate a wide variety of habitat conditions.

In the U.S. alone, over 4,500 invasive species have been identified, and 15% cause severe ecological or economic harm. For example, 142 species of introduced vertebrates have self-sustaining populations in the wild. These include ring-necked pheasants (*Phasianus colchicus*), which were brought over by hunters, and Burmese pythons (*Python molorus*), which were introduced by pet owners. Of the 300 most invasive weeds in the U.S., over half were brought in for gardening, horticulture, or landscape purposes. These include purple loosestrife (*Lythrum salicaria*) and Japanese honeysuckle (*Lonicera japonica*) in the Northeast, kudzu (*Pueraria lobata*) in the Southeast, Chinese tallow (*Sapium sebiferum*) in the South, and leafy spurge (*Euphorbia esula*) in the Great Plains. In this section, we will examine how the interactions between invasive and native species may involve competition, predation, and parasitism.

### Invasive Species May Compete Against Native Species

Ecologists have discovered many examples in which invasive species outcompete native species. For example, placental mammals have been particularly effective competitors in Australia, where introduced feral dogs (*Canis lupus,* ssp. *dingo*) are thought to have outcompeted the thylacine (*Thylacine cynocephalus*), a native marsupial, wolflike animal, in mainland Australia (**Figure 59.29a**). Sheep introduced for the wool industry appear to successfully compete with a variety of kangaroo species, especially the brush-tailed rock wallaby and larger species, such as the red kangaroo and the western gray kangaroo.

Introduced vertebrates are a problem in aquatic environments, too. In California, 48 of 137 species of freshwater fish are nonnative. Of these, 24 are known to have a negative impact on native fish.

Invasive plants are major competitors of native plants throughout the world. In some cases, invasive plants have become so successful that they have produced near monocultures—areas where they exist as the sole plant species. In the U.S. Northeast, purple loosestrife is a strong competitor, choking out other plant species in wetlands (**Figure 59.29b**). In south Florida, three invasives—Brazilian pepper (*Schinus terebinthifolius*), Australian pine (*Casuarina equisetifolia*), and, especially punk-tree (*Melaleuca quinquenervia*)—are outcompeting native vegetation in the Everglades.

### Invasive Species May Act as Predators

Many striking examples of the powerful effects of invasive species have been provided by predator introductions. Sea lampreys (*Petromyzon marinus*) are primitive, jawless fish that feed by rasping holes in the sides of larger fish species and feeding on their blood. They spawn in freshwater streams, and the juveniles return to salt water (or one of the Great Lakes as a substitute) to develop. Sea lampreys found their way into Lake Ontario in the mid-1800s by way of the Erie Canal (**Figure 59.30a**). Improvements to the Welland Canal allowed

**(a) Feral dog**

**(b) Purple loosestrife**

**Figure 59.29** **Introduced species as competitors.** **(a)** Feral dogs, also called dingoes, are thought to have outcompeted the thylacine in Australia. **(b)** Introduced plants can choke out native vegetation, as has occurred with purple loosestrife in the U.S. Northeast. a: ©kongsak sumano/123RF; b: ©Reimar Gaertner/Alamy Stock Photo

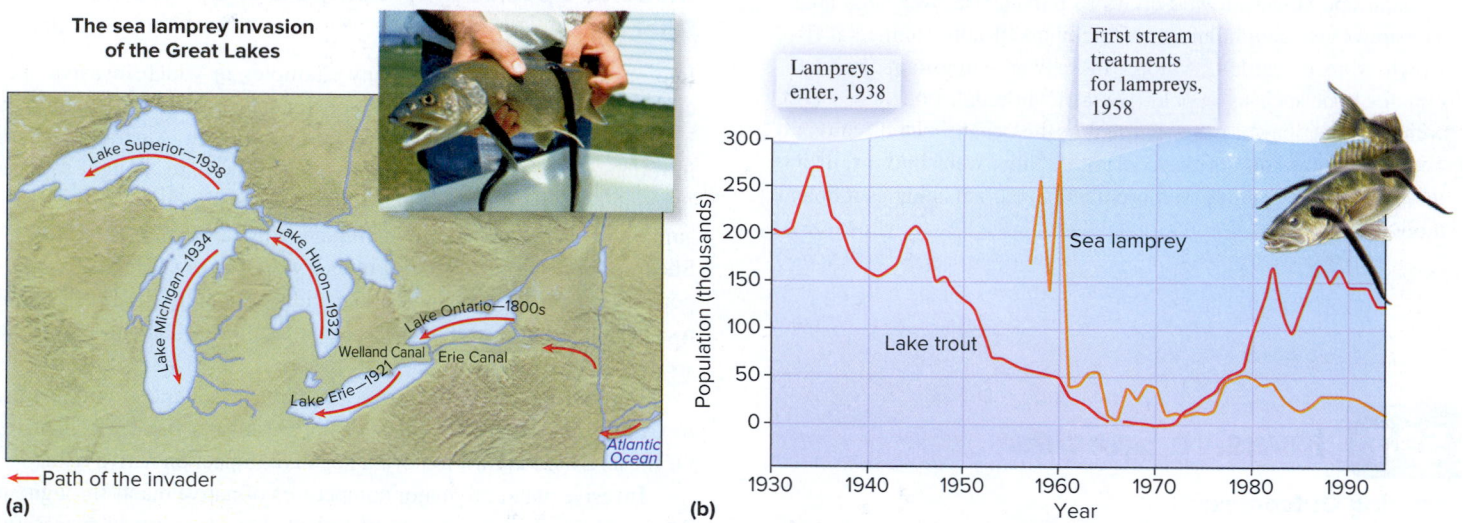

The sea lamprey invasion of the Great Lakes

Lake Superior—1938
Lake Huron—1934
Lake Michigan—1934
Lake Huron—1932
Lake Ontario—1800s
Welland Canal
Erie Canal
Lake Erie—1921
Atlantic Ocean

← Path of the invader

**(a)**

Lampreys enter, 1938

First stream treatments for lampreys, 1958

Sea lamprey

Lake trout

**(b)**

**Figure 59.30** **Effects of invasive sea lampreys on the lake trout population in the Great Lakes.** **(a)** Historical passage of lampreys into the Great Lakes from the Atlantic. **(b)** Effects of lampreys on the trout population in Lake Superior. Note that detailed records of sea lamprey populations were not available before 1956. a: Source: U.S. Fish and Wildlife Service, Bugwood.org

lampreys to bypass Niagara Falls and colonize the rest of the Great Lakes. In 1921, lampreys were found in Lake Erie and by 1938 they had been found all the way up into Lake Superior. Lake trout (*Salvelinus namaycush*) were the lampreys' preferred prey in the Great Lakes. The lake trout fishing industry declined somewhat in the 1940s due to overfishing, but the decline was hastened by the arrival of lampreys (**Figure 59.30b**). By the 1960s, lake trout catches had been reduced by over 90% of their historic averages, and the fishery collapsed.

In the late 1940s, the state of Michigan began control measures for lampreys. Barriers were erected across the mouths of streams to prevent the return of spawning adults, but they were difficult to maintain and were not 100% effective. Attention turned to chemical control of juvenile lampreys, which spend 3–5 years buried in stream

sediments, filter feeding. The chemical TFM (trifluoromethyl nitrophenol) proved effective in controlling juvenile lampreys in streams and was first applied in 1958 to streams feeding into Lake Superior (see Figure 59.30b). These treatments reduced lamprey densities by 90%, and the population of lake trout increased in the 1980s, aided by an active program of restocking. To maintain lamprey control, streams are treated every 3–5 years. Removal of invasive species is time-consuming and costly but may lead to recovery of native prey species.

A second example of an invasive species acting as a predator is the Burmese python in the Florida Everglades (see the chapter opening photo). Pythons were likely introduced prior to 1985 and the population has grown exponentially since then. Pythons in Florida

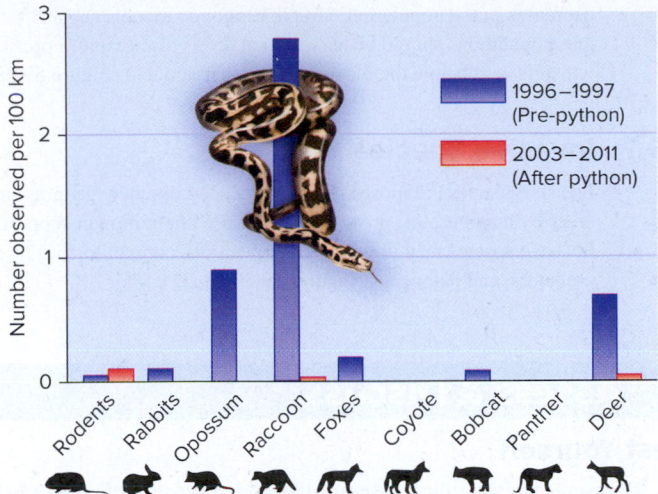

**Figure 59.31** **Effect of invasive Burmese pythons on wildlife abundance in the Everglades National Park, Florida.** According to roadside surveys, a decline in mammal abundance occurred after pythons became common (2003–2011).

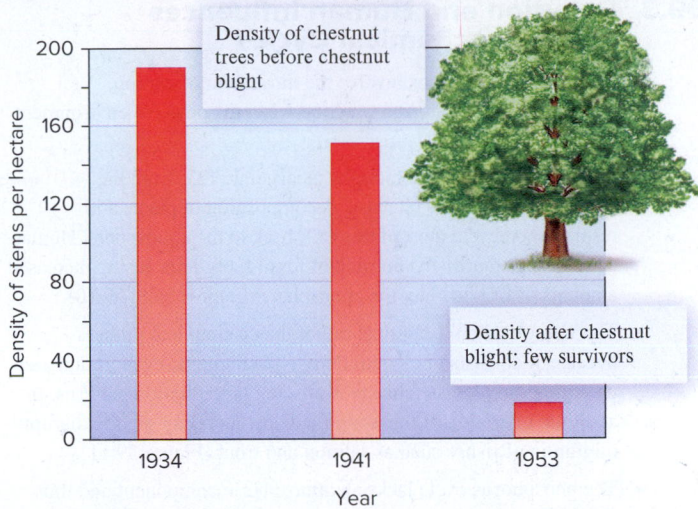

**Figure 59.32** **Effects of an introduced fungus on American chestnut trees.** The reduction in density of American chestnut trees in North Carolina following the introduction of the fungus *Cryphonectria parasitica* from Asia shows the severe effect that pathogens can have on their hosts. By the 1950s, this once prevalent species had been virtually eliminated.

consume a wide range of vertebrate prey, including endangered species such as the wood stork (*Mycteria americana*) and the Key Largo woodrat (*Neotoma floridana*). Survey data gathered by counting animals and roadkills showed that raccoons, opossums, and rabbits were often observed during the period 1993–1999, before pythons became common. In the period 2003–2011, when pythons were common, the decline in observations was 99.3% for raccoons, 98.9% for opossums, 94.1% for white-tailed deer, and 87.5% for bobcats (**Figure 59.31**). No rabbits or foxes were even observed. In south Florida today, these species exist only in locations peripheral to the Everglades, where pythons are rare.

### Some Invasive Species Are Lethal Pathogens

A **pathogen** is an agent that causes disease symptoms in its host. Some pathogens become invasive species when moved to a new geographical location. For example, Chestnut blight (*Cryphonectria parasitica*) is a fungus from Asia that was accidentally introduced to New York around 1904 from imported Asian chestnut trees (*Castanea crenata*). At that time, the American chestnut tree (*Castanea dentata*) was one of the most common trees in the eastern U.S. It was said that a squirrel could jump from one chestnut tree to another all the way from Maine to Georgia without touching the ground. The densest populations of chestnuts occurred in the Appalachian Mountains. Here, humans and wildlife alike feasted on the bountiful supply of nuts. The tree was ingrained in American culture, as in "chestnuts roasting on an open fire." By the 1950s, however, chestnut blight had significantly reduced the density of American chestnut trees in all areas of the United States, including North Carolina (**Figure 59.32**). Because at least one tree in four in these forests was a chestnut, the effect was very noticeable, although oaks and hickories replaced chestnuts in the canopy. Eventually, the fungus eliminated nearly all chestnut trees across North America.

## Summary of Key Concepts

### 59.1 Human Population Growth

- Human population growth has skyrocketed over the last 250 years (Figure 59.1).

- Differences in the age structure of a population, the relative numbers of individuals in defined age groups, can influence future patterns of population growth (Figure 59.2).

- The total fertility rate (TFR) is the average number of live births a woman has during her lifetime. TFRs are used to predict the growth of human populations (Figures 59.3, 59.4).

- The ecological footprint refers to the amount of productive land needed to support each person on Earth (Figure 59.5).

### 59.2 Global Warming and Climate Change

- Global warming is an increase in the average surface temperature on Earth. It is expected to result in climate change.

- The greenhouse effect is the process by which short-wave solar radiation passes through the atmosphere to warm the Earth and is radiated back into the atmosphere as long-wave infrared radiation. Much of this radiation is directed back to Earth's surface a second time, causing the surface temperature to rise (Figures 59.6).

- An increase in greenhouse gases, particularly $CO_2$, is intensifying the greenhouse effect, causing climate change (Table 59.1; Figure 59.7).

- Two consequences of climate change are rising sea levels and changes in precipitation patterns.

- Ecologists predict that climate change will have a large impact on the distribution of the world's organisms (Figure 59.8).

## 59.3 Pollution and Human Influences on Biogeochemical Cycles

- Biogeochemical cycles involve the movement of carbon, phosphorus, nitrogen, and water between the physical environment and organisms.

- In the carbon cycle, phototrophs incorporate $CO_2$ from the atmosphere into their biomass; decomposition of plants and respiration recycle most of this $CO_2$ back to the atmosphere. Human activities, primarily the burning of fossil fuels, are causing increased amounts of $CO_2$ to enter the atmosphere (Figures 59.9, 59.10).

- The water cycle is a physical rather than a chemical process because it consists of essentially two phenomena: evaporation and precipitation. Climate change is altering precipitation patterns. In addition, alteration of the water cycle by dams can greatly disrupt migration of fishes such as salmon and trout (Figure 59.11).

- The phosphorus cycle lacks an atmospheric component and thus is a local cycle. An overabundance of phosphorus can cause the overgrowth of algae and subsequent depletion of oxygen levels, called eutrophication (Figures 59.12, 59.13, 59.14).

- The nitrogen cycle has five parts: nitrogen fixation, nitrification, assimilation, ammonification, and denitrification. The activities of humans, including the production and use of fertilizers, have altered the nitrogen cycle (Figure 59.15).

## 59.4 Pollution and Biomagnification

- Biomagnification is the tendency of certain chemicals to concentrate in higher trophic levels in food chains (Figures 59.16, 59.17).

## 59.5 Habitat Destruction

- Habitat destruction involves the conversion of natural habitat to agricultural land, urbanization, the draining of swamps, strip mining, dam destruction, and many other forms of land modification. It is the primary cause of species extinction and is threatening many current species (Figure 59.18).

- Deforestation—in particular, of tropical forests—threatens many species with extinction because of the high numbers of species that live in forests and the high rate of forest loss (Figures 59.19, 59.20).

- The planting of crops and the grazing of livestock are among the most important drivers of ecological change. Habitat destruction has changed huge areas of the planet into agricultural areas and rangelands for livestock (Figure 59.21; Table 59.2).

## 59.6 Overexploitation

- Overexploitation is the practice in which humans harvest a particular species at a rate that is unsustainable. The overexploitation of plants and animals occurs via hunting, fishing, and removing species from their native habitat. On average, humans decrease prey populations much more than natural predators do (Figure 59.22).

- Overexploitation has decreased the populations of many species and driven some of them to extinction. Examples include whales, the passenger pigeon, and many species of fishes and plants (Figures 59.23–59.27).

- The maximum sustainable yield is the largest number of individuals that can be harvested without causing long-term

decreases in the population. Simple economic models suggest that populations should be harvested at levels of maximum profit, which is well below the maximum sustainable yield (Figure 59.28).

## 59.7 Invasive Species

- Invasive species are those species introduced into new geographic areas by humans and spread on their own without human support. Invasive species may impact native species as competitors, predators, and pathogens (Figures 59.29–59.32).

## Assess & Discuss

### Test Yourself

1. The average total fertility rate needed for zero population change across the world is
   - a. 1.7.
   - b. 1.9.
   - c. 2.0.
   - d. 2.3.
   - e. 2.5.

2. Which anthropogenic gas contributes most to global warming?
   - a. methane
   - b. chlorofluorocarbons
   - c. carbon dioxide
   - d. nitrogen oxides
   - e. sulfur dioxide

3. The data show that atmospheric carbon dioxide levels have increased by what percentage over the past 58 years?
   - a. 1%
   - b. 5%
   - c. 12%
   - d. 25%
   - e. 52%

4. Eutrophication is
   - a. caused by an overabundance of nitrogen, which leads to a decrease in bacteria populations.
   - b. caused by an overabundance of nutrients, which leads to an increase in algal populations.
   - c. usually caused by oil spills.
   - d. normally seen in dry, hot regions of the world.
   - e. accelerated by habitat loss.

5. Human production of fertilizers first impacts which part of the nitrogen cycle?
   - a. nitrogen fixation
   - b. nitrification
   - c. assimilation
   - d. ammonification
   - e. denitrification

6. The concentration of certain chemicals, such as DDT, in higher trophic levels is known as
   - a. eutrophication.
   - b. biomagnification.
   - c. biogeochemical cycling.
   - d. energy transfer.
   - e. turnover.

7. Which is *not* an example of habitat destruction?
   - a. deforestation
   - b. agriculture
   - c. urbanization
   - d. draining of wetlands
   - e. overharvesting

8. The passenger pigeon, once the most common bird in North America, was driven to extinction by
   - a. pollution.
   - b. overexploitation.
   - c. habitat destruction.
   - d. invasive species.
   - e. all of the above.

9. Most recorded extinctions have been caused by
   a. invasive species.
   b. habitat destruction.
   c. overexploitation.
   d. a and b equally.
   e. a, b, and c equally.

10. Invasive species enter an area through
    a. agricultural introductions.
    b. accidental transportation via ships.
    c. landscape plants and their pests.
    d. a and b.
    e. a, b, and c.

## Conceptual Questions

1. The Earth's atmosphere consists of 78% nitrogen. Why is nitrogen a limiting nutrient?

2. Why does maximum sustainable yield occur at the midpoint of the logistic curve and not where the population is at carrying capacity?

3. **Core Skill: Science and Society** In one family, parents, who were born in 1900, have twins at age 20 but then have no more children. Their children, grandchildren, and so on behave in the same way. In another family, parents, who were also born in 1900, delay reproduction until age 33 but have triplets. Their children and grandchildren behave in the same way. Which family has the most descendants by 2000? What can you conclude?

## Collaborative Questions

1. Discuss what might limit human population growth in the future.

2. As a group, try to predict what effects an atmospheric concentration of 700 ppm of $CO_2$ might have on the environment.

# Biodiversity and Conservation Biology

# 60

**Spix's macaw (*Cyanopsitta spixii*).** Less than 100 individuals of this species are known to exist in the rain forests of Brazil.

©Patrick Pleul/dpa/picture-alliance/Newscom

I n 2009, Jeff Corwin, an American conservationist and host for programs on Animal Planet and other television networks, published a book entitled *100 Heartbeats: The Race to Save the Earth's Most Endangered Species*. The Hundred Heartbeat Club had been created earlier by biologist E. O. Wilson to highlight the plight of animal species, such as Spix's macaw (*Cyanopsitta spixii*) in Brazil (see the chapter opening photo), the Chinese river dolphin (*Lipotes vexillifer*), and the Philippine eagle (*Pithecophaga jefferyi*), that have 100 or fewer individuals left alive (and hence are that number of heartbeats away from extinction). Saving species from extinction is important in its own right, but, as we will see, conservation of biological diversity also has great economic and social value to humankind.

Biological diversity, or **biodiversity**, encompasses the genetic diversity of species, the variety of species, and the different ecosystems they form. The field of **conservation biology** uses principles and knowledge from molecular biology, genetics, and ecology to protect and sustain biodiversity. Because it draws from nearly all chapters of this textbook, a discussion of conservation biology is a fitting way to conclude our study of biology. In this chapter, we will begin by examining the questions of what biodiversity is and why it should be conserved and exploring how much biodiversity is needed for ecosystems to function properly.

Later, we will consider what is being done to help conserve the world's endangered plant and animal life. This includes identifying global areas rich in species and establishing parks and refuges of the appropriate size, number, and connectivity. We will also discuss conservation of particularly important types of species and how to restore damaged habitats to a more natural condition. We then examine how captive-breeding programs help to build populations of rare species prior to their release back into the wild. Some programs have also used modern genetic techniques such as cloning to help breed and perhaps eventually increase populations of endangered species.

## 60.1 Genetic, Species, and Ecosystem Diversity

### Learning Outcome:

1. List and describe the three levels of biodiversity.

Biodiversity can be examined on three levels: genetic diversity, species diversity, and ecosystem diversity. Each level of biodiversity provides valuable benefits to humanity.

**Genetic diversity** is the amount of genetic variation occurring within and between populations. Without such variation, populations may be unable to respond quickly to changes in environmental conditions, resulting in population decline and even extinction. Maintaining genetic variation in the wild relatives of crops may be vital to the continued success of crop-breeding programs. For example, the café marron (*Ramosmania rodriguesii*), a wild relative of the coffee plant that is native to a tiny island off the coast of

**Figure 60.1** **Café marron being cultured in London's Kew Gardens.** These plants are derived from just one surviving individual found in Mauritius. ©Florapix/Alamy Stock Photo

Mauritius, was assumed to be extinct until 1979, when one surviving tree was identified. Today, cuttings from the tree are being cultured in London's Kew Gardens (**Figure 60.1**). The plant may carry genes that could allow coffee to be grown in a wider range of soils and elevations.

A second level of biodiversity is **species diversity**, the number and relative abundance of species in a community (refer back to Section 58.1). In 1973, the U.S. Endangered Species Act (ESA) was enacted, which was designed to protect both endangered and threatened species. **Endangered species** are those species that are in danger of extinction throughout all or a significant portion of their range. **Threatened species** are those species likely to become endangered in the foreseeable future. Many species are currently threatened. According to the International Union for Conservation of Nature and Natural Resources (IUCN), more than 25% of the fish species that live on coral reefs and 22% of all mammals, 12% of birds, and 31% of amphibians are threatened with extinction. In 2000, the World Wildlife Foundation placed Atlantic cod (*Gadus morhua*) on the endangered species list as a result of overfishing. Nine of 17 populations of commercially important Chinook salmon (*Oncorhynchus tshawytscha*) in California and Oregon are listed as endangered or threatened.

The last level of biodiversity is **ecosystem diversity**, the diversity of structure and function within an ecosystem. Conservation has largely focused attention on species-rich ecosystems such as tropical rain forests. Over 120 prescription drugs used to treat malaria, cancer, and other diseases were developed from rain forest plants, yet less than 1% of such plants have been tested for medicinal properties. However, some ecologists have argued that relatively species-poor ecosystems such as prairies are also highly threatened and in equal need of conservation. More than 99% of the original tallgrass prairie in the United States has been converted to agricultural land.

## 60.2 Biodiversity and Ecosystem Function

### Learning Outcomes:

1. **CoreSKILLS »** Create graphical representations of possible relationships between biodiversity level and ecosystem function.

2. Describe experimental evidence that shows how species richness and ecosystem function are linked.

Because biodiversity affects the health of ecosystems, ecologists have explored the question of how much diversity is needed for ecosystems to function properly. In this section, we will examine several models that explore the relationship between ecosystem function and species richness and examine an experimental approach used to study this relationship.

### Ecologists Have Proposed Several Models for the Relationship Between Ecosystem Function and Species Richness

Because biodiversity affects the health of ecosystems, ecologists have explored the question of how many species are needed for ecosystems to function properly. In doing so, they have described several possible relationships between ecosystem function and species richness. In the 1950s, ecologist Charles Elton proposed the **diversity-stability hypothesis**, which suggests that species-rich communities are more stable than those with fewer species (refer back to Section 58.2). If we use stability as a measure of ecosystem function, Elton's hypothesis indicates a linear correlation between ecosystem function and species richness; as the number of different species increases, ecosystem function increases proportionately (**Figure 60.2a**). Australian ecologist Brian Walker proposed an alternative to this idea, termed the **redundancy hypothesis** (**Figure 60.2b**). According to this hypothesis, ecosystem function increases rapidly at fairly low levels of species richness, but then levels off because most additional species are functionally redundant.

Two other alternative ideas relating species richness and ecosystem function have been proposed. The **keystone hypothesis** (**Figure 60.2c**) proposes that ecosystem function dramatically rises as species richness approaches its natural level. Finally, the **idiosyncratic hypothesis** suggests that although ecosystem function can change as the number of species increases or decreases, the amount and direction of change are unpredictable (**Figure 60.2d**).

Determining the correct model(s) for the relationship between ecosystem function and species richness is very important, as our understanding of the effect of species loss on ecosystem function can greatly affect the way we manage our environment. Only relatively recently has the link between ecosystem function and species richness been studied. An early and influential study, conducted by Shahid Naeem and his colleagues in laboratories in England, is described next.

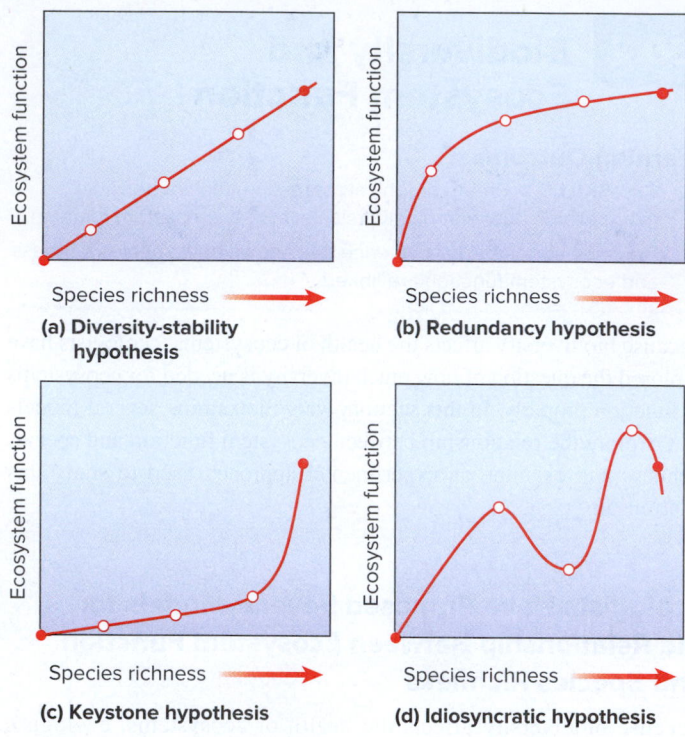

(a) Diversity-stability hypothesis

(b) Redundancy hypothesis

(c) Keystone hypothesis

(d) Idiosyncratic hypothesis

**Figure 60.2** **Graphical representations of possible relationships between ecosystem function and species richness.** The two solid

dots represent the end points of a continuum of species richness. The first dot is at the origin, where there are no species and no community services. The second dot represents a natural level of species richness. The relationship is strongest in **(a)** and weakest in **(d)**.

 **Core Skill: Modeling** The goal of this modeling challenge is to draw a graph showing a different relationship between ecosystem function and species richness, one based on the rivet hypothesis, which was proposed in 1981.

**Modeling Challenge:** In 1981, Paul and Ann Ehrlich proposed a relationship between ecosystem function and species richness called the rivet hypothesis. In this model, species are like the rivets on an airplane. Some species play a small but critical role in keeping the plane, the ecosystem, airborne, while other species do not, and we cannot tell beforehand which species affect the ecosystem function the most. The loss of a single rivet will probably not weaken the plane. However, the loss of few rivets would impair airworthiness. The plane could still function but not at maximum efficiency. Continuing on, the loss of a few more rivets could again be tolerated, but the loss of yet more rivets would prove critical to the airplane's function. Thus, ecosystem function declines with decreased species richness in a stepwise fashion. Draw a graph that depicts a model for the rivet hypothesis. Use the same format as those in Figure 60.2.

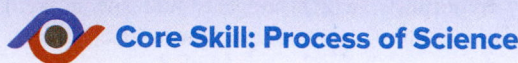 **Core Skill: Process of Science**

# Feature Investigation | Ecotron Experiments Analyzed the Relationship Between Ecosystem Function and Species Richness

In the early 1990s, American ecologist Shahid Naeem and colleagues used a series of 14 environmental chambers in a facility termed the Ecotron, at Silwood Park, England, to determine how species richness affects ecosystem function. These chambers contained terres-

trial communities that differed only in their level of species richness (**Figure 60.3**). The number of species in each chamber was manipulated to create high-, medium-, and low-richness ecosystems, each with four trophic levels. The trophic levels consisted of primary

**Figure 60.3** **Ecotron experiments comparing species richness and ecosystem function.** (3): ©Pete Manning, Ecotron Facility, NERC Centre for Population Biology

**Concept Check:** *What is one of the dangers in interpreting these results?*

**HYPOTHESIS** Reduced species richness can lead to reduced ecosystem functioning.

**STARTING LOCATION** Ecotron, a controlled environment facility at the Natural Environment Research Council (NERC) Centre for Population Biology, in Silwood Park, England.

| Experimental level | Conceptual level |
|---|---|
| 1 Construct 14 identical experimental chambers. | Temperature- and humidity-controlled chambers are used to control environmental conditions and allow identical starting conditions in all chambers. |

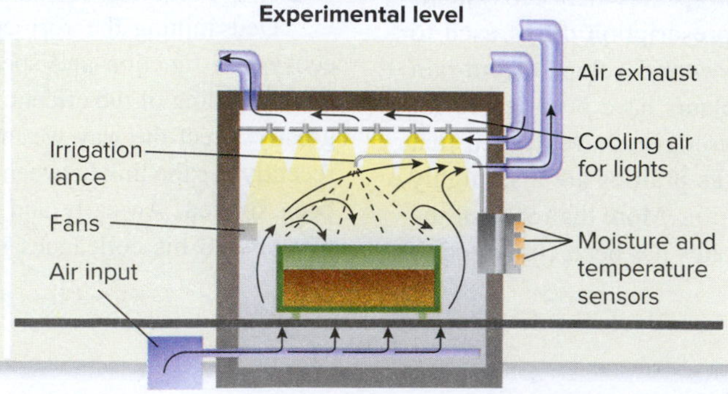

Air exhaust

Cooling air for lights

Irrigation lance

Fans

Air input

Moisture and temperature sensors

**2** Add different combinations of species to the 14 chambers. The species added were based on 3 types of model communities (food webs), each with 4 trophic levels but with varying degrees of species richness.

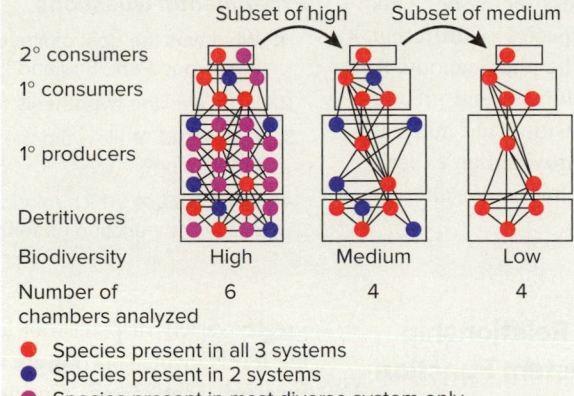

Subset of high    Subset of medium

2° consumers
1° consumers
1° producers
Detritivores

Biodiversity    High    Medium    Low
Number of chambers analyzed    6    4    4

● Species present in all 3 systems
● Species present in 2 systems
● Species present in most diverse system only

The three diagrams to the left each represent a single chamber. Circles represent species, and lines connecting them represent biotic interactions among the species. Note that each lower-diversity community is a subset of its higher-diversity counterpart and that all community types have 4 trophic levels.

**3** Measure and analyze a range of processes, including vegetation cover and nutrient uptake.

Measurements help determine how each different type of community functions.

**4** **THE DATA**

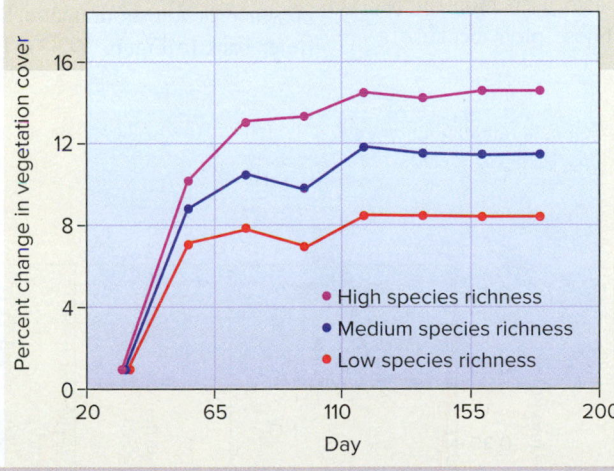

Plant productivity is linked to community diversity as measured by the percent change in vegetation cover from initial conditions.

Data reveal that low-diversity communities have lower vegetation cover and thus are less productive than high-diversity communities.

**5** **CONCLUSION** Increases in species richness lead to increases in ecosystem function. In this case, increased plant species richness results in greater vegetation cover.

**6** **SOURCE** Naeem, S., et al. 1994. Declining biodiversity can alter the performance of ecosystems. *Nature* 368: 734–737.

producers (annual plants), primary consumers (insects, snails, and slugs), secondary consumers (parasitoids that fed on the herbivores), and detritivores (earthworms and soil insects). The experiment ran for just over 6 months, and species were added only after the trophic level below them was established. For example, parasitoids were not added until herbivores were abundant.

Researchers monitored and analyzed a range of measures of ecosystem function, including community respiration, decomposition, nutrient retention rates, and community productivity. The data shown in step 4 of Figure 60.3 focus only on community productivity. The result was that community productivity, expressed as percent change in vegetation cover (the amount of ground covered by leaves of plants), increased as species richness increased. This productivity increase occurred because plant species of different heights could utilize light at different levels of the plant canopy. A larger ground cover also meant a larger plant biomass and greater community productivity, and increased decomposition and nutrient uptake rates. For the first time, ecologists had provided an experimental demonstration that a loss of species richness can alter or impair the functioning of an ecosystem.

### Experimental Questions

1. What was the goal of Naeem and colleagues in their experiment at Silwood Park, England?

2. What was the hypothesis tested by the researchers?

3. **CoreSKILL »** How did the researchers test for ecosystem functioning?

4. **CoreSKILL »** Which relationship between species richness and ecosystem function do these results support? (Refer to Figure 60.2.)

## Field Experiments Have Explored the Relationship Between Species Richness and Ecosystem Function

In the mid-1990s, David Tilman and colleagues performed experiments in the field to determine how species richness affected proper ecosystem functioning. Tilman's previous experiments had suggested that species-rich grasslands were more stable (that is, they were more resistant to the ravages of drought and recovered from drought more quickly) than species-poor grasslands (refer back to Figure 58.4). In subsequent experiments, Tilman's group sowed multiple plots, each 3 m by 3 m and having comparable soils, with seeds of 1, 2, 4, 6, 8, 12, or 24 species of prairie plants. Exactly which species were sown into each plot was determined randomly from a pool of 24 native species. The treatments were replicated 21 times, for a total of 147 plots.

The results showed that plots with more species had increased productivity, expressed as a percentage of plant cover (the amount of ground covered by leaves of plants) than plots with fewer species (**Figure 60.4a**). More species-rich plots also used more nutrients, such as nitrate ($NO_3^-$), than lower richness plots because a

greater variety of plant root lengths could utilize nutrients at different levels of the soil (**Figure 60.4b**). Furthermore, the frequency of invasive plant species (species not originally planted in the plots) decreased with increased plant species richness (**Figure 60.4c**).

Although Tilman's results show a relationship between species richness and ecosystem function, they also suggest that most of the advantages of increasing richness come with the first 5–10 species, beyond which adding more species appears to have little to no effect. These data support the redundancy hypothesis (see Figure 60.2b).

Confirmation of the redundancy hypothesis is also observed on a larger scale. The productivity of temperate forests on different continents is roughly the same, despite different numbers of tree species being present: 729 species in East Asia, 253 in North America, and 124 in Europe. The presence of more tree species may ensure a supply of "backups," should some of the most productive species die off from insect attack or disease. This effect was seen following the demise of the American chestnut tree. Disease devastated this species, and its presence in forests dramatically decreased by the mid-20th century (refer back to Figure 59.32). The forests filled in with other species

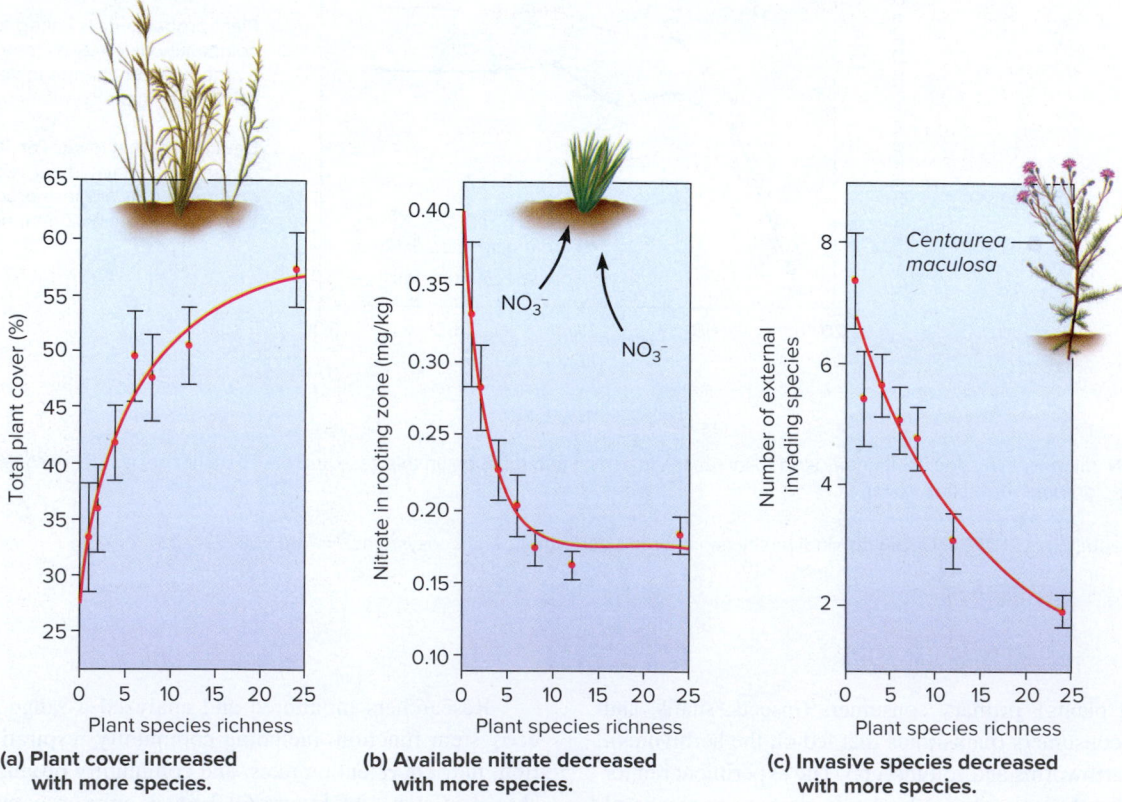

(a) **Plant cover increased with more species.**

(b) **Available nitrate decreased with more species.**

(c) **Invasive species decreased with more species.**

**Figure 60.4** **The relationship of species richness to ecosystem function.**

and continued to function as before in terms of nutrient cycling and gas exchange. However, although the forests continued to function without the American chestnuts, some important changes occurred. For example, the loss of chestnuts deprived bears and other animals of an important source of food and may have affected their reproductive capacity and hence the size of their populations.

## 60.3 Value of Biodiversity to Human Welfare

### Learning Outcomes:

1. Outline the benefits of biological diversity to human welfare.

Why should biodiversity be a concern? American biologists Paul Ehrlich and E. O. Wilson have suggested that the loss of biodiversity should be an area of great concern for at least three reasons:

1. Humans depend on plants, animals, and microorganisms for a wide range of foods, medicines, and industrial products.

2. Ecosystems provide an array of essential services, such as clean air and water.

3. Humans have an ethical responsibility to protect what are our only known living companions in the universe.

In this section, we examine some of the primary reasons why it is so important to preserve biodiversity.

### Society Benefits Economically from Biodiversity

Many different sectors of society benefit from biodiversity. Three key sectors are the pharmaceutical, agricultural, and the natural products industries.

***Pharmaceutical Industry*** The pharmaceutical industry is heavily dependent on plant, animal, fungal, and bacterial products for source material. Worldwide prescription drug sales are forecast to be $1 trillion in 2020.

An estimated 50,000–70,000 plant species are used in traditional and modern medicine. About 25% of the prescription drugs in the U.S. alone are derived from plants, and the 2015 market value of such drugs was estimated to be $374 billion, accounting for a little less than half the global pharmaceutical market. Many medicines come from plants found only in tropical rain forests. These include quinine, a drug from the bark of the cinchona tree (*Cinchona officinalis*), which is used for treating malaria, and vincristine, a drug derived from rosy periwinkle (*Catharanthus roseus*), which is a treatment for leukemia and Hodgkin's disease (**Figure 60.5a**).

Animal products have also been important to the pharmaceutical industry. For example, the venom of the gila monster (*Heloderma suspectum*), one of only two venomous lizards in the world, is being used to treat people who are resistant to conventional treatment for type 2 diabetes, a disease that may affect 30% of Americans at some point in their lives. A protein from the South American pit viper (*Bothrops jacara*) may help control human blood pressure. Tarantula venom may be helpful in treating neurological disorders such as Parkinson's disease.

Fungi and soil bacteria have provided an important reservoir of drugs. In 1928, Scottish scientist Alexander Fleming discovered that a fungus of the genus *Penicillium* produces penicillin, one of the first and most widely used antibiotics. In 1964, a group of Canadian scientists traveled to Easter Island and identified a drug produced by a soil bacterium that suppresses immune reactions in human. It is used to prevent organ rejection in transplants and as a coating on heart stents. The drug was named rapamycin after the indigenous name of Easter Island, which is Rapa Nui.

***Agriculture*** Although humans depend on only about 20 plant species to provide 90% of the world's food, wild relatives of these crops provide a useful reservoir of genetic material for developing pest-resistant varieties or strains that can grow in marginal areas. A gene from longstamen rice (*Oryza longistaminata*), a grass species from Africa, has been integrated into the genome of commercially grown rice (*O. sativa*) to confer resistance to rice blight disease (**Figure 60.5b**). In the 1970s, infusion of genetic material from wild corn in Mexico

(a)    (b)    (c)

**Figure 60.5 Societal benefits from biodiversity. (a)** Bark of the cinchona tree (*Cinochona officinalis*), a plant found only in tropical rain forests, is used to produce quinine, an effective treatment for malaria. **(b)** A gene from longstamen rice (*Oryza longistaminata*), native to Africa, was used to reduce rice blight disease in rice, *O. sativa,* native to Asia. **(c)** Commercial fishing for salmon is an important part of the economy in the U.S. Pacific Northwest. a: ©Heather Angel/Natural Visions; b: ©Arterra Picture Library lamy Stock Photo; c: ©Joshua Roper/Alamy Stock Photo

was used to protect commercially grown U.S. corn from a leaf fungus, which had killed 15% of the crop. Lake Placid mint (*Dicerandra frutescens*), native only in central Florida, produces a powerful insect-repelling chemical that may have benefits for crop protection. Another endangered species, buffalo clover (*Trifolium stoloniferum*), has high protein content and is a perennial, making it of high potential value as a forage crop.

***Natural Products Industry***  Many products that are used by humans are harvested from plant species growing in their native environments. Biodiversity is important for the maintenance of those native environments. Natural plant products include wood, rubber, and dietary supplements. The forest products industry manufactures over $200 billion in products in the U.S. annually. Maple syrup, nuts, blueberries and algae are all harvested in addition to the trees themselves. Interestingly, a relatively new commercial crop is made by the flowering shrub *Parthenium argentartum*. It produces high amounts of natural rubber and grows in the deserts of the Southwest U.S., adding economic value to marginal lands.

Animals, too, have great commercial value in the natural products industry. Salmon fishing in the Pacific Northwest supports over 60,000 jobs and injects over $1 billion into the economy (**Figure 60.5c**). In the U.K., the sea fish catch is worth over $500 million annually. Exploitation of wild animals, often known as bushmeat, in Africa is a major source of animal protein for more than a billion of the world's poorest people. Game birds, especially pheasants and ducks, are often shot in North America and Europe.

## Natural Ecosystems Provide Essential Services to Humans

Beyond the direct economic gains from biodiversity, humans benefit enormously from the essential services that natural ecosystems provide (**Table 60.1**). For example, forests soak up carbon dioxide, maintain soil fertility, and retain water, helping to prevent or minimize flooding. Estuaries provide water filtration and protect rivers and coastal shores from excessive erosion. *Prochlorococcus*, an abundant ocean-dwelling genus of cyanobacteria, was discovered only in 1986, yet it is estimated to produce about 20% of the oxygen we breathe. Other ecosystem functions include the maintenance of populations of natural predators to regulate pest outbreaks and of reservoirs of pollinators to pollinate crops and other plants. In addition, approximately 75% of the 100,000 chemicals released into the environment can be degraded by living organisms. The loss of biodiversity can disrupt an ecosystem's ability to carry out such functions.

In the 1990s, farmers in India began using the anti-inflammatory drug diclofenac to reduce pain and fever in their livestock. They could hardly anticipate that vultures scavenging on dead carcasses would accumulate large doses of the drug and die of renal failure. But the consequences did not stop there. Following a 97% reduction in vulture numbers over a 14-year period, the population of feral dogs exploded, because of the greater availability of uneaten carcasses. The incidence of rabies in humans increased, with estimates of an additional 48,000 people dying over the 14-year time span. The loss of the scavenging services of the vultures was estimated to cost India $24 billion.

A 2014 paper in the journal *Global Environmental Change* by economist Robert Costanza and colleagues made an attempt to calculate the monetary value of ecosystems to various economies (in 2007

| Table 60.1 | Examples of the World's Ecosystem Services |
|---|---|
| **Service** | **Example** |
| Atmospheric gas supply | Regulation of carbon dioxide, ozone, and oxygen levels |
| Climate regulation | Regulation of carbon dioxide, nitrogen dioxide, and methane levels |
| Disturbance regulation | Storm protection; flood control |
| Water regulation | Regulation of hydrological flows |
| Water supply | Irrigation; water for industry |
| Erosion control | Retention of topsoil; reduction of accumulation of sediments in lakes |
| Soil formation | Soil formation processes |
| Nutrient cycling | Nitrogen, phosphorus, and carbon cycles |
| Waste treatment | Sewage purification |
| Pollination | Pollination of crops |
| Biological control | Pest population regulation |
| Wilderness and refuges | Habitat for wildlife |
| Food production | Crops; livestock |
| Raw materials | Fossil fuels; timber |
| Genetic resources | Medicines; genes for plant resistance |
| Recreation | Ecotourism; outdoor recreation |
| Cultural | Aesthetic and educational value |

dollars). They came to the conclusion that, at the time, the world's ecosystems were worth more than $124 trillion a year, nearly twice the gross national product of the world's economies combined ($75.2 trillion in 2007 dollars).

## Ethical Reasons Underlie the Conservation of Biodiversity

Arguments can also be made against the loss of biodiversity on ethical grounds. As only one of many species on Earth, many people have argued that we have no right to destroy other species and the environment around us. John Muir, the founder of the Sierra Club, thought that natural areas had spiritual value and should be preserved rather than used as a source of natural products. This idea, which became known as the preservationist ethic, contrasts with the resource conservation ethic, which focuses on management of natural areas to allow the wisest current and future use of natural resources. In the United States, California condors (*Gymnogyps californianus*) and grizzly bears (*Ursus arctos*) have little economic value to people yet we assign them enough value that we preserve the habitats they live in. American philosopher Tom Regan suggests that animals should be treated with respect because they have a life of their own and therefore have value apart from anyone else's interests. American law professor Christopher Stone, in an influential 1972 article titled "Should Trees Have Standing?" has argued that entities such as nonhuman natural objects, like trees or lakes, should be given legal rights just as corporations are treated as individuals for certain purposes. In 1984, E. O. Wilson proposed the concept known as biophilia: Humans have innate attachments with other species and natural habitats because of our close association with them over millions of years.

## 60.4 Conservation Strategies

**Learning Outcomes:**

1. List and describe the criteria that conservation biologists use to identify areas for protection.
2. Explain how the principles of island biogeography and landscape ecology are used to create nature preserves.
3. Describe different approaches that conservation biologists use to protect individual species.
4. Define restoration ecology, and describe the approaches used to restore degraded ecosystems and populations of species.
5. Describe the advantages and disadvantages of using cloning as a conservation strategy to help save endangered species.

As discussed in Chapter 59, the activities of humans have had many negative effects that threaten biodiversity and ecosystems. The goal of conservation biologists is to manage natural resources with the aim of protecting species, their habitats, and ecosystems. In their efforts to maintain the diversity of life on Earth, these biologists are active on many fronts.

We begin this section by discussing how conservation biologists identify the global areas most in need of conservation. We will then examine some of the factors that must be considered when designing nature preserves, also called parks. These factors include the size and shape of the preserve and the ability of species to move from one nature preserve to another. In some cases, efforts are aimed at protecting species that have a great ecological impact; in others, they are aimed at protecting species, such as the panda bear, which are recognizable and garner support from humans. We will then explore the topic of **habitat restoration**—the full or partial repair or replacement of biological habitats and/or their populations that have been degraded or destroyed. Such restoration may involve captive breeding programs to reestablish populations of threatened species in the wild. Finally, we will discuss how genetic cloning may be used as an additional tool to help conserve endangered species.

### Conservation Seeks to Establish Protected Areas

Currently, about 15.4% of global land area is under some form of environmental protection. More than 217,155 separate areas and 3.4% of the global ocean area are protected, with more being added daily. Conservation biologists often must make decisions regarding which habitats should be protected. Many conservation efforts have focused on saving habitats in so-called megadiversity countries, because they often have the greatest number of species. However, more recent strategies have promoted preservation of certain key areas with the highest levels of unique species or the preservation of representative areas of all types of habitat, even relatively species-poor areas.

**Megadiversity Countries**   One strategy of targeting areas for conservation is to identify **megadiversity countries**, those countries with the greatest numbers of species. Using the number of plants, vertebrates, and selected groups of insects as criteria, American biologist Russell Mittermeier and colleagues determined that just 17 countries are home to nearly 70% of all known species. Brazil, Indonesia, and Colombia top the list, followed by Australia, Peru, Mexico, Madagascar, China, and nine other countries. The megadiversity country approach suggests that conservation efforts should be focused on the most biologically rich countries. However, although megadiversity areas may contain the most species, they do not necessarily contain the greatest number of unique species. The mammal species list for Peru is 344, and for Ecuador, it is 271; of these, however, 208 species are common to both countries.

**Areas Rich in Endemic Species**   Another approach for setting conservation priorities—one adopted by the organization Conservation International—takes into account the number of species that are **endemic**; that is, they are found only in a particular place or region. This approach suggests that conservationists focus their efforts on **biodiversity hot spots**, regions that are biologically diverse and under threat of destruction. To qualify as a biodiversity hot spot, a region must meet two criteria: (1) It must contain at least 1,500 species of vascular plants as endemic species, and (2) it must have lost at least 70% of its original habitat. Vascular plants were chosen as the primary group of organisms to determine whether or not an area qualifies as a hot spot, mainly because most other terrestrial organisms depend on them to some extent.

Conservationists Norman Myers, Russell Mittermeier, and colleagues identified 34 biodiversity hot spots that together occupy a mere 2.3% of the Earth's surface but contain 150,000 endemic plant species, or 50% of the world's total (**Figure 60.6**). This approach proposes that protecting such hot spots will prevent the extinction of a larger number of endemic species than would protecting areas of a similar size elsewhere. The main argument against using hot spots as the criterion for targeting conservation efforts is that the areas richest in endemic species—tropical forests—would receive the majority of attention and funding, perhaps at the expense of protecting other areas.

**Representative Habitats and Crisis Ecoregions**   In a third approach to prioritizing areas for conservation, some conservation biologists have argued that we need to conserve representatives of all major habitats. Prairies are a case in point. An example is the Pampas region of South America, which is arguably the most threatened habitat on the continent because of rapid conversion of its natural grasslands to ranch land and agriculture (**Figure 60.7**). The Pampas does not compare well in richness or endemics with the rain forests, but it is a unique area that, without preservation efforts, could disappear. By selecting habitats that are most distinct from those already preserved, many areas that are threatened but not biologically rich may be preserved in addition to the less immediately threatened, but richer, tropical forests.

With regard to representative habitats, habitat loss has been most extensive in temperate grasslands and tropical and temperate deciduous forests, where, in each case, over 50% of the land has been converted to other uses, such as agriculture. At the same time, very little, less than 10%, of these habitats have been protected. Such habitats are thus termed **crisis ecoregions**.

**Last of the Wild**   The final approach to conservation involves preservation of regions of the world relatively untouched by humans.

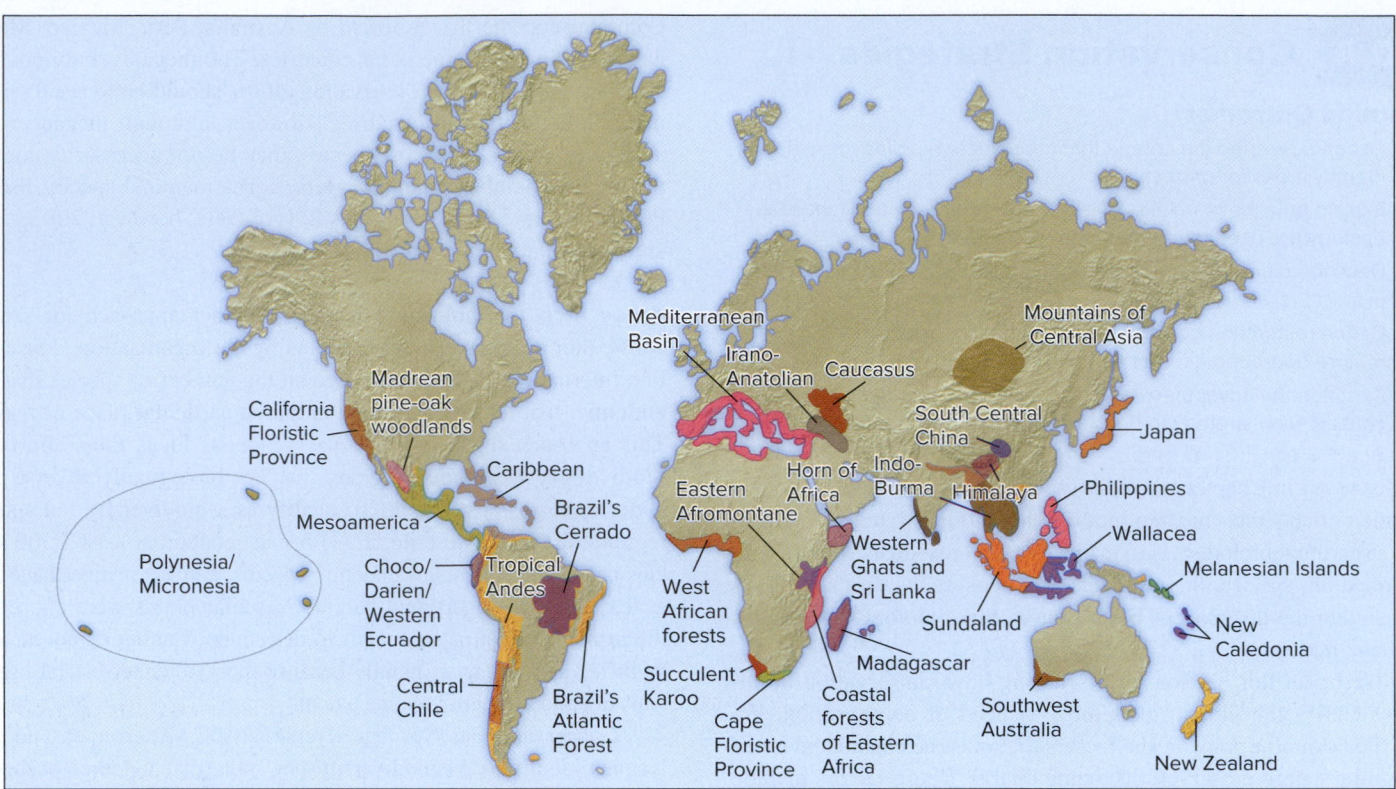

**Figure 60.6** **Location of the major biodiversity hot spots.** Hot spots have high numbers of endemic species. Different colors distinguish the biodiversity hot spots.

**Figure 60.7** **The Pampas, Argentina.** This habitat is not rich in species but is threatened due to conversion to ranch land and agriculture. The guanaco, shown here, is a characteristic grazer of pampas grass. ©Vicki Fisher/Alamy Stock Photo

Scientists have mapped out the extent of the human footprint on the globe. The areas of the Earth that fall within the lowest 10% of the human-affected areas have been termed the "last of the wild." Such areas, because they are relatively pristine, offer a great opportunity for conservationists because of their relatively intact communities. These wild areas include the tundra and boreal forests of Russia and Canada as well as some desert biomes and tropical forests.

## Preserve Design Incorporates Principles of Island Biogeography and Landscape Ecology

After identifying areas to preserve, conservationists must determine the size, arrangement, and management of the protected land. Among the questions conservationists ask is whether one large preserve is preferable to an equivalent area composed of smaller preserves. Ecologists also need to determine whether nature preserves should be close together or far apart and whether or not they should be connected by strips of suitable habitat to allow the movement of plants and animals between them.

***The Role of Island Biogeography*** According to the equilibrium model of island biogeography (refer back to Section 58.4), the biodiversity on an island tends toward an equilibrium number that is determined by the balance between two factors: immigration rates and extinction rates. This theory has been applied to nature preserves because, in essence, they are islands in a sea of human-altered habitat.

One question for conservationists is how large a protected area should be (**Figure 60.8a**). According to island biogeography, the number of species should increase with increasing area (the species-area effect). Thus, a larger area would mean that a larger number of species would be protected. In addition, larger preserves have other benefits. For example, they are beneficial for organisms that require

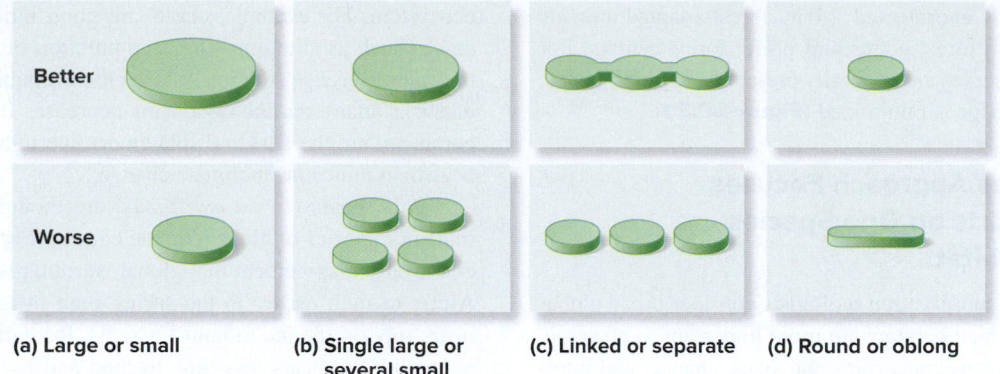

(a) Large or small  (b) Single large or several small  (c) Linked or separate  (d) Round or oblong

**Figure 60.8** **The theoretical design of nature preserves.** **(a)** A larger preserve holds more species and has low extinction rates. **(b)** A given area should be fragmented into as few pieces as possible. **(c)** Maintaining or creating corridors between fragments may also enhance dispersal. **(d)** Circular-shaped areas minimize the amount of edge effects. The labels "Better" and "Worse" refer to theoretical principles generated by the equilibrium model of island biogeography, but empirical data have not supported all the predictions.

*Concept Check:* *What are some of the potential risks in connecting preserved areas with movement corridors?*

large spaces, including migrating species and species with extensive territories, such as lions and tigers.

A related question is whether it is preferable to protect a single, large preserve or several smaller ones (**Figure 60.8b**). This question is called the **SLOSS debate** (for single large or several small). Proponents of a single, large preserve claim that a larger preserve is better able to protect more species and larger populations than an equal area divided into small areas. According to island biogeography, a larger block of habitat should support more species than several smaller blocks.

However, many empirical studies suggest that multiple small sites of equivalent area will contain more species, because a series of small sites is more likely to contain a broader variety of habitats than one large site. Looking at a variety of sites, American conservationists Jim Quinn and Susan Harrison concluded that animal life was richer in collections of small preserves than in a smaller number of larger ones. In their study, having more habitat types outweighed the effect of larger area size on species richness. In addition, another benefit of a series of smaller preserves is a reduction of extinction risk by a single event such as a wildfire or the spread of disease.

*The Role of Landscape Ecology* Landscape ecology is an area of ecology that examines the spatial arrangement of communities and ecosystems in a geographic area. Landscape ecologists have suggested that small preserves should be linked together by **movement corridors**, thin strips of land that permit the movement of species between preserved patches (**Figure 60.8c**). Such corridors ideally facilitate movements of organisms that are vulnerable to predation outside their natural habitat or have poor powers of dispersal between habitat patches. In this way, if a population in one small preserve experiences a disaster, immigrants from neighboring populations can more easily recolonize it. This avoids the need for humans to physically move new plants or animals into an area.

Several types of habitat function as movement corridors, including hedgerows (a linear patch of shrubs and small trees), which facilitate movement and dispersal of species between forest fragments (**Figure 60.9**). In China, corridors of habitat have been established to

link small, adjacent populations of giant pandas. However, some disadvantages are associated with movement corridors. Corridors also can facilitate the spread of disease, invasive species, and fire between small preserves.

Finally, nature preserves are often designed to minimize **edge effects**, the special physical conditions that exist at the boundaries, or edges, of ecosystems. Habitat edges, particularly those between a natural habitat such as a forest and developed land, are often different in physical characteristics from the habitat core. For example, the center of a forest is shaded by trees and has less wind and light than

Hedgerows

**Figure 60.9** **Movement corridors.** ©Andrew Parker/Alamy Stock Photo

*Concept Check:* *How do these European hedgerows act as movement corridors?*

the forest edge, which is unprotected. Many forest-adapted animals therefore shy away from forest edges and prefer forest centers. For this reason, circular preserves are generally preferable to oblong ones, because the amount of edge is minimized (**Figure 60.8d**).

## The Single-Species Approach Focuses Conservation Efforts on One Species in a Particular Habitat

As we have seen, many conservation biologists consider broad global issues when deciding which habitats are more important to preserve, as well as more local issues concerning the sizes, shapes, and interconnectedness of preserves. When establishing and managing nature preserves, they may take a single-species approach to their conservation efforts, which focuses on saving species that are deemed particularly important. As with habitat conservation, different approaches are used to identify the species that are most important.

*Indicator Species*   Some conservation biologists have suggested that certain organisms can be monitored as **indicator species**, those species whose status provides information on the overall health of an

ecosystem. For example, corals are good indicators of marine processes such as siltation—the accumulation of sediments transported by water. Because siltation reduces the availability of light, the abundance of many marine organisms decreases in such situations, with corals among the first to display a decline in health. Coral bleaching is also an indicator of climate change.

Polar bears (*Ursus maritimus*) are thought to be a mammalian indicator species of global climate change (**Figure 60.10a**). Most scientists are in agreement that global warming is causing the ice in the Arctic to melt earlier in the spring than in the past. Because polar bears rely on the ice to hunt for seals, the earlier breakup of the ice is leaving the bears less time to feed and build the fat that enables them to sustain themselves and their young. A U.S. Geological Survey study concluded that future reduction of Arctic ice could result in a loss of two-thirds of the world's polar bear population within 50 years. In May 2008, the polar bear was listed as a threatened species under the U.S. Endangered Species Act (ESA).

*Umbrella Species*   **Umbrella species** are those whose habitat requirements are so large that protecting them would protect many other species existing in the same habitat. The Northern spotted owl

(a) Indicator species: polar bear

(b) Umbrella species:
    northern spotted owl

(c) Flagship species: Florida panther

(d) Keystone species: American beaver

**Figure 60.10** Indicator, umbrella, flagship, and keystone species.  **(a)** Polar bears have been called an indicator species with respect to global climate change. **(b)** The northern spotted owl is considered an umbrella species for the old-growth forest in the Pacific Northwest. **(c)** The Florida panther has become a flagship species for Florida. **(d)** The American beaver, a keystone species, creates large dams across streams, and the resultant lakes provide habitats for a great diversity of species. a: ©Robert E. Barber/Alamy Stock Photo; b: Source: John & Karen Hollingsworth/U.S. Fish & Wildlife Service; c: Source: Larry Richardson, US Fish & Wildlife Service; d: ©Wildlife/Alamy Stock Photo

(*Strix occidentalis*) of the Pacific Northwest is considered to be an important umbrella species (**Figure 60.10b**). A pair of birds needs at least 800 hectares of old-growth forest for survival and reproduction, so maintaining healthy owl populations helps to ensure survival of many other forest-dwelling species. In the southeastern U.S., the red-cockaded woodpecker (*Picoides borealis*) is often seen as that region's equivalent of the Northern spotted owl because it requires large tracts of old-growth long-leaf pine (*Pinus palustris*), including old diseased trees in which it can excavate its nests.

***Flagship Species*** In the past, conservation resources were often allocated to a **flagship species**, a single large or instantly recognizable species. Such species were typically chosen because they were attractive and thus more readily engendered support from the public for their conservation. The concept of the flagship species, typically a charismatic vertebrate such as the American buffalo (*Bison bison*), has often been used to raise awareness for conservation in general. The giant panda (*Ailuropoda melanoleuca*) is the World Wildlife Fund's emblem for endangered species, and the Florida panther (*Puma concolor*) has become a symbol of the state's conservation campaign (**Figure 60.10c**).

***Keystone Species*** A different conservation strategy focuses on **keystone species**, species within a community that have a role out of proportion to their abundance or biomass. The beaver, a relatively small animal, can have a dramatic impact on a community by building a dam and flooding an entire river valley (**Figure 60.10d**). The resultant lake may become a home to fish species, wildfowl, and aquatic vegetation. A decline in the number of beavers may have serious ramifications for the remaining community members, promoting fish die-offs, waterfowl loss, and the death of vegetation adapted to waterlogged soil.

American ecologist John Terborgh considers tropical palm nuts and figs to be keystone plant species because they produce fruit during otherwise fruitless times of the year and are thus critical resources for tropical forest fruit-eating animals, including primates, rodents, and many birds. Together, these fruit eaters account for as much as three-quarters of the tropical forest animal biomass. Without the fruit trees, widespread extinction of these animals could occur.

## Habitat Restoration Attempts to Improve Degraded Habitats

Another aspect of conservation efforts is improving the quality of a previously degraded habitat. As noted at the beginning of this section, habitat restoration is the full or partial repair or replacement of biological habitats and/or their populations that have been degraded or destroyed. It can focus on complete restoration, rehabilitation, or replacement of a habitat, and it can involve returning species to the wild following captive breeding.

***Types of Habitat Restoration*** In complete restoration, conservation biologists attempt to return a habitat to its composition and condition prior to the disturbance. For example, under the leadership of American ecologist Aldo Leopold, the University of Wisconsin pioneered the restoration of prairie habitats as early as 1935, converting agricultural land back to species-rich prairies (**Figure 60.11a**).

The second approach of habitat restoration aims to return the habitat to something similar to, but a little less than, full restoration, a goal called rehabilitation. In Florida, phosphate mining involves removing a layer of topsoil, mining the phosphate-rich layers, returning the topsoil, and then replanting the area. Unfortunately, species such as cogongrass (*Imperata cylindrica*), an invasive Southeast Asian species, may invade these disturbed areas. Even so, the restoration serves to revegetate the area (**Figure 60.11b**).

The third approach to repairing a habitat, termed replacement, makes no attempt to restore what was originally present but instead replaces the original ecosystem with a different one. Ecosystem replacement is particularly useful for places in which the terrain has been substantially altered by past human activities. It would be nearly impossible to re-create the original landscape of an area that was mined for stone or gravel. In these situations, however, wetlands or lakes may be created in the open pits (**Figure 60.11c**).

**(a) Complete restoration**

**(b) Rehabilitation**

**(c) Ecosystem replacement**

**Figure 60.11** **Habitat restoration. (a)** The University of Wisconsin pioneered the practice of complete restoration of agricultural land to native prairies. **(b)** In Florida, complete restoration after phosphate mining is not usually possible. After topsoil is replaced, invasive species such as cogongrass often grow, resulting in habitat rehabilitation rather than complete restoration. **(c)** This old limestone mine in Crimea has been flooded to provide a valuable freshwater habitat, replacing the ecosystem that was originally present. a: ©University of Wisconsin-Madison Arboretum; b: Courtesy of DL Rockwood, School of Forest Resources and Conservation, University of Florida, Gainesville, FL; c: ©Vladimir Mulder/Shutterstock

 **Core Skill: Science and Society** Restoration of human-degraded habitats can lead to recovery of habitat and increased biodiversity.

***Reintroductions and Captive Breeding***   Reintroducing species into areas where they previously existed is a valuable conservation strategy that can re-establish populations in areas where they once occurred (look back at the opening photograph in Chapter 56). Captive breeding, the propagation of animals and plants outside their natural habitat to produce stock for subsequent release into the wild, has proved valuable in re-establishing breeding populations following extinction or near extinction. Zoos, aquariums, and botanical gardens often play a key role in captive breeding, propagating species that are highly threatened in the wild.

Several classic programs illustrate the value of captive breeding and reintroduction. The peregrine falcon (*Falco peregrinus*) had become extinct in nearly all of the eastern U.S. by the mid-1940s, a decline that was linked to the effects of DDT (refer back to Figure 59.16). In 1970, American biologist Tom Cade gathered falcons from other parts of the country to start a captive breeding program at Cornell University. Since then, the program has released thousands of birds into the wild, and in 1999, the peregrine falcon was removed from the list of endangered species.

A captive breeding program is also helping to save the California condor (*Gymnogyps californicus*) from extinction. At a cost of $35 million, this species conservation project is the most expensive one ever undertaken in the U.S. In the 1980s, the population had dropped to only 22 known condors, some in captivity and some in the wild. Scientists made the decision to capture the remaining wild birds in order to protect and breed them (**Figure 60.12a**). By 2016, the captive population numbered 167 individuals, and 268 birds were living in the coastal mountains of California; northern Baja California, Mexico; the Grand Canyon area of Arizona; and Zion National Park, Utah (**Figure 60.12b**). A milestone was reached in 2003, when a pair of captive-reared California condors bred in the wild.

## Can Cloning Save Endangered Species?

The use of reproductive cloning to clone endangered species is a new method that may eventually help bolster populations of captive-bred species. In 1997, geneticist Ian Wilmut and colleagues at Scotland's Roslin Institute announced to the world that they had cloned a sheep, the now-famous Dolly, from mammary cells of an adult ewe. Since then, interest has arisen among conservation biologists about whether the same technology might be used to save species on the verge of extinction.

Scientists were encouraged that in January 2001, an Iowa farm cow called Bessie gave birth to a cloned Asian gaur (*Bos gaurus*), an endangered species. The gaur, an oxlike animal native to the jungles of India and Burma, was cloned from a single skin cell taken from a dead animal. To clone the gaur, scientists removed the nucleus from a cow's egg and replaced it with a nucleus from the gaur's cell. The treated egg was then placed into the cow's uterus. Unfortunately, the gaur died from dysentery 2 days after birth, although scientists believe this was unrelated to the cloning procedure.

More recently, other endangered or extinct species have been cloned. In 2003, another type of endangered wild cattle, the Javan banteng (*Bos javanicus*), was successfully cloned (**Figure 60.13**). In 2005, clones of the African wildcat (*Felis libyca*) successfully produced wildcat kittens. This is the first time that clones of a wild species have bred. In 2009, a cloned Pyrenean ibex (*Capra pyrenaica*) was born but lived only 7 minutes due to physical defects in the lungs. The last wild Pyrenean ibex died in Spain in 2000. Other candidates for cloning include the Sumatran tiger (*Panthera tigris*) and the giant panda. Brazil plans to clone eight of its endangered species. Cloning long-extinct animals such as the woolly mammoth (*Mammuthus primigenius*) or Tasmanian tiger (*Thylacinus cynocephalus*) would be more difficult due to a lack of fully preserved DNA.

Despite the promise of cloning, a number of issues remain unresolved:

1. Scientists would have to develop an intimate knowledge of different species' reproductive cycles. For sheep and cows, this was routine, based on the vast experience in breeding these species, but eggs of different species, even if they could be

**(a) A condor chick being fed using a puppet**     **(b) A released captive-bred condor**

**Figure 60.12  Captive breeding programs.**  The California condor (*Gymnogyps californicus*), the largest bird in the U.S., with a wingspan of nearly 3 m, has been bred in captivity in California. **(a)** A researcher at the San Diego Wild Animal Park feeds a chick with a puppet so that the birds will not become habituated to the presence of humans. **(b)** This captive-bred condor soars over the Grand Canyon. Note the tag on the underside of its wing. a: ©Charlie Neuman/San Diego Union-Tribune/ZUMA Press/Alamy Stock Photo; b: ©Simpson/Photri Images/Alamy Stock Photo

**Figure 60.13 Cloning an endangered species.** In 2004, this 8-month-old cloned Javan banteng (*Bos javanicus*) made its public debut at the San Diego Zoo. ©Yvette Cardozo/Alamy Stock Photo

 **Core Concept: Information** Reproductive cloning utilizes a somatic cell, which contains the complete DNA or genetic blueprint of the animal that is to be cloned. The somatic cell is fused with an oocyte that has had its nucleus removed.

harvested, often require different nutritive media in laboratory cultures.

2. Because it is desirable to use natural mothers for gestation of cloned animals, scientists will have to identify surrogate females of similar but more common species that can carry the fetus to term.

3. Some argue that cloning does not address the root causes of species loss, such as habitat fragmentation or poaching, and that resources would be better spent elsewhere—for example, in preserving the remaining habitat of endangered species.

4. Cloning from a limited number of sources might not be able to increase the genetic variation of a species. However, if cells were obtained from many different deceased animals—for example, from their skin—these clones could theoretically reintroduce lost genes back into the population.

Many biologists believe that while cloning may have a role in conservation, it is only part of the solution and we need to address what made the species go extinct before attempting to restore it.

### Preserving Biodiversity Is an Important Goal for Human Societies

As we saw in Section 60.3, biodiversity is of great value to human welfare. Therefore, conservation is clearly a matter of great importance, and a failure to value and protect our natural resources adequately could be a grave mistake. Some authors, most recently American ecologist and geographer Jared Diamond, have investigated why many societies of the past—including Angkor Wat, Easter Island,

and the Mayans—collapsed or vanished, leaving behind monumental ruins. Diamond has concluded that the collapse of these societies occurred partly because people inadvertently destroyed the ecological resources on which their societies depended.

Modern nations such as Rwanda face similar issues. The country's population density is the highest in Africa, and it has a limited amount of land that can be used for growing crops. By the late 1980s, the need to feed a growing population had led to the wholesale clearing of Rwanda's forests and wetlands, with the result that little additional land was available to farm. Increased population pressure, along with food shortages fueled by environmental scarcity, were likely contributing factors in igniting the genocide of 1994.

As we hope you have seen throughout this textbook, an understanding of biology is vital to learning and helping to solve many of society's problems. The study of biology has a huge potential for improving people's lives and society at large. Biology offers the opportunity to unlock new diagnoses and treatments for diseases, to improve nutrition and food production, and to maintain biological diversity.

## Summary of Key Concepts

### 60.1 Genetic, Species, and Ecosystem Diversity

- Biodiversity represents diversity at three levels: genetic diversity, species diversity, and ecosystem diversity. Conservation biology uses knowledge from molecular biology, genetics, and ecology to protect biological diversity (Figure 60.1).

### 60.2 Biodiversity and Ecosystem Function

- Four models describe the relationship between ecosystem function and species richness: the diversity-stability, redundancy, keystone, and idiosyncratic hypotheses (Figure 60.2).

- Laboratory and field experiments have shown that increased species richness results in increased ecosystem function and support the redundancy hypothesis (Figures 60.3, 60.4).

### 60.3 Value of Biodiversity to Human Welfare

- The preservation of biodiversity has been justified because of its economic value, because of the value of services that ecosystems provide, and on ethical grounds (Figure 60.5, Table 60.1).

### 60.4 Conservation Strategies

- Habitat conservation strategies commonly target megadiversity countries, countries with the largest number of species; biodiversity hot spots, areas with the largest number of endemic species, or those unique to the area; representative habitats including crisis ecoregions, areas that represent the major habitats; and "last of the wild," or pristine areas that have not been severely affected by human activities (Figures 60.6, 60.7).

- Conservation biologists employ many strategies in protecting biodiversity. Principles of the equilibrium model of island biogeography and landscape ecology are used in the theory and practice of preserve design (Figure 60.8, 60.9).

- The single-species approach focuses conservation efforts on indicator species, umbrella species, flagship species, or keystone species (Figure 60.10).

- Habitat restoration seeks to repair or replace populations and their habitats. Three basic approaches to habitat restoration are complete restoration, rehabilitation, and ecosystem replacement (Figure 60.11).

- Captive breeding is propagating animals or plants outside their natural habitat and reintroducing them to the wild. Cloning of endangered species has been accomplished on a very small scale and, despite its limitations, may eventually have a role in conservation biology (Figures 60.12, 60.13).

## Assess & Discuss

### Test Yourself

1. Which of the following best describes an endangered species?
   a. a species that is likely to become extinct in a portion of its range
   b. a species that has disappeared in a particular community but is present in other natural environments
   c. a species that is extinct
   d. a species that is in danger of becoming extinct throughout all or a significant portion of its range
   e. Both b and d are true of endangered species.

2. In 1977, Rafael Guzman, a Mexican biologist, discovered a previously unknown wild relative of corn that is resistant to many of the viral diseases that infect domestic corn. Agriculturalists believe that crossbreeding the wild corn with domestic corn could improve current corn crops. In this case, biodiversity is important at which level?
   a. ecosystem
   b. species
   c. genetic
   d. community
   e. both a and b

3. The idea that humans have an innate attachment to other life-forms, put forth by E. O. Wilson, is known as
   a. biodiversity.
   b. biophilia.
   c. the call of the wild.
   d. biotheology.
   e. the "last of the wild."

4. The research conducted by Tilman and colleagues demonstrated that
   a. as diversity increases, productivity increases.
   b. as diversity decreases, productivity increases.
   c. areas with higher diversity demonstrate less efficient use of nutrients.
   d. increases in species richness lead to an increase in invasive species.
   e. increased diversity results in increased susceptibility to disease.

5. Which hypothesis best describes the idea that a small decline in species richness results in a large drop in ecosystem function?
   a. Diversity-stability hypothesis
   b. Redundancy hypothesis
   c. Keystone hypothesis
   d. Idiosyncratic hypothesis
   e. Community hypothesis

6. According to the equilibrium model of island biogeography, which type of preserve would contain the most species?
   a. One large park
   b. Several small parks with a combined area equal to that of a large park
   c. Several small parks connected by a movement corridor
   d. A circular park
   e. All of the above

7. Saving endangered habitats, such as the Argentine Pampas, focuses on
   a. saving genetic diversity.
   b. saving keystone species.
   c. conservation in a megadiversity country.
   d. preserving an area rich in endemic species.
   e. preserving a representative habitat.

8. Biodiversity hot spots are those areas rich in
   a. species.
   b. habitats.
   c. rare species.
   d. biodiversity.
   e. endemic species.

9. Over time, dark forms of the peppered moth (*Biston betularia*) became more common in polluted environments because predators were less able to detect them on trees darkened by soot. These moths are regarded by many as a(n)
   a. keystone species.
   b. indicator species.
   c. umbrella species.
   d. flagship species.
   e. endangered species.

10. After being used for mining, what was once a deciduous forest is replaced by grassland to be used for public recreation. This process is known as
    a. complete restoration.
    b. rehabilitation.
    c. ecosystem replacement.
    d. bioremediation.
    e. habitat repair.

### Conceptual Questions

1. What is the value of increased biodiversity for human society?

2. Distinguish among the following as bases for strategies to conserve biodiversity: megadiversity countries, biodiversity hot spots, crisis ecoregions, and "last of the wild."

3.  **Core Concept: Systems** Describe how the following hypotheses link species richness with ecosystem function: diversity-stability, redundancy, keystone, and idiosyncratic.

### Collaborative Questions

1. Discuss the differences between indicator species, umbrella species, flagship species, and keystone species.

2. You are called upon to design a park with maximum biodiversity in a tropical country. What are your recommendations?

## Periodic Table of the Elements

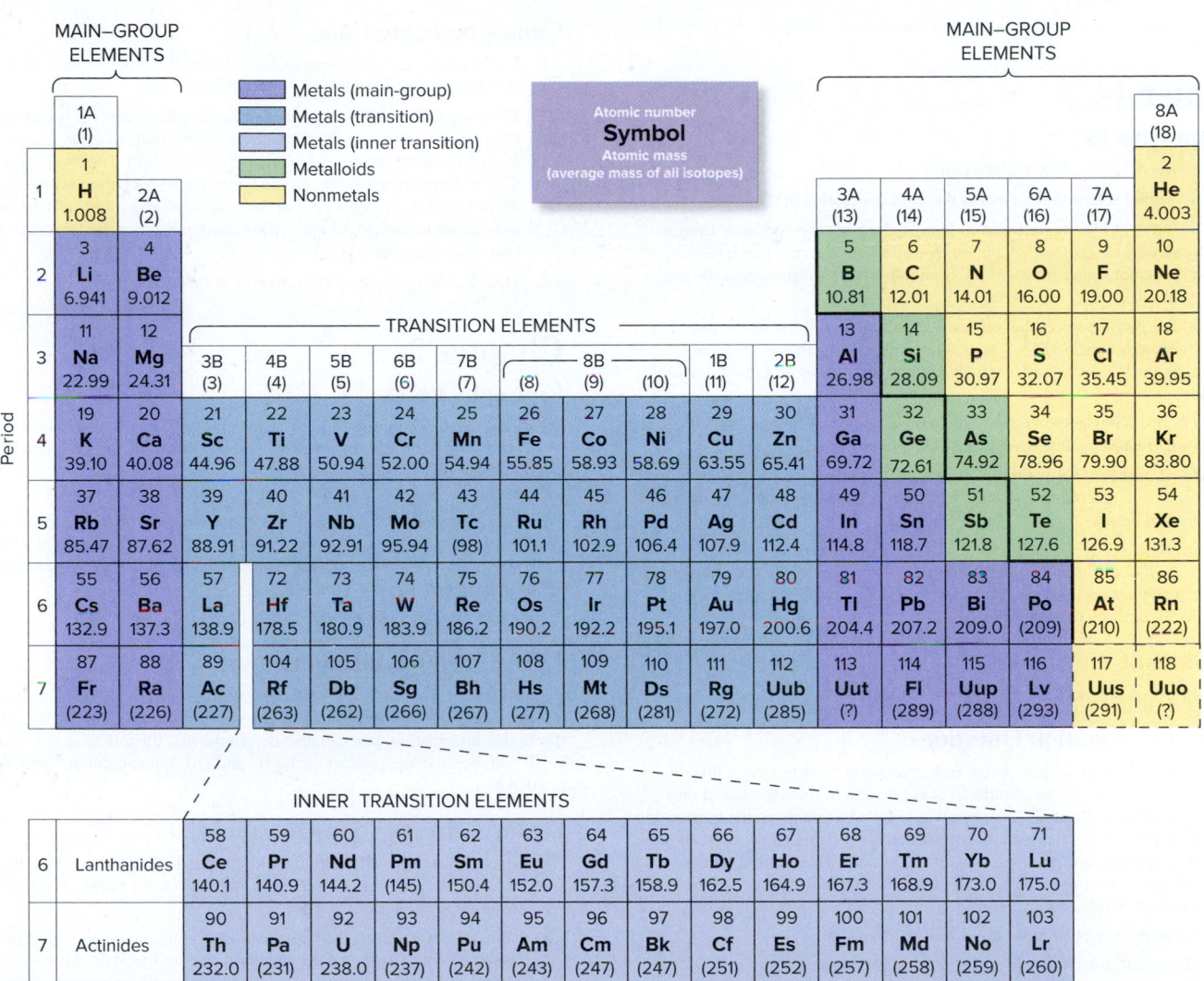

**The complete Periodic Table of the Elements.** Group numbers are different in some cases from those presented in Figure 2.5, because of the inclusion of transition elements. In some cases, the average atomic mass has been rounded to one or two decimal places, and in others only an estimate is given in parentheses due to the short-lived nature or rarity of those elements. The symbols and names of some of the elements between 112–118 are temporary until the chemical characteristics of these elements become better defined. Element 117 is currently not confirmed as a true element, and little is known about element 118. The International Union of Pure and Applied Chemistry (IUPAC) has recently proposed adopting the name copernicium (Cp) for element 112 in honor of scientist and astronomer Nicolaus Copernicus.

# Appendix B

## Answer Key

Answers to Collaborative Questions can be found on the website.

## Chapter 1

### Concept Checks

*Figure 1.3* The herd is at the population level.

*Figure 1.6* Natural selection is a process that leads to evolution.

*Figure 1.8* A tree of life suggests that all living organisms evolved from a single ancestor by vertical descent with mutation. A web of life assumes that both vertical descent and horizontal gene transfer have been important mechanisms in the evolution of new species.

*Figure 1.9* The genome stores the information used to make proteins. In and of itself, the genome is merely DNA. The traits of cells and organisms are largely determined by the structures and functions of the thousands of different proteins they produce.

*Figure 1.11* Taxonomy helps us appreciate the unity and diversity of life. Organisms that are closely related evolutionarily are placed in smaller groups.

*Figure 1.14* A researcher can compare the results from the experimental group and the control group to determine if a single variable is causing a particular outcome in the experimental group.

*Figure 1.15* After the *CFTR* gene was identified by discovery-based science, researchers realized that this gene was similar to other genes that encoded proteins that were already known to be transport proteins. This provided an important clue about the likely function of the *CFTR* gene.

### Core Skills: Connections

*Figure 1.10* Fungi are more closely related to animals.

### Feature Investigation Questions

1. In discovery-based science, a researcher does not need to have a preconceived hypothesis. Experimentation is conducted in the hope that it may have practical applications or may provide new information that will lead to a hypothesis. By comparison, hypothesis testing occurs when a researcher forms a hypothesis that makes certain predictions. Experiments are conducted to see if those predictions are correct. In this way, the hypothesis may be accepted or rejected.

2. This strategy may be described as a five-stage process:
   1. Observations are made regarding natural phenomena.
   2. These observations lead to a hypothesis that tries to explain the phenomena. A useful hypothesis is one that is testable because it makes specific predictions.
   3. Experimentation is conducted to determine if the predictions are correct.
   4. The data from the experiment are analyzed.
   5. The hypothesis is accepted or rejected.

3. In an ideal experiment, the control and experimental groups differ by only one factor. Biologists apply statistical analyses to their data to determine if the outcomes for the control and experimental groups are likely to differ because of the single variable that is different between the two groups. This provides an objective way to accept or reject a hypothesis.

### Test Yourself

1. d  2. a  3. b  4. c  5. d  6. b  7. d  8. d  9. a  10. b

### Conceptual Questions

1. Evolution applies only to populations, whereas the other four core concepts could apply to individuals and to populations.

2. The unity among different species occurs because modern species have evolved from a group of related ancestors. Some of the traits in those ancestors are also found in modern species, which thereby unites them. The diversity is due to the variety of environments on the Earth. Each species has evolved to occupy its own unique environment. For every species, many traits are evolutionary adaptations to survival in a specific environment. For this reason, evolution also promotes diversity.

3. Students should rephrase the concepts in their own words.

## Chapter 2

### Concept Checks

*Figure 2.4* An electron shell is a region outside the nucleus of an atom occupied by electrons of a given energy level. More than one orbital can be found within an electron shell. An orbital may be spherical or dumbbell-shaped and contains up to two electrons.

*Figure 2.11* Strand separation requires energy, because the DNA strands are held together by a large number of hydrogen bonds. Although each hydrogen bond is weak, the collective strength of such bonds in a molecule of DNA adds up to a considerable amount.

*Figure 2.17* The oil will be in the centers of the soap micelles.

### Core Skills: Connections

*Figure 2.20* At a pH of 5.0, the $H^+$ concentration would be $10^{-5}$ M, as can be seen in Figure 2.20 and can also be calculated with the equation $pH = -\log_{10}[H^+]$. (From this information, you can also determine that the $OH^-$ concentration must be $10^{-9}$ M, because the product of the $H^+$ and $OH^-$ concentrations must be equal to $10^{-14}$ M.)

### Feature Investigation Questions

1. Scientists were aware that atoms contained charged particles. Many believed that the positive charges and mass were evenly distributed throughout the atom.

2. Rutherford was testing the hypothesis that atoms are composed of positive charges evenly distributed throughout the atom. Based on this model of the structure of the atom, α particles, which are positively charged nuclei of helium atoms, should be slightly deflected as they pass through gold foil, due to the presence of positive charges spread throughout the foil.

3. Instead of showing slight deflection as they passed through the gold foil, the majority, 98%, of the α particles passed directly through the foil without deflection. Less than 2% of them showed deflection, and a much smaller percentage bounced back from the gold foil. Rutherford suggested that since most of the α particles passed unimpeded through the gold foil, most of the volume of atoms is empty space. Rutherford also proposed that the bouncing back of some of the α particles indicated that most of the positively charged particles in an atom were concentrated in a compact area. These results ran counter to the hypothesized model.

### Test Yourself

1. b  2. b  3. b  4. d  5. e  6. e  7. e  8. c  9. c  10. e

## Conceptual Questions

1. Covalent bonds are bonds in which atoms share electrons. A nonpolar covalent bond is one between two atoms of similar electronegativities, such as two carbon atoms. A hydrogen bond is a weak interaction that forms when a hydrogen atom in a polar molecule becomes electrically attracted to an electronegative atom. The van der Waal dispersion forces are temporary, weak interactions, resulting from random electrical forces generated by the changing distributions of electrons in the outer shells of nearby atoms. The strong attraction between two oppositely charged atoms forms an ionic bond.

2. Within limits, bonds within molecules can rotate and thereby change the molecules' shapes. This is important because it is the shape of a molecule that determines, in part, the ability of that molecule to interact with other molecules. Also, when two molecules do interact through such forces as hydrogen bonds, the shape of one or both molecules may change as a consequence. The change in shape is often part of the mechanism by which signals are sent within and between cells.

3. When two or more atoms react with each other to form a new substance, that new substance often has emergent properties that are quite different from the starting materials. A good example of emergent properties at the molecular level occurs in the formation of sodium chloride (NaCl), a solid white crystalline compound that is very important for most living organisms. In their elemental states, sodium is a soft, highly reactive metal and chlorine is a toxic gas. When they combine through ionic bonds, the two elements produce a completely new and harmless substance found in all the world's oceans and soils. Another example from this chapter is water, a liquid that is vital for all life but which is formed from two gases, hydrogen and oxygen, with very different properties.

# Chapter 3

## Concept Checks

*Figure 3.5* One reason is that the binding of a molecule to an enzyme depends on the spatial arrangements of the atoms in that molecule. Enantiomers have different spatial relationships and are mirror images of each other. Therefore, one may bind very tightly to an enzyme and the other may not be recognized at all.

*Figure 3.6* Recall from Figure 3.4 that the reverse of a dehydration reaction is a hydrolysis reaction, in which a molecule of water is added to the molecule being broken down, resulting in the formation of monomers.

*Figure 3.10* Hydrogenation is the addition of hydrogens to double-bonded carbon atoms, changing them from unsaturated to saturated. This causes the resulting fat to have a higher melting point and thus be solid at room temperature.

*Figure 3.14* The process would produce 71 water molecules, one less than the number of amino acids in the polypeptide.

*Figure 3.18* If the primary structure of protein 1 were altered in this way, the changes would, in turn, most likely alter the secondary and tertiary structures of protein 1. Therefore, it is possible that the precise fit between proteins 1 and 2 would no longer be possible and the two proteins would lose the ability to interact.

*Figure 3.23* Yes. The opposite strand must be complementary to the first strand, because A must be paired with T, and G with C. For instance, if a portion of the first strand has the sequence AATGCA, the opposite strand for that portion will be TTACGT.

## Core Skills: Connections

*Figure 3.7* Cellulose is believed to be the most abundant organic molecule on Earth. In addition to being part of plant cells, it is also found in many other organisms, including many protists.

## Feature Investigation Questions

1. Anfinsen was testing the hypothesis that the information necessary for determining the three-dimensional shape of a protein is contained within the protein itself. In other words, the chemical characteristics of the amino acids that make up a protein determine the protein's three-dimensional shape.

2. The urea disrupts hydrogen and ionic bonds that are necessary for protein folding. The β-mercaptoethanol breaks the S−S bonds that form between certain amino acids of the same polypeptide. Both substances cause the polypeptide to unfold, disrupting the three-dimensional shape.

3. Anfinsen removed the urea and β-mercaptoethanol from the ribonuclease by size-exclusion chromatography. After removing these substances, Anfinsen discovered that the protein refolded into its proper three-dimensional shape and became functional again. This was important because the solution at that point contained only the protein and lacked any other cellular material that could possibly assist in protein folding. This demonstrated that the protein could refold itself into the functional conformation.

## Test Yourself

1. b   2. e   3. e   4. a   5. c   6. b   7. e   8. d   9. b   10. b

## Conceptual Questions

1. Isomers are molecules with the same chemical formula but with different structures and arrangements of their atoms. There are two major types of isomers: structural isomers and stereoisomers. Because many chemical reactions in biology depend on the actions of enzymes, which are often highly specific with respect to the spatial arrangement of atoms in a molecule, one isomer of a pair may have biological functions, and the other may not.

2. Saturated fatty acids are saturated with hydrogens and have only single (C−C) bonds, whereas unsaturated fatty acids have one or more double (C=C) bonds. The double bonds in unsaturated fatty acids alters the shape, resulting in one or more kinks in the chain. Saturated fatty acids are unkinked and are better able to stack tightly together. Fats containing saturated fatty acids have a higher melting point than those containing mostly unsaturated fatty acids; consequently, saturated fats tend to be solids at room temperatures, and unsaturated fats are usually liquids at room temperature.

3. The structures of all these macromolecules determine their functions. For example, the structure of a protein determines its three-dimensional shape. This, in turn, allows a protein to interact specifically with certain other molecules. Also, protein domains have specific structures that determine their functions. A single protein may have multiple domains, allowing that protein to perform a fairly complex function, such as activating genes in response to hormone binding. Likewise, the structures of different lipids determine such functional characteristics as male/female differences, cellular membrane formation, and energy storage. The different structures of polysaccharides determine their usefulness as energy stores or as components of plant cell walls.

# Chapter 4

## Concept Checks

*Figure 4.2* These vents release hot gases from the interior of the Earth. Organic molecules can form in the temperature gradient between the extremely hot vent water and the cold water that surrounds the vent.

*Figure 4.3* A liposome is more similar to today's cells, which are surrounded by a membrane that is composed of a phospholipid bilayer.

*Figure 4.4* Chemical selection occurs when certain molecules, such as RNAs, have properties that provide advantages and therefore cause them to increase in number relative to other molecules in the same environment.

*Figure 4.5* You would use an electron microscope. A light microscope does not have good enough resolution.

*Figure 4.7* The primary advantage is that it gives an image of the 3-D surface of an object.

*Figure 4.11* Centrioles: Not found in plant cells; their role is not entirely clear, but they are found in the centrosome, which is where microtubules are anchored. Chloroplasts: Not found in animal cells; function in photosynthesis. Cell wall: Not found in animal cells; important in cell shape. Central vacuole: Not found in animal cells; site that provides storage and regulates cell volume.

*Figure 4.16* Both dynein and microtubules are anchored in place. Using ATP as a source of energy, dynein tugs on microtubules. Because the microtubules are anchored, they bend in response to the force exerted by dynein.

*Figure 4.19* The nuclear matrix, located inside the nucleus, serves to organize the chromosomes into chromosome territories.

*Figure 4.22* The protein is synthesized into the ER and then travels through the *cis*, medial, and *trans* Golgi before being secreted.

*Figure 4.26* Of these three functions, membrane transport is probably the most important to metabolism because it determines which molecules can enter the cell and participate in metabolism and which products of metabolism are exported from the cell.

*Figure 4.28* The invaginations increase the surface area where ATP synthesis takes place, thereby increasing the amount of ATP synthesis.

*Figure 4.33* The signal sequence of such a protein is recognized by SRP, which halts translation. The growing polypeptide and its ribosome are then transferred to the ER membrane, where translation resumes.

*Figure 4.34* If chaperone proteins were not present in the cytosol, the protein destined for the mitochondrial matrix would start to fold, which might prevent it from passing through the channels in the outer and inner mitochondrial membranes. Normally, a protein is threaded through these channels in an unfolded state.

## Core Skills: Connections

*Figure 4.3* Phospholipids are amphipathic molecules; they have a polar end (the head) and a nonpolar end (the two fatty acid tails). Phospholipids form a bilayer such that the heads interact with water, whereas the tails are shielded from the water. This is an energetically favorable structure.

*Figure 4.10* Alternative splicing produces proteins with slightly different structures, because they have certain regions that have different amino acid sequences. The functions of such proteins are often similar, but specialized for the cell type in which they are expressed.

*Figure 4.12* The surfaces of structures involved with gas exchange are highly convoluted. This provides them with much greater surface areas, thereby facilitating the movement of gases through the surfaces.

*Figure 4.15* The type of movement shown in part (b) of Figure 4.15 occurs during muscle contraction.

*Figure 4.20* During cell division, the chromosomes condense and form more compact structures.

*Figure 4.30* These processes are similar in that DNA replication occurs and then the mitochondrion or bacterial cell splits in two. They are different in that bacteria have cell walls and such a wall must form between the two daughter cells, which does not occur when mitochondria divide.

## Feature Investigation Questions

1. In a pulse-chase experiment, radioactive material is provided to cells in a single administration. This is referred to as the pulse. After a few minutes, a large amount of nonradioactive material is provided to the cells to "chase away" the ability to use the radioactive material. The researchers were attempting to determine the movement of proteins through the different compartments of a cell. Radioactive amino acids were used to label the proteins and enable the researchers to visualize where the proteins were at different times.

2. Pancreatic cells produce large numbers of proteins that they then secrete. Thus, these cells provide researchers with an ideal system for studying protein movement through a cell.

3. Using electron microscopy, the researchers found that the proteins, labeled with radioactivity, were first found in the ER of the cells. Later the radiolabeled proteins moved to the Golgi and then into vesicles near the plasma membrane.

   The researchers concluded that proteins move through several cellular compartments before they are secreted from a cell. Also, the movement of proteins through these compartments is not random but follows a particular pathway: ER, Golgi, secretory vesicles, plasma membrane, and, finally, extracellular environment.

## Test Yourself

1. d 2. d 3. a 4. b 5. e 6. a 7. e 8. e 9. a The protein would go there first, because targeting to the ER occurs cotranslationally. 10. c It is true that they all carry out metabolism, but so do eukaryotic cells.

## Conceptual Questions

1. Stage 1: Nucleotides and amino acids were produced prior to the existence of cells.

   Stage 2: Nucleotides and amino acids became polymerized to form DNA, RNA, and proteins.

   Stage 3: Polymers became enclosed in membranes.

   Stage 4: Polymers enclosed in membranes evolved cellular properties.

2. If the motor protein is bound to a cargo and can walk along a cytoskeletal filament that is fixed in place, this will cause movement of the cargo when the

motor protein is activated. If the motor protein is fixed in place and the filament is free to move, this will cause the filament to move when the motor protein is activated. If both the motor protein and the filament are fixed in place, the activation of the motor protein will cause the filament to bend.

3. ATP synthesis occurs along the inner mitochondrial membrane. The invaginations of this membrane greatly increase its surface area, thereby allowing for a greater amount of ATP synthesis.

# Chapter 5

## Concept Checks

*Figure 5.6* Phospholipids are transferred to the other leaflet of the ER membrane via enzymes called flippases.

*Figure 5.7* The most common feature causing a transmembrane segment to form is a stretch (about 20) of amino acids that mostly have hydrophobic (nonpolar) side chains.

*Figure 5.13* Water will move from outside to inside, from the lower to the higher concentration.

*Figure 5.15* The purpose of gating is to regulate the function of channels, allowing them to be open or closed.

*Figure 5.21* The protein coat allows the budding process at the surface of the Golgi membrane to form a vesicle.

## Core Skills: Connections

*Figure 5.5* Transmembrane proteins called cell adhesion molecules bind to each other to promote cell-to-cell adhesion. In addition, they can bind to filaments in the extracellular matrix, such as collagen fibers, thereby causing a cell to adhere to the extracellular matrix.

*Figure 5.10* Leucine would be more likely to cross an artificial phospholipid bilayer because it is more hydrophobic than lysine.

*Figure 5.11* Gradients of sodium and potassium ions are important for the conduction of action potentials.

## Feature Investigation Questions

1. Most cells allow movement of water across the cell membrane by passive diffusion. However, it was noted that certain cell types had a much higher rate of water movement, indicating that something different was occurring in these cells.

2. The researchers identified water channels by characterizing proteins that are present in red blood cells and kidney cells but not other types of cells. Red blood cells and kidney cells have a faster rate of water movement across the membrane than other cell types. These cells are more likely to have water channels. By identifying proteins that are found in both of these types of cells but not in other cells, the researchers were identifying possible candidate proteins that function as water channels. In addition, CHIP28 had a structure that resembled other known channel proteins.

   Agre and his associates experimentally created multiple copies of the gene that produces the CHIP28 protein and then artificially transcribed the genes to produce many mRNAs. The mRNAs were injected into frog oocytes where they could be translated to make the CHIP28 proteins. After altering the frog oocytes by introducing the CHIP28 mRNAs, the researchers compared the rate of water transport in the altered oocytes versus normal frog oocytes. This procedure allowed them to introduce the candidate protein into a cell type that normally does not have the protein present.

3. After artificially introducing the candidate protein into the frog oocytes, the researchers found that the experimental oocytes took up water at a much faster rate in a hypotonic solution as compared to the control oocytes. The results indicated that the presence of the CHIP28 protein did increase water transport into cells.

## Test Yourself

1. c   2. c   3. b   4. d   5. b   6. e   7. d   8. e   9. e   10. c

## Conceptual Questions

1. See Figure 5.1 for the type of drawing you should have made. The membrane is considered a mosaic of lipid, protein, and carbohydrate molecules. The membrane exhibits properties that resemble a fluid because lipids and proteins can move relative to each other within the plane of the membrane.

2. Integral membrane proteins can contain transmembrane segments that cross the membrane, or they may contain lipid anchors. Peripheral membrane proteins are noncovalently bound to integral membrane proteins or to the polar heads of phospholipids.

3. The lipid bilayer, channels, and transporters cause the plasma membrane to be selectively permeable. This allows a cell to take up needed nutrients from its extracellular environment and to export waste products into the environment.

# Chapter 6

## Concept Checks

*Figure 6.2* The solution of dissolved $Na^+$ and $Cl^-$ has more entropy. A salt crystal is very ordered, whereas the ions in solution are much more disordered.

*Figure 6.4* If a large amount of ADP was broken down, the cell would not be able to synthesize as much ATP, which is made by the attaching a phosphate to ADP. The ATP cycle would be inhibited.

*Figure 6.5* Lowering $E_A$ speeds up the rate. When the activation energy is lower, it takes less time for reactants to reach the transition state, where a reaction can occur. Lowering $E_A$ does not affect the direction of a reaction.

*Figure 6.7* At a substrate concentration of 0.5 mM, the reaction catalyzed by enzyme A would have the higher velocity. That reaction would be very near its $V_{max}$, whereas enzyme B's reaction would be well below its $V_{max.}$

*Figure 6.14* Protein degradation eliminates proteins that are worn out, misfolded, or no longer needed by the cell. Such proteins could interfere with normal cell function. In addition, the recycling of amino acids saves the cell energy.

## Core Skills: Connections

*Figure 6.9* If RNase P did not function properly, ptRNAs could not be converted to their mature forms. The ptRNAs would be too large to fit into the sites on the ribosome. Therefore, translation would be inhibited.

*Figure 6.13* Feedback inhibition prevents the excessive breakdown of carbohydrates and thereby prevents the excessive synthesis of ATP. Cells don't waste energy making ATP if they don't need it.

## Feature Investigation Questions

1. RNase P has both a protein and an RNA subunit. To determine which subunit has catalytic function, it was necessary to purify them individually and then see which one was able to cleave ptRNA.

2. The experimental strategy was to incubate RNase P or subunits of RNase P with ptRNA and then do gel electrophoresis to determine if ptRNA had been cleaved to a mature tRNA and a 5′ fragment. The control without the protein subunit was used to determine if the RNA alone could catalyze the cleavage. The control without the RNA subunit was used to determine if some other factor in the experiment (for instance, $Mg^{2+}$ or protein) was able to cleave the ptRNA.

3. The critical results occurred when the researchers incubated the purified RNA subunit at high $Mg^{2+}$ concentrations with the ptRNA. Under these conditions, the ptRNA was cleaved. These results indicated that the RNA subunit has catalytic activity. A high $Mg^{2+}$ concentration is needed to keep it catalytically active in the absence of the protein subunit.

## Test Yourself

1. e  2. b  3. b  4. c  5. a  6. c  7. b  8. e  9. e  10. a

## Conceptual Questions

1. Exergonic reactions are spontaneous. They proceed in a particular direction. An exergonic reaction could be slow or fast. By comparison, an endergonic reaction is not spontaneous. It will not proceed in a particular direction unless free energy is supplied. An endergonic reaction can be fast or slow.

2. During feedback inhibition, the product of a metabolic pathway binds to an allosteric site on an enzyme that acts earlier in the pathway. The product inhibits this enzyme, thereby preventing the overaccumulation of the product.

3. Recycling of amino acids and nucleotides conserves a great deal of energy. Cells don't have to remake these building blocks, which would require a large amount of energy.

# Chapter 7

## Concept Checks

*Figure 7.3* The molecules that donate phosphates are 1,3-bisphosphoglycerate and phosphoenolpyruvate.

*Figure 7.6* For each acetyl group that is oxidized, the main products are 2 $CO_2$, 3 NADH, 1 $FADH_2$, and 1 GTP.

*Figure 7.8* The protein complex is called cytochrome oxidase because it removes electrons from (oxidizes) cytochrome $c$.

*Figure 7.10* No. The role of the electron transport chain is to produce an $H^+$ electrochemical gradient, which is what drives ATP synthase. If the $H^+$ electrochemical is made another way, such as by bacteriorhodopsin, ATP synthase can still make ATP.

*Figure 7.14* The advantage is that cells can use the same enzymes to metabolize different kinds of organic molecules. This saves energy because it would require a lot of energy to make many different enzymes, which are composed of proteins.

## Core Skills: Connections

*Figure 7.2* Glycolytic muscle fibers rely on glycolysis for their ATP needs. Because glycolysis does not require oxygen, such muscle fibers can function without oxygen.

*Figure 7.4* FDG is radiolabeled so it can be specifically detected by a PET scan.

## Feature Investigation Questions

1. The researchers attached an actin filament to the γ subunit of ATP synthase. The actin filament was fluorescently labeled, so the researchers could determine if the actin filament moved when viewed under a fluorescence microscope.

2. When functioning in the hydrolysis of ATP, the actin filament was seen to rotate. The actin filament was attached to the γ subunit of ATP synthase. The rotational movement of the filament was the result of the rotational movement of the γ subunit. In the control part of the experiment, no ATP was added to stimulate enzyme activity. In the absence of ATP, no movement was observed.

3. No, the counterclockwise rotation observed by the researchers is the opposite of what would be expected inside mitochondria. During the experiment, the ATP synthase was not functioning to synthesize ATP but instead was running backward and hydrolyzing ATP.

## Test Yourself

1. a  2. b  3. b  4. c  5. a  6. c  7. b  8. d  9. d  10. b

## Conceptual Questions

1. The purpose of the electron transport chain is to pump $H^+$ across the inner mitochondrial membrane to establish an $H^+$ electrochemical gradient. When $H^+$ flow back across the membrane through ATP synthase, ATP is synthesized.

2. The movement of $H^+$ through the contact site between the $c$ and $a$ subunits causes the γ subunit to rotate. As it rotates, it sequentially alters the conformation of the β subunits, where ATP is made. This causes (1) ADP and $P_i$ to bind with moderate affinity, (2) ADP and $P_i$ to bind very tightly such that ATP is made, and (3) ATP to be released.

3. The phases of glucose metabolism are regulated in a variety of ways. For example, key enzymes in glycolysis and the citric acid cycle are regulated by the availability of their substrates and by feedback inhibition. The electron transport chain is regulated by the ATP/ADP ratio. Such regulation ensures that a cell does not waste energy making ATP when it is in sufficient supply. Also, the production of too much NADH is potentially harmful because at high levels it has the potential to haphazardly donate its electrons to other molecules and promote the formation of free radicals, highly reactive chemicals that damage DNA and cellular proteins.

# Chapter 8

## Concept Checks

*Figure 8.3* The Calvin cycle can occur in the dark as long as sufficient $CO_2$, ATP, and NADPH are present in the stroma.

*Figure 8.4* Gamma rays have higher energy than radio waves.

*Figure 8.5* To become more stable by dropping down to a lower energy level, a photoexcited electron can release energy in the form of heat, release energy in the form of light, or transfer energy to another electron by resonance energy transfer.

*Figure 8.7* Having different pigment molecules allows plants to absorb a wider range of wavelengths of light.

*Figure 8.8* ATP and NADPH are produced in the stroma. $O_2$ is produced in the thylakoid lumen.

*Figure 8.9* Linear electron flow produces equal amounts of ATP and NADPH. However, plants usually need more ATP than NADPH. Cyclic photophosphorylation allows plants to make just ATP, thereby increasing the relative amount of ATP.

*Figure 8.10* Because these two proteins are homologous, this means that the genes that encode them were derived from the same ancestral gene. Therefore, the amino acid sequences of these two proteins are expected to be very similar, though not identical. Because the amino acid sequence of a protein determines its structure, two proteins with similar amino acid sequences would be expected to have similar structures.

*Figure 8.12* An electron has the highest amount of energy just after its energy has been boosted by light in photosystem I.

*Figure 8.13* NADPH reduces organic molecules and makes them more able to form C—C and C—H bonds.

*Figure 8.16* The arrangement of cells in $C_4$ plants makes the level of $CO_2$ high and the level of $O_2$ low in the bundle-sheath cells.

*Figure 8.17* When there is plenty of moisture and it is not too hot, $C_3$ plants are more efficient. However, under hot and dry conditions, $C_4$ and CAM plants have the advantage because they lose less water and avoid photorespiration.

## Core Skills: Connections

*Figure 8.2* Two guard cells make up one stoma.

## Feature Investigation Questions

1. The researchers were attempting to determine the biochemical pathway of the process of carbohydrate synthesis via photosynthesis. The researchers wanted to identify different molecules produced over time to determine the steps of the biochemical pathway.

2. The purpose for using $^{14}C$-labeled $CO_2$ was to label the different carbon molecules produced during the biochemical pathway. The researchers could "follow" the carbon molecules from the radiolabeled $CO_2$ that were incorporated into the organic molecules during photosynthesis. The radioactive isotope provided the researchers with a method of labeling the different molecules.

   The purpose of the experiment was to determine the steps in the biochemical pathway of photosynthesis. By examining samples from different times after the introduction of the labeled carbon source, the researchers were able to determine which molecules were produced first and, thus, distinguish products of the earlier steps of the pathway from products of later steps of the pathway.

   The researchers used two-dimensional paper chromatography to separate the different molecules from each other. After being separated, the different molecules were identified by various chemical methods.

3. Calvin and his colleagues were able to determine the biochemical process that incorporates $CO_2$ into organic molecules during photosynthesis. The researchers were able to identify the biochemical steps and the molecules produced at these steps of what is now called the Calvin cycle.

## Test Yourself

1. c   2. c   3. c   4. a   5. b   6. b   7. c   8. b   9. e   10. c

## Conceptual Questions

1. The two stages of photosynthesis are the light reactions and the Calvin cycle. The key products of the light reactions are ATP, NADPH, and $O_2$. The key product of the Calvin cycle is carbohydrates. The initial product is G3P, which is used to make sugars and other organic molecules.

2. NADPH is used during the reduction phase of the Calvin cycle. It donates its electrons to 1,3-BPG.

3. At the level of the biosphere, the role of photosynthesis is to incorporate carbon dioxide into organic molecules. These organic molecules can then be broken down, by autotrophs and by heterotrophs, to make ATP. The organic molecules made during photosynthesis are also used as starting materials to synthesize a wide variety of organic molecules and macromolecules that are made by cells.

# Chapter 9

## Concept Checks

*Figure 9.1* It is glucose.

*Figure 9.3* Endocrine signals are more likely to exist for a longer period of time. Their longer existence is necessary because these signals, called hormones, travel relatively long distances to reach their target cells. Therefore, they must exist long enough to get to their destinations.

*Figure 9.4* The effect of a signaling molecule is to cause a cellular response. Most signaling molecules do not enter the cell. Therefore, to exert an effect, they must alter the conformation of a receptor protein, which, in turn, stimulates an intracellular signal transduction pathway that leads to a cellular response.

*Figure 9.7* The α subunit has to hydrolyze its bound GTP to GDP and $P_i$. This changes the conformation of the α subunit so that it can reassociate with the β/γ dimer.

*Figure 9.12* The signal transduction pathway begins with activation of the G protein and ends with the activated subunits of protein kinase A . The cellular response involves the phosphorylation of target proteins, which changes their function in some way.

*Figure 9.13* Depending on the protein involved, phosphorylation can activate or inhibit protein function. Phosphorylation of phosphorylase kinase and glycogen phosphorylase activates their function, whereas phosphorylation inhibits glycogen synthase.

*Figure 9.14* Signal amplification allows a single signaling molecule to affect many proteins within a cell, thereby amplifying a cellular response.

*Figure 9.18* The initiator caspase is part of the death-inducing signaling complex. It is directly activated when a cell receives a death signal. The initiator caspase then activates the executioner caspase, which cleaves various cellular proteins and thereby causes the destruction of the cell.

## Core Skills: Connections

*Figure 9.2* Light is sensed by the cells on the illuminated side of the shoot tip and they send auxin to cells on the nonilluminated side, which accumulate more auxin, causing them to elongate.

*Figure 9.5* Most receptors and enzymes bind their ligands noncovalently and with high specificity. Enzymes, however, convert their ligands (which are reactants) into products, whereas receptors undergo a conformational change due to the binding of their ligands.

*Figure 9.10* The GTP-bound form of Ras is active and promotes cell division. To turn the signal transduction pathway off, Ras hydrolyzes GTP to GDP and $P_i$. If this cannot occur due to a mutation of Ras, the pathway will be continuously on, and uncontrolled cell division (cancer) will result.

## Feature Investigation Questions

1. Compared with control rats, those injected with prednisolone alone would be expected to have a decrease in the number of adrenal cortex cells because prednisolone suppresses ACTH synthesis. Therefore, apoptosis would be higher. By comparison, rats injected with prednisolone + ACTH would have a normal number of cells because the addition of ACTH would compensate for the effects of prednisolone. The rats injected with ACTH alone would be expected to have a greater number of cells; apoptosis would be inhibited.

2. Yes, injection of both prednisolone and ACTH probably inhibited the ability of the rats to make their own ACTH. Even so, they were given ACTH by the injection, so they didn't need to make their own ACTH to prevent apoptosis.

3. The lowest level of apoptosis would occur in the rats given ACTH alone, because they could make their own ACTH plus they were given ACTH. With such high levels of ACTH, they probably had the lowest level of apoptosis; it was already known that ACTH promotes cell division.

## Test Yourself

1. d   2. c   3. d   4. e   5. a   6. c   7. e   8. e   9. e   10. b

## Conceptual Questions

1. Cells need to respond to a changing environment, and cells need to communicate with each other.

2. In the first stage, a signaling molecule binds to a receptor, causing receptor activation. In the second stage, one type of signal is transduced, or converted to a different signal inside the cell. In the third stage, the cell responds in some way to the signal, possibly by altering the activity of enzymes, structural

proteins, or transcription factors. When the estrogen receptor is activated, the second stage, signal transduction, does not occur because the estrogen receptor is an intracellular receptor that directly activates the transcription of genes to elicit a cellular response.

3. Cell signaling allows cells to respond to environmental changes. For example, if a yeast cell is exposed to glucose, cell signaling will allow it to adapt to that change and utilize glucose more readily. Likewise, cell signaling allows plants to grow toward light. In addition, cells in a multicellular organism respond to changes in signaling molecules, such as hormones, and thereby coordinate their activities.

2. Cadherins and integrins are both integral membrane proteins that function as cell adhesion molecules. They also can function in cell signaling. Cadherins bind one cell to another cell, whereas integrins bind a cell to the extracellular matrix. Cadherins require calcium ions to function, but integrins do not.

3. Cell junctions are important in the proper arrangement of cells in a multicellular organism. In animals, for example, cells junctions allow cells to recognize and bind to each other. This is very important during embryonic development. In addition, cell junctions adhere cells to the extracellular matrix. Likewise, in plants, the cell wall and middle lamella are important in forming connections between plant cells that gives plants their correct morphology and function.

# Chapter 10

## Concept Checks

*Figure 10.1* The four functions of the ECM in animals are strength, structural support, organization, and cell signaling.

*Figure 10.2* The extension sequences of procollagen prevent large fibers from forming intracellularly.

*Figure 10.3* The proteins would become more linear, and the fiber would come apart.

*Figure 10.4* GAGs are highly negatively charged molecules that tend to attract positively charged ions and water. The high water content gives the ECM a gel-like character, which makes it difficult to compress.

*Figure 10.7* Adherens junctions and desmosomes are cell-to-cell junctions, whereas hemidesmosomes and focal adhesions are cell-to-ECM junctions.

*Figure 10.10* In contrast to the result shown in Figure 10.10, the lanthanum would be on the side of the cell layer facing the digestive tract. You would see it on that side of the cell layer up to the tight junction, but not on the other side of the tight junction.

*Figure 10.13* Middle lamellae are similar to anchoring junctions and desmosomes in that they also function in cell-to-cell adhesion. However, their structures are quite different. Middle lamellae are composed primarily of negatively charged polysaccharides that interact with divalent cations. By comparison, anchoring junctions and desmosomes hold cells together via proteins such as cadherins and integrins.

*Figure 10.15* Connective tissue has the most extensive ECM.

*Figure 10.17* Dermal tissue is found on the surfaces of leaves, stems, and roots.

## Core Skills: Connections

*Figure 10.9* If tight junctions did not exist, substances in the lumen of your intestine might directly enter your blood. This could be potentially harmful if you consumed something with a toxic molecule in it. Likewise, materials from blood could be lost by diffusing into the lumen of your small intestine.

*Figure 10.14* Plasmodesmata facilitate the movement of nutrients in a cell-to-cell manner. This is called symplastic transport.

## Feature Investigation Questions

1. The purpose of this study was to determine the sizes of molecules that can move through gap junctions from one cell to another.

2. The researchers used fluorescent dyes to visibly monitor the movement of material from one cell to an adjacent cell through the gap junctions. First, single layers of rat liver cells were cultured. Next, fluorescent dyes with molecules of various masses were injected into particular cells. The researchers then used fluorescence microscopy to determine whether or not the dyes were transferred from one cell to the next.

3. The researchers found that molecules with masses less than 1,000 daltons (Da) could pass through the gap junction channels. Molecules with masses larger than 1,000 Da could not pass through the gap junctions. Further experimentation revealed variation in gap-junction channel size among different cell types. However, the upper limit of the gap-junction channel size was determined to be around 1,000 Da.

## Test Yourself

1. e   2. c   3. b   4. e   5. e   6. d   7. e   8. d   9. e   10. a

## Conceptual Questions

1. The primary cell wall is synthesized first between the two newly made daughter cells. It is relatively thin and allows cells to expand and grow. The secondary cell wall is made in layers by the deposition of cellulose fibrils and other components. In many cell types, it is relatively thick.

# Chapter 11

## Concept Checks

*Figure 11.4* Cytosine is found in both DNA and RNA.

*Figure 11.5* The phosphate is attached to the 5′ carbon in a DNA nucleotide.

*Figure 11.11* The expected fractions would be 1/8 half-heavy and 7/8 light.

*Figure 11.15* The oxygen in a new phosphoester bond comes from the sugar.

*Figure 11.17* The lagging strand is made discontinuously in the direction opposite to the movement of the replication fork.

*Figure 11.19* When primase is synthesizing a primer in the lagging strand, it moves from left to right in this figure. After it is done making a primer, it needs to hop to the opening of the replication fork to make a new primer. This movement is from right to left in this figure.

*Figure 11.21* Telomerase uses an RNA sequence that it contains as a template to make the DNA repeat sequence.

*Figure 11.24* Proteins hold the bases of the radial loop domains in place.

## Core Skills: Connections

*Figure 11.1* When a bacterium dies, it may release some of its DNA into the environment. Such DNA can be taken up via transformation by living bacteria, even bacteria of other species. If the DNA that is taken up encodes an antibiotic resistance gene, the gene may be incorporated into the genome of the living bacterium and make it resistant to an antibiotic.

*Figure 11.25* If chromosomes did not become very compact, they might get tangled up with each other during cell division, which would prevent their even distribution into the two daughter cells.

## Feature Investigation Questions

1. Previous studies had indicated that mixing different strains could lead to transformation, or the changing of a strain into a different one. Griffith had shown that mixing heat-killed type S with living type R bacteria would result in the transformation of the type R to type S. Though mutations could cause a change of identity of certain strains, the type R to type S transformation was not due to mutation but was more likely due to the transmission of a biochemical substance between the two strains. Griffith recognized this and referred to the biochemical substance as the "transformation principle." If Avery, MacCleod, and McCarty could determine the biochemical identity of this "transformation principle," they could identify the genetic material for the bacteria.

2. A DNA extract contains DNA that has been purified from a sample of cells.

3. The researchers could not verify that the DNA extract was completely pure, that is, lacking small amounts of contaminating molecules, such as proteins and RNA. The researchers were able to treat the extract with enzymes to degrade proteins (using protease), RNA (using RNase), or DNA (using DNase). Eliminating the proteins or RNA did not alter the transformation of the type R to type S strains. Only the enzymatic degradation of DNA disrupted the transformation, indicating that DNA is the genetic material.

## Test Yourself

1. a   2. b   3. d   4. d   5. b   6. c   7. b   8. d   9. d   10. c

## Conceptual Questions

1. The genetic material must contain the information necessary to construct an entire organism. The genetic material must be accurately copied and transmitted from parent to offspring and from cell to cell during cell division in multicellular organisms. The genetic material must contain differences that can account for the known variation within each species and among different species.

Griffith discovered something called the transformation principle, and his experiments showed the existence of biochemical genetic information. In addition, he showed that this genetic information can move from one individual to another of the same species. In his experiments, Griffith took heat-killed type S bacteria and mixed them with living type R bacteria and injected them into a live mouse, which died after the injection. By themselves, these two strains would not kill the mouse, but when they were put together, the genetic information from the heat-killed type S bacteria was transferred into the living type R bacteria, thus transforming the type R bacteria into type S.

2. Since DNA is double stranded, 560 nucleotides form 280 bp. With 10 bp per turn, this DNA double helix will have 28 complete turns.

3. In a DNA double helix, the two strands hydrogen-bond with each other according to the AT/GC rule. This provides the basis for DNA replication. In addition, as described in later chapters, hydrogen bonding between complementary bases is the basis for the transcription of RNA, which is needed for gene expression. The sequence of bases within DNA strands have the function of storing information such as the sequence of amino acids within a polypeptide.

# Chapter 12

## Concept Checks

*Figure 12.1* A person with two defective copies of phenylalanine hydroxylase would have phenylketonuria.

*Figure 12.2* The ability to convert ornithine into citrulline is missing.

*Figure 12.3* The usual direction of flow of genetic information is from DNA to RNA to protein, though exceptions occur.

*Figure 12.9* The ends of protein-encoding genes do not have a poly T region that acts as a template for the synthesis of a poly A tail. Instead, the poly A tail is added to the pre-mRNA after transcription by an enzyme that attaches many adenine nucleotides in a row.

*Figure 12.11* A protein-encoding gene would still be transcribed into RNA if the start codon was missing. However, it would not be translated properly into a polypeptide.

*Figure 12.19* A region near the 5′ end of the mRNA is complementary to a region of rRNA in the small subunit. These complementary regions hydrogen bond with each other to promote the binding of the mRNA to the small ribosomal subunit.

## Core Skills: Connections

*Figure 12.5* Both DNA and RNA polymerase use a DNA strand as a template and connect nucleotides to each other in a 5′ to 3′ direction based on the complementarity of base pairing. One difference is that DNA polymerase needs a pre-existing strand, such as a RNA primer, to begin DNA replication, whereas RNA polymerase can begin the synthesis of RNA on a bare template strand. Another key difference is that DNA polymerase connects deoxyribonucleotides, whereas RNA polymerase connects ribonucleotides.

*Figure 12.15* The attachment of an amino acid to a tRNA is an endergonic reaction. ATP provides the energy to drive this reaction.

## Feature Investigation Questions

1. A triplet mimics mRNA because it can cause a specific tRNA to bind to the ribosome. This was useful to Nirenberg and Leder because it allowed them to correlate the binding of a tRNA carrying a specific amino acid with a triplet sequence.

2. The researchers were attempting to match codons with appropriate amino acids. By radiolabeling one amino acid in each of the 20 tubes for each codon, the researchers were able to identify the correct relationship by detecting which tube produced radioactivity on the filter.

3. For the AUG triplet, the filter would have shown radioactivity when methionine was radiolabeled. Even though AUG acts as the start codon, it also codes for the amino acid methionine. The other three codons act as stop codons and do not code for an amino acid. In these cases, the researchers would not have detected radioactivity on any of the filters.

## Test Yourself

1. b   2. a   3. d   4. e   5. e   6. e   7. d   8. d   9. d   10. b

## Conceptual Questions

1. Confirmation of their one gene/one enzyme hypothesis came from studies involving arginine biosynthesis. Biochemists had already established that particular enzymes are involved in a pathway to produce arginine. Intermediates in this pathway are ornithine and citrulline. Mutants in single genes disrupted the ability of cells to catalyze just one reaction in this pathway, thereby suggesting that a single gene encodes a single enzyme.

2. Each of these 20 enzymes catalyzes the attachment of a specific amino acid to a specific tRNA molecule.

3. During transcription, a DNA strand is used as a template for the synthesis of RNA. Most genes encode mRNAs, which contain the information to make polypeptides. During translation, an mRNA binds to a ribosome and a polypeptide is made, which becomes a unit within a functional protein.

# Chapter 13

## Concept Checks

*Figure 13.1* The binding of an ncRNA to DNA or another RNA could be inhibited by the formation of a stem-loop, because it would interfere with base pairing.

*Figure 13.4* HOTAIR binds to the target gene because a segment of the RNA in HOTAIR is complementary to the target gene.

*Figure 13.6* RISC binds to an mRNA because the miRNA is complementary to the mRNA. The miRNA and mRNA hydrogen bond to each other.

## Core Skills: Connections

*Figure 13.7* In eukaryotes, proteins that are destined for the ER, Golgi, vacuoles, lysosomes, or secretion need SRP to reach their proper locations. In bacteria and archaea, proteins that are secreted need SRP.

## Feature Investigation Questions

1. The *mex-3* mRNA corresponds to the sense strand.

2. The researchers began with a copy of the *mex-3* gene in a plasmid. When RNA polymerase and nucleotides were added, the sense strand of *mex-3* mRNA was made. They also modified this plasmid by placing the promoter on the opposite side of the *mex-3* coding sequence. When RNA polymerase and nucleotides were added, the antisense strand of *mex-3* mRNA was made.

3. The mixture of sense and antisense *mex-3* mRNA was the most effective at causing the degradation of *mex-3* mRNA.

## Test Yourself

1. e   2. a   3. e   4. c   5. d   6. d   7. e   8. a   9. c   10. d

## Conceptual Questions

1. HOTAIR: scaffold, guide

   SRP RNA: scaffold, alterer of protein function

   miRNA: guide

   crRNA: guide

2. RNA interference is the phenomenon in which an miRNA or an siRNA silences the expression of an mRNA. The double-stranded pre-miRNA or pre-siRNA is cleaved by dicer into a smaller double-stranded RNA that associates with RISC. One of the RNA strands is then degraded, so that only a short single-stranded miRNA or siRNA that recognizes an mRNA is found within RISC. This miRNA or siRNA guides RISC to the mRNA and causes it to be silenced.

3. The structure of HOTAIR provides a scaffold for the binding of two histone-modifying complexes. The structure is also complementary to GA-rich sequences next to certain target genes, such as the *HoxD* gene, and HOTAIR thereby guides the histone-modifying complexes to those genes. Such genes are then covalently modified, which causes them to be repressed.

# Chapter 14

## Concept Checks

*Figure 14.2* Gene regulation causes each type of cell to express its own unique set of proteins, which, in turn, are largely responsible for the morphology and function of that type of cell.

*Figure 14.6* The *lacZ*, *lacY*, and *lacA* genes are under the control of the *lac* promoter.

*Figure 14.7* Negative control refers to the action of a repressor protein, which inhibits transcription when it binds to the DNA. Inducible refers to the action of a small effector molecule. When it is present, it promotes transcription.

*Figure 14.11* In the case of bacterial metabolism of sugars, the repressor keeps the *lac* operon turned off unless lactose is present in the environment. The activator allows the bacterium to choose between glucose and lactose.

*Figure 14.16* When an activator interacts with mediator, it causes RNA polymerase to proceed to the elongation stage of transcription.

*Figure 14.18* Some histone modifications enhance transcription, whereas others inhibit it.

*Figure 14.21* The advantage of alternative splicing is that it allows a single gene to encode two or more polypeptides. This enables organisms to have smaller genomes, which are more efficient and easier to package into a cell.

*Figure 14.22* When iron levels rise in the cell, the iron binds to IRP and removes it from the mRNA that encodes ferritin. This results in the rapid translation of ferritin protein, which can store excess iron. Unfortunately, ferritin storage does have limits, so iron poisoning can still occur if too much is ingested.

## Core Skills: Connections

*Figure 14.10* In eukaryotic cells, cAMP acts as a second messenger in signal transduction pathways.

*Figure 14.19* A nucleosome is composed of DNA wrapped around an octamer of histone proteins.

## Feature Investigation Questions

1. The first observation was the identification of rare bacterial strains that showed constitutive expression of the *lac* operon. Normally, the *lac* genes are expressed only when lactose is present. These mutant strains expressed the genes continuously. The researchers also observed that some of these strains had mutations in the *lacI* gene. These two observations were key to the development of hypotheses explaining the relationship between the *lacI* gene and the regulation of the *lac* operon.

2. The correct hypothesis is that the *lacI* gene encodes a repressor protein that inhibits the operon.

3. The researchers introduced an F′ factor into the mutant strain that carried the wild-type *lacI* gene. In this merozygote, the cells that contained the F′ factor had both a mutant and a normal copy of the *lacI* gene. In the merozygote with an F′ factor with a normal copy of the *lacI* gene, regulation of the *lac* operon was restored. The researchers concluded that the normal *lacI* gene produced adequate amounts of a diffusible protein that could interact with the operator site on the chromosomal DNA as well as the F′ factor DNA and regulate transcription.

## Test Yourself

1. d   2. b   3. c   4. c   5. c   6. d   7. c   8. c   9. d   10. c

## Conceptual Questions

1. In an inducible operon, the presence of a small effector molecule causes transcription to occur. An example is the *lac* operon, which is induced with allolactose. In repressible operons, a small effector molecule inhibits transcription. An example is the *trp* operon, which is repressed by high levels of tryptophan. The effects of these small molecules are mediated through regulatory proteins that bind to the DNA.

2. a. regulatory protein; b. small effector molecule; c. segment of DNA; d. small effector molecule; and e. regulatory protein.

3. Gene regulation offers key advantages, such as (1) proteins are made only when they are needed; (2) proteins are made in the correct cell type; and (3) proteins are made at the correct stage of development. These advantages are important for reproduction and sustaining life.

# Chapter 15

## Concept Checks

*Figure 15.1* At neutral pH, glutamic acid is negatively charged. Perhaps the negative charges repel each other and prevent hemoglobin proteins from aggregating into fiber-like structures.

*Figure 15.3* This trait is due to a mutation that occurred in a somatic cell, so it cannot be transmitted to the individual's offspring.

*Figure 15.6* A thymine dimer is harmful because it can cause an error in DNA replication that results in a mutation.

*Figure 15.8* UvrC and UvrD are responsible for removing the damaged DNA. UvrC makes cuts on both sides of the damage, and then UvrD removes the damaged region.

*Figure 15.9* The Sun produces UV light and other harmful radiation that can damage DNA. This person has a defect in the nucleotide excision repair system. Therefore, her DNA is more likely to suffer mutations, which can cause pigmentation changes and other effects on the skin.

*Figure 15.11* Growth factors turn on a signal transduction pathway that ultimately leads to cell division.

*Figure 15.14* The type of cancer associated with this translocation is leukemia, which is a cancer of blood cells. The fused gene is expressed in white blood cells because it has the *bcr* promoter. The abnormal fusion protein promotes cancer in these cells.

*Figure 15.15* Checkpoints prevent cell division if a genetic abnormality is detected. This mechanism helps to properly maintain the genome by minimizing the possibility that a cell harboring a mutation will divide to produce two daughter cells.

*Figure 15.16* Cancer would not occur if both copies of the *Rb* gene and both copies of the *E2F* gene were rendered inactive due to mutations. An activated copy of the *E2F* gene is needed to promote cell division.

## Core Skills: Connections

*Figure 15.11* Drugs that inhibit protein kinases may be used to combat cancer if they target the protein kinases that are overactive in certain forms of cancer.

## Feature Investigation Questions

1. Some biologists believed that heritable traits may be altered by physiological events. This view suggests that mutations may be stimulated by certain needs of the organism. Others believed that mutations are random. If a mutation had a beneficial effect that improved survival and/or reproductive success, it would be more likely to be maintained in the population through natural selection.

2. The Lederbergs were testing the hypothesis that mutations are random events.

3. When the researchers looked at the locations on the secondary plates of colonies that were resistant to the bacteriophages, the pattern was the same on the two plates. This indicates that the mutation that allowed the colonies to be resistant to the virus occurred on the master plate. Thus, the mutation occurred randomly and was not caused by exposure to the virus.

## Test Yourself

1. d   2. d   3. d   4. e   5. d   6. b   7. b   8. b   9. c   10. e

## Conceptual Questions

1. Random mutations are more likely to be harmful than beneficial. The genes within each species have evolved to work properly. They have functional promoters, coding sequences, terminators, and so on, that are all needed for expression. Mutations are more likely to disrupt these sequences. For example, mutations within the coding sequence may produce early stop codons, frameshift mutations, and missense mutations that result in a nonfunctional polypeptide. On rare occasions, however, mutations are beneficial; they may produce a gene that is expressed better than the original gene or produce a polypeptide that functions better.

2. A spontaneous mutation originates within a living cell. It may be due to spontaneous changes in nucleotide structure, errors in DNA replication, or alterations of the structure of DNA by products of normal metabolism. The causes of induced mutations originate from outside the cell. They may be physical agents, such as UV light or X-rays, or chemicals that act as mutagens. Both spontaneous and induced mutations may cause a harmful phenotype such as a cancer. In many cases, induced mutations are avoidable if the individual can prevent exposure to the environmental agent that acts as a mutagen.

3. Mutations may alter the expression of a gene and/or alter the function of a protein encoded by a gene. In many cases, such changes are harmful because a gene may not be expressed at the correct level, or the protein may not function as well as the normal (nonmutant) protein.

# Chapter 16

## Concept Checks

*Figure 16.1* Chromosomes are readily seen when they are compacted in a dividing cell. By adding such a drug, the researchers increase the percentage of cells that are actively dividing.

*Figure 16.2* Interphase consists of the $G_1$, S, and $G_2$ phases of the cell cycle.

**Figure 16.8** The astral microtubules, which extend away from the chromosomes, are important for positioning the spindle apparatus within the cell. The polar microtubules project into the region between the two poles. Polar microtubules that overlap with each other play a role in the separation of the two poles. Kinetochore microtubules are attached to kinetochores at the centromeres and are involved in sorting the chromosomes.

**Figure 16.9** Cytokinesis in both animal and plant cells follows mitosis and separates a mother cell into two daughter cells. In animal cells, cytokinesis involves the formation of a cleavage furrow, which constricts like a drawstring to separate the cells. In plants, the two daughter cells are separated by a cell plate, which forms a cell wall between them.

**Figure 16.14** The purpose of meiosis in animals is to produce gametes. These gametes combine during fertilization to produce a diploid organism. Following fertilization, the purpose of mitosis is to produce a multicellular organism.

**Figure 16.16** Inversions and reciprocal translocations do not affect the total amount of genetic material.

## Core Skills: Connections

**Figure 16.3** Checkpoint proteins prevent cancer by checking the integrity of the genome. If abnormalities in DNA structure are detected or if a chromosome is not properly attached to the spindle apparatus, the checkpoint protein will delay cell division until the problem is fixed. If it cannot be fixed, the checkpoint protein will initiate the process of apoptosis, thereby killing a cell that may harbor mutations. This prevents the proliferation of cells that have the potential to be cancerous.

## Feature Investigation Questions

1. Researchers had demonstrated that the binding of progesterone to receptors in oocytes caused the cells to advance from the $G_2$ phase of the cell cycle to mitosis. It appeared that progesterone acted as a signaling molecule for advancement through the cell cycle.

2. The researchers proposed that progesterone acted as a signaling molecule that led to the synthesis of molecules that cause the oocyte to advance through the cell cycle and achieve maturation.

   To test their hypothesis, donor eggs were exposed to progesterone for either 2 or 12 hours. Control donor oocytes were not exposed to progesterone. Cytosol from each treatment was then transferred to recipient oocytes. The researchers recorded whether or not the recipient oocytes underwent maturation.

3. The oocytes that were exposed to the progesterone for only 2 hours did not induce maturation in the recipient oocytes, whereas the oocytes that were exposed to progesterone for 12 hours did induce maturation in the recipient oocytes. The researchers suggested that a time span greater than 2 hours is needed to accumulate the proteins that are necessary to promote maturation.

## Test Yourself

1. b   2. e   3. b   4. e   5. c   6. a   7. c   8. d   9. b   10. c

## Conceptual Questions

1. In diploid species, chromosomes are present in pairs, one from each parent, and contain similar gene arrangements. Such chromosomes are homologous. When DNA is replicated, two identical copies are created, and these are sister chromatids.

2. There are four copies. A karyotype shows homologous chromosomes that come in pairs. Each member of the pair has replicated to form a pair of sister chromatids. Therefore, four copies of each gene are present. See the inset to Figure 16.1.

3. Mitosis is a process that produces two daughter cells with the same genetic material as the original daughter cell. In the case of plants and animals, this allows a fertilized egg to develop into a multicellular organism composed of many, genetically identical cells.

# Chapter 17

## Concept Checks

**Figure 17.4** The stamens are removed from the purple flower to prevent self-fertilization.

**Figure 17.5** The reason why offspring of the $F_1$ generation exhibit only one variant of each character is because one trait is dominant over the other.

**Figure 17.6** The ratio of alleles ($T$ to $t$) is 1:1. The reason why the phenotypic ratio is 3:1 is because $T$ is dominant to $t$.

**Figure 17.7** The genotype was $Pp$. To produce white offspring, which are $pp$, the original plant had to have at least one copy of the $p$ allele. Because it had purple flowers, it also had to have one copy of the $P$ allele. So, its genotype must have been $Pp$.

**Figure 17.8** If the linked assortment hypothesis had been correct, the ratio would have been 3 yellow, round to 1 green, wrinkled.

**Figure 17.11** There are four possible ways that the chromosomes can align, and eight different types of gametes ($ABC$, $abc$, $ABc$, $abC$, $Abc$, $aBC$, $AbC$, $aBc$) can be produced.

**Figure 17.12** No. If two parents are affected with the disease, they would have to be homozygous for the mutant allele if it's recessive. Two homozygous parents would produce all affected offspring, barring rare mutations. If they produce an unaffected offspring, then the mutant allele is not recessive.

**Figure 17.13** All affected offspring having at least one affected parent suggests a dominant pattern of inheritance.

**Figure 17.14** The person is a female. In mammals, the presence of the Y chromosome causes maleness. Therefore, lacking a Y chromosome, a person with a single X chromosome develops into a female.

**Figure 17.18** No. You need a genetically homogenous population to study the norm of reaction. A wild population of squirrels is not genetically homogenous, so it could not be used.

**Figure 17.19** The recessive allele is the result of a loss-of-function mutation. In a $Ccpp$ individual, the enzyme encoded by the $P$ gene is defective.

## Core Skills: Connections

**Figure 17.9** Sexual reproduction is the process in which two haploid gametes (for example, sperm and egg) combine with each other to begin the life of a new individual. Each gamete contributes one set of chromosomes. The resulting zygote has chromosomes that occur in pairs (one from each parent). The members of each pair are called homologs of each other; they carry the same types of genes.

**Figure 17.10** Alleles segregate, or go into separate cells, during the process of meiosis. Meiosis begins with a diploid mother cell that has pairs of genes, which may be found in different alleles. During meiosis, these pairs of genes separate and end up in different haploid cells. Therefore, each haploid cell has only one copy of each gene. In other words, each haploid cell has only one allele of a given gene.

## Feature Investigation Questions

1. Morgan was testing the hypothesis of use and disuse. This hypothesis suggests that if a structure is not used, over time, it will diminish and/or disappear. Morgan was originally testing to see if flies reared in the dark would lose some level of eye development.

2. When the $F_1$ individuals were crossed, only male $F_2$ offspring expressed the white eye color. At this time, Morgan was aware of sex chromosome differences between male and female flies. He realized that because males only possess one copy of X-linked genes, this would explain why only $F_2$ males exhibited the recessive trait.

3. In a cross between a white-eyed male and a female that is heterozygous for the white and red alleles, 1/2 of the female offspring would have white eyes. Also, a cross between a white-eyed male and a white-eyed female would yield all offspring with white eyes.

## Test Yourself

1. c 2. b. Mendel's law of segregation refers to the separation of the two alleles into separate cells. Meiosis is the nuclear division process that produces haploid cells. During the first meiotic division, a diploid cell divides to produce haploid cells. This is the phase in which the two alleles segregate, or separate, from each other. 3. d   4. c   5. e   6. d. Half of the males will be affected, but only half of the children will be males, so you multiply: 0.5 x 0.5 = 0.25, or 25%.   7. d   8. d   9. d   10. c

## Conceptual Questions

1. Two affected parents having an unaffected offspring would rule out recessive inheritance. If two unaffected parents have an affected offspring, dominant inheritance is ruled out. However, it should be noted that this answer assumes that no new mutations are happening. In rare cases, a new mutation could cause or alter these results. For recessive inheritance, two affected parents could have an unaffected offspring if the offspring had a new mutation that converted the recessive allele to the dominant allele. Similarly for dominant inheritance, two unaffected parents could have an affected offspring if the offspring inherited a new mutation that was dominant. Note: New mutations are expected to be very rare.

2. The individual probabilities are as follows: $AA = 0.25$; $bb = 0.5$; $CC = 0.5$; and $Dd = 0.5$. These are determined by making small Punnett squares. We use the product rule to calculate the probability of $AAbbCCDd = (0.25)(0.5)(0.5)(0.5) = 0.03125$, or 3.125%.

3. The environment is necessary for the expression of genes. For example, organic molecules and energy are needed for transcription and translation. In addition, environmental factors influence the outcomes of traits. For example, sunlight can cause a tanning response, thereby affecting the darkness of the skin.

# Chapter 18

## Concept Checks

*Figure 18.1* Only the allele inherited from the father would be expressed in the offspring. Because he is heterozygous, half of the offspring would be normal size and half would be dwarf.

*Figure 18.3* The Barr body is much more compact than the other X chromosome in the cell. This compaction prevents most of the genes on the Barr body from being expressed.

*Figure 18.7* The gene is located in the chloroplast DNA. In this species, chloroplasts are transmitted from parent to offspring via eggs but not via sperm.

*Figure 18.10* Crossing over occurred during oogenesis in the heterozygous female of the $F_1$ generation to produce the recombinant offspring of the $F_2$ generation.

*Figure 18.11* One strategy would be to begin with two true-breeding parental strains: $alal\ dpdp$ and $al^+al^+\ dp^+dp^+$ and cross them to get $F_1$ heterozygotes, $al^+al\ dp^+dp$. Then testcross female $F_1$ heterozygotes to male $alal\ dpdp$ homozygotes. In the $F_2$ generation, the recombinant offspring would be $al^+al\ dpdp$ and $alal\ dp^+dp$, and the nonrecombinants would be $al^+al\ dp^+dp$, and $alal\ dpdp$.

## Core Skills: Connections

*Figure 18.6* The evolutionary origin of these organelles is an ancient endosymbiotic relationship. Mitochondria are derived from purple bacteria, and chloroplasts are derived from cyanobacteria.

## Feature Investigation Questions

1. Bateson and Punnett were testing the hypothesis that the gene pairs that influence flower color and pollen shape would assort independently of each other. The two traits were expected to show a pattern consistent with Mendel's law of independent assortment.

2. The expected results were a phenotypic ratio of 9:3:3:1. The researchers expected 9/16 of the offspring would have purple flowers and long pollen, 3/16 of the offspring would have purple flowers and round pollen, 3/16 of the offspring would have red flowers and long pollen, and 1/16 of the offspring would have red flowers and round pollen.

3. Though all four of the expected phenotypes were seen, they were not in the predicted ratio of 9:3:3:1. The number of individuals with the phenotypes found in the parental generation (purple flowers and long pollen or red flowers and round pollen) was much higher than expected. Bateson and Punnett suggested that the gene controlling flower color was somehow coupled with the gene that controls pollen shape. This would explain why these traits did not always assort independently.

## Test Yourself

1. e    2. d    3. c    4. a    5. a    6. e    7. e    8. d    9. c    10. c

## Conceptual Questions

1. Epigenetics is the study of mechanisms that lead to changes in gene expression that can be passed from cell to cell and are reversible, but do not involve a change in the sequence of DNA. Not all epigenetic changes are passed from parent to offspring. For example, a cigarette smoker could acquire an epigenetic change in a lung cell; this change would not be passed to offspring.

2. A Barr body is an X chromosome in the somatic cells of female mammals that has undergone X-chromosome inactivation. It is highly compacted. This compaction prevents the expression of most X-linked genes.

3. Epigenetics alters the way that genes are expressed, and thereby can affect an individual's traits. Such traits, in turn, may affect reproduction.

# Chapter 19

## Concept Checks

*Figure 19.5* The advantage of the lytic cycle is that the phage can make many copies of itself and proliferate. However, sometimes the growth conditions may not be favorable to make new phages. The advantage of the lysogenic cycle is that the prophage can remain latent until conditions become favorable to make new phages.

*Figure 19.10* The loop domains are held in place by proteins that bind to the DNA at the bottoms of the loops. The proteins also bind to each other.

*Figure 19.11* Bacterial chromosomes and plasmids are similar in that they are both circular DNA molecules. However, bacterial chromosomes are usually much longer than plasmids and carry many more genes. Also, bacterial chromosomes tend to be more compacted due to the formation of loop domains and DNA supercoiling.

*Figure 19.12* In 16 hours, there will be 32 doublings. So, $2^{32} = 4,294,967,296$. (The actual number would be much less because the cells would deplete the growth media and grow more slowly than the maximal rate.)

*Figure 19.15* Yes. The two strains would have mixed together, allowing them to conjugate. Therefore, there would have been colonies on the plates.

*Figure 19.16* During conjugation, only one strand of the DNA from an F factor is transferred from the donor to the recipient cell. The single-stranded DNA in both cells is then used as a template to create double-stranded F factor DNA in both cells.

*Figure 19.18* Transduction is not a normal part of the phage life cycle. It is a mistake in which a piece of the bacterial chromosome is packaged into a phage coat and is then transferred to another bacterial cell.

## Core Skills: Connections

*Figure 19.4* Viral release occurs via a budding process in which a membrane vesicle is formed that surrounds the capsid. Similarly, exocytosis involves the formation of a membrane vesicle that encloses some type of cargo.

*Figure 19.9* A nucleoid is not a membrane-bound organelle. It is simply a site where a bacterial chromosome is found. A cell nucleus in a eukaryotic cell has an envelope with a double membrane.

*Figure 19.17* Griffiths was able to show that genetic material was transferred to type R bacteria, which converted them to type S. This occurred via transformation. Later, Avery, MacLeod, and McCarty determined that DNA was the material that was being transferred.

## Feature Investigation Questions

1. Lederberg and Tatum were testing the hypothesis that genetic material could be transferred from one bacterial strain to another.

2. The experimental growth medium lacked particular amino acids and biotin. The mutant strains were unable to synthesize these particular substances. Therefore, they were unable to grow due to the lack of the necessary nutrients. The two strains used in the experiment each lacked the ability to make two essential nutrients necessary for growth. The appearance of colonies growing on the experimental growth medium indicated that some bacterial cells had acquired the functional genes in place of the two mutations they carried. By acquiring these functional genes, they gained the ability to synthesize the essential nutrients.

3. Davis placed samples of the two bacterial strains in different arms of a U-tube. A filter allowed the free movement of the liquid in which the bacterial cells were suspended, but prevented actual contact between the bacterial cells. After incubating the strains taken from the U-tube, Davis found that gene transfer did not take place. He concluded that physical contact between cells of the two strains was required for gene transfer.

## Test Yourself

1. c    2. e    3. c    4. b    5. e    6. a    7. d    8. d    9. b    10. c

## Conceptual Questions

1. Viruses are similar to living cells in that they contain a genetic material that provides a blueprint to make new viruses. However, viruses are not composed of cells, and by themselves, they do not carry out metabolism, use energy, maintain homeostasis, or even reproduce. A virus or its genetic material must be taken up by a living cell to replicate.

2. Conjugation involves direct physical contact between two bacterial cells in which a donor cell transfers a strand of DNA to a recipient cell.

Transformation occurs when a bacterium takes up a DNA fragment from the environment, which may have come from a dead bacterium.

Transduction occurs when a bacteriophage that has infected a bacterial cell breaks up the chromosome, and a fragment of bacterial chromosomal DNA is incorporated into a newly made bacteriophage. It then transfers this fragment of DNA to a recipient bacterial cell.

3. Horizontal gene transfer is the transfer of genes from an organism to another organism that is not the offspring of the first organism. These acquired genes sometimes promote survival and therefore may have an evolutionary advantage. Such genes may even lead to the formation of new species. From a medical perspective, an important example of horizontal gene transfer is when one bacterium acquires a resistance gene from another bacterium and becomes resistant to an antibiotic. This phenomenon is making it increasingly difficult to treat a wide variety of bacterial diseases.

# Chapter 20

## Concept Checks

*Figure 20.3* Cell division and cell migration are common in the early stages of development, whereas cell differentiation and apoptosis are more common as tissues and organs start to form.

*Figure 20.4* If apoptosis did not occur, the fingers would be webbed.

*Figure 20.8* The larva would have anterior structures at both ends and would lack posterior structures such as a spiracle.

*Figure 20.9* The Bicoid protein functions as a transcription factor that promotes the formation of anterior structures. Its function is highest in the anterior end of the zygote.

*Figure 20.14* The last abdominal segment would have legs!

*Figure 20.18* Stem cells can divide, and the daughter cells can differentiate into specific cell types.

*Figure 20.20* Hematopoietic stem cells are multipotent.

*Figure 20.24* The pattern would be sepal, petal, stamen, stamen.

*Figure 20.23* Stem cells in plants are found in meristems, which are located at the tips of roots and shoots.

## Core Skills: Connections

*Figure 20.5* Some cell surface receptors recognize signals that convey positional information. After the signal binds to this type of receptor, a signal transduction pathway is activated that may cause a cell to divide, migrate, differentiate, or undergo apoptosis.

*Figure 20.17* As the number of *Hox* genes increases, the body plan of the animal becomes more complex.

## Feature Investigation Questions

1. The researchers were interested in the factors that cause cells to differentiate. In this particular study, they were attempting to identify the gene(s) involved in the differentiation of muscle cells.

2. Using genetic technology, the researcher compared the gene expression in cells that could differentiate into muscle cells with the gene expression in cells that could not differentiate into muscle cells. Though many genes were expressed in both, the researchers were able to identify three genes that were expressed in muscle cell lines that were not expressed in nonmuscle cell lines.

3. Again, using genetic technology, each of the candidate genes was introduced into a cell that normally did not give rise to skeletal muscle. This procedure was used to test whether or not the gene played a key role in muscle cell differentiation. If the genetically engineered cell gave rise to muscle cells, the researchers would have evidence that that particular candidate gene was involved in muscle cell differentiation. Of the three candidate genes, only one was shown to be involved in muscle cell differentiation. When the *MyoD* gene was expressed in fibroblasts, these cells differentiated into skeletal muscle cells.

## Test Yourself

1. c  2. d  3. e  4. b  5. c  6. e  7. a  8. a  9. b  10. d

## Conceptual Questions

1. a. This abnormality is consistent with a mutation in a segmentation gene, such as a gap gene.

   b. This abnormality is consistent with a mutation in a homeotic gene because the characteristics of a particular segment have been changed.

2. Both types of genes encode transcription factors that bind to the DNA and regulate the expression of other genes. The effects of *Hox* genes determine the characteristics of certain regions of the body, whereas the *myoD* gene is cell-specific—it causes a cell to become a skeletal muscle cell.

3. Maternal effect genes control the formation of body axes, such as the antero-posterior and dorsoventral axes. Next, the segmentation genes divide the embryo into segments, though visible segments are lost in many animal species at later stages of development. Finally, the homeotic genes determine the characteristics of each segment.

# Chapter 21

## Concept Checks

*Figure 21.3* The insertion of chromosomal DNA into the vector disrupts the *lacZ* gene, thereby preventing the expression of β-galactosidase. The functionality of *lacZ* can be determined by providing the growth medium with a colorless compound, X-Gal, which is cleaved by β-galactosidase into a blue dye. Bacterial colonies containing recircularized vectors form blue colonies, whereas colonies containing recombinant vectors carrying a segment of chromosomal DNA will be white.

*Figure 21.5* The 600-bp fragment will be closer to the bottom. Smaller pieces travel faster through the gel.

*Figure 21.6* The primers are complementary to sequences at each end of the DNA region to be amplified.

*Figure 21.8* If a ddNTP is added to a growing DNA strand, the strand can no longer grow because the 3′ —OH group, the site of attachment for the next nucleotide, is missing.

*Figure 21.9* A fluorescent spot identifies a cDNA that is complementary to a particular DNA sequence. Because the cDNA was generated from mRNA, the fluorescence identifies a gene that has been transcribed in a particular cell type under a given set of conditions.

*Figure 21.10* The sgRNA is composed of two different components of the bacterial defense system, crRNA and tracrRNA, which have been linked together.

*Figure 21.12* One reason is that more complex species tend to have more genes. A second reason is that species vary with regard to the amount of repetitive DNA sequences in their genomes.

*Figure 21.18* Retrotransposons. A single retrotransposon can be transcribed into multiple copies of RNA, which can be converted to DNA by reverse transcriptase, and inserted into multiple sites in the genome.

## Core Skills: Connections

*Figure 21.2* Plasmids are small, circular DNA molecules that exist independently of the bacterial chromosome. They have their own origin of replication. Many plasmids carry genes that convey some type of selective advantage to the host cell, such as antibiotic resistance.

*Figure 21.14* The proteins produced by family members at early stages of development (embryonic and fetal stages) have a higher affinity for oxygen than the proteins produced in an adult. This allows the embryo and fetus to obtain oxygen from the mother's bloodstream.

## Feature Investigation Questions

1. The goal of the experiment was to sequence the entire genome of *Haemophilus influenzae*. By conducting this experiment, the researchers would have information about genome size and the types of genes the bacterium has.

2. If you divide 20 by 4.1, this equals 4.9. The value of 4.9 represents $m$ in the equation: $P = e^{-m}$. If you substitute 4.1 for $m$ into the equation and solve for $P$, the probability $P$ equals 0.0074 or 0.74%. Therefore, only 0.74% would be left unsequenced, which is less than 1%.

3. The researchers were successful in sequencing the entire genome of the bacterium. The genome size was determined to be 1,830,137 base pairs, with a predicted 1,743 structural genes. The researchers were also able to predict the function of many of these genes. More importantly, the results were the first complete genomic sequence of a living organism.

## Test Yourself

1. d  2. b  3. b  4. b  5. c  6. e  7. e  8. a  9. c  10. e

## Conceptual Questions

1. A ddNTP is missing an oxygen at the 3′ position. This prevents the further growth of a DNA strand, thereby causing chain termination.

2. a. yes

   b. No, it's only one chromosome in the nuclear genome.

   c. Yes, it's corn's nuclear genome. Corn also has a mitochondrial genome and a chloroplast genome.

   d. yes

3. The genome contains the information for the production of cellular proteins; it is a blueprint. The production of proteins is largely responsible for determining cellular characteristics, which, in turn, are largely responsible for determining an organism's traits. In addition, the genome also encodes many non-coding RNAs that perform a variety of different functions.

# Chapter 22

## Concept Checks

*Figure 22.2* A single organism does not evolve. Populations may evolve from one generation to the next due to differences in reproductive success.

*Figure 22.7* Due to a changing global climate, the island fox became isolated from the mainland species. Over time, natural selection resulted in adaptations for the population on the island and eventually resulted in a new species with characteristics that are somewhat different from those of the mainland species.

*Figure 22.8* Many answers are possible. One example is the wing of a bird and the wing of a bat.

*Figure 22.11* The magnitudes of traits are changing. For example, in the breeds of dogs, the lengths of legs, body size, and so on, are quite different. Artificial selection is often aimed at changing the relative sizes of body parts or the amount of something, such as oil content.

*Figure 22.14* Orthologs have similar gene sequences because they are derived from the same ancestral gene. The sequences are not identical because after the species diverged, each one accumulated different random mutations that changed their sequences.

*Figure 22.16* Humans have one large chromosome 2, but this chromosome is divided into two separate chromosomes in the other three species. In chromosome 3, the banding patterns among humans, chimpanzees, and gorillas are very similar, but the orangutan has a large inversion that flips the arrangement of bands in the centromeric region.

## Core Skills: Connections

*Figure 22.15* The three mechanisms of horizontal gene transfer between bacterial species are conjugation, transformation and transduction.

## Feature Investigation Questions

1. The island has a moderate level of isolation but is located near enough to the mainland to have some migrants. The island is an undisturbed habitat, so the researchers would not have to consider the effects of human activity on the study. Finally, the island had an existing population of ground finches that would serve as the subjects of the study over many generations.

2. First, the researchers were able to show that beak depth is a genetic trait that has variation in the population. Second, the depth of the beak is an indicator of the types of seeds the birds can eat. The birds with larger beaks can eat larger and drier seeds; therefore, changes in the types of seeds available could act as a selective force on the bird population.

   During the study period, annual changes in rainfall occurred, which affected the seed sizes produced by the plants on the island. In the drier year, fewer small seeds were produced, so the birds would have to eat larger, drier seeds.

3. The researchers found that following the drought in 1977, the average beak depth in the finch population increased. This indicated that birds with larger beaks were better able to adapt to the environmental changes due to the drought and produce more offspring. This is direct evidence of the phenomenon of natural selection.

## Test Yourself

1. d  2. d  3. b  4. b  5. b  6. d  7. c  8. d  9. d  10. e

## Conceptual Questions

1. Some random mutations result in a phenotype with greater reproductive success. If this occurs, natural selection results in a greater proportion of such individuals in succeeding generations.

2. The process of convergent evolution produces two different species from different lineages that show similar characteristics because they occupy similar environments. An example is the long snout and tongue of both the giant anteater, found in South America, and the echidna, found in Australia. These structures enable these animals to feed on ants, but the two structures evolved independently. These observations support the idea that evolution results in adaptations to particular environments.

3. Homologous structures are two or more structures that are similar because they are derived from a common ancestor. An example is the set of bones that is found in the human arm, turtle arm, bat wing, and whale flipper. The forearms in these species have been modified to perform different functions. This supports the idea that all of these animals evolved from a common ancestor by descent with modification.

# Chapter 23

## Concept Checks

*Figure 23.3* Over the short run, alleles that confer better fitness in the warmer climate would be favored and increase in frequency, perhaps enhancing diversity. Over the long run, however, an allele that confers high fitness in the homozygous state may become monomorphic, thereby reducing genetic diversity.

*Figure 23.4* Stabilizing selection eliminates alleles that result in phenotypes that deviate significantly from the average phenotype. For this reason, it tends to decrease genetic diversity.

*Figure 23.6* If malaria was eradicated, there would be no selective advantage for the heterozygote. The $H^S$ allele would eventually be eliminated because the $H^S H^S$ homozygote has a lower fitness. Directional selection would occur.

*Figure 23.7* Courtship songs are likely to be part of intersexual selection. Such traits are likely to be involved in mate choice.

*Figure 23.11* The bottleneck effect tends to decrease genetic diversity. This may eliminate adaptations that promote survival and reproductive success. Therefore, the bottleneck effect makes it more difficult for a population to survive.

*Figure 23.13* Migration results in gene flow, which tends to make the allele frequencies in neighboring populations more similar to each other. It also promotes genetic diversity by introducing new alleles into populations.

## Core Skills: Connections

*Figure 23.12* There are lots of possibilities. The idea is that you are changing one codon to another codon that specifies the same amino acid. For example, changing a codon from GGA to GGG is likely to be neutral because both codons specify glycine.

## Feature Investigation Questions

1. The males of the two species of cichlids used in the experiment are distinguishable by coloration, and the researchers were testing the hypothesis that the females make mate choices based on this variable.

2. Individual females were placed in an aquarium that also contained one male from each species. The males were held in separate enclosures to limit their movement but allow the female to see each of the males. The researchers recorded the courtship behavior between the female and males and the number of positive encounters between the female and each of the different males. This procedure was conducted under normal lighting and under monochromatic lighting that obscured the coloration differences between the males of the two species. Comparing the behavior of the females under normal light conditions and monochromatic light conditions allowed the researchers to determine the importance of coloration in mate choice.

3. The researchers found that the female was more likely to select a mate from her own species in normal light conditions. However, under monochromatic light conditions, the species-specific mate choice was not observed. Females were as likely to choose males of the other species as they were males of their own species. This indicated that coloration is an important factor in mate choice in these species of fish.

## Test Yourself

1. d   2. c   3. c   4. e   5. b   6. c   7. b   8. d   9. b   10. a

## Conceptual Questions

1. The frequency of the disease is a genotype frequency because it represents individuals with the disease. If we let $q^2$ represent the genotype frequency, then $q$ equals the square root of 0.04, which is 0.2. If $q = 0.2$, then $p = 1 - q$, which is 0.8. The frequency of heterozygous carriers is $2pq$, which is $2(0.8)(0.2) = 0.32$, or 32%.

2. Directional selection—In this pattern of natural selection, individuals with an extreme phenotype are more likely to survive and reproduce. As a result, the extreme phenotype will become predominant in the population. In addition to selecting for a certain phenotype, the opposite extreme phenotype is removed from the gene pool.

   Stabilizing selection—This pattern of natural selection favors individuals with intermediate phenotypes, whereas individuals with extreme phenotypes are less likely to reproduce. Stabilizing selection tends to prevent major changes in the phenotypes prevalent in populations.

   Disruptive selection—This pattern of natural selection favors both extremes and removes the intermediate phenotype. It is also known as diversifying selection.

   Balancing selection—This pattern of natural selection results in a balanced polymorphism in which two or more alleles are stably maintained in a population. It can occur as a result of heterozygote advantage, as with the sickle cell allele, or negative frequency-dependent selection, as in certain prey populations.

   Sexual selection—This is a type of natural selection that is directly associated with reproductive success. It can occur by any of the previous four mechanisms. Male coloration in African cichlids is an example.

3. Genetic drift involves random changes in the genetic composition of a population from one generation to the next. Neutral changes in DNA sequences may happen randomly, and these are most likely to accumulate in a population due to genetic drift. This is evolution at the level of DNA, but it does not act upon phenotype.

# Chapter 24

## Concept Checks

*Figure 24.1* There are a lot of possibilities. Certain grass species look quite similar. Elephant species look very similar. And so on.

*Figure 24.3* Temporal isolation is an example of a prezygotic isolating mechanism. Because the two species breed at different times of the year, hybrid zygotes are not produced.

*Figure 24.5* Hybrid sterility is a type of postzygotic isolating mechanism. A hybrid is produced by the interbreeding of the two species, but it is sterile.

*Figure 24.11* The offspring would inherit 16 chromosomes from *G. tetrahit* and anywhere from 8 to 16 from the hybrid. So it would have 24 to 32 chromosomes. The hybrid parent would always pass on the 8 chromosomes that are found in pairs. With regard to the 8 chromosomes not found in pairs, it could pass on 0 to 8 of them.

*Figure 24.12* The insects on a particular type of host plant would tend to breed with each other, and natural selection would favor the development of traits that are an advantage for feeding on that host. Over time, the accumulation of genetic changes may lead to reproductive isolation between the populations of insects feeding on different hosts.

*Figure 24.14* If the *Gremlin* gene was underexpressed, less gremlin protein would be produced. Because the gremlin protein inhibits apoptosis, more cell death would occur, and the result would probably be smaller feet, possibly not webbed.

*Figure 24.17* The tip of the mouse's tail might have a mouse eye!

## Core Skills: Connections

*Figure 24.2* Female choice is a prezygotic isolating mechanism.

*Figure 24.7* The Hawaiian Islands have many different ecological niches that can be occupied by birds. The founding population evolved to occupy those niches, thereby evolving into many different species of honeycreepers.

*Figure 24.15* The *Hox* genes expressed along the anteroposterior axis during early embryonic development are homeotic genes. In insects that contain discrete body segments, each *Hox* gene determines the structures that will ultimately form in those segments. Although more complex animals such as mammals do not display discrete segments, the expression of the *Hox* genes controls what structures will form along the anteroposterior axis.

## Feature Investigation Questions

1. Podos hypothesized that the morphological changes in the beaks would also affect the birds' songs. A male bird's song is an important component for mate choice. If changes in the beak alter the song, reproductive ability will be affected. Podos suggested that changes in the beak morphology could thus lead to reproductive isolation among the finches.

2. Podos first caught male birds in the field and collected data on beak size. The birds were banded for identification and released. Later, the banded birds' songs were recorded and analyzed for range of frequencies and trill rates. The results were then compared with similar data from other species of birds to determine if beak size constrained the frequency range and trill rate of the song.

3. The results of the study did indicate that natural selection acting on beak size due to changes in diet could lead to changes in song. Considering the importance of bird song to mate choice, the changes in the songs could also lead to reproductive isolation.

   The phrase "by-product of adaptation" refers to changes in phenotype that are not directly acted on by natural selection. In the case of the Galápagos finches, the changes in beak size were directly related to diet; however, as a consequence of that selection, the song pattern was also altered. The change in song pattern was a by-product.

## Test Yourself

1. b   2. b   3. e   4. d   5. c   6. a   7. b   8. d   9. c   10. c

## Conceptual Questions

1. Prezygotic isolating mechanisms prevent the formation of a zygote. An example is mechanical isolation, the incompatibility of genitalia. Postzygotic isolating mechanisms act after the formation of the zygote. An example is inviability of the hybrid that develops. (Other examples shown in Figure 24.2 are also correct.) Postzygotic mechanisms are more costly because some energy is expended for the formation of a zygote and its subsequent growth.

2. The concept of gradualism suggests that each new species evolves continuously over long spans of time (Figure 24.13a). The principal idea is that large phenotypic differences that produce new species are due to the accumulation of many small genetic changes. According to the concept of punctuated equilibrium (Figure 24.13b), species exist relatively unchanged for many generations. During this period, the species is in equilibrium with its environment. These long periods of equilibrium are punctuated by relatively short periods during which evolution occurs at a far more rapid rate. This rapid evolution is caused by relatively few genetic changes.

3. One example involves the *Hox* genes, which control morphological features along the anteroposterior axis in animals. An increase in the number of *Hox* genes during evolution is associated with an increase in body complexity and may have spawned many different animal species.

# Chapter 25

## Concept Checks

*Figure 25.2* A phylum is broader than a family.

*Figure 25.3* Yes. They can have many common ancestors, depending on how far back you go in the tree. For example, dogs and cats have a common ancestor that gave rise to mammals, and an older common ancestor that gave rise to vertebrates. The most recent common ancestor is the point at which two species diverged from each other.

*Figure 25.4* An order is a smaller taxon that would have a more recent common ancestor.

*Figure 25.9* A hinged jaw is the character common to the salmon, lizard, and rabbit, but not to the lamprey.

*Figure 25.10* The change of G2 to A is common to species A, B, and C, but not to species G.

*Figure 25.12* The kiwis are found in New Zealand. Surprisingly, however, kiwis are more closely related to Australian and African flightless birds than they are to moas, which were found in New Zealand.

*Figure 25.15* Monophyletic groups are defined on the assumption that a particular group of species descended from a common ancestor. If horizontal gene transfer has occurred, not all of the genes carried by the species in such a group were inherited from the common ancestor, thus muddling the concept of monophyletic groups.

## Core Skills: Connections

**Figure 25.1** The domains Bacteria and Archaea have organisms with prokaryotic cells.

**Figure 25.13** There are lots of possibilities. The changes need to be ones that change one codon to another codon that specifies the same amino acid. For example, changing a codon from GGA to GGG is likely to be neutral because both codons specify glycine.

## Feature Investigation Questions

1. Molecular paleontology is the sequencing and analysis of DNA obtained from extinct species. Tissue samples from specimens of extinct species may contain DNA molecules that can be extracted, amplified, and sequenced. The DNA sequences can then be compared with those of living species to study evolutionary relationships between modern and extinct species.

   The researchers extracted DNA from tissue samples of moas, extinct flightless birds that lived in New Zealand. The DNA sequences from the moas were compared with the DNA sequences of modern species of flightless birds to determine the evolutionary relationships of this particular group of birds.

2. The researchers compared the DNA sequences of the extinct moas and modern kiwis of New Zealand to the emu and cassowary of Australia and New Guinea, the ostrich of Africa, and rheas of South America. All of the birds are flightless. By selecting these birds, the researchers could look for similarities between flightless birds over a large geographic area.

3. The DNA sequences were very similar among the different species of flightless birds. Interestingly, the sequences of the kiwis of New Zealand were more similar to those of the modern species of flightless birds found on other landmasses than they were to those of the moas found in New Zealand.

   The researchers constructed a new phylogenetic tree that suggests that kiwis are more closely related to the emu, cassowary, and ostrich than to moas. Also, based on the results of this study, the researchers suggested that New Zealand was colonized twice by ancestors of flightless birds. One ancestor gave rise to the now-extinct moas. The other ancestor gave rise to the kiwis.

## Test Yourself

1. c   2. d   3. e   4. d   5. b   6. d   7. b   8. b   9. c   10. e

## Conceptual Questions

1. The scientific name of every species has two parts, which are the genus name and the species epithet. The genus name is always capitalized, but the species name is not. Both names are italicized. An example is *Canis lupus*.

2. If neutral mutations occur at a relatively constant rate, they act as a molecular clock with which to measure evolutionary time. Genetic diversity between species that is due to neutral mutations gives an estimate of the time elapsed since the last common ancestor. A molecular clock can provide a timescale for a phylogenetic tree.

3. Morphological analysis focuses on structural features of extinct and modern species. Many traits are analyzed to obtain a comprehensive picture of how species may be related. Convergent evolution leads to similar traits that arise independently in different species as they adapt to similar environments. Convergent evolution can, therefore, cause errors if a researcher assumes that a particular trait arose only once and that all species having the trait have the same common ancestor.

# Chapter 26

## Concept Checks

**Figure 26.2** In a sedimentary rock formation, the layer at the bottom is usually the oldest.

**Figure 26.3** For this time frame, you could analyze the relative amounts of potassium-40 and argon-40, rubidium-87 and strontium-87, uranium-235 and lead-207, or uranium-238 and lead-206.

**Figure 26.9** Most animal species, including fruit flies, fishes, and humans, exhibit bilateral symmetry.

**Figure 26.16** Defining features of primates are grasping hands, forward-facing eyes to facilitate binocular vision, a large brain, digits with flat nails instead of claws, and complex social behavior including well-developed parental care.

## Core Skills: Connections

**Figure 26.7** First, the process of membrane invagination created the nuclear envelope. Second, endocytosis may have enabled an ancient archaeon to take up a bacterial cell. Over time, bacterial genes were transferred to the nucleus, which gave rise to the eukaryotic nuclear genome. An engulfed bacterial cell eventually became a mitochondrion, and an engulfed cyanobacterial cell became a chloroplast in algae and plants.

**Figure 26.13** Two key features are mammary glands and hair. Mammals also have specialized teeth, external ears, enlarged skulls that harbor highly developed brains, and four-chambered hearts. Mammals are typically endothermic.

## Test Yourself

1. c   2. e   3. a   4. d   5. e   6. c   7. b   8. a   9. d   10. d

## Conceptual Questions

1. The relative ages of fossils can be determined by their locations in a sedimentary rock formation. Older fossils are found in lower layers. A common way to determine the ages of fossils is via radiometric dating, which is often conducted on a sample of igneous rock from the vicinity of the fossil. A radioisotope is an unstable isotope of an element that decays spontaneously, releasing radiation at a constant rate. The half-life is the length of time required for a radioisotope to decay to exactly one-half of its initial quantity. To determine the age of an igneous rock (and that of a fossil found near it), scientists can measure the amounts of a given radioisotope and its decay product.

2. The process of membrane invagination led to formation of the nuclear envelope. In addition, endocytosis may have enabled an ancient archaeon to take up a bacterial cell. Over time, bacterial genes were transferred to the nucleus, which gave rise to the eukaryotic nuclear genome. An engulfed bacterial cell eventually became a mitochondrion, and an engulfed cyanobacterial cell became a chloroplast in algae and plants.

3. Several examples are described in this chapter. In some cases, catastrophic events like volcanic eruptions and glaciers caused mass extinctions, which allowed new species to evolve and flourish. In other cases, changing environmental conditions (for example, changes in temperature and moisture) played key roles. One interesting example is adaptation to terrestrial environments. Plant species evolved seeds that are desiccation resistant, whereas animal species evolved eggs with shells. Mammalian species evolved internal gestation.

# Chapter 27

## Concept Checks

**Figure 27.11** The motion of the stiff filament of a prokaryotic flagellum is more like that of a propeller shaft than the flexible arms of a human swimmer.

**Figure 27.15** Endospores allow bacterial cells to survive treatments and environmental conditions that would kill ordinary cells.

## Core Skills: Connections

**Figure 27.6** Like the bacterium *Magnetospirillum magnetotacticum*, birds such as homing pigeons and migratory fishes such as rainbow trout have the capacity to sense and respond to magnetic fields.

**Figure 27.12** The microscopic protist *Giardia intestinalis* uses flagella to move within the human small intestine.

## Feature Investigation Questions

1. Many bacteria are known to produce organic compounds that function as antibiotics, which are potential food sources for chemoheterotrophic bacteria.

2. Researchers isolated and cultivated bacteria from different types of soils, then grew the cultured bacteria on media that contained one of several common types of antibiotics as the only source of organic food.

## Test Yourself

1. c   2. b   3. c   4. d   5. a   6. a   7. b   8. e   9. d   10. d

## Conceptual Questions

1. Small cell size and simple division processes allow many bacteria to increase in number much more rapidly than eukaryotes can. This rapid population growth helps to explain why food can spoil so quickly and why infections can spread very rapidly within the body. Other factors also influence these rates.

2. Pathogen populations naturally display genetic variation in their susceptibility to antibiotics. When such populations are exposed to antibiotics, even if only a few cells are initially resistant, the number of resistant cells will eventually increase and such cells could come to dominate natural populations.

3. Humans and cyanobacteria. When humans pollute natural waters with high levels of fertilizers originating from sewage effluent or crop field runoff, cyanobacterial populations are able to grow large enough to produce harmful blooms.

# Chapter 28

## Concept Checks

*Figure 28.7*  After food particles are collected in a feeding groove, they are enclosed by membrane vesicles and then digested by enzymes.

*Figure 28.9*  The intestinal parasite *Giardia intestinalis* is transmitted from one person to another via fecal wastes, whereas the urogenital parasite *Trichomonas vaginalis* can be transmitted by sexual activity.

*Figure 28.17*  Flagellar hairs function like oars, helping to pull the cell through the water.

*Figure 28.18*  Kelps are harvested for use in the production of industrially useful materials. In addition, they nurture fishes and other wildlife of economic importance.

*Figure 28.21*  Genes that encode cell adhesion and extracellular matrix proteins are likely essential to modern choanoflagellates' ability to attach to surfaces, where they feed. Similar proteins are involved in the formation of multicellular tissues in animals. Evolutionary biologists would say that ancient choanoflagellates were preadapted for the later evolution of multicellular tissues in early animals.

*Figure 28.24*  Cysts allow protists to survive conditions that are not suitable for growth. One such condition would be the dry or cold environment outside a parasitic protist's warm, moist host tissues.

*Figure 28.29*  Gametes of *Plasmodium falciparum* undergo fusion to produce zygotes while in the mosquito host.

## Core Skills: Connections

*Figure 28.5*  In sponges, amoebocytes, which move similarly to amoeboid protists, carry food to other cells.

## Feature Investigation Questions

1. Natural growths of this alga were already known to resist microbial attack and breakdown.

2. This process is commonly used to release tough fossils from rock.

## Test Yourself

1. c  2. a  3. b  4. b  5. e  6. b  7. e  8. d  9. b  10. c

## Conceptual Questions

1. Protists are amazingly diverse, reflecting the occurrence of extensive adaptive radiation after the origin of eukaryotic cells, widespread occurrence of endosymbiosis, and adaptation to many types of moist habitats, including the tissues of animals and plants. As a result of this extensive diversity, protists cannot be classified into a single kingdom or phylum.

2. Several protists, including the apicomplexans *Cryptosporidium parvum* and *Plasmodium falciparum* and the kinetoplastids *Leishmania major* and *Trypanosoma brucei*, cause many cases of illness around the world, but few treatments are available, and organisms often evolve drug resistance. Genomic data allow researchers to identify metabolic features of these parasites that are not present in humans and are therefore good targets for development of new drugs. An example is provided by metabolic pathways of the apicoplast, a reduced plastid that is present in cells of the genus *Plasmodium*. Because the apicoplast plays essential metabolic roles in the protist but is absent from humans, drugs that disable apicoplast metabolism would kill the parasite without harming the human host.

3. Most protist cells cannot survive outside moist environments, but cysts have tough walls and dormant cytoplasm, allowing them to persist in habitats that are unfavorable for growth. While cysts play important roles in the asexual reproduction and survival of many protists, they also allow protist parasites such as *Entamoeba histolytica* (the cause of amoebic dysentery) to spread to human hosts who consume food or water that has been contaminated with cysts. Widespread contamination can cause illness or disease in thousands of people at a time.

# Chapter 29

## Concept Checks

*Figure 29.4*  Fungal hyphae growing into a substrate having a much higher solute concentration will tend to lose cell water to the substrate, a process that could inhibit fungal growth. This process explains how drying or salting foods helps to protect them from fungal degradation and thus preserves them.

*Figure 29.6*  You might filter the air entering the patient's room and limit the entry of visitors and materials that could introduce fungal spores from the outside environment.

## Core Skills: Connections

*Figure 29.7*  The *Saccharomyces cerevisiae* genome is only 12 million base pairs in size, relatively small for a eukaryote.

*Figure 29.10*  Amanitin, by interfering with the function of RNA polymerase II, inhibits transcription in eukaryotic cells, including those of humans.

## Feature Investigation Questions

1. Plants growing on soils with temperatures up to 65°C would be expected to have fungal endophytes that aid in heat stress tolerance.

2. The investigators cured some of their *Curvularia protuberata* cultures of an associated virus; then they compared the survival of plants infected with fungal endophytes that had virus versus fungal endophytes lacking virus under conditions of heat stress. Only plants infected with fungal endophytes that possessed the virus were able to survive growth on soils with a high temperature.

3. The fungus *C. protuberata* might be used to confer heat stress tolerance to crop plants, as the investigators demonstrated in tomato.

## Test Yourself

1. c  2. b  3. a  4. b  5. a  6. d  7. b  8. c  9. e  10. a

## Conceptual Questions

1. Fungi are like animals in being heterotrophic, having absorptive nutrition, and storing surplus organic compounds in their cells as glycogen. Fungi are like plants in having rigid cell walls and reproducing by means of walled spores that are dispersed by wind, water, or animals.

2. Toxic or hallucinogenic compounds likely help to protect the fungi from organisms that would consume them.

3. Many fungi degrade organic compounds, thereby contributing to decomposition, which recycles materials. Some fungi trap and consume the bodies of small animals, such as nematodes. Some fungi cause diseases of plants or animals, which also helps to control populations in nature.

# Chapter 30

## Concept Checks

*Figure 30.15*  Mycorrhizal fungi provide their plant partners with water and minerals absorbed from a much larger volume of soil than plant roots can exploit on their own.

## Feature Investigation Questions

1. Investigators fed the experimental mice germ-free food so that microbes normally found in food would not be present, allowing the effects of the fecal transplants to be more easily detected. In this way, the investigators removed a potentially confounding factor from their experiment.

2. Microbes from healthy children enabled the mice receiving them to grow larger than mice receiving microbes from stunted children. This outcome was a first clue that the microbes directly affected child growth, a connection that was established by additional studies.

## Test Yourself

1. c  2. d  3. c  4. c  5. b  6. e  7. e  8. b  9. c  10. d

## Conceptual Questions

1. Ice microbiomes include colored surface films of algae that absorb solar radiation, thereby reducing the amount reflected into space. The absorption of solar radiation has a warming effect on the physical environment, but could also influence other organisms present. Soil microbiomes contain diverse prokaryotic and eukaryotic species, including nitrogen-fixing bacteria and mycorrhizal fungi that aid plant growth.

2. One example of a host is the common, abundant nearshore green alga *Cladophora*, whose microbiota include nitrogen-fixing bacteria that aid host growth and obtain oxygen and organic molecules from the host.

3. Mycorrhizal fungi, which are components of many plant root microbiomes, mobilize and transport phosphorus and other minerals to plant roots. This improves the plant's ability to obtain minerals from the environment.

# Chapter 31

## Concept Checks

*Figure 31.5* Wind speed varies, so if the sporangium released all the spores at the same time and there were little or no wind, the spores would not travel very far and the resulting offspring might have to compete with the parent plant for scarce resources. Releasing spores gradually means that some spores may encounter strong gusts of wind that will carry them long distances, reducing competition with the parent.

*Figure 31.12* During the Carboniferous period (Coal Age), atmospheric oxygen levels reached historically high levels, enough to supply the large needs of giant insects, which obtain oxygen by diffusion.

*Figure 31.17* Larger sporophytes are able to capture more resources for use in producing larger numbers of progeny and therefore increase the fitness of ferns and seed plants.

*Figure 31.22* Although some angiosperm seeds, such as those of corn and coconut, contain abundant endosperm, many angiosperm embryos consume most or all of the nutritive endosperm during their development.

*Figure 31.24* Because the lacy integument of *R. heinzelinii* does not completely enclose the megasporangium, it probably did not function to protect the megasporangium before fertilization or act as an effective seed coat after fertilization, as do the integuments of modern seed plants. However, the lacy integument of *R. heinzelinii* might have retained the megasporangium on the parent sporophyte during the period of time when nutrients flowed from parent to developing ovule and seed. That function would prevent megasporangia from dropping off the parent plant before fertilization occurred, allow the parent plant to provide nutrients needed during embryo development, and allow seeds time to absorb and store more nutrients from the parent. Such a function would illustrate how one mutation having a positive reproductive benefit can lay the foundation for subsequent mutations that confer additional fitness. *R. heinzelinii* illustrates a first step in the multistage evolutionary process that gave rise to modern seeds.

## Core Skills: Connections

*Figure 31.23* Like the plant seed, the amniotic egg characteristic of many animals provides protection and nutrients to the developing embryo.

## Feature Investigation Questions

1. The experimental goals were to determine the rate at which organic molecules produced by gametophyte photosynthesis were able to move into sporophytes and to investigate the effect of sporophyte size on the amount of organic molecules transferred from the gametophyte.

2. The investigators shaded sporophytes with blackened glass tubing to ensure that all of the radioactive organic molecules detected in sporophytes at the end of the experiment came originally from the gametophytes.

3. The investigators measured the amount of radioactivity in gametophytes and sporophytes, and in sporophytes of different sizes. These measurements indicated the relative amounts of labeled organic compounds that were present in different plant tissues.

## Test Yourself

1. c   2. d   3. d   4. e   5. b   6. a   7. c   8. e   9. c   10. b

## Conceptual Questions

1. Streptophyte algae, particularly the complex genera *Chara* and *Coleochaete*, share many features of structure, reproduction, and biochemistry with land plants. Examples include cell division similarities, plasmodesmata, and sexual reproduction by means of flagellate sperm and eggs.

2. Bryophytes are well adapted for sexual reproduction when water is available for fertilization. Their green gametophytes efficiently transfer nutrients to developing embryos, enhancing their growth into sporophytes. Their sporophytes are able to produce many genetically diverse spores as the result of meiosis and effectively disperse these spores by means of wind.

3. Vascular tissues allow tracheophytes to effectively conduct water from roots to stems and to leaves. Waxy cuticle helps prevent loss of water by evaporation through plant surfaces. Stomata allow plants to achieve gas exchange under moist conditions and help them avoid losing excess water under arid conditions.

# Chapter 32

## Concept Checks

*Figure 32.4* The nitrogen-fixing cyanobacteria that often occur within the coralloid roots of cycads are photosynthetic organisms that require light. If coralloid roots grew underground, the cyanobacteria would not receive enough light to survive.

*Figure 32.10* Ways in which conifer leaves are adapted to resist water loss include low surface area/volume ratio, needle or scale shape, thick surface coating of waxy cuticle, and stomata that are recessed and are therefore less exposed to drying winds.

*Figure 32.12* Relatively wide vessels are commonly present in the water transport tissues of angiosperms and much less commonly in other plants. The vessels occasionally found in nonangiosperms are thought to have evolved independently from those of angiosperms.

*Figure 32.20* A large, showy perianth would not be useful to grass plants because they are wind pollinated; such a perianth would interfere with pollination in grasses. By not producing a showy perianth, grasses increase the chances of successful pollination and save resources that would otherwise be consumed during perianth development.

*Figure 32.24* The flower characteristics of *Brighamia insignis* shown in this figure (white color and deep, narrow nectar tubes) are consistent with pollination by a moth (see Table 32.2).

## Core Skills: Connections

*Figure 32.2* Modern forests are dominated by seed plants, gymnosperms and angiosperms, whereas nonseed plants dominated *Archaeopteris* forests.

*Figure 32.8* The wind-dispersed seeds of the gymnosperm pine resemble the wind-dispersed fruits of the angiosperm maple in having winglike structures that enhance transport in air.

## Feature Investigation Questions

1. The investigators obtained many samples from around the world because they wanted to increase their chances of finding as many species as possible.

2. Although cannabinoids are produced in glandular hairs that cover the plant surface, these compounds are most abundant on leaves near the flowers. Collecting such leaves reduces the chances that compounds might be missed by the analysis.

## Test Yourself

1. d   2. a   3. e   4. e   5. b   6. d   7. e   8. c   9. d   10. e

## Conceptual Questions

1. Refer to Figure 30.15 to see how plant biologists think stamens and pistils might have evolved from leaves that bore sporangia. Then consider how green leaves surrounding stamens and pistils might have been transformed into petals, sepals, or tepals.

2. Apple, strawberry, and cherry plants coevolved with animals that use the fleshy, sweet portion of the fruits as food and excrete the seeds, thereby dispersing them. Humans have sensory systems similar to those of the animal seed-dispersal agents and likewise are attracted by the same colors, odors, and tastes.

3. Unlike an apple flower, a sunflower is not a single flower, but rather is an inflorescence, a group of flowers. A pollinator visiting a compact inflorescence such as a sunflower head may be able to pollinate many flowers at the same time, thereby potentially producing many seeds per pollinator visit. Likewise, a seed disperser agent might be able to disperse more seeds per plant.

# Chapter 33

## Core Skills: Connections

*Figure 33.5* A shared derived character.

*Figure 33.10* Yellow.

## Feature Investigation Questions

1. The researchers sequenced the complete gene that encodes small subunit rRNA from a variety of representative phyla of animals to determine their phylogenetic relationships, particularly the relationships of arthropods to other phyla.

2. The results indicated a monophyletic clade containing arthropods and nematodes. This clade was called the Ecdysozoa. The results of this study indicated that nematodes are more closely related to arthropods than was previously believed.

3. The fruit fly, *Drosophila melanogaster*, and the nematode, *Caenorhabditis elegans*, have been widely studied to understand early development. Under the traditional phylogeny, these two species were not considered to be closely related, so similarities in development were assumed to have arisen early in animal evolution. With the closer relationship indicated by this study, these similarities may have evolved after the divergence of the Ecdysozoan clade. This puts into question the applicability of studies of these organisms to the understanding of human biology.

## Test Yourself

1. b    2. c    3. e    4. c    5. c    6. d    7. d    8. d    9. b    10. e

## Conceptual Questions

1. See Figure below.

2. The coelom cushioned the internal organs in fluid, preventing injury from external forces. In addition, the coelom enabled the internal organs to grow and move independently of the outer body wall. Finally, in some invertebrates, the coelom acts as a hydrostatic skeleton that supports the body and permits movement.

3. They are paraphyletic since the group contains a common ancestor but not all of its descendants.

# Chapter 34

## Concept Checks

***Figure 34.3*** Sponges aren't eaten by other organisms because they produce toxic chemicals and contain needle-like silica spicules that are hard to digest.

***Figure 34.4*** The dominant life stages are jellyfish: medusa; sea anemone: polyp; Portuguese man-of-war: polyp (in a large floating colony).

***Figure 34.5*** Cnidocytes are not reused. New ones form to replace the discharged ones.

***Figure 34.6*** Having no specialized respiratory or circulatory system, flatworms obtain oxygen by diffusion. A flattened shape ensures that no cells are too far from the body surface.

***Figure 34.10*** The lophophore functions as (1) a ciliary feeding device and (2) a respiratory organ.

***Figure 34.11*** Technically, the hearts of most mollusks pump hemolymph into vessels and then into tissues. The hemolymph collects in open, fluid-filled cavities called sinuses, which flow into the gills and then back to the heart. This is known as an open circulatory system. Only closed circulatory systems pump blood, as occurs in the cephalopods.

***Figure 34.15*** Some advantages of segmentation are organ duplication, minimization of body distortion during movement, and specialization of some segments.

***Figure 34.17*** An annelid is segmented and possesses a true coelom, whereas a nematode is unsegmented and has a pseudocoelom. In addition, nematodes molt, but annelids do not.

***Figure 34.21*** All arachnids have a body consisting of two tagmata: a cephalothorax and an abdomen. Insects have three tagmata: head, thorax, and abdomen.

***Figure 34.24*** Two other key insect adaptations are the development of wings and an exoskeleton that reduced water loss and aided in the colonization of land.

***Figure 34.30*** In embryonic development, deuterostomes show radial cleavage and indeterminate cleavage, and the blastopore becomes the anus. (In protostomes, cleavage is spiral and determinate, and the blastopore becomes the mouth.)

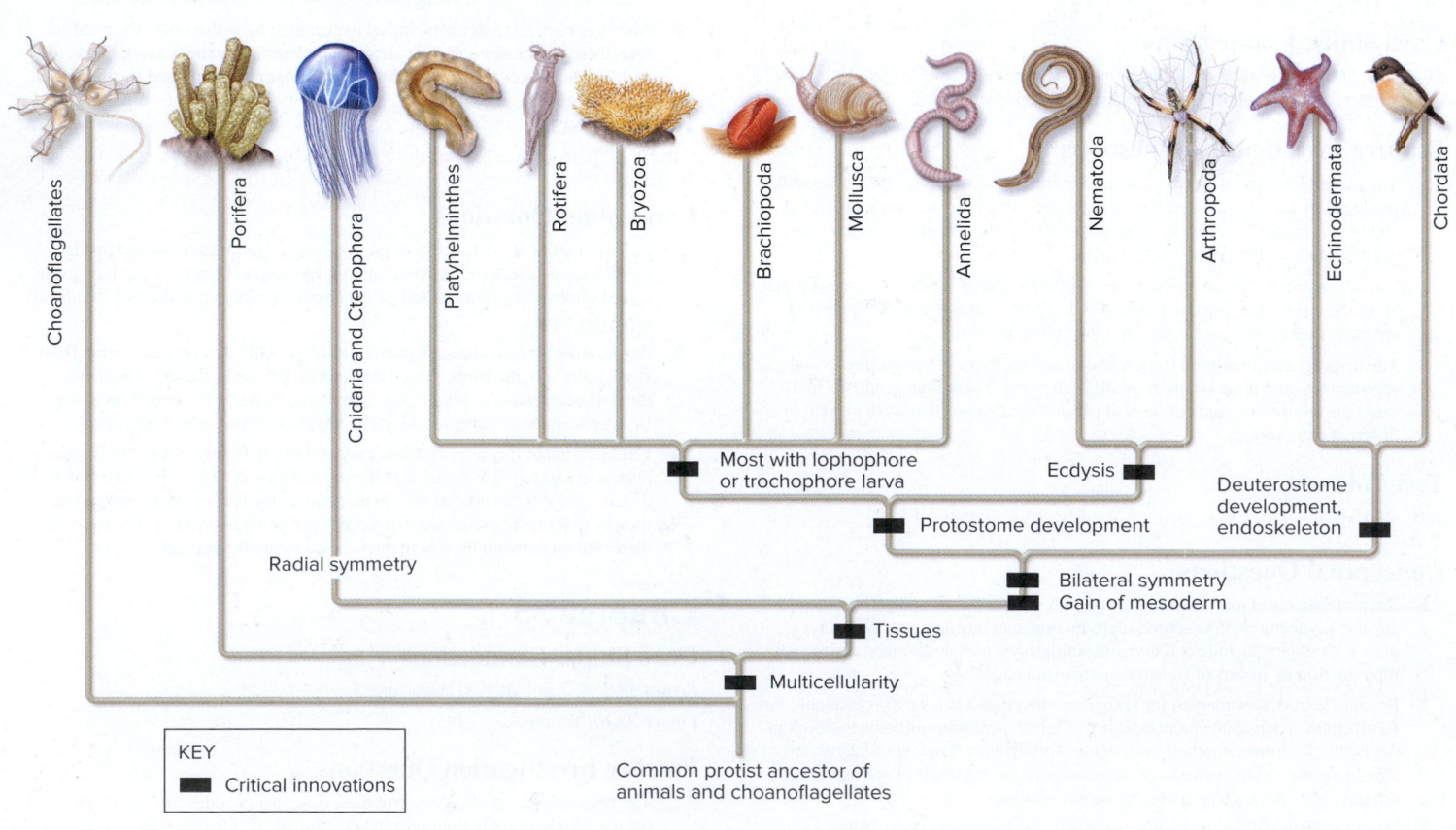

KEY

■ Critical innovations

## Core Skills: Connections

*Figure 34.19* Because most species can excrete urine that is isoosmotic or hyperosmotic to the body fluids.

*Figure 34.27* These organs, called statocysts, are located at the base of the antennules.

## Feature Investigation Questions

1. The researchers tested the hypothesis that an octopus can learn by observing the behavior of another octopus.

2. The results indicated that the observer learned by watching the training of the other octopus. The observer was much more likely to choose the same color ball that the demonstrator was trained to attack. These results support the hypothesis that octopuses can learn by observing the behavior of others.

3. The untrained octopuses had no prior exposure to the demonstrators. The results indicated that these octopuses were as likely to attack the white ball as the red ball. No preference for either color was indicated. The untrained octopuses acted as a control. This is an important factor to ensure that the results from the trials using observers indicate a response to learning and not an existing preference for a certain color.

## Test Yourself

1. b   2. d   3. d   4. d   5. b   6. c   7. b   8. a   9. c   10. b

## Conceptual Questions

1. The five main feeding methods used by invertebrate animals are (1) suspension feeding, (2) decomposition, (3) herbivory, (4) predation, and (5) parasitism. Suspension feeding is usually used to filter out food particles from the water. A great many phyla, including sponges, rotifers, bryozoans, brachiopods, some mollusks, some echinoderms and tunicates, are filter feeders. Decomposers usually feed on dead material such as animal carcasses or dead leaves. For example, many fly and beetle larvae feed on dead animals, and earthworms consume dead leaves from the surface of the Earth. Earthworms and crabs also sift through soil or mud, eating the substrate and digesting the soil-dwelling bacteria, protists, and dead organic material. Herbivores eat plants or algae and are especially common in the arthropoda. Adult moths and butterflies also consume nectar. Snails are also common plant feeders. Predators feed on other animals, killing their prey, and may be active hunters or sit-and-wait predators. Many scorpions and spiders actively pursue their prey, whereas web-spinning spiders ambush their prey using webs. Parasites also feed on other animals but do not normally kill their hosts. Endoparasites, which includes flukes, tapeworms, and nematodes, live inside their hosts. Ectoparasites (ticks and lice) live on the outside of their hosts.

2. Gametes would dry out on land, and internal fertilization prevents this from happening. Also, water facilitates the movement of gametes, reducing the need for internal fertilization.

3. Complete metamorphosis has four stages: egg, larva, pupa, and adult. The larval stage is often spent in an entirely different habitat from that of the adult, and larval and adult forms utilize different food sources. Incomplete metamorphosis has only three stages: egg, nymph, and adult. Young insects, called nymphs, look like miniature adults when they hatch from their eggs.

# Chapter 35

## Concept Checks

*Figure 35.1* Vertebrates (but not invertebrates) usually possess a (1) notochord; (2) dorsal hollow nerve chord; (3) pharyngeal slits; (4) postanal tail, exhibited by all chordates; (5) vertebral column; (6) cranium; and (7) endoskeleton of cartilage or bone.

*Figure 35.7* Ray-finned fishes (but not sharks) have a (1) bony skeleton; (2) mucus-covered skin; (3) swim bladder; and (4) operculum covering the gills.

*Figure 35.8* Both lungfishes and coelocanths are Sarcopterygii, having lobe fins.

*Figure 35.11* The advantages for animals that moved onto land included an oxygen-rich environment and a bonanza of food in the form of terrestrial plants and the insects that fed on them.

*Figure 35.13* No. Caecilians and some salamanders give birth to live young.

*Figure 35.14* Besides the amniotic egg, other critical innovations in amniotes are thoracic breathing; internal fertilization; a thicker, less permeable skin; and more efficient kidneys.

*Figure 35.16* Snakes evolved from tetrapod ancestors but subsequently lost their limbs. Some species have tiny vestigial limbs.

*Figure 35.20* Adaptations in birds to reduce body weight for flight include a lightweight skull, reduction of organ size (including smaller gonads outside of breeding season), and a lack of a urinary bladder and teeth. Also female birds have one ovary and can carry relatively few eggs.

## Core Skills: Connections

*Figure 35.5* No, an examination of Figure 35.1 shows that "fishy" organisms include an ancestral population and some of its descendants, but not all of them. The "fish" are thus a paraphyletic group. To be a monophyletic group, the fish would have to include the tetrapods as well. Therefore, there is no true monophyletic group that corresponds to the popular name "fish."

*Figure 35.17* Both crocodilians and birds and mammals have four-chambered hearts and care for their young.

*Figure 35.26* None. The bloodstreams of fetus and mother are brought into close contact in the placenta, but they do not mix.

## Feature Investigation Questions

1. The researchers were interested in determining the role of *Hox* genes in controlling limb development in mice.

2. The researchers bred mice that were homozygous for certain mutations in specific *Hox* genes. This allowed the researchers to determine the function of individual genes.

3. The researchers found that homozygous mutants developed limbs of shorter length compared to limbs of the wild-type mice. The reduced length was due to the lack of development of particular bones in the limb, specifically, the radius, ulna, and some carpels. These results indicated that simple mutations in a few genes could lead to dramatic changes in limb development.

## Test Yourself

1. b   2. e   3. d   4. e   5. a   6. d   7. d   8. c   9. c   10. a

## Conceptual Questions

1. Both vertebrates and arthropods have external limbs that move when the attached muscles contract or relax. The difference is that arthropods have external skeletons, with the muscles attached internally, whereas vertebrates have internal skeletons with the muscles attached externally.

2. Endothermy (warm-bloodedness) probably evolved independently in both birds and mammals. If the common ancestor of reptiles and birds were endothermic, the chances are that all reptiles would be endothermic.

3. Probably. Both birds and reptiles lay amniotic eggs and possess scales, though these only cover the legs in birds. Birds and crocodilians also share a four-chambered heart. Finally, birds share many skeletal similarities with certain dinosaurs.

# Chapter 36

## Concept Checks

*Figure 36.8* Having stomata located on the darker and cooler lower epidermis helps reduce water loss from the leaf.

*Figure 36.12* A twig having five sets of bud scale scars is likely to be approximately 6 years old.

*Figure 36.17* Cactus stems are green and photosynthetic; they play the role served by the leaves of most plants and make the organic compounds.

*Figure 36.21* A woody stem such as a tree trunk builds up a thicker layer of wood than inner bark in part because older tracheids and vessel element walls are not lost during shedding of bark, which is the case for secondary phloem. In addition, plants typically produce a greater volume of xylem than phloem tissue per year, in part because vessel elements are relatively wide. A large volume of water-conducting tissue helps plants maintain a large amount of internal water.

*Figure 36.24* Lateral roots are produced from internal meristematic tissue because roots do not produce axillary buds like those from which shoot branches develop. Internal production of branch roots helps to prevent them from shearing off as the root tip grows through abrasive soil.

## Core Skills: Connections

*Figure 36.5* Apical-basal polarity of the plant body resembles anterior-posterior polarity in the animal body.

*Figure 36.9* Plant cells often possess a large vacuole, whereas vacuoles in animal cells are relatively small.

*Figure 36.11* There is no difference among eukaryotes with respect to the structure of microtubules.

## Feature Investigation Questions

1. The advantages of using natural plants include the opportunity to avoid influencing plants with unnatural environmental factors, such as artificial light, and the exposure of all experimental plants to similar growth conditions. In addition, the investigators studied the leaves of some large trees, which would be hard to accommodate in a greenhouse.

2. Pinnately veined leaves were splinted to prevent their breaking, since they were cut at the single main vein, which has both support and conducting functions.

3. The researchers measured leaf water conduction at two or more places on each leaf because the effect of cutting a vein might have affected some portions of leaves more than others.

## Test Yourself

1. d  2. c  3. b  4. a  5. c  6. a  7. d  8. e  9. b  10. e

## Conceptual Questions

1. If overall plant architecture were bilaterally symmetrical, plants would be shaped like higher animals, with a distinct front (ventral surface) and back (dorsal surface). Bilaterally symmetrical plants would have a reduced ability to deploy branches and leaves in a way that would fill available lighted space and would thus be unable to take optimal photosynthetic advantage of their habitats.

2. If leaves were generally radially symmetrical (shaped like spheres or cylinders), they would not have the maximal ability to absorb sunlight, and they would not be able to optimally disperse excess heat from their surfaces.

3. Although tall herbaceous plants exist (palms and bamboo are examples), the additional support and water-conducting capacity that are provided by secondary xylem allow woody plants to grow tall.

# Chapter 37

## Concept Checks

*Figure 37.4* Auxin efflux carriers could be located on the upper sides of root cells, thereby allowing auxin to move upward in roots.

*Figure 37.6* Once a callus has been formed from a single plant having desirable characteristics using plant tissue culture, the callus can be divided into many small calluses. A grower can transfer these to separate containers having the appropriate hormone mixtures to induce root and shoot growth, then transplant the young plant clones to soil. In this way, the grower can produce many identical plants.

*Figure 37.9* The triple response of a seedling to internally produced ethylene serves to protect the delicate apical meristem from damage as the seedling pushes upward through the soil.

*Figure 37.11* The inactive conformation of phytochrome would absorb the red light in the sunlight, thereby converting the molecule into the active conformation.

*Figure 37.12* Exposing plants to brief periods of darkness during the daytime will have no effect on flowering because flowering is determined by night length.

*Figure 37.14* Yes, just as shoots exhibit negative gravitropism in upward growth, roots are capable of using negative phototropism to grow downward, because light decreases with depth in the soil.

*Figure 37.19* The immunity effect is relatively long-lasting in both cases.

## Feature Investigation Questions

1. Hypothetically, auxin enhances the rate at which cell membrane proton pumps acidify the plant cell wall, thereby allowing cells to extend. Although the evidence for acid effects on cell-wall extension is strong, the molecular basis of possible auxin effects on proton pumps is not as yet clear.

2. A small number of seedling tips could display atypical responses for a variety of reasons. The investigators actually performed the experiment with many replicate seedling tips (coleoptiles), in order to gain confidence that the responses are general.

## Test Yourself

1. c  2. c  3. a  4. e  5. d  6. d  7. c  8. d  9. d  10. d

---

**BioTIPS page 762**

**HYPOTHESIS** Light destroys auxin on lit side of shoot tips, causing unequal auxin distribution. Unlit side should grow more than lit side.

**STARTING MATERIALS** Corn seedlings.

| | Experimental level | Conceptual level |
|---|---|---|
| **1** Collect auxin into agar blocks from:<br>A  dark-grown tips<br>B  tips grown with directional light | Dark-grown tip / Auxin diffusion / Directional light-grown tip 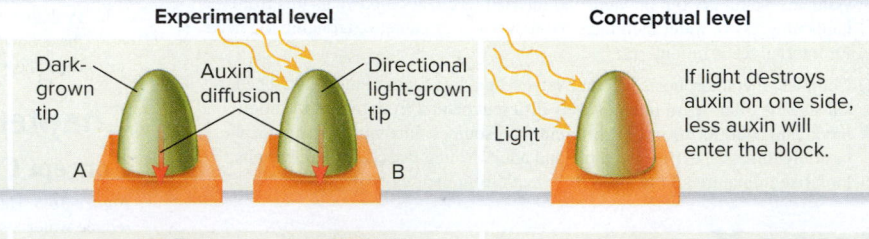 | If light destroys auxin on one side, less auxin will enter the block. Light |
| **2** Place agar blocks on right side of decapitated shoots. | Agar block — Dark-grown  Shoot  Light-grown 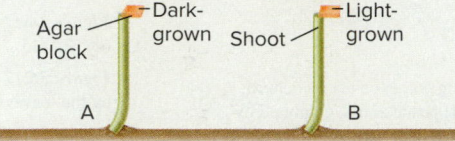 | If the block on the right side has less auxin, it will cause less bending. |

**3  THE DATA**

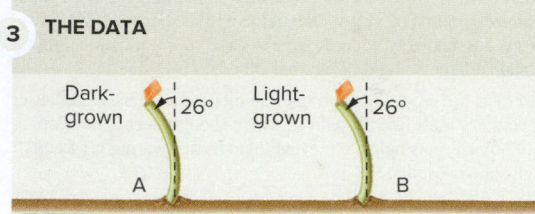
Dark-grown  26°  Light-grown  26°

A                B

**4  CONCLUSION**

Similar bending demonstrates that light did not destroy auxin in the directionally lit shoot tip. If it had, less auxin would have been present in the agar block in B, and the degree of bending would have been less. The hypothesis described under conceptual level (above) is incorrect.

## Conceptual Questions

1. Behavior is defined as the response of a living thing to a stimulus. Therefore, because plants display many kinds of responses to diverse stimuli, they display behavior.

2. Many kinds of disease-causing bacteria and fungi occur in nature, and these organisms evolve very quickly, producing diverse elicitors. Thus, plants must maintain various types of resistance genes, each having many alleles.

3. Talking implies a conversation with "listeners" who detect a message and respond to it. Thus, plants that exude volatile compounds that attract enemies of herbivores could be interpreted as "talking" to those enemies. The message is "Hey, you guys, there's food for you over here." In addition, research has revealed that some plants near those under attack respond to volatile compounds by building up defenses. "Talking" to other plants does not enhance the "talker's" fitness. But the ability to "listen" enhances the "listener's" fitness, because it can take preemptive actions to prevent attack.

# Chapter 38

## Concept Checks

**Figure 38.5**  Plastids that are clustered near the nucleus should have more rubisco than plastids at the periphery.

**Figure 38.6**  Chlorosis is not always a sign of iron deficiency; it can be a symptom of a deficiency of any of several mineral nutrients, including zinc in corn.

**Figure 38.10**  Mineral leaching occurs more readily in sandy soils than in clay soils.

**Figure 38.12**  Soil crusts containing nitrogen-fixing cyanobacteria increase soil fertility, fostering the growth of larger plants that stabilize soils against erosion and provide forage for animals.

**Figure 38.13**  Oxygen, which makes up 21% of Earth's present atmosphere, can bind to the active site of nitrogenase, thereby inactivating it.

## Core Skills: Connections

**Figure 38.20**  Both the tapeworm and the dodder obtain organic food from a host, an animal or a plant, respectively.

## Feature Investigation Questions

1. *SDQ1* expression is induced by phosphorus deficiency. This gene fosters replacement of plastid phospholipids with sulfur-containing lipids, thereby reducing the plant's phosphorus requirement.

2. They used genetic engineering techniques to place a reporter gene under the control of the *SDQ1* promoter, so that when *SDQ1* was expressed, the reporter gene was expressed also. After growing plants in nutrient solutions containing various levels of phosphorus, they removed sample leaves and treated them with a compound that turns blue when the reporter gene is expressed. When they saw blue leaves, the investigators could infer (1) that the plants from which those leaves had been taken were beginning to experience phosphorus deficiency and (2) that application of fertilizer at this point could prevent damage to the plants.

## Test Yourself

1. e   2. a   3. c   4. b   5. e   6. c   7. d   8. d   9. a   10. b

## Conceptual Questions

1. Agricultural experts are concerned that adding excess fertilizer to crop fields increases the costs of crop production. Ecologists are concerned that excess fertilizers will wash from crop fields into natural waters and cause harmful overgrowths of cyanobacteria, algae, and aquatic plants. Methods for closely monitoring crop nutrient needs so that only the appropriate amount of fertilizer is applied would help to allay both groups' concerns.

2. Use Figure 38.18 as a reference. A first arrow could be drawn from a root to rhizobia in the soil, and the arrow labeled "flavonoids." A second arrow could be drawn from rhizobia to roots and labeled "Nod factors." A third arrow from rhizobia to roots could be labeled "infection proteins." A fourth arrow from roots to rhizobia could be labeled "nodulins" and the resulting nodule environmental conditions, which influence the formation of bacterioids. A fifth arrow could represent the flow of fixed nitrogen from bacterioids to plant. A sixth arrow could represent the flow of organic compounds from plant to bacteroids.

3. Tropical plants often partner with microbes that provide essential mineral nutrients. Such partnerships include nitrogen-fixing bacteria that may live within plant roots and/or mycorrhizal fungi that harvest minerals from decaying organisms and transport the minerals to plant root tissues.

# Chapter 39

## Concept Checks

**Figure 39.5**  When placed in pure water, a turgid cell having a water potential of 1.0 will lose water, because 1 is greater than 0. When placed in pure water, a plasmolyzed cell having a water potential of –1.0 MPa will gain water. When placed in pure water, a flaccid cell having a water potential of –0.5 MPa will gain water. This is because water moves from a region of higher water potential to a region of lower water potential, and 0 is greater than –0.5.

**Figure 39.11**  You would likely see stained rings or helical ribbons extending up the insides of the walls of extensible tracheids. You would not see staining at the ends of tracheids, where they connect to form long tubes.

**Figure 39.12**  The large perforations in vessel element end walls allow an air bubble to extend from one element to another, thereby clogging a vessel and preventing water flow through it. In contrast, the much smaller pores in the end walls of tracheids do not allow water to flow as efficiently as it does through vessels, but these smaller pores also retard the movement of air bubbles. As a result, an air bubble will be confined to a single tracheid where it can do little harm.

**Figure 39.13**  Root pressure can help to reverse an embolism, thereby restoring water flow through the plant xylem.

**Figure 39.15**  The evaporation of water has a powerful cooling effect because it dissipates heat so effectively. Water has the highest heat of vaporization of any known liquid.

**Figure 39.16**  You could model a stomatal guard cell with an elongate balloon by partially inflating it, then attaching thick tape along one side to represent thickened inner walls and circles of string or thin tape to represent radial cellulose, then adding more air to the balloon. The balloon should curve as it expands, just as a guard cell does when the stomatal pore opens. Two such balloons could be used to model both guard cells and the stomatal pore.

**Figure 39.19**  In the ocotillo's desert habitat, times of drought and availability of water sufficient to support the development and photosynthetic function of leaves do not occur as predictably as in temperate forests. For this reason, ocotillo leaf abscission is not amenable to the evolution of genetic mechanisms that allow leaf drop to be precisely timed in anticipation of the onset of drought.

## Core Skills: Connections

**Figure 39.8**  Tight junctions in the intestinal epithelium of animals and Casparian strips in walls of endodermal cells of plant roots both form a tight seal, preventing movement of materials from one location to another.

**Figure 39.21**  Stomatal guard cell and phloem companion cell development both begin with an unequal cell division.

## Feature Investigation Questions

1. Replicates provide reproducibility, and controlled growth conditions reduce the effects of environment.

2. Leaf temperature indicates degree of evaporation through open stomata.

## Test Yourself

1. c   2. b   3. e   4. d   5. a   6. e   7. d   8. e   9. a   10. c

## Conceptual Questions

1. In the case of plant fertilizers, more is not better, because the ion concentration of overfertilized soil may become so high as to draw water from plant cells. In this case, the cells would be bathed in a hypertonic solution and would likely lose water to the solution. If plant cells lose too much water, they will die.

2. When the natural vegetation is removed, transpiration stops, so water is not transported from the ground to the atmosphere, where it may be an important contributor to local rainfall. Extensive removal of plants actually changes local climates in ways that reduce agricultural productivity and human survival.

3. When mineral ion concentrations in the soil are lower than those within root cells, root-hair plasma membranes take up nutrient ions by active transport, which requires energy in the form of ATP. So the plant rooted in soil of lower mineral concentration will likely expend more ATP during nutrient uptake.

# Chapter 40

## Concept Checks

*Figure 40.3*  Some flowers lack some of the major flower parts.

*Figure 40.4*  By having its stamens clustered around the pistil, the hibiscus flower increases the chance that a pollinator will both pick up pollen and deliver pollen from another hibiscus flower in the same visit.

*Figure 40.7*  The absence of showy petals often correlates with wind pollination, because large petals would interfere with the shedding of pollen in the wind.

*Figure 40.10*  The rim flowers of *Gerbera* inflorescences have bilateral symmetry, conferred by expression of a *CYCLOIDEA*-like gene. Rim flowers also possess showy petals that attract pollinators, but lack pollen-producing stamens. By contrast, central flowers display radial symmetry, lack showy petals, and possess pollen-producing stamens.

*Figure 40.12*  The maximum number of cells in a mature male gametophyte of a flowering plant is three: a tube cell and two sperm cells.

*Figure 40.14*  Female gametophytes are not photosynthetic and cannot produce their own organic food. Enclosed within ovules, female gametophytes lack direct access to the outside environment. Carpels contain veins of vascular tissue that bring nutrients from sporophytic tissue to ovules.

*Figure 40.16*  An embryo in which the TOPLESS genes were nonfunctional would have two roots and no shoots.

*Figure 40.17*  During their maturation, the cotyledons of eudicot seeds absorb the nutrients originally present in the endosperm.

## Core Skills: Connections

*Figure 40.1*  The gametophyte generation of a flowering plant is small, non-photosynthetic, and haploid.

*Figure 40.15*  The plant pollen tube is analogous in function to a human penis in that both structures accomplish internal fertilization. The plant pollen tube grows long enough to deposit sperm at the micropyle within the body of the female gametophyte, much as a male animal's penis deposits sperm within the female's body. In both cases, sperm are more likely to survive and accomplish fertilization than if they were deposited outside the female body.

## Feature Investigation Questions

1. The large flowers of this lily enabled investigators to more easily mark petals and record the positions of marks over time.

2. Time-lapse video reduced the amount of time investigators would have to spend recording changes in the positions of petal marks.

## Test Yourself

1. a   2. b   3. d   4. b   5. d   6. e   7. b   8. c   9. d   10. c

## Conceptual Questions

1. Pollen grains are vulnerable to mechanical damage and microbial attack during the journey through the air from the anthers of a flower to a stigma. Sporopollenin is an extremely tough polymer that helps to protect pollen cells from these dangers.

2. The embryos within seeds are vulnerable to mechanical damage and microbial attack after they are dispersed. Seed coats protect embryos from these dangers and also help to prevent seeds from germinating until conditions are favorable for seedling survival and growth.

3. Flower diversity is an evolutionary response to diverse pollination circumstances. For example, plants such as oak and corn that are wind-pollinated produce flowers having a poorly developed perianth. If the flowers of such plants had large, showy perianths, they would get in the way of pollen dispersal or acquisition. On the other hand, flowers that are pollinated by animals often have diverse shapes and attractive petals of differing colors or fragrances that have coevolved with different types of animal pollinators.

# Chapter 41

## Concept Checks

*Figure 41.8*  Surface area is important to any living organism that needs to exchange materials with the environment. A good example of a high surface area/volume ratio is that of most tree leaves. This high ratio makes leaves ideally suited for such processes as light absorption (required for photosynthesis; see Chapter 8) and the exchange of gases and water with the environment.

*Figure 41.13*  In nature, an animal such as a horse would have the same type of response shown here if threatened by a predator. Upon sensing the presence of the predator, the horse's respiratory and circulatory systems would begin increasing their activities in preparation for the possibility that the horse might have to flee or defend itself. The increase would occur even before the horse began to flee.

*Figure 41.14*  Blood, including plasma and blood cells, would leak out of the blood vessel into the interstitial space. The fluid level of the bloodstream would decrease, and that of the interstitial space near the site of the injury would increase. Eventually the blood that entered the interstitial space would be degraded by enzymes, resulting in the characteristic skin appearance of a bruise. If the injury were very severe, the fluid level in the blood could decrease to a point where the various tissues and organs of the body would not receive sufficient nutrients and oxygen to function normally.

*Figure 41.15*  A decrease in intracellular fluid volume, like that shown in the middle panel in Figure 41.15, will result in an increase in intracellular solute concentration (likewise, an increase in intracellular fluid volume will decrease intracellular solute concentrations). Such a change may have drastic consequences for cell function. For example, some solutes, like $Ca^{2+}$ and certain other ions, are toxic to cells at high concentrations.

*Figure 41.16*  No, obligatory exchanges must always occur, but animals can minimize obligatory losses through modifications in behavior. For example, terrestrial animals that seek shade on a hot, sunny day reduce evaporative water loss. As another example, reducing activity minimizes water loss due to respiration.

*Figure 41.18*  Humans cannot survive by drinking seawater because we do not possess specialized salt glands to rid ourselves of the excess sodium and other ions we would ingest with the seawater. The human kidneys cannot eliminate that much salt. The high blood concentrations of sodium and other ions would cause changes in cellular membrane potentials, disrupting vital functions of electrically excitable tissue such as cardiac muscle and nerve tissue.

## Core Skills: Connections

*Figure 41.6*  Note that both the stomach and the small intestine depicted in Figures 41.6 and 46.6 contain layers of muscle wrapped around a lumen. Although you will learn later that the stomach and intestine have many different functions, this similarity in anatomy suggests that both of these organs may perform the functions of mechanically breaking apart chunks of food and propelling the food from one region to another.

*Figure 41.11*  Negative feedback may occur at the molecular level. It is a common feature of some enzymatic pathways, such as the metabolic one depicted in Figure 6.13.

## Feature Investigation Questions

1. Symptoms of prolonged, heavy exercise include fatigue, muscle cramps, and even occasionally seizures. Fatigue results from the reduction in blood flow to muscles and other organs. Muscle cramps and seizures are the results of imbalances in plasma ion concentrations. Cade and his colleagues hypothesized that maintaining a normal balance of water and ions would prevent these problems from occurring, and that if water and ion concentrations were maintained, athletic performance should not decrease as rapidly with prolonged exercise.

2. To test their hypothesis, the researchers created a drink that would restore the correct proportions of water and ions lost by the athletes during exercise. If the athletes consumed the drink, they should not experience as much fatigue or muscle cramping, and thus their performance should be enhanced compared with that of a control group of athletes that drank only water.

3. Gatorade and many similar sports drinks are beneficial in replacing the ions and water lost in perspiration during exercise. These drinks also typically contain sugar that is meant to provide energy for someone who is being very active. Given that a person at rest or one who is exercising lightly would not be expected to perspire very much, a logical hypothesis is that sports drinks are not beneficial in such cases. A small degree of perspiration will not create a significant ion imbalance. Also, as a rule, consuming sugary drinks should be avoided because of their links to weight gain and dental disease. Thus, to a person at rest or engaging in only light exercise, the benefits of such drinks are minimal and may be outweighed by their potential negative effects.

## Test Yourself

1. c   2. c   3. d   4. b   5. c   6. c   7. d   8. b   9. e   10. e

## Conceptual Questions

1. Structure and function are related in that the function of a given organ, for example, depends in part on the organ's size, shape, and cellular and tissue arrangement. Clues about a physical structure's function can often be obtained by examining the structure's form. For example, the extensive surface area of a moth's antennae suggests that the antennae are important in detecting the presence of airborne

chemicals. Likewise, any structure that contains a large surface area for its volume is likely involved in some aspect of signal detection, cell-cell communication, or transport of materials within the animal or between the animal and the environment. Surface area increases by a power of 2, and volume increases by a power of 3 as an object enlarges; this means that in order to greatly increase surface area of a structure such as an antenna, without occupying enormous volumes, specializations must be present (such as folds) to package the structure in a small space.

2. Homeostasis is the ability of animals to maintain a stable internal environment by adjusting physiological processes, despite changes in the external environment. Examples include maintenance of ion and water balance, pH of body fluids, and body temperature. Some animals conform to their external environment to achieve homeostasis, but others regulate their internal environment themselves.

3. Maintaining homeostasis requires continual supplies of energy. Animals consume food, and the energy from that food helps sustain activities that maintain physiological variables such as body temperature, water and ion homeostasis, and synthesis of complex molecules. Without this energy, it would be difficult or impossible for animals to maintain many important biological processes within a narrow range despite changes in the environment.

# Chapter 42

## Concept Checks

*Figure 42.4* Many reflexes, such as the knee-jerk reflex, cannot be prevented once started. Others, however, can be controlled to an extent. Open your eyes widely and gently touch your eyelashes. A reflex that protects your eye will tend to make you close your eyelid. However, you can overcome this reflex with a bit of difficulty if you need to, for example, when you are putting in contact lenses.

*Figure 42.9* When the $K^+$ channels opened (at 1 msec), the $Na^+$ channels would still be opened, so the part of the curve that slopes downward would not occur as rapidly, and perhaps the cell would not be able to restore its resting potential.

*Figure 42.11* Saltatory conduction allows an action potential to move faster down an axon. This effect is especially important for long axons, such as those that carry signals from the spinal cord to distant muscles.

## Core Skills: Connections

*Figure 42.6* Electrochemical ion gradients function in neuronal signaling, ATP synthesis, muscle contraction, osmotic regulation, and transport of organic molecules across membranes.

*Figure 42.7* Water molecules move through membrane channels called aquaporins.

## Feature Investigation Questions

1. Loewi was aware that electrical stimulation of the vagus nerve associated with heart muscle slowed down the rate of heart contractions in a frog. Also, he knew that electrical stimulation of other nerves associated with the frog heart produced the opposite result. If the effects of the different nerves on heart muscle were mediated directly by electrical activity only, the heart muscle cells would have no way to distinguish between stimulatory and inhibitory signals. Loewi hypothesized that nerves released chemicals onto heart muscle cells and that it was these different chemicals that produced the varied effects on the heart.

2. The results could also have been explained as a chemical having been released into the bathing medium from heart 1. The stimulated vagus nerve might have caused the muscle cells to release a chemical that was then capable of inhibiting heart 2. This may have been unlikely, but Loewi's experiment did not rule it out. A control experiment might have been to electrically stimulate heart 1 directly, by applying current to the muscle itself and not indirectly via the vagus nerve. If Loewi's hypothesis was correct, this control experiment would result in no effect on heart 2.

3. If the vagus nerve released acetylcholine, and it was this neurotransmitter that slowed the rate of beating of heart 2 in Loewi's experiment, then simply adding acetylcholine to the medium of a resting heart should produce the same effect as Loewi observed in his experiment. Alternatively, repeating Loewi's experiment but with the addition of an acetylcholine receptor blocker to heart 2, should decrease or eliminate the response.

## Test Yourself

1. c   2. d   3. d   4. e   5. b   6. d   7. b   8. e   9. a   10. e

## Conceptual Questions

1. In a graded potential, a weak stimulus causes a small change in the membrane potential, whereas a strong stimulus produces a greater change. Graded

potentials occur along the dendrites and cell body. If a graded potential reaches the threshold potential at the axon hillock, an action potential results. This is a change in the membrane potential that is of a constant value and is propagated from the axon hillock to the axon terminal.

2. An increase in extracellular $Na^+$ concentration would slightly depolarize neurons, thereby changing the resting membrane potential. This effect would be minimal, however, because the resting membrane is not very permeable to $Na^+$. However, the shape of the action potentials in such neurons would be a little steeper, and the peak a little higher, because the electrochemical gradient favoring $Na^+$ entry into the cell through voltage-gated channels would be greater.

3. Neurons are among the most highly complex cells in an animal's body, with numerous extensions of the cell body. These extensions provide considerable surface area that allows for an extraordinary number of cell-to-cell contacts with other neurons and other types of cells, making them ideally suited for intercellular communication. In addition, myelin sheaths provide a structure that speeds up electric signaling along the axon, further facilitating intercellular communication.

# Chapter 43

## Concept Checks

*Figure 43.4* Not necessarily. Brain mass is not the sole determinant of intelligence or the ability to perform complex tasks. The degree of folding of the cerebral cortex is also important.

*Figure 43.6* A spinal nerve is composed of axons of both afferent and efferent neurons.

*Figure 43.7* The major symptom experienced by patients undergoing a lumbar puncture is headache, in part because the brain is no longer cushioned adequately by CSF. Within 24–48 hours, however, the CSF is replenished to normal levels.

*Figure 43.11* It was in her right hand.

## Core Skills: Connections

*Figure 43.1* As defined in Chapter 33, animals are multicellular heterotrophs (cannot make their own food) whose cells lack a cell wall. Most animals have a nervous system, muscles, and the abilities to move about during at least some phase of their life cycle and to reproduce sexually.

## Feature Investigation Questions

1. Gaser and Schlaug hypothesized that repeated exposure to musical training would increase the size of certain areas of the brain associated with motor, auditory, and visual skills. All three skills are commonly used in reading and performing musical pieces. The researchers used MRI to examine the areas of the brain associated with motor, auditory, and visual skills in three groups of individuals: professional musicians, amateur musicians, and nonmusicians. The researchers found that certain areas of the brain were larger in the professional musicians compared to the other groups, and larger in the amateur musicians compared to the nonmusicians.

2. Schmithorst and Holland found that, when exposed to music, certain regions of the brains of musicians were activated differently compared with those regions in the brains of nonmusicians. This study supports the hypothesis that there is a difference in the brains of musicians versus nonmusicians.

   The experiment conducted by Gaser and Schlaug compared the size of certain regions of the brain among professional musicians, amateur musicians, and nonmusicians. Schmithorst and Holland, however, were also able to detect functional differences between musicians and nonmusicians.

3. A trained athlete might be expected to show greater development of brain areas devoted to motor skills. For example, athletes who engage in repetitive activities such as tennis and badminton have been found to have greater volumes of motor and balance regions including the cerebellum. Someone deaf from birth might be predicted to have decreased development of auditory regions such as those in the temporal lobe. However, it is not necessarily this simple. For example, MRI studies sometimes but not always reveal differences in cell number in the temporal lobes of deaf versus hearing subjects, but there is generally significantly less development of fiber tracts (white matter) in this region in the brains of deaf persons. Therefore, the size of a brain structure (that is, the number of neurons) by itself may not always be predictive of function.

## Test Yourself

1. b   2. a   3. c   4. c   5. c   6. b   7. e   8. b   9. a   10. d

## Conceptual Questions

1. All animals with nervous systems have reflexes, which allow rapid behavioral responses to changes in the environment. When a cnidarian senses a tactile stimulus, its nerve net responds immediately and the animal reflexively contracts nearly all of its muscles, making the animal a smaller target. This behavior protects the animal from predators. When you hear a loud, unexpected, and frightening sound (such as a firecracker), you hunch your shoulders and slightly lower your head; this reflex protects you from danger by minimizing exposure of your neck and head to danger. Dilation of the pupils of the eyes in darkness, and constriction of the pupils in bright light, are reflexes that help us see in the dark and protect our retinas in bright light. Reflexes are particularly adaptive because they occur rapidly, typically with very few synapses involved, and without the need for conscious thought.

2. White matter consists of the myelinated axons that are bundled together in large tracts in the central nervous system and which connect different CNS regions. The lipid-rich myelin gives the tracts a whitish appearance. In contrast, gray matter consists of the cell bodies, dendrites, and some unmyelinated axons of neurons in the CNS.

3. The activities of animal nervous systems are replete with examples of new properties emerging from complex interactions. For example, you learned about reflexes in this chapter, which are behaviors that emerge from interactions between individual neurons that form communication circuits between the peripheral and central nervous systems. You also learned about "higher" functions of nervous systems, such as conscious thought, which also emerges from the interactions between many individual cells, each of which is in communication with up to hundreds of thousands of other cells. Individually, the cells cannot "think," but networked together in elaborate ways, they enable a person to think, remember, plan ahead, and interpret the environment.

# Chapter 44

## Concept Checks

**Figure 44.3** To think about the touch sensations you are aware of, let's take the example of sitting in a chair reading this textbook while holding it on your lap. You are aware of the constant weight of the book, the brush of the pages on your fingertips as you turn a page, a gentle breeze that may be circulating in your environment, the deep pressure from regularly adjusting your posture in your chair, an itch you may have on your skin, and the heat or cold of the room. Even a simple exercise such as this one is filled with stimuli of numerous types and durations.

**Figure 44.7** Waves of fluid (perilymph) movement caused by sound pressure on the tympanic membrane result in vibration of the basilar membrane.

**Figure 44.9** The orientation of the canals permits animals to detect circular or angular movement of the head in three different planes. The fluid in a canal that is oriented in the same plane as the plane of movement will respond maximally to the movement. For example, the canal that is oriented horizontally will show the greatest response to horizontal movements, while the other two canals will not respond as much. Overall, by comparing the signals from the three canals, the brain can interpret the motion in three dimensions.

**Figure 44.11** By comparing the intensity of signals from both sides of its head, a pit viper can determine the direction from which the heat is coming and thus localize its prey.

**Figure 44.18** Because red-green color blindness is caused by a sex-linked recessive gene, males require only a single defective allele on an X chromosome, whereas females require two defective alleles, one on each X chromosome.

**Figure 44.21** Photoreceptors synapse with bipolar cells, which in turn form synapses with ganglion cells.

**Figure 44.26** Salt is a vital nutrient needed to maintain plasma membrane potentials and fluid balance in animals' bodies. Sugar provides glucose and other monosaccharides, important energy-yielding compounds. Sour (acidic) foods, like citrus fruits, provide nutrients and important antioxidants (vitamin C, for example) that protect against disease. Bitter substances are often toxic, and their bad taste discourages animals from eating them.

## Core Skills: Connections

**Figure 44.1** The term sensory receptor refers to a type of cell that can respond to a particular type of stimulus. The term membrane receptor refers to a protein within a cell membrane that binds a ligand, thereby generating signals that initiate a cellular response.

**Figure 44.4** Cilia are cell extensions that contain in their internal structure microtubules and motor proteins that cause them to beat, or move, in a coordinated fashion. Stereocilia are membrane projections that are not motile, but instead are deformed by the movements of surrounding fluids.

**Figure 44.8** Statoliths are also found in the roots and shoots of plants. They serve as a gravity-detection mechanism that results in roots growing downward and shoots upward.

**Figure 44.16** The microvilli of the photoreceptor in the compound eye of insects and the outer segments of the rod and cone in the vertebrate eye share membrane adaptations that increase their surface area, making them better able to detect photons.

## Feature Investigation Questions

1. One possibility is that many different types of odor molecules might bind to one or just a few types of receptor proteins, with the brain responding differently depending on the number or distribution of the activated receptors. The second hypothesis is that organisms can make a large number of receptor proteins, each type binding a particular odor molecule or group of odor molecules. According to this hypothesis, it is the *type* of receptor protein, and not the number or distribution of receptors, that is important for olfactory sensing.

   The researchers extracted mRNA molecules from the olfactory receptor cells of the nasal epithelium of rats. They then used this mRNA to identify genes that encoded G-protein-coupled receptors.

2. The results of the experiment conducted by Buck and Axel support the hypothesis that animals discriminate between different odors based on having a variety of receptor proteins that recognize different odor molecules. Current research suggests that each olfactory receptor cell has a single type of receptor protein that is specific to particular odor molecules. Because most odors are due to multiple chemicals that activate many different types of odor receptor proteins, the brain detects odors based on the combination of the activated receptor proteins. Odor seems to be discriminated by many olfactory receptor proteins, which are in the membrane of separate olfactory receptor cells.

3. Among other things, animals use odors to detect potential sources of food, shelter, mates, and danger. Discerning between the odors of food versus potentially toxic material is key to the survival of many animals (think, for example, how we recognize fresh milk and spoiled milk). Many animals also use olfaction in finding shelter; some birds, for example, can distinguish the odors produced by particular beneficial plants that they use to build their nests. These plants have natural pathogen-killing properties and help maintain a healthier nest site. Odor preference also plays an important role in animals finding mates via the release of pheromones, particularly animals that live in dark environments such as nocturnal animals. The ability to discern the odors of different predators even when the predator may not be close enough to hear, or may not be in sight, is a huge survival advantage for many animals, who generally can differentiate such odors from those of nonthreatening species.

## Test Yourself

1. d   2. c   3. b   4. c   5. e   6. e   7. b   8. b   9. d   10. b

## Conceptual Questions

1. Sensory transduction is the process by which incoming stimuli are converted into neural signals. An example is the generation of signals in the retina when a photon of light strikes a photoreceptor.

   Perception is an awareness of the sensations that are experienced. An example is realizing what a particular visual image is.

2. Structurally, single-lens and compound eyes both have a lens, photoreceptors, light-absorbing pigmented cells, and structures with extensive surface areas. Functionally, both types of eye are capable of sensing the intensity of light, as well as distinguishing between wavelengths of light (color). In both cases, also, information from the eye is sent to the brain.

3. Of the various senses, the sense of olfaction (smell) is least important for the survival of humans. As diurnal animals, we rely largely on our visual sense. Sounds are a critical way to learn about impending danger, such as a car horn, but also provide our major means of communication. Other senses, such as the ability to sense pain, have acutely important functions from time to time. Olfaction, though often a pleasurable sense and at times a protective one (think of the smell of spoiled food), nonetheless provides little survival advantage to us. In fact, many people spend much of their lives with greatly diminished olfactory abilities, whether from chronic allergies or other problems, and are not hindered in any significant way. The story is very different for animals such as nocturnal mammals, which rely very heavily on olfaction to find food, locate mates, and avoid predators.

# Chapter 45

## Concept Checks

**Figure 45.1** In addition to not needing to shed their skeletons periodically, animals with endoskeletons can use their skin as an efficient means of heat transfer (and, to an extent in amphibians, water transfer). In addition, the body surface of such animals is often a highly sensitive sensory organ.

*Figure 45.3* If a tendon is torn, the attachment of a muscle to a bone is reduced or lost. Therefore, when that muscle contracts, it will not be able to move the bone, at least not as much as usual.

## Core Skills: Connections

*Figure 45.10* Voltage-gated $Ca^{2+}$ channels exist in the terminals of all axons that communicate by chemical signaling (neurotransmitter release). In those cases, depolarization of the axon terminal opens $Ca^{2+}$ channels, allowing $Ca^{2+}$ to enter the terminal and trigger exocytosis of stored vesicles containing neurotransmitter molecules.

## Feature Investigation Questions

1. PPAR-$\delta$ is a transcription factor that regulates the expression of genes that enable cells to more efficiently burn fat instead of glucose for energy.

2. Evans hypothesized that if PPAR-$\delta$ were highly activated in mice, the mice would lose weight because of the high level of fat metabolism. Evans and his coworkers developed transgenic mice with highly activated PPAR-$\delta$. Then they fed the transgenic mice and a group of wild-type mice high-fat diets. They then compared the weights of the two groups of mice to determine if the change in PPAR-$\delta$ activity affected weight. The weights of the transgenic mice were considerably lower than those of the wild-type mice. These results supported the hypothesis that highly activated PPAR-$\delta$ would lead to lower weight gain due to fat metabolism. Interestingly, the researchers also discovered that the transgenic mice could perform prolonged exercise for a much longer time than the wild-type mice. The muscle tissue of the transgenic mice was more specialized for long-term exercise.

3. Based on the mean body weight prior to the switch in diet, the wild-type and transgenic mice should weigh approximately 49–50 g and 33–34 g, respectively, at the conclusion of the study. Current dietary recommendations suggest that people consume no more than about 30% of their daily calories from fat, and that most of that fat should be in the form of unsaturated fat. According to the Centers for Disease Control and Prevention, the average consumption of fat in the U.S. amounts to roughly 33% of daily calories; however, that is an average and many individuals consume far more fat than that each day.

## Test Yourself

1. b   2. e   3. d   4. b   5. e   6. c   7. a   8. e   9. c   10. e

## Conceptual Questions

1. Exoskeletons are on the outside of an animal's body, and endoskeletons are inside the body. Both function in support and protection, but only exoskeletons protect an animal's outer surface. Exoskeletons must be shed when an animal grows, whereas endoskeletons grow with an animal.

2. Animals can fly, glide, swim, walk, hop, jump, crawl, and be moved by water or air currents. For animals adapted to it, swimming and flying are energetically less costly than moving on land. In all cases, however, friction due to contact with land or drag due to the resistance of water or air requires energy to overcome.

3. The use of energy released by the hydrolysis of ATP is fundamental to skeletal muscle function and locomotion. Recall that ATP must be hydrolyzed during the cross-bridge cycle for skeletal muscle cells to shorten. Energy is also used to maintain calcium ion balance in the sarcoplasmic reticulum and is expended in all forms of locomotion. The amount of energy expended by animals during locomotion reflects how well they are adapted to the environment in which they must move.

# Chapter 46

## Concept Checks

*Figure 46.4* Sauropod dinosaurs were herbivores that probably had a gizzard-type structure in which stones helped to grind up coarse vegetation. Such stones would have become smooth after months or even years of tumbling around in the gizzard. Some of these sauropods are known to have lacked the sort of grinding teeth characteristic of modern mammalian herbivores, and thus a gizzard would have functioned in their digestion much as it does in modern birds.

*Figure 46.7* A gallbladder stores bile and releases it precisely when it is needed, in response to a meal, which is particularly useful for animals that consume large or infrequent meals. In the absence of a gallbladder, bile flows into the intestine continuously and cannot be increased to match the amount or timing of food intake.

## Core Skills: Connections

*Figure 46.8* Transmembrane transport processes such as facilitated diffusion and secondary active transport are not unique to animals, and one or more types occur in virtually all cells.

*Figure 46.11* CCK inhibits stomach activity. This is an example of negative feedback. The arrival of chyme in the small intestine stimulates secretion of CCK, which promotes digestion. At the same time, CCK inhibits contraction of the smooth muscles of the stomach so that the entry of chyme into the small intestine is slowed down. This allows time for controlled digestion and absorption of nutrients in the small intestine, and prevents the intestine becoming overfilled with chyme. Simultaneously, CCK inhibits acid production by the stomach so that the pH of the small intestine does not become dangerously low before bicarbonate ions are able to neutralize it.

## Feature Investigation Questions

1. The surprising observations that some people with gastritis or ulcers have living bacteria (*H. pylori*) in their stomachs and that administering bacteria-killing compounds provided some relief from the symptoms led to the hypothesis that *H. pylori* infection is a cause of ulcers in humans.

2. The results did support the hypothesis. However, the results also clearly indicated that not all ulcers are due to *H. pylori* infection.

3. A combined treatment with bismuth and an antibiotic seems to be the most effective treatment. It is apparent, however, that even in individuals with continued *H. pylori* infections, some ulcers will heal on their own. In the absence of bismuth-antibiotic therapy, though, the likelihood of a recurrence of a new ulcer is much greater.

## Test Yourself

1. d   2. e   3. d   4. c   5. c   6. a   7. d   8. c   9. e   10. e

## Conceptual Questions

1. Digestion is the breaking down of large molecules into smaller ones by the action of enzymes and acid. Absorption is the transport of ions and small molecules, including those that do not require digestion, across the epithelial cells of the alimentary canal and from there into the extracellular fluid of an animal.

2. The crop is a dilation of the esophagus, which stores and softens food. The gizzard contains swallowed pebbles that help pulverize food. Both of these functions are adaptations that assist digestion in birds, which do not have teeth and therefore do not chew food. Humans, like many animals, can chew food before swallowing.

3. The small intestine contains layers of smooth muscle whose contractions help mix the contents of the intestine; this mixing facilitates digestion by bringing enzymes and food molecules into contact. Tight junctions along the apical membranes of the epithelium prevent enzymes from leaking out of the intestine. The surface area available for absorption is greatly increased by the folding of the epithelium to form villi and by the presence of microvilli, making up the brush border. The epithelial cells have digestive enzymes on their apical membranes; this ensures that the final steps of digestion will release products at the cell surface, where they can be quickly absorbed. The small intestine also contains neurons that, when activated by stretching of the intestine, signal the stomach to temporarily relax so that the intestine can process chyme in small amounts.

# Chapter 47

## Concept Checks

*Figure 47.6* Nearly all animals today show a similar relationship between body mass and metabolic rate, and there is no reason why that should not always have been true. Thus, the tiny 1-foot-tall ancestral horse *Eohippus* most likely had a higher BMR than do today's larger horses.

*Figure 47.9* Humans are homeothermic endotherms. We maintain our body temperature within a very narrow range, and we supply our own body heat.

## Core Skills: Connections

*Figure 47.3* Exocytosis involves the fusion of intracellular vesicles with the plasma membrane, resulting in the release of the vesicle contents into the extracellular fluid. See Figure 5.21 for a general description and Figure 42.13 for a specific example unique to animal cells.

## Feature Investigation Questions

1. Scientists were interested in knowing how animals regulate their body mass around a particular level, even though many animals experience changes in food supply throughout the year. This observation seemed to indicate that a mechanism existed within the body that monitored when fuel stores were higher or lower than normal and that initiated changes in behavior and metabolism to compensate.

2. The ob mice lost weight and ate less during the experimental procedure. This confirmed that something in the bloodstream of the wild-type mice was regulating body weight but was missing in the ob mice. When the unknown factor crossed into the bloodstream of the ob mice, it caused them to lose weight. In another group of parabiosed mice, however, the wild-type mice lost weight, but the db mice did not. Coleman concluded that these obese mice were not able to respond to the chemical signal that regulates body weight, even though they made the signal themselves and it was active in their parabiosed wild-type partners. It is now known that db mice produce leptin and ob mice do not; however, db mice are not sensitive to leptin and consequently they failed to lose weight in Coleman's experiment.

3. When parabiosed, leptin from db mice will enter the circulation of ob mice. The ob mice will lose weight as a consequence. The db mice will continue to gain weight because they remain insensitive to leptin.

## Test Yourself

1. c  2. a  3. d  4. c  5. a  6. c  7. e  8. c  9. c  10. e

## Conceptual Questions

1. Insulin acts on adipose and skeletal muscle cells to facilitate the diffusion of glucose from extracellular fluid into the cells' cytosol. This is accomplished by increasing the translocation of proteins called glucose transporters (GLUTs) from the cytosol to sites within the plasma membrane of insulin-sensitive cells. Insulin also inhibits glycogenolysis and gluconeogenesis in the liver, which decreases the amount of glucose secreted into the blood by the liver. Insulin is required for glucose transport because like many other polar molecules, glucose cannot move across the lipid bilayer of a plasma membrane by simple diffusion. The inhibitory effects of insulin on liver function help to ensure that liver glycogen stores will be spared for the postabsorptive period.

2. Appetite is controlled by a satiety center in the brain that receives signals from the stretched stomach and intestines after a meal. When digestion and absorption are complete, the stomach and intestines return to their original size, and the brain no longer perceives that the animal feels "full." In addition, appetite is controlled by leptin, a hormone secreted by adipose cells in direct proportion to the amount of fat stored in an animal's body. When leptin concentrations in the blood are high, appetite is suppressed. When leptin concentrations are low, as occurs when an animal is losing weight, appetite is increased. The presence of a hormone that is released into the blood in proportion to fat mass in the body allows the brain to monitor the amount of energy stored in the body. A decrease in the concentration of leptin in the blood, for example, is the mechanism that communicates to the brain that fat stores are lower than normal. This initiates the sensation of hunger, which encourages an animal to seek food.

3. Countercurrent exchange is a mechanism for retaining body heat. The physical arrangement (structure) of arteries and veins in an animal's body can contribute to the very important function of thermoregulation. As warm blood travels through arteries down a bird's leg, for example, heat moves by conduction from the artery to adjacent veins carrying cooler blood in the other direction, toward the heart. By the time the arterial blood reaches the tip of the leg, its temperature has dropped considerably, decreasing the amount of heat loss to the environment, while the heat is returned to the body's core via the warmed veins.

# Chapter 48

## Concept Checks

**Figure 48.1** Open circulatory systems evolved prior to closed systems. However, this does not mean that open systems are in some way "primitive" compared to closed circulatory systems. It is better to think of open systems as being ideally suited to the needs of those animals that have them. Arthropods are an incredibly successful order of animals, having the greatest number of species and inhabiting virtually every ecological niche on Earth. Clearly, their type of circulatory system has not hindered the success of arthropods.

**Figure 48.6** The aorta and all arteries branching from it carry oxygenated blood.

**Figure 48.8** The left and right ventricles pump blood through the semilunar valves into the aorta and the pulmonary trunk, respectively.

**Figure 48.12** No, erythrocytes never leave the blood vessels unless a vessel is cut.

**Figure 48.18** Several factors probably limit insect body size, but the respiratory system most likely is one of them. If an insect grew to the size of a human, for example, the trachea and tracheoles would be so large and extensive that there would be little room for any other internal organs in the body! Also, the mass of the animal's body

and the forces generated during locomotion would probably collapse the tracheoles. Finally, diffusion of oxygen from the surface of the body to the deepest regions of a human-sized insect would take far too long to support the metabolic demands of internal structures.

**Figure 48.19** As the lungs expand, the pressure within them decreases, as defined by Boyle's law. This permits air to flow into the lungs.

**Figure 48.25** An increase in the blood concentration of $HCO_3^-$ would favor the reaction $HCO_3^- + H^+ \rightarrow H_2CO_3 \rightarrow CO_2 + H_2O$. This would reduce the $H^+$ concentration of the blood, thereby raising the pH; the $CO_2$ formed as a result would be exhaled. These changes would shift the hemoglobin curve to the left of the usual position.

## Core Skills: Connections

**Figure 48.4** Immune defenses are found in most living organisms. Many bacteria produce antibacterial secretions that kill other bacteria. Plants, as shown in Figures 37.17 through 37.19, have a wide array of pathogen-fighting mechanisms.

**Figure 48.17** Countercurrent exchange is an efficient means of heat transfer between arteries and veins, such as those near the skin surface of the legs of a wading bird. Heat from the descending arteries is transferred to surrounding veins, which return the warm blood to the heart, preventing heat loss through the skin to the water.

**Figure 48.26** The brainstem includes the midbrain, pons, and medulla oblongata. See Figure 43.9 for an illustration of the major parts of the human brain.

## Feature Investigation Questions

1. Since glucocorticoids stimulate production of surfactant by lung tissues, injections of glucocorticoids to a pregnant woman should induce surfactant production by the lungs of the fetus. This indeed happens; the glucocorticoids cross into the fetal circulation from the mother, and stimulate surfactant in the fetus. Once born, even though the baby is premature, its lungs are relatively more mature and are able to remain open.

2. The oxygen levels achieved after surfactant therapy were all improved compared to the pre-treatment levels, but some babies actually had far higher values than is normal for a healthy baby. This was because at first all the babies were being ventilated with a mixture of gas that was very high in oxygen (much higher than is found in air); when their lung function improved, this mixture resulted in very high blood oxygen levels until the oxygen mixture was reduced. The differences in outcomes between babies likely reflected the degree of overall health of the individual babies; some were born slightly less prematurely than others.

3. Blood $CO_2$ levels were increased in the sick infants because lung ventilation is important not just for obtaining $O_2$, but for eliminating $CO_2$. If ventilation is compromised in any person, for any reason, blood levels of $CO_2$ inevitably increase.

## Test Yourself

1. b  2. c  3. b  4. a  5. d  6. a  7. a  8. b  9. b  10. b

## Conceptual Questions

1. *Closed circulatory system:* In a closed circulatory system, the blood is contained within tubes called blood vessels and is transported by a pump called the heart. All of the nutrients and oxygen that tissues require are delivered directly to them by the blood vessels. Advantages of a closed circulatory system are that different parts of an animal's body can receive blood flow in proportion to each part's metabolic requirements at any given time. Due to its efficiency, a closed circulatory system allows organisms to become larger.

   *Open circulatory system:* In an open circulatory system, the organs are bathed in hemolymph that ebbs and flows into and out of the heart(s) and body cavity, rather than having blood directed to all cells by increasingly smaller vessels. As in a closed circulatory system, there are a pump and blood vessels, but these two structures are less developed and less complex than those in a closed circulatory system. Partly as a result, the sizes of organisms such as mollusks and arthropods are generally relatively small, although exceptions do exist.

2. Carbon dioxide, hydrogen ions, and heat are produced by metabolism; the more active a cell is, the more of these products it generates. Because these products, in turn, reduce the ability of hemoglobin to bind oxygen (in other words facilitate the unloading of oxygen), more active regions of an animal's body obtain more oxygen in proportion to their metabolic demand at that time.

3. Hemoglobin is a protein with quaternary structure (see Chapter 3) in which the different subunits cooperate to bind up to a total of four oxygen molecules. It is the structure of the subunits and their relationship to each other that contributes to their ability to bind $O_2$ and to the nonlinear relationship of the oxygen-hemoglobin dissociation curve. In addition, however, interactions

of hemoglobin with other molecules, such as $CO_2$, change the structure of hemoglobin in such a way that its properties change. Under such conditions, hemoglobin is less able to bind $O_2$ and consequently it releases the gas. Any molecule that binds to hemoglobin will alter its structure and change its properties; these revert to the original state once the bound molecules are released. A particularly dramatic example of the relationship between the structure and function of hemoglobin is that which occurs in sickle cell disease, due to a mutation that changes the structure of the protein.

# Chapter 49

## Concept Checks

**Figure 49.2** Secretion of substances into the tubules is advantageous because it increases the amounts of the substances that are removed from the body by the excretory organs. The increase in amounts is important, because many substances that get secreted are potentially toxic. Filtration, though efficient, is limited by the volume of fluid that can leave the capillaries and enter the excretory tubule.

## Core Skills: Connections

**Figure 49.8** A brush border composed of microvilli is also present along the epithelial cell layer of the vertebrate small intestine (as shown in Figure 46.6). In the intestine, the brush border serves to increase the absorption of nutrients. In both the intestine and the proximal tubules of nephrons, therefore, a brush border provides extensive surface area for the transport of substances between a lumen and the epithelial cells (and from there to extracellular fluid).

**Figure 49.10** Epithelial cells like those in the distal tubule and cortical collecting duct can distribute proteins between the apical and basolateral sides of the plasma membrane. In this way, the $Na^+/K^+$-ATPase pumps that are stimulated by aldosterone are present and active only on one side of the cells, the basolateral surfaces. If the pumps were activated on the apical surfaces of the cells, aldosterone would not be able to promote reabsorption of $Na^+$ and water, because $Na^+$ would also be transported from the cells into the lumen.

**Figure 49.13** Countercurrent exchange is important in heat regulation in endotherms, in oxygen diffusion from the water into the blood across the gills of fishes, and in solute and water reabsorption in the loop of Henle in the mammalian kidney.

## Test Yourself

1. e   2. e   3. a   4. c   5. c   6. e   7. d   8. a   9. e   10. b

## Conceptual Questions

1. Nitrogenous wastes are the breakdown products of the metabolism of proteins and nucleic acids. They can be ammonia and ammonium ions, urea, or uric acid. The predominant type of waste excreted depends in part on an animal's environment. For example, aquatic animals typically excrete ammonia and ammonium ions, whereas many terrestrial animals excrete primarily urea and uric acid. Urea and uric acid are less toxic than the other types but require energy to be synthesized. Urea and uric acid also result in less water being excreted, an adaptation that is especially useful for organisms that must conserve water, such as many terrestrial species.

2. The three processes are filtration, reabsorption, and secretion. During filtration, an organ acts like a sieve or filter, removing some of the water and small solutes from the blood, interstitial fluid, or hemolymph, while excluding blood cells and large solutes such as proteins. Reabsorption is the process whereby epithelial cells of an excretory organ recapture useful solutes that were filtered. Secretion is the process whereby epithelial cells of an excretory organ transport unneeded or harmful solutes from the blood to the excretory tubules for elimination. Some substances such as glucose and amino acids are reabsorbed but not secreted, while some other substances such as toxic compounds are not reabsorbed and are secreted. Still other substances, namely proteins, are not filtered at all.

3. The respiratory system eliminates $CO_2$, the major waste product of metabolism produced by animals. The digestive system eliminates certain solid wastes from ingested food. The urinary system eliminates soluble wastes other than $CO_2$.

# Chapter 50

## Concept Checks

**Figure 50.10** Not all mammals use the energy of sunlight to synthesize vitamin D. Many animals, such as those that inhabit caves or that are strictly nocturnal, rarely are exposed to sunlight. Some of these animals get their vitamin D from dietary sources.

How others maintain $Ca^{2+}$ balance without dietary or sunlight-derived active vitamin D remains uncertain.

**Figure 50.12** $Na^+$ and $K^+$ balance is of vital importance for most animals because of the critical role these ions play in nervous system and muscle function. It is more the rule than the exception that such important physiological variables are under multiple layers of control. This control grants a high degree of fine-tuning capability so that these ions—and other similarly important substances—rarely exceed or fall below the normal range of concentration for a given animal.

**Figure 50.13** The great height of the twin on the left clearly indicates that his condition arose prior to puberty.

**Figure 50.15** Because 20-hydroxyecdysone is a steroid hormone, you would predict that its receptor would be intracellular. All steroid hormones interact with receptors located either in the cytosol or, more commonly, in the nucleus. The hormone-receptor complex then acts to promote or inhibit transcription of one or more genes. The receptor for 20-hydroxyecdysone is indeed found in cell nuclei.

## Core Skills: Connections

**Figure 50.3** When dopamine is secreted from an axon terminal into a synapse, from which it diffuses into a postsynaptic cell, it is considered a neurotransmitter. When it is secreted from an axon terminal into the extracellular fluid, from which it diffuses into the blood, it is considered a hormone.

**Figure 50.7** In addition to the pancreas, certain other organs in an animal's body may contain both exocrine and endocrine tissue or cells. For example, you learned in Chapters 46 and 47 that the vertebrate alimentary canal is composed of several types of secretory cells. Some of these cells release hormones into the blood that regulate the activities of the pancreas and other structures, such as the gallbladder. Other cells of the alimentary canal secrete exocrine products such as acids or mucus into the lumen of the canal that directly aid in digestion or act as a protective coating, respectively.

## Feature Investigation Questions

1. Banting and Best based their procedure on a condition that results when the pancreatic duct is blocked. The exocrine cells will deteriorate in a pancreas that has an obstructed duct; however, the islet cells are not affected. The researchers proposed to experimentally replicate the condition to isolate the cells suspected of secreting the glucose-lowering factor. From these cells, they assumed they would be able to extract the substance of interest without contamination or degradation due to exocrine products.

2. The extracts obtained by Banting and Best did contain insulin, the glucose-lowering factor, but were of low strength and purity. Collip developed a procedure to obtain a more purified extract with a higher concentration of insulin.

3. Normally, the concentration of glucose in a mammal's blood is never high enough to exceed the ability of the kidneys to reabsorb it all from the filtrate (refer back to Chapter 49 for details about filtration). However, like all transport processes, reabsorption of glucose from the kidney filtrate has a finite capacity that depends on the number of transporter molecules and their inherent rate of activity. In untreated diabetes, the blood concentration of glucose becomes so high that it exceeds the capacity of the kidney nephrons to fully reabsorb it from the filtrate. Consequently, some glucose appears in the urine.

## Test Yourself

1. b   2. e   3. b   4. d   5. b   6. e   7. c   8. d   9. b   10. d

## Conceptual Questions

1. Leptin acts in the hypothalamus to reduce appetite and increase metabolic rate. Because adipose tissue is typically the most important and abundant source of stored energy in an animal's body, the ability to relay information to the appetite and metabolism centers of the brain about the amount of available adipose tissue is a major benefit. In this way, the brain's centers can indirectly monitor the minute-to-minute energy status in the body. A decrease in leptin, for example, indicates a decrease in adipose tissue—as might occur during a fast. Removal of the leptin signal causes appetite to increase and metabolism to decrease, thereby conserving energy. The presence of an appetite and the subjective sensations associated with hunger motivate an animal to seek food at the expense of other activities, such as seeking shelter, finding a mate, and so on.

2. Type 1 diabetes mellitus is characterized by insufficient production of insulin due to the immune system destroying the insulin-producing cells of the pancreas. In type 2 diabetes mellitus, insulin is still produced by the pancreas (at least for a time), but adipose and muscle cells do not respond normally to the insulin.

3. Insulin acts to lower blood glucose concentrations, for example, after a meal, whereas glucagon elevates blood glucose, for example, during fasting. Insulin acts by stimulating the insertion of glucose transporters (GLUTs) into the cell membrane of muscle and fat cells. Glucagon acts by stimulating glycogenolysis in the liver. If a high dose of glucagon were injected into an mammal, including a human, the blood concentration of glucagon would increase rapidly. This would stimulate increased glycogenolysis, resulting in blood glucose concentrations that were above normal.

# Chapter 51

## Concept Checks

*Figure 51.12* Pregnancy and subsequent lactation require considerable energy and, therefore, nutrient ingestion. Consuming the placenta provides the female with a rich supply of protein and other important nutrients.

## Core Skills: Connections

*Figure 51.10* In addition to its other functions, the placenta must serve the function of the lungs for the fetus. Arteries always carry blood away from the heart; veins carry blood to the heart. Blood leaving the heart of the fetus and traveling through arteries to the placenta is deoxygenated. As blood leaves the placenta and returns to the heart, the blood has become oxygenated as oxygen diffuses from the maternal blood into fetal blood. That oxygenated blood then gets pumped from the fetal heart through other arteries to the rest of the fetus's body.

*Figure 51.11* Positive feedback also occurs during ovulation in the ovarian cycle (see Figure 51.9). Stimulation of an ovarian follicle by LH causes growth of the follicle and the release of estradiol, which further stimulates LH, which causes more follicle activity, and so on until ovulation occurs.

## Feature Investigation Questions

1. Using *Daphnia pulex*, Paland and Lynch compared the accumulation of mitochondrial mutations between sexually reproducing populations and asexually reproducing populations.

2. Of all the mutations observed in the populations, 17.7% were either moderately or mildly deleterious and all of these persisted in the asexually reproducing species, whereas only 4.4% persisted in sexually reproducing species. Thus, populations of asexually reproducing species were about four times more likely to retain a deleterious mutation.

3. Sexual reproduction allows for mixing of the different alleles of genes with each generation, thereby increasing genetic variation within the population. This increased variation could prevent the accumulation of deleterious alleles in the population.

## Test Yourself

1. d  2. c  3. e  4. a  5. d  6. c  7. b  8. c  9. c  10. e

## Conceptual Questions

1. External fertilization results in exposure of gametes to predation and other environmental dangers. Many animals have evolved the ability to lay enormous numbers of eggs to compensate for these dangers.

2. Cells of the hypothalamus produce two important hormones that regulate reproduction. GnRH stimulates the anterior pituitary gland to release two gonadotropic hormones, LH and FSH. These two hormones regulate the production of gonadal hormones and development of gametes in both sexes. In addition, increased secretion of GnRH contributes to the initiation of puberty. The mammalian hypothalamus also produces oxytocin, a hormone that is stored in the posterior pituitary gland and that acts to stimulate uterine contractions during labor and milk release during lactation.

3. Sexual reproduction requires that males and females of a species produce different gametes and that these gametes come into contact with each other. Thus, males and females must expend energy to locate mates. Also, the production of very large numbers of gametes may be necessary to increase the likelihood that the eggs are fertilized. These costs are outweighed by the genetic diversity afforded by sexual reproduction.

# Chapter 52

## Concept Checks

*Figure 52.2* Although swelling is one of the most obvious manifestations of inflammation, it has no significant adaptive value of its own. It is a consequence of fluid leaking out of blood vessels into the interstitial space. It can, however, contribute to pain sensations, because the buildup of fluid may cause distortion of connective tissue structures such as tendons and ligaments. Pain, while obviously unpleasant, is an important signal that alerts many animals to an injury and serves as a reminder to protect the injured site.

*Figure 52.10* Both B-cell and T-cell receptors have transmembrane domains, a constant region, and a variable region that binds a specific antigen.

*Figure 52.14* Because an animal may encounter the same type of pathogen many times during its life, having a secondary immune response means that future infections will be fought off much more efficiently.

## Feature Investigation Questions

1. The amino acid sequence of Toll protein shared similarities with a portion of a protein known to be involved in immune responses in vertebrates. In addition, activation of Toll protein and the vertebrate immune protein (a cytokine receptor) resulted in the generation of some of the same intracellular signals. These findings suggested that in addition to its characterized role in embryonic development, Toll may also be important in immune functions in flies.

2. No, Toll protein is not a receptor that recognizes pathogen-associated molecular patterns (PAMPs) expressed on microbial surfaces, and thus it is distinguishable from Toll-like receptors in vertebrates. Toll is, however, a transmembrane protein that binds to extracellular signals; these signals arise, however, not from the microbes themselves but rather from proteins that are endogenous to flies and that are generated during infections.

3. Yes, the results of the survival study clearly implicated Toll as a protein required for the induction of antimicrobial proteins and the ability to withstand fungal infection. Thus, the investigators' hypothesis was supported.

## Test Yourself

1. e  2. b  3. c  4. c  5. a  6. e  7. b  8. a  9. d  10. b

## Conceptual Questions

1. Innate immunity is present at birth and is found in all animals. These defenses recognize general, conserved features common to a wide array of pathogens and include internal defenses involving phagocytes and other cells. Adaptive immunity develops *after* an animal has been exposed to a *particular* antigen. The responses include humoral and cell-mediated defenses. Adaptive immunity appears to be largely restricted to vertebrates. Unlike innate immunity, in adaptive immunity, the response to an antigen is greatly increased if an animal is exposed to that antigen again at some future time.

2. Bacteria are single-celled prokaryotes that lack a true nucleus but are capable of reproducing on their own. Viruses are nucleic acids packaged in a protein coat; they require a host cell to reproduce. Eukaryotic parasites include certain fungi, protists and worms.

3. An immunoglobulin consists of four interlinked polypeptides, two heavy chains and two light chains, held together by disulfide bonds. Each immunoglobulin contains within its structure a constant region that is the same from one molecule to another within a given immunoglobulin class, and a variable region. The amino acid sequence of the variable region is what distinguishes one immunoglobulin from another and allows that region to specifically bind a particular antigen.

# Chapter 53

## Concept Checks

*Figure 53.6* Most of these receptors are located in or associated with blood vessels supplying the brain, a vital organ that among other functions controls many of the compensatory responses to changes in blood pressure or blood oxygen levels. Other receptors are located in the aorta, the first major vessel to leave the heart. Thus, blood pressure and gases are monitored in the general circulation and also specifically in the circulation entering the brain.

## Core Skills: Connections

*Figure 53.3* Stretch-sensitive receptors are widespread in animal bodies. The familiar knee-jerk response is a reflex triggered by the stretch of receptors located in tendons in the knee. Other examples include stretch receptors in muscles that provide feedback information on an animal's posture and movement, receptors in the stomach that relay a sense of fullness when the stomach is stretched after eating, and receptors in the urinary bladder that signal when the bladder is full.

*Figure 53.11* Positive feedback also occurs during the ovarian cycle in mammals at the time of ovulation (see Figure 51.9) and during the process of birth in mammals (see Figure 51.11).

## Feature Investigation Questions

1. Muscular movements help propel blood from veins in the limbs back to the heart. The increased venous return helps restore blood pressure by providing additional blood for the heart to pump. It is also possible that the unusually large decrease in pressure that occurred in the denervated dogs upon standing activated other, slower mechanisms that increased blood pressure independently of the two sets of baroreceptors described here. For example, although it was not described in the chapter, most animals including mammals have additional sets of baroreceptors in other organs that appear to play a smaller role in the control of blood pressure.

2. By testing the animals in a quiet, isolated room, the investigators reduced the possibility of other complicating variables that might cause a change in blood pressure. For example, stressful sounds or smells, or the sight of unfamiliar investigators might activate neural pathways associated with fight-or-flight responses, and these could raise blood pressure independently of baroreceptor input.

3. To ascertain the relative contributions of different baroreceptors to the responses shown by the animals in this experiment, the investigators could denervate the carotid or aortic baroreceptors independently, leaving the other set intact. This would allow a direct comparison of the effectiveness of each set of baroreceptors.

## Test Yourself

1. b   2. d   3. a   4. c   5. a   6. c   7. e   8. e   9. c   10. e

## Conceptual Questions

1. Animals in nature are confronted with many types of homeostatic challenges that often require integrated responses by multiple organ systems. For example, a bird flying at a high altitude over a mountain range faces the challenge of obtaining sufficient oxygen from the air. In such a situation, a bird might increase its breathing rate, adjust its cardiac output, or both; indeed, both of these changes and several others do occur. Similarly, fish that migrate between fresh and salt water, such as salmon, alter the function of their respiratory and urinary systems to help compensate for the changes in ion and water movement across the gills due to moving from one environment to another. Yet another common example of a challenge to homeostasis is starvation, during which the nervous, endocrine, urinary, and digestive systems will help maintain glucose homeostasis by processes such as gluconeogenesis.

2. Light-headedness can occur in some people when donating blood, which is essentially a carefully controlled hemorrhage. Initially, as the homeostatic processes described in this chapter are just beginning, there is a period of instability with respect to blood pressure control. While lying or sitting down, this is rarely noticeable, but when the person stands, gravity counteracts the movement of blood through limb veins back to the heart, causing a sudden decrease in pressure. Fainting does not usually occur, however, because the baroreceptor reflex responds immediately to this sudden change in pressure, and within seconds the heart rate and cardiac output are increased due to the actions of the sympathetic nervous system. The phenomenon of light-headedness can actually happen any time a person stands up after reclining, even under ordinary circumstances, but it is more noticeable when a person's blood volume is reduced such as after donating blood.

3. All of the homeostatic responses to hemorrhage described in this chapter require energy. Increasing the activity of any muscle, including that of the heart and the respiratory muscles, requires a considerable increase in expenditure of energy (ATP). Activity of the nervous system, so vital to the compensatory response to hemorrhage, requires continual ATP production and hydrolysis to maintain ion concentration gradients across the plasma membrane of neurons; without the maintenance of these gradients, there could be no flow of current along a neuron. Many of the transport processes in the kidneys also require ATP to drive the ion pumps that move ions across membranes and that create osmotic gradients for the movement of water.

# Chapter 54

## Concept Checks

*Figure 54.6* Cold water suppresses the ability of the coral-building organisms to secrete their calcium carbonate shell.

*Figure 54.9* In some areas when fire is prevented, fuel, in the form of old leaves and branches, can accumulate. When a fire eventually occurs, it can be so large and hot that it destroys everything in its path, even reaching high into the tree canopy.

*Figure 54.14* Acid soils are low in essential plant and animal nutrients such as calcium and nitrogen and are lethal to some soil microorganisms that are important in decomposition and nutrient cycling.

*Figure 54.16* This band is due to increasing cloudiness and rain in the tropics, which maintain fairly constant temperatures across a relatively wide latitudinal range.

*Figure 54.22* Soil conditions can also influence biome type. Nutrient-poor soils, for example, may support vegetation different from that of the surrounding area.

*Figure 54.23* Taiga.

## Core Skills: Connections

*Figure 54.13* Plants cannot readily absorb salty water because of its highly negative water potential.

## Feature Investigation Questions

1. Most believed that invasive species succeed in new environments because of the lack of natural enemies and that diseases and predators present in the original environment controlled the growth of the population. When a species is introduced into a novel environment, the natural enemies are usually absent. This allows for an unchecked increase in the population of the invasive species.

2. Callaway and Aschehoug were able to demonstrate through a controlled experiment that the presence of *C. diffusa*, an invasive species, reduced the biomass of three other native species of grasses by releasing allelochemicals. Similar experiments using species of grasses that are found in the native region of *C. diffusa* indicated that these species have evolved defenses against the allelochemicals.

3. The activated charcoal helped to remove the allelochemical(s) from the soil. The researchers conducted this experiment to provide further evidence that the chemical(s) released by *C. diffusa* was reducing the biomass of the native Montana grasses. With the removal of the chemical(s) by the addition of the charcoal, the researchers observed an increase in biomass of the native Montana grasses compared with the experiments lacking the charcoal.

4. Researchers could measure the biomass of *C. diffusa*. Because *C. diffusa* is invasive in the U.S., effects of native North American grasses on *C. diffusa* would be expected to be weak. Conversely, *C. diffusa* is not invasive in Eurasia so strong effects of Eurasian grasses on *C. diffusa* should be observed, and this is what researchers found.

## Test Yourself

1. b   2. a   3. b   4. b   5. a   6. a   7. d   8. d   9. a   10. c

## Conceptual Questions

1. Mountains are cooler than valleys because of adiabatic cooling. Air at higher altitudes expands because of decreased pressure. As it expands, air cools, at a rate of 10°C for every 1,000 m in elevation. As a result, mountain tops can be much cooler than the plains or valleys that surround them.

2. First, lightning strikes from electrical storms are usually more frequent in prairies than in deserts. Second, the vegetation in a prairie is more continuous and the biomass more extensive than in a desert, so fires burn more frequently and for longer periods.

3. Florida is a peninsula that is surrounded by the Atlantic Ocean and the Gulf of Mexico. Differential heating between the land and the sea creates onshore sea breezes on both the east and west coasts. These breezes often drive clouds across the whole peninsula, bringing heavy rain.

# Chapter 55

## Concept Checks

*Figure 55.3* In operant conditioning a behavior is reinforced by a reward or punishment. In classical conditioning, an involuntary response comes to be associated with a stimulus that did not originally elicit the response, as with Pavlov's dogs salivating at the sound of a metronome.

*Figure 55.5* The ability to sing the same distinctive song must be considered innate behavior because the cuckoo has had no opportunity to learn its song from its parents.

*Figure 55.7* Tinbergen manipulated pinecones, but not all digger wasp nests are surrounded by pinecones. You could manipulate branches, twigs, stones, and leaves to determine the necessary size and dimensions of objects that digger wasps use as landmarks.

*Figure 55.8* Monarch migration is an unusual example because the return trip involves several different generations: One generation overwinters in Mexico, but these individuals lay eggs and die on the return journey, and their offspring continue the return trip.

*Figure 55.15* The individuals in the center of the group are less likely to be attacked than those on the edge of the group. This advantage is referred to as the geometry of the selfish herd.

*Figure 55.18* All the larvae in the group are likely to be the progeny of one egg mass from one adult female moth. The death of one caterpillar in the cluster teaches a predator to avoid preying on caterpillars with that warning pattern and thus benefits the caterpillar's close kin.

## Core Skills: Connections

*Figure 55.3* Prey species converge on the color patterns displayed by toxic, bad-tasting, or dangerous species to reinforce predators' avoidance of them.

*Figure 55.4* According to studies of humans and other animals, learning a task increases the size of brain regions that are associated with learning and memory.

## Feature Investigation Questions

1. Tinbergen observed the activity of digger wasps as they prepared to leave the nest. Each time, the wasp hovered and flew around the nest for a period of time before leaving. Tinbergen suggested that during this time, the wasp was making a mental map of the nest site. He hypothesized that the wasp was using characteristics of the nest site, particularly landmarks, to help relocate it.

2. Tinbergen placed pinecones around the nest of the wasps. When the wasps left the nest, he removed the pinecones from the nest site and set them up in the same pattern a distance away, constructing a sham nest. For each trial, the wasps would go directly to the sham nest, which had the pinecones around it. This indicated to Tinbergen that the wasps identified the nest based on the pinecone landmarks.

3. No, but Tinbergen also conducted an experiment to determine if the wasps were responding to the visual cue of the pinecones or the chemical cue of the pinecone scent. The results of this experiment indicated that the wasps responded to the visual cue of the pinecones and not their scent.

## Test Yourself

1. d   2. d   3. d   4. c   5. c   6. d   7. b   8. c   9. a   10. c

## Conceptual Questions

1. The donation of the male's body to the female is the ultimate nuptial gift. It is possible that this meal enables the females to produce more eggs. In this way, the male benefits because its genes will be passed on to future generations.

2. Certainty of paternity influences degree of parental care. With internal fertilization, certainty of paternity is relatively low. With external fertilization, eggs and sperm are deposited together, and paternity is more certain. This explains why males of some species, such as mouth-breeding cichlid fish, are more likely to engage in parental care.

3. Alarm calling calls attention to the caller, so if no relatives are present, females bolt into their warren to escape a predator. However, daughters and sisters are kin and may pass on copies of a female's genes, so alarm calls are frequently made when these relatives are present.

# Chapter 56

## Concept Checks

*Figure 56.3* In a half-empty classroom, the distribution is often clumped because friends sit together.

*Figure 56.6* (a) type III, (b) type II

*Figure 56.12* Only density-dependent factors have this stabilizing tendency.

## Core Skills: Connections

*Figure 56.3* Uniform. Territorial marking is likely to keep cheetahs well separated from each other.

*Figure 56.13* It has lost the ability to produce viable seeds but it makes thousands of fully formed plantlets, borne on its leaves.

## Feature Investigation Questions

1. It became apparent that the sheep population was declining. Some individuals thought that the decline in the population was due to the negative effect of increased wolf predation on population growth. This idea led to the suggestion of culling the wolf population to reduce the level of predation on the sheep.

2. The survivorship curve is similar to a typical type I survivorship curve, which would suggest that survival is high among young and reproductively active members of the population and that mortality rates are higher for older members of the population. One difference between the actual survivorship curve and a typical type I curve is that the mortality rate of very young sheep was higher in the actual curve and then leveled off after the second year. This suggests that very young and older sheep are at greater risk from predation.

3. It was concluded that wolf predation was not the primary reason for the drop in the sheep population. It appeared that wolves prey on the vulnerable members of the population and not on the healthy, reproductively active members. The Park Service determined that several cold winters may have had a more important effect on the sheep population than wolf predation did. Based on these conclusions, the Park Service ended its wolf-control program.

## Test Yourself

1. b   2. e   3. b   4. c   5. c   6. b   7. b   8. c   9. d   10. c

## Conceptual Questions

1. It increases. Instead of recapturing 5 tagged fish, you will recapture only 4. Population size is now estimated as $50 \times 40/4 = 2,000/4 = 500$. Your population size estimate has increased to 500 when, in fact, it is more likely that 400 fish occur in the lake.

2. When population sizes are low ($N = 100$), $(K - N)/K$ is so small that growth is low.

$$\frac{dN}{dt} = (0.1)(100) \times \frac{(1,000 - 100)}{1,000}$$

$$\frac{dN}{dt} = 9$$

At medium values of $N$, $(K - N)/K$ is closer to a value of 1, and population growth is relatively large. If $K = 1,000$, $N = 500$, and $r = 0.1$, then

$$\frac{dN}{dt} = (0.1)(500) \times \frac{(1,000 - 100)}{1,000}$$

$$\frac{dN}{dt} = 25$$

By comparing these two examples with that shown in Section 56.3, we see that growth is small at high and low values of $N$ and is greatest at immediate values of $N$. Growth is greatest when $N = K/2$. However, when expressed as a percentage, growth is greatest at low population sizes. Where $N = 100$, percentage growth = $9/100 = 9\%$. Where $N = 500$, percentage growth = $25/500 = 5\%$, and where $N = 900$, percentage growth = $9/100 = 1\%$.

3. In the ponds that dry out, species would tend to be semelparous, producing all their offspring in a single reproductive event while water is present. In the permanently wet ponds, species would be iteroparous, reproducing repeatedly over the course of a lifetime.

## Collaborative Questions

2. a) $\lambda = 1,200 / 1,000 = 1.2$
   b) After 5 years, $N_5 = N_0 \lambda^5$
   $\qquad\qquad\qquad\quad = (1,000)(1.2)^5$
   $\qquad\qquad\qquad\quad = 2,488$

# Chapter 57

## Concept Checks

*Figure 57.2* Individual vultures often fight one another over small carcasses. These interactions constitute intraspecific interference competition.

*Figure 57.6* In 1974, Tom Schoener examined segregation in a more wide-ranging literature review of over 80 species, including slime molds, mollusks, and insects, as well as birds. He found segregation by habitat occurred in the majority of the examples, 55%. The second most common form of segregation was by food type, 40%.

**Figure 57.8** Omnivores, such as bears, can feed on both plant material, such as berries, and animals, such as salmon. Thus, omnivores may act as either predators or herbivores, depending on what they are feeding on.

**Figure 57.9** Batesian mimicry has a positive effect for the mimic, and the model is unaffected, so it is a +/0 relationship, like commensalism. Müllerian mimicry has a positive effect on both species, so it is a +/+ relationship, like mutualism.

**Figure 57.11** Invertebrate herbivores can eat around mechanical defenses; therefore, chemical defenses are probably most effective against these herbivores.

**Figure 57.16** It's an example of facultative mutualism, because both species can live without the other.

**Figure 57.19** Fertilizer increases plant quality and hence herbivore density, which, in turn, increases the density of spiders. This is bottom-up control.

## Core Skills: Connections

**Figure 57.9** Most mollusks are heavily armored. However, sea slugs have lost their shells. These species are aposematically colored, advertising a poisonous body. In addition, some octopuses are poisonous, and most can eject an inky chemical "smokescreen."

**Figure 57.11** Red hot chili peppers.

**Figure 57.13** Dodder, *Cuscuta pentagona,* is another important parasitic plant.

**Figure 57.17** Red.

## Feature Investigation Questions

1. The two species of barnacles can be found in the same intertidal zone, but there is a distinct difference between the realized niches of these species. *Chthamalus stellatus* is found only in the upper intertidal zone. *Semibalanus balanoides* is found only in the lower tidal zone.

2. Connell moved rocks with young *Chthamalus* from the upper intertidal zone into the lower intertidal zone to allow *Semibalanus* to colonize the rocks. After the rocks were colonized by *Semibalanus*, he removed *Semibalanus* from one side of each rock and returned the rocks to the lower intertidal zone. This procedure allowed Connell to observe the growth of *Chthamalus* in the presence and the absence of *Semibalanus*.

3. Connell observed that *Chthamalus* was more resistant to desiccation than *Semibalanus*. Though *Semibalanus* was the better competitor in the lower intertidal zone, that species was at a disadvantage in the upper intertidal zone when water levels were low. This fact allowed *Chthamalus* to flourish and outcompete *Semibalanus* in a different region of the intertidal zone.

## Test Yourself

1. d   2. c   3. b   4. d   5. c   6. b   7. d   8. b   9. b   10. c

## Conceptual Questions

1. Yes, it is possible that by removing parasites from a neighbor, a primate may be reducing the likelihood of the parasite spreading to infect it. You scratch my back, I'll scratch yours, and together we will both be better off.

2. There are at least three factors that might limit losses due to pest damage to crops. First, plants possess an array of defensive chemicals, including alkaloids, phenolics, and terpenes. Second, many herbivore populations are reduced by the action of natural enemies. Third, the low nutritive value of plants ensures that herbivore populations remain low and unlikely to affect plant populations. While we can't easily increase the levels of defensive chemicals in many crop plants or reduce their nutritive value, we can introduce more natural enemies of plant pests. We see evidence for this in the use of biological controls.

# Chapter 58

## Concept Checks

**Figure 58.3** The species richness of trees doesn't increase in the mountainous areas of the West because rainfall in the western U.S. is low compared to that in the East.

**Figure 58.5** Walking from the current edge of the glacier to the mouth of the inlet, an ecologist is walking backward in ecological time to communities that originated hundreds of years ago.

**Figure 58.8** Competition features more prominently. Although early colonists tend to make the habitat more favorable for later colonists, it is the later colonists who outcompete the earlier ones, and this fuels species change.

**Figure 58.10** At first glance, the change looks small, but the data are plotted on a log scale. On this scale, an increase in bird richness from 1.2 to 1.6 equals an increase from 16 to 40 species, a change of over 100%.

**Figure 58.14** It depends on the trophic level of their food, whether dead vegetation or dead animals. Many decomposers feed at multiple trophic levels.

## Core Skills: Connections

**Figure 58.9** The model helps conservationists design the best shaped and optimally placed nature reserves in a "sea" of developed land.

**Figure 58.13** Cyanobacteria.

## Feature Investigation Questions

1. Simberloff and Wilson were testing the three predictions of the equilibrium theory of island biogeography. One prediction suggested that the number of species should increase with increasing island size. Another prediction suggested that the number of species should decrease with increasing distance of the island from the source pool. Finally, the researchers were testing the prediction that the turnover of species on islands should be considerable.

2. Simberloff and Wilson used the information gathered from the species survey to determine whether the same types of species recolonized the islands or the colonizing species were random.

3. Island distance to the mainland affected species richness with near islands having higher numbers of species than distant islands. However, species turnover was low on all islands and was unaffected by island distance from the mainland.

## Test Yourself

1. c   2. a   3. c   4. a   5. e   6. c   7. e   8. a   9. d   10. c

## Conceptual Questions

1. The value of the Shannon diversity index is 1.609 for both forests. By this measure, diversity is equal in the two forests. The index is unable to discriminate between communities that have different species abundances but the same relative proportions of species. An observer would be more likely to encounter a variety of trees in forest A than in forest B.

2. Carrion beetles are decomposers. They feed on dead animals such as mice, at trophic level 3 or 4. Mice generally feed on vegetative material (trophic level 1) or crawling arthropods (trophic level 2), so mice themselves feed at trophic level 2 or 3.

3. *C. vulgaris* litter enriches the soil with nitrogen, facilitating the growth of the grasses. Adding fertilizer also increases soil nitrogen. The mechanism of succession operating in this case is facilitation.

# Chapter 59

## Concept Checks

**Figure 59.2** The conclusion would be that were very few juveniles in the population and many mature adults. The population would be in decline.

**Figure 59.5** Many different ecological footprint calculators are available on the Internet. Does altering inputs such as type of transportation, amount of meat eaten, or amount of waste generated make a difference?

**Figure 59.7** Hawaii lies north of the equator. The Northern Hemisphere has greater land area and plant biomass than the Southern Hemisphere, so in the northern summer more $CO_2$ is used up by plants and atmospheric $CO_2$ level declines slightly. In the northern winter, less $CO_2$ is absorbed and atmospheric $CO_2$ levels increase.

**Figure 59.9** The greatest stores are in rocks and fossil fuels.

## Core Skills: Connections

**Figure 59.9** The other three elements are oxygen, hydrogen, and nitrogen. Therefore, the water and nitrogen cycles are very important to humans.

## Feature Investigation Questions

1. The researchers wanted to learn the effects of increased carbon dioxide levels on the forest ecosystem: the effects on primary production as well as on other trophic levels in the ecosystem.

2. By increasing the carbon dioxide levels in only half of the chambers, the researchers were maintaining the control treatment necessary in all scientific studies.

By maintaining equal numbers of control and experimental treatments, the researchers could compare data to determine what effects the experimental treatment had on the ecosystem.

3.   $t_{14} = 5.667$, $P < 0.001$; $x_1 = 10.00$, s.d $= 2.93$; $x_2 = 3.20$, s.d $= 1.72$

## Test Yourself

1. d    2. c    3. d    4. b    5. a    6. b    7. e    8. b    9. b    10. e

## Conceptual Questions

1.   Nitrogen molecules have a triple bond, making them hard to break apart. Only a few species of bacteria can break apart atmospheric nitrogen and fix nitrogen. The excess ammonia, $NH_3$, or ammonium, $NH_4^+$, that they produce in this way gradually accumulates and can be used by plants.

2.   Maximum sustainable yield represents the number of individuals that can be removed from a population without affecting population growth. This is rather like removing the interest from a bank account and not touching the principal. Maximal sustainable yield occurs at the steepest point of the growth curve, which is at the midpoint of the logistic curve.

3.   The family whose members have triplets has 27 descendants in 2000, compared to 32 for the family whose members have twins. Delaying reproduction can slow population growth.

# Chapter 60

## Concept Checks

*Figure 60.3*  It is possible that the results are driven by what is known as a sampling effect. As the numbers of species in the community increase, so does the likelihood of including a "superspecies," a species with exceptionally large individuals that would use up resources. In communities with higher diversity, care has to be taken that increased species richness is driving the results, not the increased likelihood of including a superspecies.

*Figure 60.8*  Corridors might also promote the movement of invasive species or the spread of fire between areas.

*Figure 60.9*  The hedgerows act as habitat corridors because they permit species movement between forest fragments.

## Feature Investigation Questions

1.   The researchers hoped to replicate terrestrial communities that differed only in their level of species richness. This would allow the researchers to determine the relationship between species richness and ecosystem function.

2.   The hypothesis was that ecosystem function was directly related to species richness. If species richness increased, the hypothesis suggested that ecosystem function should increase.

3.   The researchers tested for ecosystem functioning by monitoring community respiration, decomposition, nutrient retention rates, and productivity. All of these indicate the efficiency of nutrient production and use in the ecosystem.

4.   The redundancy hypothesis.

## Test Yourself

1. d    2. c    3. b    4. a    5. c    6. a    7. e    8. e    9. b    10. c

## Conceptual Questions

1.   Increased species diversity increases ecosystem function. Ecosystem functions such as nutrient cycling, regulation of atmospheric gases, pollination of crops, pest regulation, water purity, storm protection, and sewage purification are all likely to be increased by increased species diversity. In addition, increased plant species diversity increases likely availability of new medicines for humans.

2.   Megadiversity countries are those with the greatest number of species. Biodiversity hot spots conserve the greatest numbers of endemic species. Crisis ecoregions are those areas of the Earth which represent distinct biome types, such as temperate grasslands and tropical deciduous forests, but have undergone substantial habitat loss. "Last of the wild" areas are relatively pristine areas such as much tundra and taiga, deserts, and some tropical rainforests.

3.   The diversity-stability hypothesis suggests a linear correlation between species richness and ecosystem function; as diversity increases, ecosystem function increases proportionately. The redundancy hypothesis suggests that ecosystem function increases rapidly at lower levels of species richness but then levels off, as additional species are functionally redundant. The keystone hypothesis suggests that ecosystem function is low at low levels of species richness and only rises substantially as species richness approaches high levels. The idiosyncratic hypothesis suggests that there is no predictable relationship between species richness and ecosystem function.

**1000 Genomes Project**   An international research effort to establish the level of human genetic variation.

**20-hydroxyecdysone**   A hormone produced by the prothoracic glands of arthropods that stimulates molting.

**30-nm fiber**   Nucleosome units organized into a more compact structure that is 30 nm in diameter.

**5′ cap**   The 7-methylguanosine structure at the 5′ end of most mature mRNAs in eukaryotes.

## A

**ABC model**   A model for flower development in which three classes of genes, called *A*, *B*, and *C*, govern the formation of sepals, petals, stamens, and carpels. More recently, a fourth class, called the *E* genes, was found to be required for this process.

**abiotic**   The term used to describe interactions between organisms and their nonliving environment.

**abscisic acid**   One of several plant hormones that help a plant cope with environmental stress.

**absolute refractory period**   The period during an action potential when the inactivation gate of the voltage-gated sodium channel is closed; during this time, it is impossible to generate another action potential.

**absorption spectrum**   A diagram that depicts the wavelengths of electromagnetic radiation that are absorbed by a pigment.

**absorption**   The process in which ions, water, and small molecules diffuse or are transported out of the alimentary canal into an animal's body fluids.

**absorptive nutrition**   The process whereby an organism uses enzymes to digest organic materials and absorbs the resulting small food molecules into its cells.

**absorptive state**   One of two alternating phases in the utilization of nutrients; occurs when ingested nutrients enter the blood from the gastrointestinal tract. The other phase is the postabsorptive state.

**accommodation**   In the vertebrate eye, the process in which contraction and relaxation of the ciliary muscles adjust the lens according to the angle at which light enters the eye.

**acetylcholinesterase**   An enzyme located on membranes of postsynaptic cells that respond to the neurotransmitter acetylcholine, such as in muscle fibers in a neuromuscular junction; breaks down excess acetylcholine released into the synaptic cleft.

**acid hydrolase**   A hydrolytic enzyme found in lysosomes that functions at acidic pH and uses a molecule of water to break a covalent bond.

**acid rain**   Precipitation with a pH of less than 5.6; results from the burning of fossil fuels.

**acid**   A molecule that releases hydrogen ions (H⁺) in solution.

**acidic**   A solution that has a pH below 7.

**acoelomate**   An animal that lacks a fluid-filled body cavity.

**acquired antibiotic resistance**   The common phenomenon in which a previously susceptible strain of bacteria becomes resistant to a specific antibiotic.

**acquired immunodeficiency syndrome (AIDS)**   A disease caused by the human immunodeficiency virus (HIV) that weakens the immune system of infected individuals.

**acrocentric**   A chromosome in which the centromere is near one end.

**acromegaly**   A condition in which a person's growth hormone level is abnormally elevated after puberty, causing many bones to thicken and enlarge.

**acrosomal reaction**   An event in fertilization in which enzymes released from a sperm's acrosome break down the outer layers of an egg cell, allowing the entry of the sperm cell's nucleus into the egg cell.

**acrosome**   A special structure at the tip of a sperm's head containing proteolytic enzymes that help break down the protective outer layers of the egg cell at fertilization.

**actin filament**   A thin type of protein filament composed of actin proteins that forms part of the cytoskeleton and supports the plasma membrane; plays a key role in cell strength, shape, and movement.

**actin**   A cytoskeletal protein, found in the thin filaments of myofibrils.

**action potential**   An electrical signal along a cell's plasma membrane; occurs in animal neuron axons and muscle cells and in some plant cells.

**action spectrum**   The rate of photosynthesis plotted as a function of the wavelength of light.

**activation energy**   An initial input of energy in a chemical reaction that allows the molecules to get close enough to cause a rearrangement of bonds.

**activator**   A transcription factor that binds to DNA and increases the rate of transcription.

**active immunity**   An animal's ability to fight off a pathogen to which it has been previously exposed. Active immunity can develop as a result of natural infection or artificial immunization.

**active site**   The location in an enzyme where a chemical reaction takes place.

**active transport**   The transport of a substance across a membrane from an area of low concentration to one of higher concentration with the aid of a transport protein; requires an input of energy.

**adaptations**   Changes in populations of living organisms that are the result of natural selection and that increase their ability to survive and reproduce in their environment.

**adaptive immunity**   A specific immune defense that develops only after an animal is exposed to a foreign substance; believed to be unique to vertebrates.

**adaptive radiation**   The process whereby a single ancestral species evolves into a wide array of descendant species that differ greatly in their habitat, form, or behavior.

**adenine (A)**   A purine base found in DNA and RNA.

**adenosine triphosphate (ATP)**   A molecule that is a common energy source for all cells.

**adenylyl cyclase**   An enzyme in the plasma membrane that synthesizes cAMP from ATP.

**adherens junction**   A mechanically strong type of cell junction between animal cells that is organized into bands. The cells are connected to each other via cadherins, and the cadherins are linked to actin filaments on the inside of the cells.

**adhesion**   The ability of two different substances to bind to each other; the ability of water to be attracted to, and thereby adhere to, a surface that is not electrically neutral.

**adiabatic cooling**   The process in which increasing elevation produces a decrease in air temperature due to lowered air pressure.

**adventitious root**   A root that is produced on the surfaces of stems (and sometimes leaves) of vascular plants; also, roots that develop at the bases of stem cuttings.

**aerenchyma**   Spongy plant tissue with large air spaces.

**aerobic respiration**   A type of cellular respiration in which O₂ is consumed and CO₂ is released.

**aerotolerant anaerobe**   A microorganism that does not use oxygen but is not poisoned by it either.

**afferent arteriole**   Blood vessel that carries blood into a glomerulus of the vertebrate kidney.

**affinity**   The degree of attraction between an enzyme and its substrate(s).

**aflatoxins**   Fungal toxins that cause liver cancer and are a major health concern worldwide.

**age structure**   The relative numbers of individuals of each defined age group in a population.

**age-specific fertility rate**   The rate of offspring production for females of a certain age; used to calculate how a population grows.

**akinete**   A thick-walled, food-filled cell produced by certain bacteria or protists that enables them to survive unfavorable conditions in a dormant state.

**aldosterone**   A steroid hormone made by the adrenal glands that regulates salt and water balance in vertebrates.

**algae** (singular, **alga**)   A term that applies to about 10 phyla of protists, including mostly photosynthetic and some nonphotosynthetic species; often also includes cyanobacteria.

**alimentary canal**   In animals, the single elongated tube of a digestive system, with an opening at either end through which food and eventually wastes pass from one end to the other.

**alkaline**   A solution with a pH above 7.

**allantois**   One of the four extraembryonic membranes in the amniotic egg. It serves as a disposal sac for metabolic wastes.

**allele frequency**   The number of copies of a particular allele in a population divided by the total number of alleles for that gene in that population.

**allele**   A variant form of a gene.

**allelochemical**   A powerful plant chemical, often a root exudate, that kills other plant species.

**allelopathy**   The suppression of growth of one species due to the release of toxic chemicals by another species.

**allergy**   Hypersensitivity reaction to an environmental antigen (an allergen) that is otherwise a harmless or relatively harmless substance.

**allopatric speciation**   A form of speciation that occurs when a population becomes geographically isolated from other populations and evolves into one or more new species.

**allopatric**   The term used to describe species occurring in different geographic areas.

**alloploid**   An organism having at least one set of chromosomes from two or more different species.

**allosteric site**   A site on an enzyme where a molecule can bind noncovalently and affect the enzyme's function.

**alternation of generations**   The phenomenon that occurs in plants and some protists in which the life cycle alternates between multicellular diploid organisms, called sporophytes, and multicellular haploid organisms, called gametophytes.

**alternative splicing**   The splicing of pre-mRNA in more than one way to allow the production of two or more different polypeptides from the same gene.

**altruism**   Behavior that appears to benefit others at a cost to oneself.

**alveolus** (plural, **alveoli**)   1. Saclike structures in the lungs where gas exchange occurs. 2. Saclike cellular features of the protists known as alveolates.

**Alzheimer's disease (AD)**   The leading worldwide cause of dementia; characterized by a loss of memory and intellectual and emotional function.

**GLOSSARY**

**AM**   *See* arbuscular mycorrhizae.

**amensalism**   One-sided competition between species, in which the interaction is detrimental to one species but not to the other.

**Ames test**   A test that helps ascertain whether or not an agent is a mutagen by using a strain of a bacterium, *Salmonella typhimurium*.

**amino acid**   Any of the monomers that are linked to form a protein. Amino acids have a common structure in which a carbon atom, called the α-carbon, is linked to an amino group (—NH$_2$) and a carboxyl group (—COOH), as well as to a hydrogen atom and a side chain that distinguishes the particular amino acid.

**aminoacyl site (A site)**   One of three sites for tRNA binding in the ribosome during translation; the other two are the peptidyl site (P site) and the exit site (E site). The A site is where incoming tRNA molecules bind to the mRNA (except for the initiator tRNA).

**aminoacyl tRNA**   *See* charged tRNA.

**aminoacyl-tRNA synthetase**   An enzyme that catalyzes the attachment of amino acids to tRNA molecules.

**ammonification**   The conversion of organic nitrogen to NH$_3$ and NH$_4^+$ during the nitrogen cycle.

**amnion**   The innermost of the four extraembryonic membranes in the amniotic egg. It protects the developing embryo in a fluid-filled sac called the amniotic cavity.

**amniotes**   A group of tetrapods with amniotic eggs that includes turtles, lizards, snakes, crocodiles, birds, and mammals.

**amniotic egg**   A type of egg produced by amniotes that contains the developing embryo and the four separate extraembryonic membranes that it produces: the amnion, the yolk sac, the allantois, and the chorion.

**amoeba (plural, amoebae)**   A protist that moves by pseudopodia, which involves extending cytoplasm into filaments or lobes.

**amoebocyte**   A mobile cell within a sponge's mesophyl that absorbs food from choanocytes, digests it, and carries the nutrients to other cells.

**amphibian**   An ectothermic, vertebrate animal that metamorphoses from a water-breathing to an air-breathing form but must return to the water to reproduce.

**amphipathic**   Refers to molecules containing a hydrophobic (water-fearing) region and a hydrophilic (water-loving) region.

**amplicon analysis**   A comparison of amplicons (amplified sequences) present in a particular DNA sample to reference sequences in a database.

**amplicon**   Any gene region for which many copies have been made (amplified) from a DNA sample with the use of specific primer sequences in polymerase chain reaction (PCR).

**ampulla (plural, ampullae)**   1. A muscular sac at the base of each tube foot of an echinoderm; used to store water. 2. A bulge in the walls of the semicircular canals of the mammalian inner ear; important for sensing circular motions of the head.

**amygdala**   An area of the limbic system of the vertebrate forebrain known to be critical for understanding and remembering emotional situations.

**amylase**   A digestive enzyme in saliva and the pancreas involved in the digestion of carbohydrates.

**anabolic reaction**   A metabolic pathway that involves the synthesis of larger molecules from smaller precursor molecules. Such reactions usually require an input of energy.

**anabolism**   A metabolic pathway that results in the synthesis of cellular molecules and macromolecules; requires an input of energy.

**anaerobic respiration**   The breakdown of organic molecules in the absence of oxygen by using a final electron acceptor that is something other than oxygen.

**anaerobic**   Refers to an environment that lacks oxygen or a process that occurs in the absence of oxygen; a form of metabolism that does not require oxygen.

**anagenesis**   The pattern of speciation in which a single species is transformed into a different species over the course of many generations.

**analogous structure**   A structure that is the result of convergent evolution. Such structures have arisen independently, two or more times, because species have occupied similar types of environments on Earth.

**anaphase**   The phase of mitosis during which the sister chromatids separate from each other and move to opposite poles; the poles themselves also move farther apart.

**anatomy**   The study of the structures of living things.

**anchoring junction**   A type of junction between animal cells that attaches cells to each other and to the extracellular matrix (ECM).

**androgens**   Steroid hormones produced by the male testes (and, to a lesser extent, the adrenal glands) that affect most aspects of male reproduction.

**aneuploidy**   Alteration of the number of a particular chromosome present in an organism or cell, so the total number of chromosomes is not an exact multiple of a set.

**angiosperm**   A flowering plant. The term means enclosed seed, which reflects the presence of seeds within fruits.

**animal pole**   In triploblast organisms, the pole of the egg with less yolk and more cytoplasm.

**Animalia**   A eukaryotic kingdom of the domain Eukarya.

**animals**   Multicellular heterotrophs with cells that lack cell walls. Most animals have nerves, muscles, the capacity to move at some point in their life cycle, and the ability to reproduce sexually, with sperm fusing directly with eggs.

**anion**   An ion that has a net negative charge.

**annual**   A plant that dies after producing seed during its first year of life.

**antagonist**   Two or more muscles that produce oppositely directed movements at a joint.

**anterior**   Refers to the end of an animal where the head is found.

**anteroposterior axis**   In bilateral animals, one of the three axes along which the adult body pattern is organized; the others are the dorsoventral axis and the left-right axis.

**anther**   The uppermost part of a flower stamen, consisting of a cluster of four sporangia that produce and release pollen.

**antheridia**   Spherical or elongate gametangia that produce sperm in plants.

**anthropogenic**   Caused by humans or their activities; the term comes from the Greek *anthropogenes*, meaning "born of man," and is often applied to environmental pollution originating from human activity.

**anthropoidea**   A group of primates that includes the monkeys and the hominoidea; these species are larger-brained, diurnal, and have opposable thumbs.

**antibiotic**   A chemical, usually made by microorganisms, that inhibits the growth of certain other microorganisms.

**antibody**   A protein secreted by plasma cells that is part of the immune response; antibodies travel all over the body to reach antigens identical to those that stimulated their production, combine with these antigens, and then guide an attack that eliminates the antigens or the cells bearing them.

**anticodon**   A three-base sequence in tRNA that is complementary to a codon in mRNA.

**antidiuretic hormone (ADH)**   A polypeptide hormone secreted by the posterior pituitary that acts on kidney cells to decrease urine production.

**antigen**   Any foreign molecule that the host does not recognize as self and that triggers a specific immune response.

**antigen-presenting cells (APCs)**   Cells of a vertebrate's acquired immune system that complex antigen with class II MHC proteins, leading to helper T cell activation.

**antiparallel**   The arrangement in DNA where one strand runs in the 5′ to 3′ direction and the other strand is oriented in the 3′ to 5′ direction.

**antiporter**   A type of transporter that binds two or more ions or molecules and transports them in opposite directions across a membrane.

**anus**   The opening at the posterior end of the alimentary canal through which solid wastes are expelled.

**aorta**   In vertebrates, a large blood vessel that exits a ventricle of the heart and leads to the systemic circulation.

**apical meristem**   In plants, a group of actively dividing cells at a growing tip.

**apical region**   The region of a plant seedling that produces the leaves and flowers.

**apical-basal polarity**   An architectural feature of plants in which they display an upper, apical pole and a lower, basal pole; the shoot apical meristem occurs at the apical pole, and the root apical meristem occurs at the basal pole.

**apical-basal-patterning genes**   A category of genes that are important in early stages of plant development during which the apical and basal axes are formed.

**apomixis**   A natural asexual reproductive process in which plant fruits and seeds are produced within flowers in the absence of fertilization.

**apoplast**   The continuum of water-filled cell walls and intercellular spaces in a plant.

**apoplastic transport**   The movement of solutes along cell walls and in the spaces between cells.

**apoptosis**   Programmed cell death.

**apoptosome**   A complex of proteins that promotes apoptosis via the intrinsic pathway by activating caspases.

**aposematic coloration**   Warning coloration that advertises an organism's unpalatable taste.

**aquaporin**   A transport protein in the form of a channel that allows the rapid diffusion of water across the cell membrane.

**aqueous solution**   A solution made with water.

**aquifer**   An underground water supply.

**arbuscular mycorrhizae**   Associations between plant cells, often root cells of vascular plants, and fungi that form highly branched hyphae.

**Archaea**   One of the three domains of life; the other two are Bacteria and Eukarya.

**archaea**   When not capitalized, refers to a species within the domain Archaea.

**archegonia (singular, archegonium)**   Flask-shaped gametangia that each enclose a single egg cell in plants.

**archenteron**   A cavity formed in an animal embryo during gastrulation that will become the organism's digestive tract.

**area hypothesis**   The proposal that larger areas contain more species than smaller areas because they can support larger populations and a greater range of habitats.

**artery**   A blood vessel that carries blood away from the heart.

**artificial selection**   *See* selective breeding.

**asci (singular, ascus)**   Fungal sporangia shaped like sacs that produce and release sexual ascospores.

**ascocarp**   The type of fruiting body produced by ascomycete fungi.

**ascomycetes**   A phylum of fungi that produce sexual spores in saclike asci located at the surfaces of fruiting bodies known as ascocarps.

**ascospore**   The type of sexual spore produced by fungi in the phylum Ascomycota.

**aseptate**   The condition of not being partitioned into smaller cells; usually refers to fungal cells.

**asexual reproduction**   A reproductive strategy that occurs when offspring are produced from a single parent, without the fusion of gametes from two parents. The offspring are therefore clones of the parent.

**assimilation**   During the nitrogen cycle, the process by which plants and animals incorporate the $NH_3$, $NH_4^+$, and $NO_3^-$ formed through nitrogen fixation and nitrification.

**assisted reproductive technologies (ART)**   A collection of procedures used to produce a pregnancy by artificial mechanisms.

**association**   A statistical result in which changes in two variables follow a pattern.

**associative learning**   A change in behavior due to the development of an association between a stimulus and a response.

**asthma**   A disease in which the smooth muscles around the bronchioles contract more than usual, decreasing airflow in the lungs.

**AT/GC rule**   Refers to the phenomenon that an A in one DNA strand always hydrogen-bonds with a T in the opposite strand, and a G in one strand always hydrogen-bonds with a C.

**atherosclerosis**   The condition in which plaques cause the arteries to narrow and harden and large plaques may occlude (block) the lumen of an artery.

**atmospheric pressure**   The pressure exerted by the gases in the air on the body surfaces of animals.

**atom**   The smallest functional unit of matter that forms all chemical substances and cannot be further broken down into other substances by ordinary chemical or physical means.

**atomic mass**   An atom's mass relative to the mass of other atoms. By convention, the most common form of carbon, which has six protons and six neutrons, is assigned an atomic mass of exactly 12.

**atomic nucleus**   The center of an atom; contains protons and neutrons.

**atomic number**   The number of protons in an atom.

**ATP synthase**   An enzyme that utilizes the energy stored in a $H^+$ electrochemical gradient for the synthesis of ATP via chemiosmosis.

**ATP-dependent chromatin remodeling complex**   A collection of proteins that alters chromatin structure.

**atrial natriuretic peptide (ANP)**   A polypeptide hormone secreted from the atria of the heart whenever blood levels of sodium increase; ANP causes a loss of $Na^+$ in the urine (natriuresis) by decreasing sodium reabsorption in the renal tubules.

**atrioventricular (AV) node**   Specialized cardiac cells in most vertebrates that sit near the junction of the atria and ventricles and conduct the electrical events from the atria to the ventricles.

**atrioventricular (AV) valve**   A one-way valve into a ventricle of the vertebrate heart through which blood moves from an atrium.

**atrium**   In the heart, a chamber to collect blood from the tissues.

**atrophy**   A reduction in the size of a structure, such as a muscle.

**audition**   The ability to detect and interpret sound waves; present in vertebrates and arthropods.

**autoimmune disease**   In humans and many other vertebrates, a disorder in which the body's normal state of immune tolerance breaks down, with the result that immune responses are directed against the body's own cells and tissues.

**autonomic nervous system**   The division of the peripheral nervous system that regulates homeostasis and organ function.

**autophagosome**   A double-membrane structure enclosing cellular material destined to be degraded; produced by the process of autophagy.

**autophagy**   A process whereby cellular material, such as a worn-out organelle, becomes enclosed in a double membrane and is degraded.

**autosomes**   All of the chromosomes found in the cell nucleus of eukaryotes except for the sex chromosomes.

**autotomy**   In echinoderms, the ability to detach a body part, such as a limb, that will later regenerate.

**autotroph**   An organism that has metabolic pathways that use energy from either inorganic molecules or light to make organic molecules.

**auxin efflux carrier**   One of several types of PIN proteins, which transport auxin out of plant cells.

**auxin influx carrier (AUX1/LAX)**   A plasma membrane protein that transports auxin into plant cells.

**auxin-responsive genes**   Plant genes that are regulated by the hormone auxin.

**auxins**   A group of plant hormones; considered to be "master" plant hormones because they influence plant structure, development, and behavior in many ways.

**avirulence gene (*Avr* gene)**   A gene in a plant pathogen that encodes a virulence-enhancing elicitor, which causes plant disease.

**axillary bud**   A bud that occurs in the axil, the upper angle where a twig or leaf emerges from a stem.

**axillary meristem**   A meristem produced in the axil, the upper angle where a twig or leaf emerges from a stem. Axillary meristems generate axillary buds, which can produce flowers or branches.

**axon hillock**   The part of the axon closest to the cell body; typically where an action potential begins.

**axon terminal**   The end of an axon, which conveys electrical or chemical messages to other cells.

**axon**   An extension of the plasma membrane of a neuron that is involved in sending signals to neighboring cells.

**axoneme**   An internal structure of eukaryotic flagella and cilia that contains microtubules, the motor protein dynein, and linking proteins.

# B

**B cell**   A type of lymphocyte that participates in acquired immune responses.

**bacilli** (singular, **bacillus**)   Rod-shaped prokaryotic cells.

**backbone**   The linear arrangement of phosphates and sugar molecules in a DNA or RNA strand.

**Bacteria**   One of the three domains of life; the other two are Archaea and Eukarya.

**bacterial colony**   A clone of genetically identical cells formed from a single bacterium by repeated cell divisions.

**bacteriophage**   A virus that infects bacteria.

**bacteroid**   A modified bacterial cell of the type known as rhizobia present in mature root nodules of some plants.

**balanced polymorphism**   The phenomenon in which two or more alleles are kept in balance and maintained in a population over the course of many generations.

**balancing selection**   A type of natural selection that maintains genetic diversity in a population.

**balloon angioplasty**   A common treatment to restore blood flow through an artery. A thin tube with a tiny, inflatable balloon at its tip is threaded through the artery to the diseased area; inflating the balloon compresses the plaque against the arterial wall, widening the lumen.

**baroreceptor reflex**   The rapid, involuntary compensatory response of vertebrates to a change in blood pressure; the pressure is detected by *baroreceptors*, which signal the brainstem to initiate changes in the activity of autonomic neurons. This, in turn, influences the function of structures of the circulatory system in such a way as to correct for a deviation of blood pressure beyond the normal range.

**baroreceptor**   A pressure-sensitive region within the walls of certain arteries that contains the endings of nerve cells; these regions sense and help to maintain blood pressure in the normal range for an animal.

**Barr body**   A highly condensed X chromosome present in the cells of female mammals.

**basal body**   A site at the base of flagella or cilia from which microtubules grow. Basal bodies are anchored on the cytosolic side of the plasma membrane.

**basal metabolic rate (BMR)**   The metabolic rate of an animal under resting conditions, in a postabsorptive state, and at a standard temperature.

**basal nuclei**   Clusters of neuronal cell bodies in the vertebrate forebrain that surround the thalamus and lie beneath the cerebral cortex; involved in planning and learning movements.

**basal region**   The region of a plant seedling that produces the roots.

**basal transcription**   A low level of transcription resulting from the action of the core promoter alone.

**base pair**   The structure in which two bases in opposite strands of DNA are held together by hydrogen bonding to each other.

**base substitution**   A mutation that involves the substitution of a single base in the DNA for another base.

**base**   1. A molecule that when dissolved in water lowers the $H^+$ concentration. 2. A component of nucleotides that is a single or double ring of carbon and nitrogen atoms.

**basidia**   Club-shaped cells that produce sexual spores in the fruiting bodies of basidiomycete fungi.

**basidiocarp**   The type of fruiting body produced by fungi in the phylum Basidiomycota.

**basidiomycetes**   A phylum of fungi whose sexual spores are produced on the surfaces of club-shaped structures (basidia).

**basidiospore**   A sexual spore of fungi in the phylum Basidiomycota.

**basilar membrane**   A component of the mammalian ear that vibrates back and forth in response to sound and bends the stereocilia in one direction and then the other.

**basophil**   A type of leukocyte that secretes the anticlotting factor heparin at the site of an infection, which helps flush out the infected site; basophils also secrete histamine, which attracts infection-fighting cells and proteins.

**Batesian mimicry**   The mimicry of an unpalatable species (the model) by a palatable one (the mimic).

**behavior**   The observable response of an organism to an external or internal stimulus.

**behavioral ecology**   A subdiscipline of organismal ecology that focuses on how the behavior of an individual organism contributes to its survival and reproductive success, which, in turn, eventually affects the population density of the species.

**benign tumor**   A precancerous mass of abnormal cells.

**bidirectional replication**   The process in which DNA replication proceeds outward from the origin in opposite directions.

**biennial**   A plant that does not reproduce during the first year of life but may reproduce within the following year.

**Bilateria**   Bilaterally symmetric animals.

**bile salts**   A group of substances produced in the liver that solubilize dietary fat and increase its accessibility to digestive enzymes.

**bile**   A substance produced by the liver that contains bicarbonate ions, cholesterol, phospholipids, a number of organic wastes, and a group of substances derived from cholesterol and collectively termed bile salts. Bile emulsifies fats so that they can be absorbed by the small intestine.

**binary fission**   The process of cell division in bacteria and archaea in which one cell divides into two cells.

**binocular (or stereoscopic) vision**   A type of vision in animals having two eyes located at the front of the head; the overlapping images coming into both eyes are processed together in the brain to form one perception. Binocular vision enables depth perception.

**binomial nomenclature**   The standard format for scientific naming of species. Each species has a genus name and a specific epithet.

**biochemistry**   The study of the chemistry of living organisms.

**biodiversity crisis**   The idea that there is currently an elevated loss of species on Earth, far beyond the normal historical extinction rate of species.

**biodiversity hot spots**   Regions that are biologically diverse and under threat of destruction.

**biodiversity**   The variety of life-forms that exist now and existed in the past.

**biofilm**   An aggregation of microorganisms that secrete adhesive mucilage, thereby gluing themselves to surfaces.

**biogeochemical cycle**   The continuous movement of a nutrient such as nitrogen, carbon, sulfur, or phosphorus from the physical environment to organisms and back.

**biogeographic region**   One of six geographic regions into which the world's biota can be divided: Nearctic, Palearctic, Neotropical, Ethiopian, Oriental, and Australian.

**biogeography**   The study of the geographic distribution of extinct and living species.

**biological control**   The use of an introduced species' natural enemies to control its proliferation.

**biological diversity**   *See* biodiversity.

**biological evolution**   A heritable change in a population of organisms from one generation to the next.

**biological membrane**   Any membrane made by living cells; can be the plasma membrane or an internal membrane that surrounds an organelle.

**biological nitrogen fixation**   Nitrogen fixation that is performed in nature by certain prokaryotes.

**biological species concept**   An approach used to distinguish species, which states that a species is a group of individuals whose members have the potential to interbreed with one another in nature to produce viable, fertile offspring but cannot successfully interbreed with members of other species.

**biology**   The study of life.

**bioluminescence**   A phenomenon in living organisms in which chemical reactions give off light rather than heat.

**biomagnification**   The increase in the concentration of a substance in living organisms from lower to higher trophic levels in a food chain.

**biomass**   A quantitative estimate of the total mass of living matter in a given area, usually measured in grams or kilograms per square meter.

**biome**   A major type of habitat characterized by distinctive plant and animal life.

**bioremediation**   The use of living organisms, usually microbes or plants, to detoxify polluted habitats such as dump sites or oil spills.

**biosphere**   The regions on the surface of the Earth and in the atmosphere where living organisms exist.

**biosynthetic reaction**   Also called an anabolic reaction; a chemical reaction in which small molecules are used to synthesize larger molecules.

**biotic**   The term used to describe interactions among organisms.

**biparental inheritance**   An inheritance pattern in which both the male and female gametes contribute organellar genes to the offspring.

**bipedal**   Having the ability to walk on two feet.

**bipolar cells**   Cells in the vertebrate eye that make synapses with photoreceptors and relay responses to the ganglion cells.

**bivalent**   Homologous pairs of sister chromatids that are associated with each other, lying side by side.

**blade**   The flattened portion of a leaf.

**blastocoel**   A cavity formed in a cleavage-stage vertebrate embryo (blastula); provides a space into which cells of the future digestive tract will migrate.

**blastocyst**   The mammalian counterpart of a blastula.

**blastoderm**   A flattened disc of dividing cells in the embryo of animals that undergo incomplete cleavage; occurs in birds and some fishes.

**blastomeres**   The two half-size daughter cells produced by each cell division during cleavage.

**blastopore**   A small opening created when a band of tissue invaginates during gastrulation. It forms the primary opening of the archenteron to the outside.

**blastula**   An animal embryo at the stage where it has an outer epithelial layer and an inner cavity, forming a hollow sphere of cells.

**blood pressure**   The force exerted by blood on the walls of blood vessels; blood pressure is responsible for moving blood through the vessels.

**blood**   A fluid connective tissue in animals consisting of cells and (in mammals) cell fragments suspended in a solution of water containing dissolved nutrients, proteins, gases, and other molecules.

**body mass index (BMI)**   A method of assessing body fat and health risk that involves calculating the ratio of weight compared with height; weight in kilograms is divided by the square of the height in meters.

**Bohr effect**   The effect of $CO_2$ and $H^+$ on the affinity of hemoglobin for oxygen (that is, on the oxygen-hemoglobin dissociation curve).

**bone**   A relatively hard component of the vertebrate skeleton; a living, dynamic tissue composed of organic materials and minerals.

**bottleneck effect**   A change in allele frequencies due to genetic drift in a population that has been dramatically reduced in size; this effect can reduce the genetic diversity of the population.

**Bowman's capsule**   A saclike structure that houses the glomerulus at the beginning of the tubular component of a nephron in the mammalian kidney.

**brain**   Organ of the central nervous system of animals that functions to process and integrate information.

**brainstem**   The part of the vertebrate brain composed of the medulla oblongata, the pons, and the midbrain.

**brassinosteroid**   One of several plant hormones that help a plant to cope with environmental stress.

**bronchi** (singular, **bronchus**)   Tubes branching from the trachea and leading into the lungs.

**bronchiole**   A thin-walled, small tube branching from the bronchi and leading to the alveoli in mammalian lungs.

**bronchodilator**   A compound that binds to receptors on the plasma membranes of smooth muscle cells of the bronchioles of the lung and causes the muscle cells to relax, thereby widening the bronchioles and easing breathing.

**brown adipose tissue**   A specialized tissue in small mammals such as hibernating bats, small rodents living in cold environments, and many newborn mammals, including humans, that can help to generate heat and maintain body temperature.

**brush border**   The collective name for the microvilli in the small intestine and the proximal tubules of the kidneys in vertebrates.

**bryophytes**   Liverworts, mosses, and hornworts, the modern nonvascular land plants.

**buccal pumping**   A form of breathing in which animals take in water or air into their mouths, then raise the floor of the mouth, creating a positive pressure that pumps water or air across the gills or into the lungs; found in fishes and amphibians.

**bud**   A miniature plant shoot having a dormant shoot apical meristem.

**budding**   A form of asexual reproduction in which a portion of the parent organism pinches off to form a complete new individual.

**buffer**   An acid-base pair that minimizes pH fluctuations in the fluids of living organisms. Buffers can raise or lower pH as needed.

**bulk flow**   The mass movement of liquid in a plant caused by pressure, gravity, or both.

# C

**$C_3$ plant**   A plant that adds $CO_2$ to RuBP to produce 3PG, a three-carbon molecule.

**$C_4$ plant**   A plant that uses PEP carboxylase to initially fix $CO_2$ into a four-carbon molecule and later uses rubisco to fix $CO_2$ into simple sugars; this mechanism is an adaptation to hot, dry environments.

**cadherin**   A cell adhesion molecule found in animal cells that promotes cell-to-cell adhesion.

**calcitonin**   A hormone that plays a role in $Ca^{2+}$ homeostasis in some vertebrates.

**calorie**   The amount of heat required to raise the temperature of 1 gram of water 1°C. The Calorie (dietary unit) is equivalent to a kilocalorie, or 1,000 calories.

**Calvin cycle**   The second stage in the process of photosynthesis. During this cycle, ATP is used as a source of energy and NADPH is used as a source of high-energy electrons, driving the synthesis of carbohydrates using $CO_2$.

**CAM (crassulacean acid metabolism) plants**   $C_4$ plants that open their stomata at night to take up $CO_2$.

**Cambrian explosion**   An event during the Cambrian period (543–490 mya) in which there was an abrupt increase (on a geological scale) in the diversity of animal species.

**camouflage**   The blending of an organism with the background of its habitat.

**cAMP**   *See* cyclic adenosine monophosphate.

**canopy**   The uppermost layer of tree foliage in a forest.

**CAP site**   One of two regulatory sites near the *lac* promoter; this site is a DNA sequence recognized by the catabolite activator protein (CAP).

**capillary**   A tiny thin-walled vessel that is the site of gas and nutrient exchange between the blood and interstitial fluid.

**capping**   The process in which 7-methylguanosine is covalently attached at the 5′ end of pre-mRNAs of eukaryotes.

**capsid**   A protein coat enclosing a virus's genome.

**capsule**   A very thick, gelatinous glycocalyx produced by certain strains of bacteria that may help them avoid being destroyed by an animal's immune (defense) system.

**carapace**   The hard protective cuticle covering the cephalothorax of a crustacean.

**carbohydrate**   A carbon-containing organic molecule often represented by the general formula, $C_n(H_2O)_n$; carbohydrates include starches, sugars, and cellulose.

**carbon fixation**   A process in which carbon from inorganic $CO_2$ is incorporated into an organic molecule such as a carbohydrate.

**carcinogen**   An agent that increases the likelihood of developing cancer, usually a mutagen.

**carcinoma**   A cancer of epithelial cells.

**cardiac angiography**   A medical procedure used to visualize the coronary arteries and check for the presence of disease.

**cardiac cycle**   The events that produce a single heartbeat, which can be divided into two phases: diastole and systole.

**cardiac muscle**   A type of muscle tissue, found only in hearts, in which physical and electrical connections between individual cells enable many of the cells to contract simultaneously.

**cardiac output (CO)**   The amount of blood the heart pumps per unit time, usually expressed in units of L/min.

**carnivore**   An animal that consumes animal flesh or fluids.

**carotenoid**   A type of photosynthetic or protective pigment found in plastids that imparts a color that ranges from yellow to orange to red.

**carpel**   A flower shoot organ that produces ovules that contain female gametophytes.

**carrying capacity ($K$)**   The upper boundary for a population size in a given environment.

**Casparian strips**   Ribbon-like structures in the walls of endodermal cells of plant roots, composed of suberin and phenolic polymers; prevent apoplastic transport of solutes into vascular tissues.

**caspase**   An enzyme that functions as a protease when it is activated during apoptosis.

**catabolic reaction**   A metabolic pathway in which a molecule is broken down into smaller components, usually releasing energy.

**catabolism**   A metabolic pathway that results in the breakdown of larger molecules into smaller molecules. Such reactions are often exergonic.

**catabolite activator protein (CAP)**   An activator protein for the *lac* operon.

**catabolite repression**   In bacteria, a process whereby transcriptional regulation is influenced by the presence of a preferred energy source (glucose).

**catalase**   An enzyme within peroxisomes that breaks down hydrogen peroxide to water and oxygen gas.

**catalyst**   An agent that speeds up the rate of a chemical reaction without being permanently changed or consumed during the reaction.

**cataract**   An accumulation of protein in the lens of the eye; causes blurring and poor night vision.

**cation exchange**   With regard to soil, the process in which hydrogen ions replace mineral cations on the surfaces of clay particles.

**cation**   An ion that has a net positive charge.

**cDNA library**   A type of DNA library in which the inserts are derived from cDNA.

**cDNA**   *See* complementary DNA.

**cecum**   The first portion of a vertebrate's large intestine.

**cell adhesion molecule (CAM)**   A membrane protein found in animal cells that promotes cell adhesion.

**cell adhesion**   A vital function of the cell membrane that allows cells to bind to each other. Cell adhesion is critical in the formation of multicellular organisms and provides a way to convey positional information between neighboring cells.

**cell biology**   The study of individual cells and their interactions with each other.

**cell body**   A part of a neuron that contains the cell nucleus and other organelles.

**cell communication**   The process by which cells can detect, interpret, and respond to signals in their environment. In multicellular organisms, cell communication is also needed to coordinate cellular activities within the whole organism.

**cell cycle**   A series of events that leads to cell division. For eukaryotes, it involves a series of phases in which a cell divides by mitosis or meiosis.

**cell differentiation**   The process by which cells become specialized into particular types.

**cell division**   The process of cell reproduction, in which one cell divides into two cells.

**cell junctions**   Specialized structures that adhere cells to each other and to the ECM.

**cell nucleus**   The membrane-bound area of a eukaryotic cell in which the genetic material is found.

**cell plate**   In plant cells, a structure that forms a cell wall between the two daughter cells during cytokinesis.

**cell signaling**   A vital function of the plasma membrane in which cells sense changes in their environment and communicate with each other.

**cell surface receptor**   A receptor found in the plasma membrane that enables a cell to respond to different kinds of extracellular signaling molecules.

**cell theory**   A theory that states that all organisms are made of cells, cells are the smallest units of living organisms, and new cells come from pre-existing cells by cell division.

**cell wall**   A relatively rigid, porous structure located outside the plasma membrane of prokaryotic, plant, fungal, and certain protist cells; provides support and protection.

**cell**   The simplest unit of a living organism.

**cell-free translation system**   *See* in vitro translation system.

**cell-mediated immunity**   A type of acquired immunity in which cytotoxic T cells directly attack and destroy infected body cells, cancer cells, or transplanted cells.

**cell-to-cell communication**   A form of cell communication that occurs between two different cells.

**cellular respiration**   A process by which living cells obtain energy from organic molecules and release waste products.

**cellular response**   Adaptation at the cellular level that involves a cell responding to signals in its environment.

**cellulose**   The main macromolecule of the cell wall of plants and many algae; a linear polymer made of thousands of glucose monomers.

**central cell**   In the female gametophyte of a flowering plant, a large cell that contains two nuclei; after double fertilization, it forms the first cell of the nutritive endosperm tissue.

**central dogma**   Refers to the steps of gene expression at the molecular level: DNA is transcribed into mRNA, and mRNA is translated into a polypeptide.

**central nervous system (CNS)**   In vertebrates, the brain and spinal cord.

**central region**   The region of a plant seedling that produces stem tissue.

**central vacuole**   An organelle that often occupies 80% or more of the volume of a plant cell and stores a large amount of water, enzymes, and inorganic ions.

**central zone**   The area of a plant shoot meristem where undifferentiated stem cells are maintained.

**centrioles**   A pair of structures within the centrosome of animal cells. Most plant cells and many protists lack centrioles.

**centromere**   The region where the two sister chromatids are tightly associated; the centromere is an attachment site for kinetochore proteins.

**centrosome**   A single structure often near the nucleus of a eukaryotic cell that forms a nucleating site for the growth of microtubules; also called a microtubule-organizing center.

**cephalization**   The localization of sensory structures at the anterior end of an animal's body.

**cerebellum**   The part of the vertebrate hindbrain, along with the pons, responsible for monitoring and coordinating body movements.

**cerebral cortex**   The surface layer of gray matter that forms the outer part of the cerebrum of the vertebrate brain.

**cerebral ganglia**   A paired structure in the head of invertebrates that receives input from sensory cells and controls motor output.

**cerebrospinal fluid**   Fluid that exists in ventricles within the central nervous system and surrounds the exterior of the brain and spinal cord; it absorbs physical shocks to the brain resulting from sudden movements or blows to the head.

**cerebrum**   A region of the vertebrate forebrain that is responsible for the higher functions of conscious thought, planning, and emotion, as well as control of motor function.

**channel**   A transmembrane protein that forms an open passageway for the facilitated diffusion of ions or molecules across a membrane.

**chaperone**   A protein that keeps another protein in an unfolded state during the process of post-translational sorting.

**character displacement**   The tendency for two species to diverge in morphology and thus resource use because of competition.

**character state**   A particular variant of a given character.

**character**   A characteristic of an organism, such as the appearance of seeds, pods, flowers, or stems in the garden pea.

**charged tRNA**   A tRNA with its attached amino acid; also called aminoacyl tRNA.

**checkpoint protein**   A protein that senses if a cell is in the proper condition to divide and prevents it from progressing through the cell cycle if it is not.

**checkpoint**   One of three critical regulatory points found in the cell cycle of eukaryotic cells. At these checkpoints, a variety of proteins act as sensors to determine if a cell is in the proper condition to divide.

**chemical equilibrium**   A state of a chemical reaction in which the rate of formation of products equals the rate of formation of reactants.

**chemical evolution**   The process by which a population of molecules changes over time to become a new population with a different chemical composition.

**chemical mutagen**   A chemical that causes mutations.

**chemical potential energy**   The potential energy contained within atoms and the bonds between atoms.

**chemical reaction**   A process in which one or more substances are changed into other substances.

**chemical selection**   The process that occurs when a chemical within a mixture has special properties or advantages that cause it to increase in amount. May have played a key role in the formation of an RNA world.

**chemical synapse**   A synapse in which a chemical called a neurotransmitter is released from an axon terminal and acts as a signal from the presynaptic to the postsynaptic cell.

**chemiosmosis**   A process for making ATP in which energy stored in an ion electrochemical gradient is used to make ATP from ADP and $P_i$.

**chemoautotroph**   An organism able to use energy obtained by chemical modifications of inorganic compounds to synthesize organic compounds.

**chemoheterotroph**   An organism that must obtain organic molecules both for energy and as a carbon source.

**chemoreceptor**   A sensory receptor in animals that responds to specific chemical compounds.

**chitin**   A tough, nitrogen-containing, polysaccharide polymer that forms the external skeleton of many insects and crustaceans and is found in the cell walls of fungi.

**chlorophyll $a$**   A type of chlorophyll pigment found in plants, algae, and cyanobacteria.

**chlorophyll $b$**   A type of chlorophyll pigment found in plants, green algae, and some other photosynthetic organisms.

**chlorophyll** A photosynthetic green pigment found in the chloroplasts of plants, algae, and some bacteria.

**chloroplast genome** The chromosome found in chloroplasts.

**chloroplast** A semiautonomous organelle found in plant and algal cells that carries out photosynthesis.

**chlorosis** The yellowing of plant leaves caused by any of various mineral deficiencies.

**choanocyte** A specialized cell of sponges that functions to trap and eat particulate matter and plankton.

**cholecystokinin (CCK)** A hormone released by cells of the small intestine in vertebrates; stimulates release of pancreatic enzymes into the small intestine and contraction of the gallbladder.

**chondrichthyans** Members of the clade Chondrichthyes, including sharks, skates, and rays.

**chorion** One of the four extraembryonic membranes in the amniotic egg. It provides gas exchange between the embryo and the surrounding air.

**chorionic gonadotropin** An LH-like hormone made by a layer of cells around the blastocyst and by the placenta that maintains the corpus luteum.

**chromatin** The complex of DNA and proteins that makes up eukaryotic chromosomes.

**chromosome territory** A distinct area where each chromosome is located within the cell nucleus of eukaryotic cells; chromosome territories do not overlap.

**chromosome theory of inheritance** An explanation of how the steps of meiosis account for the inheritance patterns observed by Mendel.

**chromosome** A discrete unit of genetic material composed of DNA and associated proteins. Eukaryotes have chromosomes in their cell nuclei and in plastids and mitochondria.

**chylomicron** Large fat droplet coated with amphipathic proteins that perform an emulsifying function similar to that of bile salts; chylomicrons are formed in intestinal epithelial cells from absorbed fats in the diet.

**chyme** A solution of water, ions, molecular fragments of proteins, nucleic acids, and carbohydrates, droplets of fat, and various other small molecules produced in the vertebrate stomach.

**cilia** (singular, **cilium**) Cell appendages that have the same internal structure as flagella and function like flagella to facilitate cell movement; cilia are shorter and more numerous than are flagella.

**ciliate** A protist that moves by means of cilia, which are tiny hairlike extensions that occur on the outside of cells and have the same internal structure as flagella.

**circadian rhythm** Internal biological clock that occurs in plants, animals, and other organisms.

**circulatory system** All of the structures in an animal's body that contribute to the movement of blood or hemolymph throughout the body; in vertebrates includes the heart, blood vessels, and blood.

**cis/trans isomers** Organic molecules with the same chemical composition but existing in two different configurations determined by the positions of hydrogen atoms on the two carbons of a C=C double bond. When the hydrogen atoms are on the same side of the double bond, it is called a *cis* isomer; when on the opposite sides of the double bond, it is a *trans* isomer.

**cis-acting element** A DNA segment that must be adjacent to the gene(s) that it regulates

**cis-effect** The effect on gene regulation that is mediated by a *cis*-acting element.

**cisternae** Flattened, fluid-filled tubules of the endoplasmic reticulum.

**citric acid cycle** A cycle that results in the breakdown of carbohydrates to $CO_2$; also known as the Krebs cycle.

**clade** A group of species consisting of a common ancestral species and all of its descendant species.

**cladistic approach** An approach used to construct a phylogenetic tree by comparing shared primitive and shared derived characters among different species.

**cladistics** The classification of species based on evolutionary relationships.

**cladogenesis** The splitting or diverging of one species into two or more species.

**cladogram** A phylogenetic tree constructed by using a cladistic approach.

**clasper** An extension of the pelvic fin of a chondrichthyan, used by the male to transfer sperm to the female.

**class** In taxonomy, a subdivision of a phylum.

**classical conditioning** A type of associative learning in which an involuntary response comes to be associated positively or negatively with a stimulus that did not originally elicit the response.

**cleavage furrow** In animal cells, an area that constricts like a drawstring to separate the cells during cytokinesis.

**cleavage** A succession of rapid cell divisions with no significant growth that produces a hollow sphere of cells called a blastula.

**climate change** A long-term change in Earth's climate or change in climate in a particular region.

**climate** The prevailing weather pattern of a given region.

**climax community** A distinct end point of succession.

**clonal deletion** One of two mechanisms that explain why normal individuals lack active lymphocytes that respond to self components; in this case, T cells with receptors capable of binding self proteins are destroyed by apoptosis.

**clonal inactivation** One of two mechanisms that explain why normal individuals lack active lymphocytes that respond to self components; in this case, the process occurs outside the thymus and causes potentially self-reacting T cells to become nonresponsive.

**clonal selection** The process by which an antigen-stimulated lymphocyte divides and forms a clone of cells, each of which recognizes that particular antigen.

**closed circulatory system** A circulatory system in which blood flows throughout an animal entirely within a series of vessels and is kept separate from the interstitial fluid.

**closed conformation** Chromatin that cannot be transcribed into RNA.

**clumped** The term used to refer to the most common pattern of dispersion within a population, in which individuals are gathered in small groups.

**cnidocil** On the surface of a cnidocyte, a hairlike trigger that detects stimuli.

**cnidocyte** A characteristic feature of cnidarians; a stinging cell that functions in defense or the capture of prey.

**coacervates** Droplets that form spontaneously from the association of charged polymers such as proteins, carbohydrates, or nucleic acids surrounded by water.

**coactivator** A protein that increases the rate of transcription but does not directly bind to the DNA itself.

**cocci** Sphere-shaped prokaryotic cells.

**cochlea** A coiled structure in the inner ear of mammals that contains the auditory receptors (organ of Corti).

**coding sequence** The region of a gene or a DNA molecule that encodes the information for the amino acid sequence of a polypeptide.

**coding strand** The DNA strand opposite to the template (or noncoding strand).

**codominance** The phenomenon in which a single individual expresses two alleles.

**codon** A sequence of three nucleotide bases that specifies a particular amino acid or a stop codon; codons function during translation.

**coefficient of relatedness (r)** The probability that any two individuals will share a copy of a particular gene.

**coelom** A fluid-filled body cavity in an animal.

**coelomate** An animal with a true coelom.

**coenzyme** An organic molecule that temporarily binds to an enzyme and participates in the chemical reaction that the enzyme catalyzes, but is left unchanged when the reaction is completed.

**coevolution** The process by which two or more species of organisms influence each other's evolutionary pathway.

**cofactor** Usually an inorganic ion that temporarily binds to the surface of an enzyme and promotes a chemical reaction.

**cognitive learning** The ability to solve problems with conscious thought and without direct environmental feedback.

**cohesion** The ability of like molecules to noncovalently bind to each other; the attraction of water molecules for each other.

**cohesion-tension theory** The explanation for long-distance water transport in plants as the combined effect of the cohesion of water and evaporative tension.

**cohort** A group of organisms of the same age.

**coleoptile** A protective sheath that encloses the first bud of the epicotyl in a mature monocot embryo.

**coleorhiza** A protective envelope that encloses the young root of a monocot.

**colinearity rule** The phenomenon whereby the order of homeotic genes along the chromosome correlates with their expression along the anteroposterior axis of the body.

**collagen** A protein secreted from animal cells that forms large fibers in the extracellular matrix.

**collenchyma cell** A type of flexible plant cell that makes up the tissue called collenchyma.

**collenchyma** A plant ground tissue that provides support to plant organs.

**colligative property** A property of a solution that depends only on the total number of dissolved solute particles.

**colon** A part of a vertebrate's large intestine consisting of three relatively straight segments—the ascending, transverse, and descending portions. The terminal portion of the descending colon is S-shaped, forming the sigmoid colon, which empties into the rectum.

**combinatorial control** The phenomenon whereby a combination of many factors determines the expression of any given gene.

**commensalism** An interaction that benefits one species and leaves the other unaffected.

**communication** The use of specially designed visual, chemical, auditory, or tactile signals to modify the behavior of others.

**community ecology** The study of how populations of species interact and form functional communities.

**community** An assemblage of populations of different species that live in the same place at the same time.

**compartmentalization** A characteristic of eukaryotic cells, in which many membrane-bound organelles separate the cell into different regions. Cellular compartmentalization allows a cell to carry out specialized chemical reactions in different places.

**competent** The term used to describe bacterial strains that have the ability to take up DNA from the environment.

**competition** An interaction that affects two or more species negatively, as they compete over food or other resources.

**competitive exclusion principle** The idea that two species with the same resource requirements cannot occupy the same niche.

**competitive inhibitor** A molecule that binds noncovalently to the active site of an enzyme and inhibits the ability of the substrate to bind.

**complement** The family of plasma proteins that provides a means for extracellular killing of microbes without prior phagocytosis.

**complementary DNA (cDNA)** DNA molecules that are made using mRNA as a starting material.

**complementary** The characteristic of the two strands of DNA that is due to the specific base pairing that occurs between nucleic acids: A pairs only with T (in DNA) or U (in RNA), and G pairs only with C.

**complete flower** A eudicot flower that possesses all four types of flower whorls or a monocot flower that has tepals, androecium, and gynoecium.

**complete metamorphosis** During development in the majority of insects, a dramatic change in body form from larva to a very different looking adult.

**compound eye** A type of image-forming visual organ in arthropods and some annelids consisting of several hundred to several thousand light detectors called ommatidia.

**compound** A molecule composed of two or more different elements.

**concentration gradient** See transmembrane gradient.

**concentration** The amount of a solute dissolved in a unit volume of solution.

**condensation reaction** A chemical reaction in which two or more molecules are combined into one larger molecule by covalent bonding, with the loss of a small molecule.

**condom** A sheathlike membrane worn over the penis; in addition to their contraceptive function, condoms significantly reduce the risk of contracting and transmitting sexually transmitted diseases.

**conduction** The process in which the body surface loses or gains heat through direct contact with cooler or warmer substances.

**cones** 1. Photoreceptors found in the vertebrate eye; they are less sensitive to low levels of light but can detect color. 2. The reproductive structures of coniferous plants.

**congenital hypothyroidism** A condition characterized by poor differentiation of the central nervous system due to a failure of neurons to become myelinated in fetal development; results in profound mental defects.

**conidia** A type of asexual reproductive cell produced by many fungi.

**conifers** A phylum of gymnosperm plants, Coniferophyta.

**conjugation** A type of gene transfer between bacteria that involves a direct physical interaction between two bacterial cells.

**connective tissues** Groups of cells that connect, anchor, and support the structures of an animal's body; include blood, adipose (fat-storing) tissue, bone, cartilage, loose connective tissue, and dense connective tissue.

**connexon** A channel that forms gap junctions in vertebrates, consisting of six connexin proteins in one cell aligned with six connexin proteins in an adjacent cell.

**conservation biology** The study that uses principles and knowledge from molecular biology, genetics, and ecology to protect and sustain the biological diversity of life.

**conservative mechanism** In this incorrect model for DNA replication, both parental strands of DNA remain together (are conserved) following DNA replication. The two newly made daughter strands are also joined together.

**constant region** The portion of the amino acid sequence in the heavy and light chains that is identical in all immunoglobulins of a given class.

**constitutive gene** An unregulated gene that has a relatively constant level of expression in all conditions over time.

**contig** A contiguous sequence of DNA fragments that consists of overlapping pieces of chromosomal DNA.

**continental drift** The process by which, over the course of billions of years, the major landmasses, known as the continents, have shifted their positions, changed their shapes, and, in some cases, become separated from each other.

**contraception** The use of methods to prevent fertilization or implantation of a fertilized egg.

**contractile vacuole** A small, membrane-enclosed, water-filled compartment that eliminates excess liquid from the cells of certain protists.

**contrast** In microscopy, relative differences in the lightness, darkness, or color between adjacent regions in a sample.

**control group** The sample in an experiment that is treated just like an experimental group except that it is not subjected to one particular variable.

**convection** The transfer of heat by the movement of air or water next to the body.

**convergent evolution** The process whereby two different species from different lineages independently develop similar characteristics because they occupy similar environments.

**core promoter** Refers to the TATA box and the transcriptional start site of a eukaryotic protein-encoding gene.

**corepressor** A small effector molecule that binds to a repressor protein to inhibit transcription.

**cork cambium** A secondary meristem in a plant that produces cork tissue.

**cornea** A thin, clear layer on the front of the vertebrate eye; also part of the lens of ommatidia in compound eyes.

**corona** The ciliated crown of members of the phylum Rotifera.

**coronary artery bypass** A common treatment to restore blood flow through a coronary artery. A small piece of healthy blood vessel is removed from one part of the body and surgically grafted onto the coronary circulation in order to bypass the diseased artery.

**coronary artery disease** A condition that occurs when plaques form in the coronary arteries.

**corpus callosum** The major tract that connects the two hemispheres of the cerebrum.

**corpus luteum** A structure that develops from a ruptured follicle following ovulation; it is responsible for secreting hormones that stimulate the development of the uterus during pregnancy.

**cortex** The area of a plant stem or root beneath the epidermis that is largely composed of parenchyma tissue.

**cortical reaction** An event in fertilization in which $IP_3$ and $Ca^{2+}$ produce barriers to more than one sperm cell binding to and uniting with an egg.

**cortisol** A glucocorticoid hormone made in the adrenal cortex.

**cotranslational sorting** The sorting process in which the synthesis of certain eukaryotic proteins begins in the cytosol and then halts temporarily until the ribosome has become bound to the ER membrane.

**cotyledon** An embryonic seed leaf.

**countercurrent exchange** 1. An arrangement of water and blood flow in which water enters a fish's mouth and flows between the lamellae of the gills in the opposite direction to blood flowing through the lamellar capillaries. 2. The transfer of heat between blood flowing in opposite directions in arteries and veins under the skin of vertebrates; regulates heat loss to the environment.

**countercurrent multiplication system** The mechanism by which the loop of Henle in the vertebrate kidney reabsorbs salts and water along it length.

**covalent bond** A chemical bond in which two atoms share a pair of electrons.

**CpG island** A cluster of CpG sites. C and G refer to the bases cytosine and guanine in DNA, and p refers to a phosphodiester linkage between the nucleotides containing those bases.

**cranial nerve** A nerve in the peripheral nervous system that is directly connected to the brain.

**cranium** A protective bony or cartilaginous housing that encases the brain of a craniate.

**crisis ecoregions** Representative habitats that are at greatest risk because of extensive habitat loss and lack of conservation or protection.

**CRISPR-Cas system** A system found in bacteria and archaea composed of noncoding RNAs and proteins that provides defense against bacteriophages, viruses, and transposons.

**CRISPR-Cas technology** An experimental technique to introduce mutations into genes.

**cristae** Projections of the highly invaginated inner membrane of a mitochondrion.

**critical innovations** New features that foster the diversification of phyla.

**critical period** A limited period of time during development in which many animals acquire species-specific patterns of behavior.

**crop** A storage organ that is a dilation of the lower esophagus; found in most birds and many invertebrates, including insects and some worms.

**cross-bridge cycle** During muscle contraction, the sequence of events that occurs between the time when a cross-bridge binds to a thin filament and when it is set to repeat the process.

**cross-bridge** A region of myosin molecules that extends from the surface of each thick filament toward a thin filament in skeletal muscle.

**cross-fertilization** Fertilization that involves the union of a female gamete and a male gamete from different individuals.

**crossing over** The exchange of genetic material between homologous chromosomes during meiosis; allows for increased variation in the genetic information that each parent may pass to the offspring.

**cross-pollination** The process in which a stigma receives pollen from a different plant of the same species.

**cryosphere** Earth's icy environments.

**cryptochrome** A type of blue-light receptor in plants and protists.

**CT scan** Computerized tomography, which is an X-ray technique used to examine the structure of bones and soft tissues, including the brain.

**C-terminus** The location of the last amino acid in a polypeptide; also known as the carboxyl end.

**cupula** A gelatinous structure within the lateral line organ of fishes that detects changes in water movement.

**cuticle** A waxy surface coating that helps to reduce water loss from plant surfaces. Also, a nonliving covering that serves to both support and protect an animal.

**cycads** A phylum of gymnosperm plants, referred to formally as Cycadophyta.

**cyclic adenosine monophosphate (cAMP)** A small molecule that is produced from ATP and acts as a second messenger.

**cyclic AMP (cAMP)** See cyclic adenosine monophosphate.

**cyclic electron flow** See cyclic photophosphorylation.

**cyclic photophosphorylation** During photosynthesis, a pattern of electron flow in the thylakoid membrane that is cyclic and generates only ATP.

**cyclin** A protein responsible for advancing a cell through the phases of the cell cycle by binding to a cyclin-dependent kinase.

**cyclin-dependent kinase (cdk)** A protein responsible for advancing a cell through the phases of the cell cycle. Its function is dependent on the binding of a cyclin.

**cyst**    A unicellular or multicellular structure that often has a thick, protective wall and can remain dormant through periods of unfavorable climate or low food availability.

**cytogenetics**    The field of genetics that involves the microscopic examination of chromosomes.

**cytokines**    A family of proteins that function in both innate and acquired immune defenses by providing a chemical communication network that synchronizes the components of the immune response.

**cytokinesis**    The division of the cytoplasm to produce two distinct daughter cells.

**cytokinin**    A type of plant hormone that promotes cell division.

**cytosine (C)**    A pyrimidine base found in DNA and RNA.

**cytoskeleton**    In eukaryotes, a network within the cytosol consisting of three different types of protein filaments called microtubules, intermediate filaments, and actin filaments.

**cytosol**    The region of a eukaryotic cell that is inside the plasma membrane and outside the organelles.

**cytotoxic T cell**    A type of lymphocyte that travels to the location of its target, binds to the target by combining with an antigen on it, and directly kills the target via secreted chemicals.

# D

**dalton (Da)**    A measure of atomic mass. One dalton equals one-twelfth the mass of a carbon atom.

**daughter strand**    The newly made strand in DNA replication.

**day-neutral plant**    A plant that flowers regardless of the night length, as long as day length meets the minimal requirements for plant growth.

**deafness**    Hearing loss, usually caused by damage to the hair cells within the cochlea.

**death receptor**    A type of cell surface receptor found in eukaryotic cells that can promote apoptosis when it becomes activated.

**decomposer**    An organism that gets its energy from the remains and waste products of other organisms.

**defecation**    The expulsion of feces that occurs through the anus of an animal's digestive canal.

**defensive mutualism**    A mutually beneficial interaction often involving an animal defending a plant or herbivore in return for food or shelter.

**deforestation**    The conversion of forested areas by humans to nonforested land.

**degenerate**    The characteristic of the genetic code that more than one codon can specify the same amino acid.

**dehydration reaction**    A type of condensation reaction in which a molecule of water is lost.

**deletion**    A type of mutation in which a segment of chromosomal material has been removed.

**demographic transition**    The shift in birth and death rates accompanying human societal development.

**demography**    The study of birth rates, death rates, age distributions, and the sizes of populations.

**dendrite**    A treelike extension of the plasma membrane of a neuron that receives electrical signals from other neurons.

**dendritic cell**    A type of cell derived from bone marrow stem cells that plays an important role in innate immunity; these cells are scattered throughout most tissues, where they perform various macrophage-like functions.

**denitrification**    The reduction of nitrate ($NO_3^-$) to gaseous nitrogen ($N_2$).

**density-dependent factor**    A mortality factor whose influence increases with the density of the population.

**density-independent factor**    A mortality factor whose influence is not affected by changes in population density.

**deoxynucleoside triphosphates**    Individual nucleotides with three phosphate groups.

**deoxyribonucleic acid (DNA)**    One of two types of nucleic acids; the other is ribonucleic acid (RNA). A DNA molecule consists of two strands of nucleotides coiled around each other to form a double helix, held together by hydrogen bonds according to the AT/GC rule.

**deoxyribose**    A five-carbon sugar found in DNA.

**depolarization**    The change in the membrane potential that occurs when a cell membrane becomes less polarized, that is, less negative inside the cell relative to the surrounding fluid.

**dermal tissue**    The covering on various parts of a plant.

**desertification**    The process by which an area becomes more desert-like, usually as a result of overstocking with domestic animals that can greatly reduce grass coverage through overgrazing.

**desmosome**    A mechanically strong type of cell junction between animal cells that typically occurs in spotlike rivets.

**determinate cleavage**    In animals, a characteristic of protostome development in which the fate of each embryonic cell is determined very early.

**determinate growth**    Growth that is of limited duration, such as the growth of flowers or of animal bodies.

**determined**    Refers to a cell that has committed to become a particular cell type.

**detritivore**    An organism that gets its energy from consuming detritus.

**detritus**    Unconsumed plants that die and decompose, along with the dead remains of animals and animal waste products.

**deuterostome**    An animal whose development exhibits radial, indeterminate cleavage and in which the blastopore becomes the anus; includes echinoderms and vertebrates.

**development**    In biology, a series of changes in the state of a cell, tissue, organ, or organism; the underlying process that gives rise to the structures and functions of living organisms.

**developmental genetics**    A field of study aimed at understanding how gene expression controls the process of development.

**diaphragm**    A large muscle that subdivides the thoracic cavity from the abdomen in mammals; contraction of the diaphragm enlarges the thoracic cavity during inhalation.

**diarrhea**    A common intestinal disorder arising from ingested microbes or other causes; usually runs its course within one or two days but, in serious cases, can require hospitalization.

**diastole**    The first phase of the cardiac cycle, in which the ventricles fill with blood coming from the atria through the open AV valves.

**dideoxy chain-termination method**    A method for determining the sequence of bases in DNA that utilizes dideoxynucleotide triphosphates as reagents.

**dideoxy sequencing**    *See* dideoxy chain-termination method.

**differential gene regulation**    The phenomenon in which the expression of genes differs under various environmental conditions and in specialized cell types.

**digestion**    The process of breaking down nutrients in food into smaller molecules that can be absorbed across the intestinal epithelia and directly used by cells.

**digestive system**    In animals, the long tube through which food is processed. In a vertebrate, this system consists of the alimentary canal plus several associated structures.

**dihybrid**    Refers to an offspring that is a hybrid with respect to two traits.

**dikaryotic mycelium**    A fungal body that is made of cells that each possess two genetically distinct nuclei.

**dimorphic fungi**    Fungi that exist in two different morphological forms.

**dinosaur**    A term, meaning terrible lizard, used to describe some of the extinct reptiles preserved as fossils.

**dioecious**    Refers to plants that produce staminate and carpellate flowers on separate plants.

**diploblastic**    Having two distinct germ layers—ectoderm and endoderm—but not mesoderm.

**diploid**    Containing two sets of chromosomes; designated as $2n$.

**diploid-dominant species**    Species in which the diploid organism is the multicellular organism in the life cycle. Animals are an example.

**direct repair**    A type of DNA repair in which an enzyme finds an incorrect structure in the DNA and directly restores the correct structure.

**directional selection**    A pattern of natural selection that favors individuals at one extreme of a phenotypic distribution.

**directionality**    In a DNA or RNA strand, refers to the orientation of the sugar molecules within that strand. Can be $5'$ to $3'$ or $3'$ to $5'$.

**disaccharide**    A carbohydrate composed of two monosaccharides.

**discovery science**    The collection and analysis of data without the need for a preconceived hypothesis; also called discovery-based science.

**discovery-based science**    The collection and analysis of data without the need for a preconceived hypothesis; also called discovery science.

**discrete trait**    A trait with clearly defined phenotypic variants.

**dispersion**    The extent to which individuals in a population are clustered together or spread out.

**dispersive mechanism**    In this incorrect model for DNA replication, segments of parental DNA and newly made DNA are interspersed in both strands following the replication process.

**dispersive mutualism**    A mutually beneficial interaction often involving plants and pollinators that disperse their pollen, and plants and fruit eaters that disperse the plant's seeds.

**dissociation constant**    An equilibrium constant for the formation and dissociation of a ligand and a protein, such as a receptor or an enzyme.

**distal tubule**    The segment of the renal tubule through which fluid flows into one of the many collecting ducts in the kidney.

**disulfide bridge**    Covalent chemical bond formed between two sulfhydryl groups on cysteine side chains in a protein; important in the tertiary structure of proteins.

**diversifying selection**    A pattern of natural selection that favors the survival of two or more different genotypes that produce different phenotypes.

**diversity-stability hypothesis**    The proposal that species-rich communities are more stable than those with fewer species.

**DNA (deoxyribonucleic acid)**    The genetic material that provides a blueprint for the organization, development, and function of living things.

**DNA helicase**    An enzyme that uses ATP to separate DNA strands during DNA replication.

**DNA library**    A collection of recombinant vectors, each containing a particular fragment of DNA from a given organism.

**DNA ligase**    An enzyme that catalyzes the formation of a covalent bond between nucleotides in adjacent DNA fragments to complete the replication process.

**DNA methylation**   A process in which methyl groups are attached to cytosines in DNA.

**DNA methyltransferase**   An enzyme that attaches methyl groups to bases in DNA.

**DNA microarray**   A technology used to monitor the expression of thousands of genes simultaneously.

**DNA polymerase**   An enzyme responsible for covalently linking nucleotides together during DNA replication.

**DNA primase**   An enzyme that synthesizes a primer for DNA replication.

**DNA replication**   The process by which DNA is copied. The original DNA strands are used as templates for the synthesis of new DNA strands

**DNA sequencing**   A procedure used to determine the base sequence of DNA.

**DNA supercoiling**   A process that compacts a chromosome through twisting of the DNA molecule to form additional coils.

**DNA topoisomerase**   An enzyme that alleviates DNA supercoiling during DNA replication.

**DNA transposon**   A type of transposable element that moves as a DNA molecule.

**DNase**   An enzyme that digests DNA.

**domain**   1. A defined region of a protein with a distinct structure and function. 2. One of the three major categories of life: Bacteria, Archaea, and Eukarya.

**domestication**   A process that involves artificial selection of plants or animals for traits desirable to humans.

**dominant species**   A species that has a large effect in a community because of its high abundance or high biomass.

**dominant**   Refers to the trait that is displayed in a heterozygote.

**dormancy**   A phase of metabolic slowdown in a plant or a seed.

**dorsal hollow nerve cord**   A hollow tract of nervous tissue that lies dorsal to the notochord and alimentary canal and which in vertebrates develops into the spinal cord and brain.

**dorsal**   Refers to the upper side of an animal.

**dorsoventral axis**   In bilateral animals, one of the three axes along which the adult body pattern is organized; the others are the anteroposterior axis and the left-right axis.

**dosage compensation**   The phenomenon in which the expression of X-linked genes is equalized between males and females; in mammals, the inactivation of one X chromosome in the female reduces the number of expressed copies (doses) of X-linked genes from two to one.

**double bond**   A bond that occurs when the atoms of a molecule share two pairs of electrons.

**double fertilization**   In angiosperms, the process in which two different fertilization events occur, producing both a zygote and the first cell of a nutritive endosperm tissue.

**double helix**   Two strands of DNA hydrogen-bonded with each other. In a DNA double helix, two DNA strands are twisted together to form a structure that resembles a spiral staircase.

**Down syndrome**   A human disorder caused by the inheritance of three copies of chromosome 21.

**droplet organelle**   An organelle that is not surrounded by a membrane but exists as a droplet formed by liquid-liquid phase separation.

**Duchenne muscular dystrophy**   An inherited, X-linked recessive disorder of humans causing muscle weakness and muscle degeneration.

**duodenum**   The first part of the vertebrate small intestine, arising from the stomach.

**duplication**   A type of mutation in which a section of a chromosome occurs two or more times.

**dynamic instability**   The oscillation of a single microtubule between growing and shortening phases; important in many cellular activities, including the sorting of chromosomes during cell division.

# E

**Ecdysis**   The process by which an animal molts, or breaks out of its old exoskeleton, and secretes a newer, larger one.

**Ecdysozoa**   A clade of molting animals that encompasses the arthropods and nematodes.

**echolocation**   The phenomenon in which certain species listen for echoes of high-frequency sound waves in order to determine the distance and location of an object.

**ecological footprint**   The amount of productive land needed to support each person on Earth.

**ecological species concept**   An approach used to distinguish species; considers a species within its native environment and states that each species occupies its own ecological niche.

**ecology**   The study of interactions among organisms and between organisms and their environments.

**ecosystem diversity**   The diversity of structure and function within an ecosystem.

**ecosystem**   The biotic community of organisms in an area, as well as the abiotic environment affecting that community.

**ecosystems ecology**   The study of the flow of energy and cycling of nutrients among organisms within a community and between organisms and the environment.

**ecotypes**   Genetically distinct populations adapted to their local environments.

**ectoderm**   In animals, the outermost layer of cells formed during gastrulation that covers the surface of the embryo and differentiates into the epidermis and nervous system.

**ectomycorrhizae**   Beneficial associations between temperate forest trees and soil fungi that coat their roots.

**ectoparasite**   A parasite that lives on the outside of the host's body.

**ectotherm**   An animal that largely depends on the environment to warm its body.

**ectothermic**   Dependent on external heat as the main source of body heat.

**edge effect**   A special physical condition that exists at the boundary, or edge, of an area of habitat.

**effective number of species**   The number of equally abundant species needed to obtain the same diversity index as that observed in a dataset of interest in which all species are not equally abundant. It is a measure of diversity that converts values from species diversity indices into equivalent numbers of species.

**effective population size**   The number of individuals that contribute genes to future populations, often smaller than the actual population size.

**effector cell**   A cloned lymphocyte that carries out the attack response during an acquired immune response.

**effector**   A molecule that directly influences a cellular response; in a homeostatic control system in animals, a structure that compensates for a deviation of a physiological variable from its set point.

**efferent arteriole**   A blood vessel that carries blood away from a glomerulus of the vertebrate kidney.

**egestion**   In animals, the process of eliminating undigested material from the body.

**egg cells**   Female haploid gametes in species that reproduce sexually.

**egg**   The mature female gamete; also called an ovum.

**ejaculation**   The movement of semen through the urethra by contraction of muscles at the base of the penis.

**elastin**   A protein that makes up elastic fibers in the extracellular matrix of animals.

**electrical synapse**   A synapse that directly passes electric current from the presynaptic to the postsynaptic cell via gap junctions.

**electrocardiogram (ECG or EKG)**   A record of the electrical impulses generated by the cells of the heart during the cardiac cycle.

**electrochemical gradient**   The combined effect of both an electrical and a chemical gradient across a membrane; determines the direction in which ions will move.

**electrogenic pump**   A pump that generates an electrical gradient across a membrane.

**electromagnetic receptor**   A sensory receptor in animals that detects radiation within a wide range of the electromagnetic spectrum, including visible, ultraviolet, and infrared light, as well as electrical and magnetic fields in some animals.

**electromagnetic spectrum**   All possible wavelengths of electromagnetic radiation, from relatively short wavelengths (gamma rays) to much longer wavelengths (radio waves).

**electron microscope**   A microscope that uses an electron beam for illumination.

**electron shell**   The region around an atom's nucleus where electrons reside; larger atoms have more electron shells than smaller atoms.

**electron transport chain (ETC)**   A group of protein complexes and small organic molecules within the inner membranes of mitochondria and chloroplasts and the plasma membrane of prokaryotes. The components accept and donate electrons to each other in a linear manner and produce an $H^+$ electrochemical gradient.

**electron**   A negatively charged particle found in orbitals around an atomic nucleus.

**electronegativity**   A measure of an atom's ability to attract electrons in a bond with another atom.

**element**   A pure substance composed of only one kind of atom.

**elicitor**   A protein produced by bacterial and fungal pathogens that promotes virulence.

**elongation factor**   A protein that is needed for polypeptide synthesis during the elongation stage of translation.

**elongation**   The second stage in transcription or translation, where RNA strands or polypeptides are made, respectively.

**embryo**   The early stages of development in a multicellular organism during which the organization of the organism is largely established.

**embryogenesis**   The process by which embryos develop from single-celled zygotes by mitotic divisions.

**embryonic development**   The process by which a fertilized egg is transformed into an organism with distinct physiological systems and body parts.

**embryonic germ cell (EG cell)**   A cell in the early mammalian embryo that later gives rise to sperm or egg cells. These cells are pluripotent.

**embryonic stem cell (ES cell)**   A cell in the early mammalian embryo that can differentiate into almost every cell type of the body. These cells are pluripotent.

**embryophytes**   The land plants.

**emerging virus**   A virus that has arisen recently or has recently shown a greater probability of causing infection.

**emphysema**   A progressive disease characterized by a loss of elastic recoil ability of the lungs, usually resulting from chronic tobacco smoking.

**empirical thought**   Thought that relies on observation to form an idea or hypothesis, rather than trying to understand life from a nonphysical or spiritual point of view.

**emulsification** A process during digestion that disrupts large lipid droplets into many tiny droplets, thereby increasing their total surface area and providing greater exposure to lipase action.

**enantiomer** One of a pair of stereoisomers that exist as mirror images.

**endangered species** Those species that are in danger of extinction throughout all or a significant portion of their range.

**endemic** Refers to species that are naturally found only in a particular place or region.

**endergonic** Refers to chemical reactions that require an addition of free energy and do not proceed spontaneously.

**endocrine disruptor** A chemical found in polluted water and soil that resembles a natural hormone; a common example are chemicals that resemble estrogen and can bind to estrogen receptors in animals.

**endocrine glands** Structures that contain epithelial cells that secrete hormones into the bloodstream, where they circulate throughout the body.

**endocrine system** All the endocrine glands and other organs containing hormone-secreting cells.

**endocytosis** A process in which the plasma membrane invaginates, or folds inward, to form a vesicle that brings substances into the cell.

**endoderm** In animals, the innermost layer of cells formed during gastrulation; lines the gut and gives rise to many internal organs.

**endodermis** In vascular plants, a thin cylinder of root tissue that forms a barrier between the root cortex and the central core of vascular tissue.

**endomembrane system** A network of membranes that includes the nuclear envelope, the endoplasmic reticulum, Golgi apparatus, lysosomes, vacuoles, peroxisomes, and plasma membrane.

**endomycorrhizae** Partnerships between plants and fungi in which the fungal hyphae grow into the spaces between root cell walls and plasma membranes.

**endoparasite** A parasite that lives inside the host's body.

**endoplasmic reticulum (ER)** A convoluted network of membranes in a cell's cytoplasm that forms flattened, fluid-filled tubules, or cisternae.

**endoskeleton** An internal hard skeleton covered by soft tissue; present in echinoderms and vertebrates.

**endosperm** A nutritive tissue that increases the efficiency with which food is stored and used in the seeds of flowering plants.

**endospore** A structure with a tough coat that is produced inside of certain bacteria and then released when the enclosing bacterial cell dies and breaks down.

**endosporic gametophyte** A plant gametophyte that grows within the confines of microspore or megaspore walls.

**endosymbiosis theory** A theory that mitochondria and chloroplasts originated from bacteria that took up residence within primordial eukaryotic cells.

**endosymbiosis** A symbiotic relationship in which the smaller species (the symbiont) lives inside the larger species (the host).

**endosymbiotic** Describes a relationship in which one organism (the endosymbiont) lives inside another (the host).

**endotherm** An animal that generates most of its body heat by metabolic processes.

**endothermic** Capable of generating body heat through metabolism.

**energy flow** The movement of energy through an ecosystem.

**energy intermediate** A molecule such as ATP or NADH that stores energy and is used to drive endergonic reactions in cells.

**energy** The ability to promote change or do work.

**enhancer** A regulatory element in eukaryotes that increases the rate of transcription when bound by an activator protein.

**enthalpy ($H$)** The total energy of a system.

**entomology** The scientific study of insects.

**entropy** The degree of disorder of a system.

**environmental science** The application of ecology to real-world problems.

**enzyme** A protein that acts as a catalyst to speed up a chemical reaction in a cell.

**enzyme-linked receptor** A receptor found in all living species that typically has two important domains: an extracellular domain, which binds a signaling molecule, and an intracellular domain, which has a catalytic function.

**enzyme-substrate complex** The structure produced by binding an enzyme and its substrate(s).

**eosinophil** A type of phagocyte found in large numbers in mucosal surfaces lining the gastrointestinal, respiratory, and urinary tracts, where they fight off parasitic infections.

**epicotyl** The portion of an embryonic plant stem with two tiny leaves in a first bud; located above the point of attachment of the cotyledons.

**epidermis** A layer of dermal tissue on the surfaces of leaves, stems, and roots that helps protect a plant from damage.

**epigenetic inheritance** An epigenetic change that is passed from parent to offspring. An example is genomic imprinting.

**epigenetics** The study of mechanisms that lead to changes in gene expression that can be passed from cell to cell and are reversible, but do not involve a change in the sequence of DNA.

**epilimnion** The upper layer of water in a lake, usually warm and containing high levels of dissolved oxygen.

**epimutation** A heritable change in gene expression that does not alter the sequence of DNA.

**episome** A plasmid or viral genome that can integrate into a host chromosome or can replicate independently.

**epistasis** A gene interaction in which the alleles of one gene mask the expression of the alleles of another gene.

**epithalamus** A region of the vertebrate forebrain that includes the pineal gland.

**epithelial tissue** In animals, a sheet of densely packed cells that covers the body, covers individual organs, or lines the walls of various cavities inside the body.

**epitope** An antigenic determinant; the polypeptide fragment of an antigen that is complexed to an MHC protein and presented to a helper T cell.

**equilibrium model of island biogeography** A model that explains the process of succession on new islands, proposing that the number of species on an island tends toward an equilibrium number that is determined by the balance between immigration rates and extinction rates.

**equilibrium potential** The membrane potential at which the flow of an ion is at equilibrium, with no net movement in either direction.

**equilibrium** 1. In a chemical reaction, occurs when the rate of the forward reaction is balanced by the rate of the reverse reaction. 2. In a population, the situation in which the population size stays the same.

**ER lumen** A single compartment enclosed by the ER membrane.

**ER signal sequence** A sorting signal in a polypeptide that is usually located near the N-terminus and is recognized by SRP (signal recognition particle), allowing the polypeptide to be directed to the ER membrane.

**erythrocyte** A cell that serves the critical function of transporting oxygen throughout an animal's body; also known as a red blood cell.

**erythropoietin (EPO)** A hormone made by the liver and kidneys in response to any situation where additional red blood cells are required.

**esophagus** In animals, the tubular structure that forms a pathway from the pharynx to the stomach.

**essential amino acid** An amino acid that is required in the diet of many animals.

**essential element** In plants, a chemical element that is required for metabolism, sometimes by functioning as an enzyme cofactor.

**essential fatty acid** An unsaturated fatty acid, such as linoleic acid, that cannot be synthesized by animal cells and must therefore be consumed in the diet.

**essential nutrient** In animals, a compound that cannot be synthesized from any ingested or stored precursor molecule and so must be obtained in the diet in its complete form. In plants, those substances needed to complete reproduction while avoiding the symptoms of nutrient deficiency.

**estradiol** The major estrogen in many vertebrates, including humans.

**estrogens** Steroid hormones produced by the ovaries that affect most aspects of female reproduction.

**ethology** Scientific studies of animal behavior.

**ethylene** A plant hormone that is particularly important in coordinating plant developmental and stress responses.

**euchromatin** The less condensed regions of a chromosome; areas that are capable of gene transcription.

**eudicots** One of the two largest lineages of flowering plants in which the embryo possesses two seed leaves.

**Eukarya** One of the three domains of life; the other two are Bacteria and Archaea.

**eukaryote** A member of the domain Eukarya. The distinguishing feature of eukaryotes is cell compartmentalization, including a cell nucleus; eukaryotes include protists, fungi, plants, and animals.

**eukaryotic** Refers to organisms having cells with internal compartments that serve various functions; includes all members of the domain Eukarya.

**euphyll** A leaf with branched veins.

**euploid** Refers to an organism that has a chromosome number that is a multiple of a chromosome set ($1n$, $2n$, $3n$, etc.).

**eusociality** An extreme form of altruism in social insects in which the vast majority of females, known as workers, do not reproduce. Instead, they help one reproductive female (the queen) raise offspring.

**eustele** In plants, a ring of vascular tissue arranged around a central pith of nonvascular tissue; typical of progymnosperms, gymnosperms, and angiosperms.

**eutherian** A placental mammal; a member of the clade Eutheria.

**eutrophic** Waters that contain relatively high levels of nutrients such as phosphate or nitrogen and typically exhibit high levels of primary productivity and low levels of biodiversity.

**eutrophication** The process by which elevated nutrient levels in a body of water lead to an overgrowth of algae or aquatic plants and a subsequent depletion of water oxygen concentrations when these photosynthesizers decay.

**evaporation** The conversion of water from the liquid to the gaseous state at normal temperatures. Animals use evaporation as a means of losing excess body heat.

**evapotranspiration rate** The rate at which water moves into the atmosphere through the processes of evaporation from the soil and other surfaces and transpiration of plants.

**evo-devo** *See* evolutionary developmental biology.

**evolution** A heritable change in a population of organisms from one generation to the next.

**evolutionarily conserved** The term used to describe homologous DNA sequences that are very similar or identical between different species.

**evolutionary developmental biology**   The field of biology that compares the development of different organisms in an attempt to understand relationships between organisms and the mechanisms that bring about evolutionary change; referred to as evo-devo.

**evolutionary lineage concept**   An approach used to distinguish species; states that a species is derived from a single distinct lineage and has its own evolutionary tendencies and historical fate.

**excitation-contraction coupling**   The sequence of events by which an action potential in the plasma membrane of a muscle fiber leads to cross-bridge activity.

**excitatory postsynaptic potential (EPSP)**   The response to an excitatory neurotransmitter that depolarizes the postsynaptic membrane; the depolarization brings the membrane potential closer to the threshold potential that would trigger an action potential.

**excretion**   In animals, the process of expelling waste or harmful materials from the body.

**excretory system**   All of an animal's organs (such as gills, lungs, kidneys, and, in some animals, the body surface) that function to remove soluble wastes generated from metabolism.

**excurrent siphon**   A structure in a tunicate used to expel water from the body

**exercise**   Any physical activity that increases an animal's metabolic rate.

**exergonic**   Refers to chemical reactions that release free energy and occur spontaneously.

**exocrine gland**   A gland in which epithelial cells secrete chemicals into a duct, which carries those molecules directly to another structure or to the outside surface of the body.

**exocytosis**   A process in which material inside a cell is packaged into vesicles and excreted into the extracellular environment.

**exon**   A portion of RNA that is found in the mature mRNA molecule after splicing is finished.

**exoskeleton**   An external skeleton made primarily of chitin that surrounds and protects most of the body surface of animals such as insects.

**expansin**   A protein that occurs in the plant cell wall and fosters cell enlargement.

**experimental group**   The sample in an experiment that is subjected to some type of variation that does not occur for the control group.

**exploitation competition**   Competition in which organisms compete indirectly through the consumption of a limited resource.

**exponential growth**   Rapid population growth that occurs when the per capita growth rate remains above zero.

**extant**   Refers to a species that is still in existence.

**extensor**   A muscle that straightens a limb at a joint.

**external fertilization**   Fertilization that occurs in aquatic environments, when eggs and sperm are released into the water in close enough proximity for fertilization to occur.

**extinct**   Refers to a species that existed in the past, but has died out.

**extinction**   The end of the existence of a species or a group of species.

**extracellular fluid**   The fluid in an organism's body that is outside of the cells.

**extracellular matrix (ECM)**   A network of material that is secreted from animal cells and forms a complex meshwork outside of cells. The ECM provides strength, support, and organization.

**extranuclear inheritance**   In eukaryotes, the transmission of genes that are located outside the cell nucleus.

**extremophile**   An organism that occurs primarily in extreme habitats.

**extrinsic pathway**   One of two different pathways for apoptosis that involves the activation of death receptors.

**eye**   The visual organ in animals that detects light and sends signals to the brain.

**eyecup**   A visual organ in flatworms that detects light and its direction but does not form an image.

# F

**F factor**   A fertility factor, or type of bacterial plasmid that plays a role in bacterial conjugation.

**$F_1$ generation**   The first generation of offspring in a genetic cross.

**$F_2$ generation**   The second generation of offspring in a genetic cross.

**facilitated diffusion**   A mechanism of passive transport in which a transport protein provides a passageway for a substance to cross a membrane from an area of higher concentration to one of lower concentration.

**facilitation**   A mechanism for succession in which a species facilitates or makes the local environment more suitable for subsequent species.

**facultative anaerobe**   A microorganism that can use oxygen in aerobic respiration, obtain energy via anaerobic fermentation, or use inorganic chemical reactions to obtain energy.

**facultative mutualism**   An interaction between mutualistic species that is beneficial but not essential to the survival and reproduction of either species.

**falsifiable**   Refers to a hypothesis that can be shown to be incorrect based on additional observations or experimentation.

**family**   In taxonomy, a subdivision of an order.

**fast fiber**   A skeletal muscle fiber containing myosin with higher ATPase activity.

**fast-glycolytic fiber**   A skeletal muscle fiber that has high myosin ATPase activity but cannot make as much ATP as an oxidative fiber because its source of ATP is glycolysis; best suited for rapid, intense actions.

**fast-oxidative fiber**   A skeletal muscle fiber that has high myosin ATPase activity and can make large amounts of ATP; used for long-term actions.

**fate**   The ultimate morphological features that a cell or a group of cells will adopt.

**feedback inhibition**   A type of regulation in which the product of a metabolic pathway inhibits an enzyme that acts early in the pathway, thus preventing the overaccumulation of the product.

**feedforward regulation**   The process by which an animal's body begins preparing for a change in some variable before it even occurs.

**female gametophyte**   A haploid multicellular plant generation that produces one or more eggs but does not produce sperm cells.

**female-enforced monogamy hypothesis**   The hypothesis that a male is monogamous due to various actions employed by his female mate.

**fermentation**   The breakdown of organic molecules to produce energy without any net oxidation of an organic molecule.

**fertilization**   The union of two gametes, such as an egg cell with a sperm cell, to form a zygote.

**fertilizer**   A soil addition that enhances plant growth by providing essential elements.

**fetus**   The maturing embryo after the eighth week of gestation in humans.

**fever**   An increase in an animal's body temperature, typically due to infection.

**fiber**   A type of tough-walled plant cell that provides support.

**fibrous root system**   The root system of monocots, which consists of multiple adventitious roots that grow from the stem base.

**fight-or-flight response**   The response of vertebrates to real or perceived danger; associated with increased activity of the sympathetic branch of the autonomic nervous system.

**filament**   1. The elongate portion of a flower's stamen; contains vascular tissue that delivers nutrients from parental sporophytes to anthers. 2. In fishes, a part of the gills.

**filtrate**   In the process of filtration in an excretory system, the material that passes through the filter and enters the excretory organ for either further processing or excretion.

**filtration**   The passive removal of water and small solutes from the blood during the production of urine.

**finite rate of increase**   In ecology, the ratio of a population size from one year to the next.

**fitness**   The relative likelihood that a genotype will contribute to the gene pool of the next generation as compared with other genotypes.

**fixed action pattern (FAP)**   An animal behavior that, once initiated, will continue until completed.

**fixed nitrogen**   Atmospheric nitrogen that has been combined with another element into a substance that can be used by plants. An example is ammonia, $NH_3$.

**flaccid**   Refers to a plant cell in which the concentration of solutes is the same as that in the external fluid environment. A flaccid cell has a water content higher than a plasmolyzed cell, but lower than a turgid cell.

**flagella** (singular, **flagellum**)   Relatively long cell appendages that facilitate cellular movement or the movement of extracellular fluids.

**flagellate**   A protist that uses one or more flagella to move in water or cause water motions useful in feeding.

**flagship species**   A single large or instantly recognizable species.

**flame cell**   A cell that exists primarily to maintain osmotic balance between an organism's body and surrounding fluids; present in flatworms.

**flexor**   A muscle that bends a limb at a joint.

**flower**   A reproductive shoot; a short stem branch bearing reproductive organs instead of leaves.

**flowering plants**   The angiosperms, which produce ovules within the protective ovaries of flowers. The ovules develop into seeds, and the ovaries develop into fruits, which function in seed dispersal.

**fluidity**   A property of biological membranes in which individual molecules remain in close association yet have the ability to move rotationally or laterally within the plane of the membrane. Membranes are semifluid.

**fluid-mosaic model**   The accepted model of a biological membrane; its basic framework is the semifluid phospholipid bilayer with a mosaic of proteins. Carbohydrates may be attached to the lipids or proteins.

**focal adhesion**   A mechanically strong type of cell junction that connects an animal cell to the extracellular matrix (ECM).

**follicle-stimulating hormone (FSH)**   A gonadotropin that stimulates follicle development.

**follicular phase**   The first half of the human ovarian cycle, in which a cohort of immature follicles begin to grow and differentiate.

**food chain**   A linear depiction of energy flow between organisms, with each organism feeding on and deriving energy from the preceding organism.

**food web**   A complex model of interconnected food chains in which there are multiple links among different species.

**food-induced thermogenesis**   A rise in an animal's metabolic rate and heat production for a few hours after eating.

**foot**   In mollusks, a muscular structure usually used for movement.

**forebrain** One of three major divisions of the vertebrate brain; the other two divisions are the midbrain and the hindbrain.

**fossil fuel** A fuel formed in the Earth from protist, plant, or animal remains, such as coal, petroleum, or natural gas.

**fossil** Recognizable preserved remains of past life on Earth.

**founder effect** Genetic drift that occurs when a small group of individuals separates from a larger population and establishes a colony in a new location.

**fovea** A small area on the retina directly behind the lens, where an image is most sharply focused.

**frameshift mutation** A mutation that involves the addition or deletion of a number of nucleotides that is not a multiple of three and alters the reading frame of a protein-encoding gene.

**Frank-Starling law of the heart** A law that states the strength of a heart contraction (beat) is related to the amount of blood that had filled the heart chambers prior to the contraction. Increased return of blood to the heart will increase the subsequent strength of a heart contraction.

**free energy (G)** The amount of a system's energy that is available and can be used to promote change or do work.

**free radical** A molecule containing an atom with a single, unpaired electron in its outer shell. A free radical is unstable and interacts with other molecules by removing electrons from their atoms.

**frequency** For a sound wave, the number of complete wavelengths that occur in 1 second, measured in hertz (Hz).

**frontal lobe** One of four lobes of the cerebral cortex of the human brain; important in a variety of functions, including judgment and conscious thought.

**fruit** A structure that develops from flower parts, encloses seeds, and fosters seed dispersal in the environment.

**fruiting bodies** The visible fungal reproductive structures that are composed of densely packed hyphae that typically grow out of the substrate.

**functional genomics** Genomic methods aimed at studying the expression of a genome.

**functional group** A group of atoms with a characteristic chemical structure that exhibits particular properties. Each functional group exhibits the same properties in all molecules in which it occurs.

**functional magnetic resonance imaging (fMRI)** A technique used to determine changes in brain activity while a person is performing specific tasks.

**fundamental niche** The optimal range of conditions in which a particular species functions best.

**Fungi** A eukaryotic kingdom of the domain Eukarya.

**fungus-like protists** Heterotrophic protists that often resemble true fungi in having threadlike, filamentous bodies and absorbing nutrients from their environment.

# G

**G protein** An intracellular protein that binds guanosine triphosphate (GTP) and guanosine diphosphate (GDP) and participates in intracellular signaling pathways.

**$G_1$ phase** The first gap phase of the cell cycle.

**$G_2$ phase** The second gap phase of the cell cycle.

**gallbladder** In many vertebrates, a small sac underneath the liver that is a storage site for bile; allows the release of large amounts of bile to be precisely timed to the consumption of fats.

**gametangia** Specialized structures produced by some fungi and many land plants in which developing gametes are protected by a jacket of tissue.

**gamete** A haploid cell that is involved with sexual reproduction, such as a sperm or egg cell.

**gametogenesis** The formation of gametes.

**gametophyte** In plants and many multicellular protists, the haploid stage that produces gametes by mitosis.

**ganglion cells** Cells in the vertebrate eye whose axons extend into the optic nerve.

**gap gene** A type of segmentation gene; a mutation in this type of gene may cause several adjacent segments to be missing in the larva.

**gap junction** A type of junction between animal cells that provides a passageway for intercellular transport.

**gas exchange** The process of moving oxygen and carbon dioxide in opposite directions between the environment and blood and between blood and cells.

**gastrin** A hormone secreted by cells of the vertebrate stomach that stimulates smooth muscle contraction and acid production by stomach cells.

**gastrovascular cavity** In certain invertebrates such as cnidarians, a body cavity with a single opening to the outside; it functions as both a digestive system and circulatory system.

**gastrula** A stage of an animal embryo that is the result of gastrulation and has three cellular layers: the ectoderm, endoderm, and mesoderm.

**gastrulation** In animals, a process in which an area in the blastula invaginates and folds inward, creating different embryonic cell layers called germ layers.

**gated** A property of many channels that allows them to open and close to control the movement of solutes across a membrane.

**gel electrophoresis** A technique used to separate macromolecules by applying an electric field that causes them to migrate through a gel matrix.

**gene cloning** The process of making multiple copies of a particular gene.

**gene expression** Gene function either at the level of traits or at the molecular level.

**gene family** A group of homologous genes within a single species that carry out related functions.

**gene flow** A transfer of alleles into or out of a population that occurs when fertile individuals migrate between populations having different allele frequencies.

**gene interaction** The phenomenon in which a single trait is controlled by two or more genes, each of which has two or more alleles.

**gene pool** All of the alleles for every gene in a population.

**gene regulation** The ability of cells to control the expression of their genes.

**gene transfer** The process by which genetic material is transferred from one bacterial cell to another.

**gene** A unit of heredity. At the molecular level, a gene is an organized unit of base sequences in a DNA strand that can transcribed into RNA and ultimately results in the formation of a functional product.

**general lineage concept** A widely accepted approach used to distinguish species; states that each species is a population of an independently evolving lineage.

**general transcription factors (GTFs)** Five different proteins that play a role in initiating transcription at the core promoter of protein-encoding genes in eukaryotes.

**generative cell** In a seed plant, one of the cells resulting from the division of a microspore; a generative cell divides to produce two sperm cells.

**genetic code** A code that specifies the relationship between the sequence of bases in the codons found in mRNA and the sequence of amino acids in a polypeptide.

**genetic diversity** The amount of genetic variation occurring within and between populations.

**genetic drift** The random change in a population's allele frequencies from one generation to the next that is attributable to chance. It occurs more quickly in small populations.

**genetic map** A chart that shows the linear arrangement of genes along a chromosome.

**genetic mapping** The use of genetic crosses to determine the linear order of genes that are linked to each other along the same chromosome.

**genetics** The branch of biology that deals with inheritance.

**genome centric metagenomics** The process of assembling whole microbial genomes from metagenomic DNA sequences.

**genome** The complete genetic material of an organism or species.

**genomic imprinting** A phenomenon in which a segment of DNA is imprinted, or marked, during egg or sperm formation in a way that affects gene expression throughout the life of the individual who inherits that DNA.

**genomic library** A type of DNA library in which the inserts are derived from chromosomal DNA.

**genomics** Techniques that are used to analyze the DNA sequence of the entire genome of a species.

**genotype frequency** In a population, the number of individuals with a given genotype divided by the total number of individuals.

**genotype** The genetic composition of an individual.

**genus (plural, genera)** In taxonomy, a subdivision of a family.

**geological timescale** A timeline of the Earth's history and major events from its origin about 4.55 bya to the present.

**germ layer** An embryonic cell layer such as ectoderm, mesoderm, or endoderm.

**germ line** Cells that give rise to gametes, such as egg and sperm cells.

**germination** In plants, the process in which an embryo absorbs water, becomes metabolically active, and grows out of the seed coat, producing a seedling.

**giant axon** A very large axon in certain species such as squids that facilitates high-speed neuronal conduction and rapid responses to stimuli.

**gibberellic acid** A type of gibberellin, a plant hormone.

**gibberellin** A plant hormone that stimulates both cell division and cell elongation.

**gills** Specialized filamentous organs in aquatic animals that are used to obtain oxygen and eliminate wastes.

**ginkgos** A phylum of gymnosperms; formally known as Ginkgophyta.

**gizzard** The muscular portion of the stomach of birds and some reptiles that is capable of grinding food into smaller fragments.

**glaucoma** A condition in which drainage of aqueous humor in the eye becomes blocked and the pressure inside the eye increases. If untreated, this pressure damages cells in the retina and leads to irreversible loss of vision.

**glia** Cells that surround the neurons; a major class of cells in nervous systems that perform various functions.

**global warming** A gradual elevation of the Earth's average surface temperature caused by an increasing greenhouse effect.

**glomerular filtration rate (GFR)** The rate at which a filtrate from plasma is formed in all the glomeruli of the vertebrate kidneys.

**glomerulus** A cluster of interconnected, fenestrated capillaries in the renal corpuscle of the kidney; the site of filtration in the kidney.

**glucagon** A hormone secreted by the pancreas of an animal that stimulates the processes of glycogenolysis, gluconeogenesis, and the synthesis of ketones in the liver.

**gluconeogenesis** A mechanism for maintaining blood glucose level in which enzymes in the liver convert noncarbohydrate precursors into glucose, which is then secreted into the blood.

**glucose sparing** A metabolic adjustment that reserves the glucose produced by the liver for use by the nervous system.

**glycocalyx** 1. An outer viscous covering surrounding a bacterium that traps water and helps protect the bacterium from drying out. 2. A carbohydrate-rich zone on the surface of animal cells; also called a cell coat.

**glycogen** A polysaccharide found in animal cells (especially liver and skeletal muscle) and sometimes called animal starch; also, the major carbohydrate storage of fungi.

**glycogenolysis** A mechanism for maintaining blood glucose level by breaking down stored glycogen by hydrolysis, yielding molecules of glucose, which are then secreted into the blood.

**glycolipid** A lipid that has carbohydrate attached to it.

**glycolysis** A metabolic pathway that breaks down glucose to pyruvate.

**glycolytic fiber** A skeletal muscle fiber that has relatively few mitochondria but possesses both a high concentration of glycolytic enzymes and a large store of glycogen.

**glycoprotein** A protein that has a carbohydrate attached to it.

**glycosaminoglycan** The most abundant type of polysaccharide in the extracellular matrix (ECM) of animals, consisting of repeating disaccharide units that give a gel-like character to the ECM.

**glycosidic bond** A covalent bond formed between two sugar molecules via a dehydration reaction.

**glycosylation** The covalent attachment of a carbohydrate to a protein or lipid, producing a glycoprotein or glycolipid, respectively.

**glyoxysome** A specialized organelle within plant seeds that contains enzymes needed to convert fats to sugars.

**gnathostomes** All vertebrate species that possess jaws.

**Golgi apparatus** A stack of flattened, membrane-bound compartments that performs three overlapping functions: secretion, processing, and protein sorting.

**gonads** The testes in males and the ovaries in females, where the gametes are formed.

**G-protein-coupled receptor (GPCR)** A common type of receptor found in the cells of eukaryotic species that interacts with G proteins to initiate a cellular response.

**graded potential** A depolarization or hyperpolarization of a neuron's plasma membrane that varies with the strength of a stimulus.

**gradualism** A concept suggesting that species evolve continuously over long spans of time.

**grain** The characteristic single-seeded fruit of cereal grasses such as rice, corn, barley, and wheat.

**granum** A structure composed of stacked membrane-bound thylakoids within a chloroplast.

**gravitropism** Plant growth in response to the force of gravity.

**gray matter** Brain tissue that consists of neuronal cell bodies, dendrites, and some unmyelinated axons.

**gross primary production (GPP)** The measure of biomass production by photosynthetic organisms; equivalent to the carbon fixed during photosynthesis.

**ground tissue** Type of tissue that makes up most of the body of a plant and has a variety of functions, including photosynthesis, storage of carbohydrates, and support. Ground tissue can be subdivided into three types: parenchyma, collenchyma, and sclerenchyma.

**group selection** A premise that attempts to explain altruism by claiming that natural selection produces outcomes beneficial for the whole group or species rather than for individuals.

**growth factors** A signaling molecule in animals that promotes cell division.

**growth hormone (GH)** A hormone produced in vertebrates by the anterior pituitary gland; GH acts on the liver to produce insulin-like growth factor-1 (IGF-1).

**growth** An increase in weight or size.

**guanine (G)** A purine base found in DNA and RNA.

**guard cell** A specialized plant cell that allows epidermal pores (stomata) to close when conditions are too dry and to open under moist conditions, allowing the entry of $CO_2$ needed for photosynthesis.

**gustation** The sense of taste.

**guttation** The phenomenon in which droplets of water form at the edges of leaves as the result of root pressure.

**gymnosperm** A plant that produces seeds that are exposed rather than seeds enclosed in fruits.

# H

**H⁺ electrochemical gradient** A transmembrane gradient for $H^+$ composed of both an electrical gradient and a concentration difference for $H^+$ across a membrane.

**habitat destruction** A usually human-driven process in which a natural habitat is altered in a way that prevents it from supporting the species that were originally present.

**habitat restoration** The full or partial repair or replacement of biological habitats and/or their populations that have been damaged.

**habituation** The form of nonassociative learning in which an organism learns to ignore a repeated stimulus.

**hair cell** A mechanoreceptor in animals that is a specialized epithelial cell with deformable stereocilia.

**half-life** 1. In the case of organic molecules in a cell, refers to the time it takes for 50% of the molecules to be broken down and recycled. 2. In the case of radioisotopes, the time it takes for 50% of the atoms to decay and emit radiation.

**halophyte** A plant that can tolerate higher than normal salt concentrations and can occupy coastal salt marshes or saline deserts.

**Hamilton's rule** The proposal that an altruistic gene will be favored by natural selection when $rB > C$, where $r$ is the coefficient of relatedness of the donor (the altruist) to the recipient, $B$ is the benefit received by the recipient, and $C$ is the cost incurred by the donor.

**haplodiploid system** A genetic system in which females develop from fertilized eggs and are diploid but males develop from unfertilized eggs and are haploid.

**haploid** Containing one set of chromosomes; designated as $1n$.

**haploid-dominant species** Species in which the haploid organism is the multicellular organism in the life cycle. Examples include fungi and some protists.

**haplorrhini** Larger-brained diurnal species of primates, including tarsiers, monkeys, gibbons, orangutans, gorillas, chimpanzees, and humans.

**Hardy-Weinberg equation** An equation ($p^2 + 2pq + q^2 = 1$) that relates allele and genotype frequencies; the equation predicts an equilibrium if no new mutations are formed, no natural selection occurs, the population size is very large, the population does not migrate, and mating is random.

**heartburn** More properly called gastroesophageal reflux; the movement of acidic stomach contents upward into the esophagus, typically causing pain.

**heat capacity** The amount of heat required to raise the temperature of an entire object or amount of substance.

**heat of fusion** The amount of heat energy that must be withdrawn or released from a substance to cause it to change from the liquid to the solid state.

**heat of vaporization** The heat required to vaporize 1 mole of any substance at its boiling point under standard pressure.

**helper T cell** A type of lymphocyte that assists in the activation and function of B cells and cytotoxic T cells.

**hematocrit** The volume of blood that is composed of red blood cells, usually between 35% and 65% in vertebrates.

**hemidesmosome** A mechanically strong type of cell junction that connects an animal cell to the extracellular matrix (ECM).

**hemiparasite** A parasitic organism that photosynthesizes, but lacks a root system to draw water and thus depends on its host for that function.

**hemispheres** The two halves of the cerebrum.

**hemizygous** Having only one copy of a particular gene; a male mammal is hemizygous for an X-linked gene.

**hemocyanin** A copper-containing pigment that binds oxygen and gives blood or hemolymph a bluish tint.

**hemodialysis** A medical procedure used to artificially perform the kidneys' function.

**hemoglobin** An iron-containing protein that reversibly binds to oxygen and is found within the cytosol of red blood cells.

**hemolymph** Mixed fluid found in one body compartment (hemocoel) in an animal; present in many invertebrates.

**hemorrhage** A loss of blood from a ruptured blood vessel.

**herbaceous plant** A plant that produces little or no wood and is composed mostly of primary vascular tissues.

**herbivore** An animal that eats only plants.

**herbivory** A form of species interaction in which herbivores feed on plants.

**heritable** A property of DNA, which means that it can be passed from cell to cell and from parent to offspring.

**hermaphrodites** In animals, individuals that possess both ovaries and testes.

**heterochromatin** The highly compacted regions of chromosomes that are usually transcriptionally inactive because of their tight conformation.

**heterochrony** Differences among species in the rate or timing of developmental events.

**heterospory** In plants, the production of two different types of spores: microspores and megaspores; microspores produce male gametophytes, and megaspores produce female gametophytes.

**heterotherm** An animal that has a body temperature that varies widely with environmental temerature; both ectotherms and endotherms may be heterotherms.

**heterotroph** Organisms that cannot produce their own organic molecules by using energy from inorganic sources or light; they must obtain one or more organic compounds from their environment.

**heterotrophic** Requiring organic food from the environment.

**heterozygote advantage** A phenomenon in which a heterozygote has a higher fitness than either corresponding homozygote.

**heterozygous** An individual with two different alleles of the same gene.

**highly repetitive sequence** A DNA sequence that is repeated tens of thousands or even millions of times throughout a genome.

**hindbrain** One of three major divisions of the vertebrate brain; the other two divisions are the midbrain and the forebrain.

**hippocampus** The area of the limbic system of the vertebrate forebrain that functions in establishing memories for spatial locations, facts, and the sequence of events.

**GLOSSARY**

**histone acetyltransferase** An enzyme that attaches acetyl groups to the amino terminal tails of histone proteins.

**histone code hypothesis** The idea that the pattern of histone modification is recognized by proteins and thus plays a role in the expression of eukaryotic genes.

**histone variant** A histone protein that has a slightly different amino acid sequence than that of a standard histone protein.

**histones** A group of proteins involved in the formation of nucleosomes, which aid in the compaction of eukaryotic DNA.

**holobiont** The combination of a host organism and its microbiome.

**holoblastic cleavage** A complete type of cell cleavage in certain animals in which the entire zygote is bisected into two equal-sized blastomeres.

**hologenome** The combined genomes of a host organism and its microbiome.

**holoparasite** A parasitic organism that lacks chlorophyll and is totally dependent on a host plant for its water and nutrients.

**homeobox** A 180-bp sequence within the coding sequence of a homeotic gene.

**homeodomain** A region of a homeotic protein that functions in binding to the DNA.

**homeostasis** The process whereby living organisms regulate their cells and bodies to maintain relatively stable internal conditions.

**homeostatic control system** A system designed to regulate particular variables in an animal's body, such as body temperature; consists of a set point, sensor, integrator, and effectors.

**homeotherm** An animal that maintains its body temperature within a narrow range.

**homeotic gene** A gene that specifies the developmental fate of a particular segment or region of an animal's body.

**hominin** A species of humans, whether extant or extinct.

**hominoidea (hominoid)** A group of primates that includes gibbons, orangutans, gorillas, chimpanzees, and humans, as well as all of their recent ancestors.

**homolog** A member of a pair of chromosomes in a diploid organism.

**homologous genes** Genes derived from the same ancestral gene that have accumulated random mutations that make their sequences slightly different.

**homologous structures** Structures that are similar to each other because they are derived from the same ancestral structure.

**homology** A similarity that occurs due to descent from a common ancestor.

**homozygous** An individual with two identical copies of an allele.

**horizontal gene transfer** A process in which an organism incorporates genetic material from another organism without being the offspring of that organism.

**hormone** In animals, a chemical signal that is produced in a gland or other structure and released into the blood or hemolymph, where it acts on distant target cells. In plants, a signaling molecule that is important in coordination of plant development or plant response to the environment.

**hornworts** A phylum of bryophytes; formally known as Anthocerophyta.

**host cell** 1. A cell that is infected by a virus, fungus, or bacterium. 2. A eukaryotic cell that contains photosynthetic or nonphotosynthetic endosymbionts.

**host plant resistance** The ability of plants to prevent herbivory via either chemical or mechanical defenses.

**host range** The number of species and cell types that a virus or bacterium can infect.

**host** A larger organism that provides an environment that is occupied by one or more smaller organisms or viruses.

**hot spot** A human-impacted geographic area with a large number of endemic species. To qualify as a hot spot, a region must contain at least 1,500 species of endemic vascular plants and have lost at least 70% of its original habitat.

**Hox genes** In animals, a class of homeotic genes involved in pattern formation in early embryos.

**Human Genome Project** A 13-year international effort coordinated by the U.S. Department of Energy and the National Institutes of Health that characterized and sequenced the entire human genome.

**human immunodeficiency virus (HIV)** A retrovirus that is the causative agent of acquired immune deficiency syndrome (AIDS).

**humoral immunity** A type of acquired immunity in which plasma cells secrete antibodies that bind to antigens.

**hybrid zone** An area where two populations can interbreed.

**hybridization** The process in which two individuals of the same species with different characteristics are bred or crossed to each other; the offspring are referred to as hybrids.

**hydrocarbon** Molecules with predominantly hydrogen–carbon bonds.

**hydrogen bond** A weak chemical attraction between a hydrogen atom in a polar molecule and an electronegative atom in another polar molecule.

**hydrolysis reaction** A chemical reaction that utilizes water to break apart other molecules.

**hydrophilic** Refers to molecules that contain ionic and/or polar covalent bonds and will dissolve in water.

**hydrophobic** Refers to molecules that do not have partial charges and therefore are not attracted to water molecules. Such molecules are composed predominantly of carbon and hydrogen and are relatively insoluble in water.

**hydrostatic skeleton** A fluid-filled body cavity in certain soft-bodied invertebrates that is surrounded by muscles and provides support and shape.

**hydroxide ion** An anion with the formula $OH^-$.

**hyperpolarization** The change in the membrane potential that occurs when the cell membrane becomes more polarized.

**hypersensitive response (HR)** A plant's local defensive response to pathogen attack.

**hypertension** High blood pressure.

**hyperthermophile** An organism that thrives in extremely hot temperatures.

**hypertonic** When the concentration of solutes outside the cell is higher and causes a cell to shrink due to osmosis of water out of the cell.

**hypha** A microscopic, branched filament of the body of a fungus.

**hypocotyl** The portion of an embryonic plant stem located below the point of attachment of the cotyledons.

**hypolimnion** The layer of cold, dense water at the bottom of a lake, often with low levels of dissolved oxygen.

**hypothalamus** A part of the vertebrate brain located ventral to the thalamus; it controls functions related to growth, reproduction, and metabolism among others, and regulates many basic behaviors such as eating and drinking.

**hypothesis testing** Also known as the scientific method, a strategy for formulating and testing the validity of a hypothesis.

**hypothesis** In biology, a proposed explanation for a natural phenomenon that is based on previous observations or experimental studies.

**hypotonic** When the concentration of solutes outside the cell is lower and causes a cell to swell due to the uptake of water via osmosis.

**I**

**idiosyncratic hypothesis** The idea that ecosystem function changes as the number of species increases or decreases but that the amount and direction of change are unpredictable.

**immune system** The cells and organs within an animal's body that contribute to immune defenses.

**immune tolerance** The process by which the body distinguishes between self and nonself components.

**immunity** The ability of an animal to ward off internal threats, including harmful microorganisms, foreign molecules, and abnormal cells such as cancer cells.

**immunoglobulin (Ig)** A Y-shaped protein with two heavy chains and two light chains that provides immunity to foreign substances; antibodies are a type of immunoglobulin.

**immunological memory** The immune system's ability to produce a secondary immune response.

**imperfect flower** A flower that lacks either stamens or carpels.

**implantation** The first event of pregnancy, when the blastocyst embeds within the uterine endometrium.

**imprinting** 1. Learning that occurs during a brief critical period and establishes a long-lasting behavioral response to a specific object or individual, such as recognition and bonding to a parent. 2. In genetics, the marking of DNA that occurs differently between males and females.

**in vitro translation system** A mixture of components isolated from cells that is capable of synthesizing polypeptides if mRNA is added.

**inactivation gate** A string of amino acids that juts out from a channel protein into the cytosol and blocks the movement of ions through the channel.

**inborn error of metabolism** A genetic defect that produces an inability to metabolize a certain compound.

**inbreeding depression** The phenomenon whereby inbreeding produces homozygotes that are less fit, thereby decreasing the reproductive success of a population.

**inbreeding** Mating between genetically related individuals.

**inclusive fitness** The term used to designate the total number of copies of genes passed on through one's relatives, as well as one's own reproductive output.

**incomplete dominance** The phenomenon in which a heterozygote that carries two different alleles exhibits a phenotype that is intermediate between the phenotypes of the corresponding homozygous individuals.

**incomplete flower** A flower that lacks one or more of the four flower whorls.

**incomplete metamorphosis** During development in some insects, a gradual change in body form from a nymph into a adult.

**incurrent siphon** A structure in a tunicate used to draw water through the mouth.

**indeterminate cleavage** In animals, a characteristic of deuterostome development in which each cell produced by early cleavage retains the ability to develop into a complete embryo.

**indeterminate growth** Growth in which plant shoot apical meristems continuously produce new stem tissues and leaves, as long as conditions remain favorable.

**indicator species** A species whose status provides information on the overall health of an ecosystem.

**indirect calorimetry** A method of determining metabolic rate in which the rate at which an animal uses oxygen is measured.

**individualistic model** A view of the nature of a community that considers it to be an assemblage of species coexisting primarily because of similarities in their physiological requirements and tolerances.

**induced fit**  The phenomenon that occurs when a substrate(s) binds to an enzyme and the enzyme undergoes a conformational change that causes the substrate(s) to bind more tightly to the enzyme.

**induced mutation**  A mutation brought about by environmental agents that enter the cell and then alter the structure of DNA.

**inducer**  A small effector molecule that increases the rate of transcription of a gene.

**inducible operon**  A type of operon for which the presence of a small effector molecule causes transcription to increase.

**induction**  1. In development, the process by which a cell or group of cells governs the developmental fate of neighboring cells. 2. In molecular genetics, refers to the process by which transcription has been turned on by the presence of a small effector molecule.

**industrial nitrogen fixation**  The human activity of producing nitrogen fertilizers.

**infertility**  The inability to produce viable offspring.

**inflammation**  An innate local response to infection or injury characterized by local redness, swelling, heat, and pain.

**inflorescence**  A tightly clustered group of flowers produced by a plant.

**infundibular stalk**  The thin piece of tissue that physically connects the hypothalamus to the pituitary gland.

**ingestion**  In animals, the act of taking food into the body.

**ingroup**  In a cladogram, the group whose evolutionary relationships are of interest.

**inheritance**  The transmission of characteristics from parent to offspring.

**inhibition**  A mechanism for succession in which early colonists exclude subsequent colonists.

**inhibitory postsynaptic potential (IPSP)**  The response to an inhibitory neurotransmitter that hyperpolarizes the postsynaptic membrane; this hyperpolarization reduces the likelihood of an action potential.

**initiation factor**  A protein that facilitates the interactions between mRNA, the first tRNA, and the ribosomal subunits during the initiation stage of translation.

**initiation**  The first stage in the process of transcription or translation.

**innate immunity**  The body's defenses that are present at birth and act against any foreign material in much the same way, regardless of that material's specific identity; these defenses include the skin and mucous membranes, plus various cellular and chemical defenses.

**innate**  The term used to describe behaviors that seem to be genetically programmed.

**inner bark**  The thin layer of secondary phloem that transports watery solutions of organic compounds in a woody stem.

**inner segment**  The part of the vertebrate photoreceptors (rods and cones) that contains the cell nucleus and cytoplasmic organelles.

**inorganic chemistry**  The study of the nature of atoms and molecules, with the exception of those that contain rings or chains of carbon.

**insulin**  A hormone found in animals that regulates metabolism in several ways, primarily by regulating the blood glucose concentration.

**insulin-like growth factor-1 (IGF-1)**  A hormone in mammals that stimulates the elongation of bones, especially during puberty.

**integral membrane protein**  A protein such as a transmembrane protein or a lipid-anchored protein that cannot be released from the membrane unless the membrane is dissolved with an organic solvent or detergent.

**integrase**  An enzyme, sometimes encoded by viruses, that catalyzes the integration of the viral genome into a host-cell chromosome.

**integrator**  The part of a homeostatic control system, typically a nucleus in the brain, in which the value of a variable is compared to a set point.

**integrin**  A cell adhesion molecule found in animal cells that connects cells to the extracellular matrix.

**integument**  In plants, a leaflike structure that encloses the sporangium to form an ovule.

**interference competition**  Competition in which organisms interact directly with one another by physical force or intimidation.

**interferon**  A protein that generally inhibits viral replication inside host cells.

**intermediate filament**  A type of protein filament of the cytoskeleton of animal cells that helps maintain cell shape and rigidity.

**intermediate-disturbance hypothesis**  The proposal that moderately disturbed communities are more diverse than undisturbed or highly disturbed communities.

**internal fertilization**  Fertilization that occurs in terrestrial animals, in which sperm are deposited within the reproductive tract of the female during copulation.

**interneuron**  A type of neuron that forms interconnections between other neurons.

**internode**  The region of a plant stem between adjacent nodes.

**interphase**  The portion of the cell cycle consisting of the $G_1$, S, and $G_2$ phases, during which the chromosomes are decondensed and found in the nucleus.

**intersexual selection**  Sexual selection between members of the opposite sex.

**interspecies hybrid**  The offspring resulting from the interbreeding of members of two different species.

**interspecific competition**  Competition between individuals of different species.

**interstitial fluid**  The fluid that surrounds cells.

**intertidal zone**  The area where the land meets the sea, which is alternately submerged and exposed by the daily cycle of tides.

**intracellular fluid**  The fluid inside cells.

**intrasexual selection**  Sexual selection that occurs via competition between members of the same sex for the opportunity to mate with individuals of the opposite sex.

**intraspecific competition**  Competition between individuals of the same species.

**intrauterine device (IUD)**  A small object that is placed in the uterus and interferes with the endometrial preparation required for acceptance of the blastocyst; used as a form of contraception.

**intrinsic pathway**  A pathway leading to apoptosis that initially involves internal DNA damage.

**intrinsic rate of increase**  The maximum value of the per capita growth rate, which is attained when conditions are optimal for population growth.

**introduced species**  A species moved by humans from a native location to another location.

**intron**  Intervening DNA sequences that are found in between the coding sequences of genes.

**invasive species**  Introduced species that spread on their own, often outcompeting native species for space and resources.

**inverse density-dependent factor**  A mortality factor whose influence decreases as population size increases.

**inversion**  A type of mutation that involves a change in the direction of the genetic material along a single chromosome.

**invertebrate**  An animal that lacks a backbone (vertebrae).

**iodine-deficient goiter**  An overgrown thyroid gland that is incapable of making thyroid hormone due to a lack of dietary iodine.

**ion**  An atom or molecule that gains or loses one or more electrons and acquires a net electric charge.

**ionic bond**  The bond that occurs when a cation binds to an anion.

**ionotropic receptor**  One of two types of postsynaptic receptors, the other being a metabotropic receptor. Consists of a ligand-gated ion channel that opens in response to binding of a neurotransmitter.

**iris**  The circle of pigmented smooth muscle and connective tissue that is responsible for eye color.

**iron regulatory element (IRE)**  A response element within the ferritin mRNA to which the iron regulatory protein binds.

**iron regulatory protein (IRP)**  An RNA-binding protein that regulates the translation of the mRNA that encodes ferritin.

**islets of Langerhans**  Spherical clusters of endocrine cells that are scattered throughout the pancreas; the cells secrete insulin or glucagon, among other hormones.

**isomers**  Two or more molecules with the same chemical formula but different structures and characteristics.

**isotonic**  Condition in which the solute concentrations on both sides of a plasma membrane are equal, which does not cause a cell to shrink or swell.

**isotope**  A form of an element that contains a different number of neutrons from the element's other isotopes.

**iteroparity**  The pattern of repeated reproduction at intervals throughout an organism's life cycle.

# J

**joint**  The juncture where two or more bones of a vertebrate endoskeleton come together.

**juvenile hormone**  A hormone made in arthropods that inhibits maturation from a larva into a pupa.

# K

**karyogamy**  The process of nuclear fusion.

**karyotype**  A photographic representation of the chromosomes from an actively dividing cell.

**kcal (kilocalorie)**  One thousand calories; the amount of heat energy required to raise the temperature of 1 kg of water by 1°C.

**ketones**  Small compounds generated from fatty acids. Ketones are made in the liver and released into the blood during prolonged fasting to provide an important energy source for many tissues, including the brain.

**keystone hypothesis**  The idea that ecosystem function plummets as soon as biodiversity declines from its natural levels.

**keystone species**  A species within a community that has a role out of proportion to its abundance or biomass.

**kidney**  The major excretory organ found in all vertebrates.

**kin selection**  Selection for behavior that lowers an individual's own fitness but enhances the reproductive success of a relative.

**kinesis**  A movement in response to a stimulus, but one that is not directed toward or away from the source of the stimulus.

**kinetic energy**  Energy associated with movement.

**kinetic skull**  A characteristic of lizards and snakes in which the joints between various parts of the skull are extremely mobile.

**kinetochore**  A group of proteins that bind to a centromere and are necessary for sorting the chromosomes.

**kingdom**  A taxonomic group; the second largest division after domain.

**$K_M$**  The substrate concentration at which an enzyme-catalyzed reaction is half of its maximal value.

**knowledge** The awareness and understanding of information.

**Koch's postulates** A series of steps used to determine whether a particular organism causes a specific disease.

*K*-selected species A species whose life history strategy shows a low rate of per capita population growth but good competitive ability.

# L

**labor** The strong rhythmic contractions of the uterus that serve to deliver a fetus during childbirth.

**lac repressor** A repressor protein that regulates the *lac* operon.

*lac* operon An operon in the genome of *E. coli* that contains the genes for the proteins that allow the bacterium to metabolize lactose.

**lacteal** A lymphatic vessel in the center of each intestinal villus; lipids are absorbed by the lacteals, which eventually empty into the circulatory system.

**lagging strand** During DNA replication, a DNA strand made as a series of small Okazaki fragments that are subsequently connected to each other to form a continuous strand.

**lamellae** (singular, **lamella**) Platelike structures in the internal gills of fishes that branch from structures called filaments; gas exchange occurs here.

**larva** A free-living organism that is morphologically very different from the embryo and adult.

**larynx** The segment of the respiratory tract that contains the vocal cords.

**latent** The term used to describe a prophage or provirus that remains inactive for a long time.

**lateral line system** Microscopic sensory organs in fishes and some toads that allows them to detect movement in surrounding water.

**lateral meristem** *See* secondary meristem.

**leaching** The dissolution and removal of inorganic ions as water percolates through materials such as soil.

**leading strand** During DNA replication, a DNA strand made in the same direction that the replication fork is moving. The strand is synthesized as one long continuous molecule.

**leaf abscission** The process by which a leaf drops after the formation of an abscission zone.

**leaf primordia** Small outgrowths that occur at the sides of a shoot apical meristem and develop into young leaves.

**leaf vein** In plants, a bundle of vascular tissue in a leaf.

**leaflet** 1. Half of a phospholipid bilayer. 2. A portion of a compound leaf.

**learning** The ability of an animal to make modifications to a behavior based on previous experience; the process by which new information is acquired.

**leaves** Flattened plant organs that emerge from stems and typically function in photosynthesis.

**left-right axis** In bilaterally symmetric animals, the left and right sides of the body.

**leghemoglobin** A protein found in legume plants that helps to regulate local oxygen concentrations around rhizobial bacteroids in root nodules.

**legume** A member of the pea (bean) family; also the distinctive fruit of such a plant.

**lek** A designated communal courting area used by certain species of animals.

**lens** 1. A structure of the eye that focuses light. 2. The glass components of a light microscope or the electromagnetic parts of an electron microscope that allow the production of magnified images of microscopic structures.

**lentic** Refers to a freshwater habitat characterized by standing water.

**lenticels** Passages in the outer bark of a woody plant stem that allow inner stem tissues to accomplish gas exchange.

**leptin** A hormone produced by adipose cells in proportion to fat mass; controls appetite and metabolic rate.

**leukocyte** A cell that develops from the marrow of certain bones of vertebrates; all leukocytes (also known as white blood cells) perform vital functions that defend the body against infection and disease.

**lichen** A complex mutualistic association between fungi and other microbes, including photosynthetic green algae or cyanobacteria.

**Liebig's law of the minimum** The principle that states that species' biomass or abundance is limited by the scarcest factor.

**life cycle** The sequence of events that characterize the steps of development of the individuals of a given species.

**life table** A table that provides data on the numbers of living individuals in various age classes in a population and their relative fertilities.

**ligand** An ion or molecule that binds to a protein, such as an enzyme or a receptor.

**ligand·receptor complex** The structure formed when a ligand and its receptor bind noncovalently to each other.

**ligand-gated ion channel** A type of cell surface receptor that binds a ligand and functions as an ion channel. Ligand binding either opens or closes a channel.

**light microscope** A microscope that utilizes light for illumination.

**light reactions** The first of two stages in the process of photosynthesis. During the light reactions, photosystem II and photosystem I absorb light energy and produce ATP, NADPH, and $O_2$.

**light-harvesting complex** A component of photosystem II and photosystem I composed of several dozen pigment molecules that are anchored to proteins in the thylakoid membrane of a chloroplast. The role of these complexes is to absorb photons of light.

**lignin** A tough polymer that adds strength and decay resistance to cell walls of tracheids, vessel elements, and other cells of plants.

**limbic system** In the vertebrate forebrain, the areas involved in the formation and expression of emotions; also plays a role in learning, memory, and the perception of smells.

**limiting factor** A factor whose amount or concentration limits the rate of a biological process or a chemical reaction.

**line transect** A sampling technique used by plant ecologists in which the number of plants located along a length of string are counted.

**lineage** A series of species that forms a line of descent.

**linear electron flow** In the light reactions of photosynthesis, the movement of electrons from PSII to PSI and ultimately to $NADP^+$ to form NADPH.

**linkage group** A group of genes that usually stay together during meiosis.

**linkage** The phenomenon in which two genes close together on the same chromosome are transmitted as a unit.

**lipase** The major fat-digesting enzyme, secreted by the pancreas.

**lipid raft** In a membrane, a group of lipids, sometimes including associated proteins, that float together as a unit in a larger sea of lipids.

**lipid** A molecule composed predominantly of hydrogen and carbon atoms. Lipids are nonpolar and therefore very insoluble in water. They include fats (triglycerides), phospholipids, waxes, and steroids.

**lipid-anchored protein** A type of integral membrane protein that is attached to the membrane via a lipid molecule.

**lipid-exchange protein** A protein that extracts a lipid from one membrane, diffuses through the cell, and inserts the lipid into another membrane.

**lipolysis** The enzymatic breakdown of triglycerides into fatty acids and either monoglycerides or glycerol.

**lipopolysaccharides** Lipids with covalently bound polysaccharides; prevalent in the thin, outer envelope that encloses the cell walls of Gram-negative bacteria.

**liposome** A vesicle surrounded by a phospholipid bilayer.

**liquid-liquid phase separation** The phenomenon in which aggregated solutes, such as proteins and RNA molecules, separate from the bulk solvent, and form a droplet.

**liver** An organ in vertebrates that performs diverse metabolic functions and is the site of bile production.

**liverworts** A phylum of bryophytes; formally called Hepatophyta.

**loam** A type of soil that contains a mixture of sand, silt, and clay and is ideal for plant cultivation.

**lobe-finned fishes** Fishes in which the fins are part of the body; the fins are supported by skeletal extensions of the pectoral and pelvic areas.

**locomotion** The movement of an animal from place to place.

**locus** The physical location of a gene on a chromosome.

**logistic equation** An equation that relates the growth of a population to the carrying capacity, *K*, of its environment.

**logistic growth** The pattern in which the growth of a population typically slows down as the population size approaches the carrying capacity.

**long-day plant** A plant that flowers in spring or early summer, when the night period is shorter (and thus the day length is longer) than a defined period.

**long-term potentiation (LTP)** The long-lasting strengthening of the connection between neurons that is believed to be part of the mechanism of learning and memory.

**loop domain** In bacteria, a chromosomal segment that is folded into a loop by attachment to proteins; a means of compacting a bacterial chromosome.

**loop of Henle** A segment of the renal tubule of the kidney containing a sharp hairpin-like loop that contributes to reabsorption of ions and water and that consists of a descending limb coming from the proximal tubule and an ascending limb leading to the distal tubule.

**lophophore** A horseshoe-shaped crown of tentacles used for feeding in several invertebrate species.

**Lophotrochozoa** A clade of animals that encompasses the mollusks, annelids, and several other phyla; they are distinguished by two morphological features: the lophophore, a crown of tentacles used for feeding, and the trochophore larva, a distinct larval stage.

**lotic** Refers to a freshwater habitat characterized by running water.

**lumen** The internal space or hollow cavity of an organelle or an organ, such as the endoplasmic reticulum, the stomach, or a blood vessel.

**lung** In terrestrial vertebrates, internal paired structures used to bring $O_2$ into the circulatory system and remove $CO_2$.

**lungfishes** The Dipnoi; fish with primitive lungs that live in oxygen-poor freshwater swamps and ponds.

**luteal phase** The phase of the human ovarian cycle that begins after ovulation and during which a corpus luteum is formed.

**luteinizing hormone (LH)** A gonadotropin that controls the production of sex steroids in both males and females.

**lycophyll** A relatively small leaf having a single unbranched vein; produced by lycophytes.

**lycophytes** Members of a phylum (Lycopodiophyta) of vascular land plants whose leaves are lycophylls.

**lymphatic system**   A system of vessels along with a group of organs and tissues where most leukocytes reside. The lymphatic vessels collect excess interstitial fluid and return it to the blood.

**lymphocyte**   A type of leukocyte that is responsible for specific immunity; may be either a B cell or a T cell.

**lysogenic cycle**   A type of viral reproductive cycle consisting of integration of phage DNA into that of a bacterium, prophage replication, and excision.

**lysosome**   A small organelle found in animal cells that contains acid hydrolases that degrade molecules and macromolecules.

**lytic cycle**   A type of viral reproductive cycle in which the production and release of new phages lyses the host cell.

## M

**M phase**   The phase of the cell cycle consisting of the sequential events of mitosis and cytokinesis.

**macroalgae**   Photosynthetic protists that can be seen with the unaided eye; also known as seaweeds.

**macroevolution**   Evolutionary changes that produce new species and groups of species.

**macromolecule**   Many molecules bonded together to form a polymer. Carbohydrates, proteins, and nucleic acids (for example, DNA and RNA) are important macromolecules found in living organisms.

**macronutrient**   An element required by plants in amounts of at least 1 g/kg of plant dry matter.

**macroparasite**   A parasite that lives in a host but releases infective juvenile stages outside the host's body.

**macrophage**   A large phagocyte capable of engulfing viruses and bacteria.

**macular degeneration**   An eye condition in which photoreceptor cells in and around the macula (which contains the fovea of the retina) are lost; one of the leading causes of blindness in the U.S.

**madreporite**   A sievelike plate on the surface of an echinoderm through which water enters the water vascular system.

**magnetic resonance imaging (MRI)**   An imaging method that relies on the use of magnetic fields and radio waves to visualize the internal structure of an organism's body.

**magnification**   The ratio between the size of an image produced by a microscope and the object's actual size.

**major depressive disorder**   A neurological disorder characterized by feelings of despair and sadness, resulting from an imbalance in neurotransmitter levels in the brain.

**major groove**   A wider groove that spirals around the DNA double helix; provides a location where a protein can bind to a particular sequence of bases and affect the expression of a gene.

**major histocompatibility complex (MHC)**   A gene family that encodes the plasma membrane self proteins that must be complexed with an antigen for T-cell recognition to occur.

**male gametophyte**   A haploid multicellular plant life cycle stage that produces sperm.

**male-assistance hypothesis**   A hypothesis to explain the existence of monogamy that maintains that males remain with females to help them rear their offspring.

**malignant tumor**   A growth of cells that has progressed to the cancerous stage.

**Malpighian tubules**   Delicate projections from the digestive tract of insects and some other taxa that function as an excretory organ.

**mantle cavity**   The chamber in a mollusk mantle that houses delicate gills.

**mantle**   In mollusks, a fold of skin draped over the visceral mass that secretes a shell in those species that form shells.

**many-eyes hypothesis**   The idea that increased group size decreases predators' success because of increased detection of predators.

**map distance**   The distance between genes along chromosomes, which is calculated as the number of recombinant offspring divided by the total number of offspring times 100.

**map unit (mu)**   A unit of distance on a chromosome equivalent to a 1% recombination frequency.

**marker genes**   Genes that mark, or indicate, the occurrence of particular metabolic functions in organisms.

**mark-recapture technique**   The capture and tagging of animals so they can be released and recaptured, allowing an estimate of population size.

**marsupial**   A member of a group of seven mammalian orders and about 280 species found in the clade Metatheria.

**mass extinction**   An event in which many species become extinct at the same time.

**mass-specific BMR**   The amount of energy expended per gram of body mass in the resting condition.

**mast cell**   A type of cell derived from bone marrow stem cells that occurs throughout connective tissues and plays an important role in innate immunity.

**mastax**   The circular muscular pharynx in the mouth of rotifers.

**mate-guarding hypothesis**   The hypothesis that a male is monogamous to prevent his mate from being fertilized by other males.

**maternal effect gene**   A gene for which only the mother's gene product affects the phenotype of the resulting offspring.

**maternal inheritance**   A phenomenon in which offspring inherit particular genes only from the female parent (through the egg).

**matrotrophy**   In plants, the phenomenon in which zygotes remain enclosed within gametophyte tissues, where they are sheltered and fed.

**matter**   Anything that has mass and takes up space.

**mature mRNA**   In eukaryotes, transcription produces a longer RNA, called pre-mRNA, which undergoes certain modifications before it exits the nucleus; mature mRNA is the final functional product.

**maximum likelihood**   One method used to evaluate a phylogenetic tree based on an evolutionary model.

**maximum sustainable yield (MSY)**   The largest number of individuals that can be removed without causing long-term decreases in the population.

**mean fitness of the population**   The average reproductive success of members of a population.

**mechanoreceptor**   A sensory receptor in animals that transduces mechanical energy such as pressure, touch, stretch, movement, and sound.

**mediator**   A large protein complex that plays a role in initiating transcription at the core promoter of protein-encoding genes in eukaryotes.

**medulla oblongata**   The part of the vertebrate hindbrain that coordinates many processes that maintain homeostasis, such as breathing.

**medusa**   A type of cnidarian body form that is motile and usually floats mouth down.

**megadiversity countries**   Those countries with the greatest numbers of species; used in targeting areas for conservation.

**megaspore**   In seed plants and some seedless plants, a large spore that produces a female gametophyte within the spore wall.

**meiosis I**   The first division of meiosis in which homologous chromosomes are separated into different cells.

**meiosis II**   The second division of meiosis in which sister chromatids are separated into different cells.

**meiosis**   The process by which haploid cells are produced from a cell that was originally diploid.

**Meissner corpuscle**   A specialized receptor that senses touch and light pressure and lies just beneath the skin surface of an animal.

**membrane attack complex (MAC)**   A multi-unit protein formed by the activation of complement proteins; the complex creates water channels in the microbial plasma membrane and causes the microbe to swell and burst.

**membrane potential**   The difference between the electric charges outside and inside a cell; also called a potential difference (or voltage).

**membrane transport**   The movement of ions or molecules across a biological membrane.

**membrane vesicle**   A small sphere enclosed by a membrane.

**memory cell**   A cloned lymphocyte that remains poised to recognize a returning antigen; a component of acquired immunity.

**memory**   The ability to retain, retrieve, and use previously learned information.

**Mendelian inheritance**   The inheritance patterns of genes that segregate and assort independently.

**Mendel's law of independent assortment**   The alleles of different genes assort independently of each other during the process that gives rise to gametes.

**Mendel's law of segregation**   The two alleles of a gene separate (segregate) from each other during the process that gives rise to gametes, so every gamete receives only one allele.

**meninges**   Three layers of sheathlike membranes that cover and protect the brain and spinal cord.

**meningitis**   A potentially life-threatening infectious disease in which the meninges become inflamed.

**menopause**   The event during which a woman permanently stops having ovarian cycles.

**menstrual cycle**   The cyclical changes that occur in the uterus in parallel with the ovarian cycle in a female mammal; also called the uterine cycle.

**menstruation**   A period of bleeding at the beginning of the uterine cycle in a female mammal.

**meristem**   In plants, a region of undifferentiated cells (stem cells) that produces new tissue by cell division.

**meroblastic cleavage**   An incomplete type of cell cleavage in which only the region of the egg containing cytoplasm at the animal pole undergoes cell division; occurs in birds and some fishes.

**merozygote**   A bacterial cell that contains an F′ factor.

**mesoderm**   In animals, a layer of cells formed during gastrulation that develops between the ectoderm and endoderm; gives rise to the skeleton, muscles, and much of the circulatory system.

**mesoglea**   A gelatinous substance between the epidermis and the gastrodermis in cnidarians.

**mesohyl**   A gelatinous, protein-rich matrix in between the choanocytes and the epithelial cells of a sponge.

**mesophyll**   The internal tissue of a plant leaf; the site of photosynthesis.

**messenger RNA (mRNA)**   RNA that contains the information to specify a polypeptide with a particular amino acid sequence.

**metabolic cycle**   A biochemical cycle in which particular molecules enter while others leave; the process is cyclical because it involves a series of organic molecules that are regenerated with each turn of the cycle.

**metabolic pathway**   In living cells, a coordinated series of chemical reactions in which each step is catalyzed by a specific enzyme.

**metabolic rate**   The total energy expenditure of an organism per unit of time.

**metabolism**   The sum of all bodily activities and chemical reactions that occur within an organism. Also, a specific set of chemical reactions occurring at the cellular level.

**metabolome** Collection of information about the types and abundances of molecules, such as sugars and fatty acids, produced by metabolism in a single organism.

**metabotropic receptor** A G-protein-coupled receptor that initiates a signaling pathway in response to a neurotransmitter. One of two types of postsynaptic receptors, the other being an ionotropic receptor.

**metacentric** Refers to a chromosome in which the centromere is near the middle.

**metagenome** The genomes of all the organisms present in a sample.

**metagenomics** A field of study that seeks to identify and analyze the collective microbial genomes contained in a community of organisms, including those not easily cultured in the laboratory.

**meta-metabolome** Collection of information about the types and abundances of molecules, such as sugars and fatty acids, produced by metabolism by an entire microbiome.

**metamorphosis** The process in which a pupal or juvenile organism changes into a mature adult with very different characteristics.

**metanephridia** Excretory filtration organs found in a variety of invertebrates, including annelids.

**metaphase plate** A plane halfway between the poles of the spindle apparatus on which the sister chromatids align during metaphase in mitosis.

**metaphase** The phase of mitosis during which the chromosomes are aligned along the metaphase plate.

**metaproteome** All the proteins produced by all the members of a microbiome.

**metastasis** The process by which cancer cells spread from their original location to distant parts of the body.

**metatranscriptome** A collection of all the mRNAs present in an environmental sample, that is, all of the mRNAs produced by all of the organisms sampled from a particular place at a particular time.

**methanogens** Several groups of anaerobic archaea that convert $CO_2$, methyl groups, or acetate to methane, and release it from their cells.

**methanotroph** An aerobic bacterium that consumes methane.

**methyl-CpG-binding protein** A protein that binds methylated DNA sequences and inhibits transcription.

**micelle** A sphere formed from the aggregation of long amphipathic molecules when they are mixed with water. The polar regions are on the surface and in contact with water and the nonpolar regions are in the center.

**microbiome** A particular assemblage of microbes (including their genes) that are associated with a defined environment.

**microbiota** Collections of microbial life cataloged by amplicon analysis.

**microcystin** A type of toxin produced by some cyanobacteria, including the genus *Microcystis*.

**microevolution** Changes in a population's gene pool, such as changes in allele frequencies, from generation to generation.

**microfilament** *See* actin filament.

**micrograph** An image taken with the aid of a microscope.

**micronutrient** An element required by plants in amounts equal to or less than 0.1 g/kg of plant dry matter; also known as a trace element.

**microparasite** A parasite that multiplies within its host, usually within the cells.

**micropyle** A small opening in the integument of a seed plant ovule through which a pollen tube grows.

**microRNAs (miRNAs)** Small RNA molecules, typically 22 nucleotides in length, that are transcribed from endogenous eukaryotic genes and silence the expression of specific mRNAs by inhibiting translation.

**microscope** A magnification tool that enables researchers to visualize the structures and inner workings of cells.

**microspore** In seed plants and some seedless plants, a relatively small spore that produces a male gametophyte within the spore wall.

**microtubule** A type of hollow protein filament composed of tubulin proteins that is part of the cytoskeleton and is important for cell shape, organization, and movement.

**microvilli** Small projections of the plasma membranes of epithelial cells in the small intestine and many other absorptive cells.

**midbrain** One of three major divisions of the vertebrate brain; the other two divisions are the hindbrain and the forebrain.

**middle lamella** An extracellular layer in plants composed primarily of carbohydrate; cements adjacent plant cell walls together.

**migration** Long-range seasonal movement of animals in order to feed or breed.

**mimicry** The resemblance of an organism (the mimic) to another organism (the model).

**mineral** An inorganic ion required by a living organism for normal cellular functioning.

**mineralization** The general process by which phosphorus, nitrogen, $CO_2$, and other minerals are released from organic compounds.

**minor groove** A groove that spirals around the DNA double helix but is smaller than the major groove.

**missense mutation** A base substitution that changes a single amino acid in a polypeptide.

**mitochondrial genome** The chromosome found in mitochondria.

**mitochondrial matrix** A compartment inside the inner membrane of a mitochondrion.

**mitochondrial pathway** *See* intrinsic pathway.

**mitochondrion** A semiautonomous organelle found in eukaryotic cells that supplies most of a cell's ATP.

**mitosis** In eukaryotes, the process in which nuclear division results in two nuclei, each of which receives the same complement of chromosomes.

**mitotic cell division** A process whereby a eukaryotic cell divides to produce two new cells that are genetically identical to the original cell.

**mitotic spindle** The structure responsible for organizing and sorting the chromosomes during mitosis; also called the mitotic spindle apparatus.

**mixotroph** An organism that is able to use photoautotrophy as well as phagotrophy or osmotrophy to obtain organic nutrients.

**model organism** An organism studied by many different researchers so that they can compare their results and determine scientific principles that apply more broadly to other species.

**model-based learning** An educational approach in which students evaluate or generate models as a way to enhance their understanding of scientific concepts and improve their critical-thinking skills.

**moderately repetitive sequence** A DNA sequence that is repeated a few hundred to several thousand times in a genome.

**molar** A term used to describe a solution's molarity; a 1 molar solution contains 1 mole of a solute dissolved in enough water to make 1 L of solution.

**molarity** The number of moles of a solute dissolved in 1 L of water.

**mole** The amount of any substance that contains the same number of particles as there are atoms in exactly 12 g of carbon.

**molecular biology** A field of study spawned largely by genetic technology that allows researchers to study the structure and function of the molecules of life.

**molecular clock** A method for estimating evolutionary time; based on the observation that neutral mutations occur at a relatively constant rate.

**molecular evolution** The process of evolution at the level of genes and proteins.

**molecular formula** A representation of a molecule that consists of the chemical symbols for all of the atoms present and subscripts that indicate how many of those atoms are present.

**molecular homologies** Similarities at the molecular level that indicate that living species evolved from a common ancestor or interrelated group of ancestors.

**molecular mass** The sum of the atomic masses of all atoms in a molecule.

**molecular systematics** A field of study that involves the analysis of genetic data, such as DNA and amino acid sequences, to identify and study genetic homologies and construct phylogenetic trees.

**molecule** Two or more atoms that are connected by chemical bonds.

**monoclonal antibodies** Antibodies of a specific type that are derived from a single clone of cells.

**monocots** One of the two largest lineages of flowering plants in which the embryo possesses a single seed leaf.

**monocular vision** A type of vision in animals that have eyes on the sides of the head; the animal sees a wide area at one time, though depth perception is reduced.

**monocyte** A type of phagocyte that circulates in the blood for only a few days, after which it enters an organ or tissue and develops into a macrophage.

**monoecious** Refers to plants that produce carpellate and staminate flowers on the same plant.

**monogamy** A mating system in which each individual mates exclusively with one partner over at least a single breeding cycle.

**monohybrid** The $F_1$ offspring, also called single-trait hybrids, of true-breeding parents that differ with regard to a single character.

**monomer** An organic molecule that can be used to form a larger molecule (polymer) consisting of many repeating units of the monomer.

**monomorphic gene** A gene that exists predominantly as a single allele in a population.

**monophagous** The term used to describe parasites that feed on one or a few closely related species.

**monophyletic group** A group of species, a taxon, that is a clade.

**monosaccharide** A simple sugar, such as a pentose or hexose.

**monotreme** A member of the mammalian order Monotremata, which consists of five species found in Australia and New Guinea: the duck-billed platypus and four species of echidna.

**morphogen** A molecule that imparts positional information and promotes developmental changes at the cellular level.

**morphology** The structure or form of a body part or an entire organism.

**morula** An early stage in a mammalian embryo in which physical contact between cells is maximized by compaction.

**mosaic** An individual with somatic regions that are genetically different from each other.

**mosses** A phylum of bryophytes; formally called Bryophyta.

**motility** The ability of a cell to move or change position within its environment.

**motor end plate** The region of a skeletal muscle cell that lies beneath an axon terminal at the neuromuscular junction.

**motor neuron** A neuron that sends signals away from the central nervous system and elicits some type of response from a gland, muscle or other structure.

**motor protein**   A type of cellular protein that uses ATP as a source of energy to promote movement; consists of three domains called the head, hinge, and tail.

**movement corridor**   Thin strips of habitat that may permit the movement of individuals between larger habitat patches.

**mRNA**   *See* messenger RNA.

**Müllerian mimicry**   A type of mimicry in which two or more noxious species converge to look the same, thus reinforcing the basic distasteful design.

**multicellular**   Describes an organism consisting of more than one cell, particularly when cell-to-cell adherence and signaling processes and cellular specialization can be demonstrated.

**multiple alleles**   The phenomenon in which a gene has three or more alleles in a natural population.

**multiple sclerosis (MS)**   A disease in which the patient's own body attacks and destroys myelin as if it were a foreign substance; impairs the function of myelinated neurons that control movement, speech, memory, and emotion.

**multipotent**   Refers to a stem cell that can differentiate into several cell types, but far fewer than a pluripotent cell can.

**muscle fiber**   Individual cell within a muscle.

**muscle tissue**   Bundles of muscle fibers (cells) that are specialized to contract when stimulated and thus generate a force that facilitates movement or exerts pressure .

**muscular dystrophy**   A group of diseases associated with progressive degeneration of skeletal and cardiac muscle fibers.

**mutagen**   An agent known to cause mutation.

**mutant allele**   An allele that has been altered by mutation.

**mutation**   A heritable change in the genetic material of an organism.

**mutation**   A heritable change in the genetic material.

**mutualism**   A symbiotic interaction in which both species benefit.

**mycelium**   A fungal body composed of microscopic branched filaments known as hyphae.

**mycorrhizae** (singular, **mycorrhiza**)   Associations between the hyphae of certain fungi and the roots of seed plants.

**myelin sheath**   In the nervous system, an insulating layer made up of specialized glial cells wrapped around the axons.

**myocardial infarction (MI)**   The death of cardiac muscle cells, which can occur if a region of the heart is deprived of blood for an extended time.

**myofibril**   Rodlike collection of myofilaments within a muscle fiber (cell); contains thick and thin filaments.

**myoglobin**   An oxygen-binding protein that provides an intracellular reservoir of oxygen for muscle fibers.

**myosin**   A motor protein found abundantly in muscle cells and also in other cell types.

# N

**NAD⁺**   Nicotinamide adenine dinucleotide; an organic molecule that functions as an energy intermediate. It combines with two electrons and H⁺ to form NADH.

**NADPH**   Nicotinamide adenine dinucleotide phosphate; an energy intermediate that provides the energy and electrons to drive the Calvin cycle during photosynthesis.

**natural killer (NK) cell**   A type of leukocyte that participates in both innate and acquired immunity; recognizes general features on the surface of cancer cells or any virus-infected cells.

**natural selection**   The process that eliminates those individuals that are less likely to survive and reproduce in a particular environment, while allowing other individuals with traits that confer greater reproductive success to increase in numbers.

**nauplius**   The first larval stage in a crustacean.

**navigation**   A mechanism of migration that involves the ability not only to follow a compass bearing but also to set or adjust it.

**negative control**   Transcriptional regulation by repressor proteins.

**negative feedback loop**   A homeostatic mechanism in animals in which a change in the variable being regulated brings about responses that move the variable in the opposite direction.

**negative frequency-dependent selection**   A pattern of natural selection in which the fitness of a genotype decreases when its frequency becomes higher; the result is balanced polymorphism.

**negative pressure filling**   The process by which reptiles, birds, and mammals ventilate their lungs.

**nekton**   Free-swimming animals in the open ocean that can swim against currents to locate food.

**nematocyst**   In a cnidarian, a powerful capsule with an inverted coiled and barbed thread that functions to immobilize small prey.

**nephron**   One of several million single-cell-thick tubules that are the functional units of the mammalian kidney.

**Nernst equation**   The formula that gives the equilibrium potential for an ion at any given concentration gradient: $E = 60 \text{ mV} \log_{10}([X_{\text{extracellular}}]/[X_{\text{intracellular}}])$.

**nerve net**   Interconnected neurons with no central control organ.

**nerve**   A structure found in the peripheral nervous system that is composed of multiple myelinated axons bound by connective tissue; carries information to or from the central nervous system.

**nervous system**   Coordinated circuits of cells that sense internal and environmental changes and transmit signals that enable an animal to respond in an appropriate way.

**nervous tissue**   Networks of cells (neurons) that receive, generate, and conduct electrical signals throughout an animal's body.

**net primary production (NPP)**   Gross primary production minus the energy lost in plant cellular respiration.

**net reproductive rate**   The population growth rate per generation.

**neural crest**   In vertebrates, a group of embryonic cells derived from ectoderm that disperse throughout the embryo and contribute to the development of the skeleton and other structures, including peripheral nerves.

**neural tube**   In chordates, a structure formed from ectoderm located dorsal to the notochord; all neurons and their supporting cells in the central nervous system originate from neural precursor cells derived from the neural tube.

**neurogenesis**   The production of new neurons by cell division.

**neurohormone**   A hormone made in and secreted by neurons whose cell bodies are in the hypothalamus.

**neuromodulator**   Another term for a neuropeptide, which is a neurotransmitter that can alter or modulate the response of a postsynaptic neuron to other neurotransmitters.

**neuromuscular junction**   The contact point between an axon terminal of a motor neuron and a skeletal or cardiac muscle fiber.

**neuron**   A highly specialized cell found in nervous systems of animals that communicates with other cells by electrical or chemical signals.

**neuroparasitology**   The study of how parasites control the nervous systems of their hosts.

**neuroscience**   The scientific study of nervous systems.

**neurulation**   The embryological process responsible for initiating central nervous system formation.

**neutral theory of evolution**   Theory proposing that most genetic variation in a population is due to the accumulation of neutral mutations that have attained high frequencies in the population via genetic drift.

**neutral variation**   Changes in genes and proteins that result from genetic drift and do not have an effect on reproductive success.

**neutron**   A neutral particle found in the nucleus of an atom.

**neutrophil**   A phagocyte that is the most abundant type of leukocyte; neutrophils engulf bacteria by endocytosis.

**niche**   The unique set of habitat resources a species requires as well as its effect on the ecosystem.

**nitrification**   The conversion by soil bacteria of ammonia ($NH_3$) or ammonium ($NH_4^+$) to nitrate ($NO_3^-$), a form of nitrogen commonly used by plants.

**nitrogen fixation**   A specialized metabolic process in which certain prokaryotes use the enzyme nitrogenase to convert inert atmospheric nitrogen gas ($N_2$) into ammonia ($NH_3$); also, the industrial process by which humans produce $NH_3$ fertilizer from $N_2$.

**nitrogenase**   An enzyme used in the biological process of fixing nitrogen.

**nitrogen-limitation hypothesis**   The proposal that organisms select food based on its nitrogen content.

**nociceptor**   A sensory receptor in animals that responds to extreme heat, cold, and pressure, as well as to certain molecules such as acids; also known as a pain receptor.

**Nod factor**   Nodulation factor; a substance produced by nitrogen-fixing bacteria in response to flavonoids secreted from the roots of potential host plants. Nod factors bind to receptors in plant root membranes, starting a process that allows the bacteria to invade roots.

**node**   1. The region of a plant stem from which one or more leaves, branches, or buds emerge. 2. The branch points in a phylogenetic tree.

**nodes of Ranvier**   Exposed areas along the axons of myelinated neurons that contain many voltage-gated $Na^+$ channels and are the sites of regeneration of action potentials.

**nodule**   A small swelling on a plant root that contains nitrogen-fixing bacteria.

**nodulin**   One of several plant proteins that foster root nodule development.

**non-coding RNA (ncRNA)**   An RNA molecule that does not encode the amino acid sequence of a polypeptide.

**noncompetitive inhibitor**   A molecule that binds noncovalently to an enzyme at a location that is outside the active site and inhibits the enzyme's function.

**non-Darwinian evolution**   The idea that much of the modern variation in gene sequences is explained by neutral variation rather than adaptive variation.

**nondisjunction**   An event in which the chromosomes do not separate properly during cell division.

**nonpolar covalent bond**   A strong bond formed between two atoms of similar electronegativities in which electrons are shared between the atoms.

**nonrandom mating**   The phenomenon that occurs when individuals choose their mates based on their genotypes or phenotypes.

**nonrecombinant**   An offspring whose combination of traits has not changed from the true-breeding parental generation.

**nonsense mutation**   A mutation that changes a normal codon into a stop codon; this causes translation to be terminated earlier than expected, producing a truncated polypeptide.

**GLOSSARY**

**nonshivering thermogenesis** An increase in an animal's metabolic rate that is not due to increased muscle activity; occurs primarily in brown adipose tissue.

**norm of reaction** The phenotype range that individuals with a particular genotype exhibit under differing environmental conditions.

**notochord** A defining characteristic of all chordate embryos; a flexible rod that lies between the digestive tract and the nerve cord.

**N-terminus** The location of the first amino acid in a polypeptide; also known as the amino end.

**nuclear envelope** A double-membrane structure that encloses the cell's nucleus.

**nuclear genome** The chromosomes found in the nucleus of a eukaryotic cell.

**nuclear lamina** A collection of protein fibers that line the inner nuclear membrane; part of the nuclear matrix.

**nuclear matrix** A filamentous network of proteins that is found inside the nucleus and lines the inner nuclear membrane. The nuclear matrix serves to organize the chromosomes.

**nuclear pore** A passageway for the movement of molecules and macromolecules into and out of the nucleus; formed where the inner and outer nuclear membranes make contact with each other.

**nucleic acid** An organic macromolecule composed of nucleotides. The two types of nucleic acids are deoxyribonucleic acid (DNA) and ribonucleic acid (RNA).

**nucleoid** The site in a bacterial cell where the genetic material (DNA) is located.

**nucleolus** A droplet organelle in the nucleus of nondividing cells where ribosome assembly occurs.

**nucleosome** A structural unit of eukaryotic chromosomes composed of an octamer of histones (eight histone proteins) wrapped with DNA.

**nucleosome-free region (NFR)** In eukaryotes, a site in the chromatin in which the DNA is not wrapped around histone proteins to form nucleosomes.

**nucleotide excision repair (NER)** A common type of DNA repair system that removes (excises) and repairs a region of the DNA where damage has occurred.

**nucleotide** An organic molecule having three components: one or more phosphate groups, a five-carbon sugar (either deoxyribose or ribose), and a single or double ring of carbon and nitrogen atoms known as a base.

**nucleus** (plural, **nuclei**) 1. In cell biology, an organelle found in eukaryotic cells that contains most of the cell's genetic material. 2. In chemistry, the region of an atom that contains protons and neutrons. 3. In neurobiology, a group of neuronal cell bodies in the brain that are involved in a particular function.

**nutrient** Any substance that is taken in by a living organism and is required for survival, growth, development, tissue repair, or reproduction.

**nutrition** The process of consuming and using nutrients.

# O

**obligate aerobes** Organisms that require oxygen for survival.

**obligatory mutualism** An interaction in which two mutualistic species cannot live without each other.

**occipital lobe** One of four lobes of the cerebral cortex of the human brain; controls aspects of vision and color recognition.

**ocelli** Photosensitive organs found in some animal species.

**octet rule** The observation that many atoms are most stable when their outermost shell is full, with eight electrons.

**Okazaki fragments** Short segments of DNA synthesized in the lagging strand during DNA replication.

**olfaction** The sense of smell.

**olfactory bulbs** Part of the limbic system of the forebrain of vertebrates; the olfactory bulbs carry information about odors to the brain.

**oligodendrocytes** Glial cells that produce the myelin sheath around neurons in the central nervous system.

**oligotrophic** The term used to describe aquatic systems that are low in nutrients such as phosphate and combined nitrogen and are consequently low in primary productivity and biomass, but typically high in species diversity.

**ommatidium** (plural, **ommatidia**) A visual unit in the compound eye of arthropods and some annelids that functions as a separate photoreceptor capable of forming an independent image.

**omnivore** An animal that consumes both plants and animals for food.

**oncogene** A type of mutant gene derived from a proto-oncogene. An oncogene is overactive, thus contributing to uncontrolled cell growth and promoting cancer.

**one-gene/one-enzyme hypothesis** An early hypothesis by Beadle and Tatum that suggested that one gene encodes one enzyme. It was later modified.

**oocyte** In animals, a cell that undergoes meiosis to produce an egg cell.

**oogenesis** Gametogenesis in a female animal, resulting in the production of an egg cell.

**oogonium** (plural, **oogonia**) In animals, the diploid germ cell that gives rise to the female gamete, the egg.

**open circulatory system** In animals, a circulatory system in which hemolymph, which does not differ from the interstitial fluid, flows throughout the body and is not confined to special vessels.

**open complex** A separation between the two DNA strands that occurs near the promoter during transcription; also called a transcription bubble.

**open conformation** Chromatin that can be transcribed into RNA.

**operant conditioning** A form of behavior modification; a type of associative learning in which an animal's behavior is reinforced by a consequence, either a reward or a punishment.

**operator** A DNA sequence in bacteria that is recognized by activator or repressor proteins that regulate the level of gene transcription.

**operculum** A protective flap that covers the gills of a bony fish.

**operon** A set of two or more genes in bacteria that are under the transcriptional control of a single promoter.

**opsin** A protein that is a component of visual pigments in the vertebrate eye.

**opsonization** The process by which an antibody binds to a pathogen and provides a means to link the pathogen with a phagocyte.

**optic disc** In vertebrates, the point on the retina where the optic nerve leaves the eye.

**optic nerve** A structure of the vertebrate eye that carries electrical signals to the brain.

**optimal foraging** The concept that in a given circumstance, an animal seeks to obtain the most energy possible with the least expenditure of energy.

**optimality theory** The theory that predicts that an animal should behave in a way that maximizes the benefits of a behavior minus its costs.

**orbital** The region surrounding the nucleus of an atom where the probability is high of finding a particular electron.

**order** In taxonomy, a subdivision of a class.

**organ of Corti** Coiled structure in the vertebrate ear responsible for detecting sound.

**organ system** Different organs that work together to perform an overall function or functions in an organism.

**organ** A collection of two or more tissues that performs a specific function or set of functions.

**organelle** A subcellular structure or membrane-bound compartment with its own unique structure and function.

**organic chemistry** The study of carbon-containing molecules.

**organic farming** The production of crops without the use of commercial inorganic fertilizers, growth substances, and pesticides.

**organic molecule** A carbon-containing molecule, so named because such molecules were first discovered in living organisms.

**organism** A living thing that maintains an internal order that is separated from the environment.

**organismal ecology** The study of the ways in which individual organisms meet the challenges of their biotic and abiotic interactions within their environments.

**organismic model** A view of the nature of a community that considers it to be equivalent to a superorganism; individuals, populations, and communities have a relationship to each other that resembles the associations found between cells, tissues, and organs.

**organizing center** A group of cells in a plant shoot meristem that ensures the proper organization of the meristem and preserves the correct number of actively dividing stem cells.

**organogenesis** The developmental stage during which cells and tissues form organs in animal embryos.

**orientation** A mechanism of migration in which animals have the ability to follow a compass bearing and travel in a straight line.

**origin of replication** A site within a chromosome that serves as a starting point for DNA replication.

**orthologs** Homologous genes found in different species.

**osmoconformer** An animal whose osmolarity conforms to that of its environment.

**osmolarity** The solute concentration of an aqueous solution, expressed as milliosmoles/liter (mOsm/L).

**osmoregulator** An animal that maintains stable internal ion concentrations and osmolarities, even when living in water whose osmolarity is very different from that of its body fluids or living on land.

**osmosis** The movement of water across a membrane to balance solute concentrations. Water diffuses from a solution that is hypotonic (lower solute concentration) into a solution that is hypertonic (higher solute concentration).

**osmotic adjustment** The process by which a plant cell modifies the solute concentration of its cytosol.

**osmotroph** An organism that relies on osmotrophy (uptake of small organic molecules) as a mechanism of nutrition.

**osmotrophy** The process of feeding by absorbing small organic molecules.

**osteichythan** A clade that includes all vertebrates with a bony skeleton.

**osteomalacia** Bone deformation in adults due to inadequate mineral intake or absorption from the small intestine.

**osteoporosis** A disease in which the mineral and organic components of bone are reduced.

**otolith** Granules of calcium carbonate found in the gelatinous substance that embeds hair cells in the vertebrate ear.

**outer bark** Protective layers of mostly dead cork cells that cover the outside of woody stems and roots.

**outer segment** The highly convoluted plasma membranes found in the rods and cones of the eye.

**outgroup** In a cladogram, a species or group of species that does not exhibit one or more shared derived characters found in the ingroup.

**ovarian cycle** The events beginning with the development of an ovarian follicle, followed by release of a secondary oocyte, and concluding with formation and subsequent degeneration of a corpus luteum.

**ovary** 1. In animals, the female gonad where eggs are formed. 2. In plants, the lowermost portion of the pistil that encloses and protects the ovules.

**overexploitation** The practice in which humans harvest a particular species at a rate that is unsustainable, based on its natural rate of mortality and capacity for reproduction.

**oviduct** A thin tube with undulating fimbriae (finger-like structures) that is connected to the uterus and extends out to the ovary; also called the Fallopian tube.

**oviparous** Refers to an animal whose young hatch from eggs laid outside the mother's body.

**ovoviparous** Refers to an animal that retains fertilized eggs covered by a protective sheath or other structure within the body, where the young hatch.

**ovulation** The process by which a mature oocyte is released from an ovary.

**ovule** In a seed plant, a megaspore-producing sporangium with enclosing structures known as integuments.

**oxidation** The removal of one or more electrons from an atom or molecule; occurs during the breakdown of small organic molecules.

**oxidative fiber** A skeletal muscle fiber that contains numerous mitochondria and has a high capacity for oxidative phosphorylation.

**oxidative phosphorylation** A process during which NADH and $FADH_2$ are oxidized to make more ATP via the phosphorylation of ADP.

**oxygen-hemoglobin dissociation curve** A curve that represents the relationship between the partial pressure of oxygen and the binding of oxygen to hemoglobin proteins.

**oxytocin** A hormone secreted by the posterior pituitary gland that stimulates contractions of the smooth muscles in the uterus of a pregnant mammal, facilitating the birth process; after birth, it is important in milk secretion.

# P

**P generation** The parental generation in a genetic cross.

**P protein** Phloem protein; the proteinaceous material used by plant phloem as a response to wounding.

**pacemaker** *See* sinoatrial (SA) node.

**Pacinian corpuscle** A specialized receptor that is located deep beneath the surface of an animal's skin and responds to deep pressure or vibration.

**pair-rule gene** A type of segmentation gene; a mutation in this gene may cause alternating segments or parts of segments to be absent.

**paleontologist** A scientist who studies fossils.

**palisade parenchyma** Photosynthetic ground tissue of the plant leaf mesophyll that consists of closely packed, elongate cells adapted to efficiently absorb sunlight.

**palmate** A type of leaf vein pattern in which veins radiate outward, resembling an open hand.

**pancreas** In vertebrates, an elongated gland located behind the stomach that secretes digestive enzymes and a fluid rich in bicarbonate ions.

**paracrine signaling** A type of cellular communication in which molecules are released into the interstitial fluid and act on nearby cells.

**paralogs** Homologous genes within a single species.

**paraphyletic group** A group of species that contains a common ancestor and some, but not all, of its descendants.

**parapodia** Fleshy, footlike structures in the Errantia, a type of Annelid worm, that are pushed into the substrate to provide traction during movement.

**parasite** An organism that feeds on another organism, called the host, for a relatively long time, but does not normally kill it outright.

**parasitism** A symbiotic association in which one organism feeds off another but does not normally kill it.

**parasympathetic division** The division of the autonomic nervous system that is involved in maintaining and restoring body functions.

**parathyroid hormone (PTH)** A hormone that acts on bone to stimulate the activity of cells that dissolve the mineral part of bone.

**parenchyma cell** A type of plant cell that is thin-walled and alive at maturity.

**parenchyma** A plant ground tissue that is composed of parenchyma cells.

**parental strand** The original strand in DNA replication.

**parietal lobe** One of four lobes of the cerebral cortex of the human brain; receives and interprets sensory input from visual and somatic pathways.

**parthenogenesis** An asexual process in which an offspring develops from an unfertilized egg.

**partial pressure** The individual pressure of each gas in the air; the sum of these pressures is known as atmospheric pressure.

**particulate inheritance** The idea that the determinants of hereditary traits are transmitted in discrete units, or particles, from one generation to the next.

**passive immunity** A type of acquired immunity that confers protection against disease through the direct transfer of antibodies from one individual to another.

**passive transport** The diffusion of a solute across a membrane in a process that is energetically favorable and does not require an input of energy.

**paternal inheritance** A pattern in which only the male gamete contributes particular organellar genes to the offspring.

**pathogen** A virus or microorganism that causes disease symptoms in its host.

**pathogen-associated molecular pattern (PAMP)** Any general molecular feature common to many pathogens that triggers an innate immune response.

**pattern formation** The process that gives rise to a plant or animal with a particular body structure.

**pedal glands** Glands in the foot of a rotifer that secrete a sticky substance that aids in attachment to the substrate.

**pedicel** 1. A flower stalk. 2. A narrow, waistlike point of attachment between the body parts of spiders and some insects.

**pedigree analysis** An examination of human traits over several generations in a family as a way to deduce the pattern of inheritance.

**peer-review process** A procedure in which experts in a particular area evaluate papers submitted to scientific journals.

**pelagic zone** The open ocean, where the water depth averages 4,000 m and nutrient concentrations are typically low.

**penis** A male external accessory sex organ found in many animals that is involved in copulation.

**pentadactyl limb** A limb ending in five digits.

**pepsin** An active enzyme in the stomach that begins the digestion of protein.

**peptide bond** The covalent bond between a carboxyl and amino group that links amino acids in a polypeptide.

**peptidoglycan** A polymer composed of carbohydrates crosslinked with peptides that is an important component of the cell walls of most bacteria.

**peptidyl site (P site)** One of three sites for tRNA binding in the ribosome during translation; the other two are the aminoacyl site (A site) and the exit site (E site). The P site holds the tRNA carrying the growing polypeptide chain.

**peptidyl transfer reaction** During translation, the transfer of the polypeptide from the tRNA in the P site to the amino acid at the A site.

**per capita growth rate** The per capita birth rate minus the per capita death rate; the rate that determines how a population grows over any time period.

**perception** An awareness of the sensations that are experienced.

**perennial** A plant that lives for more than 2 years, often producing seeds each year after it reaches reproductive maturity.

**perfect flower** A flower that has both stamens and carpels.

**perianth** The term that refers to flower petals and sepals collectively.

**pericarp** The wall of a plant's fruit.

**pericycle** A cylinder of plant tissue that has cell division (meristematic) capacity and encloses the root vascular tissue.

**periderm** The outer layers of a woody stem, composed of cork cambium, layers of cork tissue produced by the cork cambium, and associated parenchyma cells, together forming outer bark.

**peripheral membrane protein** A protein that is noncovalently bound to a region of an integral membrane protein that projects out from the membrane or noncovalently bound to the polar head group of phospholipids.

**peripheral nervous system (PNS)** In vertebrates, all nerves and ganglia outside the brain and spinal cord.

**peripheral zone** The area of a plant shoot meristem that contains dividing cells that will eventually differentiate into plant structures.

**periphyton** Communities of microorganisms that are attached by mucilage to underwater surfaces such as rocks, sand, and plants.

**peristalsis** In animals, the rhythmic, spontaneous waves of muscle contractions that propel food through the digestive system.

**peritubular capillaries** A capillary near the junction of the cortex and medulla that surrounds the renal tubule of the mammalian kidney.

**permafrost** A layer of permanently frozen soil found in tundra.

**peroxisome** A relatively small organelle that is found in all eukaryotic cells and that catalyzes detoxifying reactions.

**personalized medicine** A medical practice in which information about a patient's genotype is used to individualize the person's medical care.

**petal** A flower organ that usually serves to attract insects or other animals for pollen transport.

**petiole** A stalk that connects a leaf to the stem of a plant.

**pH** The mathematical expression of a solution's hydrogen ion ($H^+$) concentration, defined as the negative logarithm to the base 10 of the $H^+$ concentration.

**phage** *See* bacteriophage.

**phagocyte** A cell capable of phagocytosis; phagocytes provide nonspecific defense against pathogens that enter the body.

**phagocytosis** A type of endocytosis that involves the formation of a membrane vesicle, called a phagosome, or phagocytic vacuole, which engulfs a particle such as a bacterium.

**phagotroph** An organism that specializes in phagotrophy (particle feeding) by means of phagocytosis as a mechanism of nutrition.

**pharyngeal slit** A defining characteristic of all chordate embryos. In early-diverging chordates, pharyngeal slits develop into a filter-feeding device, and in some advanced chordates, they form gills.

**pharynx** A portion of the vertebrate alimentary canal; also known as the throat.

**phenotype**   The characteristics of an organism that are the result of the expression of its genes.

**pheromone**   A powerful chemical attractant used to manipulate the behavior of others.

**phloem loading**   The process of conveying sugars to sieve-tube elements for long-distance transport.

**phloem**   A specialized conducting tissue in a plant's stem.

**phoresy**   A form of commensalism in which individuals of one species use individuals of a second species for transportation.

**phosphodiester linkage**   Refers to a double linkage (two phosphoester bonds) that holds together adjacent nucleotides in DNA and RNA strands.

**phosphodiesterase**   An enzyme that breaks a bond in cAMP to form AMP.

**phospholipid bilayer**   The basic framework of a biological membrane, consisting of two layers of phospholipids.

**phospholipid**   A type of lipid that is similar in structure to a triglyceride, but with the third hydroxyl group of glycerol linked to a phosphate group instead of a fatty acid; a key component of biological membranes.

**phosphorylation**   The attachment of a phosphate to a molecule.

**photic zone**   A fairly narrow zone close to the surface of an aquatic environment, where light is sufficient to allow photosynthesis to occur.

**photoautotroph**   An organism that uses the energy from light to make organic molecules from inorganic sources.

**photoheterotroph**   An organism that is able to use light energy to generate ATP but must take in organic compounds from the environment as a source of carbon.

**photon**   One of the discrete particles that make up light. A photon is massless and travels in a wavelike pattern.

**photoperiodism**   A plant's ability to measure and respond to the amount of light and day length; used as a way of detecting seasonal change.

**photophosphorylation**   The process by which the light reactions of photosynthesis produce ATP.

**photopsin**   Any of several types of visual pigments in the cone cells of the vertebrate eye.

**photoreceptor**   A specialized sensory receptor in an animal that responds to visible light energy; in plants, molecules that respond to light.

**photorespiration**   The metabolic process occurring in $C_3$ plants when the enzyme rubisco combines with $O_2$ instead of $CO_2$ and produces only one molecule of 3PG instead of two, thereby reducing photosynthetic efficiency.

**photosynthesis**   The process whereby light energy is captured by plant, algal, or photosynthetic bacterial cells and is used to synthesize organic molecules from $CO_2$ and $H_2O$ (or $H_2S$).

**photosystem I (PSI)**   A distinct complex of proteins and pigment molecules in chloroplasts that absorbs light during the light reactions of photosynthesis.

**photosystem II (PSII)**   A distinct complex of proteins and pigment molecules in chloroplasts that absorbs light and also generates oxygen from water during the light reactions of photosynthesis.

**phototropin**   The main blue-light sensor involved in positive phototropism in plants.

**phototropism**   The tendency of a plant to grow toward a light source.

**phylogenetic tree**   A diagram that describes the evolutionary relationships among various species, based on the information available to and gathered by systematists.

**phylogeny**   The evolutionary history of a species or group of species.

**phylum** (plural, **phyla**)   In taxonomy, a subdivision of a kingdom.

**physical mutagen**   A physical agent, such as ultraviolet light, that causes mutations.

**physical systems**   Physical environments, such as oceans, ice, fresh waters, and soils, that serve as habitats for microbiomes.

**physiological ecology**   A subdiscipline of organismal ecology that investigates how organisms are physiologically adapted to their environment and how the environment impacts the distribution of species.

**physiology**   The study of the functions of living things.

**phytochrome**   A red-light and far-red-light receptor in plants.

**phytoplankton**   Microscopic photosynthetic protists that float in water or actively move through it.

**pigment**   A molecule that can absorb light energy.

**pili** (singular, **pilus**)   Threadlike surface appendages that allow bacteria to attach to each other during conjugation or to move across surfaces.

**piloting**   A mechanism of migration in which an animal moves from one familiar landmark to the next.

**pinnate**   A type of leaf vein pattern in which veins appear feather-like.

**pinocytosis**   A type of endocytosis that involves the formation of membrane vesicles from the plasma membrane as a way for cells to internalize the extracellular fluid.

**pistil**   A flower structure that may consist of a single carpel or multiple, fused carpels and is differentiated into stigma, style, and ovary.

**pit**   A thin-walled circular area in a plant cell wall where secondary wall materials such as lignin are absent and through which water moves.

**pituitary giant**   A person who has a tumor of the GH-secreting cells of the anterior pituitary gland and thus produces excess GH during childhood and, if untreated, during adulthood; the person can grow very tall before growth ceases after puberty.

**pituitary gland**   A multilobed endocrine gland sitting directly below the hypothalamus of the brain.

**placenta**   A structure in humans and other eutherian mammals that connects the developing young to the mother's uterine wall and allows the transfer of nutrients and gases.

**placental transfer tissues**   In plants, nutritive tissues that aid in the transfer of nutrients from maternal parent to embryo.

**plant tissue culture**   A laboratory process to produce thousands of identical plants having the same desirable characteristics.

**plant**   A multicellular eukaryotic organism that is usually photosynthetic (having plastids), primarily lives on land, and has cells with a cell wall containing cellulose.

**Plantae**   A eukaryotic kingdom of the domain Eukarya.

**plaques**   1. Deposits of lipids, fibrous tissue, and smooth muscle cells that may develop inside arterial walls. 2. A bacterial biofilm that may form on the surfaces of teeth.

**plasma cell**   A cell that synthesizes and secretes antibodies.

**plasma membrane**   The biological membrane that separates the internal contents of a cell from its external environment.

**plasma**   The fluid part of blood that contains water and dissolved solutes.

**plasmid**   A small circular piece of DNA that exists separately from the bacterial chromosome in many strains of bacteria; plasmids are also found in some eukaryotic cells, such as yeast, and can be used as vectors in cloning experiments.

**plasmodesma** (plural, **plasmodesmata**)   A membrane-lined, ER-containing channel that connects the cytoplasm of adjacent plant cells.

**plasmogamy**   The fusion of the cytoplasm between two gametes.

**plasmolysis**   The shrinkage of algal or plant cytoplasm that occurs when water leaves the cell by osmosis, with the result that the plasma membrane no longer presses on the cell wall.

**plastid**   A general name given to organelles found in plant and algal cells that are bound by two or more membranes and contain DNA and large amounts of either chlorophyll (in chloroplasts), carotenoids (in chromoplasts), or starch (in amyloplasts).

**platelets**   Cell fragments in the blood of mammals that play a crucial role in the formation of blood clots.

**pleiotropy**   The phenomenon in which a mutation in a single gene can have multiple effects on an individual's phenotype.

**pleural sac**   A double layer of thin, moist connective tissue that encases each lung.

**pluripotent**   Refers to the ability of embryonic stem cells to differentiate into almost every cell type of the body.

**point mutation**   A mutation that affects only a single base pair within DNA or that involves the addition or deletion of a single base pair to a DNA sequence.

**polar covalent bond**   A covalent bond between two atoms that have different electronegativities; the shared electrons are closer to the nucleus of the atom of higher electronegativity than to the nucleus of the atom of lower electronegativity. This distribution of the shared electrons around the atoms creates a polarity, or difference in electric charge, across the molecule.

**polar transport**   The process whereby auxin flows primarily downward in shoots.

**pole**   A structure of the spindle apparatus defined by each centrosome.

**pollen grain**   The immature male gametophyte of a seed plant.

**pollen tube**   In seed plants, a long, thin tube produced by a pollen grain that delivers sperm to the ovule.

**pollen**   In seed plants, tiny male gametophytes enclosed by sporopollenin-containing microspore walls.

**pollination syndromes**   The pattern of coevolved traits between particular types of flowers and their specific pollinators.

**pollination**   The process in which pollen grains are transported to an angiosperm flower or a gymnosperm cone primarily by means of wind or animal pollinators.

**pollinator**   An animal that carries pollen between angiosperm flowers or cones of gymnosperms.

**poly A tail**   A string of adenine nucleotides at the 3′ end of most mature mRNAs in eukaryotes.

**polyandry**   A mating system in which one female mates with several males, but males mate with only one female.

**polycistronic mRNA**   An mRNA that contains the coding sequences for two or more protein-encoding genes.

**polygenic**   Refers to a trait for which several or many genes contribute to the outcome.

**polygyny**   A mating system in which one male mates with several females in a single breeding season, but females mate with only one male.

**polymer**   A large molecule formed by linking many smaller molecules called monomers.

**polymerase chain reaction (PCR)**   A technique to make many copies of a gene in vitro; primers are used that flank the region of DNA to be amplified.

**polymorphic gene**   A gene that commonly exists as two or more alleles in a population.

**polymorphism**   The presence of two or more variations of a character (trait) within a population.

**polyp**   A type of cnidarian body form that is sessile and occurs mouth up.

**polypeptide**   A molecule consisting of a linear sequence of amino acids; the term denotes structure.

**polyphagous**   Parasites that feed on many host species.

**polyphyletic group**   A group of species that consists of members of several evolutionary lines and does not include the most recent common ancestor of the included lineages.

**polyploid**   Refers to an organism or cell that has three or more sets of chromosomes.

**polyploidy**   The condition in which a cell or organism has three or more sets of chromosomes.

**polysaccharide**   A long carbohydrate polymer formed of many monosaccharides linked together.

**polysome**   The complex of a single mRNA and multiple ribosomes.

**pons**   The part of the vertebrate hindbrain that, along with the cerebellum, has an integrative motor function; it also plays a role in regulating breathing.

**population density**   The number of organisms of a given species in a given unit area or volume.

**population ecology**   The study of how populations grow and what factors promote or limit growth.

**population genetics**   The study of genes and genotypes in a population.

**population**   A group of individuals of the same species that occupy the same environment and (for sexually reproducing organisms) can interbreed with one another.

**portal vein**   A vein that not only collects blood from capillaries—like all veins—but also forms another set of capillaries, as opposed to returning the blood directly to the heart.

**positional information**   Information regarding a cell's location relative to other cells that is conveyed by morphogens and cell-to-cell contacts.

**positive control**   Transcriptional regulation by activator proteins.

**positive feedback loop**   In animals, a mechanism that accelerates or amplifies a process, leading to what is sometimes called an explosive system.

**postabsorptive state**   One of two alternating phases in the utilization of nutrients; occurs when the gastrointestinal tract is empty of nutrients and the body's own stores must supply energy. The other phase is the absorptive state.

**postanal tail**   A defining characteristic of all chordate embryos; a tail of variable length that extends posterior to the anal opening.

**posterior**   Refers to the rear (tail) end of an animal.

**postsynaptic cell**   The cell that receives the electrical or chemical signal sent from a neuron.

**post-translational sorting**   The uptake of proteins into the nucleus, mitochondria, chloroplasts, or peroxisomes that occurs after the protein is completely made in the cytosol (that is, completely translated).

**postzygotic isolating mechanism**   A mechanism that prevents interbreeding by blocking the development of a viable and fertile individual after fertilization has taken place.

**potential energy**   The energy that a substance possesses due to its structure or location.

**power stroke**   In muscle, a conformation change in the myosin cross-bridge that results in binding between myosin and actin and the movement of the actin filament.

**prebiotic soup**   The medium formed by the slow accumulation of organic molecules in the early oceans over a long period of time prior to the existence of life.

**predation**   An interaction in which the action of a predator results in the death of its prey.

**prediction**   An expected outcome based on a hypothesis that can be shown to be correct or incorrect through observation or experimentation.

**pregnancy**   The time during which a developing embryo or fetus grows within the uterus of the mother; also known as gestation.

**preinitiation complex**   The assembled structure consisting of RNA polymerase II and transcription factors (GTFs) at the TATA box prior to transcription of eukaryotic protein-encoding genes.

**pre-mRNA**   In eukaryotes, the mRNA transcript before any biochemical modifications are made to it.

**pressure potential**   The component of water potential due to hydrostatic pressure.

**pressure-flow hypothesis**   Explains sugar translocation in plants as a process driven by differences in turgor pressure between cells of a sugar source, where sugar is produced, and cells of a sugar sink, where sugar is consumed.

**presynaptic cell**   The neuron that sends an electrical or chemical signal to another cell.

**prezygotic isolating mechanism**   A mechanism that blocks interbreeding by preventing the formation of a zygote.

**primary active transport**   A type of transport that involves pumps that directly use energy to transport a solute against a gradient.

**primary cell wall**   In plants, a relatively thin and flexible cell wall that is synthesized first between two newly made daughter cells.

**primary consumer**   An organism that obtains its food by eating primary producers; also called a herbivore.

**primary electron acceptor**   The molecule to which a high-energy electron from an excited pigment molecule such as P680* is transferred during photosynthesis.

**primary endosymbiosis**   The process by which a eukaryotic host cell acquires prokaryotic endosymbionts. Mitochondria and the plastids of green and red algae are examples of organelles that originated via primary endosymbiosis.

**primary growth**   Plant growth that occurs from primary meristems and produces primary tissues and organs of diverse types.

**primary immune response**   The immune response to an initial exposure to an antigen.

**primary meristem**   A meristematic tissue that increases plant length and produces new organs.

**primary oocyte**   In animals, a cell that undergoes meiosis to begin the process of egg production.

**primary plastid**   A plastid that originated from a prokaryote as the result of primary endosymbiosis.

**primary producers**   Autotrophs such as plants, algae, and photosynthetic bacteria that use sunlight and form the basis of the food chain.

**primary production**   Production by autotrophs, normally green plants.

**primary spermatocyte**   In animals, a cell that undergoes meiosis to begin the process of sperm production.

**primary structure**   The linear sequence of amino acids of a polypeptide; one of four levels of protein structure.

**primary succession**   Succession on newly exposed sites that were not previously occupied by soil and vegetation.

**primary vascular tissue**   Plant tissue composed of xylem and phloem; a conducting tissue of nonwoody plants.

**principle of parsimony**   The concept that the preferred hypothesis is the one that is the simplest.

**principle of species individuality**   A view of the nature of a community in which each species is distributed according to its physiological needs and population dynamics; most communities intergrade continuously, and competition does not create distinct vegetational zones.

**prion**   An infectious protein that causes disease by inducing the abnormal folding of other protein molecules.

**probability**   The chance that an event will have a particular outcome.

**probiotic treatment**   The introduction of one or more microbial strains into the microbiome of a host organism, usually to improve health.

**proboscis**   The coiled tongue of a butterfly or moth, which can be uncoiled, enabling the insect to drink nectar from flowers.

**processivity**   The characteristic of DNA polymerase that keeps it from falling off the template stand as it is synthesizing a new daughter strand.

**producer**   An organism that synthesizes the organic compounds used by other organisms for food.

**product rule**   The probability that two or more independent events will occur is equal to the product of their individual probabilities.

**product**   The end result of a chemical reaction.

**production efficiency**   The percentage of energy assimilated by an organism that becomes incorporated into new biomass.

**productivity hypothesis**   The proposal that greater production by plants results in greater overall species richness.

**progesterone**   A hormone secreted by the female ovaries that plays a key role in pregnancy.

**progymnosperms**   An extinct group of plants having wood but not seeds, which evolved before the gymnosperms.

**prokaryotic cell**   Refers to a cell lacking a membrane-enclosed nucleus and cell compartmentalization; includes the cells from all members of the domains Bacteria and Archaea.

**prokaryotic**   Refers to organisms having cells lacking a membrane-enclosed nucleus and cell compartmentalization; includes all members of the domains Bacteria and Archaea.

**prometaphase**   The phase of mitosis during which the nuclear envelope completely fragments into vesicles and the mitotic spindle is fully formed.

**promiscuous**   In ecology, a term for animals that have many different sexual mates.

**promoter**   A sequence of DNA within a gene that controls when and where transcription begins.

**proofreading**   The ability of DNA polymerase to identify a mismatched nucleotide and remove it from the daughter strand.

**prophage**   The DNA of a phage that has become integrated into a bacterial chromosome.

**prophase**   The phase of mitosis during which the chromosomes condense and the nuclear membrane begins to vesiculate.

**proplastid**   A type of unspecialized structure from which a plastid is derived.

**prosthetic group**   A small molecule that is permanently attached to the surface of an enzyme and aids in the enzyme's function.

**protease**   An enzyme that cuts proteins into smaller polypeptides.

**proteasome**   A protein complex that provides the primary pathway for protein degradation in archaea and eukaryotic cells.

**protein kinase cascade**   The sequential activation of multiple protein kinases.

**protein kinase**   An enzyme that transfers a phosphate group from ATP to a specific amino acid in a protein.

**protein phosphatase**   An enzyme responsible for removing phosphate groups from proteins.

**protein subunit**   An individual polypeptide within a functional protein; most functional proteins are composed of two or more polypeptides.

**protein**   A functional unit composed of one or more polypeptides. Each polypeptide is composed of a linear sequence of amino acids.

**protein-encoding gene**   A gene that serves as a template to make an mRNA molecule that contains the information to specify a polypeptide with a particular amino acid sequence; most genes are protein-encoding genes.

**protein-protein interactions** The specific interactions between proteins that occur during many critical cellular processes.

**proteoglycan** A long, linear core protein with many GAGs attached to it; found in the ECM.

**proteolysis** A processing event within a cell in which enzymes called proteases cut proteins into smaller polypeptides.

**proteome** The complete complement of proteins that a cell is currently making or an organism can make.

**proteomics** Techniques used to analyze and compare proteomes.

**prothoracicotropic hormone (PTTH)** A hormone produced in certain invertebrates that stimulates a pair of endocrine glands called the prothoracic glands.

**Protista** Formerly a eukaryotic kingdom. Protists are now placed into seven eukaryotic supergroups.

**protobiont** The term used to describe the first nonliving structure that could have evolved into a living cell.

**proton** A positively charged particle found in the nucleus of an atom. The number of protons in an atom is called the atomic number and defines each type of element.

**protonephridia** Simple excretory organs found in flatworms that are used to filter out wastes and excess water.

**proto-oncogene** A normal gene that, if mutated, can become an oncogene.

**protostome** An animal whose development exhibits spiral determinate cleavage and in which the blastopore becomes the mouth; includes mollusks, annelid worms, and arthropods.

**protozoa** A term commonly used to describe diverse heterotrophic protists.

**proventriculus** The glandular portion of the stomach of a bird.

**provirus** Viral DNA that has become integrated into a chromosome of the host cell.

**proximal tubule** The segment of the renal tubule in the kidney that drains Bowman's capsule.

**proximate cause** A specific genetic and physiological mechanism that underlies a behavior.

**pseudocoelomate** An animal with a pseudocoelom.

**pteridophytes** A phylum of vascular plants having euphylls, but not seeds; formally called Pteridophyta.

**pulmonary arteries** The arteries that bring blood from the right ventricle to the lungs in animals with a double circulation.

**pulmonary circulation** The circuit in a double circulation in which blood is pumped from the right side of the heart to the lungs to pick up oxygen from the atmosphere and release carbon dioxide.

**pump** A transporter that directly couples its conformational changes to an energy source, such as ATP hydrolysis.

**punctuated equilibrium** A concept that suggests that the tempo of evolution is more sporadic than gradual. Species rapidly evolve into new species followed by long periods of equilibrium with little evolutionary change.

**Punnett square** A common method for predicting the outcome of simple genetic crosses.

**pupa** A developmental stage in some insects that undergo metamorphosis; occurs between the larval and adult stages.

**pupil** A small opening in the eye of a vertebrate that transmits different patterns of light emitted from or reflected off objects in the animal's field of view.

**purine** Either of the bases adenine (A) and guanine (G), which have a fused double ring of carbon and nitrogen atoms.

**pyramid of biomass** A graphic representation of trophic levels in a food web in which the organisms at each trophic level are weighed.

**pyramid of energy** A graphic represention of trophic levels in a food web in which rates of energy production are used rather than biomass.

**pyramid of numbers** A graphic representation of trophic levels in a food web in which the number of individuals decreases at each trophic level, with a huge number of individuals at the base and fewer individuals at the top.

**pyrimidine** Any of the bases thymine (T), cytosine (C), and uracil (U), which have a single ring of carbon and nitrogen atoms.

# Q

**quadrat** A sampling device used by plant ecologists consisting of a square frame that often encloses an area of 0.25 m².

**quantitative trait** A trait that shows continuous variation over a range of phenotypes.

**quaternary structure** The association of two or more polypeptides to form a protein; one of four levels of protein structure.

**quorum sensing** A mechanism by which prokaryotic cells are able to communicate by chemical means when they reach a critical population size.

# R

**radial cleavage** A mechanism of animal development in which the cleavage planes are either parallel or perpendicular to the vertical axis of the embryo.

**radial loop domain** A loop of chromatin, often 25,000 to 200,000 base pairs in size, that is anchored to the nuclear matrix.

**radial pattern** A characteristic of the body plan of plants in which root and shoot cells form concentric rings.

**radial symmetry** 1. In plants, an architectural feature in which embryos display a cylindrical shape, which is retained in the stems and roots of seedlings and mature plants. In addition, new leaves or flower parts are produced in circular whorls, or spirals, around shoot tips. 2. In animals, an architectural feature in which the body can be divided into symmetrical halves by many different longitudinal planes along a central axis.

**radiation** The emission of electromagnetic waves from the surfaces of objects; a means of heat exchange in animals.

**radicle** An embryonic root, which extends from the hypocotyl in eudicot seeds.

**radioisotope** An isotope found in nature that is inherently unstable and usually does not exist for long periods of time. Such isotopes decay and emit energy in the form of radiation.

**radiometric dating** A process for estimating the age of a fossil by analyzing the decay of radioisotopes within the accompanying rock.

**radula** A unique, protrusible, tonguelike organ in a mollusk that has many teeth and is used to eat plants, scrape food particles off rocks, or bore into shells of other species.

**rain shadow** An area on the side of a mountain that is sheltered from the wind and experiences less precipitation.

**random sampling error** The deviation between the observed and the expected outcomes due to random chance.

**random** The rarest pattern of dispersion within a population, in which the location of individuals lacks a pattern.

**rate-limiting step** The slowest step in a reaction pathway.

**ray-finned fishes** The Actinopterygii, which includes all bony fishes except the coelacanths and lungfishes.

**reabsorption** In the production of urine, the process in which useful solutes in the filtrate are recaptured and transported back into the body fluids of an animal.

**reactant** A substance that participates in a chemical reaction and becomes changed by that reaction.

**reading frame** Refers to the way in which codons are read during translation, in groups of three bases beginning with the start codon.

**realized niche** The actual range of an organism in nature.

**receptacle** The enlarged region at the tip of a flower peduncle to which flower parts are attached.

**receptor potential** The membrane potential in a sensory receptor cell of an animal.

**receptor tyrosine kinase** A type of enzyme-linked receptor found in animals and choanoflagellates that can attach phosphate groups to tyrosines in the receptor itself or in other cellular proteins.

**receptor** 1. A cellular protein that recognizes a signaling molecule and becomes activated or inhibited in response to it. 2. A structure capable of detecting changes in the environment of an animal, such as a touch receptor.

**receptor-mediated endocytosis** A common type of endocytosis in which a receptor in the membrane is specific for a given cargo.

**recessive** Refers to a trait that is masked by the presence of a dominant trait in a heterozygote.

**reciprocal cross** A cross in which the sexes and phenotypes are reversed compared to another cross.

**reciprocal translocation** A type of mutation in which two different types of chromosomes exchange pieces, thereby producing two abnormal chromosomes carrying translocations.

**recombinant DNA technology** The use of laboratory techniques to bring together fragments of DNA from multiple sources.

**recombinant vector** A vector containing a piece of chromosomal DNA.

**recombinant** An offspring that has a different combination of traits from the true-breeding parental generation.

**recombination frequency** The frequency of crossing over between two genes.

**rectum** The last segment of the large intestine of vertebrates that ends at the anus, the posterior opening of the alimentary canal to the external environment.

**redox reaction** A type of reaction in which an electron that is removed during the oxidation of an atom or molecule is transferred to another atom or molecule, which becomes reduced; short for a reduction-oxidation reaction.

**reduction** The addition of one or more electrons to an atom or molecule.

**reductionism** An approach that involves reducing complex systems to simpler components as a way to understand how the system works. In biology, reductionists study the parts of a cell or organism as individual units.

**redundancy hypothesis** The proposal that ecosystem function increases rapidly at fairly low levels of species richness, but then levels off because most additional species are redundant.

**reflex arc** A simple circuit that allows an organism to respond rapidly to inputs from sensory neurons and consists of only a few neurons.

**regeneration** A form of asexual reproduction in which a complete organism forms from a fragment of an animal's body.

**regulatory element** In eukaryotes, a DNA sequence that is recognized by regulatory transcription factors and regulates the expression of genes.

**regulatory sequence**   In the regulation of transcription, a DNA sequence that functions as a binding site for regulatory transcription factor proteins, which influence the rate of transcription.

**regulatory transcription factor**   A protein that binds to DNA, usually in the vicinity of a promoter, and affects the rate of transcription of one or more nearby genes.

**relative abundance**   The frequency of occurrence of the species in a community.

**relative refractory period**   The period near the end of an action potential when voltage-gated potassium channels are still open; during this time a new action potential can be generated if a stimulus is sufficiently strong to raise the membrane potential to threshold.

**relative water content (RWC)**   The property often used to gauge the water content of a plant organ or entire plant; RWC integrates the water potential of all cells within an organ or plant and is thus a measure of relative turgidity.

**release factor**   A protein that recognizes a stop codon in the termination stage of translation and promotes the termination of translation.

**renal corpuscle**   A filtering component in the nephron of the mammalian kidney.

**renal tubule**   The major portion of the mammalian nephron, consisting of three major segments with specialized functions.

**repeatable**   A characteristic of an experiment that yields similar results when conducted on multiple occasions.

**repetitive sequence**   DNA sequences that are present in many copies in a genome.

**replica plating**   A technique in which a replica of bacterial colonies is transferred from one petri plate to a new petri plate.

**replication fork**   The area where two DNA strands have separated and new strands are being synthesized.

**replication**   1. The copying of DNA strands. 2. The performing of experiments several or many times.

**repressible operon**   A type of operon for which a small effector molecule inhibits transcription.

**repressor**   A transcription factor that binds to DNA and inhibits transcription.

**reproduction**   The generation of offspring by sexual or asexual means.

**reproductive isolating mechanisms**   Mechanisms that prevent interbreeding between different species.

**reproductive isolation**   A criterion for identifying a species; the circumstances and mechanisms that collectively prevent a species from interbreeding with other species.

**reproductive success**   The likelihood that an individual will contribute fertile offspring to the next generation.

**resistance (R)**   The tendency of blood vessels to slow the flow of blood through their lumens.

**resistance gene (R gene)**   A plant gene that has evolved as part of a defense system in response to pathogen attack.

**resolution**   In microscopy, the ability to observe two adjacent objects as distinct from one another; a measure of the clarity of an image.

**resonance energy transfer**   The process by which energy (not an electron itself) can be transferred to adjacent pigment molecules during photosynthesis.

**resource partitioning**   The differentiation of niches, both in space and time, that enables similar species to coexist in a community.

**resource-based mutualism**   A mutually beneficial interaction in which both species receive a benefit in the form of resource transfer of energy and nutrients.

**respiration**   *See* gas exchange.

**respiratory centers**   Several nuclei in the brainstem in vertebrates that initiate expansion of the lungs.

**respiratory chain**   *See* electron transport chain.

**respiratory pigment**   A large protein that contains one or more metal atoms that bind to oxygen.

**respiratory system**   All components of the body that contribute to the exchange of gas between the external environment and cells of the body; in mammals, includes the nose, mouth, airways, and lungs and the muscles and connective tissues that encase these structures within the thoracic (chest) cavity.

**resting potential**   The difference in charges across the plasma membrane in an unstimulated neuron.

**rest-or-digest response**   The response of vertebrates to situations associated with nonstressful states, such as feeding; mediated by the parasympathetic branch of the autonomic nervous system.

**restriction enzyme**   An enzyme that recognizes a particular DNA sequence and cleaves the DNA backbone at two sites.

**restriction point**   A point in the cell cycle in which a cell has become committed to divide.

**reticular formation**   An array of nuclei in the brainstem of vertebrates that plays a major role in controlling states such as sleep and arousal.

**retina**   A sheetlike layer of photoreceptors at the back of the vertebrate eye.

**retinal**   A derivative of vitamin A that is capable of absorbing light energy; a component of visual pigments in the vertebrate eye.

**retrotransposon**   A type of transposable element that moves via an RNA intermediate.

**retrovirus**   An RNA virus that utilizes reverse transcription to produce viral DNA that can be integrated into a chromosome of the host cell.

**reverse transcriptase**   A viral enzyme that catalyzes the synthesis of viral DNA starting with viral RNA as a template.

**rhizobia**   A general term for nitrogen-fixing bacteria associated with plant roots.

**rhodopsin**   The visual pigment in the rods of the vertebrate eye.

**ribonucleic acid (RNA)**   One of two classes of nucleic acids; the other is deoxyribonucleic acid (DNA). RNA consists of a single strand of nucleotides.

**ribose**   A five-carbon sugar found in RNA.

**ribosomal RNA (rRNA)**   An RNA that forms part of ribosomes, which provide the site where translation occurs.

**ribosome**   A structure composed of proteins and rRNA that is the site where translation of mRNAs and synthesis of polypeptides occurs.

**ribozyme**   A biological catalyst that is an RNA molecule.

**rickets**   A condition in children characterized by bone deformations due to inadequate mineral intake or malabsorption in the small intestine.

**ring canal**   A structure in the central disc in the water vascular system of echinoderms.

**RNA interference (RNAi)**   Refers to a type of mRNA silencing; miRNA or siRNA interferes with the proper expression of an mRNA.

**RNA modification**   The biochemical modification of an RNA.

**RNA polymerase**   The enzyme that synthesizes strands of RNA during gene transcription.

**RNA splicing**   *See* splicing.

**RNA world**   A hypothetical period on primitive Earth when both the information needed for life and the catalytic activity of living cells were contained solely in RNA molecules.

**RNA**   *See* ribonucleic acid.

**RNA-induced silencing complex (RISC)**   A complex consisting of miRNA or siRNA and proteins; mediates RNA interference.

**RNase**   An enzyme that digests RNA.

**rod**   Type of photoreceptor found in the vertebrate eye that is very sensitive to low-intensity light but does not readily discriminate different colors. Rods are useful mostly at night, and they send signals to the brain that generate a black-and-white visual image.

**root apical meristem (RAM)**   The region of rapidly dividing cells at the tip of a plant root.

**root hair**   A specialized, long, thin root epidermal cell that functions to absorb water and minerals, usually from soil.

**root pressure**   Osmotic pressure within roots that causes water to rise for some distance through a plant stem, under conditions of high soil moisture or low transpiration.

**root system**   The collection of roots and root branches produced by root apical meristems.

**root**   A plant organ that provides anchorage in the soil and also fosters efficient uptake of water and minerals.

**root-shoot axis**   The general body pattern of plants in which the root grows downward and the shoot grows upward.

**rough endoplasmic reticulum (rough ER)**   The part of the ER whose outer surface is studded with ribosomes; this region plays a key role in the initial synthesis and sorting of proteins that are destined for the ER, Golgi apparatus, lysosomes, vacuoles, plasma membrane, or outside of the cell.

**r-selected species**   A species whose life history strategy shows a high rate of per capita population growth but poor competitive ability.

**rubisco**   The enzyme that catalyzes the first step in the Calvin cycle, in which $CO_2$ is incorporated into an organic molecule.

**Ruffini corpuscle**   A specialized receptor that is located beneath the surface of the skin of mammals and responds to deep pressure and vibration.

**ruminants**   Animals such as sheep, goats, llamas, and cows that have complex stomachs consisting of several chambers.

# S

**S phase**   The DNA synthesis phase of the cell cycle.

**saltatory conduction**   The conduction of an action potential along an axon in which the action potential is regenerated at each node of Ranvier instead of along the entire length of the axon.

**sarcoma**   A tumor of connective tissue such as bone or cartilage.

**sarcomere**   One complete unit of the repeating pattern of thick and thin filaments within a myofibril.

**sarcoplasmic reticulum**   A specialized form of endoplasmic reticulum in a muscle fiber that is the source of the increase in cytosolic calcium involved in muscle contraction.

**satiety**   A feeling of fullness.

**saturated fatty acid**   A fatty acid in which all the carbons are linked by single covalent bonds.

**scanning electron microscopy (SEM)**   A type of microscopy that utilizes an electron beam to produce an image of the three-dimensional surface of a biological sample.

**Schwann cells**   The glial cells that form myelin on axons that travel outside the brain and spinal cord.

**science**   In biology, the observation, identification, experimental investigation, and theoretical explanation of natural phenomena.

**scientific method**   A series of steps to test the validity of a hypothesis. This approach often involves a comparison between control and experimental groups.

**scientific model**   In biology, a conceptual, mathematical, or physical depiction of a real-world phenomenon.

**sclereid**   A star- or stone-shaped plant cell having a tough, lignified cell wall.

**sclerenchyma** A rigid plant ground tissue composed of tough-walled fibers and sclereids.

**sea level** The average level for the surface of one or more of Earth's oceans.

**second messengers** Small molecules or ions that relay signals inside the cell after an extracellular signaling molecule (a first messenger) has activated a cell surface receptor.

**secondary active transport** A type of membrane transport that involves the utilization of a pre-existing gradient to drive the active transport of another solute.

**secondary cell wall** A thick rigid plant cell wall that is synthesized and deposited between the plasma membrane and the primary cell wall after a plant cell matures and has stopped increasing in size.

**secondary consumer** An organism that eats primary consumers; also called a carnivore.

**secondary endosymbiosis** A process that occurs when a eukaryotic host cell acquires a eukaryotic endosymbiont having a primary plastid.

**secondary growth** Plant growth that occurs from secondary meristems and increases the girth of woody plant stems and roots.

**secondary immune response** An immediate and heightened production of additional specific antibodies against the particular antigen that previously elicited a primary immune response.

**secondary meristem** A meristem in woody plants forming a ring of actively dividing cells that encircle the stem.

**secondary metabolite** Molecules that are produced by secondary metabolism.

**secondary oocyte** In animals, the large haploid cell that is produced when a primary oocyte completes meiosis I during oogenesis.

**secondary phloem** The inner bark of a woody plant.

**secondary plastid** A plastid that has originated by the endosymbiotic incorporation of a eukaryotic cell containing a primary plastid into a eukaryotic host cell.

**secondary production** The measure of production of heterotrophs and decomposers.

**secondary spermatocytes** In animals, the haploid cells produced when a primary spermatocyte undergoes meiosis I during spermatogenesis.

**secondary structure** The bending or twisting of a region of a protein into an α helix or β sheet; one of four levels of protein structure.

**secondary succession** Succession on a site that has previously supported life but has undergone a disturbance.

**secondary xylem** In plants, a type of secondary vascular tissue that is also known as wood.

**secretin** A hormone released by cells of the small intestine in vertebrates; stimulates release of bicarbonate ions from the pancreas into the small intestine.

**secretion** 1. The export of a substance from a cell. 2. In the production of urine, the process in which some solutes are actively transported into the tubules of the excretory organ; this supplements the amount of a solute that would normally be removed by filtration alone.

**secretory pathway** A pathway for the movement of larger substances, such as carbohydrates and proteins, from the ER to the outside of a cell.

**secretory vesicle** A membrane vesicle carrying different types of materials that fuses with the cell's plasma membrane to release the contents extracellularly.

**seed coat** A hard, tough covering that develops from the ovule's integuments and protects a plant embryo.

**seed plant** The informal name for gymnosperms and angiosperms.

**seed** A reproductive structure having specialized tissues that enclose a plant embryo; produced by gymnosperms and flowering plants, usually as the result of sexual reproduction.

**segmentation gene** A gene that controls the segmentation pattern of an animal embryo.

**segmentation** The organization of an animal's body into clearly defined regions.

**segment-polarity gene** A type of segmentation gene; a mutation in this gene causes portions of segments to be missing and cause adjacent regions to become mirror images of each other.

**selectable marker** A gene whose presence can allow organisms (such as bacteria) to grow under a certain set of conditions. For example, an antibiotic-resistance gene is a selectable marker that allows bacteria to grow in the presence of the antibiotic.

**selective breeding** Programs and procedures designed to modify traits in domesticated species.

**selective permeability** The property of membranes that allows the passage of certain ions or molecules but not others.

**selective serotonin reuptake inhibitors (SSRIs)** Drugs used to treat major depressive disorder that act by increasing concentrations of serotonin in the brain.

**self-compatibility (SC)** The reproductive state of plants that can serve as both mother and father to their progeny.

**self-fertilization** Fertilization that involves the union of a female gamete and male gamete from the same individual.

**self-incompatibility (SI)** Reproductive state of a plant that prevents the germination of pollen that is genetically too similar to its pistil.

**self-pollination** The process in which pollen from the anthers of a flower is transferred to the stigma of the same flower or between flowers of the same plant.

**self-splicing** The phenomenon in which an rRNA or a tRNA catalyzes the removal of its own intron(s).

**semelparity** A reproductive pattern in which organisms produce all of their offspring in a single reproductive event.

**semen** A mixture containing fluid and sperm that is released during ejaculation.

**semicircular canal** Structures of the vertebrate ear that can detect a range of motions of the head.

**semiconservative mechanism** The correct model for DNA replication; double-stranded DNA is half conserved following replication, resulting in new double-stranded DNA containing one parental strand and one daughter strand.

**semifluid** A property of biological membranes in which movement of membrane components occurs only in two dimensions.

**semilunar valve** One-way valve into the aorta or pulmonary trunk through which blood is pumped from the left or right ventricle, respectively.

**seminiferous tubule** A tightly packed tubule in the testis, where spermatogenesis takes place.

**sense** A system in an animal that consists of specialized cells that respond to a specific type of chemical or physical stimulus and send signals to the central nervous system, where the signals are received and interpreted.

**sensor** A structure such as a sensory receptor or a nucleus in the brain that detects a signal in a homeostatic control system.

**sensory neuron** A neuron that detects or senses information from the outside world, such as light, sound, touch, and heat; sensory neurons also detect internal body conditions such as blood pressure and body temperature.

**sensory receptor** In animals, a specialized cell whose function is to receive sensory inputs.

**sensory transduction** The process by which incoming stimuli are converted into neural signals.

**sepal** A flower structure that is often green and is part of the outer layer of a bud.

**septum (plural, septa)** A cross wall; examples include the cross walls that divide the hyphae of most fungi into many small cells and the structure that separates the old and new chambers of a nautilus.

**set point** The normal value for a controlled variable, such as blood pressure, in an animal.

**setae** Chitinous bristles in the integument of many invertebrates.

**sex chromosomes** A distinctive pair of chromosomes that are different in males and females of some species and determine the sex of an individual.

**sex pili** Hairlike structures made by bacterial $F^+$ cells that bind specifically to $F^-$ cells.

**sex-linked gene** A gene that is found on one sex chromosome but not on the other.

**sexual dimorphism** A pronounced difference in the morphologies of the two sexes within a species.

**sexual reproduction** A process in which two haploid gametes unite in a fertilization event to produce a cell called a zygote.

**sexual selection** A type of natural selection that is directed at certain traits of sexually reproducing species that make it more likely for individuals to find or choose a mate and/or engage in successful reproduction.

**Shannon diversity index ($HS$)** A means of measuring the diversity of a community, using the formula $H_S = -\Sigma p_i \ln p_i$.

**shared derived character** A character that is shared by two or more species or taxa and originated in their most recent common ancestor.

**shared primitive character** A character that is shared by two or more different taxa and inherited from ancestors older than their last common ancestor.

**shattering** The process by which ears of wild grain crops break apart and disperse seeds.

**shell** A tough, protective covering on an amniotic egg that is impermeable to water and prevents the embryo from drying out.

**shivering thermogenesis** Rapid muscle contractions in an animal, without any locomotion, in order to raise body temperature.

**shock** A condition in which the vertebrate circulatory system cannot provide sufficient delivery of blood containing oxygen and nutrients to the vital organs of the body, resulting in cell death and decreased function of those organs.

**shoot apical meristem (SAM)** The region of rapidly dividing cells at the tip of a plant shoot.

**shoot system** The collection of plant organs produced by shoot apical meristems.

**shoot** The portion of a plant consisting of stems and leaves.

**short stature** A condition characterized by stunted growth; formerly called pituitary dwarfism.

**short-day plant** A plant that flowers only when the night length is longer than a defined period.

**shotgun DNA sequencing** A strategy for sequencing an entire genome by randomly sequencing many different DNA fragments.

**sickle cell disease** A disease due to a mutation in a hemoglobin gene that results in sickle-shaped red blood cells that are less able to move smoothly through capillaries and can block blood flow, resulting in pain and cell death of the surrounding tissue.

**sieve plate pore** One of many perforations in a plant's sieve plate.

**sieve plate** The perforated end wall of a mature sieve-tube element.

**sieve-tube elements** A component of the phloem tissues of flowering plants; thin-walled cells arranged end to end to form pipes.

**sigma factor**  A protein that recognizes the promoter in a bacterial gene and binds RNA polymerase to the promoter.

**sign stimulus**  In animals, a trigger that initiates a fixed action pattern of behavior.

**signal recognition particle (SRP)**  A protein-RNA complex that recognizes the ER signal sequence of a polypeptide, pauses translation, and directs the ribosome to the ER to complete translation.

**signal transduction pathway**  A group of proteins that convert an initial signal to a different signal inside a cell.

**signal**  Regarding cell communication, an agent that influences the properties of cells.

**silencer**  A regulatory element in eukaryotes that prevents transcription of a given gene when bound by a repressor protein.

**silent mutation**  A gene mutation that does not alter the amino acid sequence of the polypeptide, even though the base sequence has changed.

**simple diffusion**  When a substance moves across a membrane from a region of high concentration to a region of lower concentration by passing directly through the phospholipid bilayer.

**simple Mendelian inheritance**  The inheritance pattern of traits affected by a single gene that is found in two variants, one of which is completely dominant over the other.

**simple translocation**  A type of mutation in which a single piece of chromosome is attached to another chromosome.

**single-factor cross**  A cross in which an experimenter follows the variants of only one character.

**single-lens eye**  Type of eye found in vertebrates and some invertebrates that has only one lens, as opposed to compound eyes with many lenses.

**single-nucleotide polymorphism (SNP)**  A type of genetic variation in a population in which a particular gene sequence varies at a single nucleotide.

**single-strand binding protein**  A protein that binds to both of the single strands of parental DNA and prevents them from re-forming a double helix during DNA replication.

**sinoatrial (SA) node**  A collection of modified cardiac cells in the right atrium of most vertebrates that spontaneously and rhythmically generates action potentials that spread across the entire atria; also known as the pacemaker of the heart.

**sister chromatids**  The two duplicated chromatids that are still joined to each other after DNA replication.

**skeletal muscle**  A type of muscle tissue that is attached by tendons to bones in vertebrates and to the exoskeleton of invertebrates.

**skeleton**  A structure that serves one or more functions related to support, protection, and locomotion.

**sliding filament mechanism**  The process by which a muscle fiber shortens during muscle contraction.

**SLOSS debate**  In conservation biology, the debate over whether it is preferable to protect one single, large preserve or several smaller ones.

**slow fiber**  A skeletal muscle fiber containing myosin with low ATPase activity.

**slow-oxidative fiber**  A skeletal muscle fiber that has a low rate of myosin ATPase activity but has the ability to make large amounts of ATP; used for prolonged, regular movement.

**small effector molecule**  A molecule that affects gene transcription by binding to a regulatory transcription factor, causing a conformational change in that protein.

**small intestine**  In vertebrates, the part of the alimentary canal that leads from the stomach to the large intestine (or to the anus or cloaca in animals that lack a large intestine) and that carries out nearly all digestion of food and absorption of food nutrients and water.

**small-interfering RNAs (siRNAs)**  RNA molecules that are usually from outside sources and are processed to a small size (22 nucleotides). They are usually a perfect match to pre-existing mRNAs and cause their degradation.

**smooth endoplasmic reticulum (smooth ER)**  The part of the ER whose outer surface is not studded with ribosomes. This region is continuous with the rough ER and functions in diverse metabolic processes such as detoxification, carbohydrate metabolism, accumulation of calcium ions ($Ca^{2+}$), and synthesis and modification of lipids.

**smooth muscle**  A type of muscle tissue that surrounds and forms part of the lining of hollow organs and tubes in vertebrate bodies; it is not under conscious control.

**solute potential**  The component of water potential due to the presence of solute molecules.

**solute**  A substance dissolved in a liquid.

**solution**  A liquid that contains one or more dissolved solutes.

**solvent**  The liquid in which a solute is dissolved.

**somatic cell**  The type of cell that constitutes all cells of an animal or plant body except those that give rise to gametes.

**somatic embryogenesis**  The production of plant embryos from body (somatic) cells.

**somatic nervous system**  The division of the peripheral nervous system that senses external environmental conditions and controls skeletal muscles.

**sorting signal**  A short amino acid sequence in a protein that directs the protein to its correct location in a cell; also known as a traffic signal.

**source pool**  The pool of species on the mainland that is available to colonize an island.

**spatial summation**  Occurs when two or more postsynaptic potentials are generated at one time along different regions of the dendrites and their depolarizations and hyperpolarizations sum together to initiate an action potential.

**speciation**  The formation of new species.

**species concept**  A way of defining the concept of a species and/or of providing an approach to distinguish one species from another.

**species diversity**  A measure of biological diversity that incorporates both the number of species in an area and the relative distribution of individuals among species.

**species interactions**  The various ways in which a species can interact with other species, such as predation, competition, parasitism, mutualism, and commensalism; part of the study of population ecology.

**species richness**  The numbers of species in a community.

**species**  In taxonomy, a subdivision of a genus. Each species is a group of related organisms that share a distinctive set of attributes in nature and (for sexually reproducing species) are capable of interbreeding.

**species-area effect**  The relationship between the amount of available area and the number of species present.

**species-area hypothesis**  The proposal that larger areas contain more species than smaller ones because they can support larger populations and a greater range of habitats.

**species-productivity hypothesis**  The proposal that greater production by plants results in greater overall species richness.

**species-time hypothesis**  The proposal that temperate regions have less species-rich communities than tropical ones because they are younger.

**specific heat**  The amount of energy required to raise the temperature of 1 gram of a substance by 1°C.

**sperm**  Refers to a male gamete that is generally smaller than the female gamete (egg).

**spermatids**  In animals, haploid cells produced when the secondary spermatocytes undergo meiosis II; these cells eventually differentiate into sperm cells.

**spermatogenesis**  Gametogenesis in a male animal, resulting in the production of sperm.

**spermatogonium (plural, spermatogonia)**  In animals, a diploid germ cell that gives rise to the male gamete, the sperm.

**spermatophytes**  All of the living and extinct phyla of seed-producing plants.

**spicules**  Needle-like structures that are made of protein, calcium carbonate, or silica and form lattice-like skeletons in sponges, possibly helping to reduce predation.

**spinal cord**  In chordates, the structure that connects the brain to all areas of the body and together with the brain constitutes the central nervous system.

**spinal nerve**  A nerve that connects the peripheral nervous system and the spinal cord.

**spiracle**  One of several pairs of pores on the body surface of insects through which air enters and exits the body.

**spiral cleavage**  A mechanism of animal development in which the planes of cell cleavage are oblique to the axis of the embryo.

**spirilli**  Rigid, spiral-shaped prokaryotic cells.

**spirochaetes**  Flexible, spiral-shaped prokaryotic cells.

**spliceosome**  A complex of several subunits known as snRNPs that removes introns from eukaryotic pre-mRNA.

**splicing**  The process in which introns are removed from an RNA molecule, such as a pre-mRNA, and the remaining exons are connected to each other.

**spongin**  A tough protein that lends skeletal support to a sponge.

**spongocoel**  A central cavity in the body of a sponge.

**spongy parenchyma**  Photosynthetic ground tissue of the plant leaf mesophyll that contains rounded cells separated by abundant air spaces.

**spontaneous mutation**  A mutation resulting from some abnormality in a biological process.

**sporangia (singular, sporangium)**  Structures that produce and disperse the spores of plants, fungi, or protists.

**spore**  A haploid, typically single-celled reproductive structure of fungi and plants. A spore is able to grow into a new fungal mycelium or plant gametophyte in a suitable location.

**sporophyte**  The diploid generation of plants or multicellular protists that have a sporic life cycle; this generation produces haploid spores by the process of meiosis.

**sporopollenin**  A tough material found in the walls of plant spores that helps to prevent cellular damage during transport in air.

**spring overturn**  The mixing of lake water as ice melts and storms churn up water from the bottom.

**stabilizing selection**  A pattern of natural selection that favors the survival of individuals with intermediate phenotypes.

**stamen**  A flower organ that produces the male gametophytes, pollen.

**standing crop**  The total biomass in an ecosystem at any one point in time.

**starch**  A polysaccharide composed of repeating glucose units that is produced by the cells of plants and some algal protists.

**Starling forces**  The forces that influence the movement of water into and out of capillaries in animals; they are the osmotic imbalance between the plasma and the interstitial fluid and the hydrostatic pressure of the blood entering and exiting the capillary.

**start codon**   A three-base sequence—usually AUG—that specifies the first amino acid in a polypeptide.

**statocyst**   An organ of equilibrium found in many invertebrate species.

**statolith**   1. Tiny granules of sand or other dense objects located in a statocyst that aid equilibrium in many invertebrates. 2. In plants, a starch-heavy plastid that allows both roots and shoots to detect gravity.

**stem cell**   A cell that divides in such a way that one daughter cell can remain a stem cell and the other can differentiate into a specialized cell type. Stem cells supply the cells that constitute the bodies of all animals and plants.

**stem**   A plant organ that contains vascular tissue and produces buds, leaves, branches, and reproductive structures.

**stereocilia**   Deformable projections from epithelial cells called hair cells that are bent by movements of fluid or other stimuli.

**stereoisomers**   Isomers with identical bonding relationships, but different spatial positioning of their atoms.

**sternum**   The breastbone of a vertebrate.

**steroid**   A lipid containing four interconnected rings of carbon atoms; functions as a hormone in animals and plants.

**stigma**   In a flower, the topmost portion of the pistil, which receives and recognizes pollen of the appropriate species or genotype.

**stomach**   A saclike organ in some animals that most likely evolved as a means of storing food; it partially digests some of the macromolecules in food and regulates the rate at which its contents empty into the small intestine.

**stomata** (singular, **stoma or stomate**)   Pores on plant surfaces that can be closed to retain water or open to allow the entry of $CO_2$ (needed for photosynthesis) and the exit of $O_2$ and water vapor.

**stop codon**   One of three three-base sequences—UAA, UAG, and UGA—that signals the end of translation; also called a termination codon or codon.

**strain**   Within a given species, a lineage that has genetic differences compared to another lineage.

**strand**   A structure of DNA (or RNA) formed by the covalent linkage of nucleotides in a linear manner.

**strepsirrhini**   Smaller species of primates, including bush babies, lemurs, and pottos.

**streptophyte algae**   The green algae that are closely related to land plants (embryophytes).

**streptophyte**   Land pants (embryophytes) and their close relatives among the green algae.

**stretch receptor**   A type of mechanoreceptor found widely in an animal's organs and muscle tissues that can be distended.

**stroke volume (SV)**   The amount of blood ejected with each beat, or stroke, of the heart.

**stroma**   The fluid-filled region of the chloroplast between the thylakoid membrane and the inner membrane.

**stromatolite**   A layered calcium carbonate structure generally produced by cyanobacteria living in an aquatic environment.

**strong acid**   An acid that completely dissociates into ions when added to water.

**structural formula**   A type of chemical formula for molecules in which each covalent bond is represented by a line indicating a pair of shared electrons.

**structural isomers**   Isomers that contain the same atoms but in different bonding relationships.

**style**   In a flower, the elongate portion of the pistil through which the pollen tube grows.

**stylet**   A sharp, piercing organ in the mouth of nematodes and some insects.

**submetacentric**   Refers to a chromosome in which the centromere is off center.

**subsidence zones**   Areas of high pressure that are the sites of the world's tropical deserts because the subsiding air is relatively dry, having released all of its moisture over the equator.

**subspecies**   A subdivision of a species; this designation is used when two or more geographically restricted groups of the same species differ, but not enough to warrant their placement into separate species.

**substrate**   1. The reactant molecules that bind to an enzyme at the active site and participate in a chemical reaction. 2. The organic compounds such as soil or rotting wood that fungi use as food.

**substrate-level phosphorylation**   A method of synthesizing ATP that occurs when an enzyme directly transfers a phosphate from an organic molecule to ADP.

**succession**   The gradual and continuous change in species composition of a community over time.

**sugar sink**   The plant tissues or organs in which more sugar is consumed than is produced by photosynthesis.

**sugar source**   The plant tissues or organs that produce more sugar than they consume in respiration.

**supergroup**   One of the seven subdivisions of the domain Eukarya.

**surface area-to-volume (SA/V) ratio**   The ratio between a structure's surface area and the volume in which the structure is contained.

**surface tension**   A measure of how difficult it is to break the interface between a liquid and air.

**surfactant**   A mixture of proteins and amphipathic lipids produced in type II alveolar cells that prevents the collapse of the alveoli by reducing surface tension in the lungs.

**survivorship curve**   A graphical plot of the numbers of surviving individuals for each age class in a population.

**suspensor**   A short chain of cells at the base of an early angiosperm embryo that provides anchorage and nutrients.

**swim bladder**   A gas-filled, balloon-like structure that helps a fish remain buoyant in the water even when it is completely stationary.

**symbiosis**   An intimate association between two or more organisms of different species.

**sympathetic division**   The division of the autonomic nervous system that is responsible for rapidly activating body systems to provide immediate energy in response to danger or stress.

**sympatric speciation**   A form of speciation that occurs when members of a species that initially occupy the same habitat within the same range diverge into two or more different species even though there are no physical barriers to interbreeding.

**sympatric**   The term used to describe species occurring in the same geographic area.

**symplast**   All of a plant's protoplasts (the cell contents without the cell walls) and plasmodesmata.

**symplastic transport**   The movement of a substance from the cytosol of one cell to the cytosol of an adjacent cell via membrane-lined channels called plasmodesmata.

**symplesiomorphy**   *See* shared primitive character.

**symporter**   A type of transporter that binds two or more ions or molecules and transports them in the same direction across a membrane; also called a cotransporter.

**synapomorphy**   *See* shared derived character.

**synapse**   A junction where an axon terminal meets another neuron, a muscle cell, or a gland cell and through which an electrical or chemical signal passes.

**synapsis**   The process of forming a bivalent.

**synaptic cleft**   The extracellular space between a neuron and a receiving cell.

**synaptic plasticity**   A change in strength of the connection between two neurons, which occurs as a result of learning.

**synergids**   In the female gametophyte of a flowering plant, the two cells adjacent to the egg cell that help to import nutrients from maternal sporophyte tissues.

**synthetic microbiomes**   Microbiomes generated by mixing cultures of beneficial microbial species.

**systematics**   The study of biological diversity and evolutionary relationships among species, both extant and extinct.

**systemic acquired resistance (SAR)**   A whole-plant defensive response to a pathogen attack.

**systemic circulation**   The circuit of a double circulation in which blood is pumped from the left side of an animal's heart to the body to drop off $O_2$ and nutrients and pick up $CO_2$ and wastes. The blood then returns to the right side of the heart.

**systems biology**   A field of study in which researchers investigate living organisms in terms of their underlying networks—groups of structural and functional connections—rather than their individual molecular components.

**systole**   The second phase of the cardiac cycle, in which the ventricles contract and eject the blood through the open semilunar valves.

# T

**T cells**   A type of lymphocyte that directly kills infected, mutated, or transplanted cells.

**tagmata**   Functional units composed of fused body segments.

**taproot system**   The root system of eudicots, consisting of one main root with many branch roots.

**taste buds**   Structures located in the mouth and tongue of vertebrates that contain the sensory cells, supporting cells, and associated neuronal endings that contribute to taste sensation.

**TATA box**   One of three features found in promoters of many protein-encoding genes in eukaryotes; the others are the transcriptional start site and regulatory elements.

**taxis**   A directed movement in response to a stimulus, either toward or away from the stimulus.

**taxon**   A group of species that are evolutionarily related to each other. In taxonomy, each species is placed into several taxons that form a hierarchy from large (domain) to small (genus).

**taxonomy**   The field of biology that is concerned with the theory, practice, and rules of classifying living and extinct species and also viruses.

**telocentric**   Refers to a chromosome in which the centromere is at the end.

**telomerase**   An enzyme that catalyzes the replication of the telomere.

**telomere**   A region at the ends of eukaryotic chromosomes where a specialized form of DNA replication occurs.

**telophase**   The phase of mitosis during which the chromosomes decondense and the nuclear membrane re-forms.

**temperate phage**   A bacteriophage that can follow either a lysogenic or a lytic cycle.

**template strand**   The DNA strand that is used as a template for RNA synthesis or DNA replication.

**temporal lobe**   One of four lobes of the cerebral cortex of human brain; necessary for language, hearing, and some types of memory.

**temporal summation**   Occurs when two or more postsynaptic potentials arrive at the same location in a dendrite in quick succession and their depolarizations and hyperpolarizations sum together to initiate an action potential.

**tepal**   A flower perianth part that cannot be distinguished by appearance as a petal or a sepal.

**termination codon**   *See* stop codon.

**termination**   The final stage of transcription, in which the RNA dissociates from the DNA, or of translation, in which the polypeptide is released from the ribosome.

**terminator**   A sequence of DNA within a gene that specifies the end of transcription.

**territory**   A fixed area in which an individual or group excludes other members of its own species, and sometimes other species, by aggressive behavior or territory marking.

**tertiary consumer**   An organism that feeds on secondary consumers; also called a secondary carnivore.

**tertiary endosymbiosis**   The acquisition by eukaryotic protist host cells of plastids from cells that possess secondary plastids.

**tertiary plastid**   A plastid acquired by the incorporation into a host cell of an endosymbiont having a secondary plastid.

**tertiary structure**   The three-dimensional shape of a single polypeptide; one of four levels of protein structure.

**testable**   Refers to a hypothesis that can be accepted or rejected based on experimentation.

**testcross**   (1) A cross to determine if an individual with a dominant phenotype is a homozygote or a heterozygote. (2) Also, a cross to determine if two different genes are linked.

**testis** (plural, **testes**)   In animals, the male gonad, where sperm are produced.

**testosterone**   The primary androgen in many vertebrates, including humans.

**tetraploid**   Refers to an organism or cell that has four sets of chromosomes.

**tetrapod**   A vertebrate animal having four legs or leglike appendages.

**thalamus**   A region of the vertebrate forebrain that plays a major role in relaying sensory information to appropriate parts of the cerebrum and, in turn, sending outputs from the cerebrum to other parts of the brain.

**theory**   In biology, a broad explanation of some aspect of the natural world that is substantiated by a large body of evidence. Biological theories incorporate observations, hypothesis testing, and the laws of other disciplines such as chemistry and physics. A theory makes valid predictions.

**thermocline**   The thin transitional zone in a lake that separates the epilimnion from the hypolimnion

**thermodynamics**   The study of energy interconversions.

**thermoreceptor**   A sensory receptor in animals that responds to cold and heat.

**theropods**   A group of bipedal saurischian dinosaurs.

**thick filament**   A section of the repeating pattern in a myofibril composed almost entirely of the motor protein myosin.

**thigmotropism**   Touch responses in plants.

**thin filament**   A section of the repeating pattern in a myofibril that contains the cytoskeletal protein actin, as well as two other proteins—troponin and tropomyosin—that play important roles in regulating contraction.

**thoracic breathing**   Breathing in which coordinated contractions of muscles expand the rib cage, creating a negative pressure to suck air in and then forcing it out later; used by amniotes.

**thoraco-abdominal pump**   In mammals, the mechanical effect of breathing movements on the return of blood from veins in the abdomen to the thorax where the

heart is located; increased depth and rate of breathing results in pressure differences between the two body compartments that favors flow of blood into the chest; also called the respiratory pump.

**threatened species**   Those species that are likely to become endangered in the foreseeable future.

**threshold concentration**   The concentration above which a morphogen exerts its effects but below which it is ineffective.

**threshold potential**   The membrane potential, typically around –55 to –50 mV, which is sufficient to trigger an action potential in an electrically excitable cell such as a neuron.

**thylakoid lumen**   The fluid-filled compartment within a thylakoid.

**thylakoid membrane**   A membrane within the chloroplast that forms many flattened, fluid-filled tubules that enclose a single, convoluted compartment. It contains chlorophyll and is the site where the light-dependent reactions of photosynthesis occurs.

**thylakoid**   A flattened, fluid-filled tubule found in cyanobacterial cells and the chloroplasts of photosynthetic protists and plants; the location of the light reactions of photosynthesis.

**thymine (T)**   A pyrimidine base found in DNA.

**thymine dimer**   A site in DNA where two adjacent thymine bases become covalently crosslinked to each other; may cause a mutation when the DNA strand is replicated.

**thyroxine (T$_4$)**   A weakly active thyroid hormone that contains iodine and helps regulate metabolic rate; it is converted by cells into the more active triiodothyronine (T$_3$).

**tidal ventilation**   A type of breathing in mammals in which the lungs are inflated with air and then the chest muscles and diaphragm relax and recoil back to their original positions as an animal exhales. During exhalation, air leaves via the same route that it entered during inhalation, and no new oxygen is delivered to the airways at that time.

**tight junction**   A type of junction between animal cells that forms a tight seal between adjacent cells and thereby prevents material from leaking between the cells.

**tissue**   A part of an animal or plant consisting of a group of cells having a similar structure and function, for example, muscle tissue.

**tolerance**   A mechanism for succession in which any species can start the succession, but the eventual climax community is reached in a somewhat orderly fashion; early species neither facilitate nor inhibit subsequent colonists.

**Toll-like receptors (TLRs)**   Receptor proteins that recognize nonspecific antigens in microbes; a key part of the innate immune response.

**topsoil**   The uppermost layer of a soil.

**torus** (plural, **tori**)   The nonporous, flexible central region of a conifer pit that functions like a valve.

**total fertility rate**   The average number of live births a female has during her lifetime.

**total peripheral resistance (TPR)**   The sum of all the resistance in all the arterial vessels of the systemic circulation.

**totipotent**   Refers to the ability of a fertilized egg to produce all of the cell types in the adult organism or the ability of unspecialized plant cells to regenerate an adult plant.

**toxins**   Compounds that have adverse physiological effects in living organisms; often produced by various protist and plant species.

**trace element**   An element that is essential for normal growth and function of living organisms but is required in extremely small quantities.

**trachea**   1. A sturdy tube arising from the spiracles of an insect's body; involved in respiration. 2. The name of the tube leading to the lungs of air-breathing vertebrates.

**tracheal system**   The respiratory system of insects, consisting of a series of finely branched air tubes called tracheae; air enters and exits the tracheae through spiracles, which are pores on the body surface.

**tracheary elements**   Water-conducting cells in plants that, when mature, are always dead and empty of cytosol; include tracheids and vessel elements.

**tracheid**   A type of dead, lignified plant cell in xylem that conducts water, along with dissolved minerals and hormones; also provides structural support.

**tracheophytes**   A term used to describe vascular plants.

**tract**   A parallel bundle of myelinated axons in the central nervous system.

**traffic signal**   *See* sorting signal.

**trait**   An identifiable characteristic; usually refers to a variant.

**transcription factor**   A protein that influences the ability of RNA polymerase to transcribe genes.

**transcription**   The process that produces an RNA copy of a gene.

**transcriptional start site**   The site in a eukaryotic promoter where transcription begins.

**transcriptome**   A collection of all the mRNA sequences produced by a single organism under defined conditions.

**transduction**   A type of gene transfer between bacteria in which a virus infects a bacterial cell and then a newly made virus subsequently transfers some of that cell's DNA to another bacterium.

**trans-effect**   In both prokaryotes and eukaryotes, a form of genetic regulation that can occur even though two DNA segments are not physically adjacent. The action of the lac repressor on the *lac* operon is a *trans*-effect.

**transepithelial transport**   The process of moving molecules across an epithelium, such as across the intestinal cells of animals.

**transfer RNA (tRNA)**   An RNA that carries amino acids and is used to translate mRNA into polypeptides.

**transformation**   A type of gene transfer between bacteria in which a segment of DNA from the environment is taken up by a competent cell and incorporated into the bacterial chromosome.

**transition state**   In a chemical reaction, a state in which the original bonds have stretched to their limit; once this state is reached, the reaction can proceed to the formation of products.

**transitional form**   An organism that provides a link between earlier and later forms in evolution.

**translation**   The process of synthesizing a specific polypeptide on a ribosome.

**translocation**   1. A type of mutation in which one segment of a chromosome becomes attached to a different chromosome. 2. A process in plants in which phloem transports substances from a source to a sink.

**transmembrane gradient**   A situation in which the concentration of a solute is higher on one side of a membrane than on the other.

**transmembrane protein**   A protein that has one or more regions that are physically embedded in the hydrophobic region of a membrane's phospholipid bilayer.

**transmembrane transport**   The export of material from one cell into the intercellular space and then into an adjacent cell.

**transmission electron microscopy (TEM)**   A type of microscopy in which a beam of electrons is transmitted through a biological sample to form an image on a photographic plate or screen.

**GLOSSARY**

**transpiration** The process by which water evaporates from the aerial parts of plants, usually via the stomata of leaves.

**transport protein** A transmembrane protein that provides a passageway for the movement of ions and hydrophilic molecules across the phospholipid bilayer.

**transporter** A transmembrane protein that binds a solute and undergoes a conformational change to allow the movement of the solute across a membrane; also called a carrier.

**transposable element (TE)** A segment of DNA that can move from one site to another within a genome.

**transposase** An enzyme that facilitates transposition.

**transposition** The process in which a short segment of DNA, a transposable element, moves to a new site in a genome.

**transverse tubules (T-tubules)** Invaginations of the plasma membrane of skeletal muscle fibers that open to the extracellular fluid and conduct action potentials from the outer surface of the fibers to the myofibrils.

**trichome** A projection, often hairlike, from the epidermal tissue of a plant that offers protection from excessive light, ultraviolet radiation, extreme air temperature, or attack by herbivores.

**triglyceride** A molecule composed of three fatty acids linked by ester bonds to a molecule of glycerol; also known as a triacylglycerol.

**triiodothyronine ($T_3$)** A thyroid hormone that contains iodine and helps regulate metabolic rate.

**triplet** A group of three bases that functions as a codon.

**triploblastic** Having three distinct germ layers: endoderm, ectoderm, and mesoderm.

**triploid** Refers to an organism or cell that has three sets of chromosomes.

**trochophore larva** A distinct larval stage of many invertebrate phyla.

**trophectoderm** The outer layer of cells in a developing mammalian blastocyst; continuous with the ectoderm layer.

**trophic level** Each feeding level in a food chain.

**trophic-level transfer efficiency** The amount of energy at a trophic level that is acquired by the trophic level above and incorporated into biomass.

**tropism** In plants, a growth response that is dependent on a stimulus that occurs in a particular direction.

**tropomyosin** A rod-shaped protein that plays an important role in regulating muscle contraction.

**troponin** A small globular-shaped protein that plays an important role in regulating muscle contraction through its ability to bind $Ca^{2+}$.

**trp operon** An operon of *E. coli* that encodes enzymes required to make the amino acid tryptophan, a building block of cellular proteins.

**true-breeding line** A strain that continues to exhibit the same trait after several generations of self-fertilization or inbreeding.

**trypsin** A protease involved in the breakdown of proteins in the small intestine.

**tubal ligation** A means of contraception that involves the cutting and sealing of a woman's oviducts, thereby preventing movement of a fertilized egg into the uterus.

**tube cell** In a seed plant, one of the cells resulting from the division of a microspore; produces the pollen tube to deliver sperm to the female gametophyte.

**tube feet** Echinoderm structures that function in movement, gas exchange, feeding, and excretion.

**tumor** An abnormal overgrowth of cells.

**tumor-suppressor gene** A gene that when normal (that is, not mutant) encodes a protein that prevents cancer; however, when a mutation eliminates its function, cancer may occur.

**tunic** A nonliving structure that encloses a tunicate, made of protein and a cellulose-like material called tunicin.

**turgid** Refers to a plant cell whose cytosol is so full of water that the plasma membrane presses right up against the cell wall; as a result, a turgid cell is firm or swollen.

**turgor pressure** *See* osmotic pressure.

**two-factor cross** A cross in which an experimenter simultaneously follows the inheritance of two different characters.

**type 1 diabetes mellitus (T1DM)** A disease in which the pancreas does not produce sufficient insulin; as a result, extracellular glucose cannot cross plasma membranes, and glucose accumulates to very high concentrations in the blood.

**type 2 diabetes mellitus (T2DM)** A disease in which the pancreas produces sufficient insulin, but the cells of the body lose much of their ability to respond to insulin.

## U

**ubiquitin** A small protein in eukaryotic cells that is covalently attached to an unwanted protein, which directs the protein to a proteasome for degradation.

**ulcer** An erosion of the wall of the alimentary canal; typically occurs in the lower esophagus, stomach, or duodenum.

**ultimate cause** The reason a particular behavior evolved, in terms of its effect on reproductive success.

**umbrella species** A species whose habitat requirements are so large that protecting it would protect many other species existing in the same habitat.

**uniform** A pattern of dispersion within a population in which individuals maintain a certain minimum distance between themselves to produce an evenly spaced distribution.

**uniporter** A type of transporter that binds a single ion or molecule and transports it across a membrane.

**unipotent** Refers to a stem cell found in an adult organism that can produce daughter cells that differentiate into only one cell type.

**unsaturated fatty acid** A fatty acid that contains one or more C=C double bonds.

**unsaturated** The property of certain lipids that contain one or more C=C double bonds.

**upwelling** In the ocean, a process that carries mineral nutrients from the bottom waters to the surface.

**uracil (U)** A pyrimidine base found in RNA.

**urea** A nitrogenous waste commonly produced in many terrestrial species, including mammals.

**uremia** A condition characterized by the retention of urea and other waste products in the blood; typically results from kidney disease.

**uric acid** A nitrogenous waste produced by birds, insects, and most reptiles.

**urine** The part of the filtrate formed in the kidney that remains after all reabsorption of solutes and water is complete.

**uterine cycle** *See* menstrual cycle.

**uterus** A small, pear-shaped organ capable of enlarging and specialized for carrying a developing fetus in female mammals.

## V

**vaccination** The injection into the body of small quantities of weakened or dead pathogens, resulting in the development of immunity to those pathogens without causing disease.

**vacuole** Specialized compartments found in eukaryotic cells that function in storage, the regulation of cell volume, and degradation.

**vagina** The birth canal of female mammals; also functions to receive sperm during copulation.

**vaginal diaphragm** A barrier method of preventing fertilization in which a diaphragm is placed in the upper part of the vagina just prior to intercourse; blocks movement of sperm to the cervix.

**valence electron** An electron in the outermost shell of an atom that is available to share with other atoms. Such electrons allow atoms to form chemical bonds with each other.

**van der Waals dispersion forces** Attractive forces between molecules in close proximity to each other, caused by the variations in the distribution of electron density around individual atoms.

**variable region** A unique domain within an immunoglobulin that serves as the antigen-binding site.

**vasa recta capillaries** Capillaries surrounding the renal tubules in the medulla of the kidney.

**vascular bundle** A cluster of primary plant vascular tissues that appears round or oval in cross section.

**vascular cambium** A secondary meristematic tissue of seed plants that produces both wood and inner bark.

**vascular plants** A broad category of plants distinguished by internal water and nutrient-conducting (vascular) tissues that also provide structural support

**vascular tissue** A complex plant tissue composed of interconnected cells that form conducting vessels for water, minerals, and organic compounds.

**vasectomy** A surgical procedure in men that severs the vas deferens, thereby preventing the release of sperm at ejaculation.

**vasoconstriction** A decrease in blood vessel radius; an important mechanism for directing blood flow away from specific regions of the body.

**vasodilation** An increase in blood vessel radius; an important mechanism for directing blood flow to specific regions of the body.

**vasotocin** A peptide hormone that is responsible for regulating salt and water balance in the blood of nonmammalian vertebrates.

**vector** A type of DNA that acts as a carrier of a DNA segment that is to be cloned.

**vegetal pole** In triploblast organisms, the pole of the egg where the yolk is most concentrated.

**vegetative growth** The production of new nonreproductive tissues by the shoot apical meristem and root apical meristem during seedling development and growth of mature plants.

**vegetative reproduction** A form of asexual reproduction that involves nonreproductive parts of plants.

**vein** 1. In animals, a blood vessel that returns blood to the heart. 2. In plants, a bundle of vascular tissue in a leaf.

**veliger** In mollusks, a free-swimming larva that has a rudimentary foot, shell, and mantle.

**venation** Leaf vein patterns.

**ventilation** The process of bringing oxygenated water or air into contact with a respiratory organ such as gills or lungs.

**ventral** Refers to the lower side of an animal.

**ventricle** In the heart, a chamber that pumps blood out of the heart.

**vernalization** The process in which certain species of plants require an exposure to cold temperatures in order to flower.

**vertebrae** Interlocking bony or cartilaginous structures forming a column that provides support and also protects the nerve cord, which lies within the column's tubelike structure.

**vertebrate** An animal with a backbone.

**vertical evolution** A type of evolution in which genetic changes occur in a series of related species that form a lineage; species evolve from pre-existing species by the accumulation of mutations.

**vessel element**   A type of plant cell in xylem that conducts water, along with dissolved minerals, hormones, and certain organic compounds.

**vessel**   In a plant, a water-conducting structure made up of aligned vessel elements.

**vestibular system**   The organ of balance in vertebrates, located in the inner ear next to the cochlea.

**vestigial structure**   An anatomical feature that has no current function but resembles a structure of a presumed ancestor.

**vibrios**   Comma-shaped prokaryotic cells.

**villi** (singular, **villus**)   Finger-like projections extending from the inner surface into the lumen of the vertebrate small intestine; these are specializations that aid in digestion and absorption.

**viral envelope**   A structure enclosing a viral capsid that consists of a membrane derived from the plasma membrane of the host cell and embedded with virally encoded spike glycoproteins.

**viral genome**   The genetic material of a virus.

**viral reproductive cycle**   The series of steps that results in the production of new viruses during a viral infection.

**viral vector**   A type of vector used in cloning experiments that is derived from a virus.

**viroid**   An RNA molecule that infects plant cells.

**virulence**   The ability of a pathogen to infect its host and cause disease.

**virulent phage**   A bacteriophage that follows only the lytic cycle.

**virus**   A small infectious particle that consists of nucleic acid enclosed in a protein coat. Some viruses also have a viral envelope derived from the plasma membrane of the host cell.

**visceral mass**   In mollusks, a structure that rests atop the foot and contains the internal organs.

**vitamin D**   A vitamin that is converted into a hormone in the body; regulates the calcium level in the blood through an effect on intestinal transport of calcium ions.

**vitamin**   An organic nutrient that serves as a coenzyme for a metabolic or biosynthetic reaction.

**viviparous**   Refers to an animal whose embryos develop within the uterus, receiving nourishment from the mother via a placenta.

**$V_{max}$**   The maximal velocity, or rate, of an enzyme-catalyzed reaction.

**volt**   A unit of measurement of potential difference in charge (electrical force), such as the difference between the interior and exterior of a cell.

**voltage-gated ion channels**   Ion channels that open and close in response to changes in voltage across a cell membrane.

# W

**water potential**   The potential energy of water.

**water vascular system**   A network of canals in which water pressure generated by the contraction of muscles enables extension and contraction of the tube feet, allowing echinoderms to move slowly.

**water**   The liquid form of $H_2O$.

**wavelength**   The distance from one peak to the next in a sound wave or light wave.

**weak acid**   An acid that only partially dissociates into ions when added to water.

**weathering**   The physical and chemical breakdown of rock.

**white matter**   Brain tissue that consists of myelinated axons that are bundled together in large numbers to form tracts.

**whole metagenomic sequencing (WMS)**   Base sequencing of all of the DNA present in a sample.

**whorls**   In a flower, concentric rings of sepals and petals (or tepals), stamens, and carpels.

**wild-type allele**   A prevalent allele in a population.

**wood**   A secondary plant tissue composed of numerous pipelike arrays of empty, water-conducting cells whose walls are strengthened by an exceptionally tough polymer known as lignin.

**woody plant**   A type of plant that produces both primary and secondary vascular tissues.

# X

**X inactivation center (Xic)**   A short region on the X chromosome known to play a critical role in X inactivation.

**X-chromosome inactivation (XCI)**   A process that causes an X chromosome to become highly compacted and silences the genes that it carries.

**X-linked gene**   A gene found on the X chromosome but not on the Y.

**xylem**   A specialized conducting tissue in plants that transports water, minerals, and some organic compounds.

# Y

**yeast**   A unicellular fungus that may reproduce by budding.

**yield**   The number of individuals harvested in a given unit of time.

**yolk sac**   One of the four extraembryonic membranes in the amniotic egg. The yolk sac encloses a stockpile of nutrients, in the form of yolk, for the developing embryo.

# Z

**Z scheme**   A model depicting the series of energy changes of an electron during the light reactions of photosynthesis. The electron absorbs light energy twice, resulting in an energy curve with a zigzag shape.

**zero population growth**   The situation in which no changes in population size occur.

**zombie parasite**   A parasite that infects its host and is then able to control that host's behavior.

**zona pellucida**   The glycoprotein covering that surrounds a mature oocyte.

**zone of elongation**   The area above the root apical meristem of a plant where cells extend by water uptake, thereby dramatically increasing root length.

**zone of maturation**   The area above the zone of elongation in a plant root where most root cell differentiation and tissue specialization occur.

**zooplankton**   Aquatic organisms drifting in the open ocean or fresh water; includes minute animals consisting of some worms, copepods, tiny jellyfish, and the small larvae of invertebrates and fishes.

**zygospore**   A dark-pigmented, thick-walled, multinucleate spore that matures within the zygosporangium of zygomycete fungi during sexual reproduction.

**zygote**   A diploid cell formed by the fusion of two haploid gametes.

# Index

Page numbers followed by *f* denote figures; those followed by *t* denote tables.

## A

A. *See* adenine.
A band, 955
ABA receptor protein, 831–32*f*, 831–33
ABC model, of flower development, 431
abiotic factors:
  effects of, on organisms, 1155*t*
  in environment, 1152
abiotic interactions, 1149
abiotic synthesis, 70
*abl* gene, 316
ABO blood groups, 365, 365*t*
abomasum, 978
aboveground plant microbiomes, 633
abscisic acid:
  and plant behavior, 786*t*, 792, 797
  in plant reproduction, 852
  water stress and, 831–32
abscission, 792
absolute refractory periods, 890
absorption:
  of amino acids, 982, 982*f*
  into blood and lymph, 979
  defined, 970, 971*f*
  of digested carbohydrates, 980–81, 981*f*
  of fat, 983–84, 983*f*
  of nutrients, 974–75, 975*f*, 984, 992–93, 992*f*
  in small intestine, 978–79, 979*f*
  of vitamins, minerals, and water, 984
absorption spectra, 169, 169*f*
absorptive nutrition, 606, 608*f*
absorptive state:
  blood glucose concentration in, 996
  defined, 992
  energy sources in, 994*t*
  nutrients in, 992–93, 992*f*
  regulation of, 994–97, 995*f*, 997*f*
*Acacia collinsii*, 1230*f*
acacia trees, 1*f*, 1230, 1230*f*
*Acetabularia*, 583*f*
acetaldehyde, 161
acetylcholine (ACh):
  and baroreceptors, 1135
  and cell communication, 196

discovery of, 896–98, 897*f*
and muscular-skeletal systems, 960
in response to hemorrhage, 1134
size and structure of, 895
acetylcholinesterase, 960
acetyl CoA:
  in citric acid cycle, 151, 152*f*
  in pyruvate oxidation, 150–51, 150*f*
*N*-Acetylgalactosamine, 365
acetyl groups:
  in citric acid cycle, 152*f*
  in pyruvate oxidation, 150–51, 150*f*
acetylsalicylic acid, 127
*Acheta domesticus* (house cricket), 2*t*
acid hydrolases, 94
acidic soils, 42, 1159, 1159*f*
acidic solutions, 42, 42*f*
acidity, 1159–60
acid rain, 1160
acid reflux, 986
acids:
  defined, 41
  and hydrogen ion concentration, 41
  plant nutrition, 806
  stomach, 986
*Acinonyx jubatus*, 1190, 1190*f*
acoelomates, 692, 692*f*
acquired antibiotic resistance, 411
acquired immune deficiency syndrome (AIDS), 399–400, 637, 1128–29, 1128*f*
acquired immunity, 1108
  *See also* adaptive immunity
acrocentric chromosomes, 342, 342*f*
acromegaly, 1077
acrosomal reaction, 1089, 1089*f*
acrosomes, 1089
ACTH (adrenocorticotropic hormone), 197
actin, in skeletal muscles, 955–57
actin filaments:
  and cell communication, 186
  and cell junctions, 209
  in cytoskeleton, 84–85, 85*t*
Actinistia and actinisians, 741, 742*f*
Actinobacteria, 568, 812
actinomorphic flowers, 845
Actinomycetes, 574–75
Actinopterygii, 738, 741, 741*f*
action potentials, 889–92
  all-or-none operation of, 889–90, 890*f*
  in baroreceptors, 1133, 1134
  changes occurring during, 890*f*
  depolarization phase, 889
  electrical properties of neurons, 885
  in hearing, 930, 931*f*
  in heart, 1017–18
  peak of, 890*f*
  and plant behavior, 796, 796*f*
  and plant nutrition, 815
  propagation of, 891–92, 891*f*, 892*f*
  refractory periods, 890
  repolarization phase, 889–90
  and sensory stimulus strength, 926

in skeletal muscle, 954, 958
action spectra, 169, 169*f*
*Actitis macularia*, 1198
activation, of cellular receptors, 186–91*f*
activation domain, 64
activation energy, 131, 132*f*
activators, 285
  eukaryotic, of transcription, 294–95, 296*f*, 298, 299*f*
  of *lac* operon, 291–93, 292*f*
active carbon, 1151
active immunity, 1126
active sites, of enzymes, 132, 139
active transport:
  in animal bodies, 873
  defined, 113, 114*f*
  in digestive systems, 974–75, 975*f*
  in plant cells, 820, 820*f*
  and transmembrane gradient, 115
  transport proteins for, 120–21, 121*f*
active trapping mechanisms, 815
acyl transferase, 112*f*
*Acyrthosiphon pisum*, 507
AD (Alzheimer disease), 279, 922–23, 923*f*
adaptation phase (CRISPR-Cas system), 275–76, 277*f*, 278
adaptations:
  to local environments, 506–8
  and natural selection, 482
  *See also* evolution
adaptive immunity, 1115–26
  cell-mediated immunity, 1117–18, 1117*f*, 1121–23, 1122*f*, 1124*f*
  defined, 1108
  example, 1123–24, 1125*f*
  humoral immunity, 1117–21, 1117*f*, 1119–21*f*
  immunological memory in, 1125–26, 1126*f*
  lymphocytes, 1116–17, 1116*f*
  lymphoid organs, tissues, and cells in, 1115–16, 1116*f*
  and self molecules, 1124–25
adaptive radiation, 502–3
adenine (A):
  in DNA, 224
  molecular structure of, 64, 65
adenosine diphosphate. *See* ADP.
adenosine monophosphate (AMP). *See* cAMP.
adenosine triphosphate. *See* ATP.
adenoviruses, 393, 394, 394*f*
adenylyl cyclase, 192, 193*f*
ADH. *See* antidiuretic hormone.
adherens junctions, 208, 209*f*
adhesion:
  cell, 96, 96*f*
  in long-distance plant transport, 828*f*
  of water, 40*f*, 41
adhesive proteins:
  in connective tissue, 863
  in extracellular matrix, 204, 204*t*

adiabatic cooling, 1162
adipose tissue, 999
  as connective tissue, 863*f*
  in endocrine system, 1060*f*
  leptin secretion in, 1073
ADP (adenosine diphosphate):
  from ATP hydrolysis, 129–30, 129*f*
  in citric acid cycle, 153
  in cross-bridge cycle, 957
  in oxidative phosphorylation, 153, 155
adrenal cortex, 197, 198*f*, 1060*f*, 1067, 1068
adrenal glands, 1060*f*, 1067–68, 1068*f*, 1069*t*
adrenaline. *See* epinephrine.
adrenal medulla, 1060*f*, 1067, 1068
adult-onset diabetes. *See* type 2 diabetes mellitus.
adults, structures of, 418, 419*f*
adventitious roots, 767, 785
*Aedes* genus, 399
aeration, soil, 806–7, 806*f*, 807*f*
aerenchyma, 797
aerobic exercise, 962
aerobic respiration, 146
  *See also* citric acid cycle; glycolysis; oxidative phosphorylation
aerotolerant anaerobes, 573
afferent arterioles, 1047–49, 1141*f*
afferent neurons, 883–84
afferent vessels, 1026
affinity:
  of enzymes, 133
  of GLUTs for glucose, 995
  of hemoglobin for oxygen, 1034–35, 1035*f*, 1036*f*
  of receptors, 196
aflatoxins, 610
African apes, 630, 631*f*
African cichlids, 487–89, 508
African clawed tree frog, 414
African forest elephants, 516
African savanna elephants, 516
African violets, 855
African wildcat, 1292
Afrotheria, 757
*Agalychnis callidryas*, 746*f*
agar, 787
agave plants, 1205*f*
*Agave shawii*, 1205*f*
age, and sexual selection, 487
age classes, 1205
Age of Amphibians, 745
Age of Dinosaurs, 546
Age of Fishes, 546
Age of Mammals, 547
age-specific fertility, 1208–9
age structure, of populations, 1258, 1258*f*
aggregation, and flower symmetry, 845, 846
aggression, in visual communication, 1191, 1191*f*
aging, 790

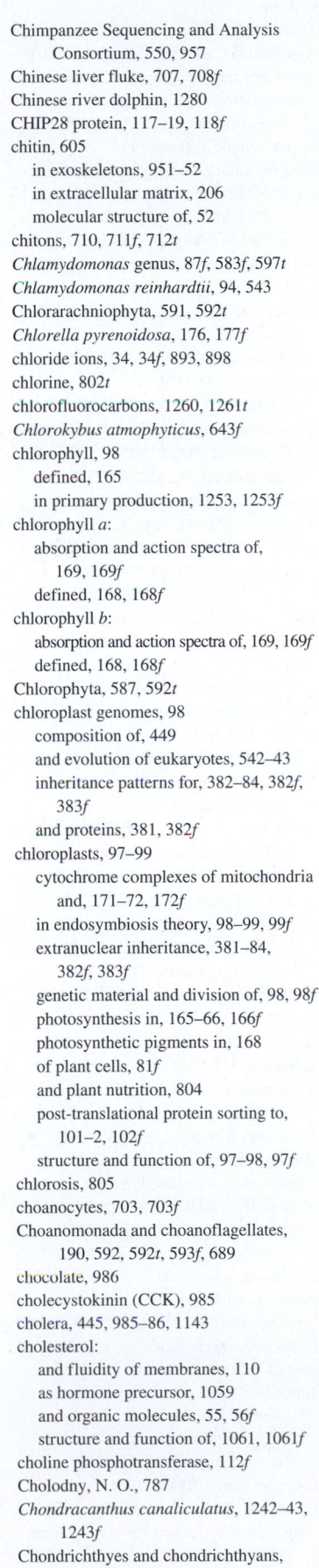

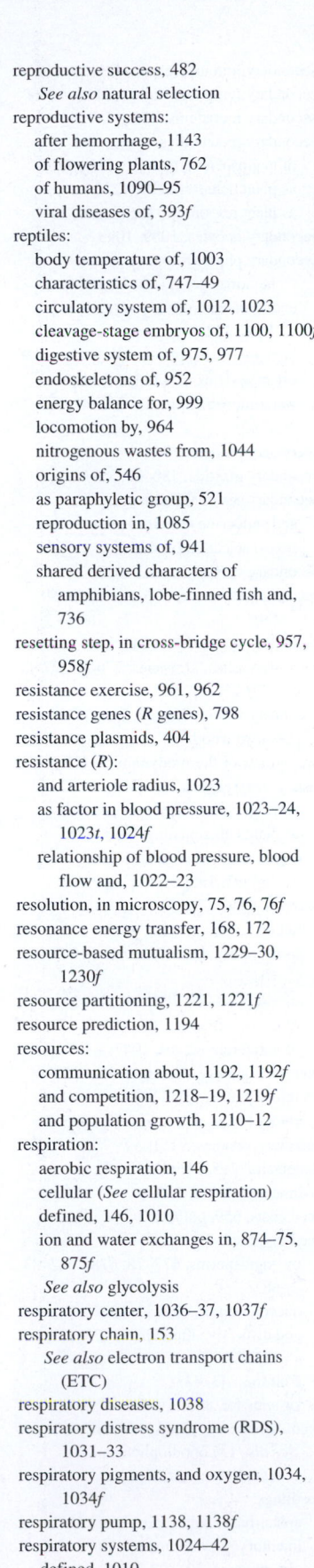

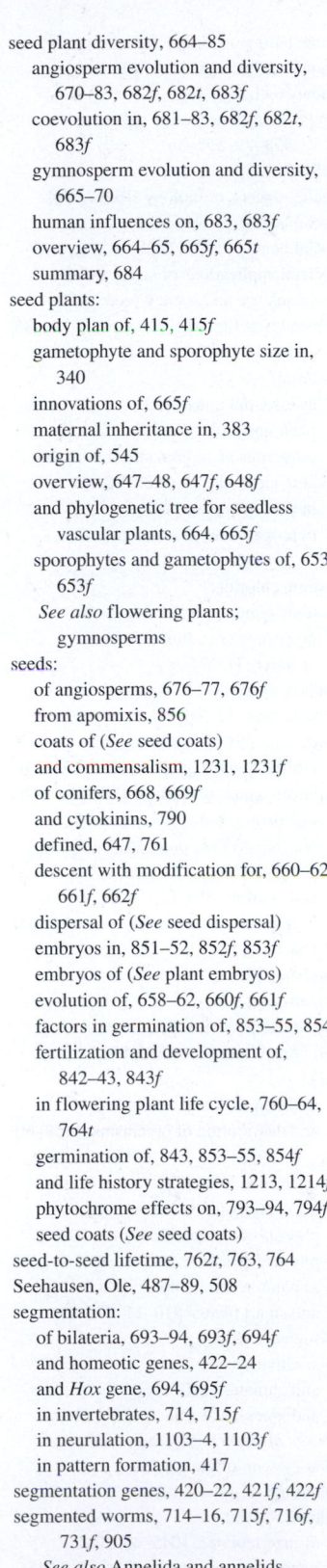

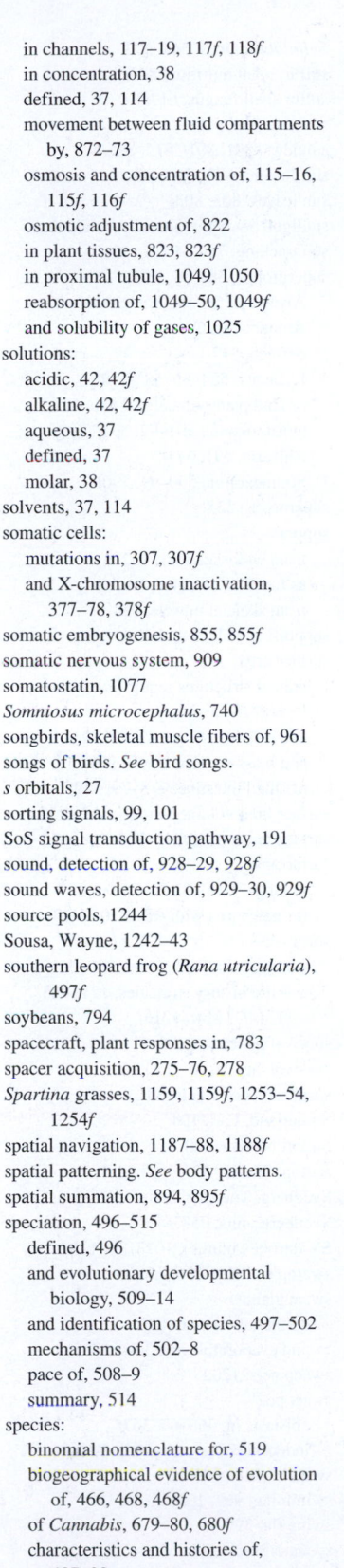

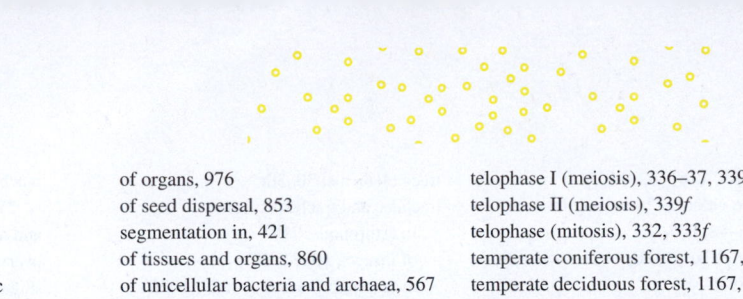

Walker, Brian, 1281
Walker, John, 157
*Wallabia bicolor*, 756*f*
Wallace, Alfred Russel, 461, 482, 1176
Wallin, Ivan, 98
wall pores, of peat moss, 627, 627*f*
walruses, 754*f*, 1004
warblers, 1221, 1221*f*
Warburg, Otto, 149
Warburg effect, 149
Warren, J. Robin, 987–88
wasps, 2*t*, 724, 882
waste removal:
　exchanges of ions and water in, 874
　processes for, 1044–45, 1045*f*
wastes, water in elimination of, 40–41, 40*f*
water, 36–41
　absorption of, 768, 769*f*, 975, 978, 984
　absorption of, in plant cells, 768, 769*f*
　and aldosterone, 1051, 1052
　and amphipathic molecules, 37, 37*f*
　balance of ions and water in internal
　　fluids, 873–74
　bulk flow of, 825–26, 825*f*, 826*f*
　cell-based transport of, in plants,
　　820–22, 821*f*, 822*f*
　in chemical reactions, 36
　circulation of, through sponges, 703,
　　703*f*
　cohesion-tension theory and transport
　　of, 828–29, 828*f*, 829*f*
　concentration of solute in, 38–39
　defined, 36
　dissolution of gases in, 1025
　and distribution of organisms, 1158–60,
　　1159*f*
　exercise and homeostatic balance of,
　　876–79
　functions of, 40–41, 40*f*
　hydrogen and hydroxide ions in, 41
　hydrogen bonds in, 33*f*
　ions and polar molecules in, 37, 37*f*
　leaf abscission and loss, 833–34
　leaf vein damage and flow of, 773
　in lycophytes and pteridophytes,
　　646–47, 647*f*
　in mass of living organisms, 29
　molecular shape of, 35*f*
　movement between fluid compartments
　　by, 873
　movement detection in, 933, 933*f*
　movement of, in osmosis, 115–16, 116*f*
　obligatory exchanges between body and
　　environment of, 874–76
　in photosynthesis, 165, 173*f*
　and plant nutrition, 802*f*, 804–5
　polar covalent bonds, 32, 32*f*
　in proximal tubule, 1049, 1050
　proximity to, and climate, 1162–63,
　　1163*f*
　reabsorption of, 1050–51, 1050*f*,
　　1052–53, 1053*f*
　resistance of, for aquatic animals, 964
　in sandy soil, 806, 807, 807*f*
　and secondary response to hemorrhage,
　　1140, 1142, 1142*f*
　states of, 39–40, 39*f*

stomata and loss of, 829–31, 830*f*
　summary, 43
　in tracheophytes, 647*f*
　and waxes, 55
　in xylem, 826–28, 827*f*, 828*f*
water availability, 1157, 1158*f*
water-breathing animals:
　ion and water balance for, 874–75,
　　875*f*
　respiratory systems of, 1025–27, 1025*f*
water-conducting cells, of xylem, 775,
　　775*f*
water cycles, 1265, 1266*f*
water deficit, 829
water-holding capacity, of soil, 805–7,
　　806*f*, 807*f*
Waterland, Robert, 379
water potential, 820–22, 822*f*
Waters, Colin, 1257
water-soluble hormones, 1059, 1061–62
water-soluble vitamins, 972, 973*t*, 984
water stress, 829–32
water vapor, 39
　as greenhouse gas, 1260, 1261*t*
　from transpiration, 818, 829, 829*f*
　*See also* transpiration
water vascular systems, 727–28, 727*f*,
　　728*f*, 728*t*
Watson, James, 226–27, 451
wavelength, 167, 929
waxes, 55
waxy cuticles of plants, 215, 216
weak acids, 41
weaponry, as predation defense, 1225*f*,
　　1226
weathering, 806
web of life, 7, 8*f*, 532–33
web spinning, 720–21, 721*f*
weight, mass vs., 29
weight, and relative water content,
　　822
Weinberg, Wilhelm, 479
Weintraub, Harold, 427–29
Weismann, August, 221, 355
*Welwitschia mirabilis*, 670, 671*f*
Went, Frits, 787
western hemlocks, 1241–42, 1242*f*
western meadowlark, 500
wetlands, 1174, 1174*f*
whales:
　evolution of, 465, 467*f*
　sensory systems of, 925, 930
　threatened due to overexploitation,
　　1272, 1272*f*
wheat, 179, 180*f*, 365–66, 366*f*, 790
wheat rust, 616, 617*f*
whisk fern *(Psilotum nudum)*, 646*f*
White, Philip, 810
white blood cells, 1015
　*See also* leukocytes
white clover, 1155, 1156*f*
white-crowned sparrows, 1183, 1184*f*
white-handed gibbon, 549*f*
white-handed gibbons. *See* gibbons.
white matter, 908, 909*f*
whitemouth moray eel, 741*f*
white muscle fibers, 961

　*See also* glycolytic fibers
white-nose syndrome, 1108
white of eyes. *See* sclera.
White pelican, 752*f*
Whittaker, Robert, 517, 1164
whole-genome duplications, in flowering
　　plant evolution, 674, 675*f*
whole metagenomic sequencing (WMS),
　　of microbiome, 624*f*, 625–26, 627*f*
whooping cough, 569
whorls, flower, 844, 844*f*
Wieschaus, Eric, 421, 1112, 1113
wild mustard plant, 470, 470*f*
wild-type alleles, 363
Wilkins, Maurice, 226, 227
Wilkinson, Gerald, 1196
Williams, George C., 1194
willow trees, 127, 1232, 1233*f*
Wilmut, Ian, 1292
Wilson, Allan, 527, 554
Wilson, E. O., 497, 1244, 1244*f*, 1246–47,
　　1280–82, 1286
Wilson, Henry V., 417
wind:
　and anemotaxis, 1185
　and Coriolis force, 1162
　and effects temperature, 1157, 1158*f*
　plant responses to, 795–96
wings, 751*f*, 965
wintercreeper, 468, 469*f*
winter wheat, flowering of, 844
witchweeds, 815
WMS (whole metagenomic sequencing),
　　of microbiome, 624*f*, 625–26, 627*f*
Woese, Carl, 517
Wöhler, Friedrich, 45, 46
Wolf, Larry, 1190
wolves, 1232, 1233*f*
wood:
　defined, 665
　in gynmosperm evolution, 665–66, 666*f*
　and roots, 777
　and secondary xylem, 776
　structure of, 776, 776*f*, 777
Woodcock, Christopher, 238
wood frog, 745, 745*f*
wood lice, 725*f*, 726
woodpeckers, 1270, 1270*f*
wood pigeon, 1193, 1193*f*
woody plants:
　as eudicots, 763
　phloem loading in, 835, 836
　plant-bacterial symbioses, 812
　secondary growth in, 766
　vascular systems of, 775
　*See also* flowering plants
　　(angiosperms); gymnosperms
woody progymnosperms. *See*
　　progymnosperms.
woody roots, 778
work ethic, 1182
World Health Organization, 985
worms, 1085
　*See also* Annelida and annelids;
　　flatworms; Nematoda and
　　nematodes
wounds, plant responses to, 835, 835*f*

wound sealing, after hemorrhage, 1132
Wright, Sewall, 477
Wright, Stephen, 683
*Wuchereria bancrofti*, 717
Wyllie, Andrew, 197, 198*f*
Wynne-Edwards, V. C., 1193–94
Wyoming, 1168, 1168*f*

# X

Xanax, 899
X-chromosome inactivation, 377–79,
　　377*f*, 378*f*, 378*t*
X chromosomes:
　of chimpanzees vs. humans, 550
　inactivation of, 377–79, 377*f*, 378*f*,
　　378*t*
　and muscular-skeletal systems, 967
　number of, 378, 378*t*
Xenartha, 757
*Xenopus laevis*, 414
xeroderma pigmentosum (XP), 304,
　　313–14, 314*f*
X-Gal, 437
X inactivation center (Xic), 378, 378*t*
*Xist* gene, 379
X-linked genes, 360–61
X-O system, 359, 360*f*
X-ray diffraction pattern:
　of DNA, 226, 226*f*
　of ribosomes, 259, 259*f*
xylem, 645, 646
　bulk flow in, 825, 825*f*
　and Casparian strips, 824
　and cytokinins, 792
　long-distance plant transport in, 818,
　　825*f*, 826–28, 827*f*
　as plant tissue, 216
　and primary stem structure
　　development, 766, 767
　secondary, 776–77, 779 (*See also*
　　wood)
　and transpiration, 828–29
　water-conducting cells of, 775, 775*f*
X-Y system, 359, 360*f*

# Y

Yamanaka, Shinya, 427
Yasumura, Yuki, 791
yeast:
　fermentation in, 161
　genome size of, 449
yeasts, 609, 620
　asexual reproduction by, 609, 610*f*
　cell division in, 334
　cellular response by, 184, 184*f*
*Yersinia pestis*, 577
yew tree, 669*f*
yolk, 1100
yolk plug, 1103
yolk sac, 746, 747*f*
Yoshida, Masasuke, 157–59
Young, Larry, 1197
*Yucca constricta*, 498
*Yucca pallida*, 498
yucca species, 498

# Z